CHILTON'S AUTO SERVICE MANUAL 2002 Edition

CHILTON AUTOMOTIVE INFORMATION

PUBLISHED BY **W. G. NICHOLS, INC.**

ISBN 0-8019-9346-6
Library of Congress Catalog Card No. 2001-135290
1234567890 8765432109

www.chiltonsonline.com

Table of Contents

Unit Repair Sections

Table of Contents

Car Sections

Model Index

EDITORIAL POLICY

Manufacturer and Model Coverage

This Manual does not seek to cover every make and model that is currently available on the market. Rather, the Chilton Editorial Staff makes judicious decisions as to which makes and models warrant coverage. Those that are included herein represent Chilton's judgement as to the makes and models that make up 90% of vehicles that will be presented to the average technician for diagnosis and repair. In general, this Manual does not cover:

- Exotics (e.g. Rolls-Royce, Dodge Viper; Alfa Romeo, etc.)
- OEM's with no U.S. presence (e.g. Fiat)
- OEM's that have not sold enough units to be a factor in the repair market.

Model Year Information

Every effort is made to gather current data for use in this Manual. Data is acquired from manufacturers at the time when each OEM chooses to release it. Different manufacturers choose to release their new model information at different times of the year. Indeed, the same manufacturer can be early one season with information, and then late the next season. As a result, not all models are equally current when each edition of this Manual goes to press. You will note that the Editorial Staff has taken care to indicate the currency of coverage for each model.

Safety Notice

Proper service and repair procedures are vital to the safe, reliable operation of all motor vehicles, as well as the personal safety of those performing the repairs. This manual outlines procedures for servicing and repairing vehicles using safe effective methods. The procedures contain many NOTES, WARNINGS and CAUTIONS which should be followed along with standard safety procedures to eliminate the possibility of personal injury or improper service which could damage the vehicle or compromise its safety.

It is important to note that repair procedures and techniques, tools and parts for servicing vehicles, as well as the skill and experience of the individual performing the work vary widely. It is not possible to anticipate all of the conceivable ways or conditions under which vehicles may be serviced, or to provide cautions as to all of the possible hazards that may result. Standard and accepted safety precautions and equipment should be used when handling toxic or flammable fluids, and safety goggles or other protection should be used during cutting, grinding, chiseling, prying, or any other process that can cause material removal or projectiles.

Some procedures require the use of tools specially designed for a specific purpose. Before substituting another tool or procedure, you must be completely satisfied that neither your personal safety, nor the performance of the vehicle will be endangered.

Although information in this manual is based on industry sources and is as complete as possible at the time of publication, the possibility exists that some vehicle manufacturers made later changes which could not be included here. Information on very late models may not be available in some circumstances. While striving for total accuracy, Nichols Publishing cannot assume responsibility for any errors, changes, or omissions that may occur in the compilation of this data.

LOCATING AND USING INFORMATION

Organization

The Table of Contents, located at the front of the book, lists each Unit Repair Section and Model Specific section in this manual.

To find where a particular model specific section is located in the book, you need only look in the Table of Contents. Once you have found the proper section, you may wish to find where specific procedures are located in that section. Turn to the Index at the front of the model specific section. At the upper left-hand side is a listing of the main topics within that section and the page number on which they may be found. Following the main topics is an alphabetical listing of all of the procedures within the section and their page numbers.

The Model Index, located just after the Table of Contents in the beginning of this manual, may also be used to locate the specific section for any vehicle model covered in this manual.

Specifications

Specifications charts for all models covered in this book are located in Chapter 1. They include: Vehicle and Engine Identification, General Engine Specifications, Engine Tune-Up Specifications, Capacities, Valve Specifications, Crankshaft & Connecting Rod Specifications, Piston & Ring Specifications, Engine Fastener Torque Specifications, Brake Specifications, Maintenance Interval Specifications, Ball Joint Specifications, Wheel & Tire Specifications, and Wheel Alignment Specifications.

Unit Repair Sections

The three Unit Repair Sections are written to cover all applicable models for the specific system or component, unless specifically noted otherwise. The procedures covered in the URS are not repeated in the model specific sections, therefore, refer to the URS for the service procedures for the applicable systems or components. Refer to the Table of Contents for URS coverage.

Model Specific Sections

The model specific sections are grouped by manufacturer and arranged in alphabetical order. The text and illustrations that comprise the service procedures in each model specific section are arranged in the following order of systems and components: Engine Repair (Gasoline, then Diesel if applicable), Fuel System (Gasoline, then Diesel if applicable), Drive Train, Steering and Suspension.

All illustrations are located as close as possible to the applicable procedure. Procedures are for all models in the particular section unless specifically noted otherwise.

Part Numbers

Part numbers listed in this book are not recommendations by Nichols Publishing for any product by brand name. They are references that can be used with interchanges manuals and aftermarket supplier catalogs to locate each brand supplier's discrete part number.

Special Tools

Special tools are recommended by the vehicle manufacturer to perform their specific job. Use has been kept to a minimum, but where absolutely necessary, they are referred to in the text by the part number of the tool manufacturer. These tools may be purchased, under the appropriate part number, from your local dealer or regional distributor, or an equivalent tool can be purchased locally from a tool supplier or parts outlet. Before substituting any tool for the one recommended, read the previous Safety Notice.

ACKNOWLEDGMENTS

This publication contains material that is reproduced and distributed under a license from Ford Motor Company. No further reproduction or distribution of the Ford Motor Company material is allowed without the expressed written permission from Ford Motor Company.

Portions of the material contained herein have been reprinted with permission of General Motors Corporation, Service Technology Group.

Nichols Publishing would like to express thanks to all of the fine companies who participate in the production of our books. Hand tools supplied by Craftsman are used during all phases of our vehicle teardown and photography. Many of the fine specialty tools used in our procedures were provided courtesy of Lisle Corporation. Lincoln Automotive Products (1 Lincoln Way, St. Louis, MO 63120) has provided their industrial shop equipment, including jacks (engine, transmission and floor), engine stands, fluid and lubrication tools, as well as shop presses. Rotary Lifts, the largest automobile lift manufacturer in the world, offering the biggest variety of surface and in-ground lifts available (1-800-640-5438 or www.Rotary-Lift.com), have fulfilled our shop's lift needs. Much of our shop's electronic testing equipment was supplied by Universal Enterprises Inc. (UEI).

SPECIFICATIONS

1

CHRYSLER CORP.

Chrysler Sebring Coupe • Dodge Avenger

ENGINE AND VEHICLE IDENTIFICATION

Engine						
Code ①	Liters (cc)	Cu. In.	Cyl.	Fuel Sys.	Engine Type	Eng. Mfg.
N	2.5 (2497)	152	V6	MFI	SOHC	Mitsubishi
Y	2.0 (1996)	122	I4	MFI	DOHC	Chrysler
H	3.0 (2972)	181	V6	MFI	DOHC	Chrysler

Model Year	
Code ②	Year
W	1998
X	1999
Y	2000
1	2001
2	2002

MFI: Multi-port Fuel Injection

DOHC: Double Overhead Camshaft

SOHC: Single Overhead Camshaft

① 8th position of the Vehicle Identification Number (VIN).

② 10th position of VIN.

93461C01

GENERAL ENGINE SPECIFICATIONS

All measurements are given in inches.

Year	Model	Engine Displacement Liters (cc)	Engine Series (ID/VIN)	Fuel System	Net Horsepower @ rpm	Net Torque @ rpm (ft. lbs.)	Bore x Stroke (in.)	Com-pression Ratio	Oil Pressure @ rpm
1998	Avenger	2.5 (2497)	N	MFI	155@5500	161@4400	3.29 x 2.99	9.4:1 ①	35-75@3000
		2.0 (1996)	Y	MFI	140@6000	130@4800	3.44 x 3.27	9.6:1	25-80@3000
	Sebring Coupe	2.5 (2497)	N	MFI	155@5500	161@4400	3.29 x 2.99	9.4:1 ①	35-75@3000
		2.0 (1996)	Y	MFI	140@6000	130@4800	3.44 x 3.27	9.6:1	25-80@3000
1999	Avenger	2.5 (2497)	N	MFI	155@5500	161@4400	3.29 x 2.99	9.4:1 ①	35-75@3000
		2.0 (1996)	Y	MFI	140@6000	130@4800	3.44 x 3.27	9.6:1	25-80@3000
	Sebring Coupe	2.5 (2497)	N	MFI	155@5500	161@4400	3.29 x 2.99	9.4:1 ①	35-75@3000
		2.0 (1996)	Y	MFI	140@6000	130@4800	3.44 x 3.27	9.6:1	25-80@3000
2000	Avenger	2.5 (2497)	N	MFI	155@5500	161@4400	3.29 x 2.99	9.4:1 ①	35-75@3000
	Sebring Coupe	2.5 (2497)	N	MFI	155@5500	161@4400	3.29 x 2.99	9.4:1 ①	35-75@3000
2001-02	Avenger	2.5 (2497)	N	MFI	155@5500	161@4400	3.29 x 2.99	9.4:1 ①	35-75@3000
	Sebring Coupe	2.5 (2497)	N	MFI	155@5500	161@4400	3.29 x 2.99	9.4:1 ①	35-75@3000
	Sebring Avenger Coupe	3.0 (2972)	H	MFI	200@5500	205@4400	3.59 x 2.99	9.0:1	43-100@3500

MFI: Multi-port Fuel Injection

① California - 9.0:1

93461C02

ENGINE TUNE-UP SPECIFICATIONS

Year	Engine Displacement Liters (cc)	Engine ID/VIN	Spark Plug Gap (in.)	Ignition Timing (deg.)	Fuel Pump (psi)	Idle Speed (rpm)		Valve Clearance	
						MT	AT	In.	Ex.
1998	2.0 (1996)	Y	0.033-0.038	①	47-50	800	800	HYD	HYD
	2.5 (2497)	N	0.039-0.043	①	47-50	—	750	HYD	HYD
1999	2.0 (1996)	Y	0.033-0.038	①	47-50	800	800	HYD	HYD
	2.5 (2497)	N	0.039-0.043	①	47-50	—	750	HYD	HYD
2000	2.5 (2497)	N	0.039-0.043	①	47-50	—	750	HYD	HYD
2001-02	2.5 (2497)	N	0.039-0.043	①	47-50	—	750	HYD	HYD
	3.0 (972)	H	0.039-0.043	①	47-50	—	750	HYD	HYD

NOTE: The Vehicle Emission Control Information label often reflects specification changes made during production. The label figures must be used if they differ from those in this chart.

HYD: Hydraulic

① Basic ignition timing is not adjustable.

93461C03

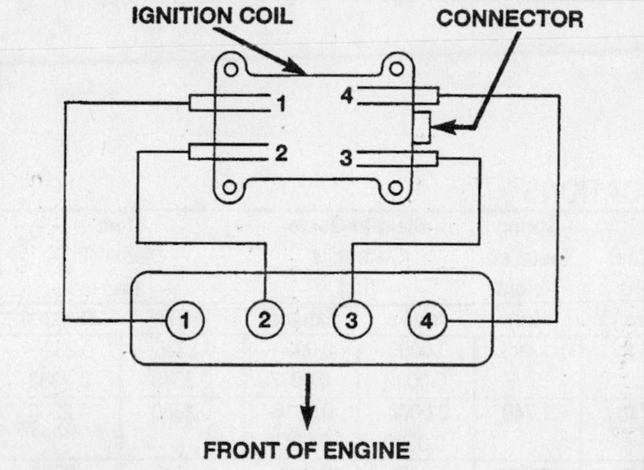

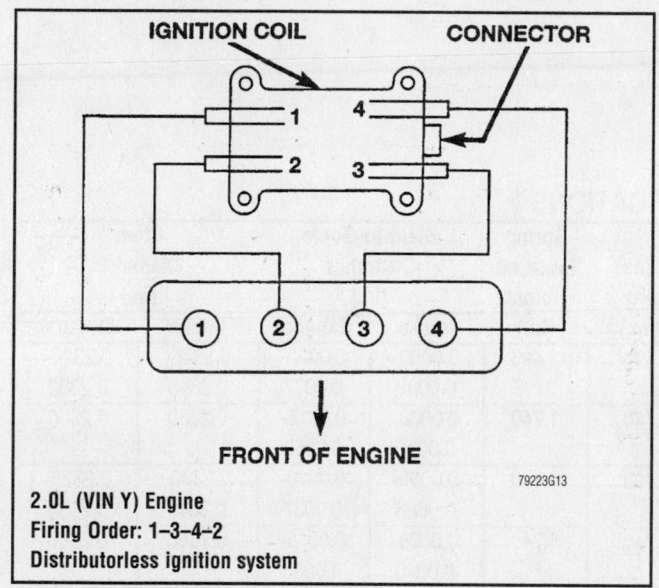

2.0L (VIN Y) Engine
Firing Order: 1–3–4–2
Distributorless ignition system

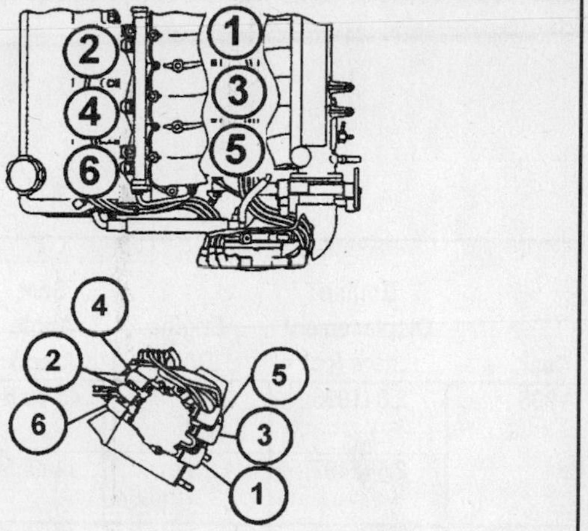

2.5L (VIN N) Engine
Firing Order: 1–2–3–4–5–6
Distributor Rotation: Clockwise

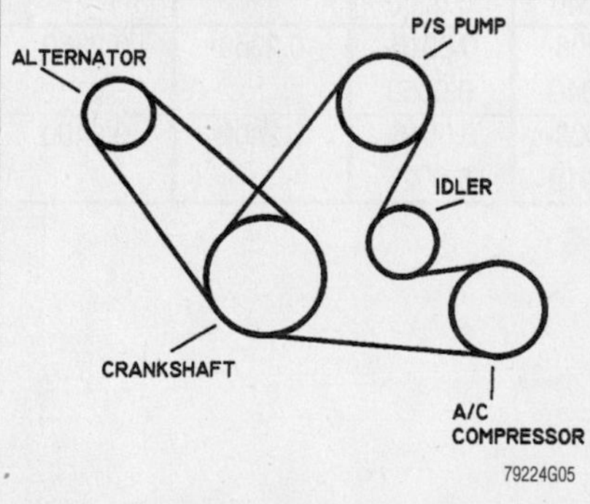

Accessory drive belt routing—Chrysler 2.0L (VIN Y)

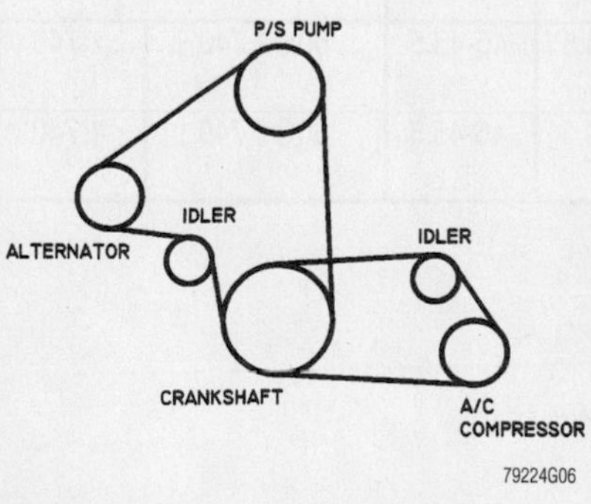

Accessory drive belt routing—Chrysler 2.5L engine

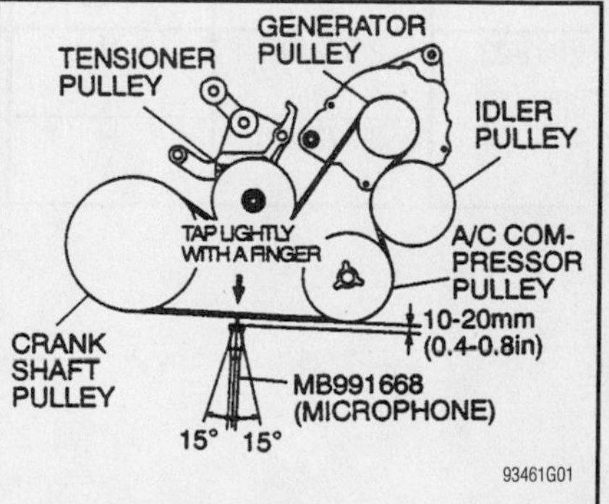

Accessory drive belt routing—Chrysler 3.0L engine

CAPACITIES

Year	Model	Engine Displacement Liters (cc)	Engine ID/VIN	Engine Oil with Filter (qts.)	Transmission (pts.) Manual	Transmission (pts.) Auto.①	Fuel Tank (gal.)	Cooling System (qts.)
1998	Avenger	2.0 (1996)	Y	4.5	4.2	18.2	16.9	7.4
		2.5 (2497)	N	4.5	—	18.2	16.9	7.4
	Sebring Coupe	2.0 (1996)	Y	4.5	4.2	18.2	16.9	7.4
		2.5 (2497)	N	4.5	—	18.2	16.9	7.4
1999	Avenger	2.0 (1996)	Y	4.5	4.2	18.2	16.9	7.4
		2.5 (2497)	N	4.5	—	18.2	16.9	7.4
	Sebring Coupe	2.0 (1996)	Y	4.5	4.2	18.2	16.9	7.4
		2.5 (2497)	N	4.5	—	18.2	16.9	7.4
2000	Avenger	2.5 (2497)	N	4.5	—	18.2	16.9	7.4
	Sebring Coupe	2.5 (2497)	N	4.5	—	18.2	16.9	7.4
2001-02	Avenger	2.5 (2497)	N	4.5	—	18.2	16.9	7.4
	Sebring Coupe	3.0 (2972)	H	4.5	4.2	8	16.3	8.5

NOTE: All capacities are approximate. Add fluid gradually and ensure a proper fluid level is obtained.

① Overhaul fill capacity with torque converter empty

93461C04

VALVE SPECIFICATIONS

Year	Engine Displacement Liters (cc)	Engine ID/VIN	Seat Angle (deg.)	Face Angle (deg.)	Spring Test Pressure (lbs. @ in.)	Spring Installed Height (in.)	Stem-to-Guide Clearance (in.) Intake	Stem-to-Guide Clearance (in.) Exhaust	Stem Diameter (in.) Intake	Stem Diameter (in.) Exhaust
1998	2.0 (1996)	Y	45-45.5	44.5-45	60@1.496	1.496	0.0009-0.0026	0.0020-0.0037	0.2336-0.2343	0.2325-0.2332
	2.5 (2497)	N	44-44.5	45-45.5	60@1.740	1.740	0.0008-0.0040	0.0016-0.0060	0.2360	0.2360
1999	2.0 (1996)	Y	45-45.5	44.5-45	60@1.496	1.496	0.0009-0.0026	0.0020-0.0037	0.2336-0.2343	0.2325-0.2332
	2.5 (2497)	N	44-44.5	45-45.5	60@1.740	1.740	0.0008-0.0040	0.0016-0.0060	0.2360	0.2360
2000	2.5 (2497)	N	44-44.5	45-45.5	60@1.740	1.740	0.0008-0.0040	0.0016-0.0060	0.2360	0.2360
2001-02	2.5 (2497)	N	44-44.5	45-45.5	60@1.740	1.740	0.0008-0.0040	0.0016-0.0060	0.2360	0.2360
	3.0 (2972)	H	44-44.5	45-45.5	60@1.740	1.740	0.0008-0.0019	0.0016-0.0027	0.2400	0.2400

93461C05

WHEEL ALIGNMENT

			Caster		Camber			
Year	Model		Range (+/-Deg.)	Preferred Setting (Deg.)	Range (+/-Deg.)	Preferred Setting (Deg.)	Toe-in (in.)	Steering Axis Inclination (Deg.)
1998	ALL	F	+1.50	+4.31	+0.50	0	0 +/- 0.13	7.31
		R	—	—	+0.50	-1.34	0.13 +/- 0.13	—
1999	ALL	F	+1.50	+4.31	+0.50	0	0 +/- 0.13	7.31
		R	—	—	+0.50	-1.34	0.13 +/- 0.13	—
2000	ALL	F	+1.50	+4.31	+0.50	0	0 +/- 0.13	7.31
		R	—	—	+0.50	-1.34	0.13 +/- 0.13	—
2001	ALL	F	+1.50	+4.31	+0.50	0	0 +/- 0.13	7.31
		R	—	—	+0.50	-1.34	0.13 +/- 0.13	—
2002	ALL	F	+1.50	+4.31	+0.50	0	0 +/- 0.13	7.31
		R	—	—	+0.50	-1.34	0.13 +/- 0.13	—

93461C10

TIRE, WHEEL AND BALL JOINT SPECIFICATIONS

		OEM Tires		Tire Pressures (psi)			
Year	Model	Standard	Optional	Front	Rear	Wheel Size	Ball Joint Inspection
1998	Avenger, base	P195/70HR14	P205/55HR16	32	29	5.5-J	①
	Avenger ES	P215/50HR17	None	32	29	6.5-JJ	①
	Sebring JX	P205/65R15	P215/55R16	32	32	6-JJ	①
	Sebring Jxi	P215/55R16	None	30	30	6-JJ	①
	Sebring LX	P195/70R14	None	30	30	5.5-J	①
	Sebring Lxi	P205/55R16	None	32	29	6-JJ	①
1999	Avenger, base	P195/70HR14	P205/55HR16	32	29	5.5-J	①
	Avenger ES	P215/50HR17	None	32	29	6.5-JJ	①
	Sebring JX	P205/65R15	P215/55R16	30	30	6-JJ	①
	Sebring Jxi	P215/55R16	None	30	30	6-JJ	①
	Sebring LX	P195/70R14	P205/55HR16	32	29	5.5-J	①
	Sebring Lxi	P215/50R17	None	32	29	6.5-JJ	①
2000	Avenger, base	P195/70HR14	P205/55HR16	32	29	5.5-J	①
	Avenger ES	P215/50HR17	None	32	29	6.5-JJ	①
	Sebring JX	P205/65R15	P215/55R16	30	30	6-JJ	①
	Sebring Jxi	P215/55R16	None	30	30	6-JJ	①
	Sebring LX	P195/70R14	P205/55HR16	32	29	5.5-J	①
	Sebring Lxi	P215/50R17	None	32	29	6.5-JJ	①
2001-02	Avenger, base	P195/70HR14	P205/55HR16	32	29	5.5-J	①
	Avenger ES	P215/50HR17	None	32	29	6.5-JJ	①
	Sebring JX	P205/65R15	P215/55R16	30	30	6-JJ	①
	Sebring Jxi	P215/55R16	None	30	30	6-JJ	①
	Sebring LX	P195/70R14	P205/55HR16	32	29	5.5-J	①
	Sebring Lxi	P215/50R17	None	32	29	6.5-JJ	①

OEM: Original Equipment Manufacturer

PSI: Pounds Per Square Inch

STD: Standard

OPT: Optional

L: Lower

U: Upper

① Replace if any measurable movement is found

93461C11

SCHEDULED MAINTENANCE INTERVALS
Chrysler—Sebring, Dodge—Avenger

TO BE SERVICED	TYPE OF SERVICE	VEHICLE MILEAGE INTERVAL (x1000)												
		7.5	15	22.5	30	37.5	45	52.5	60	67.5	75	82.5	90	97.5
Engine oil & filter	R	✓	✓	✓	✓	✓	✓	✓	✓	✓	✓	✓	✓	✓
Coolant level, hoses & clamps	S/I	✓	✓	✓	✓	✓	✓	✓	✓	✓	✓	✓	✓	✓
Rotate tires	S/I	✓	✓	✓	✓	✓	✓	✓	✓	✓	✓	✓	✓	✓
Automatic transaxle fluid level	S/I		✓		✓		✓		✓		✓		✓	
Brake hoses & disc brake pads	S/I		✓		✓		✓		✓		✓		✓	
Driveshaft boots & front suspension components	S/I		✓		✓		✓		✓		✓		✓	
Air filter element	R				✓				✓				✓	
Engine coolant	R				✓				✓				✓	
Spark plugs (DOHC)	R				✓				✓				✓	
Spark plugs (SOHC) ①	R				✓				✓				✓	
Accessory drive belts	S/I				✓				✓				✓	
Ball joints & steering linkage seals	S/I				✓				✓				✓	
Exhaust system	S/I				✓				✓				✓	
Fuel hoses	S/I				✓				✓				✓	
Manual transaxle oil	S/I				✓				✓				✓	
PCV valve	S/I				✓				✓				✓	
Rear drum brake lining & rear wheel cylinders	S/I				✓				✓				✓	
Camshaft timing belt	R								✓					
Ignition cables	R								✓					
Distributor cap & rotor	S/I								✓					
EVAP system	S/I								✓					
Fuel system	S/I								✓					

R: Replace S/I: Service or Inspect
① Spark plugs: replace every 100,000 miles.

FREQUENT OPERATION MAINTENANCE (SEVERE SERVICE)

If a vehicle is operated under any of the following conditions it is considered severe service:

- Extremely dusty areas.

- 50% or more of the vehicle operation is in 32°C (90°F) or higher temperatures, or constant operation in temperatures below 0°C (32°F).

- Prolonged idling (vehicle operation in stop and go traffic).

- Frequent short running periods (engine does not warm to normal operating temperatures).

- Police, taxi, delivery usage or trailer towing usage.

Oil & oil filter change: change every 3000 miles.

Disc brake pads: check every 6000 miles.

Air filter element: change every 15,000 miles.

Automatic transaxle fluid: change every 15,000 miles.

Rear drum brake linings & rear wheel cylinders: check every 15,000 miles.

Spark plugs: change every 15,000 miles.

93461C12

SCHEDULED MAINTENANCE INTERVALS
DAIMLERCHRYSLER CORPORATION
CHRYSLER SEBRING (through 2000)
DODGE AVENGER

The following should be used as a guide when determining the amount of work required for a particular service. In estimating how long a particular Scheduled Maintenance Service should take, please observe the following:

- Labor Time is time based on field research and data supplied by the vehicle manufacturer.
- Labor time operations are given in hours and tenths of an hour.
- All labor operations are to be used as a guide.

Mechanic Skill Level Codes:
(A) PRECISION: Highly skilled with multiple certification.
(B) GENERAL: Normally skilled with certification.
(C) MAINTENANCE: Semi-skilled working on certification.

	LABOR TIME
7500 Mile Service (C)	
All Models	.3
15000 Mile Service (C)	
All Models	1.5
22500 Mile Service (C)	
All Models	.3
30000 Mile Service (B)	
All Models	2.1
37500 Mile Service (C)	
All Models	.3
45000 Mile Service (B)	
All Models	1.5
52500 Mile Service (C)	
All Models	.3
60000 Mile Service (B)	
All Models	2.1
67500 Mile Service (B)	
All Models	1.3
Replace timing belt add	1.8
75000 Mile Service (B)	
All Models	1.5
82500 Mile Service (C)	
All Models	.3
90000 Mile Service (B)	
All Models	2.1
97500 Mile Service (C)	
All Models	.3

93461C13

Heater Core replacement is covered in Section 2 of this manual

CHRYSLER CORP.

Chrysler Cirrus • Sebring Convertible • Dodge Stratus • Plymouth Breeze

ENGINE AND VEHICLE IDENTIFICATION

Engine						
Code ①	Liters (cc)	Cu. In.	Cyl.	Fuel Sys.	Engine Type	Eng. Mfg.
C	2.0 (1996)	122	I4	MFI	SOHC	Chrysler
N	2.5 (2497)	152	V6	MFI	SOHC	Mitsubishi
G	2.4 (2351)	148	I4	MFI	SOHC	Chrysler
X	2.4 (2429)	143	I4	MFI	DOHC	Chrysler
U	2.7 (2736)	167	V6	MFI	DOHC	Chrysler
H	3.0 (2972)	181	V6	MFI	SOHC	Chrysler

Model Year	
Code ②	Year
W	1998
X	1999
Y	2000
1	2001
2	2002

MFI: Multi-point Fuel Injection

SOHC: Single Overhead Camshaft

DOHC: Double Overhead Camshaft

① 8th position of VIN

② 10th position of VIN

93461C14

GENERAL ENGINE SPECIFICATIONS

All measurements are given in inches.

Year	Model	Engine Displacement Liters (cc)	Engine Series (ID/VIN)	Fuel System	Net Horsepower @ rpm	Net Torque @ rpm (ft. lbs.)	Bore x Stroke (in.)	Com-pression Ratio	Oil Pressure @ rpm
1998	Breeze	2.0 (1996)	C	MFI	132@6000	129@5000	3.44x3.26	9.8:1	25-80@3000
	Cirrus	2.4 (2429)	X	MFI	150@5200	167@4000	3.44x3.98	9.4:1	25-80@3000
	Cirrus	2.5 (2497)	N	MFI	168@5800	170@4350	3.29x2.99	9.4:1	25-80@3000
	Sebring Convertible	2.4 (2429)	X	MFI	150@5200	167@4000	3.44x3.98	9.4:1	25-80@3000
	Sebring Convertible	2.5 (2497)	N	MFI	168@5800	170@4350	3.29x2.99	9.4:1	25-80@3000
	Stratus	2.0 (1996)	C	MFI	132@6000	129@5000	3.44x3.26	9.8:1	25-80@3000
	Stratus	2.4 (2429)	X	MFI	150@5200	167@4000	3.44x3.98	9.4:1	25-80@3000
	Stratus	2.5 (2497)	N	MFI	168@5800	170@4350	3.29x2.99	9.4:1	25-80@3000
1999	Breeze	2.0 (1996)	C	MFI	132@6000	129@5000	3.44x3.26	9.8:1	25-80@3000
	Cirrus	2.4 (2429)	X	MFI	150@5200	167@4000	3.44x3.98	9.4:1	25-80@3000
	Cirrus	2.5 (2497)	N	MFI	168@5800	170@4350	3.29x2.99	9.4:1	25-80@3000
	Sebring Convertible	2.4 (2429)	X	MFI	150@5200	167@4000	3.44x3.98	9.4:1	25-80@3000
	Sebring Convertible	2.5 (2497)	N	MFI	168@5800	170@4350	3.29x2.99	9.4:1	25-80@3000
	Stratus	2.0 (1996)	C	MFI	132@6000	129@5000	3.44x3.26	9.8:1	25-80@3000
	Stratus	2.4 (2429)	X	MFI	150@5200	167@4000	3.44x3.98	9.4:1	25-80@3000
	Stratus	2.5 (2497)	N	MFI	168@5800	170@4350	3.29x2.99	9.4:1	25-80@3000
2000	Breeze	2.4 (2429)	X	MFI	150@5200	167@4000	3.44x3.98	9.4:1	25-80@3000
	Cirrus	2.4 (2429)	X	MFI	150@5200	167@4000	3.44x3.98	9.4:1	25-80@3000
	Cirrus	2.5 (2497)	N	MFI	168@5800	170@4350	3.29x2.99	9.4:1 ①	25-80@3000
	Sebring Convertible	2.5 (2497)	N	MFI	168@5800	170@4350	3.29x2.99	9.4:1 ①	25-80@3000
	Stratus	2.4 (2429)	X	MFI	150@5200	167@4000	3.44x3.98	9.4:1	25-80@3000
	Stratus	2.5 (2497)	N	MFI	168@5800	170@4350	3.29x2.99	9.4:1 ①	25-80@3000
2001-02	Breeze	2.4 (2429)	X	MFI	150@5200	167@4000	3.44x3.98	9.4:1	25-80@3000
	Cirrus	2.4 (2429)	X	MFI	150@5200	167@4000	3.44x3.98	9.4:1	25-80@3000
	Cirrus	2.5 (2497)	N	MFI	168@5800	170@4350	3.29x2.99	9.4:1 ①	25-80@3000
	Sebring Convertible	2.5 (2497)	N	MFI	168@5800	170@4350	3.29x2.99	9.4:1 ①	25-80@3000
	Stratus	2.4 (2429)	X	MFI	150@5200	167@4000	3.44x3.98	9.4:1	25-80@3000
	Stratus	2.5 (2497)	N	MFI	168@5800	170@4350	3.29x2.99	9.4:1 ①	25-80@3000
	Sebring Sedan	2.7 (2672)	U	MFI	200@5900	192@4300	3.39x3.09	9.6:1	45-105@3000
	Sebring Coupe	2.4 (2351)	G	MFI	142@5500	158@4000	3.41x3.94	9.5:1	43-100@3500
	Stratus Coupe	2.4 (2351)	G	MFI	142@5500	158@4000	3.41x3.94	9.5:1	43-100@3500
	Sebring Coupe	3.0 (2972)	H	MFI	200@5500	205@4500	3.59x2.99	9.0:1	43-100@3500
	Stratus Coupe	3.0 (2972)	H	MFI	200@5500	205@4500	3.59x2.99	9.0:1	43-100@3500

MFI: Multi-point Fuel Injection

① California: 9.0:1

93461C15

GASOLINE ENGINE TUNE-UP SPECIFICATIONS

Year	Engine Displacement Liters (cc)	Engine ID/VIN	Spark Plug Gap (in.)	Ignition Timing (deg.) MT	Ignition Timing (deg.) AT	Fuel Pump (psi) ①	Idle Speed (rpm) MT	Idle Speed (rpm) AT	Valve Clearance In.	Valve Clearance Ex.
1998	2.0 (1996)	C	0.035	②	②	48	③	③	HYD	HYD
	2.4 (2429)	X	0.050	—	②	49	—	③	HYD	HYD
	2.5 (2497)	N	0.038-0.043	—	②	49	—	③	HYD	HYD
1999	2.0 (1996)	C	0.035	②	②	48	③	③	HYD	HYD
	2.4 (2429)	X	0.050	—	②	49	—	③	HYD	HYD
	2.5 (2497)	N	0.038-0.043	—	②	49	—	③	HYD	HYD
2000	2.4 (2429)	X	0.050	—	②	49	③	③	HYD	HYD
	2.5 (2497)	N	0.038-0.043	—	②	49	—	③	HYD	HYD
2001-02	2.4 (2429)	X	0.050	—	②	49	③	③	HYD	HYD
	2.4 (2351)	G	0.039-0.043	—	②	47-50	③	③	HYD	HYD
	2.5 (2497)	N	0.038-0.043	—	②	49	—	③	HYD	HYD
	2.7 (2672)	U	0.048-0.053	②	②	58	③	③	HYD	HYD
	3.0 (2972)	H	0.039-0.043	②	②	47-50	③	③	HYD	HYD

NOTE: The Vehicle Emission Control Information label often reflects specification changes made during production. The label figures must be used if they differ from those in this chart.

HYD: Hydraulic

① This reading measured with vacuum hose disconnected from fuel pressure regulator.

② Ignition timing cannot be adjusted. Base engine timing is set at TDC during assembly.

③ Refer to the Vehicle Emission Control Information label for correct specifications.

93461C16

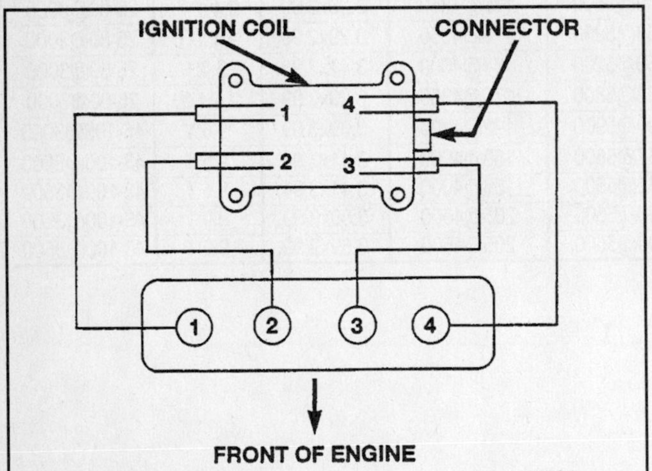

79223G13

2.0L (VIN C) and 2.4L (VIN X) Engines
Firing Order: 1–3–4–2
Distributorless ignition system

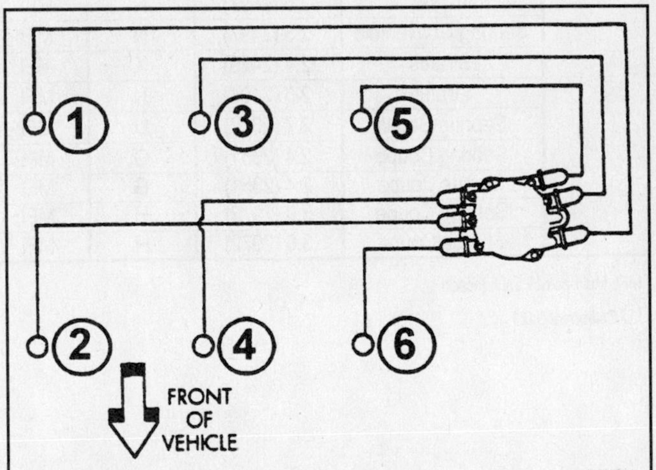

79223G04

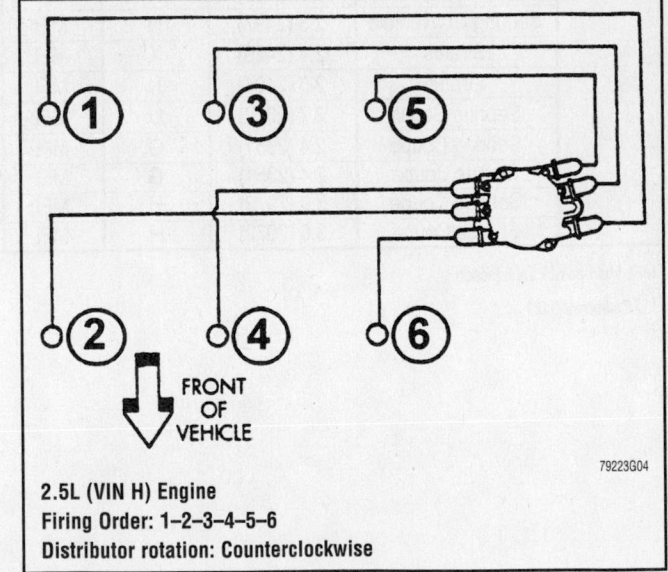

2.5L (VIN H) Engine
Firing Order: 1–2–3–4–5–6
Distributor rotation: Counterclockwise

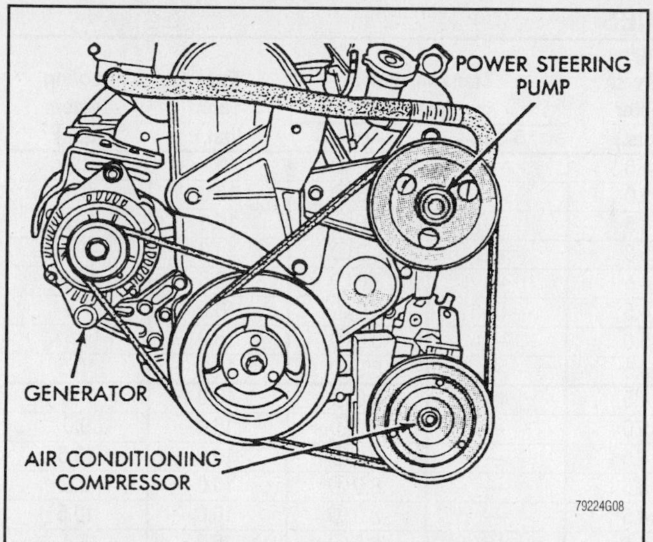

Accessory drive belt routing—Chrysler 2.0L (VIN C) engine

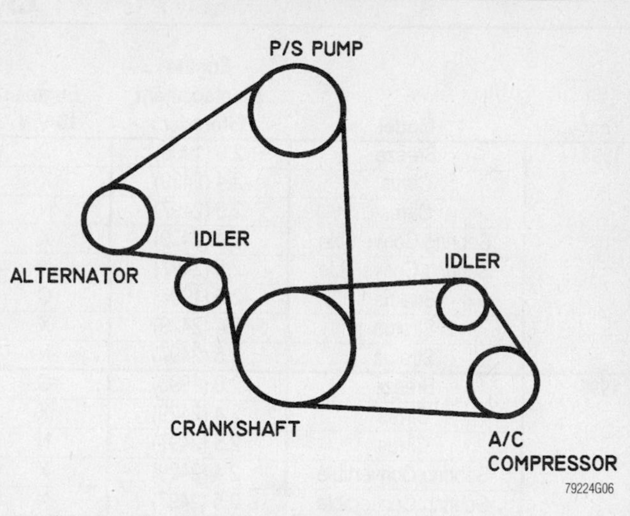

Accessory drive belt routing—Chrysler 2.5L engine

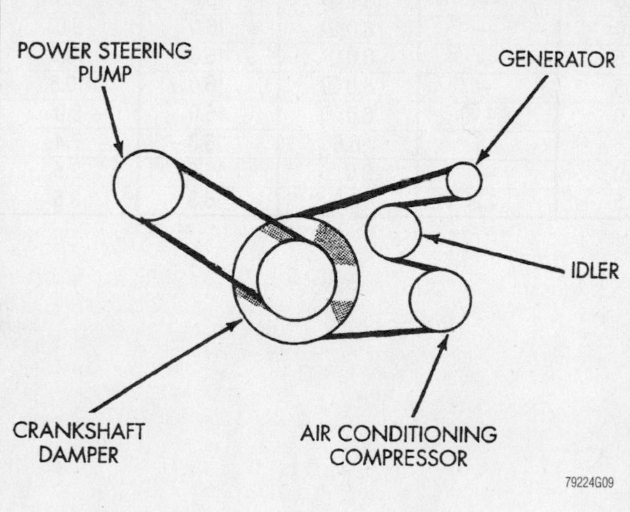

Accessory drive belt routing—Chrysler 2.4L engine

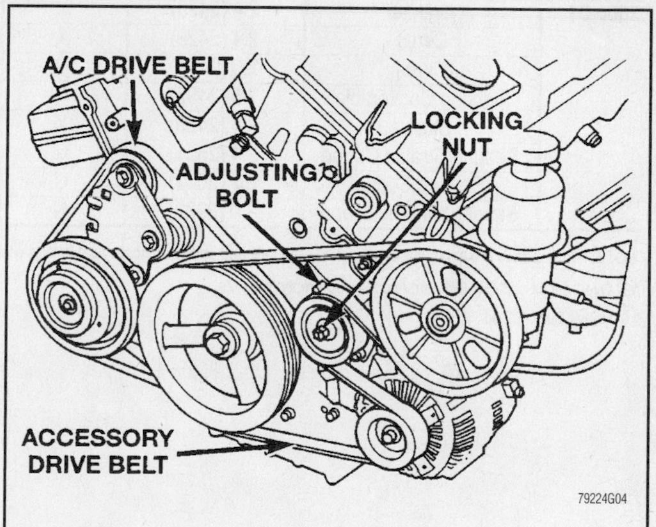

Accessory drive belt routing—Chrysler 2.7L engine

CAPACITIES

Year	Model	Engine Displacement Liters (cc)	Engine ID/VIN	Engine Oil with Filter (qts.)	Transmission (pts.)		Fuel Tank (gal.)	Cooling System (qts.)
					5-Spd	Auto.		
1998	Breeze	2.0 (1996)	C	4.5	4.4	8.0 ①	16.0	7.4
	Cirrus	2.4 (2429)	X	5.0	—	8.0 ①	16.0	9.0
	Cirrus	2.5 (2497)	N	4.5	—	8.0 ①	16.0	10.5
	Sebring Convertible	2.4 (2429)	X	5.0	—	8.0 ①	16.0	9.0
	Sebring Convertible	2.5 (2497)	N	4.5	—	8.0 ①	16.0	10.5
	Stratus	2.0 (1996)	C	4.5	4.4	8.0 ①	16.0	7.4
	Stratus	2.4 (2429)	X	5.0	—	8.0 ①	16.0	9.0
	Stratus	2.5 (2497)	N	4.5	—	8.0 ①	16.0	10.5
1999	Breeze	2.0 (1996)	C	4.5	4.4	8.0 ①	16.0	7.4
	Cirrus	2.4 (2429)	X	5.0	—	8.0 ①	16.0	9.0
	Cirrus	2.5 (2497)	N	4.5	—	8.0 ①	16.0	10.5
	Sebring Convertible	2.4 (2429)	X	5.0	—	8.0 ①	16.0	9.0
	Sebring Convertible	2.5 (2497)	N	4.5	—	8.0 ①	16.0	10.5
	Stratus	2.0 (1996)	C	4.5	4.4	8.0 ①	16.0	7.4
	Stratus	2.4 (2429)	X	5.0	—	8.0 ①	16.0	9.0
	Stratus	2.5 (2497)	N	4.5	—	8.0 ①	16.0	10.5
2000-01	Breeze	2.4 (2429)	X	5.0	—	8.0 ①	16.0	9.0
	Cirrus	2.4 (2429)	X	5.0	—	8.0 ①	16.0	9.0
	Cirrus	2.5 (2497)	N	4.5	—	8.0 ①	16.0	10.5
	Sebring Convertible	2.5 (2497)	N	4.5	—	8.0 ①	16.0	10.5
	Stratus	2.4 (2429)	X	5.0	4.4 ②	8.0 ①	16.0	9.0
	Sebring/Stratus Coupe	2.4 (2351)	G	4.5	3.1	6.6	16.3	7.4
	Sebring Sedan	2.7 (2672)	U	5.0	—	8.0 ①	16.0	9.5
	Sebring/Stratus Coupe	3.0 (2972)	N	4.5	4.2	8.0	16.3	8.5

NOTE: All capacities are approximate. Add fluid gradually and ensure a proper fluid level is obtained.

① Overhaul fill capacity with torque converter empty: 18.2 pts.

② Heavy Duty: 5.2 pts.

93461C17

VALVE SPECIFICATIONS

Year	Engine Displacement Liters (cc)	Engine ID/VIN	Seat Angle (deg.)	Face Angle (deg.)	Spring Test Pressure (lbs. @ in.)	Spring Installed Height (in.)	Stem-to-Guide Clearance (in.)		Stem Diameter (in.)	
							Intake	Exhaust	Intake	Exhaust
1998	2.0 (1996)	C	44.5-45	45-45.5	75@1.54	1.540	0.0018-0.0025	0.0029-0.0037	0.2340	0.2330
	2.4 (2429)	X	45	44.5-45	76@1.50	1.496	0.0018-0.0025	0.0029-0.0037	0.2340	0.2330
	2.5 (2497)	N	45-44.5	45-45.5	60@1.74	1.740	0.0008-0.0020	0.0016-0.0028	0.2360	0.2360
1999	2.0 (1996)	C	44.5-45	45-45.5	75@1.54	1.540	0.0018-0.0025	0.0029-0.0037	0.2340	0.2330
	2.4 (2429)	X	45	44.5-45	76@1.50	1.496	0.0018-0.0025	0.0029-0.0037	0.2340	0.2330
	2.5 (2497)	N	45-44.5	45-45.5	60@1.74	1.740	0.0008-0.0020	0.0016-0.0028	0.2360	0.2360
2000	2.4 (2429)	X	45	44.5-45	76@1.50	1.496	0.0018-0.0025	0.0029-0.0037	0.2340	0.2330
	2.5 (2497)	N	45-44.5	45-45.5	60@1.74	1.740	0.0008-0.0020	0.0016-0.0028	0.2360	0.2360
2001-02	2.4 (2429)	X	45	44.5-45	76@1.50	1.496	0.0018-0.0025	0.0029-0.0037	0.2340	0.2330
	2.4 (2351)	G	45-44.5	45-45.5	60@1.74	1.740	0.0008-0.0019	0.0012-0.0027	0.2360	0.2360
	2.5 (2497)	N	45-44.5	45-45.5	60@1.74	1.740	0.0008-0.0020	0.0016-0.0028	0.2400	0.2300
	2.7 (2672)	U	44.5-45	45-45.5	148@1.14	1.496	0.0009-0.0026	0.0020-0.0033	0.2337-0.2344	0.2326-0.2333
	3.0 (2972)	H	44.5-45	45-45.5	60@1.74	1.740	0.0008-0.0019	0.0016-0.0027	0.2400	0.2400

93461C18

CRANKSHAFT AND CONNECTING ROD SPECIFICATIONS

All measurements are given in inches.

Year	Engine Displacement Liters (cc)	Engine ID/VIN	Crankshaft				Connecting Rod		
			Main Brg. Journal Dia.	Main Brg. Oil Clearance	Shaft End-play	Thrust on No.	Journal Diameter	Oil Clearance	Side Clearance
1998	2.0 (1996)	C	2.0469-2.0475	0.0008-0.0024	0.0035-0.0094	3	1.8894-1.8900	0.0010-0.0023	0.0050-0.0150
	2.4 (2429)	X	2.3610-2.3625	0.0007-0.0023	0.0035-0.0095	3	1.9670-1.9685	0.0009-0.0027	0.0051-0.0150
	2.5 (2497)	N	2.3620	0.0008-0.0016	0.0020-0.0160	3	1.9690	0.0008-0.0016	0.0040-0.0160
1999	2.0 (1996)	C	2.0469-2.0475	0.0008-0.0024	0.0035-0.0094	3	1.8894-1.8900	0.0010-0.0023	0.0050-0.0150
	2.4 (2429)	X	2.3610-2.3625	0.0007-0.0023	0.0035-0.0095	3	1.9670-1.9685	0.0009-0.0027	0.0051-0.0150
	2.5 (2497)	N	2.3620	0.0008-0.0016	0.0020-0.0160	3	1.9690	0.0008-0.0016	0.0040-0.0160
2000	2.4 (2429)	X	2.3610-2.3625	0.0007-0.0023	0.0035-0.0095	3	1.9670-1.9685	0.0009-0.0027	0.0051-0.0150
	2.5 (2497)	N	2.3620	0.0008-0.0016	0.0020-0.0120	3	1.9690	0.0008-0.0016	0.0040-0.0160
2001-02	2.4 (2429)	X	2.3610-2.3625	0.0007-0.0023	0.0035-0.0095	3	1.9670-1.9685	0.0009-0.0027	0.0051-0.0150
	2.4 (2351)	G	1.77-2.2400	0.0008-0.0015	0.002-0.0090	3	1.7709-1.7714	0.0008-0.0019	0.0012-0.0270
	2.5 (2497)	N	2.3620	0.0008-0.0016	0.0020-0.0120	3	1.9690	0.0008-0.0016	0.0040-0.0160
	2.7 (2736)	U	2.4997-2.5004	0.0014-0.0021	0.017 max.	3	2.1067-2.1060	0.0010-0.0026	0.0052-0.0150
	3.0 (2972)	H	2.0-2.4000	0.0008-0.0016	0.0020-0.0120	3	1.7709-1.7714	0.0008-0.0019	0.0012-0.0270

93461C19

PISTON AND RING SPECIFICATIONS

All measurements are given in inches.

Year	Engine Displacement Liters (cc)	Engine ID/VIN	Piston Clearance	Ring Gap Top Compression	Ring Gap Bottom Compression	Ring Gap Oil Control	Ring Side Clearance Top Compression	Ring Side Clearance Bottom Compression	Ring Side Clearance Oil Control
1998	2.0 (1996)	C	0.0004-0.0017 ①	0.009-0.020	0.019-0.031	0.009-0.026	0.0010-0.0026	0.0010-0.0026	0.0002-0.007
	2.4 (2429)	X	0.0009-0.0022 ②	0.010-0.020	0.009-0.018	0.010-0.025	0.0011-0.0031	0.0011-0.0031	0.0004-0.0070
	2.5 (2497)	N	0.0008-0.0016	0.010-0.015	0.005-0.021	0.006-0.020	0.0012-0.0028	0.0008-0.0024	NA
1999	2.0 (1996)	C	0.0004-0.0017 ①	0.009-0.020	0.019-0.031	0.009-0.026	0.0010-0.0026	0.0010-0.0026	0.0002-0.007
	2.4 (2429)	X	0.0009-0.0022 ②	0.010-0.020	0.009-0.018	0.010-0.025	0.0011-0.0031	0.0011-0.0031	0.0004-0.0070
	2.5 (2497)	N	0.0008-0.0016	0.010-0.015	0.005-0.021	0.006-0.020	0.0012-0.0028	0.0008-0.0024	NA
2000	2.4 (2429)	X	0.0009-0.0022 ②	0.010-0.020	0.009-0.018	0.010-0.025	0.0011-0.0031	0.0011-0.0031	0.0004-0.0070
	2.5 (2497)	N	0.0008-0.0016	0.010-0.015	0.005-0.021	0.006-0.020	0.0012-0.0028	0.0008-0.0024	NA
2001-02	2.4 (2429)	X	0.0009-0.0022 ②	0.010-0.020	0.009-0.018	0.010-0.025	0.0011-0.0031	0.0011-0.0031	0.0004-0.0070
	2.4 (2351)	G	0.004-0.009	0.012-0.027	0.010-0.017	0.010-0.025	0.0012-0.0027	0.0012-0.0027	0.0004-0.0015
	2.5 (2497)	N	0.0008-0.0016	0.010-0.015	0.005-0.021	0.006-0.020	0.0012-0.0028	0.0008-0.0024	NA
	2.7 (2672)	U	0.0003-0.0016	0.008-0.014	0.0146-0.0249	0.010-0.030	0.0013-0.0032	0.0016-0.0031	0.0022-0.008
	3.0 (2972)	H	0.0003-0.0090	0.012-0.017	0.018-0.023	0.008-0.0019	0.0012-0.0017	0.0018-0.0023	0.0008-0.019

NA: Not Available

① Clearance at 11/16 inch from bottom of skirt

② Clearance at 9/16 inch from bottom of skirt

93461C20

TORQUE SPECIFICATIONS

All readings in ft. lbs.

Year	Engine Displacement Liters (cc)	Engine ID/VIN	Cylinder Head Bolts	Main Bearing Bolts	Rod Bearing Bolts	Crankshaft Damper Bolts	Flywheel Bolts	Manifold Intake	Manifold Exhaust	Spark Plug	Lug Nut
1998	2.0 (1996)	C	①	②	③	105	70	17	17	20	95
	2.4 (2429)	X	①	②	③	100	70	17	17	20	95
	2.5 (2497)	N	80	69	37	134	70	16	33	18	95
1999	2.0 (1996)	C	①	②	③	105	70	17	17	20	95
	2.4 (2429)	X	①	②	③	100	70	17	17	20	95
	2.5 (2497)	N	80	69	37	134	70	16	33	18	95
2000	2.4 (2429)	X	①	②	③	100	70	17	17	20	95
	2.5 (2497)	N	80	69	37	134	70	16	33	18	95
2001-02	2.4 (2429)	X	①	②	③	100	70	17	17	20	95
	2.4 (2351)	G	④	②	③	102	70	16	25	18	80
	2.5 (2497)	N	80	69	37	134	70	16	33	18	95
	2.7 (2672)	U	⑤	⑥	20	125	—	8	17	15	100
	3.0 (2972)	H	80	69	37	125	55	17	33	18	80

① Step 1: 25 ft. lbs.
Step 2: 50 ft. lbs.
Step 3: 50 ft. lbs.
Step 4: Plus 1/4 turn

② Step 1: 30 ft. lbs
Step 2: Plus 1/4 turn

③ Step 1: 20 ft. lbs.
Step 2: Plus 1/4 turn

④ Step 1:58 ft. lbs.
Step 2: Fully loosen
Step 3: 15 ft. lbs.
Step 4: Plus 1/4 turn
Step 5: Plus 1/4 turn

⑤ Step 1:35 ft. lbs.
Step 2: 55 ft. lbs.
Step 3: 55 ft. lbs.
Step 4: Plus 1/4 turn
Step 5: 21 ft. lbs.

⑥ Main cap inside bolts: 15 ft. lbs. Plus 1/4 turn
Main cap outside bolts: 20 ft. lbs. Plus 1/4 turn
Main cap tie bolts: 21 ft. lbs.

93461C21

WHEEL ALIGNMENT

Year	Model		Caster Range (+/-Deg.)	Caster Preferred Setting (Deg.)	Camber Range (+/-Deg.)	Camber Preferred Setting (Deg.)	Toe-in (in.)	Steering Axis Inclination (Deg.)
1998	All	F	+1.00	+3.31	+0.59	0	0.06 +/- 0.10	7.00
		R	—	—	+0.38	-0.19	0.06 +/- 0.10	—
1999	All	F	+1.00	+3.31	+0.59	0	0.06 +/- 0.10	7.00
		R	—	—	+0.38	-0.19	0.06 +/- 0.10	—
2000	All	F	+1.00	+3.31	+0.59	0	0.06 +/- 0.10	7.00
		R	—	—	+0.38	-0.19	0.06 +/- 0.10	—
2001-02	All	F	+1.00	+3.31	+0.59	0	0.06 +/- 0.10	7.00
		R	—	—	+0.38	-0.19	0.06 +/- 0.10	—

93461C22

TIRE, WHEEL AND BALL JOINT SPECIFICATIONS

Year	Model	OEM Tires		Tire Pressures (psi)		Wheel	Ball Joint
		Standard	Optional	Front	Rear	Size	Inspection
1998	Breeze	P195/70R14	None	31	31	6-J	①
	Cirrus	P195/65R15	None	30	30	6-J	①
	Sebring Conv.	P195/65R15	None	32	29	6-JJ	②
	Stratus	P195/70R14	P195/65HR15	31	31	6-JJ	①
1999	Breeze	P195/70R14	None	31	31	6-J	①
	Cirrus	P195/65R15	None	30	31	6-J	①
	Sebring Conv.	P205/65R15	P215/55R16	32	32	6-JJ	②
	Stratus	P195/70R14	P195/65HR15	31	31	6-JJ	①
2000	Breeze	P195/70R14	None	31	31	6-J	①
	Cirrus	P195/65R15	None	30	31	6-J	①
	Sebring Conv.	P205/65R15	P215/55R16	32	32	6-JJ	②
	Stratus	P195/70R14	P195/65HR15	31	31	6-JJ	①
2001-02	Breeze	P195/70R14	None	31	31	6-J	①
	Cirrus	P195/65R15	None	30	31	6-J	①
	Sebring/Stratus Coupe	P205/65R16	P215/55R17	32	32	6-JJ	②
	Sebring Conv.	P205/65R15	P215/55R16	32	32	6-JJ	②
	Stratus	P195/70R14	P195/65HR15	31	31	6-JJ	①
	Sebring/Stratus Sedan	P205/65R15	P205/60HR16	32	32	6-JJ	①

OEM: Original Equipment Manufacturer

PSI: Pounds Per Square Inch

STD: Standard

OPT: Optional

L: Lower

U: Upper

① Do not lift car. Grasp the grease fitting and attempt to move or rotate. Replace if any movement is found.

② Replace if any measurable movement is found

93461C23

For Wheel Alignment specifications, see Section 1 of this manual

BRAKE SPECIFICATIONS

All measurements in inches unless noted

			Brake Disc			Brake Drum Diameter				Brake Caliper
Year	Model		Original Thickness	Minimum Thickness	Maximum Run-out	Original Inside Diameter	Max. Wear Limit	Maximum Machine Diameter	Minimum Lining Thickness	Guide Pin Bolts (ft. lbs.)
1998	Breeze	F	0.911	0.843	0.003	—	—	—	0.035	16
		R	0.360	0.285	0.005	NA	①	①	②	16
	Cirrus	F	0.911	0.843	0.005	—	—	—	0.035	16
		R	0.360	0.285	0.005	NA	①	①	②	16
	Sebring Convertible	F	0.911	0.843	0.005	—	—	—	0.035	16
		R	0.360	0.285	0.005	NA	①	①	②	16
	Stratus	F	0.911	0.843	0.003	—	—	—	0.035	16
		R	0.360	0.285	0.005	NA	①	①	②	16
1999	Breeze	F	0.911	0.843	0.003	—	—	—	③	16
		R	0.360	0.285	0.005	NA	①	①	②	16
	Cirrus	F	0.911	0.843	0.005	—	—	—	③	16
		R	0.360	0.285	0.005	NA	①	①	②	16
	Sebring Convertible	F	0.911	0.843	0.005	—	—	—	③	16
		R	0.360	0.285	0.005	NA	①	①	②	16
	Stratus	F	0.911	0.843	0.003	—	—	—	③	16
		R	0.360	0.285	0.005	NA	①	①	②	16
2000	Breeze	F	0.911	0.843	0.003	—	—	—	③	16
		R	0.360	0.285	0.005	NA	①	①	②	16
	Cirrus	F	0.911	0.843	0.005	—	—	—	③	16
		R	0.360	0.285	0.005	NA	①	①	②	16
	Sebring Convertible	F	0.911	0.843	0.005	—	—	—	③	16
		R	0.360	0.285	0.005	NA	①	①	②	16
	Stratus	F	0.911	0.843	0.003	—	—	—	③	16
		R	0.360	0.285	0.005	NA	①	①	②	16
2001-02	Breeze	F	0.911	0.843	0.003	—	—	—	③	16
		R	0.360	0.285	0.005	NA	①	①	②	16
	Cirrus	F	0.911	0.843	0.005	—	—	—	③	16
		R	0.360	0.285	0.005	NA	①	①	②	16
	Sebring Convertible	F	0.911	0.843	0.005	—	—	—	③	16
		R	0.400	0.330	0.005	9.80	①	①	0.040	31
	Sebring/Stratus Coupe	F	0.911	0.843	0.005	—	—	—	③	16
		R	0.360	0.285	0.005	NA	①	①	②	16
	Stratus	F	0.911	0.843	0.003	—	—	—	③	16
		R	0.360	0.285	0.005	NA	①	①	②	16
	Sebring/Stratus Sedan	F	0.911	0.843	0.005	—	—	—	③	26
		R	0.360	0.285	0.005	NA	①	①	②	26

NA: Not Available

① Maximum diameter is stamped on drum

② Disc brake lining and backing total thickness: 9/32 inch
Drum brake shoe lining and backing total thickness: 1/8 inch

③ Disc brake lining and backing: 3/8 inch

93461C24

SCHEDULED MAINTENANCE INTERVALS
CHRYSLER CIRRUS & SEBRING CONVERTIBLE, DODGE STRATUS & PLYMOUTH BREEZE

TO BE SERVICED	TYPE OF SERVICE	VEHICLE MILEAGE INTERVAL (x1000)												
		7.5	15	22.5	30	37.5	45	52.5	60	67.5	75	82.5	90	97.5
Engine oil & filter	R	✓	✓	✓	✓	✓	✓	✓	✓	✓	✓	✓	✓	✓
Brake hoses	S/I	✓	✓	✓	✓	✓	✓	✓	✓	✓	✓	✓	✓	✓
Coolant level, hoses & clamps	S/I	✓	✓	✓	✓	✓	✓	✓	✓	✓	✓	✓	✓	✓
CV joints & front suspension components	S/I	✓	✓	✓	✓	✓	✓	✓	✓	✓	✓	✓	✓	✓
Exhaust system	S/I	✓	✓	✓	✓	✓	✓	✓	✓	✓	✓	✓	✓	✓
Rotate tires	S/I	✓	✓	✓	✓	✓	✓	✓	✓	✓	✓	✓	✓	✓
Accessory drive belts	S/I		✓		✓		✓		✓		✓		✓	
Brake linings	S/I			✓			✓			✓			✓	
Air filter element	R				✓				✓				✓	
Spark plugs ①②	R													
Lubricate front & rear ball joints	S/I				✓				✓				✓	
Engine coolant	R						✓				✓			
PCV valve	S/I								✓				✓	
Ignition cables ②③	R													
Camshaft timing belt ④	R													

R: Replace S/I: Service or Inspect

① 4-cylinder: every 30,000 miles.

② 6-cylinde: 100,000 miles.

③ 4-cylinder: 60,000 miles.

④ Replace at 105,000 miles for normal service; replace at 102,000 miles for severe service

FREQUENT OPERATION MAINTENANCE (SEVERE SERVICE)

If a vehicle is operated under any of the following conditions it is considered severe service:

- Extremely dusty areas.

- 50% or more of the vehicle operation is in 32°C (90°F) or higher temperatures, or constant operation in temperatures below 0°C (32°F).

- Prolonged idling (vehicle operation in stop and go traffic).

- Frequent short running periods (engine does not warm to normal operating temperatures).

- Police, taxi, delivery usage or trailer towing usage.

Oil & oil filter change: change every 3000 miles.

Rotate tires every 6000 miles.

Brake linings: check every 12,000 miles.

Air filter element: change every 15,000 miles.

Automatic transaxle fluid: service or inspect every 15,000 miles.

PCV valve: check every 30,000 miles.

Engine coolant, replace at 36,000, 51,000 & 81,000 miles.

93461C25

For Maintenance Interval recommendations, see Section 1 of this manual

SCHEDULED MAINTENANCE INTERVALS
DAIMLERCHRYSLER CORPORATION
CHRYSLER CIRRUS, SEBRING CONVERTIBLE, COUPE, SEDAN
DODGE STRATUS, PLYMOUTH BREEZE

The following should be used as a guide when determining the amount of work required for a particular service. In estimating how long a particular Scheduled Maintenance Service should take, please observe the following:

- Labor Time is time based on field research and data supplied by the vehicle manufacturer.
- Labor time operations are given in hours and tenths of an hour.
- All labor operations are to be used as a guide.

Mechanic Skill Level Codes:
(A) PRECISION: Highly skilled with multiple certification.
(B) GENERAL: Normally skilled with certification.
(C) MAINTENANCE: Semi-skilled working on certification.

	LABOR TIME
7500 Mile Service (C)	
All Models	.7
15000 Mile Service (C)	
All Models	.9
22500 Mile Service (C)	
All Models	.8
30000 Mile Service (C)	
All Models	1.0
37500 Mile Service (C)	
All Models	.7
45000 Mile Service (C)	
All Models	1.0
52500 Mile Service (C)	
All Models	1.0
60000 Mile Service (B)	
All Models	2.4
67500 Mile Service (C)	
All Models	.8
75000 Mile Service (C)	
All Models	.9
82500 Mile Service (C)	
All Models	1.0
90000 Mile Service (B)	
All Models	2.2
97500 Mile Service (C)	
All Models	.7

93461C26

CHRYSLER CORP.

Chrysler 300M • Concorde • LHS • Dodge Intrepid • Eagle Vision

ENGINE AND VEHICLE IDENTIFICATION

Engine							Model Year	
Code ①	Liters (cc)	Cu. In.	Cyl.	Fuel Sys.	Engine Type	Eng. Mfg.	Code ②	Year
G	3.5 (3518)	215	V6	MFI	SOHC	Chrysler	W	1998
J	3.2 (3231)	195	V6	MFI	SOHC	Chrysler	X	1999
R	2.7 (2736)	167	V6	MFI	DOHC	Chrysler	Y	2000
U	2.7 (2736) ③	167	V6	MFI	DOHC	Chrysler	1	2001
V	3.5 (3518) ④	215	V6	MFI	DOHC	Chrysler	2	2002

MFI: Multi-point Fuel Injection

DOHC: Double Overhead Camshafts

SOHC: Single Overhead Camshaft

① 8th position of the Vehicle Identification Number (VIN)

② 10th position of VIN

③ 120 amp alternator

④ Magnum

93461C27

GENERAL ENGINE SPECIFICATIONS

All measurements are given in inches.

Year	Model	Engine Displacement Liters (cc)	Engine Series (ID/VIN)	Fuel System	Net Horsepower @ rpm	Net Torque @ rpm (ft. lbs.)	Bore x Stroke (in.)	Com-pression Ratio	Oil Pressure @ rpm
1998	Concorde	2.7 (2736)	R	MFI	200@5800	190@4850	3.39x3.09	9.6:1	25-105@3000
		3.2 (3231)	J	MFI	225@6300	225@3800	3.62x3.19	9.5:1	25-105@3000
	Intrepid	2.7 (2736)	R	MFI	200@5800	190@4850	3.39x3.09	9.6:1	25-105@3000
		3.2 (3231)	J	MFI	225@6300	225@3800	3.62x3.19	9.5:1	25-105@3000
1999	300M	3.5 (3518)	G	MFI	253@6400	255@3950	3.78x3.19	9.6:1	25-105@3000
	Concorde	2.7 (2736)	R	MFI	200@5800	190@4850	3.39x3.09	9.6:1	25-105@3000
		3.2 (3231)	J	MFI	225@6300	225@3800	3.62x3.19	9.5:1	25-105@3000
	Intrepid	2.7 (2736)	R	MFI	200@5800	190@4850	3.39x3.09	9.6:1	25-105@3000
		3.2 (3231)	J	MFI	225@6300	225@3800	3.62x3.19	9.5:1	25-105@3000
	LHS	3.5 (3518)	G	MFI	253@6400	255@3950	3.78x3.19	9.6:1	25-105@3000
2000	300M	3.5 (3518)	G	MFI	253@6400	255@3950	3.78x3.19	9.6:1	25-105@3000
	Concorde	2.7 (2736)	R	MFI	200@5800	190@4850	3.39x3.09	9.6:1	25-105@3000
		2.7 (2736)	U	MFI	200@5800	190@4850	3.39x3.09	9.6:1	25-105@3000
		3.5 (3518)	V	MFI	253@6400	255@3950	3.78x3.19	9.6:1	25-105@3000
		3.2 (3231)	J	MFI	225@6300	225@3800	3.62x3.19	9.5:1	25-105@3000
	Intrepid	2.7 (2736)	R	MFI	200@5800	190@4850	3.39x3.09	9.6:1	25-105@3000
		2.7 (2736)	U	MFI	200@5800	190@4850	3.39x3.09	9.6:1	25-105@3000
		3.5 (3518)	V	MFI	253@6400	255@3950	3.78x3.19	9.6:1	25-105@3000
		3.2 (3231)	J	MFI	225@6300	225@3800	3.62x3.19	9.5:1	25-105@3000
	LHS	3.5 (3518)	G	MFI	253@6400	255@3950	3.78x3.19	9.6:1	25-105@3000
2001-02	300M	3.5 (3518)	G	MFI	253@6400	255@3950	3.78x3.19	9.6:1	25-105@3000
	Concorde	2.7 (2736)	R	MFI	200@5800	190@4850	3.39x3.09	9.6:1	25-105@3000
		2.7 (2736)	U	MFI	200@5800	190@4850	3.39x3.09	9.6:1	25-105@3000
		3.5 (3518)	V	MFI	253@6400	255@3950	3.78x3.19	9.6:1	25-105@3000
		3.2 (3231)	J	MFI	225@6300	225@3800	3.62x3.19	9.5:1	25-105@3000
	Intrepid	2.7 (2736)	R	MFI	200@5800	190@4850	3.39x3.09	9.6:1	25-105@3000
		2.7 (2736)	U	MFI	200@5800	190@4850	3.39x3.09	9.6:1	25-105@3000
		3.5 (3518)	V	MFI	253@6400	255@3950	3.78x3.19	9.6:1	25-105@3000
		3.2 (3231)	J	MFI	225@6300	225@3800	3.62x3.19	9.5:1	25-105@3000
	LHS	3.5 (3518)	G	MFI	253@6400	255@3950	3.78x3.19	9.6:1	25-105@3000

MFI: Multi-point Fuel Injection

93461C28

ENGINE TUNE-UP SPECIFICATIONS

Year	Engine Displacement Liters (cc)	Engine ID/VIN	Spark Plug Gap (in.)	Ignition Timing (deg.)	Fuel Pump (psi)	Idle Speed (rpm)	Valve Clearance Intake	Valve Clearance Exhaust
1998	2.7 (2736)	R	0.048-0.053	①	49	①	HYD	HYD
	3.2 (3231)	J	0.048-0.053	①	49	①	HYD	HYD
1999	2.7 (2736)	R	0.048-0.053	①	49	①	HYD	HYD
	3.2 (3231)	J	0.048-0.053	①	49	①	HYD	HYD
	3.5 (3518)	G	0.048-0.053	①	49	①	HYD	HYD
2000-01	2.7 (2736)	R	0.048-0.053	①	58	①	HYD	HYD
	2.7 (2736)	U	0.048-0.053	①	58	①	HYD	HYD
	3.5 (3518)	V	0.048-0.053	①	58	①	HYD	HYD
	3.2 (3231)	J	0.048-0.053	①	58	①	HYD	HYD
	3.5 (3518)	G	0.048-0.053	①	58	①	HYD	HYD
2001-02	2.7 (2736)	R	0.048-0.053	①	58	①	HYD	HYD
	2.7 (2736)	U	0.048-0.053	①	58	①	HYD	HYD
	3.5 (3518)	V	0.048-0.053	①	58	①	HYD	HYD
	3.2 (3231)	J	0.048-0.053	①	58	①	HYD	HYD
	3.5 (3518)	G	0.048-0.053	①	58	①	HYD	HYD

NOTE: The Vehicle Emission Control Information label often reflects specification changes made during production. The label figures must be used if they differ from those in this chart.

HYD: Hydraulic

① Controlled by the Powertrain Control Module (PCM)

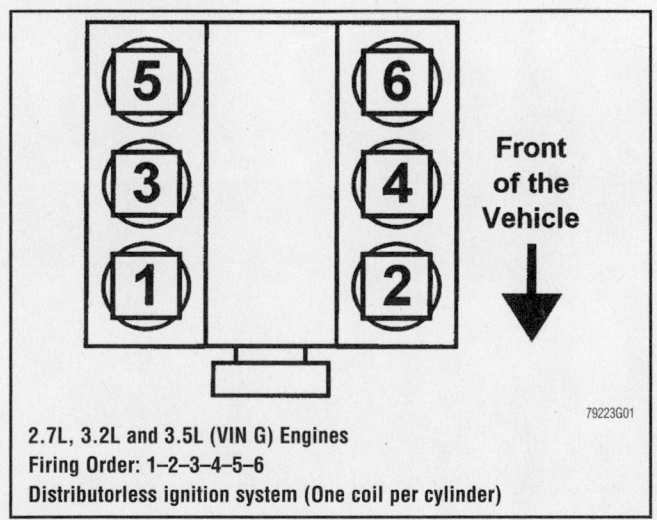

93461C29

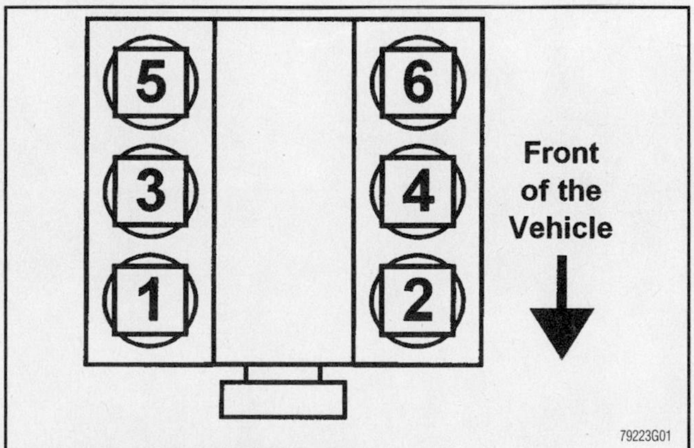

2.7L, 3.2L and 3.5L (VIN G) Engines
Firing Order: 1-2-3-4-5-6
Distributorless ignition system (One coil per cylinder)

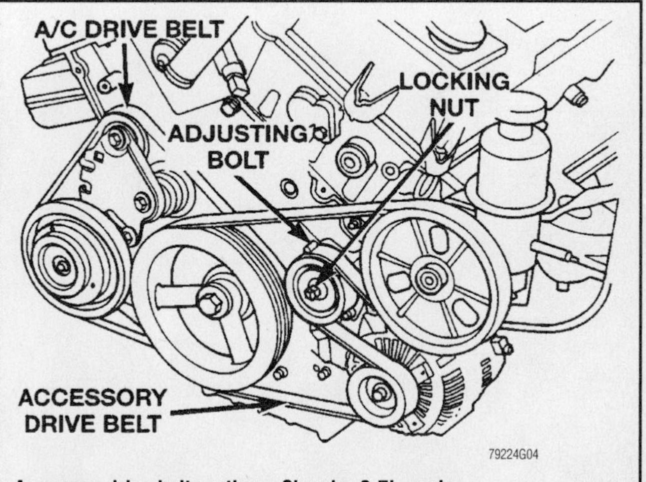

Accessory drive belt routing—Chrysler 2.7L engine

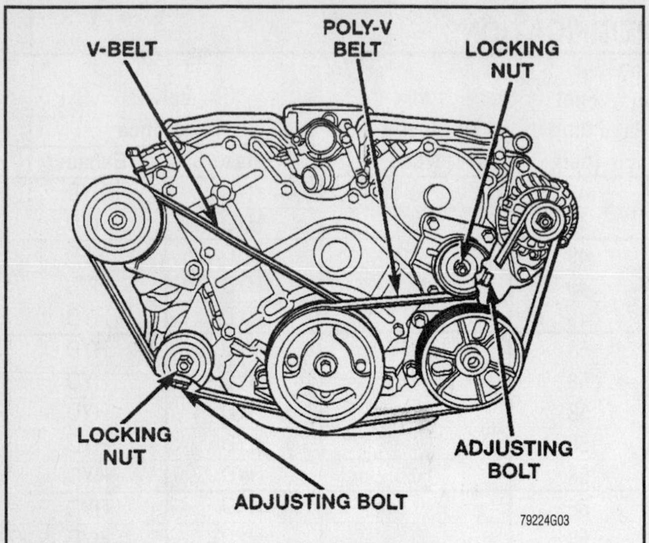

Accessory drive belt routing—Chrysler 3.2L engine

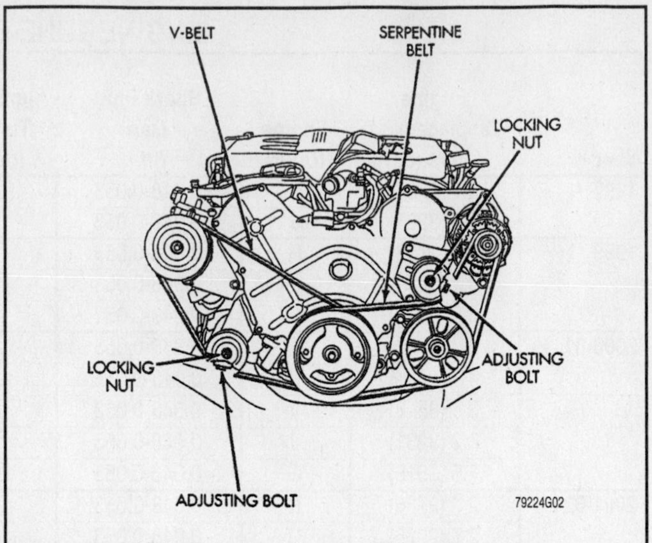

Accessory drive belt routing—Chrysler 3.5L engine

CAPACITIES

Year	Model	Engine Displacement Liters (cc)	Engine ID/VIN	Engine Oil with Filter (qts.)	Auto. Transmission (pts.) ①	Front Drive Axle (pts.)	Fuel Tank (gal.)	Cooling System (qts.)
1998	Concorde	2.7 (2736)	R	5.0	19.8	2.0	18.0	8.0
		3.2 (3231)	J	5.0	19.8	2.0	18.0	8.0
	Intrepid	2.7 (2736)	R	5.0	19.8	2.0	18.0	8.0
		3.2 (3231)	J	5.0	19.8	2.0	18.0	8.0
1999	300M	3.5 (3518)	G	5.0	19.8	2.0	18.0	9.4
	Concorde	2.7 (2736)	R	5.0	19.8	2.0	18.0	9.4
		3.2 (3231)	J	5.0	19.8	2.0	18.0	9.4
	Intrepid	2.7 (2736)	R	5.0	19.8	2.0	18.0	9.4
		3.2 (3231)	J	5.0	19.8	2.0	18.0	9.4
	LHS	3.5 (3518)	G	5.0	19.8	2.0	18.0	9.4
2000	300M	3.5 (3518)	G	5.0	19.8	2.0	18.0	9.4
	Concorde	2.7 (2736)	R	5.0	19.8	2.0	18.0	9.4
		2.7 (2736)	U	5.0	19.8	2.0	18.0	9.4
		3.5 (3518)	V	5.0	19.8	2.0	18.0	9.4
		3.2 (3231)	J	5.0	19.8	2.0	18.0	9.4
	Intrepid	2.7 (2736)	R	5.0	19.8	2.0	18.0	9.4
		2.7 (2736)	U	5.0	19.8	2.0	18.0	9.4
		3.5 (3518)	V	5.0	19.8	2.0	18.0	9.4
		3.2 (3231)	J	5.0	19.8	2.0	18.0	9.4
	LHS	3.5 (3518)	G	5.0	19.8	2.0	18.0	9.4
2001-02	300M	3.5 (3518)	G	5.0	19.8	2.0	18.0	9.4
	Concorde	2.7 (2736)	R	5.0	19.8	2.0	18.0	9.4
		2.7 (2736)	U	5.0	19.8	2.0	18.0	9.4
		3.5 (3518)	V	5.0	19.8	2.0	18.0	9.4
		3.2 (3231)	J	5.0	19.8	2.0	18.0	9.4
	Intrepid	2.7 (2736)	R	5.0	19.8	2.0	18.0	9.4
		2.7 (2736)	U	5.0	19.8	2.0	18.0	9.4
		3.5 (3518)	V	5.0	19.8	2.0	18.0	9.4
		3.2 (3231)	J	5.0	19.8	2.0	18.0	9.4
	LHS	3.5 (3518)	G	5.0	19.8	2.0	18.0	9.4

NOTE: All capacities are approximate. Add fluid gradually and ensure a proper fluid level is obtained.

① Overhaul fill capacity with torque converter empty
Estimated service fill: 9 pts.

93461C30

VALVE SPECIFICATIONS

Year	Engine Displacement Liters (cc)	Engine ID/VIN	Seat Angle (deg.)	Face Angle (deg.)	Spring Test Pressure (lbs. @ in.)	Spring Installed Height (in.)	Stem-to-Guide Clearance (in.)		Stem Diameter (in.)	
							Intake	Exhaust	Intake	Exhaust
1998	3.2 (3231)	J	45-45.5	44.5-45	①	1.496	0.0009-0.0026	0.0020-0.0037	0.2730-0.2737	0.2719-0.2726
	2.7 (2736)	R	44.5-45	45-45.5	②	1.496	0.0009-0.0026	0.0020-0.0033	0.2337-0.2344	0.2326-0.2333
1999	3.5 (3518)	G	45-45.5	44.5-45	①	1.496	0.0009-0.0026	0.0020-0.0037	0.2730-0.2737	0.2719-0.2726
	3.2 (3231)	J	45-45.5	44.5-45	①	1.496	0.0009-0.0026	0.0020-0.0037	0.2730-0.2737	0.2719-0.2726
	2.7 (2736)	R	44.5-45	45-45.5	②	1.496	0.0009-0.0026	0.0020-0.0033	0.2337-0.2344	0.2326-0.2333
2000	3.5 (3518)	G	45-45.5	44.5-45	③	1.496	0.0009-0.0026	0.0020-0.0037	0.2730-0.2737	0.2719-0.2726
	3.2 (3231)	J	45-45.5	44.5-45	③	1.496	0.0009-0.0026	0.0020-0.0037	0.2730-0.2737	0.2719-0.2726
	2.7 (2736)	R	44.5-45	45-45.5	②	1.496	0.0009-0.0026	0.0020-0.0033	0.2337-0.2344	0.2326-0.2333
	2.7 (2736)	U	44.5-45	45-45.5	②	1.496	0.0009-0.0026	0.0020-0.0033	0.2337-0.2344	0.2326-0.2333
	3.5 (3518)	V	45-45.5	44.5-45	③	1.496	0.0009-0.0026	0.0020-0.0037	0.2730-0.2737	0.2719-0.2726
2001-02	3.5 (3518)	G	45-45.5	44.5-45	③	1.496	0.0009-0.0026	0.0020-0.0037	0.2730-0.2737	0.2719-0.2726
	3.2 (3231)	J	45-45.5	44.5-45	③	1.496	0.0009-0.0026	0.0020-0.0037	0.2730-0.2737	0.2719-0.2726
	2.7 (2736)	R	44.5-45	45-45.5	②	1.496	0.0009-0.0026	0.0020-0.0033	0.2337-0.2344	0.2326-0.2333
	2.7 (2736)	U	44.5-45	45-45.5	②	1.496	0.0009-0.0026	0.0020-0.0033	0.2337-0.2344	0.2326-0.2333
	3.5 (3518)	V	45-45.5	44.5-45	③	1.496	0.0009-0.0026	0.0020-0.0037	0.2730-0.2737	0.2719-0.2726

① Intake: 69.5-80.5 lbs.@1.496 in. valve closed
Intake: 188.0-204.0 lbs.@1.1594 in. valve opened
Exhaust: 56-64 lbs.@1.4961 in. valve closed
Exhaust: 124-136 lbs.@1.239 in. valve opened

② 56-64 lbs.@ 1.496 in. valve closed
Intake: 147.9-162.1 lbs.@1.1417 in. valve opened
Exhaust: 138.0-150.8 lbs.@1.811 in. valve opened

③ Intake: 69.5-80.5 lbs.@1.496 in. valve closed
Intake: 188.0-204.0 lbs.@1.1594 in. valve opened
Exhaust: 71-79 lbs.@1.4961 in. valve closed
Exhaust: 130-144 lbs.@1.239 in. valve opened

93461C31

CRANKSHAFT AND CONNECTING ROD SPECIFICATIONS

All measurements are given in inches.

Year	Engine Displacement Liters (cc)	Engine ID/VIN	Crankshaft				Connecting Rod		
			Main Brg. Journal Dia.	Main Brg. Oil Clearance	Shaft End-play	Thrust on No.	Journal Diameter	Oil Clearance	Side Clearance
1998	2.7 (2736)	R	2.4997-2.5004	0.0014-0.0021	0.017 max.	3	2.1067-2.1060	0.0010-0.0026	0.0052-0.0150
	3.5 (3518)	J	2.5190-2.5200	0.0007-0.0028	0.0040-0.0120	2	2.2820-2.2830	0.0008-0.0034	0.0050-0.0157
1999	2.7 (2736)	R	2.4997-2.5004	0.0014-0.0021	0.017 max.	3	2.1067-2.1060	0.0010-0.0026	0.0052-0.0150
	3.2 (3231)	J	2.5190-2.5200	0.0007-0.0034	0.0040-0.0120	2	2.2830-2.2840	0.0008-0.0034	0.0050-0.0150
	3.5 (3518)	G	2.5190 2.5200	0.0004-0.0022	0.0040-0.0120	2	2.2830-2.2840	0.0008-0.0034	0.0050-0.0150
2000	2.7 (2736)	R	2.4997-2.5004	0.0014-0.0021	0.017 max.	3	2.1067-2.1060	0.0010-0.0026	0.0052-0.0150
	2.7 (2736)	U	2.4997-2.5004	0.0014-0.0021	0.017 max.	3	2.1067-2.1060	0.0010-0.0026	0.0052-0.0150
	3.5 (3518)	V	2.5190 2.5200	0.0004-0.0022	0.0040-0.0120	2	2.2830-2.2840	0.0008-0.0034	0.0050-0.0150
	3.2 (3231)	J	2.5190-2.5200	0.0007-0.0034	0.0040-0.0120	2	2.2830-2.2840	0.0008-0.0034	0.0050-0.0150
	3.5 (3518)	G	2.5190 2.5200	0.0004-0.0022	0.0040-0.0120	2	2.2830-2.2840	0.0008-0.0034	0.0050-0.0150
2001-02	2.7 (2736)	R	2.4997-2.5004	0.0014-0.0021	0.017 max.	3	2.1067-2.1060	0.0010-0.0026	0.0052-0.0150
	2.7 (2736)	U	2.4997-2.5004	0.0014-0.0021	0.017 max.	3	2.1067-2.1060	0.0010-0.0026	0.0052-0.0150
	3.5 (3518)	V	2.5190 2.5200	0.0004-0.0022	0.0040-0.0120	2	2.2830-2.2840	0.0008-0.0034	0.0050-0.0150
	3.2 (3231)	J	2.5190-2.5200	0.0007-0.0034	0.0040-0.0120	2	2.2830-2.2840	0.0008-0.0034	0.0050-0.0150
	3.5 (3518)	G	2.5190 2.5200	0.0004-0.0022	0.0040-0.0120	2	2.2830-2.2840	0.0008-0.0034	0.0050-0.0150

Max: Maximum

93461C32

Timing belt service is covered in Section 3 of this manual

PISTON AND RING SPECIFICATIONS

All measurements are given in inches.

Year	Engine Displacement Liters (cc)	Engine ID/VIN	Piston Clearance	Ring Gap Top Compression	Ring Gap Bottom Compression	Ring Gap Oil Control	Ring Side Clearance Top Compression	Ring Side Clearance Bottom Compression	Ring Side Clearance Oil Control
1998	2.7 (2736)	R	0.0001-0.0016	0.008-0.014	0.0146-0.0249	0.010-0.030	0.0016-0.0031	0.0016-0.0031	0.0025-0.0082
	3.5 (3518)	J	0.0003-0.0018	0.008-0.014	0.0087-0.0193	0.010-0.030	0.0016-0.0031	0.0016-0.0031	0.0015-0.0073
1999	2.7 (2736)	R	0.0001-0.0016	0.008-0.014	0.0146-0.0249	0.010-0.030	0.0016-0.0031	0.0016-0.0031	0.0025-0.0082
	3.2 (3231)	J	0.0003-0.0018	0.008-0.014	0.0087-0.0193	0.010-0.030	0.0016-0.0031	0.0016-0.0031	0.0015-0.0073
	3.5 (3518)	G	0.0003-0.0018	0.008-0.014	0.0087-0.0193	0.010-0.030	0.0016-0.0031	0.0016-0.0031	0.0015-0.0073
2000	2.7 (2736)	R	0.0001-0.0016	0.008-0.014	0.0146-0.0249	0.010-0.030	0.0016-0.0031	0.0016-0.0031	0.0025-0.0082
	2.7 (2736)	U	0.0003-0.0018	0.008-0.014	0.0146-0.0249	0.010-0.030	0.0016-0.0031	0.0016-0.0031	0.0025-0.0082
	3.5 (3518)	V	0.0003-0.0018	0.008-0.014	0.0087-0.0193	0.010-0.030	0.0016-0.0031	0.0016-0.0031	0.0015-0.0073
	3.2 (3231)	J	0.0003-0.0018	0.008-0.014	0.0087-0.0193	0.010-0.030	0.0016-0.0031	0.0016-0.0031	0.0015-0.0073
	3.5 (3518)	G	0.0003-0.0018	0.008-0.014	0.0087-0.0193	0.010-0.030	0.0016-0.0031	0.0016-0.0031	0.0015-0.0073
2001-02	2.7 (2736)	R	0.0001-0.0016	0.008-0.014	0.0146-0.0249	0.010-0.030	0.0016-0.0031	0.0016-0.0031	0.0025-0.0082
	2.7 (2736)	U	0.0003-0.0018	0.008-0.014	0.0146-0.0249	0.010-0.030	0.0016-0.0031	0.0016-0.0031	0.0025-0.0082
	3.5 (3518)	V	0.0003-0.0018	0.008-0.014	0.0087-0.0193	0.010-0.030	0.0016-0.0031	0.0016-0.0031	0.0015-0.0073
	3.2 (3231)	J	0.0003-0.0018	0.008-0.014	0.0087-0.0193	0.010-0.030	0.0016-0.0031	0.0016-0.0031	0.0015-0.0073
	3.5 (3518)	G	0.0003-0.0018	0.008-0.014	0.0087-0.0193	0.010-0.030	0.0016-0.0031	0.0016-0.0031	0.0015-0.0073

93461C33

TORQUE SPECIFICATIONS

All readings in ft. lbs.

Year	Engine Displacement Liters (cc)	Engine ID/VIN	Cylinder Head Bolts	Main Bearing Bolts	Rod Bearing Bolts	Crankshaft Damper Bolts	Flywheel Bolts	Manifold		Spark Plugs	Lug Nuts
								Intake	Exhaust		
1998	3.2 (3231)	J	①	②	③	75	75	④	200⑤	20	100
	2.7 (2736)	R	⑥	②	⑦	125	75	105⑤	200⑤	15	100
1999	3.5 (3518)	G	①	②	③	70	75	106⑤	200⑤	20	100
	3.2 (3231)	J	①	②	③	70	75	107⑤	200⑤	20	100
	2.7 (2736)	R	⑥	②	⑦	125	75	108⑤	200⑤	15	100
2000	3.5 (3518)	G	①	②	③	70	70	109⑤	200⑤	20	100
	3.2 (3231)	J	①	②	③	70	70	110⑤	200⑤	20	100
	2.7 (2736)	R	⑥	②	⑦	125	70	111⑤	200⑤	15	100
	2.7 (2736)	U	⑥	②	⑦	125	70	112⑤	200⑤	15	100
	3.5 (3518)	V	①	②	③	70	70	113⑤	200⑤	20	100
2001-02	3.5 (3518)	G	①	②	③	70	70	109⑤	200⑤	20	100
	3.2 (3231)	J	①	②	③	70	70	110⑤	200⑤	20	100
	2.7 (2736)	R	⑥	②	⑦	125	70	111⑤	200⑥	15	100
	2.7 (2736)	U	⑥	②	⑦	125	70	112⑤	200⑥	15	100
	3.5 (3518)	V	①	②	③	70	70	113⑤	200⑤	20	100

① Step 1: 45 ft. lbs.
Step 2: 65 ft. lbs.
Step 3: 65 ft. lbs.
Step 4: Plus 1/4 turn
Final torque should be over 90 ft. lbs.

② Main cap inside bolts: 15 ft. lbs. plus 1/4 turn
Main cap outside bolts: 20 ft. lbs. plus 1/4 turn
Main cap tie bolts: 250 inch lbs.

③ Step 1: 40 ft. lbs.
Step 2: Plus 1/4 turn

④ M8 bolts: 250 inch lbs.
M6 bolts: 105 inch lbs.

⑤ Inch lbs.

⑥ Step 1: 35 ft. lbs.
Step 2: 55 ft. lbs.
Step 3: 55 ft. lbs.
Step 4: Plus 90 degrees
Step 5: M8 bolts (3 front) 250 inch lbs.

⑦ Step 1: 20 ft. lbs.
Step 2: Plus 1/4 turn

93461C34

Heater Core replacement is covered in Section 2 of this manual

BRAKE SPECIFICATIONS

All measurements in inches unless noted

Year	Model		Brake Disc Original Thickness	Brake Disc Minimum Thickness	Brake Disc Maximum Runout	Brake Drum Diameter Original Inside Diameter	Brake Drum Diameter Max. Wear Limit	Brake Drum Diameter Maximum Machine Diameter	Minimum Lining Thickness	Brake Caliper Mounting Bolts (ft. lbs.)
1998	Concorde	F	1.024	0.960	0.0035	—	—	—	0.250	17
		R	0.468	0.409	0.0035	—	—	—	0.280	17
	Intrepid	F	1.024	0.960	0.0035	—	—	—	0.250	17
		R	0.468	0.409	0.0035	—	—	—	0.280	17
1999	300M	F	1.024	0.960	0.0035	—	—	—	0.250	17
		R	0.468	0.409	0.0035	—	—	—	0.280	17
	Concorde	F	1.024	0.960	0.0035	—	—	—	0.250	17
		R	0.468	0.409	0.0035	—	—	—	0.280	17
	Intrepid	F	1.024	0.960	0.0035	—	—	—	0.250	17
		R	0.468	0.409	0.0035	—	—	—	0.280	17
	LHS	F	1.024	0.960	0.0035	—	—	—	0.250	17
		R	0.468	0.409	0.0035	—	—	—	0.280	17
2000	300M	F	1.024	0.960	0.0035	—	—	—	0.250	17
		R	0.468	0.409	0.0035	—	—	—	0.280	17
	Concorde	F	1.024	0.960	0.0035	—	—	—	0.250	17
		R	0.468	0.409	0.0035	—	—	—	0.280	17
	Intrepid	F	1.024	0.960	0.0035	—	—	—	0.250	17
		R	0.468	0.409	0.0035	—	—	—	0.280	17
	LHS	F	1.024	0.960	0.0035	—	—	—	0.250	17
		R	0.468	0.409	0.0035	—	—	—	0.280	17
2001-02	300M	F	1.024	0.960	0.0035	—	—	—	0.250	17
		R	0.468	0.409	0.0035	—	—	—	0.280	17
	Concorde	F	1.024	0.960	0.0035	—	—	—	0.250	17
		R	0.468	0.409	0.0035	—	—	—	0.280	17
	Intrepid	F	1.024	0.960	0.0035	—	—	—	0.250	17
		R	0.468	0.409	0.0035	—	—	—	0.280	17
	LHS	F	1.024	0.960	0.0035	—	—	—	0.250	17
		R	0.468	0.409	0.0035	—	—	—	0.280	17

F: Front

R: Rear

93461C39

WHEEL ALIGNMENT

Year	Model		Caster Range (+/-Deg.)	Caster Preferred Setting (Deg.)	Camber Range (+/-Deg.)	Camber Preferred Setting (Deg.)	Toe-in (in.)	Steering Axis Inclination (Deg.)
1998	300M	F	+1.00	+3.00	+0.60	0	0.10 +/- 0.20	—
		R	—	—	+0.50	-0.20	0.10 +/- 0.30	—
	Concorde	F	+1.00	+3.00	+0.56	0	0.09 +/- 0.10	14.00
		R	—	—	+0.50	-0.20	0.09 +/- 0.15	—
	LHS	F	+1.00	+3.00	+0.60	0	0.09 +/- 0.10	14.0
		R	—	—	+0.50	-0.20	0.05 +/- 0.15	—
	Intrepid	F	+1.00	+3.00	+0.56	0	0.05 +/- 0.10	14.00
		R	—	—	+0.50	-0.20	0.05 +/- 0.15	—
1999	300M	F	+1.00	+3.00	+0.60	0	0.10 +/- 0.20	—
		R	—	—	+0.50	-0.20	0.10 +/- 0.30	—
	Concorde	F	+1.00	+3.00	+0.56	0	0.09 +/- 0.10	14.00
		R	—	—	+0.50	-0.20	0.09 +/- 0.15	—
	LHS	F	+1.00	+3.00	+0.60	0	0.09 +/- 0.10	14.0
		R	—	—	+0.50	-0.20	0.05 +/- 0.15	—
	Intrepid	F	+1.00	+3.00	+0.56	0	0.05 +/- 0.10	14.00
		R	—	—	+0.50	-0.20	0.05 +/- 0.15	—
2000	300M	F	+1.00	+3.00	+0.60	0	0.10 +/- 0.20	—
		R	—	—	+0.50	-0.20	0.10 +/- 0.30	—
	Concorde	F	+1.00	+3.00	+0.56	0	0.09 +/- 0.10	14.00
		R	—	—	+0.50	-0.20	0.09 +/- 0.15	—
	LHS	F	+1.00	+3.00	+0.60	0	0.09 +/- 0.10	14.0
		R	—	—	+0.50	-0.20	0.05 +/- 0.15	—
	Intrepid	F	+1.00	+3.00	+0.56	0	0.05 +/- 0.10	14.00
		R	—	—	+0.50	-0.20	0.05 +/- 0.15	—
2001-02	300M	F	+1.00	+3.00	+0.60	0	0.10 +/- 0.20	—
		R	—	—	+0.50	-0.20	0.10 +/- 0.30	—
	Concorde	F	+1.00	+3.00	+0.56	0	0.09 +/- 0.10	14.00
		R	—	—	+0.50	-0.20	0.09 +/- 0.15	—
	LHS	F	+1.00	+3.00	+0.60	0	0.09 +/- 0.10	14.0
		R	—	—	+0.50	-0.20	0.05 +/- 0.15	—
	Intrepid	F	+1.00	+3.00	+0.56	0	0.05 +/- 0.10	14.00
		R	—	—	+0.50	-0.20	0.05 +/- 0.15	—

93461C40

TIRE, WHEEL AND BALL JOINT SPECIFICATIONS

Year	Model	OEM Tires		Tire Pressures (psi)		Wheel	Ball Joint
		Standard	Optional	Front	Rear	Size	Inspection
1998	Concorde	P225/60R16	None	30	31	7-JJ	①
	LHS	P225/60R16	None	32	32	7-JJ	①
1999	300M	P225/55/R17	None	30	30	7-JJ	①
	Concorde	P205/70R15	P225/60R16	32	32	6-JJ	①
	LHS	P225/55/R17	None	30	30	7-JJ	①
2000	300M	P225/55/R17	None	30	30	7-JJ	①
	Concorde	P205/70R15	P225/60R16	32	32	6-JJ	①
	LHS	P225/55/R17	None	30	30	7-JJ	①
2001-02	300M	P225/55/R17	None	30	30	7-JJ	①
	Concorde	P205/70R15	P225/60R16	32	32	6-JJ	①
	LHS	P225/55/R17	None	30	30	7-JJ	①

OEM: Original Equipment Manufacturer

PSI: Pounds Per Square Inch

① Do not lift car. Grasp the grease fitting and attempt to move or rotate. Replace if any movement is found.

93461C41

SCHEDULED MAINTENANCE INTERVALS
Chrysler—300M, Concorde, LHS Dodge—Intrepid

TO BE SERVICED	TYPE OF SERVICE	VEHICLE MILEAGE INTERVAL (x1000)												
		7.5	15	22.5	30	37.5	45	52.5	60	67.5	75	82.5	90	97.5
Engine oil & filter	R	✓	✓	✓	✓	✓	✓	✓	✓	✓	✓	✓	✓	✓
Exhaust system	S/I	✓	✓	✓	✓	✓	✓	✓	✓	✓	✓	✓	✓	✓
Brake hoses	S/I	✓	✓	✓	✓	✓	✓	✓	✓	✓	✓	✓	✓	✓
CV joints & front suspension components	S/I	✓	✓	✓	✓	✓	✓	✓	✓	✓	✓	✓	✓	✓
Rotate tires	S/I	✓	✓	✓	✓	✓	✓	✓	✓	✓	✓	✓	✓	✓
Coolant level, hoses & clamps	S/I	✓	✓	✓	✓	✓	✓	✓	✓	✓	✓	✓	✓	✓
Accessory drive belts	S/I		✓		✓		✓		✓		✓		✓	
Brake linings	S/I		✓	✓				✓	✓				✓	
Spark plugs	R				✓				✓				✓	
Air filter element	R				✓				✓				✓	
Lubricate steering linkage & tie rod ends	S/I				✓				✓				✓	
Engine coolant	R						✓				✓			
PCV valve	S/I								✓				✓	
Ignition cables	R								✓					
Camshaft timing belt	R								✓					

R: Replace S/I: Service or Inspect

FREQUENT OPERATION MAINTENANCE (SEVERE SERVICE)

If a vehicle is operated under any of the following conditions it is considered severe service:

- Extremely dusty areas

- 50% or more of the vehicle operation is in 32°C (90°F) or higher temperatures, or constant operation in temperatures below 0°C (32°F)

Prolonged idling (vehicle operation in stop and go traffic)

- Frequent short running periods (engine does not warm to normal operating temperatures)

- Police, taxi, delivery usage or trailer towing usage

CV joints & front suspension components: check every 3000 miles

Oil & oil filter change: change every 3000 miles

Rotate tires: every 3000 miles

Brake linings: check every 9000 miles

Air filter element: change every 15,000 miles

Automatic transaxle fluid: change every 15,000 miles

Differential fluid: change every 15,000 miles

Tie rod ends & steering linkage: lubricate every 15,000 miles

PCV valve: check every 30,000 miles

93461C42

SCHEDULED MAINTENANCE INTERVALS
DAIMLERCHRYSLER CORPORATION
CHRYSLER 300M, CONCORDE, LHS
DODGE INTREPID, EAGLE VISION

The following should be used as a guide when determining the amount of work required for a particular service. In estimating how long a particular Scheduled Maintenance Service should take, please observe the following:

- Labor Time is time based on field research and data supplied by the vehicle manufacturer.
- Labor time operations are given in hours and tenths of an hour.
- All labor operations are to be used as a guide.

Mechanic Skill Level Codes:
(A) PRECISION: Highly skilled with multiple certification.
(B) GENERAL: Normally skilled with certification.
(C) MAINTENANCE: Semi-skilled working on certification.

	LABOR TIME
7500 Mile Service (C)	
All Models	.7
15000 Mile Service (C)	
All Models	.9
22500 Mile Service (C)	
All Models	.7
30000 Mile Service (B)	
All Models	1.5
37500 Mile Service (C)	
All Models	.7
45000 Mile Service (C)	
All Models	.9
52500 Mile Service (C)	
All Models	.9
60000 Mile Service (B)	
All Models	2.4
Add for timing belt replacement	
67500 Mile Service (C)	
All Models	.7
75000 Mile Service (C)	
All Models	.9
82500 Mile Service (C)	
All Models	.7
90000 Mile Service (B)	
All Models	1.6
97500 Mile Service (C)	
All Models	.7

93461C43

CHRYSLER CORP.
Dodge/Plymouth Neon

ENGINE AND VEHICLE IDENTIFICATION

Engine						
Code ①	Liters	Cu. In. (cc)	Cyl.	Fuel Sys.	Engine Type	Eng. Mfg.
C	2.0	122 (1996)	4	MFI	SOHC	Chrysler
Y	2.0	122 (1996)	4	MFI	DOHC	Chrysler

MFI: Multi-point Fuel Injection

SOHC: Single Overhead Camshaft

DOHC: Double Overhead Camshaft

① 8th position of VIN

② 10th position of VIN

Model Year	
Code ②	Year
W	1998
X	1999
Y	2000
1	2001
2	2002

93461C44

GENERAL ENGINE SPECIFICATIONS

All measurements are given in inches.

Year	Model	Engine Displacement Liters (cc)	Engine Series (ID/VIN)	Fuel System	Net Horsepower @ rpm	Net Torque @ rpm (ft. lbs.)	Bore x Stroke (in.)	Com-pression Ratio	Oil Pressure @ rpm
1998	Neon	2.0 (1996)	C	MFI	132@6000	129@5000	3.44x3.26	9.8:1	25-80@3000
	Neon	2.0 (1996)	Y	MFI	150@4400	133@5500	3.44x3.26	9.6:1	25-80@3000
1999	Neon	2.0 (1996)	C	MFI	132@6000	129@5000	3.44x3.26	9.8:1	25-80@3000
	Neon	2.0 (1996)	Y	MFI	150@4400	133@5500	3.44x3.26	9.6:1	25-80@3000
2000	Neon	2.0 (1996)	C	MFI	132@6000	129@5000	3.44x3.26	9.8:1	25-80@3000
2001-02	Neon	2.0 (1996)	C	MFI	132@6000	129@5000	3.44x3.26	9.8:1	25-80@3000

NA: Not available

MFI: Multi-point Fuel Injection

93461C45

For Accessory Drive Belt illustrations, see Section 1 of this manual

ENGINE TUNE-UP SPECIFICATIONS

Year	Engine Displacement Liters (cc)	Engine ID/VIN	Spark Plugs Gap (in.)	Ignition Timing (deg.) MT	Ignition Timing (deg.) AT	Fuel Pump (psi)	Idle Speed (rpm) MT	Idle Speed (rpm) AT	Valve Clearance In.	Valve Clearance Ex.
1998	2.0 (1996)	C	0.035	①	①	48	②	②	HYD	HYD
	2.0 (1996)	Y	0.035	①	①	48	②	②	HYD	HYD
1999	2.0 (1996)	C	0.035	①	①	48	②	②	HYD	HYD
	2.0 (1996)	Y	0.035	①	①	48	②	②	HYD	HYD
2000	2.0 (1996)	C	0.035	①	①	48	②	②	HYD	HYD
2001-02	2.0 (1996)	C	0.035	①	①	48	②	②	HYD	HYD

NOTE: The Vehicle Emission Control Information label often reflects specification changes made during production. The label figures must be used if they differ from those in this chart.

HYD : Hydraulic

① Refer to the Vehicle Emission Control Information label for correct timing specifications with a range of +/- 2 degrees. Ignition timing cannot be adjusted. Base engine timing is set at TDC during assembly.

② Refer to Vehicle Emissions Control Information label for proper specification

93461C46

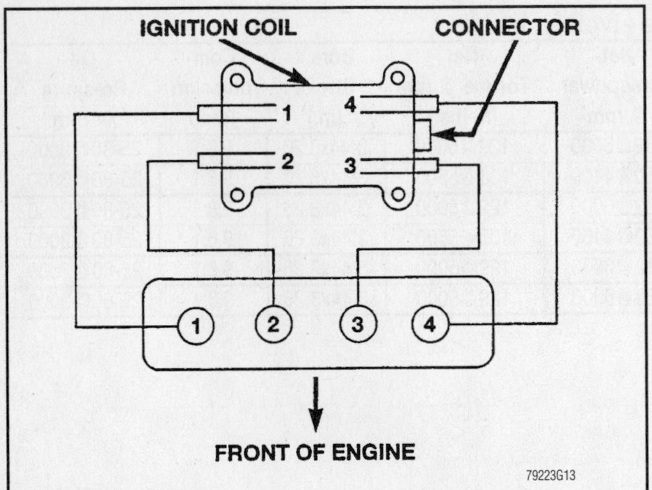

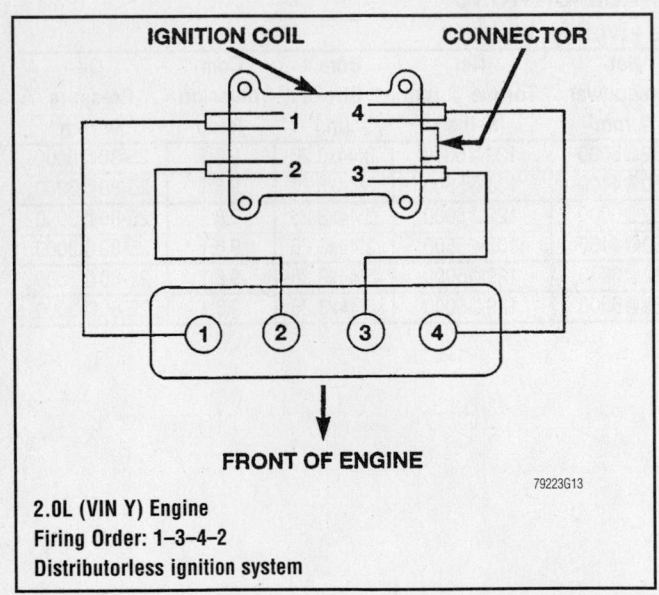

2.0L (VIN Y) Engine
Firing Order: 1–3–4–2
Distributorless ignition system

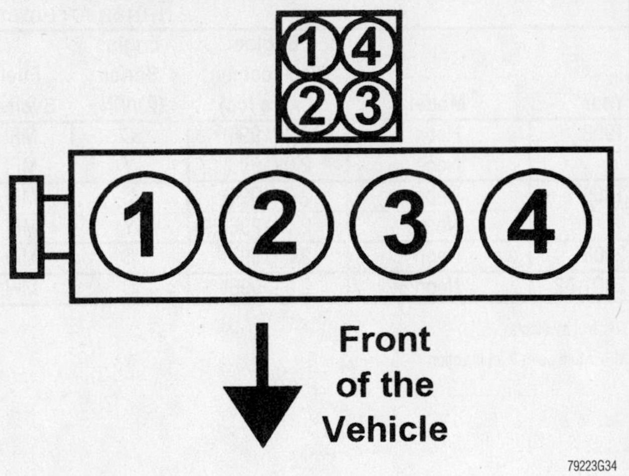

2.0L (VIN C) Engine
Firing Order: 1–3–4–2
Distributorless ignition system

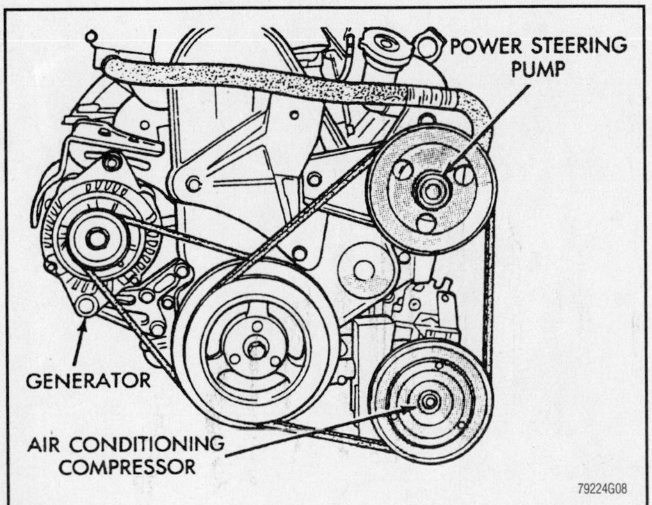

Accessory drive belt routing—Chrysler Neon 2.0L (VIN C,Y) engine

CAPACITIES

Year	Model	Engine Displacement Liters (cc)	Engine ID/VIN	Engine Oil with Filter (qts.)	Transmission (pts.) 5-Spd	Transmission (pts.) Auto.	Fuel Tank (gal.)	Cooling System (qts.)
1998	Neon	2.0 (1996)	C	4.5	4.0-4.6	8.0	12.5	7.4
	Neon	2.0 (1996)	Y	4.5	4.0-4.6	8.0	12.5	7.4
1999	Neon	2.0 (1996)	C	4.5	4.0-4.6	8.0	12.5	7.4
	Neon	2.0 (1996)	Y	4.5	4.0-4.6	8.0	12.5	7.4
2000	Neon	2.0 (1996)	C	4.5	4.0-4.6	8.0	12.5	6.5
2001-02	Neon	2.0 (1996)	C	4.5	4.0-4.6	8.0	12.5	6.5

NOTE: All capacities are approximate. Add fluid gradually and ensure a proper fluid level is obtained.

93461C47

VALVE SPECIFICATIONS

Year	Engine Displacement Liters (cc)	Engine ID/VIN	Seat Angle (deg.)	Face Angle (deg.)	Spring Test Pressure (lbs. @ in.)	Spring Installed Height (in.)	Stem-to-Guide Clearance (in.) Intake	Stem-to-Guide Clearance (in.) Exhaust	Stem Diameter (in.) Intake	Stem Diameter (in.) Exhaust
1998	2.0 (1996)	C	45	45-45.5	70@1.57	1.580	0.0018-0.0025	0.0029-0.0037	0.2340	0.2330
	2.0 (1996)	Y	44.5-45	45-45.5	55-60@1.49	1.490	0.0018-0.0025	0.0029-0.0037	0.2340	0.2330
1999	2.0 (1996)	C	45	45-45.5	70@1.57	1.580	0.0018-0.0025	0.0029-0.0037	0.2340	0.2330
	2.0 (1996)	Y	44.5-45	45-45.5	55-60@1.49	1.490	0.0018-0.0025	0.0029-0.0037	0.2340	0.2330
2000	2.0 (1996)	C	45	45-45.5	70@1.57	1.580	0.0018-0.0025	0.0029-0.0037	0.2340	0.2330
2001-02	2.0 (1996)	C	45	45-45.5	70@1.57	1.580	0.0018-0.0025	0.0029-0.0037	0.2340	0.2330

93461C48

For Tire, Wheel and Ball Joint specifications, see Section 1 of this manual

CRANKSHAFT AND CONNECTING ROD SPECIFICATIONS

All measurements are given in inches.

Year	Engine Displacement Liters (cc)	Engine ID/VIN	Crankshaft Main Brg. Journal Dia.	Crankshaft Main Brg. Oil Clearance	Crankshaft Shaft End-play	Crankshaft Thrust on No.	Connecting Rod Journal Diameter	Connecting Rod Oil Clearance	Connecting Rod Side Clearance
1998	2.0 (1996)	C	2.0469-2.0475	0.0008-0.0024	0.0035-0.0094	3	1.8894-1.8900	0.0010-0.0023	0.0050-0.0150
	2.0 (1996)	Y	2.0469-2.0475	0.0009-0.0024	0.0035-0.0094	3	1.8894-1.8900	0.0010-0.0023	0.0051-0.0150
1999	2.0 (1996)	C	2.0469-2.0475	0.0008-0.0024	0.0035-0.0094	3	1.8894-1.8900	0.0010-0.0023	0.0050-0.0150
	2.0 (1996)	Y	2.0469-2.0475	0.0009-0.0024	0.0035-0.0094	3	1.8894-1.8900	0.0010-0.0023	0.0051-0.0150
2000	2.0 (1996)	C	2.0469-2.0475	0.0008-0.0024	0.0035-0.0094	3	1.8894-1.8900	0.0010-0.0023	0.0050-0.0150
2001-02	2.0 (1996)	C	2.0469-2.0475	0.0008-0.0024	0.0035-0.0094	3	1.8894-1.8900	0.0010-0.0023	0.0050-0.0150

93461C49

PISTON AND RING SPECIFICATIONS

All measurements are given in inches.

Year	Engine Displacement Liters (cc)	Engine ID/VIN	Piston Clearance	Ring Gap Top Compression	Ring Gap Bottom Compression	Ring Gap Oil Control	Ring Side Clearance Top Compression	Ring Side Clearance Bottom Compression	Ring Side Clearance Oil Control
1998	2.0 (1996)	C	0.0004-0.0017	0.0090-0.0200	0.0190-0.0310	0.009-0.026	0.0010-0.0026	0.0010-0.0026	0.0002-0.0070
	2.0 (1996)	Y	0.0007-0.0020	0.0090-0.0200	0.0190-0.0310	0.009-0.026	0.0010-0.0026	0.0010-0.0026	0.0002-0.0070
1999	2.0 (1996)	C	0.0004-0.0017	0.0090-0.0200	0.0190-0.0310	0.009-0.026	0.0010-0.0026	0.0010-0.0026	0.0002-0.0070
	2.0 (1996)	Y	0.0007-0.0020	0.0090-0.0200	0.0190-0.0310	0.009-0.026	0.0010-0.0026	0.0010-0.0026	0.0002-0.0070
2000	2.0 (1996)	C	0.0003-0.0006	0.0090-0.0200	0.0190-0.0310	0.009-0.026	0.0010-0.0026	0.0010-0.0026	0.0002-0.0070
2001-02	2.0 (1996)	C	0.0003-0.0006	0.0090-0.0200	0.0190-0.0310	0.009-0.026	0.0010-0.0026	0.0010-0.0026	0.0002-0.0070

93461C50

TORQUE SPECIFICATIONS

All readings in ft. lbs.

Year	Engine Displacement Liters (cc)	Engine ID/VIN	Cylinder Head Bolts	Main Bearing Bolts	Rod Bearing Bolts	Crankshaft Damper Bolts	Flywheel Bolts	Manifold Intake	Manifold Exhaust	Spark Plugs	Lug Nut
1998	2.0 (1996)	C	①	②	③	105	70	8.5	17	20	100
	2.0 (1996)	Y	④	②	③	105	70	21	17	20	100
1999	2.0 (1996)	C	①	②	③	105	70	8.5	17	20	100
	2.0 (1996)	Y	④	②	③	105	70	21	17	20	100
2000	2.0 (1996)	C	①	②	③	100	70	8.5	17	20	95
2001-02	2.0 (1996)	C	①	②	③	100	70	8.5	17	20	95

① Step 1: 25 ft. lbs.
Step 2: 50 ft. lbs.
Step 3: 50 ft. lbs.
Step 4: plus 90 degrees

② M8: 22 ft. lbs.
M11:60 ft. lbs.

③ Step 1: 20 ft. lbs.
Step 2: plus 90 degrees

④ Step 1:
Bolts 1-6: 25 ft. lbs.
Bolts 7-10: 20 ft. lbs.
Step 2:
Bolts 1-6: 50 ft. lbs.
Bolts 7-10: 20 ft. lbs.
Step 3:
Bolts 1-6: 50 ft. lbs.
Bolts 7-10: 20 ft. lbs.
Step 4: plus 90 degrees

93461C51

BRAKE SPECIFICATIONS

All measurements in inches unless noted

Year	Model		Brake Disc Original Thickness	Brake Disc Minimum Thickness	Brake Disc Maximum Runout	Brake Drum Diameter Original Inside Diameter	Brake Drum Diameter Max. Wear Limit	Brake Drum Diameter Maximum Machine Diameter	Minimum Lining Thickness	Brake Caliper Bracket Bolts (ft. lbs.)	Brake Caliper Mounting Bolts (ft. lbs.)
1998	Neon	F	0.792	0.724	0.005	—	—	—	0.300	55	16
		R	0.364	0.285	0.005	7.88	NA	①	②	55	16
1999	Neon	F	0.792	0.724	0.005	—	—	—	0.300	55	16
		R	0.364	0.285	0.005	7.88	NA	①	②	55	16
2000	Neon	F	0.866	0.803	0.003	—	—	—	0.300	55	16
		R	0.364	0.285	0.003	7.88	NA	①	②	55	16
2001-02	Neon	F	0.866	0.803	0.003	—	—	—	0.300	55	16
		R	0.364	0.285	0.003	7.88	NA	①	②	55	16

NA: Not Available
F: Front
R: Rear
① Stamped on the outer edge of drum
② Disc brake pad total thickness: 0.281
Brake shoe lining thickness: 0.0625

93461C52

For Wheel Alignment specifications, see Section 1 of this manual

WHEEL ALIGNMENT

Year	Model		Caster Range (+/-Deg.)	Caster Preferred Setting (Deg.)	Camber Range (+/-Deg.)	Camber Preferred Setting (Deg.)	Toe-in (in.)	Steering Axis Inclination (Deg.)
1998	Neon	F	+1.00	+2.60	+0.40	0	0.05 +/- 0.10	13.19
		R	—	—	+0.40	-0.25	0.15 +/- 0.10	—
1999	Neon	F	+1.00	+2.60	+0.40	0	0.05 +/- 0.10	13.19
		R	—	—	+0.40	-0.25	0.15 +/- 0.10	—
2000	Neon	F	+1.00	+2.60	+0.40	0	0.05 +/- 0.10	13.19
		R	—	—	+0.40	-0.25	0.15 +/- 0.10	—
2001-02	Neon	F	+1.00	+2.60	+0.40	0	0.05 +/- 0.10	13.19
		R	—	—	+0.40	-0.25	0.15 +/- 0.10	—

93461C53

TIRE, WHEEL AND BALL JOINT SPECIFICATIONS

Year	Model	OEM Tires Standard	OEM Tires Optional	Tire Pressures (psi) Front	Tire Pressures (psi) Rear	Wheel Size	Ball Joint Inspection
1998	Neon, base	P175/70R14	P185/65R14	32	32	6-JJ	①
	Neon 2dr Competition	P185/65HR14	None	32	32	6J	①
	Neon 4dr Competition	P175/65HR14	None	32	32	6J	①
	Neon Sport, RT	P185/65R14	None	32	32	6J	①
1999	Neon, base	P185/65R14	None	30	30	5.5-JJ	①
	Neon RT, Competition	P185/65HR14	None	32	32	6-JJ	①
2000	Neon, base	P185/65R14	P185/60R15	30	30	5.5-JJ	①
	Neon ES	P185/60R15	None	32	32	6-JJ	①
2001-02	Neon, base	P185/65R14	P185/60R15	30	30	5.5-JJ	①
	Neon ES	P185/60R15	None	32	32	6-JJ	①

OEM: Original Equipment Manufacturer

PSI: Pounds Per Square Inch

① Replace if any measurable movement is found.

93461C54

SCHEDULED MAINTENANCE INTERVALS

Dodge—Neon

TO BE SERVICED	TYPE OF SERVICE	VEHICLE MILEAGE INTERVAL (x1000)												
		7.5	15	22.5	30	37.5	45	52.5	60	67.5	75	82.5	90	97.5
Engine oil & filter	R	✓	✓	✓	✓	✓	✓	✓	✓	✓	✓	✓	✓	✓
Brake hoses	S/I	✓	✓	✓	✓	✓	✓	✓	✓	✓	✓	✓	✓	✓
Coolant level, hoses & clamps	S/I	✓	✓	✓	✓	✓	✓	✓	✓	✓	✓	✓	✓	✓
CV joints & front suspension components	S/I	✓	✓	✓	✓	✓	✓	✓	✓	✓	✓	✓	✓	✓
Exhaust system	S/I	✓	✓	✓	✓	✓	✓	✓	✓	✓	✓	✓	✓	✓
Manual transaxle oil	S/I	✓	✓	✓	✓	✓	✓	✓	✓	✓	✓	✓	✓	✓
Rotate tires	S/I	✓	✓	✓	✓	✓	✓	✓	✓	✓	✓	✓	✓	✓
Accessory drive belts	S/I		✓		✓		✓		✓		✓		✓	
Brake linings	S/I			✓			✓			✓			✓	
Air filter element	R				✓				✓				✓	
Spark plugs	R				✓				✓				✓	
Lubricate ball joints	S/I				✓				✓				✓	
Engine coolant	R						✓				✓			
PCV valve	S/I								✓				✓	
Ignition cables	R								✓					
Camshaft timing belt ①	R													

R: Replace S/I: Service or Inspect

① Camshaft timing belt: replace at 105,000 miles

FREQUENT OPERATION MAINTENANCE (SEVERE SERVICE)

If a vehicle is operated under any of the following conditions it is considered severe service:

- Extremely dusty areas

- 50% or more of the vehicle operation is in 32°C (90°F) or higher temperatures, or constant operation in temperatures below 0°C (32°F)

Prolonged idling (vehicle operation in stop and go traffic)

- Frequent short running periods (engine does not warm to normal operating temperatures)

- Police, taxi, delivery usage or trailer towing usage

Oil & oil filter change: change every 3000 miles

Rotate tires every 6000 miles

Brake linings: inspect every 12,000 miles

Air filter element: service or inspect every 15,000 miles

Automatic transaxle: change fluid & adjust bands every 15,000 miles

Manual transaxle fluid: replace every 15,000 miles

Engine coolant: replace at 36,000 miles and every 30,000 miles thereafter

93461C55

SCHEDULED MAINTENANCE INTERVALS
DAIMLERCHRYSLER CORPORATION
DODGE NEON
PLYMOUTH NEON

The following should be used as a guide when determining the amount of work required for a particular service. In estimating how long a particular Scheduled Maintenance Service should take, please observe the following:

- Labor Time is time based on field research and data supplied by the vehicle manufacturer.
- Labor time operations are given in hours and tenths of an hour.
- All labor operations are to be used as a guide.

Mechanic Skill Level Codes:
(A) PRECISION: Highly skilled with multiple certification.
(B) GENERAL: Normally skilled with certification.
(C) MAINTENANCE: Semi-skilled working on certification.

	LABOR TIME
7500 Mile Service (C)	
All Models	.7
15000 Mile Service (C)	
All Models	.9
22500 Mile Service (C)	
All Models	.8
30000 Mile Service (C)	
All Models	1.5
37500 Mile Service (C)	
All Models	.7
45000 Mile Service (C)	
All Models	1.0
52500 Mile Service (C)	
All Models	1.4
60000 Mile Service (B)	
All Models	1.9
67500 Mile Service (C)	
All Models	.8
75000 Mile Service (C)	
All Models	1.0
82500 Mile Service (C)	
All Models	.9
90000 Mile Service (B)	
All Models	1.9
97500 Mile Service (C)	
All Models	.7

93461C56

CHRYSLER CORP.
Chrysler PT Cruiser

ENGINE AND VEHICLE IDENTIFICATION

Code ①	Liters (cc)	Cu. In.	Engine Cyl.	Fuel Sys.	Engine Type	Eng. Mfg.
B	2.4 (2429)	148	L4	SMFI	DOHC	Chrysler

Model Year	
Code ②	Year
1	2001
2	2002

DOHC: Double Overhead Camshaft

① 8th position of VIN

② 10th position of VIN

93461C57

GENERAL ENGINE SPECIFICATIONS

Year	Model	Engine Displacement Liters (cc)	Engine Series (ID/VIN)	Fuel System	Net Horsepower @ rpm	Net Torque @ rpm (ft. lbs.)	Bore x Stroke (in.)	Com-pression Ratio	Oil Pressure @ rpm
2001	PT Cruiser	2.4 (2429)	B	SMFI	150@5200	164@4000	3.44x3.98	9.4:1	25-80@3000
2002	PT Cruiser	2.4 (2429)	B	SMFI	150@5200	164@4000	3.44x3.98	9.4:1	25-80@3000

SMFI: Sequential Multi-port Fuel Injection

93461C58

For Tune-up, Capacities and Firing orders, see Section 1 of this manual

ENGINE TUNE-UP SPECIFICATIONS

Year	Engine Displacement Liters (cc)	Engine ID/VIN	Spark Plug Gap (in.)	Ignition Timing (deg.)	Fuel Pump (psi)	Idle Speed (rpm)	Valve Clearance	
							In.	Ex.
2001	2.4 (2429)	B	0.048-0.053	①	49	②	HYD	HYD
2002	2.4 (2429)	B	0.048-0.053	①	49	②	HYD	HYD

NOTE: The Vehicle Emission Control Information label often reflects specification changes made during production. The label figures must be used if they differ from those in this chart.

HYD: Hydraulic

① Ignition timing is regulated by the Powertrain Control Module (PCM), and cannot be adjusted.

② Idle speed is controled by the Powertrain Control Module (PCM), and cannot be adjusted.

93461C59

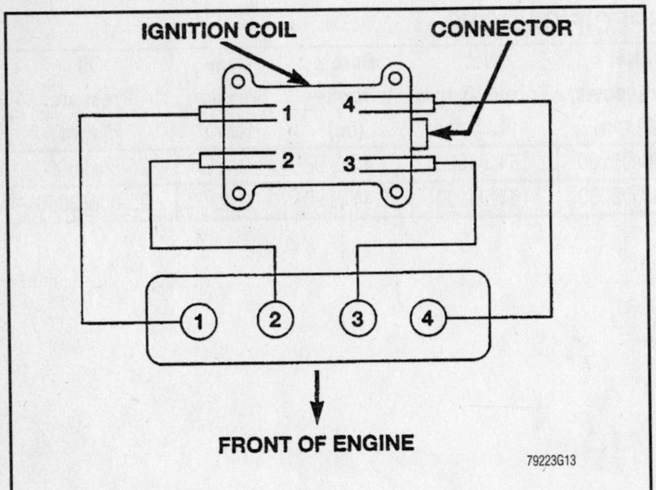

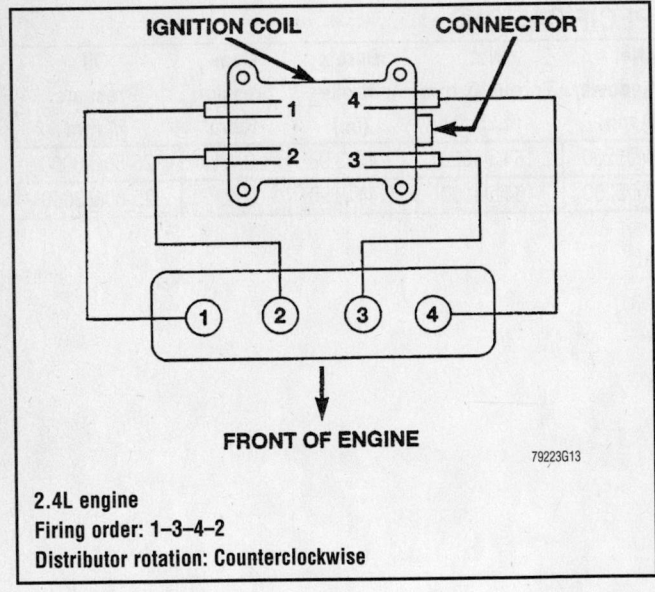

2.4L engine
Firing order: 1–3–4–2
Distributor rotation: Counterclockwise

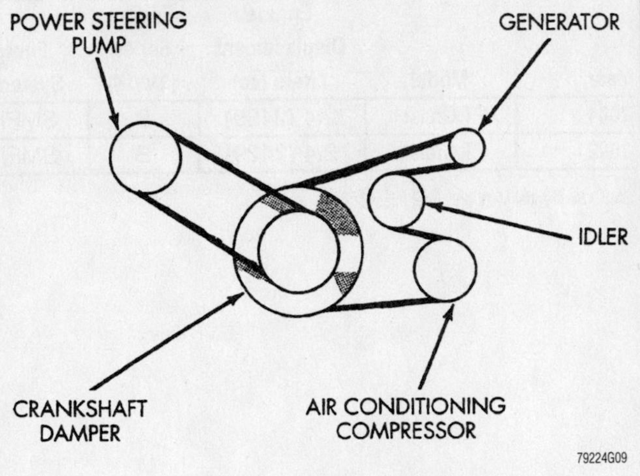

Accessory drive belt routing—Chrysler 2.4L engine

CAPACITIES

Year	Model	Engine Displacement Liters (cc)	Engine ID/VIN	Engine Oil with Filter (qts.)	Automatic Transaxle (qts.)	Power Transfer Unit (qts.)	Rear Drive Axle (pts.)	Fuel Tank (gal.)	Cooling System (qts.)
2001	PT Cruiser	2.4 (2429)	B	4.5	①	1.22	②	20.0	9.5
2002	PT Cruiser	2.4 (2429)	B	4.5	①	1.22	②	20.0	9.5

NOTE: All capacities are approximate. Add fluid gradually and check to be sure a proper fluid level is obtained.

① 31TH overhaul fill capacity with torque converter empty: 8.5 qts.
41TE overhaul fill capacity with torque converter empty: 9.1 qts.

② Overrunning clutch: 0.75 pts.

93461C60

VALVE SPECIFICATIONS

Year	Engine Displacement Liters (cc)	Engine ID/VIN	Seat Angle (deg.)	Face Angle (deg.)	Spring Test Pressure (lbs. @ in.)	Spring Installed Height (in.)	Stem-to-Guide Clearance (in.)		Stem Diameter (in.)	
							Intake	Exhaust	Intake	Exhaust
2001	2.4 (2429)	B	45	44.5-45.0	129-143@ 1.17	1.50	0.0018-0.0025	0.0029-0.0037	0.2340	0.2330
2002	2.4 (2429)	B	45	44.5-45.0	129-143@ 1.17	1.50	0.0018-0.0025	0.0029-0.0037	0.2340	0.2330

① Intake valve: 44.5 degrees
Exhaust valve: 45 degrees

93461C61

CRANKSHAFT AND CONNECTING ROD SPECIFICATIONS

All measurements are given in inches.

Year	Engine Displacement Liters (cc)	Engine ID/VIN	Crankshaft: Main Brg. Journal Dia.	Crankshaft: Main Brg. Oil Clearance	Crankshaft: Shaft End-play	Crankshaft: Thrust on No.	Connecting Rod: Journal Diameter	Connecting Rod: Oil Clearance	Connecting Rod: Side Clearance
2001	2.4 (2429)	B	2.3610-2.3625	0.0007-0.0023	0.0035-0.0094	2	1.9670-1.9685	0.0009-0.0027	0.0051-0.0150
2002	2.4 (2429)	B	2.3610-2.3625	0.0007-0.0023	0.0035-0.0094	2	1.9670-1.9685	0.0009-0.0027	0.0051-0.0150

93461C62

PISTON AND RING SPECIFICATIONS

All measurements are given in inches.

Year	Engine Displacement Liters (cc)	Engine ID/VIN	Piston Clearance	Ring Gap: Top Compression	Ring Gap: Bottom Compression	Ring Gap: Oil Control	Ring Side Clearance: Top Compression	Ring Side Clearance: Bottom Compression	Ring Side Clearance: Oil Control
2001	2.4 (2429)	B	0.0009-0.0022	0.0098-0.0200	0.0090-0.0180	0.0098-0.0250	0.0011-0.0031	0.0011-0.0031	0.0004-0.0070
2002	2.4 (2429)	B	0.0009-0.0022	0.0098-0.0200	0.0090-0.0180	0.0098-0.0250	0.0011-0.0031	0.0011-0.0031	0.0004-0.0070

① Oil control ring side rails must be free to rotate after assembly

93461C63

TORQUE SPECIFICATIONS

All readings in ft. lbs.

Year	Engine Displacement Liters (cc)	Engine ID/VIN	Cylinder Head Bolts	Main Bearing Bolts	Rod Bearing Bolts	Crankshaft Damper Bolts	Flywheel Bolts	Manifold Intake	Manifold Exhaust	Spark Plugs	Lug Nuts
2001	2.4 (2429)	B	①	②	③	100	70	20	17	20	100
2002	2.4 (2429)	B	①	②	③	100	70	20	17	20	100

① Step 1: 25 ft. lbs.
Step 2: 50 ft. lbs.
Step 3: 50 ft. lbs.
Step 4: Plus 1/4 turn

② M8 bolts: 20 ft. lbs.
M11 bolts: 30 ft. lbs. plus 1/4 turn

③ Step 1: 20 ft. lbs.
Step 2: Plus 1/4 turn

93461C64

BRAKE SPECIFICATIONS

All measurements in inches unless noted

Year	Model		Brake Disc Original Thickness	Brake Disc Minimum Thickness	Brake Disc Maximum Run-out	Brake Drum Diameter Original Inside Diameter	Brake Drum Diameter Max. Wear Limit	Brake Drum Diameter Maximum Machine Diameter	Min. Lining Thick-ness	Caliper Guide Pin Bolts (ft. lbs.)
2001	PT Cruiser	F	0.902-0.909	0.803	0.005	—	—	—	0.313	26
		R	0.344-0.364	0.285	0.005	8.63-8.65	NA	NA	①	16
2002	PT Cruiser	F	0.902-0.909	0.803	0.005	—	—	—	0.313	26
		R	0.344-0.364	0.285	0.005	8.63-8.65	NA	NA	①	16

NA: Not Available
F: Front
R: Rear

① Rear Disc: 0.350 in.
Rear Bonded Shoes: 0.062 in.
Rear Riveted Shoes: 0.031 in.

93461C65

WHEEL ALIGNMENT

			Caster		Camber			Steering
				Preferred		Preferred		Axis
			Range	Setting	Range	Setting	Toe-in	Inclination
Year	Model		(+/-Deg.)	(Deg.)	(+/-Deg.)	(Deg.)	(in.)	(Deg.)
2001	PT Cruiser	F	+1.00	+2.45	+0.40	0.00	0.00 + 0.40	N/A
		R	—	—	+0.40	+0.20	+0.00 + 0.40	N/A
2002	PT Cruiser	F	+1.00	+2.45	+0.40	0.00	0.00 + 0.40	N/A
		R	—	—	+0.40	+0.20	+0.00 + 0.40	N/A

N/A: Not Available

93461C66

TIRE, WHEEL AND BALL JOINT SPECIFICATIONS

		OEM Tires		Tire Pressures (psi)		Wheel	Ball Joint
Year	Model	Standard	Optional	Front	Rear	Size	Inspection
2001	PT Cruser	P205/55R16	None	32	32	5.5-J	①
2002	PT Cruser	P205/55R16	None	32	32	5.5-J	①

OEM: Original Equipment Manufacturer

PSI: Pounds Per Square Inch

① Replace if any measurable movement is found

93461C67

SCHEDULED MAINTENANCE INTERVALS
Chrysler—PT Cruiser

TO BE SERVICED	TYPE OF SERVICE	VEHICLE MILEAGE INTERVAL (x1000)												
		7.5	15	22.5	30	37.5	45	52.5	60	67.5	75	82.5	90	97.5
Engine oil & filter	R	✓	✓	✓	✓	✓	✓	✓	✓	✓	✓	✓	✓	✓
Brake hoses	S/I	✓	✓	✓	✓	✓	✓	✓	✓	✓	✓	✓	✓	✓
Coolant level, hoses & clamps	S/I	✓	✓	✓	✓	✓	✓	✓	✓	✓	✓	✓	✓	✓
CV joints & front suspension components	S/I	✓	✓	✓	✓	✓	✓	✓	✓	✓	✓	✓	✓	✓
Exhaust system	S/I	✓	✓	✓	✓	✓	✓	✓	✓	✓	✓	✓	✓	✓
Manual transaxle oil	S/I	✓	✓	✓	✓	✓	✓	✓	✓	✓	✓	✓	✓	✓
Rotate tires	S/I	✓	✓	✓	✓	✓	✓	✓	✓	✓	✓	✓	✓	✓
Accessory drive belts	S/I		✓		✓		✓		✓		✓		✓	
Brake linings	S/I			✓			✓			✓			✓	
Air filter element	R				✓				✓				✓	
Spark plugs	R				✓				✓				✓	
Lubricate ball joints	S/I				✓				✓				✓	
Engine coolant ①	R													
PCV valve	S/I								✓				✓	
Ignition cables	R								✓					
Camshaft timing belt ②	R													

R: Replace   S/I: Service or Inspect

① Engine coolant: flush and replace at 100,000 miles.

② Camshaft timing belt: replace at 120,000 miles.

FREQUENT OPERATION MAINTENANCE (SEVERE SERVICE)

If a vehicle is operated under any of the following conditions it is considered severe service:

- Extremely dusty areas.

- 50% or more of the vehicle operation is in 32°C (90°F) or higher temperatures, or constant operation in temperatures below 0°C (32°F).

- Prolonged idling (vehicle operation in stop and go traffic).

- Frequent short running periods (engine does not warm to normal operating temperatures).

- Police, taxi, delivery usage or trailer towing usage.

Oil & oil filter change: change every 3000 miles.

Rotate tires every 6000 miles.

Brake linings: inspect every 12,000 miles.

Air filter element: service or inspect every 15,000 miles.

Automatic transaxle: change fluid & adjust bands every 15,000 miles.

Manual transaxle fluid: replace every 15,000 miles.

Engine coolant: replace at 36,000 miles and every 30,000 miles thereafter.

93461C68

Timing belt service is covered in Section 3 of this manual

SCHEDULED MAINTENANCE INTERVALS
DAMILERCHRYSLER CORPORATION
CHRYSLER PT CRUISER

The following should be used as a guide when determining the amount of work required for a particular service. In estimating how long a particular Scheduled Maintenance Service should take, please observe the following:

- Labor Time is time based on field research and data supplied by the vehicle manufacturer.
- Labor time operations are given in hours and tenths of an hour.
- All labor operations are to be used as a guide.

Mechanic Skill Level Codes:
(A) PRECISION: Highly skilled with multiple certification.
(B) GENERAL: Normally skilled with certification.
(C) MAINTENANCE: Semi-skilled working on certification.

	LABOR TIME
7500 Mile Service (C)	
All Models	.7
15000 Mile Service (C)	
All Models	.9
22500 Mile Service (C)	
All Models	.8
30000 Mile Service (B)	
All Models	1.5
37500 Mile Service (C)	
All Models	.7
45000 Mile Service (C)	
All Models	1.0
52500 Mile Service (C)	
All Models	.9
60000 Mile Service (B)	
All Models	1.9
67500 Mile Service (C)	
All Models	.8
75000 Mile Service (C)	
All Models	.9
82500 Mile Service (C)	
All Models	.9
90000 Mile Service (B)	
All Models	1.9
97500 Mile Service (C)	
All Models	.7

93461C69

CHRYSLER CORP.
Eagle Talon

ENGINE AND VEHICLE IDENTIFICATION

Engine						
Code ①	Liters (cc)	Cu. In.	Cyl.	Fuel Sys.	Engine Type	Eng. Mfg.
F	2.0 (1997)	122	4	MFI Turbo	DOHC	Mitsubsihi
Y	2.0 (1996)	122	4	MFI	DOHC	Chrysler

Model Year	
Code ②	Year
W	1998

MFI: Multi-port Fuel Injection

DOHC: Double Overhead Camshaft

① 8th digit of the Vehicle Identification Number (VIN)

② 10th digit of the VIN

93461C70

GENERAL ENGINE SPECIFICATIONS
All measurements are given in inches.

Year	Engine Displacement Liters (cc)	Engine Series (ID/VIN)	Fuel System	Net Horsepower @ rpm	Net Torque @ rpm (ft. lbs.)	Bore x Stroke (in.)	Com-pression Ratio	Oil Pressure @ rpm
1998	2.0 (1997)	F	MFI-Turbo	①	②	3.35x3.46	8.5:1	③
	2.0 (1996)	Y	MFI	140@6000	131@4800	3.44x3.27	9.6:1	④

MFI: Multi-port Fuel Injection

① Automatic: 205@6000
Manual: 210@6000

② Automatic: 220@3000
Manual: 214@3000

③ 11.4 psi or more at curb idle speed

④ 40 psi or more at curb idle speed

93461C71

Heater Core replacement is covered in Section 2 of this manual

ENGINE TUNE-UP SPECIFICATIONS

Year	Engine Displacement Liters (cc)	Engine ID/VIN	Spark Plug Gap (in.)	Ignition Timing (deg.)		Fuel Pump (psi) ①	Idle Speed (rpm)		Valve Clearance	
				MT	AT		MT	AT	In.	Ex.
1998	2.0 (1997)	F	0.028-0.030	5B	5B	33	750	750	HYD	HYD
	2.0 (1996)	Y	0.033-0.038	②	②	38	800	800	HYD	HYD

NOTE: The Vehicle Emission Control Information label often reflects specification changes made during production. The label figures must be used if they differ from those in this chart.

HYD: Hydraulic

① Pressure at idle with vacuum applied to fuel pressure regulator

② Basic ignition timing is not adjustable

93461C72

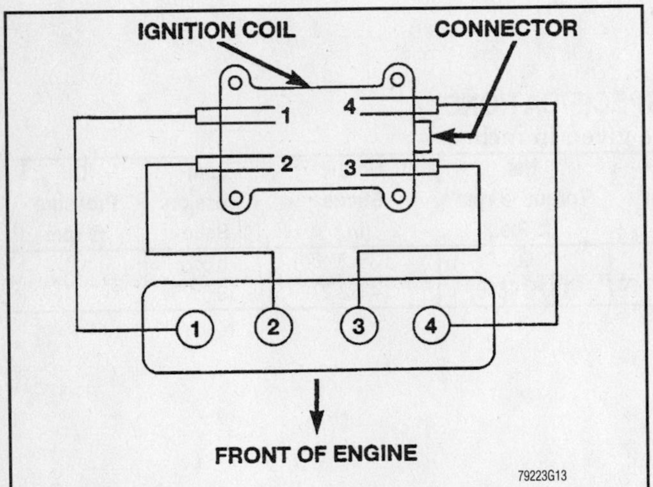

2.0L (VIN Y) Engine
Firing Order: 1–3–4–2
Distributorless ignition system

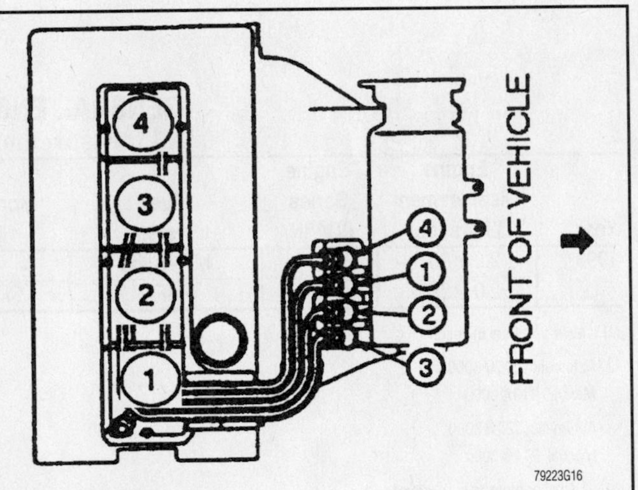

2.0L (VIN F) Engine
Firing Order: 1–3–4–2
Distributorless ignition system

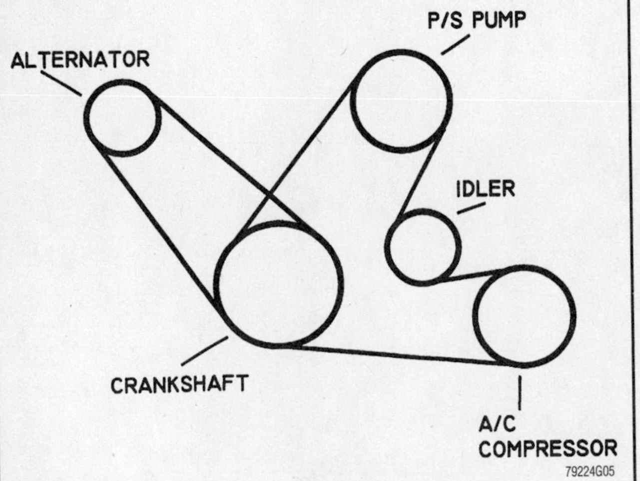

Accessory drive belt routing—Chrysler 2.0L (VIN Y) engine

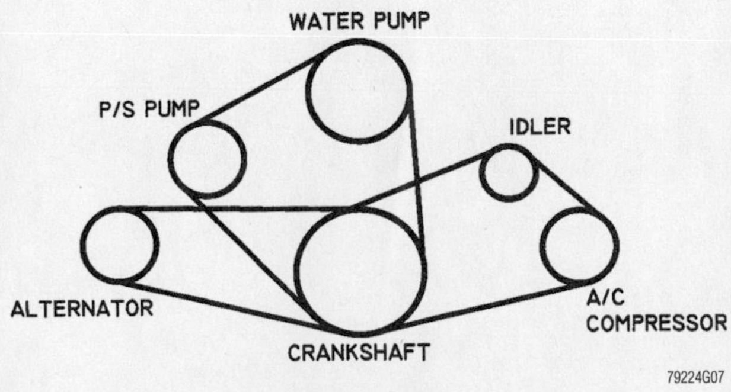

Accessory drive belt routing—Chrysler 2.0L (VIN F) turbo engine

CAPACITIES

Year	Engine Displacement Liters (cc)	Engine ID/VIN	Engine Oil with Filter (qts.)	Transmission (pts.) 5-Spd	Transmission (pts.) Auto.	Drive Axle (qts.)	Fuel Tank	Cooling System
1998	2.0 (1997)	F	4.6	①	14.2	1.8	16.0	7.4
	2.0 (1996)	Y	4.5	4.2	18.2	—	16.0	7.4

NOTE: All capacities are approximate. Add fluid gradually and ensure a proper fluid level is obtained.

① 2WD: 4.6
4WD: 4.8

93461C73

VALVE SPECIFICATIONS

Year	Engine Displacement Liters (cc)	Engine ID/VIN	Seat Angle (deg.)	Face Angle (deg.)	Spring Test Pressure (lbs. @ in.)	Spring Installed Height (in.)	Stem-to-Guide Clearance (in.) Intake	Stem-to-Guide Clearance (in.) Exhaust	Stem Diameter (in.) Intake	Stem Diameter (in.) Exhaust
1998	2.0 (1997)	F	44-44.5	45-45.5	54@1.570	1.570	0.0008-0.0040	0.0020-0.0060	0.2600	0.2560
	2.0 (1996)	Y	45	44.5-45	123-137@ 1.153	1.496	0.0019-0.0030	0.0029-0.0040	0.2336-0.2343	0.2325-0.2332

93461C74

Brake service is covered in Section 4 of this manual

CRANKSHAFT AND CONNECTING ROD SPECIFICATIONS

All measurements are given in inches.

Year	Engine Displacement Liters (cc)	Engine ID/VIN	Crankshaft				Connecting Rods		
			Main Brg. Journal Dia.	Main Brg. Oil Clearance	Shaft End-play	Thrust on No.	Journal Diameter	Oil Clearance	Side Clearance
1998	2.0 (1997)	F	2.2400	0.0008-0.0016	0.0020-0.0071	3	1.7700	0.0008-0.0020	0.0039-0.0098
	2.0 (1996)	Y	2.0469-2.0475	0.0009-0.0024	0.0035-0.0094	3	1.8894-1.8900	0.0010-0.0023	0.0051-0.0150

93461C75

PISTON AND RING SPECIFICATIONS

All measurements are given in inches.

Year	Engine Displacement Liters (cc)	Engine ID/VIN	Piston Clearance	Ring Gap			Ring Side Clearance		
				Top Compression	Bottom Compression	Oil Control	Top Compression	Bottom Compression	Oil Control
1998	2.0 (1997)	F	0.0012-0.0020	0.0098-0.0138	0.0157-0.0217	0.0039-0.0157	0.0016-0.0031	0.0008-0.0024	0.0005-0.0089
	2.0 (1996)	Y	0.0005-0.0017	0.0090-0.0200	0.0190-0.0310	0.0090-0.0260	0.0010-0.0026	0.0010-0.0026	0.0002-0.0070

93461C76

TORQUE SPECIFICATIONS

All readings in ft. lbs.

Year	Engine Displacement Liters (cc)	Engine ID/VIN	Cylinder Head Bolts	Main Bearing Bolts	Rod Bearing Bolts	Crankshaft Damper Bolts	Flywheel Bolts	Manifold Intake	Manifold Exhaust	Spark Plugs	Lug Nuts
1998	2.0 (1997)	F	①	②	③	94	94-101	14	18-22	18	65-80
	2.0 (1996)	Y	④	55	⑤	105	—	17	17	20	65-80

① Step 1: 58 ft. lbs.
Step 2: Fully loosen
Step 3: 15 ft. lbs.
Step 4: Plus 90 degrees
Step 5: Repeat Step 4

② Step 1: 18 ft. lbs.
Step 2: Plus 90 degrees

③ 14.5 ft. lbs. plus 90 degrees

④ Step 1:
Bolts 1-6: 24 ft. lbs.
Bolts 7-10: 20 ft. lbs.
Step 2:
Bolts 1-6: 49 ft. lbs.
Bolts 7-10: 20 ft. lbs.
Step 3: Plus 90 degrees

⑤ 20 ft. lbs. plus 90 degrees

93461C77

BRAKE SPECIFICATIONS

All measurements in inches unless noted

Year	Model		Brake Disc Original Thickness	Brake Disc Minimum Thickness	Brake Disc Maximum Runout	Minimum Lining Thickness	Bake Caliper Bracket Bolts (ft.bs.)	Bake Caliper Mounting Bolts (ft.bs.)
1998	Talon	F	0.940	0.882	0.003	0.080	65	54
		R	0.390	0.331	0.003	0.080	—	38

F: Front
R: Rear

93461C78

For complete Engine Mechanical specifications, see Section 1 of this manual

WHEEL ALIGNMENT

Year	Model		Caster Range (+/-Deg.)	Caster Preferred Setting (Deg.)	Camber Range (+/-Deg.)	Camber Preferred Setting (Deg.)	Toe-in (in.)	Steering Axis Inclination (Deg.)
1998	Talon ①②	F	1.50	+4.69	0.50	-0.08	0 +/- 0.12	—
		R	—	—	0.50	+1.30	—	—
	Talon ①③	F	1.50	+4.69	0.50	-0.30	0 +/- 0.13	
		R	—	—	0.50	-1.69	0.13 +/- 0.13	—
	Talon ④	F	1.50	+4.69	0.50	-0.08	0 +/- 0.12	—
		R	—	—	0.50	+1.30	—	—

① Front wheel drive
② With 14 inch wheels
③ With 16 inch wheels
④ All wheel drive

93461C79

TIRE, WHEEL AND BALL JOINT SPECIFICATIONS

Year	Model	OEM Tires Standard	OEM Tires Optional	Tire Pressures (psi) Front	Tire Pressures (psi) Rear	Wheel Size	Ball Joint Inspection
1998	Talon ESi	P195/70HR14	None	32	29	5.5-JJ	①
	Talon TSi 2wd	P205/55VR16	None	32	29	6-JJ	①
	Talon TSi 4wd	P215/50VR17	None	32	29	6.5-JJ	①

OEM: Original Equipment Manufacturer
PSI: Pounds Per Square Inch
① Torque required in inch lbs. to rotate ball joint when removed from the knuckle

93461C80

SCHEDULED MAINTENANCE INTERVALS
Eagle—Talon

TO BE SERVICED	TYPE OF SERVICE	VEHICLE MILEAGE INTERVAL (x1000)												
		7.5	15	22.5	30	37.5	45	52.5	60	67.5	75	82.5	90	97.5
Engine oil & filter (Non-turbo) ①	R	✓	✓	✓	✓	✓	✓	✓	✓	✓	✓	✓	✓	✓
Coolant level, hoses & clamps	S/I	✓	✓	✓	✓	✓	✓	✓	✓	✓	✓	✓	✓	✓
Rotate tires	S/I	✓	✓	✓	✓	✓	✓	✓	✓	✓	✓	✓	✓	✓
Automatic transaxle fluid level	S/I		✓		✓		✓		✓		✓		✓	
Brake hoses & disc brake pads	S/I		✓		✓		✓		✓		✓		✓	
Driveshaft boots & front suspension components	S/I		✓		✓		✓		✓		✓		✓	
Air filter element	R				✓				✓				✓	
Automatic transaxle fluid & filter ②	R				✓				✓				✓	
Engine coolant	R				✓				✓				✓	
Spark plugs	R				✓				✓				✓	
Accessory drive belts	S/I				✓				✓				✓	
Ball joints & steering linkage seals	S/I				✓				✓				✓	
Exhaust system	S/I				✓				✓				✓	
Fuel hoses	S/I				✓				✓				✓	
Manual transaxle oil (including transfer)	S/I				✓				✓				✓	
Rear axle oil (AWD)	S/I				✓				✓				✓	
Camshaft timing belt	R								✓					
Ignition cables	R								✓					
EVAP & fuel system	S/I								✓					

R: Replace S/I: Service or Inspect

① Engine oil & filter (Turbo): change every 5000 miles

② Turbo w/ A/T only

FREQUENT OPERATION MAINTENANCE (SEVERE SERVICE)

If a vehicle is operated under any of the following conditions it is considered severe service:

- Extremely dusty areas

- 50% or more of the vehicle operation is in 32°C (90°F) or higher temperatures, or constant operation in temperatures below 0°C (32°F)

- Frequent short running periods (engine does not warm to normal operating temperatures).

- Police, taxi, delivery usage or trailer towing usage.

Oil & filter change: change every 3000 miles

Air filter element: service or inspect every 7500 miles

Automatic transaxle fluid: change every 15,000 miles

Spark plugs: change every 15,000 miles

Disc brake pads: check more frequently than every 7500 miles

93461C81

For Accessory Drive Belt illustrations, see Section 1 of this manual

SCHEDULED MAINTENANCE INTERVALS
DAIMLERCHRYSLER CORPORATION
EAGLE TALON

The following should be used as a guide when determining the amount of work required for a particular service. In estimating how long a particular Scheduled Maintenance Service should take, please observe the following:

- Labor Time is time based on field research and data supplied by the vehicle manufacturer.
- Labor time operations are given in hours and tenths of an hour.
- All labor operations are to be used as a guide.

Mechanic Skill Level Codes:
(A) PRECISION: Highly skilled with multiple certification.
(B) GENERAL: Normally skilled with certification.
(C) MAINTENANCE: Semi-skilled working on certification.

	LABOR TIME
7500 Mile Service (C)	
All models	.6
15000 Mile Service (B)	
All models	1.9
22500 Mile Service (C)	
All models	.6
30000 Mile Service (B)	
All models	2.1
37500 Mile Service (C)	
All models	.6
45000 Mile Service (B)	
All models	1.9
52500 Mile Service (C)	
All models	.6
60000 Mile Service (B)	
All models	2.4
67500 Mile Service (C)	
All models	.6
75000 Mile Service (B)	
All models	1.9
82500 Mile Service (C)	
All models	.6
90000 Mile Service (B)	
All models	2.1
97500 Mile Service (C)	
All models	.6

93461C82

FORD MOTOR CO.
Lincoln Continental

ENGINE AND VEHICLE IDENTIFICATION

Engine						
Code ①	Liters (cc)	Cu. In.	Cyl.	Fuel Sys.	Engine Type	Eng. Mfg.
V	4.6 (4593)	281	8	SFI	DOHC	Ford

OHV: Overhead Valves

DOHC: Double Overhead Camshafts

SFI: Sequential Fuel Injection

① 8th digit of the Vehicle Identification Number (VIN)

② 10th digit of the Vehicle Identification Number (VIN)

Model Year	
Code ②	Year
W	1998
X	1999
Y	2000
1	2001
2	2002

93461C83

GENERAL ENGINE SPECIFICATIONS

Year	Model	Engine Displacement Liters (cc)	Engine ID/VIN	Fuel System Type	Net Horsepower @ rpm	Net Torque @ rpm (ft. lbs.)	Bore x Stroke (in.)	Com-pression Ratio	Oil Pressure @ rpm
1998	Continental	4.6 (4593)	V	SFI	260@5750	270@3000	3.55x3.54	9.8:1	33@1500
1999	Continental	4.6 (4593)	V	SFI	275@5750	275@4750	3.55x3.54	9.8:1	33@1500
2000	Continental	4.6 (4593)	V	SFI	275@5750	275@4750	3.55x3.54	9.8:1	33@1500
2001	Continental	4.6 (4593)	V	SFI	275@5750	275@4750	3.55x3.54	9.8:1	33@1500

SFI: Sequential Fuel Injection

93461C84

For Tire, Wheel and Ball Joint specifications, see Section 1 of this manual

ENGINE TUNE-UP SPECIFICATIONS

Year	Engine Displacement Liters (cc)	Engine ID/VIN	Spark Plug Gap (in.)	Ignition Timing (deg.)	Fuel Pump (psi) ①	Idle Speed (rpm)	Valve Clearance	
							Intake	Exhaust
1998	4.6 (4593)	V	0.052-0.056	10B	30-45	②	HYD	HYD
1999	4.6 (4593)	V	0.052-0.056	10B	30-45	②	HYD	HYD
2000	4.6 (4593)	V	0.052-0.056	10B	30-45	②	HYD	HYD
2001	4.6 (4593)	V	0.052-0.056	10B	30-45	②	HYD	HYD

NOTE: The Vehicle Emission Control Information label often reflects specification changes made during production. The label figures must be used if they differ from those in this chart.

B: Before Top Dead Center

HYD: Hydraulic

NA: Not Adjustable

① Fuel pressure with engine running, pressure regulator vacuum hose connected

② Refer to Vehicle Emission Control Information label

93461C85

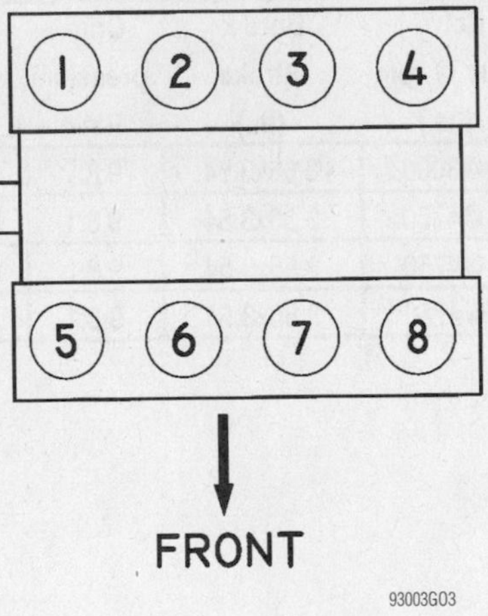

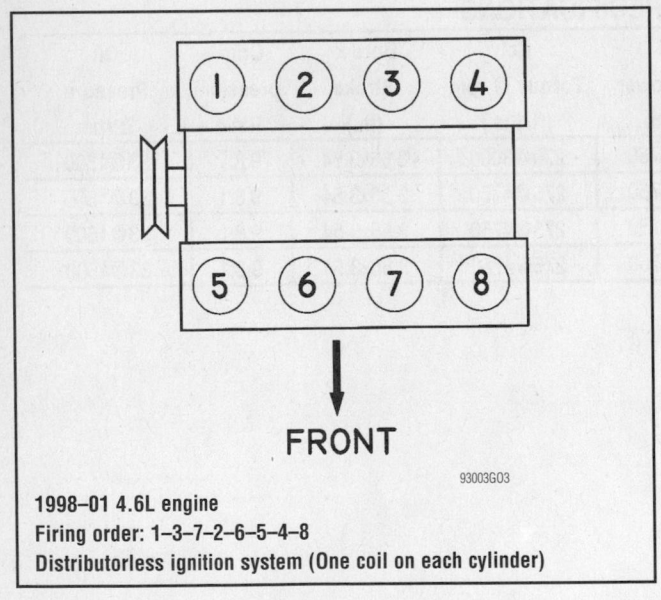

1998–01 4.6L engine
Firing order: 1–3–7–2–6–5–4–8
Distributorless ignition system (One coil on each cylinder)

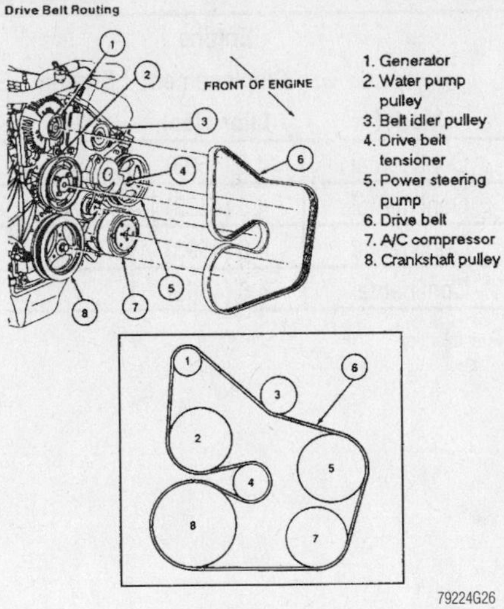

Serpentine accessory drive belt routing— 4.6L (VIN V) engine

CAPACITIES

Year	Model	Engine Displacement Liters (cc)	Engine ID/VIN	Engine Oil with Filter (qts.)	Transaxle (pts.) Auto. ①	Drive Axle (pts.)	Fuel Tank (gal.)	Cooling System (qts.)
1998	Continental	4.6 (4593)	V	6.0	27.4	②	20.0	14.3
1999	Continental	4.6 (4593)	V	6.0	27.4	②	20.0	14.3
2000	Continental	4.6 (4593)	V	6.0	27.4	②	20.0	14.3
2001	Continental	4.6 (4593)	V	6.0	27.4	②	20.0	14.3

NOTE: All capacities are approximate. Add fluid gradually and ensure a proper fluid level is obtained.

① Includes torque converter

② Included in transaxle capacity

93461C86

VALVE SPECIFICATIONS

Year	Engine Displacement Liters (cc)	Engine ID/VIN	Seat Angle (deg.)	Face Angle (deg.)	Spring Test Pressure (lbs. @ in.)	Spring Installed Height (in.)	Stem-to-Guide Clearance (in.)		Stem Diameter (in.)	
							Intake	Exhaust	Intake	Exhaust
1998	4.6 (4593)	V	45	45.5	160@1.103	1.425	0.0008-0.0027	0.0018-0.0037	0.2746-0.2754	0.2736-0.2744
1999	4.6 (4593)	V	45	45.5	160@1.103	1.425	0.0008-0.0027	0.0018-0.0045	0.2746-0.2754	0.2736-0.2744
2000	4.6 (4593)	V	45	45.5	160@1.103	1.425	0.0008-0.0027	0.0018-0.0045	0.2746-0.2754	0.2736-0.2744
2001	4.6 (4593)	V	45	45.5	160@1.103	1.425	0.0008-0.0027	0.0018-0.0045	0.2746-0.2754	0.2736-0.2744

93461C87

For Wheel Alignment specifications, see Section 1 of this manual

CRANKSHAFT AND CONNECTING ROD SPECIFICATIONS

All measurements are given in inches.

Year	Engine Displacement Liters (cc)	Engine ID/VIN	Crankshaft: Main Brg. Journal Dia.	Crankshaft: Main Brg. Oil Clearance	Crankshaft: Shaft End-play	Crankshaft: Thrust on No.	Connecting Rod: Journal Diameter	Connecting Rod: Oil Clearance	Connecting Rod: Side Clearance
1998	4.6 (4593)	V	2.6580-2.6576	0.0010-0.0018	0.0051-0.0119	5	2.0859-2.0867	0.0011-0.0027	0.0059-0.0177
1999	4.6 (4593)	V	2.6580-2.6576	0.0010-0.0018	0.0051-0.0119	5	2.0859-2.0867	0.0011-0.0027	0.0059 0.0177
2000	4.6 (4593)	V	2.6580-2.6576	0.0010-0.0018	0.0051-0.0119	5	2.0859-2.0867	0.0011-0.0027	0.0059 0.0177
2001	4.6 (4593)	V	2.6580-2.6576	0.0010-0.0018	0.0051-0.0119	5	2.0859-2.0867	0.0011-0.0027	0.0059 0.0177

93461C88

PISTON AND RING SPECIFICATIONS

All measurements are given in inches.

Year	Engine Displacement Liters (cc)	Engine ID/VIN	Piston Clearance	Ring Gap: Top Compression	Ring Gap: Bottom Auto. ①	Ring Gap: Oil Control	Ring Side Clearance: Top Compression	Ring Side Clearance: Bottom Compression	Ring Side Clearance: Oil Control
1998	4.6 (4593)	V	0.0007-0.0018	0.010-0.020	0.010-0.020	0.006-0.026	0.0004-0.0009	0.0012-0.0032	SNUG
1999	4.6 (4593)	V	0.0007-0.0018	0.010-0.020	0.010-0.020	0.006-0.026	0.0004-0.0009	0.0012-0.0032	SNUG
2000	4.6 (4593)	V	0.0007-0.0018	0.010-0.020	0.010-0.020	0.006-0.026	0.0004-0.0009	0.0012-0.0032	SNUG
2001	4.6 (4593)	V	0.0007-0.0018	0.010-0.020	0.010-0.020	0.006-0.026	0.0004-0.0009	0.0012-0.0032	SNUG

93461C89

TORQUE SPECIFICATIONS

All readings in ft. lbs.

Year	Engine Displacement Liters (cc)	Engine ID/VIN	Cylinder Head Bolts	Main Bearing Bolts	Rod Bearing Auto.	Crankshaft Damper Bolts	Flywheel Bolts	Manifold Intake	Manifold Exhaust	Spark Plugs	Lug Nuts
1998	4.6 (4593)	V	①	②	③	④	54-64	⑤	⑥	7-15	95
1999	4.6 (4593)	V	①	②	③	④	54-64	⑤	⑥	7-15	95
2000	4.6 (4593)	V	①	②	③	④	54-64	⑤	⑥	7-15	95
2001	4.6 (4593)	V	①	②	③	④	54-64	⑤	⑥	7-15	95

① Step 1: 28-31 ft. lbs.
Step 2: Plus 85-95 degrees
Step 3: Loosen all bolts 360 degrees
Step 4: 28-31 ft. lbs.
Step 5: Plus 85-95 degrees
Step 6: Plus 85-95 degrees

② Step 1: Main bearing cap bolts: 6-9 ft. lbs.
Step 2: Main bearing cap bolts, outer: 16-21 ft. lbs.
Step 3: Main bearing cap bolts, inner: 27-32 ft. lbs.
Step 4: Rotate main bearing cap bolts 85-95 degrees
Step 5: Main cap adjusting screws 4 ft. lbs. then 7.5 ft. lbs.
Step 6: Main cap side bolts: 7 ft. lbs. then 14-17 ft. lbs.

③ Step 1: 5 ft. lbs.
Step 2: 10 ft. lbs.
Step 3: 18-25 ft. lbs.
Step 4: Plus 85-95 degrees

④ Step 1: 77-99 ft. lbs.
Step 2: Loosen 360 degrees
Step 3: 35-39 ft. lbs.
Step 4: Plus 85-95 degrees

⑤ Step 1: Four inside short bolts: 9-11 ft. lbs.
Step 2: All other bolts: 13-16 ft. lbs.
Step 3: All bolts plus 85-95 degrees

⑥ Studs: 8-9 ft. lbs.
Nuts: 14-16 ft .lbs.

93461C90

BRAKE SPECIFICATIONS

All measurements in inches unless noted

			Brake Disc			Brake Drum				Brake Caliper	
Year	Model		Original Thickness	Minimum Thickness	Maximum Run-out	Original Inside Diameter	Max. Wear Limit	Maximum Machine Diameter	Minimum Lining Thickness	Bracket Bolts (ft. lbs.)	Mounting Bolts (ft. lbs.)
1998	Continental	F	1.020	0.974	0.003	—	—	—	0.039	65-87	25
		R	0.550	0.502	0.001	—	—	—	0.039	—	25
1999	Continental	F	1.020	0.974	0.003	—	—	—	0.039	65-87	25
		R	0.550	0.502	0.001	—	—	—	0.039	—	25
2000	Continental	F	1.020	0.974	0.003	—	—	—	0.039	65-87	25
		R	0.550	0.502	0.001	—	—	—	0.039	—	25
2001	Continental	F	1.020	0.974	0.003	—	—	—	0.039	65-87	25
		R	0.550	0.502	0.001	—	—	—	0.039	—	25

NOTE: Follow specifications stamped on the rotor if figures differ from those in this chart.

F: Front

R: Rear

93461C91

WHEEL ALIGNMENT

			Caster		Camber			
Year	Model		Range (+/-Deg.)	Preferred Setting (Deg.)	Range (+/-Deg.)	Preferred Setting (Deg.)	Toe-in (in.)	Steering Axis Inclination (Deg.)
1998	Continental	F	0.75	0	0.70	0	-0.20 +/- 0.25	—
		R	—	—	0.70	0	0.20 +/- 0.25	—
1999	Continental	F	0.75	0	0.70	0	-0.20 +/- 0.25	—
		R	—	—	0.70	0	0.20 +/- 0.25	—
2000	Continental	F	0.75	0	0.70	0	-0.20 +/- 0.25	—
		R	—	—	0.70	0	0.20 +/- 0.25	—
2001	Continental	F	0.75	0	0.70	0	-0.20 +/- 0.25	—
		R	—	—	0.70	0	0.20 +/- 0.25	—

93461C92

TIRE, WHEEL AND BALL JOINT SPECIFICATIONS

Year	Model	OEM Tires		Tire Pressures (psi)		Wheel	Ball Joint
		Standard	Optional	Front	Rear	Size	Inspection
1998	Continental	P225/60HR16	P225/60VR16	30	30	7-JJ	①
1999	Continental	P225/60HR16	P225/60VR16	30	30	7-JJ	①
2000-01	Continental	P225/60HR16	P225/60VR16	30	30	7-JJ	①

OEM: Original Equipment Manufacturer

PSI: Pounds Per Square Inch

① Replace if any measurable movement is found.

93461C93

SCHEDULED MAINTENANCE INTERVALS

Lincoln Continental

TO BE SERVICED	TYPE OF SERVICE	VEHICLE MILEAGE INTERVAL (X1000)																		
		5	10	15	20	25	30	35	40	45	50	55	60	65	70	75	80	85	90	95
Engine oil & filter	R	✓	✓	✓	✓	✓	✓	✓	✓	✓	✓	✓	✓	✓	✓	✓	✓	✓	✓	✓
Rotate tires	S/I	✓		✓		✓		✓		✓		✓		✓		✓		✓		✓
Engine coolant protection, hoses & clamps	S/I			✓			✓			✓			✓			✓			✓	
Passenger compartment air filter	R			✓			✓			✓			✓			✓			✓	
Air cleaner filter	R						✓						✓						✓	
Automatic transaxle fluid & filter	R						✓						✓						✓	
Brake lines & connections	S/I						✓						✓						✓	
Exhaust heat shields	S/I						✓						✓						✓	
Front and rear disc brake pads & rotors	S/I						✓						✓						✓	
Accessory drive belt(s)	S/I												✓							
Engine coolant ①	R										✓						✓			
Spark plugs	R																			
PCV valve	R												✓							

① Engine coolant - change initially at 50,000 miles & thereafter every 30,000 miles.

R - Replace S/I - Service and Inspect

FREQUENT OPERATION MAINTENANCE (SEVERE SERVICE)

If a vehicle is operated under any of the following conditions it is considered severe service:

- Extremely dusty areas.

- 50% or more of the vehicle operation is in 32°C (90°F) or higher temperatures, or constant operation in temperatures below 0°C (32°F).

- Prolonged idling (vehicle operation in stop and go traffic)..

- Frequent short running periods (engine does not warm to normal operating temperatures).

- Police, taxi, delivery usage or trailer towing usage.

Oil & oil filter - change every 3000 miles.

Rotate tires at 6000 miles & every 9000 miles thereafter.

Air cleaner element service or inspect every 15,000 miles.

Automatic transaxle fluid & filter - change every 21,000 miles.

93461C94

SCHEDULED MAINTENANCE INTERVALS
FORD MOTOR COMPANY
LINCOLN CONTINENTAL

The following should be used as a guide when determining the amount of work required for a particular service. In estimating how long a particular Scheduled Maintenance Service should take, please observe the following:

- Labor Time is time based on field research and data supplied by the vehicle manufacturer.
- Labor time operations are given in hours and tenths of an hour.
- All labor operations are to be used as a guide.

Mechanic Skill Level Codes:
(A) PRECISION: Highly skilled with multiple certification.
(B) GENERAL: Normally skilled with certification.
(C) MAINTENANCE: Semi-skilled working on certification.

	LABOR TIME
5000 Mile Service (C)	
All Models	.9
10000 Mile Service (C)	
All Models	.5
15000 Mile Service (C)	
All Models	1.0
20000 Mile Service (C)	
All Models	.7
25000 Mile Service (C)	
All Models	.9
30000 Mile Service (B)	
All Models	2.0
Replace auto. trans. fluid, filter add	.5
35000 Mile Service (C)	
All Models	.9
40000 Mile Service (C)	
All Models	.7
45000 Mile Service (C)	
All Models	1.0
50000 Mile Service (C)	
All Models	.9
Replace engine coolant add	.5
55000 Mile Service (C)	
All Models	.9
60000 Mile Service (B)	
All Models	2.2
Replace auto. trans. fluid filter add	.5
65000 Mile Service (C)	
All Models	.9
70000 Mile Service (C)	
All Models	.5
75000 Mile Service (C)	
All Models	.7
80000 Mile Service (C)	
All Models	.9
85000 Mile Service (C)	
All Models	.9
Replace engine coolant add	.5
90000 Mile Service (B)	
All Models	2.0
Replace auto. trans. fluid, filter add	.5
95000 Mile Service (C)	
All Models	.9
100000 Mile Service (C)	
All Models	1.2

93461C95

FORD MOTOR CO.
Ford Contour • Mercury Cougar & Mystique

ENGINE AND VEHICLE IDENTIFICATION

Engine						
Code ①	Liters (cc)	Cu. In.	Cyl.	Fuel Sys.	Engine Type	Eng. Mfg.
3	2.0 (1999)	122	4	SFI	DOHC	Ford
L	2.5 (2507)	153	6	SFI	DOHC	Ford
G	2.5 (2507)	153	6	SFI	DOHC	Ford

Model Year	
Code ②	Year
W	1998
X	1999
Y	2000
1	2001
2	2002

SFI: Sequential Fuel Injection

DOHC: Double Overhead Camshaft

① 8th digit of VIN

② 10th digit of VIN

93461C96

GENERAL ENGINE SPECIFICATIONS

Year	Model	Engine Displacement Liters (cc)	Engine ID/VIN	Fuel System Type	Net Horsepower @ rpm	Net Torque @ rpm (ft. lbs.)	Bore x Stroke (in.)	Com-pression Ratio	Oil Pressure @ rpm
1998	Contour	2.0 (1999)	3	SFI	125@6000	130@4500	3.39x3.46	9.6:1	20-45@1500
		2.5 (2507)	L	SFI	170@6200	165@4200	3.25x3.13	9.7:1	25-45@1500
	Contour SVT	2.5 (2507)	G	SFI	196@6750	168@5500	3.25x3.13	10.25:1	20-45@1500
	Mystique	2.0 (1999)	3	SFI	125@5500	130@4000	3.39x3.46	9.6:1	20-45@1500
		2.5 (2507)	L	SFI	170@6200	165@4200	3.25x3.13	9.7:1	20-45@1500
1999	Contour	2.0 (1999)	3	SFI	125@6000	130@4500	3.39x3.46	9.6:1	20-45@1500
		2.5 (2507)	L	SFI	170@6200	165@4200	3.25x3.13	9.7:1	25-45@1500
	Contour SVT	2.5 (2507)	G	SFI	196@6750	168@5500	3.25x3.13	10.25:1	20-45@1500
	Cougar	2.0 (1999)	3	SFI	125@6000	130@4500	3.39x3.46	9.6:1	20-45@1500
		2.5 (2507)	L	SFI	170@6200	165@4200	3.25x3.13	9.7:1	25-45@1500
	Mystique	2.0 (1999)	3	SFI	125@5500	130@4000	3.39x3.46	9.6:1	20-45@1500
		2.5 (2507)	L	SFI	170@6200	165@4200	3.25x3.13	9.7:1	20-45@1500
2000	Contour	2.5 (2507)	L	SFI	170@6200	165@4200	3.25x3.13	9.7:1	25-45@1500
	Cougar	2.0 (1999)	3	SFI	125@6000	130@4500	3.39x3.46	9.6:1	20-45@1500
		2.5 (2507)	L	SFI	170@6200	165@4200	3.25x3.13	9.7:1	25-45@1500
	Mystique	2.0 (1999)	3	SFI	125@5500	130@4000	3.39x3.46	9.6:1	20-45@1500
		2.5 (2507)	L	SFI	170@6200	165@4200	3.25x3.13	9.7:1	20-45@1500
2001	Cougar	2.0 (1999)	3	SFI	125@6000	130@4500	3.39x3.46	9.6:1	20-45@1500
		2.5 (2507)	L	SFI	170@6200	165@4200	3.25x3.13	9.7:1	25-45@1500
	Cougar S	2.5 (2507)	G	SFI	196@6750	168@5500	3.25x3.13	10.25:1	20-45@1500

SFI: Sequential Fuel Injection

93461C97

ENGINE TUNE-UP SPECIFICATIONS

Year	Engine Displacement Liters (cc)	Engine ID/VIN	Spark Plug Gap (in.)	Ignition Timing (deg.) MT	Ignition Timing (deg.) AT	Fuel Pump (psi) ①	Idle Speed (rpm) MT	Idle Speed (rpm) AT	Valve Clearance Intake	Valve Clearance Exhaust
1998	2.0 (1999)	3	0.050	10B	10B	37-41	②	②	HYD	HYD
	2.5 (2507)	G	0.054	10B	—	37-41	②	—	HYD	HYD
	2.5 (2507)	L	0.054	10B	10B	37-41	②	②	HYD	HYD
1999	2.0 (1999)	3	0.050	10B	10B	37-41	②	②	HYD	HYD
	2.5 (2507)	G	0.054	10B	—	37-41	②	—	HYD	HYD
	2.5 (2507)	L	0.054	10B	10B	37-41	②	②	HYD	HYD
2000	2.0 (1999)	3	0.050	10B	10B	37-41	②	②	HYD	HYD
	2.5 (2507)	L	0.054	10B	10B	37-41	②	②	HYD	HYD
2001	2.0 (1999)	3	0.050	10B	10B	37-41	②	②	HYD	HYD
	2.5 (2507)	G	0.054	10B	—	37-41	②	—	HYD	HYD
	2.5 (2507)	L	0.054	10B	10B	37-41	②	②	HYD	HYD

NOTE: The Vehicle Emission Control Information label often reflects specification changes made during production. The label figures must be used if they differ from those in this chart.

B: Before Top Dead Center

HYD: Hydraulic

① Fuel pressure with engine running, pressure regulator vacuum hose connected

② Refer to Vehicle Emission Control Information (VECI) label

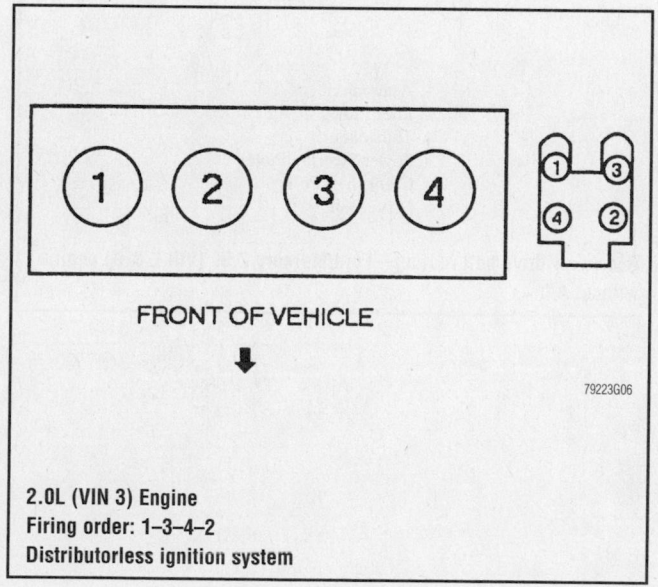

93461C98

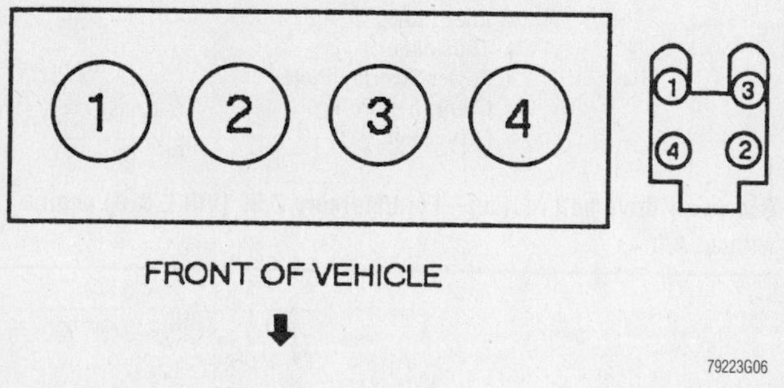

2.0L (VIN 3) Engine
Firing order: 1–3–4–2
Distributorless ignition system

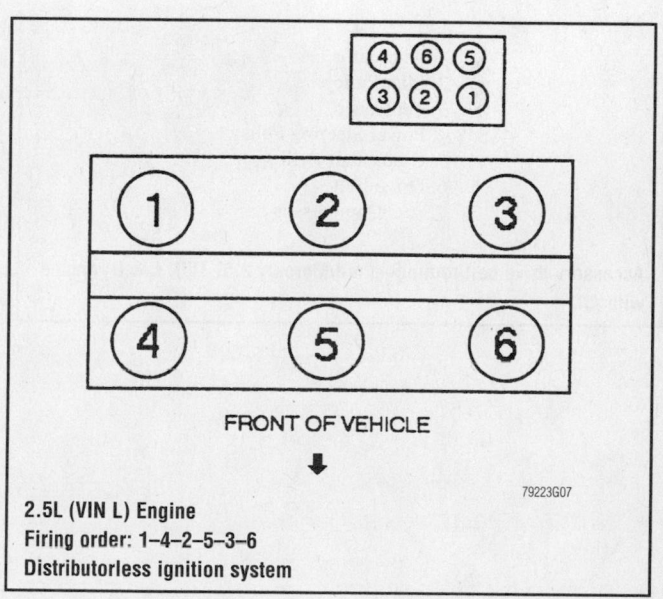

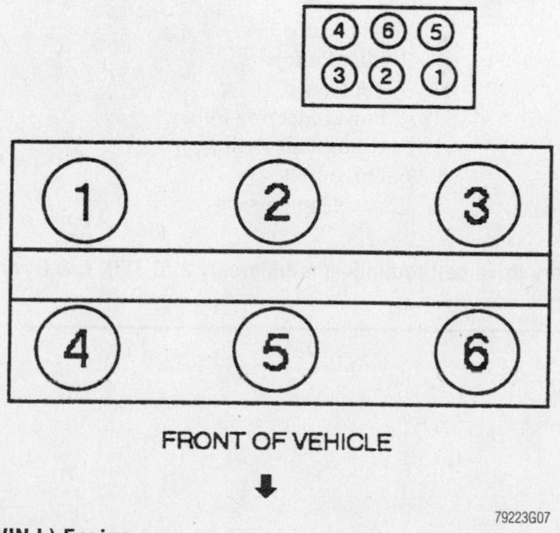

2.5L (VIN L) Engine
Firing order: 1–4–2–5–3–6
Distributorless ignition system

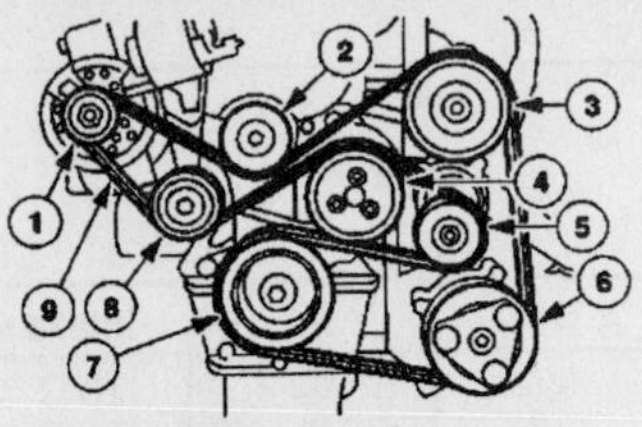

1	Generator pulley
2	Belt idler pulley
3	Power steering pump pulley
4	Water pump pulley
5	Belt tensioner
6	A/C clutch pulley
7	Crankshaft pulley damper
8	Belt idler pulley
9	Drive belt

93004G05

Accessory drive belt routing —Ford/Mercury 2.0L (VIN 3) DOHC engine

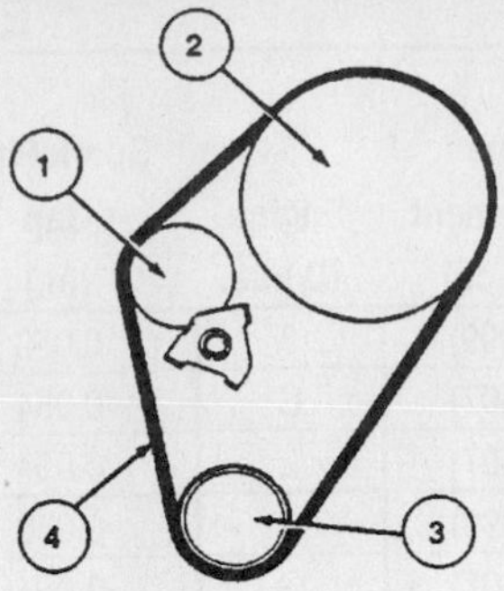

1. Drive Belt Tensioner
2. Water Pump Drive Pulley
3. Water Pump
4. Water Pump Drive Belt

79224G15

Water pump drive belt routing—Ford/Mercury 2.5L (VIN L & G) engine

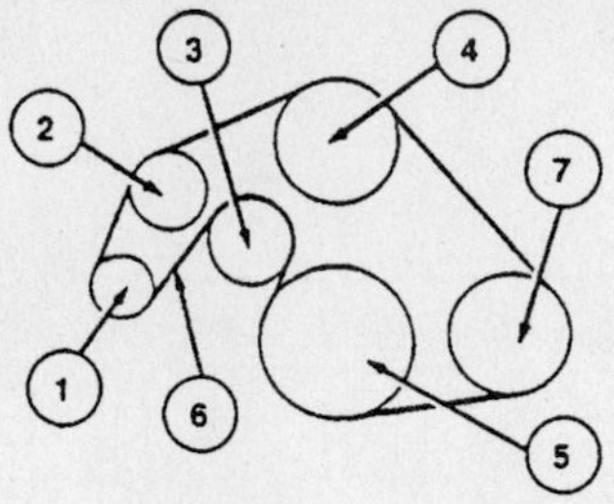

1. Alternator
2. Idler Pulley
3. Tensioner
4. Power Steering Pulley
5. Crankshaft Pulley
6. Drive Belt
7. A/C Compressor

79224G16

Accessory drive belt routing—Ford/Mercury 2.5L (VIN L & G) engine with A/C

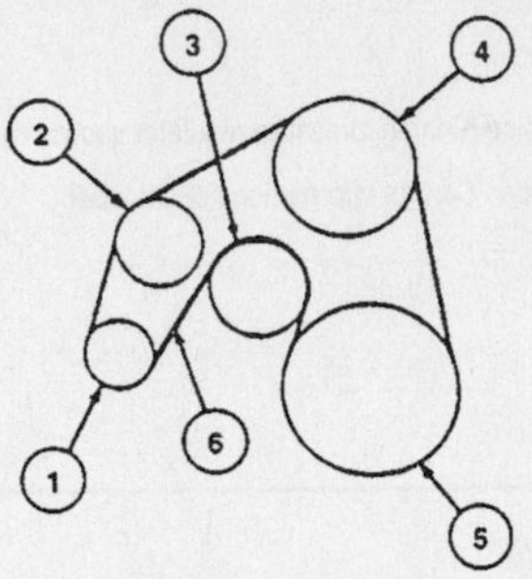

1. Alternator
2. Idler Pulley
3. Tensioner
4. Power Steering Pulley
5. Crankshaft Pulley
6. Drive Belt

79224G17

Accessory drive belt routing—Ford/Mercury 2.5L (VIN L & G) engine without A/C

CAPACITIES

Year	Model	Engine Displacement Liters (cc)	Engine ID/VIN	Engine Oil with Filter (qts.)	Transaxle (pts.) Manual	Transaxle (pts.) Auto. ①	Front Drive Axle (pts.)	Fuel Tank (gal.)	Cooling System (qts.)
1998	Contour	2.0 (1999)	3	4.5	5.5	18.0	②	14.5	③
		2.5 (2507)	L	5.8	5.5	20.6	②	14.5	④
	Contour SVT	2.5 (2507)	G	5.8	5.5	—	②	14.5	④
	Mystique	2.0 (1999)	3	4.5	5.5	18.0	②	14.5	③
		2.5 (2507)	L	5.8	5.5	20.6	②	14.5	④
1999	Contour	2.0 (1999)	3	4.5	5.5	18.0	②	14.5	③
		2.5 (2507)	L	5.8	5.5	20.6	②	14.5	④
	Contour SVT	2.5 (2507)	G	5.8	5.5	—	②	14.5	④
	Cougar	2.0 (1999)	3	4.5	5.5	18.0	②	14.5	④
		2.5 (2507)	L	5.8	5.5	20.6	②	14.5	④
	Mystique	2.0 (1999)	3	4.5	5.5	18.0	②	14.5	③
		2.5 (2507)	L	5.8	5.5	20.6	②	14.5	④
2000	Contour	2.5 (2507)	L	5.8	5.5	20.6	②	14.5	④
	Cougar	2.0 (1999)	3	4.5	5.5	18.0	②	14.5	④
		2.5 (2507)	L	5.8	5.5	20.6	②	14.5	④
	Mystique	2.0 (1999)	3	4.5	5.5	18.0	②	14.5	③
		2.5 (2507)	L	5.8	5.5	20.6	②	14.5	④
2001	Cougar	2.0 (1999)	3	4.5	5.5	18.0	②	14.5	④
		2.5 (2507)	L	5.8	5.5	20.6	②	14.5	④
	Cougar S	2.5 (2507)	G	5.8	5.5	—	②	14.5	④

NOTE: All capacities are approximate. Add fluid gradually and ensure a proper fluid level is obtained.

① Includes torque converter

② Included in transaxle capacity

③ Automatic transaxle: 7.5 qts.
Manual transaxle: 7.0 qts.

④ Automatic transaxle: 9.1 qts.
Manual transaxle: 8.9 qts.

93461C99

VALVE SPECIFICATIONS

Year	Engine Displacement Liters (cc)	Engine ID/VIN	Seat Angle (deg.)	Face Angle (deg.)	Spring Test Pressure (lbs. @ in.)	Spring Installed Height (in.)	Stem-to-Guide Clearance (in.)		Stem Diameter (in.)	
							Intake	Exhaust	Intake	Exhaust
1998	2.0 (1999)	3	45	45	NA	1.346	0.0007-0.0025	0.0014-0.0032	0.2373-0.2379	0.2366-0.2372
	2.5 (2507)	G	44.75	45.5	153@1.18	1.570	0.0007-0.0027	0.0017-0.0037	0.2350-0.2358	0.2343-0.2350
	2.5 (2507)	L	44.75	45.5	153@1.18	1.570	0.0007-0.0027	0.0017-0.0037	0.2350-0.2358	0.2343-0.2350
1999	2.0 (1999)	3	45	45	NA	1.346	0.0007-0.0025	0.0014-0.0032	0.2373-0.2379	0.2366-0.2372
	2.5 (2507)	G	44.75	45.5	153@1.18	1.570	0.0007-0.0027	0.0017-0.0037	0.2350-0.2358	0.2343-0.2350
	2.5 (2507)	L	44.75	45.5	153@1.18	1.570	0.0007-0.0027	0.0017-0.0037	0.2350-0.2358	0.2343-0.2350
2000	2.0 (1999)	3	45	45	NA	1.346	0.0007-0.0025	0.0014-0.0032	0.2373-0.2379	0.2366-0.2372
	2.5 (2507)	L	44.75	45.5	153@1.18	1.570	0.0007-0.0027	0.0017-0.0037	0.2350-0.2358	0.2343-0.2350
2001	2.0 (1999)	3	45	45	NA	1.346	0.0007-0.0025	0.0014-0.0032	0.2373-0.2379	0.2366-0.2372
	2.5 (2507)	G	44.75	45.5	153@1.18	1.570	0.0007-0.0027	0.0017-0.0037	0.2350-0.2358	0.2343-0.2350
	2.5 (2507)	L	44.75	45.5	153@1.18	1.570	0.0007-0.0027	0.0017-0.0037	0.2350-0.2358	0.2343-0.2350

NA: Not Available

93461CA1

CRANKSHAFT AND CONNECTING ROD SPECIFICATIONS

All measurements are given in inches.

Year	Engine Displacement Liters (cc)	Engine ID/VIN	Crankshaft Main Brg. Journal Dia.	Crankshaft Main Brg. Oil Clearance	Crankshaft Shaft End-play	Crankshaft Thrust on No.	Connecting Rod Journal Diameter	Connecting Rod Oil Clearance	Connecting Rod Side Clearance
1998	2.0 (1999)	3	2.2827-2.2835	0.0008-0.0017	0.0035-0.0100	3	1.8425-1.8504	0.0006-0.0028	0.0035-0.0126
	2.5 (2507)	G	2.4670-2.4790	0.0009-0.0019	0.0040-0.0090	4	1.9670-1.9680	0.0010-0.0025	0.0039-0.0118
	2.5 (2507)	L	2.4670-2.4790	0.0009-0.0019	0.0040-0.0090	4	1.9670-1.9680	0.0010-0.0025	0.0039-0.0118
1999	2.0 (1999)	3	2.2827-2.2835	0.0008-0.0017	0.0035-0.0100	3	1.8425-1.8504	0.0006-0.0028	0.0035-0.0126
	2.5 (2507)	G	2.4670-2.4790	0.0009-0.0019	0.0040-0.0090	4	1.9670-1.9680	0.0010-0.0025	0.0039-0.0118
	2.5 (2507)	L	2.4670-2.4790	0.0009-0.0019	0.0040-0.0090	4	1.9670-1.9680	0.0010-0.0025	0.0039-0.0118
2000	2.0 (1999)	3	2.2827-2.2835	0.0008-0.0017	0.0035-0.0100	3	1.8425-1.8504	0.0006-0.0028	0.0035-0.0126
	2.5 (2507)	L	2.4670-2.4790	0.0009-0.0019	0.0040-0.0090	4	1.9670-1.9680	0.0010-0.0025	0.0039-0.0118
2001	2.0 (1999)	3	2.2827-2.2835	0.0008-0.0017	0.0035-0.0100	3	1.8425-1.8504	0.0006-0.0028	0.0035-0.0126
	2.5 (2507)	G	2.4670-2.4790	0.0009-0.0019	0.0040-0.0090	4	1.9670-1.9680	0.0010-0.0025	0.0039-0.0118
	2.5 (2507)	L	2.4670-2.4790	0.0009-0.0019	0.0040-0.0090	4	1.9670-1.9680	0.0010-0.0025	0.0039-0.0118

93461CA2

Heater Core replacement is covered in Section 2 of this manual

PISTON AND RING SPECIFICATIONS

All measurements are given in inches.

	Engine			Ring Gap			Ring Side Clearance		
Year	Displacement Liters (cc)	Engine ID/VIN	Piston Clearance	Top Compression	Bottom Compression	Oil Control	Top Compression	Bottom Compression	Oil Control
1998	2.0 (1999)	3	0.0008-0.0016	0.008-0.010	0.012-0.020	0.016-0.055	0.0016-0.0028	0.0008-0.0021	Snug
	2.5 (2507)	G	0.0005-0.0009	0.004-0.010	0.011-0.017	0.006-0.026	0.0015-0.0029	0.0015-0.0033	Snug
	2.5 (2507)	L	0.0005-0.0009	0.004-0.010	0.011-0.017	0.006-0.026	0.0015-0.0029	0.0015-0.0033	Snug
1999	2.0 (1999)	3	0.0008-0.0016	0.008-0.010	0.012-0.020	0.016-0.055	0.0016-0.0028	0.0008-0.0021	Snug
	2.5 (2507)	G	0.0005-0.0009	0.004-0.010	0.011-0.017	0.006-0.026	0.0015-0.0029	0.0015-0.0033	Snug
	2.5 (2507)	L	0.0005-0.0009	0.004-0.010	0.011-0.017	0.006-0.026	0.0015-0.0029	0.0015-0.0033	Snug
2000	2.0 (1999)	3	0.0008-0.0016	0.008-0.010	0.012-0.020	0.016-0.055	0.0016-0.0028	0.0008-0.0021	Snug
	2.5 (2507)	L	0.0005-0.0009	0.004-0.010	0.011-0.017	0.006-0.026	0.0015-0.0029	0.0015-0.0033	Snug
2001	2.0 (1999)	3	0.0008-0.0016	0.008-0.010	0.012-0.020	0.016-0.055	0.0016-0.0028	0.0008-0.0021	Snug
	2.5 (2507)	G	0.0005-0.0009	0.004-0.010	0.011-0.017	0.006-0.026	0.0015-0.0029	0.0015-0.0033	Snug
	2.5 (2507)	L	0.0005-0.0009	0.004-0.010	0.011-0.017	0.006-0.026	0.0015-0.0029	0.0015-0.0033	Snug

93461CA3

TORQUE SPECIFICATIONS

All readings in ft. lbs.

Year	Engine Displacement Liters (cc)	Engine ID/VIN	Cylinder Head Bolts	Main Bearing Bolts	Rod Bearing Bolts	Crankshaft Damper Bolts	Flywheel Bolts	Manifold Intake	Manifold Exhaust	Spark Plugs	Lug Nuts
1998	2.0 (1999)	3	①	55-66	②	81-89	80-87	12-15	13-16	9-13	63
	2.5 (2507)	G	③	④	⑤	⑥	54-64	6-9	13-16	7-15	63
	2.5 (2507)	L	③	④	⑤	⑥	54-64	6-9	13-16	7-15	63
1999	2.0 (1999)	3	①	55-66	②	81-89	80-87	12-15	13-16	9-13	63
	2.5 (2507)	G	③	④	⑤	⑥	54-64	6-9	13-16	7-15	63
	2.5 (2507)	L	③	④	⑤	⑥	54-64	6-9	13-16	7-15	63
2000	2.0 (1999)	3	①	55-66	②	81-89	80-87	12-15	13-16	9-13	63
	2.5 (2507)	L	③	④	⑤	⑥	54-64	6-9	13-16	7-15	63
2001-02	2.0 (1999)	3	①	55-66	②	81-89	80-87	12-15	13-16	9-13	63
	2.5 (2507)	G	③	④	⑤	⑥	54-64	6-9	13-16	7-15	63
	2.5 (2507)	L	③	④	⑤	⑥	54-64	6-9	13-16	7-15	63

NOTE: Always follow proper torque patterns. Stretch bolts are used in all procedures that require rotating the fastener a certain number of degrees. The bolts stretch and cannot be reused. For reassembly, replace with new fastners.

① Step 1: 15-22 ft. lbs.
Step 2: 30-37 ft. lbs.
Step 3: Tighten 90-120 degrees

② Step 1: 22-25 ft. lbs.
Step 2: Tighten each bolt 85-95 degrees

③ Step 1: 27-32 ft. lbs.
Step 2: Tighten 85-95 degrees
Step 3: Loosen bolts then repeat Step 1
Step 4: Tighten 85-95 degrees
Step 5: Repeat Step 4

④ Step 1: 2.0-3.6 ft. lbs.
Step 2: Push crankshaft rearward.
Lightly seat crankshaft washer forward
Step 3: Outer cap bolts: 16-21 ft. lbs.
Step 4: Inner cap bolts: 27-32 ft. lbs.
Step 5: Tighten inner and outer cap bolts 85-95 degrees
Step 6: Remaining bolts: 15-22 ft. lbs.

⑤ 26-33 ft. lbs. plus 90-120 degrees

⑥ Step 1: 89 ft. lbs.
Step 2: Loosen bolt
Step 3: 35-39 ft. lbs.
Step 4: Tighten 85-95 degrees

93461CA4

BRAKE SPECIFICATIONS
FORD CONTOUR, MERCURY MYSTIQUE, COUGAR
All measurements in inches unless noted

			Brake Disc			Brake Drum				Brake Caliper
Year	Model		Original Thickness	Minimum Thickness	Maximum Run-out	Original Inside Diameter	Max. Wear Limit	Maximum Machine Diameter	Minimum Lining Thickness	Mounting Bolts (ft. lbs.)
1998	Contour	F	0.950	0.870	0.006	—	—	—	0.125	20
		R	0.790	0.710	0.006	8.00	NA	8.04	0.125	30
	Mystique	F	0.950	0.870	0.006	—	—	—	0.125	20
		R	0.790	0.710	0.006	8.00	NA	8.04	0.125	30
1999	Contour	F	0.950	0.870	0.006	—	—	—	0.125	20
		R	0.790	0.710	0.006	8.00	NA	8.04	0.125	30
	Cougar	F	0.950	0.870	0.006	—	—	—	0.125	20
		R	0.790	0.710	0.006	8.00	NA	8.04	0.125	30
	Mystique	F	0.950	0.870	0.006	—	—	—	0.125	20
		R	0.790	0.710	0.006	8.00	NA	8.04	0.125	30
2000	Contour	F	0.950	0.870	0.006	—	—	—	0.125	20
		R	0.790	0.710	0.006	8.00	NA	8.04	0.125	30
	Cougar	F	0.950	0.870	0.006	—	—	—	0.125	20
		R	0.790	0.710	0.006	8.00	NA	8.04	0.125	30
	Mystique	F	0.950	0.870	0.006	—	—	—	0.125	20
		R	0.790	0.710	0.006	8.00	NA	8.04	0.125	30
2001	Cougar	F	0.950	0.870	0.006	—	—	—	0.125	20
		R	0.790	0.710	0.006	8.00	NA	8.04	0.125	30

NOTE: Follow specifications stamped on rotor or drum if figures differ from those in this chart.

NA: Not Available

F: Front

R: Rear

93461CA5

WHEEL ALIGNMENT

			Caster		Camber			Steering
Year	Model		Range (+/-Deg.)	Preferred Setting (Deg.)	Range (+/-Deg.)	Preferred Setting (Deg.)	Toe-in (in.)	Steering Axis Inclination (Deg.)
1998	Contour	F	1.00	+2.10	1.30	-0.53	0 +/- 0.13	—
		R	—	—	1.00	-0.53	0.06 +/- 0.10	—
	Mystique	F	1.00	+2.10	1.00	-0.53	0 +/- 0.05	—
		R	—	—	1.00	-0.53	0.8 +/- 0.05	—
1999	Contour	F	1.00	+2.10	1.30	-0.53	0 +/- 0.13	—
		R	—	—	1.00	-0.53	0.06 +/- 0.10	—
	Cougar	F	1.10	+2.52	1.38	-0.61	0 +/- 0.05	—
		R	—	—	1.28	-0.70	0.8 +/- 0.05	—
	Mystique	F	1.00	+2.10	1.00	-0.53	0 +/- 0.05	—
		R	—	—	1.00	-0.53	0.8 +/- 0.05	—
2000	Contour	F	1.00	+2.10	1.30	-0.53	0 +/- 0.13	—
		R	—	—	1.00	-0.53	0.06 +/- 0.10	—
	Cougar	F	1.10	+2.52	1.38	-0.61	0 +/- 0.05	—
		R	—	—	1.28	-0.70	0.8 +/- 0.05	—
	Mystique	F	1.00	+2.10	1.00	-0.53	0 +/- 0.05	—
		R	—	—	1.00	-0.53	0.8 +/- 0.05	—
2001	Contour	F	1.00	+2.10	1.30	-0.53	0 +/- 0.13	—
		R	—	—	1.00	-0.53	0.06 +/- 0.10	—
	Cougar	F	1.10	+2.52	1.38	-0.61	0 +/- 0.05	—
		R	—	—	1.28	-0.70	0.8 +/- 0.05	—
	Mystique	F	1.00	+2.10	1.00	-0.53	0 +/- 0.05	—
		R	—	—	1.00	-0.53	0.8 +/- 0.05	—
2001	Cougar	F	1.10	+2.52	1.38	-0.61	0 +/- 0.05	—
		R	—	—	1.28	-0.70	0.8 +/- 0.05	—

93461CA6

TIRE, WHEEL AND BALL JOINT SPECIFICATIONS

Year	Model	OEM Tires		Tire Pressures (psi)		Wheel	Ball Joint
		Standard	Optional	Front	Rear	Size	Inspection
1998	Contour GL, LX	P185/70R14	P195/65R14	34	34	5.5-J	①
	Contour SE	P205/60R15	None	31	34	6-JJ	①
	Mystique GS	P185/70R14	P195/65R14	34	34	5.5-JJ	①
	Mystique LS	P205/60R15	None	31	34	6-JJ	①
1999	Contour LX	P185/70R14	P195/65R14	34	34	5.5-J	①
	Contour SE	P205/60R15	None	31	34	6-JJ	①
	Cougar	P205/60R15	P215/50R16	35	35	6-JJ	①
	Mystique GS	P185/70R14	P195/65R14	34	34	5.5-JJ	①
	Mystique LS	P205/60R15	None	31	34	6-JJ	①
2000	Contour LX	P185/70R14	P195/65R14	34	34	5.5-J	①
	Contour SE	P205/60R15	None	31	34	6-JJ	①
	Cougar	P205/60R15	P215/50R16	35	35	6-JJ	①
	Mystique GS	P185/70R14	P195/65R14	34	34	5.5-JJ	①
	Mystique LS	P205/60R15	None	31	34	6-JJ	①
2001	Cougar	P205/60R15	P215/50R16	35	35	6-JJ	①

OEM: Original Equipment Manufacturer

PSI: Pounds Per Square Inch

STD: Standard

OPT: Optional

① Replace if any measurable movement is found

93461CA7

SCHEDULED MAINTENANCE INTERVALS

Ford Contour, Mercury Mystique, Cougar

TO BE SERVICED	TYPE OF SERVICE	VEHICLE MILEAGE INTERVAL (x1000)												
		5	10	15	20	25	30	35	40	45	50	55	60	65
Engine oil & filter	R	✓	✓	✓	✓	✓	✓	✓	✓	✓	✓	✓	✓	✓
Rotate tires	S/I	✓		✓		✓		✓		✓		✓		✓
Front & rear brakes	S/I		✓		✓		✓		✓		✓		✓	
Cooling system, hoses, clamps & coolant strength	S/I			✓			✓			✓			✓	
Passenger compartment air filter	R				✓				✓				✓	
Air cleaner element	R						✓						✓	
Automatic transaxle fluid & filter	S/I												✓	
Exhaust heat shields	S/I						✓						✓	
Accessory drive belt(s)	S/I						✓						✓	
Fuel lines & hoses	S/I						✓						✓	
Crankcase emission filter (2.0L)	R						✓							
Engine coolant ①	R										✓			
Spark plugs ②③	R												✓	
PCV valve	R												✓	

R: Replace S/I: Service or Inspect

① Change initially at 50,000 miles & every 30,000 miles thereafter.

② 2.0L: replace every 60,000 miles.

③ 2.5L: replace every 100,000 miles.

FREQUENT OPERATION MAINTENANCE (SEVERE SERVICE)

If a vehicle is operated under any of the following conditions it is considered severe service:

- Extremely dusty areas.

- 50% or more of the vehicle operation is in 32°C (90°F) or higher temperatures, or constant operation in temperatures below 0°C (32°F).

- Prolonged idling (vehicle operation in stop and go traffic).

- Frequent short running periods (engine does not warm to normal operating temperatures).

- Police, taxi, delivery usage or trailer towing usage.

Oil & filter change: change every 3000 miles.

Front & rear brakes: check every 9000 miles.

Rotate tires at 6000 miles & every 9000 miles thereafter.

Air cleaner element: check every 15,000 miles.

Passenger compartment air filter: change every 18,000 miles.

Automatic transaxle fluid & filter: change every 30,000 miles.

Spark plugs: replace every 60,000 miles.

93461CA8

For Accessory Drive Belt illustrations, see Section 1 of this manual

SCHEDULED MAINTENANCE INTERVALS
FORD MOTOR COMPANY
FORD CONTOUR, MERCURY MYSTIQUE
MERCURY COUGAR

The following should be used as a guide when determining the amount of work required for a particular service. In estimating how long a particular Scheduled Maintenance Service should take, please observe the following:

- Labor Time is time based on field research and data supplied by the vehicle manufacturer.
- Labor time operations are given in hours and tenths of an hour.
- All labor operations are to be used as a guide.

Mechanic Skill Level Codes:
(A) PRECISION: Highly skilled with multiple certification.
(B) GENERAL: Normally skilled with certification.
(C) MAINTENANCE: Semi-skilled working on certification.

	LABOR TIME
5000 Mile Service (C)	
All Models	.8
10000 Mile Service (C)	
All Models	.7
15000 Mile Service (C)	
All Models	.8
20000 Mile Service (C)	
All Models	,7
25000 Mile Service (C)	
All Models	.8
30000 Mile Service (C)	
All Models	1.0
35000 Mile Service (C)	
All Models	.8
40000 Mile Service (C)	
All Models	.7
45000 Mile Service (C)	
All Models	.9
50000 Mile Service (C)	
All Models	1.0
55000 Mile Service (C)	
All Models	.7
60000 Mile Service (B)	
All Models	1.7
Replace AT. fluid & filter add	.6
65000 Mile Service (C)	
All Models	.8

93461CA9

FORD MOTOR CO.
Ford Thunderbird

ENGINE AND VEHICLE IDENTIFICATION

Engine							Model Year	
Code ①	Liters (cc)	Cu. In.	Cyl.	Fuel Sys.	Type	Eng. Mfg.	Code ②	Year
A	3.9 (3947)	243	8	MFI	DOHC	Ford	2	2002

MFI: Multi-Port Fuel Injection

DOHC: Double Overhead Camshaft

① 8th digit of the VIN

② 10th digit of the VIN

93461CA0

GENERAL ENGINE SPECIFICATIONS

Year	Model	Engine Displacement Liters (cc)	Engine ID/VIN	Fuel System Type	Net Horsepower @ rpm	Net Torque @ rpm (ft. lbs.)	Bore x Stroke (in.)	Com-pression Ratio	Oil Pressure @ rpm
2002	Thunderbird	3.9 (3947)	A	SFI	252@6100	267@4300	NA	10.6:1	NA

NA: Not Available

SFI: Sequential Fuel Injection

93461CB1

For Tire, Wheel and Ball Joint specifications, see Section 1 of this manual

ENGINE TUNE-UP SPECIFICATIONS

Year	Engine Displacement Liters (cc)	Engine ID/VIN	Spark Plugs Gap (in.)	Ignition Timing (deg.) ①		Fuel Pump (psi)	Idle Speed (rpm) ①		Valve Clearance	
				MT	AT		MT	AT	In.	Ex.
2002	3.9 (3947)	A	0.039-0.043	—	10-20B	43	—	650-750	0.007-0.009	0.009-0.011

The underhood specifications sticker often reflects tune-up specification changes in production. Sticker figures must be used if they disagree with those in this chart.

① Controlled by the engine computer

93461CB2

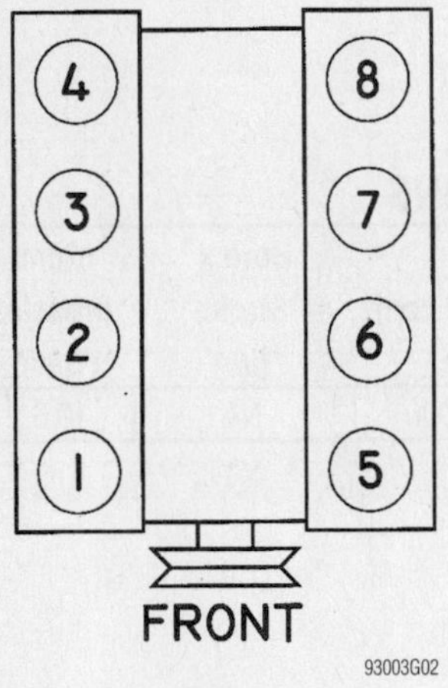

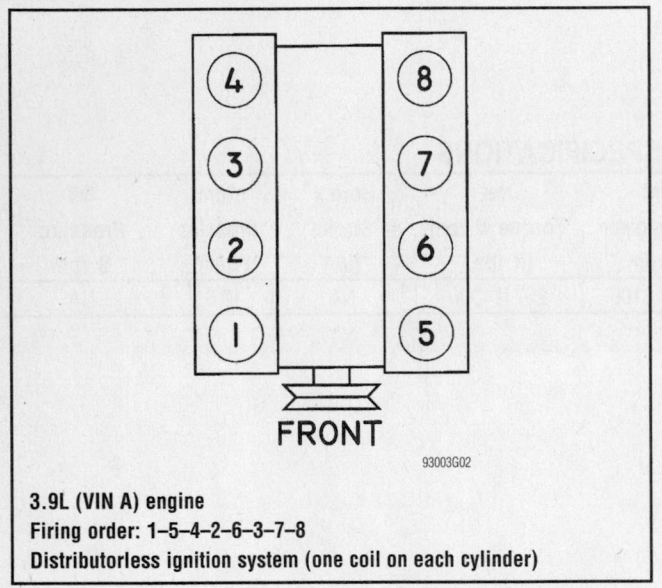

3.9L (VIN A) engine
Firing order: 1–5–4–2–6–3–7–8
Distributorless ignition system (one coil on each cylinder)

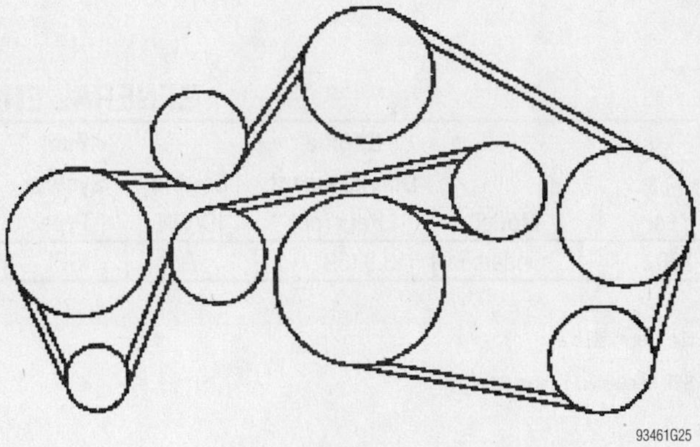

Serpentine accessory drive belt routing—Ford 3.9L (VIN A) engine

CAPACITIES

Year	Model	Engine Displacement Liters (cc)	Engine ID/VIN	Engine Oil with Filter (qts.)	Transmission (pts.)		Drive Axle Rear (pts.)	Fuel Tank (gal.)	Cooling System (qts.)
					Manual	Auto.			
2002	Thunderbird	3.9 (3947)	A	NA	—	23.8	3.0	18.0	11.3

N/A: Not Available

93461CB3

VALVE SPECIFICATIONS

Year	Engine Displacement Liters (cc)	Engine ID/VIN	Seat Angle (deg.)	Face Angle (deg.)	Spring Test Pressure (lbs. @ in.)	Spring Free Length (in.)	Stem-to-Guide Clearance (in.)		Stem Diameter (in.)	
							Intake	Exhaust	Intake	Exhaust
2002	3.9 (3947)	A	NA	NA	NA	NA	NA	NA	NA	NA

NA: Not Available

93461CB4

For Wheel Alignment specifications, see Section 1 of this manual

CRANKSHAFT AND CONNECTING ROD SPECIFICATIONS

All measurements are given in inches.

Year	Engine Displacement Liters (cc)	Engine ID/VIN	Crankshaft				Connecting Rod		
			Main Brg. Journal Dia.	Main Brg. Oil Clearance	Shaft End-play	Thrust on No.	Journal Diameter	Oil Clearance	Side Clearance
2002	3.9 (3947)	A	NA	NA	NA	NA	NA	NA	NA

NA: Not Available

93461CB5

PISTON AND RING SPECIFICATIONS

All measurements are given in inches.

Year	Engine Displacement Liters (cc)	Engine ID/VIN	Piston Clearance	Ring Gap			Ring Side Clearance		
				Top Compression	Bottom Compression	Oil Control	Top Compression	Bottom Compression	Oil Control
2002	3.9 (3947)	A	NA	NA	NA	NA	NA	NA	NA

NA: Not Available

93461CB6

TORQUE SPECIFICATIONS

All readings in ft. lbs.

Year	Engine Displacement Liters (cc)	Engine ID/VIN	Cylinder Head Bolts	Main Bearing Bolts	Rod Bearing Bolts	Crankshaft Damper Bolts	Flywheel Bolts	Manifold Intake	Manifold Exhaust	Spark Plugs	Lug Nut
2002	3.9 (3947)	A	①	NA	NA	②	③	NA	NA	NA	100

NA: Not Available

① Step 1: Tighten M10 bolts to 15 ft. lbs.
Step 2: Tighten M10 bolts to 26 ft. lbs.
Step 3: Tighten M10 bolts to 33 ft. lbs.
Step 4: Tighten M10 bolts plus 90 degrees
Step 5: Tighten M10 bolts plus 90 degrees
Step 6: Tighten M8 bolts to 15 ft. lbs.
Step 7: Tighten M8 bolts plus 90 degrees

② Step 1: 59 ft. lbs.
Step 2: Plus 80 degrees

③ Step 1: 11 ft. lbs.
Step 2: 81 ft. lbs.

93461CB7

BRAKE SPECIFICATIONS

All measurements in inches unless noted

Year	Model		Brake Disc Original Thickness	Brake Disc Minimum Thickness	Brake Disc Maximum Runout	Minimum Lining Thickness	Brake Caliper Bracket Bolts (ft. lbs.)	Brake Caliper Mounting Bolts (ft. lbs.)
2002	Thunderbird	F	NA	NA	NA	NA	NA	NA
		R	NA	NA	NA	NA	NA	NA

93461CB8

WHEEL ALIGNMENT

Year	Model		Caster Range (+/-Deg.)	Caster Preferred Setting (Deg.)	Camber Range (+/-Deg.)	Camber Preferred Setting (Deg.)	Toe-in (in.)	Steering Axis Inclination (Deg.)
2002	Thunderbird	F	0.70	0	0.70	0	0.08 +/- 0.13	—
		R	—	—	0.75	0	0.13 +/- 0.13	—

93461CB9

TIRE, WHEEL AND BALL JOINT SPECIFICATIONS

Year	Model	OEM Tires Standard	OEM Tires Optional	Tire Pressures (psi) Front	Tire Pressures (psi) Rear	Wheel Size	Ball Joint Inspection
2002	Thunderbird	P235/50VR17	—	30	30	7-J/7.5J	U ① L ① ②

OEM: Original Equipment Manufacturer

PSI: Pounds Per Square Inch

STD: Standard

OPT: Optional

L: Lower

U: Upper

① Replace if any measurable movement is found.

② Do not lift car. Inspect the boss into which the grease fitting is threaded. Replace if the boss is flush or receded below the surface of the ball joint.

93461CB0

SCHEDULED MAINTENANCE INTERVALS
Ford Thunderbird

TO BE SERVICED	TYPE OF SERVICE	VEHICLE MILEAGE INTERVAL (x1000) 5	10	15	20	25	30	35	40	45	50	55	60	65
Air cleaner filter	R						✓						✓	
Accessory drive belt	S/I												✓	
Brake system ①	S/I			✓			✓			✓			✓	
Clutch pedal operation	S/I						✓						✓	
Cooling system hoses and clamps	S/I			✓			✓			✓			✓	
CV-joint boots & axle seals	S/I						✓						✓	
Engine coolant	R	Ten years or 150,000 miles												
Engine oil & filter	R	✓	✓	✓	✓	✓	✓	✓	✓	✓	✓	✓	✓	✓
Exterior Lights	S/I	Check monthly												
PCV valve	S/I												✓	
Exhaust system & heat shields	S/I						✓						✓	
Parking brake system	S/I	Every 6 months												
Power steering fluid	S/I	Every 6 months												
Rotate tires	S/I	✓		✓		✓		✓		✓		✓		✓
Steering linkage	S/I						✓						✓	
Spark plugs	R	Change at 100,000 miles												
Suspension components	S/I						✓						✓	

R: Replace S/I: Inspect and service, if necessary L: Lubricate A: Adjust C: Clean

① Inspect the reservoir fluid level, rotor and or drum, brake lines, hoses, calipers and or wheel cylinders

FREQUENT OPERATION MAINTENANCE (SEVERE SERVICE)

If a vehicle is operated under any of the following conditions it is considered severe service:

- Extremely dusty areas.

- 50% or more of the vehicle operation is in 32°C (90°F) or higher temperatures, or constant operation in temperatures below 0°C (32°F).

- Prolonged idling (vehicle operation in stop and go traffic).

- Frequent short running periods (engine does not warm to normal operating temperatures).

- Police, taxi, delivery usage or trailer towing usage.

Oil & oil filter change: change every 3000 miles.

Air filter element: change every 15,000 miles.

93461CC1

FORD MOTOR CO.

Ford Crown Victoria • Lincoln Town Car • Mercury Grand Marquis

ENGINE AND VEHICLE IDENTIFICATION

Engine							Model Year	
Code ①	Liters (cc)	Cu. In.	Cyl.	Fuel Sys.	Type	Eng. Mfg.	Code ②	Year
W	4.6 (4593)	281	8	SFI	SOHC	Ford	W	1998
							X	1999
							Y	2000
							1	2001
							2	2002

SFI: Sequential Fuel Injection

SOHC: Single Overhead Camshaft

① 8th digit of the Vehicle Identification Number (VIN)

② 10th digit fo the Vehicle Identification Number (VIN)

93461CC2

GENERAL ENGINE SPECIFICATIONS

Year	Model	Engine Displacement Liters (cc)	Engine ID/VIN	Fuel System Type	Net Horsepower @ rpm	Net Torque @ rpm (ft. lbs.)	Bore x Stroke (in.)	Com-pression Ratio	Oil Pressure @ rpm
1998	Crown Victoria	4.6 (4593)	W	SFI	①	②	3.55x3.54	③	20-45@1500
	Grand Marquis	4.6 (4593)	W	SFI	①	②	3.55x3.54	9.0:1	20-45@1500
	Town Car	4.6 (4593)	W	SFI	210@4250	275@3250	3.55x3.54	9.0:1	20-45@1500
1999	Crown Victoria	4.6 (4593)	W	SFI	①	②	3.55x3.54	③	20-45@1500
	Grand Marquis	4.6 (4593)	W	SFI	①	②	3.55x3.54	9.0:1	20-45@1500
	Town Car	4.6 (4593)	W	SFI	210@4250	275@3250	3.55x3.54	9.0:1	20-45@1500
2000	Crown Victoria	4.6 (4593)	W	SFI	①	②	3.55x3.54	③	20-45@1500
	Grand Marquis	4.6 (4593)	W	SFI	①	②	3.55x3.54	9.0:1	20-45@1500
	Town Car	4.6 (4593)	W	SFI	210@4250	275@3250	3.55x3.54	9.0:1	20-45@1500
2001	Crown Victoria	4.6 (4593)	W	SFI	④	⑤	3.55x3.54	③	20-45@1500
	Grand Marquis	4.6 (4593)	W	SFI	④	⑤	3.55x3.54	9.4:0	20-45@1500
	Town Car	4.6 (4593)	W	SFI	④	⑤	3.55x3.54	9.4:0	20-45@1500

SFI: Sequential Fuel Injection

① Single exhaust: 190@4250
Dual exhaust: 210@4250
Crown Victoria with natural gas: 178@4500

② Single exhaust: 265@3250
Dual exhaust: 275@3250
Crown Victoria with natural gas: 237@3500

③ Gasoline engine: 9.4:1
Natural gas engine: 10.0:1

④ Single exhaust: 220@4750
Dual exhaust: 235@4000\
Crown Victoria with natural gas: 178@4500

⑤ Single exhaust: 265@4000
Dual exhaust: 275@4000
Crown Victoria with natural gas: 237@3500

93461CC3

ENGINE TUNE-UP SPECIFICATIONS

Year	Engine Displacement Liters (cc)	Engine ID/VIN	Spark Plug Gap (in.)	Ignition Timing (deg.)	Fuel Pump (psi) ①	Idle Speed (rpm)	Valve Clearance Intake	Valve Clearance Exhaust
1998	4.6 (4593)	W	0.054	10B	35-45	②	HYD	HYD
1999	4.6 (4593)	W	0.054	10B	35-45	②	HYD	HYD
2000	4.6 (4593)	W	0.054	10B	35-45	②	HYD	HYD
2001	4.6 (4593)	W	0.054	10B	35-45	②	HYD	HYD

NOTE: The Vehicle Emission Control Information label often reflects specification changes made during production. The label figures must be used if they differ from those in this chart.

B: Before Top Dead Center

HYD: Hydraulic

① Fuel pressure with engine running, pressure regulator vacuum hose connected

② Refer to Vehicle Emission Control Information label

93461CC4

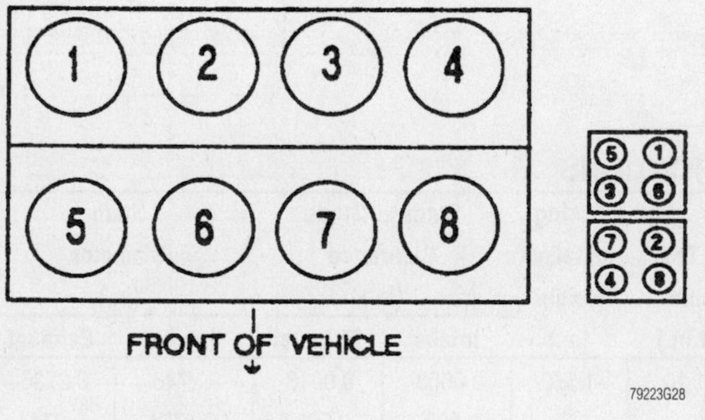

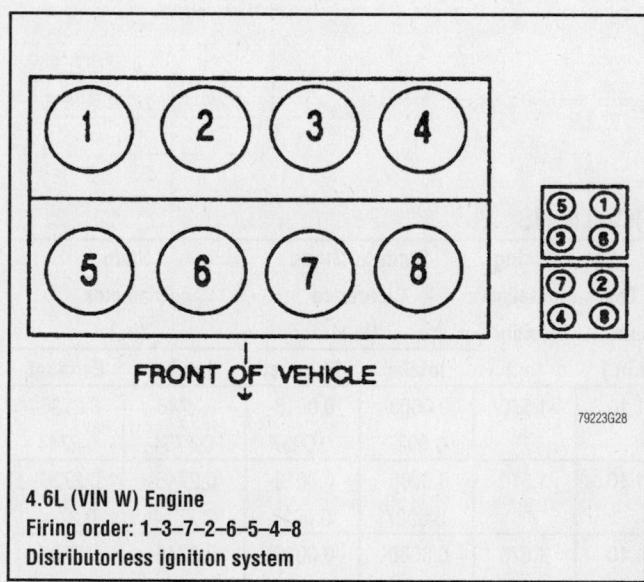

4.6L (VIN W) Engine
Firing order: 1–3–7–2–6–5–4–8
Distributorless ignition system

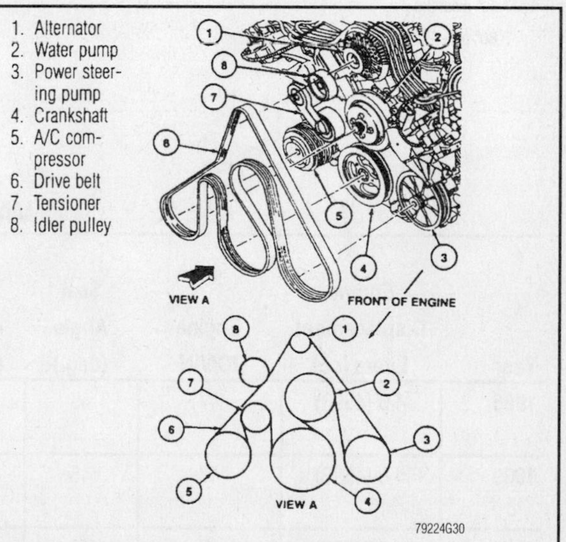

Accessory drive belt routing—Ford 4.6L (VIN W) engine

CAPACITIES

Year	Model	Engine Displacement Liters (cc)	Engine ID/VIN	Engine Oil with Filter (qts.)	Automatic Transmission (pts.) ①	Rear Drive Axle (pts.) ②	Fuel Tank (gal.)	Cooling System (qts.)
1998	Crown Victoria	4.6 (4593)	W	5.0	27.2	3.75	19.0	14.1
	Grand Marquis	4.6 (4593)	W	5.0	28.2	3.75	19.0	15.1
	Town Car	4.6 (4593)	W	5.0	28.2	3.75	19.0	15.1
1999	Crown Victoria	4.6 (4593)	W	5.0	27.2	3.75	19.0	14.1
	Grand Marquis	4.6 (4593)	W	5.0	28.2	3.75	19.0	15.1
	Town Car	4.6 (4593)	W	5.0	28.2	3.75	19.0	15.1
2000	Crown Victoria	4.6 (4593)	W	5.0	27.2	3.75	19.0	14.1
	Grand Marquis	4.6 (4593)	W	5.0	28.2	3.75	19.0	15.1
	Town Car	4.6 (4593)	W	5.0	28.2	3.75	19.0	15.1
2001	Crown Victoria	4.6 (4593)	W	5.0	27.2	3.75	19.0	14.1
	Grand Marquis	4.6 (4593)	W	5.0	28.2	3.75	19.0	15.1
	Town Car	4.6 (4593)	W	5.0	28.2	3.75	19.0	15.1

NOTE: All capacities are approximate. Add fluid gradually and ensure a proper fluid level is obtained.

① Includes torque converter

② 7.50" axle: 3.0 pts.
8.80" axle: 3.25 pts.

93461CC5

VALVE SPECIFICATIONS

Year	Engine Displacement Liters (cc)	Engine ID/VIN	Seat Angle (deg.)	Face Angle (deg.)	Spring Test Pressure (lbs. @ in.)	Spring Installed Height (in.)	Stem-to-Guide Clearance (in.) Intake	Stem-to-Guide Clearance (in.) Exhaust	Stem Diameter (in.) Intake	Stem Diameter (in.) Exhaust
1998	4.6 (4593)	W	45	45.5	132@1.10	1.570	0.0008-0.0027	0.0018-0.0037	0.2746-0.2754	0.2736-0.2744
1999	4.6 (4593)	W	45	45.5	132@1.10	1.570	0.0008-0.0027	0.0018-0.0037	0.2746-0.2754	0.2736-0.2744
2000	4.6 (4593)	W	45	45.5	132@1.10	1.570	0.0008-0.0027	0.0018-0.0037	0.2746-0.2754	0.2736-0.2744
2001	4.6 (4593)	W	45	45.5	132@1.10	1.570	0.0008-0.0027	0.0018-0.0037	0.2746-0.2754	0.2736-0.2744

93461CC6

CRANKSHAFT AND CONNECTING ROD SPECIFICATIONS

All measurements are given in inches.

Year	Engine Displacement Liters (cc)	Engine ID/VIN	Crankshaft Main Brg. Journal Dia.	Crankshaft Main Brg. Oil Clearance	Crankshaft Shaft End-play	Crankshaft Thrust on No.	Connecting Rod Journal Diameter	Connecting Rod Oil Clearance	Connecting Rod Side Clearance
1998	4.6 (4593)	W	2.6500-2.6570	0.0009-0.0026	0.0051-0.0119	5	2.0870-2.8670	0.0009-0.0026	0.0006-0.0177
1999	4.6 (4593)	W	2.6500-2.6570	0.0009-0.0026	0.0051-0.0119	5	2.0870-2.8670	0.0009-0.0026	0.0006-0.0177
2000-01	4.6 (4593)	W	2.6500-2.6570	0.0009-0.0026	0.0051-0.0119	5	2.0870-2.8670	0.0009-0.0026	0.0006-0.0177
2000-01	4.6 (4593)	W	2.6500-2.6570	0.0009-0.0026	0.0051-0.0119	5	2.0870-2.8670	0.0009-0.0026	0.0006-0.0177

93461CC7

PISTON AND RING SPECIFICATIONS

All measurements are given in inches.

Year	Engine Displacement Liters (cc)	Engine ID/VIN	Piston Clearance ①	Ring Gap Top Compression	Ring Gap Bottom Compression	Ring Gap Oil Control	Ring Side Clearance Top Compression	Ring Side Clearance Bottom Compression	Ring Side Clearance Oil Control
1998	4.6 (4593)	W	0.0002-0.0010	0.005-0.012	0.012-0.022	0.006-0.026	②	0.008-0.0024	0.0010-0.0077
1999	4.6 (4593)	W	0.0002-0.0010	0.005-0.012	0.012-0.022	0.006-0.026	②	0.008-0.0024	0.0010-0.0077
2000	4.6 (4593)	W	0.0002-0.0010	0.005-0.012	0.012-0.022	0.006-0.026	②	0.008-0.0024	0.0010-0.0077
2001	4.6 (4593)	W	0.0002-0.0010	0.005-0.012	0.012-0.022	0.006-0.026	②	0.008-0.0024	0.0010-0.0077

① Measured 1.96 in. (43mm) from the top

② On 10:1 engines: 0.0012 - 0.0028 in.
On 9:1 engines: 0.0008 - 0.0024 in.

93461CC8

TORQUE SPECIFICATIONS

All readings in ft. lbs.

Year	Engine Displacement Liters (cc)	Engine ID/VIN	Cylinder Head Bolts	Main Bearing Bolts	Rod Bearing Bolts	Crankshaft Damper Bolts	Flywheel Bolts	Manifold Intake	Manifold Exhaust	Spark Plugs	Lug Nuts
1998	4.6 (4593)	W	①	②	③	114-121	54-64	15-22	13-16	7-15	95
1999	4.6 (4593)	W	①	②	③	114-121	54-64	15-22	13-16	7-15	95
2000	4.6 (4593)	W	①	②	③	114-121	54-64	15-22	13-16	7-15	95
2001	4.6 (4593)	W	①	②	③	114-121	54-64	15-22	13-16	7-15	95

NOTE: Stretch bolts are used in all procedures that require rotating the fastener a certain number of degrees. The bolts stretch and cannot be reused. For reassembly, replace with new fasteners.

① Step 1: 22-30 ft. lbs.
Step 2: Plus 85-95 degrees
Step 3: Plus 85-95 degrees

② Step 1: Main bearing cap bolts: 22-25 ft. lbs.
Step 2: Rotate each bolt 85-95 degrees
Step 3: Main bearing cap adjusting screws: 44 inch lbs. then 80-90 inch lbs.
Step 4: Main bearing cap side bolts: 7 ft. lbs. then 14-17 ft. lbs.

③ Step 1: 8 ft. lbs.
Step 2: 12 ft. lbs.
Step 3: 25-34 ft. lbs.
Step 4: Rotate 85-95 degrees

93461CC9

BRAKE SPECIFICATIONS

All measurements in inches unless noted

		Front Brake Disc			Rear Brake Disc			Minimum	Brake Caliper	
		Original	Minimum	Maximum	Original	Minimum	Maximum	Lining	Bracket Bolts	Mounting Bolts
Year	Model	Thickness	Thickness	Run-out	Thickness	Thickness	Run-out	Thickness	(ft. lbs.)	(ft. lbs.)
1998	Crown Victoria	1.024	0.974	0.002	0.550	0.510	0.002	0.125	125-169	21-26
	Grand Marquis	1.024	0.974	0.002	0.550	0.510	0.002	0.125	125-169	21-26
	Town Car	1.024	0.974	0.002	0.550	0.510	0.002	0.125	125-169	21-26
1999	Crown Victoria	1.024	0.974	0.002	0.550	0.510	0.002	0.125	125-169	21-26
	Grand Marquis	1.024	0.974	0.002	0.550	0.510	0.002	0.125	125-169	21-26
	Town Car	1.024	0.974	0.002	0.550	0.510	0.002	0.125	125-169	21-26
2000	Crown Victoria	1.024	0.974	0.002	0.550	0.510	0.002	0.125	125-169	21-26
	Grand Marquis	1.024	0.974	0.002	0.550	0.510	0.002	0.125	125-169	21-26
	Town Car	1.024	0.974	0.002	0.550	0.510	0.002	0.125	125-169	21-26
2001	Crown Victoria	1.024	0.974	0.002	0.550	0.510	0.002	0.125	125-169	21-26
	Grand Marquis	1.024	0.974	0.002	0.550	0.510	0.002	0.125	125-169	21-26
	Town Car	1.024	0.974	0.002	0.550	0.510	0.002	0.125	125-169	21-26

NOTE: Follow specifications stamped on rotor or drum if figures differ from those in this chart.

93461CC0

WHEEL ALIGNMENT

Year	Model		Caster Range (+/-Deg.)	Caster Preferred Setting (Deg.)	Camber Range (+/-Deg.)	Camber Preferred Setting (Deg.)	Toe-in (in.)	Steering Axis Inclination (Deg.)
1998	Crown Victoria	F	0.75	+0.50	0.75	0	-0.25 +/- 0.25	—
		R	—	—	—	—	—	—
	Town Car	F	0.75	0	0.75	0	-0.12 +/- 0.25	—
		R	—	—	—	—	—	—
	Grand Marquis	F	0.75	+0.50	0.75	0	-0.13 +/- 0.25	—
		R	—	—	—	—	—	—
1999	Crown Victoria	F	0.75	+0.50	0.75	0	-0.25 +/- 0.25	—
		R	—	—	—	—	—	—
	Town Car	F	0.75	0	0.75	0	-0.12 +/- 0.25	—
		R	—	—	—	—	—	—
	Grand Marquis	F	0.75	+0.50	0.75	0	-0.13 +/- 0.25	—
		R	—	—	—	—	—	—
2000	Crown Victoria	F	0.75	+0.50	0.75	0	-0.25 +/- 0.25	—
		R	—	—	—	—	—	—
	Town Car	F	0.75	0	0.75	0	-0.12 +/- 0.25	—
		R	—	—	—	—	—	—
	Grand Marquis	F	0.75	+0.50	0.75	0	-0.13 +/- 0.25	—
		R	—	—	—	—	—	—
2001	Crown Victoria	F	0.75	+0.50	0.75	0	-0.25 +/- 0.25	—
		R	—	—	—	—	—	—
	Town Car	F	0.75	0	0.75	0	-0.12 +/- 0.25	—
		R	—	—	—	—	—	—
	Grand Marquis	F	0.75	+0.50	0.75	0	-0.13 +/- 0.25	—
		R	—	—	—	—	—	—

93461CD1

TIRE, WHEEL AND BALL JOINT SPECIFICATIONS

Year	Model	OEM Tires		Tire Pressures (psi)		Wheel	Ball Joint
		Standard	Optional	Front	Rear	Size	Inspection
1998	Crown Victoria	P225/60SR16	P225/60TR16	32	32	7-JJ	U① L①②
	Crown Victoria Police Special	P225/60VR16	None	32	32	7J	U① L①②
	Grand Marquis	P225/60SR16	P225/60TR16	35	35	6-JJ	U① L①②
	Town Car	P215/70R15	P225/60R16	30	30	Std: 6.5-JJ Opt: 7-JJ	U① L①②
1999	Crown Victoria	P225/60SR16	P225/60TR16	32	32	7-JJ	U① L①②
	Crown Victoria Police Special	P225/60VR16	None	32	32	7J	U① L①②
	Grand Marquis	P225/60SR16	P225/60TR16	35	35	6-JJ	U① L①②
	Town Car	P215/70R15	P225/60R16	30	30	Std: 6.5-JJ Opt: 7-JJ	U① L①②
2000	Crown Victoria	P225/60SR16	P225/60TR16	32	32	7-JJ	U① L①②
	Crown Victoria Police Special	P225/60SR16	P225/60VR16	35	35	7J	U① L①②
	Grand Marquis	P225/60SR16	None	32	32	6-JJ	U① L①②
	Grand Marquis w/Handling package	P225/60TR16	P225/60VR16	35	35	6-JJ	U① L①②
	Town Car	P225/70R15	P235/60R16	30	30	Std: 6.5-JJ Opt: 7-JJ	U① L①②
2001	Crown Victoria	P225/60SR16	P225/60TR16	32	32	7-JJ	U① L①②
	Crown Victoria Police Special	P225/60SR16	P225/60VR16	35	35	7J	U① L①②
	Grand Marquis	P225/60SR16	None	32	32	6-JJ	U① L①②
	Grand Marquis w/Handling package	P225/60TR16	P225/60VR16	35	35	6-JJ	U① L①②
	Town Car	P225/70R15	P235/60R16	30	30	Std: 6.5-JJ Opt: 7-JJ	U① L①②

OEM: Original Equipment Manufacturer

PSI: Pounds Per Square Inch

STD: Standard

OPT: Optional

① Replace if any measurable movement is found.

② Do not lift car. Inspect the boss into which the grease fitting is threaded. Replace if the boss is flush or receded below the surface of the ball joint.

93461CD2

Heater Core replacement is covered in Section 2 of this manual

SCHEDULED MAINTENANCE INTERVALS

Ford Crown Victoria, Mercury Grand Marquis, Lincoln Town Car

TO BE SERVICED	TYPE OF SERVICE	VEHICLE MILEAGE INTERVAL (x1000)												
		5	10	15	20	25	30	35	40	45	50	55	60	65
Engine oil & filter	R	✓	✓	✓	✓	✓	✓	✓	✓	✓	✓	✓	✓	✓
Rotate tires	S/I	✓		✓		✓		✓		✓		✓		✓
Cooling system, hoses, clamps & coolant strength	S/I			✓			✓			✓			✓	
Lubricate steering linkage	S/I			✓			✓			✓			✓	
Air cleaner element	R						✓						✓	
Automatic transaxle fluid & filter	R						✓						✓	
Spark plugs ①	R													
Exhaust heat shields	S/I						✓						✓	
Fuel filter (NGV Crown Victoria) ②	R					✓					✓			
Front & rear brakes	S/I						✓						✓	
Lubricate suspension (Town Car)	S/I						✓						✓	
Engine coolant ③	R										✓			
PCV valve	R												✓	
Accessory drive belt	S/I												✓	

R: Replace S/I: Service or Inspect

① Replace every 100,000 miles.

② Also drain coalescer assembly. Perform every 24,000 miles for severe service.

③ Change initially at 50,000 miles & thereafter every 30,000 miles.

FREQUENT OPERATION MAINTENANCE (SEVERE SERVICE)

If a vehicle is operated under any of the following conditions it is considered severe service:

- **Extremely dusty areas.**
- **50% or more of the vehicle operation is in 32°C (90°F) or higher temperatures, or constant operation in temperatures below 0°C (32°F).**
- **Prolonged idling (vehicle operation in stop and go traffic).**
- **Frequent short running periods (engine does not warm to normal operating temperatures).**
- **Police, taxi, delivery usage or trailer towing usage.**

Oil & filter change: change every 3000 miles.

Rotate tires at 6000 miles & every 9000 miles thereafter.

Automatic transmission fluid & filter: change every 21,000 miles.

93461CD3

SCHEDULED MAINTENANCE INTERVALS
FORD MOTOR COMPANY
FORD CROWN VICTORIA, LINCOLN TOWN CAR
MERCURY GRAND MARQUIS

The following should be used as a guide when determining the amount of work required for a particular service. In estimating how long a particular Scheduled Maintenance Service should take, please observe the following:

- Labor Time is time based on field research and data supplied by the vehicle manufacturer.
- Labor time operations are given in hours and tenths of an hour.
- All labor operations are to be used as a guide.

Mechanic Skill Level Codes:
(A) PRECISION: Highly skilled with multiple certification.
(B) GENERAL: Normally skilled with certification.
(C) MAINTENANCE: Semi-skilled working on certification.

	LABOR TIME
5000 Mile Service (C)	
All Models	.8
10000 Mile Service (C)	
All Models	.4
15000 Mile Service (C)	
All Models	.9
20000 Mile Service (C)	
All Models	.4
25000 Mile Service (C)	
All Models	.8
30000 Mile Service (B)	
All Models	1.9
Replace auto. trans. fluid, filter add	.5
35000 Mile Service (C)	
All Models	.8
40000 Mile Service (C)	
All Models	.4
45000 Mile Service (C)	
All Models	.9
50000 Mile Service (B)	
All Models	1.7
Replace engine coolant add	.5
55000 Mile Service (C)	
All Models	.8
60000 Mile Service (B)	
All Models	2.1
Replace auto. trans. fluid, filter add	.5
65000 Mile Service (C)	
All Models	.8

93461CD4

Brake service is covered in Section 4 of this manual

FORD MOTOR CO.

Ford Escort • Escort ZX2 • Mercury Tracer

ENGINE AND VEHICLE IDENTIFICATION

Engine							Model Year	
Code ①	Liters (cc)	Cu. In.	Cyl.	Fuel Sys.	Type	Eng. Mfg.	Code ②	Year
3	2.0 (1990)	121	4	SFI	DOHC	Ford	W	1998
P	2.0 (1999)	121	4	SFI	SOHC	Ford	X	1999
							Y	2000
							1	2001
							2	2002

SFI: Sequential Fuel Injection

DOHC: Double Overhead Camshafts

SOHC: Single Overhead Camshaft

① 8th digit of the VIN

② 10th digit of the VIN

93461CD5

GENERAL ENGINE SPECIFICATIONS

Year	Model	Engine Displacement Liters (cc)	Engine ID/VIN	Fuel System Type	Net Horsepower @ rpm	Net Torque @ rpm (ft. lbs.)	Bore x Stroke (in.)	Com-pression Ratio	Oil Pressure @ rpm
1998	Escort	2.0 (1999)	P	SFI	110@5000	125@3750	3.34x3.46	9.2:1	35-65@2000
	Escort ZX2	2.0 (1999)	3	SFI	125@5500	130@4000	3.34x3.46	10.0:1	35-65@2000
	Tracer	2.0 (1999)	P	SFI	110@5000	125@3750	3.34x3.46	9.2:1	35-65@2000
1999	Escort	2.0 (1999)	P	SFI	110@5000	125@3750	3.34x3.46	9.2:1	35-65@2000
	Escort ZX2	2.0 (1999)	3	SFI	125@5500	130@4000	3.34x3.46	10.0:1	35-65@2000
	Tracer	2.0 (1999)	P	SFI	110@5000	125@3750	3.34x3.46	9.2:1	35-65@2000
2000	Escort	2.0 (1999)	P	SFI	110@5000	125@3750	3.34x3.46	9.2:1	35-65@2000
	Escort ZX2	2.0 (1999)	3	SFI	125@5500	130@4000	3.34x3.46	10.0:1	35-65@2000
2001	Escort ZX2	2.0 (1999)	3	SFI	125@5500	130@4000	3.34x3.46	10.0:1	35-65@2000

SFI: Sequential Fuel Injection

93461CD6

ENGINE TUNE-UP SPECIFICATIONS

Year	Engine Displacement Liters (cc)	Engine ID/VIN	Spark Plug Gap (in.)	Ignition Timing (deg.) MT	Ignition Timing (deg.) AT	Fuel Pump (psi)	Idle Speed (rpm) MT	Idle Speed (rpm) AT	Valve Clearance In.	Valve Clearance Ex.
1998	2.0 (1999)	3	0.052-0.056	10B	10B	31-38 ①	②	②	HYD	HYD
	2.0 (1999)	P	0.052-0.056	10B	10B	38-45 ①	②	②	HYD	HYD
19990	2.0 (1999)	3	0.052-0.056	10B	10B	31-38 ①	②	②	HYD	HYD
	2.0 (1999)	P	0.052-0.056	10B	10B	38-45 ①	②	②	HYD	HYD
2000	2.0 (1999)	3	0.052-0.056	10B	10B	31-38 ①	②	②	HYD	HYD
	2.0 (1999)	P	0.052-0.056	10B	10B	38-45 ①	②	②	HYD	HYD
2001	2.0 (1999)	3	0.052-0.056	10B	10B	31-38 ①	②	②	HYD	HYD

NOTE: The Vehicle Emission Control Information label often reflects specification changes made during production. The label figures must be used if they differ from those in this chart.

B: Before Top Dead Center

HYD: Hydraulic

① Fuel pressure with engine running, pressure regulator vacuum hose connected

② Refer to Vehicle Emission Control Information label

93461CD7

2.0L (VIN P) Engine
Firing order: 1–3–4–2
Distributorless ignition system

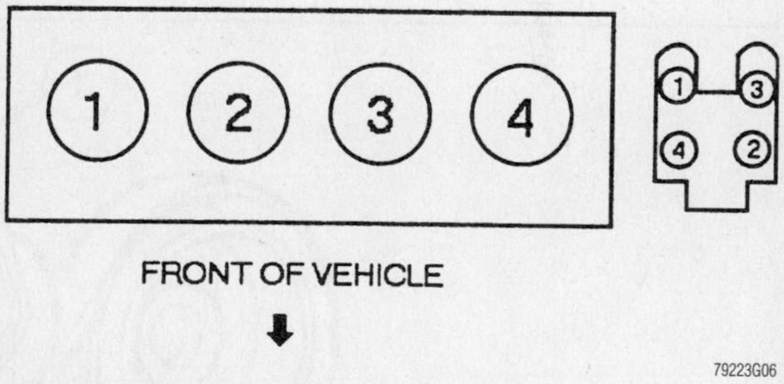

2.0L (VIN 3) Engine
Firing order: 1–3–4–2
Distributorless ignition system

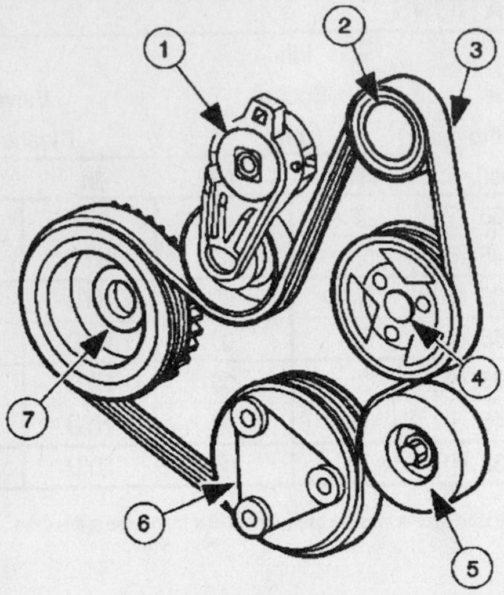

1	Belt tensioner
2	Generator pulley
3	Belt
4	Power steering pump pulley
5	Belt idler pulley
6	A/C clutch pulley
7	Crankshaft damper

93004G03

Accessory drive belt routing —Ford/Mercury 2.0L (VIN P) SOHC engine with A/C

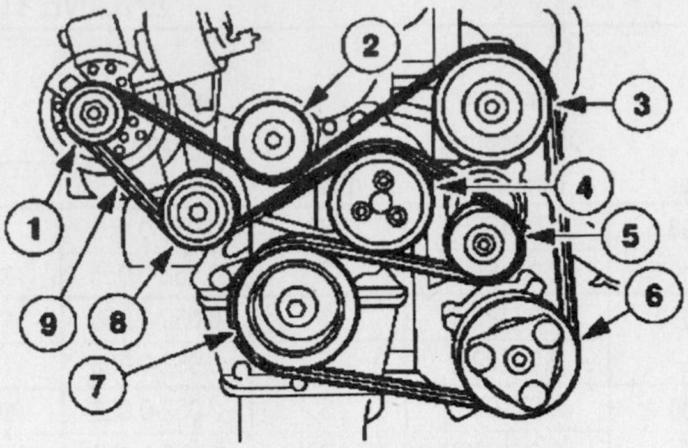

1	Generator pulley
2	Belt idler pulley
3	Power steering pump pulley
4	Water pump pulley
5	Belt tensioner
6	A/C clutch pulley
7	Crankshaft pulley damper
8	Belt idler pulley
9	Drive belt

93004G05

Accessory drive belt routing —Ford/Mercury 2.0L (VIN 3) DOHC engine

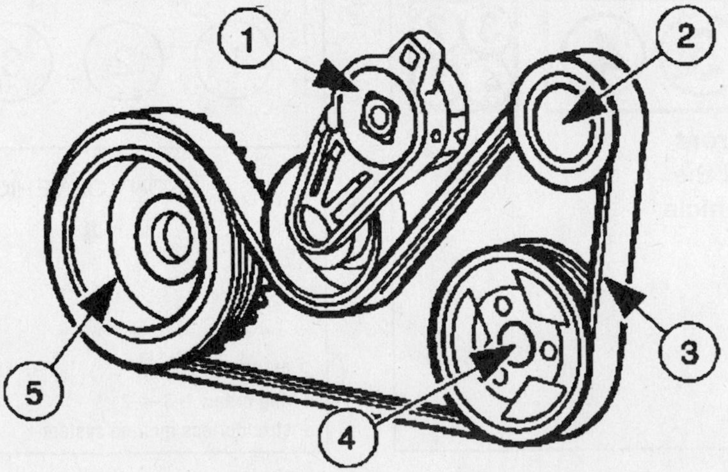

1	Belt tensioner
2	Generator pulley
3	Belt
4	Power steering pump pulley
5	Crankshaft damper

93004G04

Accessory drive belt routing —Ford/Mercury 2.0L (VIN P) SOHC engine without A/C

CAPACITIES

Year	Model	Engine Displacement Liters (cc)	Engine ID/VIN	Oil with Filter (qts.)	Transmission (pts.) Manual	Transmission (pts.) Auto. ①	Front Axle (pts.)	Fuel Tank (gal.)	Cooling System (qts.)
1998	Escort	2.0 (1999)	P	4.0	5.7	13.4	②	11.9	③
	Escort ZX2	2.0 (1999)	3	4.0	6.7	13.4	②	13.2	④
	Tracer	2.0 (1999)	P	4.0	5.7	13.4	②	11.9	③
1999	Escort	2.0 (1999)	P	4.0	5.7	13.4	②	11.9	③
	Escort ZX2	2.0 (1999)	3	4.0	6.7	13.4	②	13.2	④
	Tracer	2.0 (1999)	P	4.0	5.7	13.4	②	11.9	③
2000	Escort	2.0 (1999)	P	4.0	5.7	13.4	②	11.9	③
	Escort ZX2	2.0 (1999)	3	4.0	6.7	13.4	②	13.2	④
	Tracer	2.0 (1999)	P	4.0	5.7	13.4	②	11.9	③
2001	Escort ZX2	2.0 (1999)	3	4.0	6.7	13.4	②	13.2	④

Note: All capacities are approximates. Add fluid gradually and ensure a proper fluid level is obtained.

① Includes torque converter

② Included in transaxle capacity

③ Manual transaxle 7.9

Automatic transaxle 5.8

④ Manual transaxle 7.0

Automatic transaxle 7.5

93461CD8

For Accessory Drive Belt illustrations, see Section 1 of this manual

VALVE SPECIFICATIONS

Year	Engine Displacement Liters (cc)	Engine ID/VIN	Seat Angle (deg.)	Face Angle (deg.)	Spring Test Pressure (lbs. @ in.)	Spring Installed Height (in.)	Stem-to-Guide Clearance (in.)		Stem Diameter (in.)	
							Intake	Exhaust	Intake	Exhaust
1998	2.0 (1999)	3	45	45	NA	1.346	0.0007-0.0025	0.0014-0.0032	0.2373-0.2379	0.2366-0.2372
	2.0 (1999)	P	45	45.6	200@1.09	1.420-1.540	0.0008-0.0027	0.0018-0.0037	0.3159-0.3167	0.3149-0.3156
1999	2.0 (1999)	3	45	45	NA	1.346	0.0007-0.0025	0.0014-0.0032	0.2373-0.2379	0.2366-0.2372
	2.0 (1999)	P	45	45.6	200@1.09	1.420-1.540	0.0008-0.0027	0.0018-0.0037	0.3159-0.3167	0.3149-0.3156
2000	2.0 (1999)	3	45	45	NA	1.346	0.0007-0.0025	0.0014-0.0032	0.2373-0.2379	0.2366-0.2372
	2.0 (1999)	P	45	45.6	200@1.09	1.420-1.540	0.0008-0.0027	0.0018-0.0037	0.3159-0.3167	0.3149-0.3156
2001	2.0 (1999)	3	45	45	NA	1.346	0.0007-0.0025	0.0014-0.0032	0.2373-0.2379	0.2366-0.2372

93461CD9

CRANKSHAFT AND CONNECTING ROD SPECIFICATIONS

All measurements are given in inches.

Year	Engine Displacement Liters (cc)	Engine ID/VIN	Crankshaft				Connecting Rod		
			Main Brg. Journal Dia.	Main Brg. Oil Clearance	Shaft End-play	Thrust on No.	Journal Diameter	Oil Clearance	Side Clearance
1998	2.0 (1999)	3	2.2827-2.2835	0.0008-0.0026	0.0040-0.0120	3	1.8461-1.8500	0.0006-0.0028	0.0040-0.0110
	2.0 (1999)	P	2.2827-2.2835	0.0008-0.0026	0.0040-0.0120	3	1.7279-1.7287	0.0008-0.0026	0.0040-0.0110
1999	2.0 (1999)	3	2.2827-2.2835	0.0008-0.0026	0.0040-0.0120	3	1.8461-1.8500	0.0006-0.0028	0.0040-0.0110
	2.0 (1999)	P	2.2827-2.2835	0.0008-0.0026	0.0040-0.0120	3	1.7279-1.7287	0.0008-0.0026	0.0040-0.0110
2000	2.0 (1999)	3	2.2827-2.2835	0.0008-0.0026	0.0040-0.0120	3	1.8461-1.8500	0.0006-0.0028	0.0040-0.0110
	2.0 (1999)	P	2.2827-2.2835	0.0008-0.0026	0.0040-0.0120	3	1.7279-1.7287	0.0008-0.0026	0.0040-0.0110
2001	2.0 (1999)	3	2.2827-2.2835	0.0008-0.0026	0.0040-0.0120	3	1.8461-1.8500	0.0006-0.0028	0.0040-0.0110

93461C00

PISTON AND RING SPECIFICATIONS

All measurements are given in inches.

Year	Engine Displacement Liters (cc)	Engine ID/VIN	Piston Clearance	Ring Gap Top Compression	Ring Gap Bottom Compression	Ring Gap Oil Control	Ring Side Clearance Top Compression	Ring Side Clearance Bottom Compression	Ring Side Clearance Oil Control
1998	2.0 (1999)	3	0.0004-0.0012	0.0118-0.0197	0.0118-0.0197	0.0158-0.0551	0.0015-0.0032	0.0015-0.0035	—
	2.0 (1999)	P	0.0008-0.0027	0.010-0.030	0.010-0.030	0.016-0.066	0.0015-0.0032	0.0015-0.0035	—
1999	2.0 (1999)	3	0.0010-0.0022	0.012-0.022	0.012-0.022	0.010-0.039	0.0015-0.0032	0.0015-0.0035	—
	2.0 (1999)	P	0.0008-0.0027	0.010-0.030	0.010-0.030	0.016-0.066	0.0015-0.0032	0.0015-0.0035	—
2000	2.0 (1999)	3	0.0010-0.0022	0.012-0.022	0.012-0.022	0.010-0.039	0.0015-0.0032	0.0015-0.0035	—
	2.0 (1999)	P	0.0008-0.0027	0.010-0.030	0.010-0.030	0.016-0.066	0.0015-0.0032	0.0015-0.0035	—
2001	2.0 (1999)	3	0.0010-0.0022	0.012-0.022	0.012-0.022	0.010-0.039	0.0015-0.0032	0.0015-0.0035	—

93461CE1

TORQUE SPECIFICATIONS

All readings in ft. lbs.

Year	Engine Displacement Liters (cc)	Engine ID/VIN	Cylinder Head Bolts	Main Bearing Bolts	Rod Bearing Bolts	Crankshaft Damper Bolts	Flywheel Bolts	Manifold Intake	Manifold Exhaust	Spark Plugs	Lug Nut
1998	2.0 (1999)	P	①	66-79	26-30	80-87	71-76	15-22	15-17	12-15	76
	2.0 (1999)	3	②	55-65	③	80-87	71-76	11-12	10-12	10-12	76
1999	2.0 (1999)	3	②	55-65	③	80-87	71-76	11-12	10-12	10-12	76
	2.0 (1999)	P	①	66-79	26-30	80-87	71-76	15-22	15-17	12-15	76
2000	2.0 (1999)	3	②	55-65	③	80-87	71-76	11-12	10-12	10-12	76
	2.0 (1999)	P	①	66-79	26-30	80-87	71-76	15-22	15-17	12-15	76
2001	2.0 (1999)	3	②	55-65	③	80-87	71-76	11-12	10-12	10-12	76

NOTE: Always follow proper torque patterns

NOTE: Stretch bolts are used in all procedures that require rotating the fastener a certain number of degrees. The bolts stretch and cannot be reused. For reassembly, replace with new fastners.

① Do not reuse cylinder head bolts.

Step 1: Tighten bolts, in sequence, to 44 ft. lbs.

Step 2: Loosen bolts approx. two turns, then retighten in sequence to 44 ft. lbs.

Step 3: Turn all bolts, in sequence, plus an additional 90 degrees

Step 4: Repeat Step 3

② Step 1: 15-22 ft. lbs.

Step 2: 30-37 ft. lbs.

Step 3: Rotate 90-120 degrees

③ Step 1: 22-25 ft. lbs.

Step 2: Rotate each bolt 85-95 degrees

93461CE2

For Wheel Alignment specifications, see Section 1 of this manual

BRAKE SPECIFICATIONS

All measurements in inches unless noted

			Brake Disc			Brake Drum Diameter				Brake Caliper
Year	Model		Original Thickness	Minimum Thickness	Maximum Runout	Original Inside Diameter	Max. Wear Limit	Maximum Machine Diameter	Minimum Lining Thickness	Mounting Bolts (ft. lbs.)
1998	Escort	F	0.870	0.790	0.004	—	—	—	0.080	36-43
		R	0.350	0.280	0.004	7.87	7.95	7.91	0.040	—
	Escort ZX2	F	0.870	0.790	0.004	—	—	—	0.080	36-43
		R	0.350	0.280	0.004	7.87	7.95	7.91	0.040	—
	Tracer	F	0.870	0.790	0.004	—	—	—	0.080	36-43
		R	0.350	0.280	0.004	7.87	7.95	7.91	0.040	—
1999	Escort	F	0.870	0.790	0.004	—	—	—	0.080	36-43
		R	0.350	0.280	0.004	7.87	7.95	7.91	0.040	—
	Escort ZX2	F	0.870	0.790	0.004	—	—	—	0.080	36-43
		R	0.350	0.280	0.004	7.87	7.95	7.91	0.040	—
	Tracer	F	0.870	0.790	0.004	—	—	—	0.080	36-43
		R	0.350	0.280	0.004	7.87	7.95	7.91	0.040	—
2000	Escort	F	0.870	0.790	0.004	—	—	—	0.080	36-43
		R	0.350	0.280	0.004	7.87	7.95	7.91	0.040	—
	Escort ZX2	F	0.870	0.790	0.004	—	—	—	0.080	36-43
		R	0.350	0.280	0.004	7.87	7.95	7.91	0.040	—
2001	Escort ZX2	F	0.870	0.790	0.004	—	—	—	0.080	36-43
		R	0.350	0.280	0.004	7.87	7.95	7.91	0.040	—

NOTE: Follow specifications stamped on rotor or drum if figures differ from those in this chart.

NA: Not Available

F: Front

R: Rear

93461CE3

WHEEL ALIGNMENT

Year	Model		Caster Range (+/-Deg.)	Caster Preferred Setting (Deg.)	Camber Range (+/-Deg.)	Camber Preferred Setting (Deg.)	Toe-in (in.)	Steering Axis Inclination (Deg.)
1998	Escort	F	1.00	+2.20	1.00	+0.40	0.08 +/- 0.10	12.85
		R	—	—	1.00	+1.18	0.08 +/- 0.10	—
	Tracer	F	0.75	+2.30	0.75	-0.40	0.08 +/- 0.11	12.85
		R	—	—	1.00	-1.18	0.08 +/- 0.11	—
1999	Escort	F	1.00	+2.20	1.00	+0.40	0.08 +/- 0.10	12.85
		R	—	—	1.00	+1.18	0.08 +/- 0.10	—
	Tracer	F	0.75	+2.30	0.75	-0.40	0.08 +/- 0.11	12.85
		R	—	—	1.00	-1.18	0.08 +/- 0.11	—
2000	Escort	F	1.00	+2.20	1.00	+0.40	0.08 +/- 0.10	12.85
		R	—	—	1.00	+1.18	0.08 +/- 0.10	—
2001	Escort	F	1.00	+2.20	1.00	+0.40	0.08 +/- 0.10	12.85
		R	—	—	1.00	+1.18	0.08 +/- 0.10	—

93461CE4

TIRE, WHEEL AND BALL JOINT SPECIFICATIONS

		OEM Tires		Tire Pressures (psi)		Wheel	Ball Joint
Year	Model	Standard	Optional	Front	Rear	Size	Inspection
1998	Escort	P185/65R14	P185/60R15	32	32	5.5-JJ	0.030 in. ①
	Tracer	P185/65SR14	P185/60R15	32	32	5.5-JJ	②
1999	Escort	P185/65R14	P185/60R15	32	32	5.5-JJ	0.030 in. ①
	Tracer	P185/65SR14	P185/60R15	32	32	5.5-JJ	②
2000	Escort	P185/65R14	P185/60R15	32	32	5.5-JJ	0.030 in. ①
2001	Escort	P185/65R14	P185/60R15	32	32	5.5-JJ	0.030 in. ①

OEM: Original Equipment Manufacturer

PSI: Pounds Per Square Inch

STD: Standard

OPT: Optional

① Maximum radial tolerance in inches

② Replace if any measurable movement is found.

93461CE5

SCHEDULED MAINTENANCE INTERVALS

Ford Escort, ZX2, Mercury Tracer

TO BE SERVICED	SERVICE	VEHICLE MILEAGE INTERVAL (x1000)																			
		5	10	15	20	25	30	35	40	45	50	55	60	65	70	75	80	85	90	95	100
Engine oil & filter	R	✓	✓	✓	✓	✓	✓	✓	✓	✓	✓	✓	✓	✓	✓	✓	✓	✓	✓	✓	✓
Tires ①	S/I	✓		✓		✓		✓		✓		✓		✓		✓		✓		✓	
Air Cleaner	R						✓						✓						✓		
Spark Plugs	R																				✓
Drive Belts	S/I												✓								
Cooling system	S/I			✓			✓			✓			✓			✓			✓		
Engine coolant	R										✓						✓				
PCV valve	R												✓								
Exhaust heat shields	S/I						✓						✓						✓		
Brake linings & drums	S/I						✓						✓						✓		
Brake line hoses & connections	S/I						✓						✓						✓		
Front ball joints	S/I						✓						✓						✓		
Bolts & nuts on chassis body	S/I						✓						✓						✓		
Steering linkage operation	S/I						✓						✓						✓		
Brake pads & rotor	S/I						✓						✓						✓		
Clutch pedal operation	S/I						✓						✓						✓		
Halfshaft dust boots	S/I						✓						✓						✓		

R: Replace S/I: Inspect and service, if needed

① Rotate, inspect the tire tread for wear, and adjust air pressure.

FREQUENT OPERATION MAINTENANCE (SEVERE SERVICE)

If a vehicle is operated under any of the following conditions it is considered severe service:

- Extremely dusty areas.

- 50% or more of the vehicle operation is in 32°C (90°F) or higher temperatures, or constant operation in temperatures below 0°C (32°F).

- Prolonged idling (vehicle operation in stop and go traffic).

- Frequent short running periods (engine does not warm to normal operating temperatures).

- Police, taxi, delivery usage or trailer towing usage.

Oil & filter change: change every 3000 miles.

Rotate tires at 6000 miles & every 9000 miles thereafter.

Automatic transmission fluid & filter: change every 21,000 miles.

93461CE6

SCHEDULED MAINTENANCE INTERVALS
FORD MOTOR COMPANY
FORD ESCORT, ZX2, MERCURY TRACER

The following should be used as a guide when determining the amount of work required for a particular service. In estimating how long a particular Scheduled Maintenance Service should take, please observe the following:

- Labor Time is time based on field research and data supplied by the vehicle manufacturer.
- Labor time operations are given in hours and tenths of an hour.
- All labor operations are to be used as a guide.

Mechanic Skill Level Codes:
(A) PRECISION: Highly skilled with multiple certification.
(B) GENERAL: Normally skilled with certification.
(C) MAINTENANCE: Semi-skilled working on certification.

	LABOR TIME
5000 Mile Service (C)	
All Models	.9
10000 Mile Service (C)	
All Models	.5
15000 Mile Service (B)	
All Models	1.4
20000 Mile Service (C)	
All Models	.5
25000 Mile Service (C)	
All Models	.9
30000 Mile Service (B)	
All Models	1.6
35000 Mile Service (C)	
All Models	.9
40000 Mile Service (C)	
All Models	.5
45000 Mile Service (B)	
All Models	1.2
50000 Mile Service (C)	
All Models	.7
55000 Mile Service (C)	
All Models	.9
60000 Mile Service (B)	
All Models	1.7
65000 Mile Service (C)	
All Models	.9
70000 Mile Service (C)	
All Models	.5
75000 Mile Service (B)	
All Models	1.3
80000 Mile Service (C)	
All Models	.7
85000 Mile Service (C)	
All Models	.8
90000 Mile Service (B)	
All Models	1.6
95000 Mile Service (C)	
All Models	.9
100000 Mile Service (C)	
All Models	.5

93461CE7

FORD MOTOR CO.
Ford Focus

ENGINE AND VEHICLE IDENTIFICATION

Engine							Model Year	
Code ①	Liters (cc)	Cu. In.	Cyl.	Fuel Sys.	Type	Eng. Mfg.	Code ②	Year
3	2.0 (1999)	121	4	SFI	DOHC	Ford	Y	2000
P	2.0 (1999)	121	4	SFI	SOHC	Ford	1	2001
							2	2002

SFI: Sequential Fuel Injection

DOHC: Double Overhead Camshafts

SOHC: Single Overhead Camshaft

① 8th digit of the VIN

② 10th digit of the VIN

93461CE8

GENERAL ENGINE SPECIFICATIONS

Year	Model	Engine Displacement Liters (cc)	Engine ID/VIN	Fuel System Type	Net Horsepower @ rpm	Net Torque @ rpm (ft. lbs.)	Bore x Stroke (in.)	Com-pression Ratio	Oil Pressure @ rpm
2000	Focus LX	2.0 (1999)	P	SFI	110@5000	125@3750	3.34x3.46	9.2:1	35-65@2000
	Focus SE	2.0 (1999)	P	SFI	110@5000	125@3750	3.34x3.46	9.2:1	35-65@2000
	Focus ZTS	2.0 (1999)	3	SFI	130@5500	130@4000	3.34x3.46	10.0:1	35-65@2000
	Focus ZX3	2.0 (1999)	3	SFI	130@5500	130@4000	3.34x3.46	10.0:1	35-65@2000
2001	Focus LX	2.0 (1999)	P	SFI	110@5000	125@3750	3.34x3.46	9.2:1	35-65@2000
	Focus SE	2.0 (1999)	P	SFI	110@5000	125@3750	3.34x3.46	9.2:1	35-65@2000
	Focus SE	2.0 (1999)	3	SFI	130@5500	130@4000	3.34x3.46	10.0:1	35-65@2000
	Focus ZTS	2.0 (1999)	3	SFI	130@5500	130@4000	3.34x3.46	10.0:1	35-65@2000
	Focus ZX3	2.0 (1999)	3	SFI	130@5500	130@4000	3.34x3.46	10.0:1	35-65@2000

SFI: Sequential Fuel Injection

93461CE9

ENGINE TUNE-UP SPECIFICATIONS

Year	Engine Displacement Liters (cc)	Engine ID/VIN	Spark Plug Gap (in.)	Ignition Timing (deg.)		Fuel Pump (psi)	Idle Speed (rpm)		Valve Clearance (in.)	
				MT	AT		MT	AT	In.	Ex.
2000	2.0 (1999)	3	0.052-0.056	10B	10B	31-38 ①	②	②	0.004-0.007	0.010-0.013
	2.0 (1999)	P	0.052-0.056	10B	10B	31-38 ①	②	②	HYD	HYD
2001	2.0 (1999)	3	0.052-0.056	10B	10B	31-38 ①	②	②	0.004-0.007	0.010-0.013
	2.0 (1999)	P	0.052-0.056	10B	10B	31-38 ①	②	②	HYD	HYD

NOTE: The Vehicle Emission Control Information label often reflects specification changes made during production. The label figures must be used if they differ from those in this chart.

B: Before Top Dead Center

HYD: Hydraulic

① Fuel pressure with engine running, pressure regulator vacuum hose connected

② Refer to Vehicle Emission Control Information label

93461CE0

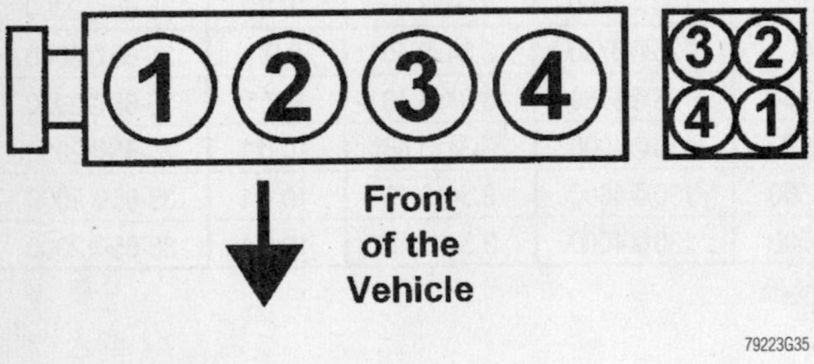

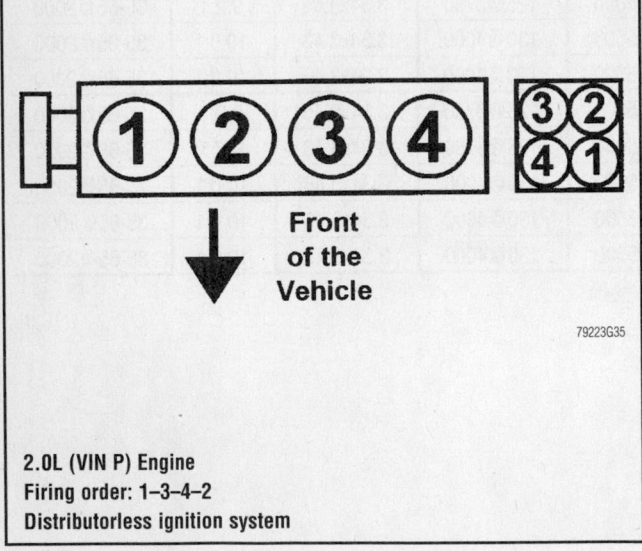

2.0L (VIN P) Engine
Firing order: 1–3–4–2
Distributorless ignition system

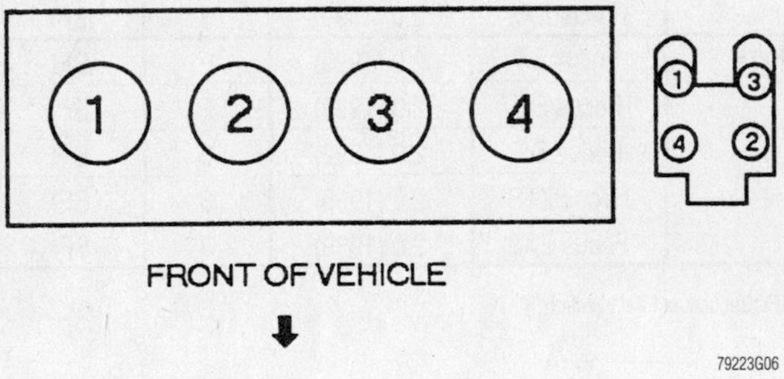

2.0L (VIN 3) Engine
Firing order: 1–3–4–2
Distributorless ignition system

CAPACITIES

Year	Model	Engine Displacement Liters (cc)	Engine ID/VIN	Engine Oil with Filter (qts.)	Transmission (pts.) Manual	Transmission (pts.) Auto.	Drive Front Axle (pts.)	Fuel Tank (gal.)	Cooling System (qts.)
2000	Focus LX	2.0 (1999)	P	4.0	4.8	14.0	—	13.2	①
	Focus SE	2.0 (1999)	P	4.0	4.8	14.0	—	13.2	①
	Focus ZTS	2.0 (1999)	3	4.5	4.0	14.0	—	13.2	②
	Focus ZX3	2.0 (1999)	3	4.5	4.0	14.0	—	13.2	②
2001	Focus LX	2.0 (1999)	P	4.0	4.8	14.0	—	13.2	①
	Focus SE	2.0 (1999)	P	4.0	4.8	14.0	—	13.2	①
	Focus SE	2.0 (1999)	3	4.5	4.0	14.0	—	13.2	②
	Focus ZTS	2.0 (1999)	3	4.5	4.0	14.0	—	13.2	②
	Focus ZX3	2.0 (1999)	3	4.5	4.0	14.0	—	13.2	②

Note: All capacities are approximates. Add fluid gradually and ensure a proper fluid level is obtained.

① Manual transaxle: 7.9
Automatic transaxle: 5.8

② Manual transaxle: 7.0
Automatic transaxle: 7.5

93461CF1

VALVE SPECIFICATIONS

Year	Engine Displacement Liters (cc)	Engine ID/VIN	Seat Angle (deg.)	Face Angle (deg.)	Spring Test Pressure (lbs. @ in.)	Spring Installed Height (in.)	Stem-to-Guide Clearance (in.) Intake	Stem-to-Guide Clearance (in.) Exhaust	Stem Diameter (in.) Intake	Stem Diameter (in.) Exhaust
2000	2.0 (1999)	3	45	45	NA	1.346	0.0007-0.0025	0.0014-0.0032	0.2373-0.2379	0.2366-0.2372
	2.0 (1999)	P	45	45.6	200@1.09	1.420-1.540	0.0008-0.0027	0.0018-0.0037	0.3159-0.3167	0.3149-0.3156
2001	2.0 (1999)	3	45	45	NA	1.346	0.0007-0.0025	0.0014-0.0032	0.2373-0.2379	0.2366-0.2372
	2.0 (1999)	P	45	45.6	200@1.09	1.420-1.540	0.0008-0.0027	0.0018-0.0037	0.3159-0.3167	0.3149-0.3156

NA: Not Available

93461CF2

CRANKSHAFT AND CONNECTING ROD SPECIFICATIONS

All measurements are given in inches.

Year	Engine Displacement Liters (cc)	Engine ID/VIN	Crankshaft Main Brg. Journal Dia.	Crankshaft Main Brg. Oil Clearance	Crankshaft Shaft End-play	Crankshaft Thrust on No.	Connecting Rod Journal Diameter	Connecting Rod Oil Clearance	Connecting Rod Side Clearance
2000	2.0 (1999)	3	2.2827-2.2835	0.0008-0.0026	0.0040-0.0120	3	1.8461-1.8500	0.0006-0.0028	0.0040-0.0110
	2.0 (1999)	P	2.2827-2.2835	0.0008-0.0026	0.0040-0.0120	3	1.7279-1.7287	0.0008-0.0026	0.0040-0.0110
2001	2.0 (1999)	3	2.2827-2.2835	0.0008-0.0026	0.0040-0.0120	3	1.8461-1.8500	0.0006-0.0028	0.0040-0.0110
	2.0 (1999)	P	2.2827-2.2835	0.0008-0.0026	0.0040-0.0120	3	1.7279-1.7287	0.0008-0.0026	0.0040-0.0110

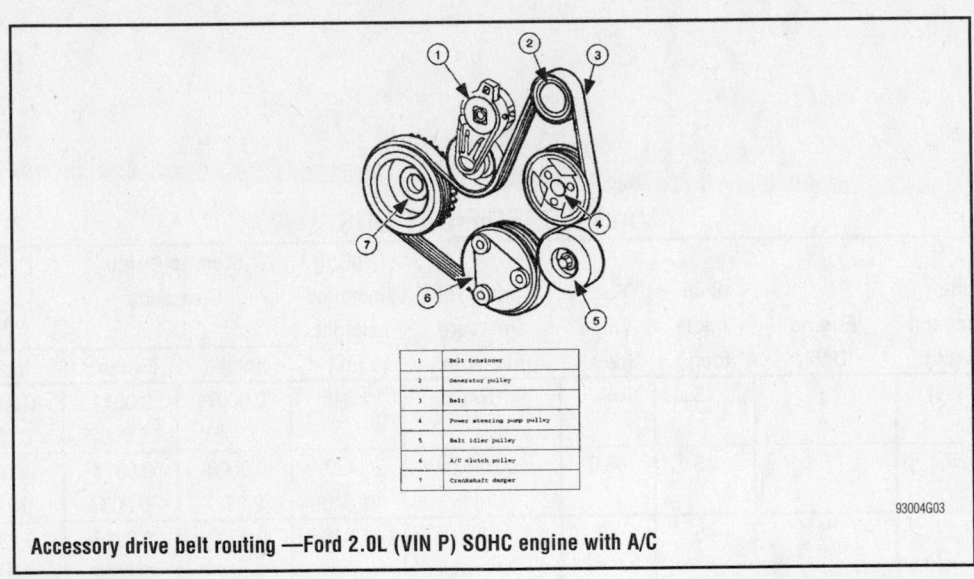

93461CF3

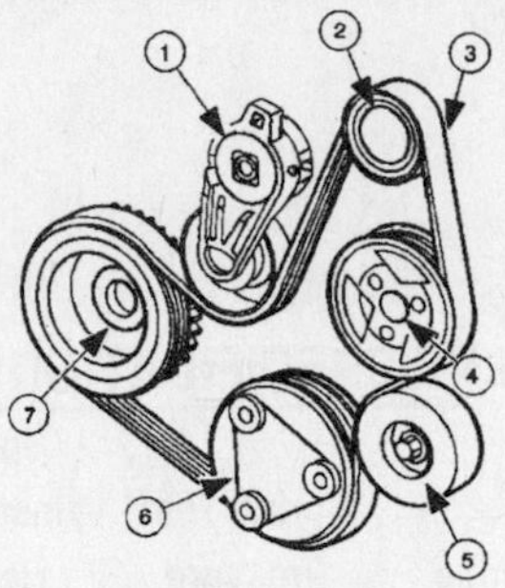

1	Belt tensioner
2	Generator pulley
3	Belt
4	Power steering pump pulley
5	Belt idler pulley
6	A/C clutch pulley
7	Crankshaft damper

93004G03

Accessory drive belt routing —Ford 2.0L (VIN P) SOHC engine with A/C

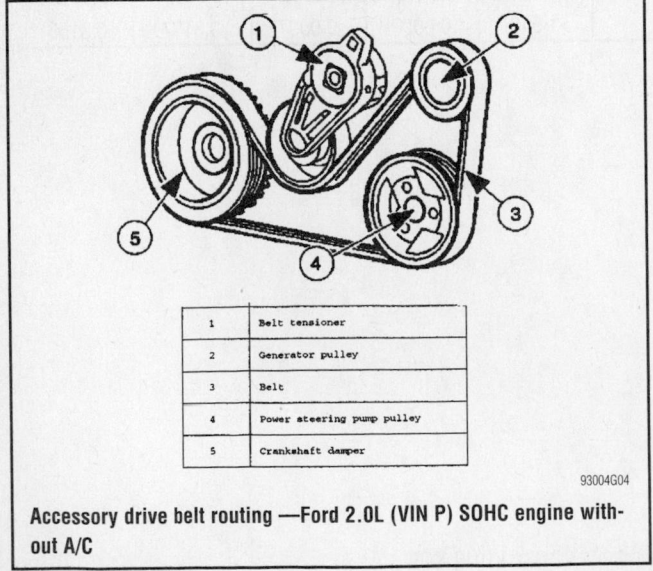

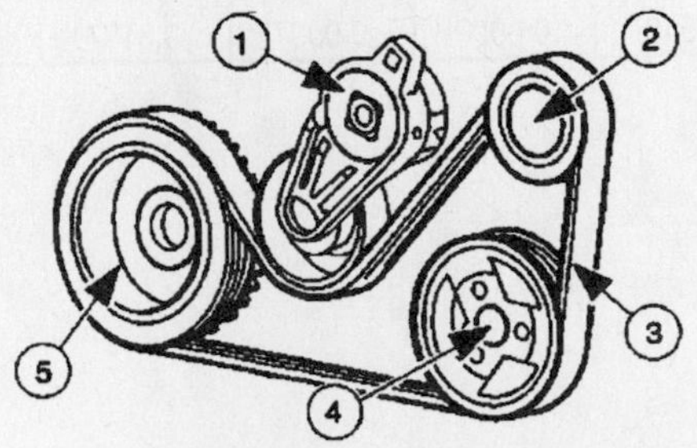

1	Belt tensioner
2	Generator pulley
3	Belt
4	Power steering pump pulley
5	Crankshaft damper

93004G04

Accessory drive belt routing —Ford 2.0L (VIN P) SOHC engine without A/C

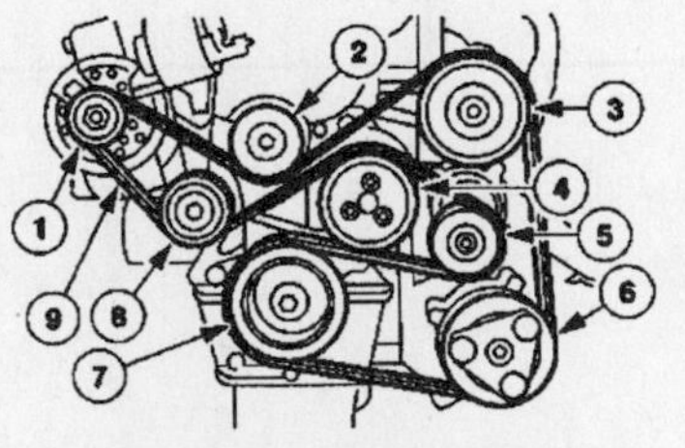

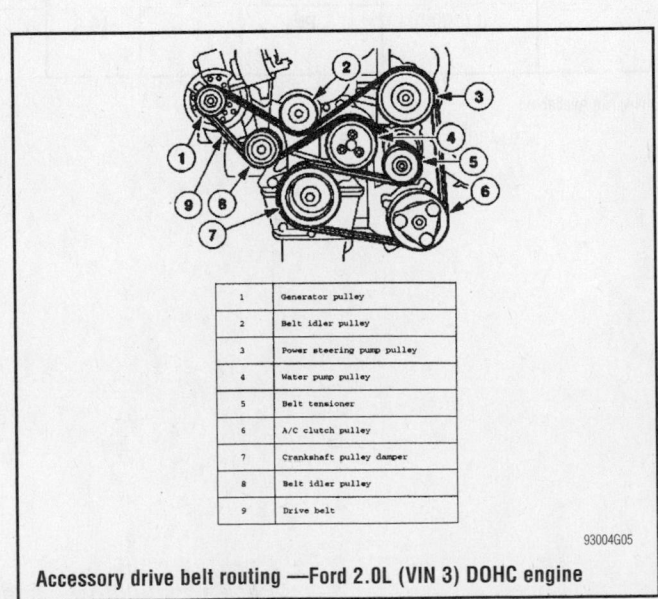

1	Generator pulley
2	Belt idler pulley
3	Power steering pump pulley
4	Water pump pulley
5	Belt tensioner
6	A/C clutch pulley
7	Crankshaft pulley damper
8	Belt idler pulley
9	Drive belt

93004G05

Accessory drive belt routing —Ford 2.0L (VIN 3) DOHC engine

PISTON AND RING SPECIFICATIONS

All measurements are given in inches.

Year	Engine Displacement Liters (cc)	Engine ID/VIN	Piston Clearance	Ring Gap Top Compression	Ring Gap Bottom Compression	Ring Gap Oil Control	Ring Side Clearance Top Compression	Ring Side Clearance Bottom Compression	Ring Side Clearance Oil Control
2000	2.0 (1999)	3	0.0010-0.0022	0.012-0.022	0.012-0.022	0.010-0.039	0.0015-0.0032	0.0015-0.0035	—
	2.0 (1999)	P	0.0008-0.0027	0.010-0.030	0.010-0.030	0.016-0.066	0.0015-0.0032	0.0015-0.0035	—
2001	2.0 (1999)	3	0.0010-0.0022	0.012-0.022	0.012-0.022	0.010-0.039	0.0015-0.0032	0.0015-0.0035	—
	2.0 (1999)	P	0.0008-0.0027	0.010-0.030	0.010-0.030	0.016-0.066	0.0015-0.0032	0.0015-0.0035	—

93461CF4

TORQUE SPECIFICATIONS

All readings in ft. lbs.

Year	Engine Displacement Liters (cc)	Engine ID/VIN	Cylinder Head Bolts	Main Bearing Bolts	Rod Bearing Bolts	Crankshaft Damper Bolts	Flywheel Bolts	Manifold Intake	Manifold Exhaust	Spark Plugs	Lug Nut
2000	2.0 (1999)	P	①	66-79	26-30	80-87	71-76	15-22	15-17	12-15	76
	2.0 (1999)	3	②	55-65	③	80-87	71-76	11-12	10-12	10-12	76
2001	2.0 (1999)	P	①	66-79	26-30	80-87	71-76	15-22	15-17	12-15	76
	2.0 (1999)	3	②	55-65	③	80-87	71-76	11-12	10-12	10-12	76

① Step 1: 37 ft. lbs.
Step 2: Loosen bolts 1/2 turn
Step 3: 37 ft. lbs.
Step 4: Plus 90 degrees
Step 5: Plus 90 degrees

② Step 1: 15 ft. lbs.
Step 2: 30 ft. lbs.
Step 3: Plus 90 degrees

③ Step 1: 22-25 ft. lbs.
Step 2: Plus 90 degrees

93461CF5

BRAKE SPECIFICATIONS

All measurements in inches unless noted

		Brake Disc			Brake Drum Diameter			Minimum Lining Thickness		Brake Caliper Mounting Bolts
Year	Model	Original Thickness	Minimum Thickness	Maximum Runout	Original Inside Diameter	Max. Wear Limit	Maximum Machine Diameter	Front	Rear	(ft. lbs.)
2000	Focus LX	0.870	0.790	0.002	7.99	8.03	—	0.080	0.040	21
	Focus SE	0.870	0.790	0.002	7.99	8.03	—	0.080	0.040	21
	Focus ZTS	0.870	0.790	0.002	7.99	8.03	—	0.080	0.040	21
	Focus ZX3	0.870	0.790	0.002	7.99	8.03	—	0.080	0.040	21
2001	Focus LX	0.870	0.790	0.002	7.99	8.03	—	0.080	0.040	21
	Focus SE	0.870	0.790	0.002	7.99	8.03	—	0.080	0.040	21
	Focus ZTS	0.870	0.790	0.002	7.99	8.03	—	0.080	0.040	21
	Focus ZX3	0.870	0.790	0.002	7.99	8.03	—	0.080	0.040	21

93461CF6

WHEEL ALIGNMENT

			Caster		Camber			Steering Axis Inclination
Year	Model		Range (+/-Deg.)	Preferred Setting (Deg.)	Range (+/-Deg.)	Preferred Setting (Deg.)	Toe-in (in.)	(Deg.)
2000	Sedan	F	1.00	+2.93	1.00	-0.36	0 +/- 0.04	—
		R	—	—	0.30	-0.93	0.08 +/- 0.05	—
	Wagon	F	1.30	+2.46	1.25	-0.62	0 +/- 0.04	—
		R	—	—	0.75	-0.62	0.12 +/- 0.05	—
2001	Sedan	F	1.00	+2.93	1.00	-0.36	0 +/- 0.04	—
		R	—	—	0.30	-0.93	0.08 +/- 0.05	—
	Wagon	F	1.30	+2.46	1.25	-0.62	0 +/- 0.04	—
		R	—	—	0.75	-0.62	0.12 +/- 0.05	—

93461CF7

TIRE, WHEEL AND BALL JOINT SPECIFICATIONS

		OEM Tires		Tire Pressures (psi)		Wheel	Ball Joint
Year	Model	Standard	Optional	Front	Rear	Size	Inspection
2000	Focus	P185/65R14	P195/65R15	32	32	5.5	①
2000	Focus	P185/65R14	P195/65R15	32	32	5.5	①

OEM: Original Equipment Manufacturer

PSI: Pounds Per Square Inch

① Replace if any measurable movement is found

93461CF8

Heater Core replacement is covered in Section 2 of this manual

SCHEDULED MAINTENANCE INTERVALS

Ford Focus

TO BE SERVICED	TYPE OF SERVICE	VEHICLE MILEAGE INTERVAL (x1000)												
		5	10	15	20	25	30	35	40	45	50	55	60	65
Air cleaner filter	R						✓						✓	
Accessory drive belt	S/I												✓	
Brake system ①	S/I			✓			✓			✓			✓	
Clutch pedal operation	S/I						✓						✓	
Cooling fan operation	S/I		✓		✓		✓		✓		✓		✓	
Cooling system hoses and clamps	S/I			✓			✓			✓			✓	
CV-joint boots & axle seals	S/I						✓						✓	
Engine coolant	R	Ten years or 150,000 miles												
Engine oil & filter	R	✓	✓	✓	✓	✓	✓	✓	✓	✓	✓	✓	✓	✓
Exterior Lights	S/I	Check monthly												
PCV valve	S/I												✓	
Exhaust system & heat shields	S/I						✓						✓	
Parking brake system	S/I	Every 6 months												
Power steering fluid	S/I	Every 6 months												
Rotate tires	S/I	✓		✓		✓		✓		✓		✓		✓
Steering linkage	S/I						✓						✓	
Spark plugs	R	Change at 100,000 miles												
Suspension components	S/I						✓						✓	

R: Replace   S/I: Inspect and service, if necessary   L: Lubricate   A: Adjust   C: Clean

① Inspect the reservoir fluid level, rotor and or drum, brake lines, hoses, calipers and or wheel cylinders

FREQUENT OPERATION MAINTENANCE (SEVERE SERVICE)

If a vehicle is operated under any of the following conditions it is considered severe service:

- Extremely dusty areas.

- 50% or more of the vehicle operation is in 32°C (90°F) or higher temperatures, or constant operation in temperatures below 0°C (32°F).

- Prolonged idling (vehicle operation in stop and go traffic).

- Frequent short running periods (engine does not warm to normal operating temperatures).

- Police, taxi, delivery usage or trailer towing usage.

Oil & oil filter change: change every 3000 miles.

Air filter element: change every 15,000 miles.

93461CF9

SCHEDULED MAINTENANCE INTERVALS
FORD MOTOR COMPANY
FORD FOCUS

The following should be used as a guide when determining the amount of work required for a particular service. In estimating how long a particular Scheduled Maintenance Service should take, please observe the following:

- Labor Time is time based on field research and data supplied by the vehicle manufacturer.
- Labor time operations are given in hours and tenths of an hour.
- All labor operations are to be used as a guide.

Mechanic Skill Level Codes:
(A) PRECISION: Highly skilled with multiple certification.
(B) GENERAL: Normally skilled with certification.
(C) MAINTENANCE: Semi-skilled working on certification.

Service	LABOR TIME
5000 Mile Service (C)	
All Models	.9
10000 Mile Service (C)	
All Models	.5
15000 Mile Service (B)	
All Models	1.2
20000 Mile Service (C)	
All Models	.5
25000 Mile Service (C)	
All Models	.9
30000 Mile Service (B)	
All Models	1.6
35000 Mile Service (C)	
All Models	.9
40000 Mile Service (C)	
All Models	.5
45000 Mile Service (B)	
All Models	1.2
50000 Mile Service (C)	
All Models	.7
55000 Mile Service (C)	
All Models	.9
60000 Mile Service (B)	
All Models	1.7
65000 Mile Service (C)	
All Models	.9

93461CF0

FORD MOTOR CO.
Lincoln LS

ENGINE AND VEHICLE IDENTIFICATION

Engine							Model Year	
Code ①	Liters (cc)	Cu. In.	Cyl.	Fuel Sys.	Type	Eng. Mfg.	Code ②	Year
S	3.0 (3049)	182	6	MFI	DOHC	Ford	Y	2000
A	3.9 (3947)	243	8	MFI	DOHC	Ford	1	2001
							2	2002

MFI: Multi-Port Fuel Injection

DOHC: Double Overhead Camshaft

① 8th digit of the VIN

② 10th digit of the VIN

93461CG1

GENERAL ENGINE SPECIFICATIONS

Year	Model	Engine Displacement Liters (cc)	Engine ID/VIN	Fuel System Type	Net Horsepower @ rpm	Net Torque @ rpm (ft. lbs.)	Bore x Stroke (in.)	Com-pression Ratio	Oil Pressure @ rpm
2000	LS6	3.0 (3049)	S	SFI	210@6500	205@4750	3.50x3.13	10.5:1	20-45@1500
	LS8	3.9 (3947)	A	SFI	252@6100	267@4300	NA	10.6:1	NA
2001	LS6	3.0 (3049)	S	SFI	210@6500	205@4750	3.50x3.13	10.5:1	20-45@1500
	LS8	3.9 (3947)	A	SFI	252@6100	267@4300	NA	10.6:1	NA

NA: Not Available

SFI: Sequential Fuel Injection

93461CG2

ENGINE TUNE-UP SPECIFICATIONS

Year	Engine Displacement Liters (cc)	Engine ID/VIN	Spark Plugs Gap (in.)	Ignition Timing (deg.) ① MT	Ignition Timing (deg.) ① AT	Fuel Pump (psi)	Idle Speed (rpm) ① MT	Idle Speed (rpm) ① AT	Valve Clearance In.	Valve Clearance Ex.
2000	3.0 (3049)	S	0.051-0.057	12-17B	12-17B	26-45	650-750	650-750	0.007-0.009	0.013-0.015
	3.9 (3947)	A	0.039-0.043	—	10-20B	43	—	650-750	0.007-0.009	0.009-0.011
2001	3.0 (3049)	S	0.051-0.057	12-17B	12-17B	26-45	650-750	650-750	0.007-0.009	0.013-0.015
	3.9 (3947)	A	0.039-0.043	—	10-20B	43	—	650-750	0.007-0.009	0.009-0.011

The underhood specifications sticker often reflects tune-up specification changes in production. Sticker figures must be used if they disagree with those in this chart.

① Controlled by the engine computer

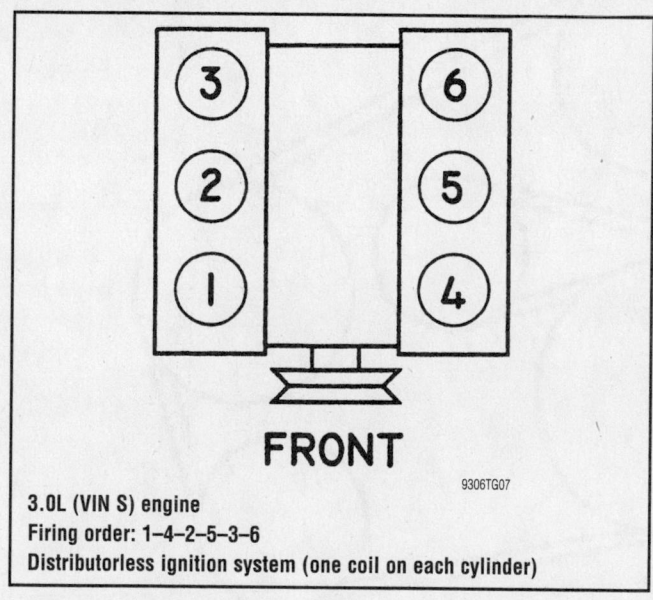

93461CG3

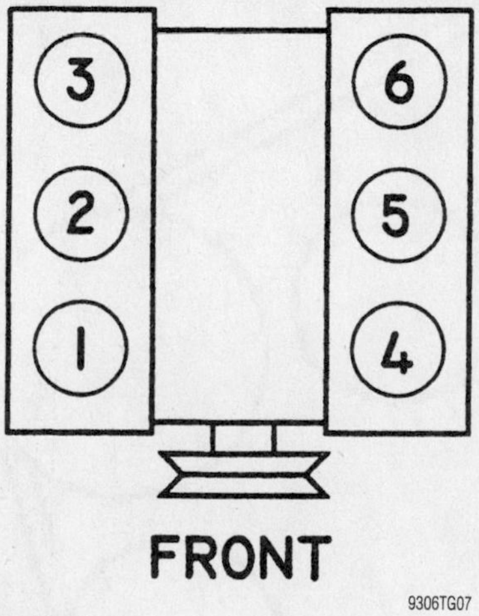

3.0L (VIN S) engine
Firing order: 1–4–2–5–3–6
Distributorless ignition system (one coil on each cylinder)

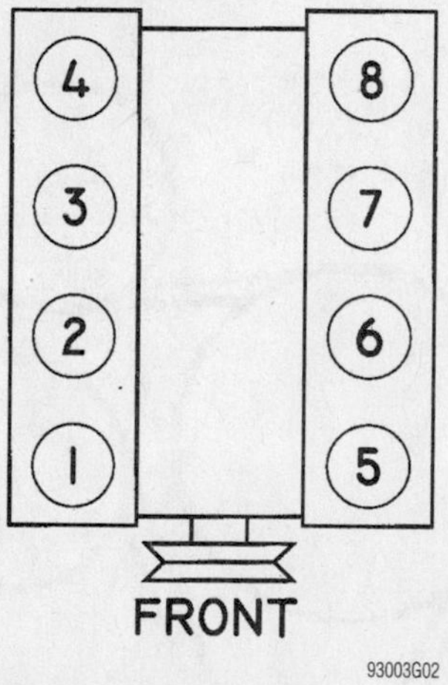

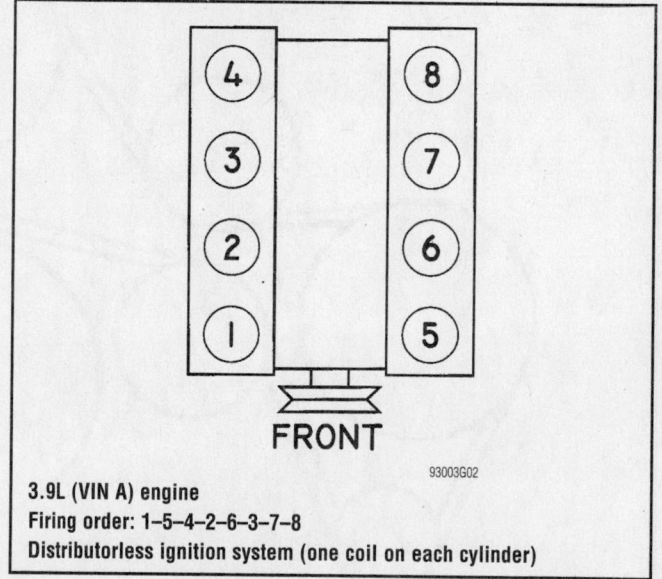

3.9L (VIN A) engine
Firing order: 1–5–4–2–6–3–7–8
Distributorless ignition system (one coil on each cylinder)

For complete Engine Mechanical specifications, see Section 1 of this manual

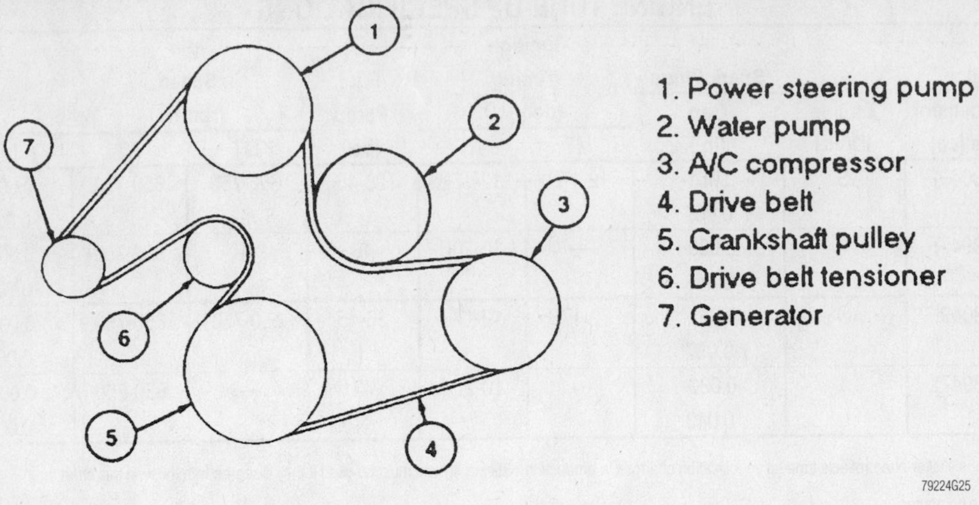

Serpentine accessory drive belt routing—Ford 3.0L (VIN S) engine

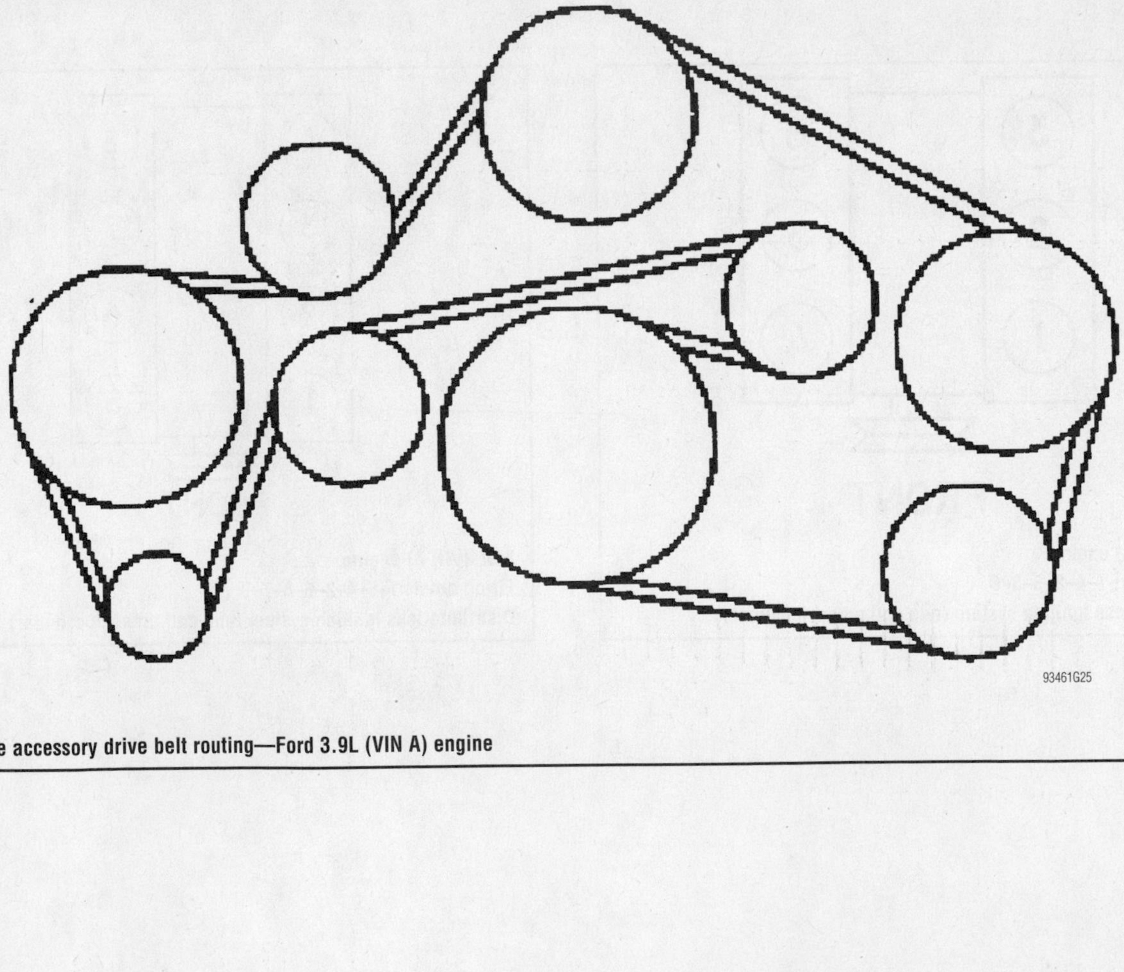

Serpentine accessory drive belt routing—Ford 3.9L (VIN A) engine

CAPACITIES

Year	Model	Engine Displacement Liters (cc)	Engine ID/VIN	Engine Oil with Filter (qts.)	Transmission (pts.) Manual	Transmission (pts.) Auto.	Drive Axle Rear (pts.)	Fuel Tank (gal.)	Cooling System (qts.)
2000	LS6	3.0 (3049)	S	6.9	NA	23.8	3.0	18.0	10.6
	LS8	3.9 (3947)	A	NA	—	23.8	3.0	18.0	11.3
2001	LS6	3.0 (3049)	S	6.9	NA	23.8	3.0	18.0	10.6
	LS8	3.9 (3947)	A	NA	—	23.8	3.0	18.0	11.3

N/A: Not Available

93461CG4

VALVE SPECIFICATIONS

Year	Engine Displacement Liters (cc)	Engine ID/VIN	Seat Angle (deg.)	Face Angle (deg.)	Spring Test Pressure (lbs. @ in.)	Spring Free Length (in.)	Stem-to-Guide Clearance (in.) Intake	Stem-to-Guide Clearance (in.) Exhaust	Stem Diameter (in.) Intake	Stem Diameter (in.) Exhaust
2000	3.0 (3049)	S	44.75	45.5	153@1.18	1.570	0.0007-0.0027	0.0017-0.0037	0.2350-0.2358	0.2343-0.2350
	3.9 (3947)	A	NA	NA	NA	NA	NA	NA	NA	NA
2001	3.0 (3049)	S	44.75	45.5	153@1.18	1.570	0.0007-0.0027	0.0017-0.0037	0.2350-0.2358	0.2343-0.2350
	3.9 (3947)	A	NA	NA	NA	NA	NA	NA	NA	NA

NA: Not Available

93461CG5

For Accessory Drive Belt illustrations, see Section 1 of this manual

CRANKSHAFT AND CONNECTING ROD SPECIFICATIONS

All measurements are given in inches.

			Crankshaft				Connecting Rod		
Year	Engine Displacement Liters (cc)	Engine ID/VIN	Main Brg. Journal Dia.	Main Brg. Oil Clearance	Shaft End-play	Thrust on No.	Journal Diameter	Oil Clearance	Side Clearance
2000	3.0 (3049)	S	2.4670-2.4790	0.0009-0.0018	0.0040 0.0090	4	1.9670-1.9680	0.0010 0.0025	0.0039-0.0118
	3.9 (3947)	A	NA	NA	NA	NA	NA	NA	NA
2001	3.0 (3049)	S	2.4670-2.4790	0.0009-0.0018	0.0040 0.0090	4	1.9670-1.9680	0.0010 0.0025	0.0039-0.0118
	3.9 (3947)	A	NA	NA	NA	NA	NA	NA	NA

NA: Not Available

93461CG6

PISTON AND RING SPECIFICATIONS

All measurements are given in inches.

				Ring Gap			Ring Side Clearance		
Year	Engine Displacement Liters (cc)	Engine ID/VIN	Piston Clearance	Top Compression	Bottom Compression	Oil Control	Top Compression	Bottom Compression	Oil Control
2000	3.0 (3049)	S	0.0005-0.0009	0.004-0.010	0.011-0.017	0.005-0.026	0.0015-0.0029	0.0015-0.0033	SNUG
	3.9 (3947)	A	NA	NA	NA	NA	NA	NA	NA
2001	3.0 (3049)	S	0.0005-0.0009	0.004-0.010	0.011-0.017	0.005-0.026	0.0015-0.0029	0.0015-0.0033	SNUG
	3.9 (3947)	A	NA	NA	NA	NA	NA	NA	NA

NA: Not Available

93461CG7

TORQUE SPECIFICATIONS

All readings in ft. lbs.

Year	Engine Displacement Liters (cc)	Engine ID/VIN	Cylinder Head Bolts	Main Bearing Bolts	Rod Bearing Bolts	Crankshaft Damper Bolts	Flywheel Bolts	Manifold Intake	Manifold Exhaust	Spark Plugs	Lug Nut
2000	3.0 (3049)	S	①	②	③	④	54-64	⑤	13-16	7-15	100
	3.9 (3947)	A	⑥	NA	NA	⑦	⑧	NA	NA	NA	100
2001	3.0 (3049)	S	①	②	③	④	54-64	⑤	13-16	7-15	100
	3.9 (3947)	A	⑥	NA	NA	⑦	⑧	NA	NA	NA	100

NA: Not Available

① Step 1: 28-31 ft. lbs.
Step 2: Plus 85-95 degrees
Step 3: Loosen one turn
Step 4: 28-31 ft. lbs.
Step 5: Plus 85-95 degrees
Step 6: Plus 85-95 degrees

② Step 1: Cap bolts 1-8 (outer) 17-20 ft. lbs.
Step 2: Cap bolts 9-16 (inner) 28-31 ft. lbs.
Step 3: Rotate bolts 1-16, 85-95 degrees
Step 4: Bolts 17-22; 15-22 ft. lbs.

③ Step 1: 30-33 ft. lbs.
Step 2: Plus 90-120 degrees

④ Step 1: 77-99 ft. lbs.
Step 2: Loosen 360 degrees
Step 3: Tighten to 35-39 ft. lbs.
Step 4: Plus 85-95 degrees

⑤ 71-106 inch lbs.

⑥ Step 1: Tighten M10 bolts to 15 ft. lbs.
Step 2: Tighten M10 bolts to 26 ft. lbs.
Step 3: Tighten M10 bolts to 33 ft. lbs.
Step 4: Tighten M10 bolts plus 90 degrees
Step 5: Tighten M10 bolts plus 90 degrees
Step 6: Tighten M8 bolts to 15 ft. lbs.
Step 7: Tighten M8 bolts plus 90 degrees

⑦ Step 1: 59 ft. lbs.
Step 2: Plus 80 degrees

⑧ Step 1: 11 ft. lbs.
Step 2: 81 ft. lbs.

93461CG8

For Tire, Wheel and Ball Joint specifications, see Section 1 of this manual

BRAKE SPECIFICATIONS

All measurements in inches unless noted

			Brake Disc			Minimum	Brake Caliper	
Year	Model		Original Thickness	Minimum Thickness	Maximum Runout	Lining Thickness	Bracket Bolts (ft. lbs.)	Mounting Bolts (ft. lbs.)
2000	LS6	F	1.180	1.120	0.004	0.079	76	26
		R	0.810	0.740	0.004	0.039	76	25
	LS8	F	1.180	1.120	0.004	0.079	76	26
		R	0.810	0.740	0.004	0.039	76	25
2001	LS6	F	1.180	1.120	0.004	0.079	76	26
		R	0.810	0.740	0.004	0.039	76	25
	LS8	F	1.180	1.120	0.004	0.079	76	26
		R	0.810	0.740	0.004	0.039	76	25

93461CG9

WHEEL ALIGNMENT

			Caster		Camber			Steering
Year	Model		Range (+/-Deg.)	Preferred Setting (Deg.)	Range (+/-Deg.)	Preferred Setting (Deg.)	Toe-in (in.)	Axis Inclination (Deg.)
2000	LS	F	0.70	0	0.70	0	0.08 +/- 0.13	—
		R	—	—	0.75	0	0.13 +/- 0.13	—
2001	LS	F	0.70	0	0.70	0	0.08 +/- 0.13	—
		R	—	—	0.75	0	0.13 +/- 0.13	—

93461CG0

TIRE, WHEEL AND BALL JOINT SPECIFICATIONS

		OEM Tires		Tire Pressures (psi)		Wheel	Ball Joint
Year	Model	Standard	Optional	Front	Rear	Size	Inspection
2000	LS6 and LS8	P215/60R16	P235/50R17	30	30	7-J/7.5J	U ① L ① ②
2001	LS6 and LS8	P215/60R16	P235/50R17	30	30	7-J/7.5J	U ① L ① ②

OEM: Original Equipment Manufacturer

PSI: Pounds Per Square Inch

STD: Standard

OPT: Optional

L: Lower

U: Upper

① Replace if any measurable movement is found.

② Do not lift car. Inspect the boss into which the grease fitting is threaded. Replace if the boss is flush or receded below the surface of the ball joint.

93461CH1

For Wheel Alignment specifications, see Section 1 of this manual

SCHEDULED MAINTENANCE INTERVALS
Lincoln LS

TO BE SERVICED	TYPE OF SERVICE	VEHICLE MILEAGE INTERVAL (x1000)												
		5	10	15	20	25	30	35	40	45	50	55	60	65
Air cleaner filter	R						✓						✓	
Accessory drive belt	S/I												✓	
Brake system ①	S/I			✓			✓			✓			✓	
Clutch pedal operation	S/I						✓						✓	
Cooling system hoses and clamps	S/I			✓			✓			✓			✓	
CV-joint boots & axle seals	S/I						✓						✓	
Engine coolant	R	Ten years or 150,000 miles												
Engine oil & filter	R	✓	✓	✓	✓	✓	✓	✓	✓	✓	✓	✓	✓	✓
Exterior Lights	S/I	Check monthly												
PCV valve	S/I												✓	
Exhaust system & heat shields	S/I						✓						✓	
Parking brake system	S/I	Every 6 months												
Power steering fluid	S/I	Every 6 months												
Rotate tires	S/I	✓		✓		✓		✓		✓		✓		✓
Steering linkage	S/I						✓						✓	
Spark plugs	R	Change at 100,000 miles												
Suspension components	S/I						✓						✓	

R: Replace S/I: Inspect and service, if necessary L: Lubricate A: Adjust C: Clean

① Inspect the reservoir fluid level, rotor and or drum, brake lines, hoses, calipers and or wheel cylinders

FREQUENT OPERATION MAINTENANCE (SEVERE SERVICE)

If a vehicle is operated under any of the following conditions it is considered severe service:

- Extremely dusty areas.

- 50% or more of the vehicle operation is in 32°C (90°F) or higher temperatures, or constant operation in temperatures below 0°C (32°F).

- Prolonged idling (vehicle operation in stop and go traffic).

- Frequent short running periods (engine does not warm to normal operating temperatures).

- Police, taxi, delivery usage or trailer towing usage.

Oil & oil filter change: change every 3000 miles.

Air filter element: change every 15,000 miles.

93461CH2

SCHEDULED MAINTENANCE INTERVALS
FORD MOTOR COMPANY
LINCOLN LS

The following should be used as a guide when determining the amount of work required for a particular service. In estimating how long a particular Scheduled Maintenance Service should take, please observe the following:

- Labor Time is time based on field research and data supplied by the vehicle manufacturer.
- Labor time operations are given in hours and tenths of an hour.
- All labor operations are to be used as a guide.

Mechanic Skill Level Codes:
(A) PRECISION: Highly skilled with multiple certification.
(B) GENERAL: Normally skilled with certification.
(C) MAINTENANCE: Semi-skilled working on certification.

	LABOR TIME
5000 Mile Service (C)	
All Models	.9
10000 Mile Service (C)	
All Models	.5
15000 Mile Service (B)	
All Models	1.6
20000 Mile Service (C)	
All Models	.5
25000 Mile Service (C)	
All Models	.9
30000 Mile Service (B)	
All Models	1.9
35000 Mile Service (C)	
All Models	.9
40000 Mile Service (C)	
All Models	.5
45000 Mile Service (B)	
All Models	1.6
50000 Mile Service (C)	
All Models	.7
55000 Mile Service (C)	
All Models	.9
60000 Mile Service (B)	
All Models	1.9
65000 Mile Service (C)	
All Models	.9

93461CH3

For Maintenance Interval recommendations, see Section 1 of this manual

FORD MOTOR CO.
Ford Mustang • Lincoln Mark VIII

ENGINE AND VEHICLE IDENTIFICATION

Engine							Model Year	
Code ①	Liters (cc)	Cu. In.	Cyl.	Fuel Sys.	Engine Type	Eng. Mfg.	Code ②	Year
4	3.8 (3802)	232	6	SFI	OHV	Ford	W	1998
V	4.6 (4593)	281	8	SFI	DOHC	Ford	X	1999
X	4.6 (4593)	281	8	SFI	SOHC	Ford	Y	2000
							1	2001
							2	2002

OHV: Overhead Valve

SOHC: Single Overhead Camshaft

DOHC: Double Overhead Camshaft

SFI: Sequential Fuel Injection

① 8th position of VIN

② 10th position of VIN

93461CH4

GENERAL ENGINE SPECIFICATIONS
All measurements are given in inches.

Year	Model	Engine Displacement Liters (cc)	Engine Series (ID/VIN)	Fuel System	Net Horsepower @ rpm	Net Torque @ rpm (ft. lbs.)	Bore x Stroke (in.)	Com-pression Ratio	Oil Pressure @ rpm
1998	Mark VIII	4.6 (4593)	V	SFI	①	②	3.55x3.54	9.85:1	20-45@1500
	Mustang	3.8 (3802)	4	SFI	150@4000	215@2750	3.81x3.39	9.0:1	40-60@2500
		4.6 (4593)	V	SFI	305@5800	300@4800	3.55x3.54	9.5:1	20-45@1500
		4.6 (4593)	X	SFI	215@4400	285@3500	3.55x3.54	9.0:1	20-45@1500
1999	Mustang	3.8 (3802)	4	SFI	190@5250	220@3000	3.81x3.39	9.0:1	40-60@2500
		4.6 (4593)	V	SFI	320@6000	317@4750	3.55x3.54	9.5:1	20-45@1500
		4.6 (4593)	X	SFI	260@5000	302@4000	3.55x3.54	9.0:1	20-45@1500
2000	Mustang	3.8 (3802)	4	SFI	190@5250	220@3000	3.81x3.39	9.0:1	40-60@2500
		4.6 (4593)	X	SFI	260@5000	302@4000	3.55x3.54	9.0:1	20-45@1500
2001	Mustang	3.8 (3802)	4	SFI	190@5250	220@3000	3.81x3.39	9.0:1	40-60@2500
		4.6 (4593)	V	SFI	320@6000	317@4750	3.55x3.54	9.5:1	20-45@1500
		4.6 (4593)	X	SFI	260@5000	302@4000	3.55x3.54	9.0:1	20-45@1500

SFI: Sequential Fuel Injection

① Mark VIII without LSC package: 280@5500
Mark VIII with LSC package: 290@5750

② Mark VIII without LSC package: 285@4500
Mark VIII with LSC package: 292@4500

93461CH5

ENGINE TUNE-UP SPECIFICATIONS

Year	Engine Displacement Liters (cc)	Engine ID/VIN	Spark Plug Gap (in.)	Ignition Timing (deg.) MT	Ignition Timing (deg.) AT	Fuel Pump (psi)①	Idle Speed (rpm) MT	Idle Speed (rpm) AT	Valve Clearance Intake	Valve Clearance Exhaust
1998	3.8 (3802)	4	0.054	②	②	28-54	②	②	HYD	HYD
	4.6 (4593)	V	0.054	10B	—	35-45	②	—	HYD	HYD
	4.6 (4593)	X	0.054	10B	10B	35-45	②	②	HYD	HYD
1999	3.8 (3802)	4	0.054	②	②	28-54	②	②	HYD	HYD
	4.6 (4593)	V	0.054	10B	—	35-45	②	—	HYD	HYD
	4.6 (4593)	X	0.054	10B	10B	35-45	②	②	HYD	HYD
2000	3.8 (3802)	4	0.054	②	②	28-54	②	②	HYD	HYD
	4.6 (4593)	X	0.054	10B	10B	35-45	②	②	HYD	HYD
2001	3.8 (3802)	4	0.054	②	②	28-54	②	②	HYD	HYD
	4.6 (4593)	V	0.054	10B	—	35-45	②	—	HYD	HYD
	4.6 (4593)	X	0.054	10B	10B	35-45	②	②	HYD	HYD

NOTE: The Vehicle Emission Control Information label often reflects specification changes made during production. The label figures must be used if they differ from those in this chart.

B: Before Top Dead Center

HYD: Hydraulic

① Fuel pressure with engine running, pressure regulator vacuum hose connected

② Refer to Vehicle Emission Control Information label

93461CH6

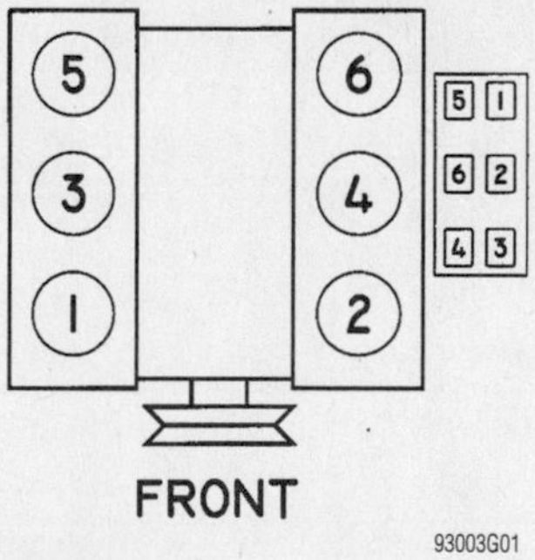

93003G01

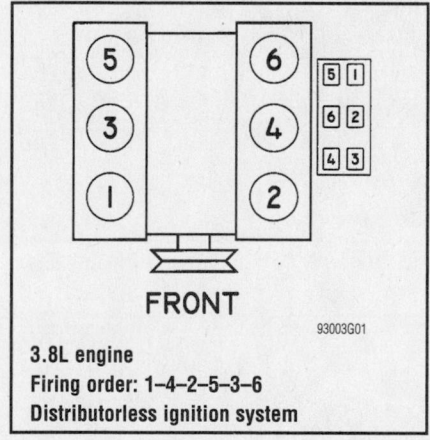

3.8L engine
Firing order: 1–4–2–5–3–6
Distributorless ignition system

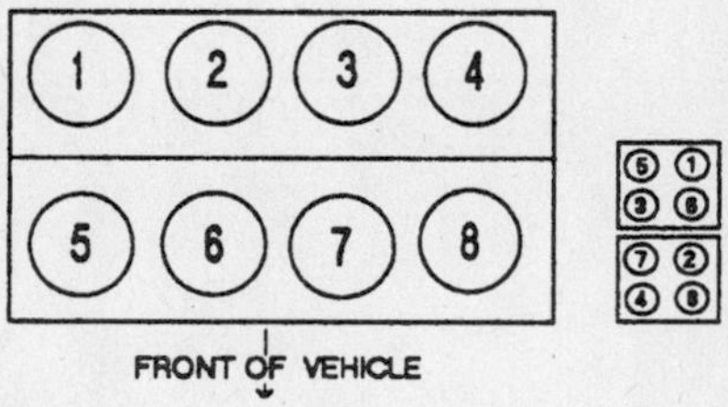

79223G28

1998 4.6L engines
Firing order: 1–3–7–2–6–5–4–8
Distributorless ignition system

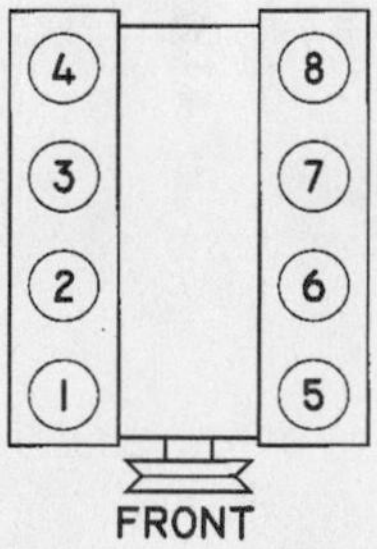

93003G02

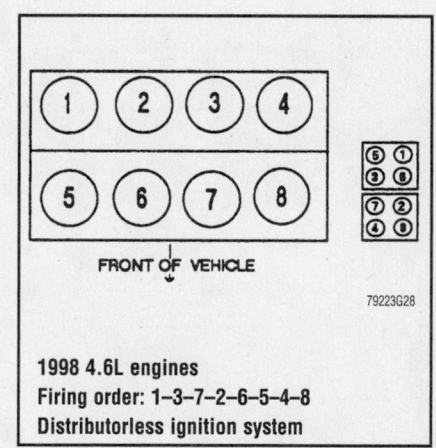

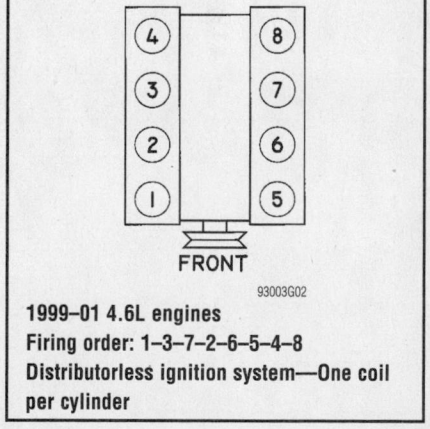

1999–01 4.6L engines
Firing order: 1–3–7–2–6–5–4–8
Distributorless ignition system—One coil per cylinder

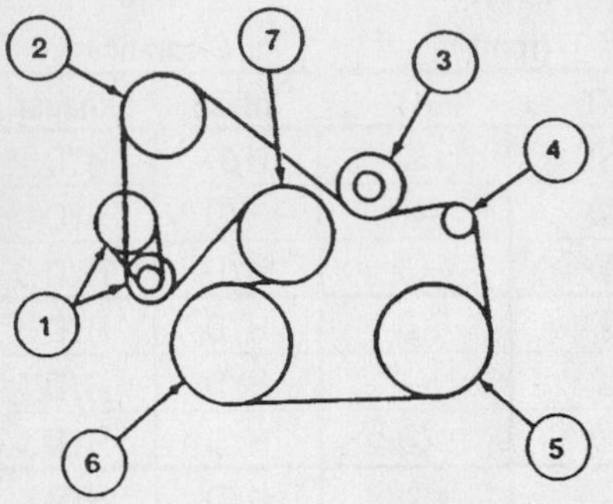

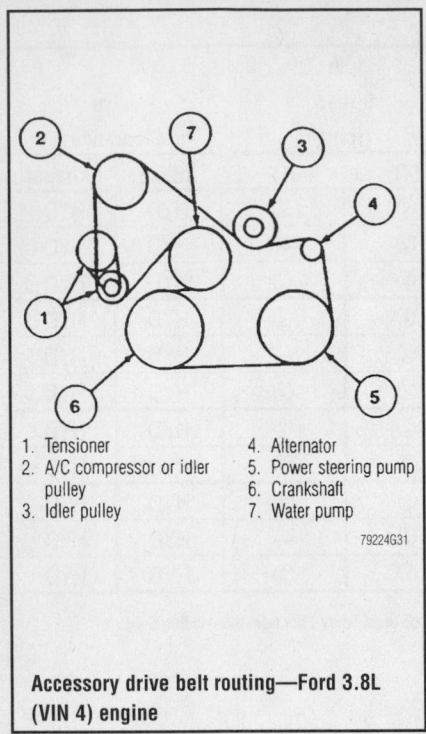

1. Tensioner
2. A/C compressor or idler pulley
3. Idler pulley
4. Alternator
5. Power steering pump
6. Crankshaft
7. Water pump

79224G31

Accessory drive belt routing—Ford 3.8L (VIN 4) engine

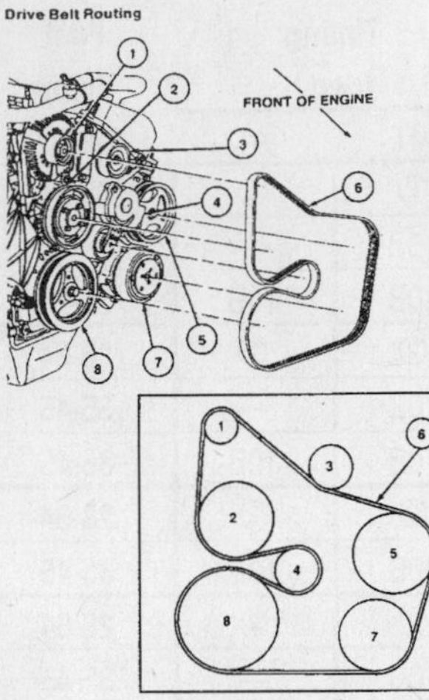

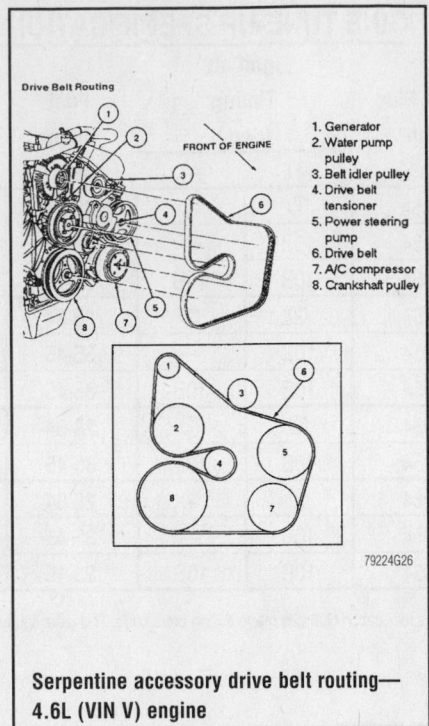

1. Generator
2. Water pump pulley
3. Belt idler pulley
4. Drive belt tensioner
5. Power steering pump
6. Drive belt
7. A/C compressor
8. Crankshaft pulley

79224G26

Serpentine accessory drive belt routing—4.6L (VIN V) engine

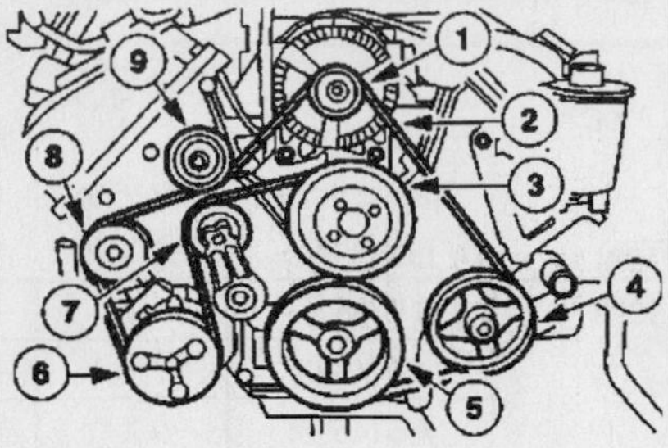

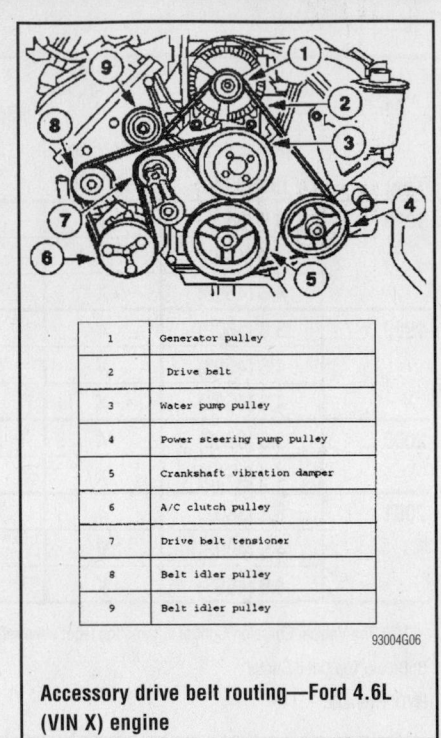

1	Generator pulley
2	Drive belt
3	Water pump pulley
4	Power steering pump pulley
5	Crankshaft vibration damper
6	A/C clutch pulley
7	Drive belt tensioner
8	Belt idler pulley
9	Belt idler pulley

93004G06

Accessory drive belt routing—Ford 4.6L (VIN X) engine

CAPACITIES

Year	Model	Engine Displacement Liters (cc)	Engine ID/VIN	Engine Oil with Filter (qts.)	Transmission (pts) Manual	Transmission (pts) Auto. ①	Drive Axle Rear (pts.)	Fuel Tank (gal.)	Cooling System (qts.)
1998	Mark VIII	4.6 (4593)	V	6.0	—	25.6	3.00	18.0	16.0
	Mustang	3.8 (3802)	4	5.0	5.6	27.8	3.50	15.4	11.8
		4.6 (4593)	V	6.0	6.5	27.8	3.75	15.4	14.1
		4.6 (4593)	X	②	6.5	25.6	3.75	15.4	14.1
1999	Mustang	3.8 (3802)	4	5.0	5.6	27.8	3.50	15.7	11.8
		4.6 (4593)	V	6.0	6.5	27.8	3.75	15.7	14.1
		4.6 (4593)	X	②	6.5	25.6	3.75	15.7	14.1
2000	Mustang	3.8 (3802)	4	5.0	5.6	27.8	3.50	15.7	11.8
		4.6 (4593)	X	②	6.5	25.6	3.75	15.7	14.1
2001	Mustang	3.8 (3802)	4	5.0	5.6	27.8	3.50	15.7	11.8
		4.6 (4593)	V	6.0	6.5	27.8	3.75	15.7	14.1
		4.6 (4593)	X	②	6.5	25.6	3.75	15.7	14.1

NOTE: All capacities are approximate. Add fluid gradually and ensure a proper fluid level is obtained.

① Includes torque converter

② If equipped with an automatic transmission: 6.7 qts.
If equipped with a manual transmission: 6.4 qts.

93461CH7

VALVE SPECIFICATIONS

Year	Engine Displacement Liters (cc)	Engine ID/VIN	Seat Angle (deg.)	Face Angle (deg.)	Spring Test Pressure (lbs. @ in.)	Spring Installed Height (in.)	Stem-to-Guide Clearance (in.) Intake	Stem-to-Guide Clearance (in.) Exhaust	Stem Diameter (in.) Intake	Stem Diameter (in.) Exhaust
1998	3.8 (3802)	4	44.5	45.8	220@1.18	1.650	0.0010-0.0027	0.0015-0.0032	0.3415-0.3423	0.3410-0.3418
	4.6 (4593)	V	45	45.5	160@1.10	1.425	0.0008-0.0027	0.0018-0.0037	0.2746-0.2754	0.2736-0.2744
	4.6 (4593)	X	45	45.5	132@1.10	1.570	0.0008-0.0027	0.0018-0.0037	0.2746-0.2754	0.2736-0.2744
1999	3.8 (3802)	4	44.7	45.7	224@1.16	1.620	0.0450-0.0900	0.0015-0.0033	0.2738-0.2751	0.2728-0.2741
	4.6 (4593)	V	45	45.5	160@1.03	1.660	0.0008-0.0027	0.0018-0.0037	0.2746-0.275	0.2736-0.2744
	4.6 (4593)	X	45	45.5	161@1.03	1.570	0.0008-0.0027	0.0018-0.0037	0.2746-0.2754	0.2736-0.2744
2000	3.8 (3802)	4	44.7	45.7	224@1.16	1.620	0.0450-0.0900	0.0015-0.0033	0.2738-0.2751	0.2728-0.2741
	4.6 (4593)	X	45	45.5	161@1.03	1.570	0.0008-0.0027	0.0018-0.0037	0.2746-0.2754	0.2736-0.2744
2001	3.8 (3802)	4	44.7	45.7	224@1.16	1.620	0.0450-0.0900	0.0015-0.0033	0.2738-0.2751	0.2728-0.2741
	4.6 (4593)	V	45	45.5	160@1.03	1.660	0.0008-0.0027	0.0018-0.0037	0.2746-0.275	0.2736-0.2744
	4.6 (4593)	X	45	45.5	161@1.03	1.570	0.0008-0.0027	0.0018-0.0037	0.2746-0.2754	0.2736-0.2744

93461CH8

TORQUE SPECIFICATIONS

All readings in ft. lbs.

Year	Engine Displacement Liters (cc)	Engine ID/VIN	Cylinder Head Bolts	Main Bearing Bolts	Rod Bearing Bolts	Crankshaft Damper Bolts	Flywheel Bolts	Manifold Intake	Manifold Exhaust	Spark Plugs	Lug Nuts
1998	3.8 (3802)	4	①	65-81	31-36	103-132	54-64	②	15-22	7-15	95
	4.6 (4593)	V	③	④	⑤	114-121	54-64	⑥	13-16	7-15	95
	4.6 (4593)	X	③	⑦	⑧	114-121	54-64	15-22	15-22	7-15	95
1999	3.8 (3802)	4	①	⑨	⑩	118	54-64	②	15-22	7-15	95
	4.6 (4593)	V	⑪	④	⑤	114-121	54-64	⑥	13-16	7-15	95
	4.6 (4593)	X	③	⑦	⑧	114-121	54-64	15-22	15-22	7-15	95
2000	3.8 (3802)	4	①	⑨	⑩	118	54-64	②	15-22	7-15	95
	4.6 (4593)	X	③	⑦	⑧	114-121	54-64	15-22	15-22	7-15	95
2001	3.8 (3802)	4	①	⑨	⑩	118	54-64	②	15-22	7-15	95
	4.6 (4593)	V	⑪	④	⑤	114-121	54-64	⑥	13-16	7-15	95
	4.6 (4593)	X	③	⑦	⑧	114-121	54-64	15-22	15-22	7-15	95

① Do not reuse cylinder head bolts.
Step 1: 15 ft. lbs.
Step 2: 29 ft. lbs.
Step 3: 37 ft. lbs.
Step 4: Loosen bolts one at a time and retorque as follows:
Long bolts: 11-18 ft. lbs.
Short bolts: 7-15 ft. lbs.
Step 5: Tighten 85-95 degrees

② Upper intake manifold bolts:
Step 1: 8 ft. lbs.
Step 2: 15 ft. lbs.
Step 3: 24 ft. lbs
Lower intake manifold bolts:
Step 1: 13 ft. lbs.
Step 2: 16 ft. lbs.

③ Do not reuse cylinder head bolts.
Step 1: 27-32 ft. lbs.
Step 2: Tighten each bolt 85-95 degrees
Step 3: Repeat Step 2

④ Step 1: Main bearing cap bolts: 6-9 ft. lbs.
Step 2: Main bearing cap bolts, outer: 16-21 ft. lbs.
Step 3: Main bearing cap bolts, inner: 27-32 ft. lbs.
Step 4: Tighten main bearing cap bolts 85-95 degrees
Step 5: Main cap adjusting screws: 4 ft. lbs. then 7.5 ft. lbs.
Step 6: Main cap side bolts: 7 ft. lbs. then 14-17 ft. lbs.

⑤ Step 1: 5ft. lbs.
Step 2: 10 ft. lbs.
Step 3: 18-25 ft. lbs.
Step 4: Tighten 85-95 degrees

⑥ Step 1: Four inside short bolts: 9-11 ft. lbs.
Step 2: All other bolts: 13-16 ft. lbs.
Step 3: Tighten 85-95 degrees

⑦ Do not reuse main cap bolts.
Step 1: Main bearing cap bolts: 22-25 ft. lbs.
Step 2: Tighten each bolt 85-95 degrees
Step 3: Main bearing cap adjust screws: 4 ft. lbs. then 6-8 ft. lbs.
Step 4: Main bearing cap side bolts: 7 ft. lbs. then 14-17 ft. lbs.

⑧ Do not reuse rod bolts.
Step 1: 12 ft. lbs.
Step 2: Tighten 85-95 degrees

⑨ Step 1: 37 ft. lbs.
Step 2: Tighten 115-125 degrees

⑩ Step 1: 18 ft. lbs.
Step 2: 33 ft. lbs.
Step 3: Tighten 90-120 degrees

⑪ Step 1: 30 ft. lbs.
Step 2: Tighten 90 degrees
Step 3: Loosen all bolts one (1) turn
Step 4: 30 ft. lbs.
Step 5: Tighten 90 degrees
Step 6: Tighten an additional 90 degrees

93461C11

Timing belt service is covered in Section 3 of this manual

BRAKE SPECIFICATIONS

All measurements in inches unless noted

Year	Model		Brake Disc Original Thickness	Brake Disc Minimum Thickness	Brake Disc Maximum Run-out	Brake Drum Diameter Original Inside Diameter	Brake Drum Diameter Max. Wear Limit	Brake Drum Diameter Maximum Machine Diameter	Minimum Lining Thickness	Brake Caliper Mounting Bolts (ft. lbs.)
1998	Mark VIII	F	1.024	0.974	0.003	—	—	—	0.125	60
		R	0.709	0.657	0.002	—	—	—	—	26
	Mustang	F	1.030	0.970	0.001	—	—	—	0.125	64
		R	0.550	0.500	0.002	—	—	—	0.123	26
	Mustang	F	1.100	1.040	0.001	—	—	—	0.125	64
	Cobra	R	0.710	0.660	0.002	—	—	—	0.123	26
1999	Mustang	F	1.030	0.970	0.001	—	—	—	0.125	64
		R	0.550	0.500	0.002	—	—	—	0.123	26
	Mustang	F	1.100	1.040	0.001	—	—	—	0.125	64
	Cobra	R	0.710	0.660	0.002	—	—	—	0.123	26
2000	Mustang	F	1.030	0.970	0.001	—	—	—	0.125	64
		R	0.550	0.500	0.002	—	—	—	0.123	26
2001	Mustang	F	1.030	0.970	0.001	—	—	—	0.125	64
		R	0.550	0.500	0.002	—	—	—	0.123	26
	Mustang	F	1.100	1.040	0.001	—	—	—	0.125	64
	Cobra	R	0.710	0.660	0.002	—	—	—	0.123	26

NOTE: Follow specifications stamped on rotor or drum if figures differ from those in this chart.

NA: Not Available

F: Front

R: Rear

93461C12

WHEEL ALIGNMENT

			Caster		Camber			Steering
Year	Model		Range (+/-Deg.)	Preferred Setting (Deg.)	Range (+/-Deg.)	Preferred Setting (Deg.)	Toe-in (in.)	Axis Inclination (Deg.)
1998	Mustang	F	0.75	0	0.75	0	0.13 +/- 0.13	—
		R	—	—	0.50	0	0.13 +/- 0.07	—
	Mark VIII	F	0.75	0	0.75	0	0.13 +/- 0.13	—
		R	—	—	0.50	0	0.06 +/- 0.13	—
1999	Mustang	F	0.75	0	0.75	0	0.13 +/- 0.13	—
		R	—	—	0.50	0	0.13 +/- 0.07	—
2000	Mustang	F	0.75	0	0.75	0	0.13 +/- 0.13	—
		R	—	—	0.50	0	0.13 +/- 0.07	—
2001	Mustang	F	0.75	0	0.75	0	0.13 +/- 0.13	—
		R	—	—	0.50	0	0.13 +/- 0.07	—

93461CI3

Heater Core replacement is covered in Section 2 of this manual

TIRE, WHEEL AND BALL JOINT SPECIFICATIONS

		OEM Tires		Tire Pressures (psi)		Wheel	Ball Joint
Year	Model	Standard	Optional	Front	Rear	Size	Inspection
1998	Mustang GT	P225/55ZR16	P245/45ZT17	30	30	Std: 7.5-JJ Opt: 8-JJ	①
	Mustang Cobra	P245/45ZR17	None	30	30	8-JJ	①
	Mark VIII	P225/60VR16	None	30	30	7-JJ	①
1999	Mustang, base	P205/65HR15	None	32	32	7-JJ	①
	Mustang GT	P225/55HR16	P245/45ZR17	30	30	8-JJ	①
	Mark VIII	P225/60VR16	None	30	30	7-JJ	①
2000	Mustang, base	P205/65HR15	None	30	30	7-JJ	①
	Mustang GT	P225/55HR16	P245/45ZR17	30	30	8-JJ	①

OEM: Original Equipment Manufacturer

PSI: Pounds Per Square Inch

STD: Standard

OPT: Optional

① Replace if any measurable movement is found

93461CI4

SCHEDULED MAINTENANCE INTERVALS

Lincoln Mk VIII, Ford Mustang

TO BE SERVICED	TYPE OF SERVICE	VEHICLE MILEAGE INTERVAL (x1000)												
		5	10	15	20	25	30	35	40	45	50	55	60	65
Engine oil & filter	R	✓	✓	✓	✓	✓	✓	✓	✓	✓	✓	✓	✓	✓
Adjust clutch pedal by lifting pedal	S/I	✓	✓	✓	✓	✓	✓	✓	✓	✓	✓	✓	✓	✓
Rotate tires	S/I	✓		✓		✓		✓		✓		✓		✓
Cooling system, hoses, clamps & coolant strength	S/I			✓			✓			✓			✓	
Air cleaner element	R						✓						✓	
Automatic transmission fluid & filter	R						✓						✓	
Engine coolant ①	R						✓						✓	
Spark plugs (Mark VIII) ②	R													
Spark plugs (Mustang)	R													
Accessory drive belt(s)	S/I						✓						✓	
Brake lines, hoses & connections	S/I						✓						✓	
Exhaust heat shields	S/I						✓						✓	
Front & rear brakes	S/I						✓						✓	
PCV valve	R												✓	
Rear axle lubricant ②	R													

R: Replace S/I: Service or Inspect

① Engine coolant: change engine coolant at 48,000 to 50,000 miles and thereafter every 30,000 miles.

② Replace every 100,000 miles.

FREQUENT OPERATION MAINTENANCE (SEVERE SERVICE)

If a vehicle is operated under any of the following conditions it is considered severe service:

- Extremely dusty areas.

- 50% or more of the vehicle operation is in 32°C (90°F) or higher temperatures, or constant operation in temperatures below 0°C (32°F).

- Prolonged idling (vehicle operation in stop and go traffic).

- Frequent short running periods (engine does not warm to normal operating temperatures).

- Police, taxi, delivery usage or trailer towing usage.

Oil & filter change: change every 3000 miles.

Rotate tires at 6000 miles & every 9000 miles thereafter.

Automatic transmission fluid & filter: change every 21,000 miles.

93461C15

Brake service is covered in Section 4 of this manual

SCHEDULED MAINTENANCE INTERVALS
FORD MOTOR COMPANY
FORD MUSTANG,
LINCOLN MARK VIII

The following should be used as a guide when determining the amount of work required for a particular service. In estimating how long a particular Scheduled Maintenance Service should take, please observe the following:

- Labor Time is time based on field research and data supplied by the vehicle manufacturer.
- Labor time operations are given in hours and tenths of an hour.
- All labor operations are to be used as a guide.

Mechanic Skill Level Codes:
(A) PRECISION: Highly skilled with multiple certification.
(B) GENERAL: Normally skilled with certification.
(C) MAINTENANCE: Semi-skilled working on certification.

	LABOR TIME
5000 Mile Service (C)	
All Models	1.0
10000 Mile Service (C)	
All Models	.5
15000 Mile Service (C)	
All Models	1.2
20000 Mile Service (C)	
All Models	.5
25000 Mile Service (C)	
All Models	1.0
30000 Mile Service (B)	
AT	2.1
MT	1.2
35000 Mile Service (C)	
All Models	1.0
40000 Mile Service (C)	
All Models	.5
45000 Mile Service (C)	
All Models	1.2
50000 Mile Service (C)	
All Models	1.0
55000 Mile Service (C)	
All Models	1.0
60000 Mile Service (B)	
AT	2.1
MT	1.2
65000 Mile Service (C)	
All Models	1.0

93461CI6

FORD MOTOR CO.

Ford Taurus • Taurus SHO • Mercury Sable

ENGINE AND VEHICLE IDENTIFICATION

Engine						
Code ①	Liters (cc)	Cu. In.	Cyl.	Fuel Sys.	Engine Type	Eng. Mfg.
N	3.4 (3393)	207	8	SFI	DOHC	Yamaha
S	3.0 (3049)	182	6	SFI	DOHC	Ford
U	3.0 (2982)	181	6	SFI	OHV	Ford

Model Year	
Code ②	Year
W	1998
X	1999
Y	2000
1	2001
2	2002

OHV: Overhead Valves

DOHC: Double Overhead Camshafts

SFI: Sequential Fuel Injection

① 8th digit of the Vehicle Identification Number (VIN)

② 10th digit of the Vehicle Identification Number (VIN)

93461C17

GENERAL ENGINE SPECIFICATIONS

Year	Model	Engine Displacement Liters (cc)	Engine ID/VIN	Fuel System Type	Net Horsepower @ rpm	Net Torque @ rpm (ft. lbs.)	Bore x Stroke (in.)	Com-pression Ratio	Oil Pressure @ rpm
1998	Sable	3.0 (2982)	U	SFI	145@5250	170@3250	3.50x3.15	9.3:1	40-60@2500
		3.0 (2998)	S	SFI	185@5750	200@4500	3.50x3.13	10.0:1	20-45@1500
	Taurus	3.0 (2982)	U	SFI	145@5250	170@3250	3.50x3.15	9.3:1	40-60@2500
		3.0 (2998)	S	SFI	185@5750	200@4500	3.50x3.13	10.0:1	20-45@1500
	Taurus SHO	3.4 (3393)	N	SFI	235@6100	230@4800	3.25x3.13	10.0:1	20-45@1500
1999	Sable	3.0 (2982)	U	SFI	145@5250	170@3250	3.50x3.15	9.3:1	40-60@2500
		3.0 (2998)	S	SFI	185@5750	200@4500	3.50x3.13	10.0:1	20-45@1500
	Taurus	3.0 (2982)	U	SFI	145@5250	170@3250	3.50x3.15	9.3:1	40-60@2500
		3.0 (2998)	S	SFI	185@5750	200@4500	3.50x3.13	10.0:1	20-45@1500
	Taurus SHO	3.4 (3393)	N	SFI	235@6100	230@4800	3.25x3.13	10.0:1	20-45@1500
2000	Sable	3.0 (2982)	U	SFI	155@4900	185@3950	3.50x3.15	9.3:1	40-60@2500
		3.0 (2998)	S	SFI	200@5750	200@4500	3.50x3.13	10.0:1	20-45@1500
	Taurus	3.0 (2982)	U	SFI	155@4900	185@3950	3.50x3.15	9.3:1	40-60@2500
		3.0 (2998)	S	SFI	200@5750	200@4500	3.50x3.13	10.0:1	20-45@1500
2001	Sable	3.0 (2982)	U	SFI	155@4900	185@3950	3.50x3.15	9.3:1	40-60@2500
		3.0 (2998)	S	SFI	200@5750	200@4500	3.50x3.13	10.0:1	20-45@1500
	Taurus	3.0 (2982)	U	SFI	155@4900	185@3950	3.50x3.15	9.3:1	40-60@2500
		3.0 (2998)	S	SFI	200@5750	200@4500	3.50x3.13	10.0:1	20-45@1500

SFI: Sequential Fuel Injection

93461C18

For complete Engine Mechanical specifications, see Section 1 of this manual

ENGINE TUNE-UP SPECIFICATIONS

Year	Engine Displacement Liters (cc)	Engine ID/VIN	Spark Plug Gap (in.)	Ignition Timing (deg.)	Fuel Pump (psi) ①	Idle Speed (rpm)	Valve Clearance Intake	Valve Clearance Exhaust
1998	3.0 (2982)	U	0.042-0.046	10B	30-45	②	HYD	HYD
	3.0 (2998)	S	0.052-0.056	10B	30-45	②	HYD	HYD
	3.4 (3393)	N	0.042-0.046	10B	35-45	②	0.006-0.010	0.010-0.014
1999	3.0 (2982)	U	0.042-0.046	10B	26-45	②	HYD	HYD
	3.0 (2998)	S	0.052-0.056	10B	26-45	②	HYD	HYD
	3.4 (3393)	N	0.042-0.046	10B	35-45	②	0.006-0.010	0.010-0.014
2000	3.0 (2982)	U	0.042-0.046	10B	26-45	②	HYD	HYD
	3.0 (2998)	S	0.052-0.056	10B	26-45	②	HYD	HYD
2001	3.0 (2982)	U	0.042-0.046	10B	26-45	②	HYD	HYD
	3.0 (2998)	S	0.052-0.056	10B	26-45	②	HYD	HYD

NOTE: The Vehicle Emission Control Information label often reflects specification changes made during production. The label figures must be used if they differ from those in this chart.

B: Before Top Dead Center

HYD: Hydraulic

① Fuel pressure with engine running, pressure regulator vacuum hose connected

② Refer to Vehicle Emission Control Information label

93461C19

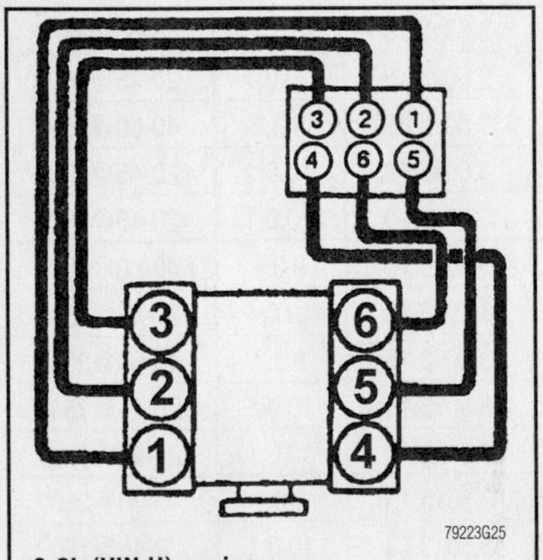

3.0L (VIN U) engine
Firing order: 1–4–2–5–3–6
Distributorless ignition system

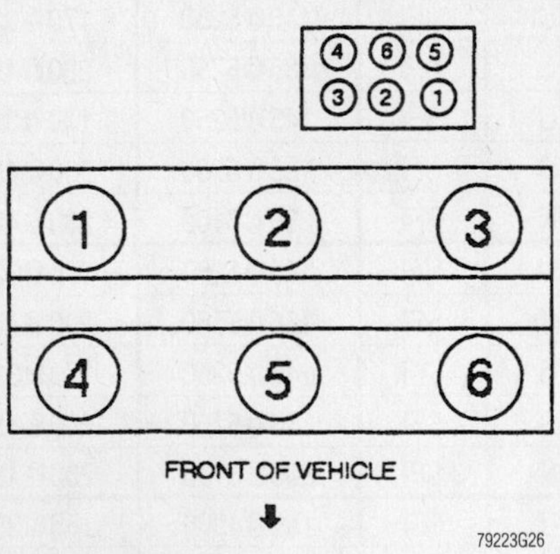

3.0L (VIN S) engine
Firing order: 1–4–2–5–3–6
Distributorless ignition system

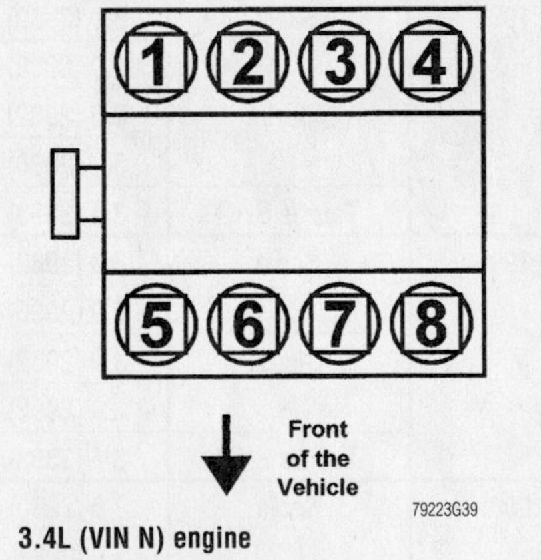

3.4L (VIN N) engine
Firing order: 1–5–4–2–6–3–7–8
Distributorless ignition system (one coil on each cylinder)

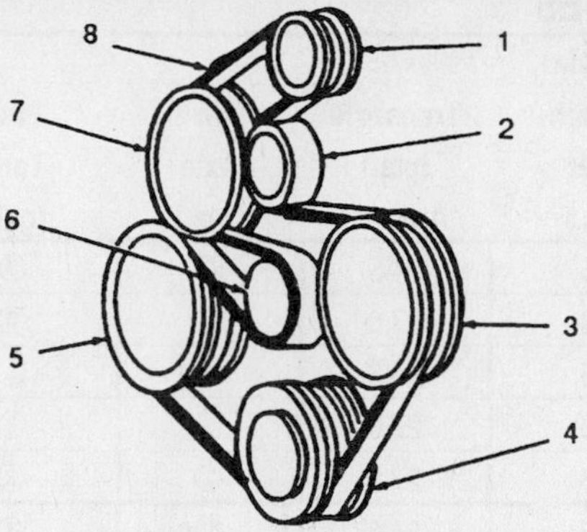

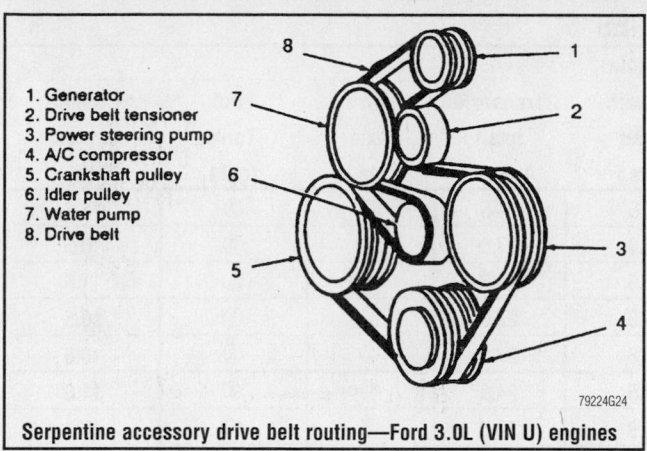

1. Generator
2. Drive belt tensioner
3. Power steering pump
4. A/C compressor
5. Crankshaft pulley
6. Idler pulley
7. Water pump
8. Drive belt

79224G24

Serpentine accessory drive belt routing—Ford 3.0L (VIN U) engines

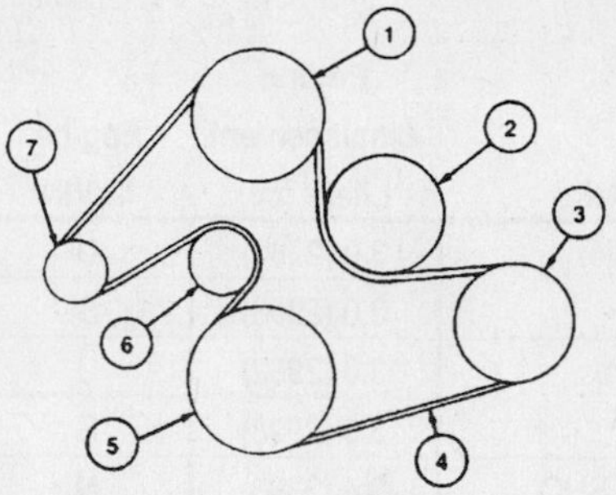

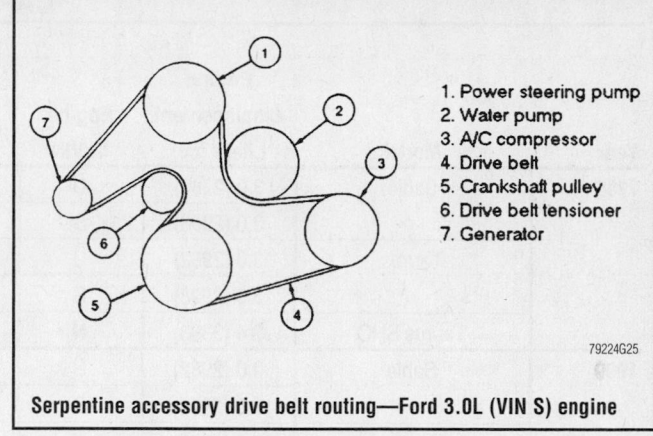

1. Power steering pump
2. Water pump
3. A/C compressor
4. Drive belt
5. Crankshaft pulley
6. Drive belt tensioner
7. Generator

79224G25

Serpentine accessory drive belt routing—Ford 3.0L (VIN S) engine

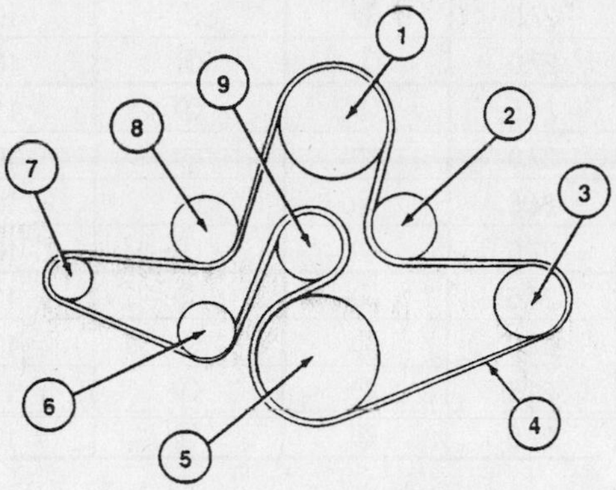

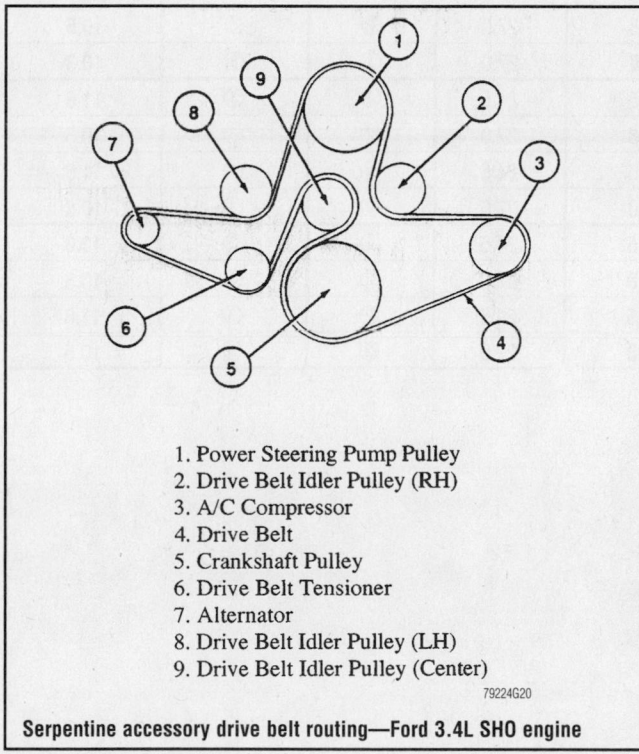

1. Power Steering Pump Pulley
2. Drive Belt Idler Pulley (RH)
3. A/C Compressor
4. Drive Belt
5. Crankshaft Pulley
6. Drive Belt Tensioner
7. Alternator
8. Drive Belt Idler Pulley (LH)
9. Drive Belt Idler Pulley (Center)

79224G20

Serpentine accessory drive belt routing—Ford 3.4L SHO engine

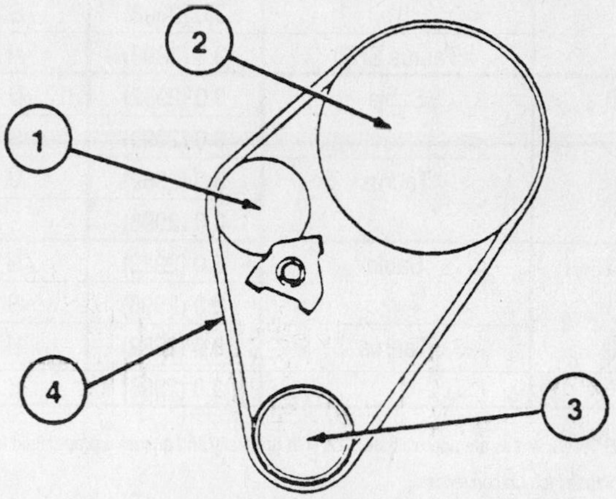

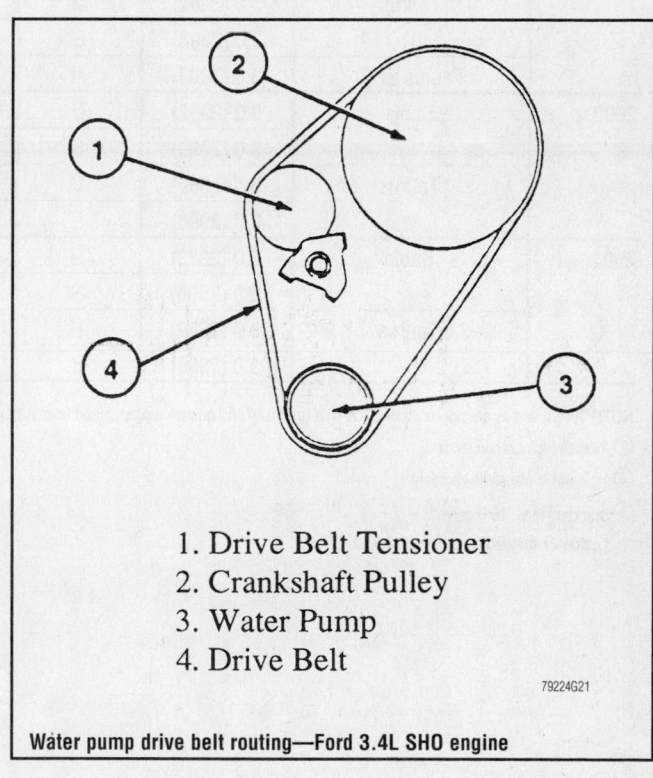

1. Drive Belt Tensioner
2. Crankshaft Pulley
3. Water Pump
4. Drive Belt

79224G21

Water pump drive belt routing—Ford 3.4L SHO engine

For Accessory Drive Belt illustrations, see Section 1 of this manual

CAPACITIES

Year	Model	Engine Displacement Liters (cc)	Engine ID/VIN	Engine Oil with Filter (qts.)	Transaxle (pts.) Auto. ①	Drive Axle (pts.)	Fuel Tank (gal.)	Cooling System (qts.)
1998	Sable	3.0 (2982)	U	4.5	24.5	②	③	11.6
		3.0 (2998)	S	5.8	27.0	②	③	10.5
	Taurus	3.0 (2982)	U	4.5	24.5	②	③	11.6
		3.0 (2998)	S	5.8	27.0	②	③	10.5
	Taurus SHO	3.4 (3393)	N	5.8	27.0	②	③	10.8
1999	Sable	3.0 (2982)	U	4.5	24.5	②	③	11.6
		3.0 (2998)	S	5.8	27.0	②	③	10.5
	Taurus	3.0 (2982)	U	4.5	24.5	②	③	11.6
		3.0 (2998)	S	5.8	27.0	②	③	10.5
	Taurus SHO	3.4 (3393)	N	6.7	27.0	②	③	10.8
2000	Sable	3.0 (2982)	U	4.5	24.5	②	③	11.6
		3.0 (2998)	S	5.8	27.0	②	③	10.5
	Taurus	3.0 (2982)	U	4.5	24.5	②	③	11.6
		3.0 (2998)	S	5.8	27.0	②	③	10.5
2001	Sable	3.0 (2982)	U	4.5	24.5	②	③	11.6
		3.0 (2998)	S	5.8	27.0	②	③	10.5
	Taurus	3.0 (2982)	U	4.5	24.5	②	③	11.6
		3.0 (2998)	S	5.8	27.0	②	③	10.5

NOTE: All capacities are approximate. Add fluid gradually and ensure a proper fluid level is obtained.

① Includes torque converter

② Included in transaxle capacity

③ Standard tank: 16.0 gals.
Optional extended range tank: 18.6 gals.

93461C10

VALVE SPECIFICATIONS

Year	Engine Displacement Liters (cc)	Engine ID/VIN	Seat Angle (deg.)	Face Angle (deg.)	Spring Test Pressure (lbs. @ in.)	Spring Installed Height (in.)	Stem-to-Guide Clearance (in.) Intake	Stem-to-Guide Clearance (in.) Exhaust	Stem Diameter (in.) Intake	Stem Diameter (in.) Exhaust
1998	3.0 (2982)	U	45	44	180@1.16	1.580	0.0001-0.0027	0.0015-0.0033	0.3126-0.3134	0.3121-0.3129
	3.0 (2998)	S	44.75	45.5	153@1.18	1.570	0.0007-0.0027	0.0017-0.0037	0.2350-0.2358	0.2343-0.2350
	3.4 (3393)	N	45	45.5	89@1.00	1.360	0.0010-0.0023	0.0012-0.0025	0.2346-0.2352	0.2344-0.2350
1999	3.0 (2982)	U	45	44	180@1.16	1.580	0.0010 0.0028	0.0015-0.0033	0.3126-0.3134	0.3121-0.3129
	3.0 (2998)	S	44.75	45.5	153@1.18	1.570	0.0007-0.0027	0.0017-0.0037	0.2350-0.2358	0.2343-0.2350
	3.4 (3393)	N	45	45.5	89@1.00	1.360	0.0010-0.0023	0.0012-0.0025	0.2346-0.2352	0.2344-0.2350
2000	3.0 (2982)	U	45	44	180@1.16	1.580	0.0010 0.0028	0.0015-0.0033	0.3126-0.3134	0.3121-0.3129
	3.0 (2998)	S	44.75	45.5	153@1.18	1.570	0.0007-0.0027	0.0017-0.0037	0.2350-0.2358	0.2343-0.2350
2001	3.0 (2982)	U	45	44	180@1.16	1.580	0.0010 0.0028	0.0015-0.0033	0.3126-0.3134	0.3121-0.3129
	3.0 (2998)	S	44.75	45.5	153@1.18	1.570	0.0007-0.0027	0.0017-0.0037	0.2350-0.2358	0.2343-0.2350

93461CJ1

CRANKSHAFT AND CONNECTING ROD SPECIFICATIONS

All measurements are given in inches.

Year	Engine Displacement Liters (cc)	Engine ID/VIN	Crankshaft Main Brg. Journal Dia.	Crankshaft Main Brg. Oil Clearance	Crankshaft Shaft End-play	Crankshaft Thrust on No.	Connecting Rod Journal Diameter	Connecting Rod Oil Clearance	Connecting Rod Side Clearance
1998	3.0 (2982)	U	2.5190-2.5198	0.0009 0.0030	0.0040 0.0080	3	2.1253-2.1261	0.0009 0.0030	0.0060 0.0140
	3.0 (2998)	S	2.4670-2.4790	0.0009-0.0018	0.0040-0.0090	4	1.9670-1.9680	0.0010-0.0025	0.0039-0.0118
	3.4 (3393)	N	2.4794-2.4803	0.0004-0.0012	0.0024-0.0103	3	1.9765-1.9685	0.0009-0.0023	0.0060-0.0120
1999	3.0 (2982)	U	2.5190-2.5198	0.0009 0.0027	0.0040 0.0080	3	2.1253-2.1261	0.0009 0.0027	0.0060 0.0140
	3.0 (2998)	S	2.4670-2.4790	0.0009-0.0018	0.0040 0.0090	4	1.9670-1.9680	0.0010 0.0025	0.0039-0.0118
	3.4 (3393)	N	2.4794-2.4803	0.0004-0.0012	0.0024-0.0103	3	1.9675 1.9685	0.0009-0.0023	0.0060 0.0120
2000	3.0 (2982)	U	2.5190-2.5198	0.0009 0.0027	0.0040 0.0080	3	2.1253-2.1261	0.0009 0.0027	0.0060 0.0140
	3.0 (2998)	S	2.4670-2.4790	0.0009-0.0018	0.0040 0.0090	4	1.9670-1.9680	0.0010 0.0025	0.0039-0.0118
2001	3.0 (2982)	U	2.5190-2.5198	0.0009 0.0027	0.0040 0.0080	3	2.1253-2.1261	0.0009 0.0027	0.0060 0.0140
	3.0 (2998)	S	2.4670-2.4790	0.0009-0.0018	0.0040 0.0090	4	1.9670-1.9680	0.0010 0.0025	0.0039-0.0118

93461CJ3

PISTON AND RING SPECIFICATIONS

All measurements are given in inches.

Year	Engine Displacement Liters (cc)	Engine ID/VIN	Piston Clearance	Ring Gap Top Compression	Ring Gap Bottom Compression	Ring Gap Oil Control	Ring Side Clearance Top Compression	Ring Side Clearance Bottom Compression	Ring Side Clearance Oil Control
1998	3.0 (2982)	U	0.0014-0.0022	0.010-0.020	0.010-0.020	0.010-0.049	0.0012 0.0031	0.0012 0.0031	SNUG
	3.0 (2998)	S	0.0005-0.0009	0.004-0.010	0.011-0.017	0.005-0.026	0.0015-0.0029	0.0015-0.0033	SNUG
	3.4 (3393)	N	0.0008-0.0016	0.008-0.014	0.008-0.014	0.008-0.028	0.0018-0.0031	0.0012-0.0028	0.0024-0.0059
1999	3.0 (2982)	U	0.0014-0.0022	0.010-0.020	0.010-0.020	0.010-0.049	0.0012 0.0031	0.0012 0.0031	SNUG
	3.0 (2998)	S	0.0005-0.0009	0.004-0.010	0.011-0.017	0.006-0.026	0.0015-0.0029	0.0015-0.0033	SNUG
	3.4 (3393)	N	0.0008-0.0016	0.008-0.014	0.008-0.014	0.008-0.028	0.0018-0.0031	0.0012-0.0028	0.0024-0.0059
2000	3.0 (2982)	U	0.0014-0.0022	0.010-0.020	0.010-0.020	0.010-0.049	0.0012 0.0031	0.0012 0.0031	SNUG
	3.0 (2998)	S	0.0005-0.0009	0.004-0.010	0.011-0.017	0.006-0.026	0.0015-0.0029	0.0015-0.0033	SNUG
2001	3.0 (2982)	U	0.0014-0.0022	0.010-0.020	0.010-0.020	0.010-0.049	0.0012 0.0031	0.0012 0.0031	SNUG
	3.0 (2998)	S	0.0005-0.0009	0.004-0.010	0.011-0.017	0.006-0.026	0.0015-0.0029	0.0015-0.0033	SNUG

93461CJ2

TORQUE SPECIFICATIONS

All readings in ft. lbs.

Year	Engine Displacement Liters (cc)	Engine ID/VIN	Cylinder Head Bolts	Main Bearing Bolts	Rod Bearing Bolts	Crankshaft Damper Bolts	Flywheel Bolts	Manifold Intake	Manifold Exhaust	Spark Plugs	Lug Nuts
1998	3.0 (2982)	U	①	56-62	23-28	93-121	54-64	②	15-18	7-15	95
	3.0 (2998)	S	③	④	⑤	⑥	54-64	⑦	13-16	7-15	95
	3.4 (3393)	N	⑧	⑨	⑩	⑥	54-64	14-20	11-19	11-15	95
1999	3.0 (2982)	U	⑪	56-62	23-28	93-121	54-64	②	15-18	7-15	95
	3.0 (2998)	S	③	④	⑤	⑥	54-64	⑦	13-16	7-15	95
	3.4 (3393)	N	⑦	⑧	⑨	⑤	55-61	14-20	11-19	11-15	95
2000	3.0 (2982)	U	⑪	56-62	23-28	93-121	54-64	②	15-18	7-15	95
	3.0 (2998)	S	③	④	⑤	⑥	54-64	⑦	13-16	7-15	95
2001	3.0 (2982)	U	⑪	56-62	23-28	93-121	54-64	②	15-18	7-15	95
	3.0 (2998)	S	③	④	⑤	⑥	54-64	⑦	13-16	7-15	95

① Step 1: 52-66 ft.lbs
Step 2: Loosen one turn
Step 3: 34-40 ft. lbs.
Step 4: 63-73 ft. lbs.

② Upper intake: 15-22 ft. lbs.
Lower Intake:
Step 1: 15-22 ft. lbs.
Step 2: 20-23 ft. lbs.

③ Step 1: 28-31 ft. lbs.
Step 2: Rotate 85-95 degrees
Step 3: Loosen one turn
Step 4: 28-31 ft. lbs.
Step 5: Rotate 85-95 degrees
Step 6: Repeat Step 5

④ Step 1: Cap bolts 1-8 (outer) 17-20 ft. lbs.
Step 2: Cap bolts 9-16 (inner) 28-31 ft. lbs.
Step 3: Rotate bolts 1-16, 85-95 degrees
Step 4: Bolts 17-22; 15-22 ft. lbs.

⑤ Step 1: 30-33 ft. lbs.
Step 2: Rotate 90-120 degrees

⑥ Step 1: 77-99 ft. lbs.
Step 2: Loosen 360 degrees
Step 3: Tighten to 35-39 ft. lbs.
Step 4: Rotate 85-95 degrees

⑦ 71-106 inch lbs.

⑧ Step 1: 20-23 ft. lbs.
Step 2: Rotate 85-95 degrees

⑨ Step 1: Cap bolts 1-10 (outer) 17-20 ft. lbs.
Step 2: Cap bolts 11-20 (inner) 28-31 ft. lbs.
Step 3: Rotate bolts 1-20, 85-95 degrees
Step 4: Bolts 21-31, 15-22 ft. lbs.

⑩ Step 1: 30-33 ft. lbs.
Step 2: Rotate 90-120 degrees

⑪ Step 1: 35-39 ft. lbs.
Step 2: Loosen one turn
Step 3: 20-24 ft. lbs.
Step 4: Rotate 85-95 degrees
Step 5: Repeat Step 4

93461CJ4

BRAKE SPECIFICATIONS
Ford Taurus, Mercury Sable
All measurements in inches unless noted

			Brake Disc			Brake Drum				Brake Caliper	
Year	Model		Original Thickness	Minimum Thickness	Maximum Run-out	Original Inside Diameter	Max. Wear Limit	Maximum Machine Diameter	Minimum Lining Thickness	Bracket Bolts (ft. lbs.)	Mounting Bolts (ft. lbs.)
1998	Sable	F	1.020	0.974	0.002	—	—	—	0.039	65-85	23-28
		R	0.550	0.500	0.004	8.86	0.06	8.92	0.039	65-87	23-25
	Taurus	F	1.020	0.974	0.002	—	—	—	0.039	65-85	23-28
		R	0.940	0.500	0.002	8.86	0.06	8.92	0.039	65-87	23-25
1999	Sable	F	1.020	0.974	0.002	—	—	—	0.039	65-85	23-28
		R	0.550	0.500	0.004	8.85	0.05	8.92	0.039	65-87	23-25
	Taurus	F	1.020	0.974	0.002	—	—	—	0.039	65-85	23-28
		R	0.940	0.500	0.002	8.85	0.05	8.92	0.039	65-87	23-25
2000	Sable	F	1.020	0.974	0.002	—	—	—	0.039	65-85	23-28
		R	0.550	0.500	0.004	8.85	0.05	8.92	0.039	65-87	23-25
	Taurus	F	1.020	0.974	0.002	—	—	—	0.039	65-85	23-28
		R	0.940	0.500	0.002	8.85	0.05	8.92	0.039	65-87	23-25
2001	Sable	F	1.020	0.974	0.002	—	—	—	0.039	65-85	23-28
		R	0.550	0.500	0.004	8.85	0.05	8.92	0.039	65-87	23-25
	Taurus	F	1.020	0.974	0.002	—	—	—	0.039	65-85	23-28
		R	0.940	0.500	0.002	8.85	0.05	8.92	0.039	65-87	23-25

NOTE: Follow specifications stamped on rotor or drum if figures differ from those in this chart.

F: Front

R: Rear

93461CJ5

WHEEL ALIGNMENT

Year	Model		Caster Range (+/-Deg.)	Caster Preferred Setting (Deg.)	Camber Range (+/-Deg.)	Camber Preferred Setting (Deg.)	Toe-in (in.)	Steering Axis Inclination (Deg.)
1998	Taurus Sedan	F	0.70	0	0.70	0	0.10 +/- 0.13	—
		R	—	—	0.70	0	0.18 +/- 0.13	—
	Taurus Wagon	F	0.70	0	0.70	0	0.10 +/- 0.13	—
		R	—	—	1.20	0	0.18 +/- 0.13	—
	Taurus SHO	F	0.70	0	0.70	0	0.10 +/- 0.13	—
		R	—	—	1.20	0	0.18 +/- 0.13	—
	Sable	F	0.70	0	0.70	0	0.20 +/- 0.25	—
		R	—	—	0.70	0	0.36 +/- 0.25	—
	Sable Wagon	F	0.70	0	0.70	0	0.20 +/- 0.25	—
		R	—	—	0.12	0	0.36 +/- 0.25	—
1999	Taurus Sedan	F	0.70	0	0.70	0	0.10 +/- 0.13	—
		R	—	—	0.70	0	0.18 +/- 0.13	—
	Taurus Wagon	F	0.70	0	0.70	0	0.10 +/- 0.13	—
		R	—	—	1.20	0	0.18 +/- 0.13	—
	Taurus SHO	F	0.70	0	0.70	0	0.10 +/- 0.13	—
		R	—	—	1.20	0	0.18 +/- 0.13	—
	Sable	F	0.70	0	0.70	0	0.20 +/- 0.25	—
		R	—	—	0.70	0	0.36 +/- 0.25	—
	Sable Wagon	F	0.70	0	0.70	0	0.20 +/- 0.25	—
		R	—	—	0.12	0	0.36 +/- 0.25	—
2000	Taurus Sedan	F	0.70	0	0.70	0	0.10 +/- 0.13	—
		R	—	—	0.70	0	0.18 +/- 0.13	—
	Taurus Wagon	F	0.70	0	0.70	0	0.10 +/- 0.13	—
		R	—	—	1.20	0	0.18 +/- 0.13	—
	Sable	F	0.70	0	0.70	0	0.20 +/- 0.25	—
		R	—	—	0.70	0	0.36 +/- 0.25	—
	Sable Wagon	F	0.70	0	0.70	0	0.20 +/- 0.25	—
		R	—	—	0.12	0	0.36 +/- 0.25	—
2001	Taurus Sedan	F	0.70	0	0.70	0	0.10 +/- 0.13	—
		R	—	—	0.70	0	0.18 +/- 0.13	—
	Taurus Wagon	F	0.70	0	0.70	0	0.10 +/- 0.13	—
		R	—	—	1.20	0	0.18 +/- 0.13	—
	Sable	F	0.70	0	0.70	0	0.20 +/- 0.25	—
		R	—	—	0.70	0	0.36 +/- 0.25	—
	Sable Wagon	F	0.70	0	0.70	0	0.20 +/- 0.25	—
		R	—	—	0.12	0	0.36 +/- 0.25	—

93461CJ7

TIRE, WHEEL AND BALL JOINT SPECIFICATIONS

		OEM Tires		Tire Pressures (psi)		Wheel	Ball Joint
Year	Model	Standard	Optional	Front	Rear	Size	Inspection
1998	Sable	P205/65R15	None	35	35	6-JJ	0.030 in. ①
	Taurus, exc. SHO	P205/65R15	None	35	35	6-JJ	0.030 in. ①
	Taurus SHO	P225/55VR16	P225/55ZR16	30	30	6.5-JJ	0.030 in. ①
1999	Sable	P205/65R15	P225/55ZR16	35	35	6-JJ	0.030 in. ①
	Taurus, exc. SHO	P205/65R15	None	30	30	6-JJ	0.030 in. ①
	Taurus SHO	P225/55ZR16	None	33	33	6.5-JJ	0.030 in. ①
2000	Sable GS, LS Sedan	P205/65R16	None	33	33	7J	0.030 in. ①
	Sable LS Premium Sedan	P215/60R16	None	33	33	7J	0.030 in. ①
	Sable Wagon	P215/60R16	None	31	31	7J	0.030 in. ①
	Taurus	P215/60R16	None	33	33	6.5-JJ	0.030 in. ①
2001	Sable GS, LS Sedan	P205/65R16	None	33	33	7J	0.030 in. ①
	Sable LS Premium Sedan	P215/60R16	None	33	33	7J	0.030 in. ①
	Sable Wagon	P215/60R16	None	31	31	7J	0.030 in. ①
	Taurus	P215/60R16	None	33	33	6.5-JJ	0.030 in. ①

OEM: Original Equipment Manufacturer

PSI: Pounds Per Square Inch

① Maximum radial tolerance in inches

93461CJ8

SCHEDULED MAINTENANCE INTERVALS
Ford Taurus, Mercury Sable

TO BE SERVICED	TYPE OF SERVICE	VEHICLE MILEAGE INTERVAL (X1000)																			
		5	10	15	20	25	30	35	40	45	50	55	60	65	70	75	80	85	90	95	100
Engine oil & filter	R	✓	✓	✓	✓	✓	✓	✓	✓	✓	✓	✓	✓	✓	✓	✓	✓	✓	✓	✓	✓
Rotate tires	S/I	✓		✓		✓		✓		✓		✓		✓		✓		✓		✓	
Engine coolant protection, hoses & clamps	S/I			✓			✓			✓			✓			✓			✓		
Passenger compartment air filter	R				✓				✓				✓				✓				✓
Air cleaner filter	R						✓						✓						✓		
Automatic transaxle fluid & filter	R						✓						✓						✓		
Brake lines & connections	S/I						✓						✓						✓		
Exhaust heat shields	S/I						✓						✓						✓		
Front and rear disc brake pads & rotors	S/I						✓						✓						✓		
Accessory drive belt(s)	S/I												✓								
Engine coolant ①	R										✓						✓				
Spark plugs (exc. 3.0L FF) ②	R																				✓
Spark plugs (3.0L FF)	R						✓						✓								
PCV valve (except 3.0L 4-valve)	R												✓								
PCV valve (3.0L 4-valve)	R																				✓

① Engine coolant: change initially at 50,000 miles & thereafter every 30,000 miles

② Platinum tip spark plugs: change every 100,000 miles

R: Replace S/I: Service and Inspect

FREQUENT OPERATION MAINTENANCE (SEVERE SERVICE)

If a vehicle is operated under any of the following conditions it is considered severe service:

- Extremely dusty areas.

- 50% or more of the vehicle operation is in 32°C (90°F) or higher temperatures, or constant operation in temperatures below 0°C (32°F).

- Prolonged idling (vehicle operation in stop and go traffic)..

- Frequent short running periods (engine does not warm to normal operating temperatures).

- Police, taxi, delivery usage or trailer towing usage.

Oil & oil filter: change every 3000 miles

Rotate tires at 6000 miles & every 9000 miles thereafter

Air cleaner element service or inspect every 15,000 miles

Automatic transaxle fluid & filter: change every 21,000 miles

93461CJ9

SCHEDULED MAINTENANCE INTERVALS
FORD MOTOR COMPANY
FORD TAURUS, MERCURY SABLE

The following should be used as a guide when determining the amount of work required for a particular service. In estimating how long a particular Scheduled Maintenance Service should take, please observe the following:

- Labor Time is time based on field research and data supplied by the vehicle manufacturer.
- Labor time operations are given in hours and tenths of an hour.
- All labor operations are to be used as a guide.

Mechanic Skill Level Codes:
(A) PRECISION: Highly skilled with multiple certification.
(B) GENERAL: Normally skilled with certification.
(C) MAINTENANCE: Semi-skilled working on certification.

	LABOR TIME
5000 Mile Service (C)	
All Models	.9
10000 Mile Service (C)	
All Models	.5
15000 Mile Service (C)	
All Models	1.0
20000 Mile Service (C)	
All Models	.7
25000 Mile Service (C)	
All Models	.9
30000 Mile Service (B)	
All Models	2.0
Replace auto. trans. fluid, filter add	.5
35000 Mile Service (C)	
All Models	.9
40000 Mile Service (C)	
All Models	.7
45000 Mile Service (C)	
All Models	1.0
50000 Mile Service (C)	
All Models	.9
Replace engine coolant add	.5
55000 Mile Service (C)	
All Models	.9
60000 Mile Service (B)	
All Models	2.2
Replace auto. trans. fluid filter add	.5
65000 Mile Service (C)	
All Models	.9
70000 Mile Service (C)	
All Models	.5
75000 Mile Service (C)	
All Models	.7
80000 Mile Service (C)	
All Models	.9
85000 Mile Service (C)	
All Models	.9
Replace engine coolant add	.5
90000 Mile Service (B)	
All Models	2.0
Replace auto. trans. fluid, filter add	.5
95000 Mile Service (C)	
All Models	.9
100000 Mile Service (C)	
All Models	1.2

93461CJ0

GENERAL MOTORS C- & H-BODIES

Buick LeSabre • Park Avenue • Oldsmobile Eighty Eight • Ninety Eight • LSS • Regency • Pontiac Bonneville

ENGINE AND VEHICLE IDENTIFICATION

Engine						
Code ①	Liters (cc)	Cu. In.	Cyl.	Fuel Sys.	Engine Type	Eng. Mfg.
1 ③	3.8 (3785)	231	6	MFI	OHV	BOC
K	3.8 (3785)	231	6	MFI	OHV	BOC

Model Year	
Code ②	Year
W	1998
X	1999
Y	2000
1	2001
2	2002

MFI: Multi-point Fuel Injection

BOC: Buick/Oldsmobile/Cadillac

OHV: Overhead Valves

① 8th position of VIN

② 10th position of VIN

③ Supercharged engine

93461CK1

GENERAL ENGINE SPECIFICATIONS

Year	Model	Engine Displacement Liters (cc)	Engine Series (ID/VIN)	Fuel System	Net Horsepower @ rpm	Net Torque @ rpm (ft. lbs.)	Bore x Stroke (in.)	Com-pression Ratio	Oil Pressure @ rpm
1998	Bonneville	3.8 (3785)	1 ①	MFI	240@5000	275@3200	3.80x3.40	9.0:1	60@1850
	Bonneville	3.8 (3785)	K	MFI	205@5200	230@4000	3.80x3.40	9.4:1	60@1850
	Eighty-Eight	3.8 (3785)	K	MFI	205@5200	230@4000	3.80x3.40	9.4:1	60@1850
	LeSabre	3.8 (3785)	K	MFI	205@5200	230@4000	3.80x3.40	9.4:1	60@1850
	LSS	3.8 (3785)	1 ①	MFI	240@5000	275@3200	3.80x3.40	9.0:1	60@1850
	LSS	3.8 (3785)	K	MFI	205@5200	230@4000	3.80x3.40	9.4:1	60@1850
	Park Avenue	3.8 (3785)	1 ①	MFI	240@5000	275@3200	3.80x3.40	9.0:1	60@1850
	Park Avenue	3.8 (3785)	K	MFI	205@5200	230@4000	3.80x3.40	9.4:1	60@1850
	Regency	3.8 (3785)	K	MFI	205@5200	230@4000	3.80x3.40	9.4:1	60@1850
1999	Bonneville	3.8 (3785)	1 ①	MFI	240@5000	275@3200	3.80x3.40	9.0:1	60@1850
	Bonneville	3.8 (3785)	K	MFI	205@5200	230@4000	3.80x3.40	9.4:1	60@1850
	Eighty-Eight	3.8 (3785)	K	MFI	205@5200	230@4000	3.80x3.40	9.4:1	60@1850
	LeSabre	3.8 (3785)	K	MFI	205@5200	230@4000	3.80x3.40	9.4:1	60@1850
	LSS	3.8 (3785)	K	MFI	205@5200	230@4000	3.80x3.40	9.4:1	60@1850
	Park Avenue	3.8 (3785)	1 ①	MFI	240@5000	275@3200	3.80x3.40	9.0:1	60@1850
	Park Avenue	3.8 (3785)	K	MFI	205@5200	230@4000	3.80x3.40	9.4:1	60@1850
2000	Bonneville	3.8 (3785)	1 ①	MFI	240@5000	275@3200	3.80x3.40	9.0:1	60@1850
	Bonneville	3.8 (3785)	K	MFI	205@5200	230@4000	3.80x3.40	9.4:1	60@1850
	LeSabre	3.8 (3785)	K	MFI	200@5200	230@4000	3.80x3.40	9.4:1	60@1850
	Park Avenue	3.8 (3785)	1 ①	MFI	240@5000	280@3600	3.80x3.40	9.0:1	60@1850
	Park Avenue	3.8 (3785)	K	MFI	200@5200	230@4000	3.80x3.40	9.4:1	60@1850
2001	Bonneville	3.8 (3785)	1 ①	MFI	240@5000	275@3200	3.80x3.40	9.0:1	60@1850
	Bonneville	3.8 (3785)	K	MFI	205@5200	230@4000	3.80x3.40	9.4:1	60@1850
	LeSabre	3.8 (3785)	K	MFI	200@5200	230@4000	3.80x3.40	9.4:1	60@1850
	Park Avenue	3.8 (3785)	1 ①	MFI	240@5000	280@3600	3.80x3.40	9.0:1	60@1850
	Park Avenue	3.8 (3785)	K	MFI	200@5200	230@4000	3.80x3.40	9.4:1	60@1850

MFI: Multi-point Fuel Injection

① Supercharged engine

93461CK2

ENGINE TUNE-UP SPECIFICATIONS

Year	Engine Displacement Liters (cc)	Engine ID/VIN	Spark Plug Gap (in.)	Ignition Timing (deg.)	Fuel Pump (psi)	Idle Speed (rpm)	Valve Clearance	
							Intake	Exhaust
1998	3.8 (3786)	1	0.060	①	41-47 ②	③	HYD	HYD
	3.8 (3785)	K	0.060	①	41-47 ②	③	HYD	HYD
1999	3.8 (3786)	1	0.060	①	41-47 ②	③	HYD	HYD
	3.8 (3785)	K	0.060	①	41-47 ②	③	HYD	HYD
2000	3.8 (3786)	1	0.060	①	41-47 ②	③	HYD	HYD
	3.8 (3785)	K	0.060	①	41-47 ②	③	HYD	HYD
2001	3.8 (3786)	1	0.060	①	41-47 ②	③	HYD	HYD
	3.8 (3785)	K	0.060	①	41-47 ②	③	HYD	HYD

NOTE: The Vehicle Emission Control Information label often reflects specification changes made during production. The label figures must be used if they differ from those in this chart.

HYD: Hydraulic

① DIS Ignition System timing not adjustable

② Pressure at fuel pump

③ Idle speed maintained by ECM. There is no recommended adjustment procedure

93461CK3

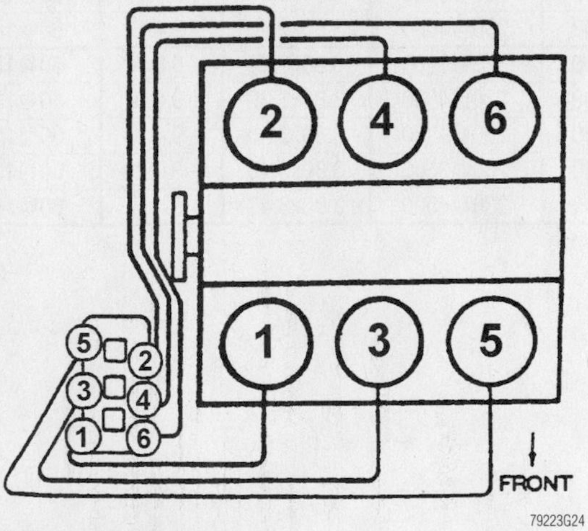

3.8L (VIN 1 and L) Engines
Firing order: 1-6-5-4-3-2
Distributorless ignition system

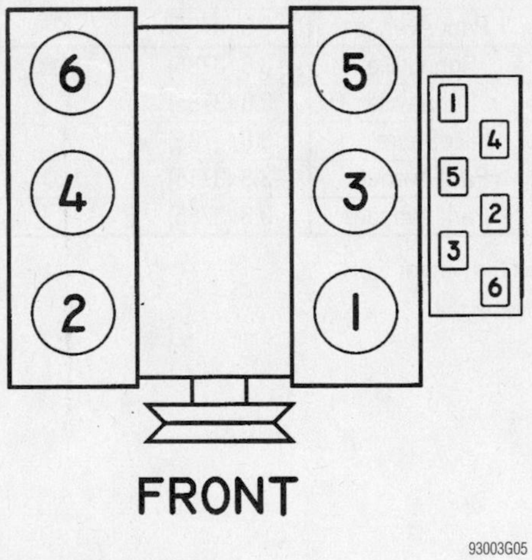

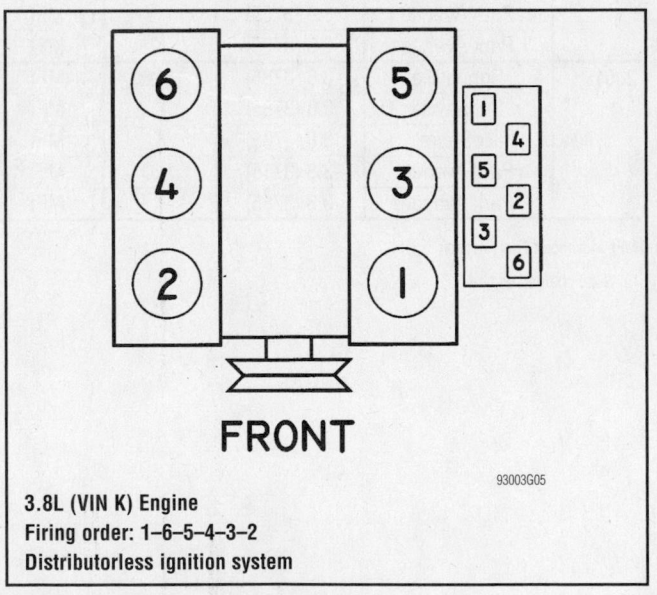

3.8L (VIN K) Engine
Firing order: 1-6-5-4-3-2
Distributorless ignition system

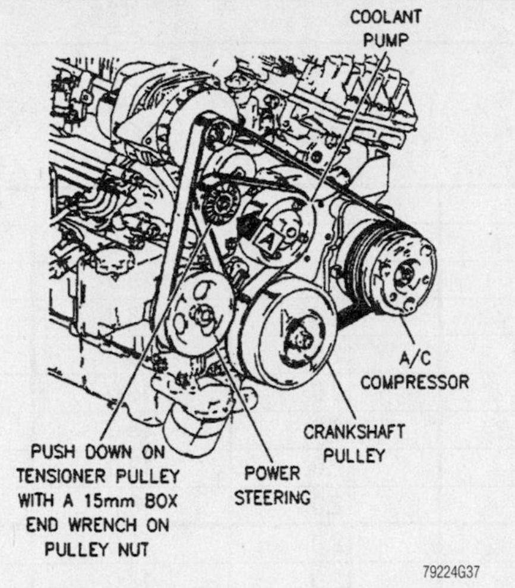

Serpentine drive belt routing—GM "C and H" Bodies 3.8L (VIN K) engine

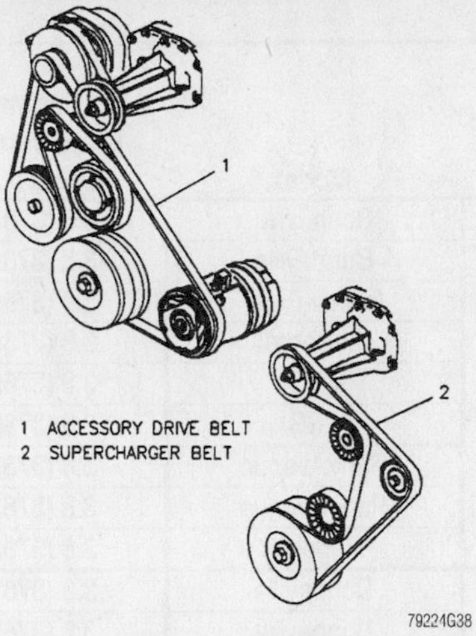

Serpentine drive belt routing—GM "C and H" Bodies 3.8L (VIN 1) engine

Timing belt service is covered in Section 3 of this manual

CAPACITIES

Year	Model	Engine Displacement Liters (cc)	Engine ID/VIN	Engine Oil with Filter (qts.)	Transmission (pts.)	Fuel Tank (gal.)	Cooling System (qts.)
1998	Bonneville	3.8 (3785)	1	4.5	12.0	18.0	13.0
	Bonneville	3.8 (3785)	K	4.5	12.0	18.0	13.0
	Eighty-Eight	3.8 (3785)	K	4.5	12.0	18.0	13.0
	LeSabre	3.8 (3785)	K	4.5	12.0	18.0	13.0
	LSS	3.8 (3785)	1	4.5	12.0	18.0	13.0
	LSS	3.8 (3785)	K	4.5	12.0	18.0	13.0
	Park Avenue	3.8 (3785)	1	4.5	12.0	18.0	13.0
	Park Avenue	3.8 (3785)	K	4.5	12.0	18.0	13.0
	Regency	3.8 (3785)	K	4.5	12.0	18.0	13.0
1999	Bonneville	3.8 (3785)	1	4.5	12.0	18.0	13.0
	Bonneville	3.8 (3785)	K	4.5	12.0	18.0	13.0
	Eighty-Eight	3.8 (3785)	K	4.5	12.0	18.0	13.0
	LeSabre	3.8 (3785)	K	4.5	12.0	18.0	13.0
	LSS	3.8 (3785)	K	4.5	12.0	18.0	13.0
	Park Avenue	3.8 (3785)	1	4.5	12.0	18.0	13.0
	Park Avenue	3.8 (3785)	K	4.5	12.0	18.0	13.0
2001	Bonneville	3.8 (3785)	1	4.5	12.0	18.0	13.0
	Bonneville	3.8 (3785)	K	4.5	12.0	18.0	13.0
	LeSabre	3.8 (3785)	K	4.5	12.0	18.0	13.0
	Park Avenue	3.8 (3785)	1	4.5	12.0	18.0	13.0
	Park Avenue	3.8 (3785)	K	4.5	12.0	18.0	13.0

NOTE: All capacities are approximate. Add fluid gradually and ensure a proper fluid level is obtained.

93461CK4

VALVE SPECIFICATIONS

Year	Engine Displacement Liters (cc)	Engine ID/VIN	Seat Angle (deg.)	Face Angle (deg.)	Spring Test Pressure (lbs. @ in.)	Spring Installed Height (in.)	Stem-to-Guide Clearance (in.)		Stem Diameter (in.)	
							Intake	Exhaust	Intake	Exhaust
1998	3.8 (3785)	1	45	45	80@1.750	1.690-1.720	0.0015-0.0032	0.0015-0.0032	NA	NA
	3.8 (3785)	K	45	45	80@1.750	1.690-1.720	0.0015-0.0032	0.0015-0.0032	NA	NA
1999	3.8 (3785)	1	45	45	80@1.750	1.690-1.720	0.0015-0.0032	0.0015-0.0032	NA	NA
	3.8 (3785)	K	45	45	80@1.750	1.690-1.720	0.0015-0.0032	0.0015-0.0032	NA	NA
2000	3.8 (3785)	1	45	45	80@1.750	1.690-1.720	0.0015-0.0032	0.0015-0.0032	NA	NA
	3.8 (3785)	K	45	45	80@1.750	1.690-1.720	0.0015-0.0032	0.0015-0.0032	NA	NA
2001	3.8 (3785)	1	45	45	80@1.750	1.690-1.720	0.0015-0.0032	0.0015-0.0032	NA	NA
	3.8 (3785)	K	45	45	80@1.750	1.690-1.720	0.0015-0.0032	0.0015-0.0032	NA	NA

NA: Not Available

93461CK5

Heater Core replacement is covered in Section 2 of this manual

CRANKSHAFT AND CONNECTING ROD SPECIFICATIONS

All measurements are given in inches.

Year	Engine Displacement Liters (cc)	Engine ID/VIN	Crankshaft				Connecting Rod		
			Main Brg. Journal Dia.	Main Brg. Oil Clearance	Shaft End-play	Thrust on No.	Journal Diameter	Oil Clearance	Side Clearance
1998	3.8 (3786)	1	2.4988-2.4998	①	0.0030-0.0110	2	2.2487-2.2499	0.0005-0.0026	0.0040-0.0200
	3.8 (3786)	K	2.4988-2.4998	①	0.0030-0.0110	2	2.2487-2.2499	0.0005-0.0026	0.0040-0.0200
1999	3.8 (3786)	1	2.4988-2.4998	①	0.0030-0.0110	2	2.2487-2.2499	0.0005-0.0026	0.0040-0.0200
	3.8 (3786)	K	2.4988-2.4998	①	0.0030-0.0110	2	2.2487-2.2499	0.0005-0.0026	0.0040-0.0200
2000	3.8 (3786)	1	2.4988-2.4998	①	0.0030-0.0110	2	2.2487-2.2499	0.0005-0.0026	0.0040-0.0200
	3.8 (3786)	K	2.4988-2.4998	①	0.0030-0.0110	2	2.2487-2.2499	0.0005-0.0026	0.0040-0.0200
2001	3.8 (3786)	1	2.4988-2.4998	①	0.0030-0.0110	2	2.2487-2.2499	0.0005-0.0026	0.0040-0.0200
	3.8 (3786)	K	2.4988-2.4998	①	0.0030-0.0110	2	2.2487-2.2499	0.0005-0.0026	0.0040-0.0200

① Journal 1: 0.0007 - 0.0016
Journals 2 and 3: 0.0010 - 0.0020
Journal 4: 0.0009 - 0.0018

93461CK6

PISTON AND RING SPECIFICATIONS

All measurements are given in inches.

Year	Engine Displacement Liters (cc)	Engine ID/VIN	Piston Clearance	Ring Gap Top Compression	Ring Gap Bottom Compression	Ring Gap Oil Control	Ring Side Clearance Top Compression	Ring Side Clearance Bottom Compression	Ring Side Clearance Oil Control
1998	3.8 (3786)	1	0.0004-0.0020	0.012-0.022	0.030-0.040	0.010-0.030	0.0013-0.0031	0.0013-0.0031	0.0009-0.0079
	3.8 (3786)	K	0.0004-0.0020	0.012-0.022	0.030-0.040	0.010-0.030	0.0013-0.0031	0.0013-0.0031	0.0009-0.0079
1999	3.8 (3786)	1	0.0004-0.0020	0.012-0.022	0.030-0.040	0.010-0.030	0.0013-0.0031	0.0013-0.0031	0.0009-0.0079
	3.8 (3786)	K	0.0004-0.0020	0.012-0.022	0.030-0.040	0.010-0.030	0.0013-0.0031	0.0013-0.0031	0.0009-0.0079
2000	3.8 (3786)	1	0.0004-0.0020	0.012-0.022	0.030-0.040	0.010-0.030	0.0013-0.0031	0.0013-0.0031	0.0009-0.0079
	3.8 (3786)	K	0.0004-0.0020	0.012-0.022	0.030-0.040	0.010-0.030	0.0013-0.0031	0.0013-0.0031	0.0009-0.0079
2001	3.8 (3786)	1	0.0004-0.0020	0.012-0.022	0.030-0.040	0.010-0.030	0.0013-0.0031	0.0013-0.0031	0.0009-0.0079
	3.8 (3786)	K	0.0004-0.0020	0.012-0.022	0.030-0.040	0.010-0.030	0.0013-0.0031	0.0013-0.0031	0.0009-0.0079

93461CK7

TORQUE SPECIFICATIONS

All readings in ft. lbs.

Year	Engine Displacement Liters (cc)	Engine ID/VIN	Cylinder Head Bolts	Main Bearing Bolts	Rod Bearing Bolts	Crankshaft Damper Bolts	Flywheel Bolts	Manifold Intake	Manifold Exhaust	Spark Plugs	Lug Nuts
1998	3.8 (3786)	1	①	②	③	④	⑤	⑥	22	11	100
	3.8 (3786)	K	①	⑦	③	④	⑤	11	38	11	100
1999	3.8 (3786)	1	①	②	③	④	⑤	⑥	22	11	100
	3.8 (3786)	K	①	⑦	③	④	⑤	11	38	11	100
2000	3.8 (3786)	1	①	②	③	④	⑤	⑥	22	11	100
	3.8 (3786)	K	①	⑦	③	④	⑤	11	38	11	100
2001	3.8 (3786)	1	①	②	③	④	⑤	⑥	22	11	100
	3.8 (3786)	K	①	⑦	③	④	⑤	11	38	11	100

① Step 1: Tighten all bolts to 35 ft. lbs.
Step 2: Turn all bolts 130 degrees
Step 3: Rotate four center bolts an additional 30 degrees

② Step 1: Tighten caps in equal increments to 52 ft. lbs.
Step 2: Loosen 360 degrees
Step 3: 15 ft. lbs.
Step 4: 54 ft. lbs.
Step 5: Plus three turns of 35 degrees for a total of 105 degrees

③ 20 ft. lbs. plus 50 degrees

④ 111 ft. lbs. plus 76 degrees

⑤ 11 ft. lbs. plus 50 degrees

⑥ Upper manifold: 8 ft. lbs.
Lower manifold: 11 ft. lbs.

⑦ 26 ft. lbs. plus 50 degrees

93461CK8

BRAKE SPECIFICATIONS

All measurements in inches unless noted

						Brake Drum Diameter			Minimum		Brake Caliper
			Brake Disc			Original	Max.	Maximum	Lining		Mounting
			Original	Minimum	Maximum	Inside	Wear	Machine	Thickness		Bolts
Year	Model		Thickness	Thickness	Runout	Diameter	Limit	Diameter	Front	Rear	(ft. lbs.)
1998	Bonneville		1.260	1.209	0.002	8.860	8.909	8.920	0.030	0.030	38
	Eighty-Eight		1.276	1.209	0.004	8.860	8.909	8.800	0.030	①	38
	LeSabre		1.260	1.209	0.002	8.863	8.909	8.920	0.030	①	38
	LSS		1.276	1.209	0.004	8.860	8.909	8.800	0.030	①	38
	Park Avenue		1.260	1.209	0.002	8.863	8.909	8.920	0.030	①	38
	Regency		1.276	1.209	0.004	8.860	8.909	8.800	0.030	①	38
1999	Bonneville		1.260	1.209	0.002	8.860	8.909	8.920	0.030	0.030	38
	Eighty-Eight		1.276	1.209	0.004	8.860	8.909	8.800	0.030	①	38
	LeSabre		1.260	1.209	0.002	8.863	8.909	8.920	0.030	①	38
	LSS		1.276	1.209	0.004	8.860	8.909	8.800	0.030	①	38
	Park Avenue		1.260	1.209	0.002	8.863	8.909	8.920	0.030	①	38
2000	Bonneville	F	1.267	1.209	0.002	—	—	—	0.030	—	63
		R	0.433	0.374	0.002	—	—	—	—	0.030	20
	LeSabre	F	1.267	1.200	0.002	—	—	—	0.030	—	63
		R	0.433	0.374	0.002	—	—	—	—	0.030	20
	Park Avenue	F	1.267	1.200	0.002	—	—	—	0.030	—	63
		R	0.433	0.374	0.002	—	—	—	—	0.030	20
2001	Bonneville	F	1.267	1.209	0.002	—	—	—	0.030	—	63
		R	0.433	0.374	0.002	—	—	—	—	0.030	20
	LeSabre	F	1.267	1.200	0.002	—	—	—	0.030	—	63
		R	0.433	0.374	0.002	—	—	—	—	0.030	20
	Park Avenue	F	1.267	1.200	0.002	—	—	—	0.030	—	63
		R	0.433	0.374	0.002	—	—	—	—	0.030	20

① 0.030 over rivet head; If bonded lining, use 0.062 from shoe

93461CK9

WHEEL ALIGNMENT
PONTIAC BONNEVILLE

Year	Model		Caster Range (Deg.)	Caster Preferred Setting (Deg.)	Camber Range (Deg.)	Camber Preferred Setting (Deg.)	Toe-in (in.)	Steering Axis Inclination (Deg.)
1998	Bonneville	F	0.50	3.00	0.50	0.19	0 +/- 0.09	12.50
		R	—	—	0.50	-0.31	0.06 +/- 0.12	—
1999	Bonneville	F	0.50	3.00	0.50	0.19	0 +/- 0.09	12.50
		R	—	—	0.50	-0.31	0.06 +/- 0.12	—
2000	Bonneville	F	0.50	6.00	0.50	-0.20	0.10 +/- 0.10	—
		R	—	—	0.50	-0.30	0.10 +/- 0.10	—
2001	Bonneville	F	0.50	6.00	0.50	-0.20	0.10 +/- 0.10	—
		R	—	—	0.50	-0.30	0.10 +/- 0.10	—

93461CK0

WHEEL ALIGNMENT
OLDSMOBILE 88, 98, LSS

Year	Model		Caster Range (Deg.)	Caster Preferred Setting (Deg.)	Camber Range (Deg.)	Camber Preferred Setting (Deg.)	Toe-in (in.)	Steering Axis Inclination (Deg.)
1998	88, LSS	F	0.50	3.00	0.50	0.19	0 +/- 0.09	12.50
		R	—	—	0.50	-0.31	0.06 +/- 0.12	—
1999	88, LSS	F	0.50	3.00	0.50	0.19	0 +/- 0.09	12.50
		R	—	—	0.50	-0.31	0.06 +/- 0.12	—
2000	88	F	0.50	3.00	0.50	0.19	0 +/- 0.09	12.50
		R	—	—	0.50	-0.31	0.06 +/- 0.12	—
2001	88	F	0.50	3.00	0.50	0.19	0 +/- 0.09	12.50
		R	—	—	0.50	-0.31	0.06 +/- 0.12	—

93461CL1

WHEEL ALIGNMENT
BUICK PARK AVENUE

Year	Model		Caster Range (Deg.)	Caster Preferred Setting (Deg.)	Camber Range (Deg.)	Camber Preferred Setting (Deg.)	Toe-in (in.)	Steering Axis Inclination (Deg.)
1998	Park avenue	F	+0.50	+6.00	+0.50	-0.20	0.20 +/- 0.20	—
		R	—	—	+0.50	-0.30	0.20 +/- 0.20	—
1999	Park avenue	F	+0.50	+6.00	+0.50	-0.20	0.20 +/- 0.20	—
		R	—	—	+0.50	-0.30	0.20 +/- 0.20	—
2000	Park avenue	F	+0.50	+6.00	+0.50	-0.20	0.20 +/- 0.20	—
		R	—	—	+0.50	-0.30	0.20 +/- 0.20	—
2001	Park avenue	F	+0.50	+6.00	+0.50	-0.20	0.20 +/- 0.20	—
		R	—	—	+0.50	-0.30	0.20 +/- 0.20	—

93461CL2

TIRE, WHEEL AND BALL JOINT SPECIFICATIONS
Buick
LeSabre

Year	Model	OEM Tires Standard	OEM Tires Optional	Tire Pressures (psi) Front	Tire Pressures (psi) Rear	Wheel Size	Ball Joint Inspection
1998	LeSabre	P205/70R15	P215/60R16	30	30	6-JJ	①
1999	LeSabre	P205/70R15	P215/60R16	30	30	6-JJ	①
2000	LeSabre	P205/70R15	P215/60R16	30	30	6-JJ	①
2001	LeSabre	P205/70R15	P215/60R16	30	30	6-JJ	①

OEM: Original Equipment Manufacturer

PSI: Pounds Per Square Inch

① Do not lift car. Inspect the boss into which the grease fitting is threaded. Replace if the boss is flush or receded below the surface of the ball joint.

93461CL3

For Accessory Drive Belt illustrations, see Section 1 of this manual

TIRE, WHEEL AND BALL JOINT SPECIFICATIONS
Oldsmobile
88,98 and LSS

Year	Model	OEM Tires		Tire Pressures (psi)		Wheel	Ball Joint
		Standard	Optional	Front	Rear	Size	Inspection
1998	Eighty-eight	P205/70R15	P215/65R15 P225/60R16	30	30	6-JJ	①
	LSS	P205/70R15	P215/65R15 P225/60R16	30	30	6-JJ	①
	Ninety-eight	P205/70R15	None	30	30	6-JJ	①
1999	Eighty-eight	P205/70R15	P215/65R15 P225/60R16	30	30	6-JJ	①
	LSS	P205/70R15	P215/65R15 P225/60R16	30	30	6-JJ	①
	Ninety-eight	P205/70R15	None	30	30	6-JJ	①
2000	Eighty-eight	P205/70R15	P215/65R15 P225/60R16	30	30	6-JJ	①
	LSS	P205/70R15	P215/65R15 P225/60R16	30	30	6-JJ	①
	Ninety-eight	P205/70R15	None	30	30	6-JJ	①
2001	Eighty-eight	P205/70R15	P215/65R15 P225/60R16	30	30	6-JJ	①
	LSS	P205/70R15	P215/65R15 P225/60R16	30	30	6-JJ	①
	Ninety-eight	P205/70R15	None	30	30	6-JJ	①

OEM: Original Equipment Manufacturer

PSI: Pounds Per Square Inch

① Replace if any measurable movement is found.

93461CL4

TIRE, WHEEL AND BALL JOINT SPECIFICATIONS
Buick
LeSabre

Year	Model	OEM Tires		Tire Pressures (psi)		Wheel	Ball Joint
		Standard	Optional	Front	Rear	Size	Inspection
1998	LeSabre	P205/70R15	P215/60R16	30	30	6-JJ	①
1999	LeSabre	P205/70R15	P215/60R16	30	30	6-JJ	①
2000	LeSabre	P205/70R15	P215/60R16	30	30	6-JJ	①
2001	LeSabre	P205/70R15	P215/60R16	30	30	6-JJ	①

OEM: Original Equipment Manufacturer

PSI: Pounds Per Square Inch

① Do not lift car. Inspect the boss into which the grease fitting is threaded. Replace if the boss is flush or receded below the surface of the ball joint.

93461CL5

TIRE, WHEEL AND BALL JOINT SPECIFICATIONS
Pontiac
Bonneville

Year	Model	OEM Tires		Tire Pressures (psi)		Wheel	Ball Joint
		Standard	Optional	Front	Rear	Size	Inspection
1998	Bonneville SE	P215/65R15	P225/60R16	30	30	6-JJ	①
	Bonneville SSE	P225/60R16	None	30	30	7-JJ	①
1999	Bonneville SE	P215/65R15	P225/60R16	30	30	6-JJ	①
	Bonneville SSE	P225/60R16	None	30	30	7-JJ	①
2000	Bonneville SE	P215/65R15	P225/60R16	30	30	6-JJ	①
	Bonneville SSE	P225/60R16	None	30	30	7-JJ	①
2001	Bonneville SE	P215/65R15	P225/60R16	30	30	6-JJ	①
	Bonneville SSE	P225/60R16	None	30	30	7-JJ	①

OEM: Original Equipment Manufacturer

PSI: Pounds Per Square Inch

① Do not lift car. Inspect the boss into which the grease fitting is threaded. Replace if the boss is flush or receded below the surface of the ball joint.

93461CL6

For Tire, Wheel and Ball Joint specifications, see Section 1 of this manual

SCHEDULED MAINTENANCE INTERVALS
GM C & H BODIES—BUICK LESABRE, PARK AVENUE, OLDSMOBILE LSS, REGENCY & EIGHTY-EIGHT & PONTIAC BONNEVILLE

TO BE SERVICED	TYPE OF SERVICE	VEHICLE MILEAGE INTERVAL (x1000)												
		7.5	15	22.5	30	37.5	45	52.5	60	67.5	75	82.5	90	97.5
Engine oil & filter	R	✓	✓	✓	✓	✓	✓	✓	✓	✓	✓	✓	✓	✓
Exhaust system & brake hoses	S/I	✓	✓	✓	✓	✓	✓	✓	✓	✓	✓	✓	✓	✓
Driveshaft boots & front suspension components	S/I	✓	✓	✓	✓	✓	✓	✓	✓	✓	✓	✓	✓	✓
Lubricate chassis, suspension, steering linkage, transaxle shift linkage, parking brake cable guides, underbody contact points & linkage	S/I	✓	✓	✓	✓	✓	✓	✓	✓	✓	✓	✓	✓	✓
Coolant level, hoses & clamps	S/I	✓	✓	✓	✓	✓	✓	✓	✓	✓	✓	✓	✓	✓
Throttle linkage	S/I	✓	✓	✓	✓	✓	✓	✓	✓	✓	✓	✓	✓	✓
Brake linings & rotate tires	S/I	✓		✓		✓		✓		✓		✓		✓
Accessory drive belts supercharger oil	S/I				✓				✓				✓	
Engine coolant ①	R													
Spark plugs ②	R				✓				✓				✓	
Air filter element	R				✓				✓				✓	
PCV filter	R				✓				✓				✓	
Ignition cables	S/I				✓				✓				✓	
EGR & fuel systems	S/I				✓				✓				✓	
Automatic transaxle fluid & filter	R													✓
Throttle body mount bolt torque	S/I	✓												

R: Replace   S/I: Service or Inspect

① Engine coolant: replace every 100,000 miles. Use O.E. specified (DEX-COOL™) coolant only. If any silicate coolant is used, the service interval is every 30,000 miles.

② Platinum tip spark plugs: replace every 100,000 miles.

FREQUENT OPERATION MAINTENANCE (SEVERE SERVICE)

If a vehicle is operated under any of the following conditions it is considered severe service:

- Extremely dusty areas.
- 50% or more of the vehicle operation is in 32°C (90°F) or higher temperatures, or constant operation in temperatures below 0°C (32°F).
- Prolonged idling (vehicle operation in stop and go traffic).
- Frequent short running periods (engine does not warm to normal operating temperatures).
- Police, taxi, delivery usage or trailer towing usage.

CV joints & front suspension components: service or inspect every 3000 miles.
Engine oil & filter change: change every 3000 miles.
Brake linings: check every 6000 miles.
Chassis lubrication: lubricate every 6000 miles.
Suspension, steering linkage, transaxle shift linkage, parking cable guides, underbody contact points: lubricate every 6000 miles.
Throttle body mount bolt torque: tighten at 6000 miles.
Air filter element: service or inspect every 15,000 miles.
Automatic transaxle fluid: change every 50,000 miles (1997).
Inspect throttle body bore & throttle plate for deposits: clean as required every 15,000 miles.
Rotate tires at 6000 miles, then every 15,000 miles.

93461CL7

SCHEDULED MAINTENANCE INTERVALS
GENERAL MOTORS CORPORATION
C & H BODIES
BUICK LESABRE, PARK AVENUE
OLDSMOBILE LSS, REGENCY, EIGHTY EIGHT
PONTIAC BONNEVILLE

The following should be used as a guide when determining the amount of work required for a particular service. In estimating how long a particular Scheduled Maintenance Service should take, please observe the following:

- Labor Time is time based on field research and data supplied by the vehicle manufacturer.
- Labor time operations are given in hours and tenths of an hour.
- All labor operations are to be used as a guide.

Mechanic Skill Level Codes:
(A) PRECISION. Highly skilled with multiple certification.
(B) GENERAL: Normally skilled with certification.
(C) MAINTENANCE: Semi-skilled working on certification.

	LABOR TIME
7500 Mile Service (C)	
All Models	1.4
15000 Mile Service (C)	
All Models	.9
22500 Mile Service (C)	
All Models	1.4
30000 Mile Service (B)	
All Models	3.0
37500 Mile Service (C)	
All Models	1.4
45000 Mile Service (C)	
All Models	.9
52500 Mile Service (C)	
All Models	1.4
60000 Mile Service (B)	
All Models	3.0
67500 Mile Service (C)	
All Models	1.4
75000 Mile Service (C)	
All Models	.9
82500 Mile Service (C)	
All Models	1.4
90000 Mile Service (B)	
All Models	2.9
97500 Mile Service (B)	
All Models	1.9

93461CL8

GENERAL MOTORS E- & K-BODIES

Cadillac DeVille • DeVille Concours • Eldorado • Seville

ENGINE AND VEHICLE IDENTIFICATION

Engine						
Code ①	Liters (cc)	Cu. In.	Cyl.	Fuel Sys.	Engine Type	Eng. Mfg.
9	4.6 (4565)	279	8	MFI	DOHC	Cadillac
Y	4.6 (4565)	279	8	MFI	DOHC	Cadillac

Model Year	
Code ②	Year
W	1998
X	1999
Y	2000
1	2001
2	2002

DOHC: Double Overhead Camshafts

MFI: Multi-point Fuel Injection

① 8th position of VIN

② 10th position of VIN

93461CL9

GENERAL ENGINE SPECIFICATIONS

Year	Model	Engine Displacement Liters (cc)	Engine Series (ID/VIN)	Fuel System	Net Horsepower @ rpm	Net Torque @ rpm (ft. lbs.)	Bore x Stroke (in.)	Com-pression Ratio	Oil Pressure @ rpm
1998	DeVille	4.6 (4565)	Y	MFI	275@5600	300@4000	3.66x3.31	10.3:1	35@2000
	DeVille Concours	4.6 (4565)	9	MFI	300@6000	290@4400	3.66x3.31	10.3:1	35@2000
	Eldorado	4.6 (4565)	Y	MFI	275@5600	300@4000	3.66x3.31	10.3:1	35@2000
	Eldorado ETC	4.6 (4565)	9	MFI	300@6000	290@4400	3.66x3.31	10.3:1	35@2000
	Seville SLS	4.6 (4565)	Y	MFI	275@5600	300@4000	3.66x3.31	10.3:1	35@2000
	Seville STS	4.6 (4565)	9	MFI	300@6000	290@4400	3.66x3.31	10.3:1	35@2000
1999	DeVille	4.6 (4565)	Y	MFI	275@5600	300@4000	3.66x3.31	10.3:1	35@2000
	DeVille Concours	4.6 (4565)	9	MFI	300@6000	290@4400	3.66x3.31	10.3:1	35@2000
	Eldorado	4.6 (4565)	Y	MFI	275@5600	300@4000	3.66x3.31	10.3:1	35@2000
	Eldorado ETC	4.6 (4565)	9	MFI	300@6000	290@4400	3.66x3.31	10.3:1	35@2000
	Seville SLS	4.6 (4565)	Y	MFI	275@5600	300@4000	3.66x3.31	10.3:1	35@2000
	Seville STS	4.6 (4565)	9	MFI	300@6000	290@4400	3.66x3.31	10.3:1	35@2000
2000	DeVille	4.6 (4565)	Y	MFI	275@5600	300@4000	3.66x3.31	10.0:1	35@2000
	DeVille DTS	4.6 (4565)	9	MFI	300@6000	295@4400	3.66x3.31	10.0:1	35@2000
	DeVille DHS	4.6 (4565)	Y	MFI	275@5600	300@4000	3.66x3.31	10.0:1	35@2000
	Eldorado	4.6 (4565)	Y	MFI	275@5600	300@4000	3.66x3.31	10.0:1	35@2000
	Eldorado ETC	4.6 (4565)	9	MFI	300@6000	295@4400	3.66x3.31	10.0:1	35@2000
	Seville SLS	4.6 (4565)	Y	MFI	275@5600	300@4000	3.66x3.31	10.0:1	35@2000
	Seville STS	4.6 (4565)	9	MFI	300@6000	295@4400	3.66x3.31	10.0:1	35@2000
2001	DeVille	4.6 (4565)	Y	MFI	275@5600	300@4000	3.66x3.31	10.0:1	35@2000
	DeVille DTS	4.6 (4565)	9	MFI	300@6000	295@4400	3.66x3.31	10.0:1	35@2000
	DeVille DHS	4.6 (4565)	Y	MFI	275@5600	300@4000	3.66x3.31	10.0:1	35@2000
	Eldorado	4.6 (4565)	Y	MFI	275@5600	300@4000	3.66x3.31	10.0:1	35@2000
	Eldorado ETC	4.6 (4565)	9	MFI	300@6000	295@4400	3.66x3.31	10.0:1	35@2000
	Seville SLS	4.6 (4565)	Y	MFI	275@5600	300@4000	3.66x3.31	10.0:1	35@2000
	Seville STS	4.6 (4565)	9	MFI	300@6000	295@4400	3.66x3.31	10.0:1	35@2000

MFI: Multi-point Fuel Injection

93461CL0

ENGINE TUNE-UP SPECIFICATIONS

Year	Engine Displacement Liters (cc)	Engine ID/VIN	Spark Plug Gap (in.)	Ignition Timing (deg.)	Fuel Pump (psi)	Idle Speed (rpm)	Valve Clearance Intake	Valve Clearance Exhaust
1998	4.6 (4565)	9	0.050	①	40-50	①	HYD	HYD
	4.6 (4565)	Y	0.050	①	40-50	①	HYD	HYD
1999	4.6 (4565)	9	0.050	①	40-50	①	HYD	HYD
	4.6 (4565)	Y	0.050	①	40-50	①	HYD	HYD
2000	4.6 (4565)	9	0.050	①	40-50	①	HYD	HYD
	4.6 (4565)	Y	0.050	①	40-50	①	HYD	HYD
2001	4.6 (4565)	9	0.050	①	40-50	①	HYD	HYD
	4.6 (4565)	Y	0.050	①	40-50	①	HYD	HYD

NOTE: The Vehicle Emission Control Information label often reflects specification changes made during production. The label figures must be used if they differ from those in this chart.

HYD: Hydraulic

① Refer to Vehicle Emission Control Information label

93461CM1

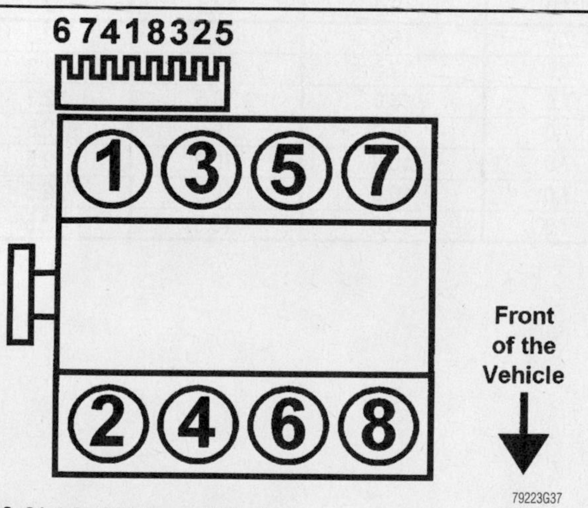

1998–01 4.6L (VIN 9 and Y) Engines
Firing order: 1–2–7–3–4–5–6–8
Distributorless ignition system

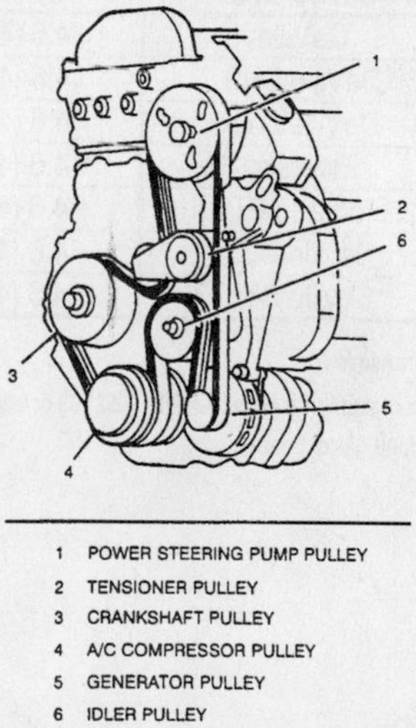

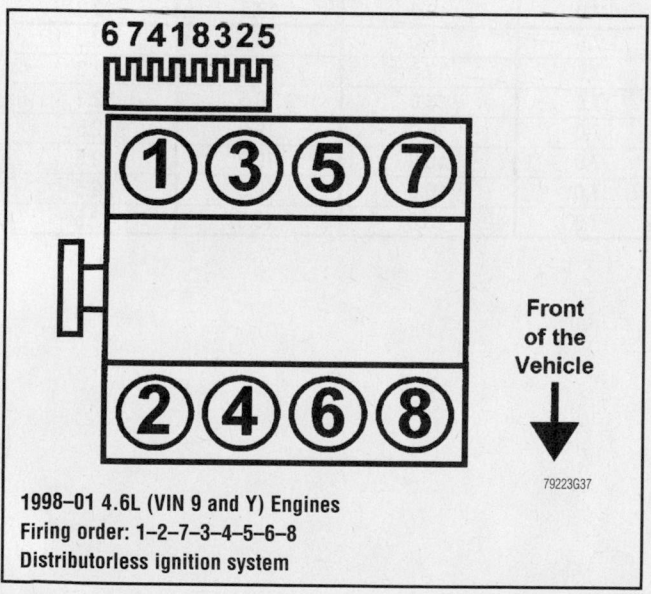

Serpentine drive belt routing—GM "E and K" Bodies 4.6L engine

CAPACITIES

Year	Model	Engine Displacement Liters (cc)	Engine ID/VIN	Engine Oil with Filter (qts.)	Transmission (pts.) ①	Fuel Tank (gal.)	Cooling System (qts.) ②
1998	DeVille	4.6 (4565)	Y	7.5	16.0	20.0	12.3
	DeVille Concours	4.6 (4565)	9	7.5	16.0	18.0	12.3
	Eldorado	4.6 (4565)	Y	7.5	16.0	20.0	12.3
	Eldorado ETC	4.6 (4565)	9	7.5	16.0	20.0	12.3
	Seville SLS	4.6 (4565)	Y	7.5	16.0	20.0	12.3
	Seville STS	4.6 (4565)	9	7.5	16.0	20.0	12.3
1999	DeVille	4.6 (4565)	Y	7.5	16.0	20.0	12.3
	DeVille Concours	4.6 (4565)	9	7.5	16.0	18.0	12.3
	Eldorado	4.6 (4565)	Y	7.5	16.0	20.0	12.3
	Eldorado ETC	4.6 (4565)	9	7.5	16.0	20.0	12.3
	Seville SLS	4.6 (4565)	Y	7.5	16.0	20.0	12.3
	Seville STS	4.6 (4565)	9	7.5	16.0	20.0	12.3
2000	DeVille	4.6 (4565)	Y	7.5	16.0	18.5	12.3
	DeVille DHS	4.6 (4565)	9	7.5	16.0	18.5	12.3
	DeVille DTS	4.6 (4565)	9	7.5	16.0	18.5	12.3
	Eldorado	4.6 (4565)	Y	7.0	16.0	19.0	12.5
	Eldorado ETC	4.6 (4565)	9	7.0	16.0	19.0	12.5
	Seville SLS	4.6 (4565)	Y	7.0	16.0	18.5	12.5
	Seville STS	4.6 (4565)	9	7.0	16.0	18.5	12.5
2001	DeVille	4.6 (4565)	Y	7.5	16.0	18.5	12.3
	DeVille DHS	4.6 (4565)	9	7.5	16.0	18.5	12.3
	DeVille DTS	4.6 (4565)	9	7.5	16.0	18.5	12.3
	Eldorado	4.6 (4565)	Y	7.0	16.0	19.0	12.5
	Eldorado ETC	4.6 (4565)	9	7.0	16.0	19.0	12.5
	Seville SLS	4.6 (4565)	Y	7.0	16.0	18.5	12.5
	Seville STS	4.6 (4565)	9	7.0	16.0	18.5	12.5

① Total capacity of dry transmission

② Dex-Cool engine coolant and three pellets of P/N 1052753 or equivalent
(Do not mix with ethylene glycol base)

93461CM2

VALVE SPECIFICATIONS

Year	Engine Displacement Liters (cc)	Engine ID/VIN	Seat Angle (deg.)	Face Angle (deg.)	Spring Test Pressure (lbs. @ in.)	Spring Installed Height (in.)	Stem-to-Guide Clearance (in.)		Stem Diameter (in.)	
							Intake	Exhaust	Intake	Exhaust
1998	4.6 (4565)	9	46	45	53@1.190	1.190	0.0010-0.0030	0.0020-0.0040	0.2331-0.2339	0.2331-0.2339
	4.6 (4565)	Y	46	45	46@1.190	1.190	0.0010-0.0030	0.0020-0.0040	0.2331-0.2339	0.2331-0.2339
1999	4.6 (4565)	9	46	45	53@1.190	1.190	0.0010-0.0030	0.0020-0.0040	0.2331-0.2339	0.2331-0.2339
	4.6 (4565)	Y	46	45	46@1.190	1.190	0.0010-0.0030	0.0020-0.0040	0.2331-0.2339	0.2331-0.2339
2000	4.6 (4565)	9	46	45	53@1.190	1.190	0.0010-0.0030	0.0020-0.0040	0.2331-0.2339	0.2331-0.2339
	4.6 (4565)	Y	46	45	46@1.190	1.190	0.0010-0.0030	0.0020-0.0040	0.2331-0.2339	0.2331-0.2339
2001	4.6 (4565)	9	46	45	53@1.190	1.190	0.0010-0.0030	0.0020-0.0040	0.2331-0.2339	0.2331-0.2339
	4.6 (4565)	Y	46	45	46@1.190	1.190	0.0010-0.0030	0.0020-0.0040	0.2331-0.2339	0.2331-0.2339

93461CM3

CRANKSHAFT AND CONNECTING ROD SPECIFICATIONS

All measurements are given in inches.

	Engine		Crankshaft				Connecting Rod		
Year	Displacement Liters (cc)	Engine ID/VIN	Main Brg. Journal Dia.	Main Brg. Oil Clearance	Shaft End-play	Thrust on No.	Journal Diameter	Oil Clearance	Side Clearance
1998	9	4.6 (4565)	2.5335-2.5337	0.0006-0.0025	0.0020-0.0200	3	2.1239-2.1235	0.001-0.003	0.0080-0.0200
	Y	4.6 (4565)	2.5335-2.5337	0.0006-0.0025	0.0020-0.0200	3	2.1239-2.1235	0.001-0.003	0.0080-0.0200
1999	9	4.6 (4565)	2.5335-2.5337	0.0006-0.0025	0.0020-0.0200	3	2.1239-2.1235	0.001-0.003	0.0080-0.0200
	Y	4.6 (4565)	2.5335-2.5337	0.0006-0.0025	0.0020-0.0200	3	2.1239-2.1235	0.001-0.003	0.0080-0.0200
2000	9	4.6 (4565)	2.5335-2.5337	0.0006-0.0025	0.0020-0.0200	3	2.1239-2.1235	0.001-0.003	0.0080-0.0200
	Y	4.6 (4565)	2.5335-2.5337	0.0006-0.0025	0.0020-0.0200	3	2.1239-2.1235	0.001-0.003	0.0080-0.0200
2001	9	4.6 (4565)	2.5335-2.5337	0.0006-0.0025	0.0020-0.0200	3	2.1239-2.1235	0.001-0.003	0.0080-0.0200
	Y	4.6 (4565)	2.5335-2.5337	0.0006-0.0025	0.0020-0.0200	3	2.1239-2.1235	0.001-0.003	0.0080-0.0200

93461CM4

PISTON AND RING SPECIFICATIONS

All measurements are given in inches.

Year	Engine Displacement Liters (cc)	Engine ID/VIN	Piston Clearance	Ring Gap Top Compression	Ring Gap Bottom Compression	Ring Gap Oil Control	Ring Side Clearance Top Compression	Ring Side Clearance Bottom Compression	Ring Side Clearance Oil Control
1998	4.6 (4565)	9	0.0008-0.0020	0.010-0.016	0.014-0.020	0.010-0.030	0.0016-0.0037	0.0016-0.0037	①
	4.6 (4565)	Y	0.0008-0.0020	0.010-0.016	0.014-0.020	0.010-0.030	0.0016-0.0037	0.0016-0.0037	①
1999	4.6 (4565)	9	0.0008-0.0020	0.010-0.016	0.014-0.020	0.010-0.030	0.0016-0.0037	0.0016-0.0037	①
	4.6 (4565)	Y	0.0008-0.0020	0.010-0.016	0.014-0.020	0.010-0.030	0.0016-0.0037	0.0016-0.0037	①
2000	4.6 (4565)	9	0.0008-0.0020	0.010-0.016	0.014-0.020	0.010-0.030	0.0016-0.0037	0.0016-0.0037	①
	4.6 (4565)	Y	0.0008-0.0020	0.010-0.016	0.014-0.020	0.010-0.030	0.0016-0.0037	0.0016-0.0037	①
2001	4.6 (4565)	9	0.0008-0.0020	0.010-0.016	0.014-0.020	0.010-0.030	0.0016-0.0037	0.0016-0.0037	①
	4.6 (4565)	Y	0.0008-0.0020	0.010-0.016	0.014-0.020	0.010-0.030	0.0016-0.0037	0.0016-0.0037	①

① Side sealing

93461CM5

TORQUE SPECIFICATIONS

All readings in ft. lbs.

Year	Engine Displacement Liters (cc)	Engine ID/VIN	Cylinder Head Bolts	Main Bearing Bolts	Rod Bearing Bolts	Crankshaft Damper Bolts	Flywheel Bolts	Manifold Intake	Manifold Exhaust	Spark Plugs	Lug Nuts
1998	4.6 (4565)	9	①	②	③	④	⑤	⑥	18	11	100
	4.6 (4565)	Y	①	②	③	④	⑤	⑥	18	11	100
1999	4.6 (4565)	9	①	②	③	④	⑤	⑥	18	11	100
	4.6 (4565)	Y	①	②	③	④	⑤	⑥	18	11	100
2000	4.6 (4565)	9	①	②	③	④	⑤	⑥	18	11	100
	4.6 (4565)	Y	①	②	③	④	⑤	⑥	18	11	100
2001	4.6 (4565)	9	①	②	③	④	⑤	⑥	18	11	100
	4.6 (4565)	Y	①	②	③	④	⑤	⑥	18	11	100

① M11 bolts:
Step 1: 30 ft. lbs. plus an additional 70 degrees
Step 2: Plus 60 degrees
Step 3: Plus 60 degrees
M6 bolts:
106 inch lbs.

② M 10 bolts:
Step 1: 15 ft. lbs.
Step 2: Plus 65 degrees
M8 bolts:
22 ft. lbs.

③ Step 1: 22 ft. lbs.
Step 2: Loosen completely
Step 3: 18 ft. lbs.
Step 4: Plus 110 degrees

④ Step 1: 37 ft. lbs.
Step 2: Plus 120 degrees

⑤ Step 1: 11 ft. lbs.
Step 2: Plus 50 degrees

⑥ 89 inch lbs.

93461CM6

BRAKE SPECIFICATIONS
CADILLAC DEVILLE, ELDORADO AND SEVILLE

All measurements in inches unless noted

Year	Model		Brake Disc Original Thickness	Brake Disc Minimum Thickness	Brake Disc Maximum Runout	Minimum Lining Thickness	Brake Caliper Mounting Bolts (ft. lbs.)
1998	DeVille	F	1.268	1.209	0.002	0.030	38
		R	0.433	0.374	0.002	0.030	20
	DeVille Concours	F	1.268	1.209	0.002	0.030	38
		R	0.433	0.374	0.002	0.030	20
	Eldorado	F	1.268	1.209	0.002	0.030	38
		R	0.433	0.374	0.002	0.030	20
	Eldorado ETC	F	1.268	1.209	0.002	0.030	38
		R	0.433	0.374	0.002	0.030	20
	Seville SLS	F	1.268	1.209	0.002	0.030	38
		R	0.433	0.374	0.002	0.030	20
	Seville STS	F	1.268	1.209	0.002	0.030	38
		R	0.433	0.374	0.002	0.030	20
1999	DeVille	F	1.268	1.209	0.002	0.030	38
		R	0.433	0.374	0.002	0.030	20
	DeVille Concours	F	1.268	1.209	0.002	0.030	38
		R	0.433	0.374	0.002	0.030	20
	Eldorado	F	1.268	1.209	0.002	0.030	38
		R	0.433	0.374	0.002	0.030	20
	Eldorado ETC	F	1.268	1.209	0.002	0.030	38
		R	0.433	0.374	0.002	0.030	20
	Seville SLS	F	1.268	1.209	0.002	0.030	38
		R	0.433	0.374	0.002	0.030	20
	Seville STS	F	1.268	1.209	0.002	0.030	38
		R	0.433	0.374	0.002	0.030	20
2000	DeVille	F	1.268	1.209	0.002	0.030	38
		R	0.433	0.374	0.002	0.030	20
	DeVille DHS	F	1.268	1.209	0.002	0.030	38
		R	0.433	0.374	0.002	0.030	20
	DeVille DTS	F	1.268	1.209	0.002	0.030	38
		R	0.433	0.374	0.002	0.030	20
	Eldorado	F	1.268	1.209	0.002	0.030	38
		R	0.433	0.374	0.002	0.030	20
	Eldorado ETC	F	1.268	1.209	0.002	0.030	38
		R	0.433	0.374	0.002	0.030	20
	Seville SLS	F	1.268	1.209	0.002	0.030	38
		R	0.433	0.374	0.002	0.030	20
	Seville STS	F	1.268	1.209	0.002	0.030	38
		R	0.433	0.374	0.002	0.030	20
2001	DeVille	F	1.268	1.209	0.002	0.030	38
		R	0.433	0.374	0.002	0.030	20
	DeVille DHS	F	1.268	1.209	0.002	0.030	38
		R	0.433	0.374	0.002	0.030	20
	DeVille DTS	F	1.268	1.209	0.002	0.030	38
		R	0.433	0.374	0.002	0.030	20
	Eldorado	F	1.268	1.209	0.002	0.030	38
		R	0.433	0.374	0.002	0.030	20
	Eldorado ETC	F	1.268	1.209	0.002	0.030	38
		R	0.433	0.374	0.002	0.030	20
	Seville SLS	F	1.268	1.209	0.002	0.030	38
		R	0.433	0.374	0.002	0.030	20
	Seville STS	F	1.268	1.209	0.002	0.030	38
		R	0.433	0.374	0.002	0.030	20

93461CM7

WHEEL ALIGNMENT
CADILLAC DeVille

			Caster		Camber			
Year	Model		Range (Deg.)	Preferred Setting (Deg.)	Range (Deg.)	Preferred Setting (Deg.)	Toe-in (in.)	Steering Axis Inclination (Deg.)
1998	DeVille	F	+1.00	+2.30	+0.50	0	0.09 +/- 0.09	—
		R	—	—	+0.50	0	0.09 +/- 0.09	—
1999	DeVille	F	+1.00	+2.30	+0.50	0	0.09 +/- 0.09	—
		R	—	—	+0.50	0	0.09 +/- 0.09	—
2000	DeVille	F	0.50	+6.00	+0.50	-0.20	0.09 +/- 0.09	—
		R	—	—	+0.50	-0.30	0.09 +/- 0.09	—
2001	DeVille	F	+0.50	+6.00	+0.50	-0.20	0.09 +/- 0.09	—
		R	—	—	+0.50	-0.30	0.09 +/- 0.09	—

93461CM8

WHEEL ALIGNMENT
CADILLAC ELDORADO

			Caster		Camber			
Year	Model		Range (Deg.)	Preferred Setting (Deg.)	Range (Deg.)	Preferred Setting (Deg.)	Toe-in (in.)	Steering Axis Inclination (Deg.)
1998	Eldorado	F	+1.00	+2.30	+0.50	0	0.20 +/- 0.20	—
		R	—	—	+0.50	0	0.20 +/- 0.20	—
1999	Eldorado	F	+1.00	+2.30	+0.50	0	0.20 +/- 0.20	—
		R	—	—	+0.50	0	0.20 +/- 0.20	—
2000	Eldorado	F	+1.00	+2.30	+0.50	0	0.20 +/- 0.20	—
		R	—	—	+0.50	0	0.20 +/- 0.20	—
2001	Eldorado	F	+1.00	+2.30	+0.50	0	0.20 +/- 0.20	—
		R	—	—	+0.50	0	0.20 +/- 0.20	—

93461CM9

WHEEL ALIGNMENT
CADILLAC SEVILLE

Year	Model		Caster Range (Deg.)	Caster Preferred Setting (Deg.)	Camber Range (Deg.)	Camber Preferred Setting (Deg.)	Toe-in (in.)	Steering Axis Inclination (Deg.)
1998	Seville	F	+0.50	+6.00	+0.50	-0.20	0.09 +/- 0.09	—
		R	—	—	+0.50	-0.30	0.09 +/- 0.09	—
1999	Seville	F	+0.50	+6.00	+0.50	-0.20	0.09 +/- 0.09	—
		R	—	—	+0.50	-0.30	0.09 +/- 0.09	—
2000	Seville	F	+0.50	+6.00	+0.50	-0.20	0.09 +/- 0.09	—
		R	—	—	+0.50	-0.30	0.09 +/- 0.09	—
2001	Seville	F	+0.50	+6.00	+0.50	-0.20	0.09 +/- 0.09	—
		R	—	—	+0.50	-0.30	0.09 +/- 0.09	—

93461CM0

TIRE, WHEEL AND BALL JOINT SPECIFICATIONS
Cadillac

Year	Model	OEM Tires		Tire Pressures (psi)		Wheel Size	Ball Joint Inspection
		Standard	Optional	Front	Rear		
1998	deVille	P225/60R16	None	30	30	7-JJ	①
	Eldorado	P225/60R16	P225/60ZR16	Std: 28 Opt: 29	Std: 26 Opt: 29	7-JJ	①
	Fleetwood	P225/60R16	None	30	30	7-JJ	①
	Seville	P225/60R16	P225/60ZR16	Std: 28 Opt: 29	Std: 26 Opt: 29	7-JJ	①
1999	deVille	P225/60R16	None	30	30	7-JJ	①
	Eldorado	P225/60R16	P225/60ZR16	Std: 28 Opt: 29	Std: 26 Opt: 29	7-JJ	①
	Fleetwood	P225/60R16	None	30	30	7-JJ	①
	Seville	P225/60R16	P225/60ZR16	Std: 28 Opt: 29	Std: 26 Opt: 29	7-JJ	①
2000	deVille	P225/60R16	None	30	30	7-JJ	①
	Eldorado	P225/60R16	P225/60ZR16	Std: 28 Opt: 29	Std: 26 Opt: 29	7-JJ	①
	Fleetwood	P225/60R16	None	30	30	7-JJ	①
	Seville	P225/60R16	P225/60ZR16	Std: 28 Opt: 29	Std: 26 Opt: 29	7-JJ	①
2001	deVille	P225/60R16	None	30	30	7-JJ	①
	Eldorado	P225/60R16	P225/60ZR16	Std: 28 Opt: 29	Std: 26 Opt: 29	7-JJ	①
	Fleetwood	P225/60R16	None	30	30	7-JJ	①
	Seville	P225/60R16	P225/60ZR16	Std: 28 Opt: 29	Std: 26 Opt: 29	7-JJ	①

OEM: Original Equipment Manufacturer

PSI: Pounds Per Square Inch

STD: Standard

OPT: Optional

L: Lower

U: Upper

① Replace if any measurable movement is found.

93461CN1

SCHEDULED MAINTENANCE INTERVALS

GM E & K BODIES—CADILLAC DEVILLE, DEVILLE CONCOURS, ELDORADO & SEVILLE

TO BE SERVICED	TYPE OF SERVICE	VEHICLE MILEAGE INTERVAL (x1000)												
		7.5	15	22.5	30	37.5	45	52.5	60	67.5	75	82.5	90	97.5
Engine oil & filter	R	✓	✓	✓	✓	✓	✓	✓	✓	✓	✓	✓	✓	✓
Coolant level, hoses & clamps	S/I	✓	✓	✓	✓	✓	✓	✓	✓	✓	✓	✓	✓	✓
Drive shaft boots & front suspension components	S/I	✓	✓	✓	✓	✓	✓	✓	✓	✓	✓	✓	✓	✓
Exhaust system, brake hoses & throttle linkage	S/I	✓	✓	✓	✓	✓	✓	✓	✓	✓	✓	✓	✓	✓
Lubricate chassis, suspension, steering linkage, transaxle shift linkage, parking brake cable guides, underbody contact points & linkage	S/I	✓	✓	✓	✓	✓	✓	✓	✓	✓	✓	✓	✓	✓
Brake linings	S/I	✓		✓		✓		✓		✓		✓		✓
Rotate tires	S/I	✓		✓		✓		✓		✓		✓		✓
Inspect throttle body bore & throttle plate for deposits	S/I		✓				✓				✓		✓	
Air filter element	R				✓				✓				✓	
Engine coolant ①	R													
PCV valve	R				✓				✓				✓	
Spark plugs ②	R				✓				✓				✓	
Accessory drive belt(s)	S/I				✓				✓				✓	
Automatic transaxle fluid & filter	S/I				✓				✓				✓	
EGR & fuel systems	S/I				✓				✓				✓	
Ignition cables	S/I				✓				✓				✓	

R: Replace S/I: Service or Inspect

① Engine coolant: replace every 100,000 miles. Use O.E. specified (DEX-COOL™) coolant only. If any silicate coolant is used, the service interval is every 30,000 miles.

② Platinum tip spark plugs: replace every 100,000 miles.

FREQUENT OPERATION MAINTENANCE (SEVERE SERVICE)

If a vehicle is operated under any of the following conditions it is considered severe service:

- Extremely dusty areas.

- 50% or more of the vehicle operation is in 32°C (90°F) or higher temperatures, or constant operation in temperatures below 0°C (32°F).

- Prolonged idling (vehicle operation in stop and go traffic).

- Frequent short running periods (engine does not warm to normal operating temperatures).

- Police, taxi, delivery usage or trailer towing usage.

CV joints & front suspension components: service or inspect every 3000 miles.

Engine oil & filter change: change every 3000 miles.

Brake linings: check every 6000 miles.

Chassis lubrication: lubricate every 6000 miles.

Suspension, steering linkage, transaxle shift linkage, parking cable guides, underbody contact points: lubricate every 6000 miles.

Air filter element: service or inspect every 15,000 miles.

Automatic transaxle fluid: change every 50,000 miles (1997 only).

Inspect throttle body bore & throttle plate for deposits: clean as required every 15,000 miles.

Rotate tires at 6000 miles, then every 15,000 miles.

93461CN2

Heater Core replacement is covered in Section 2 of this manual

SCHEDULED MAINTENANCE INTERVALS
GENERAL MOTORS CORPORATION
E & K BODIES
CADILLAC DEVILLE, DEVILLE CONCOURS, ELDORADO, SEVILLE

The following should be used as a guide when determining the amount of work required for a particular service. In estimating how long a particular Scheduled Maintenance Service should take, please observe the following:

- Labor Time is time based on field research and data supplied by the vehicle manufacturer.
- Labor time operations are given in hours and tenths of an hour.
- All labor operations are to be used as a guide.

Mechanic Skill Level Codes:
(A) PRECISION: Highly skilled with multiple certification.
(B) GENERAL: Normally skilled with certification.
(C) MAINTENANCE: Semi-skilled working on certification.

	LABOR TIME
7500 Mile Service (C)	
All Models	1.9
15000 Mile Service (C)	
All Models	.9
22500 Mile Service (C)	
All Models	1.7
30000 Mile Service (B)	
All Models	4.2
37500 Mile Service (C)	
All Models	1.6
45000 Mile Service (C)	
All Models	1.1
52500 Mile Service (C)	
All Models	1.7
60000 Mile Service (B)	
All Models	4.1
67500 Mile Service (C)	
All Models	1.7
75000 Mile Service (C)	
All Models	1.3
82500 Mile Service (C)	
All Models	1.8
90000 Mile Service (B)	
All Models	4.2
97500 Mile Service (C)	
All Models	1.8

93061CL4

GENERAL MOTORS F-BODY
Chevrolet Camaro • Pontiac Firebird

ENGINE AND VEHICLE IDENTIFICATION

Engine						
Code ①	Liters (cc)	Cu. In.	Cyl.	Fuel Sys.	Engine Type	Eng. Mfg.
G	5.7 (5665)	350	8	SFI	OHV	CPC
K	3.8 (3785)	231	6	SFI	OHV	CPC

SFI: Sequential Fuel Injection
CPC: Chevrolet/Pontiac/Canada
OHV: Overhead Valve
① 8th position of VIN
② 10th position of VIN

Model Year	
Code ②	Year
W	1998
X	1999
Y	2000
1	2001
2	2002

93461CN3

GENERAL ENGINE SPECIFICATIONS

Year	Model	Engine Displacement Liters (cc)	Engine Series (ID/VIN)	Fuel System	Net Horsepower @ rpm	Net Torque @ rpm (ft. lbs.)	Bore x Stroke (in.)	Com-pression Ratio	Oil Pressure @ rpm
1998	Camaro	3.8 (3785)	K	SFI	200@5200	225@4000	3.80x3.40	9.4:1	60@1850
		5.7 (5665)	G	SFI	305@5200	335@4000	3.89x3.62	10.0:1	18@2000
	Firebird	3.8 (3785)	K	SFI	200@5200	225@4000	3.80x3.40	9.4:1	60@1850
		5.7 (5665)	G	SFI	305@5200	335@4000	3.89x3.62	10.0:1	18@2000
1999	Camaro	3.8 (3785)	K	SFI	200@5200	225@4000	3.80x3.40	9.4:1	60@1850
		5.7 (5665)	G	SFI	305@5200	335@4000	3.89x3.62	10.0:1	18@2000
	Firebird	3.8 (3785)	K	SFI	200@5200	225@4000	3.80x3.40	9.4:1	60@1850
		5.7 (5665)	G	SFI	305@5200	335@4000	3.89x3.62	10.0:1	18@2000
2000	Camaro	3.8 (3785)	K	SFI	200@5200	225@4000	3.80x3.40	9.4:1	60@1850
		5.7 (5665)	G	SFI	305@5200	335@4000	3.89x3.62	10.0:1	18@2000
	Firebird	3.8 (3785)	K	SFI	200@5200	225@4000	3.80x3.40	9.4:1	60@1850
		5.7 (5665)	G	SFI	305@5200	335@4000	3.89x3.62	10.0:1	18@2000
2001	Camaro	3.8 (3785)	K	SFI	200@5200	225@4000	3.80x3.40	9.4:1	60@1850
		5.7 (5665)	G	SFI	305@5200	335@4000	3.89x3.62	10.0:1	18@2000
	Firebird	3.8 (3785)	K	SFI	200@5200	225@4000	3.80x3.40	9.4:1	60@1850
		5.7 (5665)	G	SFI	305@5200	335@4000	3.89x3.62	10.0:1	18@2000

SFI: Sequential Fuel Injection

93461CN4

Brake service is covered in Section 4 of this manual

ENGINE TUNE-UP SPECIFICATIONS

Year	Engine Displacement Liters (cc)	Engine ID/VIN	Spark Plug Gap (in.)	Ignition Timing (deg.) MT	Ignition Timing (deg.) AT	Fuel Pump (psi)	Idle Speed (rpm) MT	Idle Speed (rpm) AT	Valve Clearance Intake	Valve Clearance Exhaust
1998	3.8 (3785)	K	0.045	①	①	41-47	①	①	HYD	HYD
	5.7 (5665)	G	0.060	①	①	48-55	①	①	HYD	HYD
1999	3.8 (3785)	K	0.045	①	①	41-47	①	①	HYD	HYD
	5.7 (5665)	G	0.060	①	①	48-55	①	①	HYD	HYD
2000	3.8 (3785)	K	0.045	①	①	41-47	①	①	HYD	HYD
	5.7 (5665)	G	0.060	①	①	48-55	①	①	HYD	HYD
2001	3.8 (3785)	K	0.045	①	①	41-47	①	①	HYD	HYD
	5.7 (5665)	G	0.060	①	①	48-55	①	①	HYD	HYD

NOTE: The Vehicle Emission Control Information label often reflects specification changes made during production. The label figures must be used if they differ from those in this chart.

HYD: Hydraulic

① Refer to Vehicle Emission Control Information label

93461CN5

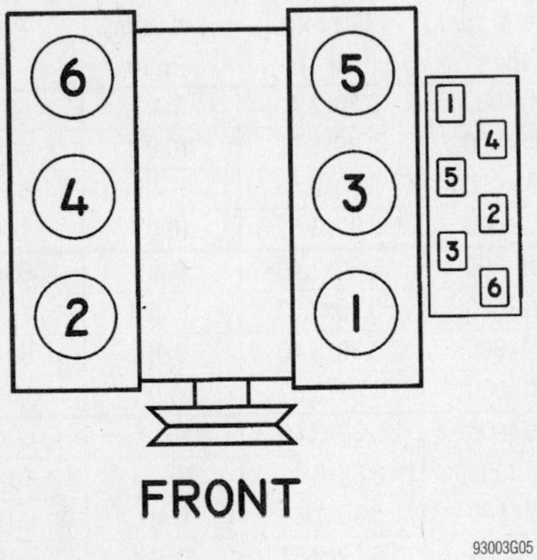

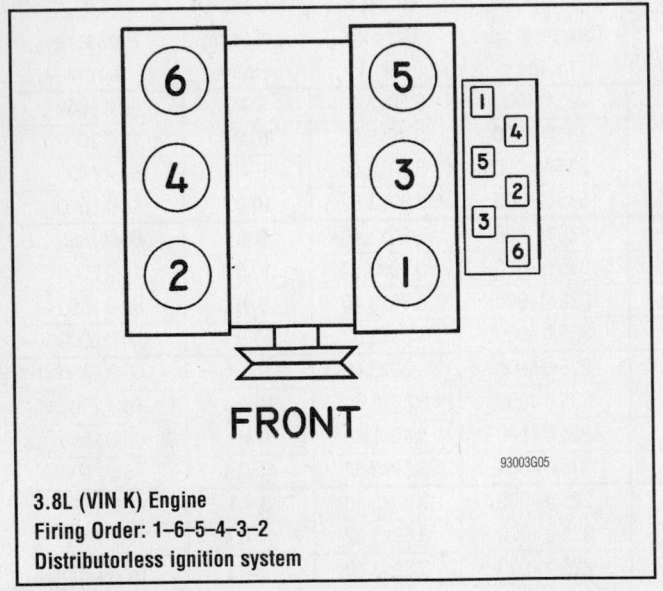

3.8L (VIN K) Engine
Firing Order: 1–6–5–4–3–2
Distributorless ignition system

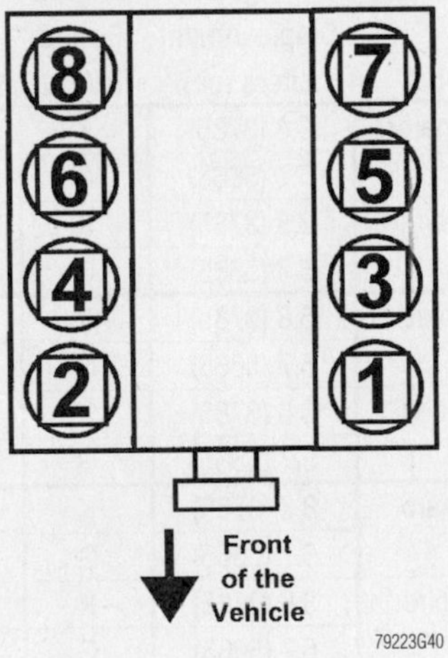

5.7L (VIN G) Engine
Firing Order: 1–8–7–2–6–5–4–3
Distributorless ignition system (one coil on each cylinder)

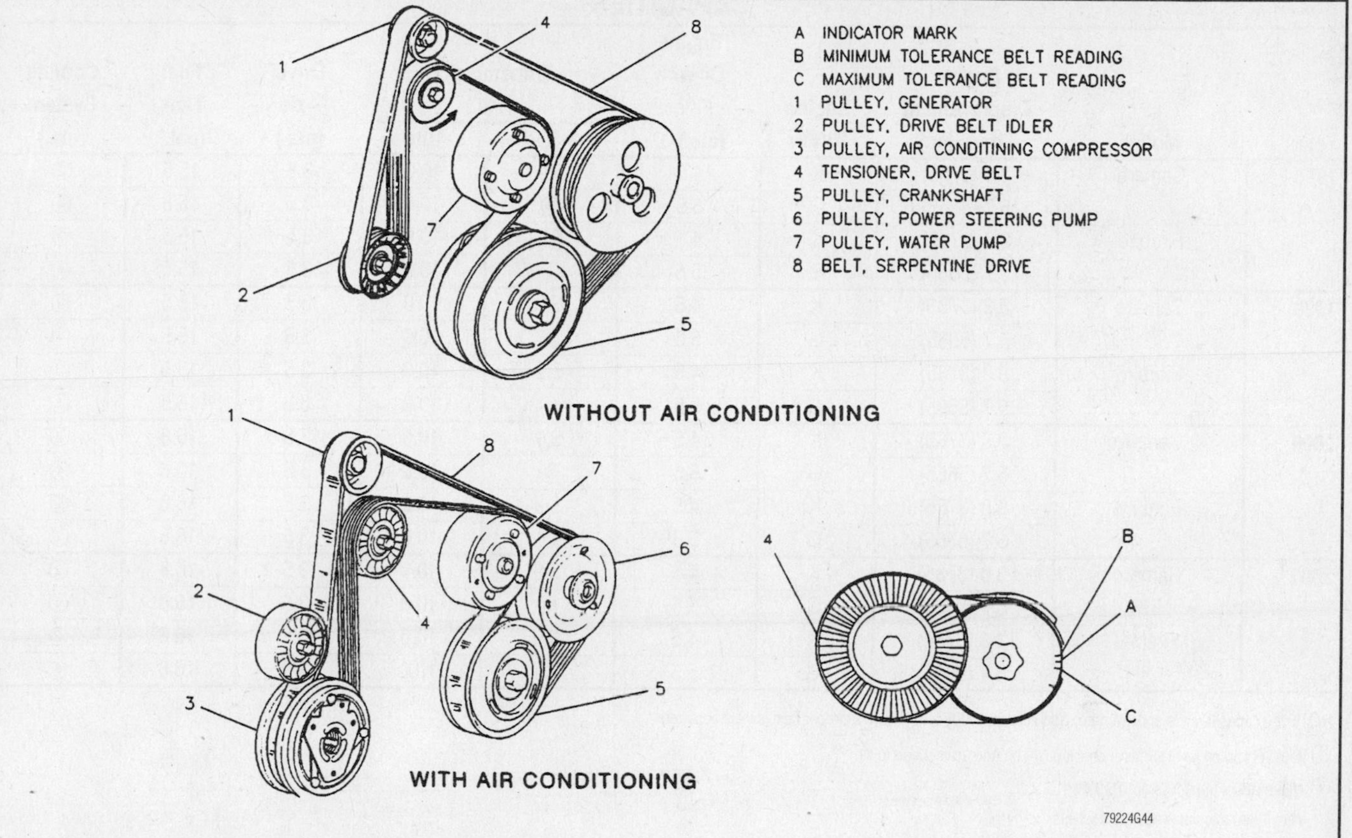

Serpentine drive belt routing—GM "F" Bodies 3.8L engine

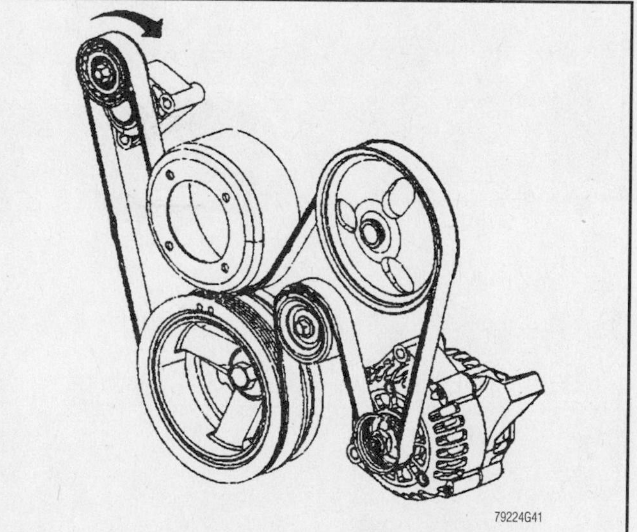

Serpentine drive belt routing—GM "F" Bodies 5.7L engine

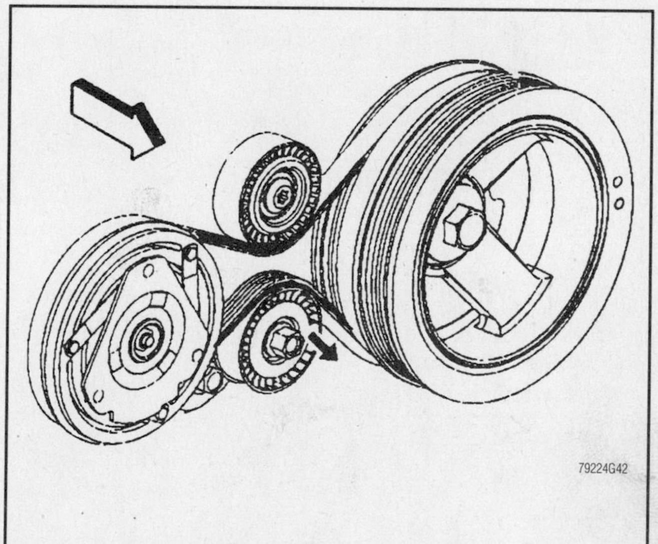

A/C drive belt routing—GM "F" Bodies 5.7L engine

CAPACITIES

Year	Model	Engine Displacement Liters (cc)	Engine ID/VIN	Engine Oil with Filter (qts.) ①	Transmission (pts.) 5-Spd	Transmission (pts.) Auto.	Drive Axle (pts.)	Fuel Tank (gal.)	Cooling System (qts.)
1998	Camaro	3.8 (3785)	K	4.5	5.9	10.0	3.5	15.5	②
		5.7 (5665)	G	5.5	③	10.0	3.5	15.5	④
	Firebird	3.8 (3785)	K	4.5	5.9	10.0	3.5	15.5	②
		5.7 (5665)	G	5.5	③	10.0	3.5	15.5	④
1999	Camaro	3.8 (3785)	K	4.5	5.9	10.0	3.5	15.5	②
		5.7 (5665)	G	5.5	③	10.0	3.5	15.5	④
	Firebird	3.8 (3785)	K	4.5	5.9	10.0	3.5	15.5	②
		5.7 (5665)	G	5.5	③	10.0	3.5	15.5	④
2000	Camaro	3.8 (3785)	K	4.5	5.9	10.0	3.5	16.8	②
		5.7 (5665)	G	5.5	③	10.0	3.5	16.8	④
	Firebird	3.8 (3785)	K	4.5	5.9	10.0	3.5	16.8	②
		5.7 (5665)	G	5.5	③	10.0	3.5	16.8	④
2001	Camaro	3.8 (3785)	K	4.5	5.9	10.0	3.5	16.8	②
		5.7 (5665)	G	5.5	③	10.0	3.5	16.8	④
	Firebird	3.8 (3785)	K	4.5	5.9	10.0	3.5	16.8	②
		5.7 (5665)	G	5.5	③	10.0	3.5	16.8	④

NOTE: All capacities are approximate. Add fluid gradually and ensure a proper fluid level is obtained.

① With vehicle on level surface, check oil level. Add as required to fill.

② With manual transmission: 15.3 qts.
With automatic transmission: 15.1 qts.

③ ZF 6 speed transmission: 4.4 pts.

④ With manual transmission: 12.5 qts.
With automatic transmission: 12.3 qts.

93461CN6

VALVE SPECIFICATIONS

Year	Engine Displacement Liters (cc)	Engine ID/VIN	Seat Angle (deg.)	Face Angle (deg.)	Spring Test Pressure (lbs. @ in.) ①	Spring Installed Height (in.)	Stem-to-Guide Clearance (in.) Intake	Stem-to-Guide Clearance (in.) Exhaust	Stem Diameter (in.) Intake	Stem Diameter (in.) Exhaust
1998	3.8 (3785)	K	45	45	228@1.277	1.69-1.72	0.0015-0.0035	0.0015-0.0032	NA	NA
	5.7 (5665)	G	46	45	220@1.320	1.80	0.0010-0.0026	0.0010-0.0026	0.0313-0.0314	0.0313-0.0314
1999	3.8 (3785)	K	45	45	228@1.277	1.69-1.72	0.0015-0.0035	0.0015-0.0032	NA	NA
	5.7 (5665)	G	46	45	220@1.320	1.80	0.0010-0.0026	0.0010-0.0026	0.0313-0.0314	0.0313-0.0314
2000	3.8 (3785)	K	45	45	228@1.277	1.69-1.72	0.0015-0.0035	0.0015-0.0032	NA	NA
	5.7 (5665)	G	46	45	220@1.320	1.80	0.0010-0.0026	0.0010-0.0026	0.0313-0.0314	0.0313-0.0314
2001	3.8 (3785)	K	45	45	228@1.277	1.69-1.72	0.0015-0.0035	0.0015-0.0032	NA	NA
	5.7 (5665)	G	46	45	220@1.320	1.80	0.0010-0.0026	0.0010-0.0026	0.0313-0.0314	0.0313-0.0314

NA: Not Available

① With valve open

93461CN7

For Accessory Drive Belt illustrations, see Section 1 of this manual

CRANKSHAFT AND CONNECTING ROD SPECIFICATIONS

All measurements are given in inches.

Year	Engine Displacement Liters (cc)	Engine ID/VIN	Crankshaft Main Brg. Journal Dia.	Main Brg. Oil Clearance	Shaft End-play	Thrust on No.	Connecting Rod Journal Diameter	Oil Clearance	Side Clearance
1998	3.8 (3786)	K	2.4988-2.4998	①	0.0030-0.0110	2	2.2487-2.2499	0.0005-0.0026	0.0040-0.0200
	5.7 (5665)	G	2.5580-2.5590	0.0007-0.0021	0.0015-0.0078	3	2.0991-2.0999	0.0006-0.0030	0.0043-0.0200
1999	3.8 (3786)	K	2.4988-2.4998	①	0.0030-0.0110	2	2.2487-2.2499	0.0005-0.0026	0.0040-0.0200
	5.7 (5665)	G	2.5580-2.5590	0.0007-0.0021	0.0015-0.0078	3	2.0991-2.0999	0.0006-0.0030	0.0043-0.0200
2000	3.8 (3786)	K	2.4988-2.4998	①	0.0030-0.0110	2	2.2487-2.2499	0.0005-0.0026	0.0040-0.0200
	5.7 (5665)	G	2.5580-2.5590	0.0007-0.0021	0.0015-0.0078	3	2.0991-2.0999	0.0006-0.0030	0.0043-0.0200
2001	3.8 (3786)	K	2.4988-2.4998	①	0.0030-0.0110	2	2.2487-2.2499	0.0005-0.0026	0.0040-0.0200
	5.7 (5665)	G	2.5580-2.5590	0.0007-0.0021	0.0015-0.0078	3	2.0991-2.0999	0.0006-0.0030	0.0043-0.0200

① Journal 1: 0.0007 - 0.0016
Journals 2 and 3: 0.0010 - 0.0020
Journal 4: 0.0009 - 0.0018

93461CN8

PISTON AND RING SPECIFICATIONS

All measurements are given in inches.

Year	Engine Displacement Liters (cc)	Engine ID/VIN	Piston Clearance	Ring Gap Top Compression	Bottom Compression	Oil Control	Ring Side Clearance Top Compression	Bottom Compression	Oil Control
1998	3.8 (3786)	K	0.0004-0.0020	0.012-0.022	0.030-0.040	0.010-0.030-	0.0013-0.0031	0.0013-0.0031	0.0009-0.0079
	5.7 (5665)	G	0.0007-0.0021	0.009-0.015	0.017-0.025	0.007-0.027	0.0016-0.0034	0.0016-0.0032	0.0004-0.0087
1999	3.8 (3786)	K	0.0004-0.0020	0.012-0.022	0.030-0.040	0.010-0.030-	0.0013-0.0031	0.0013-0.0031	0.0009-0.0079
	5.7 (5665)	G	0.0007-0.0021	0.009-0.015	0.017-0.025	0.007-0.027	0.0016-0.0034	0.0016-0.0032	0.0004-0.0087
2000	3.8 (3786)	K	0.0004-0.0020	0.010-0.018	0.023-0.033	0.010-0.030-	0.0013-0.0031	0.0013-0.0031	0.0009-0.0079
	5.7 (5665)	G	0.0007-0.0021	0.009-0.015	0.017-0.025	0.007-0.027	0.0016-0.0034	0.0016-0.0032	0.0004-0.0087
2001	3.8 (3786)	K	0.0004-0.0020	0.010-0.018	0.023-0.033	0.010-0.030-	0.0013-0.0031	0.0013-0.0031	0.0009-0.0079
	5.7 (5665)	G	0.0007-0.0021	0.009-0.015	0.017-0.025	0.007-0.027	0.0016-0.0034	0.0016-0.0032	0.0004-0.0087

93461CN9

TORQUE SPECIFICATIONS

All readings in ft. lbs.

Year	Engine Displacement Liters (cc)	Engine ID/VIN	Cylinder Head Bolts	Main Bearing Bolts	Rod Bearing Bolts	Crankshaft Damper Bolts	Flywheel Bolts	Manifold Intake	Manifold Exhaust	Spark Plugs	Lug Nuts
1998	3.8 (3785)	K	①	②	20	③	④	⑤	17	23	100
	5.7 (5665)	G	⑥	⑦	⑧	⑨	⑩	⑪	⑫	12	100
1999	3.8 (3785)	K	①	②	20	③	④	⑤	17	23	100
	5.7 (5665)	G	⑥	⑦	⑧	⑨	⑩	⑪	⑫	12	100
2000	3.8 (3785)	K	①	②	20	③	④	⑤	17	23	100
	5.7 (5665)	G	⑥	⑦	⑧	⑨	⑩	⑪	⑫	12	100
2001	3.8 (3785)	K	①	②	20	③	④	⑤	17	23	100
	5.7 (5665)	G	⑥	⑦	⑧	⑨	⑩	⑪	⑫	12	100

NA: Not Available

① Step 1: 37 ft. lbs. plus 120 degrees

② Step 1: Tighten to 52 ft. lbs.
Step 2: Loosen bearing cap 360 degrees counter-clockwise
Step 3: Tighten caps to 15 ft. lbs.
Step 4: Tighten caps to 30 ft. lbs.
Step 5: Turn bolts an additional 35 degrees.
Step 6: Turn bolts an additional 35 degrees
Step 7: Turn bolts an additional 40 degrees -- for a total of 110 degrees

③ 111 ft. lbs. plus 76 degrees

④ 11 ft. lbs. plus 50 degrees

⑤ Upper manifold:
Bolts 1-10, in sequence: 11 ft. lbs. (15 Nm)
Coolant outlet bolts: 20 ft. lbs. (27 Nm)
Side bolts: 22 ft. lbs. (30 Nm)
Lower manifold bolt/nut: 11 ft. lbs.

⑥ M11 bolts:
Step 1: Bolts 1-10 22 ft. lbs.
Step 2: Rotate bolts 1-10 90 degrees
Step 3: Rotate bolts 1-8 90 degrees
Step 4: Rotate bolts 9-10 50 degrees
Step 5: Tighten inner (M8) bolts to 22 ft. lbs.

⑦ Step 1: Inner bolts: 15 ft. lbs.
Step 2: Inner bolts: Rotate 80 degrees
Step 3: Outer side bolts: 18 ft. lbs.
Step 4: Outer side studs: 15 ft. lbs. plus 53 degrees

⑧ Step 1: 15 ft. lbs.
Step 2: Rotate 60 degrees

⑨ Step 1: Tighten bolt to 240 ft. lbs. to seat balancer
Step 2: Tighten new bolt to 37 ft. lbs.
Step 3: Rotate an additional 140 degrees

⑩ Step 1: 15 ft. lbs.
Step 2: 37 ft. lbs.
Step 3: 74 ft. lbs.

⑪ Step 1: 44 inch lbs.
Step 2: 89 inch lbs.

⑫ Step 1: 11 ft. lbs.
Step 2: 18 ft. lbs.

93461CN0

BRAKE SPECIFICATIONS

All measurements in inches unless noted

			Brake Disc			Brake Drum Diameter				Brake Caliper	
						Original	Max.	Maximum	Minimum	Bracket	Mounting
			Original	Minimum	Maximum	Inside	Wear	Machine	Lining	Bolt	Bolt
Year	Model		Thickness	Thickness	Runout	Diameter	Limit	Diameter	Thickness	(ft. lbs.)	(ft. lbs.)
1998	Camaro	F	1.260	1.223	0.005	—	—	—	0.030	160	23
		R	NA	0.985	0.005	—	—	—	0.030	160	27
	Firebird	F	1.260	1.223	0.005	—	—	—	0.030	160	23
		R	NA	0.985	0.005	—	—	—	0.030	160	27
1999	Camaro	F	1.260	1.223	0.005	—	—	—	0.030	160	23
		R	NA	0.985	0.005	—	—	—	0.030	160	27
	Firebird	F	1.260	1.223	0.005	—	—	—	0.030	160	23
		R	NA	0.985	0.005	—	—	—	0.030	160	27
2000	Camaro	F	1.260	1.223	0.005	—	—	—	0.030	160	23
		R	NA	0.985	0.005	—	—	—	0.030	160	27
	Firebird	F	1.260	1.223	0.005	—	—	—	0.030	160	23
		R	NA	0.985	0.005	—	—	—	0.030	160	27
2001	Camaro	F	1.260	1.223	0.005	—	—	—	0.030	160	23
		R	NA	0.985	0.005	—	—	—	0.030	160	27
	Firebird	F	1.260	1.223	0.005	—	—	—	0.030	160	23
		R	NA	0.985	0.005	—	—	—	0.030	160	27

NA: Not Available

F: Front

R: Rear

93461C01

WHEEL ALIGNMENT

			Caster		Camber			Steering
Year	Model		Range (+/-Deg.)	Preferred Setting (Deg.)	Range (+/-Deg.)	Preferred Setting (Deg.)	Toe-in (in.)	Axis Inclination (Deg.)
1998	Camaro	F	1.00	+5.00	1.00	+0.40	0 +/- 0.20	—
		R	—	—	0.60	0	0 +/- 0.30	—
	Firebird	F	0.50	+5.00	0.50	+0.41	0 +/- 0.09	—
		R	—	—	0.56	0	0 +/- 0.16	—
1999	Camaro	F	1.00	+5.00	1.00	+0.80	-0.20 +/- 0.20	—
		R	—	—	0.60	0	0 +/- 0.30	—
	Firebird	F	0.50	+5.00	0.50	+0.41	0 +/- 0.09	—
		R	—	—	0.56	0	0 +/- 0.16	—
2000	Camaro	F	1.00	+5.00	1.00	+0.80	-0.20 +/- 0.20	—
		R	—	—	0.60	0	0 +/- 0.30	—
	Firebird	F	0.50	+5.00	0.50	+0.41	0 +/- 0.09	—
		R	—	—	0.56	0	0 +/- 0.16	—
2001	Camaro	F	1.00	+5.00	1.00	+0.80	-0.20 +/- 0.20	—
		R	—	—	0.60	0	0 +/- 0.30	—
	Firebird	F	0.50	+5.00	0.50	+0.41	0 +/- 0.09	—
		R	—	—	0.56	0	0 +/- 0.16	—

93461C02

For Tire, Wheel and Ball Joint specifications, see Section 1 of this manual

TIRE, WHEEL AND BALL JOINT SPECIFICATIONS

Year	Model	OEM Tires		Tire Pressures (psi)		Wheel	Ball Joint
		Standard	Optional	Front	Rear	Size	Inspection
1998	Camaro base	215/60R16	P235/55R16	30	30	7.5-JJ	U: 0.125 in.
							L: 0.047 in.
	Camaro RS	P235/55R16	None	30	30	8-J	U: 0.125 in.
							L: 0.047 in.
	Camaro Z28	P235/55R16	P245/50ZR16	30	30	8-J	U: 0.125 in.
			P275/40ZR17	30	30	9-J	L: 0.047 in.
	Firebird, base	P215/60R16	P235/55R16	30	30	7.5-JJ	0.125 in.
	Firebird Formula	P235/55R16	P245/50ZR16	30	30	8-J	0.125 in.
	Trans Am Coupe	P235/55R16	None	30	30	8-JJ	0.125 in.
	Trans Am Conv.	P245/50ZR16	P275/40ZR17	30	30	8-JJ	0.125 in.
1999	Camaro RS	P235/55R16	None	30	30	8-J	U: 0.125 in.
							L: 0.047 in.
	Camaro Z28	P235/55R16	P245/50ZR16	30	30	8-J	U: 0.125 in.
			P275/40ZR17	30	30	9-J	L: 0.047 in.
	Firebird, base	P215/60R16	P235/55R16	30	30	7.5-JJ	0.125 in.
	Firebird Formula	P235/55R16	P245/50ZR16	30	30	8-J	0.125 in.
	Trans Am Coupe	P235/55R16	None	30	30	8-JJ	0.125 in.
	Trans Am Conv.	P245/50ZR16	P275/40ZR17	30	30	8-JJ	0.125 in.
2000	Camaro RS	P215/60TR16	P215/60HR16	30	30	8-J	U: 0.125 in.
			P235/55R16				L: 0.047 in.
	Camaro Z28	P235/55R16	P245/50ZR16	30	30	8-J	U: 0.125 in.
			P275/40ZR17	30	30	9-J	L: 0.047 in.
	Firebird, base	P215/60R16	P235/55R16	30	30	7.5-JJ	0.125 in.
	Firebird Formula	P245/50R16	P275/40ZR17	30	30	8-J	0.125 in.
	Trans Am Coupe	P245/50ZR16	P275/40ZR17	30	30	8-JJ	0.125 in.
2001	Camaro RS	P215/60TR16	P215/60HR16	30	30	8-J	U: 0.125 in.
			P235/55R16				L: 0.047 in.
	Camaro Z28	P235/55R16	P245/50ZR16	30	30	8-J	U: 0.125 in.
			P275/40ZR17	30	30	9-J	L: 0.047 in.
	Firebird, base	P215/60R16	P235/55R16	30	30	7.5-JJ	0.125 in.
	Firebird Formula	P245/50R16	P275/40ZR17	30	30	8-J	0.125 in.
	Trans Am Coupe	P245/50ZR16	P275/40ZR17	30	30	8-JJ	0.125 in.

OEM: Original Equipment Manufacturer

PSI: Pounds Per Square Inch

STD: Standard

OPT: Optional

L: Lower

U: Upper

Fr: Front

Rr: Rear

93461C03

SCHEDULED MAINTENANCE INTERVALS
GM F BODY—CAMARO AND FIREBIRD

TO BE SERVICED	TYPE OF SERVICE	VEHICLE MILEAGE INTERVAL (x1000)												
		7.5	15	22.5	30	37.5	45	52.5	60	67.5	75	82.5	90	97.5
Engine oil & filter	R	✓	✓	✓	✓	✓	✓	✓	✓	✓	✓	✓	✓	✓
Coolant level, hoses & clamps	S/I	✓	✓	✓	✓	✓	✓	✓	✓	✓	✓	✓	✓	✓
Exhaust system, brake hoses & throttle linkage	S/I	✓	✓	✓	✓	✓	✓	✓	✓	✓	✓	✓	✓	✓
Lubricate chassis and suspension	S/I	✓	✓	✓	✓	✓	✓	✓	✓	✓	✓	✓	✓	✓
Lubricate steering linkage and transaxle shift linkage	S/I	✓	✓	✓	✓	✓	✓	✓	✓	✓	✓	✓	✓	✓
Lubricate parking brake cable guides, underbody contact points & linkage	S/I	✓	✓	✓	✓	✓	✓	✓	✓	✓	✓	✓	✓	✓
Brake hoses & brake linings	S/I	✓		✓		✓		✓		✓		✓		✓
Rotate tires ①	S/I	✓		✓		✓		✓		✓		✓		✓
Automatic transmission fluid & filter ②	S/I													
Air filter element & PCV filter	R				✓				✓				✓	
Engine coolant ③	R													
Spark plugs ④	R				✓				✓				✓	
Ignition cables, EGR & fuel systems	S/I				✓				✓				✓	
Serpentine drive belt	S/I				✓				✓				✓	
Rear axle oil (Limited slip) ⑤	R	✓												

R: Replace

S/I: Service or Inspect

① For models with P245/50ZR16 tires, rotate front-to-rear only, & be sure that the tires roll in the direction indicated by the arrows on the side walls

② Automatic transmission fluid & filter: replace every 100,000 miles

③ Engine coolant: replace every 100,000 miles. Use O.E. specified (DEX-COOL™) coolant only. If any silicate coolant is used, the service interval is every 30,000 miles

④ Platinum tip spark plugs: replace every 100,000 miles

③ Engine coolant: replace every 100,000 miles. Use O.E. specified (DEX-COOL™) coolant only. If any silicate coolant is used, the service interval is every 30,000 miles

⑤ If the vehicle is used to tow a trailer, change the rear axle fluid every 7500 miles in either type of differential

FREQUENT OPERATION MAINTENANCE (SEVERE SERVICE)

If a vehicle is operated under any of the following conditions it is considered severe service:

- Extremely dusty areas.

- 50% or more of the vehicle operation is in 32°C (90°F) or higher temperatures, or constant operation in temperatures below 0°C (32°F).

- Prolonged idling (vehicle operation in stop and go traffic).

- Frequent short running periods (engine does not warm to normal operating temperatures).

- Police, taxi, delivery usage or trailer towing usage.

Oil & oil filter: change every 3000 miles

Chassis lubrication: lubricate every 6000 miles

Automatic transmission fluid & filter: change every 15,000 miles

Air filter element: service or inspect every 15,000 miles

Rotate tires at 6000 miles, then every 15,000 miles

93461C04

For Wheel Alignment specifications, see Section 1 of this manual

SCHEDULED MAINTENANCE INTERVALS
GENERAL MOTORS CORPORATION
F BODY
CHEVROLET CAMARO, PONTIAC FIREBIRD

The following should be used as a guide when determining the amount of work required for a particular service. In estimating how long a particular Scheduled Maintenance Service should take, please observe the following:

- Labor Time is time based on field research and data supplied by the vehicle manufacturer.
- Labor time operations are given in hours and tenths of an hour.
- All labor operations are to be used as a guide.

Mechanic Skill Level Codes:
(A) PRECISION: Highly skilled with multiple certification.
(B) GENERAL: Normally skilled with certification.
(C) MAINTENANCE: Semi-skilled working on certification.

	LABOR TIME
7500 Mile Service (C)	
All Models	1.4
15000 Mile Service (C)	
All Models	.9
22500 Mile Service (C)	
All Models	1.4
30000 Mile Service (B)	
All Models	3.1
37500 Mile Service (C)	
All Models	1.4
45000 Mile Service (C)	
All Models	.9
52500 Mile Service (C)	
All Models	1.4
60000 Mile Service (B)	
All Models	3.1
67500 Mile Service (C)	
All Models	1.4
75000 Mile Service (C)	
All Models	.9
82500 Mile Service (C)	
All Models	1.4
90000 Mile Service (B)	
All Models	3.1
97500 Mile Service (C)	
All Models	1.4

93461C05

GENERAL MOTORS G-BODY
Buick Riviera • Oldsmobile Aurora

ENGINE AND VEHICLE IDENTIFICATION

Engine							Model Year	
Code ①	Liters (cc)	Cu. In.	Cyl.	Fuel Sys.	Engine Type	Eng. Mfg.	Code ②	Year
H	3.5 (3475)	212	6	SFI	DOHC	BOC	W	1998
1 ③	3.8 (3785)	231	6	MFI	OHV	BOC	X	1999
K	3.8 (3785)	231	6	MFI	OHV	BOC	1	2001
C	4.0 (3995)	244	8	MFI	DOHC	BOC	2	2002

NOTE: Oldsmobile did not market an Aurora for 2000.

DOHC: Double Overhead Camshafts

OHV: Overhead Valves

MFI: Multi-point Fuel Injection

SFI: Sequential Fuel Injection

BOC: Buick/Oldsmobile/Cadillac

① 8th position of VIN

② 10th position of VIN

③ Supercharged Engine

93461C06

GENERAL ENGINE SPECIFICATIONS

Year	Model	Engine Displacement Liters (cc)	Engine Series (ID/VIN)	Fuel System	Net Horsepower @ rpm	Net Torque @ rpm (ft. lbs.)	Bore x Stroke (in.)	Com-pression Ratio	Oil Pressure @ rpm
1998	Aurora	4.0 (3995)	C	MFI	250@5600	245@4400	3.43x3.31	10.2:1	30@2000
	Riviera	3.8 (3785) ①	1	MFI	225@5000	275@3200	3.80x3.40	9.0:1	60@1850
		3.8 (3785)	K	MFI	205@5200	230@4000	3.80x3.40	9.4:1	60@1850
1999	Aurora	4.0 (3995)	C	MFI	250@5600	245@4400	3.43x3.31	10.2:1	30@2000
	Riviera	3.8 (3785)	1	MFI	225@5000	275@3200	3.80x3.40	9.0:1	60@1850
2001	Aurora	3.5 (3475)	H	SFI	215@5600	230@4400	3.52x3.62	9.3:1	30@2000
		4.0 (3995)	C	SFI	250@5600	260@4000	3.43x3.31	10.2:1	30@2000
2002	Aurora	3.5 (3475)	H	SFI	215@5600	230@4400	3.52x3.62	9.3:1	30@2000
		4.0 (3995)	C	SFI	250@5600	260@4000	3.43x3.31	10.2:1	30@2000

MFI: Multi-point Fuel Injection

SFI: Sequential Fuel Injection

① Supercharged engine

93461C07

For Maintenance Interval recommendations, see Section 1 of this manual

ENGINE TUNE-UP SPECIFICATIONS

Year	Engine Displacement Liters (cc)	Engine ID/VIN	Spark Plug Gap (in.)	Ignition Timing (deg.)	Fuel Pump (psi)	Idle Speed (rpm)	Valve Clearance Intake	Valve Clearance Exhaust
1998	3.8 (3785)	1	0.060	①	40-47	②	HYD	HYD
	3.8 (3785)	K	0.060	①	40-47	②	HYD	HYD
	4.0 (3995)	C	0.050	①	41-47	②	HYD	HYD
1999	3.8 (3785)	1	0.060	①	40-47	②	HYD	HYD
	4.0 (3995)	C	0.050	①	41-47	②	HYD	HYD
2001	3.5 (3475)	H	0.050	①	41-47	②	HYD	HYD
	4.0 (3995)	C	0.050	①	41-47	②	HYD	HYD
2002	3.5 (3475)	H	0.050	①	41-47	②	HYD	HYD
	4.0 (3995)	C	0.050	①	41-47	②	HYD	HYD

NOTE: The Vehicle Emission Control Information label often reflects specification changes made during production. The label figures must be used if they differ from those in this chart.

HYD: Hydraulic

① DIS Ignition System timing is not adjustable

② Idle speed is maintained by the ECM. There is no recommended adjustment procedure

93461C08

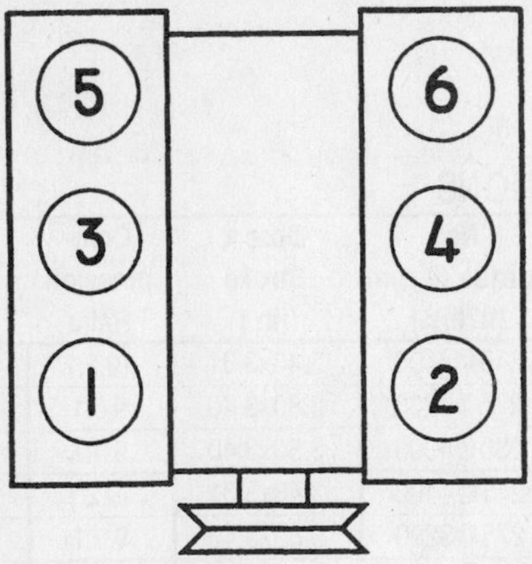

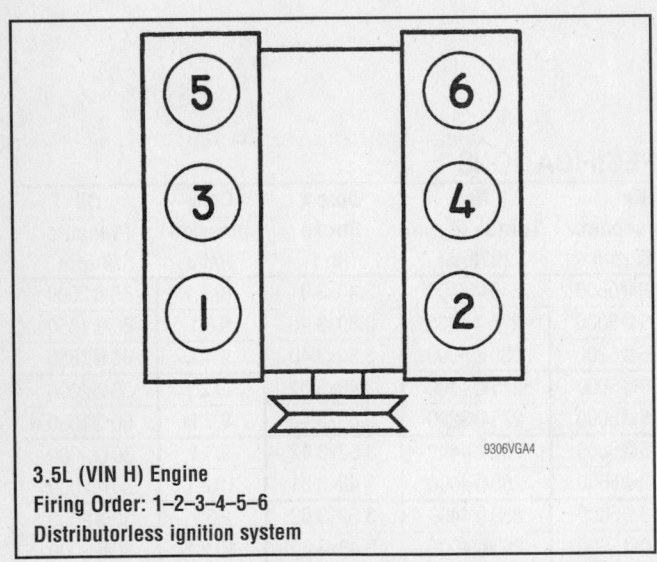

3.5L (VIN H) Engine
Firing Order: 1-2-3-4-5-6
Distributorless ignition system

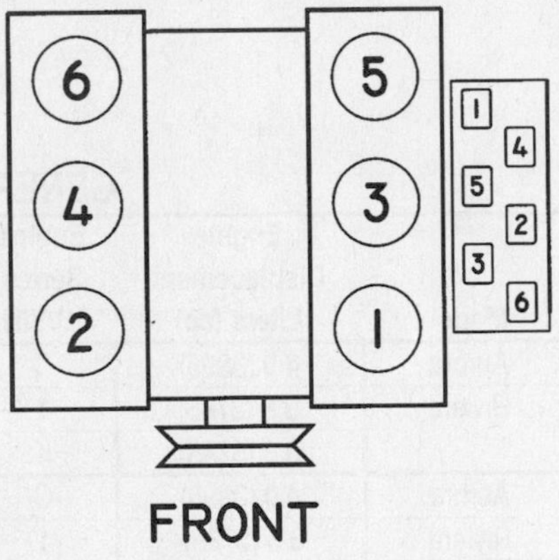

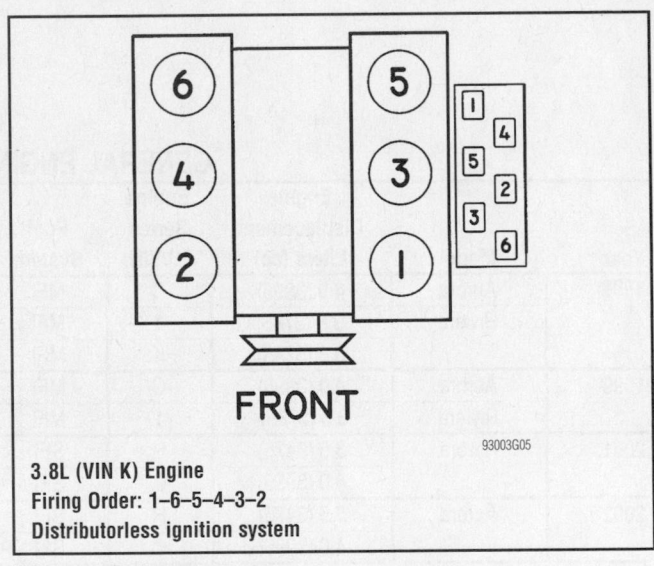

3.8L (VIN K) Engine
Firing Order: 1-6-5-4-3-2
Distributorless ignition system

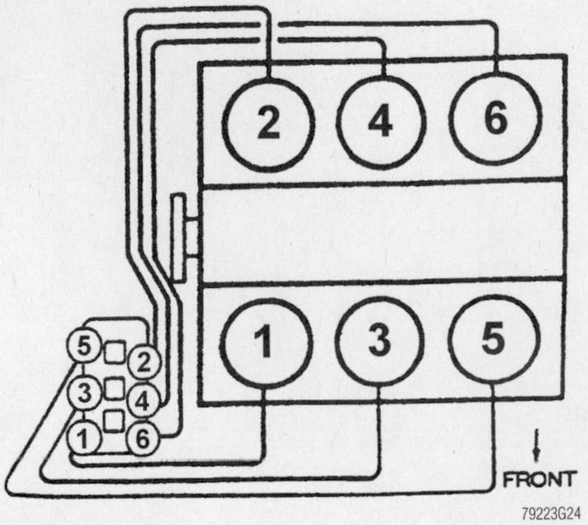

3.8L (VIN 1) Engine
Firing Order: 1-6-5-4-3-2
Distributorless ignition system

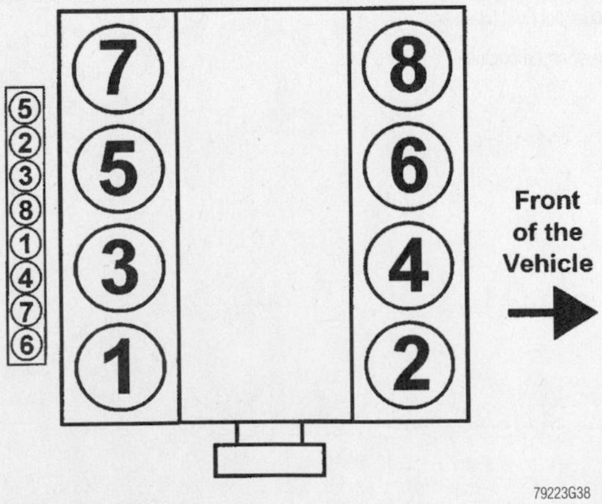

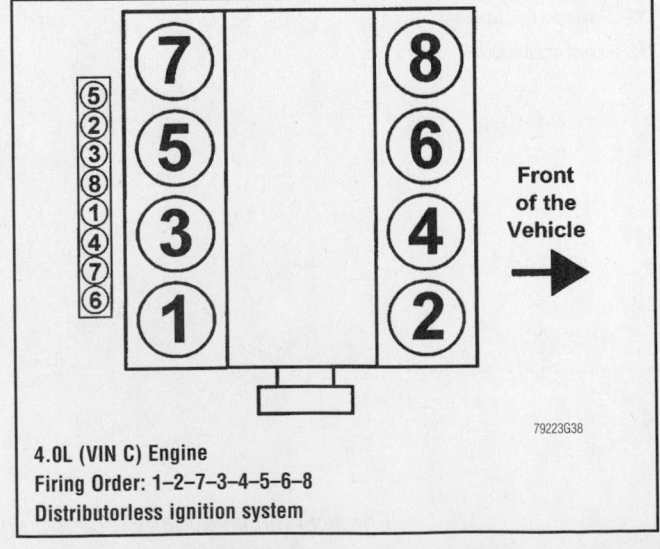

4.0L (VIN C) Engine
Firing Order: 1-2-7-3-4-5-6-8
Distributorless ignition system

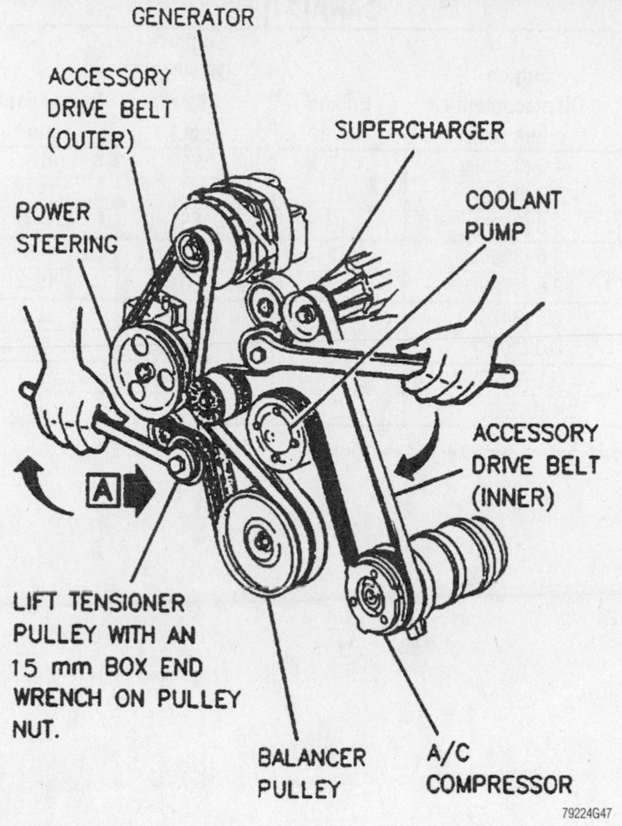

Serpentine drive belt routing—GM "G" Bodies 3.8L (VIN 1) engine

Serpentine drive belt routing—GM "G" Bodies 3.8L (VIN K) engine

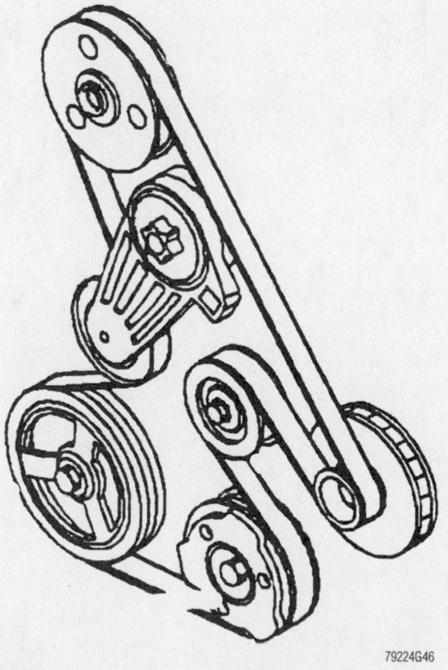

Serpentine drive belt routing—GM "G" Body 4.0L engine

CAPACITIES

Year	Model	Engine Displacement Liters (cc)	Engine ID/VIN	Engine Oil with Filter (qts.)	Transmission (pts.)	Fuel Tank (gal.)	Cooling System (qts.)
1998	Aurora	4.0 (3995)	C	7.5	6.5	20.0	13.0
	Riviera	3.8 (3785)	1	5.0	12.0	20.0	13.0
		3.8 (3785)	K	4.5	12.0	20.0	13.0
1999	Aurora	4.0 (3995)	C	7.5	6.5	20.0	13.0
	Riviera	3.8 (3785)	1	5.0	12.0	20.0	13.0
2001	Aurora	3.5 (3475)	H	6.0	12.0	18.5	13.0
		4.0 (3995)	C	7.5	6.5	17.5	13.0
2002	Aurora	3.5 (3475)	H	6.0	12.0	18.5	13.0
		4.0 (3995)	C	7.5	6.5	17.5	13.0

NOTE: All capacities are approximate. Add fluid gradually and ensure a proper fluid level is obtained.

93461C09

VALVE SPECIFICATIONS

Year	Engine Displacement Liters (cc)	Engine ID/VIN	Seat Angle (deg.)	Face Angle (deg.)	Spring Test Pressure (lbs. @ in.)	Spring Installed Height (in.)	Stem-to-Guide Clearance (in.)		Stem Diameter (in.)	
							Intake	Exhaust	Intake	Exhaust
1998	3.8 (3785)	1	45	45	210@1.315	1.690-1.720	0.0015-0.0035	0.0015-0.0032	NA	NA
	3.8 (3785)	K	45	45	210@1.315	1.690-1.720	0.0015-0.0035	0.0015-0.0032	NA	NA
	4.0 (3995)	C	46	45	92@0.854	1.190	0.0010-0.0030	0.0020-0.0040	NA	NA
1999	3.8 (3785)	1	45	45	210@1.315	1.690-1.720	0.0015-0.0035	0.0015-0.0032	NA	NA
	4.0 (3995)	C	46	45	92@0.854	1.190	0.0010-0.0030	0.0020-0.0040	NA	NA
2001	3.5 (3475)	H	45.75	45	136@0.964	1.377	0.0010-0.0030	0.0020-0.0040	0.233-0.234	0.233-0.234
	4.0 (3995)	C	46	45	92@0.854	1.190	0.0010-0.0030	0.0020-0.0040	NA	NA
2002	3.5 (3475)	H	45.75	45	136@0.964	1.377	0.0010-0.0030	0.0020-0.0040	0.233-0.234	0.233-0.234
	4.0 (3995)	C	46	45	92@0.854	1.190	0.0010-0.0030	0.0020-0.0040	NA	NA

NA: Not Available

93461C00

CRANKSHAFT AND CONNECTING ROD SPECIFICATIONS

All measurements are given in inches.

Year	Engine Displacement Liters (cc)	Engine ID/VIN	Crankshaft Main Brg. Journal Dia.	Crankshaft Main Brg. Oil Clearance	Crankshaft Shaft End-play	Crankshaft Thrust on No.	Connecting Rod Journal Diameter ①	Connecting Rod Oil Clearance	Connecting Rod Side Clearance
1998	3.8 (3785)	1	2.4988-2.4998	0.0008-0.0022	0.0030-0.0110	2	2.3738-2.3745	0.0005-0.0026	0.0030-0.0150
	3.8 (3785)	K	2.4988-2.4998	0.0008-0.0022	0.0030-0.0110	2	2.3738-2.3745	0.0005-0.0026	0.0030-0.0150
	4.0 (3995)	C	2.5335-2.5337	0.0006-0.0025	0.0020-0.0200	2	2.2490	0.0010-0.0030	0.0080-0.0200
1999	3.8 (3785)	1	2.4988-2.4998	0.0008-0.0022	0.0030-0.0110	2	2.3738-2.3745	0.0005-0.0026	0.0030-0.0150
	4.0 (3995)	C	2.5335-2.5337	0.0006-0.0025	0.0020-0.0200	2	2.2490	0.0010-0.0030	0.0080-0.0200
2001	3.5 (3475)	H	2.7550-2.7560	0.0006-0.0021	0.0050-0.0200	3	2.1829-2.1835	0.0009-0.0025	0.0040-0.0130
	4.0 (3995)	C	2.5335-2.5337	0.0006-0.0025	0.0020-0.0200	2	2.2490	0.0010-0.0030	0.0080-0.0200
2002	3.5 (3475)	H	2.7550-2.7560	0.0006-0.0021	0.0050-0.0200	3	2.1829-2.1835	0.0009-0.0025	0.0040-0.0130
	4.0 (3995)	C	2.5335-2.5337	0.0006-0.0025	0.0020-0.0200	2	2.2490	0.0010-0.0030	0.0080-0.0200

① Large end of connecting rod ID

93461CP1

PISTON AND RING SPECIFICATIONS

All measurements are given in inches.

Year	Engine ID/VIN	Engine Displacement Liters (cc)	Piston Clearance	Ring Gap Top Compression	Ring Gap Bottom Compression	Ring Gap Oil Control	Ring Side Clearance Top Compression	Ring Side Clearance Bottom Compression	Ring Side Clearance Oil Control
1998	3.8 (3785)	1	0.0004-0.0020	0.012-0.022	0.030-0.040	0.010-0.030	0.0013-0.0031	0.0013-0.0031	0.0009-0.0079
	3.8 (3785)	K	0.0004-0.0020	0.012-0.022	0.030-0.040	0.010-0.030	0.0013-0.0031	0.0013-0.0031	0.0009-0.0079
	4.0 (3995)	C	0.0008-0.0020	0.010-0.016	0.014-0.020	0.010-0.030	0.0016-0.0037	0.0016-0.0037	side-sealing
1999	3.8 (3785)	1	0.0004-0.0020	0.012-0.022	0.030-0.040	0.010-0.030	0.0013-0.0031	0.0013-0.0031	0.0009-0.0079
	4.0 (3995)	C	0.0008-0.0020	0.010-0.016	0.014-0.020	0.010-0.030	0.0016-0.0037	0.0016-0.0037	Side-sealing
2001	3.5 (3475)	H	0.0010-0.0025	0.008-0.018	0.014-0.020	0.010-0.030	0.0016-0.0037	0.0016-0.0037	Side-sealing
	4.0 (3995)	C	0.0008-0.0020	0.010-0.016	0.014-0.020	0.010-0.030	0.0016-0.0037	0.0016-0.0037	Side-sealing
2002	3.5 (3475)	H	0.0010-0.0025	0.008-0.018	0.014-0.020	0.010-0.030	0.0016-0.0037	0.0016-0.0037	Side-sealing
	4.0 (3995)	C	0.0008-0.0020	0.010-0.016	0.014-0.020	0.010-0.030	0.0016-0.0037	0.0016-0.0037	Side-sealing

93461CP2

TORQUE SPECIFICATIONS

All readings in ft. lbs.

Year	Engine Displacement Liters (cc)	Engine ID/VIN	Cylinder Head Bolts	Main Bearing Bolts	Rod Bearing Bolts	Crankshaft Damper Bolts	Flywheel Bolts	Manifold		Spark Plugs	Lug Nuts
								Intake	Exhaust		
1998	3.8 (3786)	1	①	②	③	④	⑤	⑥	22	11	100
	3.8 (3786)	K	①	②	③	④	⑤	⑦	22	11	100
	4.0 (3995)	C	⑧	⑨	⑩	⑪	⑤	7.5	18	⑫	100
1999	3.8 (3786)	1	①	②	③	④	⑤	⑥	22	11	100
	4.0 (3995)	C	⑧	⑨	⑩	⑪	⑤	7.5	22	⑫	100
2001	3.5 (3475)	H	⑬	⑭	⑮	⑯	⑤	5	18	11	100
	4.0 (3995)	C	⑧	⑨	⑩	⑪	⑤	7.5	18	⑫	100
2002	3.5 (3475)	H	⑬	⑭	⑮	⑯	⑤	5	18	11	100
	4.0 (3995)	C	⑧	⑨	⑩	⑪	⑤	7.5	18	⑫	100

① Step 1: 37 ft. lbs.
Step 2: 130 degrees
Step 3: Plus 30 degrees additional on four center bolts
(NOTE: Must use new bolts)

② Step 1: Tighten caps in equal increments to 30 ft. lbs.
Step 2: Plus 110 degrees
Side bolts:
Step 1: 11 ft. lbs.
Step 2: Plus 45 degrees

③ Step 1: 20 ft. lbs.
Step 2: 50 degrees

④ Step 1: 110 ft. lbs.
Step 2: 76 degrees

⑤ Step 1: 11 ft. lbs.
Step 2: 50 degrees

⑥ Supercharger: 17 ft. lbs.
Lower manifold: 11 ft. lbs.

⑦ Upper intake: 89 inch lbs.
Lower manifold: 11 ft. lbs.

⑧ M11 bolts:
Step 1: 30 ft. lbs.
Step 2: 70 degrees
Step 3: 60 degrees
Step 4: 60 degrees
M6 bolts: 106 inch lbs.

⑨ Step 1: 15 ft. lbs.
Step 2: 65 degrees

⑩ Step 1: 22 ft. lbs.
Step 2: Loosen completely
Step 3: 18 ft. lbs.
Step 4: 110 degrees

⑪ Step 1: 37 ft. lbs.
Step 2: 150 degrees

⑫ New cylinder head 1st time installation: 20 ft. lbs.
All others: 11 ft. lbs.

⑬ M11 bolts:
Step 1: 22 ft. lbs.
Step 2: 60 degrees
Step 3: 60 degrees
Step 4: 60 degrees
M6 bolts: 22 ft. lbs.

⑭ Cap Bolts
Step 1: 15 ft. lbs.
Step 2: 70 degrees
Perimeter bolts: 22 ft. lbs.

⑮ Step 1: 22 ft. lbs.
Step 2: Loosen completely
Step 3: 18 ft. lbs.
Step 4: 110 degrees

⑯ Step 1: 37 ft. lbs.
Step 2: 120 degrees

93461CP3

BRAKE SPECIFICATIONS

All measurements in inches unless noted

Year	Model		Brake Disc Original Thickness	Brake Disc Minimum Thickness	Brake Disc Maximum Run-out	Minimum Lining Thickness	Brake Caliper Mounting Bolt (ft. lbs.)
1998	Aurora	F	1.260	1.209	0.002	0.030	38
		R	0.433	0.374	0.002	0.030	20
	Riviera	F	1.260	1.209	0.002	0.030	38
		R	0.433	0.374	0.002	0.030	20
1999	Aurora	F	1.260	1.209	0.002	0.030	38
		R	0.433	0.374	0.002	0.030	20
	Riviera	F	1.260	1.209	0.002	0.030	38
		R	0.433	0.374	0.002	0.030	20
2001	Aurora	F	1.260	1.209	0.002	0.030	38
		R	0.433	0.374	0.002	0.030	20
2002	Aurora	F	1.260	1.209	0.002	0.030	38
		R	0.433	0.374	0.002	0.030	20

F: Front

R: Rear

93461CP4

WHEEL ALIGNMENT

Year	Model		Caster Range (+/-Deg.)	Caster Preferred Setting (Deg.)	Camber Range (+/-Deg.)	Camber Preferred Setting (Deg.)	Toe-in (in.)	Steering Axis Inclination (Deg.)
1998	Aurora	F	+0.50	+6.00	+0.50	+0.20	0 +/- 0.09	12.50
		R	—	—	+0.50	-0.31	0.09 +/- 0.09	—
	Riviera	F	+0.50	+6.00	+0.50	+0.20	0 +/- 0.20	12.50
		R	—	—	+0.50	-0.30	0.20 +/- 0.20	—
1999	Aurora	F	+0.50	+6.00	+0.50	+0.20	0 +/- 0.09	12.50
		R	—	—	+0.50	-0.31	0.09 +/- 0.09	—
	Riviera	F	+0.50	+6.00	+0.50	+0.20	0 +/- 0.20	12.50
		R	—	—	+0.50	-0.30	0.20 +/- 0.20	—
2001	Aurora	F	+0.50	+6.00	+0.50	+0.20	0 +/- 0.09	12.50
		R	—	—	+0.50	-0.31	0.09 +/- 0.09	—
2002	Aurora	F	+0.50	+6.00	+0.50	+0.20	0 +/- 0.09	12.50
		R	—	—	+0.50	-0.31	0.09 +/- 0.09	—

F: Front

R: Rear

93461CP5

TIRE, WHEEL AND BALL JOINT SPECIFICATIONS

Year	Model	OEM Tires		Tire Pressures (psi)		Wheel	Ball Joint
		Standard	Optional	Front	Rear	Size	Inspection
1998	Aurora	P235/60R16	P235/60VR16	30	30	7-J	①
	Riviera	P225/60R16	None	30	30	6.5-JJ	①
1999	Aurora	P235/60R16	P235/60VR16	30	30	7-J	①
	Riviera	P225/60R16	None	30	30	6.5-JJ	①
2001	Aurora	P235/60R16	P235/60VR16	30	30	7-J	①
2002	Aurora	P235/60R16	P235/60VR16	30	30	7-J	①

OEM: Original Equipment Manufacturer

PSI: Pounds Per Square Inch

① Replace if any measurable movement is found.

93461CP6

Heater Core replacement is covered in Section 2 of this manual

SCHEDULED MAINTENANCE INTERVALS
GM G BODY—BUICK RIVIERA & OLDSMOBILE AURORA

TO BE SERVICED	TYPE OF SERVICE	VEHICLE MILEAGE INTERVAL (x1000)												
		7.5	15	22.5	30	37.5	45	52.5	60	67.5	75	82.5	90	97.5
Engine oil & filter	R	✓	✓	✓	✓	✓	✓	✓	✓	✓	✓	✓	✓	✓
Coolant level, hoses & clamps	S/I	✓	✓	✓	✓	✓	✓	✓	✓	✓	✓	✓	✓	✓
Driveshaft boots & front suspension components	S/I	✓	✓	✓	✓	✓	✓	✓	✓	✓	✓	✓	✓	✓
Exhaust system & brake hoses	S/I	✓	✓	✓	✓	✓	✓	✓	✓	✓	✓	✓	✓	✓
Lubricate chassis and suspension	S/I	✓	✓	✓	✓	✓	✓	✓	✓	✓	✓	✓	✓	✓
Lubricate steering linkage and transaxle linkage	S/I	✓	✓	✓	✓	✓	✓	✓	✓	✓	✓	✓	✓	✓
Lubricate parking brake cable guides, underbody contact points & linkage	S/I	✓	✓	✓	✓	✓	✓	✓	✓	✓	✓	✓	✓	✓
Throttle linkage	S/I	✓	✓	✓	✓	✓	✓	✓	✓	✓	✓	✓	✓	✓
Brake linings	S/I	✓		✓		✓		✓		✓		✓		✓
Rotate tires	S/I	✓		✓		✓		✓		✓		✓		✓
Air filter element	R				✓				✓				✓	
Engine coolant ①	R													
Spark plugs ②	R				✓				✓				✓	
Accessory drive belt(s)	S/I				✓				✓				✓	
Automatic transaxle fluid & filter	S/I				✓				✓				✓	
Fuel system	S/I				✓				✓				✓	
Ignition cables	R				✓				✓				✓	
Inspect throttle body bore & throttle plate for deposits	S/I		✓				✓				✓			
Supercharger oil	S/I				✓				✓				✓	

R: Replace

S/I: Service or Inspect

① Engine coolant: replace every 100,000 miles. Use O.E. specified (DEX-COOL™) coolant only. If any silicate coolant is used, the service interval is every 30,000 miles.

② Platinum tip spark plugs: replace every 100,000 miles.

FREQUENT OPERATION MAINTENANCE (SEVERE SERVICE)

If a vehicle is operated under any of the following conditions it is considered severe service:

- Extremely dusty areas.

- 50% or more of the vehicle operation is in 32°C (90°F) or higher temperatures, or constant operation in temperatures below 0°C (32°F).

- Prolonged idling (vehicle operation in stop and go traffic).

- Frequent short running periods (engine does not warm to normal operating temperatures).

- Police, taxi, delivery usage or trailer towing usage.

CV joints & front suspension components: service or inspect every 3000 miles.

Engine oil & filter change: change every 3000 miles.

Brake linings: check every 6000 miles.

Chassis lubrication: lubricate every 6000 miles.

Suspension, steering linkage, transaxle shift linkage, parking cable guides, underbody contact points: lubricate every 6000 miles.

Throttle body mount bolt torque: tighten at 6000 miles.

Air filter element: service or inspect every 15,000 miles.

Inspect throttle body bore & throttle plate for deposits: clean as required every 15,000 miles.

Rotate tires at 6000 miles, then every 15,000 miles.

93461CP7

SCHEDULED MAINTENANCE INTERVALS
GENERAL MOTORS CORPORATION
G BODY
BUICK RIVIERA
OLDSMOBILE AURORA

The following should be used as a guide when determining the amount of work required for a particular service. In estimating how long a particular Scheduled Maintenance Service should take, please observe the following:

- Labor Time is time based on field research and data supplied by the vehicle manufacturer.
- Labor time operations are given in hours and tenths of an hour.
- All labor operations are to be used as a guide.

Mechanic Skill Level Codes:
(A) PRECISION: Highly skilled with multiple certification.
(B) GENERAL: Normally skilled with certification.
(C) MAINTENANCE: Semi-skilled working on certification.

	LABOR TIME
7500 Mile Service (C)	
All Models	1.4
15000 Mile Service (C)	
All Models	.9
Inspect throttle body for deposits add	.1
22500 Mile Service (C)	
All Models	1.4
30000 Mile Service (B)	
All Models	2.5
37500 Mile Service (C)	
All Models	1.4
45000 Mile Service (C)	
All Models	.9
Inspect throttle body for deposits add	.1
52500 Mile Service (C)	
All Models	1.4
60000 Mile Service (B)	
All Models	2.5
67500 Mile Service (C)	
All Models	1.4
75000 Mile Service (C)	
All Models	.9
Inspect throttle body for deposits add	.1
82500 Mile Service (C)	
All Models	1.4
90000 Mile Service (B)	
All Models	2.4
97500 Mile Service (B)	
All Models	1.8

93461CP8

Brake service is covered in Section 4 of this manual

GENERAL MOTORS J-BODY
Chevrolet Cavalier • Pontiac Sunfire

ENGINE AND VEHICLE IDENTIFICATION CHART

Engine						
Code ①	Liters (cc)	Cu. In.	Cyl.	Fuel Sys.	Engine Type	Eng. Mfg.
4	2.2 (2180)	133	4	MFI	OHV	CUS
T	2.4 (2392)	146	4	MFI	DOHC	CUS

Model Year	
Code ②	Year
W	1998
X	1999
Y	2000
1	2001
2	2002

CUS: Chevrolet/United States
MFI: Multi-point Fuel Injection
OHV: Overhead Valves
DOHC: Double Overhead Camshafts
① 8th position of VIN
② 10th position of VIN

93461CP9

GENERAL ENGINE SPECIFICATIONS

Year	Model	Engine Displacement Liters (cc)	Engine Series (ID/VIN)	Fuel System	Net Horsepower @ rpm	Net Torque @ rpm (ft. lbs.)	Bore x Stroke (in.)	Com-pression Ratio	Oil Pressure @ rpm
1998	Cavalier	2.2 (2180)	4	MFI	120@5200	130@3200	3.50x3.46	9.0:1	56@3000
		2.4 (2392)	T	MFI	150@6000	155@4400	3.54x3.70	9.5:1	30@3000
	Sunfire	2.2 (2180)	4	MFI	120@5200	130@3200	3.50x3.46	9.0:1	56@3000
		2.4 (2392)	T	MFI	150@6000	155@4400	3.54x3.70	9.5:1	30@3000
1999	Cavalier	2.2 (2180)	4	MFI	120@5200	130@3200	3.50x3.46	9.0:1	56@3000
		2.4 (2392)	T	MFI	150@6000	155@4400	3.54x3.70	9.5:1	30@3000
	Sunfire	2.2 (2180)	4	MFI	120@5200	130@3200	3.50x3.46	9.0:1	56@3000
		2.4 (2392)	T	MFI	150@6000	155@4400	3.54x3.70	9.5:1	30@3000
2000	Cavalier	2.2 (2180)	4	MFI	115@5000	136@3600	3.50x3.46	9.0:1	56@3000
		2.4 (2392)	T	MFI	150@5600	155@4400	3.54x3.70	9.5:1	30@3000
	Sunfire	2.2 (2180)	4	MFI	115@5000	136@3600	3.50x3.46	9.0:1	56@3000
		2.4 (2392)	T	MFI	150@5600	155@4400	3.54x3.70	9.5:1	30@3000
2001	Cavalier	2.2 (2180)	4	MFI	115@5000	136@3600	3.50x3.46	9.0:1	56@3000
		2.4 (2392)	T	MFI	150@5600	155@4400	3.54x3.70	9.5:1	30@3000
	Sunfire	2.2 (2180)	4	MFI	115@5000	136@3600	3.50x3.46	9.0:1	56@3000
		2.4 (2392)	T	MFI	150@5600	155@4400	3.54x3.70	9.5:1	30@3000

MFI: Multi-point Fuel Injection

93461CP0

ENGINE TUNE-UP SPECIFICATIONS

Year	Engine Displacement Liters (cc)	Engine ID/VIN	Spark Plug Gap (in.)	Ignition Timing (deg.)	Fuel Pump (psi)	Idle Speed (rpm)	Valve Clearance Intake	Valve Clearance Exhaust
1998	2.2 (2180)	4	0.060	①	41-47	①	HYD	HYD
	2.4 (2392)	T	0.060	①	41-47	①	HYD	HYD
1999	2.2 (2180)	4	0.060	①	41-47	①	HYD	HYD
	2.4 (2392)	T	0.060	①	41-47	①	HYD	HYD
2000	2.2 (2180)	4	0.060	①	41-47	①	HYD	HYD
	2.4 (2392)	T	0.060	①	41-47	①	HYD	HYD
2001	2.2 (2180)	4	0.060	①	41-47	①	HYD	HYD
	2.4 (2392)	T	0.060	①	41-47	①	HYD	HYD

NOTE: The Vehicle Emission Control Information label often reflects specification changes made during production. The label figures must be used if they differ from those in this chart.

HYD: Hydraulic

① Refer to Vehicle Emission Control Information label

93461CQ1

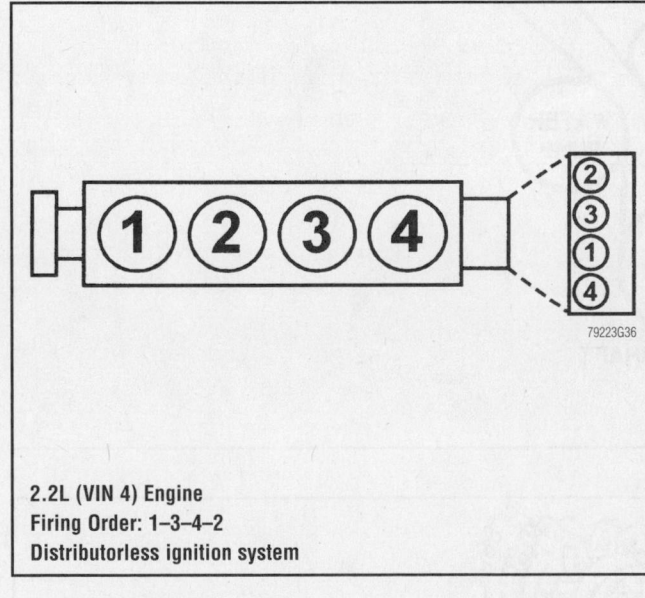

2.2L (VIN 4) Engine
Firing Order: 1–3–4–2
Distributorless ignition system

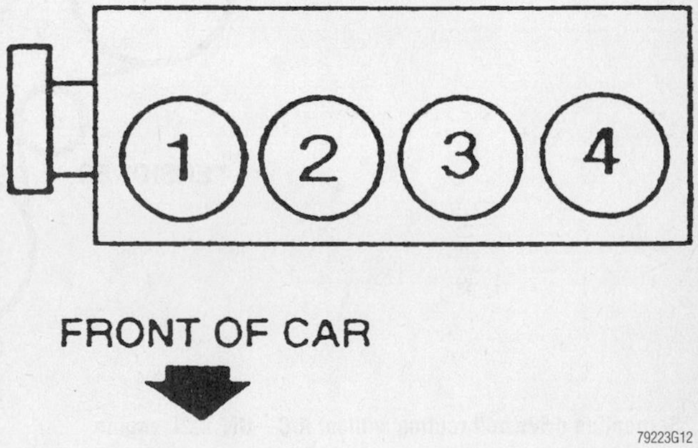

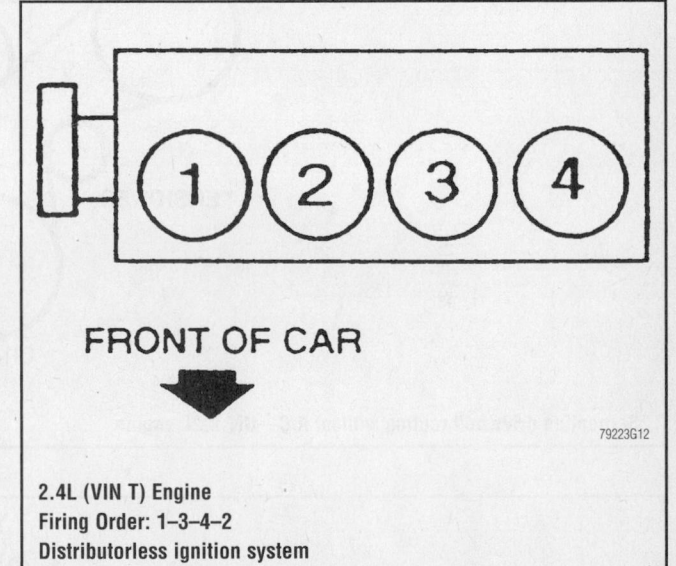

2.4L (VIN T) Engine
Firing Order: 1–3–4–2
Distributorless ignition system

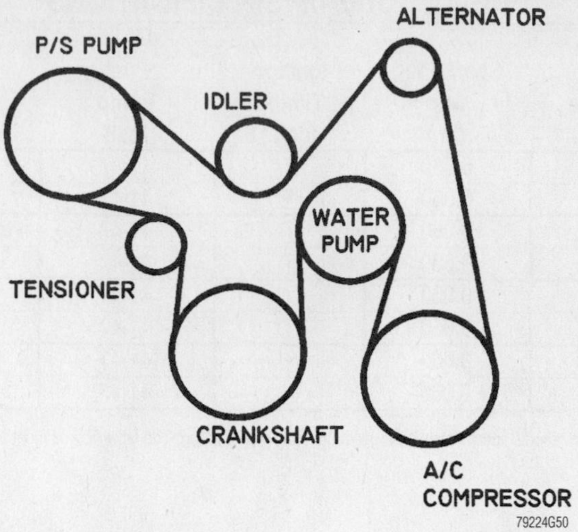

Serpentine drive belt routing with A/C—GM 2.2L engine

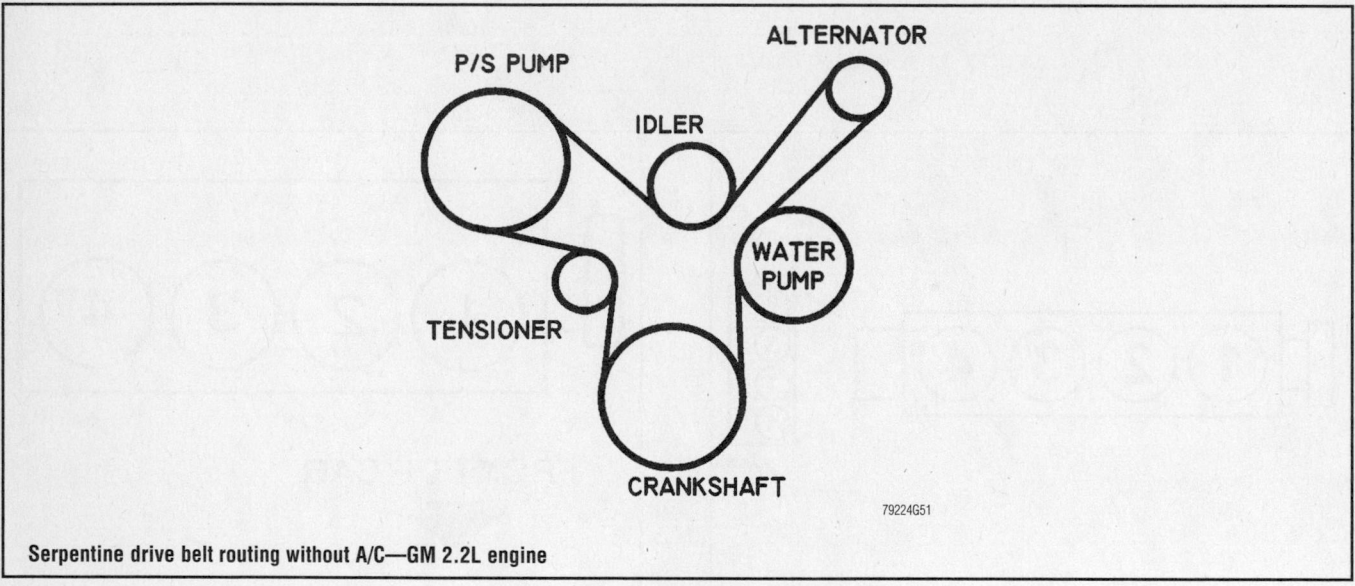

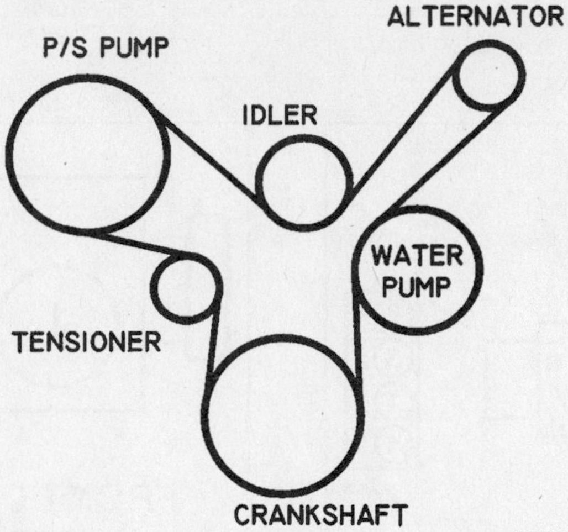

Serpentine drive belt routing without A/C—GM 2.2L engine

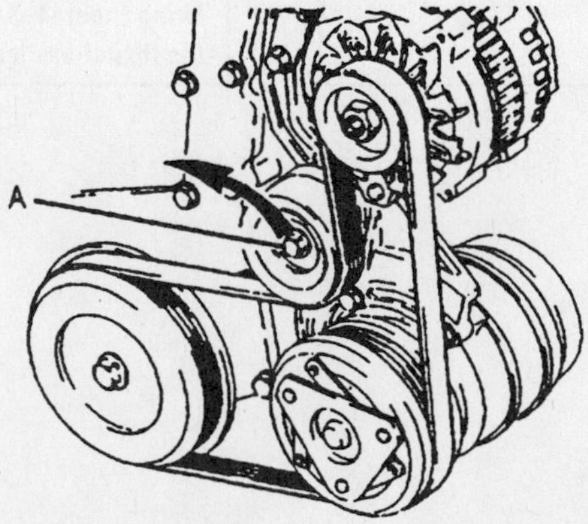

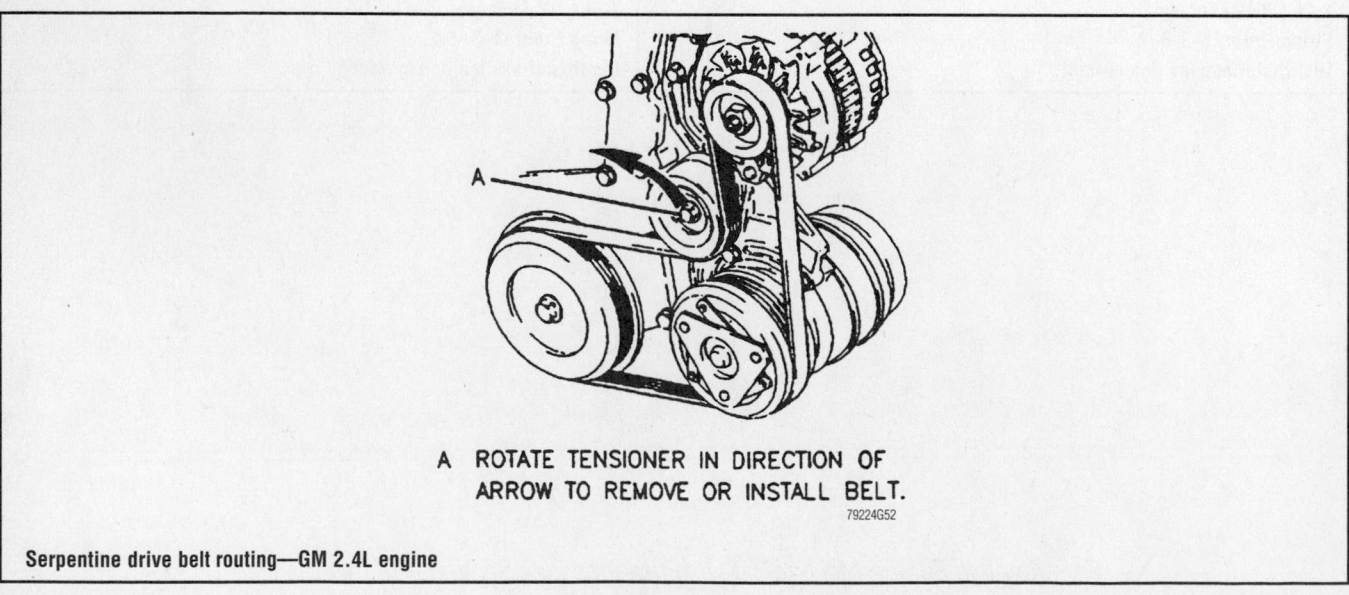

A ROTATE TENSIONER IN DIRECTION OF ARROW TO REMOVE OR INSTALL BELT.

79224G52

Serpentine drive belt routing—GM 2.4L engine

CAPACITIES

Year	Model	Engine Displacement Liters (cc)	Engine ID/VIN	Engine Oil with Filter (qts.)	Transmission (pts.) Manual	Transmission (pts.) Auto.	Fuel Tank (gal.)	Cooling System (qts.)
1998	Cavalier	2.2 (2180)	4	4.0	4.0	①	15.2	10.5
		2.4 (2392)	T	4.5	4.0	①	15.2	10.5
	Sunfire	2.2 (2180)	4	4.0	4.0	①	15.2	10.5
		2.4 (2392)	T	4.5	4.0	①	15.2	10.5
1999	Cavalier	2.2 (2180)	4	4.0	4.0	①	15.2	10.5
		2.4 (2392)	T	4.5	4.0	①	15.2	10.5
	Sunfire	2.2 (2180)	4	4.0	4.0	①	15.2	10.5
		2.4 (2392)	T	4.5	4.0	①	15.2	10.5
2000	Cavalier	2.2 (2180)	4	4.0	4.0	①	15.2	9.6
		2.4 (2392)	T	4.5	4.0	①	15.2	9.9
	Sunfire	2.2 (2180)	4	4.0	4.0	①	15.2	9.6
		2.4 (2392)	T	4.5	4.0	①	15.2	9.9
2001	Cavalier	2.2 (2180)	4	4.0	4.0	①	15.2	9.6
		2.4 (2392)	T	4.5	4.0	①	15.2	9.9
	Sunfire	2.2 (2180)	4	4.0	4.0	①	15.2	9.6
		2.4 (2392)	T	4.5	4.0	①	15.2	9.9

NOTE: All capacities are approximate. Add fluid gradually and ensure a proper fluid level is obtained.

① 3 Speed: 8.0 pts.
4 Speed: 14.8 pts.

93461CQ2

VALVE SPECIFICATIONS

Year	Engine Displacement Liters (cc)	Engine ID/VIN	Seat Angle (deg.)	Face Angle (deg.)	Spring Test Pressure (lbs. @ in.)	Spring Installed Height (in.)	Stem-to-Guide Clearance (in.)		Stem Diameter (in.)	
							Intake	Exhaust	Intake	Exhaust
1998	2.2 (2180)	4	46	45	72-81@1.60	1.600	0.0007-0.0020	0.0014-0.0029	0.2740-0.2743	0.2731-0.2736
	2.4 (2392)	T	45	46	50-55@1.44	1.437	0.0009-0.0025	0.0016-0.0032	0.2331-0.2339	0.2326-0.2334
1999	2.2 (2180)	4	46	45	72-81@1.60	1.600	0.0007-0.0020	0.0014-0.0029	0.2740-0.2743	0.2731-0.2736
	2.4 (2392)	T	45	46	50-55@1.44	1.437	0.0009-0.0025	0.0016-0.0032	0.2331-0.2339	0.2326-0.2334
2000	2.2 (2180)	4	46	45	72-81@1.60	1.600	0.0007-0.0020	0.0014-0.0029	0.2740-0.2743	0.2731-0.2736
	2.4 (2392)	T	45	46	50-55@1.44	1.437	0.0009-0.0025	0.0016-0.0032	0.2331-0.2339	0.2326-0.2334
2001	2.2 (2180)	4	46	45	72-81@1.60	1.600	0.0007-0.0020	0.0014-0.0029	0.2740-0.2743	0.2731-0.2736
	2.4 (2392)	T	45	46	50-55@1.44	1.437	0.0009-0.0025	0.0016-0.0032	0.2331-0.2339	0.2326-0.2334

93461CQ3

CRANKSHAFT AND CONNECTING ROD SPECIFICATIONS

All measurements are given in inches.

	Engine		Crankshaft				Connecting Rod		
Year	Displacement Liters (cc)	Engine ID/VIN	Main Brg. Journal Dia.	Main Brg. Oil Clearance	Shaft End-play	Thrust on No.	Journal Diameter	Oil Clearance	Side Clearance
1998	2.2 (2180)	4	2.4945-2.4954	0.0006-0.0019	0.0020-0.0070	4	1.9983-1.9994	0.0010-0.0030	0.0039-0.0149
	2.4 (2392)	T	2.3612-2.3631	0.0004-0.0023	0.0034-0.0095	3	1.8887-1.8897	0.0004-0.0026	0.0059-0.0177
1999	2.2 (2180)	4	2.4945-2.4954	0.0006-0.0019	0.0020-0.0070	4	1.9983-1.9994	0.0010-0.0030	0.0039-0.0149
	2.4 (2392)	T	2.3612-2.3631	0.0004-0.0023	0.0034-0.0095	3	1.8887-1.8897	0.0004-0.0026	0.0059-0.0177
2000	2.2 (2180)	4	2.4945-2.4954	0.0006-0.0019	0.0020-0.0070	4	1.9983-1.9994	0.0010-0.0030	0.0039-0.0149
	2.4 (2392)	T	2.3612-2.3631	0.0004-0.0023	0.0034-0.0095	3	1.8887-1.8897	0.0004-0.0026	0.0059-0.0177
2001	2.2 (2180)	4	2.4945-2.4954	0.0006-0.0019	0.0020-0.0070	4	1.9983-1.9994	0.0010-0.0030	0.0039-0.0149
	2.4 (2392)	T	2.3612-2.3631	0.0004-0.0023	0.0034-0.0095	3	1.8887-1.8897	0.0004-0.0026	0.0059-0.0177

93461C04

For Tire, Wheel and Ball Joint specifications, see Section 1 of this manual

PISTON AND RING SPECIFICATIONS

All measurements are given in inches.

Year	Engine Displacement Liters (cc)	Engine ID/VIN	Piston Clearance	Ring Gap			Ring Side Clearance		
				Top Compression	Bottom Compression	Oil Control	Top Compression	Bottom Compression	Oil Control
1998	2.2 (2180)	4	0.0006-0.0018	0.0100-0.0200	0.0100-0.0200	0.0100-0.0500	0.0011-0.0019	0.0011-0.0019	0.0019-0.0082
	2.4 (2392)	T	0.0006-0.0015	0.0060-0.0120	0.0098-0.0157	0.0098-0.0299	0.0016-0.0031	0.0012-0.0028	NA
1999	2.2 (2180)	4	0.0006-0.0018	0.0100-0.0200	0.0100-0.0200	0.0100-0.0500	0.0011-0.0019	0.0011-0.0019	0.0019-0.0082
	2.4 (2392)	T	0.0006-0.0015	0.0060-0.0120	0.0098-0.0157	0.0098-0.0299	0.0016-0.0031	0.0012-0.0028	NA
2000	2.2 (2180)	4	0.0006- ① 0.0018	0.0100-0.0200	0.0120-0.0177	0.0100-0.0300	0.0020-0.0035	0.0016-0.0031	0.0005-0.0087
	2.4 (2392)	T	0.0006- ② 0.0009	0.0060-0.0120	0.0098-0.0157	0.0098-0.0299	0.0016-0.0031	0.0012-0.0028	NA
2001	2.2 (2180)	4	0.0006- ① 0.0018	0.0100-0.0200	0.0120-0.0177	0.0100-0.0300	0.0020-0.0035	0.0016-0.0031	0.0005-0.0087
	2.4 (2392)	T	0.0006- ② 0.0009	0.0060-0.0120	0.0098-0.0157	0.0098-0.0299	0.0016-0.0031	0.0012-0.0028	NA

① 42.7mm from top of piston

② 42.7mm from top of piston

93461CQ5

TORQUE SPECIFICATIONS

All readings in ft. lbs.

Year	Engine Displacement Liters (cc)	Engine ID/VIN	Cylinder Head Bolts	Main Bearing Bolts	Rod Bearing Bolts	Crankshaft Damper Bolts	Flywheel Bolts	Manifold		Spark Plugs	Lug Nuts
								Intake	Exhaust		
1998	2.2 (2180)	4	①	70	38	77 ②	55	③	10	13	100
	2.4 (2392)	T	④	⑤	⑥	⑦	⑧	⑨	⑩	13	100
1999	2.2 (2180)	4	①	70	38	77 ②	55	③	10	13	100
	2.4 (2392)	T	④	⑤	⑥	⑦	⑧	⑨	⑩	13	100
2000	2.2 (2180)	4	①	70	38	77 ②	55	③	10	13	100
	2.4 (2392)	T	④	⑤	⑥	⑦	⑧	⑨	⑩	13	100
2001	2.2 (2180)	4	①	70	38	77 ②	55	③	10	13	100
	2.4 (2392)	T	④	⑤	⑥	⑦	⑧	⑨	⑩	13	100

NA: Not Available

① Step 1: Long bolts: 46 ft. lbs.
Step 2: Short bolts: 43 ft. lbs.
Step 3: Long bolts an additional 90 degree turn
Step 4: Short bolts an additional 90 degree turn

② Center bolt spec shown; Pulley-to-hub bolts: 37 ft. lbs.

③ Bolts: 17 ft. lbs.
Nuts: 17 ft. lbs.
Studs: 9 ft. lbs.

④ Step 1: bolts 1-8; 40 ft. lbs.
Step 2: bolts 9-10: 30 ft. lbs.
Step 3: An additional 90 degree turn

⑤ 15 ft. lbs. plus 90 degrees

⑥ 18 ft. lbs. plus 80 degrees

⑦ 129 ft. lbs. plus 90 degrees

⑧ 15 ft. lbs. plus 45 degrees

⑨ Nuts: 18 ft. lbs.
Studs: 97 inch lbs.

⑩ Nuts: 31 ft. lbs.
Studs: 97 inch lbs.

93461CQ6

For Wheel Alignment specifications, see Section 1 of this manual

BRAKE SPECIFICATIONS

All measurements in inches unless noted

Year	Model	Brake Disc Original Thickness	Brake Disc Minimum Thickness	Brake Disc Maximum Run-out	Brake Drum Original Inside Diameter	Brake Drum Max. Wear Limit	Brake Drum Maximum Machine Diameter	Minimum Lining Thickness Front	Minimum Lining Thickness Rear	Brake Caliper Mounting Bolts (ft. lbs.)
1998	Cavalier	0.786	0.736	0.003	7.880	7.930	7.900	0.030	0.030	38
	Sunfire	0.786	0.736	0.003	7.880	7.930	7.900	0.030	0.030	38
1999	Cavalier	0.786	0.736	0.003	7.880	7.930	7.900	0.030	0.030	38
	Sunfire	0.786	0.736	0.003	7.880	7.930	7.900	0.030	0.030	38
2000	Cavalier	0.786	0.736	0.003	NA	8.909	8.879	0.030	0.030	38
	Sunfire	0.786	0.736	0.003	NA	8.909	8.879	0.030	0.030	38
2001	Cavalier	0.786	0.736	0.003	NA	8.909	8.879	0.030	0.030	38
	Sunfire	0.786	0.736	0.003	NA	8.909	8.879	0.030	0.030	38

NA: Not Available

93461C07

WHEEL ALIGNMENT

Year	Model		Caster Range (+/-Deg.)	Caster Preferred Setting (Deg.)	Camber Range (+/-Deg.)	Camber Preferred Setting (Deg.)	Toe-in (in.)	Steering Axis Inclination (Deg.)
1998	All	F	+1.00	+1.45	+1.00	0	0 +/- 0.25	—
		R	—	—	+0.75	-0.40	0.20 +/- 0.30	—
1999	All	F	+1.00	+1.45	+1.00	0	0 +/- 0.25	—
		R	—	—	+0.75	-0.40	0.20 +/- 0.30	—
2000	All	F	+1.00	+4.30	+1.00	0	0 +/- 0.25	—
		R	—	—	+0.75	-0.40	0.20 +/- 0.30	—
2001	All	F	+1.00	+4.30	+1.00	0	0 +/- 0.25	—
		R	—	—	+0.75	-0.40	0.20 +/- 0.30	—

93461CQ8

TIRE, WHEEL AND BALL JOINT SPECIFICATIONS

Year	Model	OEM Tires		Tire Pressures (psi)		Wheel	Ball Joint
		Standard	Optional	Front	Rear	Size	Inspection
1998	Cavalier base	P195/70R14	None	30	30	6-JJ	①
	Cavalier LS, RS	P195/65R15	None	30	30	6-JJ	①
	Cavalier Z24	P205/55R16	None	30	30	6-JJ	①
	Sunfire SE	P195/70R14	P195/65R15	30	30	6-JJ	0.125 in.
	Sunfire GT	P205/55R16	None	30	30	6-JJ	0.125 in.
1999	Cavalier base	P195/70R14	None	30	30	6-JJ	①
	Cavalier LS, RS	P195/65R15	None	30	30	6-JJ	①
	Cavalier Z24	P205/55R16	None	30	30	6-JJ	①
	Sunfire SE	P195/70R14	P195/65R15	30	30	6-JJ	0.125 in.
	Sunfire GT	P205/55R16	None	30	30	6-JJ	0.125 in.
2000	Cavalier base	P195/70R14	None	30	30	6-JJ	①
	Cavalier LS, RS	P195/65R15	None	30	30	6-JJ	①
	Cavalier Z24	P205/55R16	None	30	30	6-JJ	①
	Sunfire SE	P195/70R14	P195/65R15	30	30	6-JJ	0.125 in.
	Sunfire GT	P205/55R16	None	30	30	6-JJ	0.125 in.

OEM: Original Equipment Manufacturer

PSI: Pounds Per Square Inch

① Replace if any measurable movement is found

93461CQ9

SCHEDULED MAINTENANCE INTERVALS
GM J BODY—CHEVROLET CAVALIER & PONTIAC SUNFIRE

TO BE SERVICED	TYPE OF SERVICE	7.5	15	22.5	30	37.5	45	52.5	60	67.5	75	82.5	90	97.5
Engine oil & filter	R	✓	✓	✓	✓	✓	✓	✓	✓	✓	✓	✓	✓	✓
Exhaust system & brake hoses	S/I	✓	✓	✓	✓	✓	✓	✓	✓	✓	✓	✓	✓	✓
Drive shaft boots & front suspension components	S/I	✓	✓	✓	✓	✓	✓	✓	✓	✓	✓	✓	✓	✓
Coolant level, hoses & clamps	S/I	✓	✓	✓	✓	✓	✓	✓	✓	✓	✓	✓	✓	✓
Throttle linkage	S/I	✓	✓	✓	✓	✓	✓	✓	✓	✓	✓	✓	✓	✓
Lubricate chassis and suspension	S/I	✓	✓	✓	✓	✓	✓	✓	✓	✓	✓	✓	✓	✓
Lubricate steering linkage and transaxle shift linkage	S/I	✓	✓	✓	✓	✓	✓	✓	✓	✓	✓	✓	✓	✓
Lubricate parking brake cable guides, underbody contact points and linkage	S/I	✓	✓	✓	✓	✓	✓	✓	✓	✓	✓	✓	✓	✓
Brake linings & rotate tires	S/I	✓		✓		✓		✓		✓		✓		✓
Automatic transmission fluid & filter ①	S/I													
Air filter element & PCV filter	R				✓				✓				✓	
Engine coolant ②	R													
Spark plugs ③	R				✓				✓				✓	
Accessory drive belt(s)	R				✓				✓				✓	
EGR & fuel systems	S/I				✓				✓				✓	
Ignition cables	S/I				✓				✓				✓	

R: Replace

S/I: Service or Inspect

① Automatic transaxle fluid & filter: replace at 100,000 miles (if not changed previously).

② Engine coolant: replace every 100,000 miles. Use O.E. specified (DEX-COOL™) coolant only. If any silicate coolant is used, the service interval is every 30,000 miles.

③ Platinum tip spark plugs: replace every 100,000 miles.

FREQUENT OPERATION MAINTENANCE (SEVERE SERVICE) ADDITIONS

If a vehicle is operated under any of the following conditions it is considered severe service:

- Towing a trailer or using a camper or car-top carrier.

- Extensive idling or low-speed driving for long distances as in heavy commercial use, such as delivery, taxi or police cars.

- Operating on rough, muddy or salt-covered roads.

- Operating on unpaved or dusty roads.

- 50% or more of the vehicle operation is in 32°C (90°F) or higher temperatures, or constant operation in temperatures below 0°C (32°F).

Engine oil and filter: change every 3000 miles or 3 months, whichever occurs first.

Wheels and tires: inspect and rotate every 6000 miles.

Air cleaner element: inspect every 15,000 miles and replace or clean as needed. Replace it at least every 30,000 miles.

Automatic transaxle fluid & filter: replace every 50,000 miles.

93461CQ0

SCHEDULED MAINTENANCE INTERVALS
GENERAL MOTORS CORPORATION
J BODY
CHEVROLET CAVALIER
PONTIAC SUNFIRE

The following should be used as a guide when determining the amount of work required for a particular service. In estimating how long a particular Scheduled Maintenance Service should take, please observe the following:

- Labor Time is time based on field research and data supplied by the vehicle manufacturer.
- Labor time operations are given in hours and tenths of an hour.
- All labor operations are to be used as a guide.

Mechanic Skill Level Codes:
(A) PRECISION: Highly skilled with multiple certification.
(B) GENERAL: Normally skilled with certification.
(C) MAINTENANCE: Semi-skilled working on certification.

	LABOR TIME
7500 Mile Service (C)	
All Models	1.4
15000 Mile Service (C)	
All Models	.9
22500 Mile Service (C)	
All Models	1.4
30000 Mile Service (B)	
All Models	3.0
37500 Mile Service (C)	
All Models	1.4
45000 Mile Service (C)	
All Models	.9
52500 Mile Service (C)	
All Models	1.4
60000 Mile Service (B)	
All Models	3.0
67500 Mile Service (C)	
All Models	1.4
75000 Mile Service (C)	
All Models	.9
82500 Mile Service (C)	
All Models	1.4
90000 Mile Service (B)	
All Models	3.0
97500 Mile Service (C)	
All Models	1.4

93461CR1

GENERAL MOTORS L/N-BODY
Chevrolet Malibu • Oldsmobile Cutlass

ENGINE AND VEHICLE IDENTIFICATION

Engine							Model Year	
Code ①	Liters (cc)	Cu. In.	Cyl.	Fuel Sys.	Engine Type	Eng. Mfg.	Code ②	Year
M	3.1 (3130)	191	6	SFI	OHV	BOC	W	1998
T	2.4 (2392)	146	4	SFI	DOHC	CUS	X	1999
							Y	2000
							1	2001
							2	2002

BOC: Buick/Oldsmobile/Cadillac

CUS: Chevrolet/United States

SFI: Sequential Fuel Injection

① 8th position of VIN

② 10th position of VIN

93461CR2

GENERAL ENGINE SPECIFICATIONS

Year	Model	Engine Displacement Liters (cc)	Engine Series (ID/VIN)	Fuel System	Net Horsepower @ rpm	Net Torque @ rpm (ft. lbs.)	Bore x Stroke (in.)	Com-pression Ratio	Oil Pressure @ rpm
1998	Cutlass	3.1 (3130)	M	SFI	160@5200	185@4000	3.50x3.31	9.5:1	15@1100
	Malibu	3.1 (3130)	M	SFI	160@5200	185@4000	3.50x3.31	9.5:1	15@1100
	Malibu	2.4 (2392)	T	SFI	150@6000	150@5600	3.54x3.70	9.5:1	30@3000
1999	Cutlass	3.1 (3130)	M	SFI	160@5200	185@4000	3.50x3.31	9.5:1	15@1100
	Malibu	3.1 (3130)	M	SFI	160@5200	185@4000	3.50x3.31	9.5:1	15@1100
	Malibu	2.4 (2392)	T	SFI	150@6000	150@5600	3.54x3.70	9.5:1	30@3000
2000	Malibu	3.1 (3130)	M	SFI	170@5200	190@4000	3.50x3.31	9.6:1	15@1100
2001	Malibu	3.1 (3130)	M	SFI	170@5200	190@4000	3.50x3.31	9.6:1	15@1100

SFI: Multi-point Fuel Injection

93461CR3

ENGINE TUNE-UP SPECIFICATIONS

Year	Engine Displacement Liters (cc)	Engine ID/VIN	Spark Plug Gap (in.)	Ignition Timing (deg.) MT	Ignition Timing (deg.) AT	Fuel Pump (psi)	Idle Speed (rpm) MT	Idle Speed (rpm) AT	Valve Clearance In.	Valve Clearance Ex.
1998	2.4 (2392)	T	0.035	①	①	41-47	①	①	HYD	HYD
	3.1 (3130)	M	0.060	①	①	41-47	①	①	HYD	HYD
1999	2.4 (2392)	T	0.035	①	①	41-47	①	①	HYD	HYD
	3.1 (3130)	M	0.060	①	①	41-47	①	①	HYD	HYD
2000	3.1 (3130)	M	0.060	①	①	41-47	①	①	HYD	HYD
2001	3.1 (3130)	M	0.060	①	①	41-47	①	①	HYD	HYD

NOTE: The Vehicle Emission Control Information label often reflects specification changes made during production. The label figures must be used if they differ from those in this chart.

HYD: Hydraulic

① Refer to Vehicle Emission Control Information label

93461CR4

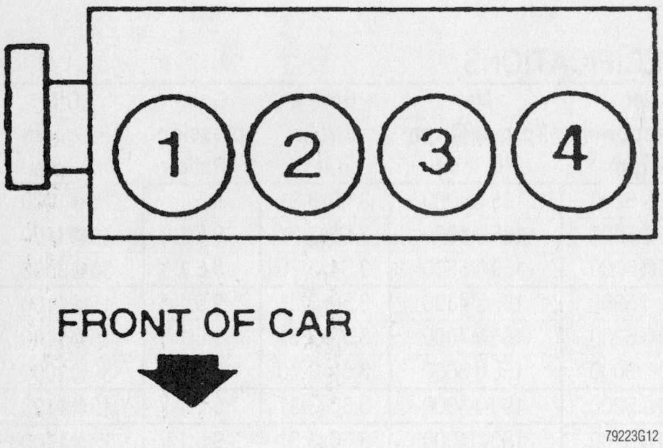

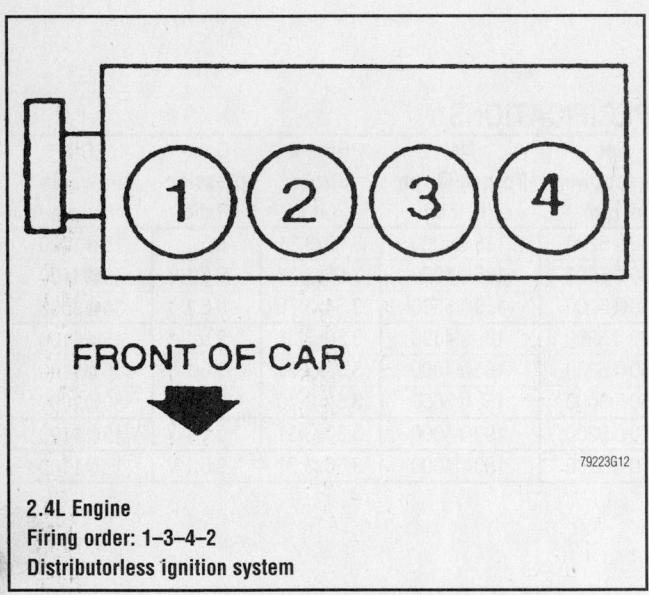

2.4L Engine
Firing order: 1–3–4–2
Distributorless ignition system

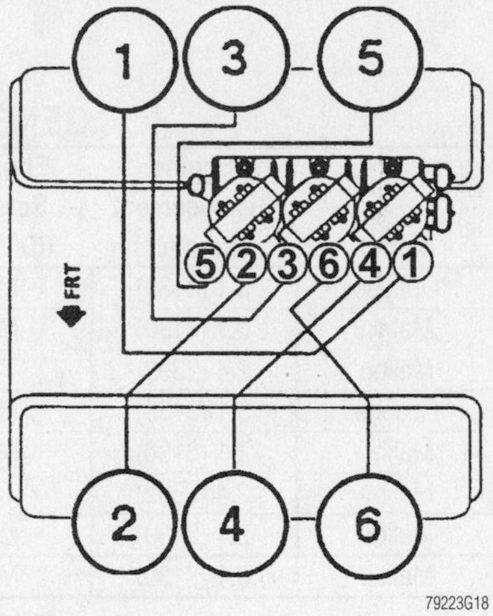

3.1L Engine
Firing order: 1–2–3–4–5–6
Distributorless ignition system

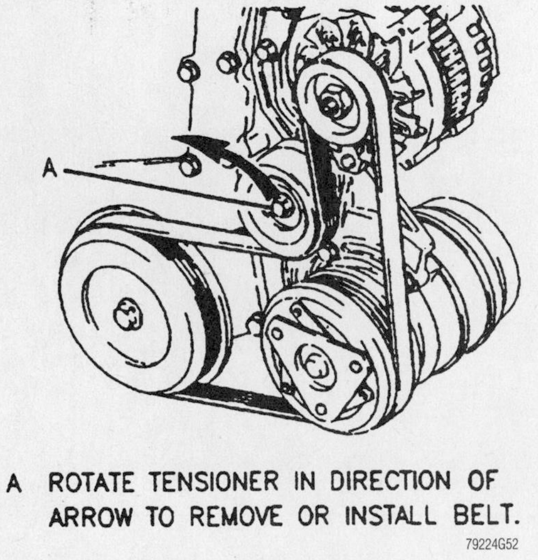

Serpentine drive belt routing—GM 2.4L engine

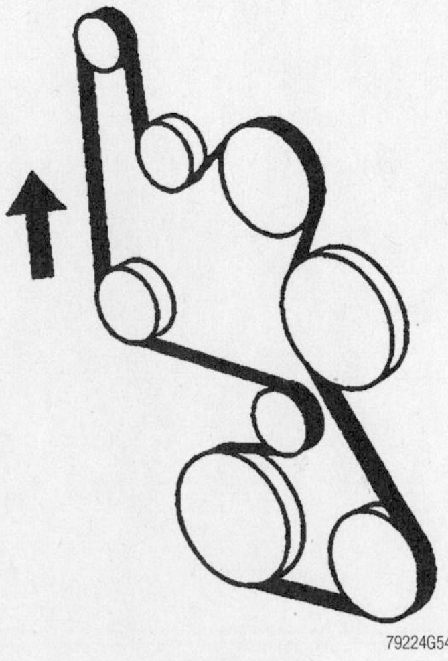

Serpentine drive belt routing—GM 3.1L engine L/N bodies

CAPACITIES

Year	Model	Engine Displacement Liters (cc)	Engine ID/VIN	Engine Oil with Filter (qts.)	Transmission (pts.)	Fuel Tank (gal.)	Cooling System (qts.)
1998	Cutlass	3.1 (3130)	M	4.0	14.8	15.2	13.6
	Malibu	3.1 (3130)	M	4.0	14.8	15.2	13.6
	Malibu	2.4 (2392)	T	4.0	14.8	15.2	11.3
1999	Cutlass	3.1 (3130)	M	4.0	14.8	15.2	13.6
	Malibu	3.1 (3130)	M	4.0	14.8	15.2	13.6
	Malibu	2.4 (2392)	T	4.0	14.8	15.2	11.3
2000	Malibu	3.1 (3130)	M	4.5	13.8	15.0	13.6
2001	Malibu	3.1 (3130)	M	4.5	13.8	15.0	13.6

NOTE: All capacities are approximate. Add fluid gradually and ensure a proper fluid level is obtained.

93461CR5

VALVE SPECIFICATIONS

Year	Engine Displacement Liters (cc)	Engine ID/VIN	Seat Angle (deg.)	Face Angle (deg.)	Spring Test Pressure (lbs. @ in.)	Spring Installed Height (in.)	Stem-to-Guide Clearance (in.)		Stem Diameter (in.)	
							Intake	Exhaust	Intake	Exhaust
1998	2.4 (2392)	T	45	46	50-55@ 1.437	1.437	0.0009-0.0025	0.0016-0.0032	0.2331-0.2339	0.2326-0.2334
	3.1 (3130)	M	45	45	230@1.260	1.701	0.0010-0.0027	0.0010-0.0027	NA	NA
1999	2.4 (2392)	T	45	46	50-55@ 1.437	1.437	0.0009-0.0025	0.0016-0.0032	0.2331-0.2339	0.2326-0.2334
	3.1 (3130)	M	45	45	230@1.260	1.701	0.0010-0.0027	0.0010-0.0027	NA	NA
2000	3.1 (3130)	M	45	45	230@1.260	1.701	0.0010-0.0027	0.0010-0.0027	NA	NA
2001	3.1 (3130)	M	45	45	230@1.260	1.701	0.0010-0.0027	0.0010-0.0027	NA	NA

NA: Not Available

93461CR6

CRANKSHAFT AND CONNECTING ROD SPECIFICATIONS

All measurements are given in inches.

Year	Engine Displacement Liters (cc)	Engine ID/VIN	Crankshaft Main Brg. Journal Dia.	Crankshaft Main Brg. Oil Clearance	Crankshaft Shaft End-play	Crankshaft Thrust on No.	Connecting Rod Journal Diameter	Connecting Rod Oil Clearance	Connecting Rod Side Clearance
1998	2.4 (2392)	T	2.3622-2.3631	0.0004-0.0023	0.0034-0.0095	3	1.8887-1.8897	0.0004-0.0026	0.0059-0.0177
	3.1 (3130)	M	2.6473-2.6383	0.0008- ① 0.0025	0.0024-0.0083	3	1.9987-1.9994	0.0007-0.0024	0.0070-0.0170
1999	2.4 (2392)	T	2.3622-2.3631	0.0004-0.0023	0.0034-0.0095	3	1.8887-1.8897	0.0004-0.0026	0.0059-0.0177
	3.1 (3130)	M	2.6473-2.6383	0.0008- ① 0.0025	0.0024-0.0083	3	1.9987-1.9994	0.0007-0.0024	0.0070-0.0170
2000	3.1 (3130)	M	2.6473-2.6383	0.0008- ① 0.0025	0.0024-0.0083	3	1.9987-1.9994	0.0007-0.0024	0.0070-0.0170
2001	3.1 (3130)	M	2.6473-2.6383	0.0008- ① 0.0025	0.0024-0.0083	3	1.9987-1.9994	0.0007-0.0024	0.0070-0.0170

① Thrust bearing: 0.0012 - 0.0030

93461CR7

Timing belt service is covered in Section 3 of this manual

PISTON AND RING SPECIFICATIONS

All measurements are given in inches.

Year	Engine Displacement Liters (cc)	Engine ID/VIN	Piston Clearance	Ring Gap Top Compression	Ring Gap Bottom Compression	Ring Gap Oil Control	Ring Side Clearance Top Compression	Ring Side Clearance Bottom Compression	Ring Side Clearance Oil Control
1998	2.4 (2392)	T	0.0006-0.0015	0.006-0.012	0.010-0.016	0.010-0.030	0.0016-0.0031	0.0012-0.0028	0.0005-0.0089
	3.1 (3130)	M	0.0013-0.0027	0.006-0.014	0.020-0.028	0.010-0.050	0.0020-0.0033	0.0020-0.0035	0.0080
1999	2.4 (2392)	T	0.0006-0.0015	0.006-0.012	0.010-0.016	0.010-0.030	0.0016-0.0031	0.0012-0.0028	0.0005-0.0089
	3.1 (3130)	M	0.0013-0.0027	0.006-0.014	0.020-0.028	0.010-0.050	0.0020-0.0033	0.0020-0.0035	0.0080
2000	3.1 (3130)	M	0.0013-0.0027	0.006-0.014	0.020-0.028	0.010-0.050	0.0020-0.0033	0.0020-0.0035	0.0080
2001	3.1 (3130)	M	0.0013-0.0027	0.006-0.014	0.020-0.028	0.010-0.050	0.0020-0.0033	0.0020-0.0035	0.0080

93461CR8

TORQUE SPECIFICATIONS

All readings in ft. lbs.

Year	Engine Displacement Liters (cc)	Engine ID/VIN	Cylinder Head Bolts	Main Bearing Bolts	Rod Bearing Bolts	Crankshaft Damper Bolts	Flywheel Bolts	Manifold Intake	Manifold Exhaust	Spark Plug	Lug Nut
1998	2.4 (2392)	T	①	②	③	④	⑤	19	11	13	100
	3.1 (3130)	M	⑥	⑦	⑧	76	52	⑨	12	⑩	100
1999	2.4 (2392)	T	①	②	③	④	⑤	19	11	13	100
	3.1 (3130)	M	⑥	⑦	⑧	76	52	⑨	12	⑩	100
2000	3.1 (3130)	M	⑥	⑦	⑧	76	52	⑨	12	⑩	100
2001	3.1 (3130)	M	⑥	⑦	⑧	76	52	⑨	12	⑩	100

NA: Not Available

① Bolts 1-8: 40 ft. lbs.
Bolts 9-10: 30 ft. lbs.
Tighten all bolts an additional 90 degrees

② 15 ft. lbs. plus 90 degrees

③ 18 ft. lbs. plus 80 degrees

④ 129 ft. lbs. plus 90 degrees

⑤ 22 ft. lbs. plus 45 degrees

⑥ Coat threads with sealer and torque to 37 ft. lbs., then turn 1/4 turn (90 degrees)

⑦ 37 ft. lbs. plus 75 degrees

⑧ 15 ft. lbs. plus 75 degrees

⑨ Lower intake manifold bolts: 10 ft. lbs.
Upper intake manifold bolts: 18 ft. lbs.

⑩ New cylinder first-time installation: 20 ft. lbs.
All others: 11 ft. lbs.

03461CR9

Heater Core replacement is covered in Section 2 of this manual

BRAKE SPECIFICATIONS

All measurements in inches unless noted

					Brake Drum Diameter			Minimum Lining Thickness		Brake Caliper	
		Brake Disc			Original	Max.	Maximum			Bracket	Mounting
		Original	Minimum	Maximum	Inside	Wear	Machine			Bolts	Bolts
Year	Model	Thickness	Thickness	Runout	Diameter	Limit	Diameter	Front	Rear	(ft. lbs.)	(ft. lbs.)
1998	Cutlass	1.031	0.987	0.003	8.900-8.909	8.920	8.909	0.030	①	85	23
	Malibu	1.031	0.987	0.003	8.900-8.909	8.920	8.909	0.030	①	85	23
1999	Cutlass	1.031	0.987	0.003	8.900-8.909	8.920	8.909	0.030	①	85	23
	Malibu	1.031	0.987	0.003	8.900-8.909	8.920	8.909	0.030	①	85	23
2000	Malibu	1.031	0.980	0.003	8.900-8.909	8.920	8.909	0.030	①	85	23
2001	Malibu	1.031	0.980	0.003	8.900-8.909	8.920	8.909	0.030	①	85	23

① 0.030 over rivet head; If bonded lining, use 0.062 from shoe

93461CR0

WHEEL ALIGNMENT

Year	Model		Caster Range (+/-Deg.)	Caster Preferred Setting (Deg.)	Camber Range (+/-Deg.)	Camber Preferred Setting (Deg.)	Toe-in (in.)	Steering Axis Inclination (Deg.)
1998	Malibu	F	+1.00	+4.10	+0.80	-0.20	0.06 +/- 0.12	—
		R	—	—	+0.50	-0.20	0.03 +/- 0.05	—
	Cutlass	F	+1.00	+4.10	+1.00	-0.19	0.50 +/- 0.12	+13.13
		R	—	—	+0.50	-0.19	0.03 +/- 0.04	—
1999	Malibu	F	+1.00	+4.10	+0.80	-0.20	0.06 +/- 0.12	—
		R	—	—	+0.50	-0.20	0.03 +/- 0.05	—
	Cutlass	F	+1.00	+4.10	+1.00	-0.19	0.50 +/- 0.12	+13.13
		R	—	—	+0.50	-0.19	0.03 +/- 0.04	—
2000	Malibu	F	+1.00	+4.10	+0.80	-0.20	0.06 +/- 0.12	—
		R	—	—	+0.50	-0.20	0.03 +/- 0.05	—
	Cutlass	F	+1.00	+4.10	+1.00	-0.19	0.50 +/- 0.12	+13.13
		R	—	—	+0.50	-0.19	0.03 +/- 0.04	—
2001	Malibu	F	+1.00	+4.10	+0.80	-0.20	0.06 +/- 0.12	—
		R	—	—	+0.50	-0.20	0.03 +/- 0.05	—
	Cutlass	F	+1.00	+4.10	+1.00	-0.19	0.50 +/- 0.12	+13.13
		R	—	—	+0.50	-0.19	0.03 +/- 0.04	—

F: Front

R: Rear

93461CS1

TIRE, WHEEL AND BALL JOINT SPECIFICATIONS

Year	Model	OEM Tires Standard	OEM Tires Optional	Tire Pressures (psi) Front	Tire Pressures (psi) Rear	Wheel Size	Ball Joint Inspection
1998	Cutlass	P215/60R15	None	30	26	6.5-JJ	①
	Malibu	P215/60TR15	P215/60HR15	30	26	6-JJ	①
1999	Cutlass	P215/60R15	None	30	26	6.5-JJ	①
	Malibu	P215/60TR15	P215/60HR15	30	26	6-JJ	①
2000	Malibu	P215/60TR15	P215/60HR15	30	26	6-JJ	①
2001	Malibu	P215/60TR15	P215/60HR15	30	26	6-JJ	①

OEM: Original Equipment Manufacturer

PSI: Pounds Per Square Inch

① Replace if any measurable movement is found.

93461CS2

SCHEDULED MAINTENANCE INTERVALS
GM L/N BODY—CHEVROLET MALIBU & OLDSMOBILE CUTLASS

TO BE SERVICED	TYPE OF SERVICE	VEHICLE MILEAGE INTERVAL (x1000)																	
		7.5	15	22.5	30	37.5	45	52.5	60	67.5	75	82.5	90	97.5	100	105	113	120	150
Accessory drive belt(s)	S/I								✓									✓	
Air cleaner element	R				✓				✓				✓					✓	
Cooling system ①	S/I																		✓
Cooling system hoses	S/I																		✓
Engine coolant	R																		✓
Engine oil and filter	R	✓	✓	✓	✓	✓	✓	✓	✓	✓	✓	✓	✓	✓		✓	✓	✓	
Fuel tank, cap & lines	S/I				✓				✓				✓					✓	
Radiator, condenser, pressure cap and neck	C																		✓
Spark plug wires	S/I														✓				
Spark plugs	R														✓				
Tires and wheels ②	S/I	✓	✓	✓	✓	✓	✓	✓	✓	✓	✓	✓	✓	✓		✓	✓	✓	

R: Replace

S/I: Service or Inspect

C: Clean

① Pressure test the cooling system and pressure cap.

② This includes rotating the tires.

FREQUENT OPERATION MAINTENANCE (SEVERE SERVICE) ADDITIONS

If a vehicle is operated under any of the following conditions it is considered severe service:

- Towing a trailer or using a camper or car-top carrier.

- Extensive idling or low-speed driving for long distances as in heavy commercial use, such as delivery, taxi or police cars.

- Operating on rough, muddy or salt-covered roads.

- Operating on unpaved or dusty roads.

- 50% or more of the vehicle operation is in 32°C (90°F) or higher temperatures, or constant operation in temperatures below 0°C (32°F).

Engine oil and filter: change every 3000 miles or 3 months, whichever occurs first.

Wheels and tires: inspect and rotate every 6000 miles.

Air cleaner element: inspect every 15,000 miles and replace or clean as needed. Replace it at least every 30,000 miles.

Automatic transaxle fluid & filter: replace every 50,000 miles.

93461CS3

SCHEDULED MAINTENANCE INTERVALS
GENERAL MOTORS CORPORATION
L/N BODY
CHEVROLET MALIBU
OLDSMOBILE CUTLASS

The following should be used as a guide when determining the amount of work required for a particular service. In estimating how long a particular Scheduled Maintenance Service should take, please observe the following:

- Labor Time is time based on field research and data supplied by the vehicle manufacturer.
- Labor time operations are given in hours and tenths of an hour.
- All labor operations are to be used as a guide.

Mechanic Skill Level Codes:
(A) PRECISION: Highly skilled with multiple certification.
(B) GENERAL: Normally skilled with certification.
(C) MAINTENANCE: Semi-skilled working on certification.

	LABOR TIME
8000 Mile Service (C)	
All Models	1.0
15000 Mile Service (C)	
All Models	1.0
22500 Mile Service (C)	
All Models	1.0
30000 Mile Service (C)	
All Models	1.3
37500 Mile Service (C)	
All Models	1.0
45000 Mile Service (C)	
All Models	1.0
52500 Mile Service (C)	
All Models	1.0
60000 Mile Service (C)	
All Models	1.4
67500 Mile Service (C)	
All Models	1.0
75000 Mile Service (C)	
All Models	1.0
82500 Mile Service (C)	
All Models	1.0
90000 Mile Service (C)	
All Models	1.3
97500 Mile Service (C)	
All Models	1.0
100000 Mile Service (C)	
4 Cyl.	.8
V6	1.0
105000 Mile Service (C)	
All Models	1.0
113000 Mile Service (C)	
All Models	1.0
120000 Mile Service (C)	
All Models	1.4
150000 Mile Service (C)	
All Models	.8

93461CS4

For complete Engine Mechanical specifications, see Section 1 of this manual

GENERAL MOTORS N-BODY

Buick Skylark • Oldsmobile Achieva • Alero • Pontiac Grand Am

ENGINE AND VEHICLE IDENTIFICATION

Engine						
Code ①	Liters (cc)	Cu. In.	Cyl.	Fuel Sys.	Engine Type	Eng. Mfg.
M	3.1 (3130)	191	6	SFI	OHV	BOC
T	2.4 (2392)	146	4	SFI	DOHC	CUS

Model Year	
Code ②	Year
W	1998
X	1999
Y	2000
1	2001
2	2002

BOC: Buick/Oldsmobile/Cadillac

CUS: Chevrolet/United States

SFI: Sequential Fuel Injection

① 8th position of VIN

② 10th position of VIN

93461CS5

GENERAL ENGINE SPECIFICATIONS

Year	Model	Engine Displacement Liters (cc)	Engine Series (ID/VIN)	Fuel System	Net Horsepower @ rpm	Net Torque @ rpm (ft. lbs.)	Bore x Stroke (in.)	Com-pression Ratio	Oil Pressure @ rpm
1998	Cutlass	3.1 (3130)	M	SFI	160@5200	185@4000	3.50x3.31	9.5:1	15@1100
	Malibu	3.1 (3130)	M	SFI	160@5200	185@4000	3.50x3.31	9.5:1	15@1100
	Malibu	2.4 (2392)	T	SFI	150@6000	150@5600	3.54x3.70	9.5:1	30@3000
1999	Cutlass	3.1 (3130)	M	SFI	160@5200	185@4000	3.50x3.31	9.5:1	15@1100
	Malibu	3.1 (3130)	M	SFI	160@5200	185@4000	3.50x3.31	9.5:1	15@1100
	Malibu	2.4 (2392)	T	SFI	150@6000	150@5600	3.54x3.70	9.5:1	30@3000
2000	Malibu	3.1 (3130)	M	SFI	170@5200	190@4000	3.50x3.31	9.6:1	15@1100
2001	Malibu	3.1 (3130)	M	SFI	170@5200	190@4000	3.50x3.31	9.6:1	15@1100

SFI: Multi-point Fuel Injection

93461CS6

ENGINE TUNE-UP SPECIFICATIONS

Year	Engine Displacement Liters (cc)	Engine ID/VIN	Spark Plugs Gap (in.)	Ignition Timing (deg.)		Fuel Pump (psi)	Idle Speed (rpm)		Valve Clearance	
				MT	AT		MT	AT	In.	Ex.
1998	2.4 (2392)	T	0.060	①	①	41-47	①	①	HYD	HYD
	3.1 (3130)	M	0.060	①	①	41-47	①	①	HYD	HYD
1999	2.4 (2392)	T	0.060	①	①	41-47	①	①	HYD	HYD
	3.1 (3130)	M	0.060	①	①	41-47	①	①	HYD	HYD
	3.4 (3934)	E	0.060	①	①	41-47	①	①	HYD	HYD
2000	2.4 (2392)	T	0.050	①	①	41-47	①	①	HYD	HYD
	3.4 (3934)	E	0.060	①	①	41-47	①	①	HYD	HYD
2001	2.4 (2392)	T	0.050	①	①	41-47	①	①	HYD	HYD
	3.4 (3934)	E	0.060	①	①	41-47	①	①	HYD	HYD

NOTE: The Vehicle Emission Control Information label often reflects specification changes made during production. The label figures must be used if they differ from those in this chart.

HYD: Hydraulic

① Refer to Vehicle Emission Control Information label

93461CS7

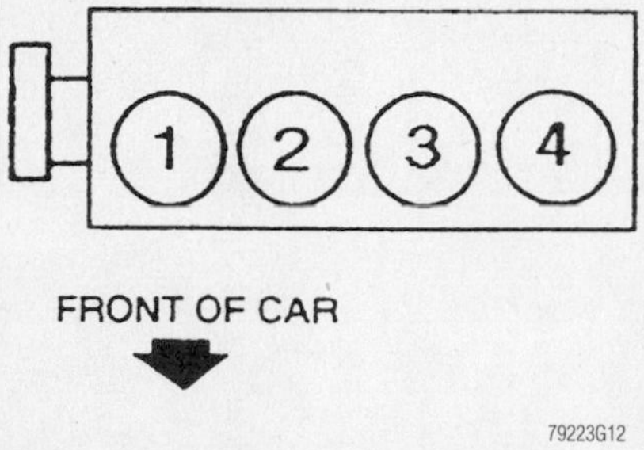

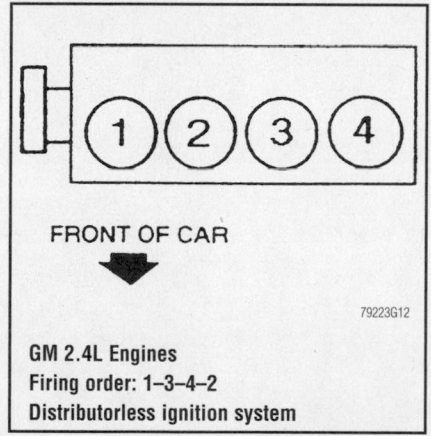

GM 2.4L Engines
Firing order: 1–3–4–2
Distributorless ignition system

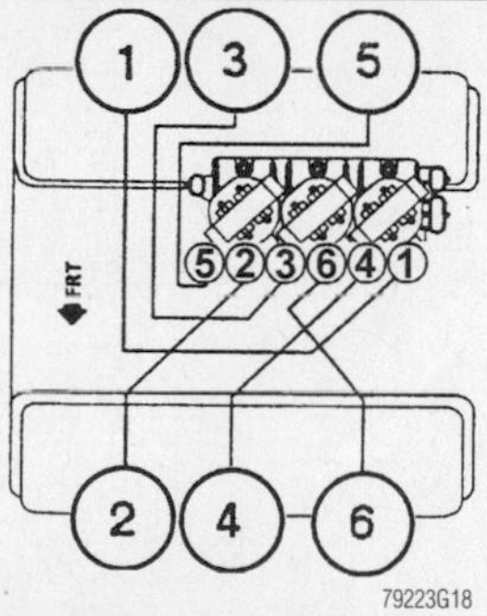

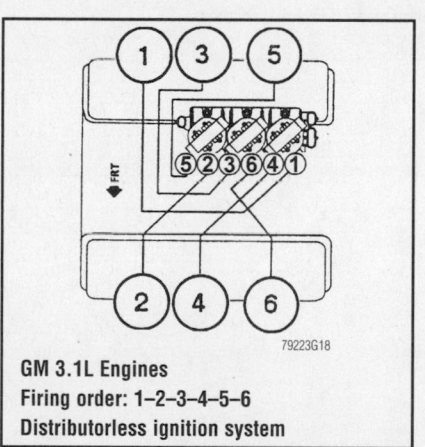

GM 3.1L Engines
Firing order: 1–2–3–4–5–6
Distributorless ignition system

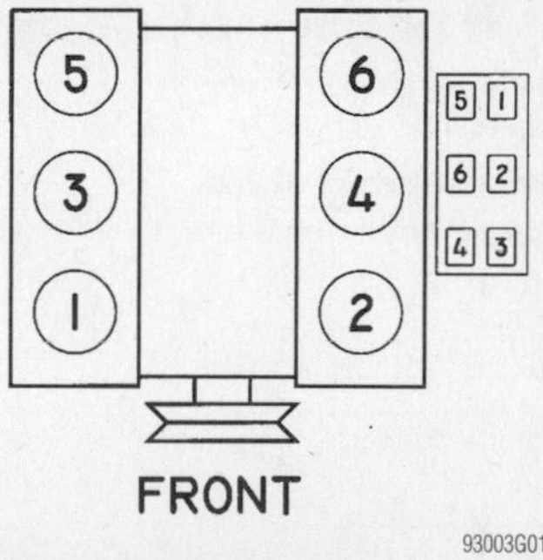

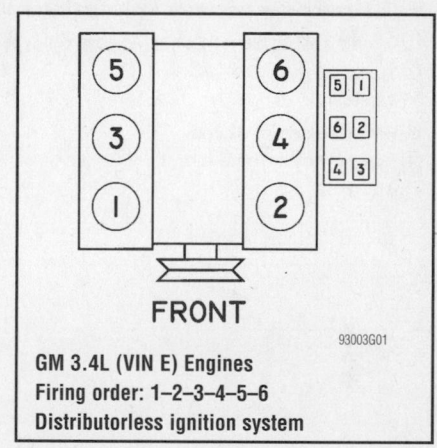

GM 3.4L (VIN E) Engines
Firing order: 1–2–3–4–5–6
Distributorless ignition system

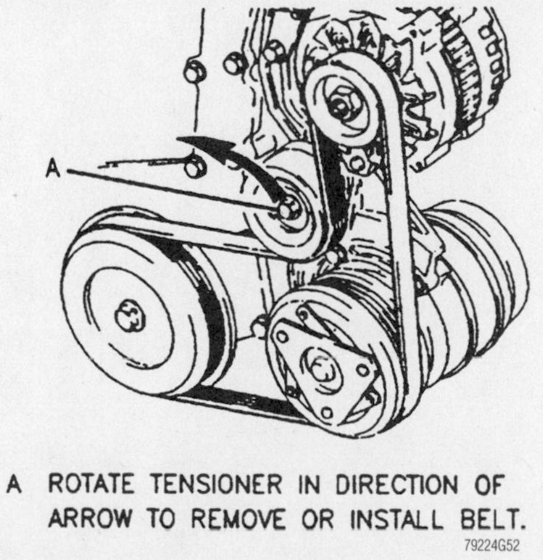

Serpentine drive belt routing—GM 2.4L engine

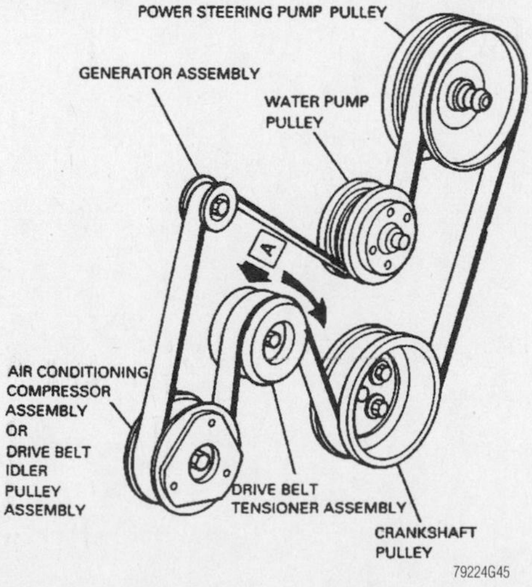

Serpentine drive belt routing—GM 3.4L (VIN E) engine

For Accessory Drive Belt illustrations, see Section 1 of this manual

CAPACITIES

Year	Model	Engine Displacement Liters (cc)	Engine ID/VIN	Engine Oil with Filter (qts.)	Transmission (pts.) Manual	Transmission (pts.) Auto.	Fuel Tank (gal.)	Cooling System (qts.)
1998	Achieva	2.4 (2392)	T	4.0 ①	4.0	②	15.2	10.4
	Achieva	3.1 (3130)	M	4.0 ①	4.0	④	15.2	10.8
	Grand Am	2.4 (2392)	T	4.5	4.0	8.0 ③	15.2	10.4
	Grand Am	3.1 (3130)	M	4.5	4.0	8.0 ③	15.2	13.1
1999	Alero	2.4 (2392)	T	4.0 ①	4.0	②	14.3	11.3
	Alero	3.4 (3934)	E	4.0 ①	4.0	②	14.3	13.6
	Grand Am	2.4 (2392)	T	4.5	4.0	8.0 ③	15.2	10.4
	Grand Am	3.1 (3130)	M	4.5	4.0	8.0 ③	15.2	11.3
	Grand Am	3.4 (3934)	E	4.0 ①	4.0	②	15.2	13.6
2000	Alero	2.4 (2392)	T	4.0 ①	4.0	②	14.3	11.3
	Alero	3.4 (3934)	E	4.0 ①	4.0	②	14.3	13.6
	Grand Am	2.4 (2392)	T	4.5	4.0	8.0 ③	15.2	11.3
	Grand Am	3.4 (3934)	E	4.0 ①	4.0	②	15.2	13.6
2001	Alero	2.4 (2392)	T	4.0 ①	4.0	②	14.3	11.3
	Alero	3.4 (3934)	E	4.0 ①	4.0	②	14.3	13.6
	Grand Am	2.4 (2392)	T	4.5	4.0	8.0 ③	15.2	11.3
	Grand Am	3.4 (3934)	E	4.0 ①	4.0	②	15.2	13.6

NOTE: All capacities are approximate. Add fluid gradually and ensure a proper fluid level is obtained.

① Capacity is without filter replacement; Additional oil may be required

② 3 speed: 8.0
4 speed: 12

③ With 4T60E transaxle: 12.0 pts.

④ 3 speed: 8.0
4 speed: 12

93461CS8

VALVE SPECIFICATIONS

Year	Engine Displacement Liters (cc)	Engine ID/VIN	Seat Angle (deg.)	Face Angle (deg.)	Spring Test Pressure (lbs. @ in.)	Spring Installed Height (in.)	Stem-to-Guide Clearance (in.)		Stem Diameter (in.)	
							Intake	Exhaust	Intake	Exhaust
1998	2.4 (2392)	T	45	46	50-55@ 1.437	1.437	0.0009-0.0025	0.0016-0.0032	0.2331-0.2339	0.2326-0.2334
	3.1 (3130)	M	45	45	250@ 1.239	1.710	0.0010-0.0027	0.0010-0.0027	NA	NA
1999	2.4 (2392)	T	45	46	50-55@ 1.437	1.437	0.0009-0.0025	0.0016-0.0032	0.2331-0.2339	0.2326-0.2334
	3.1 (3130)	M	45	45	250@ 1.239	1.710	0.0010-0.0027	0.0010-0.0027	NA	NA
	3.4 (3934)	E	45	45	75@ 1.701	1.701	0.0010-0.0027	0.0010-0.0027	NA	NA
2000	2.4 (2392)	T	45	46	50-55@ 1.437	1.437	0.0009-0.0025	0.0016-0.0032	0.2331-0.2339	0.2326-0.2334
	3.4 (3934)	E	45	45	75@ 1.701	1.701	0.0010-0.0027	0.0010-0.0027	NA	NA
2001	2.4 (2392)	T	45	46	50-55@ 1.437	1.437	0.0009-0.0025	0.0016-0.0032	0.2331-0.2339	0.2326-0.2334
	3.4 (3934)	E	45	45	75@ 1.701	1.701	0.0010-0.0027	0.0010-0.0027	NA	NA

NA: Not available

93461CS9

CRANKSHAFT AND CONNECTING ROD SPECIFICATIONS

All measurements are given in inches.

	Engine		Crankshaft				Connecting Rod		
Year	Displacement Liters (cc)	Engine ID/VIN	Main Brg. Journal Dia.	Main Brg. Oil Clearance	Shaft End-play	Thrust on No.	Journal Diameter	Oil Clearance	Side Clearance
1998	2.4 (2392)	T	2.3622-2.3631	0.0004-0.0023	0.0034-0.0095	3	1.8887-1.8897	0.0004-0.0026	0.0059-0.0177
	3.1 (3130)	M	2.6473-2.6483	0.0008-0.0025 ①	0.0024-0.0083	3	1.9987-1.9994	0.0007-0.0024	0.0070-0.0170
1999	2.4 (2392)	T	2.3622-2.3631	0.0004-0.0023	0.0034-0.0095	3	1.8887-1.8897	0.0004-0.0026	0.0059-0.0177
	3.1 (3130)	M	2.6473-2.6483	0.0008-0.0025 ①	0.0024-0.0083	3	1.9987-1.9994	0.0007-0.0024	0.0070-0.0170
	3.4 (3934)	E	2.6473-2.6483	0.0008-0.0025 ①	0.0024-0.0083	3	1.9987-1.9994	0.0007-0.0024	0.0070-0.0170
2000	2.4 (2392)	T	2.3622-2.3631	0.0004-0.0023	0.0034-0.0095	3	1.8887-1.8897	0.0004-0.0026	0.0059-0.0177
	3.4 (3934)	E	2.6473-2.6483	0.0008-0.0025 ①	0.0024-0.0083	3	1.9987-1.9994	0.0007-0.0024	0.0070-0.0170
2001	2.4 (2392)	T	2.3622-2.3631	0.0004-0.0023	0.0034-0.0095	3	1.8887-1.8897	0.0004-0.0026	0.0059-0.0177
	3.4 (3934)	E	2.6473-2.6483	0.0008-0.0025 ①	0.0024-0.0083	3	1.9987-1.9994	0.0007-0.0024	0.0070-0.0170

① Thrust bearing: 0.0012 - 0.0030

93461CS0

PISTON AND RING SPECIFICATIONS

All measurements are given in inches.

Year	Engine Displacement Liters (cc)	Engine ID/VIN	Piston Clearance	Ring Gap Top Compression	Ring Gap Bottom Compression	Ring Gap Oil Control	Ring Side Clearance Top Compression	Ring Side Clearance Bottom Compression	Ring Side Clearance Oil Control
1998	2.4 (2392)	T	0.0006-0.0015	0.006-0.012	0.010-0.016	0.010-0.030	0.0016-0.0031	0.0012-0.0028	0.0005-0.0089
	3.1 (3130)	M	0.0013-0.0027	0.006-0.014	0.0197-0.0280	0.0098-0.0500	0.0020-0.0033	0.0020-0.0035	0.0080
1999	2.4 (2392)	T	0.0006-0.0015	0.006-0.012	0.010-0.016	0.010-0.030	0.0016-0.0031	0.0012-0.0028	0.0005-0.0089
	3.1 (3130)	M	0.0013-0.0027	0.006-0.014	0.0197-0.0280	0.0098-0.0500	0.0020-0.0033	0.0020-0.0035	0.0080
	3.4 (3934)	E	0.0013-0.0027	0.006-0.014	0.0197-0.0280	0.0098-0.0500	0.0020-0.0033	0.0020-0.0035	0.0080
2000	2.4 (2392)	T	0.0006-0.0015	0.006-0.012	0.010-0.016	0.010-0.030	0.0016-0.0031	0.0012-0.0028	0.0005-0.0089
	3.4 (3934)	E	0.0013-0.0027	0.006-0.014	0.0197-0.0280	0.0098-0.0500	0.0020-0.0033	0.0020-0.0035	0.0080
2001	2.4 (2392)	T	0.0006-0.0015	0.006-0.012	0.010-0.016	0.010-0.030	0.0016-0.0031	0.0012-0.0028	0.0005-0.0089
	3.4 (3934)	E	0.0013-0.0027	0.006-0.014	0.0197-0.0280	0.0098-0.0500	0.0020-0.0033	0.0020-0.0035	0.0080

93461CT1

For Wheel Alignment specifications, see Section 1 of this manual

TORQUE SPECIFICATIONS

All readings in ft. lbs.

Year	Engine Displacement Liters (cc)	Engine ID/VIN	Cylinder Head Bolts	Main Bearing Bolts	Rod Bearing Bolts	Crankshaft Damper Bolts	Flywheel Bolts	Manifold Intake	Manifold Exhaust	Spark Plug	Lug Nut
1998	2.4 (2392)	T	①	②	③	④	⑤	⑥	⑦	13	100
	3.1 (3130)	M	⑧	⑨	⑩	76	52	⑪	12	20	100
1999	2.4 (2392)	T	①	②	③	④	⑤	⑥	⑦	13	100
	3.4 (3934)	E	⑫	⑬	⑩	76	52	⑪	12	20	100
2000	2.4 (2392)	T	①	②	③	④	⑤	⑥	⑦	13	100
	3.4 (3934)	E	⑫	⑬	⑩	76	52	⑪	12	20	100
2001	2.4 (2392)	T	①	②	③	④	⑤	⑥	⑦	13	100
	3.4 (3934)	E	⑫	⑬	⑩	76	52	⑪	12	20	100

① Nos. 1-8: 40 ft. lbs.
Nos. 9-10: 30 ft. lbs.
Tighten all bolts an additional 90 degrees

② 15 ft. lbs. plus 90 degrees

③ 18 ft. lbs. plus 80 degrees

④ 129 ft. lbs. plus 90 degrees

⑤ 22 ft. lbs. plus 45 degrees

⑥ Nuts: 19 ft. lbs.
Studs: 97 inch lbs.

⑦ Nuts: 11 ft. lbs.
Studs: 97 inch lbs.

⑧ Coat threads with sealer torque to 33 ft. lbs., then turn 1/4 turn (90 degrees)

⑨ 37 ft. lbs. plus 77 degrees

⑩ 15 ft. lbs. plus 75 degrees

⑪ Lower: 115 inch lbs.
Upper: 18 ft. lbs.

⑫ 37 ft. lbs. plus 90 degrees

⑬ 37 ft. lbs. plus 77 degrees

93461CT2

BRAKE SPECIFICATIONS

All measurements in inches unless noted

			Brake Disc			Brake Drum Diameter			Minimum Lining Thickness		Brake Caliper	
Year	Model		Original Thickness	Minimum Thickness	Maximum Runout	Original Inside Diameter	Max. Wear Limit	Maximum Machine Diameter	Front	Rear	Bracket Bolts (ft. lbs.)	Mounting Bolts (ft. lbs.)
1998	Achieva	F	0.806	0.736	0.003	—	—	—	0.030	—	—	38
		R	—	—	—	7.874-7.890	7.930	7.899	—	①	—	—
	Skylark	F	0.806	0.736	0.003	—	—	—	0.030	—	—	38
		R	—	—	—	7.874-7.890	7.930	7.899	—	①	—	—
	Grand Am	F	0.806	0.736	0.003	—	—	—	0.030	—	—	38
		R	—	—	—	7.874-7.890	7.930	7.899	—	①	—	—
1999	Alero	F	1.031	0.972	0.003	—	—	—	0.030	—	85	23
		R	0.430	0.410	—	8.863	8.920	8.909	—	①	85	81
	Grand Am	F	1.031	0.972	0.003	—	—	—	0.030	—	85	23
		R	0.430	0.410	—	8.863	8.920	8.909	—	①	85	81
2000	Alero	F	1.031	0.972	0.003	—	—	—	0.030	—	85	23
		R	0.430	0.410	—	8.863	8.920	8.909	—	①	85	81
	Grand Am	F	1.031	0.972	0.003	—	—	—	0.030	—	85	23
		R	0.430	0.410	—	8.863	8.920	8.909	—	①	85	81
2001	Alero	F	1.031	0.972	0.003	—	—	—	0.030	—	85	23
		R	0.430	0.410	—	8.863	8.920	8.909	—	①	85	81
	Grand Am	F	1.031	0.972	0.003	—	—	—	0.030	—	85	23
		R	0.430	0.410	—	8.863	8.920	8.909	—	①	85	81

NA: Not available

① 0.030 over rivet head; If bonded lining, use 0.030 from shoe

93461CT3

WHEEL ALIGNMENT

Year	Model		Caster Range (+/-Deg.)	Caster Preferred Setting (Deg.)	Camber Range (+/-Deg.)	Camber Preferred Setting (Deg.)	Toe-in (in.)	Steering Axis Inclination (Deg.)
1998	Skylark	F	1.00	+1.44	1.00	0	0 +/- 0.12	—
		R	—	—	0.75	-0.40	0.20 +/- 0.30	—
	Achieva	F	1.00	+1.75	1.00	0	0 +/- 0.13	—
		R	—	—	0.75	-0.40	0.10 +/- 0.15	—
	Grand Am	F	1.00	+1.44	1.00	0	0 +/- 0.13	13.20
		R	—	—	0.75	-0.41	0.09 +/- 0.15	—
1999	Alero	F	1.00	+4.10	1.00	-0.20	0.05 +/- 0.12	—
		R	—	—	0.50	-0.20	0.03 +/- 0.04	—
	Grand Am	F	1.00	+4.10	1.00	-0.20	0.05 +/- 0.13	—
		R	—	—	0.50	-0.20	0.03 +/- 0.10	—
2000	Alero	F	1.00	+4.10	1.00	-0.20	0.05 +/- 0.12	—
		R	—	—	0.50	-0.20	0.03 +/- 0.04	—
	Grand Am	F	1.00	+4.10	1.00	-0.20	0.05 +/- 0.13	—
		R	—	—	0.50	-0.20	0.03 +/- 0.10	—
2001	Alero	F	1.00	+4.10	1.00	-0.20	0.05 +/- 0.12	—
		R	—	—	0.50	-0.20	0.03 +/- 0.04	—
	Grand Am	F	1.00	+4.10	1.00	-0.20	0.05 +/- 0.13	—
		R	—	—	0.50	-0.20	0.03 +/- 0.10	—

93461CT4

TIRE, WHEEL AND BALL JOINT SPECIFICATIONS

Year	Model	OEM Tires		Tire Pressures (psi)		Wheel	Ball Joint
		Standard	Optional	Front	Rear	Size	Inspection
1998	Achieva	P195/70R14	P195/65R15	30	30	6-J	①
	Grand AM SE	P195/70R14	P195/65R15 P205/55R16	30	30	6-JJ	0.125 in.
	Grand Am GT, GTE	P195/70R14	P195/65R15 P205/55R16	30	30	6-JJ	0.125 in.
	Skylark, exc.GS	P195/70R14	P195/65R15	30	30	6-JJ	①
	Skylark GS	P205/55R16	None	30	30	6.5-JJ	①
1999	Alero	P215/60R15	P225/50R16	30	30	7-J	①
	Grand AM SE	P195/70R14	P195/65R15 P205/55R16	30	30	6-JJ	0.125 in.
	Grand Am GT, GTE	P195/70R14	P195/65R15 P205/55R16	30	30	6-JJ	0.125 in.
	Skylark, exc.GS	P195/70R14	P195/65R15	30	30	6-JJ	①
	Skylark GS	P205/55R16	None	30	30	6.5-JJ	①
2000	Alero GX, GL	P215/60R15	P225/50HR15	30	30	6-JJ	①
	Alero GL1, GL2, GL3	P225/50VR16	None	30	30	6-JJ	①
	Alero GLS	P225/50R16	None	30	30	6-JJ	①
	Grand AM SE/SE1	P215/60R15	None	30	30	6-JJ	0.125 in.
	Grand Am SE2	P225/50R16	None	30	30	6-JJ	0.125 in.
	Grand Am GT, GT1	P225/50R16	None	30	30	6-JJ	0.125 in.
2001	Alero GX, GL	P215/60R15	P225/50HR15	30	30	6-JJ	①
	Alero GL1, GL2, GL3	P225/50VR16	None	30	30	6-JJ	①
	Alero GLS	P225/50R16	None	30	30	6-JJ	①
	Grand AM SE/SE1	P215/60R15	None	30	30	6-JJ	0.125 in.
	Grand Am SE2	P225/50R16	None	30	30	6-JJ	0.125 in.
	Grand Am GT, GT1	P225/50R16	None	30	30	6-JJ	0.125 in.

OEM: Original Equipment Manufacturer
PSI: Pounds Per Square Inch
STD: Standard
OPT: Optional
① Replace if any measurable movement is found.

93461CT5

SCHEDULED MAINTENANCE INTERVALS

GM N BODY—BUICK SKYLARK, OLDSMOBILE ACHIEVA, ALERO AND PONTIAC GRAND AM

TO BE SERVICED	TYPE OF SERVICE	VEHICLE MILEAGE INTERVAL (x1000)												
		7.5	15	22.5	30	37.5	45	52.5	60	67.5	75	82.5	90	97.5
Engine oil & filter	R	✓	✓	✓	✓	✓	✓	✓	✓	✓	✓	✓	✓	✓
Automatic transaxle fluid & filter ①	S/I	✓	✓	✓	✓	✓	✓	✓	✓	✓	✓	✓	✓	✓
Brake hoses	S/I	✓	✓	✓	✓	✓	✓	✓	✓	✓	✓	✓	✓	✓
Chassis lubrication	S/I	✓	✓	✓	✓	✓	✓	✓	✓	✓	✓	✓	✓	✓
Coolant level, hoses & clamps	S/I	✓	✓	✓	✓	✓	✓	✓	✓	✓	✓	✓	✓	✓
Driveshaft boots & front suspension components	S/I	✓	✓	✓	✓	✓	✓	✓	✓	✓	✓	✓	✓	✓
Exhaust system	S/I	✓	✓	✓	✓	✓	✓	✓	✓	✓	✓	✓	✓	✓
Lubricate chassis & suspension	S/I	✓	✓	✓	✓	✓	✓	✓	✓	✓	✓	✓	✓	✓
Lubricate steering linkage & transaxle shift linkage	S/I	✓	✓	✓	✓	✓	✓	✓	✓	✓	✓	✓	✓	✓
Lubricate parking brake cable guides, underbody contact points & linkage	S/I	✓	✓	✓	✓	✓	✓	✓	✓	✓	✓	✓	✓	✓
Manual transaxle oil	S/I	✓	✓	✓	✓	✓	✓	✓	✓	✓	✓	✓	✓	✓
Throttle linkage	S/I	✓	✓	✓	✓	✓	✓	✓	✓	✓	✓	✓	✓	✓
Brake linings	S/I	✓		✓		✓		✓		✓		✓		✓
Rotate tires	S/I	✓		✓		✓		✓		✓		✓		✓
Air filter element & PCV filter	R				✓				✓				✓	
Engine coolant ②	R													
Spark plugs ③	R				✓				✓				✓	
Accessory drive belt(s)	S/I				✓				✓				✓	
EGR & fuel systems	S/I				✓				✓				✓	
Ignition cables	S/I				✓				✓				✓	

R: Replace

S/I: Service or Inspect

① Automatic transaxle fluid & filter: replace at 100,000 miles (if not changed previously).

② Engine coolant: replace every 100,000 miles. Use O.E. specified (DEX-COOL™) coolant only. If any silicate coolant is used, the service interval is every 30,000 miles.

③ Platinum tip spark plugs: replace every 100,000 miles.

FREQUENT OPERATION MAINTENANCE (SEVERE SERVICE) ADDITIONS

If a vehicle is operated under any of the following conditions it is considered severe service:

- Towing a trailer or using a camper or car-top carrier.

- Extensive idling or low-speed driving for long distances as in heavy commercial use, such as delivery, taxi or police cars.

- Operating on rough, muddy or salt-covered roads.

- Operating on unpaved or dusty roads.

- 50% or more of the vehicle operation is in 32°C (90°F) or higher temperatures, or constant operation in temperatures below 0°C (32°F).

Engine oil and filter: change every 3000 miles or 3 months, whichever occurs first.

Wheels and tires: inspect and rotate every 6000 miles.

Air cleaner element: inspect every 15,000 miles and replace or clean as needed. Replace it at least every 30,000 miles.

Automatic transaxle fluid & filter: replace every 50,000 miles.

93461CT6

SCHEDULED MAINTENANCE INTERVALS
GENERAL MOTORS CORPORATION
N BODY
BUICK SKYLARK
OLDSMOBILE ACHIEVA, ALERO
PONTIAC GRAND AM

The following should be used as a guide when determining the amount of work required for a particular service. In estimating how long a particular Scheduled Maintenance Service should take, please observe the following:

- Labor Time is time based on field research and data supplied by the vehicle manufacturer.
- Labor time operations are given in hours and tenths of an hour.
- All labor operations are to be used as a guide.

Mechanic Skill Level Codes:
(A) PRECISION: Highly skilled with multiple certification.
(B) GENERAL: Normally skilled with certification.
(C) MAINTENANCE: Semi-skilled working on certification.

	LABOR TIME
7500 Mile Service (C)	
All Models	1.4
15000 Mile Service (C)	
All Models	.9
22500 Mile Service (C)	
All Models	1.4
30000 Mile Service (B)	
All Models	2.4
37500 Mile Service (C)	
All Models	1.4
45000 Mile Service (C)	
All Models	.9
52500 Mile Service (C)	
All Models	1.4
60000 Mile Service (B)	
All Models	2.4
67500 Mile Service (C)	
All Models	1.4
75000 Mile Service (C)	
All Models	.9
82500 Mile Service (C)	
All Models	1.4
90000 Mile Service (B)	
All Models	2.4
97500 Mile Service (C)	
All Models	1.4

93461CT7

GENERAL MOTORS V-BODY
Cadillac Catera

ENGINE AND VEHICLE IDENTIFICATION

Engine						
Code ①	Liters (cc)	Cu. In.	Cyl.	Fuel Sys.	Engine Type	Eng. Mfg.
R	3.0 (2972)	181	V6	SMFI	DOHC	Opel

Model Year	
Code ②	Year
W	1998
X	1999
Y	2000
1	2001
2	2002

SMFI: Sequential Multi-port Fuel Injection
DOHC: Double Overhead Camshaft
① 8th position of VIN
② 10th position of VIN

93461CT8

GENERAL ENGINE SPECIFICATIONS

Year	Model	Engine Displacement Liters (cc)	Engine Series (ID/VIN)	Fuel System	Net Horsepower @ rpm	Net Torque @ rpm (ft. lbs.)	Bore x Stroke (in.)	Com-pression Ratio	Oil Pressure @ rpm
1998	Catera	3.0 (2972)	R	MFI	200@6000	192@3600	3.38x3.34	10.0:1	22@900
1999	Catera	3.0 (2972)	R	MFI	200@6000	192@3600	3.38x3.34	10.0:1	22@900
2000	Catera	3.0 (2972)	R	MFI	200@6000	192@3600	3.38x3.34	10.0:1	22@900
2001	Catera	3.0 (2972)	R	MFI	200@6000	192@3600	3.38x3.34	10.0:1	22@900

MFI: Multi-point Fuel Injection

93461CT9

ENGINE TUNE-UP SPECIFICATIONS

Year	Engine Displacement Liters (cc)	Engine ID/VIN	Spark Plug Gap (in.)	Ignition Timing (deg.) AT	Fuel Pump (psi)	Idle Speed (rpm) AT	Valve Clearance In.	Valve Clearance Ex.
1998	3.0 (2972)	R	0.034 - 0.043	①	41-47	①	HYD	HYD
1999	3.0 (2972)	R	0.034 - 0.043	①	41-47	①	HYD	HYD
2000	3.0 (2972)	R	0.034 - 0.043	①	41-47	①	HYD	HYD
2001	3.0 (2972)	R	0.034 - 0.043	①	41-47	①	HYD	HYD

NOTE: The Vehicle Emission Control Information label often reflects specification changes made during production. The label figures must be used if they differ from those in this chart.

HYD: Hydraulic

① Refer to Vehicle Emission Control Information label

93461CT0

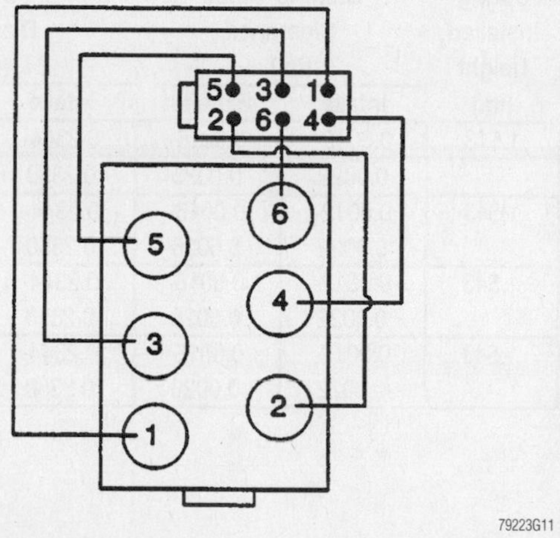

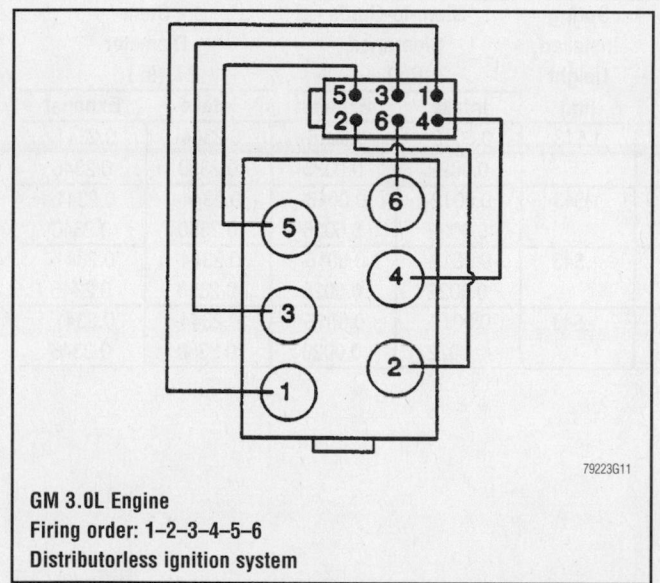

GM 3.0L Engine
Firing order: 1–2–3–4–5–6
Distributorless ignition system

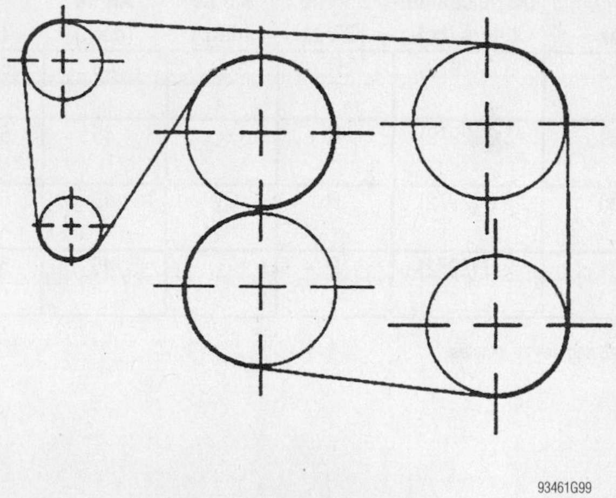

Serpentine drive belt routing—GM 3.0L engine

CAPACITIES

Year	Model	Engine Displacement Liters (cc)	Engine ID/VIN	Engine Oil with Filter (qts.)	Transmission (pts.)	Drive Axle Rear (pts.)	Fuel Tank (gal.)	Cooling System (qts.)
1998	Catera	3.0 (2972)	R	6.0	14	3.5	18	10
1999	Catera	3.0 (2972)	R	6.0	14	3.5	18	10
2000	Catera	3.0 (2972)	R	6.0	14	3.5	18	10
2001	Catera	3.0 (2972)	R	6.0	14	3.5	18	10

NOTE: All capacities are approximate. Add fluid gradually and ensure a proper fluid level is obtained.

93461CU1

VALVE SPECIFICATIONS

Year	Engine Displacement Liters (cc)	Engine ID/VIN	Seat Angle (deg.)	Face Angle (deg.)	Spring Test Pressure (lbs. @ in.)	Spring Installed Height (in.)	Stem-to-Guide Clearance (in.)		Stem Diameter (in.)	
							Intake	Exhaust	Intake	Exhaust
1998	3.0 (2972)	R	①	45	56.6@1.338	1.543	0.0012-0.0022	0.0016-0.0026	0.2344-0.2350	0.2341-0.2346
1999	3.0 (2972)	R	①	45	56.6@1.338	1.543	0.0012-0.0022	0.0016-0.0026	0.2344-0.2350	0.2341-0.2346
2000	3.0 (2972)	R	①	45	56.6@1.338	1.543	0.0012-0.0022	0.0016-0.0026	0.2344-0.2350	0.2341-0.2346
2001	3.0 (2972)	R	①	45	56.6@1.338	1.543	0.0012-0.0022	0.0016-0.0026	0.2344-0.2350	0.2341-0.2346

① 45 degrees 20 minutes

93461CU2

CRANKSHAFT AND CONNECTING ROD SPECIFICATIONS

All measurements are given in inches.

Year	Engine Displacement Liters (cc)	Engine ID/VIN	Crankshaft: Main Brg. Journal Dia.	Crankshaft: Main Brg. Oil Clearance	Crankshaft: Shaft End-play	Crankshaft: Thrust on No.	Connecting Rod: Journal Diameter	Connecting Rod: Oil Clearance	Connecting Rod: Side Clearance
1998	3.0 (2972)	R	①	0.0006-0.0017	0.0004-0.0300	2	②	0.0005-0.0024	0.0027-0.0110
1999	3.0 (2972)	R	①	0.0006-0.0017	0.0004-0.0300	2	②	0.0005-0.0024	0.0027-0.0110
2000	3.0 (2972)	R	①	0.0006-0.0017	0.0004-0.0300	2	②	0.0005-0.0024	0.0027-0.0110
2001	3.0 (2972)	R	①	0.0006-0.0017	0.0004-0.0300	2	②	0.0005-0.0024	0.0027-0.0110

① Green 2.6763 - 2.6766
Brown 26766 - 2.6770
Green/Blue 2.6665 - 2.6668
Brown/Blue 2.6668 - 2.6671
Green/White 2.6566 - 2.6570
Brown/White 2.6570 - 2.6573

② Undersize 0.25; 1.918 - 1.919
Undersize 0.50; 1.908 - 1.909

93461CU3

PISTON AND RING SPECIFICATIONS

All measurements are given in inches.

Year	Engine Displacement Liters (cc)	Engine ID/VIN	Piston Clearance	Ring Gap: Top Compression	Ring Gap: Bottom Compression	Ring Gap: Oil Control	Ring Side Clearance: Top Compression	Ring Side Clearance: Bottom Compression	Ring Side Clearance: Oil Control
1998	3.0 (2972)	R	0.0010-0.0018	0.0118-0.0196	0.0118-0.0196	0.0157-0.0551	0.0008-0.0015	0.0008-0.0015	0.0004-0.0012
1999	3.0 (2972)	R	0.0010-0.0018	0.0118-0.0196	0.0118-0.0196	0.0157-0.0551	0.0008-0.0015	0.0008-0.0015	0.0004-0.0012
2000	3.0 (2972)	R	0.0010-0.0018	0.0118-0.0196	0.0118-0.0196	0.0157-0.0551	0.0008-0.0015	0.0008-0.0015	0.0004-0.0012
2001	3.0 (2972)	R	0.0010-0.0018	0.0118-0.0196	0.0118-0.0196	0.0157-0.0551	0.0008-0.0015	0.0008-0.0015	0.0004-0.0012

93461CU4

Timing belt service is covered in Section 3 of this manual

TORQUE SPECIFICATIONS

All readings in ft. lbs.

Year	Engine Displacement Liters (cc)	Engine ID/VIN	Cylinder Head Bolts	Main Bearing Bolts	Rod Bearing Bolts	Crankshaft Damper Bolts	Flywheel Bolts	Manifold Intake	Manifold Exhaust	Spark Plugs	Lug Nut
1998	3.0 (2972)	R	①	②	③	15	④	15	15	18	100
1999	3.0 (2972)	R	①	②	③	15	④	15	15	18	100
2000	3.0 (2972)	R	①	②	③	15	④	15	15	18	100
2001	3.0 (2972)	R	①	②	③	15	④	15	15	18	100

① Step 1: 18 ft. lbs.
Step 2: Rotate 90 degrees
Step 3: Rotate 90 degrees
Step 4: Rotate 90 degrees
Step 5: Rotate 15 degrees

② Step 1: 37 ft. lbs.
Step 2: Rotate 60 degrees
Step 3: Rotate 15 degrees

③ Step 1: 26 ft. lbs.
Step 2: Rotate 45 degrees
Step 3: Rotate 15 degrees

④ Step 1: 48 ft. lbs.
Step 2: Rotate 30 degrees
Step 3: Rotate 15 degrees

93461CU5

BRAKE SPECIFICATIONS
GM V BODY

All measurements in inches unless noted

Year	Model		Brake Disc Original Thickness	Brake Disc Minimum Thickness	Maximum Runout	Minimum Lining Thickness	Brake Caliper Bracket Bolts (ft. lbs.)	Brake Caliper Mounting Bolts (ft. lbs.)
1998	Catera	F	1.102	0.984	0.001	0.315 ①	137	63
		R	0.472	0.393	0.004	NA	92	32
1999	Catera	F	1.102	0.984	0.001	0.315 ①	137	63
		R	0.472	0.393	0.004	NA	92	32
2000	Catera	F	1.102	0.984	0.001	0.315 ①	137	63
		R	0.472	0.393	0.004	NA	92	32
2001	Catera	F	1.102	0.984	0.001	0.315 ①	137	63
		R	0.472	0.393	0.004	NA	92	32

F: Front
R: Rear
NA: Not available
① Total thickness of lining and backing plate

93461CU6

WHEEL ALIGNMENT

Year	Model		Caster Range (+/-Deg.)	Caster Preferred Setting (Deg.)	Camber Range (+/-Deg.)	Camber Preferred Setting (Deg.)	Toe-in (in.)	Steering Axis Inclination (Deg.)
1998	Catera	F	1.00	+5.00	0.55	-0.60	0.06 +/- 0.05	—
		R	—	—	0.55	-1.75	0 +/- 0.05	—
1999	Catera	F	1.00	+5.00	0.55	-0.60	0.06 +/- 0.05	—
		R	—	—	0.55	-1.75	0 +/- 0.05	—
2000	Catera	F	1.00	+5.00	0.55	-0.60	0.06 +/- 0.05	—
		R	—	—	0.55	-1.75	0 +/- 0.05	—
2001	Catera	F	1.00	+5.00	0.55	-0.60	0.06 +/- 0.05	—
		R	—	—	0.55	-1.75	0 +/- 0.05	—

93461CU7

Heater Core replacement is covered in Section 2 of this manual

TIRE, WHEEL AND BALL JOINT SPECIFICATIONS

Year	Model	OEM Tires		Tire Pressures (psi)		Wheel	Ball Joint
		Standard	Optional	Front	Rear	Size	Inspection
1998	Catera	P225/55HR16	None	32	32	7-JJ	0.125 in. ①
1999	Catera	P225/55HR16	None	32	32	7-JJ	0.125 in. ①
2000	Catera	P225/55HR16	None	32	32	7-JJ	0.125 in. ①
2001	Catera	P225/55HR16	None	32	32	7-JJ	0.125 in. ①

OEM: Original Equipment Manufacturer

PSI: Pounds Per Square Inch

① Replace if any measurable movement is found.

93461CU8

SCHEDULED MAINTENANCE INTERVALS
GM V BODY—CADILLAC CATERA

TO BE SERVICED	TYPE OF SERVICE	VEHICLE MILEAGE INTERVAL (x1000)																			
		5	10	15	20	25	30	35	40	45	50	55	60	65	70	75	80	85	90	95	100
Engine oil & filter	R	✓	✓	✓	✓	✓	✓	✓	✓	✓	✓	✓	✓	✓	✓	✓	✓	✓	✓	✓	✓
Rotate tires	S/I	✓		✓		✓		✓		✓		✓		✓		✓		✓		✓	
Brake hoses	S/I	✓		✓		✓		✓		✓		✓		✓		✓		✓		✓	
Passenger Compartment Air Filter (Pollen Filter)	R			✓			✓			✓			✓			✓			✓		
Air filter element	S/I			✓			✓			✓			✓			✓			✓		
	R						✓						✓						✓		
Fuel tank, cap and lines	S/I						✓						✓								
Rear axle fluid level	S/I						✓						✓								
Automatic transaxle fluid & filter ①	R										✓										✓
Accessory drive belt(s)	S/I												✓								
Spark plugs ②	R																				✓
Ignition cables	S/I																				✓
Camshaft timing belt ③	R												✓								✓
Fuel filter	R																				✓
Engine coolant ④	R																				

R: Replace S/I: Service or Inspect

① Automatic transaxle fluid & filter: replace at 50,000 miles (83,000 km) if the vehicle has experienced severe service usage.

② Platinum tip spark plugs: replace every 100,000 miles.

③ Replace at 60,000 miles (96,000 km) if the engine was driven without an engine coolant heater being used and where temperatures fall below -20°F (-28°C). Otherwise, replace the belt at 100,000 miles (160,000 km).

④ Engine coolant: replace every 150,000 miles. Use O.E. specified (DEX-COOL™) coolant only. If any silicate coolant is used, the service interval is every 30,000 miles.

FREQUENT OPERATION MAINTENANCE (SEVERE SERVICE)

If a vehicle is operated under any of the following conditions it is considered severe service:

- Extremely dusty areas.

- 50% or more of the vehicle operation is in 32°C (90°F) or higher temperatures, or constant operation in temperatures below 0°C (32°F).

- Prolonged idling (vehicle operation in stop and go traffic).

- Frequent short running periods (engine does not warm to normal operating temperatures).

- Police, taxi, delivery usage or trailer towing usage.

Oil & oil filter: change every 5000 miles

Rotate tires at 5000 miles, then every 10,000 miles.

Air filter element: service or inspect every 15,000 miles.

Camshaft timing belt: change every 60,000 miles for severe service.

93461CU9

SCHEDULED MAINTENANCE INTERVALS
GENERAL MOTORS CORPORATION
V BODY
CADILLAC CATERA

The following should be used as a guide when determining the amount of work required for a particular service. In estimating how long a particular Scheduled Maintenance Service should take, please observe the following:

- Labor Time is time based on field research and data supplied by the vehicle manufacturer.
- Labor time operations are given in hours and tenths of an hour.
- All labor operations are to be used as a guide.

Mechanic Skill Level Codes:
(A) PRECISION: Highly skilled with multiple certification.
(B) GENERAL: Normally skilled with certification.
(C) MAINTENANCE: Semi-skilled working on certification.

	LABOR TIME
5000 Mile Service (C)	
All Models	.9
10000 Mile Service (C)	
All Models	.3
15000 Mile Service (C)	
All Models	1.5
20000 Mile Service (B)	
All Models	.3
25000 Mile Service (C)	
All Models	.9
30000 Mile Service (C)	
All Models	1.2
35000 Mile Service (C)	
All Models	.9
40000 Mile Service (B)	
All Models	.3
45000 Mile Service (C)	
All Models	1.5
50000 Mile Service (C)	
All Models	1.3
55000 Mile Service (C)	
All Models	.9
60000 Mile Service (B)	
All Models	3.9
65000 Mile Service (C)	
All Models	.9
70000 Mile Service (B)	
All Models	.3
75000 Mile Service (C)	
All Models	1.5
80000 Mile Service (B)	
All Models	.3
85000 Mile Service (C)	
All Models	.9
90000 Mile Service (B)	
All Models	.9
95000 Mile Service (C)	
All Models	.9
100000 Mile Service (B)	
All Models	4.4
Replace fuel filter add in line	.5

93061C08

GENERAL MOTORS W-BODY

Buick Regal • 1998-01 Century • Chevrolet Impala • Lumina • Monte Carlo • Oldsmobile Cutlass Supreme • Intrigue • Pontiac Grand Prix

ENGINE AND VEHICLE IDENTIFICATION

Engine Code							Model Year	
Code ①	Liters (cc)	Cu. In.	Cyl.	Fuel Sys.	Engine Type	Eng. Mfg.	Code ②	Year
1	3.8 (3785)	231	6	MFI	OHV	BOC	W	1998
K	3.8 (3785)	231	6	MFI	OHV	CPC	X	1999
H	3.5 (3475)	212	6	MFI	DOHC	BOC	Y	2000
M	3.1 (3130)	191	6	MFI	OHV	BOC	1	2001
J	3.1 (3130)	191	6	SFI	OHV	BOC	2	2002
E	3.4 (3393)	207	6	MFI	OHV	CPC		

MFI: Multi-point Fuel Injection

BOC: Buick/Oldsmobile/Cadillac

CPC: Chevrolet/Pontiac/Canada

DOHC: Dual Overhead Camshafts

OHV: Overhead Valves

① 8th position of VIN

② 10th position of VIN

93461CU0

For complete Engine Mechanical specifications, see Section 1 of this manual

GENERAL ENGINE SPECIFICATIONS

Year	Model	Engine Displacement Liters (cc)	Engine Series (ID/VIN)	Fuel System	Net Horsepower @ rpm	Net Torque @ rpm (ft. lbs.)	Bore x Stroke (in.)	Com-pression Ratio	Oil Pressure @ rpm
1998	Century	3.1 (3130)	M	MFI	160@5200	185@4000	3.50x3.31	9.5:1	15@1100
	Century	3.8 (3785)	1	MFI	240@5200	280@3200	3.80x3.40	9.0:1	60@1850
	Century	3.8 (3785)	K	MFI	205@5200	230@4000	3.80x3.40	9.4:1	60@1850
	Grand Prix	3.8 (3785)	1	MFI	240@5200	280@3200	3.80x3.40	9.0:1	60@1850
	Grand Prix	3.8 (3785)	K	MFI	205@5200	230@4000	3.80x3.40	9.4:1	60@1850
	Grand Prix	3.1 (3130)	M	MFI	160@5200	185@4000	3.50x3.31	9.5:1	15@1100
	Intrigue	3.8 (3785)	K	MFI	205@5200	230@4000	3.80x3.40	9.4:1	60@1850
	Lumina	3.1 (3130)	M	MFI	160@5200	185@4000	3.50x3.31	9.5:1	15@1100
	Monte Carlo	3.1 (3130)	M	MFI	160@5200	185@4000	3.50x3.31	9.5:1	15@1100
	Regal	3.8 (3785)	1	MFI	240@5200	280@3200	3.80x3.40	9.0:1	60@1850
	Regal	3.8 (3785)	K	MFI	205@5200	230@4000	3.80x3.40	9.4:1	60@1850
	Regal	3.1 (3130)	M	MFI	160@5200	185@4000	3.50x3.31	9.5:1	15@1100
1999	Century	3.1 (3130)	M	MFI	160@5200	185@4000	3.50x3.31	9.5:1	15@1100
	Century	3.8 (3785)	1	MFI	240@5200	280@3200	3.80x3.40	9.0:1	60@1850
	Century	3.8 (3785)	K	MFI	205@5200	230@4000	3.80x3.40	9.4:1	60@1850
	Grand Prix	3.8 (3785)	1	MFI	240@5200	280@3200	3.80x3.40	9.0:1	60@1850
	Grand Prix	3.8 (3785)	K	MFI	205@5200	230@4000	3.80x3.40	9.4:1	60@1850
	Grand Prix	3.1 (3130)	M	MFI	160@5200	185@4000	3.50x3.31	9.5:1	15@1100
	Intrigue	3.8 (3785)	K	MFI	205@5200	230@4000	3.80x3.40	9.4:1	60@1850
	Intrigue	3.5 (3475)	H	MFI	215@5600	230@4400	3.52x3.62	9.3:1	29@2000
	Lumina	3.8 (3785)	K	MFI	205@5200	230@4000	3.80x3.40	9.4:1	60@1850
	Lumina	3.1 (3130)	M	MFI	160@5200	185@4000	3.50x3.31	9.5:1	15@1100
	Monte Carlo	3.8 (3785)	K	MFI	205@5200	230@4000	3.80x3.40	9.4:1	60@1850
	Monte Carlo	3.1 (3130)	M	MFI	160@5200	185@4000	3.50x3.31	9.5:1	15@1100
	Regal	3.8 (3785)	1	MFI	240@5200	280@3200	3.80x3.40	9.0:1	60@1850
	Regal	3.8 (3785)	K	MFI	205@5200	230@4000	3.80x3.40	9.4:1	60@1850
	Regal	3.1 (3130)	M	MFI	160@5200	185@4000	3.50x3.31	9.5:1	15@1100
2000	Century	3.1 (3130)	M	MFI	160@5200	185@4000	3.50x3.31	9.5:1	15@1100
	Century	3.8 (3785)	1	MFI	240@5200	280@3200	3.80x3.40	9.0:1	60@1850
	Century	3.8 (3785)	K	MFI	205@5200	230@4000	3.80x3.40	9.4:1	60@1850
	Grand Prix	3.8 (3785)	1	MFI	240@5200	280@3200	3.80x3.40	9.0:1	60@1850
	Grand Prix	3.8 (3785)	K	MFI	205@5200	230@4000	3.80x3.40	9.5:1	60@1850
	Grand Prix	3.1 (3130)	J	SFI	160@5200	185@4000	3.50x3.31	9.5:1	15@1100
	Impala	3.4 (3393)	E	MFI	210@5200	215@4000	3.62x3.31	9.5:1	15@1100
	Impala	3.8 (3785)	K	MFI	205@5200	230@4000	3.80x3.40	9.4:1	60@1850
	Intrigue	3.5 (3475)	H	MFI	215@5600	230@4400	3.52x3.62	9.3:1	29@2000
	Lumina	3.1 (3130)	J	SFI	160@5200	185@4000	3.50x3.31	9.5:1	15@1100
	Monte Carlo	3.8 (3785)	K	MFI	205@5200	230@4000	3.80x3.40	9.4:1	60@1850
	Monte Carlo	3.4 (3393)	E	MFI	210@5200	215@4000	3.62x3.31	9.5:1	15@1100
	Regal	3.8 (3785)	1	MFI	240@5200	280@3200	3.80x3.40	9.0:1	60@1850
	Regal	3.8 (3785)	K	MFI	205@5200	230@4000	3.80x3.40	9.4:1	60@1850
	Regal	3.1 (3130)	M	MFI	160@5200	185@4000	3.50x3.31	9.5:1	15@1100
2001	Century	3.1 (3130)	M	MFI	160@5200	185@4000	3.50x3.31	9.5:1	15@1100
	Century	3.8 (3785)	1	MFI	240@5200	280@3200	3.80x3.40	9.0:1	60@1850
	Century	3.8 (3785)	K	MFI	205@5200	230@4000	3.80x3.40	9.4:1	60@1850
	Grand Prix	3.8 (3785)	1	MFI	240@5200	280@3200	3.80x3.40	9.0:1	60@1850
	Grand Prix	3.8 (3785)	K	MFI	205@5200	230@4000	3.80x3.40	9.5:1	60@1850
	Grand Prix	3.1 (3130)	J	SFI	160@5200	185@4000	3.50x3.31	9.5:1	15@1100
	Impala	3.4 (3393)	E	MFI	210@5200	215@4000	3.62x3.31	9.5:1	15@1100
	Impala	3.8 (3785)	K	MFI	205@5200	230@4000	3.80x3.40	9.4:1	60@1850

93461CV1

GENERAL ENGINE SPECIFICATIONS

Year	Model	Engine Displacement Liters (cc)	Engine Series (ID/VIN)	Fuel System	Net Horsepower @ rpm	Net Torque @ rpm (ft. lbs.)	Bore x Stroke (in.)	Com-pression Ratio	Oil Pressure @ rpm
2001	Intrigue	3.5 (3475)	H	MFI	215@5600	230@4400	3.52x3.62	9.3:1	29@2000
(Cont.)	Lumina	3.1 (3130)	J	SFI	160@5200	185@4000	3.50x3.31	9.5:1	15@1100
	Monte Carlo	3.8 (3785)	K	MFI	205@5200	230@4000	3.80x3.40	9.4:1	60@1850
	Monte Carlo	3.4 (3393)	E	MFI	210@5200	215@4000	3.62x3.31	9.5:1	15@1100
	Regal	3.8 (3785)	1	MFI	240@5200	280@3200	3.80x3.40	9.0:1	60@1850
	Regal	3.8 (3785)	K	MFI	205@5200	230@4000	3.80x3.40	9.4:1	60@1850
	Regal	3.1 (3130)	M	MFI	160@5200	185@4000	3.50x3.31	9.5:1	15@1100

MFI: Multi-point Fuel Injection

93461CV2

For Accessory Drive Belt illustrations, see Section 1 of this manual

ENGINE TUNE-UP SPECIFICATIONS

Year	Engine Displacement Liters (cc)	Engine ID/VIN	Spark Plug Gap (in.)	Ignition Timing (deg.)	Fuel Pump (psi)	Idle Speed (rpm)	Valve Clearance	
							In.	Ex.
1998	3.1 (3130)	M	0.060	①	41-47	②	HYD	HYD
	3.8 (3785)	1	0.060	①	41-47	②	HYD	HYD
	3.8 (3785)	K	0.060	①	41-47	②	HYD	HYD
1999	3.1 (3130)	M	0.060	①	41-47	②	HYD	HYD
	3.5 (3475)	H	0.050	①	41-47	②	HYD	HYD
	3.8 (3785)	1	0.060	①	41-47	②	HYD	HYD
	3.8 (3785)	K	0.060	①	41-47	②	HYD	HYD
2000	3.1 (3130)	M	0.060	①	41-47	②	HYD	HYD
	3.1 (3130)	J	0.060	①	41-47	②	HYD	HYD
	3.4 (3393)	E	0.045	①	41-47	②	HYD	HYD
	3.5 (3475)	H	0.050	①	41-47	②	HYD	HYD
	3.8 (3785)	1	0.060	①	41-47	②	HYD	HYD
	3.8 (3785)	K	0.060	①	41-47	②	HYD	HYD
2001	3.1 (3130)	M	0.060	①	41-47	②	HYD	HYD
	3.1 (3130)	J	0.060	①	41-47	②	HYD	HYD
	3.4 (3393)	E	0.045	①	41-47	②	HYD	HYD
	3.5 (3475)	H	0.050	①	41-47	②	HYD	HYD
	3.8 (3785)	1	0.060	①	41-47	②	HYD	HYD
	3.8 (3785)	K	0.060	①	41-47	②	HYD	HYD

NOTE: The Vehicle Emission Control Information label often reflects specification changes made during production. The label figures must be used if they differ from those in this chart.

HYD: Hydraulic

① Distributorless Ignition System (DIS) timing is not adjustable

② Idle speed is maintained by the Engine Control Module (ECM). There is no recommended adjustment procedure.

93461CV3

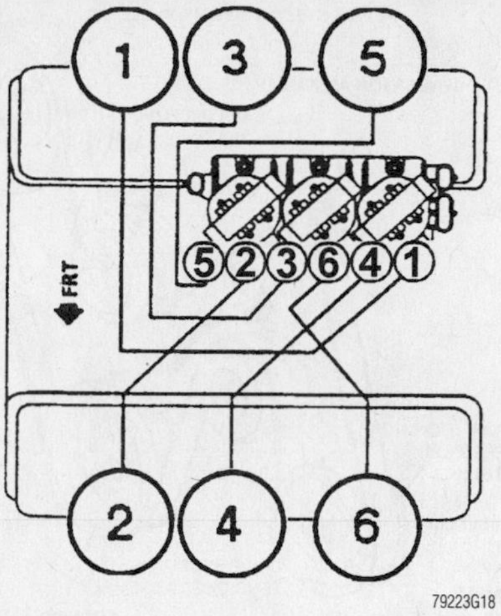

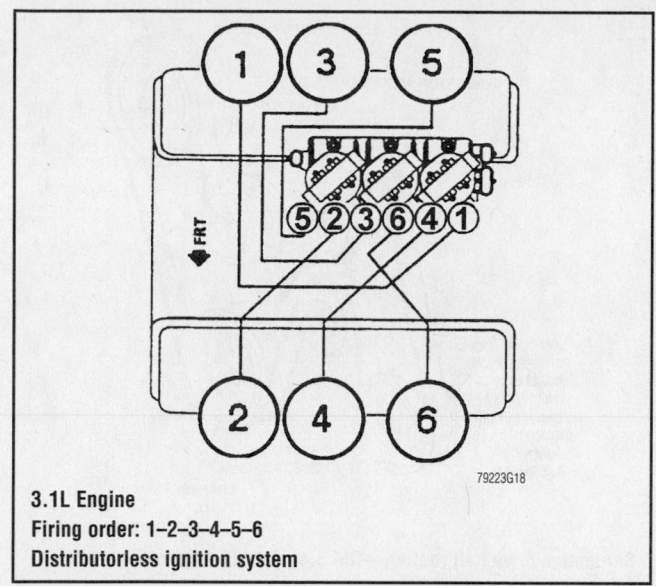

3.1L Engine
Firing order: 1–2–3–4–5–6
Distributorless ignition system

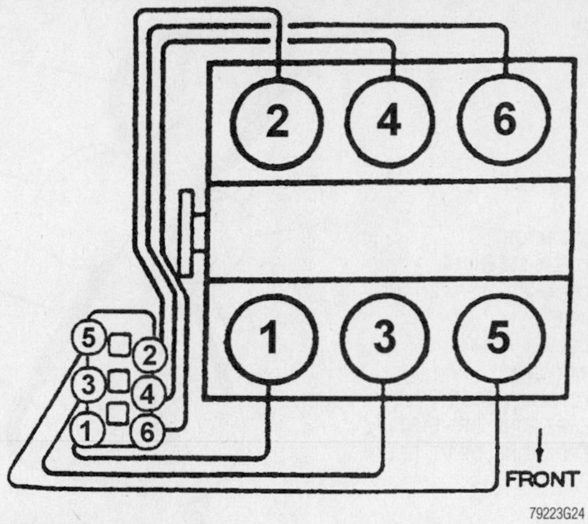

3.8L (VIN 1) Engine
Firing order: 1–6–5–4–3–2
Distributorless ignition system

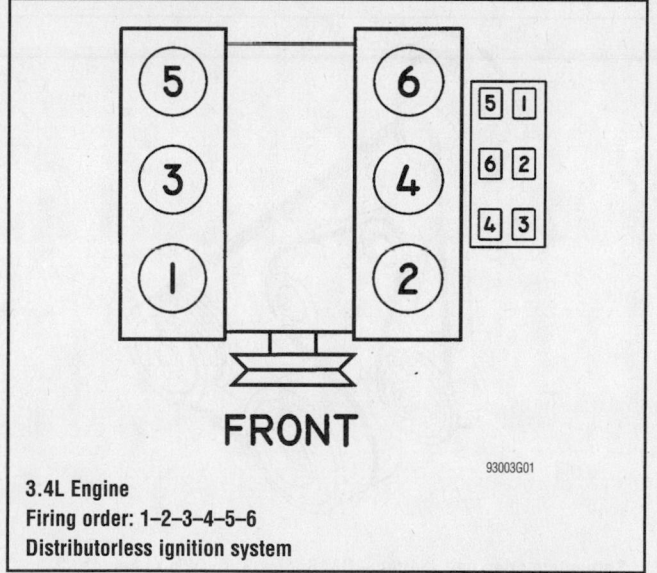

3.4L Engine
Firing order: 1–2–3–4–5–6
Distributorless ignition system

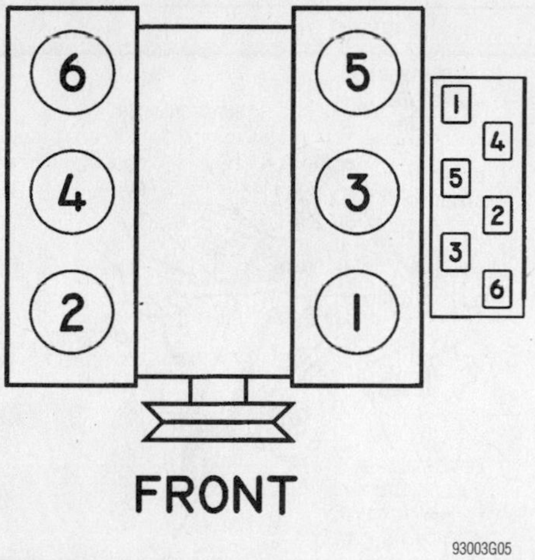

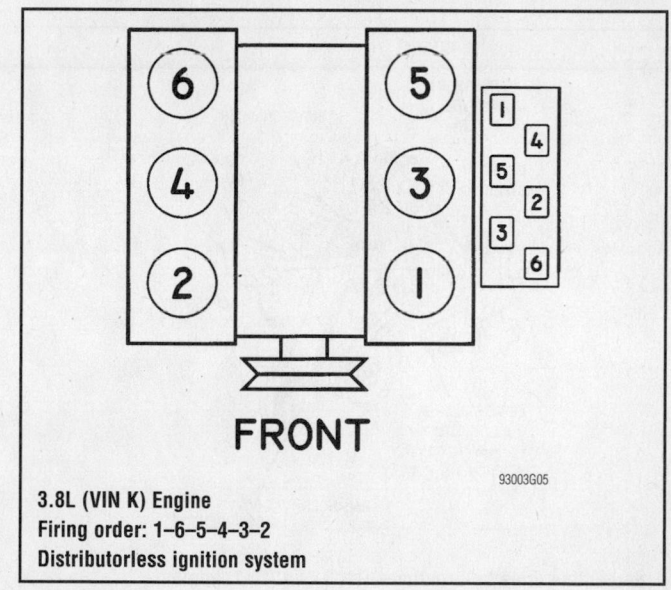

3.8L (VIN K) Engine
Firing order: 1–6–5–4–3–2
Distributorless ignition system

3.5L Engine
Firing order: 1–2–3–4–5–6
Distributorless ignition system

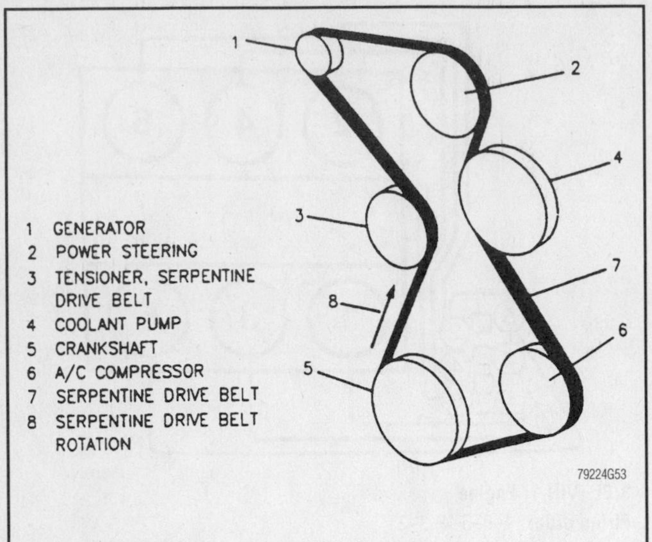

Serpentine drive belt routing—GM 3.1L engine

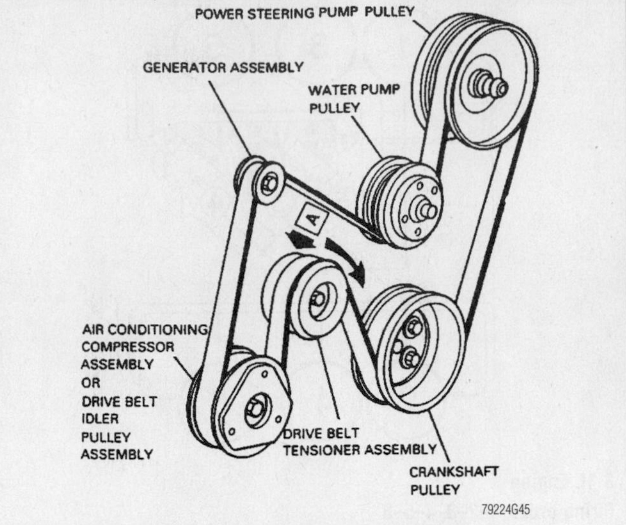

Serpentine drive belt routing—GM 3.4L (VIN E) engine

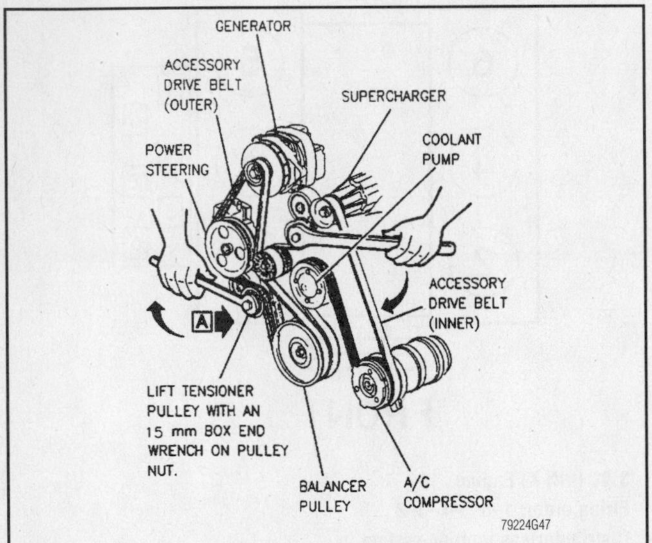

Serpentine drive belt routing—GM 3.8L (VIN 1) engine

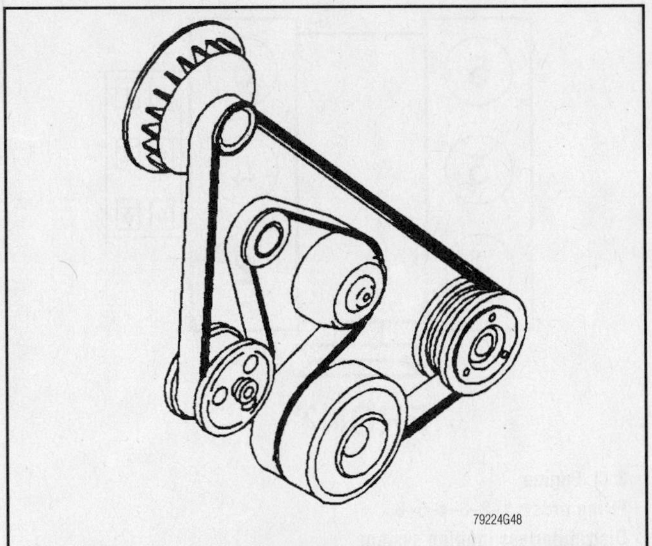

Serpentine drive belt routing—GM Bodies 3.8L (VIN K) engine

CAPACITIES

Year	Model	Engine Displacement Liters (cc)	Engine ID/VIN	Engine Oil with Filter (qts.)	Transmission (pts.)	Fuel Tank (gal.)	Cooling System (qts.)
1998	Century	3.1 (3130)	M	4.5	①	16.0	12.6
	Century	3.8 (3785)	1	5.0	①	16.5	12.7
	Century	3.8 (3785)	K	5.0	①	16.5	12.7
	Grand Prix	3.8 (3785)	1	5.0	①	16.5	12.7
	Grand Prix	3.8 (3785)	K	5.0	①	16.5	12.7
	Grand Prix	3.1 (3130)	M	4.5	①	16.0	12.6
	Intrigue	3.8 (3785)	K	5.0	①	16.5	12.7
	Lumina	3.1 (3130)	M	4.5	①	16.0	12.6
	Monte Carlo	3.1 (3130)	M	4.5	①	16.5	12.6
	Regal	3.8 (3785)	1	5.0	①	16.5	12.7
	Regal	3.8 (3785)	K	5.0	①	17.1	11.1
	Regal	3.1 (3130)	M	4.5	①	17.1	12.5
1999	Century	3.1 (3130)	M	4.5	①	16.0	12.6
	Century	3.8 (3785)	1	5.0	①	16.5	12.7
	Century	3.8 (3785)	K	5.0	①	16.5	12.7
	Grand Prix	3.8 (3785)	1	5.0	①	16.5	12.7
	Grand Prix	3.8 (3785)	K	5.0	①	16.5	12.7
	Grand Prix	3.1 (3130)	M	4.5	①	16.0	12.6
	Intrigue	3.5 (3475)	H	6.0	①	18.0	9.6
	Lumina	3.8 (3785)	K	5.0	①	16.5	12.7
	Lumina	3.1 (3130)	M	4.5	①	16.0	12.6
	Monte Carlo	3.8 (3785)	K	5.0	①	16.5	12.7
	Monte Carlo	3.1 (3130)	M	4.5	①	16.5	12.6
	Regal	3.8 (3785)	1	5.0	①	16.5	12.7
	Regal	3.8 (3785)	K	5.0	①	17.1	11.1
	Regal	3.1 (3130)	M	4.5	①	17.1	12.5
2000	Century	3.1 (3130)	M	4.5	①	16.0	12.6
	Century	3.8 (3785)	1	5.0	①	16.5	12.7
	Century	3.8 (3785)	K	5.0	①	16.5	12.7
	Grand Prix	3.1 (3130)	J	5.0	①	16.5	12.7
	Grand Prix	3.8 (3785)	K	5.0	①	16.5	12.7
	Grand Prix	3.1 (3130)	M	4.5	①	16.0	12.6
	Impala	3.4 (3393)	E	5.0	①	16.5	12.7
	Impala	3.8 (3785)	K	5.0	①	16.5	12.7
	Intrigue	3.5 (3475)	H	6.0	①	18.0	9.6
	Lumina	3.1 (3130)	J	4.5	①	16.0	12.6
	Monte Carlo	3.8 (3785)	K	5.0	①	16.5	12.7
	Monte Carlo	3.4 (3393)	E	5.0	①	16.5	12.7
	Regal	3.8 (3785)	1	5.0	①	16.5	12.7
	Regal	3.8 (3785)	K	5.0	①	17.1	11.1
	Regal	3.1 (3130)	M	4.5	①	17.1	12.5
2001	Century	3.1 (3130)	M	4.5	①	16.0	12.6
	Century	3.8 (3785)	1	5.0	①	16.5	12.7
	Century	3.8 (3785)	K	5.0	①	16.5	12.7
	Grand Prix	3.1 (3130)	J	5.0	①	16.5	12.7
	Grand Prix	3.8 (3785)	K	5.0	①	16.5	12.7
	Grand Prix	3.1 (3130)	M	4.5	①	16.0	12.6
	Impala	3.4 (3393)	E	5.0	①	16.5	12.7
	Impala	3.8 (3785)	K	5.0	①	16.5	12.7
	Intrigue	3.5 (3475)	H	6.0	①	18.0	9.6
	Lumina	3.1 (3130)	J	4.5	①	16.0	12.6
	Monte Carlo	3.8 (3785)	K	5.0	①	16.5	12.7

93461CV4

For Wheel Alignment specifications, see Section 1 of this manual

CAPACITIES

Year	Model	Engine Displacement Liters (cc)	Engine ID/VIN	Engine Oil with Filter (qts.)	Transmission (pts.)	Fuel Tank (gal.)	Cooling System (qts.)
2001	Monte Carlo	3.4 (3393)	E	5.0	①	16.5	12.7
(Cont.)	Regal	3.8 (3785)	1	5.0	①	16.5	12.7
	Regal	3.8 (3785)	K	5.0	①	17.1	11.1
	Regal	3.1 (3130)	M	4.5	①	17.1	12.5

NOTE: All capacities are approximate. Add fluid gradually and ensure a proper fluid is obtained.

① Capacity is without filter replacement; Additional oil may be required

93461CV5

VALVE SPECIFICATIONS

Year	Engine Displacement Liters (cc)	Engine ID/VIN	Seat Angle (deg.)	Face Angle (deg.)	Spring Test Pressure (lbs. @ in.)	Spring Installed Height (in.)	Stem-to-Guide Clearance (in.) Intake	Stem-to-Guide Clearance (in.) Exhaust	Stem Diameter (in.) Intake	Stem Diameter (in.) Exhaust
1998	3.1 (3130)	M	45	45	250@1.239	1.710	0.0001-0.0027	0.0010-0.0027	NA	NA
	3.8 (3785)	1	45	45	80@1.750	1.690-1.720	0.0015-0.0032	0.0015-0.0032	NA	NA
	3.8 (3785)	K	45	45	210@1.32	1.690-1.720	0.0015-0.0035	0.0015-0.0032	NA	NA
1999	3.1 (3130)	M	45	45	250@1.239	1.710	0.0001-0.0027	0.0010-0.0027	NA	NA
	3.5 (3475)	H	45.75	45	130-142@0.964	1.377	0.0010-0.0030	0.0020-0.0040	0.233-0.234	0.233-0.234
	3.8 (3785)	1	45	45	80@1.750	1.690-1.720	0.0015-0.0032	0.0015-0.0032	NA	NA
	3.8 (3785)	K	45	45	210@1.32	1.690-1.720	0.0015-0.0035	0.0015-0.0032	NA	NA
2000	3.1 (3130)	M	45	45	250@1.239	1.710	0.0001-0.0027	0.0010-0.0027	NA	NA
	3.4 (3393)	E	46	45	75@1.40	1.400	0.0011-0.0026	0.0014-0.0031	NA	NA
	3.5 (3475)	H	45.75	45	130-142@0.964	1.377	0.0010-0.0030	0.0020-0.0040	0.233-0.234	0.233-0.234
	3.8 (3785)	J	45	45	80@1.750	1.690-1.720	0.0015-0.0032	0.0015-0.0032	NA	NA
	3.8 (3785)	1	45	45	80@1.750	1.690-1.720	0.0015-0.0032	0.0015-0.0032	NA	NA
	3.8 (3785)	K	45	45	210@1.32	1.690-1.720	0.0015-0.0035	0.0015-0.0032	NA	NA
2001	3.1 (3130)	M	45	45	250@1.239	1.710	0.0001-0.0027	0.0010-0.0027	NA	NA
	3.4 (3393)	E	46	45	75@1.40	1.400	0.0011-0.0026	0.0014-0.0031	NA	NA
	3.5 (3475)	H	45.75	45	130-142@0.964	1.377	0.0010-0.0030	0.0020-0.0040	0.233-0.234	0.233-0.234
	3.8 (3785)	J	45	45	80@1.750	1.690-1.720	0.0015-0.0032	0.0015-0.0032	NA	NA
	3.8 (3785)	1	45	45	80@1.750	1.690-1.720	0.0015-0.0032	0.0015-0.0032	NA	NA
	3.8 (3785)	K	45	45	210@1.32	1.690-1.720	0.0015-0.0035	0.0015-0.0032	NA	NA

NA: Not Available

93461CV6

CRANKSHAFT AND CONNECTING ROD SPECIFICATIONS

All measurements are given in inches.

Year	Engine Displacement Liters (cc)	Engine ID/VIN	Crankshaft Main Brg. Journal Dia.	Crankshaft Main Brg. Oil Clearance	Crankshaft Shaft End-play	Crankshaft Thrust on No.	Connecting Rod Journal Diameter	Connecting Rod Oil Clearance	Connecting Rod Side Clearance
1998	3.1 (3130)	M	2.6473-2.6383	0.0008-0.0025 ①	0.0024-0.0083	3	1.9987-1.9994	0.0007-0.0024	0.0070-0.017
	3.8 (3785)	1	2.4988-2.4998	0.0008-0.0022	0.0030-0.011	2	2.3738-2.3745	0.0005-0.0026	0.0030-0.0150
	3.8 (3785)	K	2.4988-2.4998	0.0008-0.0022	0.0030-0.0110	2	2.3738-2.3745	0.0005-0.0026	0.0030-0.0150
1999	3.1 (3130)	M	2.6473-2.6383	0.0008-0.0025 ①	0.0024-0.0083	3	1.9987-1.9994	0.0007-0.0024	0.0070-0.0170
	3.5 (3475)	H	2.7550-2.7560	0.0006-0.0021	0.0050-0.0200	3	2.1829-2.1835	0.0009-0.0025	0.0040-0.0130
	3.8 (3785)	1	2.4988-2.4998	0.0008-0.0022	0.0030-0.0110	2	2.3738-2.3745	0.0005-0.0026	0.0030-0.0150
	3.8 (3785)	K	2.4988-2.4998	0.0008-0.0022	0.0030-0.0110	2	2.3738-2.3745	0.0005-0.0026	0.0030-0.0150
2000	3.1 (3130)	M	2.6473-2.6383	0.0008-0.0025 ①	0.0024-0.0083	3	1.9987-1.9994	0.0007-0.0024	0.0070-0.0170
	3.4 (3393)	E	2.6472-2.6479	0.0008-0.0025	0.0024-0.0083	3	1.9987-1.9994	0.0007-0.0024	0.0070-0.0170
	3.5 (3475)	H	2.7550-2.7560	0.0006-0.0021	0.0050-0.0200	3	2.1829-2.1835	0.0009-0.0025	0.0040-0.0130
	3.8 (3785)	J	2.4988-2.4998	0.0008-0.0022	0.0030-0.0110	2	2.3738-2.3745	0.0005-0.0026	0.0030-0.0150
	3.8 (3785)	1	2.4988-2.4998	0.0008-0.0022	0.0030-0.0110	2	2.3738-2.3745	0.0005-0.0026	0.0030-0.0150
	3.8 (3785)	K	2.4988-2.4998	0.0008-0.0022	0.0030-0.0110	2	2.3738-2.3745	0.0005-0.0026	0.0030-0.0150
2001	3.1 (3130)	M	2.6473-2.6383	0.0008-0.0025 ①	0.0024-0.0083	3	1.9987-1.9994	0.0007-0.0024	0.0070-0.0170
	3.4 (3393)	E	2.6472-2.6479	0.0008-0.0025	0.0024-0.0083	3	1.9987-1.9994	0.0007-0.0024	0.0070-0.0170
	3.5 (3475)	H	2.7550-2.7560	0.0006-0.0021	0.0050-0.0200	3	2.1829-2.1835	0.0009-0.0025	0.0040-0.0130
	3.8 (3785)	J	2.4988-2.4998	0.0008-0.0022	0.0030-0.0110	2	2.3738-2.3745	0.0005-0.0026	0.0030-0.0150
	3.8 (3785)	1	2.4988-2.4998	0.0008-0.0022	0.0030-0.0110	2	2.3738-2.3745	0.0005-0.0026	0.0030-0.0150
	3.8 (3785)	K	2.4988-2.4998	0.0008-0.0022	0.0030-0.0110	2	2.3738-2.3745	0.0005-0.0026	0.0030-0.0150

① Thrust bearing: 0.0012 - 0.0030

93461CV7

PISTON AND RING SPECIFICATIONS

All measurements are given in inches.

Year	Engine Displacement Liters (cc)	Engine ID/VIN	Piston Clearance	Ring Gap Top Compression	Ring Gap Bottom Compression	Ring Gap Oil Control	Ring Side Clearance Top Compression	Ring Side Clearance Bottom Compression	Ring Side Clearance Oil Control
1998	3.1 (3130)	M	0.0010-0.0018	0.0118-0.0196	0.0118-0.0196	0.0157-0.0551	0.0008-0.0015	0.0008-0.0015	0.0004-0.0012
	3.8 (3785)	1	0.0004-0.0020	0.0120-0.0220	0.0300-0.0400	0.0100-0.0300	0.0013-0.0031	0.0013-0.0031	0.0009-0.0079
	3.8 (3785)	K	0.0004-0.0020	0.0120-0.0220	0.0300-0.0400	0.0100-0.0300	0.0013-0.0031	0.0013-0.0031	0.0009-0.0079
1999	3.1 (3130)	M	0.0010-0.0018	0.0118-0.0196	0.0118-0.0196	0.0157-0.0551	0.0008-0.0015	0.0008-0.0015	0.0004-0.0012
	3.5 (3475)	H	0.0010-0.0025	0.0080-0.0180	0.0140-0.0200	0.0100-0.0300	0.0016-0.0037	0.0016-0.0037	side-sealing
	3.8 (3785)	1	0.0004-0.0020	0.0120-0.0220	0.0300-0.0400	0.0100-0.0300	0.0013-0.0031	0.0013-0.0031	0.0009-0.0079
	3.8 (3785)	K	0.0004-0.0020	0.0120-0.0220	0.0300-0.0400	0.0100-0.0300	0.0013-0.0031	0.0013-0.0031	0.0009-0.0079
2000	3.1 (3130)	M	0.0010-0.0018	0.0118-0.0196	0.0118-0.0196	0.0157-0.0551	0.0008-0.0015	0.0008-0.0015	0.0004-0.0012
	3.4 (3393)	E	0.0008-0.0020	0.0080-0.0180	0.0220-0.0320	0.0098-0.0229	0.0013-0.0031	0.0013-0.0031	0.0011-0.0081
	3.5 (3475)	H	0.0010-0.0025	0.0080-0.0180	0.0140-0.0200	0.0100-0.0300	0.0016-0.0037	0.0016-0.0037	side-sealing
	3.8 (3785)	J	0.0004-0.0020	0.0100-0.0160	0.0300-0.0400	0.0100-0.0300	0.0013-0.0031	0.0013-0.0031	0.0009-0.0079
	3.8 (3785)	1	0.0004-0.0020	0.0100-0.0160	0.0300-0.0400	0.0100-0.0300	0.0013-0.0031	0.0013-0.0031	0.0009-0.0079
	3.8 (3785)	K	0.0004-0.0020	0.0120-0.0220	0.0300-0.0400	0.0100-0.0300	0.0013-0.0031	0.0013-0.0031	0.0009-0.0079
2001	3.1 (3130)	M	0.0010-0.0018	0.0118-0.0196	0.0118-0.0196	0.0157-0.0551	0.0008-0.0015	0.0008-0.0015	0.0004-0.0012
	3.4 (3393)	E	0.0008-0.0020	0.0080-0.0180	0.0220-0.0320	0.0098-0.0229	0.0013-0.0031	0.0013-0.0031	0.0011-0.0081
	3.5 (3475)	H	0.0010-0.0025	0.0080-0.0180	0.0140-0.0200	0.0100-0.0300	0.0016-0.0037	0.0016-0.0037	side-sealing
	3.8 (3785)	J	0.0004-0.0020	0.0100-0.0160	0.0300-0.0400	0.0100-0.0300	0.0013-0.0031	0.0013-0.0031	0.0009-0.0079
	3.8 (3785)	1	0.0004-0.0020	0.0100-0.0160	0.0300-0.0400	0.0100-0.0300	0.0013-0.0031	0.0013-0.0031	0.0009-0.0079
	3.8 (3785)	K	0.0004-0.0020	0.0120-0.0220	0.0300-0.0400	0.0100-0.0300	0.0013-0.0031	0.0013-0.0031	0.0009-0.0079

93461CV8

TORQUE SPECIFICATIONS

All readings in ft. lbs.

Year	Engine Displacement Liters (cc)	Engine ID/VIN	Cylinder Head Bolts	Main Bearing Bolts	Rod Bearing Bolts	Crankshaft Damper Bolts	Flywheel Bolts	Manifold Intake	Manifold Exhaust	Spark Plug	Lug Nut
1998	3.1 (3130)	M	①	②	③	76	61	④	10	⑤	100
	3.8 (3785)	1	⑧	⑨	20	⑩	⑪	⑬	22	11	100
	3.8 (3785)	K	⑧	⑨	20	⑩	⑪	⑫	38	11	100
1999	3.1 (3130)	M	①	②	③	76	61	④	10	⑭	100
	3.5 (3475)	H	⑮	⑯	⑰	⑱	⑲	5	18	15	100
	3.8 (3785)	1	⑧	⑨	⑳	⑩	⑪	⑬	22	11	100
	3.8 (3785)	K	⑧	⑨	⑳	⑩	⑪	⑫	38	11	100
2000	3.1 (3130)	M	①	②	③	76	61	④	10	⑭	100
	3.1 (3130)	J	①	②	③	76	61	④	10	⑭	100
	3.4 (3393)	E	⑥	②	39	78	61	18	⑦	11	100
	3.5 (3475)	H	⑮	⑯	⑰	⑱	⑲	5	18	15	100
	3.8 (3785)	1	⑧	⑨	⑳	⑩	⑪	⑬	22	11	100
	3.8 (3785)	K	⑧	⑨	⑳	⑩	⑪	⑫	38	11	100
2001	3.1 (3130)	M	①	②	③	76	61	④	10	⑭	100
	3.1 (3130)	J	①	②	③	76	61	④	10	⑭	100
	3.4 (3393)	X	⑥	②	39	78	61	18	⑦	11	100
	3.5 (3475)	H	⑮	⑯	⑰	⑱	⑲	5	18	15	100
	3.8 (3785)	1	⑧	⑨	⑳	⑩	⑪	⑬	22	11	100
	3.8 (3785)	K	⑧	⑨	⑳	⑩	⑪	⑫	38	11	100

① Coat threads with sealer torque to 33 ft. lbs., then turn 1/4 turn (90 degrees)

② 37 ft. lbs. plus 77 degrees

③ 15 ft. lbs. plus 75 degrees

④ Torque all bolts to 15 ft. lbs. Retorque to 24 ft. lbs.

⑤ New cylinder head:
1st-time installation: 20 ft. lbs.
All other installations: 11 ft. lbs.

⑥ 37 ft. lbs. plus 90 degrees

⑦ 115 inch lbs.

⑧ Step 1: 35 ft. lbs.
Step 2: plus 130 degrees
Step 3: Rotate four center bolts an additional 30 degrees

⑨ 30 ft. lbs. plus 110 degrees
Side bolts: 11 ft. lbs. plus 45 degrees

⑩ 110 ft. lbs. plus 76 degrees

⑪ 11 ft. lbs. plus 50 degrees

⑫ Upper manifold: 18 ft. lbs.
Lower manifold bolt/nut: 22 ft. lbs.
Upper manifold studs: 89 inch lbs.

⑬ Upper manifold: 8 ft. lbs.
Lower manifold: 11 ft. lbs.

⑭ 20 ft. lbs. plus 50 degrees

⑮ Step 1: 22 ft. lbs.
Step 2: plus 60 degrees
Step 3: plus 60 degrees
Step 4: plus 80 degrees

⑯ Cap bolts
Step 1: 15 ft. lbs.
Step 2: plus 70 degrees
Perimeter bolts: 22 ft. lbs.

⑰ Step 1: 22 ft. lbs.
Step 2: Loosen completely
Step 3: 18 ft. lbs.
Step 4: plus 110 degrees

⑱ Step 1: 37 ft. lbs.
Step 2: plus 150 degrees

⑲ Step 1: 11 ft. lbs.
Step 2: plus 50 degrees

⑳ 20 ft. lbs. plus 50 degrees

93461CV9

BRAKE SPECIFICATIONS

GM W BODY

All measurements in inches unless noted

			Brake Disc			Brake Drum Diameter			Minimum Lining Thickness		Brake Caliper	
Year	Model		Original Thickness	Minimum Thickness	Maximum Runout	Original Inside Diameter	Max. Wear Limit	Maximum Machine Diameter	Front	Rear	Bracket Bolts (ft. lbs.)	Mounting Bolts (ft. lbs.)
1997	Cutlass	F	1.039	0.972	0.003	—	—	—	0.030	0.030	148	80
	Supreme	R	0.492	0.429	0.003	8.863	8.920	8.909	0.030	0.030	81	20
	Grand Prix	F	1.039	0.972	0.004	—	—	—	0.030	0.030	148	80
		R	0.492	0.429	0.004	8.863	8.920	8.909	0.030	0.030	81	20
	Lumina	F	1.040	0.972	0.004	—	—	—	0.030	0.030	148	80
		R	0.492	0.429	0.004	8.863	8.920	8.909	0.030	0.030	81	20
	Monte Carlo	F	1.040	0.972	0.004	—	—	—	0.030	0.030	148	80
		R	0.492	0.429	0.004	8.863	8.920	8.909	0.030	0.030	81	20
	Regal	F	1.039	0.972	0.003	—	—	—	0.030	0.030	148	80
		R	0.492	0.429	0.003	8.863	8.920	8.909	0.030	0.030	81	20
1998	Century	F	1.039	0.972	0.003	—	—	—	0.030	0.030	148	80
		R	0.492	0.429	0.003	8.863	8.920	8.909	0.030	0.030	81	20
	Grand Prix	F	1.039	0.972	0.004	—	—	—	0.030	0.030	137	63
		R	0.492	0.429	0.004	8.863	8.920	8.909	0.030	0.030	92	33
	Intrigue	F	1.039	0.972	0.003	—	—	—	0.030	0.030	137	63
		R	0.492	0.429	0.003	—	—	—	0.030	0.030	92	33
	Lumina	F	1.040	0.972	0.004	—	—	—	0.030	0.030	148	80
		R	0.492	0.429	0.004	8.863	8.920	8.909	0.030	0.030	81	20
	Monte Carlo	F	1.040	0.972	0.004	—	—	—	0.030	0.030	148	80
		R	0.492	0.429	0.004	8.863	8.920	8.909	0.030	0.030	81	20
	Regal	F	1.039	0.972	0.003	—	—	—	0.030	0.030	137	63
		R	0.492	0.429	0.003	8.863	8.920	8.909	0.030	0.030	92	33
1999	Century	F	1.039	0.972	0.003	—	—	—	0.030	0.030	137	63
		R	0.492	0.429	0.003	8.863	8.920	8.909	0.030	0.030	92	33
	Grand Prix	F	1.039	0.972	0.003	—	—	—	0.030	0.030	137	63
		R	0.492	0.429	0.003	—	—	—	0.030	0.030	92	33
	Intrigue	F	1.039	0.972	0.003	—	—	—	0.030	0.030	137	63
		R	0.492	0.429	0.003	—	—	—	0.030	0.030	92	33
	Lumina	F	1.040	0.972	0.004	—	—	—	0.030	0.030	148	80
		R	0.492	0.429	0.004	8.863	8.920	8.909	0.030	0.030	81	20
	Monte Carlo	F	1.040	0.972	0.004	—	—	—	0.030	0.030	148	80
		R	0.492	0.429	0.004	8.863	8.920	8.909	0.030	0.030	81	20
	Regal	F	1.039	0.972	0.003	—	—	—	0.030	0.030	137	63
		R	0.492	0.429	0.003	8.863	8.920	8.909	0.030	0.030	92	33
2000-01	Century	F	1.039	0.972	0.003	—	—	—	0.030	0.030	137	63
		R	0.492	0.429	0.003	8.863	8.920	8.909	0.030	0.030	92	33
	Grand Prix	F	1.039	0.972	0.003	—	—	—	0.030	0.030	137	63
		R	0.492	0.429	0.003	—	—	—	0.030	0.030	92	33
	Impala	F	1.040	0.972	0.004	—	—	—	0.030	0.030	148	80
		R	0.492	0.429	0.004	8.863	8.920	8.909	0.030	0.030	81	20
	Intrigue	F	1.039	0.972	0.003	—	—	—	0.030	0.030	137	63
		R	0.492	0.429	0.003	—	—	—	0.030	0.030	92	33
	Lumina	F	1.040	0.972	0.004	—	—	—	0.030	0.030	148	80
		R	0.492	0.429	0.004	8.863	8.920	8.909	0.030	0.030	81	20
	Monte Carlo	F	1.040	0.972	0.004	—	—	—	0.030	0.030	148	80
		R	0.492	0.429	0.004	8.863	8.920	8.909	0.030	0.030	81	20
	Regal	F	1.039	0.972	0.003	—	—	—	0.030	0.030	137	63
		R	0.492	0.429	0.003	8.863	8.920	8.909	0.030	0.030	92	33

F: Front

R: Rear

93461CV0

WHEEL ALIGNMENT
BUICK CENTURY

Year	Model		Caster Range (Deg.)	Caster Preferred Setting (Deg.)	Camber Range (Deg.)	Camber Preferred Setting (Deg.)	Toe-in (in.)	Steering Axis Inclination (Deg.)
1998	Century	F	+0.50	+1.80	+0.50	+0.70	0 +/- 0.20	—
		R	—	—	+0.50	-0.15	0.10 +/- 0.10	—
1999	Century	F	+0.50	+1.80	+0.50	+0.70	0 +/- 0.20	—
		R	—	—	+0.50	-0.15	0.10 +/- 0.10	—
2000	Century	F	+0.50	+3.00	+0.50	-0.90	0.10 +/- 0.20	—
		R	—	—	+0.50	-0.90	0 +/- 0.20	—
2001	Century	F	+0.50	+3.00	+0.50	-0.90	0.10 +/- 0.20	—
		R	—	—	+0.50	-0.90	0 +/- 0.20	—

93461CW1

WHEEL ALIGNMENT
BUICK REGAL

Year	Model		Caster Range (Deg.)	Caster Preferred Setting (Deg.)	Camber Range (Deg.)	Camber Preferred Setting (Deg.)	Toe-in (in.)	Steering Axis Inclination (Deg.)
1998	Regal	F	+0.50	+1.80	+0.50	+0.70	0 +/- 0.20	—
		R	—	—	+0.50	-0.15	0.10 +/- 0.15	—
1999	Regal	F	+0.50	+1.80	+0.50	+0.70	0 +/- 0.20	—
		R	—	—	+0.50	-0.15	0.10 +/- 0.15	—
2000	Regal	F	+0.50	+3.00	+0.50	-0.90	0.10 +/- 0.20	—
		R	—	—	+0.50	-0.90	0 +/- 0.20	—
2001	Regal	F	+0.50	+3.00	+0.50	-0.90	0.10 +/- 0.20	—
		R	—	—	+0.50	-0.90	0 +/- 0.20	—

93461CW2

WHEEL ALIGNMENT
CHEVROLET IMPALA

Year	Model		Caster Range (Deg.)	Caster Preferred Setting (Deg.)	Camber Range (Deg.)	Camber Preferred Setting (Deg.)	Toe-in (in.)	Steering Axis Inclination (Deg.)
2000	Impala	F	+0.50	+3.20 ①	+0.50	+0.85	0.10 +/- 0.20	0
		R	—	—	+0.50	0	0.10 +/- 0.20	—
2001	Impala	F	+0.50	+3.20 ①	+0.50	+0.85	0.10 +/- 0.20	0
		R	—	—	+0.50	0	0.10 +/- 0.20	—

① Not adjustable, for reference only

93461CW3

WHEEL ALIGNMENT
CHEVROLET LUMINA

Year	Model		Caster Range (Deg.)	Caster Preferred Setting (Deg.)	Camber Range (Deg.)	Camber Preferred Setting (Deg.)	Toe-in (in.)	Steering Axis Inclination (Deg.)
1998	Lumina	F	+0.50	+1.80	+0.50	+0.70	0 +/- 0.09	—
		R	—	—	+0.50	-0.15	0.06 +/- 0.06	—
1999	Lumina	F	+0.50	+1.80	+0.50	+0.70	0 +/- 0.09	—
		R	—	—	+0.50	-0.15	0.06 +/- 0.06	—
2000	Lumina	F	+0.50	+1.80	+0.50	+0.70	0 +/- 0.09	—
		R	—	—	+0.50	-0.15	0.06 +/- 0.06	—
2001	Lumina	F	+0.50	+1.80	+0.50	+0.70	0 +/- 0.09	—
		R	—	—	+0.50	-0.15	0.06 +/- 0.06	—

93461CW4

WHEEL ALIGNMENT
CHEVROLET MONTE CARLO

Year	Model		Caster Range (Deg.)	Caster Preferred Setting (Deg.)	Camber Range (Deg.)	Camber Preferred Setting (Deg.)	Toe-in (in.)	Steering Axis Inclination (Deg.)
1998	Monte Carlo	F	+0.50	+1.80	+0.50	+0.70	0 +/- 0.20	—
		R	—	—	+0.50	-0.15	0.10 +/- 0.10	—
1999	Monte Carlo	F	0.50	+1.80	+0.50	+0.70	0 +/- 0.20	—
		R	—	—	+0.50	-0.15	0.10 +/- 0.10	—
2000	Monte Carlo	F	0.50	+1.80	+0.50	+0.70	0 +/- 0.20	—
		R	—	—	+0.50	-0.15	0.10 +/- 0.10	—
2001	Monte Carlo	F	0.50	+1.80	+0.50	+0.70	0 +/- 0.20	—
		R	—	—	+0.50	-0.15	0.10 +/- 0.10	—

93461CW5

WHEEL ALIGNMENT
OLDSMOBILE CUTLASS

Year	Model		Caster Range (Deg.)	Caster Preferred Setting (Deg.)	Camber Range (Deg.)	Camber Preferred Setting (Deg.)	Toe-in (in.)	Steering Axis Inclination (Deg.)
1998	Cutlass	F	1.00	4.10	1.00	-0.19	0.50 +/- 0.12	13.13
		R	—	—	0.50	-0.19	0.03 +/- 0.04	—
1999	Cutlass	F	1.00	4.10	1.00	-0.19	0.50 +/- 0.12	13.13
		R	—	—	0.50	-0.19	0.03 +/- 0.04	—
2000	Cutlass	F	1.00	4.10	1.00	-0.19	0.50 +/- 0.12	13.13
		R	—	—	0.50	-0.19	0.03 +/- 0.04	—
2001	Cutlass	F	1.00	4.10	1.00	-0.19	0.50 +/- 0.12	13.13
		R	—	—	0.50	-0.19	0.03 +/- 0.04	—

93461CW6

WHEEL ALIGNMENT
OLDSMOBILE INTRIGUE

Year	Model		Caster Range (Deg.)	Caster Preferred Setting (Deg.)	Camber Range (Deg.)	Camber Preferred Setting (Deg.)	Toe-in (in.)	Steering Axis Inclination (Deg.)
1998	Intrigue	F	0.50	3.00	0.50	-0.90	0.05 +/- 0.10	—
		R	—	—	0.50	-0.90	0 +/- 0.10	—
1999	Intrigue	F	0.50	3.00	0.50	-0.90	0.05 +/- 0.10	—
		R	—	—	0.50	-0.90	0 +/- 0.10	—
2000	Intrigue	F	0.50	3.00	0.50	-0.90	0.05 +/- 0.10	—
		R	—	—	0.50	-0.90	0 +/- 0.10	—
2001	Intrigue	F	0.50	3.00	0.50	-0.90	0.05 +/- 0.10	—
		R	—	—	0.50	-0.90	0 +/- 0.10	—

93461CW7

WHEEL ALIGNMENT
PONTIAC GRAND PRIX

Year	Model		Caster Range (Deg.)	Caster Preferred Setting (Deg.)	Camber Range (Deg.)	Camber Preferred Setting (Deg.)	Toe-in (in.)	Steering Axis Inclination (Deg.)
1998	Grand Prix	F	0.50	3.00	0.50	-0.90	0.05 +/- 0.10	14.00
		R	—	—	0.50	-0.90	0 +/- 0.10	—
1999	Grand Prix	F	0.50	3.00	0.50	-0.90	0.05 +/- 0.10	14.00
		R	—	—	0.50	-0.90	0 +/- 0.10	—
2000	Grand Prix	F	0.50	3.00	0.50	-0.90	0.05 +/- 0.10	14.00
		R	—	—	0.50	-0.90	0 +/- 0.10	—
2001	Grand Prix	F	0.50	3.00	0.50	-0.90	0.05 +/- 0.10	14.00
		R	—	—	0.50	-0.90	0 +/- 0.10	—

93461CW8

TIRE, WHEEL AND BALL JOINT SPECIFICATIONS
Buick Century and Regal

Year	Model	OEM Tires		Tire Pressures (psi)		Wheel	Ball Joint
		Standard	Optional	Front	Rear	Size	Inspection
1998	Regal, exc GS	P205/70R15	P215/70R15 P225/60R16	30	30	6-JJ	①
	Regal GS	P225/60R16	None	30	30	6.5-JJ	①
	Century 2dr	P225/60R16	None	30	30	6.5-JJ	①
	Century 4dr	P205/70R15	P225/60R16	30	30	Std: 6-JJ Opt: 6.5-JJ	①
1999	Regal, exc GS	P205/70R15	P215/70R15 P225/60R16	30	30	6-JJ	①
	Regal GS	P225/60R16	None	30	30	6.5-JJ	①
	Century 2dr	P225/60R16	None	30	30	6.5-JJ	①
	Century 4dr	P205/70R15	P225/60R16	30	30	Std: 6-JJ Opt: 6.5-JJ	①
2000	Regal, exc GS	P205/70R15	P215/70R15 P225/60R16	30	30	6-JJ	①
	Regal GS	P225/60R16	None	30	30	6.5-JJ	①
	Century 2dr	P225/60R16	None	30	30	6.5-JJ	①
	Century 4dr	P205/70R15	P225/60R16	30	30	Std: 6-JJ Opt: 6.5-JJ	①
2001	Regal, exc GS	P205/70R15	P215/70R15 P225/60R16	30	30	6-JJ	①
	Regal GS	P225/60R16	None	30	30	6.5-JJ	①
	Century 2dr	P225/60R16	None	30	30	6.5-JJ	①
	Century 4dr	P205/70R15	P225/60R16	30	30	Std: 6-JJ Opt: 6.5-JJ	①

OEM: Original Equipment Manufacturer

PSI: Pounds Per Square Inch

STD: Standard

OPT: Optional

L: Lower

U: Upper

① Replace if any measurable movement is found.

93461CW9

TIRE, WHEEL AND BALL JOINT SPECIFICATIONS
Chevrolet Impala, Lumina and Monte Carlo

Year	Model	OEM Tires		Tire Pressures (psi)		Wheel Size	Ball Joint Inspection
		Standard	Optional	Front	Rear		
1998	Monte Carlo	P195/70R14	None	30	30	6-JJ	①
			P225/60R16N	30	30	6-JJ	
	Impala	P225/60R16	P225/60R16	30	30	6-JJ	①
			P225/60R16N	30	30	6-JJ	
	Lumina 3.1L	P205/70R15	P215/65R16	30	30	6-JJ	①
			P225/60R16	30	30	6.5-JJ	
	Lumina 3.4L	P225/60R16	None	30	30	6.5-JJ	①
1999	Monte Carlo	P195/70R14	None	30	30	6-JJ	①
			P225/60R16N	30	30	6-JJ	
	Impala	P225/60R16	P225/60R16	30	30	6-JJ	①
			P225/60R16N	30	30	6-JJ	
	Lumina 3.1L	P205/70R15	P215/65R16	30	30	6-JJ	①
			P225/60R16	30	30	6.5-JJ	
	Lumina 3.4L	P225/60R16	None	30	30	6.5-JJ	①
2000-01	Monte Carlo	P195/70R14	None	30	30	6-JJ	①
			P225/60R16N	30	30	6-JJ	
	Impala	P225/60R16	P225/60R16	30	30	6-JJ	①
			P225/60R16N	30	30	6-JJ	
	Lumina 3.1L	P205/70R15	P215/65R16	30	30	6-JJ	①
			P225/60R16	30	30	6.5-JJ	
	Lumina 3.4L	P225/60R16	None	30	30	6.5-JJ	①
2001	Monte Carlo	P195/70R14	None	30	30	6-JJ	①
			P225/60R16N	30	30	6-JJ	
	Impala	P225/60R16	P225/60R16	30	30	6-JJ	①
			P225/60R16N	30	30	6-JJ	
	Lumina 3.1L	P205/70R15	P215/65R16	30	30	6-JJ	①
			P225/60R16	30	30	6.5-JJ	
	Lumina 3.4L	P225/60R16	None	30	30	6.5-JJ	①

OEM: Original Equipment Manufacturer

PSI: Pounds Per Square Inch

① Replace if any measurable movement is found

93461CW0

TIRE, WHEEL AND BALL JOINT SPECIFICATIONS
Oldsmobile Intrigue and Cutlass Supreme

		OEM Tires		Tire Pressures (psi)		Wheel	Ball Joint
Year	Model	Standard	Optional	Front	Rear	Size	Inspection
1998	Cutlass Supreme	P215/60R16	None	30	30	6.5-JJ	①
	Intrigue	P225/60R16	None	30	30	6.5-JJ	①
1999	Cutlass Supreme	P215/60R16	None	30	30	6.5-JJ	①
	Intrigue	P225/60R16	None	30	30	6.5-JJ	①
2000	Cutlass Supreme	P215/60R16	None	30	30	6.5-JJ	①
	Intrigue	P225/60R16	None	30	30	6.5-JJ	①
2001	Intrigue	P225/60R16	None	30	30	6-JJ	①

OEM: Original Equipment Manufacturer

PSI: Pounds Per Square Inch

STD: Standard

OPT: Optional

① Replace if any measurable movement is found.

93461CX1

TIRE, WHEEL AND BALL JOINT SPECIFICATIONS
Pontiac Grand Prix

		OEM Tires		Tire Pressures (psi)		Wheel	Ball Joint
Year	Model	Standard	Optional	Front	Rear	Size	Inspection
1998	Grand Prix	P205/70R15	P225/60R16	30	30	7-JJ	0.125 in.
1999	Grand Prix	P205/70R15	P225/60R16	30	30	7-JJ	0.125 in.
2000-01	Grand Prix	P205/70R15	P225/60R16	30	30	7-JJ	0.125 in.
2000-01	Grand Prix	P205/70R15	P225/60R16	30	30	7-JJ	0.125 in.

OEM: Original Equipment Manufacturer

PSI: Pounds Per Square Inch

93461CX2

SCHEDULED MAINTENANCE INTERVALS

GM W BODY—BUICK CENTURY, REGAL, CHEVROLET IMPALA, LUMINA, MONTE CARLO, OLDSMOBILE INTRIGUE, CUTLASS SUPREME & PONTIAC GRAND PRIX

TO BE SERVICED	TYPE OF SERVICE	VEHICLE MILEAGE INTERVAL (x1000)												
		7.5	15	22.5	30	37.5	45	52.5	60	67.5	75	82.5	90	97.5
Engine oil & filter	R	✓	✓	✓	✓	✓	✓	✓	✓	✓	✓	✓	✓	✓
Automatic transaxle fluid & filter ①	S/I	✓	✓	✓	✓	✓	✓	✓	✓	✓	✓	✓	✓	✓
Brake hoses	S/I	✓	✓	✓	✓	✓	✓	✓	✓	✓	✓	✓	✓	✓
Coolant level, hoses & clamps	S/I	✓	✓	✓	✓	✓	✓	✓	✓	✓	✓	✓	✓	✓
Drive shaft boots & front suspension components	S/I	✓	✓	✓	✓	✓	✓	✓	✓	✓	✓	✓	✓	✓
Exhaust system & throttle linkage	S/I	✓	✓	✓	✓	✓	✓	✓	✓	✓	✓	✓	✓	✓
Lubricate chassis, suspension, steering linkage, transaxle shift linkage, parking brake cable guides, underbody contact points & linkage	S/I	✓	✓	✓	✓	✓	✓	✓	✓	✓	✓	✓	✓	✓
Rotate tires	S/I	✓		✓		✓		✓		✓		✓		✓
Air filter element	R				✓				✓				✓	
Engine coolant ②	R				✓				✓				✓	
PCV filter	R				✓				✓				✓	
Spark plugs ③	R				✓				✓				✓	
Accessory drive belt(s)	S/I				✓				✓				✓	
Ignition cables, EGR & fuel systems	S/I				✓				✓				✓	
Camshaft timing belt	R								✓					

R: Replace S/I: Service or Inspect

① Automatic transaxle fluid & filter: replace at 100,000 miles (if not changed previously).

② Engine coolant: replace every 100,000 miles. Use O.E. specified (DEX-COOL™) coolant only. If any silicate coolant is used, the service interval is every 30,000 miles

③ Platinum tip spark plugs: replace every 100,000 miles.

FREQUENT OPERATION MAINTENANCE (SEVERE SERVICE)

If a vehicle is operated under any of the following conditions it is considered severe service:

- Extremely dusty areas.

- 50% or more of the vehicle operation is in 32°C (90°F) or higher temperatures, or constant operation in temperatures below 0°C (32°F).

- Prolonged idling (vehicle operation in stop and go traffic).

- Frequent short running periods (engine does not warm to normal operating temperatures).

- Police, taxi, delivery usage or trailer towing usage.

Oil & oil filter: change every 3000 miles

Chassis lubrication: lubricate every 6000 miles.

Rotate tires at 6000 miles, then every 12,000 miles.

Air filter element: service or inspect every 15,000 miles.

Camshaft timing belt: change every 60,000 miles.

93461CX3

SCHEDULED MAINTENANCE INTERVALS
GENERAL MOTORS CORPORATION
W BODY
BUICK CENTURY, REGAL
CHEVROLET IMPALA, LUMINA, MONTE CARLO
OLDSMOBILE INTRIGUE, CUTLASS SUPREME
PONTIAC GRAND PRIX

The following should be used as a guide when determining the amount of work required for a particular service. In estimating how long a particular Scheduled Maintenance Service should take, please observe the following:

- Labor Time is time based on field research and data supplied by the vehicle manufacturer.
- Labor time operations are given in hours and tenths of an hour.
- All labor operations are to be used as a guide.

Mechanic Skill Level Codes:
(A) PRECISION: Highly skilled with multiple certification.
(B) GENERAL: Normally skilled with certification.
(C) MAINTENANCE: Semi-skilled working on certification.

	LABOR TIME
7500 Mile Service (C)	
All Models	1.3
15000 Mile Service (C)	
All Models	.8
22500 Mile Service (C)	
All Models	1.3
30000 Mile Service (B)	
All Models	3.0
37500 Mile Service (C)	
All Models	1.3
45000 Mile Service (C)	
All Models	.8
52500 Mile Service (C)	
All Models	1.3
60000 Mile Service (B)	
All Models	3.0
Add for timing belt R&R.	
67500 Mile Service (C)	
All Models	1.3
75000 Mile Service (C)	
All Models	.8
82500 Mile Service (C)	
All Models	1.3
90000 Mile Service (B)	
All Models	3.0
97500 Mile Service (C)	
All Models	1.3

93461CX4

Heater Core replacement is covered in Section 2 of this manual

GENERAL MOTORS Y-BODY
Chevrolet Corvette

ENGINE AND VEHICLE IDENTIFICATION

Engine						
Code ①	Liters (cc)	Cu. In.	Cyl.	Fuel Sys.	Engine Type	Eng. Mfg.
G	5.7 (5665)	350	8	SFI	OHV	CPC

Model Year	
Code ②	Year
W	1998
X	1999
Y	2000
1	2001
2	2002

CPC: Chevrolet/Pontiac/Canada

MFI: Multi-point Fuel Injection

OHV: Over Head Valves

① 8th position of VIN

② 10th position of VIN

93461CX5

GENERAL ENGINE SPECIFICATIONS

Year	Model	Engine Displacement Liters (cc)	Engine Series (ID/VIN)	Fuel System	Net Horsepower @ rpm	Net Torque @ rpm (ft. lbs.)	Bore x Stroke (in.)	Com-pression Ratio	Oil Pressure @ rpm
1998	Corvette	5.7 (5665)	G	SFI	345@5600	350@4400	3.89x3.62	10.1:1	18@2000
1999	Corvette	5.7 (5665)	G	SFI	345@5600	350@4400	3.89x3.62	10.1:1	18@2000
2000	Corvette	5.7 (5665)	G	SFI	345@5600	350@4400	3.89x3.62	10.1:1	18@2000
2001	Corvette	5.7 (5665)	G	SFI	345@5600	350@4400	3.89x3.62	10.1:1	18@2000

SFI: Sequential Fuel Injection

93461CX6

ENGINE TUNE-UP SPECIFICATIONS

Year	Engine Displacement Liters (cc)	Engine ID/VIN	Spark Plug Gap (in.)	Ignition Timing (deg.)		Fuel Pump (psi)	Idle Speed (rpm)		Valve Clearance	
				MT	AT		MT	AT	In.	Ex.
1998	5.7 (5665)	G	0.060	①	①	48-55	①	①	HYD	HYD
1999	5.7 (5665)	G	0.060	①	①	48-55	①	①	HYD	HYD
2000	5.7 (5665)	G	0.060	①	①	48-55	①	①	HYD	HYD
2001	5.7 (5665)	G	0.060	①	①	48-55	①	①	HYD	HYD

NOTE: The Vehicle Emission Control Information label often reflects specification changes made during production. The label figures must be used if they differ from those in this chart.

HYD: Hydraulic

① Refer to Vehicle Emission Control Information label

93461CX7

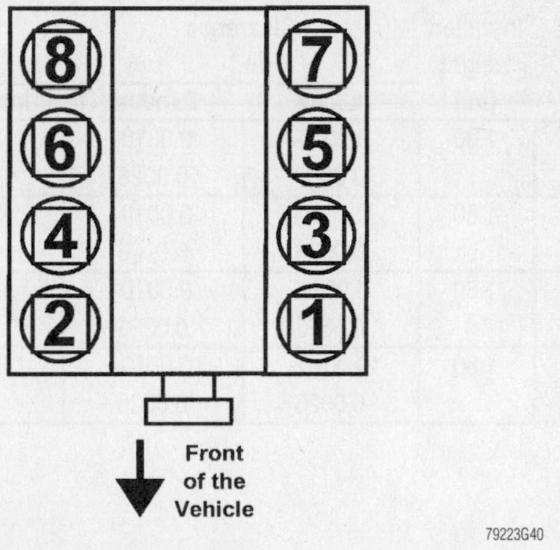

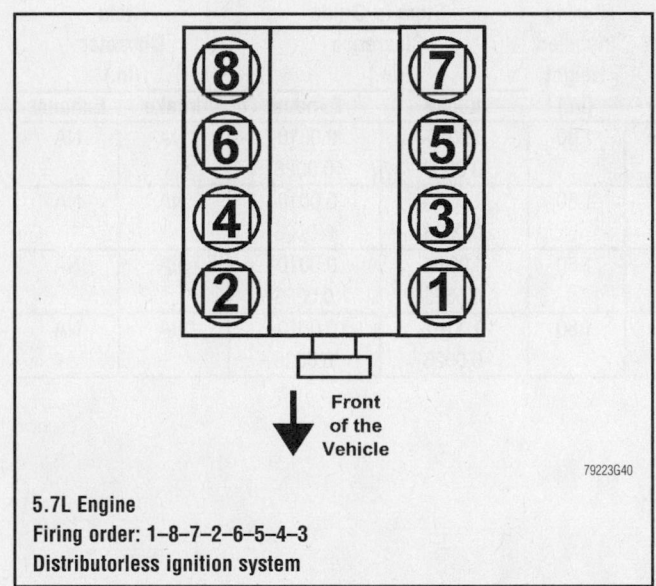

5.7L Engine
Firing order: 1–8–7–2–6–5–4–3
Distributorless ignition system

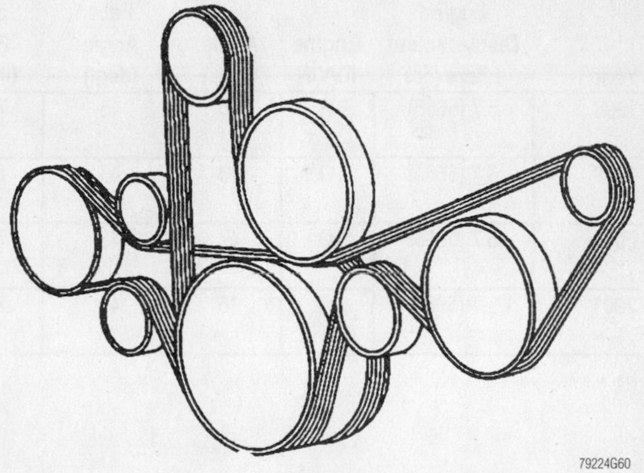

Serpentine drive belt routing—GM 5.7L (VIN G) engine

CAPACITIES

Year	Model	Engine Displacement Liters (cc)	Engine ID/VIN	Engine Oil with Filter (qts.)	Transmission (pts.) 6-Spd	Transmission (pts.) Auto.	Drive Axle (pts.)	Fuel Tank (gal.)	Cooling System (qts.)
1998	Corvette	5.7 (5737)	G	6.5	①	②	3.8	20.0	14.7
1999	Corvette	5.7 (5737)	G	6.5	①	②	3.8	20.0	14.7
2000	Corvette	5.7 (5737)	G	6.5	①	②	3.8	20.0	14.7
2001	Corvette	5.7 (5737)	G	6.5	①	②	3.8	20.0	14.7

NOTE: All capacities are approximate. Add fluid gradually and ensure a proper fluid level is obtained.

① MM 6 speed trans: 4.1 pts.

② 4L60E trans: 10.0 pts.

93461CX8

VALVE SPECIFICATIONS

Year	Engine Displacement Liters (cc)	Engine ID/VIN	Seat Angle (deg.)	Face Angle (deg.)	Spring Test Pressure (lbs. @ in.)	Spring Installed Height (in.)	Stem-to-Guide Clearance (in.) Intake	Stem-to-Guide Clearance (in.) Exhaust	Stem Diameter (in.) Intake	Stem Diameter (in.) Exhaust
1998	5.7 (5665)	G	46	45	76@1.80	1.80	0.0010-0.0026	0.0010-0.0026	NA	NA
1999	5.7 (5665)	G	46	45	76@1.80	1.80	0.0010-0.0026	0.0010-0.0026	NA	NA
2000	5.7 (5665)	G	46	45	76@1.80	1.80	0.0010-0.0026	0.0010-0.0026	NA	NA
2001	5.7 (5665)	G	46	45	76@1.80	1.80	0.0010-0.0026	0.0010-0.0026	NA	NA

NA: Not Available

93461CX0

CRANKSHAFT AND CONNECTING ROD SPECIFICATIONS

All measurements are given in inches.

Year	Engine Displacement Liters (cc)	Engine ID/VIN	Crankshaft Main Brg. Journal Dia.	Crankshaft Main Brg. Oil Clearance	Crankshaft Shaft End-play	Crankshaft Thrust on No.	Connecting Rod Journal Diameter	Connecting Rod Oil Clearance	Connecting Rod Side Clearance
1998	5.7 (5665)	G	2.558-2.559	0.0007-0.0021	0.0015-0.0078	5	2.0987	0.0006-0.0025	0.0043-0.0200
1999	5.7 (5665)	G	2.558-2.559	0.0007-0.0021	0.0015-0.0078	5	2.0987	0.0006-0.0025	0.0043-0.0200
2000	5.7 (5665)	G	2.558-2.559	0.0007-0.0021	0.0015-0.0078	5	2.0987	0.0006-0.0025	0.0043-0.0200
2001	5.7 (5665)	G	2.558-2.559	0.0007-0.0021	0.0015-0.0078	5	2.0987	0.0006-0.0025	0.0043-0.0200

93461CX9

PISTON AND RING SPECIFICATIONS

All measurements are given in inches.

Year	Engine Displacement Liters (cc)	Engine ID/VIN	Piston Clearance	Ring Gap Top Compression	Ring Gap Bottom Compression	Ring Gap Oil Control	Ring Side Clearance Top Compression	Ring Side Clearance Bottom Compression	Ring Side Clearance Oil Control
1998	5.7 (5665)	G	0.0007-0.0021	0.009-0.015	0.017-0.025	0.007-0.027	0.0016-0.0033	0.0016-0.0031	0.0004-0.0087
1999	5.7 (5665)	G	0.0007-0.0021	0.009-0.015	0.017-0.025	0.007-0.027	0.0016-0.0033	0.0016-0.0031	0.0004-0.0087
2000	5.7 (5665)	G	0.0007-0.0021	0.009-0.015	0.017-0.025	0.007-0.027	0.0016-0.0033	0.0016-0.0031	0.0004-0.0087
2001	5.7 (5665)	G	0.0007-0.0021	0.009-0.015	0.017-0.025	0.007-0.027	0.0016-0.0033	0.0016-0.0031	0.0004-0.0087

93461CY1

TORQUE SPECIFICATIONS

All readings in ft. lbs.

Year	Engine Displacement Liters (cc)	Engine ID/VIN	Cylinder Head Bolts	Main Bearing Bolts	Rod Bearing Bolts	Crankshaft Damper Bolts	Flywheel Bolts	Manifold		Spark Plugs	Lug Nut
								Intake	Exhaust		
1998	5.7 (5665)	G	①	②	③	④	⑤	⑥	⑦	11	100
1999	5.7 (5665)	G	①	②	③	④	⑤	⑥	⑦	11	100
2000	5.7 (5665)	G	①	②	③	④	⑤	⑥	⑦	11	100
2001	5.7 (5665)	G	①	②	③	④	⑤	⑥	⑦	11	100

① M11 bolts: 22 ft. lbs.
M11 bolts: plus 90 degrees
M11 bolts 1<en dash>8: plus 90 degrees
M11 bolts 9<en dash>10: plus 50 degrees
M8 bolts (11<en dash>15): 22 ft. lbs.

② Inner Bolts:
Step 1: 15 ft. lbs.
Step 2: 80 degrees
Side bolts: 18 ft. lbs.
Outer studs:
Step 1: 15 ft. lbs.
Step 2: 53 degrees

③ Step 1: 15 ft. lbs.
Step 2: Rotate 75 degrees

④ Step 1: 240 ft. lbs.
Step 2: Install a new bolt and tighten to 37 ft. lbs.
Step 3: Rotate 140 degrees

⑤ Step 1: 15 ft. lbs.
Step 2: 37 ft. lbs.
Step 3: 74 ft. lbs.

⑥ Step 1: 44 inch lbs.
Step 2: 89 inch lbs.

⑦ Step 1: 11 ft. lbs.
Step 2: 18 ft. lbs.

93461CY2

BRAKE SPECIFICATIONS

All measurements in inches unless noted

Year	Model	Brake Disc			Minimum Lining Thickness		Brake Caliper	
		Original Thickness	Minimum Thickness	Maximum Runout	Front	Rear	Bracket Bolts (ft. lbs.)	Mounting Bolts (ft. lbs.)
1998	Corvette	①	②	0.006	0.030	0.030	—	③
1999	Corvette	①	②	0.006	0.030	0.030	—	③
2000	Corvette	①	②	0.006	0.030	0.030	—	③
2001	Corvette	①	②	0.006	0.030	0.030	—	③

① Heavy duty: 1.110; Std.: 0.795

② Heavy duty: 1.059; Std.: 0.744

③ Front: not available, Rear: upper 26 ft. lbs., lower 16 ft. lbs.

93461CY3

WHEEL ALIGNMENT

			Caster		Camber			Steering
				Preferred		Preferred		Axis
			Range	Setting	Range	Setting	Toe-in	Inclination
Year	Model		(+/-Deg.)	(Deg.)	(+/-Deg.)	(Deg.)	(in.)	(Deg.)
1998	Corvette	F	0.50	+6.90	0.50	-0.20	0.05 +/- 0.10	—
		R	—	—	0.50	-0.18	0 +/- 0.10	—
1999	Corvette	F	0.50	+6.90	0.50	-0.20	0.05 +/- 0.10	—
		R	—	—	0.50	-0.18	0 +/- 0.10	—
2000	Corvette	F	0.50	+6.90	0.50	-0.20	0.05 +/- 0.10	—
		R	—	—	0.50	-0.18	0 +/- 0.10	—
2001	Corvette	F	0.50	+6.90	0.50	-0.20	0.05 +/- 0.10	—
		R	—	—	0.50	-0.18	0 +/- 0.10	—

93461CY4

TIRE, WHEEL AND BALL JOINT SPECIFICATIONS

		OEM Tires		Tire Pressures (psi)		Wheel	Ball Joint
Year	Model	Standard	Optional	Front	Rear	Size	Inspection
1998	Corvette	Fr: P245/45ZR17 Rr: P275/40ZR18	None	30	30	8.5	U: 0.125 in. L: 0.047 in.
1999	Corvette	Fr: P245/45ZR17 Rr: P275/40ZR18	None	30	30	8.5	U: 0.125 in. L: 0.047 in.
2000	Corvette	Fr: P245/45ZR17 Rr: P275/40ZR18	None	30	30	8.5	U: 0.125 in. L: 0.047 in.
2001	Corvette	Fr: P245/45ZR17 Rr: P275/40ZR18	None	30	30	8.5	U: 0.125 in. L: 0.047 in.

OEM: Original Equipment Manufacturer
PSI: Pounds Per Square Inch
L:Lower
U:Upper
Fr: Front
Rr: Rear

93461CY5

For Tire, Wheel and Ball Joint specifications, see Section 1 of this manual

SCHEDULED MAINTENANCE INTERVALS
GM Y BODY—CHEVROLET CORVETTE

TO BE SERVICED	TYPE OF SERVICE	VEHICLE MILEAGE INTERVAL (x1000)												
		7.5	15	22.5	30	37.5	45	52.5	60	67.5	75	82.5	90	97.5
Engine oil & filter ①	R	✓	✓	✓	✓	✓	✓	✓	✓	✓	✓	✓	✓	✓
Brake hoses & brake lining	S/I	✓	✓	✓	✓	✓	✓	✓	✓	✓	✓	✓	✓	✓
Coolant level, hoses & clamps	S/I	✓	✓	✓	✓	✓	✓	✓	✓	✓	✓	✓	✓	✓
Exhaust system & throttle linkage	S/I	✓	✓	✓	✓	✓	✓	✓	✓	✓	✓	✓	✓	✓
Lubricate chassis, suspension, steering linkage, transaxle shift linkage, parking brake cable guides, underbody contact points & linkage	S/I	✓	✓	✓	✓	✓	✓	✓	✓	✓	✓	✓	✓	✓
Rear axle fluid level	S/I	✓	✓	✓	✓	✓	✓	✓	✓	✓	✓	✓	✓	✓
Automatic transaxle fluid & filter ②	S/I		✓		✓		✓		✓		✓		✓	
Air filter element	R				✓				✓				✓	
Engine coolant ③	R				✓				✓				✓	
Ignition cables, EGR & fuel systems	S/I				✓				✓				✓	
Serpentine drive belt	S/I				✓				✓				✓	
Spark plugs ④	R													

R: Replace S/I: Service or Inspect

① Corvette engines require a special oil meeting GM Standard 4718M.

② Automatic transaxle fluid & filter: replace at 100,000 miles (if not changed previously).

③ Engine coolant: replace every 100,000 miles. Use O.E. specified (DEX-COOL™) coolant only. If any silicate coolant is used, the service interval is every 30,000 miles.

④ Platinum tip spark plugs: replace every 100,000 miles.

FREQUENT OPERATION MAINTENANCE (SEVERE SERVICE)

If a vehicle is operated under any of the following conditions it is considered severe service:

- Extremely dusty areas.

- 50% or more of the vehicle operation is in 32°C (90°F) or higher temperatures, or constant operation in temperatures below 0°C (32°F).

- Prolonged idling (vehicle operation in stop and go traffic).

- Frequent short running periods (engine does not warm to normal operating temperatures).

- Police, taxi, delivery usage or trailer towing usage.

Engine oil & oil filter: change every 3000 miles

Chassis lubrication: lubricate every 6000 miles.

Lubricate suspension, parking brake cable guides, underbody contact points & linkage: lubricate every 6000 miles.

Air filter element: service or inspect every 15,000 miles.

Automatic transmission fluid & filter: change every 15,000 miles.

93461CAL

SCHEDULED MAINTENANCE INTERVALS
GENERAL MOTORS CORPORATION
Y BODY
CHEVROLET CORVETTE

The following should be used as a guide when determining the amount of work required for a particular service. In estimating how long a particular Scheduled Maintenance Service should take, please observe the following:

- Labor Time is time based on field research and data supplied by the vehicle manufacturer.
- Labor time operations are given in hours and tenths of an hour.
- All labor operations are to be used as a guide.

Mechanic Skill Level Codes:
(A) PRECISION: Highly skilled with multiple certification.
(B) GENERAL: Normally skilled with certification.
(C) MAINTENANCE: Semi-skilled working on certification.

	LABOR TIME
7500 Mile Service (C)	
All Models	1.2
15000 Mile Service (C)	
All Models	1.3
22500 Mile Service (C)	
All Models	1.2
30000 Mile Service (B)	
All Models	2.5
37500 Mile Service (C)	
All Models	1.2
45000 Mile Service (C)	
All Models	1.3
52500 Mile Service (C)	
All Models	1.2
60000 Mile Service (B)	
All Models	2.5
67500 Mile Service (C)	
All Models	1.2
75000 Mile Service (C)	
All Models	1.3
82500 Mile Service (C)	
All Models	1.2
90000 Mile Service (B)	
All Models	2.5
97500 Mile Service (C)	
All Models	1.2

93461CY6

GEO/CHEVROLET
Metro • Prism

ENGINE AND VEHICLE IDENTIFICATION

Engine						
Code ①	Liters (cc)	Cu. In.	Cyl.	Fuel Sys.	Engine Type	Eng. Mfg.
6	1.0 (993)	(61)	3	MFI	SOHC	Suzuki
8	1.8 (1803)	(110)	4	MFI	DOHC	Toyota
9	1.3 (1300)	(79)	4	MFI	DOHC	Suzuki

Model Year	
Code ②	Year
W	1998
X	1999
Y	2000
1	2001
2	2002

MFI: Multi-point Fuel Injection

DOHC: Dual Overhead Camshaft

SOHC: Single Overhead Camshaft

① 8th position of VIN

② 10th position of VIN

93461CY7

GENERAL ENGINE SPECIFICATIONS

Year	Model	Engine Displacement Liters (cc)	Engine Series (ID/VIN)	Fuel System	Net Horsepower @ rpm	Net Torque @ rpm (ft. lbs.)	Bore x Stroke (in.)	Com-pression Ratio	Oil Pressure @ rpm
1998	Metro	1.0 (993)	6	MFI	55@5700	58@3300	2.91x3.03	9.5:1	54@3000
	Metro	1.3 (1300)	9	MFI	70@5500	74@3500	2.91x3.03	9.5:1	54@3000
	Prizm	1.8 (1803)	8	MFI	115@5200	117@2800	3.20x3.40	9.5:1	36-71@3000
1999	Metro	1.0 (993)	6	MFI	55@5700	58@3300	2.91x3.03	9.5:1	54@3000
	Metro	1.3 (1300)	9	MFI	70@5500	74@3500	2.91x3.03	9.5:1	54@3000
	Prizm	1.8 (1803)	8	MFI	115@5200	117@2800	3.20x3.40	9.5:1	36-71@3000
2000	Metro	1.0 (993)	6	MFI	55@5700	58@3300	2.91x3.03	9.5:1	54@3000
	Metro	1.3 (1300)	9	MFI	70@5500	74@3500	2.91x3.03	9.5:1	54@3000
	Prizm	1.8 (1803)	8	MFI	115@5200	117@2800	3.20x3.40	9.5:1	36-71@3000
2001	Metro	1.0 (993)	6	MFI	55@5700	58@3300	2.91x3.03	9.5:1	54@3000
	Metro	1.3 (1300)	9	MFI	70@5500	74@3500	2.91x3.03	9.5:1	54@3000
	Prizm	1.8 (1803)	8	MFI	115@5200	117@2800	3.20x3.40	9.5:1	36-71@3000

MFI: Multi-point Fuel Injection

93461CY8

TUNE-UP SPECIFICATIONS

Year	Engine Displacement Liters (cc)	Engine ID/VIN	Spark Plug Gap (in.)	Ignition Timing (deg.) MT	Ignition Timing (deg.) AT	Fuel Pump (psi)	Idle Speed (rpm) MT	Idle Speed (rpm) AT	Valve Clearance Intake	Valve Clearance Exhaust
1998	1.0 (993)	6	0.041	5B ①	5B ①	23-30	800	850	HYD	HYD
	1.3 (1300)	9	0.041	5B ①	5B ①	23-30	800	850	HYD	HYD
	1.8 (1803)	8	0.031	10B ②	10B ②	31-37	700-750	700-750	0.0060-0.0100	0.0100-0.0140
1999	1.0 (993)	6	0.041	5B ①	5B ①	23-30	800	850	HYD	HYD
	1.3 (1300)	9	0.041	5B ①	5B ①	23-30	800	850	HYD	HYD
	1.8 (1803)	8	0.031	10B ②	10B ②	31-37	700-750	700-750	0.0060-0.0100	0.0100-0.0140
2000	1.0 (993)	6	0.041	5B ①	5B ①	23-30	800	850	HYD	HYD
	1.3 (1300)	9	0.041	5B ①	5B ①	23-30	800	850	HYD	HYD
	1.8 (1803)	8	0.031	10B ②	10B ②	31-37	700-750	700-750	0.0060-0.0100	0.0100-0.0140
2001	1.0 (993)	6	0.041	5B ①	5B ①	23-30	800	850	HYD	HYD
	1.3 (1300)	9	0.041	5B ①	5B ①	23-30	800	850	HYD	HYD
	1.8 (1803)	8	0.031	10B ②	10B ②	31-37	700-750	700-750	0.0060-0.0100	0.0100-0.0140

NOTE: The Vehicle Emission Control Information label often reflects specification changes made during production. The label figures must be used if they differ from those in this chart.

B: Before top dead center

HYD: Hydraulic

① Connect a fused jumper from Duty Check cavity 4 to cavity 5 for fixed timing (DLC connector located at left strut tower)

② Insert jumper wire between terminals in DLC connector E1 and TE1

93461CY9

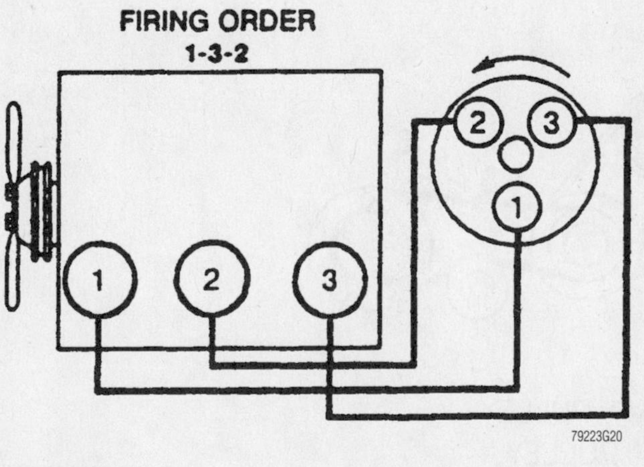

1.0L Engine
Firing order: 1–3–2
Distributor rotation: Counterclockwise

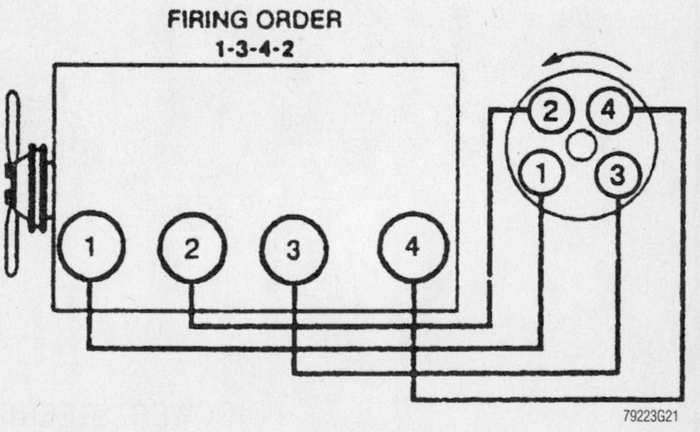

1.3L Engine
Firing order: 1–3–4–2
Distributor rotation: Counterclockwise

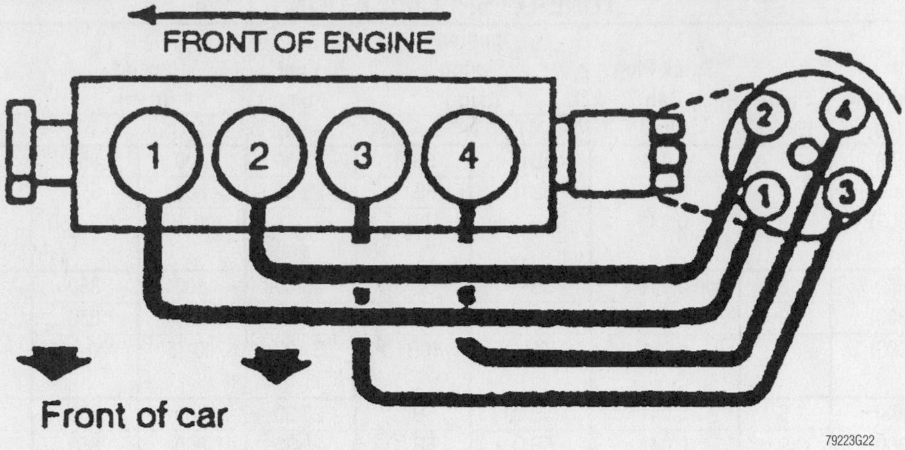

GEO 1.8L Engines
Firing order: 1–3–4–2
Distributor rotation: Counterclockwise

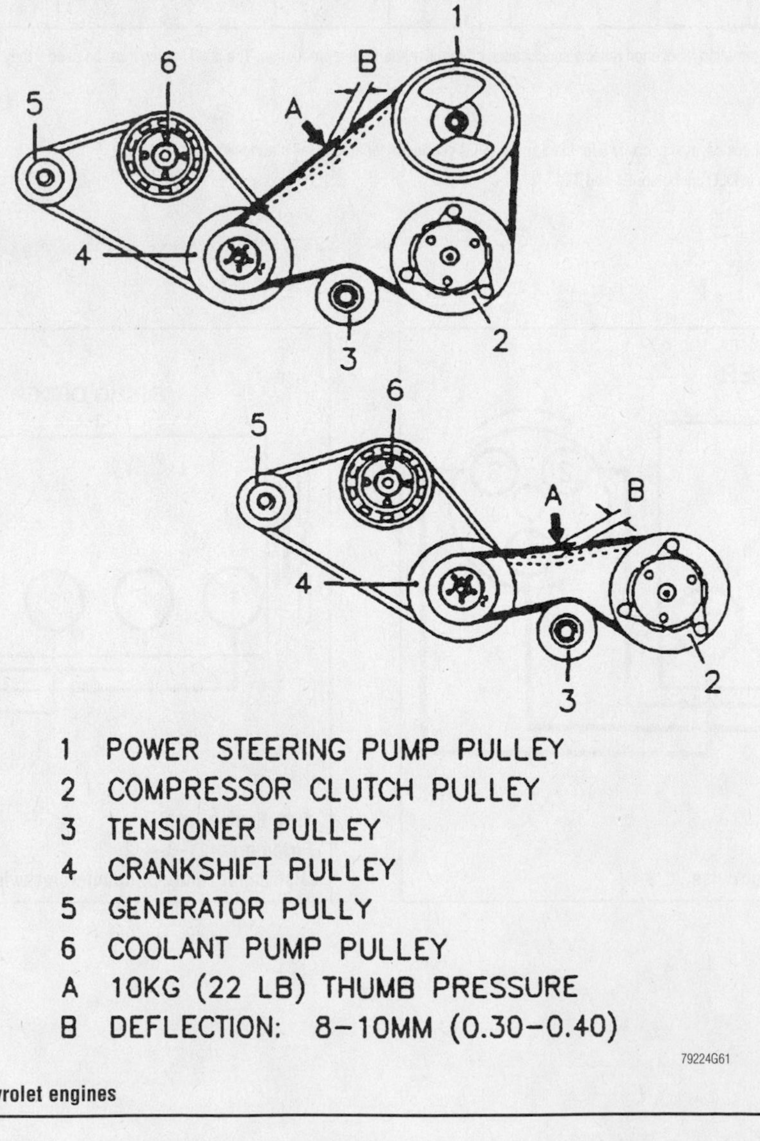

Accessory drive belt—GEO/Chevrolet engines

CAPACITIES

Year	Model	Engine ID/VIN	Engine Displacement Liters (cc)	Engine Oil with Filter	Transmission (pts.)		Drive Axle (pts.)	Fuel Tank (gal.)	Cooling System (qts.)
					5-Spd	Auto.			
1998	Metro	6	1.0 (993)	3.7	5.0	10.1 ①	—	10.6	4.2
	Metro	9	1.3 (1300)	3.7	5.0	10.1 ①	—	10.6	4.9
	Prizm	8	1.8 (1803)	3.9	4.0	6.6	3.0 ②	13.2	6.7
1999	Metro	6	1.0 (993)	3.7	5.0	10.1 ①	—	10.6	4.2
	Metro	9	1.3 (1300)	3.7	5.0	10.1 ①	—	10.6	4.9
	Prizm	8	1.8 (1803)	3.9	4.0	6.6	3.0 ②	13.2	6.7
2000	Metro	6	1.0 (993)	3.7	5.0	10.1 ①	—	10.6	4.2
	Metro	9	1.3 (1300)	3.7	5.0	10.1 ①	—	10.6	4.9
	Prizm	8	1.8 (1803)	3.9	4.0	6.6	3.0 ②	13.2	6.7
2001	Metro	6	1.0 (993)	3.7	5.0	10.1 ①	—	10.6	4.2
	Metro	9	1.3 (1300)	3.7	5.0	10.1 ①	—	10.6	4.9
	Prizm	8	1.8 (1803)	3.9	4.0	6.6	3.0 ②	13.2	6.7

NOTE: All capacities are approximate. Add fluid gradually and ensure a proper fluid level is obtained.

① Automatic transmission: Specification is after complete overhaul. Drain and fill will be less

② 3 speed automatic only

93461CY0

VALVE SPECIFICATIONS

Year	Engine Displacement Liters (cc)	Engine ID/VIN	Seat Angle (deg.)	Face Angle (deg.)	Spring Test Pressure (lbs. @ in.)	Spring Installed Height (in.)	Stem-to-Guide Clearance (in.)		Stem Diameter (in.)	
							Intake	Exhaust	Intake	Exhaust
1998	1.0 (993)	6	45	45	46.1-51.8@ 1.28	1.28	0.0008-0.0022	0.0018-0.0028	0.2148-0.2157	0.2142-0.2148
	1.3 (1300)	9	45	45	54.7-64.3@ 1.63	1.63	0.0008-0.0019	0.0014-0.0025	0.2742-0.2748	0.2737-0.2742
	1.8 (1803)	8	45	45.5	37.3@1.25	1.25	0.0010-0.0024	0.0012-0.0026	0.2350-0.2356	0.2348-0.2354
1999	1.0 (993)	6	45	45	46.1-51.8@ 1.28	1.28	0.0008-0.0022	0.0018-0.0028	0.2148-0.2157	0.2142-0.2148
	1.3 (1300)	9	45	45	54.7-64.3@ 1.63	1.63	0.0008-0.0019	0.0014-0.0025	0.2742-0.2748	0.2737-0.2742
	1.8 (1803)	8	45	45.5	37.3@1.25	1.25	0.0010-0.0024	0.0012-0.0026	0.2350-0.2356	0.2348-0.2354
2000	1.0 (993)	6	45	45	46.1-51.8@ 1.28	1.28	0.0008-0.0022	0.0018-0.0028	0.2148-0.2157	0.2142-0.2148
	1.3 (1300)	9	45	45	54.7-64.3@ 1.63	1.63	0.0008-0.0019	0.0014-0.0025	0.2742-0.2748	0.2737-0.2742
	1.8 (1803)	8	45	45.5	37.3@1.25	1.25	0.0010-0.0024	0.0012-0.0026	0.2350-0.2356	0.2348-0.2354
2001	1.0 (993)	6	45	45	46.1-51.8@ 1.28	1.28	0.0008-0.0022	0.0018-0.0028	0.2148-0.2157	0.2142-0.2148
	1.3 (1300)	9	45	45	54.7-64.3@ 1.63	1.63	0.0008-0.0019	0.0014-0.0025	0.2742-0.2748	0.2737-0.2742
	1.8 (1803)	8	45	45.5	37.3@1.25	1.25	0.0010-0.0024	0.0012-0.0026	0.2350-0.2356	0.2348-0.2354

93461CYZ1

CRANKSHAFT AND CONNECTING ROD SPECIFICATIONS

All measurements are given in inches.

Year	Engine Displacement Liters (cc)	Engine ID/VIN	Crankshaft Main Brg. Journal Dia.	Crankshaft Main Brg. Oil Clearance	Crankshaft Shaft End-play	Crankshaft Thrust on No.	Connecting Rod Journal Diameter	Connecting Rod Oil Clearance	Connecting Rod Side Clearance
1998	1.0 (993)	6	①	0.00008-0.00015	0.0044-0.0122	NA	1.6529-1.6535	0.0012-0.0019	0.0039-0.0078
	1.3 (1300)	9	①	0.00008-0.00015	0.0044-0.0122	NA	1.6529-1.6535	0.0012-0.0019	0.0039-0.0078
	1.8 (1803)	8	②	0.0006-0.0013	0.0006-0.0087	3	1.8893-1.8898	0.0008-0.0019	0.0050-0.0150
1999	1.0 (993)	6	①	0.00008-0.00015	0.0044-0.0122	NA	1.6529-1.6535	0.0012-0.0019	0.0039-0.0078
	1.3 (1300)	9	①	0.00008-0.00015	0.0044-0.0122	NA	1.6529-1.6535	0.0012-0.0019	0.0039-0.0078
	1.8 (1803)	8	②	0.0006-0.0013	0.0006-0.0087	3	1.8893-1.8898	0.0008-0.0019	0.0050-0.0150
2000	1.0 (993)	6	①	0.00008-0.00015	0.0044-0.0122	NA	1.6529-1.6535	0.0012-0.0019	0.0039-0.0078
	1.3 (1300)	9	①	0.00008-0.00015	0.0044-0.0122	NA	1.6529-1.6535	0.0012-0.0019	0.0039-0.0078
	1.8 (1803)	8	②	0.0006-0.0013	0.0006-0.0087	3	1.8893-1.8898	0.0008-0.0019	0.0050-0.0150
2001	1.0 (993)	6	①	0.00008-0.00015	0.0044-0.0122	NA	1.6529-1.6535	0.0012-0.0019	0.0039-0.0078
	1.3 (1300)	9	①	0.00008-0.00015	0.0044-0.0122	NA	1.6529-1.6535	0.0012-0.0019	0.0039-0.0078
	1.8 (1803)	8	②	0.0006-0.0013	0.0006-0.0087	3	1.8893-1.8898	0.0008-0.0019	0.0050-0.0150

NA: Not Available

① 1 Stamping: 1.7714-1.7716
2 Stamping: 1.7712-1.7714
3 Stamping: 1.7710-1.7712

② 0 Stamping: 1.8895-1.8898
1 Stamping: 1.8893-1.8895
2 Stamping: 1.8891-1.8893

93461CYZ2

PISTON AND RING SPECIFICATIONS

All measurements are given in inches.

Year	Engine Displacement Liters (cc)	Engine ID/VIN	Piston Clearance	Ring Gap			Ring Side Clearance		
				Top Compression	Bottom Compression	Oil Control	Top Compression	Bottom Compression	Oil Control
1998	1.0 (993)	6	0.0008-0.0015	0.0079-0.0118	0.0079-0.0118	0.0079-0.0275	0.0012-0.0027	0.0008-0.0023	—
	1.3 (1300)	9	0.0008-0.0015	0.0079-0.0118	0.0079-0.0012	0.0079-0.0275	0.0014-0.0027	0.0008-0.0023	—
	1.8 (1803)	8	0.0033-0.0041	0.0098-0.0138	0.0138-0.0197	0.0039-0.0157	0.0018-0.0033	0.0012-0.0028	—
1999	1.0 (993)	6	0.0008-0.0015	0.0079-0.0118	0.0079-0.0118	0.0079-0.0275	0.0012-0.0027	0.0008-0.0023	—
	1.3 (1300)	9	0.0008-0.0015	0.0079-0.0118	0.0079-0.0012	0.0079-0.0275	0.0014-0.0027	0.0008-0.0023	—
	1.8 (1803)	8	0.0033-0.0041	0.0098-0.0138	0.0138-0.0197	0.0039-0.0157	0.0018-0.0033	0.0012-0.0028	—
2000	1.0 (993)	6	0.0008-0.0015	0.0079-0.0118	0.0079-0.0118	0.0079-0.0275	0.0012-0.0027	0.0008-0.0023	—
	1.3 (1300)	9	0.0008-0.0015	0.0079-0.0118	0.0079-0.0012	0.0079-0.0275	0.0014-0.0027	0.0008-0.0023	—
	1.8 (1803)	8	0.0033-0.0041	0.0098-0.0138	0.0138-0.0197	0.0039-0.0157	0.0018-0.0033	0.0012-0.0028	—
2001	1.0 (993)	6	0.0008-0.0015	0.0079-0.0118	0.0079-0.0118	0.0079-0.0275	0.0012-0.0027	0.0008-0.0023	—
	1.3 (1300)	9	0.0008-0.0015	0.0079-0.0118	0.0079-0.0012	0.0079-0.0275	0.0014-0.0027	0.0008-0.0023	—
	1.8 (1803)	8	0.0033-0.0041	0.0098-0.0138	0.0138-0.0197	0.0039-0.0157	0.0018-0.0033	0.0012-0.0028	—

93461CYZ3

TORQUE SPECIFICATIONS

All readings in ft. lbs.

Year	Engine Displacement Liters (cc)	Engine ID/VIN	Cylinder Head Bolts	Main Bearing Bolts	Rod Bearing Bolts	Crankshaft Damper Bolt	Flywheel Bolts	Manifold		Spark Plugs	Lug Nuts
								Intake	Exhaust		
1998	1.0 (993)	6	54	40	26	96 ①	55	17	17	21	44
	1.3 (1300)	9	49	40	26	94 ①	55	17	17	21	44
	1.8 (1803)	8	②	③	④	105 ①	⑤	13	36	21	76
1999	1.0 (993)	6	54	40	26	96 ①	55	17	17	21	44
	1.3 (1300)	9	49	40	26	94 ①	55	17	17	21	44
	1.8 (1803)	8	②	③	④	105 ①	⑤	13	36	21	76
2000	1.0 (993)	6	54	40	26	96 ①	55	17	17	21	44
	1.3 (1300)	9	49	40	26	94 ①	55	17	17	21	44
	1.8 (1803)	8	②	③	④	105 ①	⑤	13	36	21	76
2001	1.0 (993)	6	54	40	26	96 ①	55	17	17	21	44
	1.3 (1300)	9	49	40	26	94 ①	55	17	17	21	44
	1.8 (1803)	8	②	③	④	105 ①	⑤	13	36	21	76

① Crankshaft timing belt sprocket

② Step 1: 36 ft. lbs.
Step 2: Plus 90 degrees

③ Step 1: 16 ft. lbs.
Step 2: 32 ft. lbs.
Step 3: Plus 90 degrees

④ 15 ft. lbs. plus 90 degrees

⑤ Manual transaxle: 36 ft. lbs. Plus 90 degrees
Automatic transaxle: 61 ft. lbs.

93461CYZ4

BRAKE SPECIFICATIONS

All measurements in inches unless noted

		Brake Disc			Brake Drum Diameter			Minimum Lining Thickness ①		Brake Caliper	
Year	Model	Original Thickness	Minimum Thickness	Maximum Runout	Original Inside Diameter	Max. Wear Limit	Maximum Machine Diameter	Front	Rear	Bracket Bolts (ft. lbs.)	Mounting Bolts (ft. lbs.)
1998	Metro	0.670	0.590	0.004	②	②	②	0.236	0.111	—	18
	Prizm	0.866	0.787	0.003	7.87	7.91	7.91	0.390	0.390	—	③
1999	Metro	0.670	0.590	0.004	②	②	②	0.236	0.111	—	18
	Prizm	0.866	0.787	0.003	7.87	7.91	7.91	0.390	0.390	—	③
2000	Metro	0.670	0.590	0.004	②	②	②	0.236	0.111	—	18
	Prizm	0.866	0.787	0.003	7.87	7.91	7.91	0.390	0.390	—	③
2001	Metro	0.670	0.590	0.004	②	②	②	0.236	0.111	—	18
	Prizm	0.866	0.787	0.003	7.87	7.91	7.91	0.390	0.390	—	③

① Minimum lining thickness includes pad/shoe backing

② 2 door: 7.09
4 door: 7.87

③ Front: 18 ft. lbs.
Rear: 14 ft. lbs.

93461CYZ5

WHEEL ALIGNMENT

			Caster		Camber			
Year	Model		Range (+/-Deg.)	Preferred Setting (Deg.)	Range (+/-Deg.)	Preferred Setting (Deg.)	Toe-in (in.)	Steering Axis Inclination (Deg.)
1998	Metro	F	2.00	+3.00 ①	0.75	+0.75	0.08 +/- 0.08	—
		R	—	—	0.55	+0.45	0.20 +/- 0.10	—
	Prizm	F	0.75	+1.32 ①	0.75	-0.18	0.05 +/- 0.10	—
		R	—	—	0.75	-0.92 ①	0.20 +/- 0.10	—
1999	Metro	F	2.00	+3.00 ①	1.00	+0.50	0.08 +/- 0.08	—
		R	—	—	1.00	0 ①	0.23 +/- 0.07	—
	Prizm	F	0.75	+1.32 ①	0.75	-0.18	0.05 +/- 0.10	—
		R	—	—	0.75	-0.92 ①	0.20 +/- 0.10	—
2000	Metro	F	2.00	+3.00 ①	1.00	+0.50	0.08 +/- 0.08	—
		R	—	—	1.00	0 ①	0.23 +/- 0.07	—
	Prizm	F	0.75	+1.32 ①	0.75	-0.18	0.05 +/- 0.10	—
		R	—	—	0.75	-0.92 ①	0.20 +/- 0.10	—
2001	Metro	F	2.00	+3.00 ①	1.00	+0.50	0.08 +/- 0.08	—
		R	—	—	1.00	0 ①	0.23 +/- 0.07	—
	Prizm	F	0.75	+1.32 ①	0.75	-0.18	0.05 +/- 0.10	—
		R	—	—	0.75	-0.92 ①	0.20 +/- 0.10	—

① Not adjustable, for reference only

93461CYZ6

TIRE, WHEEL AND BALL JOINT SPECIFICATIONS

		OEM Tires		Tire Pressures (psi)		Wheel	Ball Joint
Year	Model	Standard	Optional	Front	Rear	Size	Inspection
1998	Metro	P155/80R13	None	32	32	4.5-B	①
	Prizm	P175/65R14	P185/65R14	30	30	5.5J	①
	Prizm Lsi	P185/65R14	None	30	30	5.5J	①
1999	Metro	P155/80R13	None	32	32	4.5-B	①
	Prizm	P175/65R14	P185/65R14	30	30	5.5J	①
	Prizm Lsi	P185/65R14	None	30	30	5.5J	①
2000	Metro	P155/80R13	None	32	32	4.5-B	①
	Prizm	P175/65R14	P185/65R14	30	30	5.5J	①
	Prizm Lsi	P185/65R14	None	30	30	5.5J	①
2001	Metro	P155/80R13	None	32	32	4.5-B	①
	Prizm	P175/65R14	P185/65R14	30	30	5.5J	①
	Prizm Lsi	P185/65R14	None	30	30	5.5J	①

OEM: Original Equipment Manufacturer

PSI: Pounds Per Square Inch

STD: Standard

OPT: Optional

① Replace if any measurable movement is found.

93461CYZ7

SCHEDULED MAINTENANCE INTERVALS
GEO METRO

TO BE SERVICED	TYPE OF SERVICE	VEHICLE MILEAGE INTERVAL (x1000)												
		7.5	15	22.5	30	37.5	45	52.5	60	67.5	75	82.5	90	97.5
Engine oil & filter	R	✓	✓	✓	✓	✓	✓	✓	✓	✓	✓	✓	✓	✓
Chassis lubrication	S/I	✓	✓	✓	✓	✓	✓	✓	✓	✓	✓	✓	✓	✓
Locking front hubs	S/I	✓	✓	✓	✓	✓	✓	✓	✓	✓	✓	✓	✓	✓
Lubricate parking brake cable guides, underbody contact points & linkage	S/I	✓	✓	✓	✓	✓	✓	✓	✓	✓	✓	✓	✓	✓
Rotate tires	S/I	✓	✓	✓	✓	✓	✓	✓	✓	✓	✓	✓	✓	✓
Brake system	S/I	✓		✓		✓		✓		✓		✓		✓
Exhaust system	S/I	✓		✓		✓		✓		✓		✓		✓
Fuel tank, cap & lines	S/I		✓		✓		✓		✓		✓		✓	
Air cleaner filter	R				✓				✓				✓	
Engine coolant	R				✓				✓				✓	
Fuel filter	R				✓				✓				✓	
Manual transaxle oil	R				✓				✓				✓	
Spark plugs	R				✓				✓				✓	
Accessory drive belt(s)	S/I				✓				✓				✓	
Cooling system	S/I				✓				✓				✓	
Automatic transaxle fluid & filter	S/I				✓				✓				✓	
EGR system	S/I				✓				✓				✓	
Engine timing	S/I								✓					
Ignition cables	S/I								✓					
PCV system	S/I				✓				✓				✓	
Brake fluid	R								✓					
PCV valve ①	R													
Timing belt	R								✓					
Throttle body unit mount bolt torque ②	S/I													

R: Replace S/I: Service or Inspect

① PCV valve: replace at 50,000 miles.

② Torque to 16 ft. lbs. (22 Nm) at 6000 miles.

FREQUENT OPERATION MAINTENANCE (SEVERE SERVICE)

If a vehicle is operated under any of the following conditions it is considered severe service:

- Extremely dusty areas.

- 50% or more of the vehicle operation is in 32°C (90°F) or higher temperatures, or constant operation in temperatures below 0°C (32°F).

- Prolonged idling (vehicle operation in stop and go traffic).

- Frequent short running periods (engine does not warm to normal operating temperatures).

- Police, taxi, delivery usage or trailer towing usage.

Oil & oil filter: change every 3000 miles.

Chassis lubrication: lubricate every 6000 miles.

Throttle body mount bolt torque: torque to 16 ft. lbs. (22 Nm) at 6000 miles.

Air cleaner filter: service or inspect every 15,000 miles.

Rotate tires: rotate at 6000 miles and then every 15,000 miles thereafter.

Automatic transaxle fluid & filter: replace every 50,000 miles.

93461CYZ8

SCHEDULED MAINTENANCE INTERVALS
GEO PRISM

TO BE SERVICED	TYPE OF SERVICE	VEHICLE MILEAGE INTERVAL (x1000)												
		7.5	15	22.5	30	37.5	45	52.5	60	67.5	75	82.5	90	97.5
Engine oil & filter	R	✓	✓	✓	✓	✓	✓	✓	✓	✓	✓	✓	✓	✓
Lubricate parking brake cable guides, underbody contact points & linkage	S/I	✓	✓	✓	✓	✓	✓	✓	✓	✓	✓	✓	✓	✓
Rotate tires	S/I	✓	✓	✓	✓	✓	✓	✓	✓	✓	✓	✓	✓	✓
Brake system	S/I	✓		✓		✓		✓		✓		✓		✓
Engine idle speed	S/I	✓		✓		✓		✓		✓		✓		✓
Exhaust system	S/I	✓		✓		✓		✓		✓		✓		✓
Chassis lubrication	S/I		✓		✓		✓		✓		✓		✓	
Fuel tank, cap & lines	S/I				✓				✓				✓	
Valve clearance	S/I								✓					
Air cleaner filter	R				✓				✓				✓	
Engine coolant	R				✓				✓				✓	
Fuel tank cap gasket	R				✓				✓				✓	
Manual transaxle oil	R				✓				✓				✓	
Spark plugs	R				✓				✓				✓	
Accessory drive belt(s)	S/I								✓		✓		✓	✓
Cooling system	S/I						✓				✓			
Automatic transaxle fluid & filter	S/I				✓				✓				✓	
EGR system	S/I				✓				✓				✓	
Ignition cables	S/I								✓					
PCV system	S/I				✓				✓				✓	
Timing belt	R								✓					
EVAP canister	S/I								✓					
Throttle body unit mount bolt torque ①	S/I													

R: Replace S/I: Service or Inspect

① Torque to 16 ft. lbs. (22 Nm) at 6000 miles.

FREQUENT OPERATION MAINTENANCE (SEVERE SERVICE)

If a vehicle is operated under any of the following conditions it is considered severe service:

- Extremely dusty areas.

- 50% or more of the vehicle operation is in 32°C (90°F) or higher temperatures, or constant operation in temperatures below 0°C (32°F).

- Prolonged idling (vehicle operation in stop and go traffic).

- Frequent short running periods (engine does not warm to normal operating temperatures).

- Police, taxi, delivery usage or trailer towing usage.

Oil & oil filter: change every 3000 miles.

Chassis lubrication: lubricate every 6000 miles.

Throttle body mount bolt torque: torque to 16 ft. lbs. (22 Nm) at 6000 miles.

Air cleaner filter: service or inspect every 15,000 miles.

Differential fluid: replace every 15,000 miles.

93461CYZZ

Heater Core replacement is covered in Section 2 of this manual

SCHEDULED MAINTENANCE INTERVALS
GENERAL MOTORS CORPORATION
CHEVROLET/GEO METRO, PRIZM

The following should be used as a guide when determining the amount of work required for a particular service. In estimating how long a particular Scheduled Maintenance Service should take, please observe the following:

- Labor Time is time based on field research and data supplied by the vehicle manufacturer.
- Labor time operations are given in hours and tenths of an hour.
- All labor operations are to be used as a guide.

Mechanic Skill Level Codes:
(A) PRECISION: Highly skilled with multiple certification.
(B) GENERAL: Normally skilled with certification.
(C) MAINTENANCE: Semi-skilled working on certification.

	LABOR TIME
7500 Mile Service (C)	
All Models	1.8
15000 Mile Service (C)	
All Models	.7
Rotate tires add	.5
22500 Mile Service (C)	
All Models	1.6
30000 Mile Service (B)	
All Models	2.7
Rotate tires add	.5
37500 Mile Service (C)	
All Models	1.6
45000 Mile Service (C)	
All Models	.7
Rotate tires add	.5
60000 Mile Service (B)	
All Models	4.9
67500 Mile Service (C)	
All Models	1.6
75000 Mile Service (C)	
All Models	.7
Rotate tires add	.5
82500 Mile Service (C)	
All Models	1.6
90000 Mile Service (B)	
All Models	2.8
97500 Mile Service (C)	
All Models	1.6

93061CW5

SATURN

LS • LS1 • LS2 • LW1 • LW2 • SC1 • SC2 • SL • SL1 • SL2 • SW1 • SW2

ENGINE AND VEHICLE IDENTIFICATION

Engine						
Code ①	Liters (cc)	Cu. In.	Cyl.	Fuel Sys.	Engine Type	Eng. Mfg.
7	1.9 (1901)	116	4	MFI	DOHC	Saturn
8	1.9 (1901)	116	4	MFI	SOHC	Saturn
F	2.2 (2199)	134	4	SFI	DOHC	Saturn
R	3.0 (3000)	183	6	SFI	DOHC	Saturn

Model Year	
Code ②	Year
W	1998
X	1999
Y	2000
1	2001
2	2002

MFI: Multi-point Fuel Injection

SFI: Sequential Fuel Injection

DOHC: Double Overhead Camshafts

SOHC: Single Overhead Camshaft

① 8th digit of VIN

② 10th digit of VIN

93401CYZ9

Brake service is covered in Section 4 of this manual

GENERAL ENGINE SPECIFICATIONS

Year	Model	Engine Displacement Liters (cc)	Engine ID/VIN	Fuel System Type	Net Horsepower @ rpm	Net Torque @ rpm (ft. lbs.)	Bore x Stroke (in.)	Com-pression Ratio	Oil Pressure @ rpm
1998	Coupe	1.9 (1901)	7	MFI	124@5600	122@4800	3.23x3.54	9.5:1	29@2000
	Coupe	1.9 (1901)	8	MFI	100@5000	114@2400	3.23x3.54	9.3:1	36@2000
	Sedan	1.9 (1901)	7	MFI	124@5600	122@4800	3.23x3.54	9.5:1	29@2000
	Sedan	1.9 (1901)	8	MFI	100@5000	114@2400	3.23x3.54	9.3:1	36@2000
	Wagon	1.9 (1901)	7	MFI	124@5600	122@4800	3.23x3.54	9.5:1	29@2000
	Wagon	1.9 (1901)	8	MFI	100@5000	114@2400	3.23x3.54	9.3:1	36@2000
1999	Coupe	1.9 (1901)	7	MFI	124@5600	122@4800	3.23x3.54	9.5:1	29@2000
	Coupe	1.9 (1901)	8	MFI	100@5000	114@2400	3.23x3.54	9.3:1	36@2000
	Sedan	1.9 (1901)	7	MFI	124@5600	122@4800	3.23x3.54	9.5:1	29@2000
	Sedan	1.9 (1901)	8	MFI	100@5000	114@2400	3.23x3.54	9.3:1	36@2000
	Wagon	1.9 (1901)	7	MFI	124@5600	122@4800	3.23x3.54	9.5:1	29@2000
	Wagon	1.9 (1901)	8	MFI	100@5000	114@2400	3.23x3.54	9.3:1	36@2000
2000	Sedan	1.9 (1901)	7	MFI	124@5600	122@4800	3.23x3.54	9.5:1	29@2000
	Sedan	1.9 (1901)	8	MFI	100@5000	114@2400	3.23x3.54	9.3:1	36@2000
	Wagon	1.9 (1901)	7	MFI	124@5600	122@4800	3.23x3.54	9.5:1	29@2000
	Wagon	1.9 (1901)	8	MFI	100@5000	114@2400	3.23x3.54	9.3:1	36@2000
	Sedan	2.2 (1901)	F	SFI	137@5800	147@4400	3.38x3.5	9.5:1	50@1000
	Wagon	2.2 (1901)	F	SFI	137@5800	147@4400	3.38x3.50	9.5:1	50@1000
	Sedan	3.0 (3000)	R	SFI	182@6000	190@3600	3.38x3.50	10.0:1	22@1000
	Wagon	3.0 (3000)	R	SFI	182@6000	190@3600	3.38x3.50	10.0:1	22@1000
2001	Sedan	1.9 (1901)	7	MFI	124@5600	122@4800	3.23x3.54	9.5:1	29@2000
	Sedan	1.9 (1901)	8	MFI	100@5000	114@2400	3.23x3.54	9.3:1	36@2000
	Wagon	1.9 (1901)	7	MFI	124@5600	122@4800	3.23x3.54	9.5:1	29@2000
	Wagon	1.9 (1901)	8	MFI	100@5000	114@2400	3.23x3.54	9.3:1	36@2000
	Sedan	2.2 (1901)	F	SFI	137@5800	147@4400	3.38x3.5	9.5:1	50@1000
	Wagon	2.2 (1901)	F	SFI	137@5800	147@4400	3.38x3.50	9.5:1	50@1000
	Sedan	3.0 (3000)	R	SFI	182@6000	190@3600	3.38x3.50	10.0:1	22@1000
	Wagon	3.0 (3000)	R	SFI	182@6000	190@3600	3.38x3.50	10.0:1	22@1000

MFI: Multi-port Fuel Injection

SFI: Sequential Fuel Injection

93461CYZ0

ENGINE TUNE-UP SPECIFICATIONS

Year	Engine Displacement Liters (cc)	Engine ID/VIN	Spark Plug Gap (in.)	Ignition Timing (deg.) MT	Ignition Timing (deg.) AT	Fuel Pump (psi) ①	Idle Speed (rpm) MT ②	Idle Speed (rpm) AT ②	Valve Clearance In.	Valve Clearance Ex.
1998	1.9 (1901)	7	0.040	③	③	31-36	850	750	HYD	HYD
	1.9 (1901)	8	0.040	③	③	31-36	750	650	HYD	HYD
1999	1.9 (1901)	7	0.040	③	③	31-36	850	750	HYD	HYD
	1.9 (1901)	8	0.040	③	③	31-36	750	650	HYD	HYD
2000	1.9 (1901)	7	0.040	③	③	31-36	850	750	HYD	HYD
	1.9 (1901)	8	0.040	③	③	31-36	750	650	HYD	HYD
	2.2 (2199)	F	0.043	③	③	55-65	④	④	HYD	HYD
	3.0 (3000)	R	0.043	③	③	39-49	④	④	HYD	HYD
2001	1.9 (1901)	7	0.040	③	③	31-36	850	750	HYD	HYD
	1.9 (1901)	8	0.040	③	③	31-36	750	650	HYD	HYD
	2.2 (2199)	F	0.043	③	③	55-65	④	④	HYD	HYD
	3.0 (3000)	R	0.043	③	③	39-49	④	④	HYD	HYD

NOTE: The Vehicle Emission Control Information label often reflects specification changes made during production. The label figures must be used if they differ from those in this chart.

HYD: Hydraulic

① Pressure measured at idle

② Idle speed measured with manual transmission in Neutral; automatic transmission in D (drive)

③ Engines equipped with Distributorless Ignition System (DIS). Ignition timing is not adjustable

④ Refer to the Vehicle Emission Control Information label

93461CYAA

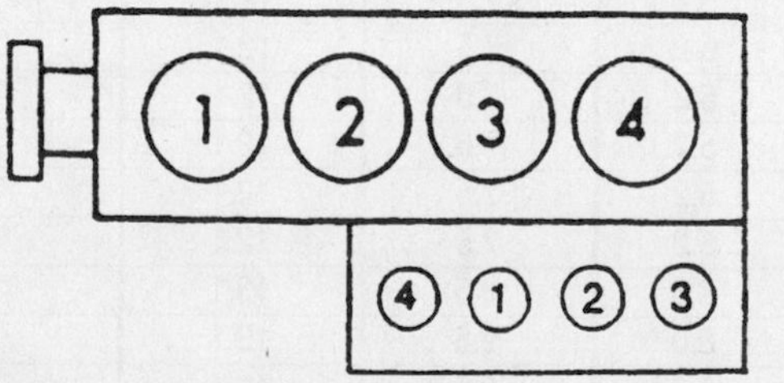

79223G14

1.9L Engine
Firing order: 1–3–4–2
Distributorless ignition system

79223G36

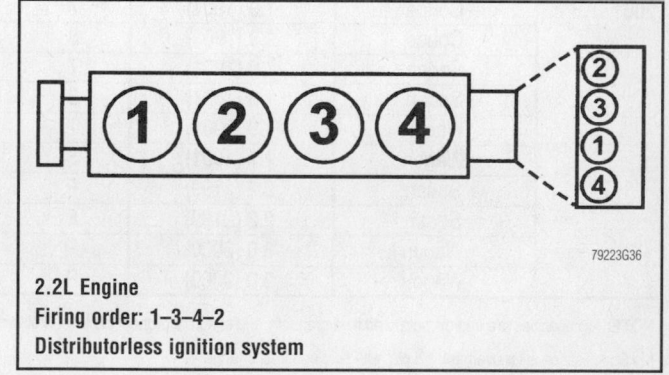

2.2L Engine
Firing order: 1–3–4–2
Distributorless ignition system

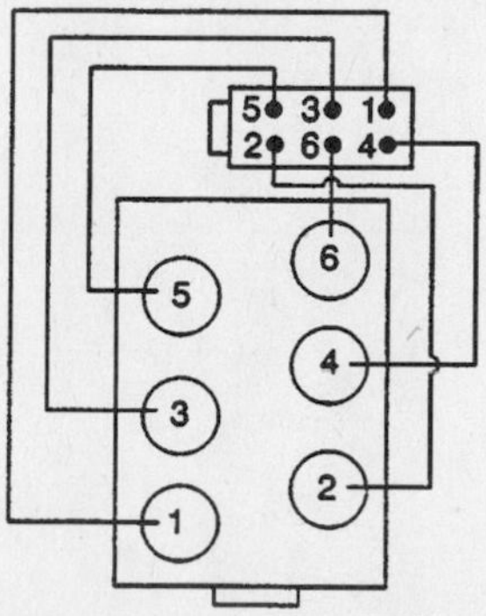

79223G11

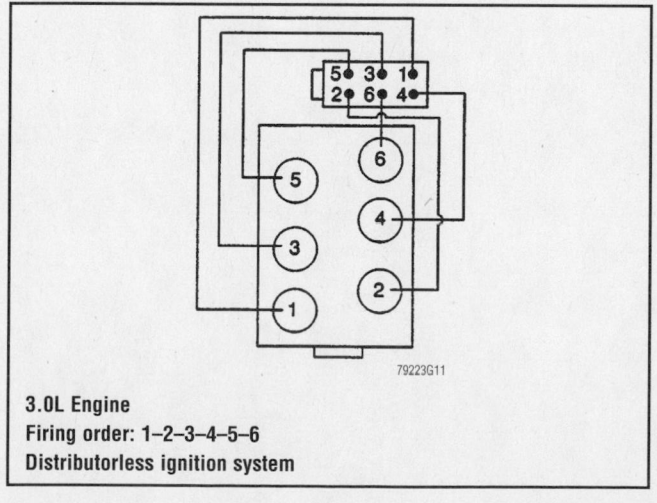

3.0L Engine
Firing order: 1–2–3–4–5–6
Distributorless ignition system

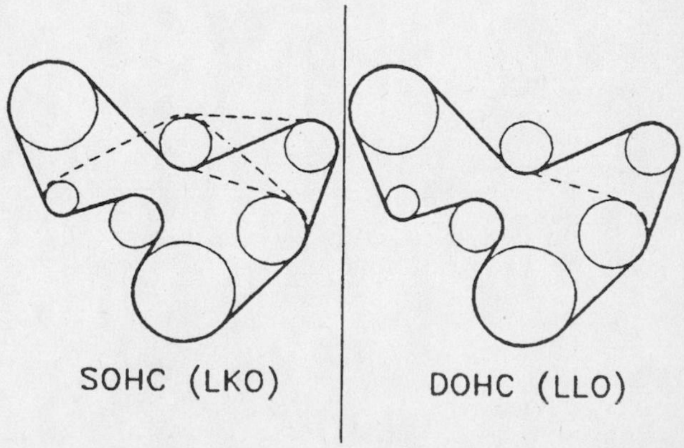

79224G62

Serpentine drive belt routing—Saturn engines

CAPACITIES

Year	Model	Engine Displacement Liters (cc)	Engine ID/VIN	Engine Oil with Filter (qts.)	Transaxle (pts.) Manual	Transaxle (pts.) Auto. ①	Fuel Tank (gal.)	Cooling System (qts.)
1998	Coupe	1.9 (1901)	7	4.0	5.2	7.5	12.8	7.0
	Coupe	1.9 (1901)	8	4.0	5.2	7.5	12.8	7.0
	Sedan	1.9 (1901)	7	4.0	5.2	7.5	12.8	7.0
	Sedan	1.9 (1901)	8	4.0	5.2	7.5	12.8	7.0
	Wagon	1.9 (1901)	7	4.0	5.2	7.5	12.8	7.0
	Wagon	1.9 (1901)	8	4.0	5.2	7.5	12.8	7.0
1999	Coupe	1.9 (1901)	7	4.0	5.2	7.5	12.8	7.0
	Coupe	1.9 (1901)	8	4.0	5.2	7.5	12.8	7.0
	Sedan	1.9 (1901)	7	4.0	5.2	7.5	12.8	7.0
	Sedan	1.9 (1901)	8	4.0	5.2	7.5	12.8	7.0
	Wagon	1.9 (1901)	7	4.0	5.2	7.5	12.8	7.0
	Wagon	1.9 (1901)	8	4.0	5.2	7.5	12.8	7.0
2000	Coupe	1.9 (1901)	7	4.0	5.2	7.5	12.1	7.0
	Coupe	1.9 (1901)	8	4.0	5.2	7.5	12.1	7.0
	Sedan	1.9 (1901)	7	4.0	5.2	7.5	12.1	7.0
	Sedan	1.9 (1901)	8	4.0	5.2	7.5	12.1	7.0
	Wagon	1.9 (1901)	7	4.0	5.2	7.5	12.1	7.0
	Wagon	1.9 (1901)	8	4.0	5.2	7.5	12.1	7.0
	Sedan	2.2 (2199)	F	5.0	8.0	9.5	13.1	7.3
	Sedan	2.2 (2199)	F	5.0	8.0	9.5	13.1	7.3
	Wagon	3.0 (3000)	R	5.0	8.0	9.5	13.1	7.4
	Wagon	3.0 (3000)	R	5.0	8.0	9.5	13.1	7.4
2001	Coupe	1.9 (1901)	7	4.0	5.2	7.5	12.1	7.0
	Coupe	1.9 (1901)	8	4.0	5.2	7.5	12.1	7.0
	Sedan	1.9 (1901)	7	4.0	5.2	7.5	12.1	7.0
	Sedan	1.9 (1901)	8	4.0	5.2	7.5	12.1	7.0
	Wagon	1.9 (1901)	7	4.0	5.2	7.5	12.1	7.0
	Wagon	1.9 (1901)	8	4.0	5.2	7.5	12.1	7.0
	Sedan	2.2 (2199)	F	5.0	8.0	9.5	13.1	7.3
	Sedan	2.2 (2199)	F	5.0	8.0	9.5	13.1	7.3
	Wagon	3.0 (3000)	R	5.0	8.0	9.5	13.1	7.4
	Wagon	3.0 (3000)	R	5.0	8.0	9.5	13.1	7.4

NOTE: All capacities are approximate. Add fluid gradually and ensure a proper fluid level is obtained.

① Specification is for overhaul. 8.4 pts. with fluid and filter change

93461CYAB

VALVE SPECIFICATIONS

Year	Engine Displacement Liters (cc)	Engine ID/VIN	Seat Angle (deg.)	Face Angle (deg.)	Spring Test Pressure (lbs. @ in.)	Spring Free-Length (in.)	Stem-to-Guide Clearance (in.) Intake	Stem-to-Guide Clearance (in.) Exhaust	Stem Diameter (in.) Intake	Stem Diameter (in.) Exhaust
1998	1.9 (1901)	7	44.5-45.4	45-45.5	163-180@ 0.984	1.6100	0.0010- 0.0025	0.0015- 0.0032	0.2736- 0.2740	0.2729- 0.2736
	1.9 (1901)	8	44.5-45.4	45-45.25	202-211@ 1.280	1.8898- 1.9134	0.0010- 0.0025	0.0015- 0.0032	0.2736- 0.2741	0.2736- 0.2740
1999	1.9 (1901)	7	44.5-45.4	45-45.5	163-180@ 0.984	1.6100	0.0010- 0.0025	0.0015- 0.0032	0.2736- 0.2740	0.2729- 0.2736
	1.9 (1901)	8	44.5-45.4	45-45.25	202-211@ 1.280	1.8898- 1.9134	0.0010- 0.0025	0.0015- 0.0032	0.2736- 0.2741	0.2736- 0.2740
2000	1.9 (1901)	7	44.5-45.4	45-45.5	163-180@ 0.984	1.6100	0.0010- 0.0025	0.0015- 0.0032	0.2736- 0.2740	0.2729- 0.2736
	1.9 (1901)	8	44.5-45.4	45-45.25	202-211@ 1.280	1.8898- 1.9134	0.0010- 0.0025	0.0015- 0.0032	0.2736- 0.2741	0.2736- 0.2740
	2.2 (2199)	F	44.5-45.4	45-45.5	① ②	1.6100	0.0012 0.0022	0.0016 0.0026	0.2344 0.2355	0.2341 0.2347
	3.0 (3000)	R	44.5-45.4	45-45.25	56.6@1.338	NA	0.0012 0.0022	0.0016 0.0026	0.2344 0.2350	0.2341 0.2346
2001	1.9 (1901)	7	44.5-45.4	45-45.5	163-180@ 0.984	1.6100	0.0010- 0.0025	0.0015- 0.0032	0.2736- 0.2740	0.2729- 0.2736
	1.9 (1901)	8	44.5-45.4	45-45.25	202-211@ 1.280	1.8898- 1.9134	0.0010- 0.0025	0.0015- 0.0032	0.2736- 0.2741	0.2736- 0.2740
	2.2 (2199)	F	44.5-45.4	45-45.5	① ②	1.6100	0.0012 0.0022	0.0016 0.0026	0.2344 0.2355	0.2341 0.2347
	3.0 (3000)	R	44.5-45.4	45-45.25	56.6@1.338	NA	0.0012 0.0022	0.0016 0.0026	0.2344 0.2350	0.2341 0.2346

NA: Not available

① Valve spring load closed: 245-271 N

② Valve spring load open: 525-575 N

93461CYAC

For Accessory Drive Belt illustrations, see Section 1 of this manual

CRANKSHAFT AND CONNECTING ROD SPECIFICATIONS

All measurements are given in inches.

Year	Engine Displacement Liters (cc)	Engine ID/VIN	Crankshaft: Main Brg. Journal Dia.	Crankshaft: Main Brg. Oil Clearance	Crankshaft: Shaft End-play	Crankshaft: Thrust on No.	Connecting Rod: Journal Diameter	Connecting Rod: Oil Clearance	Connecting Rod: Side Clearance
1998	1.9 (1901)	7	2.2438-2.2444	0.0002-0.0020	0.0020-0.0079	3	1.9761-1.9767	0.0001-0.0021	0.0065-0.0171
	1.9 (1901)	8	2.2438-2.2444	0.0002-0.0020	0.0020-0.0079	3	1.9761-1.9767	0.0001-0.0021	0.0065-0.0171
1999	1.9 (1901)	7	2.2438-2.2444	0.0002-0.0020	0.0020-0.0079	3	1.9761-1.9767	0.0001-0.0021	0.0065-0.0171
	1.9 (1901)	8	2.2438-2.2444	0.0002-0.0020	0.0020-0.0079	3	1.9761-1.9767	0.0001-0.0021	0.0065-0.0171
2000	1.9 (1901)	7	2.2438-2.2444	0.0002-0.0020	0.0020-0.0079	3	1.9761-1.9767	0.0001-0.0021	0.0065-0.0171
	1.9 (1901)	8	2.2438-2.2444	0.0002-0.0020	0.0020-0.0079	3	1.9761-1.9767	0.0001-0.0021	0.0065-0.0171
	2.2 (2199)	F	2.2045-2.2050	0.0040 0.0040	0.0012-0.0150	NA	1.9291-1.9297	0.0001-0.0021	0.0028-0.0146
	3.0 (3000)	R	2.6763-2.6766	0.0040 0.0040	0.0004-0.0300	NA	1.927-1.9280	0.0001-0.0021	0.0027-0.0110
2001	1.9 (1901)	7	2.2438-2.2444	0.0002-0.0020	0.0020-0.0079	3	1.9761-1.9767	0.0001-0.0021	0.0065-0.0171
	1.9 (1901)	8	2.2438-2.2444	0.0002-0.0020	0.0020-0.0079	3	1.9761-1.9767	0.0001-0.0021	0.0065-0.0171
	2.2 (2199)	F	2.2045-2.2050	0.0040 0.0040	0.0012-0.0150	NA	1.9291-1.9297	0.0001-0.0021	0.0028-0.0146
	3.0 (3000)	R	2.6763-2.6766	0.0040 0.0040	0.0004-0.0300	NA	1.927-1.9280	0.0001-0.0021	0.0027-0.0110

NA: Not available

93461CYAD

PISTON AND RING SPECIFICATIONS

All measurements are given in inches.

Year	Engine Displacement Liters (cc)	Engine ID/VIN	Piston Clearance	Ring Gap Top Compression	Ring Gap Bottom Compression	Ring Gap Oil Control	Ring Side Clearance Top Compression	Ring Side Clearance Bottom Compression	Ring Side Clearance Oil Control
1998	1.9 (1901)	7	①	0.0098-0.0157	0.0098-0.0197	0.0098-0.0429	0.0016-0.0032	0.0012-0.0031	SNUG
	1.9 (1901)	8	①	0.0098-0.0157	0.0098-0.0197	0.0098-0.0429	0.0016-0.0032	0.0012-0.0031	SNUG
1999	1.9 (1901)	7	①	0.0098-0.0157	0.0098-0.0197	0.0098-0.0429	0.0016-0.0032	0.0012-0.0031	SNUG
	1.9 (1901)	8	①	0.0098-0.0157	0.0098-0.0197	0.0098-0.0429	0.0016-0.0032	0.0012-0.0031	SNUG
2000	1.9 (1901)	7	①	0.0098-0.0157	0.0098-0.0197	0.0098-0.0429	0.0016-0.0032	0.0012-0.0031	SNUG
	1.9 (1901)	8	①	0.0098-0.0157	0.0098-0.0197	0.0098-0.0429	0.0016-0.0032	0.0012-0.0031	SNUG
	2.2 (2199)	F	0.0001-0.0005	0.0060-0.015	0.0012 0.0027	NA	0.0028-0.0146	0.0005-0.0024	SNUG
	3.0 (3000)	R	0.0027-0.0110	0.0008-0.0015	0.0118 0.0196	NA	0.0027-0.0110	0.0005-0.0024	SNUG
2001	1.9 (1901)	7	①	0.0098-0.0157	0.0098-0.0197	0.0098-0.0429	0.0016-0.0032	0.0012-0.0031	SNUG
	1.9 (1901)	8	①	0.0098-0.0157	0.0098-0.0197	0.0098-0.0429	0.0016-0.0032	0.0012-0.0031	SNUG
	2.2 (2199)	F	0.0001-0.0005	0.0060-0.015	0.0012 0.0027	NA	0.0028-0.0146	0.0005-0.0024	SNUG
	3.0 (3000)	R	0.0027-0.0110	0.0008-0.0015	0.0118 0.0196	NA	0.0027-0.0110	0.0005-0.0024	SNUG

NA: Not available

① Piston No. 1, 2 and 3: 0.0002-0.0017
Piston No. 4: 0.0002-0.0021

93461CYAE

TORQUE SPECIFICATIONS

All readings in ft. lbs.

Year	Engine Displacement Liters (cc)	Engine ID/VIN	Cylinder Head Bolts	Main Bearing Bolts	Rod Bearing Bolts	Crankshaft Damper Bolts	Flywheel Bolts	Manifold Intake	Manifold Exhaust	Spark Plugs	Lug Nuts
1998	1.9 (1901)	7	①	37	33	159	59②	22③	19③	20	103
	1.9 (1901)	8	①	37	33	159	59②	22③	16③	20	103
1999	1.9 (1901)	7	①	37	33	159	59②	22③	19③	20	103
	1.9 (1901)	8	①	37	33	159	59②	22③	16③	20	103
2000	1.9 (1901)	7	①	37	33	159	59②	22③	19③	20	103
	1.9 (1901)	8	①	37	33	159	59②	22③	16③	20	103
	2.2 (2199	F	④	37	18	⑤	59②	7.5	13	15	138
	3.0 (3000)	R	⑥	37	26	15	59②	15	18	18	139
2001	1.9 (1901)	7	①	37	33	159	59②	22③	19③	20	103
	1.9 (1901)	8	①	37	33	159	59②	22③	16③	20	103
	2.2 (2199	F	④	37	18	⑤	59②	7.5	13	15	138
	3.0 (3000)	R	⑥	37	26	15	59②	15	18	18	139

① Step 1: 22 ft. lbs.
Step 2: 33 ft. lbs.
Step 3: 90 degrees

② Flexplate specification: 44 ft. lbs.

③ Studs: 106 inch lbs.

④ Step 1: 22 ft. lbs.
Step 2: 155 degrees

⑤ 74 ft. lbs. Plus 75 degrees

⑥ Refer to the torque sequence chart for specifications

93461CYAF

BRAKE SPECIFICATIONS

All measurements in inches unless noted

			Brake Disc			Brake Drum Diameter				Brake Caliper	
Year	Model		Original Thickness	Minimum Thickness	Maximum Runout	Original Inside Diameter	Max. Wear Limit	Maximum Machine Diameter	Minimum Lining Thickness	Bracket Bolt (ft. lbs.)	Mounting Bolt (ft. lbs.)
1998	Coupe	F	0.710	0.633	0.0024	—	—	—	0.080	81	27
		R	0.430	0.370	0.0024	7.87	7.93	7.91	0.040	63	27
	Sedan	F	0.710	0.633	0.0024	—	—	—	0.080	81	27
		R	0.430	0.370	0.0024	7.87	7.93	7.91	0.040	63	27
	Wagon	F	0.710	0.633	0.0024	—	—	—	0.080	81	27
		R	0.430	0.370	0.0024	7.87	7.93	7.91	0.040	63	27
1999	Coupe	F	0.710	0.633	0.0024	—	—	—	0.080	81	27
		R	0.430	0.370	0.0024	7.87	7.93	7.91	0.040	63	27
	Sedan	F	0.710	0.633	0.0024	—	—	—	0.080	81	27
		R	0.430	0.370	0.0024	7.87	7.93	7.91	0.040	63	27
	Wagon	F	0.710	0.633	0.0024	—	—	—	0.080	81	27
		R	0.430	0.370	0.0024	7.87	7.93	7.91	0.040	63	27
2000	Coupe ①	F	0.710	0.633	0.0024	—	—	—	0.080	81	27
		R	0.430	0.370	0.0024	7.87	7.93	7.91	0.040	63	27
	Sedan ①	F	0.710	0.633	0.0024	—	—	—	0.080	81	27
		R	0.430	0.370	0.0024	7.87	7.93	7.91	0.040	63	27
	Wagon ①	F	0.710	0.633	0.0024	—	—	—	0.080	81	27
		R	0.430	0.370	0.0024	7.87	7.93	7.91	0.040	63	27
	Coupe ②	F	0.430	0.008	0.006	—	—	—	0.080	81	27
		R	0.160	0.420	0.0024	7.87	7.93	7.91	0.040	63	27
	Sedan ②	F	0.550	0.008	0.005	—	—	—	0.900	70	27
		R	0.420	0.004	0.0024	9.05	9.09	9.08	0.035	63	22
	Wagon ②	F	0.550	0.008	0.005	—	—	—	0.900	70	27
		R	0.420	0.004	0.0024	9.05	9.09	9.08	0.035	63	22
2001	Coupe ①	F	0.710	0.633	0.0024	—	—	—	0.080	81	27
		R	0.430	0.370	0.0024	7.87	7.93	7.91	0.040	63	27
	Sedan ①	F	0.710	0.633	0.0024	—	—	—	0.080	81	27
		R	0.430	0.370	0.0024	7.87	7.93	7.91	0.040	63	27
	Wagon ①	F	0.710	0.633	0.0024	—	—	—	0.080	81	27
		R	0.430	0.370	0.0024	7.87	7.93	7.91	0.040	63	27
	Coupe ②	F	0.430	0.008	0.006	—	—	—	0.080	81	27
		R	0.160	0.420	0.0024	7.87	7.93	7.91	0.040	63	27
	Sedan ②	F	0.550	0.008	0.005	—	—	—	0.900	70	27
		R	0.420	0.004	0.0024	9.05	9.09	9.08	0.035	63	22
	Wagon ②	F	0.550	0.008	0.005	—	—	—	0.900	70	27
		R	0.420	0.004	0.0024	9.05	9.09	9.08	0.035	63	22

NA: Not Available

F: Front

R: Rear

① S series

② L series

93461CYAG

For Wheel Alignment specifications, see Section 1 of this manual

WHEEL ALIGNMENT

Year	Model		Caster Range (+/-Deg.)	Caster Preferred Setting (Deg.)	Camber Range (+/-Deg.)	Camber Preferred Setting (Deg.)	Toe-in (in.)	Steering Axis Inclination (Deg.)
1998	SC,SL,SW	F	0.60	+1.69	0.69	-0.50	0.09 +/- 0.05	—
		R	—	—	0.69	-0.69	0.09 +/- 0.05	—
1999	SC,SL,SW	F	0.60	+1.69	0.69	-0.50	0.09 +/- 0.05	—
		R	—	—	0.69	-0.69	0.09 +/- 0.05	—
2000	S Series	F	0.60	+1.70	0.70	-0.50	0.20 +/- 0.10	—
		R	—	—	0.70	-0.70	0.20 +/- 0.10	—
	L series	F	1.00	+3.70	0.50	-1.00	0.20 +/- 0.15	—
		R	—	—	0.60	-1.00	0.15 +/- 0.07	—
2001	L series	F	1.00	+3.70	0.50	-1.00	0.20 +/- 0.15	—
		R	—	—	0.60	-1.00	0.15 +/- 0.07	—
	S Series	F	0.60	+1.70	0.70	-0.50	0.20 +/- 0.10	—
		R	—	—	0.70	-0.70	0.20 +/- 0.10	—

93461CYAH

TIRE, WHEEL AND BALL JOINT SPECIFICATIONS

Year	Model	OEM Tires Standard	OEM Tires Optional	Tire Pressures (psi) Front	Tire Pressures (psi) Rear	Wheel Size	Ball Joint Inspection
1998	SC1, SL, SL1, SW1	P175/70R14	None	30	28	5-JJ	NS
	SL2, SW2	P185/65R15	None	30	28	5-JJ	NS
	SC2	P195/60R15	None	30	28	5-JJ	NS
1999	SC1, SL, SL1, SW1	P175/70R14	None	30	28	5-JJ	NS
	SL2, SW2	P185/65R15	None	30	28	5-JJ	NS
	SC2	P195/60R15	None	30	28	5-JJ	NS
2000	LS, LS1	P195/65R15	None	30	26	6J	NS
	LS2	P205/65R15	None	30	30	6J	NS
	LW1	P195/65R15	None	30	26	6J	NS
	LW2	P205/65R15	None	30	30	6J	NS
	SC1, SL, SL1	P185/65R14	None	30	26	6J	NS
	SC2	P195/60R15	None	30	26	6J	NS
	SL2, SW2	P185/65R15	None	30	26	6J	NS
2001	LS, LS1	P195/65R15	None	30	26	6J	NS
	LS2	P205/65R15	None	30	30	6J	NS
	LW1	P195/65R15	None	30	26	6J	NS
	LW2	P205/65R15	None	30	30	6J	NS
	SC1, SL, SL1	P185/65R14	None	30	26	6J	NS
	SC2	P195/60R15	None	30	26	6J	NS
	SL2, SW2	P185/65R15	None	30	26	6J	NS

OEM: Original Equipment Manufacturer

PSI: Pounds Per Square Inch

STD: Standard

OPT: Optional

NS: Not specified by manufacturer

93461CYAI

SCHEDULED MAINTENANCE INTERVALS
SATURN LS, LS1, LW1, SC, SC1, SC2, SL, SL1, SL2, SW1 & SW2

		VEHICLE MILEAGE INTERVAL (x1000)												
TO BE SERVICED	TYPE OF SERVICE	3	6	9	12	15	18	21	24	27	30	33	36	39
Engine oil & filter	R		✓		✓		✓		✓		✓		✓	
Lubricate chassis, suspension, steering linkage, transaxle shift linkage, parking brake cable guides, underbody contact points & linkage	S/I		✓		✓		✓		✓		✓		✓	
Driveshaft boots, suspension bushings & ball joint seals	S/I		✓				✓		✓		✓		✓	
Exhaust system & throttle linkage	S/I		✓		✓		✓		✓		✓		✓	
Rotate tires	S/I		✓		✓		✓				✓			
Brake hoses & brake lining	S/I		✓				✓				✓			
Accessory drive belt(s)	S/I						✓						✓	
Engine coolant level, hoses & clamps	S/I						✓						✓	
Air filter element	R										✓			
Engine coolant	R												✓	
Manual transaxle oil	R		✓											
Spark plugs ①	R										✓			
Automatic transaxle fluid & filter	S/I										✓			
Ignition cables & fuel systems	S/I										✓			
Vacuum line/hose	S/I										✓			
Fuel filter ②	R													

R: Replace S/I: Service or Inspect

① Platinum tip spark plugs: replace every 100,000 miles.

② Replace every 60,000 miles.

FREQUENT OPERATION MAINTENANCE (SEVERE SERVICE)

If a vehicle is operated under any of the following conditions it is considered severe service:

- Extremely dusty areas.

- 50% or more of the vehicle operation is in 32°C (90°F) or higher temperatures, or constant operation in temperatures below 0°C (32°F).

- Prolonged idling (vehicle operation in stop and go traffic).

- Frequent short running periods (engine does not warm to normal operating temperatures).

- Police, taxi, delivery usage or trailer towing usage.

Engine oil & oil filter: change every 3000 miles

93461CYAJ

SCHEDULED MAINTENANCE INTERVALS
SATURN
L SERIES
S SERIES

The following should be used as a guide when determining the amount of work required for a particular service. In estimating how long a particular Scheduled Maintenance Service should take, please observe the following:

- Labor Time is time based on field research and data supplied by the vehicle manufacturer.
- Labor time operations are given in hours and tenths of an hour.
- All labor operations are to be used as a guide.

Mechanic Skill Level Codes:
(A) PRECISION: Highly skilled with multiple certification.
(B) GENERAL: Normally skilled with certification.
(C) MAINTENANCE: Semi-skilled working on certification.

	LABOR TIME
6000 Mile Service (B)	
All Models	1.7
12000 Mile Service (C)	
All Models	.9
18000 Mile Service (B)	
All Models	1.9
24000 Mile Service (C)	
All Models	.9
30000 Mile Service (B)	
All Models	2.8
36000 Mile Service (B)	
All Models	2.3

93461CYAK

HEATER CORES

2

1998–02 CHRYSLER CORPORATION

Concorde, Intrepid, LHS and Vision

REMOVAL & INSTALLATION

1. Disconnect the negative battery cable from the remote battery post located near the right strut tower.

CAUTION

After disconnecting the negative battery cable, wait 2 minutes for the driver's/passenger's air bag system capacitor to discharge before attempting to do any work around the steering column or instrument

2. Remove the instrument panel by removing or disconnecting the following:
   - Shifter knob Allen screw
   - Both instrument panel end covers

➡On the LHS or 300M, remove the center bezel prior to the shifter bezel

3. Using a trim stick tool, gently pry upward on the shifter bezel; then, discon-

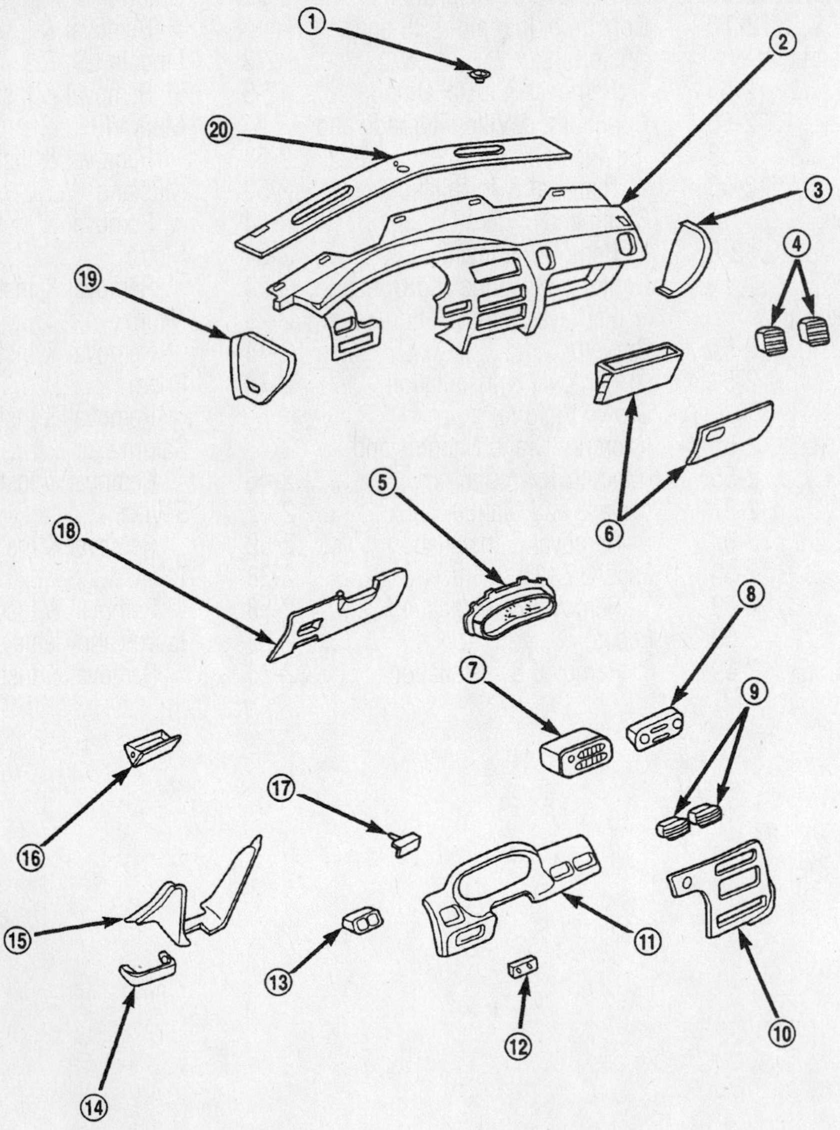

1 – SPEAKER, INSTRUMENT PANEL CENTER
2 – INSTRUMENT PANEL ASSEMBLY
3 – BEZEL, INSTRUMENT PANEL END CAP
4 – LOUVER, AIR OUTLET
5 – HOUSING INSTRUMENT CLUSTER
6 – GLOVE BOX ASSEMBLY
7 – RADIO
8 – CONTROL ASSEMBLY, INSTRUMENT PANEL
9 – LOUVER, AIR OUTLET
10 – BEZEL, INSTRUMENT PANEL TRIM-CENTER
11 – BEZEL INSTRUMENT CLUSTER
12 – SWITCH HEADLAMP
13 – LOUVER, AIR OUTLET
14 – COVER, INSTRUMENT PANEL CENTER SUPPORT/BIN (6 PASS. ONLY)
15 – 6 PASS. ONLY
16 – ASH RECEIVER
17 – LEVER, PARKING BRAKE
18 – COVER, LOWER INSTRUMENT PANEL
19 – BEZEL, INSTRUMENT PANEL END CAP
20 – COVER, UPPER INSTRUMENT PANEL

93111G81

Exploded view of the instrument panel—Intrepid

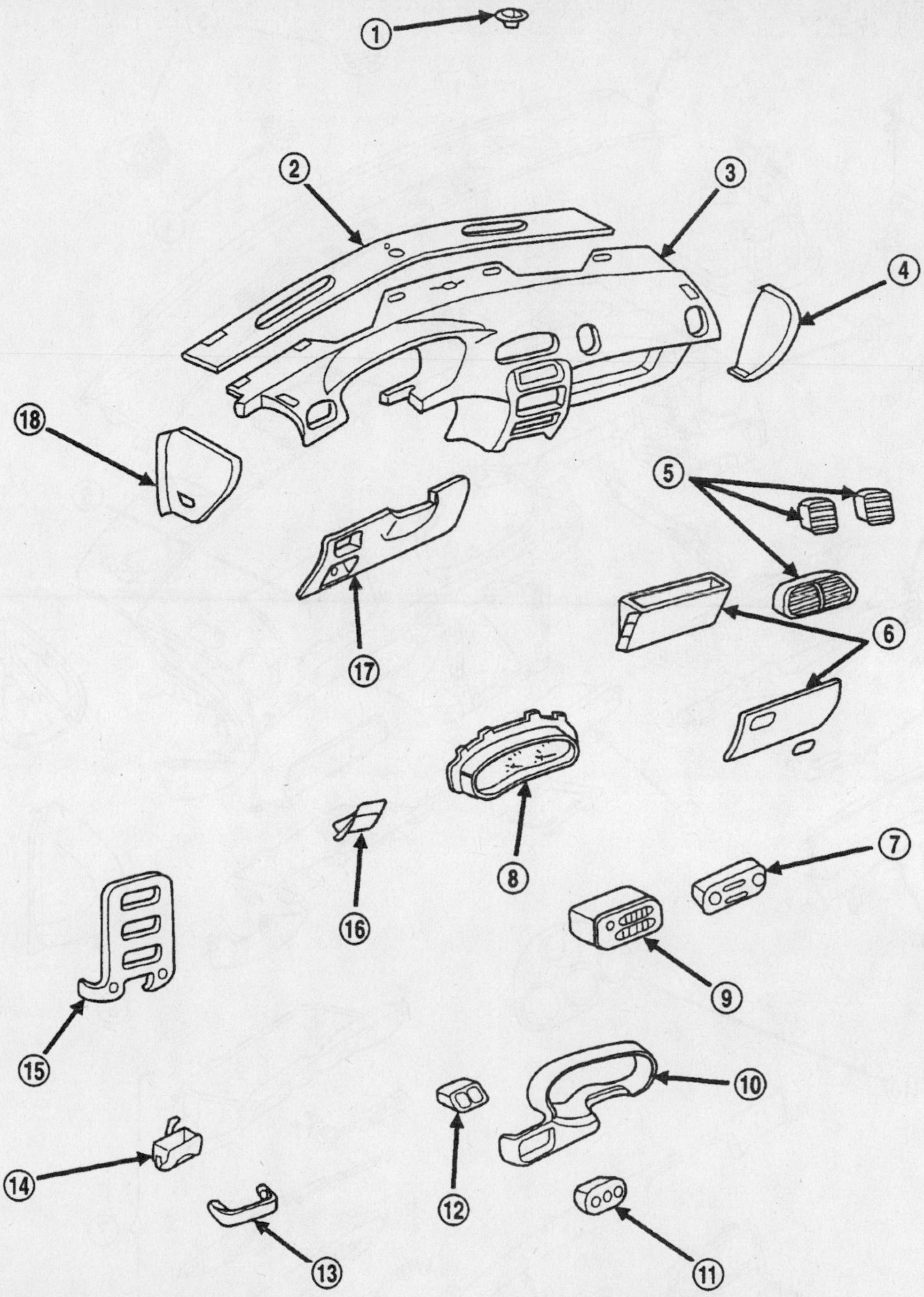

1 – SPEAKER, INSTRUMENT PANEL CENTER
2 – COVER, UPPER INSTRUMENT PANEL
3 – INSTRUMENT PANEL ASSEMBLY
4 – BEZEL, INSTRUMENT PANEL END CAP
5 – LOUVER, AIR OUTLET
6 – GLOVE BOX ASSEMBLY
7 – CONTROL ASSEMBLY, INSTRUMENT PANEL
8 – HOUSING, INSTRUMENT CLUSTER
9 – RADIO
10 – BEZEL, INSTRUMENT CLUSTER
11 – SWITCH, HEADLAMP
12 – LOUVER, AIR OUTLET
13 – COVER, INSTRUMENT PANEL CENTER SUPPORT/BIN (6 PASS. ONLY)
14 – ASH RECEIVER
15 – BEZEL, INSTRUMENT PANEL TRIM-CENTER
16 – LEVER, PARKING BRAKE
17 – COVER, LOWER INSTRUMENT PANEL
18 – BEZEL, INSTRUMENT PANEL END CAP

93111G82

Exploded view of the instrument panel—Concorde

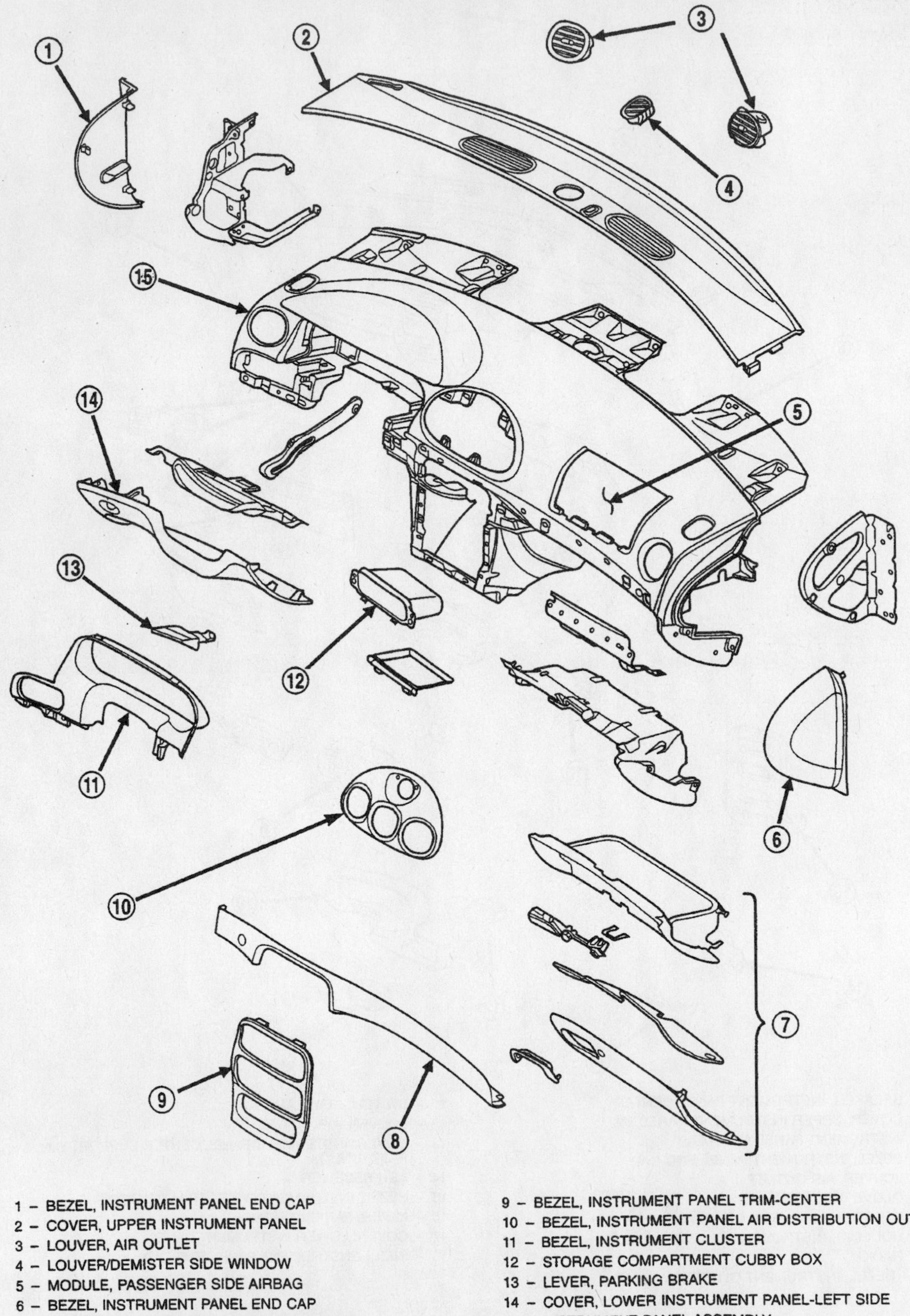

1 – BEZEL, INSTRUMENT PANEL END CAP
2 – COVER, UPPER INSTRUMENT PANEL
3 – LOUVER, AIR OUTLET
4 – LOUVER/DEMISTER SIDE WINDOW
5 – MODULE, PASSENGER SIDE AIRBAG
6 – BEZEL, INSTRUMENT PANEL END CAP
7 – GLOVE BOX ASSEMBLY, INSTRUMENT PANEL
8 – BEZEL, INSTRUMENT PANEL UPPER RIGHT TRIM
9 – BEZEL, INSTRUMENT PANEL TRIM-CENTER
10 – BEZEL, INSTRUMENT PANEL AIR DISTRIBUTION OUTLET
11 – BEZEL, INSTRUMENT CLUSTER
12 – STORAGE COMPARTMENT CUBBY BOX
13 – LEVER, PARKING BRAKE
14 – COVER, LOWER INSTRUMENT PANEL-LEFT SIDE
15 – INSTRUMENT PANEL ASSEMBLY

93111G83

Exploded view of the instrument panel—LHS and 300M

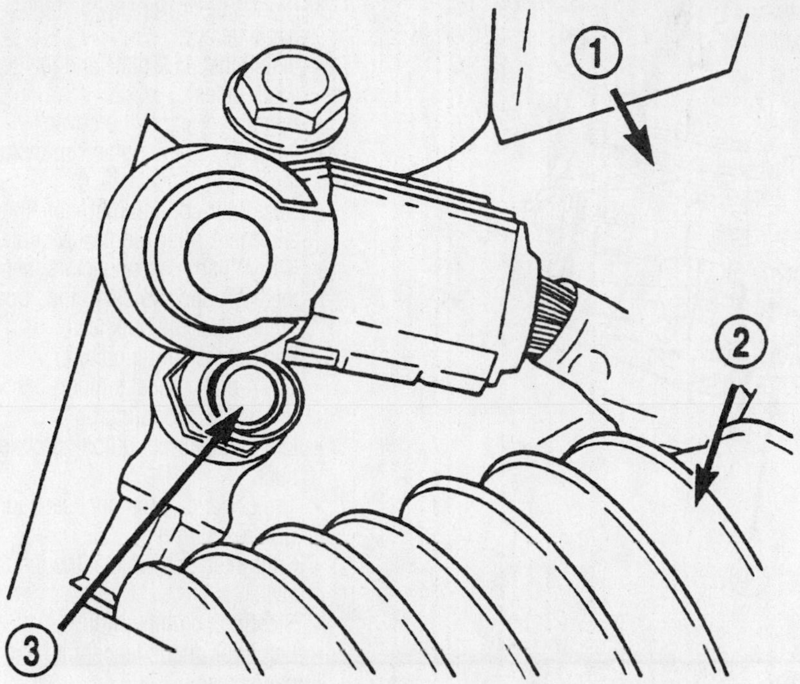

1. Right strut tower
2. Air cleaner inlet tube
3. Remote terminal

93111G84

View of the remote battery cable terminal—LHS, 300M, Concorde and Intrepid

nect both wiring connectors and a light bulb socket connector; then, remove the socket.

4. Remove the 2 lower instrument panel cover screws (outside end) and disconnect the deck lid release switch wiring connector. Pull the lower instrument panel cover rearward releasing the clips. Remove the cable-to-brake release handle. Pull the cover rearward and remove it.

5. If equipped for 6 passengers, remove the lower floor bin.

6. If equipped for 5 passengers, remove or disconnect the following items:

- Center bezel using the trim stick
- Heater/air conditioning housing assembly control switch and the traction control switch wiring harnesses
- Bezel from the vehicle
- Two left side console side cover screws, pull the cover outward and remove from the vehicle
- Lower glove box door; the 2 screws are located at the right console side cover, pull outward and remove it
- 2 bracket screws and the 2 inside the console bin; then, remove the console
- 2 center lower instrument panel nuts
- 4 steel reinforcement-to-lower instrument panel cover bottom bolts
- 16-way DLC from the reinforcement
- left floor duct/silencer pad screw and the pad
- Steering column shrouds
- Shift interlock cable at the ignition switch
- Column wiring
- The under column duct section
- Left panel air conditioning outlet duct
- The 4 brake pedal support bracket-to-column mounting bolts
- Lower the steering column
- Heater/air conditioning housing and the Air bag Control Module (ACM) harness connectors then, the 2 ground eyelets from the floor tunnel near the bulkhead

7. Using a trim stick, carefully, pry out on the left and right A-pillar trim moldings and slide them rearward to remove them.

8. Using a trim stick, carefully pry upward on the instrument panel top cover, and remove it.

9. From both sides, remove the scuff plate screw. Using a trim stick, carefully, pry out the scuff plates and remove them.

10. Remove or disconnect the following:

- 3 right side cowl panel screws and the panel
- 3 left side cowl panel screws and the panel
- Remove the right side under dash silencer and pad
- 2 right side radio antenna and amplifier harness connectors
- Left side junction block and Body Control Module (BCM) harness connectors
- 8 instrument panel-to-chassis screws
- Pull the instrument panel rearward
- Instrument panel from the vehicle with the help of an assistant

11. Discharge and recover the air conditioning system refrigerant.

12. Drain the cooling system into a clean container for reuse.

13. Place shop cloths in the interior to catch any excess coolant.

14. Remove or disconnect the following:

- Air cleaner hose and the air distribution duct
- Heater hoses-to-dash panel spring type fasteners and the heater hoses from the heater core. Plug the openings to prevent coolant spillage
- Refrigerant lines-to-expansion valve nut and separate the refrigerant lines from the expansion valve. Plug the openings to prevent contamination

⁂ WARNING

The air conditioning lubricant used in this system absorbs moisture readily. Do not leave any portion of the system open for any long period of time.

- 3 heater/air conditioning housing assembly-to-chassis nuts from the engine compartment
- 2 defroster duct screws and the duct
- Heater/air conditioning housing assembly-to-dash panel nuts (2) and screws (2)
- 4 rear seat heat duct nuts and the duct
- Rear seat heat duct elbow push-pin fastener
- Electrical connector from the

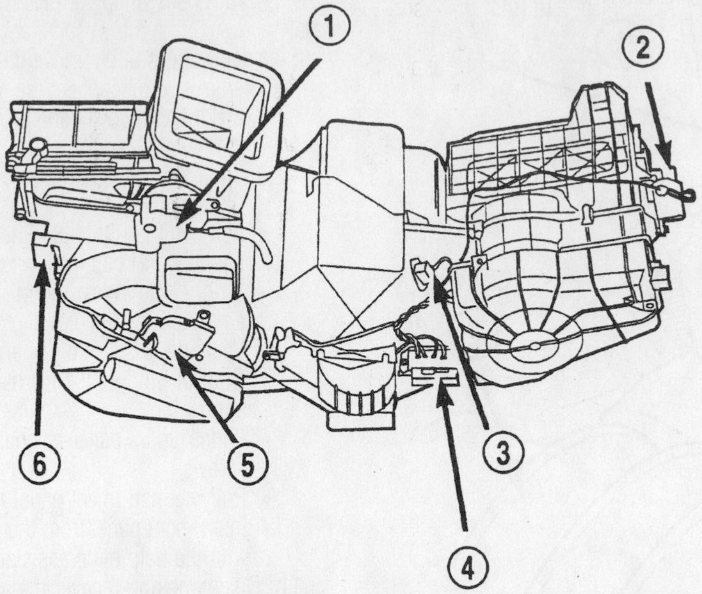

1 – MODE DOOR ACTUATOR
2 – RECIRCULATION DOOR ACTUATOR
3 – EVAPORATOR TEMPERATURE SENSOR
4 – POWER MODULE OR BLOWER RESISTOR
5 – BLEND DOOR ACTUATOR
6 – HVAC PLENUM CONNECTOR

93111G85

View of the heater/air conditioning housing assembly—LHS, 300M, Concorde and Intrepid

heater/air conditioning housing assembly
- Pull the heater/air conditioning housing assembly rearward and remove it from the vehicle
- 2 heater core-to-heater/air conditioning housing assembly screws
- Heater core from the heater/air conditioning housing assembly

To install:

15. Install or connect the following:
    - Heater core to the heater/air conditioning housing assembly
    - 2 heater core-to-heater/air conditioning housing assembly screws
    - Push the heater/air conditioning housing assembly forward (carefully) and install it to the vehicle
    - Electrical connector to the heater/air conditioning housing assembly
    - Rear seat heat duct elbow push pin fastener
    - Duct and the 4 rear seat heat duct nuts
    - Heater/air conditioning housing assembly-to-dash panel nuts (2) and screws (2)
    - 2 defroster duct and the duct screws
    - In the engine compartment, the 3 heater/air conditioning housing assembly-to-chassis nuts
    - Refrigerant lines to the expansion valve
    - Refrigerant lines-to-expansion valve nut
    - Heater hoses to the heater core and the heater hoses-to-dash panel spring type fasteners
    - Air cleaner hose and the air distribution duct
16. Refill the cooling system.
17. Evacuate, charge and leak-test the air conditioning system refrigerant.
18. Install the instrument panel by installing or connecting the following:
    - Instrument panel to the vehicle
    - Push the instrument panel forward
    - 8 instrument panel-to-chassis screws
    - Left side junction block and Body Control Module (BCM) harness connectors
    - 2 right side radio antenna and amplifier harness connectors
    - Right side under dash silencer and pad
    - Left side cowl panel and the 3 panel screws
    - Right side cowl panel and the 3 panel screws
    - Install the scuff plate and the screw (both sides)
    - Instrument panel top cover
    - Left and right A-pillar trim moldings
    - Heater/air conditioning housing and the Air bag Control Module (ACM) harness connectors then, the 2 ground eyelets to the floor tunnel near the bulkhead
    - Raise the steering column
    - The 4 brake pedal support bracket-to-column mounting bolts
    - Left panel air conditioning outlet duct
    - The under column duct section
    - Column wiring
    - Shift interlock cable at the ignition switch
    - Steering column shrouds
    - Left floor duct/silencer pad and the pad screw
    - 4 steel reinforcement-to-lower instrument panel cover bottom bolts
    - 6-way Diagnostic Link Connector (DLC) to the reinforcement
    - 2 center lower instrument panel nuts
    - The 2 bracket screws and the 2 inside the console bin; then, the console (5 passenger models only)
    - 2 screws are located at the right console side cover (5 passenger models only)
    - Two left side console side cover screws (5 passenger models only)
    - Bezel to the vehicle (5 passenger models only).
    - Heater/air conditioning housing assembly control switch and the traction control switch wiring harnesses.
    - Center bezel using the trim stick.
    - Lower floor bin (6 passenger models only).
    - 2 lower instrument panel cover screws (outside end) and connect the deck lid release switch wiring connector. Install the cable-to-brake release handle
    - Light bulb socket. Both wiring connectors and a light bulb socket connector
    - Shifter bezel
    - Center bezel prior to the shifter bezel (LHS or 300M models)
    - Both instrument panel end covers
    - Shifter knob Allen screw
19. Connect the negative battery cable to

the remote battery post located near the right strut tower.

20. Operate the engine to normal operating temperatures; then, check the climate control operation and check for leaks.

PT Cruiser

2001

1. Before servicing the vehicle, refer to the precautions in the beginning of this section.
2. Remove the A/C refrigerant by using approved equipment.
3. Drain the cooling system.
4. Disconnect the negative battery cable.
5. Remove the instrument panel by removing or disconnecting the following:
   - Left and right **A**pillar trim moldings using a trim stick
   - Front power window switch
   - Center bezel retaining screw
   - Heating, Ventilation and Air Conditioning (HVAC) control knobs
   - Center bezel using a trim stick
   - Top cover retaining screws and pull it rearward to remove
   - HVAC control unit from the instrument panel
   - Retaining screws from the upper and lower steering column shrouds
   - Left lower instrument panel bezel
   - Steering column wire connectors
   - Steering column
   - Brake pedal support bracket
   - Left side instrument panel end cap from the wiring connectors
   - Rear power window switch from the console and pull the parking handle all the way up
   - Auxiliary power outlet wire connector
   - Center console
   - Glove box assembly
   - Right side instrument panel end cap
   - 5 right side instrument panel wire connectors
   - Instrument panel from the vehicle, with the help of an assistant
6. Remove or disconnect the following:
   - Heater hoses from the heater core
   - Reposition the vehicle speed control servo, if equipped
   - Suction and liquid lines from the evaporator
   - Drain tube
   - A/C vacuum harness connector
   - Defroster duct from the heater—A/C unit housing
   - Heater—A/C unit housing from the vehicle
   - Heater core from the housing

To install:

7. Install or connect the following:
   - Heater core to the housing
   - Heater—A/C unit housing to the vehicle
   - Defroster duct to the housing
   - A/C vacuum harness
   - Drain tube
   - Suction and liquid lines to the evaporator
   - Reposition the vehicle speed control servo, if equipped
   - Heater hoses to the heater core
8. Install the instrument panel by installing or connecting the following:
   - Instrument panel
   - Right side instrument panel wire connectors
   - Right side instrument panel end cap
   - Glove box assembly
   - Center console
   - Auxiliary power outlet wire connector
   - Rear power window switch
   - Left side instrument panel end cap
   - Left side wire connectors
   - Brake pedal support bracket
   - Steering column with a new pinch bolt
   - Steering column wire connectors
   - Left lower instrument panel bezel
   - HVAC control unit
   - Center bezel
   - Front power window switch
   - Left and right **A** pillar trim moldings
   - Negative battery cable
9. Fill the cooling system.
10. Recharge the A/C system.
11. Start the vehicle and verify proper system operations.
12. Roadtest the vehicle and check for any rattles, repair if necessary.

2002

1. Before servicing the vehicle, refer to the precautions in the beginning of this section.
2. Remove the A/C refrigerant by using approved equipment.

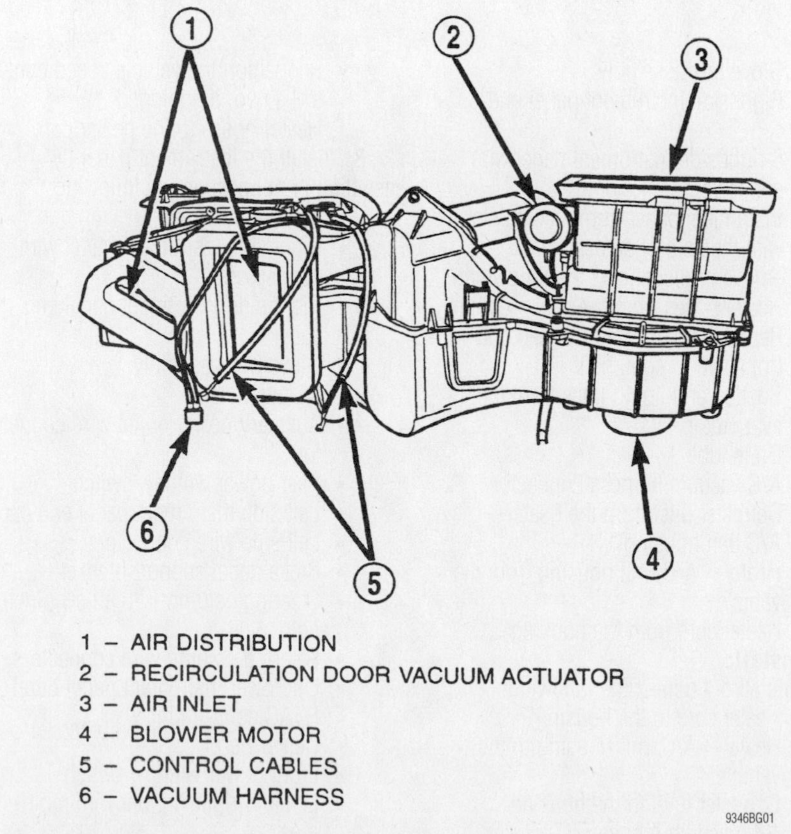

Heater-A/C unit housing foam seals—PT Cruiser

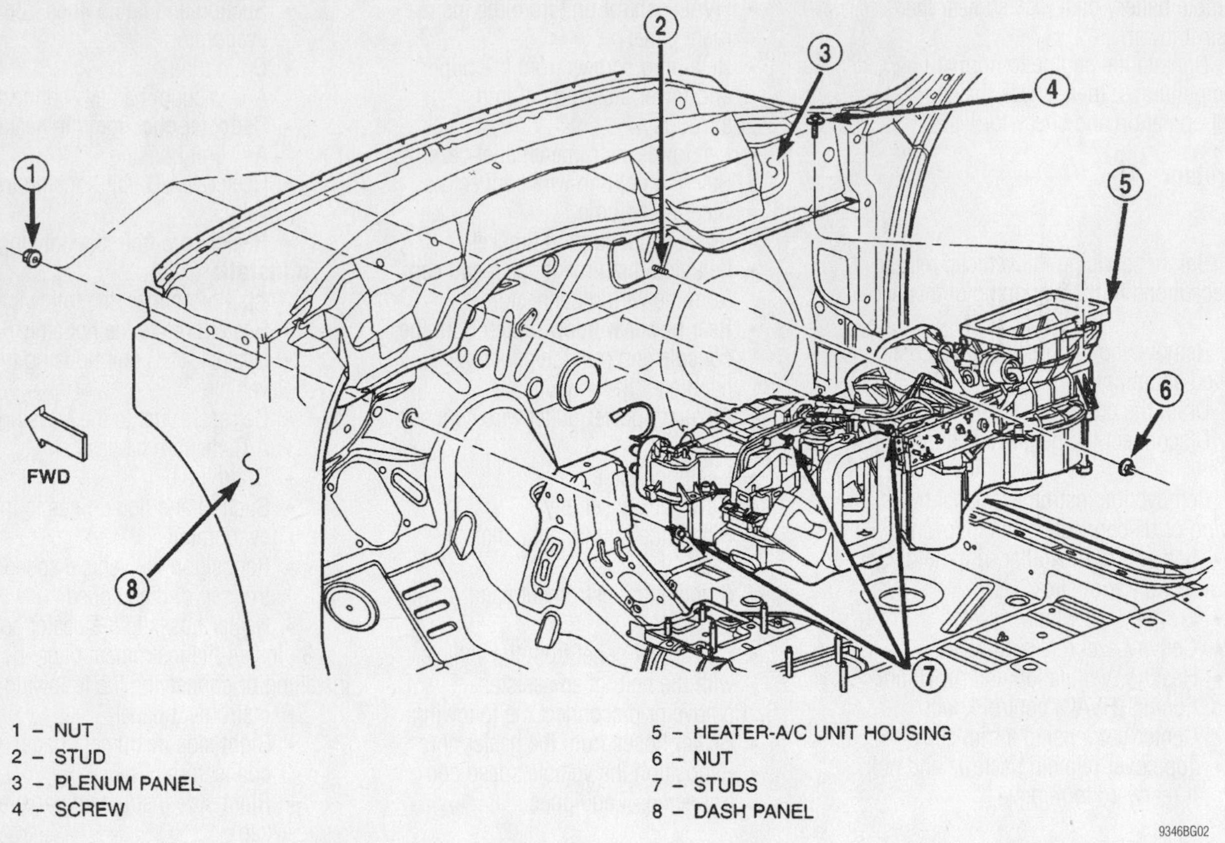

Heater-A/C unit housing removal—PT Cruiser

3. Drain the cooling system.
4. Disconnect the negative battery cable.
5. Remove the instrument panel by removing or disconnecting the following:
   - Left and right **A**pillar trim moldings using a trim stick
   - Front power window switch
   - Center bezel retaining screw
   - Heating, Ventilation and Air Conditioning (HVAC) control knobs
   - Center bezel using a trim stick
   - Top cover retaining screws and pull it rearward to remove
   - HVAC control unit from the instrument panel
   - Retaining screws from the upper and lower steering column shrouds
   - Left lower instrument panel bezel
   - Steering column wire connectors
   - Steering column
   - Brake pedal support bracket
   - Left side instrument panel end cap from the wiring connectors
   - Rear power window switch from the console and pull the parking handle all the way up
   - Auxiliary power outlet wire connector
   - Center console
   - Glove box assembly
   - Right side instrument panel end cap
   - 5 right side instrument panel wire connectors
   - Instrument panel from the vehicle, with the help of an assistant
6. Remove or disconnect the following:
   - Heater hoses from the heater core
   - Reposition the vehicle speed control servo, if equipped
   - Suction and liquid lines from the evaporator
   - Drain tube
   - A/C vacuum harness connector
   - Defroster duct from the heater—A/C unit housing
   - Heater—A/C unit housing from the vehicle
   - Heater core from the housing

To install:

7. Install or connect the following:
   - Heater core to the housing
   - Heater—A/C unit housing to the vehicle
   - Defroster duct to the housing
   - A/C vacuum harness
   - Drain tube
   - Suction and liquid lines to the evaporator
   - Reposition the vehicle speed control servo, if equipped
   - Heater hoses to the heater core
8. Install the instrument panel by installing or connecting the following:
   - Instrument panel
   - Right side instrument panel wire connectors
   - Right side instrument panel end cap
   - Glove box assembly
   - Center console
   - Auxiliary power outlet wire connector
   - Rear power window switch
   - Left side instrument panel end cap
   - Left side wire connectors
   - Brake pedal support bracket
   - Steering column with a new pinch bolt
   - Steering column wire connectors
   - Left lower instrument panel bezel
   - HVAC control unit
   - Center bezel
   - Front power window switch
   - Left and right **A** pillar trim moldings
   - Negative battery cable
9. Fill the cooling system.
10. Recharge the A/C system.

11. Start the vehicle and verify proper system operations.

12. Roadtest the vehicle and check for any rattles, repair if necessary.

Avenger and Sebring Coupe

REMOVAL & INSTALLATION

1. Disconnect the negative battery cable. Properly drain the cooling system.
2. Disconnect the heater hoses from the heater core.
3. Remove the floor console by removing or disconnecting the following:
   - Center console panel
   - Shift knob
   - Accessory box or ashtray
   - Floor console panel assembly
   - Shift lever cover assembly
   - Floor console assembly
4. From under the steering wheel, remove the hood lock release handle, the driver's side under cover and the lap cooler grille.
5. Remove the steering wheel by removing or disconnecting the following:
   - Air bag module-to-steering wheel screws (rear of the steering wheel) and the air bag module

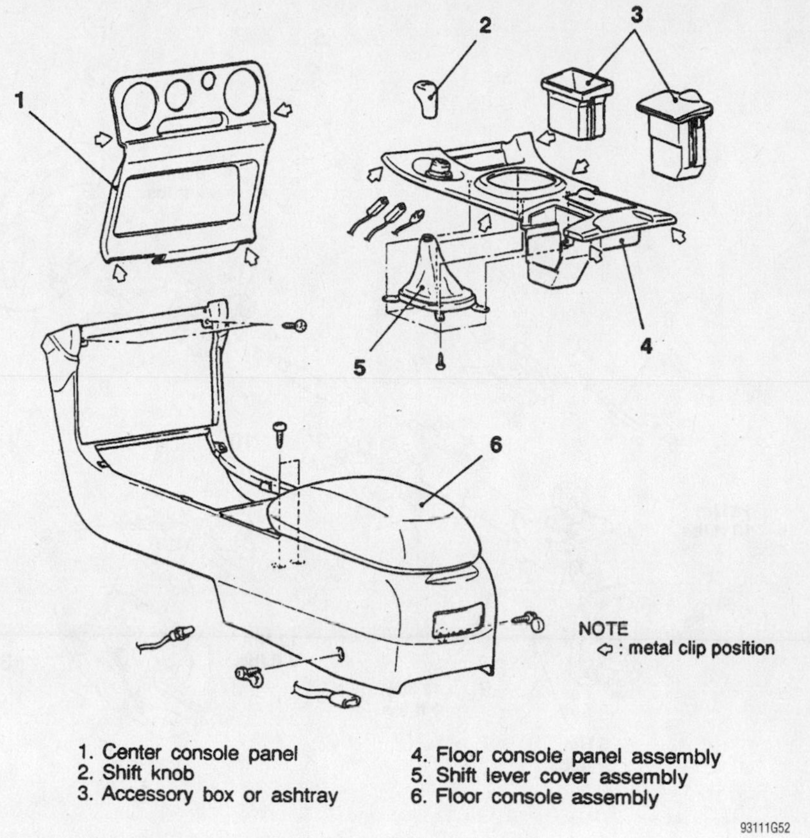

Exploded view of the center console—Avenger and Sebring coupe

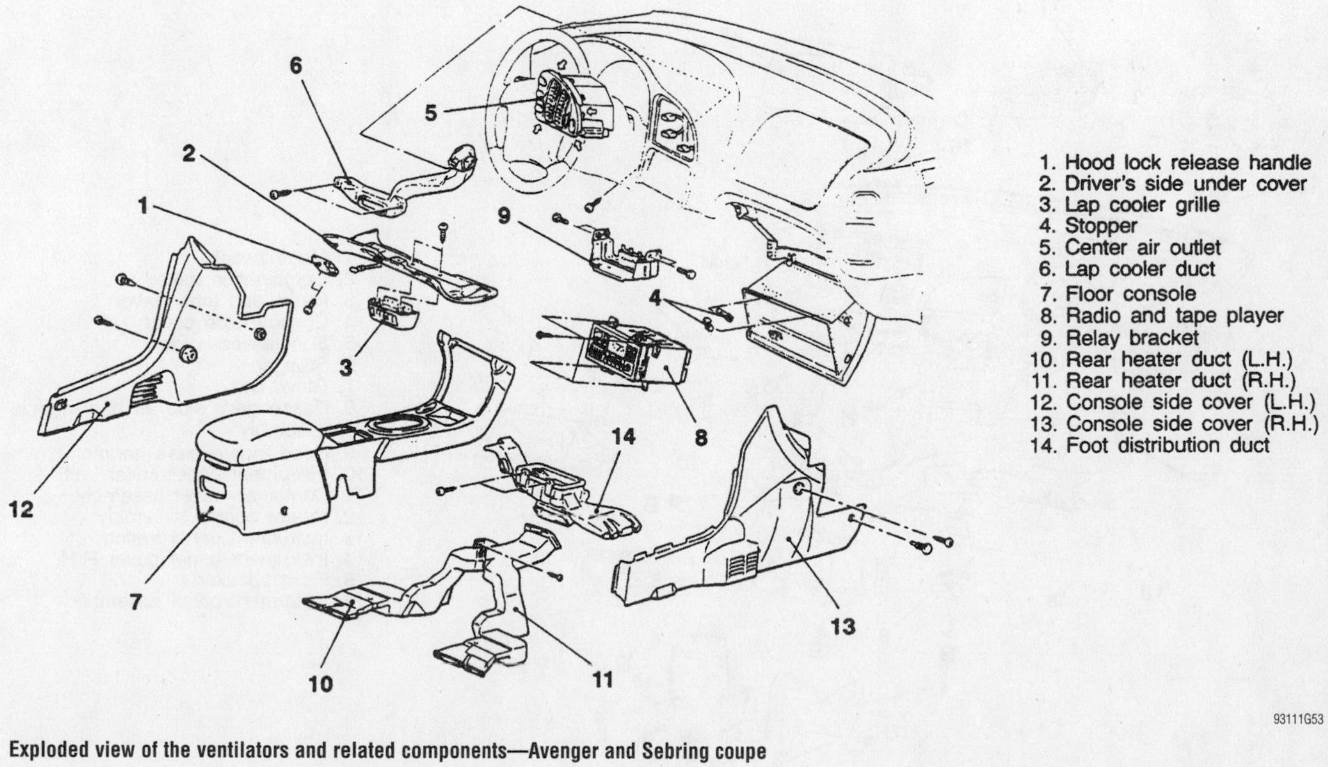

Exploded view of the ventilators and related components—Avenger and Sebring coupe

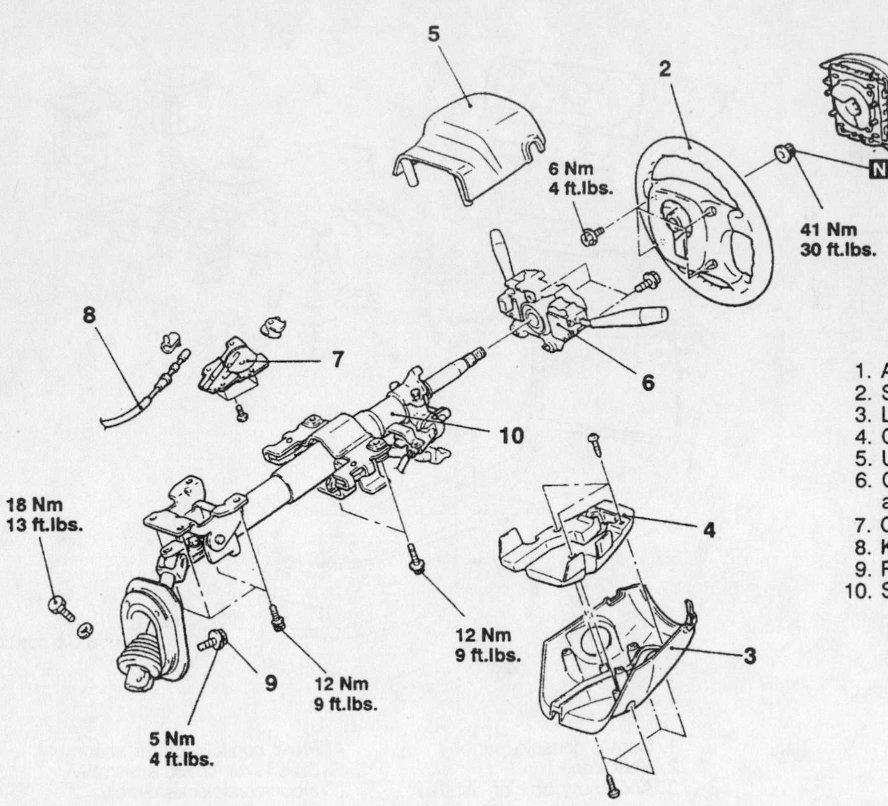

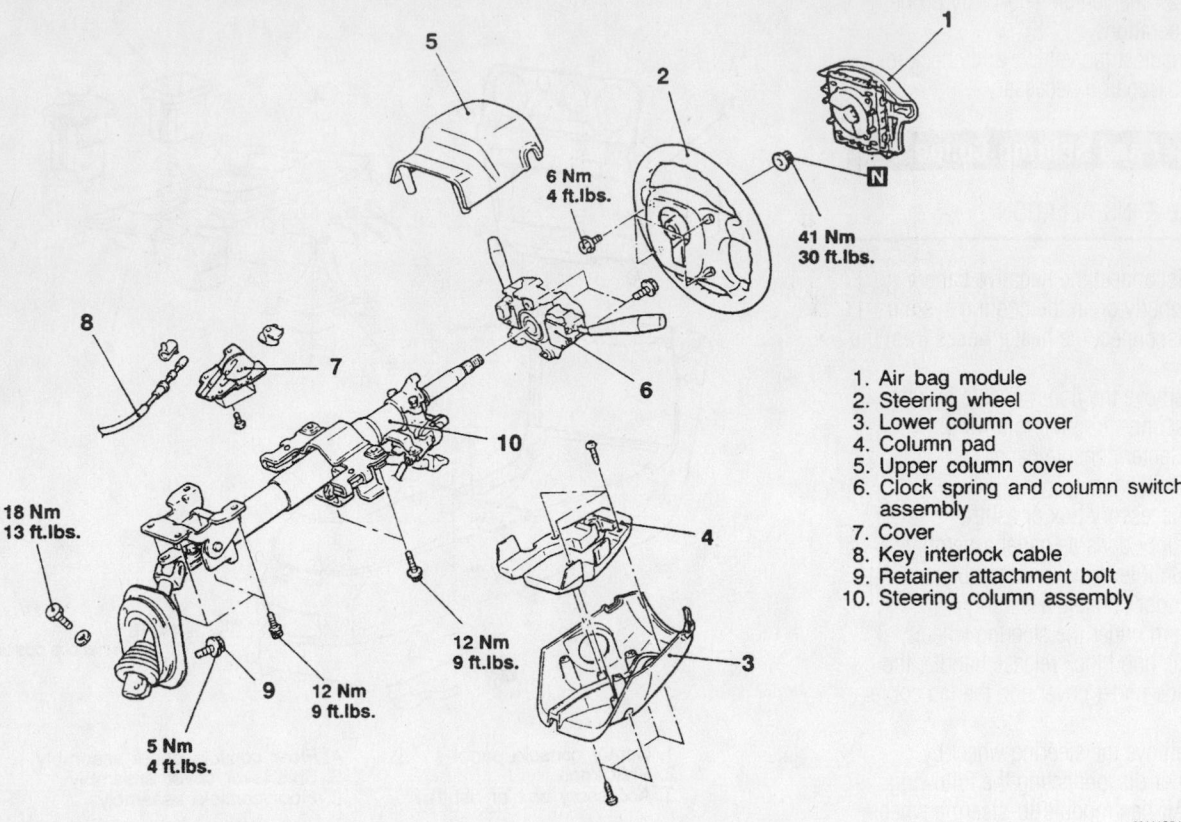

1. Air bag module
2. Steering wheel
3. Lower column cover
4. Column pad
5. Upper column cover
6. Clock spring and column switch assembly
7. Cover
8. Key interlock cable
9. Retainer attachment bolt
10. Steering column assembly

93111G54

Exploded view of the steering column and related components—Avenger and Sebring coupe

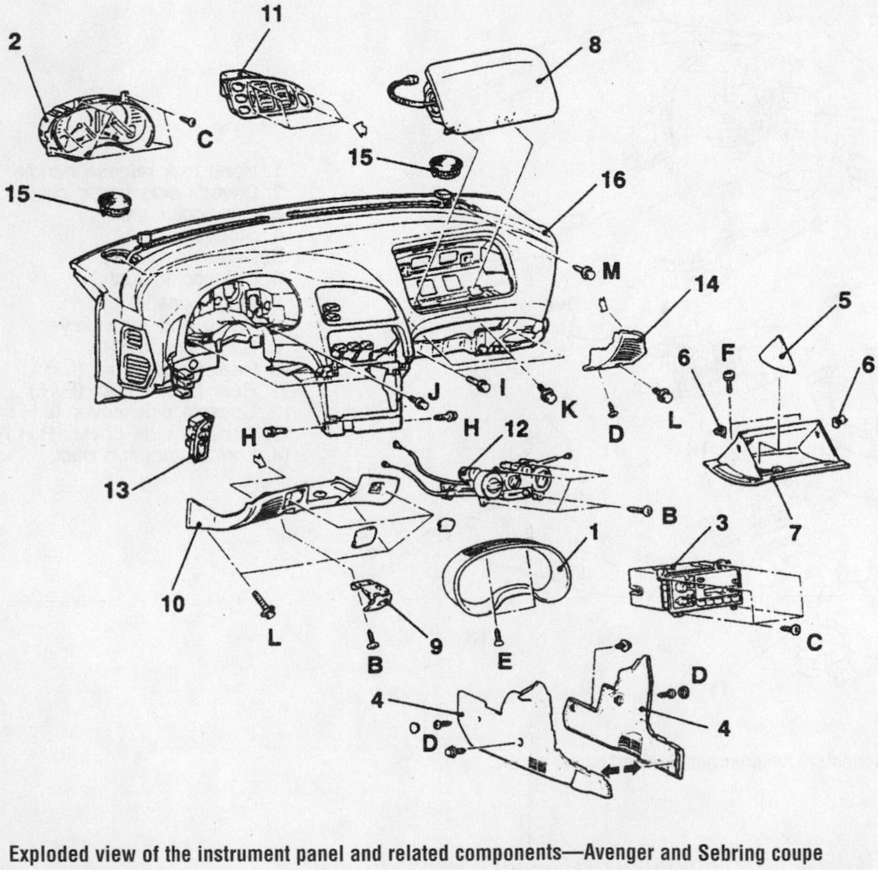

1. Meter bezel
2. Combination meter
3. Radio and tape player
4. Console side cover
5. Sunglasses holder
6. Stopper
7. Glove box
8. Passenger's side air bag module assembly
9. Hood lock release handle
10. Instrument under cover L.H.
11. Center air outlet assembly
12. Heater control assembly
13. Instrument panel switch
14. Instrument under cover R.H.
15. Front speaker
16. Instrument panel assembly

93111G55

Exploded view of the instrument panel and related components—Avenger and Sebring coupe

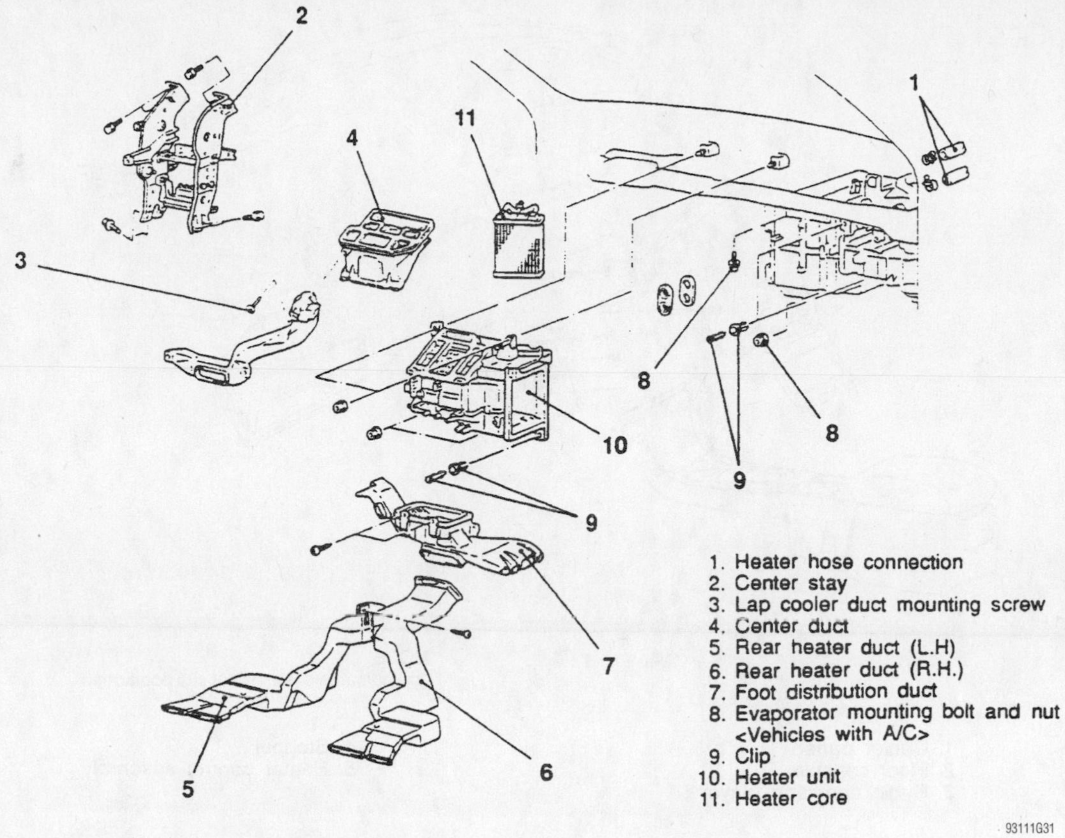

Exploded view of the heater core housing and related components—Avenger and Sebring coupe

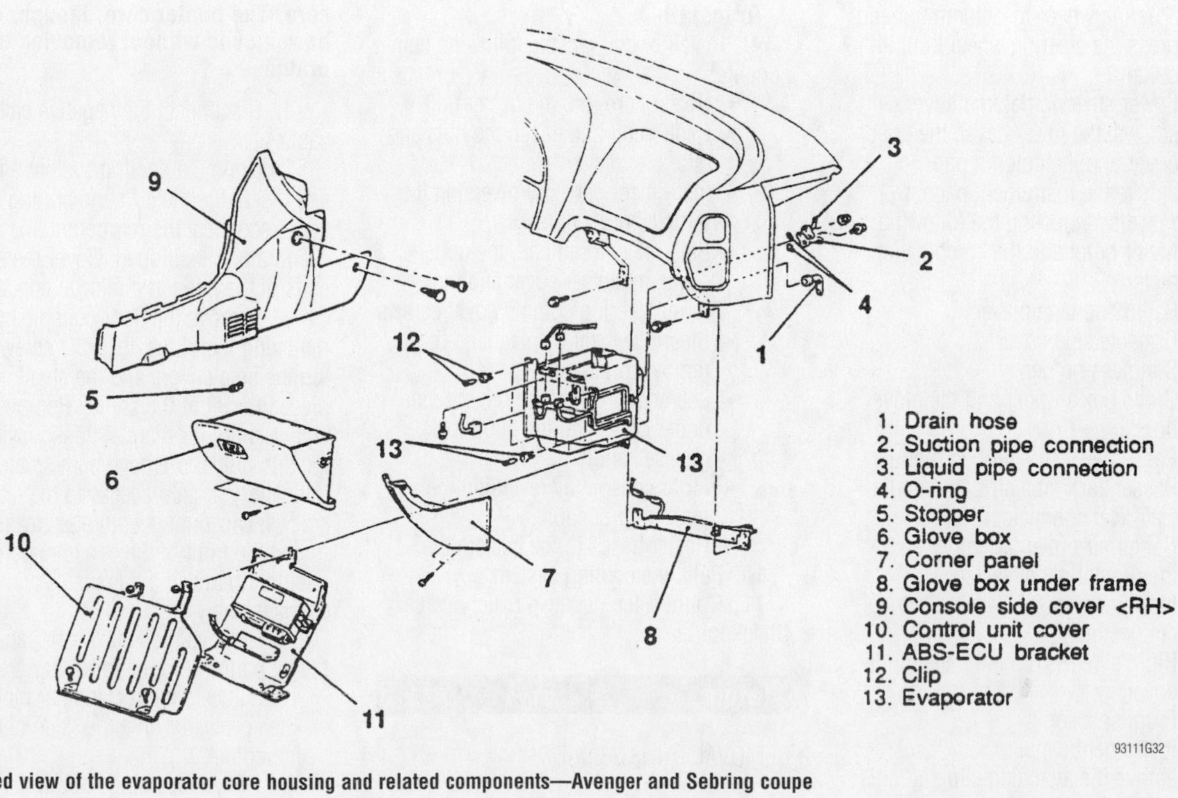

Exploded view of the evaporator core housing and related components—Avenger and Sebring coupe

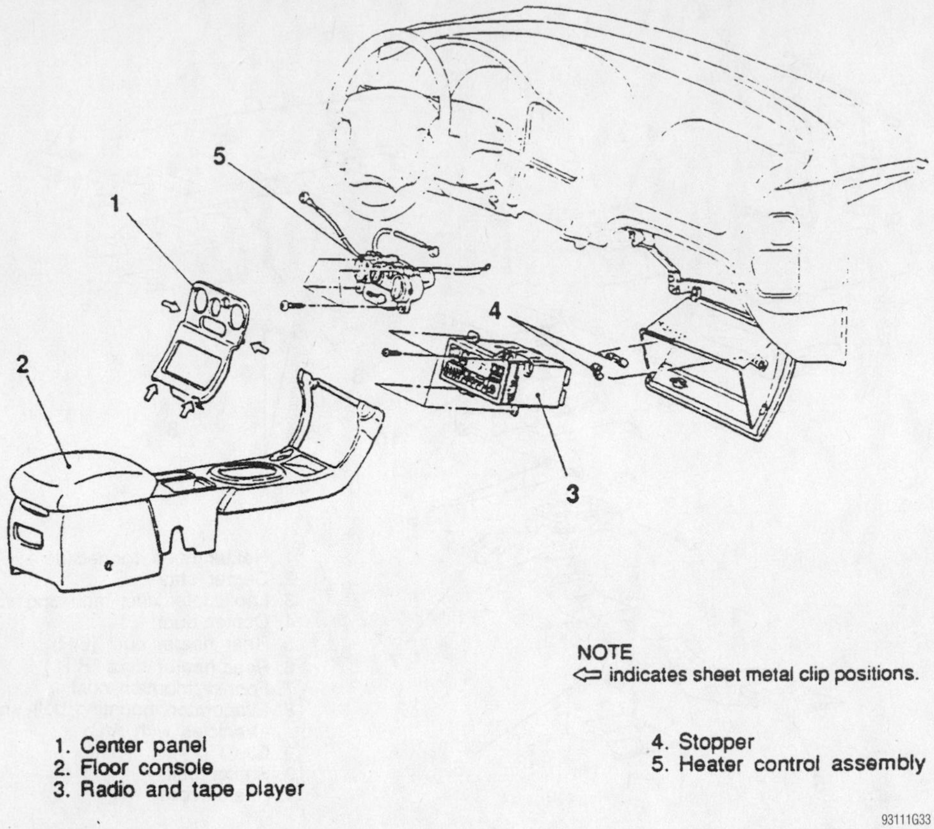

Exploded view of the manual control head and related components—Avenger and Sebring coupe

- Steering wheel-to-column nut and press the steering wheel from the column
- Lower steering column cover screws, the lower cover, the upper cover and the column pad

6. Remove the instrument panel by removing or disconnecting the following:

- Meter bezel and the combination meter
- Radio and tape player
- Console side cover
- Sunglass holder
- Glove box stopper and the glove box
- Passenger's side air bag module
- Passenger's side air bag module electrical connector
- Center air outlet assembly
- Instrument panel switch
- Instrument panel switch electrical connectors
- Right side instrument panel under cover
- Front speaker
- Instrument panel

7. Remove the attaching clip and nuts and remove the heater unit.

8. Remove the heater core from the heater unit.

To install:

9. Install or connect the following components:

- Heater core into the heater unit
- Heater unit and attach the nuts and clip
- Instrument panel by reversing the removal procedures
- Steering column pad, the upper cover, the lower cover and the lower steering column cover screws
- Steering wheel by reversing the removal procedure
- Lap cooler grille, the driver's side under cover and the hood lock release handle
- Floor console by reversing the removal procedure
- Heater hoses to the heater core

10. Refill the cooling system.

11. Connect the negative battery cable. Check for leaks.

Talon

REMOVAL & INSTALLATION

➡The evaporator housing can be removed by it self, without removing the console, instrument panel or heater core. The heater core, though, cannot be removed without removing the evaporator.

1. Disconnect the negative battery cable.

2. Drain the cooling system and properly discharge the air conditioning system and disconnect the refrigerant lines from the evaporator, if equipped. Cover the exposed ends of the lines to minimize contamination.

3. Remove the floor console by first removing the plugs, then the screws retaining the side covers and the small cover piece in front of the shifter. Remove the shifter knob, for manual transmission, and the cup holder. Remove both small pieces of upholstery to gain access to the retainer screws. Disconnect both electrical connectors at the front of the console. Remove the shoulder harness guide plates and the console assembly.

4. Remove the instrument panel assembly by performing the following procedure:

a. Locate the rectangular plugs in the knee protector on either side of the steering column.

b. Pry these plugs out, and remove the screws.

c. Remove the screws from the hood lock release lever and the knee protector.

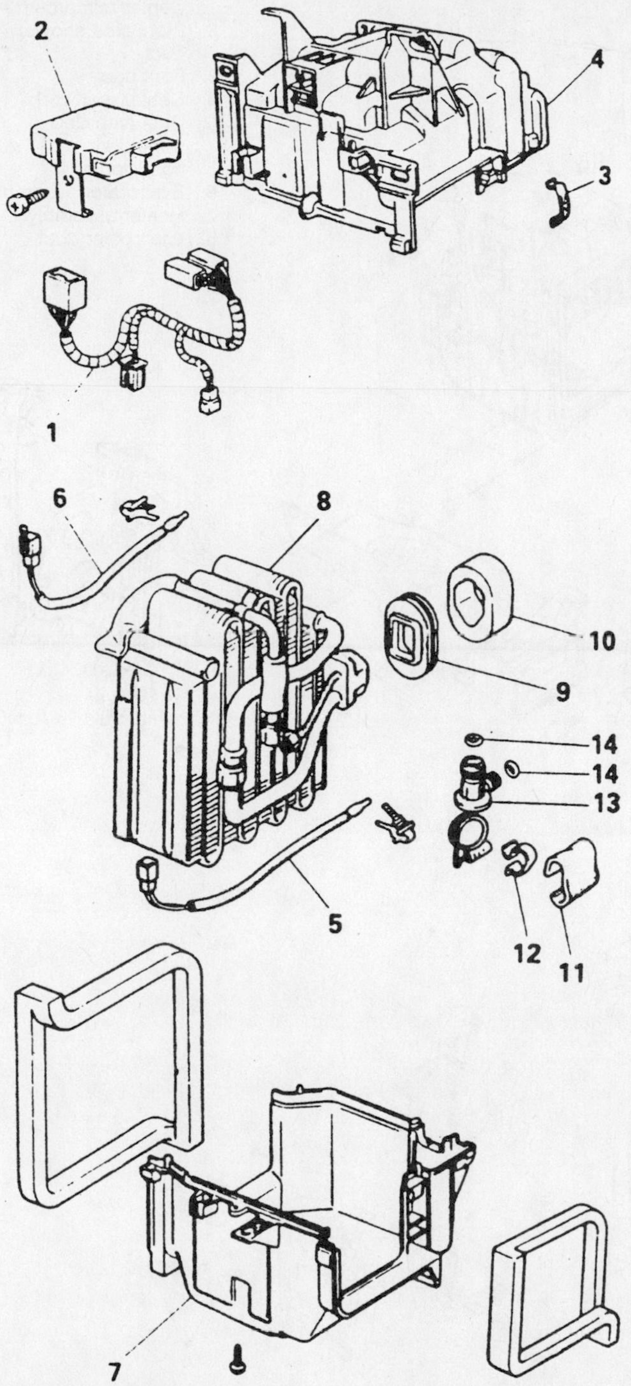

1. Wiring harness
2. Air conditioning control unit
3. Clips
4. Upper evaporator case
5. Air inlet sensor
6. Air thermo sensor
7. Lower evaporator case
8. Evaporator assembly
9. Grommet
10. Insulator
11. Rubber insulator
12. Clip
13. Expansion valve
14. O-ring

93111G26

Exploded view of the evaporator and related components

5. Remove or disconnect the following:
   - Upper and lower column covers
   - Narrow panel covering the instrument cluster cover screws and the cover
   - Radio panel and the radio
   - Center air outlet assembly by reaching through the grille and pushing the side clips out with a small flat-tipped tool while carefully prying the outlet free
   - Heater control knobs (pull off) and the heater control panel assembly
   - Open the glove box, remove the plugs from the sides and the glove box assembly
   - Instrument gauge cluster and the speedometer adapter by disconnecting the speedometer cable from the transaxle, pulling the cable slightly towards the vehicle interior, then giving a slight twist on the adapter to release it
   - Left and right speaker covers from the top of the instrument panel
   - Center plate below the heater controls
   - Lower air ducts
   - Steering column bolts and lower the column
   - Instrument panel mounting screws, bolts and the instrument panel assembly
   - Both stamped steel reinforcement pieces
   - Lower duct work from the heater box
   - Upper center duct

6. Vehicles without air conditioning will have a square duct in place of the evaporator. Remove this duct if present. If equipped with air conditioning, remove the evaporator assembly by removing or disconnecting the following:
   - Wiring harness connectors and the electronic control unit
   - Drain hose and lift out the evaporator unit

7. If servicing the assembly, disassemble the housing and remove the expansion valve and evaporator.

8. With the evaporator removed, remove the heater unit. To prevent bolts from falling inside the blower assembly, set the inside/outside air-selection damper to the position that permits outside air introduction.

9. Remove the cover plate around the heater tubes and remove the core fastener

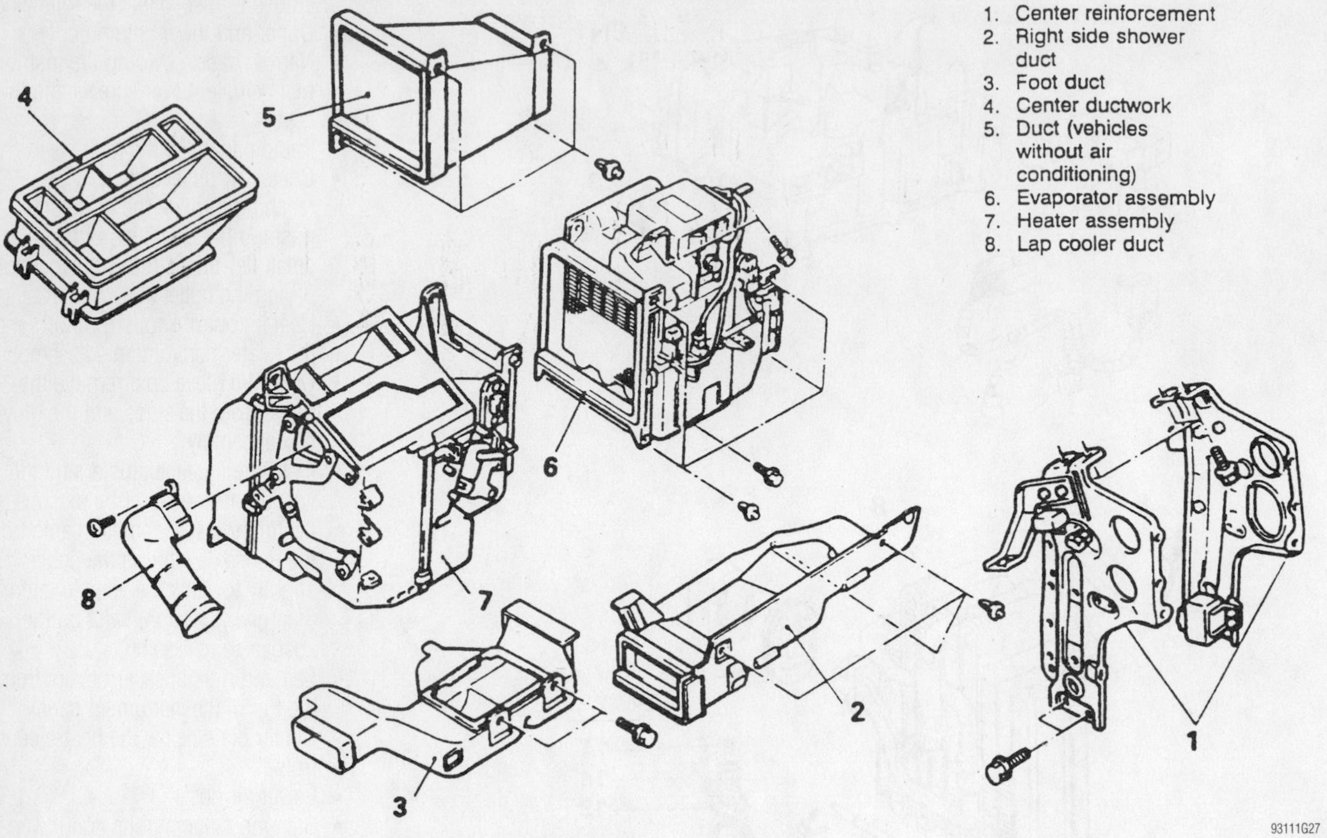

1. Center reinforcement
2. Right side shower duct
3. Foot duct
4. Center ductwork
5. Duct (vehicles without air conditioning)
6. Evaporator assembly
7. Heater assembly
8. Lap cooler duct

Exploded view of the heater and air conditioning housing—Talon

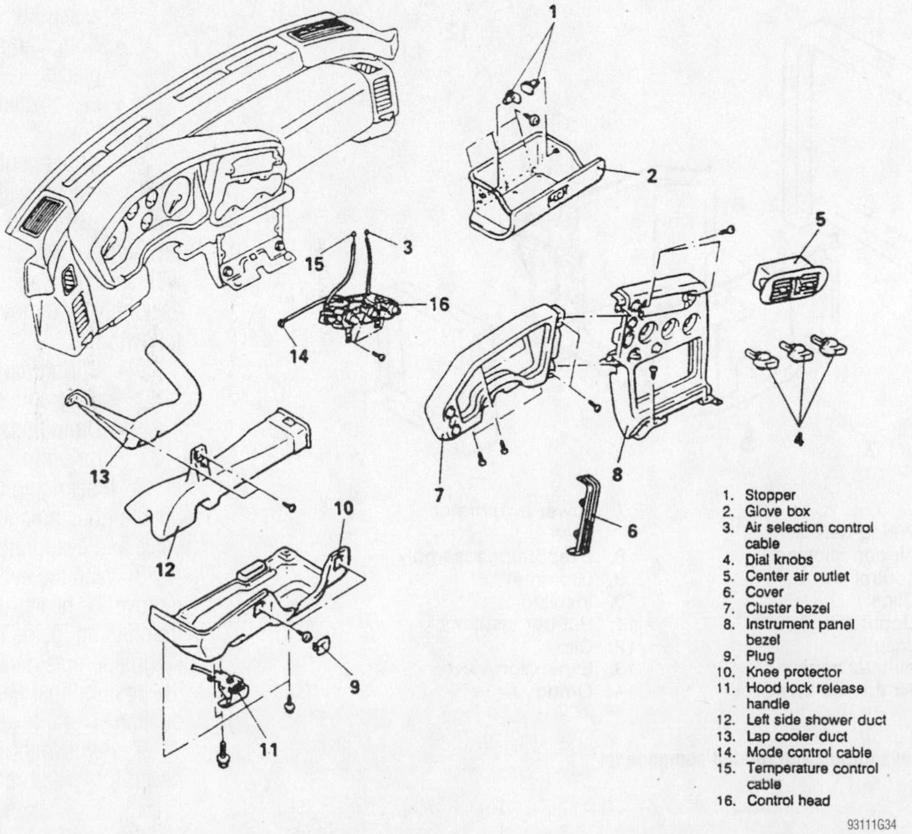

1. Stopper
2. Glove box
3. Air selection control cable
4. Dial knobs
5. Center air outlet
6. Cover
7. Cluster bezel
8. Instrument panel bezel
9. Plug
10. Knee protector
11. Hood lock release handle
12. Left side shower duct
13. Lap cooler duct
14. Mode control cable
15. Temperature control cable
16. Control head

Exploded view of the manual control head and related components—Talon

clips. Pull the heater core from the heater box, being careful not to damage the fins or tank ends.

To install:

10. Install the heater core to the heater box. Install the clips and cover.
11. Install the heater box and connect the duct work.
12. Assemble the housing, evaporator and expansion valve, making sure the gaskets are in good condition. Install the evaporator housing.
13. Using new lubricated O-rings, connect the refrigerant lines to the evaporator.
14. Install or connect the following:
    - Electronic transmission ELC box
    - All wire and control cables
    - Instrument panel assembly and the console by reversing their removal procedures
15. Evacuate, charge and leak-test the air conditioning system. If the evaporator was replaced, add 2 oz. of refrigerant oil during the recharge.
16. Refill the cooling system.
17. Connect the negative battery cable and check the entire climate control system for proper operation. Check the system for leaks.

Neon

REMOVAL & INSTALLATION

1. Disconnect the negative battery cable.
2. Discharge and recover the air conditioning system refrigerant.
3. Remove the instrument panel from the vehicle by removing or disconnecting the following:
    - Push the seats back all the way
    - Pry out the left and right A-pillar trim moldings and remove (using a trim stick tool)
    - Upper instrument panel cover
    - Pull up on the cluster bezel (carefully), and remove it from the vehicle
    - Pull the instrument panel cover rearward (carefully), and remove it from the vehicle

WARNING

Lock the steering wheel in the straight-ahead position; this will prevent damage to the clockspring

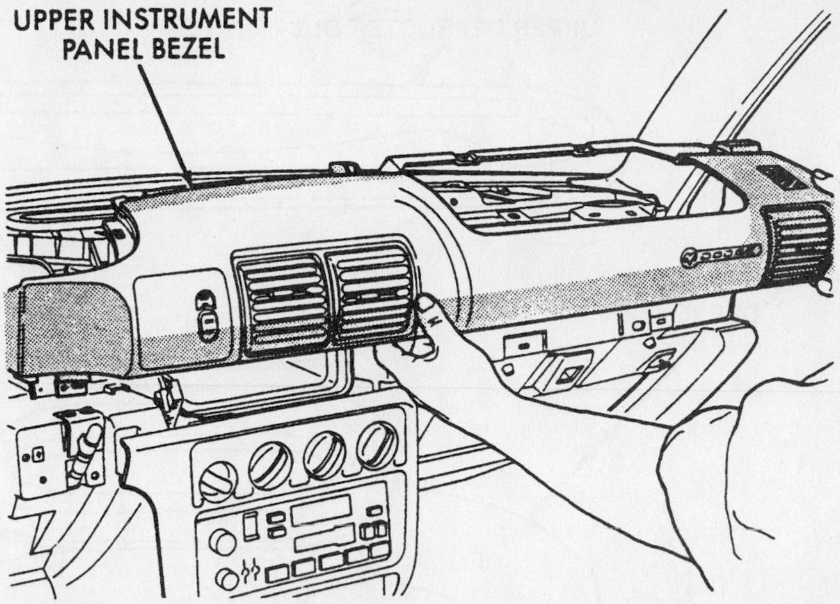

Remove the right side upper instrument panel bezel—Neon

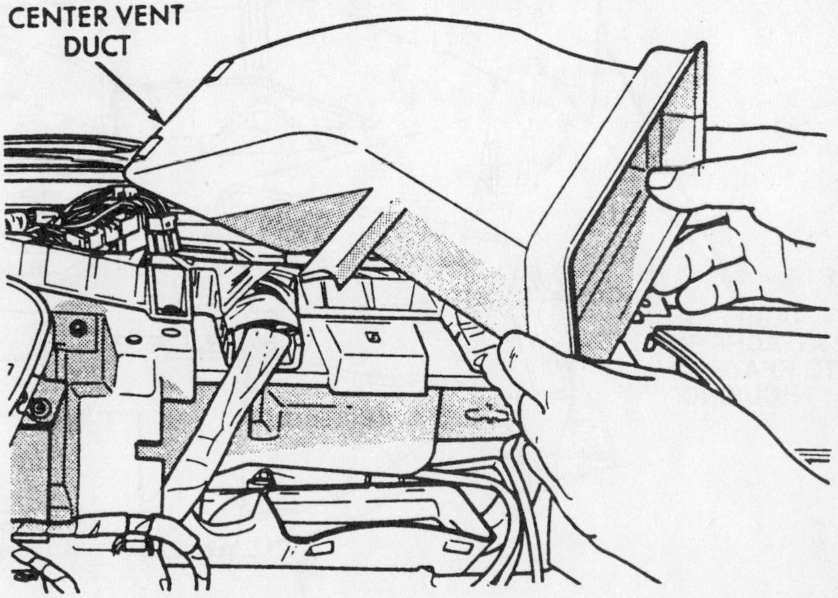

You must remove the instrument panel center vent duct—Neon

- Steering column as an assembly
- Left and right instrument panel end caps
- Center console
- Depress the Data Link Connector (DLC) sides and remove the DLC from the instrument panel reinforcement
- 4 bulkhead instrument panel screws
- 2 brake pedal support bracket bolts
- 2 center support mounting bolts
- Left and right A-pillar bolts; there are 2 on each side
- Antenna connector from the right side
- Left and right A-pillar door harness connectors
- 2 HVAC wiring harness connectors from the top right of the instrument panel
- Left side wiring harness connector

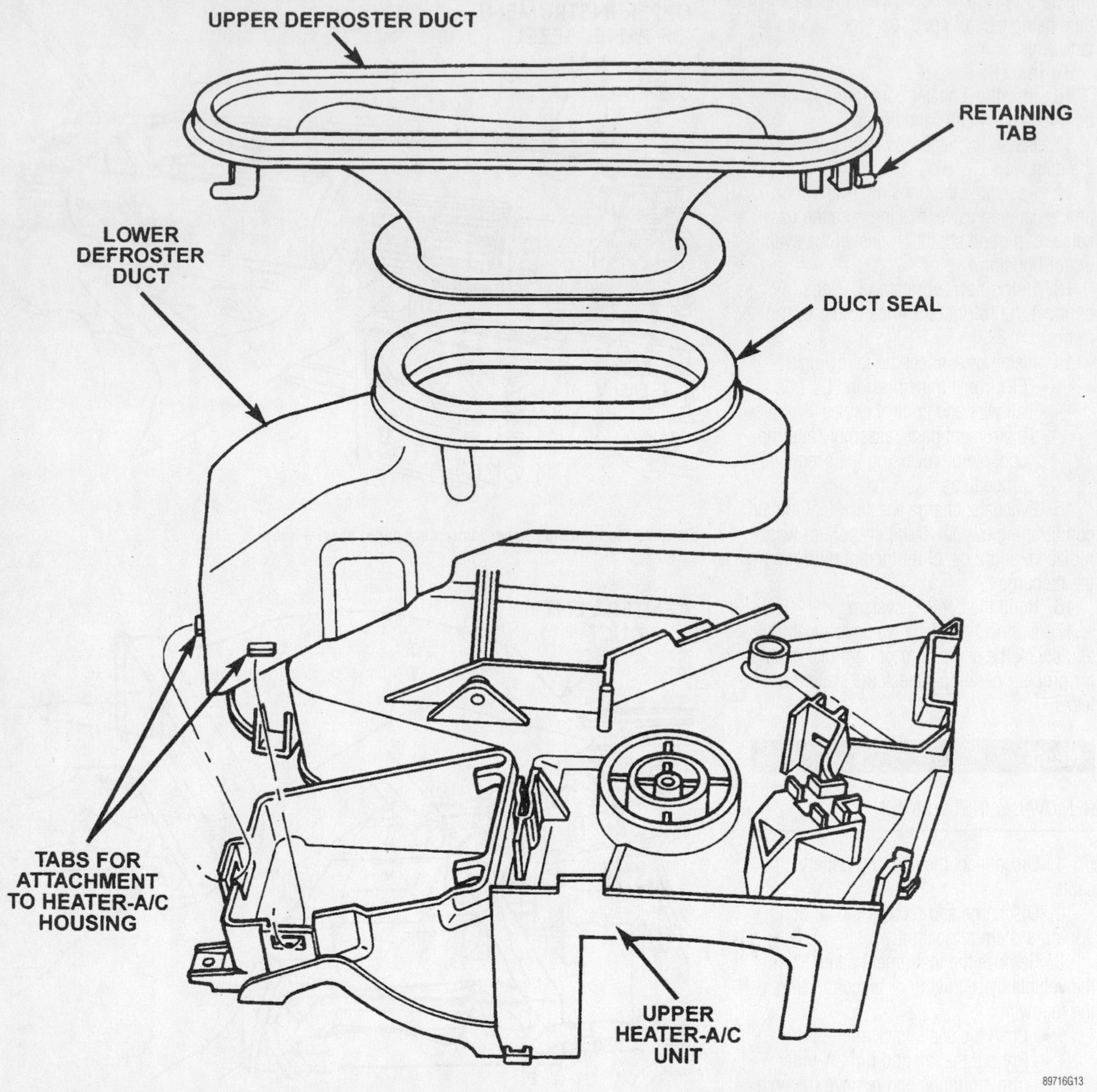

The upper defrost duct is secured with retaining tabs—Neon

from the top left of the instrument panel for the vanity and rear view mirrors

- Pull off the HVAC control head knobs
- 2 top front center bezel screws
- Using a trim stick, carefully pry out the instrument panel center bezel and remove it
- 2 HVAC control head screws
- Instrument panel wiring harness connector
- Vacuum harness connector

4. Pull the HVAC control head out of the instrument panel, twist it 90 degrees and push it back through the opening; do not disconnect the control cables.

5. Remove or disconnect the following:

- Air bag Control Module (ACM) from the center console
- Parking Brake Warning Lamp Switch from the center console
- Transmission Range Indicator Lamp from the center console

6. Using an assistant, pull the instrument panel rearward and remove it from the vehicle.

7. Drain the cooling system and remove the heater hoses at the dash panel. Place plugs in the heater core outlets to prevent coolant spillage during the unit housing removal.

8. Remove the suction line at the expansion valve. Place a piece of tape over the open refrigerant line to prevent moisture and/or dirt from entering the line.

9. Remove the expansion valve from the evaporator fitting to prevent moisture and/or dirt from entering the evaporator.

10. Remove or disconnect the following:

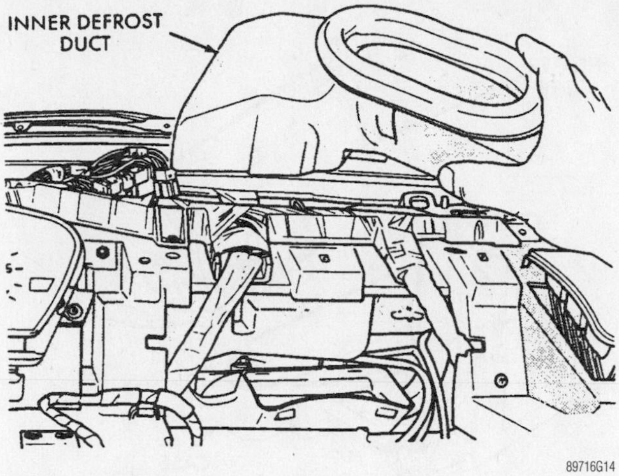

Remove the inner defrost duct—Neon

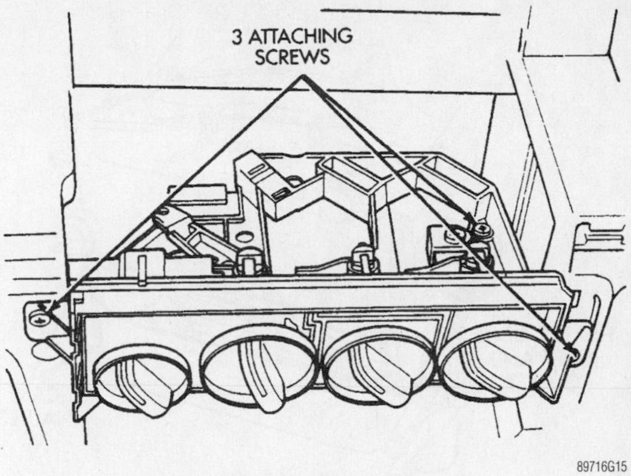

Exploded view of the control panel retainer locations—Neon

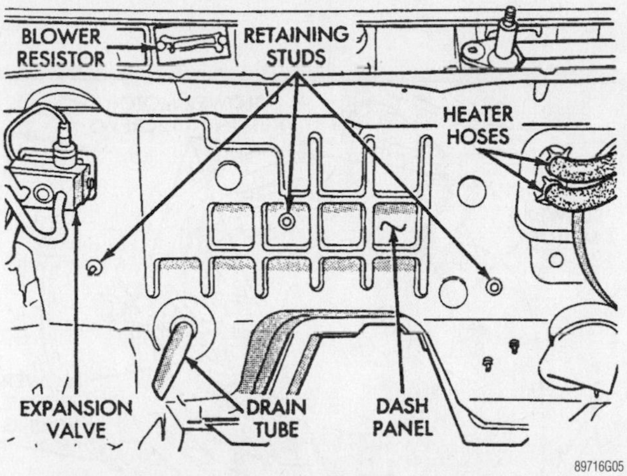

Location of the 3 dash panel retaining studs—Neon

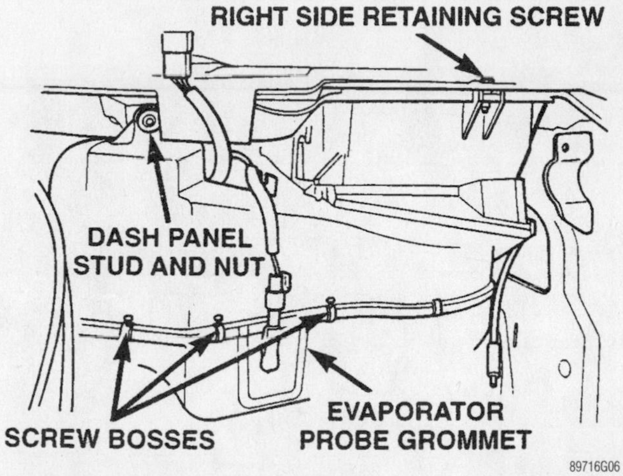

Unit housing retaining screw location—Neon

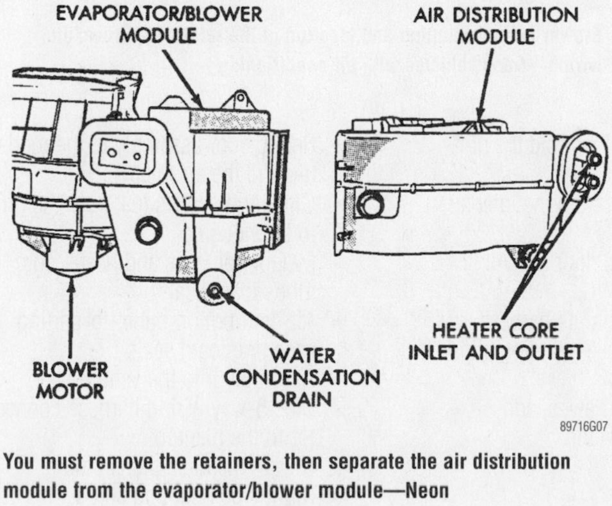

You must remove the retainers, then separate the air distribution module from the evaporator/blower module—Neon

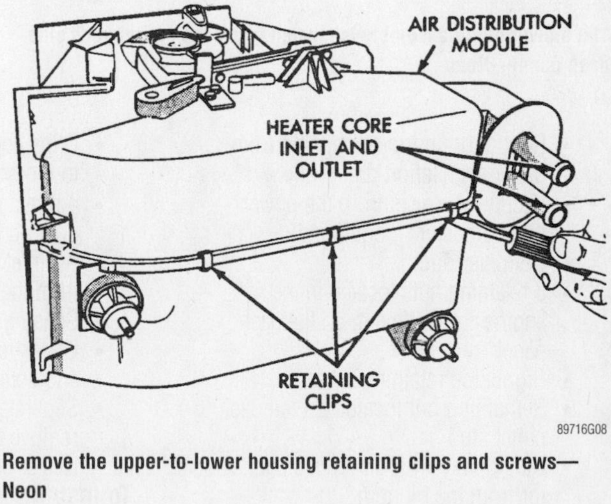

Remove the upper-to-lower housing retaining clips and screws—Neon

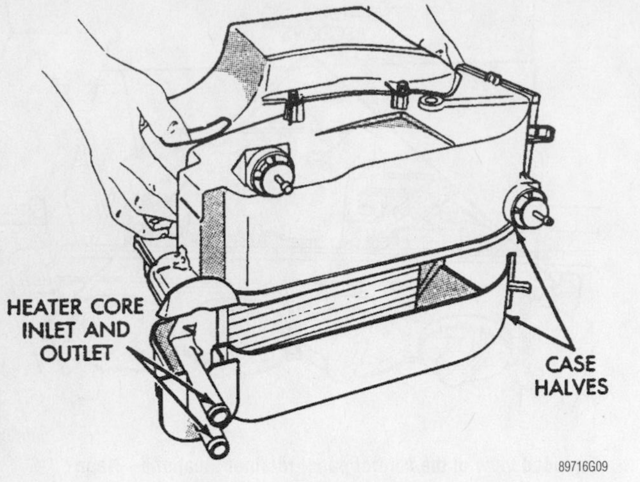

Then separate the 2 halves of the module—Neon

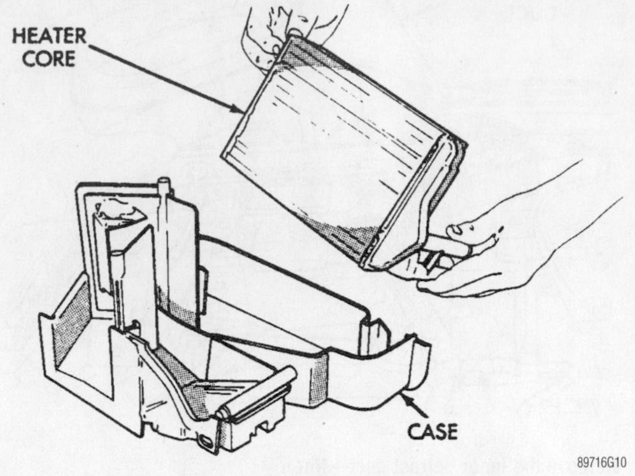

Remove the heater core from the case—Neon

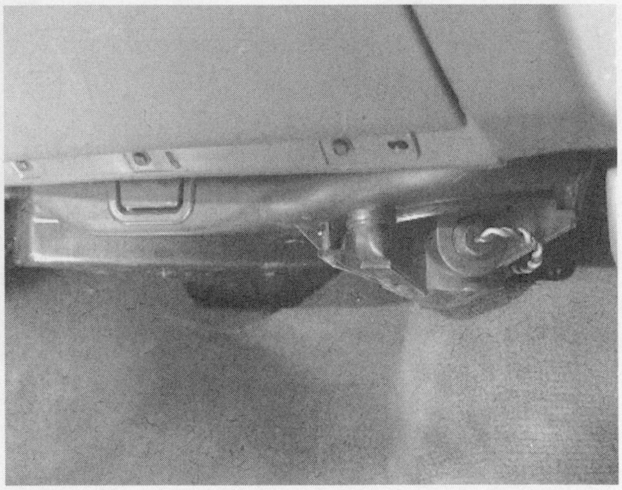

The blower motor assembly is located under the passenger's side dash panel—Neon

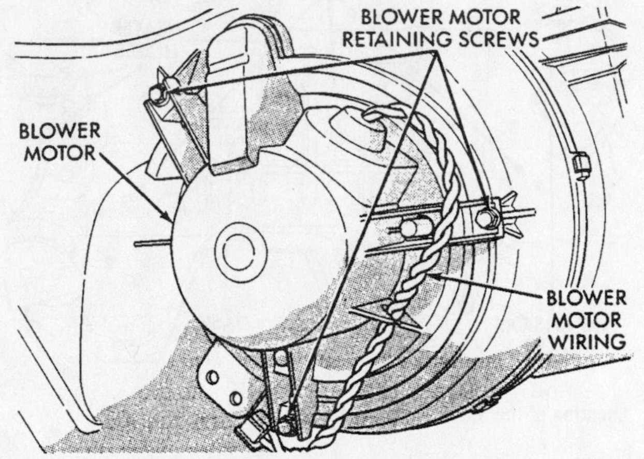

Blower motor mounting and location of the retaining screws and wiring—Neon vehicles with air conditioning

- Rubber drain tube extension from the condensation drain tube
- Vacuum harness from the power brake booster (if equipped)
- Defroster duct
- 3 retaining nuts located in the engine compartment, on the dash panel
- Right side retaining screw
- Remaining nut located on the dash panel stud
- Blue 5-way wiring harness connector from the plenum
- Heater/air conditioning assembly from the vehicle
- Separate the air distribution outlet-to-parting case line foam seals
- Evaporator lines and heater core tubes foam seals
- Clips and screws that hold the unit to the housing
- 4 retaining screws from the inlet air duct on the module
- Air inlet and recirc door assembly from the module
- Sensing switch from its harness
- Upper-to-lower case retaining clips and screws
- Separate the case halves, and remove the evaporator
- Heater core

To install:

11. Install or connect the following:
    - Heater core into the heater/air conditioning unit.
    - Evaporator and the upper-to-lower case retaining clips and screws
    - Sensing switch to its harness
    - Recirc door assembly to the module and the air inlet
    - Clips and screws that hold the unit to the housing
    - Evaporator lines and heater core tubes foam seals.
    - Air distribution outlet-to-parting case line foam seals
    - Assembly into the vehicle
    - Blue 5-way wiring harness connector to the plenum
    - Nut on the dash panel stud
    - Right side retaining screw
    - 3 retaining nuts located in the engine compartment, on the dash panel
    - Defroster duct
    - If equipped, the vacuum harness to the power brake booster

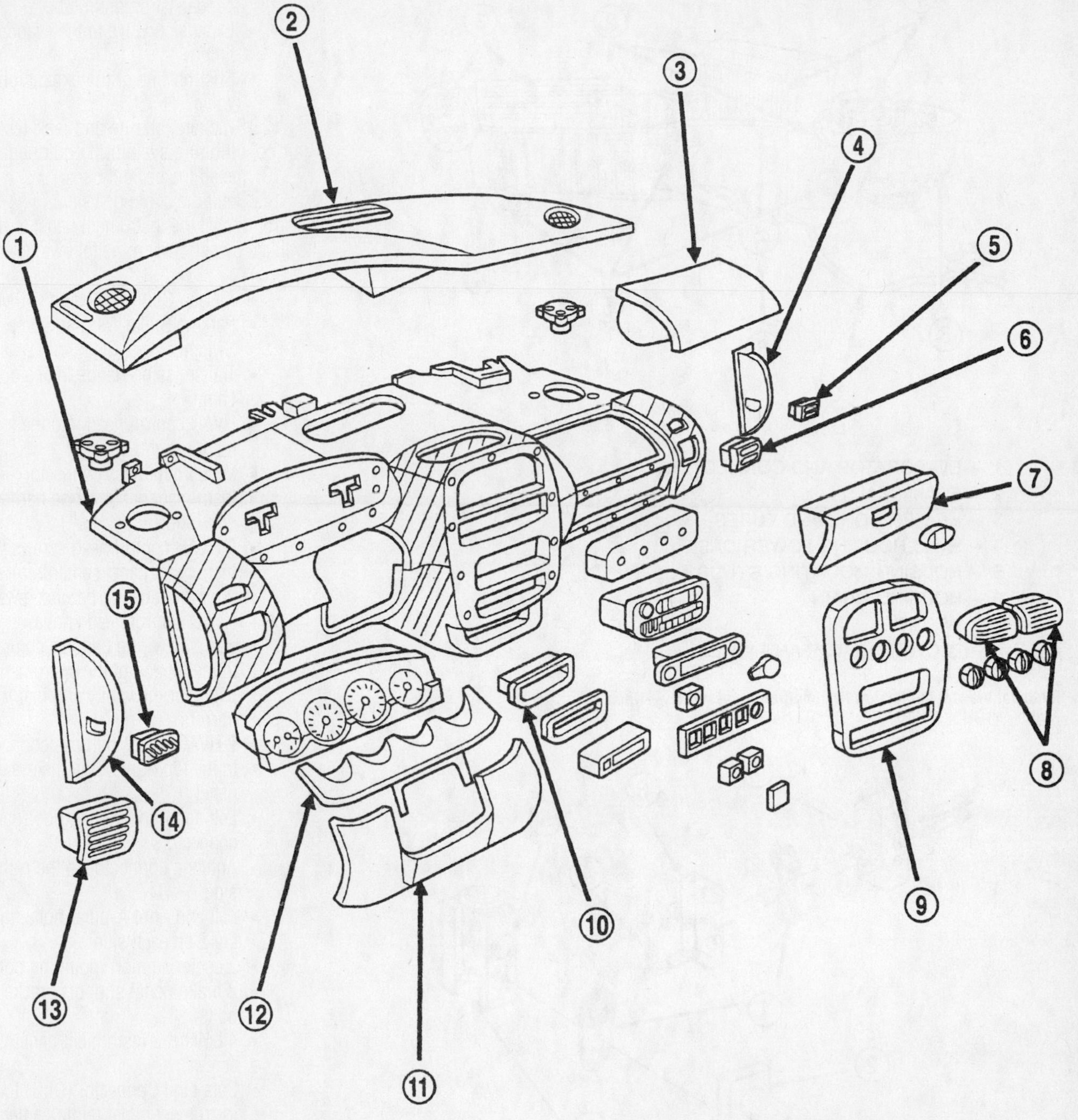

1 – INSTRUMENT PANEL ASSEMBLY
2 – UPPER COVER INSTRUMENT PANEL
3 – MODULE, PASSENGER SIDE AIRBAG
4 – END CAP, RIGHT
5 – DEMISTER GRILLE, RIGHT
6 – LOUVER, AIR OUTLET, RIGHT
7 – DOOR, GLOVE BOX
8 – LOUVER, AIR OUTLET, CENTER
9 – BEZEL INSTRUMENT PANEL, CENTER
10 – BIN, LOWER STORAGE
11 – COVER, LOWER INSTRUMENT PANEL
12 – CLUSTER BEZEL
13 – LOUVER, AIR OUTLET, LEFT
14 – END CAP, LEFT
15 – DEMISTER GRILLE, LEFT

93111G86

Exploded view of the instrument panel—Neon

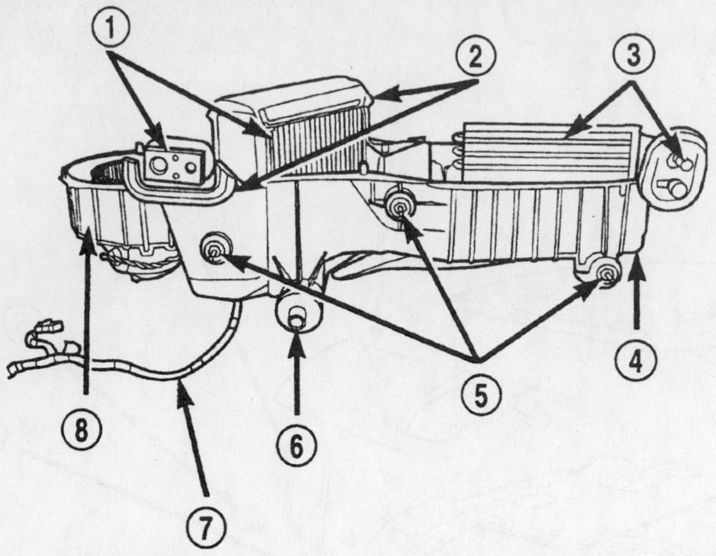

Sectional view of the heater/air conditioning assembly—Neon

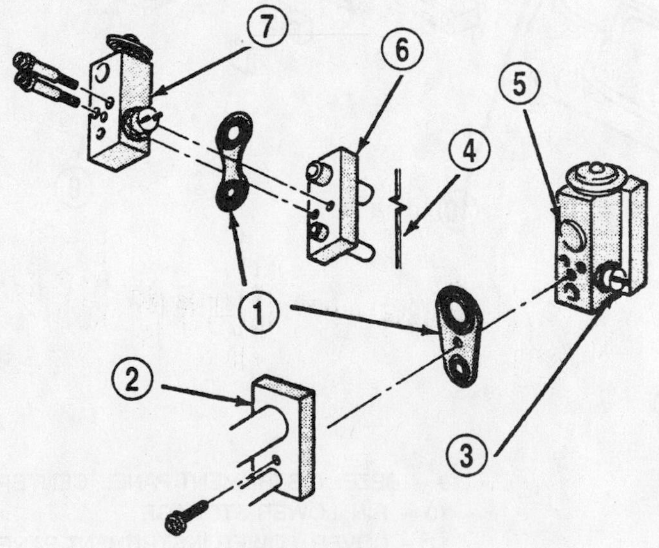

Exploded view of the expansion valve—Neon

- Rubber drain tube extension to the condensation drain tube
- Expansion valve to the evaporator fitting
- Suction line at the expansion valve

12. Evacuate, charge and leak-test the air conditioning system, if equipped.
13. Install the heater hoses.
14. Refill the cooling system.
15. Install the instrument panel to the vehicle by performing the following procedure:
    - Air bag Control Module (ACM)
    - Parking Brake Warning Lamp Switch
    - Transmission Range Indicator Lamp
    - HVAC control head into the instrument panel
    - Vacuum harness connector
    - Instrument panel wiring harness connector
    - 2 HVAC control head screws
    - Instrument panel center bezel
    - 2 top front center bezel screws
    - HVAC control head knobs
    - Left side wiring harness connector to the top left of the instrument panel for the vanity and rear view mirrors
    - 2 HVAC wiring harness connectors to the top right of the instrument panel
    - Left and right A-pillar door harness connectors
    - Antenna connector to the right side
    - Left and right A-pillar bolts; there are 2 on each side
    - 2 center support mounting bolts
    - 2 brake pedal support bracket bolts
    - 4 bulkhead instrument panel screws
    - Data Link Connector (DLC) to the instrument panel reinforcement
    - Center console
    - Left and right instrument panel end caps
    - Steering column as an assembly
    - Instrument panel cover
    - Cluster bezel
    - Upper instrument panel cover
    - Left and right A-pillar trim moldings
    - Move seats forward
    - Negative battery cable
16. Operate the engine to normal operating temperatures; then, check the climate control operation and check for leaks.

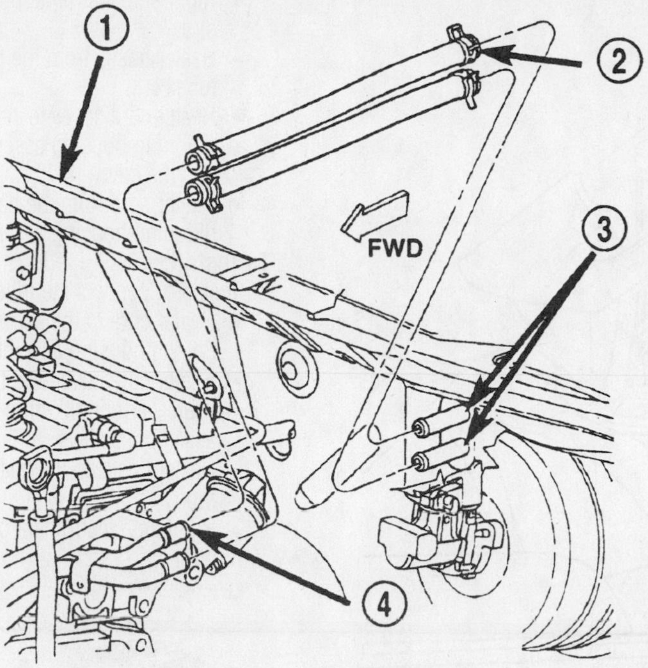

1 – COWL PANEL
2 – HEATER HOSE AND CLAMPS
3 – HEATER CORE TUBES
4 – HEATER HOSE SUPPLY AND RETURN TUBES

93111G89

View of the heater hoses and clamps—Neon

Cirrus, Sebring Convertible, Breeze, Stratus

REMOVAL & INSTALLATION

1. Disconnect the negative battery cable.
2. Drain the cooling system into a clean container for reuse.
3. From the center of the instrument panel, grasp and pull to remove the radio/control module bezel.
4. At the right side of the instrument panel, remove the side trim.
5. Remove or disconnect the following:
   - 2 lower right side support beam screws
   - Instrument panel support-to-A-pillar bolt
   - On left side of the instrument panel, remove the side trim
   - Upper instrument panel bezel
   - Lower knee bolster
   - Console-to-instrument panel screws
   - Gearshift knob and the shifter bezel
   - Rear console screws and the rear console half
   - Front console screws and the front console half

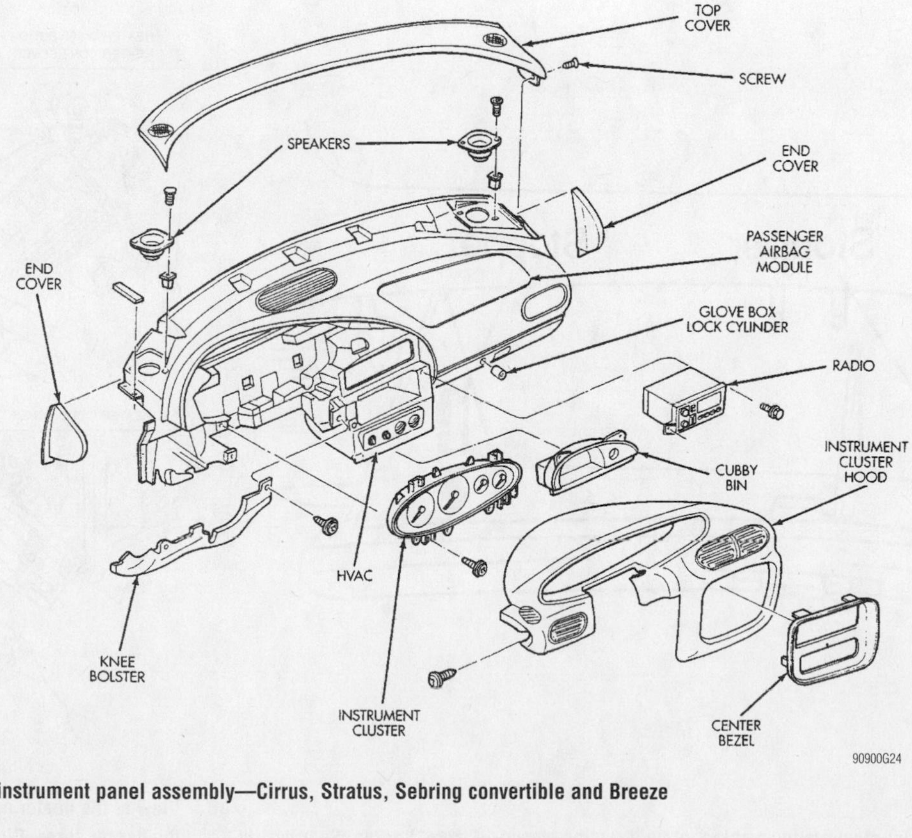

90900G24

Exploded view of the instrument panel assembly—Cirrus, Stratus, Sebring convertible and Breeze

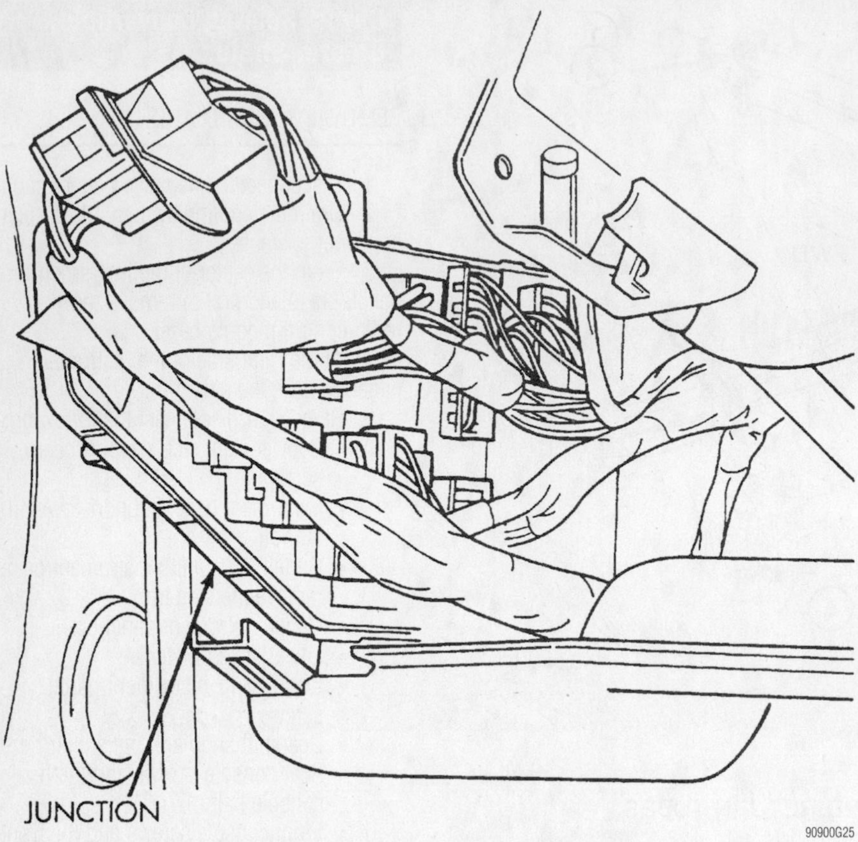

Junction block location—Cirrus, Stratus, Sebring convertible and Breeze

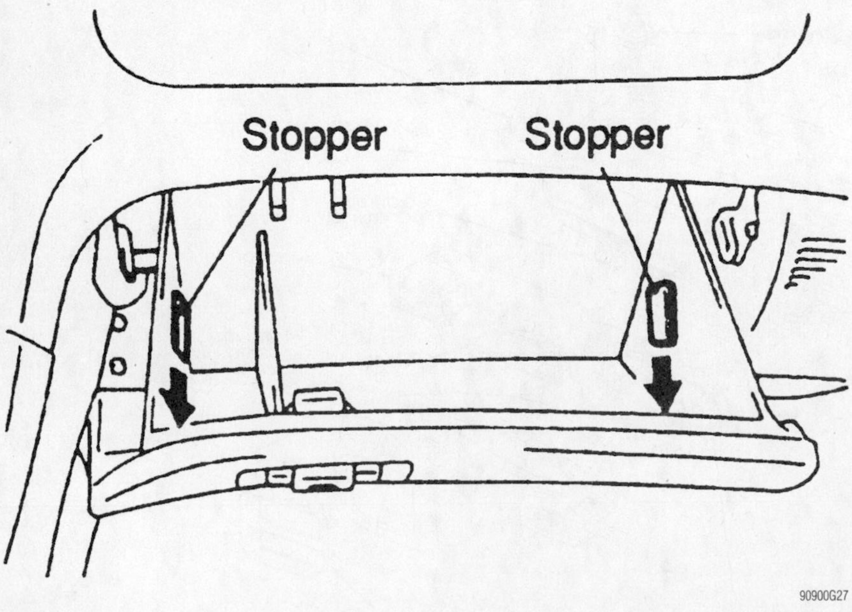

Unlock the glove box door stoppers by pushing in the direction shown—Cirrus, Stratus, Sebring convertible and Breeze

- Right side instrument panel support strut
- Heater hoses from the heater core tubes
- Heater core cover-to-heater/air conditioning housing assembly screws and the cover
- Heater core from the heater/air conditioning housing assembly

To install:

6. Install or connect the following:
   - Heater core to the heater/air conditioning housing assembly
   - Heater core cover and the cover-to-heater/air conditioning housing assembly screws
   - Heater hoses to the heater core tubes
   - Right side instrument panel support strut

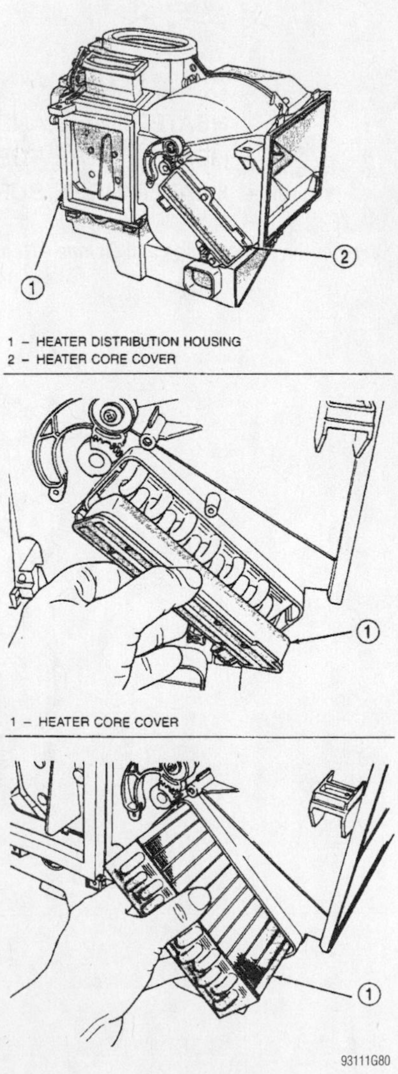

View of the heater housing assembly and the heater core—Cirrus, Stratus and Breeze

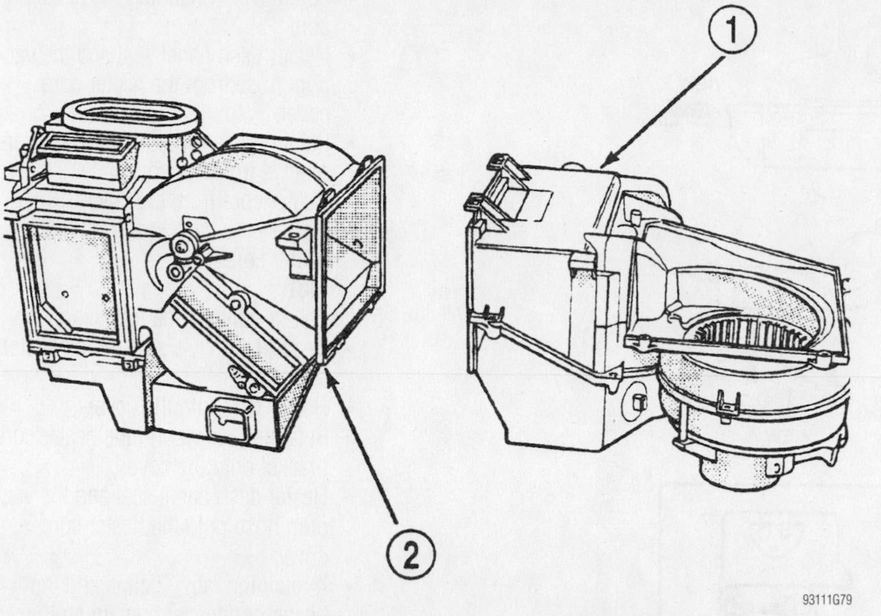

View of the heater/air conditioning housing assembly—Cirrus, Stratus and Breeze

- Front console half and the front console screws
- Rear console half and the rear console screws
- Shifter bezel and the gearshift knob
- Console-to-instrument panel screws
- Lower knee bolster
- Upper instrument panel bezel
- On left side of the instrument panel, install the side trim
- Instrument panel support-to-A-pillar bolt
- 2 lower right side support beam screws
- Right side of the instrument panel, install the side trim
- On center of the instrument panel, install the radio/control module bezel

7. Refill the cooling system.
8. Connect the negative battery cable.
9. Operate the engine to normal operating temperatures; then, check the climate control operation and check for leaks.

1998–01 FORD MOTOR COMPANY

Contour, Mystique and Cougar

REMOVAL & INSTALLATION

1. Disconnect the negative battery cable.
2. Remove the center console.
3. Disarm the air bag system and remove the air bag diagnostic monitor and bracket.
4. Remove or disconnect the following:
   - Screw retaining the air transfer duct to the heater outlet floor duct. Push the transfer duct inside of the heater outlet floor duct
   - 3 screws retaining the heater outlet floor duct to the heater core cover and release the retaining tabs on each side of the duct and remove the duct
   - Heater hoses from the heater core, and plug the heater core tubes and the hoses to prevent coolant loss.
   - Vacuum supply hose (the black hose) from the vacuum source in the engine compartment
   - Vacuum supply hose (the black hose) from the air conditioning vacuum reservoir tank
   - Release the 4 retaining tabs and remove the 2 clips and the heater

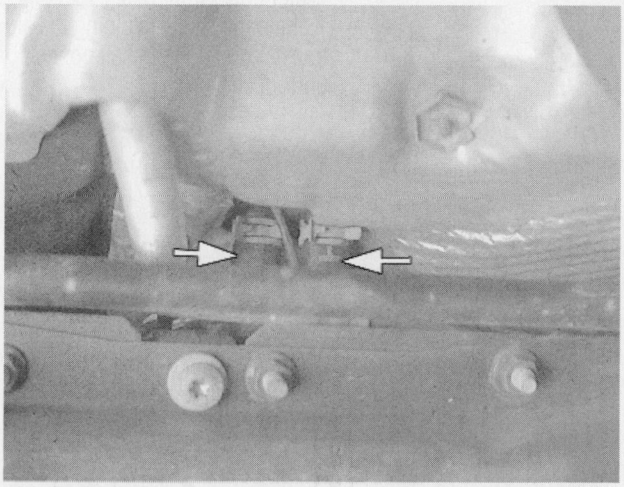

The heater hoses are best accessed from underneath the vehicle—Contour and Mystique

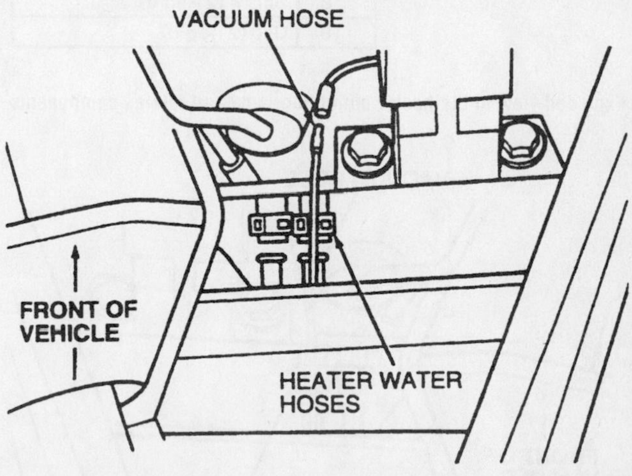

Disconnect the heater hose and vacuum line from underneath the vehicle—Contour and Mystique

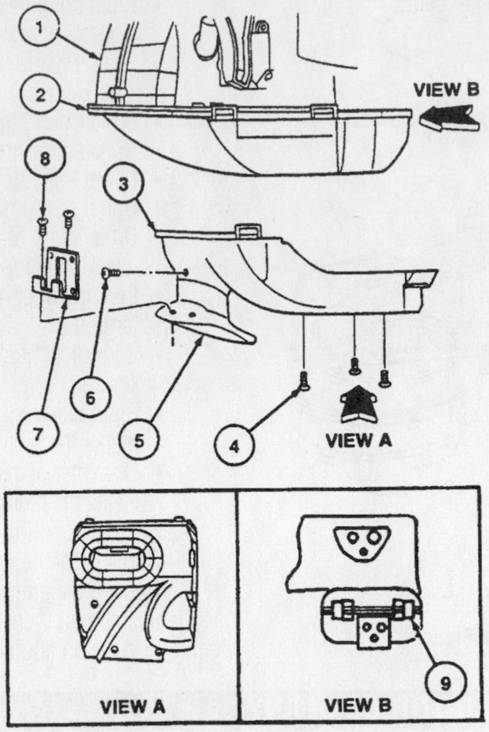

Item	Description
1	A/C Evaporator Housing
2	Heater Core Cover (Part of 19850)
3	Heater Outlet Floor Duct
4	Screw (3 Req'd)
5	Air Transfer Duct (Part of 18C433)
6	Screw (1 Req'd)
7	Air Bag Diagnostic Monitor Bracket
8	Screw (2 Req'd)
9	Clip (2 Req'd)

91056G26

Exploded view of the heater outlet floor duct and related components—Contour and Mystique

core cover containing the heater core
- Heater dash panel seal and the vacuum hose from the heater core cover
- Retaining screw and the heater core bracket from the cover
- Heater core from the cover
- Heater core case seal from the heater core

To install:

5. Install or connect the following:
   - Heater core case seal on the heater core
   - Heater core into the cover
   - Retaining screw and the heater core bracket onto the cover
   - Heater dash panel seal and the vacuum hose onto the heater core cover
   - 4 retaining tabs, 2 clips and the heater core cover containing the heater core
   - Vacuum supply hose (the black hose) to the air conditioning vacuum reservoir tank
   - Vacuum supply hose (the black hose) to the vacuum source in the engine compartment
   - Unplug the hoses and the core tubes (if installing old core)
   - Heater hoses onto the heater core
   - Heater outlet floor duct, engage the retaining tabs and tighten the 3 screws retaining the heater outlet floor duct to the heater core cover
   - Push the transfer duct out of the heater outlet floor duct
   - Tighten the screw retaining the air transfer duct to the heater outlet floor duct

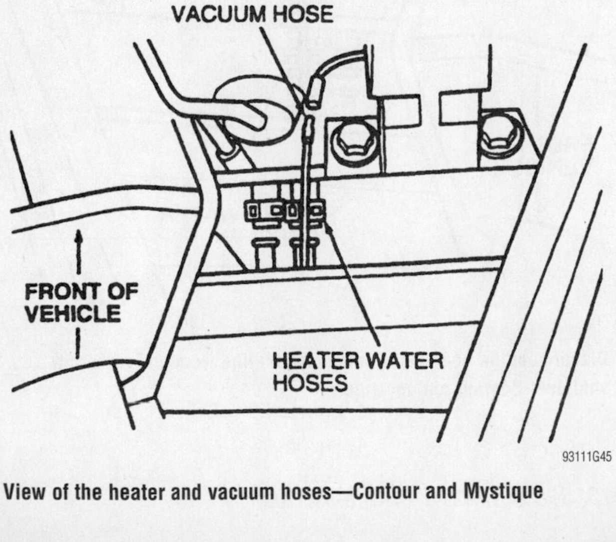

93111G45

View of the heater and vacuum hoses—Contour and Mystique

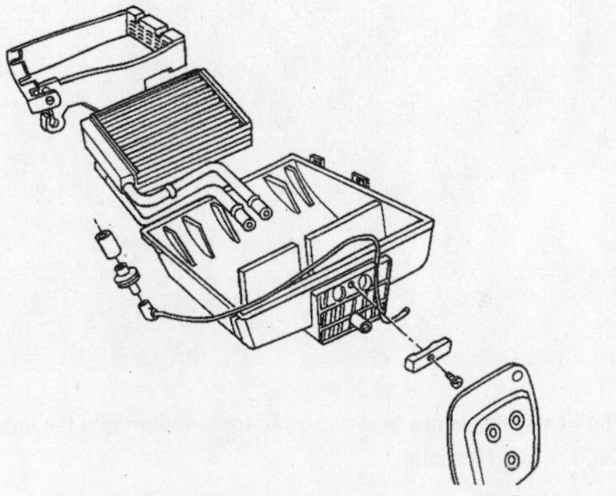

93111G46

Exploded view of the heater core and related components—Contour and Mystique

- Air bag diagnostic monitor and bracket
- Center console

6. Connect the negative battery cable.

Focus

REMOVAL & INSTALLATION

1. Before servicing the vehicle, refer to the precautions in the beginning of this section.
2. Drain the cooling system.
3. Recover the A/C refrigerant, if equipped.
4. Remove or disconnect the following:

- Negative battery cable
- Radio
- Left and right lower panels
- Data link electrical connector
- Steering column upper shroud
- Audio unit control switch, if equipped
- Steering column lower shroud
- Steering column
- Instrument cluster
- Glove compartment
- Passenger air bag module
- Climate control and audio trim panel
- Ash tray
- Heater core housing assembly panel
- Center console
- Instrument panel upper retaining screws
- Instrument panel side trim panels
- Instrument panel side bolts
- Instrument panel bolts behind the instrument cluster, climate control and audio panel and passenger air bag module
- Defroster tubes
- Instrument panel
- Heater hoses
- A/C evaporator refrigerant lines
- Cross vehicle beam
- Rear footwell ventilation hoses
- Heater housing wiring harness
- Heater housing. Refer to the illustrations for housing fastener locations.

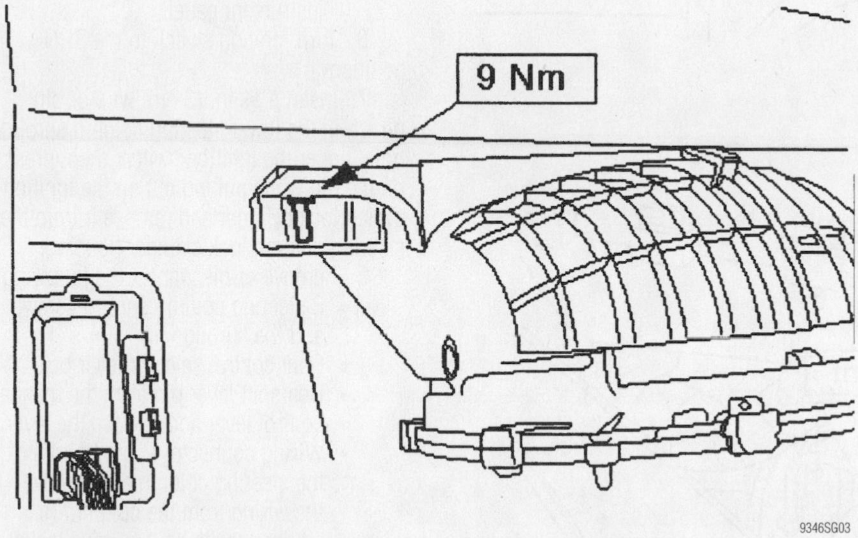

Heater housing fastener location

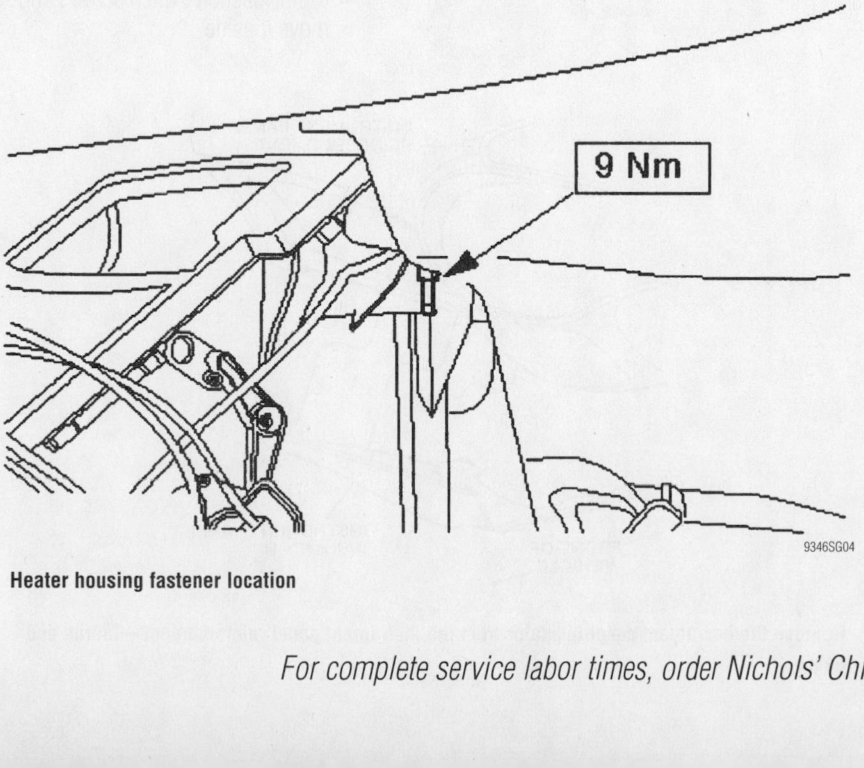

Heater housing fastener location

To install:

5. Install or connect the following:

- Heater housing. Refer to the illustrations for housing fastener locations and torque specifications.
- Heater housing wiring harness
- Rear footwell ventilation hoses
- Cross vehicle beam. Tighten the beam fasteners to 15 ft. lbs. (20 Nm) and the beam bracket bolts to 12 ft. lbs. (16 Nm).
- A/C evaporator refrigerant lines
- Heater hoses
- Instrument panel
- Defroster tubes
- Instrument panel bolts behind the instrument cluster, climate control and audio panel and passenger air bag module
- Instrument panel side bolts
- Instrument panel side trim panels
- Instrument panel upper retaining screws
- Center console
- Heater core housing assembly panel
- Ash tray
- Climate control and audio trim panel
- Passenger air bag module
- Glove compartment
- Instrument cluster
- Steering column. Tighten the nuts to 13 ft. lbs. (18 Nm) and the Torx screw to 17 ft. lbs. (23 Nm)
- Steering column lower shroud
- Audio unit control switch, if equipped
- Steering column upper shroud
- Data link electrical connector

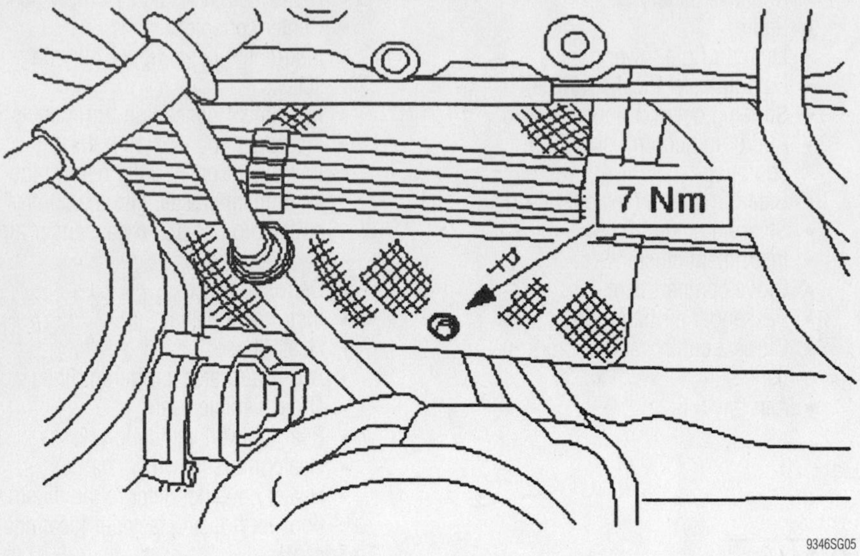

Heater housing fastener location

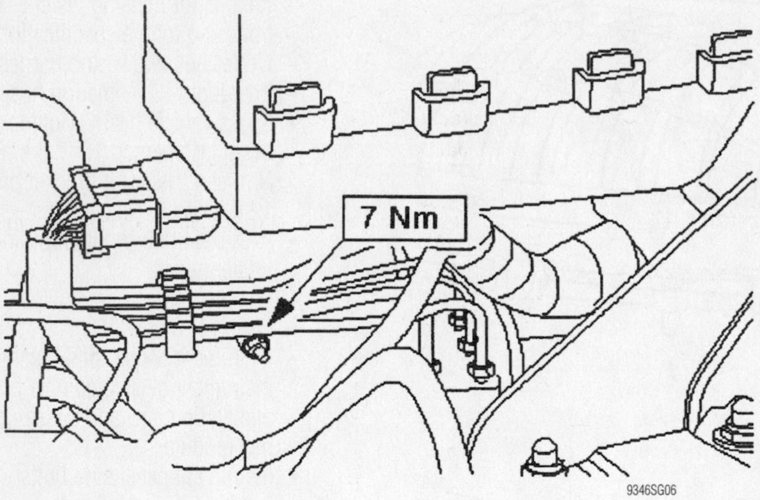

Heater housing fastener location

- Left and right lower panels
- Radio
- Negative battery cable

6. Fill the cooling system.
7. Recharge the A/C system, if equipped.
8. Start the engine and check for leaks.

Taurus and Sable

REMOVAL & INSTALLATION

1. Disconnect the negative battery cable.

CAUTION

After disconnecting the negative battery cable, wait for at least 1 minute for the SRS or air bag module to deplete its energy.

2. Place the front wheels in the straight-ahead position.
3. Lock the steering column.
4. Remove or disconnect the following:
   - SRS bolt covers from both sides of the steering wheel
   - SRS module-to-steering wheel bolts
   - SRS module and disconnect the electrical connector
   - Steering wheel bolt and discard it
5. Press the steering wheel from the steering column.
   - Lower steering column shaft bolt
   - 2 lower instrument panel cover-to-instrument panel screws and unsnap the lower cover from the instrument panel
6. Turn ignition switch to the RUN position
7. Insert a ⅛ in. (3mm) wire or pin punch in the lower steering column shroud hole, under the ignition switch; then, press on the pin while pulling out on the ignition switch lock cylinder and remove it from the steering column lock cylinder housing.
8. Remove or disconnect the following:
   - 3 steering column shroud screws and the shrouds
   - Shift control selector lever boot
   - Gearshift lever pin from the manual control lever and remove the lever
   - Wiring connector at the bottom of the steering column and remove the wiring from the column, (if equipped with an overdrive lockout switch on the manual control lever)
   - Electrical connectors
   - Multi-function switch screws and move it aside

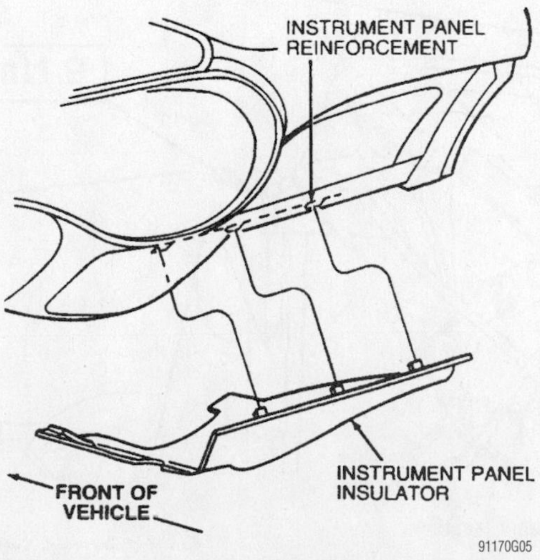

Remove the instrument panel insulator from the instrument panel reinforcement—Taurus and Sable

- Shift indicator cable from the shifter tube
- Shifter indicator-to-column adjustment cable screw
- Interlock cable and actuator (if equipped with a column shift)
- 4 steering column-to-instrument panel bracket nuts and steering column
- Push pins and the lower instrument cover from the instrument panel reinforcement (at the passenger's side)
- Console finish panel

9. Under the steering column, 2 instrument panel brace screws, the courtesy lamp socket, the 2 Diagnostic Link Connector (DLC) screws and the instrument panel brace.

10. If not equipped with an Electronic Automatic Temperature Control (EATC), disconnect the electrical connectors and the vacuum harness from the evaporator housing and the blower motor.

11. If equipped with an Electronic Automatic Temperature Control (EATC), remove the sensor hose/elbow and disconnect the electrical harness connectors and the vacuum hose harness from the evaporator housing.

12. Insert the Radio Removing tools 415-001into the integrated control panel faceplate; then, push the tools in approximately 1½ in. (38mm) to release the retaining clips. Spread the tools slightly and pull the integrated control panel from the instrument panel.

13. Remove or disconnect the following:
- Electrical connectors and the automatic temperature control sensor hose and elbow, if equipped
- 6 console center finish panel screws and the panel, (if equipped with a floor shift)
- 4 instrument panel finish panel screws and panel, located at the right side of the steering column, (if equipped with a column shift)
- Instrument cluster finish panel-to-instrument panel clips and pull the finish panel straight out
- 4 instrument cluster-to-instrument panel screws, the electrical connectors and the instrument cluster
- Upper finish panel (located at the top of the instrument panel)
- Light sensor amplifier connector, (if equipped with an autolamp)
- Instrument panel finish end panels on both sides of the instrument panel

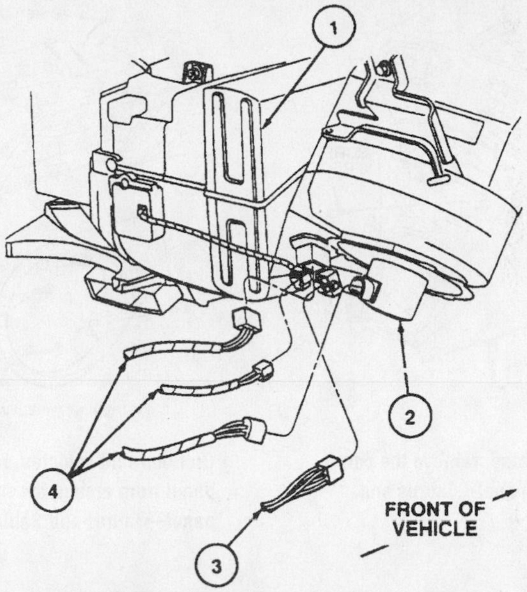

Item	Description
1	A/C Evaporator Housing
2	Blower Motor
3	Vacuum Hose Harness
4	Main Wiring

91170G06

Manual air conditioning/heater equipped vehicles disconnection points—Taurus and Sable

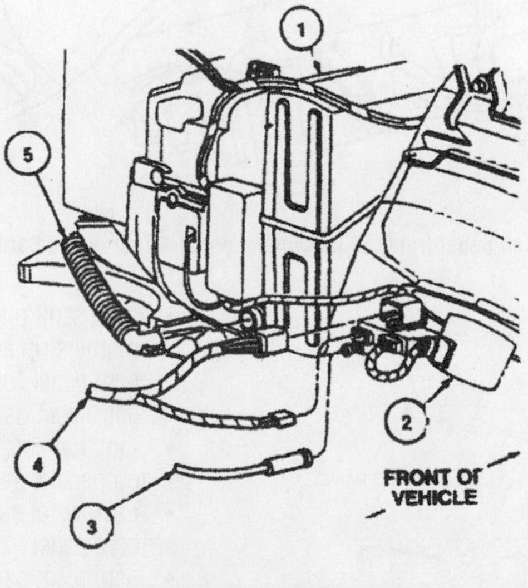

Item	Description
1	A/C Evaporator Housing
2	Blower Motor
3 & 4	Main Wiring
5	Automatic Temperature Control Sensor Hose and Elbow

91170G07

EATC equipped vehicles disconnection points—Taurus and Sable

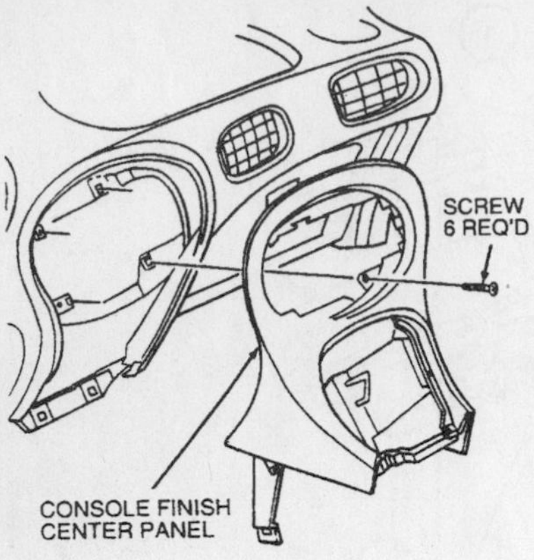

On floor shift vehicles, remove the console center finish panel—Taurus and Sable

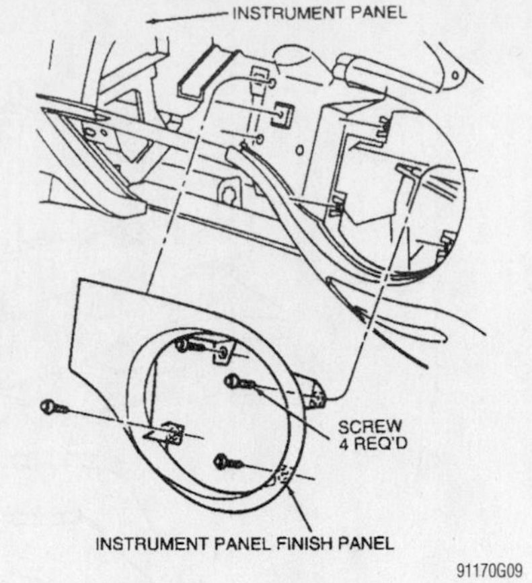

On floor shift vehicles, remove the finish panel from around the integrated control panel—Taurus and Sable

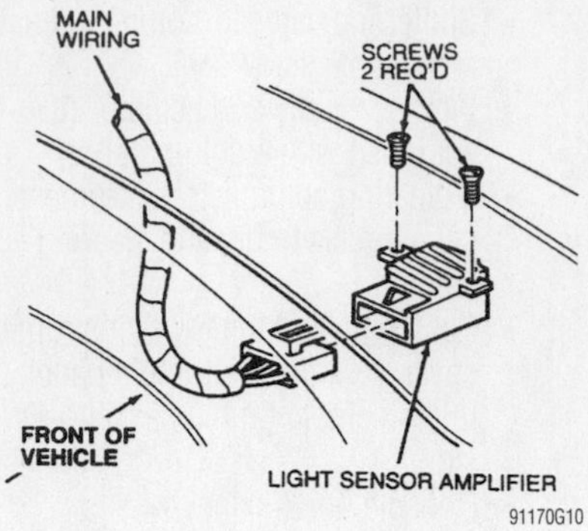

If equipped with autolamps, detach the light sensor amplifier—Taurus and Sable

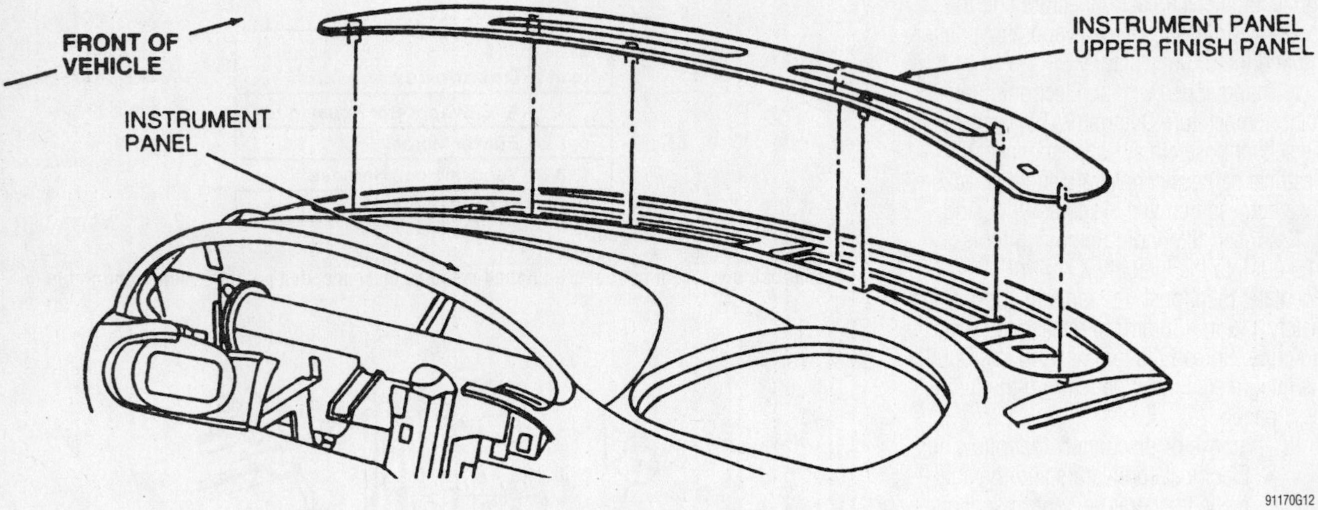

Unsnap the upper finish panel from the instrument panel—Taurus and Sable

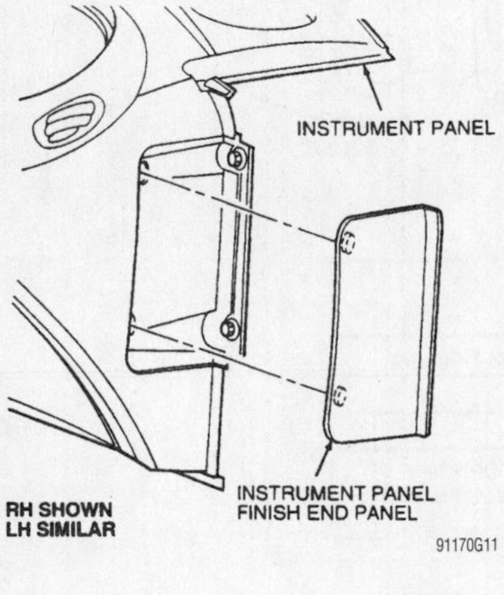

Remove the instrument panel finish end panel—Taurus and Sable

- Front scuff plates and pull the door weatherstrip away from the instrument panel (on both sides of the instrument panel)
- 3 instrument panel-to-upper cowl top panel screws
- 2 screws at each end of the instrument panel
- Instrument panel electrical connectors and the parking brake switch
- Instrument panel

14. Drain the cooling system into a clean container for reuse

15. Remove or disconnect the following:

- Heater hoses from the heater core and plug the openings
- 4 air conditioning electronic blend door actuator-to-heater/air conditioning housing assembly screws and the actuator
- Metal cover, disengage the spring from the heater core cover and from the lever
- Gently, depress the locking ramp and remove the lever from the secondary air temperature control door end

WARNING

Do not attempt to bend any part of the lever, for it is brittle and will break.

16. Rotate the primary air conditioning air temperature control door shaft downward, swing the metal link counterclockwise and remove it from the pin.

17. Remove the 3 heater core cover-to-heater/air conditioning housing assembly screws, the cover and the seal.

18. Press on the heater core tubes and

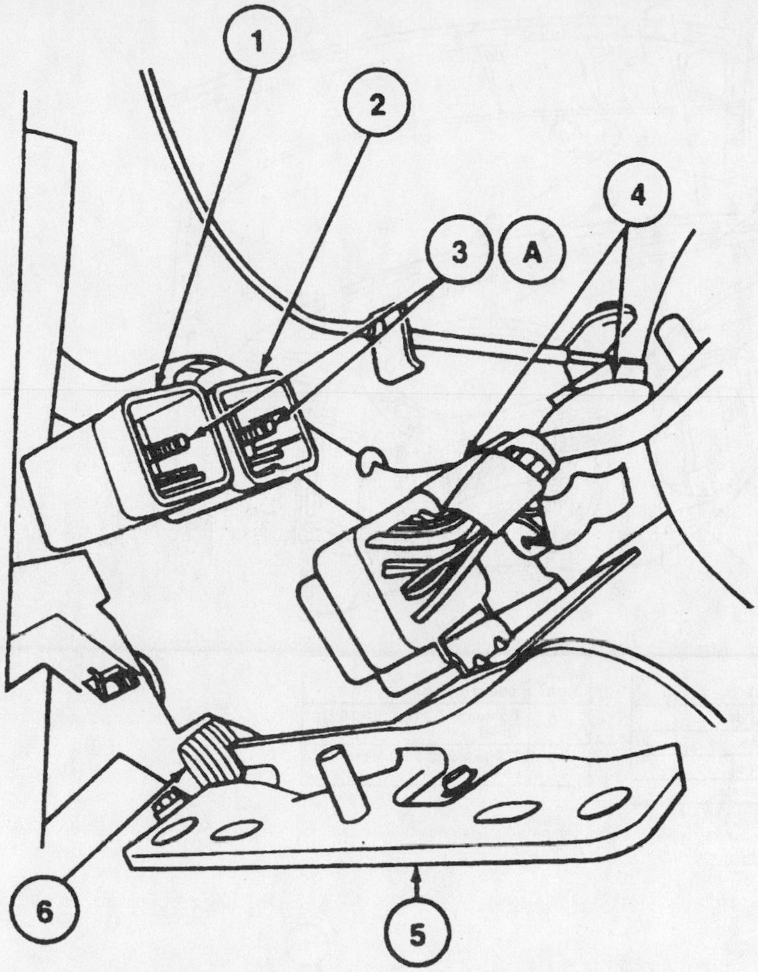

Item	Description
1	Wiring Assy Electrical Connector
2	Wiring Assy Electrical Connector
3	Bolt (Part of 14290 and 14A005)
4	Wiring Assy Electrical Connector
5	Instrument Panel
6	Parking Brake Control
A	Tighten to 4-6 N·m (36-53 Lb-In)

91170G13

Detach the electrical connectors—Taurus and Sable

remove the heater core from the heater/air conditioning housing assembly.

To install:

19. Install or connect the following:
    - Heater core to the heater/air conditioning housing assembly
    - Heater core cover, the seal and the 3 heater core cover-to-heater/air conditioning housing assembly screws
    - Lever to the secondary air conditioning air temperature control door end
    - Spring to the lever, engage the spring to the heater core cover and install the metal cover
    - Air conditioning electronic blend door actuator and the 4 actuator-to-heater/air conditioning housing assembly screws
    - Heater hoses to the heater core
    - Instrument panel with the help of an assistant
    - Instrument panel electrical connectors and the parking brake switch
    - 2 screws at each end of the instrument panel with the help of an assistant
    - 3 instrument panel-to-upper cowl top panel screws
    - Front scuff plates and the door weatherstrip (on both sides of the instrument panel)
    - Instrument panel finish end panels (on both sides of the instrument panel)
    - Electrical connector to the light sensor amplifier (if equipped with an autolamp)
    - Upper finish panel (at the top of the instrument panel)
    - Instrument cluster, electrical connectors and the 4 instrument cluster-to-instrument panel screws
    - Instrument cluster finish panel and engage the finish panel-to-instrument panel clips
    - Instrument panel finish panel and 4 panel screws, located at the right side of the steering column (if equipped with a column shift)
    - Console center finish panel and the 6 panel screws (if equipped with a floor shift)
    - Electrical connectors and the automatic temperature control sensor hose and elbow, if equipped
    - Integrated control panel faceplate

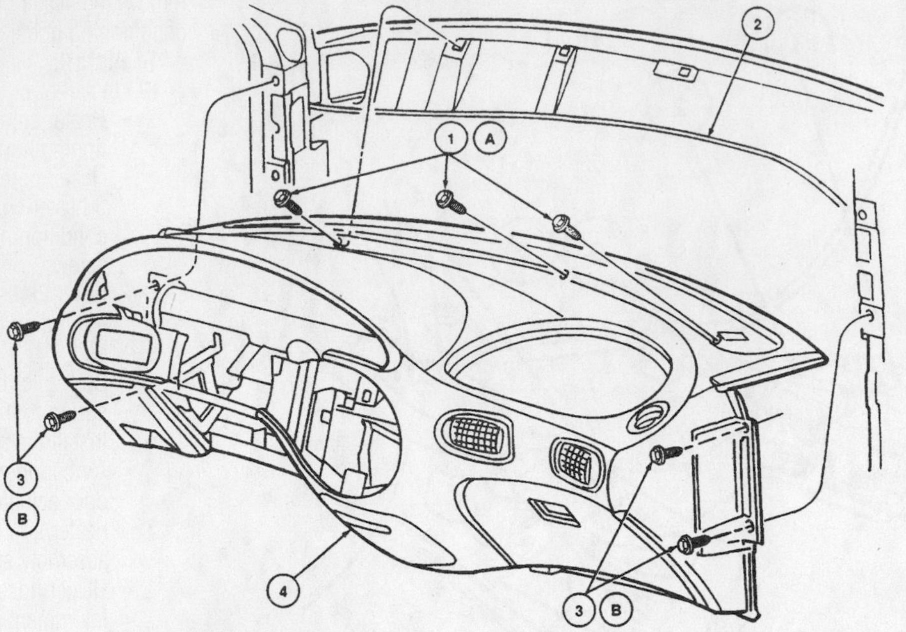

Item	Description
1	Screw (3 Req'd)
2	Cowl Top Panel
3	Screw (4 Req'd)
4	Instrument Panel

Item	Description
A	Tighten to 7-9 N·m (62-79 Lb-In)
B	Tighten to 19-25 N·m (15-18 Lb-Ft)

91170G14

Instrument panel-to-cowl panel mounting—Taurus and Sable

**** WARNING**

Do not use excessive force when installing the radio, the retaining clips can become damaged.

- Sensor hose/elbow, the electrical harness connectors and the vacuum hose harness to the evaporator housing (if equipped with an Electronic Automatic Temperature Control (EATC)

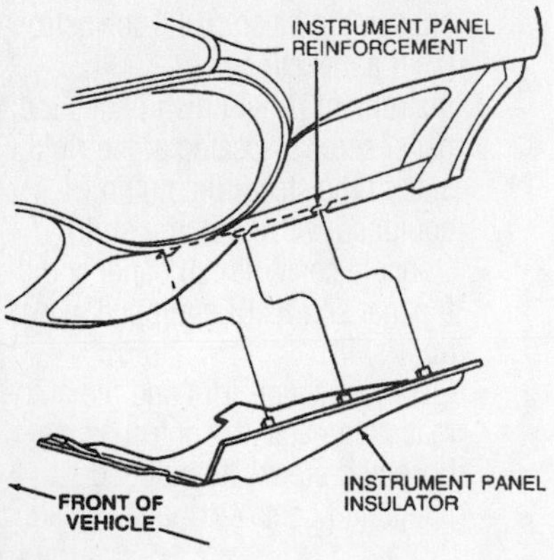

Remove the 4 blend door actuator retaining screws—Taurus and Sable

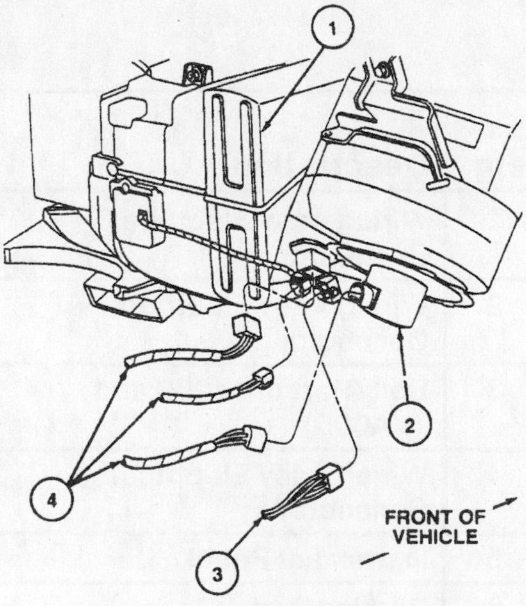

Item	Description
1	A/C Evaporator Housing
2	Blower Motor
3	Vacuum Hose Harness
4	Main Wiring

91176G06

Secondary air temperature door connections—Taurus and Sable

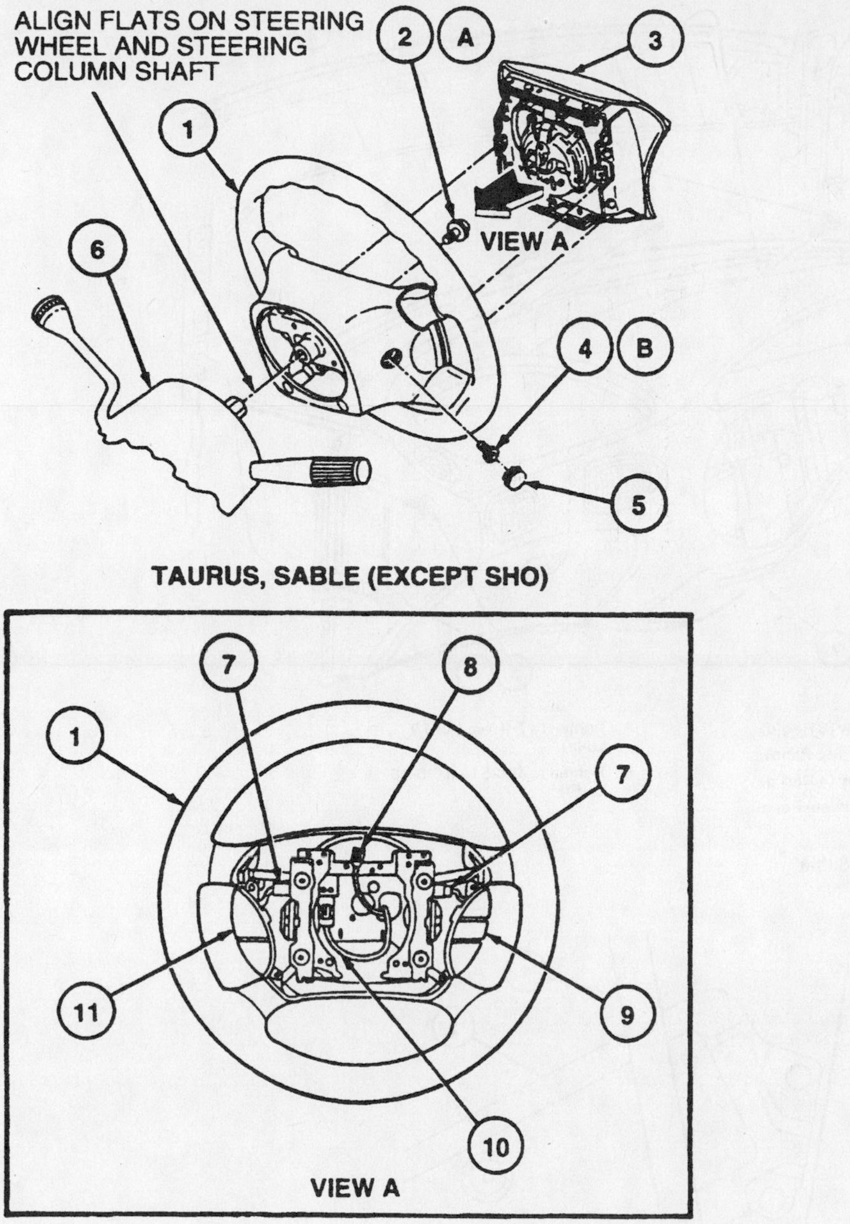

1	Steering Wheel	10	Speed Control / Horn Wire Connector
2	Bolt	11	Speed Control Actuator Switch (Left Hand)
3	Driver Side Air Bag Module	A	Tighten to 34-46 N·m (26-33 Lb-Ft)
4	Screw (2 Req'd)	B	Tighten to 10-14 N·m (89-123 Lb-In)
5	Steering Wheel Spoke Cover		
6	Steering Column Tube		
7	Electrical Connector		
8	Air Bag Electrical Connector		
9	Speed Control Actuator Switch (Right Hand)		

93111G90

Exploded view of the SRS module and the steering wheel assembly—Taurus and Sable

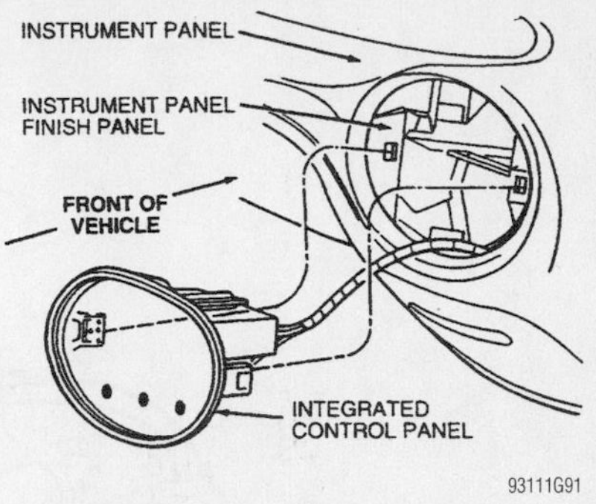

View of the integrated control panel—Taurus and Sable

- Connect the electrical connectors and the vacuum harness to the evaporator housing and the blower motor (if not equipped with an EATC)
- Instrument panel brace and the 2 Diagnostic Link Connector (DLC) screws
- Courtesy lamp socket and install the 2 instrument panel brace screws
- Console finish panel (if equipped with a floor shift)
- Lower instrument cover to the instrument panel reinforcement (located on the passenger's side)
- Steering column and torque the 4 steering column-to-instrument panel bracket nuts to 10–13 ft. lbs. (13–17 Nm)
- Interlock cable and actuator (if equipped with a column shift)
- Shifter indicator-to-column adjustment cable screw
- Shift indicator cable to the shifter tube
- Multi-function switch and the screws
- Electrical connectors
- Wiring connector at the bottom of the steering column and install the wiring to the column (if equipped with an overdrive lockout switch on the manual control lever)
- Gearshift lever and the lever pin to the manual control lever
- 3 steering column shroud screws and the shrouds
- Ignition switch to the steering column lock cylinder housing
- Lower instrument panel cover and the 2 lower cover-to-instrument panel screws
- Lower steering column shaft bolt

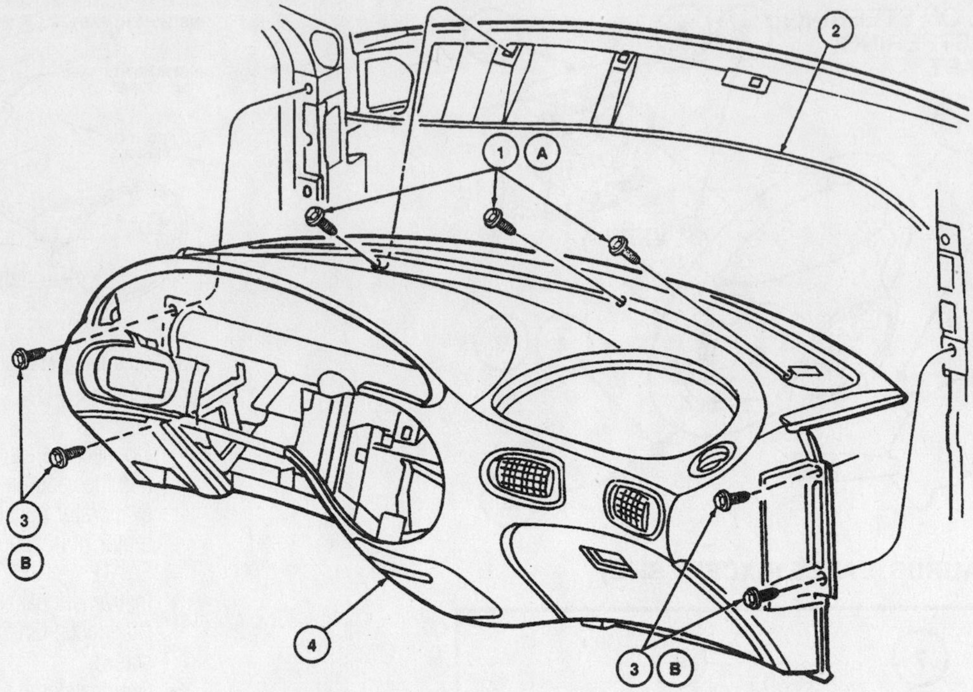

1 Screw (3 Req'd)
2 Cowl Top Panel
3 Screw (4 Req'd)
4 Instrument Panel

A Tighten to 7-9 N·m (62-79 Lb-In)
B Tighten to 19-25 N·m (15-18 Lb-Ft)

93111G92

Exploded view of the instrument panel—Taurus and Sable

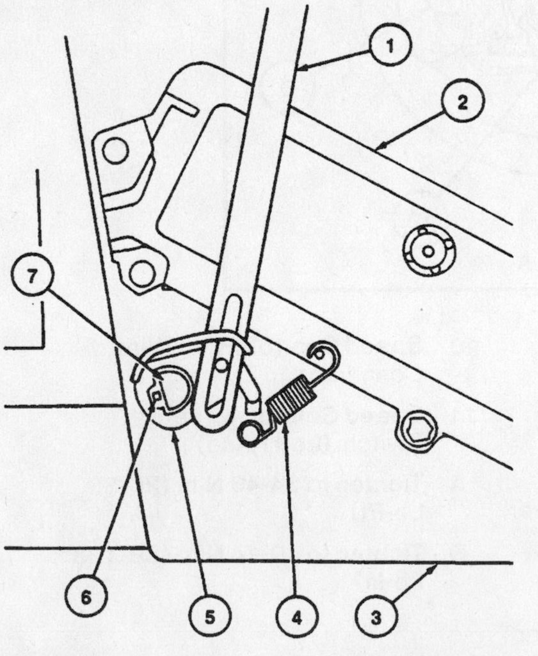

1 Metal Link
2 Heater Core Cover
3 A/C Evaporator Housing
4 Spring
5 Lever
6 Locking Ramp (Part of Secondary A/C Air Temperature Control Door Shaft)
7 Secondary A/C Air Temperature Control Door Shaft

93111G93

View of the temperature control mechanism—Taurus and Sable

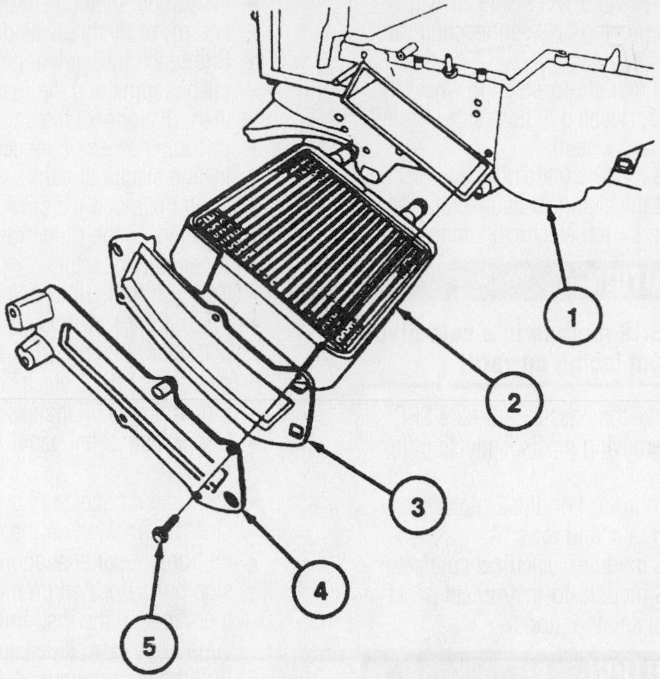

1 A/C Evaporator Housing
2 Heater Core
3 Heater Core Cover Seal
4 Heater Core Cover
5 Screw

Exploded view of the heater core—Taurus and Sable

and torque it to 17–20 ft. lbs. (22–26 Nm)
- Steering wheel to the steering column
- New steering wheel bolt and torque it to 26–33 ft. lbs. (34–46 Nm)
- SRS module and connect the electrical connector
- SRS module-to-steering wheel bolts and torque the bolts to 89–123 inch lbs. (10–14 Nm)
- SRS bolt covers
- Negative battery cable

20. Fill the cooling system.
21. Operate the engine to normal operating temperatures; then, check the climate control operation and check for leaks.

Lincoln LS

REMOVAL & INSTALLATION

1. Before servicing the vehicle, refer to the precautions in the beginning of this section.
2. Drain the cooling system.
3. Recover the A/C refrigerant.
4. Remove or disconnect the following:
   - Negative battery cable
   - Heater hose assembly
   - Cabin air filter plenum
   - Thermostatic expansion valve manifold and tube assembly
   - Driver's side air bag module
   - Floor console and A/C duct
   - Shift lever assembly, if equipped with automatic transmission
   - Left and right instrument panel insulators
   - Left and right door sill scuff plates
   - Left and right door weatherstrips
   - Left and right A pillar lower trim panels
   - Left and right windshield side garnish moldings
   - Instrument panel defroster opening grille assembly
   - Instrument panel cowl top screws
   - Instrument panel upper reinforcement bolts
   - Left and right instrument panel side finish panels
   - Hood release handle and cable
   - Upper right bulkhead electrical connector
   - Right instrument panel electrical connectors
   - Passenger side tunnel electrical connector
   - Steering column intermediate shaft
   - Left junction box electrical connectors
   - Left instrument panel electrical connectors
   - Ignition shift interlock connector, if equipped
   - 4 instrument panel tunnel brace bolts
   - Left outer instrument panel cowl side cover and reinforcement bolt
   - Left instrument panel cowl side bolt and nut
   - Right instrument panel cowl side bolt and nut
   - Instrument panel
   - Evaporator housing electrical connector
   - Cowl top attachment bolt
   - Evaporator housing attachment bolt
   - 3 evaporator nuts and washers in the engine compartment
   - Rear seat floor ducts
   - Evaporator core housing
   - Evaporator core housing air inlet
   - Bypass door harness connector
   - Heater core

To install:

5. Install or connect the following:
   - Heater core
   - Bypass door harness connector
   - Evaporator core housing air inlet
   - Evaporator core housing
   - Rear seat floor ducts
   - 3 evaporator nuts and washers in the engine compartment
   - Evaporator housing attachment bolt
   - Cowl top attachment bolt
   - Evaporator housing electrical connector
   - Instrument panel
   - Instrument panel cowl top screws. Tighten to 27 inch lbs. (3 Nm).
   - Right instrument panel cowl side bolt and nut. Tighten the fasteners to 15 ft. lbs. (20 Nm).
   - Left instrument panel cowl side bolt and nut. Tighten the fasteners to 15 ft. lbs. (20 Nm).
   - Left outer instrument panel cowl side cover and reinforcement bolt. Tighten the bolt to 15 ft. lbs. (20 Nm).
   - Instrument panel upper reinforce-

ment bolts. Tighten the bolts to 15 ft. lbs. (20 Nm).
- 4 instrument panel tunnel brace bolts. Tighten the bolts to 15 ft. lbs. (20 Nm).
- Ignition shift interlock connector, if equipped
- Left instrument panel electrical connectors
- Left junction box electrical connectors
- Steering column intermediate shaft. Tighten the pinch bolt to 26 ft. lbs. (35 Nm).
- Passenger side tunnel electrical connector
- Right instrument panel electrical connectors
- Upper right bulkhead electrical connector
- Hood release handle and cable
- Left and right instrument panel side finish panels
- Instrument panel defroster opening grille assembly
- Left and right windshield side garnish moldings
- Left and right A pillar lower trim panels
- Left and right door weatherstrips
- Left and right door sill scuff plates
- Left and right instrument panel insulators
- Shift lever assembly, if equipped with automatic transmission
- Floor console and A/C duct
- Driver's side air bag module
- Thermostatic expansion valve manifold and tube assembly
- Cabin air filter plenum
- Heater hose assembly
- Negative battery cable

6. Fill the cooling system.
7. Recharge the A/C system.
8. Start the engine and check for leaks.

Continental

REMOVAL & INSTALLATION

1. Disconnect the negative battery cable.

CAUTION

After disconnecting the negative battery cable, wait for at least 1 minute for the SRS or air bag module to deplete its energy.

2. Drain the cooling system into a clean container for reuse.

3. Remove the driver's side air bag module by removing or disconnecting the following:
   - SRS module-to-steering wheel bolts, (located at both sides of the steering wheel)
   - SRS module (carefully) and disconnect the electrical connector
   - Horn switch electrical connector

CAUTION

Place the SRS module in a safe place with the front facing upward.

4. Remove the passenger's side SRS module by removing or disconnecting the following:
   - Push inward on the 2 glove box door tabs and lower it
   - SRS module's electrical connector
   - SRS module-to-instrument panel bolts and the module

CAUTION

Place the SRS module in a safe place with the front facing upward

5. Remove the instrument panel by removing or disconnecting the following:
   - Floor or mini console (if equipped)
   - Rear seat climate control air duct sleeve (if equipped)
   - Left side instrument panel insulator pushpins and the insulator; then, disconnect the courtesy lamp
   - Instrument panel steering column cover screws and the cover
   - Pull the hood release handle, remove the screws and move it aside
   - Pull the parking brake handle, remove the bolts, the parking brake release handle and the cable
   - Steering column opening cover reinforcement-to-instrument panel bolts and the cover reinforcement
   - Release the cable and the conduit from the parking brake actuator
   - Steering column
   - Loosen the bolt and disconnect the left side outboard bulkhead electrical connector
   - Loosen the bolt and disconnect the left side inboard bulkhead electrical connector
   - Heated seat switch electrical connectors (if equipped)
   - 2 steering column mounting support-to-instrument cowl brace bolts
   - Instrument panel dash brace-to-instrument panel bolt
   - Instrument panel dash brace nut and move the brace aside
   - Right side instrument panel insulator pushpins and the insulator; then, disconnect the courtesy lamp
   - Vacuum harness connector
   - In-line electrical harness connector
   - Scuff plate and the cowl trim panel (located on the right side)
   - Antenna connector
   - Brake shift interlock actuator cable
   - Pry out the instrument panel defroster opening grille assembly, disconnect the 2 electrical connectors and remove the assembly
   - Upper instrument panel-to-cowl screws
   - Instrument panel support-to-cowl side nut (located on the right side)
   - Instrument panel support-to-cowl side bolts (located on the left side); then, loosen the instrument panel support-to-cowl side captive bolt
   - Pull the instrument panel rearward; then, disconnect the Electronic Air Temperature Control (EATC) hose from the heater plenum and make sure all electrical connectors are disconnected
   - Instrument panel with the help of an assistant

➡Check the upper cowl clips for damage; if necessary, replace them.

6. Loosen the screw and disconnect the PCM electrical connector.
7. Disconnect the heater hoses from the heater core.
8. Remove or disconnect the following:
   - 2 metal cover-to-heater/air conditioning housing screws and the metal cover (located below the heater core cover)
   - Electrical harness connector (at the heater core cover); then, remove the air conditioning electronic blend door actuator-to-heater/air conditioning housing screws and the actuator
   - Air conditioning air intake flue damper assist spring

WARNING

Do not bend any part of the air conditioning damper door shaft for it is brittle and will break.

9. Depress the air conditioning damper door shaft locking ramp, disconnect it from the air temperature control door shaft, swing the locking ramp counterclockwise and

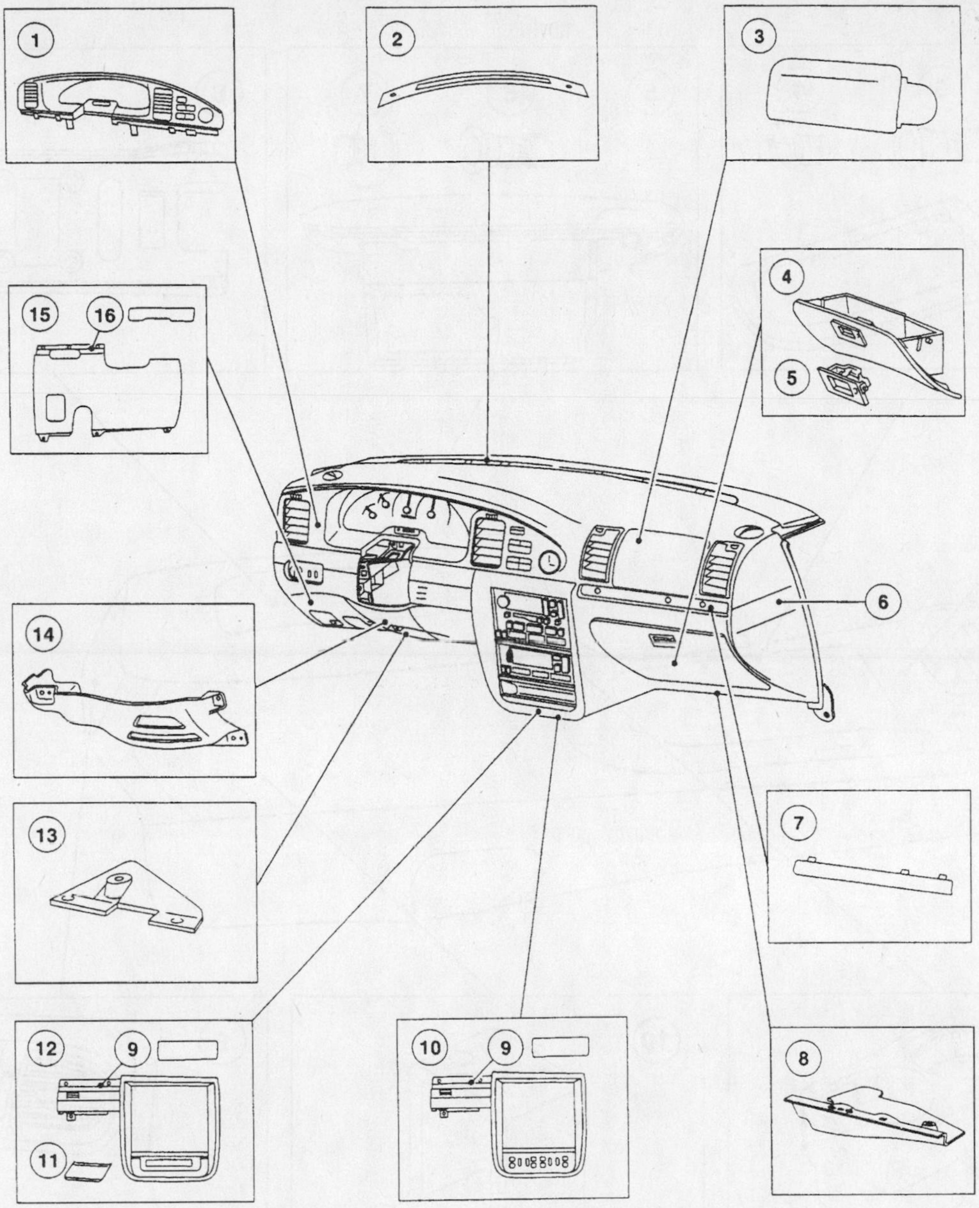

1 Instrument Panel Cluster Finish Panel
2 Instrument Panel Defroster Opening Grille
3 Passenger Side Air Bag Module
4 Glove Compartment
5 Glove Compartment Door Latch
6 Instrument Panel
7 Instrument Panel Finish Panel (RH) (Wood)
8 Instrument Panel Insulator (RH)
9 Instrument Panel Finish Panel (Wood)
10 Instrument Panel Center Finish Panel (Service, Heated Seat)
11 Utility Compartment Lower Mat
12 Instrument Panel Center Finish Panel (Base)
13 Instrument Panel Insulator (LH)
14 Instrument Panel Steering Column Opening Cover Reinforcement
15 Instrument Panel Steering Column Cover
16 Instrument Panel Finish Panel (LH) (Wood)

93111G99

Exploded view of the instrument panel and related components—Continental

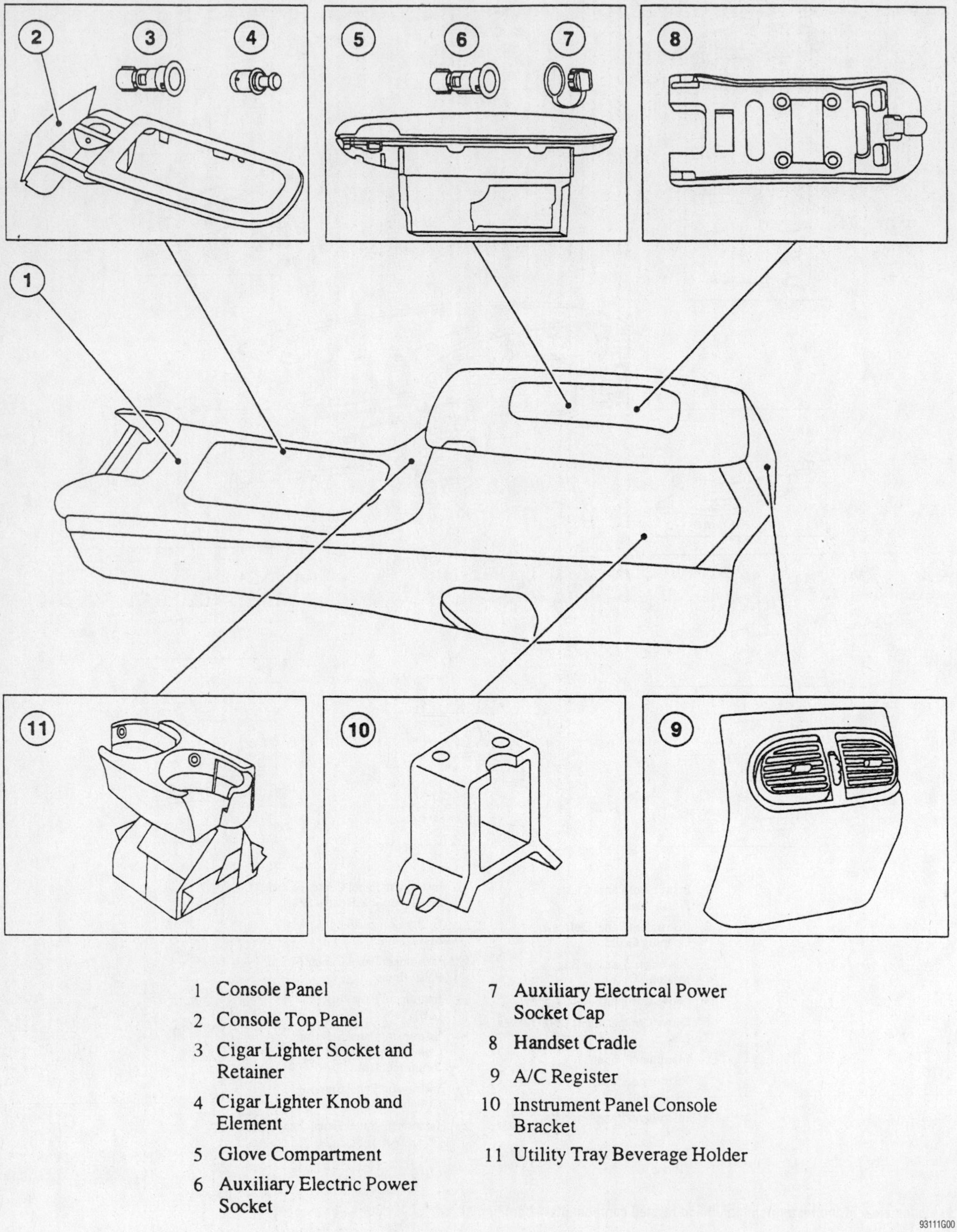

Exploded view of the floor console and related components—Continental

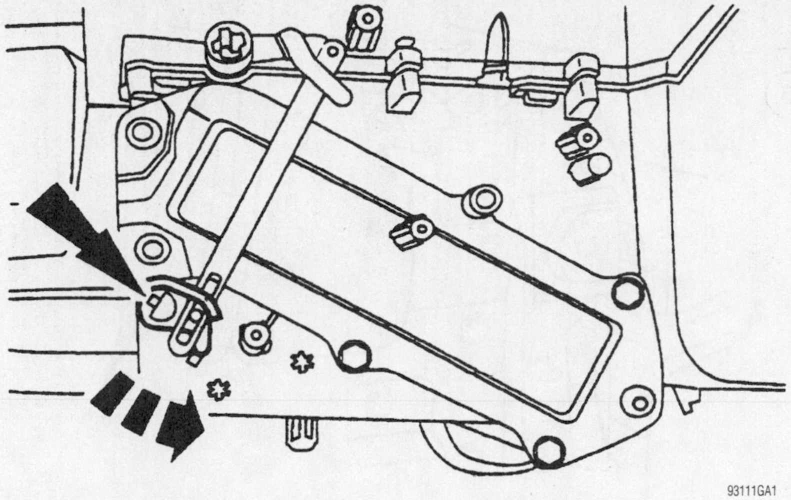

View of the air temperature control door shaft and heater core cover—Continental

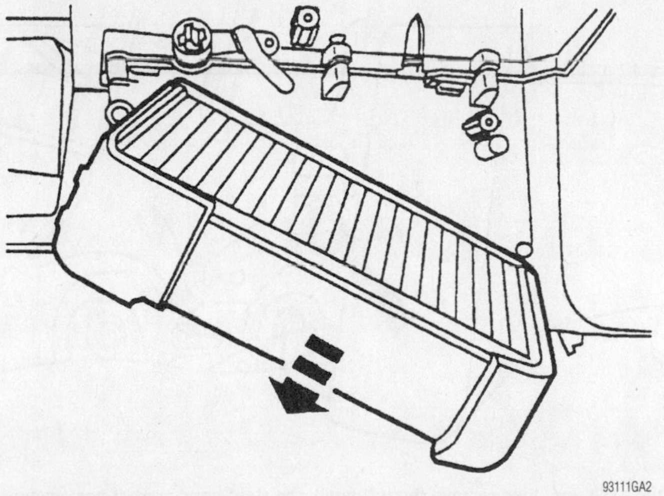

View of the heater core—Continental

remove it from the air conditioning damper door shaft.

10. Move the air conditioning damper door shaft counterclockwise and remove it.

11. Remove or disconnect the following:
- Heater core cover-to-heater/air conditioning housing screws and the cover
- Heater core cover seal
- Heater core from the heater/air conditioning housing

To install:

12. Install or connect the following:
- Heater core to the heater/air conditioning housing
- Heater core cover seal
- Heater core cover and the cover-to-heater/air conditioning housing screws
- Air conditioning damper door shaft and move it clockwise and install it
- Air conditioning damper door shaft locking ramp, swing the locking ramp clockwise and connect it to the air temperature control door shaft
- Air conditioning air intake flue damper assist spring

⁂ WARNING

Do not bend any part of the air conditioning damper door shaft for it is brittle and will break.

13. At the heater core cover, install the air conditioning electronic blend door actuator and the actuator-to-heater/air conditioning housing screws; then, connect the electrical harness connector.

14. Install or connect the following:
- Metal cover and the 2 metal cover-to-heater/air conditioning housing screws (located below the heater core cover)
- Heater hoses to the heater core
- PCM electrical connector and tighten the screw

15. Install the instrument panel by installing or connecting the following:
- Instrument panel with the help of an assistant
- Pull the instrument panel rearward; then, connect the Electronic Air Temperature Control (EATC) hose to the heater plenum and make sure all electrical connectors are connected
- Instrument panel support-to-cowl side bolts and the instrument panel support-to-cowl side captive bolt (located on the left side)
- Instrument panel support-to-cowl side nut (located on the right side)
- Upper instrument panel-to-cowl screws
- 2 electrical connectors and the instrument panel defroster opening grille assembly
- Brake shift interlock actuator cable
- Antenna connector
- Scuff plate and the cowl trim panel (located on the right side)
- In-line electrical harness connector
- Vacuum harness connector
- Right side instrument panel insulator and secure the insulator with the pushpins; then, connect the courtesy lamp
- Instrument panel dash brace and the brace nut
- Instrument panel dash brace-to-instrument panel bolt
- 2 steering column mounting support-to-instrument cowl brace bolts
- Heated seat switch electrical connectors (if equipped)
- Left side inboard bulkhead electrical connector and tighten the bolt
- Left side outboard bulkhead electrical connector and tighten the bolt
- Steering column
- Cable and the conduit to the parking brake actuator
- Steering column opening cover reinforcement and the cover reinforcement-to-instrument panel bolts

- Parking brake handle, the bolts and the cable
- Hood release handle and the screws
- Instrument panel steering column cover and the cover screws
- Left side instrument panel insulator and secure the insulator with the pushpins; then, connect the courtesy lamp
- Rear seat climate control air duct sleeve, if equipped
- Floor or mini console, if equipped

16. Install the passenger's side SRS module by installing or connecting the following:
    - SRS module and torque the module-to-instrument panel bolts to 62–97 inch lbs. (7–11 Nm)
    - SRS module's electrical connector
    - Glove box door
17. Install the driver's side air bag module by installing or connecting the following:
    - Horn switch electrical connector
    - Electrical connector and install the SRS module
    - SRS module-to-steering wheel bolts (located at both sides of the steering wheel), and torque the bolts to 90–122 inch lbs. (10–14 Nm)
18. Refill the cooling system.
19. Connect the negative battery cable.
20. Operate the engine to normal operating temperatures; then, check the climate control operation and check for leaks.

Escort, ZX2 and Tracer

REMOVAL & INSTALLATION

1. Disconnect the negative battery cable.

⁂ CAUTION

After disconnecting the negative battery cable, wait for at least 1 minute for the SRS or air bag module to deplete its energy.

2. Drain the cooling system into a clean container for reuse.
3. On the Coupe, remove the air cleaner outlet tube.
4. Disconnect the heater hoses from the heater core.
5. Place the front wheels in the straight-ahead position. Lock the steering column.

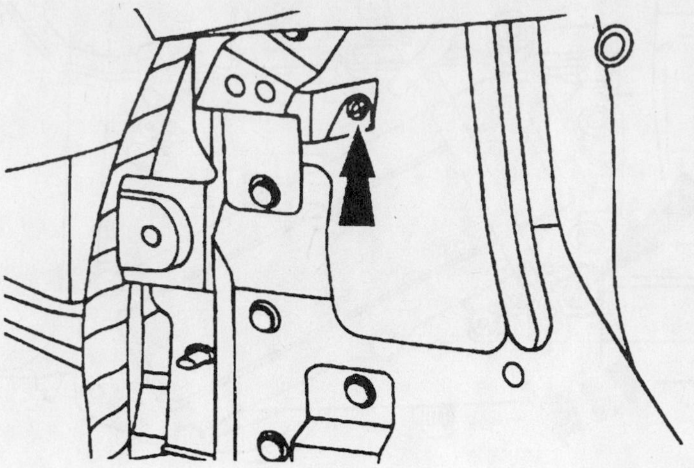

Remove the screw from the center instrument panel finish panel—Escort and Tracer

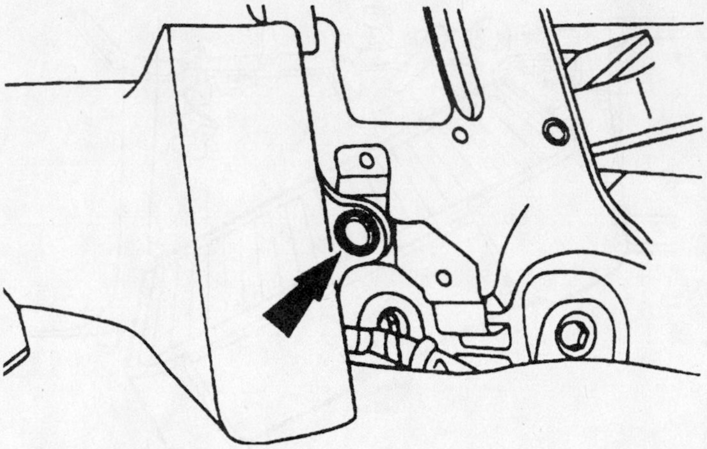

Unfasten the pushpins, then remove the left-hand and right-hand control box covers—Escort and Tracer

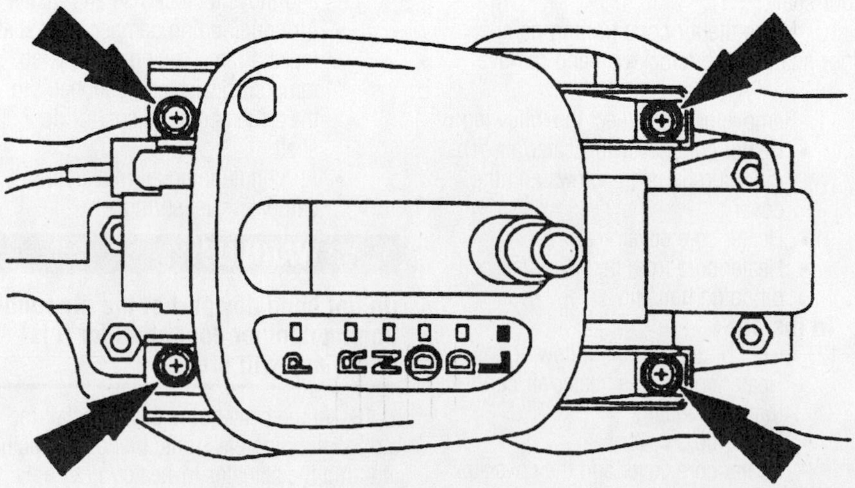

Unfasten the retaining screws and position the transaxle control selector dial bezel sideways—Escort and Tracer

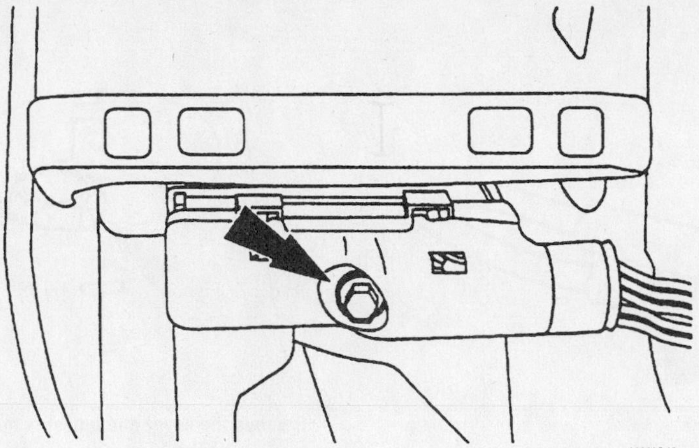

90980G43

Unfasten the PCM electrical connector bolt, unplug the connector and move it aside—Escort and Tracer

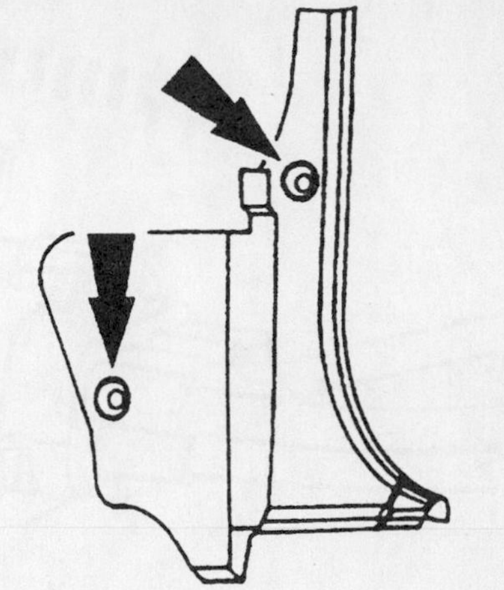

90980G48

Remove the pushpins from each cowl side trim panel, and remove the panels—Escort and Tracer

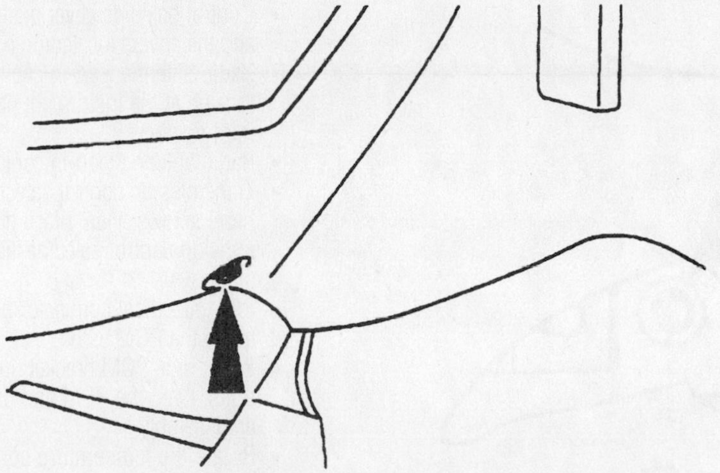

90980G46

Unfasten the screw and remove the instrument panel steering column cover—Escort and Tracer

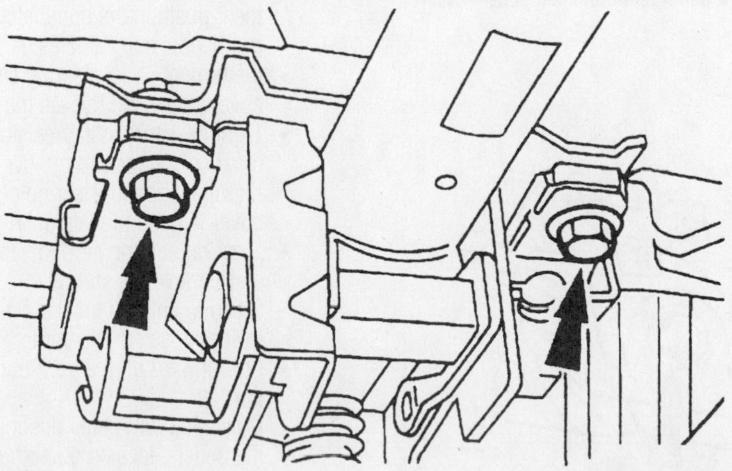

90980G47

Remove the steering column bracket bolts and lower the column—Escort and Tracer

6. Remove the steering wheel by removing or disconnecting the following:
   - SRS module-to-steering wheel bolts
   - SRS module (carefully), and disconnect the horn switch and the SRS electrical connectors

CAUTION

Place the SRS module in a safe place with the front facing upward

   - Steering wheel bolt and discard it
   - Press the steering wheel from the steering column

7. Remove the passenger's side SRS module by removing or disconnecting the following:
   - Push inward on the 2 glove box door tabs and lower it
   - SRS module's electrical connector
   - SRS module-to-instrument panel bolts and the module

CAUTION

Place the SRS module in a safe place with the front facing upward.

8. Remove the instrument panel by removing or disconnecting the following:
   - Floor console
   - Screw located at the center instrument panel finish panel
   - Pull the center instrument panel finish panel straight out from the instrument panel reinforcement (Coupe models only)
   - Power point socket electrical connectors and remove the center

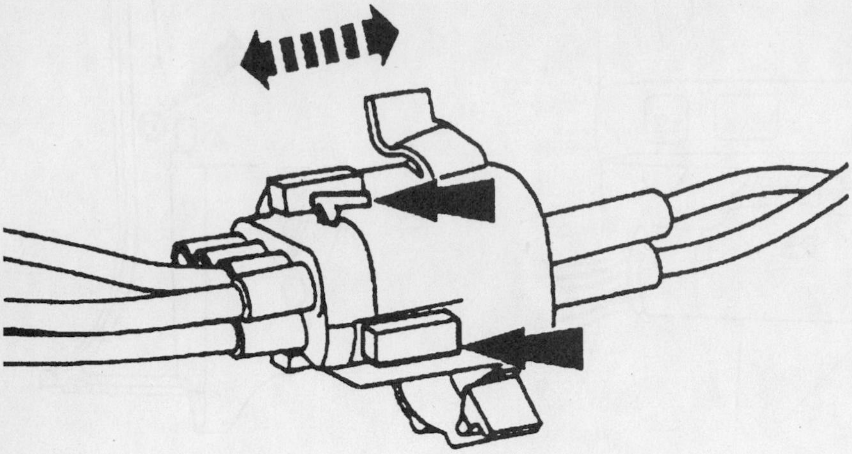

Unplug the vacuum line harness connector—Escort and Tracer

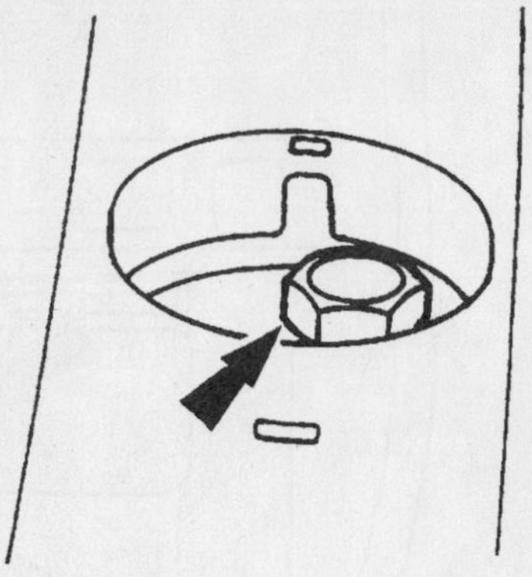

Remove the cover and unfasten the upper instrument panel reinforcement bolt—Escort and Tracer

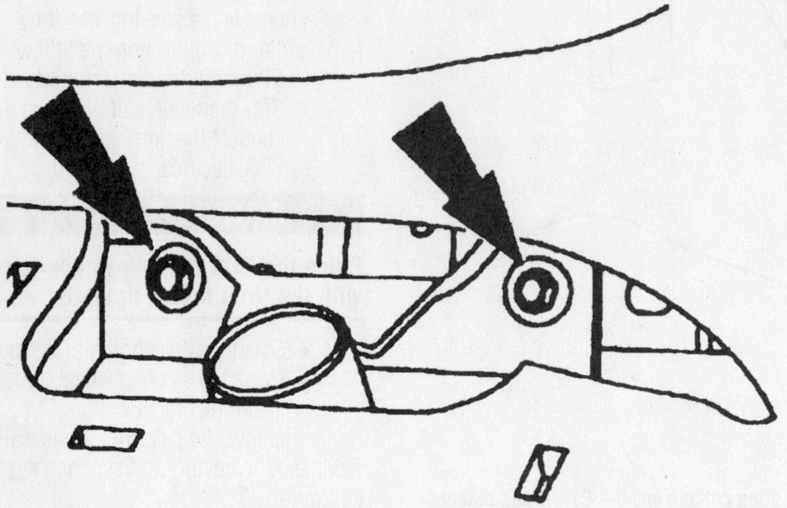

Unfasten both the left-hand and right-hand center instrument panel reinforcement bolts—Escort and Tracer

Unfasten both the left-hand and right-hand lower instrument panel reinforcement bolts—Escort and Tracer

instrument panel finish panel, if equipped

- Control box side cover pushpins and the covers located on both sides
- Radio antenna lead-in cable, located at the instrument panel reinforcement
- Radio antenna lead-in cable
- Transmission control selector dial bezel screws; then, place the transmission control selector dial bezel sideways
- PCM electrical connector bolt and move the PCM aside
- 2 rear side PCM bracket nuts, the 2 bolts; then, the PCM and bracket as an assembly
- Rotate the temperature control switch to the COOL position and disconnect the heater control cable
- Hood latch control handle nut; then, position the hood latch control handle and cable aside
- Instrument panel steering column cover screw and release the cover
- Light switch rheostat resistor electrical connector and remove the instrument panel steering column cover (Coupe models only)
- Steering column shroud screws and remove the shrouds
- Steering column bracket bolts and lower the steering column
- Pull upward on the front door scuff plates
- Cowl side trim panel pushpins and the panels, located on both sides
- Interior fuse junction panel electrical connector
- Lower instrument panel reinforcement in-line electrical connector, located on the left side

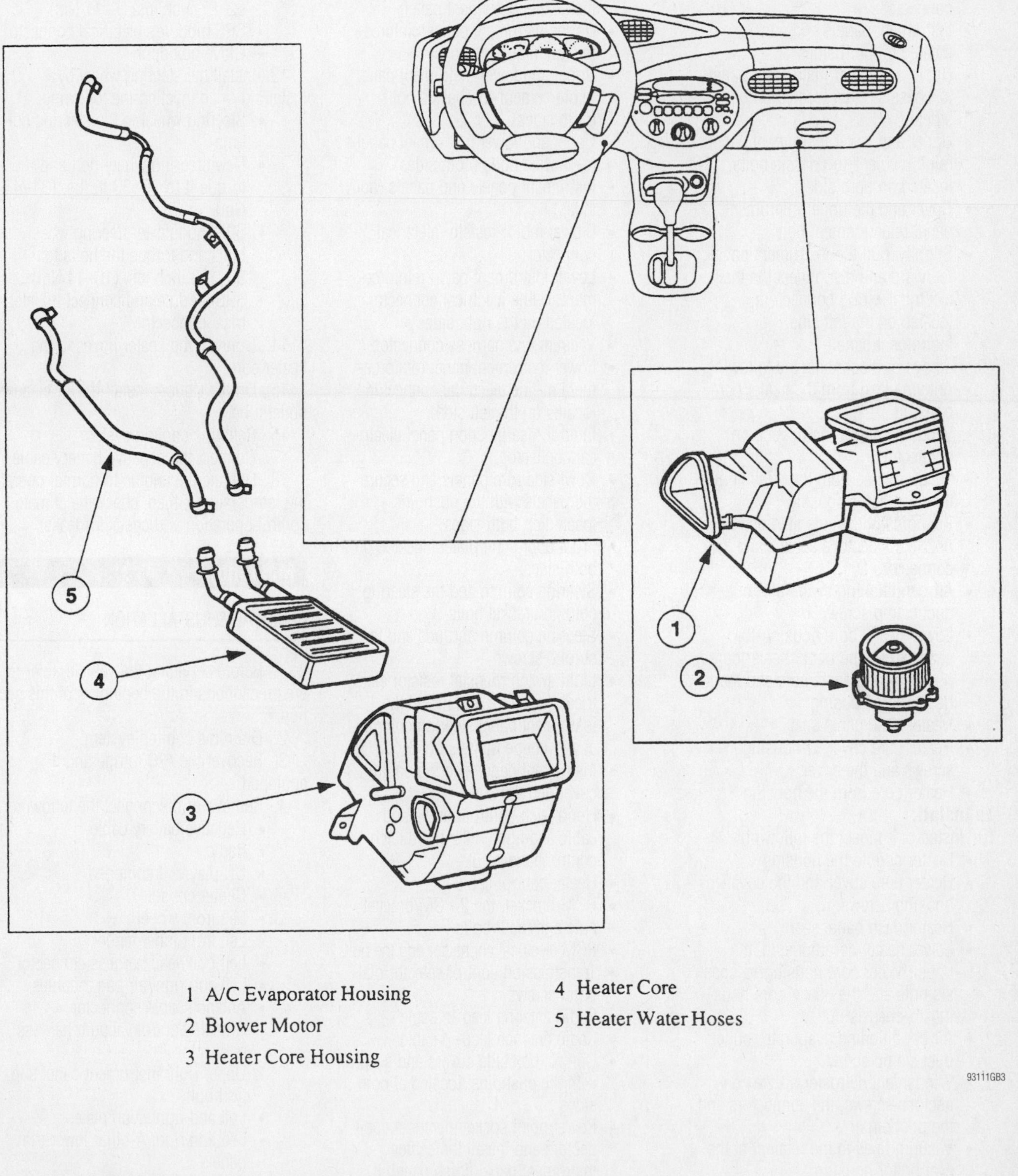

Exploded view of the heater core, heater housing and related components—Escort and Tracer

- Vacuum line harness connector
- Lower instrument panel reinforcement in-line electrical connector, located on the right side
- Blower motor resistor electrical connector
- Instrument panels end panels, located on both sides
- Upper and lower instrument panel-to-chassis bolts, located on both sides
- Upper and lower instrument panel reinforcement-to-chassis bolts, located on both sides
- Cover and the upper instrument panel reinforcement bolt
- Slightly, pull the instrument panel rearward and disconnect the main electrical wiring connectors, located on the left side
- Instrument panel

9. Remove or disconnect the following:
   - Antenna lead from the heater core housing
   - Vacuum control motor vacuum connector
   - Vacuum lines from the retainer at the evaporator housing
   - Pushpins, the windshield defroster nozzle connector's screws and the connectors
   - Air conditioning evaporator outlet duct clamp screw
   - Lower heater core housing-to-chassis nut, the upper heater core housing-to-chassis nuts and the heater core housing
   - Heater dash panel seal
   - Heater core cover-to-housing screws and the cover
   - Heater core from the housing

To install:

10. Install or connect the following:
    - Heater core to the housing
    - Heater core cover and the cover-to-housing screws
    - Heater dash panel seal
    - Lower heater core housing, the upper heater core housing-to-chassis nuts and the heater core housing-to-chassis nut
    - Air conditioning evaporator outlet duct clamp screw
    - Windshield defroster nozzle connector's screws, the connectors and the pushpins
    - Vacuum lines to the retainer at the evaporator housing
    - Vacuum control motor vacuum connector
    - Antenna lead to the heater core housing

11. Install the instrument panel by installing or connecting the following:
    - Instrument panel
    - Main electrical wiring connectors (located on the left side), and install the instrument panel
    - Upper instrument panel reinforcement bolt and the cover
    - Upper and lower instrument panel reinforcement-to-chassis bolts (both sides)
    - Upper and lower instrument panel-to-chassis bolts (both sides)
    - Instrument panels end panels (both sides)
    - Blower motor resistor electrical connector
    - Lower instrument panel reinforcement in-line electrical connector, located on the right side
    - Vacuum line harness connector.
    - Lower instrument panel reinforcement in-line electrical connector, located on the left side
    - Interior fuse junction panel electrical connector
    - Cowl side trim panels and secure the panels with the pushpins, located on both sides
    - Front door scuff plates, located on both sides
    - Steering column and the steering column bracket bolts
    - Steering column shrouds and the shroud screws
    - Light switch rheostat resistor electrical connector and install the instrument panel steering column cover (coupe models only)
    - Instrument panel steering column cover and the cover screw
    - Hood latch control handle and cable and tighten the hood latch control handle nut
    - Heater control cable
    - PCM/bracket, the 2 PCM/bracket nuts and the 2 bolts
    - PCM electrical connector and the bolt
    - Transmission control selector dial bezel screws
    - Radio antenna lead-in cable
    - Radio antenna lead-in cable
    - Control box side covers and secure with the pushpins, located at both sides
    - Power point socket electrical connectors and install the center instrument panel finish panel, if equipped
    - Center instrument panel finish panel screw
    - Floor console

12. Install the passenger's side SRS module by installing or connecting the following:
    - SRS module and torque the module-to-instrument panel bolts to 62–97 inch lbs. (7–11 Nm)
    - SRS module's electrical connector
    - Glove box door

13. Install the steering wheel by installing or connecting the following:
    - Steering wheel to the steering column
    - New steering wheel bolt and torque it to 26–33 ft. lbs. (34–46 Nm)
    - SRS module-to-steering wheel bolts and torque the bolts to 89–123 inch lbs. (10–14 Nm)
    - SRS module and connect the electrical connector

14. Connect the heater hoses to the heater core.
15. On the Coupe, install the air cleaner outlet tube.
16. Refill the cooling system.
17. Connect the negative battery cable.
18. Operate the engine to normal operating temperatures; then, check the climate control operation and check for leaks.

Mustang

REMOVAL & INSTALLATION

1. Before servicing the vehicle, refer to the precautions in the beginning of this section.
2. Drain the cooling system.
3. Recover the A/C refrigerant, if equipped.
4. Remove or disconnect the following:
   - Negative battery cable
   - Radio
   - CD player, if equipped
   - Center console
   - Left front wheel
   - Left front inner fender
   - Left bulkhead harness connector
   - Left and right air bag modules
   - Antenna cable connector
   - Climate control vacuum harness connector
   - Upper right instrument panel support bolt
   - Left and right scuff plates
   - Left and right A-pillar lower trim panels
   - Left and right windshield side garnish moldings
   - Left and right door weatherstrips
   - Right main wire harness connectors

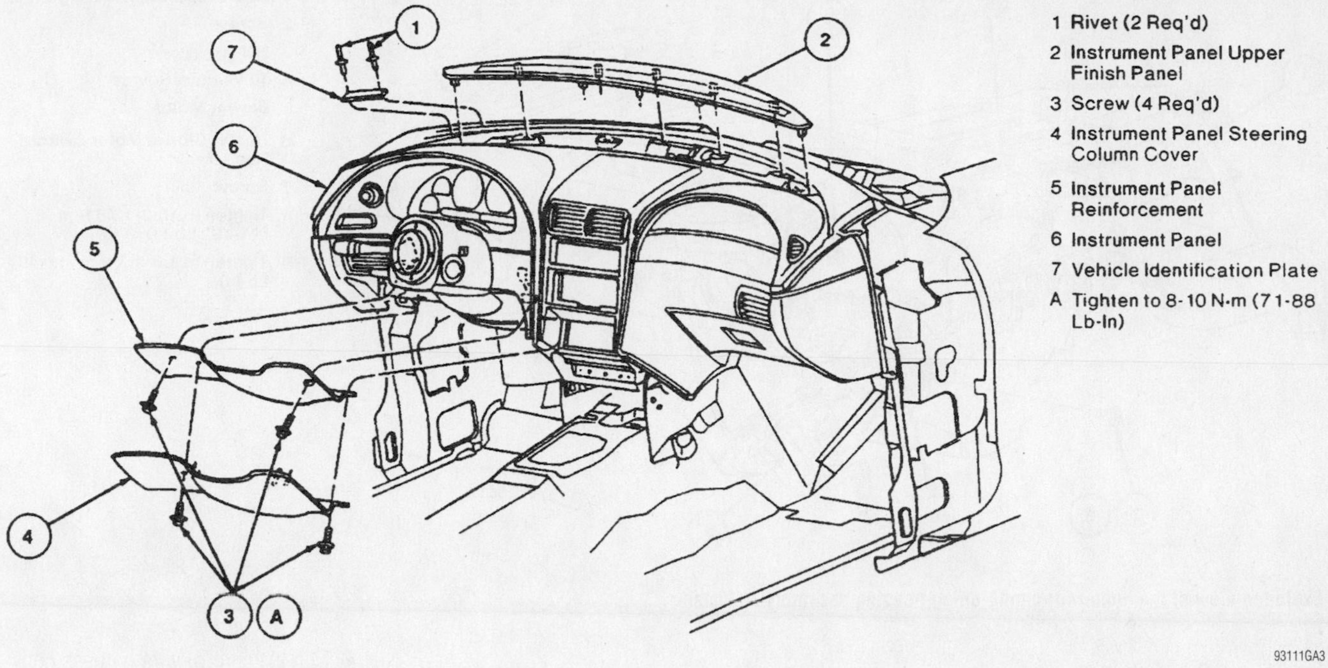

Exploded view of the instrument panel, cover and reinforcement—Mustang

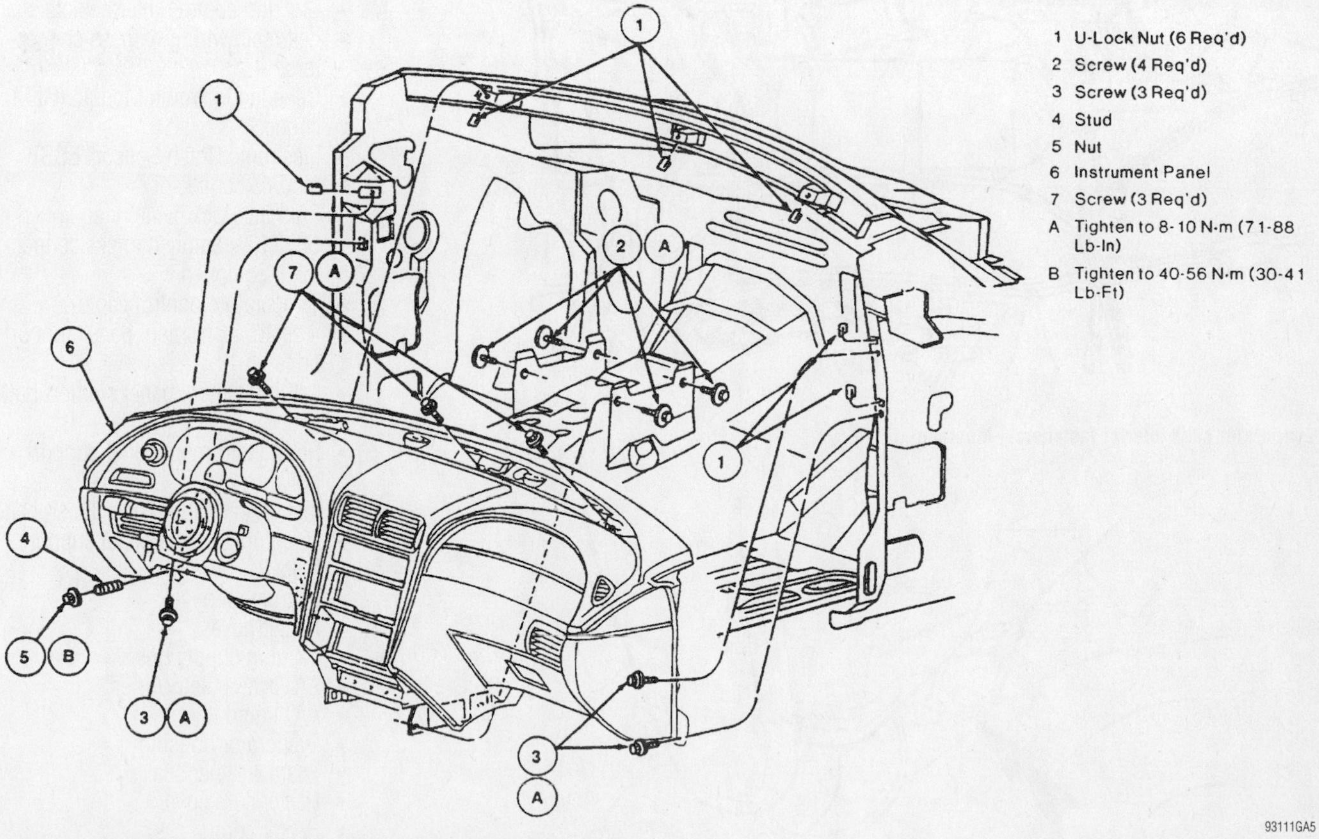

Exploded view of the instrument panel and related components—Mustang

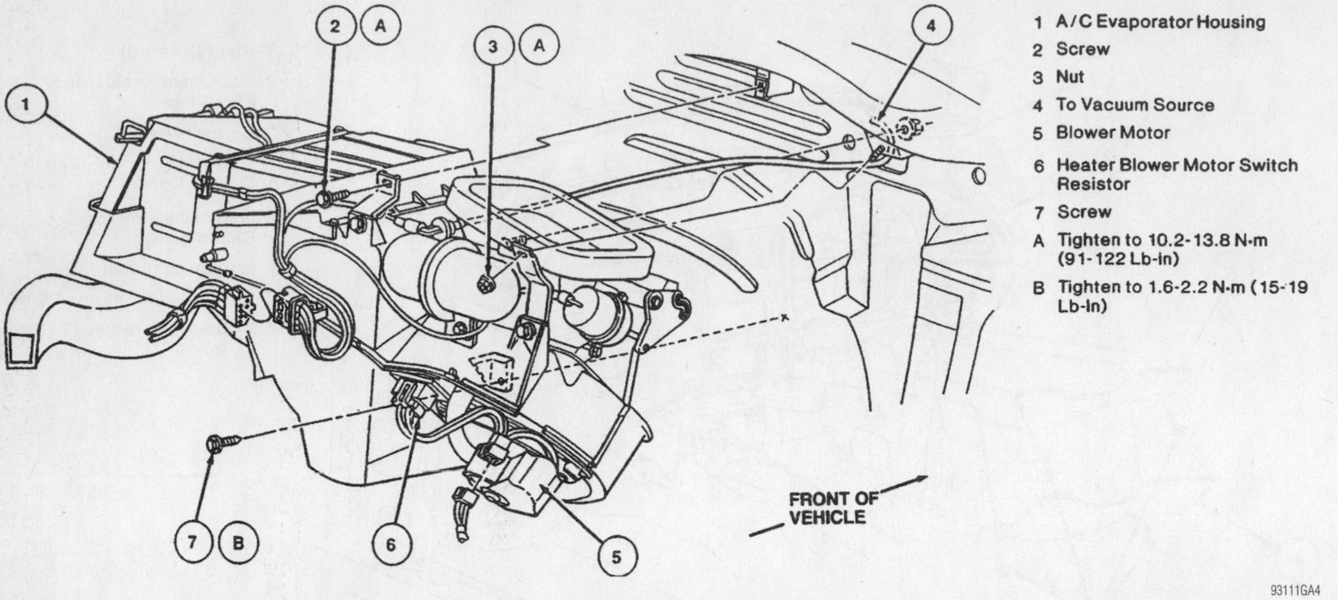

Exploded view of the heater/air conditioning housing assembly—Mustang

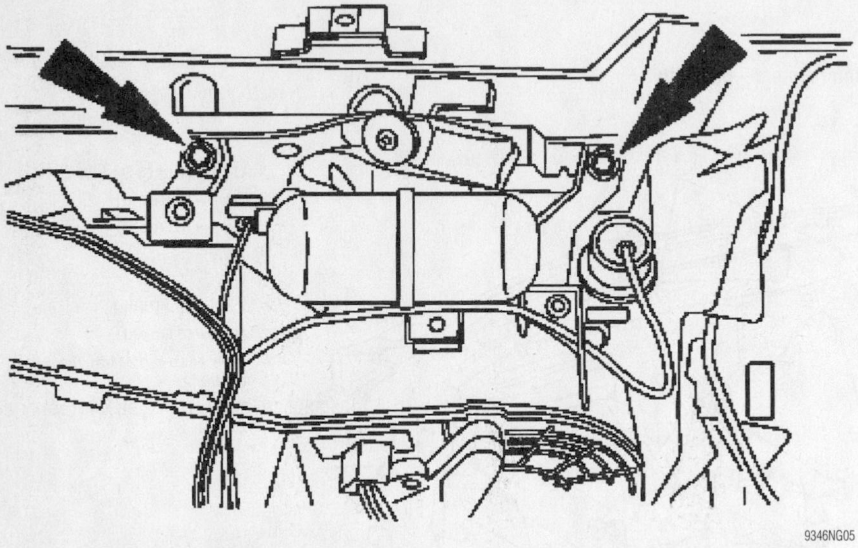

Evaporator case interior fasteners—Mustang

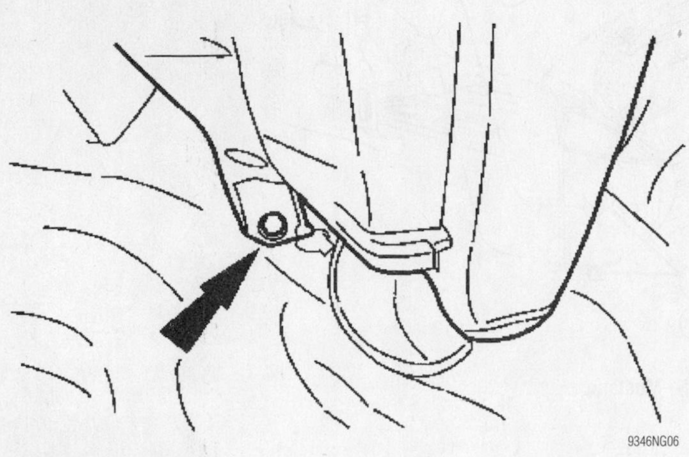

Evaporator case screw—Mustang

- Climate control wire harness connector
- Instrument panel steering column cover
- Instrument panel reinforcement
- Steering column intermediate shaft
- Left main wiring harness connectors
- Generic Electronic Module (GEM) connectors
- Electronic Crash Sensor (ECS) module connector
- Shift interlock cable, if equipped
- Shifter assembly harness connector, if equipped
- Temperature control cable
- 4 center instrument panel support bolts
- Left instrument panel support bolt and nut
- Right instrument panel support bolt
- Instrument panel upper finish panel
- Upper instrument panel support bolts
- Instrument panel
- Heater hoses
- Vacuum supply hose
- A/C accumulator/drier
- A/C liquid line
- Evaporator housing
- Foam weather seal
- Heater core cover
- Heater core

To install:

5. Install or connect the following:
   - Heater core
   - Heater core cover

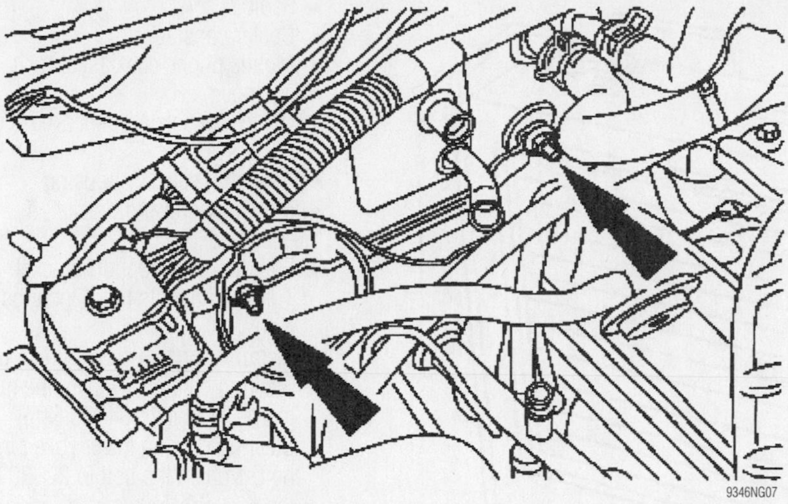

Evaporator case under-hood nuts—Mustang

- Foam weather seal
- Evaporator housing. Tighten the fasteners to 71 inch lbs. (8 Nm).
- A/C liquid line
- A/C accumulator/drier
- Vacuum supply hose
- Heater hoses
- Instrument panel
- Upper instrument panel support bolts. Tighten the bolts to 80 inch lbs. (9 Nm).
- Instrument panel upper finish panel
- Right instrument panel support bolt. Tighten the bolt to 80 inch lbs. (9 Nm).
- Left instrument panel support bolt and nut. Tighten the bolt to 80 inch lbs. (9 Nm) and the nut to 35 ft. lbs. (48 Nm).
- 4 center instrument panel support bolts. Tighten the bolts to 80 inch lbs. (9 Nm).
- Temperature control cable
- Shifter assembly harness connector, if equipped
- Shift interlock cable, if equipped
- ECS module connector
- GEM connectors
- Left main wiring harness connectors
- Steering column intermediate shaft. Tighten the pinch bolt to 35 ft. lbs. (47 Nm).
- Instrument panel reinforcement. Tighten the screws to 80 inch lbs. (9 Nm).
- Instrument panel steering column cover. Tighten the screws to 80 inch lbs. (9 Nm).
- Climate control wire harness connector
- Right main wire harness connectors
- Left and right door weatherstrips
- Left and right windshield side garnish moldings
- Left and right A-pillar lower trim panels
- Left and right scuff plates
- Upper right instrument panel support bolt. Tighten the bolt to 80 inch lbs. (9 Nm).
- Climate control vacuum harness connector
- Antenna cable connector
- Left and right air bag modules
- Left bulkhead harness connector
- Left front inner fender
- Left front wheel
- Center console
- CD player, if equipped
- Radio
- Negative battery cable

6. Fill the cooling system.
7. Recharge the A/C system.
8. Start the engine and check for leaks.

Mark VIII

REMOVAL & INSTALLATION

1. Before servicing the vehicle, refer to the precautions in the beginning of this section.
2. Drain the cooling system.
3. Recover the A/C refrigerant.
4. Remove or disconnect the following:
   - Negative battery cable
   - Transmission wiring harness at the transmission crossmember
   - Left and right bulkhead electrical connectors
   - Left air bag module
   - Steering wheel
   - Left instrument panel close-out panels
   - Left reinforcement brace
   - Left lower A/C duct
   - Steering column electrical connectors
   - Steering column lower yoke pinch bolt
   - Steering column
   - Instrument panel center finish panel
   - Floor console support bracket
   - Mobile phone and bracket, if equipped
   - Gear shift selector housing
   - Center console wiring harness connectors
   - Mobile phone coaxial cable, if equipped
   - Center console
   - Radio
   - Electronic Automatic Temperature Control (EATC) module
   - Heater outlet duct
   - Left and right door weatherstrips
   - Left and right cowl trim panels
   - Left and right roof side inner front moldings
   - Mobile phone speaker connector, if equipped
   - Instrument panel defroster grille
   - Windshield wiper control module and bracket
   - Left wiring harness connectors
   - Shift interlock cable
   - Right instrument panel close-out panel
   - Glove compartment
   - Right air bag module
   - EATC sensor hose
   - Climate control vacuum harness
   - Wiring harness in-line connector
   - Right cowl insulator panel
   - Right wiring harness connectors
   - Antenna cable
   - Transmission wiring harness grommet
   - Left instrument panel cowl side nuts
   - Instrument panel center support nuts
   - Instrument panel floor bracket
   - Right instrument panel cowl side nut
   - Instrument panel cowl top bolts
   - Instrument panel
   - A/C liquid line
   - A/C accumulator/drier and bracket

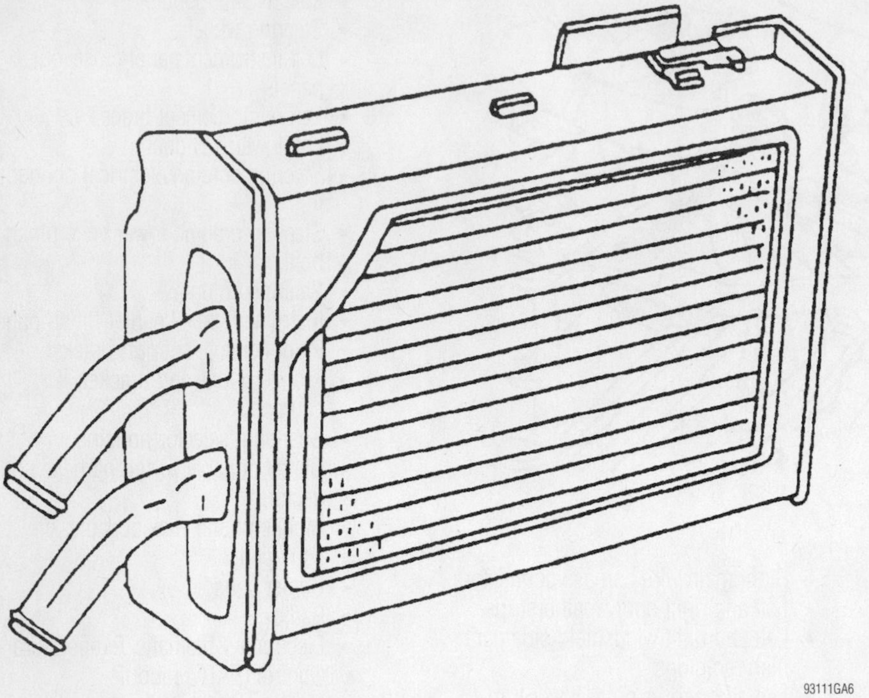

View of the heater core—Lincoln Mark VIII

- Vacuum supply hose
- Heater hoses
- Heater outlet floor ducts
- Evaporator housing from the vehicle
- A/C blend door actuator
- 13 evaporator case screws
- Heater core cover screws
- Heater core

To install:

5. Install or connect the following:
   - Heater core
   - Heater core cover screws. Tighten the screws to 18–26 inch lbs. (2–3 Nm).
   - 13 evaporator case screws. Tighten the screws to 18–26 inch lbs. (2–3 Nm).
   - A/C blend door actuator
   - Evaporator housing to the vehicle. Tighten the fasteners to 41–54 inch lbs. (5–6 Nm).
   - Heater outlet floor ducts
   - Heater hoses
   - Vacuum supply hose
   - A/C accumulator/drier and bracket
   - A/C liquid line
   - Instrument panel
   - Instrument panel cowl top bolts. Tighten the bolts to 54–70 inch lbs. (6–8 Nm).
   - Right instrument panel cowl side nut. Tighten the nut to 40–52 ft. lbs. (54–72 Nm).
   - Instrument panel floor bracket. Tighten the bolts to 20–24 ft. lbs. (26–34 Nm).
   - Instrument panel center support nuts. Tighten the nuts to 96–120 inch lbs. (10–14 Nm).
   - Left instrument panel cowl side nuts. Tighten the nuts to 40–52 ft. lbs. (54–72 Nm).
   - Transmission wiring harness grommet
   - Antenna cable
   - Right wiring harness connectors
   - Right cowl insulator panel
   - Wiring harness in-line connector
   - Climate control vacuum harness
   - EATC sensor hose
   - Right air bag module
   - Glove compartment
   - Right instrument panel close-out panel
   - Shift interlock cable
   - Left wiring harness connectors
   - Windshield wiper control module and bracket
   - Instrument panel defroster grille
   - Mobile phone speaker connector, if equipped
   - Left and right roof side inner front moldings
   - Left and right cowl trim panels
   - Left and right door weather-strips
   - Heater outlet duct
   - Electronic Automatic Temperature Control (EATC) module
   - Radio
   - Center console
   - Mobile phone coaxial cable, if equipped
   - Center console wiring harness connectors
   - Gear shift selector housing
   - Mobile phone and bracket, if equipped
   - Floor console support bracket. Tighten the bolts to 27 inch lbs. (3 Nm).
   - Instrument panel center finish panel
   - Steering column. Tighten the nuts to 24–34 ft. lbs. (34–46 Nm).
   - Steering column lower yoke pinch bolt. Tighten the bolt to 22–30 ft. lbs. (30–40 Nm).
   - Steering column electrical connectors
   - Left lower A/C duct
   - Left reinforcement brace
   - Left instrument panel close-out panels
   - Steering wheel. Tighten the bolt to 25–34 ft. lbs. (34–46 Nm).
   - Left air bag module
   - Left and right bulkhead electrical connectors
   - Transmission wiring harness at the transmission crossmember
   - Negative battery cable
6. Fill the cooling system.
7. Recharge the A/C system.
8. Start the engine and check for leaks.

Ford Crown Victoria, Mercury Grand Marquis and Lincoln Town Car

1. Before servicing the vehicle, refer to the precautions in the beginning of this section.

2. If equipped, turn the air suspension service switch to the **OFF** position before raising the vehicle.

3. Drain the cooling system.

4. Remove or disconnect the following:
   - Negative battery cable
   - Front seats
   - Carpeting
   - Rear airflow duct
   - Heater hoses

5. Remove the driver's side air bag module by removing or disconnect the following:
   - SRS module-to-steering wheel bolts, from both sides of the steering wheel.
   - SRS module and disconnect the electrical connector
   - Horn switch electrical connector

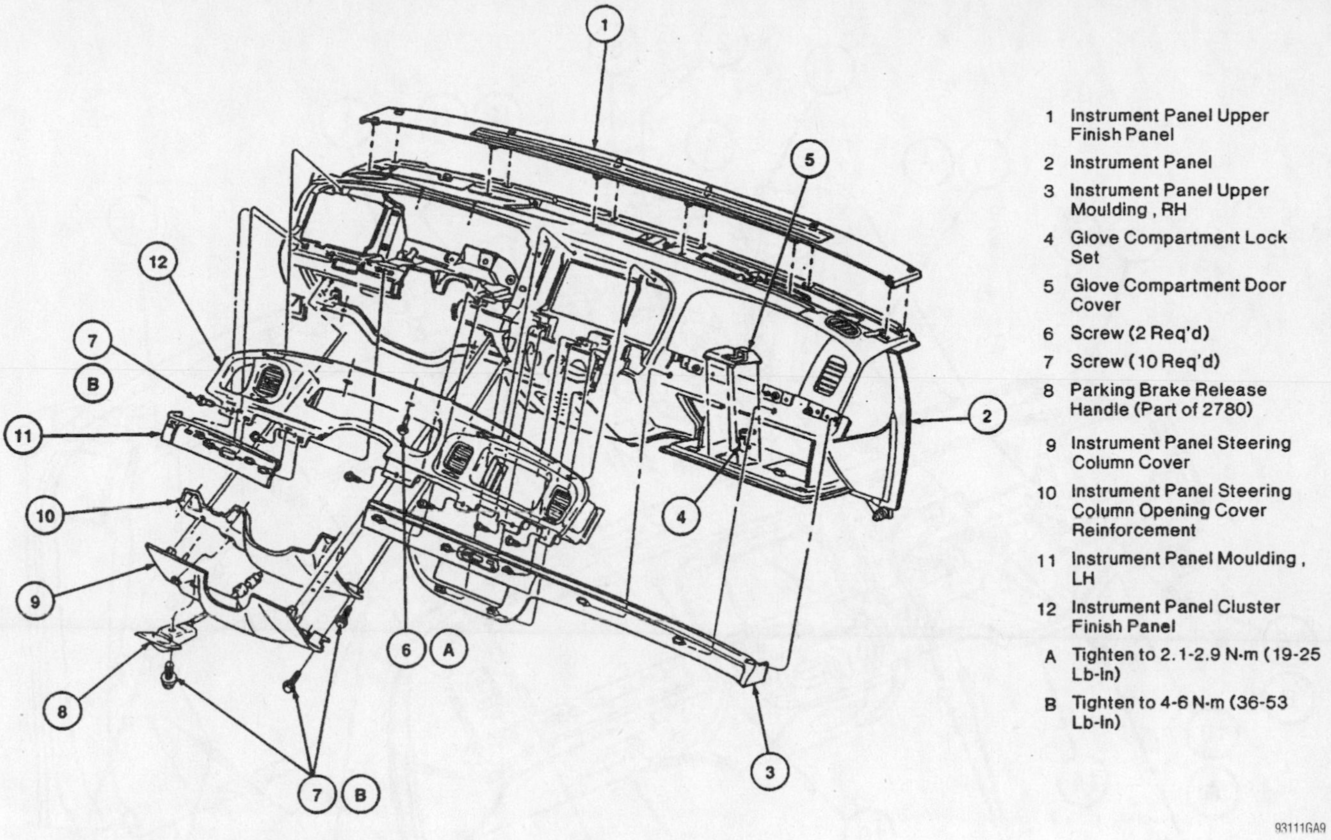

View of the instrument panel finish panels—Crown Victoria, Grand Marquis and Town Car

6. Remove the passenger's side SRS module by removing or disconnecting the following:
   - Open the glove box and disconnect the glove compartment isolator
   - Push inward on the 2 glove box door tabs and lower it
   - SRS module's electrical connector
   - SRS module-to-instrument panel bolts and the module

7. Remove the instrument panel by removing or disconnecting the following:
   - Speed control servo nuts and move the servo aside
   - Bolt and disconnect the left side bulkhead connector
   - Left side bulkhead connector from the dash panel
   - Windshield washer fluid reservoir screw, and position the reservoir aside
   - Blower motor electrical connector
   - Air conditioning pressure cut-off switch electrical connector
   - In-line electrical harness connector
   - Electronic Automatic Temperature Control (EATC) variable blower motor controller electrical connector
   - Right front wheel
   - Right front fender splash shield bolts and move the shield away from the cowl
   - Wiring harness from the evaporator case
   - Wiring harness from the cowl
   - Right side instrument panel lower insulator pushpins, disconnect the power point electrical connector, remove the courtesy lamp from the socket and remove the insulator
   - Unseat the wiring grommet and feed the wiring harness through the cowl
   - Cowl side trim panels and the windshield side garnish moldings from both sides
   - Locking clip and the Electronic Crash Sensor (ECS) module electrical connector
   - Antenna connector
   - Ground wire bolts from the right side
   - EATC hose at the evaporator housing
   - Bolt and disconnect the bulkhead connector from the right side
   - Pull the door weatherstrip seals away from the instrument panel on both sides
   - Tunnel brace trim cover
   - Climate control head vacuum harness connector
   - Instrument panel tunnel brace nuts and the brace
   - Ground wire bolts from the left side
   - Wiring harness connectors from the left side
   - Parking brake switch electrical connector

   a. Pry out the instrument panel defroster opening grille, disconnect the electrical connectors and the remove the grille.

8. Remove or disconnect the following:
   - 3 upper instrument panel screws
   - Instrument panel cowl side nut from both sides
   - Instrument panel cowl side bolt. From the left side
   - Instrument panel
   - Windshield wiper mounting arm and the pivot shaft

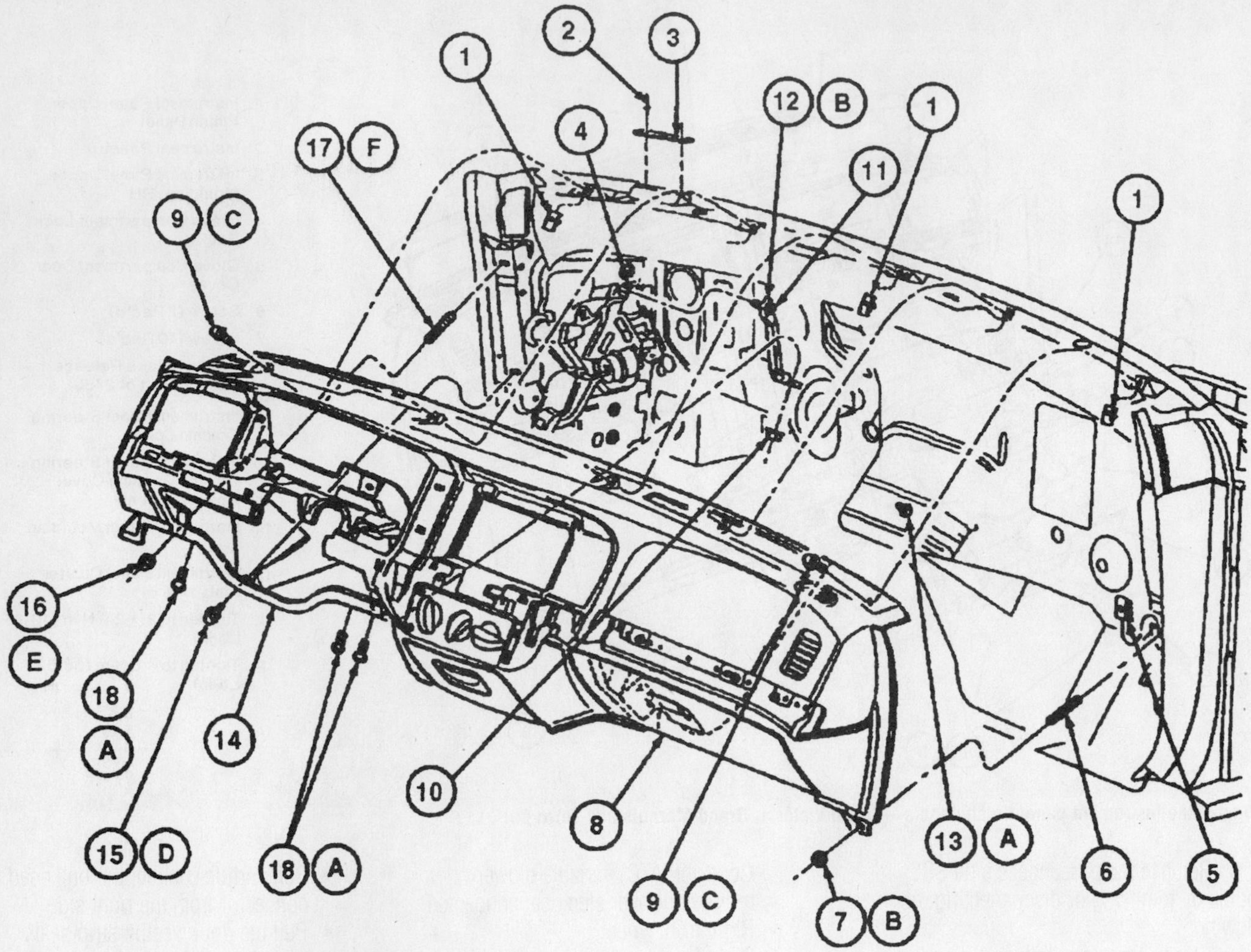

1 Nut (3 Req'd)
2 Rivet (2 Req'd)
3 Vehicle Identification Plate
4 Brake Pedal Support
5 J Nut
6 Stud
7 Nut
8 Instrument Panel
9 Screw (3 Req'd)
10 Steering Column Retaining Nut
11 Bolt (Part of 3F659)
12 Brake Pedal Support Steering Column Brace
13 Bolt (2 Req'd)
14 Instrument Panel Steering Column Opening Cover Reinforcement
15 Bolt (2 Req'd)
16 Nut
17 Stud
18 Bolt (3 Req'd)
A Tighten to 9-14 N·m (80-123 Lb-In)
B Tighten to 10-14 N·m (89-123 Lb-In)
C Tighten to 2-3 N·m (18-26 Lb-In)
D Tighten to 47-63 N·m (35-46 Lb-Ft)
E Tighten to 45-70 N·m (34-51 Lb-Ft)
F Tighten to 22-34 N·m (17-25 Lb-Ft)

93111GA0

Exploded view of the instrument panel—Crown Victoria, Grand Marquis and Town Car

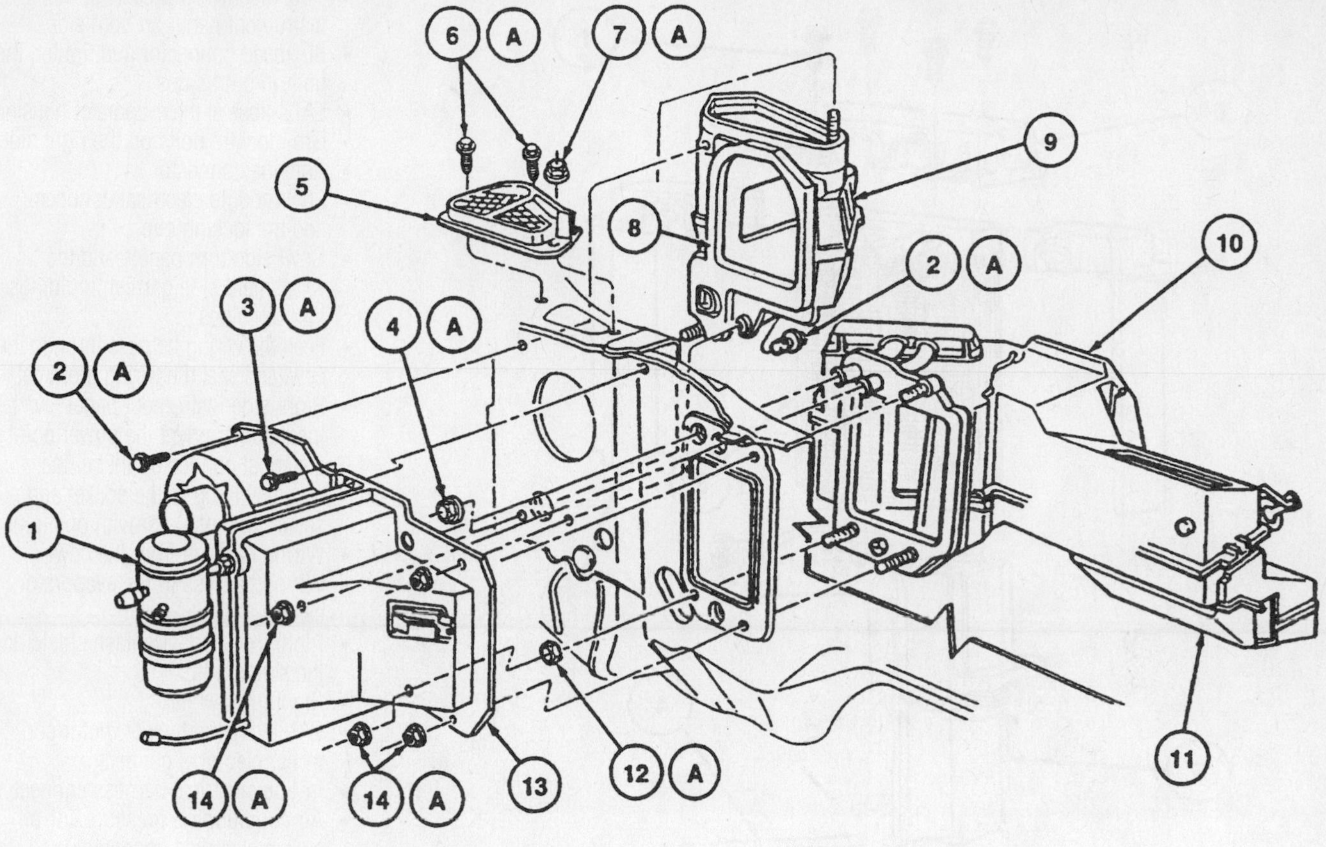

1	Suction Accumulator / Drier	9	A / C Recirculating Air Duct
2	Screw (2 Req'd)	10	Heater Air Plenum Chamber
3	Screw	11	Heater Outlet Floor Duct
4	Nut and Washer Assy	12	Nut
5	A / C Recirculating Air Duct Screen	13	A / C Evaporator Core Housing
6	Screw (2 Req'd)	14	Nut and Washer
7	Nut and Washer	A	Tighten to 2.5-3.2 N·m (23-28 Lb-In)
8	A / C Air Inlet Door Inner Seal		

Exploded view of the heater air plenum chamber, the evaporator housing and the air inlet duct—Crown Victoria, Grand Marquis

- Electrical connector and remove the upper cowl extension screw and the extension
- Vacuum hoses and the wire harness connector from the evaporative emissions canister purge valve
- Evaporative emissions purge valve nuts and the valve
- Cowl side nut
- Right side rear fender apron screws and reposition the apron
- Evaporator housing nut and screw
- Vacuum hoses and the wiring harness connectors
- Plenum chamber's lower flange nuts
- Plenum chamber's upper flange nut and the plenum chamber
- Heater core cover-to-heater plenum chamber screws and the cover
- Seal and the heater core

To install:

9. Install or connect the following:
   - Seal and the heater core
   - Heater core cover and the cover-to-heater plenum chamber screws
   - Plenum chamber and the plenum chamber's upper flange nut
   - Plenum chamber's lower flange nuts
   - Vacuum hoses and the wiring harness connectors
   - Evaporator housing nut and screw
   - Right side rear fender apron and the apron screws
   - Cowl side nut
   - Evaporative emissions purge valve and the valve nuts

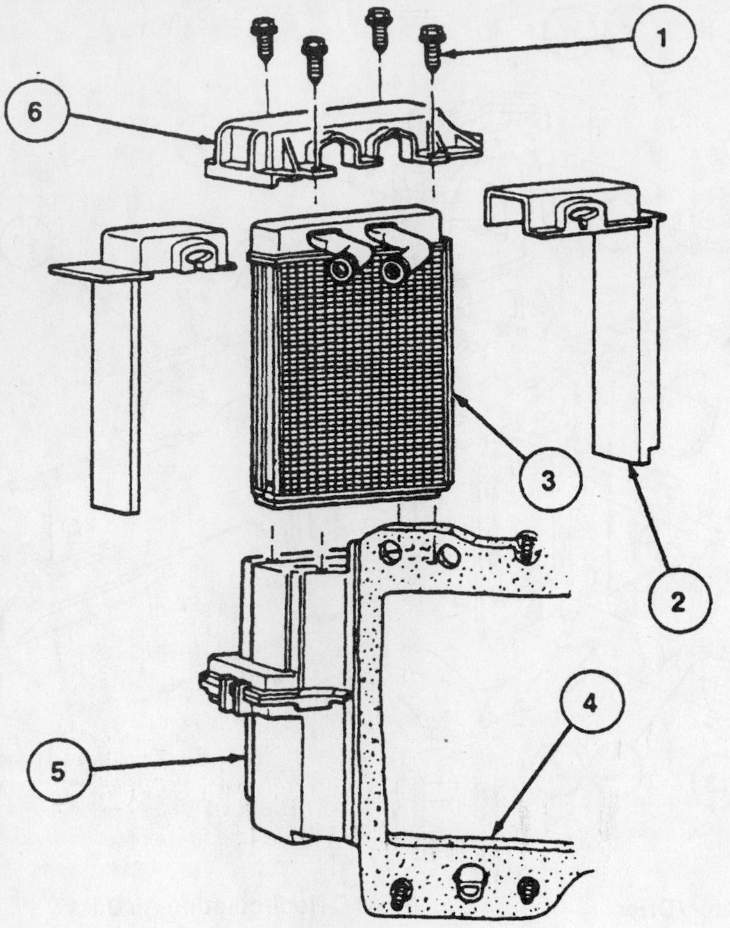

1 Screw (4 Req'd)
2 Heater Core Case Seal
3 Heater Core
4 Heater Dash Gasket
5 Heater Air Plenum Chamber
6 Heater Core Cover

93111GB2

View of the heater core—Crown Victoria, Grand Marquis and Town Car

- Vacuum hoses and the wire harness connector to the evaporative emissions canister purge valve
- Electrical connector; then, install the upper cowl extension screw and the extension
- Windshield wiper mounting arm and the pivot shaft

10. Install the instrument panel by install or connect the following:

- Instrument panel
- At the left side, install the instrument panel cowl side bolt
- Instrument panel cowl side nut on both sides
- 3 upper instrument panel screws
- Electrical connectors and the install the instrument panel defroster opening grille
- Parking brake switch electrical connector
- Wiring harness connectors on the left side
- Ground wire bolts on the left side
- Instrument panel tunnel brace and the brace nuts
- Climate control head vacuum harness connector
- Install the tunnel brace trim cover.
- Door weatherstrip seals to the instrument panel on both sides
- Bulkhead connector and tighten the bolt on both sides
- EATC hose at the evaporator housing
- Ground wire bolts on the right side
- Antenna connector
- ECS module electrical connector and the locking clip
- Cowl side trim panels and the windshield side garnish moldings on both sides
- Feed the wiring harness through the cowl and seat the wiring grommet
- Right side instrument panel lower insulator, connect the power point electrical connector, install the courtesy lamp to the socket and secure the insulator with pushpins
- Wiring harness from the cowl
- Wiring harness to the evaporator case
- Right front fender splash shield and the shield bolts
- Right front wheel
- EATC variable blower motor controller electrical connector
- In-line electrical harness connector
- Air conditioning pressure cut-off switch electrical connector
- Blower motor electrical connector
- Windshield washer fluid reservoir and the reservoir screw
- Left side bulkhead connector to the dash panel
- Left side bulkhead connector and tighten the bolt
- Speed control servo and the servo nuts
- Right SRS module and torque the module-to-instrument panel bolts to 62–97 inch lbs. (7–11 Nm)
- Right SRS module electrical connector
- Glove compartment isolator
- Glove box door
- Horn switch electrical connector
- Electrical connector and install the left SRS module
- Left SRS module-to-steering wheel bolts on both sides of the wheel, and torque the bolts to 90–122 inch lbs. (10–14 Nm)
- Rear airflow duct
- Carpeting
- Front seats
- Heater hoses

11. Refill the cooling system.
12. Connect the negative battery cable.
13. Operate the engine to normal operating temperature. Check the climate control operation and check for leaks.

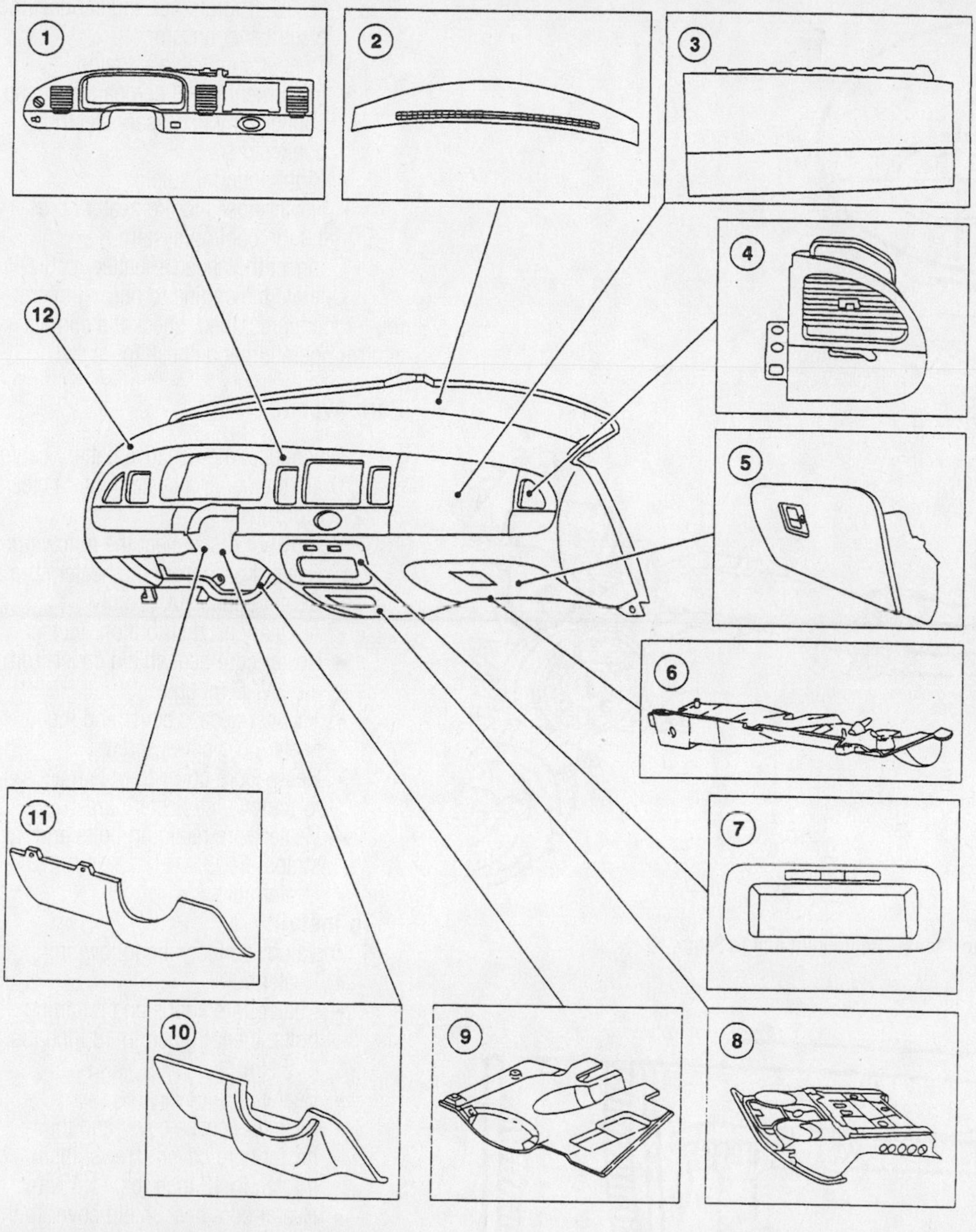

1 Instrument Panel Finish Panel
2 Instrument Panel Defroster Opening Grille
3 Passenger Side Air Bag Module
4 Instrument Panel Finish Panel
5 Glove Compartment
6 Instrument Panel Lower Insulator (RH)
7 Instrument Panel Finish Panel
8 Instrument Panel Ash Receptacle
9 Instrument Panel Lower Insulator (LH)
10 Instrument Panel Steering Column Cover
11 Instrument Panel Steering Column Opening Cover Reinforcement
12 Instrument Panel

Exploded view of the instrument panel—Town Car

1998–01 GENERAL MOTORS

Bonneville, Eighty-Eight, LeSabre, LSS, Park Avenue and Regency

REMOVAL & INSTALLATION

Bonneville and LeSabre

1. Disconnect the negative battery cable.
2. Drain the cooling system into a clean container for reuse.
3. Remove or disconnect the following:
   - Heater hoses from the heater core
   - Right sound insulator
   - All necessary electrical connectors and remove the instrument panel compartment
   - Temperature valve actuator
   - Electrical connector and remove the HVAC programmer
   - Heater core cover-to-heater assembly screws and the cover
   - Heater core-to-heater assembly screws and the heater core

To install:

4. Install or connect the following:
   - Heater core and the heater core-to-heater assembly screws, then, tighten the screws to 12 inch lbs. (1.4 Nm)
   - Heater core cover and the cover-to-heater assembly screws, then, tighten the screws to 12 inch lbs. (1.4 Nm)

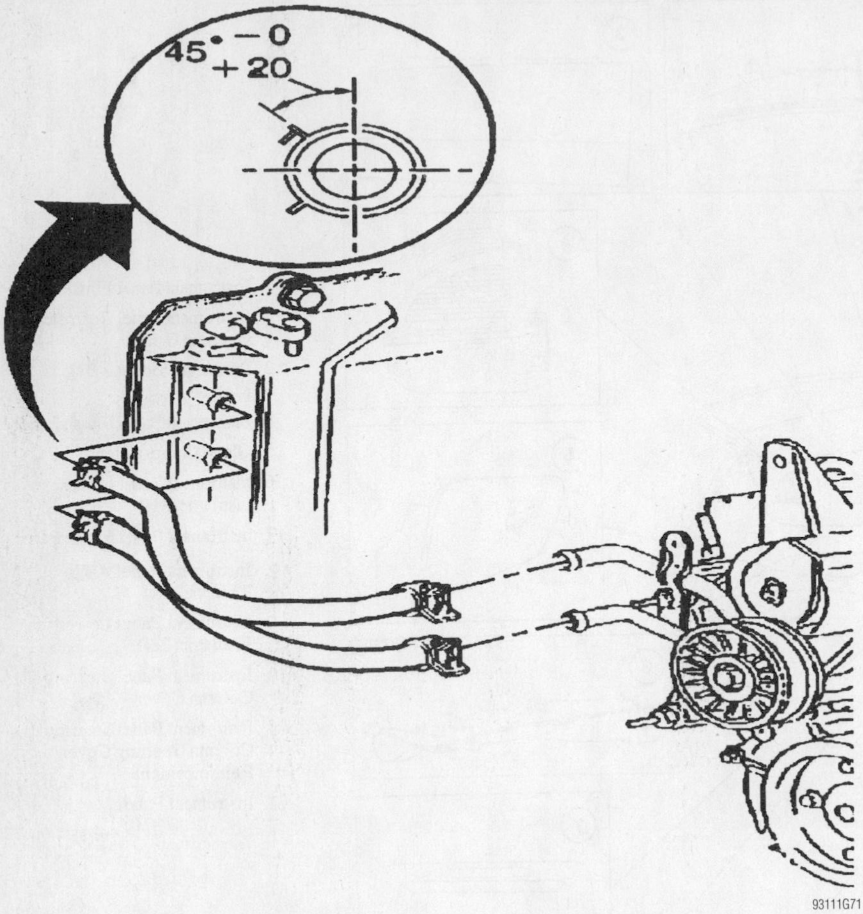

93111G71

View of the heater hoses and clamp positions—Bonneville and LeSabre

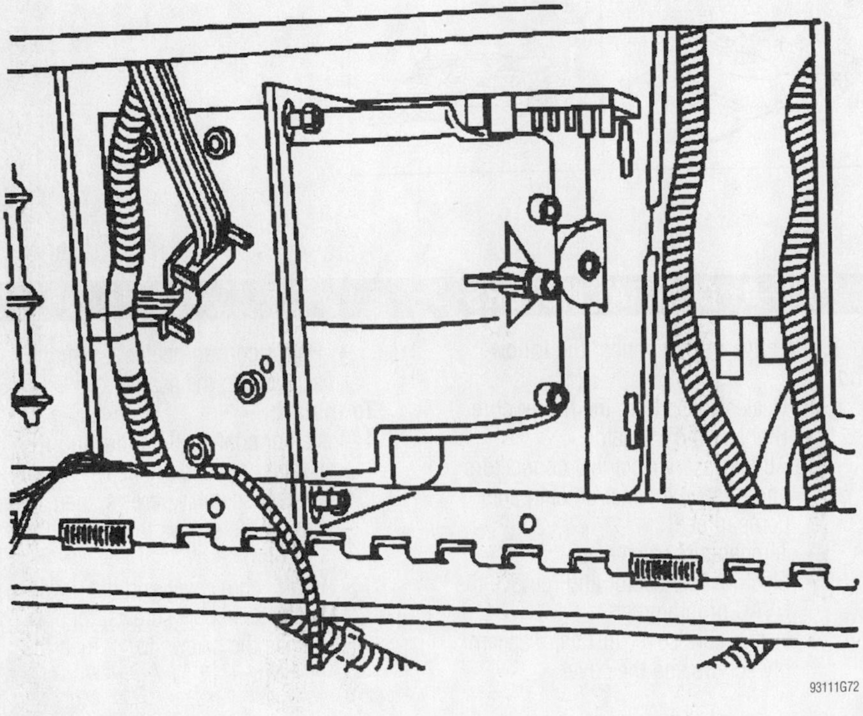

93111G72

View of the heater core cover—Bonneville and LeSabre

- HVAC programmer and connect the electrical connector
- Temperature valve actuator
- Instrument panel compartment and connect any necessary electrical connectors
- Right sound insulator
- Heater hoses to the heater core

5. Refill the cooling system.
6. Connect the negative battery cable.
7. Operate the engine to normal operating temperatures; then, check the climate control operation and check for leaks.

Park Avenue

1. Disconnect the negative battery cable.
2. Drain the cooling system into a clean container for reuse.
3. Remove or disconnect the following:
   - Heater hoses from the heater core
   - Center console assembly
   - Auxiliary air distribution duct
   - Heater core heat shield cover from the HVAC module
   - Air filter access cover and the heater core cover screws
   - Heater core cover to obtain access to remove the heater core
   - Heater core retaining bolts and straps
   - Heater core

To install:

4. Install or connect the following:
   - Heater core
   - Heater core straps and retaining bolts, then, tighten to 18 inch lbs. (1.5 Nm)
   - Heater core cover
   - Air filter access cover and the heater core cover screws, then tighten to 12 inch lbs. (1.4 Nm)
   - Heater core heat shield cover to the HVAC module

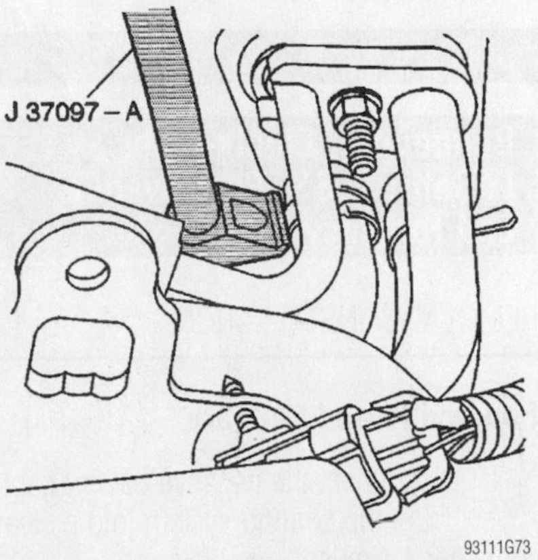

93111G73

Using Hose Clamp Pliers tool J-37097-A remove the heater hoses clamp—Park Avenue

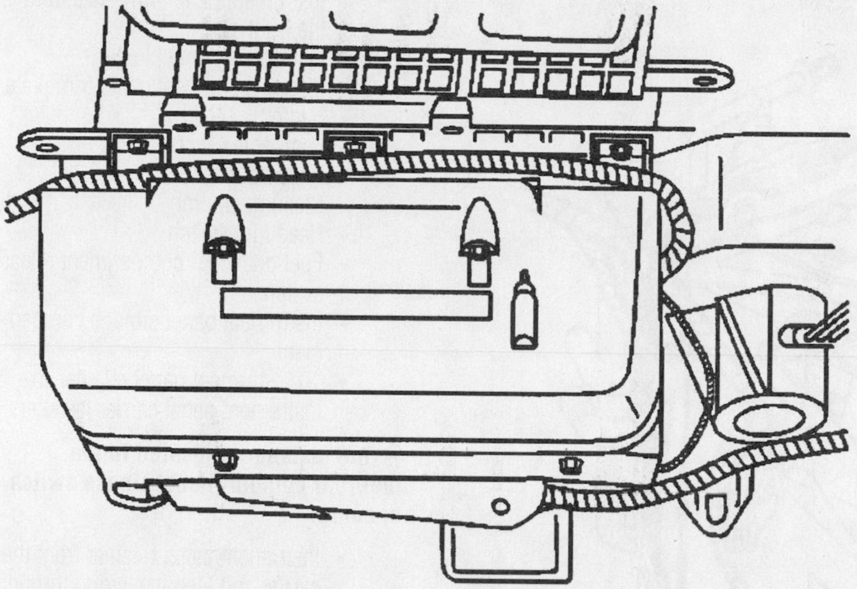

View of the heater core cover—Park Avenue

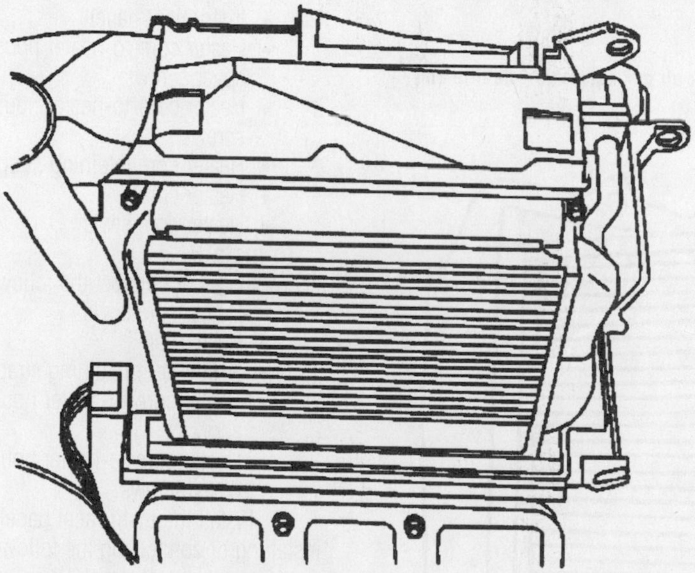

View of the heater core—Park Avenue

- Auxiliary air distribution duct
- Center console assembly
- Heater hoses to the heater core

5. Refill the cooling system.
6. Connect the negative battery cable.
7. Operate the engine to normal operating temperatures; then, check the climate control operation and check for leaks.

LSS and Regency

1. Before servicing the vehicle, refer to the precautions in the beginning of this section.
2. Drain the cooling system.
3. Remove or disconnect the following:
   - Fuel injector sight shield
   - Heater hoses from the heater core
   - Left and right sound insulators
   - Air distributor duct
   - Heater core heat shield
   - Heater core cover
   - Heater core

To install:

4. Install or connect the following:
   - Heater core. Torque the screw to 12 inch lbs. (1.5 Nm).
   - Heater core cover. Torque the screws to 12 inch lbs. (1.5 Nm).
   - Heater core heat shield. Torque the 12 inch lbs. (1.5 Nm).
   - Air distributor duct. Torque the screws to 12 inch lbs. (1.5 Nm).
   - Left and right sound insulators. Torque the fasteners to 18 inch lbs. (2 Nm).
   - Heater hoses
   - Fuel injector sight shield
5. Refill the cooling system.
6. Start the engine and check for leaks.

Concours, deVille, Eldorado and Seville

REMOVAL & INSTALLATION

1. Disconnect the negative battery cable.
2. Drain the cooling system into a clean container for reuse.
3. Remove or disconnect the following:
   - Instrument panel compartment
   - Sound insulator from the right side
   - Electrical connector and remove the heater/air conditioning programmer
   - Driver's air mix actuator from the bottom of the heater assembly
   - Passenger's air mix actuator from the bottom of the heater assembly
   - Heater core cover-to-heater assembly screws and the cover
   - Heater hoses from the heater core, located in the engine compartment
   - Heater core-to-heater assembly screws
   - Heater core

To install:

4. Install or connect the following:
   - Heater core
   - Heater core-to-heater assembly screws, then, tighten to 18 inch lbs. (2 Nm)
   - Heater hoses to the heater core
   - Heater core cover and the cover-to-heater assembly screws, then, tighten to 18 inch lbs. (2 Nm)
   - Passenger's air mix actuator at the top of the heater assembly
   - Driver's air mix actuator at the bottom of the heater assembly
   - Heater/air conditioning programmer and connect the electrical connector
   - Sound insulator on the right side
   - Instrument panel compartment
5. Refill the cooling system.
6. Connect the negative battery cable.

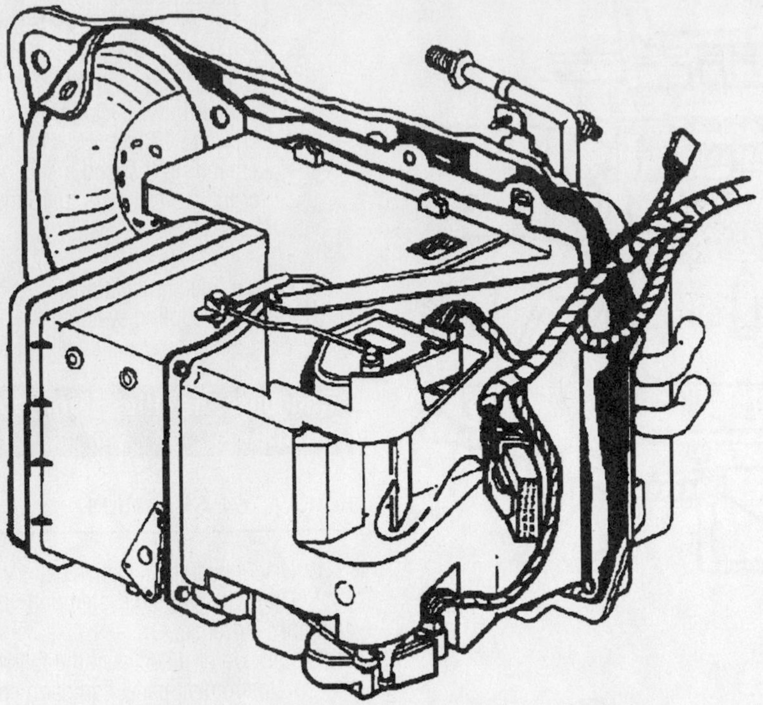

View of the heater assembly with the driver's and passenger's air mix actuators—deVille and Eldorado

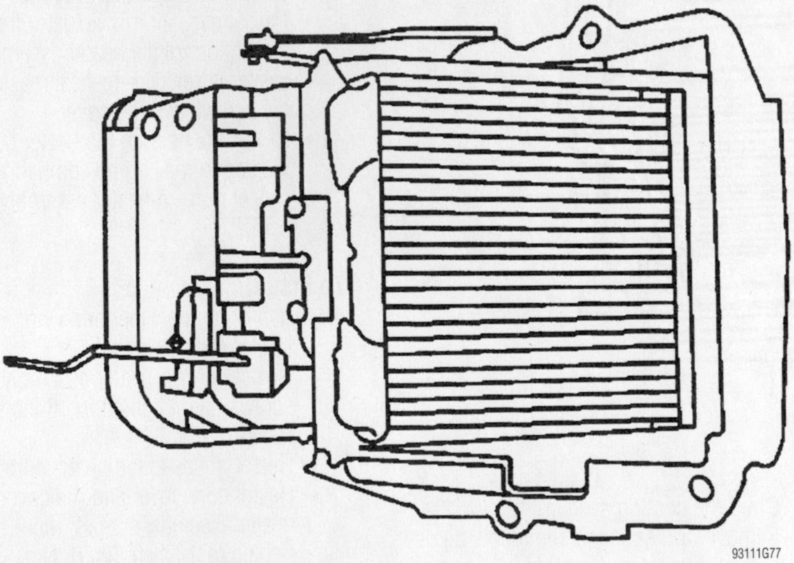

View of the heater core—deVille and Eldorado

Seville

REMOVAL & INSTALLATION

1. Disconnect the negative battery cable.
2. Drain the cooling system into a clean container for reuse.
3. Disconnect the heater hoses from the heater core.
4. Disable the SIR system.
5. Remove the instrument panel by removing or disconnecting the following:
   - Console
   - Right insulator panel
   - Instrument panel end caps from both sides
   - Side windows air outlets
   - Upper instrument panel trim pad
   - Passenger's side SIR module
   - Instrument panel trim plate from the right side
   - Instrument panel cluster
   - Center instrument panel trim plate
   - Radio
   - Lap cooler duct
   - Knee bolster.
   - Steering column
   - Headlight switch.
   - Fuel door/rear compartment release switch.
   - Instrument panel storage compartment
   - 10 instrument panel retainer-to-instrument panel carrier fasteners

➡One fastener is located in the fuel/rear compartment release switch opening.

   - Instrument panel retainer from the carrier and feed the wiring through the openings, as necessary
   - Inside air temperature sensor electrical connector
   - Instrument panel
   - Heater core-to-heater housing cover screws
   - Heater core-to-heater housing screws
   - Heater core retaining straps
   - Heater core
   - Heater core seals

To install:

6. Install or connect the following:
   - Heater core seals
   - Heater core.
   - Heater core retaining straps
   - Heater core-to-heater housing screws.
   - Heater core-to-heater housing cover screws.
7. Install the instrument panel by installing or connecting the following:
   - Instrument panel
   - Inside air temperature sensor electrical connector
   - Instrument panel retainer to the carrier and feed the wiring through the openings, as necessary
   - 10 instrument panel retainer-to-instrument panel carrier fasteners

➡One fastener is located in the fuel/rear compartment release switch opening.

   - Instrument panel storage compartment
   - Fuel door/rear compartment release switch
   - Headlight switch
8. Install or connect the following:
   - Steering column

- Knee bolster
- Lap cooler duct
- Radio
- Center instrument panel trim plate
- Instrument panel cluster
- Install the instrument panel trim plate on the right side
- Passenger's side SIR module
- Upper instrument panel trim pad
- Side windows air outlets
- Instrument panel end caps on both sides
- Right insulator panel
- Console

9. Enable the SIR system.

10. Connect the heater hoses to the heater core.

11. Refill the cooling system.

12. Connect the negative battery cable.

13. Operate the engine to normal operating temperatures; then, check the climate control operation and check for leaks.

Camaro, Z28, Firebird and Trans Am

**** CAUTION**

Some vehicles are equipped with an SIR or air bag system. The air bag system must be disabled before performing service on or around the air bag, instrument panel components, wiring and sensors. Failure to follow safety and disabling procedures could result in accidental air bag deployment, possible personal injury and unnecessary air bag system repairs.

REMOVAL & INSTALLATION

1. With the ignition key removed, disconnect the negative battery cable.

2. Disable the SIR system by removing or disconnecting the following:

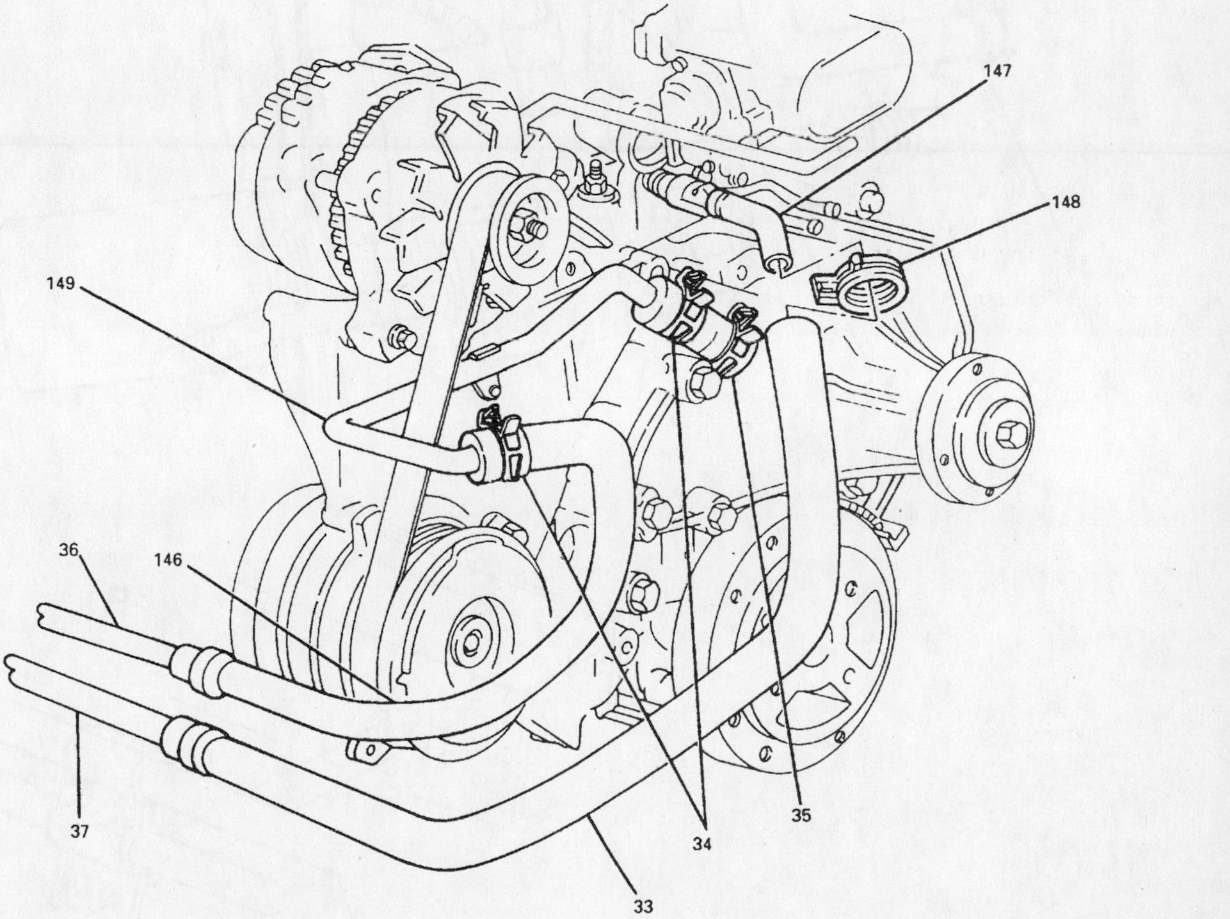

33 HOSE ASSEMBLY, HEATER INLET
34 CLAMP, HEATER OUTLET FRONT HOSE
35 CLAMP, HEATER INLET FRONT HOSE
36 PIPE, HEATER HOSE OUTLET
37 PIPE, HEATER HOSE INLET
146 HOSE ASSEMBLY, HEATER OUTLET
147 HOSE, THROTTLE BODY HEATER RETURN
148 CLAMP, THROTTLE BODY HEATER RETURN HOSE
149 PIPE, HEATER OUTLET

88146G08

Heater hose and pipe routing—Camaro/Firebird with the 3.4L engine

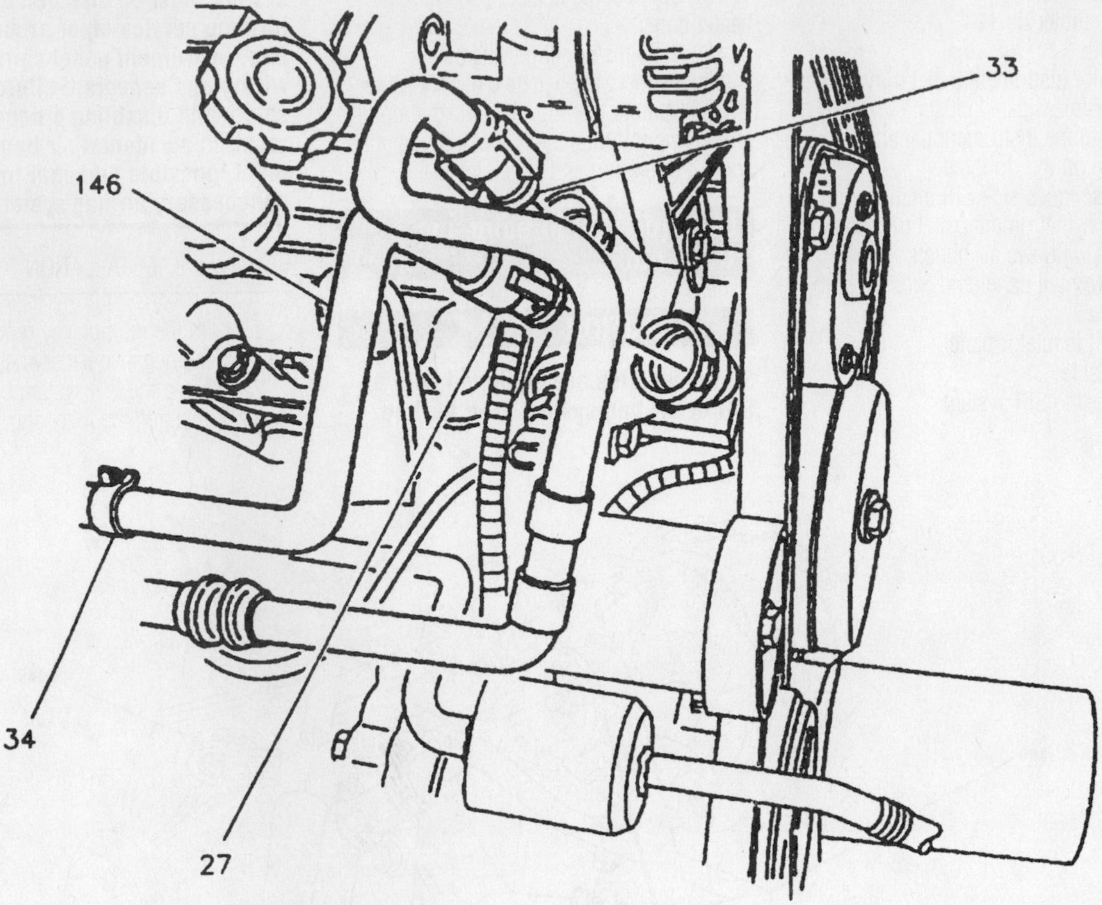

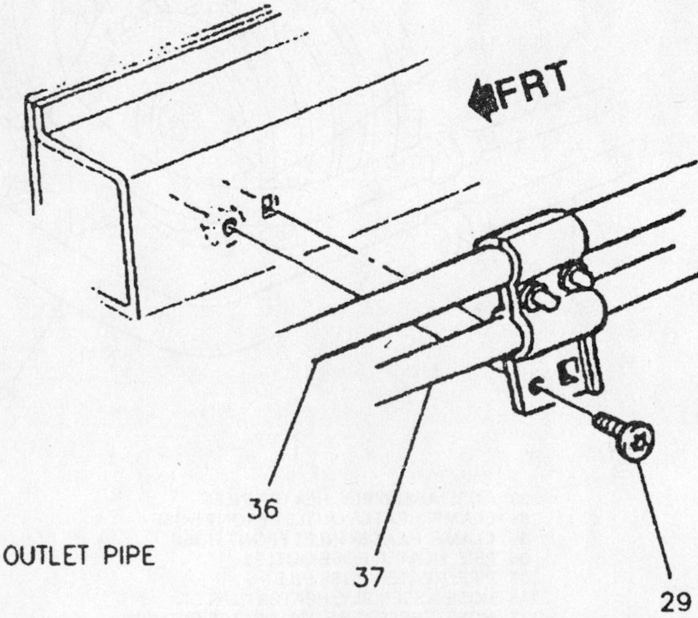

27 FITTING, QUICK CONNECT SNAP
29 BOLT/SCREW, HEATER INLET AND OUTLET PIPE
33 HOSE, HEATER INLET
34 CLAMP, HEATER OUTLET
36 PIPE, HEATER HOSE OUTLET
37 PIPE, HEATER HOSE INLET
146 HOSE, HEATER OUTLET

88146G09

Heater hose and pipe routing—Camaro/Firebird with the 3.8L engine

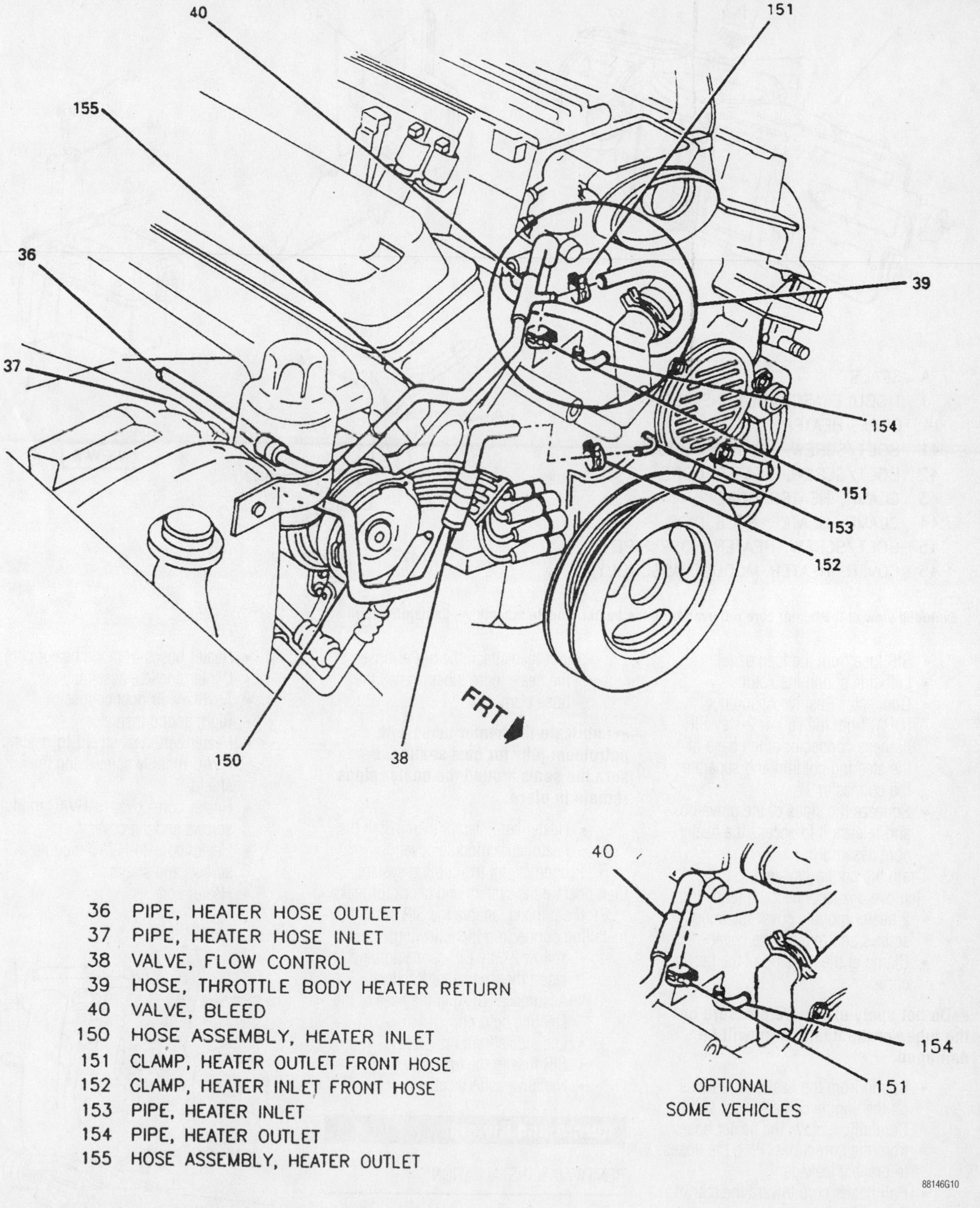

Heater hose and pipe routing—Camaro/Firebird with the 5.7L engine

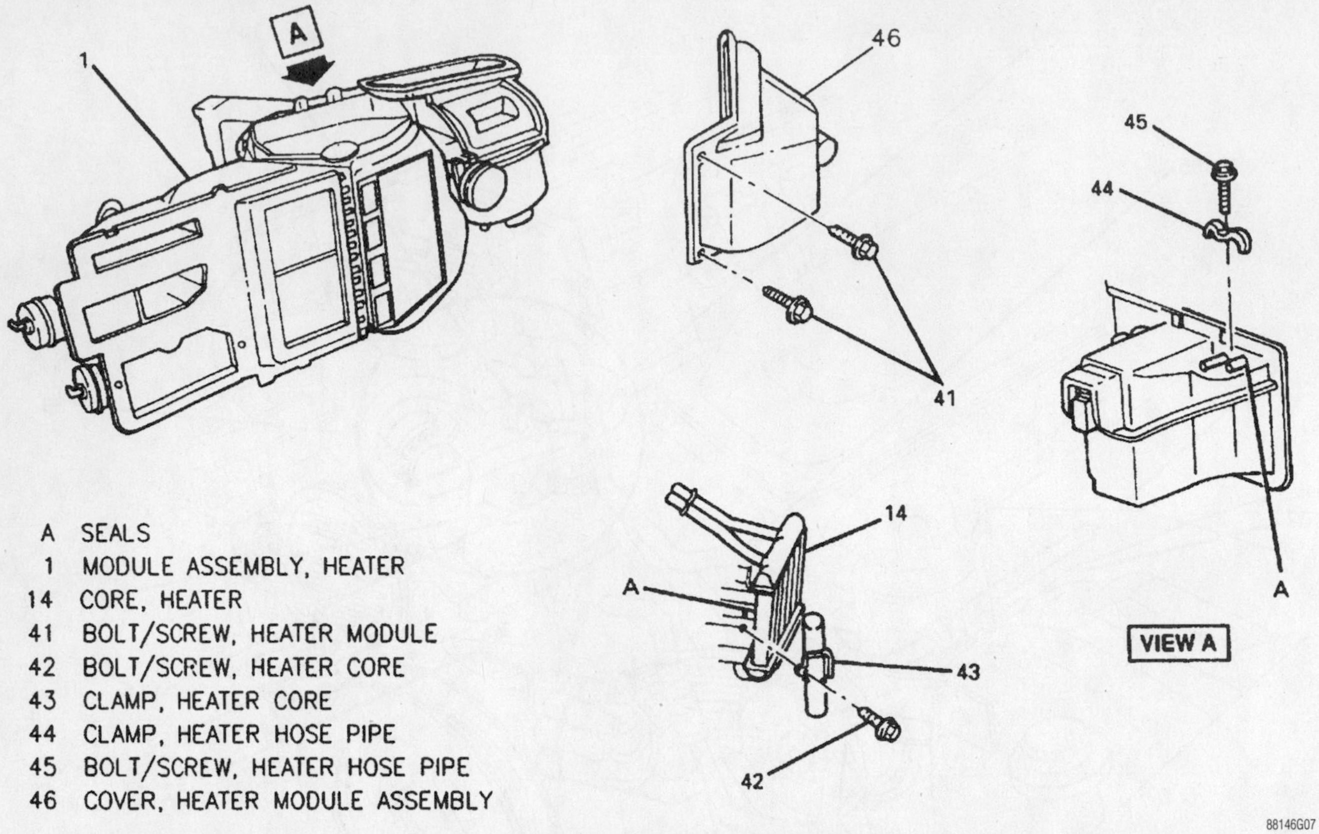

88146G07

Exploded view of the heater core removed from the heater module assembly—Camaro/Firebird

- SIR fuse from the fuse panel
- Left side sound insulator
- Connector Positive Assurance (CPA) from the yellow 2-way SIR harness connector at the base of the steering column and separate the connector
- Squeeze the sides of the glove box and release it to access the heater core assembly

3. Drain the cooling system.
4. Remove or disconnect the following:
   - 2 heater module cover retaining screws and remove the cover
   - Clamp at the left side of the heater core

➡Do not apply excessive pressure on the tubes or the heater core will be damaged.

- Clamp from the heater core tubes on the engine compartment side. Carefully, remove the heater hoses from the core tubes. Plug the hoses to prevent leakage
- Pull heater core toward the rear of the vehicle to remove it

To install:

5. Install or connect the following:
   - Position the heater core into place and then attach the heater hoses to the heater core tubes. Install the hose clamp

➡Lubricate the heater tubes with petroleum jelly for best sealing. Be sure the seals around the heater pipes remain in place.

- Heater core clamp, then install the heater core module cover

6. Properly, refill the cooling system, then operate the system and check for leaks.
7. If equipped, enable the SIR system by installing connecting the following:
   - Yellow 2-way SIR connector and insert the Connector Positive Assurance (CPA) at the base of the steering column
   - Left side sound insulator
   - SIR fuse in the fuse panel
   - Negative battery cable

Aurora and Riviera

REMOVAL & INSTALLATION

1. Disconnect the negative battery cable.
2. Drain the cooling system into a clean container for reuse.
3. Remove or disconnect the following:
   - Heater hoses from the heater core
   - Center console assembly
   - Auxiliary air duct connector
   - Right sound insulator
   - Heater core heat shield-to-heating, HVAC module screws and the shield
   - Heater core cover-to-HVAC module screws and the cover
   - Heater core-to-HVAC module screws and straps
   - Heater core

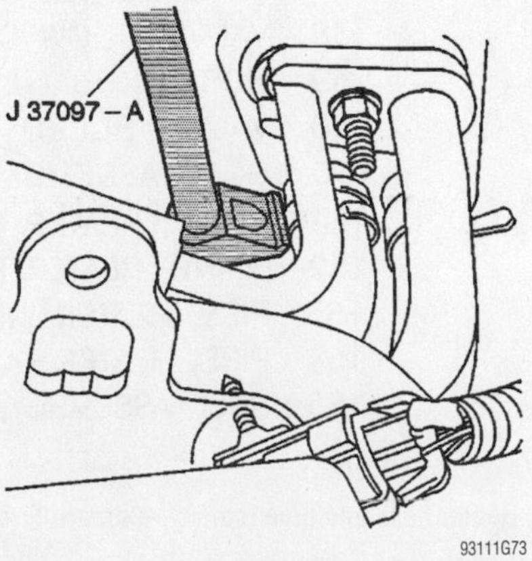

93111G73

Remove the heater hoses clamp—Aurora and Riviera

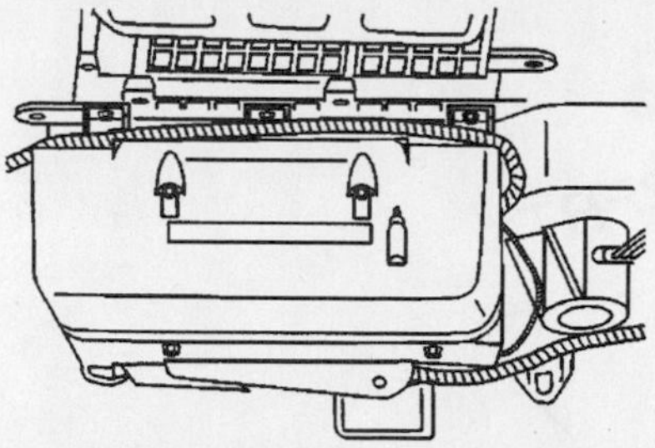

93111G74

View of the heater core cover—Aurora and Riviera

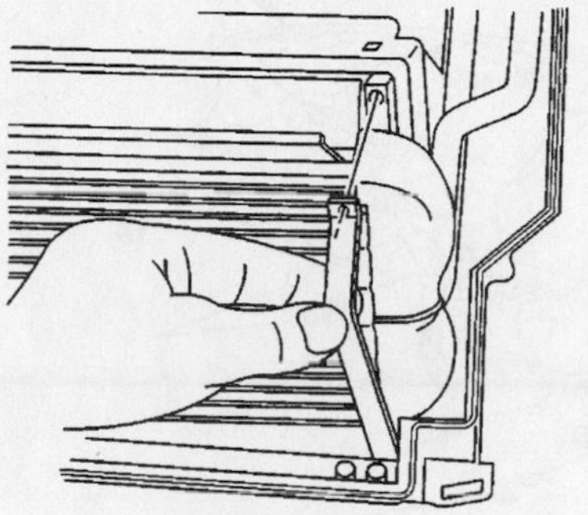

93111G78

View of the heater core retaining straps—Aurora and Riviera

To install:

4. Install or connect the following:
   - Heater core
   - Heater core-to-HVAC module screws and straps, then, tighten to 18 inch lbs. (1.5 Nm)
   - Heater core cover and the cover-to-HVAC module screws, then, tighten to 18 inch lbs. (1.5 Nm)
   - Heater core heat shield and the shield-to-heating, HVAC module screws, then, tighten to 18 inch lbs. (1.5 Nm)
   - Right sound insulator
   - Auxiliary air duct connector
   - Center console assembly
   - Heater hoses to the heater core
5. Refill the cooling system.
6. Connect the negative battery cable.

Cavalier and Sunfire

REMOVAL & INSTALLATION

1. Disable the SIR system by performing the following procedure:
   a. Point the wheel in the straight-ahead position.
   b. Turn the ignition switch to the LOCK position.
   c. Remove the AIR BAG fuse from the fuse block.
   d. At the base of the steering column, remove the left sound insulator.
   e. At the base of the steering column, disconnect the Connector Position Assurance (CPA), the yellow 2-way electrical connectors and the passenger's side module electrical connector.
2. Disconnect the negative battery cable.
3. Drain the cooling system into a clean container for reuse.

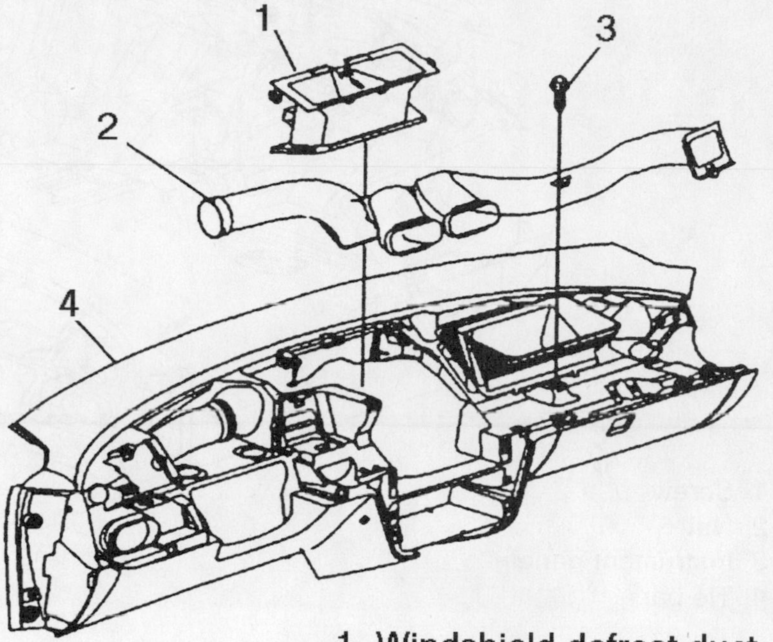

87950083

Air distribution duct mounting—Cavalier shown

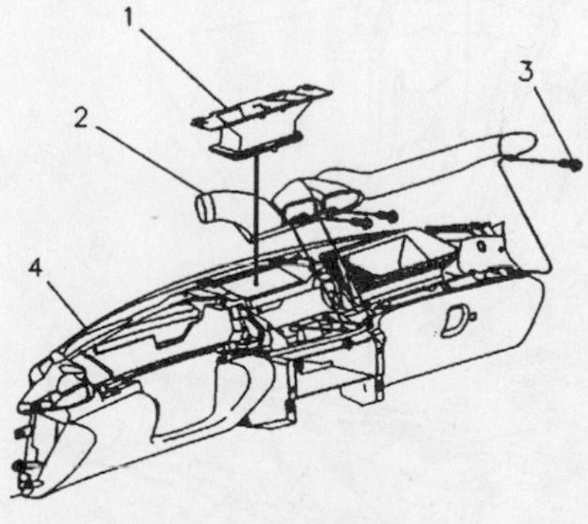

87950084

Location of the air distribution duct mounting—Sunfire shown

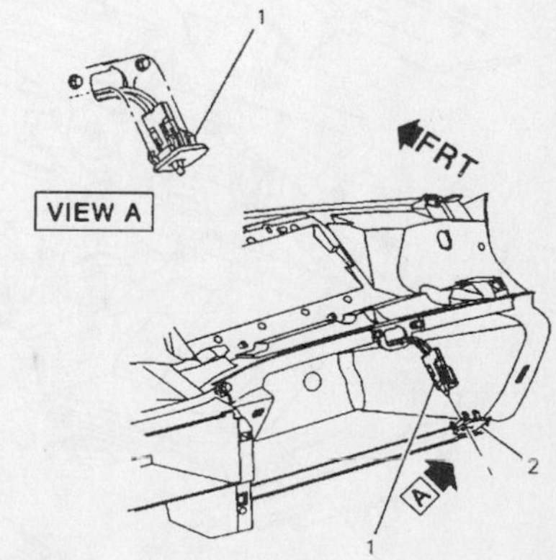

87950085

Detach the instrument panel lamp connector—Cavalier and Sunfire

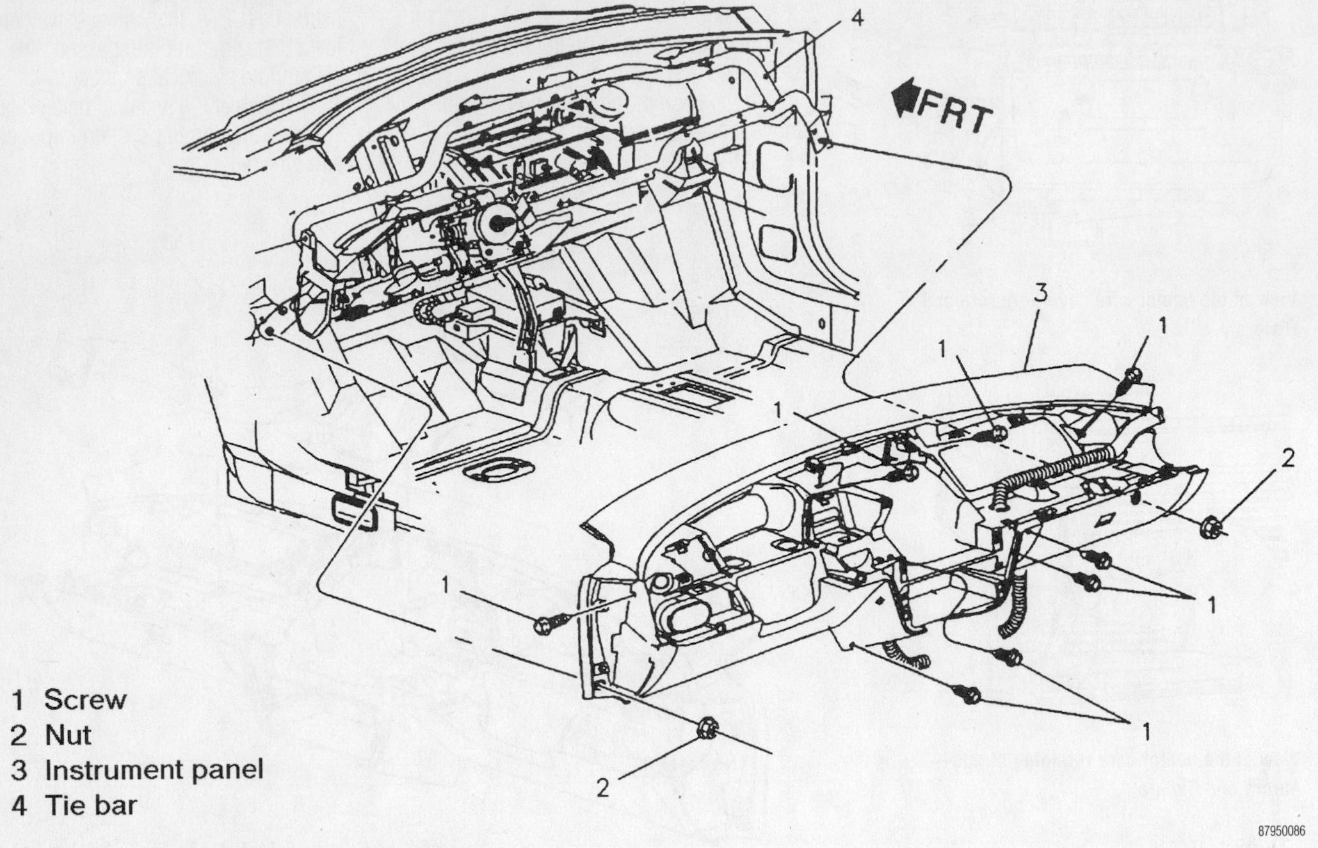

Instrument panel-to-tie bar attachments—Cavalier

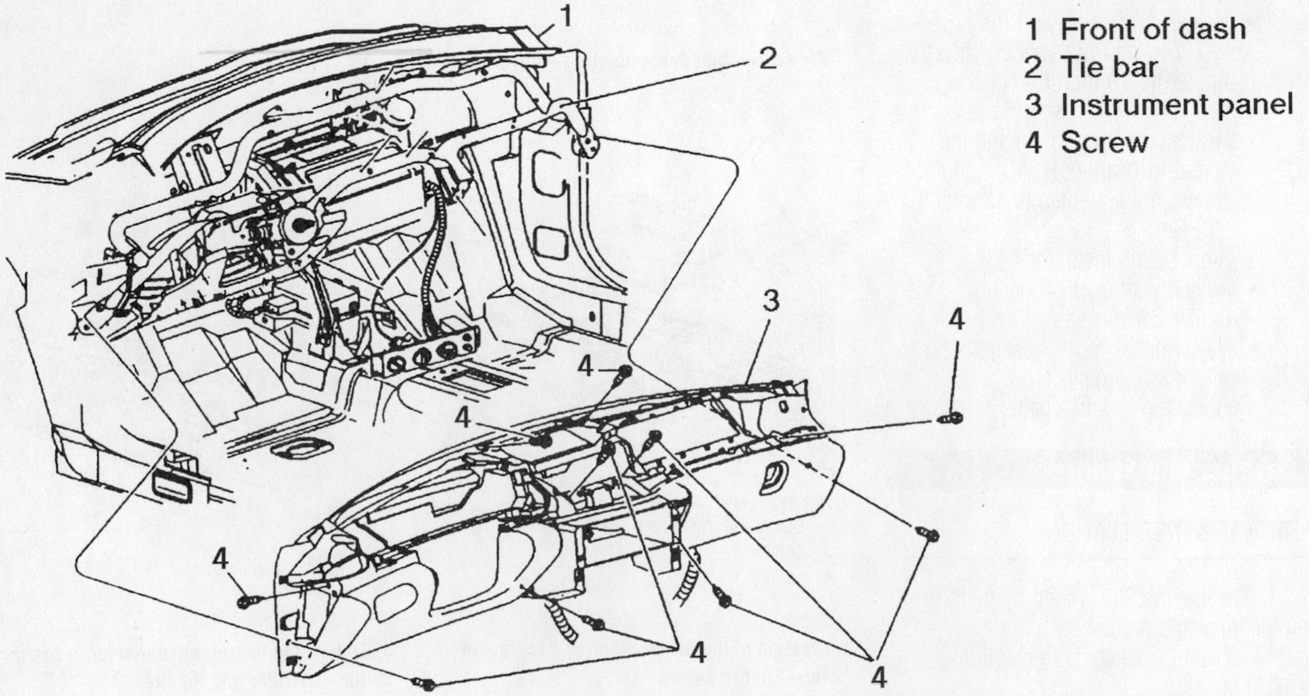

Instrument panel-to-tie bar mounting—Sunfire

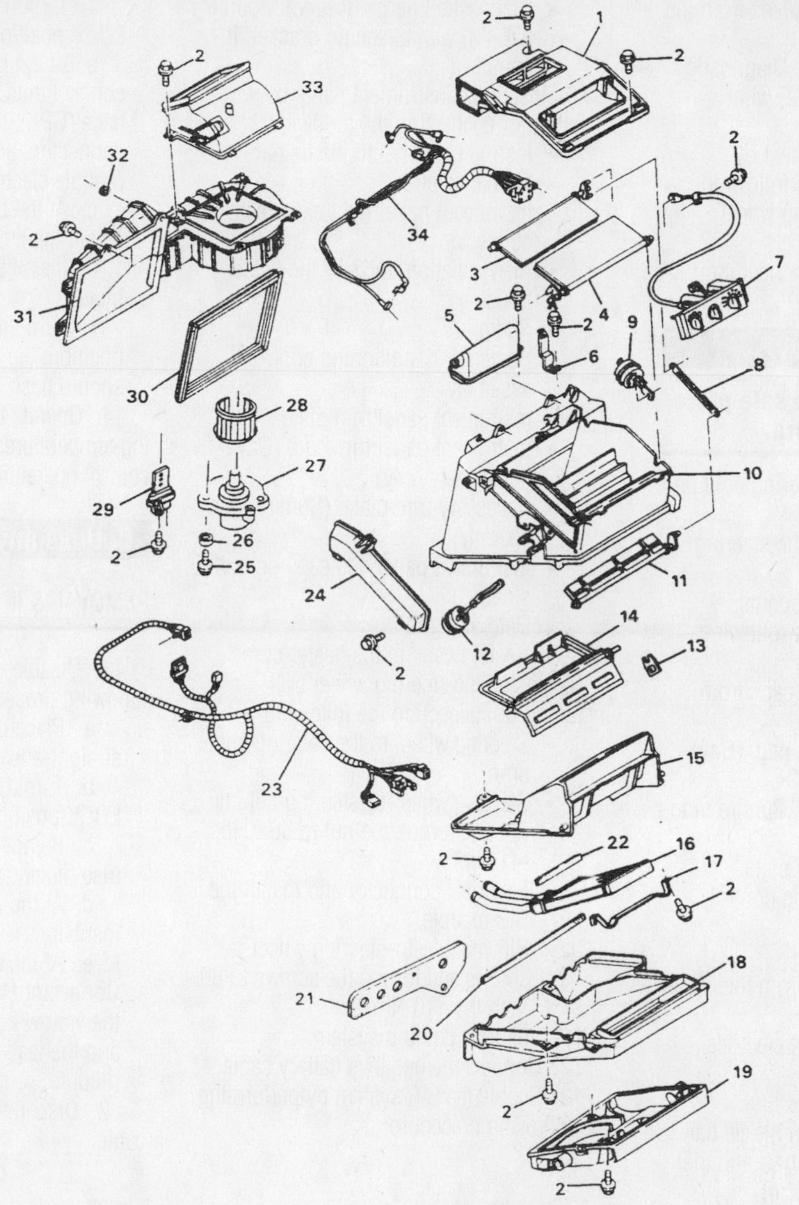

1 VALVE HOUSING COVER	18 HEATER COVER
2 HEATER/BLOWER MODULE BOLT	19 HEATER OUTLET
3 DEFROSTER VALVE	20 HEATER CORE SHROUD SEAL
4 MODE VALVE	21 HEATER CORE TUBE AND MOUNT SEAL
5 HEATER-VACUUM TANK	22 HEATER CORE SEAL
6 HEATER MODULE MOUNTING BRACKET	23 HEATER AND A/C CONTROL SWITCH HARNESS
7 HEATER-CONTROL	24 DEFROSTER DUCT
8 HEATER VALVE LEVER LINK	25 BLOWER MOTOR BOLTS
9 DEFROSTER VALVE ACTUATOR	26 BLOWER MOTOR ISOLATOR
10 HEATER CASE	27 BLOWER MOTOR
11 HEATER VALVE	28 BLOWER FAN
12 MODE VALVE ACTUATOR	29 BLOWER RESISTOR
13 TEMPERATURE VALVE CLIP	30 HEATER CASE SEAL
14 TEMPERATURE VALVE	31 BLOWER AND AIR INLET CASE
15 HEATER CORE SHROUD	32 MOUNTING SEAL
16 HEATER CORE	33 AIR INLET HOUSING
17 HEATER CORE STRAP	34 VACUUM HARNESS

93111GB4

Exploded view of the heater/evaporator housing assembly—Cavalier and Sunfire

4. Disconnect the heater hoses from the heater core.

5. If equipped, remove the Diagnostic Energy Reserve Module (DERM) with attaching brackets.

6. Remove the steering wheel by removing or disconnecting the following:

- SIR module-to-steering wheel screws
- SIR module and disconnect the electrical connector

CAUTION

Place the SIR module in a safe place with the front facing upward.

- Steering wheel-to-steering column nut
- Steering wheel from the steering column

7. Remove the instrument panel by removing or disconnecting the following:

- Defroster grille
- Instrument panel end caps from both sides
- Instrument panel trim pad, (Cavalier models only)
- Accessory trim plate, (Sunfire models only)
- Instrument panel trim plate
- Heater/air conditioning control assembly
- Radio
- Air ventilation ducts from the heater housing
- Instrument panel light electrical connector
- Tie bar screws
- Instrument panel from the tie bar
- Heater core outlet screws and the outlet
- Heater core cover-to-heater case screws and the cover.

➡The heater core mounting screw is located in the recess in the center of the heater core cover.

- Heater core clamp from the heater assembly
- Heater core from the heater assembly

To install:

8. Install or connect the following:

- Heater core to the heater assembly
- Heater core clamp to the heater assembly and torque the clamp screws to 9 inch lbs. (1 Nm)
- Heater core cover and the cover-to-heater case screw and torque the screws to 9 inch lbs. (1 Nm)
- Heater core outlet and torque the screws to 9 inch lbs. (1 Nm)
- Diagnostic Energy Reserve Module (DERM) with attaching bracket, if equipped

9. Install the instrument panel by installing or connecting the following:

- Instrument panel to the tie bar
- Tie bar screws
- Instrument panel light electrical connector
- Air ventilation ducts to the heater housing
- Radio
- Heater/air conditioning control assembly
- Instrument panel trim plate
- Instrument panel trim pad, (Cavalier models only)
- Accessory trim plate, (Sunfire models only)
- Instrument panel end caps on both sides
- Defroster grill
- Heater hoses to the heater core

10. Install the steering wheel by installing or connecting the following:

- Steering wheel to the steering column
- Steering wheel-to-steering column nut and torque the nut to 30 ft. lbs. (41 Nm)
- Electrical connector and install the SIR module
- SIR module-to-steering wheel screws and torque the screws to 89 inch lbs. (10 Nm)

11. Refill the cooling system.

12. Connect the negative battery cable.

13. Enable the SIR system by performing the following procedure:

a. Turn the ignition switch to the LOCK position.

b. At the base of the steering column, connect the Connector Position Assurance (CPA), the yellow 2-way electrical connectors and the passenger's side module electrical connector.

c. At the base of the steering column, install the left sound insulator.

d. Install the AIR BAG fuse to the fuse block.

e. Turn the ignition switch to the RUN position; the INFL REST warning light should flash 7–9 times then turn OFF.

14. Operate the engine to normal operating temperatures; then, check the climate control operation and check for leaks.

Cutlass and Malibu

REMOVAL & INSTALLATION

1. Disable the air bag by performing the following procedure:

a. Place the front wheel in the straight-ahead position.

b. Turn the ignition switch to the LOCK position.

c. Remove the air bag fuse from the fuse block.

d. At the left side, remove the sound insulator.

e. At the driver's side, disconnect the Connector Position Assurance (CPA) and the yellow 2-way electrical connectors and the lead to the passenger's side SIR module.

2. Disconnect the negative battery cable.

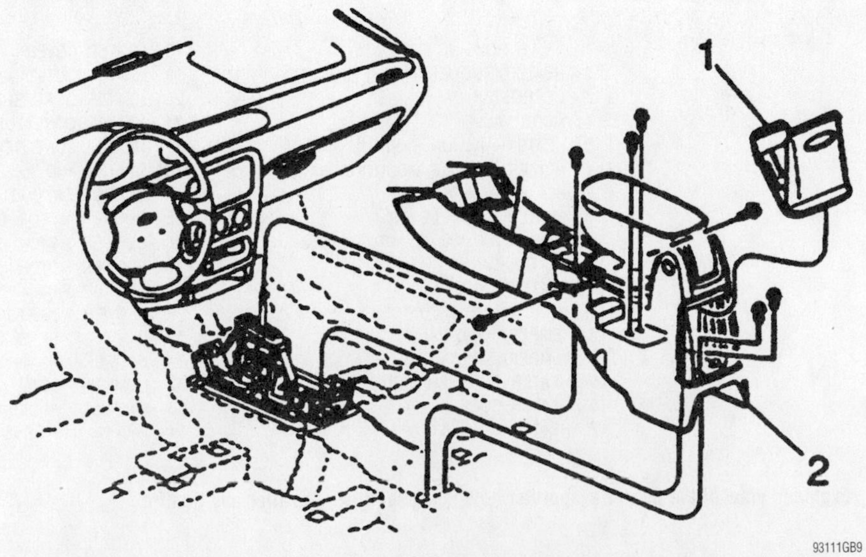

Exploded view of the console and related components—Cutlass and Malibu

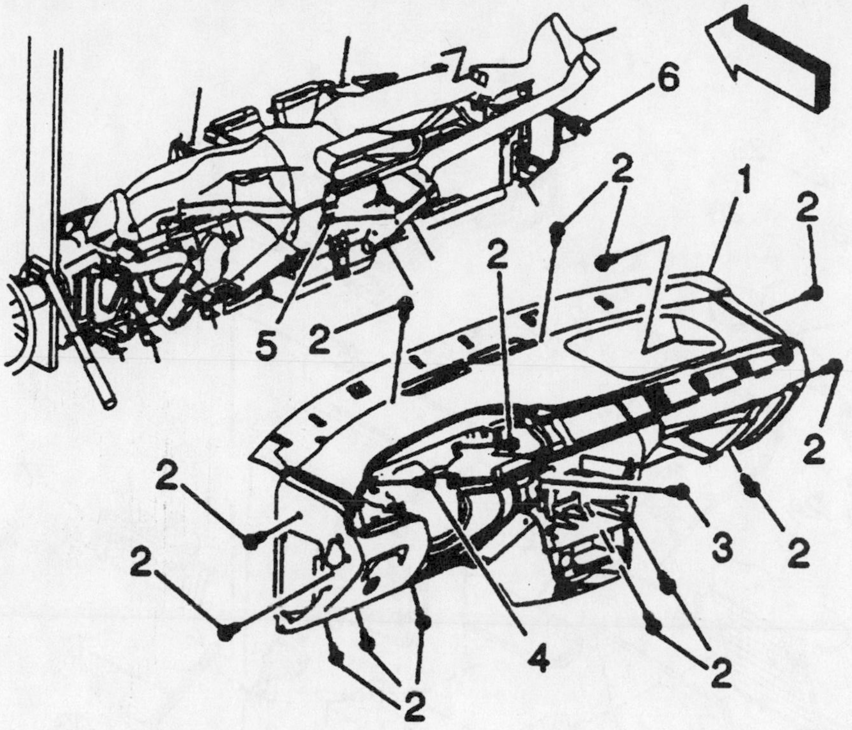

View of instrument panel—Cutlass and Malibu

3. Drain the cooling system into a clean container for reuse.

4. Disconnect the heater hoses from the heater core and remove the drain tube from the heater/air conditioning housing.

5. Remove the steering wheel by removing or disconnecting the following:

- 2 SIR module-to-steering wheel screws (located at the rear of the steering wheel), and the module
- SIR module and disconnect the electrical connector and remove the SIR module

CAUTION

Place the SIR module in a safe place with the front facing upward.

- Horn and cruise control electrical connectors
- Steering wheel-to-steering column nut
- Steering wheel from the steering column

6. Remove the console by removing or disconnecting the following:

- Front seats
- Fold the console compartment upward
- Console trim plate
- Rear cup holder
- Console-to-chassis screws and the console

7. Remove the instrument panel by removing or disconnecting the following:

- Defroster grille
- Instrument panel valance screws and the valance
- Instrument panel end caps from both sides
- Screws located under the instrument panel end caps
- Instrument panel compartment
- Steering column covers
- Steering column stalks
- Instrument cluster and accessory trim plates
- Instrument cluster fasteners, disconnect the electrical connectors and remove the cluster
- Heater/air conditioning control assembly
- Stereo/tape deck assembly
- Ignition switch
- Upper windshield side garnish molding
- Loosen the center console, pull it rearward to disengage it from the instrument panel
- Instrument panel-to-tie bar screws
- Instrument panel
- Outlet from the heater/air conditioning housing
- Heater core cover-to-heater/air conditioning housing

➡There is a screw located in the recess in the center of the cover.

- Heater core-to-heater/air conditioning housing clamps
- Heater core

To install:

8. Install or connect the following:

- Heater core
- Heater core-to-heater/air conditioning housing clamps
- Heater core cover-to-heater/air conditioning housing

➡There is a screw located in the recess in the center of the cover.

- Outlet to the heater/air conditioning housing

9. Install the instrument panel by installing or connecting the following:

- Instrument panel
- Instrument panel-to-tie bar screws
- Move the center console forward and engage it to the instrument panel
- Upper windshield side garnish molding
- Ignition switch
- Stereo/tape deck assembly
- Heater/air conditioning control assembly
- Instrument cluster, connect the electrical connectors and install the cluster fasteners
- Instrument cluster accessory trim plates
- Steering column stalks
- Steering column covers
- Instrument panel compartment
- Instrument panel end caps, install the screws
- Instrument panel end caps on both sides
- Instrument panel valance and the valance screws
- Defroster grille

10. Install the console by installing or connecting the following:

- Console and the console-to-chassis screws
- Rear cup holder
- Console trim plate
- Front seats

11. Install the steering wheel by installing or connecting the following:

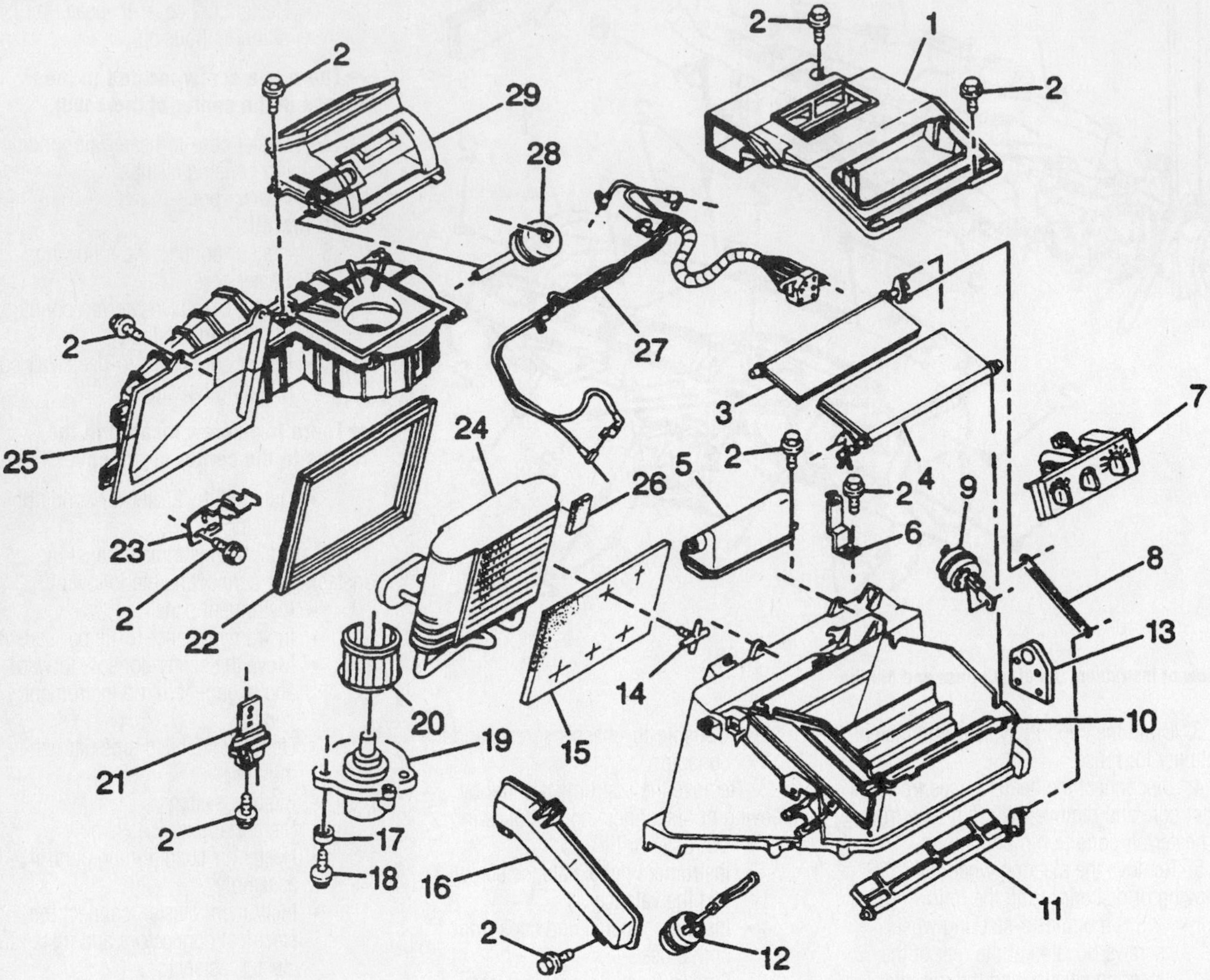

1	Valve Housing Cover	16	Defroster Case
2	Bolt	17	Blower Motor Bolt Insulator
3	Defroster Valve	18	Blower Motor Bolt
4	Mode Valve	19	Blower Motor
5	Vacuum Tank	20	Blower Motor Fan
6	HVAC Module Bracket	21	Blower Motor Resistor
7	HVAC Control Assembly	22	Evaporator Core Seal
8	Heater Valve Link	23	Evaporator Core Bracket
9	Defroster Valve Actuator	24	Evaporator Core
10	Evaporator Case	25	Blower and Air Inlet Case
11	Heater Valve	26	Evaporator Core Spacer
12	Mode Valve Actuator	27	HVAC Vacuum Harness
13	Temperature Control Motor	28	Air Inlet Valve Actuator
14	Water Filter Retainer	29	Air Inlet Case
15	Water Filter		

93111GC1

View of the evaporator core, upper heater/air conditioning housing and related components—Cutlass and Malibu

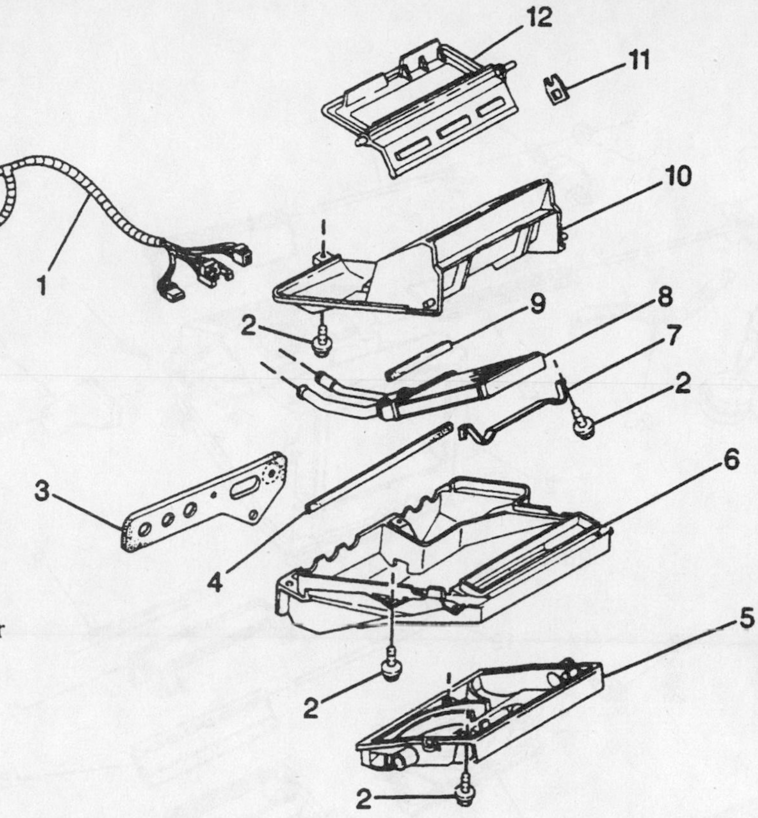

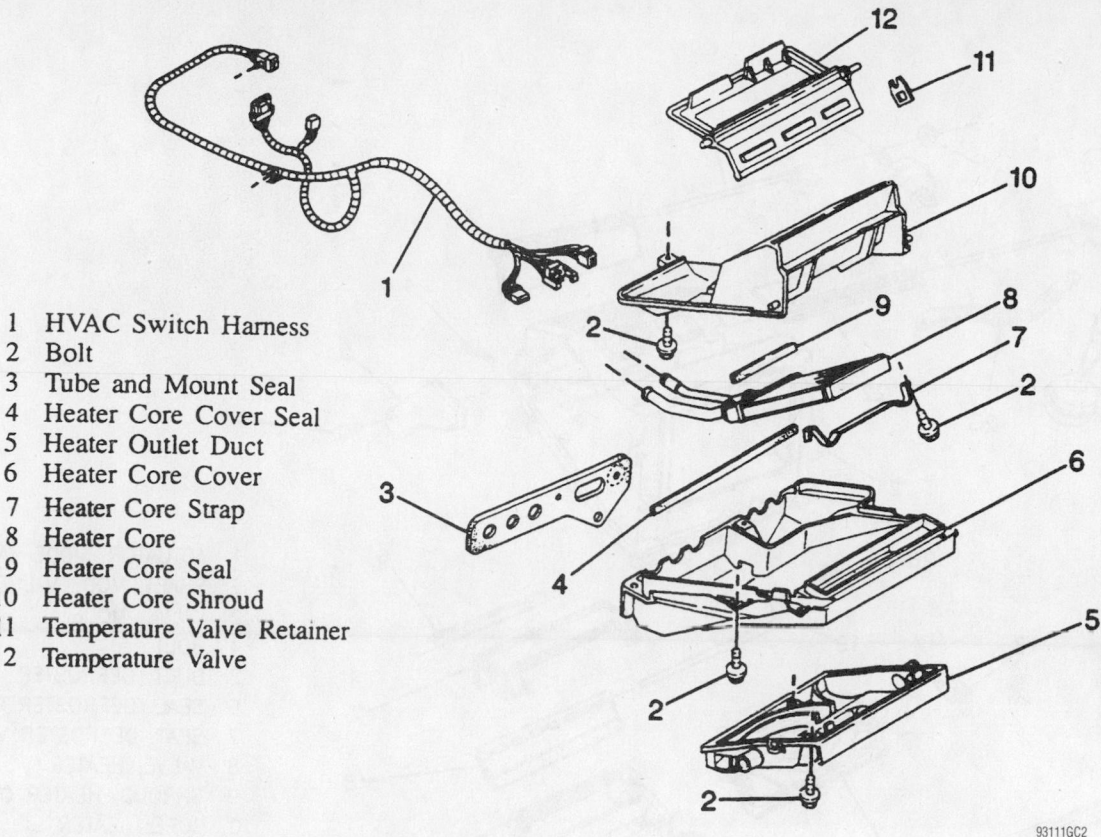

1 HVAC Switch Harness
2 Bolt
3 Tube and Mount Seal
4 Heater Core Cover Seal
5 Heater Outlet Duct
6 Heater Core Cover
7 Heater Core Strap
8 Heater Core
9 Heater Core Seal
10 Heater Core Shroud
11 Temperature Valve Retainer
12 Temperature Valve

View of the heater core, lower heater/air conditioning housing and related components—Cutlass and Malibu

- Steering wheel to the steering column
- Steering wheel-to-steering column nut and torque to 30 ft. lbs. (41 Nm)
- Horn and cruise control electrical connectors
- SIR module and connect the electrical connector
- SIR module and torque the 2 module-to-steering wheel screws to 89 inch lbs. (10 Nm)
- Heater hoses to the heater core and install the drain tube to the heater/air conditioning housing

12. Refill the cooling system.
13. Connect the negative battery cable.
14. Enable the SIR or air bag by installing or connecting the following:
    - Connector Position Assurance (CPA), the yellow 2-way electrical connectors and the lead to the passenger's side SIR module (located on the drivers' side)
    - Sound insulator, located on the left side
    - Air bag fuse to the fuse block
    a. Turn the ignition switch to RUN and verify that the Air Bag Warning light flashes 7–9 times and turns OFF.

➡If the SIR system does not operate as described, perform the SIR diagnostic system check.

15. Operate the engine to normal operating temperatures; then, check the climate control operation and check for leaks.

Achieva, Alero, Grand Am and Skylark

REMOVAL & INSTALLATION

1998–99

1. Disconnect the negative battery cable.
2. Drain the cooling system into a clean container for reuse.
3. Raise and safely support the vehicle.
4. Remove the drain tube from the heater housing.
5. Disconnect the heater hoses from the heater core.
6. Lower the vehicle.
7. Remove or disconnect the following:
    - Console, if equipped
    - Sound insulators from both sides
    - Steering column opening filler
    - Air outlet duct
    - Rear floor air outlet screws and the outlet
    - Heater core cover
    - Heater core-to-housing mounting clamps and the heater core

To install:

8. Install or connect the following:
    - Heater core and the heater core-to-housing mounting clamps
    - Heater core cover
    - Rear floor air outlet and the outlet screws
    - Air outlet duct
    - Steering column opening filler
    - Sound insulators, if equipped
    - Console, if equipped
    - Heater hoses to the heater core.
    - Drain tube to the heater housing.
9. Refill the cooling system.
10. Connect the negative battery cable.
11. Operate the engine to normal operating temperatures. Check the climate control operation and check for leaks.

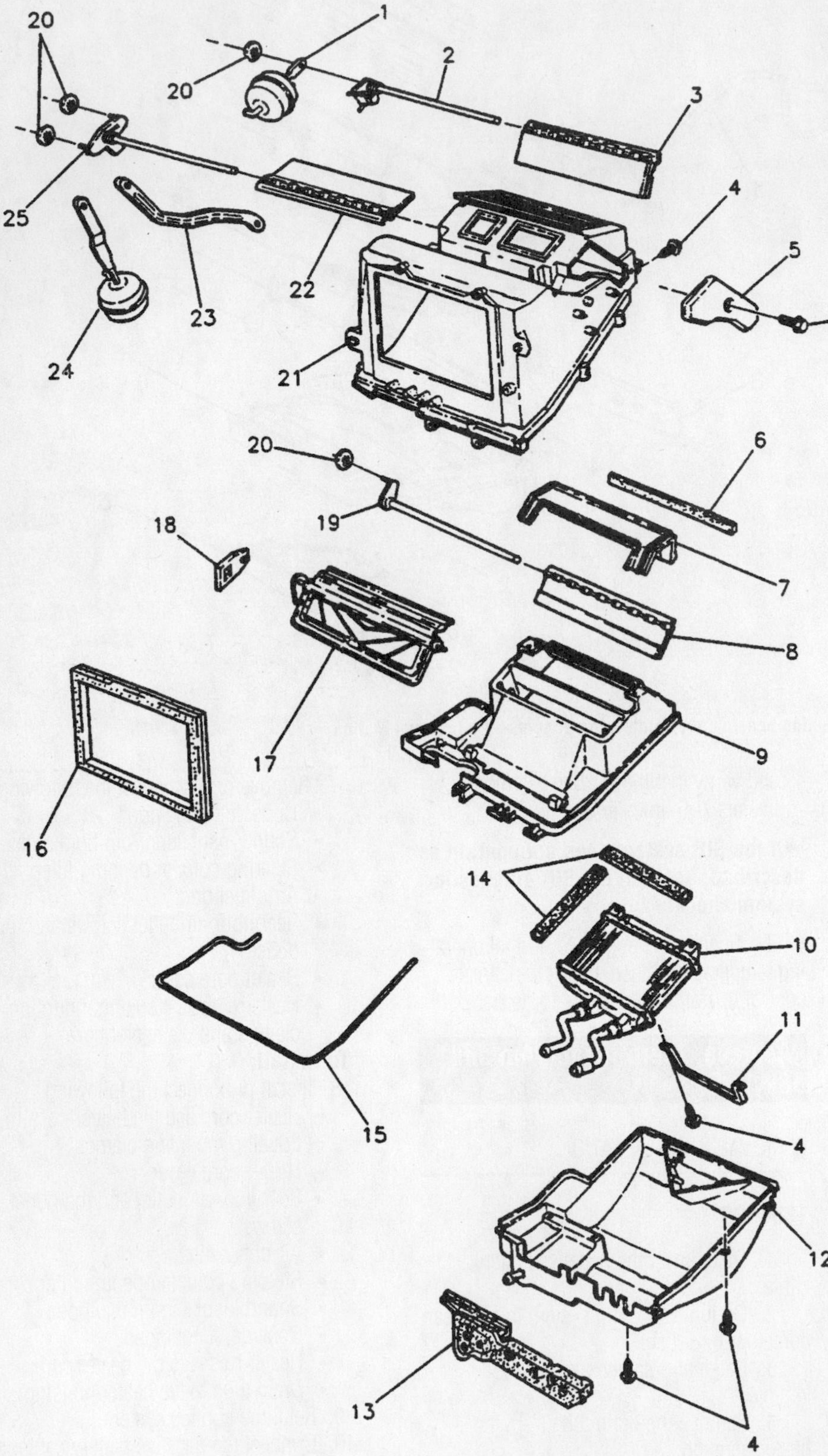

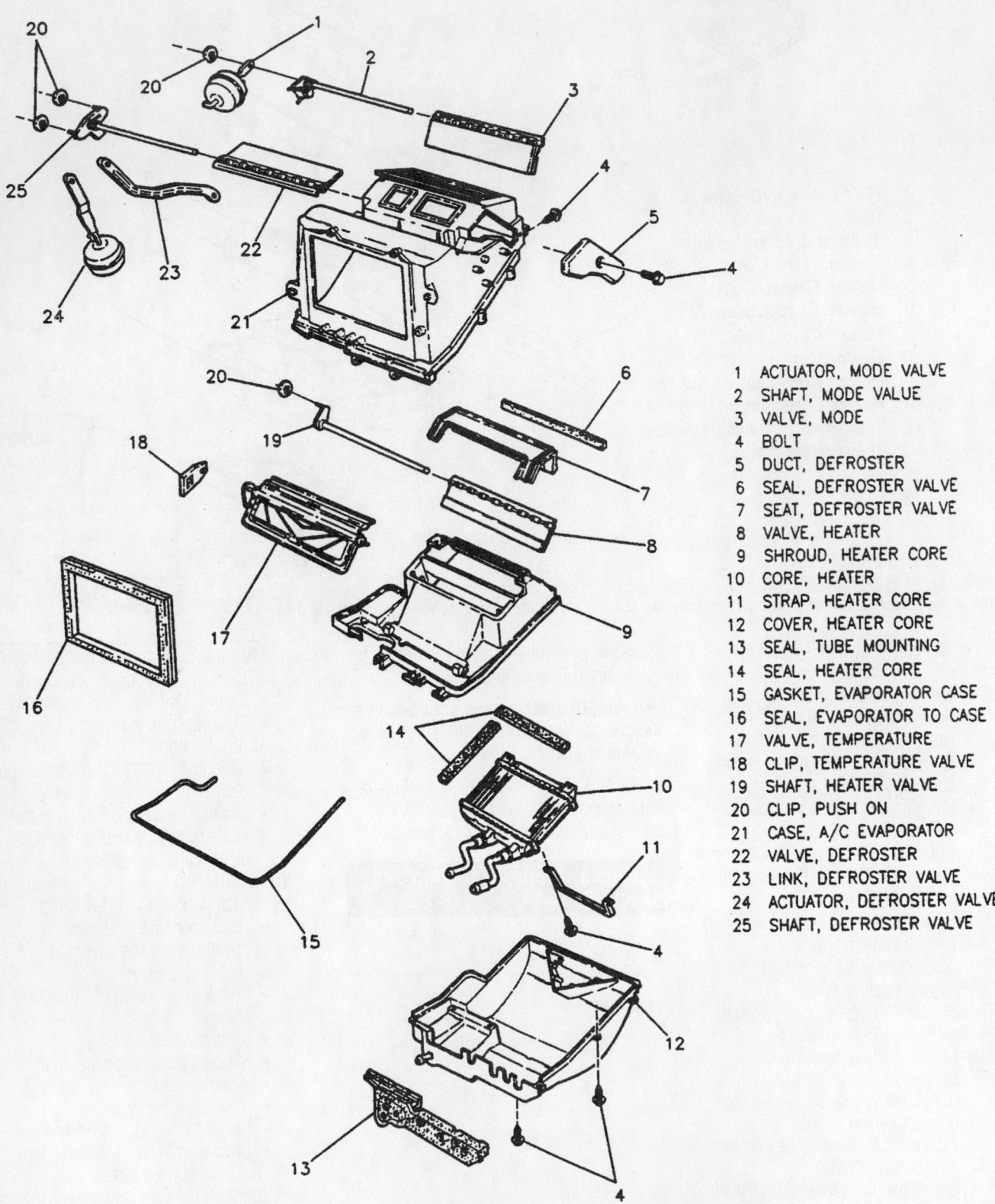

1 ACTUATOR, MODE VALVE
2 SHAFT, MODE VALUE
3 VALVE, MODE
4 BOLT
5 DUCT, DEFROSTER
6 SEAL, DEFROSTER VALVE
7 SEAT, DEFROSTER VALVE
8 VALVE, HEATER
9 SHROUD, HEATER CORE
10 CORE, HEATER
11 STRAP, HEATER CORE
12 COVER, HEATER CORE
13 SEAL, TUBE MOUNTING
14 SEAL, HEATER CORE
15 GASKET, EVAPORATOR CASE
16 SEAL, EVAPORATOR TO CASE
17 VALVE, TEMPERATURE
18 CLIP, TEMPERATURE VALVE
19 SHAFT, HEATER VALVE
20 CLIP, PUSH ON
21 CASE, A/C EVAPORATOR
22 VALVE, DEFROSTER
23 LINK, DEFROSTER VALVE
24 ACTUATOR, DEFROSTER VALVE
25 SHAFT, DEFROSTER VALVE

Exploded view of the heater core, heater housing and related components—Grand Am, Achieva and Skylark

2000–01 Models

1. Before servicing the vehicle, refer to the precautions in the beginning of this section.
2. Drain the engine cooling system.
3. Disable the Supplemental Inflatable Restraint (SIR) system.
4. Recover the A/C refrigerant.
5. Remove or disconnect the following:
    - Evaporator hose from the evaporator
    - Heater hoses
    - Drain tube elbow
    - Heater case plate for the heater pipes
    - Heater case plate for the evaporator block
    - End caps
    - Sound insulators
    - Instrument panel compartment
    - Passenger inflator module
    - Tilt steering lever
    - Upper and the lower steering column covers
    - Hazard warning switch
    - Instrument panel cluster trim plate
    - Instrument panel cluster
    - Driver inflator module
    - Steering wheel
    - Multifunction lever
    - Accessory trim plate
    - Radio
    - Cup holder
    - Gearshift handle
    - Ignition switch bolts
    - Instrument panel upper bolt covers
    - Electrical junction box connections
    - Fog lamp switch
    - Windshield side upper garnish molding
    - Console from the instrument panel
    - Instrument panel from the tie bar
    - Air distribution duct
    - Floor air duct
    - Tie bar
    - Daytime Running Lights (DRL) sensor wiring harness from the Heating, Ventilation, Air Conditioning (HVAC) module
    - Blower motor electrical connectors
    - Temperature actuator electrical connector
    - HVAC module from the vehicle
    - Heater core case cover
    - Heater core bracket
    - Heater core

To install:

6. Install or connect the following:
    - Heater core
    - Heater core bracket. Torque the screw to 9 inch lbs. (1 Nm).
    - Heater core case cover. Torque the screws to 9 inch lbs. (1 Nm).
    - HVAC module
    - Temperature actuator electrical connector
    - Blower motor electrical connectors
    - DRL wiring harness
    - Tie bar. Torque the bolts to 15 ft. lbs. (20 Nm).
    - Floor air duct. Torque the bolts to 18 inch lbs. (2 Nm).
    - Air distribution duct. Torque the bolts to 27 inch lbs. (3 Nm).
    - Instrument panel
    - Console to the instrument panel
    - Windshield side upper garnish molding
    - Fog lamp switch
    - Electrical junction box connections
    - Instrument panel upper bolt covers
    - Ignition switch bolts
    - Gearshift handle
    - Cup holder
    - Radio. Torque the bolts to 27 inch lbs. (3 Nm).
    - Accessory trim plate
    - Multifunction lever. Torque the screw to 35 inch lbs. (4 Nm).
    - Steering wheel. Torque the nut to 27 ft. lbs. (37 Nm).
    - Driver inflator module. Torque the screw to 89 inch lbs. (10 Nm).
    - Instrument panel cluster
    - Instrument panel cluster trim plate
    - Hazard warning switch
    - Upper and the lower steering column covers. Torque the screw to 18 inch lbs. (2 Nm).
    - Tilt steering lever
    - Passenger inflator module. Torque the fasteners to 89 inch lbs. (10 Nm).
    - Instrument panel compartment
    - Sound insulators
    - End caps
    - Heater case plate for the evaporator block. Torque the nuts to 89 inch lbs. (10 Nm).
    - Heater case plate for the heater pipes
    - Drain tube elbow
    - Heater hoses
    - Evaporator hose. Torque the fitting to 18 ft. lbs. (25 Nm).
7. Fill the engine cooling system.
8. Evacuate and recharge the A/C system.

Catera

REMOVAL & INSTALLATION

1. Disable the SIR system.
2. Disconnect the negative battery cable.
3. Drain the cooling system.
4. Remove the heater hose quick connects from the heater core pipes by performing the following procedure:
    a. At the passenger side, raise the air inlet screen and open the access door near the pollen filter.
    b. Unlock the quick connect collars by squeezing the tabs and carefully pulling back on the tabs to disconnect the sleeve.
    c. If green assembly marks are attached, discard them.

✱✱ WARNING

The front wheels must be maintained in the straight-ahead position and the steering column must be in the LOCK position. Failure to do so will cause improper alignment of some components during installation and may result to damage to the SIR coil assembly.

5. Remove the steering column by removing or disconnecting the following:

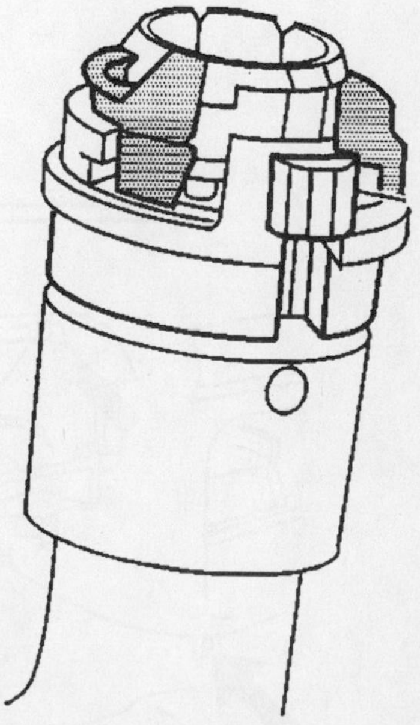

View of a heater hose connector—Catera

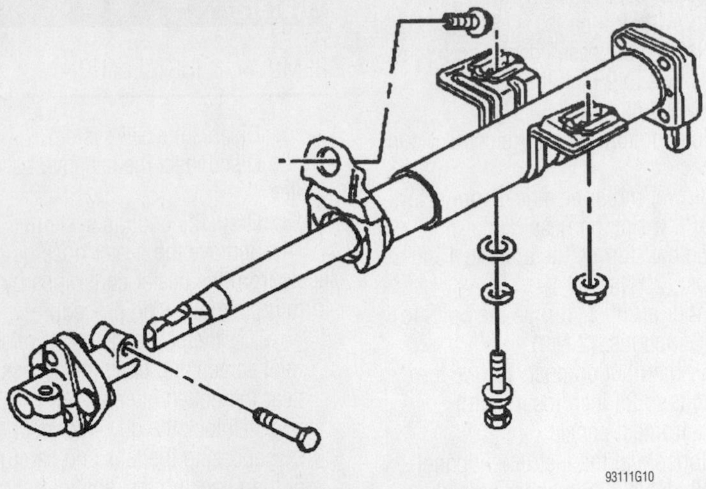

View of the steering column assembly—Catera

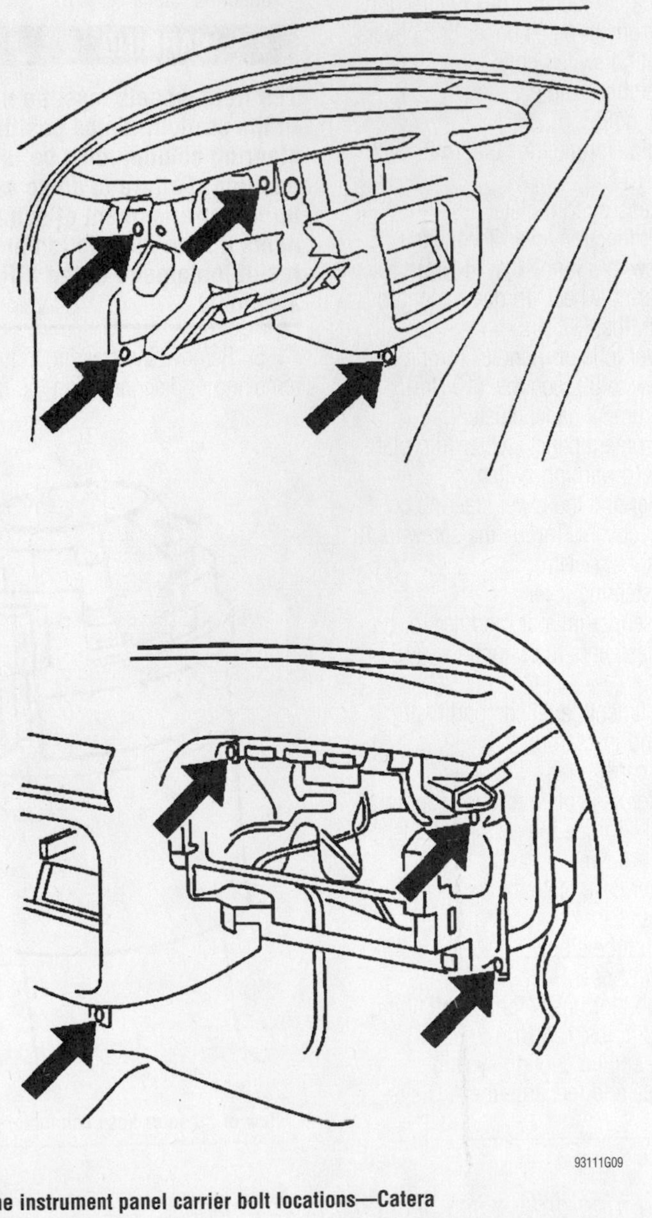

View of the instrument panel carrier bolt locations—Catera

- Instrument panel driver knee bolster energy absorber and sound insulator
- Steering column electrical connector(s)
- Coupler bolt from the lower steering column connection and slightly separate the coupler to aid in the shaft removal
- Using a chisel and a hammer, rotate the forward support strap shear nut and bolt (located under the steering column) counterclockwise in order to remove them
- Steering column-to-rear support bracket bolt
- Pull the steering column straight back and through the dash panel and remove it from the vehicle.

WARNING

Handle the steering column with care for it is very susceptible to damage. Dropping, leaning or hammering on it could cause damage to its collapsible design.

6. Remove the instrument panel carrier by removing or disconnecting the following:
- Windshield pillar moldings
- Access panel, the air deflector outlet screw, the air deflector outlet and the air outlet duct, located at the right-side of the instrument panel
- Instrument panel SIR (air bag) module cover, the instrument panel compartment and the SIR module
- Upper center console and the lower center console
- Center console air duct screw and the center console air duct
- Radio tape player bezel and the radio
- Climate control head
- Center air outlet deflector, the outlet screws and the outlet housing
- Driver's side access panel
- Driver's side air outlet deflector, the outlet screw and the outlet housing
- Driver's side lower outlet duct
- Instrument cluster
- Fuse and relay panel screws and the panels
- Instrument panel carrier bolts
- Electrical connectors from the instrument panel and/or the wiring harness clips from the instrument panel carrier, if necessary
- Instrument panel carrier

➡It is not necessary to physically remove the instrument panel but it is necessary to pull the carrier rearward

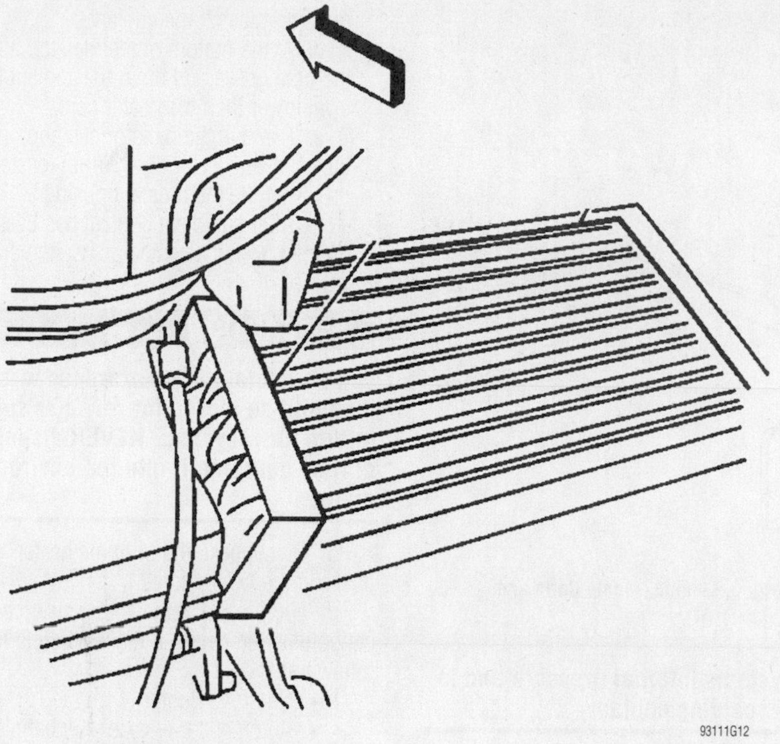

View of a heater core and seal—Catera

to enable the heater core to be removed from its housing.

- Blower motor housing screws and the housing with the motor
- Heater core pipe bracket-to-chassis screw and the bracket
- Heater core inlet/outlet pipe bracket-to-heater core screw and the inlet/outlet pipe bracket
- Instrument panel support brace bolts from the instrument panel and the transmission well, then remove the brace
- Heater core-to-housing retaining screw. Plug the heater core pipes and protect the interior from coolant spills
- Heater core and the rubber seal from the heater housing

To install:

7. Install or connect the following:
   - Rubber seal and heater core into the heater housing; be careful not to damage the fins
   - Heater core-to-housing retainer screw
   - Instrument panel support brace to the transmission well and instrument panel, then torque the bolts to 16 ft. lbs. (22 Nm)
   - Heater core inlet/outlet pipe bracket to the heater core (using a new O-ring lightly coated with coolant), and torque the screw to 44 inch lbs. (5 Nm)
   - Heater core pipe bracket with the screw to the chassis, be careful not to strip the screw
   - Blower motor/housing assembly and torque the screws to 35 inch lbs. (4 Nm)
   - Instrument panel carrier
   - Instrument panel carrier by reversing the removal procedures and torque the instrument panel carrier bolts to 16 ft. lbs. (22 Nm)
8. Remove the steering column by installing or connecting the following:
   - Steering column into the vehicle, through the dash panel and into the lower steering coupling
   - Hand start the rear support bracket bolt, the forward support strap nut and shear bolt
   - Rear support bracket bolt. Torque to 16 ft. lbs. (22 Nm)
   - Forward support strap nut. Torque to 16 ft. lbs. (22 Nm)
   - New forward support shear bolt. Torque to 15 ft. lbs. (21 Nm)
   - Lower steering column shaft bolt and torque to 16 ft. lbs. (22 Nm)
   - Steering column electrical connector(s)
   - Sound insulator and the instrument panel driver knee bolster energy absorber
9. Connect the heater hoses to the heater core pipes by performing the following procedure:
   a. If not attached to the quick connect, discard the green assembly marker(s).
   b. Push the quick connects into the pipes until they are fully seated.
   c. Squeeze the locking tabs and press the retaining sleeve into the locked position.
   d. At the passenger side, close the access door near the pollen filter and lower the air inlet screen.
10. Refill the cooling system by performing the following procedure:
   a. Add a 50/50 mixture of water and DEX-COOL® antifreeze to the KALT/COLD mark (seam) on the surge tank.
   b. Start the engine and allow it to idle for 1 min.
   c. Add more coolant to the surge tank as necessary.
   d. Install the radiator sure tank cap.
   e. Cycle the engine, from idle to 3000 rpm, in 30 second intervals, until the engine reaches normal operating temperatures.

➡The cooling system will bleed itself automatically during warm-up.

   f. Turn the engine OFF and recheck the coolant level when the engine is cool
11. Connect the negative battery cable.
12. Enable the SIR system.
13. Reprogram the necessary accessories.

Century, Lumina and Regal

REMOVAL & INSTALLATION

1. Disconnect the negative battery cable.
2. If equipped with a 3.1L engine, remove the air cleaner and duct assembly.
3. If equipped with a 3.8L engine, remove the fuel injector sight shield by performing the following procedures:
   a. Clean the area around the oil filler cap/tube assembly location.
   b. Rotate the oil filler cap/tube assembly counterclockwise from the valve cover.

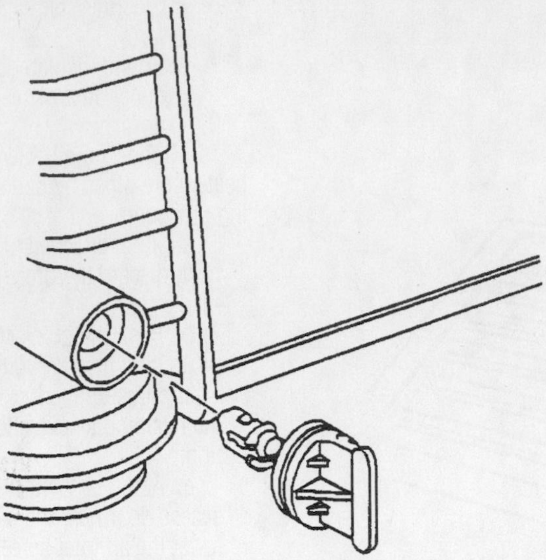

93111G01

View of the radiator drain valve—Century, Grand Prix, Intrigue, Lumina, Monte Carlo and Regal

c. Lift the fuel injector sight shield up at the front and slide it from the rear engine bracket.

d. Reinstall the oil filler cap/tube assembly in the valve cover.

CAUTION

Before draining the cooling system, allow the engine to cool to relieve the systems internal pressure and to avoid scalding coolant.

4. Drain the cooling system by performing the following procedure:

a. Raise and safely support the front of the vehicle.

b. Remove and clean the coolant recovery tank.

c. Place a 2 gallon pan under the radiator to catch the coolant.

d. At the bottom of the radiator, open the drain valve and drain the coolant to a level lower than the heater core.

e. Remove the radiator cap and open the air bleed screw (2–3 turns) located on top of the thermostat housing.

f. After sufficient coolant has been drained from the system, close the drain valve.

CAUTION

Engine coolant is a hazardous waste; it should be stored for reuse or submitted for recycling. NEVER dispose of it by dumping it into the environment.

5. Disconnect and plug the heater hoses at the heater core.

6. If equipped, remove the lower center console by removing or disconnecting the following:

- Cigarette lighter
- Automatic transaxle shift handle
- Upper console trim plate from the front floor console by unsnapping it
- Electrical connectors from the upper console trim plate and remove the trim plate
- CD storage compartment from the front floor console by unsnapping it
- Raise the front floor console armrest and remove the compartment mat
- Console-to-chassis bolts and screws
- Electrical connectors from the console and remove the console from the vehicle

7. Remove the lower instrument panel lower compartment by removing or disconnecting the following:

- Lower right instrument panel insulator
- Compartment-to-panel bolts and screws, from under the instrument panel compartment
- Instrument panel compartment door screws and the door
- Instrument panel compartment screws and plastic clips; then slide the compartment from the instrument panel
- Electrical connector from the compartment
- Ashtray and bracket, if necessary
- Lower heater duct
- Heater core cover and discard the cover seals
- Heater core mounting clip and bracket
- Heater core

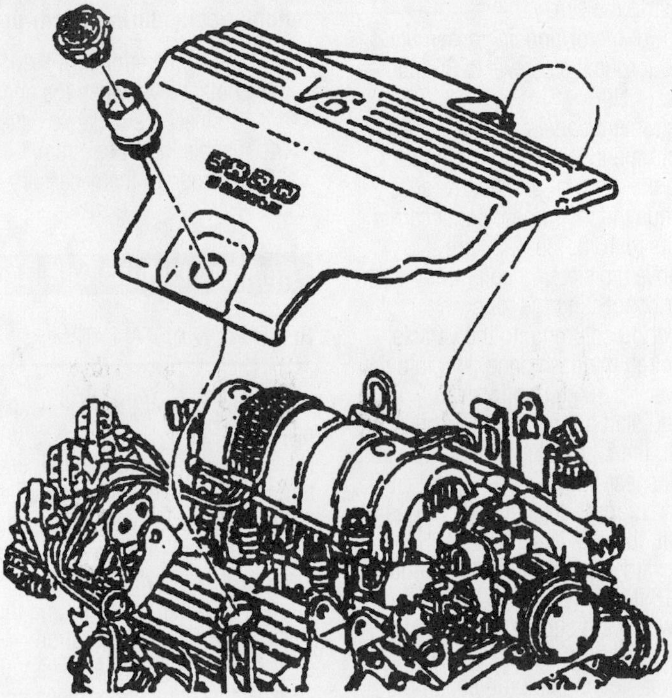

93111G02

View of the fuel injector sight shield—Century, Grand Prix, Intrigue, Lumina, Monte Carlo and Regal

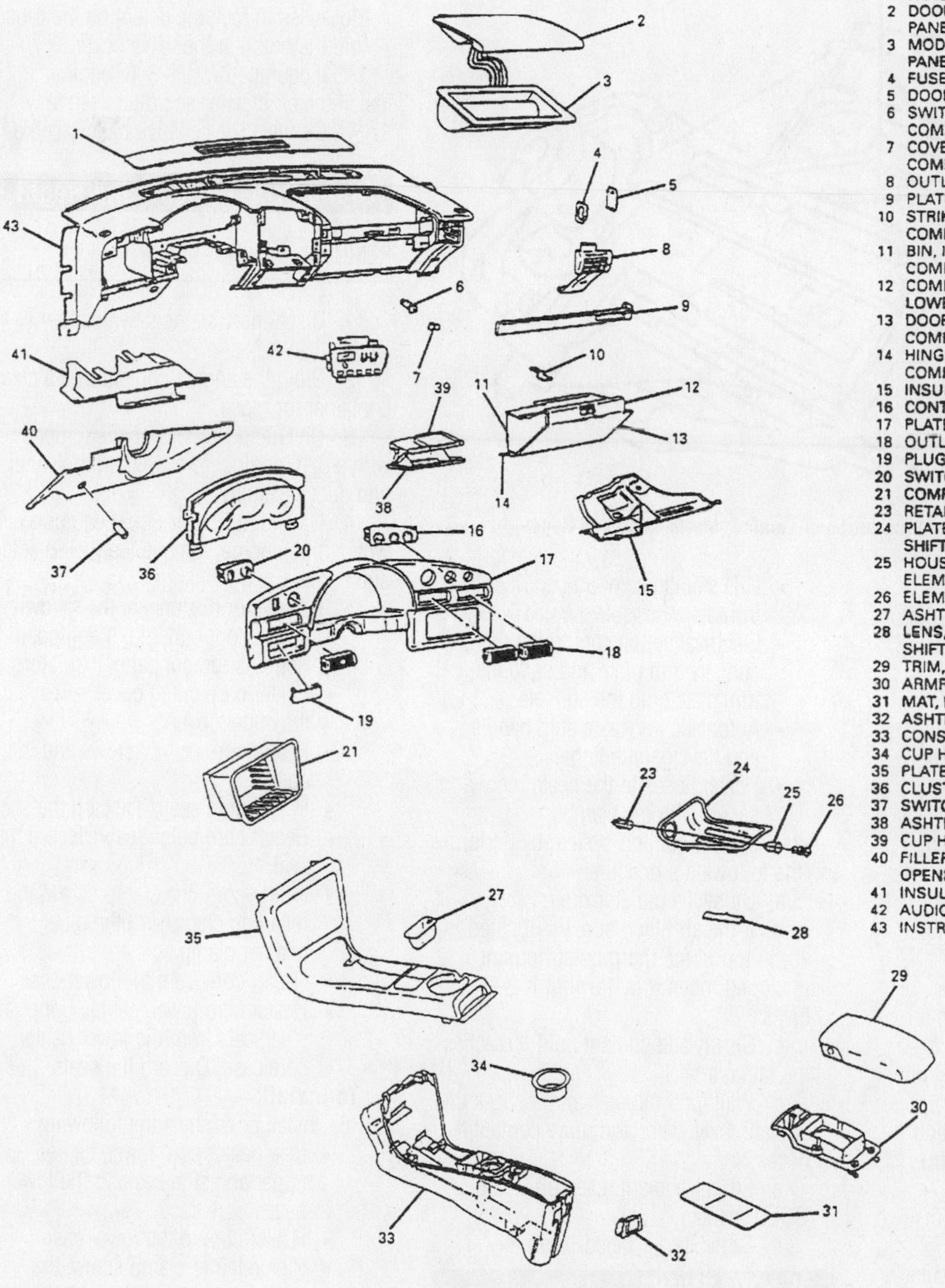

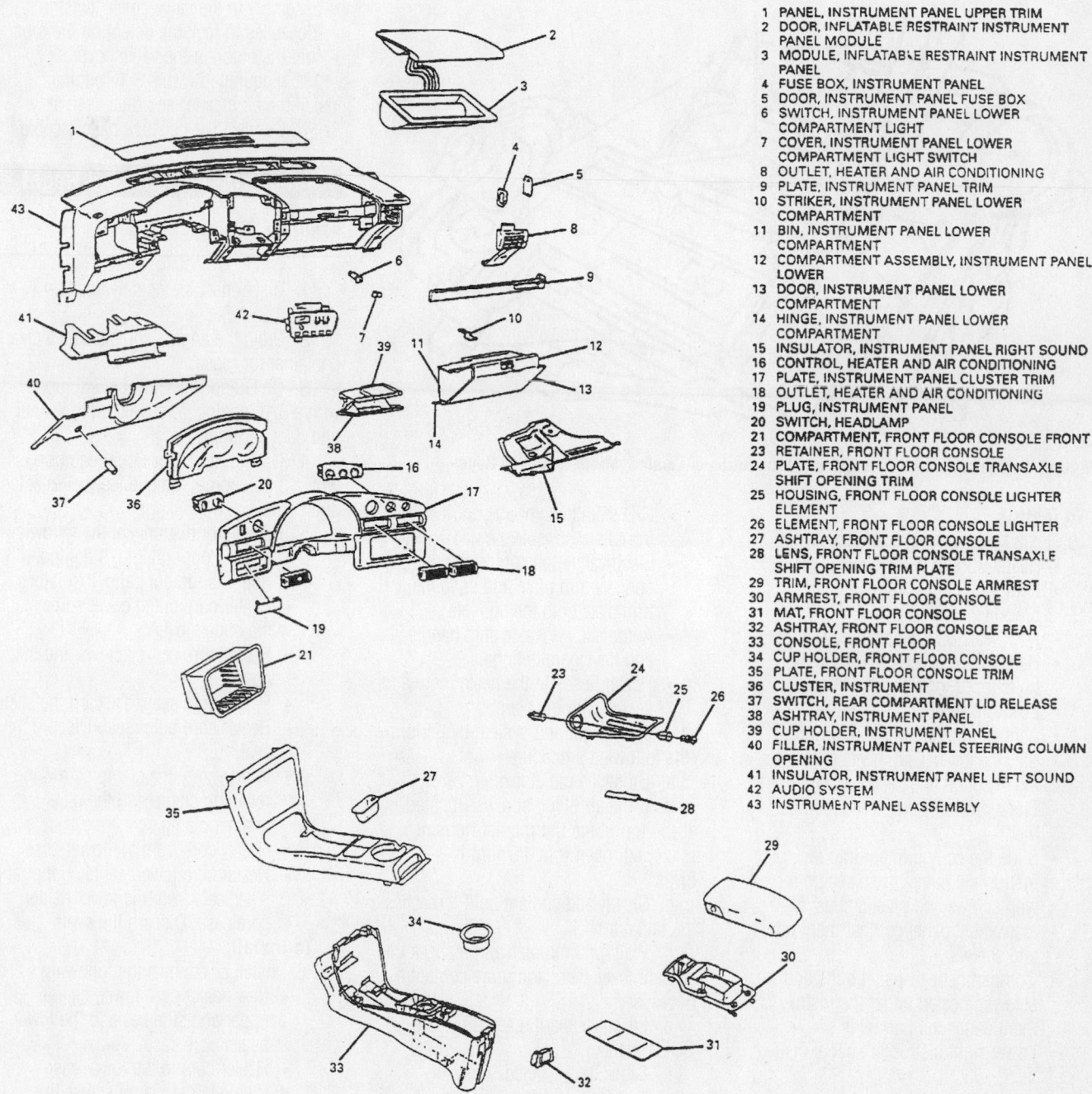

1 PANEL, INSTRUMENT PANEL UPPER TRIM
2 DOOR, INFLATABLE RESTRAINT INSTRUMENT PANEL MODULE
3 MODULE, INFLATABLE RESTRAINT INSTRUMENT PANEL
4 FUSE BOX, INSTRUMENT PANEL
5 DOOR, INSTRUMENT PANEL FUSE BOX
6 SWITCH, INSTRUMENT PANEL LOWER COMPARTMENT LIGHT
7 COVER, INSTRUMENT PANEL LOWER COMPARTMENT LIGHT SWITCH
8 OUTLET, HEATER AND AIR CONDITIONING
9 PLATE, INSTRUMENT PANEL TRIM
10 STRIKER, INSTRUMENT PANEL LOWER COMPARTMENT
11 BIN, INSTRUMENT PANEL LOWER COMPARTMENT
12 COMPARTMENT ASSEMBLY, INSTRUMENT PANEL LOWER
13 DOOR, INSTRUMENT PANEL LOWER COMPARTMENT
14 HINGE, INSTRUMENT PANEL LOWER COMPARTMENT
15 INSULATOR, INSTRUMENT PANEL RIGHT SOUND
16 CONTROL, HEATER AND AIR CONDITIONING
17 PLATE, INSTRUMENT PANEL CLUSTER TRIM
18 OUTLET, HEATER AND AIR CONDITIONING
19 PLUG, INSTRUMENT PANEL
20 SWITCH, HEADLAMP
21 COMPARTMENT, FRONT FLOOR CONSOLE FRONT
23 RETAINER, FRONT FLOOR CONSOLE
24 PLATE, FRONT FLOOR CONSOLE TRANSAXLE SHIFT OPENING TRIM
25 HOUSING, FRONT FLOOR CONSOLE LIGHTER ELEMENT
26 ELEMENT, FRONT FLOOR CONSOLE LIGHTER
27 ASHTRAY, FRONT FLOOR CONSOLE
28 LENS, FRONT FLOOR CONSOLE TRANSAXLE SHIFT OPENING TRIM PLATE
29 TRIM, FRONT FLOOR CONSOLE ARMREST
30 ARMREST, FRONT FLOOR CONSOLE
31 MAT, FRONT FLOOR CONSOLE
32 ASHTRAY, FRONT FLOOR CONSOLE REAR
33 CONSOLE, FRONT FLOOR
34 CUP HOLDER, FRONT FLOOR CONSOLE
35 PLATE, FRONT FLOOR CONSOLE TRIM
36 CLUSTER, INSTRUMENT
37 SWITCH, REAR COMPARTMENT LID RELEASE
38 ASHTRAY, INSTRUMENT PANEL
39 CUP HOLDER, INSTRUMENT PANEL
40 FILLER, INSTRUMENT PANEL STEERING COLUMN OPENING
41 INSULATOR, INSTRUMENT PANEL LEFT SOUND
42 AUDIO SYSTEM
43 INSTRUMENT PANEL ASSEMBLY

93111G03

Exploded view of the instrument panel—Century, Grand Prix, Intrigue, Lumina, Monte Carlo and Regal

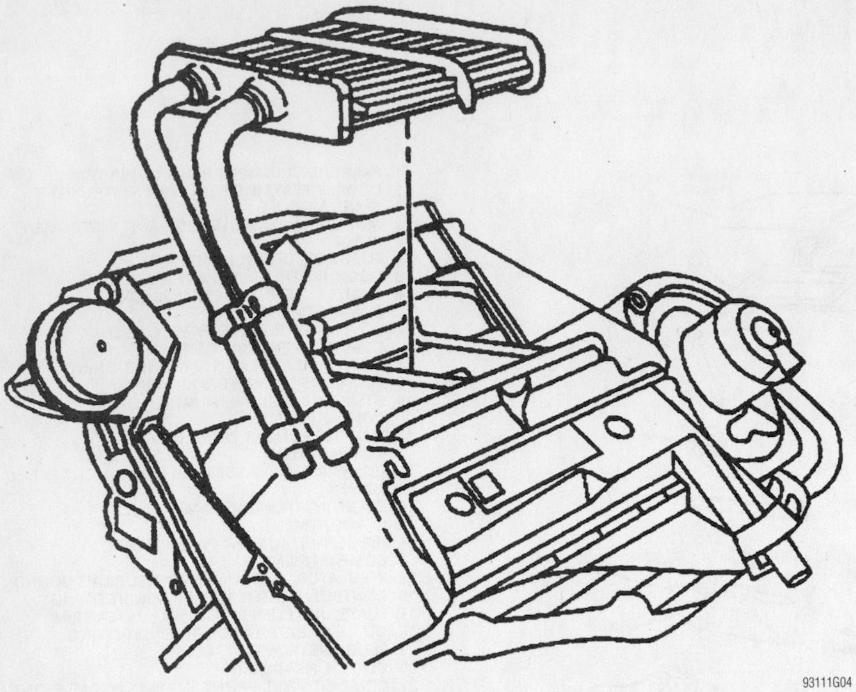

View of the heater core—Century, Grand Prix, Intrigue, Lumina, Monte Carlo and Regal

To install:

8. Install and connect the following:
   - Heater core in the vehicle
   - Heater core mounting clip and bracket
   - New seals on the heater core cover
   - Heater core cover and torque the bolts to 13 inch lbs. (1.5 Nm)
   - Lower heater duct, if removed
   - Ashtray and bracket, if removed
9. Install the lower instrument panel lower by installing or connecting the following:
   - Electrical connector to the compartment
   - Slide the compartment into the instrument panel; then, secure it with screws and plastic clips
   - Instrument panel compartment door and screws
   - Compartment-to-panel bolts and screws, located under the instrument panel compartment
   - Lower right instrument panel insulator
10. If equipped, install the lower center console install or connect the following:
   - Console in the vehicle and connect the electrical connectors
   - Console-to-chassis bolts and screws, tighten in sequence beginning at the front right and continue in a clockwise order to 106 inch lbs. (12 Nm)
   - Armrest compartment mat and lower the front floor console armrest
   - C/D storage compartment into the front floor console
   - Electrical connectors to the upper console trim plate and snap the trim plate onto the console
   - Automatic transaxle shift handle and the cigarette lighter
   - Heater hoses to the heater core and secure with the clamps
11. Refill the cooling system by performing the following procedure:
   a. Close the radiator drain valve.
   b. If the air bleed screws (located at the top of the thermostat housing) is closed, open it by turning it 2–3 turns.
   c. Slowly add coolant until it reaches the radiator neck.
   d. Wait for 2 minutes and recheck the coolant level; then, add more coolant if necessary.
   e. Fill the coolant reservoir to the COLD mark.
   f. Close the air bleed screw.

⁂ WARNING

Do not over-tighten the air bleed screw for it is made of brass.

   g. Install the radiator cap and make sure that the arrows align with the overflow tube.
12. If equipped with a 3.8L engine, install the fuel injector sight shield by performing the following procedures:
   a. Remove the oil filler cap/tube assembly from the valve cover.
   b. Slide the fuel injector sight shield into the rear engine bracket and lower it into place.
   c. Reinstall the oil filler cap/tube assembly in the valve cover. Twist it clockwise to lock the detent on the tube into the notch in the valve cover.
13. If equipped with a 3.1L engine, install the air cleaner and duct assembly.
14. Connect the negative battery cable.

Grand Prix, Intrigue and Regal

REMOVAL & INSTALLATION

1. Disconnect the negative battery cable.
2. Drain the engine coolant into a clean container for reuse.
3. On Grand Prix or Regal equipped with a 3.1L engine, remove the air cleaner and duct assembly.
4. On Grand Prix or Regal equipped with a 3.8L engine, remove the fuel injector sight shield.
5. Remove or disconnect the following:
   - Lower floor console, if equipped
   - Both instrument panel insulators
   - Heater core outlet cover screws and the outlet cover
   - Heater core cover screws and the cover
   - Heater core seals. Discard the seals
   - Heater core outer seal. Discard the seal
   - Heater core line clamp screw, the retaining clamp and the pipe retainer clamp screw
   - Heater core from the lower case
   - Heater core lower, center, upper and side seals from the lower heater core case. Discard the seals

To install:

6. Install or connect the following:
   - New heater core lower, center, upper and side seals to the lower heater core case
   - Heater core to the lower case
   - Pipe retainer clamp screw, the retaining clamp and the heater core line clamp screw
   - Heater core outer seal
   - Heater core seals
   - Heater core cover and the outlet cover screws, then, tighten the screws to 13 inch lbs. (1.5 Nm)
   - Heater core outlet cover and the outlet cover screws, then, tighten the screws to 13 inch lbs. (1.5 Nm)

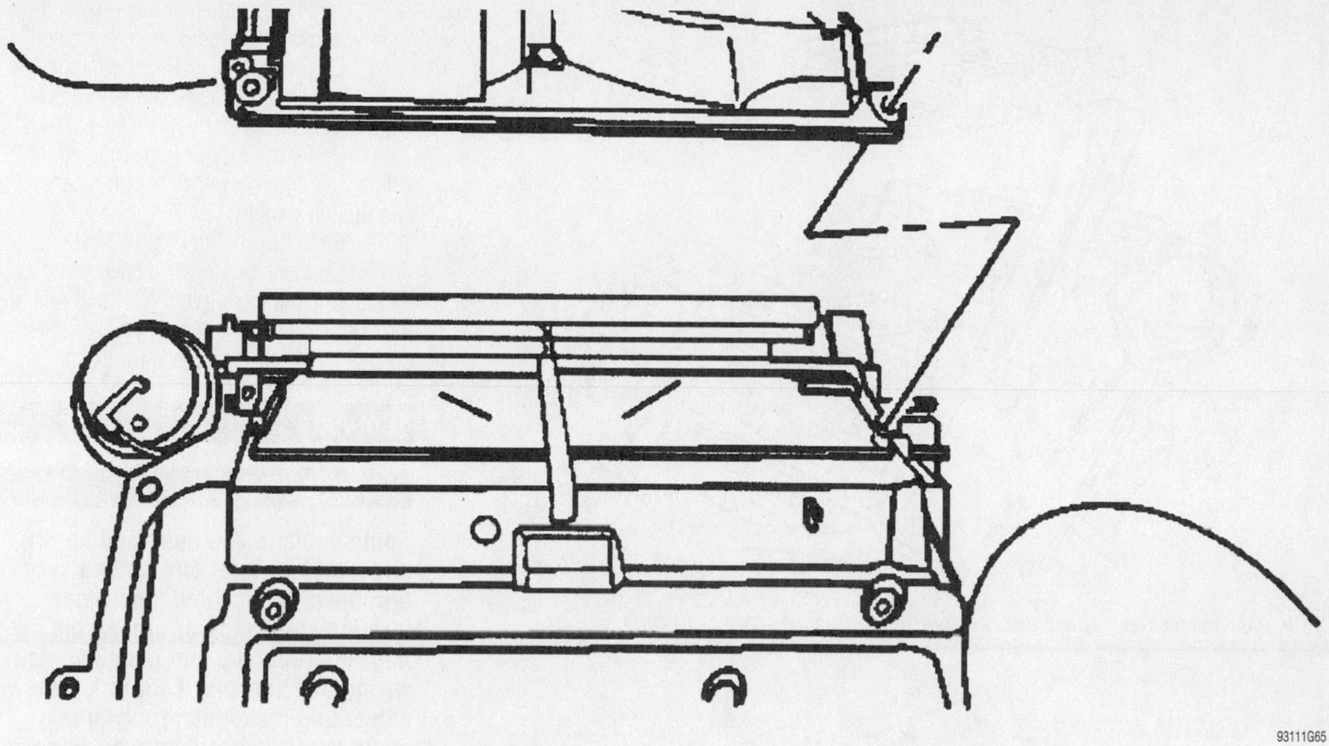

View of the heater core outlet cover—Grand Prix, Intrigue and Regal

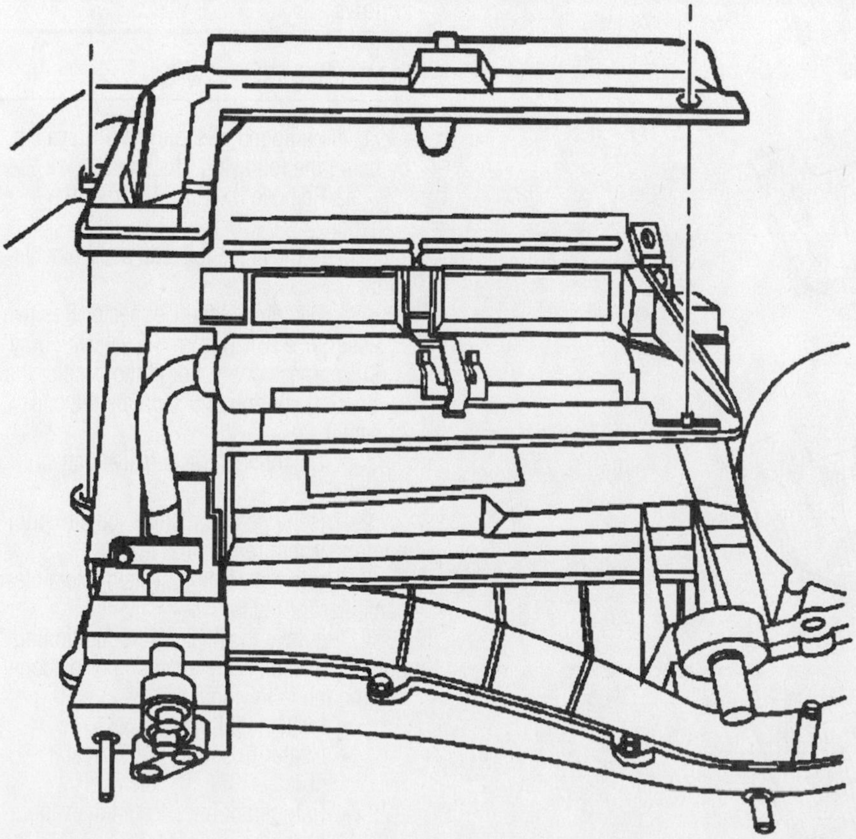

View of the heater core cover—Grand Prix, Intrigue and Regal

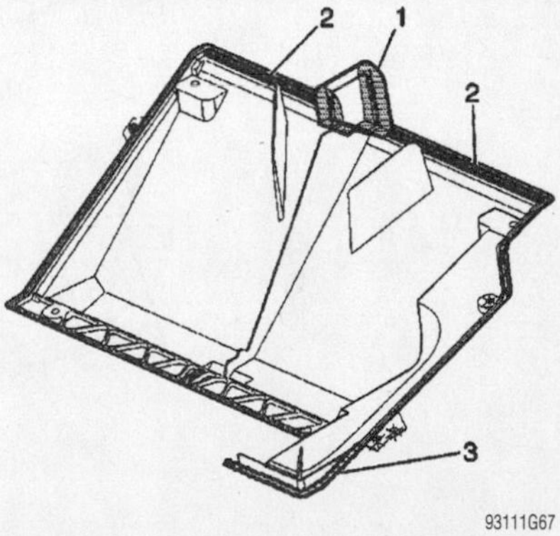

Location of the heater core cover seals—Grand Prix, Intrigue and Regal

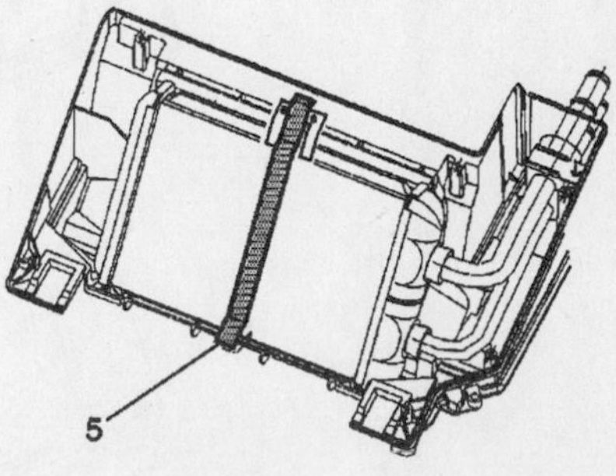

View of the heater core outer seal—Grand Prix, Intrigue and Regal

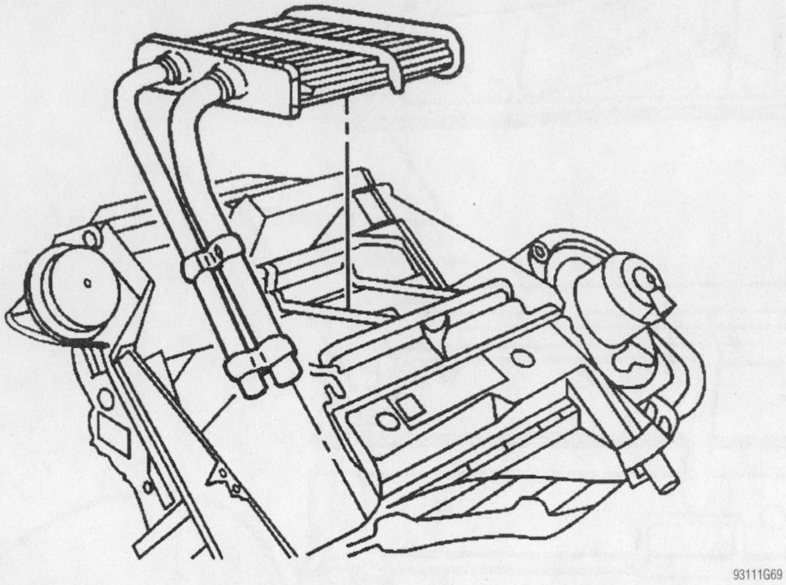

View of the heater core—Grand Prix, Intrigue and Regal

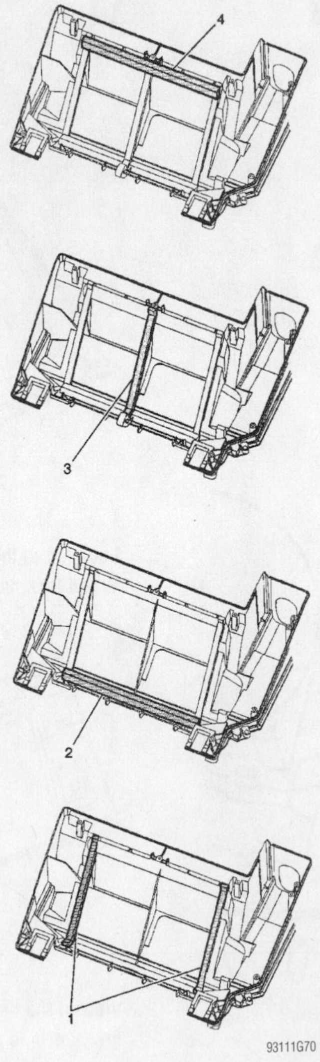

View of the heater core lower, center, upper and side seals of the lower heater core case—Grand Prix, Intrigue and Regal

- Both instrument panel insulators
- Lower floor console, if equipped

7. On Grand Prix or Regal equipped with a 3.8L engine, install the fuel injector sight shield.
8. On Grand Prix or Regal equipped with a 3.1L engine, install the air cleaner and duct assembly.
9. Refill the engine cooling system.
10. Connect the negative battery cable.
11. Operate the engine to normal operating temperatures; then, check the climate control operation and check for leaks.

Corvette

✱✱ CAUTION

Some vehicles are equipped an SIR or air bag system. The air bag system must be disabled before performing service on or around the air bag, instrument panel components, wiring and sensors. Failure to follow safety and disabling procedures could result in accidental air bag deployment, possible personal injury and unnecessary air bag system repairs.

REMOVAL & INSTALLATION

1. If equipped, disable the SIR system by using the following procedure:
   a. Remove the SIR fuse from the fuse panel.
   b. Remove the left side sound insulator.
   c. Disconnect the Connector Positive Assurance (CPA) from the yellow 2-way SIR harness connector at the base of the steering column and separate the connector.
   d. Disconnect the negative battery cable.
2. Discharge and recover the air conditioning system refrigerant.
3. Drain the cooling system into a clean container for reuse.
4. Remove the heater/air conditioning housing assembly by removing or disconnecting the following:
   - Intake manifold
   - Heater hoses from the heater core
   - Refrigerant lines from the evaporator core
   - Drain tube from the heater/air conditioning housing assembly
   - Upper instrument pad and the ventilation duct assembly

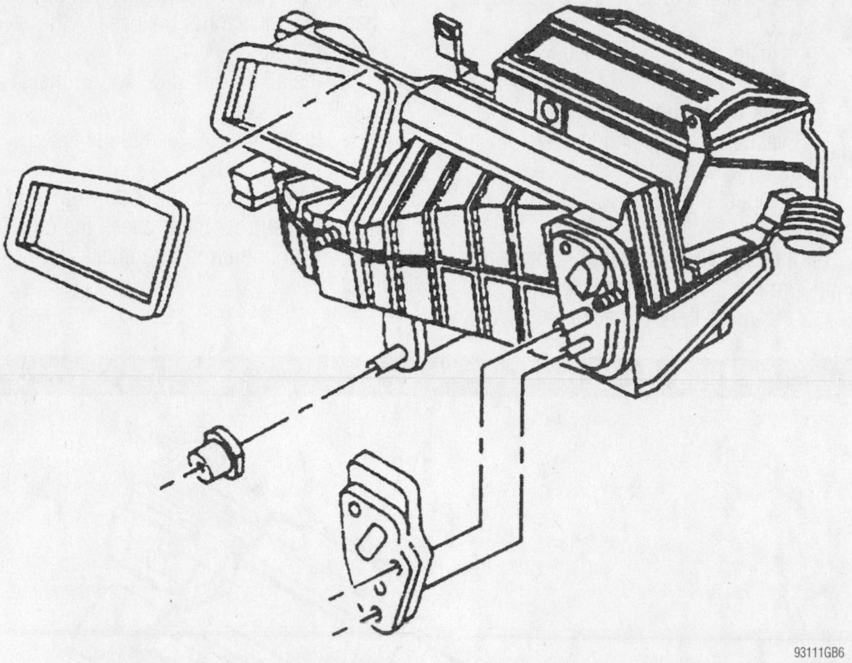

View of the heater/air conditioning housing assembly—Corvette

- Electrical connector from the instrument cluster and remove the instrument cluster
- Inboard side window glass defroster duct from the windshield defroster duct on both sides
- DRL electrical connector and the sun load temperature sensor from the windshield defroster duct, if equipped
- Windshield defroster duct from the heater/air conditioning housing assembly
- Inside air temperature sensor duct from the heater/air conditioning housing assembly
- Passenger's side SIR module electrical connector
- Passenger's side SIR bracket and module assembly
- Instrument panel center support bracket.
- Electrical and vacuum connectors from the heater/air conditioning control head
- Knee bolster
- Driver's floor hush panel
- Floor air outlet-to-heater/air conditioning housing assembly duct from the left side
- Passenger's side floor hush panel
- Lower air floor air duct from the upper air duct from the right side
- Upper floor air duct from the heater/air conditioning housing from the right side
- Blower motor electrical connector
- Blower motor from the heater/air conditioning housing
- Temperature door and the vacuum solenoid electrical connectors, if equipped with the CJ2 air conditioning system
- Vacuum source line from the instrument panel harness connection
- Temperature door control motor electrical connector
- Rear floor air ducts from the heater/air conditioning housing from both side
- Heater/air conditioning housing-to-bulkhead/heater core module and instrument panel nuts/bolts
- Heater/air conditioning housing assembly
- Air inlet, drain and plumbing seals. Discard the seals
- Heater core cover from the heater/air conditioning housing assembly
- Heater core cover seals and discard the seals
- Outer heater core seal and discard the seal
- Heater core-to-heater/air conditioning housing mounting clip screws
- Heater core pipe screw
- Heater core

To install:

5. Install or connect the following:
   - Heater core
   - Heater core pipe screw
   - Heater core-to-heater/air conditioning housing mounting clip screws
   - New outer heater core seal, if equipped with the CJ2 air conditioning system
   - New heater core cover seals
   - Heater core cover to the heater/air conditioning housing assembly
6. Install the heater/air conditioning housing assembly by installing or connecting the following:
   - New air inlet, drain and plumbing seals
   - Heater/air conditioning housing assembly
   - Heater/air conditioning housing-to-bulkhead/heater core module and instrument panel nuts/bolts
   - Rear floor air ducts to the heater/air conditioning housing on both sides
   - Temperature door control motor electrical connector
   - Vacuum source line to the instrument panel harness connection
   - Temperature door and the vacuum solenoid electrical connectors, if equipped with the CJ2 air conditioning system
   - Blower motor to the heater/air conditioning housing
   - Blower motor electrical connector
   - Upper floor air duct to the heater/air conditioning housing on the right side
   - Lower air floor air duct to the upper air duct on the right side
   - Passenger's side floor hush panel
   - Floor air outlet-to-heater/air conditioning housing assembly duct on the left side
   - Driver's floor hush panel
   - Knee bolster
   - Electrical and vacuum connectors to the heater/air conditioning control head
   - Instrument panel center support bracket
   - Passenger's side SIR bracket and module assembly
   - Inside air temperature sensor duct
   - Windshield defroster duct to the heater/air conditioning housing assembly
   - DRL electrical connector and the sun load temperature sensor, if

equipped with a windshield defroster duct
- Inboard side window glass defroster duct to the windshield defroster duct on both sides
- Electrical connector to the instrument cluster and install the instrument cluster
- Upper instrument pad and the ventilation duct assembly
- Drain tube to the heater/air conditioning housing assembly
- Refrigerant lines to the evaporator core
- Heater hoses to the heater core
- Intake manifold

7. Refill the cooling system.
8. Evacuate, charge and leak test the air conditioning system refrigerant.
9. Connect the negative battery cable.
10. If equipped, enable the SIR system by performing the following procedure:
    a. Connect the Connector Positive Assurance (CPA) to the yellow 2-way SIR harness connector at the base of the steering column.
    b. Install the left side sound insulator.
    c. Install the SIR fuse to the fuse panel.
11. Operate the engine to normal operating temperatures; then, check the climate control operation and check for leaks.

1998–01 GEO

Metro

REMOVAL & INSTALLATION

1. Disconnect the negative battery cable.
2. Properly drain the cooling system into a clean container for reuse.
3. Remove the instrument panel.
4. Remove or disconnect the following:
   - Heater hoses from the heater core tubes at the firewall
   - Temperature and mode control cables from the heater housing
   - Air duct from the heater housing to the blower motor assembly (on models not equipped with A/C); otherwise, detach the heater housing from the evaporator assembly
   - 2 bolts and nuts and remove the heater case assembly
   - Heater case halves
   - Heater core from the heater case assembly

To install:

5. Install or connect the following:
   - Heater core into the heater case assembly
   - Assemble the heater case assembly
   - Heater assembly and install the 2 bolts and nuts
   - Heater case assembly to the evaporator case assembly, if equipped with A/C
   - Mode control and the temperature cables to the heater housing
   - Heater hoses to the heater core
6. Install the instrument panel.
7. Refill the cooling system.
8. Connect the negative battery cable.
9. Operate the engine to normal operating temperatures; then, check for leaks and the heater operation.

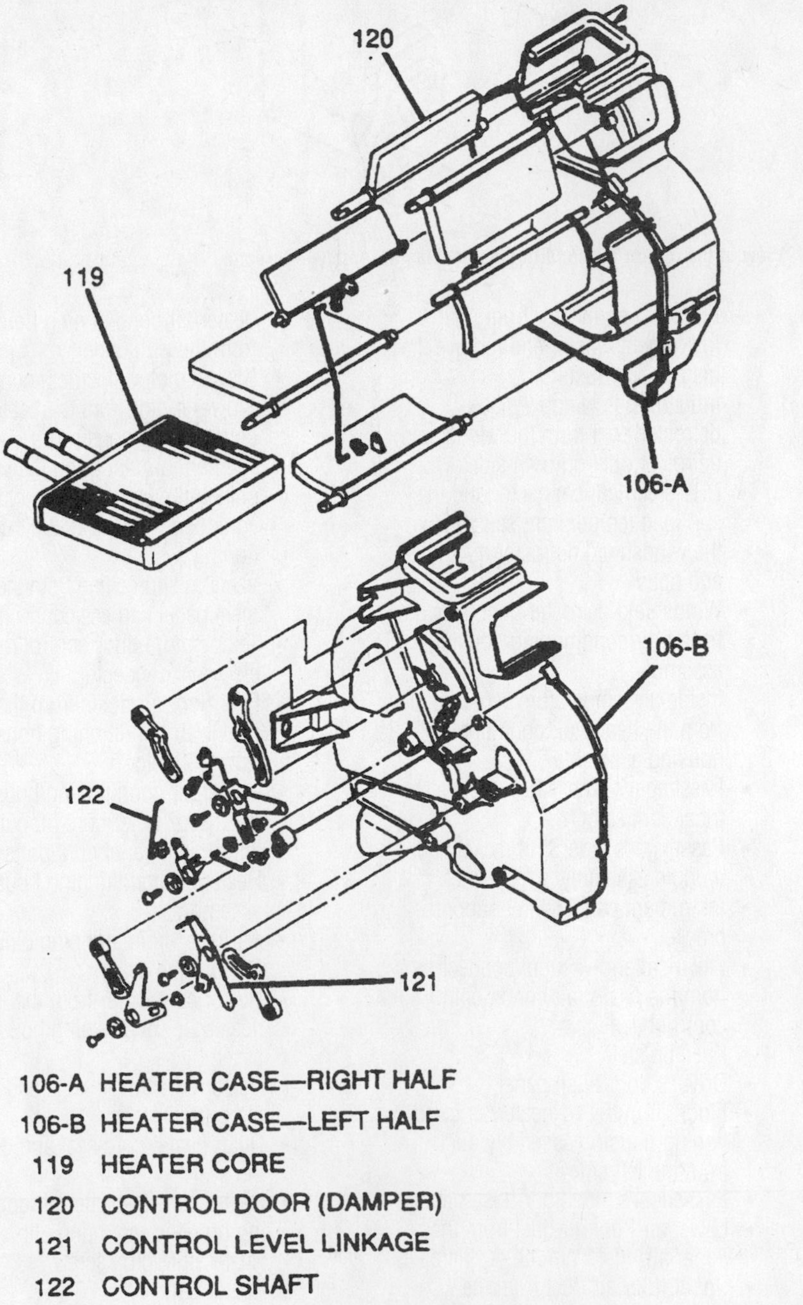

Exploded view of the heater core, heater case and related components—Metro

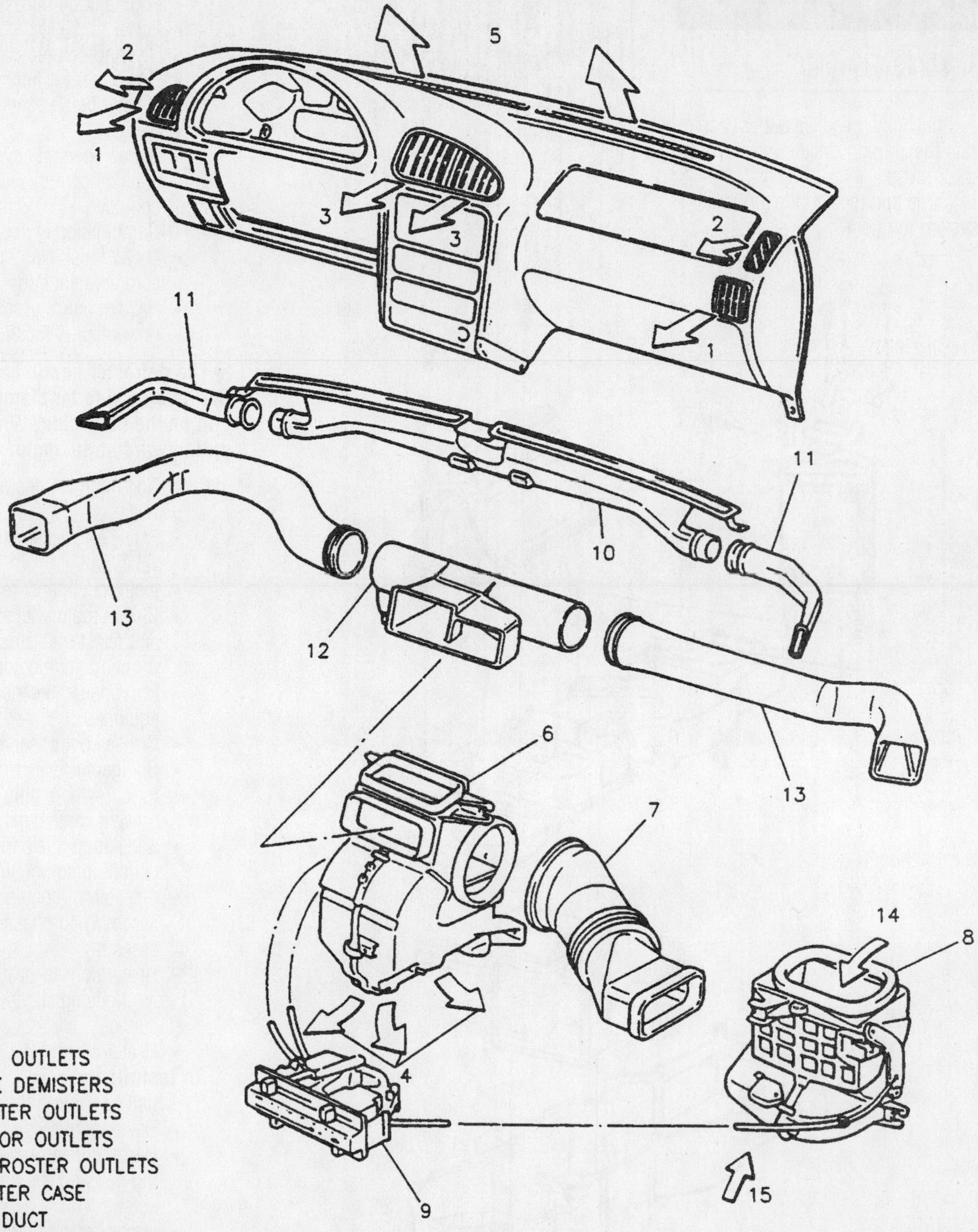

1 SIDE OUTLETS
2 SIDE DEMISTERS
3 CENTER OUTLETS
4 FLOOR OUTLETS
5 DEFROSTER OUTLETS
6 HEATER CASE
7 AIR DUCT
8 BLOWER MOTOR CASE
9 HEATER CONTROL UNIT
10 DEFROSTER DUCT
11 SIDE DEMISTER DUCTS
12 CENTER VENT DUCT
13 SIDE VENT DUCTS
14 OUTSIDE AIR
15 RECIRCULATED AIR

93111GC3

Exploded view of the instrument panel, heater housing, ventilation ducts and related components—Metro

Prizm

REMOVAL & INSTALLATION

1. Disconnect the negative battery cable.
2. Drain the cooling system into a clean container for reuse.
3. Discharge and recover the air conditioning system refrigerant.
4. Remove the heater hoses from the heater core.
5. Remove the instrument panel by removing or disconnecting the following:
   - Steering wheel-disarm air bag system if equipped
   - Right and left front pillar garnish trim
   - Floor console bin
   - Engine hood release lever
   - Lower finish No. 1 trim panel
   - Steering column cover
   - Center cluster finish panel
   - Cluster finish panel
   - Radio
   - Stereo opening cover or center differential control switch if so equipped
   - Combination meter assembly
   - Lower finish No. 2 panel with glove compartment door
   - Heater control assembly
   - Lower center finish panel

➡The defroster nozzle has a boss on the reverse side for clamping onto the clip on the body side. When removing, pull upward at an angle.

   - No. 1 and No. 2 side defroster nozzles
   - Safety pad assembly from the vehicle
   - Instrument panel reinforcement
   - Blower motor electrical connector and resistor electrical connector
   - Air conditioning compressor control module electrical connector, if equipped
   - Cruise control servo
   - Refrigerant lines from the evaporator core; then, plug the openings to prevent contamination
   - Ventilation ducts from the heater/air conditioning housing
   - 6 heater/air conditioning housing assembly-to-chassis nuts and the assembly
   - Heater core-to-heater/air conditioning housing clamp screws and the clamps
   - Heater core

To install:

6. Install or connect the following:
   - Heater core
   - Heater core-to-heater/air conditioning housing clamps and the clamp screws
   - Heater/air conditioning housing assembly and the 6 assembly-to-chassis nuts
   - Ventilation ducts to the heater/air conditioning housing
   - Refrigerant lines to the evaporator core
   - Cruise control servo, if equipped
   - Air conditioning compressor control module electrical connector
   - Blower motor electrical connector and resistor electrical connector
   - Instrument panel reinforcement

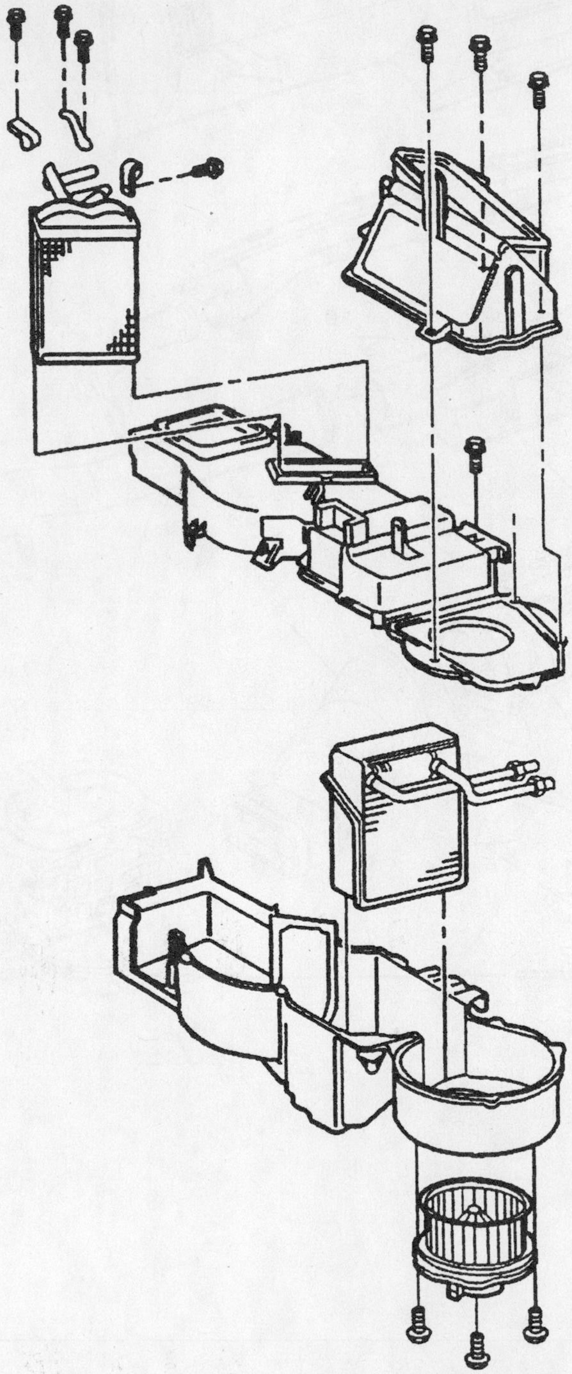

Exploded view of the heater core, the evaporator core, the heater/air conditioning housing and related components—Prizm

- Safety pad assembly in the vehicle
- No. 1 and No. 2 side defroster nozzles
- Lower center finish panel
- Heater control assembly
- Lower finish No. 2 panel with glove compartment door
- Combination meter assembly
- Stereo opening cover or center differential control switch if so equipped
- Radio
- Cluster finish panel
- Center cluster finish panel
- Steering column covers
- No. 1 lower finish panel
- Engine hood release lever
- Floor console box
- Right and left front pillar garnish trim
- Steering wheel
- Heater hoses to the heater core

7. Evacuate, charge and leak test the air conditioning system refrigerant.
8. Refill the cooling system.
9. Connect the negative battery cable.

1998–01 SATURN

REMOVAL & INSTALLATION

1. Disconnect the negative battery cable.
2. Drain the cooling system into a clean container for reuse.
3. Raise and safely support the vehicle.
4. Heater hoses from the heater core

➡Carefully, use compressed air to blow the remainder of the coolant from the heater core to prevent spillage of coolant in the interior when removing the heater core.

5. Remove or disconnect the following
   - Lower trim panel extensions (located at both sides of the console-to-center instrument panel), by pulling outward at the dual lock locations; then, rotate the panels outward to disengage the hinges at the console
   - Lower instrument panel closeout panel (located on the right side of the heater/air conditioning housing), by pulling it out at the top edge and rotating the top downward
   - Lower heater duct-to-heater/air conditioning housing 2 screws and 2 clips (located at the bottom of the heater/air conditioning housing); then, drop the duct down and carefully, slide it out sideways

✲✲ WARNING

Be careful not to damage the heater duct-to-rear floor heater duct seal.

   - Release the retaining clip and disconnect the temperature control cable from the heater door hook
   - Heater core side cover
   - 4 lower heater core cover-to-heater/air conditioning housing assembly screws and the cover
   - Heater core clamp and the heater core

To install:

6. Install or connect the following:
   - Heater core, the clamp and the heater core-to-heater/air conditioning housing assembly screw
   - Lower heater core cover and the 4 cover-to-heater/air conditioning housing assembly screws
   - Heater core side cover and the cover screw, located at the left side of the heater/air conditioning housing assembly
   - Temperature control cable to the heater door hook and secure with the retaining clip

✲✲ WARNING

Be careful not to damage the heater duct-to-rear floor heater duct seal

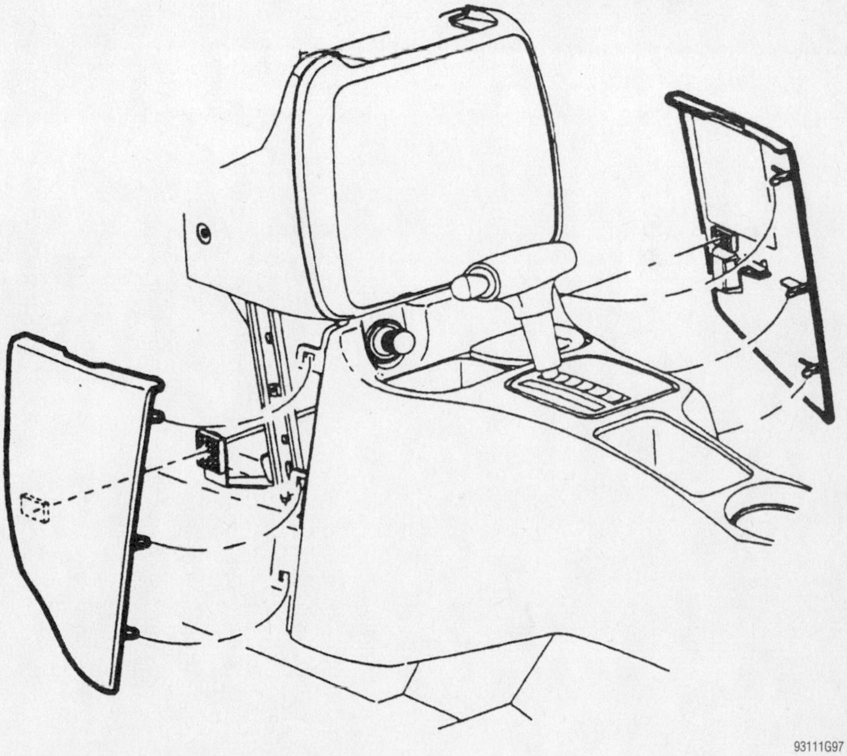

View of the lower trim panel extensions—Saturn

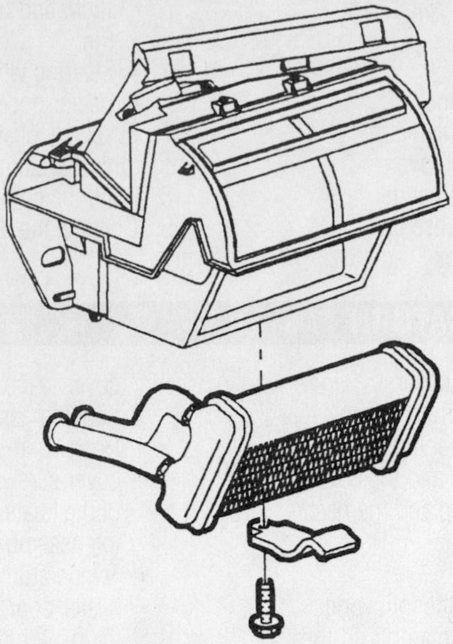

Exploded view of the heater core and heater case—Saturn

- Lower heater duct; then, the lower heater duct-to-heater/air conditioning housing 2 screws and 2 clips, located at the bottom of the heater/air conditioning housing
- Lower instrument panel closeout panel, located at the right side of the heater/air conditioning housing
- Console-to-center instrument panel location (on both sides), install the lower trim panel extensions
- Heater hose-to-heater core clamps

7. Lower the vehicle.
8. Refill the cooling system.
9. Connect the negative battery cable.
10. Operate the engine to normal operating temperatures; then, check the climate control operation and check for leaks.

TIMING BELTS

3

GENERAL INFORMATION

Timing belts are typically only used on overhead camshaft engines. Timing belts are used to synchronize the crankshaft with the camshaft, similar to a timing chain on an overhead valve (pushrod) engine. Unlike a timing belt, a timing chain will normally last the life of the engine without needing service or replacement. Timing belts use raised teeth to mesh with sprockets to operate the valvetrain of an overhead camshaft engine.

Whenever a vehicle with an unknown service history comes into your repair facility or is recently purchased, here are some points that should be asked to help prevent costly engine damage:

- Does the owner know if or when the belt was replaced?
- If the vehicle purchased is used or the condition and mileage of the last timing belt replacement are unknown, it is recommended to inspect, replace or at least inform the owner that the vehicle is equipped with a timing belt.
- Note the mileage of the vehicle. The average replacement interval for a timing belt is approximately 60,000 miles (96,000 km).

Interference Engines

Engines, chain- or belt-driven, can be classified as either free-running or interference, depending on what would happen if the piston-to-valve timing is disrupted. A free-running engine is designed with enough clearance between the pistons and valves to allow the crankshaft to rotate (pistons still moving) while the camshaft stays in one position (several valves fully open). If this condition occurs normally, no internal engine damage will result. In an interference engine, there is not enough clearance between the pistons and valves to allow the crankshaft to turn without the camshaft being in time.

An interference engine can suffer extensive internal damage if a timing belt fails. The piston design does not allow clearance for the valve to be fully open and the piston to be at the top of its stroke. If the belt fails, the piston will collide with the valve and will bend or break the valve, damage the piston, and/or bend a connecting rod. When this type of failure occurs, the engine will need to be replaced or disassembled for further internal inspection; either choice costing many times that of replacing the timing belt.

TIMING BELT SERVICE

Inspection

➡For manufacturer's recommended service interval, refer to the maintenance interval chart located in this manual.

The average replacement interval for a timing belt is approximately 60,000 miles (96,000km). If, however, the timing belt is inspected earlier or more frequently than suggested, and shows signs of wear or defects, the belt should be replaced at that time.

⁂ WARNING

Never allow antifreeze, oil or solvents to come into with a timing belt. If this occurs immediately wash the solution from the timing belt. Also, never excessive bend or twist the timing belt; this can damage the belt so that its lifetime is severely shortened.

Inspect both sides of the timing belt. Replace the belt with a new one if any of the following conditions exist:

- Hardening of the rubber—back side is glossy without resilience and leaves no indentation when pressed with a fingernail
- Cracks on the rubber backing
- Cracks or peeling of the canvas backing
- Cracks on rib root
- Cracks on belt sides
- Missing teeth or chunks of teeth
- Abnormal wear of belt sides—the sides are normal if they are sharp, as if cut by a knife.

If none of these conditions exist, the belt does not need replacement unless it is at the recommended interval. The belt MUST be replaced at the recommended interval.

TCCS1242

Never bend or twist a timing belt excessively, and do not allow solvents, antifreeze, gasoline, acid or oil to come into contact with the belt

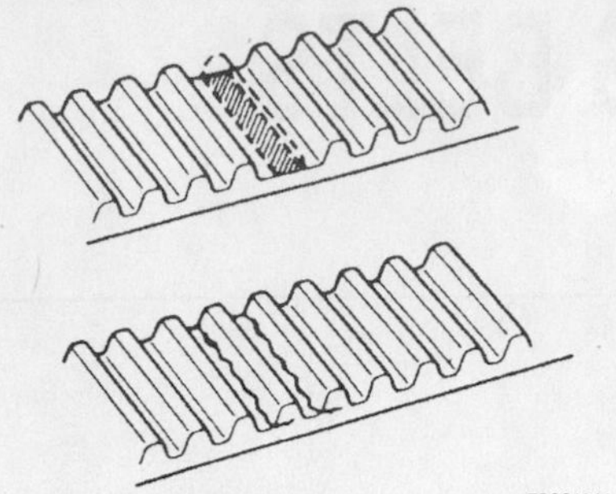

TCCS1244

Broken tooth may be due to a damaged pulley

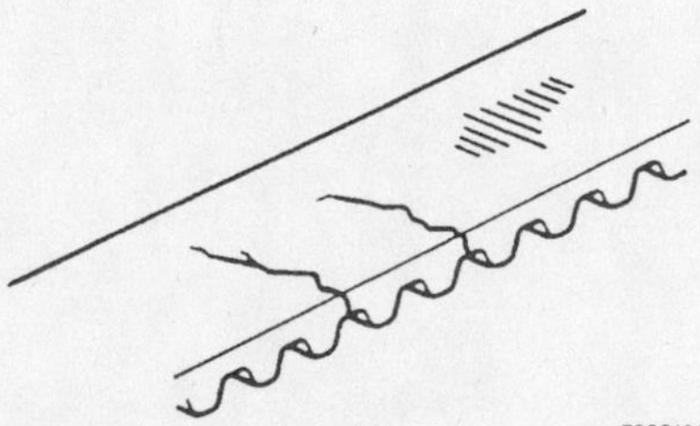

TCCS1245

Back surface worn or cracked from a possible overheated engine or interference with the belt cover

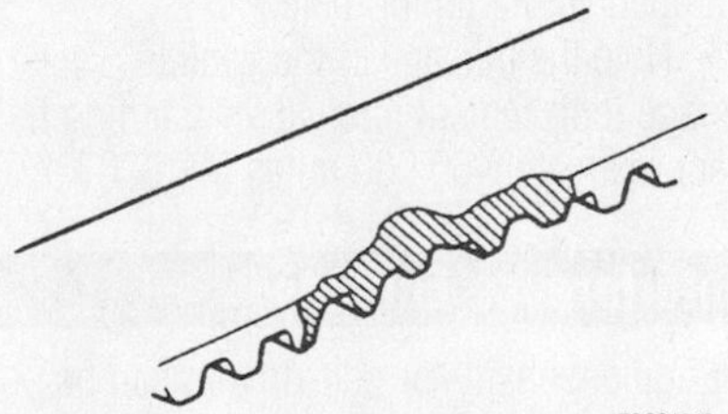

TCCS1246

Side wear from improper installation

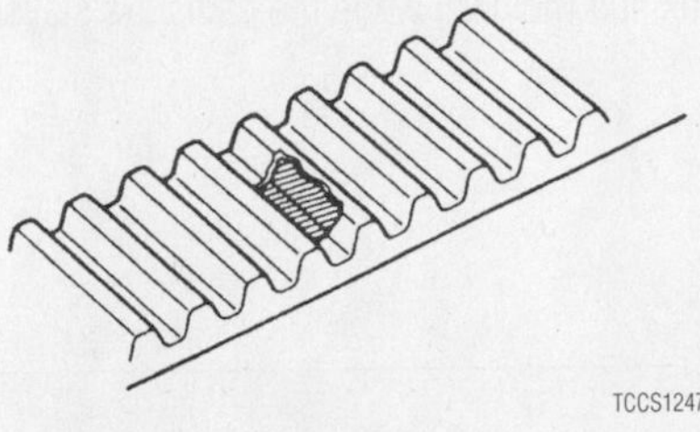

TCCS1247

Worn teeth from excessive belt tension, camshaft or distributor not turning properly or fluid leaking on the belt

⁂ WARNING

On interference engines, it is very important to replace the timing belt at the recommended intervals, otherwise expensive engine damage will likely result if the belt fails.

Removal & Installation

CHRYSLER CORPORATION

2.0L (VIN C) SOHC Engine

1. On Cirrus, Stratus, Breeze, and Sebring Convertible models, disconnect the negative battery cable from the left strut tower. The ground cable is equipped with a insulator grommet which should be placed on the stud to prevent the negative battery cable from accidentally grounding.
2. On Neon models, disconnect the negative battery cable.
3. Remove the drive belts and accessories.
4. Remove the right inner splashshield.
5. Remove the crankshaft damper.
6. Remove the right engine mount.
7. Place a support under the engine.
8. Remove the engine mount bracket.
9. Remove the timing belt cover.
10. Loosen the timing belt tensioner bolts.
11. Remove the timing belt and the tensioner.

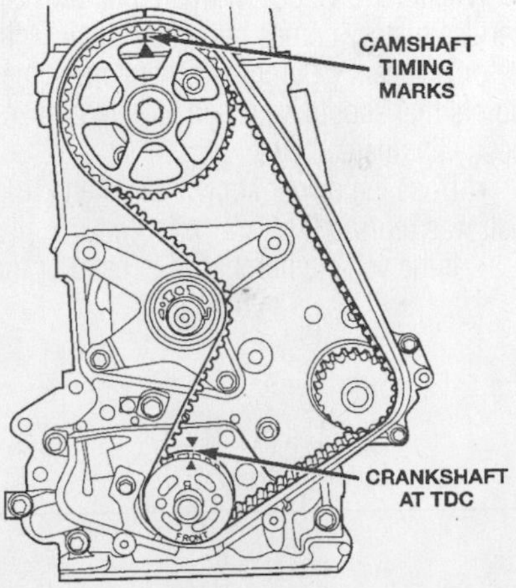

79225G99

TDC alignment for timing belt installation—2.0L (VIN C) SOHC Engine

➡When tensioner is removed from the engine it is necessary to compress the plunger into the tensioner body.

12. Place the tensioner in a soft-jawed vise to compress the tensioner.
13. After compressing the tensioner place a pin (a 5/64 in. Allen wrench will work) into the plunger side hole to retain the plunger until installation.

To install:

14. Set the crankshaft sprocket to Top Dead Center (TDC) by aligning the notch on the sprocket with the arrow on the oil pump housing, then back off the sprocket three notches before TDC.
15. Set the camshaft to align the timing marks.
16. Move the crankshaft to ½ notch before TDC.
17. Install the timing belt starting at the crankshaft, around the water pump, then around the camshaft last.
18. Move the crankshaft to TDC to take up the belt slack.
19. Reinstall the tensioner to the block but do not tighten it.
20. Using a torque wrench apply 250 inch lbs. (28 Nm) of torque to the tensioner pulley.
21. With torque being applied to the tensioner pulley, move the tensioner up against the tensioner bracket and tighten the fasteners to 275 inch lbs. (31 Nm).
22. Remove the tensioner plunger pin, the tension is correct when the plunger pin can be removed and replaced easily.
23. Rotate the crankshaft two revolutions and recheck the timing marks.
24. Reinstall the timing belt cover.
25. Reinstall the engine mount bracket.
26. Reinstall the right engine mount.
27. Remove the engine support.
28. Reinstall the crankshaft damper and tighten to 105 ft. lbs. (142 Nm).
29. Reinstall the drive belts and accessories.
30. Reinstall the right inner splash-shield.
31. Perform the crankshaft and camshaft relearn alignment procedure using the DRB scan tool or equivalent.

2.0L (VIN F) Engine

1. Disconnect the negative battery cable.
2. Remove the engine undercover.
3. Remove the engine mount bracket.
4. Remove the drive belts.

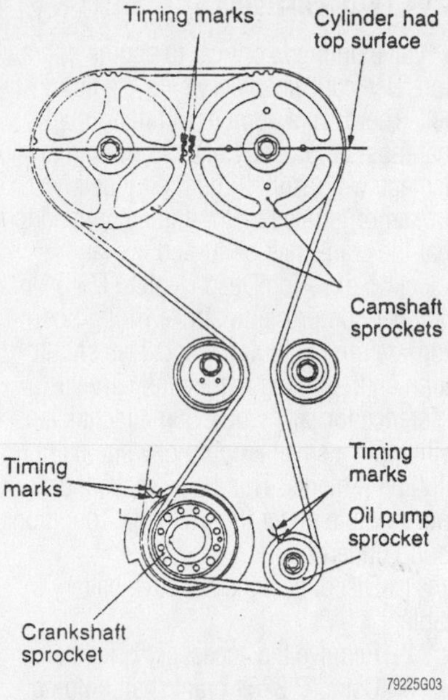

Notice the timing mark on the oil pump drive sprocket—Chrysler 2.0L (VIN F) engines

5. Remove the belt tensioner pulley.
6. Remove the water pump pulleys.
7. Remove the crankshaft pulley.
8. Remove the stud bolt from the engine support bracket and remove the timing belt covers.
9. Rotate the crankshaft clockwise to align the camshaft timing marks. Always turn the crankshaft in the forward direction only.
10. Loosen the tension pulley center bolt.

➡If the timing belt is to be reused, mark the direction of rotation on the flat side of the belt with an arrow.

11. Move the tension pulley towards the water pump and remove the timing belt.
12. Remove the crankshaft sprocket center bolt using special tool MB990767 to hold the crankshaft sprocket while removing the center bolt. Then, use MB998778 puller to remove the sprocket.
13. Mark the direction of rotation on the timing belt "B" with a arrow.
14. Loosen the center bolt on the tensioner and remove the belt.
15. To remove the camshaft sprocket, remove the cylinder head cover. Use a wrench to hold the hexagonal part of the camshaft and remove the sprocket mounting bolt.

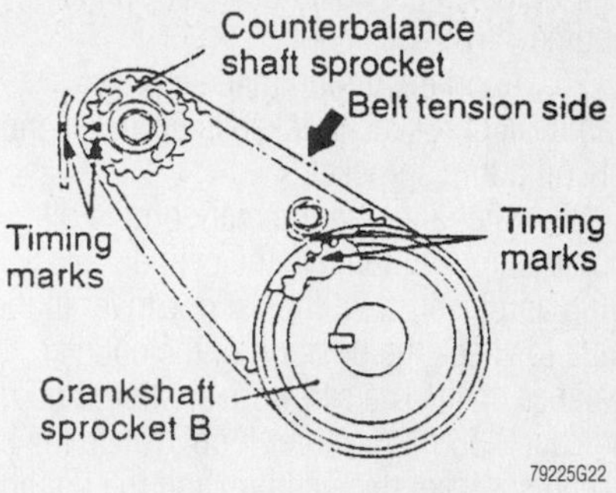

Timing belt "B" timing marks locations—Chrysler 2.0L (VIN F) engines

⁂ WARNING

Do not rotate the camshafts or the crankshaft while the timing belt is removed.

To install:

16. Use a wrench to hold the camshaft, then install the sprocket and mounting bolt. Tighten the bolt(s) to 65 ft. lbs. (88 Nm).
17. Install the cylinder head cover.
18. Place the crankshaft sprocket on the crankshaft. Use tool MB990767 to hold the crankshaft sprocket while tightening the center bolt. Tighten the center bolt to 80–94 ft. lbs. (108–127 Nm).
19. Align the timing marks on the crankshaft sprocket "B" and the balance shaft.
20. Install timing belt "B" on the sprockets. Position the center of the tensioner pulley to the left and above the center of the mounting bolt.
21. Push the pulley clockwise toward the crankshaft to apply tension to the belt and tighten the mounting bolt to 14 ft. lbs. (19 Nm). Do not let the pulley turn when tightening the bolt because it will cause excessive tension on the belt. The belt should deflect 0.20–0.28 in. (5–7mm) when finger pressure is applied between the pulleys.
22. Install the crankshaft sensing blade and the crankshaft sprocket. Apply engine oil to the mounting bolt and tighten the bolt to 80–94 ft. lbs. (108–127 Nm).
23. Use a press or vise to compress the auto tensioner pushrod. Insert a set pin when the holes are aligned.

⁂ WARNING

Do not compress the pushrod too quickly, damage to the pushrod can occur.

24. Install the auto tensioner on the engine.

25. Align the timing marks on the camshaft sprocket, crankshaft sprocket and the oil pump sprocket.

26. After aligning the mark on the oil pump sprocket, remove the cylinder block plug and insert a Phillips screwdriver in the hole to check the position of the counter balance shaft. The screwdriver should go in at least 2.36 in. or more. if not, rotate the oil pump sprocket once and realign the timing mark so the screwdriver goes in. Do not remove the screwdriver until the timing belt is installed.

27. Install the timing belt on the intake camshaft and secure it with a clip.

28. Install the timing belt on the exhaust camshaft. Align the timing marks with the cylinder head top surface using two wrenches. Secure the belt with another clip.

29. Install the belt around the idler pulley, oil pump sprocket, crankshaft sprocket and the tensioner pulley.

30. Turn the tension pulley so the pinholes are at the bottom. Press the pulley lightly against the timing belt.

31. Screw the special tool into the left engine support bracket until it contacts the tensioner arm, then screw the tool in a little more and remove the pushrod pin from the auto tensioner. Remove the special tool and tighten the center bolt to 35 ft. lbs. (48 Nm).

32. Turn the crankshaft ¼ turn counterclockwise, then clockwise until the timing marks are aligned.

33. Loosen the center bolt. Install special tool MD998767 on the tension pulley. Turn the tension pulley counterclockwise with a torque of 2.6 ft. lbs. (3.5 Nm) and tighten the center bolt to 35 ft. lbs. (48 Nm). Do not let the tension pulley turn with the bolt.

34. Turn the crankshaft two revolutions to the right and align the timing marks. After 15 minutes, measure the protrusion of the pushrod on the auto tensioner. The standard measurement is 0.150–0.177 in. (3.8–4.5mm). If the protrusion is out of specification, loosen the tension pulley, apply the proper torque to the belt and retighten the center bolt.

35. Install the crankshaft pulley. Tighten the mounting bolts to 18 ft. lbs. (25 Nm).

36. Install the water pump. Tighten the mounting bolts to 6.5 ft. lbs. (8.8 Nm).

37. Install and adjust the drive belts.

38. Install the engine mount bracket.

39. Install the engine undercover.

40. Connect the negative battery cable.

2.0L (VIN Y) Engine

Valve timing is critical to engine operation. Use care when servicing the timing belt. There are a number of timing marks that must be properly aligned or engine damage will result. If the timing belt has not broken or jumped teeth, it is recommended that the crankshaft be turned by hand (clockwise) to Top Dead Center (TDC) No.1 cylinder compression stroke (firing position) before beginning work. This should align all the timing marks and serve as a reference for later work. Some technicians will apply a small amount of white paint to all timing marks. This helps make them more visible under the low-light conditions found underhood.

1. Disconnect the negative battery cable.
2. Remove the accessory drive belts.
3. Using C 3281 crankshaft holding tool, remove the crankshaft pulley center retaining bolt.
4. Using puller tools 1026 and 6827, remove the crankshaft pulley.
5. Remove the power steering pump from the bracket and position it aside. Do not disconnect the hoses.
6. Remove the power steering pump bracket from the engine.
7. Use a floor jack with a piece of wood on it and raise the engine to take the weight off of the engine mount.
8. Remove the engine mount and bracket.
9. Remove the front timing belt cover.
10. If not done so previously, align the timing marks. Loosen the timing belt tensioner and remove the belt.

⁂ WARNING

Do not rotate the crankshaft or camshafts after removing the timing belt or valvetrain components may be damaged. Always align the timing marks before removing the timing belt.

11. For 1997 models:

    a. If the timing belt tensioner is to be replaced, remove the retaining bolts and remove the timing belt tensioner. When the timing belt tensioner is removed from the engine, it is necessary to compress the plunger into the tensioner body.

    b. Place the tensioner in a vise and slowly compress the plunger.

➡Position the tensioner in the vise the same way it will be installed on the engine. This is to ensure proper pin orientation for when the tensioner is installed on the engine.

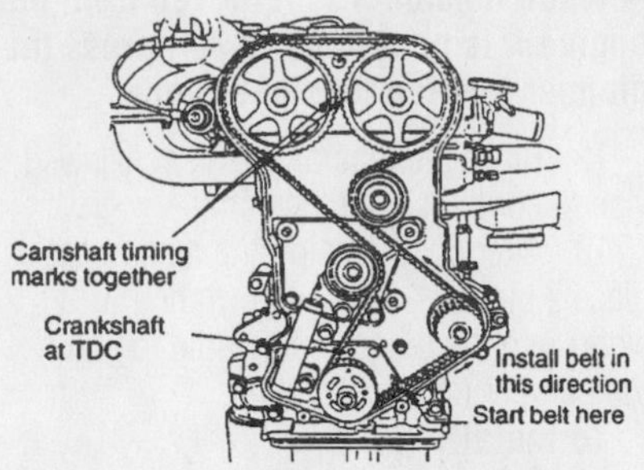

Install the timing belt by starting at the crankshaft sprocket and working around the other pulleys in a counterclockwise direction—Chrysler 2.0L (VIN Y) engine

    c. When the plunger is compressed into the tensioner body, install a pin through the body and plunger to hold the plunger in place until the tensioner is installed.

12. For 1998–01 models:

    a. Place an 8mm Allen wrench into the belt tensioner, then using the long end of a ⅛ in. (3mm) Allen wrench, rotate the tensioner counterclockwise until it slides into the locking hole.

13. Remove the timing belt.

To install:

14. Check that all timing marks are still aligned. Bring the crankshaft sprocket to ½ a notch before TDC.
15. Install the timing belt. Starting at the crankshaft, route the belt around the water pump sprocket, idler pulley, camshaft sprockets, then around the tensioner pulley.
16. Move the crankshaft to TDC to take up the slack in the belt. Install the tensioner

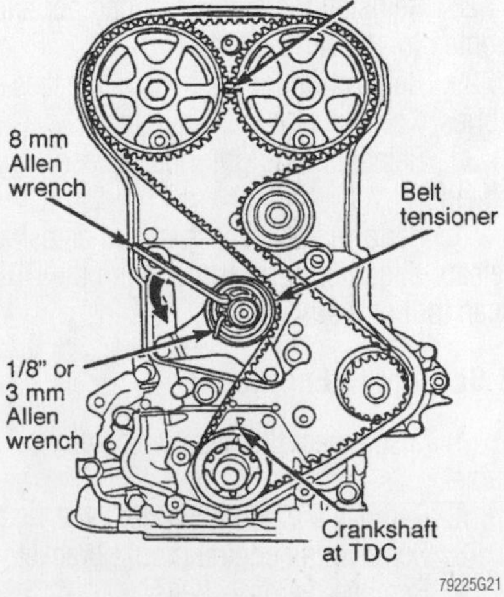

Wrench position for properly locking the tensioner on 1998–01 Chrysler 2.0L (VIN Y) engine

to the block but do not tighten the retaining bolts.

17. For 1997 models:

a. Using a torque wrench on the tensioner pulley, apply 21 ft. lbs. (28 Nm) of torque to the pulley.

b. With the torque being applied to the tensioner pulley, move the tensioner up against the tensioner pulley bracket and tighten the retaining bolts to 23 ft. lbs. (31 Nm).

c. Remove the tensioner plunger pin. Pretension is correct when the pin can be removed and installed.

18. For 1998–01 models, remove the wrenches from the tensioner.

19. Rotate the crankshaft two revolutions and check the timing marks. If the timing marks are not properly aligned, remove the belt and reinstall it as described.

20. Install the front timing belt cover.

21. Lower the engine enough to install the engine mount bracket.

22. Install the bracket and remove the floor jack.

23. Install the power steering pump bracket and pump.

24. Install the crankshaft pulley using C-4685-C pulley installer.

25. Tighten the mounting bolt to 105 ft. lbs. (142 Nm).

26. Install the accessory drive belts.

27. Connect the negative battery cable.

28. Start the engine. Check for leaks and proper engine operation.

2.4L (VIN X) Engine

1. Disconnect the negative battery cable from the left strut tower. The ground cable is equipped with a insulator grommet which should be placed on the stud to prevent the negative battery cable from accidentally grounding.

2. Remove the right inner splash-shield.

3. Remove the accessory drive belts.

4. Remove the crankshaft damper.

5. Remove the right engine mount.

6. Place a suitable floor jack under the vehicle to support the engine.

7. Remove the engine mount bracket.

8. Remove the timing belt cover.

➡Do not rotate the crankshaft or the camshafts after the timing belt has been removed. Damage to the valve components may occur. Before removing the timing belt, always align the timing marks.

9. Align the timing marks of the timing belt sprockets to the timing marks on the rear timing belt cover and oil pump cover. Loosen the timing belt tensioner bolts.

10. Remove the timing belt and the tensioner.

11. Remove the camshaft timing belt sprockets.

12. Remove the crankshaft timing belt sprocket using special removal tool No. 6793.

13. Place the tensioner into a soft-jawed vise to compress the tensioner.

14. After compressing the tensioner, place a pin (a 5/64 in. Allen wrench will work) into the plunger side hole to retain the plunger until installation.

To install:

15. Using special tool No. 6792, install the crankshaft timing belt sprocket onto the crankshaft.

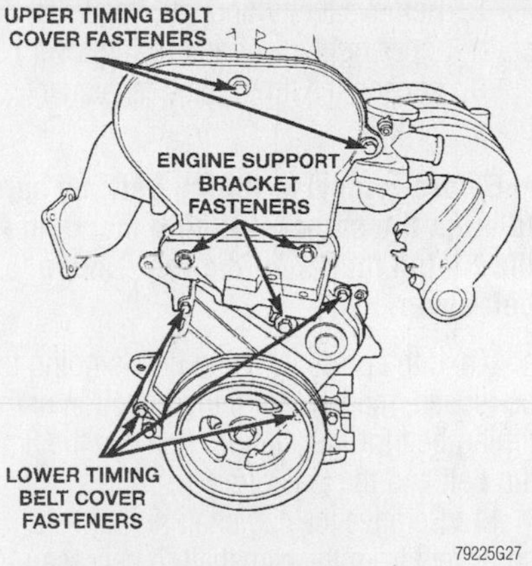

Timing belt cover bolt locations—2.4L (vin x) Engine

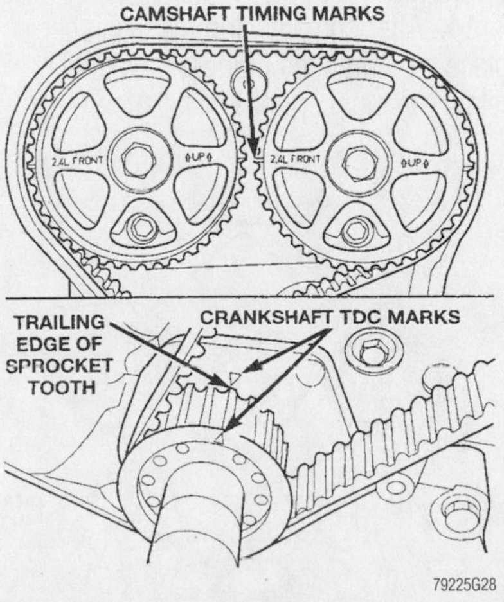

Crankshaft and camshaft alignment marks—2.4L (vin x) Engine

16. Install the camshaft sprockets onto the camshafts. Install and tighten the camshaft sprocket bolts to 75 ft. lbs. (101 Nm).

17. Set the crankshaft sprocket to Top Dead Center (TDC) by aligning the notch on the sprocket with the arrow on the oil pump housing.

18. Set the camshafts to align the timing marks on the sprockets.

19. Move the crankshaft to ½ notch before TDC.

20. Install the timing belt starting at the crankshaft, then around the water pump sprocket, idler pulley, camshaft sprockets and the tensioner pulley.

21. Move the crankshaft sprocket to TDC to take up the belt slack.

22. Reinstall the tensioner to the block but do not tighten it at this time.

23. Using a torque wrench on the tensioner pulley, apply 250 inch lbs. (28 Nm) of torque to the tensioner pulley.

24. With torque being applied to the tensioner pulley, move the tensioner up against the tensioner pulley bracket and tighten the fasteners to 275 inch lbs. (31 Nm).

25. Remove the tensioner plunger pin, the tension is correct when the plunger pin can be removed and replaced easily.

26. Rotate the crankshaft two revolutions and recheck the timing marks. Wait several minutes, then recheck that the plunger pin can easily be removed and installed.

27. Reinstall the front timing belt cover.

28. Reinstall the engine mount bracket.

29. Reinstall the right engine mount.

30. Remove the floor jack from under the vehicle.

31. Install the crankshaft damper and tighten to 105 ft. lbs. (142 Nm).

32. Install and adjust the accessory drive belts.

33. Install the right inner splash-shield.

34. Reconnect the negative battery cable.

35. Perform the crankshaft and camshaft relearn alignment procedure using the DRB scan tool or equivalent.

2.4L (VIN B) Engine

1. Before servicing the vehicle, refer to the precautions in the beginning of this section.

2. Remove the A/C refrigerant using approved equipment.

3. Remove or disconnect the following:

- Negative battery cable
- Right front wheel and inner splash shield

- Accessory drive belts
- Crankshaft damper
- Lower torque strut
- Exhaust system from the manifold
- A/C compressor switch
- Upper torque strut and bracket
- Upper radiator support crossmember
- Power steering pump and bracket without disconnecting the lines
- Right engine mount through bolt after supporting the engine
- Engine support bracket

4. Rotate the crankshaft until the Top Dead Center (TDC) mark on the oil pump housing aligns with the TDC mark on the crankshaft sprocket.
5. Install an allen wrench into the tensioner. Rotate the tensioner counterclockwise while pushing on the wrench until it slides into the locking hole.
6. Remove the timing belt.

To install:

7. Set the crankshaft sprocket at TDC by aligning the sprocket with the arrow on the oil pump housing.
8. Set the camshafts timing marks so that the exhaust camshaft sprocket is a ½notch below the intake camshaft sprocket.
9. Install the timing belt by starting at the crankshaft. Go around the water pump sprocket idler pulley, camshaft sprockets and the tensioner.
10. Move the exhaust camshaft sprocket counterclockwise to align the marks and to remove any slack.
11. Remove the wrench from the belt tensioner.
12. Rotate the crankshaft two full revolutions and verify that the TDC marks are properly aligned.
13. Install or connect the following:
    - Lower timing belt cover and torque the bolts to 40 inch (4.5 Nm)
    - Upper timing belt cover and torque the bolts to 40 inch lbs. (4.5 Nm)
    - Right engine support bracket and reposition the power steering pump. Torque the bracket bolts to 45 ft. lbs. (61 Nm).
    - Right engine mount through bolt and torque it to 87 ft. lbs. (118 Nm)
    - Upper radiator support crossmember
    - Torque strut bracket
    - Upper torque strut
    - A/C lines and pressure switch
    - Exhaust system to the manifold
    - Crankshaft damper
    - Accessory drive belts
    - Lower torque strut
    - Right splash shield and wheel
    - Negative battery cable
14. Recharge the A/C system.
15. Perform the camshaft/crankshaft synchronization procedure.

2.5L (VIN H) Engine

1. Disconnect the negative battery cable from the left strut tower. The ground cable is equipped with a insulator grommet which should be placed on the stud to prevent the negative battery cable from accidentally grounding.
2. Remove the right inner splash-shield.
3. Remove the accessory drive belts.
4. Remove the crankshaft damper.
5. Remove the right engine mount.
6. Place a suitable floor jack under the vehicle to support the engine.
7. Remove the right engine mount bracket.
8. Remove the timing belt upper left cover, upper right cover and lower cover.
9. Loosen the timing belt tensioner bolts.

➡Before removing timing belt, be sure to align the sprocket timing marks to the timing marks on the rear timing belt cover.

10. If the present timing belt is going to be reused, mark the running direction of the timing belt for installation. Remove the timing belt and the tensioner.
11. Remove the camshaft timing belt sprockets from the camshaft, if necessary.
12. Remove the crankshaft timing belt sprocket and key.
13. Place the tensioner into a soft-jawed vise to compress the tensioner.
14. After compressing the tensioner place a pin into the plunger side hole to retain the plunger until installation.

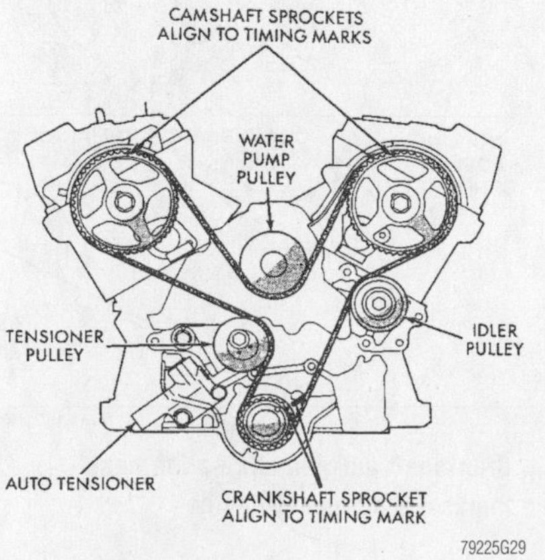

Timing belt engine sprocket timing—2.5L (vin h) Engine

To install:

15. If removed, reinstall the camshaft sprockets onto the camshaft. Install the camshaft sprocket bolt and tighten to 65 ft. lbs. (88 Nm).
16. If removed, reinstall the crankshaft timing belt sprocket and key onto the crankshaft.
17. Set the crankshaft sprocket to Top Dead Center (TDC) by aligning the notch on the sprocket with the arrow on the oil pump housing, then back off the sprocket three notches before TDC.
18. Set the camshafts to align the timing marks on the sprockets with the marks on the rear timing belt cover.
19. Install the belt on the rear camshaft sprocket first.
20. Install a binder clip on the belt to the sprocket so it won't slip out of position.
21. Keeping the belt taut, install it under the water pump pulley and around the front camshaft sprocket.
22. Install a binder on the front sprocket and belt.
23. Rotate the crankshaft to TDC.
24. Continue routing the belt by the idler pulley and around the crankshaft sprocket to the tensioner pulley.
25. Move the crankshaft sprocket clockwise to TDC to take up the belt slack. Check that all timing marks are in alignment.
26. Reinstall the tensioner to the block but do not tighten it at this time.
27. Using special tool No. MD998767 and a torque wrench on the tensioner pulley, apply 39 inch lbs. (4.4 Nm) of torque to the tensioner. Tighten the tensioner pulley bolt to 35 ft. lbs. (48 Nm).
28. With torque being applied to the tensioner pulley, move the tensioner up against the tensioner bracket and tighten the fasteners to 17 ft. lbs. (23 Nm).
29. Remove the tensioner plunger pin, the tension is correct when the plunger pin can be removed and replaced easily.
30. Rotate the crankshaft two revolutions clockwise and recheck the timing marks. Check to be sure the tensioner plunger pin can be easily installed and removed. If the pin does not remove and install easily, perform the procedure again.
31. Reinstall the timing belt cover.
32. Reinstall the engine mount bracket.
33. Reinstall the right engine mount.
34. Remove the engine support.
35. Install the crankshaft damper and tighten to 134 ft. lbs. (182 Nm).
36. Reinstall the accessory drive belts and adjust them.
37. Reinstall the right inner splash-shield.

38. Perform the crankshaft and camshaft relearn alignment procedure using the DRB scan tool or equivalent.

2.5L (VIN N) Engine

1. Disconnect the negative battery cable.
2. Remove the accessory drive belts.
3. Using crankshaft holding tools MB990767 and MB998754, remove the crankshaft bolt and remove the pulley.
4. Remove the heated oxygen sensor connection.
5. Remove the power steering pump with the hose attached and position it aside.
6. Remove the power steering pump bracket.
7. Place a floor jack under the engine oil pan, with a block of wood in between, and raise the engine so that the weight of the engine is no longer being applied to the engine support bracket.
8. Remove the upper engine mount. Spraying lubricant, slowly remove the reamer (alignment) bolt and remaining bolts and remove the engine support bracket.

➡The reamer bolt is sometimes heat-seized on the engine support bracket.

9. Remove the front timing belt covers.
10. If the timing belt is to be reused, draw an arrow indicating the direction of rotation on the back of the belt for reinstallation.
11. Align the timing marks by turning the crankshaft with MD998769 crankshaft turning tool. Loosen the center bolt on the timing belt tensioner pulley and remove the belt.

⁂ WARNING

Do not rotate the crankshaft or camshaft after removing the timing belt or valvetrain components may be damaged. Always align the timing marks before removing the timing belt.

12. Check the belt tensioner for leaks and check the pushrod for cracks.
13. If the timing belt tensioner is to be replaced, remove the retaining bolts and remove the timing belt tensioner. When the timing belt tensioner is removed from the engine, it is necessary to compress the plunger into the tensioner body.
14. Place the tensioner in a vise and slowly compress the plunger. Take care not to damage the pushrod.

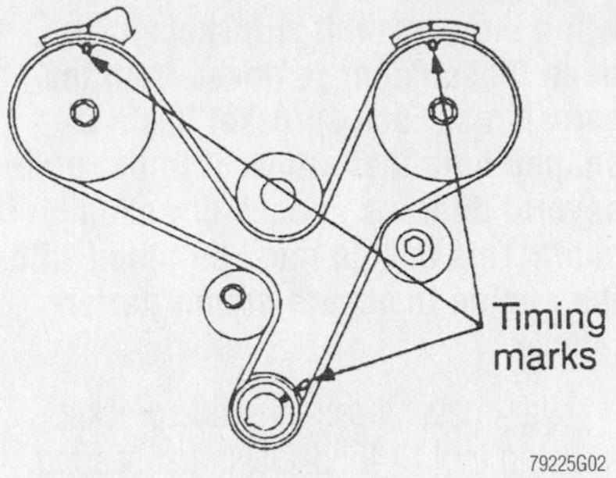

Camshaft and crankshaft alignment marks—Chrysler 2.5L (VIN N) engines

➡Position the tensioner in the vise the same way it will be installed on the engine. This is to ensure proper pin orientation for when the tensioner is installed on the engine.

15. When the plunger is compressed into the tensioner body, install a pin through the body and plunger to hold the plunger in place until the tensioner is installed.

To install:

16. Install the timing belt tensioner and tighten the retaining bolts to 17 ft. lbs. (24 Nm), but do not remove the pin at this time.
17. Check that all timing marks are still aligned.
18. Use bulldog clips (large paper binder clips) or other suitable tool to secure the timing belt and to prevent it from slacking. Install the timing belt. Starting at the crankshaft, go around the idler pulley, then the front camshaft sprocket, the water pump pulley, the rear camshaft sprocket and the tensioner pulley.
19. Be sure the belt is tight between the crankshaft and front camshaft sprocket, between the camshaft sprockets and the water pump. Gently raise the tensioner pulley, so that the belt does not sag, and temporarily tighten the center bolt.
20. Move the crankshaft ¼ turn counterclockwise, then turn it clockwise to the position where the timing marks are aligned.
21. Loosen the center bolt of the tensioner pulley. Using MD998767 tensioner tool, and a torque wrench apply 3.3 ft. lbs. (4.4 Nm) tensional torque to the timing belt and tighten the center bolt to 35 ft. lbs. (48 Nm). When tightening the bolt, be sure that the tensioner pulley shaft does not rotate with the bolt.
22. Remove the tensioner plunger pin. Pretension is correct when the pin can be removed and installed easily. If the pin cannot be easily removed and installed it is still satisfactory as long as it is within its standard value.
23. Check that the tensioner pushrod is within the standard value. When the tensioner is engaged the pushrod should measure 0.149–0.177 in. (3.8–4.5mm).
24. Rotate the crankshaft two revolutions and check the timing marks. If the timing marks are not properly aligned remove the belt and repeat Steps 17 through 23.
25. Install the timing belt covers.
26. Install the engine mounting bracket.
27. Lower the engine enough to install the engine mount onto bracket and remove the floor jack.
28. Install the power steering pump bracket and pump.
29. Install the crankshaft pulley and tighten the retaining bolt to 13 ft. lbs. (18 Nm).
30. Install the accessory drive belts.
31. Properly fill the cooling system.
32. Connect the negative battery cable.
33. Check for leaks and proper engine and cooling system operation.

3.2L (VIN J) and 3.5L (VIN F) Engines

Use care when servicing a timing belt. Valve timing is absolutely critical to engine performance. If the valve timing marks on all drive sprockets are not properly aligned, engine damage will result. If only the belt and tensioner are being serviced, do not loosen the camshaft drive sprockets unless they are to be replaced. The sprockets have oversized openings and can be rotated several degrees in each direction on their shafts. This means the sprockets must be retimed, requiring some special tools.

⁂ CAUTION

Fuel injection systems remain under pressure, even after the engine has been turned off. The fuel system pressure must be relieved before disconnecting any fuel lines. Failure to do so may result in fire and/or personal injury.

1. Disconnect the negative battery cable.
2. Rotate the engine to Top Dead Center (TDC) on the compression stroke for cylinder No. 1.
3. Release the fuel system pressure using the recommended procedure.
4. Place a pan under the radiator and drain the coolant.

5. Remove the radiator and cooling fan assemblies.
6. Remove the accessory drive belts.
7. Remove the upper radiator hose.
8. Remove the crankshaft damper with a quality puller tool gripping the inside of the pulley.
9. Remove the stamped steel cover. Do not remove the sealer on the cover; it may be reusable.
10. Remove the left side cast cover. If necessary, remove the lower belt cover, located behind the crankshaft damper.
11. If the timing belt is to be reused, mark the timing belt with the running direction for installation.
12. Align the camshaft sprockets with the marks on the rear covers.
13. Remove the timing belt and tensioner.
14. If it is necessary to service the camshaft sprockets, use the following procedure:
    a. Hold the camshaft sprocket with a 36mm box end wrench, loosen and remove the sprocket retaining bolt and washer.

➡To remove the camshaft sprocket retainer bolt while the engine is in the vehicle, it may be necessary to raise that side of the engine due to the length of the retainer bolt. The right bolt is 8 ⅜ in. (213mm) long, while the left bolt is 10 in. (254mm) long. These bolts are not interchangeable and their original location during removal should be noted.

    b. Remove the camshaft sprocket from the camshaft. The camshaft sprockets are not interchangeable from side-to-side.
    c. Remove the crankshaft sprocket using Puller L-4407A.

To install:

15. If it was necessary to remove the camshaft sprockets, use the following procedure:

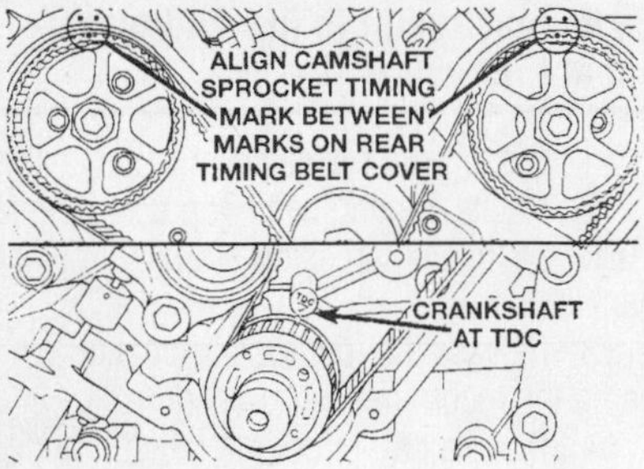

Timing belt alignment marks—Chrysler 3.2L and 3.5L engines

➡This procedure can only be used when the camshaft sprockets have been loosened or removed from the camshafts. Each sprocket has a D shaped hole that allows it to be rotated several degrees in each direction on its shaft. This design must be timed with the engine to ensure proper performance.

    a. Install the crankshaft sprocket, using tool C-4685-C1, thrust bearing, washer and 12mm bolt.
    b. When the camshaft sprockets are loosened or removed, the camshafts must be timed to the engine. Install the Camshaft Alignment tools 6642-A, to the rear of the cylinder heads. These tools lock the camshafts in the proper position.
16. Preload the belt tensioner as follows:
    a. Place the tensioner in a vise the same way it is mounted on the engine.
    b. Slowly compress the plunger into the tensioner body.
    c. When the plunger is compressed into the tensioner body, install a pin through the body and plunger to retain the plunger in place until the tensioner is installed.
17. Install both camshaft sprockets to the appropriate shafts. The left camshaft sprocket has the Distributorless Ignition System (DIS) pick-up as part of the sprocket.

➡The right bolt is 8 ⅜ in. (213mm) long, while the left bolt is 10 in. (254mm) long. These bolts are not interchangeable.

18. Apply Loctite®271, to the threads of the camshaft sprocket retainer bolts and install to the appropriate shafts. Do not tighten the bolts at this time.
19. Align the camshaft sprockets between the marks on the rear belt covers.
20. Align the crankshaft sprocket with the TDC mark on the oil pump cover.
21. Install the timing belt, starting at the crankshaft sprocket and going in a counterclockwise direction. After the belt is installed on the right sprocket, keep tension on the belt until it is past the tensioner pulley.
22. Holding the tensioner pulley against the belt, install the timing belt tensioner into the housing and tighten to 21 ft. lbs. (28 Nm).
23. When the tensioner is in place, pull the retainer pin out to allow tensioner to extend to the pulley bracket.

➡Be sure that the timing marks on the cam sprockets are still between the marks on the rear cover.

24. Remove the spark plug in the No.1 cylinder and install a dial indicator to check for TDC of the piston. Rotate the crankshaft until the piston is exactly at TDC.
25. Hold the camshaft sprocket hex with a 36mm wrench and tighten the right camshaft sprocket bolt to 75 ft. lbs. (102 Nm) plus an additional 90 degree turn. Tighten the left camshaft sprocket bolt to 85 ft. lbs. (115 Nm) plus an additional 90 degree turn.
26. Remove the dial indicator. Install the spark plug and tighten to 20 ft. lbs. (28 Nm).
27. Remove the camshaft alignment tools from the back of the cylinder heads and install the cam covers with new O-rings.
28. Tighten the fasteners to 20 ft. lbs. (27 Nm). Repeat this procedure on the other camshaft.
29. Rotate the crankshaft sprocket two revolutions and check for proper alignment of the timing marks on the camshaft and the crankshaft. If the timing marks do not align, repeat the procedure.
30. Before installing, inspect the sealer on the stamped steel cover. If some sealer is missing, use MOPAR Silicone Rubber Adhesive sealant to replace the missing sealer.
31. Install the lower belt cover behind the crankshaft damper, if necessary.
32. Install the stamped steel cover and the left side cast cover. Tighten the 6mm bolts to 105 inch lbs. (12 Nm), the 8mm bolts to 250 inch lbs. (28 Nm) and the 10mm bolts to 40 ft. lbs. (54 Nm).
33. Install the crankshaft damper using special tool L-4524, a 5.9 in. long bolt, thrust bearing and washer. Tighten the center bolt to 85 ft. lbs. (115 Nm).
34. Install the upper radiator hose.
35. Install the accessory drive belts and adjust them to the proper tension.
36. Install the radiator and cooling fan assemblies.
37. Refill and bleed the cooling system.
38. Connect the negative battery cable.
39. With the radiator cap off so coolant can be added, run the engine. Watch for leaks and listen for unusual engine noises.

FORD MOTOR COMPANY

1.3L (VIN H) Engine

1. Disconnect the negative battery cable.
2. Remove the accessory drive belts.
3. Remove the three water pump pulley attaching bolts and remove the water pump pulley.

4. Remove the right front wheel and tire assembly and the right inner fender panel.
5. Remove the four attaching bolts and the screws from the crankshaft pulley. Remove the spacer and outer pulley, if equipped. Remove the inner spacer, inner pulley and the baffle or guide plates, as required.
6. Remove the attaching bolts and the upper and lower covers.
7. Rotate the crankshaft until the sprocket timing marks are aligned.
8. Remove the timing belt tensioner spring, spring cover and timing belt tensioner bolt. Remove the timing belt.

➡If the timing belt is to be reused, mark the direction of rotation on the belt, using a crayon, so the belt can be reinstalled in the same direction.

9. If the camshaft sprocket requires removal, proceed as follows:
   a. Hold the camshaft stationary with an open end wrench and remove the camshaft sprocket retaining bolt.
   b. Pull the camshaft sprocket with the dowel pin off of the camshaft. Use care not to drop the dowel pin.
10. If the crankshaft sprocket requires removal, proceed as follows:
   a. Remove the crankshaft pulley retaining bolt.
   b. Pull the crankshaft pulley hub, sprocket and key from the crankshaft. Be sure not to drop the crankshaft key.

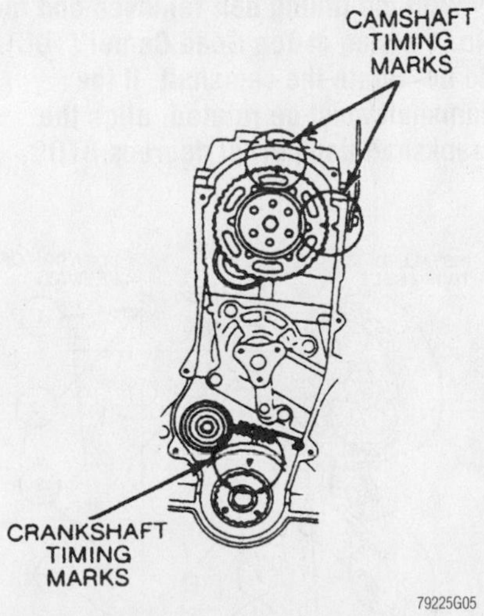

Before removing or installing the timing belt, be sure the crankshaft and both camshaft matchmarks are aligned as shown—ford 1.3L (VIN H) engine

To install:

11. If removed, install the crankshaft sprocket as follows:
   a. Install the sprocket with the key onto the crankshaft.
   b. Install the crankshaft pulley hub.
   c. Clean the threads of the crankshaft pulley bolt and coat with a non-hardening sealer.
   d. Install the bolt and tighten to 80–85 ft. lbs. (108–118 Nm).
12. If removed, install the camshaft sprocket as follows:
   a. Position the sprocket and dowel pin to the camshaft and install the retaining bolt.
   b. Hold the camshaft stationary with an open end wrench and tighten the retaining bolt to 36–45 ft. lbs. (49–61 Nm).
13. Align the camshaft and crankshaft timing marks with the marks located on the cylinder head and oil pump housing.
14. If reusing the original timing belt, install the timing belt with the mark made indicating the direction of rotation.
15. Install the timing belt tensioner spring and cover on the tensioner. Position the tensioner and spring assembly on the engine and install the attaching bolt. Do not tighten the bolt at this time.
16. Rotate the crankshaft two turns in the direction of normal rotation and align the timing marks. Ensure all marks are still correctly aligned.
17. Reconnect the free end of the spring to the spring anchor. Tighten the tensioner bolt to 14–19 ft. lbs. (19–26 Nm).
18. Install the upper and lower covers. Install the attaching bolts and tighten to 71–97 inch lbs. (8–11 Nm).
19. Install the crankshaft pulley baffle with the curved lip facing outward or install the large guide plate, then the small guide plate, as required.
20. Install the inner pulley with the deep recess facing outward. Install the spacer, then the outer pulley, spacer and screws. Install the pulley bolts and tighten to 109–152 inch lbs. (12–17 Nm).
21. Install the inner fender panel.
22. Install the wheel and tire assembly. Tighten the lug bolts to 65–87 ft. lbs. (88–118 Nm).
23. Install the water pump pulley and tighten the bolts to 36–45 ft. lbs. (49–61 Nm).
24. Install the accessory drive belts.
25. Connect the negative battery cable.
26. Run the engine and check for proper operation.

1.8L (VIN 8) Engine

1. Disconnect the negative battery cable.
2. Remove the timing belt upper cover and gasket.
3. Remove the accessory drive belts.
4. Remove the water pump pulley bolts and remove the pulley.
5. Remove the right front wheel and tire assembly.
6. Remove the right upper and lower splash-shields.
7. Remove the timing belt middle and lower covers along with the gaskets.
8. Remove the crankshaft pulley hub bolt and hub.
9. Rotate the crankshaft and align the timing marks located on the camshaft sprockets and seal plate.
10. Check that the crankshaft sprocket and the oil pump are aligned.

➡If the timing belt is to be reused, mark an arrow on the belt to indicate it's rotational direction for installation reference.

11. Loosen the timing belt tensioner bolt.
12. Turn the timing belt tensioner counterclockwise and hand-tighten the tensioner bolt to relieve the tension on the timing belt.
13. Remove the timing belt.
14. If the camshaft sprockets are to be removed, continue as follows:
   a. Disconnect and tag the ignition wires and vacuum lines blocking the removal of the cylinder head cover.
   b. Remove the cylinder head cover

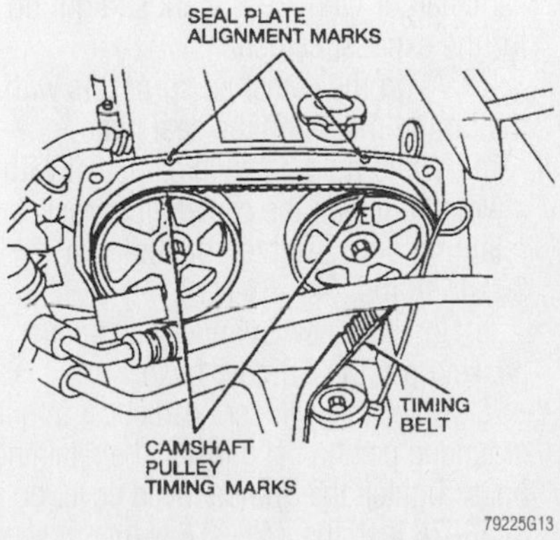

Camshaft timing alignment marks—Ford 1.8L (VIN 8) engines

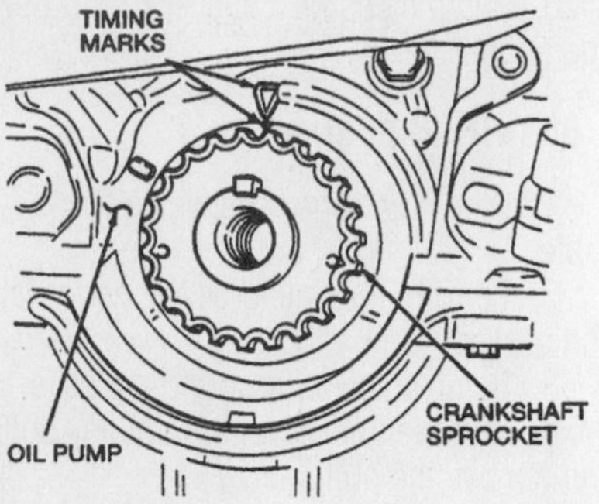

Crankshaft timing mark position—Ford 1.8L (VIN 8) engines

retaining bolts and remove the cover and gasket.

c. While holding the camshaft with a wrench, remove the camshaft sprocket retaining bolt.

d. Remove the camshaft sprocket.

e. If removing both camshaft sprockets, tag the sprockets for identification at reassembly.

15. If removing the crankshaft sprocket, remove the crankshaft pulley bolt and hub, if not already done. Slide the crankshaft sprocket off the crankshaft.

16. Inspect the timing belt tensioner and spring, replace if necessary.

To install:

17. If the crankshaft sprocket was removed, install the crankshaft key with the tapered end facing the oil pump. Install the crankshaft sprocket onto the crankshaft while making sure to match the alignment grooves.

18. If removed, install the camshaft sprockets as follows

a. Turn the camshaft until the dowel pins face straight up.

b. Install the camshaft sprocket with the **I** mark straight up for the intake camshaft or with the **E** mark straight up for the exhaust camshaft.

c. Align the camshaft sprockets with the timing marks on the seal plate.

d. While holding each camshaft with a wrench, install the camshaft sprocket retaining bolts. Tighten the bolts to 36–45 ft. lbs. (49–61 Nm).

e. Install a new cylinder head cover gasket onto the cylinder head.

f. Place the cylinder head cover into its mounting position and install the retaining bolts. Tighten the cylinder head cover bolts to 43–78 inch lbs. (4.9–8.8 Nm).

g. Install the ignition wires to the spark plugs and connect the vacuum hoses to the cylinder head cover.

19. Temporarily secure the timing belt tensioner in the far left position.

20. Verify that the timing marks on the crankshaft sprocket and the oil pump are aligned.

21. Verify that the timing marks on the camshaft sprockets and the seal plate are aligned.

22. Install the timing belt in a counterclockwise motion. Be sure there is no looseness on the idler side of the timing belt or between the camshaft sprockets.

➡If using the old timing belt, be sure to install the belt in the same direction of travel as it was removed.

23. Loosen the timing belt tensioner bolt. Allow the tensioner spring to apply tension to the timing belt.

24. Rotate the crankshaft 1 5⁄6 turns clockwise and align the timing belt pulley mark with the tension set mark which is located at approximately the 10 o'clock position.

25. Turn the crankshaft two turns clockwise and align the crankshaft sprocket with the tension set mark on the oil pump.

26. Verify that all timing marks are aligned. If not, remove the timing belt and repeat the installation procedures.

27. Apply tension to the timing belt tensioner and tighten the tensioner lockbolt to 27–38 ft. lbs. (37–52 Nm).

28. Rotate the crankshaft 2 1⁄6 (780 degrees) turns clockwise and verify that the camshaft and crankshaft timing marks are aligned.

29. Measure the timing belt deflection by applying 22 lbs. (98 N) of pressure on the timing belt between the camshaft sprockets. The timing belt deflection should be 0.35–0.45 in. (9–11.5mm). If necessary to adjust the timing belt deflection, rotate the crankshaft two turns clockwise and ensure that the timing marks are still aligned. If the timing marks are not aligned, repeat the installation procedure.

30. Install the crankshaft pulley hub and tighten the retaining bolt to 80–87 ft. lbs. (108–118 Nm).

31. Install the crankshaft pulley and washer. Install the retaining bolts and tighten to 109–152 inch lbs. (12–17 Nm).

32. Install the timing belt middle and lower covers with the gaskets. Tighten the middle and lower timing belt cover retaining bolts to 65–95 inch lbs. (7.8–11 Nm).

33. Install the power steering drive belt.

34. Install the water pump pulley and retaining bolts. Tighten the bolts to 69–95 inch lbs. (7.8–11.0 Nm).

35. Install the alternator/water pump drive belt.

36. Install the splash-shields. Tighten the bolts to 69–95 inch lbs. (7.8–11.0 Nm).

37. Install the right wheel and tire assembly. Tighten the lug nuts to 65–87 ft. lbs. (88–118 Nm).

38. Install the timing belt upper cover and gasket. Tighten the bolts to 69–95 inch lbs. (7.8–11.0 Nm).

39. Connect the negative battery cable.

40. Run the engine and check for leaks.

41. Road test the vehicle and check for proper engine operation.

1.9L (VIN J) Engine

1. Disconnect the negative battery cable.
2. Remove the accessory drive belt automatic tensioner and the accessory drive belt.
3. Remove the timing belt cover.
4. Align the timing mark on the camshaft sprocket with the timing mark on the cylinder head.
5. Confirm that the timing mark on the crankshaft sprocket is aligned with the timing mark on the oil pump housing.
6. Loosen the belt tensioner attaching bolt, pry the tensioner away from the timing belt and retighten the bolt.
7. Remove the spark plugs. Remove the right engine mount.
8. Remove the right side splash-shield.
9. Remove the flywheel inspection shield.
10. Use a suitable tool to hold the flywheel in place.
11. Remove the crankshaft damper bolt and washer and remove the bolt.
12. Remove the timing belt.

➡With the timing belt removed and the No. 1 piston at Top Dead Center (TDC), do not rotate the camshaft. If the camshaft must be rotated, align the crankshaft damper 90 degrees BTDC.

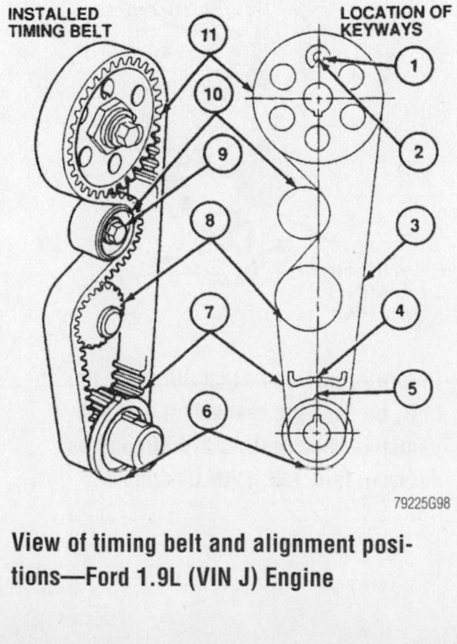

View of timing belt and alignment positions—Ford 1.9L (VIN J) Engine

To install:

13. Install the timing belt over the sprockets in a counterclockwise direction starting at the crankshaft. Keep the belt span from the crankshaft to the camshaft tight while installing over the remaining sprocket.

14. Loosen the belt tensioner attaching bolt, allowing the tensioner to snap against the belt.

15. Rotate the crankshaft clockwise two complete revolutions, stopping at TDC. This will allow the tensioner spring to load the timing belt.

➡Do not turn the engine counterclockwise to align the timing marks. Do not rotate the crankshaft with the spark plugs installed.

16. Recheck the camshaft and crankshaft timing marks for alignment, to be sure the timing belt has not skipped a tooth during rotation. Repeat the procedure if the timing marks are not aligned.

17. Tighten the tensioner attaching bolt to 17–22 ft. lbs. (23–30 Nm).

18. Install the crankshaft dampener and the bolt and washer. Tighten the bolt to 81–96 ft. lbs. (110–130 Nm).

19. Install the flywheel inspection shield.

20. Install the splash-shield and lower the vehicle.

21. Install the right engine mount. Install the spark plugs.

22. Install the timing belt cover.

23. Install the accessory drive belt automatic tensioner and the accessory drive belt.

24. Connect the negative battery cable.

2.0L (VIN 3) Engine

1. Remove or disconnect the following:
   - Negative battery cable
   - Spark plugs
   - Catalytic converter
   - Right front wheel
   - Right inner splash shield
   - Accessory drive belt and idler pulley

2. Rotate the crankshaft to Top Dead Center (TDC) and install Crankshaft TDC Timing Peg T97P-6000-A.

3. Rotate the crankshaft clockwise against the peg.

4. Remove or disconnect the following:
   - Water pump pulley
   - Valve cover
   - Crankshaft pulley
   - Timing belt covers

5. Install Camshaft Alignment Timing Tool T94P-6256-CH into the slots of both camshafts at the rear of the cylinder head to lock the camshafts into position.

6. Loosen the timing belt tensioner bolt and relieve the tension on the timing belt by disconnecting the tensioner tab from the timing cover back plate.

7. Remove the timing belt.

To install:

➡Always replace the timing belt after loosening or removing it.

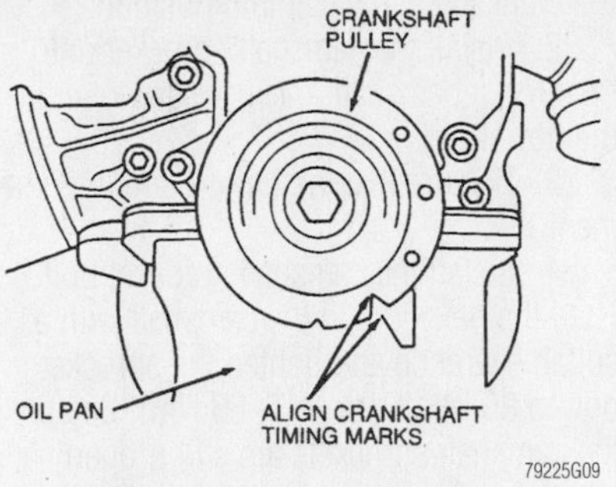

Crankshaft alignment position—Ford 2.0L (VIN 3) engines

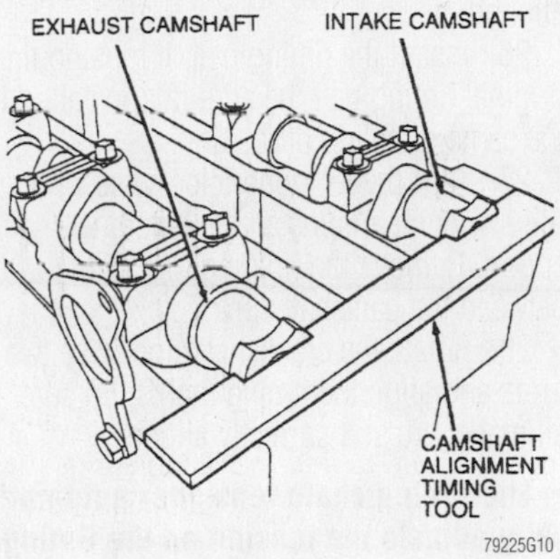

Placement of camshaft alignment timing tool T94P-6256-CH—Ford 2.0L (VIN 3) engines

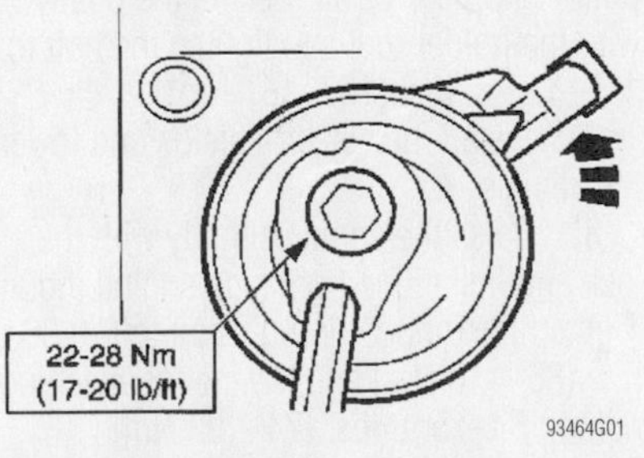

Timing belt tensioner index marks—Ford 2.0L (VIN 3) engines

8. Install the timing belt.

9. Engage the timing belt tensioner tab into the timing cover back plate.

10. Adjust the timing belt tensioner until the index marks are aligned. Tighten the tensioner bolt to 17–20 ft. lbs. (22–28 Nm).

11. Remove the camshaft alignment tool and the crankshaft TDC peg.

12. Install or connect the following:
   - Timing belt covers
   - Crankshaft pulley. Tighten the bolt to 81–89 ft. lbs. (110–120 Nm).
   - Valve cover
   - Water pump pulley
   - Accessory drive belt and idler pulley
   - Right inner splash shield
   - Right front wheel
   - Catalytic converter
   - Spark plugs
   - Negative battery cable

13. Start the engine and check for proper operation.

2.0L (VIN A) Engine

1. Disconnect the negative battery cable.

2. Label and disconnect the spark plug wires and clips from the cylinder head cover. Remove the ignition distributor with wiring and set it aside.

3. Remove the power steering hose brackets from the cylinder head cover. If necessary disconnect the crankshaft position sensor.

4. Disconnect the breather tube and PCV valve from the cylinder head cover.

5. Loosen the cylinder head cover bolts in 2–3 steps. Remove the cylinder head cover.

6. Remove the power steering belt shield. Loosen the power steering adjusting bolt, lockbolt and through-bolt and remove the power steering belt.

7. Loosen the alternator adjusting bolt and upper mounting bolt. Remove the alternator belt.

8. Support the engine with engine support tool 014–00750. Raise the engine slightly with a jack and remove the right side engine support insulator (mount).

9. Remove the oil level indicator bolt and four upper timing belt cover bolts and remove the upper timing belt cover.

10. Remove the splash-shields. Using holder tool T92C-6316-AH, hold the crankshaft pulley and remove the pulley bolt. Use a suitable puller to remove the pulley, then remove the guide plate.

11. Remove the four lower timing belt cover bolts and remove the lower timing belt cover.
12. Temporarily install the crankshaft pulley bolt.
13. Turn the crankshaft until the timing mark on the crankshaft sprocket aligns with the timing mark on the oil pump and the camshaft sprocket timing marks, **E** and **I**, align on the camshaft sprockets.
14. Insert camshaft sprocket holding tool T92C-6256-AH, between the camshaft sprockets.
15. Turn the timing belt tensioner with an Allen wrench and remove the tensioner spring from the tensioner spring pin.
16. If the timing belt is to be reused, mark the direction of rotation on the timing belt. Remove the timing belt.
17. If necessary to remove the sprockets, remove the camshaft sprocket holding tool. Hold the camshaft by placing a suitable wrench on the hexagon which is cast into the camshaft. Place another wrench onto the camshaft sprocket retaining bolt and loosen the bolt.

➡Before removing the camshaft sprocket(s), be sure that the camshafts are still in alignment and tag each sprocket to the camshaft from which it was removed.

18. Remove the camshaft sprocket bolt and the camshaft sprocket from the camshaft.
19. Repeat the camshaft sprocket removal procedure for the opposite camshaft if required.
20. Slide off the crankshaft sprocket and remove the crankshaft key.

To install:

21. Install the crankshaft key and slide the crankshaft sprocket into position.
22. Install the camshaft sprocket onto the proper camshaft, making sure to align the dowel pin.
23. Be sure that the **I** and **E** are in alignment.
24. Install the camshaft sprocket bolt. Hold the hexagon on the camshaft with a suitable wrench and tighten the sprocket bolt to 35–48 ft. lbs. (47–65 Nm). Be sure the camshaft sprockets are still properly aligned and reinstall the sprocket holding tool.
25. Be sure the timing marks on the camshaft and crankshaft sprockets are still aligned.
26. Install the timing belt. If reusing the original timing belt, be sure it is installed in the same direction of rotation.
27. Turn the tensioner clockwise with an Allen wrench and install the tensioner spring. Remove the holding tool from between the camshaft sprockets.
28. Rotate the crankshaft clockwise two turns and align the timing marks. Be sure all marks are still correctly aligned.

➡The timing chain tensioner automatically adjusts the tension on the timing belt.

29. Install the timing belt lower cover and tighten the four bolts to 71–88 inch lbs. (8–10 Nm).
30. Install the guide plate, crankshaft pulley and pulley bolt. Secure the pulley with the holder tool and tighten the bolt to 116–123 ft. lbs. (157–167 Nm).
31. Install the splash-shields and lower the vehicle.
32. Raise the engine slightly with the jack and install the right side engine mount. Tighten the mount through-bolt to 63–86 ft. lbs. (86–116 Nm) and the mount attaching nuts to 54–75 ft. lbs. (74–103 Nm). Remove the engine support tool.
33. Install the upper timing belt cover and tighten the bolts to 71–88 inch lbs. (8–10 Nm).
34. Clean the cylinder head and valve cover mating surfaces thoroughly.
35. Apply silicone sealant to the cylinder head surface in the area adjacent to the front camshaft bearing caps. Apply sealant to a new gasket and install it on the cylinder head cover.
36. Install the cylinder head cover and tighten the bolts.
37. Install the power steering hose brackets and tighten the bolts to 71–88 inch lbs. (8–10 Nm). Connect the spark plug wires and wire clips. Connect the breather tube and PCV valve. If necessary, connect the crankshaft position sensor.
38. Install the alternator belt and adjust the tension. Tighten the upper mounting bolt to 14–18 ft. lbs. (19–25 Nm) and the lower through-bolt to 27–38 ft. lbs. (37–52 Nm).
39. Install the power steering belt and adjust the tension. Tighten the through-bolt to 32–45 ft. lbs. (43–61 Nm) and the lock-bolt to 23–34 ft. lbs. (31–46 Nm). Install the power steering belt shield and tighten the bolts to 61–86 inch lbs. (7–9 Nm).
40. Connect the negative battery cable.
41. Run the engine and check for leaks and proper engine operation.

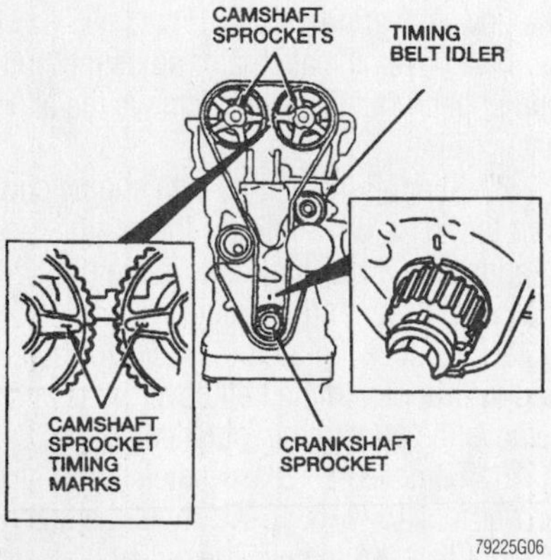

Be sure that the intake and exhaust camshaft timing marks align so that they face each other—Ford 2.0L (VIN A) engines

2.0L (VIN P) Engine

1. Remove or disconnect the following:
   - Negative battery cable
   - Accessory drive belt and tensioner
   - Timing belt cover
   - Right front wheel
   - Right inner splash shield
   - Crankshaft pulley
2. Align the timing marks as shown.
3. Refer to the illustration and remove the timing belt as follows:
   a. Loosen the timing belt tensioner bolt (1).
   b. Use an 8mm Allen wrench, and turn the tensioner (2) counterclockwise ¼ turn.
   c. Insert a ⅛ inch drill bit in the hole (3) to lock the belt tensioner in place.
   d. Remove the timing belt (4).

To install:

4. Install the timing belt in a counter-

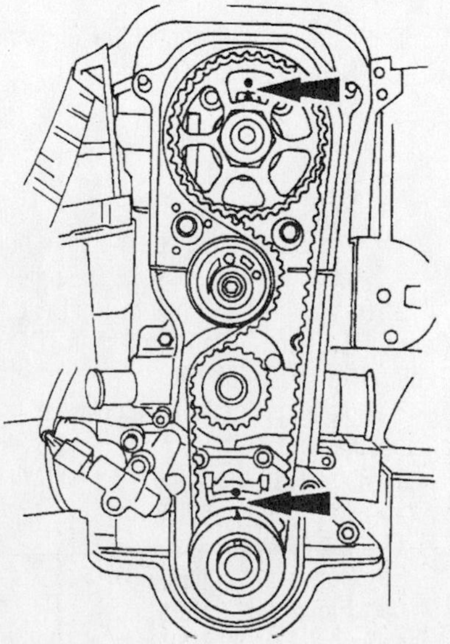

Align the timing marks before removing or installing the timing belt—2.0L (VIN P) SOHC engines

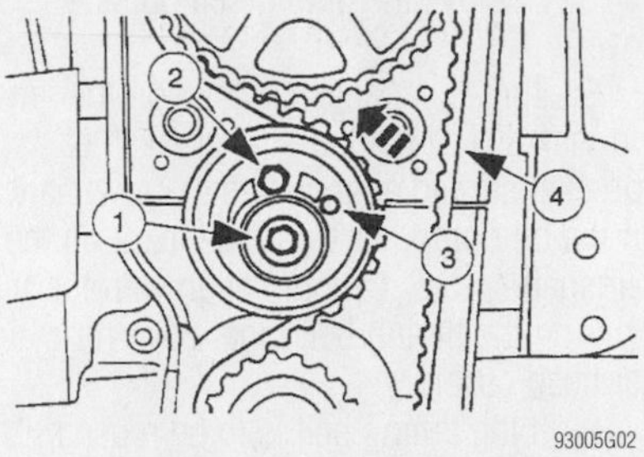

Remove the timing belt by following these 4 numbered steps (refer to the text for an explanation)—2.0L (VIN P) SOHC engine

clockwise direction starting at the crankshaft.

5. Remove the drill bit to unlock the timing belt tensioner.

6. Rotate the crankshaft two complete turns and check that the timing marks align.

7. Tighten the tensioner bolt to 15–22 ft. lbs. (20–30 Nm).

8. Install or connect the following:

- Crankshaft pulley. Tighten the bolt to 81–96 ft. lbs. (110–130 Nm).
- Right inner splash shield
- Right front wheel
- Timing belt cover
- Accessory drive belt and tensioner
- Negative battery cable

9. Start the engine and check for proper operation.

2.5L (VIN B) Engine

1. Disconnect the negative battery cable.

2. Label and disengage the electrical connectors from the coolant elbow. Label and remove the electrical connectors from the Knock Sensor (KS), if required and the Crankshaft Position (CKP) sensor.

3. Loosen the drive belt tensioner locknuts and adjusting bolts. Remove the accessory drive belts.

4. Remove the lower bolt from the A/C and alternator tensioner bracket.

5. Remove the right wheel and splashshields.

6. Hold the crankshaft pulley (damper) with holder tool T92C-6316-AH, and remove the crankshaft pulley bolt. Remove the crankshaft pulley, using a puller if needed.

7. Remove the 5 front timing belt cover bolts.

8. Hold the water pump pulley with holder tool T92C-6312-AH, remove the four bolts and the water pump pulley.

9. Hold the power steering pump pulley with strap wrench D85L-6000-A and remove the power steering pump pulley nut and pulley.

10. Remove the upper bolt from the belt idler bracket and remove the bracket.

11. Remove the engine oil dipstick tube retaining bolt and the tube.

12. Remove the 8 rear timing belt cover retaining bolts and remove the timing belt covers.

13. Temporarily reinstall the crankshaft pulley bolt.

14. Remove the three nuts and through-bolt from the right-hand engine support insulator and remove the support insulator. Remove the support insulator bracket.

15. Turn the crankshaft to Top Dead Center (TDC) No. 1 cylinder in the direction of normal rotation. Be sure that the timing mark on the crankshaft sprocket aligns with the timing mark on the oil pump.

16. Remove the two bolts from the timing belt tensioner arm, removing the lower bolt first.

17. Remove the timing belt tensioner arm.

18. If the timing belt is to be reused, mark the direction of rotation on the timing belt.

19. Loosen the Allen bolt on the timing belt tensioner.

20. Remove the timing belt.

21. If the timing belt sprockets are to be removed, proceed as follows:

a. Remove the intake manifold.

b. Label and disconnect the necessary hoses from the cylinder head covers.

c. Label and disconnect the spark plug wires from the spark plugs.

d. Remove the cylinder head cover retaining bolts and remove the cylinder head covers.

e. Hold the camshaft using a suitable wrench on the hexagon cast into the

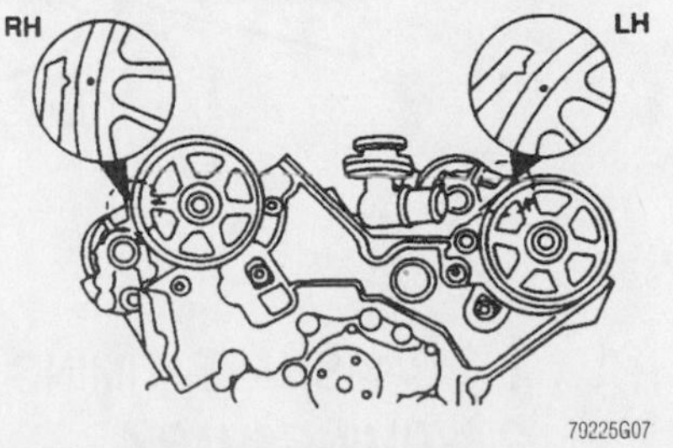

Left and right camshaft timing position—Ford 2.5L (VIN B) engines

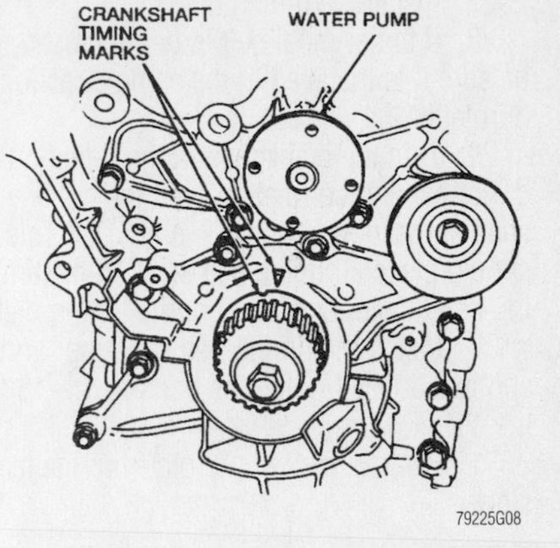

TDC alignment for the crankshaft—Ford 2.5L (VIN B) engines

camshaft. Remove the camshaft sprocket bolts and the camshaft sprockets.

f. Use crankshaft damper puller T74P-6316-A, to remove the crankshaft sprocket. Remove the crankshaft sprocket key.

To install:

22. If the timing belt sprockets were removed, proceed as follows:

a. Install the crankshaft sprocket key and crankshaft sprocket.

b. Install the camshaft sprockets on the camshafts with the retaining bolts.

c. Hold the camshaft using a suitable wrench on the hexagon cast into the camshaft. Tighten the camshaft sprocket bolts to 90–103 ft. lbs. (123–140 Nm).

d. Be sure the cylinder head cover and cylinder head contact surfaces are clean and free of dirt, oil and old sealant and gasket material.

e. Apply silicone sealant to the cylinder heads in the area adjacent to the front and rear camshaft caps. Install new gaskets on the cylinder heads.

f. Install the cylinder head covers and tighten the retaining bolts.

g. Connect the spark plug wires to the spark plugs and connect the hoses to the cylinder head covers.

h. Install the intake manifold.

23. Position the timing belt tensioner arm in a suitable press.

24. Compress the tensioner until the hole in the piston is aligned with the 2nd hole in the tensioner case. Insert a 0.060 in. (1.6mm) diameter wire or pin through the 2nd hole to keep the piston compressed.

25. Align the camshaft sprockets to TDC.

26. Turn the crankshaft counterclockwise until the crankshaft sprocket is offset from TDC by one tooth.

27. Install the timing belt.
28. If the original belt is being reused, be sure it is installed in the same direction of rotation.
29. Turn the crankshaft in the direction of normal engine rotation until the crankshaft sprocket timing mark is at TDC. This should place all of the belt slack in the timing belt tensioner portion of the timing belt.
30. Install the timing belt tensioner arm and two bolts. Tighten the bolts to 14–18 ft. lbs. (19–25 Nm).
31. Remove the wire or pin from the tensioner.

➡When properly timed, the crankshaft timing marks will align and the crankshaft sprocket timing mark will no longer be one tooth off.

32. Rotate the crankshaft two complete turns in the direction of normal rotation and align the timing marks. Be sure all marks are still correctly aligned. This will also set the timing belt tension.

➡The timing belt tensioner will automatically adjust the timing belt tension.

33. Tighten the timing belt tensioner Allen bolt to 28–32 ft. lbs. (35–51 Nm).
34. Install the right-hand engine support insulator. Tighten the three nuts to 54–76 ft. lbs. (74–103 Nm) and the through-bolt to 50–68 ft. lbs. (67–93 Nm).
35. Remove the crankshaft damper bolt.
36. Install the timing belt covers with the rear 8 bolts. Tighten to 71–88 inch lbs. (8–10 Nm).
37. Install the engine oil dipstick tube and retaining nut.
38. Install the belt idler bracket and the upper retaining bolt.
39. Install the power steering pump pulley and nut. Tighten the nut to 36–43 ft. lbs. (49–59 Nm) while holding the pulley with a strap wrench.
40. Install the water pump pulley and four bolts. Secure the pulley with the holder tool and tighten the bolts to 71–88 inch lbs. (8–10 Nm).
41. Install the 5 front timing belt cover bolts and tighten to 71–88 inch lbs. (8–10 Nm).
42. Install the crankshaft pulley (damper) with the bolt. Hold the crankshaft pulley with the holding tool and tighten to 116–122 ft. lbs. (157–166 Nm).
43. Install the splash-shields.
44. Install the wheel and tighten the lug nuts to 65–87 ft. lbs. (88–118 Nm).
45. Install the lower bolt into the A/C and alternator tensioner bracket.
46. Install the accessory drive belts and adjust the tension.
47. Engage the electrical connectors to the sensors at the coolant elbow and the KS and CKP sensors.
48. Connect the negative battery cable.
49. Run the engine and check for leaks and proper engine operation.

GENERAL MOTORS

1.0L (VIN 6) and 1.3 (VIN 9) Engines

1. Disconnect the negative battery cable.
2. Remove the clips and right side splash-shield.
3. Remove the lower alternator cover plate.
4. Remove the alternator drive belt and if equipped, the A/C drive belt.

➡It is not necessary to remove the crankshaft timing sprocket bolt (center bolt) to remove the crankshaft pulley.

5. Remove the crankshaft pulley bolts and crankshaft pulley.
6. Remove the water pump pulley.
7. Remove the retaining bolts and nut from the timing belt outside cover.
8. Remove the timing belt outside cover.
9. Turn the crankshaft to align the timing marks. The mark on the crankshaft sprocket should align with the arrow mark on the oil pump housing. The mark on the camshaft sprocket should align with the **V** mark on the timing belt inner cover or cylinder head cover.
10. If the timing belt is to be reused, mark the direction of rotation on the belt.
11. Remove the timing belt tensioner, tensioner plate, tensioner spring, spring damper and timing belt.

➡Never turn the camshaft or crankshaft independently after the timing belt has been removed. Interference may occur between the pistons and valves, and parts may be damaged.

12. Inspect the timing belt for wear or cracks, and replace as necessary. Check the tensioner for smooth rotation.
13. If the timing belt sprockets are to be removed, proceed as follows:
    a. Using a 0.39 in. (10mm) rod inserted into the camshaft, hold the camshaft and remove the retaining bolt and camshaft sprocket.

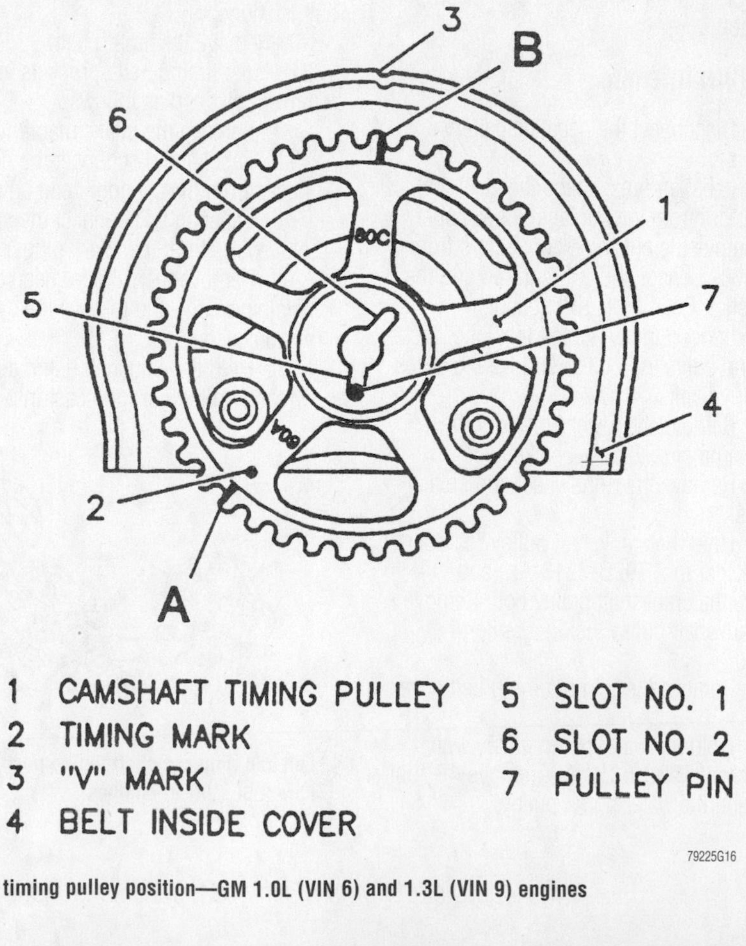

Upper timing pulley position—GM 1.0L (VIN 6) and 1.3L (VIN 9) engines

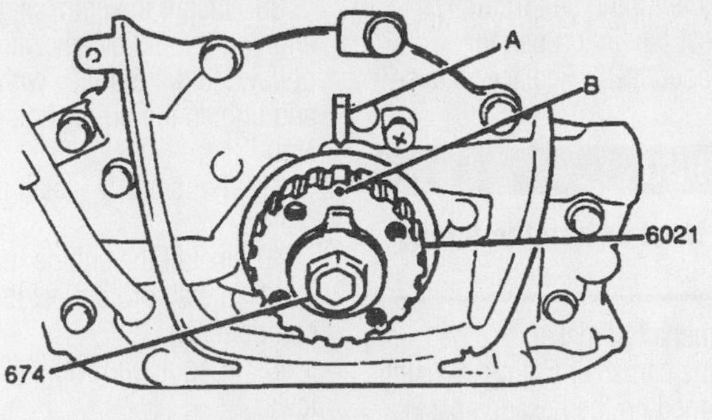

Crankshaft timing mark—GM 1.0L (VIN 6) and 1.3L (VIN 9) engines

WARNING

Be careful not to damage the cylinder head or cylinder head cover mating surfaces. Place a clean shop cloth between the rod and cylinder head. Do not bump the rod hard against the cylinder head when loosening the bolt.

b. If equipped with a manual transaxle, lock the crankshaft in position by inserting a suitable flat-bladed tool into the hole in the bottom of the bell housing to engage the flywheel teeth.

c. If equipped with an automatic transaxle, lock the crankshaft in position by inserting a suitable flat-bladed tool between the flywheel teeth and against the engine block.

d. Remove the crankshaft sprocket bolt and crankshaft sprocket.

To install:

14. If the timing belt sprockets were removed, proceed as follows:

a. Install the crankshaft sprocket, aligning the keyway. Lock the crankshaft in place and tighten the crankshaft sprocket bolt to 81 ft. lbs. (110 Nm).

b. Install the camshaft sprocket and retaining bolt. Lock the camshaft in place using the rod, and tighten the bolt to 44 ft. lbs. (60 Nm). Remove the locking rod.

15. Install the tensioner plate to the tensioner.

16. Insert the lug of the tensioner plate into the hole of the tensioner.

17. Install the tensioner, tensioner plate and spring. Do not fully tighten the tensioner bolt and stud at this time.

18. Move the tensioner plate in a counterclockwise direction. This should cause the tensioner to move in the same direction. If it does not, remove the tensioner and tensioner plate, and reinsert the tensioner plate lug in the timing plate tensioner hole.

19. Check that the camshaft timing marks are aligned. If not, align the two marks by turning the camshaft.

20. Check that the punch mark on the crankshaft timing belt sprocket is aligned with the arrow mark on the oil pump case. If not, align the two marks by turning the crankshaft.

21. With the timing marks aligned, install the timing belt on the two sprockets. If the old belt is being reused, be sure to install it running in the same direction of original rotation.

22. Install the tensioner spring and spring damper. Turn the timing belt two rotations clockwise after installing the tensioner spring and damper to remove any belt slack. Tighten the tensioner stud to 8 ft. lbs. (11 Nm), then the tensioner bolt to 20 ft. lbs. (27 Nm).

➡Confirm that both sets of timing marks are aligned properly.

23. Using a new seal, install the timing belt cover and tighten the bolts and nut to 97 inch lbs. (11 Nm).

24. Install the crankshaft pulley. Fit the keyway on the pulley to the crankshaft timing belt sprocket and tighten the bolts to 8–12 ft. lbs. (11–16 Nm).

25. Install the alternator drive belt, if equipped, A/C drive belt.

26. Install the lower alternator cover plate. Tighten the bolts to 89 inch lbs. (10 Nm).

27. Install the right side splash-shield and clips.

28. Connect the negative battery cable.

29. Run the engine. Check for leaks.

1.6L (VIN 6) and 1.8L (VIN 8) Engines

1. Disconnect the negative battery cable.

2. Remove the windshield washer reservoir from the engine compartment.

3. If equipped with cruise control, proceed as follows:

a. Remove the cruise control actuator cover.

b. Disconnect the cruise control harnesses.

c. Disconnect the control cable.

d. Remove the bolts and actuator from the vehicle.

4. Remove the right front wheel.

5. Remove the bolts and plastic clips and the right front wheel housing.

6. Remove the alternator/water pump drive belt.

7. If equipped with A/C, proceed as follows:

a. Remove the A/C compressor drive belt.

b. Disconnect the compressor harness.

c. Remove the bolts and compressor, without disconnecting the refrigerant lines. Suspend the compressor aside.

d. Remove the compressor mounting bracket.

8. Remove the power steering pump drive belt.

9. Disconnect the wiring from the alternator and oil pressure switch.

10. Remove the engine wiring harness cover.

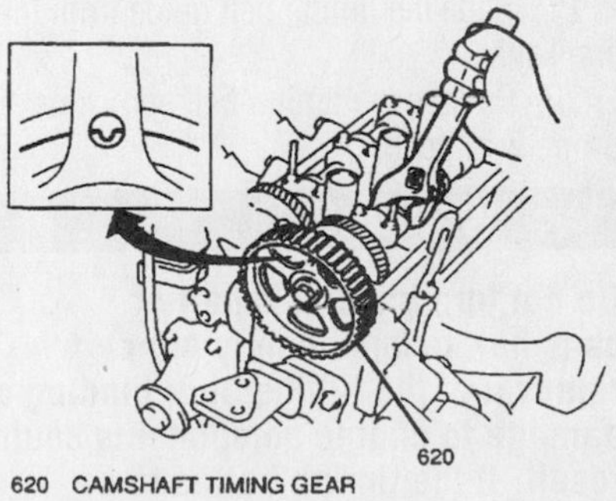

Aligning the camshaft timing marks—GM 1.6L (VIN 6) and 1.8L (VIN 8) engines

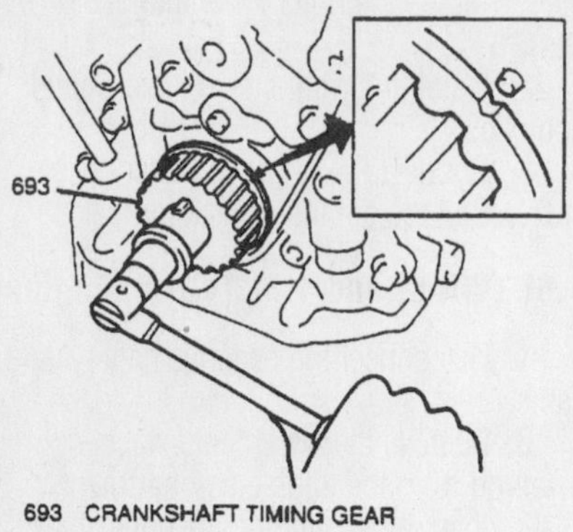

Crankshaft pulley alignment indicator—GM 1.6L (VIN 6) and 1.8L (VIN 8) engines

11. Remove the wiring harness from the cylinder head cover.
12. Disconnect the ignition wires from the spark plugs, then remove the spark plugs.
13. Remove the PCV hoses from the valve cover.
14. Remove the cap nuts, the seal washers and the cylinder head cover with the gasket.
15. Turn the crankshaft to align the timing mark on the crankshaft pulley at **0**, setting the piston in the No. 1 cylinder at Top Dead Center (TDC) on the compression stroke. Check that the valve lash adjusters on the No. 1 cylinder are loose. If not, turn the crankshaft pulley one complete revolution (360 degrees).
16. Remove the engine ground wire from the right fender apron.
17. Install a suitable support under the engine and remove the engine mount.
18. Remove the water pump pulley.
19. Remove the crankshaft pulley using a suitable puller.
20. Remove the 9 retaining bolts and the timing belt covers.
21. Slide the timing belt guide from the crankshaft.
22. Be sure the timing belt sprockets are properly aligned.

**** WARNING**

Do not turn the crankshaft or camshaft independently after removal of the timing belt; binding or damage to engine components could result. If the timing belt is to be reused, mark the belt with an arrow showing the direction of engine revolution.

23. Remove the timing belt tensioner bolt, tensioner and tension spring.
24. Remove the timing belt from the sprockets. Inspect the timing belt for cracked or damaged teeth. Replace as necessary.

**** WARNING**

Do not bend, twist or turn the timing belt.

25. If the camshaft sprocket is to be removed, hold the camshaft stationary using a wrench positioned on the hexagon cast into the camshaft, and remove the sprocket retaining bolt and sprocket.

**** WARNING**

Be careful not to damage the cylinder head when holding the camshaft in place.

26. If the crankshaft sprocket is to be removed, pry it from the crankshaft using two flat-bladed prybars.

To install:

27. Align the camshaft key with the groove on the sprocket and slide the sprocket on. Hold the camshaft with the wrench at the hexagonal portion of the camshaft, and tighten the camshaft timing sprocket bolt to 43 ft. lbs. (59 Nm).
28. Be sure the sprocket is still properly aligned.
29. Install the crankshaft timing sprocket. Align the crankshaft key with the groove on the sprocket and slide it on.
30. Reinstall the timing belt tensioner and the tension spring. Pry the tensioner to the left as far as it will go and temporarily tighten the retaining bolt.
31. Install the timing belt. If installing the old belt, observe the matchmarks made during removal.
32. Loosen the retaining bolt for the timing belt tensioner and allow it to tension the belt.
33. Temporarily install the crankshaft pulley bolt and turn the crankshaft clockwise two full revolutions. Be sure each timing mark realigns exactly.
34. Tighten the timing belt tensioner bolt to 27 ft. lbs. (37 Nm).
35. Measure the timing belt deflection. Correct deflection should be 0.20–0.24 in. (5–6mm) at 4 lbs. (20 Nm) of pressure. If the deflection is not correct, adjust it with the timing belt tensioner.
36. Install the timing belt guide, with the cup side facing outward.
37. Install the timing belt covers, installing the bottom one first. Tighten the 9 cover bolts to 62 inch lbs. (7 Nm).
38. Install the crankshaft pulley after aligning the pulley key with the slot on the pulley. Hold the pulley with tool J-8614-01, and tighten the pulley bolt to 87 ft. lbs. (118 Nm).
39. Temporarily install the water pump pulley.
40. Install the engine mount.
41. Install or connect the remaining components.
42. If equipped with A/C, proceed as follows:
    a. Install the compressor mounting bracket and tighten the bolts to 35 ft. lbs. (47 Nm).
    b. Install the compressor and tighten the bolts to 18 ft. lbs. (25 Nm).
    c. engage the compressor wiring connector.
    d. Install the compressor drive belt and adjust the tension.
43. If equipped with cruise control, proceed as follows:
    a. Install the cruise control actuator and tighten the bolts to 89 inch lbs. (10 Nm).
    b. Connect the cruise control cable.
    c. Install the cruise control actuator cover.
44. Connect the negative battery cable.
45. Start the engine and check vehicle operation.

3.0L (VIN R) Engine

1. Before servicing the vehicle, refer to the precautions in the beginning of this section.
2. Remove or disconnect the following:
    - Negative battery cable
    - Intake air resonator
    - Intake plenum
    - Splash shield
    - Secondary air injection (AIR) pipe
    - Accessory drive belt
    - Water pump and power steering pump pulleys
    - Drive belt tensioner
    - Front timing belt cover
    - Crankshaft balancer
3. Rotate the crankshaft clockwise to 60 degrees before top dead center (BTDC).
4. Loosen the timing belt tensioner and idler pulleys.
5. Remove the timing belt.

To install:

6. Start at the crankshaft sprocket and install the timing belt with the double dash (TDC) aligned with the marks on the oil pump and on the belt drive gear.
7. Set the initial timing belt tension as follows:

TIMING BELT INSTALLATION AND ADJUSTMENT TABLE

Step	Action	Value	Yes	No
1	Install the timing belt and align marks on the belt with the marks on the camshaft gears and the crankshaft gear. Check the timing belt deflection between the idler pulley for camshafts 3 & 4 and camshaft number 4. Is the timing belt installed, the marks aligned and the timing belt deflection adjusted?	1 cm (0.4 in) maximum	Go to Step 2	—
2	Set the initial timing belt tension at the timing belt tensioner. Is the initial timing belt tension set?	—	Go to Step 3	—
3	Rotate the engine two complete revolutions and secure the crankshaft at Top Dead Center (TDC) with the J 42069-10. Has the engine been rotated and the crankshaft secured to TDC?	—	Go to Step 4	—
4	Starting with camshafts 3 and 4, check the alignment of the marks on the camshaft gears with the marks on the J 42069-20 checking gauge. Do the marks on the camshaft gears align exactly with the marks on J 42069-20?	—	Go to Step 5	Go to Step 6
5	Check the alignment of the marks on camshafts gears 1 and 2 with the marks on the J 42069-20 checking gauge. Do the marks on the camshaft gears align exactly with the marks on J 42069-20?	—	Go to Step 14	Go to Step 10
6	Do the camshaft gear marks line up to the left (BTDC) of the marks on the J 42069-20 checking gauge?	—	Go to Step 8	Go to Step 7
7	Do the camshaft gear marks line up to the right (ATDC) of the marks on the J 42069-20 checking gauge?	—	Go to Step 9	—
8	Turn the idler pulley eccentric, for camshafts 3 and 4, counterclockwise until the marks on the camshaft gear align exactly with the marks on J 42069-20. Rotate the engine two complete revolutions, lock the crankshaft at TDC with J 42069-10 and recheck the alignment of the camshaft gear marks to the marks on J 42069-20. Do the marks on the camshaft gears align exactly with the marks on J 42069-20?	—	Go to Step 5	Go to Step 6
9	Turn the idler pulley eccentric, for camshafts 3 and 4, clockwise until the marks on the camshaft gear align exactly with the marks on J 42069-20. Rotate the engine two complete revolutions, lock the crankshaft at TDC with J 42069-10 and recheck the alignment of the camshaft gear marks to the marks on J 42069-20. Do the marks on the camshaft gears align exactly with the marks on J 42069-20?	—	Go to Step 5	Go to Step 6

79225G35

Timing belt installation and adjustment table—GM 3.0L (VIN R) engine

a. Turn the tensioner nut COUNTERCLOCKWISE to full stop, then, turn the nut back until the reference mark is 1 mm (0.003 in) over the flange.

b. Tighten the timing belt tensioner locking nut until snug, the locking nut will be tightened to specifications after all final adjustments are made.

8. Rotate the engine in the clockwise direction two revolutions stopping at 60 degrees BTDC.

9. Inspect the alignment of the reference marks on the camshaft gears with the notches on the rear timing belt cover, as well as, the mark on the crankshaft sprocket and oil pump housing.

10. If timing belt adjustment IS NOT required, set the final timing belt tension:

a. Loosen the timing belt tensioner locking nut.

b. Turn the locking nut counterclockwise to full stop, then back until the reference mark is 2-4 mm (0.078-

Step	Action	Value	Yes	No
10	Do the camshaft gear marks line up to the left (BTDC) of the marks on the J 42069-20 checking gauge?	—	Go to Step 12	Go to Step 11
11	Do the camshaft gear marks line up to the right (ATDC) of the marks on the J 42069-20 checking gauge?	—	Go to Step 13	—
12	Turn the idler pulley eccentric, for camshafts 1 and 2, counterclockwise until the marks on the camshaft gear align exactly with the marks on J 42069-20. Rotate the engine two complete revolutions, lock the crankshaft at TDC with J 42069-10 and recheck the alignment of the camshaft gear marks to the marks on J 42069-20. Do the marks on the camshaft gears align exactly with the marks on J 42069-20?	—	Go to Step 14	Go to Step 10
13	Turn the idler pulley eccentric, for camshafts 1 and 2, clockwise until the marks on the camshaft gear align exactly with the marks on J 42069-20. Rotate the engine two complete revolutions, lock the crankshaft at TDC with J 42069-10 and recheck the alignment of the camshaft gear marks to the marks on J 42069-20. Do the marks on the camshaft gears align exactly with the marks on J 42069-20?	—	Go to Step 14	Go to Step 10
14	Set the final timing belt tension at the timing belt tensioner. Is the final timing belt tension set?	—	Go to Step 15	—
15	Again, rotate the engine two complete revolutions and lock the crankshaft at TDC. Do a final inspection of the camshaft gear marks' relationship to the J 42069-20 marks. The marks must align exactly. Do the marks on the camshaft gears align exactly with the marks on the J 42069-20?	—	Go to Step 16	Go to Step 2
16	Remove all checking tools and ensure all idler pulleys and the tensioner locking nut are tightened to specifications. Continue with re-assembly of the engine.	—	—	—

79225G36

Timing belt installation and adjustment table (continued)—GM 3.0L (VIN R) engine

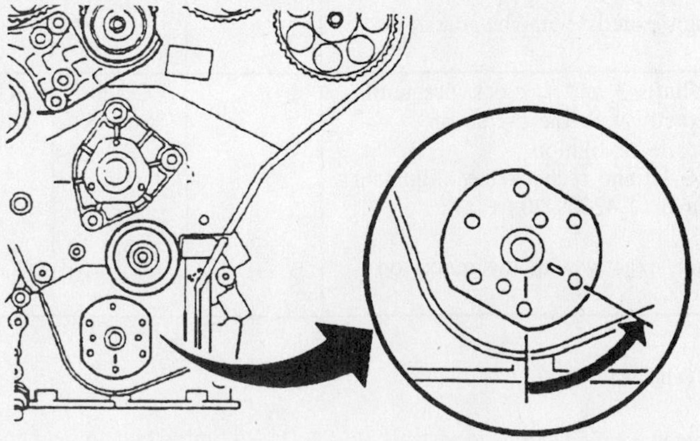

Crankshaft alignment to 60 degrees BTDC—GM 3.0L (VIN R) engine

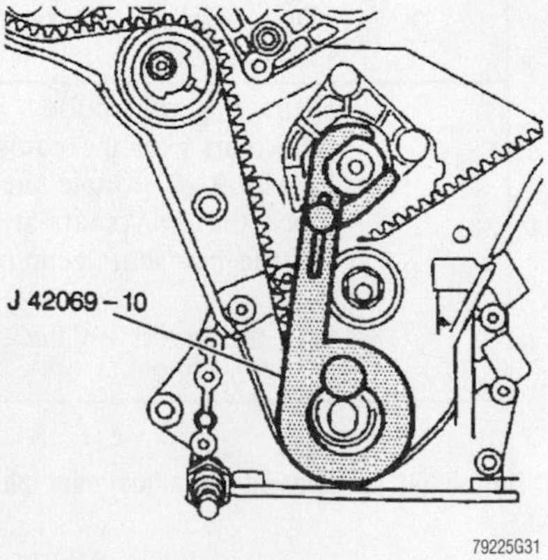

Securing the crankshaft—GM 3.0L (VIN R) engine

0.157 in) ABOVE the reference mark on the flange.

11. Tighten the timing belt tensioner locking nut 15 ft. lbs. (20 Nm).

12. Tighten the idler pulley bolts to 30 ft. lbs. (40 Nm).

13. Install or connect the following:
- Crankshaft balancer. Torque the bolts to 15 ft. lbs. (20 Nm).
- Front timing belt cover. Torque the bolts to 71 inch lbs. (8 Nm).
- Drive belt tensioner. Torque the bolts to 30 ft. lbs. (40 Nm).
- Water pump pulley. Torque the bolts to 71 inch lbs. (8 Nm).
- Power steering pump pulley. Torque the bolts to 15 ft. lbs. (20

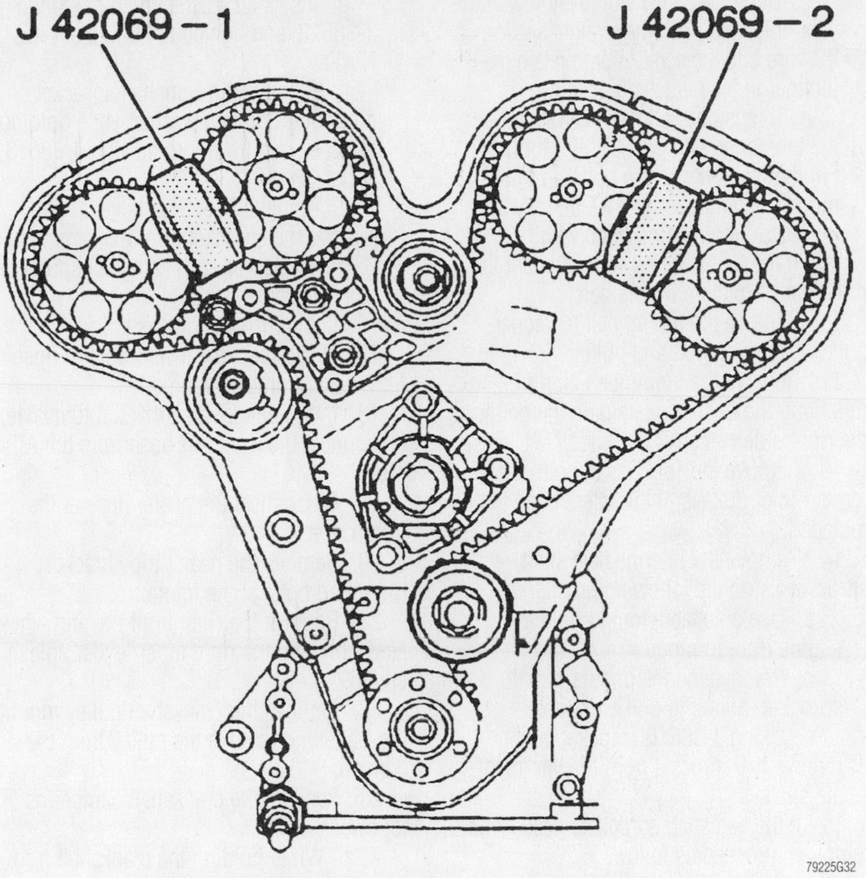

Locking the camshaft—GM 3.0L (VIN R) engine

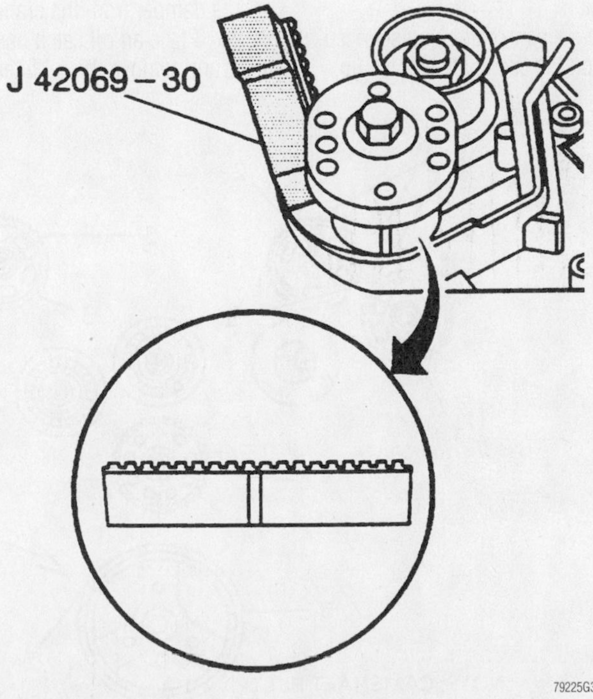

Using the tool to pin the timing belt—GM 3.0L (VIN R) engine

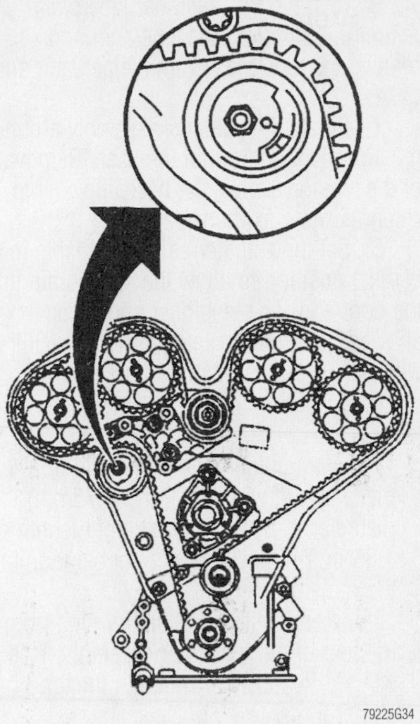

Initial timing belt tension adjustment—GM 3.0L (VIN R) engine

Nm).

- Accessory drive belt
- AIR injection pipe
- Splash shield
- Intake plenum
- Intake air resonator

3.4L (VIN X) Engine

The 3.4L (VIN X) engine uses a timing chain and camshaft timing belts.

1. Disconnect the negative battery cable.
2. Disconnect and remove the power steering pump from the pump mounting bracket.
3. Remove the left, right and center timing belt covers.
4. Rotate the engine clockwise to align the timing marks, TDC on the No.1 exhaust stroke, on the camshaft sprockets and intermediate shaft.
5. Loosely clamp the two camshaft sprockets on each side of the engine together using clamping pliers or the equivalent. Secure the belt to the right side cam sprocket with a C-clamp and a wide pad on the belt.

➡When clamping the sprockets no deflection should be noticed. If any deflection is noticed, loosen the clamping devices. DO NOT mar the camshaft sprockets with the clamping device.

6. Remove the tensioner side plate retaining bolts from the tensioner and remove the side plate from the actuator and base.

7. Rotate the actuator assembly around the arm pivot and out of the base. Removal of the tensioner from the base allows it to extend to its maximum travel.

8. Set the actuator aside on a table in a vertical position to allow the oil to drain into the boot end. The tensioner should be allowed to sit for 5 minutes prior to refilling with oil.

9. Reset the timing belt actuator as follows:

a. Straighten out a paper clip or a piece of stiff wire 0.032 in. (0.75mm) diameter to a minimum straight length of 1.85 in. (47mm). Form a double loop in the remaining end.

b. Remove the rubber end plug from the rear of the tensioner assembly. This will aid in allowing the oil in the tensioner to escape.

c. Hold the tensioner in your hand with the rubber boot end of the tensioner pointing down.

d. DO NOT remove the vent plug. Push the paper clip through the center hole in the vent plug and into the pilot hole.

e. Insert a small screwdriver into the screw slot inside the end of the tensioner.

f. Retract the tensioner by rotating the tensioner plunger in a clockwise direction while pushing the rod tip against a table top.

g. Align the screw slot to align with the vent hole, and push the straight section of the wire into the screw slot to retain the plunger in the retracted position.

h. If tensioner oil has been lost, fill the tensioner with SAE 5W30 Mobil 1®. Fill the tensioner to the bottom of the plug. The tensioner **MUST** be fully retracted before being filled with oil.

10. If the belt is being reused, mark the direction of rotation on the belt.

11. Remove the timing belt tensioner pulley mounting bolt and pulley.

12. Remove the timing belt after first removing the C-clamp retaining the belt to the right side camshaft sprocket.

13. Remove the Torx® head bolts securing the idler pulleys, if the idlers need to be replaced.

14. Remove the intermediate shaft sprocket using the following procedure:

a. Use a suitable tool to hold the engine from turning.

b. Remove the intermediate shaft sprocket mounting bolt and washer.

c. Using J-38616 sprocket puller, remove the sprocket from the intermediate shaft.

15. If the camshaft sprockets need to be removed, proceed as follows:

a. Remove the camshaft carrier cover(s).

b. Remove the camshaft sprocket clamping pliers.

c. Rotate the camshaft being serviced so the flats on the camshaft are face up.

d. Install a camshaft hold-down tool J-38613, and tighten to 22 ft. lbs. (30 Nm).

e. Remove the camshaft sprocket mounting bolt and washer while holding the camshaft from turning with J-38613 and J-38614.

f. Using J-38616 sprocket puller, remove the sprocket from the camshaft.

g. Repeat for each camshaft sprocket as necessary.

16. Drain the cooling system.

17. Disconnect the lower radiator hose from water pump inlet pipe.

18. If equipped with a manual transaxle, disconnect the front AIR hose from the AIR pipe.

19. Disconnect the heater hose at the front cover.

20. Remove the heater pipe bracket mounting bolts at the frame.

21. Remove the right front tire and wheel assembly and the right inner fender splash-shield.

22. Remove the crankshaft pulley mounting bolts and remove the pulley from the damper.

23. Remove the crankshaft damper as follows:

a. While holding the crankshaft from turning using a suitable tool, remove the damper mounting bolt and washer.

b. Install tool J-24420-B and remove the damper from the crankshaft.

24. Place an oil catch pan under the oil filter and remove the oil filter.

A LOCATION OF TIMING MARKS WITH CAM HOLD DOWN TOOLS J 38613 INSTALLED (#4 TDC COMPRESSION STROKE)
B FRONT COVER TIMING MARK
C LOCATION OF TIMING MARKS WITH DRIVE BELT INSTALLED
D LOCATION WHERE CAM HOLD DOWN TOOLS ARE INSTALLED
1 RH EXHAUST CAMSHAFT SPROCKET
2 RH INTAKE CAMSHAFT SPROCKET
3 LH INTAKE CAMSHAFT SPROCKET
4 LH EXHAUST CAMSHAFT SPROCKET
5 PERMANENT MARKS PAINTED DOTS REMOVE PREVIOUS MARKS IF TIMING IS BEING CHANGED AND MARKS AGAIN IN THESE LOCATIONS
6 CRANKSHAFT BALANCER
7 INTERMEDIATE SHAFT SPROCKET

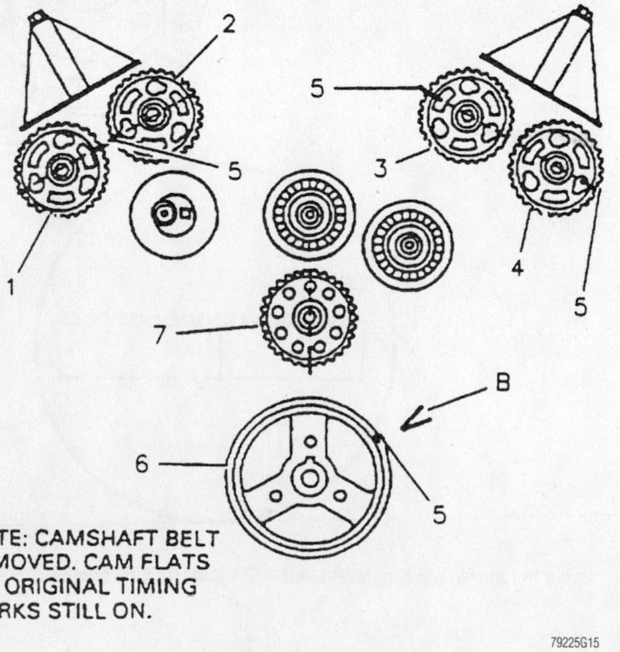

79225G15

Timing marks with hold-down tool in place—GM 3.4L (VIN X) engines

25. Remove the A/C compressor mounting bracket bolts.
26. Remove the lower front cover bolts.
27. On automatic transaxle vehicles, remove the halfshaft following the recommended procedure.
28. Remove the rear alternator bracket.
29. Disconnect and remove the starter following the recommended procedure.
30. Remove the intermediate shaft drive belt sprocket retaining bolt and remove the intermediate shaft drive belt sprocket using puller J38616.
31. Remove the upper alternator mounting bolts.
32. Remove the forward light relay center screws and position the relay center aside.
33. Disconnect the oil cooler hose from the front cover.
34. Remove the water pump pulley.
35. Remove the upper front cover bolts and remove the front cover.
36. Mark the intermediate shaft sprocket, chain link, crankshaft sprocket and cylinder block for assembly reference. The marks should be made with paint so they won't be lost when the components are removed.
37. Retract the timing chain tensioner shoe as follows:
    a. Insert tool J-33875 on both sides of the tensioner.
    b. Pull on the through-pin in the tensioner arm to retract the spring located in the tensioner arm.
    c. While compressing the spring, use a suitable tool, a cotter pin or nail, and insert the pin in the hole in the tensioner assembly to hold the tensioner compressed. The tool used must be strong enough to hold the tensioner compressed.

➡The timing chain, crankshaft sprocket and intermediate shaft sprocket will be removed at the same time. If, when removing the assembly, the intermediate shaft sprocket does not easily come off the intermediate shaft, rotate the crankshaft back and forth to loosen the intermediate shaft sprocket.

38. Install a suitable puller, J-38611 and J-8433.
39. Tighten the bolt on the puller and slowly pull the crankshaft sprocket off the crankshaft. Be sure the intermediate shaft sprocket is moving along with the crankshaft sprocket.
40. Remove the timing chain and sprockets.
41. Remove the tensioner mounting bolts and remove the tensioner assembly.

To install:

42. Install the tensioner assembly and tensioner assembly mounting bolts finger-tight first. Tighten the bolt in the slotted hole first to 18 ft. lbs. (25 Nm), then tighten the remainder of the bolts to 18 ft. lbs. (25 Nm).
43. Check to ensure that the crankshaft key is fully seated in the crankshaft cutout and the tensioner assembly is fully retracted.
44. Assemble the timing chain, intermediate shaft sprocket and crankshaft sprocket on a work bench. The timing marks made should be in alignment. The large chamfer and counterbore of the crankshaft sprocket are installed facing toward the engine and the intermediate shaft spline sockets are installed facing away from the engine.
45. Install the sprocket and chain assembly onto the engine. As the sprockets are installed, parallel alignment must be maintained.
46. The crankshaft sprocket will have to be pressed on the final 0.31 in. (8mm). This can be done using J-38612 or an equivalent puller.
47. Ensure timing was maintained.
48. Remove the retaining pin from tensioner. Clean all gasket surfaces completely.
49. Apply GM sealer 1052080 to the lower edges of the sealing surface of the front cover. Install a new gasket on the front cover.
50. Install the front cover on the engine. Apply thread sealant to the large bolts and tighten the bolts enough to pull the front cover against the engine block.
51. Install the water pump pulley.
52. Connect the oil cooler hose to the front cover.
53. Position the forward light relay center and install the mounting screws.
54. Install the upper alternator mounting bolt and tighten it to 22 ft. lbs. (30 Nm).
55. Install the intermediate shaft drive belt sprocket. The sprocket must lock into the intermediate shaft timing chain sprocket. Install the mounting bolt and washer. While holding the engine from turning, tighten the bolt to 95 ft. lbs. (130 Nm).
56. Connect and install the starter. Tighten the starter mounting bolts to 32 ft. lbs. (43 Nm).
57. Install the halfshaft following the recommended procedure.
58. Install the rear alternator bracket. Tighten the mounting bolt to 22 ft. lbs. (30 Nm) and the mounting stud to 41 ft. lbs. (55 Nm) and the lower bolt to 61 ft. lbs. (83 Nm).
59. Install the lower front cover bolts and tighten the small bolts to 18 ft. lbs. (25 Nm).
60. Install the A/C compressor mounting bolts and tighten the mounting bolts to 37 ft. lbs. (50 Nm).
61. Install a new oil filter.
62. Install the crankshaft damper as follows:
    a. Coat the seal contact area on the damper with clean engine oil.
    b. Align the notch inside the damper with the crankshaft key and slide the damper on until the key is started into the notch.
    c. Using J-29113 puller, press the damper into position on the crankshaft.
    d. Install the crankshaft pulley and pulley mounting bolts. Tighten the pulley mounting bolts to 37 ft. lbs. (50 Nm).
    e. Install the crankshaft damper mounting bolt and washer, and tighten to 78 ft. lbs. (105 Nm).
63. Install the right side inner fender splash-shield.
64. Install the tire and wheel assembly.
65. Tighten the upper front cover small bolts to 18 ft. lbs. (25 Nm) and the large bolts to 35 ft. lbs. (47 Nm).
66. Connect the heater hose to the front cover.
67. Install the heater pipe bracket mounting bolts.
68. If equipped with a manual transaxle, connect the front AIR hose to the AIR pipe.
69. Connect the lower radiator hose to the water pump.
70. Install the right side cooling fan. Install the upper radiator support.
71. Position the torque strut mounting bracket and install the mounting bolts and tighten to 52 ft. lbs. (70 Nm).
72. Install the front engine lift hook and tighten the mounting bolt to 52 ft. lbs. (70 Nm).
73. To install the camshaft sprockets, proceed as follows:
    a. Wipe the camshaft noses with clean engine oil.
    b. Install the camshaft sprocket onto the nose of the camshaft.
    c. Install the lockring and shim ring.
    d. Install, but DO NOT tighten, the camshaft sprocket mounting bolts at this time.

74. Install the intermediate shaft sprocket as follows:

a. Lubricate the seal contact area on the intermediate shaft sprocket with clean engine oil.

b. Slide the sprocket through the intermediate shaft sprocket seal and engage the locking tangs into the sockets of the chain sprocket.

c. Lightly lubricate the shaft seal and place it in position on the end of the intermediate shaft.

d. Install the intermediate shaft sprocket mounting bolt and washer. Tighten the bolt to 96 ft. lbs. (130 Nm) while holding the crankshaft from turning.

75. Install the timing belt idler pulleys and tighten the Torx® bolts to 37 ft. lbs. (50 Nm).

76. Install the actuator assembly and side plate. Tighten the actuator mounting bracket bolts to 37 ft. lbs. (50 Nm).

77. Install the belt, taking note of direction of rotation if the old belt was used.

78. Install the tensioner pulley to the mounting base. Tighten the bolt to 37 ft. lbs. (50 Nm).

79. Rotate the tensioner pulley counterclockwise into the belt using the cast square lug on the body and engage the ball end of the actuator into the socket on the pulley arm.

80. Remove the tensioner lockpin allowing the tensioner shaft to extend and the pulley to move into the belt.

81. Rotate the tensioner pulley counterclockwise, applying 14 ft. lbs. (18 Nm) of torque.

82. Rotate the engine clockwise three times to seat the belt. Align the crankshaft reference marks during the final rotation to TDC. Do not allow the crankshaft to spring back or reverse its direction of rotation.

➡The timing flats on the camshafts should be 180 degrees apart from the left side to the right side. Both camshafts on the same side should be the same.

83. To perform the camshaft timing procedure, proceed as follows:

a. Rotate the camshaft flats up on the right side camshafts and install a camshaft hold-down tool J-38613, and tighten to 22 ft. lbs. (30 Nm).

b. Seat the lockring on the right exhaust and intake camshaft sprockets by threading in the mounting bolt and washer.

c. Hold the sprocket from turning using tool J-38614.

➡Running torque of the bolts before seating should be 55 ft. lbs. (75 Nm).

d. If less torque is required, replace the shim ring and lockring.

e. If more torque is required, replace the shim ring and lockring and inspect the bolts for burrs.

f. Seating of the lockring is accomplished when the edge is flush with the sprocket hub.

g. With the lockring seated tighten the bolt to final torque of 81 ft. lbs. (110 Nm).

h. Remove J-38613.

i. Rotate the engine clockwise one full revolution or any number of odd revolutions. **DO NOT** rotate the engine backward.

j. Be sure the timing mark on the damper aligns with the mark on the front cover.

k. Repeat Substeps **a** through **i** for the left side.

84. Install the camshaft carrier covers.

85. Install timing the belt left, right and center covers and retaining bolts.

86. Connect the negative battery cable.

87. Start the engine and verify proper operation and engine performance.

BRAKES

4

1998–02 CHRYSLER CORPORATION

Caliper

REMOVAL & INSTALLATION

Chrysler Concorde, New Yorker, LHS
Dodge Intrepid
Eagle Vision

FRONT

1. Remove the appropriate wheel and tire assemblies.
2. Remove the 2 caliper guide pin bolts and remove the caliper assembly. If the caliper is not being removed from the vehicle as during brake pad renewal, simply hang the caliper with a piece of wire to take the weight off the brake hose. If the caliper is being removed for rebuild or replacement, continue to Step 5.
3. Remove the bolt retaining the brake hose to the caliper. Be sure to plug the end of the brake hose or cover it with a plastic bag to prevent contamination from entering the hydraulic system. Remove the caliper from the vehicle.

To install:

4. Reconnect the brake hose to the caliper, if removed, using new sealing washers.
5. If new linings are being installed, the caliper pistons must be pushed back into their bore to accommodate the thickness of

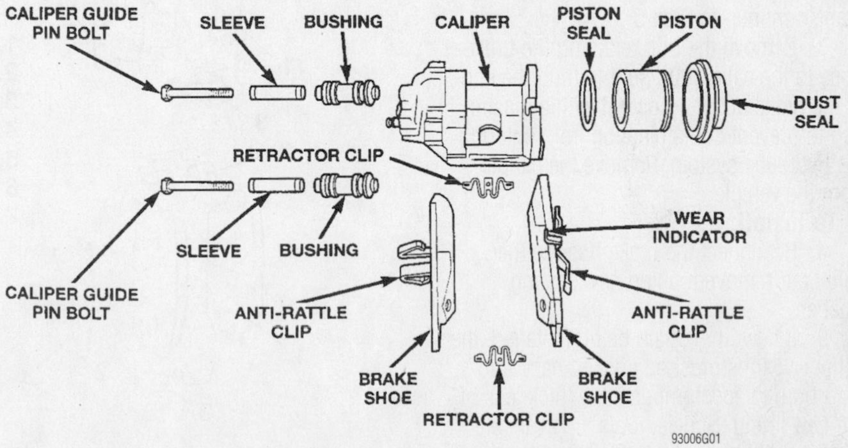

Front brake caliper and related components—Concorde, New Yorker, LHS, Intrepid, Vision

the new lining. Special tools are available for pushing the piston back although a large C-clamp can often be used. It is good practice to remove some (⅓–½) brake fluid from the master cylinder reservoir. This prevents overflow caused by brake fluid being forced through the lines as the piston is pushed back.

6. Install the brake pads into the caliper.

7. Install the caliper to the steering knuckle in the correct position. Lightly lubricate the machined areas that support the caliper with high-temperature grease.

8. Install and torque the caliper guide pin bolts. Use care not to cross the threads of the caliper pin bolts. Torque the guide pin bolts to 15 ft. lbs. (20 Nm).

9. Torque the brake hose fittings to 35 ft. lbs. (48 Nm).

10. Be sure to bleed the brake system.

11. Install the wheels and lug nuts. Torque the wheel lug nuts in a star pattern sequence to 95–100 ft. lbs. (129–135 Nm).

12. Before attempting to move the vehicle, pump the brake pedal to seat the pads against the rotors. Make sure the vehicle has a firm brake pedal. Check the level of the brake fluid and add DOT 3 brake fluid, if necessary.

13. Road test the vehicle and make several stops to wear off any foreign material on the brakes and to seat the brake linings, if replaced.

REAR

1. Remove the appropriate wheel and tire assemblies.

2. Remove the 2 caliper guide pin bolts and remove the caliper assembly. If the caliper is not being removed from the vehicle as during brake pad renewal, simply hang the caliper with a piece of wire to take the weight off the brake hose. If the caliper is being removed for rebuild or replacement, continue to Step 5.

3. Remove the bolt retaining the brake hose to the caliper. Be sure to plug the end of the brake hose or cover it with a plastic bag to prevent contamination from entering the hydraulic system. Remove the caliper from the vehicle.

To install:

4. Reconnect the brake hose to the caliper, if removed, using new sealing washers.

5. If new linings are being installed, the caliper pistons must be pushed back into their bore to accommodate the thickness of the new lining. Special tools are available for this, although a large C-clamp can often be used. It is good practice to remove some brake fluid from the master cylinder reservoir. This prevents overflow caused by brake fluid being forced through the lines as the piston is pushed back.

6. Install the brake pads into the caliper.

7. Install the brake caliper to the caliper adapter. Lightly lubricate machined areas that support the caliper with high-temperature grease.

8. Install and tighten the caliper guide pin bolts. Use care not to cross the threads of the caliper pin bolts. Torque the guide pin bolts to 17 ft. lbs. (22 Nm).

9. Torque the brake hose fittings to 35 ft. lbs. (48 Nm).

10. Be sure to bleed the brake system.

11. Install the wheels and lug nuts. Torque the wheel lug nuts in a star pattern sequence to 95–100 ft. lbs. (129–135 Nm).

12. Before attempting to move the vehicle, pump the brake pedal to seat the pads against the rotors. Make sure the vehicle has a firm brake pedal.

13. Check the level of the brake fluid and add DOT 3 brake fluid, if necessary.

14. Road test the vehicle and make several stops to wear off any foreign material on the brakes and to seat the brake linings, if replaced.

Chrysler Sebring
Dodge Avenger

FRONT

1. Remove about half of the brake fluid from the master cylinder.

2. Position a C-clamp, or other suitable tool, over the caliper. Smoothly apply pressure, forcing the caliper piston into the caliper bore until it bottoms. Remove the C-clamp, if used.

3. If the caliper is to be completely removed from the vehicle, remove the brake hose attaching bolt, disconnect the brake hose from the caliper and plug the hose to prevent fluid contamination or loss.

4. Remove the caliper mounting bolts and lift the caliper off of the support bracket.

5. Remove the caliper from the vehicle. If the caliper is only removed for access to other components, support the caliper, with the brake hose attached, so that there is no strain on the brake hose.

To install:

6. Position the caliper to the steering knuckle. Lubricate and install the mounting bolts. Torque the bolts to 54 ft. lbs. (74 Nm).

7. If removed, unplug and install the brake line hose to the caliper and torque the inlet fitting bolt to 22 ft. lbs. (29 Nm).

8. Fill the master cylinder with fresh brake fluid and, if the brake hose was removed, bleed the brake system.

9. Install the wheel assembly.

10. Depress the brake pedal 3–4 times to seat the brake linings and to restore pressure in the system.

✲✲ CAUTION

Do not move the vehicle until a firm pedal is obtained.

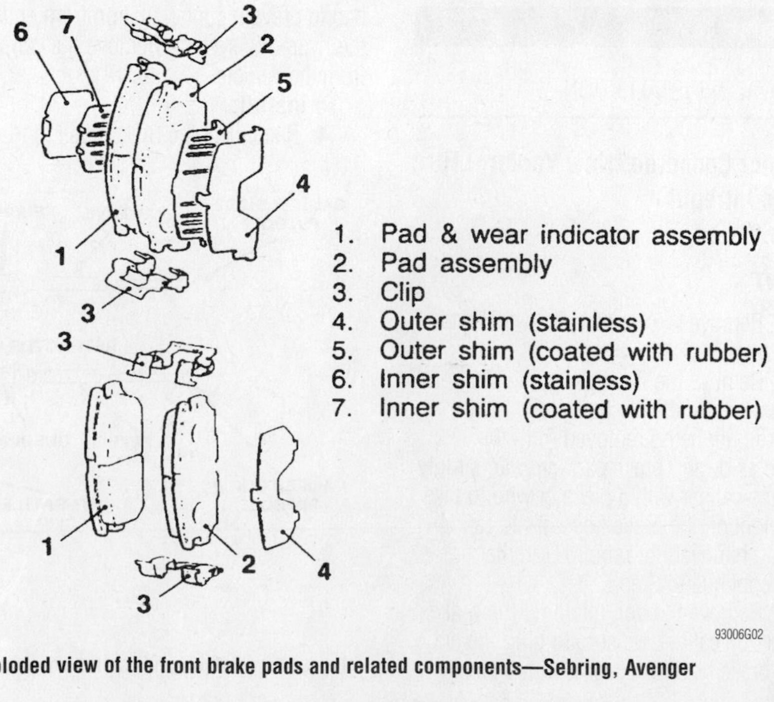

Exploded view of the front brake pads and related components—Sebring, Avenger

REAR

Unlike many rear disc brake designs, this system does not incorporate the parking brake system, into the rear brake caliper. The rear brake system is serviced the same as the front system.

1. Remove about half of the brake fluid from the master cylinder.
2. Remove the wheel assembly.
3. Position a C-clamp, or other suitable tool, over the caliper. Smoothly apply pressure, forcing the caliper piston into the caliper bore until it bottoms. Remove the C-clamp, if used.
4. If the caliper is to be completely removed from the vehicle, remove the brake hose attaching bolt, disconnect the brake hose from the caliper and plug the hose to prevent fluid contamination or loss.
5. Remove the caliper mounting bolts and lift the caliper off of the support bracket.
6. Remove the caliper from the vehicle. If the caliper is only removed for access to other components, support the caliper, with the brake hose attached, so that there is no strain on the brake hose.

To install:

7. Position the caliper on the support bracket, lubricate and install the mounting bolts. Torque the bolts to 54 ft. lbs. (74 Nm).
8. If removed, unplug and install the brake line hose to the caliper and torque the inlet fitting bolt to 22 ft. lbs. (29 Nm).
9. Fill the master cylinder with fresh brake fluid and, if the brake hose was removed, bleed the brake system.
10. Install the wheel assembly.
11. Depress the brake pedal 3–4 times to seat the brake linings and to restore pressure in the system.

⁂ CAUTION

Do not move the vehicle until a firm pedal is obtained.

Eagle Talon

FRONT

1. Remove the appropriate tire and wheel assembly.
2. To disconnect the front brake hose, hold the nut on the brake hose side and loosen the flared brake line nut. Remove the brake hose from the caliper.
3. Remove the caliper guide and lock pins and lift the caliper assembly from the caliper support.

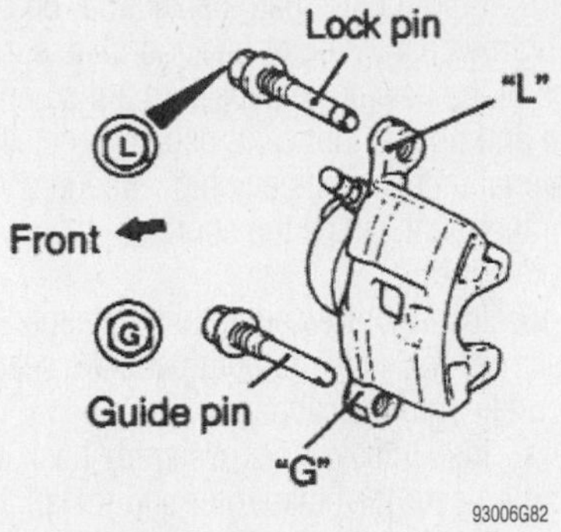

Caliper locking pin and guide pin description—Talon

To install:

4. Position caliper onto the caliper support. Install the guide pin and lock pin. Torque guide and locking pins to 54 ft. lbs. (75 Nm).
5. Reconnect the brake hose.
6. Bleed the brake system.
7. Apply brake pedal and inspect system. Ensure proper operation and no leakage.
8. Reinstall tire and wheel assembly. Torque lug nuts to 87–101 ft. lbs. (120–140 Nm).

REAR

1. Disconnect the battery negative cable.
2. Remove the appropriate tire and wheel assemblies. Loosen the parking brake cable adjustment from inside the vehicle.
3. Disconnect the parking brake cable end installed to the rear brake caliper assembly.
4. Remove the caliper lock and guide pins. Lift the caliper assembly from the caliper support.
5. Remove the rear brake hose from the caliper. Remove the caliper from the vehicle.

To install:

6. Install the rear brake hose onto the caliper with new washers in place. If equipped with brake hose retainer bolt, torque bolt to 25 ft. lbs. (35 Nm) torque. If no bolt is used, torque the brake hose fitting to 12 ft. lbs. (17 Nm).
7. Install the caliper over the brake pads. Lubricate and install the lock pin and torque to 23 ft. lbs. (32 Nm). Reinstall the guide pin and torque to 23 ft. lbs. (32 Nm).
8. Bleed the brake system.
9. Inspect the brake system for leaks and ensure proper operation.
10. Install tire and wheel assemblies. Torque wheels to 87–101 ft. lbs. (120–140 Nm).
11. Properly adjust parking brake cable.

Dodge/Plymouth Neon, PT Cruiser

1. On 1998–01 models, isolate the master cylinder as follows:
   a. Use a brake pedal holding tool and depress the brake pedal past its first one inch of travel and hold it in this position. (This will keep brake fluid from draining from the master cylinder).
2. Remove the front wheels.
3. Remove the 2 caliper guide pin bolts.
4. Disconnect the brake hose from the caliper.
5. Remove the caliper from the steering spindle.

To install:

6. Reconnect the brake hose to the caliper and torque to 35 ft. lbs. (48 Nm).
7. Reinstall the caliper to the steering spindle.
8. Reinstall the caliper guide pin bolts and torque to 16 ft. lbs. (22 Nm).
9. Properly bleed the brake system.
10. Reinstall the front wheels and lower the vehicle.
11. Pump the brake pedal to seat the front brake pads before moving the vehicle.
12. Road test the vehicle and check for proper operation.

Chrysler Cirrus, Sebring
Dodge Stratus
Plymouth Breeze

1. Remove the front wheels.
2. Remove the 2 caliper-to-steering knuckle guide pin bolts.
3. If the caliper is to be removed from the vehicle completely, for example, for overhaul perform the following procedure.
   a. Disconnect the brake hose from the caliper.
   b. Cover the opening of the brake hose so the hydraulic system does not become contaminated.
4. Remove the caliper from the steering knuckle.

To install:

5. Clean and lubricate both steering knuckle abutments with a coating of multi-purpose grease.
6. Position the caliper and brake pad assembly over the brake rotor. Be sure to properly install the caliper assembly into the abutments of the steering knuckle. Be sure the caliper guide pin bolts, rubber bushings and sleeves are clear of the steering knuckle bosses.
7. Reinstall the caliper guide pin bolts

and torque to 16 ft. lbs. (22 Nm). On Sebring and Avenger models, torque the caliper guide pin bolts to 54 ft. lbs. (74 Nm).

8. If removed, connect the brake hose to the caliper and torque to 35 ft. lbs. (48 Nm).

9. Properly bleed the brake system.

10. Reinstall the wheel and tire and torque the lug nuts in a star pattern sequence to half specification. Repeat the tightening procedure to full specified torque of 95–100 ft. lbs. (129–135 Nm).

11. Pump the brake pedal to seat the front brake pads before moving the vehicle.

12. Road test the vehicle and check for proper operation.

Disc Brake Pads

REMOVAL & INSTALLATION

Chrysler Concorde, New Yorker, LHS
Dodge Intrepid
Eagle Vision

FRONT AND REAR

1. Remove some of the fluid from the master cylinder.
2. Remove the appropriate wheels.
3. Remove the 2 caliper guide pin bolts. Remove the caliper assembly by swinging the top part of the caliper away from the brake rotor edge, then lift the caliper assembly up.
4. Prevent strain or other damage to the brake hose by supporting the caliper assembly with a strong piece of wire hanging from the strut.
5. Remove the outboard brake pad by prying the brake pad retaining clip over raised area on the caliper. Then slide the pad down and off the caliper.
6. Before removing the inboard brake pad, use a large C-clamp to press the piston back into the caliper. This will prevent possible damage to the caliper piston. It is good practice to remove some (⅓–½) of the brake fluid from the reservoir. This is because as the caliper piston is pushed back into the caliper, brake fluid will be pushed back through the lines, back into the master cylinder and fluid reservoir, possibly causing the reservoir to overflow.
7. Remove the inboard brake pad by pulling away from piston until the retainer clip is free from the cavity in the piston.

To install:

8. Lubricate both the caliper mating surface and the machined abutment surfaces with multi-purpose lubricant.
9. Before brake pad installation, be sure to lightly coat the outer backing plate surface of the new brake pads with a disc brake pad anti-squeal lubricant, usually a gel-like material that deadens any high-frequency vibration that can be the source of disc brake squeal.
10. Install brake pads into the caliper assembly making sure both pads are seated securely onto the caliper.
11. Install the caliper assembly back into position over the brake rotor and install the caliper guide pin bolts. Torque the front and rear caliper guide pin bolts to 17 ft. lbs. (23 Nm).
12. Install the wheels and lug nuts. Torque the lug nuts, in a star pattern sequence, to 95–100 ft. lbs. (129–135 Nm).
13. Top off the master cylinder to the appropriate level, using DOT 3 type brake fluid only.
14. Before moving the vehicle, pump the brakes until a firm pedal is obtained. Road test the vehicle to make sure the brake operation is normal.

Chrysler Sebring
Dodge Avenger

FRONT

1. Remove some of the brake fluid from the master cylinder reservoir. The reservoir should be no more than ½ full. When the pistons are depressed into the calipers, excess fluid will flow up into the reservoir.
2. Remove the appropriate tire and wheel assemblies.
3. Remove the caliper guide and lock pins and lift the caliper assembly from the caliper support. Tie the caliper out of the way using wire. Do not allow the caliper to hang by the brake line.

➡On some models the caliper can be flipped up by leaving the upper pin in place, using it as a pivot point.

4. Remove the brake pads, spring clip and shims. Take note of positioning to aid installation.
5. Install the wheel lug nuts onto the studs and lightly tighten. This is done to hold the disc on the hub.

To install:

6. Use a large C-clamp to compress the piston(s) back into the caliper bore.
7. Lubricate the slide points and install the brake pads, shims and spring clip onto the caliper support. Install the caliper over the brake pads.
8. Lubricate and install the caliper guide and lock pins in their original positions. Torque guide and locking pins to 54 ft. lbs. (74 Nm).
9. Install the tire and wheel assemblies.

✲✲ CAUTION

Pump brake pedal several times, until firm, before attempting to move the vehicle.

10. Road test the vehicle and check brakes for proper operation.

REAR

Unlike many rear disc brake designs, this system does not incorporate the parking brake system, into the rear brake caliper, therefore, the rear brake system is serviced the same as the front system.

1. Remove some of the brake fluid from the master cylinder reservoir. The reservoir should be no more than ½ full. When the pistons are depressed into the calipers, excess fluid will flow up into the reservoir.
2. Remove the appropriate tire and wheel assemblies.
3. Remove the caliper guide and lock pins and lift the caliper assembly from the caliper support. Tie the caliper out of the way using wire. Do not allow the caliper to hang by the brake hose.

➡On some models, the caliper can be flipped up by leaving the upper pin in place, using it as a pivot point.

4. Remove the brake pads, spring clip and shims. Take note of positioning to aid installation.
5. Install the wheel lug nuts onto the studs and lightly tighten. This is done to hold the brake disc on the hub.

To install:

6. Use a large C-clamp to compress the piston(s) back into the caliper bore.
7. Lubricate the slide points and install the brake pads, shims and spring clip onto the caliper support. Install the caliper over the brake pads.
8. Lubricate and install the caliper guide and lock pins in their original positions. Torque guide and locking pins to 54 ft. lbs. (74 Nm).
9. Install the tire and wheel assemblies.

✲✲ CAUTION

Pump brake pedal several times, until firm, before attempting to move the vehicle.

10. Road test the vehicle and check brakes for proper operation.

Eagle Talon

FRONT

1. Remove some of the brake fluid from the master cylinder reservoir. The reservoir should be no more than half full. When the pistons are pressed into the calipers, excess fluid will flow up into the reservoir.
2. Remove the appropriate tire and wheel assemblies.
3. Remove the caliper guide and lock pins and lift the caliper assembly from the caliper support. Tie the caliper out of the way using wire. Do not allow the caliper to hang by the brake line.

➡On some vehicles, the caliper can be flipped up by leaving the upper pin in place and using it as a pivot point.

4. Remove the brake pads, spring clip and shims. Take note of positioning to aid installation.
5. Install the wheel lug nuts onto the studs and lightly tighten. This is done to hold the disc on the hub.

To install:

6. Use a large C-clamp to compress piston(s) back into caliper bore. On 2 piston calipers both pistons will have to be retracted together.
7. Lubricate slide points and install the brake pads, shims and spring clip onto the caliper support. Install the caliper over the brake pads.
8. Lubricate and install the caliper guide and lock pins in their original positions. Torque guide and locking pins to 54 ft. lbs. (75 Nm).
9. Reinstall the tire and wheel assemblies.

➡Pump brake pedal several times, until firm, before attempting to move vehicle.

10. Road test the vehicle and check brakes for proper operation.

REAR

1. Remove some of the brake fluid from the master cylinder reservoir. The reservoir should be no more than half full. When the pistons are depressed into the calipers, excess fluid will flow up into the reservoir.
2. Remove the appropriate tire and wheel assemblies. Loosen the parking brake cable adjustment from inside the vehicle.
3. Disconnect the parking brake cable end installed to the rear brake caliper assembly.
4. Remove the caliper lock and guide pins and lift the caliper assembly from the caliper support. Tie the caliper out of the way using wire. Do not allow the caliper to hang by the brake line.
5. Remove the outer shim, brake pads and spring clips from the caliper support. Take note of positioning of each to aid in installation.
6. Reinstall the wheel lug nuts onto the studs and lightly tighten. This is done to hold the disc on the hub.
7. Clean the caliper piston. Using rear disc brake driver tool MB990652, thread the piston into the caliper bore. Be sure, at this point, that the stopper groove of the piston correctly fits into the projection on the replacement brake pads rear surface.

To install:

8. Lubricate all sliding and pivot points. Install the brake pads, shims and spring clip to the caliper support. Install the caliper over the brake pads.
9. Lubricate and install the caliper guide and lock pins. Torque the pins to 23 ft. lbs. (32 Nm). Attach the parking brake cable to the rear brake assembly.
10. Start the engine and forcefully depress the brake pedal 5–6 times. Apply the parking brake and make sure the adjustment is within specifications. Adjust the parking brake cable, as required.
11. Reinstall the tire and wheel assemblies.
12. Test the brakes for proper operation.

Dodge/Plymouth Neon

1. Remove the front wheels.
2. Remove the 2 caliper to steering knuckle guide pin bolts.
3. Lift the caliper away from the steering knuckle by first rotating the free end of the caliper away from the steering knuckle. Then, slide the opposite end of the caliper out from under the machined end of the steering knuckle.
4. Support the caliper from the upper control arm to prevent the weight of the caliper from being supported by the brake flex hose that will damage the hose.
5. Remove the brake pads from the caliper. Remove the outboard brake pad by prying the pad retaining clip over the raised area on the caliper. Then, slide the pad down and off the caliper. Pull the inboard brake pad away from the piston until the retaining clip is free from the cavity in the piston.
6. If required, the rotor can be removed by pulling it straight off the wheel mounting studs.

To install:

7. Clean all parts well. Inspect the caliper for piston seal leaks (brake fluid in and around the boot area and inboard lining) and for any ruptures of the piston dust boot. If the boot is damaged or fluid leak is visible, disassemble the caliper and install a new seal and boot (and piston, if scored).
8. Inspect the caliper pin bushings. Replace if damaged, dry or brittle.

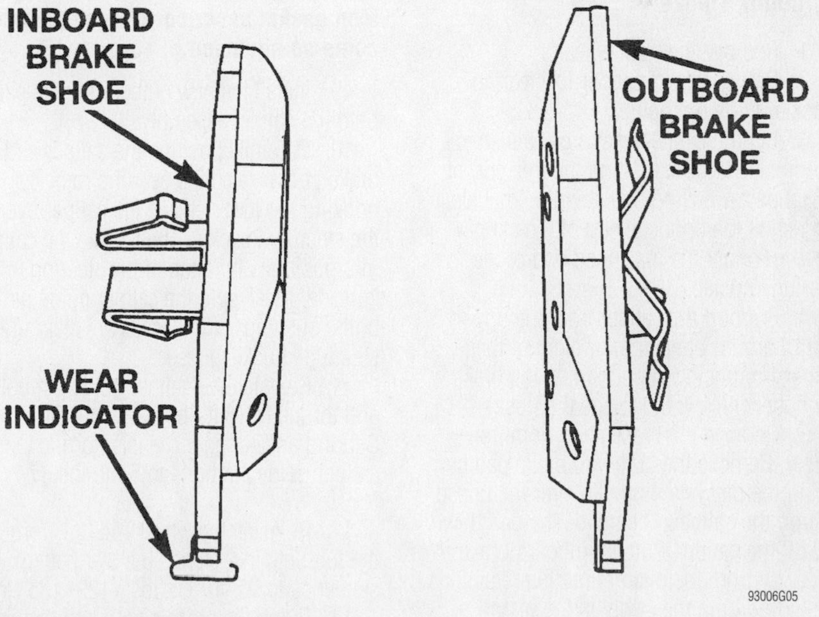

Disc brake pad identification—Neon

9. Completely compress the piston into the caliper using a large C-clamp or other suitable tool.

10. Lubricate the area on the steering knuckle where the caliper slides with high temperature grease.

11. Reinstall the rotor if removed.

12. Reinstall the brake pads into the caliper. Note that the inboard and outboard pads are different. Make sure the inboard brake shoe assembly is positioned squarely against the face of the caliper piston.

➡Be sure to remove the noise suppression gasket paper cover if the pads come so equipped.

13. Carefully position the caliper and brake shoe assemblies over the rotor by hooking the lower end of the caliper over the steering knuckle. Then, rotate the caliper into position at the top of the steering knuckle. Make sure the caliper guide pin bolts, bushings and sleeves are clear of the steering knuckle bosses.

14. Reinstall the caliper guide pin bolts and torque to 16 ft. lbs. (22 Nm).

15. Reinstall the wheels and torque the mounting bolts to 100 ft. lbs. (135 Nm).

16. Pump the brake pedal until the brake pads are seated and a firm pedal is achieved before attempting to move the vehicle.

17. Road test the vehicle for proper operation.

Chrysler Cirrus, Sebring
Dodge Stratus
Plymouth Breeze

1. Remove the front wheels.

2. Remove the 2 caliper-to-steering knuckle guide pin bolts.

3. Lift the caliper away from the steering knuckle by first rotating the free end of the caliper away from the steering knuckle. Then slide the opposite end of the caliper out from under the machined end of the steering knuckle.

4. Support the caliper from the upper control arm to prevent the weight of the caliper from being supported by the brake flex hose which will damage the hose.

5. Remove the brake pads from the caliper. Remove the outboard brake pad by prying the pad retaining clip over the raised area on the caliper. Then slide the pad down and off the caliper. Pull the inboard brake pad away from the piston until the retaining clip is free from the cavity in the piston.

To install:

6. Completely depress the piston into the caliper using a large C-clamp or other suitable tool.

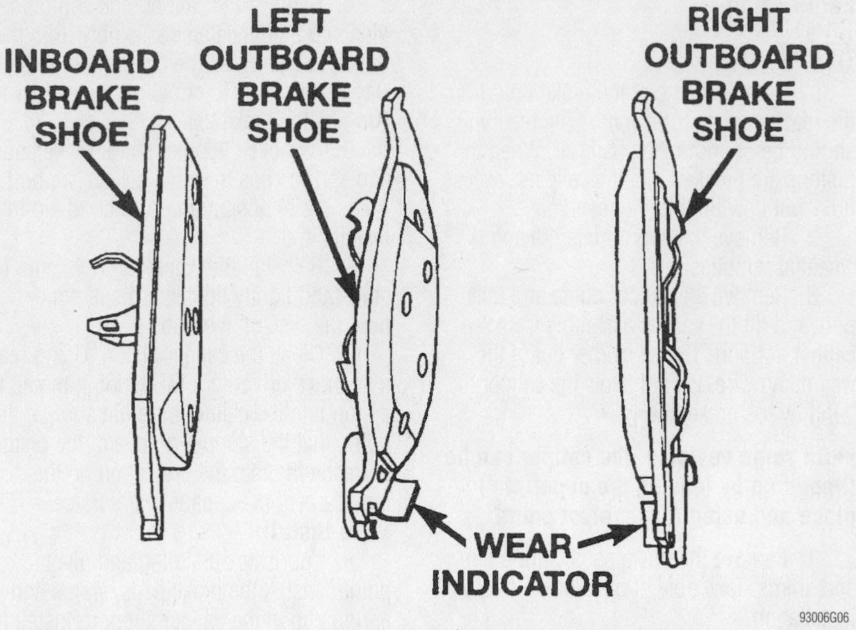

Disc brake pad identification—Cirrus, Sebring, Stratus, Breeze

7. Lubricate the area on the steering knuckle where the caliper slides with high temperature grease.

8. Install the new inboard brake pad into the caliper piston by firmly pressing into the piston bore. Install the brake pads into the caliper. Note that the inboard and outboard pads are different. Make sure the inboard brake pad assembly is positioned squarely against the face of the caliper piston.

➡Be sure to remove the noise suppression gasket paper cover if the pads come so equipped.

9. Install the new outboard brake pad onto the caliper assembly.

10. Carefully position the caliper and brake pad assemblies over the rotor by hooking the lower end of the caliper over the steering knuckle. Then rotate the caliper into position at the top of the steering knuckle. Make sure the caliper guide pin bolts, bushings and sleeves are clear of the steering knuckle bosses.

11. Reinstall the caliper guide pin bolts and torque to 16 ft. lbs. (22 Nm). On Sebring and Avenger models, torque the caliper guide pin bolts to 54 ft. lbs. (74 Nm).

12. Reinstall the wheel and tire. Torque the lug nuts in 2 steps, in a star pattern sequence to 95–100 ft. lbs. (129–135 Nm).

13. Pump the brake pedal until the brake pads are seated and a firm pedal is achieved before attempting to move the vehicle.

14. Road test the vehicle to check for proper operation.

Brake Drums

REMOVAL & INSTALLATION

Chrysler Concorde, New Yorker, LHS
Dodge Intrepid
Eagle Vision

1. Remove the rear wheels.

2. Remove the brake drum from the rear hub and bearing assembly. If the drum is difficult to remove, increase the clearance between the brake shoes and the drum as follows:
   a. Remove the rubber plug from the top of the brake support plate.
   b. Rotate the automatic shoe adjuster screw with an upward motion using a medium size flat tipped tool.

3. Remove the brake drum from the rear hub and bearing.

4. Inspect the brake shoe linings and drums for wear, contamination and scoring.

To install:

5. Install the rear brake drum onto the rear hub and bearing assembly.

6. Adjust the brake shoes as follows:
   a. Rotate the automatic shoe adjuster in a downward motion using a medium size flat tip tool. Turn the adjuster until there is a slight drag felt while turning the drum.
   b. Install the rubber plug back into the top part of the brake support plate.

7. Install the rear wheels and lug nuts. Torque the lug nuts in a star pattern sequence to 95–100 ft. lbs. (129–135 Nm).

Chrysler Sebring
Dodge Avenger

1. Remove the wheel assembly.
2. Remove the brake drum detent (retaining) screw and remove the drum from the axle.
3. If difficulty is encountered in removing the drum:
   a. Verify the parking brake is released.
   b. Loosen the parking brake cable.
   c. Remove the access hole plug from the backing plate and move the parking brake lever until the lever stop rests on the brake shoe.
   d. If necessary, turn the adjuster so that it draws in to allow more clearance between the shoe and drum.

To install:

4. Inspect all parts. Check the drum for cracks or excessive wear.
5. Turn the adjuster until it is drawn all the way in to the stop. Check that the adjuster turns freely. The nut must NOT lock at the end of the adjuster. Check that the parking brake lever stops are against the edge of the shoe web.
6. Install the brake drum and detent screw.
7. Install the wheel assembly.
8. Apply the foot brake at least 10 times until clicking of the adjustment actuator can no longer be heard. This procedure will automatically adjust the clearance between the shoe and drum.

Dodge/Plymouth Neon

1. Remove the rear wheels.
   a. Locate and remove the rubber plug from the top of the brake support plate (backing plate).
   b. Insert a small prying tool through the adjuster access hole and engage the teeth on the adjuster wheel.
   c. Rotate the adjuster wheel so it is moved toward the front of the vehicle. This will back off the adjustment of the rear brake shoes.
   d. Continue moving the adjuster wheel toward the front of the vehicle until it stops moving.
2. Remove the rear brake drum from the hub assembly.

To install:

3. Inspect the brake drums for cracks or signs of overheating. Measure the drum runout and diameter. If not to specification, resurface the drum. Runout should not exceed 0.006 inch (0.152mm). The diameter variation (oval shape) of the drum braking surface must not exceed either 0.0025 inch (0.0635mm) in 30° or 0.0035 inch (0.089mm) in 360°. All brake drums are marked with the maximum allowable brake drum diameter on the face of the drum.
4. Install the rear brake drum onto the hub assembly.
5. Reinstall the wheels and properly adjust the brakes.
6. Road test vehicle to check brake operation.

Chrysler Cirrus, Sebring
Dodge Stratus
Plymouth Breeze

All vehicles except Sebring Convertible, are equipped with rear wheel, 2-shoe leading/trailing, internal expanding type of drum brakes with automatic self-adjuster mechanisms. The automatic self-adjuster mechanisms used on these vehicles are new designs and function differently than the screw type adjusters used in the past. These new self-adjusters are still actuated each time the vehicle's service brakes are applied. The new adjusters are located directly below the wheel cylinders.

The Sebring Convertible's rear wheel drum brake is a 2-shoe leading/trailing internal expanding type with an automatic self-adjuster mechanism. The automatic self-adjuster mechanism used on this vehicle is the screw type adjuster. The self-adjuster mechanism is actuated each time the vehicle service brakes are applied. Generally, drum brakes with a self-adjusting mechanism do not require manual brake shoe adjustment. Although, in the event that the brake shoes are replaced, it is advisable to make the initial adjustment manually to speed up the initial adjustment time. The initial adjustment procedure must be done prior to driving the vehicle.

1. Remove the rear wheel assembly.
2. For all vehicles, except Sebring Convertible, use the following procedure:
   a. Locate and remove the rubber plug from the brake support plate (backing plate).
   b. Insert a brake adjuster tool or similarly shaped prytool through the automatic adjuster access hole and engage the teeth on the adjuster wheel. Rotate the adjuster wheel so it is moved toward the front of the vehicle. Continue moving the adjuster until it stops; this will back off the adjustment of the rear brake shoes.
3. For the Sebring Convertible, use the following procedure for releasing the self-adjusting mechanism:
   a. Locate and remove the rubber plug from the brake support plate (backing plate).
   b. Insert a brake adjuster tool or similarly shaped prytool through the automatic adjuster access hole and carefully push the adjuster actuating lever out of engagement with the adjuster starwheel. While holding the lever away from the starwheel, insert a second prytool through the access hole and engage the

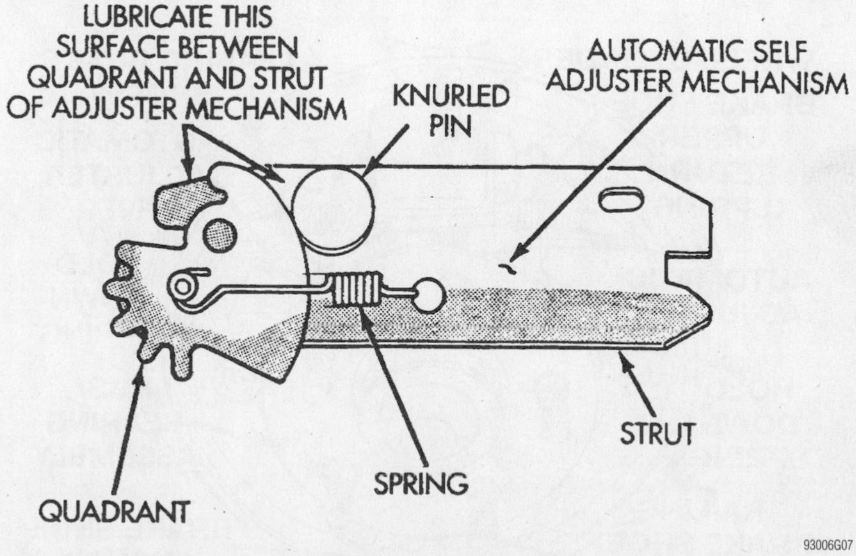

Automatic self-adjuster mechanism—Cirrus, Sebring Coupe, Stratus, Breeze

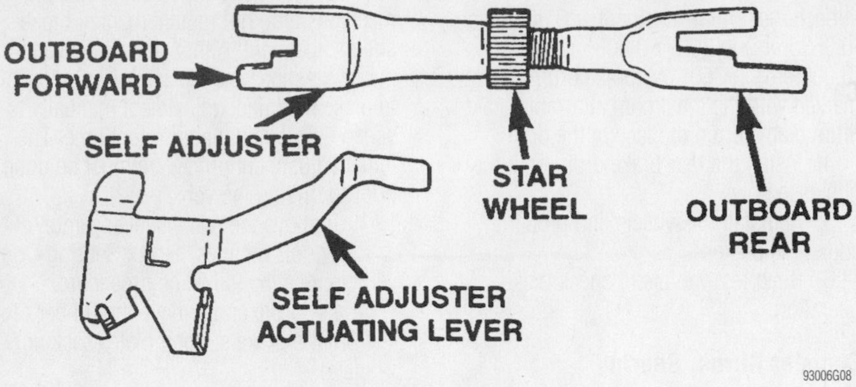

Rear brake shoe automatic self-adjuster mechanism and actuating lever—Sebring Convertible

teeth on the adjuster wheel. Rotate the adjuster wheel upward away from the ground; this will back off the adjustment of the rear brake shoes.

4. Remove the rear brake drum from the hub assembly.

To install:

5. Inspect the brake drums for cracks or signs of overheating. Measure the drum run-out and diameter. If not to specification, resurface the drum. Run-out should not exceed 0.006 in. (0.15mm). The diameter variation (oval shape) of the drum braking surface must not exceed either 0.0025 in. (0.064mm) in 30 degrees rotation, or 0.0035 in. (0.089mm) in 360 degrees rotation. All brake drums are marked with the maximum allowable brake drum diameter on the face of the drum.

6. Install the rear brake drum onto the hub assembly.

7. Reinstall the wheel and tire. Torque the lug nuts in a star pattern sequence to about 45 ft. lbs. (61 Nm); then, repeat the pattern and final torque to 95–100 ft. lbs. (129–135 Nm).

8. Properly adjust the rear brakes.

9. Road test vehicle to check for proper brake operation.

Brake Shoes

REMOVAL & INSTALLATION

Chrysler Concorde, New Yorker, LHS Dodge Intrepid Eagle Vision

1. Remove the rear wheels and brake drums.

2. Remove the dust cap from the rear hub and bearing assembly.

3. Remove the cotter pin, nut retainer and wave washer. Discard the old cotter pin.

4. Remove the rear hub and bearing assembly retainer nut and washer. Remove the rear hub and bearing assembly from the spindle.

5. Remove the automatic adjuster spring from the adjuster lever.

6. Rotate the automatic adjuster star-wheel enough so both shoes move out far enough to be free of the wheel cylinder boots.

7. Disconnect the parking brake cable from the actuating lever. Disconnect parking brake cable one side at a time.

8. Remove the both lower brake shoe to anchor springs.

9. Remove the 2 brake shoe hold-down springs from the brake shoes.

10. Remove the brake shoes, upper shoe-to-shoe return spring, automatic adjuster and automatic adjuster lever from the backing plate as an assembly.

11. Separate the brake shoes from the automatic adjuster mechanism.

12. Remove the brake shoe automatic adjuster lever from the leading brake shoe.

To install:

13. Thoroughly clean and dry the backing plate. To prepare the backing plate, lubricate the bosses, anchor pin and parking brake actuating lever pivot surface lightly with lithium based grease.

14. Remove, clean and dry all parts still on the old shoes. Lubricate the starwheel shaft threads with anti-seize lubricant.

15. Assemble both brake shoes, the top shoe to shoe return spring, automatic adjuster and automatic adjuster lever before mounting on vehicle. Make sure the ends of the automatic adjusters are positioned above the extruded pins in the webbing of the brake shoes prior to installation.

16. Install the brake shoe assembly onto the brake support plate and install the hold-down springs.

17. Install the lower anchor springs and reconnect the parking brake cable to the park brake lever of the trailing brake shoe.

18. Rotate the serrated adjuster nut to remove the free-play from the adjuster assembly.

19. Install the automatic adjuster lever spring on the lead brake shoe assembly and the automatic adjuster lever.

20. Install the rear hub and bearing assembly. Install washer and retainer nut and torque to 124 ft. lbs. (168 Nm).

21. Install the wave washer, nut retainer and a new cotter pin onto the spindle. Install dust cap.

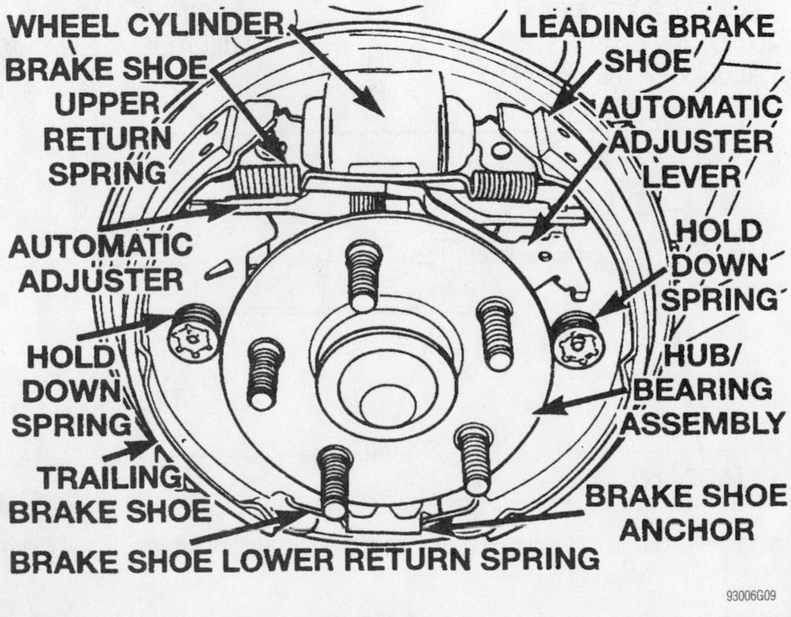

Kelsey Hayes rear brake assembly—Sebring Convertible

22. Adjust brake shoes so not to interfere with brake drum installation. Install the rear brake drum.

➡After installing the brake drums, pump the brake pedal several times to partially adjust the brake shoes. To verify proper operation of the self-adjusting parking brake, be sure that both rear brakes are not dragging when the parking brake pedal is released.

23. Install the rear wheels and lug nuts. Torque the lug nuts, in a star pattern sequence, to 95 ft. lbs. (129 Nm).

24. Road test the vehicle. The automatic adjusters will continue brake adjustment during the road test of the vehicle.

Chrysler Sebring Dodge Avenger

1. Remove the wheel assembly.
2. Remove the detent screw and brake drum. If difficulty is encountered in removing the drum:
   a. Verify the parking brake is released.
   b. Loosen the parking brake cable.
   c. Remove the access hole plug from the backing plate and move the parking brake lever until the lever stop rests on the brake shoe.
   d. If necessary, turn the adjuster so that it draws in to allow more clearance between the shoe and drum.

➡Note the location of all springs and clips for proper reassembly.

3. Remove the shoe-to-lever spring and remove the adjuster lever.
4. Remove the auto adjuster assembly.
5. Remove the retainer spring.

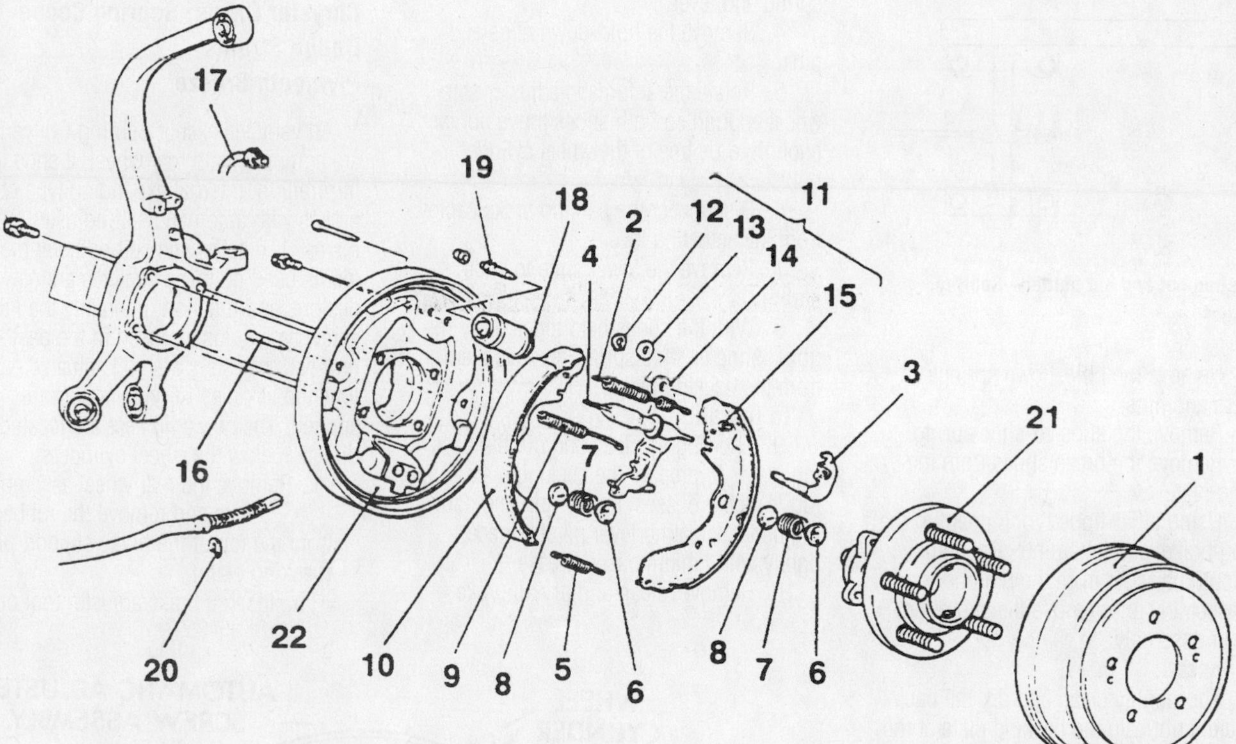

1. Brake drum
2. Shoe-to-lever spring
3. Adjuster lever
4. Auto adjuster assembly
5. Retainer spring
6. Shoe hold-down cup
7. Shoe hold-down spring
8. Shoe hold-down cup
9. Shoe-to-shoe spring
10. Shoe and lining assembly
11. Shoe and lever assembly
12. Retainer
13. Wave washer
14. Parking lever
15. Shoe and lining assembly
16. Shoe hold-down pin
17. Brake pipe connection
18. Wheel cylinder
19. Bleeder screw
20. Snap ring
21. Rear hub assembly
22. Backing plate

93006G10

Exploded view of the drum brake assembly—Sebring, Avenger

Heater Core replacement is covered in Section 2 of this manual

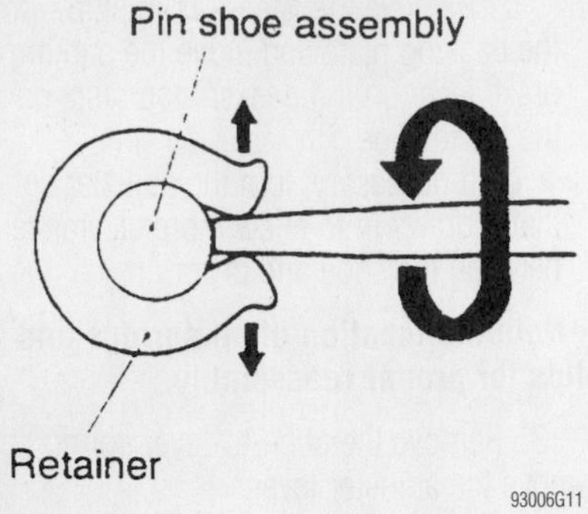

Opening the retainer clip—Sebring, Avenger

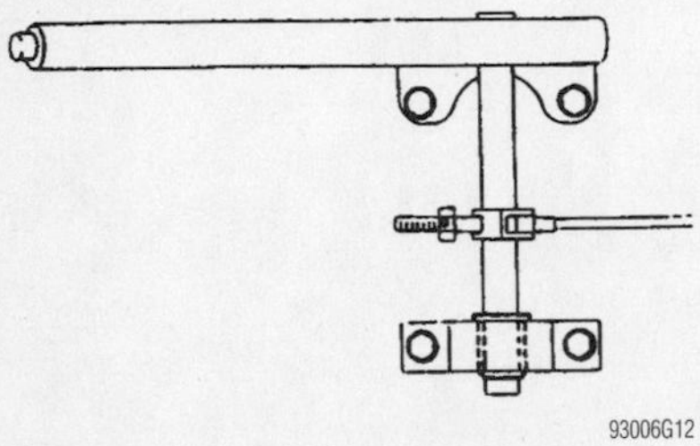

Adjusting nut and nut holder—Sebring, Avenger

6. Remove the hold-down springs, washers and pins.
7. Remove the shoe-to-shoe spring.
8. Remove the brake shoes from the backing plate.
9. Using a flat-tipped tool, open up the parking brake lever retaining clip. Remove the clip and washer from the pin on the shoe assembly and remove the shoe from the lever assembly.

To install:

10. Thoroughly clean and dry the backing plate. Lubricate the backing plate at the brake shoe contact points.
11. Lubricate backing plate bosses, anchor pin, and parking brake actuating mechanism with a lithium-based grease.
12. Install the parking brake lever assembly on the lever pin. Install the wave washer and a new retaining clip. Use pliers or the like to install the retainer on the pin. If removed, connect the parking brake lever to the parking brake cable and verify that the cable is properly routed.
13. Clean and lubricate the adjuster assembly. Make sure the nut-adjuster is drawn all the way to the stop, but the nut must NOT lock firmly at the end of the assembly.
14. Install the brake shoes on the backing plate with the hold-down springs, washers and pins.
15. Install the shoe-to-shoe spring.
16. Install the retainer spring.
17. Install the auto adjuster assembly and install the adjuster lever and the shoe-to-lever spring.
18. Pre-adjust the shoes so the drum slides on with a light drag and install the brake drum.
19. Adjust the rear brake shoes and install the rear wheels.
20. Adjust the parking brake cable.
21. Check for proper brake operation.

Dodge/Plymouth Neon

1. Remove the rear wheels.
2. Remove the drums.
3. Remove the automatic adjuster spring and lever.
4. Remove the hold-down clips and pins.
5. Rotate the automatic adjuster starwheel enough so both shoes move out far enough to be free of the wheel cylinder boots.
6. Disconnect the parking brake cable from the actuating lever.
7. Remove the lower shoe to shoe spring.
8. With the shoes held together by the upper shoe to shoe spring, remove them from the backing plate.

To install:

9. Thoroughly clean and dry the backing plate. To prepare the backing plate, lubricate the bosses, anchor pin and parking brake actuating lever pivot surface lightly with lithium based grease.
10. Remove, clean and dry all brake components. Lubricate the starwheel shaft threads with anti-seize lubricant and transfer all parts to their proper locations on the new shoes.
11. Reinstall the lower spring.
12. Reconnect the parking brake cable.
13. Reinstall the automatic adjuster lever and spring.
14. Adjust the starwheel.
15. Remove any grease from the linings and install the drum.
16. Reinstall the wheels and properly adjust the brakes.
17. Check for proper brake system operation.

Chrysler Cirrus, Sebring Coupe
Dodge Stratus
Plymouth Breeze

All vehicles except Sebring Convertible, are equipped with rear wheel, 2 shoe leading/trailing, internal expanding type of drum brakes with automatic self-adjuster mechanisms. The automatic self-adjuster mechanisms used on these vehicles are new designs and function differently than the screw type adjusters used in the past. These new self-adjusters are still actuated each time the vehicles' service brakes are applied. The new adjusters are located directly below the wheel cylinders.

1. Remove the rear wheel assembly.
   a. Locate and remove the rubber plug from the top of the brake support plate (backing plate).
   b. Insert a brake adjuster tool or simi-

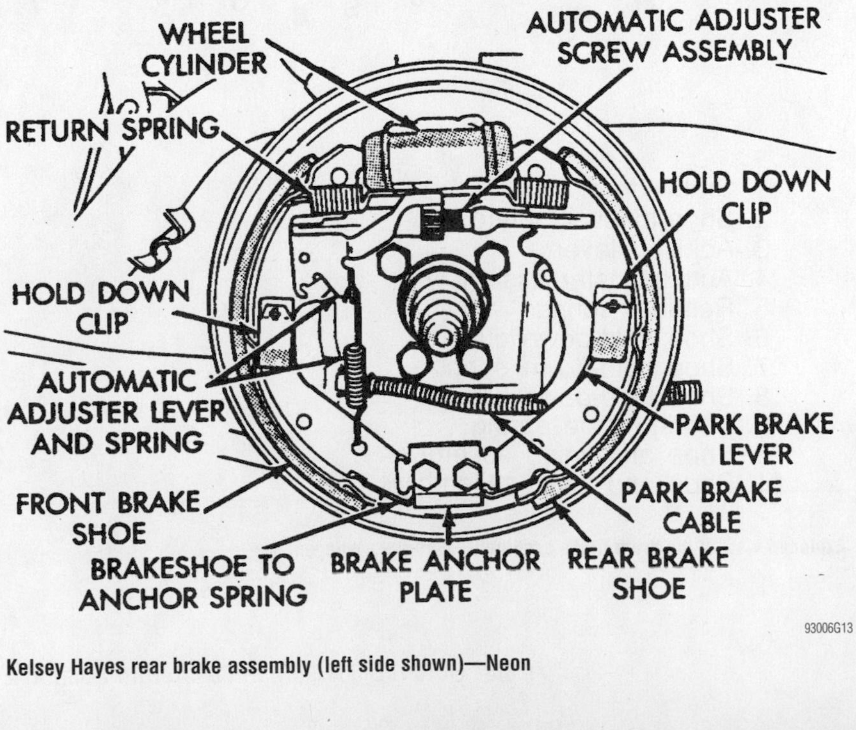

Kelsey Hayes rear brake assembly (left side shown)—Neon

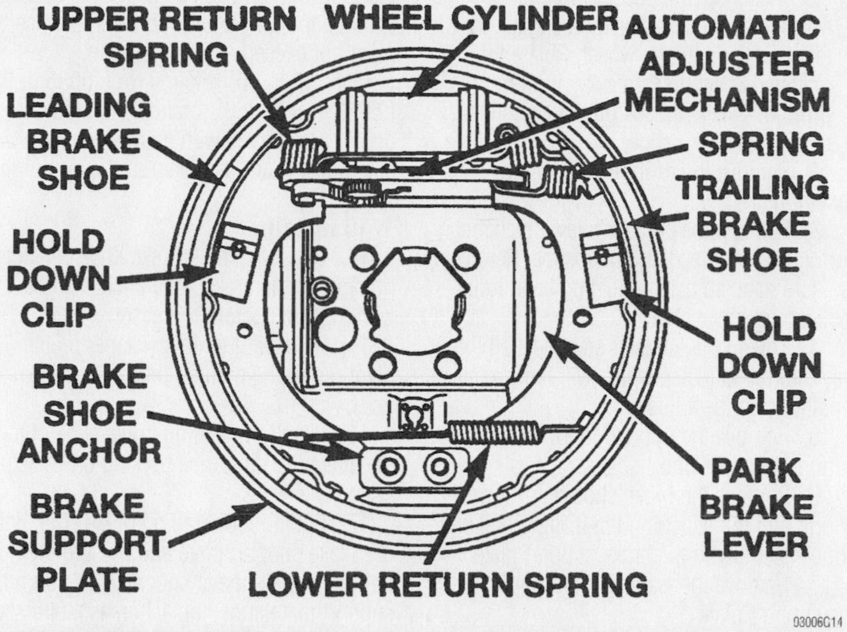

Varga rear wheel brake assembly (left side shown)—Cirrus, Sebring Coupe, Stratus, Breeze

larly shaped prytool through the automatic adjuster access hole and engage the teeth on the adjuster quadrant. Then rotate the quadrant so the teeth of the quadrant are moved toward the front of the vehicle. This will back off the adjustment of the rear brake shoes.

c. Continue moving the quadrant toward the front of the vehicle until it stops moving.

2. Remove the drum from the hub assembly.
3. Remove the actuating spring from the adjuster mechanism and trailing brake shoe.
4. Remove the upper return spring from the brake shoes.
5. Remove the lower return spring from the brake shoes.
6. Remove the brake shoe retainer and pin attaching the leading brake shoe assembly to the brake support plate.
7. Remove the leading brake shoe and the adjuster mechanism as an assembly from the rear brake support plate. The adjuster mechanism cannot be separated from the leading brake shoe until the brake shoe and the adjuster mechanism is removed from the support plate.
8. Remove the trailing brake shoe retainer and pin attaching the trailing brake shoe assembly to the brake support plate. Remove the trailing brake shoe assembly.

➡On this vehicle, the parking brake actuating lever is permanently attached to the trailing brake shoe assembly. Do not attempt to remove it from the original brake shoe assembly or reuse the original actuating lever on a replacement brake shoe assembly. All replacement brake shoe assemblies for this vehicle must have the actuating lever as part of the trailing brake shoe assembly.

9. Remove the parking brake cable from the parking brake lever. Do not remove the lever from the brake shoe.
10. Remove the automatic adjuster mechanism from the brake shoe by fully extending the adjuster and rotate the adjuster out to release from the brake shoe.

To inspect:

11. Thoroughly clean all parts. The brake lining should show contact across the entire width and from heel to toe; otherwise, replace. Clean and inspect the brake support plate and the automatic adjuster mechanism. Be sure the quadrant (toothed part) of the adjuster is free to rotate throughout its entire tooth contact range and is free to slide the full length of its mounting slot. Check the knurled pin. It should be securely attached to the adjuster mechanism and its teeth should be in good condition. If the adjuster is worn or damaged, replace it. If the adjuster is serviceable, lubricate lightly with high-temperature grease between the strut and the quadrant. Check the brake springs. Overheating indications are paint discoloration or distorted end coils. Replace parts as required.
12. Inspect the brake drums for cracks or signs of overheating. Measure the drum run-out and diameter. If not to specification, reface the drum. Run-out should not exceed 0.006 in. (0.15mm). The diameter variation (oval shape) of the drum braking surface must not exceed either 0.0025 in. (0.064mm) in 30 degrees rotation, or 0.0035 in. (0.089mm) in 360 degrees rotation. All brake drums are marked with the maximum allowable brake drum diameter on the face of the drum.

To install:

13. Lubricate the 8 brake shoe contact points with high-temperature grease.

➡The trailing brake shoe assemblies used on the rear brakes of this vehicle are unique to the left and right side of the vehicle. Care must be taken to ensure the brake shoes are properly installed in their correct side of the vehicle. When the trailing shoes are properly installed on their correct side of the vehicle, the park brake actuating lever will be positioned under the brake shoe web.

14. Reinstall the parking brake cable onto the parking brake lever and install the trailing brake shoe and attaching pin.
15. Reinstall the automatic self-adjuster on the leading brake shoe by rotating it inward to attach. Install the leading shoe and adjuster assembly to the brake support plate.
16. Make sure the leading brake shoe is squarely seated on the brake support plate shoe contact areas, and install the brake retainer on the retainer pin.
17. Reinstall the lower return spring.

➡The upper brake shoe return spring and adjuster mechanism actuating spring are unique to the side of the vehicle they are used on. The springs are colored for identification. The left side springs are green and the right side springs are blue.

18. Reinstall the upper return spring (blue, right side; green, left side) on the leading brake shoe first, then on the trailing brake shoe.
19. Reinstall the self-adjuster spring on the trailing brake shoe first then attach it to the adjuster.
20. Reinstall the rear brake drums.
21. Reinstall the wheel and tire. Torque the lug nuts in a star pattern sequence to

about 45 ft. lbs. (61 Nm), then repeat the pattern and final torque to 95–100 ft. lbs. (129–135 Nm).

22. Adjust the rear brakes by depressing the brake pedal. Brake shoe adjustment will occur the first time the brake pedal is depressed, pushing the rear brake shoes against the braking surface of the rear brake drums. Brake shoes should now be correctly adjusted and will not require any type of manual adjustment.

23. Road test vehicle to check for proper brake operation.

Sebring Convertible

The Sebring Convertible's rear wheel drum brake is a 2-shoe leading/trailing internal expanding type with an automatic self-adjuster mechanism. The automatic self-adjuster mechanism used on this vehicle is the screw type adjuster. The self-adjuster mechanism is actuated each time the vehicle service brakes are applied. Generally, drum brakes with a self-adjusting mechanism do not require manual brake shoe adjustment. Although, in the event that the brake shoes are replaced, it is advisable to make the initial adjustment manually to speed up the initial adjustment time. The initial adjustment procedure must be done prior to driving the vehicle.

➡When removing the rear brake shoes, replace the brake shoes from only one side of the vehicle at a time. This is due to the automatic adjustment feature of the parking brake system. If the brake shoes are removed from both sides of the vehicle at the same time, the automatic adjuster will remove all slack from the parking brake cables, which will make brake shoe installation extremely difficult.

1. Remove the rear wheel assembly.
   a. Locate and remove the rubber plug from the brake support plate (backing plate).
   b. Insert a brake adjuster tool or similarly shaped prytool through the automatic adjuster access hole and carefully push the adjuster actuating lever out of engagement with the adjuster starwheel. While holding the lever away from the starwheel, insert a second prytool through the access hole and engage the teeth on the adjuster wheel. Rotate the adjuster wheel upward away from the ground. This will back off the adjustment of the rear brake shoes.
2. Remove the drum from the hub assembly.
3. Remove the adjusting lever actuating spring from the leading brake shoe. Remove the automatic adjuster actuating lever from the leading brake shoe.
4. Thread the adjuster starwheel all the way into the adjuster, which will remove all tension from the adjuster.
5. Remove the upper and lower return springs from the brake shoes.
6. Remove the brake shoe hold-down spring and pin attaching the leading brake shoe assembly to the brake support plate.
7. Remove the leading brake shoe from the support plate.
8. Remove the automatic adjuster from the parking brake actuating lever and trailing brake shoe.
9. Remove the retaining clip securing the parking brake actuating lever to the trailing brake shoe.
10. Remove the trailing brake shoe hold-down spring and pin attaching the trailing brake shoe assembly to the brake support plate.
11. Remove the trailing brake shoe from the brake support plate and separate the shoe from the parking brake actuating lever.

To inspect:

12. Clean all parts well. The brake lining should show contact across the entire width and from heel to toe; otherwise, replace. Clean and inspect the brake support plate and the automatic adjuster mechanism. Be sure the adjuster is free to rotate throughout its entire range. If the adjuster is worn or damaged, replace it. If the adjuster is serviceable, lightly lubricate the threaded portion with high-temperature grease. Check the brake springs. Overheating indications are paint discoloration or distorted end coils. Replace parts as required.

13. Inspect the brake drums for cracks or signs of overheating. Measure the drum run-out and diameter. If not to specification, reface the drum. Run-out should not exceed 0.006 in. (0.15mm). The diameter variation (oval shape) of the drum braking surface must not exceed either 0.0025 in. (0.064mm) in 30° rotation, or 0.0035 in. (0.089mm) in 360° rotation. All brake drums are marked with the maximum allowable brake drum diameter on the face of the drum.

To install:

14. Lubricate the 6 brake shoe contact points and the brake shoe anchor points with high-temperature grease.

15. Reinstall the wave washer on the pivot pin of the parking brake actuating lever.

16. Install the trailing brake shoe onto the attaching pin of the parking brake actuating lever.

17. Position the trailing brake shoe onto the brake support plate and be sure the trailing brake shoe is squarely seated on the support plate shoe contact areas and install the brake shoe hold-down spring on the hold-down pin.

18. Reinstall the parking brake actuating lever-to-trailing brake shoe retaining clip.

19. Reinstall the automatic adjuster on the trailing brake shoe and the parking brake actuating lever.

20. Place the leading brake shoe onto the brake support plate in proper position and install the attaching pin and hold-down spring.

21. Reinstall the lower and upper return springs.

22. Reinstall the automatic adjuster actuating lever and spring onto the leading brake shoe.

23. Manually adjust the brake shoes to the furthest adjusted position but not so far as to interfere with the installation of the brake drum.

24. Reinstall the rear brake drums. Check and adjust the brake shoes as necessary.

25. Reinstall the wheel and tire. Torque the lug nuts in a star pattern sequence to about 45 ft. lbs. (61 Nm), then repeat the pattern and final torque to 100 ft. lbs. (135 Nm).

26. Road test vehicle to check for proper brake operation.

1998–01 FORD MOTOR COMPANY

Brake Caliper

REMOVAL & INSTALLATION

Ford Focus

1. Before servicing the vehicle, refer to the precautions in the beginning of this section.
2. Remove or disconnect the following:
   - Front wheel
   - Brake hose from the support bracket
   - Outer brake pad retaining clip
   - Brake caliper from the steering knuckle
   - Brake pads
   - Brake hose from the caliper

To install:

3. Install or connect the following:
   - Brake hose to the caliper. Tighten the fitting to 11 ft. lbs. (15 Nm).
   - Brake pads
   - Brake caliper to the steering knuckle. Tighten the bolts to 21 ft. lbs. (28 Nm).
   - Outer brake pad retaining clip
   - Brake hose to the support bracket
   - Front wheel
4. Bleed the brakes and check for proper operation.

Ford Contour, Mercury Mystique

FRONT

1. Remove the wheel and tire assembly.
2. Remove the outer disc brake pad spring clip (anti-rattle clip).
3. Remove the 2 locator pin covers and remove the locator pins.
4. Free the hose from its mounting on the strut.

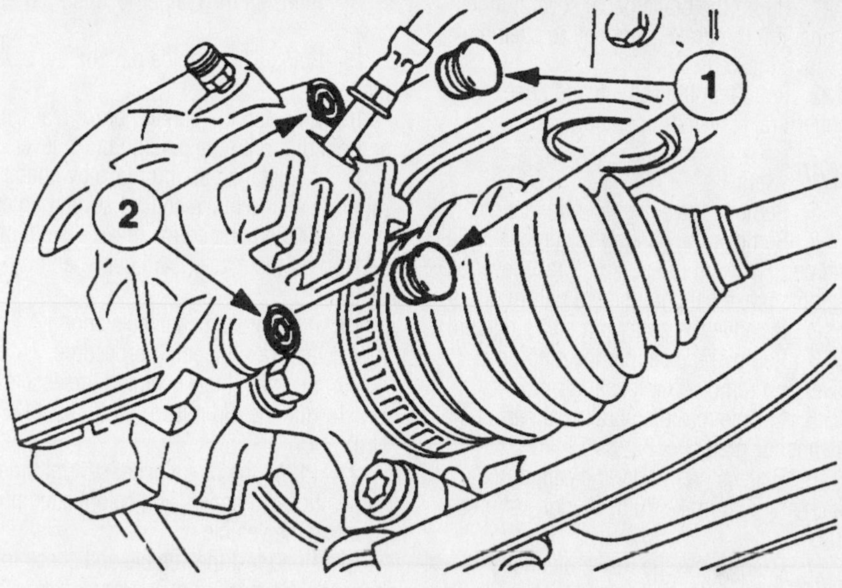

93006G18

View of disc brake caliper locating pins—Contour, Mystique

5. Lift the caliper off of the brake rotor.
6. Remove the inboard disc brake pad from the caliper.
7. If the brake caliper is to be removed from the vehicle, place a pan under the caliper to catch the brake fluid for proper disposal. Disconnect the brake hose at the caliper and allow to drain.
8. Remove the brake caliper.
9. If the caliper is not to be serviced, tie off the caliper to prevent strain on the brake hose.

To install:

10. If removed, install the brake hose onto the caliper. Torque the fitting to 10 ft. lbs. (14 Nm).
11. Position the inboard disc brake pad into the caliper.
12. Make sure that the outboard disc brake pad is positioned properly.
13. Reinstall the caliper over the brake rotor and position onto the caliper anchor plate.
14. Reinstall the 2 caliper locator pins and torque to 20 ft. lbs. (28 Nm).
15. Reinstall the caliper locator pin covers.
16. Reinstall the outer disc brake pad spring clip.
17. Reconnect the brake hose to the front strut.
18. Bleed the brake system of air. Top off the master cylinder when complete.
19. If the brake pedal feels spongy, repeat the brake bleeding procedure.
20. Reinstall the wheel and tire assem-

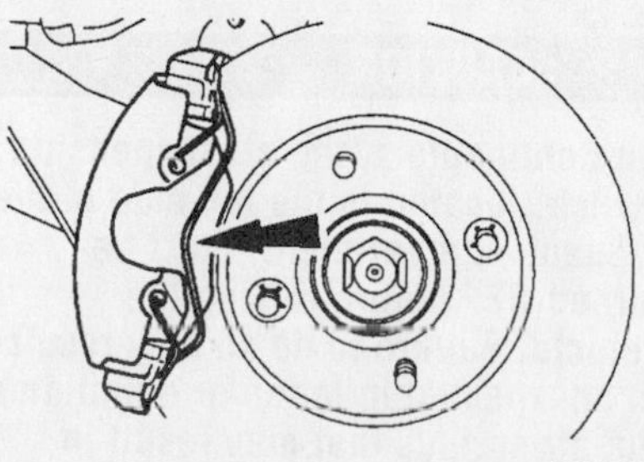

93006G17

Position of disc brake pad (brake shoe and lining) spring clip—Contour, Mystique

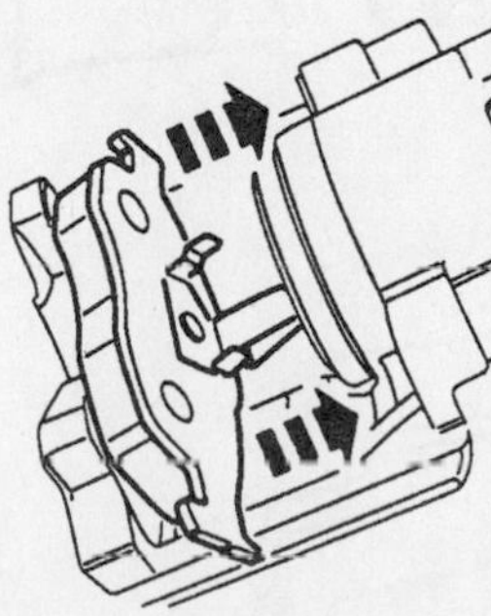

93006G19

View of inboard disc brake pad (brake shoe and lining)—Contour, Mystique

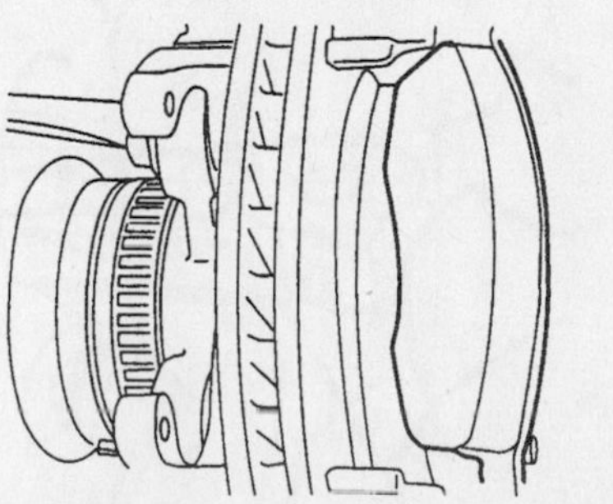

93006G20

Positioning the outboard disc brake pad (brake shoe and lining)—Contour, Mystique

bly. Torque the lug nuts to 62 ft. lbs. (85 Nm).

21. Pump the brake pedal several times to position the brake pads before attempting to move the vehicle.

22. Road test the vehicle and check for proper brake system operation.

REAR

1. Remove the wheel and tire assembly.
2. Remove the parking brake rear cable and conduit from the parking brake lever at the disc brake caliper, using a pair of pliers.
3. Remove the cotter pin and guide pin.
4. Remove the caliper locating pin cover and remove the locating pin.
5. Lift the rear disc brake caliper off of the anchor plate.
6. Place a pan under the caliper to catch any lost brake fluid. Dispose of properly.
7. Crack open the rear brake hose fitting and allow the brake fluid to drain.
8. Finish removing the rear brake hose and washers from the caliper. Discard the washers.
9. Remove the disc brake caliper.

To install:

10. Retract the piston fully into the caliper using service tool T87P–2588–A.
11. Reconnect the rear brake hose to the caliper using new washers.
12. Fit the rear disc brake caliper over the rotor and position onto the anchor plate.
13. Reinstall the caliper locating pin and torque to 30 ft. lbs. (41 Nm).
14. Reinstall the caliper locating pin cover.
15. Reinstall the guide pin and the cotter pin.
16. Reinstall the parking brake rear cable and conduit onto the parking brake lever.
17. Adjust the parking brake by operating the parking brake control several times.
18. Properly bleed the brake system of air. Top off the master cylinder when complete.
19. If the brake pedal feels spongy, repeat the brake bleeding procedure.
20. Reinstall the wheel and tire assembly. Torque the lug nuts to 94 ft. lbs. (128 Nm).
21. Pump the brake pedal several times to position the brake pads before attempting to move the vehicle.
22. Road test the vehicle and check for proper brake system operation.

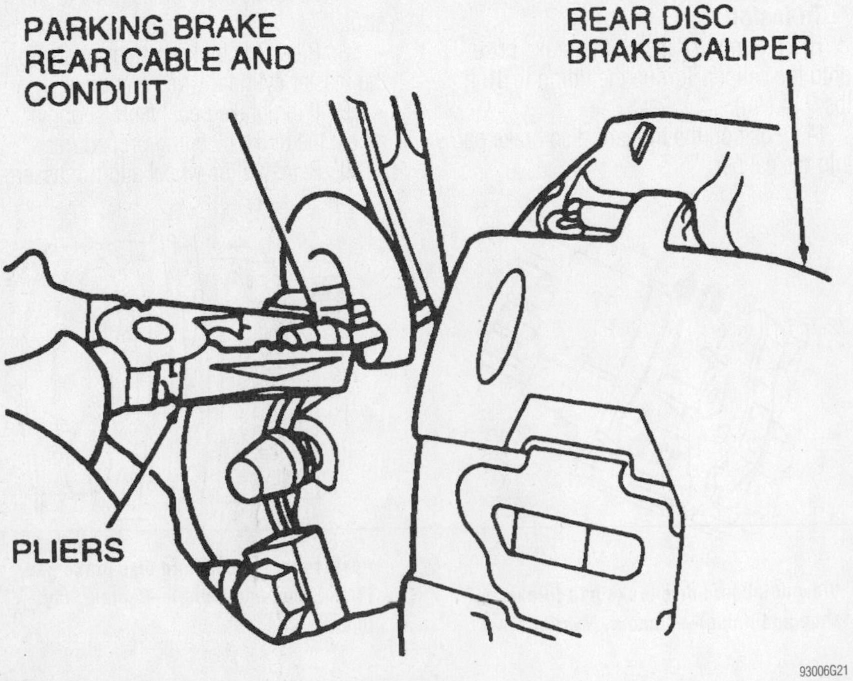

Method of removing the parking brake cable—Contour, Mystique

Lincoln LS

FRONT

1. Before servicing the vehicle, refer to the precautions in the beginning of this section.
2. Remove or disconnect the following:
   - Front wheel
   - Brake fluid hose
   - Caliper mounting bolts
   - Brake caliper

To install:

3. Install or connect the following:
   - Brake caliper. Tighten the mounting bolts to 26 ft. lbs. (35 Nm).
   - Brake fluid hose. Use new copper washers and tighten the bolt to 35 ft. lbs. (47 Nm).
   - Front wheel
4. Bleed the brake system.
5. Before attempting to move the vehicle, pump the brake pedal to seat the pads against the rotors. Make sure the vehicle has a firm brake pedal. Check the level of the brake fluid and add DOT 3 or 4 brake fluid if necessary.

REAR

1. Before servicing the vehicle, refer to the precautions in the beginning of this section.
2. Remove or disconnect the following:
   - Rear wheel
   - Parking brake cable
   - Caliper mounting bolts
   - Brake fluid hose
   - Brake caliper

To install:

3. Install or connect the following:
   - Brake fluid hose. Use new copper washers and tighten the bolt to 36 ft. lbs. (48 Nm).
   - Brake caliper. Tighten the mounting bolts to 25 ft. lbs. (33 Nm).
   - Parking brake cable
   - Rear wheel
4. Bleed the brake system.
5. Before attempting to move the vehicle, pump the brake pedal to seat the pads against the rotors. Make sure the vehicle has a firm brake pedal. Check the level of the brake fluid and add DOT 3 or 4 brake fluid if necessary.

Ford Taurus
Lincoln Continental
Mercury Sable

CAUTION

On Continentals, the air suspension switch, located in the left side of the luggage compartment, must be turned OFF before raising the vehicle. Failure to do so may result in unexpected inflation or deflation of the air springs that may result in shifting of the vehicle during service.

FRONT

1. Remove brake fluid from the brake master cylinder reservoir until the reservoir is ½ full.

2. For Continentals, turn the air suspension switch, located in the left side of the luggage compartment, to the **OFF** position.

3. Remove the wheel and tire assembly.

4. Mark the disc brake caliper to ensure that it is reinstalled in the correct location.

5. Remove the hollow bolt connecting the brake hose to the disc brake caliper and plug the brake hose. Discard the 2 copper sealing washers.

6. Remove the caliper locating pins and lift the caliper off the rotor using a rotating motion.

To install:

7. Retract the disc brake caliper piston fully in the piston bore, using an old brake pad or block of wood and a C-clamp.

➡Make sure the clip-on insulators are attached to the brake pads.

8. Install the disc brake pads to the caliper. Make sure the brake pad insulators are correctly attached to the brake pad plate.

9. Position the disc brake caliper and pad assembly above the rotor and install it with a rotating motion. Make sure the inner and outer pads are properly positioned and the outer anti-rattle spring is properly positioned.

10. Lubricate the locating pins and the inside of the insulators with silicone grease. Torque the locating pins to 25 ft. lbs. (34 Nm).

11. Remove the plug and install the brake hose to the disc brake caliper. Use 2 new copper washers and torque the hollow bolt to 30–40 ft. lbs. (41–54 Nm).

12. Bleed the brake system, filling the master cylinder as required.

13. Install the wheel and tire assembly; torque the nuts to 85–104 ft. lbs. (115–142 Nm).

14. Pump the brake pedal several times to position the brake pads prior to moving the vehicle.

15. For Continentals, turn the air suspension service switch to the **ON** position.

16. Road test the vehicle and check for proper brake system operation.

REAR

1. Remove brake fluid from the brake master cylinder reservoir until the reservoir is ½ full.

2. For Continentals, turn the air suspension switch, located in the left side of the luggage compartment, to the **OFF** position.

3. Remove the wheel and tire assembly.

4. Remove the retaining bolt and disconnect the brake hose from the caliper assembly. Discard the copper sealing washers.

5. Remove the retaining clip from the parking brake at the caliper. Disengage the parking brake cable end from the lever arm.

6. Lift the rear disc brake caliper away from the rear disc support bracket.

7. Remove the disc brake caliper locating pins and boots from the rear disc support bracket.

To install:

8. Using rear caliper piston adjuster tool T87P-2588-A, rotate the rear disc brake piston and adjuster clockwise until fully seated.

➡Make sure one of the 2 slots in the rear disc brake piston and adjuster face is positioned so it will engage the nib on the disc brake pad.

9. Apply silicone dielectric compound to the inside of the slider pin boots and the slider pins.

10. Position the slider pins and boots in the support bracket. Position the caliper assembly on the support bracket. Make sure the brake pads are installed correctly.

11. Remove the residue from the pin retainer threads and apply 1 drop of threadlock and sealer. Install the pin retainers and torque to 23–26 ft. lbs. (31–35 Nm).

12. Attach the cable end to the parking brake lever. Install the cable retaining clip on the caliper assembly.

13. Using new washers, connect the brake flex hose to the caliper. Torque the retaining bolt to 40 ft. lbs. (54 Nm).

14. Bleed the brake system, filling the master cylinder as required.

15. Install the wheel and tire assembly; torque the nuts to 85–104 ft. lbs. (115–142 Nm).

16. Pump the brake pedal several times to position the brake pads prior to moving the vehicle.

17. For Continentals, turn the air suspension service switch to the **ON** position.

18. Road test the vehicle and check for proper brake system operation.

Ford Escort
Mercury Tracer

FRONT

1. Remove the wheel and tire assembly.

2. Remove the disc brake pads.

3. Remove the banjo bolt securing the brake hose to the brake caliper. Discard 2 copper sealing washers.

4. Remove 2 brake caliper retaining bolts and remove the brake caliper.

To install:

5. Place the brake caliper in position and install 2 brake caliper retaining bolts. Torque the bolts to 29–36 ft. lbs. (39–49 Nm).

6. Install the brake hose to the brake caliper and install the banjo bolt with 2 new copper sealing washers. Torque the banjo bolt to 16–22 ft. lbs. (22–29 Nm).

7. Install the disc brake pads.

8. Bleed the brake system.

9. Install the wheel and tire assembly. Torque the lug nuts to 65–87 ft. lbs. (88–118 Nm).

10. Pump the brake pedal several times to position the brake pads to the brake rotor.

11. Road test the vehicle and check the brake system for proper operation.

REAR

1. Remove the wheel and tire assembly.

2. Remove the disc brake pads.

3. Remove the parking brake cable bracket bolt and position the bracket aside.

4. Remove the parking brake cable from the operating lever.

5. Remove the banjo bolt securing the brake hose to the brake caliper. Discard 2 copper sealing washers.

6. Remove the upper brake caliper retaining bolt.

7. Slide the brake caliper off the mounting bracket and remove from the vehicle.

To install:

8. Place the brake caliper on the mounting bracket and slide into position.

9. Install the upper brake caliper retaining bolt. Torque the bolt to 33–43 ft. lbs. (45–59 Nm).

10. Install the brake hose to the brake caliper and secure with the banjo bolt using 2 new copper sealing washers. Torque the banjo bolt to 16–22 ft. lbs. (22–29 Nm).

11. Connect the parking brake cable to the operating lever. Position the bracket and install the bracket bolt.

12. Install the disc brake pads.

13. Bleed the brake system.

14. Install the wheel and tire assembly. Torque the lug nuts to 65–87 ft. lbs. (88–118 Nm).

15. Pump the brake pedal several times

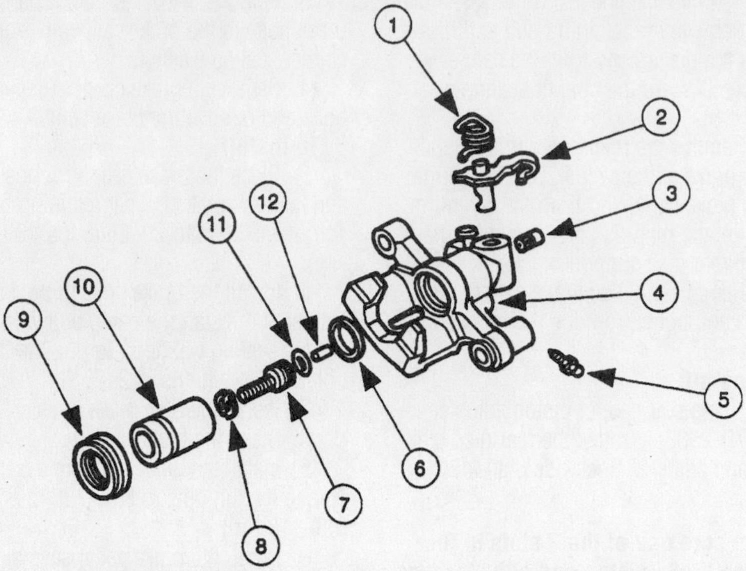

1	Rear Wheel Brake Caliper Spring	7	Rear Disc Brake Adjuster Spindle
2	Rear Brake Operating Lever	8	Snap Ring
3	Brake Adjuster Screw	9	Dust Seal
4	Rear Disc Brake Caliper	10	Piston
5	Wheel Cylinder Bleeder Screw	11	O-Ring
6	Piston Seal	12	Brake Connecting Link

93006G22

Rear disc brake assembly—Escort, Tracer

to position the brake pads to the brake rotor.

16. Road test the vehicle and check the brake system for proper operation.

Ford Mustang

FRONT

1. Remove the wheel and tire assembly.
2. Mark the disc brake caliper to ensure that it will be installed to the same side if both calipers are being removed.
3. Remove the banjo bolt that connects the brake hose fitting to the disc brake caliper. Remove and plug the brake hose. Discard 2 copper sealing washers.
4. On single piston calipers, remove the lower brake caliper locating pin.
5. Rotate the brake caliper approximately 90° away from the anchor assembly.
6. Slide the brake caliper away from anchor assembly until the brake caliper disengages from the upper locating pin and remove from the vehicle.
7. On dual piston calipers, remove the clip and washer with the locating pin and remove the disc brake caliper.

To install:

8. If required, compress the caliper piston(s) using a C-clamp and an old disc brake pad or block of wood.
9. On single piston calipers, engage the upper disc brake caliper locating pin into the insulator.
10. Make sure the anti–rattle is properly installed on caliper.
11. Make sure the brake pads are properly installed.
12. Rotate the brake caliper assembly into place on the anchor plate.

➡New locating pins are recommended due to the thread locking compound on the threads.

13. Install the lower disc brake caliper locating pin and torque to 23 ft. lbs. (31 Nm).
14. On dual piston calipers, place the disc brake caliper over the brake rotor with the brake pads installed and lower the caliper onto the brake rotor.
15. Install the disc brake caliper locating pin and clip by pressing the caliper down to compress the bias springs and the locating pin into position and secure with the washer and clip.
16. Unplug and install the brake hose to the disc brake caliper using 2 new copper sealing washers, 1 on each side of the fitting. Install the banjo bolt and torque to 30 ft. lbs. (40 Nm).
17. Bleed the brake system and check for leaks. Replace the bleeder screw rubber cap.
18. Install the wheel and tire assembly. Torque the lug nuts in a star pattern to 85–105 ft. lbs. (115–142 Nm).
19. Pump the brake pedal several times to position the brake pads before the vehicle is moved.
20. Road test the vehicle and check for proper brake system operation.

REAR

1. Remove the wheel and tire assembly.
2. Remove the banjo bolt securing the brake hose to the disc brake caliper. Separate the brake hose from the caliper and plug the hose. Discard 2 copper sealing washers.
3. Remove the retaining clip from the parking brake rear cable at the disc brake caliper. Release the tension from the parking brake cable and disengage the cable end from the parking brake lever on the brake caliper.
4. Hold the disc brake caliper locating pin hex-heads with an open-end wrench and remove 2 brake pin retainer bolts.
5. Lift the disc brake caliper assembly away from the disc support bracket and remove the disc brake caliper from the vehicle.
6. Remove the brake caliper locating pins and boots from support bracket.

To install:

7. If required, use Caliper Piston Adjuster T87P-2588-A to rotate the caliper piston into the caliper bore. Align the tool tips with the slots in the piston and rotate the tool clockwise until the piston is seated.
8. Apply a suitable silicone dielectric compound to the inside of the locating pin boots and the caliper locating pins.
9. Install the brake caliper locating pins and boots in the disc support bracket.
10. Properly install the disc brake pads and anti-rattle clips.
11. Apply a drop of suitable thread locking compound to the threads of the brake pin retainer bolts and install the brake pin retainers. Torque the brake pin retainer bolts to 23–26 ft. lbs. (31–35 Nm) while holding the brake caliper locating pins with an open-end wrench.
12. Attach the end of the parking brake cable to the parking brake lever on the disc brake caliper.
13. Unplug and install the brake hose to the disc brake caliper using 2 new copper sealing washers on the fitting. Insert the banjo bolt through both washers and the

brake line fitting and torque to 30 ft. lbs. (40 Nm).

14. Properly bleed the brake system and check for leaks.

15. Install the wheel and tire assembly. Torque the lug nuts in a star pattern to 85–105 ft. lbs. (115–142 Nm).

16. Pump the brake pedal several times to position the brake pads before moving the vehicle.

17. Road test the vehicle and check the brake system for proper operation.

Lincoln Mark VIII

FRONT

1. Remove the wheel and tire assembly.

2. Loosen the brake line fitting that connects the brake hose to the brake line at the frame bracket. Plug the brake line. Remove the retaining clip from the hose and bracket and disengage the hose from the bracket.

3. Remove the hollow bolt attaching the brake hose to the caliper and remove the brake hose. Discard the sealing washers.

4. Disconnect the brake pad wear sensor connector (if equipped).

5. Remove the caliper locating pins and remove the caliper. If removing both calipers, mark the right and left sides so they may be reinstalled correctly.

To install:

6. Fully retract piston into caliper bore.

7. Install the caliper over the rotor with the outer brake pad against the rotor's braking surface. This prevents pinching the piston boot between the inner brake pad and the piston.

8. Lubricate the locating pins and the inside of the locating pin insulators with silicone dielectric grease. Install the caliper locating pins and thread them into the spindle/anchor plate assembly by hand. Torque the caliper locating pins to 45–65 ft. lbs. (61–88 Nm).

9. Connect brake pad wear sensor connector (if equipped).

10. Install new sealing washers on each side of the brake hose fitting outlet and install the hollow bolt through the hose fitting and into the caliper. Torque the bolt to 30 ft. lbs. (41 Nm).

11. Place the brake hose in the bracket and install the retaining clip. Make sure the hose is not twisted.

12. Remove the plug from the brake line, connect the brake line to the brake hose and torque the fitting nut to 10–18 ft. lbs. (13–24 Nm).

13. Bleed the brake system using clean DOT 3 brake fluid from a closed container.

14. Install the wheel and tire assembly. Torque the lug nuts to 85–105 ft. lbs. (115–142 Nm).

15. Apply the brake pedal several times before moving the vehicle, to position the brake pads.

16. Road test the vehicle and check for proper brake system operation.

REAR

1. Remove the wheel and tire assembly.

2. Disconnect brake pad wear sensor connector (if equipped).

3. Remove the brake fitting retaining bolt from the disc brake caliper and disconnect the flexible brake hose from the caliper. Plug the hose and the caliper fitting.

4. Release the tension from the parking brake cable. Disconnect the cable from the caliper.

5. Remove the disc brake caliper locating pins. Lift the caliper off the brake rotor and anchor plate using a rotating motion.

6. Remove the disc brake pads.

To install:

7. Place the Rear Caliper Piston Adjuster T87P-2588-A in the caliper. Be sure the pins on the tool are meshed with the slots on the piston. Rotate the tool clockwise until the piston is seated in the caliper bore.

8. Install the disc pads in the support bracket. Make sure that the pads are on the correct side.

9. Place the disc brake caliper assembly above the brake rotor with the anti-rattle spring located on the lower adapter support arm. Install the caliper over the rotor with a rotating motion. Make sure the inner pad is properly positioned with the nub in the piston slot.

10. Apply a suitable thread-locking compound to the caliper locating pin threads and start them in by hand. Torque the locating pins to 19–26 ft. lbs. (26–35 Nm).

11. Install the brake hose on the disc brake caliper with a new gasket on each side of the fitting outlet. Insert the retaining bolt and torque to 30–40 ft. lbs. (40–54 Nm).

12. Connect the parking brake cable to the caliper and adjust the tension.

13. Connect the brake pad wear sensor connector (if equipped).

14. Bleed the brake system using clean DOT 3 brake fluid from a closed container.

15. Install the wheel and tire assembly. Torque the lug nuts to 85–105 ft. lbs. (115–142 Nm).

16. Apply the brake pedal several times before moving the vehicle, to position the brake pads.

17. Verify proper parking brake system operation.

18. Road test the vehicle and check for proper brake system operation.

Ford Crown Victoria
Lincoln Town Car
Mercury Grand Marquis

FRONT

➡Before continuing with this procedure, make sure to have available, 2 new disc brake caliper guide pin bolts and 2 banjo bolt sealing washers, per caliper. Once removed, these parts lose their torque holding ability or retention capability and must not be reused.

1. If equipped with air suspension, the air suspension switch, located on the right-hand side of the luggage compartment, must be turned to the **OFF** position before raising the vehicle.

2. Remove the front wheel and tire assembly.

3. Loosen and remove the banjo bolt securing the brake hose from the disc brake caliper. Plug the brake hose. Discard the sealing washers.

4. Remove 2 disc brake caliper guide pin bolts and discard. If removing both calipers, mark the right and left sides so they may be reinstalled correctly.

5. Lift the disc brake caliper off of the anchor plate.

To install:

6. Retract the disc brake caliper piston fully in the piston bore, using an old brake pad or block of wood and a C-clamp.

7. Install the disc brake pads to the caliper. Make sure that the brake pad insulators are correctly attached to the brake pad plate.

8. Position the disc brake caliper onto the anchor plate. Make sure the inner and outer pads are properly positioned and the anti-rattle spring is properly positioned. The caliper bleed screw should be positioned on top of the caliper when assembled on the vehicle.

9. Install 2 new caliper guide pin bolts. Torque the bolts to 21–26 ft. lbs. (28–26 Nm).

10. Unplug and install the brake hose to the disc brake caliper using 2 new copper

sealing washers on the banjo bolt. Torque the bolt to 30–40 ft. lbs. (41–54 Nm).

11. Bleed the brake system, filling the master cylinder as required. Only use clean DOT 3 brake fluid from a sealed container.

12. Install the wheel and tire assembly. Torque the lug nuts in a star pattern to 85–104 ft. lbs. (115–142 Nm).

13. If equipped with air suspension, turn the air suspension switch to the **ON** position.

14. Pump the brake pedal several times to position the brake pads prior to moving the vehicle.

15. Road test the vehicle and check for proper brake system operation.

REAR

➡Before continuing with this procedure, make sure to have available, 2 banjo bolt sealing washers, per caliper. Once removed, these parts lose their torque holding ability or retention capability and must not be reused.

1. If equipped with air suspension, the air suspension switch, located on the right-hand side of the luggage compartment, must be turned to the **OFF** position before raising the vehicle.

2. Remove the rear wheel and tire assembly.

3. Loosen and remove the banjo bolt securing the brake hose to the disc brake caliper. Plug the brake hose and discard both sealing washers.

4. Remove 2 disc brake caliper locating bolts. Lift the disc brake caliper off the rotor and anchor plate using a rotating motion.

To install:

5. Retract the disc brake caliper piston fully in the piston bore, using an old brake pad or block of wood and a C-clamp.

6. Install the disc brake pads on the caliper. Make sure that the pads are on the correct side.

7. Position the disc brake caliper assembly above the rotor with the anti-rattle spring located on the lower adapter support arm. Install the caliper over the rotor with a rotating motion.

8. Clean the inner surface of the caliper bushings and locating bolts. Lubricate the caliper locating bolts with a suitable silicone dielectric compound. Install and start the locating bolts by hand only. Torque both bolts to 16–20 ft. lbs. (22–27 Nm).

9. Unplug and install the brake hose to the disc brake caliper using 2 new copper sealing washers on the banjo bolt. Torque the bolt to 30–40 ft. lbs. (41–54 Nm).

10. Bleed the brake system, filling the master cylinder as required. Only use clean DOT 3 brake fluid from a sealed container.

11. Replace the rubber rear disc brake bleeder screw cap.

12. Install the wheel and tire assembly. Torque the lug nuts in a star pattern to 85–104 ft. lbs. (115–142 Nm).

13. If equipped with air suspension, turn the air suspension switch to the **ON** position.

14. Pump the brake pedal several times to position the brake pads prior to moving the vehicle.

15. Road test the vehicle and check for proper brake system operation.

Disc Brake Pads

REMOVAL & INSTALLATION

Focus

1. Remove or disconnect the following:
   - Front wheel
   - Brake hose from the support bracket
   - Outer brake pad retaining clip
   - Brake caliper
   - Inner and outer brake pads

To install:

2. Compress the caliper piston into the caliper bore.
3. Install or connect the following:
   - Inner and outer brake pads
   - Brake caliper. Tighten the bolts to 21 ft. lbs. (28 Nm).
   - Brake hose to the support bracket
   - Front wheel

Ford Contour
Mercury Mystique
Mercury Cougar (1999–01 model)

FRONT

1. Remove ½ of the brake fluid from the master cylinder reservoir.

2. Remove the wheel and tire assembly.

3. Remove the outer disc brake pad spring clip (anti-rattle clip).

4. Remove the 2 locating pin covers and remove the locating pins.

5. Free the hose from its mounting on the strut.

6. Lift the caliper off of the brake rotor and tie it off to prevent damage to the brake hose.

7. Remove the outboard disc brake pad from the anchor plate.

8. Remove the inboard disc brake pad from the brake caliper.

To install:

9. If installing new disc brake pads, use a C-clamp or similar tool to push the caliper piston into the caliper bore. This will allow room for the new pads.

10. Place the inboard disc brake pad into the caliper.

11. Place the outboard disc brake pad into position in the anchor plate.

12. Position the disc brake caliper onto the rotor.

13. Reinstall the 2 locating pins and torque to 20 ft. lbs. (28 Nm).

14. Reinstall the caliper locating pin covers.

15. Secure the brake hose to its support on the strut.

16. Reinstall the disc brake pad spring clip.

17. Reinstall the wheel and tire assembly. Torque the lug nuts to 62 ft. lbs. (85 Nm).

18. Pump the brake pedal several times to achieve a good pedal before attempting to move the vehicle.

19. Check the brake fluid level in the master cylinder fluid reservoir and add fluid as necessary.

20. Road test the vehicle and check for proper brake system operation.

REAR

1. Remove ½ of the brake fluid from the master cylinder reservoir.

2. Remove the wheels.

3. Remove the cotter pin and guide pin.

4. Remove the caliper locating pin cover and remove the locating pin.

5. Swing the rear disc brake caliper away from the brake rotor and anchor plate. There is no need to remove the parking brake cable or brake hose.

6. Remove the inner and outer disc brake pads.

To install:

7. If installing new disc brake pads, use rear caliper piston adjuster T87P-2588-A or similar tool, to rotate the rear disc brake piston clockwise, retracting the caliper piston. This will allow room for the new brake pads.

8. Install the inner and outer disc brake pads.

9. Swing the rear disc brake caliper back into position over the disc brake pads.

10. Clean the locating pin threads and apply 1 drop of a thread locking agent or similar sealer.

11. Apply a small amount of disc brake caliper slide grease to the shaft of the locating pin.

12. Reinstall the locating pin and torque to 30 ft. lbs. (41 Nm).

13. Reinstall the guide pin and the cotter pin.

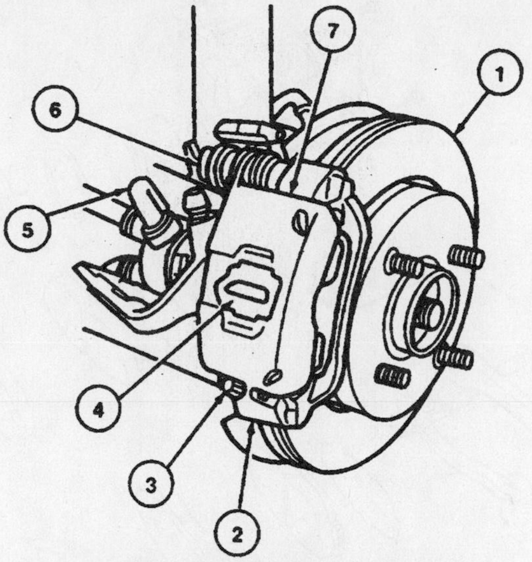

1	Rear Disc Brake Rotor
2	Rear Disc Brake Caliper Anchor Plate
3	Guide Pin
4	Anti-Rattle Clip
5	Parking Brake Lever

93006G26

Rear disc brake components—Contour, Mystique

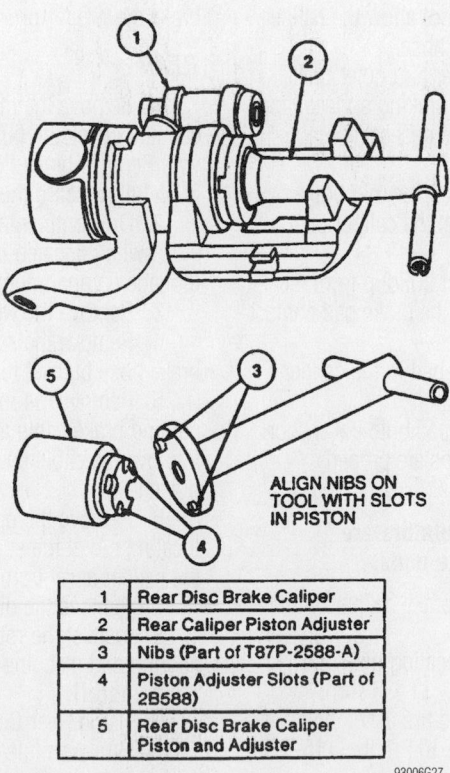

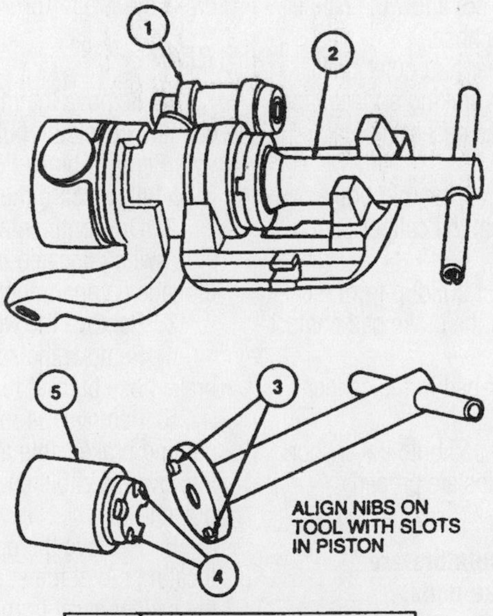

1	Rear Disc Brake Caliper
2	Rear Caliper Piston Adjuster
3	Nibs (Part of T87P-2588-A)
4	Piston Adjuster Slots (Part of 2B588)
5	Rear Disc Brake Caliper Piston and Adjuster

93006G27

Diagram of rear caliper piston and the adjuster tool—Contour, Mystique

14. Adjust the parking brake by operating the parking brake control several times.
15. Reinstall the wheel and tire assembly. Torque the lug nuts to 62 ft. lbs. (85 Nm).
16. Adjust the parking brake by operating the parking brake control several times.
17. Pump the brake pedal several times to achieve a good pedal before attempting to move the vehicle.
18. Check the brake fluid level in the master cylinder fluid reservoir and add fluid as necessary.
19. Road test the vehicle and check for proper brake system operation.

Lincoln LS

FRONT

1. Remove the master cylinder reservoir cap and check the fluid level in the reservoir. Remove brake fluid until the reservoir is ½ full. Discard the removed fluid.
2. Remove the wheel and tire assembly.
3. Remove the disc brake caliper locating pins. Lift the caliper assembly from the anchor plate and rotor.
4. Suspend the caliper inside the fender housing with wire. Do not allow the caliper to hang from the brake hose.
5. Remove the inner and outer brake pads. Inspect the rotor braking surfaces for scoring and machine as necessary.

To install:

6. Use a C-clamp and an old brake pad or block of wood to seat the caliper piston in its bore.
7. Remove any rust buildup from the inside of the caliper in the brake pad contact area.
8. Install the inner pad in the caliper piston.
9. Install the outer pad onto the anchor plate. Make sure the clips are properly seated.
10. Install the disc brake caliper onto the anchor plate.
11. Install caliper locating pins and torque to 26 ft. lbs. (35 Nm).
12. Install wheel and tire assembly and torque lugs nuts to 85–104 ft. lb. (115-142 Nm).
14. Pump the brake pedal several times prior to moving the vehicle to position the brake pads to the rotor.
15. Refill the master cylinder reservoir as necessary, using only clean DOT 3 or 4 brake fluid from a closed container.

REAR

1. Remove the master cylinder reservoir cap and check the fluid level in the reservoir. Remove brake fluid until the reservoir is ½ full. Discard the removed fluid.
2. Remove the wheel and tire assembly.
3. Remove the disc brake caliper locating pins at the support bracket.
4. Remove the caliper.
5. Remove the disc brake pads.
6. Inspect the rotor braking surfaces for scoring and machine as necessary.

To install:

7. Using Rear Caliper Piston Adjuster T87P-2588-A, rotate the piston clockwise until it is fully seated. Make sure one of the slots in the piston face is positioned so it will engage the nib on the brake pad.
8. Install the brake pads in the support bracket. Place the caliper assembly over the rotor into position on the support bracket. Make sure the brake pads are installed correctly.
9. Install and torque the disc brake caliper locating pin to 25 ft. lbs. (33 Nm).
10. Install the wheel and tire assembly and torque the lug nuts to 85–104 ft. lbs. (115–142 Nm).
11. Pump the brake pedal several times prior to moving the vehicle, to position the brake pads to the rotor.
12. Refill the master cylinder reservoir if necessary, using only clean DOT 3 or 4 brake fluid from a closed container.
13. Road test the vehicle and check the brake system for proper operation.

Ford Taurus
Lincoln Continental
Mercury Sable

CAUTION

For Continentals, the air suspension switch, located in the left side of the luggage compartment, must be turned OFF before raising the vehicle. Failure to do so may result in unexpected inflation or deflation of the air springs that may result in shifting of the vehicle during service.

FRONT

1. Remove the master cylinder reservoir cap and check the fluid level in the reservoir. Remove brake fluid until the reservoir is ½ full. Discard the removed fluid.
2. On Continentals, turn the air suspension switch, located in the left side of the luggage compartment, to the **OFF** position.
3. Remove the wheel and tire assembly.

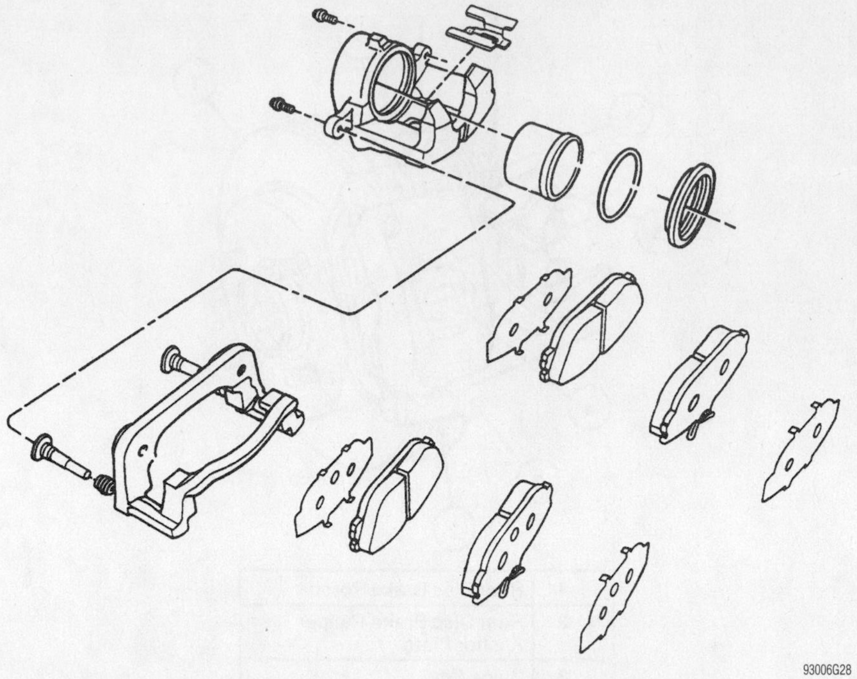

Front disc brake caliper, pads and related components—Taurus, Continental, Sable

4. Remove the disc brake caliper locating pins. Lift the caliper assembly from the anchor plate and rotor using a rotating motion.
5. Suspend the caliper inside the fender housing with wire. Do not allow the caliper to hang from the brake hose.
6. Remove the inner and outer brake pads. Inspect the rotor braking surfaces for scoring and machine as necessary.

To install:

7. Use a C-clamp and an old brake pad or block of wood to seat the caliper piston in its bore.
8. Remove any rust buildup from the inside of the caliper in the brake pad contact area.
9. Install the inner pad in the caliper piston.
10. Install the outer pad onto the anchor plate. Make sure the clips are properly seated.

➡Make sure the insulators are installed on the brake pads.

11. Install the disc brake caliper onto the anchor plate.
12. Install caliper locating pins and torque to 23–28 ft. lbs. (31–38 Nm).
13. Install wheel and tire assembly and torque lugs nuts to 85–104 ft. lb. (115-142 Nm)
14. Pump the brake pedal several times prior to moving the vehicle to position the brake pads to the rotor.
15. Refill the master cylinder reservoir as necessary, using only clean DOT 3 brake fluid from a closed container.
16. On Continentals, turn the air suspension service switch to the **ON** position.
17. Road test the vehicle and check the brake system for proper operation.

REAR

1. Remove the master cylinder reservoir cap and check the fluid level in the reservoir. Remove brake fluid until the reservoir is ½ full. Discard the removed fluid.
2. On Continentals, turn the air suspension switch, located in the left side of the luggage compartment, to the **OFF** position.
3. Remove the wheel and tire assembly.
4. Remove the screw retaining the brake hose bracket to the frame side rail.
5. Remove the retaining clip from the parking brake cable at the disc brake caliper. Remove the cable end from the parking brake lever.
6. Remove the upper disc brake caliper locating pin at the support bracket. Rotate the caliper away from the rotor.
7. Remove the disc brake pads.
8. Inspect the rotor braking surfaces for scoring and machine as necessary.

To install:

9. Using Rear Caliper Piston Adjuster T87P-2588-A, rotate the piston clockwise until it is fully seated. Make sure one of the slots in the piston face is positioned so it will engage the nib on the brake pad.
10. Install the brake pads in the support

bracket. Rotate the caliper assembly over the rotor into position on the support bracket. Make sure the brake pads are installed correctly.

11. Remove the residue from the rear brake pin retainer bolt threads and apply 1 drop of a suitable threadlock sealer. Install and torque the disc brake caliper locating pin to 23–26 ft. lbs. (31–35 Nm).

12. Attach the cable end to the parking brake lever. Install the cable retaining clip on the caliper assembly. Position the brake flex hose and bracket assembly to the side rail, and install the retaining screw. Torque to 11 ft. lbs. (16 Nm).

13. Install the wheel and tire assembly and torque lug nuts to 85–104 ft. lbs. (115–142 Nm).

14. Pump the brake pedal several times prior to moving the vehicle, to position the brake pads to the rotor.

15. Refill the master cylinder reservoir if necessary, using only clean DOT 3 brake fluid from a closed container.

16. On Continentals, turn the air suspension service switch to the **ON** position.

17. Road test the vehicle and check the brake system for proper operation.

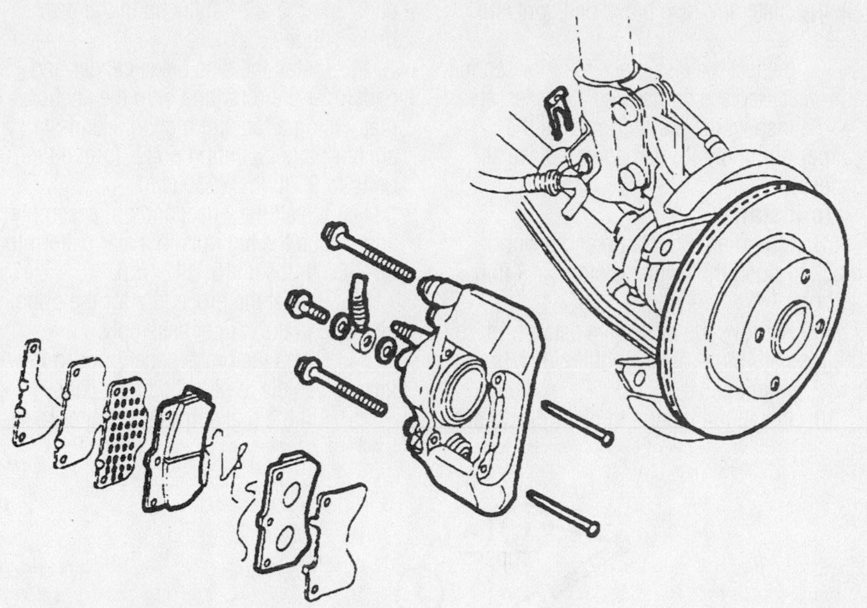

Front disc brake pads—Escort, Tracer

Ford Escort
Mercury Tracer

FRONT

1. Remove the master cylinder reservoir cap and check the fluid level in the reservoir. Remove brake fluid until the reservoir is ½ full. Discard the removed fluid.
2. Remove the wheel and tire assembly.
3. Remove the W-spring.
4. Remove 2 disc brake pad locating pins and remove the M-spring.
5. Remove the brake pads and shims from the brake caliper.
6. Inspect the brake rotor. Resurface or replace if needed.

To install:

7. Use a suitable tool to push the piston into the brake caliper bore.
8. Apply a suitable grease between the shims and the disc brake pad guide plates and position the brake pads and shims to the brake caliper.
9. Install the W-spring and 2 brake pad locating pins.
10. Install the M-spring.
11. Install the wheel and tire assembly. Torque the lug nuts to 65–87 ft. lbs. (88–118 Nm).
12. Pump the brake pedal several times prior to moving the vehicle to position the brake pads.
13. Check the fluid level in the master cylinder reservoir and fill as needed.
14. Road test the vehicle and check for proper brake system operation.

REAR

1. Remove the master cylinder reservoir cap and check the fluid level in the reservoir. Remove brake fluid until the reservoir is ½ full. Discard the removed fluid.
2. Remove the wheel and tire assembly.
3. If necessary, remove the screw plug and turn the adjustment gear counterclockwise with an Allen® wrench to pull the piston fully inward.
4. Remove the lower brake caliper bolt.
5. Using a small prybar, pivot the caliper on its mounting bracket to access the brake pads. If the upper lock bolt requires lubrication or service, remove it and suspend the caliper with mechanics wire.
6. Remove the brake pads, shims, spring and guides.
7. Inspect the brake rotor. Resurface or replace if needed.

To install:

8. Apply an appropriate brake pad grease between the shims and the brake pads.
9. Pivot the caliper on its mounting bracket and position the brake pads, shims, spring and guides to the brake rotor.
10. Lubricate and install the lower caliper bolt. Tighten the bolt to 33–43 ft. lbs. (45–59 Nm).
11. Turn the adjustment gear clockwise with an Allen wrench until the brake pads just touch the rotor, then loosen the gear ⅓ of a turn. Install the screw plug and torque to 12 ft. lbs. (16 Nm).
12. Install the wheel and tire assembly. Torque the lug nuts to 65–87 ft. lbs. (88–118 Nm).
13. Pump the brake pedal several times prior to moving the vehicle to position the brake pads.
14. Check the brake fluid level in the master cylinder reservoir and fill if needed.
15. Road test the vehicle and check for proper brake system operation.

Mustang

FRONT WITH SINGLE PISTON CALIPER

1. Remove and discard ½ of the brake fluid from the brake master cylinder reservoir. Properly dispose of the used brake fluid.
2. Remove the wheel and tire assembly.
3. Remove 2 anchor plate mounting bolts and lift the disc brake caliper from the anchor plate and rotor using a rotating motion. Hang the disc brake caliper from the body with wire. Do not let the disc brake caliper hang by the brake hose.
4. Remove the outer and inner disc brake pads and the anti-rattle clip from the disc brake caliper.
5. Clean any residue from the caliper

anchor plate and disc brake pad contact areas.

6. Inspect the disc brake rotor for scoring and wear. Replace or machine, as necessary.

7. Inspect the piston boot and the caliper pin boots for damage. Replace as necessary.

To install:

8. Use a large C-clamp and a wood block to push the caliper piston back into its bore.

9. Remove the protective paper from the insulators on the back of the disc brake pads, if equipped.

10. Install the inner and outer disc brake pads and the anti-rattle clip in the disc brake caliper.

11. Install the disc brake caliper and pads over the rotor and onto the anchor plate using a rotating motion. Hand-start 2 anchor plate mounting bolts. Torque the bolts to 95 ft. lbs. (130 Nm).

12. Install the wheel and tire assembly and torque the lug nuts in a star pattern to 85–105 ft. lbs. (115–142 Nm).

13. Repeat the procedure for the opposite disc brake caliper assembly.

14. Pump the brake pedal prior to moving the vehicle to seat the brake pads.

15. Fill the brake master cylinder reservoir with clean DOT 3 brake fluid from a closed container.

16. If the disc brake calipers were replaced or repaired, be sure to bleed the system.

17. Road test the vehicle and check the brake system for proper operation.

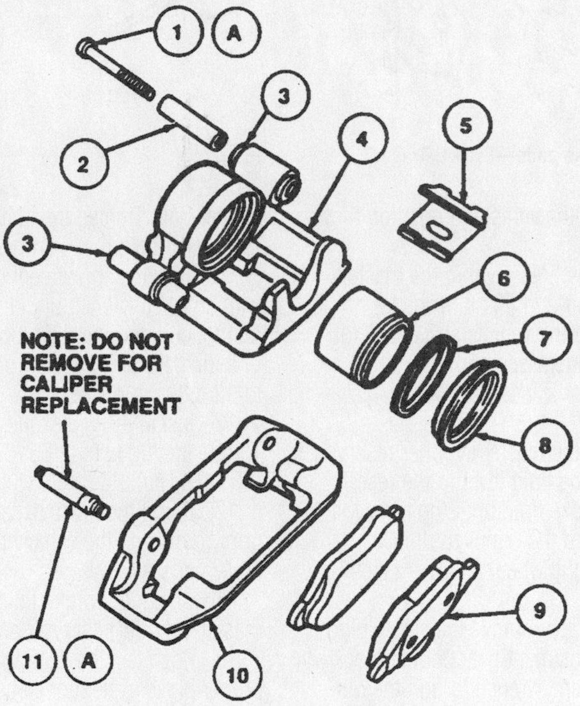

1	Upper Caliper Brake Bolt
2	Front Caliper Sleeve
3	Insulator
4	Caliper Housing (Part of 2B119)
5	Disc Brake Pad Anti-Rattle Clip
6	Caliper Piston
7	Piston Seal
8	Dust Boot
9	Brake Shoe and Lining
10	Front Disc Brake Caliper Anchor Plate
11	Disc Brake Caliper Locating Pin
A	Tighten to 88 N·m (65 Lb-Ft)

93006G30

Disc brake pads and related components—Mustang with single piston caliper

FRONT WITH DUAL PISTON CALIPER

1. Remove and discard ½ of the brake fluid from the brake master cylinder reservoir. Properly dispose of the used brake fluid.

2. Remove the wheel and tire assembly.

3. Remove the clip, washer and disc brake caliper locating pin.

4. Remove the disc brake caliper from the brake rotor and the caliper anchor plate. Hang the disc brake caliper from the body with wire. Do not let the disc brake caliper hang by the brake hose.

5. Remove the outer and inner disc brake pads from the disc brake caliper.

6. Clean any residue from the caliper anchor plate and disc brake pad contact areas.

7. Inspect the disc brake rotor for scoring and wear. Replace or machine, as necessary.

8. Inspect the piston boots and the caliper pin boots for damage. Replace as necessary.

To install:

9. Use a large C-clamp and a wood block to push the caliper pistons back into their bores.

10. Remove the protective paper from the insulators on the back of the disc brake pads, if equipped.

11. Install the inner and outer disc brake pads in the disc brake caliper. Make sure the pads are seated properly.

12. Install the disc brake caliper and pads over the rotor and onto the anchor plate using a rotating motion. Make sure that the guiding surfaces of the disc brake pads and the anchor bracket are seating correctly.

13. Press the disc brake caliper down by hand to compress the bias spring and slide the disc brake caliper locating pin into position.

14. Install the washer and clip.

15. Install the wheel and tire assembly. Torque the lug nuts in a star pattern to 85–105 ft. lbs. (115–142 Nm).

16. Repeat the procedure for the opposite disc brake caliper assembly.

17. Pump the brake pedal prior to moving the vehicle to position the brake pads.

18. Fill the brake master cylinder reservoir with clean DOT 3 brake fluid from a closed container.

19. If the disc brake calipers were replaced or repaired, be sure to bleed the system.

20. Road test the vehicle and check the brake system for proper operation.

REAR

1. Remove and discard ½ of the brake fluid from the brake master cylinder reservoir. Properly dispose of the used brake fluid.
2. Remove the wheel and tire assembly.
3. Remove the screw retaining the brake hose bracket to the shock absorber bracket.
4. Remove the retaining clip from the parking brake cable at the disc brake caliper.
5. Release the tension from the parking brake cable and disengage the parking brake cable end from the parking brake lever on the disc brake caliper.
6. Hold the disc brake caliper locating pin hex-head with an open-end wrench. Remove the upper brake pin retainer.
7. Rotate the disc brake caliper away from the brake rotor.
8. Remove the inner and outer disc brake pads and the anti-rattle clip from the disc support bracket.
9. Inspect the disc brake rotor for scoring and wear. Replace or machine, as necessary.

To install:

10. Using Rear Caliper Piston Adjuster T87P-2588-A, rotate the caliper piston clockwise until the piston is fully seated.

➡Ensure that one of the 2 slots in the rear brake piston face is positioned so it will engage the nib on the back of the inner brake pad.

11. Install the disc brake pad anti-rattle clips and install the inner and outer disc brake pads on the rear disc support bracket.
12. Rotate the disc brake caliper over the brake rotor into position on the disc support bracket. Make sure the disc brake pads and anti-rattle clips are correctly installed.
13. Apply a suitable thread-locking compound to the threads of the upper brake pin retainer.
14. Install the brake pin retainer and tighten to 23–26 ft. lbs. (31–35 Nm) while holding the disc brake caliper locating pin with an open-end wrench.
15. Install the parking brake cable end to the parking brake lever on the disc brake caliper. Install the retaining clip to the disc brake caliper.
16. Install the wheel and tire assembly. Torque the lug nuts in a star pattern to 85–105 ft. lbs. (115–142 Nm).
17. Repeat the procedure for the opposite disc brake caliper assembly.
18. Pump the brake pedal prior to moving the vehicle to position the brake pads.
19. Fill the brake master cylinder reservoir with clean DOT 3 brake fluid from a closed container.
20. If the disc brake calipers were replaced or repaired, be sure to bleed the system.
21. Road test the vehicle and check the brake system for proper operation.

Mercury Cougar (1998 Model)

FRONT

1. If equipped with air suspension, the air suspension switch, located on the right-hand side of the luggage compartment, must be turned to the **OFF** position before raising the vehicle.
2. Remove and properly dispose ½ of the brake fluid from the master cylin-der.
3. Remove the front wheel and tire assembly.
4. Remove 2 disc brake caliper locating pins and remove the caliper from the anchor plate and rotor. Suspend the caliper inside the fender housing with a length of wire. Do not let the caliper hang by the brake hose.
5. Remove the disc brake pads from the caliper.
6. Inspect the disc brake rotor for scoring and wear. Replace or machine as necessary. Observe the minimum thickness specification stamped on the rotor if machining is required.

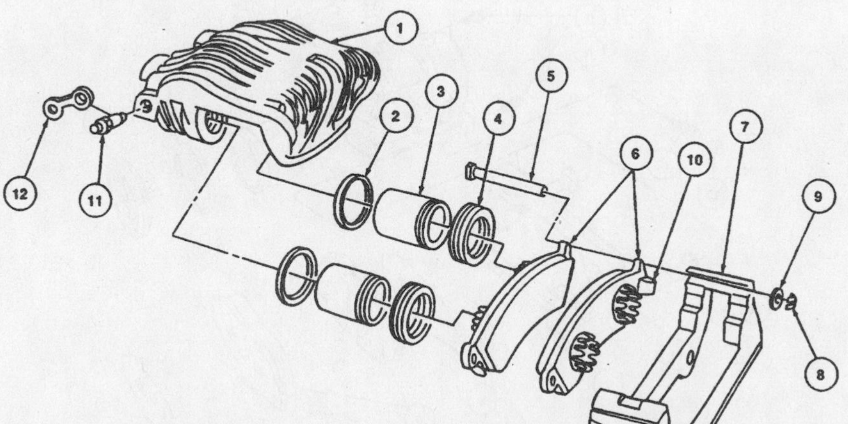

1	Caliper Housing
2	Piston Seal (2 Req'd)
3	Caliper Piston (2 Req'd)
4	Dust Boot (2 Req'd)
5	Front Brake Locating Pin
6	Brake Shoe and Lining
7	Front Disc Brake Caliper Anchor Plate
8	Clip
9	Washer
10	Anti-Rattle Clip and Insulator
11	Wheel Cylinder Bleeder Screw
12	Cover

93006G31

Disc brake pads and related components—Mustang with dual piston caliper

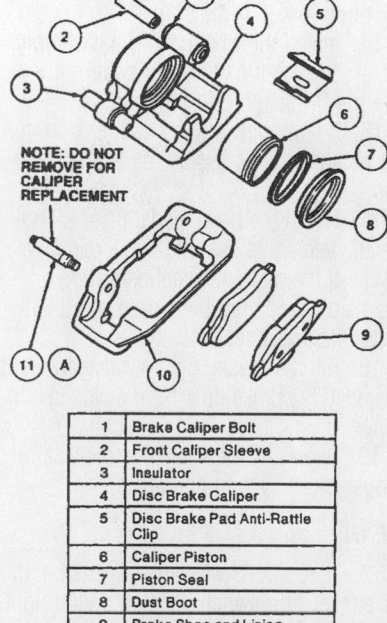

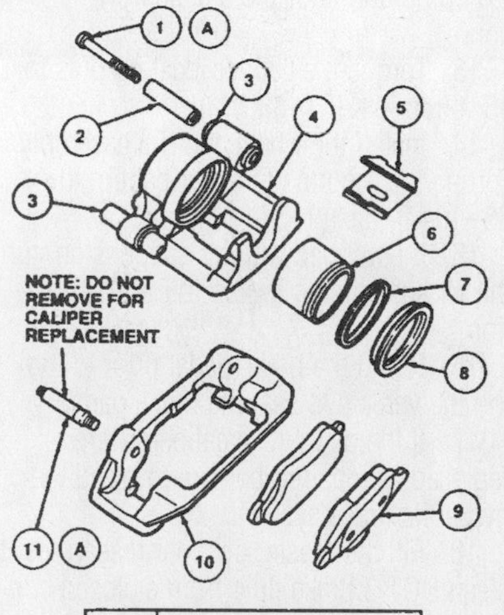

1	Brake Caliper Bolt
2	Front Caliper Sleeve
3	Insulator
4	Disc Brake Caliper
5	Disc Brake Pad Anti-Rattle Clip
6	Caliper Piston
7	Piston Seal
8	Dust Boot
9	Brake Shoe and Lining
10	Front Disc Brake Caliper Anchor Plate
11	Disc Brake Caliper Locating Pin
A	Tighten to 81 N-m (60 Lb-Ft)

93006G32

Front Disc brake pads and related components (15 inch wheel)—Cougar

To install:

7. Use a C-clamp and an old brake pad, wood block to seat the caliper piston in its bore. Do not allow metal or sharp objects to come into direct contact with the caliper piston surface or damage will result.

8. Install the correct inner disc brake pad to the caliper piston using care not to bend the anti-rattle clips.

9. Install the correct outer disc brake pad to the caliper making sure to properly install the anti-rattle clip. The outer pads are marked for left-hand or right-hand installation.

10. Install the disc brake caliper over the rotor with the outer brake pad against the rotor's braking surface. This prevents pinching the piston boot between the inner brake pad and the piston.

11. If worn, install 2 new locating pin insulators in the disc brake caliper by fabricating a tool for this purpose.

12. Lubricate the locating pins and the inside of the locating pin insulators with silicone dielectric grease. Install the caliper locating pins and thread them into the spindle/anchor plate assembly by hand.

13. Torque the caliper locating pins to 45–65 ft. lbs. (61–88 Nm).

14. Install the wheel and tire assembly. Torque the lug nuts in a star pattern to 85–104 ft. lbs. (115–142 Nm).

15. If equipped with air suspension, turn the air suspension switch to the **ON** position.

16. Pump the brake pedal prior to moving the vehicle to seat the brake pads.

17. If the disc brake calipers were replaced or repaired be sure to bleed the hydraulic brake system.

18. Fill the master cylinder reservoir with clean DOT 3 brake fluid from a closed container.

19. Road test the vehicle and check for proper brake system operation.

REAR

1. If equipped with air suspension, the air suspension switch, located on the right-hand side of the luggage compartment, must be turned to the **OFF** position before raising the vehicle.

2. Remove and properly dispose ½ of the brake fluid from the master cylin-der.

3. Remove the rear wheel and tire assembly.

4. Release the tension from the parking brake cable and disconnect the parking brake cable from the disc brake caliper.

5. Remove 2 caliper locating pins. Lift the disc brake caliper off the rotor and anchor plate using a rotating motion.

6. Remove the inner and outer disc brake pads.

7. Inspect the disc brake rotor for scoring and wear. Replace or machine, as necessary. Observe the minimum thickness specification stamped on the rotor if machining is required.

To install:

8. Use a C-clamp and an old brake pad, wood block to seat the caliper piston in its bore. Do not allow metal or sharp objects to come into direct contact with the plastic caliper piston surface or damage will result.

9. Position Caliper Piston Tool T87P-2588-A on the rear disc brake caliper. Be sure the pins on the tool are meshed with the slots on the piston. Rotate the tool clockwise until the piston is seated in the caliper.

10. Install the inner and outer disc brake pads in the support bracket. Make sure the pads are on the correct sides.

11. Position the disc brake caliper above the rotor with the anti-rattle spring located on the lower adapter support arm. Install the caliper over the rotor with a rotating motion. Make sure the inner pad is properly positioned.

12. Insert 2 disc brake caliper locating pins and thread them in by hand. Torque the pins to 19–26 ft. lbs. (26–35 Nm).

13. Install the wheel and tire assembly. Torque the lug nuts in a star pattern to 85–104 ft. lbs. (115–142 Nm).

14. If equipped with air suspension, turn the air suspension switch to the **ON** position.

15. Pump the brake pedal prior to moving the vehicle to seat the brake pads.

16. If the disc brake caliper was repaired or replaced, make sure to bleed the hydraulic brake system.

17. Fill the master cylinder reservoir with clean DOT 3 brake fluid from a closed container.

18. Road test the vehicle and check for proper brake system operation.

Lincoln Mark VIII

FRONT

1. Remove and discard ½ of the brake fluid from the brake master cylinder.

2. Turn the air suspension service switch to the **OFF** position.

3. Remove the wheel and tire assembly.

4. Remove the disc brake caliper pin retainers from the locating pins and remove the caliper from the anchor plate and disc

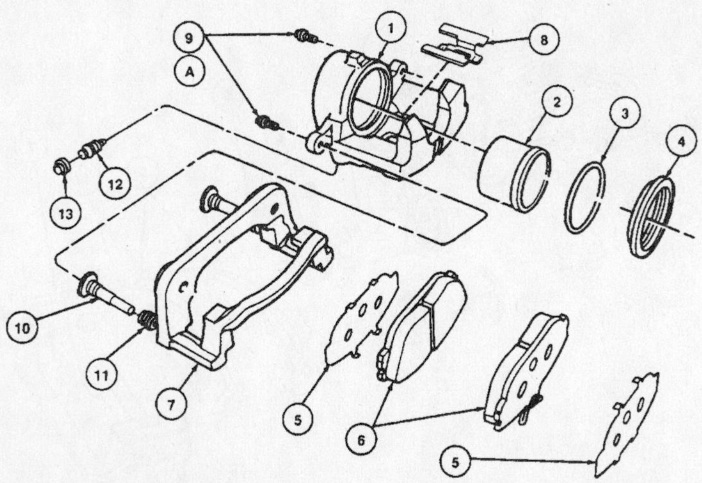

1	Disc Brake Caliper Housing
2	Caliper Piston
3	Brake Piston Seal
4	Piston Boot
5	Front Wheel Disc Brake Shoe Insulator (2 Req'd)
6	Brake Shoe and Lining
7	Front Disc Brake Caliper Anchor Plate
8	Brake Shoe Hold-Down Spring

9	Rear Brake Pin Retainer (2 Req'd)
10	Disc Brake Caliper Locating Pin
11	Caliper Guide Pin Excluder
12	Wheel Cylinder Bleeder Screw
13	Bleed Screw Cap
A	Tighten to 31-38 N·m (23-28 Lb-Ft)

93006G33

Front disc brake pads and related components—Mark VIII

brake rotor. Suspend the caliper inside the fender housing with a length of wire. Do not let the caliper hang by the brake hose.

5. Remove the outer, then inner disc brake pads from the caliper.

6. Inspect the disc brake rotor for scoring, runout and wear. Replace or machine as necessary. Do not exceed the minimum allowable thickness stamped on each disc brake rotor.

To install:

7. Using a C-clamp and a block of wood press the disc brake caliper piston into the caliper housing to allow for the new disc brake pads to fit the brake rotor.

8. Install the inner, then outer disc brake pads in the caliper.

9. Install the disc brake caliper over the rotor with the outer brake pad against the rotor's braking surface. This prevents pinching the piston boot between the inner brake pad and the piston.

10. Install the disc brake caliper pin retainers to the locating pins, by hand. Torque the caliper pin retainers to 16–23 ft. lbs. (22–32 Nm).

11. Install the wheel and tire assembly. Torque the lug nuts in a star pattern to 85–105 ft. lbs. (115–142 Nm).

12. Turn the air suspension service switch to the **ON** position.

13. Pump the brake pedal prior to moving the vehicle to seat the disc brake pads.

14. If the disc brake calipers were replaced or repaired, be sure to bleed the brake system.

15. Fill the brake master cylinder as needed with clean DOT 3 brake fluid from a closed container.

16. Road test the vehicle and check for proper brake system operation.

REAR

1. Remove and discard ½ of the brake fluid from the brake master cylinder.

2. Verify that the parking brake control is released.

3. Turn the air suspension service switch to the **OFF** position.

4. Remove the wheel and tire assembly.

5. Remove the retaining clip from the parking brake cable at the disc brake caliper.

6. Release the parking brake cable tension and disengage the cable end from the caliper lever.

7. Remove the disc brake caliper pin retainers securing the caliper to the locating pins.

8. Rotate the caliper away from the disc brake rotor.

9. Remove the inner and outer disc brake pads from the disc brake support bracket.

10. Inspect the disc brake rotor for scoring, runout and wear. Replace or machine as necessary. Do not exceed the minimum allowable thickness stamped on each disc brake rotor.

To install:

11. Using Caliper Piston Adjuster T87P-2588-A, rotate the disc brake caliper piston until it is fully seated in the caliper. Be sure the pins on the tool are meshed with the slots on the piston.

12. Install the inner and outer disc brake pads in the rear disc support bracket. Check to be sure the pads are on the correct sides, and that the nib on the inner disc brake pad is correctly installed to the slot in the caliper piston.

13. Rotate the disc brake caliper over the rotor into position on the support bracket. Install the caliper pin retainers to the caliper locating pins and thread them in by hand. Torque the pin retainers to 23–26 ft. lbs. (31–35 Nm).

14. Install the wheel and tire assembly. Torque the lug nuts in a star pattern to 85–105 ft. lbs. (115–142 Nm).

15. Turn the air suspension service switch to the **ON** position.

16. Pump the brake pedal prior to moving the vehicle to seat the disc brake pads.

17. If the disc brake calipers were replaced or repaired, be sure to bleed the brake system.

18. Fill the brake master cylinder as needed with clean DOT 3 brake fluid from a closed container.

19. Road test the vehicle and check for proper brake system operation.

Ford Crown Victoria
Lincoln Town Car
Mercury Grand Marquis

➡Before continuing with this procedure, make sure to have available, 2 new disc brake caliper anchor bracket mounting bolts, per caliper. Once removed, these parts lose their torque holding ability or retention capability and must not be reused.

FRONT

1. If equipped with air suspension, the air suspension switch, located on the right-hand side of the luggage compartment, must be turned to the **OFF** position before raising the vehicle.

2. Remove ½ of the brake fluid from the brake master cylinder reservoir. Properly dispose of the used brake fluid.

3. Remove the front wheel and tire assembly.

4. Remove 2 disc brake caliper anchor bracket mounting bolts and discard. Lift the caliper assembly from the disc brake rotor using a rotating motion. Suspend the caliper inside the fender housing with wire. Do not allow the caliper to hang from the brake hose.

5. Remove the inner and outer disc brake pads. Inspect the rotor braking surfaces for scoring and machine as necessary. Refer to the minimum rotor thickness specification when machining. If machining is not necessary, hand-sand the glaze from the braking surfaces with medium grit sandpaper. Make sure to wear an approved respirator.

To install:

6. Use a C-clamp and an old brake pad, wood block to seat the caliper piston in its bore. Do not allow metal or sharp objects to come into direct contact with the plastic caliper piston surface or damage will result.

7. Remove all rust buildup from the inside of the caliper legs.

8. Make sure the anti-rattle spring is seated in the caliper lining inspection opening and that it is installed from the lining side.

9. Install the inner disc brake pad to the caliper piston. Do not bend the pad clips during installation in the piston or distortion and rattles can occur. Install the outer disc brake pad. Make sure the clips are properly seated.

10. Install the disc brake caliper over the rotor and install 2 new anchor bracket mounting bolts. Torque the bolts to 126–169 ft. lbs. (170–230 Nm).

11. Install the wheel and tire assembly. Torque the lug nuts in a star pattern to 85–104 ft. lbs. (115–142 Nm).

12. If equipped with air suspension, turn the air suspension switch to the **ON** position.

13. Pump the brake pedal prior to moving the vehicle to seat the brake pads.

14. Fill the master cylinder reservoir with clean DOT 3 brake fluid from a closed container.

15. If the disc brake calipers were

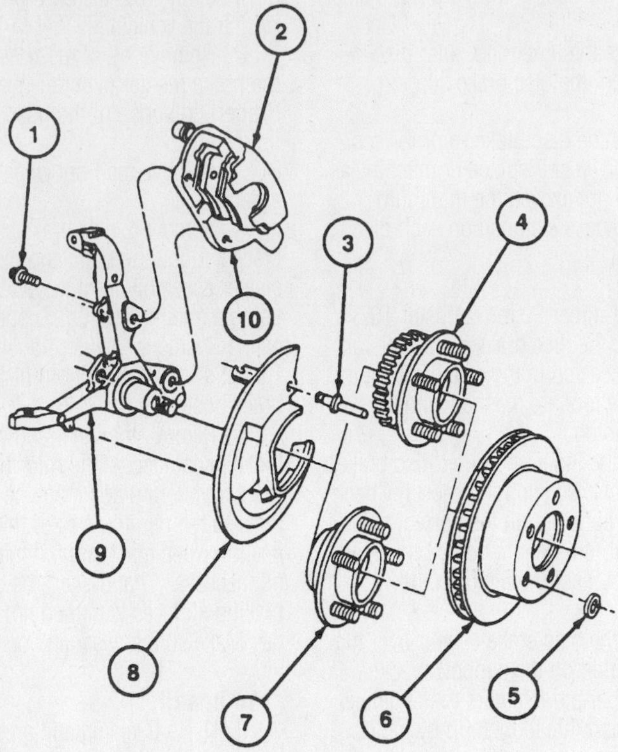

Item	Description
1	Bolt (2 Req'd)
2	Disc Brake Caliper
3	Rivet (3 Req'd)
4	Wheel Hub with ABS
5	Washer (2 Req'd)
6	Front Disc Brake Rotor
7	Wheel Hub Without ABS
8	Front Disc Brake Rotor Shield
9	Front Wheel Spindle
10	Front Disc Brake Caliper Anchor Plate Assembly

93006G83

Exploded view of the front disc assembly—Crown Victoria, Town Car, Grand Marquis

replaced or repaired be sure to bleed the system.

16. Road test the vehicle and check for proper brake system operation.

REAR

1. If equipped with air suspension, the air suspension switch, located on the right-hand side of the luggage compartment, must be turned to the **OFF** position before raising the vehicle.

2. Remove ½ of the brake fluid from the brake master cylinder reservoir. Properly dispose of the used brake fluid.

3. Remove the rear wheel and tire assembly.

4. Remove 2 disc brake caliper retaining bolts.

5. Lift the disc brake caliper off the disc brake rotor and anchor plate using a rotating motion.

6. Remove the inner and outer disc brake pads.

7. Inspect the disc brake rotor for scoring and wear. Inspect the rotor braking surfaces for scoring and machine as necessary. Refer to the minimum rotor thickness specification when machining. If machining is not necessary, hand-sand the glaze from the braking surfaces with medium grit sandpaper. Make sure to wear an approved respirator.

To install:

8. Use a C-clamp and an old brake pad, wood block to seat the caliper piston in its bore. Do not allow metal or sharp objects to come into direct contact with the plastic caliper piston surface or damage will result.

9. Remove all rust buildup from the inside of the caliper legs.

10. Install the inner disc brake pad to the caliper piston. Do not bend the pad clips during installation in the piston or distortion and rattles can occur. Install the outer disc brake pad. Make sure the clips are properly seated.

➡Make sure the insulators are installed on the brake pads.

11. Install the disc brake caliper over the disc brake rotor and install 2 caliper retaining bolts. Torque to 16–20 ft. lbs. (22–27 Nm).

12. Install the wheel and tire assembly. Torque the lug nuts in a star pattern to 85–104 ft. lbs. (115–142 Nm).

13. If equipped with air suspension, turn the air suspension switch to the **ON** position.

14. Pump the brake pedal prior to moving the vehicle to seat the disc brake pads.

15. Fill the master cylinder reservoir with clean DOT 3 brake fluid from a closed container.

16. If the disc brake calipers were replaced or repaired be sure to bleed the system.

17. Road test the vehicle and check for proper brake system operation.

Brake Drums

REMOVAL & INSTALLATION

Focus

1. Remove or disconnect the following:
   - Rear wheel
   - Wheel speed sensor, if equipped
   - Brake drum, bearing and spindle assembly.

To install:

2. Install or connect the following:
   - Brake drum, bearing and spindle assembly. Tighten the bolts to 49 ft. lbs. (66 Nm).
   - Wheel speed sensor, if equipped
   - Rear wheel

Ford Contour
Mercury Mystique

1. Remove the wheel and tire assembly.
2. Remove the brake drum retainers, if installed.
3. Grasp the brake drum and remove.
4. If the drum will not slide off with light force, then the brake shoes will need to be backed off as follows:
   a. Remove the rubber plug on the backing plate and insert a screwdriver or small brake adjusting tool into the slot to contact the brake strut and quadrant.
   b. A forward motion of the tool will separate the quadrant from the knurled wheel and allow the brake shoes to retract.
   c. Remove the brake drum.

To install:

5. Make sure the brake drum and shoes are clean of any oils or protective coatings.
6. Position the brake drum onto the wheel hub.
7. Install new brake drum retainers if available.
8. Reinstall the wheel and tire assembly. Torque the lug nuts to 94 ft. lbs. (128 Nm).
9. Work the parking brake control several times to adjust the rear brake shoes.
10. Pump the brake pedal several times to assure a good pedal before attempting to move the vehicle.
11. Road test the vehicle and check for proper brake system operation.

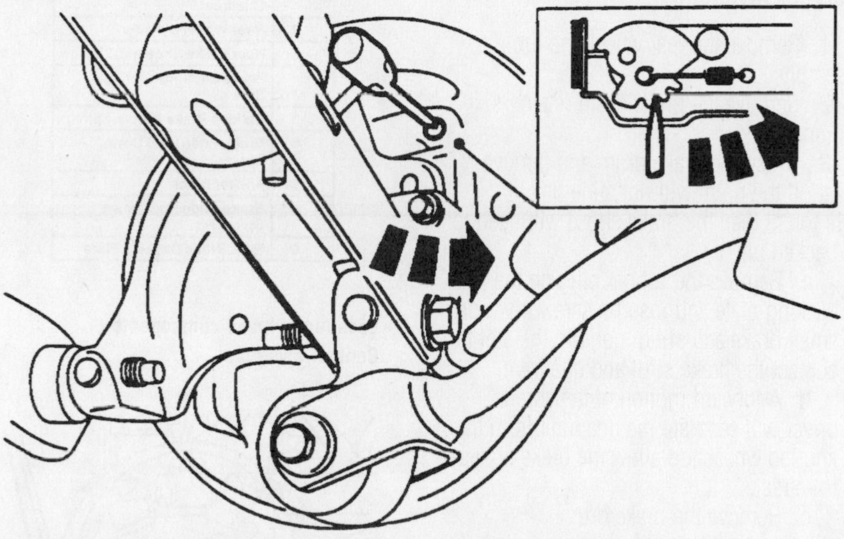
93006G38

Positioning a screwdriver to release the brake shoes—Contour, Mystique

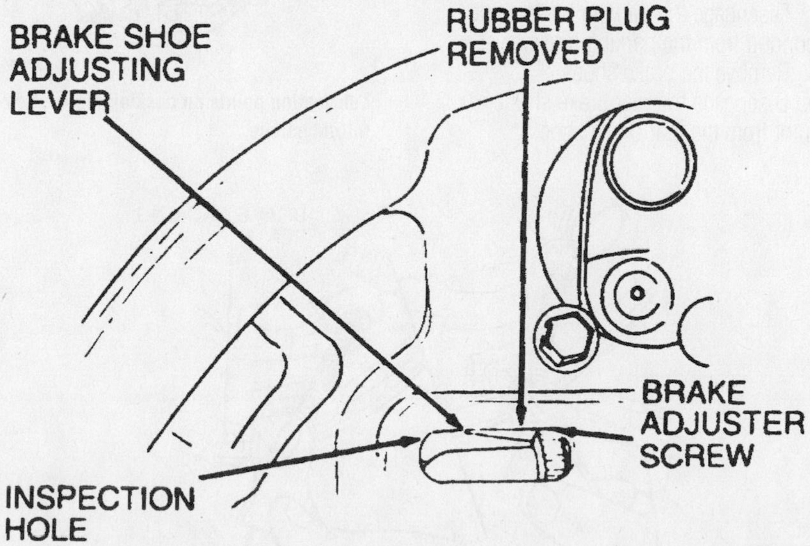

93006G39

Retracting the brake shoes to allow drum removal—Taurus and Sable

Ford Taurus
Mercury Sable

1. Remove the wheel and tire assembly.
2. Remove the brake drum.

➡If the brake drum cannot be removed easily, remove the brake tube-to-axle retention bracket and pry rubber plug from rear brake backing plate inspection hole. This will allow sufficient room for insertion of a screwdriver and brake tools to disengage brake shoe adjusting lever and back off the brake adjuster screw.

3. Inspect the drum for scoring and/or other wear. Machine or replace, as necessary.

To install:

4. Measure the brake drum inside diameter using D81L-1103-A brake adjustment gauge.
5. Using the brake adjustment gauge, adjust the brake shoes to the same dimensions as the brake drum.
6. Position the brake drum over the brake shoes on the axle hub.
7. Install the wheel and tire assembly. Torque the lug nuts to 85–104 ft. lbs. (115–141 Nm).
8. Pump the brake pedal several times to position the brake shoes and complete the adjustment.
9. Road test the vehicle and check for proper brake system operation.

Ford Escort
Mercury Tracer

1. Remove the wheel and tire assembly.
2. Remove 2 brake drum retaining screws.
3. Pull the brake drum from the hub. Inspect the drum and refinish or replace, as necessary. If refinishing, check the maximum inside diameter specification.

To install:

4. If needed, adjust the brake shoes to fit the brake drum.
5. Place the brake drum on the hub.
6. Install 2 brake drum retaining screws. Torque to 89–123 inch lbs. (10–14 Nm).
7. Install the wheel and tire assembly.

Torque the lug nuts to 73–100 ft. lbs. (100–135 Nm).

8. Check the brake system for proper operation.

Ford Mustang

1. Remove the wheel and tire assembly.
2. Remove the drum retaining nuts, if equipped, and remove the brake drum.
3. Inspect the brake drum for scoring and wear. Check the drums for conditioning and wear. Replace or machine as necessary. If machining, observe the maximum diameter specification.

To install:

4. If a new brake drum is being installed, remove the protective coating from the drum using brake cleaner.
5. Adjust the brake shoes to the brake drum using Brake Adjusting Gauge D81L-1103-A.
6. Install the brake drum.
7. Install the wheel and tire assembly.
8. Complete the brake adjustment by carefully backing the vehicle up several times while applying the brakes with a minimum force of 25 lbs. (111 N).
9. Road test the vehicle and check the brake system for proper operation.

Brake Shoes

REMOVAL & INSTALLATION

Focus

1. Remove or disconnect the following:
   - Rear wheel
   - Wheel speed sensor, if equipped
   - Brake drum, bearing and spindle assembly
   - Brake shoe hold-down springs and pins
   - Brake shoe assembly from the wheel cylinder and anchor block
   - Parking brake cable
   - Lower return spring
   - Upper return spring
   - Primary shoe from the strut and brake shoe adjuster
   - Parking brake return spring
   - Secondary shoe from the strut support

To install:

2. Install or connect the following:
   - Secondary shoe to the strut support
   - Parking brake return spring
   - Primary shoe to the strut and brake shoe adjuster. Rotate the adjuster fully clockwise.
   - Upper return spring
   - Lower return spring
   - Parking brake cable
   - Brake shoe assembly to the wheel cylinder and anchor block
   - Brake shoe hold-down springs and pins
   - Brake drum, bearing and spindle assembly. Tighten the bolts to 49 ft. lbs. (66 Nm).
   - Wheel speed sensor, if equipped
   - Rear wheel
3. Operate the brake pedal to adjust the rear brakes.

Ford Contour
Mercury Mystique

1. Remove the rear wheel and tire assembly.
2. Remove the brake drum retainers, if equipped.
3. Grasp the brake drum and remove.
4. If the drum will not slide off with light force, then the brake shoes will need to be backed off:
   a. Remove the rubber plug on the backing plate and insert a screwdriver or small brake adjusting tool into the slot to contact the brake strut and quadrant.
   b. A forward motion of the screwdriver will separate the quadrant from the knurled wheel and allow the brake shoes to retract.
   c. Remove the brake drum.
5. Remove the brake shoe hold down springs and the brake shoe hold down pins.
6. Remove the brake shoe retracting springs.
7. Disengage the parking brake cable and conduit from the parking brake lever.
8. Remove the brake shoes.
9. Disengage the rear brake strut and quadrant from the rear brake shoe.

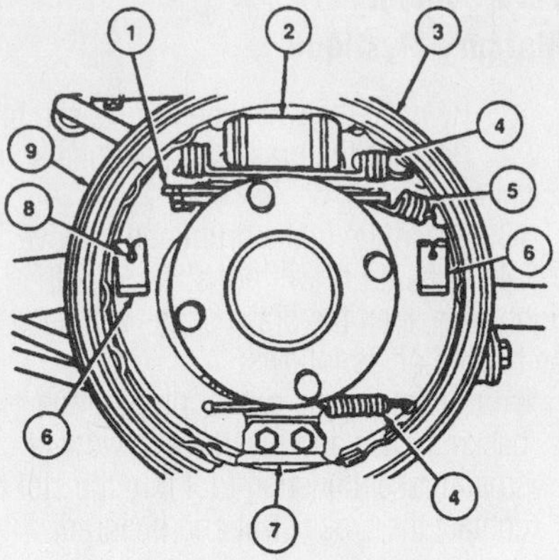

1	Rear Brake Strut and Quadrant
2	Rear Wheel Cylinder
3	Rear Brake Shoe and Lining
4	Brake Shoe Retracting Spring
5	Parking Brake Return Spring
6	Brake Shoe Hold Down Spring
7	Anchor Block
8	Brake Shoe Hold Down Spring Pin
9	Rear Brake Backing Plate

93006G50

Location of brake components—Contour/Mystique

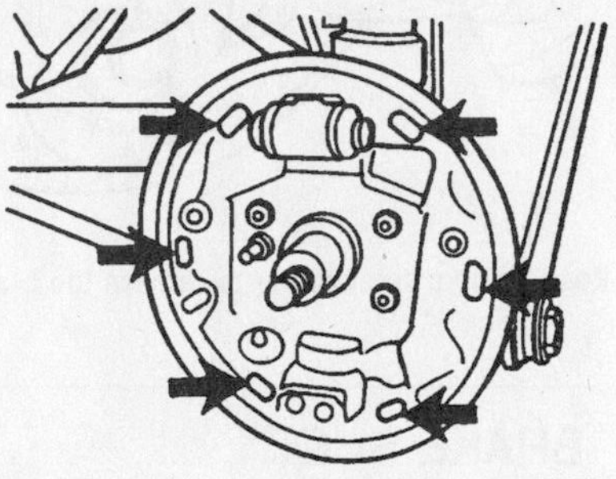

93006G52

Lubricating points on backing plate—Contour/Mystique

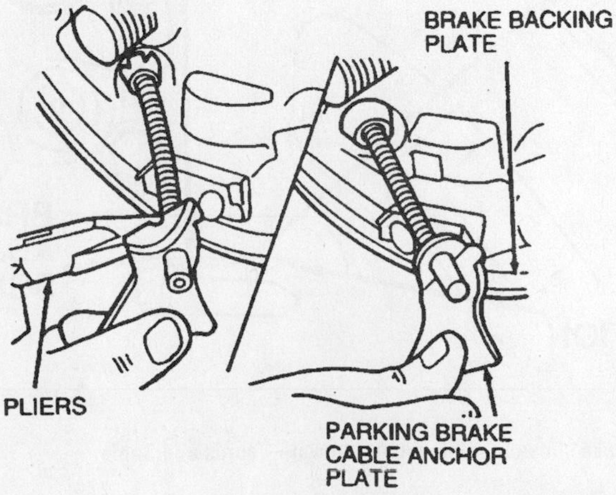

93006G51

Removal of parking brake cable—Contour/Mystique

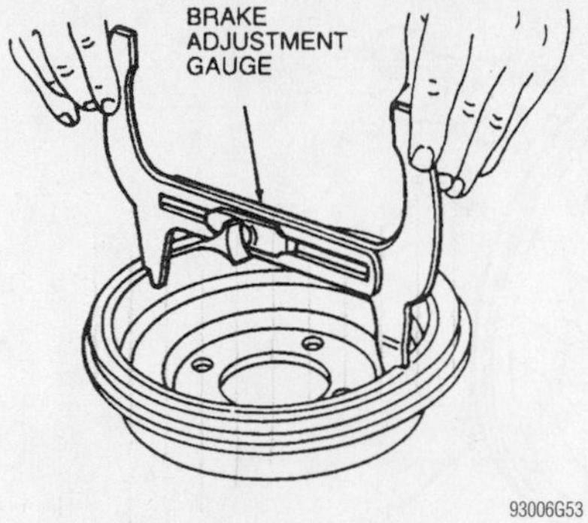

Measuring the brake drum inner diameter—Contour/Mystique

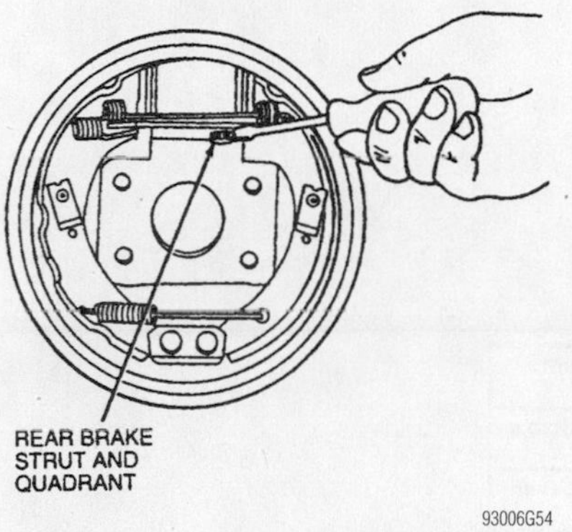

Adjusting the brake shoes to fit the drum—Contour/Mystique

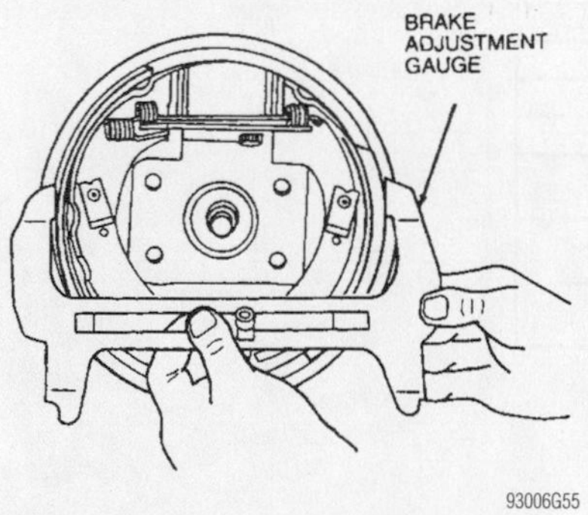

Measuring the brake shoe adjustment—Contour/Mystique

10. Remove the parking brake rear cable and conduit from the parking brake cable anchor on the trailing brake shoe.

To install:

11. Lubricate the rear brake shoe contact points on the backing plate with an appropriate grease.
12. Engage the parking brake rear cable and conduit into the parking brake cable anchor on the trailing brake shoe.
13. Position the trailing brake shoe on the backing plate.
14. Engage the rear brake strut and quadrant.
15. Reinstall the parking brake rear spring.
16. Reinstall the brake shoe hold down spring pin and the hold down spring.
17. Insert the leading brake shoe into the slot on the rear brake strut and quadrant.
18. Reinstall the brake shoe hold down pin and hold down spring.
19. Reinstall the brake shoe retracting springs.
20. Adjust the brake shoes by first measuring the inside drum diameter with an appropriate brake adjustment gauge.
21. Insert a small screwdriver or similar tool into the knurled quadrant of the brake strut and quadrant to adjust the brake shoes to the same measurement of the brake drum by expanding the brake strut and quadrant.
22. Trial fit the brake drum. The shoes should just contact the drum surface when properly adjusted.
23. Make sure that the brake drum and brake shoes are clean of any oils or protective coatings.
24. Reinstall the brake drum.
25. Reinstall new drum retainers, if available.
26. Reconnect the parking brake rear cable and conduit to the parking brake lever. It may be necessary to back off on the parking brake cable adjustment to allow for the new brake shoes.
27. Reinstall the wheel and tire assembly. Torque the lug nuts to 62 ft. lbs. (85 Nm).
28. Work the parking brake control several times to complete the brake shoe adjustment and to check the parking brake adjustment as well.
29. Pump the brake pedal several times to assure a good pedal.
30. Road test the vehicle and check for proper brake system operation.

Ford Taurus
Mercury Sable

1. Remove the wheel and tire assembly.
2. Remove the brake drum.
3. Remove the parking brake cable from the parking brake lever.
4. Remove the 2 brake shoe hold-down springs and pins.
5. Lift the brake shoes, springs and adjuster assembly off the backing plate and wheel cylinder assembly. When removing the assembly, be careful not to bend the adjusting lever.
6. Remove the retracting springs from the lower brake attachments and upper shoe-to-adjusting lever attachment points.
7. Remove the horseshoe retaining clip and spring washer and slide the lever off the parking brake lever pin on the trailing shoe. Discard the horseshoe clip.

To install:

8. Apply a light coating of disc brake caliper slide grease at the points where the brake shoes contact the backing plate.
9. Apply a thin coat of lubricant to the adjuster screw threads and socket end of the adjusting screw. Install the stainless steel washer over the socket end of the adjusting screw and install the socket. Turn the adjusting screw into the adjusting pivot nut to the limit of the threads and then back off ½ turn.
10. Assemble the parking brake lever to the trailing shoe by installing the spring washer and a new horseshoe retaining clip. Crimp the clip until it retains the lever to the shoe securely.
11. Position the trailing shoe on the backing plate and attach the rear parking brake cable.
12. Position the leading shoe on the backing plate and attach the lower brake shoe adjusting spring to the brake shoes.
13. Install the adjuster assembly in the slots on the brake shoes. The wide slot on the dual slotted end must fit into the leading shoe. The narrow slot on the dual slotted end fits into the shoe adjusting lever. The single slotted side of the adjuster assembly must fit into the slots on the trailing shoe and the rear parking brake cable bracket.

➡The adjuster socket blade is marked R for the right or L for the left brake assemblies. The adjuster blade must be installed with the letter R or L in the upright position, facing the wheel cylinder. Make sure the adjuster socket fits into the parking brake lever.

14. Complete the installation by reversing the removal procedures.
15. Pump the brake pedal several times to position the brake shoes and finish the brake shoe adjustment.

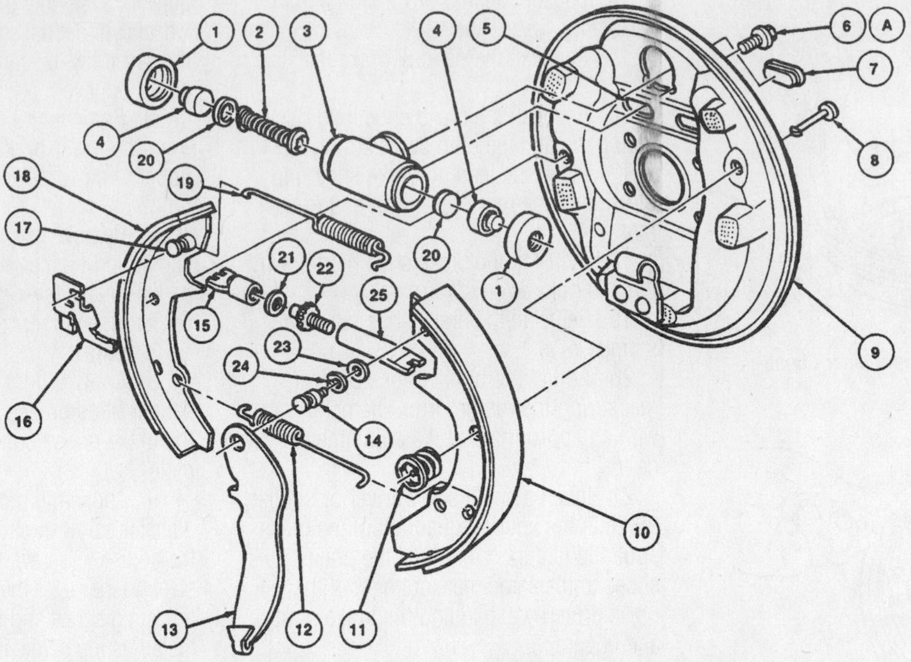

1	Boot
2	Spring Expander
3	Rear Wheel Cylinder
4	Piston and Insert
5	Shoe Adjustment Access Hole
6	Wheel Cylinder Retaining Bolt (2 Req'd)
7	Brake Adjusting Hole Cover
8	Brake Shoe Hold-Down Spring Pin
9	Rear Brake Backing Plate
10	Trailing Shoe and Lining
11	Brake Shoe Hold-Down Spring
12	Brake Shoe Retracting Spring
13	Parking Brake Lever

14	Parking Brake Lever Pin (Inner)
15	Brake Shoe Adjusting Screw Socket
16	Brake Shoe Adjusting Lever
17	Parking Brake Lever Pin
18	Leading Shoe and Lining
19	Brake Shoe Adjusting Screw Spring
20	Cup
21	Washer
22	Brake Adjuster Screw
23	Washer
24	Parking Brake Lever Pin Retainer
25	Adjusting Pivot Nut
A	Tighten to 12-18 N·m (107-159 Lb-In)

93006G56

Brake shoes and related components—Taurus and Sable

16. Road test the vehicle and check the brake system for proper operation.

Ford Escort
Mercury Tracer

1. Remove the wheel and tire assembly.
2. Remove 2 brake drum retaining screws and remove the brake drum.
3. Remove 2 brake shoe return springs.
4. Remove the right-hand anti-rattle spring.
5. Push and turn to release 2 brake shoe hold-down springs and remove the springs.
6. Remove the leading and trailing shoes from the brake backing plate.

To install:

7. Use a suitable high temperature grease to lightly lubricate the brake shoe contact points on the backing plate.
8. Position the trailing brake shoe on the backing plate and install one of the brake shoe hold-down springs.
9. Position the leading brake shoe on the backing plate and install the other brake shoe hold-down spring.
10. Install the right-hand anti-rattle spring.
11. Install 2 brake shoe return springs.
12. Using a brake adjusting gauge, measure the inside diameter of the brake drum.
13. Compare the brake drum measurement to the brake shoes.
14. Adjust the brake shoes by inserting a screwdriver into the knurled quadrant of the rear quad operating lever and adjust the shoes to the same measurement as the brake drum.
15. Install the brake drum.
16. Install 2 brake drum retaining screws.
17. Install the wheel and tire assembly. Torque the lug nuts to 65–87 ft. lbs. (88–118 Nm).
18. Complete the brake shoe adjustment by sharply applying the brakes several times while driving the vehicle alternating between forward and reverse gears.
19. Check the brake system operation by making several stops while driving forward.

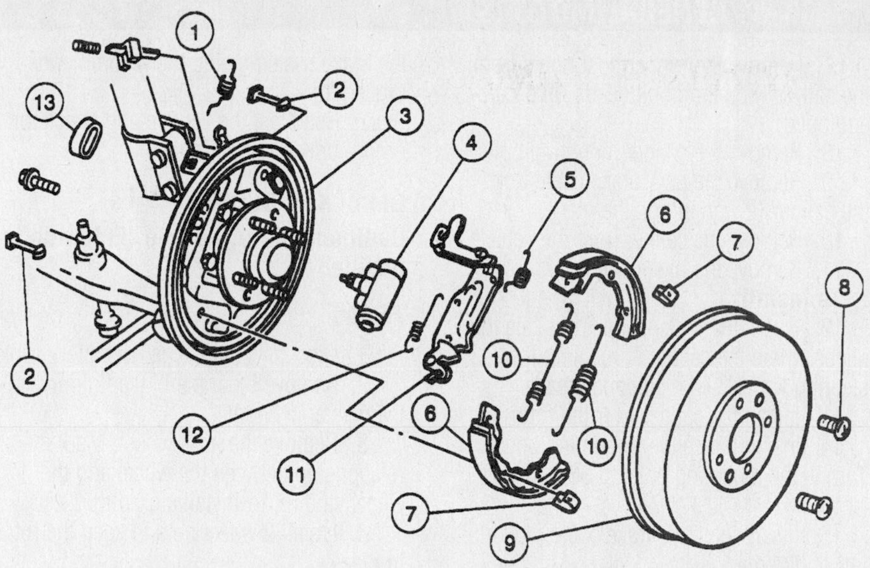

1	Parking Brake Link Spring	7	Brake Shoe Hold-Down Spring
2	Brake Shoe Hold-Down Spring Pin	8	Brake Drum Screw (2 Req'd)
3	Rear Brake Backing Plate	9	Brake Drum
4	Rear Wheel Cylinder	10	Brake Shoe Retracting Spring
5	Right Hand Anti-Rattle Spring	11	Parking Brake Lever
6	Rear Brake Shoe and Lining	12	Parking Brake Return Spring
		13	Brake Adjusting Hole Cover

93006G57

Brake shoes and related components—Escort/Tracer

Mercury Cougar (1998 Model)

1. Remove the wheel and tire assembly.
2. Remove the brake drum.
3. Disconnect the parking brake rear cable from the parking brake lever.
4. Remove 2 brake shoe hold-down retainers, brake shoe hold-down springs and hold-down pins.
5. Lift the brake shoes, springs and adjuster off of the brake backing plate as an assembly using care not to bend the brake shoe adjusting lever.
6. Remove the brake shoe adjusting screw spring.

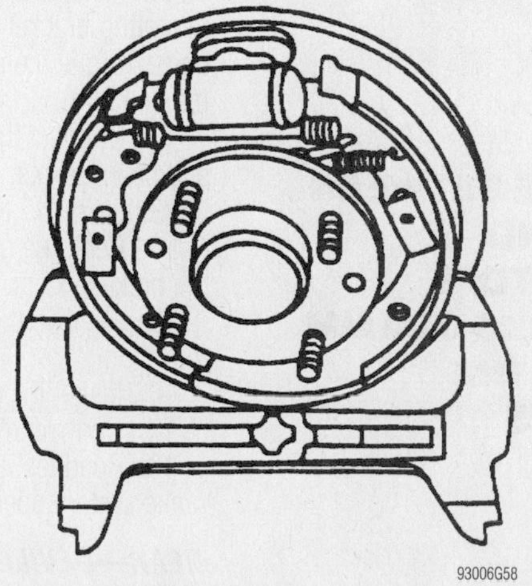

93006G58

Checking the adjustment of the brake shoes—Escort/Tracer

7. Separate the brake shoes by removing the brake shoe retracting springs.
8. Remove the parking brake lever pin retainer and spring washer, then remove the brake shoe adjusting lever from the adapter.

To install:

9. Apply a light coating of caliper slide grease to the brake backing plate brake shoe contact areas as well as the adjuster screw threads.
10. Assemble the adjuster screw and the brake shoe adjusting screw socket and washer. Turn the brake shoe adjusting screw socket all the way down, then back off ½ turn.
11. Install the parking brake lever to the trailing shoe with the spring washer and a new parking brake lever pin retainer. Crimp the retainer to securely retain the parking brake lever.
12. Place the trailing shoe on the brake backing plate and attach the parking brake cable.
13. Install the brake shoe hold-down pin, spring and retainers on the trailing brake shoe.
14. Place the leading shoe on the brake backing plate and attach the lower brake shoe retracting spring between both brake shoes.
15. Install the leading brake shoe hold-down pin, spring and retainers.
16. Install the brake adjuster assembly to the slots between the brake shoes. The brake shoe adjusting screw socket end must fit into the slot in the leading shoe and the adjuster nut end must fit into the slots of the trailing brake shoe and parking brake lever.
17. Install the brake shoe adjusting lever on the pin of the leading shoe and to the slot in the brake shoe adjusting screw socket.
18. Install the upper brake shoe retracting spring in the slot on the trailing shoe and the slot in the brake shoe adjusting lever. Verify that the brake shoe adjusting lever is contacting the starwheel on the adjuster assembly.
19. Adjust the brake shoes to the brake drum.
20. Install the brake drum.
21. Install the wheel and tire assembly. Using a torque wrench, torque the lug nuts in a star pattern to 85–105 ft. lbs. (115–142 Nm).
22. Apply the brakes several times while backing up the vehicle. After each stop, the vehicle must be moved forward.
23. Road test the vehicle and check for proper brake system operation by making several stops from varying forward speeds.

1998–01 GENERAL MOTORS

Brake Caliper

REMOVAL & INSTALLATION

GM C- & H-BODIES:
Buick LeSabre, Park Avenue
Oldsmobile Eighty Eight Royale, Ninety Eight, Regency
Pontiac Bonneville

FRONT & REAR

➡The Bosch 2U ABS system cannot increase brake pressure above master cylinder pressure applied by during braking. There is no need to depressurize the system prior to service.

1. Remove brake fluid from the master cylinder reservoir until the reservoir is approximately ⅓ full.
2. Remove the front wheel. Mark the position of the wheel to the wheel studs, prior to removal, for installation reference.
3. Install 2 lug nuts to retain the rotor once the caliper is removed.
4. Using a large C-clamp, bottom the piston in the caliper bore by positioning the C-clamp on the outboard pad and on the round portion of the brake caliper where the piston is housed.
5. Remove the banjo bolt that fastens the brake hose to the brake caliper. Discard the gaskets.
6. Cap the brake line to avoid excessive fluid loss or fluid contamination.
7. Remove the rubber dust boots from the caliper mounting bolt heads (if equipped).
8. Remove the caliper mounting bolts.
9. Remove the park brake cable from the caliper (if working on the rear).
10. Remove the caliper from the vehicle.
11. Remove the brake pads.

To install:

12. Install the brake pads. Lubricate the slides where the caliper mounts on the steering knuckle with silicone grease.
13. Install the caliper over the rotor.
14. Install the caliper mounting bolts. Torque the mounting bolts to 38 ft. lbs. (51 Nm).
15. Install the park brake cable (if working on the rear).
16. Install the rubber dust boots over the caliper mounting bolt heads (if equipped).
17. Check the clearance between the brake caliper and caliper bracket stops. If the clearance is too tight, check the caliper leading and trailing edges for build up. File down as necessary.
18. Connect the brake hose to the caliper. Install the brake hose banjo bolt, using new gaskets, and torque to 33 ft. lbs. (45 Nm).
19. Remove the lug nuts used to secure the rotor.
20. Install the wheel, aligning the reference marks made during removal, and torque the lug nuts to 100 ft. lbs. (140 Nm).
21. Refill the master cylinder and bleed the brake system using the recommended procedure.
22. Road test the vehicle and check for proper braking performance

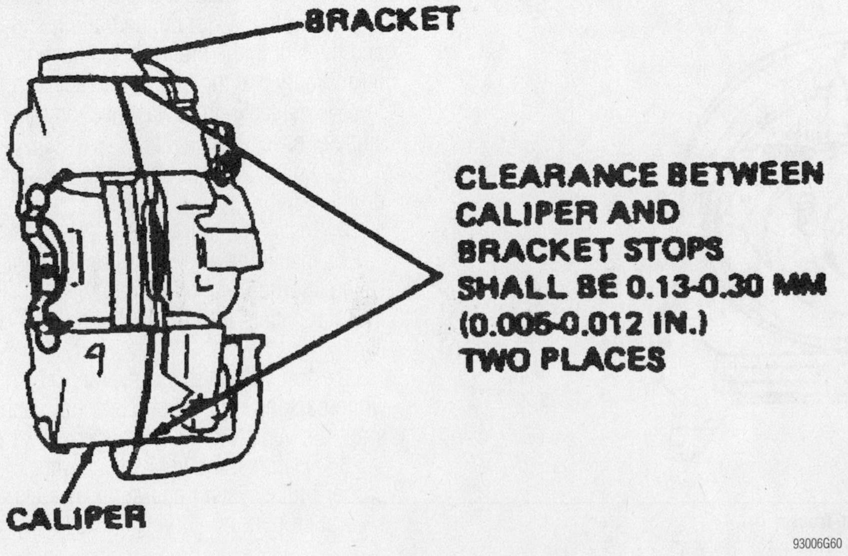

Measuring the caliper clearance—LeSabre, Park Avenue, Eighty Eight Royale, Ninety Eight, Regency, Bonneville

GM E- K- V-BODIES:
Cadillac Catera, deVille, Eldorado, Seville

FRONT

1. Disconnect the negative battery cable.
2. Remove ⅔ of the brake fluid from the master cylinder.
3. Remove the front wheel. Mark the relationship between the wheel and the wheel stud for re-installation purposes.
4. Install 2 wheel nuts to keep the rotor in place.
5. Using a large C-clamp against the inboard pad, compress the caliper piston into the caliper to provide clearance during removal.
6. Place a catch pan under the caliper.
7. Disconnect the brake hose from the caliper. Cap the line to prevent excessive fluid loss or contamination.
8. Remove the caliper mounting bolts and remove the caliper from the vehicle.
9. Inspect the mounting bolts; sleeves and boots for wear and/or damage. Replace parts as necessary.

To install:

10. Before installing the caliper, make sure the piston is fully seated in the bore and the brake pads are properly seated.
11. Lubricate the mounting bolt shafts and inner diameter of the sleeves with silicone grease.
12. Install the caliper in the caliper mounting bracket and install the mounting bolts. Torque the mounting bolts to 38 ft. lbs. (51 Nm).
13. Connect the brake hose with the bolt and new gaskets. Torque the brake hose bolt to 33 ft. lbs. (45 Nm).
14. Refill the master cylinder and bleed the brake system.
15. Remove the 2 wheel nuts securing the rotor.
16. Install the wheel and tire assembly.
17. Connect the negative battery cable.
18. Road test the vehicle for proper brake system operation.

REAR—DEVILLE, ELDORADO, SEVILLE

1. Disconnect the negative battery cable.
2. Remove ⅔ of the brake fluid from the master cylinder.

3. Remove the rear wheel.

4. Install 2 wheel nuts to keep the rotor in place.

5. Place a catch pan under the caliper.

6. Disconnect the brake hose from the caliper. Cap the line to prevent fluid loss or contamination.

7. Loosen the tension on the parking brake at the equalizer.

8. Remove the parking brake cable mounting lever, and remove the cable end by lifting up and disengaging the end.

9. Remove the caliper sleeve bolt.

10. Lift the caliper up and slide the caliper inboard off of the pin sleeve to remove the caliper from the vehicle.

11. Use a suitable tool in the caliper piston slots to turn the piston and thread it into the caliper. After bottoming the piston, lift the inner edge of the boot next to the piston and press out any trapped air the boot must lay flat.

To install:

12. Inspect the pin boot, bolt boot and sleeve boot for cuts, tears or deterioration and replace as necessary.

13. Inspect the bolt sleeve and pin sleeve for corrosion or damage. Pull the boots to gain access to the sleeves for inspection or replacement. Replace corroded or damaged sleeves; do not try to polish away corrosion.

14. If not replaced, remove the pin boot from the caliper and install the small end over the pin sleeve (installed on caliper support) until the boot seats in the pin groove. This prevents cutting the pin boot when sliding the caliper onto the pin sleeve.

15. Hold the caliper in the position as removed and start it over the end of the pin sleeve. As the caliper approaches the pin boot, work the large end of the pin boot in the caliper groove, then push the caliper fully onto the pin.

16. Pivot the caliper down, being careful not to damage the piston boot on the inboard disc brake pad. Compress the sleeve boot by hand as the caliper moves into position to prevent boot damage.

17. After the caliper is in position, recheck the position of the pad clips. If necessary, use a small prybar to reseat or enter the pad clips on the bracket abutments.

18. Install the sleeve bolt and torque to 20 ft. lbs. (27 Nm).

19. Install the parking brake cable bracket, with the cable attached, and torque the bolt to 32 ft. lbs. (43 Nm).

20. Install the parking brake cable onto the parking brake lever and the retaining clip onto the parking brake cable.

21. Connect the brake hose with the bolt and new gaskets and torque the bolt to 32 ft. lbs. (43 Nm).

22. Adjust the parking brake cable.

23. Refill the master cylinder and bleed the brake system.

24. Remove the wheel nuts retaining the rotor and install the wheel.

25. Connect the negative battery cable.

26. Road test the vehicle for proper brake system operation.

REAR—CATERA

1. Remove the wheel(s).
2. Attach a hose to the bleeder screw on the caliper.
3. Open the bleeder screw.
4. Compress the pistons into the caliper housing to provide clearance.
5. Use a punch to drive the caliper retaining pins out of the caliper from the outside inward.
6. Remove or disconnect the following:
   - Brake caliper spring retainer
   - Brake pads
   - Brake caliper pipe
   - Caliper mounting bolts
   - Brake caliper

To install:

7. Install or connect the following:
   - Brake caliper. Torque the bolts to 59 ft. lbs. (80 Nm).
   - Brake caliper pipe. Torque the fitting to 12 ft. lbs. (16 Nm).
   - Brake pads
8. Install one caliper retaining pin.
9. Install the caliper spring retainer.
10. Install the second retaining pin.
11. Fill and bleed the brake system.
12. Install the wheel(s).

GM F-BODY:
Chevrolet Camaro
Pontiac Firebird

1. Remove ⅔ of the brake fluid from the master cylinder.

2. Matchmark the relationship of the wheel and hub, then remove the wheel and tire assembly.

3. If completely removing the caliper from the vehicle for replacement or service, remove the bolt, inlet fitting and 2 gaskets from the caliper housing. Plug the openings in the caliper housing and inlet fitting to prevent system contamination or excessive fluid loss.

4. Remove the circlip and retainer pin.

5. Remove the caliper housing from the rotor and mounting bracket. If the caliper is not being completely removed, it must be suspended from the vehicle using a wire hook in order to prevent damage to the brake lines.

To install:

6. Check the inlet fitting bolt for blockage, clear or replace as necessary.

7. Install the caliper housing over the rotor and onto the mounting bracket. Ensure the guiding surfaces on the inboard and outboard disc brake pads and mounting bracket are seated correctly.

8. Press the caliper housing down to compress the bias springs. Slide a new retainer pin into position and install a new circlip.

9. If removed, install the inlet fitting, bolt and 2 new gaskets. Torque the bolt to 30 ft. lbs. (40 Nm).

10. Fill the master cylinder and bleed the brake system.

11. Align the matchmarks made earlier and install the wheel and tire assembly.

12. Lower the vehicle, then with the engine running pump the brake pedal slowly and firmly 3 times to seat then brake pads.

REAR

1. Matchmark the relationship between the wheel and hub, then remove the wheel and tire assembly. Install 2 wheel nuts to retain the rotor.

2. If completely removing the caliper from the vehicle for replacement or service, remove the brake hose bolt inlet fitting and 2 gaskets from the caliper housing.

3. Plug the openings in the caliper housing and inlet fitting to prevent system contamination or excessive fluid loss.

4. Remove the 2 caliper guide pin bolts.

5. Remove the caliper housing from the rotor and mounting bracket. If the caliper is not being completely removed, it must be suspended from the vehicle using a wire hook in order to prevent damage to the brake lines.

To install:

6. Inspect the guide pins and boots and replace if corroded, worn or damaged. Check the inlet fitting bolt for blockage, clear or replace as necessary.

7. Install the caliper housing over the rotor and into the mounting bracket.

8. Install 2 caliper guide pin bolts start-

ing with the upper pin and torque the bolts to 27 ft. lbs. (37 Nm).

9. Attach the brake hose to the caliper using 2 new copper gaskets. Torque the bolt to 22 ft. lbs. (30 Nm).

10. If the hose fitting was removed, bleed the entire brake system.

11. Lower the vehicle sufficiently and cycle the parking brake.

12. Raise and safely support the vehicle.

13. Inspect the caliper parking brake levers and ensure they are against the stops on the caliper housing. If the levers are not on their stops, check the parking brake adjustment.

14. Remove the 2 nuts securing the rotor, then align the matchmarks made earlier and install the wheel assembly.

15. Lower the vehicle, then with the engine running pump the brake pedal slowly and firmly 3 times to seat then brake pads.

16. Check the hydraulic system for leaks.

GM G-BODY:
Buick Riviera
Oldsmobile Aurora

FRONT

1. Siphon ⅔ of the brake fluid out of the master cylinder reservoir.

2. Remove the tire and wheel assembly.

3. Install 2 wheel nuts loosely to secure the rotor when the caliper is removed.

4. Remove the bolts securing the brake hose to the caliper and disconnect the brake hose from the caliper.

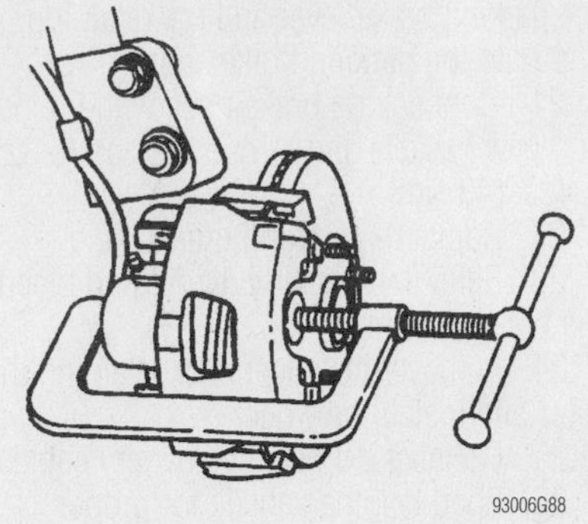

93006G88

Compressing the front caliper piston—Riviera, Aurora

5. Plug the hose to prevent excessive fluid loss and possible fluid contamination.

6. Compress the piston into the caliper bore to provide clearance for removal.

7. Remove the caliper mounting bolts.

8. Remove the caliper from the anchor bracket.

9. If the caliper is being replaced remove the brake pads from the caliper or anchor bracket.

To install:

10. Seat the caliper piston fully in its bore.

11. Install the brake pads in the caliper or anchor bracket.

12. Install the caliper on the anchor bracket and install the mounting bolts.

13. Torque the caliper mounting bolts to 38 ft. lbs. (51 Nm). Torque the mounting bolts to 63 ft. lbs. (85 Nm).

14. Reconnect the brake hose to the caliper using new washers. Torque the fitting bolt to 33 ft. lbs. (45 Nm).

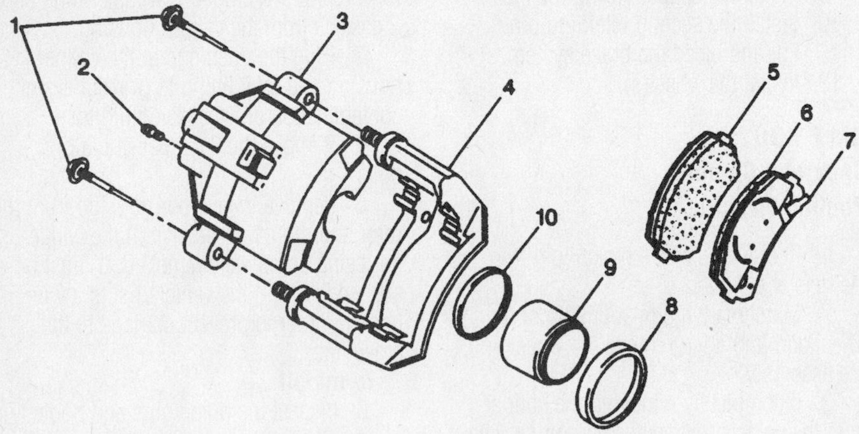

(1) Pin Bolts
(2) Outboard Pad
(3) Caliper Boot
(4) Inboard Pad
(5) Piston Seal
(6) Piston
(7) Anchor bracket
(8) Bleeder Valve
(9) Caliper Housing
(10) Caliper Anchor Bracket

93006G87

Front caliper and brake pad components—Riviera, Aurora

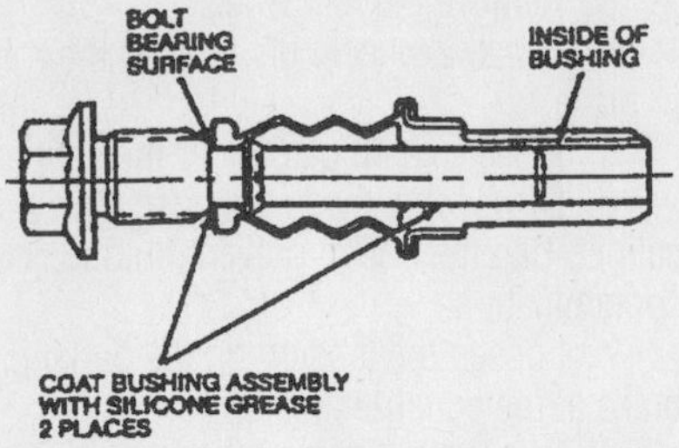

93006G89

Caliper mounting bolt and sleeve lubrication points—Riviera, Aurora

15. Remove the 2 lug nuts securing the brake rotor.

16. Install the wheel and tire assembly and torque the wheel nuts to 100 ft. lbs. (140 Nm).

17. Refill the master cylinder with fluid and bleed the brake system.

18. Pump the brake pedal several times to seat the brake pads against the rotor. Verify no hydraulic leaks and a good firm brake pedal.

REAR

1. Siphon ⅔ of the brake fluid out of the master cylinder reservoir.

2. Remove the tire and wheel assembly.

3. Install 2 wheel nuts loosely to secure the rotor when the caliper is removed.

4. Remove the bolt securing the brake hose to the caliper and disconnect the brake hose from the caliper.

5. Plug the hose to prevent excessive fluid loss and possible fluid contamination.

6. Remove the park brake cable

7. Remove upper and lower caliper bolts and lift caliper from mounting bracket.

To install:

8. Seat the caliper piston fully in its bore.

9. Make sure the notches in the caliper piston are at 6 and 12 o'clock.

10. Mount caliper onto anchor bracket. Install the 2 mounting bolts and torque bolts to 63 ft. lbs. (85 Nm)

11. Reconnect the brake hose to the caliper using new copper washers. Torque the mounting bolt to 33 ft. lbs. (45 Nm).

12. Install the park brake cable.

13. Remove the wheel nuts securing the rotor.

14. Reinstall the tire and wheel assembly and torque the wheel nuts to 100 ft. lbs. (140 Nm).

15. Refill the master cylinder with fluid and bleed the brake system.

16. Pump the brake pedal several times to seat the brake pads against the rotor. Verify no hydraulic leaks and a good firm brake pedal.

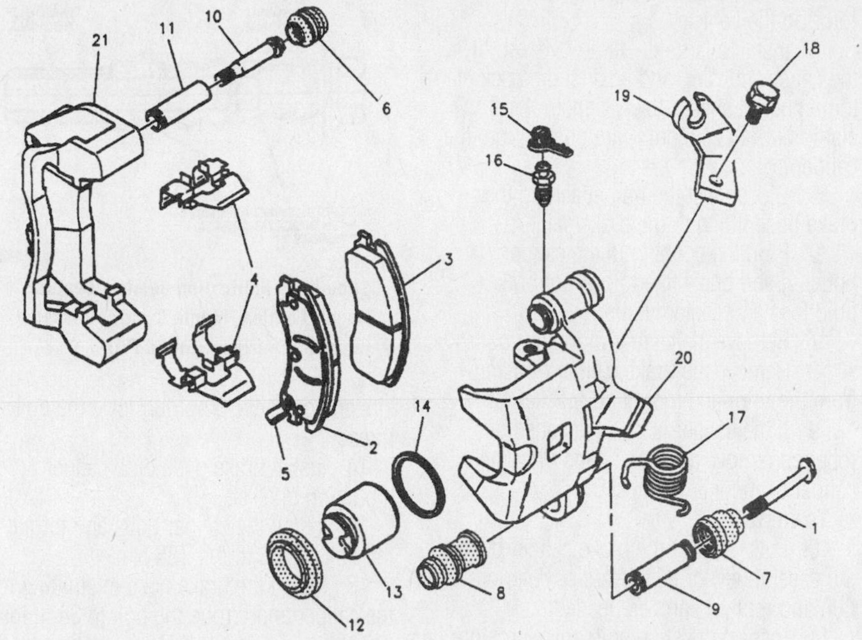

1 SLEEVE BOLT
2 OUTBOARD SHOE & LINING
3 INBOARD SHOE & LINING
4 PAD CLIP
5 WEAR SENSOR
6 PIN BOOT
7 BOLT BOOT
8 SLEEVE BOLT
9 BOLT SLEEVE
10 PIN BOLT
11 PIN SLEEVE
12 PISTON BOOT
13 PISTON ASSEMBLY
14 PISTON SEAL
15 BLEEDER VALVE CAP
16 BLEEDER VALVE
17 LEVER RETURN SPRING
18 BOLT AND WASHER
19 CABLE SUPPORT BRACKET
20 CALIPER BODY ASSEMBLY
21 CALIPER SUPPORT

93006G90

Rear caliper and brake pad components—Riviera, Aurora

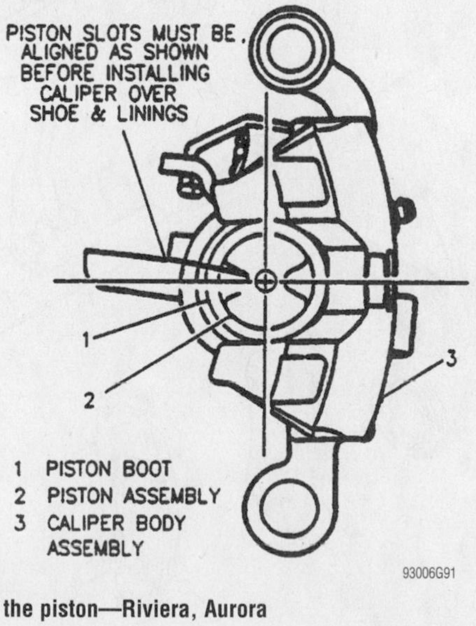

93006G91

Aligning the notches in the piston—Riviera, Aurora

GM J-BODY:
Chevrolet Cavalier
Pontiac Sunfire

1. Siphon ⅔ of the brake fluid out of the master cylinder.
2. Remove the tire and wheel assembly.
3. Compress the caliper piston back into the caliper bore using a large pair of pliers, C-clamp or special piston retracting tool.
4. If the caliper is to be completely removed from the vehicle for bench service, disconnect and cap the brake line from the caliper. Discard the old washers.
5. Remove the caliper mounting bolts and sleeves.
6. Remove the caliper from the knuckle.
7. Remove the brake pads from the caliper, if caliper is being replaced.

To install:

8. Install the brake pads in the caliper.
9. Install the caliper on the steering knuckle.
10. Install the mounting bolts and sleeves and torque to 40 ft. lbs. (51 Nm).
11. If the caliper brake line was disconnected, uncap and connect the brake hose to the caliper using new copper washers. Torque the mounting bolt to 35 ft. lbs. (44 Nm).
12. Refill the master cylinder and bleed the brake system.
13. Install the tire and wheel assembly and tighten to specification.
14. Verify correct brake operation.

GM L/N and N-BODY:
Buick Skylark
Chevrolet Malibu
Oldsmobile Achieva, Cutlass
Pontiac Grand Am

1. Siphon ⅔ of the brake fluid out of the master cylinder.
2. Remove the tire and wheel assembly.
3. Compress the caliper piston back into the caliper bore using a large pair of pliers, C-clamp or special piston retracting tool.
4. Remove the brake hose from the caliper and discard the copper washers.
5. Plug the hose to prevent excessive fluid loss and possible fluid contamination.
6. Remove the caliper mounting bolts and remove the caliper from the knuckle.
7. Remove the brake pads from the caliper, if the caliper is being replaced.

To install:

8. Inspect the condition of the caliper support for rust and corrosion that will hinder the travel of the caliper.

9. Inspect the caliper mounting hardware. New bolts are usually recommended.
10. Lubricate the mounting bushings and sleeves with silicone grease as required.
11. Install the brake pads in the caliper.
12. Install the caliper on the steering knuckle.
13. Install the mounting bolts and torque to 40 ft. lbs. (51 Nm).
14. Connect the brake hose to the caliper using new copper washers and torque the mounting bolt to 35 ft. lbs. (44 Nm).
15. Refill the master cylinder and bleed the brake system.
16. Install the tire and wheel assembly.
17. Verify correct brake operation.

GM W-BODY:
Buick Century, Regal
Chevrolet Lumina, Monte Carlo
Oldsmobile Cutlass Supreme, Intrigue
Pontiac Grand Prix

FRONT

1. Remove ⅔ of the brake fluid from the master cylinder assembly.
2. Remove the tire and wheel assembly. Mark a relationship between the wheel and the wheel stud for re-installation purposes.
3. Install 2 wheel nuts to retain the rotor on the vehicle.
4. Install a large C-clamp over top of the caliper housing and against the back of the outboard shoe. Slowly tighten the C-clamp until the piston(s) are pushed into the caliper bore.
5. Disconnect the bolt attaching the brake hose fitting to the brake caliper.
6. Plug the opening in the caliper housing and brake line to prevent brake fluid loss and contamination.
7. Remove the caliper mounting bolts.
8. Remove the brake caliper housing from the rotor and mounting bracket.
9. If the caliper is to be replaced or repaired remove the brake pads from the caliper or mounting bracket.

To install:

10. Fully inspect the brake caliper bushing assemblies for cuts, tears or deterioration and replace parts as needed.
11. Inspect the slide bolts for corrosion. If corrosion is found, replace the slide bolts and bushings before installing the brake caliper assembly.
12. Before installing the caliper, make sure the piston(s) are seated in the bore, and that the brake pads are correctly installed.
13. Lubricate the caliper slide bolts with silicone grease. Do not lubricate the bolt threads.

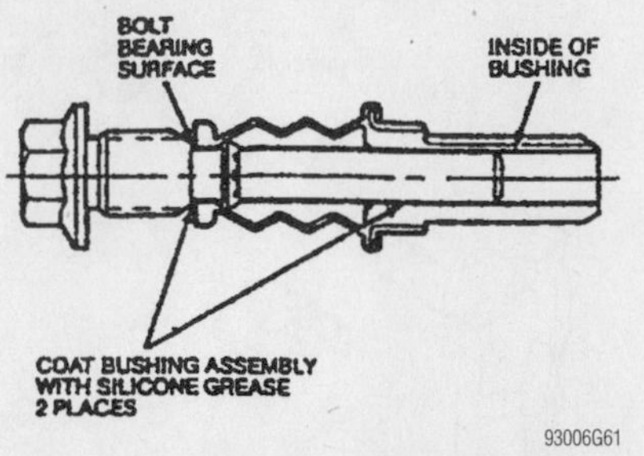

Slide bolts lubrication points—Century, Regal, Lumina, Monte Carlo, Cutlass Supreme, Intrigue, Grand Prix

14. Install brake pads onto caliper or mounting bracket.
15. Install the caliper bolts and torque the bolts to 63 ft. lbs. (85 Nm).
16. Install the brake hose inlet fitting to the caliper and torque the bolt to 24 ft. lbs. (32 Nm).
17. Remove wheel nuts securing the brake rotor to the hub.
18. Install the tire and wheel assembly.
19. Fill the master cylinder to the proper level.
20. Bleed the brake system.
21. Verify correct brake operation.

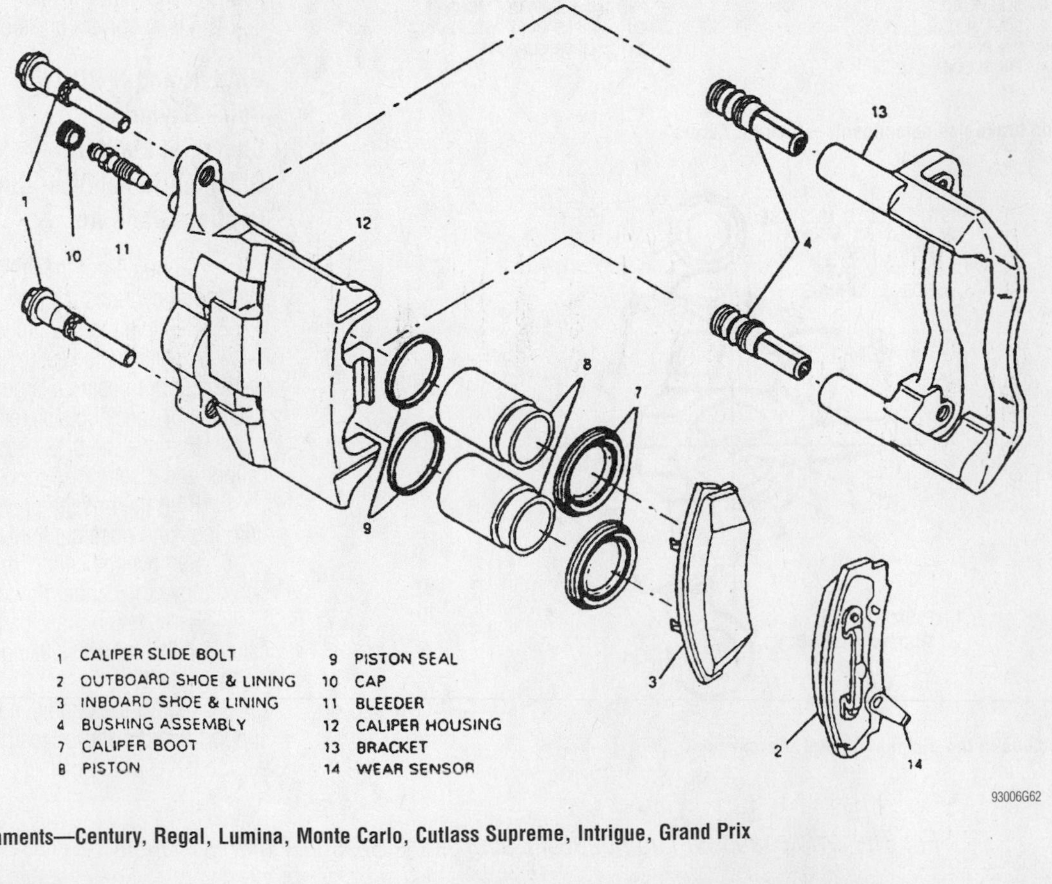

Caliper attachments—Century, Regal, Lumina, Monte Carlo, Cutlass Supreme, Intrigue, Grand Prix

REAR

1. Remove ⅔ of the brake fluid from the master cylinder assembly.
2. Remove the tire and wheel assembly. Mark a relationship between the wheel and the wheel stud for re-installation purposes.
3. Install 2 wheel nuts to retain the rotor.
4. Remove the brake hose from the caliper and discard the copper washers.
5. Plug the openings in the caliper and the brake hose to prevent brake fluid loss and contamination.
6. The following steps are for models with the park brake cable attached to the caliper:
    a. Disconnect the parking brake cable from the parking brake lever on the caliper. Lift up one end of the cable spring clip free end of the cable from the lever.
    b. Remove the bolt and washer attaching the cable support bracket to the caliper body assembly.
    c. Remove the caliper sleeve bolts.
7. For all others, remove the two caliper mounting bolts.
8. Remove the caliper body assembly from the vehicle. Pivot the caliper assembly up to clear the rotor and then slide it inboard off the pin sleeve.

To install:

9. Inspect the caliper bolt boots, pins and sleeve bolt for cuts, tears or deterioration. Replace as necessary.
10. Lubricate the mounting surfaces and the mounting sleeves.
11. Hold the caliper body assembly in the position from which it was removed, and start it over the end of the pin sleeve.
12. As the caliper body assembly approaches the pin boot, work the large end of the pin boot in the caliper body groove. Push the caliper body fully onto the pin.
13. Pivot the caliper body assembly down, using care not to damage the piston boot on the inboard shoe. Compress the sleeve boot by hand as the caliper body moves into position to prevent boot damage.
14. After installing the caliper assembly into position, recheck the installation of the pad clips. If necessary, use a small prying tool to reset or center the pad clips.
15. Install the brake caliper sleeve bolts and torque to 20 ft. lbs. (27 Nm).
16. On models without caliper mounted park brake cable, torque caliper mounting bolts to 32 ft. lbs. (43 Nm)
17. To install the park brake cable, the following steps apply (some models).
    a. Install the cable support bracket with the cable attached. Torque bolt to 32 ft. lbs. (43 Nm).
    b. Lift up on the end of the cable spring clip and work the end of the parking brake cable into the notch of the parking brake lever.
18. Connect the brake hose to the brake caliper and torque the bolt to 32 ft. lbs. (44 Nm).
19. Remove the wheel nuts securing the rotor to the hub and bearing assembly.
20. Install the tire and wheel assembly.
21. Fill the master cylinder to the proper level with clean brake fluid.
22. Bleed the brake system using the recommended procedure.
23. Apply approximately 175 lbs. (79 kg) of force, 3 times, to properly seat the brake shoe and linings against the rotor.
24. Adjust the parking brake cable as necessary (some models).

GM Y-BODY:
Chevrolet Corvette

FRONT

1. Disconnect the negative battery cable and remove ⅔ of the brake fluid from the master cylinder reservoir.

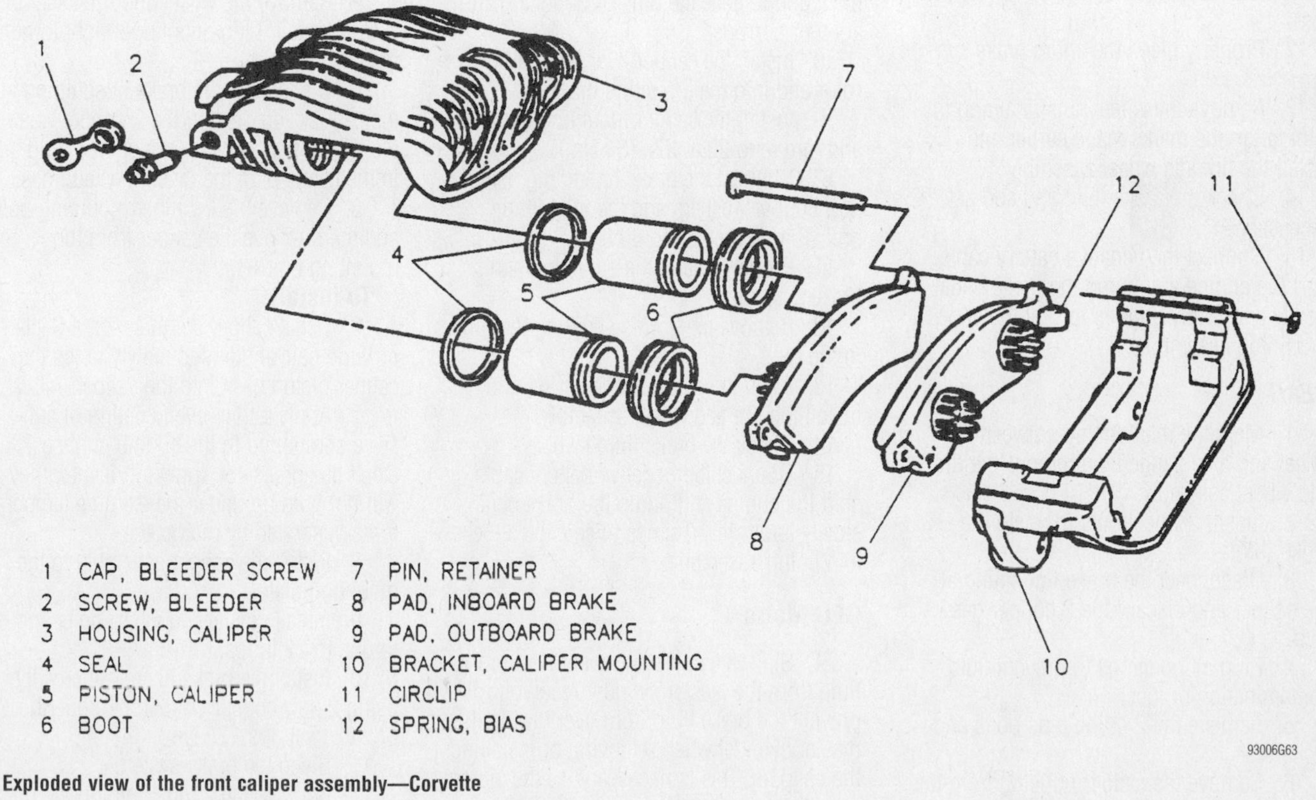

Exploded view of the front caliper assembly—Corvette

2. Mark the relationship between the wheel and axle flange, then remove the tire and wheel assembly.

3. Install 2 wheel nuts to retain the brake rotor.

4. Depress the caliper pistons into the caliper bores in order to provide clearance between the pads and the rotor.

5. Disconnect the brake hose fitting at the caliper by removing the bolt. Discard the 2 copper washers.

6. Plug all openings to prevent fluid contamination or loss.

➡Do not allow the fluid to come into contact with the front transverse spring, as damage to the spring may occur.

7. For Corvette, remove the 2 caliper mounting bolts.

8. Remove the caliper housing from the rotor and the caliper mounting bracket.

To install:

9. Install the caliper over the brake rotor and into the caliper mounting bracket. Make sure the shoe lining guiding surfaces are correctly seated in the bracket.

10. Compress the bias springs by applying pressure to the mounting bracket, then install the new retainer pin.

11. Connect the brake hose inlet fitting using 2 new copper washers and the inlet fitting bolt. Torque the bolt to 30 ft. lbs. (40 Nm).

12. Properly bleed the entire brake system.

13. Remove the wheel nuts retaining the rotor, align the marks made earlier and install the tire and wheel assembly.

14. Check the brake fluid and add as necessary.

15. Connect the negative battery cable, start the engine and pump the brake pedal slowly and firmly 3 times to seat the shoe and lining assemblies.

REAR

1. Mark the relationship between the wheel and axle flange then remove the tire and wheel assembly.

2. Install 2 wheel nuts to retain the brake rotor.

3. Disconnect the brake line fitting at the caliper and discard the 2 copper gaskets.

4. Plug all openings to prevent fluid contamination or loss.

5. Remove the 2 guide pins bolts and discard.

6. Remove the caliper housing from the brake rotor and caliper mounting bracket.

To install:

7. Inspect the guide pins for free movement and replace the pins or boots if damaged or corroded.

8. Install the caliper over the brake rotor and into the mounting bracket.

9. Install the caliper mounting bolts and torque to 23 ft. lbs. (31 Nm).

10. Connect the brake line fitting using 2 new copper washers and the inlet fitting bolt. Torque the bolt to 30 ft. lbs. (40 Nm).

11. Properly bleed the entire hydraulic brake system.

12. Remove the 2 nuts securing the rotor to the hub.

13. Align the marks made earlier and install the tire and wheel assembly.

14. Check the brake fluid level.

15. Connect the negative battery cable, start the engine and pump the brake pedal slowly and firmly 3 times to seat the shoe and lining assemblies.

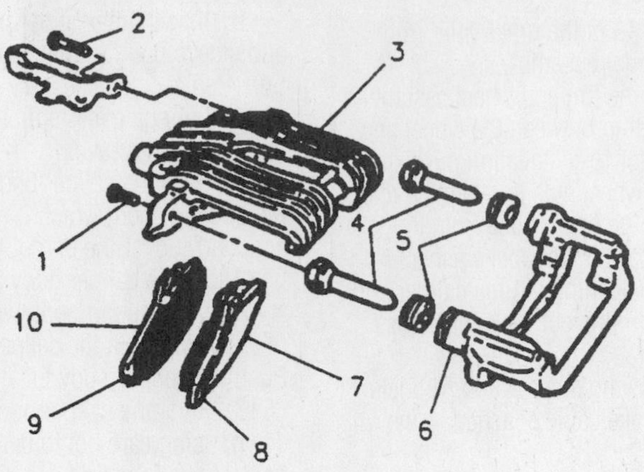

1 BOLT, UPPER GUIDE PIN
2 BOLT, LOWER GUIDE PIN
3 HOUSING, CALIPER
4 PIN, GUIDE
5 BOOT
6 BRACKET, MOUNTING
7 INSULATOR
8 PAD, OUTBOARD BRAKE
9 SENSOR, WEAR
10 PAD, INBOARD BRAKE

93006G64

Exploded view of the rear caliper assembly—Corvette

GEO Metro

1. Siphon a sufficient quantity of brake fluid from the master cylinder reservoir to prevent the brake fluid from overflowing the master cylinder when removing or installing the calipers. This is necessary, as the piston must be forced into the cylinder bore to provide sufficient clearance to install the caliper.

2. Remove the wheel and tire assembly.

3. Install 2 lug nuts finger tight to hold the disc in place.

4. Disconnect the brake hose fitting at the caliper and discard the 2 copper washers. Use a pan to catch any spilled fluid and immediately plug the disconnected hose.

5. Remove the 2 caliper mounting bolts and then remove the caliper from the mounting bracket.

To install:

6. Use a caliper compressor, a C-clamp or large pair of pliers to slowly press the caliper piston back into the caliper.

7. Apply a thin, even coating of anti-seize compound to the sliding surfaces. Don't use grease or spray lubricants; they will not hold up under the extreme temperatures generated by the brakes.

8. Install the caliper assembly to the mounting plate.

9. Install caliper mounting bolts and torque to 22 ft. lbs. (30 Nm).

10. Install the brake hose to the caliper using 2 new copper washers. Torque the bolt to 17 ft. lbs. (23 Nm).

11. Bleed the brake system.

12. Remove the 2 lugs holding the disc in place and install the wheel and tire assembly.

13. Check the level of the brake fluid in the master cylinder reservoir; it should be at least to the middle of the reservoir.

GEO Prizm

FRONT

1. Siphon a sufficient quantity of brake fluid from the master cylinder reservoir to prevent the brake fluid from overflowing the master cylinder when removing or installing the calipers. This is necessary, as the piston must be forced into the cylinder bore to provide sufficient clearance to install the caliper.
2. Remove the wheel and tire assembly.
3. Install 2 lug nuts finger tight to hold the brake rotor in place.
4. Disconnect the brake hose at the caliper and discard the 2 copper washers. Use a pan to catch any spilled fluid and immediately plug the disconnected hose.
5. Remove the 2 caliper mounting bolts and then remove the caliper from the mounting bracket.

To install:

6. Use a caliper compressor, a C-clamp or large pair of pliers to slowly press the caliper piston back into the caliper.
7. Apply a thin, even coating of anti-seize compound to the sliding surfaces. Don't use grease or spray lubricants; they will not hold up under the extreme temperatures generated by the brakes.
8. Install the caliper assembly to the mounting plate.
9. Install the caliper mounting bolts and torque to 25 ft. lbs. (34 Nm).
10. Install the brake hose to the caliper using 2 new copper washers. Torque the brake hose bolt to 22 ft. lbs. (30 Nm).
11. Bleed the brake system.
12. Remove the 2 lugs holding the disc in place and install the wheel.
13. Check the level of the brake fluid in the master cylinder reservoir; it should be at least to the middle of the reservoir.

Saturn

FRONT

1. Remove the front wheel and tire assembly.
2. Disconnect the brake hose from the caliper and discard the 2 copper washers.
3. Plug the openings to prevent system contamination or excessive fluid loss.
4. Remove the lock pin and guide pin from the caliper.
5. Remove the caliper from the support, being careful not to damage the pin boots.
6. Remove the pin boots from the caliper support and inspect for damage.

To install:

7. If necessary, bottom the caliper piston by hand, or by using a C-clamp.
8. If removed, install the brake pads and clips to the caliper support.
9. Lubricate the pin boots and guide pins with silicone grease.
10. Install the pin boots into the caliper support, using the pin to assure that the boot passes all the way through the support.
11. Position the caliper onto the support and over the brake pads.
12. Lubricate the non-threaded portion of the guide and lock pins with silicone grease.
13. Install the pins through the caliper and torque to 27 ft. lbs. (36 Nm).

➡Make sure the brake line is properly routed with loop to the rear and that the hose is not twisted.

14. Install the brake hose using 2 new copper washers. Torque the fitting bolt to 36 ft. lbs. (49 Nm).
15. Properly bleed the hydraulic brake system.
16. Install the wheel and tire assembly.

REAR

1. Remove the rear wheel and tire assembly.
2. Disconnect the brake hose from the caliper and discard the 2 copper washers.
3. Plug the openings to prevent system contamination or excessive fluid loss.
4. Slip the end of the parking cable off the parking brake lever.
5. Remove the cable outer housing from the cable bracket with SA9151BR cable release tool.
6. Remove the lock pin and guide pin.
7. Remove the caliper from the support, being careful not to damage the pin boots.
8. Remove the pin boots from the caliper support for inspection and lubrication.

To install:

9. Make sure the piston is bottomed in the bore. Do not compress the piston using a C-clamp; instead the piston must be rotated into the caliper on its threads using a piston driver tool.
10. If removed, install the brake pads and clips to the caliper support.
11. Lubricate the pin boots and guide pins with silicone grease.
12. Install the pin boots into the caliper support, using the pin to assure that the boot passes all the way through the support.
13. Position the caliper onto the caliper support.
14. Lubricate the non-threaded portion of the guide and lock pins. Install the pins and torque to 27 ft. lbs. (36 Nm).
15. Install the brake hose using new copper washers and torque the fitting bolt to 36 ft. lbs. (49 Nm).

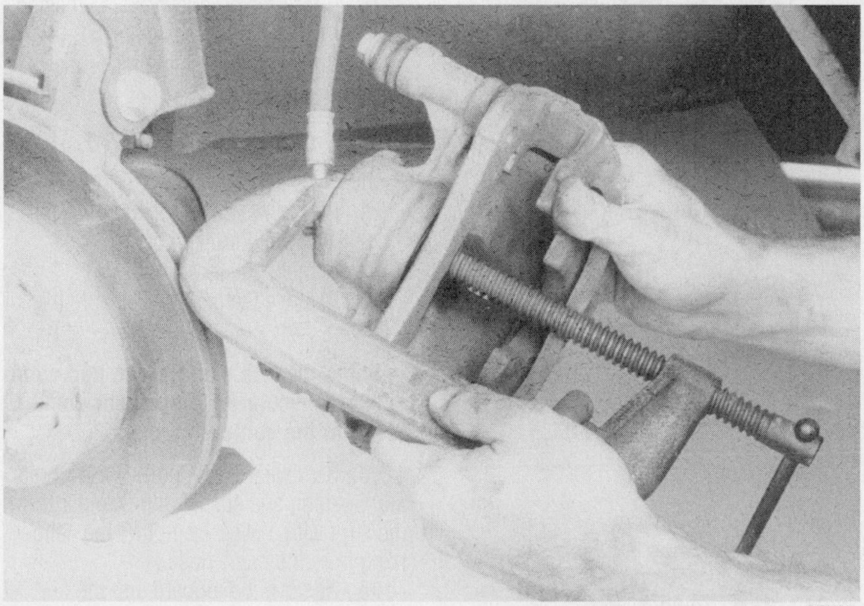

84229044

Using a large C-clamp to compress the piston back into the caliper bore

16. Connect the parking brake cable.
17. Properly bleed the hydraulic brake system.
18. Install the wheel and tire assembly.

Disc Brake Pads

REMOVAL & INSTALLATION

GM C- and H-BODIES:
Buick LeSabre, Park Avenue
Oldsmobile Eighty Eight Royale, Ninety Eight, Regency
Pontiac Bonneville

1. Remove brake fluid from the master cylinder reservoir until the reservoir is approximately ⅓ full. Discard the removed fluid.
2. Remove the front wheel. Mark the position of the wheel to the wheel studs, prior to removal, for installation reference.
3. Install 2 lug nuts to retain the rotor once the caliper is removed.
4. Remove the caliper mounting sleeve bolts. Support the caliper out of the way using wire. DO NOT disconnect the brake hose or allow the caliper to hang from the brake hose.
5. Remove the outboard pad by pushing it in toward the piston until the mounting tabs clear the holes in the caliper body. With the tabs clear of the holes, push the pad out the bottom of the caliper.
6. Remove the inboard pad from the piston by pulling the top of the pad out to disengage the retainer spring.

To install:

7. Before installing the pads in the caliper, the piston must be fully seated in the bore. A large C-clamp can be used to compress the piston.
8. Install the inboard pad in the caliper by inserting the top pad ears in first, then sliding the bottom of the pad into place until the spring clip snaps into place. Make sure the inboard pad seats flush against the caliper piston.
9. Install the outboard pad by lining up the tabs on the rear of the pad with the mounting holes in the caliper body. Press the pad firmly down into the caliper until the tabs snap into the mounting holes.
10. Clean and lubricate the caliper bolt and sleeve assemblies and install them into the caliper. Lubricate the caliper slides and mountings.
11. Install the caliper over the rotor and torque the mounting bolts to 38 ft. lbs. (51 Nm).
12. Remove the 2 lug nuts used to secure the rotor in place.
13. Install the wheel, aligning the marks made during removal, and torque the lug nuts to 100 ft. lbs. (140 Nm).
14. Pump the brake pedal several times to seat the pads against the rotor.
15. Refill the master cylinder using DOT 3 brake fluid only and road test to verify proper brake operation.

GM E-, K- and V-BODIES:
Cadillac Catera, deVille, Eldorado, Seville

FRONT

1. Remove ⅔ of the brake fluid from the master cylinder reservoir.
2. Remove the front wheel.
3. Remove the caliper mounting bolts.
4. Remove the caliper from the steering knuckle without disconnecting the brake hose.
5. Suspend the caliper from the coil spring with wire. Do not let the caliper hang from the brake hose.

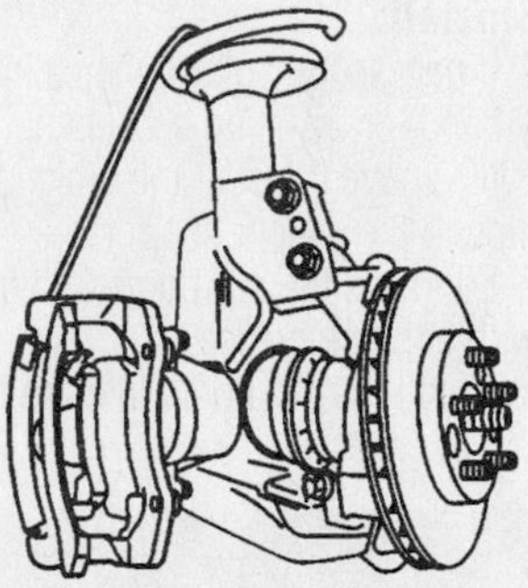

93006G67

Supporting the front caliper—Catera, deVille, Eldorado, Seville

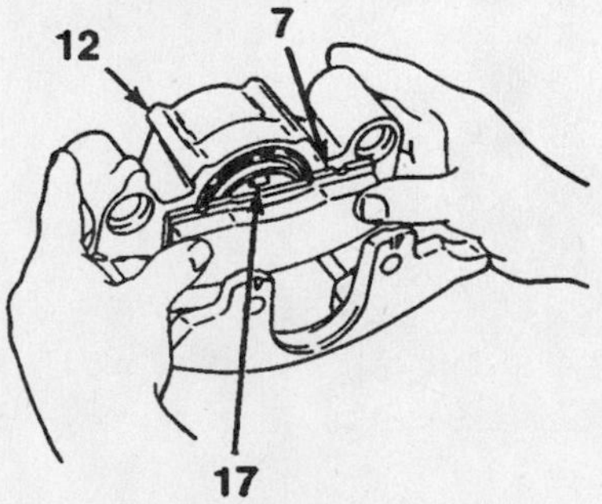

7. Inboard Shoe & Lining
12. Caliper Housing
17. Shoe Retainer Spring

93006G68

Installing the inboard disc brake pad into the front caliper—Catera, deVille, Eldorado, Seville

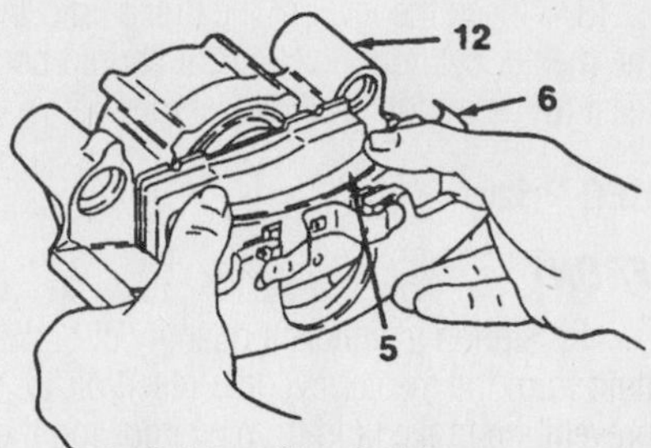

5. Outboard Shoe & Lining
6. Wear Sensor
12. Caliper Housing

93006G69

Installing the outboard disc brake pad in the front caliper—Catera, deVille, Eldorado, Seville

6. Remove brake pads from anchor bracket.

To install:

7. Fully seat the caliper piston into the bore using a large C-clamp.
8. Lubricate the brake caliper mounting surfaces.
9. Install the brake pads onto the anchor bracket using new shims and clips.
10. Place the caliper onto the anchor bracket and install the mounting bolts.
11. Torque the mounting bolts to 63 ft. lbs. (85 Nm).
12. Install the wheel and tire assembly and torque to specification.
13. Lower the vehicle.
14. Pump the brake pedal several times to seat the brake pads.
15. Check the fluid level in the master cylinder and fill as necessary.
16. Road test the vehicle for proper brake operation.

REAR—DEVILLE, ELDORADO, SEVILLE

1. Remove ⅔ of the brake fluid from the master cylinder reservoir.
2. Remove the rear wheel.
3. Remove the park brake cable from the caliper.

➡Some models use a brake pad wear sensor. Disconnect the sensor connector from the vehicle harness.

4. Remove the caliper mounting bolts and position and suspend the caliper from the strut with a wire. Do not let the caliper hang from the brake hose.
5. Remove the inboard and the outboard brake pads from the caliper mounting bracket.

To install:

6. Lubricate mounting surfaces and install new brake pad clips.

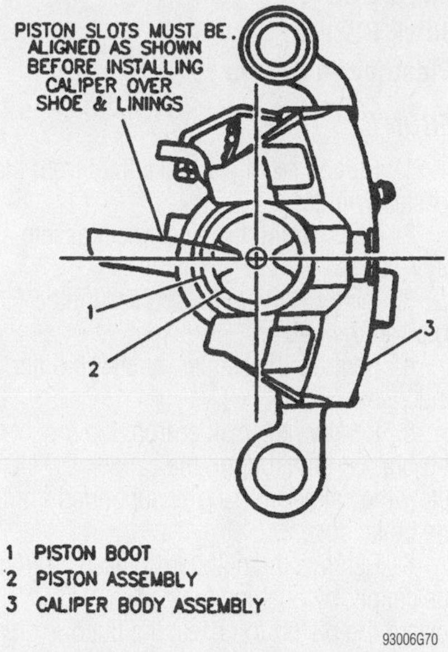

Aligning the slots on the rear caliper—Catera, deVille, Eldorado, Seville

7. The caliper piston must be fully seated in the bore. A spanner type tool can be used to bottom the piston into its bore by turning it in. Make sure the slots in the piston are straight across from each other so the notches on the brake pad will seat.

8. Install the inboard pad by inserting the pad into the straight tabs on the retainer, then pressing down and snapping the pad under the S-shaped tabs. The pad should lay flat against the rotor. Make sure the D-shaped notches are in line with the buttons on the back of the pad lining. If they are not in alignment, rotate the piston until the D-shaped notches face the caliper mounting bolt holes.

9. Install the outer pad into the brake pad clips. Make sure the wear indicator is at the leading edge of the pad during forward wheel rotation.

10. Install the caliper on the mounting bracket and install the bolts. Torque the mounting bolts to 63 ft. lbs. (85 Nm).

11. Install the parking brake cable onto the caliper.

12. Install the wheel and tire assembly and torque to specification.

13. Fill the master cylinder reservoir to the FULL mark and pump the brake pedal several times to seat the brake pads.

14. Check the fluid level in the master cylinder again and fill as necessary.

15. Road test the vehicle for proper brake operation.

REAR—CATERA

1. Remove the wheel(s).
2. Attach a hose to the bleeder screw on the caliper.
3. Open the bleeder screw.
4. Compress the pistons into the caliper housing to provide clearance.
5. Use a punch to drive the caliper retaining pins out of the caliper from the outside inward.
6. Remove or disconnect the following:
   - Brake caliper spring retainer
   - Brake pads

To install:

7. Install the brake pads.
8. Install one caliper retaining pin.
9. Install the caliper spring retainer.
10. Install the second retaining pin.
11. Fill and bleed the brake system.
12. Install the wheel(s).

GM F-BODY: Chevrolet Camaro Pontiac Firebird

FRONT

1. Remove the caliper assembly from the rotor and mounting bracket without disconnecting the brake line, then position aside. Do not allow the caliper to hang by the brake hose, suspend it with a length of wire or a fabricated hook.
2. Remove the outer pad assembly using a small prytool, if necessary, to disengage the show buttons from the caliper assembly.
3. Remove the inner pad.

To install:

4. If not done already, bottom the piston into the caliper using a large C-clamp positioned on the housing and inside the piston well.
5. With the piston bottomed in the caliper bore, lift the inner edge of the boot next to the piston and press out any trapped air. Make sure the boot is flat.
6. Install the inner pad assembly by positioning the retainer spring into the piston. The pad must lay flat against the piston and the boot must not touch the pad. If the boot and pad are in contact, remove the pad and reposition the boot.
7. Install the outer pad assembly, making sure the back of the pad is flat against the caliper. The wear sensor should be at the trailing lower edge of the outer pad during forward wheel rotation or the outer pad has been installed on the wrong side.
8. Install the caliper assembly to the rotor and mounting bracket.

REAR

1. Remove ⅔ of the brake fluid from the master cylinder reservoir.
2. Matchmark the relationship of the wheel to the axle flange, then remove the wheel and tire assembly. Install 2 wheel nuts to retain the rotor.
3. Position a C-clamp and tighten until the piston bottoms in the base of the caliper housing. Make sure 1 end of the C-clamp rests on the inlet fitting bolt and the other against the outboard disc brake pad.

➡It is not necessary to remove the parking brake caliper lever return spring to replace the disc brake pads.

4. Remove the upper caliper guide pin bolt and discard.
5. Rotate the caliper housing on the lower caliper mounting bolt. Be careful not to strain the hose or cable conduit. It may be necessary to loosen the lower caliper guide pin slightly.
6. Remove the disc brake pads and discard the shim.

To install:

7. Clean all residue from the pad guide surfaces on the mounting bracket and caliper housing. Inspect the guide pins for free movement in the mounting bracket. Replace the guide pins or boots, if they are corroded or damaged.
8. Install a new shim, then install the disc brake pads. The outboard pad with insulator is installed toward the caliper housing. The inboard pad with the wear sensor is installed nearest the caliper piston. The wear sensor must be on the leading edge with forward wheel rotation or the pad has been installed on the wrong side.
9. Rotate the caliper housing into its operating position. The springs on the outboard brake pad must not stick through the inspection hole in the caliper housing. If the springs are sticking through the inspection hole in the caliper housing, lift the caliper housing and make the necessary corrections to the outboard brake pad positions.
10. Install a new upper caliper guide pin bolt and torque to 27 ft. lbs. (37 Nm).

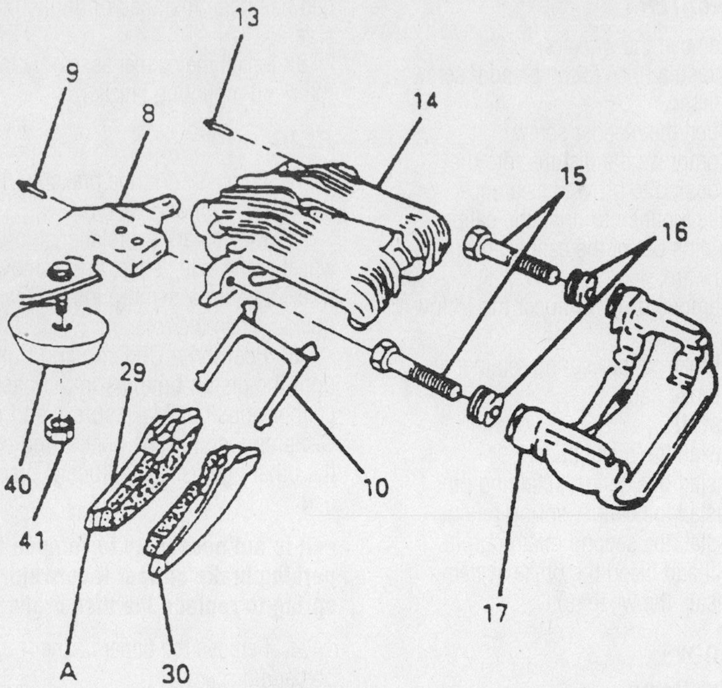

A WEAR SENSOR
8 BRACKET, PARKING BRAKE CABLE
9 BOLT/SCREW, LOWER CALIPER GUIDE PIN
10 SHIM
13 BOLT/SCREW, UPPER CALIPER GUIDE PIN
14 HOUSING, REAR BRAKE CALIPER
15 PIN, REAR BRAKE CALIPER GUIDE
16 BOOT, REAR BRAKE CALIPER GUIDE PIN
17 BRACKET, REAR BRAKE CALIPER ANCHOR
29 PAD, REAR DISC BRAKE INNER
30 PAD, REAR DISC BRAKE OUTER
40 DAMPENER, REAR BRAKE VIBRATION
41 NUT, REAR BRAKE VIBRATION DAMPENER

93006G92

Exploded view of the rear caliper assembly—Camaro, Firebird

Ensure that the lower caliper guide bolt is torqued to 27 ft. lbs. (37 Nm).

11. With the engine running, pump the brake pedal slowly and firmly to seat the brake pads.
12. Check the caliper parking brake levers to make sure they are against the stops on the caliper housing. If the levers are not on their stops, check the parking brake adjustment.
13. Remove the 2 wheel nuts from the rotor, then align the marks and install the wheel assembly.
14. Check the master cylinder fluid level and road test the vehicle.

GM G-BODY:
Buick Riviera
Oldsmobile Aurora

FRONT

1. Siphon ⅔ of the brake fluid from the master cylinder.
2. Remove the tire and wheel assembly.
3. Install 2 wheel nuts to secure the rotor on the hub.
4. Remove the caliper mounting bolts and sleeves.
5. Remove the caliper from the mounting bracket and support the caliper. DO NOT allow the caliper to hang unsupported from the brake hose.
6. Remove the outboard brake pad from the caliper by pushing the pad inward toward the piston to unseat the buttons on the back of pad from the holes in the caliper. Once the buttons are unseated push the pad out of the caliper.
7. Remove the inboard pad from the caliper by pulling the top of the pad away from the piston and disengaging the spring clip from the caliper.
8. Compress the piston back into the bore using a C-Clamp.

To install:

9. Install the inboard pad into the caliper so the spring clip seats in the piston. Make sure lower edge of the spring clip is engaged in the piston and the pad is even against the base of the piston. Push the pad flat against caliper piston.
10. Install the outboard pad into the caliper so the wear indicator is at the trailing edge of the pad during forward wheel rotation. Push the pad the straight down into the caliper so the spring clips rode along the outside of the caliper. The buttons on the pad will snap into the mounting holes when the pad is correctly installed.

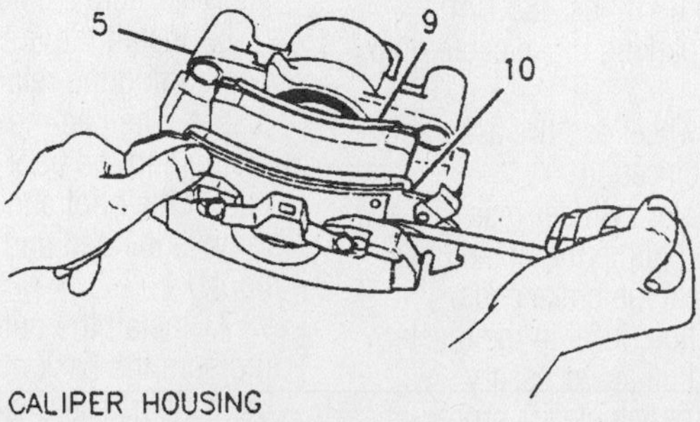

5 CALIPER HOUSING
9 INBOARD SHOE AND LINING
10 OUTBOARD SHOE AND LINING

93006G93

Removing the outboard front pad—Riviera, Aurora

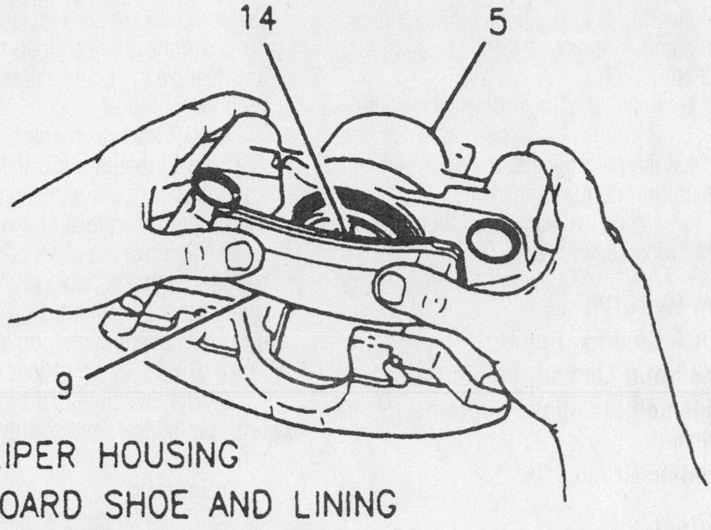

Installing the inboard front pad—Riviera, Aurora

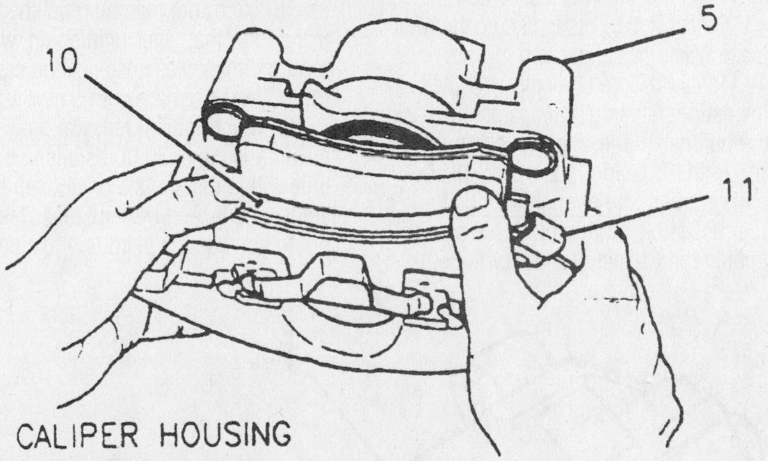

Installing the outboard front pad—Riviera, Aurora

11. Reinstall the caliper onto the steering knuckle and torque the mounting bolts to 38 ft. lbs. (51 Nm).
12. Remove the 2 wheel nuts securing the rotor.
13. Reinstall the tire and wheel assembly and torque the wheel nuts to 100 ft. lbs. (140 Nm).
14. Refill the master cylinder. Pump the brake pedal several times to seat the pads against the rotor.
15. Check the master cylinder level and add fluid as necessary.

REAR

1. Siphon ⅔ of the brake fluid from the master cylinder.
2. Remove the tire and wheel assembly.
3. Install 2 wheel nuts to secure the rotor on the hub.
4. Pivot the parking brake actuator level and disconnect the parking brake cable from the lever.
5. Remove the mounting bolt securing the parking brake cable bracket to the caliper and position the cable and bracket out of the way.
6. Remove the lower caliper mounting bolt.
7. Pivot the caliper upward and secure the caliper in a position over the mounting bracket. DO NOT remove the caliper from the upper pivot pin.
8. Remove the inboard and outboard pads from the caliper bracket.
9. Remove the brake pad clips from the mounting bracket.

To install:

10. Spin the caliper piston back into the bore. Make sure the piston notches are at 6 and 12 o'clock after the piston is compressed.
11. Install new brake pad clips in the caliper mounting bracket.
12. Install the inboard and outboard pads in the mounting bracket.
13. Pivot the caliper down over the pads.
14. Reinstall the lower mounting bolt and torque to 20 ft. lbs. (27 Nm).
15. Reinstall the parking brake cable bracket on the caliper and torque the mounting bolt to 32 ft. lbs. (43 Nm).
16. Pivot the actuator lever and connect the cable end to the actuator.
17. Remove the 2 wheel nuts securing the rotor.
18. Reinstall the tire and wheel assembly and torque the wheel nuts to 100 ft. lbs. (140 Nm).
19. Refill the master cylinder. Pump the brake pedal several times to seat the pads against the rotor.
20. Check the master cylinder level and add fluid as necessary.

GM J-BODY:
Chevrolet Cavalier
Pontiac Sunfire

1. Siphon ⅔ of the brake fluid out of the master cylinder reservoir.
2. Remove the tire and wheel assembly.
3. Compress the caliper piston back into the caliper bore using a large pair of pliers, C-clamp or special caliper piston retracting tool.
4. Remove the caliper mounting bolts and sleeves.
5. Remove the caliper from the steering knuckle without disconnecting the brake hose. Do not allow the caliper to hang from the brake hose. Support the caliper with a piece of wire.
6. Remove the outboard pad by pushing in on the outside edge of the pad to release the mounting dowel from the hole in

the caliper. When both dowels are unseated, push the pad out the bottom of the caliper.

7. Remove the inboard pad from the caliper by pulling it out of the caliper.

To install:

8. Inspect the caliper for any signs of leakage. If the caliper is leaking, new brake pads will be damaged. Also inspect the condition of the caliper support for rust and corrosion which will hinder the travel on the caliper.

9. Inspect the caliper mounting hardware. New bolts are usually recommended. Clean all parts well. Lubricate the mounting bushings and sleeves with silicone grease as required.

10. Install the inboard pad in the caliper so the spring clip on the pad back engages in the caliper piston.

11. Install the outboard pad over the caliper end until the mounting dowels snap into the mounting holes in the caliper.

12. Install the caliper over the rotor onto the steering knuckle.

13. Install the mounting bolts and sleeves and torque to 40 ft. lbs. (51 Nm).

14. Install the tire and wheel assembly.

15. Pump the brake pedal several times to seat the pads against the rotor before attempting to move the vehicle.

16. Check the master cylinder level and add fluid as necessary.

GM L/N and N-BODY:
Buick Skylark
Chevrolet Malibu
Oldsmobile Achieva, Cutlass
Pontiac Grand Am

1. Siphon ⅔ of the brake fluid out of the master cylinder reservoir.

2. Remove the tire and wheel assembly.

3. Remove the caliper from the steering knuckle without disconnecting the brake hose. DO NOT allow the caliper to hang from the brake hose. Support the caliper with a piece of wire.

4. Remove the outboard pad by pushing in on the outside edge of the pad to release the mounting dowel from the hole in the caliper. When both dowels are unseated push the pad out the bottom of the caliper.

5. Remove the inboard pad from the caliper by pulling it out of the caliper.

To install:

6. Install the inboard pad in the caliper so the spring clip on the pad back engages in the caliper piston.

7. Install the outboard pad over the caliper end until the mounting dowels snap into the mounting holes in the caliper.

8. Install the caliper over the rotor onto the steering knuckle. Install the mounting bolts and sleeves and torque to 40 ft. lbs. (51 Nm).

9. Install the tire and wheel assembly.

10. Pump the brake pedal several times to seat the pads against the rotor before attempting to move the vehicle.

11. Check the master cylinder level and add fluid as necessary.

GM W-BODY:
Buick Century, Regal
Chevrolet Lumina, Monte Carlo
Oldsmobile Cutlass Supreme, Intrigue
Pontiac Grand Prix

FRONT

1. Siphon ⅔ of the brake fluid out of the master cylinder.

2. Mark the relationship of the wheel to the wheel stud for re-installation purposes. Remove the tire and wheel assembly.

3. Install 2 lug nuts to secure the rotor in place when the caliper is removed.

4. Install a large C-clamp over the top of the caliper housing and against the back of the outboard shoe. Slowly tighten the C-clamp until the caliper pistons are pushed into the caliper bore enough to slide the caliper assembly off the rotor. Use care not to tighten the C-clamp too far or the outboard shoe retaining spring will be deformed and require replacement.

5. Remove the caliper mounting bolts and remove the brake caliper from the mounting bracket.

6. DO NOT disconnect the brake hose from the caliper or allow the brake hose to support the weight of the caliper. Support the caliper on a piece of wire out of the way.

7. Remove the outer brake pad from the caliper using a suitable prying tool to lift the outboard shoe retaining spring so that it will clear the caliper center lug and pull the brake pad out of the caliper.

8. Remove the inner brake pad by unsnapping the shoe springs from the piston.

To install:

9. Clean all parts well. If the brake pads were worn so badly that the brake rotor is damaged, it must be replaced. Light scoring of the rotor surfaces not exceeding 0.060 inch (1.5mm) in depth is not harmful to brake operation and may result from normal use. Brake rotors may be refinished. Do not use a rotor that, after refinishing, will not meet the thickness specification cast in the rotor. Always replace with a new rotor.

10. If not done at removal, now use a C-clamp and clamp both pistons at the same time with a metal plate or wooden block across the face of both pistons. Take care not to damage the pistons or caliper boots.

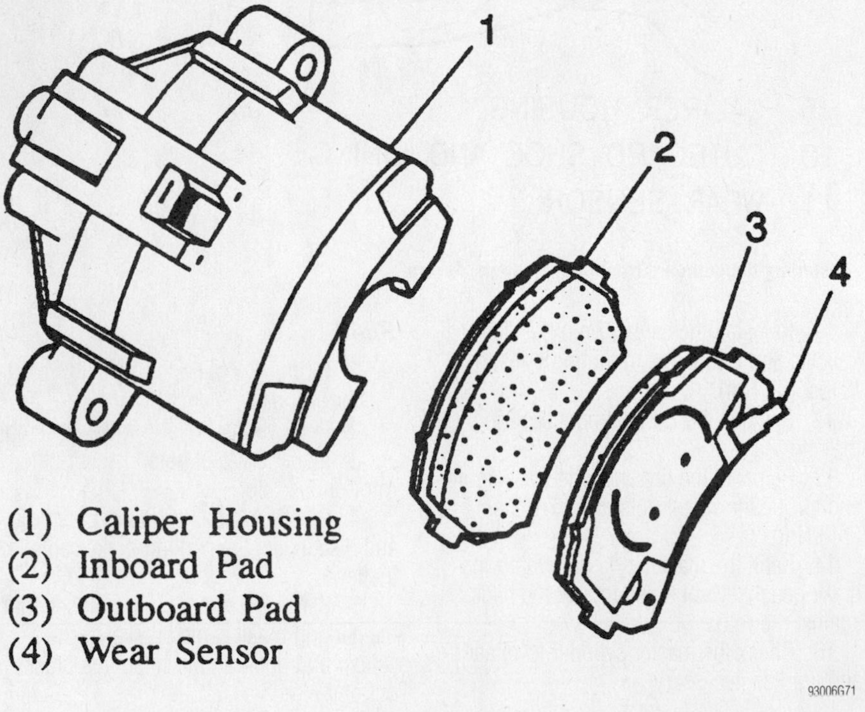

Front brake pads and caliper—Century, Regal, Lumina, Monte Carlo, Cutlass Supreme, Intrigue, Grand Prix

➡After bottoming the pistons into the caliper bore, lift the inner edge of each caliper boot next to the piston and press out any trapped air. Make sure each boot convolution is tucked back into place. Boots must lay flat.

11. Inspect the caliper bushings for wear. Replace as necessary. Carefully inspect the slide bolts for corrosion. If corrosion if found, use new parts including the bushing assemblies when installing the caliper. Do not attempt to polish away corrosion. Lubricate caliper slide bolts with silicone grease.
12. Install the new inner disc brake pad in the caliper by snapping the shoe retainer springs into the piston making sure both sets of locking tabs are seated in the caliper pistons. The pad must seat flat against the pistons.
13. Install the outer pad into the caliper by snapping the outboard shoe retaining spring over the caliper center lug and into the housing slot. The pad will slide up onto the caliper and the retaining ring will lock into place on the groove in the caliper.
14. The outer pad wear sensor should be at the trailing edge of the shoe during forward wheel rotation.
15. Install the caliper mounting bolts and torque to 80 ft. lbs. (108 Nm).
16. Remove the 2 nuts temporarily securing the rotor.
17. Install the tire and wheel assembly and tighten to specification.
18. Pump the brake pedal several times to seat the pads against the rotor.
19. Check the brake fluid level and top off as necessary.
20. Road test the vehicle to ensure the proper brake performance.

REAR

1. Siphon ⅔ of the brake fluid out of the master cylinder.
2. Remove the tire and wheel assembly.
3. Install 2 lug nuts to secure the rotor in place when the caliper is removed.
4. Remove bolt and washer attaching cable support bracket to caliper body assembly. It is not necessary to disconnect the parking brake lever or disconnect the brake hose.
5. Remove the caliper retaining bolt.
6. Pivot the caliper body assembly up from the rotor and remove from the bracket. Do not completely remove the caliper assembly body.
7. Remove the outboard and inboard shoe and linings from the caliper support assembly.
8. Remove 2 brake lining clips from the caliper support.

To install:

➡In order for the rear pads to seat in the caliper properly, the cut outs in the caliper piston must be at the 6 and 12 o'clock positions. Failure to align the piston correctly can lead to brake drag, premature brake wear and possible brake failure.

9. Using a suitable type spanner tool turn the piston in to bottom the piston fully into the caliper bore. Once the caliper is fully seated make sure the cutouts in the piston are at the 6 and 12 o'clock positions.
10. After bottoming the piston into the caliper bore, lift the inner edge of boot next to the piston assembly and press out any trapped air.
11. Install 2 pad clips in the caliper support.
12. Lubricate the inner pad where it contacts the piston and mounting surfaces.
13. Install outboard and inboard shoe and linings in caliper support. Position the wear sensors downward at the leading edge of the rotor during forward wheel rotation.
14. Hold the metal shoe edge against the spring end of clips in the caliper support. Push brake pad in towards the hub, bending spring ends slightly and engage shoe notches with support abutments.
15. Pivot the caliper body assembly down over the brake pad. Compress the sleeve boot by hand as the caliper body moves into position to prevent boot damage.

➡After the caliper body assembly is in position, recheck installation of the brake pad clips. If necessary, use a small prying tool to reseat or center the pad clip on the support abutments.

16. Install the sleeve bolts and torque bolt to 20 ft. lbs. (27 Nm).
17. Install the cable support bracket with the cable attached and bolt washer. Torque bolt to 32 ft. lbs. (43 Nm).
18. Remove the 2 lug nuts securing the rotor.
19. Install the tire and wheel assembly and tighten to specification.
20. Pump the brake pedal several times to seat the pads against the rotor.
21. Check the brake fluid level and top off as necessary.

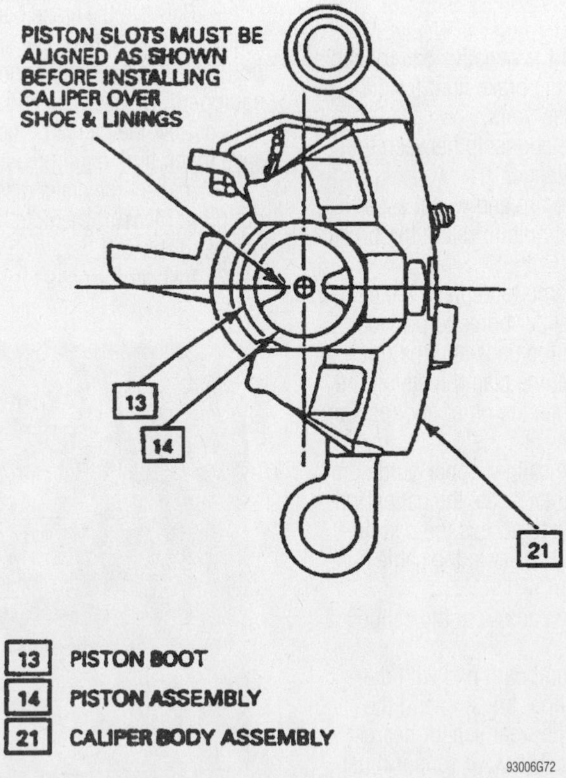

Positioning rear caliper piston slots—Century, Regal, Lumina, Monte Carlo, Cutlass Supreme, Intrigue, Grand Prix

22. Adjust parking brake as necessary.
23. Road test the vehicle for proper brake performance.

GM Y-BODY:
Chevrolet Corvette

FRONT

1. Disconnect the negative battery cable.
2. Remove the caliper from the mounting bracket but do not disconnect the brake hose and inlet fitting assembly.
3. Suspend the caliper from the upper control arm with wire to avoid damage to the brake hose.
4. Remove the pad and lining assemblies from the caliper.

To install:

5. Clean all residue from the pad and lining assembly guiding surfaces on the caliper housing and the mounting bracket.
6. Install the outboard pad with the insulator to the caliper housing and the inboard pad with the wear sensor into the caliper pistons. Press the pads firmly until they are they are fully seated.
7. Remove the support and install the caliper to the mounting bracket.
8. Connect the negative battery cable.

REAR

1. Disconnect the negative battery cable and remove ⅔ of the brake fluid from the master cylinder reservoirs.
2. Mark the relationship between the wheel to the axle flange.
3. Remove the tire and wheel assembly. Install 2 wheel nuts to retain the brake rotor.
4. Use a C-clamp to depress the caliper pistons into the caliper bores to provide clearance between the pads and the rotor. Make sure 1 end of the clamp rests on the inlet fitting bolt while the other end rests on the outboard pad.
5. Remove the caliper upper guide pin bolt and discard, then rotate the caliper on the lower guide pin to access the pad linings. Be careful not to strain the cable conduit or the hoses.
6. Remove the pads from the caliper.

To install:

7. Install the outboard pad with the insulator to the caliper housing and the inboard pad with the wear sensor nearest the caliper pistons. The wear sensor must be in the trailing position during forward wheel rotation. Press the pads firmly until they are they are fully seated.
8. Rotate the caliper housing into position, then install a new upper guide pin bolt and torque 26 ft. lbs. (35 Nm).
9. Remove the wheel nuts securing the rotor to the hub and install the tire and wheel assembly.
10. Fill the master cylinder to the proper level with clean brake fluid.
11. Connect the negative battery cable, start the engine and pump the brake pedal slowly and firmly 3 times to seat the shoe and lining assemblies.

GEO Metro

1. Set the parking brake and block the rear wheels.
2. Siphon a sufficient quantity of brake fluid from the master cylinder reservoir to prevent the brake fluid from overflowing from the master cylinder when removing or installing the brake pads. This is necessary, as the piston must be forced into the cylinder bore to provide sufficient clearance to install the pads.
3. Remove the wheel, then reinstall 2 lug nuts finger tight to hold the disc in place.
4. Remove the 2 caliper mounting bolts and then remove the caliper from the mounting bracket. Position the caliper out of the way and support it with wire so it doesn't hang by the brake line.
5. Remove the brake pads, the wear indicators, the anti-squeal shims, the support plates and the anti squeal springs (if so equipped). Disassemble slowly and take note of how the parts fit together. This will save much time during reassembly.
6. Inspect the brake disc for scoring or gouging. Measure the disc for both thickness and run-out.
7. Inspect the pads for remaining thick-

84249014

Front disc brake assembly

84249015

Removing caliper mounting bolts

84249016

Removing the caliper from the rotor

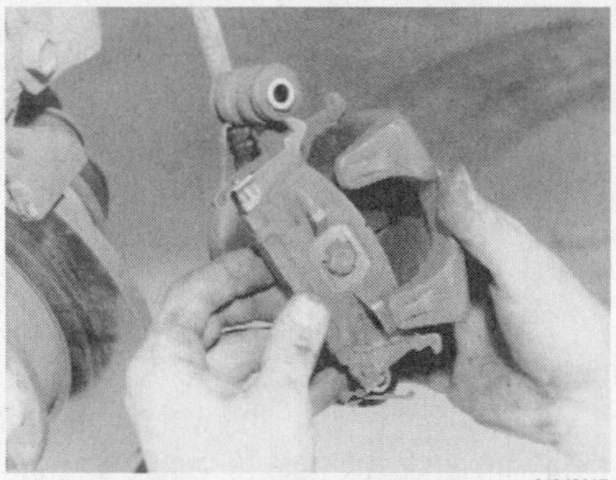

84249017

Removing the outside brake pad

84249018

Removing the inside brake pad

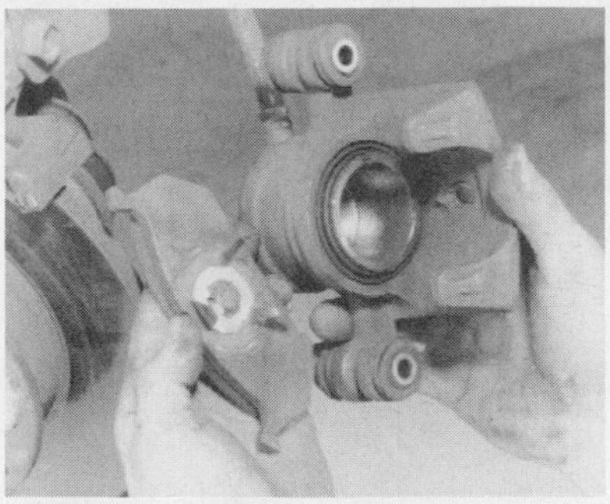

84249019

The inside brake pad is attached to the caliper piston with a spring clip

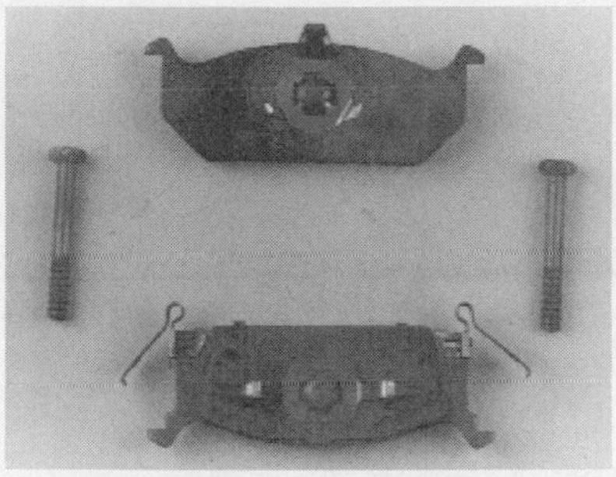

84249020

Inside pad is at top and outside pad is on bottom. Note wear indicator springs protruding from sides of outside pad

84249021

Removing the anti-squeal springs from the caliper wear indicators

ness and condition. Any sign of uneven wear, cracking, heat checking or spotting is cause for replacement. Compare the wear of the inner pad to the outer pad. While they will not wear at exactly the same rate, the remaining thickness should be about the same on both pads. If one is heavily worn and the other is not, suspect either a binding caliper piston or dirty slides in the caliper mount.

8. Examine the 2 caliper retaining bolts

84249022

Using a C-clamp to compress the caliper piston

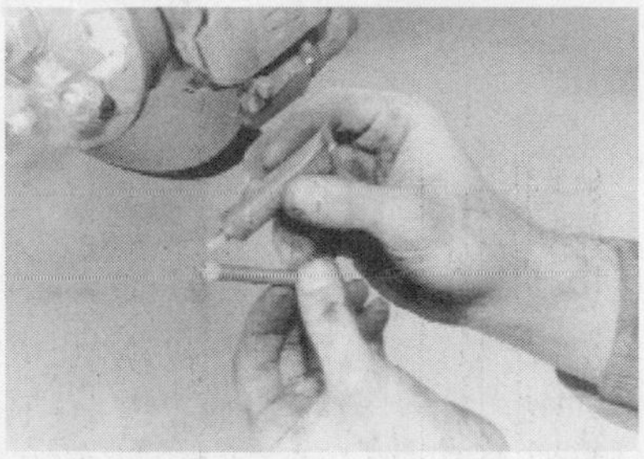

84249023

Coat the caliper mounting bolts with anti-seize during installation

and the slide bushings in which they run. Everything should be clean and dry. If cleaning is needed, use spray solvents and a clean cloth. Do not wire brush or sand the bolts. This will cause grooves in the metal which will trap more dirt. Check the condition of the rubber dust boots and replace them if damaged.

To install:

9. Install the pad support plates onto the mounting bracket.
10. Install new pad wear indicators onto each pad, making sure the arrow on the tab points in the direction of disc rotation.
11. Install new anti-squeal pads to the back of the pads.
12. Install the pads into the mounting bracket and install the anti-squeal springs.
13. Use a caliper compressor, or a C-clamp to slowly press the caliper piston back into the caliper. If the piston is frozen, or if the caliper is leaking hydraulic fluid, the caliper must be overhauled or replaced.
14. Install the caliper assembly to the mounting plate. Before installing the retaining bolts, apply a thin, even coating of anti-seize compound to the threads and slide surfaces. Don't use grease or spray lubricants; they will not hold up under the extreme temperatures generated by the brakes. Torque the bolts to specification.
15. Remove the 2 lugs holding the disc in place and install the wheel.
16. Check the level of the brake fluid in the master cylinder reservoir; it should be at least to the middle of the reservoir.
17. Depress the brake pedal several times and make sure that the movement feels normal. The first brake pedal application may result in a very "long" pedal due to the pistons being retracted. Always make several brake applications before starting the vehicle. Bleeding is not usually necessary after pad replacement.
18. Recheck the fluid level and add to the **MAX** line if necessary.

GEO Prizm

FRONT

1. Siphon a sufficient quantity of brake fluid from the master cylinder reservoir to prevent the brake fluid overflowing from the master cylinder when removing or installing the brake pads. This is necessary, as the piston must be forced into the cylinder bore to provide sufficient clearance to install the pads.
2. Remove the wheel and tire assembly.
3. Install 2 lug nuts finger tight to hold the disc in place.
4. Remove the 2 caliper mounting bolts and then remove the caliper from the mounting bracket. Position the caliper out of the way and support it with wire so it doesn't hang by the brake line.
5. Remove the 2 brake pads, the 2 wear indicators, the 4 anti-squeal shims, the 4 support plates (anti-rattle springs) and the 2 anti squeal springs, if so equipped. Disassemble slowly and take note of how the parts fit together. This will save much time during reassembly.
6. Inspect the brake disc for scoring or gouging. Measure the disc for both thickness and run-out. Complete inspection procedures are given later in this section.

7. Inspect the pads for remaining thickness and condition. Any sign of uneven wear, cracking, heat checking or spotting is cause for replacement. Compare the wear of the inner pad to the outer pad. While they will not wear at exactly the same rate, the remaining thickness should be about the same on both pads. If one is heavily worn and the other is not, suspect either a binding caliper piston or dirty slides in the caliper mount.

8. Examine the 2 caliper retaining bolts and the slide bushings in which they run. Everything should be clean and dry. If

84229034

You will need a short extension to remove the front caliper mounting bracket bolts

84229035

Carefully remove the caliper assembly from the rotor

84229036

Suspend the caliper assembly out of the way to avoid hanging from the hose

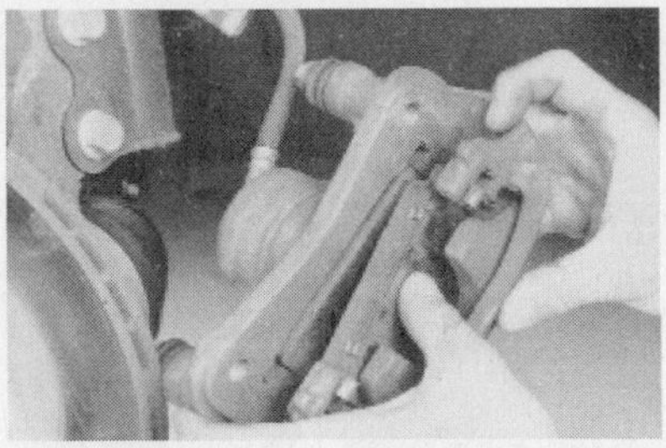
84229037

Removing the outer brake pad from the caliper assembly

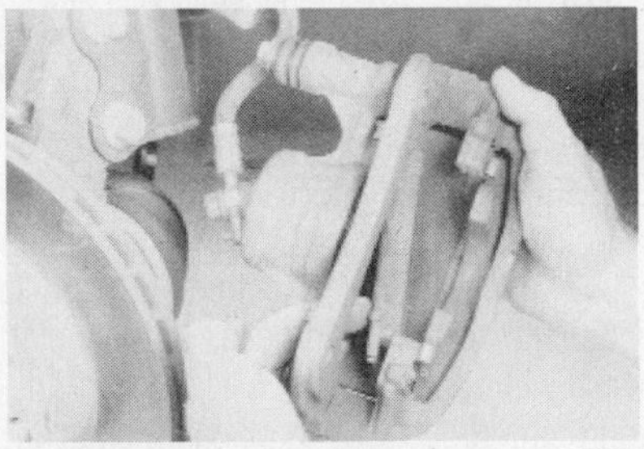
84229038

Removing the inner brake pad from the caliper assembly

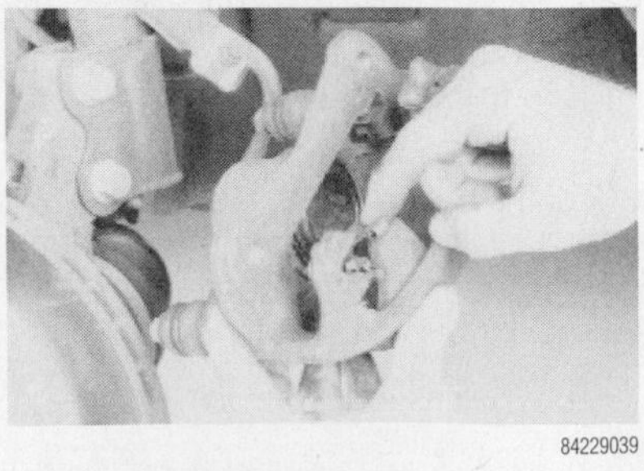
84229039

Note the correct position support plates (anti-rattle springs)

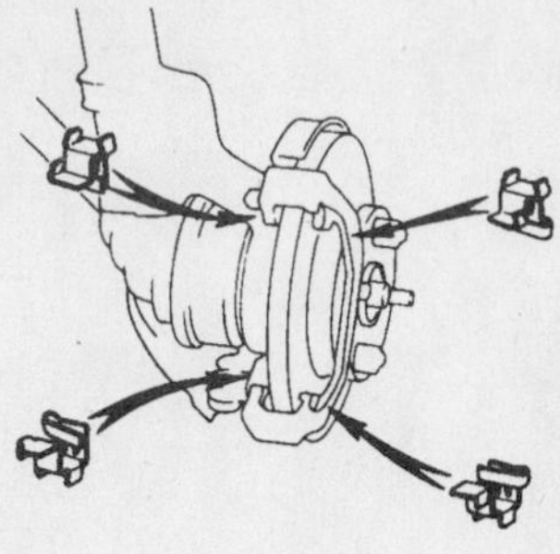
84229041

Correct placement of the support plates (anti-rattle springs)

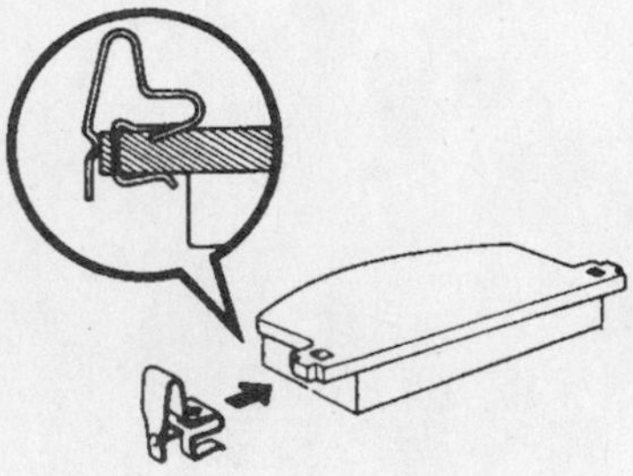
84229042

Make certain the brake wear indicators are correctly installed

cleaning is needed, use spray solvents and a clean cloth. Do not wire brush or sand the bolts. This will cause grooves in the metal which will trap more dirt. Check the condition of the rubber dust boots and replace them if damaged.

To install:

9. Install the 4 pad support plates (anti-rattle springs) onto the mounting bracket.

10. Install new pad wear indicators onto each pad, making sure the arrow on the tab points in the direction of disc rotation.

11. Install new anti-squeal pads to the back of the pads.

12. Install the pads into the mounting bracket and install the anti-squeal springs, if so equipped.

13. Use a caliper compressor, a C-clamp or large pair of pliers to slowly press the caliper piston back into the caliper. If the piston is frozen, or if the caliper is leaking hydraulic fluid, the caliper must be overhauled or replaced.

14. Install the caliper assembly to the mounting plate. Before installing the retaining bolts, apply a thin, even coating of anti-seize compound to the threads and slide surfaces. Don't use grease or spray lubricants; they will not hold up under the extreme temperatures generated by the

brakes. Torque the bolts to 25 ft. lbs. (34 Nm).

15. Remove the 2 lugs holding the disc in place and install the wheel.

16. Check the level of the brake fluid in the master cylinder reservoir; it should be at least to the middle of the reservoir.

17. Depress the brake pedal several times and make sure that the movement feels normal. The first brake pedal application may result in a very "long" pedal due to the pistons being retracted. Always make several brake applications before starting the vehicle. Bleeding is not usually necessary after pad replacement.

18. Recheck the fluid level and add to the "MAX" line if necessary.

REAR

1. Raise and safely support the rear of the vehicle on jackstands. Block the front wheels.

2. Siphon a sufficient quantity of brake fluid from the master cylinder reservoir to prevent the brake fluid from overflowing the master cylinder when removing or installing the brake pads. This is necessary, as the

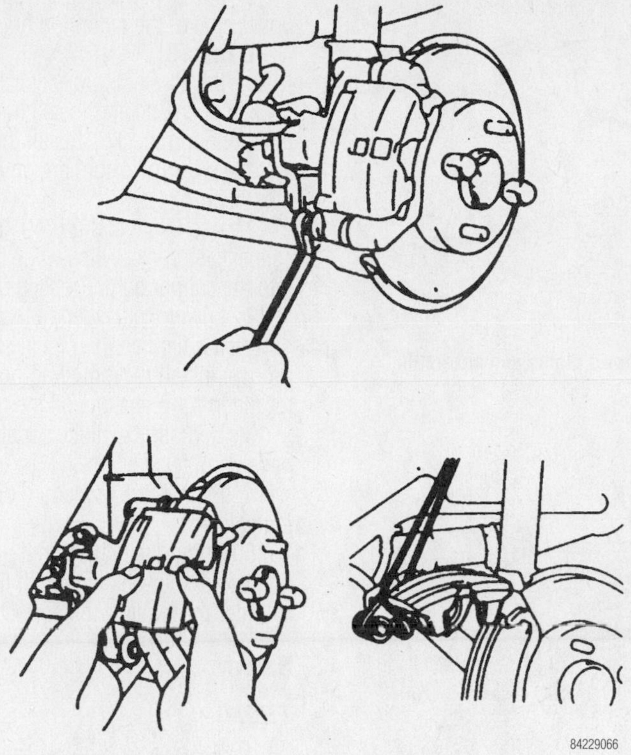

Correct method for raising the rear caliper to gain access to the pads

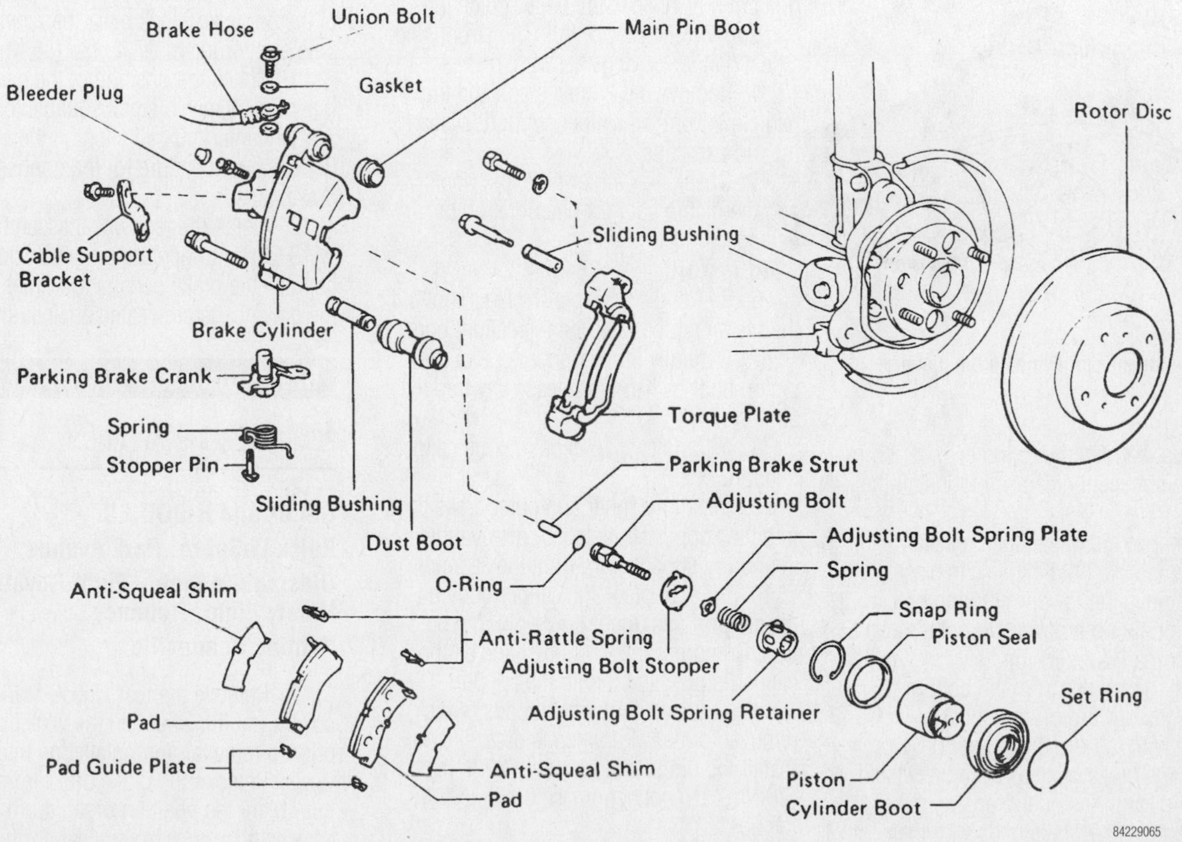

Rear disc brake components

Heater Core replacement is covered in Section 2 of this manual

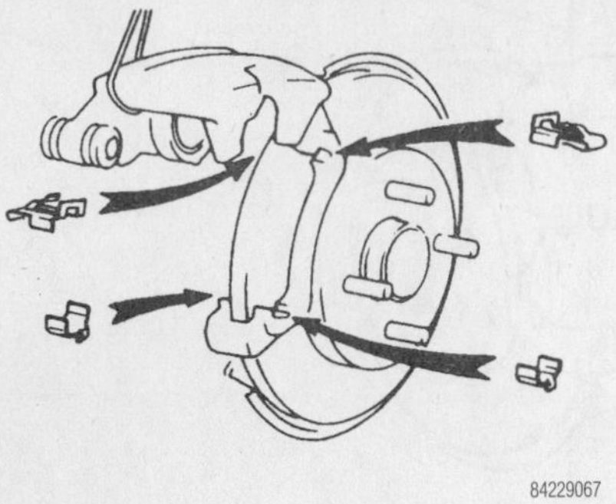

84229067

Installing support plates and anti-rattle shims

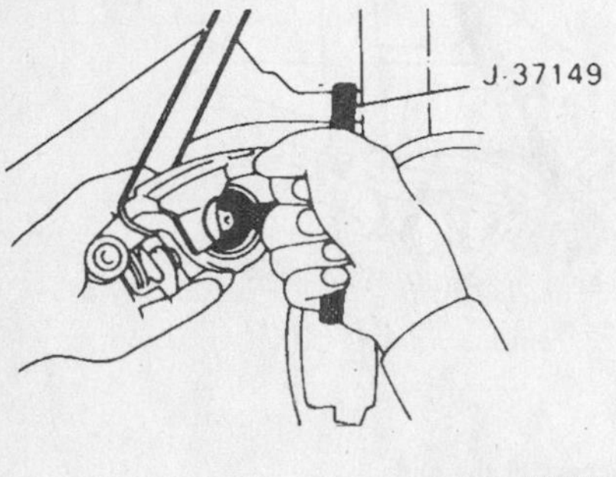

84229068

Using the GM tool to retract the rear caliper piston

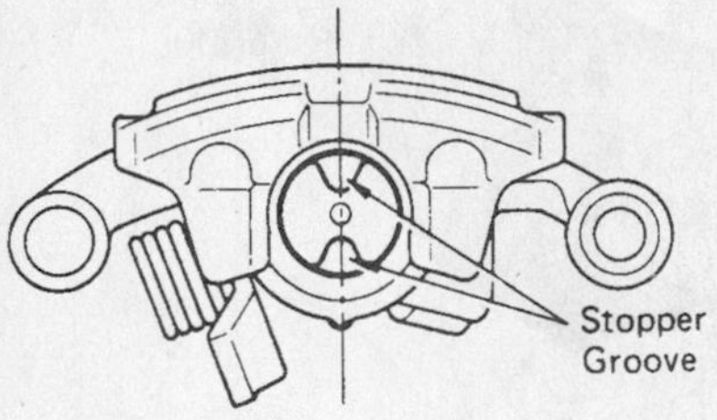

84229069

Align the stopper groove with the pad protrusion

piston must be forced into the cylinder bore to provide sufficient clearance to install the pads.

3. Remove the wheel, then reinstall 2 lug nuts finger tight to hold the disc in place.

4. Remove the lower mounting bolt from the mounting bracket. Do not remove the caliper main (upper) pin.

5. Lift the caliper from the bottom so that it hinges upward on the upper pin. Use a piece of wire to hold the caliper up. Do not allow the brake hose to become twisted or kinked during this operation.

6. Remove the brake pads with their shims, springs and support plates.

7. Check the rotor thickness.

8. Install new pad support plates to the lower sides of the mounting bracket.

9. Install new anti-rattle springs to the upper side of the mounting bracket.

To install:

10. Install a new anti-squeal shim to the back of each pad and install the pads onto the mounting bracket. Install the pads so that the wear indicators are on the uppermost side.

11. Use GM tool J-37149, to turn the caliper piston clockwise while pressing it into the caliper bore. Proceed until it locks.

12. Lower the caliper so that the pad protrusion fits into the piston stopper groove. Install the mounting bolts and torque to 14 ft. lbs. Install the rear wheel.

13. Depress the brake pedal until a firm pedal is obtained.

14. Repeat this procedure on the remaining caliper.

15. Pump the brakes until a firm pedal is felt. Lower the vehicle and fill the master cylinder reservoir to the correct level.

Saturn

FRONT

1. Remove the front wheels.

2. Remove the caliper lower lock pins.

3. Either pivot the caliper up on the guide pin or remove the upper guide pin and support the caliper from the strut using a coat hanger or length of wire.

4. Remove the 2 brake pads and the pad clips from the caliper support. Discard the old pad clips.

5. Check the caliper pins, pin boots and the piston boot for deterioration or damage.

To install:

6. By hand or using a C-clamp, bottom the piston all the way into the caliper bore.

7. Carefully lift the inner edge of the piston boot by hand to release any trapped air.

8. Install new pad clips into the caliper support.

9. Install the inner and outer brake pads into the support. If installed, remove the temporary support wire from the caliper.

10. Pivot or place the caliper body on the support and upper guide pin into position. Compress the boots by hand as the caliper is positioned onto the support.

11. Lubricate the smooth ends of the removed pin(s) with silicone grease, then install the pin(s) and torque to 27 ft. lbs. (36 Nm). Do not get grease on the pin threads.

12. Install the wheels.

13. Prior to operating the vehicle, depress the brake pedal a few times until the brake pads are seated against the rotor.

REAR

1. Remove the rear wheels.

2. Remove the caliper lock and guide pins.

3. Remove the caliper from the support, being careful not to damage the pin boots and suspend the caliper from a wire.

4. Remove the brake pads from the support.

To install:

5. Using SA91110NE piston driver tool, bottom the piston by rotating it clockwise into the caliper bore; do not use a C-clamp to press the piston into the bore.

6. Align the piston slots so they are perpendicular to the brake pads.

7. Carefully lift the inner edge of the piston boot to release any trapped air. The boot must lie flat below the level of the piston face.

8. Install new pad clips into the caliper support.

9. Install the inner and outer brake pads into the clips on the support. The pad with the wear sensor should be located outboard. The piston indentation slots should be positioned to correctly accept the brake pads.

10. Position the caliper body onto the support. Lubricate the non-threaded portion of the guide and lock pins, then install the pins and torque to 27 ft. lbs. (36 Nm).

11. Check the position of the pad clips. If necessary, use a small suitable tool to reseat or center the pad clips on the support. Repeat the procedure for the opposite side brake pads.

12. Install the rear wheel assemblies.

13. Prior to operating the vehicle, depress the brake pedal a few times until the brake pads are seated against the rotor.

Brake Drums

REMOVAL & INSTALLATION

GM C- and H-BODIES:
Buick LeSabre, Park Avenue
Oldsmobile Eighty Eight Royale, Ninety Eight, Regency
Pontiac Bonneville

1. Remove the rear wheel. Mark the position of the wheel on the wheel studs, prior to removal, for installation reference.

2. Remove the brake drum from the hub. Mark the position of the drum on the wheel studs, prior to removal, for installation reference.

3. If the brake drum is difficult to remove, check the following:

a. Make sure that the parking brake is released.

b. Back off on the parking brake cable adjustment.

c. Use a hammer and a small metal punch to bend in the backing plate knockout to provide access to the park brake lever.

d. Insert a screwdriver through the hole and press in to push the park brake lever off its stop. This lets the brake shoes retract slightly.

e. Apply a small amount of penetrating oil around the drum pilot hole.

4. Remove the brake drum from the vehicle.

5. Inspect the wheel cylinder for leakage and inspect the brake shoes for wear; replace as necessary.

6. Inspect the brake drum for scoring, cracks or other wear; machine or replace as necessary. If machining, observe the maximum diameter specification.

To install:

7. Measure the inside diameter of the brake drum using clearance gauge J 21177-A.

8. Turn the starwheel on the adjusting screw assembly until the brake shoe diameter is 0.050 in. (1.27mm) less than the drum inside diameter.

9. Install the brake drum onto the wheel hub, aligning the marks made during removal.

10. Install the wheel, aligning the marks made during removal. Torque the lug nuts to 100 ft. lbs. (140 Nm).

11. Road test the vehicle and check brake operation.

GM F-BODY:
Chevrolet Camaro
Pontiac Firebird

1. Matchmark the relationship of the wheel to the axle flange, then remove the wheel and tire assembly.

2. Matchmark the relationship of the drum to the axle flange and remove the brake drum. If the brake drum is difficult to remove, try the following:

a. Ensure the parking brake is released.

b. Back off the parking brake cable adjustment.

c. Remove the adjusting hole cover or knockout plate from the backing plate. Back off the adjustment screw, using suitable brake adjusting tools.

d. Use a rubber mallet to tap gently on the outer rim of the drum and/or around the inner drum diameter by the spindle. Be careful not to deform the drum by excessive use of force.

To install:

3. Adjust the brake shoes. The outside diameter of the shoe and linings should be 0.050 inch (1.27mm) less than the inside diameter of the brake drum on each wheel.

4. Install the drum, aligning the marks on the drum and the axle flange.

5. Install the wheel and tire assembly, aligning the marks on the wheel and axle flange.

GM J-BODY:
Chevrolet Cavalier
Pontiac Sunfire

1. Remove the wheel and tire assembly.

2. Pull the brake drum off. It may be necessary to gently tap the rear edge of the drum to start it off the studs. If extreme resistance to removal is encountered, it will be necessary to retract the brake shoe self-adjuster screw, sometimes called the star-wheel. Knock out the access hole in the brake drum and turn the adjuster to retract the linings from the drum. Install a replacement hole cover before reinstalling the drum.

➡Do not hammer on the brake drum to remove it.

To install:

3. Inspect the inside of the brake drum. If worn, heavily grooved or if the opening is distorted, the drum should be refinished or replaced. If refinishing, observe the maximum drum diameter specification.

4. Inspect the wheel cylinder for signs of brake fluid leakage. Inspect the brake shoe springs and self-adjuster mechanism. The adjuster should usually be disassembled, cleaned and lubricated when the drum is removed for brake service.

5. Install the drum over the brake shoes.

6. Install the wheel and tire assembly.

7. Adjust the brakes.

8. Check brake operation.

GM L/N and N-BODY:
Buick Skylark
Chevrolet Malibu
Oldsmobile Achieva, Cutlass
Pontiac Grand Am

1. Remove the wheel.

2. Pull the brake drum off. It may be necessary to gently tap the rear edge of the drum to start it off the studs. If extreme resistance to removal is encountered, it will be necessary to retract the brake shoe self-adjuster screw, sometimes called the star-wheel. Knock out the access hole in the brake drum and turn the adjuster to retract the linings from the drum. Install a replacement hole cover before reinstalling the drum.

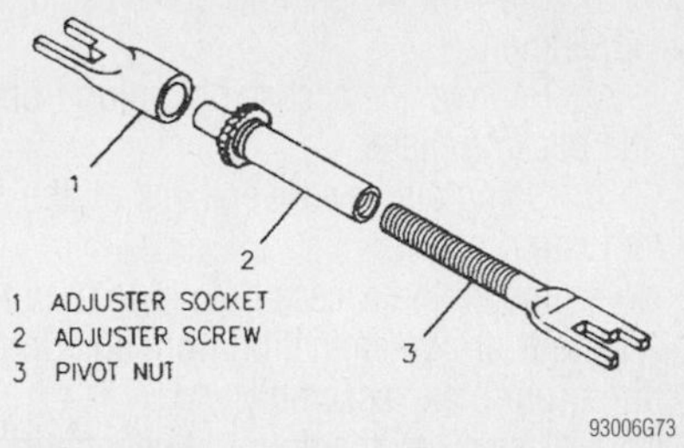

Brake adjuster—Skylark, Malibu, Achieva, Cutlass, Grand Am

➡DO NOT hammer on the brake drum to remove it.

To install:

3. Inspect the inside of the brake drum. If worn, heavily grooved or if the opening is distorted, the drum should be refinished or replaced. If refinishing, observe the maximum drum diameter specification.

4. Inspect the wheel cylinder for signs of brake fluid leakage. Inspect the brake shoe springs and self-adjuster mechanism. The adjuster should usually be disassembled, cleaned and lubricated when the drum is removed for brake service.

5. Install the drum over the brake shoes.

6. Install the wheel. Adjust the brakes.

7. Check brake operation.

GM W-BODY:
Buick Century, Regal
Chevrolet Lumina, Monte Carlo
Oldsmobile Cutlass Supreme, Intrigue
Pontiac Grand Prix

1. Mark the relationship of the wheel to the axle flange to help maintain wheel balance after assembly.

2. Remove the tire and wheel assembly.

3. Mark the relationship of the brake drum to the axle flange.

4. If difficulty is encountered in removing the brake drum, the following steps may be of assistance.

a. Make sure the parking brake is released.

b. Back off the parking brake cable adjustment.

c. Remove the access hole plug from the backing plate.

d. Using a screwdriver, back off the adjusting screw.

e. Install the access hole plug to prevent dirt or contamination from entering the drum brake assembly.

f. Use a small amount of penetrating oil applied around the brake drum pilot hole.

g. Carefully remove the brake drum from the vehicle.

5. After removing the brake drum it should be checked for the following:

a. Inspecting for cracks and deep grooves.

b. Inspect for out of round and taper.

c. Inspecting for hot spots (black in color).

To install:

6. Install the brake drum onto the vehicle aligning the reference marks on the axle flange.

7. Install the tire and wheel assembly, torquing it to specifications.

8. Road test for proper brake operation.

GEO Metro

1. Make sure the parking brake is fully released.

2. Remove the rear wheels.

3. Without ABS, perform the following:

a. Remove the dust cap. Using a hammer and punch, unstake the spindle nut.

b. Remove the spindle nut and washer. Discard the spindle nut.

4. If equipped with ABS, perform the following:

a. Remove the 2 brake drum retaining screws.

b. Install 2, 8mm screws in the 2 opposite threaded holes on the brake drum.

c. Alternately rotate each screw slowly until the 2 screws pull the drum from the hub and wheel studs.

5. Remove the drum from the vehicle. Remove the 2, 8mm screws from the drum.

6. Remove the rear brake drum assembly from the vehicle.

7. Remove the brake drum. If the drum is frozen on the hub, tap the drum with a rubber mallet or wooden hammer handle to loosen it.

8. Inspect the brake shoes for wear and/or damage; replace as necessary. Inspect the wheel cylinder for signs of brake fluid leakage; replace as necessary.

9. Inspect the drum for cracks, scoring or other wear; machine or replace as necessary. If machining, observe the maximum drum diameter specification.

To install:

10. Install the brake drum assembly onto the spindle.

11. On non-ABS equipped models;

a. Install the washer and a new spindle nut.

b. Torque the spindle nut to 74 ft. lbs. (100 Nm).

c. Stake the spindle nut.

d. Install the dust cap.

12. On ABS equipped models, install the brake drum and tighten the brake drum retaining screws.

13. Install the wheel.

14. To adjust the brakes, depress the brake pedal 3–5 times with approximately 66 lbs. (30 kg) of pressure.

15. Road test and check brake operation.

GEO Prizm

1. Make sure the parking brake is fully released.

2. Remove the rear wheel.

3. Without ABS, perform the following:

a. Remove the dust cap. Using a hammer and punch, unstake the spindle nut.

b. Remove the spindle nut and washer. Discard the spindle nut.

4. If equipped with ABS, perform the following:

a. Remove the 2 brake drum retaining screws.

b. Install 2, 8mm screws in the 2 opposite threaded holes on the brake drum.

c. Alternately rotate each screw slowly until the 2 screws pull the drum from the hub and wheel studs.

5. Remove the drum from the vehicle. Remove the 2, 8mm screws from the drum.

6. Remove the rear brake drum assembly from the vehicle.

7. Mark the position of 1 wheel lug stud in relation to a hole in the drum. This will allow the drum to be reinstalled in the original position.

8. Remove the brake drum. If the drum is frozen on the hub, tap the drum with a rubber mallet or wooden hammer handle to loosen it.

9. Inspect the brake shoes for wear and/or damage and replace as necessary. Inspect the wheel cylinder for signs of brake fluid leakage; replace as necessary.

10. Inspect the drum for cracks, scoring or other wear; machine or replace as necessary. If machining, observe the maximum drum diameter specification.

To install:

11. Install the brake drum assembly onto the spindle.

12. On non-ABS equipped models;

a. Install the washer and a new spindle nut.

b. Torque the spindle nut to 74 ft. lbs. (100 Nm).

c. Stake the spindle nut.

d. Install the dust cap.

13. On ABS equipped models, install the brake drum and tighten the brake drum retaining screws.

14. Install the drum, observing the matchmarks made earlier. Keep the drum straight while installing it, or it could damage the brake shoes.

15. Install the wheel.

16. To adjust the brakes, depress the brake pedal 3–5 times with approximately 66 lbs. (30 kg) of pressure.

17. Road test and check brake operation.

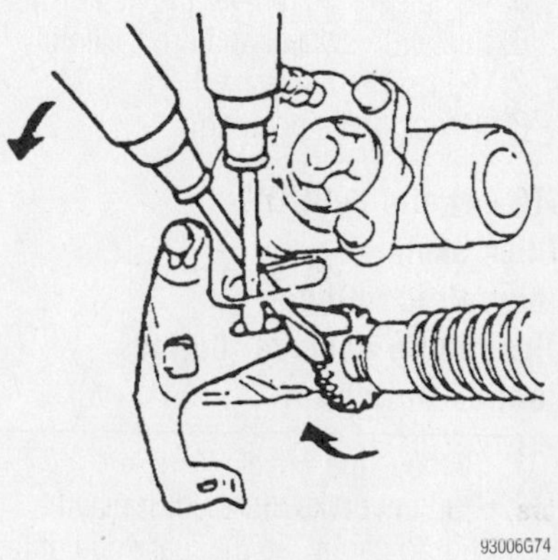

Backing off the rear brake adjuster—Prizm

Saturn

1. Remove the rear wheel and tire assembly.

2. Remove the brake drum.

3. If necessary, turn the starwheel of the brake adjuster assembly to loosen the brake shoes and allow for drum removal.

4. Remove the drum.

To install:

5. Install brake drum over brake shoes and onto hub.

6. Install tire and wheel assembly. Torque to the proper specification.

7. Adjust brakes following the proper procedure.

8. Road test for braking operation.

Brake Shoes

REMOVAL & INSTALLATION

GM C- and H-BODIES:
Buick LeSabre, Park Avenue
Oldsmobile Eighty Eight Royale, Ninety Eight, Regency
Pontiac Bonneville

1. Remove the rear wheel. Mark the position of the wheel to the wheel studs, prior to removal, for installation reference.
2. Remove the brake drum. Mark the position of the brake drum to the wheel studs, prior to removal, for installation reference.
3. Using tool J 38400 brake spanner tool and remover, remove the actuator spring from the adjuster lever. Use care not to distort the spring when removing it.
4. Lift the end of the retractor spring from the adjuster shoe assembly. Insert the hook end of the J-38400 between the retractor spring and the shoe. Pry slightly to remove the spring end from the hole in the shoe.
5. Pry the end of the retractor spring toward the axle with the flat end of the tool until the spring snaps down off the shoe web onto the backing plate.
6. Remove the one brake shoe and remove the adjuster assembly.
7. Disconnect the parking brake lever from the shoe. DO NOT remove the parking brake lever from the cable end unless it is being replaced.
8. Using J 38400, lift the end of the retractor spring from the adjuster shoe assembly. Insert the hook end of the tool between the retractor spring and the shoe. Pry slightly to remove the spring end from the hole in the shoe. Pry the end of the retractor spring toward the axle with the flat end of the tool, until the spring snaps down off the shoe web onto the backing plate. Leave the spring there. Do not remove it.
9. Remove the brake shoe.

To install:

10. Check the backing plate attaching bolts to make sure that they are tight. Use fine emery cloth to clean all rust and dirt from the shoe contact surfaces on the plate and lubricate with high temperature grease. Check the wheel cylinder for signs of leakage.
11. Clean all parts completely in brake solvent and air dry.
12. Clean the backing plate shoe contact points.
13. Inspect the inside of the brake drum. If worn, heavily grooved or if the opening is distorted, the drum should be machined or replaced. If machining, observe the maximum drum diameter specification.
14. Disassemble, clean and lubricate the adjuster screw.
15. Position the brake shoe that connects to the parking brake lever, on the backing plate. Using J-38400, pull the end of the retractor spring up to rest on the web of the shoe. Pull the end of the retractor spring up until it snaps into the slot in the brake shoe.
16. Connect the parking brake lever.
17. Install the remaining shoe and the adjuster screw assembly.
18. Position the brake shoe using J-38400 and pull the end of the retractor spring up to rest on the web of the shoe. Pull the end of the retractor spring up until it snaps into the slot in the brake shoe.
19. Using J-38400, spread the brake shoes and work the adjuster screw into position.
20. Install the actuator spring with the U-shaped end going through the web.
21. Measure the inside diameter of the brake drum using clearance gauge J 21177-A.
22. Turn the starwheel on the adjusting screw assembly until the brake shoe diameter is 0.050 in. (1.27mm) less than the drum inside diameter.
23. Install the brake drum onto the wheel hub, aligning the marks made during removal.
24. Install the wheel, aligning the marks made during removal. Torque the lug nuts to 100 ft. lbs. (140 Nm).
25. Repeat the procedure for the brake shoe assembly on the opposite side of the vehicle.
26. Apply and release the brake pedal 30–35 times with normal pedal force. Pause about 1 second between applications.
27. Road test the vehicle and check brake operation.

A ACCESS HOLE PLUG.
1 ADJUSTER SOCKET
2 ADJUSTER SCREW
3 PIVOT NUT
4 RETRACTOR SPRING
5 ADJUSTER SHOE AND LINING
6 WHEEL CYLINDER
7 BLEEDER VALVE
8 BOLT
9 BACKING PLATE
10 PARK BRAKE SHOE AND LINING
11 PARK BRAKE LEVER
12 ACTUATOR SPRING
13 ADJUSTER ACTUATOR

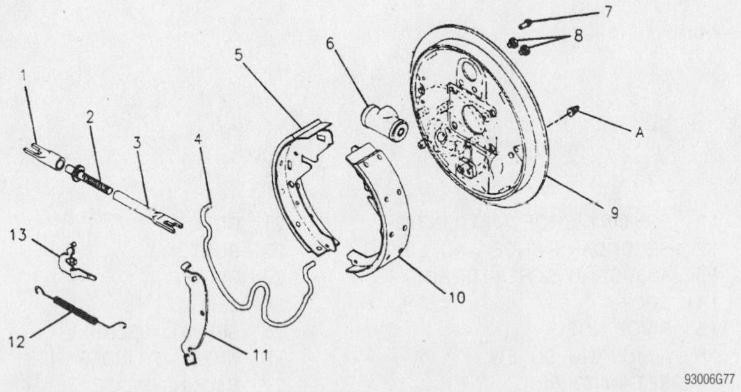

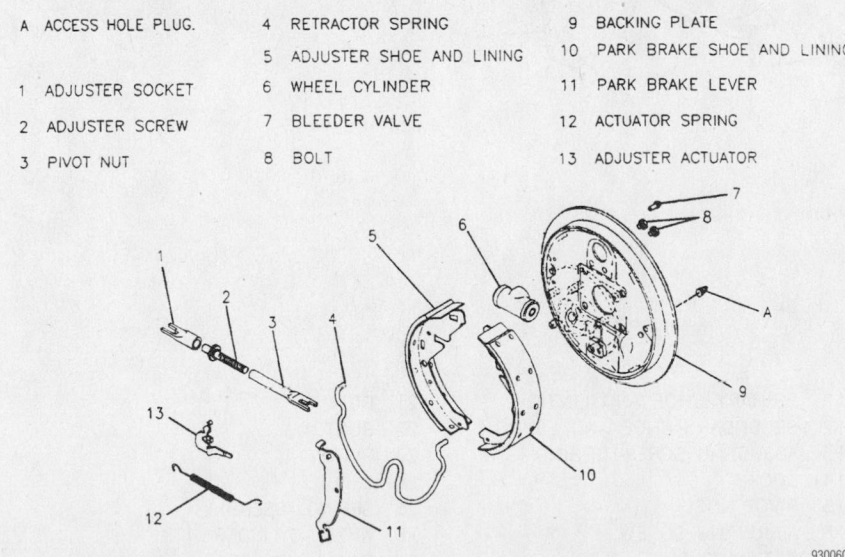

Exploded view of the drum brake components—LeSabre, Park Avenue, Eighty Eight, Royale, Ninety Eight, Regency, Bonneville

GM F-BODY:
Chevrolet Camaro
Pontiac Firebird

1. Matchmark the relationship of the wheel to the axle flange, then remove the wheel and tire assembly.
2. Matchmark the relationship of the drum to the axle flange and remove the brake drum.
3. Remove the return springs using a suitable tool.
4. Remove the hold-down springs and pins. Remove the actuator pivot.
5. Remove the actuator link while lifting up on the actuator.
6. Remove the actuator lever and lever return spring.
7. Remove the shoe guide, parking brake strut and strut spring.
8. Remove the brake shoes and discon-

nect the parking brake lever from the appropriate shoe.

9. Remove the adjusting screw assembly and spring.

To install:

10. Install the parking brake lever on the appropriate shoe by hooking the lever tab into the slot.

11. Install the adjusting screw and spring. Lubricate the adjusting screw with suitable brake grease.

12. Clean and lubricate the contact points of the backing plate, then install the brake shoe assemblies and the parking brake cable to the backing plate.

13. Install the parking brake strut and strut spring by spreading the shoes apart. The strut end with the spring engages the parking brake lever and shoe. The other end engages the opposite shoe.

14. Install the shoe guide, actuator lever and lever return spring.

15. Install the hold-down pins, actuator pivot and springs.

16. Install the actuator link on the anchor pin. Install the actuator link into the actuator lever while holding up on the lever.

17. Install the shoe return springs and adjust the brakes. When properly adjusted, the brake shoe linings will be approximately 0.050 inch (1.27mm) less than the inner diameter of the brake drum.

18. Align and install the brake drum, then the wheel and tire assembly.

GM J-BODY:
Chevrolet Cavalier
Pontiac Sunfire

1. Remove the tire and wheel assembly.

2. Remove the brake drum. It may be necessary to gently tap the rear edge of the drum to start it off the studs. If extreme resistance to removal is encountered, it will

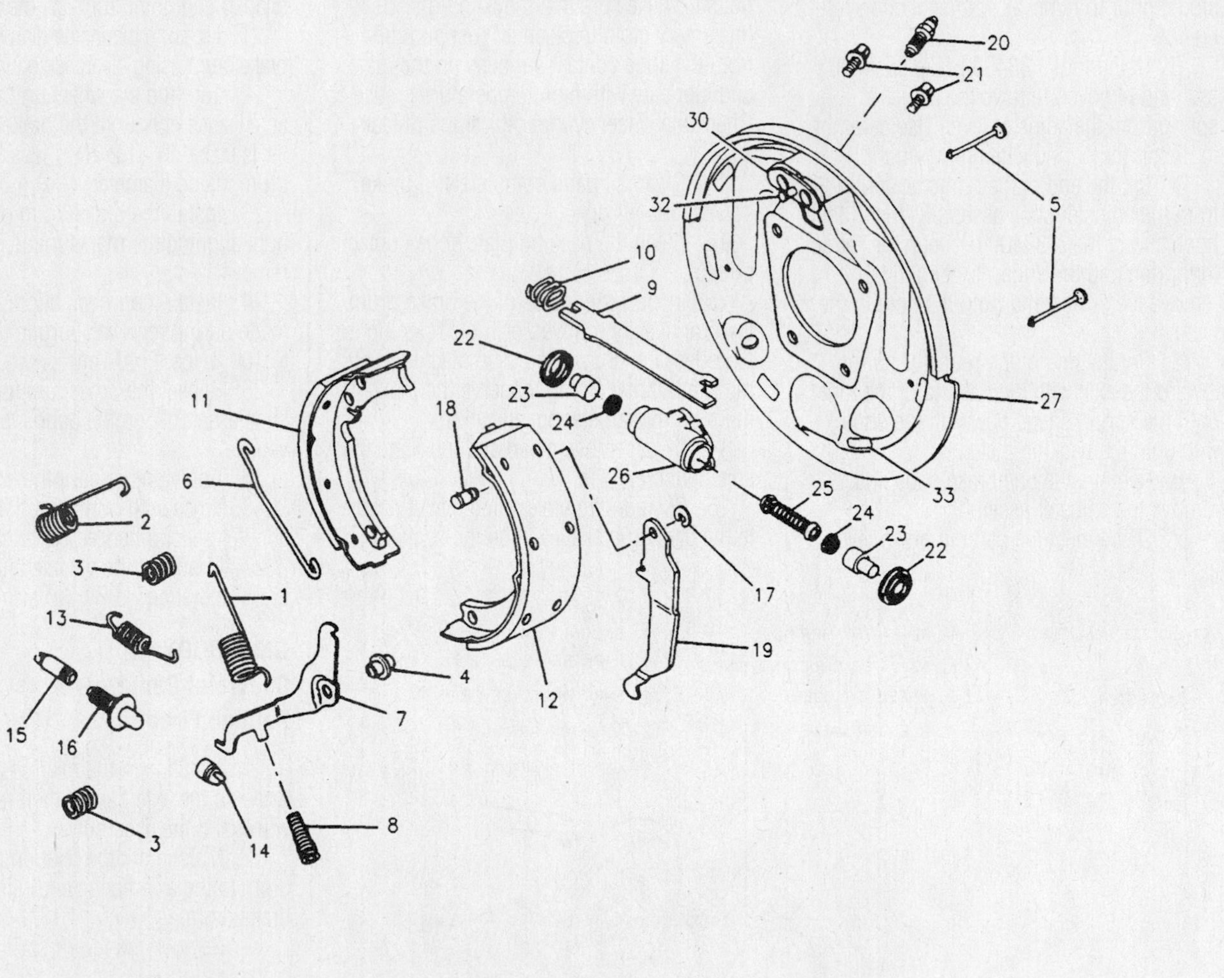

Exploded view of drum brake components—Cavalier, Sunfire

be necessary to retract the brake shoe self-adjuster screw, sometime called the star-wheel. Turn the adjuster to retract the linings from the drum.

➡Do not hammer on the brake drum to remove it.

3. Remove the upper return springs from the shoes using tool J-8057 brake spring tool.
4. Remove the hold-down springs using J-8049 brake spring tool.
5. Remove the shoe hold-down pins from behind the brake backing plate.
6. Lift up the actuator lever for the self-adjusting mechanism and remove the actuating link. Remove the actuator lever, pivot, and the pivot return spring.
7. Spread the shoes apart to clear the wheel cylinder pistons, then remove the parking brake strut and spring.
8. Disconnect the parking brake cable from the lever. Remove the shoes, still connected by their adjusting screw spring.
9. With the shoes removed, note the position of the adjusting spring and remove the spring and adjusting screw.
10. Remove the C-clip from the parking brake lever and remove the lever from the secondary shoe.
11. Use a damp cloth to remove all dirt and dust from the backing plate and brake parts.

To install:

12. Check the backing plate attaching bolts to make sure they are tight. Use fine emery cloth to clean all rust and dirt from the shoe contact surfaces on the plate and lubricate with brake grease. Check the wheel cylinder for signs of leakage.
13. Clean all parts completely in brake solvent and air dry.
14. Clean the backing plate shoe contact points.
15. Inspect the inside of the brake drum. If worn, heavily grooved or if the opening is distorted, the drum should be refinished or replaced.
16. Inspect the brake shoe springs and self-adjuster mechanism. Disassemble the adjuster mechanism and clean the threads and coat with grease. Make sure the adjuster assembly turns freely before installing in the vehicle.
17. Install the parking brake lever on the secondary shoe and secure with C-clip.
18. Install the adjusting screw and spring on the shoes, connecting them together. The coils of the spring must not be over the starwheel on the adjuster. The left and right hand springs are not interchangeable.
19. Spread the shoe assemblies apart and connect the parking brake cable. Install the shoes on the backing plate, engaging the shoes at the top temporarily with the wheel cylinder pistons. Make sure the starwheel on the adjuster is lined up with the adjusting hole in the backing plate, if equipped.
20. Spread the shoes apart slightly and install the parking brake strut and spring. Make sure the end of the strut without the spring engages the parking brake lever. The end with the spring engages the primary shoe (the one with the shorter lining).
21. Install the actuator pivot, lever and return spring. Install the actuating link in the shoe retainer. Lift up the actuator lever and hook the link into the lever.
22. Install the hold-down pins through the back of the plate using J-8057. Install the lever pivots and hold-down springs. Install the shoe return springs using J-8049. Be careful not to stretch or distort the springs.
23. Make sure the linings are in the right place, the self-adjusting mechanism is correctly installed, and the parking brake parts are hooked up.
24. Measure the distance from the edge of the primary lining to the edge secondary lining, then measure the inside width of the drum. Adjust the linings by means of the adjuster so the drum will fit onto the linings.
25. Install the hub and bearing assembly onto the axle if removed. Torque the retaining bolts to 35 ft. lbs. (55 Nm.).
26. Install the drum.
27. Install the wheel and tire assembly.
28. Adjust the brakes. Install a rubber hole cover in the adjustment knock-out hole after the adjustment is complete. Adjust the parking brake.
29. Road test the vehicle and verify proper brake operation.

GM L/N and N-BODY:
Buick Skylark
Chevrolet Malibu
Oldsmobile Achieva, Cutlass
Pontiac Grand Am

1. Remove the tire and wheel assembly.
2. Remove the brake drum.
3. Remove the upper return springs from the shoes using tool J-8057 brake spring tool.
4. Remove the hold-down springs using J-8049 brake spring tool.
5. Remove the shoe hold-down pins from behind the brake backing plate.
6. Lift up the actuator lever for the self-adjusting mechanism and remove the actuating link. Remove the actuator lever, pivot, and the pivot return spring.
7. Spread the shoes apart to clear the wheel cylinder pistons and remove the parking brake strut and spring.
8. Disconnect the parking brake cable from the lever. Remove the shoes, still connected by their adjusting screw spring.
9. With the shoes removed, note the position of the adjusting spring and remove the spring and adjusting screw.
10. Remove the C-clip from the parking brake lever and the lever from the secondary shoe.
11. Use a damp cloth to remove all dirt and dust from the backing plate and brake parts.

To install:

12. Check the backing plate attaching bolts to make sure they are tight. Use fine emery cloth to clean all rust and dirt from the shoe contact surfaces on the plate and lubricate with brake grease. Check the wheel cylinder for signs of leakage.
13. Clean all parts completely in brake solvent and air dry. Clean the backing plate shoe contact points.
14. Inspect the inside of the brake drum. If worn, heavily grooved or if the opening is distorted, the drum should be refinished or replaced.
15. Inspect the brake shoe springs and self-adjuster mechanism. Disassemble the adjuster mechanism and clean the threads and coat with grease. Make sure the adjuster assembly turns freely before installing in the vehicle.
16. Install the parking brake lever on the secondary shoe and secure with C-clip.
17. Install the adjusting screw and spring on the shoes, connecting them together. The coils of the spring must not be over the starwheel on the adjuster. The left and right hand springs are not interchangeable.
18. Spread the shoe assemblies and connect the parking brake cable. Install the shoes on the backing plate, engaging the shoes at the top temporarily with the wheel cylinder pistons. Make sure the starwheel on the adjuster is lined up with the adjusting hole in the backing plate, if equipped.
19. Spread the shoes slightly and install

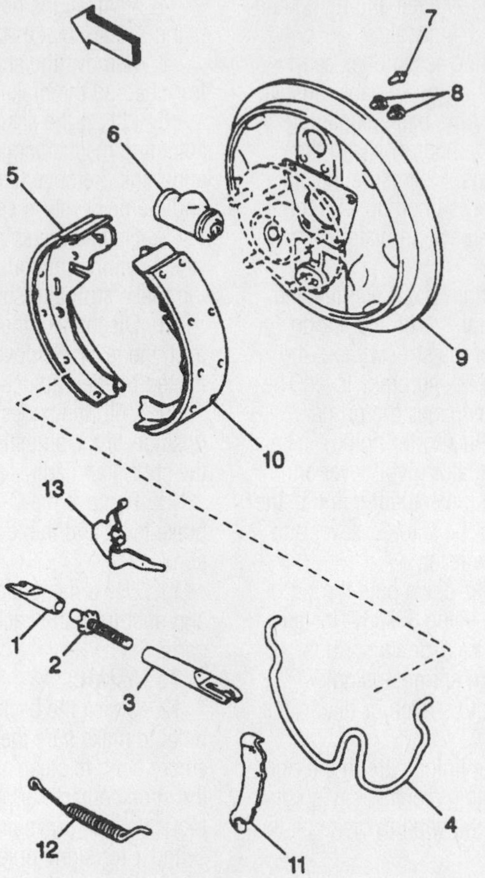

(1) Socket, Brake Adjuster
(2) Screw, Brake Adjuster
(3) Nut, Brake Pivot
(4) Spring, Retractor
(5) Brake Shoe and Lining
(6) Cylinder, Wheel Brake
(7) Valve, Bleeder
(8) Bolts, Wheel Cylinder
(9) Plate, Brake Backing
(10) Brake Shoe and Lining
(11) Lever, Park Brake
(12) Spring
(13) Adjuster Actuator

93006G78

Drum brake components, exploded view—Skylark, Malibu, Achieva, Cutlass, Grand Am

the parking brake strut and spring. Make sure the end of the strut without the spring engages the parking brake lever. The end with the spring engages the primary shoe (the one with the shorter lining).

20. Install the actuator pivot, lever and return spring. Install the actuating link in the shoe retainer. Lift up the actuator lever and hook the link into the lever.

21. Install the hold-down pins through the back of the plate using J-8057. Install the lever pivots and hold-down springs. Install the shoe return springs using J-8049. Be careful not to stretch or distort the springs.

22. Make sure the linings are in the right place, the self-adjusting mechanism is correctly installed, and the parking brake parts are hooked up.

23. Measure the distance from the edge of the primary lining to the edge secondary lining, then measure the inside width of the drum. Adjust the linings by means of the adjuster so the drum will fit onto the linings.

24. Install the hub and bearing assembly onto the axle if removed. Torque the retaining bolts to 35 ft. lbs. (55 Nm).

25. Install the drum and the wheels.

26. Adjust the brakes. Install a rubber hole cover in the adjustment knock-out hole after the adjustment is complete. Adjust the parking brake.

27. Road test the vehicle and verify proper brake operation.

GM W-BODY:
Buick Century, Regal
Chevrolet Lumina, Monte Carlo
Oldsmobile Cutlass Supreme, Intrigue
Pontiac Grand Prix

1. Remove the tire and wheel assembly.
2. Remove the brake drum.
3. Using tool J-38400 brake spanner and remover, remove the actuator spring from the adjuster lever. Use care not to distort the spring when removing it.

**** CAUTION**

During the following steps when removing the retractor spring from either shoe and lining assembly, do not over stretch the spring. This will reduce its effectiveness. Keep fingers away from retractor spring to prevent fingers from being pinched between the spring and shoe web or spring and the backing plate.

4. Lift the end of the retractor spring from the adjuster shoe assembly. Insert the hook end of the J-38400, between the retractor spring and the shoe. Pry slightly to remove the spring end from the hole in the shoe.
5. Pry the end of the retractor spring toward the axle with the flat end of the tool until the spring snaps down off the shoe web onto the backing plate.
6. Remove the one brake shoe and remove the adjuster assembly.
7. Disconnect the parking brake lever from the shoe. DO NOT remove the parking brake lever from the cable end unless it is being replaced.
8. Using J-38400, lift the end of the retractor spring from the adjuster shoe assembly. Insert the hook end of the J-38400, between the retractor spring and the shoe. Pry slightly to remove the spring end from the hole in the shoe. Pry the end of the retractor spring toward the axle with the flat end of the tool until the spring snaps down off the shoe web onto the backing plate.
9. Remove the brake shoe.

To install:

10. Clean all the brake spring completely with brake solvent and allow to air dry.
11. Disassemble, clean and lubricate the adjuster screw. Once lubricated, reassemble.
12. Clean the backing plate and after it is

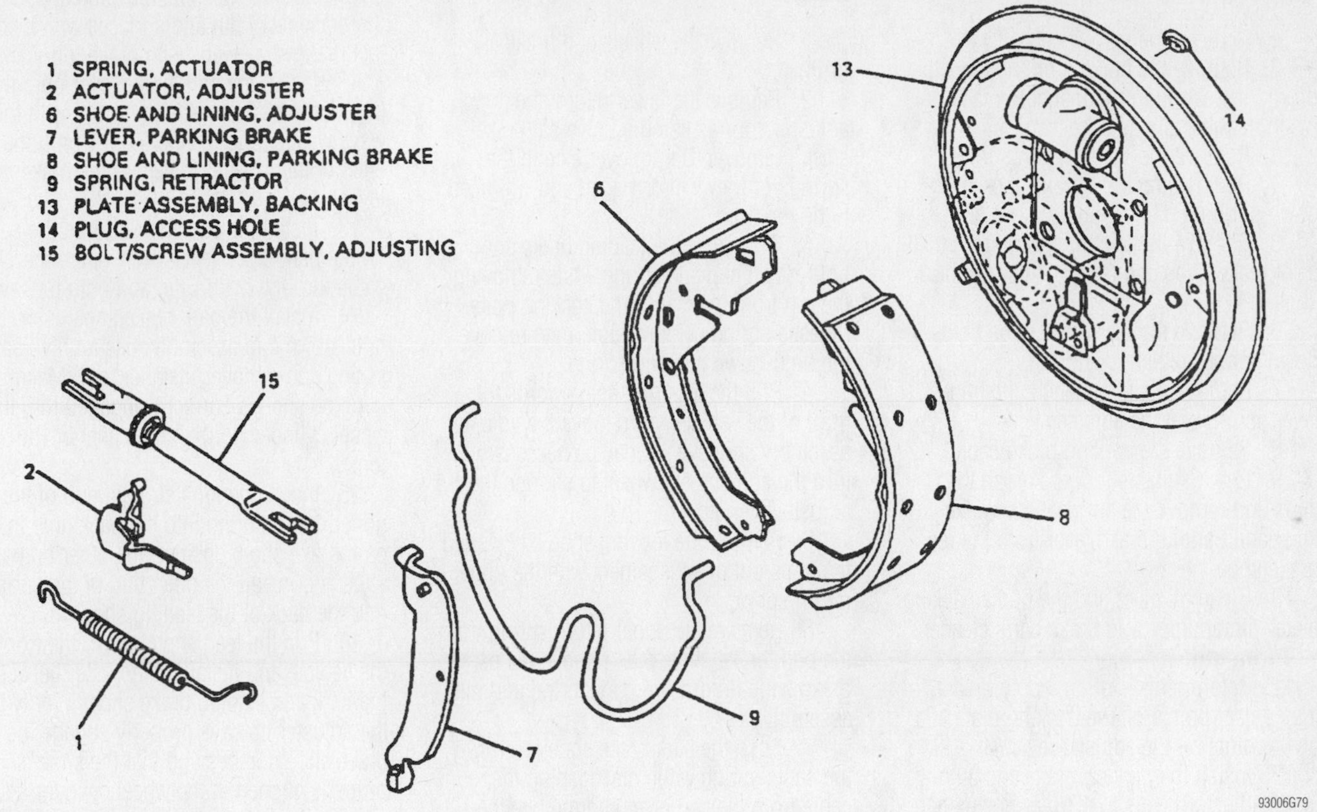

Rear brake assembly component alignment—Century, Regal, Lumina, Monte Carlo, Cutlass Supreme, Intrigue, Grand Prix

dry apply a thin coat of brake grease to the brake shoe contact points on the backing plate.

13. Position the brake shoe that connects to the parking brake lever, on the backing plate. Using J-38400, pull the end of the retractor spring up to rest on the web of the shoe. Pull the end of the retractor spring up until it snaps into the slot in the brake shoe.
14. Connect the parking brake lever.
15. Install the remaining shoe and the adjuster screw assembly.
16. Position the brake shoe, using J-38400, pull the end of the retractor spring up to rest on the web of the shoe. Pull the end of the retractor spring up until it snaps into the slot in the brake shoe.
17. Using J-38400, spread the brake shoes and work the adjuster screw into position.
18. Install the actuator spring with the U-shaped end going through the web.
19. Install the brake drum.
20. Install the tire and wheel assembly.
21. Adjust the brakes.
22. Road test for proper brake operation.

GEO Metro

1. Remove the rear wheels.
2. Remove the brake drum.
3. Remove the upper and lower return springs from the brake shoes.
4. Remove the anti-rattle spring and brake shoe adjustment strut.
5. Remove the primary and secondary brake shoe hold-down springs by pressing in and turning the hold-down pins.

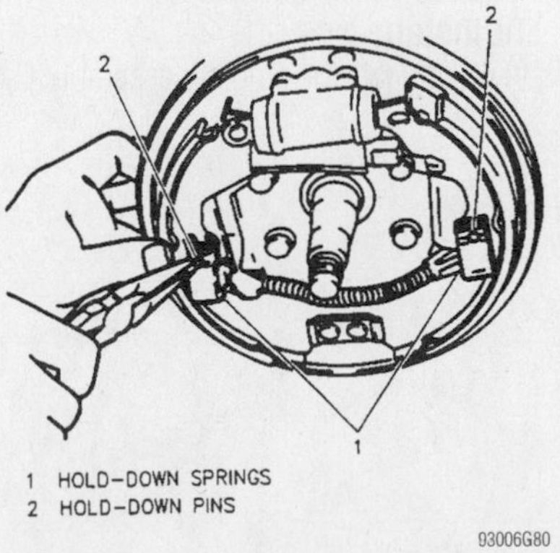

Removing the shoe hold-down springs—Metro

6. Remove the parking brake cable.
7. Inspect all parts for wear and/or damage and replace as necessary.
8. Clean the brake backing plate and all parts to be reused with suitable brake parts cleaner and allow it to dry.

To install:

9. Lubricate all contact points on the backing plate with grease.
10. Install the hold-down pins on the backing plate.
11. Install the brake cable.
12. Install the primary and secondary brake shoes to the vehicle and secure with the hold-down springs.
13. Install the brake adjustment strut and anti-rattle spring.
14. Install the upper and lower return springs to the primary and secondary brake shoes.
15. Install the brake drum.
16. Install the rear wheels.
17. Press the brake pedal 3–5 times with approximately 66 lbs. (30 kg) pressure to adjust the brake shoe clearance.
18. Check to ensure the brake drum is free from dragging and proper braking is obtained.

GEO Prizm

1. Remove the rear wheels.
2. Remove the brake drum. Inspect the drum for wear and/or damage. Replace or machine the drum as necessary.
3. Remove the upper return spring.
4. Remove the front brake shoe hold-down spring, retainers and pin.
5. Remove the anchor spring from the front brake shoe and remove the front brake shoe.
6. Remove the rear brake shoe hold-down spring, retainers and pin.
7. Remove the automatic adjusting lever spring and the adjuster.
8. Remove the parking brake cable.
9. Use a suitable tool to spread the C-clips and remove the automatic adjuster lever and parking brake lever from the rear brake shoe.
10. Clean all parts and the brake backing plate with an approved brake parts cleaner.

To install:

11. Before reinstallation, apply grease to the contact points on the backing plate and pivot points on the adjuster lever.
12. Install the parking brake and automatic adjuster levers to the rear brake shoe.
13. Install new C-clips and use pliers to install them. Do not bend more than necessary to hold in place.
14. Check the clearance between the parking brake lever and the brake shoe using a feeler gauge. The clearance must be at least 0.0138 in. (0.35mm). If the clearance is not as specified, replace the shim between the lever and shoe.
15. Connect the parking brake cable.
16. Install the adjuster and the lever spring.
17. Install the rear brake shoe on the backing plate with the end of the shoe inserted in the wheel cylinder and the adjuster in place. Install the hold-down spring, retainer and pin.
18. Install the anchor spring between the front and rear shoe.
19. Install the front brake shoe with the end of the shoe inserted in the wheel cylinder and the adjuster in place.
20. Install the hold-down spring, retainers and pin.
21. Connect the return spring.
22. Adjust the brake shoes. When adjusted properly, the outside diameter of the brake shoes will be approximately 0.024 in. (0.6mm) less than the inner diameter of the brake drum.
23. Install the brake drum and the wheel.
24. Road test and check for proper brake operation.

Saturn

1. Remove the wheels and brake drums.
2. Remove the lower return and adjuster springs using a universal brake spring remover. Do not over extend the springs or they will damaged and will need to be replaced.
3. Compress the leading brake shoe hold-down cup and spring while removing the pin from the rear of the backing plate. Release spring compression, then remove the hold-down cup and spring.
4. Pull the leading shoe towards the front of the vehicle and remove the adjuster assembly and lever. It may be necessary to turn the adjuster starwheel to shorten the adjuster's length.
5. Remove the leading shoe by twisting the shoe out of engagement with the upper return spring.
6. Remove the upper return spring from the park brake shoe, then remove the park brake shoe hold-down cup, spring and pin assembly.
7. Push the park brake shoe lever into the cable spring while disengaging the cable from the end lever and remove the parking brake shoe, lever and cable spring from the vehicle.
8. Remove the retainer and wave washer, then remove the park brake lever from the shoe.
9. Disassemble the brake adjuster socket, screw and nut, then clean the components in denatured alcohol. Inspect the assembly, making sure the screw threads smoothly into the adjusting nut over the full threaded length.
10. Inspect the wheel cylinder for signs of leakage and for cut or damaged boots. Do not attempt to repair a damaged cylinder, the assembly must be replaced.

To install:

11. Lubricate the adjuster assembly, the 6 backing plate raised shoe contact pads, the brake lever pin and surfaces which contact brake shoe webs with brake lubricant.

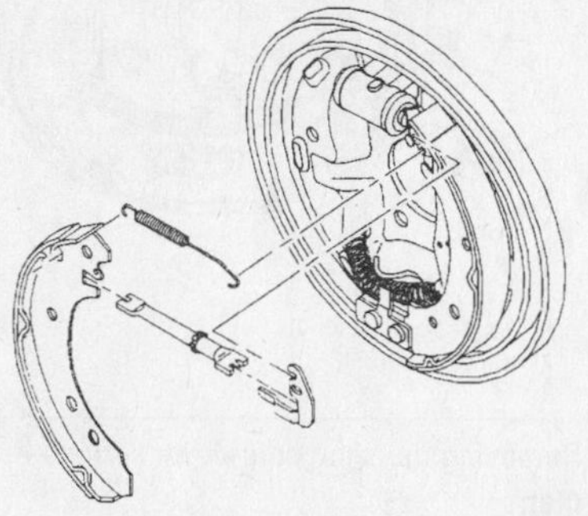
93006G81

Rear brake shoe and adjuster installation—Saturn

12. Install the park brake lever onto the pin on the brake shoe and secure with the wave washer and retainer clip. Crimp the ends of the retainer to secure the brake lever.
13. Install the cable spring into the cage on the park brake lever, then install the cable through the spring and onto the lever.
14. Install the park brake shoe using the hold-down cup assembly; use a universal spring cup remover/installer tool. Make sure the shoe is correctly engaged into the wheel cylinder (top) and the anchor (bottom).
15. Install the long straight end of the upper return spring into the back hole in the park brake shoe, position the other brake shoe and install the other end of the spring into the back of the leading shoe.
16. Pull the lead shoe toward the front of the vehicle and install the adjuster between the park and leading brake shoes. Verify that the adjuster notches properly engage the brake shoe notches and that the shoe is properly aligned in the wheel cylinder and anchor.
17. Install the adjuster lever and adjuster spring. Make sure the notch on the lever engages the pin on the park shoe and the notch on the adjusting socket. The lower leg of the lever should engage the teeth of the starwheel adjuster assembly.
18. Secure the leading brake shoe using the hold-down cup assembly.
19. Install the adjuster spring to the upper side of the brake shoes with the short end to the lead shoe and the long end to the adjuster lever. Then install the lower return spring into the lower holes of the shoes.
20. Verify the correct location of all brake components, if necessary, use the other side brake assembly for comparison.
21. Using a suitable drum clearance gauge, measure the inner diameter of the brake drum and adjust the outside diameter of the brake shoes to 0.02 inch (0.50mm) less than the inner diameter of the drum.
22. Repeat the procedure for the opposite brake shoes and install the brake drums.
23. If the wheel cylinders have been replaced, bleed the hydraulic brake system.
24. Install the rear wheels.
25. Apply and release the brake pedal 20 times to allow the adjuster to properly position the brake shoes.
26. Check and adjust the parking brake cable, as necessary.

CHRYSLER CORPORATION

1998–02

Chrysler-Concorde • LHS • 300M • **Dodge-**Intrepid • **Eagle-**Vision

5

PRECAUTIONS

Before servicing any vehicle, please be sure to read all of the following precautions, which deal with personal safety, prevention of component damage, and important points to take into consideration when servicing a motor vehicle:

- Never open, service or drain the radiator or cooling system when the engine is hot; serious burns can occur from the steam and hot coolant.
- Observe all applicable safety precautions when working around fuel. Whenever servicing the fuel system, always work in a well-ventilated area. Do not allow fuel spray or vapors to come in contact with a spark, open flame or excessive heat (a hot drop light, for example). Keep a dry chemical fire extinguisher near the work area. Always keep fuel in a container specifically designed for fuel storage; also, always properly seal fuel containers to avoid the possibility of fire or explosion. Refer to the additional fuel system precautions later in this section.
- Fuel injection systems often remain pressurized, even after the engine has been turned **OFF**. The fuel system pressure must be relieved before disconnecting any fuel lines. Failure to do so may result in fire and/or personal injury.
- Brake fluid often contains polyglycol ethers and polyglycols. Avoid contact with the eyes and wash your hands thoroughly after handling brake fluid. If you do get brake fluid in your eyes, flush your eyes with clean, running water for 15 minutes. If eye irritation persists, or if you have taken brake fluid internally, IMMEDIATELY seek medical assistance.
- The EPA warns that prolonged contact with used engine oil may cause a number of skin disorders, including cancer! You should make every effort to minimize your exposure to used engine oil. Protective gloves should be worn when changing oil. Wash your hands and any other exposed skin areas as soon as possible after exposure to used engine oil. Soap and water, or waterless hand cleaner should be used.
- All new vehicles are now equipped with an air bag system. The system must be disabled before performing service on or around system components, steering column, instrument panel components, wiring and sensors. Failure to follow safety and disabling procedures could result in accidental air bag deployment, possible personal injury and unnecessary system repairs.
- Always wear safety goggles when working with, or around, the air bag system. When carrying a non-deployed air bag, be sure the bag and trim cover are pointed away from your body. When placing a non-deployed air bag on a work surface, always face the bag and trim cover upward, away from the surface. This will reduce the motion of the module if it is accidentally deployed. Refer to the additional air bag system precautions later in this section.
- Clean, high quality brake fluid from a sealed container is essential to the safe and proper operation of the brake system. You should always buy the correct type of brake fluid for your vehicle. If the brake fluid becomes contaminated, completely flush the system with new fluid. Never reuse any brake fluid. Any brake fluid that is removed from the system should be discarded. Also, do not allow any brake fluid to come in contact with a painted surface; it will damage the paint.
- Never operate the engine without the proper amount and type of engine oil; doing so WILL result in severe engine damage.
- Timing belt maintenance is extremely important! Many models utilize an interference-type, non-freewheeling engine. If the timing belt breaks, the valves in the cylinder head may strike the pistons, causing potentially serious (also time-consuming and expensive) engine damage. Refer to the maintenance interval charts in the front of this manual for the recommended replacement interval for the timing belt, and to the timing belt section for belt replacement and inspection.
- Disconnecting the negative battery cable on some vehicles may interfere with the functions of the on-board computer system(s) and may require the computer to undergo a relearning process once the negative battery cable is reconnected.
- When servicing drum brakes, only disassemble and assemble one side at a time, leaving the remaining side intact for reference.
- Only an MVAC-trained, EPA-certified automotive technician should service the air conditioning system or its components.

ENGINE REPAIR

Alternator

REMOVAL

2.7L Engine

1. Remove or disconnect the following:
   - Negative battery cable
   - Lower plastic splash shield
   - Transmission cooler and position it aside
   - Lower radiator crossmember support
2. Loosen but do not remove:
3. Adjusting "T" bolt and the pivot bolt.
4. Drive belt adjusting bolt.
5. Remove or disconnect the following:
   - Drive belt
   - Alternator field circuit plug
   - Alternator B+ terminal nut and wire
   - Pivot bolt; be careful not to lose the spacer
   - Alternator

3.2L and 3.5L Engines

1. Remove or disconnect the following:
   - Negative battery cable
   - Lower mounting bolt and pivot bolt
   - Drive belt tension bolt
   - Drive belt
   - Bracket, lower mounting bolt and pivot bolt
   - Alternator
   - Alternator field circuit plug
   - Alternator B+ terminal nut and wire

INSTALLATION

2.7L Engine

1. Install or connect the following:
   - Alternator
   - Pivot bolt with the spacer and leave loose
   - Alternator field circuit plug
   - Alternator B+ wire and terminal nut; then, torque the nut to 89 inch lbs. (10 Nm)
   - Drive belt
2. Using a belt tension gauge, torque the adjusting "T" bolt until the tension gauge reads 120 lbs. (534 N) for a used belt or 180–200 lbs. (792–880 N) for a new belt.

- Lower radiator crossmember support
- Transmission cooler
- Lower plastic splash shield
- Negative battery cable

3. Torque the bolts to the following specifications:
   a. Pivot bolt to 40 ft. lbs. (54 Nm).
   b. 8mm mounting bolt to 30 ft. lbs. (41 Nm).
   c. 10mm mounting bolt to 40 ft. lbs. (54 Nm).

3.5L (VIN F) Engine

1. Install or connect the following:
   - Alternator B+ wire and terminal nut and torque to 89 inch lbs. (10 Nm)
   - Alternator field circuit plug
   - Alternator
   - Drive belt
   - Pivot bolt and lower mounting bolt
2. Using a belt tension gauge, torque the drive belt until the tension gauge reads 120 lbs. (534 N) for a used belt or 140–160 lbs. (623–711 N) for a new belt.
3. Torque the pivot and lower mounting bolts to 40 ft. lbs. (54 Nm).
4. Install the bracket and connect the negative battery cable
5. Connect the negative battery cable.

3.2L and 3.5L (VIN G) Engines

1. Install or connect the following:
   - Alternator B+ wire and terminal nut; then, torque the nut to 89 inch lbs. (10 Nm)
   - Alternator field circuit plug
   - Alternator
   - Drive belt
   - Pivot bolt and lower mounting bolt
2. Using a belt tension gauge, torque the "V" drive belt until the tension gauge reads 120 lbs. (534 N) for a used belt or 140–160 lbs. (623–711 N) for a new belt.
3. Using a belt tension gauge, torque the Poly "V" belt until the tension gauge reads 120 lbs. (534 N) for a used belt or 180–200 lbs. (792–880 N) for a new belt.
4. Install the bracket.
5. Torque the bolts to the following specifications:
   a. Pivot bolt to 40 ft. lbs. (54 Nm).
   b. 8mm mounting bolt to 30 ft. lbs. (41 Nm).
   c. 10mm mounting bolt to 40 ft. lbs. (54 Nm).
6. Connect the negative battery cable.

Ignition Timing

ADJUSTMENT

All models utilize a Distributorless Ignition System (DIS). It is a fixed ignition timing system, which means that basic ignition timing cannot be adjusted. All spark advance is permanently set by the Powertrain Control Module (PCM).

Engine Assembly

REMOVAL & INSTALLATION

2.7L Engine

1. Before servicing the vehicle, refer to the precautions in the beginning of this section.
2. Relieve the fuel system pressure.
3. Drain the cooling system.
4. Drain the engine oil.
5. Remove or disconnect the following:
   - Negative battery cable at the remote terminal near the right strut tower
   - Hood
   - Wiper arms, cowl covers and cowl support
   - Air intake duct and air cleaner assembly
   - Upper radiator crossmember
   - Hood release cable from the latch
   - Cooling fan assembly
   - Upper and lower radiator hoses
   - A/C condenser-to-radiator fasteners
   - Radiator
   - Accessory drive belts
   - Power steering pump and position it aside without disconnecting the lines

➡Remove the alternator with the engine.

   - A/C compressor and position it aside without disconnecting the lines
   - V-band clamps at the exhaust manifolds
   - Speed control and throttle cables from the throttle body
   - Heater hoses and the coolant hoses at the recovery tank
   - All vacuum lines, electrical connectors and ground straps from the engine
   - Catalytic converter down pipes-to-rear mount fasteners
   - Structural collar mounting bolts and collar
6. Matchmark the flexplate-to-torque converter and remove the torque converter bolts.
   - Both transaxle cooler line-to-engine brackets
   - Starter and Crankshaft Position (CKP) sensor
   - 2 lower transaxle-to-engine bolts
   - Exhaust Gas Recirculation (EGR) valve assembly

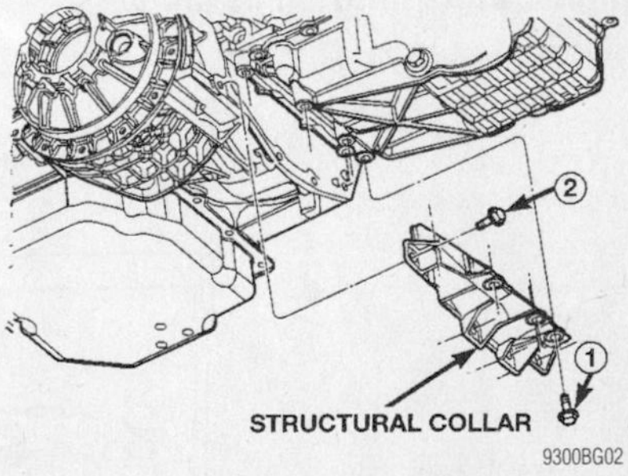

Removing the structural collar—2.7L engine

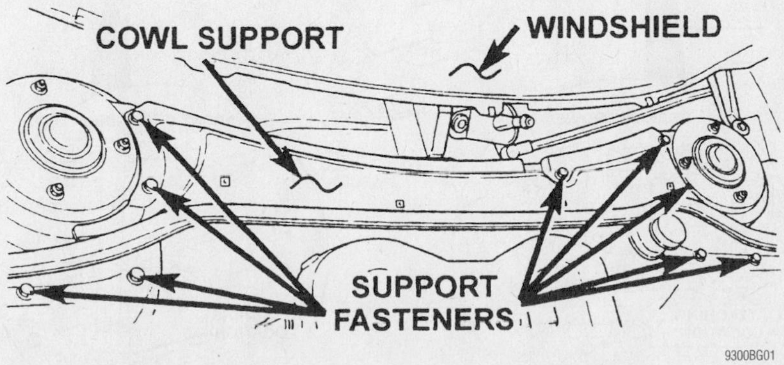

Remove the cowl support for clearance to remove the engine—2.7L engine

- Fuel line and engine harness from the throttle body bracket

7. Remove the following brackets from the double-ended transaxle-to-engine bolts:

- Wiring harness bracket
- Transaxle shift cable bracket
- Throttle body support bracket
- Upper transaxle-to-engine bolts
- Both (right and left) engine mount isolators-to-engine mount bracket fasteners

WARNING

Be careful not to damage the intake manifold or cylinder head covers when lifting the engine.

8. Attach an engine lifting fixture to the engine.
9. Support the transaxle with a jack and remove the engine.

To install:

10. Install the engine.
11. Align the engine mounts and install the fasteners, but do not tighten them at this time.

WARNING

Do not tighten the transaxle-to-engine bolts until all of the bolts have been hand-started and the engine is flush against the transaxle.

12. Install the transaxle-to-engine bolts. Torque the bolts to 75 ft. lbs. (102 Nm).
13. Remove the engine support fixture.
14. Torque the engine mount isolator-to-bracket fasteners to 45 ft. lbs. (61 Nm).
15. Align the matchmarks and install the flexplate-to-torque converter bolts. Torque the bolts to 55 ft. lbs. (75 Nm).
16. Install the starter and the CKP sensor.
17. Install the structural collar using the following sub-steps:
    a. Torque the vertical collar-to-oil pan bolts to 10 inch lbs. (1.1 Nm).
    b. Torque the collar-to-transaxle bolts to 40 ft. lbs. (55 Nm).
    c. Torque the center vertical bolt to 40 ft. lbs. (55 Nm); then, the remaining vertical bolts to the same torque.
18. Complete the installation by reversing the removal procedure. Keep the following in mind:
    - Use new V-band clamps and torque them to 100 inch lbs. (11 Nm)
    - Torque the A/C compressor and power steering pump fasteners to 21 ft. lbs. (28 Nm)
    - Negative battery cable

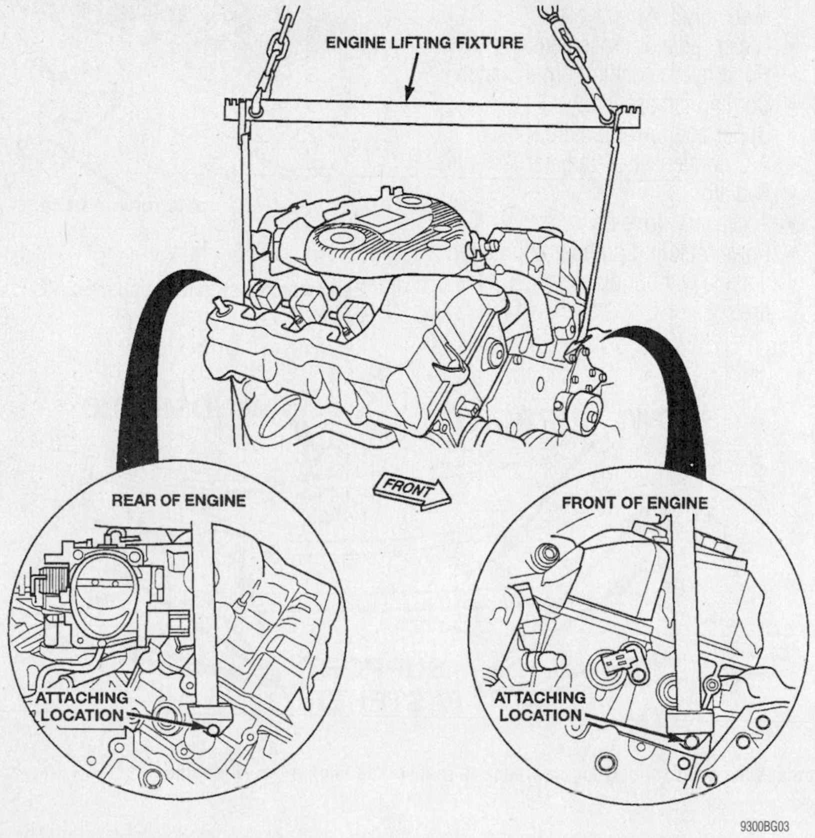

Engine support fixture attaching points—2.7L engine

19. Fill the cooling system to the proper level.
20. Fill the engine with clean oil.
21. Start the vehicle, check for leaks and repair if necessary.

3.2L and 3.5L Engines

1. Before servicing the vehicle, refer to the precautions in the beginning of this section.
2. Drain the engine oil.
3. Drain the engine coolant.
4. Properly relieve the fuel system pressure.
5. Remove or disconnect the following:

- Negative battery cable
- Hood
- Wiper arms
- Cowl covers and supports
- Air cleaner and air inlet duct
- Upper radiator support and hood release cable
- Fan module
- Accessory drive belts
- Upper and lower radiator hoses
- Engine oil and transmission cooler lines at the radiator
- Alternator, if necessary.
- A/C compressor mounting bolts and position it aside with the lines connected
- Power steering pump mounting bolts and position the pump aside
- V-band clamp at the right exhaust manifold
- Right side catalytic converter down pipe bracket fasteners
- Fuel lines
- Throttle and cruise control cables
- All vacuum hoses
- Ground straps at both cylinder heads

WARNING

The intake manifold is a composite design and should be removed before lifting the engine or it may be damaged and require replacement.

- Upper intake manifold and cover the openings
- Heater hoses
- Throttle body support brackets
- Upper transaxle-to-engine mounting bolts
- All electrical and vacuum hose connections
- Structural collar and matchmark the flexplate-to-torque converter position
- Flexplate-to-torque converter mounting bolts

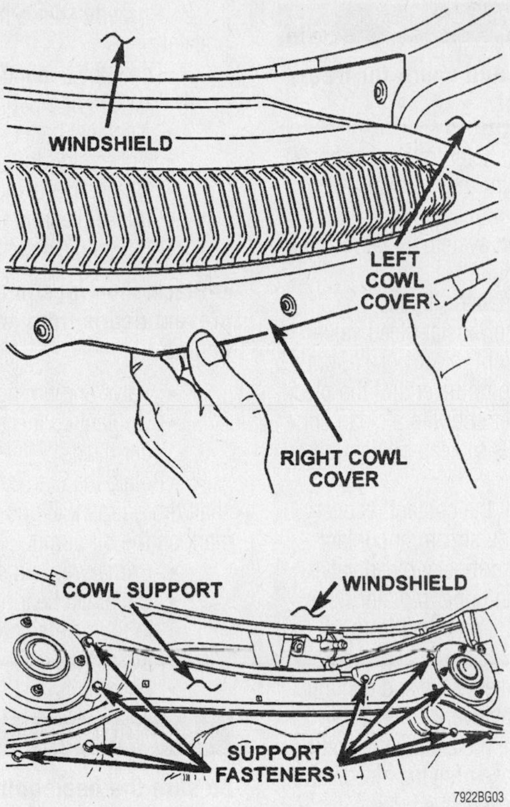

Be careful not to damage the windshield when removing the cowl covers and supports 3.2L and 3.5L models

- Left exhaust manifold V-band clamp and left catalytic converter support brackets
- Starter
- Left and right engine mounting bolts
- Crankshaft Position (CKP) sensor
- Lower engine-to-transaxle mounting bolts

6. Attach a suitable lifting device to the engine.
7. Support the transaxle using a floor jack with a small block of wood in between.
8. Slowly lift the engine from the vehicle.

To install:

9. Install the engine.
10. Align the engine mounts and install the fasteners, but do not tighten them until all of the mounting bolts have been installed.
11. Install the engine-to-transaxle mounting bolts and torque to 75 ft. lbs. (102 Nm).
12. Remove the engine lifting device.
13. Torque the engine mount nuts/bolts 45 ft. lbs. (61 Nm).
14. Align the flexplate-to-torque converter matchmarks and torque the bolts to 55 ft. lbs. (75 Nm).
15. Install or connect the following:
    - CKP sensor and the starter
    - Left-side exhaust manifold V-band clamp and torque to 100 inch lbs. (11 Nm)
    - Left-side catalytic converter mounting bracket fasteners
16. Install the structural collar using the following procedure:
    a. Install the vertical collar-to-oil pan bolts and tighten, temporarily to 10 inch lbs. (1.1 Nm).
    b. Install the collar-to-transaxle bolts and torque to 40 ft. lbs. (55 Nm).
    c. Starting with the center vertical bolt and working outward, torque the bolts to 40 ft. lbs. (55 Nm).
17. Install or connect the following:
    - Throttle body support bracket
    - Heater hoses, all ground straps and vacuum hoses
    - Upper intake manifold
    - All engine wiring harnesses
18. Adjust the throttle and cruise control cables
    - Fuel lines
    - V-band clamp on the right exhaust manifold and torque to 100 inch lbs. (11 Nm)
    - Right-side catalytic converter mounting bracket fasteners
    - A/C compressor and torque the bolts to 21 ft. lbs. (28 Nm)
    - Alternator and its wiring harness
    - Radiator
    - Power steering pump
    - Engine oil and transaxle cooler lines
    - Radiator hoses and accessory drive belt
    - Hood release cable and upper radiator support
    - Air cleaner and inlet hose
    - Cowl covers, supports and wiper arms
    - Hood
    - Negative battery cable
19. Fill the cooling system to the proper level.
20. Fill the engine with clean oil.
21. Start the vehicle, check for leaks and repair if necessary.

Water Pump

REMOVAL & INSTALLATION

The water pump has a die cast aluminum body and a stamped steel impeller. It bolts directly to the chain case cover using an O-ring for sealing. It is driven by the back side of the serpentine belt.

It is normal for a small amount of coolant to drip from the weep hole located on the water pump body (small black spot). If this condition exists, DO NOT replace the water pump. Only replace the water pump if a heavy deposit or steady flow of brown/green coolant is visible on the water pump body from the weep hole, which would indicate shaft seal failure. Before replacing the water pump, be sure to perform a thorough inspection. A defective pump will not be able to circulate heated coolant through the long heater hose.

2.7L Engine

1. Before servicing the vehicle, refer to the precautions in the beginning of this section.
2. Drain the cooling system.
3. Remove or disconnect the following:
    - Negative battery cable
    - Upper radiator crossmember
    - Fan module
    - Accessory drive belts

➡**The water pump is driven by the primary timing chain.**

- Crankshaft damper, timing chain cover, timing chain and all guides
- Water pump mounting bolts
- Water pump

4. Clean the water pump mounting surfaces.

To install:

5. Install or connect the following:
   - Water pump and gasket and torque the bolts to 105 inch lbs. (12 Nm)
   - Guides, timing chain and timing chain cover
   - Crankshaft damper and torque the center bolt to 125 ft. lbs. (170 Nm)
   - Accessory drive belts
   - Fan module and upper radiator crossmember
   - Negative battery cable
6. Fill the cooling system to the proper level.
7. Start the vehicle, check for leaks and repair if necessary.

3.2L and 3.5L Engines

1. Before servicing the vehicle, refer to the precautions in the beginning of this section.
2. Drain the cooling system.

⁂ WARNING

Do not use pliers to open the plastic drain.

3. Remove or disconnect the following:
   - Negative battery cable
   - Coolant recovery cap and open the thermostat bleed valve
   - Timing belt. Refer to the appropriate section for the removal and installation procedure.

➡**It is good practice to turn the crankshaft until the No. 1 cylinder is at Top Dead Center (TDC) of its compression stroke (firing position).**

- Water pump mounting bolts and pump. Discard the O-ring seal

4. Clean the gasket sealing surfaces, being careful not to scratch the aluminum surfaces.

To install:

5. Install or connect the following:
   - New O-ring and wet with clean coolant prior to installation. Be sure to keep the new O-ring free of any oil or grease.
   - Water pump with a new O-ring. Torque the water pump-to-engine bolts to 105 inch lbs. (12 Nm).

⁂ WARNING

Rotate the pump and check for freedom of movement.

- Timing belt. Refer to the appropriate section for the removal and installation procedure.

6. Fill the cooling system by performing the following procedure:

a. Close the radiator drain.

b. Open the thermostat bleed valve. Install a ¼ in. (6mm) clear hose about 48 in. (1.2m) long to the end of the bleed valve and the other end into a clean container. The intent is to keep coolant off of the drive belt(s).

c. Slowly, refill the coolant recovery bottle until a steady stream of coolant flows out of the thermostat bleed valve. Gently squeeze the upper radiator hose until all of the air is removed from the system.

d. Close the bleed valve and continue to fill the coolant recovery bottle to the proper level. Install the cap on the bottle and remove the hose from the bleed valve.

7. Reconnect the negative battery cable. Start the engine and allow it to reach normal operating temperatures.
8. Check the cooling system for leaks and correct coolant level. Be sure that the thermostat bleed valve is closed once the cooling system has been bled of any trapped air.

Cylinder Head

REMOVAL & INSTALLATION

2.7L Engine

1. Before servicing the vehicle, refer to the precautions in the beginning of this section.
2. Properly relieve the fuel system pressure.
3. Drain the cooling system.
4. Remove or disconnect the following:
   - Negative battery cable
   - Accessory drive belts
   - Crankshaft damper
   - Intake plenum, lower intake manifold and exhaust manifold

➡**Place shop rags in the openings to prevent debris from entering the engine.**

- Valve and timing chain covers
- Coolant connections for the cylinder heads

5. Rotate the crankshaft until the crankshaft timing mark aligns with the timing mark on the oil pump.
   - Primary timing chain
   - Camshaft bearing caps, gradually, in the reverse order of the tightening sequence
   - Camshafts

⁂ WARNING

Be sure the head bolts 9–11 are removed before attempting to remove the cylinder head, the head and/or block may be damaged.

- Cylinder head bolts in reverse order of installation starting with bolts 11–9, then 8–1
- Cylinder head(s)

To install:

6. Thoroughly clean and dry the mating surfaces of the head and block. Check the cylinder head for cracks, damage or engine coolant leakage. Remove scale, sealing compound and carbon. Clean the oil passages thoroughly.
7. Place a new head gasket on the cylinder block over the locating dowels.
8. Inspect the cylinder head bolts for necking (stretching) by holding a straight-

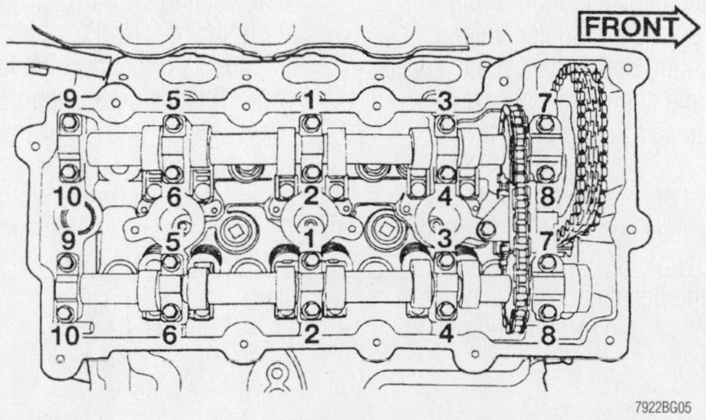

Camshaft bolt tightening sequence—2.7L engine

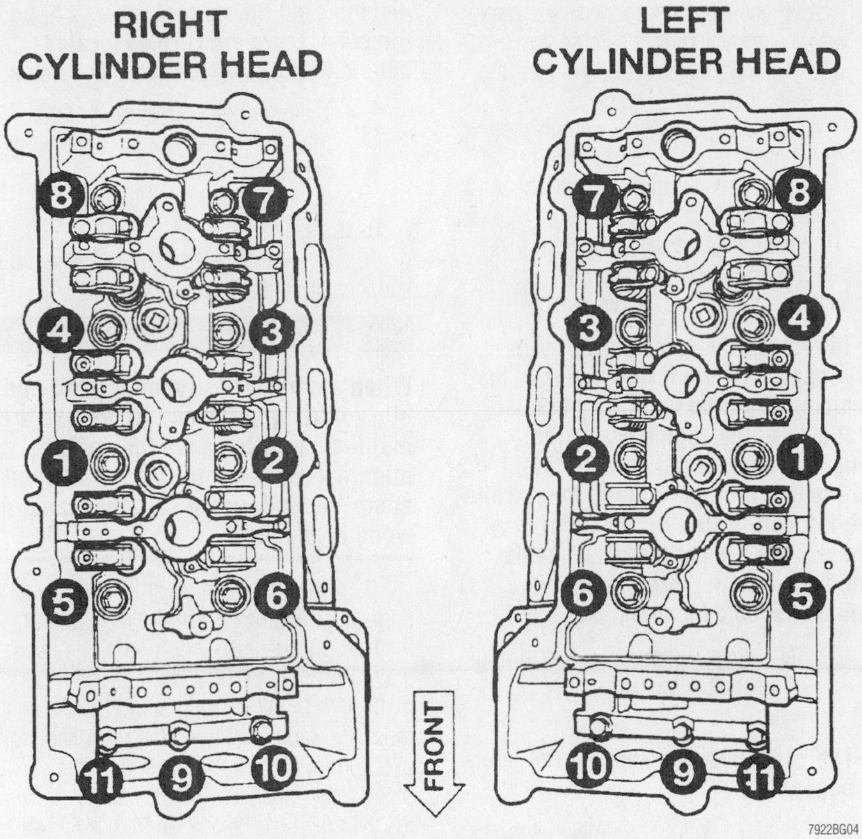

Cylinder head bolt tightening sequence —2.7L engine

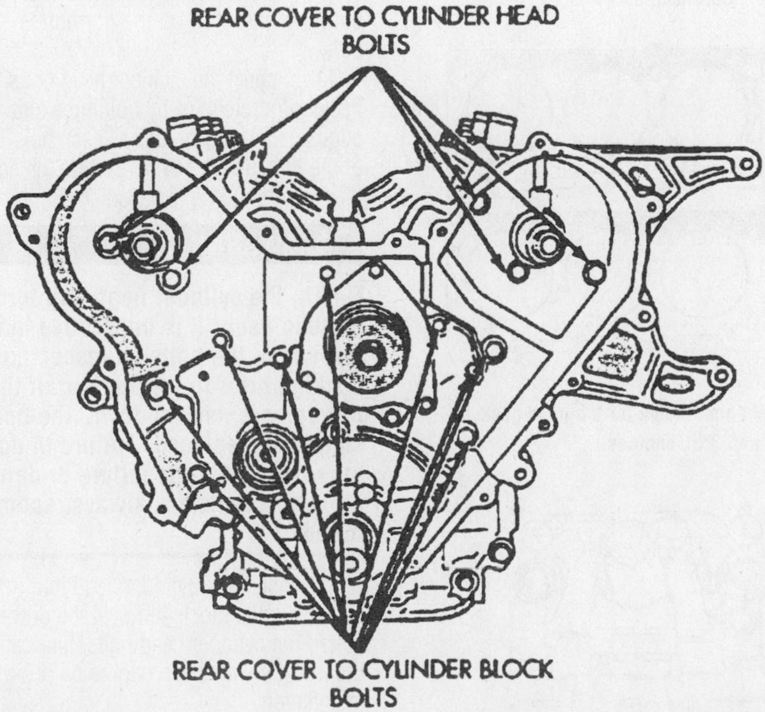

Remove the rear timing belt cover-to-cylinder head bolts, noting the bolt locations—3.2L and 3.5L engines

edge against the threads of each bolt. If all of the threads are not contacting the straightedge, the bolt should be replaced. New head bolts are recommended.

⁂ WARNING

Due to the cylinder head bolt torque method used, it is imperative that the bolt threads be inspected for necking (stretching) prior to installation. If the threads are necked down, the bolt should be replaced. Failure to do so may result in parts failure or damage.

9. Lubricate the bolt threads with clean engine oil, then install them.
10. Torque the head bolts in the sequence shown in the illustration, utilizing the following steps and tightening values:
    a. Step 1: bolts 1 through 8: 35 ft. lbs. (48 Nm).
    b. Step 2: bolts 1 through 8: 55 ft. lbs. (75 Nm).
    c. Step 3: bolts 1 through 8: 55 ft. lbs. (75 Nm).
    d. Step 4: bolts 1 through 8: plus 90 degree turn using a torque angle meter.
    e. Step 5: bolts 9 through 11: 21 ft. lbs. (28 Nm).
11. Install or connect the following:
    - Camshafts, timing chain and sprockets
    - Water connections to the cylinder head
    - Valve and timing chain covers
    - Crankshaft damper and torque the center bolt to 125 ft. lbs. (170 Nm)
    - Lower intake manifold and intake plenum
    - Exhaust manifolds
    - Accessory drive belts
    - Negative battery cable

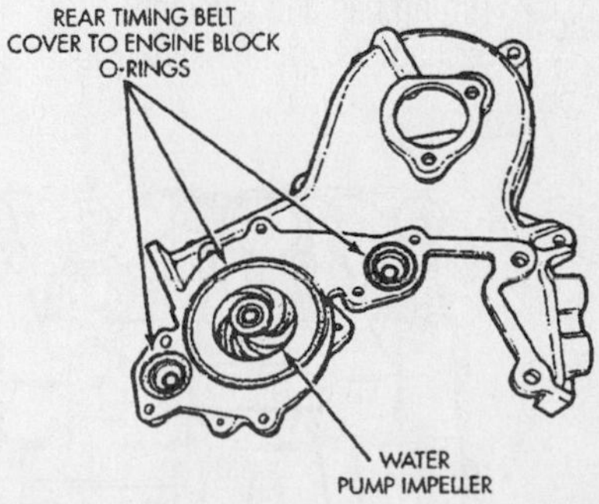

Right side belt cover, water pump and O-rings—3.2L and 3.5L engines

12. Fill the cooling system to the proper level.

13. Start the vehicle, check for leaks and repair if necessary.

3.2L and 3.5L Engines

1. Before servicing the vehicle, refer to the precautions in the beginning of this section.

This engine uses aluminum alloy cylinder heads. Use care when working with light alloy components. The heads are common to either cylinder bank, but in practice a cylinder head should be returned to the side of the engine from which it was removed. Removal of a cylinder head involves removal of the timing belt. Great care is required to install the belt, paying attention to all valve timing marks. Please note that camshaft removal on this engine does require the removal of the cylinder head.

2. Properly relieve the fuel system pressure.

3. Drain the cooling system.

4. Remove or disconnect the following:

- Negative battery cable
- Radiator and cooling fan assemblies
- Air cleaner assembly and intake manifold plenum

➡Cover the lower intake manifold during service.

- Accessory drive belts
- Crankshaft damper using the proper puller
- Engine valve covers
- Timing belt covers

➡Mark the timing belt running direction for installation. Align the camshaft sprockets with the marks on the rear covers.

- Timing belt and tensioner

5. Pre-load the timing belt tensioner as follows:

a. Place tensioner in a vise the same way it is mounted on the engine.

b. Slowly compress the plunger into the tensioner body.

c. Once the plunger is compressed, install a pin through the body and plunger to retain it in place until the tensioner is installed.

6. Hold the camshaft sprocket with a 36mm box wrench, loosen and remove the sprocket retaining bolt and washer.

➡To remove the camshaft sprocket retainer bolt while the engine is in the vehicle, it may be necessary to raise that side of the engine due to the length of the retainer bolt. The right bolt is 8.370 in. (212.6mm) long, while the left bolt is 10.0 in. (253mm) long. These bolts are not interchangeable and their original location during removal should be noted.

7. Remove or disconnect the following:

- Camshaft sprocket from the camshaft

➡The camshaft sprockets are not interchangeable.

- Intake manifold assembly using the recommended procedure
- Rear timing belt cover-to-cylinder head fasteners

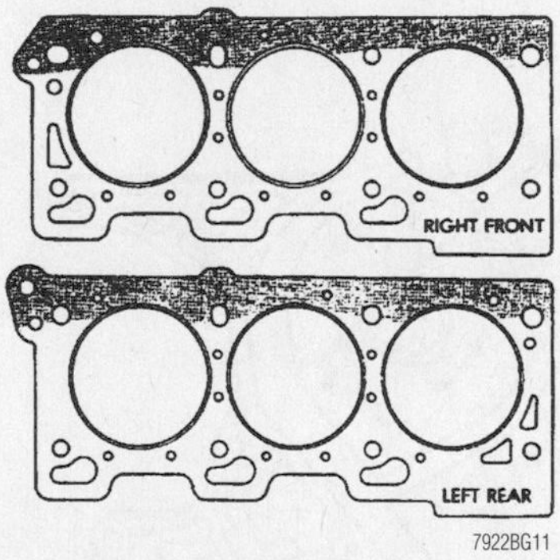

Correct positioning for the head gaskets—3.2L and 3.5L engines

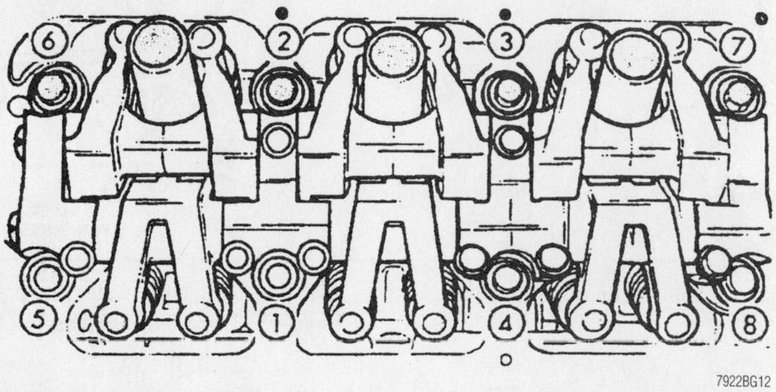

Cylinder head bolt tightening sequence —3.2L and 3.5L engines

➡If the right timing belt cover is to be removed, there are O-rings located behind it for the water pump passages.

- Cylinder head mounting bolts in the reverse order of the tightening sequence
- Cylinder head

To install:

8. Thoroughly clean and dry the mating surfaces of the head and block.

⁂ WARNING

When cleaning the cylinder head and block mating surfaces, do not use a metal scraper because the soft aluminum surfaces could be cut or damaged. Instead, use a scraper made of wood or plastic.

9. Check the cylinder head for cracks, damage or engine coolant leakage. Check the head for flatness. End-to-end, the head should be within 0.002 in. (0.051mm) normally with 0.008 in. (0.203mm) the maximum allowed out of true. The resurface limit is 0.008 in. (0.203mm) maximum, the combined total dimension of stock removal from the cylinder head, if any, and block top surface.

10. Place a new head gasket on the cylinder block locating dowels, being sure the gasket is on the correct side.

11. Inspect the cylinder head bolts for necking (stretching) by holding a straightedge against the threads of each bolt. If all of the threads are not contacting the scale, the bolt should be replaced.

⁂ WARNING

Due to the cylinder head bolt torque method used, it is imperative that the threads of the bolts be inspected for necking prior to installation. If the threads are necked down, the bolt should be replaced. Failure to do so may result in parts failure or damage. New bolts are always recommended.

12. Install the cylinder head into position on the engine block and over the dowels. Install the cylinder head bolts, lubricating the threads with clean engine oil prior to installation.

13. Torque the cylinder head bolts using the proper sequence as follows:

a. Step 1: torque in sequence to 45 ft. lbs. (61 Nm).

b. Step 2: torque in sequence to 65 ft. lbs. (88 Nm).

c. Step 3: torque in sequence to 65 ft. lbs. (88 Nm).
d. Step 4: torque in sequence an additional ¼ turn by using a torque angle meter.

➡Inspect the bolt torque after tightening. The torque should be over 90 ft. lbs. (122 Nm). If not, replace the cylinder head bolt.

14. Install the rear timing belt cover bolts and torque as follows:
a. M6 bolts: 105 inch lbs. (12 Nm).
b. M8 bolts: 21 ft. lbs. (28 Nm).
c. M10 bolts: 40 ft. lbs. (54 Nm).
15. Install the intake manifold assembly and torque the bolts following the proper sequence to 21 ft. lbs. (28 Nm).

➡The following procedure can only be used when the camshaft sprockets have been loosened or removed from the shafts.

16. When the camshaft sprockets are loosened or removed, the camshafts must be timed to the engine. Install the Camshaft Alignment tools 6642-A, to the rear of the cylinder heads.
17. Install both camshaft sprockets to the appropriate shafts. The left camshaft sprocket has the Distributorless Ignition System (DIS) pickup as part of the sprocket.
18. Apply thread locking compound to the threads of the camshaft sprocket retainer bolts and install to the appropriate shafts. The right bolt is 8.380 in. (21.3cm) long, while the left bolt is 10.0 in. (25.4cm) long. These bolts are not interchangeable. Do not tighten the bolts at this time. The camshaft marks should be positioned between the marks on the cover.
19. Place the crankshaft sprocket to the Top Dead Center (TDC) mark on the oil pump housing. Install the timing belt starting at the crankshaft sprocket and working in a counterclockwise direction.
20. After the belt is installed around the last sprocket keep tension on the belt until it is past the tensioner pulley.
21. Holding the tensioner pulley against the belt, install the tensioner housing and torque to 21 ft. lbs. (28 Nm).
22. When the tensioner is in place pull the retainer pin to allow the tensioner to extend to the pulley bracket.
23. Hold the right camshaft sprocket hex with a 36mm box wrench and torque the right camshaft sprocket bolt to 75 ft. lbs. (102 Nm). Turn the sprocket bolt an additional 90 degrees.
24. Hold the left camshaft sprocket hex with a 36mm box wrench and torque the left camshaft sprocket bolt to 85 ft. lbs. (115 Nm). Turn the sprocket bolt an additional 90 degrees.
25. Install or connect the following:
- Camshaft alignment tools from the back of the cylinder heads
- Cam covers and new O-rings. Torque the fasteners to 20 ft. lbs. (27 Nm). Repeat this procedure on the other camshaft.
- Timing belt covers and crankshaft damper and torque the crankshaft damper bolt to 85 ft. lbs. (115 Nm)
- Valve covers and torque the bolts to 105 inch lbs. (12 Nm)
- Spark plug tube nut and O-ring and torque to 60 inch lbs. (7 Nm)
- Spark plug and torque to 20 ft. lbs. (28 Nm)
- A/C compressor and torque the mounting bracket bolts to 30 ft. lbs. (41 Nm)
- Spark plug wires
- Accessory drive belts and adjust to the proper tension
- Intake manifold plenum using the recommended procedure
- Air cleaner assembly
- Radiator and cooling fan assemblies
- Negative battery cable

26. Check to be sure that all hoses, wiring connectors, cables, fluid and vacuum lines are reconnected.
27. Change the engine oil and oil filter.
28. Fill and bleed the cooling system.
29. Run the vehicle with the radiator cap off so coolant can be added as required until the thermostat opens. Watch for leaks and for unusual engine noises. Refill the radiator completely as required.
30. Once the vehicle has cooled, recheck the coolant and oil level.

Rocker Arm/Shafts

REMOVAL & INSTALLATION

2.7L Engine

1. Before servicing the vehicle, refer to the precautions in the beginning of this section.
2. Disconnect the negative battery cable.
3. Remove the valve covers.
4. Position the camshaft so that the base circle (heel) is facing the rocker arm being serviced.

✲✲ WARNING

Depress the valve spring only enough to remove the rocker arm or damage to the spring may result.

5. Using Valve Spring tool 8215 and Adapter 8216, depress the valve spring enough to release the tension on the rocker arm.

➡If the rocker arms are to be reused, identify their positions for reassembly in their original positions.

6. Repeat this procedure for each rocker arm being removed.

To install:

7. Lubricate the rocker arms with clean engine oil, prior to installation.
8. Position the camshaft so that the base circle (heel) is facing the rocker arm being installed.

✲✲ WARNING

Depress the valve spring only enough to install the rocker arm or damage to the spring may result.

9. Using Valve Spring tool 8215 and

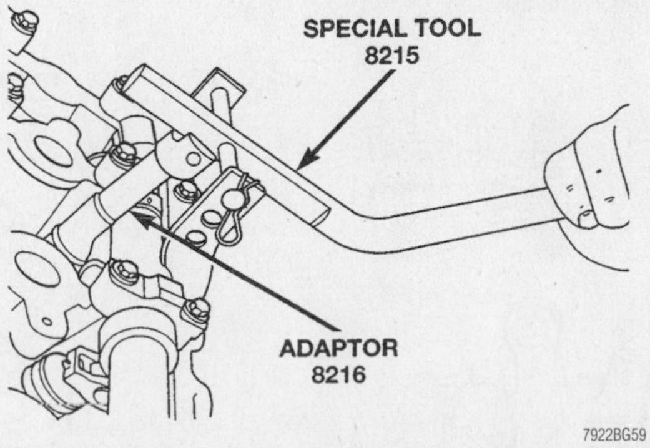

Only depress the valve spring enough to remove the rocker arm—2.7L engine

Adapter 8216, depress the valve spring enough to install the rocker arm.

10. Install the rocker arm in the original position (if reused) over the valve and lash adjuster.

➡**Inspect the rocker arm for proper engagement into the lash adjuster and valve tip.**

11. Release the tension on the valve spring and remove the tools.
12. Install the valve covers and connect the negative battery cable.

3.2L and 3.5L Engines

1. Before servicing the vehicle, refer to the precautions in the beginning of this section.
2. Relieve the fuel system pressure.
3. Remove or disconnect the following:
   - Negative battery cable
   - Air cleaner assembly and the intake manifold plenum

➡**Cover the lower intake manifold during service.**

   - Cylinder head covers
   - Rocker arm assembly
4. Inspect the rocker arms for wear or damage. Inspect the roller for scuffing or wear. Replace assembly as necessary.

⁂ WARNING

Do not remove the lash adjusters from the rocker arm assembly. The rocker arm and the adjuster are serviced as an assembly.

5. Identify the rocker arm assemblies and rocker arms and disassemble the shaft as follows:
   a. Thread a nut, washer and spacer onto a 4mm screw.
   b. Insert and tighten the 4mm screw into the dowel pin on the shaft.

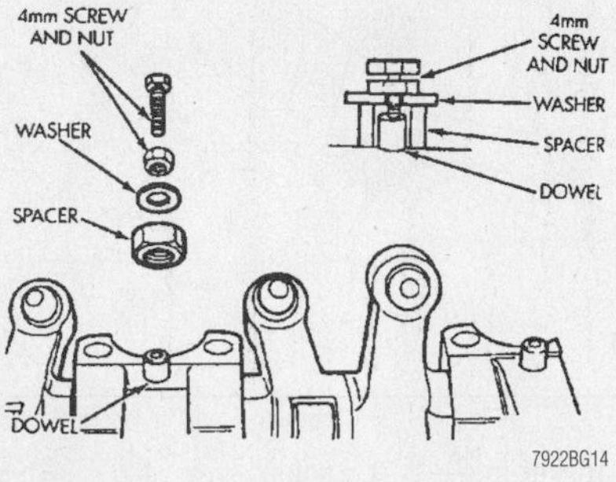

Remove the dowel pin using a 4mm screw, nut, spacer and washer installed into the pin—3.2L and 3.5L engines

   c. Loosen the nut on the screw. This will pull the dowel pin from the shaft support.
   d. Remove the rocker arms and pedestals, keeping them in order.
   e. Check the oil holes for restrictions with a small wire and clean as required.

To install:

6. Assemble the rocker shaft as follows:
   a. Install the rocker arms and pedestals onto the shaft, keeping them in the original order.
   b. Press the dowel pins into the pedestals until they bottom out in the pedestals.
7. Position the camshaft so that the timing mark on the right camshaft timing belt sprocket aligns with the timing mark on the

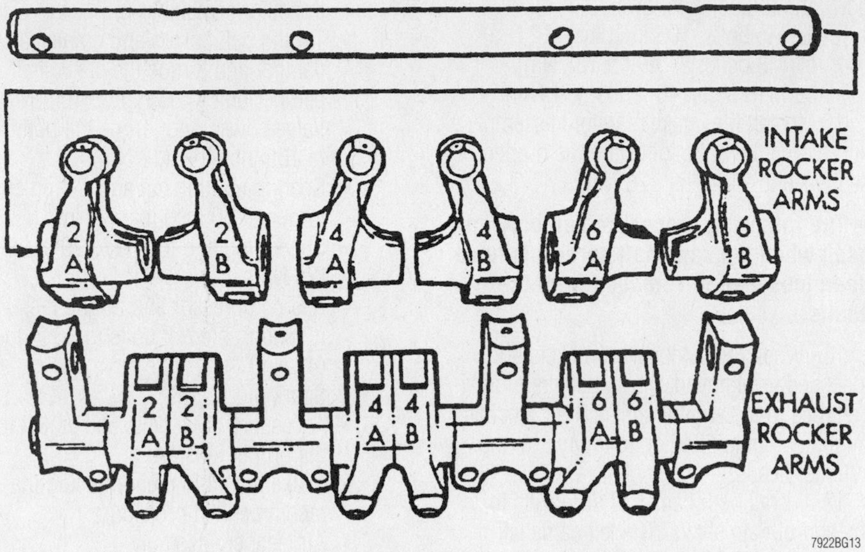

Left bank rocker arm and shaft identification—3.2L and 3.5L engines

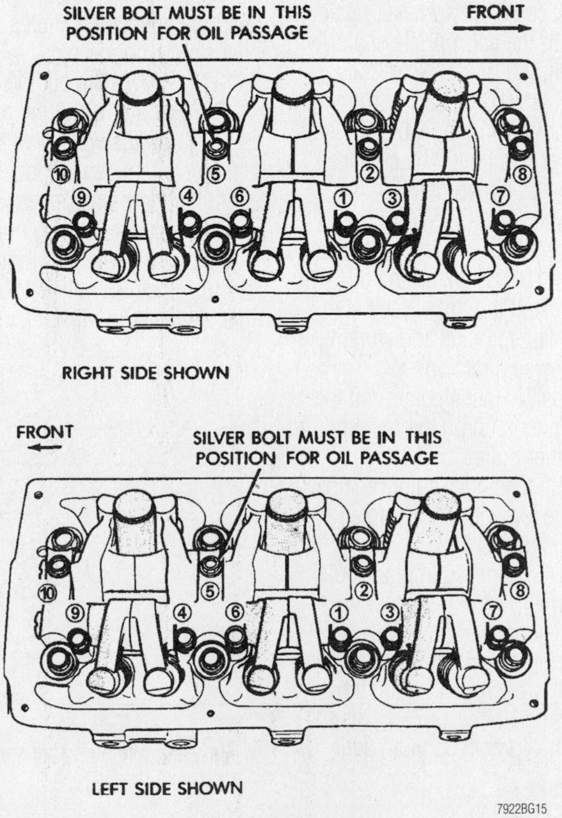

Proper torque sequence for the rocker arm and shaft assemblies—3.2L and 3.5L engines

rear timing belt cover and the timing mark on the left sprocket is 45 degrees from the mark on the rear timing belt cover. There will be no load on the shaft during installation. Install the rocker shafts so the identification marks are facing toward the front of the engine.

8. Install the oil feed bolt in the correct location on the rocker shaft retainer. Torque the bolts in proper sequence to 23 ft. lbs. (31 Nm).

9. Install or connect the following:
- Valve covers and torque the bolts to 105 inch lbs. (12 Nm)
- Intake manifold plenum
- Air cleaner assembly
- Negative battery cable

Intake Manifold

REMOVAL & INSTALLATION

2.7L Engine

1. Before servicing the vehicle, refer to the precautions in the beginning of this section.
2. Disconnect the negative battery cable.
3. Properly relieve the fuel system pressure.
4. Drain the cooling system.
5. Disconnect the air tube from the air cleaner and the throttle body.
6. Hold the throttle lever in the wide-open position and remove the throttle cable and the speed control cable from the lever. Compress the locking tabs on the cables and remove them from the mounting brackets.
7. Unplug the electrical connections from the following:
- Solenoid on the Exhaust Gas Recirculation (EGR) valve transducer
- Manifold Absolute Pressure (MAP) sensor
- Throttle Position (TP) sensor
- Idle Air Control (IAC) motor
- Vacuum hose from the Positive Crankcase Ventilation (PCV) valve as well as the power brake booster at the intake manifold nipple
- Vacuum line at the fuel pressure regulator
- Purge hose from the throttle body
- Electrical connector from the TP sensor and the IAC motor
- EGR tube-to-intake manifold plenum screws
- Intake manifold plenum (upper part of the manifold)

➡Cover the lower part of the intake manifold to prevent foreign material from entering the engine.

- Fuel supply and return tubes from the fuel rail at the rear of the intake manifold

8. Disconnect the fuel/return tubes by pushing the quick-connect fitting toward the fuel tube while depressing the built-in disconnect tool with Quick-Connect Fitting tool 6751. To disconnect the fitting from the fuel rail, slightly twist the fitting while maintaining downward pressure on tool 6751. Wrap shop towels around the fuel hoses to absorb any fuel spillage.

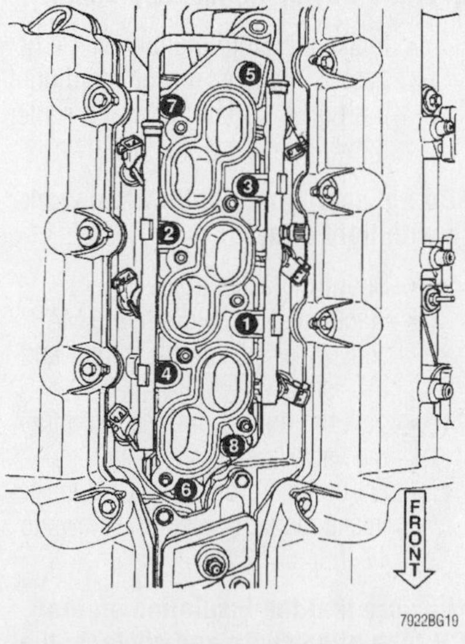

Intake manifold loosening and tightening sequence —2.7L engine

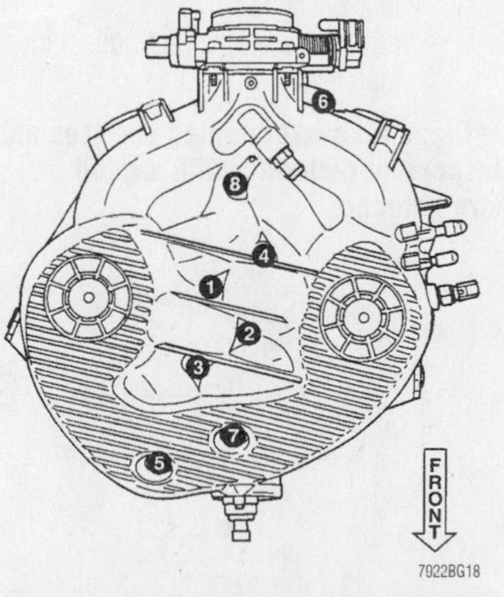

Intake plenum tightening sequence —2.7L engine

9. Plug the fuel line openings to prevent system contamination.
10. Remove or disconnect the following:
- Fuel clamp screw and tubes from the bracket
- Electrical harness from the injectors and turn toward the center of the engine
- Fuel rail mounting bolts and lift the fuel rail with the injectors attached straight up and off the engine. Cover the injector openings.
- Intake manifold bolts and the manifold
- Intake manifold seal retainer screws and intake manifold gasket

11. Clean all mating surfaces.
12. Inspect the manifold for damage, cracks or clogged passages. Repair, clean or replace the manifold as required.

To install:

13. Verify that all intake manifold and cylinder head sealing surfaces are clean. Place a drop of sealant onto each of the 4 corners of the intake manifold gasket, where the cylinder head meets the engine block.

✲✲ CAUTION

The intake manifold gasket is made of very thin metal and can cause cuts if handled carelessly.

14. Install the intake manifold and 8 mounting bolts. Snug down evenly to just 10 inch lbs. (1.1 Nm).
15. Torque the lower intake manifold bolts in the proper sequence to 105 inch lbs. (12 Nm).
16. Install the fuel injectors by performing the following procedure:
    a. Apply a light coat of clean engine oil to the O-ring on the nozzle end of each injector.
    b. Insert the fuel injector nozzles into the openings in the intake manifold. Seat the injectors in place and install the fuel rail mounting bolts, torque to 16 ft. lbs. (22 Nm).
17. Install or connect the following:
- Electrical connectors to each fuel injector. Rotate the injectors toward the cylinder head covers.
- Fuel supply and return tubes to the fuel rail. Be sure that the black plastic release ring to the quick-connect fitting is in the OUT position. Place special tool 6751 under the largest diameter of the quick-connect fitting.

18. Pull tool 6751 toward the fuel rail until the quick-connect fitting clicks into place. Place the special tool between the shoulder of the built-in disconnect tool and top of the quick-connect fitting, then inspect the security of the fitting by applying a slight downward force against the fitting. It should be locked in place.

- Intake plenum with new gasket onto the intake manifold. Loosely install the mounting bolts.
- EGR tube to the manifold with a new gasket in place. Loosely install the mounting screws.
- Torque the intake manifold plenum mounting bolts to 105 inch lbs. (12 Nm) following the outlined sequence
- Left and right support brackets to the manifold and torque the lower fasteners to 50 inch lbs. (6 Nm) and the upper fasteners to 105 inch lbs. (12 Nm)
- EGR tube mounting bolts
- PCV valve hose and power brake booster hose
- Electrical connectors to the EGR transducer solenoid, IAC motor, Map and TP sensors
- Throttle cable and speed control cable to the mounting bracket and connect to the throttle body lever while holding lever in the wide-open position
- Purge hose to the throttle body
- Reconnect the air tube to the air cleaner and the throttle body
- Negative battery cable

19. Change the engine oil and oil filter.

20. Fill the cooling system. Run the vehicle with the radiator cap removed until the thermostat opens, adding coolant as required. Watch for fuel and coolant leaks and for correct engine operation.

21. Once the vehicle has cooled, recheck the coolant level and add, if necessary.

3.2L and 3.5L Engines

1. Before servicing the vehicle, refer to the precautions in the beginning of this section.

2. Properly relieve the fuel system pressure.

3. Drain the cooling system.

4. Remove or disconnect the following:

- Negative battery cable
- Engine cover from the top of the intake manifold
- Accelerator and the speed control cable from the throttle lever
- Idle Air Control (IAC) motor
- Intake Air Temperature (IAT) sensor
- Manifold Absolute Pressure (MAP) sensor
- Ground screw from the intake manifold
- Electrical connector from the Throttle Position (TP) sensor
- Vacuum hoses from the manifold tuning valve, Positive Crankcase Ventilation (PCV) make-up air hose, IAC motor supply hose and the purge hose from the throttle bodies
- Brake booster hose, PCV hose and the remaining vacuum hoses from the intake manifold
- Exhaust Gas Recirculation (EGR) tube-to-intake manifold plenum bolts
- Plenum support bracket mounting bolts on each side of the plenum
- Intake plenum mounting bolts

➡The intake manifold plenum (upper half of the intake manifold assembly) uses 2 different length bolts. Take note of their position and be sure they are installed in the same location during installation.

5. Remove the intake manifold plenum from the intake manifold.

➡Discard the old gasket. Cover the intake manifold openings with tape to keep debris from entering the engine.

6. Remove or disconnect the following:

- Upper radiator hose from the thermostat housing
- Heater hose from the rear of the intake manifold
- Lower intake manifold bolts and manifold

➡Clean all gasket mating surfaces and inspect for distortion with a good straightedge.

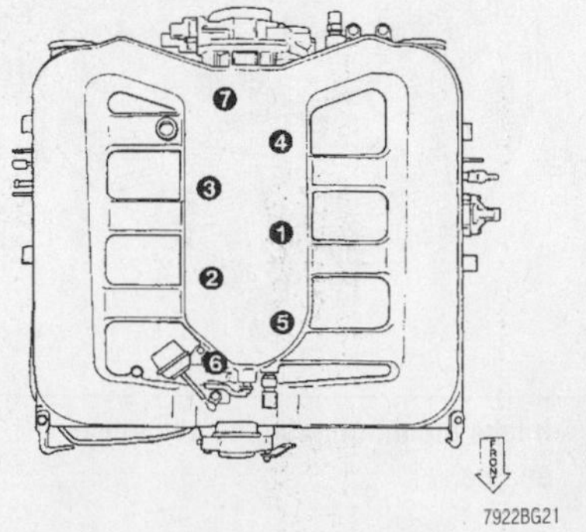

Upper intake manifold loosening and tightening sequence —3.2L and 3.5L engines

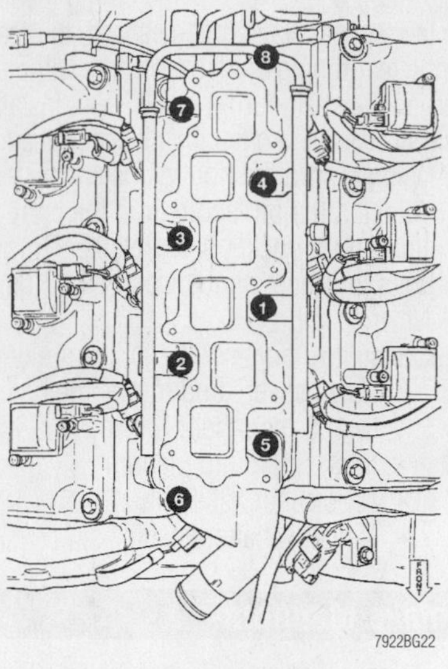

Lower intake manifold loosening and tightening sequence —3.2L and 3.5L

To install:

➡Verify that all intake manifold and cylinder head sealing surfaces are clean.

7. Install or connect the following:

- Intake manifold gasket, then the lower manifold. Torque the bolts in the proper sequence to 21 ft. lbs. (28 Nm).
- Upper radiator hose to the thermostat housing
- Heater hose to the rear of the intake manifold

➡Ensure the ignition cables are routed out of the way of the intake plenum.

- Intake manifold plenum with a new gasket in place. Torque the mounting bolts, working from the center outward, to 21 ft. lbs. (28 Nm)

➡Do not overtighten bolts when working with light alloys.

- Support bracket bolts
- Electrical connectors to the MAP sensor, TP sensor, IAC motor and IAT sensor
- Vacuum hose to the manifold tuning valve
- EGR tube. Torque the EGR tube-to-intake manifold plenum screws to 17 ft. lbs. (22 Nm).

➡Be sure that the insulation on the EGR tube aligns with and contacts the insulation on the vacuum harness at the rear of the engine. Rotate the throttle lever to the wide-open position and reconnect the speed control and throttle cables.

- PCV valve hose
- Air cleaner plenum and plenum hose
- Ground wire to the intake manifold plenum
- Brake booster hose to intake manifold plenum fitting
- Throttle body purge tubes
- Intake manifold plenum cover
- Negative battery cable

8. Fill and bleed the cooling system.
9. Change the engine oil and filter.
10. Test run the engine, check for fuel and coolant leaks and verify correct engine operation.

Exhaust Manifold

REMOVAL & INSTALLATION

2.7L Engine

RIGHT MANIFOLD

1. Before servicing the vehicle, refer to the precautions in the beginning of this section.
2. Remove or disconnect the following:
   - Negative battery cable
   - Air intake plenum and the air filter housing
   - Battery cable housing tube-to-transaxle bolt
   - Ehaust Gas Recirculation (EGR) valve and tube
   - Oxygen (O_2S) sensor
   - V-band clamp from the manifold

➡Do not reuse the V-band clamps.

   - Heat shield
   - Exhaust manifold

3. Remove all traces of the old manifold gasket and clean both gasket mating surfaces.

To install:

4. Install or connect the following:
   - Exhaust manifold and new gasket. Torque the bolts to 17 ft. lbs. (23 Nm) working from the center outward.
   - Heat shields and torque the bolts to 105 inch lbs. (12 Nm)
   - New V-band clamp and torque to 100 inch lbs. (11.3 Nm)
   - O_2S sensor
   - EGR valve and tube using new gaskets, then torque to 95 inch lbs. (11 Nm)
   - Battery cable tube-to-transaxle and torque the bolt to 75 ft. lbs. (101 Nm)
   - Air inlet plenum and air filter housing
   - Negative battery cable

LEFT MANIFOLD

1. Before servicing the vehicle, refer to the precautions in the beginning of this section.
2. Remove or disconnect the following:
   - Negative battery cable
   - Exhaust system
   - V-band clamps and left catalytic converter

➡Do not reuse the V-band clamps.

3. Loosen and rotate the transaxle dipstick tube out of the way.
4. Remove or disconnect the following:
   - Engine wiring harness support bracket from the cylinder head
   - Oxygen (O_2S) sensor
   - Engine oil dipstick tube and manifold heat shield
   - Exhaust manifold bolts and manifold

To install:

5. Remove all traces of the old manifold gasket and clean both gasket mating surfaces.
6. Install or connect the following:
   - Exhaust manifold and new gasket and torque the bolts to 17 ft. lbs. (23 Nm) working from the center outward
   - Heat shields and torque the bolts to 105 inch lbs. (12 Nm)
   - O_2S sensor
   - Transaxle dipstick tube
   - O-ring for the engine oil dipstick tube and tube
   - Catalytic converter with a new V-band clamp and torque to 89 inch lbs. (10 Nm)
   - Engine wiring harness support bracket to the cylinder head
   - Exhaust system
   - Negative battery cable

3.2L and 3.5L Engines

1. Before servicing the vehicle, refer to the precautions in the beginning of this section.
2. Remove or disconnect the following:
   - Negative battery cable
   - Exhaust pipes from the exhaust manifold
   - Heated Oxygen (HO_2S) sensor electrical wiring
   - Heat shield-to-exhaust manifold screws
   - Exhaust manifold bolts and manifold
3. Inspect the manifold for damage or cracks. Check for distortion against a straight-edge or thickness gauge. Replace manifold if required.
4. Remove all traces of the old manifold gasket and clean both gasket mating surfaces.

To install:

5. Install or connect the following:
   - New manifold gasket and exhaust manifold to the cylinder head and torque to 15 ft. lbs. (20 Nm)
   - Exhaust pipe to the exhaust manifold and torque the nuts to 21 ft. lbs. (28 Nm)
   - Heat shield and torque the retaining screws to 11 ft. lbs. (15 Nm)
   - HO_2S sensor electrical connector
   - Negative battery cable
6. Operate the vehicle and inspect for exhaust leaks.

Front Crankshaft Seal

➡The front crankshaft seal procedures are for timing belt equipped engines only. For engines that utilize timing chains, please refer to the applicable procedure later in this section.

REMOVAL & INSTALLATION

3.2L and 3.5L Engines

Note that the timing belt must be removed from the vehicle to perform this service. Use care to be sure all valve timing marks are carefully aligned both before removing the belt and after belt installation and all service has been completed. It may be good practice to set the engine to Top Dead Center (TDC) No. 1 cylinder compres-

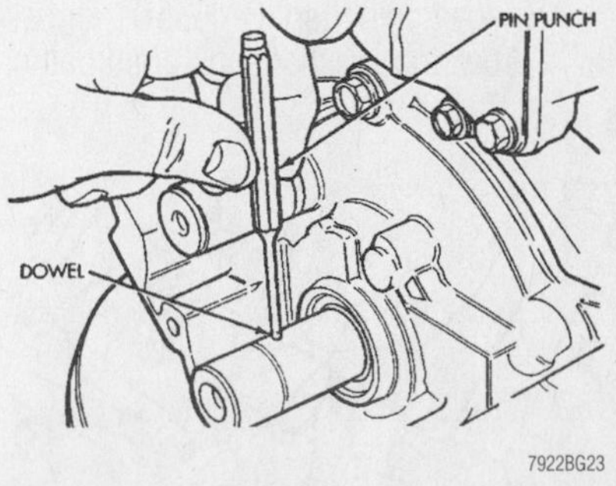

Removing the timing belt sprocket dowel pin from the crankshaft—3.2L and 3.5L engines

sion stroke (firing position) and aligning all timing marks before removing the timing belt. This serves as a reference for all work that follows.

1. Before servicing the vehicle, refer to the precautions at the beginning of this section.
2. Properly relieve the fuel system pressure.
3. Drain the cooling system.
4. Remove or disconnect the following:
   - Negative battery cable
   - Radiator and cooling fan module assembly
   - Accessory drive belts
   - Crankshaft damper bolt
   - Timing belt front cover

➡The sealer on the timing belt front cover may be reusable and should not be removed. Use silicone rubber adhesive sealant to replace any missing sealer.

   - Timing belt and tensioner. Refer to the appropriate section for the removal and installation procedure
   - Crankshaft timing belt sprocket
5. Locate the small dowel pin in the crankshaft. With a small punch, carefully tap out the dowel from the end of the crankshaft.
6. Remove the crankshaft seal using tool 6341A, taking care not to nick the shaft seal surface or seal bore during removal.

To install:

7. Inspect the crankshaft seal lip surface for varnish and dirt. Polish the area using 400 grit sandpaper to remove varnish as necessary.
8. Install or connect the following:
   - Crankshaft seal using seal installer tool 6342
   - Rear lower timing belt cover
   - Dowel into the crankshaft so that it protrudes 0.047 in. (1.2mm)
   - Timing belt sprocket at the crankshaft using tool C-4685C1, thrust bearing, washer and 12mm bolt or an equivalent setup to pull the sprocket onto crankshaft. Do not hammer on the sprocket.
9. Verify that all valve timing marks are aligned.
   - Timing belt and tensioner using the recommended procedure
10. Rotate the crankshaft 2 complete turns and recheck the timing marks on the camshafts and crankshaft. The marks must align with their respective locations. If the marks do not align, repeat the timing belt installation procedure. When correct valve timing has been verified, install the timing belt covers.
   - Crankshaft damper. Hold the crankshaft damper, using tool L-3281, and torque the bolt to 85 ft. lbs. (115 Nm).
   - Accessory drive belts and adjust to the proper tension
   - Radiator and cooling fan assemblies
   - Negative battery cable
11. Fill and bleed the cooling system.
12. Start the vehicle, check for leaks and repair if necessary.

Camshaft and Valve Lifters

REMOVAL & INSTALLATION

2.7L Engine

1. Before servicing the vehicle, refer to the precautions in the beginning of this section.

⁂ WARNING

When the timing chain is removed and the cylinder heads are installed, DO NOT turn the crankshaft or camshaft without first locating the proper crankshaft position. Failure to do so will result in piston-to-valve contact.

2. Remove or disconnect the following:
   - Primary timing chain
   - Second chain tensioner mounting bolts
3. Slowly loosen the camshaft bearing cap retaining bolts in the reverse order of the tightening sequence.
   - Bearing caps
   - Camshafts, secondary chain and tensioner as an assembly
   - Tensioner and chain from the crankshaft

To install:

4. Assemble the chain on the camshafts. Ensure the plated links are facing toward the front. Align the plated links to the dots on the camshaft sprockets.

➡There are 2 different styles of camshaft (secondary) chain tensioners . The Early Build tensioners will separate into sub-components, the Later Build tensioners will not. Compress the camshaft (secondary) chain as follows:

5. For early build vehicles:
   a. Separate the tensioner cylinder from the tensioner housing.
   b. Carefully drain the oil from the housing using care not to remove the internal tensioner components.
   c. Assemble the tensioner housing.
   d. Using hand pressure, compress and lock the tensioner using a fabricated lockpin.
6. For late build vehicles:

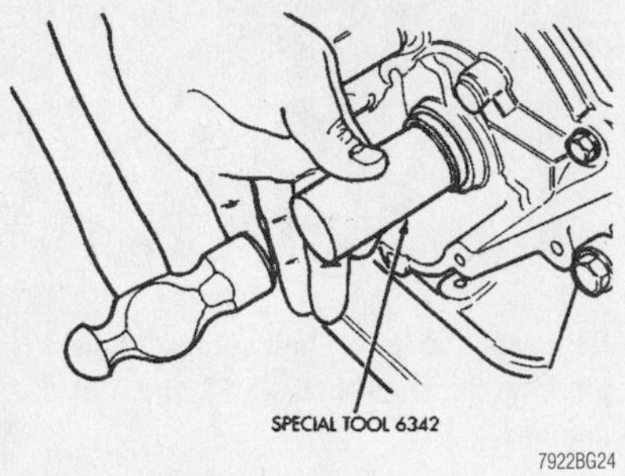

Installing the crankshaft oil seal—3.2L and 3.5L engines

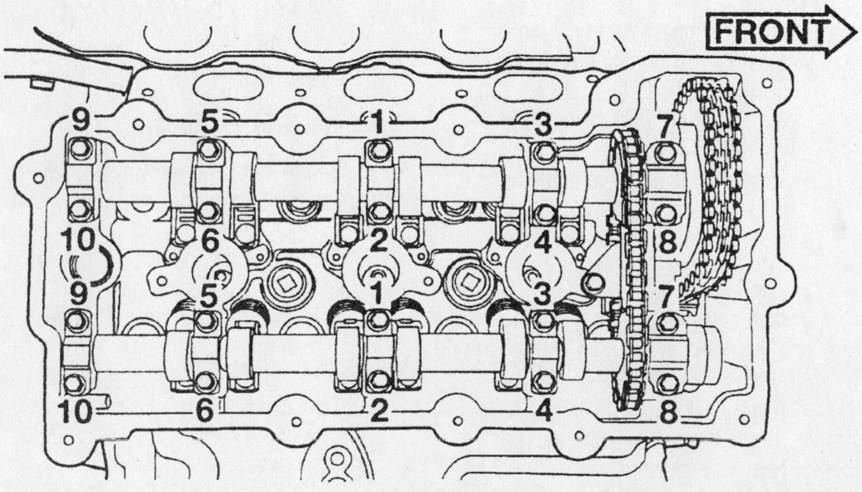

Camshaft bearing cap tightening sequence—2.7L engine

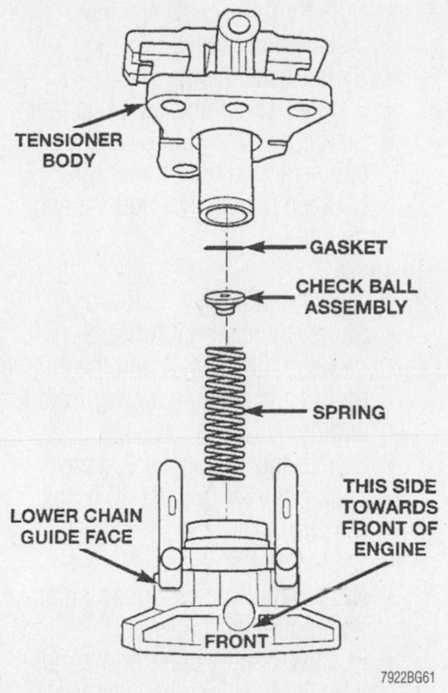

Exploded view of the camshaft (secondary) chain tensioner, early build—2.7L engine

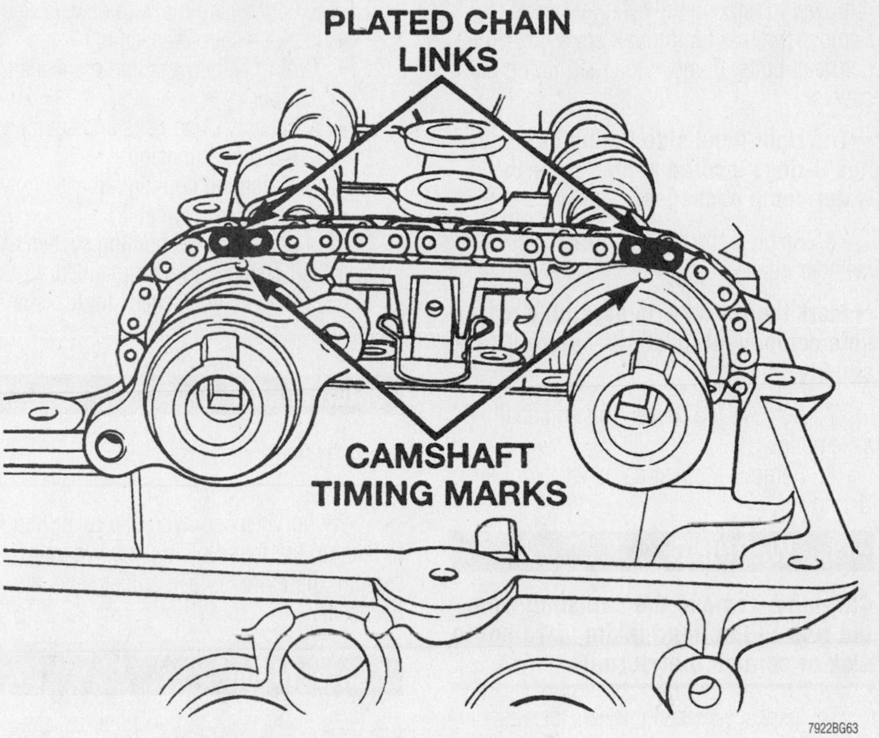

Proper camshaft (secondary) chain alignment—2.7L engine

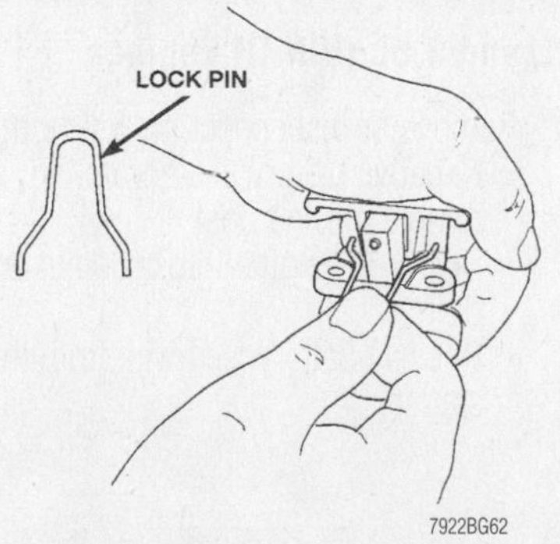

Fabricate a lockpin, as shown, to keep the tensioner compressed—2.7L engine

a. Place the tensioner in a soft-jawed vise.

b. Slowly compress the tensioner until the fabricated lockpin can be installed.

c. Remove the compressed and locked tensioner from the vise.

7. Insert the compressed and locked camshaft chain tensioner in between the camshafts and chain.

8. Position the camshafts so that the plated links and dots are facing upward.

9. Install or connect the following:

- Camshafts

⁂ WARNING

Ensure that the rocker arms are correctly seated and in proper positions.

- Camshaft bearing caps and torque the bolts gradually, in sequence, to 105 inch lbs. (12 Nm)
- Secondary chain tensioner and torque the bolts to 105 inch lbs. (12 Nm)
- Primary timing chain
- Negative battery cable

10. Remove the lockpin from the secondary chain tensioner.

3.2L and 3.5L Engines

1. Before servicing the vehicle, refer to the precautions in the beginning of this section.

Camshafts are serviced from the rear of the cylinder head. Although the engine does not need to be removed for camshaft service, the cylinder head must be removed from the vehicle. Note too, that the camshaft sprockets have a D-shaped hole that allows it to rotate several degrees in each direction on its shaft.

2. Properly relieve the fuel system pressure.

3. Drain the cooling system.

4. Remove or disconnect the following:

- Negative battery cable
- Radiator/cooling fan assemblies and the accessory drive belts
- Crankshaft damper and the timing belt covers

➡Mark the timing belt rotation direction for installation. Align the timing belt sprockets with marks on the rear timing belt covers before removing the timing belt.

- Timing belt tensioner and timing belt. Refer to the appropriate section for the removal and installation procedure
- Camshaft timing belt sprockets
- Intake manifold assembly using the recommended procedure
- Exhaust manifold

➡Be sure to clean the gasket mating surfaces between the exhaust manifold and the cylinder head.

5. The rear timing belt cover must be

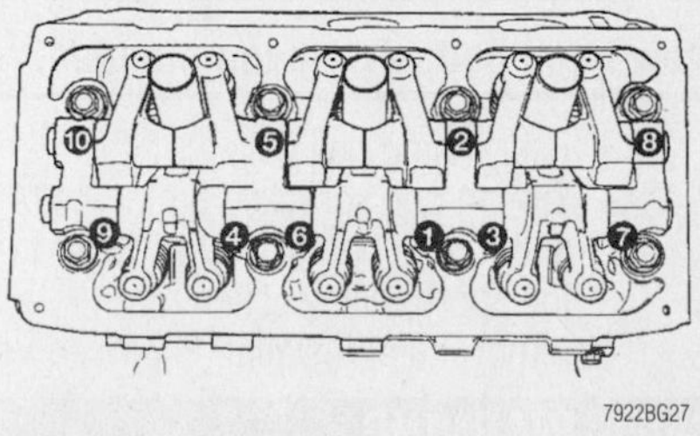

Rocker arm/shaft tightening sequence—3.2L and 3.5L engines

removed to remove the cylinder heads. Remove the rear timing belt cover-to-cylinder head bolts. Remove the rear timing belt covers.

➡The right-hand side timing belt cover has O-rings located behind it for the water pump passages.

6. Remove the cylinder head bolts and cylinder head.

➡Mark the rocker arm assembly to note component locations before disassembly.

7. Remove the rocker arm and shaft assemblies.
8. Remove the rear camshaft cover and O-ring.

**** WARNING**

Carefully, remove the camshaft from the rear of the head taking care not to nick or scratch the journals.

9. Inspect camshaft journals for wear or damage. If wear is present, inspect the cylinder head for damage. Inspect the head oil holes for clogging. Replace the camshaft as required.
10. Measure the height of the cam using a micrometer. Measure in 2 places: the unworn area and in the wear zone. Subtract the figures to get cam wear. The standard specification is 0.001 in. (0.0254mm) with the wear limit being 0.010 in. (0.254mm). Replace the camshaft if it is worn beyond this specification.

To install:

11. Lubricate the camshaft journals and lobes with clean engine oil.
12. Install or connect the following:
    - Camshaft
    - Camshaft cover and O-ring and torque the bolts to 21 ft. lbs. (28 Nm)
    - Rocker arm assemblies
    - Cylinder head assembly

**** WARNING**

New head bolts are recommended.

- Rear timing belt covers
- Timing belt sprocket, timing belt and timing belt tensioner. Refer to the appropriate section for the removal and installation procedure.

**** WARNING**

Be sure that once they are all installed, the timing of the camshaft(s) is accurate.

- Exhaust manifold with new gasket
- Intake manifold assembly
- Timing belt covers and crankshaft damper
- Accessory drive belts and set them to the proper tension
- Radiator and cooling fan assembly
- Negative battery cable

13. Fill and bleed the cooling system. An oil and filter change is recommended.
14. Start the vehicle, check for leaks and repair if necessary.

Valve Lash

ADJUSTMENT

These engines use hydraulic roller lifters to take up the free-play in the valve train system, therefore no lash adjustments are necessary.

Starter Motor

REMOVAL & INSTALLATION

2.7L Engine

1. Remove or disconnect the following:
    - Negative battery cable from the remote ground post
    - Battery feed and posi-lock connectors
    - Nuts from the catalyst support bracket and mount
    - Starter heat shield
    - 3 starter-to-engine/transaxle bolts
    - Starter by rotating it toward the engine and sliding it rearward between the catalyst and the engine mount

To install:

2. Install or connect the following:
    - Starter by sliding it forward between the catalyst and the engine mount and rotating it away from the engine
    - Starter and torque the 3 starter-to-engine/transaxle bolts 40 ft. lbs. (54 Nm)
    - Starter heat shield
    - Nuts to the catalyst support bracket and mount
    - Positive battery cable and torque the nut to 89 inch lbs. (10 Nm)
    - Posi-lock connectors
    - Negative battery cable to the remote ground post

3.2L and 3.5L (VIN G) Engines

1. Remove or disconnect the following:
    - Negative battery cable from the remote ground post
    - Starter-to-engine/transaxle nut and bolts
    - Positive battery feed wire from the starter

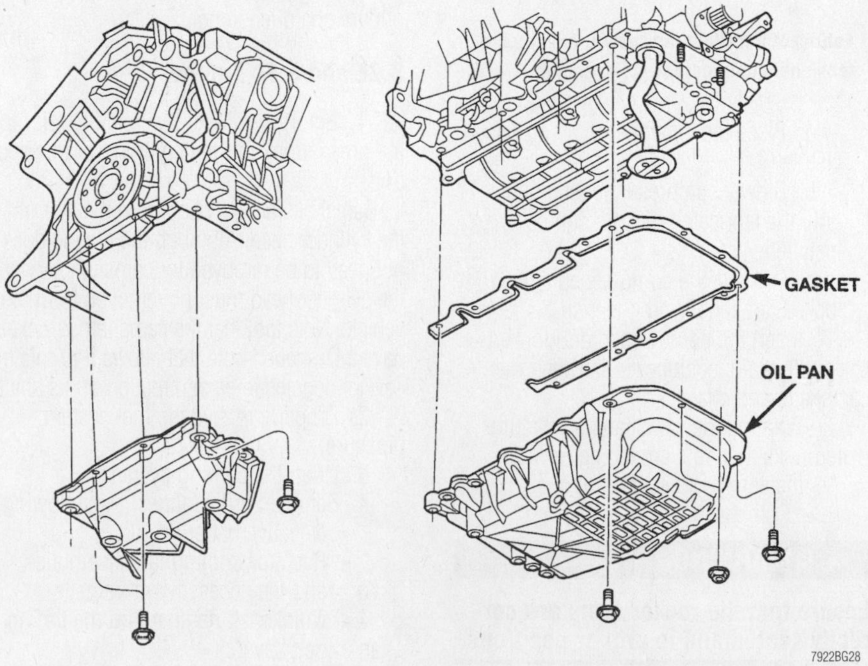

Exploded view of the oil pan removal and installation—2.7L and 3.5L (VIN G) engines

- Starter and position it to gain access to the posi-lock connector

2. Place a support under the engine and slightly relieve the pressure from the left engine mount.
3. Remove the 3 left engine mount-to-engine bolts.
4. Slightly, raise the engine to provide more room.
5. Remove the starter by sliding it rearward between the catalyst and the engine mount.
6. Remove the posi-lock connector.

To install:

7. Install or connect the following:
   - Posi-lock connector
   - Starter by sliding it forward between the catalyst and the engine mount
8. Slightly, lower the engine.
   - 3 left engine mount-to-engine bolts
   - Positive battery feed wire to the starter and torque the nut to 89 inch lbs. (10 Nm)
   - Starter and torque the starter-to-engine/transaxle bolts 40 ft. lbs. (54 Nm)
   - Negative battery cable to the remote ground post

Oil Pan

REMOVAL & INSTALLATION

1. Before servicing the vehicle, refer to the precautions in the beginning of this section.
2. Drain the engine oil and remove the oil filter.
3. Remove or disconnect the following:
   - Negative battery cable
   - Dipstick and housing

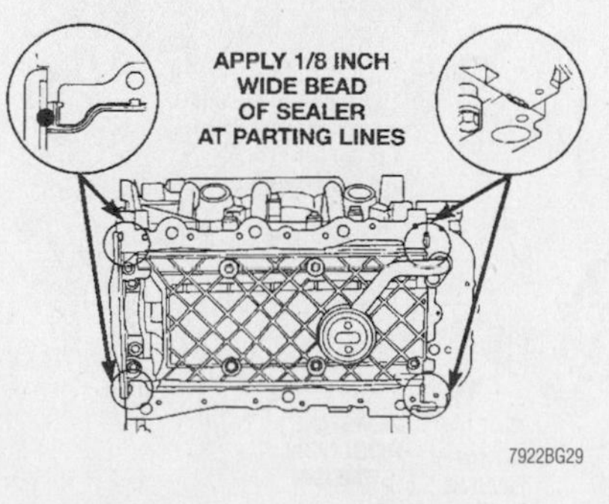

To ensure a proper seal, apply sealer as shown—2.7L and 3.5L (VIN F) engines

   - Structural collar from the rear of the oil pan and transmission housing
   - Engine oil cooler lines from the oil pan, if equipped
   - Transmission oil cooler line clips, if necessary
   - Oil pan mounting bolts, oil pan and gasket
4. Clean the oil pan and all gasket surfaces.

To install:

5. Apply a ⅛ in. (3mm) bead of sealer at the parting line of the oil pump body and the rear seal retainer.
6. Install or connect the following:
   - Oil pan and torque the M8 nuts/bolts to 21 ft. lbs. (28 Nm) and the M6 nuts/bolts to 105 inch lbs. (12 Nm)
7. Install the structural collar using the following procedure:
   a. Install the vertical collar to the oil pan mounting bolts and tighten, temporarily to 10 inch lbs. (1.1 Nm).
   b. Install the collar-to-transaxle bolts and torque to 40 ft. lbs. (55 Nm).
   c. Starting with the center vertical bolt and working outward. Torque the bolts to 40 ft. lbs. (55 Nm).
8. Install the dipstick and housing and connect the negative battery cable.
9. Fill the engine with the proper amount of clean SAE 5W-30 or SAE 10W-30 engine oil only. Do not mix the two grades of oil.
10. Start the engine, check for leaks and repair if necessary.

Oil Pump

REMOVAL & INSTALLATION

2.7L Engine

1. Before servicing the vehicle, refer to the precautions in the beginning of this section.
2. Drain the engine oil.
3. Remove or disconnect the following:
   - Crankshaft damper
   - Timing chain cover
   - Timing chain
   - Crankshaft sprocket
   - Oil pan
   - Oil pickup tube and o-ring
   - Oil pump

To install:

4. Fill the oil pump rotor cavity with clean engine oil.

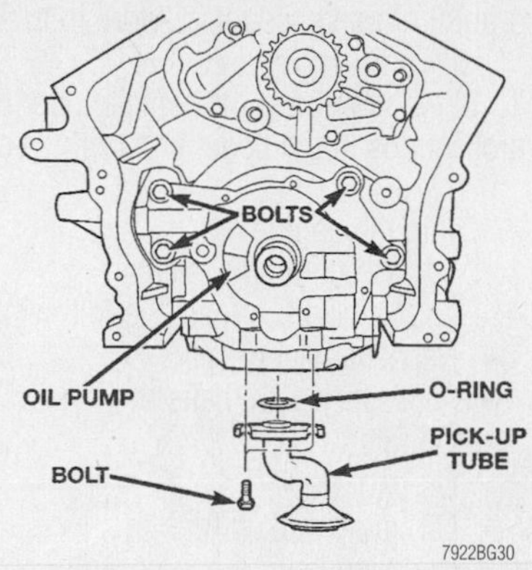

Oil pump mounting bolt locations—2.7L engine

5. Carefully, install the oil pump over the crankshaft and into position.
6. Install the oil pump mounting bolts and torque to 21 ft. lbs. (28 Nm).
7. Lubricate the new pickup tube O-ring with clean engine oil and torque the pickup tube mounting bolts to 21 ft. lbs. (28 Nm).
8. Install or connect the following:
   - Oil pan
   - Crankshaft sprocket
   - Timing chain
   - Timing chain cover
   - Crankshaft damper
9. Fill the engine with clean engine oil.
10. Start the vehicle, check for leaks and repair if necessary.

3.2L and 3.5L Engines

The timing belt must be removed to access the oil pump located behind the crankshaft drive sprocket. It is good practice to turn the crankshaft to Top Dead Center (TDC) No. 1 cylinder compression stroke (firing position) before starting disassembly. This should align all timing marks and be a

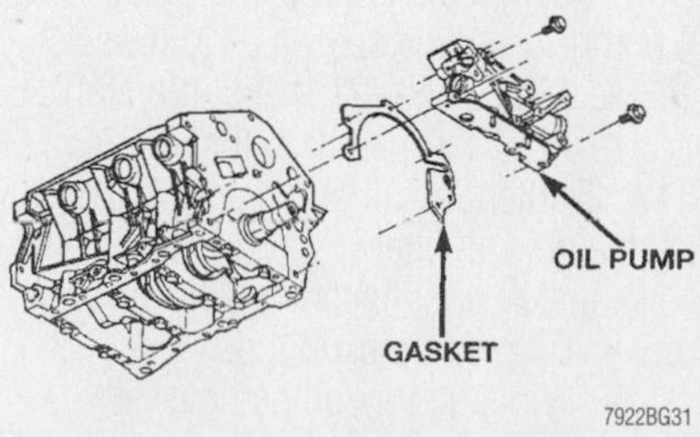

Prime the oil pump before installation, because a dry pump will wear prematurely and cause low oil pressure—3.2L and 3.5L engines

good point of reference for all work to follow.

1. Before servicing the vehicle, refer to the precautions in the beginning of this section.
2. Drain the engine oil.
3. Drain the cooling system.
4. Remove or disconnect the following:
   - Negative battery cable
   - Accessory drive belts
   - Oil filter
   - Oil pan
   - Oil pump pickup tube
   - Windage tray/oil pan gasket
   - Crankshaft damper using a suitable puller tool
   - Timing belt covers
5. Place matchmarks on the timing belt to aid installation. Align the matchmarks on the camshaft sprockets to marks on the rear timing belt covers before removing the timing.
6. Remove or disconnect the following:
   - Timing belt. Refer to the appropriate section for the removal and installation procedure.
   - Crankshaft sprocket using a suitable puller tool
   - Oil pump-to-engine screws and pump
   - Oil pump cover screws and cover
   - Oil pump rotors
7. Wash all parts in solvent and inspect carefully for damage or wear.

To install:

8. Clean all parts well. There should be no traces of old gasket/sealer on any components.
9. Assemble the oil pump with new parts as required.
10. Install the oil pump cover. Torque the fasteners to 108 inch lbs. (12 Nm).
11. Prime the oil pump prior to installation by filling the rotor cavity with clean engine oil.
12. Install the oil pump and tighten the oil pump-to-engine screws as follows:
    a. M8 screws: 21 ft. lbs. (28 Nm).
    b. M10 screws: 40 ft. lbs. (55 Nm).
13. Tighten the oil pan drain plug and install a new oil filter.
14. Install or connect the following:
    - Oil pump pickup tube
    - Windage tray/oil pan gasket
    - Oil pan and torque fasteners to 108 inch lbs. (12 Nm). Pay attention to sealing the oil pan gasket and its integral windage tray.
    - Crankshaft sprocket using tool C-4685C1, thrust bearing, washer and 12mm bolt to draw the sprocket onto the crankshaft
    - Timing belt. Refer to the appropriate section for the removal and installation procedure.
    - Timing belt covers, thrust bearing, washer plate and vibration dampe using tool L-4524
    - Accessory drive belts
    - Radiator and radiator hoses
    - Negative battery cable
15. Fill and bleed the cooling system.
16. Fill the engine with the correct amount of clean SAE 5W-30 or SAE 10W-30 engine oil only. Do not mix the two grades of oil.
17. Start the engine, check for leaks and proper oil pressure.

Rear Main Seal

REMOVAL & INSTALLATION

1. Before servicing the vehicle, refer to the precautions in the beginning of this section.
2. Drain the transaxle fluid.
3. Remove the negative battery cable.
4. Remove the transaxle, inspection cover and flywheel/flexplate.
5. Using a small prytool, carefully pry out the rear oil seal. Be careful not to nick or damage the crankshaft flange seal surface or the retainer bore.

To install:

6. Place the Seal Pilot tool C-4681 on the crankshaft.
7. Lightly coat the oil seal outside diameter with Loctite® Stud N' Bearing Mount® or the equivalent.
8. Apply a light coating of engine oil to the entire circumference of the oil seal lip.
9. Place the seal over the special tool and tap the seal in place with a plastic mallet.
10. Install the flexplate/flywheel and transaxle.
11. Connect the negative battery cable.
12. Fill the transaxle with the proper fluid.
13. Start the vehicle, check for leaks and repair if necessary.

Timing Chain, Sprockets, Front Cover and Seal

REMOVAL & INSTALLATION

2.7L Engine

**** WARNING**

When aligning the timing marks, rotate the crankshaft, not the camshafts. DO NOT rotate the

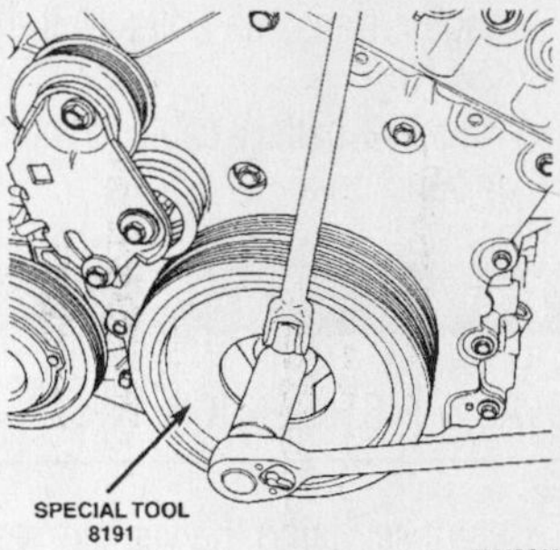

Removing the crankshaft center bolt using the Crankshaft Damper Holder tool—2.7L engine

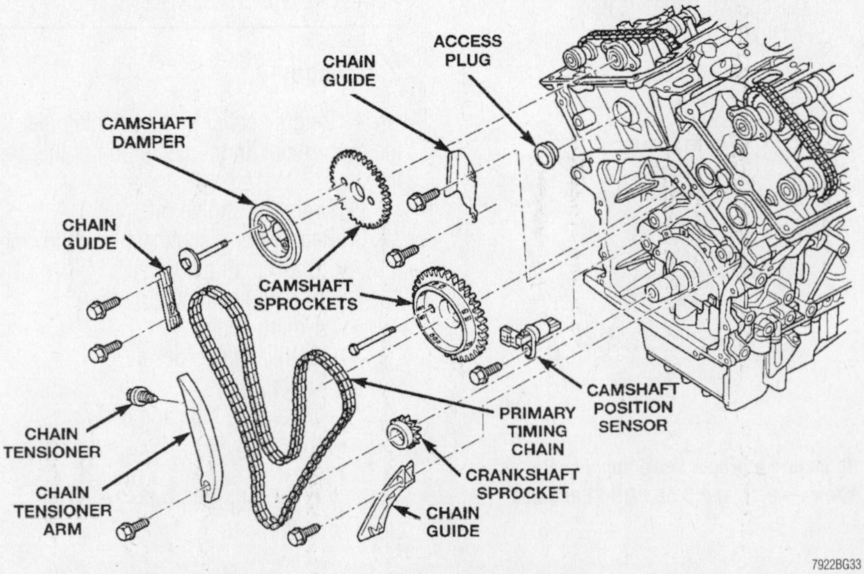

Exploded view of the timing chain drive assembly—2.7L engine

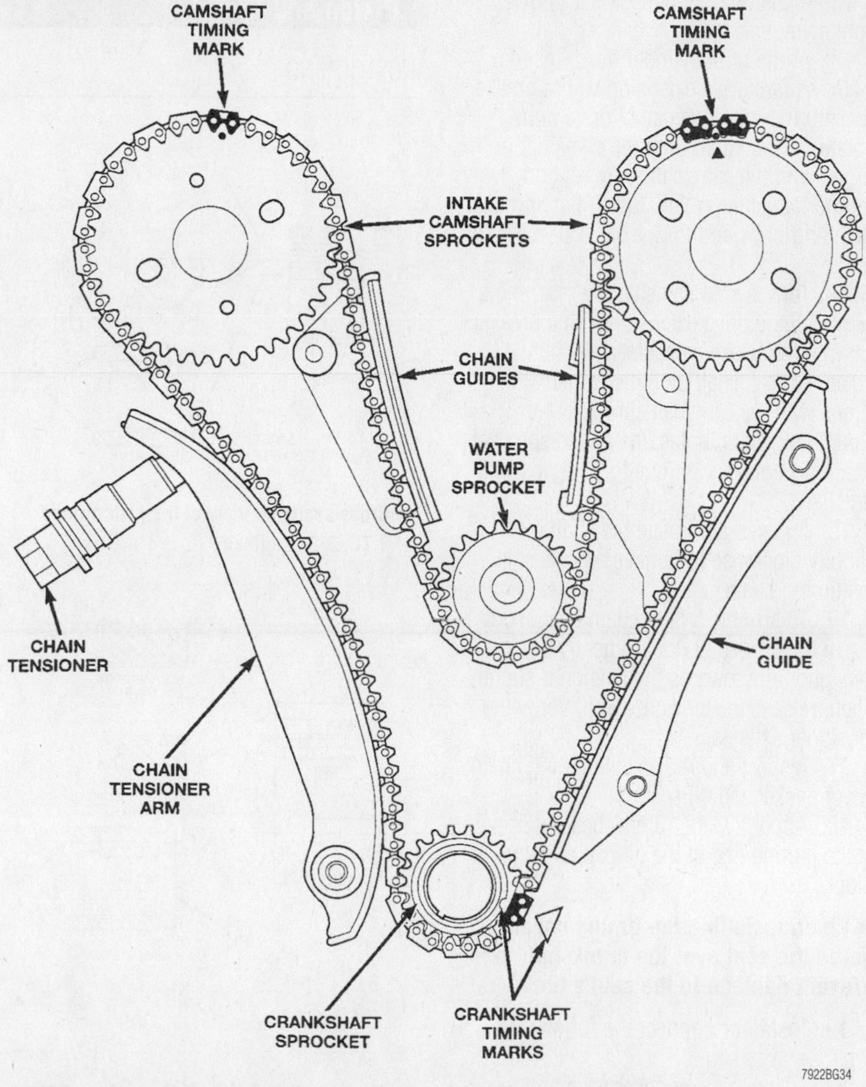

After the timing chain is installed, the timing marks should be aligned—2.7L engine

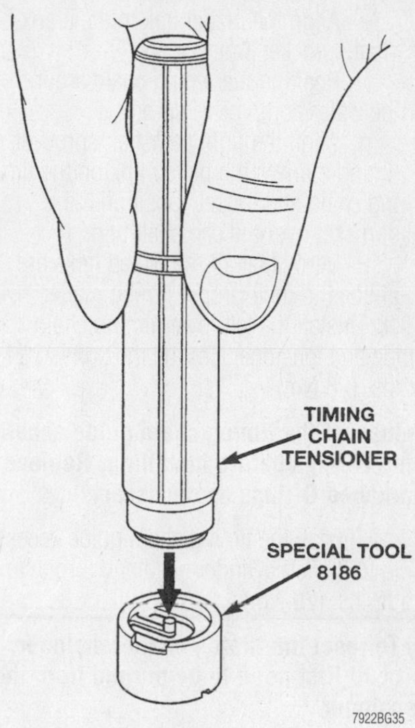

Using the Tensioner Resetting Special tool, to purge the oil from the tensioner—2.7L engine

camshafts or crankshaft with the timing chain removed without locating the crankshaft position, piston and/or valve damage may occur.

1. Before servicing the vehicle, refer to the precautions in the beginning of this section.
2. Remove or disconnect the following:
   - Upper intake manifold and valve covers
   - Upper radiator crossmember
   - Fan module
   - Accessory drive belts
3. Using Crankshaft Damper Holder tool 8191, hold the crankshaft and remove the center bolt.
   - Damper using a 3-jaw puller
   - Power steering pump and position it aside without disconnecting the hydraulic lines
   - Accessory drive belt tensioner pulley
   - Timing chain cover bolts
4. Clean and inspect the sealing surfaces.
5. Align the crankshaft sprocket timing mark with the oil pump housing mark.

➡The mark on the oil pump housing is 60 degrees at Top Dead Center (TDC).

6. Remove or disconnect the following:
   - Primary timing chain tensioner from the right cylinder head
   - Camshaft Position (CKP) sensor and timing chain access plug from the left cylinder head

➡The camshafts will rotate clockwise, when the camshaft sprocket bolts are removed.

   - Right camshaft sprocket mounting bolts, camshaft damper and sprocket
   - Left camshaft sprocket bolts and sprocket
   - Lower timing chain guide, tensioner arm and primary timing chain

To install:

➡Lubricate the timing chain and guides with clean engine oil before installation.

7. Install the timing chain by performing the following procedure:
   a. Verify that the crankshaft sprocket timing mark is aligned with the mark on the oil pump housing.
   b. Place the left side primary timing chain sprocket onto the chain, while aligning the timing mark on the sprocket to the 2 plated links on the chain.
   c. Lower the chain with the left sprocket through the left cylinder head opening.

d. Loosely position the left camshaft sprocket over the camshaft hub.

e. Align the plated link to the crankshaft sprocket timing mark.

f. Position the timing chain around the water pump drive sprocket.

g. Align the right camshaft sprocket timing mark to the plated link on the timing chain and loosely position the sprocket over the camshaft hub.

h. Verify that all the plated links are aligned to their proper timing marks.

8. Install the left lower timing chain guide and tensioner. Torque the bolts to 21 ft. lbs. (28 Nm).

➡Inspect the timing chain guide access plug O-rings before installing. Replace damaged O-rings as necessary.

9. Install the timing chain guide access plug to the left cylinder head and torque to 15 ft. lbs. (20 Nm).

➡To reset the timing chain tensioner, oil will first need to be purged from the tensioner.

10. Purge oil from the timing chain tensioner using the following procedure:

a. Remove the tensioner from the tensioner housing.

b. Place the check ball end of the tensioner into the shallow end of the Tensioner Resetting Special tool 8186.

c. Using hand pressure, slowly depress the tensioner until oil is purged from the cylinder.

d. Reinstall the tensioner into the tensioner housing.

11. Reset the timing chain tensioner using the following procedure:

a. Position the cylinder plunger into the deeper side of the Tensioner Resetting special tool 8186.

b. Apply a downward force until the tensioner is reset.

⁂ WARNING

Ensure that the tensioner is properly reset. The tensioner body must be bottomed against the top edge of the Tensioner Resetting Special tool 8186. Failure to properly perform the resetting procedure may cause tensioner jamming.

12. Install the chain tensioner into the right cylinder head.

13. At the right cylinder head, insert a ⅜ in. square drive extension with a breaker bar into the intake camshaft drive hub. Rotate the camshaft until the camshaft hub aligns with the camshaft sprocket and damper attaching holes. Install the sprocket attaching bolts and torque to 21 ft. lbs. (28 Nm).

14. Turn the left camshaft by inserting a ⅜ in. square drive extension with a breaker bar into the intake camshaft drive hub. Rotate the camshaft until the camshaft hub aligns with the camshaft sprocket and damper attaching holes. Install the sprocket attaching bolts and torque to 21 ft. lbs. (28 Nm).

15. If necessary, rotate the engine slightly clockwise to remove any slack in the timing chain.

16. To arm the timing chain tensioner: Use a flat-bladed prytool to gently pry the tensioner arm towards the tensioner slightly. Then, release the tensioner arm. Verify the tensioner extends.

17. Inspect and replace the timing chain cover gasket and oil seal.

18. Apply a ⅛ in. (3mm) bead of sealer at the parting line of the oil pan and engine block.

➡When installing the timing cover, guide the seal over the crankshaft to prevent damage to the seal's lip.

19. Install or connect the following:

- Timing cover and gasket and torque the M10 bolts to 40 ft. lbs. (54 Nm) and the M6 bolts to 105 inch lbs. (12 Nm)
- Crankshaft damper
- Accessory drive belt tensioner pulley
- Power steering pump
- Crankshaft damper
- Crankshaft center bolt and torque the bolt to 125 ft. lbs. (170 Nm) using tool 8191
- Accessory drive belts
- Fan module and electrical wiring harness
- Upper radiator crossmember
- Negative battery cable

Piston and Rings

POSITIONING

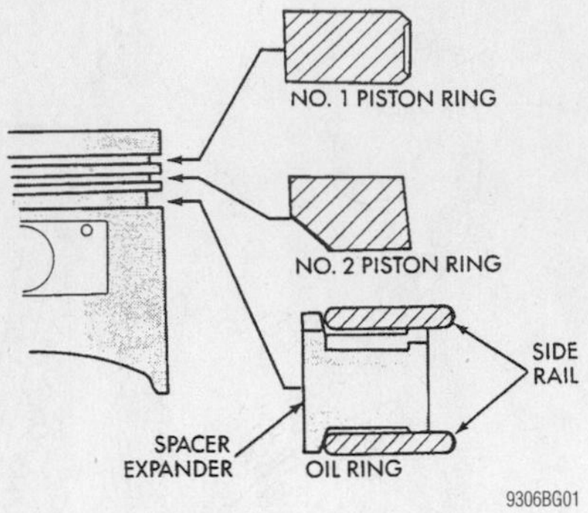

Cross-sectional view of the piston rings—2.7L, 3.2L Engines

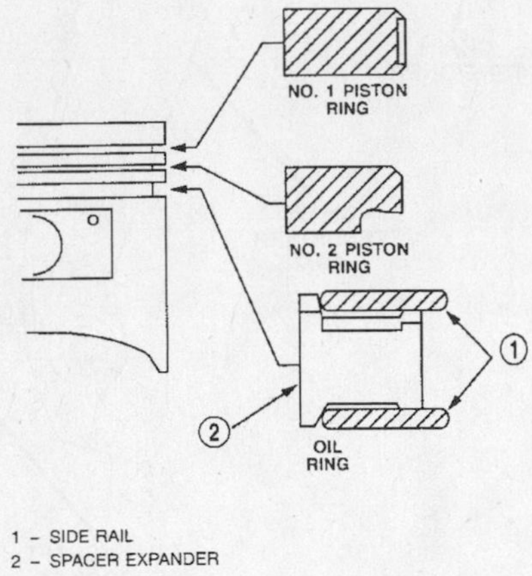

Cross-sectional view of the piston rings—3.5L Engine

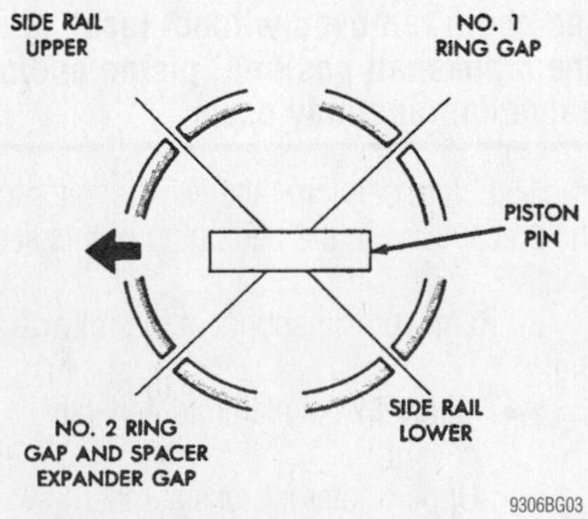

Piston ring gap positions—2.7L, 3.2L, 3.5L Engines

FUEL SYSTEM

Fuel System Service Precautions

Safety is the most important factor when performing not only fuel system maintenance but any type of maintenance. Failure to conduct maintenance and repairs in a safe manner may result in serious personal injury or death. Maintenance and testing of the vehicle's fuel system components can be accomplished safely and effectively by adhering to the following rules and guidelines:

- To avoid the possibility of fire and personal injury, always disconnect the negative battery cable unless the repair or test procedure requires that battery voltage be applied.
- Always relieve the fuel system pressure prior to disconnecting any fuel system component (injector, fuel rail, pressure regulator, etc.), fitting or fuel line connection. Exercise extreme caution whenever relieving fuel system pressure, to avoid exposing skin, face and eyes to fuel spray. Please be advised that fuel under pressure may penetrate the skin or any part of the body that it contacts.
- Always place a shop towel or cloth around the fitting or connection prior to loosening to absorb any excess fuel due to spillage. Ensure that all fuel spillage (should it occur) is quickly removed from engine surfaces. Ensure that all fuel soaked cloths or towels are deposited into a suitable waste container.
- Always keep a dry chemical (Class B) fire extinguisher near the work area.
- Do not allow fuel spray or fuel vapors to come into contact with a spark or open flame.
- Always use a back-up wrench when loosening and tightening fuel line connection fittings. This will prevent unnecessary stress and torsion to fuel line piping.
- Always replace worn fuel fitting O-rings with new. Do not substitute fuel hose, where fuel pipe is installed.

Before servicing the vehicle, also make sure to refer to the precautions in the beginning of this section as well.

Fuel System Pressure

RELIEVING

1. Before servicing the vehicle, refer to the precautions in the beginning of this section.
2. At the Power Distribution Center (PDC), remove the Fuel Pump Relay.
3. Operate the engine until the engine stalls; then, continue restarting the engine until it will no longer run.
4. Turn the ignition switch **OFF**.

**** CAUTION**

The previous steps must be performed to relieve the high pressure fuel from the fuel rail. The following steps must be performed to remove excess fuel from the fuel rail. Do not use the following steps to relieve high pressure for excessive fuel will be forced into a cylinder chamber.

5. Disconnect the electrical connector from any injector.
6. Connect 1 end of a jumper wire (with an alligator clip) to either injector terminal and the other end to the positive side of the battery.
7. Connect 1 end of a second jumper wire to the other injector terminal.

**** WARNING**

Applying power to an injector for more than a few seconds will damage the injector.

8. Momentarily, touch the other end of the jumper wire to a ground for no more than a few seconds.
9. At the Power Distribution Center (PDC), install the fuel pump relay.

➡When the fuel pump relay is removed, 1 or more Diagnostic Trouble Codes (DTC's) may be stored in the Powertrain Control Module (PCM) memory. A DRB scan tool must by used to clear the DTC's.

Fuel Filter

The fuel filter mounts to the frame rail in front of the fuel tank. The inlet and outlet ends of the filter are marked for installation purposes. Install the fuel filter making certain that it is properly orientated.

REMOVAL & INSTALLATION

➡The fuel filter is part of the fuel pressure regulator mounted on the fuel pump module.

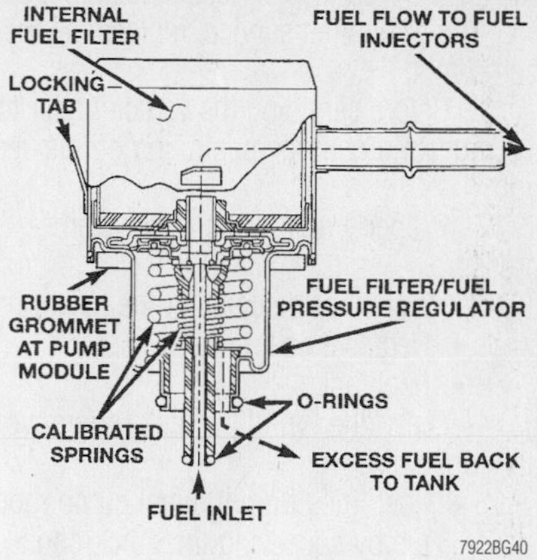

Cut away view of the fuel filter/pressure regulator

1. Before servicing the vehicle, refer to the precautions in the beginning of this section.
2. Properly relieve the fuel system pressure.
3. Lower the fuel tank.
4. Remove or disconnect the following:
   - Negative battery cable
   - Purge and vent lines
   - Fuel line from the pressure regulator
   - Filter/regulator by pushing in the locking tab, turning the regulator to unlock it and pulling the it straight up

To install:

5. Push the fuel filter/regulator into the fuel pump module and turn to lock it into position.
6. Connect the fuel lines and install the tank.
7. Start the engine, check for leaks and repair if necessary.

Fuel Pump

REMOVAL & INSTALLATION

The in-tank fuel pump module contains the fuel pump and pressure regulator which adjusts fuel system pressure. Fuel pump voltage is supplied through the fuel pump relay.

The fuel pump is serviced as part of the fuel pump module. The fuel pump module is installed in the top of the fuel tank and contains the electric fuel pump, fuel pump reservoir, inlet strainer fuel gauge sending unit, fuel supply and return line connections and the pressure regulator. The inlet strainer, fuel pressure regulator and level sensor are the only serviceable items. If the fuel pump requires service, replace the fuel pump module.

1. Before servicing the vehicle, refer to the precautions in the beginning of this section.
2. Properly relieve the fuel system pressure.
3. Remove or disconnect the following:
   - Negative battery cable
   - Fuel tank
4. Clean the top of the tank to remove any loose dirt.
   - Fuel lines from the fuel pump module by squeezing the quick-connect fitting with thumb and forefinger
   - Fuel pump module electrical connector from the top of the fuel pump module
5. Using special tool 6856, remove the fuel pump locknut by turning it counterclockwise.

CAUTION

The fuel reservoir of the fuel pump module does not empty out when the tank is drained. The fuel in the reservoir may spill out when the module is removed.

6. Remove the fuel pump and O-ring from the tank and discard the O-ring.

To install:

7. Thoroughly clean all parts. Wipe the seal area of the tank clean. Place a new O-ring on the ledge between the tank threads and the pump module opening.

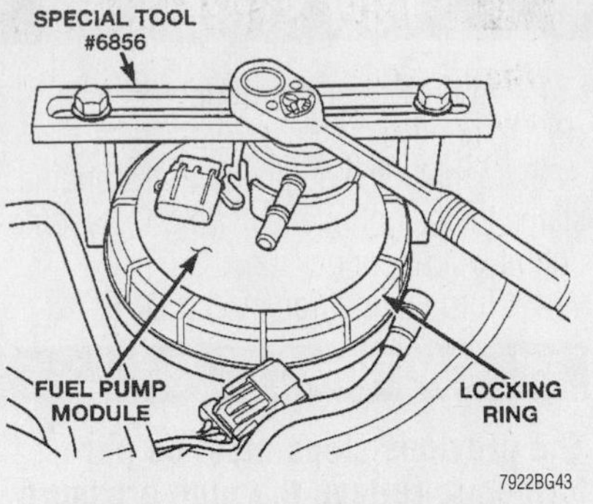

Using special tool 6856, remove the fuel pump module locknut

8. Position the fuel pump module in the tank. Be sure the alignment tab on the underside of the pump module flange sits in the corresponding notch in the fuel tank.
9. While holding the fuel pump module in place install the locking ring and torque to 40 inch lbs. (5 Nm) using special tool 6856 or a spanner-type tool.
10. Install or connect the following:
    - Fuel tank
    - Fuel pump module electrical connector
    - Negative battery cable
11. Fill the fuel tank with fuel. Install the fuel filler cap. Turn the ignition switch to the **ON** position to pressurize the system. Check the fuel system for leaks.

Fuel Injector

REMOVAL & INSTALLATION

1. Before servicing the vehicle, refer to the precautions in the beginning of this section.
2. Relieve the fuel system pressure.
3. Remove or disconnect the following:
   - Negative battery cable
   - Intake manifold plenum and cover opening with a clean cloth
4. Place a shop rag under the fuel rail's quick-connect fitting; then, squeeze the quick-connect fitting's retainer tabs together and pull the fitting assembly off of the fuel tube nipple.
   - Fuel injector electrical connectors
   - Fuel rail-to-engine bolts and fuel rail
   - Fuel injector-to-fuel rail retainer clips
   - Fuel injectors

To install:

5. Lubricate the injector O-rings with clean engine oil.
6. Install or connect the following:
   - Fuel injector and secure with retaining clips
   - Fuel rail onto cylinder head and press rail into place. Make sure that the injectors are fully seated.
   - Fuel rail-to-cylinder head bolts and torque bolts to 100 inch lbs. (11 Nm)
   - Intake plenum
7. Lubricate the quick-connect fitting's O-rings with clean engine oil; then, push the connector together until the retainer seats and a click is heard.
8. Connect the negative battery cable.

DRIVE TRAIN

Transaxle Assembly

REMOVAL & INSTALLATION

The 42LE four speed transaxle uses fully-adaptive controls. Adaptive controls are those which perform their functions based on real-time feedback sensor information. The transaxle is conventional in the use of hydraulically applied clutches to shift a planetary gear train. However, it uses electronics to control virtually all other functions. The following components are serviceable in the vehicle: valve body assembly, solenoid pack, manual valve lever position sensor, input and output speed sensors, transfer chain and sprockets, short (right side) stub shaft seal and the long (left side) stub shaft and ball bearing. Note that the factory recommends that before attempting any repair on the 42LE four-speed automatic transaxle, always check for proper shift linkage adjustment. Also, check for diagnostic trouble codes with the Chrysler DRB scan tool.

Use MOPAR Type 7176 Automatic Transmission Fluid only. Do not substitute transaxle fluid. If the differential sump requires fluid, use 80W-90 petroleum based Hypoid gear lubricant.

1. Before servicing the vehicle, refer to the precautions in the beginning of this section.
2. Remove or disconnect the following:
   - Negative battery cable
   - Engine air inlet tube
   - Crankshaft Position (CKP) sensor connector and remove the sensor from the upper right side of the transaxle bell housing
   - Transaxle wiring connector block located on the right shock tower. To free the connector from the harness, remove the wire ties.
   - Front wheels
   - Strut-to-steering knuckle bolts on both sides of the vehicle and/or tie rod ends, if required
   - Antilock Brake System (ABS) wheel speed sensor, if equipped
   - Halfshafts

➡Remove the halfshafts by inserting a prybar between the halfshaft and the transaxle case and prying the shafts from the transaxle housing. Swing the shafts out of the way, keeping the joints straight and suspend using wire. Be careful not to damage the halfshaft seals.

WARNING

Do not let the halfshafts or CV-joints hang unsupported. Internal joint damage may result if allowed to hang free.

3. Remove or disconnect the following:
   - Engine-to-transaxle brackets
   - Transaxle bell housing cover
4. Mark the driveplate to the torque converter and remove the torque converter bolts. The driveplate-to-torque converter bolts are not to be reused.
   - Starter assembly from the bell housing and allow the starter motor to sit between the engine and the frame
   - Transaxle oil cooler lines and plug the openings
   - Transaxle dipstick
   - Transaxle gear selector cable
   - Exhaust pipe from the exhaust manifold and position out of the way

➡**If the clearance will not allow for transaxle removal, remove the exhaust system from the vehicle.**

5. Support the transaxle using a transmission jack. Raise the transaxle slightly to relieve the weight off the rear transaxle mount.
   - Engine-to-transaxle brackets and transaxle mount through-bolt
   - Rear crossmember

➡**Pry the transaxle mount rearward to separate the mount from the transaxle.**

6. Lower the rear of the transaxle to gain access to the bell housing bolts. Remove the bell housing bolts.
7. Place a drain pan under the dipstick in the transaxle to catch transaxle fluid that will drain out of the case.
8. Remove or disconnect the following:
   - Transaxle dipstick tube and plug hole
   - Engine-to-transaxle bolts and transaxle

➡**The driveplate-to-torque converter bolts and the driveplate-to-crankshaft bolts must not be reused. Install new bolts whenever these bolts are removed.**

9. Inspect the driveplate for cracks. If cracks are present, replace the driveplate.

To install:

➡**Apply a light coating of grease to the pilot hole of the crankshaft if the torque converter is being replaced.**

WARNING

When installing the transaxle, be careful that the fuel tubes at the rear of the engine do not contact the following:

10. Install or connect the following:
    - Transaxle wiring harness
    - Driveplate and torque the fastener to 75 ft. lbs. (101 Nm)
    - Transaxle and torque the engine-to-transaxle case bolts to 75 ft. lbs. (101 Nm)
    - Rear transaxle case mount and rear crossmember in position and secure all fasteners
    - Transaxle dipstick tube
    - Exhaust pipe—to—engine exhaust manifold
    - Transaxle gear selector cable and oil cooler lines
    - Starter. Torque the bolts to 40 ft. lbs. (54 Nm). Be sure that the starter ground strap is installed correctly.
    - Align torque converter matchmarks and torque new torque converter-to-driveplate bolts to 60 ft. lbs. (81 Nm)
    - Bell housing cover and engine-to-transaxle brackets
11. While pulling the top of the steering knuckle outward, install the inner CV-joint, with new retainer clip in place, into the transaxle.
    - ABS wheel sensor (if removed) and strut-to-steering knuckle bolts
    - Front wheels and torque the lug nuts, in a star pattern to 95–100 ft. lbs. (129–135 Nm)
    - Transaxle dipstick
    - Transaxle wiring harness connector on the right shock tower
    - CKP sensor
    - Air inlet tube and negative battery cable
12. Start the engine and allow it to idle for 2 minutes. Apply the parking brake and move the selector through each gear position, ending in **N**. Recheck the fluid level and add if necessary. Be sure the vehicle is level when refilling the transaxle. Use Mopar Type 7176 Automatic Transmission Fluid (ATF) only. Do not substitute transaxle fluid. If the differential sump requires fluid, use 80W-90 petroleum based Hypoid gear lubricant.
13. Check the transaxle or proper operation. Adjust the shift linkage, if necessary. Be sure the reverse lamps come on when in reverse.

Halfshaft

REMOVAL & INSTALLATION

WARNING

Allowing the CV-joint assemblies to dangle unsupported, or pulling or pushing the ends, can damage boots or CV-joints. Always support both ends of the halfshaft to prevent damage or disengagement of the Tri-pot joint.

1. Before servicing the vehicle, refer to the precautions in the beginning of this section.
2. Remove or disconnect the following:
   - Negative battery cable
   - Front wheels
   - Front caliper assembly from steering knuckle
   - Front brake rotor from the hub
   - Speed sensor cable routing bracket from strut assembly
   - Hub and bearing-to-stub axle retainer nut
3. Install a puller tool onto the hub and bearing assembly and secure it into place using the wheel lug nuts.
4. Protect wheel stud threads by installing a wheel lug nut onto a wheel stud. Use a flat-bladed prying tool to prevent the hub from turning. Using the puller tool, force the halfshaft outer stub axle from the hub and bearing assembly.
5. Dislodge the inner Tri-pot joint from the stub shaft retaining snapring on the transaxle. To do this, insert a prybar

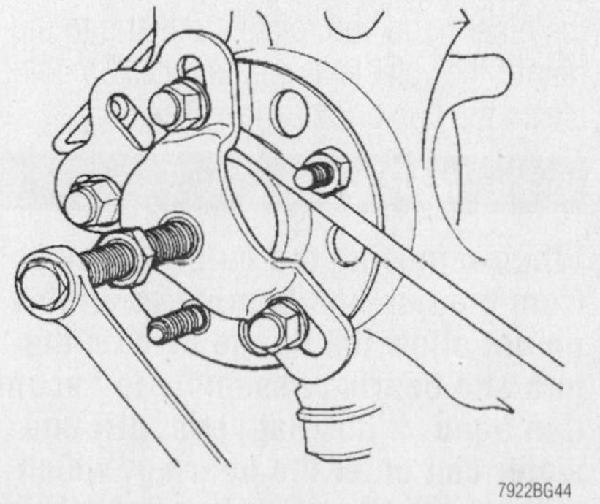

Removing the stub axle from the front hub/bearing assembly—halfshaft service

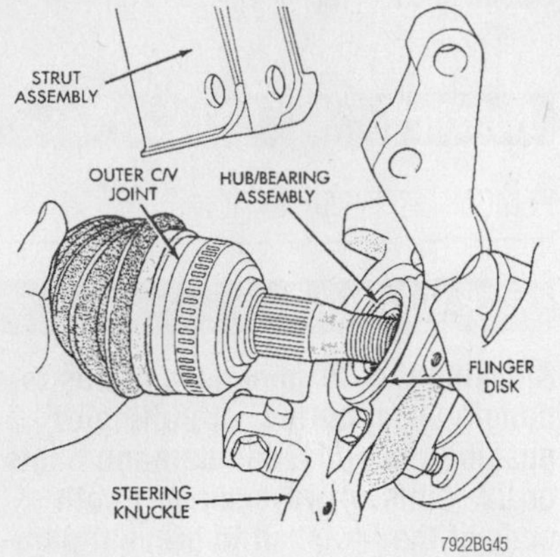

Be careful not to damage the threads for the axle nut when removing the outer CV-joint from the steering knuckle—halfshaft service

between the transaxle case and the inner Tri-pot joint and pry on Tri-pot joint.

➡**Do not try to remove the inner Tri-pot joint from the transaxle stub shaft at this time. Only disengage the inner Tri-pot joint from the retainer snapring.**

6. Remove the strut assembly-to-steering knuckle attaching bolts from the strut assembly.

⁂ WARNING

The strut assembly-to-steering knuckle bolts are serrated (toothed) where they go through the strut assembly and steering knuckle. When removing the bolts, turn the nuts off the bolt; do not turn the bolts in the steering knuckle or damage to the steering knuckle will result.

7. Separate the top of the steering knuckle from the lower end of the strut.

8. Hold the outer joint assembly with one hand. Grasp the steering knuckle with the other hand and rotate it out and to the rear of the vehicle, until the outer CV-joint clears the hub and bearing assembly.

⁂ WARNING

When removing the outer CV-joint from the hub and bearing assembly, do not allow the flange disc on the hub and bearing assembly to become damaged. If this happens, dirt and water can enter the bearing, which will cause premature bearing failure.

9. Remove the halfshaft inner joint from the transaxle stub shaft by grasping the inner Tri-pot joint and the interconnecting shaft and pulling both pieces at the same time. Take care not to pull on the interconnecting shaft to remove or separation of the spider assembly will occur.

To install:

10. Install the Tri-pot joint side of the halfshaft by performing the following procedure:

a. Replace the inner Tri-pot joint retaining circlip and O-ring seal on the transaxle stub shaft. These components are not reusable and must be replaced whenever the halfshaft is removed.

b. Apply an even coat of grease on the splines of the inner Tri-pot joint, where the O-ring seats against the Tri-pot joint.

c. Install the halfshaft through the hole in the splash shield. Grasp the inner joint in 1 hand and interconnecting shaft in the other. Align the inner Tri-pot joint spline with the stub shaft spline on the transaxle. Use a rocking motion with the inner Tri-pot joint to get it past the circlip on the transaxle stub shaft.

d. Continue pushing the Tri-pot joint onto transaxle stub shaft until it stops moving. The O-ring on the stub shaft should not be visible when the inner Tri-pot joint is fully installed. Check that the inner Tri-pot joint is locked in position by grasping the inner joint and pulling. If locked in position, the joint will not move on the stub shaft.

11. Hold the outer CV-joint assembly with one hand. Grasp the steering knuckle with the other and rotate it out and to the rear of the vehicle. Install the outer CV-joint into the hub and bearing assembly.

12. Install or connect the following:

- Top of the steering knuckle into the strut assembly. Align the steering knuckle-to-strut assembly mounting holes.
- Strut assembly-to-steering knuckle attaching bolts. Install the nuts to the attaching bolts and while holding the bolt heads, torque the nuts to 125 ft. lbs. (170 Nm). Turn the nuts on the bolts. DO NOT turn the bolts.
- New hub and bearing assembly-to-stub shaft retainer nut. Tighten but do not torque the nut at this time.
- Speed sensor cable routing bracket and screw
- Brake rotor and caliper assembly; then, tighten the caliper guide pin bolts to 30 ft. lbs. (41 Nm)
- Front wheels and lug nuts
- Negative battery cable

13. Pump the brakes until a firm pedal is obtained.

14. Apply the brakes and torque the new stub shaft-to-hub and bearing assembly retainer nut to 120 ft. lbs. (163 Nm).

⁂ WARNING

When tightening the stub shaft retaining nut, be careful not to exceed the maximum torque specification of 120 ft. lbs. (163 Nm). If this specification is exceeded, failure of the halfshaft could result.

15. Road test the vehicle to check for noise or vibration.

CV-Joints

OVERHAUL

Inner (Tri-pot) Joint

1. Before servicing the vehicle, refer to the precautions in the beginning of this section.
2. Disconnect the negative battery cable.
3. Remove the halfshaft and retaining clamps.
4. Slide the boot down the shaft away from the tri-pot housing.

➡**When separating the spider joint from the tri-pot joint housing, hold the rollers in place on the trunions to prevent the rollers and needle bearings from falling away.**

5. Carefully, slide the shaft/spider assembly from the tri-pot housing.
6. Remove the spider assembly-to-shaft snapring; then, slide the spider assembly off the shaft.

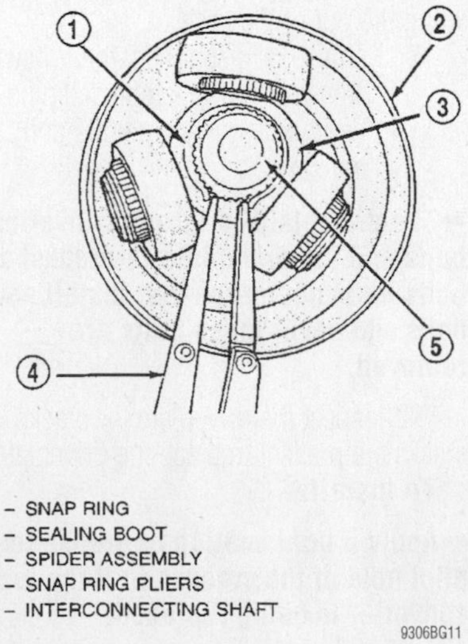

View of the halfshaft inner tri-pot joint and snapring

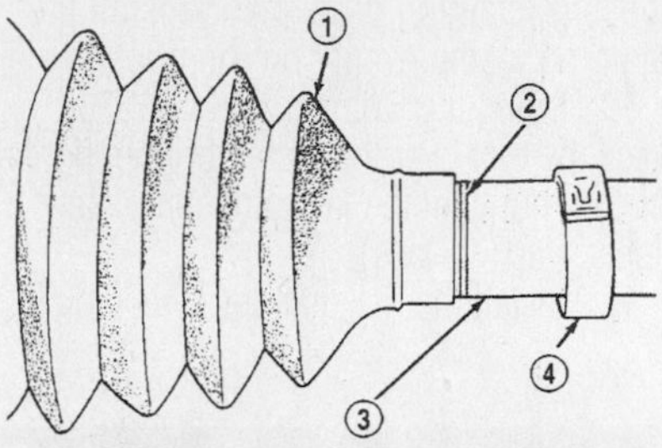

View of the halfshaft boot, shaft and boot clamp

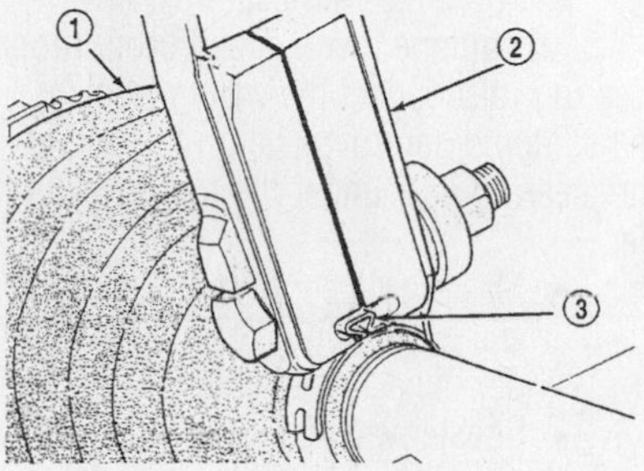

Securing the halfshaft boot clamp

✲✲ WARNING

If necessary, tap the spider assembly off the shaft using a brass drift; be careful not to hit the outer bearings.

7. Slide the boot off the shaft.
8. Throughly, inspect all parts for signs of excessive wear; if necessary, replace the halfshaft.

➡Component parts are not serviceable and must be replaced as an assembly.

To install:

9. Slide the inner tri-pot boot clamp and boot onto the shaft; then, position the boot so that only the thinnest (sight) groove is visible on the shaft.
10. Install the spider assembly onto the shaft just far enough so that the snapring can be installed.

✲✲ WARNING

If necessary, tap the spider assembly onto the shaft using a brass drift; be careful not to hit the outer bearings.

11. Install the snapring onto the shaft; make sure that the snapring is fully seated in the groove.
12. If installing a new boot, distribute ½ of the grease in the service package inside the tri-pot housing and the other ½ inside the boot.
13. Carefully, slide the spider assembly and shaft into the tri-pot housing.
14. Position the inner boot clamp evenly on the sealing boot.
15. Using the Crimper tool C-4975, place the tool over the clamp bridge, tighten the tool nut until the jaws are completely closed (face-to-face).

✲✲ WARNING

The seal must not be dimpled, stretched or out of shape. If necessary, equalize the seal pressure and shape it by hand.

16. Position the boot onto the tri-pot housing retaining groove and install the retaining clamp evenly on the boot.
17. Using the Crimper tool C-4975, place the tool over the clamp bridge, tighten the tool nut until the jaws are completely closed (face-to-face).
18. Install the halfshaft into the vehicle.

Outer CV-Joint

1. Before servicing the vehicle, refer to the precautions in the beginning of this section.
2. Disconnect the negative battery cable.
3. Remove the halfshaft and the retaining clamps.
4. Slide the boot down the shaft away from the CV-joint housing.
5. Remove the grease to expose the CV-joint-to-shaft retaining ring.
6. Spread the snapring ears apart and slide the CV-joint assembly off of the shaft.
7. Slide the boot off the shaft.
8. Throughly, clean and inspect all parts for signs of excessive wear; if necessary, replace the halfshaft.

➡Component parts are not serviceable and must be replaced as an assembly.

To install:

9. Slide the outer CV-joint boot clamp and boot onto the shaft; then, position the boot so that only the thinnest (sight) groove is visible on the shaft.
10. Slide the outer CV-joint assembly on the shaft, spread the snapring ears, position the CV-joint and verify that the snapring is fully seated in the shaft groove.
11. If installing a new boot, distribute ½ of the grease in the service package into the

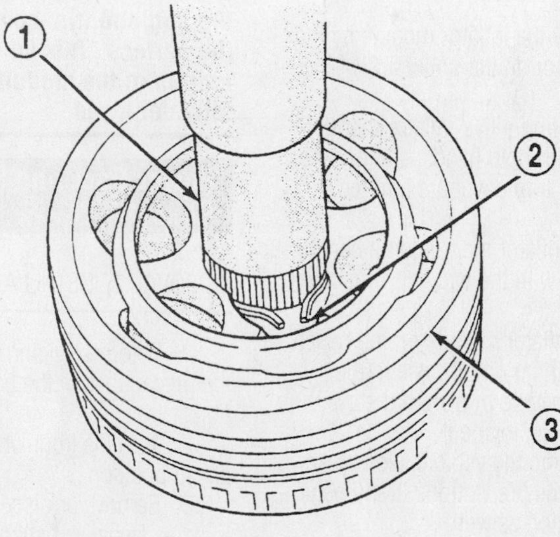

View of the halfshaft outer CV-joint and snapring

CV-joint housing and the other ½ inside the boot.

12. Position the outer boot clamp evenly on the sealing boot.

13. Using the Crimper tool C-4975, place the tool over the clamp bridge, tighten the tool nut until the jaws are completely closed (face-to-face).

⁂ WARNING

The seal must not be dimpled, stretched or out of shape. If necessary, equalize the seal pressure and shape it by hand.

14. Position the boot onto the CV-joint housing retaining groove and install the retaining clamp evenly on the boot.

15. Using the Crimper tool C-4975, place the tool over the clamp bridge, tighten the tool nut until the jaws are completely closed (face-to-face).

16. Install the halfshaft into the vehicle.

STEERING AND SUSPENSION

Air Bag

⁂ CAUTION

Some vehicles are equipped with an air bag system. The system must be disabled before performing service on or around system components, steering column, instrument panel components, wiring and sensors. Failure to follow safety and disabling procedures could result in accidental air bag deployment, possible personal injury and unnecessary system repairs.

PRECAUTIONS

Several precautions must be observed when handling the inflator module to avoid accidental deployment and possible personal injury.

- Never carry the inflator module by the wires or connector on the underside of the module.
- When carrying a live inflator module, hold securely with both hands, and ensure that the bag and trim cover are pointed away.
- Place the inflator module on a bench or other surface with the bag and trim cover facing up.
- With the inflator module on the bench, never place anything on or close to the module which may be thrown in the event of an accidental deployment.

Before servicing the vehicle, also make sure to refer to the precautions in the beginning of this section as well.

DISARMING

⁂ CAUTION

The Air Bag system must be disarmed before repair and/or removal of any component in its immediate area including the air bag itself. Failure to do so may cause accidental deployment of the air bag, resulting in unnecessary system repairs and/or personal injury.

1. Disconnect the negative battery cable and isolate the cable using an appropriate insulator (wrap with quality electrical tape).
2. Allow the system capacitor to discharge for 2 minutes before starting any repair on any air bag system or related components. This will disable the air bag system.

⁂ CAUTION

Always wear safety goggles when working with or around the air bag system. When carrying a live air bag, be sure the bag and trim cover are pointed away from the body. In the unlikely event of an accidental deployment, the bag will, then deploy with minimal chance of injury. When placing a live air bag on a bench or other surface, always face the bag and trim cover up, away from the surface. This will reduce the motion of the module if it is accidentally deployed.

Power Rack and Pinion Steering Gear

REMOVAL & INSTALLATION

1. Before servicing the vehicle, refer to the precautions in the beginning of this section.
2. Turn the front wheels to the straight-ahead position.
3. Remove or disconnect the following:
   - Negative battery cable
   - Wiper arms
   - Wiper module cover and cowl cover
   - Reinforcement from the strut towers and wiper module
   - Throttle body's intake air duct and resonator
4. Clamp the steering wheel in the center position with some type of holding device.
   - Intermediate shaft from the steering column coupler
   - Both tie rods from the steering gear and place them on the transaxle bell housing
   - Siphon the power steering fluid from the reservoir
   - Power steering gear fluid lines
5. If equipped with speed proportional steering, disconnect the wiring from the solenoid control valve on the end of the steering gear under the master cylinder.
   - Master cylinder from the booster and position it to the side without disconnecting the fluid lines
   - Booster vacuum line

➡It may be helpful to loosen the 2 bolts attaching the right mounting bracket to the steering gear in order the clear the air conditioning lines.

   - 4 steering gear mounting bolts (2 on each side)
6. Move the steering gear and intermediate shaft into the engine compartment to gain access to the roll pin.

➡Special tools are commercially available for removing roll pins.

   - Roll pin and separate the intermediate shaft from the steering gear
   - Right front wheel
   - Tie rod end from the steering arm on the right strut

➡If equipped with a 2.7L engine, turn the left front tire towards the left as far as possible to provide clearance.

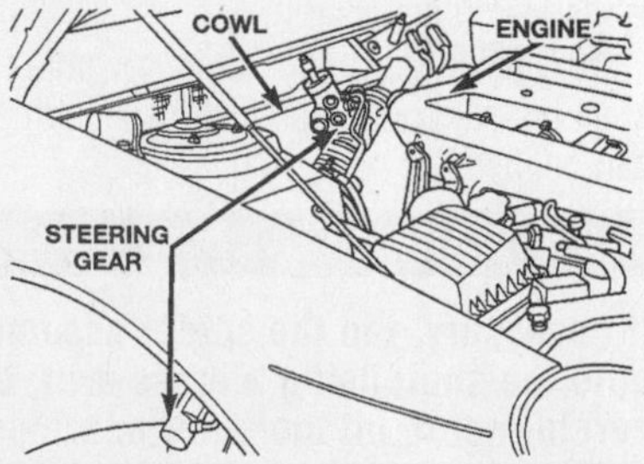

The steering gear is removed or installed through the right side of the vehicle

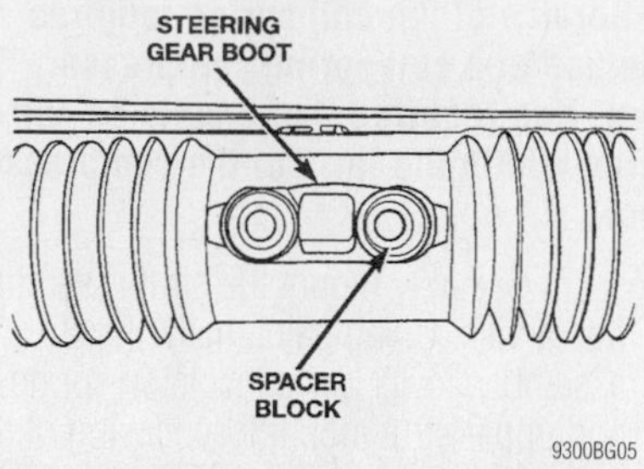

Center the spacer block in the steering gear before installing the tie rods

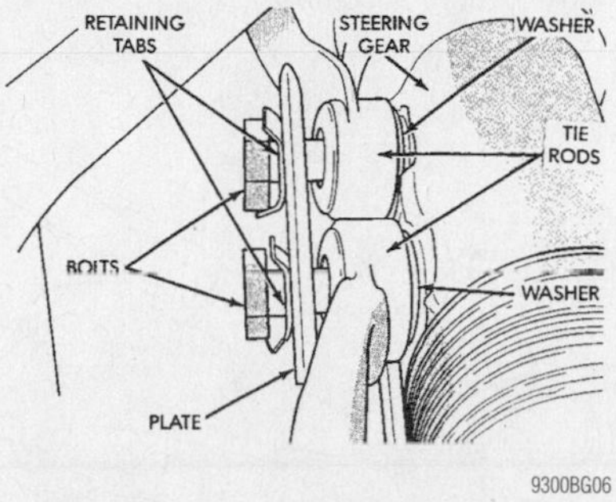

Be sure to position the washers between the tie rods and the steering gear as shown

7. Slide the right end of the steering gear out through the tie rod hole while raising the left end of the steering gear upward.

To install:

8. Position the tie rod attaching points at the center of the steering gear travel.

9. Install or connect the following:
   - Steering gear

➡If equipped with a 2.7L engine, turn the left front tire straight-ahead.

   - Right tie rod on the steering arm and torque the nut to 27 ft. lbs. (37 Nm)
   - Right front wheel. Torque the lug nuts in sequence to half of the specified torque, then tighten the nuts in sequence again to 95 ft. lbs. (129 Nm).
   - Intermediate shaft to the steering gear using the roll pin. Be sure the pin is centered in the joint.
   - 4 steering gear mounting bolts and torque the bolts to 43 ft. lbs. (58 Nm)
   - Bracket bolts; tighten them (if loosened) to 27 ft. lbs. (37 Nm)
   - Steering gear fluid lines; torque the nuts to 35 ft. lbs. (47 Nm)
   - Solenoid valve, if equipped with speed proportional steering

⁂ WARNING

Be sure the spacer block in the rack is centered before connecting the tie rods to the steering gear.

   - Tie rods to the steering gear with the washers between the rods and the steering gear. Torque the bolts to 60 ft. lbs. (82 Nm).
   - Remaining components in the reverse order of the removal steps
   - Negative battery cable

10. Add Mopar® power steering fluid to the reservoir and let it settle for at least 2 minutes. Start the engine for a few seconds and turn it **OFF**. Add fluid as necessary. Repeat this procedure until the fluid level remains constant.

11. Raise the front wheels off the floor and start the engine. Turn the steering from lock-to-lock several times. Do not hold the wheel in the locked position for more than 2 seconds at a time. Refill the reservoir as needed.

12. Lower the vehicle and repeat the procedure. If the fluid is extremely foamy, allow the vehicle to stand with the engine **OFF** for a few minutes, then repeat the procedure.

Strut

REMOVAL & INSTALLATION

Front

1. Before servicing the vehicle, refer to the precautions in the beginning of this section.

➡Service of the coil spring requires the use of a coil spring compressor tool. It is required that 5 coils be captured within the jaws of the compressor tool.

➡Do not support the vehicle by placing supports under the suspension arms. The suspension arms must hang freely.

2. Remove or disconnect the following:
   - Negative battery cable
   - Front wheel(s)
   - Stabilizer bar attaching link at the strut assembly

3. Loosen, but do not remove the outer tie rod end-to-strut assembly steering arm attaching nut. Then, remove the outer tie rod end from the steering arm using puller MB-990635.
   - Speed sensor wiring harness mounting bracket from the strut, if equipped with Antilock Brake System (ABS)
   - Brake caliper assembly. Support the caliper assembly from the vehicle frame with a strong piece of wire. Do not allow the assembly to hang by the brake hose. Remove the front brake rotor disc

⁂ WARNING

The strut assembly-to-steering knuckle bolts are serrated where they go through the strut and steering knuckle. Do not turn the bolts during removal. If the bolts are turned, damage to the steering knuckle will result.

4. The strut assembly-to-steering knuckle bolts must not be turned during strut removal. Hold the bolt head with a wrench and turn the nuts off the bolts.

5. Remove the 3 strut assembly upper mount-to-shock tower mounting nuts and washers. Remove the strut from the vehicle.

6. Disassemble the strut by performing the following procedure:
   a. Securely mount the strut assembly into a vise. Using paint, mark the strut unit, lower spring isolator, spring and upper strut mount for indexing of the parts at assembly.
   b. Position the spring compressor tool onto the strut. Compress the coil spring until all load is off the upper strut mount assembly.
   c. Install Strut Rod Socket tool L-4558A on the strut shaft nut and a 10mm socket on the end of the strut shaft to prevent it from turning. Remove the strut shaft nut.
   d. Remove the upper mount assembly, jounce bumper and seat bearing and dust shield as an assembly.
   e. Remove the coil spring and compressor as an assembly from the strut. Remove the lower spring isolator from the strut assembly lower spring seat.
   f. Inspect all components for abnormal wear, oil leakage or failure. Replace parts as required.

To install:

7. Assemble the strut by performing the following procedure:
   a. Inspect the strut assembly for signs of leakage. Actual leakage will be a stream of fluid running down the side and dripping off the lower end of the strut. A slight amount of seepage between the strut rod and strut shaft seal

is not unusual and does not affect performance of the strut assembly.

b. Install the lower spring isolator on the strut unit. Install the compressed coil spring onto the strut assembly aligning the paint marks made during removal.

c. Install the strut bearing into the bearing seat. The bearing must be installed into the seat with the notches on the bearings facing down.

d. Lower the seat bearing and dust shield onto the strut and spring assembly. Align the paint marks made during removal.

e. Install the jounce bumper and upper mount on the strut shaft, aligning the paint marks.

f. Install the strut mount-to-shaft retainer nut. Inspect all alignment marks made during removal and align as required. While holding the strut shaft from turning with a 10mm socket, torque the strut shaft nut to 70 ft. lbs. (94 Nm).

g. Equally loosen the spring compressor tool until all tension is released. Remove the spring compressor tool.

8. Install the front strut into the strut tower. Torque the 3 upper nuts to 25 ft. lbs. (33 Nm).

9. Position the steering knuckle neck into the strut assembly. Install the strut assembly-to-steering knuckle bolts. Install the nuts onto the attaching bolts and torque to 125 ft. lbs. (169 Nm). Do not turn the serrated bolt heads during installation. Turn only the nuts.

⁂ WARNING

The strut assembly-to-steering knuckle bolts are serrated (toothed) where they go through the strut and steering knuckle. Do not turn the bolts during removal. If bolts are turned, damage to the steering knuckle will result.

10. Install or connect the following:
- Brake rotor and caliper assembly to the adapter and torque the caliper bolts to 14 ft. lbs. (19 Nm)
- Front speed sensor cable routing bracket onto the front strut, if equipped
- Outer tie rod on the steering arm and torque the attaching nut to 27 ft. lbs. (37 Nm)
- Stabilizer link assembly onto the strut assembly and torque the attaching nut to 70 ft. lbs. (95 Nm)
- Front wheel and lug nuts. Torque the lug nuts, in sequence, to 95–100 ft. lbs. (129–135 Nm).
- Negative battery cable

Rear

1. Before servicing the vehicle, refer to the precautions in the beginning of this section.

The rear strut assemblies support the weight of the vehicle using coil springs positioned around the struts. The coil springs are contained between the upper mount of the strut assembly and a lower spring seat on the body of the strut assembly. The strut is attached to the spindle by a split collar on the rear spindle with a pinch bolt to hold the spindle to the strut.

2. Remove or disconnect the following:
- Rear wheel
- Caliper assembly and rotor from the hub, if equipped with rear disc brakes
- Brake flex hose from the support bracket and wheel cylinder, if equipped with rear drum brakes. Plug the brake flex hose to prevent system contamination. Do not allow the rear caliper to hang by the brake hose. Support the caliper off of the frame with a strong piece of wire.
- Speed sensor cable routing bracket and tube, if equipped with ABS
- Lateral links to the rear spindle assembly bolts
- Rear strut assembly-to-stabilizer bar attaching link at the stabilizer bar

➡Hold the hex on the attaching link stud while breaking the nut loose. The attaching link does not have to be removed from the strut.

3. Remove the rear spindle-to-strut assembly pinch bolt. Install a center punch in the hole on the spindle and tap the punch into the hole until jammed. This will spread the spindle casting allowing it to be removed from the strut.

4. Using a hammer, tap on the top surface of the spindle, driving the spindle down and off the end of the strut assembly. Let the spindle and assembled components hang from the trailing arm while the strut is being serviced.

5. From inside the trunk of the vehicle, remove the 3 upper strut mounting bolts and remove the strut from the vehicle.

6. Disassemble the strut by performing the following procedure:

➡Service of the coil spring requires the use of a coil spring compressor tool. It is required that 5 coils be captured within the jaws of the compressor tool.

a. Securely mount the strut assembly into a vise. Using paint, mark the strut assembly, lower spring isolator, spring and upper strut mount for indexing of the parts at reassembly.

b. Position a spring compressor tool onto the coil spring. Compress the coil spring until all load is off of the upper strut mount assembly.

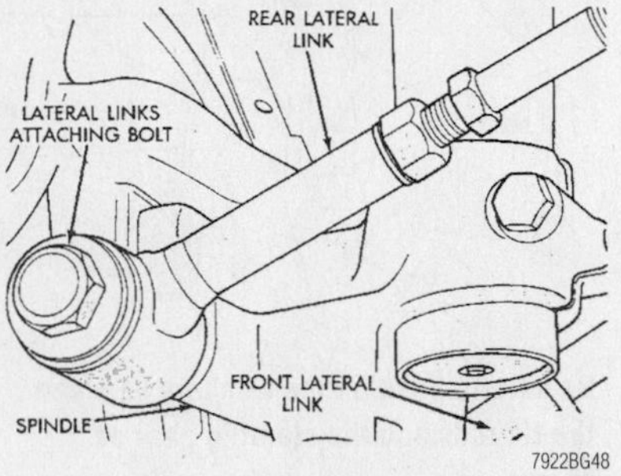

Separate the lateral links from the spindle—rear strut service

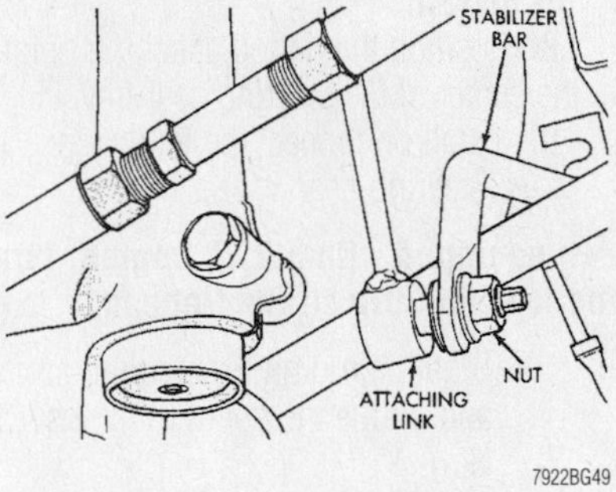

Remove the nut from the stabilizer-to-strut attaching link stud at the bar—rear strut service

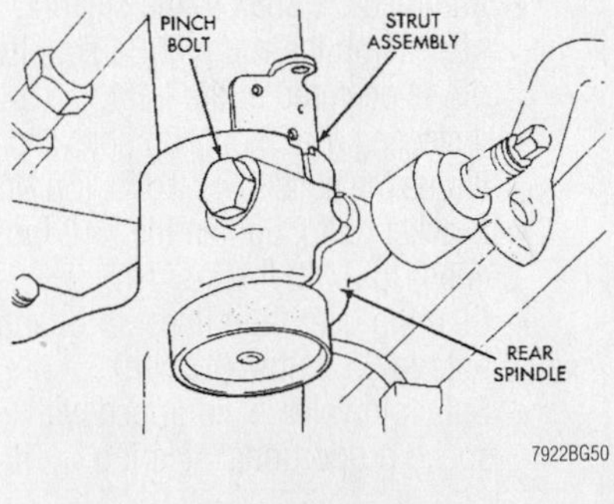

Loosen, then remove the rear spindle-to-strut pinch bolt—rear strut service

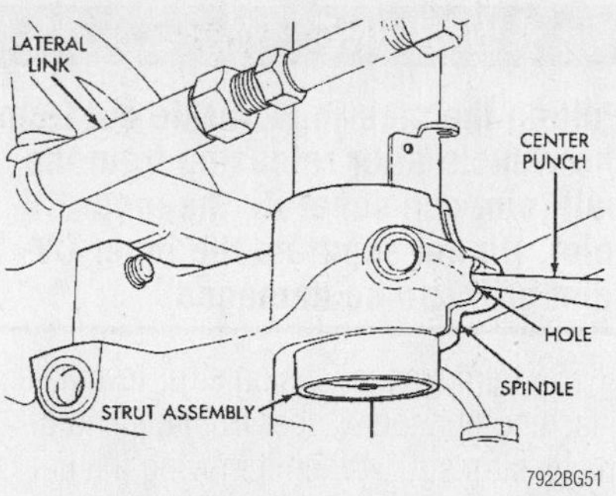

Insert a center punch into the hole on the spindle and tap until the casting is spread—rear strut service

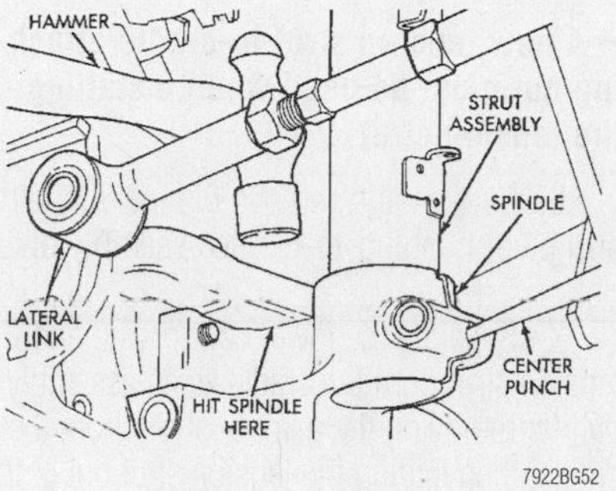

Tap with a hammer on the surface of the spindle driving it down and off the end of the strut—rear strut service

c. Install the strut rod socket tool L-4558 on the strut shaft nut and an 8mm Allen wrench on the end of the strut shaft to prevent it from turning. Remove the strut shaft nut.

d. Remove the upper strut mount assembly off of the strut shaft. Remove the coil spring and compressor tool as an assembly from the strut.

e. Remove the plate, dust shield and jounce bumper off of the strut unit.

f. Inspect all components for abnormal wear, oil leakage or failure. Replace parts as required.

To install:

7. Assemble the strut by performing the following procedure:

a. Install the lower spring isolator on the strut unit. If it is the original isolator, align the paint marks.

b. Install the jounce bumper into the dust shield. Install the plate on top of the dust shield and into the jounce bumper.

c. Install the dust shield, jounce bumper and the top plate onto the strut unit as an assembly.

d. Install the coil spring and compressor tool onto the strut unit and align the paint marks on the spring to that of the strut unit.

e. Install the upper strut mount assembly onto the strut shaft. Align the paint marks and install the strut shaft retaining nut.

f. Using the strut rod socket tool L-4558 and the 8mm Allen wrench to prevent the strut shaft from turning, torque the strut shaft nut to 70 ft. lbs. (95 Nm).

g. Equally loosen the spring compressor tool until all tension is released. Remove the spring compressor tool.

8. Position the strut in the vehicle. Torque the 3 upper mounting nuts to 20 ft. lbs. (28 Nm).

9. Install the spindle assembly onto the bottom of the strut. Push or tap the spindle assembly onto the strut, until the notch in the spindle is tightly seated against the locating tap on the strut assembly. Remove the center punch from the hole in the spindle.

10. Install or connect the following:

- Strut-to-spindle pinch bolt and torque to 40 ft. lbs. (55 Nm)
- Lateral link-to-spindle attaching bolt and torque to 105 ft. lbs. (140 Nm)
- Stabilizer bar attaching link and torque the stabilizer link-to-stabilizer bar attaching nut to 70 ft. lbs. (95

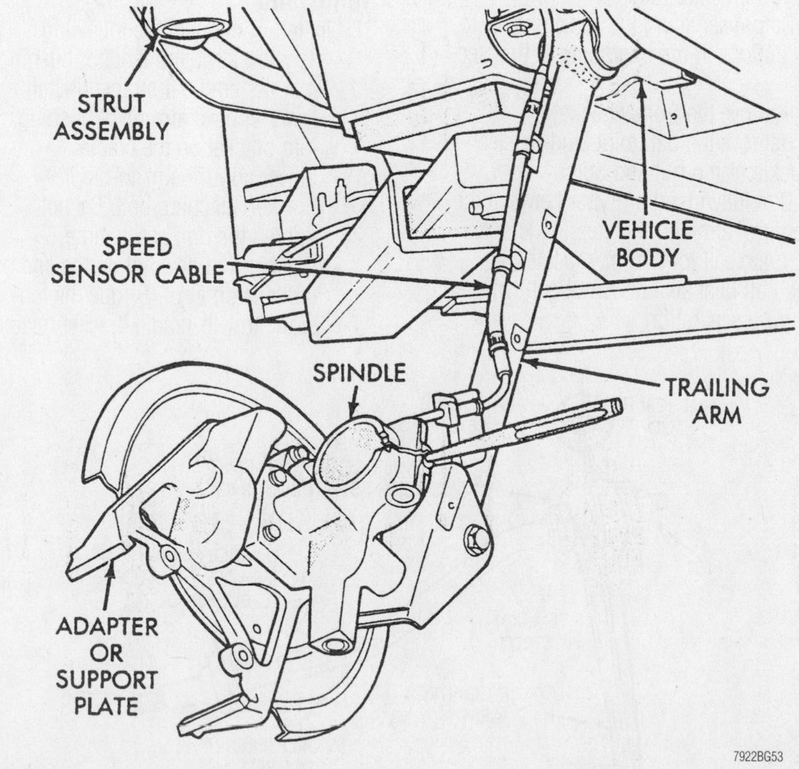

Allow the components to hang from the trailing arm as shown—rear strut service

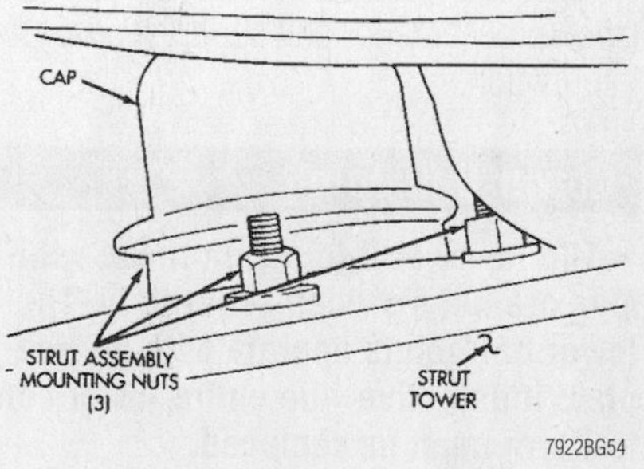

Access the 3 upper strut mounting nuts through the trunk—rear strut service

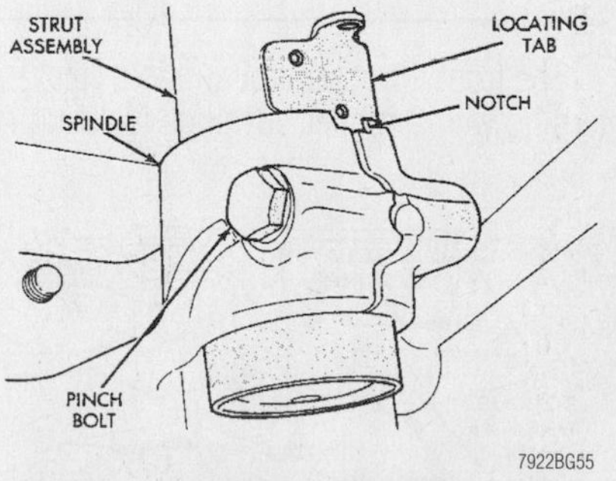

During reassembly, tap the spindle onto the strut until the notch in the spindle is tightly seated against the tab—rear strut

Nm), while holding the stabilizer link stud at the hex with a wrench

- Rear speed sensor cable routing tube and bracket, if equipped with ABS
- Rotor and caliper assembly (if equipped with rear disc brakes) and torque the bolts to 16 ft. lbs. (22 Nm)
- Rear brake flex hose to the wheel cylinder and support plate, if equipped with rear drum brakes
- Rear wheel(s) and torque the lug nuts, in sequence, to 95 ft. lbs. (129 Nm)

11. Bleed the brake system, if equipped with rear drum brakes.

12. Have the rear wheel toe set to specifications.

Coil Springs

REMOVAL & INSTALLATION

Front

Refer to the front strut removal and installation procedure for coil spring service information.

Rear

Refer to the rear strut removal and installation procedure for coil spring service information.

Lower Ball Joint

➡The lower ball joints on these vehicles are not serviced separately. The lower ball joints operate with no freeplay. If defective, the entire lower control arm must be replaced.

Lower Control Arm

REMOVAL & INSTALLATION

The front lower control arm is a steel forging with 2 rubber bushings isolating the lower control arm from the front cradle assembly. The isolator bushings consist of a metal encased pivot bushing and a solid rubber tension strut bushing. The lower control arm is bolted to the cradle assembly using a pivot bolt through the center of the rubber pivot bushing at the tension strut isolator bushing. The ball joint is built into the lower control arm and is non-serviceable. If the ball joint becomes worn, the entire lower control arm must be replaced. The ball joint seal, however, is replaceable as well as the lower control arm inner bushing. If the lower control arm is damaged, do not attempt to repair or straighten a broken or bent lower control arm.

1. Before servicing the vehicle, refer to the precautions in the beginning of this section.

2. Remove the front wheel(s).

3. Remove the ball joint stud-to-steering knuckle clamp nut and bolt.

4. Carefully insert a prybar between the lower control arm and the steering knuckle and separate ball joint from knuckle. Be sure the ball joint seal does not get damaged during separation.

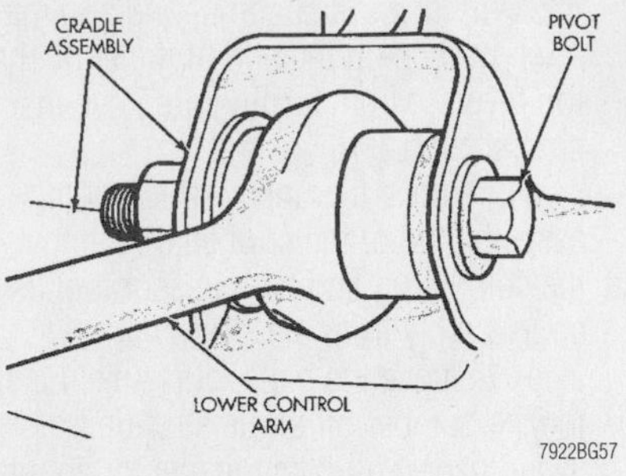

Loosen the pivot bolt and remove it—lower control arm service

⁂ WARNING

Pulling the steering knuckle out from the vehicle after releasing from the ball joint can separate the inner CV-joint. Do not separate the inner CV-joint or it can be damaged.

5. Remove the tension strut-to-cradle attaching nut and washer from the end of the tension strut. When removing the nut, keep the strut from turning by holding the tension strut at the flats using an open end wrench. Discard the tension strut-to-cradle retainer nut. A new nut must be used during installation.

➡A new tension strut-to-cradle attaching nut must be used when installing the tension strut.

6. Loosen and remove the lower control arm pivot bushing-to-cradle assembly pivot bolt.

7. Separate the lower control arm and tension strut from the cradle as an assembly by first removing the pivot bushing from the cradle, then sliding the tension strut out of the isolator bushing. Inspect the control arm and tension strut for distortion, check the rubber bushings for excessive wear or deterioration and replace these components, if necessary.

To install:

8. Install or connect the following:

- Tension strut and isolator bushing into the cradle first, then install lower control arm pivot bushing into bracket on the cradle
- Lower control arm-to-cradle bracket attaching bolt. Do not tighten the bolt at this time.
- Washer and new nut on the end of the tension strut. Torque the tension strut-to-cradle bracket retainer

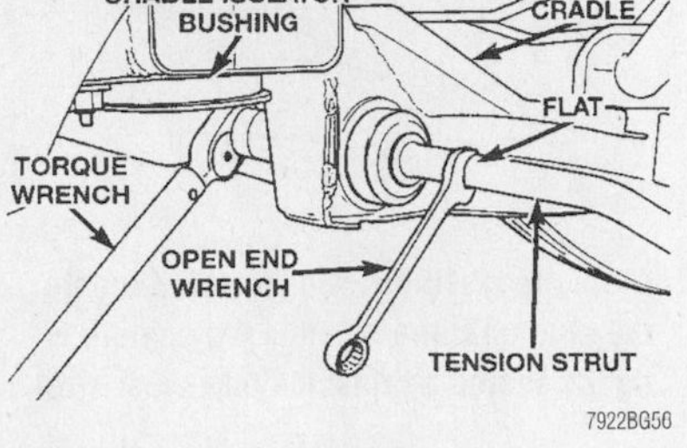

Remove the tension strut-to-cradle nut and washer—lower control arm service

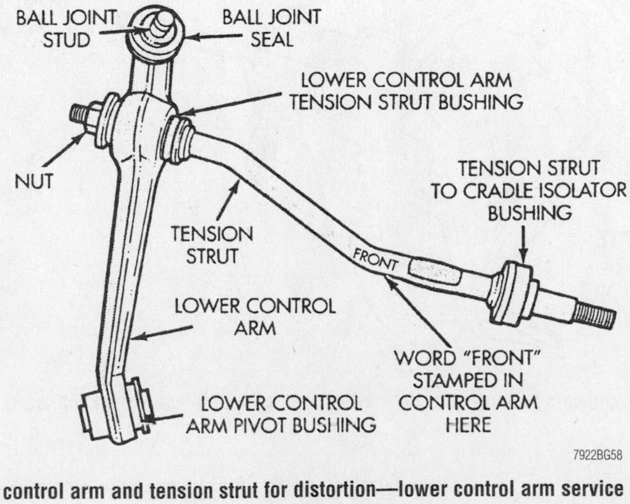

Inspect the control arm and tension strut for distortion—lower control arm service

nut to 110 ft. lbs. (150 Nm), while holding the tension strut flat with an open end wrench.

9. Inspect the ball joint seal and replace it if damaged. Install the lower ball joint stud into the steering knuckle and install the clamp bolt and nut. Torque the bolt to 40 ft. lbs. (55 Nm).

10. Install the front wheel and lug nuts. Torque the lug nuts, in a star pattern, to 95–100 ft. lbs. (129–135 Nm). Lower the vehicle so the suspension is supporting the weight of the vehicle

11. Torque the lower control arm pivot bushing-to-cradle bracket attaching bolt to 105 ft. lbs. (142 Nm).

CONTROL ARM BUSHING REPLACEMENT

1. Remove the lower control arm/tension strut assembly from the vehicle.

2. Remove the nut and separate the tensions strut from the lower control arm.

3. Using an arbor press, position the lower control arm with the large end of the pivot bushing inside the Receiver tool MB-990799.

4. Position Remover tool 6644-2 on top of the pivot bushing.

5. Using the arbor press, press the pivot bushing out of the lower control arm.

To install:

6. Turn the lower control arm over and reposition it on the Receiver tool MB-990799.

7. Using a new pivot bushing, position it in the lower control arm so that it is square with the bushing hole.

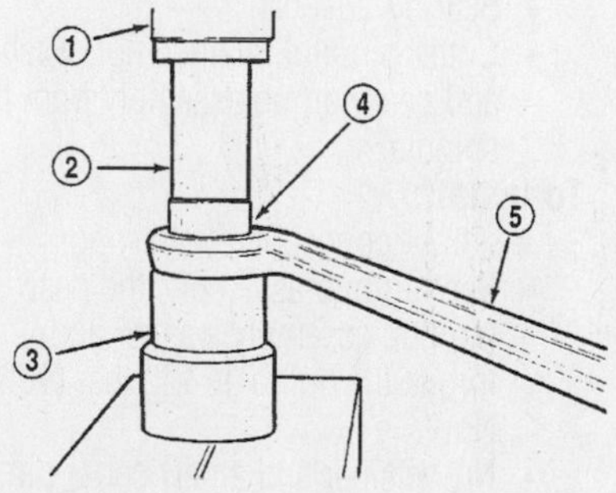

1 – ARBOR PRESS RAM
2 – SPECIAL TOOL 6644-2
3 – SPECIAL TOOL MB990799
4 – PIVOT BUSHING
5 – LOWER CONTROL ARM

9306BG25

Removing the lower control arm pivot bushing

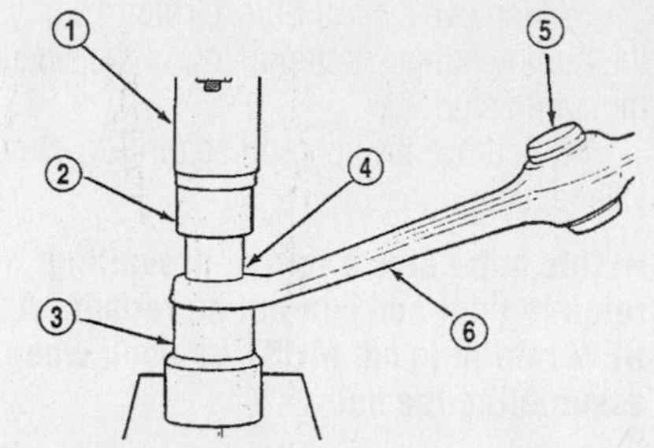

1 – ARBOR PRESS RAM
2 – SPECIAL TOOL 6644-1
3 – SPECIAL TOOL MB990799
4 – PIVOT BUSHING
5 – TENSION STRUT BUSHING
6 – LOWER CONTROL ARM

9306BG22

Installing the lower control arm pivot bushing

8. Position Installer tool 6644-1 on top of the pivot bushing with the pivot bushing setting in the recessed area of the installer tool.

9. Using the arbor press, press the pivot bushing into the lower control arm until the installer tool squarely bottoms against the surface of the lower control arm.

10. Install the tension strut into the lower control arm's tension strut bushing with the word FRONT, stamped on the tension strut, facing away from the control arm.

11. Position an open end wrench on the flat of the tension strut to keep it from turning. Torque the tension strut-to-lower control arm nut to 95 ft. lbs. (130 Nm).

TENSION STRUT BUSHING REPLACEMENT

1. Remove the lower control arm/tension strut assembly from the vehicle.

2. Remove the nut and separate the tensions strut from the lower control arm.

3. Using an arbor press, position the lower control arm with the tension strut bushing inside the Receiver tool MB-990799.

4. Position Remover tool 6644-4 on top of the tension strut bushing.

5. Using the arbor press, press the tension strut bushing out of the lower control arm.

➡As the Remover tool is press through the tension strut bushing, it will cut the bushing into 2 pieces.

6. Remove the lower control arm, both bushing pieces and the Remover tool.

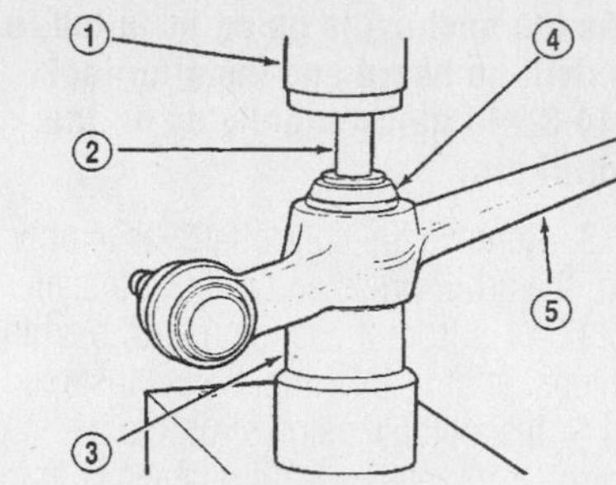

1 – ARBOR PRESS RAM
2 – SPECIAL TOOL 6644-4
3 – SPECIAL TOOL MB990799
4 – TENSION STRUT BUSHING
5 – LOWER CONTROL ARM

9306BG23

Removing the lower control arm's tension strut bushing

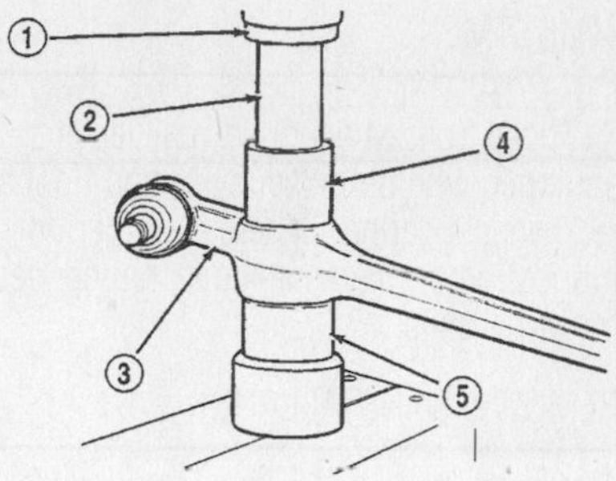

1 – ARBOR PRESS
2 – SPECIAL TOOL 6644-2
3 – LOWER CONTROL ARM
4 – SPECIAL TOOL 6644-3
5 – SPECIAL TOOL MB990799

9306BG24

Installing the lower control arm's tension strut bushing

To install:

7. Thoroughly, lubricate the new tension strut bushing, the lower control arm and the Installer tool 6644-3 with Rubber Bushing Installation Lube.

8. Press the new tension strut bushing by hand, into the large end of the Installer tool 6644-3 as far as it will go.

9. Position the lower control arm onto on the arbor press so that the tension strut hole is centered on the Receiver tool MB-990799.

10. Position Installer tool 6644-3, with the bushing installed, inside the tension strut bushing hole in the lower control arm.

11. Position Installer tool 6644-2 on top of the tension strut bushing.

12. Using the arbor press, press the tension strut bushing into the lower control arm.

Heater Core replacement is covered in Section 2 of this manual

➡As the bushing is being installed, a pop will be heard and Installer tool 6644-3 will move slightly up off the control arm.

13. Remove the control arm assembly from the arbor press and remove Installer tool 6644-3 from the tension strut bushing; the tension strut bushing is now installed.

14. Install the tension strut into the lower control arm's tension strut bushing with the word FRONT, stamped on the tension strut, facing away from the control arm.

15. Position an open end wrench on the flat of the tension strut to keep it from turning. Torque the tension strut-to-lower control arm nut to 95 ft. lbs. (130 Nm).

Wheel Bearings

ADJUSTMENT

These front wheel drive vehicles are equipped with permanently sealed front and rear wheel bearings. There is no periodic lubrication or maintenance recommended for these units.

REMOVAL & INSTALLATION

Front

1. Before servicing the vehicle, refer to the precautions in the beginning of this section.
2. Remove the front wheel.
3. Remove the front caliper assembly from the steering knuckle by removing the 2 guide pin bolts, then rotating the top of the caliper away from the knuckle and lifting the caliper off the machined abutment on the steering knuckle.
4. Suspend the caliper out of the way with a piece of wire.
5. Remove the front brake rotor from the hub by pulling it straight off of the wheel mounting stud.
6. Remove the hub and bearing-to-stub axle retainer nut.

➡This hub nut is a torque prevailing retaining nut and can not be reused. A NEW retaining nut MUST be used when assembling the hub.

7. Remove the 3 attaching bolts that mount the hub and bearing assembly to the steering knuckle assembly.

➡If the metal seal on the hub and bearing assembly is seized to the steering knuckle and becomes dislodged on the hub and bearing during removal, the hub and bearing must be replaced. If the flinger disc becomes damaged during the removal procedure, the hub and bearing assembly must be replaced.

8. Remove the hub and bearing assembly from the steering knuckle by sliding it straight out of the knuckle and off the ends of the stub shaft.
9. Gently pry the assembly out with a prybar or tap it out with a soft-faced hammer, if necessary. Be very careful not to damage the hub and bearing assembly.

To install:

10. Clean the hub and bearing mounting surfaces of dirt and be sure there are no nicks present.
11. Install or connect the following:
    - Hub and bearing squarely onto the stub shaft and the steering knuckle
    - Bearing assembly mounting bolts and tighten equally until the bearing assembly is seated squarely against the front of the steering knuckle. Torque the mounting bolts to 80 ft. lbs. (110 Nm).
    - New hub and bearing assembly-to-stub shaft retainer nut. A NEW retaining nut MUST be used when assembling the hub. Tighten but do not torque the nut at this time.
    - Brake rotor and the caliper assembly. Torque the brake caliper guide bolts to 16 ft. lbs. (22 Nm).
    - Wheel and lug nuts. Torque the lug nuts in a star pattern to 95–100 ft. lbs. (129–135 Nm).
12. Pump the brakes until a firm pedal is obtained.
13. With the weight on the vehicle on its wheels, apply the brakes to keep the vehicle from moving. Torque the hub and bearing assembly-to-stub shaft retaining nut to 120 ft. lbs. (163 Nm).

⁂ WARNING

When tightening the hub and bearing assembly to stub shaft retaining nut, do not exceed the maximum torque of 120 ft. lbs. (163 Nm). If the maximum torque is exceeded this may result in a failure of the halfshaft.

14. Inspect the toe setting on the vehicle and adjust, if necessary.

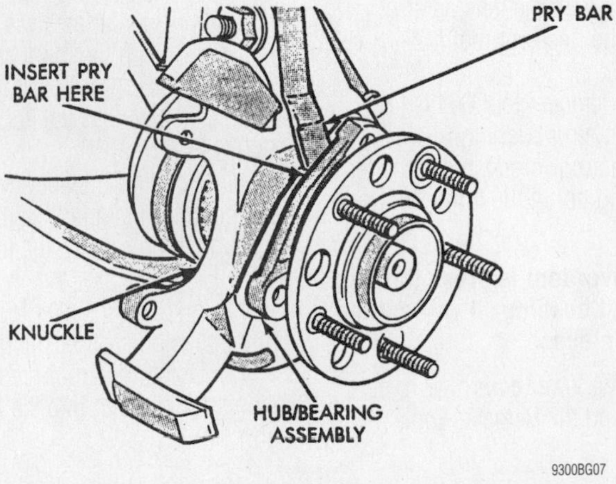

Carefully pry the front bearing/hub assembly away from the steering knuckle

Rear

1. Before servicing the vehicle, refer to the precautions in the beginning of this section.
2. Remove or disconnect the follow-ing:
    - Rear wheel
    - Brake caliper and rotor, if equipped with rear disc brakes
    - Brake drum, if equipped with drum brakes
    - Bearing dust cap
    - Cotter pin, nut retainer, nut, washer and bearing/hub assembly from the spindle

To install:

3. Install or connect the following:
    - Bearing/hub assembly, the bearing/hub assembly washer and torque the nut to 124 ft. lbs. (168 Nm)
    - Nut retainer and a new cotter pin
    - Dust cap
    - Brake drum or rotor and caliper assembly
    - Rear wheel and torque the lug nuts, in a star pattern, to 95–100 ft. lbs. (129–135 Nm)
4. Road test the vehicle to verify no excessive noise from the rear wheel bearing area.

CHRYSLER CORP.

6

1998–02

Chrysler-Sebring Coupe • **Dodge-**Avenger

PRECAUTIONS

Before servicing any vehicle, please be sure to read all of the following precautions, which deal with personal safety, prevention of component damage, and important points to take into consideration when servicing a motor vehicle:

- Never open, service or drain the radiator or cooling system when the engine is hot; serious burns can occur from the steam and hot coolant.
- Observe all applicable safety precautions when working around fuel. Whenever servicing the fuel system, always work in a well-ventilated area. Do not allow fuel spray or vapors to come in contact with a spark, open flame or excessive heat (a hot drop light, for example). Keep a dry chemical fire extinguisher near the work area. Always keep fuel in a container specifically designed for fuel storage; also, always properly seal fuel containers to avoid the possibility of fire or explosion. Refer to the additional fuel system precautions later in this section.
- Fuel injection systems often remain pressurized, even after the engine has been turned **OFF**. The fuel system pressure must be relieved before disconnecting any fuel lines. Failure to do so may result in fire and/or personal injury.
- Brake fluid often contains polyglycol ethers and polyglycols. Avoid contact with the eyes and wash your hands thoroughly after handling brake fluid. If you do get brake fluid in your eyes, flush your eyes with clean, running water for 15 minutes. If eye irritation persists, or if you have taken brake fluid internally, IMMEDIATELY seek medical assistance.
- The EPA warns that prolonged contact with used engine oil may cause a number of skin disorders, including cancer! You should make every effort to minimize your exposure to used engine oil. Protective gloves should be worn when changing oil. Wash your hands and any other exposed skin areas as soon as possible after exposure to used engine oil. Soap and water, or waterless hand cleaner should be used.
- All new vehicles are now equipped with an air bag system, often referred to as a Supplemental Restraint System (SRS) or Supplemental Inflatable Restraint (SIR) system. The system must be disabled before performing service on or around system components, steering column, instrument panel components, wiring and sensors. Failure to follow safety and disabling procedures could result in accidental air bag deployment, possible personal injury and unnecessary system repairs.
- Always wear safety goggles when working with, or around, the air bag system. When carrying a non-deployed air bag, be sure the bag and trim cover are pointed away from your body. When placing a non-deployed air bag on a work surface, always face the bag and trim cover upward, away from the surface. This will reduce the motion of the module if it is accidentally deployed. Refer to the additional air bag system precautions later in this section.
- Clean, high quality brake fluid from a sealed container is essential to the safe and proper operation of the brake system. You should always buy the correct type of brake fluid for your vehicle. If the brake fluid becomes contaminated, completely flush the system with new fluid. Never reuse any brake fluid. Any brake fluid that is removed from the system should be discarded. Also, do not allow any brake fluid to come in contact with a painted surface; it will damage the paint.
- Never operate the engine without the proper amount and type of engine oil; doing so WILL result in severe engine damage.
- Timing belt maintenance is extremely important! Many models utilize an interference-type, non-freewheeling engine. If the timing belt breaks, the valves in the cylinder head may strike the pistons, causing potentially serious (also time-consuming and expensive) engine damage. Refer to the maintenance interval charts in the front of this manual for the recommended replacement interval for the timing belt, and to the timing belt section for belt replacement and inspection.
- Disconnecting the negative battery cable on some vehicles may interfere with the functions of the on-board computer system(s) and may require the computer to undergo a relearning process once the negative battery cable is reconnected.
- When servicing drum brakes, only disassemble and assemble one side at a time, leaving the remaining side intact for reference.
- Only an MVAC-trained, EPA-certified automotive technician should service the air conditioning system or its components.

ENGINE REPAIR

Distributor

REMOVAL

2.5L Engine

The 2.5L engine is equipped with a camshaft driven mechanical distributor. This engine uses a fixed ignition timing system, in which the basic ignition timing is not adjustable. The Powertrain Control Module (PCM) determines spark advance. The Crankshaft Position (CKP) sensor and Camshaft Position (CMP) sensor are Hall effect devices. The CKP sensor is mounted remotely from the distributor, while the CMP sensor is mounted inside the distributor housing. Both sensors generate pulses that serve as inputs to the PCM; the PCM determines crankshaft position from these sensors, then calculates injector sequence and ignition timing, based on the data.

1. Before servicing the vehicle, refer to the precautions in the beginning of this section.
2. Disconnect the negative battery cable.
3. Remove or disconnect, if necessary for access:
   - Air inlet resonator to the intake manifold bolt
   - Air cleaner cover to air cleaner housing cover clamps
   - Positive Crankcase Ventilation (PCV) make-up air hose from the air inlet tube
   - Throttle body hose clamp
   - Air cleaner cover, resonator and inlet tube
   - Exhaust Gas Recirculation (EGR) tube
4. Remove or disconnect the following:
   - Spark plug wires from the distributor cap
   - Distributor cap
5. Matchmark the rotor-to-distributor and the distributor-to-engine positioning to help with installation.
   - Rotor
   - Distributor's 2 electrical harness connectors
   - Distributor hold-down nuts and washers
   - Spark plug cable mounting bracket (if necessary)
   - Transaxle dipstick tube
   - Distributor

INSTALLATION

2.5L Engine

TIMING NOT DISTURBED

➡**Before servicing the vehicle, refer to the precautions in the beginning of this section.**

1. Inspect the rotor for cracks or burned electrodes, and replace if defective. Install the rotor onto the distributor.
2. Inspect the O-ring seal. If nicked or cracked, replace with a new one. Be sure the O-ring is properly seated on the distributor.
3. Install or connect the following:
   - Distributor drive with the slotted end of the camshaft

➡**When the distributor is installed properly, the rotor will be aligned with the previously made mark.**

- Distributor hold-down nuts/washers and tighten to 108 inch lbs. (13 Nm)
- Spark plug cable bracket
- Both distributor wiring connectors
- Distributor cap
- Spark plug cables
- Transaxle dipstick tube
- EGR tube and tighten the bolts to 95 inch lbs. (11 Nm)
- Air cleaner cover, resonator and inlet tube
- Throttle body hose clamp
- PCV hose
- Air cleaner housing cover clamps
- Air inlet resonator to intake manifold bolt
- Negative battery cable

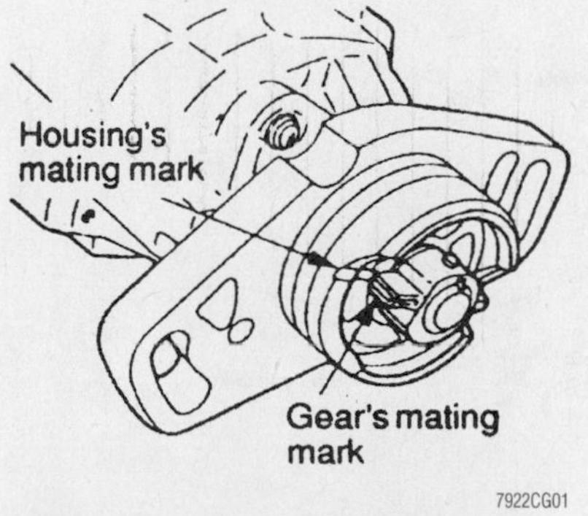

Prior to installation, align the distributor shaft with the distributor housing—2.5L engine

TIMING DISTURBED

1. Rotate the crankshaft until the No. 1 piston is at Top Dead Center (TDC) of the compression stroke.
2. Align the rotor with the distributor housing matchmark.
3. Install or connect the following:
   - Distributor

➡**With the distributor fully seated, the rotor should align the No. 1 terminal of the distributor cap.**

- Distributor hold-down nuts/washers and tighten the nuts to 108 inch lbs. (13 Nm)
- Spark plug cable bracket
- Both distributor wiring connectors
- Distributor cap
- Spark plug cables
- Transaxle dipstick tube

4. Install or connect the following, if removed:
   - EGR tube and tighten the bolts to 95 inch lbs. (11 Nm)
   - Air cleaner cover, resonator and inlet tube
   - Throttle body hose clamp
   - PCV hose
   - Air cleaner housing cover clamps
   - Air inlet resonator to intake manifold bolt
   - Negative battery cable

Alternator

REMOVAL

2.0L Engine

Remove or disconnect the following:
- Negative battery cable
- Right front wheel
- Wheel well side cover
- Auto-cruise speed control assembly
- Alternator drive belt
- Alternator electrical connectors
- Alternator bracket
- Alternator

2.5L Engine

Remove or disconnect the following:
- Negative battery cable
- Right front wheel
- Wheel well side cover
- Auto-cruise control reservoir assembly
- Power steering pump drive belt cover
- A/C compressor drive belt
- Power steering pump/alternator drive belt
- Intake manifold plenum
- Alternator bracket
- Intake manifold plenum stay
- Alternator electrical connectors
- Alternator

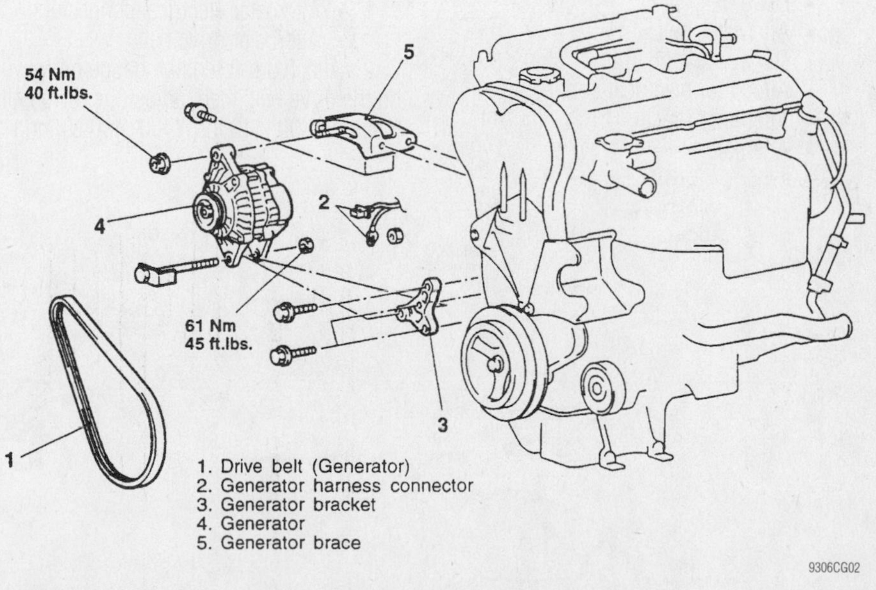

Exploded view of the alternator and related components—2.0L engine

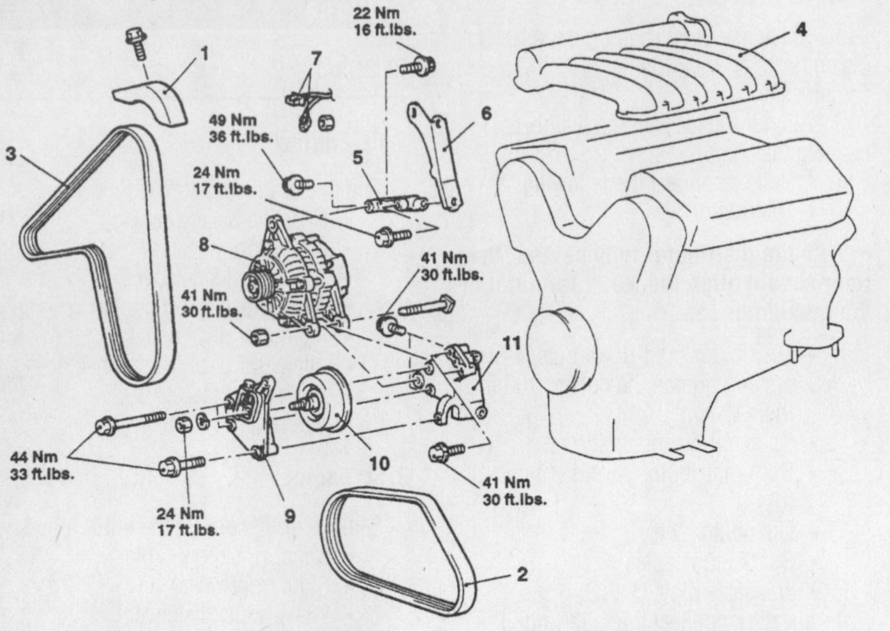

1. Pump cover
2. Drive belt (A/C compressor)
3. Drive belt (Power steering oil pump and generator)
4. Intake manifold plenum (Refer to GROUP 11 – Intake Manifold.)
5. Generator bracket
6. Intake manifold plenum stay
7. Generator harness connector
8. Generator
9. Tensioner pulley bracket
10. Tensioner pulley
11. Generator bracket

9306CG01

Exploded view of the alternator and related components—2.5L engine

3.0L Engine

1. Before servicing the vehicle, refer to the precautions in the beginning of this section.
2. Remove or disconnect the following:
   - Negative battery cable
   - Drive belt
   - Alternator belt
   - Oil level gauge unit
   - Alternator electrical connectors
   - Alternator brace
   - Alternator

INSTALLATION

2.0L Engine

1. Install or connect the following:
   - Alternator
   - Alternator bracket
   - Alternator electrical connectors
   - Alternator drive belt
2. Attach a belt tension gauge midway on the drive belt; then, adjust the drive belt tension to 90–110 lbs. (400–490 N) for a used belt or 110–160 lbs. (490–712 N) for a new belt.

WARNING

If installing a new belt, tension the belt to 70 lbs. (310 N) and operate the engine for 5 minutes or more; then, apply the final tension.

3. Torque the alternator-to-lower alternator bracket nut to 45 ft. lbs. (61 Nm) and the alternator-to-upper bracket bolt to 40 ft. lbs. (54 Nm).
4. Install or connect the following:
   - Auto-cruise speed control assembly
   - Wheel well side cover
   - Right front wheel
   - Negative battery cable

2.5L Engine

1. Install or connect the following:
   - Alternator
   - Alternator electrical connectors
   - Intake manifold plenum stay
   - Alternator bracket. Torque the alternator-to-lower alternator bracket nut to 30 ft. lbs. (41 Nm) and the alternator-to-upper bracket bolt to 16 ft. lbs. (22 Nm).
   - Intake manifold plenum
   - Power steering pump/alternator drive belt
2. Attach a belt tension gauge midway on the drive belt, between the alternator and power steering pump pulleys; then, adjust the drive belt tension to 99–121 lbs. (440–540 N) for a used belt or 143–187 lbs. (642–838 N) for a new belt.
3. Torque the tension pulley bracket nut to 17 ft. lbs. (24 Nm).
4. Install the A/C compressor's drive belt.

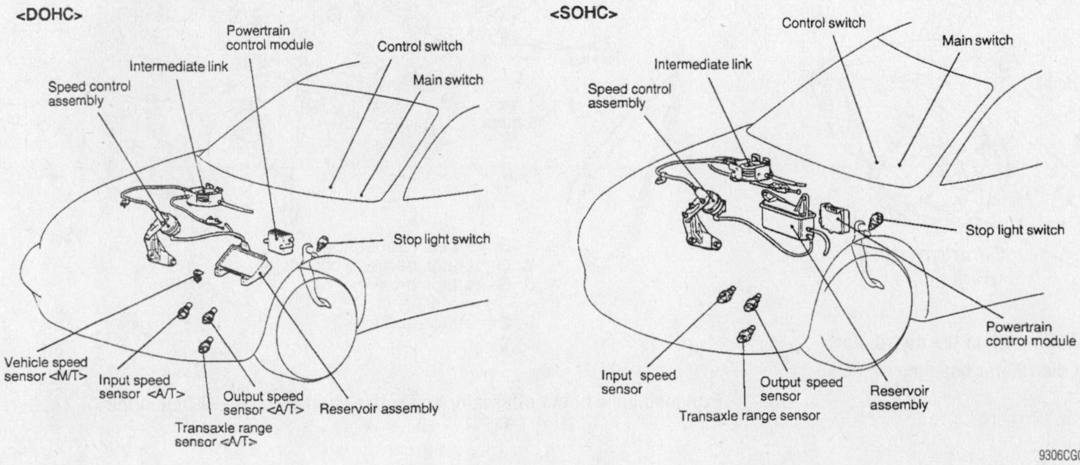

9306CG03

View of the auto-cruise speed control and related components—2.0L and 2.5L engines

5. Attach a belt tension gauge midway on the drive belt, between the crankshaft and tension pulleys; then, adjust the drive belt tension to 66–86 lbs. (295–385 N) for a used belt or 110–132 lbs. (490–590 N) for a new belt.

6. Torque the tension pulley bracket nut to 17 ft. lbs. (24 Nm).

7. Install or connect the following:

- Power steering pump drive belt cover
- Auto-cruise control reservoir assembly
- Wheel well side cover
- Right front wheel
- Negative battery cable

3.0L Engine

Install or connect the following:

- Alternator and torque the bolts to 33 ft. lbs. (44 Nm)
- Alternator brace and torque the bolts to 36 ft. lbs. (49 Nm)
- Oil level gauge unit and torque the bolt to 10 ft. lbs. (15 Nm)
- Alternator belt
- Drive belt
- Negative battery cable

Ignition Timing

Ignition timing is controlled by the Powertrain Control Module (PCM) and is not adjustable.

Engine Assembly

REMOVAL & INSTALLATION

2.0L and 2.5L Engines

1. Before servicing the vehicle, refer to the precautions in the beginning of this section.

➡The transaxle must be removed before removing the engine. They will not come out as a unit.

2. Drain the engine coolant.
3. Drain the engine oil
4. Drain the transmission oil.
5. Properly relieve the fuel system pressure.
6. Remove or disconnect the following:

- Hood
- Negative battery cable
- Engine under cover
- Transaxle assembly
- Radiator by disconnecting the hoses at the engine
- Accelerator cable and bracket
- Heater hoses
- Brake booster vacuum hose at the engine
- Vacuum hoses running to the bulkhead
- High pressure fuel line and discard the O-ring
- Fuel return hose from the fuel supply rail

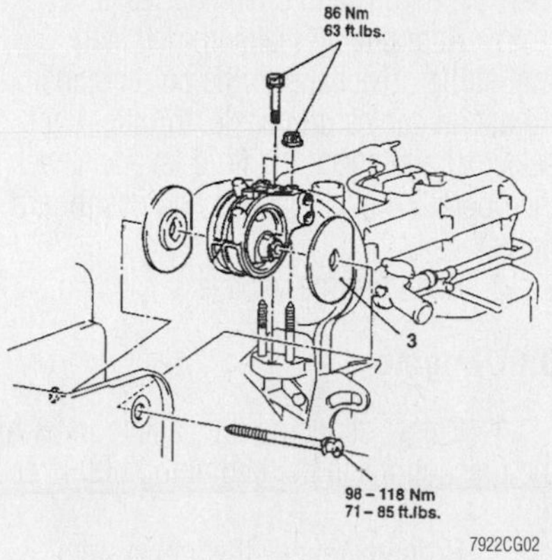

Exploded view of the right-side engine mount—2.0L engine

- Electrical connectors from the engine components

➡All wires and connectors should be labeled at the time of engine removal. This will save time during assembly.

- Accessory drive belts
- Power steering pump move aside

✲✲ WARNING

Do not disconnect the hoses or allow the pump to hang by the hoses.

- Power steering pump bracket
- A/C compressor and move aside

✲✲ CAUTION

Do not loosen or remove the A/C hoses or discharge the system.

- Exhaust system joint bolts just below the manifold and discard the gasket and nuts

7. Install the engine hoist equipment and make certain the attaching points on the engine are secure. Draw tension on the hoist just enough to support the engine's weight but no more. Do not dis-

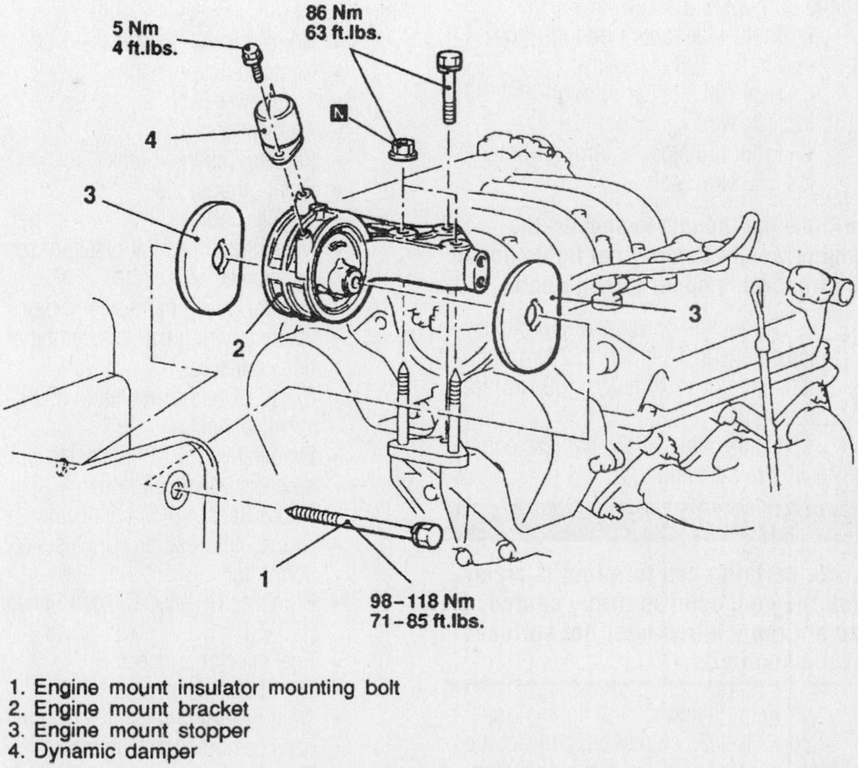

Exploded view of the right-side engine mount—2.5L engine

turb the placement of the vehicle on the stands.

- Rear (bulkhead side) roll stopper through-bolt
- Front engine roll stopper through-bolt
- Upper (right-side) engine mount-to-engine nuts/bolts, through-bolt and mount assembly
- Support bracket below the mount
- Engine

⁂ WARNING

Immediately place it on an engine stand or support it with wooden blocks. Do not allow it to rest on the oil pan or lie on its side. Do not leave the engine hanging from the hoist.

To install:

8. Installation is the reverse of the removal procedure. Please note the following important steps:

a. Connect the exhaust system to the manifold, using a new gasket. Torque the bolts to 33 ft. lbs. (44 Nm).

9. Torque the engine mount nuts and bolts as follows:

a. Right-side mount-to-engine nut and bolt: 63 ft. lbs. (86 Nm).

b. Right-side mount through-bolt: 71–85 ft. lbs. (98–118 Nm).

c. Rear roll stopper through-bolt: 32 ft. lbs. (44 Nm).

d. Front roll stopper through-bolt: 41 ft. lbs. (56 Nm).

➡Allow the mounts to support the engine weight before final tightening of the front roll stopper through-bolt.

e. Torque the power steering pump bracket bolts to 16 ft. lbs. (22 Nm) for 2.0L engine or to 29 ft. lbs. (39 Nm) for 2.5L engine.

f. Connect the wiring harness connectors to the engine.

⁂ WARNING

Make certain each terminal is clean and the connector is firmly seated. Do not route wires near hot surfaces or moving parts.

g. Using a new O-ring lightly lubricated with clean engine oil, connect the high pressure fuel line and torque the bolts to 22 inch lbs. (2.5 Nm).

h. Check the engine oil drain plug and secure it if necessary, and refill the engine.

i. Check the transaxle drain plug, tightening it if needed, and refill the transaxle.

j. Check the radiator and engine drain cocks, closing them if necessary, and refill the cooling system.

k. Double check all installation items, paying particular attention to loose hoses or hanging wires, loosened nuts, poor routing of hoses and wires.

10. Connect the negative battery cable. Start the engine and check for leaks.

11. Attend to all leaks immediately. Adjust the drive belts to the correct tension. Adjust all cables (transaxle, throttle, shift selector) and check the fluid levels. Check the operation of all gauges and dashboard lights.

12. Road test the vehicle.

3.0L Engine

1. Before servicing the vehicle, refer to the precautions in the beginning of this section.
2. Properly relieve the fuel system pressure.
3. Drain the engine oil.
4. Drain the cooling system.
5. Recover the A/C refrigerant.
6. Remove or disconnect the following:

- Hood
- Negative battery cable
- Strut tower bar
- Air cleaner
- Radiator reservoir tank
- Front exhaust pipe
- Throttle cable
- Manifold differential pressure sensor connector
- Control wiring harness and power steering wiring harness combination connector
- Exhaust Gas Recirculation (EGR) valve connector
- Evaporative (EVAP) Emission purge solenoid valve connector
- Knock Sensor (KS) connector
- Crankshaft Position (CKP) sensor connector
- Right bank Heated Oxygen Sensor (HO_2S)
- Fuel injector connector
- Distributor connector
- Control wiring/injector wiring harness combination connector
- Throttle Position Sensor (TPS) connector
- Idle Air Control (IAC) motor connector
- Ground wire
- Engine Coolant Temperature (ECT) gauge unit
- ECT sensor connector
- Left bank Heated Oxygen Sensor (HO_2S)
- Brake booster vacuum hose
- Fuel supply and return hoses
- Heater hoses
- Starter electrical connectors
- Oil Pressure switch
- Alternator wiring
- Drive belts
- A/C compressor
- Power steering pressure switch
- Power steering oil pump. Do not disconnect the hose
- Engine mount stay and install an engine Lifting Tool, such as MZ203827
- Engine mount bracket and stopper
- Engine assembly

To install:

7. Install or connect the following:

- Engine assembly
- Engine mount bracket and stopper and torque the bolts to 60 ft. lbs. (81 Nm)
- Engine mount stay and torque the bolts to 26 ft. lbs. (35 Nm). Remove the engine lifting devise
- Power steering oil pump and torque the bolts to 31 ft. lbs. (42 Nm)
- Power steering pressure switch
- A/C compressor
- Drive belts
- Alternator wiring
- Oil pressure switch
- Starter electrical connectors
- Heater hoses
- Fuel supply and return lines
- Brake booster vacuum hose
- Left bank (HO_2S)
- ECT sensor connector
- ECT gauge unit
- Ground wire
- IAC motor electrical connector
- TPS connector
- Control wiring/injector wiring harness combination connector
- Distributor connector
- Fuel injector connector
- Right bank (HO_2S)
- CKP sensor connector
- KS connector
- EVAP solenoid valve connector
- EGR valve connector
- Control wiring harness and power steering harness combination connector
- Manifold differential Pressure sensor connector
- Throttle cable

- Front exhaust pipe
- Radiator reservoir tank
- Air cleaner
- Strut tower bar
- Negative battery cable
- Hood and adjust as needed

8. Fill the engine with clean oil and replace the filter.
9. Fill the cooling system to the proper level.
10. Recharge the A/C system with the proper refrigerant.
11. Start the vehicle, check for leaks and repair if necessary.

Water Pump

REMOVAL & INSTALLATION

2.0L Engine

1. Before servicing the vehicle, refer to the precautions in the beginning of this section.
2. Drain the engine coolant.

✱✱ WARNING

This procedure requires removing the timing belt and auto-tensioner. The factory specifies that the timing marks should always be aligned before removing the timing belt. Set the engine at Top Dead Center (TDC) on the No. 1 cylinder compression stroke. This should align all timing marks on the crankshaft sprocket and both camshaft sprockets.

3. Remove or disconnect the following:
   - Negative battery cable
   - Right inner splash shield
   - Accessory drive belts
   - Properly support the engine and remove the right motor mount
   - Timing belt, tensioner and camshaft sprockets

✱✱ WARNING

With the timing belt removed, DO NOT rotate the camshaft or crankshaft or damage to the engine may occur.

   - Rear timing belt cover
   - Water pump

To install:

4. Thoroughly clean all sealing surfaces.

✱✱ WARNING

Replace the water pump if there are any cracks, signs of coolant leakage from the shaft seal, loose or rough turning bearings, damaged impeller or sprocket or loose or damaged sprocket flange.

5. Install or connect the following:
   - New water pump O-ring

✱✱ WARNING

Be sure the O-ring is properly seated in the water pump groove before tightening the screws.

   - Water pump and torque the bolts to 105 inch lbs. (12 Nm)

6. Using a cooling system pressure tester, pressurize the cooling system to 15 psi (103 kPa) and check for leaks. If okay, release the pressure and continue the engine assembly process.
7. Rotate the water pump by hand to check for freedom of movement.
8. Install or connect the following:
   - Rear timing belt cover
   - Camshaft sprocket(s), timing belt and tensioner

✱✱ WARNING

DO NOT allow the camshafts to turn while the sprocket bolts are being tightened.

✱✱ WARNING

Do not attempt to compress the tensioner plunger with the tensioner assembly installed in the engine. This will cause damage to the tensioner and other related components. The tensioner MUST be compressed in a vise.

   - Timing belt covers
   - Right engine mount bracket and mount

9. Remove the engine support.
   - Crankshaft damper
   - Right inner splash shield
   - Accessory drive belts
10. Refill and bleed the cooling system.
11. Start the engine and check for proper operation.
12. Check and top off the cooling system, if necessary.

2.5L Engine

1. Before servicing the vehicle, refer to the precautions in the beginning of this section.
2. Drain the engine coolant.

➡This procedure requires removing the timing belt and auto-tensioner. Set the engine at Top Dead Center (TDC) on

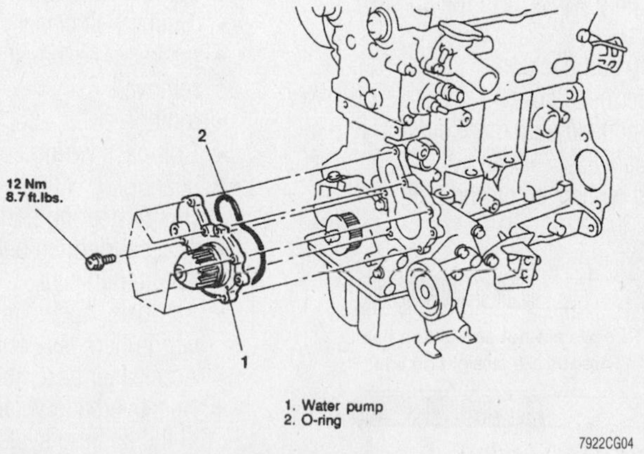

Exploded view of the water pump mounting—2.0L engine

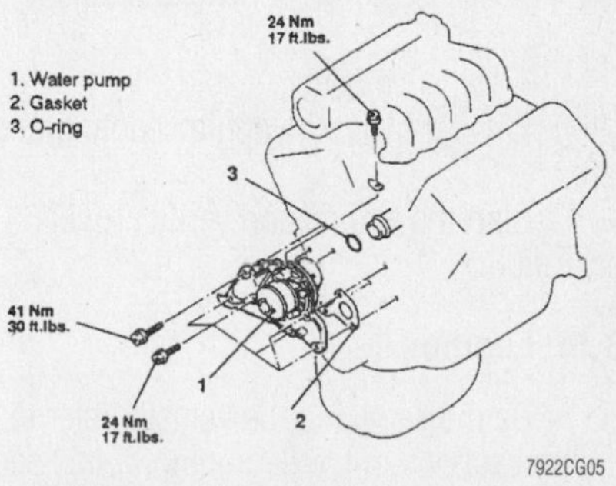

Exploded view of the water pump mounting—2.5L engine

the No. 1 cylinder compression stroke. This should align all timing marks on the crankshaft sprocket and both camshaft sprockets.

3. Remove or disconnect the following:
   - Negative battery cable
   - Accessory drive belts and crankshaft damper
   - Support the engine and remove the right engine mount
   - Timing belt covers
   - Timing belt and tensioner
   - Water pump mounting bolts
   - Water pump

To install:

4. Thoroughly clean all sealing surfaces. Inspect the pump for damage or cracks, signs of coolant leakage at the vent and excessive looseness or rough turning bearing. Any problems require a new pump.
5. Install or connect the following:
   - New water inlet pipe O-ring.

⁂ WARNING

Wet the O-ring with water to make installation easier; DO NOT use oil or grease.

   - New water pump gasket

➡Fit the pump inlet opening over the water pipe and press the assembly together to force the pipe into the water pump.

   - Water pump and torque the pump-to-engine bolts to 20 ft. lbs. (27 Nm)
   - Timing belt and timing belt tensioner. Set the timing belt tension
   - Timing belt covers
   - Right engine mount. Remove the floor jack and engine block from underneath the engine.
   - Crankshaft damper
   - Accessory drive belts and adjust tension
   - Negative battery cable
6. Fill and bleed the engine cooling system.
7. Start the engine and verify proper operation.

3.0L Engine

1. Before servicing the vehicle, refer to the precautions in the beginning of this section.
2. Drain the cooling system.
3. Remove or disconnect the following:
   - Negative battery cable
   - Timing belt
   - Thermostat and bracket
   - Water pump and discard the gasket and O-ring

To install:

4. Install or connect the following:
   - Water pump with a new O-ring and gasket and torque the bolts to 17 ft. lbs. (23 Nm)
   - Bracket and torque the bolts to 17 ft. lbs. (23 Nm)
   - Thermostat
   - Timing belt
   - Negative battery cable
5. Fill the cooling system to the proper level.
6. Start the vehicle, check for leaks and repair if necessary.

Cylinder Head

REMOVAL & INSTALLATION

2.0L Engine

1. Before servicing the vehicle, refer to the precautions in the beginning of this section.
2. Properly relieve the fuel system pressure.
3. Drain the engine coolant.
4. Remove or disconnect the following:
   - Negative battery cable
   - Air cleaner assembly
   - All vacuum hoses, lines and wiring harness connections required for cylinder head removal
   - Fuel line
   - Throttle linkage
   - Accessory drive belt(s)
   - Power steering pump and move aside
   - Coil pack wiring connector
   - Spark plug wires from the spark plugs
   - Ignition coil pack unit
   - Cylinder head cover
   - Intake and exhaust manifolds, if necessary
   - Timing belt cover, timing belt, camshaft sprocket and rear timing belt cover
   - Rocker arm/rocker arm shaft assemblies
   - Cylinder head bolts and cylinder head

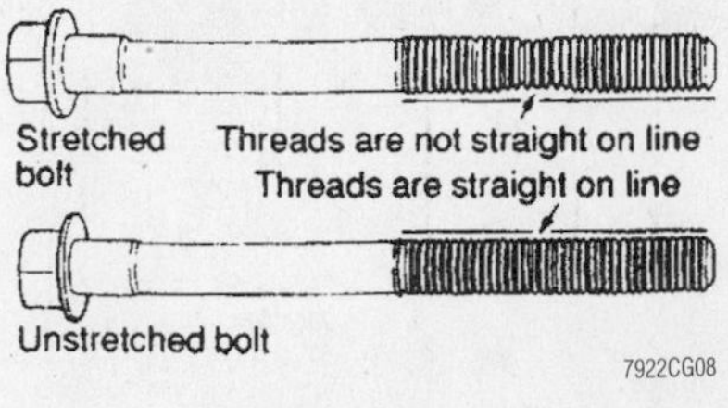

Checking the cylinder head bolts for necking (stretching)—2.0L engine

To install:

⁂ WARNING

Stretch-type bolts are used. New head bolts are recommended.

5. Thoroughly clean all parts. Clean all sealing surfaces. Use care not to scratch the aluminum cylinder head sealing surface. Check the cylinder head for flatness using a feeler gauge and a straight-edge. The cylinder head must be flat within 0.004 in. (0.1mm).
6. Check the cylinder head for cracks or other damage.
7. Install a new gasket and the cylinder head.
8. Lubricate the cylinder head bolt threads with clean engine oil.
9. Install the cylinder head bolts, the short bolts (110mm) are to be installed in positions 7, 8, 9 and 10.
10. Torque the cylinder head bolts in 4 steps as follows:
    a. Step 1: 25 ft. lbs. (34 Nm).
    b. Step 2: 50 ft. lbs. (68 Nm).
    c. Step 3: Again to 50 ft. lbs. (68 Nm).
    d. Step 4: Plus ¼ (90 degree) turn.

➡Do not use a torque wrench for the 4th step.

11. Install or connect the following:
    - Rocker arm/rocker arm shaft assemblies
    - Cylinder head cover
    - Timing belt rear cover, camshaft sprocket and timing belt
    - Timing belt cover
    - Intake and exhaust manifolds, if removed
    - Ignition coil pack
    - Coil pack wiring connector and spark plug wires
    - Power steering pump
    - Accessory drive belts
    - Throttle linkage
    - All ducts, hoses, fuel lines and wiring harness connectors
    - Air cleaner assembly
    - Negative battery cable
12. Fill the cooling system.

➡A complete oil and filter change is recommended.

13. Start the engine and check for leaks. Run the engine with the radiator cap off so

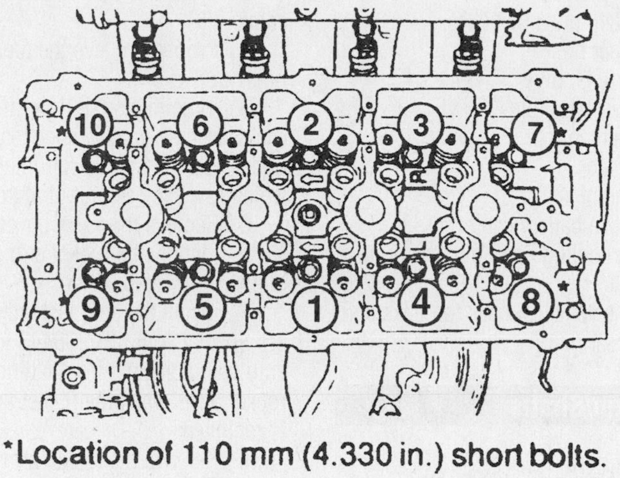

Cylinder head bolt tightening sequence—2.0L engine

as the engine warms and the thermostat opens, coolant can be added to the radiator. When the cooling system is full, shut the engine **OFF**, install the radiator cap and allow the engine to cool.

14. With the engine cool, check all fluid levels. Add coolant and oil as required. Restart the engine and test drive vehicle to check for proper operation.

2.5L Engine

1. Before servicing the vehicle, refer to the precautions in the beginning of this section.
2. Properly relieve the fuel system pressure.
3. Drain the cooling system.
4. Remove or disconnect the following:
   - Negative battery cable
   - Timing belt and camshaft sprockets
   - Wiring harnesses, vacuum hoses and lines that may inhibit cylinder head removal
   - Intake manifold assembly
   - Water pump inlet pipe retaining bolt from the inner rear of the front cylinder head
   - Valve covers and rocker arm assemblies
   - Distributor assembly
   - Ground strap from the left end of the front cylinder head
   - Exhaust manifolds and crossover pipe
   - Cylinder head bolts; keep them in numbered order
   - Cylinder head and discard the gasket

To install:

5. Thoroughly clean and dry the mating surfaces of the head and block. Check the cylinder head for cracks, damage or engine coolant leakage. Remove scale, sealing compound and carbon. Clean the oil passages thoroughly. Check the head for flatness. End-to-end, the head should be no more than 0.008 in. (0.2mm) out-of-true. If the service limit is exceeded, correct to meet specifications. Note that the maximum amount from stock allowed to be removed from the cylinder head and mating cylinder block is 0.0079 in.

(0.2mm). If the cylinder head cannot be made serviceable by removing this amount, replace it.

6. Check that the head gaskets have the proper identification marks for the engine. Lay the head gasket with the identification mark at the front top.

➡Do not apply sealant to the head gasket or mating surfaces. Stretch-type cylinder head bolts are used. All new head bolts are recommended.

7. Install the cylinder head straight down onto the block.
8. Lubricate the bolts with clean engine oil.
9. Install the bolts and the special washers by hand and just start each bolt one or 2 turns on the threads.

➡The washers must be installed correctly. The rounded shoulder of the washer denotes the face in contact with the bolt. The flat side of the washer contacts the head.

10. Torque the head bolts in the proper sequence as follows:
    a. Step1: 62 ft. lbs. (84 Nm).
    b. Step 2: 70 ft. lbs. (95 Nm).
    c. Step 3: 80 ft. lbs. (108 Nm).
11. Install or connect the following:
    - Valve cover and gasket
    - Exhaust manifolds and crossover pipe
    - Distributor
    - Intake manifold
    - All wiring harnesses, vacuum hoses and lines
    - Negative battery cable
12. Fill the cooling system. Changing the oil and filter is recommended to eliminate pollutants such as coolant in the oil.
13. With the radiator cap off, start the engine and check for leaks of fuel, vacuum, oil or coolant. Check the operation of all engine electrical systems as well as dash-

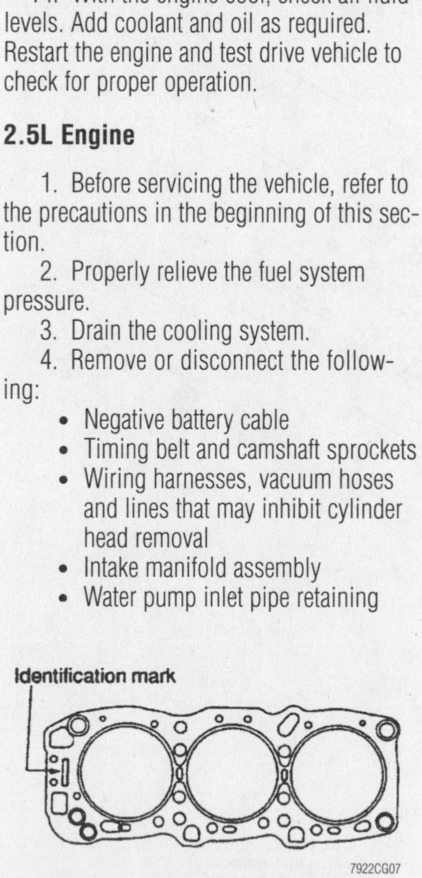

Install the head gasket with the identification mark at the front facing upward—2.5L engine

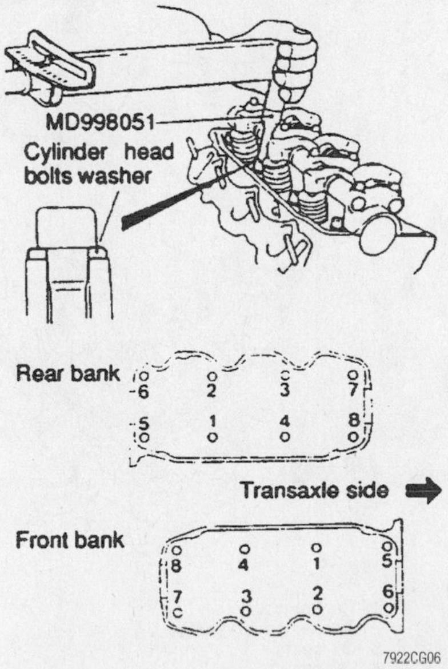

Cylinder head bolt tightening sequence—2.5L engine

Heater Core replacement is covered in Section 2 of this manual

board gauges and lights. Add coolant as the engine warms.

14. Perform necessary adjustments to the accelerator cable and drive belts. Allow the engine to cool and once again check and adjust the coolant level.

3.0L Engine

1. Before servicing the vehicle, refer to the precautions in the beginning of this section.
2. Relieve the fuel system pressure.
3. Drain the cooling system.
4. Remove or disconnect the following:
   - Negative battery cable
   - Timing belt
   - Alternator
   - Intake manifold
   - Exhaust manifold
   - Water inlet pipe
   - Breather and blow by hoses
   - Positive Crankcase Ventilation (PCV) hose
   - Spark plug cables
   - Rocker arm cover
   - Timing belt rear cover
   - Cylinder head and discard the gasket

To install:

5. Clean all mating surfaces of any residual gasket material.
6. Lubricate the threads of the cylinder head bolts with clean engine oil.
7. Install or connect the following:
   - Cylinder head with a new gasket
8. Torque the bolts, in sequence, as follows:
   a. Step 1: 80 ft. lbs. (108 Nm).
   b. Step 2: Loosen all bolts in sequence.
   c. Step 3: 80 ft. lbs. (108 Nm).
9. Install or connect the following:
   - Rear timing belt cover
   - Rocker arm cover

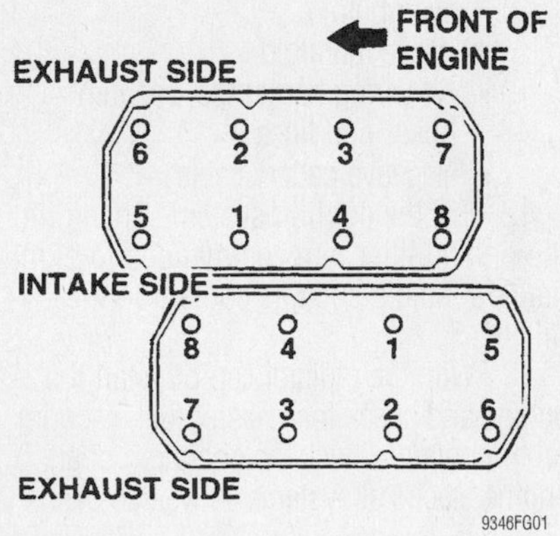

Cylinder head bolt torque sequence—3.0L engine

- Spark plug cables
- PCV and blow by hoses
- Breather hose
- Water inlet pipe
- Exhaust manifold
- Intake manifold
- Alternator
- Timing belt
- Negative battery cable

10. Fill the cooling system to the proper level.
11. Start the vehicle, check for leaks and repair if necessary.

Rocker Arm/Shafts

REMOVAL & INSTALLATION

2.5L Engine

1. Before servicing the vehicle, refer to the precautions in the beginning of this section.
2. Disconnect the negative battery cable.
3. Properly relieve the fuel system pressure.
4. If removing the right (firewall) side rocker arm/shaft assembly, remove the upper intake manifold (air intake plenum), which is a 2-piece unit of aluminum alloy.
5. Remove the valve cover(s).
6. Identify the rocker arm shaft assemblies before removal.
7. Install the auto lash adjuster retainers Special Tool MD 998443 to keep the auto lash adjusters from falling out of the rocker arms when the rocker arm assembly is removed.
8. Remove the rocker arm shaft assemblies.

➡The hydraulic automatic lash adjusters are precision units installed in the machined openings in the rocker arm units. Do not disassemble the auto lash adjusters.

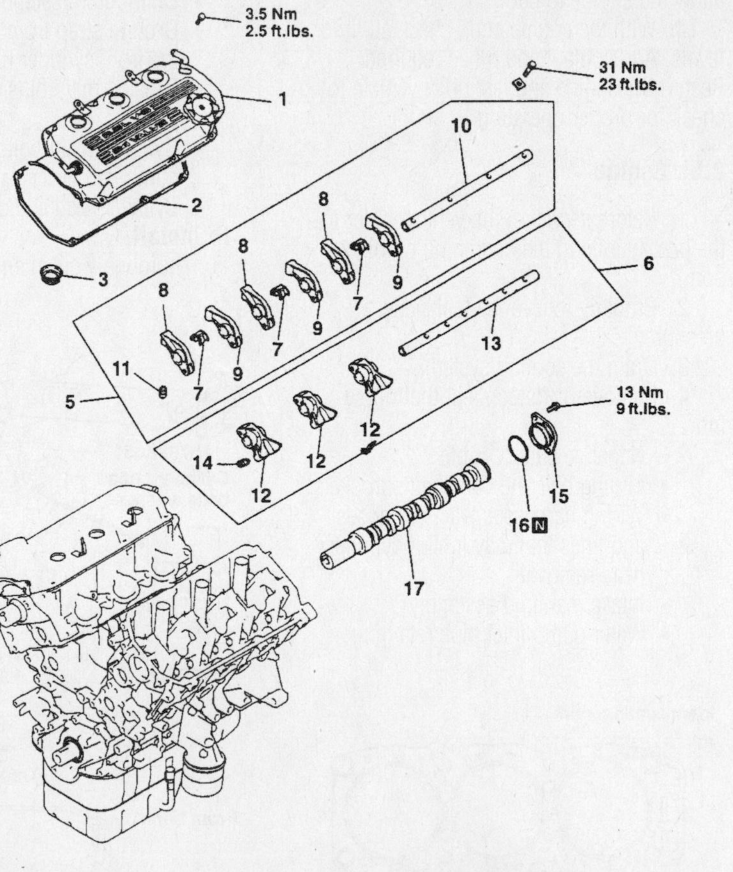

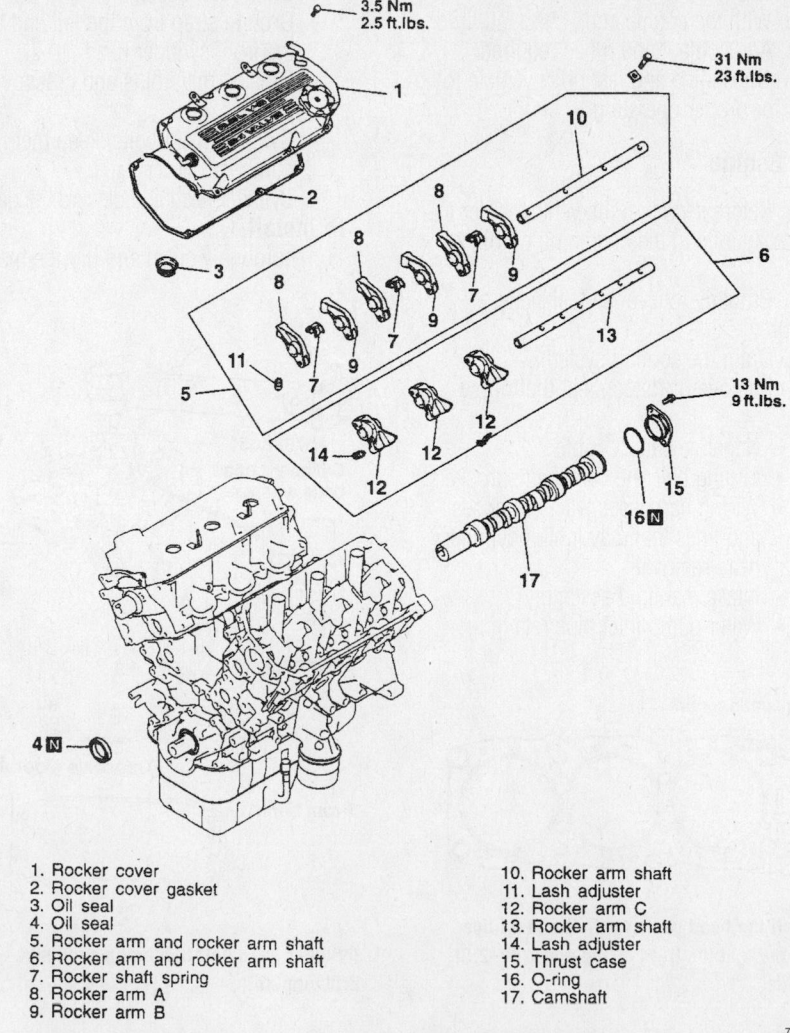

1. Rocker cover
2. Rocker cover gasket
3. Oil seal
4. Oil seal
5. Rocker arm and rocker arm shaft
6. Rocker arm and rocker arm shaft
7. Rocker shaft spring
8. Rocker arm A
9. Rocker arm B
10. Rocker arm shaft
11. Lash adjuster
12. Rocker arm C
13. Rocker arm shaft
14. Lash adjuster
15. Thrust case
16. O-ring
17. Camshaft

7922CG10

Exploded view of the cylinder head valve train assembly—2.5L engine

To install:

9. The rocker arm shafts are hollow and used as a lubrication oil duct. Ensure all valve train parts are clean. Check the rocker arm mounting portion of the shafts for wear or damage. Replace if necessary. Check all oil holes for clogging with a small wire and clean as required. If any rockers were removed, lubricate and install them on the shafts in their original positions.
10. For the right cylinder head, install the rocker arm and shaft assemblies with the FLAT in the rocker arm shafts facing toward the timing belt side of the engine.
11. For the left cylinder head, install the rocker arm and shaft assembly with the FLAT in the rocker arm shaft facing toward the transaxle side of the engine.
12. Install the retainers and spring clips in their original positions on the exhaust and intake shafts. Torque the retainer bolts to 23 ft. lbs. (31 Nm) working from the center, outward. Remove the valve lash retainer tools that may have been installed at disassembly.
13. Inspect the spark plug tube seals located on the ends of each tube. These seals slide onto each tube to seal the cylinder head cover to the spark plug tube. If these seals show signs of hardness and/or cracks, they should be replaced.
14. Install or connect the following:
    - Valve cover(s)
    - Upper intake manifold (plenum), if necessary
    - All remaining electrical connectors and tighten the air tube connections
    - Negative battery cable
15. An oil and filter change is recommended.
16. Start the engine and check for leaks, abnormal noises and vibrations.

3.0L Engine

1. Before servicing the vehicle, refer to the precautions in the beginning of this section.
2. Relieve the fuel system pressure using the recommended procedure.
3. Remove or disconnect the following:
    - Negative battery cable
    - Breather hose
    - Positive Crankcase Ventilation (PCV) hose and valve
    - Rocker arm cover
    - Rocker shaft cap
    - Rocker arms and shafts

To install:

4. Install or connect the following:
    - Rocker arms and shafts. When properly aligned, torque the rocker shaft cap to 23 ft. lbs. (31 Nm)
    - Oil seals
    - Rocker arm cover with a new gasket
    - PCV valve with a new gasket
    - PCV hose
    - Breather hose
    - Negative battery cable

Intake Manifold

REMOVAL & INSTALLATION

2.0L Engine

This engine uses a 2-piece aluminum intake manifold. A non-reusable gasket joins the 2 halves.

➡Be sure to observe all cautions and warnings in the beginning of the section that may be related to this procedure.

1. Relieve the fuel system pressure.
2. Drain the cooling system.
3. Remove or disconnect the following:
    - Negative battery cable
    - Accelerator cable, breather hose and air intake hose
    - Vacuum connection at the power brake booster and the Positive Crankcase Ventilation (PCV) valve
    - All remaining vacuum hoses and pipes, as necessary. Tag for identification, if necessary, to save time at assembly.
    - Fuel line(s), the throttle control cable and brackets

⁂ CAUTION

Do not use conventional fuel filters, hoses or clamps when servicing fuel injection systems. They are not compatible with the injection system and could fail, causing personal injury or damage to the vehicle. Use only hoses and clamps specifically designed for fuel injection.

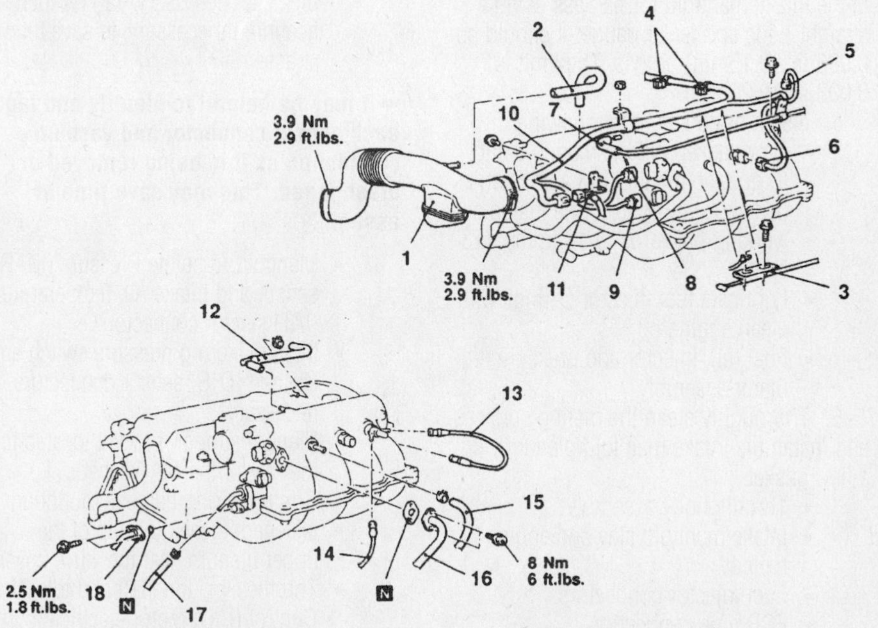

1. Air intake hose
2. Breather hose
3. Accelerator cable connection
4. Clip
5. MAP sensor connector
6. Intake air temperature sensor connector
7. Vacuum hose connection
8. TPS connector
9. Idle air control motor connector
10. Control wiring harness
11. Generator wiring harness connection
12. PCV hose assembly
13. Vacuum hose
14. Vacuum hose connection
15. Brake booster vacuum hose connection
16. EGR pipe connection
17. Fuel return hose connection
18. High-pressure fuel hose connection

N. use new components

7922CG11

Intake manifold hose, cable and wire attachment identification—2.0L engine

- Alternator wiring harness
- Manifold Absolute Pressure (MAP) sensor and Intake Air Temperature (IAT) sensor connectors
- Throttle Position (TP) sensor connector and position the engine wiring harness aside
- Exhaust Gas Recirculation (EGR) pipe connection
- Intake manifold stay and engine hanger
- Fuel injector connectors
- Throttle body assembly
- Intake manifold plenum and gasket
- Fuel rail assembly

➡Use care since the fuel injectors can drop out of the fuel rail as it is being removed.

- Intake manifold and gasket

To install:

4. Clean all gasket material from the cylinder head and intake manifold assembly. Check both surfaces for cracks or other damage. Check the intake manifold water passages and air passages for clogging. Clean if necessary. Check the gasket surface of the intake manifold for flatness using a straight-edge and feeler gauge. It should be 0.006 in. (0.15mm) or less. The limit is 0.008 in. (0.20mm).
5. Install or connect the following:
   - New gasket and the manifold and torque the manifold in a crisscross pattern, starting from the inside and working outwards to 17 ft. lbs. (23 Nm)
   - Lubricate fuel injector O-rings with clean engine oil
   - Fuel rail, injector and pressure regulator assembly
6. Thoroughly clean the mating surfaces and install the intake manifold plenum with a new gasket.
   - Throttle body assembly
   - Intake manifold stay and engine hanger
   - Fuel injector connectors
   - EGR pipe connection
   - Engine control electrical connectors
   - Alternator wiring harness connection
   - Fuel line(s)
   - Throttle control cable brackets
   - Vacuum hose at the power brake booster and the PCV valve
   - All remaining vacuum hoses and pipes
   - Accelerator cable, breather hose and air intake hose
   - Negative battery cable
7. Start the engine and check for proper operation.

2.5L Engine

1. Before servicing the vehicle, refer to the precautions in the beginning of this section.
2. Disconnect the negative battery cable.
3. Properly relieve the fuel system pressure.
4. Disconnect the fuel line(s) from the fuel rail assembly. For quick-connect fittings, squeeze the fitting retainer tabs together and separate the connection.

✲✲ CAUTION

To prevent fuel from getting in your eyes, wrap shop towels around the connection to catch any gasoline spillage.

5. Remove or disconnect the following:
   - Throttle body air inlet hose clamp
   - Air cleaner cover and inlet hose
   - Vacuum connection from the power brake booster and the Positive Crankcase Ventilation (PCV) valve
   - All remaining vacuum hoses and pipes, as necessary. Tag for identification, if necessary, to save time at assembly.

➡It may be helpful to identify and tag each sensor connector and vacuum connection as it is being removed or disengaged. This may save time at assembly.

- Manifold Absolute Pressure (MAP) sensor and Intake Air Temperature (IAT) sensor connectors
- Power steering pressure switch and Oxygen (O_2S) sensor connectors, if necessary
- Plenum support bracket located to the rear of the MAP sensor
- Control wiring harness mounting fasteners located on top of the upper plenum near the valve cover
- Throttle Position (TP) and Idle Air Control (IAC) motor electrical connections
- Throttle body assembly
- Throttle cable bracket
- Exhaust Gas Recirculation (EGR) tube from the intake manifold
- EGR valve and transducer assembly
- Plenum support bracket located to the rear of the EGR tube
- 7 upper intake plenum-to-lower manifold bolts, plenum and gasket
- Fuel injector electrical connectors
- 4 fuel rail-to-intake manifold bolts and fuel rail

➡Use care, there are spacers under each fuel rail bolt.

➡It may be necessary to remove the power steering fluid reservoir and mounting bracket to access all of the lower intake manifold fasteners.

- Lower intake manifold bolts, intake manifold and discard the gaskets

To install:

6. Clean all gasket sealing surfaces. Check both surfaces for cracks or other damage. Check the intake manifold air passages for clogging. Clean if necessary.
7. Check upper and lower manifold gasket surfaces for flatness using a straight-edge and feeler gauge.
8. The surfaces must be flat within 0.006 in. (0.15mm) per 12 in. (30.5cm) of manifold length. The limit is 0.008 in. (0.20mm).
9. Properly position the new gaskets on the heads and install the lower intake manifold. Torque the nuts as follows:
   a. Front bank: 60 inch lbs. (7 Nm).
   b. Rear bank: 14–17 ft. lbs. (20–23 Nm).
   c. Front bank: 14–17 ft. lbs. (20–23 Nm).
   d. Repeat once more.
10. Lubricate the fuel injector O-rings with clean engine oil.
11. Install or connect the following:
    - Fuel injectors
    - Seat the injectors and torque the fuel rail bolts to 105 inch lbs. (12 Nm)
    - Fuel injector electrical connectors
    - Fuel line(s) to the fuel rail assembly

➡Exert a slight tug on the fuel line away from the fuel rail to verify positive engagement.

- Upper intake plenum with new gaskets and torque the bolts to 13 ft. lbs. (18 Nm)
- Plenum support brackets and tighten to 13 ft. lbs. (18 Nm)
- EGR valve and transducer assembly
- EGR tube and torque the screws to 95 inch lbs. (11 Nm)
- Throttle cable bracket
- Throttle body assembly
- TPS and IAC electrical connections
- Control wiring harness and tighten the mounting fasteners
- Power steering pressure switch and oxygen sensor connectors, if disconnected.
- MAP sensor and IAT sensor connectors

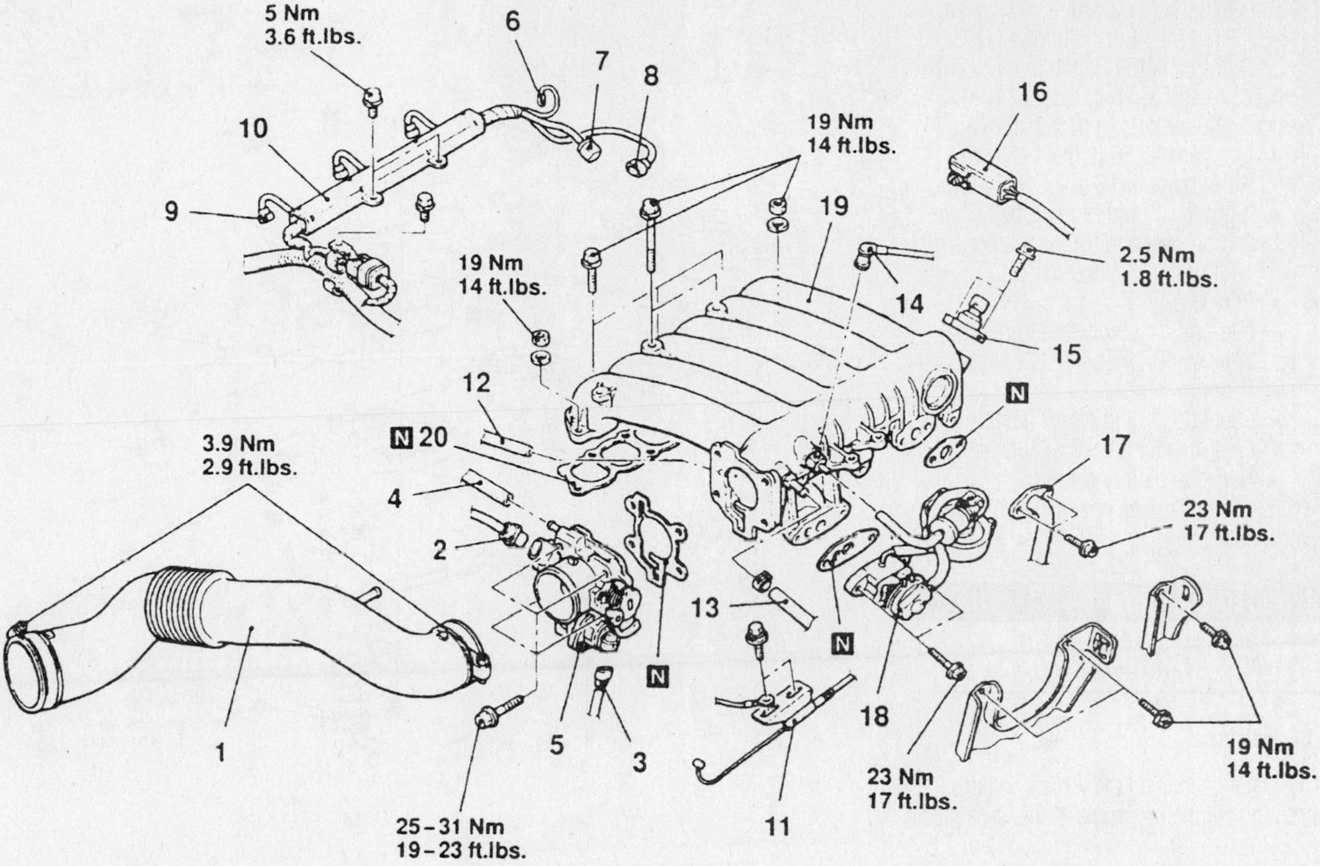

1. Air intake hose
2. TPS connector
3. Idle air control motor connector
4. Vacuum hose connection
5. Throttle body assembly
6. Power steering oil pressure switch connector
7. Heated oxygen sensor connector
8. Intake air temperature sensor connector
9. Injector connector
10. Control wiring harness
11. Accelerator cable connection
12. Vacuum hose connection
13. Brake booster vacuum hose connection
14. Vacuum hose connection
15. MAP sensor
16. Heated oxygen sensor harness
17. EGR pipe connection
18. EGR valve and EGR transducer assembly
19. Intake manifold plenum
20. Intake manifold plenum gasket

7922CG12

Exploded view of the plenum and intake manifold—2.5L engine

- Vacuum hose at the power brake booster and the PCV valve
- All remaining vacuum hoses and pipes
- Remaining engine control system electrical connectors
- Air cleaner cover and air inlet hose. Tighten the intake hose-to-throttle body hose clamp.
- Negative battery cable

12. Start the engine, check for leaks and repair if necessary.

3.0L Engine

1. Before servicing the vehicle, refer to the precautions in the beginning of this section.
2. Properly relieve the fuel system pressure.
3. Remove or disconnect the following:
   - Negative battery cable
   - Intake manifold plenum
   - Fuel injector connector
   - Fuel supply and return hoses
   - Vacuum hose
   - Fuel rail, injectors and pressure regulator as an assembly
   - Insulators
   - Positive Crankcase Ventilation (PCV) hose
   - Right front upper timing belt cover and bracket
   - Intake manifold and discard the gasket

To install:

4. Clean all mating surfaces of any residual gasket material.

5. Install or connect the following:
   - Intake manifold with a new gasket. Torque the nuts as follows:
     a. Right bank to 62 inch lbs. (7 Nm).
     b. Left bank to 16 ft. lbs. (22 Nm).
     c. Right bank to 16 ft. lbs. (22 Nm).
     d. Left bank to 16 ft. lbs. (22 Nm).
     e. Right bank to 16 ft. lbs. (22 Nm).
   - Right front upper timing belt cover and bracket and torque the bolts to 10 ft. lbs. (15 Nm)
   - PCV hose
   - Fuel rail assembly and torque the bolts to 102 inch lbs. (12 Nm)
   - Vacuum hose
   - Fuel supply and return hoses
   - Fuel injector electrical connector
   - Negative battery cable
6. Start the vehicle, check for leaks and repair if necessary.

Exhaust Manifold

REMOVAL & INSTALLATION

2.0L Engine

1. Before servicing the vehicle, refer to the precautions in the beginning of this section.
2. Properly drain the engine coolant.
3. Remove or disconnect the following:
   - Negative battery cable
   - Air intake hose and small air hose connection
   - Upper radiator hose from the thermostat housing
   - Control wiring harness connection
   - Water pipe assembly and engine oil level dipstick
   - Heat shield and engine hanger
   - Pulsed secondary air injection valve, if equipped
   - Exhaust pipe-to-exhaust manifold locknuts, separate the exhaust pipe and discard the gasket
4. Loosen the mounting fasteners and remove the exhaust manifold.

To install:

5. Clean all gasket material from the mating surfaces and check the manifold for cracks or warpage.
6. Install or connect the following:
   - New gasket and manifold and torque the fasteners, in a crisscross pattern, to 17 ft. lbs. (23 Nm)
   - Exhaust pipe to the exhaust manifold with a new gasket and new locknuts. Torque the nuts to 33 ft. lbs. (44 Nm)
   - Pulsed secondary air injection valve, if equipped

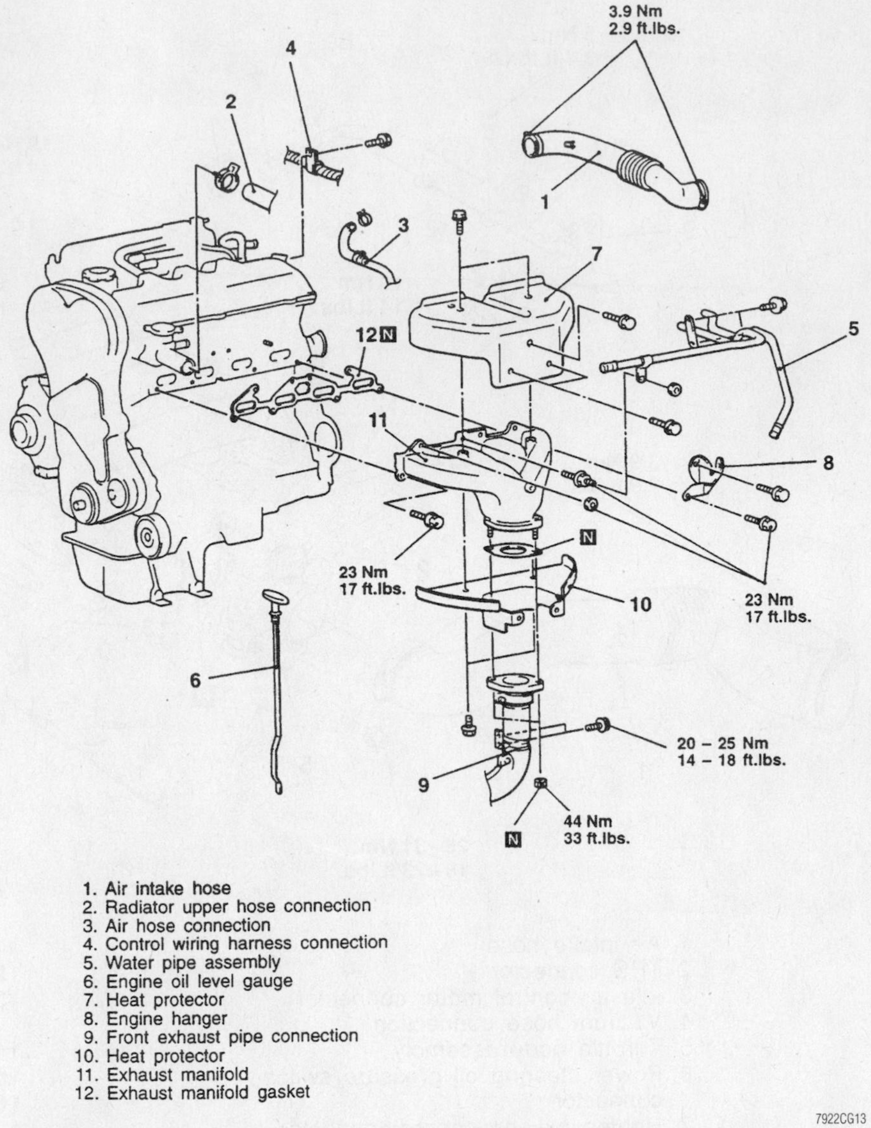

Exploded view of the exhaust manifold and related components—2.0L engine

   - Heat shield and the engine hanger
   - Control wiring harness connection
   - Upper radiator hose to the thermostat housing
   - Air intake hose and small air hose
   - Negative battery cable
7. Fill the cooling system.
8. Start the engine and check for exhaust leaks.

2.5L Engine

FRONT BANK

1. Before servicing the vehicle, refer to the precautions in the beginning of this section.
2. Remove or disconnect the following:
   - Negative battery cable
   - Cooling fan assembly
   - Engine oil level dipstick and tube
   - Engine hanger or lower heat shield, if equipped
   - Exhaust pipe-to-exhaust manifold locknuts, separate the exhaust pipe and discard the gasket
   - Exhaust manifold mounting fasteners, the exhaust manifold stay (brace), the exhaust manifold and the gasket

To install:

3. Clean all gasket material from the mating surfaces and check the manifold for cracks or warpage.
4. Install or connect the following:
   - New gasket, manifold and stay (brace) and torque the fasteners, in a crisscross pattern, to 22 ft. lbs. (29 Nm)
   - New gasket, install the exhaust pipe

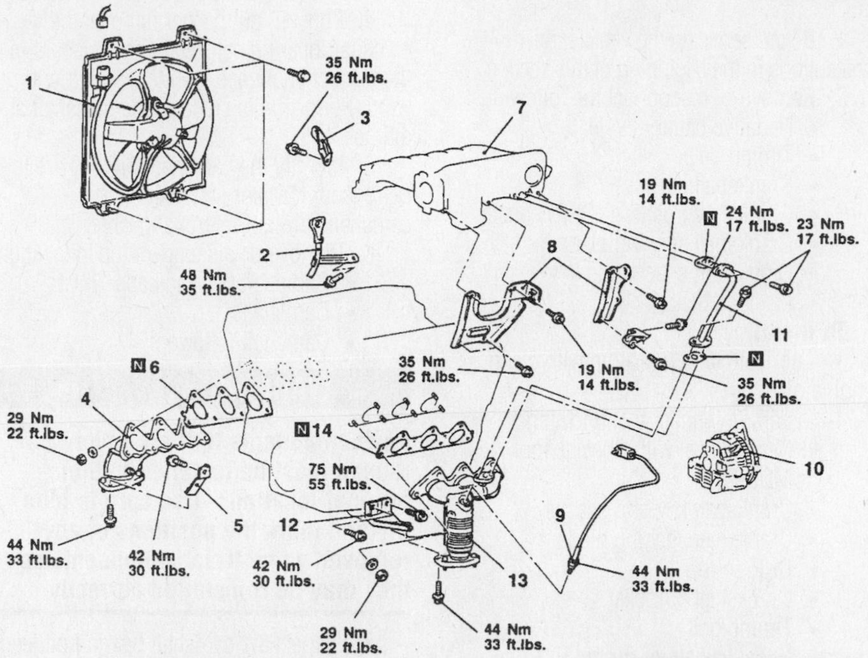

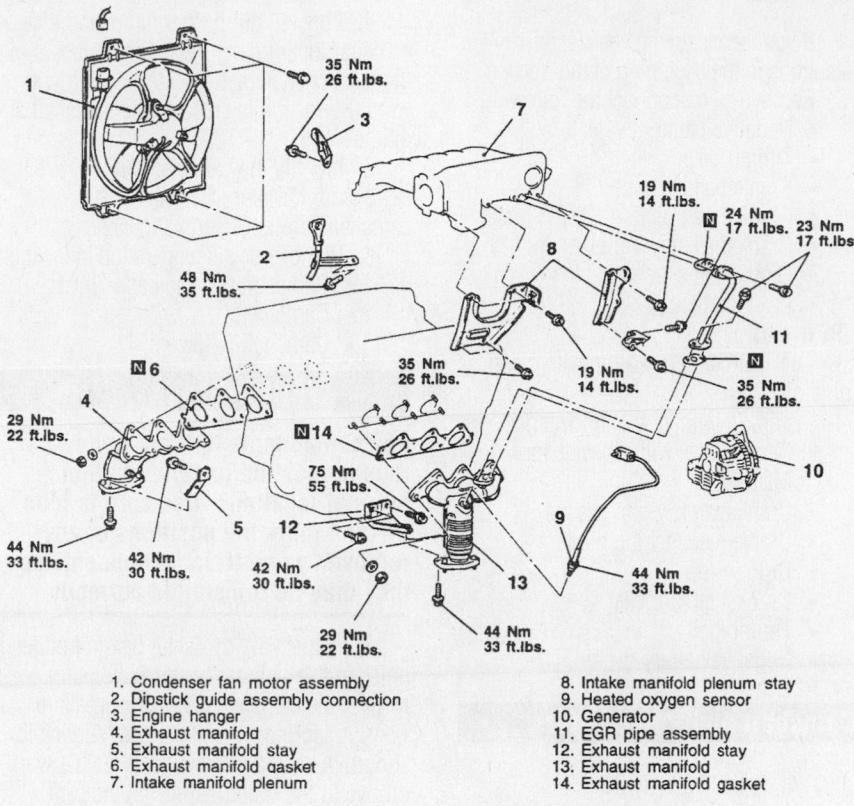

1. Condenser fan motor assembly
2. Dipstick guide assembly connection
3. Engine hanger
4. Exhaust manifold
5. Exhaust manifold stay
6. Exhaust manifold gasket
7. Intake manifold plenum
8. Intake manifold plenum stay
9. Heated oxygen sensor
10. Generator
11. EGR pipe assembly
12. Exhaust manifold stay
13. Exhaust manifold
14. Exhaust manifold gasket

7922CG14

Exploded view of the exhaust manifold assembly and related components—2.5L engine

to the exhaust manifold and torque the nuts to 33 ft. lbs. (44 Nm).
- Engine hanger or lower heat shield, if equipped and torque the lower heat shield fasteners to 10 ft. lbs. (13 Nm)
- Engine oil level dipstick and tube
- Cooling fan assembly
- Negative battery cable

5. Start the vehicle, check for leaks and repair if necessary.

REAR BANK

1. Before servicing the vehicle, refer to the precautions in the beginning of this section.
2. Remove or disconnect the following:
   - Negative battery cable
   - Intake manifold plenum and plenum stay
   - Heated Oxygen (HO_2S) sensor
   - Alternator
   - Exhaust Gas Recirculation (EGR) pipe assembly
   - Exhaust pipe-to-exhaust manifold locknuts, separate the exhaust pipe and discard the gasket
   - Manifold heat shield, if equipped
   - Exhaust manifold bolts, the exhaust manifold brace, the exhaust manifold and the gasket

To install:

3. Clean all gasket material from the mating surfaces and check the manifold for cracks or warpage.
4. Install or connect the following:
   - New gasket, the manifold and manifold brace and torque the nuts, in a crisscross pattern, to 22 ft. lbs. (30 Nm)
   - Manifold heat shield, if equipped
   - Exhaust pipe-to-exhaust manifold using a new gasket and torque the nuts to 33 ft. lbs. (44 Nm)
   - EGR pipe assembly
   - Alternator
   - HO_2S sensor
   - Intake manifold plenum and plenum stay (brace)
   - Negative battery cable
5. Start the engine, check for exhaust leaks and repair if necessary.

3.0L Engine

1. Before servicing the vehicle, refer to the precautions in the beginning of this section.
2. Remove or disconnect the following:
   - Battery and tray
   - Front exhaust pipe
   - Air cleaner assembly
   - Oil dipstick guide
   - Strut tower bar
   - Upper and lower heat shields
   - Exhaust Gas Recirculation (EGR) pipe, left side only
   - Exhaust manifold and discard the gasket

To install:

3. Clean all mating surfaces of any residual gasket material.
4. Install or connect the following:
   - Exhaust manifold with a new gasket and torque the bolts to 33 ft. lbs. (44 Nm)
   - EGR pipe, if removed. Torque the bolt to 13 ft. lbs. (18 Nm)
   - Upper and lower heat shields and torque the bolts to 10 ft. lbs. (15 Nm)
   - Strut tower bar
   - Oil dipstick guide
   - Air cleaner assembly
   - Front exhaust pipe
   - Battery and tray
5. Start the vehicle, check for leaks and repair if necessary.

Front Crankshaft Seal

REMOVAL & INSTALLATION

2.0L Engine

1. Before servicing the vehicle, refer to the precautions in the beginning of this section.
2. Drain the engine oil.
3. Remove or disconnect the following:
   - Negative battery cable
   - Accessory drive belts
   - Crankshaft damper/pulley
   - Timing belt cover
   - Timing belt
   - Crankshaft sprocket

**** WARNING**

Be careful not to nick the seal surface of the crankshaft or the seal bore.

   - Front crankshaft seal using a seal puller tool

**** WARNING**

Be careful not to damage the seal contact area of the crankshaft.

To install:

4. Lubricate the new seal with clean engine oil. Install the new front crankshaft oil seal by using oil seal installer Tool No. 6780-1.
5. Place a new oil seal into the opening with the seal spring facing the inside of the engine. Be sure the oil seal is installed flush with the front cover.
6. Install or connect the following:
   - Crankshaft timing belt sprocket
   - Timing belt
   - Timing belt cover
   - Crankshaft damper/pulley
   - Accessory drive belts
   - Negative battery cable
7. Change the oil filter and refill the crankcase.
8. Start the engine, check for leaks and repair if necessary.

2.5L Engine

1. Before servicing the vehicle, refer to the precautions in the beginning of this section.
2. Drain the engine oil.
3. Remove or disconnect the following:
   - Negative battery cable
   - Accessory drive belts
   - Crankshaft damper/pulley
   - Front timing belt covers
   - Timing belt
   - Crankshaft sprocket and key
   - Front crankshaft seal by prying it out with a flat tipped prytool; be sure to cover the end of the prytool tip with a shop towel

**** WARNING**

Be careful not to nick the seal surface of the crankshaft or the seal bore.

To install:

4. Apply a light coating of clean engine oil to the lip of the new oil seal. Install the new front crankshaft oil seal into the oil pump housing by using Oil Seal Installer Tool MD998717. Be sure the oil seal is installed flush with the oil pump cover.
5. Install or connect the following:
   - Crankshaft timing belt sprocket and key
   - Timing belt
   - Timing belt covers
   - Crankshaft damper/pulley
   - Accessory drive belts
   - Negative battery cable
6. Change the oil filter and fill the engine with clean oil.
7. Start the engine, check for leaks and repair if necessary.

3.0L Engine

1. Before servicing the vehicle, refer to the precautions in the beginning of this section.
2. Remove or disconnect the following:
   - Negative battery cable
   - Timing belt
   - Crankshaft sprocket
   - Crankshaft Position (CKP) sensor
   - Crankshaft sensing blade
   - Crankshaft spacer and key
   - Front oil seal

To install:

3. Lubricate the oil seal lip with clean engine oil.
4. Install or connect the following:
   - New oil seal with Special Tool MD998717
   - Crankshaft key and spacer
   - Crankshaft sensing blade
   - CKP sensor
   - Crankshaft sprocket
   - Timing belt
   - Negative battery cable

Camshaft and Valve Lifters

REMOVAL & INSTALLATION

2.0L Engine

1. Before servicing the vehicle, refer to the precautions in the beginning of this section.
2. Properly relieve the fuel system pressure.
3. Remove or disconnect the following:
   - Negative battery cable
   - Spark plug wires
   - Ignition coil pack with spark plug wires
   - Cylinder head cover and discard the gasket
   - Ground strap
   - Timing belt covers, timing belt and camshaft sprockets

➡The camshaft bearing caps are numbered for correct location during installation.

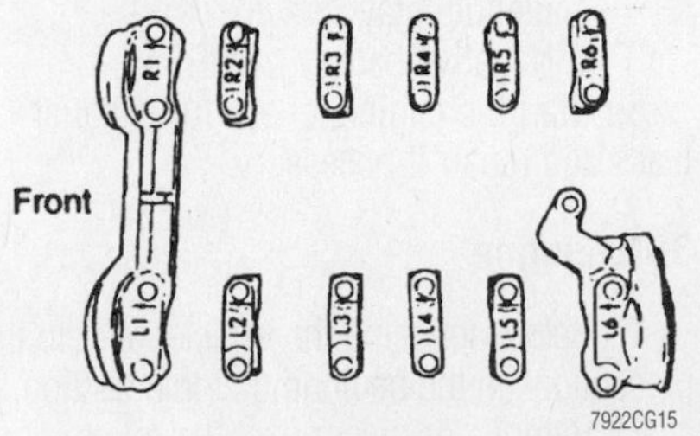

Camshaft bearing cap identification—2.0L engine

   - Outer camshaft bearing caps, first
4. Loosen, but do not remove, the camshaft bearing cap retaining fasteners in the correct sequence, inside working outward. Perform this step on one camshaft at a time.
5. Identify the camshafts, if they are to be reused, for later installation. The camshafts are not interchangeable.
6. Remove or disconnect the following:
   - Camshaft bearing caps
   - Camshafts
   - Camshaft followers

**** WARNING**

Any components that are to be reused must be installed in their original locations. Use care to identify and mark the positions of any removed valve train components so they may be reinstalled correctly.

7. Inspect the camshaft bearing oil feed holes in the cylinder head for clogging. Inspect the camshaft bearing journals for wear or scoring. Check the cam surface for abnormal wear and damage. A visible worn groove in the roller path or on the cam lobes is cause for replacement.

To install:

8. Thoroughly clean all camshaft and related parts.
9. If the fit and condition of the camshafts are acceptable, remove the camshafts for installation of the cam followers.
10. The hydraulic valve lash adjusters are inside the roller cam followers. Be sure they are clean, well lubricated with clean engine oil and properly positioned. Install the cam followers in their original positions on the hydraulic adjuster and valve stem.

**** WARNING**

To avoid valve to piston contact, be sure NONE of the pistons are at Top Dead Center (TDC) when installing the camshafts.

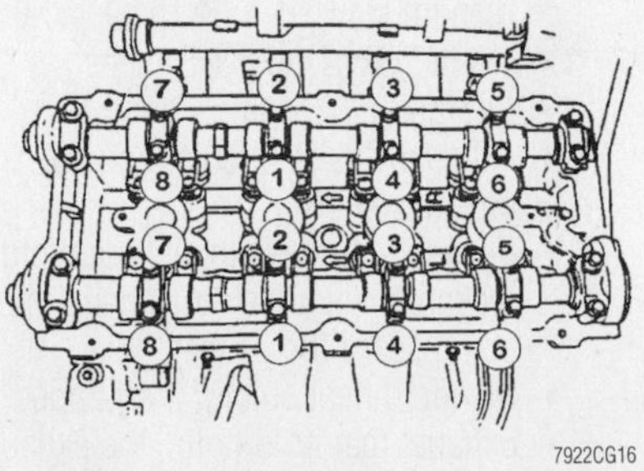

Camshaft bearing cap retaining bolt removal sequence—2.0L engine

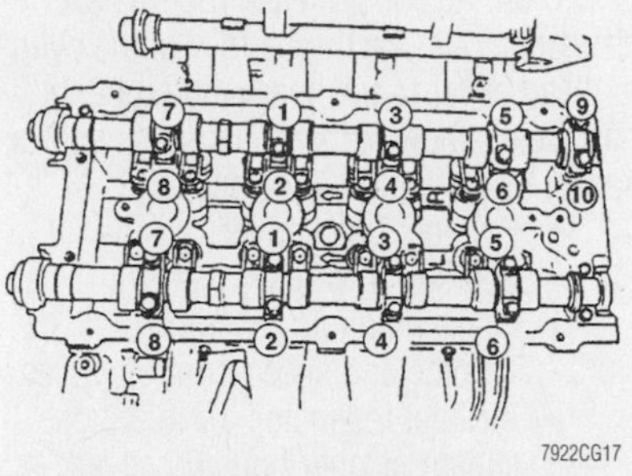

Camshaft bearing cap retaining bolt tightening sequence—2.0L engine

11. Lubricate the camshaft bearing journals and cam followers with clean engine oil.
12. Install or connect the following:
    - Camshafts
    - Right and left camshaft bearing caps No. 2 through No. 5 and right side No. 6
13. Torque the M6 fasteners to 105 inch lbs. (12 Nm) in correct sequence
14. Apply Mopar® Gasket Maker, sealer to the No. 1 and left-side No. 6 bearing caps.
    - Bearing caps and torque the M8 fasteners to 21 ft. lbs. (28 Nm)

➡**The end caps must be installed before the seals may be installed.**

- Camshaft end seals
- Camshaft sprockets, if removed

➡**Make sure all timing marks are properly aligned, using the recommended procedure.**

- Timing belt
- Timing belt covers

⁂ WARNING

Verify that all timing marks are correct. If the timing belt or sprockets are incorrectly installed, engine damage will occur.

15. Clean all sealing surfaces. Make certain the rails are flat.
    - New cylinder head cover gaskets

⁂ WARNING

DO NOT allow oil or solvents to contact the timing belt as they can deteriorate the rubber and cause tooth skipping. Apply Mopar Silicone Rubber Adhesive Sealant, at the camshaft cap corners and at the top edge of the ½ round seal.

➡**Inspect the spark plug well seals for cracking and/or swelling and replace, if necessary.**

16. Install the cylinder head cover assembly.
17. Torque the cylinder head cover fasteners, in sequence, using the following 3 steps:
    a. Step 1: 40 inch lbs. (4.5 Nm).
    b. Step 2: 80 inch lbs. (9 Nm).
    c. Step 3: 105 inch lbs. (12 Nm).
18. Install or connect the following:
    - Ignition coil pack and torque the fasteners to 105 inch lbs. (12 Nm)
    - Spark plug wires
    - Ground strap
    - All vacuum lines and remaining wiring

➡**An oil and filter change is recommended to wash out any sealant or gasket material that may have fallen into the engine.**

19. Connect the negative battery cable.
20. Test run the vehicle. Check for leaks and for proper operation.

2.5L Engine

1. Before servicing the vehicle, refer to the precautions in the beginning of this section.

➡**For camshaft service, the cylinder head must be removed.**

2. Relieve the fuel system pressure.
3. Drain the cooling system.
4. Drain the engine oil.
5. Remove or disconnect the following:
    - Negative battery cable
    - Timing belt covers, timing belt and camshaft sprockets
    - Upper intake manifold plenum
    - Lower intake manifold
    - Cylinder head bolts and cylinder head

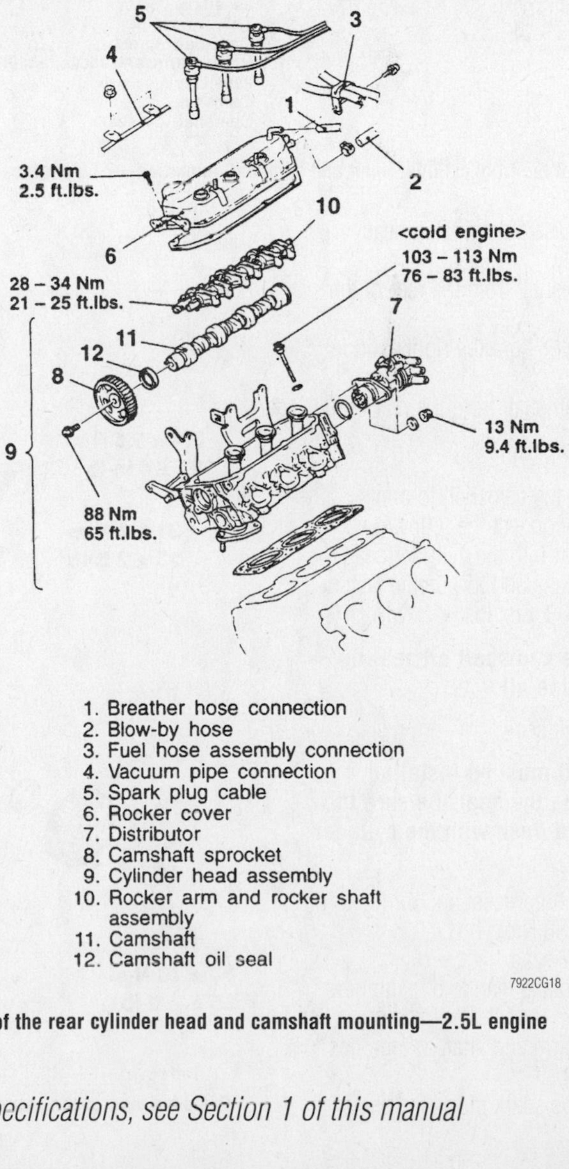

Exploded view of the rear cylinder head and camshaft mounting—2.5L engine

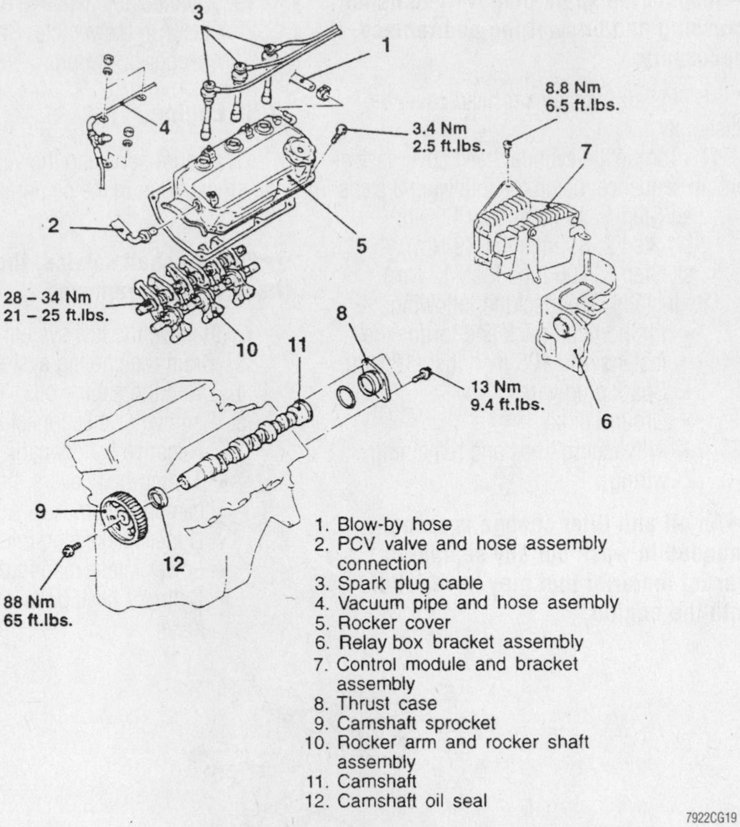

Exploded view of the front cylinder head and camshaft mounting—2.5L engine

- Thrust case from the left head assembly
- Left camshaft from the rear of the head
- Distributor from the right cylinder head
- Right camshaft from the rear of the head

To install:

6. Lubricate the camshaft journals.
7. Install or connect the following:
   - Camshaft into the cylinder head
   - Thrust case and torque the fasteners to 108 inch lbs. (13 Nm)

➡Lubricate the camshaft oil seal lip with clean engine oil.

- Camshaft seal

➡The camshaft must be installed before installing the seal. Be sure the seal is installed flush with the cylinder head surface.

- Camshaft sprocket and torque to 65 ft. lbs. (88 Nm)
- Cylinder head
- Lower intake manifold using new gaskets
- Rocker arm and shaft assemblies
- Timing belt

8. Inspect the spark plug tube seals located on the ends of each tube. These seals slide onto each tube to seal the cylinder head cover to the spark plug tube. If these seals show signs of hardness and/or cracks, they should be replaced.
   - Cylinder head cover
   - Spark plug wires
   - Intake manifold plenum
   - Throttle and speed control cables
   - Air inlet resonator, air inlet hose and air cleaner housing cover
   - All remaining electrical connectors
   - Tighten the air tube connections
   - Negative battery cable
9. Fill and bleed the cooling system.
10. Change the oil filter and fill the engine with clean oil.
11. Start the engine and check for leaks, abnormal noises and vibrations.

3.0L Engine

LEFT BANK

1. Before servicing the vehicle, refer to the precautions in the beginning of this section.
2. Remove or disconnect the following:
   - Negative battery
   - Timing belt
   - Thermostat housing
   - Blow by hose

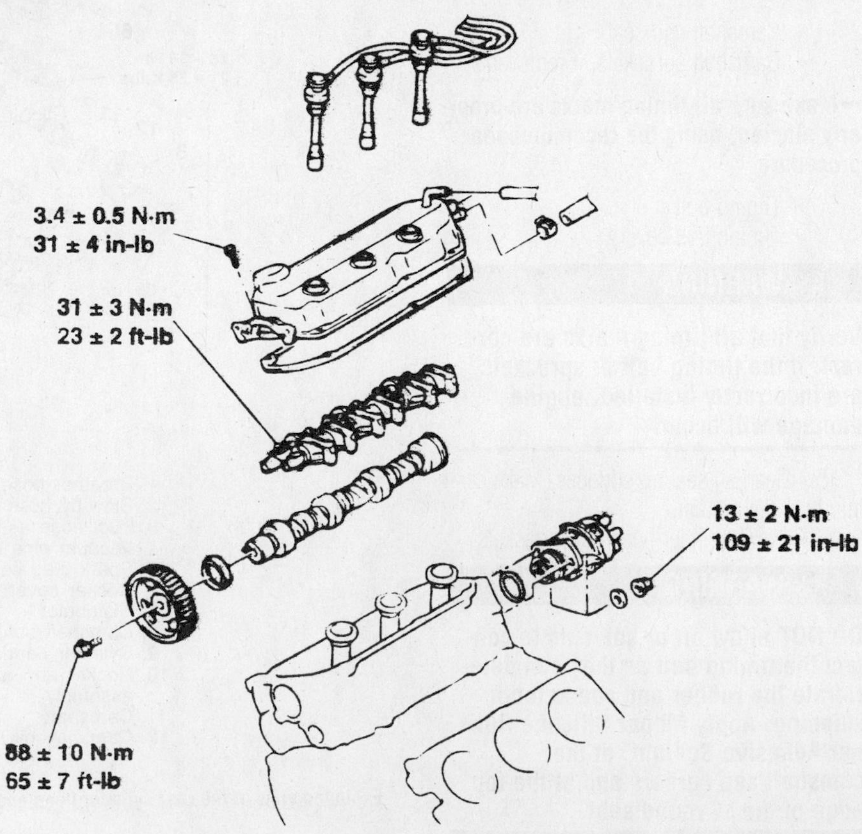

Exploded view of the rocker arm and camshaft assembly—3.0L engine

- Positive Crankcase Ventilation (PCV) hose
- Spark plug cables
- Rocker arm cover
- Rocker arm and shaft
- Camshaft sprocket
- Thrust case
- Camshaft

To install:

3. Install or connect the following:
   - Camshaft
   - Thrust case and torque the bolt to 109 inch lbs. (13 Nm)
   - Camshaft sprocket and torque the bolt to 65 ft. lbs. (88 Nm)
   - Rocker arm and shaft assembly and torque to 23 ft. lbs. (31 Nm)
   - Rocker arm cover and torque the bolts to 35 inch lbs. (4 Nm)
   - Spark plug cables
   - PCV and blow by hoses
   - Thermostat housing
   - Timing belt
   - Negative battery cable

RIGHT BANK

1. Before servicing the vehicle, refer to the precautions in the beginning of this section.
2. Remove or disconnect the following:
   - Negative battery
   - Timing belt
   - Intake manifold plenum
   - Breather hose
   - Blow by hose
   - Spark plug cables
   - Rocker arm cover
   - Rocker arm and shaft assembly with Tool MD998443
   - Distributor
   - Camshaft sprocket with Tool MB990767 and MD998715
   - Camshaft

To install:

3. Install or connect the following:
   - Camshaft and a new oil seal, if removed
   - Camshaft sprocket and torque the bolt to 65 ft. lbs. (88 Nm)
   - Distributor
4. Rotate the camshaft until the dowel pin on the front end is properly positioned.
5. Install or connect the following:
   - Rocker arm cover and torque the bolts to 35 inch lbs. (4 Nm)
   - Spark plug cables
   - Blow by and breather hose connections
   - Intake manifold plenum and torque the bolts to 13 ft. lbs. (18 Nm)
   - Timing belt
   - Negative battery cable

Valve Lash

ADJUSTMENT

The engines in these vehicles do not require periodic valve lash adjustment.

Starter Motor

REMOVAL & INSTALLATION

2.0 and 2.5L Engines

1. Before servicing the vehicle, refer to the precautions in the beginning of this section.
2. Remove or disconnect the following:
   - Negative battery cable
   - Engine under cover
   - Heated Oxygen (HO_2) sensor electrical connector
   - Front exhaust pipe
   - Starter motor electrical connec-tors
   - Starter motor

To install:

3. Install or connect the following:
   - Starter motor and torque starter-to-transaxle bolts to 40 ft. lbs. (54 Nm) for 2.0L engine or 20–25 ft. lbs. (26–33 Nm) for 2.5L engine
   - Starter motor electrical connectors

➡Use new gaskets when installing the front exhaust pipe.

   - Front exhaust pipe. Torque the front

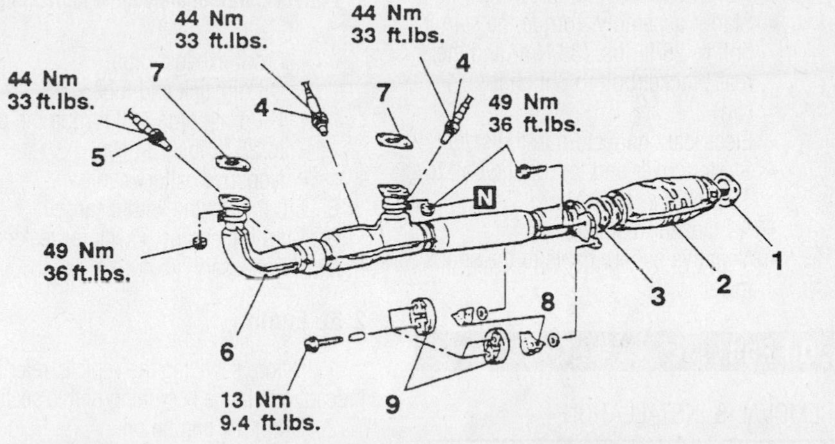

1. Gasket
2. Catalytic converter
3. Gasket
4. Heated oxygen sensor <Except vehicles for California>
5. Heated oxygen sensor
6. Front exhaust pipe
7. Gasket
8. Protector
9. Hanger

9306CG04

Exploded view of the front exhaust pipe and related components—2.5L engine

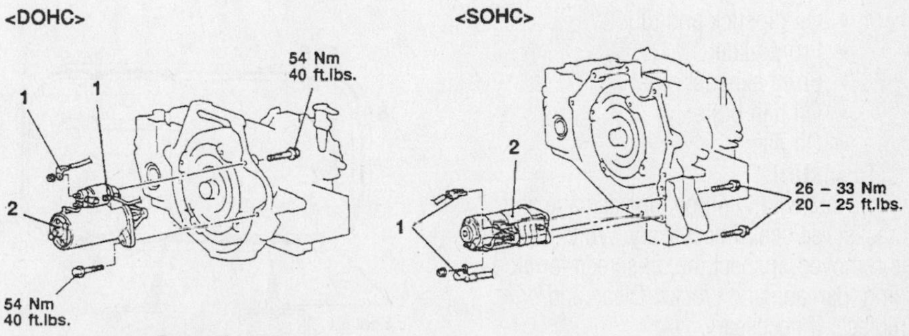

Removal steps
1. Starter terminal and connector
2. Starter motor

9306CG05

View of the starter motor—2.0L (DOHC) engine and 2.5L (SOHC) engine

exhaust pipe-to-exhaust manifold nuts to 36 ft. lbs. (49 Nm) and the front exhaust pipe-to-catalytic converter bolts to 36 ft. lbs. (49 Nm).
- HO_2sensor electrical connector
- Engine under cover
- Negative battery cable

3.0L Engine

1. Before servicing the vehicle, refer to the precautions in the beginning of this section.
2. Remove or disconnect the following:
   - Negative battery cable from the shock tower
   - Air cleaner resonator
   - Starter motor cover
3. Starter electrical connectors
   - Starter assembly

To install:

4. Install or connect the following:
   - Starter assembly. Torque the starter bolt to 25 ft. lbs. (33 Nm) and the rear bracket bolt to 8 inch lbs. (1 Nm).
   - Electrical connectors to the starter
   - Starter cover and torque the bolt to 15 inch lbs. (6 Nm)
   - Air cleaner resonator
   - Negative battery cable to the shock tower

Oil Pan

REMOVAL & INSTALLATION

2.0L Engine

1. Before servicing the vehicle, refer to the precautions in the beginning of this section.
2. Drain the engine oil.
3. Remove or disconnect the following:
   - Negative battery cable
   - Oil dipstick and tube
   - Front plate
   - Front exhaust pipe
   - Oil pan bolts
   - Oil pan

To install:

4. Inspect the oil pan for damage and cracks; replace, if necessary. While the pan is removed, inspect the oil screen for clogging, damage and cracks. Clean and/or replace, if necessary.
5. Thoroughly, clean the mating surfaces of the cylinder block and the oil pan.
6. Apply sealant to the seams between the oil pan and the engine block.
7. Install or connect the following:
   - Oil pan and torque the bolts to 108 inch lbs. (12 Nm)

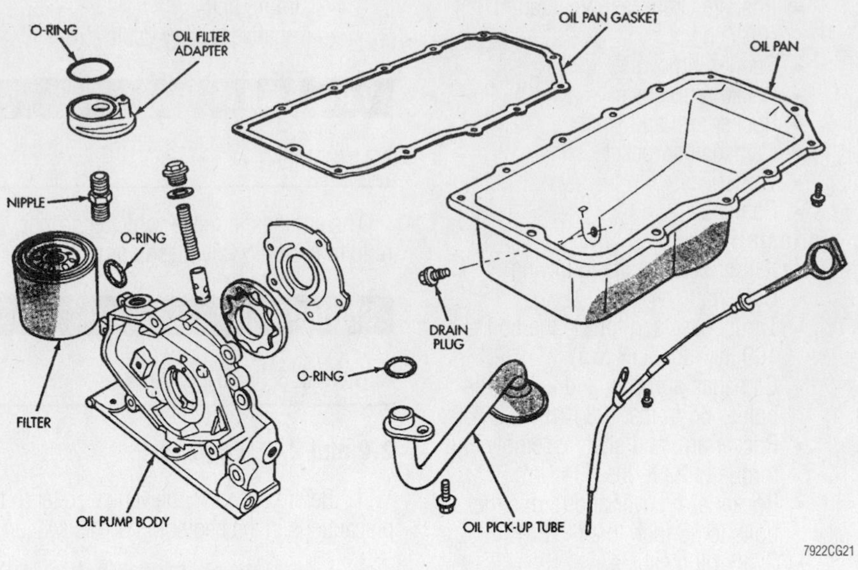

Exploded view of the engine lubricating components—2.0L engine

   - Front exhaust pipe
   - Oil dipstick and tube
   - Oil drain plug and torque the plug to 25 ft. lbs. (34 Nm)
   - Negative battery cable
8. Fill the engine with clean oil.
9. Start the engine, check for leaks and repair if necessary.

2.5L Engine

1. Before servicing the vehicle, refer to the precautions in the beginning of this section.
2. Drain the engine oil.
3. Remove or disconnect the following:
   - Negative battery cable and lower the front exhaust pipe
   - Center-member
   - Engine oil dipstick tube and dipstick
   - Starter motor
   - Front and rear plates
   - Transaxle inspection cover

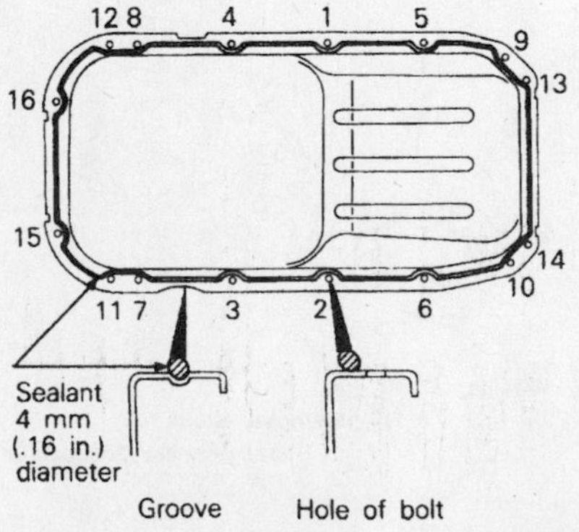

To ensure a leak-free seal, apply sealer as shown and tighten the bolts following the specified sequence—2.5L engine

   - Oil pan attaching bolts
   - Oil pan

To install:

4. Thoroughly clean and dry the oil pan, cylinder block and cylinder block bolts and bolt holes.
5. Apply a continuous 3/16 inch (4mm) bead of silicone adhesive sealant to the oil pan gasket surface. Be sure to circle all mounting bolt holes as well. Install the oil pan within a 10–15 minute period of applying the gasket material to ensure proper sealing.
6. Install or connect the following:
   - Oil pan and torque the bolts to 53 inch lbs. (6 Nm)
   - Transaxle inspection cover
   - Front and rear plates and torque the bolts to 80 ft. lbs. (108 Nm)
   - Starter motor
   - Engine oil dipstick tube and dipstick
   - Center-member and torque the bolts to 65 ft. lbs. (88 Nm)
   - Front exhaust pipe
   - Oil pan drain plug and gasket and torque the drain plug to 29 ft. lbs. (40 Nm)
   - Negative battery cable
7. Fill the engine with clean oil and install a new oil filter.
8. Start the engine, check for leaks and repair if necessary.

3.0L Engine

1. Before servicing the vehicle, refer to the precautions in the beginning of this section.
2. Drain the engine oil.
3. Remove the engine support module as follows:

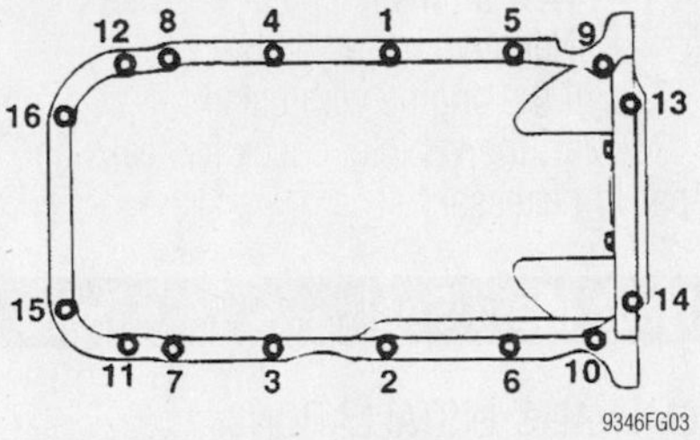

Tighten the upper oil pan bolts in sequence

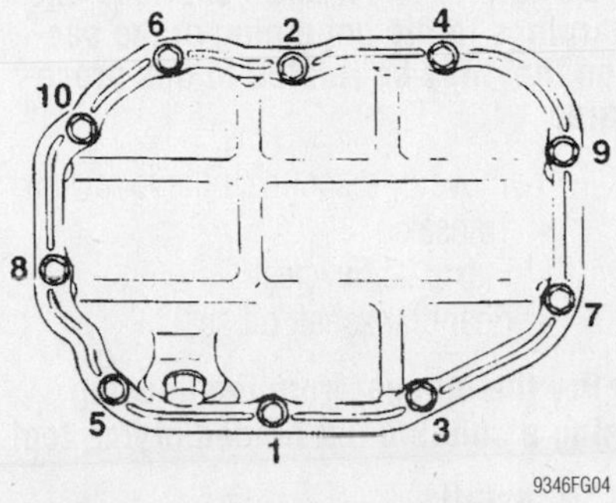

Tighten the lower oil pan bolts in sequence

- Negative battery cable
- Front exhaust pipe
- Lower oil pan and discard the gasket
- Starter electrical connectors
- Starter
- Oil dipstick and guide
- Cover
- Upper oil pan and discard the gasket

4. Clean all mating surfaces of any residual gasket material.

To install:

5. Install or connect the following:
   - Upper oil pan with a new gasket. Torque the bolts, in sequence, to 53 inch lbs. (6 Nm)
   - Cover and torque the bolts to 96 inch lbs. (11 Nm)
   - Starter
   - Starter electrical connectors
   - Oil dipstick and guide with a new O-ring and torque the bolt to 36 ft. lbs. (48 Nm)
   - Lower oil pan with a new gasket and torque the bolts to 96 inch lbs. (11 Nm)
   - Front exhaust pipe
   - Negative battery cable
6. Fill the engine with clean oil.
   - Start the vehicle, check for leaks and repair if necessary.

Oil Pump

REMOVAL & INSTALLATION

2.0L Engine

1. Before servicing the vehicle, refer to the precautions in the beginning of this section.
2. Drain the engine oil.
3. Remove or disconnect the following:
   - Negative battery cable
   - Timing belt
   - Oil pan
   - Crankshaft sprocket
   - Oil pump pick-up tube and O-ring
   - Oil pump and front crankshaft seal

⁂ WARNING

The front cover/oil pump mounting bolts may be different sizes and must be reinstalled in their original locations. Remove and tag the front cover mounting bolts.

4. Inspect the oil pump case for damage and remove the rear cover.
5. Remove the pump rotors and inspect the inside of the case for excessive wear.
6. Check that the oil relief plunger slides smoothly and check for a broken spring.

To install:

7. Clean all parts well. Be sure the block and pump surfaces are clean and free of old sealer.
8. Assemble the pump using new parts as required with clean oil. Align the marks on the inner and outer rotors when assembling.
9. Install the pump back cover and torque the screws to 88 inch lbs. (10 Nm).
10. Reinstall the pump relief valve, spring, gasket and valve cap. Torque the valve cap to 30–33 ft. lbs. (41–44 Nm).
11. Apply gasket maker to the engine block mounting surface of the oil pump body.
12. Install the oil ring into the discharge passage of the pump body.
13. Prime the oil pump before installation by filling the rotor cavity with clean engine oil.
14. Align the flats of the oil pump rotor with the flats on the crankshaft as you install the pump to the engine block.
15. Install or connect the following:
    - Oil pump-to-engine block bolts and torque to 17–21 ft. lbs. (23–28 Nm)
    - New front oil seal
    - Crankshaft sprocket
    - Oil pump pickup tube and O-ring and torque the oil pump pickup tube screw to 21 ft. lbs. (28 Nm)
    - Oil pan
    - Timing belt and covers
    - Cankshaft damper
    - New oil filter
16. Fill the engine with clean oil.
17. Test run the vehicle to check for leaks. An oil pressure gauge should be installed to verify proper engine oil pressure.

2.5L Engine

1. Before servicing the vehicle, refer to the precautions in the beginning of this section.
2. Drain the engine oil.
3. Drain the engine coolant.
4. Remove or disconnect the following:
   - Negative battery cable
   - Drive belts and accessories
   - Crankshaft damper
   - Timing belt upper and lower covers
   - Timing belt
   - Crankshaft sprocket
   - Oil pump
5. Inspect the oil pump case for damage and remove the rear cover.
6. Remove the pump rotors and inspect the inside of the case for excessive wear.
7. Check that the oil relief plunger slides smoothly and check for a broken spring.

To install:

8. Clean all parts well. Be sure the block and pump surfaces are clean and free of old sealer.
9. Assemble the pump using new parts as required with clean oil. Align the marks on the inner and outer rotors when assembling.
10. Install the pump back cover and torque the screws to 89 inch lbs. (10 Nm).
11. Reinstall the pump relief valve, spring, gasket and valve cap. Torque the valve cap to 30–33 ft. lbs. (41–44 Nm).
12. Prime the pump before installation by filling the rotor cavity with clean engine oil.
13. Apply gasket maker sealer on the pump. Install the O-ring into the counterbore on the pump body discharge passage. Position the pump onto the crankshaft until seated on the block. Torque the size M8 fasteners to 10 ft. lbs. (14 Nm) and size M10 fasteners to 30 ft. lbs. (41 Nm).
14. Install or connect the following:
    - Timing belt and crankshaft sprocket
    - Timing belt cover
    - Crankshaft damper
    - Drive belts and accessories

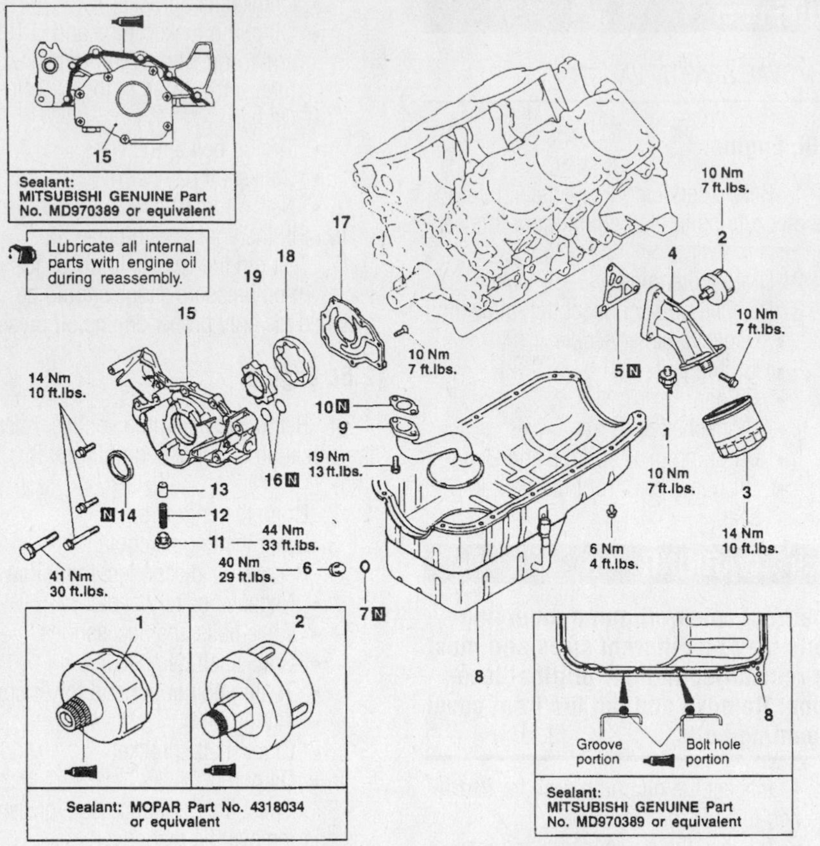

1. Oil pressure switch
2. Oil pressure gauge unit
3. Oil filter
4. Oil filter bracket
5. Oil filter bracket gasket
6. Drain plug
7. Drain plug gasket
8. Oil pan
9. Oil screen
10. Oil screen gasket
11. Plug
12. Relief spring
13. Relief plunger
14. Crankshaft oil seal
15. Oil pump case
16. O-ring
17. Oil pump cover
18. Oil pump outer rotor
19. Oil pump inner rotor

7922CG20

Exploded view of the oil pan and pump assembly—2.5L engine

15. Fill the cooling system.
16. Fill the engine with clean oil and install a new oil filter.
17. Road test the vehicle. Check for proper operation as well as leaks.

3.0L Engine

1. Before servicing the vehicle, refer to the precautions in the beginning of this section.
2. Drain the engine oil.
3. Remove the engine support module as follows:
   - Negative battery cable
   - Oil filter
   - Oil pressure switch
   - Oil filter bracket
   - Lower and upper oil pan
   - Baffle plate
   - Oil screen and gasket
   - Baffle plate
   - Relief spring and plunger
   - Front crankshaft oil seal
   - Oil pump case
   - Oil pump cover
   - Oil pump outer and inner rotors
   - Oil pump case
4. Clean all mating surfaces of any residual gasket material.

To install:

5. Prime the oil pump before installation.
6. Install or connect the following:
   - Oil pump case
   - Inner and outer rotors
   - Oil pump cover and torque the bolts to 89 inch lbs. (10 Nm)
   - Oil pump case assembly
   - New front crankshaft oil seal
   - Relief plunger and spring
   - Baffle plate
   - Oil screen with a new gasket
   - Baffle plate
   - Upper oil pan and cover
   - Lower oil pan
   - Oil filter bracket with a new gasket
   - Oil pressure switch
   - New oil filter
   - Negative battery cable
7. Fill the engine with clean oil.
8. Start the vehicle, check for leaks and repair if necessary.

Rear Main Seal

REMOVAL & INSTALLATION

2.0L and 3.0L Engines

➡Be sure to observe all cautions and warnings in the beginning of the section that may be related to this procedure.

1. Remove or disconnect the following:
   - Transaxle
   - Flexplate/flywheel
   - Rear crankshaft oil seal

➡Pry the oil seal from the housing using a suitable flat bladed prying tool.

To install:

➡When installing the new seal there is no need to lubricate sealing surface.

2. Install or connect the following:
   - Oil seal into housing using a suitable installation tool
   - Flexplate/flywheel and torque the bolts to 68 ft. lbs. (92 Nm)
   - Transaxle

2.5L Engine

1. Before servicing the vehicle, refer to the precautions in the beginning of this section.
2. Remove or disconnect the following:
   - Transaxle
   - Flexplate/flywheel
   - 5 rear seal housing-to-engine bolts
   - Oil seal housing
   - Rear crankshaft oil seal from the oil

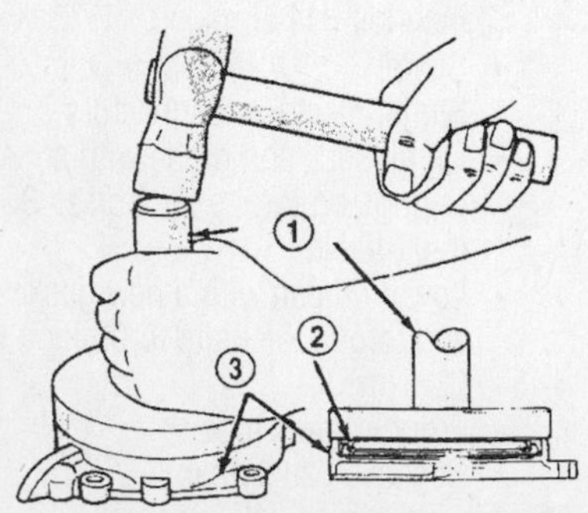

Fig. 93 Install Crankshaft Rear Oil Seal

1 – SPECIAL TOOL MD-998718
2 – SEAL
3 – HOUSING

9346FG05

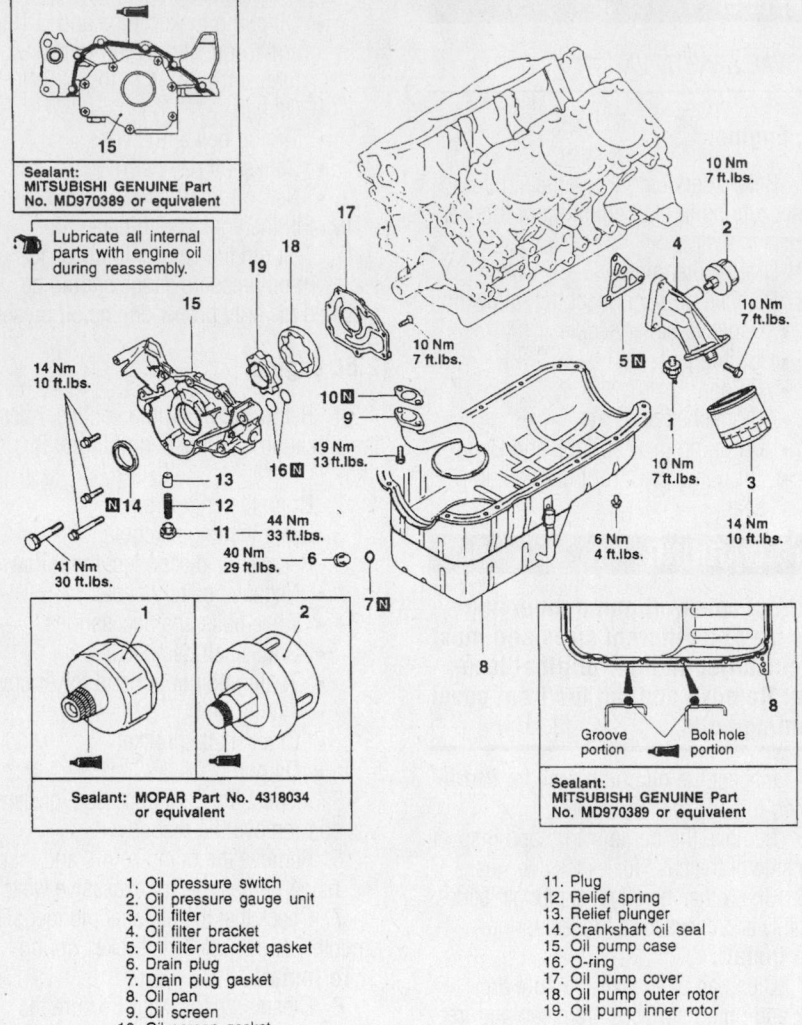

Install the new rear main seal

seal housing using a suitable flat bladed prying tool

To install:

➨When installing the seal there is no need to lubricate sealing surface.

3. Install the seal into the housing using Tool MD998718.

4. Apply silicone rubber adhesive sealant to the mating surface of the seal housing.

5. Apply a light coating of engine oil to the entire oil seal lip circumference.

6. Install or connect the following:

- Oil seal and housing and torque the mounting bolts to 96 inch lbs. (11 Nm)
- Flexplate/flywheel and torque the bolts to 68 ft. lbs. (92 Nm)
- Transaxle

Piston and Ring

POSITIONING

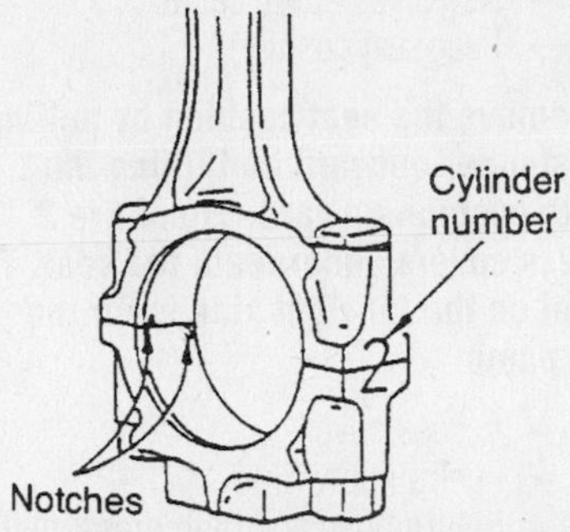

Chrysler engine connecting rod and cap installation—ensure to matchmark the cap and rod prior to disassembly

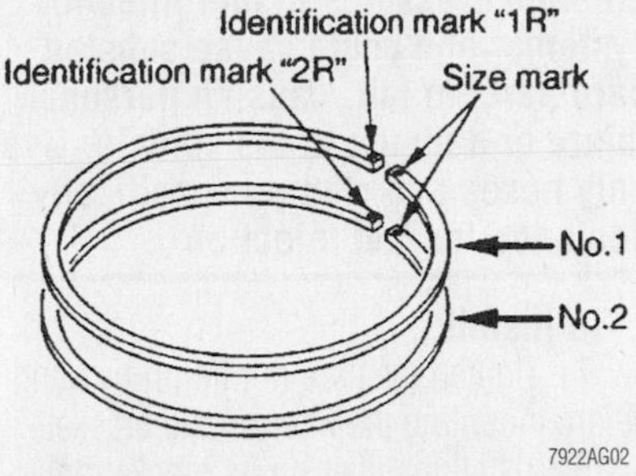

Chrysler piston ring identification mark locations

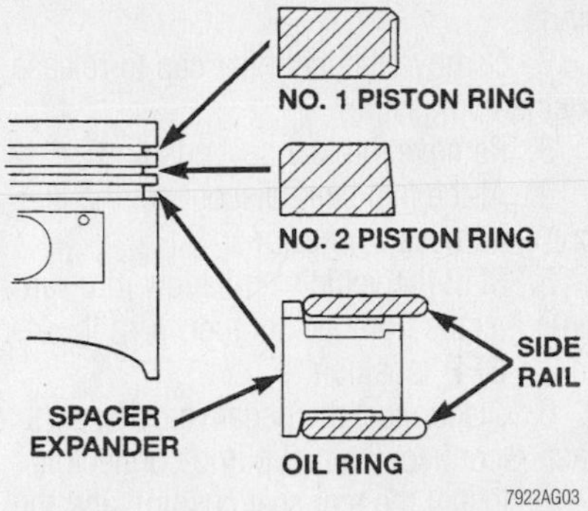

Piston ring orientation—2.0L (VIN Y) and 2.5L (VIN N) engines

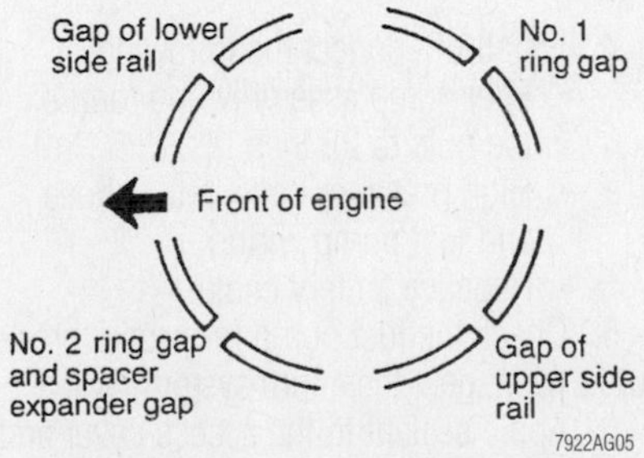

Piston ring end-gap spacing—2.0L (VIN Y) engine

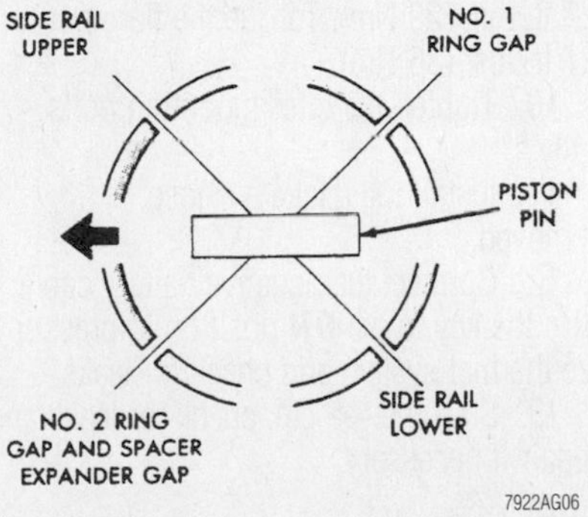

Piston ring end-gap spacing—2.5L and 3.0L engines

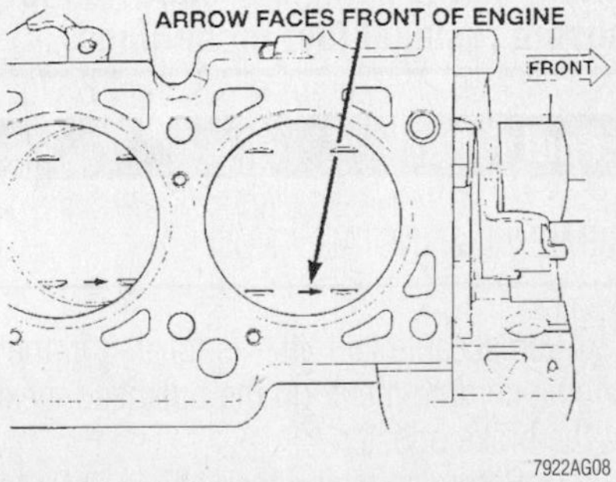

Piston positioning. The small arrows on the crown of the pistons must point toward the front of the engine—2.0L and 2.5L engines

FUEL SYSTEM

Fuel System Service Precautions

Safety is the most important factor when performing not only fuel system maintenance but any type of maintenance. Failure to conduct maintenance and repairs in a safe manner may result in serious personal injury or death. Maintenance and testing of the vehicle's fuel system components can be accomplished safely and effectively by adhering to the following rules and guidelines.

- To avoid the possibility of fire and personal injury, always disconnect the negative battery cable unless the repair or test procedure requires that battery voltage be applied.
- Always relieve the fuel system pressure before disconnecting any fuel system component (injector, fuel rail, pressure regulator, etc.), fitting or fuel line connection. Exercise extreme caution whenever relieving fuel system pressure, to avoid exposing skin, face and eyes to fuel spray. Please be advised that fuel under pressure may penetrate the skin or any part of the body that it contacts.
- Always place a shop towel or cloth around the fitting or connection prior to loosening to absorb any excess fuel due to spillage. Ensure that all fuel spillage (should it occur) is quickly removed from engine surfaces. Ensure that all fuel soaked cloths or towels are deposited into a suitable waste container.
- Always keep a dry chemical (Class B) fire extinguisher near the work area.
- Do not allow fuel spray or fuel vapors to come into contact with a spark or open flame.
- Always use a back-up wrench when loosening and tightening fuel line connection fittings. This will prevent unnecessary stress and torsion to fuel line piping.
- Always replace worn fuel fitting O-

rings with new. Do not substitute fuel hose, where fuel pipe is installed.

Fuel System Pressure

RELIEVING

1. Before servicing the vehicle, refer to the precautions in the beginning of this section.
2. Remove the fuel filler cap to release fuel tank pressure.
3. Remove the rear seat cushion.
4. At the fuel tank, disconnect the fuel pump harness connector.
5. Start the vehicle and allow it to run until it stalls from lack of fuel. Turn the key to the **OFF** position.
6. Disconnect the negative battery cable, then reconnect the fuel pump connector.
7. Install the rear seat cushion and the fuel filler cap.

CAUTION

Always wrap shop towels around a fitting that is being disconnected to absorb residual fuel in the lines.

Fuel Filter

REMOVAL & INSTALLATION

A replaceable fuel filter is located in the engine compartment, on the bulkhead, next to the brake booster.

1. Before servicing the vehicle, refer to the precautions in the beginning of this section.
2. Properly relieve the fuel system pressure.
3. Disconnect the negative battery cable and remove the air intake hose for access.
4. Hold the fuel filter housing securely with a wrench. Cover the hoses with shop towels and remove the eyebolt. Discard the gaskets.

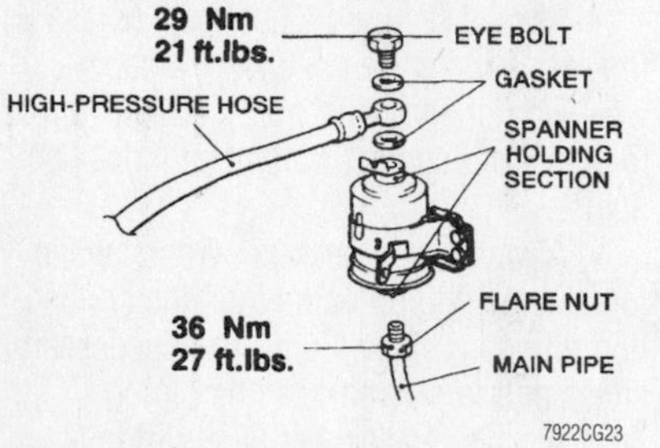

Exploded view of the fuel line-to-filter connection

5. Separate the flare nut connection at the bottom of the filter.
6. Remove the mounting bolts and the fuel filter from the vehicle.

CAUTION

Do not use conventional fuel filters, hoses or clamps when servicing fuel injection systems. They are not compatible with the injection system and the high pressures in fuel injection systems, and could cause substandard parts to fail, causing personal injury or damage to the vehicle. Use only hoses and clamps specifically designed for fuel injection.

To install:

7. Tighten the flare nut fitting by hand before mounting the filter on the bracket.
8. Install the filter on its bracket only finger-tight. Movement of the filter will ease attachment of the fuel lines.
9. Using new gaskets, connect the high pressure hose and eye bolt. While holding the fuel filter housing, torque the eye bolt to 22 ft. lbs. (29 Nm). Torque the flare nut to 27 ft. lbs. (36 Nm).
10. Tighten the filter mounting bolts fully.
11. Install the intake air hose, if removed.
12. Connect the negative battery cable, turn the key to the **ON** position to pressurize the fuel system and check for leaks.
13. Start the vehicle, check for leaks and repair if necessary.

Fuel Pump

REMOVAL & INSTALLATION

1. Before servicing the vehicle, refer to the precautions in the beginning of this section.

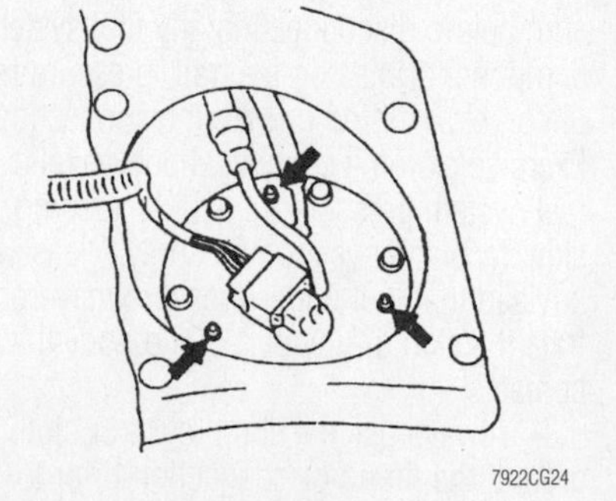

Position the fuel pump for installation by aligning the seal projections (arrows) with the fuel pump holes

Do not use conventional fuel filters, hoses or clamps when servicing fuel injection systems. They are not compatible with the injection system and could fail, causing personal injury or damage to the vehicle. Use only hoses and clamps specifically designed for fuel injection.

2. Relieve the fuel system pressure.
3. Remove or disconnect the following:
   - Negative battery cable
   - Rear seat cushion

➡Remove the seat cushion by pulling the stopper outward and lifting the lower cushion upward. There are 2 access covers underneath the seat. The panel on the far right side is for the fuel pump.

   - Access cover
   - Fuel pump wiring
   - Return hose and high pressure fuel hose
   - Fuel pump nuts and pump

To install:

➡Align the seal position projections with the holes in the fuel pump assembly

4. Install or connect the following:
   - Fuel pump assembly and torque the nuts to 22 inch lbs. (2.5 Nm)
   - High pressure hose, return hose and fuel pump wiring
   - Negative battery cable
5. Check the fuel pump for proper pressure and inspect the entire system for leaks.
6. Apply sealant to the access cover and install the cover.
7. Install the rear seat cushion.
8. Pressurize the fuel system by turning the ignition key to the **ON** position.
9. Start the engine to verify proper fuel pump performance.

Fuel Injector

REMOVAL & INSTALLATION

2.0L Engine

1. Before servicing the vehicle, refer to the precautions in the beginning of this section.
2. Relieve the fuel system pressure.
3. Remove or disconnect the following:
   - Battery
   - Air intake hose
   - High pressure fuel hose connection and discard O-ring
   - Fuel injector harness connector
   - Fuel injector electrical connectors

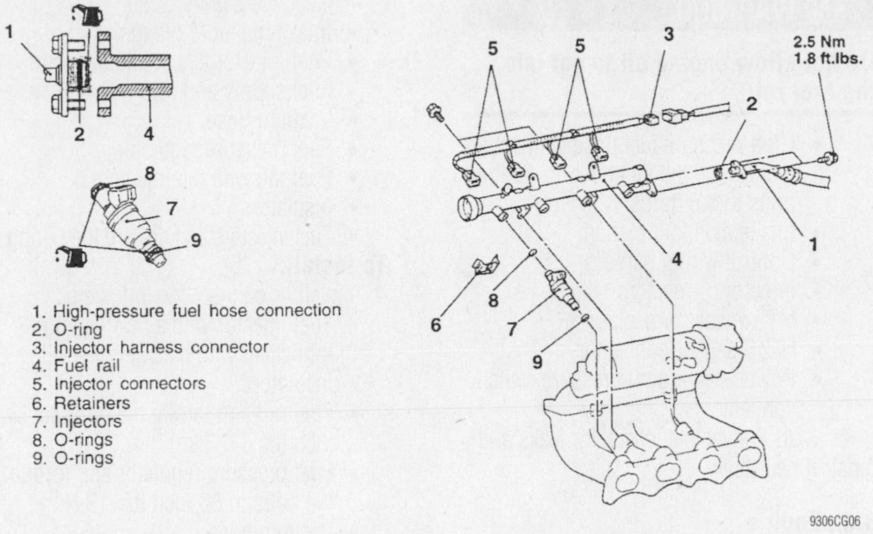

Exploded view of the fuel injector, fuel rail and related components—2.0L engine

- Fuel rail and injectors as an assembly
- Fuel injector-to-fuel rail retaining clips
- Fuel injectors and discard the O-rings

To install:

4. Install or connect the following:
   - New fuel injector O-rings
   - Fuel injectors
5. Lubricate the O-rings with clean engine oil; then, install the injectors into the fuel rail by twisting them (left and right) to make sure that they turn smoothly in the seat.

⁂ WARNING

Do not allow engine oil to get into the fuel rail.

- Fuel injector-to-fuel rail retaining clips
- Fuel injector electrical connectors
- Fuel rail
- Fuel injector harness connector
- New high-pressure fuel hose O-ring lubricated with clean engine oil

⁂ WARNING

Do not allow engine oil to get into the fuel rail.

- High pressure fuel hose connection and torque the fuel hose-to-fuel rail bolts to 1.8 ft. lbs. (2.5 Nm)
- Air intake hose
- Battery

6. Start the vehicle, check for leaks and repair if necessary.

2.5L Engine

1. Before servicing the vehicle, refer to the precautions in the beginning of this section.
2. Relieve the fuel system pressure.
3. Remove or disconnect the following:
   - Power steering oil pressure switch connector
   - Heated Oxygen (HO_2) sensor connector
   - Intake Air Temperature (IAT) sensor connector
   - Injector connectors
   - Control wiring harness
   - Intake manifold plenum
   - Fuel injector electrical connectors
   - High pressure fuel hose connection
   - O-ring from the high pressure fuel hose connection
   - Fuel rail and injectors as an assembly
   - Fuel injector retaining clips
   - Fuel injector(s) and discard the O-rings and grommet

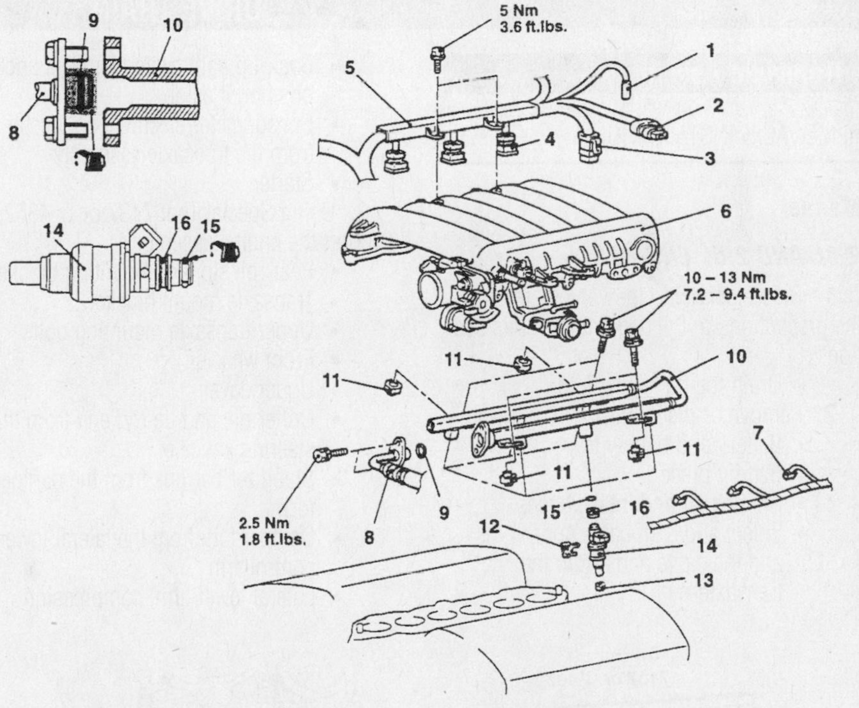

1. Power steering oil pressure switch connector
2. Heated oxygen sensor connector
3. Intake air temperature sensor connector
4. Injector connectors
5. Control wiring harness
6. Intake manifold plenum
7. Injector connectors
8. High-pressure fuel hose connection
9. O-ring
10. Fuel rail
11. Insulators
12. Injector supports
13. Insulators
14. Injectors
15. O-rings
16. Grommets

9306CG07

Exploded view of the fuel injector, fuel rail and related components—2.5L engine

To install:

4. Lubricate the O-rings and grommet(s) with clean engine oil; then, install the injectors into the fuel rail by twisting them (left and right) to make sure that they turn smoothly in the seat.

WARNING

Do not allow engine oil to get into the fuel rail.

5. Install or connect the following:
   - New fuel injector O-rings and grommet(s)
   - Fuel injectors
   - Fuel injector-to-fuel rail retaining clips
   - Fuel injector electrical connectors
   - Fuel rail and torque the fuel rail-to-intake manifold bolts to 7.2–9.4 ft. lbs. (10–13 Nm)
   - New high-pressure fuel hose O-ring
6. Lubricate the O-ring with clean engine oil.

WARNING

Do not allow engine oil to get into the fuel rail.

   - High pressure fuel hose connection and torque the fuel hose-to-fuel rail bolts to 1.8 ft. lbs. (2.5 Nm)
   - Intake manifold plenum
   - Control wiring harness
   - Injector connectors
   - IAT sensor connector
   - HO_2S sensor connector
   - Power steering oil pressure switch connector
7. Start the vehicle, check for leaks and repair if necessary.

3.0L Engine

1. Before servicing the vehicle, refer to the precautions in the beginning of this section.
2. Properly relieve the fuel system pressure.
3. Remove or disconnect the following:
   - Negative battery cable
   - Intake manifold plenum
   - Fuel injector electrical connector
   - Fuel supply and return hoses
   - Vacuum hose
   - Fuel pressure regulator
   - Fuel rail with the injectors
   - Insulators
   - Fuel injectors and discard the O-rings

To install:

4. Install or connect the following:
   - Fuel injector with a new O-ring lubricated with clean engine oil
   - Insulators
   - Fuel rail and torque the bolts to 13 inch lbs. (12 Nm)
   - Fuel pressure regulator and torque the bolts to 80 inch lbs. (9 Nm)
   - Vacuum hose
   - Fuel return and supply hoses
   - Fuel injector electrical connector
   - Intake manifold plenum
   - Negative battery cable
5. Start the vehicle, check for leaks and repair if necessary.

DRIVE TRAIN

Transaxle Assembly

REMOVAL & INSTALLATION

Manual

2.0L AND 2.5L ENGINES

1. Before servicing the vehicle, refer to the precautions in the beginning of this section.
   - Drain the transaxle fluid.
2. Remove or disconnect the following:
   - Battery and battery tray
   - Battery brace
   - Air cleaner and intake hoses
   - Select and shift cable cotter pins and the cable ends from the transaxle
   - Back-up light switch harness and position it aside
   - Speedometer electrical connector, from the transaxle assembly
   - Starter
3. Using Special Tool 7137 or C-4852, support the engine assembly.
   - Rear roll stopper mounting bracket
   - Transaxle mount bracket
   - Upper transaxle mounting bolts
   - Front wheels
   - Under cover
   - Cotter pin and tie rod end from the steering knuckle
   - Stabilizer bar link from the damper fork
   - Damper fork from the lateral lower control arm
   - Lateral lower arm, compression lower arm and lower ball joints from the steering knuckle
   - Halfshafts from the transaxle and secure aside
   - Clutch release cylinder connection and move aside without disconnecting the hydraulic line
   - Bell housing cover
   - Engine front roll stopper through-bolt
   - Centermember
4. Support the transaxle, using a transmission jack, and remove the transaxle lower coupling bolt.

➡The coupling bolt threads from the engine side, into the transaxle, and is located just above the halfshaft opening.

5. Slide the transaxle rearward and carefully lower it from the vehicle.

To install:

6. Install or connect the following:
   - Transaxle and torque the transaxle-to-engine bolts to 70 ft. lbs. (95 Nm)
   - Bell housing cover and torque the bolts to 84 inch lbs. (9 Nm)
   - Centermember and torque the front mounting bolts to 65 ft. lbs. (88 Nm) and the rear bolt to 54 ft. lbs. (73 Nm)
   - Front engine roll stopper through-

7922CG25

For transaxle removal, properly support the engine assembly as shown

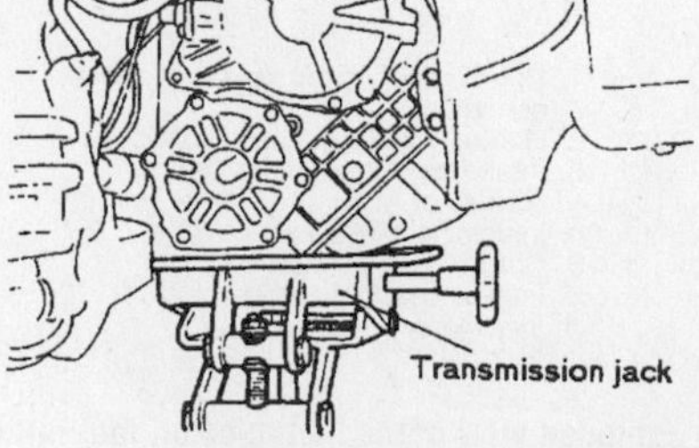

7922CG26

Also, use a transmission jack to support the transaxle assembly

bolt and lightly tighten. Once the full weight of the engine is on the mounts, torque the bolt to 42 ft. lbs. (57 Nm).
- Clutch release cylinder
- Halfshafts using new circlips

⁂ WARNING

When installing the halfshaft, keep the inboard joint straight in relation to the axle, to avoid damaging the oil seal lip of the transaxle, with the serrated part of the halfshaft.

- Tie rod and ball joints to the steering knuckle. Torque the ball joint self-locking nuts to 48 ft. lbs. (65 Nm). Torque the tie rod end nut to 21 ft. lbs. (28 Nm) and secure with a new cotter pin.
- Damper fork to the lower control arm and torque the through-bolt to 65 ft. lbs. (88 Nm)
- Stabilizer link to the damper fork and torque the self-locking nut to 29 ft. lbs. (39 Nm)
- Under cover
- Wheels
- Transaxle mount bracket to the transaxle and torque the nuts to 32 ft. lbs. (43 Nm)
- Rear roll stopper mounting bracket

7. Remove the engine support. Tighten the transaxle mount through-bolt to 51 ft. lbs. (69 Nm) and tighten the front engine roll stopper through-bolt.
- Upper transaxle mounting bolts and torque them to 35 ft. lbs. (48 Nm)
- Starter
- Back-up light switch and speedometer connector
- Select and shift cables and new cotter pins
- Air cleaner and air intake hose
- Battery tray and battery
- Battery stay

8. Be sure the vehicle is level, and refill the transaxle with Mopar® MS9417 MTX fluid, part number 4773167.

9. Check the transaxle for proper operation. Be sure the reverse lights come ON when in reverse.

3.0L ENGINE

1. Before servicing the vehicle, refer to the precautions in the beginning of this section.
2. Drain the transaxle fluid.
3. Remove or disconnect the following:
- Battery and tray
- Air cleaner assembly
- Front exhaust pipe
- Shift and selection cable connections
- Back—up light switch connection
- Vehicle Speed Sensor (VSS) connector
- Starter
- Clutch release cylinder connector
- Transaxle assembly upper coupling bolts
- Centermember
- Rear roll stopper
- Transaxle mount bracket and stopper
- Stabilizer link
- Wheel speed sensor, if equipped
- Brake hose clamp
- Tie rod end
- Lower arm
- Clutch release bearing engagement
- Driveshaft connection
- Upper oil pan bolt
- Bell housing cover
- Transaxle assembly lower coupling bolts
- Transaxle assembly

To install:

4. Install or connect the following:
- Transaxle assembly and torque the lower coupling bolts to 36 ft. lbs. (48 Nm)
- Bell housing cover and torque the cover—to—engine bolt to 80 inch lbs. (9 Nm) and the cover—to—transaxle bolt to 19 ft. lbs. (26 Nm).
- Upper oil pan bolt
- Driveshaft connection
- Clutch release bearing engagement
- Lower arm and torque the nut to 80 ft. lbs. (108 Nm)
- Tie rod end and torque the nut to 21 ft. lbs. (28 Nm)
- Wheel speed sensor connector, if equipped
- Stabilizer link and torque the nut to 33 ft. lbs. (45 Nm)
- Transaxle mount bracket and torque the nut to 42 ft. lbs. (57 Nm)
- Tranasxle mount stopper and torque the nut to 60 ft. lbs. (81 Nm)
- Rear roll stopper and torque the nut to 33 ft. lbs. (45 Nm)
- Centermember
- Clutch release cylinder and torque to 13 ft. lbs. (18 Nm)
- Starter motor and electrical connectors
- VSS connector
- Back—up light switch connector
- Shift and selection cable connections
- Front exhaust pipe
- Air cleaner assembly
- Battery and tray

5. Fill the transaxle fluid to the proper level.

6. Start the vehicle, check for leaks and repair if necessary.

Automatic

1. Before servicing the vehicle, refer to the precautions in the beginning of this section.
2. Drain the transaxle fluid.
3. Remove or disconnect the following:
- Battery and battery tray
- Battery brace
- Air cleaner and intake hoses
- Shifter lever-to-transaxle nut, cable retaining clip and cable from the transaxle
- Shifter cable mounting bracket
- Speedometer, solenoid, neutral safety switch (inhibitor switch), the pulse generator, kickdown servo switch and oil temperature sensor electrical connectors
- Transaxle oil cooler lines
- Transaxle dipstick and tube
- Starter and using Special Tool 7137 or C-4852, support the engine assembly
- Rear roll stopper mounting bracket
- Transaxle mount bracket
- Upper transaxle mounting bolts
- Front wheels
- Left side undercover
- Tie rod end from the steering knuckle
- Stabilizer bar link from the damper fork
- Damper fork from the lateral lower control arm
- Lateral lower arm and compression lower arm, lower ball joints from the steering knuckle
- Halfshafts from the transaxle and secure aside
- Bell housing cover
- Engine front roll stopper through-bolt
- Centermember
- Flexplate-to-torque converter bolts. Rotate the crankshaft to bring the bolts into a position for removal, one at a time.

➡To make installation easier, use chalk or paint to make matchmarks on the torque converter and flexplate. These marks will be used at assembly to realign the assembly, keeping these parts in balance.

⁂ WARNING

After removing the bolts, push the torque converter toward the transaxle. This will prevent the converter from remaining in contact with the engine, possibly damaging the converter.

4. Support the transaxle using a transmission jack (at the side of the case, NOT at the pan), and remove the transaxle lower coupling bolt.

➡The coupling bolt is inserted from the engine side into the transaxle and is located just above the halfshaft opening.

5. Slide the transaxle rearward and carefully lower it from the vehicle.

To install:

6. After the torque converter has been mounted on the transaxle, install the transaxle assembly to the engine. Install the mounting bolts and tighten to 70 ft. lbs. (95 Nm).
7. Align the balance matchmarks and torque the torque converter-to-flexplate bolts to 55 ft. lbs. (75 Nm).
8. Install or connect the following:
   - Bell housing cover and torque the bolts to 108 inch lbs. (12 Nm)
   - Centermember and torque the front mounting bolts to 65 ft. lbs. (88 Nm) and the rear bolts to 51–58 ft. lbs. (69–78 Nm).
   - Front engine roll stopper through-bolt and lightly tighten. Once the full weight of the engine is on the mounts, torque the bolt to 42 ft. lbs. (56 Nm).
   - Halfshafts using new circlips

⁂ WARNING

When installing the halfshaft, keep the inboard joint straight in relation to the axle to avoid damaging the oil seal lip of the transaxle with the splined part of the halfshaft.

   - Tie rod and ball joints to the steering knuckle. Torque the ball joint self-locking nuts to 48 ft. lbs. (65 Nm), the tie rod end nut to 21 ft. lbs. (28 Nm) and secure with a new cotter pin.
   - Damper fork to the lower control arm and torque the through-bolt to 65 ft. lbs. (88 Nm)
   - Stabilizer link to the damper fork and torque the self-locking nut to 29 ft. lbs. (39 Nm)
   - Left side undercover
   - Wheels
   - Transaxle mount bracket and torque the mounting nuts to 32 ft. lbs. (43 Nm)
   - Rear roll stopper mounting bracket and remove the engine support
   - Transaxle mount through-bolt to 51 ft. lbs. (69 Nm) and torque the front engine roll stopper bolt
   - Upper transaxle mounting bolts and tighten to 35 ft. lbs. (48 Nm)
   - Starter
   - Dipstick tube and dipstick
   - Shifter cable mounting bracket
   - Shifter lever and torque the nut to 14 ft. lbs. (19 Nm)
   - Oil cooler lines and secure with clamps
   - Speedometer, solenoid, neutral safety switch (inhibitor switch), the pulse generator, kickdown servo switch and oil temperature sensor electrical connectors
   - Air cleaner and air intake hose
   - Battery tray and battery

9. Refill the transaxle with MOPAR® ATF PLUS transmission fluid. Start the engine and allow it to idle for 2 minutes. Apply the parking brake and move the selector through each gear position, ending in **N**. Recheck fluid level and add if necessary. Fluid level should be between the marks in the **HOT** range on the dipstick.
10. Check the transaxle for proper operation.

Clutch

ADJUSTMENT

Pedal Height and Free-Play

1. Measure the clutch pedal height from the face of the pedal pad to the bulkhead. Compare the measured value with the desired distance of 7.0–7.09 in. (175–180mm).
2. Measure the clutch pedal clevis pin play at the face of the pedal pad. Press the pedal lightly until resistance is met, and measure this distance. The clutch pedal clevis pin play should be within 0.040–0.120 in. (1–3mm).
3. If the clutch pedal height or clevis pin

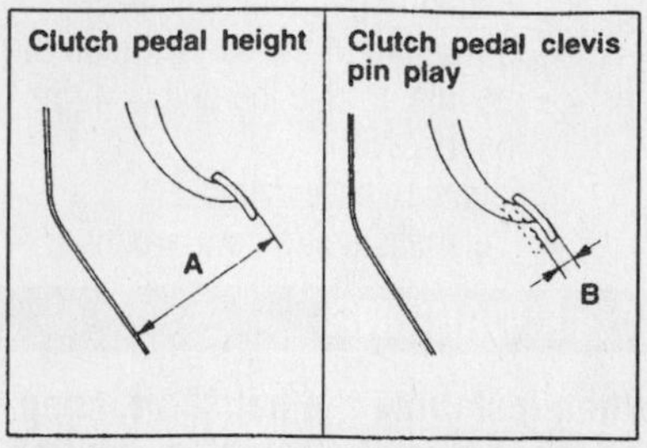

Clutch pedal height and free-play adjustment measurements

play, is not within the standard values, adjust as follows:

   a. If not equipped with cruise control, turn and adjust the stop bolt so the pedal height is the standard value, then tighten the locknut.
   b. If equipped with cruise control system, disconnect the clutch switch connector and turn the switch to obtain the standard clutch pedal height. Then, lock by tightening the locknut.
   c. Turn the pushrod to adjust the clutch pedal clevis pin play to agree with the standard value and secure the pushrod with the locknut.

➡When adjusting the clutch pedal height or the clutch pedal clevis pin play, be careful not to push the pushrod toward the master cylinder.

   d. Check that when the clutch pedal is depressed all the way, the interlock switch switches over from **ON** to **OFF**.

4. Move the clutch pedal until the resistance begins to increase; measure between this point and the pedal resting point to determine the clutch pedal free-play. The clutch pedal free-play measurement should be between 0.240–0.510 in. (6–13mm). With the pedal fully disengaged, check the distance between the bulkhead and the top of the pedal pad. The measurement should be 2.760 in. (70mm) or more.
5. If the measurements are not within specification, bleed the clutch hydraulic system. If after bleeding the measurements are still not within specified range, there is a faulty component in the system, which must be replaced.

REMOVAL & INSTALLATION

2.0L and 2.5L Engines

1. Before servicing the vehicle, refer to the precautions in the beginning of this section.
2. Remove or disconnect the following:

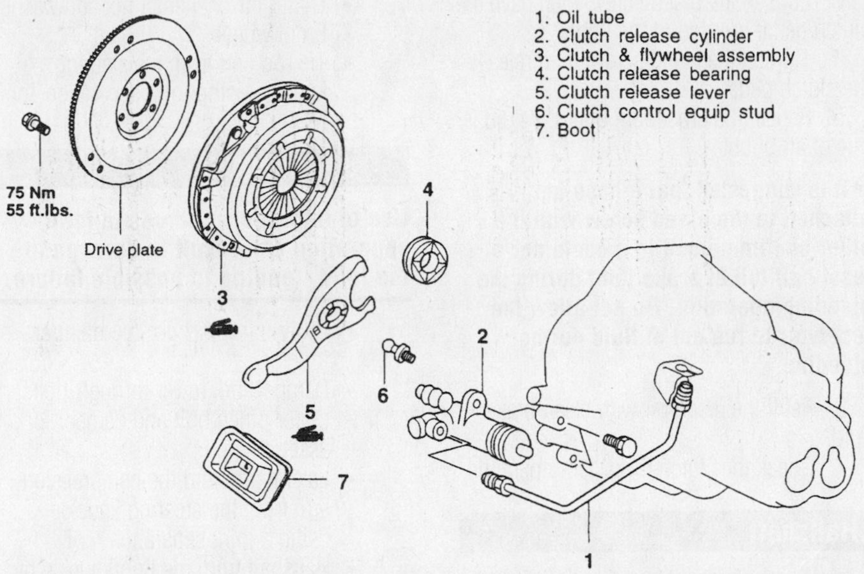

Exploded view of the clutch assembly

- Negative battery cable
- Transaxle
- Pressure plate-to-flywheel bolts, pressure plate and clutch disc

➡If the pressure plate is to be reused, loosen the bolts in a diagonal pattern, one or 2 turns at a time. This will prevent warping the clutch cover assembly.

3. Remove the return clip and the pressure plate release bearing. Do not use solvent to clean the bearing.
4. Inspect the clutch release fork and fulcrum for damage or wear. If necessary, remove the release fork and the fulcrum from the transaxle.
5. Carefully inspect the condition of the clutch components and replace any worn or damaged parts.

To install:

6. Inspect the flywheel for heat damage or cracks. Resurface or replace the flywheel as required.
7. Install or connect the following:
   - Flywheel using new bolts
   - Fulcrum, if removed, and tighten
   - Release fork

➡Apply a coating of multi-purpose grease to the point of contact with the fulcrum and the point of contact with the release bearing. Apply a coating of multi-purpose grease to the end of the release cylinder's pushrod and the pushrod hole in the release fork.

⁂ WARNING

When installing the clutch, apply grease to each part, but be careful not to apply excessive grease. Excessive grease will cause clutch slippage and shudder.

8. Apply multi-purpose grease to the clutch release bearing. Pack the bearing inner surface and the groove with grease. Do not apply grease to the resin portion of the bearing. Place the bearing in position and install the return clip.
9. Apply a coating of grease to the clutch disc splines, then use a brush to rub it in the grooves. Using a clutch disc alignment tool, position the clutch disc on the flywheel. Install the retainer bolts and tighten a little at a time, in a diagonal sequence.
10. Install the transaxle assembly and check the fluid level.
11. Connect the negative battery cable.
12. Verify proper clutch operation.

3.0L Engine

1. Before servicing the vehicle, refer to the precautions in the beginning of this section.
2. Remove or disconnect the following:
   - Negative battery cable
   - Clutch fluid line bracket
   - Insulator, washer and clutch tube
   - Union, union bolt and gasket
   - Valve and spring
   - Clutch release cylinder
   - Clutch cover and disc

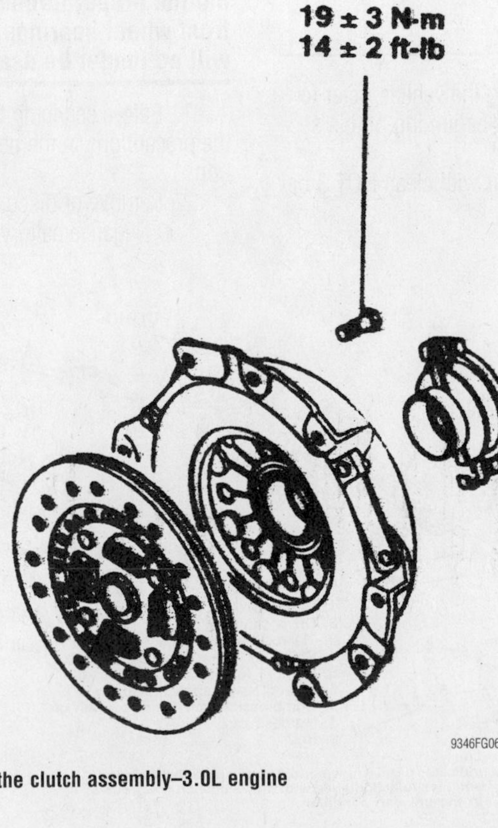

Exploded view of the clutch assembly–3.0L engine

- Release fork shaft
- Left support spring
- Release fork
- Clutch release bearing
- Release fork boot and fulcrum
- Release fork boot and maintenance cover

To install:

3. Install or connect the following:
   - Release fork boot and maintenance cover
   - Clutch release bearing
   - Release fork
   - Left support spring and release fork shaft
   - Clutch disc and cover and torque the bolts to 14 ft. lbs. (19 Nm)
   - Clutch release cylinder
   - Valve and spring
   - Union, bolt and gasket and torque the bolt to 17 ft. lbs. (23 Nm)
   - Insulator, washer and clutch tube and torque the fastener to 11 ft. lbs. (15 Nm)
   - Fluid line bracket and torque the bolt to 14 ft. lbs. (19 Nm)
   - Negative battery cable
4. Check and top off the fluid level, if necessary.

Hydraulic Clutch System

BLEEDING

1. Before servicing the vehicle, refer to the precautions in the beginning of this section.
2. Fill the reservoir with clean DOT 3 or DOT 4 brake fluid.
3. Loosen the bleed screw, then have the clutch pedal pressed to the floor.
4. Tighten the bleed screw, then release the clutch pedal.
5. Repeat the procedure until the fluid is free of air bubbles.

➡It is suggested that a hose be attached to the bleed screw with the other end immersed in a container at least half full of brake fluid during the bleeding operation. Do not allow the reservoir to run out of fluid during bleeding.

6. Refill the reservoir with clean brake fluid.
7. Check the clutch for proper operation.

Halfshaft

REMOVAL & INSTALLATION

2.0L and 2.5L Engines

**** WARNING**

If the vehicle is going to be rolled while the halfshafts are out of the vehicle, obtain 2 outer CV-joints or proper equivalent tools and install to the hubs. If the vehicle is rolled without the proper torque applied to the front wheel bearings, the bearings will no longer be usable.

1. Before servicing the vehicle, refer to the precautions in the beginning of this section.
2. Remove or disconnect the following:
   - Negative battery cable

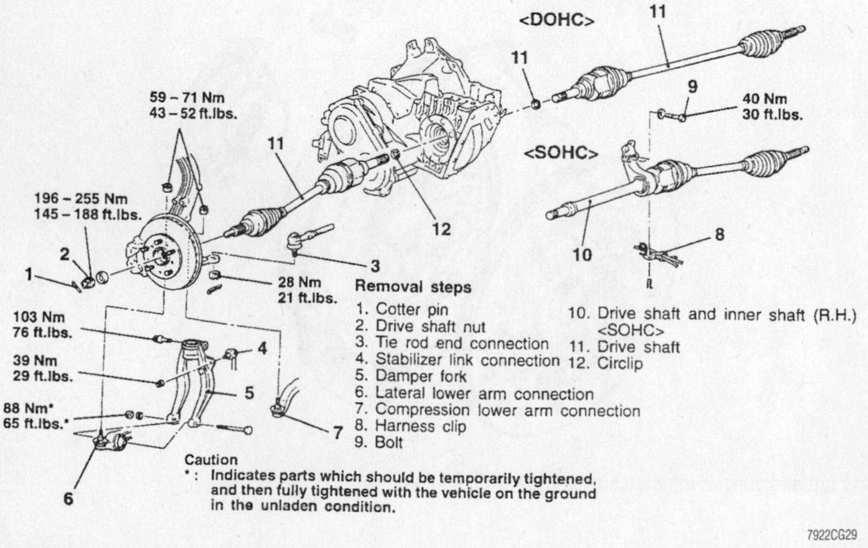

Exploded view of the left and right halfshaft assemblies' mounting

- Cotter pin, halfshaft nut and washer
- Front wheel
- Tie rod end from the steering knuckle using joint separation Tool MB991113

**** WARNING**

Use of improper methods of joint separation can result in damage to the joint, leading to possible failure.

- Sway bar link from the damper fork
- Damper fork lower through-bolts, upper pinch bolt and damper fork assembly
- Lateral arm and the compression arm from the steering knuckle using a joint separation tool
- Halfshaft from the hub/knuckle by setting up a puller on the outside wheel hub and pushing the halfshaft from the front hub
- Halfshaft from the transaxle by inserting a prybar between the transaxle case and the halfshaft

3. If equipped with a center bearing on the right halfshaft, remove the center bearing bracket-to-chassis bolts; then, tap lightly on the center bearing bracket to separate the inner shaft from the transaxle.

**** WARNING**

Do not pull on the shaft. Doing so damages the inboard joint. Do not insert the prybar too far or the oil seal in the case may be damaged.

To install:

4. Inspect the halfshaft boot for damage or deterioration. Check the ball joints and splines for wear.
5. Replace the circlips on the ends of the halfshaft(s).
6. Insert the halfshaft into the transaxle. Be sure it is fully seated.
7. If equipped with a center bearing on the right halfshaft, install the center bearing bracket-to-chassis bolts and torque to 30 ft. lbs. (40 Nm).
8. Pull the knuckle assembly outward and install the other end of the halfshaft into the hub.
9. Install the washer so the chamfered edge faces outward. Install the halfshaft nut and tighten temporarily.
10. Connect the lateral arm and the compression arm to the steering knuckle. Torque the self-locking nuts to 43–52 ft. lbs. (59–71 Nm).
11. Install or connect the following:

- Damper fork then torque the lower through-bolt to 65 ft. lbs. (88 Nm) and the upper pinch bolt to 76 ft. lbs. (103 Nm)
- Tie rod end to the steering knuckle and torque the nut to 17–25 ft. lbs. (24–33 Nm) and install a new cotter pin
- Sway bar link to the damper fork and torque the nut to 29 ft. lbs. (39 Nm)
- Lockwasher and axle nut; then, torque the axle nut with the Special Tool MB990767 to hold the hub from turning to 145–188 ft. lbs. (200–260 Nm)

➡Before securely tightening the axle nut, make sure there is no load on the wheel bearings.

- New cotter pin
- Front wheel
- Negative battery cable

12. Fill the transaxle to the proper level.
13. Test drive the vehicle and check for proper operation.

3.0L Engine

1. Before servicing the vehicle, refer to the precautions in the beginning of this section.
2. Remove or disconnect the following:
   - Negative battery cable
   - Front wheel
   - Vehicle Speed Sensor (VSS) cable, if equipped
   - Brake hose clip and cotter pin
   - Driveshaft nut
   - Lower arm ball joint connection
   - Tie rod end
   - Stabilizer link
   - Driveshaft and circlip

To install:

3. Install or connect the following:
   - Driveshaft
   - Stabilizer link and torque the nut to 33 ft. lbs. (44 Nm)
   - Tie rod end and torque the nut to 21 ft. lbs. (28 Nm)
   - New cotter pin to the tie rod end
   - Lower arm ball joint
   - Driveshaft nut and torque it to a maximum of 167 ft. lbs. (226 Nm)
   - New cotter pin to the driveshaft nut
   - Brake hose clip
   - VSS cable, if equipped
   - Front wheel
   - Negative battery cable

CV-Joints

OVERHAUL

✲✲ WARNING

The Birfield joint assembly, located on the wheel side of the halfshaft, is not to be disassembled; repair of this joint is only by replacement of the halfshaft.

Tri-Pot Joint

2.0L (DOHC) ENGINE

1. Before servicing the vehicle, refer to the precautions in the beginning of this section.
2. Remove halfshaft and place it in a soft jawed vise.
3. Disassemble or remove:
   - Tri-pot boot bands
   - Tri-pot case and remove any grease from the case
   - Halfshaft snapring
   - Tri-pot spider assembly

✲✲ WARNING

Do not disassemble the spider assembly.

   - Tri-pot boot

✲✲ WARNING

If the boot is to be reused, wrap plastic tape around the shaft splines to protect the boot from damage.

4. If necessary, remove the dynamic damper bands and slide the damper from the halfshaft.

To install:

5. If removed, install the dynamic damper by performing the following procedure:
   a. Slide the damper onto the halfshaft with new bands.
   b. Position the damper so the distance from the front of the birfield joint to

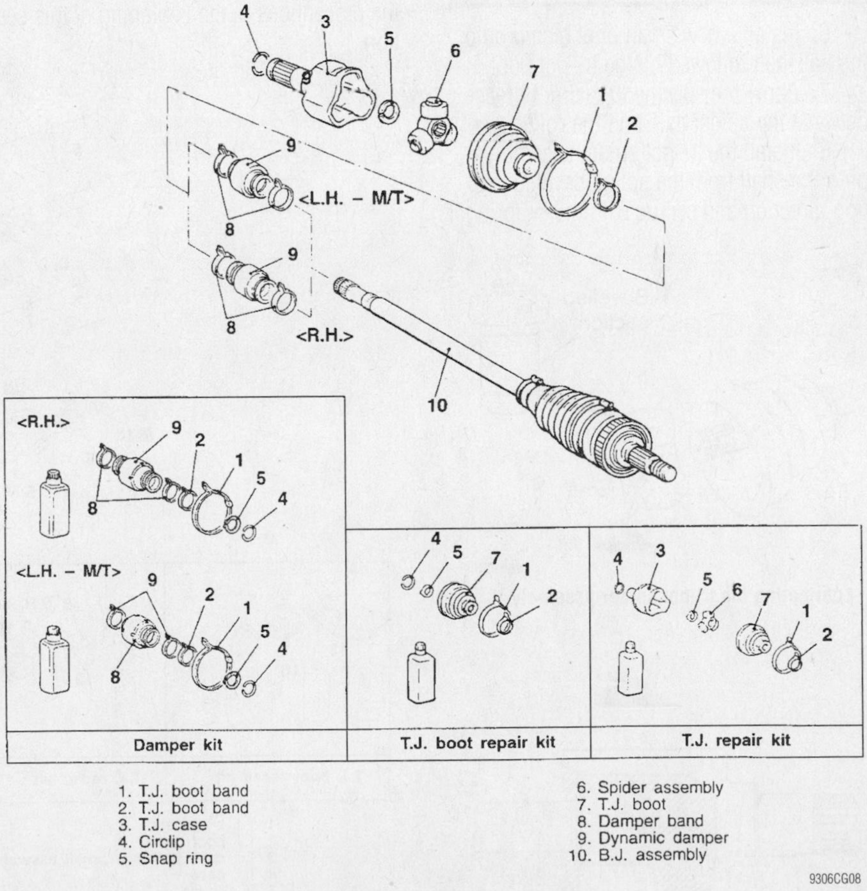

1. T.J. boot band
2. T.J. boot band
3. T.J. case
4. Circlip
5. Snap ring
6. Spider assembly
7. T.J. boot
8. Damper band
9. Dynamic damper
10. B.J. assembly

Exploded view of the halfshaft assemblies—2.0L (DOHC) Engine

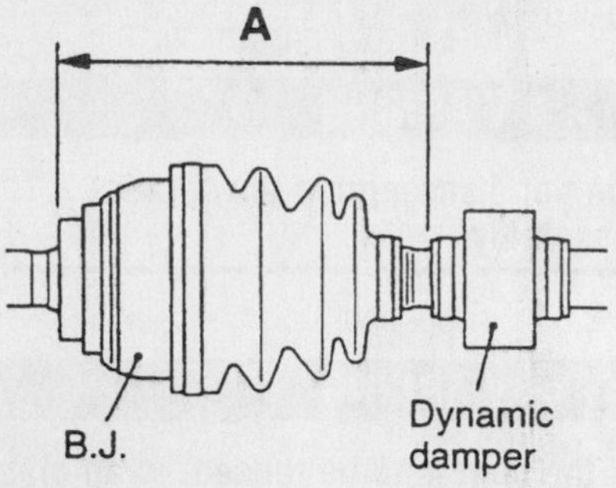

9306CG09

Positioning the dynamic damper on the halfshaft assemblies—2.0L (DOHC) Engine

the front edge of the damper assembly is 14.60–14.84 in. (371–377mm) for the right halfshaft or 7.52–7.76 in. (191–197mm) for the left halfshaft.

c. Tighten and secure the damper bands.

⁂ WARNING

Wrap plastic tape around the shaft splines to protect the boot from damage.

6. Install a new small boot clamp onto the halfshaft followed by the tri-pot boot.

7. Apply the specified repair kit grease between the spider axle and the roller.

8. Install the tri-pot spider assembly onto the shaft from the spline beveled section direction and secure with the snapring.

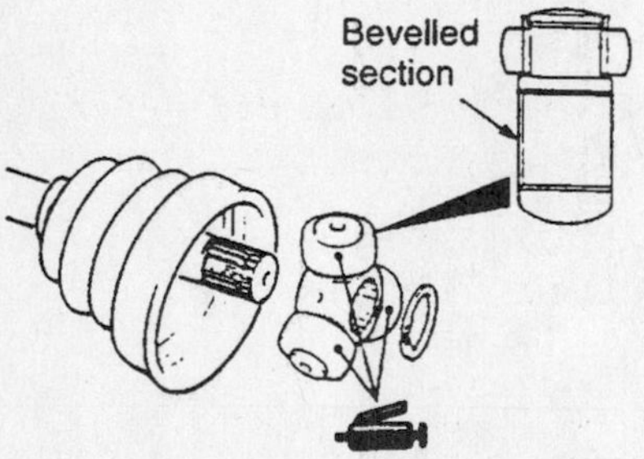

9306CG10

Lubricating the tri-pot spider assembly

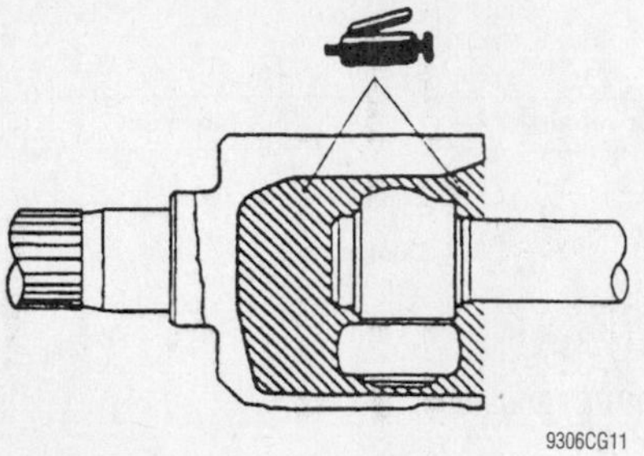

9306CG11

Lubricating the tri-pot case assembly installed

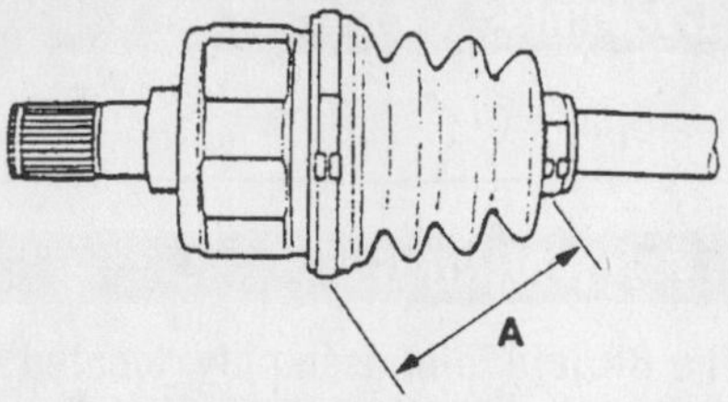

9306CG28

Measuring the tri-pot joint boot

9. Distribute a portion of the 3.7 oz. (105 g) specified repair kit grease into the tri-pot case, insert the spider assembly and add the remaining grease.

10. Install the tri-pot boot by performing the following procedure:

a. Position the boots large end on the tri-pot case.

b. Position the large and small boot clamps onto the boot.

c. Adjust the boot so that the bands are spaced at 3.03–3.27 in. (77–83mm).

d. Tighten the band securely.

2.5L (SOHC) ENGINE

1. Before servicing the vehicle, refer to the precautions in the beginning of this section.

2. Remove halfshaft and place it in a soft jawed vise.

3. Disassemble or remove:

- Tri-pot boot bands
- Tri-pot case for left halfshaft or tri-pot case/inner shaft assembly for right halfshaft

➡Wipe the grease from the tri-pot case.

- Halfshaft snapring
- Tri-pot spider assembly

⁂ WARNING

Do not disassemble the spider assembly.

- Tri-pot boot

⁂ WARNING

If the boot is to be reused, wrap plastic tape around the shaft splines to protect the boot from damage.

4. If working with the right halfshaft and it is necessary to replace the tri-pot case, press the case from the inner shaft assembly.

To install:

5. If installing a new tri-pot case onto

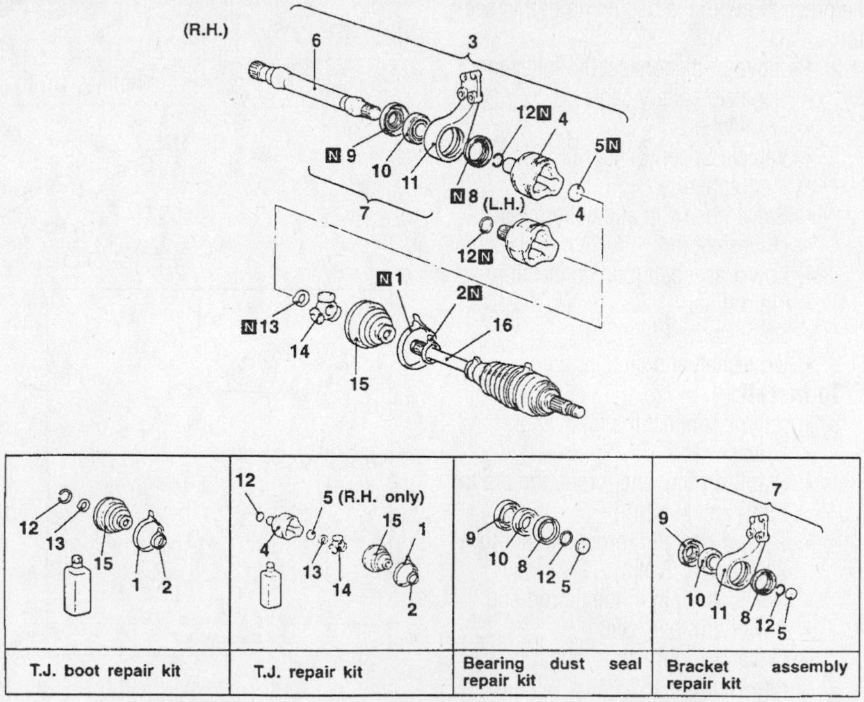

1. T.J. boot band
2. T.J. boot band
3. T.J. case and inner shaft assembly
4. T.J. case
5. Seal plate
6. Inner shaft
7. Bracket assembly
8. Dust seal outer
9. Dust seal inner
10. Center bearing
11. Center bearing bracket
12. Circlip
13. Snap ring
14. Spider assembly
15. T.J. boot
16. B.J. assembly

9306CG12

Exploded view of the halfshaft assemblies—2.5L (SOHC) Engine

the right halfshaft, perform the following procedure:

a. Lubricate the inner shaft splines with Multi-Mileage Grease No. 2525035.

b. Press the inner shaft assembly into tri-pot case.

c. Secure the tri-pot case with special Tool MB991248 on a hydraulic press with the case facing upward.

d. Position a new seal plate in the center of the tri-pot case.

e. Place a 1.18 in. (30mm) pipe on the seal plate and press the seal into the tri-pot case.

⁂ WARNING

Wrap plastic tape around the shaft splines to protect the boot from damage.

6. Install a new small boot clamp onto the halfshaft followed by the tri-pot boot.

7. Apply the specified repair kit grease between the spider axle and the roller.

8. Install the tri-pot spider assembly onto the shaft from the spline beveled section direction and secure with the snapring.

9. Distribute a portion of the 4.23 oz. (120 g) specified repair kit grease into the tri-pot case, insert the spider assembly and add the remaining grease.

10. Install the tri-pot boot by performing the following procedure:

a. Position the boots large end on the tri-pot case.

b. Position the large and small boot clamps onto the boot.

c. Adjust the boot so that the bands are spaced at 3.03–3.27 in. (77–83mm).

d. Tighten the band securely.

Center Bearing

2.5L (SOHC) ENGINE

1. Remove the right halfshaft and place it in a soft-jawed vise.

2. Remove the tri-pot spider assembly from the tri-pot case.

3. Remove the tri-pot case by performing the following procedure:

a. Position the inner shaft/tri-pot case assembly on a hydraulic press supported by Tool MB991248 with the tri-pot case facing upward.

b. Position a bar inside the tri-pot case, on the end of the inner shaft and press the case from the inner shaft assembly.

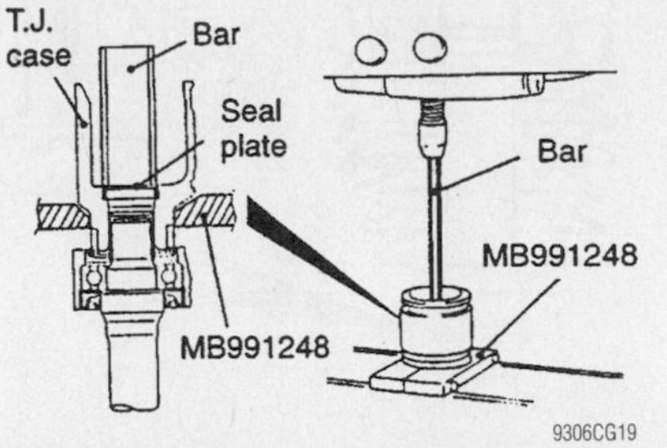

Removing the tri-pot case from the inner shaft assembly—Right halfshaft with 2.5L (SOHC) Engine

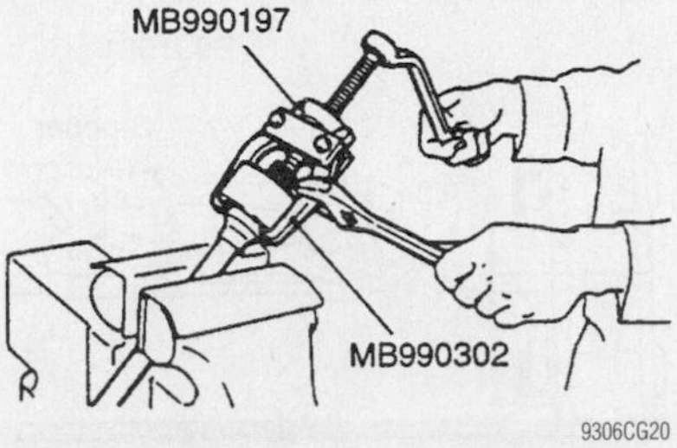

Removing the center bearing bracket from the inner shaft assembly—Right halfshaft with 2.5L (SOHC) Engine

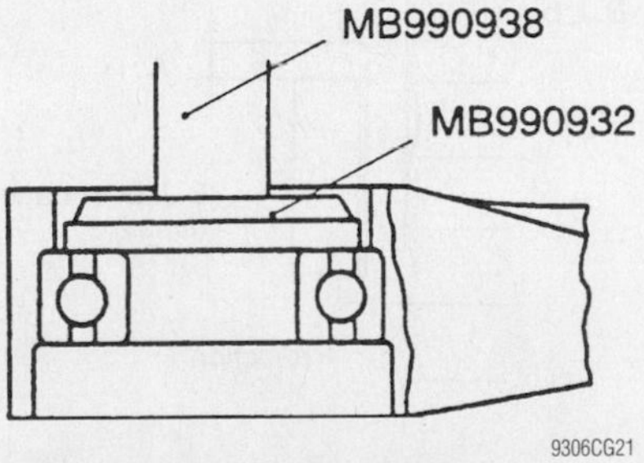

Removing the center bearing from the center bearing bracket—Right halfshaft with 2.5L (SOHC) Engine

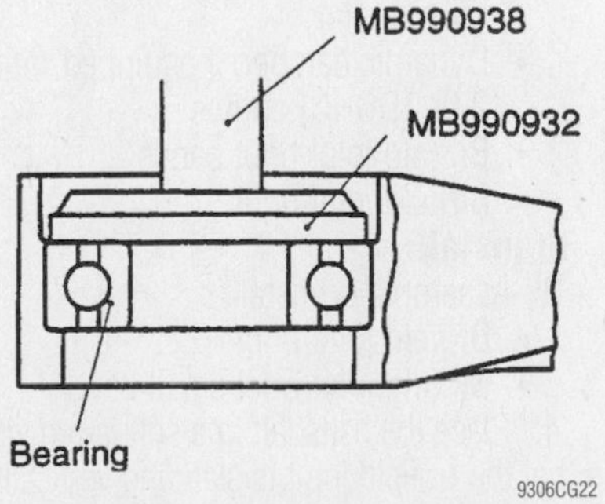

Installing the center bearing to the center bearing bracket—Right halfshaft with 2.5L (SOHC) Engine

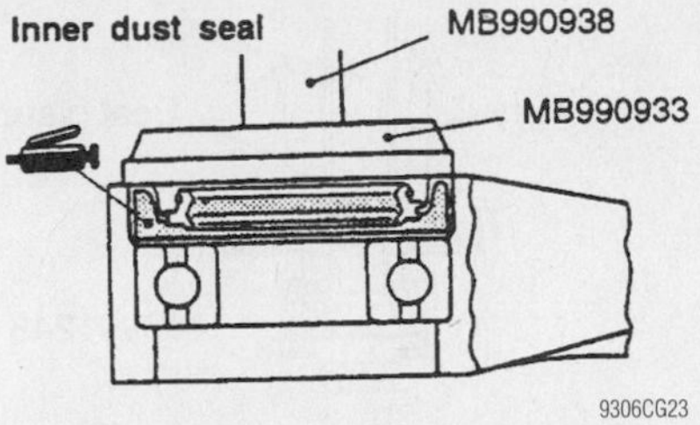

Installing the inner dust seal to the center bearing bracket—Right halfshaft with 2.5L (SOHC) Engine

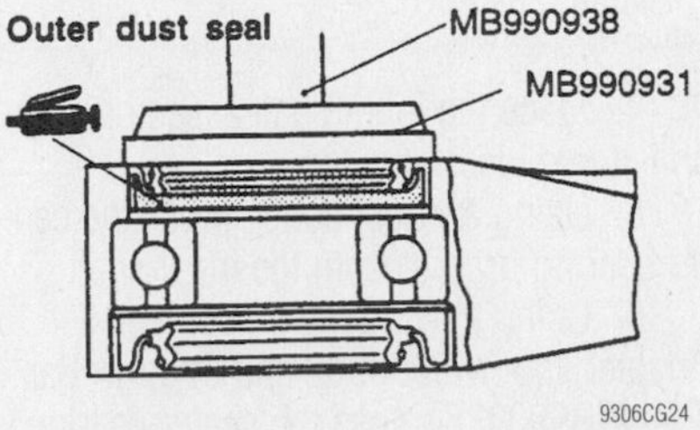

Installing the outer dust seal to the center bearing bracket—Right halfshaft with 2.5L (SOHC) Engine

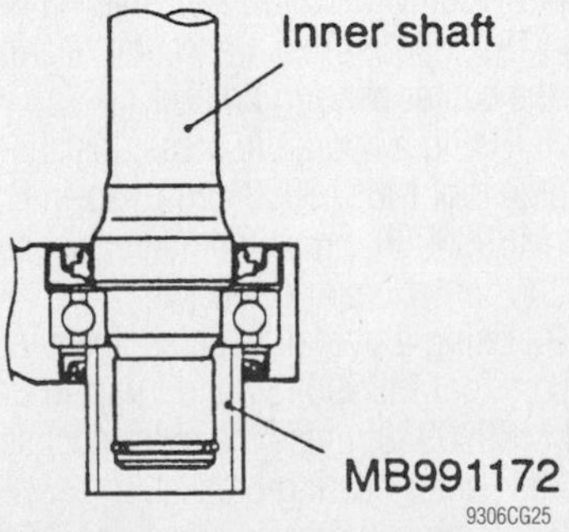

Installing the inner shaft into the center bearing bracket—Right halfshaft with 2.5L (SOHC) Engine

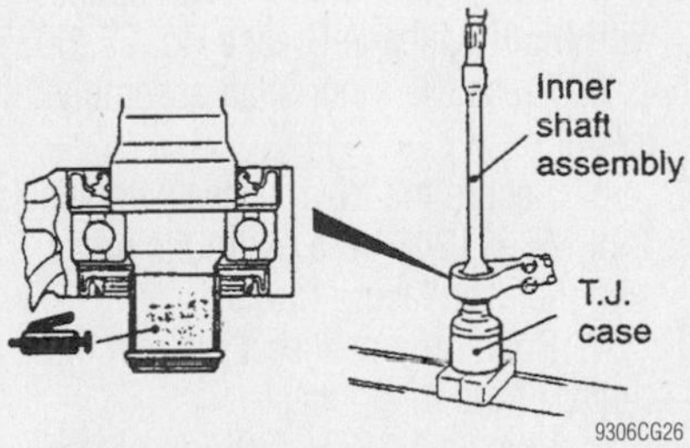

Installing the inner shaft assembly into the tri-pot case—Right halfshaft with 2.5L (SOHC) Engine

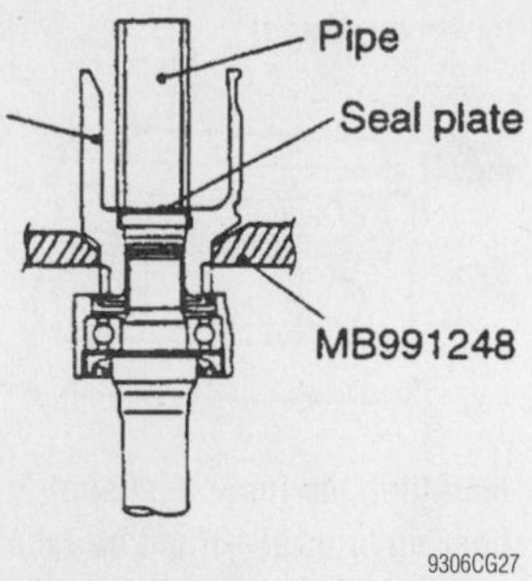

Installing the seal plate into the tri-pot case—Right halfshaft with 2.5L (SOHC) Engine

4. Place the inner shaft assembly in a soft-jawed vise.

5. Using a wheel puller, press the center bearing bracket from the inner shaft.

6. Using a hydraulic press, Installer Adapter Tool MB990932 and Snap-in Bar Tool MB990938, press the center bearing and inner dust seal from the center bearing bracket.

To install:

7. Using a hydraulic press, Installer Adapter Tool MB990932 and Snap-in Bar Tool MB990938, press the center bearing into the center bearing bracket.

8. Using a hydraulic press, Installer Adapter Tool MB990933 and Snap-in Bar Tool MB990938, press the inner duct seal into the center bearing bracket.

9. Using a hydraulic press, Installer Adapter Tool MB990931 and Snap-in Bar Tool MB990938, press the outer duct seal into the center bearing bracket.

10. Using a hydraulic press and Adapter Tool MB991172, press the inner shaft into the center bearing bracket.

11. Install the tri-pot case onto the right halfshaft by performing the following procedure:

a. Lubricate the inner shaft splines with Multi-Mileage Grease No. 2525035.

b. Press the inner shaft assembly into tri-pot case.

c. Secure the tri-pot case with special Tool MB991248 on a hydraulic press with the case facing upward.

d. Position a new seal plate in the center of the tri-pot case.

e. Place a 1.18 in. (30mm) pipe on the seal plate and press the seal into the tri-pot case.

12. Assemble the tri-pot joint assembly.

13. Install the right halfshaft.

Birfield Joint Boot

1. Remove halfshaft and place it in a soft jawed vise.

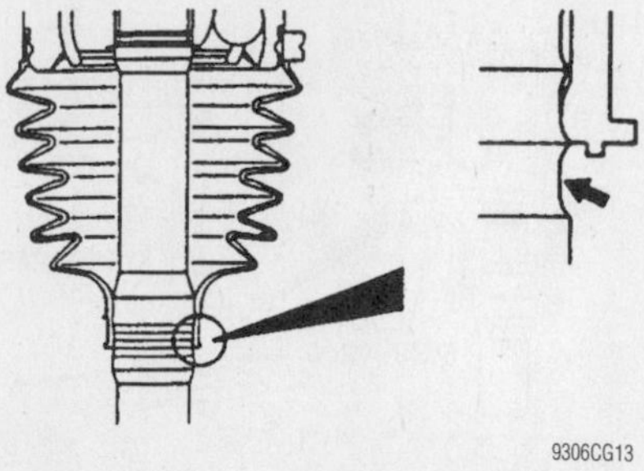

Positioning the birfield joint boot's small diameter

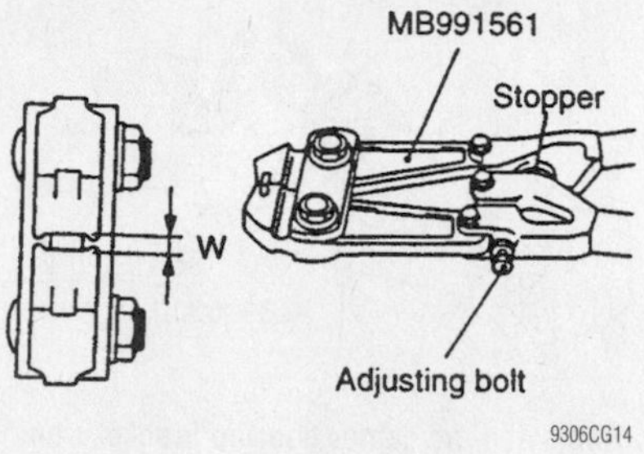

View of the band crimper tool

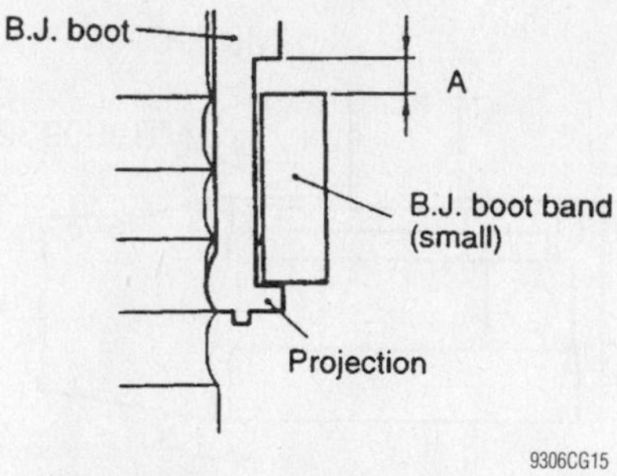

Positioning the birfield joint boot's small band

2. Disassemble or remove:

- Tri-pot joint
- Dynamic damper, if equipped with a 2.0L (DOHC) engine
- Birfield joint boot bands
- Birfield joint boot

To install:

3. Assemble or install:

- Birfield joint boot
- Birfield joint boot small band

4. Place the halfshaft in a soft jawed vise so that the birfield joint is standing vertically.

5. Position the boots small diameter so that only 1 groove is exposed on the shaft.

6. Adjust the Band Crimper Tool MB991561 so that the jaw opening is 0.114 in. (2.9mm).

7. Position the small band so there is clearance between the boot and the bands mating edge.

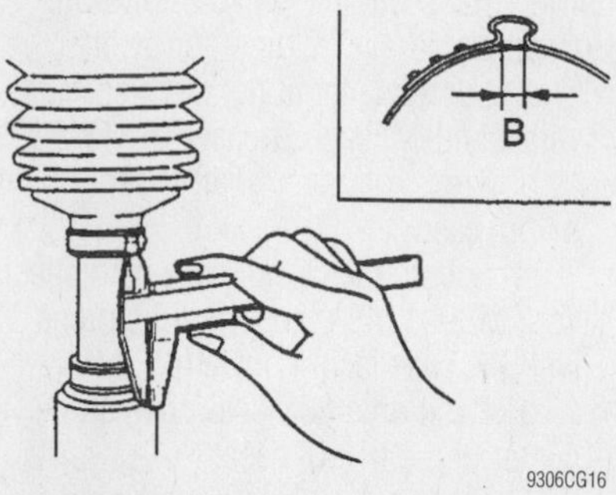

Measuring the small band's crimp

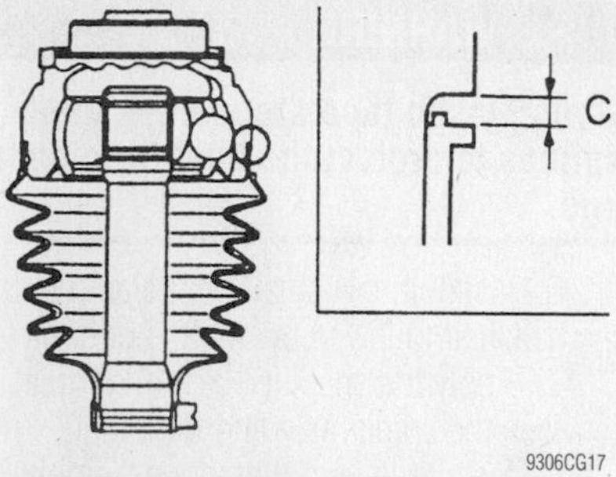

Positioning the birfield joint boot's large diameter

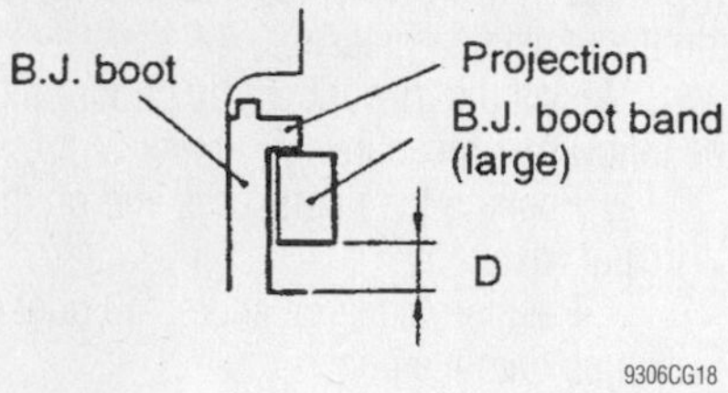

Positioning the birfield joint boot's large band

8. Using a Band Crimper Tool MB991561, crimp the small band until the internal measurement of the crimp is 0.094–0.110 in. (2.4–2.8mm).

WARNING

If the crimp dimension is not correct, remove the band install a new one.

9. Using 5.47 oz. (155g) specified amount of grease in the repair kit, pack the birfield boot.

10. Position the boot so 0.004–0.061 in. (0.1–1.55mm) of clearance exists between the boots large diameter end and the birfield joint housing shoulder.

11. Adjust the Band Crimper Tool MB991561 so that the jaw opening is 0.126 in. (3.2mm).

12. Install and position the large boot band so that it rests against the boot(s) projection and a gap exists between the clamp and the boot.

13. Using a Band Crimper Tool MB991561, crimp the small band until the internal measurement of the crimp is 0.094–0.110 in. (2.4–2.8mm).

⁂ WARNING

If the crimp dimension is not correct, remove the band install a new one.

14. If equipped with a 2.0L (DOHC) engine, install the dynamic damper by performing the following procedure:

a. Slide the damper onto the halfshaft with new bands.

b. Position the damper so the distance from the front of the birfield joint to the front edge of the damper assembly is 14.60–14.84 in. (371–377mm) for the right halfshaft or 7.52–7.76 in. (191–197mm) for the left halfshaft.

c. Tighten and secure the damper bands.

15. Install the tri-pot joint.

3.0L ENGINE

1. Before servicing the vehicle, refer to the precautions in the beginning of this section.
2. Remove halfshaft and place it in a soft jawed vise.
3. Disassemble or remove:
   - Tri-pot joint
   - Dynamic damper
   - Birfield joint boot bands
   - Birfield joint boot

To install:

4. Assemble or install:
   - Birfield joint boot
   - Birfield joint boot small band
5. Place the halfshaft in a soft jawed vise so that the birfield joint is standing vertically.
6. Position the boots small diameter so that only 1 groove is exposed on the shaft.
7. Adjust the Band Crimper Tool MB991561 so that the jaw opening is 0.114 in. (2.9mm).
8. Position the small band so there is clearance between the boot and the bands mating edge.
9. Using a Band Crimper Tool

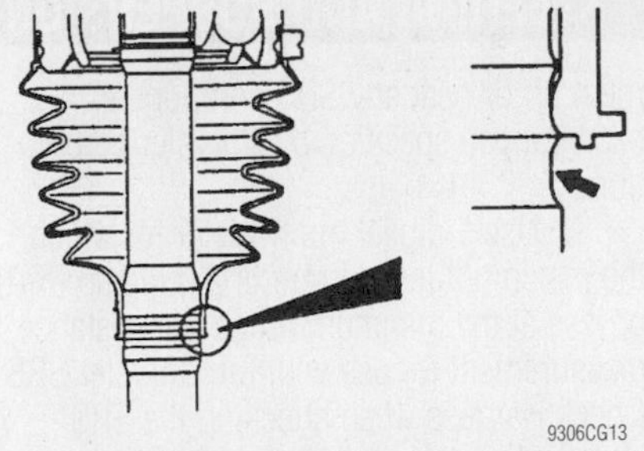

Positioning the birfield joint boot's small diameter

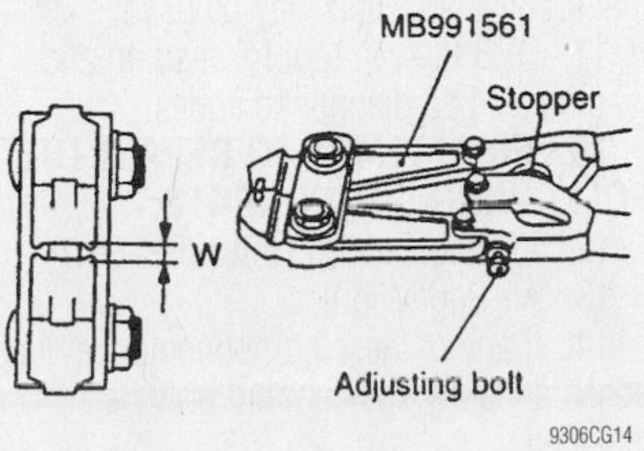

View of the band crimper tool

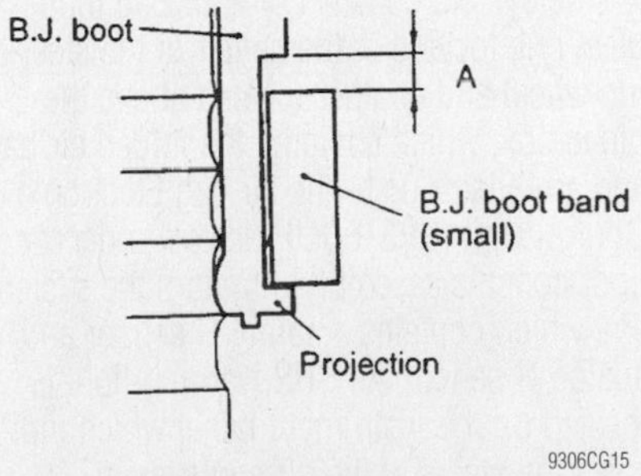

Positioning the birfield joint boot's small band

MB991561, crimp the small band until the internal measurement of the crimp is 0.094–0.110 in. (2.4–2.8mm).

⁂ WARNING

If the crimp dimension is not correct, remove the band install a new one.

10. Using 5.47 oz. (155g) specified amount of grease in the repair kit, pack the birfield boot.

11. Position the boot so 0.004–0.061 in. (0.1–1.55mm) of clearance exists between the boots large diameter end and the birfield joint housing shoulder.

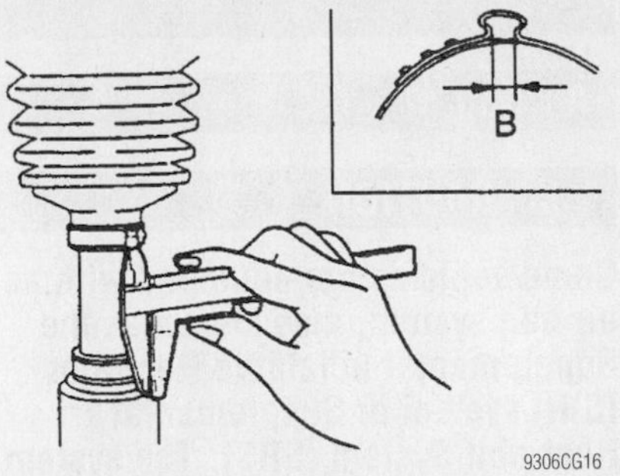

Measuring the small band's crimp

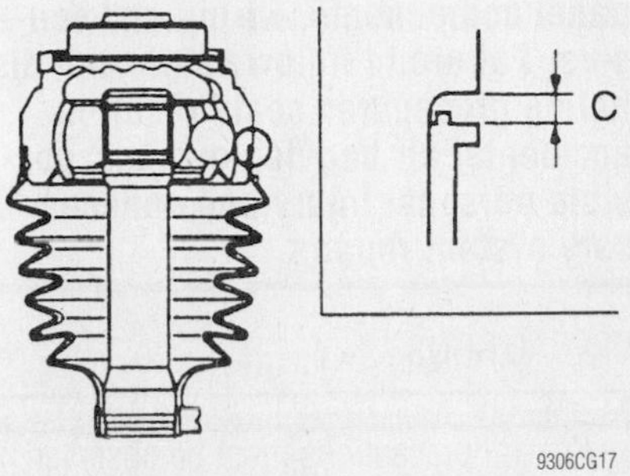

Positioning the birfield joint boot's large diameter

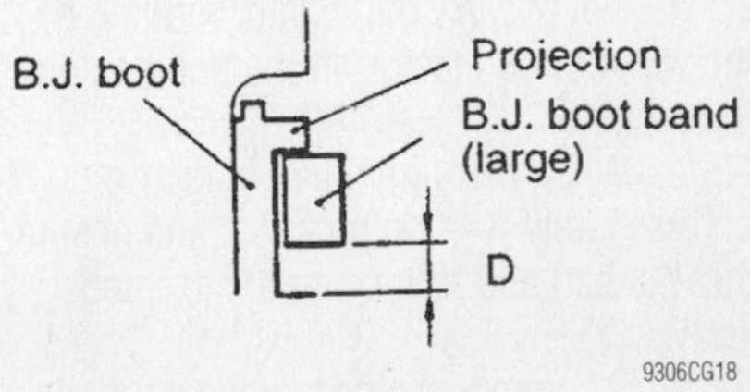

Positioning the birfield joint boot's large band

12. Adjust the Band Crimper Tool MB991561 so that the jaw opening is 0.126 in. (3.2mm).

13. Install and position the large boot band so that it rests against the boot(s) projection and a gap exists between the clamp and the boot.

14. Using a Band Crimper Tool MB991561, crimp the small band until the internal measurement of the crimp is 0.094–0.110 in. (2.4–2.8mm).

⁂ WARNING

If the crimp dimension is not correct, remove the band install a new one.

Air Bag

** CAUTION

Some vehicles are equipped with an air bag system, also known as the Supplemental Inflatable Restraint (SIR) system or Supplemental Restraint System (SRS). The system must be disabled before performing service on or around system components, steering column, instrument panel components, wiring and sensors. Failure to follow safety and disabling procedures could result in accidental air bag deployment, possible personal injury and unnecessary system repairs.

PRECAUTIONS

Several precautions must be observed when handling the inflator module to avoid accidental deployment and possible personal injury. Along with the precautions in the beginning of this section, observe the following:

1. Never carry the inflator module by the wires or connector on the underside of the module.
2. When carrying a live inflator module, hold securely with both hands, and ensure that the bag and trim cover are pointed away.
3. Place the inflator module on a bench or other surface with the bag and trim cover facing up.
4. With the inflator module on the bench, never place anything on or close to the module which may be thrown in the event of an accidental deployment.
5. Do not attempt to repair any of the air bag system wiring harness connectors. If any of the connectors or wires are faulty, replace that harness.
6. Air bag components should not be subjected to heat over 200°F (93°C). Remove the Supplemental Restraint System Air Bag Control Unit (SRS-ECU), the air bag modules themselves and the clock spring before drying or baking the vehicle after painting.
7. After air bag system service, check the SRS warning light operation to be sure that the system functions properly.
8. Make certain that the ignition switch is in the **OFF** position when a scan tool is connected or disconnected.

DO NOT use any electrical test equipment on or near any SRS components except those specified by Chrysler corporation:

9. Use a digital multi-meter for which the maximum test current is 2 miliamp (mA) or less at the minimum range of resistance measurement for use with the Chrysler SRS Check Harness when checking the SRS electrical circuitry.
10. Chrysler Special Tool MB991613 SRS Check Harness acts like a "break-out box" for checking SRS wiring. There are other factory special tool wiring adapters that are available and may be used.
11. DRB III scan tool for reading and erasing air bag diagnostic codes.

NEVER ATTEMPT TO REPAIR THE FOLLOWING COMPONENTS:

12. Air Bag Control Unit (SRS-ECU)
13. Air Bag Modules
14. If any of these components are diagnosed as faulty, they should only be replaced.

DISARMING

The system consists of 2 air bag modules, one located in the center of the steering wheel and another located above the glove box, which contains the folded air bag and an inflator unit. The air bag Electronic Control Unit (SRS-ECU) located under the floor console assembly monitors the system and which contains a safing G sensor and analog G sensor. An SRS warning light is located on the instrument panel which indicates the status of the air bag system. A clock spring interconnection is located within the steering column.

To deploy the air bags, the SRS-ECU must respond to the output signal from the analog G sensor and the safing G sensor must be ON. The SRS-ECU, then causes the air bag modules to ignite and deploy.

Service technicians should use care when working around any vehicle equipped with an air bag system, to avoid injury to the technician by inadvertent deployment of the air bag or to the driver by rendering the air bag system inoperative.

The SRS-ECU not only controls the air bag system, it can provide diagnostic information. The SRS-ECU monitors the air bag system and stores data concerning any detected faults in the system. When the ignition key is turned to the **ON** or **START** position, the SRS warning light should illuminate for about 7 seconds, then turn off. That indicates that the SRS system is in operating condition. If the SRS warning light does not illuminate as described or stays on for more than 7 seconds or if the SRS light illuminates while driving, immediate inspection is required. If the vehicle's SRS warning light is in any of these 3 conditions, the SRS system must be inspected, diagnosed and serviced.

To avoid injury from accidental deployment of the air bag during vehicle servicing, refer to all service precautions.

** CAUTION

The Air Bag system must be disarmed before removing many components. Failure to do so may cause accidental deployment of the air bag, resulting in unnecessary system repairs and/or personal injury.

1. Disarm the air bag system using the following procedure:
   a. Position the front wheels in the straight-ahead position and place the key in the **LOCK** position. Remove the key from the ignition lock cylinder.
   b. Disconnect the negative battery cable and insulate the cable end with high-quality electrical tape or similar non-conductive wrapping.
   c. Wait at least one minute before working on the vehicle. The air bag system is designed to retain enough voltage to deploy the air bag for a short period of time even after the battery has been disconnected.

Power Rack and Pinion Steering Gear

REMOVAL & INSTALLATION

2.0L and 2.5L Engines

** CAUTION

Prior to removal of the steering rack and pinion unit, center the front wheels and remove the ignition key. Failure to do so may damage the SRS (air bag system) clock spring under the steering wheel and render the Security Restraint System (SRS) inoperative, risking serious driver injury.

1. Before servicing the vehicle, refer to the precautions in the beginning of this section.
2. Drain the power steering fluid using the following procedure:

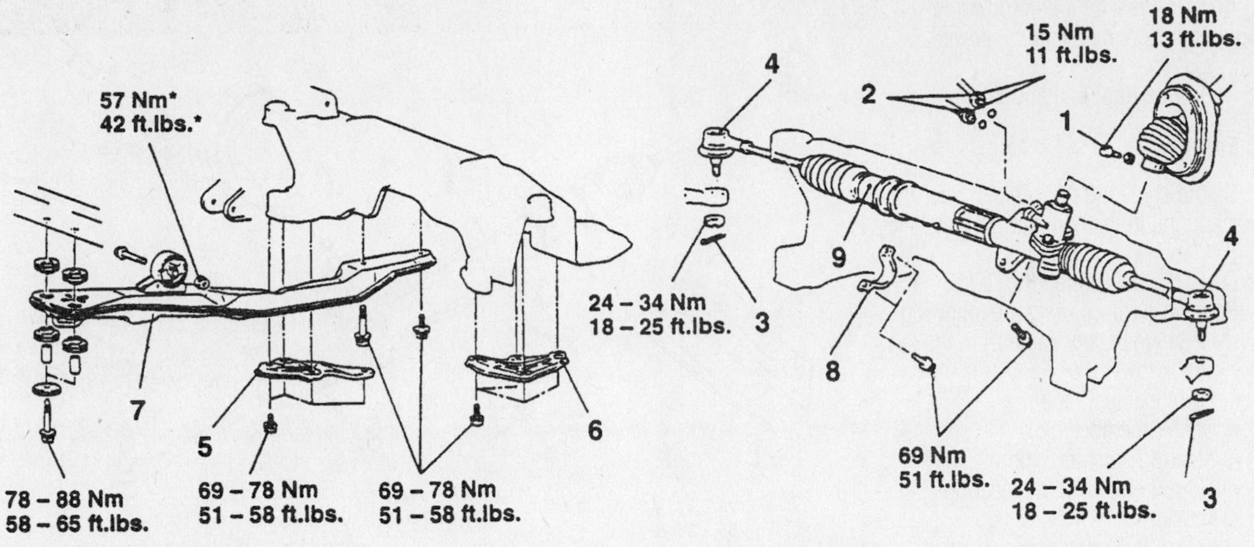

1. Joint assembly and gear box connecting bolt
2. Power steering pipe connection
3. Cotter pin
4. Tie rod end and knuckle connection
5. Stay (L.H.)
6. Stay (R.H.)
7. Center member assembly
8. Clamp
9. Gear box assembly

Caution
The fasteners marked * should be temporarily tightened before they are finally tightened once the total weight of the engine has been placed on the vehicle body.

7922CG30

Exploded view of the power rack and pinion steering gear mounting

a. Disconnect the power steering return (low side) hose.

b. Connect a suitable container to the hose.

c. Properly disable the ignition system by removing the connector from the ignition coil.

d. While cranking the engine, turn the wheels, several times, from side-to-side, until the fluid is removed.

3. Disarm the SRS system.
4. Remove or disconnect the following:
   - Both front wheels
   - Lower steering column joint-to-rack and pinion input shaft bolt
   - Stabilizer bar and washer tank
   - Cotter pins and tie rod ends from the steering knuckles
   - Solenoid's wiring harness connector, if equipped with Electronic Control Power steering (EPS)
   - Both triangular braces near the crossmember
5. Support the center crossmember. Remove the through-bolt from the front round roll stopper and remove the 3 bolts securing the center crossmember.
   - Center crossmember and properly support the engine and remove the rear roll stopper through-bolt. Lower the engine slightly

WARNING

In order to prevent damage to the engine, when supporting and jacking the engine, place a block of wood between the jack and the oil pan.

   - Power steering fluid pressure pipe and return hose from the rack fittings; then, plug the fittings to prevent excessive fluid leakage
   - Rack assembly clamp bolts and the 2 rack assembly-to-chassis bolts
   - Rack and pinion steering assembly and its rubber mounts

➡When removing the rack and pinion assembly, tilt the assembly to the inner side of the compression lower arm, and remove from the left side of the vehicle. Use caution to avoid damaging the boots.

To install:

6. Align the rack assembly so the splines are inserted into the steering column shaft.
7. Install or connect the following:
   - Steering rack and torque the mounting bolts to 51 ft. lbs. (69 Nm)
   - Pinch bolt and torque the bolt to 13 ft. lbs. (18 Nm)
   - Power steering fluid lines; torque the high side fitting to 11 ft. lbs. (15 Nm) and secure the low side hose with the clamp and raise the engine into position
   - Rear roll stopper through-bolt and torque to 32 ft. lbs. (43 Nm)
   - Crossmember and torque the front bolts to 58–65 ft. lbs. (78–88 Nm) and the rear bolt to 51–58 ft. lbs. (69–78 Nm)
   - Front roll stopper bolt and torque the nut to 32 ft. lbs. (43 Nm)
   - Both triangular braces and torque the bolts to 50–56 ft. lbs. (69–78 Nm)
   - Stabilizer bar
   - Tie rod ends and torque the nuts to 20 ft. lbs. (27 Nm)
   - Solenoid's wiring harness connector, if equipped with EPS
   - Front wheels

8. Refill the reservoir with power steering fluid and properly bleed the power steering system.
9. Perform a front end alignment.

3.0L Engine

1. Before servicing the vehicle, refer to the precautions in the beginning of this section.
2. Drain the power steering fluid.
3. Remove or disconnect the following:
   - Negative battery cable from the left shock tower
   - Front exhaust pipe
   - Center member
   - Steering shaft assembly
   - Tie rod end from the steering knuckle
   - Return hose and pressure tube connections
   - Cylinder clamp
   - Steering gear from the vehicle

To install:

4. Install or connect the following:
   - Steering gear to the vehicle
   - Cylinder clamp and torque the bolts to 51 ft. lbs. (69 Nm)
   - Pressure tube and return hoses and torque to 11 ft. lbs. (15 Nm)
   - Tie rod ends to the steering knuckle and torque the nut to 21 ft. lbs. (29 Nm)
   - Stay and torque the bolts to 55 ft. lbs. (74 Nm)
   - Steering shaft assembly and torque the bolt to 13 ft. lbs. (18 Nm)
   - Center member
   - Negative battery cable
5. Fill and bleed the power steering system.
6. Start the vehicle, check for leaks and repair if necessary.
7. Perform a front wheel alignment.

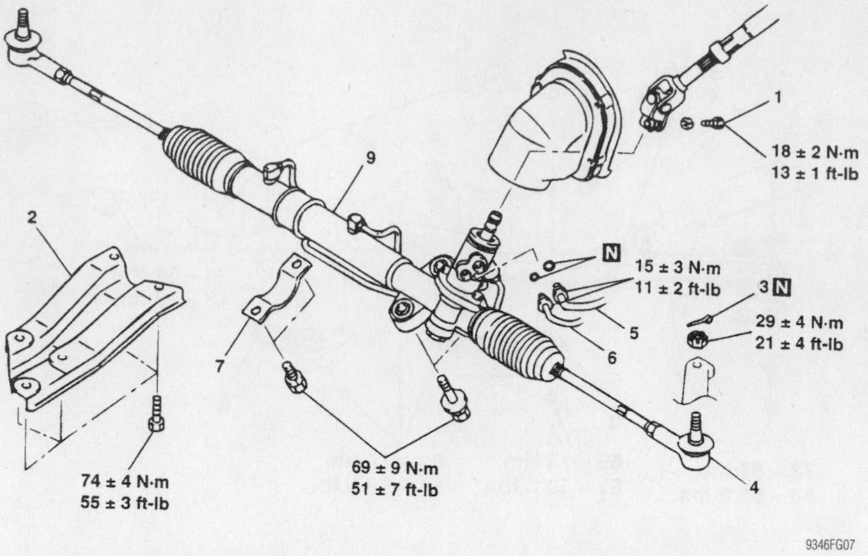

Exploded view of the steering gear assembly—3.0L

Strut

REMOVAL & INSTALLATION

3.0L Engine

1. Before servicing the vehicle, refer to the precautions in the beginning of this section.
2. Remove or disconnect the following:
   - Negative battery cable
   - Front wheel
   - Brake hose clamp
   - Front speed sensor harness clamp, if equipped
   - Sway bar link
   - Lower control arm
   - Strut from the steering knuckle
   - Upper mounting bolts
   - Strut assembly

To install:

3. Install or connect the following:
   - Strut assembly and torque the upper bolts to 33 ft. lbs. (44 Nm)
   - Strut assembly to the lower control arm and steering knuckle. Torque the bolts to 221 ft. lbs. (300 Nm)
   - Sway bar link and torque the nut to 33 ft. lbs. (44 Nm)
   - Front speed sensor harness clamp, if equipped
   - Brake hose clamp
   - Front wheel
   - Negative battery cable

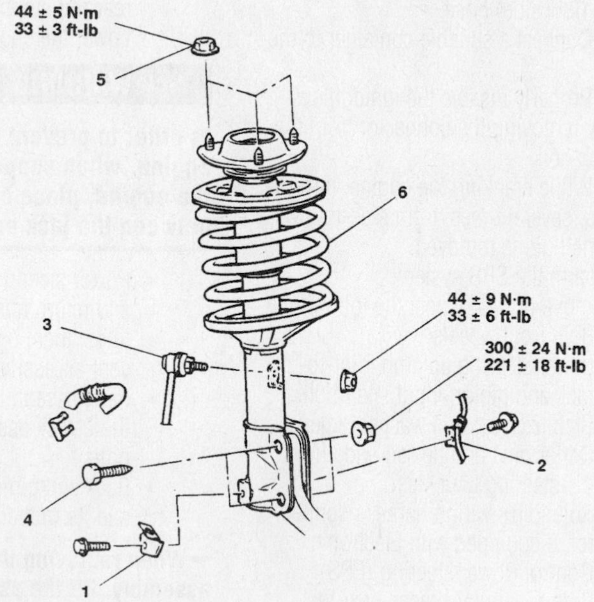

1. BRAKE HOSE CLAMP
2. FRONT SPEED SENSOR HARNESS CLAMP <VEHICLES WITH ABS>
3. STABILIZER LINK
4. BOLTS
5. NUT
6. STRUT ASSEMBLY

9346FG08

Exploded view of the front strut assembly—3.0L

4. Check and adjust the front end alignment, if needed.

Shock Absorber

REMOVAL & INSTALLATION

front

2.0L AND 2.5L ENGINES

1. Before servicing the vehicle, refer to the precautions in the beginning of this section.
2. Remove or disconnect the following:
   - Negative battery cable
   - Wheel
   - Sway bar link from the damper fork
   - Damper fork lower through-bolt, upper pinch bolt and damper fork assembly
   - Shock absorber

⁂ CAUTION

Do Not remove the large center nut.

To install:

3. Install or connect the following:
   - Shock absorber and torque the upper mounting nuts to 32 ft. lbs. (44 Nm)
   - Damper fork then torque the lower through-bolt/nut to 65 ft. lbs. (88 Nm) and the upper pinch bolt to 76 ft. lbs. (103 Nm)
   - Sway bar link and torque the link-to-damper fork nut to 29 ft. lbs. (39 Nm)
   - Wheel
4. Perform a front end alignment.

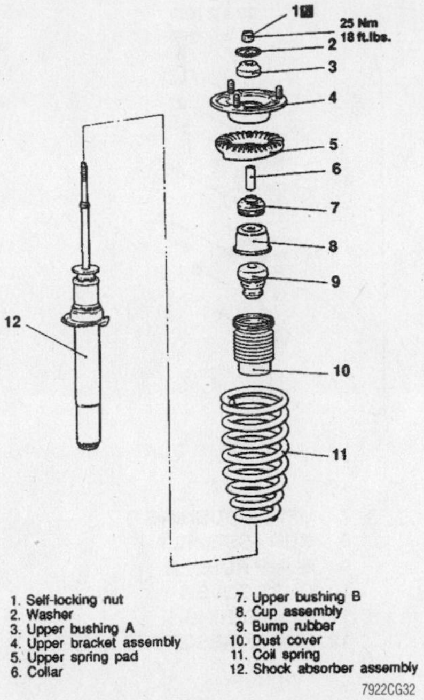

Exploded view of the front shock absorber

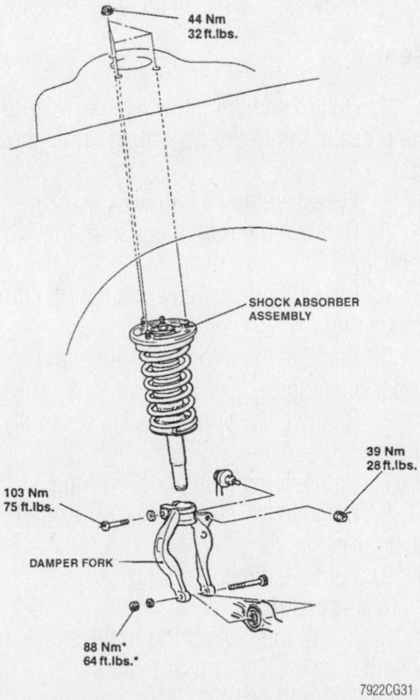

Exploded view of the front shock absorber assembly mounting

Rear

➡**The shock absorber assembly is a load bearing component; therefore, the vehicle chassis and axle weight must be supported separately, requiring the use of 2 separate lifting devices.**

The rear package shelf front cover(s) must be removed to access the top mounting nuts. Most connections are plastic clips. Use care when removing these components to avoid unnecessary damage.

1. Remove or disconnect the following:
   - Rear shelf speaker covers
   - Rear shelf top assembly
   - Front cover(s) to access the shock absorber top mounting nuts
2. Raise and support lower control arm assembly slightly.
   - Shock absorber upper mounting nuts
   - Shock absorber lower mounting bolt
   - Shock absorber

To install:

3. Position the shock absorber assembly so that the lower mounting bolt can be installed and lightly tightened.

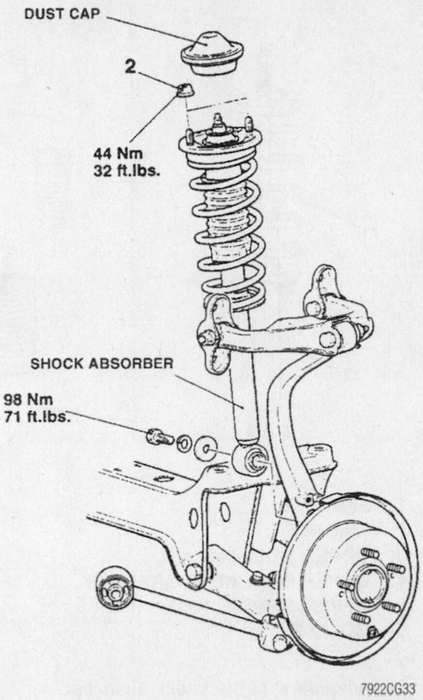

Exploded view of the rear shock absorber assembly mounting

4. Use a jack to raise or lower the lower control arm, so that the top shock absorber plate studs align through the body. Raise the jack to hold the shock absorber assembly in position.
5. Install or connect the following:

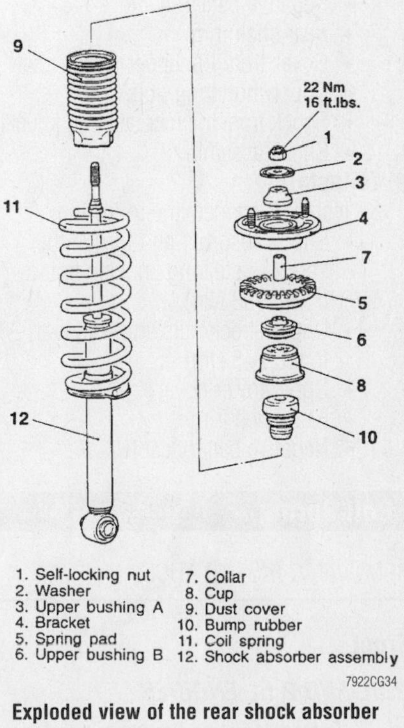

Exploded view of the rear shock absorber

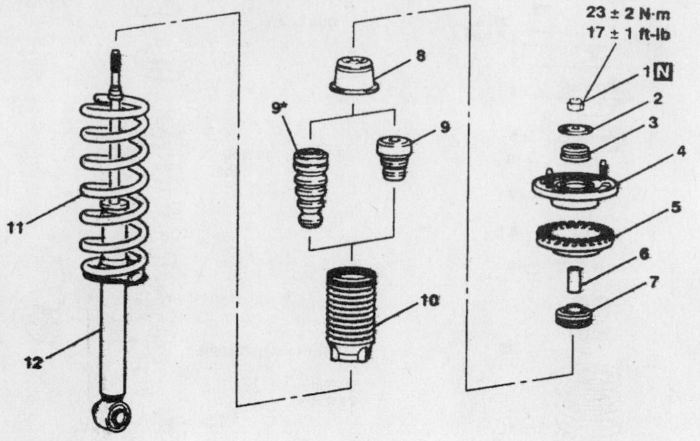

1. JAM NUT
2. WASHER
3. UPPER BUSHING A
4. UPPER BRACKET ASSEMBLY
5. UPPER SPRING PAD
6. COLLAR
7. UPPER BUSHING B
8. CUP ASSEMBLY
9. BUMP RUBBER
10. DUST COVER
11. COIL SPRING
12. SHOCK ABSORBER ASSEMBLY

9346FG09

Exploded view of the shock absorber

- Top plate nuts and torque the mounting nuts to 32 ft. lbs. (44 Nm)
- Lower mounting bolt and torque the bolt to 71 ft. lbs. (98 Nm)
- Interior trim pieces

3.0L ENGINE

1. Before servicing the vehicle, refer to the precautions in the beginning of this section.
2. Remove or disconnect the following:
   - Negative battery cable
   - Rear shelf trim
   - Cover from the upper shock tower
   - Upper mounting nuts
   - Shock from the rear steering knuckle
   - Shock assembly

To install:

3. Install or connect the following:
   - Shock absorber and torque the shock to steering knuckle bolt to 73 ft. lbs. (98 Nm)
   - Upper shock nuts and torque to 32 ft. lbs. (45 Nm)
   - Upper tower cover
   - Rear shelf trim
   - Negative battery cable

Coil Spring

REMOVAL & INSTALLATION

Front

2.0L AND 2.5L ENGINES

1. Before servicing the vehicle, refer to the precautions in the beginning of this section.
2. Place the shock absorber assembly in Compressor Tools MB991237 and MB991238 spring compressor assembly or equivalent. Tighten the compressor and compress the spring slowly. Make certain the compressor is properly engaged before tightening.
3. After tension has been removed from the shock absorber assembly and shock absorber plate, remove the piston rod nut and washer from the top of the shock absorber assembly.
4. If the spring is to be replaced, slowly release the tension on the spring compressor. Allow the spring to expand fully. If only the shock absorber is being replaced, the spring may remain in the compressor assembly.
5. By hand, remove the upper shock absorber bearing, washer, mount and shock absorber shield. Remove the upper insulator ring, the spring, the bumper and lower insulator from the spring. Take notice to each components location for proper reassembly.

To install:

6. Install the lower insulator, the shock absorber bumper and the uncompressed spring. Install the upper spring insulator, shock absorber shield, mount and washer.
7. Install or align the spring compressor. Make certain the spring is correctly positioned relative to the upper and lower insulator rings. Smoothly compress the spring.
8. Install the washer and piston rod nut. Torque the nut on the front strut to 18 ft. lbs. (25 Nm) or 16 ft. lbs. (22 Nm) on the rear strut. Install the dust cap.
9. Carefully release the spring compressor, watching the spring position as it seats. When the spring is properly seated, release/remove the compressor tools.
10. Reinstall the shock absorber assembly.

3.0L ENGINES

1. Before servicing the vehicle, refer to the precautions in the beginning of this section.
2. Disassemble as follows:
3. Remove the strut/shock assembly and place it in a suitable compressor tool.
4. Install Special Tool MB991176 to secure the strut and remove the dust cover and jam nut.
5. Remove the strut insulator.
6. Remove the upper spring seat and pad.
7. Rempve the bump rubber and dust cover.
8. Remove the coil spring.
9. Remove the lower spring pad.

To assemble:

10. Install the lower spring pad and coil spring.
11. Install the dust cover and bump rubber.
12. Install the upper spring pad and seat.
13. Carefully compress the spring and install the insulator and jam nut.
14. When properly aligned, torque the jam nut to 47 ft. lbs. (64 Nm).
15. Install the dust cover and remove the compressor tools.
16. Install the strut/shock to the vehicle.

Rear

1. Before servicing the vehicle, refer to the precautions in the beginning of this section.
2. Disassemble as follows:
3. Remove the rear shock from the vehicle.
4. Install the assembly in a spring compressor tool.
5. Remove the jam nut, washer and upper bushing.
6. Remove the upper bracket assembly and spring pad.
7. Remove the collar and bushing.
8. Remove the cup, bump rubber and dust cover.
9. Remove the coil spring

To assemble:

10. Install the coil spring to the shock absorber using Compressor Tools MB991237 and MB991239.
11. Install the dust cover and bump rubber.
12. Install the cup and bushing
13. Install the collar, supper spring pad and bracket assembly.

14. Install the upper bushing and washer.

15. When properly aligned, install the jam nut and torque to 17 ft. lbs. (23 Nm).

16. Remove the compressor tools and install the shock absorber.

Upper Ball Joint

REMOVAL & INSTALLATION

The upper ball joint is an integrated part of the upper control arm assembly for the 2.0L and 2.5L engines, and cannot be serviced separately. A worn or damaged ball joint requires replacement of the upper control arm assembly.

Lower Ball Joint

REMOVAL & INSTALLATION

The front suspension on the 2.0L and 2.5L engines is called a Multi-Link Suspension. There are 2 lower arms used in this front suspension; a curved arm called the Compression Lower Arm and also a straight arm called the Lateral Lower Arm. Both arms contain lower ball joints since there are 2 sockets in the steering knuckle. Ball joints and lower arms are removed and replaced as an assembly. A front end alignment is required after these procedures.

Upper Control Arm

REMOVAL & INSTALLATION

2.0L and 2.5L Engines

1. Before servicing the vehicle, refer to the precautions in the beginning of this section.
2. Remove or disconnect the following:
   - Front wheel
   - Ball joint stud from the steering knuckle, using the joint separation Tool MB991113
3. The ball joint can be checked using the following procedure:
   a. An adapter (MB 990326) is available that fits onto the ball joint stud and adapts to an inch-pound torque wrench. If this tool is not available, a shop-made substitute can be fabricated.
   b. Turn the ball joint stud with the torque wrench. The factory standard for breakaway torque is 3–13 inch lbs. (0.34–1.45 Nm).

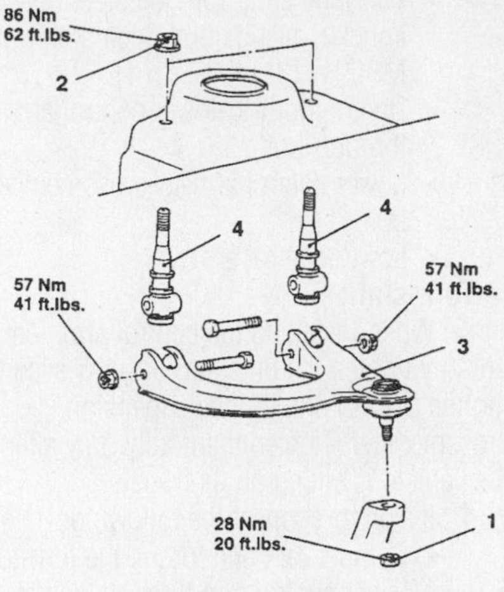

Exploded view of the upper control arm mounting

   c. If the ball joint stud is out of specification (turns too easily or is too stiff), continue with the ball joint/control arm assembly replacement.

➡The upper ball joint boot can be removed and the joint greased. A new replacement ball joint boot is recommended.

4. Inside the engine compartment, at the shock absorber tower, locate the upper control arm mounting nuts. Remove the nuts and using the joint separation tool, separate the upper arm shafts from the shock absorber tower.
5. Remove or disconnect the following:
   - Upper control arm assembly
   - Upper arm shaft assembly-to-upper control arm nuts/bolts and the shaft(s)

To install:

6. Install the upper arm shaft assemblies to the upper control arm and torque the nuts and bolts to 41 ft. lbs. (57 Nm).
7. Align the upper control arm shafts to the shock absorber tower and secure it with the mounting nuts. Torque the mounting nuts to 62 ft. lbs. (86 Nm).
8. Install or connect the following:
   - Ball joint to the knuckle and torque the locking nut to 20 ft. lbs. (28 Nm)
   - Front wheel
9. Check the wheel alignment and adjust, if necessary.

3.0L Engine

1. Before servicing the vehicle, refer to the precautions in the beginning of this section.
2. Remove or disconnect the following:
   - Rear wheel
   - Upper control arm from the steering knuckle
   - Upper control arm assembly
   - Upper control arm bracket
   - Upper control arm

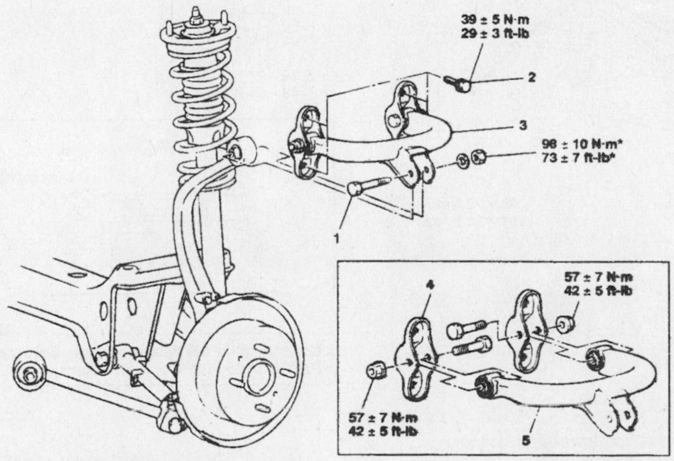

Exploded view of the upper control arm assembly

To install:

3. Install or connect the following:
   - Upper control arm and brackets and torque the bolts to 42 ft. lbs. (57 Nm)
   - Upper control arm assembly and torque the bolts to 29 ft. lbs. (39 Nm)
   - Upper control arm to the steering knuckle and torque the bolt 73 ft. lbs. (98 Nm)
   - Rear wheel

Lower Control Arm

REMOVAL & INSTALLATION

Lateral Lower Arm

1. Before servicing the vehicle, refer to the precautions in the beginning of this section.
2. Remove or disconnect the following:
   - Front wheel
   - Stay bracket from the crossmember
   - Ball joint stud from the steering knuckle, using Joint Separator Tool MB991113
   - Damper fork-to-lower control arm through-bolt
   - Lower control arm-to-crossmember bolt
   - Lower control arm

To install:

3. When installing the control arm, temporarily tighten the nuts and/or bolts securing the control arm to the suspension crossmember. Tighten them fully only after the vehicle is sitting on its wheels.
4. Install or connect the following:
   - Damper fork and torque the fork-to-lower control arm through-bolt to 64 ft. lbs. (88 Nm)
   - Ball joint stud to the knuckle and torque the nut to 65–80 ft. lbs. (88–108 Nm). Install a new cotter pin.
   - Tension rod to the control arm and torque the nuts to 80–94 ft. lbs. (108–127 Nm)

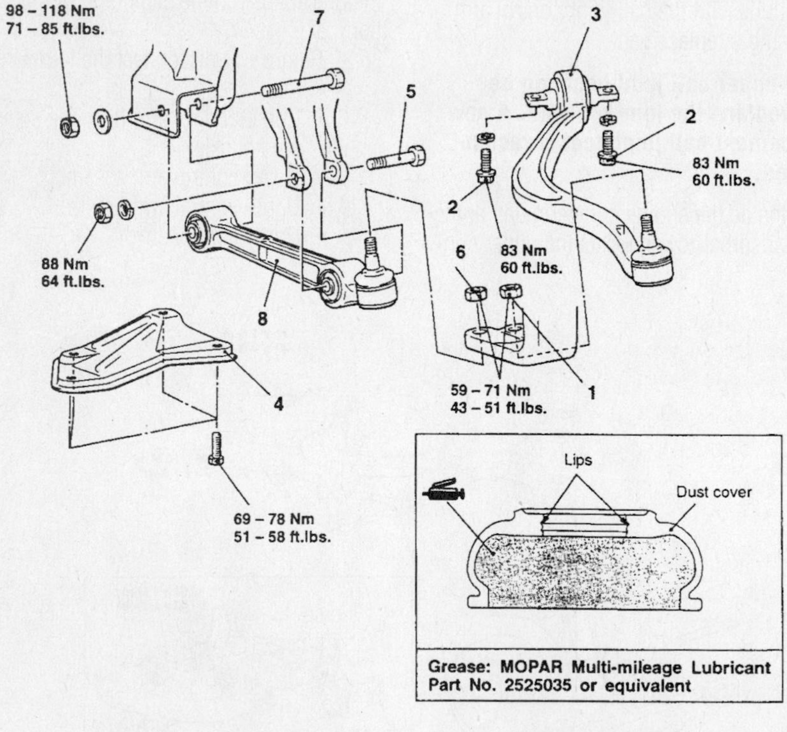

1. Compression lower arm ball joint and knuckle connection
2. Compression lower arm mounting bolt
3. Compression lower arm assembly
4. Stay
5. Shock absorber lower mounting bolt and nut
6. Lateral lower arm ball joint and knuckle connection
7. Lateral lower arm mounting bolt and nut
8. Lateral lower arm assembly

Caution
*: Indicates parts which should be temporarily tightened, and then fully tightened with the vehicle on the ground in the unladen condition.

7922CG36

Exploded view of the lateral lower arm and the compression lower arm mounting

   - Stay bracket to the crossmember and torque the mounting bolts to 50–56 ft. lbs. (69–78 Nm)
   - Front wheel

5. Once the full weight of the vehicle is on the suspension, torque the inner lower arm mounting bolt nut to 71–85 ft. lbs. (98–118 Nm).
6. Check the front end alignment and adjust as required.

Compression Lower Arm

1. Before servicing the vehicle, refer to the precautions in the beginning of this section.
2. Remove or disconnect the following:
   - Front wheel
   - Ball joint stud from the steering knuckle, using Joint Separator Tool MB991113
   - Lower control arm-to-crossmember bolt

To install:

3. Install or connect the following:
   - Control arm to the crossmember and torque the bolts to 60 ft. lbs. (83 Nm)
   - Ball joint stud to the knuckle and torque the nut to 43–51 ft. lbs. (59–71 Nm)
   - Front wheel
4. Check the front end alignment and adjust as required.

3.0L ENGINE

1. Before servicing the vehicle, refer to the precautions in the beginning of this section.
2. Remove or disconnect the following:
   - Front wheel
   - Lower control arm from the steering knuckle
   - Lower control arm mounting bolt and clamp
   - Lower control arm
3. If replacing the bushing, perform the following:
   a. Apply soapy water between the shaft and old bushing and pry the old bushing out.

To install:

4. Apply soapy water to the shaft and new bushing and install the new bushing, if removed.
5. Install or connect the following:
   - Lower control arm
   - Lower control arm clamp and torque the bolts to 60 ft. lbs. (81 Nm)
   - Lower control arm to the steering knuckle and torque the bolts to 80 ft. lbs. (108 Nm)

- Front wheel

6. Check and adjust the front toe, if needed.

Rear

1. Before servicing the vehicle, refer to the precautions in the beginning of this section.
2. Remove or disconnect the following:
   - Rear wheel
   - Stabilizer link
   - Wheel Speed Sensor (WSS), if equipped
   - Lower control arm assembly to steering knuckle
   - Lower control arm

To install:

3. Install or connect the following:
   - Lower control arm and torque the bolt to 55 ft. lbs. (74 Nm)
   - Lower control arm to the steering knuckle and torque the bolt to 80 ft. lbs. (108 Nm)
   - WSS, if equipped
   - Stabilizer link and torque the bolt to 29 ft. lbs. (39 Nm)
   - Rear wheel

Wheel Bearings

ADJUSTMENT

Front

To check hub and bearing assembly end-play, remove the caliper and rotor. Position a dial indicator to bear against the hub flange near the center ridge. Wiggle the hub back and forth. If end-play exceeds 0.002 in. (0.05mm), replace the front hub and bearing assembly.

Rear

The rear hub and wheel bearing assembly is designed for the life of the vehicle and requires no type of adjustment or periodic maintenance. The bearing is a sealed unit with the wheel hub and can only be removed and/or replaced as one unit.

REMOVAL & INSTALLATION

Front

1. Before servicing the vehicle, refer to the precautions in the beginning of this section.
2. Remove the cotter pin, halfshaft nut and washer.

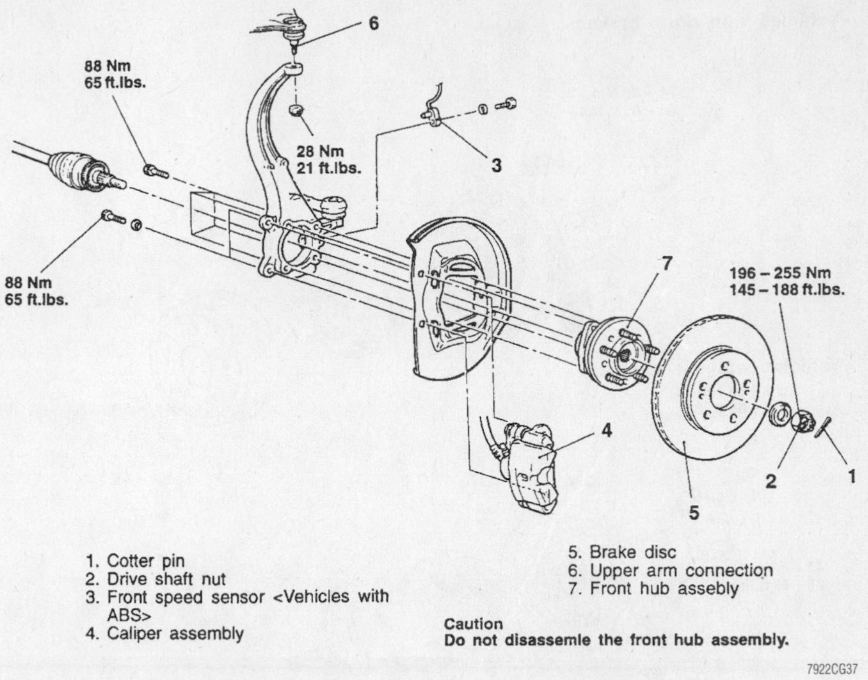

Exploded view of the front hub assembly mounting and related components

3. Remove or disconnect the following:
   - Front wheel
   - Vehicle Speed Sensor (VSS), if equipped
   - Caliper and brake pads; then, support the caliper out of the way using wire
   - Brake rotor from the hub assembly
   - Upper ball joint from the steering knuckle using a press type tool and pull the knuckle outward

⁂ WARNING

Use of improper methods of joint separation can result in damage to joint, leading to possible failure. Never use wedge-type tools or the ball joint can be damaged.

   - 4 hub-to-steering knuckle bolts
   - Hub and bearing assembly from the knuckle

➡The hub and wheel bearing assembly is not serviceable and should not be disassembled.

To install:

4. Install or connect the following:
   - Hub to the steering knuckle and torque the bolts to 65 ft. lbs. (88 Nm)
   - Upper ball joint to the steering knuckle and torque the self-locking nut to 21 ft. lbs. (28 Nm)
5. Position the rotor on the hub. Install a couple of lug nuts and lightly tighten to hold the rotor on the hub.
   - Caliper holder and place the brake pads in the holder. Slide the caliper over the brake pads and install the guide pins. Once the caliper is secured, the lug nuts can be removed.
   - VSS, if equipped
   - Front wheel
   - New cotter pin and bend to secure
6. Examine the driveshaft (halfshaft) hub washer. Locate the chamfered side. This side is installed outward, away from the hub. Install the washer and hub nut. Tighten the axle nut with the brakes applied. Torque the nut to 145–188 ft. lbs. (200–260 Nm).

⁂ WARNING

Pump the brake pedal until hard, before attempting to move the vehicle.

Rear

WITH DRUM BRAKES

1. Before servicing the vehicle, refer to the precautions in the beginning of this section.
2. Remove or disconnect the following:

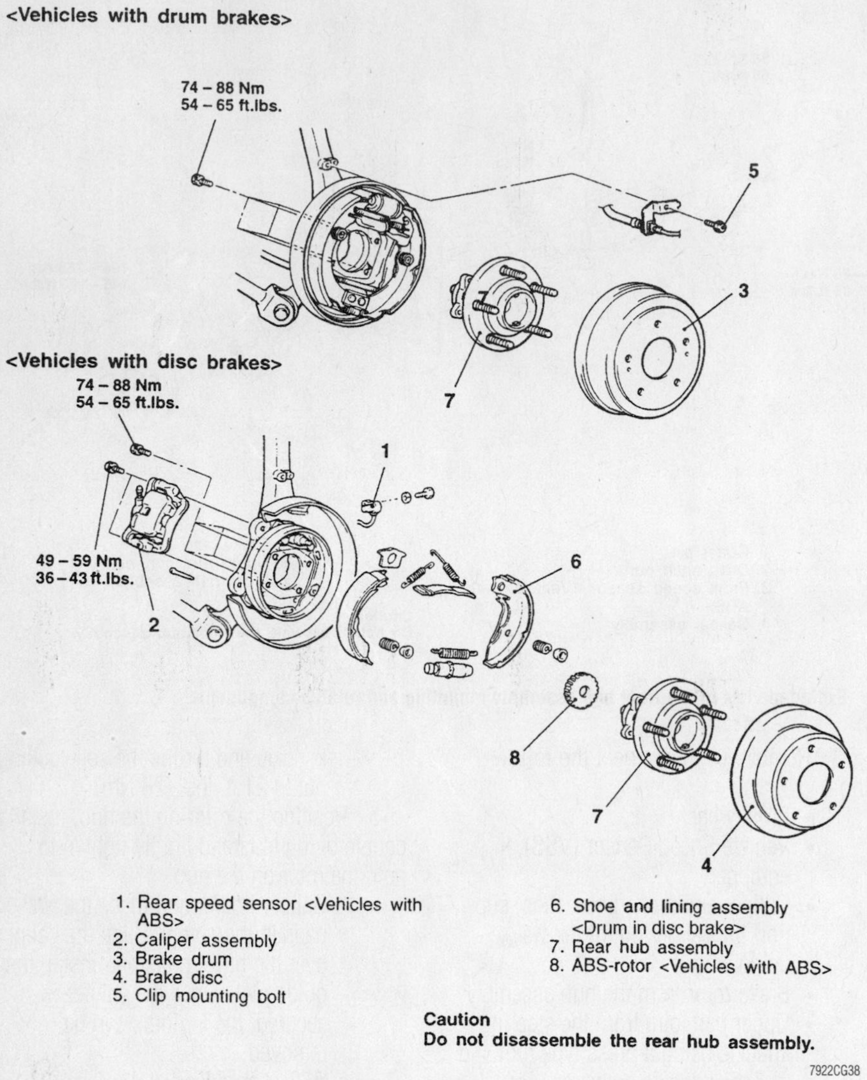

Exploded view of the rear hub assembly mounting

- Rear wheel
- Vehicle Speed Sensor (VSS), if equipped with Anti-lock Brake System (ABS)
- Brake drum
- 4 hub to the knuckle bolts
- Hub and bearing assembly from the knuckle

➡The hub assembly is not serviceable and should not be disassembled.

3. If replacing the hub, use special Socket MB991248 and a press, to remove the wheel sensor rotor from the hub.

To install:

4. Press the wheel sensor rotor onto the hub.
5. Install or connect the following:
   - Hub to the knuckle and torque the bolts to 54–65 ft. lbs. (74–88 Nm)
   - Brake drum

- VSS, if equipped with ABS
- Rear wheel and lower the vehicle

WITH DISC BRAKES

1. Before servicing the vehicle, refer to the precautions in the beginning of this section.
2. Remove or disconnect the following:
   - Rear wheel
   - Vehicle Speed Sensor (VSS), if equipped with Anti-lock Brake System (ABS)
   - Caliper and brake pads; then, support the caliper out of the way using wire
   - Brake rotor
3. Remove the parking brake shoes as follows:
   a. Upper shoe-to-anchor springs.
   b. Lower shoe-to-shoe spring.
   c. Brake shoe hold-down springs.
   d. Parking brake cable from the actuating lever.
4. Remove the 4 hub-to-knuckle bolts.
5. Remove the hub and bearing assembly from the knuckle.

➡The hub assembly is not serviceable and should not be disassembled.

6. If replacing the hub, use special Socket MB991248 and a press, to remove the wheel sensor rotor from the hub.

To install:

7. Press the wheel sensor rotor onto the hub.
8. Install the hub to the knuckle and torque the bolts to 54–65 ft. lbs. (74–88 Nm)
9. Install the parking brake shoes.
10. Position the rotor on the hub. Install a couple of lug nuts and lightly tighten to hold rotor on hub.
11. Install the caliper holder and place brake pads in holder. Slide the caliper over brake pads and install guide pins. Once caliper is secured, lug nuts can be removed.
12. Install the VSS, if equipped
13. Install the rear wheel.

CHRYSLER CORPORATION

7

1997

Eagle-Talon

PRECAUTIONS

Before servicing any vehicle, please be sure to read all of the following precautions, which deal with personal safety, prevention of component damage, and important points to take into consideration when servicing a motor vehicle:

- Never open, service or drain the radiator or cooling system when the engine is hot; serious burns can occur from the steam and hot coolant.
- Observe all applicable safety precautions when working around fuel. Whenever servicing the fuel system, always work in a well-ventilated area. Do not allow fuel spray or vapors to come in contact with a spark, open flame or excessive heat (a hot drop light, for example). Keep a dry chemical fire extinguisher near the work area. Always keep fuel in a container specifically designed for fuel storage; also, always properly seal fuel containers to avoid the possibility of fire or explosion. Refer to the additional fuel system precautions later in this section.
- Fuel injection systems often remain pressurized, even after the engine has been turned **OFF**. The fuel system pressure must be relieved before disconnecting any fuel lines. Failure to do so may result in fire and/or personal injury.
- Brake fluid often contains polyglycol ethers and polyglycols. Avoid contact with the eyes and wash your hands thoroughly after handling brake fluid. If you do get brake fluid in your eyes, flush your eyes with clean, running water for 15 minutes. If eye irritation persists, or if you have taken brake fluid internally, IMMEDIATELY seek medical assistance.
- The EPA warns that prolonged contact with used engine oil may cause a number of skin disorders, including cancer! You should make every effort to minimize your exposure to used engine oil. Protective gloves should be worn when changing oil. Wash your hands and any other exposed skin areas as soon as possible after exposure to used engine oil. Soap and water, or waterless hand cleaner should be used.
- All new vehicles are now equipped with an air bag system. The system must be disabled before performing service on or around system components, steering column, instrument panel components, wiring and sensors. Failure to follow safety and disabling procedures could result in accidental air bag deployment, possible personal injury and unnecessary system repairs.
- Always wear safety goggles when working with, or around, the air bag system. When carrying a non-deployed air bag, be sure the bag and trim cover are pointed away from your body. When placing a non-deployed air bag on a work surface, always face the bag and trim cover upward, away from the surface. This will reduce the motion of the module if it is accidentally deployed. Refer to the additional air bag system precautions later in this section.
- Clean, high quality brake fluid from a sealed container is essential to the safe and proper operation of the brake system. You should always buy the correct type of brake fluid for your vehicle. If the brake fluid becomes contaminated, completely flush the system with new fluid. Never reuse any brake fluid. Any brake fluid that is removed from the system should be discarded. Also, do not allow any brake fluid to come in contact with a painted surface; it will damage the paint.
- Never operate the engine without the proper amount and type of engine oil; doing so WILL result in severe engine damage.
- Timing belt maintenance is extremely important! Many models utilize an interference-type, non-freewheeling engine. If the timing belt breaks, the valves in the cylinder head may strike the pistons, causing potentially serious (also time-consuming and expensive) engine damage. Refer to the maintenance interval charts in the front of this manual for the recommended replacement interval for the timing belt, and to the timing belt section for belt replacement and inspection.
- Disconnecting the negative battery cable on some vehicles may interfere with the functions of the on-board computer system(s) and may require the computer to undergo a relearning process once the negative battery cable is reconnected.
- When servicing drum brakes, only disassemble and assemble one side at a time, leaving the remaining side intact for reference.
- Only an MVAC-trained, EPA-certified automotive technician should service the air conditioning system or its components.

ENGINE REPAIR

Alternator

REMOVAL & INSTALLATION

Non-Turbo

1. Before servicing the vehicle, refer to the precautions in the beginning of this section.
2. Disconnect the negative battery cable.
3. Remove or disconnect the following:
   - Right wheel
   - Under cover side panel
   - Speed control assembly
   - Alternator drive belt
   - Alternator electrical connector
   - Alternator bracket
   - Alternator

To install:

4. Install or connect the following:
   - Alternator
   - Alternator bracket
   - Alternator electrical connector
   - Alternator drive belt
5. Adjust the alternator drive belt deflection and tension to the following specifications:
   a. Used belt deflection: 0.35–0.47 in. (9.0–12.0mm)
   b. New belt deflection: 0.30–0.41 in. (7.5–10.5mm)
   c. Used belt tension: 90–110 lbs. (400–490 N)
   d. New belt tension: 110–160 lbs. (490–712 N)
6. Torque the alternator-to-upper bracket nut to 40 ft. lbs. (54 Nm) and the alternator-to-lower bracket nut to 45 ft. lbs. (61 Nm).
7. Install or connect the following:
   - Speed control assembly
   - Under cover side panel
   - Right wheel
   - Negative battery cable

Turbo

1. Before servicing the vehicle, refer to the precautions in the beginning of this section.
2. Disconnect the negative battery cable.
3. Remove or disconnect the following:
   - Right wheel
   - Under cover side panel
   - Alternator drive belt
   - Power steering pump drive belt
   - Power steering pump

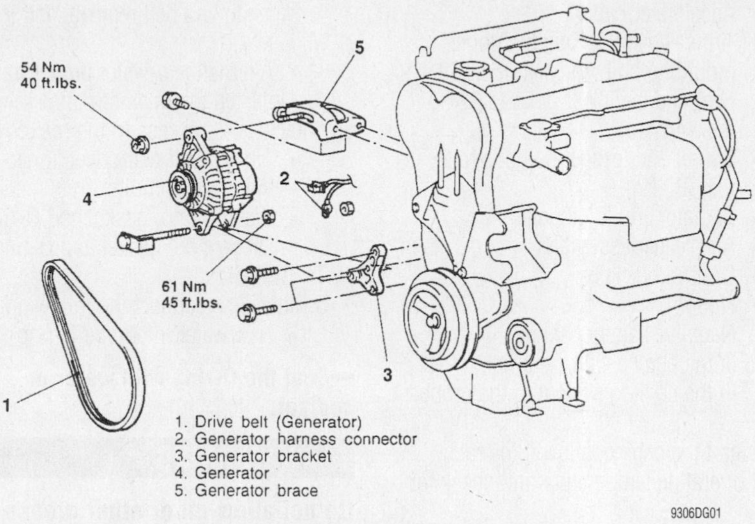

Exploded view of the alternator and related components—Non-turbo

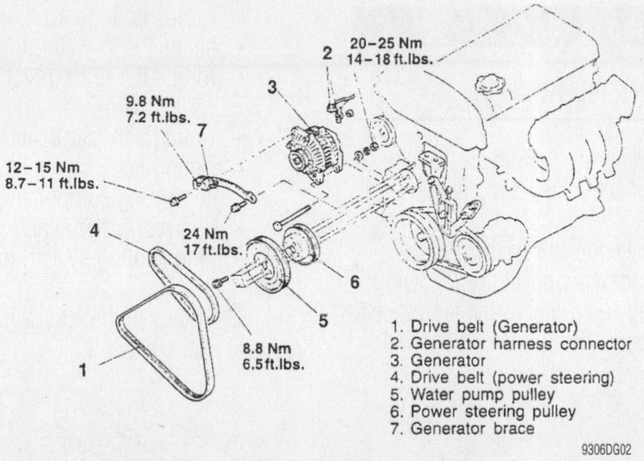

Exploded view of the alternator and related components—Turbo

➡Support the power steering pump above the engine without disconnecting the hoses.

- Alternator electrical connector
- Alternator

To install:

4. Install or connect the following:
   - Alternator
   - Alternator electrical connector
   - Alternator drive belt

5. Adjust the alternator drive belt deflection and tension to the following specifications:
   a. Used belt deflection: 0.35–0.45 in. (9–11.5mm).
   b. New belt deflection: 0.30–0.35 in. (7–9.0mm).
   c. Used belt tension: 55.1–110.2 lbs. (245–490 N).
   d. New belt tension: 110.2–154.3 lbs. (490–686 N).

6. Torque the alternator-to-upper bracket bot to 8.7–11 ft. lbs. (12–15 Nm) and the alternator-to-lower bracket nut/bolt to 14–18 ft. lbs. (20–25 Nm).

7. Adjust the power steering drive belt deflection and tension to the following specifications:
   a. Used belt deflection: 0.22–0.32 in. (5.5–8 mm).
   b. New belt deflection: 0.18–0.22 in. (4.5–5.5mm).
   c. Used belt tension: 55.1–110.2 lbs. (245–490 N).
   d. New belt tension: 110.2–154.3 lbs. (490–686 N).

8. Install or connect the following:
   - Under cover side panel
   - Right wheel
   - Negative battery cable

Ignition Timing

ADJUSTMENT

Ignition timing is controlled by the Powertrain Control Module (PCM). No adjustment is necessary or possible.

Engine Assembly

REMOVAL & INSTALLATION

The following procedure can be used on all vehicles. Slight variations may occur due to extra connections, etc., but the basic procedure should cover all models.

1. Before servicing the vehicle, refer to the precautions in the beginning of this section.
2. Relieve the fuel system pressure.
3. Drain the cooling system and the crankcase.
4. Remove or disconnect the following:
   - Negative battery cable
   - Engine undercover, if equipped
   - Hood assembly. Matchmark the hood hinges prior to removal to ease alignment during installation.
   - Air cleaner assembly and air intake ducts
   - Radiator, coolant reservoir and intercooler
   - Transaxle and transfer case, if equipped with All Wheel Drive (AWD)
   - Accelerator cable
   - Heater hoses
   - Brake vacuum hose
   - Connection for vacuum hoses
   - High pressure fuel line
   - Fuel return line
   - Oxygen (O_2S) sensor connection
   - Coolant temperature gauge connection
   - Coolant Temperature Sensor (CTS) connector
   - Connection for thermo switch sensor
   - Transaxle assembly
   - Idle speed control connection
   - Motor Position (MP) sensor connector
   - Throttle Position (TP) sensor connector
   - Exhaust Gas Recirculation (EGR) temperature sensor connection on California vehicles only
   - Fuel injector connectors

- Power transistor connector
- Ignition coil connector
- Condenser and noise filter connector
- Crankshaft Position (CKP) sensor
- Camshaft Position (CMP) sensor connectors
- Manifold differential pressure sensor connector
- Distributor and control harness
- Alternator and oil pressure switch connectors
- Drive belt and compressor and wire the compressor aside, if equipped. DO NOT discharge the system or disconnect the refrigerant lines
- Power steering pump and wire aside
- Exhaust manifold-to-pipe nuts and discard the gasket

5. Attach a hoist to the engine and support the engine weight.
   - Engine mount bracket
   - Torque control brackets (roll stoppers)

➡Some engine mount pieces have arrows on them for proper assembly. Make certain to note the orientation before removal.

6. Remove the engine assembly

To install:

7. Install or connect the following:
   - Engine and torque the engine mount bolts to 50 ft. lbs. (69 Nm)

➡The front lower mount through-bolt nut should not be tightened until the full weight of the engine is on the mount.

- Exhaust pipe and torque the nuts to 36 ft. lbs. (49 Nm)
- Vacuum hoses
- Water hoses
- Fuel return and high pressure hoses
- Brake booster vacuum hose
- Wiring harness connections
- A/C compressor electrical connectors
- CKP sensor
- CMP sensor connector
- Ignition coils and fuel injector connectors
- Manifold differential pressure sensor connector
- Capacitor
- TP sensor connector
- Ignition power transistor connector
- ECT sensor
- Engine coolant temperature gauge connector
- Oxygen (O_2S) sensor connection
- Accelerator cable
- Transaxle and torque the upper mounting bolts to 51 ft. lbs. (69 Nm) and the lower bolts to 32 ft. lbs. (44 Nm)
- Starter and torque the bolts to 22 ft. lbs. (30 Nm)
- Radiator and intercooler
- Air cleaner assembly
- Control brackets
- Hood
- Negative battery cable

8. Fill the engine with clean oil.
9. Fill the cooling system to the proper level.
10. Start the engine, allow it to reach normal operating temperature and check for leaks.
11. Road test the vehicle and check all functions for proper operation.

Water Pump

REMOVAL & INSTALLATION

1. Before servicing the vehicle, refer to the precautions in the beginning of this section.
2. Drain the cooling system.
3. Remove or disconnect the following:
   - Timing belt. Refer to the timing belt unit repair section in this manual for timing belt removal and installation.
   - Alternator-to-water pump brace
4. Unfasten the retainers, then remove any brackets for access to the rear cover.
   - Timing belt rear cover, if necessary
   - Water pump bolts
   - Water pump, gasket and O-ring. Discard the gasket and O-ring.

To install:

5. Install or connect the following:
   - New water inlet pipe O-ring

➡Coat the O-ring with water or coolant.

⁂ WARNING

Do not allow oil or other grease to contact the O-ring.

- Water pump, using a new gasket. Torque bolts to 10 ft. lbs. (13 Nm).
- Alternator-to-water pump brace. Torque the brace pivot bolt to 17 ft. lbs. (24 Nm).
- Timing belt rear cover, if removed
- Timing belt
- Remaining components
- Negative battery cable

6. Fill the cooling system to the proper level.
7. Start the engine, check for leaks and repair if necessary.

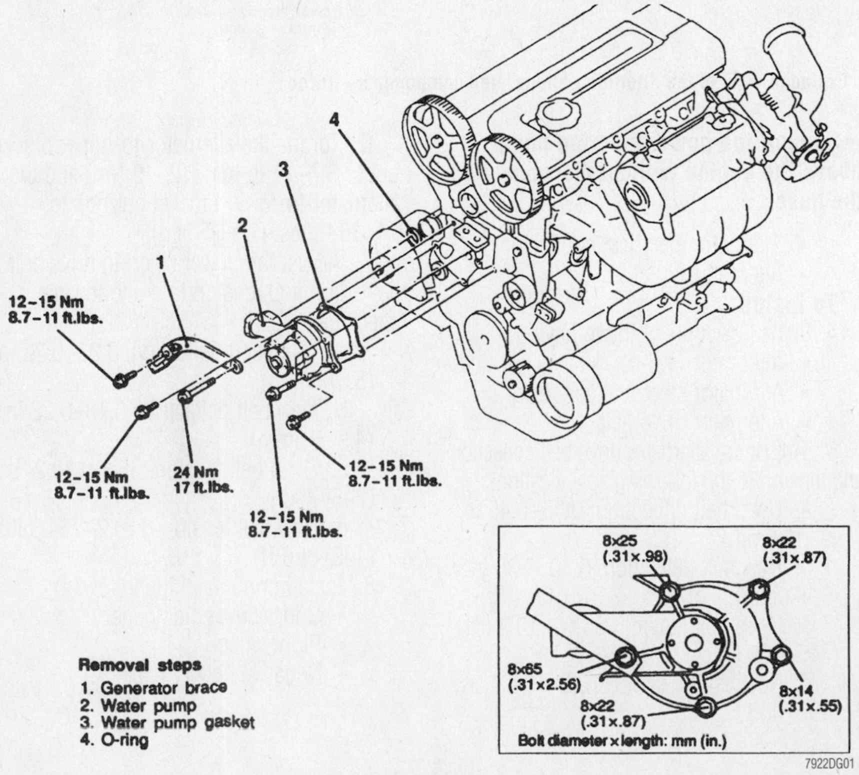

Water pump mounting and bolt locations—2.0L (VIN F) engine

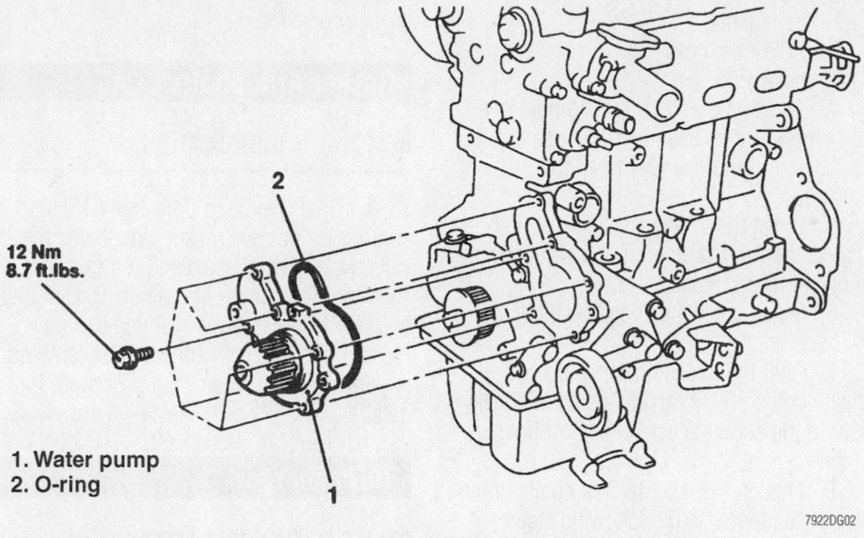

Exploded view of the water pump and O-ring mounting—2.0L (VIN Y) engine

Cylinder Head

REMOVAL & INSTALLATION

2.0l (Vin F) Engine

1. Before servicing the vehicle, refer to the precautions in the beginning of this sec tion.
2. Drain the cooling system and crankcase
3. Remove or disconnect the following:
   - Negative battery cable
   - Accelerator cable and mounting bracket
   - Intake air duct (hose) from the throttle body
4. Label and disconnect the following:
   - Idle Air Control (IAC) motor
   - Knock (ks) sensor
   - Heated Oxygen (HO_2S) sensor
   - Engine coolant temperature gauge sender
   - Engine Coolant Temperature (ECT) sensor
   - Ignition module (power transistor)
   - Throttle Position (TP) sensor
   - Condenser
   - Manifold Differential Pressure (MDP) sensor
   - Fuel injectors
   - Ignition coil
   - Camshaft Position (CMP) sensor
   - Crankshaft Position (CKP) sensor
   - Air conditioning compressor
   - Engine control wiring harness
5. Remove or disconnect the following:
   - Engine center cover
   - Spark plug wires
   - Brake booster vacuum hose
   - Fuel lines from the fuel supply rail
   - Bypass hose and water hose con-nections
   - Vacuum hoses, breather hose and Positive Crankcase Ventilation (PCV) hose
   - Timing belt
   - Power steering pump
   - Cylinder head cover and semi-cir-cular packing
   - Heat protector
   - Water and radiator hoses
   - Thermostat housing and O-ring
   - Intake manifold stay
   - Turbocharger assembly from exhaust manifold
   - Cylinder head bolts gradually, in 2–3 steps, using the specified sequence
   - Cylinder head and discard the gas-ket

To install:

6. Thoroughly, clean the deck surface of the engine block and the sealing surface of the cylinder head. Check the cylinder head for warpage.
7. Measure the length of the cylinder head bolts from below the head to the end; if the bolt measures more than 3.91 in. (99.4 mm), replace the bolt.
8. Install a new cylinder head gasket, with identification mark facing upwards.
9. Install the cylinder head. Apply clean engine oil to the bolts and install them fin-ger-tight.

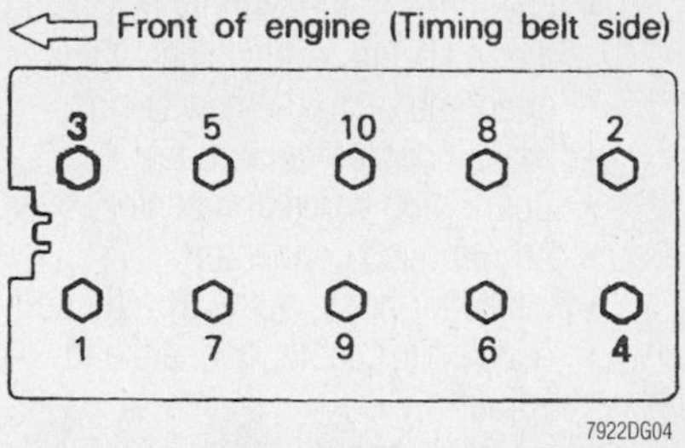

Cylinder head bolt removal sequence—2.0L (VIN F and Y) engines

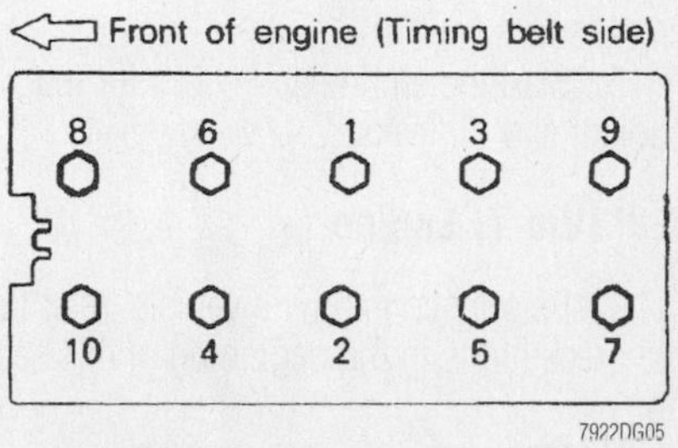

Cylinder head bolt tightening sequence—2.0L (VIN F and Y) engines

10. Tighten the bolts in sequence, using the following procedure:
    a. Tighten the bolts in sequence to 58 ft. lbs. (78 Nm).
    b. Loosen the bolts completely in the reverse order.
    c. Tighten the bolts in sequence to 15 ft. lbs. (20 Nm).
    d. Tighten the bolt ¼ turn (90 degrees) from the mark.
    e. Tighten the bolt an additional ¼ turn (90 degrees).
11. Install or connect the following:
    - Turbocharger, using a new gasket and torque the bolts to 25 ft. lbs. (35 Nm)
    - Intake manifold stay and torque the bolts to 17 ft. lbs. (23 Nm)
    - Thermostat housing and torque the bolts to 16 ft. lbs. (22 Nm)
    - Radiator and water hoses
    - Heat protector
    - Rocker arm cover and apply sealant to the semi-circular packing
12. Apply sealant at the front of the cylinder head where the camshaft oil seal retainer and the cylinder head come together
    - Power steering pump
    - Timing belt. Refer to the appropri-ate section for the removal and installation procedure.
    - PCV, breather and vacuum hoses
    - Water hose and bypass hose

- Fuel lines to fuel supply rail, using a new O-ring. Apply a small amount of engine oil to the new O-ring.
- Brake booster vacuum hose
- Spark plug wires and center cover
- All removed connectors
- Intake air hose, at the throttle body
- Accelerator cable, connect and adjust
- Negative battery cable

13. Fill the cooling system to the proper level.
14. Replace the oil filter and refill the crankcase.
15. Start the engine and check for fuel, coolant and oil leaks.

2.0l (Vin Y) Engine

1. Before servicing the vehicle, refer to the precautions in the beginning of this section.
2. Drain the cooling system and crankcase.
3. Remove or disconnect the following:
   - Negative battery cable
   - Air cleaner and air intake duct
   - Air conditioning compressor
   - Power Steering Pump (PSP) switch
   - Heated Oxygen (HO_2S) sensor
   - Engine coolant temperature gauge sender
   - Engine Coolant Temperature (ECT) sensor
   - Manifold Absolute Pressure (MAP) sensor
   - Intake Air Temperature (IAT) sensor
   - Throttle Position (TP) sensor
   - Idle Air Control (IAC) motor
   - Injector harness
   - Ignition coil
   - Camshaft Position (CMP) sensor
   - Exhaust Gas Recirculation (EGR) solenoid valve
   - Accelerator cable from the throttle body
   - Heater hoses from the rear of the engine
   - Fuel lines from the fuel supply rail
   - Purge air hose and brake booster vacuum hose connections
   - Overflow tube connection
   - Upper radiator hose and water hose connections.
   - Timing belt. Refer to the timing belt unit repair section in this manual for timing belt removal and installation.
   - Intake manifold stay
   - Intake and exhaust camshafts
   - Exhaust pipe from the exhaust manifold
   - Cylinder head bolts, in the proper sequence
   - Cylinder head

To install:

4. Thoroughly clean the cylinder head and engine block sealing surfaces. Check the deck and the cylinder head for warpage.
5. Clean the cylinder head bolts and inspect them for stretching. If a bolt appears to be stretched, replace it.
6. Install a new head gasket and the cylinder head .
7. Coat the threads of the bolts with clean engine oil and install the bolts finger-tight in the engine block. The short bolts go in the corners.
8. Tighten the cylinder head bolts in the proper sequence, in the following steps:
   a. Center bolts 1 through 6: 25 ft. lbs. (33 Nm).
   b. Outer bolts 7 through 10: 20 ft. lbs. (27 Nm).
   c. Center bolts 1 through 6: 50 ft. lbs. (67 Nm).
   d. Outer bolts 7 through 10: 20 ft. lbs. (27 Nm).
   e. Center bolts 1 through 6: 50 ft. lbs. (67 Nm).
   f. Outer bolts 7 through 10: 20 ft. lbs. (27 Nm).
   g. Turn all fasteners 1 through 10: ¼ turn (90 degrees) more in sequence. Do not use a torque wrench for this step.
9. Install or connect the following:
   - Front exhaust pipe-to-exhaust manifold using a new gasket
   - Camshafts
   - Timing belt. Refer to the timing belt unit repair section in this manual for timing belt removal and installation.
   - Intake manifold stay
   - Upper radiator hose
   - Water hose at the water pipe
   - Overflow tube
   - Brake booster vacuum hose and purge air hose
   - Fuel lines-to-fuel supply rail using a new gasket
   - Heater hose
   - All electrical connectors
   - Accelerator cable and adjust as necessary
   - Air intake duct and air cleaner assembly
   - Oil filter
10. Fill the engine with clean oil.
11. Fill the cooling system to the proper level.
12. Turn the ignition to the **ON** position and check for fuel leaks. Then, start the engine and check for coolant leaks and proper operation.

Rocker Arm/Shafts

REMOVAL & INSTALLATION

The Dual Overhead Camshaft (DOHC) engines do not use rocker arm shafts; the valves are directly actuated by rocker arms. To remove the arms, the camshaft must first be removed. It is recommended that all rocker arms and lash adjusters be replaced together. Refer the camshaft procedure for details.

Turbocharger

REMOVAL & INSTALLATION

1. Before servicing the vehicle, refer to the precautions in the beginning of this section.
2. Drain the engine oil.
3. Drain the cooling system.
4. Remove or disconnect the following:
   - Negative battery cable
   - Radiator
   - Condenser fan/radiator assembly, if equipped with A/C
   - Oxygen (O_2S) sensor
   - Oil dipstick and tube
   - Air intake bellows hose
   - Wastegate vacuum hose
   - Air outlet hose connections
   - Upper and lower heat shield
   - Power steering pump and bracket assembly. Leave the hoses connected and wire the pump aside.
   - Exhaust manifold self-locking nuts
   - Triangular engine hanger bracket
   - Oil feed line-to-turbo eyebolt and gaskets
   - Water cooling lines

➡The water line under the turbo has a threaded connection.

   - Exhaust pipe nuts and discard the gasket
   - Exhaust manifold
   - Exhaust manifold-to-turbocharger through-bolts/nuts
   - Oil return line capscrews and gasket from under the turbo
   - Turbocharger

➡Both water pipes and oil feed line can remain attached.

5. Inspect the following items:
   - Turbine (hot side) and compressor

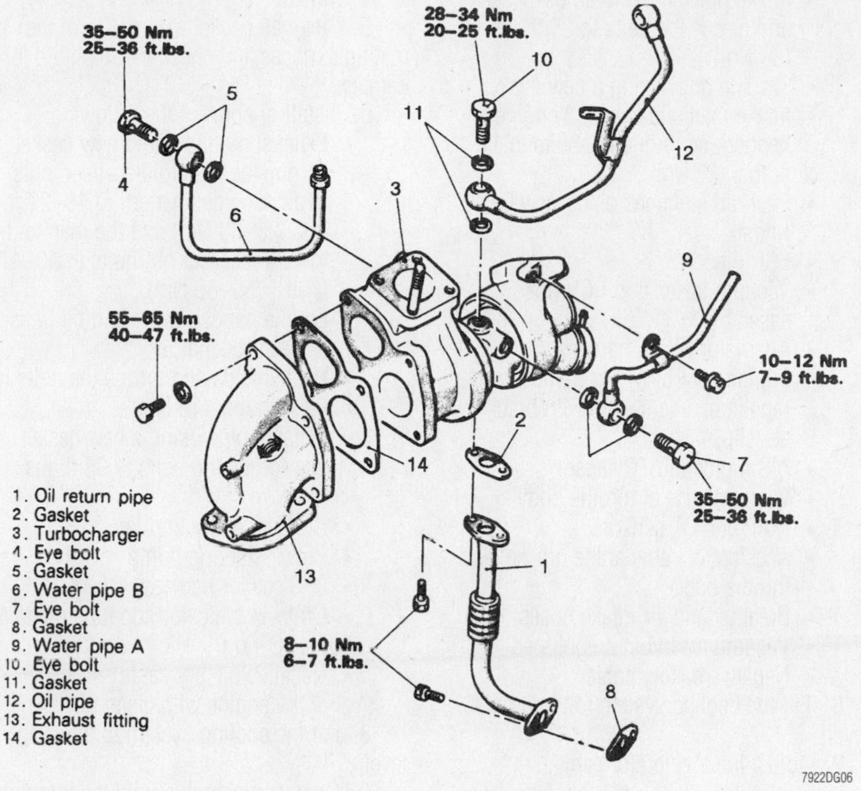

Exploded view of the turbocharger assembly—2.0L (VIN F) engine

(cold side) wheels for cracking or other damage
- Turbine and compressor wheels for ease of movement
- Oil leakage
- Wastegate valve operation. If any problem is found, replace the part.
- Oil passages for restriction or deposits, clean as required

6. Pressure test the wastegate valve by applying approximately 9 psi (62 kPa) to the actuator to make sure the rod moves.

WARNING

Do not apply more than 10 psi (71 kPa) for the wastegate diaphragm may be damaged.

➡The wastegate actuator should maintain vacuum; if not, replace it.

WARNING

Do not attempt to adjust the wastegate valve.

To install:

7. Prime the oil return line with clean engine oil. Replace all locking nuts. Before installing the threaded connection for the water inlet pipe, apply light oil to the inner pipe flange surface.
8. Install or connect the following:
   - Turbocharger
   - Exhaust manifold using a new gasket and torque the bolts to 36 ft. lbs. (49 Nm)
   - Water cooling lines and torque to 31 ft. lbs. (42 Nm)
   - Oil feed line and torque the bolts to 14 ft. lbs. (19 Nm)
   - Engine hanger and torque the bolts to 10 ft. lbs. (15 Nm)
   - Power steering pump and bracket, if removed
   - Heat shields
   - Air outlet hose
   - Wastegate hose
   - Air intake bellows
   - Oil dipstick tube and dipstick
   - O_2S sensor
   - Radiator assembly
   - Negative battery cable
9. Fill the engine with clean oil.
10. Fill the cooling system to the proper level.
11. Start the vehicle, check for leaks and repair if necessary.

Intake Manifold

REMOVAL & INSTALLATION

2.0L (VIN F) Engine

1. Before servicing the vehicle, refer to the precautions in the beginning of this section.
2. Relieve the fuel system pressure.
3. Drain the cooling system.
4. Remove or disconnect the following:
   - Negative battery cable
   - Accelerator cable, breather hose and air intake hose
   - Upper radiator hose, heater hose and water bypass hose
   - All necessary vacuum hoses and pipes
   - Brake booster vacuum line
   - High pressure fuel line and fuel return hose
5. Label and disconnect the following the electrical connectors:
   - Oxygen (O_2S) sensor
   - Engine Coolant Temperature (ECT) sensor
   - Thermo switch
   - Idle speed control
   - Exhaust Gas Recirculation (EGR) temperature sensor
   - Spark plug wires
   - Fuel rail, with injectors and pressure regulator attached
   - Intake manifold bracket
   - Throttle body water hose
   - Water inlet and heater connections
   - Thermostat housing, if necessary
   - Power brake booster and Positive Crankcase ventilation (PCV) valve vacuum lines, if connected
   - Intake manifold

To install:

6. Inspect the intake manifold's mounting surfaces for cleanliness, cracks or other damage and water passages and air passages for clogging.
7. Install or connect the following:
   - Intake manifold, using a new gasket. Torque the manifold-to-cylinder head bolts in a crisscross pattern, starting from the inside and working outwards, to 11–14 ft. lbs. (15–19 Nm).
   - Fuel rail, with injectors and pressure regulator, lubricate all seals lightly with oil. Torque the retaining bolts to 84–108 inch lbs. (10–13 Nm).

- Thermostat housing, if removed
- Intake manifold brace bracket
- Throttle body bracket
- All hoses, cables and electrical connectors
- Negative battery cable

8. Refill the cooling system
9. Run the engine until the thermostat opens, refill the radiator completely and check for leaks.
10. Adjust the accelerator cable. Check and adjust the ignition timing. Once the vehicle has cooled, recheck the coolant level.

2.0L (VIN Y) Engine

1. Before servicing the vehicle, refer to the precautions in the beginning of this section.
2. Properly relieve the fuel system pressure.
3. Drain the cooling system.
4. Remove or disconnect the following:
   - Negative battery cable
   - Vacuum reservoir, if equipped with cruise control
   - Air intake and breather hoses
   - Accelerator cable from the bracket
   - Engine harness retaining clips
   - Manifold Absolute Pressure (MAP) sensor
   - Coolant Temperature (CT) sensor connector
   - Throttle Position (TP) sensor connector
   - Air Injector Service (AIS) motor connector
   - Engine control wiring harness and move it aside
   - Alternator wiring harness
   - Pollution Control Valve (PCV) hose assembly
   - Vacuum hoses
   - Exhaust Gas Recirculation (EGR) pipe
   - Fuel lines from the fuel rail
   - Intake manifold stay and engine hanger
   - Throttle body
   - Intake manifold plenum and gasket
   - Injector connectors
   - Fuel rail with injectors
   - Intake manifold

To install:

5. Install or connect the following:
   - Intake manifold using a new gasket and torque the retainers in a crisscross pattern to 17 ft. lbs. (23 Nm)
   - Fuel rail assembly and torque the bolts to 89 inch lbs. (10 Nm)
   - Injector connectors
   - Intake plenum using a new gasket and torque the bolts to 25 ft. lbs. (34 Nm)
   - Throttle body using a new gasket
   - Intake manifold stay and engine hanger and torque the bolts to 17 ft. lbs. (23 Nm)
   - Fuel rail fuel lines using new O-rings
   - EGR pipe
   - Vacuum hoses and PCV hose assembly
   - Alternator wiring harness
   - Engine control wiring harness and reposition and secure with brackets/clips
   - AIS motor and TP sensor
   - Vacuum hose at throttle body
   - MAP and CT sensors
   - Accelerator cable at the bracket and throttle body
   - Breather and air intake hoses
   - Vacuum reservoir, if equipped
   - Negative battery cable
6. Fill the cooling system to the proper level.
7. Adjust the accelerator cable.

Exhaust Manifold

REMOVAL & INSTALLATION

2.0L (VIN F) Engine

1. Before servicing the vehicle, refer to the precautions in the beginning of this section.
2. Drain the engine oil.
3. Drain the cooling system.
4. Remove or disconnect the following:
   - Negative battery cable
   - Condenser cooling fan, if equipped with A/C
   - Power steering pump and move it aside
   - Oxygen (O_2S) sensor harness
   - Dipstick tube
   - Exhaust pipe-to-turbocharger nuts, separate the exhaust pipe and discard the gasket
   - Air intake and vacuum hose connections
   - Heat shields at exhaust manifold and turbocharger
   - Exhaust manifold-to-turbocharger bolts and nut
   - Engine hanger, water and oil lines from the turbocharger, if necessary
   - Exhaust manifold and discard gasket

To install:

5. Clean all gasket material from the mating surfaces and check the manifold for damage.
6. Install or connect the following:
   - Exhaust manifold and new gasket. Torque the manifold-to-head nuts in a crisscross pattern to 18–22 ft. lbs. (25–30 Nm) and the manifold-to-turbocharger nut/bolts to 40–47 ft. lbs. (55–65 Nm).
   - Engine hanger, water and oil lines at the turbocharger
   - Heat shields and torque the bolts to 89 inch lbs. (10 Nm)
   - Exhaust pipe using a new gasket and torque the bolts to 33 ft. lbs. (44 Nm)
   - Condenser cooling fan
   - Power steering pump
   - O_2S sensor harness
   - Oil level indicator and tube using a new O-ring
   - Negative battery cable
7. Fill the engine with clean oil.
8. Fill the cooling system to the proper level.
9. Operate the engine until the thermostat opens.
10. Check for fluid and exhaust leaks. Top off the engine coolant.

2.0L (VIN Y) Engine

1. Before servicing the vehicle, refer to the precautions in the beginning of this section.
2. Drain the cooling system.
3. Remove or disconnect the following:
   - Negative battery cable
   - Air intake hose
   - Upper radiator hose from water outlet
   - Air hose connection
   - Engine control wiring harness from the rear of the engine
   - Water pipe assembly
   - Oil dipstick
   - Upper heat shield
   - Engine hanger
   - Front exhaust pipe from the manifold
   - Lower heat shield
   - Exhaust manifold and discard gasket

To install:

4. Install or connect the following:
   - Exhaust manifold with a new gasket and torque the bolts to 17 ft. lbs. (23 Nm)
   - Lower heat shield and torque the bolts to 10 ft. lbs. (15 Nm)
   - Front exhaust pipe using a new

gasket and torque the bolts to 33 ft. lbs. (44 Nm)
- Engine hanger and torque the bolts to 10 ft. lbs. (15 Nm)
- Upper heat shield and torque the bolts to 10 ft. lbs. (15 Nm)
- Oil dipstick and water pipe
- Engine wiring harness to the rear of the engine
- Air hose and upper radiator hose
- Air intake hose
- Negative battery cable

5. Fill the cooling system to the proper level.
6. Start the vehicle, check for leaks and repair if necessary.

Front Crankshaft Seal

REMOVAL & INSTALLATION

2.0L (VIN F) Engine

1. Before servicing the vehicle, refer to the precautions in the beginning of this section.
2. Remove or disconnect the following:
   - Negative battery cable
   - Timing belt. Refer to the timing belt unit repair section in this manual for timing belt removal and installation.

➡If reusing the timing belt, be sure to mark the direction of rotation on the belt. This will ensure the same direction of rotation, thereby extending belt life.

- Crankshaft pulley
- Timing belt crankshaft sprocket

➡If the sprocket is difficult to remove, an appropriate puller may be used.

- Crankshaft seal by prying it out using a suitable prytool

To install:

3. Install or connect the following:
   - New crankshaft seal using a driver tool
   - Crankshaft sprocket
   - Timing belt. Refer to the timing belt unit repair section in this manual for timing belt removal and installation.
   - Negative battery cable

2.0L (VIN Y) Engine

1. Before servicing the vehicle, refer to the precautions in the beginning of this section.

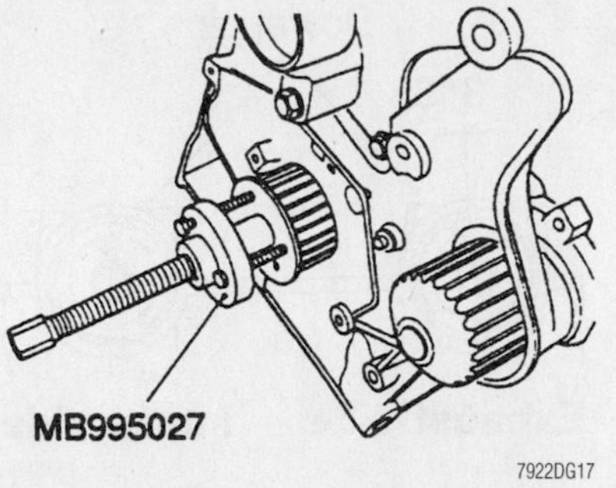

To prevent damaging the end of the crankshaft, use the proper sprocket removal tool—2.0L (VIN Y) engine

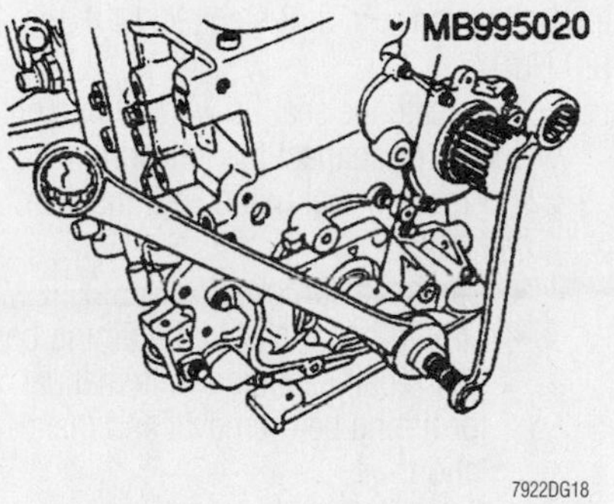

Using the special tool to remove the front crankshaft seal—2.0L (VIN Y) engine

2. Remove or disconnect the following:
   - Negative battery cable
   - Timing belt. Refer to the timing belt unit repair section in this manual for timing belt removal and installation.
   - Crankshaft sprocket using Crankshaft Sprocket Removal Tool MB995027
   - Crankshaft oil seal using Crankshaft oil seal Removal Tool MB995020

➡Be careful not the scratch the oil seal bore or the crankshaft sealing surface.

To install:

3. Apply clean engine oil to the oil seal.
4. Install or connect the following:
   - Oil seal using a seal driver
   - Crankshaft sprocket using special Tools MB995035 and MB995026
   - Timing belt. Refer to the timing belt unit repair section in this manual for timing belt removal and installation.
   - Negative battery cable

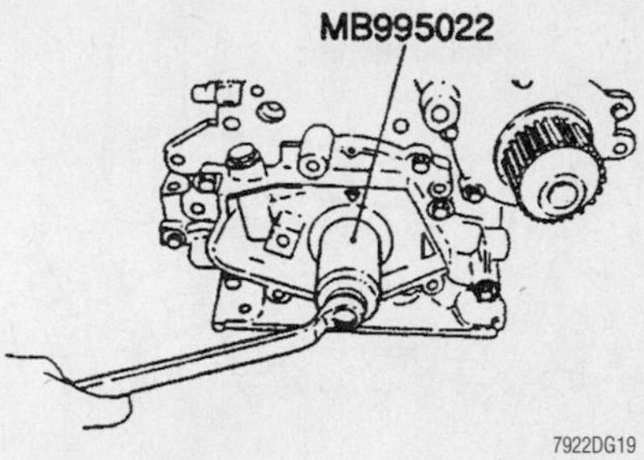

To avoid damaging the front seal, use the front oil seal installer Tool MB995022—2.0L (VIN Y) engine

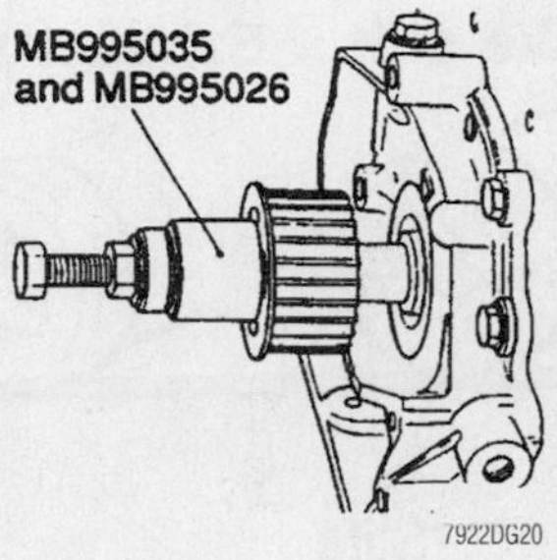

Installing the front crankshaft sprocket using the Crankshaft Sprocket Installer Tools MB995035 and MB995026—2.0L (VIN Y) engine

Camshaft and Valve Lifters

REMOVAL & INSTALLATION

2.0L (VIN F) Engine

1. Before servicing the vehicle, refer to the precautions in the beginning of this section.
2. Remove or disconnect the following:
   - Negative battery cable
   - Accelerator cable from the throttle body
   - Cable bracket from intake plenum
   - Engine center cover
   - Spark plug cables
   - Breather hose and Pollution Control Valve (PCV) hose, from the rocker cover
   - Rocker cover
3. Position the No. 1 piston at Top Dead Center (TDC) on the compression stroke.
   - Timing belt. Refer to the timing belt unit repair section in this manual for timing belt removal and installation.
   - Camshaft sprockets

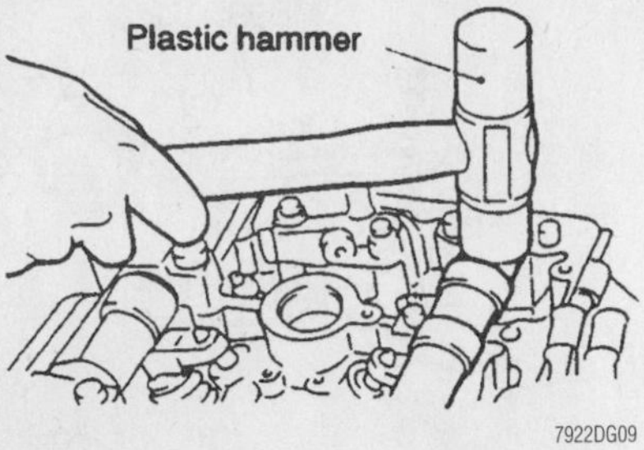

Tap the camshaft with a plastic hammer to loosen the bearing caps—2.0L (VIN F) engine

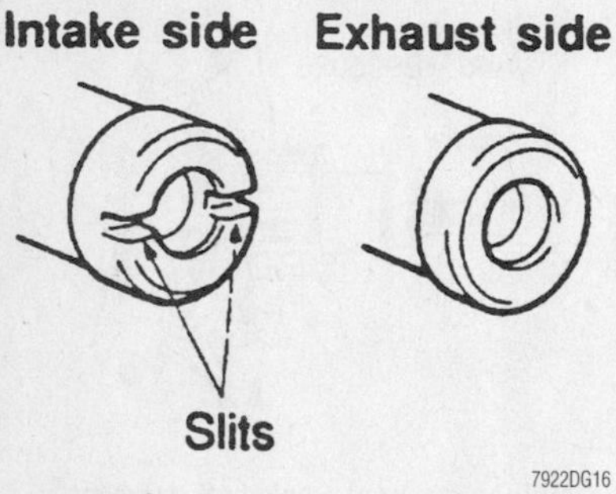

Identifying the rocker arm shafts—notice the slits in the intake side

➡Use a wrench on the hex shaped part of the camshaft to hold the cam when removing the sprockets.

- Bearing caps. Loosen the bearing cap bolts in 2–3 steps.

➡If the bearing caps are hard to remove, tap the rear of the camshaft with a plastic hammer.

⁂ WARNING

Bearing caps and rocker arms must be installed in the same location from which they were removed.

- Camshaft(s) and seals

To install:

4. Install or connect the following:
   - Camshafts, lubricate with engine oil and position with the dowels facing up

⁂ WARNING

If new camshaft(s) are being installed, remove the rocker arms and install the camshaft(s) and bearing caps. Be sure the camshaft(s) can be turned by hand. After checking, remove the camshafts and install the rocker arms.

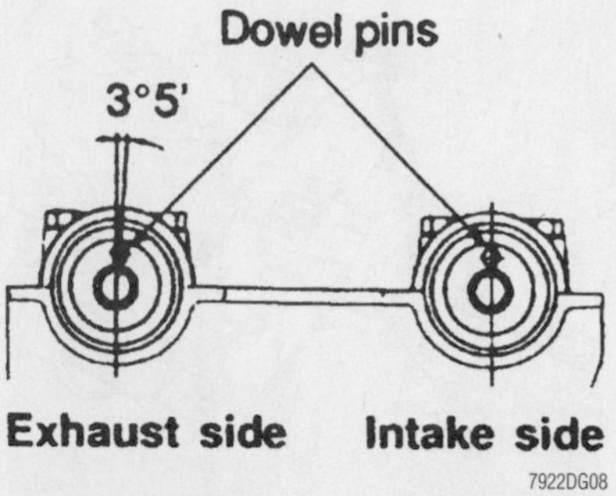

For installation, position the camshafts with the dowels facing up, as shown—2.0L (VIN F) engine

5. Install the bearing caps and torque the bolts evenly, in 2–3 steps to 14 ft. lbs. (20 Nm).
6. Lubricate the seal lip with engine oil.
7. Install or connect the following:
   - Front oil seal using Seal Installer MB998713
   - Camshaft sprockets
   - Timing belt. Refer to the timing belt unit repair section in this manual for timing belt removal and installation.
   - Cylinder head by applying sealant to the semi-circular packing
8. Apply sealant to the lower part of the front and rear bearing caps where they meet the cylinder head.
   - Rocker cover using a new gasket
   - PCV hose and breather hose
   - Spark plug wires
   - Center cover
   - Accelerator cable and adjust it
   - Negative battery cable

2.0L (VIN Y) Engine

1. Before servicing the vehicle, refer to the precautions in the beginning of this section.
2. Remove or disconnect the following:
   - Negative battery cable
   - Ignition coil pack
   - Pollution Control Valve (PCV) hose and breather hose from cylinder head cover
   - Semi-circular packing from the rear of the head
   - Camshaft Position (CMP) sensor
   - Timing belt. Refer to the timing belt unit repair section in this manual for timing belt removal and installation.
   - Camshaft sprockets

➡Use tool MB990767 and MB998719, to secure the camshaft sprockets when removing the sprocket mounting bolt.

- Rear timing belt cover and bracket
- Outside camshaft bearing cap
- Bearing caps. Gradually, loosen the camshaft bearing caps in the reverse of the tightening sequence, one camshaft at a time.

➡Keep the bearing caps in order. They must be installed in the location from which they were removed.

- Camshafts

➡The camshafts are not interchangeable. Mark the camshafts for later identification.

To install:

3. Install or connect the following:
   - Camshafts, lubricate with engine oil
   - Bearing caps and torque the bolts evenly and in sequence to 21 ft. lbs. (28 Nm)
   - Outside camshaft bearing caps, lubricate them with Loctite 518®
   - Camshaft oil seal

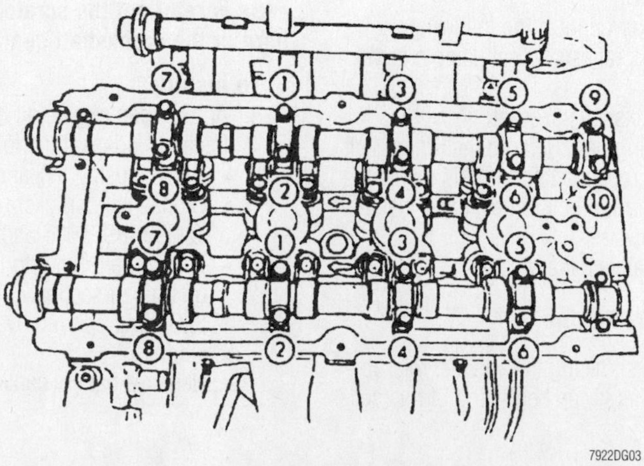

Camshaft bearing cap bolt tightening sequence—2.0L (VIN Y) engine

- Rear timing belt cover and bracket
- Camshaft sprockets
- Timing belt. Refer to the timing belt unit repair section in this manual for timing belt removal and installation.
- Semi-circular packing, lubricate with Loctite 5699®, at rear of cylinder head
- CMP sensor

4. Install the cylinder head cover. Torque the bolts in the proper sequence, evenly as follows:
    a. Step 1: 40 inch lbs. (4.5 Nm).
    b. Step 2: 80 inch lbs. (9.0 Nm).
    c. Step 3: 106 inch lbs. (12 Nm).
5. Install or connect the following:
    - Air, breather and PCV hoses
    - Coil pack
    - Negative battery cable

Valve Lash

ADJUSTMENT

Valve clearance is not adjustable. Proper valve clearance is maintained by the hydraulic lash adjusters.

Starter

REMOVAL & INSTALLATION

Non-Turbo

1. Before servicing the vehicle, refer to the precautions in the beginning of this section.
2. Remove or disconnect the following:
    - Negative battery cable
    - Starter electrical connectors
    - Starter bolts and starter

To install:

3. Install or connect the following:
    - Starter and torque the starter-to-transaxle bolts to 40 ft. lbs. (54 Nm)
    - Starter electrical connectors
    - Negative battery cable

Turbo

1. Before servicing the vehicle, refer to the precautions in the beginning of this section.
2. Remove or disconnect the following:
    - Battery
    - Air hose "C"
    - Starter electrical connectors
    - Starter bolts and starter

To install:

3. Install or connect the following:
    - Starter and torque the starter-to-transaxle bolts to 22 ft. lbs. (30 Nm)

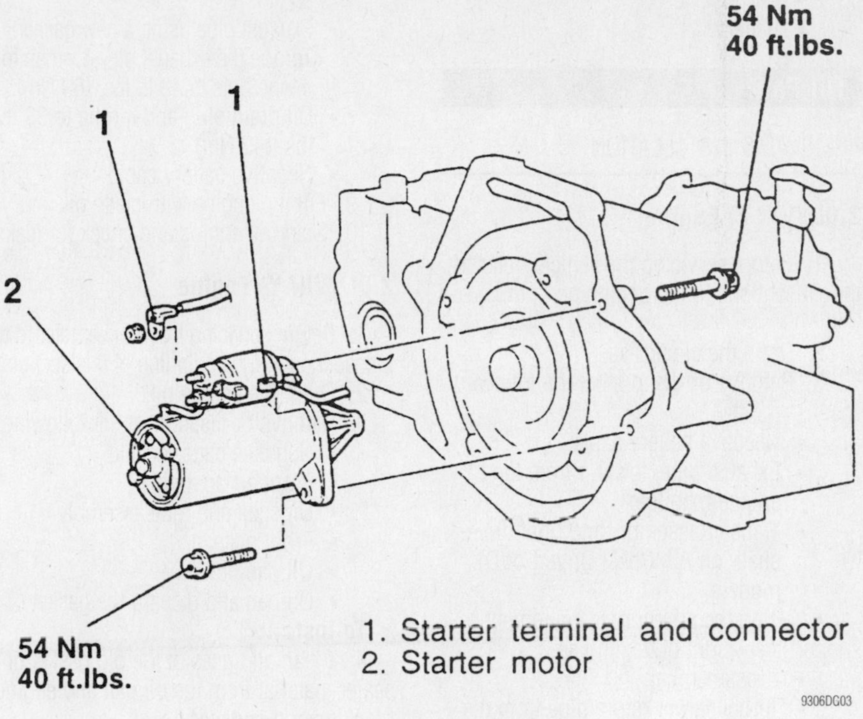

Exploded view of the starter and related components—Non-turbo

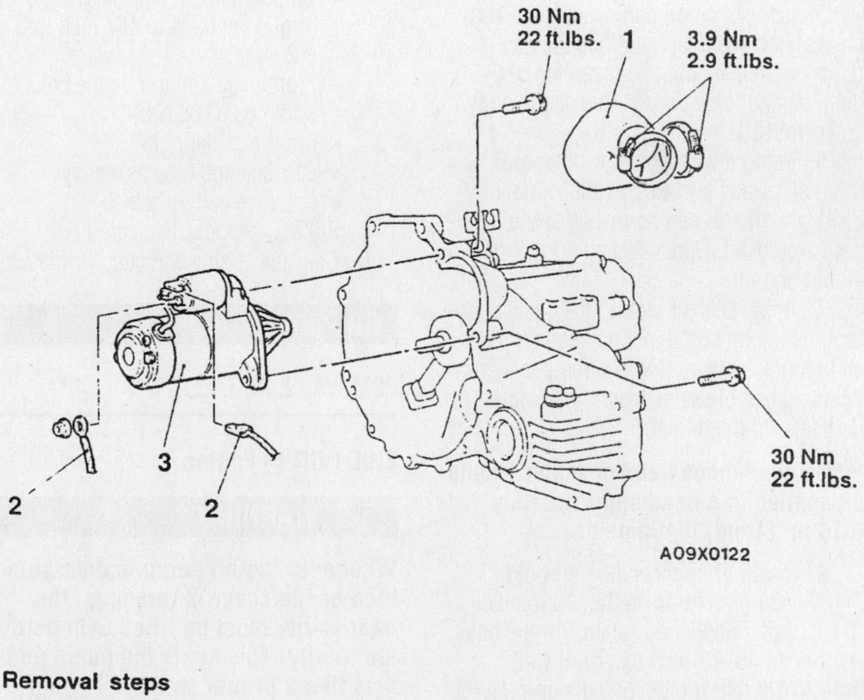

Exploded view of the starter and related components—Turbo

- Starter electrical connectors
- Air hose "C"
- Battery

Oil Pan

REMOVAL & INSTALLATION

2.0L (VIN F) Engine

1. Before servicing the vehicle, refer to the precautions in the beginning of this section.
2. Drain the crankcase.
3. Remove or disconnect the following:
   - Negative battery cable
   - Exhaust pipe, lower it from the engine manifold
   - Transfer assembly and right driveshaft, on All Wheel Drive (AWD) models
4. Using the appropriate equipment, support the weight of the engine.
   - Crossmember
   - Turbocharger return pipe from the side of the oil pan
   - Oil pan bolts. Tap a thin prytool between the engine block and oil pan to break the seal.
5. Inspect the oil pan for damage and cracks. Replace if faulty. While the pan is removed, inspect the oil screen for clogging, damage and cracks. Replace if faulty.

To install:

6. Using a wire brush or other tool, clean all gasket surfaces of the cylinder block and the oil pan so that all loose material is removed. Clean sealing surfaces of all dirt and oil.
7. Apply sealant around the gasket surfaces of the oil pan in such a manner that all bolt holes are circled and there is a continuous bead of sealer around the entire perimeter of the oil pan.

➡The continuous bead of sealer should be applied in a bead approximately 0.16 in. (4mm) in diameter.

8. Install or connect the following:
   - Oil pan. Install within 15 minutes after applying sealant. Torque bolts to 48–72 inch lbs. (6–8 Nm).
   - Oil return pipe, using a new gasket and torque the retainers to 62–88 inch lbs. (7–10 Nm)
   - Crossmember and torque the bolts to 72 ft. lbs. (100 Nm)
   - Left member and torque the forward retainer bolts to 72 ft. lbs. (100 Nm) and the rearward left member bolts to 58 ft. lbs. (80 Nm)
   - Transfer assembly and right driveshaft
   - Exhaust pipe using a new gasket. Torque the exhaust pipe-to-manifold flange nuts to 43 ft. lbs. (60 Nm).
   - Oil drain plug and torque to 33 ft. lbs. (42 Nm)
   - Negative battery cable
9. Fill the engine with clean oil.
10. Start the engine and check for leaks.

2.0L (VIN Y) Engine

1. Before servicing the vehicle, refer to the precautions in the beginning of this section.
2. Drain the engine oil.
3. Remove or disconnect the following:
   - Negative battery cable
   - Front exhaust pipe
   - Dipstick and tube assembly
   - Front plate
   - Oil pan bolts
   - Oil pan and discard the gasket

To install:

4. Clean all traces of the old gasket or sealer material from the oil pan and engine block mating surfaces.
5. Apply sealant at the point where the engine block meets the oil pump.
6. Install or connect the following:
   - Oil pan using a new gasket and torque the bolts to 107 inch lbs. (12 Nm)
   - Front plate and torque the bolts to 80 ft. lbs. (108 Nm)
   - Front exhaust pipe
   - Dipstick and tube assembly
   - Negative battery cable
7. Fill the engine with clean oil.
8. Start the engine and check for leaks.

Oil Pump

REMOVAL & INSTALLATION

2.0L (VIN F) Engine

⁂ WARNING

Whenever the oil pump is disassembled or the cover is removed, the gear cavity must be filled with petroleum jelly. This seals the pump and acts like a primer so the oil pump draws oil as soon as the engine turns. Do not use grease.

1. Before servicing the vehicle, refer to the precautions in the beginning of this section.
2. Disconnect the negative battery cable. Rotate the engine so that the No. 1 cylinder is at Top Dead Center (TDC) of its compression stroke. The timing marks should be aligned at this point.
3. Drain the engine oil.
4. Using the proper equipment, support the weight of the engine.
5. Remove or disconnect the following:
   - Front engine mount bracket and accessory drive belts
   - Timing belt upper and lower covers
   - Timing belt and crankshaft sprocket
   - Oil pressure sending unit electrical connector
   - Oil pressure sensor
   - Oil filter and bracket
   - Oil pan, oil screen and discard gasket
   - Plug cap in the engine front cover, using special plug removal Tool MD998162
   - Plug on the side of the engine block. Insert a suitable tool with a shaft diameter of 0.32 in. (8mm) into the plug hole. This will hold the silent shaft in position.
   - Oil pump driven gear-to-silent shaft bolt
   - Front cover mounting bolts

➡Note the lengths of the mounting bolts as they are removed for proper installation.

   - Front case cover and oil pump assembly

➡If necessary, the silent shaft can come out with the cover assembly.

   - Oil pump cover at the rear of the engine front cover.
   - Oil pump drive and driven gears
6. After disassembling the oil pump, clean all components and remove all residual gasket material from the mating surfaces.
7. Assemble the oil pump gears into the front case and rotate it to ensure smooth movement and no looseness. Be sure there

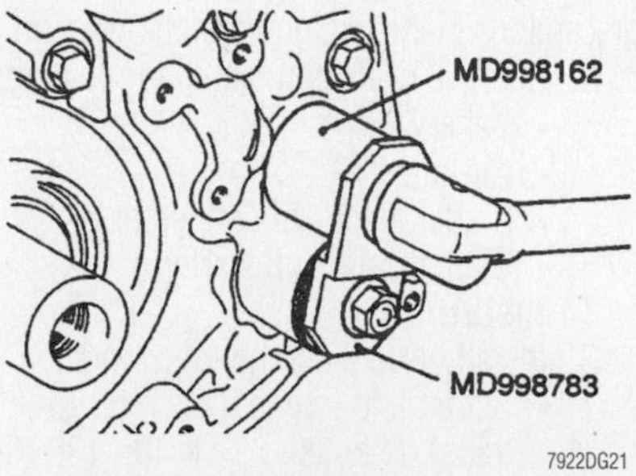

Using the special Tool MD998162 to tighten the plug cap—2.0L (VIN F) engine

is no ridge wear on the contact surface between the front case and the gear surface of the oil pump front cover.

To install:

8. Install or connect the following:
   - Oil pump, by aligning the oil pump drive gear with the driven gear timing marks. Lubricate the gears with engine oil.
   - Oil pump cover and torque the bolts to 13 ft. lbs. (18 Nm)
   - New crankshaft seal using an appropriate driver
   - New front case gasket. Position seal guide Tool MD998285, on the front end of the crankshaft to protect the seal from damage. Apply a thin coat of oil to the outer circumference of the seal pilot tool.
   - Front case assembly, through a new front case gasket and torque the bolts to 17 ft. lbs. (24 Nm)
   - Oil filter on the bracket using a new oil filter bracket gasket and torque the bolts to 25 ft. lbs. (34 Nm)
9. Insert a suitable tool into the hole in the left side of the engine block to lock the silent shaft in place.
10. Secure the oil pump drive gear onto the left silent shaft by installing and tightening the driven gear bolt to 29 ft. lbs. (40 Nm).
11. Install or connect the following:
    - Plug cap using a new O-ring in the front case groove. Using the special Tool MD998162, tighten the cap to 20 ft. lbs. (27 Nm).
    - Oil screen using a new gasket
12. Clean the oil pan and cylinder block mating surfaces. Apply sealant in the oil pan flange groove, keeping towards the inside of the bolt holes. The width of the sealant bead applied is to be about 0.16 in. (4mm) wide.

➡Install the oil pan within 15 minutes of sealant application.

13. Install or connect the following:
    - Oil pan and torque the bolts to 108 inch lbs. (12 Nm)
    - Oil pressure gauge unit and oil pressure switch
    - Oil pressure gauge unit and oil pressure switch
    - Electrical harness
    - Oil cooler and torque the bolt to 33 ft. lbs. (45 Nm)
    - Oil filter and fill the crankcase
    - Negative battery cable
14. Start the engine. Verify oil pressure and inspect for leaks.

2.0L (VIN Y) Engine

1. Before servicing the vehicle, refer to the precautions in the beginning of this section.
2. Drain the engine oil.
3. Remove or disconnect the following:
   - Negative battery cable
   - Rear plate
   - Oil filter and adapter
   - Oil pan
   - Oil pick-up tube
   - Timing belt. Refer to the timing belt unit repair section in this manual for timing belt removal and installation.
   - Crankshaft sprocket using the crankshaft sprocket Removal tool MB995027

⁂ WARNING

Do not nick the crankshaft sealing surface or the seal bore.

   - Crankshaft oil seal using crankshaft oil seal tool MB995020
   - Oil pump mounting bolts
   - Oil pump

To install:

4. Apply a bead of sealant to the sealing surface of the oil pump and install a new O-ring into the counterbore on the oil pump discharge passage.
5. Install or connect the following:
   - Oil pump and torque the bolts to 17 ft. lbs. (23 Nm)
   - New crankshaft oil seal in the oil pump
   - Crankshaft sprocket using the proper installation tools
   - Timing belt. Refer to the timing belt unit repair section in this manual for timing belt removal and installation
   - Oil pickup tube
6. Apply Loctite® 18718, at the point where the oil pump meets the engine block.
   - Oil pan using a new gasket and torque the bolts to 108 inch lbs. (12 Nm)
   - Oil filter adapter using a new O-ring. Align the roll pin with the hole and torque the assembly to 40 ft. lbs. (55 Nm).
   - New oil filter
   - Rear plate
   - Negative battery cable
7. Fill the engine with clean oil.
8. Start the engine and check for leaks.

Rear Main Seal

REMOVAL & INSTALLATION

1. Before servicing the vehicle, refer to the precautions in the beginning of this section.
2. Remove or disconnect the following:
   - Negative battery cable
   - Transaxle
   - Transfer case, if equipped
   - Drive plate, if equipped with an automatic transaxle
   - Bell housing cover and flywheel, if equipped with a manual transaxle
   - Crankshaft rear oil seal case, if leaking; otherwise, the oil seal.

➡Some engines have a separator that should also be removed.

To install:

3. Lubricate the inner diameter of the new seal with clean engine oil.
4. Install or connect the following:
   - Oil seal in the crankshaft rear oil seal case using Seal Installer Tool MD998376

➡Press the seal all the way in without tilting it.

5. If removed, drive the oil separator into the oil seal case so the its oil hole is facing downward.
6. Install or connect the following:
   - Seal case using a new gasket, if removed
   - Flywheel or drive plate
   - Transfer case, if equipped
   - Transaxle
   - Negative battery cable
7. Refill the crankcase. Start the engine and check for leaks.

Piston and Ring

POSITIONING

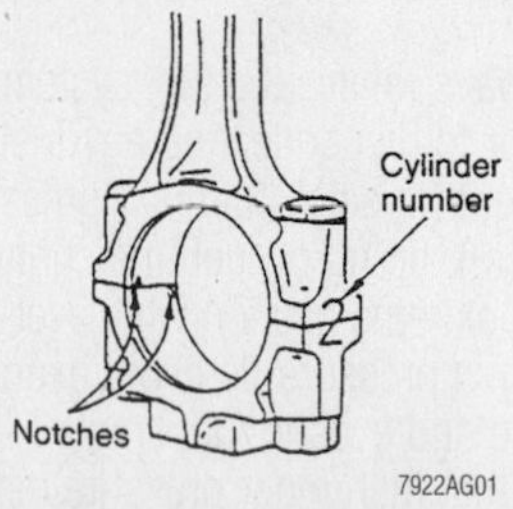

Chrysler engine connecting rod and cap installation—ensure to matchmark the cap and rod prior to disassembly

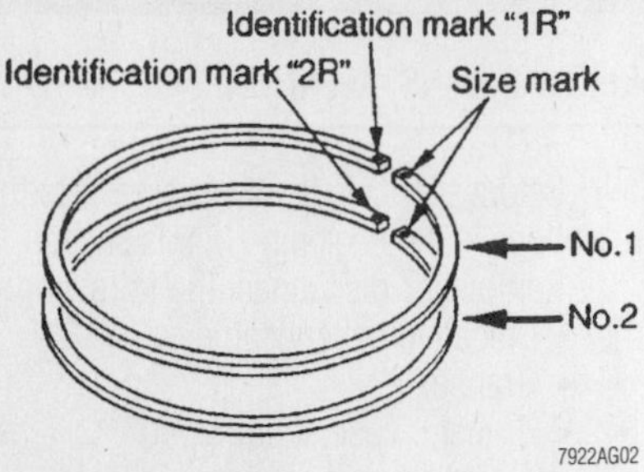

Chrysler piston ring identification mark locations

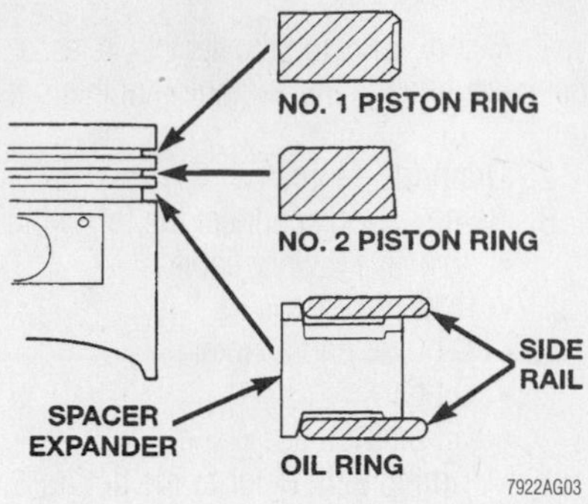

Piston ring orientation—2.0L Engine

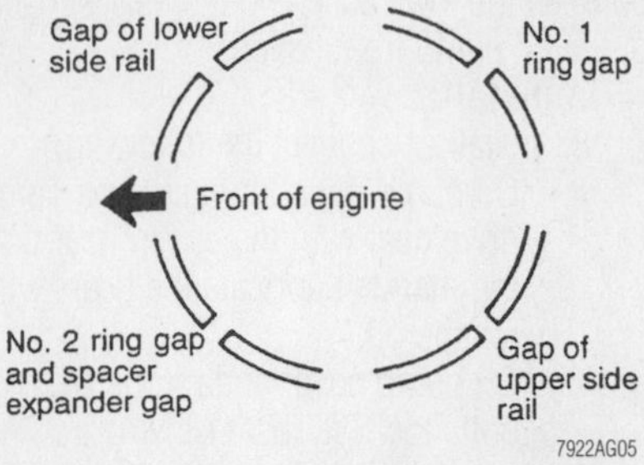

Piston ring end-gap spacing—2.0L (VIN Y) engine

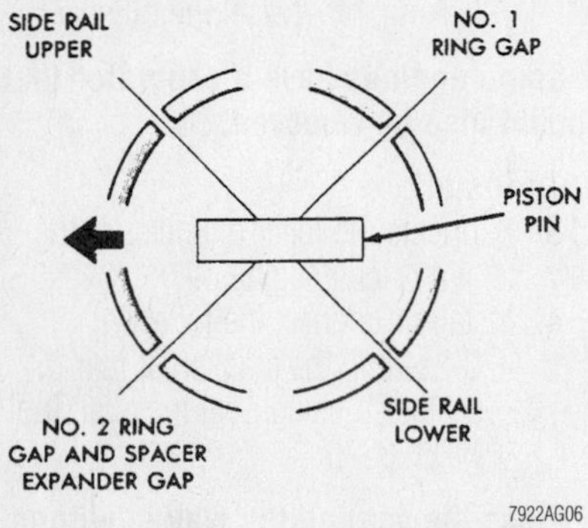

Piston ring end-gap spacing—2.0L (VIN F) engine

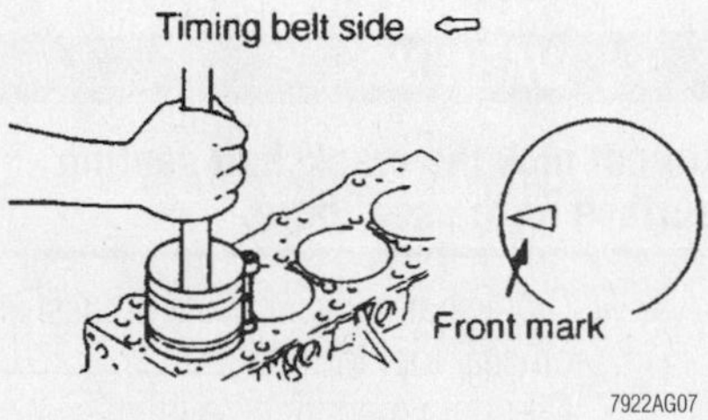

Piston positioning—2.0L (VIN F) engine

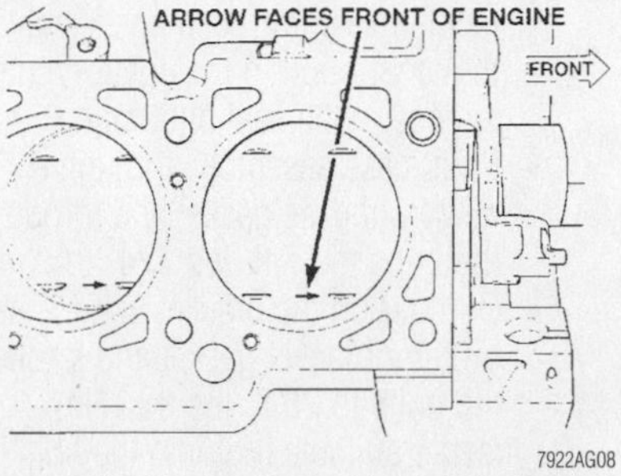

Piston positioning—2.0L (VIN Y) engine. The small arrows on the crown of the pistons must point toward the front of the engine

FUEL SYSTEM

Fuel System Service Precautions

Safety is an important factor when servicing the fuel system. Failure to conduct maintenance and repairs in a safe manner may result in serious personal injury. Maintenance and testing of the vehicle's fuel system components can be accomplished safely and effectively by adhering to the following rules and guidelines.

- To avoid the possibility of fire and personal injury, always disconnect the negative battery cable unless the repair or test procedure requires that battery voltage be applied.
- Always relieve the fuel system pressure prior to disconnecting any fuel system component (injector, fuel rail, pressure regulator, etc.), fitting or fuel line connection. Exercise extreme caution whenever relieving fuel system pressure, to avoid exposing skin, face and eyes to fuel spray. Please be advised that fuel under pressure may penetrate the skin or any part of the body that it contacts.
- Always place a shop towel or cloth around the fitting or connection prior to loosening to absorb any excess fuel due to spillage. Ensure that all fuel spillage is quickly removed from engine surfaces. Ensure that all fuel soaked cloths or towels are deposited into a suitable waste container.
- Always keep a dry chemical (Class B) fire extinguisher near the work area.
- Do not allow fuel spray or fuel vapors to come into contact with a spark or open flame.
- Always use a back-up wrench when loosening and tightening fuel line connection fittings. This will prevent unnecessary stress and torsion to fuel line piping.
- Always replace worn fuel fitting O-rings. Do not substitute fuel hose where fuel pipe is installed.

Fuel System Pressure

RELIEVING

1. Be sure to observe all cautions and warnings in the beginning of the section that may be related to this procedure.
2. Remove or disconnect the following:
   - Rear seat cushion
   - Protector
   - Fuel pump connector
3. Start the engine and allow it to run until it stops, due to lack of fuel. Turn the ignition switch to the **OFF** position.
4. Disconnect the negative battery cable.

After relieving fuel pressure:

5. Install or connect the following:
   - Fuel pump connector
   - Protector
   - Rear seat cushion

Fuel Filter

REMOVAL & INSTALLATION

On most vehicles, the fuel filter is located in the engine compartment, mounted on the firewall. On some non-turbo engines, the fuel filter is mounted under the vehicle near the fuel tank.

CAUTION

Do not use conventional fuel filters, hoses or clamps when servicing fuel injection systems. They are not compatible with the injection system and could fail, causing personal injury or damage to the vehicle. Use only

hoses and clamps specifically designed for fuel injection systems.

1. Properly relieve the fuel system pressure.
2. On non-turbo engines, raise and safely support the vehicle to gain access to the filter.
3. Before servicing the vehicle, refer to the precautions in the beginning of this section.

➡Wrap shop towels around the fitting that is being disconnected to absorb residual fuel in the lines.

4. Cover the hose connection with shop towels to prevent any splash of fuel that could be caused by residual pressure in the fuel pipe line.
5. Remove or disconnect the following:
   - Negative battery cable
   - Eye bolt, by holding the fuel filter nut securely with a back-up wrench
   - High-pressure fuel line from the filter and discard gaskets

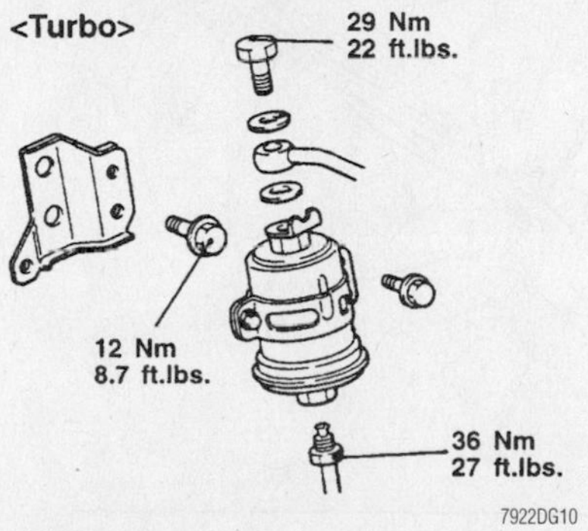

Exploded view of the fuel filter mounting—2.0L (VIN F) engine

   - Main pipe flare nut, loosen it while holding the fuel filter nut securely with a back-up wrench
   - Flare nut connection from the filter and discard gaskets
6. If equipped with a fuel tank mounted filter, remove or disconnect the following:
   - Eyebolt, gasket and connector
   - Pressure regulator
   - Fuel filter bolts and filter
   - Fuel filter bracket, if necessary

To install:

7. Install or connect the following:
   - Fuel filter in its bracket and finger-tighten at this time

➡Movement of the filter will ease attachment of the fuel lines.

⁂ CAUTION

Be sure new O-rings are installed prior to assembly.

   - Main pipe at the filter. Manually, screw in the main pipe's flare nut.
   - Fuel filter nut. Using a back-up wrench, torque the eyebolts to 22 ft. lbs. (30 Nm) and the flare nut to 25 ft. lbs. (35 Nm).
   - Fuel filter mounting bolts and torque to 10 ft. lbs. (14 Nm)
   - Negative battery cable
8. Turn the key to the **ON** position to pressurize the fuel system and check for leaks.

⁂ CAUTION

If repairs of a leak are required, remember to release the fuel pressure before opening the fuel system.

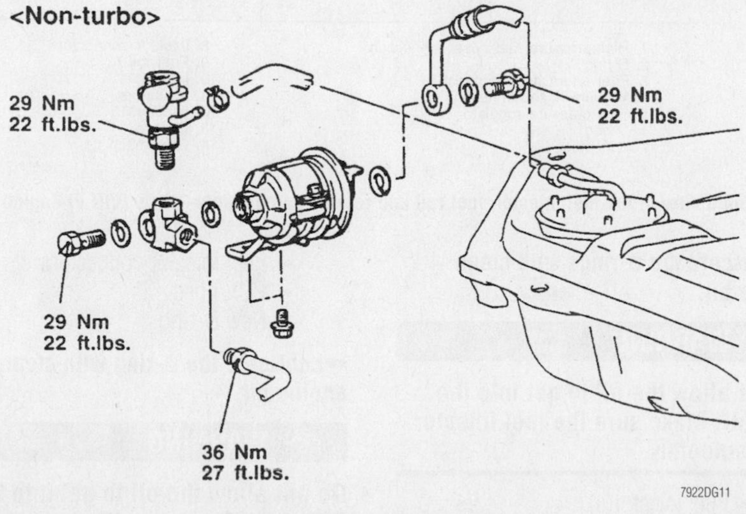

Exploded view of the fuel filter mounting—2.0L (VIN Y) engine

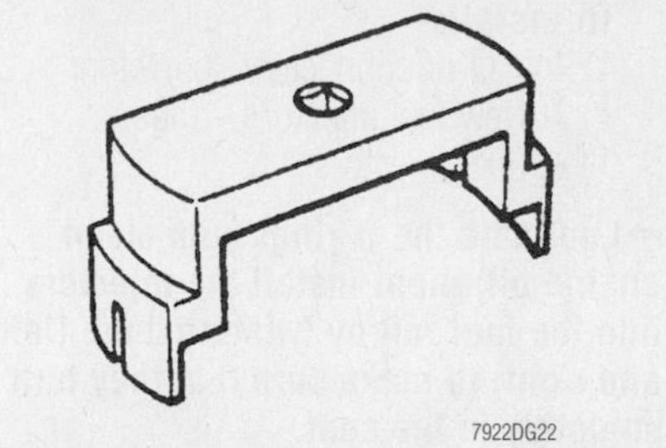

To service the fuel pump on AWD vehicles, a Fuel Pump Locking Ring Removal tool MB991480 is needed

Fuel Pump

REMOVAL & INSTALLATION

1. Before servicing the vehicle, refer to the precautions in the beginning of this section.
2. Relieve the fuel system pressure.
3. Remove or disconnect the following:
   - Negative battery cable
   - Rear seat cushion, by pulling the seat stopper near the floor and lifting the cushion up
   - Harness connector
   - Fuel lines
   - Fuel pump assembly from the tank
4. Use Fuel Pump Locking Ring Removal Tool MB991480, to remove the locking ring on the All Wheel Drive (AWD) model.

To install:

5. Install or connect the following:
   - Fuel pump in the tank
   - Fuel hoses
   - Harness connector
   - Rear seat cushion
   - Negative battery cable

Fuel Injector

REMOVAL & INSTALLATION

2.0L (VIN Y) Engine

1. Before servicing the vehicle, refer to the precautions in the beginning of this section.
2. Relieve the fuel system pressure.
3. Remove or disconnect the following:
   - Battery
   - Air intake hose
   - High pressure fuel hose from the fuel rail and discard the O-ring
   - Fuel injector harness connector
   - Fuel rail
   - Fuel injector electrical connectors
   - Fuel injector-to-fuel rail retainer(s)
   - Fuel injectors and discard the O-rings

To install:

4. Install or connect the following:
   - New fuel injector O-rings
   - Fuel injectors

➡Lubricate the O-rings with clean engine oil; then, install the injectors into the fuel rail by twisting them (left and right) to make sure that they turn smoothly in the seat.

⁂ WARNING

Do not allow engine oil to get into the fuel rail.

   - Fuel injector-to-fuel rail retainer(s)
   - Fuel injector electrical connectors
   - Fuel rail and torque the bolts to 2 ft. lbs. (3 Nm)
   - Fuel injector harness connector
   - New high-pressure fuel hose O-ring

➡Lubricate the O-ring with clean engine oil.

⁂ WARNING

Do not allow engine oil to get into the fuel rail.

   - High pressure fuel hose connection and torque the fuel hose-to-fuel rail bolts to 1.8 ft. lbs. (2.5 Nm).
   - Air intake hose
   - Battery

2.0L (VIN F) Engine

1. Before servicing the vehicle, refer to the precautions in the beginning of this section.
2. Relieve the fuel system pressure.
3. Remove or disconnect the following:
   - Negative battery cable
   - Spark plug cables
   - High pressure fuel line connection from the fuel rail
   - O-ring and discard it
   - Fuel return line connection
   - Vacuum hose connection
   - Fuel pressure regulator
   - O-ring and discard it
   - Positive Crankcase Ventilation (PCV) hose
   - Fuel injector connectors
   - Fuel rail
   - Insulators
   - Fuel injector(s)
   - O-rings and discard them
   - Grommets and discard them

To install:

4. Install or connect the following:
   - New grommets
   - New O-rings

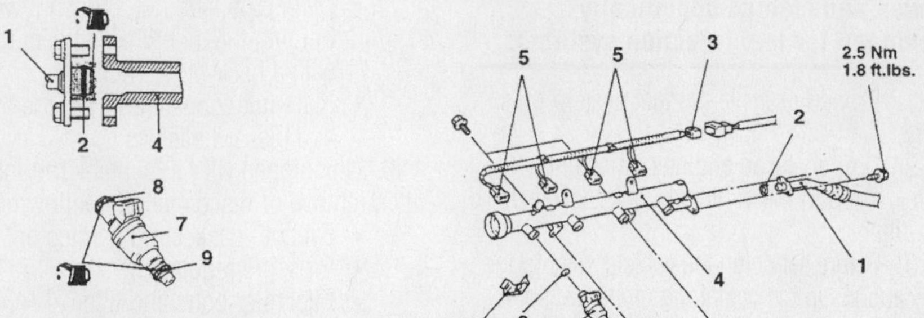

1. High-pressure fuel hose connection
2. O-ring
3. Injector harness connector
4. Fuel rail
5. Injector connectors
6. Retainers
7. Injectors
8. O-rings
9. O-rings

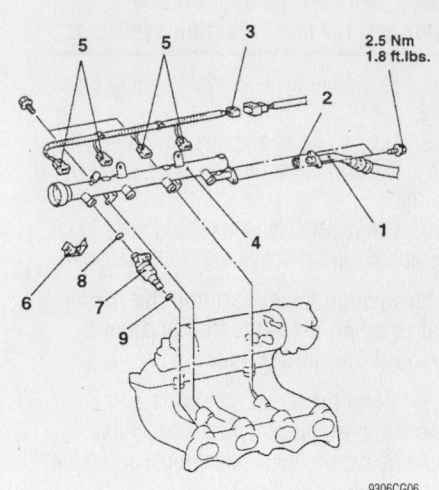

Exploded view of the fuel injector, fuel rail and related components—2.0L (VIN Y) engine

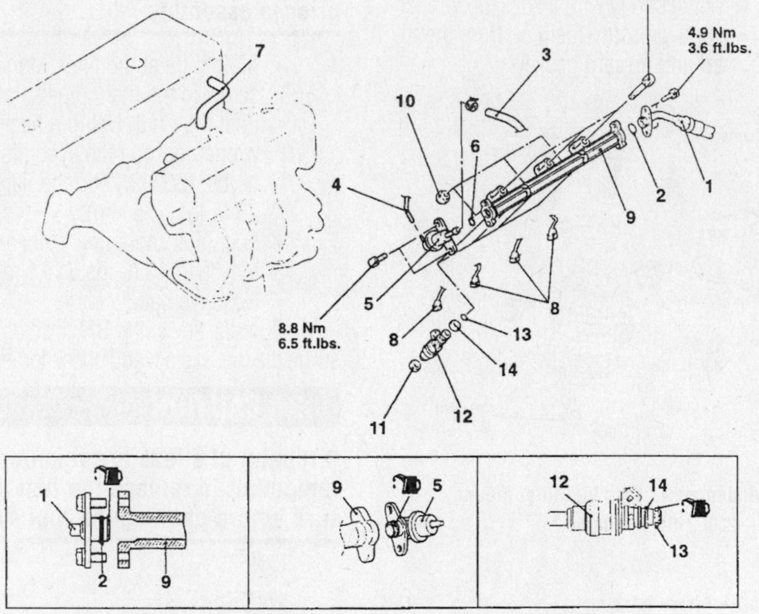

1. High-pressure fuel hose connection
2. O-ring
3. Fuel return hose connection
4. Vacuum hose connection
5. Fuel pressure regulator
6. O-ring
7. PCV hose
8. Injector connectors
9. Fuel rail
10. Insulators
11. Insulators
12. Injectors
13. O-rings
14. Grommets

9306DG05

Exploded view of the fuel injector, fuel rail and related components—2.0L (VIN F) engine

➡Lubricate the O-rings with clean engine oil.

⁂ WARNING

Do not allow the oil to get into the fuel rail. Make sure the fuel injector turns smoothly.

   - Fuel injector(s)
   - New insulators
   - Fuel rail and torque the bolts to 2 ft. lbs. (3 Nm)
   - Fuel injector connectors
   - PCV hose
   - New O-ring

➡Lubricate the O-ring with clean engine oil.

⁂ WARNING

Do not allow the oil to get into the fuel rail. Make sure the fuel pressure regulator turns smoothly.

- Fuel pressure regulator and torque to 7 ft. lbs. (9 Nm)
- Vacuum hose connection
- Fuel return line connection
- New O-ring

➡Lubricate the O-ring with clean engine oil.

⁂ WARNING

Do not allow the oil to get into the fuel rail.

- High pressure fuel line connection and torque the bolt to 1.8 ft. lbs. (2.5 Nm)
- Spark plug cables
- Negative battery cable

DRIVE TRAIN

Transaxle Assembly

REMOVAL & INSTALLATION

Manual

1. Before servicing the vehicle, refer to the precautions in the beginning of this section.
2. Drain the transaxle and transfer case fluid, if equipped.
3. Remove or disconnect the following:
   - Battery
   - Air intake hoses
   - Battery tray and support
   - Auto-cruise actuator and bracket, if equipped with cruise control
   - Charcoal canister and bracket
   - Shift and select cables from the transaxle
   - Back-up light switch
   - Vehicle Speed Sensor (VSS)
   - Starter
4. Attach a support fixture to the engine.
   - Transaxle mounting bolts
   - Rear roll stopper bracket mounting bolts
   - Transaxle mounting bracket nuts
   - Engine undercover
   - Transfer case assembly, if equipped with All Wheel Drive (AWD)

⁂ WARNING

Do not remove or install the axle shaft nut when the vehicle is on the floor or damage to the bearings will occur.

   - Halfshafts
   - Slave cylinder from the bell housing. Do not disconnect the fluid line and position it aside.
   - Bell housing cover
   - Right-hand center member stay (support)
   - Center member
5. Place a transmission jack under the transaxle and support it.
   - Transaxle mounting bolt
   - Transaxle mount and transaxle

To install:

6. Install or connect the following:
   - Transaxle and mount and torque the through-bolt to 50 ft. lbs. (69 Nm) and the transaxle assembly bolts to 22–25 ft. lbs. (30–34 Nm)
   - Center member assembly and right-hand stay and torque the bolts to 58 ft. lbs. (78 Nm)
   - Bell housing cover and slave cylinder and torque the bolts to 9 ft. lbs. (12 Nm)
   - Halfshafts

➡Be sure to install the washer in the proper direction.

   - Engine undercover and lower the vehicle
   - Transfer case assembly, if removed
   - Transaxle mounting bracket nuts and torque the bolts to 70 ft. lbs. (95 Nm)
   - Rear roll stopper bracket mounting bolts and torque to 54 ft. lbs. (74 Nm)
   - Transaxle assembly mounting bolts and torque to 35 ft. lbs. (48 Nm)
7. Remove the engine support fixture.
   - Starter
   - VSS and back-up light connectors
   - Cruise control actuator, if removed
   - Battery tray support and tray
   - Charcoal canister bracket and canister
   - Air duct and air cleaner assembly
8. Refill the transaxle. On turbocharged models, fill with gear oil of classification GL-4 or higher, SAE 75W-85W or 75W-90. On non-turbocharged models, fill with Mopar MS9417 MTX Fluid P/N 4773167.
9. Refill the transfer case with gear oil of classification GL-4 or higher, SAE 75W-85W or 75W-90.

Automatic

1. Before servicing the vehicle, refer to the precautions in the beginning of this section.
2. Drain the transaxle fluid.
3. Remove or disconnect the following:
   - Negative battery cable
   - Battery and battery tray
   - Control actuator and bracket, if equipped with auto-cruise
   - Air cleaner assembly, intercooler and air hose, as required
4. Mark the shift cable's location.
   - Adjusting nut and shift cable
   - Dipstick and tube assembly
   - Solenoid connector
   - Neutral safety switch (inhibitor switch) connector
   - Pulse generator kickdown servo switch connector
   - Oil temperature sensor connector
   - Speedometer cable and oil cooler lines
   - Starter
   - Upper transaxle-to-engine bolts
   - Transaxle mounting bracket supporting the transaxle
   - Sheet metal undercover
   - Tie rod ends and ball joints from steering knuckle
   - Halfshafts
   - Exhaust pipe and transfer case, on All Wheel Drive (AWD) vehicles
   - Lower bell housing cover
   - Flexplate-to-torque converter special bolts.

➡To remove the bolts, turn the engine crankshaft with a box wrench and bring the bolts into a position appropriate for removal, one at a time.

⁂ WARNING

After removing the bolts, push the torque converter toward the transaxle so it doesn't stay on the engine, allowing oil to pour out of the converter hub or cause damage to the converter.

   - Lower transaxle-to-engine bolts and transaxle

To install:

5. Install or connect the following:
   - Transaxle and torque the transaxle-to-engine bolts to 35 ft. lbs. (48 Nm) and the torque converter-to-driveplate bolts to 34–38 ft. lbs. (46–53 Nm)
   - Bell housing cover and torque the bolt to 4 ft. lbs. (5 Nm)
   - Transfer case and exhaust pipe, using a new gasket, for AWD models and torque the bolts to 51 ft. lbs. (69 Nm)
   - Circlips and halfshafts
   - Tie rods and ball joint and torque the castle nut to 25 ft. lbs. (34 Nm)
   - Transaxle mounting bracket and torque the bolts to 35 ft. lbs. (48 Nm)
   - Under-guard
   - Starter
   - Speedometer cable and oil cooler lines
   - Solenoid connector
   - Neutral safety switch (inhibitor switch) connector
   - Pulse generator kickdown servo switch connector
   - Oil temperature sensor connector
   - All remaining components

6. Refill with Dexron®II, Mopar ATF Plus type 7176, Mitsubishi Plus ATF automatic transaxle fluid. If equipped with AWD, check and fill the transfer case.

7. Start the engine and allow it to idle for 2 minutes. Apply the parking brake and move the selector through each gear position, ending in **N** (neutral). Recheck the fluid level and add, if necessary.

➡If the vehicle has run for less than 15 minutes but more than one minute, the fluid is considered warm and should be above the ADD mark on the dipstick. Do not add fluid unless the level is at or below the ADD mark.

Clutch

ADJUSTMENTS

Clutch Pedal Free-Play

1. Before servicing the vehicle, refer to the precautions in the beginning of this section.
2. Measure the clutch pedal height from the face of the pedal pad to the firewall.
3. Compare the measured value with the proper distance of 6.93–7.17 in. (176–182mm).
4. Measure the clutch pedal clevis pin play at the face of the pedal pad. Press the pedal lightly until resistance is met, and measure this distance. The clutch pedal clevis pin play should be within 0.04–0.12 in. (1–3mm).
5. If the clutch pedal height or clevis pin play are not within the standard values, adjust as follows:

   a. If not equipped with cruise control, turn and adjust the stop bolt so the pedal height is the standard value, then tighten the locknut.

   b. If equipped with cruise control, detach the clutch switch connector and turn the switch to obtain the standard clutch pedal height. Then, lock by tightening the locknut.

   c. Turn the pushrod to adjust the clutch pedal clevis pin play to agree with the standard value and secure the pushrod with the locknut.

➡When adjusting the clutch pedal height or the clutch pedal clevis pin play, be careful not to push the pushrod toward the master cylinder.

   d. Check that when the clutch pedal is depressed all the way, the interlock switch changes from ON to OFF.

REMOVAL & INSTALLATION

2.0l (VIN F) Engine

1. Before servicing the vehicle, refer to the precautions in the beginning of this section.
2. Remove or disconnect the following:
   - Transaxle
   - Clutch oil tubes by unscrewing the fittings
   - Clutch oil fluid chamber
   - Clutch release (slave) cylinder union bolt, gaskets and union
   - Valve plate and valve plate spring
   - Clutch release (slave) cylinder, without disconnecting the fluid line
   - Pressure plate
   - Clutch disc

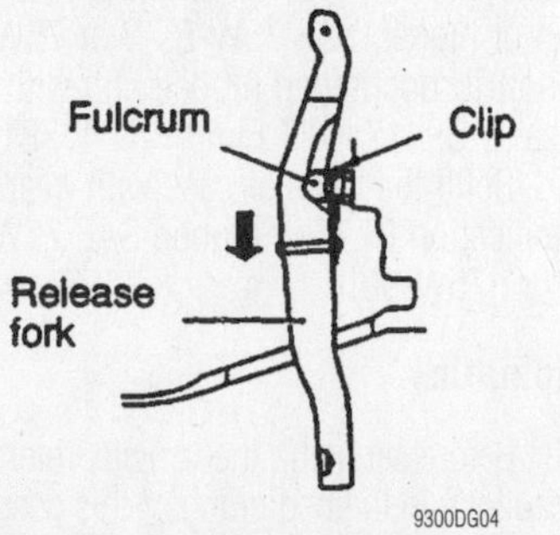

9300DG04

Slide the release fork in the direction of the arrow to disengage it from the pivot ball

   - Clutch release bearing by unfastening the return clip
   - Release fork by sliding it away from the fulcrum

⁂ WARNING

Be careful not to cause damage to the clip by pushing the fork in any other direction or removing it with force.

   - Release fork boot and fulcrum

To install:

3. Install or connect the following:
   - Fulcrum and release fork boot
   - Clutch release fork
   - Return clip
4. Apply grease to the clutch release fork contact areas.
5. Lubricate the clutch release bearing.
6. Lightly apply multi-purpose grease to the clutch disc splines.
7. Install or connect the following:
   - Clutch disc using a suitable guide to position the disc on the flywheel
   - Pressure plate
   - Clutch release (slave) cylinder and torque the bolts to 13 ft. lbs. (18 Nm)

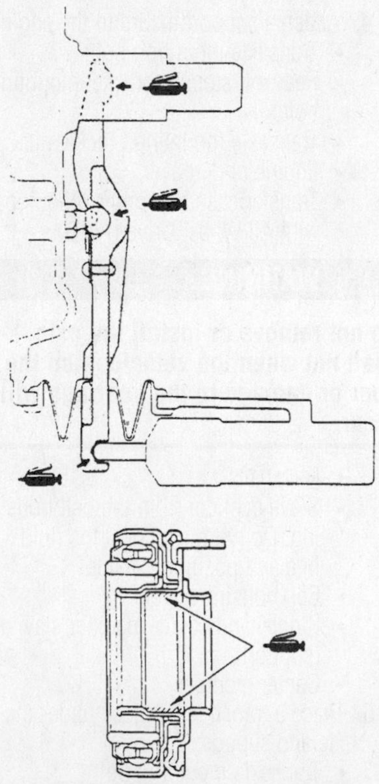

7922DG15

Lubrication points for the clutch release lever and bearing

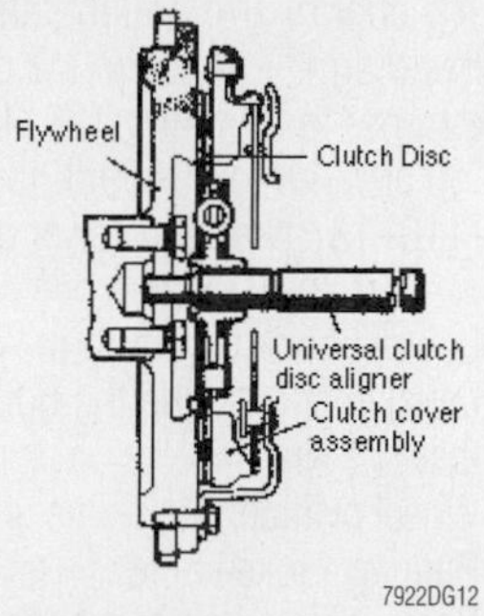

Cross-sectional view of proper clutch disc and pressure plate alignment, showing the alignment tool

- Valve plate spring and valve plate
- Gasket, union, gasket and union bolt, on the release cylinder
- Clutch oil fluid chamber. Uncap or unplug the clutch oil tubes, then attach them and tighten the fittings to 11 ft. lbs. (15 Nm).
- Transaxle

2.0L (Vin Y) Engine

1. Before servicing the vehicle, refer to the precautions in the beginning of this section.
2. Remove or disconnect the following:
   - Transaxle
   - Clutch oil tube by unscrewing the fitting

➡**Plug or cap the ends to prevent contamination from entering the line.**

- Release (slave) cylinder
- Pressure plate and clutch disc
- Clutch release bearing
- Clutch release lever
- Boot

To install:

3. Apply a suitable grease to the clutch release lever.
4. Lightly apply multi-purpose grease to the clutch disc splines.
5. Install or connect the following:
   - Boot assembly
   - Clutch release lever
   - Clutch release bearing
   - Clutch disc and pressure plate assembly
   - Clutch release (slave) cylinder and torque the bolt to 14 ft. lbs. (19 Nm)
   - Clutch oil tube and torque to 11 ft. lbs. (15 Nm)
   - Transaxle

Hydraulic Clutch System

BLEEDING

➡**Do not allow the reservoir to run out of fluid during bleeding, otherwise the entire procedure must be repeated.**

1. Before servicing the vehicle, refer to the precautions in the beginning of this section.
2. Fill the reservoir with clean brake fluid meeting DOT 3 specifications.
3. Attach a hose to the bleeder valve on the slave cylinder with the other end of the hose submerged in a container at least half full of fresh DOT 3 brake fluid from a sealed container.
4. Have an assistant press the clutch pedal to the floor, then loosen the bleed screw on the slave cylinder.
5. Tighten the bleed screw and have your helper release the clutch pedal.
6. Repeat the procedure until the fluid is free of air bubbles.

Transfer Case Assembly

REMOVAL & INSTALLATION

1. Before servicing the vehicle, refer to the precautions in the beginning of this section.
2. Drain the transfer oil.
3. Remove or disconnect the following:
   - Negative battery cable
   - Front exhaust pipe
   - Transfer case assembly and torque the bolts to 44 ft. lbs. (59 Nm)

➡**Be careful not to damage the transfer case's output housing oil seal. Do not let the rear driveshaft hang; suspend it from the body with a piece of wire. Cover the opening in the transaxle and transfer case to keep oil from dripping and to keep dirt out.**

To install:

4. Lubricate the driveshaft sleeve yoke and oil seal lip on the transfer extension housing with clean engine oil.
5. Install or connect the following:
   - Transfer case assembly and torque the transfer case-to-transaxle bolts to 40–43 ft. lbs. (55–60 Nm) on manual transaxle or 43–58 ft. lbs. (60–80 Nm) on automatic transaxle.

➡**Use care when installing the rear driveshaft to the transfer case output shaft.**

- Front exhaust pipe using a new gasket and torque the bolts to 33 ft. lbs. (44 Nm)

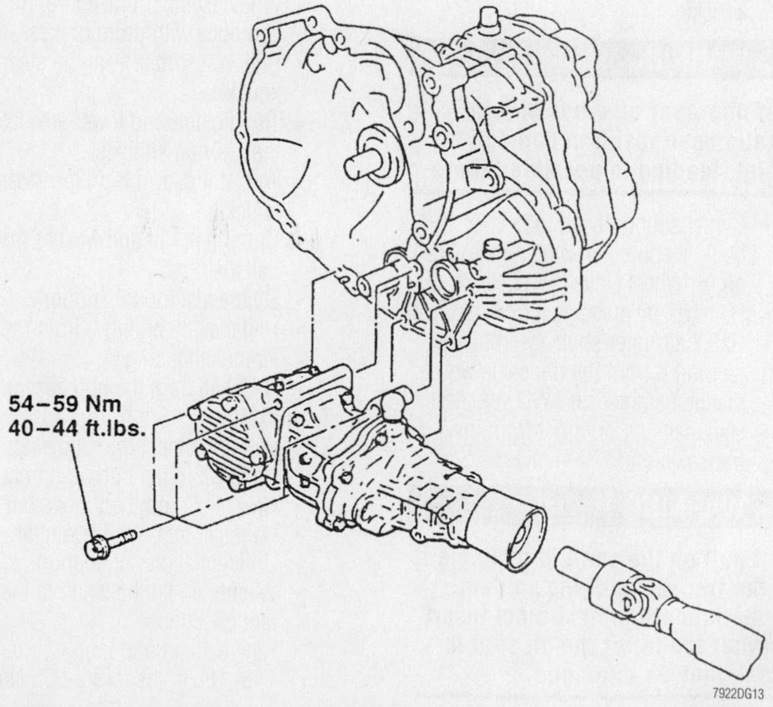

Exploded view of the transfer case assembly mounting—Vehicles with manual transaxle

6. Fill the transfer case with gear oil of classification GL-4 or higher, SAE 75W-85W or 75W-90. Check the fluid level in the transaxle and add, as required.

Halfshaft

REMOVAL & INSTALLATION

Front

1. Before servicing the vehicle, refer to the precautions in the beginning of this section.

WARNING

If the vehicle is going to be rolled with the halfshafts removed, install 2 outer CV-joints tools in the hubs. Also, if proper torque is not applied to the front wheel bearings, the bearings will no longer be usable.

2. Remove or disconnect the following:
   - Negative battery cable.
   - Cotter pin, halfshaft nut and washer
   - Speedometer drive from the right extension housing if removing the right halfshaft
   - Tie rod from the steering knuckle
   - Stabilizer link from the damper fork
   - Damper fork from the lateral lower arm
   - Lateral lower arm from the steering knuckle

WARNING

Use of improper methods of joint separation can result in damage to the joint, leading to possible failure.

   - Center support bearing bracket bolts, if equipped with an inner shaft on All Wheel Drive (AWD) vehicles
   - Halfshaft by pressing it from the hub
   - Halfshaft/inner shaft assembly by tapping it from the transaxle with a plastic hammer on AWD vehicles
   - Halfshafts by prying it from the transaxle on FWD vehicles

WARNING

Do not pull on the shaft to dislodge it from the transaxle; doing so damages the inboard joint. Do not insert the prybar too far or the oil seal in the case may be damaged.

To install:

3. Inspect the halfshaft boot for damage or deterioration. Check the ball joints and splines for wear.
4. Install or connect the following:
   - Circlips on the halfshaft ends
   - Halfshaft into the transaxle

➡Be sure it is fully seated.

   - Halfshaft into the hub
   - Center bearing bracket, if equipped and torque the bolts to 33 ft. lbs. (45 Nm)
   - Washer, face the chamfered edge outward
   - Halfshaft nut, temporarily tighten
   - Tie rod end and ball joint to the steering knuckle and torque the nut to 21 ft. lbs. (28 Nm) and the tie rod nut to 17–25 ft. lbs. (24–33 Nm)
   - Wheel and torque the axle nut to 145–188 ft. lbs. (200–260 Nm) with the brakes applied
   - New cotter pin

Rear

1. Before servicing the vehicle, refer to the precautions in the beginning of this section.
2. Install or connect the following:
   - Rear wheel(s)
   - Wheel speed sensor, if equipped with Anti-Lock Brake System (ABS)
   - Brake caliper and rotor or brake drum
   - Parking brake shoes (disc brakes) or shoe/lever assembly (drum brakes)
   - Parking brake cable from the backing plate
   - Wheel cylinder brake line, if equipped with drum brakes
   - Shock absorber from the steering knuckle
   - Trailing arm and lower arm from the steering knuckle
   - Toe control arm from the steering knuckle
   - Cotter pin, nut and washer from the halfshaft
   - Differential mount support.
   - Halfshaft by prying it from the differential housing
   - Halfshaft from the hub assembly

To install:

3. Install or connect the following:
   - Halfshaft in the hub assembly
   - New circlip on the inner shaft
   - Halfshaft into the differential
   - Differential mount support
   - Washer on the halfshaft in the correct direction
   - Halfshaft nut and torque it to 145–188 ft. lbs. (196–255 Nm)

➡If the cotter pin hole does not align, tighten the nut to 188 ft. lbs. (255 Nm) and install the cotter pin in the first hole that aligns.

   - Toe control arm, trailing arm and lower arm. Torque the toe control arm nut to 20 ft. lbs. (28 Nm), trailing arm nut to 85–99 ft. lbs. (118–137 Nm) and lower arm nut to 71 ft. lbs. (98 Nm)
   - Lower shock mount to the steering knuckle and torque the bolt to 52 ft. lbs. (71 Nm)
   - Wheel cylinder brake line, if removed
   - Parking brake cable
   - Remaining brake components
   - ABS wheel speed sensor, if removed
   - Rear wheel
   - Brake drums, if equipped
4. Bleed the brake system.

CV-Joints

OVERHAUL

WARNING

The Birfield joint assembly, located on the wheel side of the halfshaft, is not to be disassembled; repair of this joint is only by replacement of the halfshaft.

Tri-Pot Joint

FRONT WHEEL DRIVE— (FWD)

1. Before servicing the vehicle, refer to the precautions in the beginning of this section.
2. Remove halfshaft and place it in a soft jawed vise.
3. Disassemble or remove the following:
   - Tri-pot boot bands
   - Tri-pot case

➡Wipe the grease from the tri-pot case.

   - Halfshaft snapring
   - Tri-pot spider assembly

WARNING

Do not disassemble the spider assembly.

   - Tri-pot boot

WARNING

If the boot is to be reused, wrap plastic tape around the shaft splines to protect the boot from damage.

   - Dynamic damper bands from the halfshaft, if necessary

To install:

4. If removed, install the dynamic damper by performing the following procedure:

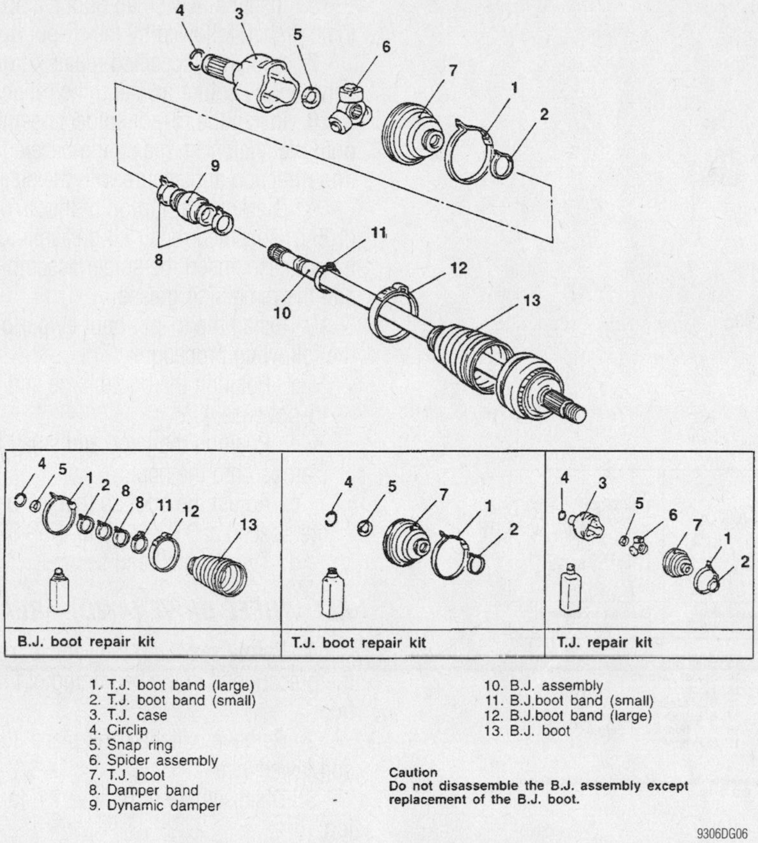

Exploded view of the front halfshaft assemblies—Front wheel drive

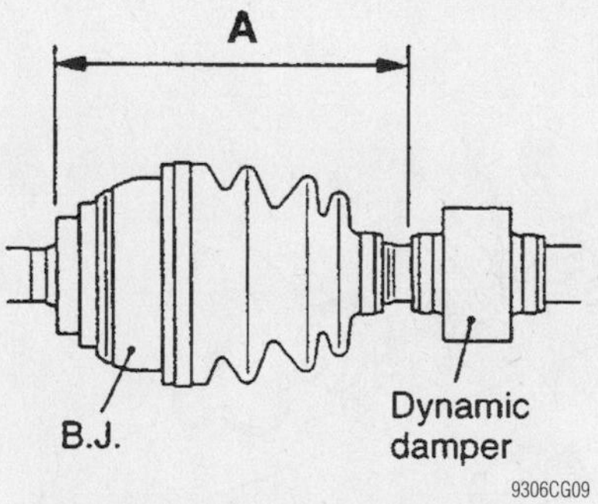

Positioning the dynamic damper on the front halfshaft assemblies—Front wheel drive

a. Slide the damper onto the halfshaft with new bands.

b. Position the damper so the distance from the front of the Birfield joint to the front edge of the damper assembly is 14.60–14.84 in. (371–377mm) for the right halfshaft (non-turbo)/left halfshaft (turbo) or 7.52–7.76 in. (191–197mm) for the left halfshaft (non-turbo, M/T).

c. Tighten and secure the damper bands.

⁂ WARNING

Wrap plastic tape around the shaft splines to protect the boot from damage.

5. Install a new small boot clamp onto the halfshaft followed by the tri-pot boot.
6. Apply the specified repair kit grease between the spider axle and the roller.
7. Install the tri-pot spider assembly onto the shaft from the spline beveled section direction and secure with the snapring.
8. Distribute a portion of the 3.7 oz.

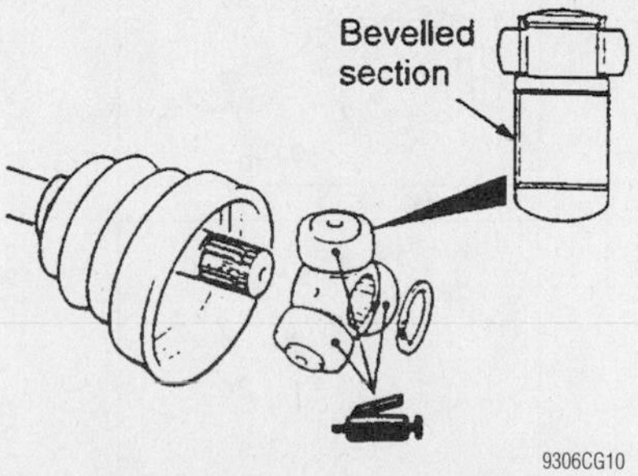

Lubricating the tri-pot spider assembly

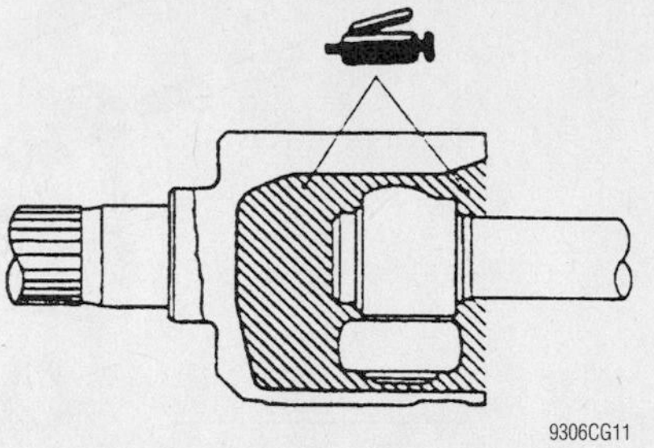

Lubricating the tri-pot case assembly installed

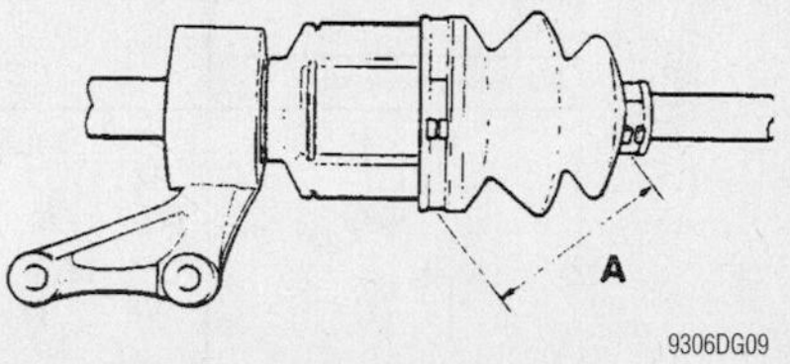

Measuring the tri-pot joint boot

(105 g) for non-turbo or 4.23 oz. (120 g) for turbo specified repair kit grease into the tri-pot case, insert the spider assembly and add the remaining grease.

9. Install the tri-pot boot by performing the following procedure:
   a. Position the boots large end on the tri-pot case.
   b. Position the large and small boot clamps onto the boot.
   c. Adjust the boot so that the bands are spaced at 3.03–3.27 in. (77–83mm).
   d. Tighten the band securely.

ALL WHEEL DRIVE (AWD)— FRONT

1. Before servicing the vehicle, refer to the precautions in the beginning of this section.
2. Remove halfshaft and place it in a soft jawed vise.
3. Disassemble or remove the following:
   - Tri-pot boot bands
   - Tri-pot case (right halfshaft) or tri-pot case/inner shaft assembly (left halfshaft)

➡Wipe the grease from the tri-pot case.

   - Halfshaft snapring
   - Tri-pot spider assembly

⁂ WARNING

Do not disassemble the spider assembly.

   - Tri-pot boot

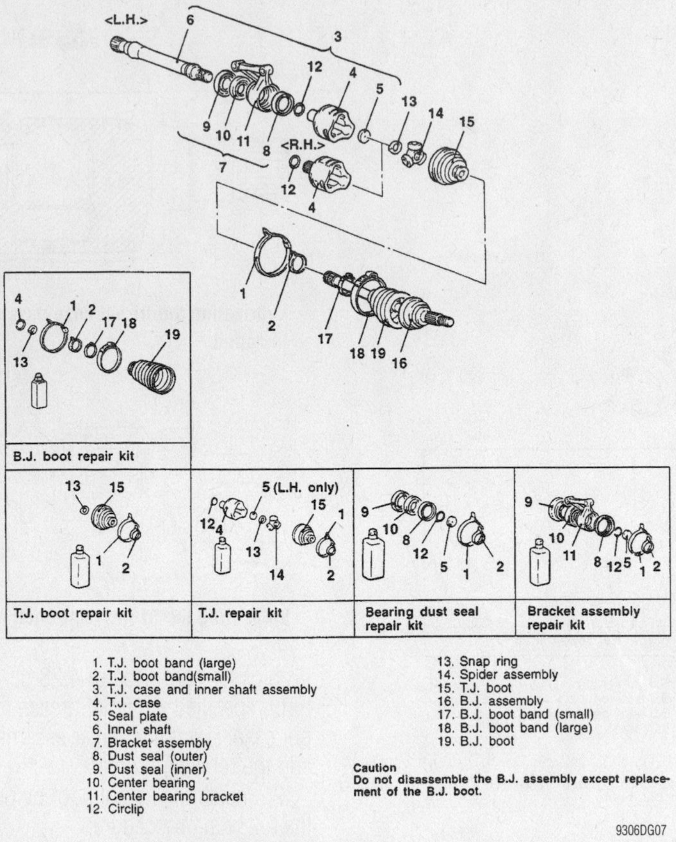

Exploded view of the front halfshaft assemblies—AWD

⁂ WARNING

If the boot is to be reused, wrap plastic tape around the shaft splines to protect the boot from damage.

4. If working with the left halfshaft and it is necessary to replace the tri-pot case, press the case from the inner shaft assembly.

To install:

5. If installing a new tri-pot case onto the left halfshaft, perform the following procedure:
   a. Lubricate the inner shaft splines with Multi-Mileage Grease No. 2525035.
   b. Press the inner shaft assembly into tri-pot case.
   c. Secure the tri-pot case with special tool MB991248 on a hydraulic press with the case facing upward.
   d. Position a new seal plate in the center of the tri-pot case.
   e. Place a 1.18 in. (30mm) pipe on the seal plate and press the seal into the tri-pot case.

⁂ WARNING

Wrap plastic tape around the shaft splines to protect the boot from damage.

6. Install a new small boot clamp onto the halfshaft followed by the tri-pot boot.
7. Apply the specified repair kit grease between the spider axle and the roller.
8. Install the tri-pot spider assembly onto the shaft from the spline beveled section direction and secure with the snapring.
9. Distribute a portion of the 3.70 oz. (105 g) specified repair kit grease into the tri-pot case, insert the spider assembly and add the remaining grease.
10. Install the tri-pot boot by performing the following procedure:
   a. Position the boots large end on the tri-pot case.
   b. Position the large and small boot clamps onto the boot.
   c. Adjust the boot so that the bands are spaced at 3.03–3.27 in. (77–83mm).
   d. Tighten the band securely.

ALL WHEEL DRIVE (AWD)—REAR

1. Before servicing the vehicle, refer to the precautions in the beginning of this section.
2. Remove halfshaft and place it in a soft jawed vise.
3. Disassemble or remove the following:
   - Tri-pot boot bands

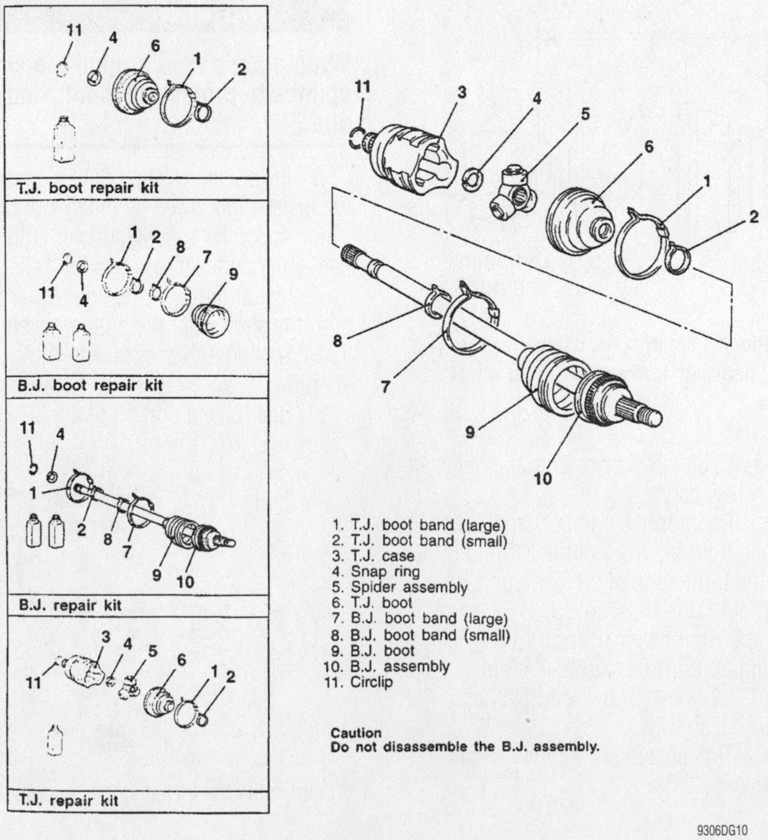

Exploded view of the rear halfshaft assemblies—AWD

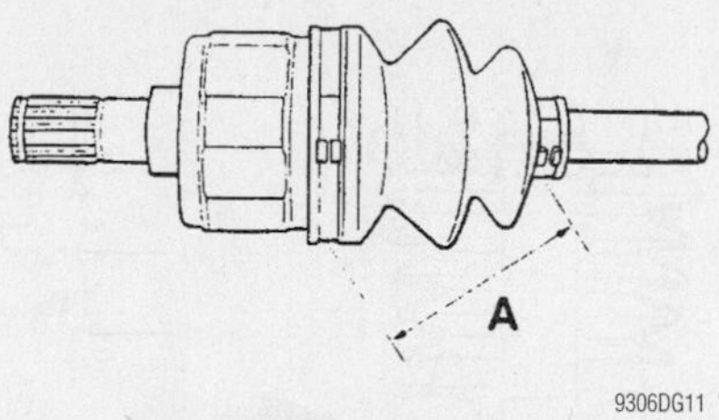

Positioning the tri-pot boot bands—Rear halfshaft assemblies with AWD

- Tri-pot case and clean the case thoroughly
- Halfshaft snapring
- Tri-pot spider assembly

⁂ WARNING

Do not disassemble the spider assembly.

- Tri-pot boot

⁂ WARNING

If the boot is to be reused, wrap plastic tape around the shaft splines to protect the boot from damage.

To install:

⁂ WARNING

Wrap plastic tape around the shaft splines to protect the boot from damage.

4. Install a new small boot clamp onto the halfshaft followed by the tri-pot boot.
5. Apply the specified repair kit grease between the spider axle and the roller.
6. Install the tri-pot spider assembly onto the shaft from the spline beveled section direction and secure with the snapring.
7. Distribute a portion of the specified repair kit grease into the tri-pot case, insert the spider assembly and add the remaining grease.
8. Conventional type of grease: 3.35 oz. (95 g).
9. Limited slip type of grease: 3.70 oz. (105 g).
10. Install the tri-pot boot by performing the following procedure:
    a. Position the boots large end on the tri-pot case.
    b. Position the large and small boot clamps onto the boot. Be sure to check the boot band identification numbers:
    - Large band: 20-98 No. BJ 82
    - Small band: 20-83 No. BJ 82

    c. Adjust the boot so that the bands are spaced at:
    - Conventional: 2.99–3.23 in. (76–82mm)
    - Limited slip: 3.19–3.43 in. (81–87mm)

    d. Tighten the band securely.

Center Bearing

ALL WHEEL DRIVE (AWD)

1. Before servicing the vehicle, refer to the precautions in the beginning of this section.
2. Remove the left halfshaft and place it in a soft-jawed vise.
3. Remove the tri-pot spider assembly from the tri-pot case.
4. Remove the tri-pot case by performing the following procedure:
    a. Position the inner shaft/tri-pot case assembly on a hydraulic press supported by Tool MB991248 with the tri-pot case facing upward.
    b. Position a bar inside the tri-pot case on the end of the inner shaft and press the case from the inner shaft assembly.
5. Remove and disconnect the following:
    - Inner shaft assembly, place it in a soft-jawed vise
    - Center bearing bracket, press it from the inner shaft with a wheel puller

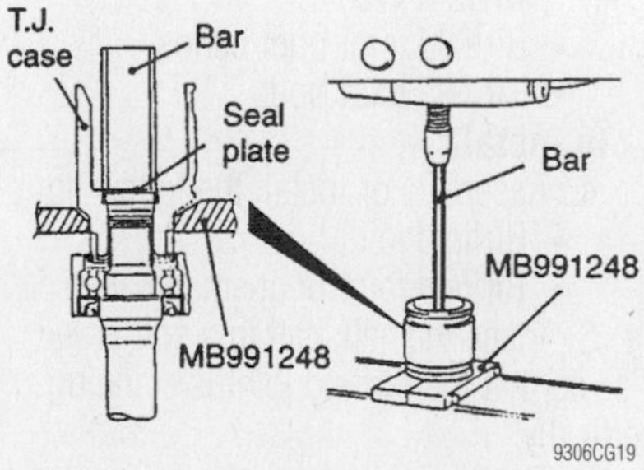

Removing the tri-pot case from the inner shaft assembly—left front halfshaft with AWD

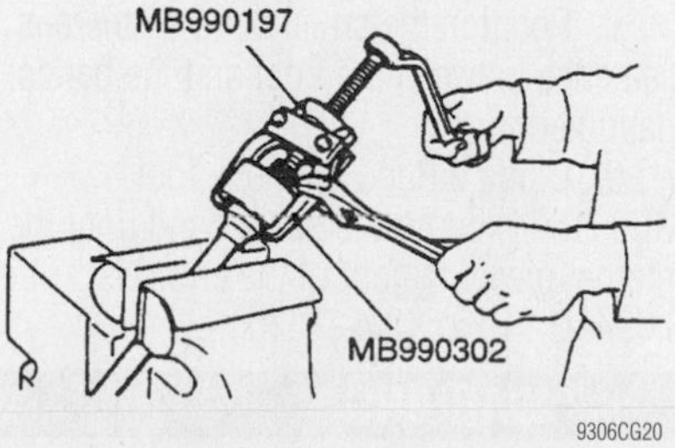

Removing the center bearing bracket from the inner shaft assembly—left front halfshaft with AWD

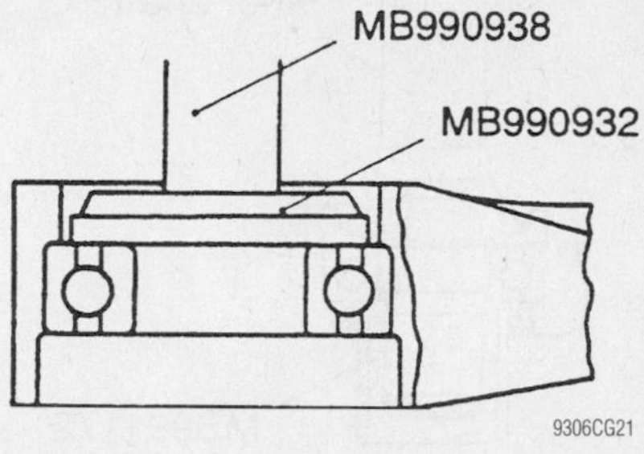

Removing the center bearing from the center bearing bracket—left front halfshaft with AWD

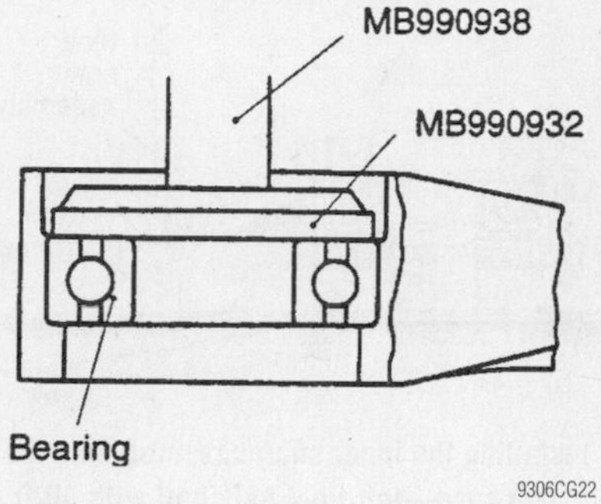

Installing the center bearing to the center bearing bracket—left front halfshaft with AWD

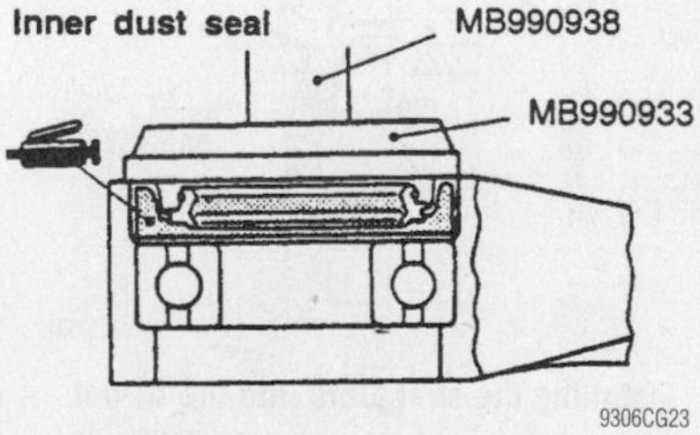

Installing the inner dust seal to the center bearing bracket—left front halfshaft with AWD

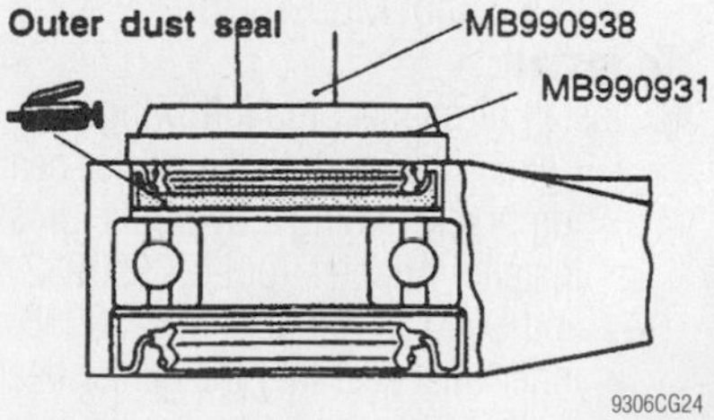

Installing the outer dust seal to the center bearing bracket—left front halfshaft with AWD

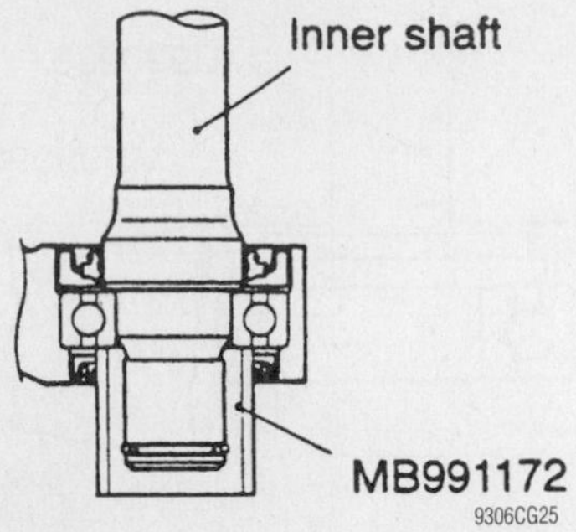

Installing the inner shaft into the center bearing bracket—left front halfshaft with AWD

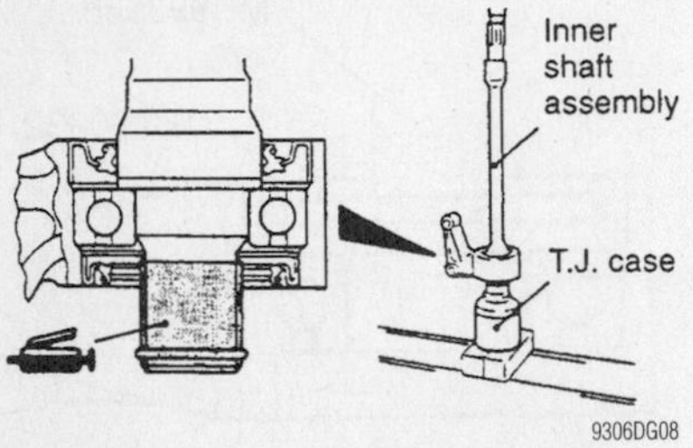

Installing the inner shaft assembly into the tri-pot case—left front halfshaft with AWD

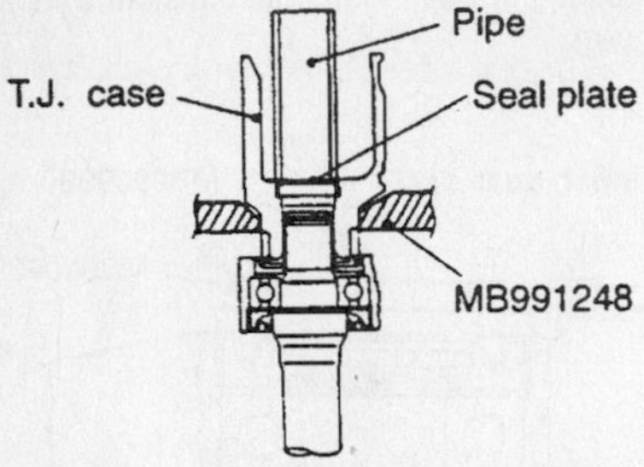

Installing the seal plate into the tri-pot case—left front halfshaft with AWD

- Center bearing and inner dust seal from the center bearing bracket using a hydraulic press, Installer Adapter Tool MB990932 and Snap-in Bar Tool MB990938

To install:

6. Install or connect the following:
   - Center bearing into the center bearing bracket using a hydraulic press, Installer Adapter Tool MB990932 and Snap-in Bar Tool MB990938
   - Inner duct seal into the center bearing bracket using a hydraulic press, Installer Adapter Tool MB990933 and Snap-in Bar Tool MB990938
   - Outer duct seal into the center bearing bracket using a hydraulic press, Installer Adapter Tool MB990931 and Snap-in Bar Tool MB990938
   - Inner shaft into the center bearing bracket using a hydraulic press and Adapter Tool MB991172

7. Install the tri-pot case onto the left halfshaft by performing the following procedure:
   a. Lubricate the inner shaft splines with Multi-Mileage Grease No. 2525035.
   b. Press the inner shaft assembly into tri-pot case.
   c. Secure the tri-pot case with special Tool MB991248 on a hydraulic press with the case facing upward.
   d. Position a new seal plate in the center of the tri-pot case.
   e. Place a 1.18 in. (30mm) pipe on the seal plate and press the seal into the tri-pot case.
8. Install or connect the following:
   - Tri-pot joint assembly
   - Left halfshaft

Birfield Joint Boot

FRONT

1. Before servicing the vehicle, refer to the precautions in the beginning of this section.
2. Remove the halfshaft and place it in a soft jawed vise.
3. Disassemble or remove the following:
   - Tri-pot joint
   - Dynamic damper, for Front Wheel Drive (FWD)
   - Birfield joint boot bands
   - Birfield joint boot

To install:

4. Assemble or install the following:
   - Birfield joint boot
   - Birfield joint boot small band
5. Place the halfshaft in a soft jawed vise so that the Birfield joint is standing vertically.
6. Position the boots small diameter so that only 1 groove is exposed on the shaft.
7. Adjust the Band Crimper Tool MB991561 so that the jaw opening is 0.114 in. (2.9mm).
8. Position the small band so there is clearance between the boot and the bands mating edge.
9. Using a Band Crimper Tool MB991561, crimp the small band until the internal measurement of the crimp is 0.094–0.110 in. (2.4–2.8mm).

⁂ WARNING

If the crimp dimension is not correct, remove the band install a new one.

10. Using 3.88 oz. (110 g) for non-turbo (FWD), 4.59 oz. (130 g) for turbo (FWD) or 3.35 oz. (95 g) for AWD, specified amount of grease in the repair kit, pack the Birfield boot.
11. Position the boot so 0.004–0.061 in. (0.1–1.55mm) of clearance exists between the boots large diameter end and the Birfield joint housing shoulder.
12. Adjust the Band Crimper Tool

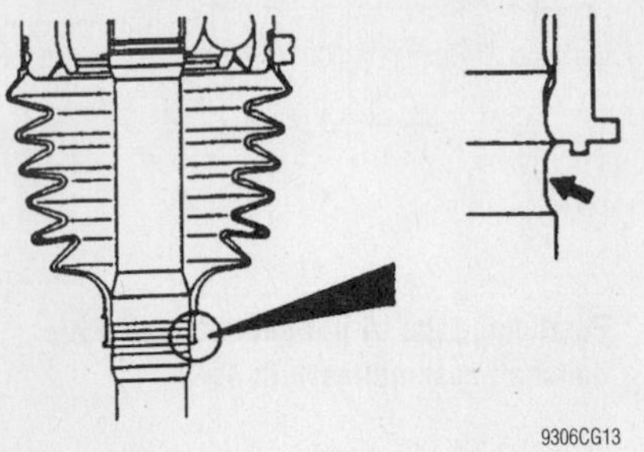

Positioning the Birfield joint boot's small diameter

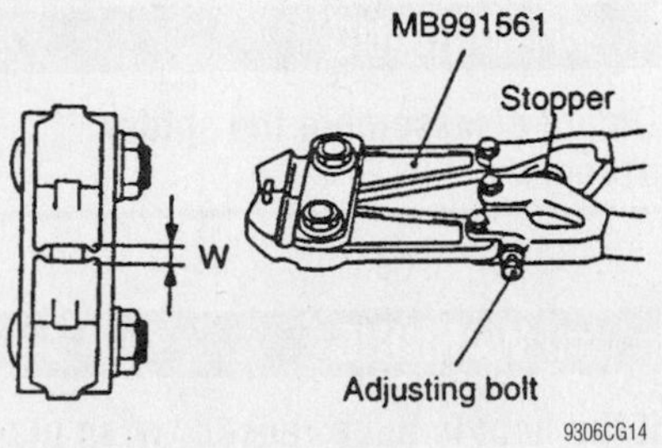

View of the band crimper tool

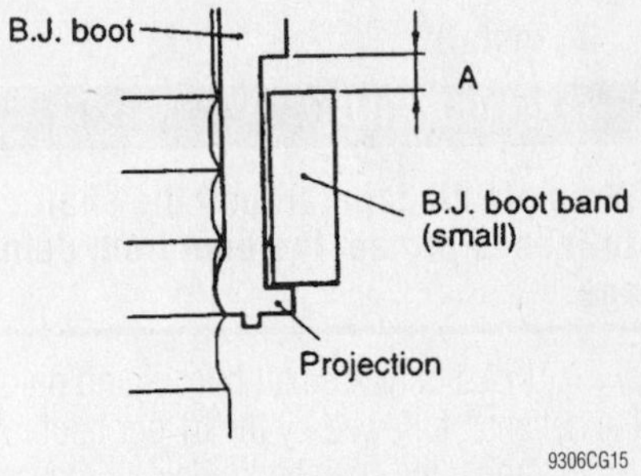

Positioning the Birfield joint boot's small band

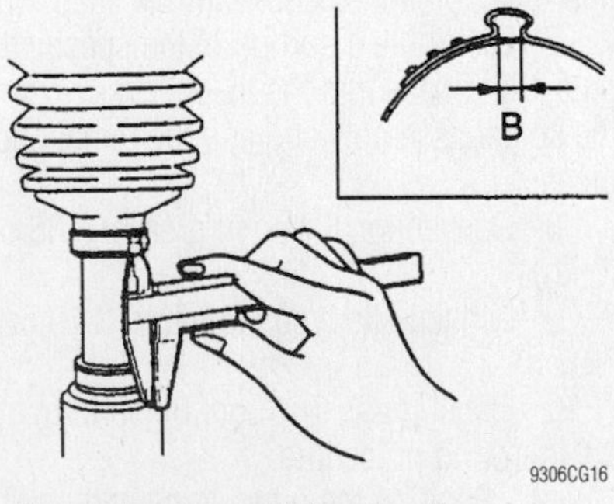

Measuring the small band's crimp

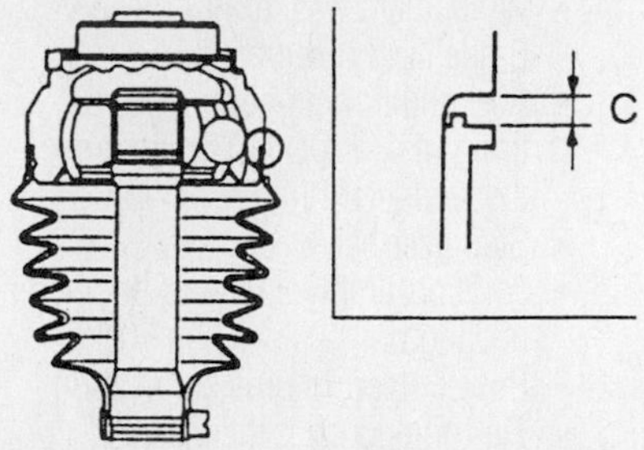

9306CG17

Positioning the Birfield joint boot's large diameter

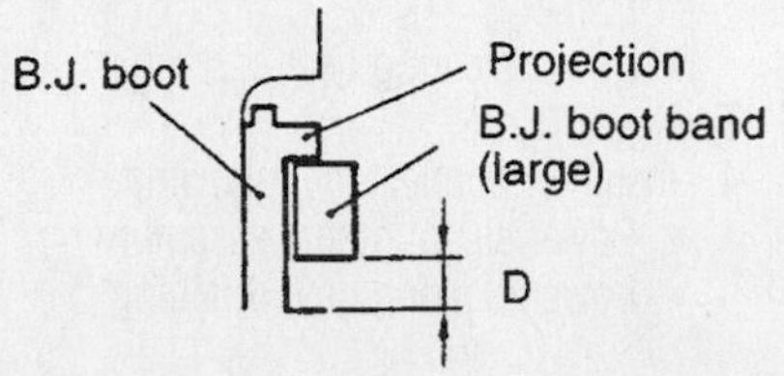

9306CG18

Positioning the Birfield joint boot's large band

MB991561 so that the jaw opening is 0.126 in. (3.2mm).

13. Install and position the large boot band so that it rests against the boot(s) projection and a gap exists between the clamp and the boot.
14. Using a Band Crimper Tool MB991561, crimp the small band until the internal measurement of the crimp is 0.094–0.110 in. (2.4–2.8mm).

⁂ WARNING

If the crimp dimension is not correct, remove the band install a new one.

15. If equipped with FWD, install the dynamic damper by performing the following procedure:
    a. Slide the damper onto the halfshaft with new bands.
    b. Position the damper so the distance from the front of the Birfield joint to the front edge of the damper assembly is 14.60–14.84 in. (371–377mm) for the right halfshaft (non-turbo)/left halfshaft (turbo) or 7.52–7.76 in. (191–197mm) for the left halfshaft (non-turbo, M/T).
    c. Tighten and secure the damper bands.
16. Install the tri-pot joint.

REAR

1. Before servicing the vehicle, refer to the precautions in the beginning of this section.
2. Remove the halfshaft and place it in a soft jawed vise.
3. Disassemble or remove the following:
    - Tri-pot joint
    - Birfield joint boot bands
    - Birfield joint boot

To install:

4. Assemble or install the following:
    - Birfield joint boot
    - Birfield joint boot small band
5. Position the large and small boot clamps onto the boot. Be sure to check the boot band identification numbers:
    - Large band: 20-110 No. BJ 87
    - Small band: 20-83 No. BJ 82
6. Install the tri-pot joint.

Pinion Seal

REMOVAL & INSTALLATION

1. Before servicing the vehicle, refer to the precautions in the beginning of this section.
2. Remove or disconnect the follow-ing:
    - Driveshaft
    - Companion flange nut. Use End Yoke Holder Tool MB990850, to secure the companion flange.

➡Matchmark the companion flange to the drive pinion.

- Companion flange. Use a bearing puller to pull the companion flange from the drive pinion.
- Pinion oil seal

To install:

3. Install or connect the following:
    - New pinion oil seal. Use an Oil Seal Installer tool MB995020 to drive the new seal into the axle housing
    - Grease oil seal lip and companion flange. Lubricate with Multi-Mileage grease No. 2525035.
    - Align the companion flange-to-drive pinion matchmarks
    - New companion flange-to-drive pinion nut. Use an End Yoke Holder tool MB990850 to secure the flange and torque the nut to 137 ft. lbs. (186 Nm)
    - Driveshaft and torque the bolts to 22–25 ft. lbs. (29–34 Nm)

Differential Carrier

REMOVAL & INSTALLATION

1. Before servicing the vehicle, refer to the precautions in the beginning of this section.
2. Remove or disconnect the following:
    - Halfshafts
    - Driveshaft

➡Support the differential carrier with a jack.

- Differential carrier mount-to-rear crossmember bolt
- Differential carrier

To install:

3. Raise the carrier into position.
4. Install or connect the following:
    - Differential carrier and torque the mount-to-rear crossmember bolt to 72 ft. lbs. (98 Nm)
    - Driveshaft
    - Halfshafts

STEERING AND SUSPENSION

Air Bag

⁂ CAUTION

Some vehicles are equipped with an air bag system, also known as the Supplemental Inflatable Restraint (SIR) or Supplemental Restraint System (SRS). The system must be disabled before performing service on or around system components, steering column, instrument panel components, wiring and sensors. Failure to follow safety and disabling procedures could result in accidental air bag deployment, possible personal injury and unnecessary system repairs.

PRECAUTIONS

Several precautions must be observed when handling the inflator module to avoid accidental deployment and possible personal injury.

- Never carry the inflator module by the wires or connector on the underside of the module.

• When carrying a live inflator module, hold securely with both hands, and ensure that the bag and trim cover are pointed away from you.

• Place the inflator module on a bench or other surface with the bag and trim cover facing up.

• With the inflator module on the bench, never place anything on or close to the module that may be thrown in the event of an accidental deployment.

DISARMING

1. Position the front wheels in the straight-ahead position and place the ignition key in the **LOCK** position. Remove the key from the ignition lock cylinder.
2. Disconnect the negative battery cable and insulate the cable end with high-quality electrical tape or similar non-conductive wrapping.
3. Wait at least one minute before working on the vehicle. The air bag system is designed to retain enough voltage to deploy the air bag for a short period of time after the battery has been disconnected.

To arm:

4. Reconnect the negative battery cable, turn the ignition switch to the **ON** position and check the air bag warning light for proper operation.

Power Rack and Pinion Steering Gear

REMOVAL & INSTALLATION

1. Before servicing the vehicle, refer to the precautions in the beginning of this section.
2. Drain the power steering fluid into a suitable container.
3. Remove or disconnect the following:
   - Negative battery cable
   - Windshield washer fluid reservoir
   - Brake fluid reservoir
   - Air conditioning compressor. Position it aside; DO NOT disconnect the refrigerant lines
   - Stabilizer bar
   - Joint assembly and gear body connecting bolt
   - Power steering pipe connection
   - Tie rod end from the steering knuckle
   - Left and right side stays
   - Center member assembly retainers
   - Center member
   - Power steering gear retaining clamp
   - Power steering gear assembly

To install:

4. Install or connect the following:
   - Power steering gear assembly
   - Power steering gear retaining

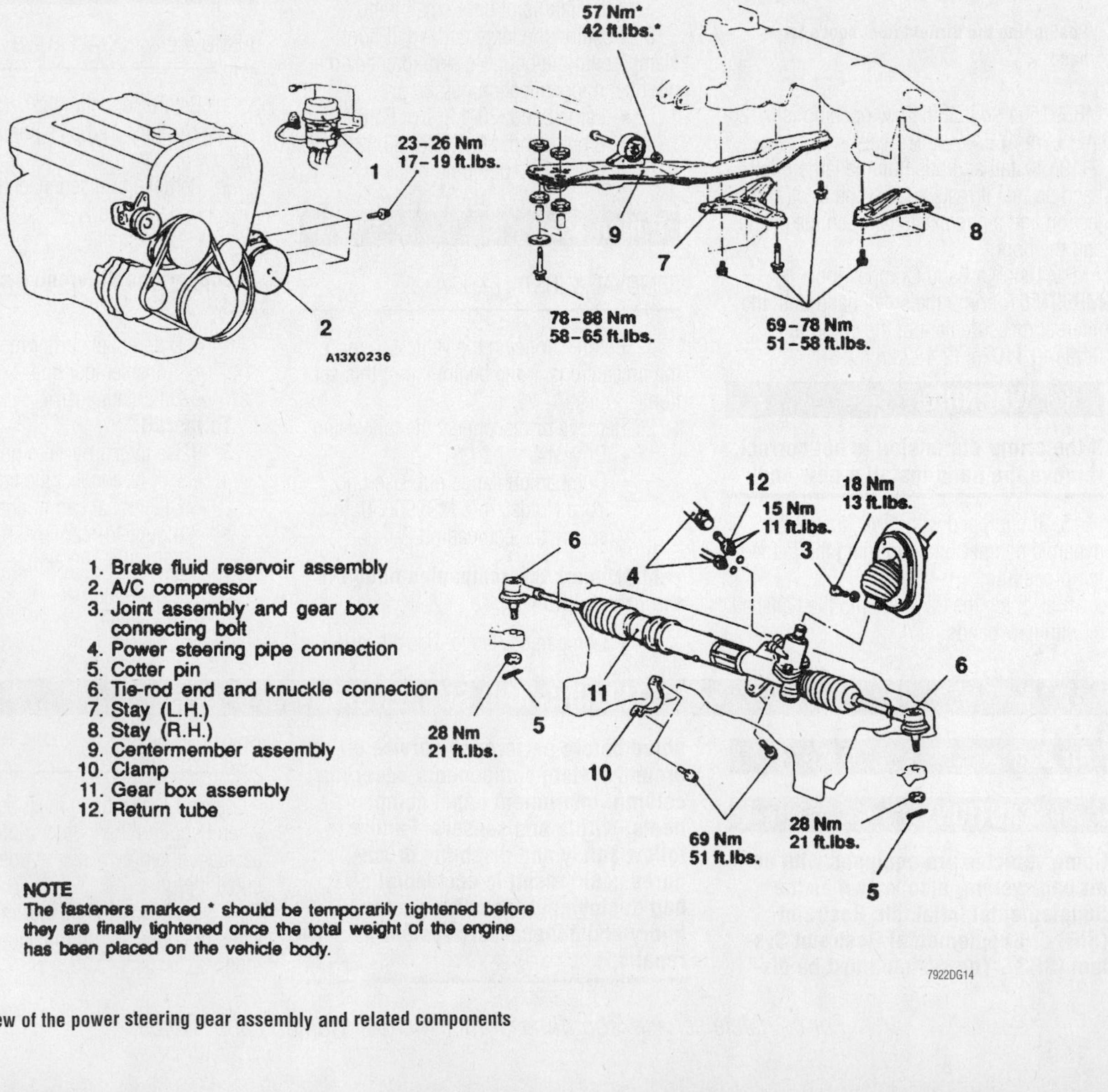

Exploded view of the power steering gear assembly and related components

clamp and torque the bolts to 51 ft. lbs. (69 Nm)
- Center member
- Center member assembly retainers and torque the bolts to 58–65 ft. lbs. (78–88 Nm)
- Left and right side stays and torque the bolts to 51–58 ft. lbs. (69–78 Nm)
- Tie rod end to the steering knuckle and torque the nut to 21 ft. lbs. (28 Nm)
- Power steering pipe connection and torque the nuts to 13 ft. lbs. (18 Nm)
- Joint assembly and gear body connecting bolt and torque the bolt to 11 ft. lbs. (15 Nm)
- Stabilizer bar
- Air conditioning compressor
- Brake fluid reservoir
- Windshield washer fluid reservoir
- Negative battery cable

5. Fill the reservoir with power steering fluid and bleed the system.

6. Check the front end alignment and adjust, if necessary.

Strut

REMOVAL & INSTALLATION

Front

1. Before servicing the vehicle, refer to the precautions in the beginning of this section.
2. Remove or disconnect the following:
   - Upper strut assembly-to-chassis nuts, from the engine compartment
   - Front wheel
   - Stabilizer link nut
   - Strut assembly lower mounting bolts
   - Damper fork
   - Strut

To install:

3. Install or connect the following:
   - Strut
   - Damper fork and torque the damper fork-to-lower arm nut/bolt to 64 ft. lbs. (88 Nm)
   - Strut assembly lower mounting bolts and torque the strut-to-damper fork bolt to 65–87 ft. lbs. (88–118 Nm)
   - Stabilizer link nut and torque the nut to 28 ft. lbs. (39 Nm)
   - Front wheel
   - Upper strut assembly-to-chassis nuts, in the engine compartment and torque to 32 ft. lbs. (44 Nm)
4. Check and adjust the front alignment, if necessary.

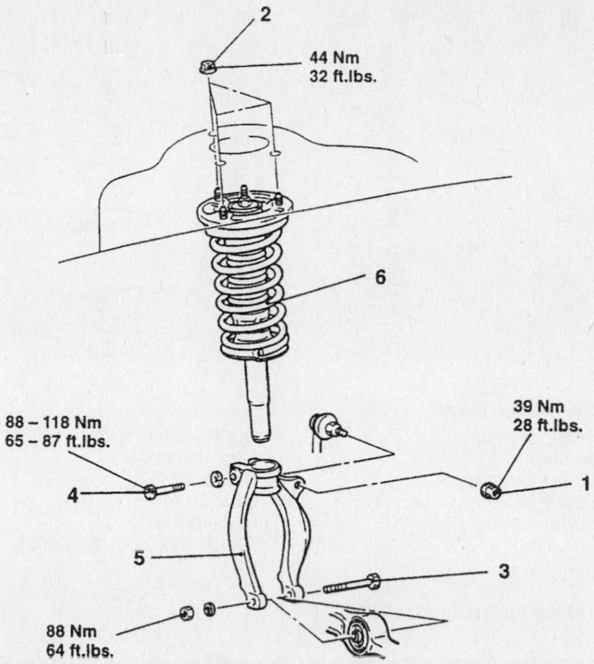

Exploded view of the front strut mounting

Rear

1. Before servicing the vehicle, refer to the precautions in the beginning of this section.
2. Remove or disconnect the following:
   - Service lid in the luggage compartment
   - Upper mounting bracket-to-body cap and flange nuts

➡Do not remove the large nut in the center of the strut assembly.

   - Strut-to-steering knuckle bolt
   - Strut

To install:

3. Install or connect the following:
   - Strut and torque the upper strut-to-chassis nuts to 32 ft. lbs. (44 Nm)
4. Raise the suspension with a jack or adjustable stand to align the strut lower mounting holes.
5. Install or connect the following:

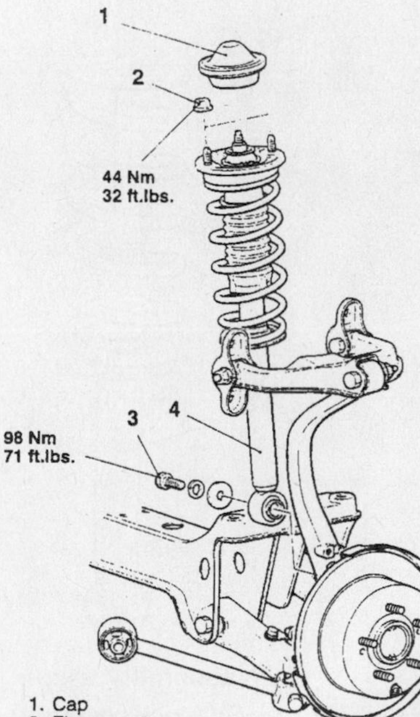

Exploded view of the rear strut mounting

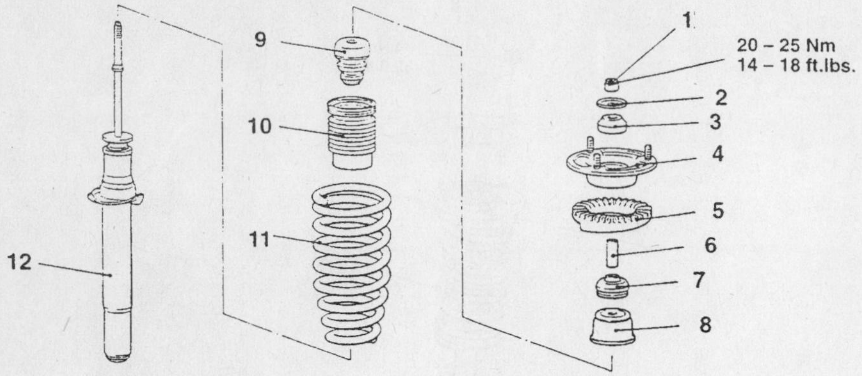

Disassembly steps

1. Self-locking nut
2. Washer
3. Upper bushing A
4. Upper bracket assembly
5. Upper spring pad
6. Collar
7. Upper bushing B
8. Cup assembly
9. Bump rubber
10. Dust cover
11. Coil spring
12. Shock absorber assembly

7922DG28

Exploded view of the front strut assembly

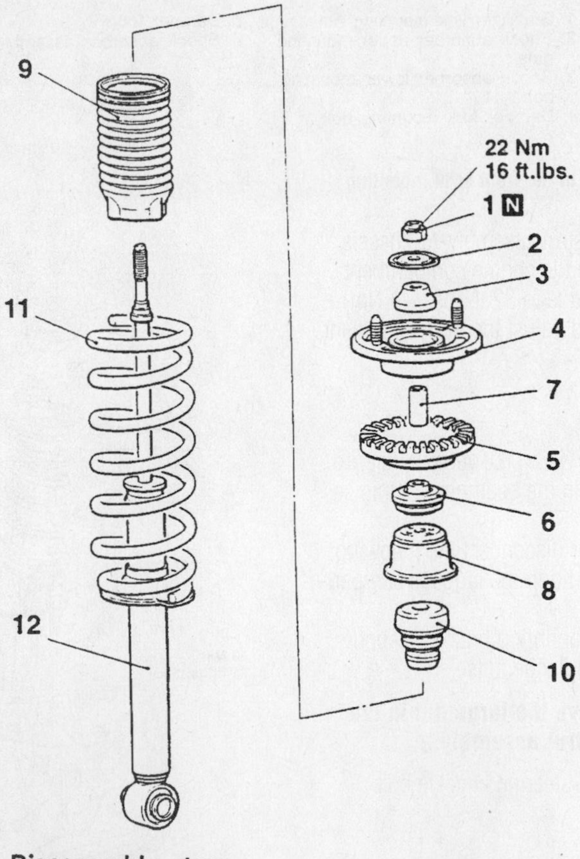

Disassembly steps

1. Self-locking nut
2. Washer
3. Upper bushing A
4. Upper bracket assembly
5. Upper spring pad
6. Upper bushing B
7. Collar
8. Cup assembly
9. Dust cover
10. Bump rubber
11. Coil spring
12. Shock absorber assembly

7922DG27

Exploded view of the rear strut assembly

- Lower strut mounting bolt and torque to 71 ft. lbs. (97 Nm)
- Cap and service lid

Coil Spring

REMOVAL & INSTALLATION

Front

1. Before servicing the vehicle, refer to the precautions in the beginning of this section.
2. Remove the strut assembly.

**** CAUTION**

Do not use air tools to tighten the compressor tool bolt.

3. Compress the coil spring, using Spring Compressing tools MB991237 and MB991239.

**** WARNING**

Be sure to install the tools evenly so the maximum length will be attained within the installation range.

4. Remove or disconnect the following:
   - Self-locking nut, by holding the piston rod
   - Washer
   - Upper bushing **A**
   - Upper bracket assembly
   - Upper spring pad
   - Collar
   - Upper bushing **B**
   - Cup assembly
   - Rubber bumper
   - Dust cover
   - Coil spring

To assemble:

5. Use the compressor tools to compress the coil spring, then install it to the strut. Align the edge of the coil spring to the stepped portion of the strut spring seat.
6. Install or assemble the following:
   - Dust cover
   - Rubber bumper
   - Cup
   - Upper bushing **B**
   - Collar
   - Upper spring pad
   - Upper bracket assembly

➡Install the assembly so the position of the 3 bolts are in the proper orientation with the damper fork.

   - Upper bushing **A**
   - Washer and self-locking nut. Temporarily, tighten the self-locking nut, then remove the spring com-

pressor tools and tighten the nut to 14–18 ft. lbs. (20–25 Nm) using a torque wrench.

WARNING

Do not use air tools to tighten the self-locking nut!

- Strut

7. Check and adjust the front alignment, if necessary.

Rear

1. Before servicing the vehicle, refer to the precautions in the beginning of this section.
2. Remove or disassemble the following:
   - Strut
   - Compress the spring using a coil spring compressor
   - Self-locking nut by holding the piston rod
   - Upper bracket assembly and spring pad
   - Collar
   - Upper bushing
   - Cup assembly
   - Bump rubber
   - Dust cover
   - Coil spring

To assemble:

3. Align the end of the coil spring with the stepped part of the spring seat and install the compressed coil spring on the strut.
4. Install or assemble the following:
   - Dust cover
   - Bump rubber
   - Cup assembly
   - Upper bushing
   - Collar
   - Upper spring pad
   - Bracket assembly
   - Upper bushing and washer
   - New self-locking nut on the piston rod and temporarily tighten the nut
5. Carefully, remove the spring compressor from the spring. Tighten the self-locking nut to 16 ft. lbs. (25 Nm).
6. Install the strut.

Upper Ball Joint

REMOVAL & INSTALLATION

The upper ball joint is an integral part of the upper control arm. If the upper ball joint is to be serviced, the upper control arm will have to be replaced.

Lower Ball Joint

REMOVAL & INSTALLATION

The lower ball joint is an integral part of the lower control arm assembly, and cannot be serviced separately. A worn or damaged ball joint, requires replacement of the lower control arm assembly.

Upper Control Arm

REMOVAL & INSTALLATION

1. Before servicing the vehicle, refer to the precautions in the beginning of this section.
2. Remove or disconnect the following:
   - Front wheel(s)
   - Upper ball joint from the steering knuckle using special tool MB991113
   - Upper shaft mounting nuts from the body
   - Upper arm
   - Upper arm-to-shafts through-bolts

To install:

3. Assemble the upper arm to the shafts at the proper angle. Tighten the through-bolts and nuts to 41 ft. lbs. (57 Nm). The proper angle is 84–86 degrees. After the arm and the shafts are connected at the correct angle, measure dimensions A and B to insure correct assembly.
   - A—11.8 in. (299.9mm)
   - B—9.2 in. (234.0mm)

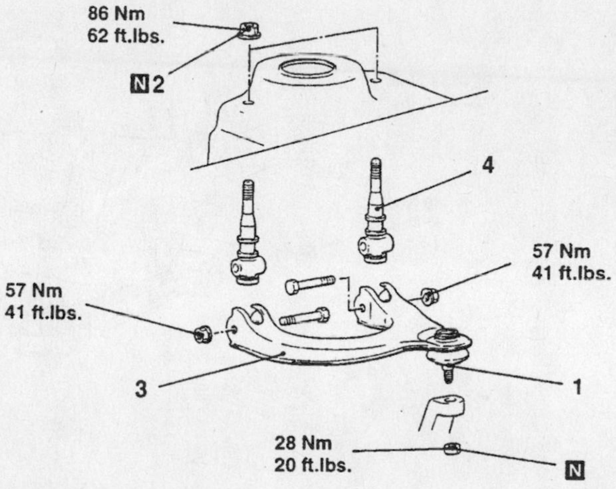

Removal steps
1. Upper arm ball joint and knuckle connection
2. Upper arm self-locking nut
3. Upper arm assembly
4. Upper arm shaft assembly

9300DG01

Exploded view of the upper control arm and related components

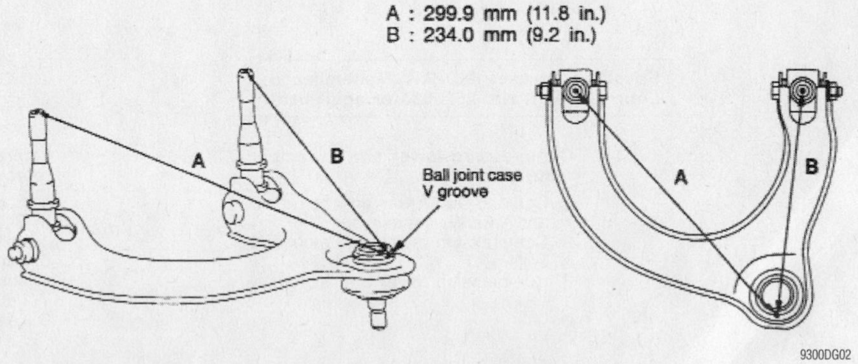

9300DG02

Be sure to install the shafts on the control arm at the correct angle

4. Install or connect the following:
   - Control arm assembly and torque the self-locking nuts to 62 ft. lbs. (86 Nm)
   - Upper ball joint-to-steering knuckle and torque the new locking nut to 20 ft. lbs. (28 Nm)
   - Front wheel(s)
5. Perform a front wheel alignment and adjust, if necessary.

Lower Control Arm

REMOVAL & INSTALLATION

Compression Lower Arm

1. Before servicing the vehicle, refer to the precautions in the beginning of this section.
2. Remove or disconnect the following:
   - Wheel
   - Compression lower arm ball joint at the steering knuckle with special tool MB991113
   - Both compression lower arm bolts
   - Compression lower arm

To install:

3. Install or connect the following:
   - Compression lower arm and torque the 2 mounting bolts to 60 ft. lbs. (81 Nm)
   - Ball joint to the knuckle assembly and torque the new self-locking nut to 43–51 ft. lbs. (59–71 Nm)
   - Wheel
4. Check and adjust the front wheel alignment, if necessary.

Lateral Lower Arm

1. Before servicing the vehicle, refer to the precautions in the beginning of this section.
2. Remove or disconnect the following:
   - Wheel
   - Stay (bracket)
   - Strut lower mounting bolts
   - Lateral lower arm and ball joint at the knuckle assembly
   - Lateral lower arm mounting bolts
   - Lateral arm

To install:

3. Install or connect the following:
   - Lateral lower arm and temporarily, install the mounting bolts; do not tighten the bolt until the vehicle is on the floor at normal riding height

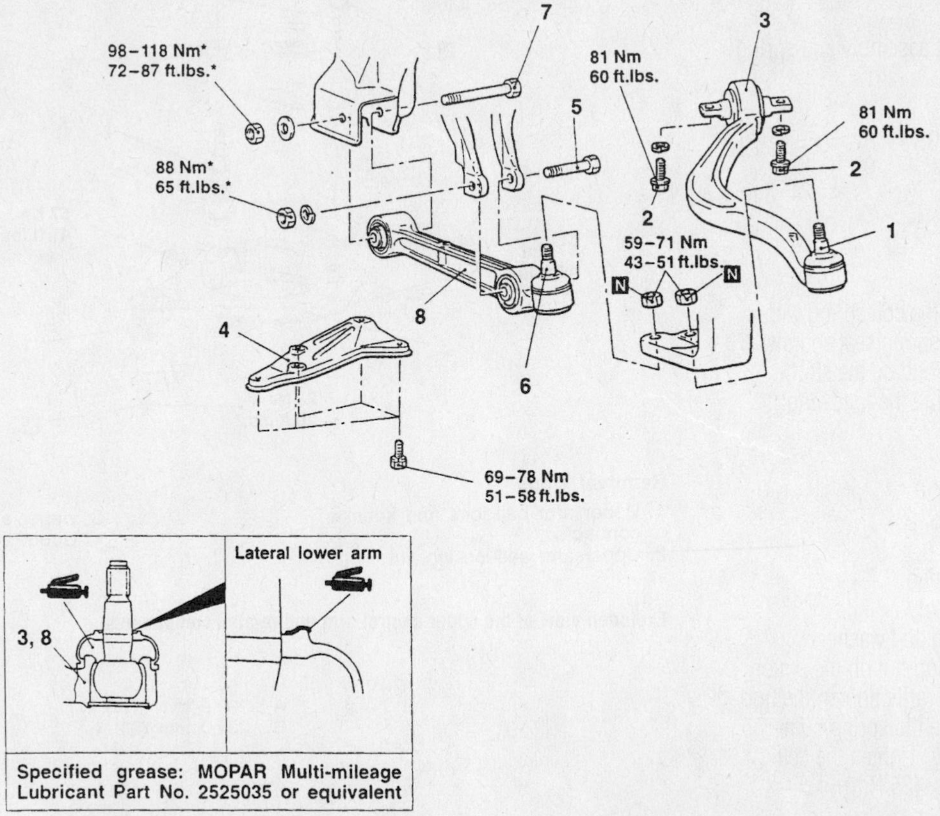

Compression lower arm assembly removal steps

1. Compression lower arm ball joint and knuckle connection
2. Compression lower arm mounting bolts
3. Compression lower arm assembly

Lateral lower arm assembly removal steps

4. Stay
5. Shock absorber lower mounting bolt
6. Lateral lower arm ball joint and knuckle connection
7. Lateral lower arm mounting bolt
8. Lateral lower arm assembly

Caution
*: **Indicates parts which should be temporarilly tightened, and then fully tightened with the vehicle on the ground in the unladen condition.**

9300DG03

Exploded view of the lower arms and related components

- Ball joint to the knuckle assembly and torque the new self-locking nut to 43–51 ft. lbs. (59–71 Nm)
- Strut lower mounting bolts and torque to 64 ft. lbs. (88 Nm)
- Stay and torque the bolts to 51–58 ft. lbs. (69–78 Nm)
- Wheel

4. Torque the lateral lower arm through-bolt and nut to 71–85 ft. lbs. (97–118 Nm).
5. Check and adjust the front wheel alignment, if necessary.

Wheel Bearings

ADJUSTMENT

Front

The front wheel bearing is sealed and not adjustable. If the wheel bearing shows signs of play, the hub assembly must be replaced.

Rear

The wheel bearing play is not adjustable. If the wheel bearing play is not within specifications, the hub assembly must be replaced.

FRONT WHEEL DRIVE (FWD)

1. Before servicing the vehicle, refer to the precautions in the beginning of this section.

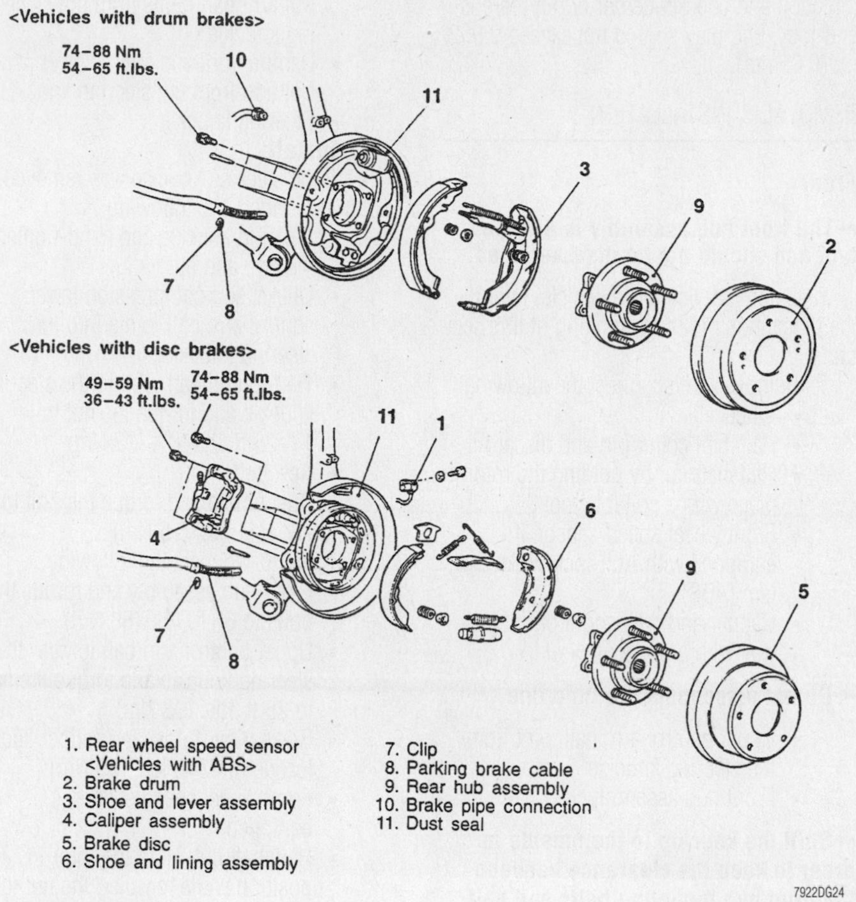

Exploded view of the rear hub assembly

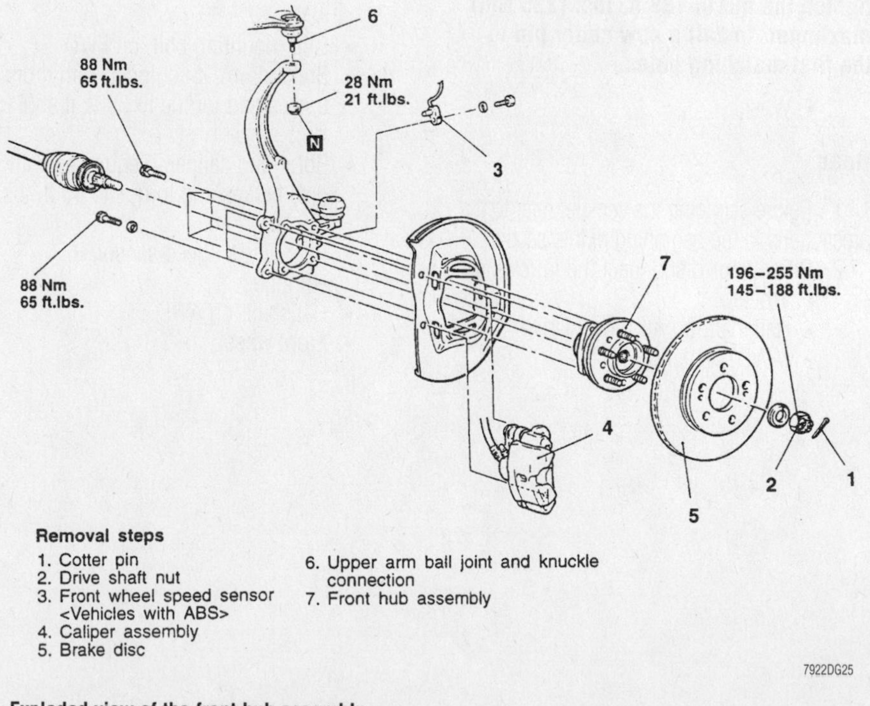

Exploded view of the front hub assembly

2. Release the parking brake and if equipped, remove the brake drum.
3. If equipped with rear disc brakes, remove the caliper assembly and the brake disc (rotor).
4. Place a dial gauge against the hub surface, then move the hub in the axial direction and check whether or not there is end-play. Specification is 0.002 in (0.05mm).
5. To check for rotary sliding resistance, turn the hub a few times to seat the bearing. Wind a rope around the hub bolts and turn the hub by pulling at a 90° angle with a spring scale. Measure to determine whether or not the rotary sliding resistance of the rear hub is at the limit value. Specification is 3.9 lbs. (1.8 kg) or less.

ALL WHEEL DRIVE (AWD)

1. Before servicing the vehicle, refer to the precautions in the beginning of this section.
2. Remove the caliper assembly and the brake disc (rotor).
3. Place a dial gauge against the hub surface, then move the hub in the axial

direction and check whether or not there is end-play. End play should not exceed 0.002 in. (0.05mm).

REMOVAL & INSTALLATION

Front

➡The front hub assembly is a sealed unit and should not be disassembled.

1. Before servicing the vehicle, refer to the precautions in the beginning of this section.
2. Remove or disconnect the following:
   - Wheel
   - Halfshaft cotter pin and discard it
   - Halfshaft nut by holding the rotor in place with a suitable tool
   - Front wheel speed sensor, if equipped with Anti-lock Brake System (ABS)
   - Caliper and suspend it out of the way with a piece of wire.

➡Do not disconnect the fluid line.

   - Upper control arm ball joint from the steering knuckle
   - Front hub assembly

➡Shift the knuckle to the outside in order to keep the clearance between the front hub mounting bolts and halfshaft.

⁂ WARNING

Be careful not to damage the ball joint boot. If equipped with ABS, be careful not to damage the rotor.

3. If necessary to remove the steering knuckle, remove or disconnect the following:
   - Dust shield
   - Tie rod end from the steering knuckle and discard the cotter pin
   - Compression and lateral lower control arm ball joints from the steering knuckle
   - Damper fork and lateral lower control arm from the steering knuckle
   - Steering knuckle

To install:

4. If the steering knuckle was removed, install or connect the following:
   - Steering knuckle and hand-tighten the bolt and nut
   - Lateral and compression lower control arm ball joints into the steering knuckle
   - Tie rod end ball joint to the steering knuckle and torque the nut to 17–25 ft. lbs. (24–33 Nm)
   - New cotter pin
   - Dust shield and torque the bolt to 78 inch lbs. (8.5 Nm)
5. Install or connect the following:
   - Front hub assembly and torque the bolts to 65 ft. lbs. (88 Nm)
   - Upper control arm ball joint to the steering knuckle and torque the nut to 25 ft. lbs. (33 Nm)
   - Brake disc (rotor) and caliper and torque to 43 ft. lbs. (59 Nm)
   - Front wheel speed sensor, if equipped
   - Halfshaft nut. Hold the rotor in position while torquing the nut to 145–188 ft. lbs. (196–255 Nm).

➡If the cotter pin holes do not align, tighten the nut to 188 ft. lbs. (225 Nm) maximum. Install a new cotter pin in the first matching holes.

   - Wheel

Rear

1. Before servicing the vehicle, refer to the precautions in the beginning of this section.
2. Remove or disconnect the following:
   - Wheel
   - Halfshaft, on All Wheel Drive (AWD) vehicles
   - Rear wheel speed sensor, if equipped with Anti-lock Brake System (ABS)
   - Caliper, if equipped with rear disc brakes and suspend the caliper on a suitable piece of wire

➡Do not disconnect the fluid line.

   - Brake drum, if equipped with drum brakes
   - Clip mounting bolt, on Front Wheel Drive (FWD)
   - Brake shoe and lining assembly, on AWD
   - Parking brake shoe and lining assembly, if equipped with disc brakes
   - Parking brake cable clip and cable on AWD
   - Rear hub and bearing assembly
3. On FWD vehicles equipped with ABS, use a suitable inner shaft remover to press off the ABS rotor.

To install:

4. If removed, install the ABS rotor.
5. Install or connect the following:
   - Rear hub and bearing and torque to 65 ft. lbs. (88 Nm)
   - Parking brake cable and retaining clip, on AWD
   - Parking brake shoe and lining assembly, if equipped with disc brakes
   - Shoe and lining assembly, on AWD
   - Clip mounting bolt, on FWD
   - Brake drum, if equipped with drum brakes and torque to 60 ft. lbs. (81 Nm)
   - Rotor and caliper, if equipped with disc brakes and torque to 43 ft. lbs. (59 Nm)
   - Rear wheel speed sensor, if equipped
   - Halfshaft, on AWD
   - Front wheel

CHRYSLER CORPORATION

8

1998–02
Dodge/Plymouth-Neon

PRECAUTIONS

Before servicing any vehicle, please be sure to read all of the following precautions, which deal with personal safety, prevention of component damage, and important points to take into consideration when servicing a motor vehicle:

- Never open, service or drain the radiator or cooling system when the engine is hot; serious burns can occur from the steam and hot coolant.
- Observe all applicable safety precautions when working around fuel. Whenever servicing the fuel system, always work in a well-ventilated area. Do not allow fuel spray or vapors to come in contact with a spark, open flame, or excessive heat (a hot drop light, for example). Keep a dry chemical fire extinguisher near the work area. Always keep fuel in a container specifically designed for fuel storage; also, always properly seal fuel containers to avoid the possibility of fire or explosion. Refer to the additional fuel system precautions later in this section.
- Fuel injection systems often remain pressurized, even after the engine has been turned **OFF**. The fuel system pressure must be relieved before disconnecting any fuel lines. Failure to do so may result in fire and/or personal injury.
- Brake fluid often contains polyglycol ethers and polyglycols. Avoid contact with the eyes and wash your hands thoroughly after handling brake fluid. If you do get brake fluid in your eyes, flush your eyes with clean, running water for 15 minutes. If eye irritation persists, or if you have taken brake fluid internally, IMMEDIATELY seek medical assistance.
- The EPA warns that prolonged contact with used engine oil may cause a number of skin disorders, including cancer! You should make every effort to minimize your exposure to used engine oil. Protective gloves should be worn when changing oil. Wash your hands and any other exposed skin areas as soon as possible after exposure to used engine oil. Soap and water, or waterless hand cleaner should be used.
- All new vehicles are now equipped with an air bag system, often referred to as a Supplemental Restraint System (SRS) or Supplemental Inflatable Restraint (SIR) system. The system must be disabled before performing service on or around system components, steering column, instrument panel components, wiring and sensors. Failure to follow safety and disabling procedures could result in accidental air bag deployment, possible personal injury and unnecessary system repairs.
- Always wear safety goggles when working with, or around, the air bag system. When carrying a non-deployed air bag, be sure the bag and trim cover are pointed away from your body. When placing a non-deployed air bag on a work surface, always face the bag and trim cover upward, away from the surface. This will reduce the motion of the module if it is accidentally deployed. Refer to the additional air bag system precautions later in this section.
- Clean, high quality brake fluid from a sealed container is essential to the safe and proper operation of the brake system. You should always buy the correct type of brake fluid for your vehicle. If the brake fluid becomes contaminated, completely flush the system with new fluid. Never reuse any brake fluid. Any brake fluid that is removed from the system should be discarded. Also, do not allow any brake fluid to come in contact with a painted surface; it will damage the paint.
- Never operate the engine without the proper amount and type of engine oil; doing so WILL result in severe engine damage.
- Timing belt maintenance is extremely important! Many models utilize an interference-type, non-freewheeling engine. If the timing belt breaks, the valves in the cylinder head may strike the pistons, causing potentially serious (also time-consuming and expensive) engine damage. Refer to the maintenance interval charts in the front of this manual for the recommended replacement interval for the timing belt, and to the timing belt section for belt replacement and inspection.
- Disconnecting the negative battery cable on some vehicles may interfere with the functions of the on-board computer system(s) and may require the computer to undergo a relearning process once the negative battery cable is reconnected.
- When servicing drum brakes, only disassemble and assemble one side at a time, leaving the remaining side intact for reference.
- Only an MVAC-trained, EPA-certified automotive technician should service the air conditioning system or its components.

ENGINE REPAIR

Alternator

REMOVAL & INSTALLATION

1998–99 Models

1. Before servicing the vehicle, refer to the precautions in the beginning of this section.
2. Remove or disconnect the following:
   - Negative battery cable
   - Alternator's adjustment nut, loosen only
   - Plastic lower splash shield
   - Alternator field circuit wiring connector, push the **RED** locking tab to release
   - Alternator pivot bolt, loosen only
   - Alternator drive belt

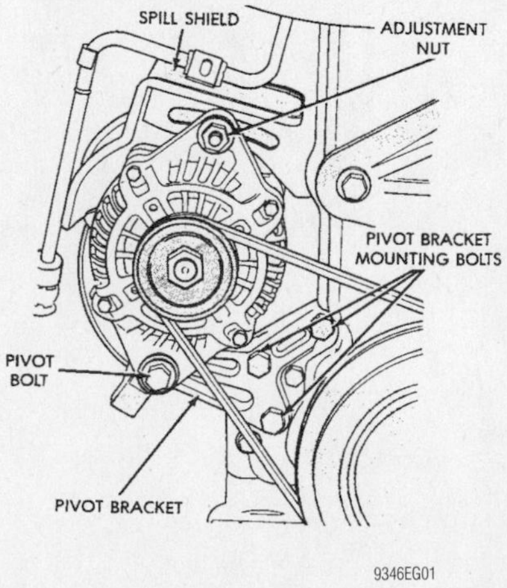

Exploded view of the alternator for 1998–99 models

➡The alternator spill shield does not need to be removed.

- 3 pivot bracket-to-engine bolts
- Pivot bracket
- Alternator, by sliding it off the "T" bolt and out through the wheel well

➡It is not necessary to remove the "T" bolt.

To install:

3. Install or connect the following:
   - Alternator, by moving it through the wheel well and sliding it onto the "T" bolt
   - Pivot bracket
   - 3 pivot bracket-to-engine bolts and torque to 40 ft. lbs. (54 Nm)
   - Alternator drive belt
   - Alternator pivot bolt

- Alternator field circuit wiring connector, push the**RED**locking tab in until it snaps in place
- Plastic lower splash shield
- Tension the drive belt to 100 lbs. (used) or 135 lbs. (new)
- Alternator adjustment nut and torque all nuts/bolts to 40 ft. lbs. (54 Nm)
- Negative battery cable

2000–02 Models

1. Before servicing the vehicle, refer to the precautions in the beginning of this section.
2. Remove or disconnect the following:
   - Negative battery cable
   - Alternator jam nut, loosen only
   - Alternator adjustment nut, loosen only
   - Accessory drive splash shield
   - Alternator lower bolt, loosen only
   - Alternator drive belt
   - Alternator field circuit wiring connector, push the**RED**locking tab to release
   - Battery positive terminal
   - Upper and lower bolts, move alternator off pivot bracket
   - Pivot bracket
   - Alternator

To install:

3. Install or connect the following:
   - Alternator
   - Pivot bracket and torque the bolts to 40 ft. lbs. (54 Nm)
   - Alternator, move it onto pivot bracket
   - Alternator field circuit wiring connector, push **RED** locking tab in until it snaps in place
   - Battery positive terminal
   - Alternator drive belt
   - Tension the drive belt to 100 lbs. (used) or 135 lbs. (new)
   - Alternator adjustment bolt
   - Alternator jam nut and torque nut to 40 ft. lbs. (54 Nm)
   - Alternator mount bolts and torque the bolts to 40 ft. lbs. (54 Nm)
   - Accessory drive splash shield
   - Negative battery cable

Ignition Timing

ADJUSTMENT

Ignition timing is controlled by the Powertrain Control Module (PCM). No adjustment is necessary or possible.

Engine Assembly

REMOVAL & INSTALLATION

1998–99 Models

1. Before servicing the vehicle, refer to the precautions in the beginning of this section.

➡After all components are installed on the engine a DRB scan tool is necessary to perform the camshaft and crankshaft timing relearn procedure.

2. Properly recover the air conditioning system refrigerant.
3. Properly relieve the fuel system pressure.
4. Drain the cooling system.
5. Drain the engine oil.
6. Remove or disconnect the following:
   - Battery and battery tray
   - Powertrain Control Module (PCM) and move it aside
   - Upper radiator hose
   - Radiator
   - Fan module assembly
   - Lower radiator hose
   - Automatic transaxle cooler lines and plug them, if equipped
   - Clutch cable, if equipped with a manual transaxle
   - Transaxle shift linkage
   - Throttle body linkage
   - Engine wiring harness
   - Heater hoses
   - Right inner splash shield
   - Accessory drive belts
   - Halfshafts
   - Exhaust pipe from the manifold and support the engine and transaxle assembly with a jack
   - Front engine mount
   - Power hop damper, if equipped with manual transaxle
   - Air cleaner assembly
   - Power steering pump/reservoir assembly and move them aside
   - A/C compressor, if equipped
   - Chassis ground straps
7. Raise the vehicle enough to allow an engine dolly and cradle to be placed under the engine.
8. Loosen the engine support posts in

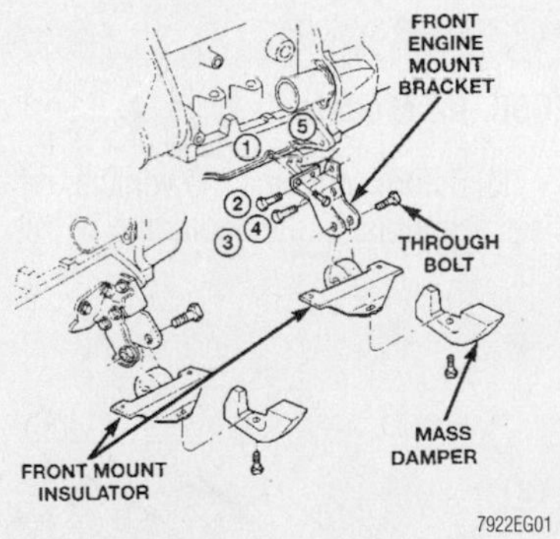

Front engine mount location and bolt identification—1998–99 models

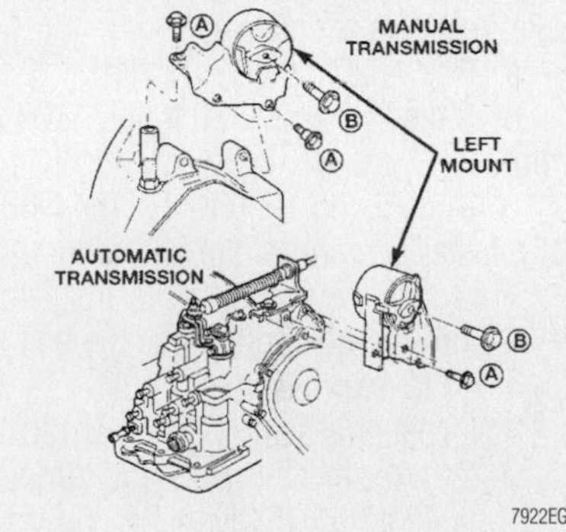

Exploded view of the left engine mount—1998–99 models

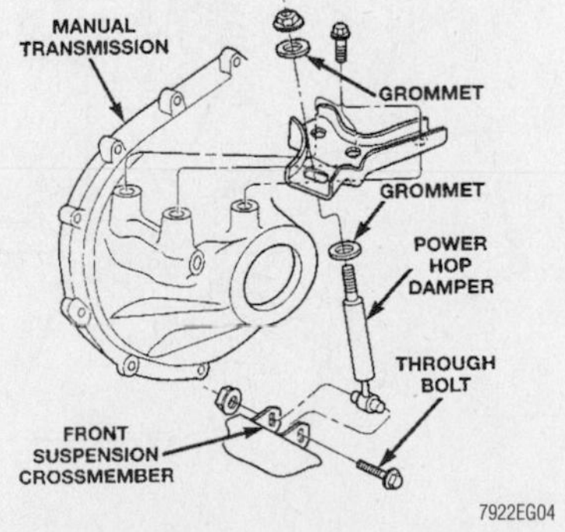

View of the power hop damper—1998–99 models—manual transaxle

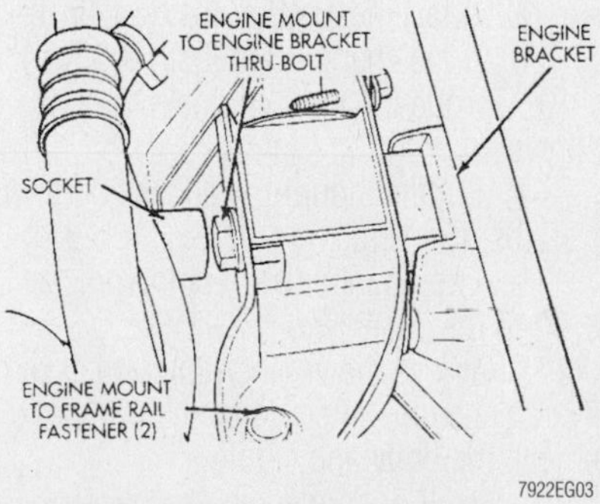

Location of the right engine mount—1998–99 models

order to allow movement for positioning onto the engine locating holes and flange on the engine bedplate. Lower the vehicle and position the cradle until the engine is resting on the support posts. Tighten the mounts to the cradle frame. This will keep the support posts from moving when removing or installing the engine and transaxle.

9. Install safety straps around the engine to the cradle; tighten the straps and lock them into position.

10. Raise the vehicle enough to see if the straps are tight enough to hold the cradle assembly to the engine.

11. Lower the vehicle so the weight of the engine and transaxle ONLY is on the cradle.

12. Remove the engine and transaxle mount through-bolts.

13. Raise the vehicle slowly, it might be necessary to move the engine/transaxle assembly with the cradle to allow removal around the body flanges.

To install:

14. Position the engine/transaxle assembly under the vehicle and lower the vehicle over the assembly.

15. Install the engine and transaxle mounts.

16. Torque the front engine mount bracket retainers as follows:

a. Bolt 1: 20 inch lbs. (3 Nm).

b. Bolts 2, 3 and 4: 80 ft. lbs. (108 Nm).

c. Bolts 5 and 1: 40 ft. lbs. (54 Nm).

17. Install or connect the following:

- Engine mount bracket-to-insulator through-bolt and torque the bolt to 40 ft. lbs. (54 Nm)
- Insulator assembly-to-lower radiator crossmember nuts and torque to 40 ft. lbs. (54 Nm)
- Mass damper bolt and torque to 40 ft. lbs. (54 Nm)

18. Torque the left engine mount as follows:

a. **A** fasteners: 40 ft. lbs. (54 Nm).

b. **B** fasteners: 80 ft. lbs. (108 Nm).

19. Torque the right engine mount as follows:

a. Engine mount-to-rail fasteners: 40 ft. lbs. (54 Nm).

b. Engine mount-to-engine bracket: 80 ft. lbs. (108 Nm).

20. Remove the engine/transaxle assembly safely straps; then, raise the vehicle to remove the dolly and cradle.

21. Install or connect the following:

- Power hop damper, if equipped and torque the retainers to 40 ft. lbs. (54 Nm)
- Exhaust pipe to the manifold
- Halfshafts
- Accessory drive belts
- Right inner splash shield
- Heater hoses
- Engine wiring harness
- Throttle body linkage
- Transaxle shift linkage
- Clutch cable if equipped with a manual transaxle
- Automatic transaxle cooler lines, if equipped
- Lower radiator hose
- Fan module assembly
- Radiator
- Upper radiator hose
- PCM
- Battery and battery tray

22. Fill the engine with clean oil.

23. Fill the cooling system.

24. Perform the camshaft and crankshaft timing relearn procedure as follows:

a. Connect a DRB scan tool to the DLC (located under the instrument panel, near the steering column).

b. Turn the ignition switch **ON** and access the "miscellaneous" screen.

c. Select "re-learn cam/crank" option and follow the directions on the scan tool screen.

25. If equipped with air conditioning, recharge the system.

2000–02 Models

1. Before servicing the vehicle, refer to the precautions in the beginning of this section.

➡After all components are installed on the engine, a DRB scan tool is necessary to perform the camshaft and crankshaft timing relearn procedure.

2. Properly recover the air conditioning system refrigerant.

3. Properly relieve the fuel system pressure.

4. Drain the engine oil.

5. Drain the cooling system.

6. Remove or disconnect the following:

- Battery and battery tray
- Air intake duct from the intake manifold
- Throttle cables
- Electrical connectors
- Air cleaner assembly
- Upper radiator hose
- Fan module assembly
- Lower radiator hose
- Cooler lines if equipped with an automatic transaxle
- Clutch cable if equipped with a manual transaxle
- Shift linkage from the transaxle
- Engine wiring harness
- Powertrain Control Module (PCM)
- Positive cable from the Power Distribution Center (PDC)
- Ground wires
- Heater hoses
- Brake booster vacuum hose
- Coolant recovery hose
- Accessory drive belts
- Power steering pump and reservoir, move them aside

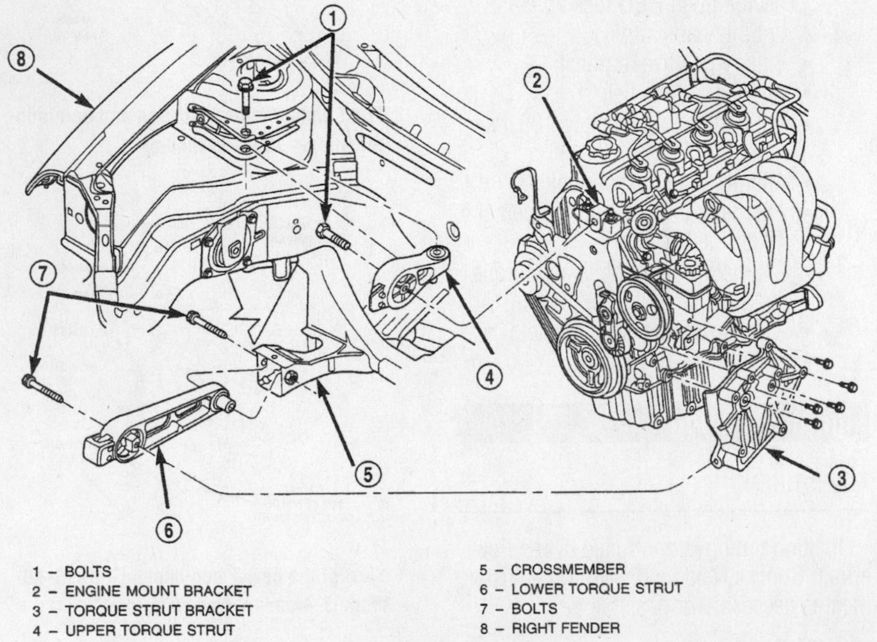

Exploded view of the torque struts and bracket—2000–02 models

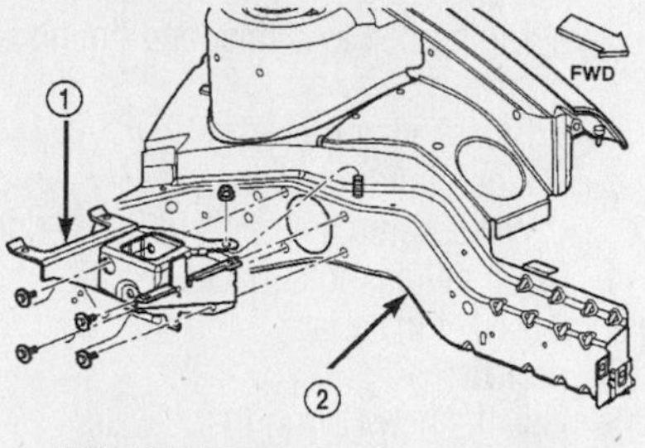

1 - MOUNT BRACKET
2 - BODY FRAME RAIL

9306EG14

Exploded view of the left mount bracket—2000–02 models

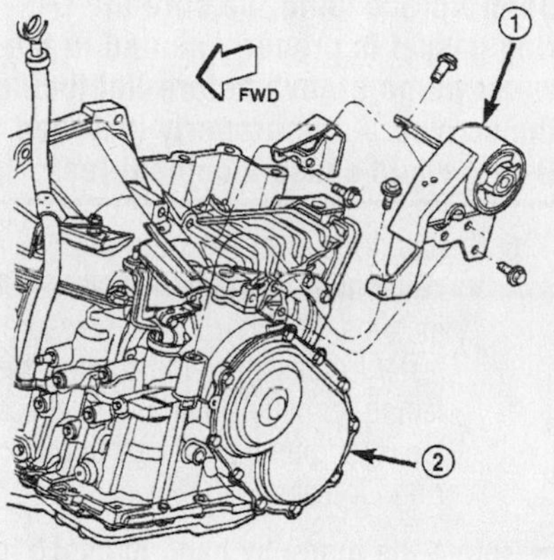

1 - MOUNT
2 - TRANSAXLE

9306EG15

Exploded view of the left mount—Automatic transaxle 2000–02 models

- Right inner splash shield
- Front wheels
- Halfshafts
- Exhaust pipe from the manifold
- Downstream Oxygen (O_2S) sensor connector
- Lower engine torque strut
- Structural collar
- Air conditioning compressor, if equipped

7. Raise the vehicle enough to allow an Engine Dolly Tool 6135 and Cradle Tool 6710 to be placed under the engine.

8. Loosen the engine support posts in order to allow movement for positioning onto the engine locating holes and flange on the engine bedplate. Carefully lower the vehicle and position the cradle until the engine is resting on the support posts. Tighten the mounts to the cradle frame. This will keep the support posts from moving when removing or installing the engine and transaxle.

9. Install safety straps around the engine to the cradle; tighten the straps and lock them into position.

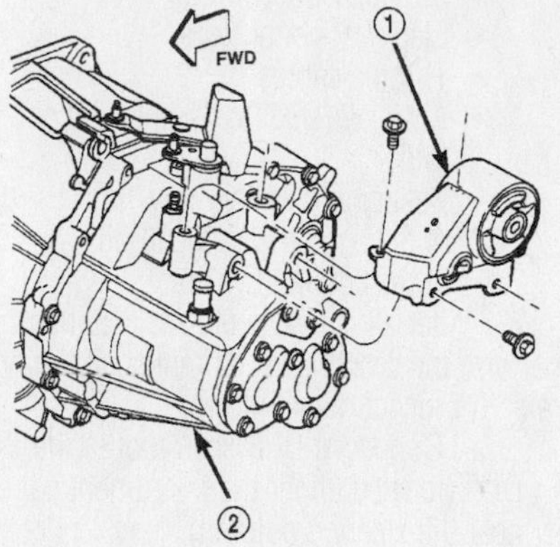

1 - MOUNT
2 - TRANSAXLE

9306EG16

Exploded view of the left mount—Manual transaxle 2000–02 models

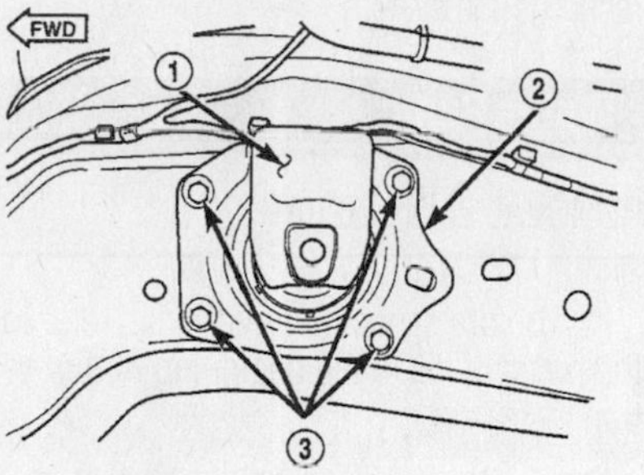

1 - SNUBBER PAD
2 - RIGHT ENGINE MOUNT
3 - BOLTS

9306EG17

Exploded view of the right mount—2000–02 models

10. Raise the vehicle enough to see if the straps are tight enough to hold the cradle assembly to the engine.

11. Lower the vehicle so the weight of the engine and transaxle ONLY is on the cradle.

12. Remove the upper engine torque strut.

13. Remove the right and left engine/transaxle mount through-bolts.

14. Raise the vehicle slowly, until it is about 6 in. (15cm) above normal engine mount locations.

15. Remove the alternator.

16. Lower bracket and upper mounting bolt.

17. Continue raising the vehicle until the engine/transaxle assembly clears the compartment.

➡It may be necessary to move the engine/transaxle assembly with the cradle to clear the body flanges.

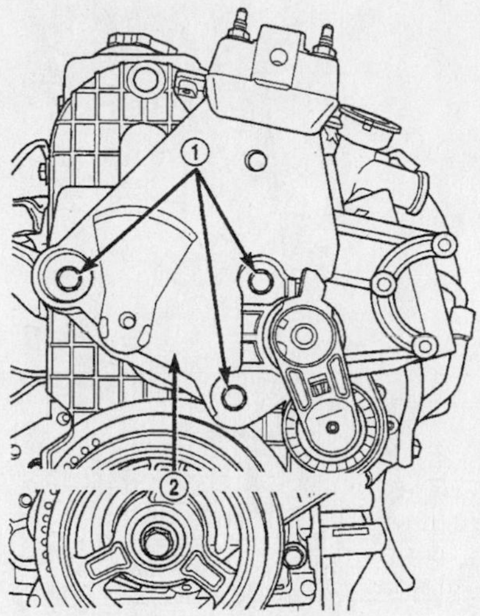

1 - BOLTS
2 - ENGINE MOUNT BRACKET ASSEMBLY

9306EG19

Exploded view of the right mount bracket—2000–02 models

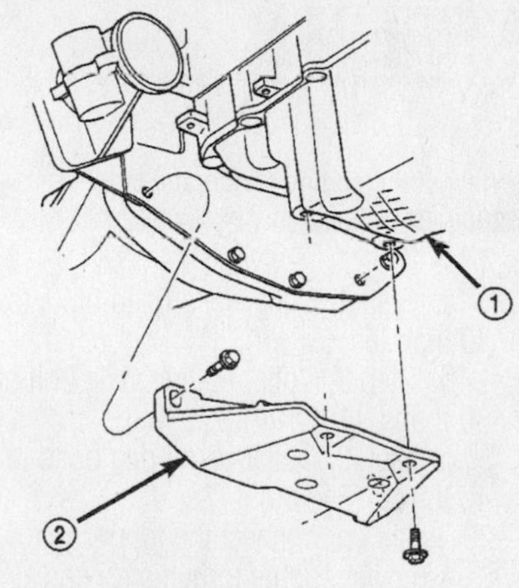

1 - OIL PAN
2 - STRUCTURAL COLLAR

9306EG18

Exploded view of the structural collar—2000–02 models

To install:

18. Install or connect the following:
    - Engine/transmission assembly to the vehicle and torque the engine mount bolts to 87 ft. lbs. (118 Nm)
    - Upper engine torque strut and remove the engine safety straps
    - Lower bracket and upper mounting bolt
    - Alternator
    - Halfshafts

19. Install the structural collar and torque the bolts, in 3 steps, using the following procedure:

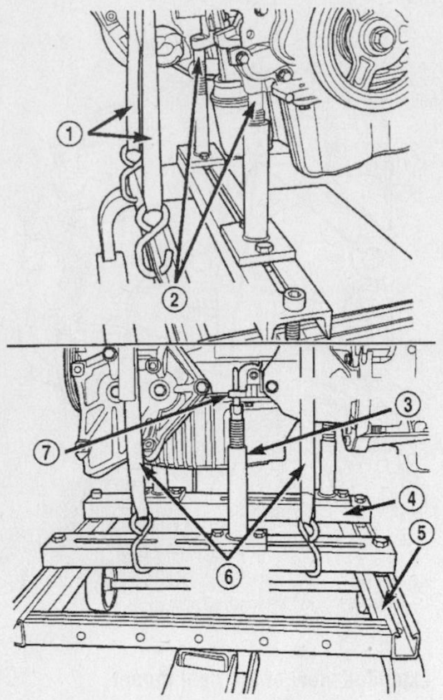

View of the engine cradle and dolly—2000–02 models

a. Step 1: Collar-to-oil pan bolts to 30 inch lbs. (3 Nm).
b. Step 2: Collar-to-transaxle bolts to 80 ft. lbs. (108 Nm).
c. Step 3: Collar-to-oil pan bolts to 40 ft. lbs. (54 Nm).

20. Install or connect the following:
- Lower engine torque strut and torque the through-bolts to 87 ft. lbs. (118 Nm)
- Downstream O_2S sensor connector
- Exhaust pipe to the manifold
- A/C compressor, if equipped
- Drive belts
- Right inner splash shield
- Front wheels
- Power steering pump and reservoir
- Clutch cable if equipped with a manual transaxle
- Shift linkage to the transaxle
- Automatic transaxle cooler lines, if equipped
- Fuel lines
- Heater hoses
- Ground wires
- PCM
- Engine wire harness
- Lower radiator hose
- Fan module assembly
- Upper radiator hose
- Air cleaner assembly
- Electrical connectors
- Throttle cables
- Air intake duct to the intake manifold
- Battery and battery tray

21. Fill the engine with clean oil.
22. Fill the cooling system.
23. After all components are installed, perform the camshaft and crankshaft timing relearn procedure as follows:
a. Connect a DRB scan tool to the DLC (located under the instrument panel, near the steering column).
b. Turn the ignition switch **ON**, and access the "miscellaneous" screen.
c. Select "re-learn cam/crank" option and follow the directions on the scan tool screen.
24. If equipped with air conditioning, recharge the system.

Water Pump

REMOVAL & INSTALLATION

1. Before servicing the vehicle, refer to the precautions in the beginning of this section.

➡After all components are installed on the engine, a DRB scan tool is necessary to perform the camshaft and crankshaft timing relearn procedure.

2. Drain the cooling system.
3. Remove or disconnect the following:
- Negative battery cable
- Right inner splash shield
- Accessory drive belts
- Power steering pump and support the engine from the bottom with a jack
- Right engine mount
- Power steering pump bracket bolts, move the pump/bracket assembly aside

➡It is not necessary to disconnect the power steering lines.

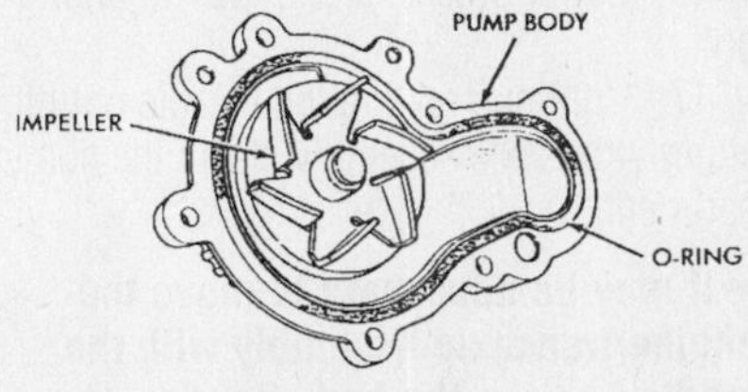

Be sure the O-ring is seated in the groove before installing the new pump

- Right engine mount bracket
- Timing belt tensioner and timing belt
- Camshaft sprocket(s)
- Inner timing belt cover
- Water pump and discard the O-ring

4. Clean all mating surfaces of any residual gasket material.

To install:

5. Install a new O-ring in the water pump groove.

➡Hold the O-ring in place with a few small dabs of silicone sealant.

⁂ WARNING

Before proceeding, be sure the O-ring gasket is properly seated in the water pump groove before tightening the screws. An improperly installed O-ring could cause a coolant leak.

6. Install or connect the following:
- Water pump, torque the bolt to 108 inch lbs. (12 Nm). Use a pressure tester to pressurize the cooling system to 15 psi (103 kPa) and check the water pump shaft seal and O-ring for leaks.

➡Rotate the pump by hand to check for freedom of movement.

- Inner timing belt cover
- Timing belt and tensioner
- Right engine mount bracket and engine mount
- Power steering pump
- Accessory drive belts
- Negative battery cable

7. Refill the cooling system.
8. Use a DRB scan tool to perform the camshaft and crankshaft timing relearn procedure, as follows:
a. Connect the scan tool to the DLC, located under the instrument panel near the steering column.
b. Turn the ignition switch **ON**, and access the "miscellaneous" screen.
c. Select the "re-learn cam/crank" option, then follow the instructions on the scan tool screen.

Cylinder Head

REMOVAL & INSTALLATION

1998–99 Models

1. Before servicing the vehicle, refer to the precautions in the beginning of this section.

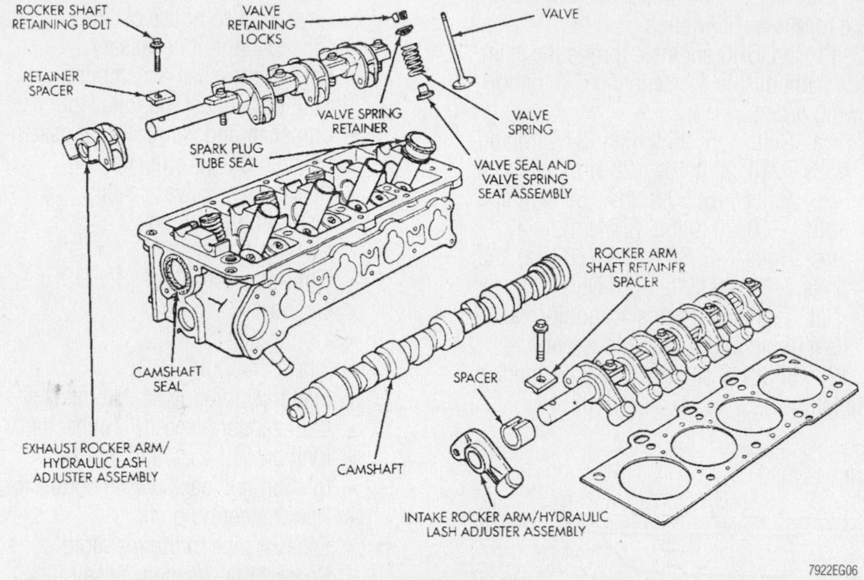

Exploded view of the cylinder head and valvetrain components—SOHC engine

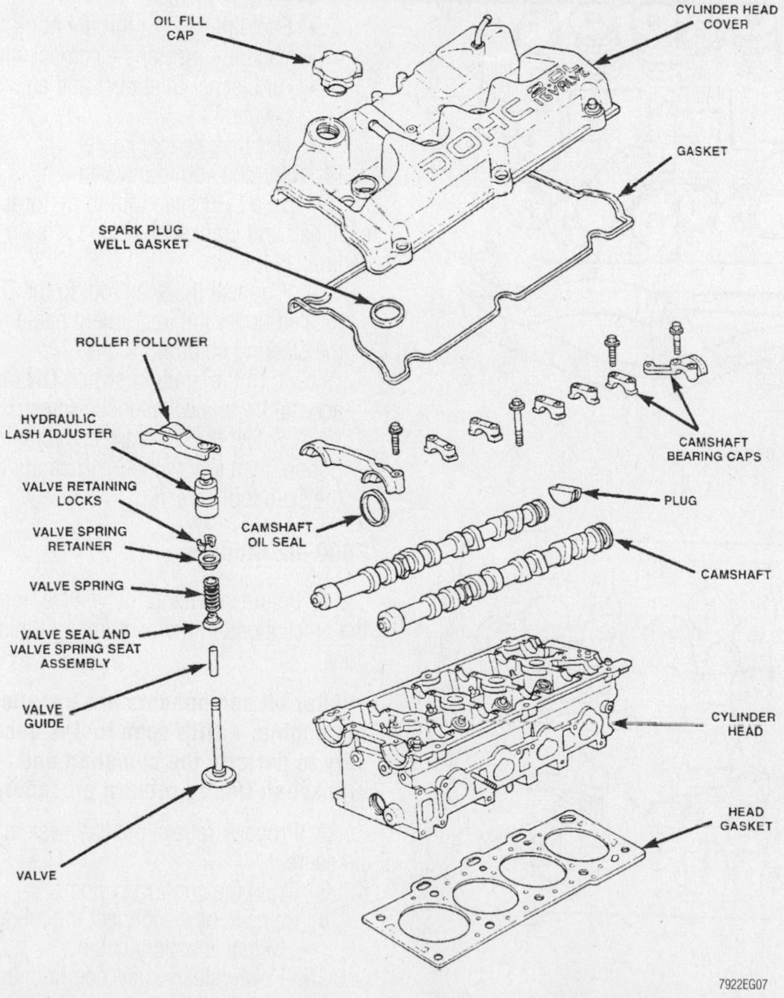

Exploded view of the cylinder head and valvetrain components—DOHC engine

➡After all components are installed on the engine, a DRB scan tool is necessary to perform the camshaft and crankshaft timing relearn procedure.

2. Properly relieve the fuel system pressure.
3. Drain the cooling system.
4. Remove or disconnect the following:
   - Negative battery cable
   - Air cleaner inlet duct and air cleaner
   - Vacuum lines and electrical wiring
   - Fuel lines from the throttle body
   - Throttle linkage
   - Accessory drive belts
   - Power brake vacuum hose from the intake manifold
   - Exhaust pipe from the manifold
   - Power steering pump, move it aside. DO NOT disconnect the fluid lines.
   - Ignition coil pack wiring connector
   - Ignition coil pack and bracket
   - Camshaft Position (CMP) sensor electrical connector
   - Fuel injector electrical connector
   - Intake manifold
   - Timing belt
   - Timing belt tensioner, for DOHC engine
   - Camshaft sprocket
   - Inner timing belt cover
   - Rocker arm cover
   - Camshaft and cam follower assemblies on DOHC engine
5. On SOHC engine, remove or disconnect the following:
   - Rocker arm shaft assemblies
   - Oil separator, if necessary
   - Heater hoses
   - Any remaining hoses or lines
6. Remove or disconnect the following:
   - Cylinder head bolts, working from the center outward
   - Cylinder head

➡The cylinder head bolts must be inspected before they can be reused. If the threads of bolts are stretched, they must be replaced. Check for thread stretching by holding a scale or other straightedge against the threads. If all the threads do not contact the scale, the bolts must be replaced.

⁂ WARNING

Use only a plastic scraper to clean the mating surfaces. NEVER use metal, as this may gouge the surfaces and cause leaks!

7. Cover the combustion chambers, then use a plastic scraper to thoroughly and carefully clean the engine block and cylinder head mating surfaces.

To install:

8. Install the cylinder head with a new gasket.

9. Lubricate the cylinder head bolt threads. The 4 short bolts (110mm for 1998 or 164mm for 1999) are installed in positions 7, 8, 9 and 10.

10. On SOHC engines, torque the cylinder head bolts, in the sequence, to:
  a. Step 1: 25 ft. lbs. (34 Nm).
  b. Step 2: 50 ft. lbs. (68 Nm).
  c. Step 3: 50 ft. lbs. (68 Nm).
  d. Step 4: An additional ¼ turn using a torque angle wrench.

11. On DOHC engines, torque the cylinder head bolts, in the sequence, in the following order:
  a. Bolts 1–6: 25 ft. lbs. (34 Nm) and bolts 7–10: 20 ft. lbs. (28 Nm).
  b. Bolts 1–6: 50 ft. lbs. (68 Nm) and bolts 7–10: 20 ft. lbs. (28 Nm).
  c. Bolts 1–6: 50 ft. lbs. (68 Nm) and bolts 7–10: 20 ft. lbs. (28 Nm).
  d. Tighten all bolts an additional ¼ turn using a torque angle wrench.

12. On SOHC engine, install or connect the following:
  - Heater hoses
  - Any removed hoses or lines
  - Oil separator, if necessary
  - Rocker arm shaft assemblies

13. Install or connect the following:
  - Camshaft and cam follower assemblies on DOHC engines
  - Rocker arm (valve) cover
  - Inner timing belt cover
  - Camshaft sprocket
  - Timing belt tensioner, for DOHC engine
  - Timing belt
  - Intake manifold
  - Fuel injector electrical connector
  - Cam sensor electrical connector
  - Ignition coil pack and bracket
  - Ignition coil pack wiring connector
  - Power steering pump
  - Exhaust pipe to the manifold
  - Power brake vacuum hose to the intake manifold
  - Accessory drive belts
  - Throttle linkage
  - Fuel lines to the throttle body
  - Vacuum lines and electrical wiring
  - Air cleaner inlet duct and air cleaner
  - Negative battery cable

14. Refill the cooling system.

15. Use a DRB scan tool to perform the camshaft and crankshaft timing relearn procedure, as follows:
  a. Connect the scan tool to the DLC, located under the instrument panel near the steering column.
  b. Turn the ignition switch **ON**, and access the "miscellaneous" screen.
  c. Select the "re-learn cam/crank" option, then follow the instructions on the scan tool screen.

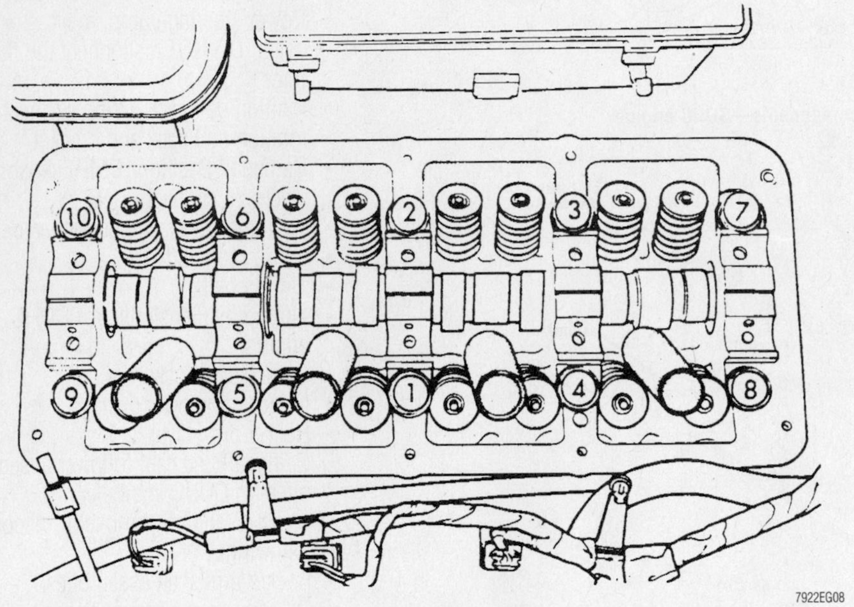

Cylinder head bolt torque sequence—2.0L (VIN C) SOHC engine

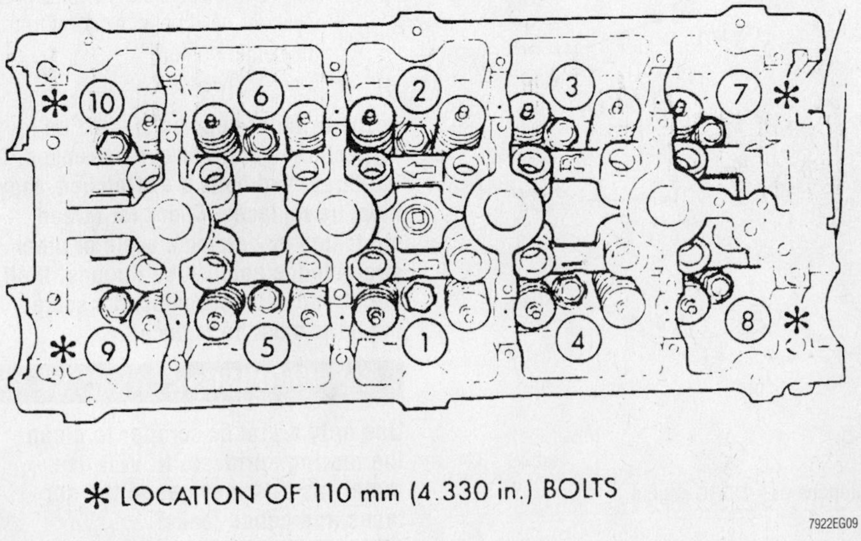

Cylinder head bolt torque sequence—2.0L (VIN Y) DOHC engine

2000–02 Models

1. Before servicing the vehicle, refer to the precautions in the beginning of this section.

➡After all components are installed on the engine, a DRB scan tool is necessary to perform the camshaft and crankshaft timing relearn procedure.

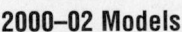

2. Properly relieve the fuel system pressure.
3. Drain the cooling system.
4. Remove or disconnect the following:
  - Negative battery cable
  - Power steering/air conditioning drive belt
  - Exhaust pipe from the manifold
  - Right front wheel
  - Right side splash shield

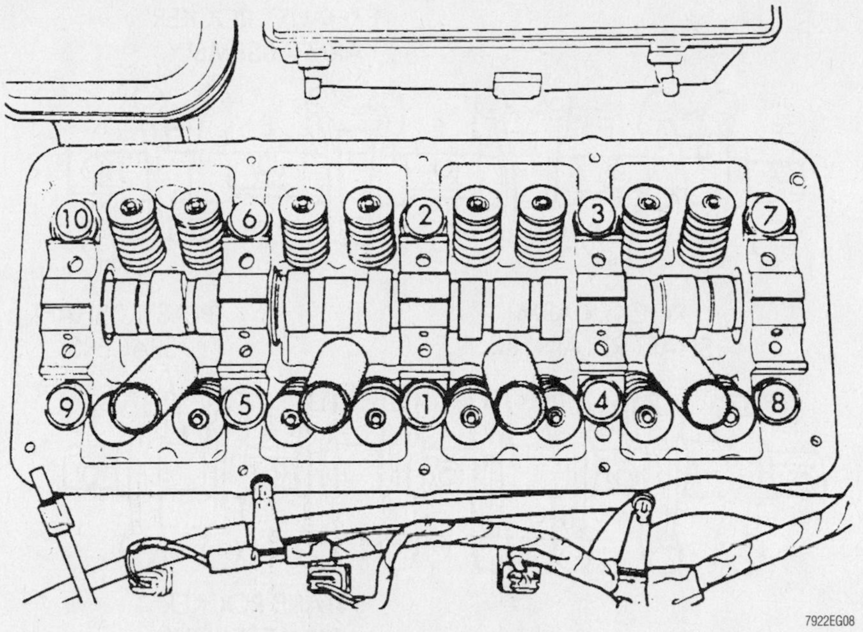

Cylinder head bolt torque sequence—2.0L (VIN C) SOHC engine

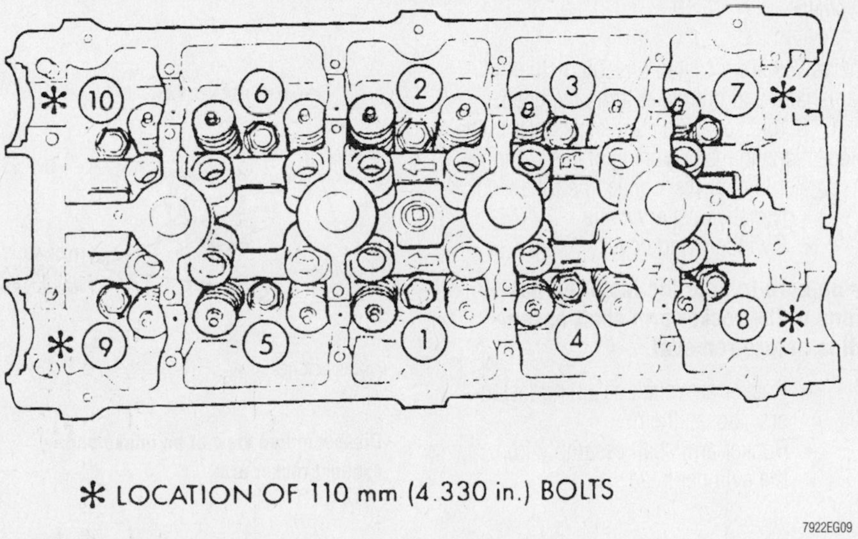

Cylinder head bolt torque sequence—2.0L (VIN Y) DOHC engine

- Alternator drive belt
- Crankshaft damper
- Lower torque strut
- Upper torque strut
- Ground strap from the engine mount bracket
- Power steering hose support clip from the engine mount bracket
- Power steering pump, move it asidc

5. Place a jack under the engine and support it.
   - Right side engine mount-to-bracket through-bolt
   - Lower engine mount bracket bolt
6. Slightly raise the engine.
   - Upper engine mount bracket bolts
   - Engine mount bracket

➡It may be necessary to raise and lower the engine until the bracket clears the engine components.

- Front timing belt cover

7. Rotate the crankshaft and align the timing marks.
   - Timing belt and tensioner
   - Camshaft sprocket
   - Rear timing belt cover
   - Fuel line from the fuel rail
   - Coolant recovery container
   - Ground wire and from the cylinder head
   - Upper radiator hose
   - Intake manifold
   - Ignition coil electrical connector
   - Coil pack
   - Spark plug wires
   - Crankcase Closed Ventilation (CCV) hose from the valve cover
   - Cam sensor electrical connector
   - Coolant Temperature (CT) sensor electrical connector
   - Heater tube from the cylinder head
   - Heater hose from the thermostat housing connector
   - Cylinder head cover
   - Cylinder head bolts, working from the center out
   - Cylinder head and discard gasket

➡The cylinder head bolts must be inspected before they can be reused. If the threads of bolts are stretched, they must be replaced. Check for thread stretching by holding a scale or other straightedge against the threads. If all the threads do not contact the scale, the bolts must be replaced.

⁂ WARNING

Use only a plastic scraper to clean the mating surfaces. NEVER use metal, as this may gouge the surfaces and cause leaks!

8. Cover the combustion chambers, then use a plastic scraper to thoroughly and carefully clean the engine block and cylinder head mating surfaces.

To install:

9. Install the cylinder head with a new gasket after applying Mopar Gasket Sealant to both sides of the gasket.
10. Lubricate the cylinder head bolt threads. The 4 short bolts (164mm) are installed in positions 7, 8, 9 and 10.
11. Torque the cylinder head bolts, in the sequence, to:
    a. Step 1: 25 ft. lbs. (34 Nm).
    b. Step 2: 50 ft. lbs. (68 Nm).
    c. Step 3: 50 ft. lbs. (68 Nm).
    d. Step 4: An additional ¼ turn using a torque angle wrench.
12. Install or connect the following:
    - Cylinder head cover
    - Heater hose to the thermostat housing connector

- Heater tube to the cylinder head
- CT sensor electrical connector
- Cam sensor electrical connector
- CCV hose to the valve cover
- Coil pack
- Spark plug wires
- Ignition coil electrical connector
- Intake manifold
- Upper radiator hose
- Ground wire to the cylinder head
- Coolant recovery container
- Fuel line to the fuel rail
- Rear timing belt cover
- Camshaft sprocket
- Timing belt and tensioner
- Front timing belt cover
- Engine mount bracket
- Upper engine mount bracket bolts
- Lower engine mount bracket bolt
- Right side engine mount-to-bracket through-bolt and torque it to 87 ft. lbs. (118 Nm)

13. Remove the jack from under the engine.
    - Power steering pump
    - Power steering hose support clip to the engine mount bracket
    - Ground strap to the engine mount bracket
    - Upper torque strut
    - Lower torque strut
    - Crankshaft damper and torque the bolt to 100 ft. lbs. (136 Nm)
    - Alternator drive belt
    - Right side splash shield
    - Right front wheel
    - Exhaust pipe to the manifold
    - Power steering/air conditioning drive belt
    - Negative battery cable
14. Refill the cooling system.
15. Use a DRB scan tool to perform the camshaft and crankshaft timing relearn procedure, as follows:
    a. Connect the scan tool to the DLC, located under the instrument panel near the steering column.
    b. Turn the ignition switch **ON**, and access the "miscellaneous" screen.
    c. Select the "re-learn cam/crank" option, then follow the instructions on the scan tool screen.

Rocker Arms/Shafts

REMOVAL & INSTALLATION

This procedure applies to Single Overhead Camshaft (SOHC) engines only. On Dual Overhead Camshaft (DOHC) engines, the valves are actuated directly by the camshafts and no rocker arms are used.

1. Before servicing the vehicle, refer to the precautions in the beginning of this section.
2. Remove or disconnect the following:
    - Negative battery cable
    - Cylinder head cover

➡Be sure to note the installed positions of the rocker arm shaft assemblies before removal.

- Rocker arm shaft attaching fasteners, loosen them
- Rocker arm shaft assembly from the cylinder head

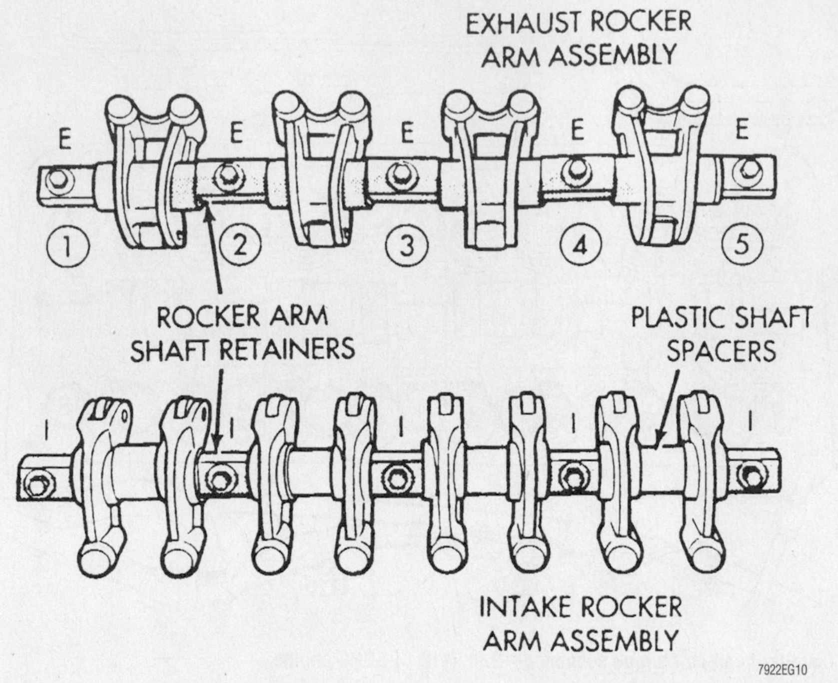

Be sure to note the rocker arm shaft component positions before disassembling the rocker arm shafts

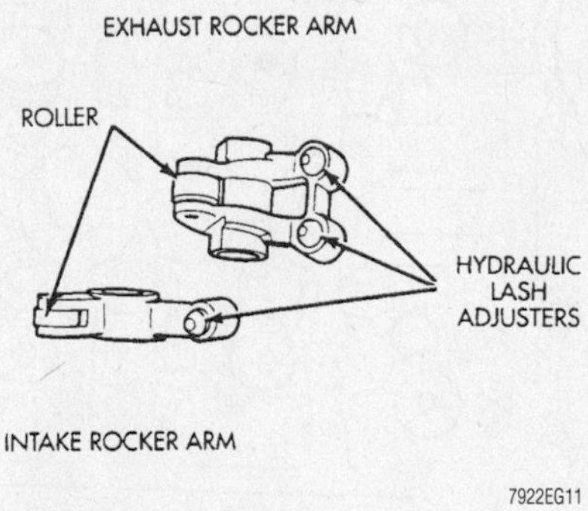

Disassembled view of an intake and exhaust rocker arm

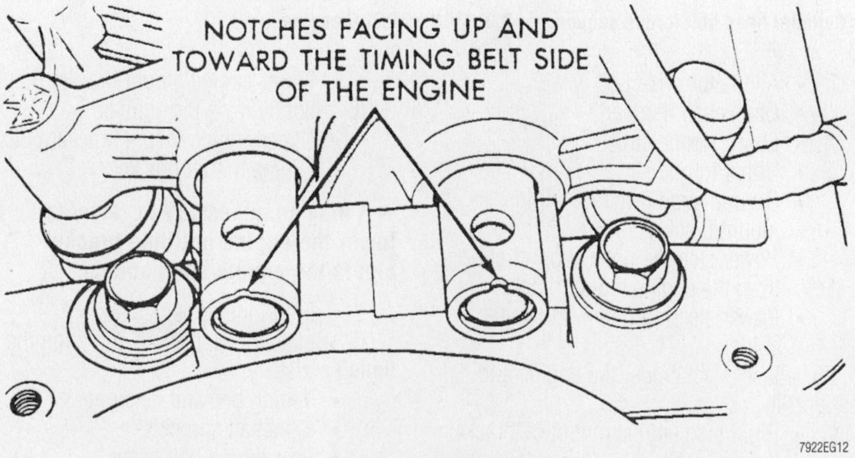

Be sure the notches in the rocker arm shafts point up and toward the timing belt side of the engine

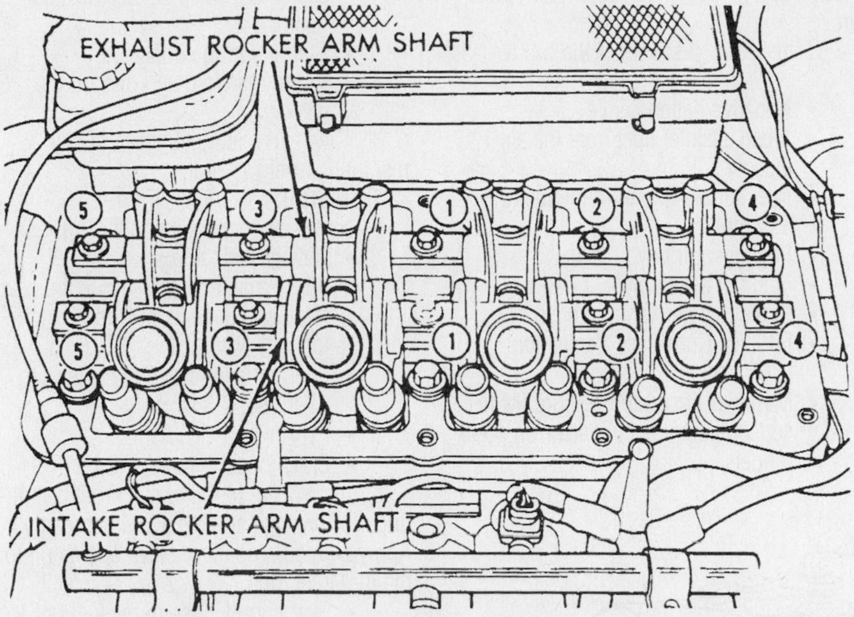

Rocker arm shaft assembly bolt tightening sequence

- Rocker arm assemblies, slide the rocker arms and spacers off the shaft

➡**Be sure to keep the spacers and rocker arms in their original locations for installation.**

3. Inspect the rocker arm for scoring, wear on the roller or damage to the rocker arm; replace any components showing damage. Check the location where the rocker arms mount to the shafts for wear or damage. Replace if damaged or worn. The rocker arm shaft is hollow and is used as a lubrication oil duct. Check the oil holes for clogs with a small piece of wire, and clean as required. Lubricate the rocker arms and spacers. Be sure to install in their original locations.

To install:

⁂ WARNING

Set the crankshaft to 3 notches before Top Dead Center (TDC) before installing the rocker arm shafts, otherwise valvetrain damage may occur when the mounting bolts are tightened.

4. Set the crankshaft sprocket to TDC by aligning the mark on the sprocket with the arrow on the oil pump housing, then back off to 3 notches before TDC.
5. Install the rocker arm/hydraulic lash adjuster assembly making sure that the adjusters are at least partially full of oil. This is indicated by little or no plunger travel when the lash adjuster is depressed. If there is excessive plunger travel, submerge the rocker arm assembly into clean engine oil and pump the plunger until the lash adjuster travel is taken up. If travel is not reduced, replace the assembly. The hydraulic lash adjuster and rocker arm are serviced as an assembly.
6. Install or connect the following:
   - Rocker arm and shaft assemblies. Position the rocker arm shafts with the **NOTCH FACING UP** and toward the timing belt side of the engine.
   - Retainers in their original positions on the exhaust and intake shafts. Tighten the bolts, in the sequence, to 21 ft. lbs. (28 Nm).

⁂ WARNING

When installing the intake rocker arm shaft assembly, be sure the plastic spacers do not interfere with the spark plug tubes. If the spacers do interfere, rotate them until they are at the proper angle. To avoid damaging the spark plug tubes, do not try to rotate the spacers by forcing the shaft down.

7. Install the rocker arm (valve) cover.
8. Connect the negative battery cable.

Intake Manifold

REMOVAL & INSTALLATION

SOHC Engine —Except High Output Engines

1. Before servicing the vehicle, refer to the precautions in the beginning of this section.

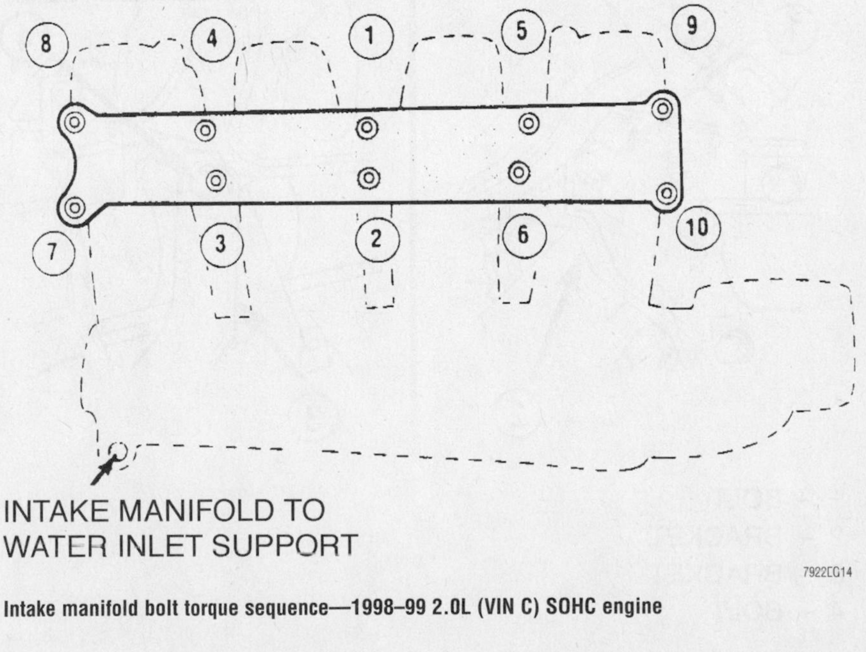

Intake manifold bolt torque sequence—1998–99 2.0L (VIN C) SOHC engine

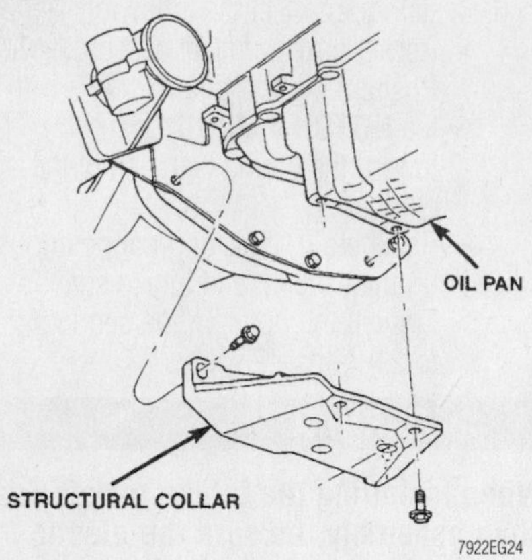

Intake manifold bolt torque sequence—2000–02 2.0L (VIN C) SOHC engine

2. Properly relieve the fuel system pressure.

3. Remove or disconnect the following:

- Negative battery cable
- Fresh air inlet duct from the air cleaner
- Fuel supply line from the fuel tube assembly
- Fuel injector cover
- Fuel rail assembly
- Brake booster vacuum hose
- Positive Crankcase Ventilation (PCV) hose
- Manifold Absolute Pressure (MAP) sensor electrical connector on 1999 models
- Knock Sensor (KS) electrical connector
- Starter wiring connectors
- Intake manifold and discard the gasket

4. Clean all mating surfaces of any residual gasket material.

To install:

5. Install or connect the following:

- New gaskets and seals
- Intake manifold and torque the new retainers, in sequence, to 95 inch lbs. (11 Nm)
- Fuel rail and torque the screws to 17 ft. lbs. (23 Nm)
- PCV hose
- Brake booster hose

6. Inspect the fuel line quick-connect fittings for damage and replace if necessary. Apply a small amount of clean engine oil to the fuel inlet tube.

- Fuel supply hose to the fuel rail assembly. Ensure the connection is fastened securely by pulling on the connector.
- Fuel injector cover
- MAP sensor wiring
- KS sensor connector
- Air inlet duct
- Negative battery cable

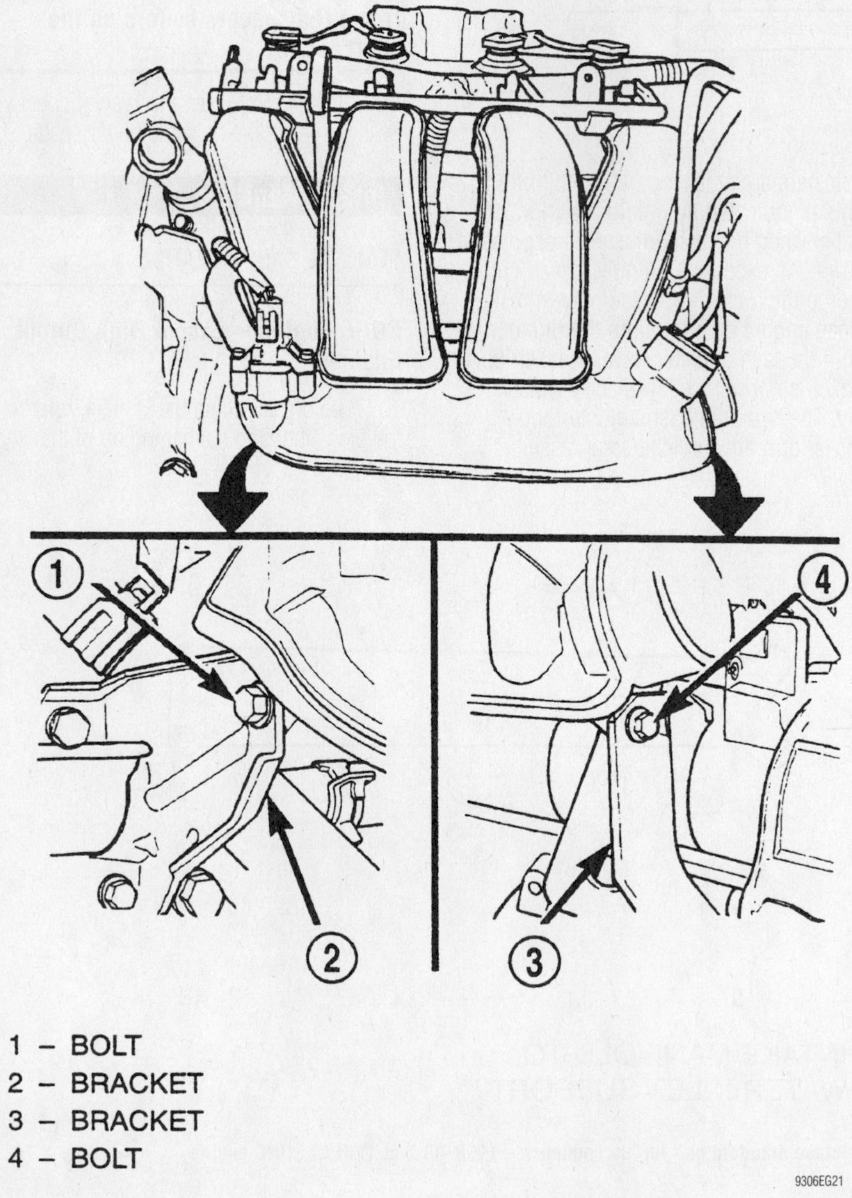

View of the intake manifold lower supports—2000–02 SOHC engines

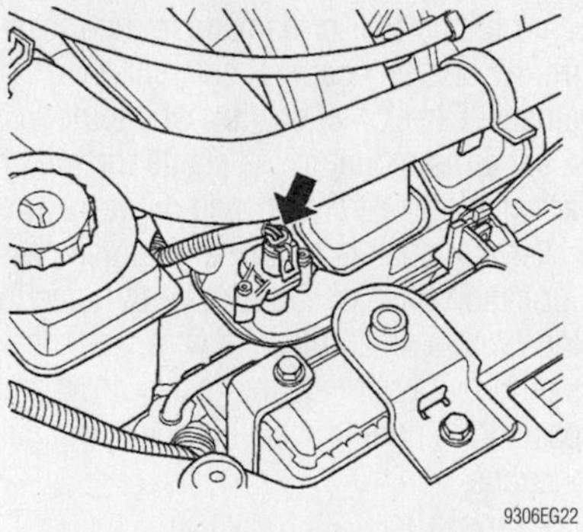

View of the Manifold Absolute Pressure (MAP) sensor—2000–02 SOHC engines

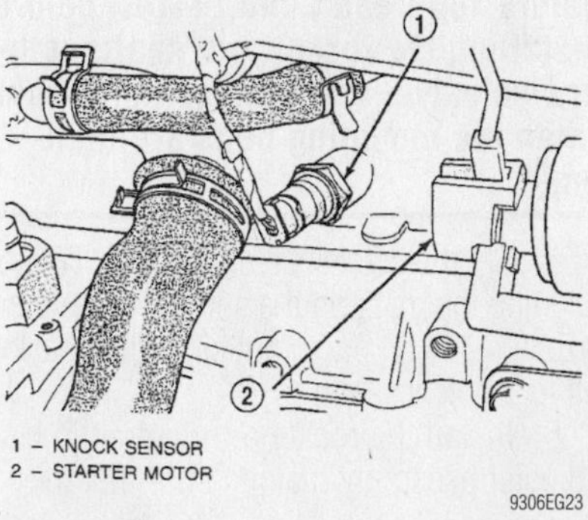

View of the Knock Sensor (KS)—2000–02 SOHC engines

7. Start the vehicle, check for leaks and repair if necessary.

High Output Engine

1. Before servicing the vehicle, refer to the precautions in the beginning of this section.
2. Properly relieve the fuel system pressure.
3. Remove or disconnect the following:
   - Negative battery cable
   - Throttle body-to-intake manifold air duct

✱✱ CAUTION

Wrap towels around the fitting to catch any spilled fuel.

- Fuel supply line quick-connect from the fuel tube assembly
- Fuel rail cover from the engine
- Fuel injector connector and harness from the injectors and fuel rail
- Positive Crankcase Ventilation (PCV) hose
- Lower intake manifold support bracket bolt
- Oil dipstick
- Manifold Absolute Pressure (MAP) sensor wiring
- Manifold Tuning Valve (MTV) actuator wiring
- Intake manifold-to-cylinder head bolts and move the manifold forward
- Knock (KS) Sensor wiring
- Brake booster hose
- Intake manifold and discard the gasket

To install:

4. Clean the mating surfaces of any residual gasket material.
5. Install or connect the following:
   - Lower manifold to the upper manifold with a new gasket and torque the bolts to 105 inch lbs. (12 Nm).
   - Intake manifold on the cylinder head with a new gasket
   - Engine wiring harness between the middle intake runners
   - KS wiring connector
   - Brake booster hose and torque the intake manifold bolts, in sequence, to 105 inch lbs. (12 Nm)
   - Lower manifold support bracket bolt and torque the bolt to 21 ft. lbs. (28 Nm)
   - MAP sensor connector
   - MTV actuator connector
   - Oil dipstick
   - Inlet air duct to the manifold and the throttle body
   - PCV hose
   - Fuel injector connectors and harness to the fuel rail
   - Fuel rail cover
   - Fuel supply line
   - Negative battery cable
6. Start the vehicle, check for leaks and repair if necessary.

DOHC Engine

1. Before servicing the vehicle, refer to the precautions in the beginning of this section.
2. Relieve the fuel system pressure.
3. Remove or disconnect the following:
   - Negative battery cable
   - Air inlet duct from the intake manifold

✱✱ CAUTION

Wrap towels around the fitting to catch any spilled fuel.

- Fuel supply line quick-connect from the fuel rail
- Engine Coolant Temperature (ECT) sensor connector
- Heater hose from the intake manifold
- Heater tube from the bottom of the intake manifold
- Upper radiator and coolant recovery hoses
- Fuel rail assembly

✱✱ WARNING

Do not allow the injectors to rest on their tips as this may damage them.

➡Cover the injector openings to prevent debris from entering the ports.

- Accelerator, kickdown and cruise control cables, if equipped, from the throttle lever and bracket
- Idle Air Control (IAC) motor electrical connector
- Throttle Position (TP) sensor electrical connector
- Throttle body vacuum hoses
- Throttle body
- Manifold Absolute Pressure/Idle Air Temperature (TMAP) sensor electrical connector on 1998 models
- Manifold Absolute Pressure (MAP) sensor electrical connector on 1999 models
- Knock Sensor (KS) electrical connector. Detach the wiring harness from the tab located on the heater tube

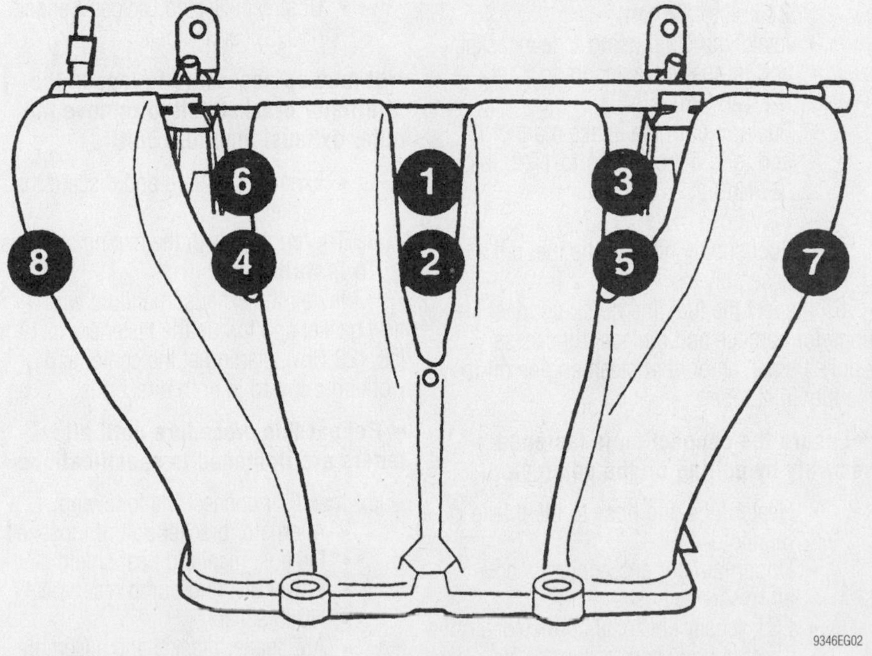

Intake manifold bolt torque sequence—High Output 2000–02 2.0L (VIN C) SOHC engine

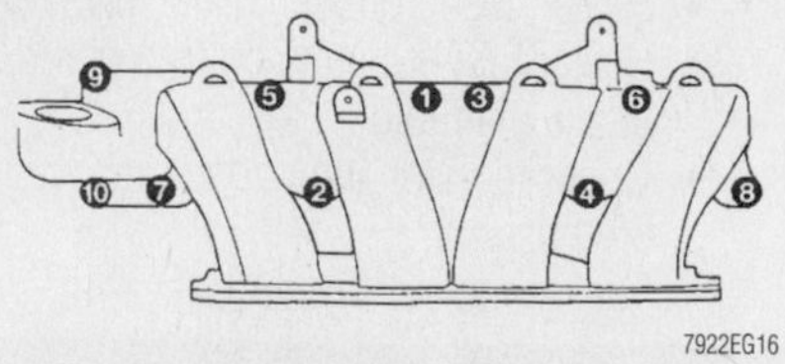

7922EG16

Intake manifold bolt torque sequence—2.0L (VIN Y) DOHC engine

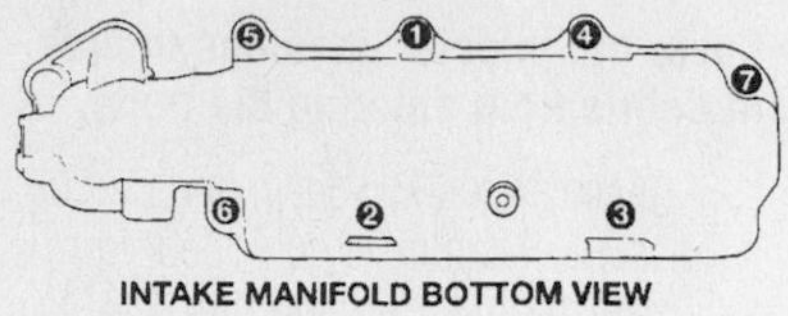

7922EG15

Intake manifold bolt torque sequence—2.0L (VIN Y) DOHC engine

- Starter electrical connectors
- Exhaust Gas Recirculation (EGR) tube
- Upper and lower intake manifold assemblies and discard the gaskets

➡If necessary, separate the upper and lower manifolds.

4. Thoroughly clean all of the gasket mating surfaces.

To install:

5. Install or connect the following:
   - Upper and lower manifolds with new gaskets and torque the bolts to 21 ft. lbs. (28 Nm)
   - Intake manifold using a new gasket and torque the fasteners to 21 ft. lbs. (28 Nm)
   - Fuel rail onto the intake manifold and torque the screws to 17 ft. lbs. (23 Nm)
   - PCV hose
   - Fuel supply hose to the fuel rail assembly
6. Inspect the fuel line quick-connect fittings for damage and replace, if necessary. Apply a small amount of clean engine oil to the fuel inlet tube.

➡Ensure the connection is fastened securely by pulling on the connector.

- Heater tube and hose to the intake manifold
- Upper radiator and coolant recovery reservoir hoses
- ECT sensor electrical connector
- Throttle body and torque the fastener to 16 ft. lbs. (22 Nm)
- TMAP sensor electrical connector on 1998 models
- MAP sensor electrical connector on 1999 models
- IAC sensor electrical connector on 1999 models
- KS and starter electrical connectors
- Wiring harness to the heater tube tab
- TP sensor electrical connectors
- Throttle body vacuum hoses
- Accelerator, kickdown and speed control cables, if equipped, to their bracket and throttle lever
- EGR tube and torque to 95 inch lbs. (11 Nm)
- Intake manifold side retainers and torque to 95 inch lbs. (11 Nm)
- Air duct to the air filter housing and torque the clamp to 25 inch lbs. (3 Nm)
- Negative battery cable

Exhaust Manifold

REMOVAL & INSTALLATION

1998 Models

1. Before servicing the vehicle, refer to the precautions in the beginning of this section.
2. Remove or disconnect the following:
   - Negative battery cable
   - Air cleaner assembly and bracket
   - Power steering pump reservoir, move it aside. DO NOT disconnect the fluid lines.
   - Exhaust manifold heat shield
   - Upstream Heated Oxygen Sensor (HO_2S) connector

➡It may be necessary to loosen the alternator bracket bolt to remove the outer exhaust manifold bolt.

- Exhaust manifold and discard the gasket

3. Thoroughly clean the mating surfaces.

To install:

4. Install the exhaust manifold with a new gasket and torque the fasteners to 17 ft. lbs. (23 Nm), starting at the center and working outward in both directions

➡Repeat this procedure until all fasteners are tightened to specifications.

5. Install or connect the following:
   - Alternator bracket bolt, if loosened
   - Exhaust manifold heat shield
   - Power steering pump reservoir
   - HO_2S sensor
   - Air cleaner bracket and assembly
   - Exhaust pipe to the manifold and torque the fasteners to 21 ft. lbs. (28 Nm)
   - Negative battery cable

1999 Models

1. Before servicing the vehicle, refer to the precautions in the beginning of this section.
2. Release the fuel pressure.
3. Remove or disconnect the following:
   - Negative battery cable
   - Fuel supply line from the fuel rail
   - Air cleaner assembly and bracket
   - Power steering reservoir bracket and move the reservoir aside
   - Upper heat shield from the exhaust manifold
   - Upstream Heated Oxygen (HO_2S) sensor connector
   - Right side splash shield
   - Alternator drive belt
   - Alternator spill shield
   - Exhaust pipe from the manifold. If Low Emission Vehicle (LEV) equipped, discard the manifold-to-flex joint gasket.
   - Lower heat shield from the exhaust manifold
   - Lower exhaust manifold fasteners
   - Upper exhaust manifold fasteners
   - Exhaust manifold and discard the gasket
4. Thoroughly clean the mating surfaces.

To install:

5. Install or connect the following:
   - New gasket
   - Exhaust manifold with a new gasket and torque to 17 ft. lbs. (23 Nm), starting at the center and working outward in both directions. Repeat this procedure until all fasteners are tightened to specifications.
   - Lower heat shield to the exhaust manifold
   - Exhaust pipe to the manifold with a new manifold-to-flex joint gasket, if LEV equipped. Torque the fasteners to 21 ft. lbs. (28 Nm).
   - Alternator spill shield
   - Alternator drive belt
   - Right side splash shield
   - HO_2S sensor connector
   - Upper heat shield to the exhaust manifold
   - Power steering reservoir and bracket
   - Air cleaner assembly and bracket
   - Fuel supply line to the fuel rail
   - Negative battery cable

2000–02 Models

1. Before servicing the vehicle, refer to the precautions in the beginning of this section.

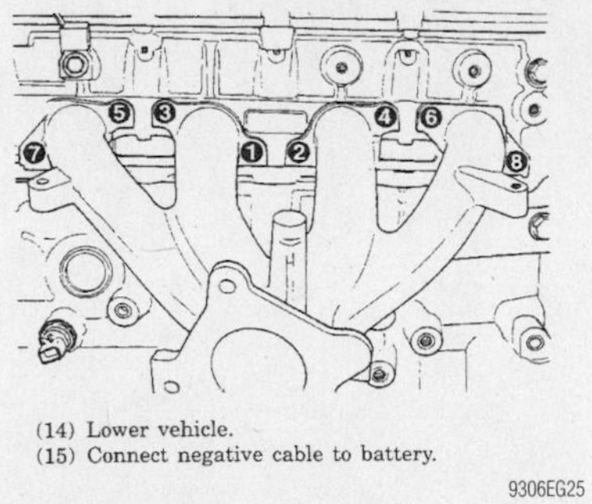

Exhaust manifold tightening sequence–2000–02 models and High Output Engines

2. Remove or disconnect the following:
   - Negative battery cable
   - Wiring harness heat shield-to-exhaust manifold support bracket bolt, if Low Emission Vehicle (LEV) equipped
   - Wiring harness heat shield-to-exhaust manifold, if Ultra Low Emission Vehicle (ULEV) equipped
   - Exhaust manifold support bracket bolt, if LEV equipped
   - Make up air hose
   - Speed control vacuum reservoir, if equipped
   - Oxygen (O_2S) sensor connector and harness clip
   - Upper heat shield
   - Exhaust manifold bolts
   - Cylinder head cover, if ULEV equipped
   - Exhaust manifold and discard gasket

To install:

3. Install or connect the following:
   - Exhaust manifold with a new gasket and torque the bolts, in sequence, to 16 ft. lbs. (23 Nm)
   - Upper heat shield, if ULEV equipped and torque bolts to 16 ft. lbs. (23 Nm)
   - Upper and lower heat shields and torque the bolts to 95 inch lbs. (11 Nm)
   - O_2S sensor connector and harness clip
   - Cylinder head cover, if ULEV equipped
   - Speed control vacuum reservoir, if equipped
   - Make up air hose
   - Exhaust manifold support bracket bolt, if LEV equipped. Torque the M10 bolt to 40 ft. lbs. (54 Nm), M12 bolt to 70 ft. lbs. (95 Nm) and the nut to 21 ft. lbs. (28 Nm).
   - Cylinder head-to-exhaust manifold support bracket bolt, if ULEV equipped. Torque the bolts to 40 ft. lbs. (54 Nm).
   - Wiring harness heat shield-to-exhaust manifold, if ULEV equipped
   - Wiring harness heat shield-to-exhaust manifold support bracket bolt, if LEV equipped
   - Negative battery cable

Front Crankshaft Seal

REMOVAL & INSTALLATION

SOHC Engine

1. Before servicing the vehicle, refer to the precautions in the beginning of this section.
2. Remove or disconnect the following:
   - Negative battery cable
   - Accessory drive belts
   - Crankshaft damper, using Puller Tool 1026 and Insert Tool 6827-A
   - Timing belt
   - Crankshaft sprocket using special Tools 6793 and C-4685-C2
   - Crankshaft sprocket key
   - Crankshaft oil seal, using a seal puller Tool 6771

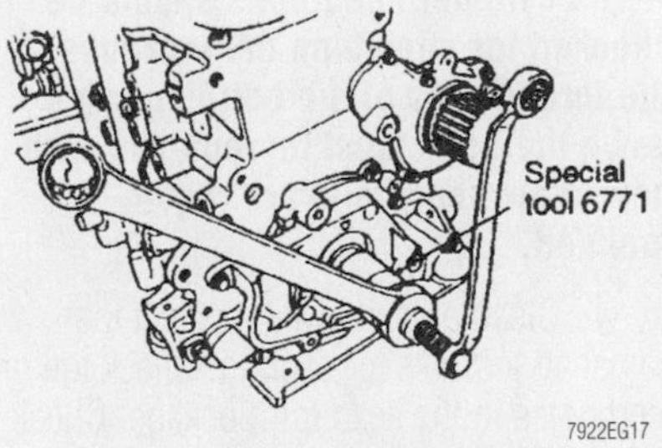

Removing the front crankshaft oil seal

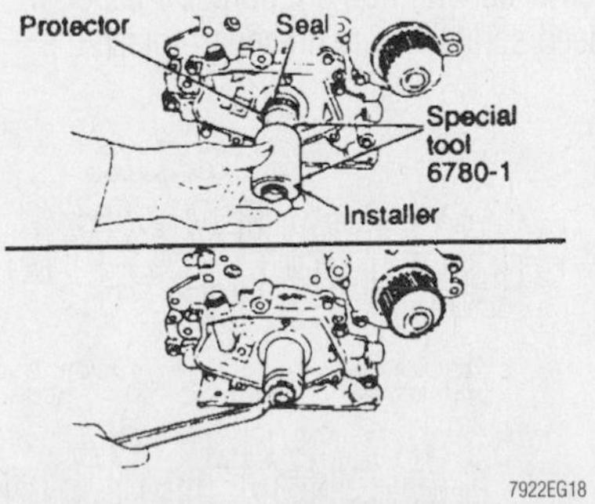

Installing a new seal using seal installer 6780–1; proceed with caution if using substitute tools

To install:

3. Install or connect the following:
   - New front crankshaft oil seal, using Crankshaft Installer Tool 6780–1
   - Crankshaft key
   - Crankshaft sprocket using Installer Tool 6792

➡**Make sure the word FRONT on the crankshaft sprocket is facing outward.**

   - Timing belt
   - Crankshaft damper using thrust bearing washer and bolt from Installer Tool 6792. Torque the damper bolt to 105 ft. lbs. (142 Nm) for 1998–99 models or 100 ft. lbs. (136 Nm) for 2000–02 models.
   - Accessory drive belts
   - Negative battery cable

DOHC Engine

1. Before servicing the vehicle, refer to the precautions in the beginning of this section.
2. Remove or disconnect the following:
   - Negative battery cable
   - Crankshaft damper with special Tools 1026 and 6287-A
   - Timing belt
   - Crankshaft sprocket with special Tools 6793 and insert C-4685-C2
   - Crankshaft oil seal with special Tool 6771

To install:

3. Install or connect the following:
   - Front crankshaft oil seal with special Tool 6780-1
   - Crankshaft sprocket with special Tool 6792

➡**Make sure the word FRONT on the crankshaft sprocket is facing outward.**

   - Timing belt
   - Crankshaft damper using thrust bearing washer and bolt from Installer Tool 6792 and torque the damper bolt to 105 ft. lbs. (142 Nm)
   - Negative battery cable

Camshaft and Lifters

REMOVAL & INSTALLATION

SOHC Engine

1998–1999 MODELS

This engine uses a camshaft running in an aluminum cylinder head. Rocker arm shafts mount directly to the cylinder head.

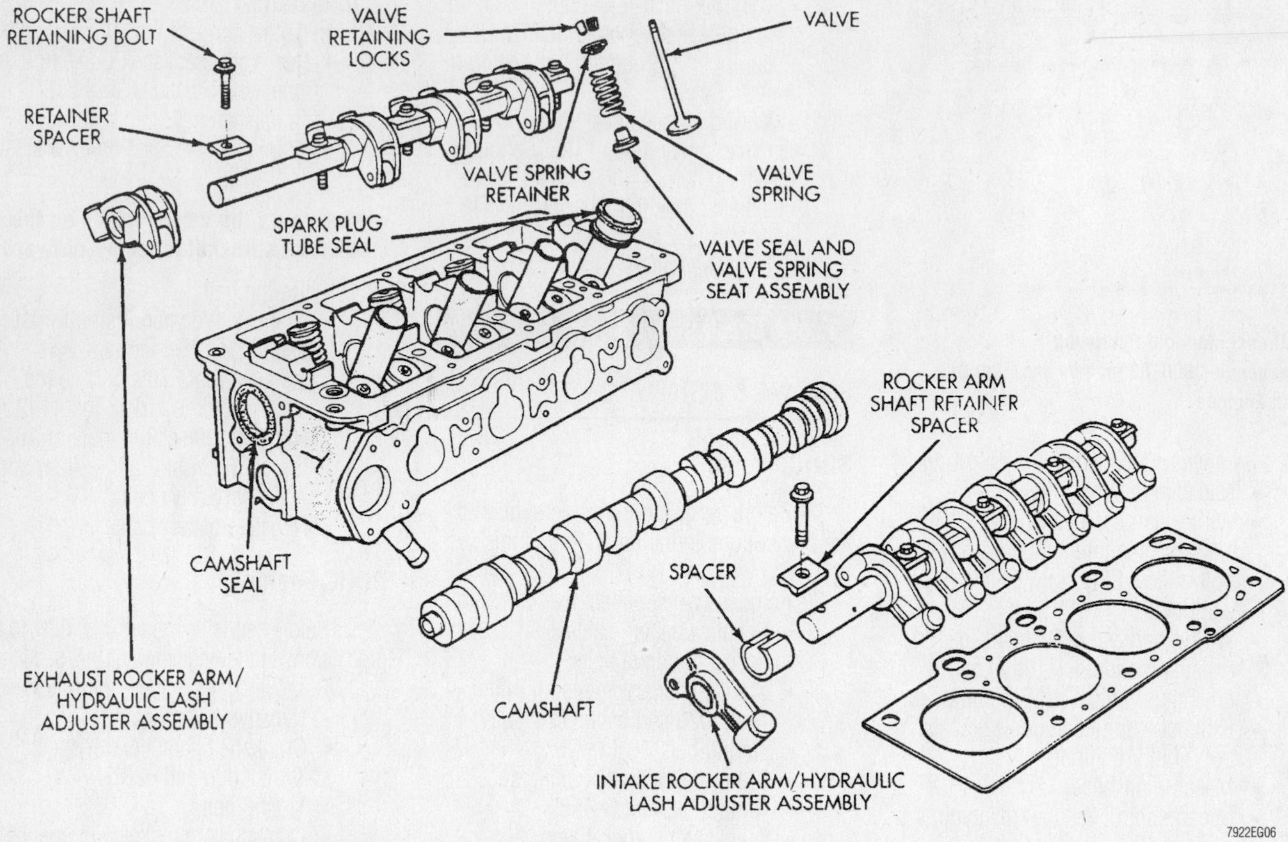

Exploded view of the cylinder head and valvetrain assembly—SOHC engine

Care must be taken to ensure all valve timing marks align after cylinder head and valvetrain service. Please note that the cylinder head must be removed from the vehicle to service the camshaft.

➡After all components are installed on the engine, a DRB scan tool is necessary to perform the camshaft and crankshaft timing relearn procedure.

1. Before servicing the vehicle, refer to the precautions in the beginning of this section.
2. Relieve the fuel system pressure.
3. Remove or disconnect the following:
   - Negative battery cable
   - Cylinder head cover

➡Mark the rocker arm shaft assemblies to identify them for later installation.

- Rocker arm shaft bolts
- Rocker arm assemblies
- Timing belt and tensioner
- Camshaft sprocket
- Rear timing belt cover
- Cylinder head
- Camshaft Position (CMP) sensor
- Camshaft from the rear of the cylinder head

To install:

➡The cylinder head bolts should be checked for stretching before reuse. If the thread area of the bolt is necked down the bolts must be replaced with new. New head bolts are recommended.

4. Clean all parts well. Inspect the camshaft journals for scoring. Check the oil feed holes in the head for blockage. Check the camshaft bearing journals for scoring. If light scratches are present, they may be removed with 400 grit abrasive paper. If deep scratches are present, replace the camshaft and check the cylinder head for damage. Replace the cylinder head if worn or damaged.

5. If the camshaft lobes show signs of wear, check the corresponding rocker arm roller for wear or damage. Replace rocker arms/hydraulic lash adjuster if worn or damaged. If the camshaft lobes show signs of pitting on the nose, flank or base circle, replace the camshaft.

6. If the rocker arms and shaft are to be serviced, mark the rocker arms so any that are to be returned to service will be installed in their original locations. Slide the rocker arms off the shaft. Keep the spacers and rocker arms in the same location for reassembly.

7. Inspect the rocker arms for scoring,

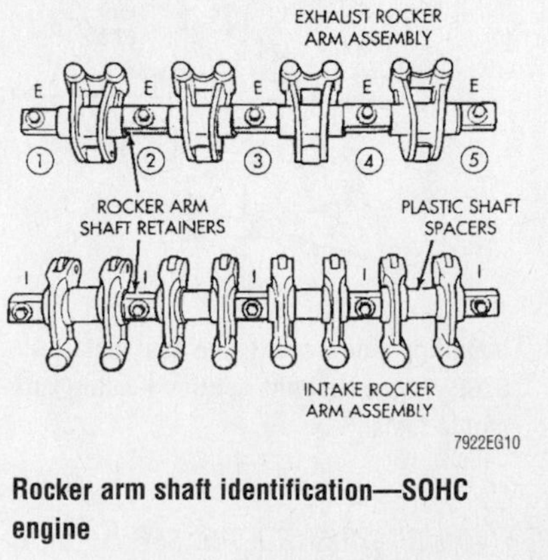

Rocker arm shaft identification—SOHC engine

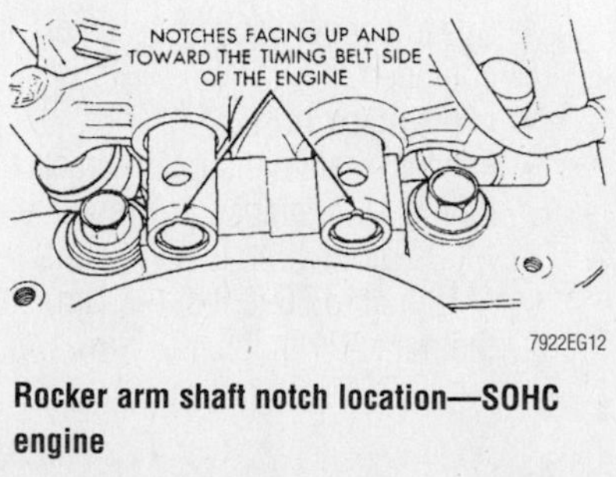

Rocker arm shaft notch location—SOHC engine

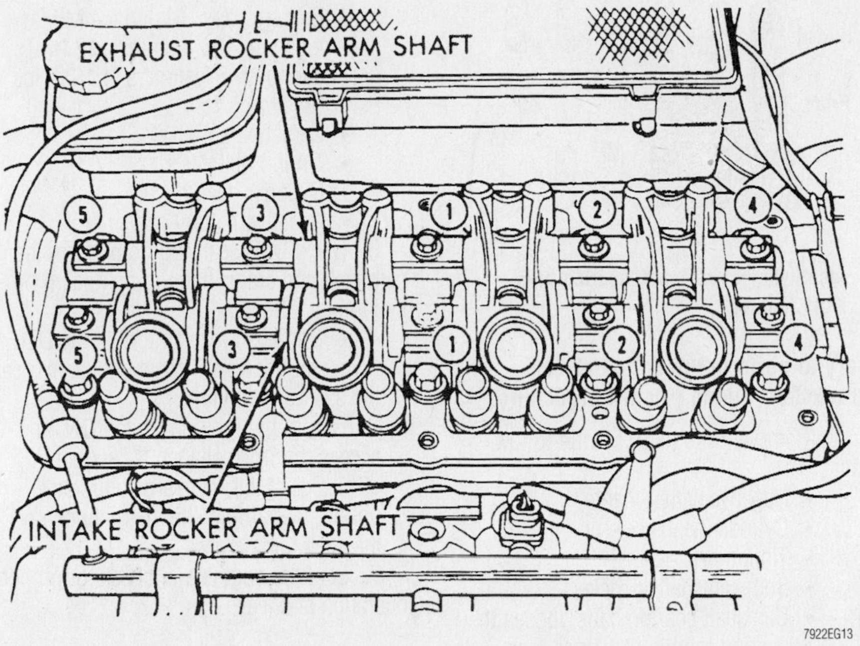

Rocker arm shaft tightening sequence—SOHC engine

wear on the roller or damage to the shaft. Replace parts as necessary.

8. The rocker arm shaft is hollow and used as a lubrication oil duct. Check that the shaft is clean inside and out.

9. Check all oil holes for clogging with a small wire and clean as required.

10. To assemble, thoroughly lubricate all rocker arm components and spacers, then install them on the rocker arm shaft in their original locations.

11. If the vehicle exhibited a tappet-like noise, the valve lash adjusters built into the rocker arms should be cleaned and checked. Lash adjusters removed from a rocker arm should be returned to their original locations. Replace worn or defective lash adjusters.

12. To install a lash adjuster, use the following procedure:

a. Lubricate the lash adjuster thoroughly with clean engine oil.

b. Install the adjuster into the rocker arm making sure the adjuster is at least partially filled with oil.

c. Submerge the rocker arm in clean engine oil and pump the plunger until the lash adjuster travel is taken up. If travel is not reduced, replace the adjuster.

d. Install the rocker arm on the rocker arm shaft.

13. Camshaft end-play can be checked, using the following procedure:

a. Install the camshaft target magnet on the end of the camshaft. Tighten the screw to 30 inch lbs. (3.4 Nm). Be sure to fit the dowels on the magnet in the holes on the camshaft.

b. Setup a dial indicator to touch on the nose of the camshaft.

c. Using a prybar, move the camshaft as far rearward as it will go. Be sure the dial indicator probe is in contact with the camshaft.

d. Zero the dial indicator.

e. Move the camshaft as far forward as it will go.

f. Read the end-play on the dial indicator. Specification is 0.005–0.013 inch (0.13–0.33mm).

14. Install or connect the following:

- Camshaft, lubricate the bearing journals thoroughly

➡Be sure the camshaft turns freely.

- CMP sensor and torque the screws to 85 inch lbs. (9.6 Nm)
- Camshaft seal flush with the cylinder head
- Camshaft sprocket and torque the bolt to 85 ft. lbs. (115 Nm)
- Cylinder head

15. Before installing the rocker arm and shaft assemblies, set the crankshaft to 3 notches before Top Dead Center (TDC) on the crankshaft sprocket.

- Rocker arm and shaft assemblies. Position the rocker shafts pointing up and toward the timing belt side of the engine
- Rocker arm shaft retainers, in their original positions and torque the bolts to 17 ft. lbs. (23 Nm)
- Timing belt and tensioner
- All electrical, vacuum and fluid connections
- Negative battery cable.

➡An oil and filter change are recommended.

16. Use a DRB scan tool to perform the camshaft and crankshaft timing relearn procedure, as follows:

a. Connect the scan tool to the DLC (located under the instrument panel, near the steering column).

b. Turn the ignition switch **ON**, and access the "miscellaneous" screen.

c. Select the "re-learn cam/crank" option, then follow the instructions on the scan tool screen.

17. Start the engine and check for leaks. Run the engine with the radiator cap off so as the engine warms and the thermostat opens, coolant can be added to the radiator. Test drive vehicle to check for proper operation.

2000–02 MODELS

➡After all components are installed on the engine, a DRB scan tool is necessary to perform the camshaft and crankshaft timing relearn procedure.

1. Before servicing the vehicle, refer to the precautions in the beginning of this section.
2. Relieve the fuel system pressure.
3. Drain the cooling system.
4. Remove or disconnect the following:
   - Battery and tray
   - Inlet Air Temperature (IAT) sensor electrical connector
   - Air intake duct from the intake manifold
   - Throttle cables
   - Throttle Position (TP) sensor
   - Idle Air Control (IAC) valve from the throttle body
   - Air make-up hose from the air cleaner
   - Air cleaner housing
   - Timing belt
   - Camshaft sprocket
   - Cylinder head cover and mark the rocker arm shaft assemblies to ease installation
   - Rocker arm shaft assemblies
   - Engine Coolant Temperature (ECT) sensor electrical connector

- Heater supply and return hoses
- Heater tube support bracket
- Power Distribution Center (PDC) screws aside to ease the camshaft removal
- Camshaft Position (CMP) sensor and camshaft target magnet
- Camshaft from the rear of the cylinder head

To install:

5. Lubricate the camshaft journals with clean engine oil.
6. Install or connect the following:
   - Camshaft and reposition the PDC and heater tubes
   - Camshaft target magnet and torque the screw to 30 inch lbs. (3 Nm)
   - CMP sensor
   - Front camshaft seal, if removed
   - Timing belt rear cover
   - Camshaft sprocket and torque to 85 ft. lbs. (115 Nm)
   - Timing belt and tensioner
   - Rocker arms in the proper order and torque to 21 ft. lbs. (28 Nm)
   - Cylinder head cover
   - Ignition coil and spark plug cables and torque the fasteners to 105 inch lbs. (12 Nm)
   - Heater tube bracket bolt
   - Heater tubes
   - ECT sensor electrical connector
   - PDC attaching screws
   - Battery and tray
   - Air cleaner housing
   - Throttle body connections
   - IAT sensor
7. Fill the cooling system to the proper level.
8. Use a DRB scan tool to perform the camshaft and crankshaft timing relearn procedure, as follows:
   a. Connect the scan tool to the DLC (located under the instrument panel, near the steering column).
   b. Turn the ignition switch **ON**, and access the "miscellaneous" screen.
   c. Select the "re-learn cam/crank" option, then follow the instructions on the scan tool screen.
9. Start the engine and check for leaks. Run the engine with the radiator cap off so as the engine warms and the thermostat opens, coolant can be added to the radiator. Test drive vehicle to check for proper operation.

DOHC Engine

1. Be sure to observe all precautions in the beginning of this section.

➡**After all components are installed on the engine, a DRB scan tool is necessary to perform the camshaft and crankshaft timing relearn procedure.**

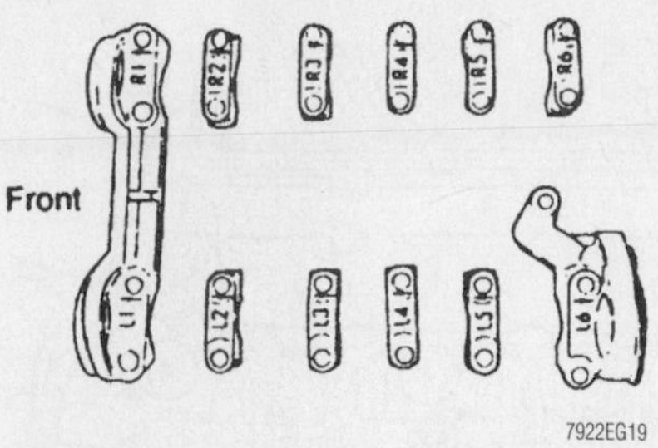

Identifying the camshaft bearing caps—DOHC engine

2. Remove or disconnect the following:
   - Negative battery cable
   - Cylinder head cover
   - Timing belt
   - Both camshaft sprockets
   - Camshaft bearing caps, loosen the bearing caps in sequence, one camshaft at a time.

➡**The bearing caps are identified for location. Remove the outside bearing caps first. If the bearing caps are difficult to remove, use a plastic hammer to gently tap the rear part of the camshaft.**

   - Intake and exhaust camshafts

To install:

3. Before installation, clean the cylinder head and cover mating surfaces. Make certain that the rails are flat.
4. Lubricate the bearing journals and cam followers with clean engine oil.
5. Install or connect the following:
   - Camshafts
   - Bearing caps. Torque the M6 fasteners to 105 inch lbs. (12 Nm) and the M8 fasteners to 18 ft. lbs. (24 Nm).
   - Camshaft sprockets
   - Timing belt
   - Cylinder head cover
   - Negative battery cable
6. Use a DRB scan tool to perform the camshaft and crankshaft timing relearn procedure, as follows:
   a. Connect the scan tool to the DLC (located under the instrument panel, near the steering column).
   b. Turn the ignition switch **ON**, and access the "miscellaneous" screen.
   c. Select the "re-learn cam/crank" option, then follow the instructions on the scan tool screen.
7. Start the engine and check for proper operation and leaks.

Valve Lash

ADJUSTMENT

The engines in these vehicles do not require periodic valve lash adjustment.

Starter Motor

REMOVAL & INSTALLATION

1998–99 Models

1. Remove the negative battery cable.
2. If equipped with A/C, perform the following procedure:

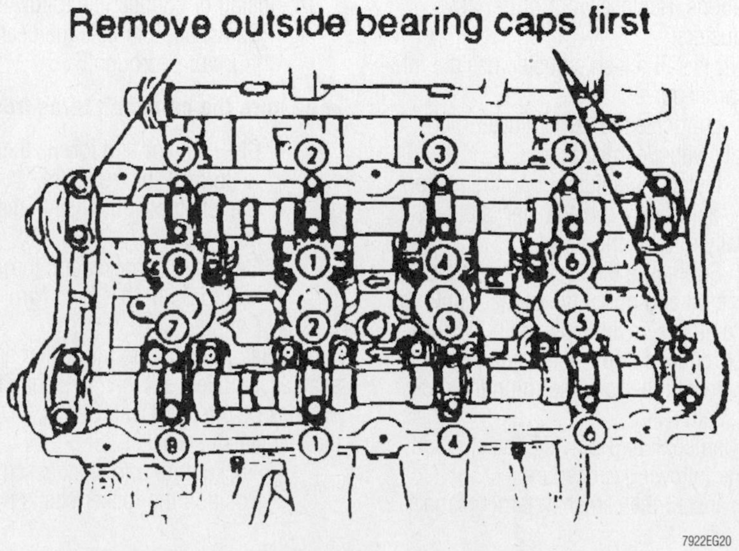

Bearing cap removal sequence—DOHC engine

a. Support the engine/transaxle assembly with a jack.

b. Remove the front engine mount bolt from the mount/crossmember bracket.

c. Lower and rotate the engine/transaxle assembly forward to provide starter clearance.

➡**Do not remove the wiring at this time.**

3. Remove or disconnect the following:
- Both starter-to-transaxle bolts
- Starter electrical connectors
- Starter by positioning it vertically and lowering it

➡**If equipped, move the A/C lines aside.**

To install:

4. Install or connect the following:
- Starter
- Starter electrical connectors
- Starter-to-transaxle bolts and torque the bolts to 40 ft. lbs. (54 Nm)
- Negative battery cable

5. If equipped with A/C, perform the following steps:

a. Raise and support the engine/transaxle assembly.

b. Install the front engine mount-to mount/crossmember bracket bolt and torque to 40 ft. lbs. (54 Nm)

2000–02 Models

1. Remove or disconnect the following:
- Negative battery cable
- Intake manifold on high output engines only
- Starter electrical connectors
- Starter-to-engine bolts
- Starter

To install:

2. Install or connect the following:
- Starter and torque bolts to 40 ft. lbs. (54 Nm)
- Starter electrical connectors
- Intake manifold on high output engines
- Negative battery cable

Oil Pan

REMOVAL & INSTALLATION

1998–99 Models

1. Before servicing the vehicle, refer to the precautions in the beginning of this section.

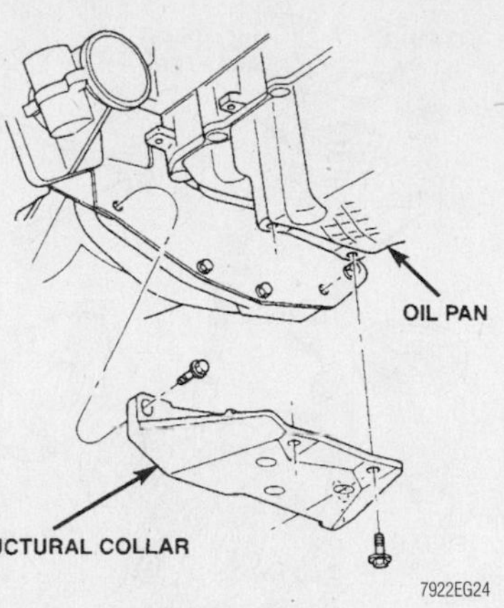

Exploded view of the structural collar mounting—1998–99 models

2. Drain the engine oil.

3. Properly support the engine and transaxle assembly.

4. Remove or disconnect the following:
- Front engine mount bracket
- Powertrain bending strut
- Oil pan-to-transaxle structural collar
- Transaxle lower dust cover
- Oil filter and adapter, if equipped with air conditioning
- Oil pan

5. Thoroughly, clean the gasket mating surfaces.

To install:

6. Apply silicone sealer to the oil pump-to-engine block parting line.

7. Install or connect the following:
- New oil pan gasket
- Oil pan and torque the retainers to 105 inch lbs. (12 Nm)
- Oil filter and adapter, if removed
- Transaxle lower dust cover
- Power train bending strut
- Front engine mount and bracket

8. Torque the front engine mount retainers as follows:

a. Engine mount bracket bolt 1: 20 inch lbs. (3 Nm).

b. Engine mount bracket bolts 2, 3 and 4: 80 ft. lbs. (108 Nm).

c. Engine mount bracket bolts 5 and 1: 40 ft. lbs. (54 Nm).

d. Engine mount bracket-to-insulator assembly through-bolt: 40 ft. lbs. (54 Nm).

e. Insulator-to-lower radiator crossmember nuts: 40 ft. lbs. (54 Nm).

f. Mass damper bolt: 40 ft. lbs. (54 Nm).

✻✻ WARNING

Follow the proper tightening sequence for the structural collar or damage to the collar or oil pan may occur.

9. Install the structural collar and torque the retainers as follows:

a. Collar-to-oil pan bolts: 30 inch lbs. (3 Nm).

b. Collar-to-transaxle bolts: 80 ft. lbs. (108 Nm).

c. Collar-to-oil pan bolts: 40 ft. lbs. (54 Nm), final torque.

10. Fill the engine with clean oil.

11. Start the engine and check for leaks; then, recheck the fluid level and add as necessary.

2000–02 Models

1. Before servicing the vehicle, refer to the precautions in the beginning of this section.
2. Raise and safely support the vehicle.
3. Drain the engine oil.
4. Remove or disconnect the following:
- Negative battery cable
- Oil filter
- Oil filter adapter
- Structural collar
- Lateral bending brace
- Transaxle lower dust cover
- Oil pan

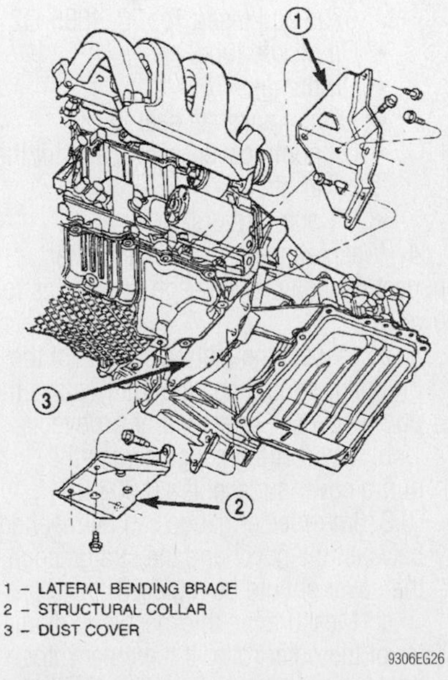

Exploded view of the bending brace, structural collar and dust cover—2000–02 models

5. Thoroughly, clean the gasket mating surfaces.

To install:

6. Apply silicone sealer to the oil pump-to-engine block parting line.
7. Install or connect the following:
   - New oil pan gasket
   - Oil pan and torque the screws to 105 inch lbs. (12 Nm)
   - Transaxle lower dust cover
   - Lateral bending brace
   - Structural collar
   - Oil filter adapter and torque the bolts to 60 ft. lbs. (80 Nm)
   - Oil filter
8. Fill the engine with clean oil
9. Start the engine, check for leaks and repair if necessary.

Oil Pump

REMOVAL & INSTALLATION

1. Before servicing the vehicle, refer to the precautions in the beginning of this section.
2. Drain the engine oil.
3. Remove or disconnect the following:
   - Negative battery cable
   - Crankshaft damper
   - Timing belt
   - Timing belt tensioner, on 2000–02 models
   - Oil pan
   - Crankshaft sprocket, using Tool 6795 and Insert Tool C-4685-C2
   - Oil pickup tube
   - Oil pump
   - Front crankshaft seal
   - Oil pump cover screws and lift the cover off
   - Oil pump rotors
4. Wash all parts in a solvent; then, inspect carefully for damage or wear, as follows:

   a. Inspect the mating surface of the oil pump should be smooth. Replace the pump cover, if scratched or grooved.

   b. Lay a straightedge across the pump cover surface. If a 0.003 in. (0.076mm) feeler gauge can be inserted between the cover and the straightedge, the cover should be replaced.

   c. Measure the thickness and diameter of the outer rotor. If the outer rotor thickness measures 0.301 in. (7.64mm) or less, or if the diameter is 3.148 in. (79.95mm) or less, replace the outer rotor.

   d. If the inner rotor measures 0.301 in. (7.64mm) or less, replace the inner rotor.

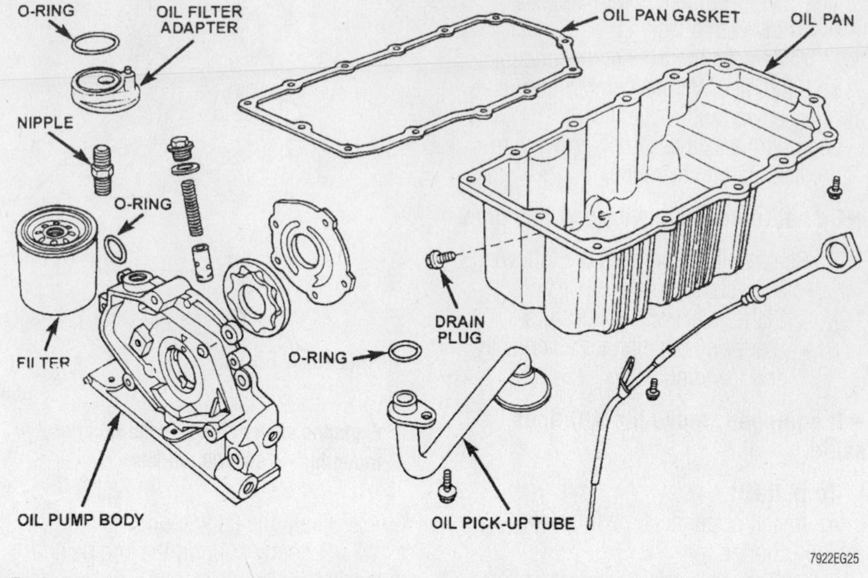

Exploded view of the oil pump and related component mounting

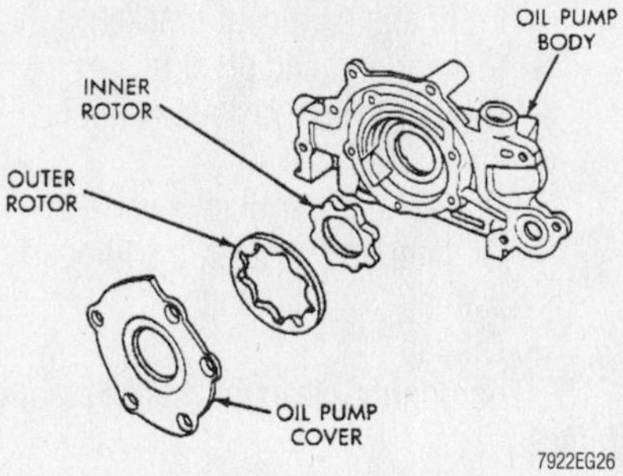

Exploded view of the oil pump assembly

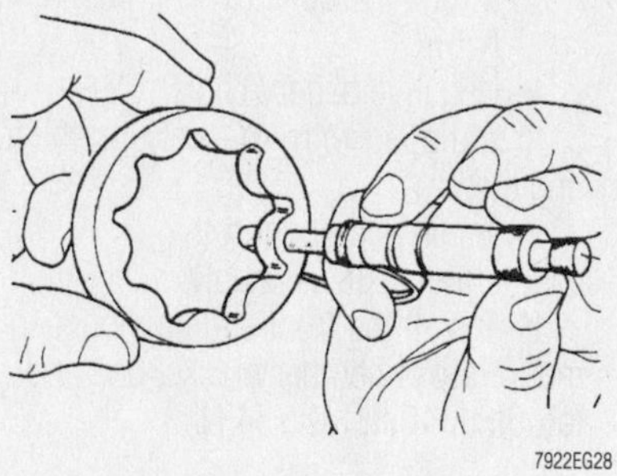

Use calipers to measure the outer rotor thickness

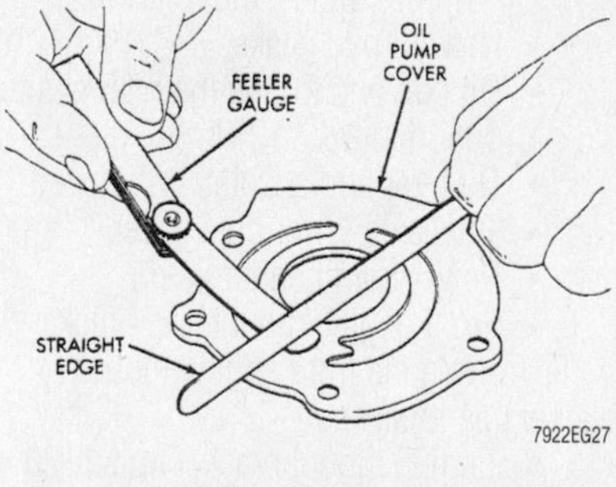

Use a feeler gauge and straightedge to check the oil pump cover for warpage

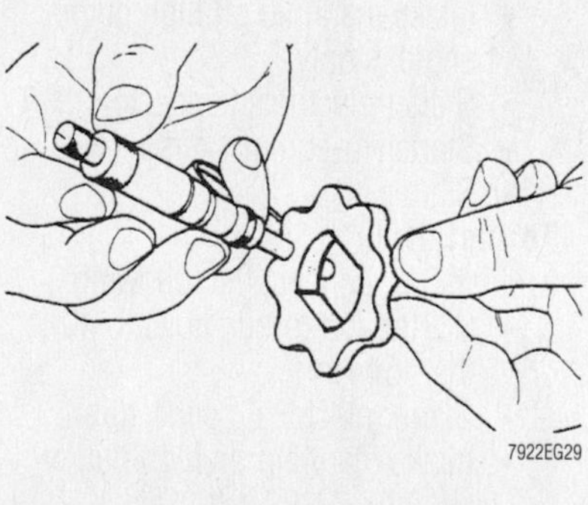

and the inner rotor thickness

   e. Slide the outer rotor into the pump housing, press to one side with your fingers and measure the clearance between the rotor and the housing. If the measurement is 0.015 in. (0.39mm) or more, replace the housing only if the outer rotor is within specification.

   f. Install the inner rotor into the pump housing, If the clearance between the inner and outer rotors is 0.008 in. (0.203mm) or more, replace both rotors.

   g. Place a straightedge across the face of the pump housing, between the bolt holes. If a feeler gauge of 0.004 in. (0.102mm) or more can be inserted between the rotors and the straightedge, replace the pump assembly.

   h. Inspect the oil pressure relief valve plunger for scoring and free operation in its bore. Small marks may be removed with 400 grit wet or dry sandpaper.

   i. The relief valve spring has a free length of about 2.39 in. (60.7mm) and should test between 18–19 lbs. (8.1–8.6

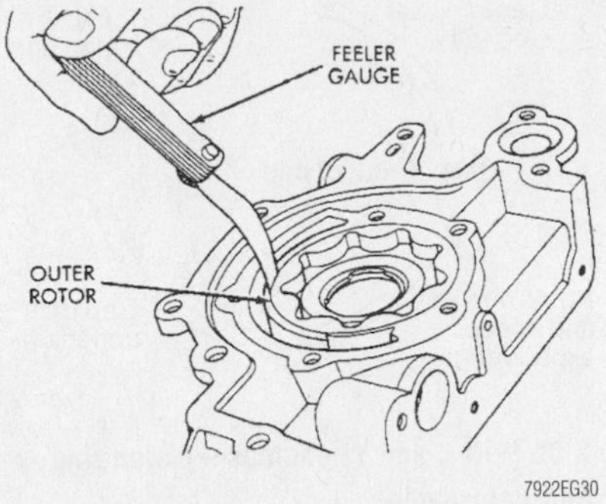

Measure the outer rotor clearance in the housing

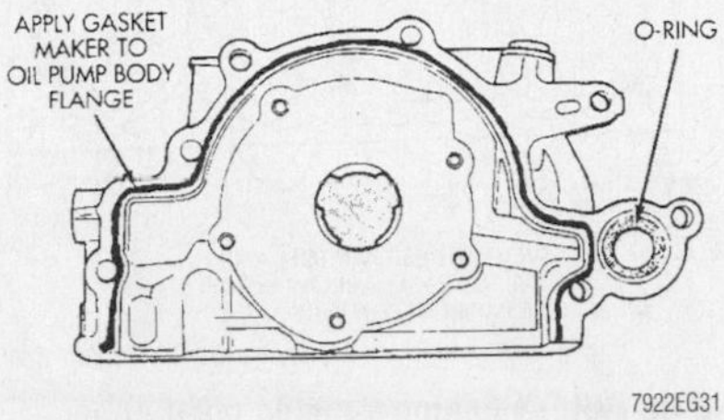

Apply a small amount of gasket maker to the pump body cover mounting surface

kg) when compressed to 1.60 in. (40.6mm). Replace the spring, if it falls outside of specifications.

j. If the oil pressure is low and the pump is within specifications, inspect for worn engine bearings or for other reasons for oil pressure loss.

To install:

5. Assemble the pump, using new parts as required, as follows:

a. Install the inner rotor with the chamfer facing the cast iron oil pump cover.

b. Apply Mopar gasket maker to the oil pump.

c. Install the oil ring into the oil pump body discharge passage.

6. Prime the oil pump before installation by filling the rotor cavity with engine oil.

7. Install or connect the following:

- Oil pump, align the rotor flats with the crankshaft flats. Torque the pump bolts to 21 ft. lbs. (28 Nm).

⁂ WARNING

The front crankshaft seal MUST be out of the pump to align or damage may result.

- New front crankshaft seal, using Seal Driver Tool 6780
- Crankshaft sprocket, using a Crankshaft Sprocket Installer Tool 6792
- Oil pump pickup tube
- Oil pan
- Timing belt tensioner, on 2000–02 models
- Timing belt and front cover
- Crankshaft damper
- Accessory drive belts
- Negative battery cable

8. Fill the engine with clean oil.

9. Start the engine and check for leaks; then, recheck the fluid level and add as necessary.

Rear Main Seal

REMOVAL & INSTALLATION

1. Before servicing the vehicle, refer to the precautions in the beginning of this section.

2. Remove or disconnect the following:

- Transaxle
- Flexplate/flywheel
- Rear main seal. Insert a seal remover between the dust lip and the metal case of the crankshaft seal.
- Angle the tool through the dust lip against the metal case of the seal. Pry out the seal.

⁂ WARNING

DO NOT let the prytool contact the crankshaft seal surface. Contact of the tool blade against the crankshaft edge (chamfer) is permitted.

To install:

⁂ WARNING

If the crankshaft edge (chamfer) has any burrs or scratches on the, clean it up with 400 grit sand paper to prevent seal damage during installation of the new seal.

➡No lubrication is necessary when installing the seal.

3. Place Crankcasc Seal Pilot Tool 6926–1 on the crankshaft; this is a pilot tool with a magnetic base.

4. Position the seal over the Pilot Tool;

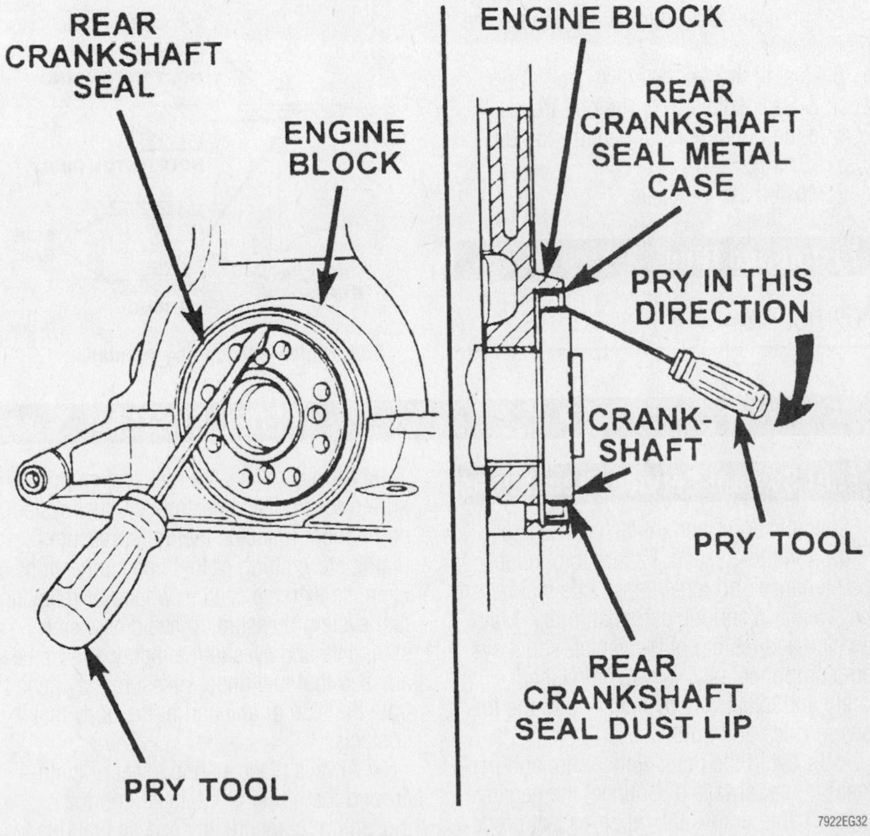

When prying the seal out, be sure to use the prytool at the proper angle

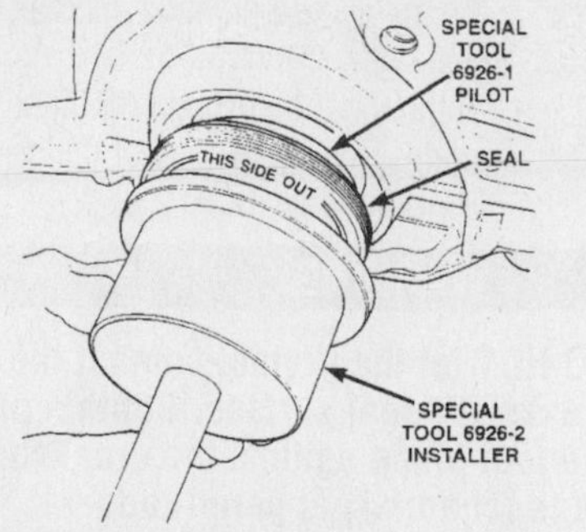

Place a proper size pilot tool with a magnetic base on the crankshaft

be sure the words THIS SIDE OUT on the seal can be read.

➡**The pilot tool should stay on the crankshaft during installation of the seal. Be sure the seal lip faces the crankcase during installation.**

✲✲ WARNING

If the seal is driven in the block past flush, this may cause an oil leak.

5. Drive the seal into the block, using Crankshaft Seal Tool 6926-2 and Handle C-4171, until the tool bottoms out against the block.
6. Install the flexplate/flywheel. Apply Lock & Seal Adhesive to the bolt treads. Torque the bolts in a star pattern, to 70 ft. lbs. (95 Nm).
7. Install the transaxle.

Piston and Ring

POSITIONING

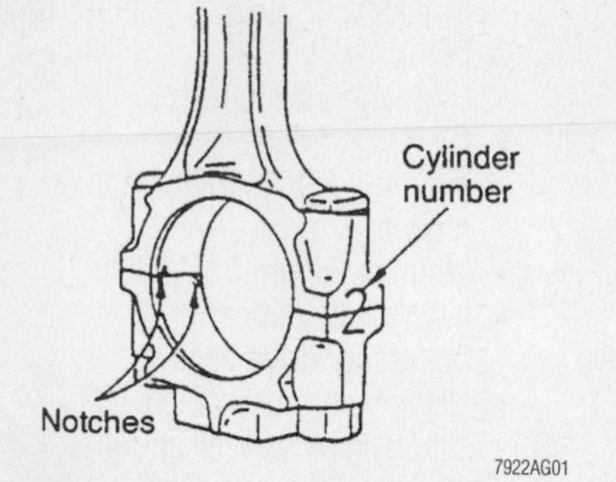

Chrysler engine connecting rod and cap installation—ensure to matchmark the cap and rod prior to disassembly

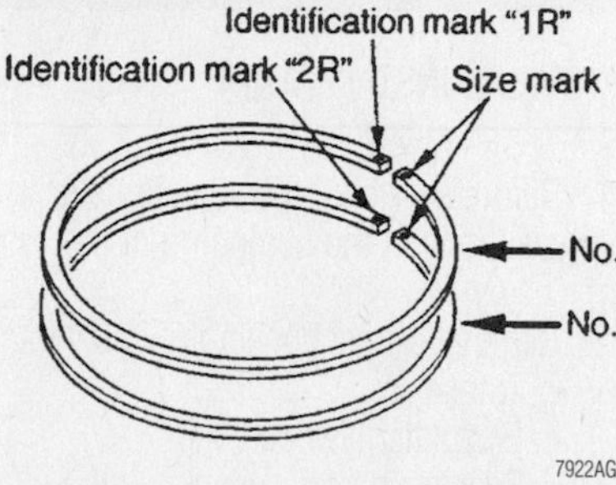

Common Chrysler piston ring identification mark locations

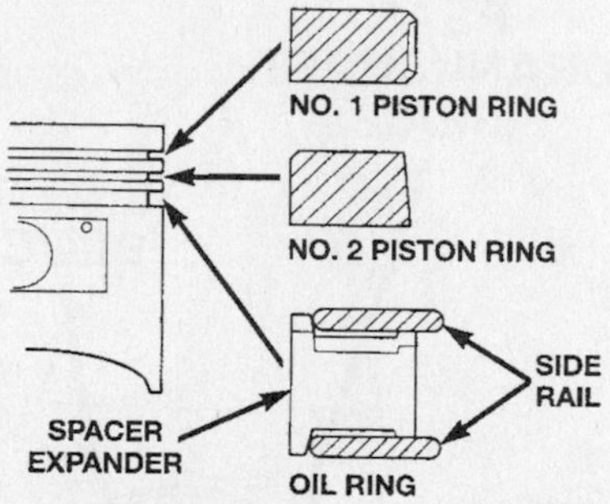

2.0L Engine—piston ring orientation

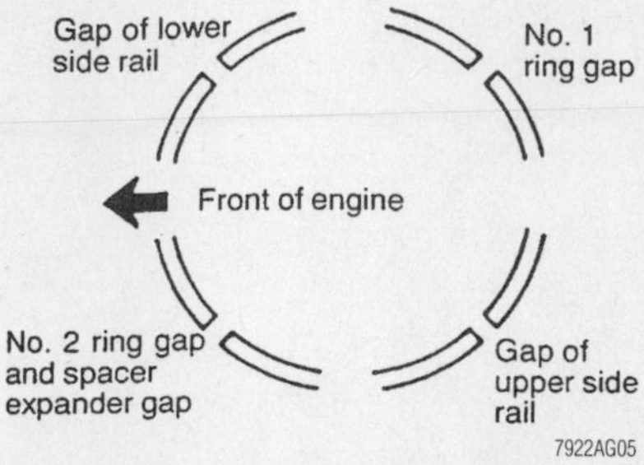

2.0L (VIN C and Y) engines—piston ring end-gap spacing

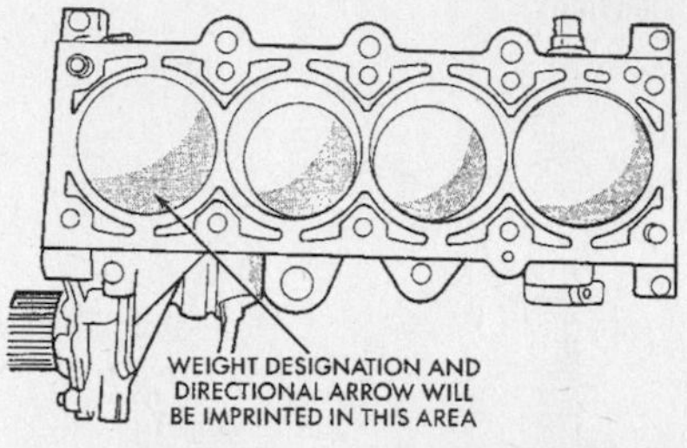

2.0L (VIN C) Engine—piston positioning. The arrow or weight marking (L or H) must face the timing belt side of the engine

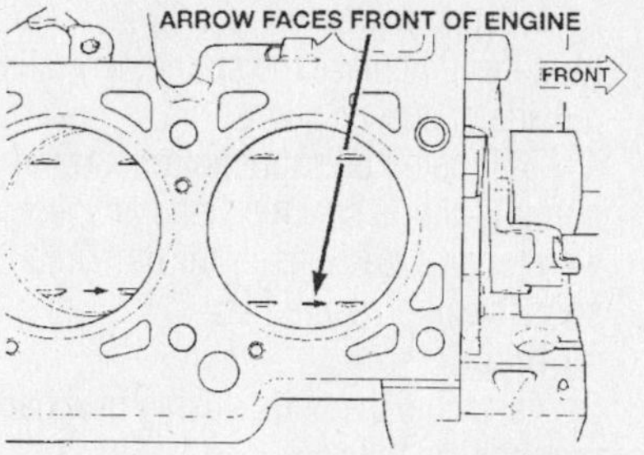

2.0L (VIN Y) Engine—piston positioning. The small arrows on the crown of the pistons must point toward the front of the engine

FUEL SYSTEM

Fuel System Service Precautions

Safety is an important factor when servicing the fuel system. Failure to conduct maintenance and repairs in a safe manner may result in serious personal injury. Maintenance and testing of the vehicle's fuel system components can be accomplished safely and effectively by adhering to the following rules and guidelines:

- To avoid the possibility of fire and personal injury, always disconnect the negative battery cable unless the repair or test procedure requires that battery voltage be applied.
- Always relieve the fuel system pressure prior to disconnecting any fuel system component (injector, fuel rail, pressure regulator, etc.), fitting or fuel line connection. Exercise extreme caution whenever relieving fuel system pressure, to avoid exposing skin, face and eyes to fuel spray. Please be advised that fuel under pressure may penetrate the skin or any part of the body that it contacts.
- Always place a shop towel or cloth around the fitting or connection prior to loosening to absorb any excess fuel due to spillage. Ensure that all fuel spillage is quickly removed from engine surfaces. Ensure that all fuel soaked cloths or towels are deposited into a suitable waste container.
- Always keep a dry chemical (Class B) fire extinguisher near the work area.
- Do not allow fuel spray or fuel vapors to come into contact with a spark or open flame.
- Always use a back-up wrench when loosening and tightening fuel line connection fittings. This will prevent unnecessary stress and torsion to fuel line piping.
- Always replace worn fuel fitting O-rings. Do not substitute fuel hose where fuel pipe is installed.

Fuel System Pressure

RELIEVING

⁂ CAUTION

Relieve the fuel system pressure before servicing any components of the fuel system. Service vehicles in well ventilated areas and avoid ignition sources. NEVER smoke while servicing the vehicle!

1. Before servicing the vehicle, refer to the precautions in the beginning of this section.
2. Remove or disconnect the following:
   - Negative battery cable
   - Fuel filler cap
   - Protective cap from the fuel rail's pressure port
3. Place the open end of a fuel pressure release hose (tool C-4799–1) into an approved gasoline container. Connect the other end of the hose to the fuel pressure test port. The fuel pressure will bleed off through the hose into the gasoline container.

➡Fuel pressure gauge kit C-4799-B contains hose C-4799–1.

4. The vehicle is now safe for servicing.
5. When finished working on the fuel system, install the fuel filler cap.

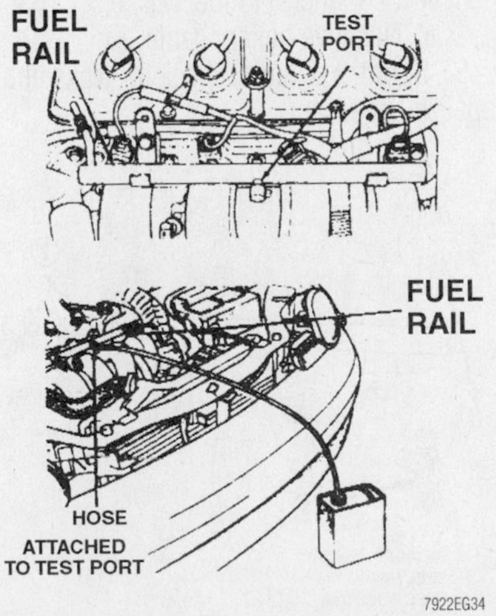

Relieve the fuel system, allowing the pressure to bleed off through the hose into the container

Fuel Filter

A combination fuel filter/pressure regulator assembly is used, which is located on the top of the fuel pump module.

REMOVAL & INSTALLATION

⁂ CAUTION

Do not allow fuel spray or fuel vapors to come in contact with a spark or open flame. Keep a dry chemical fire extinguisher nearby. Never store fuel in an open container due to risk of fire or explosion.

1. Before servicing the vehicle, refer to the precautions in the beginning of this section.
2. Properly relieve the fuel system pressure.
3. Disconnect the negative battery cable.
4. Disconnect the quick-connect fuel supply line from the filter/regulator nipple.
5. Depress the locking spring tab, located on the side of the fuel filter/regulator, then rotate 90 degrees and pull out. Be sure the upper and lower O-rings are still on the filter assembly.

To install:

6. Lightly coat the filter O-rings with clean engine oil. Insert the filter into the opening in the fuel pump module, then align the 2 hold-down tabs with the flange.
7. While applying downward pressure, rotate the filter clockwise until the spring tab catches in the locating slot.
8. Attach the fuel line to the filter/regulator assembly.
9. Connect the negative battery cable.

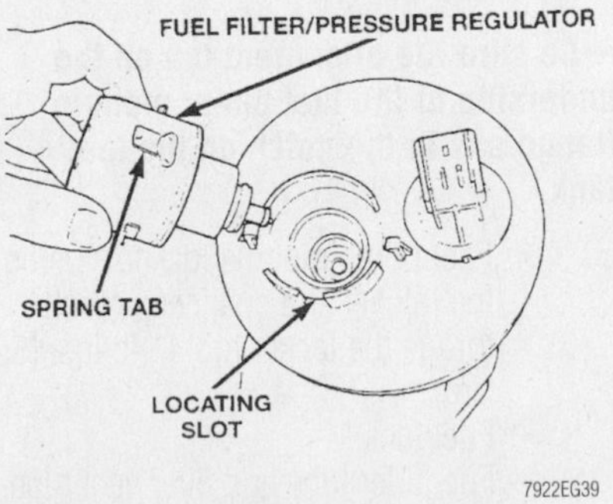

Depress the spring tab, then rotate and pull the fuel filter assembly out

Fuel Pump

The fuel pump is integral with the pump module, which also contains the fuel reservoir, level sensor, inlet strainer and fuel pressure regulator. The inlet strainer, fuel pressure regulator and level sensor are the only serviceable items. If the fuel pump requires service, replace the entire fuel pump module.

REMOVAL & INSTALLATION

1998–99 Models

1. Before servicing the vehicle, refer to the precautions in the beginning of this section.
2. Properly relieve the fuel system pressure.
3. Drain the fuel tank.

⁂ WARNING

The fuel reservoir of the fuel pump module does not empty out when the tank is drained. The fuel in the reservoir will spill out when the module is removed.

4. Remove or disconnect the following:
   - Negative battery cable
   - Fuel lines from the fuel pump module by depressing the quick-connect retainers
   - Fuel pump module electrical lock by sliding it to unlock it
   - Fuel pump module electrical connector by pushing down on the connector retainer and pulling the connector off of the module
5. Support the fuel tank with a transmission jack.
   - Fuel tank strap bolts; then, lower

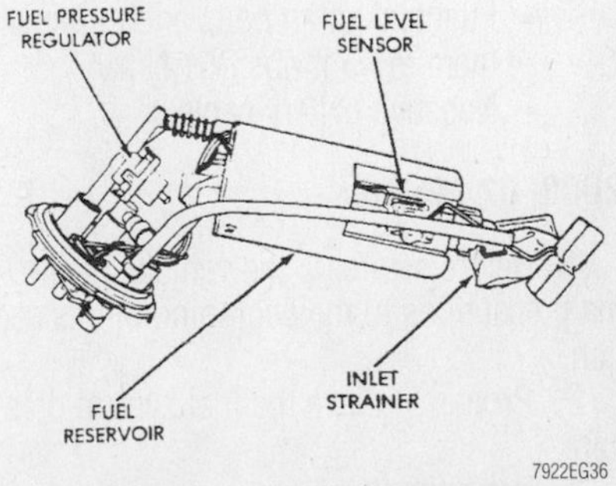

The fuel module assembly contains the pump, pressure regulator, reservoir, inlet strainer and level sensor

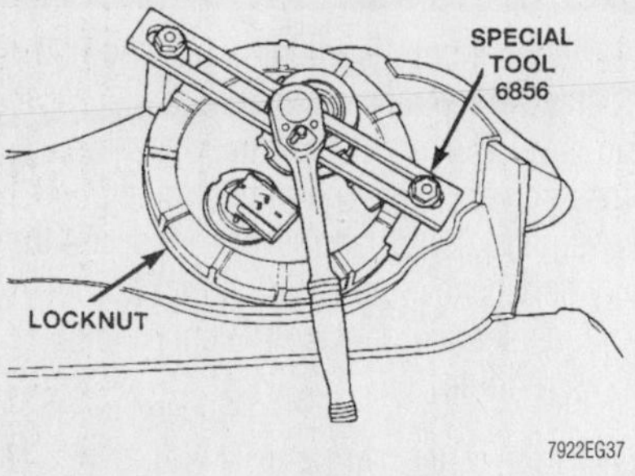

To loosen the fuel pump module locknut, use a ratchet and spanner wrench

the tank slightly for access to the module
- Fuel pump module locknut using a ratchet and spanner wrench
- Fuel pump and discard the O-ring seal

To install:

⁂ WARNING

Take care to avoid wiping dirt or debris into the fuel tank opening. Such contaminants may clog or damage the new fuel pump.

6. Wipe the fuel pump mounting flange and surrounding area of the tank clean.
7. Install or connect the following:
- New O-ring seal in the tank opening
- Fuel pump

➡Be sure the alignment tab on the underside of the fuel pump module flange sits in the notch on the fuel tank.

- Fuel pump module locknut. Using the ratchet and spanner wrench, torque the locknut to 41 ft. lbs. (55 Nm).
- Fuel tank strap bolts and torque them to 45 ft. lbs. (61 Nm)
- Negative battery cable

2000–02 Models

1. Before servicing the vehicle, refer to the precautions in the beginning of this section.
2. Properly relieve the fuel system pressure.
3. Drain the fuel tank.
4. Remove or disconnect the following:
- Negative battery cable
- Vapor line from the Evaporative Emissions (EVAP) canister tube
- EVAP canister

⁂ WARNING

The fuel reservoir of the fuel pump module does not empty out when the tank is drained. The fuel in the reservoir will spill out when the module is removed.

- Fuel line from the fuel pump module by depressing the quick-connect retainers with your thumb and forefinger
- Fuel pump module electrical connector by pushing down on the connector retainer and pulling the connector off of the module
- Fuel filler tube and filler vent tube from the fuel tank's filler hose

5. Support the fuel tank with a transmission jack.
- Fuel tank strap bolts; then, lower the tank slightly for access to the module
- Fuel pump module locknut using a ratchet and spanner wrench
- Fuel pump and discard the O-ring seal

To install:

⁂ WARNING

Take care to avoid wiping dirt or debris into the fuel tank opening. Such contaminants may clog or damage the new fuel pump.

6. Wipe the fuel pump mounting flange and surrounding area of the tank clean.
7. Install or connect the following:
- New O-ring seal in the tank opening
- Fuel pump

➡Be sure the alignment tab on the underside of the fuel pump module flange sits in the notch on the fuel tank.

- Fuel pump module locknut. Using the ratchet and spanner wrench, torque the locknut to 41 ft. lbs. (55 Nm).
- Fuel tank
- Fuel filler tube and filler vent tube to the fuel tank's filler hose
- Fuel pump module electrical connector
- Fuel line, to the fuel pump module
- EVAP canister
- Vapor line to the EVAP canister tube
- Negative battery cable

8. Start the vehicle, check for leaks and repair if necessary.

Fuel Injector

REMOVAL & INSTALLATION

1. Before servicing the vehicle, refer to the precautions in the beginning of this section.
2. Release the fuel system pressure.
3. Remove or disconnect the following:
- Negative battery cable
- Fuel line(s) from fuel rail
- Fuel injector electrical connectors
- Fuel rail with the fuel injectors
- Fuel injector retainers
- Fuel injector(s) and discard the O-rings

To install:

4. Install or connect the following:
- Fuel injector(s) using new O-rings

➡Lubricate the O-rings with clean engine oil.

- Fuel injector retainers
- Fuel rail with the fuel injectors and torque the bolts to 170–230 inch lbs. (20–25 Nm)
- Fuel injector electrical connectors
- Fuel line(s) to fuel rail
- Negative battery cable

5. Start the vehicle, check for leaks and repair if necessary.

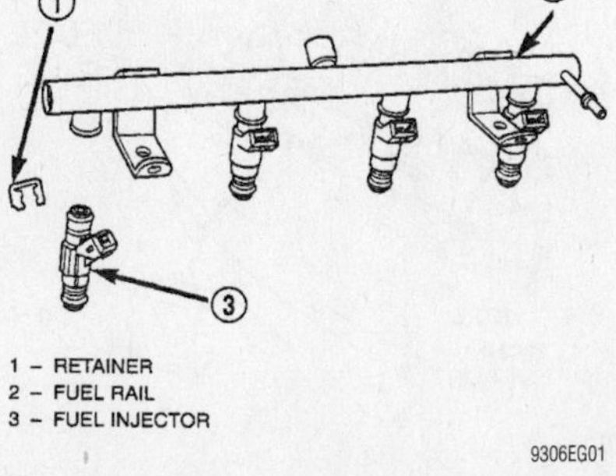

Exploded view of the fuel rail assembly

DRIVE TRAIN

Transaxle Assembly

REMOVAL & INSTALLATION

Manual Transaxle

1998–99 MODELS

1. Before servicing the vehicle, refer to the precautions in the beginning of this section.
2. Drain the transaxle fluid.
3. Remove or disconnect the following:
   - Battery cables
   - Power Distribution Center (PDC), pull it up and out of its bracket and move it aside
   - Battery heat shield
   - Battery and tray
   - Cruise control, if equipped
   - Vehicle Speed Sensor (VSS) wire
   - Back-up lamp switch wiring from the transaxle

✲✲ WARNING

Pry with equal amounts of force on both sides of the shifter cable isolator bushing to avoid damaging the cable isolator bushing.

   - Both gearshift cable ends from the transaxle shift levers, using 2 pry tools
   - Clutch housing vent cap, exposing the clutch cable end and clutch release lever
   - Clutch cable from the bell housing
   - Shift cable (linkage) mounting bracket
   - Accelerator cable shield, if equipped
   - Intake manifold support bracket
   - Upper starter bolt
   - Upper bell housing bolt
4. Install an engine bridge fixture and support the engine securely.
   - Front wheels
   - Halfshafts

✲✲ WARNING

When installing the halfshafts, new retaining clips MUST be used. NEVER reuse the old clips. If old clips are used, the inner CV-joints may disengage.

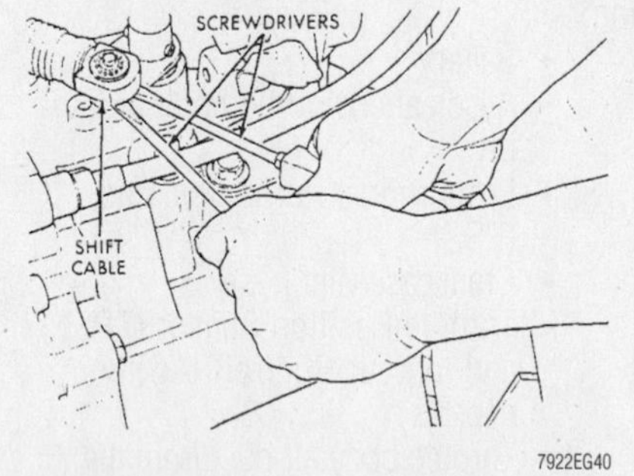

Use 2 prytools to disconnect the gearshift cable ends from the shift levers—1998–99 manual transaxle

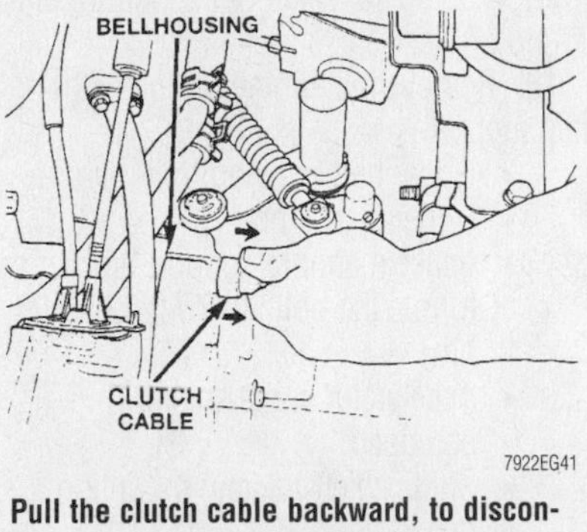

Pull the clutch cable backward, to disconnect the clutch cable from the bell housing—1998–99 manual transaxle

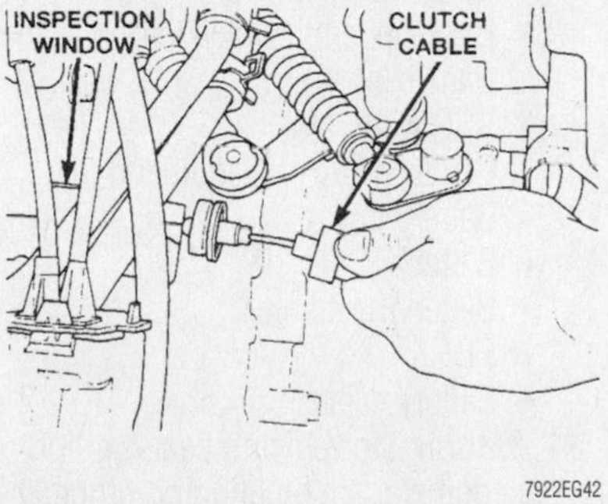

Remove the clutch cable from the lever—1998–99 manual transaxle

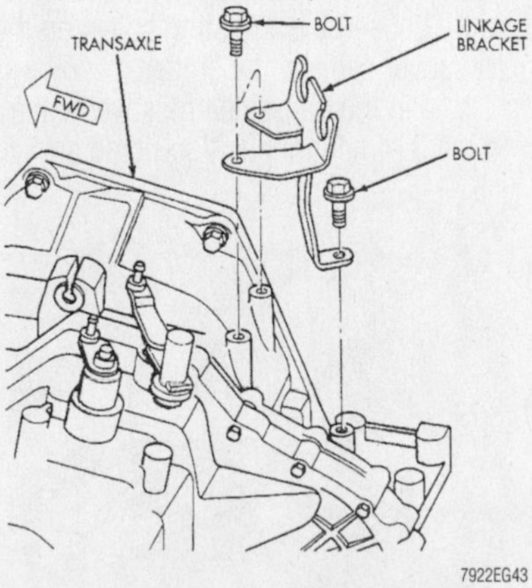

Exploded view of the linkage mounting bracket—1998–99 manual transaxle

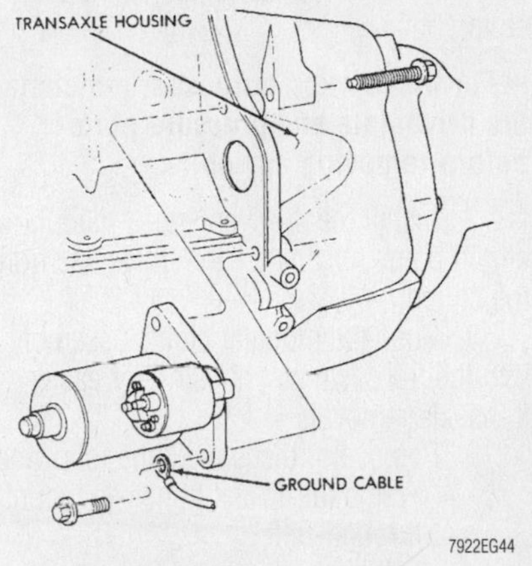

Removing the lower starter mounting bolts—1998–99 manual transaxle

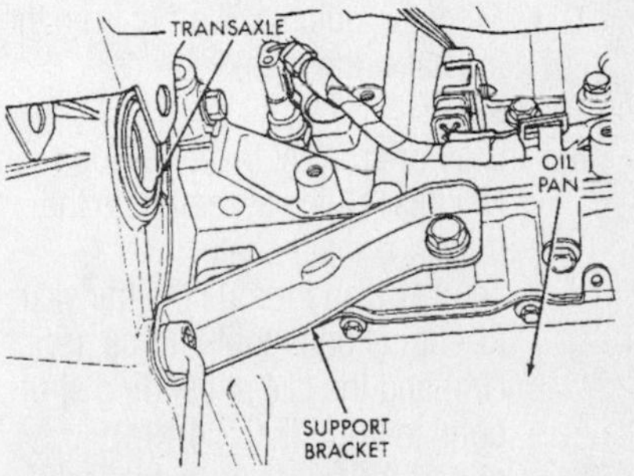

Unfasten the 2 bolts, then remove the bracket (strut) from the engine and transaxle—1998–99 manual transaxle

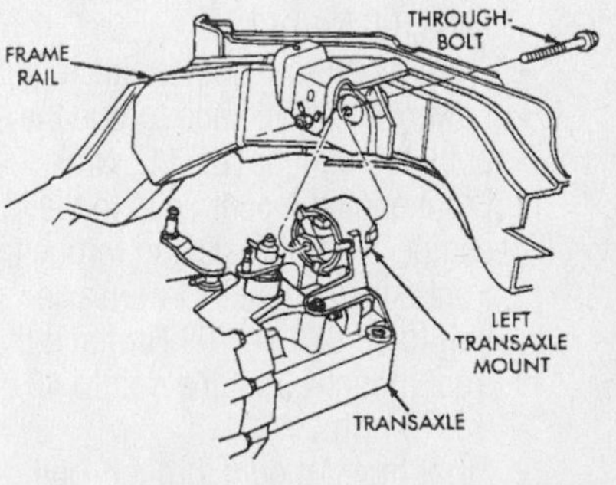

Exploded view of the left manual transaxle mount through-bolt—1998–99 models

- Power hop damper and bracket
- Lower starter mounting bolt
- Transaxle-to-rear lateral bending strut from the engine and transaxle

5. Support the transaxle with a jack.
   - Front motor mount through-bolt
   - Front motor mount bolts from the engine and transaxle
   - Lower dust shield

6. Rotate the crankshaft clockwise in order to access the driveplate-to-modular clutch bolts.

➡For installation purposes, matchmark the driveplate and pressure plate before removing any bolts.

7. Install the 4 driveplate-to-modular clutch bolts, to separate the driveplate from the clutch.

8. Push the modular clutch assembly into the transaxle bell housing for easier transaxle removal.

9. Remove or disconnect the following:
   - Frame rail-to-left transaxle mount through-bolt
   - Left transaxle mount from the transaxle, then push the mount up to provide clearance
   - Transaxle
   - Modular clutch assembly from the transaxle input shaft

To install:

10. Install or connect the following:
    - Modular clutch assembly to the transaxle input shaft
    - Transaxle and torque the transaxle-to-engine bolts to 70 ft. lbs. (95 Nm) and the lateral bending strut bolts to 40 ft. lbs. (54 Nm)

11. Support the transaxle with a jack.
    - Left transaxle mount to the transaxle and torque the bolt to 40 ft. lbs. (54 Nm)
    - Frame rail-to-left transaxle mount through-bolt and torque to 80 ft. lbs. (108 Nm)
    - 4 driveplate-to-modular clutch bolts
    - Lower dust shield and torque the bolts to 108 inch lbs. (12 Nm)
    - Front engine mount bolts to the engine and transaxle and torque the front engine mount-to-transaxle bolt to 80 ft. lbs. (108 Nm) and the front mount-to-engine bolt to 40 ft. lbs. (54 Nm)
    - Front motor mount through-bolt and torque the through-bolt to 45 ft. lbs. (61 Nm)
    - Lower starter mounting bolt
    - Power hop damper and bracket and torque the damper bolts to 40 ft. lbs. (54 Nm)

⁂ WARNING

When installing the halfshafts, new retaining clips MUST be used. NEVER reuse the old clips. If old clips are used, the inner cv joints may disengage.

- Halfshafts
- Front wheels

12. Fill the transaxle to the bottom of the fill plug hole.

13. Remove the engine bridge fixture and support.
    - Upper bell housing bolt
    - Upper starter bolt
    - Intake manifold support bracket and torque the bolts to 70 ft. Lbs. (95 nm)
    - Accelerator cable shield, if equipped
    - Shift cable (linkage) mounting bracket
    - Clutch cable to the bell housing
    - Clutch housing vent cap
    - Both gearshift cable ends to the transaxle shift levers
    - Back-up lamp switch wiring to the transaxle
    - VSS wire
    - Cruise control, if equipped
    - Battery tray
    - Battery
    - Battery heat shield
    - PDC
    - Battery cables

14. Be sure the vehicle's back-up lights and speedometer are functioning properly.

15. Adjust the crossover cable in order to ensure proper shifter adjustment as follows:

    a. Remove the floor shift console.

    b. Loosen the adjusting screw on the crossover cable at the shifter.

    c. Pin the transaxle crossover cable in the 3–4 neutral position using a ¼ in. drill bit. Align the hole in the crossover lever with the hole in the boss on the transaxle case. Be sure the drill bit goes into the transaxle case at least ½ in. (12mm).

    d. The shifter is spring loaded and self-centering. Allow the shifter to rest in its neutral position. Torque the adjustment screw to 70 inch lbs. (8 nm). Be careful to avoid moving the shift mechanism off center during screw tightening.

    e. Remove the drill bit from the transaxle case and perform a functional check by shifting the transaxle into all gears.

    f. Reinstall the center shift console. Blouse the boot out around the console. Seat the boot lip on the top of the console.

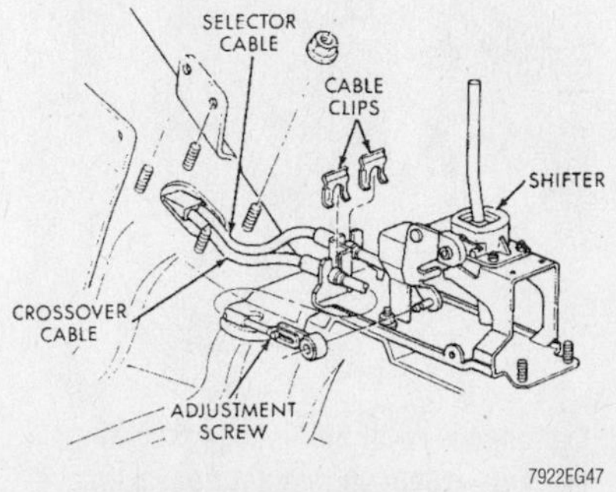

Loosen the adjustment screw on the crossover cable at the shifter—manual transaxle

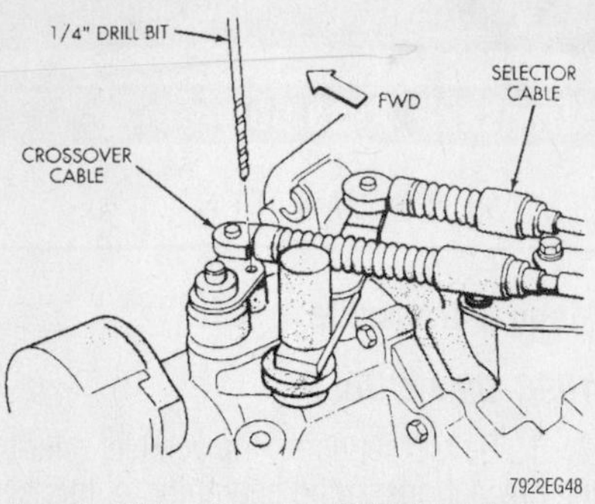

Insert a ¼ in. diameter drill bit to pin the crossover lever in the 3–4 neutral position—1998–99 manual transaxle

16. Road test the vehicle to be sure the transaxle is operating properly.

2000–02 MODELS

1. Before servicing the vehicle, refer to the precautions in the beginning of this section.

2. Drain the transaxle fluid.

3. Remove or disconnect the following:
   - Battery
   - Air cleaner/throttle body assembly
   - Proportional Purge Solenoid (PSS)
   - Crankcase vent hose
   - Throttle Position Sensor (TPS) and Idle Air Control (IAC) connectors
   - Throttle body air duct from the intake manifold
   - Accelerator cable
   - Cruise control cable, if equipped
   - Battery tray from the bracket
   - Ground cable from the battery tray bracket

- Back-up light switch connector
- Bell housing cap
- Clutch cable from the fork and transaxle
- Shift cable-to-bracket clips
- Shift selector and crossover cables from the levers and move aside
- Halfshafts
- Structural collar
- Left engine-to-transaxle lateral bending brace
- Bell housing dust cover
- Right engine-to-transaxle lateral bending brace
- Starter
- Driveplate-to-clutch module bolts

4. Place a jack and block of wood under the engine's oil pan and support the engine. Remove the transaxle's upper mount through-bolt.
5. Lower the engine/transaxle assembly to provide clearance.
   - Transaxle-to-engine bolts, using an assistant to support the transaxle
   - Transaxle

To install:

6. If installing a new or replacement transaxle, transfer the upper mount to the new transaxle; then, torque the upper mount-to-transaxle bolts to 50 ft. lbs. (68 Nm).
7. Install or connect the following:
   - Transaxle
   - Transaxle-to-engine bolts, using an assistant to support the transaxle. Torque the transaxle-to-engine bolts to 70 ft. lbs. (95 Nm).
8. Raise the engine/transaxle assembly until the upper mount aligns with the mount bracket hole. Torque the upper mount-to-mount bracket through-bolt to 80 ft. lbs. (108 Nm).
9. Remove the jack and wooden block.
   - Driveplate-to-clutch module bolts and torque to 65 ft. lbs. (88 Nm)
   - Starter and torque the bolts to 40 ft. lbs. (54 Nm)
   - Starter electrical connectors and torque the positive cable bolt to 89 inch lbs. (10 Nm)

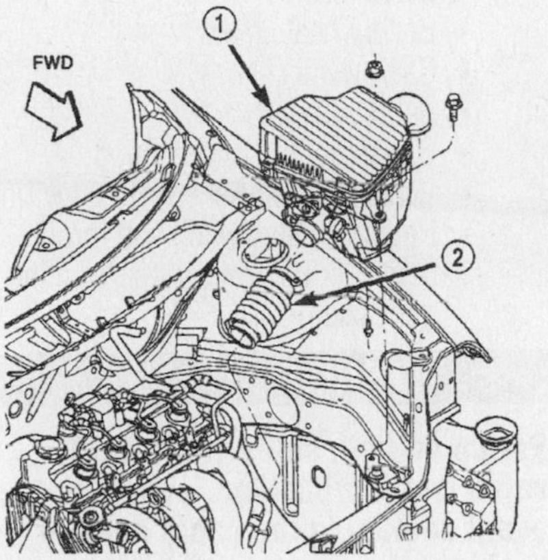

View of the air cleaner assembly and throttle body duct—2000–02 models

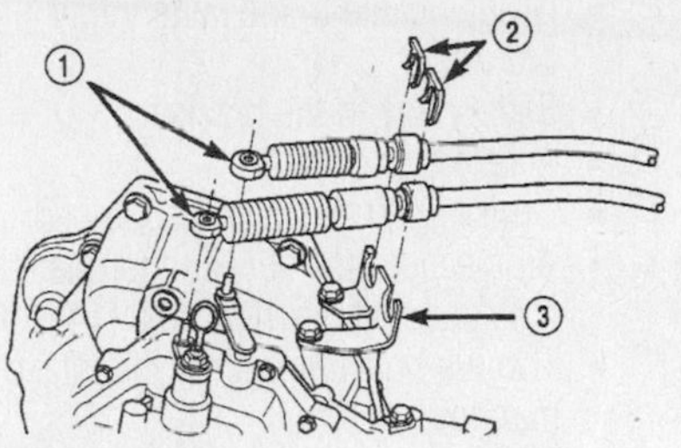

View of the shift cables—2000–02 manual transaxle

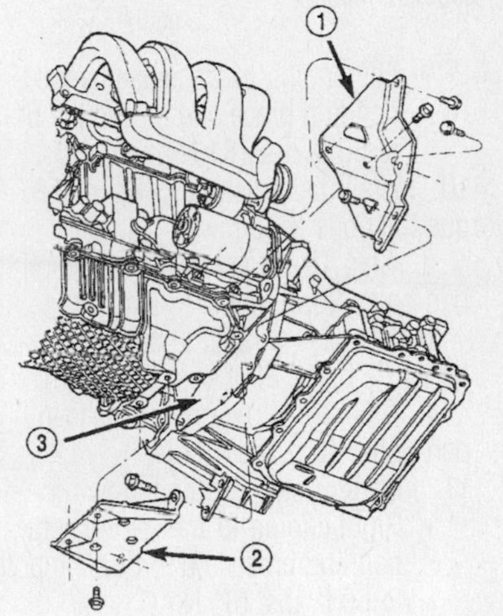

View of the left lateral bending brace and structural collar—2000–02 models

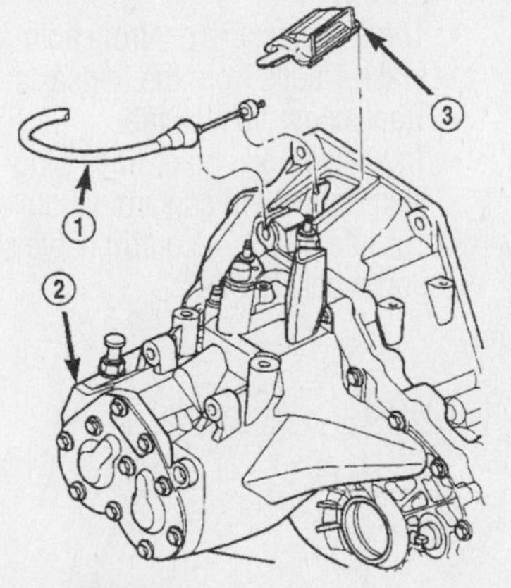

View of the clutch cable and bell housing cap—2000–02 manual transaxle

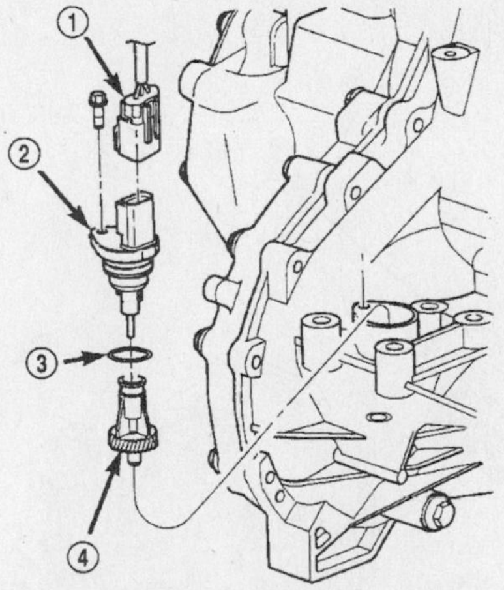

View of the Vehicle Speed Sensor (VSS)—2000–02 manual transaxle

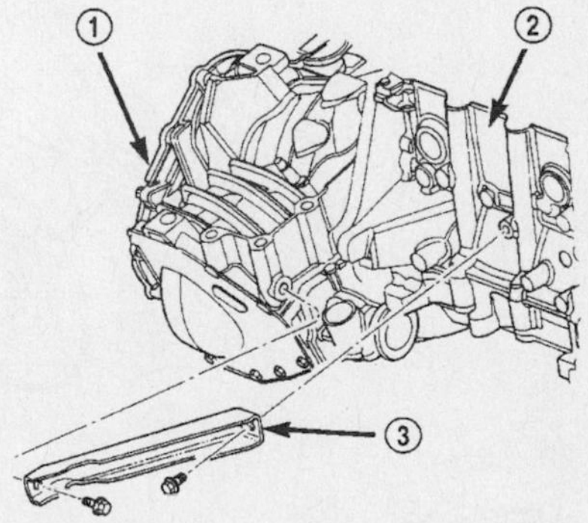

View of the right lateral bending brace—2000–02 models

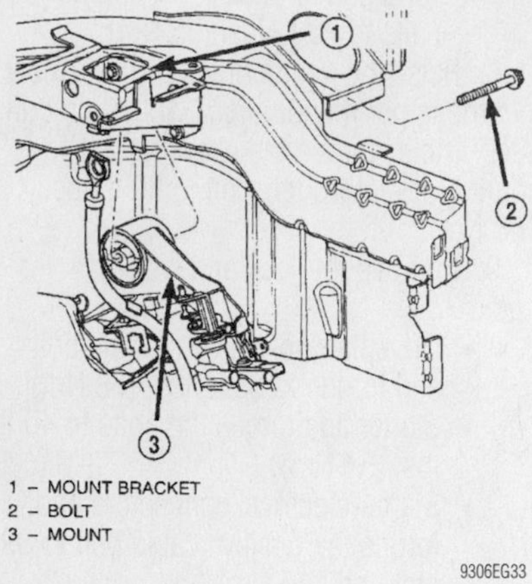

View of the upper mount through-bolt—2000–02 models

- Left engine-to-transaxle lateral bending brace and torque the bolts to 60 ft. lbs. (81 Nm)

10. Install the structural collar and torque the bolts as follows:

a. Step 1—Structural collar-to-oil pan bolts: 30 inch lbs. (3 Nm).

b. Step 2—Structural collar-to-transaxle bolts: 80 ft. lbs. (108 Nm).

c. Step 3—Structural collar-to-oil pan bolts: 40 ft. lbs. (54 Nm).

11. Install or connect the following:

- Right engine-to-transaxle lateral bending brace and torque the bolts to 60 ft. lbs. (81 Nm)
- Halfshafts
- Bell housing dust cover
- VSS electrical connector
- Shift selector and crossover cables to the levers
- Shift cable-to-bracket clips

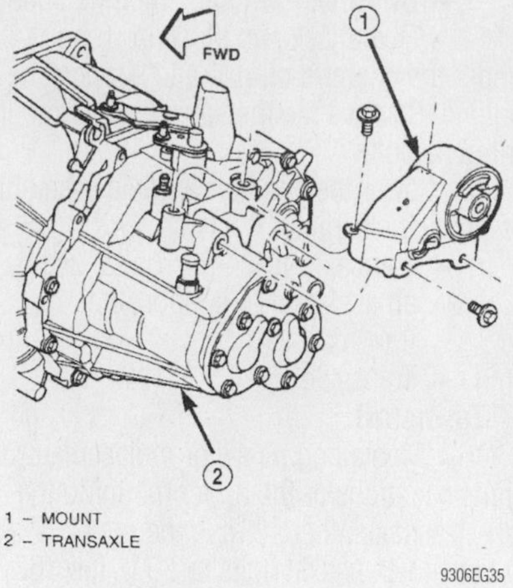

View of the upper mount—2000–02 manual transaxle

- Clutch cable to the fork and transaxle
- Bell housing cap
- Back-up light switch connector
- Ground cable to the battery tray bracket
- Battery tray to the bracket
- Accelerator cable
- Cruise control cable, if equipped
- Air cleaner assembly and torque the nuts and bolts to 10 ft. lbs. (14 Nm)
- Throttle body air duct to the intake manifold
- TPS and IAC connectors
- Crankcase vent hose to the throttle body
- PSS to the throttle body
- Battery

12. Fill the transaxle to the proper level.

13. Road test the vehicle and inspect for leaks.

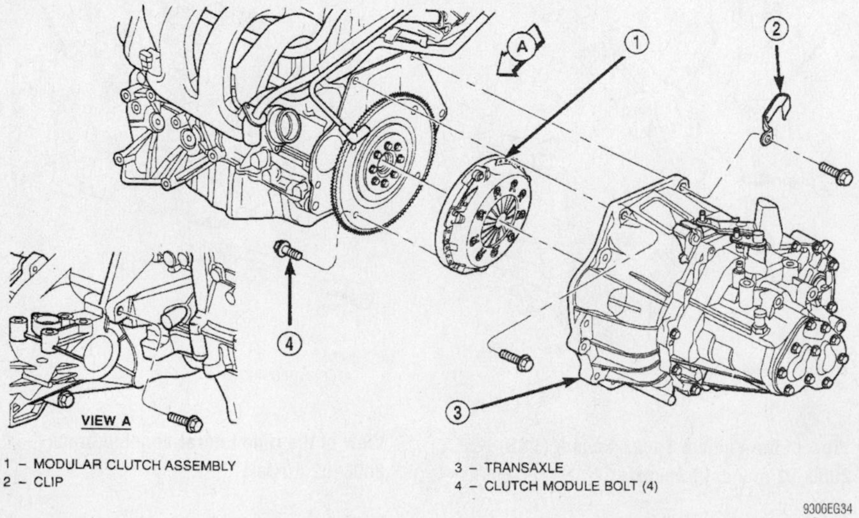

Exploded view of the transaxle and related components—2000–02 manual transaxle

Automatic Transaxle

1998–99 MODELS

1. Before servicing the vehicle, refer to the precautions in the beginning of this section.

The transaxle and torque converter must be removed as an assembly; otherwise the torque converter driveplate, pump bushing or oil seal may be damaged. The driveplate will not support a load; therefore, none of the weight of the transaxle should be allowed to rest on the plate during removal.

2. Remove or disconnect the following:

- Battery cables
- Power Distribution Center (PDC), pull it up and out of its bracket and move it aside
- Battery heat shield
- Battery
- Battery tray
- Cruise control, if equipped
- Vehicle Speed Sensor (VSS) wiring
- Neutral safety switch and torque converter control wiring, from the transaxle

**** WARNING**

Pry up on both sides of the shift cable isolator bushing, evenly, to avoid damaging the cable isolator bushing.

- Gearshift cable end from the transaxle shift lever
- Gearshift cable bracket bolt from the transaxle
- Throttle pressure control cable from the lever
- Throttle pressure control cable bracket bolts from the transaxle
- Transaxle's dipstick tube
- Transaxle's oil cooler lines plug them to prevent contamination
- Throttle pressure control cable support bracket bolts

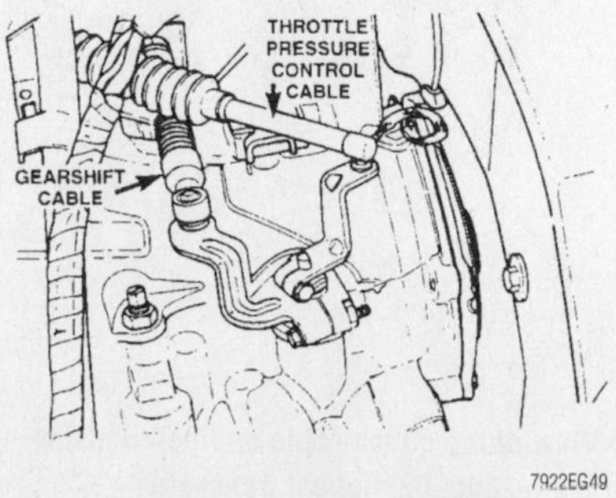

Disconnect the gearshift cable end from the transaxle shift lever—1998–99 automatic transaxle

- Upper bell housing bolts
- Upper starter bolt

3. Install an engine bridge fixture and support the engine.

- Front wheels
- Halfshafts

⁂ WARNING

When installing the halfshafts, new retaining clips must be used. DO NOT reuse the old clips. Failure to use new clips could cause the inner CV-joints to disengage.

⁂ WARNING

The exhaust flex joint must be disconnected from the exhaust manifold anytime the engine is lowered. If the engine is lowered with the flex pipe attached, damage will occur.

- Exhaust flex joint-to-exhaust manifold bolts
- Exhaust pipe from the manifold
- Transaxle-to-rear lateral bending strut, from the engine and transaxle
- Front engine bracket through-bolt
- Front engine bracket bolts
- Lower starter bolt
- Lower dust shield screws

4. Rotate the engine clockwise and remove the converter bolts.

➡Matchmark the converter to the flexplate for alignment during installation.

5. Support the transaxle with a transaxle jack.
6. Remove or disconnect the following:

- Left mount through-bolt
- Left mount-to-transaxle bolts
- Left mount
- Rear engine bolt from the transaxle
- Transaxle

7. Carefully work the transaxle and torque converter assembly rearward off the engine block dowels. Disengage the converter hub from the end of the crankshaft. Attach a small C-clamp to the edge of the bell housing. This will hold the torque converter in place during transaxle removal.

To install:

8. Install the transaxle and torque the transaxle-to-cylinder block bolt: 70 ft. lbs. (95 Nm)
9. Support the transaxle with a transaxle jack.
10. Install or connect the following:

- Rear engine bolt to the transaxle
- Left mount
- Left mount-to-transaxle bolts and torque to 40 ft. lbs. (54 Nm)
- Left mount through-bolt

11. If the torque converter was removed from the transaxle, be sure to align the pump inner gear pilot flats with the torque converter impeller hub flats. Torque the flexplate-to-crankshaft bolts to 70 ft. lbs. (95 Nm) and the flexplate-to-torque converter bolts: 50 ft. lbs. (68 Nm).
12. Rotate the engine clockwise to get access to the converter bolts.
13. Install or connect the following:

- Lower dust shield screws and torque the bell housing cover bolts to 108 inch lbs. (12 Nm)
- Lower starter bolt
- Front engine bracket bolts
- Front engine bracket through-bolt
- Transaxle-to-rear lateral bending strut to the engine and transaxle
- Exhaust pipe to the manifold
- Exhaust flex joint-to-exhaust manifold bolts
- Halfshafts

⁂ WARNING

When installing the halfshafts, new retaining clips must be used. DO NOT reuse the old clips. Failure to use new clips could cause the inner CV-joints to disengage.

- Front wheels

14. Remove the engine bridge fixture and support.

- Upper starter bolt
- Upper bell housing bolts
- Throttle pressure control cable support bracket bolts
- Transaxle's oil cooler lines. Torque the oil cooler line-to-radiator fitting to 108 inch lbs. (12 Nm) and transaxle oil cooler line fitting to 21 ft. lbs. (28 Nm).
- Transaxle's dipstick tube
- Throttle pressure control cable bracket bolts to the transaxle
- Throttle pressure control cable to the lever
- Gearshift cable bracket bolt to the transaxle
- Gearshift cable end to the transaxle shift lever

15. Adjust the gearshift and throttle cables.

- Neutral safety switch and torque converter control wiring to the transaxle
- VSS wiring
- Cruise control, if equipped
- Battery tray
- Battery
- Battery heat shield
- PDC
- Battery cables

16. Fill the transaxle to the proper level.
17. Be sure the vehicle's back-up lights and speedometer are working properly.

2000–2002 MODELS

1. Before servicing the vehicle, refer to the precautions in the beginning of this section.

The transaxle and torque converter must be removed as an assembly; otherwise the torque converter driveplate, pump bushing or oil seal may be damaged. The driveplate will not support a load; therefore, none of the weight of the transaxle should be allowed to rest on the plate during removal.

2. Drain the transaxle fluid.
3. Remove or disconnect the following:

- Battery
- Air cleaner/throttle body assembly
- Proportional Purge Solenoid (PSS)
- Crankcase vent hose
- Throttle Position Sensor (TPS) and Idle Air Control (IAC) connectors
- Throttle body air duct from the intake manifold
- Accelerator cable
- Transaxle kickdown cable
- Cruise control cable, if equipped
- Air cleaner assembly
- Battery tray from the bracket
- Torque converter clutch solenoid connector
- Neutral safely/back-up light switch connector
- Transaxle oil cooler lines and plug the lines to prevent contamination
- Shift cable and bracket from the transaxle
- Kickdown cable and bracket from the transaxle
- Transaxle oil pan
- Halfshafts
- Structural collar
- Left engine-to-transaxle lateral bending brace
- Bell housing duct cover
- Starter
- Torque converter-to-driveplate bolts

4. Using a jack and a block of wood, support the engine/transaxle assembly.
5. Remove the transaxle upper mount through-bolt.

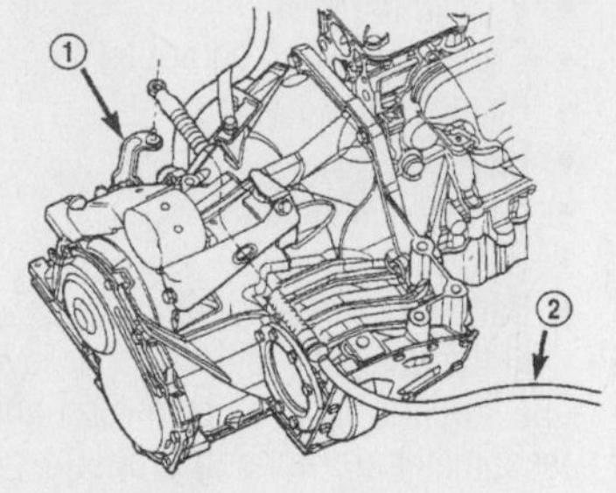

1 – SHIFT LEVER
2 – SHIFT CABLE

9306EG36

View of the gear shift cable—2000–02 automatic transaxle

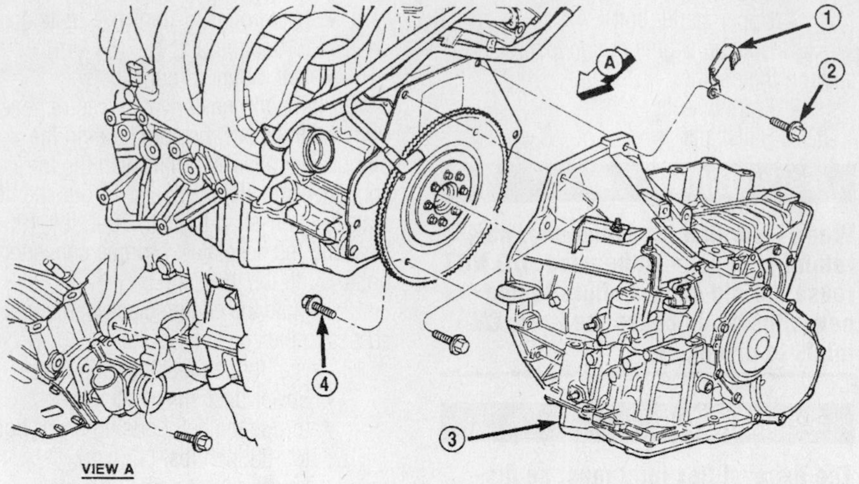

1 – CLIP
2 – BOLT (3)
3 – TRANSAXLE
4 – CONVERTER BOLT (4)

9306EG39

Exploded view of the transaxle and related components—2000–02 automatic transaxle

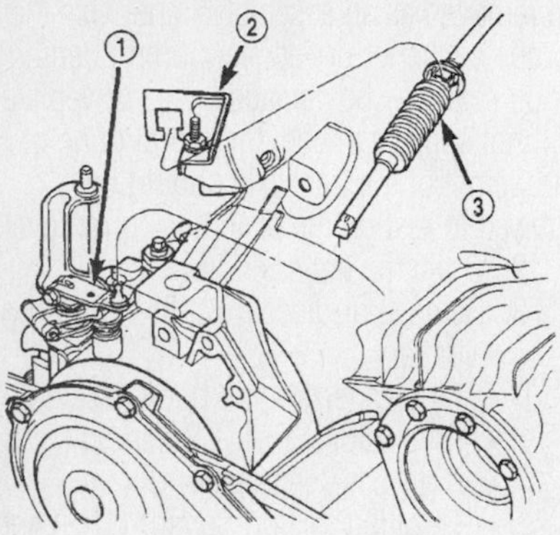

1 – LEVER
2 – BRACKET
3 – KICKDOWN CABLE

9306EG37

View of the kickdown cable—2000–02 automatic transaxle

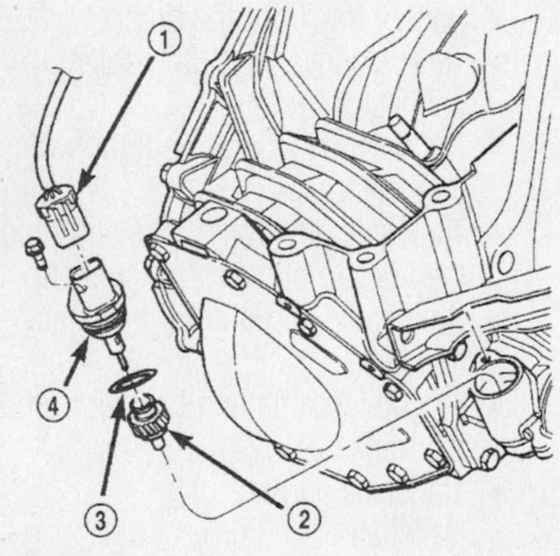

1 – CONNECTOR
2 – SPEEDO PINION
3 – O-RING
4 – SENSOR

9306EG38

View of the Vehicle Speed Sensor (VSS)—2000–02 automatic transaxle

➡The upper mount through-bolt may be accessed through the left side wheel house.

6. Lower the engine/transaxle assembly for clearance.
7. Remove or disconnect the following:

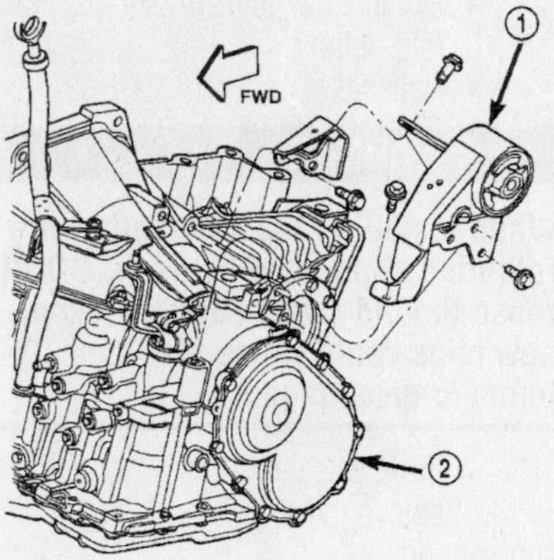

1 – MOUNT
2 – TRANSAXLE

9306EG40

View of the upper mount—2000–02 automatic transaxle

- Transaxle-to-engine bolts, using an assistant to support the transaxle
- Transaxle
- Torque converter from the front pump

To install:

8. If installing a new or replacement transaxle, transfer the upper mount to the new transaxle; then, torque the upper mount-to-transaxle bolts to 50 ft. lbs. (68 Nm).
9. Install transaxle and tighten the transaxle-to-engine bolts, using an assistant to support the transaxle. Torque the transaxle-to-engine bolts to 70 ft. lbs. (95 Nm).
10. Raise the engine/transaxle assembly until the upper mount aligns with the mount bracket hole. Torque the upper mount-to-mount bracket through-bolt to 80 ft. lbs. (108 Nm).
11. Remove the jack and wooden block.
12. Install or connect the following:
    - Torque converter-to-driveplate bolts and torque the bolts to 65 ft. lbs. (88 Nm)
    - Starter and torque the bolts to 40 ft. lbs. (54 Nm)
    - Starter electrical connectors and torque the positive cable bolt to 90 inch lbs. (10 Nm)
    - Bell housing dust cover
    - Left engine-to-transaxle bending brace and torque the bolts to 60 ft. lbs. (81 Nm)
13. Install the structural collar and torque the bolts as follows:
    a. Step 1—Structural collar-to-oil pan bolts: 30 inch lbs. (3 Nm).
    b. Step 2—Structural collar-to-transaxle bolts: 80 ft. lbs. (108 Nm).
    c. Step 3—Structural collar-to-oil pan bolts: 40 ft. lbs. (54 Nm).
14. Install or connect the following:
    - Halfshafts
    - Speed control cable, if equipped
    - Right engine-to-transaxle lateral bending brace and torque the bolts to 60 ft. lbs. (81 Nm)
    - Transaxle oil cooler lines and torque the oil cooler line-to-radiator fitting to 108 inch lbs. (12 Nm) and transaxle oil cooler line fitting to 21 ft. lbs. (28 Nm)
    - Torque converter clutch solenoid and neutral safely/back-up light switch connectors
    - Transaxle dipstick tube
    - Gearshift cable bracket bolt to the transaxle

- Gearshift cable end to the transaxle shift lever
- Transaxle kickdown cable to the lever and bracket
- Battery tray
- Battery
- Accelerator cable
- Transaxle kickdown cable
- Cruise control cable, if equipped
- Air cleaner assembly and torque the nuts and bolts to 10 ft. lbs. (14 Nm)
- TPS and IAC connectors
- Crankcase vent hose to the throttle body
- PSS to the throttle body

15. Fill the transaxle and check for leaks.
16. Be sure the vehicle's back-up lights and speedometer are working properly.

Clutch

ADJUSTMENT

1998–99 Models

The manual transaxle clutch release system has a unique self-adjusting mechanism to compensate for clutch disc wear. This adjuster mechanism is located with the clutch cable assembly. The preload spring maintains tension on the cable. This tension keeps the clutch release bearing continuously loaded against the fingers of the clutch cover assembly. No manual adjustment is obtainable.

When servicing this vehicle or if removing and installing the clutch cable, do not pull on the clutch cable housing to remove it from the dash panel. Damage to the cable self-adjuster may occur.

To check the function of the adjuster mechanism, use the following procedure:

1. With slight pressure, pull the clutch release lever end of the cable to draw the cable taut.
2. Push the clutch cable housing toward the dash panel. With less than 25 lbs. (11kg) of effort the cable housing should move 1.2–2.0 in. (30–50mm). This indicates proper adjuster mechanism function.
3. If the cable does not adjust, determine if the mechanism is properly seated on the bracket.

REMOVAL & INSTALLATION

➡Vehicles made at the Toluca, Mexico assembly plant have conventional clutch and flywheel assembles. Vehicles made at the Belvidere assembly plant have modular clutch assemblies. If the 11th digit of the VIN code is "D", the vehicle was produced at the Belvidere assembly plant; if the VIN code is "T", the vehicle was produced at the Toluca assembly plant.

1998–99 Models

TOLUCA BUILT VEHICLES

1. Before servicing the vehicle, refer to the precautions in the beginning of this section.
2. Remove the transaxle assembly.
3. Matchmark the position of the clutch cover and flywheel for proper alignment during installation.
4. Install a clutch alignment tool through the clutch disc hub to prevent the clutch disc from falling and damaging the facings.
5. Loosen the clutch cover attaching bolts, 1–2 turns at a time, in a crisscross pattern. This releases the spring pressure gradually, avoiding cover damage.

➡DO NOT touch the clutch disc facing with oily or dirty hands. Oil or dirt transferred from your hands onto the clutch disc may cause clutch chatter.

6. Remove the clutch pressure plate.
7. Remove the clutch cover assembly and disc from the flywheel.
8. Remove the flywheel, if necessary.
   a. Inspect the clutch assembly as follows:

➡Handle the components carefully to avoid contaminating the friction surfaces.

   b. Inspect for oil leakage through the engine rear main bearing oil seal and transaxle input shaft seal. If there is leakage, it should be fixed at this time.

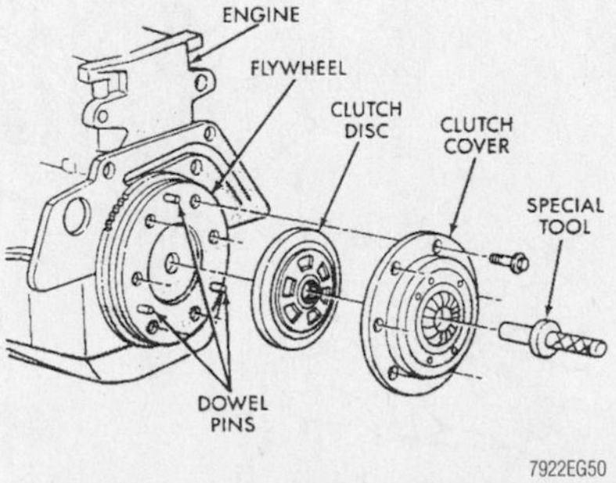

Exploded view of the conventional clutch components—1998–99 vehicles built in Toluca

   c. The friction faces of the flywheel and pressure plate should not have excessive discoloration, burned areas, cracks, deep grooves or ridges. Replace parts as required.
   d. Clean the flywheel face with medium sandpaper, then wipe the surface with mineral spirits. If the surface is severely scored, heat checked, cracked or warped, replace the flywheel.
   e. The clutch disc should be handled without touching the facings. Replace the disc if the facings show grease or oil soakage, or wear to within less than 0.008 in. (0.20mm) of the rivet heads. The splines on the disc hub and transaxle input shaft should be a snug fit without signs of excessive wear. Metallic portions of the disc assembly should be dry, clean and not discolored from excessive heat. Each of the arched springs between the facings should be tight.
   f. Wipe the friction surface of the pressure plate with mineral spirits.
   g. Using a straightedge, check the pressure plate for flatness. The pressure plate friction area should be flat-to-slightly concave, with the inner diameter 0.000–0.0039 in. (0.0–0.1mm) below the outer diameter. It should also be free from discoloration, burned areas, cracks, grooves or ridges.
   h. Using a surface plate, test the cover for flatness. All sections around the attaching bolt holes should be in contact with the surface plate within 0.015 in. (0.381mm).
   i. The cover should be a snug fit on the flywheel dowels. If the clutch assembly does not meet these requirements, it should be replaced.

To install:

➡The heavy side of the flywheel is indicated by a white paint mark, near the outside diameter.

9. To minimize the effects of flywheel unbalance, perform the following installation procedure:
   a. Loosely assembly the flywheel to the crankshaft. If available, use new flywheel attaching bolts which have sealant on the threads. If new bolts are not available, apply Loctite® sealant to the threads of the original bolts. This sealant is required to prevent engine oil leakage.
   b. Rotate the flywheel and crankshaft until the white paint (heavy side) is at the 12 o'clock position.

c. Tighten the flywheel attaching bolts, in a crisscross pattern, to 70 ft. lbs. (95 Nm).

10. Mount the clutch assembly on the flywheel with the disc centered on the alignment tool, being careful to properly align the dowels and the alignment marks made before removal. The flywheel side of the clutch disc is marked for proper installation. If the new clutch or flywheel is installed, align the orange cover balance spot as close as possible to the orange flywheel balance spot. Apply pressure to the alignment tool. Center the tip of the tool into the crankshaft and the sliding cone into the clutch fingers. Tighten the clutch attaching bolts sufficiently to hold the disc in position.

11. To avoid distorting the clutch cover, tighten the bolts gradually, a few turns at a time. Use a crisscross pattern until all bolts are seated. Tighten the bolts to a final torque of 21 ft. lbs. (28 Nm).

12. Remove the clutch alignment tool.

13. Install the transaxle.

BELVIDERE BUILT VEHICLES

1. Before servicing the vehicle, refer to the precautions in the beginning of this section.
2. Remove or disconnect the following:
   - Negative battery cable
   - Starter motor assembly
   - Rear and front transaxle brackets
   - Modular clutch-driveplate bolts and discard the bolts
   - Transaxle

➡The transaxle and modular clutch will come out as an assembly.

- Modular clutch assembly from the transaxle input shaft.

➡Handle the components carefully to avoid contaminating the friction surface.

3. Inspect for oil leakage through the engine rear main bearing oil seal and transaxle input shaft seal. If any leakage is noted, it should be fixed at this time.

To install:

➡Always use new bolts when mounting the modular clutch assembly to the driveplate.

4. Install or connect the following:
   - Modular clutch assembly onto the input shaft
   - Transaxle
5. To avoid distorting the driveplate, tighten the bolts gradually a few turns at a time. Use a crisscross pattern, until all bolts are seated. Tighten the bolts to a final torque of 55 ft. lbs. (75 Nm).
   - Clutch inspection cover
   - Transaxle lower support brackets
   - Starter assembly
   - Negative battery cable

2000–02 Models

1. Before servicing the vehicle, refer to the precautions in the beginning of this section.
2. Remove or disconnect the following:
   - Negative battery cable
   - Modular clutch-to-driveplate bolts and discard the bolts
   - Transaxle assembly
   - Modular clutch assembly from the transaxle input shaft

To install:

3. Install or connect the following:
   - Modular clutch assembly onto the transaxle input shaft
   - Transaxle assembly and torque the transaxle-to-engine bolts to 70 ft. lbs. (95 Nm)
   - New modular clutch-to-driveplate bolts and torque the bolts to 65 ft. lbs. (88 Nm)
   - Negative battery cable

Halfshafts

REMOVAL & INSTALLATION

1. Before servicing the vehicle, refer to the precautions in the beginning of this section.
2. On 1998–99 vehicles, remove the cotter pin (discard it), locknut and spring washer, from the end of the outer CV-joint stub axle.
3. With the vehicle on the ground and brakes applied, loosen, but do not remove the stub axle-to-hub and bearing retaining nut.

➡The front hub and driveshaft are splined together and retained by the hub nut.

4. On 2000–02 vehicles equipped with Anti-lock Brake System (ABS), disconnect the front wheel speed sensor.
5. Remove or disconnect the following:
   - Front wheel
   - Front caliper-to-steering knuckle bolts
   - Caliper from the steering knuckle
6. Support the caliper out of the way by suspending it with a piece of wire from the strut. DO NOT allow the caliper to hang by the brake hose.
   - Rotor from the hub
   - Outer tie rod end-to-steering knuckle nut by holding the tie rod end stud with a 11/32 in. socket while loosening the nut

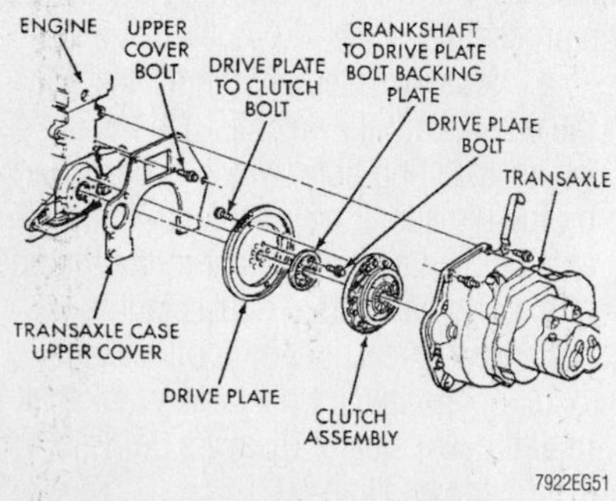

The transaxle and modular clutch are removed as an assembly—1998–99 vehicles

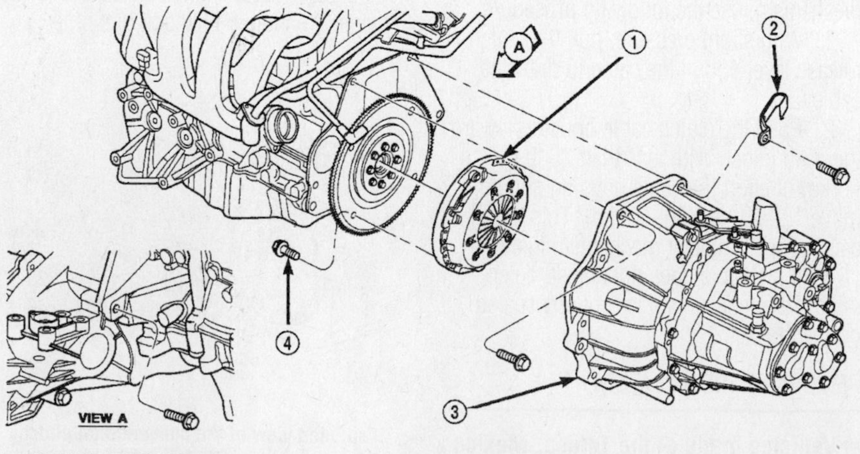

Exploded view of the manual transaxle, clutch module and related components—2000–02 vehicles

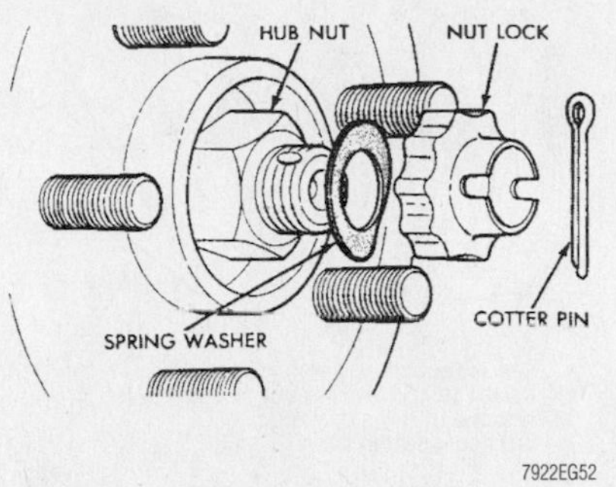

Exploded view of the cotter pin, locknut and washer—halfshaft service

- Tie rod end stud from the steering knuckle
- Ball joint stud-to-steering knuckle nut and bolt

WARNING

Be careful when separating the ball joint stud from the steering knuckle, so the ball joint seal does not get damaged.

- Ball joint stud from the steering knuckle by prying down on the lower control arm

7. For 1998–00 vehicles, perform the following steps:

a. Remove the hub and bearing-to-stub axle retaining nut.

b. Install a puller on the hub and bearing assembly, using the lug nuts to hold it in place.

c. Install a wheel lug nut on wheel stud to protect the threads on the stud. Install a flat-bladed prytool to keep the hub from turning. Using the puller, force the outer stub axle from the hub and bearing.

WARNING

Be careful when separating the inner CV-joint during this operation. Do not let the driveshaft hang by the inner CV-joint, the driveshaft must be supported.

8. Pull the steering knuckle assembly out and away from the outer CV-joint of the driveshaft assembly. Support the outer end of the driveshaft assembly.

9. Insert a prybar between the inner tripod joint and the transaxle case. Pry against the inner tripod joint until the joint retaining snapring is disengaged from the transaxle side gear.

➡Inner tripod joint removal is easier by applying outward pressure on the joint while hitting the punch with a hammer.

10. Remove the inner tripod joints from the transaxle side gears using a punch to dislodge the inner tripod joint retaining ring from the transaxle side gear. If removing the right side inner tripod joint, position the punch against the inner tripod joint. Hit the punch sharply with a hammer to dislodge the right inner joint from the side gear. If removing the left side inner tripod joint, position the punch in the groove of the inner tripod joint. Hit the punch sharply with a hammer to dislodge the left inner tripod joint from the side gear.

11. Hold the inner tripod joint and interconnecting shaft of the driveshaft assembly. Remove the inner tripod joint from the transaxle by pulling it straight out of the transaxle side gear and transaxle oil seal. When removing the tripod joint, do not let the spline or snapring drag across the sealing lip of the transaxle-to-tripod joint oil seal.

WARNING

The driveshaft, when installed, acts as a bolt which secures the front hub and bearing assembly. If the vehicle is to be supported or moved on its wheels with a driveshaft removed, install a proper-sized bolt and nut through the front hub. Tighten the bolt and nut to 135 ft. lbs. (183 Nm). This will ensure that the hub bearing cannot loosen.

To install:

12. Thoroughly clean the spline and oil seal sealing surface on the tripod joint. Lightly lubricate the oil seal sealing surface on the tripod joint with fresh, clean transmission fluid.

13. Holding the driveshaft assembly by the tripod joint and interconnecting shaft, install the tripod joint into the transaxle side gear as far as possible by hand.

14. Carefully align the tripod joint with the transaxle side gears. Then, grasp the driveshaft interconnecting shaft and push the tripod joint into the transaxle side gear until fully seated. Be sure the snapring is fully engaged with the side gear by trying to remove the tripod joint from the transaxle by hand. If the snapring is fully seated with the side gear, the tripod joint will not be removable by hand.

15. Clean all debris and moisture out of the steering knuckle.

16. Be sure that the outer CV-joint, which fits into the steering knuckle, has no debris or moisture on it before installing into the steering knuckle.

17. Slide the driveshaft back into the front hub. Install the steering knuckle into the ball joint stud.

18. Install a NEW steering knuckle-to-ball joint stud bolt and nut. Tighten the nut and bolt to 70 ft. lbs. (95 Nm).

19. Insert the tie rod end into the steering knuckle. Start the tie rod end-to-steering knuckle nut onto the stud of the tie rod end. While holding the stud of the tie rod end stationary, tighten the nut. Then, using a crow's foot and 11⁄32 in socket, tighten the tie rod end nut to 45 ft. lbs. (61 Nm).

20. Install the rotor back onto the hub and bearing assembly.

21. Position the caliper on the steering knuckle. Slide the top of the caliper under the top abutment on the steering knuckle, then install the bottom of the caliper against the bottom abutment of the steering knuckle.

22. Install the caliper-to-knuckle bolts and tighten to 23 ft. lbs. (31 Nm).

23. On 2000–02 vehicles equipped with ABS, connect the front wheel speed sensor.

24. Clean all foreign matter from the threads of the outer CV-joint stub axle. Install hub nut and washer onto the threads of the stub axle and tighten the nut.

25. With the vehicle's brakes applied to prevent the axle shaft from turning, tighten the hub nut to 150 ft. lbs. (203 Nm) for 1998–99 models or 180 ft. lbs. (244 Nm) for 2000–02 models.

26. On 1998–99 vehicles, install the spring washer, locknut and new cotter pin into the outer CV-joint stub axle.

27. Install the front wheel and tire assembly. Install the lug nuts and tighten to 100 ft. lbs. (135 Nm) for 1998–99 models or 95 ft. lbs. (128 Nm) for 2000–02 models.

28. Check the transaxle fluid level, lowering the vehicle as necessary.

CV-Joints

OVERHAUL

Tri-Pot (Inner) Joint

1. Remove or disassemble the following:
   - Halfshaft and place it in a soft-jawed vise

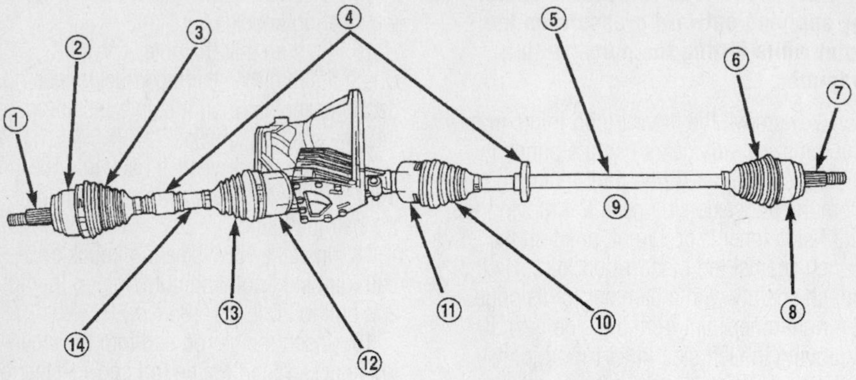

1 – STUB AXLE
2 – OUTER C/V JOINT
3 – OUTER C/V JOINT BOOT
4 – TUNED RUBBER DAMPER WEIGHT
5 – INTERCONNECTING SHAFT
6 – OUTER C/V JOINT BOOT
7 – STUB AXLE
8 – OUTER C/V JOINT
9 – RIGHT DRIVESHAFT
10 – INNER TRIPOD JOINT BOOT
11 – INNER TRIPOD JOINT
12 – INNER TRIPOD JOINT
13 – INNER TRIPOD JOINT BOOT
14 – INTERCONNECTING SHAFT LEFT DRIVESHAFT

9306EG02

View of the halfshaft assemblies—Typical

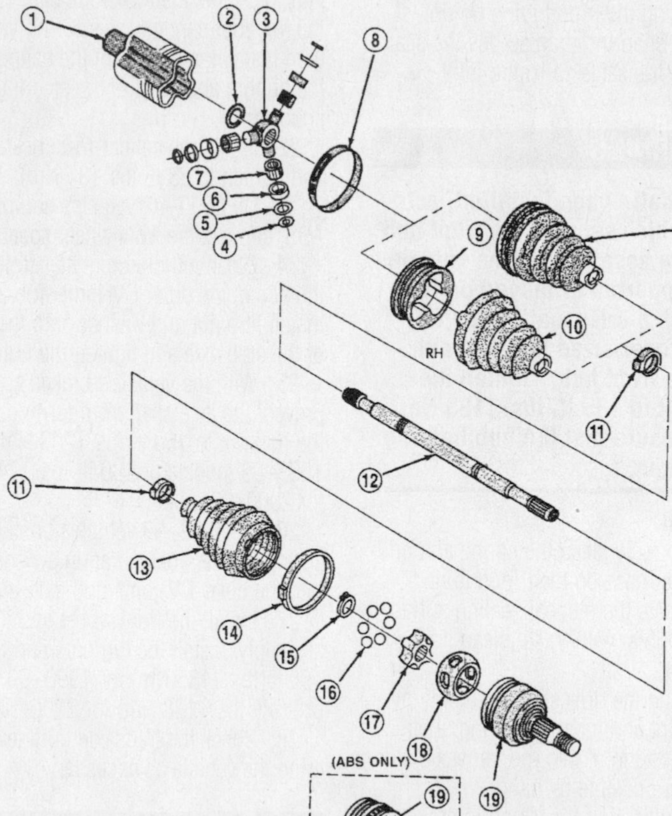

1 – HOUSING ASM, RETAINER &
2 – RING, SPACER
3 – SPIDER, TRIPOTD JOINT
4 – RING, RETAINING
5 – RETAINER, BALL & ROLLER
6 – BALL, TRIPOD JOINT
7 – ROLLER, NEEDLE
8 – CLAMP, SEAL RETAINING
9 – BUSHING, TRILOBAL TRIPOD
10 – SEAL, DRIVE AXLE INBOARD
11 – CLAMP, SEAL RETAINING
12 – SHAFT, AXLE (RH SHOWN, LH SIMILAR)
13 – SEAL, DRIVE AXLE OUTBOARD
14 – CLAMP, SEAL RETAINING
15 – RING, RACE RETAINING
16 – BALL, CHROME ALLOY
17 – RACE, C/V JOINT INNER
18 – CAGE, C/V JOINT
19 – RACE, C/V JOINT OUTER

9306EG03

Exploded view of the halfshaft assembly—Typical

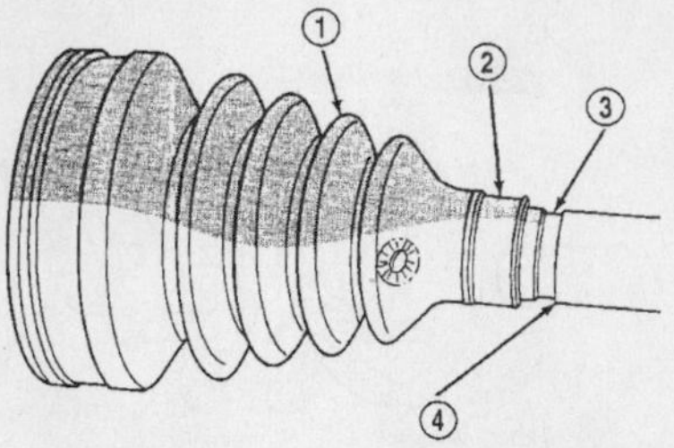

1 – SEALING BOOT
2 – RAISED BEAD IN THIS AREA OF SEALING BOOT
3 – GROOVE
4 – INTERCONNECTING SHAFT

9306EG04

Positioning the boot on the shaft—Tri-pot and Rzeppa joints

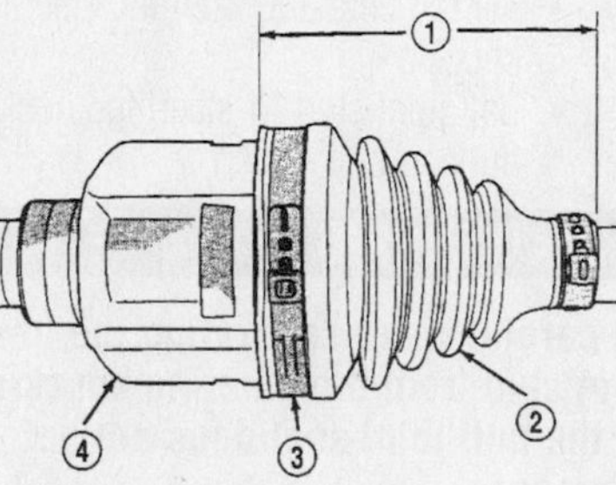

1 – 107 MILLIMETERS
2 – HYTREL SEALING BOOT
3 – SEALING BOOT CLAMP
4 – INNER TRIPOD JOINT

9306EG05

Adjusting the Tri-pot boot length—Hytrel plastic boot

- Tri-pot joint boot clamps and the boot down the shaft

⁂ WARNING

When removing the spider joint, hold the rollers in place on the trunions to keep the rollers and needle bearings in place.

- Spider/shaft assembly from the tri-pot housing
- Snapring from the shaft
- Spider assembly

➡If necessary, tap the spider assembly from the shaft with a brass drift.

⁂ WARNING

When removing the spider assembly, do not hit the outer bearings.

- Boot by sliding it off the shaft

⁂ WARNING

If any parts show excessive wear, replace the halfshaft assembly; the component parts are not serviceable.

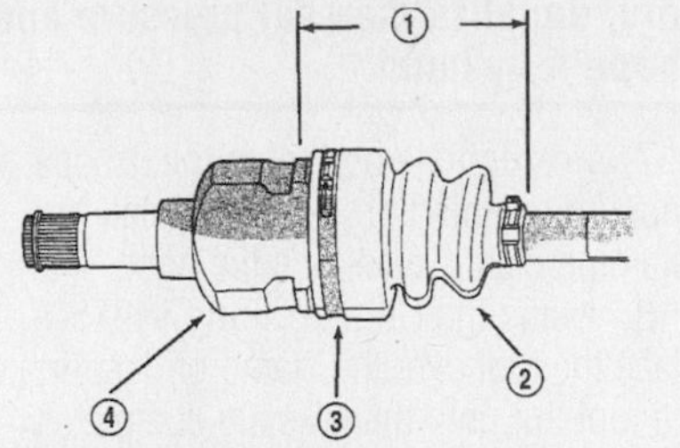

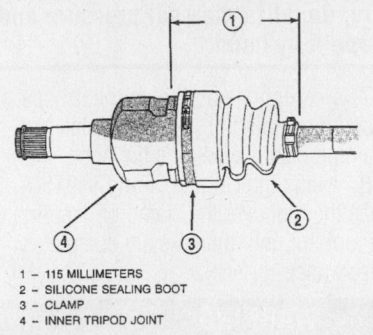

1 - 115 MILLIMETERS
2 - SILICONE SEALING BOOT
3 - CLAMP
4 - INNER TRIPOD JOINT

9306EG06

Adjusting the Tri-pot boot length—Silicone rubber boot

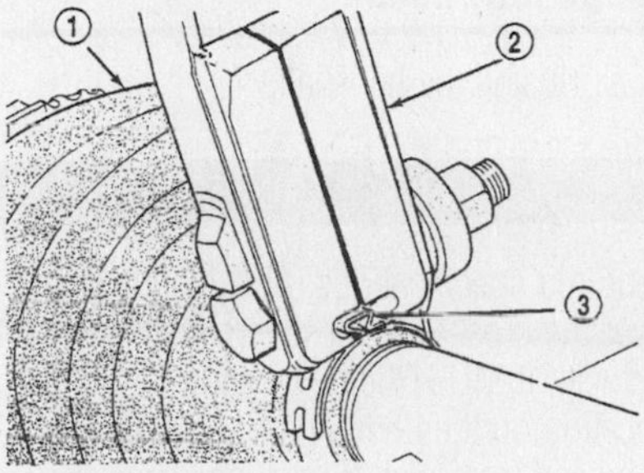

1 - SEALING BOOT
2 - SPECIAL TOOL C-4975
3 - CLAMP BRIDGE

93060G13

Tightening the high profile boot clamp—Tri-pot joint (Hytrel plastic) boot and Rzeppa joint boots

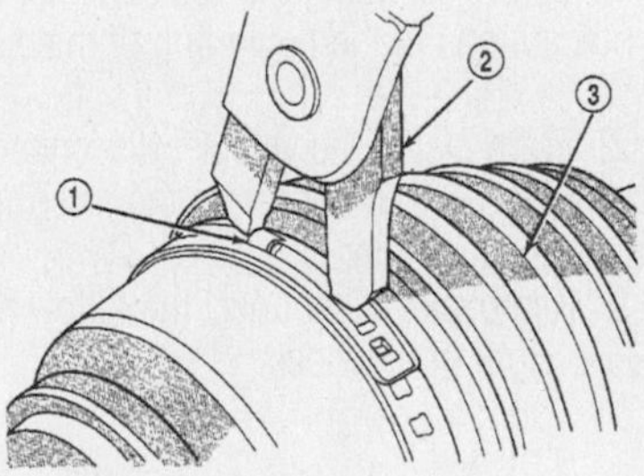

1 - CLAMP
2 - SPECIAL TOOL YA3050
3 - SEALING BOOT

9306EG07

Tightening the low profile boot clamp—Silicone rubber Tri-pot boot

To assemble:

⁂ WARNING

The Tri-pot sealing boots are made of 2 different types of material; silicon rubber (high temperature) which is soft and pliable or hytrel plastic (standard temperature) which is stiff and rigid. Be sure to replace the boot made of the correct material.

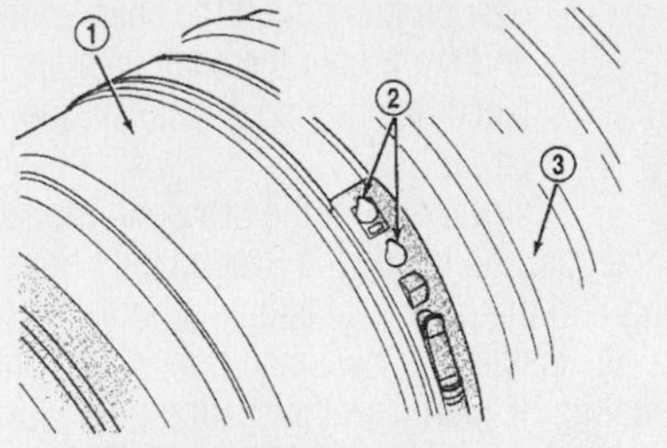

1 - INNER TRIPOD JOINT HOUSING
2 - TOP BANK OF CLAMP MUST BE RETAINED BY TABS AS SHOWN HERE TO CORRECTLY LATCH BOOT CLAMP
3 - SEALING BOOT

9306EG08

Secured the low profile boot clamp—Silicone rubber Tri-pot boot

2. Install a new small boot clamp and slide it on the shaft.
3. Position the boot so that the raised bead on the inside the boot seal is in the shaft groove.
4. If installing a new boot, distribute ½ of the grease in the service package inside the tri-pot housing and the other ½ inside the boot.
5. Install or connect the following:
   - Spider assembly, face the chamfered side toward the shaft

⁂ WARNING

If necessary, tap the spider assembly onto the shaft using a brass drift; be careful not to hit the outer bearings.

   - Snapring making sure it is fully seated in the groove
   - Spider/shaft assembly into the tri-pot housing
   - New inner boot clamp and position it evenly on the sealing boot
6. Using a trim stick, adjust the boot length to 107mm (hytrel plastic) or 115mm (silicone rubber).
7. If installing a high profile boot clamp, perform the following procedure:
   a. Using the Crimper Tool C-4975-A, place the tool over the clamp bridge, tighten the tool nut until the jaws are completely closed (face-to-face).

⁂ WARNING

The seal must not be dimpled, stretched or out of shape. If necessary, equalize the seal pressure and shape it by hand.

   b. Position the boot onto the tri-pot housing retaining groove and install the retaining clamp evenly on the boot.
   c. Using the Crimper Tool C-4975-A, place the tool over the clamp bridge, tighten the tool nut until the jaws are completely closed (face-to-face).
8. If installing a low profile latching type boot clamp, position Snap-On® Clamp Locking Tool YA3050 prongs in the clamp holes and squeeze the tool until the upper clamp band is latched behind the 2 tabs on the lower clamp band.
9. Install the halfshaft.

Rzeppa (Outer) Joint

1. Remove or disassemble the following:
   - Halfshaft and place it in a soft-jawed vise
   - Rzeppa joint boot clamps and slide the boot down the shaft
   - Rzeppa joint housing, by sharply hitting it with a soft-faced hammer to drive it off the shaft
   - Circlip from the shaft
   - Boot by sliding it off the shaft

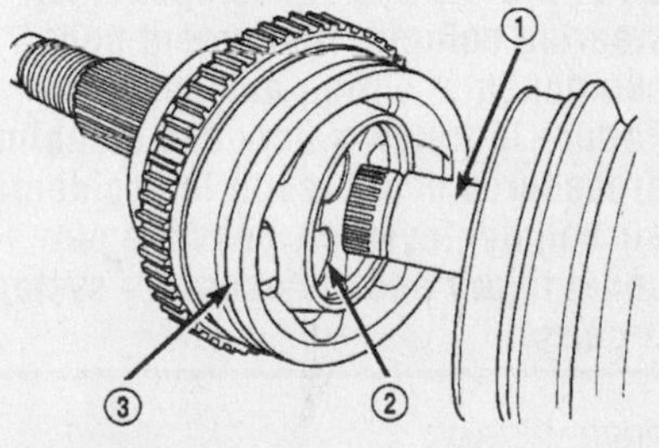

1 - INTERCONNECTING SHAFT
2 - CROSS
3 - OUTER C/V JOINT ASSEMBLY

9306EG09

Aligning the cross splines with the shaft splines—Rzeppa joint

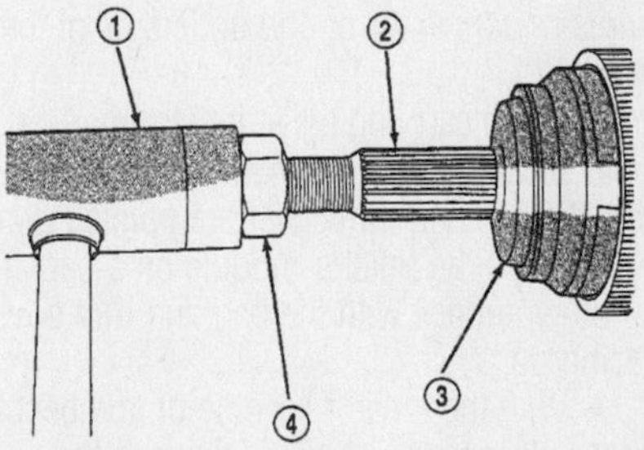

1 - SOFT FACED HAMMER
2 - STUB AXLE
3 - OUTER C/V JOINT
4 - NUT

9306EG10

Driving the Rzeppa joint onto the shaft

⁂ WARNING

If any parts show excessive wear, replace the halfshaft assembly; the component parts are not serviceable.

To assemble:

2. Install or assemble the following:
   - New small boot clamp and slide it onto the shaft
   - Boot and slide it onto the shaft
   - Circlip, if removed
3. Position the boot so that the raised bead on the inside the boot seal is in the shaft groove.
   - Halfshaft hub nut onto the Rzeppa joint threaded shaft so it is flush with the end
   - Rzeppa joint, align the shaft splines and tap it onto the shaft with a soft-faced hammer so it locks on the circlip
4. Distribute ½ of the grease in the service package inside the Rzeppa joint housing and the other ½ inside the boot.
5. Install the new small boot clamp and position it evenly on the sealing boot
6. Using the Crimper Tool C-4975-A, place the tool over the clamp bridge, tighten the tool nut until the jaws are completely closed (face-to-face).

⁂ WARNING

The seal must not be dimpled, stretched or out of shape. If necessary, equalize the seal pressure and shape it by hand.

7. Position the boot onto the Rzeppa housing retaining groove and install the retaining clamp evenly on the boot.
8. Using the Crimper Tool C-4975-A, place the tool over the clamp bridge, tighten the tool nut until the jaws are completely closed (face-to-face).

⁂ WARNING

The seal must not be dimpled, stretched or out of shape. If necessary, equalize the seal pressure and shape it by hand.

9. Install the halfshaft.

STEERING AND SUSPENSION

Air Bag

⁂ CAUTION

Some vehicles are equipped with an air bag system. The system MUST BE disabled before performing service on or around system components, steering column, instrument panel components, wiring and sensors. Failure to follow safety and disabling procedures could result in accidental air bag deployment, possible personal injury and unnecessary system repairs.

PRECAUTIONS

Several precautions must be observed when handling the inflator module to avoid accidental deployment and possible personal injury:

- Never carry the inflator module by the wires or connector on the underside of the module.
- When carrying a live inflator module, hold securely with both hands, and ensure that the bag and trim cover are pointed away.
- Place the inflator module on a bench or other surface with the bag and trim cover facing up.
- With the inflator module on the bench, never place anything on or close to the module which may be thrown in the event of an accidental deployment.

DISARMING

Proper Supplemental Restraint Ssystem (SRS) disarming can be obtained by disconnecting and isolating the negative battery cable. Allow the air bag system capacitor at least 2 minutes to discharge before removing any air bag system components.

Rack and Pinion Steering Gear

These vehicles are designed and assembled using NET BUILD front suspension alignment settings. This means that the alignment settings are determined as the vehicle is designed by the location of the front suspension components in relation to the body. This is carried out when building the vehicle, by precisely locating the front crossmember to meter gauge holes located in the underbody of the vehicle. With this method of designing and building a vehicle, it is no longer possible to adjust a vehicle's front suspension alignment settings to the required specifications. As a result, whenever the crossmember is removed from a vehicle, it MUST be replaced in the same location on the body of the vehicle it was removed from. The front suspension toe settings can still be adjusted by the outer tie rod ends.

REMOVAL & INSTALLATION

1998–99 Models

1. Before servicing the vehicle, refer to the precautions in the beginning of this section.
2. From inside the vehicle, disconnect the steering gear coupler from the steering column shaft coupler.
3. Remove or disconnect the following:
   - Both front wheels

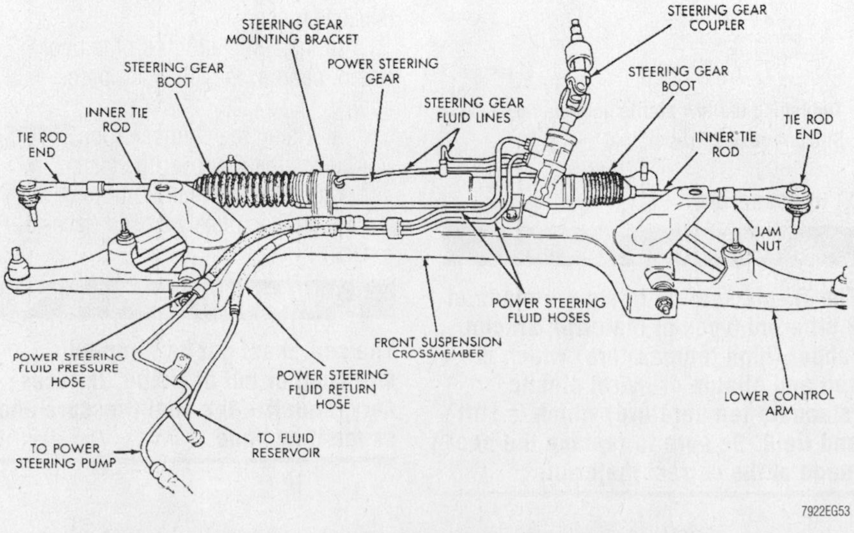

Identification of the rack and pinion steering gear components—1998–99 vehicles

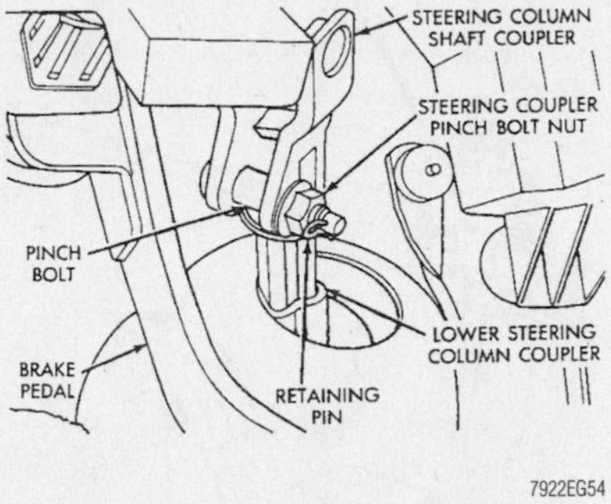

Disconnect the steering gear coupler from inside the vehicle—1998–99 vehicles

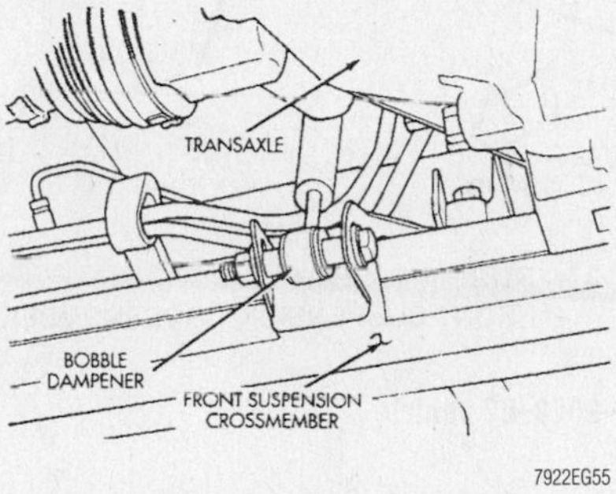

Some vehicles are equipped with a bobble damper that must be removed from the crossmember during steering gear service—1998–99 vehicles

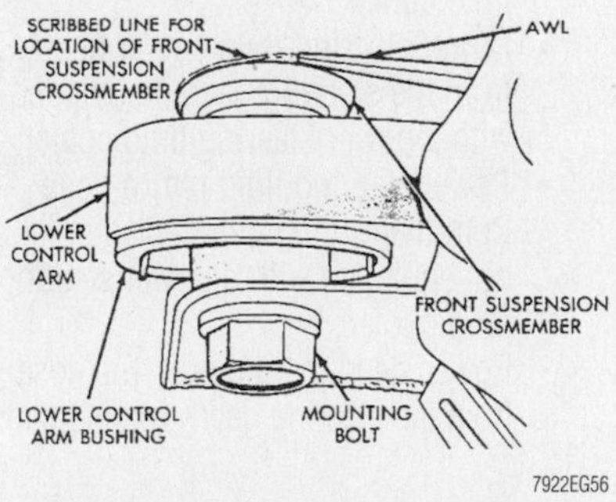

Because of the method used for aligning the vehicle, matchmark the crossmember installed position prior to steering gear removal—1998–99 vehicles

- Engine/transaxle bobble damper from the front suspension crossmember, if equipped

➡The bobble strut does not have to be removed from the transaxle.

- Outer tie rod end-to-steering knuckle nut

➡The nut is removed by holding the tie rod end stud with an 11⁄32 in. socket while loosening the nut with a wrench.

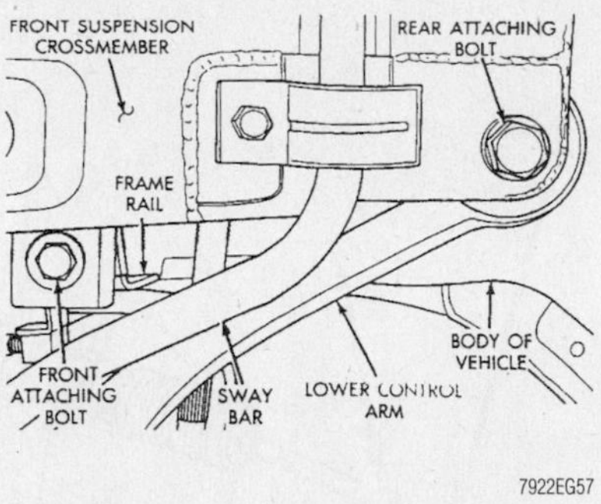

Location of the front crossmember mounting bolts—1998–99 vehicles

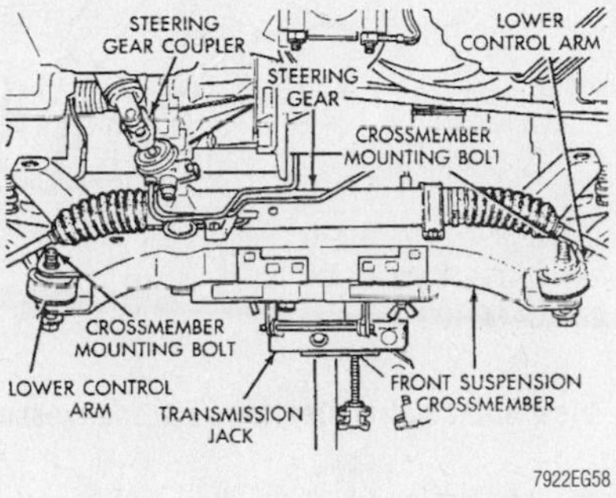

Lower the crossmember for access to the steering gear—1998–99 vehicles

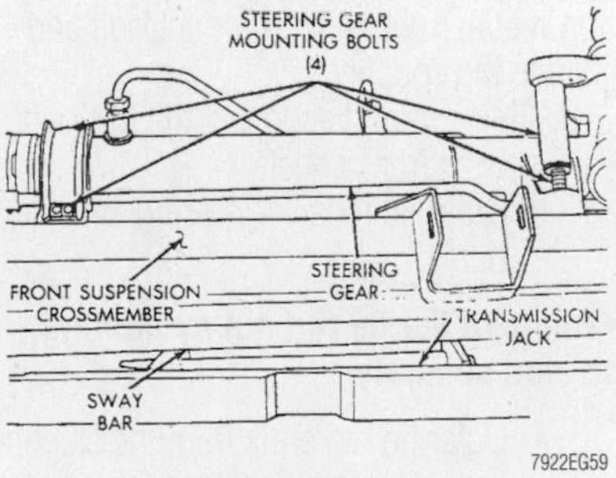

View of the steering gear mounting bolts—1998–99 vehicles

- Tie rod ends from the steering knuckles, using a Side Puller Tool MB-991113.
- Power steering fluid pressure switch connector
- Power steering pressure/return hose routing bracket from the front crossmember

➡The bracket does not have to be removed from the power steering pressure and return hoses.

- Power steering fluid pressure and return hoses from the power steering gear

⁂ WARNING

Before removing the crossmember from the vehicle, the location of the crossmember MUST be scribed or marked on the vehicle. This must be done so the crossmember can be reinstalled in its exact location. If this is not done, the proper NET BUILD alignment specifications will not be obtained and may lead to handling and/or tire wear problems.

4. Matchmark the crossmember to the chassis.
5. Place a transmission jack under the center of the crossmember. The jack will be used to lower, support and raise the crossmember when removing the steering gear.
6. Remove both front crossmember-to-frame rails bolts. Then, loosen both rear crossmember/lower control arm to the body bolts. Lower the crossmember while loosening the rear bolts.
7. Using the transmission jack, lower the crossmember enough to allow the steering gear to be removed from the crossmember. When lowering the crossmember, do not let the crossmember hang from the lower control arms. The weight should be supported by the jack.
8. Remove or disconnect the following:
   - 4 steering gear-to-crossmember bolts
   - Steering gear

➡If a new steering gear is being installed, transfer any necessary parts from the old steering gear to the new steering gear.

To install:

9. Install the steering gear on the crossmember and torque the bolts to 50 ft. lbs. (68 Nm).
10. Using the transmission jack, raise the crossmember/steering gear assembly against the chassis frame rails. Start the 2 rear crossmember-to-chassis bolts; then, the 2 front crossmember-to-frame rail volts. Tighten the 4 bolts until the crossmember is aligned with the mounting points. Torque the bolts to 20 inch lbs. (2 Nm).

➡When the crossmember is installed, it MUST align with the removal matchmarks; the purpose is maintain NET BUILD front suspension alignment settings.

11. Tap the crossmember into position, with a rubber mallet, until it is aligned with the matchmarks. Once positioned, torque both rear crossmember/lower control arm bolts to 120 ft. lbs. (163 Nm); then, both front bolts to 120 ft. lbs. (163 Nm).

12. Install or connect the following:

- Power steering fluid pressure/return hoses into the steering gear ports and torque the line-to-steering gear tube nuts to 23 ft. lbs. (31 Nm)
- Power steering fluid pressure/return hose routing bracket and screw on the cross-member and torque the bracket-to-attaching bolt to 17 ft. lbs. (23 Nm)
- Wiring harness connector onto the power steering gear's fluid pressure switch and securely latch the wiring harness connector's locking tab
- Tie rod end, into the steering knuckle. While holding the tie rod stud, torque the tie rod end-to-steering knuckle nut, using a crow's foot wrench and 11/32 in. socket, to 40 ft. lbs. (55 Nm).
- Engine/transaxle bobble strut, if equipped, onto the crossmember bracket
- Both front wheels and torque the lug nuts, in a crisscross pattern, to 100 ft. lbs. (135 Nm)
- Steering gear coupler to the steering column shaft coupler, under the dash, and torque the bolt to 21 ft. lbs. (28 Nm)
- Upper-to-lower steering coupler retaining bolt retention pin

WARNING

When refilling and bleeding the power steering system, always use the proper type of fluid. NEVER substitute automatic transmission fluid for the specified fluid.

13. Refill the power steering pump fluid reservoir to the FULL-COLD level with the proper type and amount of fluid. Bleed the system.

2000–02 Models

1. Before servicing the vehicle, refer to the precautions in the beginning of this section.

2. Place the steering wheel in the straight-ahead position. Lock the steering wheel in place, using a steering wheel holder.

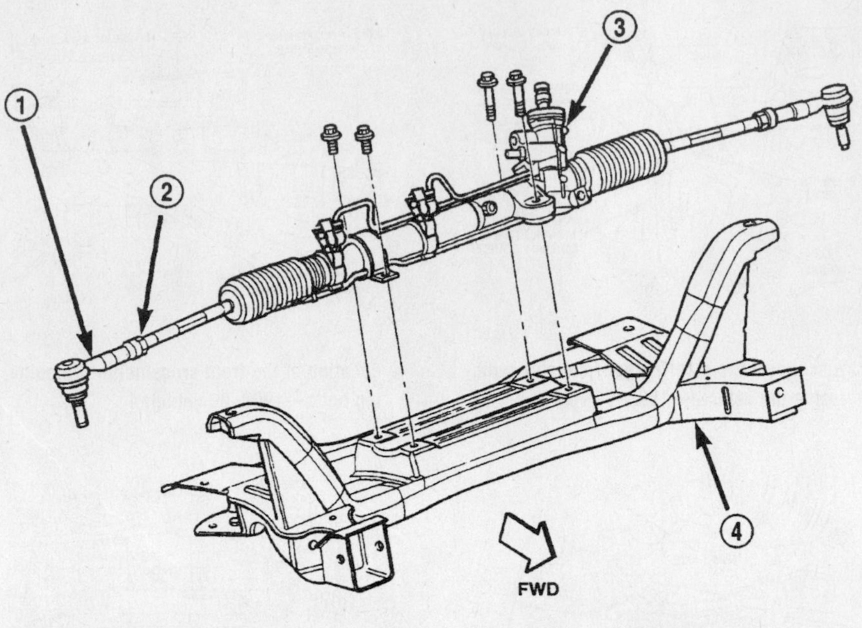

View of the power steering gear and crossmember—2000–02 vehicle

➡Locking the steering wheel keeps the clockspring in alignment position.

3. Working inside the passenger compartment, remove the steering column coupling retainer pin, pinch bolt nut/bolt and separate the couplings.

4. Remove or disconnect the following:

- Both front wheels
- Outer tie rods-to-steering knuckle nuts

➡Remove the tie rod nut by holding the stud securely.

- Outer tie rod ends from the steering knuckle using Remover Tool MB991113
- Tie rod heat shield
- Power steering fluid pressure switch wiring connector by releasing the locking tab
- Power steering fluid pressure hose from the steering gear
- Power steering fluid return hose from the steering gear, if not equipped with a power steering fluid cooler
- Power steering fluid cooler hose from the steering gear, if equipped with a power steering fluid cooler
- Power steering fluid return hose from the routing clip C-clamps, if not equipped with a power steering fluid cooler
- Power steering fluid pressure hose from the steering gear's routing clips

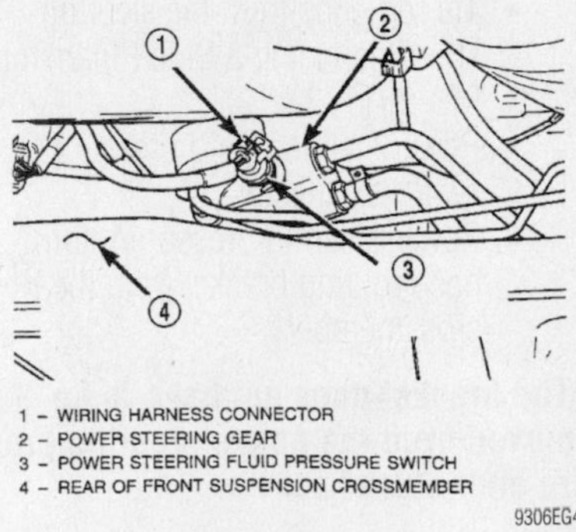

View of the power steering gear fluid pressure switch—2000–02 vehicle

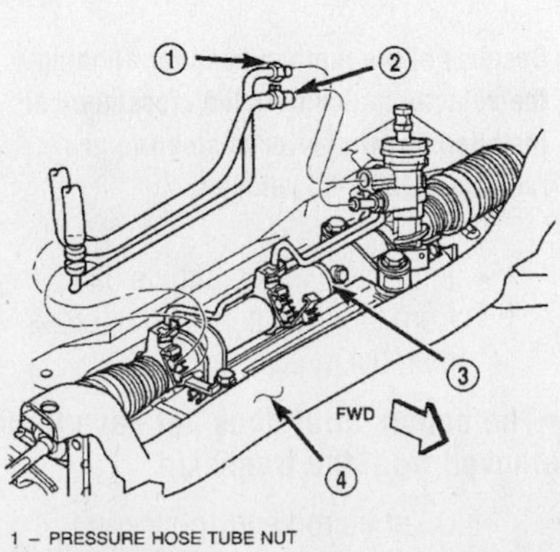

View of the power steering gear hoses and routing clips—2000–02 vehicle

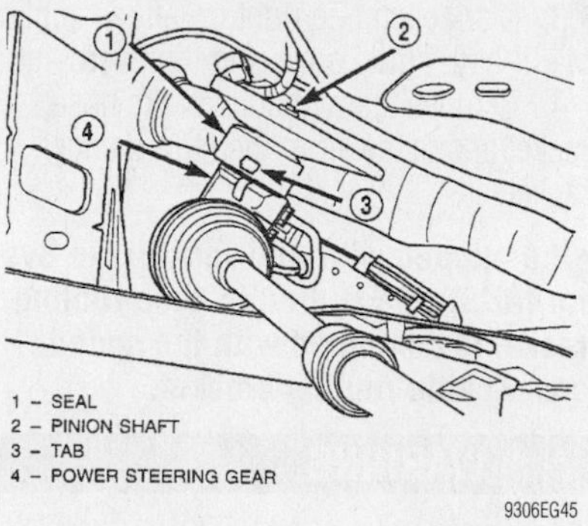

View of the pinion shaft dash cover seal—2000–02 vehicle

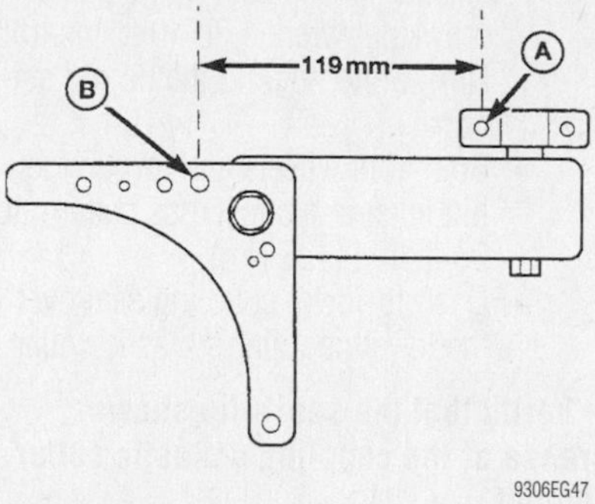

Measuring the engine torque bracket—2000–02 vehicle

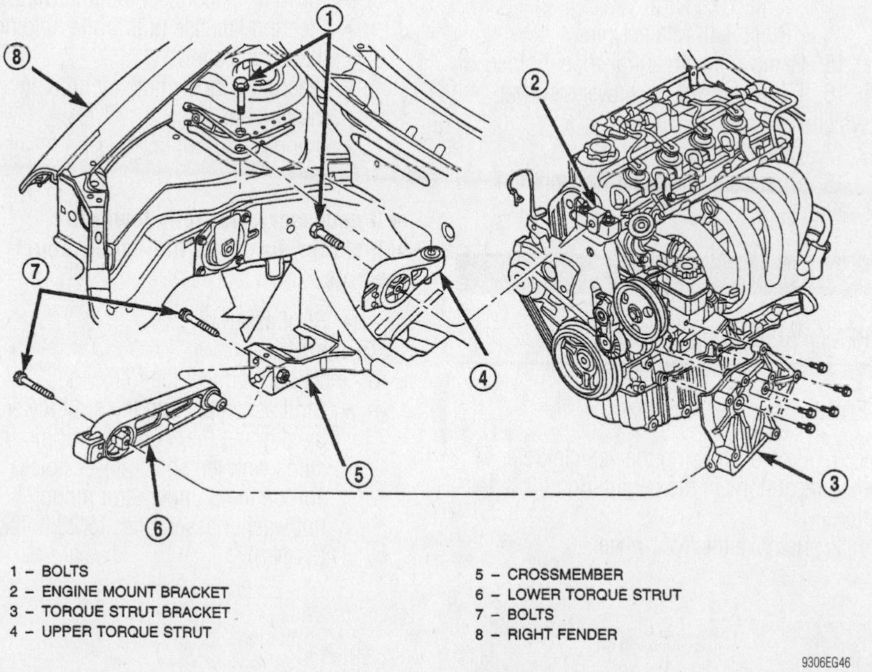

View of the engine torque struts and related components—2000–02 vehicle

- Power steering cooler hose from the steering gear's right routing clip, if equipped
- Both power steering cooler screws from the front suspension crossmember, if equipped, and move the cooler aside

➡The screws are located behind the cooler and can be accessed from above.

- Power steering cooler, if equipped, and move it aside
- Engine torque strut-to-front suspension crossmember bolt from the right forward corner of the crossmember

5. Matchmark the front suspension crossmember-to-chassis location.

⁂ WARNING

If the front suspension crossmember-to-chassis location is not matchmarked, the front wheel alignment setting will be lost.

6. Place a transmission jack under the front crossmember and support it.
7. Remove both front suspension crossmember-to-frame rail bolts, one located at each side.
8. Loosen both rear suspension crossmember-to-frame rail bolts, one located at each side, until they release from the threaded tapping plates in the bolt.

⁂ WARNING

Do not completely remove the rear bolts for they are designed to disengage from the body threads and will stay within the lower control arm rear isolator bushing.

➡The threaded tapping plates allow the lower control arm to stay in place on the crossmember.

9. Using the transmission jack, lower the front suspension crossmember enough to allow the power steering gear to be removed form the rear of the crossmember. Use the jack to support the crossmember's weight.
10. Remove or disconnect the following:
    - Lower steering column coupling-to-power steering gear pinion shaft's roll pin, using a roll pin punch
    - Lower steering column coupling from the power steering column pinion shaft
    - Pinion shaft dash cover seal from the tabs cast into the power steering gear housing
    - Power steering gear from the front suspension crossmember

To install:

11. Install or connect the following:
    - Power steering gear onto the front suspension crossmember and torque the bolts to 45 ft. lbs. (61 Nm)
    - Pinion shaft dash cover seal over the shaft and onto the power steering gear housing

➡Align the seal holes with the tabs cast into the power steering gear housing.

- Lower steering column coupling by aligning the coupling and steering gear pinion shaft flats
- Lower steering column coupling-to-pinion shaft's roll pin until it is centered

12. Center the power steering gear rack's travel.
    - Front suspension crossmember/power steering gear assembly by raising it with the jack until is aligns with its matchmarks
    - Lower steering column coupling, guide it through the dash panel hole as it is raised
    - Both rear crossmember-to-tapping plate bolts
    - Both front crossmember-to-frame rail bolts and torque them to 20 inch lbs. (2 Nm)

WARNING

Be sure to align the front suspension crossmember-to-chassis matchmarks; otherwise, the front wheel alignment setting will be lost.

- Once aligned, torque both rear crossmember-to-rear lower control arm bolts to 150 ft. lbs. (203 Nm) and both front crossmember bolts to 105 ft. lbs. (142 Nm)
- Engine torque strut to the right forward corner of the front suspension crossmember

13. Adjust the engine torque strut by performing the following procedure:
   a. Loosen the upper torque strut at the shock tower bracket.
   b. Position a floor jack on the forward edge of the bell housing to prevent the least amount of upward lifting of the engine.
   c. Slowly, lift the assembly, allowing the engine to rotate rearward so the distance between center of the engine mount bracket's rearmost attaching stud (point A) and the center of the shock tower bracket's washer hose clip hole (point B) is 4.70 in. (119mm).
   d. Torque the upper and lower torque strut bolts to 87 ft. lbs. (118 Nm).
   e. Remove the floor jack.
14. Install or connect the following:
   - Power steering hose-to-power steering gear using a new O-ring lubricated with power steering oil, if not equipped with a power steering cooler
   - Power steering fluid cooler line-to-power steering gear, if equipped with a power steering cooler
   - Power steering fluid return hose to the routing clip C-clamps, if not equipped with a power steering fluid cooler
   - Power steering fluid pressure hose to the steering gear's routing clips
   - Power steering cooler hose to the steering gear's right routing clip, if equipped
   - Torque the power steering pressure hose-to-power steering gear nut to 25 ft. lbs. (34 Nm)
   - Both power steering cooler screws to the front suspension crossmember, if equipped
   - Power steering fluid pressure switch wiring connector be sure the locking tab is secure latched
   - Tie rod heat shield, facing outboard
   - Outer tie rod ends to the steering knuckle. Torque the nut, using a crowsfoot wrench, to 40 ft. lbs. (55 Nm), while holding the tie rod stationary
   - Both front wheels and torque the lug nuts, in a crisscross pattern, to 95 ft. lbs. (128 Nm)
   - Dash-to-lower coupling seal over the lower coupling's plastic collar

➡Verify that the seal's lip shows grease at the coupling's plastic collar contact.

   - Steering column lower coupling-to-steering column upper coupling pinch bolt and torque the nut to 21 ft. lbs. (28 Nm)
   - Pinch bolt retainer pin

15. Remove the steering wheel holder.
16. Fill and bleed the power steering system.
17. Check for leaks.
18. Check and/or adjust the front toe setting.

Strut

REMOVAL & INSTALLATION

Front

1. Before servicing the vehicle, refer to the precautions in the beginning of this section.
2. Remove the front wheels.
3. Mark each one right or left, as applicable, if both struts are being removed.
4. Remove the hydraulic brake hose bracket and screw from the strut damper bracket.

➡If equipped with Anti-lock Brake System (ABS), the hydraulic hose routing bracket is combined with the speed sensor cable routing bracket.

WARNING

The steering knuckle-to-strut assembly attaching bolts are serrated and must not be turned during removal.

5. Remove or disconnect the following:
   - Steering knuckle nuts while holding the bolts stationary
   - Steering knuckle nuts by holding the bolts in place
   - 3 upper strut mount-to-strut tower nuts

➡If necessary, partially lower the vehicle for access to the upper mounting nuts.

   - Strut assembly

To install:

6. Install or connect the following:
   - Strut assembly into the strut tower by aligning the 3 upper strut mount studs with the shock tower holes. Torque the 3 upper strut mount nut/washer assemblies to 23 ft. lbs. (31 Nm).

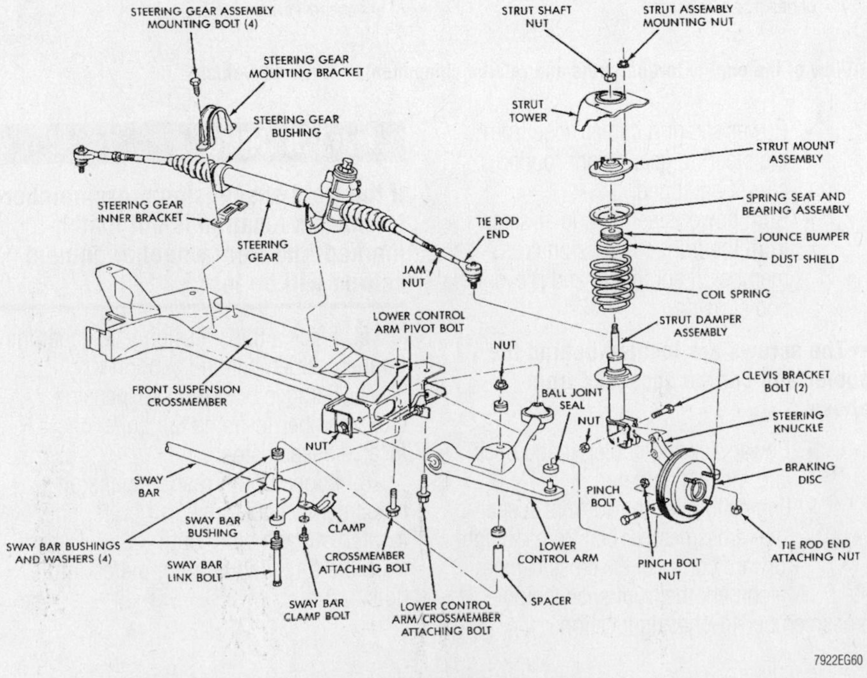

Exploded view of the front suspension—1998–99 models

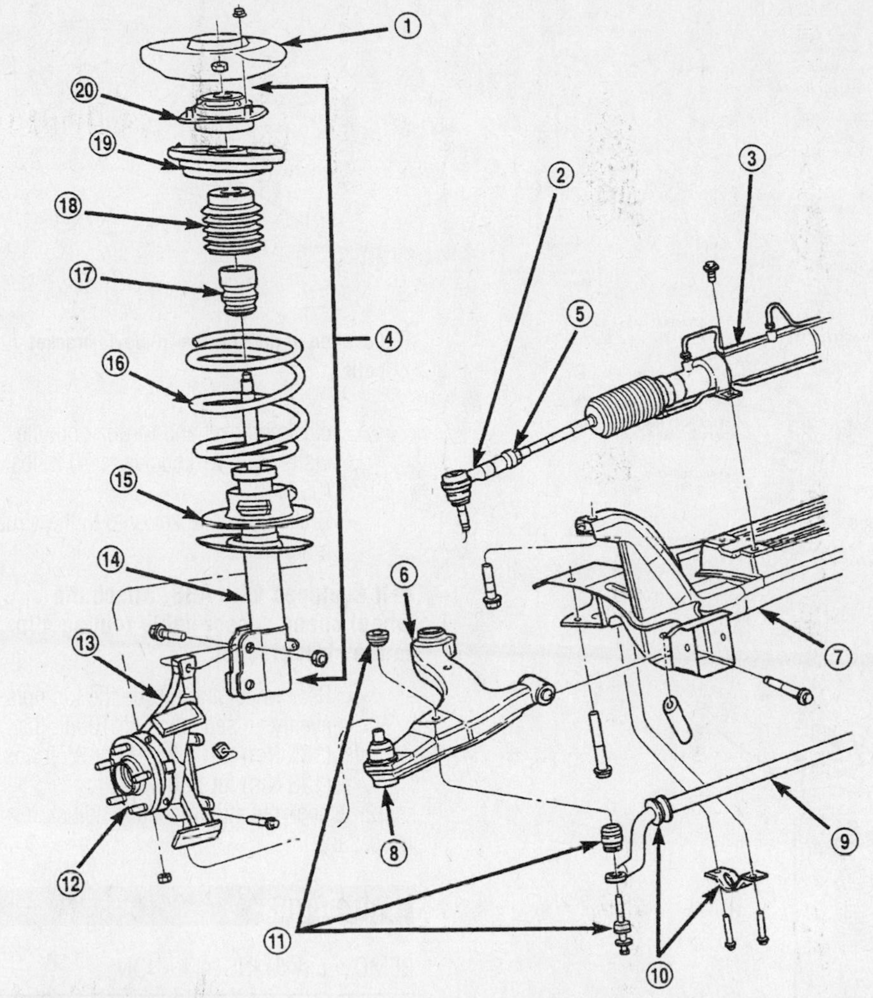

1 – VEHICLE STRUT TOWER
2 – OUTER TIE ROD
3 – STEERING GEAR
4 – STRUT ASSEMBLY
5 – JAM NUT
6 – LOWER CONTROL ARM
7 – CROSSMEMBER
8 – BALL JOINT
9 – STABILIZER BAR
10 – STABILIZER BAR CUSHION AND RETAINER
11 – STABILIZER BAR LINK
12 – HUB
13 – KNUCKLE
14 – STRUT
15 – LOWER SPRING ISOLATOR
16 – COIL SPRING
17 – JOUNCE BUMPER
18 – DUST SHIELD
19 – SPRING SEAT AND BEARING
20 – UPPER MOUNT

9306EG11

Exploded view of the front suspension—2000–02 models

⁂ WARNING

The steering knuckle-to-strut assembly attaching bolts are serrated and must not be turned during installation.

- Steering knuckle nuts while holding the bolts stationary
- Steering knuckle arm and position it into the strut assembly by aligning the strut assembly-to-steering knuckle holes
- Both strut-to-steering knuckle bolts and torque both bolts to 40 ft. lbs. (53 Nm), plus an additional 1/4 turn after the specified torque is met

➡The bolts should be installed with the nuts facing the front of the vehicle.

- Hydraulic brake hose routing bracket and screw onto the strut damper bracket and torque the bracket bolts to 10 ft. lbs. (13 Nm)

➡If equipped with ABS, the hydraulic hose routing bracket is combined with the speed sensor cable routing bracket.

- Front wheels and torque the lug nuts, in a criss-cross pattern, to 100 ft. lbs. (135 Nm) for 1998–99 or 95 ft. lbs. (128 Nm) for 2000–02

Rear

1. Before servicing the vehicle, refer to the precautions in the beginning of this section.
2. Remove the rear wheel.
3. Remove the hydraulic flex hose bracket from the strut bracket.

➡If equipped with Anti-lock Brake System (ABS), the wheel speed sensor cable routing clip is also attached to the strut assembly bracket.

4. Remove the clevis bracket-to-knuckle bolts by supporting the rear knuckle, suspension and brake components.

➡DO NOT allow the weight of the knuckle and related components to hang without support when the strut is removed.

⁂ WARNING

The knuckle-to-strut attaching bolts are serrated and must not be turned during removal. Remove the nuts while holding the bolts stationary in the knuckle.

5. Remove both strut-to-knuckle clevis bracket nuts.

➡Access to the rear upper strut mount-to-strut tower attaching bolts is through the trunk of the vehicle.

6. Remove the carpet from the top of the strut tower, if necessary.
7. Remove the rubber dust shield from the top of the strut tower.
8. Loosen, but do not remove the 4 upper strut mounting nuts.
9. Remove the 4 strut-to-chassis mount nuts, by supporting it.
10. Remove the strut from the knuckle by sliding the knuckle out of the clevis bracket.

To install:

11. Install or connect the following:
    - Strut and torque the 4 strut mount-to-body nuts to 25 ft. lbs. (34 Nm)
    - Dust shield onto the top of the strut tower opening
    - Carpeting on top of the strut tower
    - Knuckle into the strut assemblies

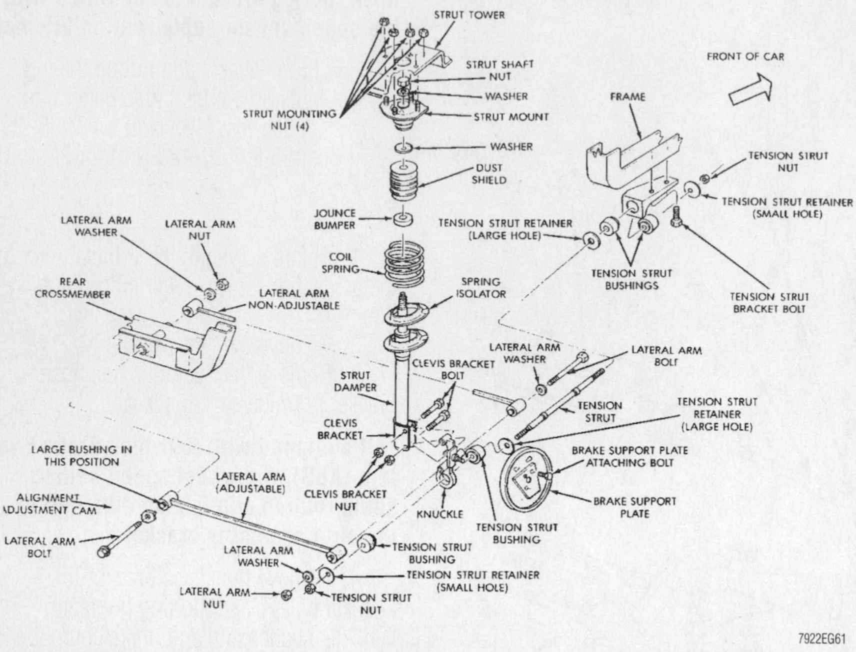

Exploded view of the rear suspension—1998–99

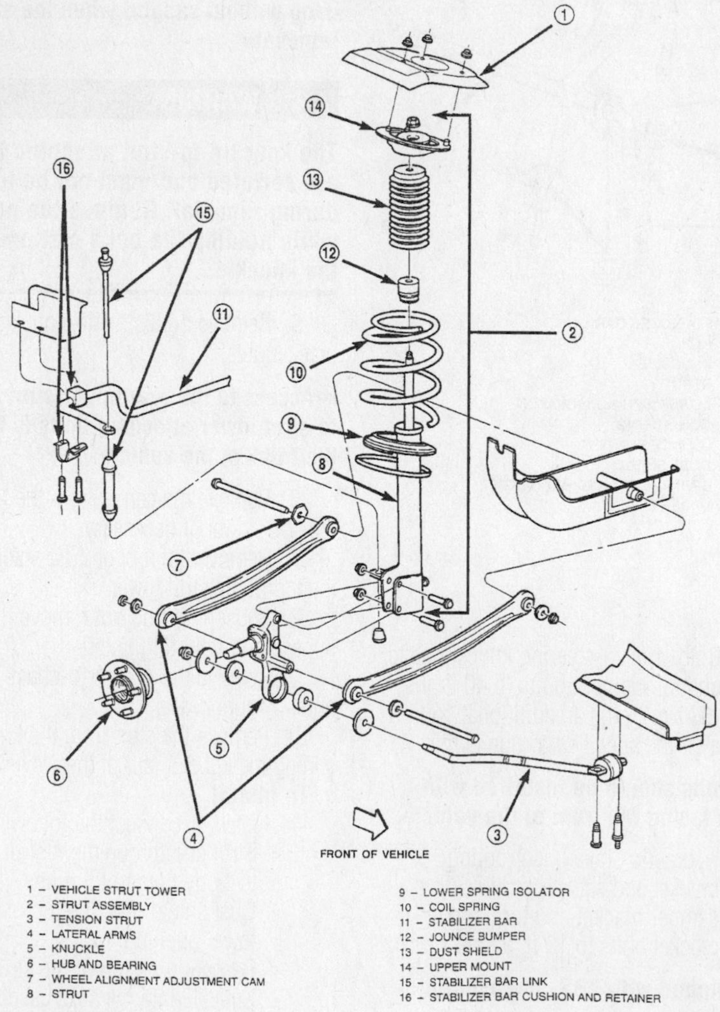

Exploded view of the rear suspension—2000–02

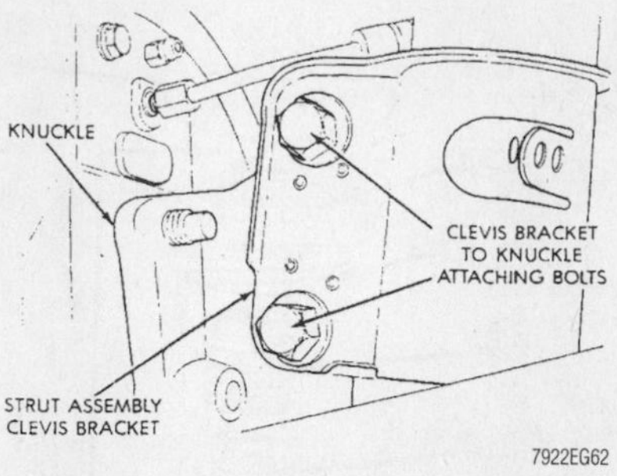

Location of the knuckle-to-clevis bracket bolts

clevis bracket and torque both clevis bracket-to-knuckle to 70 ft. lbs. (95 Nm)

- Brake hose bracket onto to the strut bracket

➡If equipped with ABS, attach the wheel speed sensor cable routing clip to the strut bracket.

- Rear wheel and torque the lug nuts evenly, in sequence, to 100 ft. lbs. (135 Nm) for 1998–99 or 95 ft. lbs. (128 Nm) for 2000–02

12. Check the alignment and adjust, if necessary.

Coil Spring

REMOVAL AND INSTALLATION

1. Before servicing the vehicle, refer to the precautions in the beginning of this section.
2. Remove the strut assembly.
3. Install the strut assembly in a spring compressor.
4. Set the lower hooks first then the upper hooks.
5. Position the strut clevis bracket straight outward away from the compressor
6. Place a clamp on the lower end of the coil spring so the strut is held in place.
7. Compress the spring until all tension is removed from the upper mount.
8. Install a strut nut socket, Special Tool 6864, on the shaft retaining nut.
9. Install socket on the hex end of the strut shaft.
10. Hold the shaft from turning and remove the nut from the shaft.
11. Remove or disconnect the following:
    - Upper mount from the strut shaft
    - Upper spring seat, bearing and upper spring isolator as an assembly
    - Dust shield

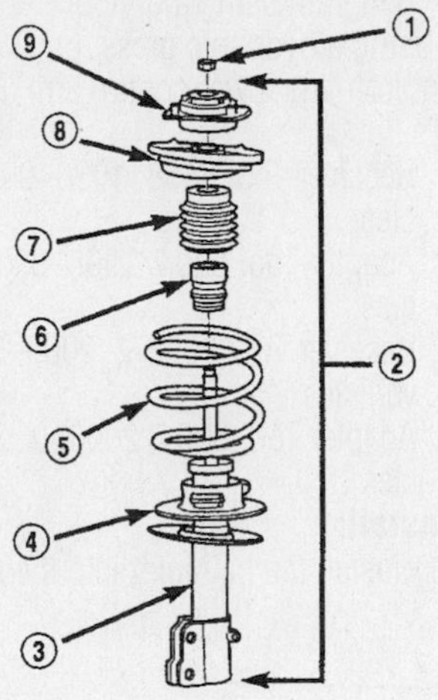

1 - NUT
2 - STRUT ASSEMBLY
3 - STRUT
4 - LOWER SPRING ISOLATOR
5 - COIL SPRING
6 - JOUNCE BUMPER
7 - DUST SHIELD
8 - SPRING SEAT AND BEARING (WITH SPRING ISOLATOR)
9 - UPPER MOUNT

9346EG03

Exploded view of the strut assembly

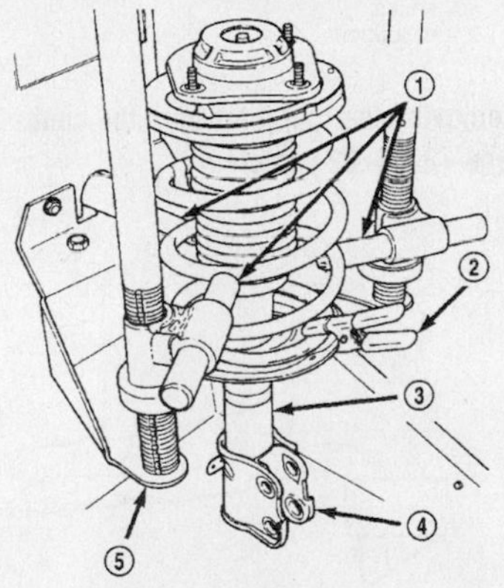

1 - LOWER HOOKS
2 - CLAMP
3 - STRUT ASSEMBLY
4 - CLEVIS BRACKET
5 - SPRING COMPRESSOR

9346EG04

Assemble the coil spring as shown

- Jounce bumper
- Clamp from the bottom of the coil spring
- Strut assembly through the bottom of the coil spring
- Lower spring isolator
- Coil spring by releasing the tension on the compressor drive

To install:

12. Place the coil spring in a compressor and rotate the spring so that the end of the top coil is directly in back.

13. Slowly compress the coil spring until enough room is available for reassembly.

14. Install or connect the following:
    - Lower spring isolator on the lower spring seat
    - Strut through the bottom of the coil spring until the spring seat contacts the lower end of the coil spring

15. Rotate the strut, if necessary, until the clevis bracket is positioned straight outward and away from the compressor.
    - Clamp on the lower end of the coil spring and strut
    - Jounce bumper on the strut shaft with the small end facing down
    - Dust shield on the strut shaft
    - Upper spring isolator on the upper spring seat and bearing
    - Upper spring seat and bearing on the top of the coil spring
    - Strut upper mount over the strut shaft and on top of the upper spring seat and bearing
    - Retaining nut on the strut shaft, loosely
    - Strut nut socket Tool 6864 on the retaining nut
    - Socket on the hex end of the shaft and torque the nut to 55 ft. lbs. (75 Nm)

16. Slowly release the tension from the from the coil spring and make certain that the coil spring is properly aligned.

17. Remove the clamp from the lower end of the coil spring and strut.

18. Remove the upper and lower hooks from the compressor tool and remove the tool.

19. Install the strut assembly.

Tension Strut

REMOVAL & INSTALLATION

1998–99 Models

1. Before servicing the vehicle, refer to the precautions in the beginning of this section.
2. Remove or disconnect the following:
    - Rear wheels
    - Front tension strut-to-knuckle nut by holding the strut with a wrench
    - Forward tension strut retainer
    - Rear tension strut-to-knuckle nut by holding the strut with a wrench
    - Rear tension strut retainer
    - Tension strut

To install:

3. Install or connect the following:
    - Tension strut
    - Rear tension strut bushings making sure the stepped area faces toward the bracket
    - Rear tension strut retainer

✻✻ WARNING

When installing the tension strut retainers, be sure to face the cupped surfaces away from the bracket.

- Rear tension strut-to-knuckle nut by holding the strut with a wrench and torque the nut to 70 ft. lbs. (95 Nm)
- Forward tension strut bushings making sure that the stepped area faces toward the knuckle
- Forward tension strut retainer

✻✻ WARNING

When installing the tension strut retainers, be sure to face the cupped surfaces away from the knuckle.

- Forward tension strut-to-knuckle nut by holding the strut with a wrench and torque the nut to 70 ft. lbs. (95 Nm)
- Rear wheels and torque the lug nuts to 100 ft. lbs. (135 Nm).

4. Check and/or align the rear wheels.

2000–02 Models

1. Remove or disconnect the following:
    - Rear wheels
    - Tension strut-to-knuckle nut by holding the strut with a wrench
    - Forward tension strut retainer
    - Rear tension strut bayonet bushing
    - Parking brake cable from tension strut bolt
    - Tension strut from the chassis

To install:

2. Install or connect the following:
    - Tension strut to the chassis and torque the bolts to 70 ft. lbs. (95 Nm)
    - Parking brake cable to tension strut nut and torque the bolts to 21 ft. lbs. (28 Nm)

➡The mounting bolt with the stud on the head is installed on the inboard side.

- Rear tension strut bayonet bushing making sure that the stepped area faces toward the knuckle

- Tension strut retainer
- Tension strut-to-knuckle nut by holding the strut with a wrench and torque the nut to 70 ft. lbs. (95 Nm)
- Rear wheels and torque the lug nuts to 100 ft. lbs. (135 Nm)

3. Check and/or align the rear wheels.

Lower Ball Joint

REMOVAL & INSTALLATION

The front suspension ball joints operate with no free-play. The ball joints are replaceable ONLY as an assembly. Do not attempt any type of repair on the ball joint assembly. The ball joint is a press fit into the lower control arm with the joint stud retained in the steering knuckle by the clamp bolt. To check the ball joint, with the weight of the vehicle resting on the road wheels, grasp the grease fitting and without using any tools, attempt to move the grease fitting. If the ball joint is worn the grease fitting will move easily. If movement is noted, replacement of the ball joint is recommended.

1. Before servicing the vehicle, refer to the precautions in the beginning of this section.
2. Remove or disconnect the following:
   - Wheel
   - Steering knuckle-to-ball joint stud's pinch bolt and nut
   - Stabilizer bar-to-lower control arm links
3. Loosen, but do not remove the bolts holding the stabilizer bar retainers to the crossmember. Then, rotate the stabilizer bar and attaching links away from the lower control arms.

✲✲ WARNING

Pulling the steering knuckle outward after releasing the ball joint can separate the inner CV-joint.

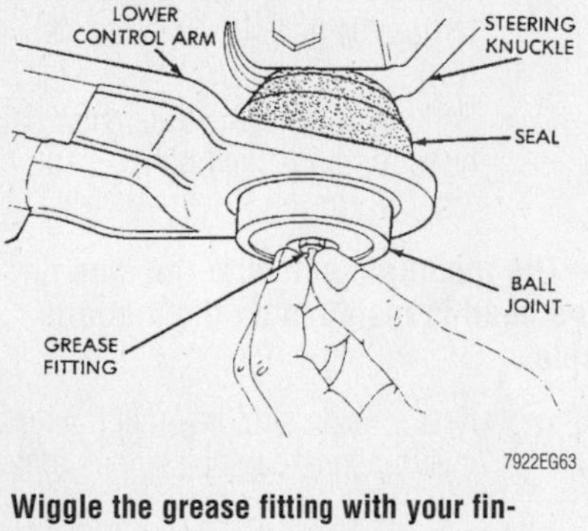

Wiggle the grease fitting with your fingers—if it moves, the ball joint should be replaced

4. Remove or disconnect the following:
   - Ball joint from the steering knuckle using a prybar

✲✲ WARNING

Be careful when separating the ball joint stud from the knuckle, so the seal does not become damaged.

   - Front lower control arm bushing-to-crossmember nut and bolt
   - Rear lower control arm-to-crossmember bolt
   - Lower control arm

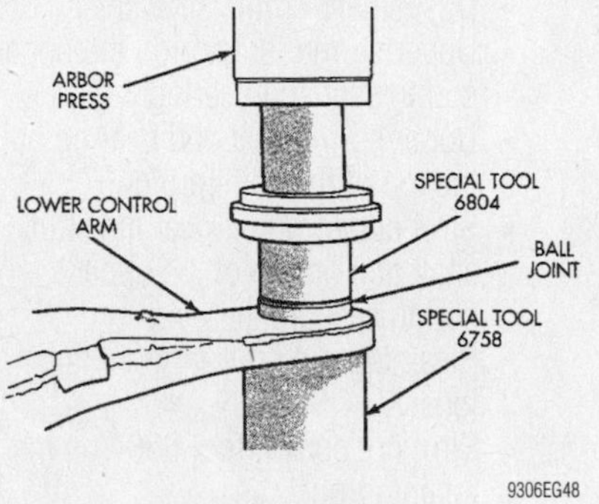

Removing the ball joint from the control arm—1998–99 vehicles

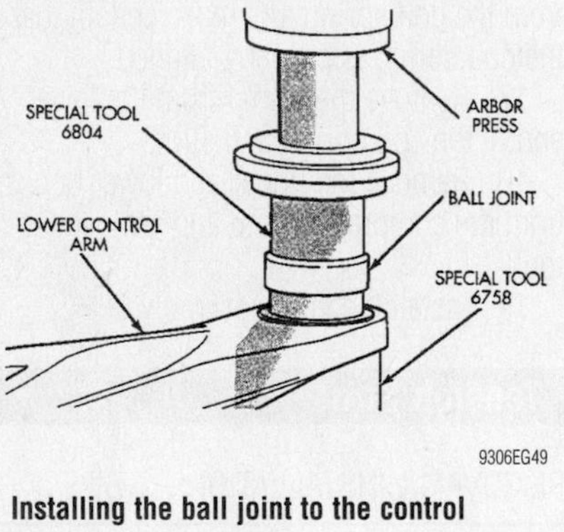

Installing the ball joint to the control arm—1998–99 vehicles

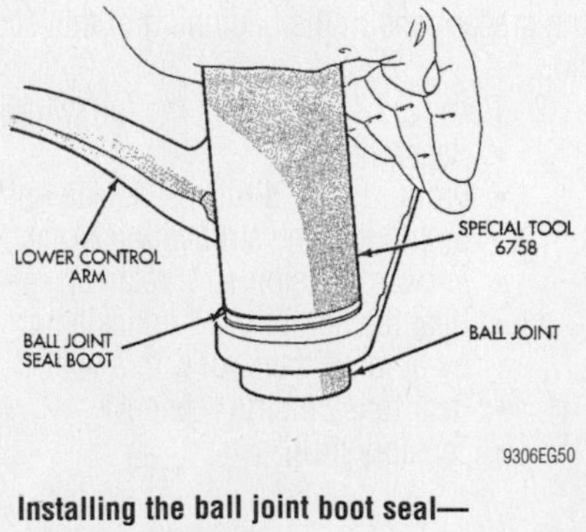

Installing the ball joint boot seal—1998–99 vehicles

   - Ball joint using a prytool

5. Using a hydraulic press, press the ball joint from the lower control arm using tools:
   - Receiver Tool 6758: 1998–99 vehicles
   - Adapter Tool 6804: 1998–99 vehicles
   - Receiver Tool 6908-2: 2000–02 vehicles
   - Adapter Tool 6804: 2000–02 vehicles

To install:

6. Reinstall the ball joint into the lower

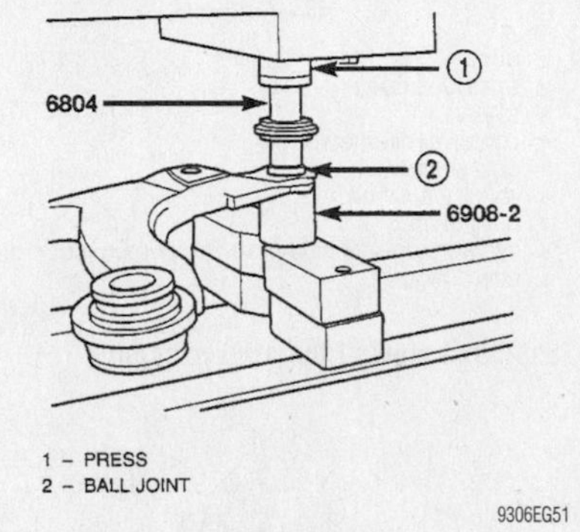

Removing the ball joint from the control arm—2000–02 vehicles

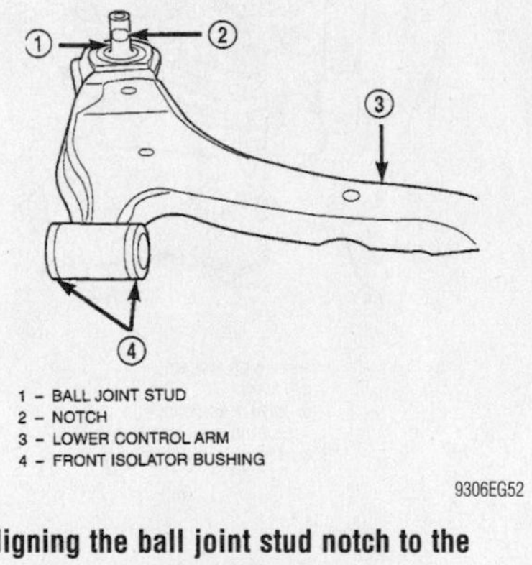

Aligning the ball joint stud notch to the control arm—2000–02 vehicles

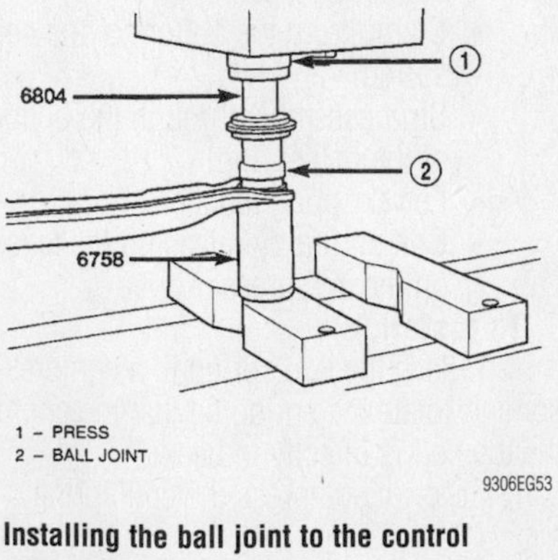

Installing the ball joint to the control arm—2000–02 vehicles

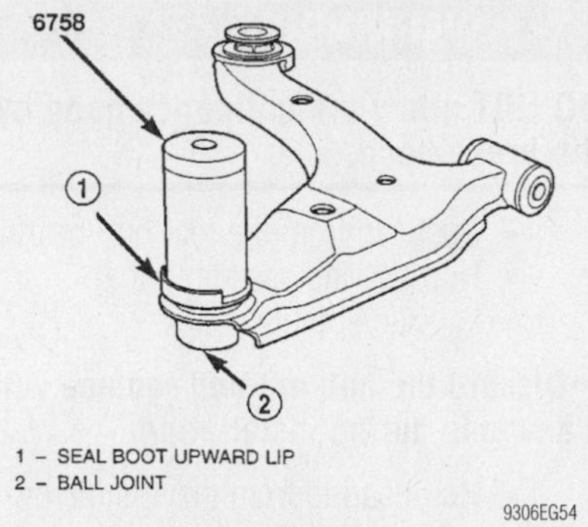

Installing the ball joint boot seal—2000–02 vehicles

control arm with the notch in the ball joint stud facing the front lower control arm bushing.

7. Using a hydraulic press, press the ball joint into the lower control arm using tools:
   a. Receiver Tool 6758 and Adapter Tool 6804.
8. Install or connect the following:
   - Ball joint boot seal using a driver tool such as a large socket or suitable sized piece of pipe

✲✲ WARNING

Do not use a shop press that was used to install the ball joint, for the press exerts too much force.

   - Lower control arm into the front crossmember
   - Rear lower control arm-to-crossmember and frame rail bolt

➡DO NOT tighten the rear bolt at this time.

   - Front lower control arm-to-crossmember nut and bolt
9. Torque the lower control arm fasteners to:
   a. Front control arm nut/bolt: 120 ft. lbs. (163 Nm) for 1998–99 models.
   b. Rear control arm nut/bolt: 120 ft. lbs. (163 Nm) for 1998–99 models.
   c. Rear pivot bolt: 150 ft. lbs. (203 Nm) for 2000–02 models.
   d. Front pivot bolt: 120 ft. lbs. (163 Nm) for 2000–02 models.
10. Install the ball joint stud into the steering knuckle. Torque the steering knuckle-to-ball joint stud pinch bolt and nut to 70 ft. lbs. (95 Nm).
11. Assemble the stabilizer bar-to-lower control arm link assemblies and bushings.
12. Rotate the stabilizer bar into position, installing the stabilizer bar links into the lower control arms. Install the top stabilizer bar link bushings and nuts. DO NOT tighten the link yet.
13. Install the wheel.
14. Lower the vehicle so the suspension is supporting the total weight of the vehicle.
15. Torque the stabilizer bar-to-lower control arm links to 21 ft. lbs. (28 Nm) for 1998 models and 17 ft. lbs. (23 Nm) for 1999–02 models.
16. Torque the stabilizer bar bushing retainer-to-crossmember bolts to 21 ft. lbs. (28 Nm).
17. Check and/or adjust the toe, as necessary.

Wheel Bearings

ADJUSTMENT

Neons are equipped with sealed hub and bearing assemblies. The hub and bearing assembly is non-serviceable. If the assembly is damaged, the complete unit must be replaced.

REMOVAL & INSTALLATION

Front

1. Before servicing the vehicle, refer to the precautions in the beginning of this section.
2. Remove the steering knuckle and hub and bearing assembly.
3. Remove a wheel lug stud from the hub flange using a C-clamp and Adapter Tool 4150A.
4. Rotate the hub to align the removed lug stud with the notch in the bearing retainer plate.
5. Rotate the hub so the stud hole is facing away from the brake caliper's lower rail on the steering knuckle.

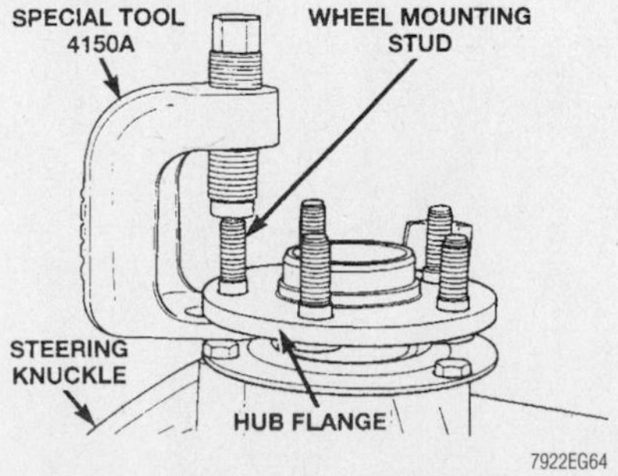

Use a proper C-clamp and adapter tool to press out one of the lug studs

6. Install ½ of a Bearing Splitter Tool 1130, between the hub and the bearing retainer plate. The threaded hole in this ½ is to be aligned with the caliper rail on the steering knuckle.
7. Install the remaining pieces of the bearing splitter on the steering knuckle. Hand-tighten the nuts to hold the splitter in place on the knuckle.
8. When the bearing splitter is installed, be sure the 3 bolts attaching the bearing retainer plate to the knuckle are contacting the bearing splitter. The bearing retainer plate should not support the knuckle or contact the splitter.

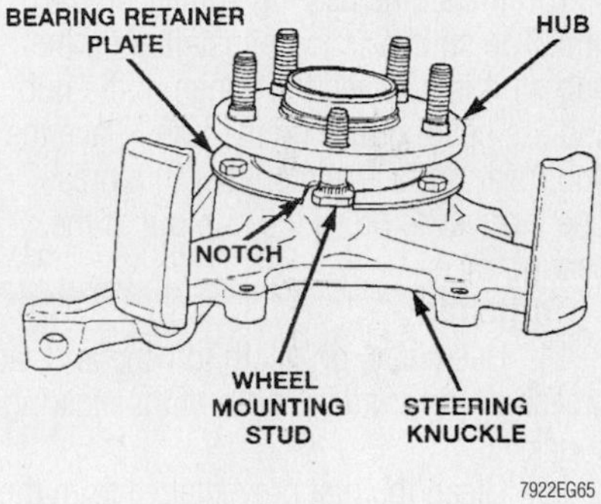

Rotate the hub in order to remove the lug stud

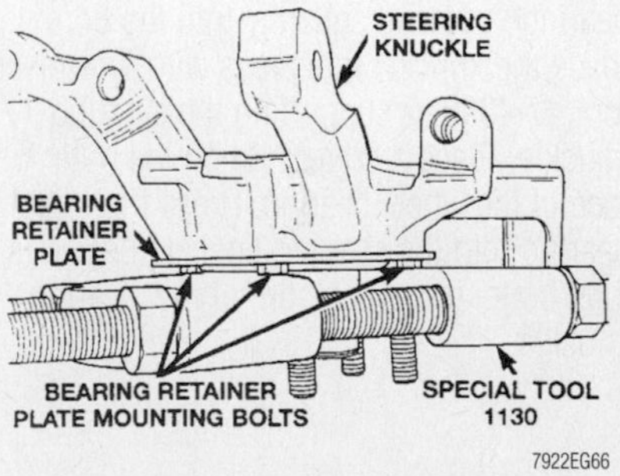

Proper installation of the bearing splitter

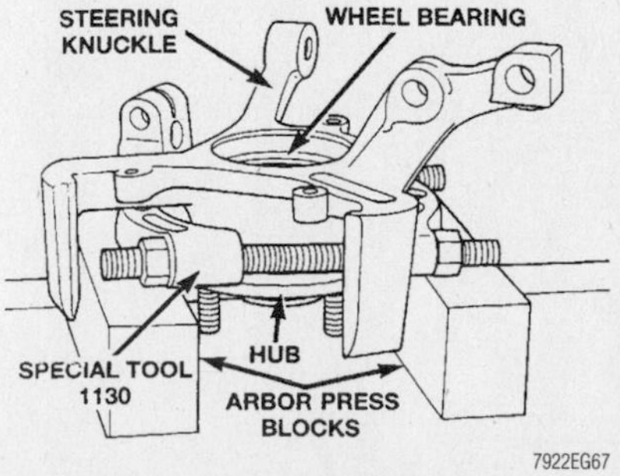

Properly support the steering knuckle for hub and bearing removal

9. Place the steering knuckle in a hydraulic press, supported by the bearing splitter.

10. Position a driver on the small end of the hub. Using the press, remove the hub from the wheel bearing. The outer bearing race will come out of the wheel bearing when the hub is pressed out of the bearing.

11. Remove the bearing splitter tool from the knuckle.

12. Place the knuckle in a press supported by the press block. The blocks must not obstruct the bore in the steering knuckle so the wheel bearing can be pressed out of the knuckle. Place a driver on the outer race of the wheel bearing, then press the bearing out of the knuckle.

13. Install the bearing splitter on the hub. The splitter is to be installed on the hub so it is between the flange of the hub and the bearing race on the hub. Place the hub, bearing race and splitter in a press. Use a driver to press the hub out of the bearing race.

To install:

14. Use clean, dry cloth to wipe and grease or dirt from the bore of the steering knuckle.

15. Clean the rust preventative from the replacement wheel bearing using a clean, dry towel.

16. Place the new wheel bearing into the bore of the steering knuckle. Be sure the bearing is placed squarely into the bore. Place the knuckle in a press with a receiver tool, C-4698–2 supporting the steering knuckle. Place a driver tool on the outer race of the wheel bearing. Press the wheel bearing into the steering knuckle until it is fully bottomed in the bore of the steering knuckle.

➡Only the original or original equipment replacement bolts should be used to mounting the bearing retainer to the knuckle. If a bolt requires replacement when installing the bearing retainer plate, be sure to get the proper type of replacement.

17. Install the bearing retainer plate on the steering knuckle. Install the 3 bearing retainer mounting bolts. Tighten the bolts to 21 ft. lbs. (28 Nm).

18. Install the removed wheel lug stud into the hub flange.

19. Place the hub with the lug stud installed, in a press supported by Adapter Tool C-4698–1. Press the wheel lug stud into the hub flange until it is fully seated against the back side on the hub flange.

20. Place the steering knuckle with the wheel bearing installed, in a press with special Receiver Tool MB-990799 supporting the inner race of the wheel bearing. Place the hub in the wheel bearing, making sure it is square with the bearing. Press the hub into the wheel bearing until it is fully bottomed in the wheel bearing.

21. Install the steering knuckle and the wheel.

22. Check and/or adjust the front alignment.

Rear

1. Before servicing the vehicle, refer to the precautions in the beginning of this section.
2. Remove or disconnect the following:
   - Wheel
   - Rear brake drum, if equipped
   - Caliper (suspend on a wire) and the rotor, if equipped with rear disc brakes

⁂ WARNING

DO NOT allow the caliper to hang by the brake hose.

- Dust cap from the rear hub/bearing
- Hub/bearing assembly-to-knuckle/spindle nut

➡Discard the hub nut and replace with a new one during installation.

- Hub/bearing from the spindle by pulling it off the end of the spindle by hand

To install:

⁂ WARNING

The hub/bearing nut must be tightened to, but NOT over, its specified torque value. The proper specification is crucial to the life of the hub bearing.

3. Position the hub/bearing assembly on the rear spindle/knuckle. Install a NEW hub nut and tighten to 160 ft. lbs. (217 Nm).
4. Install or connect the following:
   - Dust cap and seat it using a soft face hammer to carefully tap it into place
   - Brake drum, if equipped with drum brakes
   - Rotor, if equipped with disc brakes
   - Caliper and 2 guide pin bolts, if equipped with disc brakes. Tighten the bolts to 16 ft. lbs. (22 Nm).
5. Install the wheel and tire assembly. Tighten the lug nuts in a crisscross pattern, to 100 ft. lbs. (135 Nm) for 1998–99 or 95 ft. lbs. (128 Nm) for 2000–02.

CHRYSLER CORP.

9

1998–02

Chrysler-Cirrus • Sebring Convertible • **Dodge-**Stratus • **Plymouth-**Breeze

PRECAUTIONS

Before servicing any vehicle, please be sure to read all of the following precautions. The following precautions deal with personal safety, preventing of component damage, and important points to take into consideration when servicing a motor vehicle:

- Never open, service or drain the radiator or cooling system when the engine is hot; serious burns can occur from the steam and hot coolant.
- Observe all applicable safety precautions when working around fuel. Whenever servicing the fuel system, always work in a well-ventilated area. Do not allow fuel spray or vapors to come in contact with a spark, open flame or excessive heat (a hot drop light, for example). Keep a dry chemical fire extinguisher near the work area. Always keep fuel in a container specifically designed for fuel storage; also, always properly seal fuel containers to avoid the possibility of fire or explosion. Refer to the additional fuel system precautions later in this section.
- Fuel injection systems often remain pressurized, even after the engine has been turned **OFF**. The fuel system pressure must be relieved before disconnecting any fuel lines. Failure to do so may result in fire and/or personal injury.
- Brake fluid often contains polyglycol ethers and polyglycols. Avoid contact with the eyes and wash your hands thoroughly after handling brake fluid. If you do get brake fluid in your eyes, flush your eyes with clean, running water for 15 minutes. If eye irritation persists, or if you have taken brake fluid internally, IMMEDIATELY seek medical assistance.
- The EPA warns that prolonged contact with used engine oil may cause a number of skin disorders, including cancer! You should make every effort to minimize your exposure to used engine oil. Protective gloves should be worn when changing the oil. Wash your hands and any other exposed skin areas as soon as possible after exposure to used engine oil. Soap and water, or waterless hand cleaner should be used.
- All vehicles are equipped with an air bag system, often referred to as a Supplemental Restraint System (SRS) or as a Supplemental Inflatable Restraint (SIR) system. The system must be disabled before performing service on or around system components, steering column, instrument panel components, wiring and sensors. Failure to follow safety and disabling procedures could result in accidental air bag deployment, possible personal injury and unnecessary system repairs.
- Always wear safety goggles when working with, or around, the air bag system. When carrying a non-deployed air bag, be sure the bag and trim cover are pointed away from your body. When placing a non-deployed air bag on a work surface, always face the bag and trim cover upward, away from the surface. This will reduce the motion of the module if it is accidentally deployed.
- Clean, high quality brake fluid from a sealed container is essential for the safe and proper operation of the brake system. You should always buy the grade of fluid recommended for your vehicle. If the brake fluid becomes contaminated, drain and flush the system, then refill the master cylinder with new fluid. Never reuse any brake fluid. Any brake fluid that is removed from the system should be discarded. Also, do not allow any brake fluid to come in contact with a painted surface; it will damage the paint.
- Never operate the engine without the proper amount and type of engine oil; doing so WILL result in severe engine damage.
- Timing belt maintenance is extremely important! Many models may utilize an interference-type, non-free-wheeling engine. If the timing belt breaks, the valves in the cylinder head may strike the pistons, causing potentially serious (also time-consuming and expensive) engine damage. Refer to the maintenance interval charts in the front of this manual for the recommended replacement interval for the timing belt, and to the timing belt section for belt replacement and inspection.
- Disconnecting the negative battery cable on some vehicles may interfere with the functions of the on board computer system(s) and may require the computer to undergo a relearning process once the negative battery cable is reconnected.
- When servicing drum brakes, only disassemble and assemble one side at a time, leaving the remaining side intact for reference.
- Only an MVAC-trained, EPA-certified, automotive technician should service the air conditioning system or its components.

ENGINE REPAIR

Distributor

REMOVAL

2.5L Engine

The 2.5L engine is equipped with a camshaft driven mechanical distributor. This engine uses a fixed ignition timing system. The basic ignition timing is not adjustable. The Powertrain Control Module (PCM) determines spark advance. The Crankshaft Position (CKP) sensor and Camshaft Position (CMP) sensor are Hall Effect devices. The CKP sensor is mounted remotely from the distributor while the CMP sensor is mounted inside the distributor housing. Both sensors generate pulses that are inputs to the PCM. The PCM determines crankshaft position from these sensors. The PCM calculates injector sequence and ignition timing. There is a resistor built into the distributor cap. An ohmmeter connected between the center button and ignition coil terminal should read 5000 ohms.

1. Before servicing the vehicle, refer to the precautions in the beginning of this section.
2. Remove or disconnect the following:
   - Negative battery cable from the left shock tower

➡The ground cable is equipped with an insulator grommet which should be placed on the stud to prevent the negative battery cable from accidentally grounding.

   - Air inlet resonator-to-intake manifold bolt
   - Air cleaner cover-to-air cleaner housing clamp
   - Positive Crankcase Ventilation (PCV) air hose from the air inlet tube
   - Exhaust Gas Recirculation (EGR) tube
   - Spark plug wires from the distributor cap
   - Distributor cap

➡Mark the rotor position. A scribe mark indicates where to position the rotor when reinstalling the distributor.

   - Rotor
   - Both electrical connectors from the distributor

- Distributor hold-down nuts and washers
- Spark plug cable mounting bracket
- Transaxle dipstick tube
- Distributor

INSTALLATION

Timing Not Disturbed

1. Inspect the rotor for cracks or a burned electrode. Replace if defective.
2. Install or connect the following:
   - Rotor onto the distributor
   - New O-ring on the distributor
3. Carefully, engage the distributor drive with the slotted end of the camshaft. When the distributor is installed properly, the rotor will be aligned with the previously made mark.
4. Verify proper rotor alignment with mark made at disassembly.
5. Install or connect the following:
   - Distributor hold-down nuts and washers. Torque the nuts to 108 inch lbs. (13 Nm).
   - Spark plug cable bracket
   - Both distributor wiring connectors
   - Distributor cap
   - Spark plug cables
   - Transaxle dipstick tube
   - EGR tube. Torque the bolts to 95 inch lbs. (11 Nm).
   - PCV hose
   - Air cleaner
   - Air inlet resonator
   - Negative battery cable

Timing Disturbed

1. Rotate the crankshaft until No. 1 piston is at Top Dead Center (TDC) of its compression stroke.
2. Rotate the rotor to the No. 1 terminal position on the distributor cap.
3. Lower the distributor into place, engaging the distributor drive with the drive on the camshaft. With the distributor fully seated on the engine, the rotor should be under the No. 1 terminal.
4. Install or connect the following:
   - Distributor hold-down nuts and washers. Torque the nuts to 108 inch lbs. (13 Nm).
   - Spark plug cable bracket
   - Both distributor wiring connectors
   - Distributor cap
   - Spark plug cables
   - Transaxle dipstick tube
   - EGR tube. Torque the bolts to 95 inch lbs. (11 Nm).
   - PCV hose
   - Air cleaner
   - Air inlet resonator
   - Negative battery cable

Alternator

REMOVAL

2.0L Engine

1. Before servicing the vehicle, refer to the precautions in the beginning of this section.
2. Remove or disconnect the following:
   - Negative battery cable from the shock tower
   - Belt cover
   - Alternator electrical connectors
   - Alternator adjusting and pivot bolts, loosen them
   - Alternator drive belt
   - Alternator adjusting and pivot bolts, do not drop the pivot bolt's spacer
   - Alternator, by moving it toward the head light bucket area

2.4L Engine (VIN G)

1. Before servicing the vehicle, refer to the precautions in the beginning of this section.
2. Remove or disconnect the following:
   - Negative battery cable from the shock tower
   - Alternator electrical connectors
   - Alternator adjusting and pivot bolts, loosen them
   - Alternator drive belt
   - Alternator adjusting and pivot bolts
   - Anti-lock Brake System (ABS) braking unit, by removing the 2 lower plate mounting bolts
   - Coolant overflow bottle
   - Alternator, by moving it under the refrigerant lines toward the passenger's side

2.4L Engine (VIN X)

1. Before servicing the vehicle, refer to the precautions in the beginning of this section.
2. Remove or disconnect the following:
   - Negative battery cable from the shock tower
   - Accessory belt cover
   - Alternator electrical connectors
   - Accessory belt splash shield
   - Accessory belt
   - Manifold Absolute Pressure (MAP) sensor from the intake manifold
   - Alternator

2.5L Engine

1. Before servicing the vehicle, refer to the precautions in the beginning of this section.
2. Remove or disconnect the following:
   - Negative battery cable from the shock tower
   - Alternator electrical connectors
   - Upper mounting ear, pivot and idler adjusting bolts, loosen them

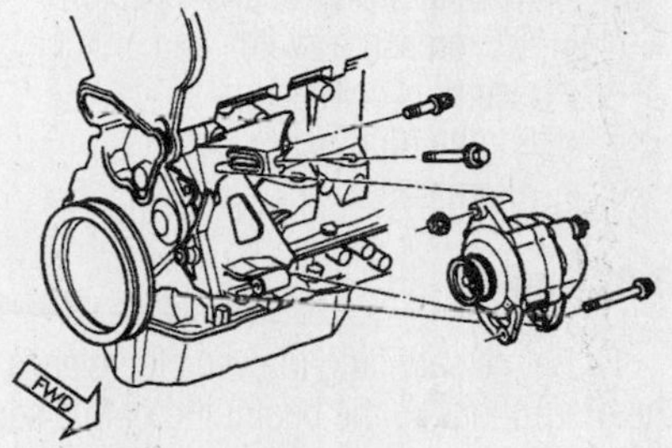

9306FG04

Exploded view of the alternator and related components—2.0L engine

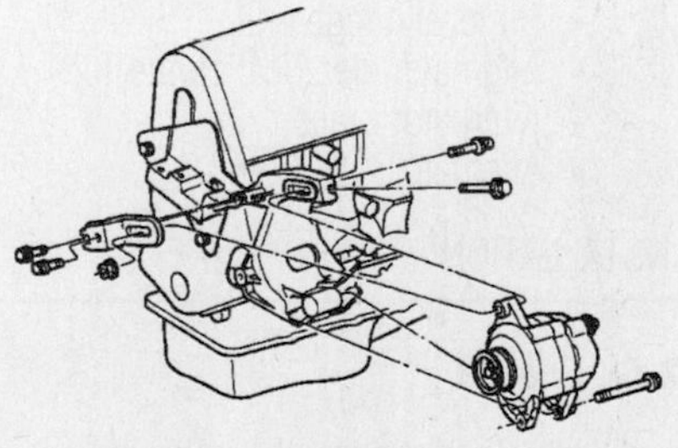

9306FG05

Exploded view of the alternator and related components—2.4L engine

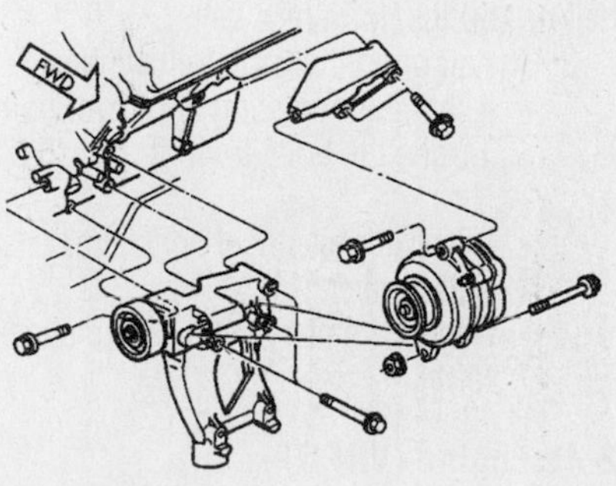

9306FG06

Exploded view of the alternator and related components—2.5L engine

- Alternator drive belt
- Alternator pivot bolt, do not drop the spacer
- Upper mounting ear bolt
- Upper alternator bracket
- Alternator

2.7L Engine

1. Before servicing the vehicle, refer to the precautions in the beginning of this section.
2. Remove or disconnect the following:
   - Negative battery cable from the shock tower
   - Accessory belt splash shield and loosen the belt
   - Alternator electrical connectors
   - A/C pressure switch and clutch electrical connectors
   - Engine oil dipstick
   - Alternator

3.0L Engine

1. Before servicing the vehicle, refer to the precautions in the beginning of this section.
2. Remove or disconnect the following:
   - Negative battery cable
   - Drive belt
   - Alternator belt
   - Oil level gauge unit
   - Alternator electrical connectors
   - Alternator brace
   - Alternator

INSTALLATION

2.0L Engine

1. Install or connect the following:
   - Alternator
   - Alternator drive belt
2. Using a Belt Tension Gauge, adjust the drive belt tension to 80 lbs. for a used belt or 150 lbs. for a new belt.
3. Install or connect the following:
   - Alternator adjusting and pivot bolts. Torque the bolts to 40 ft. lbs. (54 Nm).
   - Alternator electrical connectors
   - Belt cover
   - Negative battery cable to the shock tower

2.4L Engine (VIN G)

1. Install or connect the following:
   - Alternator
   - Coolant overflow bottle
   - ABS)braking unit. Torque the 2 lower plate mounting bolts to 21 ft. lbs. (28 Nm)
   - Alternator drive belt
2. Using a Belt Tension Gauge, adjust the drive belt tension to 80 lbs. for a used belt or 150 lbs. for a new belt.
3. Install or connect the following:
   - Alternator adjusting and pivot bolts. Torque the bolts to 40 ft. lbs. (54 Nm).
   - Alternator electrical connectors
   - Negative battery cable to the shock tower

2.4L Engine (VIN X)

1. Install or connect the following:
   - Alternator and torque the bolts to 33 ft. lbs. (44 Nm)
   - MAP sensor
   - Drive belt and splash shield
   - Alternator electrical connectors
   - Accessory belt cover
   - Negative battery cable

2.5L Engine

1. Install or connect the following:
   - Alternator and torque the bolts to 25 ft. lbs. (33 Nm)
   - Upper alternator bracket
   - Upper mounting ear and pivot bolt, do not tighten
   - Alternator drive belt
2. Using a Belt Tension Gauge, adjust the drive belt tension to 80 lbs. for a used belt or 150 lbs. for a new belt.
3. Install or connect the following:
   - Upper mounting ear and pivot bolts. Torque the bolts to 40 ft. lbs. (54 Nm).
   - Alternator electrical connectors
   - Negative battery cable to the shock tower

2.7L Engine

1. Install or connect the following:
   - Alternator and torque the bolts to 40 ft. lbs. (54 Nm)
   - Engine oil dipstick tube
   - A/C pressure switch and clutch electrical connectors
   - Alternator electrical connectors
   - Drive belt and splash shield
   - Negative battery cable

3.0L Engine

1. Install or connect the following:
   - Alternator and torque the bolts to 33 ft. lbs. (44 Nm)
   - Alternator brace and torque the bolts to 36 ft. lbs. (49 Nm)
   - Oil level gauge unit and torque the bolt to 10 ft. lbs. (15 Nm)
   - Alternator belt
   - Drive belt
   - Negative battery cable

Ignition Timing

ADJUSTMENT

These engines use a fixed ignition system. The Powertrain Control Module (PCM) regulates the ignition timing. Basic ignition timing is not adjustable.

Engine Assembly

REMOVAL & INSTALLATION

2.0L, 2.4L (VIN G) and 2.5L Engines

1. Before servicing the vehicle, refer to the precautions in the beginning of this section.
2. Relieve the fuel system pressure using the recommended procedure. Disconnect the fuel line quick-connect fitting from the fuel rail by squeezing the retainer tabs together and pulling the fuel tube/quick-connect fitting assembly off the fuel tube nipple.
3. Drain the engine oil.
4. Drain the cooling system.
5. Recover the A/C refrigerant.
6. Place the ignition switch in the **OFF** unlocked position and turn **OFF** all accessories.
7. Turn the steering wheel to the extreme left position.
8. Remove or disconnect the battery and battery tray as follows:
   - Negative battery cable from the left shock tower

➡The ground cable is equipped with an insulator grommet which should be placed on the stud to prevent the nega-

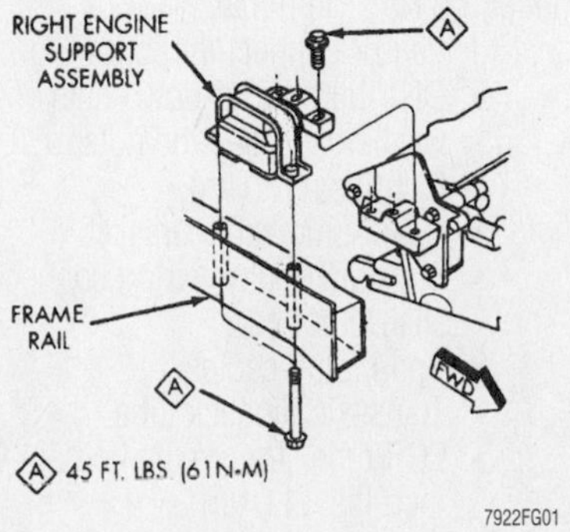

Exploded view of the right side engine mount—all engines

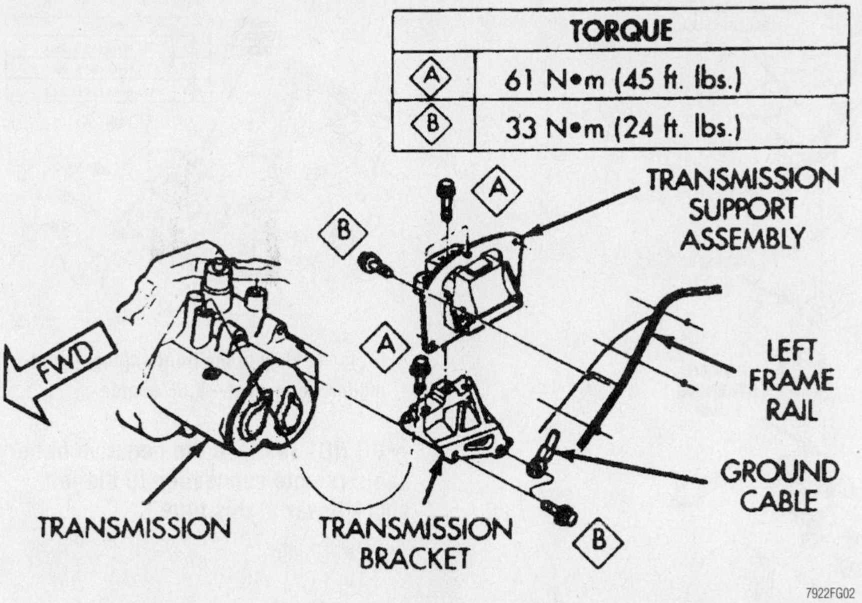

Exploded view of the left side engine mount—Type 1

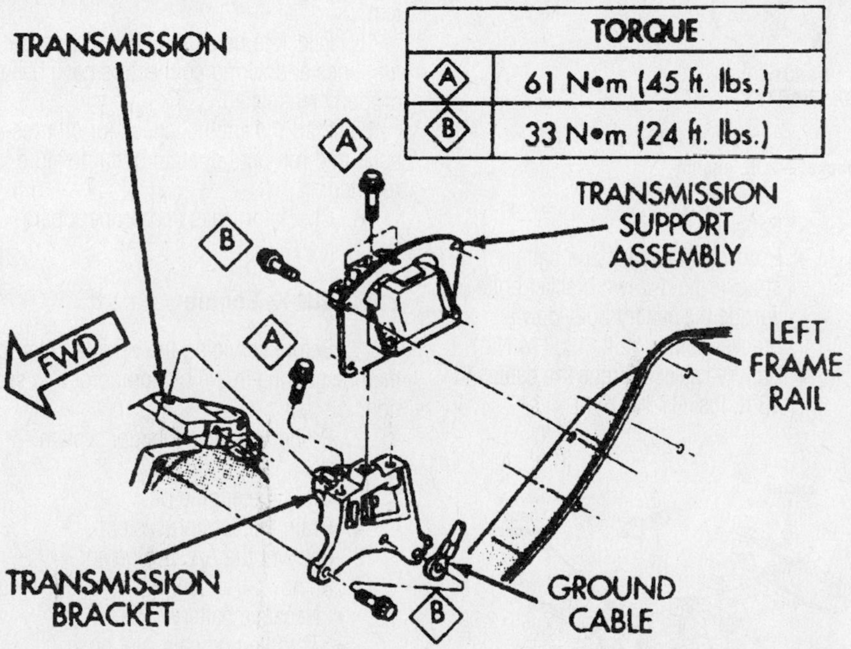

Exploded view of the left side engine mount—Type 2

tive battery cable from accidentally grounding.

- Shield, by twisting the 4 plastic screws ¼ turn
- Battery blanket heater, if equipped
- Battery cables
- Battery strap-to-battery hold-down bracket and hold-down bracket bolts
- Battery
- Battery tray and battery strap
- Air cleaner and inlet duct assembly
- Powertrain Control Module (PCM), move it aside
- Upper/lower radiator hose, radiator and cooling fan
- Automatic transaxle cooler lines, if equipped, plug the lines
- Clutch cable and transaxle shift linkage, if equipped
- Throttle body linkage and the engine wiring harness
- Heater hoses
- Front wheels
- Right side inner splash shield
- Accessory drive belts
- Both halfshafts
- Exhaust pipe from the exhaust manifold
- Front/rear engine mount brackets, from the body
- Power steering pump and reservoir; move them aside
- Compressor clutch wire lead
- Refrigerant lines from the compressor
- Compressor mounting bolts
- Compressor, plug all openings to prevent moisture contamination
- Engine ground straps and raise the vehicle and install an engine dolly under the vehicle and support engine
- Transaxle and engine mount through-bolts
- Transaxle/engine assembly

9. Raise the vehicle slowly allowing the engine/transaxle assembly to remain on the dolly.

To install:

10. Position the engine/transaxle assembly under the vehicle and lower the vehicle onto the engine assembly.
11. Install or connect the following:

- Right and left mount bolts
- Transaxle mount
- Both halfshafts
- Transaxle and engine braces
- Splash shields
- Exhaust pipe to the exhaust manifold
- Power steering pump and reservoir
- A/C compressor and torque the mounting bolts to 30 ft. lbs. (41 Nm)
- Refrigerant hoses using new seals
- A/C compressor clutch wire
- Accessory drive belts and adjust them
- Front engine mount
- Inner splash shield
- Front wheels. Torque the lug nuts to 95–100 ft. lbs. (129–135 Nm)
- Clutch cable and linkages, if equipped with a manual transaxle
- Shifter and kickdown linkages, if equipped with an automatic transaxle
- Fuel lines and heater hoses

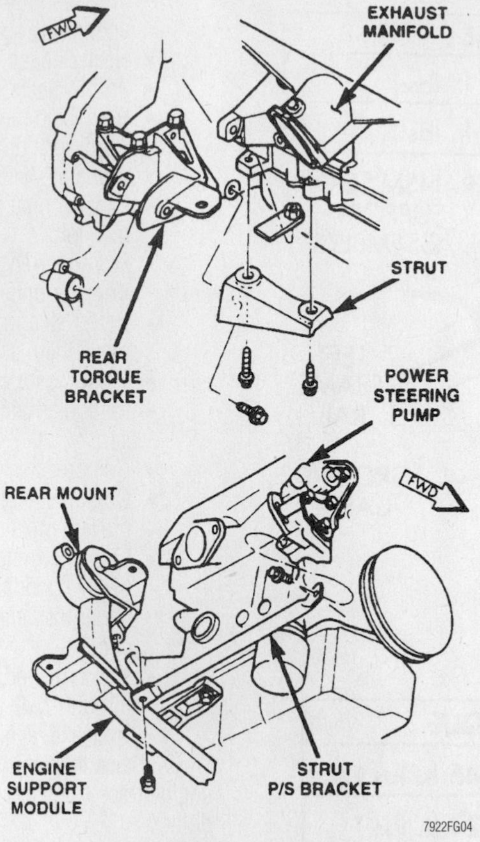

Exploded view of the rear engine mounting torque bracket—2.0L engine

- Ground straps
- Throttle body electrical connectors
- Throttle body linkage
- Radiator, cooling fan/shroud assembly and hoses
- Transaxle cooler lines to the radiator, if equipped

12. Fill the cooling system.
13. Install a new oil filter and refill the crankcase, if drained.
14. Install or connect the battery tray and the battery as follows:
    - Battery strap and battery tray
    - Battery blanket heater, if equipped
    - Battery
    - Hold-down bracket and battery strap-to-hold-down bracket bolts. Torque the battery hold-down bracket bolt to 10 ft. lbs. (14 Nm)
    - Battery cables. Torque the cables to 13 ft. lbs. (17 Nm)

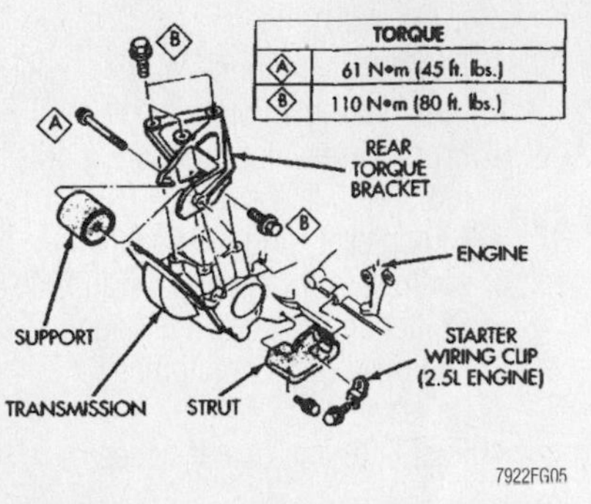

Exploded view of the rear engine mounting torque bracket—2.4L and 2.5L engines

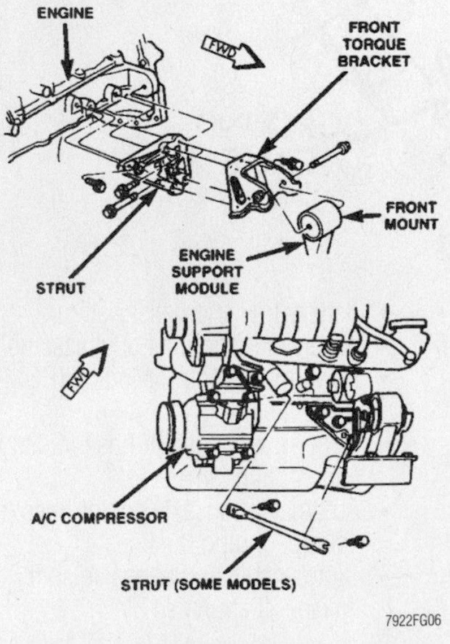

Exploded view of the front engine mounting torque bracket—2.0L and 2.4L

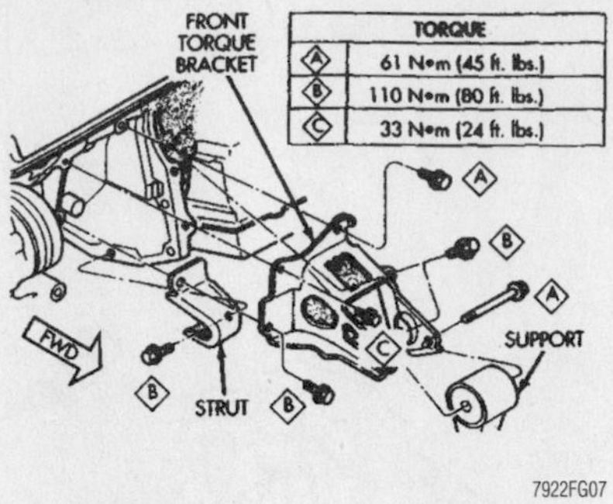

Exploded view of the front engine mounting torque bracket—2.5L engine

➡DO NOT reattach the negative battery cable remote connection to the left shock tower at this time.

- Shield
- PCM
- Air cleaner
- Air inlet duct

15. Recharge the air conditioning system.
16. Check to be sure all ducts, hoses, fuel lines and wiring connectors have been properly reattached.
17. Start the engine, check for oil pressure, and run until operating temperature is reached.
18. Check for leaks and proper operation.

2.4L (VIN X) Engine

1. Before servicing the vehicle, refer to the precautions in the beginning of this section.
2. Properly relieve the fuel system pressure.
3. Drain the engine oil.
4. Drain the cooling system.
5. Recover the A/C refrigerant.
6. Remove or disconnect the following:
    - Negative battery cable
    - Throttle body air inlet hose
    - Air cleaner assembly
    - Upper radiator crossmember
    - Upper and lower radiator hoses
    - Transmission cooler lines
    - A/C lines at the condenser
    - Radiator, fan module and A/C condenser
    - Transmission electrical harness connectors
    - Transmission shift cable
    - Engine electrical harness from the Powertrain Control Module (PCM) and bulkhead
    - Both front wheels and splash shields

- Both halfshafts
- Drive belts
- Power steering pump from the bracket and move it aside
- Heater return hose from the right front frame rail
- A/C compressor electrical connectors
- Exhaust pipe from the manifold
- Through bolts from the front and rear engine mounts
- Rear mount bracket from the transmission
- Structural collar and torque reaction bracket
- Torque converter bolts after matchmarking them
- Positive battery cable from the battery and the Power Distribution Center (PDC)
- Ground cable from the left side transmission mount bracket
- Throttle and speed control cables
- Coolant overflow hose
- Heater hose from the thermostat housing
- Engine ground straps
- Brake booster and vacuum purge hoses
- Fuel lines from the rail
- Intake manifold
- Alternator
- A/C suction line from the compressor and place an engine dolly and cradle in position

7. Loosen the cradle engine mounts and lower the engine/transmission assembly onto the dolly.

8. Remove the vertical engine mount bolts

9. Slowly raise the vehicle and make certain that the cradle is positioned properly.

10. Secure the engine/transmission assembly to the cradle and remove the assembly.

To install:

11. Position the engine/transmission assembly under the vehicle and slowly lower the vehicle into position.

12. Continue to the lower the vehicle until the right side engine mount and left side transaxle mount are properly aligned.

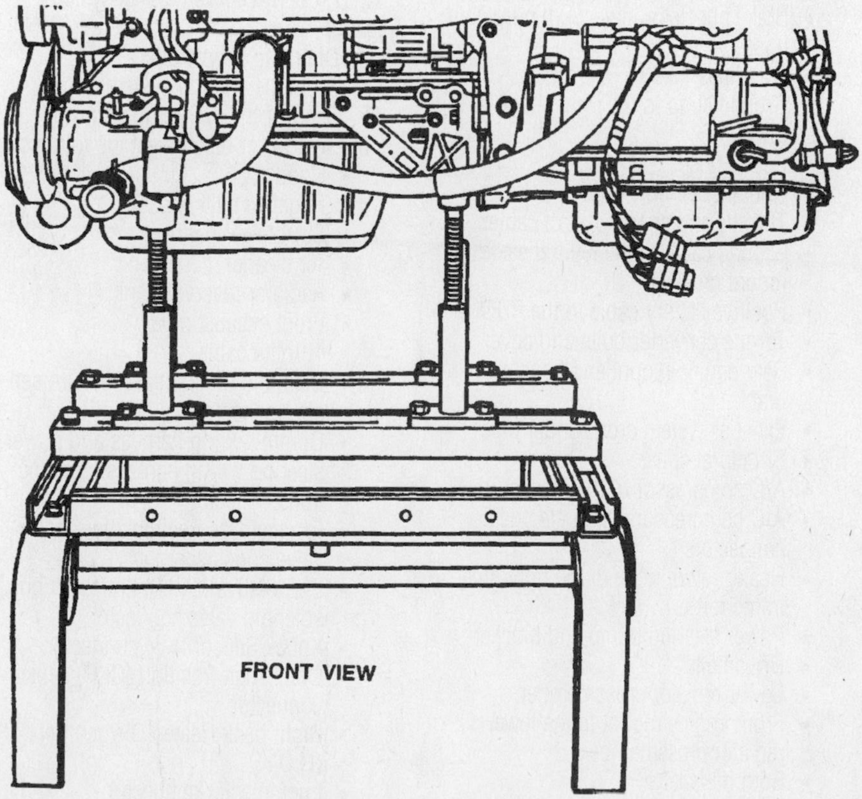

Place the engine/transmission assembly on the cradle as shown

13. Install or connect the following:
    - Mounting bolts and torque them to 45 ft. lbs. (61 Nm). Raise the vehicle and remove the cradle
    - A/C compressor and suction line
    - Alternator
    - Intake manifold and torque the bolts to 105 inch lbs. (12 Nm)
    - Alternator electrical connectors
    - Fuel line to the rail
    - Brake booster and vacuum purge hoses
    - Engine ground straps
    - Heater hose to the thermostat housing
    - Coolant overflow hose
    - Throttle and speed control cables
    - Ground cable to the left side transmission mount bracket
    - Positive battery cable to the PDC
    - Torque converter bolts
    - Structural collar and the torque reaction bracket
    - Rear mount bracket to the transmission
    - Front and rear engine mount through bolts. Torque the bolts to 45 ft. lbs. (61 Nm)
    - Exhaust pipe to the manifold and torque the bolts to 21 ft. lbs. (28 Nm)
    - A/C compressor electrical connectors
    - Heater return hose to the right front frame rail
    - Power steering pump to the bracket
    - Drive belts
    - Both halfshafts
    - Splash shields and front wheels
    - Engine electrical harness to the PCM and the bulkhead connectors
    - Transmission shift cable
    - Transmission electrical connectors
    - Radiator, fan module and A/C condenser
    - A/C lines to the condenser
    - Transmission cooler lines
    - Upper and lower radiator hoses
    - Upper radiator crossmember
    - Air cleaner assembly
    - Throttle body air inlet hose
    - Negative battery cable

14. Fill the engine with clean oil.

15. Fill the cooling system to the proper level.

16. Recharge the A/C system.

17. Start the vehicle, check for leaks and repair if necessary.

Timing belt service is covered in Section 3 of this manual

2.7L Engine

1. Before servicing the vehicle, refer to the precautions in the beginning of this section.
2. Properly relieve the fuel system pressure.
3. Drain the engine oil.
4. Drain the cooling system.
5. Recover the A/C refrigerant.
6. Remove or disconnect the following:
   - Negative battery cable
   - Throttle body air inlet hose
   - Air cleaner assembly
   - Both front wheels and splash shields
   - Lower front fascia from the crossmember
   - Lower air shield from the crossmember
   - Front fascia
   - Upper radiator crossmember
   - Upper and lower radiator hoses
   - Transmission oil cooler lines
   - A/C lines from the condenser
   - Radiator, fan and A/C condenser
   - Transmission electrical harness connectors
   - Transmission shift cable
   - Engine electrical harness from the Powertrain Control Module (PCM) and bulkhead connectors
   - Antilock Brake System (ABS) brake module from the lower radiator crossmember
   - Brake line from the lower radiator crossmember
   - Both halfshafts
   - Front engine mount through bolt
   - Front engine mount from the lower radiator crossmember
   - Lower radiator crossmember
   - Drive belts
   - Power steering pump and bracket. Do not disconnect the lines
   - Heater return hose from the right front frame rail
   - A/C compressor electrical connectors and reposition the compressor
   - Structural collar
   - Exhaust system cross under pipe
   - Rear engine mount and transaxle bracket
   - Torque converter housing cover and match mark the bolts
   - Positive battery cable from the Power Distribution Center (PDC)
   - Ground cable from the left side transaxle mount bracket
   - Throttle and speed control cables
   - Coolant overflow hose
   - Heater hose
   - Ground strap from the right shock tower
   - Fuel lines
   - Brake booster and vacuum purge hoses
   - Engine ground straps
7. Loosen the cradle engine mounts and lower the engine/transmission assembly onto the dolly.
8. Remove the vertical engine mount bolts
9. Slowly raise the vehicle and make certain that the cradle is positioned properly.
10. Secure the engine/transmission assembly to the cradle and remove the assembly.

To install:

11. Position the engine/transmission assembly under the vehicle and slowly lower the vehicle into position.
12. Continue to the lower the vehicle until the right side engine mount and left side transaxle mount are properly aligned.
13. Install or connect the following:
    - Mounting bolts and torque them to 45 ft. lbs. (61 Nm). Raise the vehicle and remove the cradle
    - Engine ground straps
    - Brake booster and vacuum purge hoses
    - Fuel lines
    - Ground strap to the right shock tower
    - Heater hose
    - Coolant overflow hose
    - Throttle and speed control cables
    - Ground cable to the left transaxle mount bracket
    - Positive battery cable to the PDC
    - Torque converter bolts and cover
    - Rear engine mount and transaxle bracket
    - Exhaust system cross under pipe
    - Structural collar
    - A/C compressor to the bracket
    - A/C compressor clutch electrical connectors
    - Heater return hose to the right side frame rail
    - Power steering pump and bracket
    - Drive belts
    - Lower radiator crossmember
    - Front engine mount to the lower radiator crossmember
    - Both halfshafts
    - Brake line to the lower radiator crossmember
    - ABS module to the lower radiator crossmember
    - Engine electrical harness to the PCM and bulkhead connectors
    - Transmission shift cable
    - Transmission electrical harness connectors
    - Radiator, fan and A/C condenser
    - A/C lines to the condenser
    - Transmission oil cooler lines
    - Upper and lower radiator hoses
    - Upper radiator crossmember
    - Front bumper fascia and lower air shield
    - Splash shields and both front wheels
    - Air cleaner assembly and throttle body air inlet hose
    - Negative battery cable
14. Fill the engine with clean oil and replace the filter.
15. Fill the cooling system to the proper level.
16. Recharge the A/C system with the proper refrigerant.
17. Start the vehicle, check for leaks and repair if necessary.

3.0L Engine

1. Before servicing the vehicle, refer to the precautions in the beginning of this section.
2. Properly relieve the fuel system pressure.
3. Drain the engine oil.
4. Drain the cooling system.
5. Recover the A/C refrigerant.
6. Remove or disconnect the following:
   - Hood
   - Negative battery cable
   - Strut tower bar
   - Air cleaner
   - Radiator reservoir tank
   - Front exhaust pipe
   - Throttle cable
   - Manifold differential pressure sensor connector
   - Control wiring harness and power steering wiring harness combination connector
   - Exhaust Gas Recirculation (EGR) valve connector
   - Evaporative (EVAP) Emission purge solenoid valve connector
   - Knock Sensor (KS) connector
   - Crankshaft Position (CKP) sensor connector
   - Right bank Heated Oxygen Sensor (HO_2S)
   - Fuel injector connector
   - Distributor connector
   - Control wiring/injector wiring harness combination connector
   - Throttle Position Sensor (TPS) connector

- Idle Air Control (IAC) motor connector
- Ground wire
- Engine Coolant Temperature (ECT) gauge unit
- ECT sensor connector
- Left bank Heated Oxygen Sensor (HO_2S)
- Brake booster vacuum hose
- Fuel supply and return hoses
- Heater hoses
- Starter electrical connectors
- Oil Pressure switch
- Alternator wiring
- Drive belts
- A/C compressor
- Power steering pressure switch
- Power steering oil pump. Do not disconnect the hose
- Engine mount stay and install an engine lifter, such as MZ203827
- Engine mount bracket and stopper
- Engine assembly

To install:

7. Install or connect the following:
   - Engine assembly
   - Engine mount bracket and stopper. Torque the bolts to 60 ft. lbs. (81 Nm)
   - Engine mount stay and torque the bolts to 26 ft. lbs. (35 Nm). Remove the engine lifting devise
   - Power steering oil pump and torque the bolts to 31 ft. lbs. (42 Nm)
   - Power steering pressure switch
   - A/C compressor
   - Drive belts
   - Alternator wiring
   - Oil pressure switch
   - Starter electrical connectors
   - Heater hoses
   - Fuel supply and return lines
   - Brake booster vacuum hose
   - Left bank (HO_2S)
   - ECT sensor connector
   - ECT gauge unit
   - Ground wire
   - IAC motor electrical connector
   - TPS connector
   - Control wiring/injector wiring harness combination connector
   - Distributor connector
   - Fuel injector connector
   - Right bank (HO_2S)
   - CKP sensor connector
   - KS connector
   - EVAP solenoid valve connector
   - EGR valve connector
   - Control wiring harness and power steering harness combination connector
   - Manifold differential Pressure sensor connector
   - Throttle cable
   - Front exhaust pipe
   - Radiator reservoir tank
   - Air cleaner
   - Strut tower bar
   - Negative battery cable
   - Hood and adjust as needed
8. Fill the engine with clean oil and replace the filter
9. Fill the cooling system to the proper level.
10. Recharge the A/C system with the proper refrigerant.
11. Start the vehicle, check for leaks and repair if necessary.

Water Pump

REMOVAL & INSTALLATION

2.0L and 2.4L (VIN X) Engines

This engine uses a die-cast aluminum body water pump with a stamped steel impeller. The water pump bolts directly to the block. Cylinder block-to-water pump sealing is provided by a large rubber O-ring. The water pump is driven by the timing belt, that must be removed to service the water pump.

1. Before servicing the vehicle, refer to the precautions in the beginning of this section.
2. Disconnect the negative battery cable from the left shock tower.

➡The ground cable is equipped with an insulator grommet which should be placed on the stud to prevent the negative battery cable from accidentally grounding.

➡This procedure requires removing the engine timing belt and the autotensioner. The factory specifies that the timing marks should always be aligned before removing the timing belt. Set the engine at Top Dead Center (TDC) on No. 1 compression stroke. This should align all timing marks on the crankshaft sprocket and both camshaft sprockets.

3. Drain the cooling system.
4. Remove or disconnect the following:
   - Right inner splash shield
   - Accessory drive belts and support the engine
   - Right motor mount
   - Power steering pump bracket bolts and move the pump/bracket assembly aside

➡Do not disconnect the power steering fluid lines.

   - Right engine mount bracket
   - Timing belt front covers
   - Timing belt tensioner and timing belt by loosening the tensioner screws

⁂ WARNING

With the timing belt removed, DO NOT rotate the camshaft or crankshaft or damage to the engine could occur.

   - Camshaft sprockets

➡Do not allow the camshafts to turn when the camshaft sprockets are being removed.

   - Rear timing belt cover
   - Water pump

To install:

5. Thoroughly clean all sealing surfaces. Replace the water pump if there are any cracks, signs of coolant leakage from the shaft seal, loose or rough turning bearings, damaged impeller or sprocket or sprocket flange loose or damaged.
6. Install or connect the following:
   - New rubber O-ring into the water pump

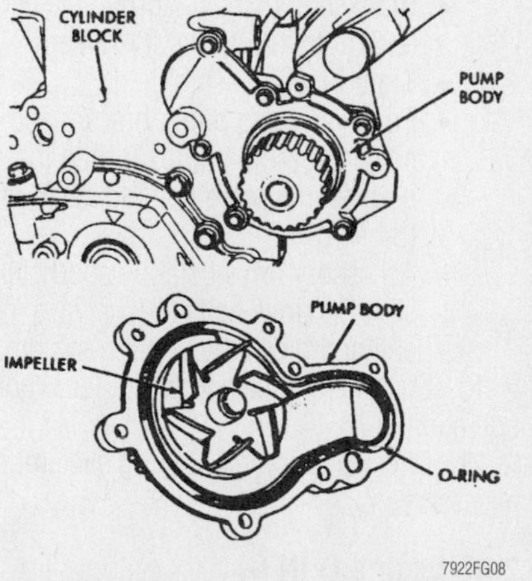

When installing the water pump, properly install the O-ring to ensure a tight seal—2.0L and 2.4L engines

⁂ WARNING

Be sure the O-ring is properly seated in the water pump groove before tightening the screws. An improperly located O-ring may cause damage to the O-ring and cause a coolant leak.

- Water pump. Torque the bolts to 105 inch lbs. (12 Nm)

7. Pressurize the cooling system to 15 psi (103.4 kPa) and check for leaks. If okay, release the pressure and continue the engine assembly process.
8. Install or connect the following:
   - Rear timing belt cover
   - Camshaft sprockets. Torque the bolts to 75 ft. lbs. (101 Nm)

➡DO NOT allow the camshafts to turn while the sprockets bolts are being tightened to maintain timing mark alignment.

⁂ WARNING

Do not attempt to compress the tensioner plunger with the tensioner assembly installed in the engine. This will cause damage to the tensioner and other related components. The tensioner MUST be compressed in a vise.

- Timing belt tensioner and timing belt. Properly tension the timing belt.
- Front upper and lower timing belt covers
- Right engine mount bracket and engine mount
- Crankshaft damper. Torque the center bolt to 105 ft. lbs. (142 Nm)
- Right inner splash shield
- Power steering pump bracket and power steering pump. Torque the bracket mounting bolts to 40 ft. lbs. (54 Nm)
- Accessory drive belts. Properly tension the drive belts

9. Refill and bleed the cooling system.
10. Start the engine and check for proper operation.
11. Check and top off cooling system, if necessary.

2.4L Engine (VIN G)

1. Before servicing the vehicle, refer to the precautions in the beginning of this section.
2. Drain the cooling system.
3. Remove or disconnect the following:
   - Negative battery cable
   - Timing belt tensioner pulley
   - Alternator brace
   - Water pump and discard the gasket and O-ring

To install:

4. Clean all mating surfaces of any residual gasket material
5. Install or connect the following:
   - New O-ring gasket
   - Water pump and torque the bolts to 117 inch lbs. (14 Nm)
   - Alternator brace and torque the bolts to 17 ft. lbs. (23 Nm)
   - Timing belt tensioner pulley
   - Negative battery cable
6. Fill the cooling system to the proper level.
7. Start the vehicle, check for leaks and repair if necessary.

2.5L Engine

The water pump bolts directly to the engine block using a gasket for pump-to-block sealing. The pump is serviced as a unit. The 2.5L engine uses metal piping beyond the lower radiator hose to route coolant to the suction side of the water pump, located in the "V" of the cylinder banks. These pipes also have connections for thermostat bypass and heater return coolant hoses. The pipes use O-rings for sealing.

The water pump is driven by the timing belt which must be removed to service the water pump. Timing belt covers must be removed to access the timing belt.

1. Before servicing the vehicle, refer to the precautions in the beginning of this section.
2. Disconnect the negative battery cable from the left shock tower.

➡The ground cable is equipped with an insulator grommet which should be placed on the stud to prevent the negative battery cable from accidentally grounding.

3. Drain the cooling system.

➡This procedure requires removing the engine timing belt and the auto-tensioner. To help assure proper alignment at assembly, it may be helpful to set the engine at Top Dead Center (TDC) on No. 1 compression stroke. This should align all timing marks on the crankshaft sprocket and both camshaft sprockets.

4. Remove or disconnect the following:
   - Accessory drive belts
   - Crankshaft damper
   - Right engine mount by safely supporting the engine
   - Timing belt covers, in this order: upper left, upper right and lower covers
   - Timing belt and tensioner
   - Water pump by separating it from the water inlet pipe

To install:

5. Thoroughly clean all sealing surfaces. Inspect the pump for damage or cracks, signs of coolant leakage at the vent and excessive looseness or rough turning bearing. Any problems require a new pump.
6. Install or connect the following:
   - New O-ring on the water inlet pipe by wetting it with water
   - New gasket on the water pump
   - Water pump. Torque the bolts to 20 ft. lbs. (27 Nm)
   - Timing belt and timing belt tensioner by adjusting tension
   - Timing belt covers
   - Right engine mount
   - Crankshaft damper
   - Accessory drive belts by adjusting tension
   - Negative battery cable
7. Refill and bleed the cooling system.
8. Start the engine and verify proper operation.

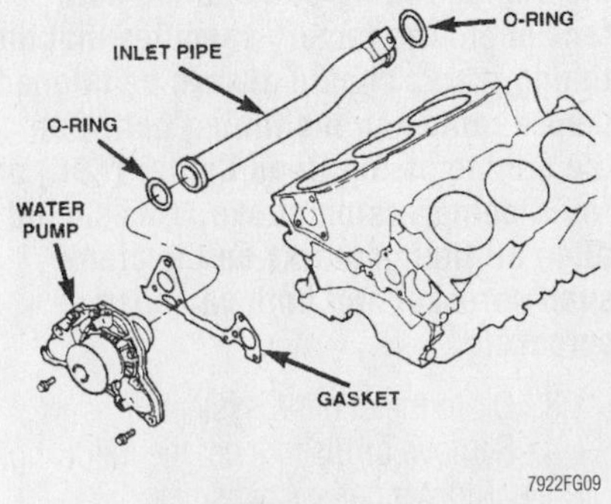

Exploded view of the water pump mounting—2.5L engine

2.7L Engine

1. Before servicing the vehicle, refer to the precautions in the beginning of this section.
2. Drain the cooling system.
3. Remove or disconnect the following:
   - Negative battery cable
   - Timing chain cover
   - Timing chain and guides
   - Water pump and discard the gasket

To install:

4. Clean all mating surfaces of any residual gasket material.

5. Install or connect the following:
   - Water pump with a new gasket and torque the bolts to 105 inch lbs. (12 Nm)
   - Timing chain guides and chain
   - Timing chain cover
   - Negative battery cable
6. Fill the cooling system to the proper level.
7. Start the vehicle, check for leaks and repair if necessary.

3.0L Engine

1. Before servicing the vehicle, refer to the precautions in the beginning of this section.
2. Drain the cooling system.
3. Remove or disconnect the following:
   - Negative battery cable
   - Timing belt
   - Thermostat and bracket
   - Water pump and discard the gasket and O-ring

To install:

4. Install or connect the following:
   - Water pump with a new O-ring and gasket. Torque the bolts to 17 ft. lbs. (23 Nm)
   - Bracket. Torque the bolts to 17 ft. lbs. (23 Nm)
   - Thermostat
   - Timing belt
   - Negative battery cable
5. Fill the cooling system to the proper level.
6. Start the vehicle, check for leaks and repair if necessary.

Cylinder Head

REMOVAL & INSTALLATION

2.0L Engine

This engine uses a Single Over Head Camshaft (SOHC) 4-valves per cylinder cross flow aluminum cylinder head. Care must be taken to be sure all valve timing marks align after cylinder head service.

1. Before servicing the vehicle, refer to the precautions in the beginning of this section.
2. Relieve the fuel system pressure using the recommended procedure.
3. Disconnect the negative battery cable from the left shock tower.

➡The ground cable is equipped with an insulator grommet which should be placed on the stud to prevent the negative battery cable from accidentally grounding.

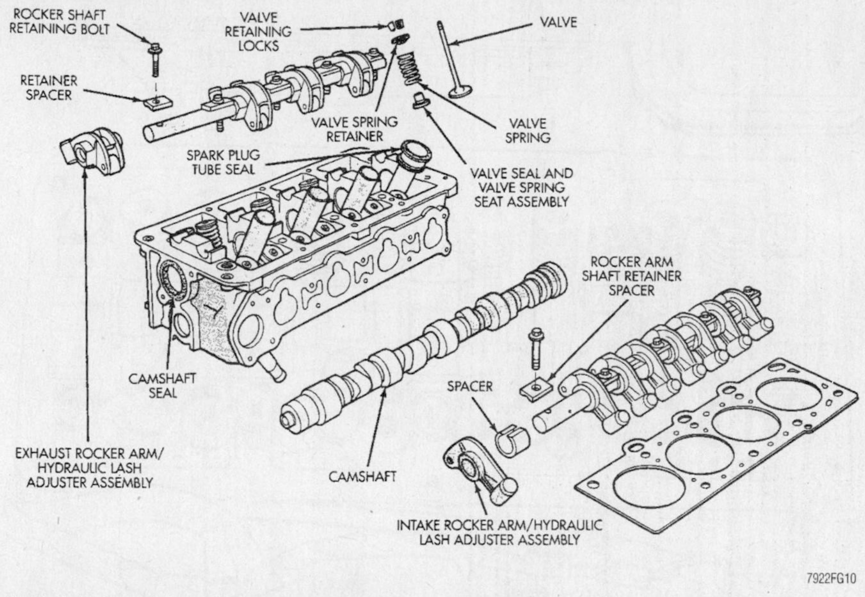

Exploded view of the cylinder head and valvetrain components—2.0L engine

4. Drain the cooling system.
5. Remove or disconnect the following:
   - Air cleaner assembly
   - All vacuum hoses and electrical connectors from the throttle body
   - Fuel line quick-connect fitting from the fuel rail, by squeezing the retainer tabs together and pulling the fuel tube/quick-connect fitting assembly off the fuel tube nipple
   - Throttle linkage
   - Accessory drive belt(s)
   - Power brake booster vacuum hose from the intake manifold
   - Exhaust pipe from the exhaust manifold
   - Power steering pump and move it aside
   - Coil pack wiring connector
   - Spark plug wires from the spark plugs
   - Ignition coil pack
   - Cylinder head cover
   - Cam sensor and fuel injector wiring
   - Intake and exhaust manifolds, if necessary
   - Timing belt cover
   - Timing belt
   - Camshaft sprocket
   - Rear timing belt cover
   - Rocker arm/rocker arm shaft assemblies
   - Cylinder head bolts
   - Cylinder head

To install:

➡The cylinder head bolts should be checked for stretching before reuse. If the thread area of the bolt is necked-down the bolts must be replaced with new. New head bolts are recommended.

6. Thoroughly clean all parts. Clean all sealing surfaces. Use care not to scratch the aluminum cylinder head sealing surface. Check the cylinder head for flatness using a feeler gauge and a straightedge. The cylinder head must be flat within 0.004 in. (0.1mm).
7. Check the cylinder head for cracks or other damage.
8. Install or connect the following:
   - New cylinder head gasket
   - Cylinder head
9. Be sure to oil the cylinder head bolt threads with clean engine oil. Install the cylinder head bolts, the 4.330 in. (110mm) short bolts are to be installed in positions 7, 8, 9 and 10. Tighten the bolts in proper sequence.
10. Torque the cylinder head bolts, in sequence, using the following steps:
    a. Step 1: 25 ft. lbs. (34 Nm).
    b. Step 2: 50 ft. lbs. (68 Nm).
    c. Step 3: Again to 50 ft. lbs. (68 Nm).
    d. Step 4: an additional ¼ turn.

➡Do not use a torque wrench for the 4th step.

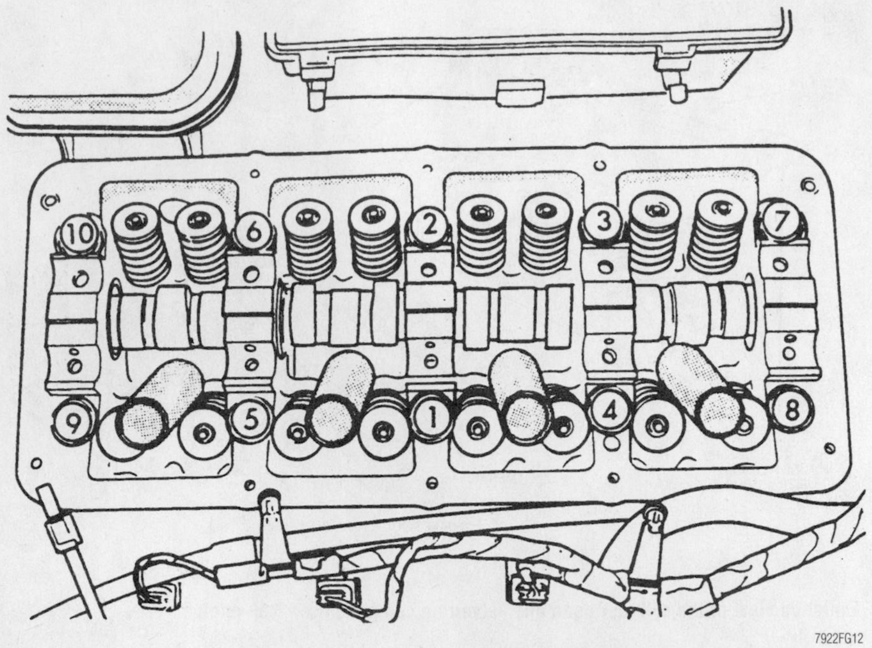

Cylinder head bolt torque sequence—2.0L engine

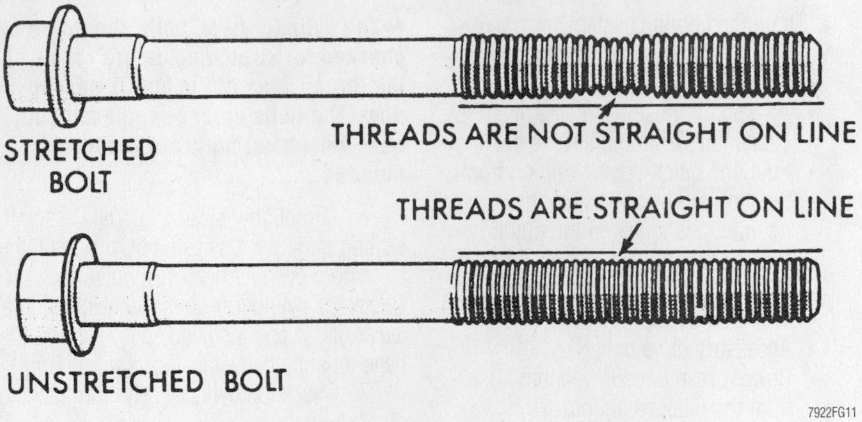

Before reusing old cylinder head bolts, check them for necking (stretching)—2.0L and 2.4L

11. Set the crankshaft to 3 notches Before Top Dead Center (BTDC) before installing the rocker arm shafts.

12. Install or connect the following:
- Rocker arm/rocker arm shaft assemblies
- Cylinder head cover using a new gasket. Torque the bolts to 105 inch lbs. (12 Nm).

➡Be sure the cover gasket mating surfaces are clean of any dirt, oil or old gasket material.

- Timing belt rear cover
- Camshaft sprocket
- Timing belt
- Timing belt cover
- Intake and exhaust manifolds, if removed
- Cam sensor and fuel injector electrical connectors
- Ignition coil pack
- Spark plug wires
- Power steering pump
- Exhaust pipe to the exhaust manifold
- Brake booster vacuum line
- Accessory drive belts and adjust them
- Throttle linkage
- All vacuum hoses and electrical connectors to the throttle body
- Fuel line to the fuel rail
- Air cleaner assembly
- Negative battery cable

13. Refill the cooling system and install a new oil filter.

14. Check to be sure all ducts, hoses, fuel lines and wiring connectors have been properly reattached.

15. Start the engine and check for leaks. Run the engine with the radiator cap off so as the engine warms and the thermostat opens, coolant can be added to the radiator. When satisfied that the cooling system is full, shut the engine **OFF**, install the radiator cap and allow the engine to cool.

16. With the engine cool, check all fluid levels. Add coolant and oil as required. Restart the engine and test drive the vehicle to check for proper operation.

2.4L Engine (VIN X)

This engine uses a Dual Over Head Camshaft (DOHC) 4-valves per cylinder cross flow aluminum cylinder head. The valves are actuated by roller cam followers which pivot on stationary hydraulic valve adjusters. Care must be taken to be sure all valve timing marks align after cylinder head and valvetrain service.

1. Before servicing the vehicle, refer to the precautions in the beginning of this section.
2. Relieve the fuel system pressure.
3. Drain the cooling system.
4. Disconnect the negative battery cable from the left shock tower.

➡The ground cable is equipped with an insulator grommet which should be placed on the stud to prevent the negative battery cable from accidentally grounding.

5. Remove or disconnect the following:
- Air cleaner assembly
- All vacuum lines, electrical connectors and fuel lines from the throttle body
- Throttle linkage
- Accessory drive belts
- Power brake vacuum hose from the intake manifold
- Exhaust pipe from the exhaust manifold
- Power steering pump and move it aside. Do not disconnect the fluid lines
- Spark plug wires
- Coil pack electrical connector
- Coil pack with the spark plug wires
- Cam sensor and fuel injector electrical connectors
- Timing belt covers
- Timing belt
- Camshaft sprockets
- Timing belt idler pulley
- Rear timing belt cover

- Cylinder head cover
- Ground strap

6. Identify the camshafts if they are to be reused for later installation. The camshafts are not interchangeable.
7. Remove or disconnect the following:
   - Camshaft bearing cap bolts in sequence
   - Camshafts

➡Refer to the camshaft removal and installation procedure for correct bolt removal/installation sequence.

- Camshaft followers

➡Any components that are to be reused must be installed in their original locations. Use care to identify and mark the positions of any removed valvetrain components so they may be reinstalled correctly.

- Intake and exhaust manifolds
- Cylinder head bolts
- Cylinder head

➡Be careful not to damage the aluminum gasket surfaces.

8. Remove all gasket material from the cylinder head and engine block. Be careful not to gouge or scratch the sealing surface of the aluminum head. The cylinder head should be checked for flatness using a good straightedge and feeler gauges. The cylinder head must be flat within 0.004 in. (0.1mm).
9. Inspect the camshaft bearing oil feed holes in the cylinder head for clogging. Inspect the camshaft bearing journals for wear or scoring. Check the cam surface for abnormal wear and damage. A visible worn groove in the roller path or on the cam lobes is cause for replacement. Valve service may be performed at this time.

To install:

10. Thoroughly clean all parts.

➡The cylinder head bolts are stretch-type. New cylinder head bolts are recommended.

11. Thoroughly clean all sealing surfaces.
12. Install or connect the following:
    - New cylinder head gasket
    - Cylinder head
13. Before installing the bolts, the threads should be oiled with clean engine oil.
14. Torque the cylinder head bolts in sequence, using the following 4 Steps:
    a. Tighten all bolts to 25 ft. lbs. (34 Nm).
    b. Tighten all bolts to 50 ft. lbs. (68 Nm).
    c. Tighten all bolts again to 50 ft. lbs. (68 Nm).
    d. Tighten all bolts and additional ¼ turn.

➡Do not use a torque wrench for the 4th step.

15. Check the camshaft end-play using the recommended procedure, then install the camshaft.
16. Apply Mopar Gasket Maker sealer to the No. 1 and No. 6 bearing caps. Install the bearing caps and tighten the M8 fasteners to 21 ft. lbs. (28 Nm). The end caps must be installed before the seals may be installed.
17. Apply a light coating of clean engine oil to the lip of the new camshaft seal. Install the camshaft seal until it fits flush with the cylinder head.
18. Install or connect the following:
    - Camshaft sprockets, if removed
    - Rear timing belt cover
    - Timing belt and properly align the timing marks
    - Timing belt cover

⁂ WARNING

Verify that all timing marks are correct. If the timing belt or sprockets are incorrectly installed, engine damage will occur. Take time to be sure all timing marks are correctly aligned.

- Intake and exhaust manifolds
- New cylinder head cover gasket
- Cylinder head cover

⁂ WARNING

DO NOT allow oil or solvents to contact the timing belt as they can deteriorate the rubber and cause tooth skipping.

➡Apply Mopar Silicone Rubber Adhesive Sealant at the camshaft cap corners and at the top edge of the ½ round seal.

19. Torque the cylinder head cover fasteners, in sequence, using the following Steps:
    a. Step 1: 40 inch lbs. (4.5 Nm).
    b. Step 2: 80 inch lbs. (9 Nm).
    c. Step 3: 105 inch lbs. (12 Nm).
20. Install or connect the following:
    - Ground strap
    - Coil pack and spark plug wiring
    - Cam sensor and fuel injector wiring
    - Power steering pump assembly
    - Exhaust pipe to the exhaust manifold
    - All vacuum lines and electrical connectors
    - Throttle linkage and fuel lines
    - Accessory drive belts and adjust them
    - Negative battery cable
21. Refill the cooling system.

➡An oil and filter change is recommended since coolant can enter the oil system when a head is removed.

22. Connect the remaining air ducting. Test run vehicle. Check for leaks and for proper operation.

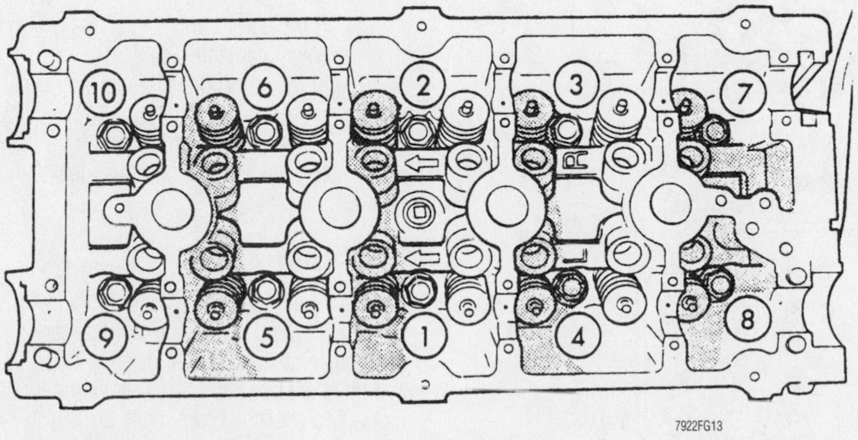

Cylinder head bolt torque sequence—2.4L (VIN X) engine

2.4L Engine (VIN G)

1. Before servicing the vehicle, refer to the precautions in the beginning of this section.
2. Relieve the fuel system pressure.
3. Drain the cooling system.
4. Drain the engine oil.
5. Remove or disconnect the following:
   - Negative battery cable
   - Strut tower bar
   - Air cleaner assembly
   - Thermostat case
   - Front exhaust pipe
   - Accelerator cable
   - Purge hose
   - Brake booster vacuum hose
   - Ignition coil
   - Fuel injector electrical connector
   - Ignition failure sensor connector
   - Manifold differential pressure sensor connector
   - Throttle Position Sensor (TPS) connector
   - Heated Oxygen Sensor (HO_2S) connector
   - Capacitor connector
   - Engine Coolant Temperature (ECT) sensor
   - Camshaft Position (CMP) sensor
   - Idle Air Control (IAC) motor connector
   - Evaporative (EVAP) emission purge solenoid valve
   - Exhaust Gas Recirculation (EGR) valve
   - Fuel supply and return hoses
   - Pressure hose
   - Oil dipstick and guide
   - Ignition coil and spark plug cables
   - Upper radiator hose
   - Positive Crankcase Ventilation (PCV) hose
   - Breather hose
   - Rocker arm cover
   - Spark plug guide oil seal
   - Water hose
   - Timing belt
   - Power steering pressure switch
   - Power steering pump and bracket
   - Exhaust manifold bracket
   - Cylinder head and discard the gasket

To install:

6. Clean all mating surfaces of any residual gasket material.
7. Lubricate the threads of the cylinder head bolts with clean engine oil.
8. Install or connect the following:
   - Cylinder head with a new gasket

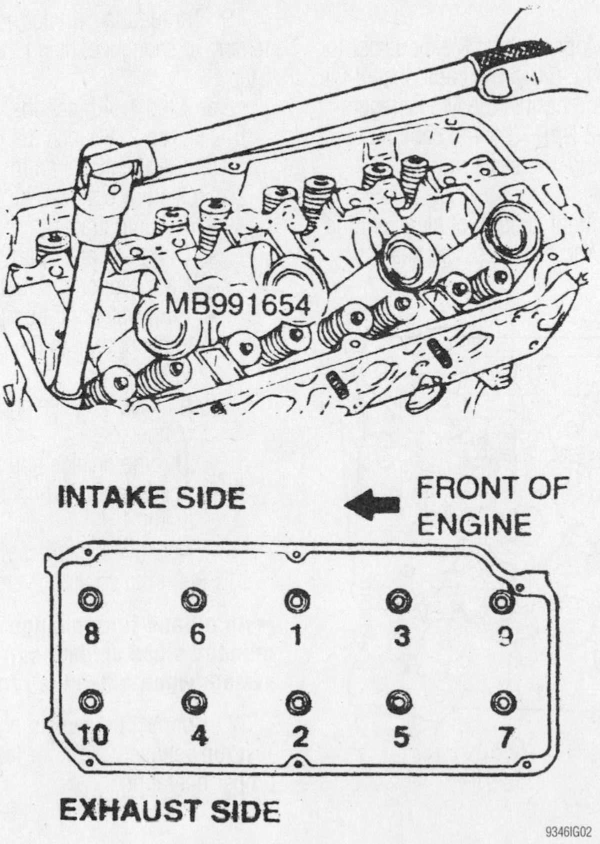

Cylinder head bolt torque sequence—2.4L (VIN G) engine

9. Torque the bolts, at a 90 degree angle with special Tool MB991654, using the following procedure:
   a. Step 1: 58 ft. lbs. (79 Nm).
   b. Step 2: Fully loosen all the bolts.
   c. Step 3: 15 ft. lbs. (20 Nm).
   d. Step 4: An additional 90 degrees.
   e. Step 5: An additional 90 degrees.
   - Exhaust manifold bracket and torque the bolts to 26 ft. lbs. (35 Nm)
   - Power steering pump and bracket
   - Power steering pressure switch
   - Timing belt
   - Water hose
   - Spark plug guide oil seal
   - Rocker arm cover
   - Breather and PCV hoses
   - Upper radiator hose
   - Ignition coil and spark plug cables
   - Pressure hose
   - Oil dipstick and tube. Torque the bolt to 177 inch lbs. (14 Nm).
   - Fuel return and supply hoses
   - EGR solenoid valve
   - EVAP purge solenoid valve
   - IAC motor connector
   - CMP sensor
   - ECT sensor
   - Capacitor
   - HO_2S connector
   - TPS connector
   - Manifold differential pressure sensor connector
   - Ignition failure sensor connector
   - Ignition coil connector
   - Brake booster vacuum hose
   - Purge hose
   - Accelerator and adjust as needed
   - Front exhaust pipe
   - Thermostat case
   - Air cleaner
   - Strut tower bar
   - Negative battery cable
10. Fill the cooling system to the proper level.
11. Fill the engine with clean oil.
12. Start the vehicle, check for leaks and repair if necessary.

2.5L Engine

This engine uses aluminum alloy cylinder heads with 4-valves per cylinder and pressed-in cast iron valve guides. The cylinders are common to either cylinder bank. Two overhead camshafts are supported by 4 bearing journals which are part of the head with the distributor driven off the right (firewall side) cylinder head. Right and left camshaft drive sprockets are inter-

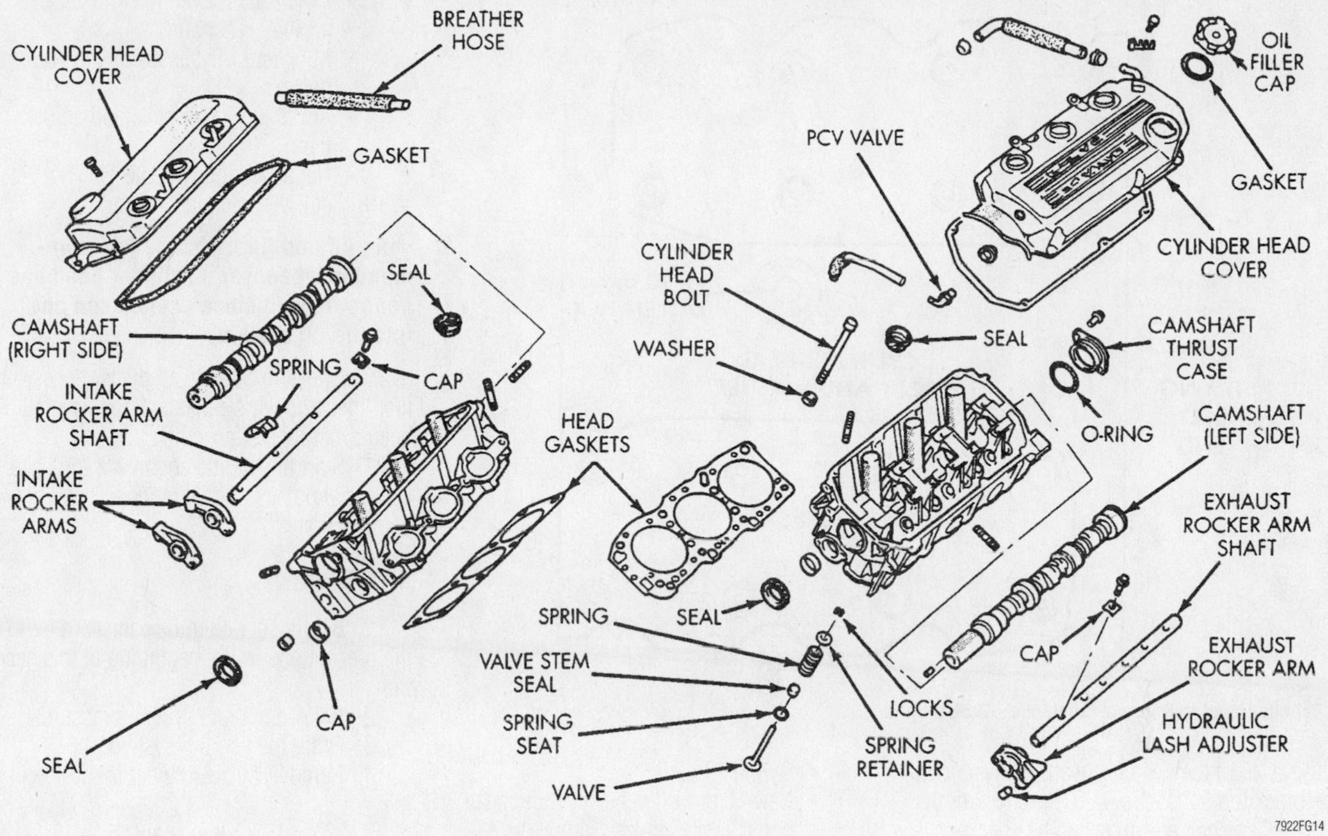

Exploded view of the cylinder head, camshafts and rocker assemblies—2.5L engine

changeable. The sprockets and engine water pump are driven by the timing belt. Care must be taken to be sure all valve timing marks align after cylinder head and valve-train service.

➡Please note that for camshaft service, the cylinder head must be removed.

1. Before servicing the vehicle, refer to the precautions in the beginning of this section.
2. Disconnect the negative battery cable from the left shock tower.

➡The ground cable is equipped with an insulator grommet which should be placed on the stud to prevent the negative battery cable from accidentally grounding.

3. Relieve the fuel system pressure using the recommended procedure.
4. Drain the cooling system.
5. Remove or disconnect the following:
   - Accessory drive belts
   - Front timing belt covers
   - Timing belt
   - Camshaft sprockets

➡The intake manifold is a 2-piece unit. The upper part is a large air intake plenum of aluminum alloy. Use care working with light alloy parts.

   - Air intake plenum
   - Manifold Absolute Pressure (MAP) and Intake Air Temperature (IAT) sensor wiring connectors
   - Air cleaner cover, inlet hoses and air inlet resonator
   - Throttle Position Sensor (TPS) and Idle Air Control (IAC) motor electrical connectors
   - Throttle cable
   - Speed control cable, if equipped
   - Exhaust Gas Recirculation (EGR) tube
   - Lower intake manifold
   - Spark plug wires and move them aside
   - Cylinder head cover
6. Identify the rocker arm shaft assemblies before removal.
7. Install the Auto Lash Adjuster Retainers MD 998443, to keep the auto lash adjusters from falling out of the rocker arms when the rocker arm assembly is removed.
8. Remove or disconnect the following:
   - Rocker arm shaft assemblies
   - Distributor assembly
   - Exhaust manifold and crossover
   - Cylinder head bolts
   - Cylinder head
9. If the camshaft(s) are to be serviced, remove the thrust case from the left head assembly and remove the camshaft from the rear of the head. If not already done, remove the distributor from the right cylinder head and remove the camshaft from the rear of the head.
10. Valve service may be performed at this time, if required.

To install:

11. Thoroughly clean all parts well. All sealing surfaces on the engine block, cylinder head(s) and both the upper and lower sections of the intake manifold must be clean. Check for cracks, signs of wear in the camshaft bores or other damage. With a straightedge and feeler gauge, check the head for flatness. It should be within 0.0012 in. (0.03mm) along its length. The service limit is 0.008 in. (0.2mm). If the head must be resurfaced, the grinding limit is 0.008 in. (0.2mm). Note that this dimension is a combined total dimension of stock material

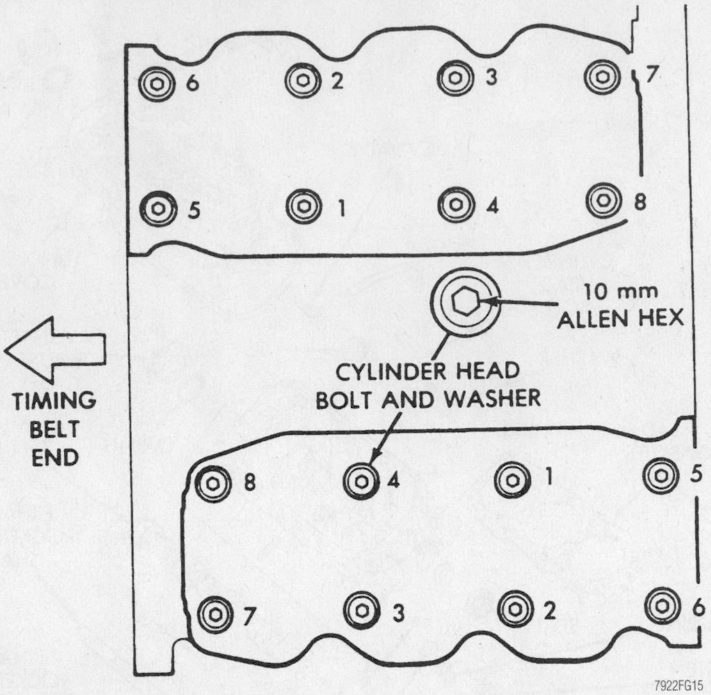

Cylinder head bolt torque sequence—2.5L engine

removal from the cylinder head, if any, and the block top surface is 0.0079 in. (0.2mm).

12. Camshaft end-play can be checked as follows:

a. Oil the camshaft journals with clean engine oil and install (if removed) the camshaft without the rocker arm assemblies.

b. Move the camshaft as far rearward as it will go.

c. Mount a dial indicator to bear on the front of the camshaft.

d. Zero the indicator.

e. Move the camshaft as far forward as it will go. End-play should be 0.004–0.008 in. (0.1–0.2mm). Maximum allowed end-play is 0.016 in. (0.4mm).

13. If the camshafts were removed, lubricate the camshaft journals and carefully reinstall the camshaft into the cylinder head. Install the thrust case and torque the thrust case mounting bolts to 108 inch lbs. (13 Nm).

14. Apply a light coating of engine oil to the lip of the camshaft seal(s).

➡Refer to Section 1 of this manual for the cylinder head torque sequence illustration. The illustration is located after the Torque Specification Chart.

15. Install or connect the following:

- Camshaft seal(s)
- Camshaft sprocket(s) and torque the bolts to 65 ft. lbs. (88 Nm)
- New cylinder head gasket
- Cylinder head
- New 10mm Allen hex cylinder head bolts, with washers. Torque the cylinder head bolts, in proper sequence, in 2–3 steps, to 80 ft. lbs. (108 Nm).
- Lower intake manifold using new gaskets
- Rocker arm assemblies
- Exhaust manifold and crossover exhaust pipe
- Timing belt, properly align the timing marks
- Timing belt covers

16. Inspect the spark plug tube seals located on the ends of each tube. These seals slide onto each tube to seal the cylinder head cover to the spark plug tube. If these seals show signs of hardness and/or cracks, they should be replaced.

17. Thoroughly clean all sealing surfaces.

18. Remove or disconnect the following:

- Cover using a new gasket. Torque bolts to 88 inch lbs. (10 Nm).
- Distributor assembly
- Spark plug wires
- Upper intake manifold (plenum) using a new gasket. Torque the bolts to 13 ft. lbs. (18 Nm).
- Plenum support bracket bolts
- EGR tube
- Speed control cable, if equipped
- Throttle cable
- TPS and IAC motor electrical connectors
- Air cleaner cover, inlet hoses and air inlet resonator
- MAP and IAT sensor wiring connectors
- Accessory drive belts and adjust them
- Negative battery cable

19. Refill the cooling system.

➡An oil and filter change is recommended whenever a cylinder head has been removed since coolant can get into the oil system.

20. Check to be sure all ducts, hoses, fuel lines and wiring connections have all been properly reattached.

21. Start the engine and check for leaks, abnormal noises and vibrations. Bleed the cooling system.

2.7L Engine

1. Before servicing the vehicle, refer to the precautions in the beginning of this section.

2. Relieve the fuel system pressure.

3. Drain the cooling system.

4. Remove or disconnect the following:

- Negative battery cable
- Drive belts
- Vibration damper
- Exhaust cross under pipe
- Catalytic converter
- Upper and lower intake manifolds
- Rocker arm covers
- Water outlet connector and rotate the crankshaft until the sprocket timing mark aligns with the timing mark on the oil pump housing
- Primary timing chain
- Camshaft bearing caps gradually
- Camshafts and valve train components
- Cylinder head and discard the gasket

To install:

5. Clean all mating surfaces of any residual gasket material.

6. Lubricate the threads of the cylinder head bolts with clean engine oil.

7. Install or connect the following:

- Cylinder head with a new gasket

8. Torque the bolts, in sequence, using the following procedure:

a. Step 1: Bolts 1–8: 35 ft. lbs. (48 Nm).

b. Step 2: Bolts 1–8: 55 ft. lbs. (75 Nm).

c. Step 3: Bolts 1–8: 55 ft. lbs. (75 Nm).

d. Step 4: Bolts 1–8: An additional 90

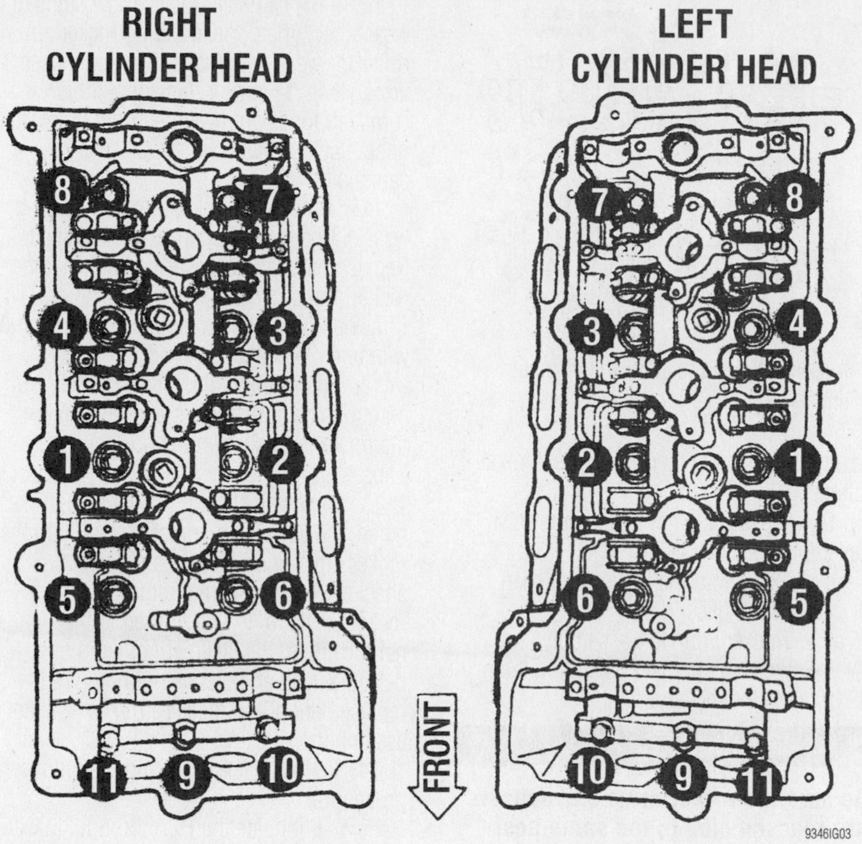

Cylinder head bolt torque sequence—2.7L engine

degrees. Do not use a torque wrench for this step.

e. Step 5: Bolts 9–11: 21 ft. lbs. (28 Nm).

- Valve train components and the camshafts
- Timing chain and sprockets
- Water outlet connector
- Rocker arm covers
- Timing chain cover
- Crankshaft vibration damper
- Lower and upper intake manifolds
- Catalytic converter
- Exhaust cross under pipe
- Drive belts
- Negative battery cable

9. Fill the cooling system to the proper level.
10. Start the vehicle, check for leaks and repair if necessary.

3.0L Engine

1. Before servicing the vehicle, refer to the precautions in the beginning of this section.
2. Relieve the fuel system pressure.
3. Drain the cooling system.
4. Remove or disconnect the following:
   - Negative battery cable
   - Timing belt
   - Alternator
   - Intake manifold
   - Exhaust manifold
   - Water inlet pipe
   - Breather and blow by hoses
   - Positive Crankcase Ventilation (PCV) hose
   - Spark plug cables
   - Rocker arm cover
   - Timing belt rear cover
   - Cylinder head and discard the gasket

To install:

5. Clean all mating surfaces of any residual gasket material.
6. Lubricate the threads of the cylinder head bolts with clean engine oil.
7. Install or connect the following:
   - Cylinder head with a new gasket
8. Torque the bolts, in sequence, as follows:
   a. Step 1: 80 ft. lbs. (108 Nm).
   b. Step 2: Loosen all bolts in sequence.
   c. Step 3: 80 ft. lbs. (108 Nm).
   - Rear timing belt cover
   - Rocker arm cover

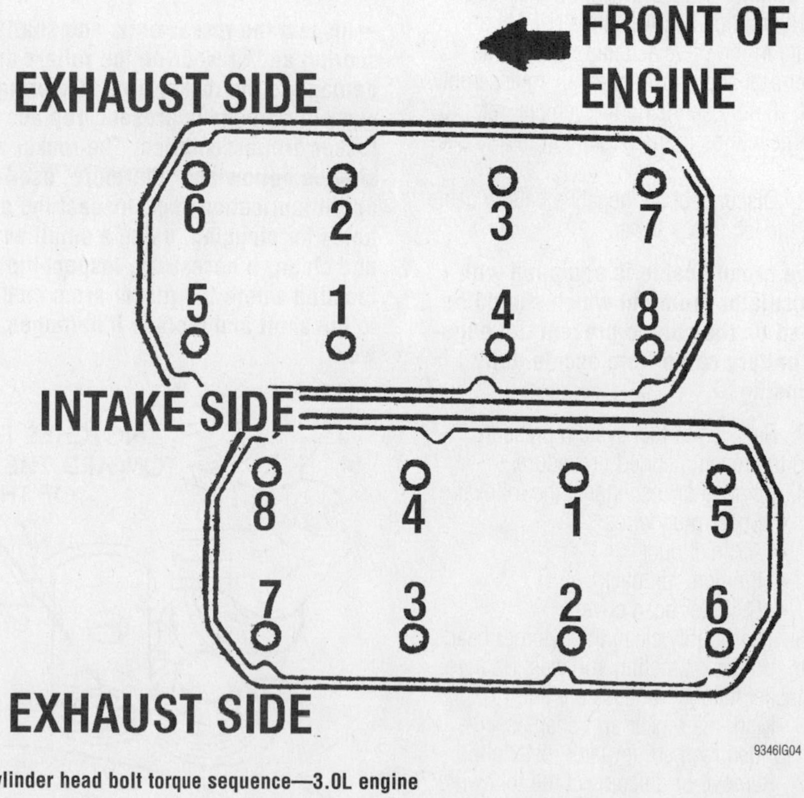

Cylinder head bolt torque sequence—3.0L engine

- Spark plug cables
- PCV and blow by hoses
- Breather hose
- Water inlet pipe
- Exhaust manifold
- Intake manifold
- Alternator
- Timing belt
- Negative battery cable

9. Fill the cooling system to the proper level.
10. Start the vehicle, check for leaks and repair if necessary.

Rocker Arm/Shaft

REMOVAL & INSTALLATION

The 2.0L and 2.5L engines are equipped with rocker arms/shafts. On the 2.4L (VIN X) engine the camshaft acts directly on the valve, therefore, no rocker arms/shafts are used.

2.0L Engine

This engine uses an Single Over Head Camshaft (SOHC) running in an aluminum cylinder head. Rocker arm shafts mount directly to the cylinder head. Care must be taken to be sure all valve timing marks align after cylinder head and valvetrain service. The Hydraulic Lash Adjusters (HLAs) are located in the valve actuating end of the rocker arm and are serviced as an assembly.

1. Before servicing the vehicle, refer to the precautions in the beginning of this section.
2. Disconnect the negative battery cable from the left shock tower.

➡The ground cable is equipped with an insulator grommet which should be placed on the stud to prevent the negative battery cable from accidentally grounding.

3. Relieve the fuel system pressure using the recommended procedure.
4. Remove or disconnect the following:
   - Spark plug wires
   - Air inlet duct
   - Ignition coil pack
   - Cylinder head cover
5. Thoroughly clean the cylinder head cover and gasket mating surfaces. Be sure the gasket mating surfaces are flat.
6. Mark the rocker arm shaft assemblies to identify them for later installation.
7. Remove or disconnect the following:
   - Rocker arm shaft bolts
   - Rocker arm assemblies
8. Mark the rocker arm spacers and retainers to identify them for correct installation.

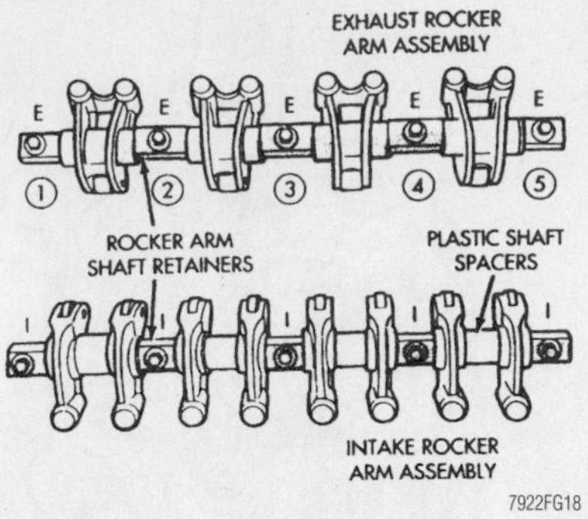

Rocker arm shaft identification—2.0L engine

9. Disassemble the rocker arm/shaft assemblies as follows:
   - Bolts from the rocker arm shaft assemblies
   - Rocker arm/HLA assembly and rocker arm spacers, sliding them off the rocker arm shaft

⁂ WARNING

Be sure the rocker arms and spacers are reassembled in the same positions they are removed.

To install:

➡Inspect the rocker arms and shaft for scoring and/or wear on the rollers or damage to the rocker arm. If scoring, wear or damage is present, replace the rocker arm assemblies. The rocker arm shaft is hollow and, therefore, used as an oil lubrication duct. Inspect the oil holes for clogging, using a small wire and clean, if necessary. Inspect the location where the rocker arms mount to the shaft and replace if damaged or worn.

10. If the camshaft lobes show signs of wear, check the corresponding rocker arm roller for wear or damage. Replace rocker arms/HLAs if worn or damaged. If the camshaft lobes show signs of pitting on the nose, flank or base circle replace the camshaft.
11. Inspect the rocker arms for scoring, wear on the roller or damage to the shaft. Replace parts as necessary. Check that the rocker arm shaft is clean inside and out. Check all oil holes for clogging with a small wire and clean and required.
12. Thoroughly lubricate all rocker arm components and spacers and reinstall on the rocker arm shaft in the original locations.
13. If the vehicle exhibited a tappet-like noise, the valve lash adjusters built into the rocker arms should be cleaned and checked. Lash adjusters removed from a rocker arm should be returned to their original locations. Replacement of worn or defective lash adjusters would require the replacement of the rocker arm/HLAs as an assembly.
14. Install a HLA, by using the following procedure:
    a. Lubricate the HLA thoroughly with clean engine oil.
    b. Reinstall the HLA into the rocker arm making sure the HLA is at least partially filled with oil.
    c. Place the rocker arm in clean engine oil and pump the plunger until the HLA travel is taken up. If travel is not reduced, replace the adjuster with the rocker arm as an assembly.
    d. Reinstall the rocker arm back on the rocker arm shaft.
15. Before installing the rocker arm and shaft assemblies, set the crankshaft to 3 notches Before Top Dead Center (BTDC) on the crankshaft sprocket.

Rocker arm shaft notch locations—2.0L engine

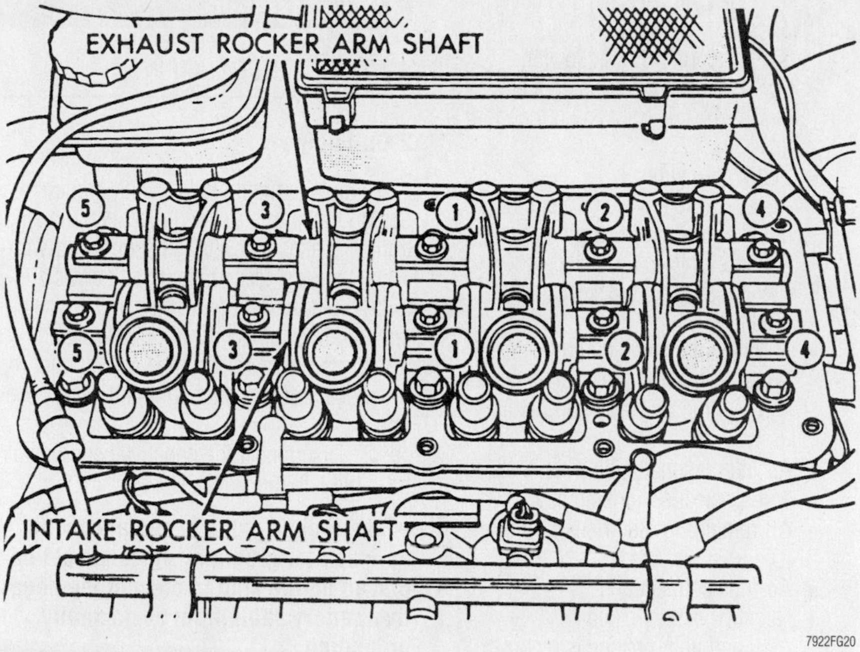

Rocker arm shaft tightening sequence—2.0L engine

➡When installing the intake rocker arm/shaft assembly, be sure the plastic rocker arm spacers do not interfere with the spark plug tubes. If there is interference, rotate the plastic spacers until they are at the proper angle. Do not rotate the spacers by forcing down on the shaft assembly or damage to the spark plug tubes will occur.

16. Reinstall the rocker arm and shaft assemblies with the small notches in the rocker shafts pointing up and toward the timing belt side of the engine. Install the retainers in their original positions on the exhaust and intake shafts. Tighten the bolts in proper sequence to 17 ft. lbs. (23 Nm).
17. Install or connect the following:
    - Cylinder head cover using a new gasket. Torque the bolts to 105 inch lbs. (12 Nm)
    - Ignition coil pack. Torque the fasteners to 17 ft. lbs. (23 Nm)
    - Spark plug wires
    - Air cleaner inlet duct
    - Negative battery cable
18. Check to be sure all electrical, vacuum and fluid connections are reattached as required.
19. An oil and filter change are recommended.
20. Start the engine and check for leaks. Test drive the vehicle to check for proper operation.

2.4L (VIN G) and 3.0L Engines

1. Before servicing the vehicle, refer to the precautions in the beginning of this section.
2. Relieve the fuel system pressure using the recommended procedure.
3. Remove or disconnect the following:
    - Negative battery cable
    - Breather hose
    - Positive Crankcase Ventilation (PCV) hose and valve
    - Rocker arm cover
    - Rocker shaft cap
    - Rocker arms and shafts

To install:

4. Install or connect the following:
    - Rocker arms and shafts. When properly aligned, torque the rocker shaft cap to 23 ft. lbs. (31 Nm)
    - Oil seals
    - Rocker arm cover with a new gasket
    - PCV valve with a new gasket
    - PCV hose
    - Breather hose
    - Negative battery cable

2.5L Engine

1. Before servicing the vehicle, refer to the precautions in the beginning of this section.
2. Disconnect the negative battery cable from the left shock tower.

➡The ground cable is equipped with an insulator grommet which should be placed on the stud to prevent the negative battery cable from accidentally grounding.

3. Relieve the fuel system pressure using the recommended procedure.
4. If removing the right (firewall) side rocker arm/shaft assembly, remove the upper intake manifold (air intake plenum), which is a 2-piece unit of aluminum alloy. Use care working with light alloy parts.
5. Remove or disconnect the air intake plenum using the following procedure:
    - Fuel supply tube from the fuel rail. Squeeze the tabs together and pull the fuel tube/quick-connect fitting assembly off the fuel tube nipple
    - Manifold Absolute Pressure (MAP) sensor and Intake Air Temperature (IAT) sensor electrical connections
    - Plenum support bracket bolt located rearward of the MAP sensor
    - Air inlet resonator-to-intake manifold bolt
    - Throttle body air inlet hose clamp
    - Air cleaner housing cover and inlet hoses
    - Throttle Position (TP) sensor electrical connector and Idle Air Control (IAC) motor electrical connections
    - Throttle cable by sliding it out of the bracket
    - Speed control cable by sliding it out of its bracket, if equipped
    - Exhaust Gas Recirculation (EGR) tube from the intake manifold
    - Plenum support bracket bolt from the rear of the EGR tube
    - Upper intake plenum
6. Remove or disconnect the following:
    - Spark plug wires
    - Cylinder head cover
7. Identify the rocker arm shaft assemblies before removal.
8. Install the auto lash adjuster retainers Special Tool MD 998443, to keep the auto lash adjusters from falling out of the rocker arms when the rocker arm assembly is removed.
9. Loosen the fasteners and remove the rocker arm shaft assemblies.

➡The hydraulic automatic lash adjusters are precision units installed in the machined openings in the rocker arm units. Do not disassemble the auto lash adjusters from the rocker arms.

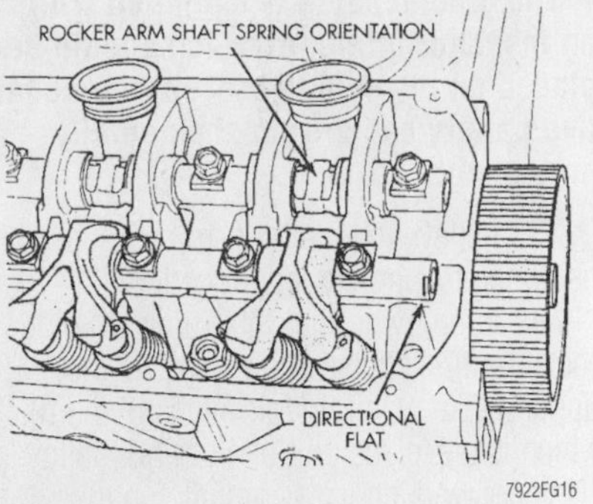

The flats on the rocker arm shaft aids proper orientation—2.5L engine

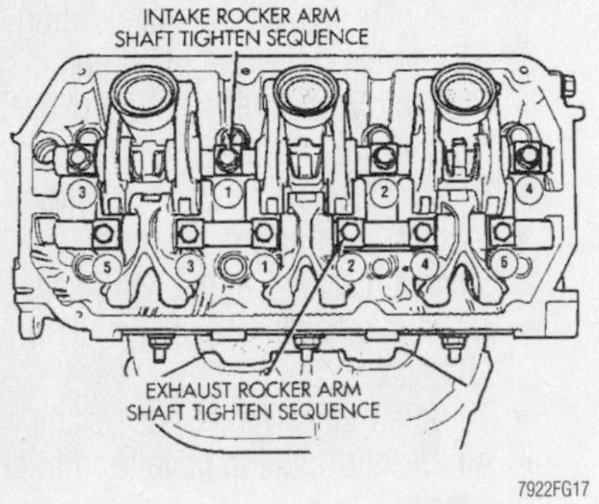

Rocker arm shaft tightening sequence—2.5L engine

To install:

10. The rocker arm shafts are hollow and used as a lubrication oil duct. Be sure all valvetrain parts are clean. Check the rocker arm mounting portion of the shafts for wear or damage. Replace if necessary. Check all oil holes for clogging with a small wire and clean as required. If any rockers were removed, lubricate and install on the shafts in their original positions.

11. For the right cylinder head, install the rocker arm and shaft assemblies with the FLAT in the rocker arm shafts facing the timing belt side of the engine. For the left cylinder head, install the rocker arm and shaft assembly with the FLAT in the rocker arm shaft facing the transaxle side of the engine. Install the retainers and spring clips in their original positions on the exhaust and intake shafts. Tighten the retainer bolts to 23 ft. lbs. (31 Nm) working from the center, outward. Remove the valve lash retainer tools that should have been installed at disassembly.

12. Inspect the spark plug tube seals located on the ends of each tube. These seals slide onto each tube to seal the cylinder head cover to the spark plug tube. If these seals show signs of hardness and/or cracking, they should be replaced.

13. Clean the cylinder head and cover mating surfaces.

14. Install or connect the following:
- Cylinder head cover using a new gasket. Torque bolts to 88 inch lbs. (10 Nm)
- Spark plug wires
- Upper intake manifold (plenum) using a new gasket. Tighten the bolts to 13 ft. lbs. (18 Nm)
- Plenum support bracket bolts
- EGR tube
- Throttle and speed control cables
- TP sensor, MAP sensor, IAT sensor and IAC motor electrical connectors
- Air inlet resonator, air inlet hose and air cleaner housing cover
- All remaining electrical connectors
- Air tube connections
- Negative battery cable

15. An oil and filter change is recommended.

✻✻ WARNING

Operating the engine without the proper amount and type of engine oil will result in severe engine damage.

16. Start the engine and check for leaks, abnormal noises and vibrations.

2.7L Engine

1. Before servicing the vehicle, refer to the precautions in the beginning of this section.
2. Relieve the fuel system pressure using the recommended procedure.
3. Remove or disconnect the following:
   - Negative battery cable
   - Rocker arm cover and rotate the engine until the cam lobe is on its base circle
4. Using special tools8215—A and 8216—A, depress the valve spring enough to release the tension on the rocker arm.
5. Remove the rocker arm from the cylinder head.

To install:

6. Lubricate the rocker arms with clean engine oil.
7. Depress the valve spring far enough to install the rocker arm.
8. Install or connect the following:
   - Rocker arm into its original position over the valve and lash adjuster.
   - Rocker arm cover
   - Negative battery cable

Intake Manifold

REMOVAL & INSTALLATION

2.0L Engine

The intake manifold is a long branch design made of a molded plastic composition. It is attached to the cylinder head with 10 fasteners. Please note that all seals are to be replaced with new seals and all fasteners are to be replaced with new fasteners.

1. Before servicing the vehicle, refer to the precautions in the beginning of this section.
2. Disconnect the negative battery cable from the left shock tower.

➡The ground cable is equipped with an insulator grommet which should be placed on the stud to prevent the negative battery cable from accidentally grounding.

3. Relieve the fuel system pressure using the recommended procedure.
4. Remove the air inlet resonator as follows:
   - Air inlet resonator-to-throttle body screw
   - Air inlet resonator-to-air inlet tube clamp, loosen it
   - Resonator
5. Remove or disconnect the following:
   - Fuel supply line quick-disconnect fitting from the fuel rail. Squeeze the retainer tabs together and pull the fuel tube fitting from the fuel tube nipple. The retainer will remain on the fuel tube.

➡Wrap shop towels around the fuel line openings to catch any spilling fuel.

   - Fuel rail

✻✻ WARNING

Use care when handling the fuel injectors. Do not set them on their tips. Cover the fuel injector openings after fuel rail removal.

   - Accelerator, kickdown and speed control cables from the throttle lever and bracket
   - Throttle Position (TP) sensor and Idle Air Control (IAC) motor electrical connections
   - Vacuum hoses from the throttle body
   - Manifold Absolute Pressure (MAP) sensor, Idle Air Control (IAC) motor and Intake Air Temperature (IAT) sensor

- Vapor and brake booster hoses
- Knock sensor electrical con-nec-tor
- Starter relay electrical connector
- Wiring harness from the intake manifold tab
- Transaxle-to-throttle body support bracket fasteners at the throttle body and loosen the fastener at the transaxle end
- Throttle body assembly
- Exhaust Gas Recirculation (EGR) tube
- Intake manifold-to-inlet water tube support fastener
- Intake manifold fasteners and discard them
- Intake manifold

To install:

6. Clean all sealing surfaces. Check upper and lower manifold gasket surfaces for flatness with a straightedge. Surface must be flat within 0.006 in. (0.15mm) per foot (30cm).

➡Replace all seals and fasteners with new ones.

7. Install or connect the following:
   - Intake manifold with new O-ring seals. Torque the fasteners, in proper sequence, to 105 inch lbs. (12 Nm)
8. Lubricate the fuel injector O-rings with engine oil.
9. Install or connect the following:
   - Fuel injector/fuel rail assembly. Torque the fuel rail bolts to 17 ft. lbs. (23 Nm)
   - Fuel injector electrical connectors
10. Lubricate the quick-connect fittings with engine oil.
11. Install or connect the following:
   - Fuel supply line to the fuel rail. Pull on the connector to insure it is locked into position.
   - PCV and brake booster hoses
   - Throttle body and torque the fasteners to 17 ft. lbs. (23 Nm)
   - Transaxle-to-throttle body support bracket. Torque the fasteners to 105 inch lbs. (12 Nm)

➡Tighten the support bracket at the throttle body first; then, at the transaxle.

- MAP and IAT sensor electrical connectors
- Knock sensor electrical and starter relay connectors
- Wiring harness to the intake manifold tab
- IAC and TP sensor electrical connectors
- Throttle body vacuum hoses
- Accelerator, kickdown and speed control cables, to their bracket(s) and throttle lever
- EGR tube. Torque the fasteners to 95 inch lbs. (11 Nm)

➡Tighten the EGR tube-to-valve fasteners first; then, at the intake manifold.

- Fresh air duct to the air filter housing
- Air inlet resonator to the throttle body
- Air inlet tube to the resonator. Torque the clamps to 20–30 inch lbs. (2–3 Nm).
- Negative battery cable

2.4L (VIN X) Engine

The intake manifold is a long branch design made of cast aluminum. It is attached to the cylinder head with 8 fasteners.

1. Before servicing the vehicle, refer to the precautions in the beginning of this section.
2. Disconnect the negative battery cable from the left shock tower.

➡The ground cable is equipped with an insulator grommet which should be placed on the stud to prevent the negative battery cable from accidentally grounding.

3. Relieve the fuel system pressure using the recommended procedure.
4. Remove the air inlet resonator as follows:
   - Both air inlet resonator-to-intake manifold bolts
   - Resonator-to-throttle body screw, loosen it
   - Air inlet resonator-to-air inlet tube clamp, loosen it
   - Resonator
5. Remove or disconnect the following:
   - Fuel supply line quick-disconnect at the fuel tube assembly

➡Squeeze the retainer tabs together and pull the fuel tube/quick-disconnect fitting assembly from the fuel tube nipple. The retainer will remain on the fuel tube. Use shop towels to catch any dripping fuel.

- Fuel rail

⁂ WARNING

Use care when handling the fuel injectors. Do not set them on their tips. Cover the fuel injector openings after fuel rail removal.

- Accelerator, kickdown and speed control cables, from the throttle lever and bracket
- Idle Air Control (IAC) motor and Throttle Position (TP) sensor electrical connections
- Throttle body vacuum hoses
- Manifold Absolute Pressure (MAP) sensor and Intake Air Temperature (IAT) sensor electrical connections
- Vapor and brake booster hoses
- Knock sensor electrical connector
- Wiring harness from the intake manifold tab
- Transaxle-to-throttle body support bracket fasteners at the throttle body and loosen the fastener at the transaxle end
- Throttle body
- Exhaust Gas Recirculation (EGR) tube
- Intake manifold support bracket
- Intake manifold

To install:

6. Thoroughly clean all parts. Clean all sealing surfaces
7. Install or connect the following:
   - New gasket
   - Intake manifold. Torque the fasteners to 17 ft. lbs. (23 Nm), in correct sequence, starting at the center and working outward

➡Make sure the fuel injector openings clean.

- Fuel rail assembly to the intake manifold and torque the screws to 17 ft. lbs. (23 Nm)

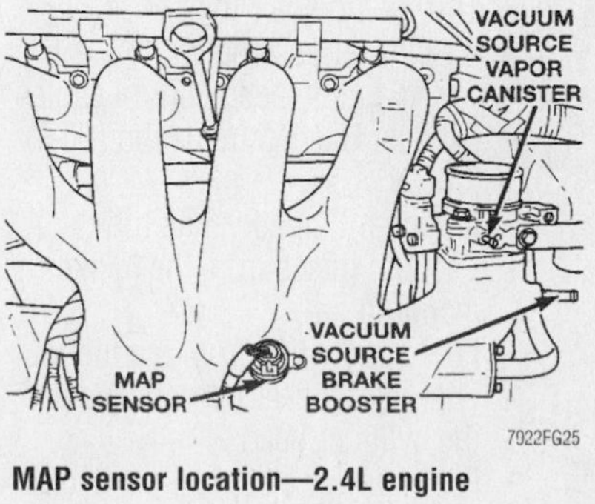

MAP sensor location—2.4L engine

- Positive Crankcase Ventilation (PCV) and brake booster hoses

8. Lubricate the fuel tube with engine oil.
9. Install or connect the following:
   - Fuel supply line to the fuel rail. Pull on the connector to insure it is locked into position
   - Throttle body and torque the fasteners to 17 ft. lbs. (23 Nm)
   - Transaxle-to-throttle body support bracket. Torque the fasteners to 105 inch lbs. (12 Nm) at the throttle body first; then, at the transaxle
   - MAP and IAT electrical connectors
   - Knock sensor electrical connector
   - Wiring harness to the intake manifold tab
   - IAC motor and TP sensor electrical connectors
   - Throttle body vacuum hoses
   - Accelerator, kickdown and speed control cables to the throttle lever and bracket
   - EGR tube. Torque the fasteners to 95 inch lbs. (11 Nm), at the EGR valve first; then, the intake manifold
   - Air inlet resonator to the throttle body
   - Air inlet tube to the resonator and torque the clamps to 20–30 inch lbs. (2.5–3.5 Nm)
   - Both air inlet resonator-to-intake manifold bolts
   - Negative battery cable

2.4L (VIN G) Engine

The intake manifold is a long branch design made of cast aluminum. It is attached to the cylinder head with 8 fasteners.

1. Before servicing the vehicle, refer to the precautions in the beginning of this section.
2. Drain the cooling system.
3. Properly relieve the fuel system pressure.
4. Remove or disconnect the following:
   - Negative battery cable
   - Throttle body air inlet hose
   - Air cleaner assembly
   - Throttle and speed control cables
   - Exhaust Gas Recirculation (EGR) tube
   - Engine oil dipstick and tube
   - Vacuum hoses from the intake manifold
   - Fuel supply line from the fuel rail
   - Fuel rail bracket from the side of the cylinder head
   - Fuel injector electrical connector
   - Knock Sensor (KS) connector
   - Engine Coolant Temperature (ECT) sensor connector
   - Idle Air Control (IAC) motor connector
   - Throttle position Sensor (TPS) connector
   - Manifold Absolute Pressure (MAP) sensor
   - A/C pressure sensor connector
   - A/C compressor clutch connector
   - Alternator electrical connectors
   - Fuel rail
   - Coolant outlet connector
   - Intake manifold and discard the gasket

To install:

5. Install or connect the following:
   - Intake manifold with a new gasket and torque the bolts, in sequence, to 105 inch lbs. (12 Nm)
   - Coolant outlet connector
   - Fuel rail and torque the bolts to 100 inch lbs. (12 Nm)
   - Alternator electrical connectors
   - A/C compressor clutch connector
   - A/C pressure sensor connector
   - MAP sensor connector
   - TPS electrical connector
   - IAC motor connector
   - ECT sensor connector
   - KS electrical connector
   - Fuel injector electrical connector
   - Fuel rail bracket to the cylinder head
   - Fuel supply hose to the fuel rail
   - Intake manifold vacuum hoses
   - Engine oil dipstick and tube
   - EGR tube
   - Throttle and speed control cables
   - Negative battery cable
6. Fill the cooling system to the proper level.
7. Start the vehicle, check for leaks and repair if necessary.

2.5L Engine

The intake manifold assembly is composed of an upper plenum and lower manifold. This aluminum alloy manifold has long runners to improve airflow inertia. The plenum chamber absorbs air pulsations created during the suction phase of each cylinder. The lower intake manifold is machined for 6 injectors and the fuel rail mounts.

1. Before servicing the vehicle, refer to the precautions in the beginning of this section.
2. Disconnect the negative battery cable from the left shock tower.

➡The ground cable is equipped with an insulator grommet which should be placed on the stud to prevent the negative battery cable from accidentally grounding.

3. Relieve the fuel system pressure using the recommended procedure.
4. Disconnect the fuel supply line quick-disconnect fitting from the fuel rail. Squeeze the fitting retainer tabs together and separate the connection.

⁂ CAUTION

Wrap shop towels around the connection to catch any gasoline spillage.

➡It may be helpful to identify and tag each sensor connector as it is being removed. This may save time at assembly.

5. Remove or disconnect the following:
   - Manifold Absolute Pressure (MAP) sensor and Intake Air Temperature (IAT) sensor electrical connections
   - Plenum support bracket at the rear of the MAP sensor
   - Air inlet resonator bolt
   - Throttle body air inlet hose clamp, loosen it
   - Air cleaner cover and inlet hoses
   - Throttle Position Sensor (TPS) and Idle Air Control (IAC) motor electrical connections
   - Throttle cable from the bracket
   - Speed control cable out of the bracket, if equipped
   - Exhaust Gas Recirculation (EGR) tube from the intake manifold
   - Plenum support bracket at the rear of the EGR tube
   - Upper intake plenum from the intake manifold
   - Fuel injector electrical connectors
   - Fuel rail

➡There are spacers under each fuel rail bolt.

   - Lower intake manifold and discard the gaskets

To install:

6. Clean all sealing surfaces.
7. Check upper and lower manifold gasket surfaces for flatness with a straightedge. Surface must be flat within 0.006 in. (0.15mm) per 12 in. (30cm) of manifold length.
8. Install or connect the following:
   - New gaskets
   - Lower intake manifold
9. For 1998–99 engines, torque the nuts, in the correct sequence, to 15 ft. lbs. (21 Nm).
10. For 2000–02 engines, torque each bank nuts in the following order:

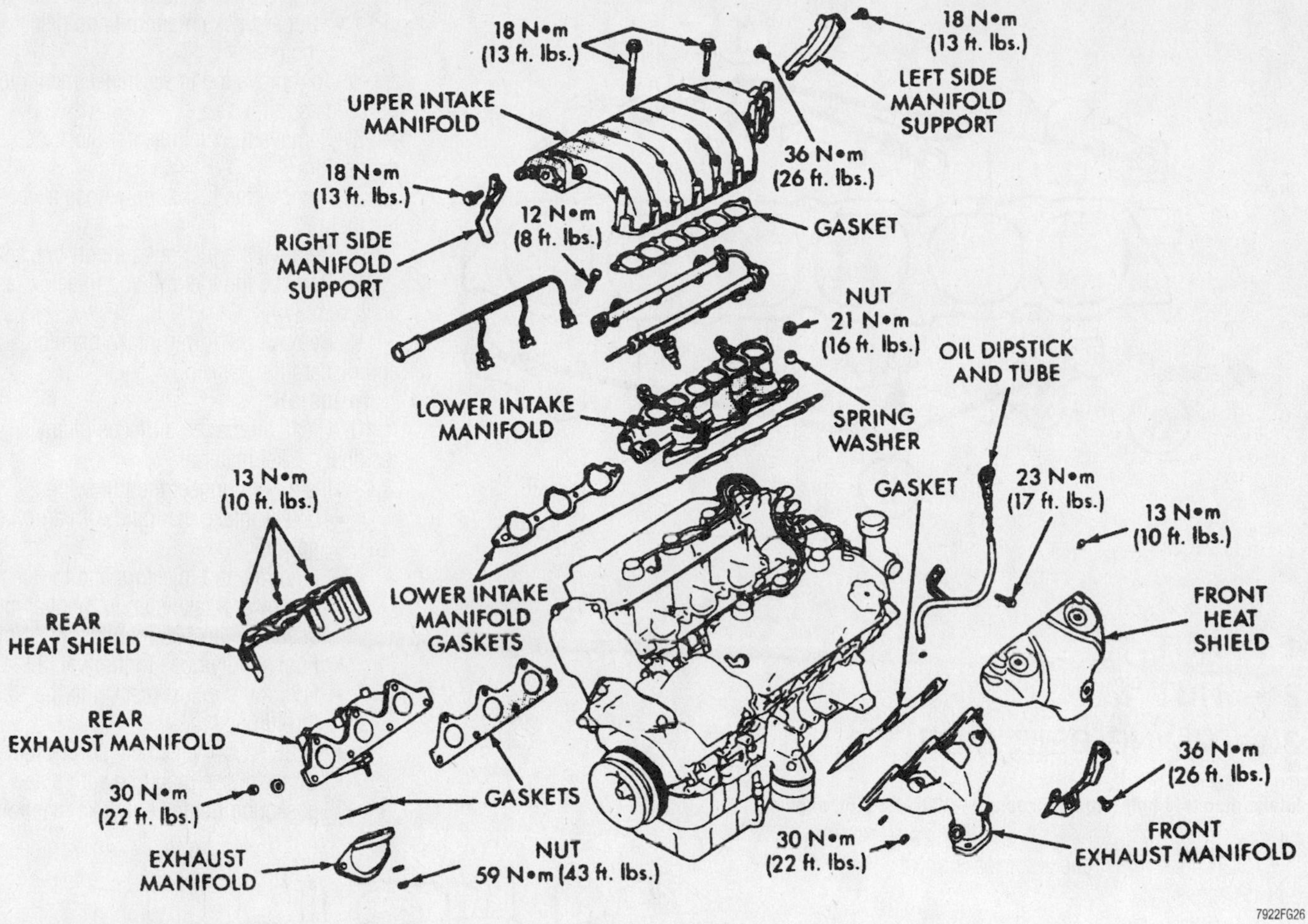

Exploded view of the intake/exhaust manifolds and related components—2.5L engine

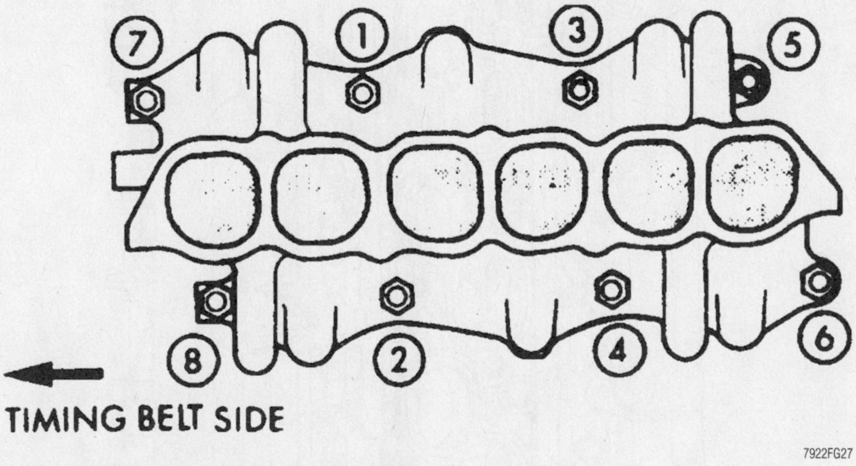

Intake manifold bolt torque sequence—1998–99 2.5L engine

a. Step 1: Right bank to 56 inch lbs. (6.4 Nm).
b. Step 2: Left bank: 15 ft. lbs. (21 Nm).
c. Step 3: Right bank: 15 ft. lbs. (21 Nm).
d. Step 4: Left bank: 15 ft. lbs. (21 Nm).
e. Step 5: Right bank: 15 ft. lbs. (21 Nm).

11. Lubricate the fuel injector O-rings with engine oil.

12. Install or connect the following:
- Fuel injectors and properly seat them
- Fuel rail and torque the bolts to 96 inch lbs. (12 Nm)
- Fuel injector electrical connectors
- Fuel supply line to the fuel rail
- Upper intake plenum using new gaskets and torque the bolts to 13 ft. lbs. (18 Nm)
- Plenum support brackets and torque the fasteners to 13 ft. lbs. (18 Nm)
- EGR tube and torque the screws to 95 inch lbs. (11 Nm)
- Throttle cables
- TP sensor, MAP sensor, IAT sensor and IAC motor electrical connectors
- Air cleaner assembly and torque the hose clamps to 25 inch lbs. (3 Nm)
- Air inlet resonator bolt
- Negative battery cable

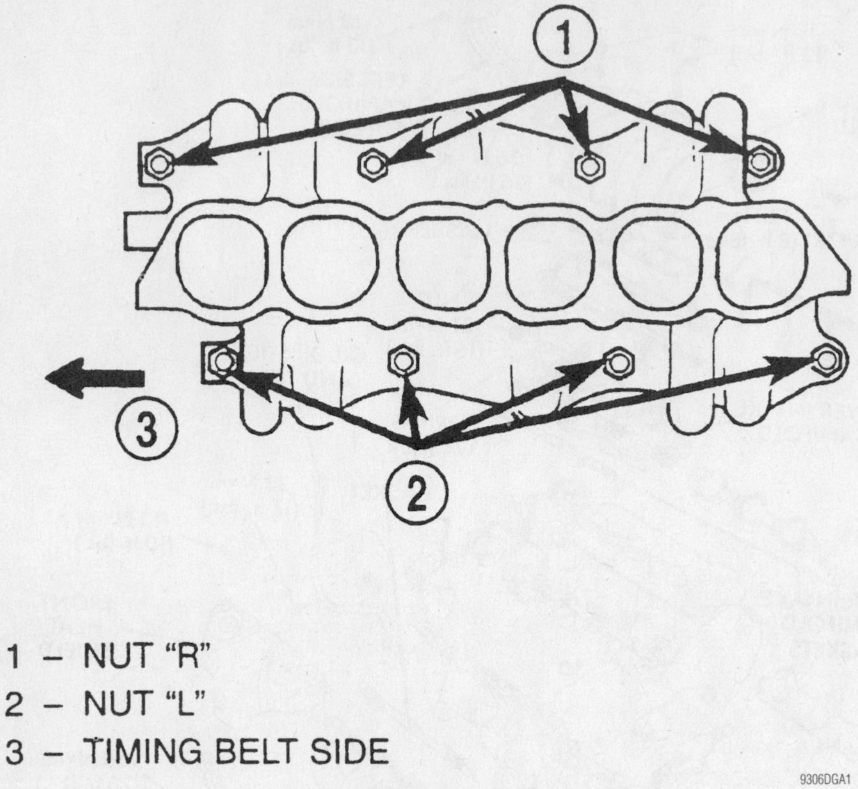

Intake manifold bolt torque sequence—2000–02 2.5L engine

13. Start the engine, check for leaks and repair if necessary.

2.7L Engine

1. Before servicing the vehicle, refer to the precautions in the beginning of this section.
2. Properly relieve the fuel system pressure.
3. Remove or disconnect the following:
   - Negative battery cable
   - Throttle body air inlet hose
   - Air cleaner assembly
   - Throttle cable shield
   - Throttle and speed control cables
   - Throttle cable bracket
   - Manifold Absolute Pressure (MAP) sensor connector
   - Throttle Position Sensor (TPS) connector
   - Idle Air Control (IAC) motor connector
   - Manifold Tuning Valve (MTV) connector
   - Vapor purge hose
   - Brake booster vacuum hose
   - Speed Control Servo
   - Positive Crankcase Ventilation (PCV) hose
   - Exhaust Gas Recirculation (EGR) upper tube
   - Upper throttle body support bracket
   - Upper intake manifold and discard the gasket
4. To remove the lower intake manifold proceed as follows:
5. Remove the fuel injector electrical connectors.
6. Remove the fuel supply hose from the fuel rail.
7. Remove the fuel rail support bracket.
8. Remove the fuel rail and injectors as an assembly.
9. Remove the lower intake manifold and discard the gasket.

To install:

10. Clean all mating surfaces of any residual gasket material.
11. Install or connect the following:
    - Lower intake manifold with a new gasket
    - Fuel rail and injectors and torque the lower intake manifold bolts, in sequence, to 105 inch lbs. (12 Nm)
    - Fuel supply hose to the fuel rail
    - Fuel rail support bracket to the throttle body
    - Fuel injector electrical connectors
    - Upper intake manifold with a new gasket on the lower intake manifold

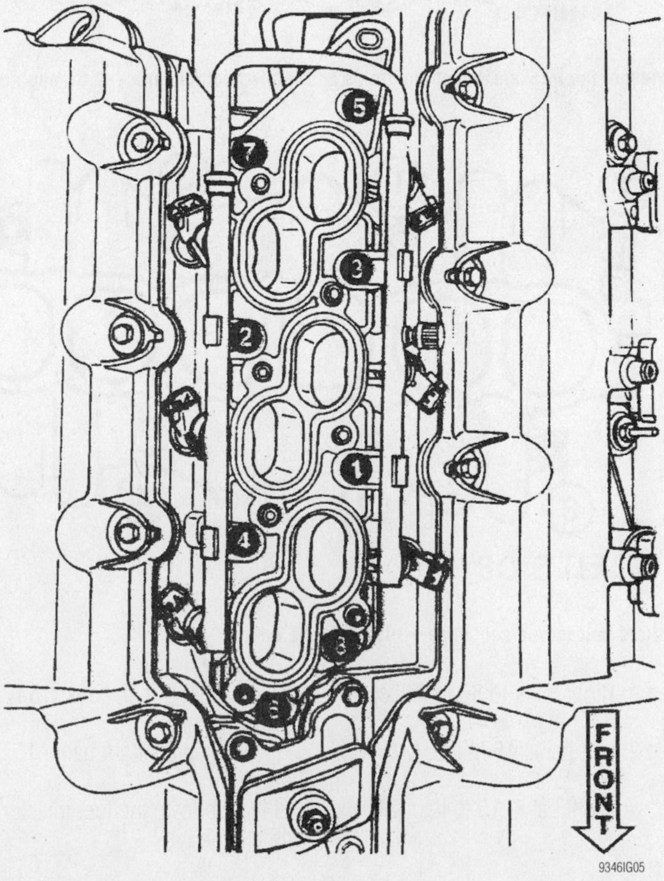

Intake manifold bolt torque sequence 2.7L engine

and torque the bolts, in sequence, to 105 inch lbs. (12 Nm)
- EGR upper tube
- Speed control servo
- PCV, brake booster and vapor purge hoses
- MAP sensor electrical connector
- TPS sensor electrical connector
- IAC motor electrical connector
- MTV electrical connector
- Throttle cable bracket
- Throttle and speed control cables
- Throttle cable shield
- Throttle body air inlet hose
- Air cleaner assembly
- Negative battery cable

12. Start the vehicle, check for leaks and repair if necessary.

3.0L Engine

1. Before servicing the vehicle, refer to the precautions in the beginning of this section.
2. Properly relieve the fuel system pressure.
3. Remove or disconnect the following:
   - Negative battery cable
   - Intake manifold plenum
   - Fuel injector connector
   - Fuel supply and return hoses
   - Vacuum hose
   - Fuel rail, injectors and pressure regulator as an assembly
   - Insulators
   - Positive Crankcase Ventilation (PCV) hose
   - Right front upper timing belt cover and bracket
   - Intake manifold and discard the gasket

To install:

4. Clean all mating surfaces of any residual gasket material.
5. Install or connect the following:
   - Intake manifold with a new gasket. Torque the nuts as follows:
     a. Right bank to 62 inch lbs. (7 Nm).
     b. Left bank to 16 ft. lbs. (22 Nm).
     c. Right bank to 16 ft. lbs. (22 Nm).
     d. Left bank to 16 ft. lbs. (22 Nm).
     e. Right bank to 16 ft. lbs. (22 Nm).
   - Right front upper timing belt cover and bracket and torque the bolts to 10 ft. lbs. (15 Nm)
   - PCV hose
   - Fuel rail assembly and torque the bolts to 102 inch lbs. (12 Nm)
   - Vacuum hose
   - Fuel supply and return hoses
   - Fuel injector electrical connector
   - Negative battery cable
6. Start the vehicle, check for leaks and repair if necessary.

Exhaust Manifold

REMOVAL & INSTALLATION

2.0L and 2.4L Engines

1. Before servicing the vehicle, refer to the precautions in the beginning of this section.
2. Disconnect the negative battery cable from the left shock tower.

➡The ground cable is equipped with an insulator grommet which should be placed on the stud to prevent the negative battery cable from accidentally grounding.

3. Remove or disconnect the following:
   - Exhaust pipe from the exhaust manifold
   - Exhaust manifold heat shield
   - Heated Oxygen Sensor (HO_2S), if necessary
   - Exhaust manifold and discard the gasket

To install:

4. Thoroughly clean all parts. Clean all sealing surfaces of the manifold and cylinder head. Check the manifold gasket surface for flatness with a straightedge and feeler gauge. The surface must be flat within 0.006 in. (0.15mm) per foot (30cm) of manifold length. Inspect the manifold for cracks or distortion. Replace if necessary.
5. Install or connect the following:
   - New gasket
   - Exhaust manifold. Torque the bolts, starting at the center and working outward, to 17 ft. lbs. (23 Nm)

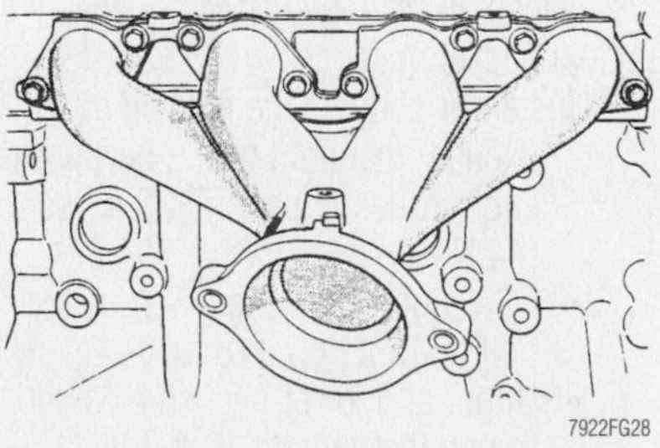

Be sure the gasket mating surfaces are clean and flat before installing the exhaust manifold—2.0L engine

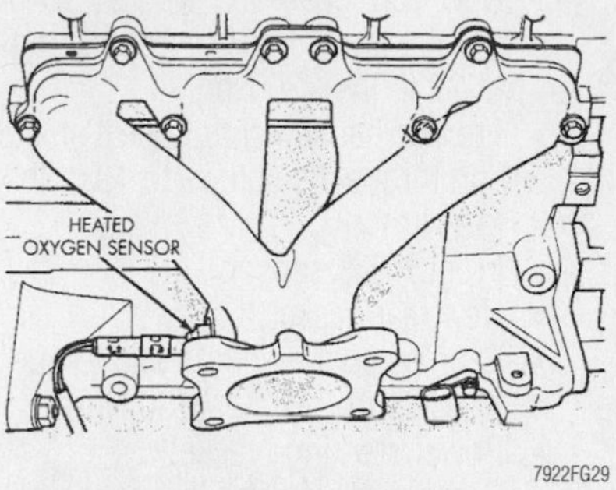

Be careful not to damage the oxygen sensor when servicing the manifold—2.4L engine

   - HO_2S sensor
   - Heat shield
   - Exhaust pipe and torque the fasteners to 21 ft. lbs. (28 Nm)
   - Negative battery cable
6. Start the engine and allow it to idle while inspecting the manifold for exhaust leaks.

2.5L Engine

1. Before servicing the vehicle, refer to the precautions in the beginning of this section.
2. Disconnect the negative battery cable from the left shock tower.

➡The ground cable is equipped with an insulator grommet which should be placed on the stud to prevent the negative battery cable from accidentally grounding.

3. Remove or disconnect the following:
   - Exhaust pipe from the rear (cowl side) exhaust manifold, at the flex joint
   - Crossover pipe assembly
   - Heated Oxygen Sensor (HO_2S) from the rear manifold
   - Power steering bracket
   - Rear exhaust manifold heat shield
   - Rear exhaust manifold and discard the gasket
   - Front HO_2S sensor
   - Front manifold heat shield
   - Front manifold

To install:

4. Thoroughly clean all parts. Inspect the exhaust manifolds for damage or cracks and check for distortion of the cylinder head sealing surface and exhaust crossover sealing surface with a straightedge and thickness gauge.
5. Install or connect the following:

- New front gasket
- Front manifold and torque the nuts to 22 ft. lbs. (30 Nm)
- Front exhaust manifold heat shield and torque the screws to 10 ft. lbs. (13 Nm)
- Front HO_2S sensor
- New rear gasket
- Rear exhaust manifold and torque the nuts to 22 ft. lbs. (30 Nm)
- Power steering bracket
- Crossover pipe and torque the nuts to 22 ft. lbs. (30 Nm)
- Rear HO_2S sensor
- Exhaust pipe to the rear manifold and torque the bolts to 21 ft. lbs. (28 Nm)
- Negative battery cable

6. Lower the vehicle.
7. Start the engine and allow the engine to idle while inspecting the vehicle for exhaust leaks at the manifold.

2.7L Engine

FRONT

1. Before servicing the vehicle, refer to the precautions in the beginning of this section.
2. Remove or disconnect the following:
   - Negative battery cable
   - Cross under pipe
   - Front catalytic converter
   - Exhaust manifold and discard the gasket

To install:

3. Clean all mating surfaces of any residual gasket material.
4. Install or connect the following:
   - Exhaust manifold with a new gasket. Torque the bolts, starting in the center and working outward, to 17 ft. lbs. (23 Nm)
   - Front catalytic converter and heat shield and torque the bolts to 21 ft. lbs. (28 Nm)
   - Cross under pipe and torque the bolts to 21 ft. lbs. (28 Nm)
   - Negative battery cable
5. Start the vehicle, check for leaks and repair if necessary.

REAR

1. Before servicing the vehicle, refer to the precautions in the beginning of this section.
2. Remove or disconnect the following:
   - Negative battery cable
   - Throttle body air inlet hose
   - Air cleaner assembly
   - Exhaust Gas Recirculation (EGR) tube from the exhaust manifold and EGR valve
   - Exhaust system
   - Cross under pipe
   - Rear catalytic converter
   - Rear Oxygen Sensor (O_2S)
   - Exhaust manifold and discard the gasket

To install:

3. Clean all mating surfaces of any residual gasket material.
4. Install or connect the following:
   - Exhaust manifold with a new gasket. Torque the bolts, starting in the center and working outward, to 17 ft. lbs. (23 Nm)
   - Rear catalytic converter and heat shield and torque the bolts to 21 ft. lbs. (28 Nm)
   - Rear Oxygen Sensor (O_2S)
   - Cross under pipe and torque the bolts to 21 ft. lbs. (28 Nm)
   - Exhaust system
   - EGR tube with new gaskets
   - Air cleaner assembly and air intake hose
   - Negative battery cable
5. Start the vehicle, check for leaks and repair if necessary.

3.0L Engine

1. Before servicing the vehicle, refer to the precautions in the beginning of this section.
2. Remove or disconnect the following:
   - Battery and tray
   - Front exhaust pipe
   - Air cleaner assembly
   - Oil dipstick guide
   - Strut tower bar
   - Upper and lower heat shields
   - Exhaust Gas Recirculation (EGR) pipe, left side only
   - Exhaust manifold and discard the gasket

To install:

3. Clean all mating surfaces of any residual gasket material.
4. Install or connect the following:
   - Exhaust manifold with a new gasket and torque the bolts to 33 ft. lbs. (44 Nm)
   - EGR pipe, if removed. Torque the bolt to 13 ft. lbs. (18 Nm)
   - Upper and lower heat shields and torque the bolts to 10 ft. lbs. (15 Nm)
   - Strut tower bar
   - Oil dipstick guide
   - Air cleaner assembly
   - Front exhaust pipe
   - Battery and tray
5. Start the vehicle, check for leaks and repair if necessary.

Front Crankshaft Seal

REMOVAL & INSTALLATION

2.0L and 2.4L Engines

The timing belt must be removed for this procedure. Use care that all timing marks are aligned after installation or the engine will be damaged.

1. Drain the engine oil.
2. Before servicing the vehicle, refer to the precautions in the beginning of this section.
3. Disconnect the negative battery cable from the left shock tower.

➡The ground cable is equipped with an insulator grommet which should be placed on the stud to prevent the negative battery cable from accidentally grounding.

4. Remove or disconnect the following:
   - Accessory drive belts
   - Crankshaft damper/pulley using a jaw puller tool
   - Timing belt
   - Crankshaft timing belt sprocket using a gear/sprocket puller

➡Be careful not to nick the seal surface of the crankshaft or the seal bore.

   - Front crankshaft seal using Seal Removal Tool No. 6771

➡Be careful not to damage the seal contact area of the crankshaft.

To install:

5. Lubricate the new oil seal lip with engine oil.
6. Install or connect the following:
   - New crankshaft oil seal with the spring facing inward using Oil Seal Installer Tool 6780–1, until it is flush with the front cover
   - Crankshaft timing belt sprocket using Tool No. 6792

➡Be sure the word "FRONT" on the timing belt sprocket is facing outward.

   - Timing belt and timing belt cover
   - Crankshaft damper/pulley using the thrust bearing/washer and 12M-1.75 x 150mm bolt from special Tool No. 6792. Torque the bolt to 105 ft. lbs. (142 Nm)
   - Accessory drive belts and adjust the tension
   - Negative battery cable

⁂ WARNING

Operating the engine without the proper amount and type of engine oil will result in severe engine damage.

7. Fill the engine with clean oil.
8. Start the engine and check for leaks.

2.5L Engine

The timing belt must be removed for this procedure. Use care to be sure all timing marks are aligned after this service or the engine will be damaged.

1. Before servicing the vehicle, refer to the precautions in the beginning of this section.
2. Disconnect the negative battery cable from the left shock tower.

➡The ground cable is equipped with an insulator grommet which should be placed on the stud to prevent the negative battery cable from accidentally grounding.

3. Remove the accessory drive belts.
4. Raise and safely support the vehicle. Drain the engine oil.
5. Remove or disconnect the following:
   - Right inner splash shield
   - Crankshaft damper/pulley
   - Timing belt covers and timing belt
   - Crankshaft timing belt sprocket and key
   - Crankshaft seal by prying it out with a flat-tipped prytool

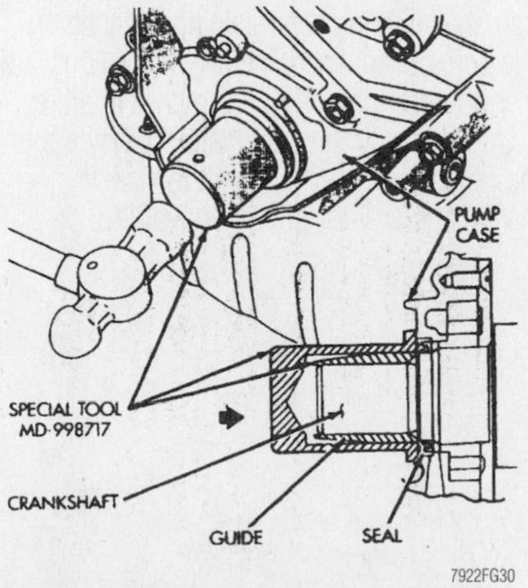

To prevent damage to the end of the crankshaft, use Oil Seal Installer MD998717, as shown—2.5L engine

⁂ WARNING

Be sure to cover the end of the pry-tool tip with a shop towel. Be careful not to nick the seal surface of the crankshaft or the seal bore.

To install:

6. Lubricate the new oil seal lip with engine oil.
7. Install or connect the following:
   - New crankshaft oil seal until it is flush with the oil pump cover, using Oil Seal Installer MD998717
   - Crankshaft timing belt sprocket and key
   - Timing belt and timing belt covers

⁂ WARNING

Verify that all timing marks are correctly aligned or the engine will be damaged.

   - Crankshaft damper/pulley and torque the bolt to 134 ft. lbs. (182 Nm)
   - Right inner splash shield
   - Accessory drive belts and adjust the tension
   - Negative battery cable

⁂ WARNING

Operating the engine without the proper amount and type of engine oil will result in severe engine damage.

8. Fill the engine with clean oil.
9. Start the engine and check for leaks.

2.7L Engine

1. Before servicing the vehicle, refer to the precautions in the beginning of this section.
2. Remove or disconnect the following:
   - Negative battery cable
   - Crankshaft damper
   - Front crankshaft seal with special Tool 6771

To install:

3. Install or connect the following:
   - New seal with special Tools 6780–2, 8179 and 6780–1
   - Crankshaft damper
   - Negative battery cable

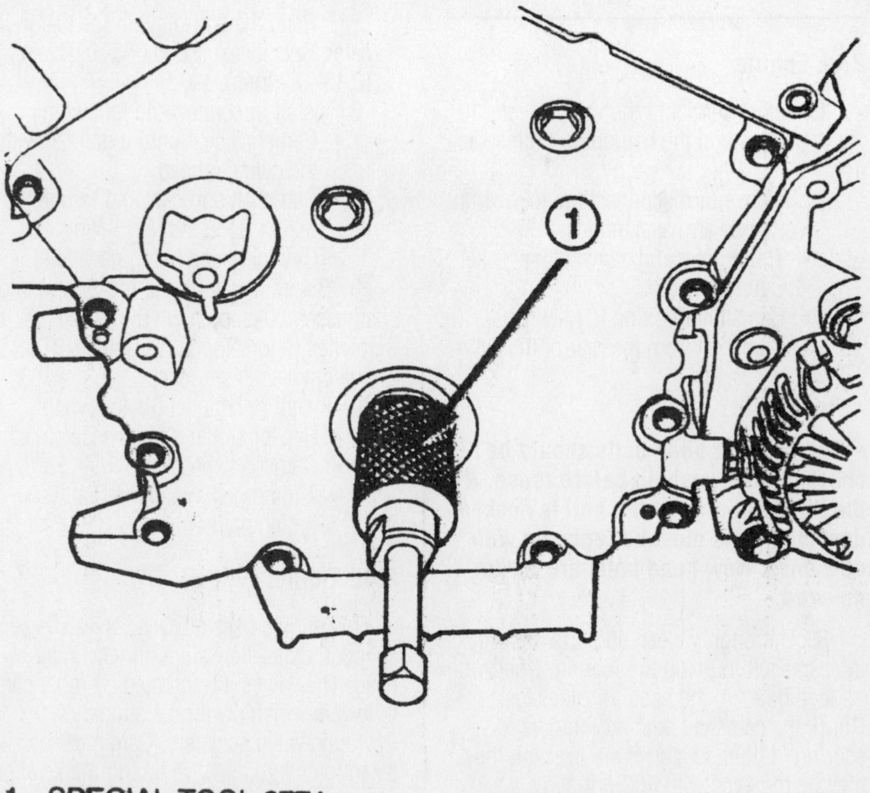

Remove the front crankshaft seal with special tool 6771

3.0L Engine

1. Before servicing the vehicle, refer to the precautions in the beginning of this section.
2. Remove or disconnect the following:
   - Negative battery cable
   - Timing belt
   - Crankshaft sprocket
   - Crankshaft Position (CKP) sensor
   - Crankshaft sensing blade
   - Crankshaft spacer and key
   - Front oil seal

To install:

3. Lubricate the oil seal lip with clean engine oil.
4. Install or connect the following:
   - New oil seal with special Tool MD998717
   - Crankshaft key and spacer
   - Crankshaft sensing blade
   - CKP sensor
   - Crankshaft sprocket
   - Timing belt
   - Negative battery cable

Camshaft

REMOVAL & INSTALLATION

2.0L Engine

1. Before servicing the vehicle, refer to the precautions in the beginning of this section.
2. Remove or disconnect the following:
   - Rocker arm assemblies
   - Timing belt and camshaft sprocket
   - Cylinder head
   - Camshaft Position (CMP) sensor
   - Camshaft from the rear of the cylinder head

To install:

➡The cylinder head bolts should be checked for stretching before reuse. If the threaded area of the bolt is necked-down the bolts must be replaced with new ones. New head bolts are recommended.

3. Thoroughly clean all parts. Inspect the camshaft journals for scoring. Check the oil feed holes in the head for blockage. Check the camshaft bearing journals for scoring. If light scratches are present, they may be removed with 400 grit abrasive paper. If deep scratches are present, replace the camshaft and check the cylinder head for damage. Replace the cylinder head if worn or damaged.
4. If the camshaft lobes show signs of wear, check the corresponding rocker arm roller for wear or damage. Replace rocker arms/Hydraulic Lash Adjuster (HLAs) if worn or damaged. If the camshaft lobes show signs of pitting on the nose, flank or base circle, replace the camshaft.
5. If the rocker arms and shaft are to be serviced, mark the rocker arms so any that are to be returned to service will be installed in their original locations.
6. Install or connect the following:
   - Rocker arm on the rocker arm shaft
   - Camshaft, be sure it turns freely
   - Cam sensor and torque the screws to 85 inch lbs. (10 Nm)
7. Camshaft end-play can be checked by using the following procedure:
   a. Lubricate the camshaft journals and install the camshaft without the rocker arm assemblies.
   b. Install the cam sensor and torque the screws to 85 inch lbs. (10 Nm).
   c. Adjust a dial indicator to touch the camshaft nose.
   d. Pry the camshaft rearward as far as it will go. Be sure the dial indicator probe is in contact with the camshaft.
   e. Zero the dial indicator.
   f. Move the camshaft forward as far as it will go.
   g. Read the end-play on the dial indicator. Specification is 0.005–0.013 in. (0.13–0.33mm).
8. Install or connect the following:
   - Camshaft seal until it is flush with the cylinder head
   - Camshaft sprocket and torque the bolt to 85 ft. lbs. (115 Nm)
   - Cylinder head using new bolts
9. Before installing the rocker arm and shaft assemblies, position the crankshaft to 3 notches Before Top Dead Center (BTDC) on the crankshaft sprocket.
10. Install or connect the following:
    - Rocker arm and shaft assemblies
    - Camshaft sprocket
    - Timing belt, by aligning all valve timing marks

2.4L Engine (VIN X)

This engine uses a DOHC, 4-valves per cylinder, cross-flow aluminum cylinder head. The valves are actuated by roller cam followers which pivot on stationary hydraulic valve adjusters. Care must be taken to ensure all valve timing marks align after cylinder head and valvetrain service.

1. Before servicing the vehicle, refer to the precautions in the beginning of this section.
2. Disconnect the negative battery cable from the left shock tower.

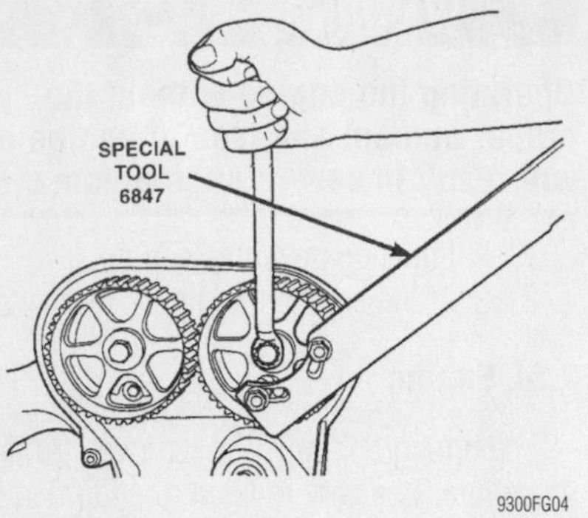

Use special tool 6847 to hold the camshaft sprocket while removing or installing the center bolt—2.4L engine

➡The ground cable is equipped with an insulator grommet which should be placed on the stud to prevent the negative battery cable from accidentally grounding.

3. Relieve the fuel system pressure using the recommended procedure.
4. Remove or disconnect the following:
   - Spark plugs cables
   - Ignition coil pack with the spark plug cables
   - Cylinder head cover and discard the gasket
   - Ground strap
   - Timing belt covers and timing belt
   - Camshaft sprockets by holding them with Tool 6847, while removing the bolt
5. Take note that the camshaft bearing caps are numbered for correct location during installation. Remove the outer bearing caps first.
6. Loosen, but do not remove, the camshaft bearing cap retaining fasteners in the correct sequence, inside working outward. Perform this step on one camshaft at a time.
7. Identify the camshafts, if they are to be reused for later installation. The camshafts are not interchangeable.

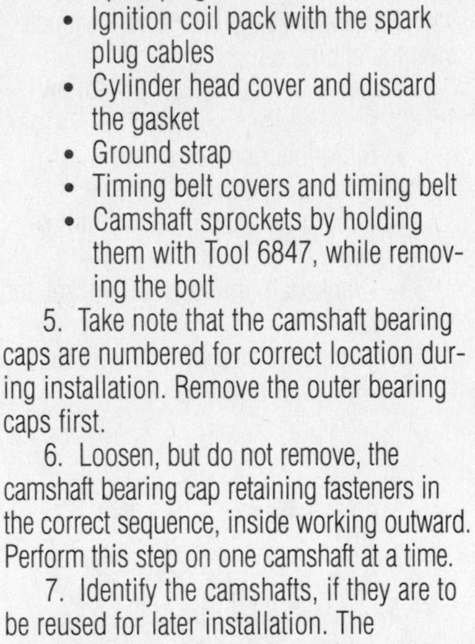

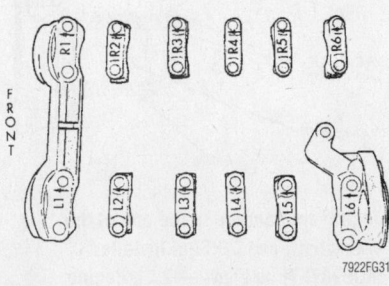

Camshaft bearing cap identification—2.4L engine

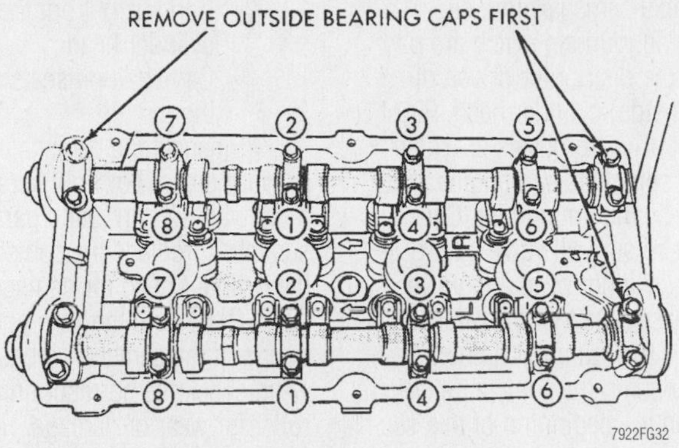

Camshaft bearing cap bolt removal sequence—2.4L engine

8. Remove or disconnect the following:
   - Camshaft bearing caps
   - Camshafts
   - Camshaft followers

9. Any components that are to be reused must be installed in their original locations. Use care to identify and mark the positions of any removed valvetrain components so they may be reinstalled correctly.

10. Inspect the camshaft bearing oil feed holes in the cylinder head for clogging. Inspect the camshaft bearing journals for wear or scoring. Check the cam surface for abnormal wear and damage. A visible worn groove in the roller path or on the cam lobes is cause for replacement.

To install:

11. Thoroughly clean all camshaft and related parts.

12. Inspect the camshaft end-play using the following procedure:
   a. Lubricate the camshaft journals and install the camshaft **WITHOUT** the cam follower assemblies. Install the rear cam caps and tighten to 21 ft. lbs. (28 Nm).
   b. Push the camshaft rearward as far as it will go.
   c. Adjust a dial indicator to rest against the front of the camshaft (the sprocket end). Zero the indicator.
   d. Move the camshaft forward as far as it will go. Read the dial indicator. End-play specification is 0.002–0.010 in. (0.05–0.15mm).
   e. If excessive end-play is present, inspect the cylinder head and camshaft for wear; replace if necessary.

13. If the fit and condition of the camshafts are acceptable, remove the camshafts for installation of the cam followers.

14. The hydraulic valve lash adjusters are inside the roller cam followers. Be sure they are clean, well lubricated with engine oil and properly positioned. Install the cam followers in their original positions on the hydraulic adjuster and valve stem.

⁂ WARNING

Be sure NONE of the pistons are at Top Dead Center (TDC) when installing the camshafts.

15. Lubricate the camshaft bearing journals and cam followers with clean engine oil

16. Install or connect the following:
   - Camshafts
   - Right/left-side camshaft bearing caps Nos. 2 through 5 and right-side No. 6. Torque the M6 fasteners to 105 inch lbs. (12 Nm) in the correct sequence.

17. Apply Mopar® Gasket Maker sealer to the No. 1 and left-side No. 6 bearing caps.

18. Install or connect the following:
   - Bearing caps and torque the M8 fasteners to 21 ft. lbs. (28 Nm)
   - Camshaft end seals
   - Camshaft sprockets, if removed, and torque the bolts to 75 ft. lbs. (101 Nm)
   - Timing belt, making sure all timing marks are aligned
   - Timing belt covers

⁂ WARNING

If the timing belt or sprockets are incorrectly installed, engine damage will occur. Take time to be sure all timing marks are correctly aligned.

19. Clean all sealing surfaces. Make certain the rails are flat.

20. Install new cylinder head cover gaskets. Apply Mopar® Silicone Rubber Adhesive Sealant, at the camshaft cap corners and at the top edge of the ½ round seal.

➡Inspect the spark plug well seals for cracking and/or swelling and replace if necessary.

21. Install the cylinder head cover and torque the fasteners, in sequence, using the following:
   a. Step 1: 40 inch lbs. (4.5 Nm).

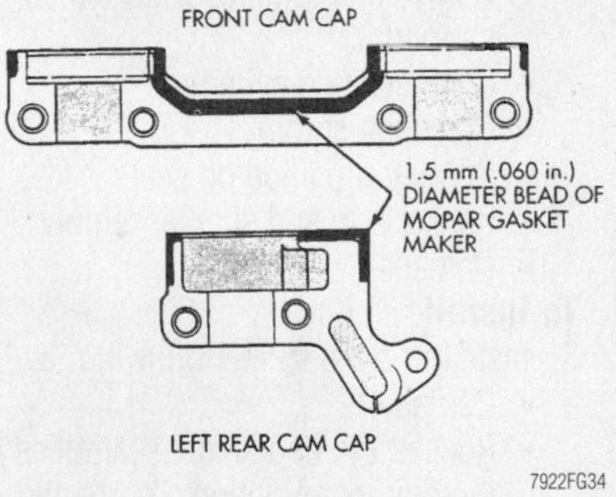

Apply sealer as shown to prevent oil leakage from the camshaft bearing end caps—2.4L engine

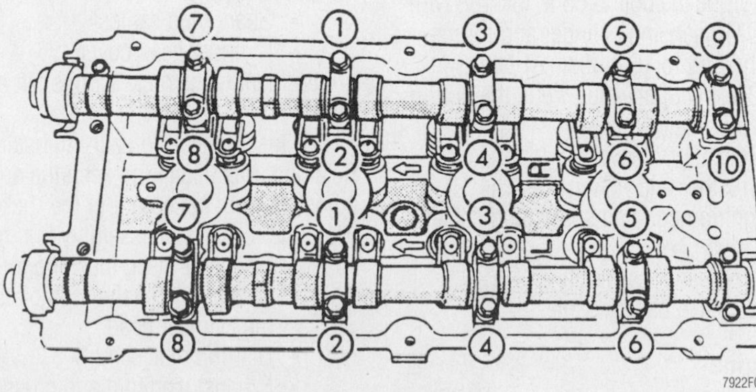

Camshaft bearing cap tightening sequence—2.4L engine

b. Step 2: 80 inch lbs. (9 Nm).
c. Step 3: 105 inch lbs. (12 Nm).

22. Install or connect the following:
- Ignition coil pack and torque the fasteners to 105 inch lbs. (12 Nm)
- Spark plug cables
- Ground strap
- All vacuum lines and wiring
- Negative battery cable

➡An oil and filter change is recommended.

23. Test run vehicle. Check for leaks and for proper operation.

2.4L Engine (VIN G)

1. Before servicing the vehicle, refer to the precautions in the beginning of this section.
2. Remove or disconnect the following:
   - Negative battery
   - Air cleaner assembly
   - Timing belt
   - Ignition coils and spark plug cables
   - Positive Crankcase Ventilation (PCV) hose
   - Breather hose
   - Rocker arm cover
   - Camshaft Position (CMP) sensor support
   - CMP sensing cylinder
   - Camshaft sprocket
   - Spark plug guide oil seal
   - Rocker arm and shaft assembly
   - Camshaft

To install:

3. Install or connect the following:
   - Camshaft
   - Rocker arm and shaft assemblies in there proper position. Torque the assemblies to 23 ft. lbs. (31 Nm)
   - Spark plug guide oil seals
   - Camshaft sprocket with special Tools MB990767 and MD998719. Torque the bolt to 65 ft. lbs. (88 Nm)
   - CMP sensing cylinder and torque the bolt to 16 ft. lbs. (21 Nm)
   - CMP support and torque the bolt to 10 ft. lbs. (15 Nm)
   - Rocker arm cover
   - Breather and PCV hoses
   - Ignition coil and spark plug cables
   - Timing belt
   - Air cleaner assembly
   - Negative battery cable

2.5L Engine

This engine uses aluminum alloy cylinder heads with 4-valves per cylinder and pressed-in cast iron valve guides. The cylinders are common to either cylinder bank. Two overhead camshafts are supported by 4 bearing journals which are part of the head with the distributor driven off the right (firewall side) cylinder head. Right and left camshaft drive sprockets are interchangeable. The sprockets and engine water pump are driven by the timing belt. Care must be taken to be sure all valve timing marks align after cylinder head and valve train service. Please note that for camshaft service, the cylinder head must be removed.

1. Before servicing the vehicle, refer to the precautions in the beginning of this section.
2. Disconnect the negative battery cable from the left shock tower.

➡The ground cable is equipped with an insulator grommet which should be placed on the stud to prevent the negative battery cable from accidentally grounding.

3. Relieve the fuel system pressure using the recommended procedure.
4. Drain the cooling system.
5. Remove or disconnect the following:
   - Timing belt covers and timing belt
   - Camshaft sprockets by holding them with Tool 6847 while removing the bolt
   - Upper intake plenum
   - Manifold Absolute Pressure (MAP) and Intake Air Temperature (IAT) sensor wiring connectors
   - Air cleaner cover, inlet hoses and air inlet resonator
   - Throttle Position Sensor (TPS) and Idle Air Control (IAC) motor electrical connectors
   - Throttle cable
   - Speed control cable, if equipped
   - Exhaust Gas Recirculation (EGR) tube
   - Lower intake manifold and discard the gasket
   - Spark plug cables
   - Cylinder head cover
6. Identify the rocker arm shaft assemblies.
7. Install the Auto Lash Adjuster Retainers MD 998443 to keep the auto lash adjusters from falling out of the rocker arms when the rocker arm assembly is removed.
8. Remove or disconnect the following:
   - Rocker arm shaft assemblies from the cylinder head
   - Distributor assembly
   - Exhaust manifold and crossover pipe
   - Cylinder head
   - Thrust case from the left cylinder head
   - Camshafts from the rear of the cylinder heads
   - Camshaft oil seals and discard them
9. Inspect the camshafts for scratches or worn areas. For light scratches, remove them with 400 grit sand paper. For deep scratches, replace the camshaft and check the cylinder head for damage. Check the oil holes to be sure they are open and free of debris. If the camshaft lobes show signs of wear, check the corresponding rocker arm roller for wear or damage. Replace the rocker arm if worn or damaged. If the camshaft shows signs of wear on the lobes, replace it.

To install:

10. Check camshaft end-play, as follows:
   a. Lubricate the camshaft journals with engine oil and install the camshaft **WITHOUT** the rocker arm assemblies.
   b. Move the camshaft rearward as far as it will go.
   c. Mount a dial indicator to rest on the front of the camshaft.
   d. Zero the indicator.
   e. Move the camshaft forward as far as it will go. End-play should be 0.004–0.008 in. (0.1–0.2mm). Maximum allowed end-play is 0.016 in. (0.4mm).
11. Lubricate the camshaft journals
12. Install or connect the following:
   - Camshafts
   - Thrust case to the left cylinder head and torque the fasteners to 108 inch lbs. (13 Nm)
   - New camshaft seal (lubricate the seal lip), until it is flush with the cylinder head surface
   - Camshaft sprocket and torque the bolt to 75 ft. lbs. (101 Nm)
   - Cylinder head(s)
   - Lower intake manifold using new gaskets
13. The rocker arm shafts are hollow and used as a lubrication oil duct. Be sure all valvetrain parts are clean. Check the rocker arm mounting portion of the shafts for wear or damage, replace if necessary. Check all oil holes for clogging with a small wire and clean as required. If any rockers were removed, lubricate and install on the shafts in their original positions.
14. Install or connect the following:
   - Rocker arm/shaft assemblies
   - Timing belt

⁂ WARNING

It is very important that all valve timing marks align properly or engine damage will result.

15. Inspect the spark plug tube seals located on the ends of each tube. These seals slide onto each tube to seal the cylinder head cover to the spark plug tube. If these seals show signs of hardness and/or cracking, replace them.
16. Clean all sealing surfaces.
17. Install or connect the following:
- Cylinder head covers using new gaskets and torque the bolts to 88 inch lbs. (10 Nm)
- Spark plug cables
- Upper intake manifold (plenum) using a new gasket and torque the bolts to 13 ft. lbs. (18 Nm)
- Plenum support brackets
- EGR tube
- Throttle and speed control cables
- TP sensor and IAC motor electrical connectors
- MAP and IAT sensor electrical connectors
- Air inlet resonator, air inlet hose and air cleaner housing cover
- All remaining electrical connectors
- Air tube connections and tighten them
- Negative battery cable

18. Refill the cooling system.

➡An oil and filter change is recommended whenever a cylinder head has been removed since coolant can get into the oil system.

⁂ WARNING

Operating the engine without the proper amount and type of engine oil will result in severe engine damage.

19. Start the engine and check for leaks, abnormal noises and vibrations. Bleed the cooling system.

2.7L Engine

1. Before servicing the vehicle, refer to the precautions in the beginning of this section.
2. Remove or disconnect the following:
- Negative battery
- Timing chain and secondary chain tensioner bolts
- Camshaft bearing caps
- Camshafts, secondary chain and tensioner as an assembly
- Tensioner and chain from the camshafts

To install:

3. Install or connect the following:

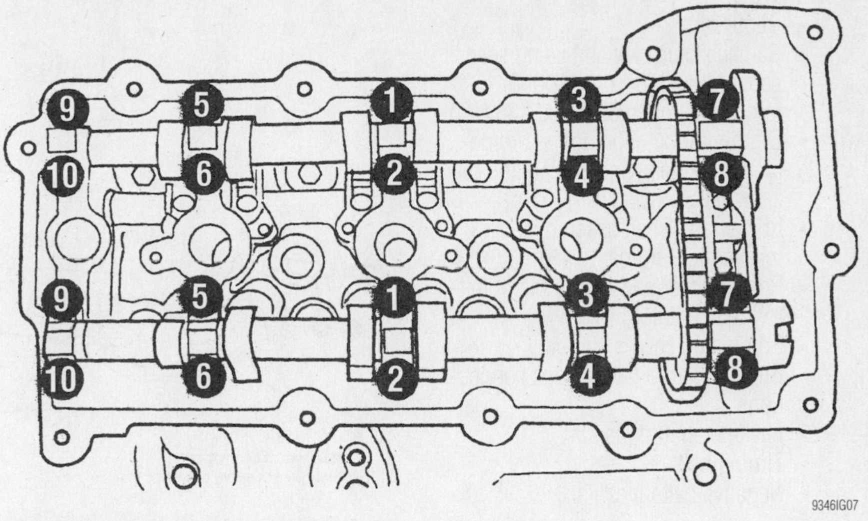

Tighten the camshaft bearing caps in the proper sequence–2.7L Engine

- Camshaft chain on the cams and make certain that the plated links are facing forward

4. Compress the chain tensioner as follows:
   a. Place the tensioner in a soft jaw vise.
   b. Slowly compress the tensioner until the fabricated lock pin can be installed in the locking holes.
- Tensioner between the camshafts and chain. Rotate the cams until the plated links and dots are in the 12 o'clock position
- Cams to the cylinder head
- Camshaft bearing caps. Slowly, and in sequence, torque the bolts to 105 inch lbs. (12 Nm)
- Secondary chain tensioner bolts and torque them to 105 inch lbs. (12 Nm). Remove the locking pin from the tensioner
- Primary timing chain
- Negative battery cable

3.0L Engine

LEFT BANK

1. Before servicing the vehicle, refer to the precautions in the beginning of this section.
2. Remove or disconnect the following:
- Negative battery
- Timing belt
- Thermostat housing
- Blow by hose
- Positive Crankcase Ventilation (PCV) hose
- Spark plug cables
- Rocker arm cover
- Rocker arm and shaft
- Camshaft sprocket
- Thrust case
- Camshaft

To install:

3. Install or connect the following:
- Camshaft
- Thrust case and torque the bolt to 109 inch lbs. (13 Nm)
- Camshaft sprocket and torque the bolt to 65 ft. lbs. (88 Nm)
- Rocker arm and shaft assembly and torque to 23 ft. lbs. (31 Nm)
- Rocker arm cover and torque the bolts to 35 inch lbs. (4 Nm)
- Spark plug cables
- PCV and blow by hoses
- Thermostat housing
- Timing belt
- Negative battery cable

RIGHT BANK

1. Before servicing the vehicle, refer to the precautions in the beginning of this section.
2. Remove or disconnect the following:
- Negative battery
- Timing belt
- Intake manifold plenum
- Breather hose connection
- Blow by hose connection
- Spark plug cables
- Rocker cover
- Rocker arm and shaft assemblies
- Distributor
- Camshaft sprocket
- Camshaft

Heater Core replacement is covered in Section 2 of this manual

To install:

- Camshaft
- Camshaft sprocket and using special Tools MB990767 and MD998715 torque the bolt to 65 ft. lbs. (88 Nm)
- Distributor by aligning the timing marks on the cylinder head and the camshaft sprocket
- Rocker arm/shaft assemblies and torque to 23 ft. lbs. (31 Nm)
- Rocker cover and torque the bolts to 35 inch lbs. (4 Nm)
- Spark plug cables
- Breather and blow by hose connections
- Intake manifold plenum
- Timing belt
- Negative battery cable

Valve Lash

ADJUSTMENT

2.0L and 2.5L Engines

The engines are equipped with Hydraulic Lash Adjusters (HLAs) which are precision units installed in machined openings in the valve actuating ends of the rocker arms. Valve clearance adjustments are not performed.

2.4L, 2.7L and 3.0L Engines

The valves are actuated by roller cam followers which pivot on stationary Hydraulic Lash Adjusters (HLAs). The HLAs are precision units installed in machined openings of the cam follower. Valve clearance adjustments are not performed.

Starter Motor

REMOVAL & INSTALLATION

2.0L Engine

MANUAL TRANSAXLE

1. Before servicing the vehicle, refer to the precautions in the beginning of this section.
2. Remove or disconnect the following:
   - Negative battery cable from the shock tower
   - Air cleaner resonator
   - Electrical connectors from the starter
   - Starter-to-transaxle bolts
   - Starter

To install:

3. Install or connect the following:

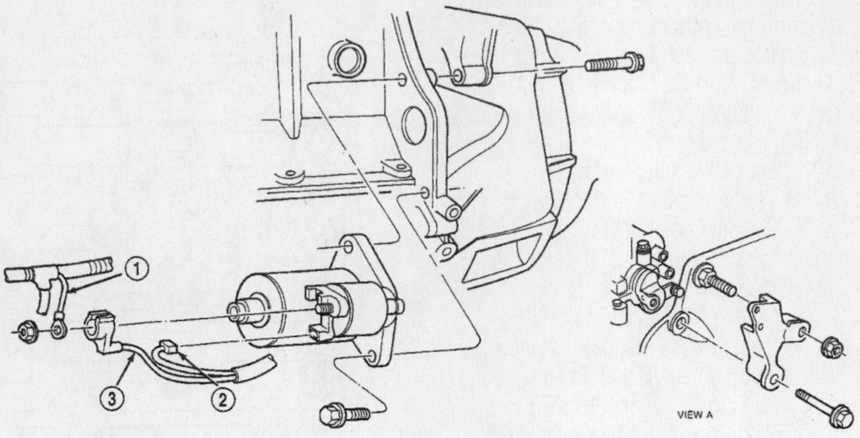

Exploded view of the starter and related components—2.0L engine with a manual transaxle

   - Starter and torque the bolts to 40 ft. lbs. (54 Nm)
   - Electrical connectors to the starter
   - Air cleaner resonator
   - Negative battery cable to the shock tower

2.0L and 2.4L (VIN G) Engines

AUTOMATIC TRANSAXLE

1. Before servicing the vehicle, refer to the precautions in the beginning of this section.
2. Remove or disconnect the following:
   - Negative battery cable from the shock tower
   - Air cleaner resonator
   - Transmission Control Module (TCM) and move it aside

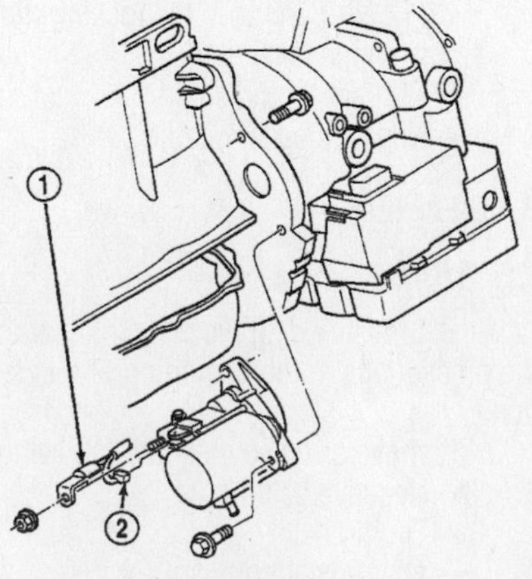

Exploded view of the starter and related components—2.0L engine with an automatic transaxle and 2.4L engine

➡Do not disconnect the TCM electrical connector(s).

   - Upper starter-to-transaxle bolt
   - Battery cable from the starter
   - Solenoid connector by unlocking the red tab and compressing it
   - Lower starter-to-transaxle bolt
   - Starter

To install:

3. Install or connect the following:
   - Starter
   - Lower starter-to-transaxle bolt and torque the bolt to 40 ft. lbs. (54 Nm)
   - Electrical connectors to the starter
   - Upper starter-to-transaxle bolt and torque the bolt to 40 ft. lbs. (54 Nm)
   - TCM
   - Air cleaner resonator
   - Negative battery cable to the shock tower

2.4L (VIN X) Engine

1. Before servicing the vehicle, refer to the precautions in the beginning of this section.
2. Remove or disconnect the following:
   - Negative battery cable
   - Air cleaner assembly
   - Starter cover
   - Starter electrical connectors
   - Starter

To install:

3. Install or connect the following:
   - Starter and torque the bolt to 23 ft. lbs. (30 Nm)
   - Electrical connectors to the starter
   - Starter cover and torque the bolt to 44 inch lbs. (5 Nm)

- Air cleaner assembly
- Negative battery cable

2.5L Engine

1. Before servicing the vehicle, refer to the precautions in the beginning of this section.
2. Remove or disconnect the following:
   - Negative battery cable from the shock tower
   - Oil filter
   - Electrical connectors from the starter
   - Starter-to-transaxle bolts
   - Starter

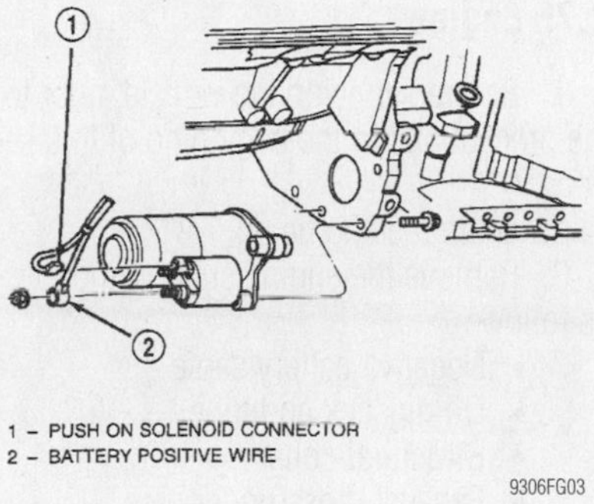

Exploded view of the starter and related components—2.5L engine

To install:

3. Install or connect the following:
   - Starter and torque the bolt to 40 ft. lbs. (54 Nm)
   - Electrical connectors to the starter
   - Oil filter
   - Negative battery cable

2.7L Engine

1. Before servicing the vehicle, refer to the precautions in the beginning of this section.
2. Remove or disconnect the following:
   - Negative battery cable
   - Oxygen Sensor (O_2S) electrical connector
   - Front engine mount through bolt
   - Starter electrical connectors
   - Starter

To install:

3. Install or connect the following:
   - Starter and hand tighten the lower bolt
   - Starter electrical connectors and torque the starter bolts to 40 ft. lbs. (54 Nm)
   - Engine mount through bolt and torque it to 45 ft. lbs. (61 Nm)
   - O_2sensor and torque it to 20 ft. lbs. (27 Nm)
   - Negative battery cable

3.0L Engine

1. Before servicing the vehicle, refer to the precautions in the beginning of this section.
2. Remove or disconnect the following:
   - Negative battery cable
   - Air cleaner assembly
   - Starter electrical connectors
   - Starter

To install:

3. Install or connect the following:
   - Starter and torque the bolts to 23 ft. lbs. (30 Nm)
   - Starter electrical connectors
   - Air cleaner assembly
   - Negative battery cable

Oil Pan

REMOVAL & INSTALLATION

2.0L Engine

1. Before servicing the vehicle, refer to the precautions in the beginning of this section.

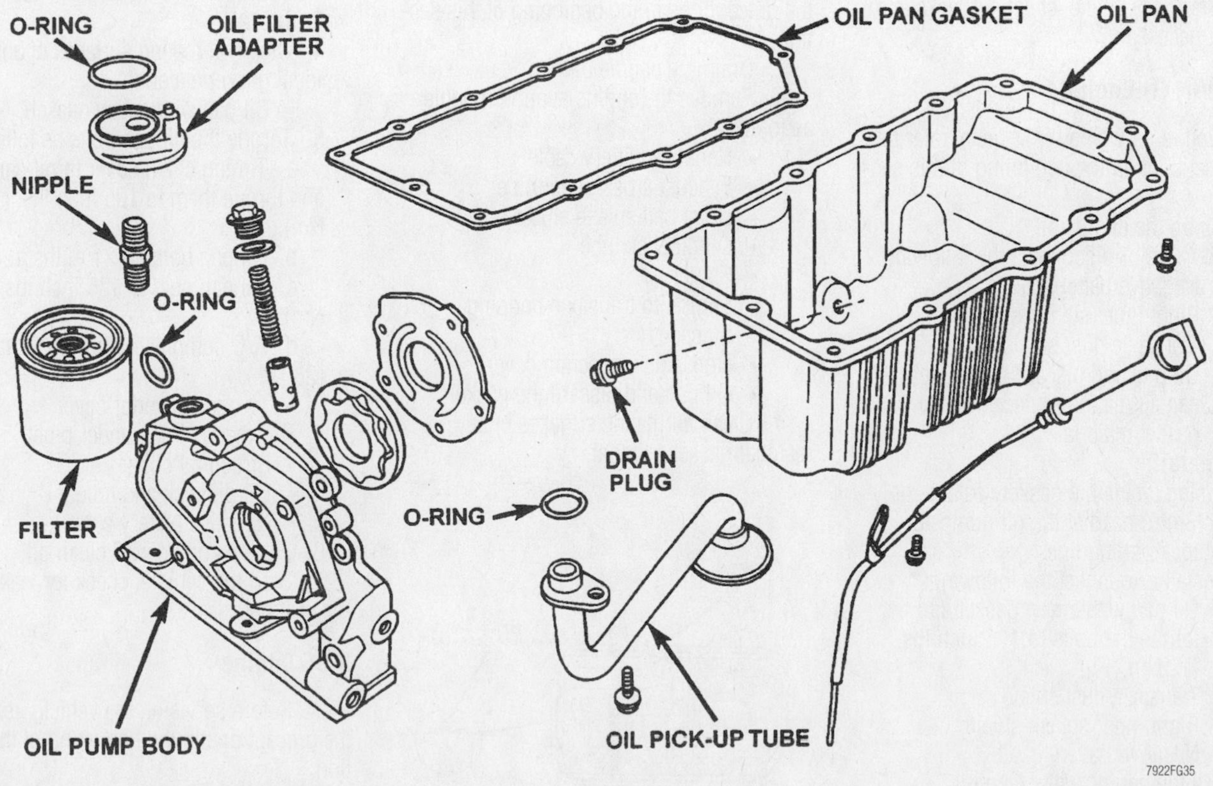

Exploded view of the oil pan and pump assembly—2.0L engine

2. Drain the engine oil.
3. Remove or disconnect the following:
   - Negative battery cable
   - Transaxle bending bracket
   - Front engine mount and bracket
   - Structural collar from the oil pan
   - Transaxle inspection cover
   - Oil filter and oil filter adapter if equipped with A/C
   - Oil pan and discard the gasket
4. Clean the oil pan and gasket sealing surfaces.

To install:

5. Using a suitable rubber adhesive gasket sealant, apply a ⅛ in. (3mm) bead at the oil pump-to-engine block parting line.
6. Install or connect the following:
   - Oil pan with a new gasket and torque the bolts 105 inch lbs. (12 Nm)
   - New oil filter and adapter, if removed
   - Transaxle lower dust cover
   - Structural collar to the oil pan
   - Front engine mount and bracket
   - Transaxle bending bracket
   - Oil pan drain plug and gasket and torque the plug to 25 ft. lbs. (34 Nm).
   - Negative battery cable
7. Fill the engine with clean oil.
8. Start the vehicle, check for leaks and repair if necessary.

2.4L (VIN G) Engine

1. Before servicing the vehicle, refer to the precautions in the beginning of this section.
2. Drain the engine oil.
3. Remove or disconnect the following:
   - Negative battery cable
   - Right inner splash shield
   - Transaxle dust shield
   - Oil pan and discard the gasket
4. Clean all mating surfaces of any residual gasket material.

To install:

5. Using a suitable gasket sealant apply a ⅛ in. (3mm) bead at the oil pump-to-engine block parting line.
6. Install or connect the following:
   - Oil pan with a new gasket and torque the bolts to 105 inch lbs. (12 Nm)
   - Transaxle dust shield
   - Right inner splash shield
   - Negative battery cable
7. Fill the engine with clean oil.
   - Start the vehicle, check for leaks and repair if necessary.

2.4L (VIN X) Engine

1. Before servicing the vehicle, refer to the precautions in the beginning of this section.
2. Drain the engine oil.
3. Remove or disconnect the following:
   - Negative battery cable
   - Oil dipstick
   - Front exhaust pipe
   - Bell housing cover
   - Oil pan and discard the gasket
4. Clean all mating surfaces of any residual gasket material.

To install:

5. Install or connect the following:
   - Oil pan with a new gasket and torque the bolts to 62 inch lbs. (7 Nm)
   - Bell housing cover and torque the upper bolt to 80 inch lbs. (9 Nm) and the lower bolt to 19 ft. lbs. (26 Nm)
   - Front exhaust pipe
   - Oil dipstick
   - Negative battery cable
6. Fill the engine with clean oil.
   - Start the vehicle, check for leaks and repair if necessary.

2.5L Engine

1. Before servicing the vehicle, refer to the precautions in the beginning of this section.
2. Drain the engine oil.
3. Remove the engine support module as follows:
   - Negative battery cable
   - Exhaust cross under pipe
   - Drive belt splash shield
   - Oil dipstick tube
   - Starter
   - Engine to transaxle bending braces
   - Transaxle inspection cover
   - Oil pan and discard the gasket
4. Clean all mating surfaces of any residual gasket material.

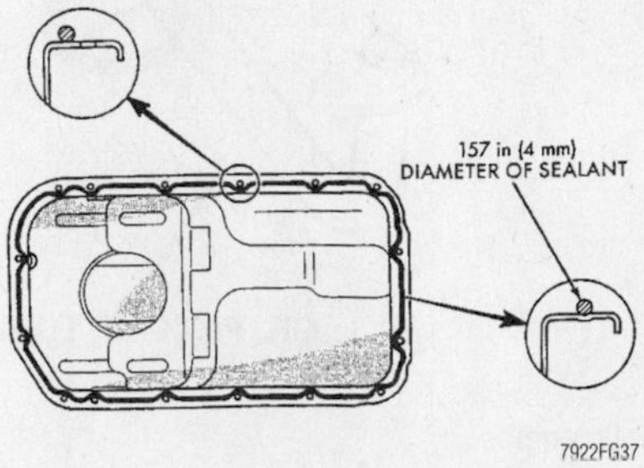

Apply sealer to the oil pan as shown—2.5L engine

To install:

5. Install or connect the following:
   - Oil pan with a new gasket and torque the bolts to 50 inch lbs. (6 Nm)
   - Transaxle inspection cover
   - Engine to transaxle bending braces
   - Exhaust cross under pipe
   - Starter
   - Drive belt splash shield
   - Oil dipstick tube
   - Negative battery cable
6. Fill the engine with clean oil.
7. Start the vehicle, check for leaks and repair if necessary.

2.7L Engine

1. Before servicing the vehicle, refer to the precautions in the beginning of this section.
2. Drain the engine oil.
3. Remove the engine support module as follows:
   - Negative battery cable
   - Oil dipstick and tube
   - Structural collar
   - Exhaust cross under pipe
   - Torque converter cover
   - Lower bolt from the A/C compressor
   - Oil pan and discard the gasket

To install:

4. Clean all mating surfaces of any residual gasket material.
   - Oil pan with a new gasket
5. Torque the oil pan bolts as follows:
   a. Timing chain cover to oil pan bolts and torque them to 105 inch lbs. (12 Nm).
   b. Oil pan bolts to 21 ft. lbs. (28 Nm).
   c. Oil pan nuts to 105 inch lbs. (12 Nm).
   d. A/C compressor bolt to 21 ft. lbs. (28 Nm).
   - Torque converter cover
   - Exhaust cross under pipe
   - Structural collar
   - Oil dipstick and tube
   - Negative battery cable
6. Fill the engine with clean oil.
7. Start the vehicle, check for leaks and repair if necessary.

3.0L Engine

1. Before servicing the vehicle, refer to the precautions in the beginning of this section.
2. Drain the engine oil.
3. Remove the engine support module as follows:
   - Negative battery cable

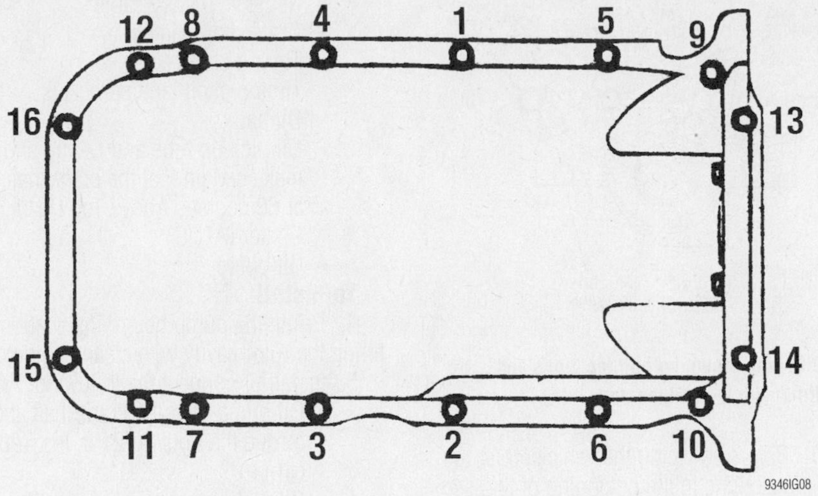

Tighten the upper oil pan bolts in sequence

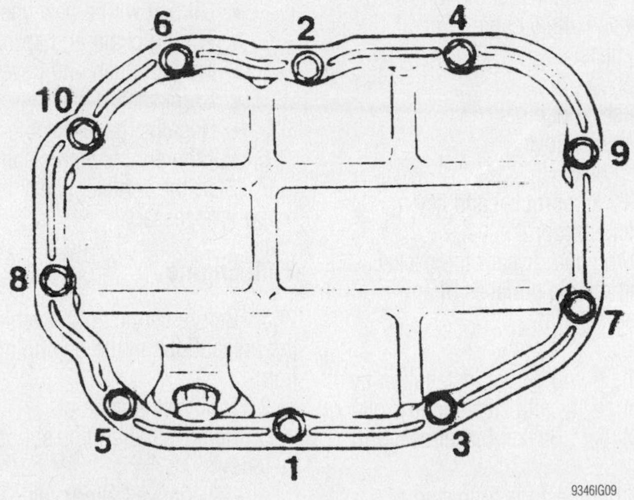

Tighten the lower oil pan bolts in sequence

- Front exhaust pipe
- Lower oil pan and discard the gasket
- Starter electrical connectors
- Starter
- Oil dipstick and guide
- Cover
- Upper oil pan and discard the gasket

4. Clean all mating surfaces of any residual gasket material.

To install:

5. Install or connect the following:
   - Upper oil pan with a new gasket and torque the bolts, in sequence, to 53 inch lbs. (6 Nm)
   - Cover and torque the bolts to 96 inch lbs. (11 Nm)
   - Starter
   - Starter electrical connectors
   - Oil dipstick and guide with a new O-ring and torque the bolt to 36 ft. lbs. (48 Nm)
   - Lower oil pan with a new gasket and torque the bolts to 96 inch lbs. (11 Nm)
   - Front exhaust pipe
   - Negative battery cable
6. Fill the engine with clean oil.
   - Start the vehicle, check for leaks and repair if necessary.

Oil Pump

REMOVAL & INSTALLATION

2.0L and 2.4L (VIN X) Engines

The oil drawn up through the pick-up tube is pressurized by the pump and routed through the full flow filter to the main oil galley running the length of the cylinder block. The oil pick-up, pump and check valve provide oil flow to the main oil gallery. A vertical hole at the No. 5 bulkhead routes pressurized oil through a restrictor up past a cylinder head bolt to an oil galley running the length of the cylinder head. The camshaft journals are slotted to allow pressurized oil to pass into the bearing cap cavities. Small holes in the bearing caps direct oil to the camshaft lobes.

1. Before servicing the vehicle, refer to the precautions in the beginning of this section.
2. Drain the engine oil.
3. Remove the engine support module as follows:
   - Negative battery cable
   - Crankshaft damper
   - Timing belt and tensioner
   - Camshaft sprocket
   - Rear timing belt cover
   - Oil pan
   - Oil pick up tube
   - Oil pump and discard the gasket
4. Clean all mating surfaces of any residual gasket material.

To install:

5. Prime the oil pump before installation.
6. Install or connect the following:
   - Oil pump with the flats aligned with the crankshaft flats and torque the bolts to 21 ft. lbs. (28 Nm)
   - New front crankshaft seal
   - Crankshaft sprocket
   - Oil pick up tube and oil pan
   - Camshaft sprocket
   - Rear timing belt cover
   - Timing belt and tensioner
   - Front timing belt cover
   - Crankcase damper
   - Negative battery cable
7. Fill the engine with clean oil.
8. Start the vehicle, check for leaks and repair if necessary.

2.4L (VIN G) Engine

1. Before servicing the vehicle, refer to the precautions in the beginning of this section.
2. Drain the engine oil.
3. Remove the engine support module as follows:
   - Negative battery cable
   - Oil filter
   - Oil pressure switch
   - Oil pan and screen

- Flange bolt
- Relief plug, spring and plunger
- Oil filter bracket
- Oil pump case
- Front case gasket
- Oil pump cover
- Oil pump driven gear
- Oil pump drive gear
- Oil pump

4. Clean all mating surfaces of any residual gasket material.

To install:

5. Prime the oil pump before installation.
6. Install or connect the following:
   - New oil seal into the front case
   - Oil pump
   - Oil pump drive gear
   - Oil pump driven gear
   - Oil pump cover
   - Front case gasket
   - Oil pump case
   - Oil filter bracket
   - Relief plunger, spring, gasket and plug
   - Flange bolt
   - Oil pan and screen
   - Oil pressure switch
   - New oil filter
   - Negative battery cable
7. Fill the engine with clean oil.
8. Start the vehicle, check for leaks and repair if necessary.

2.5L Engine

The oil pump assembly is mounted on the timing belt end of the cylinder block with the inner pump rotor indexed and installed on the crankshaft nose. The oil pump case also retains the crankshaft front oil seal and provides oil pan front end closure.

Oil pressure can be checked with a mechanical oil pressure gauge installed at the oil switch location. Oil pressure should be 6 psi (41.4 kPa) at idle and 35–75 psi (241.3–517.1 kPa) at 3000 rpm with the engine at operating temperature.

⁂ WARNING

If an oil pressure problem is suspected or if oil pressure is zero at idle, do not run the engine up to 3000 rpm in an attempt to raise oil pressure or the engine will be severely damaged.

Because the oil pump is driven off the crankshaft, the timing belt must be removed to access the pump. Use care to properly align all valve timing marks.

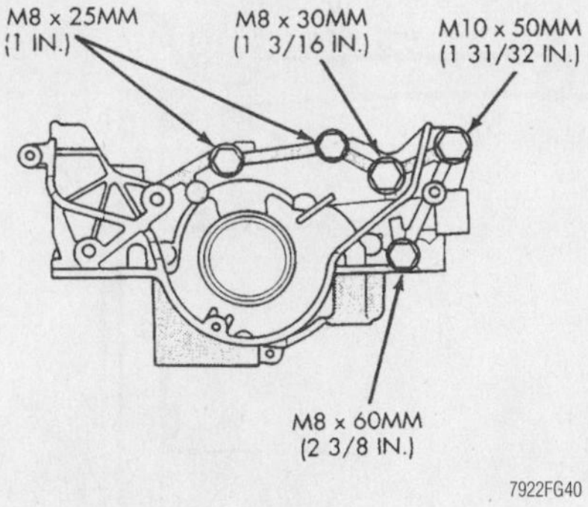

Oil pump mounting bolt locations and dimensions—2.5L engine

1. Before servicing the vehicle, refer to the precautions in the beginning of this section.
2. Drain the engine oil.
3. Remove or disconnect the following:
   - Negative battery cable
   - Drive belts
   - Oil pan
   - Crankshast damper
   - Timing belt covers
   - Timing belt
   - Crankshaft sprocket and key
   - Oil pick up tube
   - Oil pump and discard the gasket
4. Clean all mating surfaces of any residual gasket material.

To install:

5. Prime the pump before installation by filling the rotor cavity with clean engine oil.
6. Apply Mopar® Gasket Maker sealer to the pump.
7. Install or connect the following:
   - New O-ring into the pump body counterbore discharge passage
   - Oil pump and torque the M8 bolts to 10 ft. lbs. (14 Nm) and M10 bolts to 30 ft. lbs. (41 Nm)
   - Oil pick up tube and torque the bolts to 12 ft. lbs. (19 Nm)
   - Crankshaft sprocket and key
   - Timing belt and covers
   - Crankshaft damper
   - Oil pan with a new gasket
   - Drive belts
   - Negative battery cable
8. Fill the engine with clean oil.
9. Start the vehicle, check for leaks and repair if necessary.

2.7L Engine

1. Before servicing the vehicle, refer to the precautions in the beginning of this section.
2. Drain the engine oil.
3. Remove the engine support module as follows:
   - Negative battery cable
   - Crankshaft damper
   - Timing chain cover
   - Timing chain and sprockets
   - Oil pan
   - Oil pick up tube and O-ring and make certain that the crankshaft is at 60 degrees Above Top Dead Center (ATDC)
   - Oil pump

To install:

4. Prime the pump before installation by filling the rotor cavity with clean engine oil.
5. Install or connect the following:
   - Oil pump over the crankshaft and torque the bolts to 21 ft. lbs. (28 Nm)
   - Oil pick up tube with a new O-ring and torque the bolt to 21 ft. lbs. (28 Nm)
   - Oil pan with a new gasket
   - Timing chain and sprockets
   - Timing chain and cover
   - Crankshaft damper
   - Negative battery cable
6. Fill the engine with clean oil.
7. Start the vehicle, check for leaks and repair if necessary.

3.0L Engine

1. Before servicing the vehicle, refer to the precautions in the beginning of this section.
2. Drain the engine oil.
3. Remove the engine support module as follows:
   - Negative battery cable
   - Oil filter
   - Oil pressure switch
   - Oil filter bracket
   - Lower and upper oil pan
   - Baffle plate
   - Oil screen and gasket
   - Baffle plate
   - Relief spring and plunger
   - Front crankshaft oil seal
   - Oil pump case
   - Oil pump cover
   - Oil pump outer and inner rotors
   - Oil pump case
4. Clean all mating surfaces of any residual gasket material.

To install:

5. Prime the oil pump before installation.
6. Install or connect the following:
   - Oil pump case
   - Inner and outer rotors
   - Oil pump cover and torque the bolts to 89 inch lbs. (10 Nm)
   - Oil pump case assembly

- New front crankshaft oil seal
- Relief plunger and spring
- Baffle plate
- Oil screen with a new gasket
- Baffle plate
- Upper oil pan and cover
- Lower oil pan
- Oil filter bracket with a new gasket
- Oil pressure switch
- New oil filter
- Negative battery cable

7. Fill the engine with clean oil.
8. Start the vehicle, check for leaks and repair if necessary.

Rear Main Seal

REMOVAL & INSTALLATION

1. Before servicing the vehicle, refer to the precautions in the beginning of this section.
2. Remove or disconnect the following:
   - Negative battery cable from the left shock tower.

➡The ground cable is equipped with an insulator grommet which should be placed on the stud to prevent the negative battery cable from accidentally grounding.

- Transaxle
- Flexplate/flywheel
- Rear crankshaft oil seal using a flat-bladed prying tool

To install:

➡When installing the seal there is no need to lubricate the sealing surface.

3. Install or connect the following:
   - New seal using a suitable driver or seal installer
   - Flexplate/flywheel and torque the bolts to 70 ft. lbs. (95 Nm)
   - Transaxle
   - Negative battery cable
4. Start the engine and check for leaks.

Balance Shaft Chain, Sprockets And Cover

REMOVAL & INSTALLATION

2.4L (VIN X) Engine

1. Before servicing the vehicle, refer to the precautions in the beginning of this section.
2. Drain the engine oil.
3. Remove or disconnect the following:
   - Oil pan and discard the gasket
   - Chain cover, guide and tensioner
   - Gear cover and balance shaft gears
   - Balance shaft gear and crankshaft chain sprocket
   - Carrier gear cover and balance shafts

Balance Shaft Timing and Installation

➡During installation of the balance shafts to crankshaft, the proper timing procedure must be followed to prevent internal engine damage.

1. Install or connect the following:
2. Install the balance shafts and carrier on the crankshaft. Torque the bolts to 40 ft. lbs. (54 Nm)
3. Turn the balance shafts until both key ways are pointing up. Install the short hub drive gear on the sprocket driven shaft.
4. Make certain that the key ways and the gear timing marks are properly aligned.
5. Install the gear cover and torque the double ended stud and washer to 105 inch lbs. (12 Nm)
6. Align the flat of the balance shaft drive sprocket to the flat on the crankshaft
7. Install the balance shaft drive

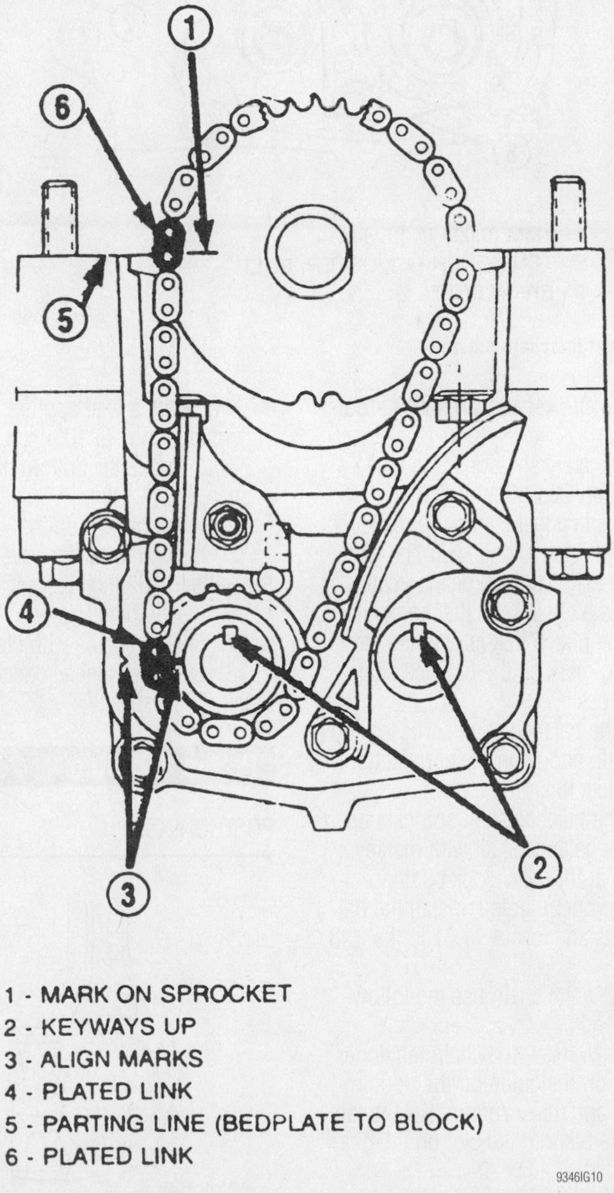

Align the key ways and the gear timing marks

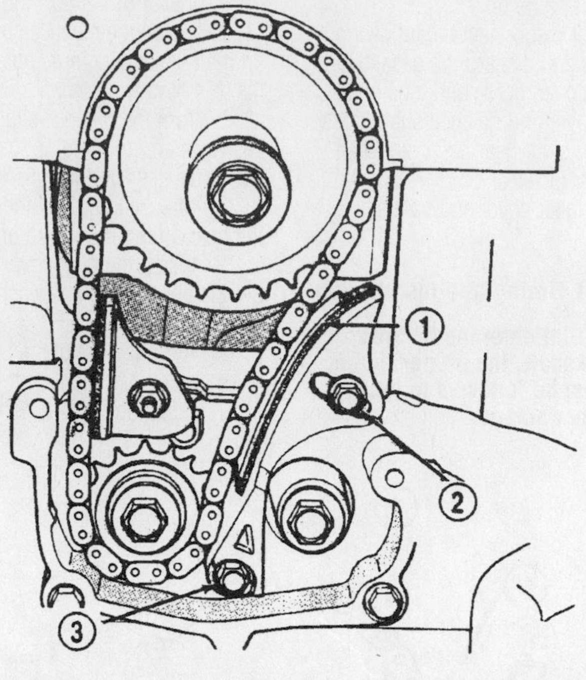

Properly adjust the chain tension

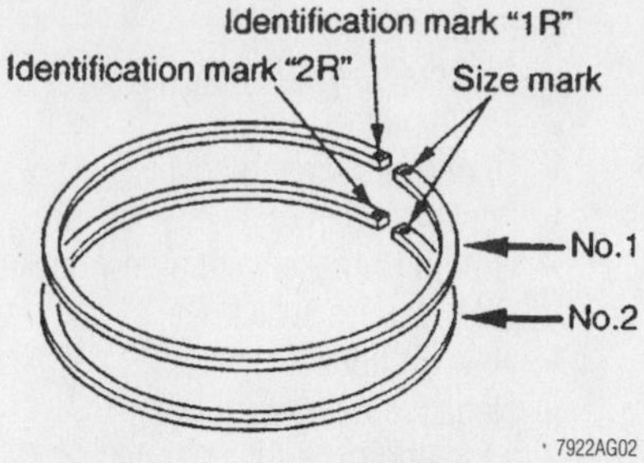

Piston ring identification mark locations

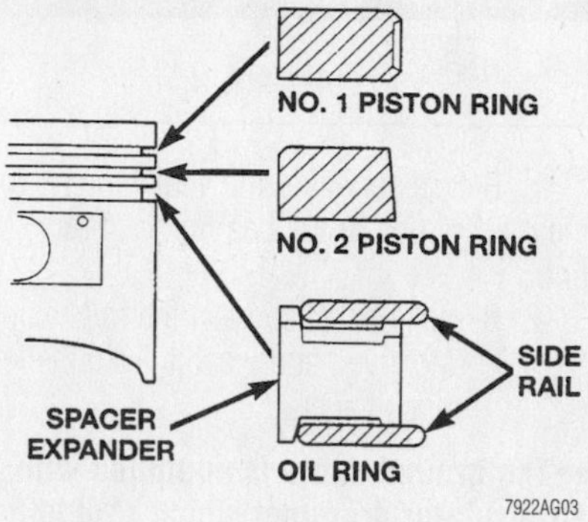

Piston ring orientation—2.0L, 2.4L, 2.5L, 2.7L and 3.0L engines

sprocket to the crankshaft with special tool 6052.

8. Turn the crankshaft until the No. 1 cylinder is at Top Dead Center (TDC).

9. Make certain that the timing marks on the chain sprocket are aligned with the parting line on the left side main bearing cap.

10. Install the chain over the crankshaft and be sure the plated link is over the No. 1 cylinder timing mark on the balance shaft crankshaft sprocket.

11. Place the balance shaft sprocket into the timing chain and align the sprocket dot to the lower plate link on the chain.

12. Slide the balance shaft sprocket on the nose of the balance shaft with the key ways pointing at the 12 o'clock position.

13. When properly timed, install the balance shaft bolts and torque to 21 ft. lbs. (28 Nm)

14. To tighten the chain use the following procedure:

a. Loosely install the chain tensioner.

b. Position the guide on the double ended stud and make certain that the tab on the guide fits into the slot on the gear cover.

c. Install and tighten the nut/washer assembly to 105 inch lbs. (12 Nm).

d. Place a 1 mm thick by 70 mm long shim between the tensioner and chain.

e. With a load applied, torque the top tensioner bolt to 105 inch lbs. (12 Nm).

f. Torque the bottom pivot bolt to 105 inch lbs. (12 Nm).

g. Remove the shim.

15. Install the carrier covers and torque the bolts to 105 inch lbs. (12 Nm).

16. Install the oil pan.

17. Fill the engine with clean oil.

18. Start the vehicle, check for leaks and repair if necessary.

Piston and Ring

POSITIONING

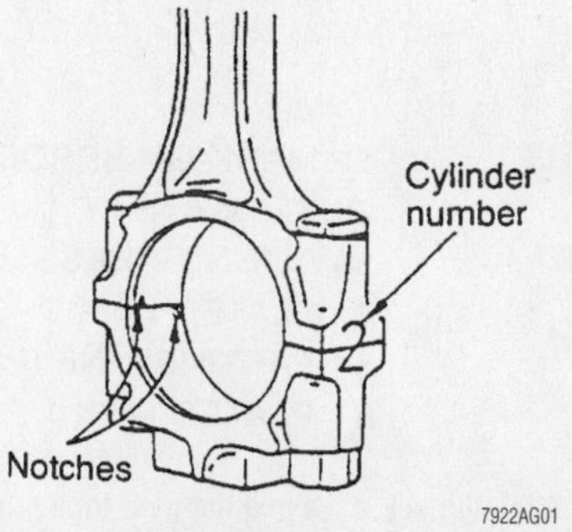

Engine connecting rod and cap installation—ensure to matchmark the cap and rod prior to disassembly

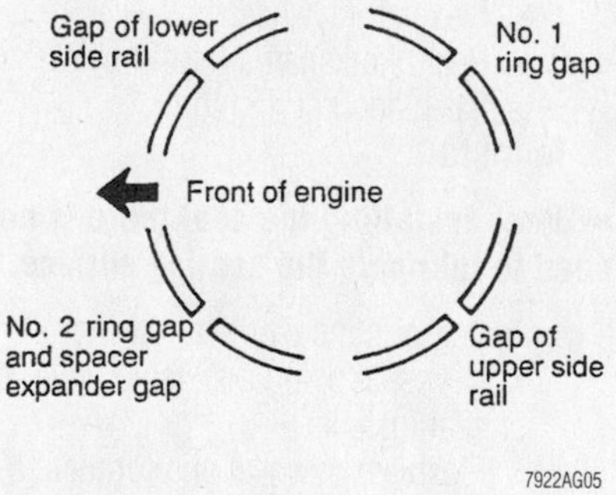

Piston ring end-gap spacing—2.0L and 2.4L (VIN X) engines

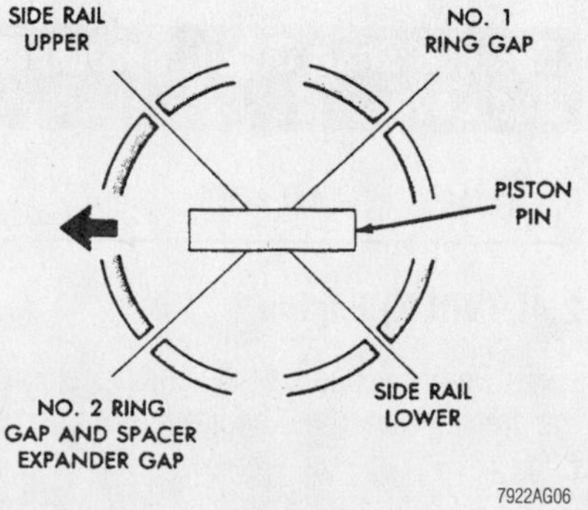

Piston ring end-gap spacing—2.4L (VIN G), 2.5L, 2.7L and 3.0L engine

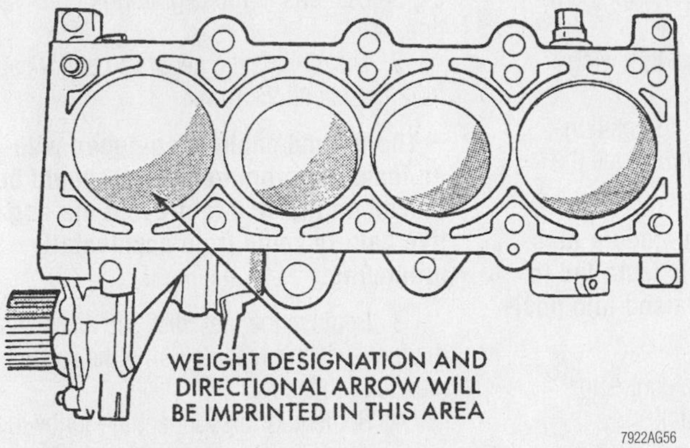

Piston positioning. The arrow or weight marking (L or H) must face toward the timing belt side of the engine—2.0L and 2.4L (VIN X) engines

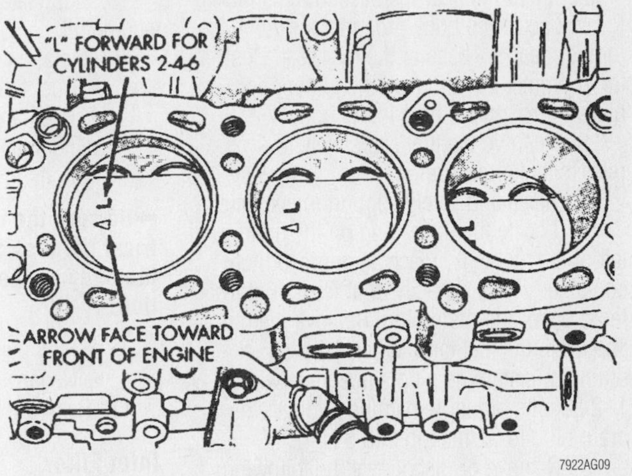

Piston positioning mark locations—2.5L engine

FUEL SYSTEM

Fuel System Service Precautions

Safety is the most important factor when performing not only fuel system maintenance but any type of maintenance. Failure to conduct maintenance and repairs in a safe manner may result in serious personal injury or death. Maintenance and testing of the vehicle's fuel system components can be accomplished safely and effectively by adhering to the following rules and guidelines.

- To avoid the possibility of fire and personal injury, always disconnect the negative battery cable unless the repair or test procedure requires that battery voltage be applied.
- Always relieve the fuel system pressure prior to disconnecting any fuel system component (injector, fuel rail, pressure regulator, etc.), fitting or fuel line connection. Exercise extreme caution whenever relieving fuel system pressure to avoid exposing skin, face and eyes to fuel spray. Please be advised that fuel under pressure may penetrate the skin or any part of the body that it contacts.
- Always place a shop towel or cloth around the fitting or connection prior to loosening to absorb any excess fuel due to spillage. Ensure that all fuel spillage (should it occur) is quickly removed from engine surfaces. Ensure that all fuel soaked cloths or towels are deposited into a suitable waste container.
- Always keep a dry chemical (Class B) fire extinguisher near the work area.
- Do not allow fuel spray or fuel vapors to come into contact with a spark or open flame.
- Always use a back-up wrench when loosening and tightening fuel line connection fittings. This will prevent unnecessary stress and torsion to fuel line piping.
- Always replace worn fuel fitting O-rings with new. Do not substitute fuel hose where fuel pipe is installed.

Fuel System Pressure

RELIEVING

1. Before servicing the vehicle, refer to the fuel system precautions and to the precautions in the beginning of this section.
2. Remove or disconnect the following:
3. Remove the fuel pump relay from the Power Distribution Center (PDC).
4. Start and run the engine until it stalls.
5. Attempt to restart the engine until it no longer runs.
6. Turn the ignition key to the off position.

Fuel Filter

REMOVAL & INSTALLATION

Frame Mounted

The fuel delivery system contains a replaceable inline filter. The fuel filter mounts to the frame above the rear of the fuel tank. The fuel tank assembly must be loosened and lowered slightly to access the filter. The inlet and outlet tubes are permanently attached to the filter. Please note that the fuel system pressure must be relieved before servicing fuel system components. In addition, quick-disconnect fittings are used on fuel line connections. When the tubes are fully connected, the locking ears and the fuel tube shoulder are visible in the windows of the connector.

1. Before servicing the vehicle, refer to the precautions in the beginning of this section.
2. Disconnect the negative battery cable from the left shock tower.

➡The ground cable is equipped with an insulator grommet which should be placed on the stud to prevent the negative battery cable from accidentally grounding.

3. Relieve the fuel system pressure using the recommended procedure.
4. From inside the trunk, disconnect the fuel pump module wiring jumper from the main body harness. The 4-pin connector is located under the trunk mat on the left

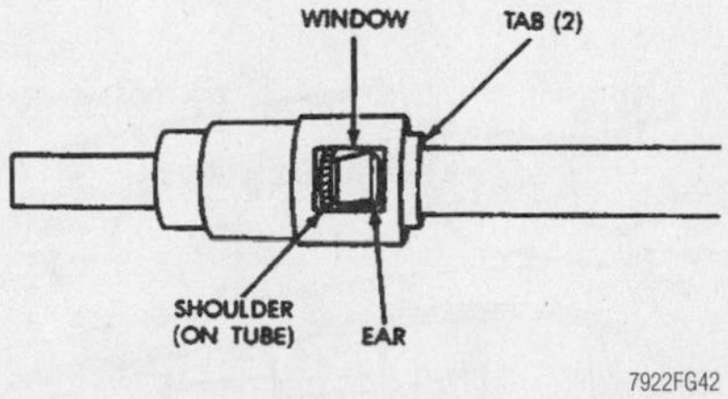

Be sure the shoulder and locking tabs are visible through the window of the connector when attached

side of the trunk near the base of the shock tower. Locate the body grommet for the jumper near the base of the rear seat. Push the grommet out and feed the jumper completely through the hole in the body.

5. Remove the fuel cap slowly to release tank pressure.
6. Raise and safely support the vehicle.
7. Locate the drain plug on the bottom left of the fuel tank. Place an approved fuel container with a capacity of at least 16 gallons, under the drain plug. Remove the plug and drain the fuel tank. When finished draining, install the plug since there will be 1–2 gallons of fuel remaining. Tighten the drain plug to 32 inch lbs. (3.6 Nm).
8. Remove or disconnect the following:
   - Driver's side fuel tank strap
   - Passenger's side fuel tank strap loosen it

➡**Allow the tank to lower until the fuel tank neck touches the rear suspension crossmember.**

⁂ CAUTION

Wrap shop towels around the fuel hoses to catch any gasoline spillage.

   - Fuel line quick-disconnect fittings from the fuel pump module. Squeeze fuel filter hose connector releasing tabs and detach the fuel lines
   - Fuel filter

To install:

➡**The fuel supply (to filter) tube and the return tube (to fuel pump module) are permanently attached to the fuel filter. The ends of the fuel supply and return tubes have different size quick-disconnect fittings. The large quick-disconnect fitting attaches to the large nipple (supply side) on the fuel pump module. The smaller quick-disconnect fitting attaches the small nipple (return side) on the fuel pump module.**

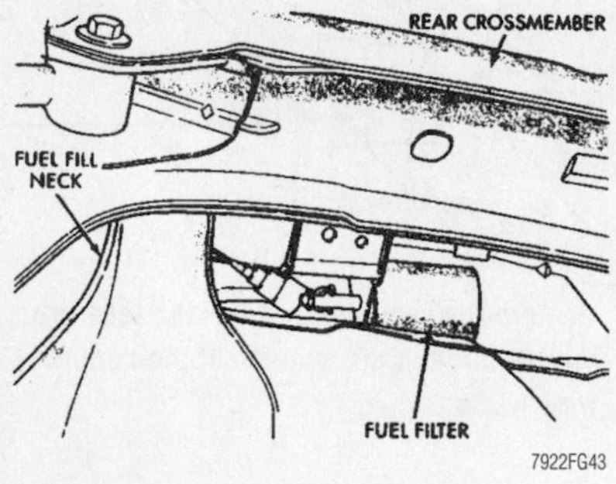

Fuel filter mounting location

9. Lubricate the fuel filter nipples with engine oil.
10. Install or connect the following:
   - Fuel tubes
   - Fuel tank raise it into position
   - Fuel tank straps and torque the bolts to 17 ft. lbs. (23 Nm)

➡**Be sure the fuel pump module electrical harness grommet is installed in the body as the tank is raised into position.**

   - Fuel pump module connector
   - Negative battery cable
11. Refill the fuel tank.

Inlet Filter

1. Before servicing the vehicle, refer to the precautions in the beginning of this section.
2. Properly relieve the fuel system pressure.
3. Remove or disconnect the following:
   - Negative battery cable
   - Fuel pump module
   - Locking tabs on the fuel pump
   - Strainer and O-ring
   - Inlet filter from the assembly

To install:

4. Install or connect the following:
   - Inlet filter
   - New O-ring and strainer
   - Locking tabs on the fuel pump module
   - Fuel pump Module
   - Negative battery cable
5. Start the vehicle, check for leaks and repair if necessary.

Fuel Pump

REMOVAL & INSTALLATION

The in-tank fuel pump module contains the fuel pump and pressure regulator which adjusts fuel system pressure to approximately 49 psi (337.8 kPa). Voltage to the fuel pump is supplied through the fuel pump relay.

The fuel pump is serviced as part of the fuel pump module. The fuel pump module is installed in the top of the fuel tank and contains the electric fuel pump, fuel pump reservoir, inlet strainer fuel gauge sending unit, fuel supply and return line connections and the pressure regulator. The inlet strainer, fuel pressure regulator and level sensor are the only serviceable items. If the fuel pump requires service, replace the fuel pump module.

1. Before servicing the vehicle, refer to the precautions in the beginning of this section.
2. Disconnect the negative battery cable from the left shock tower.

➡**The ground cable is equipped with an insulator grommet which should be placed on the stud to prevent the negative battery cable from accidentally grounding.**

3. Remove the fuel filler cap and relieve the fuel system pressure using the recommended procedure.
4. Remove or disconnect the following:
   - Fuel pump harness electrical connector on the left side of the trunk near the base of the shock tower
   - Fuel pump harness grommet and push the harness through the hole
5. Raise the vehicle and drain the fuel tank.
6. Remove or disconnect the following:
   - Fuel tank inlet and from the filler hose
   - Fuel tank straps. Support with a jack prior to loosening the straps.
7. Carefully lower the tank.
8. Clean the top of the tank to remove any loose dirt.
9. Remove or disconnect the following:
   - Fuel lines from the fuel pump module
   - Fuel pump locknut using Spanner wrench tool 6856

⁂ CAUTION

The fuel reservoir of the fuel pump module does not empty out when the tank is drained. The fuel in the reservoir may spill out when the module is removed.

   - Fuel pump
   - Fuel tank O-ring and discard it

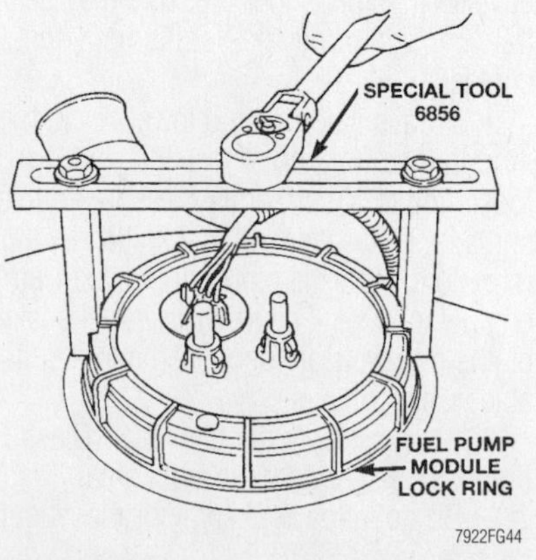

Removing the fuel pump module lock ring using special tool 6856

To install:

10. Thoroughly clean all parts. Wipe the seal area of the tank clean.
11. Install or connect the following:
    - New O-ring onto the fuel tank
    - Fuel pump module in the tank

➡**Be sure the alignment tab on the underside of the pump module flange sits in the corresponding notch in the fuel tank.**

- Locking ring and torque it to 40–45 ft. lbs. (54–61 Nm), using Spanner Wrench Tool 6856
- Fuel tank assembly
- Negative battery cable

12. Refill the fuel tank.
13. Turn the ignition switch **ON** to pressurize the system. Check the fuel system for leaks.

Fuel Injector

REMOVAL & INSTALLATION

2.0L and 2.4L Engines

1. Before servicing the vehicle, refer to the precautions in the beginning of this section.
2. Disconnect the negative battery cable from the left shock tower.

➡**The ground cable is equipped with an insulator grommet which should be placed on the stud to prevent the negative battery cable from accidentally grounding.**

3. Relieve the fuel system pressure using the recommended procedure.
4. Remove or disconnect the following:
    - Fuel supply line quick quick-connect fitting from the fuel rail
    - Fuel injector electrical connectors
    - Fuel rail from the intake manifold

➡**Cover the fuel injector holes in the intake manifold.**

- Fuel injector clip
- Fuel injector(s) from the fuel rail and discard the O-rings

To install:

5. Install or connect the following:
    - Fuel injector to the fuel rail using new O-rings lubricated with engine oil
    - Fuel injector clip
    - Fuel rail and torque the screws to 14–19 ft. lbs. (19.5–25.5 Nm)

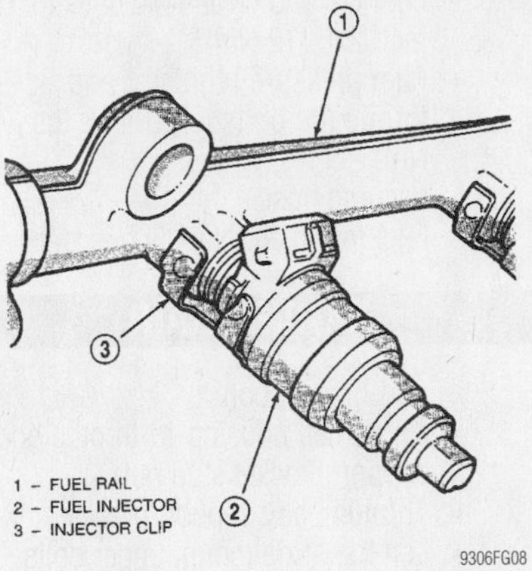

View of the fuel injector—2.0L, 2.4L and engines

- Fuel injector electrical connectors
- Fuel supply line quick quick-connect fitting to the fuel rail
- Negative battery cable

6. Use a Diagnostic Readout Box (DRB) scan tool Automatic Shutdown (ASD) fuel system test to pressurize the fuel system. Check for leaks.

2.5L and 2.7L Engines

1. Before servicing the vehicle, refer to the precautions in the beginning of this section.
2. Disconnect the negative battery cable from the left shock tower.

➡**The ground cable is equipped with an insulator grommet which should be placed on the stud to prevent the negative battery cable from accidentally grounding.**

3. Relieve the fuel system pressure using the recommended procedure.

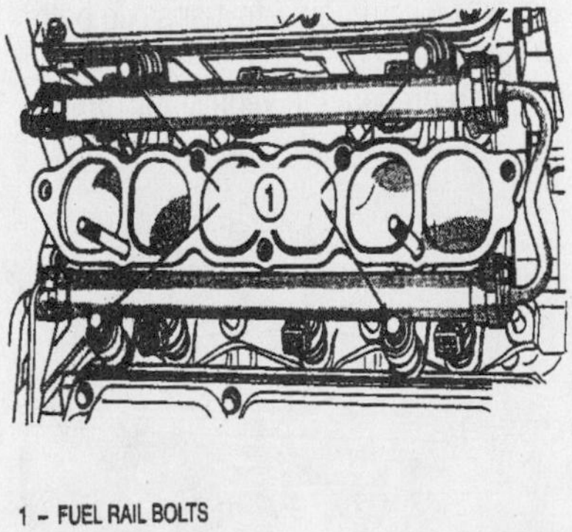

View of the fuel rail assembly—2.5L engine

4. Remove or disconnect the following:
    - Fuel supply line quick quick-connect fitting from the fuel rail
    - Upper intake manifold plenum and discard the gasket
    - Fuel injector electrical connectors
    - Fuel rail from the intake manifold

➡**Cover the intake manifold openings to keep dirt from entering the system.**

- Fuel injector clip
- Fuel injector(s) from the fuel rail and discard the O-rings

To install:

5. Install or connect the following:
    - Fuel injector to the fuel rail using new O-rings lubricated with engine oil
    - Fuel injector clip
    - Fuel rail and torque the bolts to 8 ft. lbs. (12 Nm)

➡**Be sure the spacers are located under the fuel rail**

- Fuel injector electrical connectors
- Fuel supply line quick quick-connect fitting to the fuel rail
- Upper intake manifold plenum using a new gasket and torque the bolts to 13 ft. lbs. (18 Nm)
- Throttle cables
- Sensor electrical connectors
- Negative battery cable

6. Use a Diagnostic Readout Box (DRB) scan tool Automatic Shutdown (ASD) fuel system test to pressurize the fuel system. Check for leaks.

3.0L Engine

1. Before servicing the vehicle, refer to the precautions in the beginning of this section.
2. Properly relieve the fuel system pressure.
3. Drain the cooling system.
4. Remove or disconnect the following:
    - Negative battery cable
    - Intake manifold plenum
    - Fuel injector electrical connector
    - Fuel supply and return hoses
    - Vacuum hose
    - Fuel pressure regulator
    - Fuel rail with the injectors
    - Insulators

- Fuel injectors and discard the O-rings

To install:

5. Install or connect the following:
   - Fuel injector with a new O-ring lubricated with clean engine oil
   - Insulators
   - Fuel rail and torque the bolts to 13 inch lbs. (12 Nm)
   - Fuel pressure regulator and torque the bolts to 80 inch lbs. (9 Nm)
   - Vacuum hose
   - Fuel return and supply hoses
   - Fuel injector electrical connector
   - Intake manifold plenum
   - Negative battery cable

6. Fill the cooling system to the proper level.

7. Start the vehicle, check for leaks and repair if necessary.

DRIVE TRAIN

Transaxle Assembly

REMOVAL & INSTALLATION

Manual

2.0L, 2.4L (VIN X) AND 2.5L ENGINES

WARNING

If the vehicle is going to be rolled on its wheels while the transaxle is out of the vehicle, obtain 2 outer CV-joints to install in the hubs. If the vehicle is rolled without the proper torque applied to the front wheel bearings, the bearings will no longer be usable.

1. Before servicing the vehicle, refer to the precautions in the beginning of this section.

2. Remove or disconnect the following:
   - Negative battery cable from the left shock tower.

➡The ground cable is equipped with an insulator grommet which should be placed on the stud to prevent the negative battery cable from accidentally grounding.

- Air cleaner and intake hoses
- Clutch housing vent cap
- Clutch cable from the bell housing

WARNING

Using equal force, pry up on both sides of the shifter cable isolator bushings to avoid damaging the cable isolator bushing.

- Selector lever and crossover cables from the transaxle
- Shift cable mounting bracket from the transaxle
- Accelerator cables from the throttle body
- Accelerator cable bracket from the throttle body
- Upper starter-to-throttle body support bracket bolt
- Upper bell housing-to-throttle body support bracket stud nut
- Throttle body support bracket
- Left transaxle mount upper bolts
- Upper bell housing bolts
- Vehicle Speed Sensor (VSS)
- Back-up light electrical connector from the transaxle

3. Install an engine support fixture tool and support the engine.

4. Remove or disconnect the following:
   - Front wheels
   - Halfshafts
   - Lower splash shield/battery cover from the left side
   - Lower bracket bolts, from the left transaxle mount
   - Engine-to-lower crossbar bolts
   - Front steel engine mount bracket
   - 3 front aluminum engine mount bracket bolts
   - Starter
   - Rear transaxle mount bracket
   - Transaxle-to-rear lateral bending strut from engine/transaxle assembly
   - Transaxle bell housing cover

5. Using a transaxle jack, support the transaxle.

6. Rotate the engine clockwise to gain access to the driveplate clutch bolts.
   - Driveplate clutch bolts
   - Lower engine-to-transaxle bolts
   - Transaxle

7. To prepare the vehicle for rolling, support the engine with a suitable support or reinstall the front motor mount to the engine. Then, reinstall the ball joints to the steering knuckle and install the retaining bolt. Install the obtained outer CV-joints to the hubs, install the washers and tighten the axle nuts to 180 ft. lbs. (244 Nm). The vehicle may now be safely rolled.

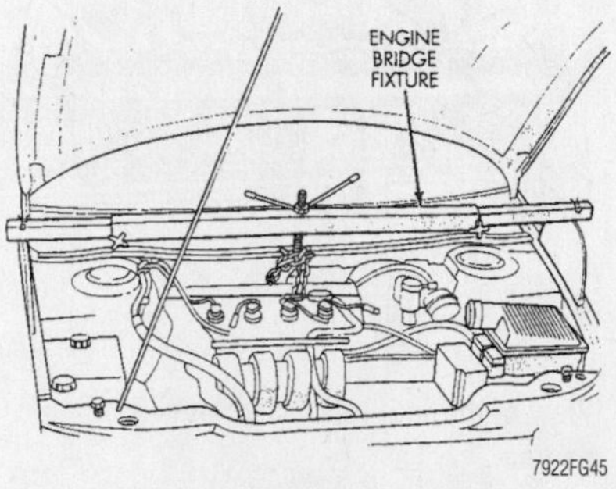

Secure the engine assembly with an appropriate support fixture

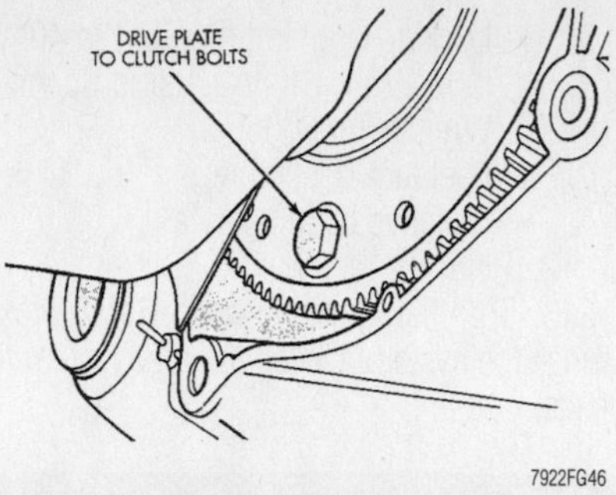

Remove the driveplate clutch bolts—rotate the engine clockwise to advance to the next bolt

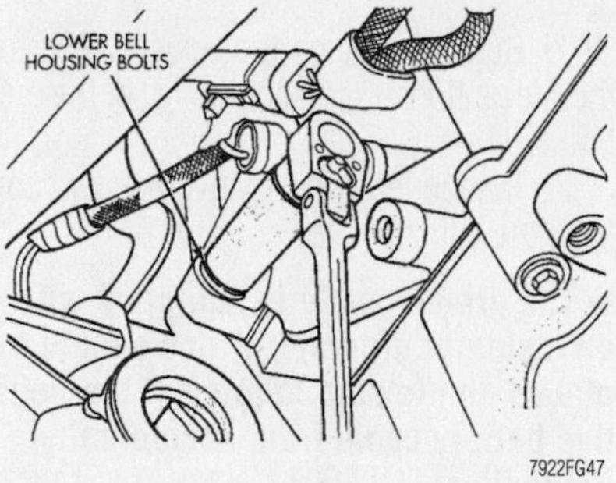

Removing the lower engine-to-transaxle mounting bolts

To install:

8. Install or connect the following:
   - Transaxle and torque the transaxle-to-engine bolts to 70 ft. lbs. (95 Nm)
   - Driveplate clutch bolts
   - Lower engine-to-transaxle bolts and torque the bolts to 70 ft. lbs. (95 Nm)
   - Transaxle bell housing cover and torque bolts to 9 ft. lbs. (12 Nm)
   - Transaxle-to-rear lateral bending

strut to the engine/transaxle assembly and torque the bolt to 40 ft. lbs. (54 Nm)
- Rear transaxle mount bracket and torque the bolts to 40 ft. lbs. (54 Nm)
- Starter
- 3 front aluminum engine mount bracket bolts and torque the bolts to 40 ft. lbs. (54 Nm)
- Front steel engine mount bracket and torque the bolt to 45 ft. lbs. (61 Nm)
- Engine-to-lower crossbar bolts
- Lower bracket bolts to the left transaxle mount and torque the bolts to 40 ft. lbs. (54 Nm)
- Lower splash shield/battery cover to the left side
- Halfshafts
- Front wheels

9. Remove engine support fixture tool.
10. Install or connect the following:
- Back-up light electrical connector
- VSS sensor
- Upper bell housing bolts and torque the bolts to 70 ft. lbs. (95 Nm)
- Left transaxle mount upper bolts and torque the bolts to 40 ft. lbs. (54 Nm)
- Throttle body support bracket
- Upper bell housing-to-throttle body support bracket stud nut and torque the nut to 32 ft. lbs. (43 Nm)
- Upper starter-to-throttle body support bracket bolt
- Accelerator cable bracket to the throttle body

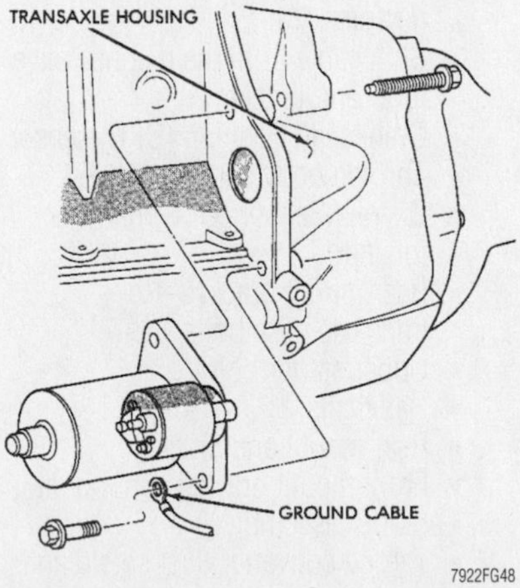

Be sure to connect the ground cable at the starter

- Accelerator cables to the throttle body
- Shift cable mounting bracket to the transaxle
- Selector lever and crossover cables to the transaxle
- Clutch cable to the bell housing
- Clutch housing vent cap
- Air cleaner and intake hoses
- Negative battery cable

11. Check to be sure that all fasteners are tightened and connections made.
12. Refill the transaxle.
13. Check the transaxle for proper operation. Be sure the reverse lights turn **ON** when in reverse.

2.4L (VIN G) AND 3.0L ENGINES

1. Before servicing the vehicle, refer to the precautions in the beginning of this section.
2. Drain the transaxle fluid.
3. Remove or disconnect the following:
- Battery and tray
- Air cleaner assembly
- Front exhaust pipe, 3.0L only
- Shift and selection cable connections
- Back—up light switch connection
- Vehicle Speed Sensor (VSS) connector
- Starter
- Clutch release cylinder connector
- Transaxle assembly upper coupling bolts
- Centermember
- Rear roll stopper
- Transaxle mount bracket and stopper
- Stabilizer link
- Wheel Speed Sensor (WSS), if equipped
- Brake hose clamp
- Tie rod end
- Lower arm
- Clutch release bearing engagement, 3.0L engine
- Driveshaft connection
- Upper oil pan bolt, 3.0L engine
- Bell housing cover
- Transaxle assembly lower coupling bolts
- Transaxle assembly

To install:

4. Install or connect the following:
- Transaxle assembly. Torque the lower coupling bolts to 36 ft. lbs. (48 Nm) 2.4L engine and to 52 ft. lbs. (71 Nm) 3.0L engine
- Bell housing cover and torque the cover—to—engine bolt to 80 inch lbs. (9 Nm) and the cover—to—transaxle bolt to 19 ft. lbs. (26 Nm)
- Upper oil pan bolt, 3.0L engine
- Driveshaft connection
- Clutch release bearing engagement
- Lower arm and torque the nut to 80 ft. lbs. (108 Nm)
- Tie rod end and torque the nut to 21 ft. lbs. (28 Nm)
- WSS, if equipped
- Stabilizer link and torque the nut to 33 ft. lbs. (45 Nm)
- Transaxle mount bracket and torque the nut to 42 ft. lbs. (57 Nm)
- Tranasxle mount stopper and torque the nut to 60 ft. lbs. (81 Nm)
- Rear roll stopper and torque the nut to 33 ft. lbs. (45 Nm)
- Centermember
- Clutch release cylinder and torque to 13 ft. lbs. (18 Nm)
- Starter motor and electrical connectors
- VSS connector
- Back—up light switch connector
- Shift and selection cable connections
- Front exhaust pipe, 3.0L engine
- Air cleaner assembly
- Battery and tray

5. Fill the transaxle fluid to the proper level.
6. Start the vehicle, check for leaks and repair if necessary.

Automatic

WARNING

If the vehicle is going to be rolled on its wheels while the transaxle is out of the vehicle, obtain 2 outer CV-joints to install to the hubs. If the vehicle is rolled without the proper torque applied to the front wheel bearings, the bearings will no longer be usable.

1. Before servicing the vehicle, refer to the precautions in the beginning of this section.
2. Drain the transaxle.
3. Disconnect the negative battery cable from the left shock tower.

➡The ground cable is equipped with an insulator grommet which should be placed on the stud to prevent the negative battery cable from accidentally grounding.

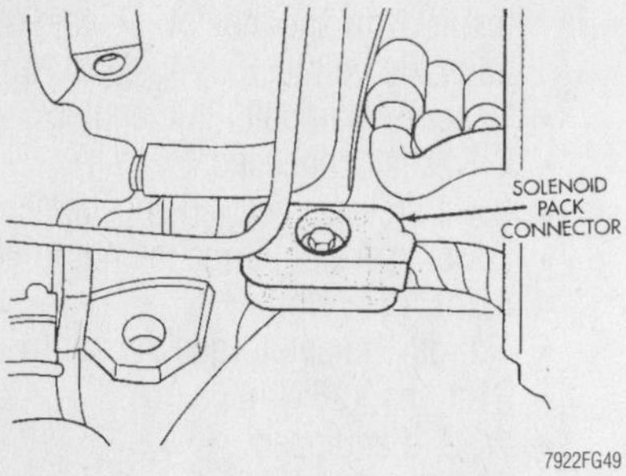

Location of the transaxle solenoid assembly 8-way connector and retaining bolt

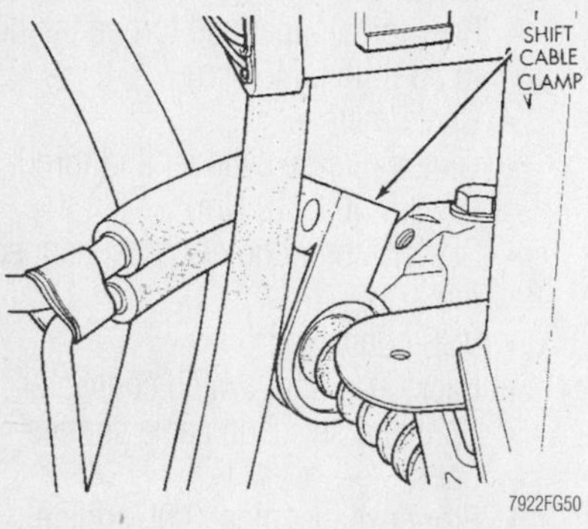

Removing the shift cable and clamp

4. Remove or disconnect the following:
   - Air cleaner duct
   - Transmission Control Module (TCM)
   - Solenoid pack electrical connector
   - Dipstick tube from the transaxle
   - Transaxle cooler lines
   - Shift lever cable from the transaxle lever
5. Using an engine support fixture tool, support the engine assembly.
6. Remove or disconnect the following:
   - Upper transaxle mount top bolts from the left side
   - Front wheels
   - Lower splash shields from both sides
   - Exhaust pipe from the exhaust manifold
   - Upper transaxle mount remaining bolts from the left side
   - Engine oil filter
   - Starter
   - Front engine mount bracket
   - Rear mount bracket through-bolt
   - Centermember bolts
   - Rear mount bracket
   - Radiator lower crossmember
   - Both lateral bending strut brackets
   - Flexplate cover
7. Matchmark the converter-to-flexplate location. Rotate the crankshaft clockwise to align the converter bolts.
8. Install or connect the following:
   - Converter-to-flexplate bolts
   - Crankshaft Position (CKP) sensor, if equipped
   - Transaxle electrical connectors
   - Right side steering gear and K-frame bolts
   - Sway bar mounts
9. Using a transmission jack and a safety chain, secure the transaxle and support it.
10. Install or connect the following:
    - Bell housing-to-engine bolts
    - Transaxle, by moving the K-frame rearward
11. To prepare the vehicle for rolling, support the engine with a suitable support or reinstall the front motor mount to the engine. Then, reinstall the ball joints to the steering knuckle and install the retaining bolt. Install the obtained outer CV-joints to the hubs, install the washers and tighten the axle nuts to 180 ft. lbs. (244 Nm). The vehicle may now be safely rolled.

To install:

12. Install or connect the following:
    - Transaxle, by moving the K-frame rearward
    - Bell housing-to-engine bolts and torque the bolts to 70 ft. lbs. (95 Nm)
    - Sway bar mounts
    - Right side steering gear and K-frame bolts
    - Transaxle electrical connectors
    - CKP sensor, if equipped
13. Align the converter-to-flexplate matchmark. Rotate the crankshaft clockwise to align the converter bolts.
14. Install or connect the following:
    - Torque converter and torque the converter-to-flexplate bolts to 55 ft. lbs. (74 Nm)
    - Flexplate cover and torque the bolts to 108 inch lbs. (12 Nm)
    - Both lateral bending strut brackets and torque the bolts to 45 ft. lbs. (61 Nm)
    - Radiator's lower crossmember and torque the bolts to 45 ft. lbs. (61 Nm)
    - Rear mount bracket
    - Centermember bolts and torque the bolts to 45 ft. lbs. (61 Nm)
    - Rear mount bracket through-bolt and torque the through-bolt to 45 ft. lbs. (61 Nm)
    - Front engine mount bracket and torque the bolts to 24 ft. lbs. (33 Nm)
    - Starter
    - New engine oil filter
    - Upper transaxle mount remaining bolts to the left side
    - Exhaust pipe to the exhaust manifold
    - Lower splash shields to both sides
    - Front wheels
    - Upper transaxle mount top bolts, to the left side
    - Shift lever cable to the transaxle lever and torque the nut to 14 ft. lbs. (19 Nm)
    - Transaxle cooler lines
    - Dipstick tube to the transaxle
    - Solenoid pack electrical connector
    - TCM
    - Air cleaner duct
    - Negative battery cable
15. Adjust the gearshift and throttle cables.
16. Refill the transaxle.
17. Check the transaxle for proper operation. Be sure the back-up lights and speedometer are working properly.

SEBRING, STRATUS SEDAN ONLY

1. Before servicing the vehicle, refer to the precautions in the beginning of this section.
2. Drain the transaxle.
3. Remove or disconnect the following:
   - Negative battery cable
   - Air cleaner assembly
   - Transaxle dipstick tube
   - Transaxle cooler lines flush with the fittings
   - Solenoid/pressure switch assembly
   - Transmission range sensor connector
   - Input and output speed sensor connectors
   - Shift cable from the manual valve lever and bracket
   - Crankshaft Position (CKP) sensor
   - Throttle body support bracket
   - Oxygen Sensor (O_2S) harness retainer
   - Rear mount bracket—to—transaxle case bolts
   - Upper starter bolt
   - Halfshafts
   - Rear mount and bracket
   - Front mount bracket and mount
   - Starter assembly
   - Torque converter dust shield and properly support the engine/transaxle assembly
   - Left mount bracket —to—transaxle bolts and lower the assembly slightly
   - Transaxle assembly

To install:

4. Install or connect the following:
   - Transaxle and torque the bolts to 70 ft. lbs. (95 Nm)
   - Upper mount. Raise the assembly into position and torque the through bolt to 55 ft. lbs. (70 Nm)
   - Torque converter bolts and remove the engine/transaxle assembly support
   - Torque converter dust shield
   - Starter and lower bolt
   - Front mount and bracket
   - Rear mount and bracket
   - Halfshafts
   - Upper starter bolt
   - O_2S harness retainer
   - Throttle body support bracket
   - CKP sensor electrical connector
   - Gearshift cable to manual valve lever
   - Output and input sensor connectors
   - Transmission range sensor connector
   - Solenoid/pressure switch connector
   - Oil cooler lines using the service splice kit
   - Transaxle dipstick tube
   - Air cleaner assembly
   - Negative battery cable

5. Fill the transaxle fluid to the proper level.
6. Start the vehicle, check for leaks and repair if necessary.

Clutch

ADJUSTMENT

2.0L, 2.4L (VIN X) and 2.5L Engines

FREE-PLAY

The manual transaxle clutch release system has a unique self-adjusting mechanism to compensate for clutch disc wear. This adjuster mechanism is located with the clutch cable assembly. The preload spring maintains tension on the cable. This tension keeps the clutch release bearing continuously loaded against the fingers of the clutch cover assembly. No manual adjustment is necessary.

When servicing this vehicle or if removing and installing the clutch cable, do not pull on the clutch cable housing to remove it from the dash panel. Damage to the cable self-adjuster may occur.

To check the function of the adjuster mechanism, use the following procedure:

1. With slight pressure, pull the clutch release lever end of the cable to draw the cable taut.
2. Push the clutch cable housing toward the dash panel. With less than 25 lbs. (11 kg) of effort, the cable housing should move 1.2–2.0 in. (30–50mm). This indicates proper adjuster mechanism function.
3. If the cable does not adjust, determine if the mechanism is properly seated on the bracket.

REMOVAL & INSTALLATION

➡**The transaxle assembly must be removed to service the clutch assembly.**

1. Before servicing the vehicle, refer to the precautions in the beginning of this section.

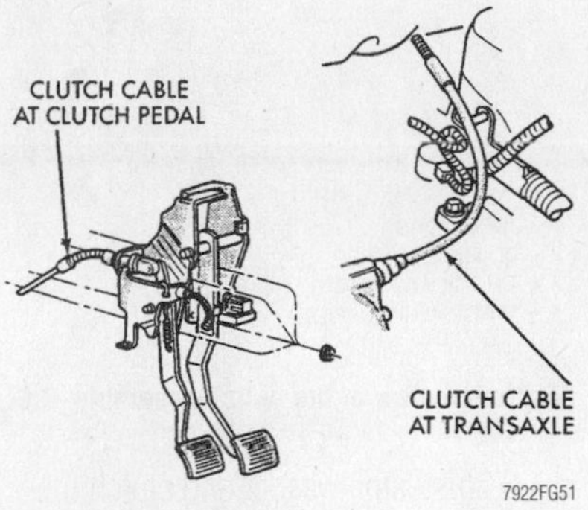

Clutch cable routing

2. Remove or disconnect the following:
   - Negative battery cable from the left shock tower

➡**The ground cable is equipped with an insulator grommet which should be placed on the stud to prevent the negative battery cable from accidentally grounding.**

   - Starter
   - Rear and front transaxle support brackets
   - Clutch inspection cover
   - Modular clutch-to-flywheel bolts
   - Transaxle assembly with the clutch as an assembly
   - Clutch assembly from the transaxle input shaft

To install:

3. Clean all parts well. Inspect for oil leakage through the engine rear crankshaft oil seal and transaxle input shaft seal. If leakage is noted, it should be corrected at this time.
4. Examine the throwout or clutch release bearing. It is pre-lubricated and sealed and should not be washed in solvent. The bearing should turn smoothly when held in the hand with a light thrust load. A light drag caused by the lubricant fill is normal. If the bearing is noisy, rough or dry, replace the complete bearing assembly. In most cases where a clutch is being serviced, the complete clutch assembly and

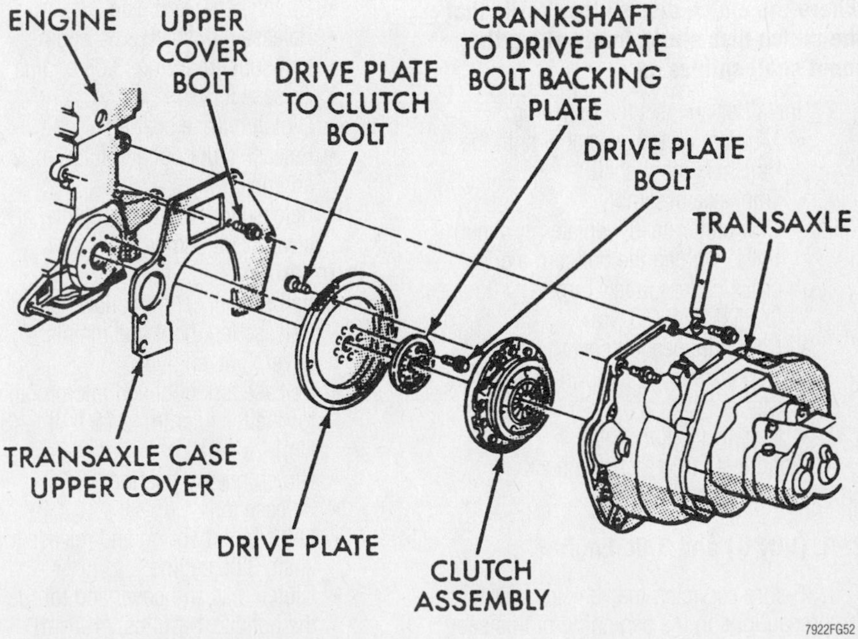

Exploded view of the clutch assembly

release bearing are usually replaced together.

5. Check the condition of the stud pivot spring clips on the back side of the clutch fork. If the clips are broken or distorted, replace the clutch fork. The pivot ball pocket in the fork is Teflon® coated and should be installed **WITHOUT** any lubricant such as grease which will break down the Teflon® coating. Be sure the ball stud and fork pocket are clean of contamination and dirt. When assembling the fork to the bearing, the small pegs on the bearing must go over the fork arms.

6. Check the flywheel for cracks, glazing or grooves. If any of these conditions exist, machine (reface) or replace the flywheel to prevent clutch chatter and premature clutch wear.

➡The manual transaxle is equipped with a reverse brake. It functions as a synchronizer, but only if the vehicle is not moving. When the clutch pedal is depressed to the floor and held for 3 seconds, and the transaxle shifts to reverse, no gear clash should be present. If there is, the input shaft should be checked. When the transaxle is removed for clutch service, check the input clutch shaft, clutch disc splines and release bearing for dry rust. If present, clean rust off and apply a light coat of high temperature bearing grease to the input shaft splines. Apply grease on the input shaft splines only where the clutch disc slides. Verify that the clutch disc slides freely along the input shaft splines.

7. Install or connect the following:
   - Modular clutch assembly onto the transaxle input shaft
   - Transaxle assembly
   - New clutch-to-driveplate (flywheel) bolts. Tighten the bolts, in a criss-cross pattern, a few turns at a time to 55 ft. lbs. (75 Nm)
   - Clutch inspection cover
   - Transaxle lower support brackets
   - Starter
   - Negative battery cable

8. Road test the vehicle to check for proper clutch operation.

2.4L (VIN G) and 3.0L Engine

1. Before servicing the vehicle, refer to the precautions in the beginning of this section.
2. Remove or disconnect the following:
   - Negative battery cable
   - Clutch fluid line bracket

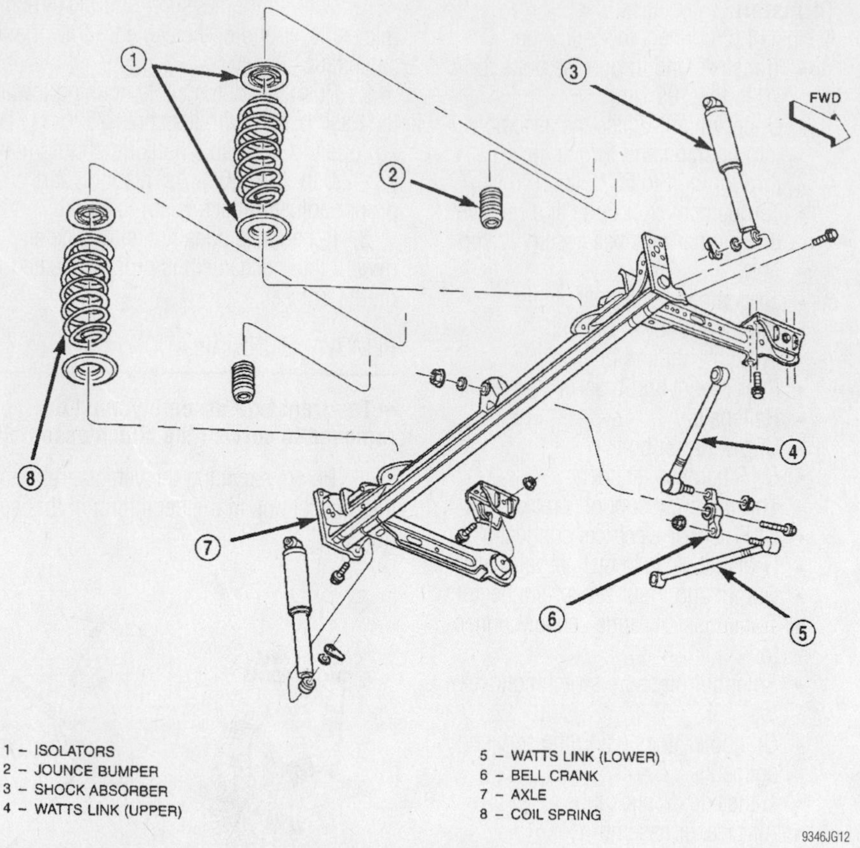

Exploded view of the clutch assembly–3.0L engine

   - Insulator, washer and clutch tube
   - Union, union bolt and gasket
   - Valve and spring, 3.0L engine
   - Clutch release cylinder
   - Clutch cover and disc
   - Release fork shaft, 3.0L engine
   - Left support spring, 3.0L engine
   - Release fork
   - Clutch release bearing
   - Release fork boot and fulcrum, 2.4L engine
   - Release fork boot and maintenance cover, 3.0L engine

To install:

3. Install or connect the following:
   - Release fork boot and maintenance cover, 3.0L engine
   - Release fork boot and fulcrum and torque the fulcrum to 26 ft. lbs. (35 Nm)
   - Clutch release bearing
   - Release fork
   - Left support spring and release fork shaft, 3.0L engine
   - Clutch disc and cover and torque the bolts to 14 ft. lbs. (19 Nm)
   - Clutch release cylinder
   - Valve and spring, 3.0L engine
   - Union, bolt and gasket and torque the bolt to 17 ft. lbs. (23 Nm)
   - Insulator, washer and clutch tube and torque the fastener to 11 ft. lbs. (15 Nm)
   - Fluid line bracket and torque the bolt to 14 ft. lbs. (19 Nm)
   - Negative battery cable

4. Check and top off the fluid level, if necessary.

Halfshafts

REMOVAL & INSTALLATION

2.0L, 2.4L (VIN X) 2.5L and 2.7L Engines

➡If the vehicle is going to be rolled while the halfshafts are out of the vehicle, obtain 2 outer CV-joints or proper equivalent tools and install to the hubs.

⁂ WARNING

If the vehicle is rolled without the proper torque applied to the front wheel bearings, the bearings will no longer be usable.

1. Before servicing the vehicle, refer to the precautions in the beginning of this section.

2. Remove or disconnect the following:

- Negative battery cable
- Cotter pin, nut lock and spring washer
- Halfshaft nut, loosen it while the vehicle is on the floor with the brakes applied
- Wheel
- Brake caliper assembly and support it on a wire
- Brake rotor
- Halfshaft nut and washer
- Tie rod end from the steering knuckle using joint separation Tool MB991113

⁂ WARNING

Use of improper methods of joint separation can result in damage to the joint, leading to possible failure.

- Speed sensor cable routing bracket, if equipped with an Anti-lock Brake System (ABS)
- Sway bar link from the damper fork, if necessary
- Damper fork assembly
- Steering knuckle from the lower control arm
- Halfshaft by pressing it from the hub

➡After pressing the outer shaft, insert a prybar between the transaxle case and the halfshaft and pry the shaft from the transaxle.

⁂ WARNING

Do not pull on the shaft. Doing so damages the inboard joint. Do not insert the prybar too far or the oil seal in the case may be damaged.

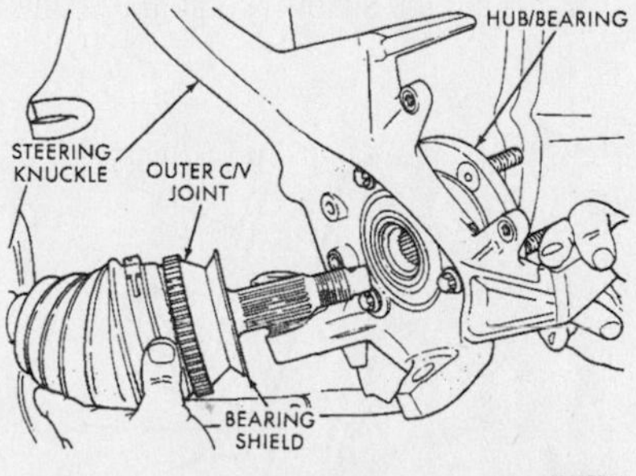

Carefully remove the outer CV-joint from the steering knuckle—be careful NOT to damage the threads or splines on the joint

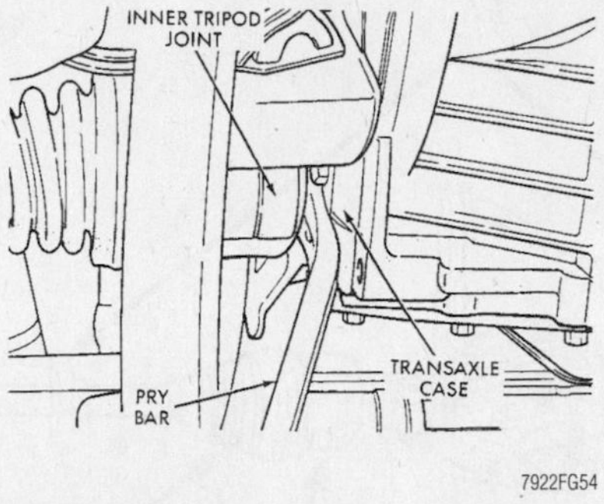

Inserting the prybar too far may damage the transaxle seal

To install:

3. Inspect the halfshaft boot for damage or deterioration. Check the ball joints and splines for wear.
4. Replace the circlips on the ends of the halfshaft(s).
5. Install or connect the following:

- Halfshaft into the transaxle until it is fully seated
- Halfshaft into the hub by pulling the knuckle assembly outward
- Washer. Make sure the chamfered edge faces outward
- Halfshaft nut and tighten temporarily
- Control arm to the steering knuckle and torque the nuts to 43–52 ft. lbs. (59–71 Nm)
- Damper fork and torque the lower through-bolt/nut to 65 ft. lbs. (88 Nm) and the upper pinch bolt to 76 ft. lbs. (103 Nm)
- Tie rod end, to the steering knuckle and torque the nut to 17–25 ft. lbs. (24–33 Nm) and install a new cotter pin
- Sway bar link to the damper fork and torque the link nut to 29 ft. lbs. (39 Nm)
- Lockwasher and axle nut and torque the nut to 145–188 ft. lbs. (200–260 Nm)

➡Before securely tightening the axle nut, be sure there is no load on the wheel bearings.

- Brake rotor and caliper assembly
- New cotter pin
- Wheel
- Negative battery cable

6. Refill the transaxle.
7. Test drive the vehicle and check for proper operation.

2.4L (VIN G) and 3.0L Engine

1. Before servicing the vehicle, refer to the precautions in the beginning of this section.
2. Remove or disconnect the following:

- Negative battery cable
- Front wheel
- Vehicle Speed Sensor (VSS) cable, if equipped
- Brake hose clip and cotter pin
- Driveshaft nut
- Lower arm ball joint connection
- Tie rod end
- Stabilizer link
- Driveshaft and circlip

To install:

3. Install or connect the following:

- Driveshaft
- Stabilizer link and torque the nut to 33 ft. lbs. (44 Nm)
- Tie rod end and torque the nut to 21 ft. lbs. (28 Nm)
- New cotter pin to the tie rod end
- Lower arm ball joint
- Driveshaft nut and torque it to a maximum of 167 ft. lbs. (226 Nm)
- New cotter pin to the driveshaft nut
- Brake hose clip
- VSS cable, if equipped
- Front wheel
- Negative battery cable

CV-Joints

OVERHAUL

Inner Tri-pot Joint

1. Remove or disassemble the following:

- Negative battery cable
- Halfshaft
- Large and small boot retaining clamps

2. Slide the boot down the shaft away from the tri-pot housing.

➡When separating the spider joint from the tri-pot joint housing, hold the rollers in place on the trunions to prevent the rollers and needle bearings from falling away.

3. Carefully slide the shaft/spider assembly from the tri-pot housing.
4. Remove the spider assembly-to-shaft snapring; then, slide the spider assembly off the shaft.

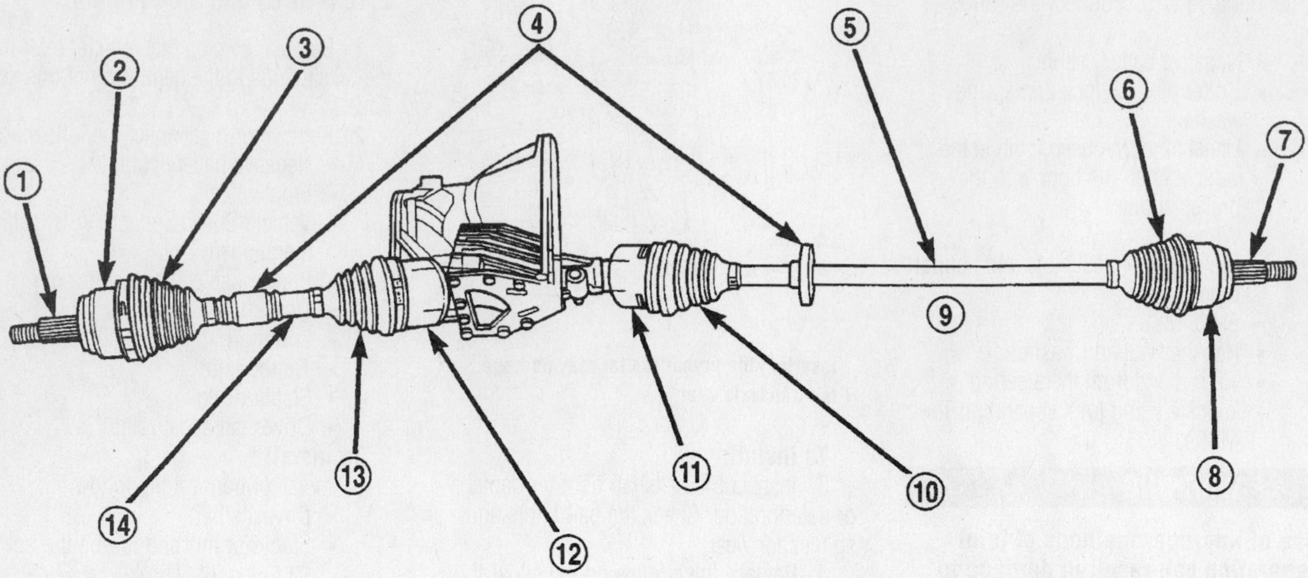

1 – STUB AXLE
2 – OUTER C/V JOINT
3 – OUTER C/V JOINT BOOT
4 – TUNED RUBBER DAMPER WEIGHT
5 – INTERCONNECTING SHAFT
6 – OUTER C/V JOINT BOOT
7 – STUB AXLE
8 – OUTER C/V JOINT
9 – RIGHT DRIVESHAFT
10 – INNER TRIPOD JOINT BOOT
11 – INNER TRIPOD JOINT
12 – INNER TRIPOD JOINT
13 – INNER TRIPOD JOINT BOOT
14 – INTERCONNECTING SHAFT LEFT DRIVESHAFT

9306EG02

View of the halfshaft assemblies—Typical

⁂ WARNING

If necessary, tap the spider assembly off the shaft using a brass drift; be careful not to hit the outer bearings.

5. Slide the boot off the shaft.
6. Throughly inspect all parts for signs of excessive wear. If necessary, replace the halfshaft.

➡Component parts are not serviceable and must be replaced as an assembly.

To assemble:

⁂ WARNING

The Tri-pot sealing boots are made of 2 different types of material; silicon rubber (high temperature) which is soft and pliable or Hytrel plastic (standard temperature) which is stiff and rigid. Be sure to replace the boot made of the correct material.

7. Slide the inner tri-pot boot clamp and boot onto the shaft; then, position the boot so that only the thinnest (sight) groove is visible on the shaft.
8. Install the spider assembly onto the shaft just far enough so that the snapring can be installed.

⁂ WARNING

If necessary, tap the spider assembly onto the shaft using a brass drift; be careful not to hit the outer bearings.

9. Install the snapring onto the shaft;

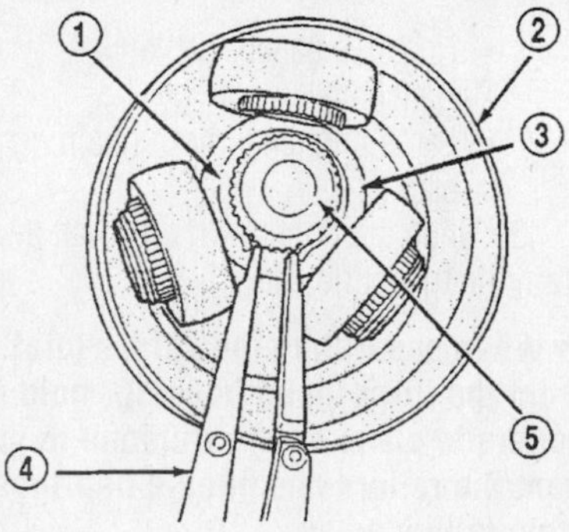

1 – SNAP RING
2 – SEALING BOOT
3 – SPIDER ASSEMBLY
4 – SNAP RING PLIERS
5 – INTERCONNECTING SHAFT

9306BG11

View of the halfshaft inner tri-pot joint and snapring

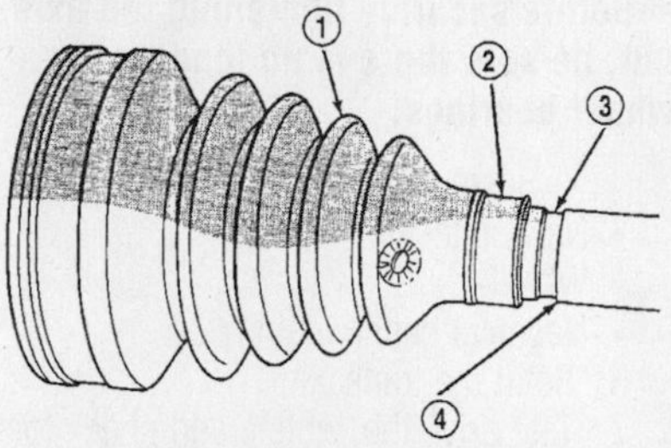

1 – SEALING BOOT
2 – RAISED BEAD IN THIS AREA OF SEALING BOOT
3 – GROOVE
4 – INTERCONNECTING SHAFT

9306EG04

View of the halfshaft boot and shaft

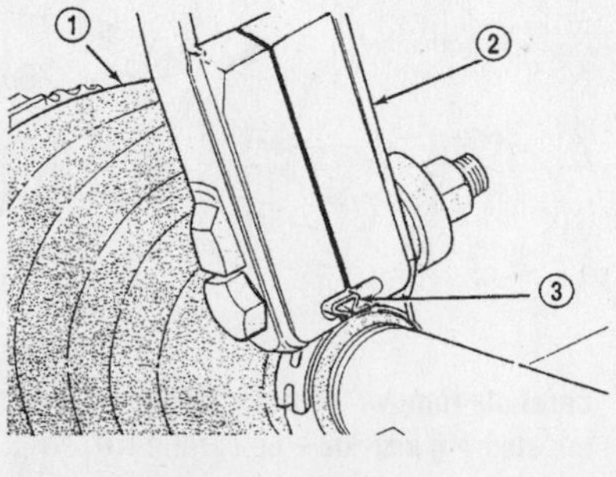

1 – SEALING BOOT
2 – SPECIAL TOOL C-4975
3 – CLAMP BRIDGE

9306BG13

Securing the halfshaft boot clamp

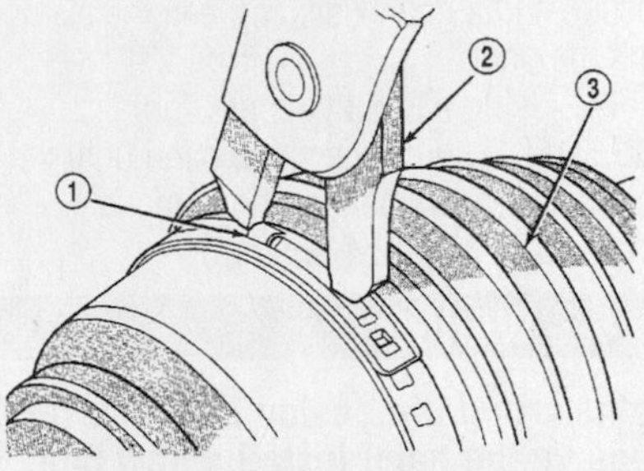

1 - CLAMP
2 - SPECIAL TOOL YA3050
3 - SEALING BOOT

9306EG07

Tightening the low profile boot clamp—Silicone rubber Tri-pot boot

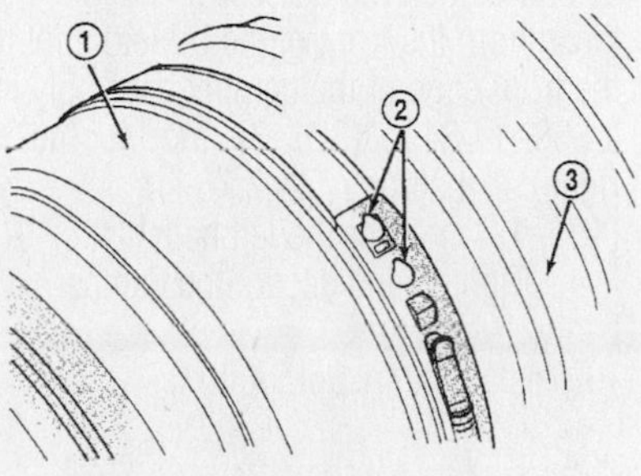

1 - INNER TRIPOD JOINT HOUSING
2 - TOP BANK OF CLAMP MUST BE RETAINED BY TABS AS SHOWN HERE TO CORRECTLY LATCH BOOT CLAMP
3 - SEALING BOOT

9306EG08

Secured the low profile boot clamp—Silicone rubber Tri-pot boot

make sure that the snapring is fully seated in the groove.

10. If installing a new boot, distribute ½ of the grease in the service package inside the tri-pot housing and the other ½ inside the boot.

11. Carefully, slide the spider assembly and shaft into the tri-pot housing.

12. Position the inner boot clamp evenly on the sealing boot.

⁂ WARNING

The seal must not be dimpled, stretched or out of shape. If necessary, use a trim stick to seat and equalize the seal pressure and shape it by hand.

➡If using a Hytrel (hard plastic) boot, be sure the stick is inserted between the soft rubber insert and the tri-pot housing, not between the hard plastic sealing boot and the soft rubber insert.

13. Position the boot onto the tri-pot housing retaining groove and install the retaining clamp evenly on the boot.

14. If using a crimp type boot clamp, perform the following procedure:

a. Using the Crimper Tool C-4975-A, place the tool over the clamp bridge, tighten the tool nut until the jaws are completely closed (face-to-face).

b. Using the Crimper Tool C-4975-A, place the tool over the clamp bridge, tighten the tool nut until the jaws are completely closed (face-to-face).

15. If using a latching type boot clamp, perform the following procedure:

a. Position Snap-On® Clamp Locking Tool YA3050 prongs in the clamp holes.

b. Squeeze the tool until the upper clamp band is latched behind the 2 tabs on the lower clamp band.

16. Install the halfshaft into the vehicle.

Rzeppa (Outer) Joint

1. Remove or disassemble the following:
- Halfshaft and place it in a soft-jawed vise
- Rzeppa joint boot clamps and slide the boot down the shaft
- Rzeppa joint housing by sharply hitting it with a soft-faced hammer to drive it off the shaft
- Circlip from the shaft
- Boot by sliding it off the shaft

⁂ WARNING

If any parts show excessive wear, replace the halfshaft assembly; the component parts are not serviceable.

To assemble:

2. Install or assemble the following:
- New small boot clamp and slide it onto the shaft

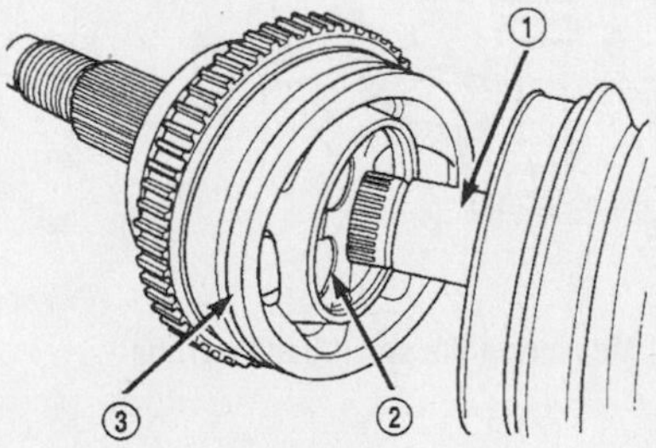

1 - INTERCONNECTING SHAFT
2 - CROSS
3 - OUTER C/V JOINT ASSEMBLY

9306EG09

Aligning the cross splines with the shaft splines—Rzeppa joint

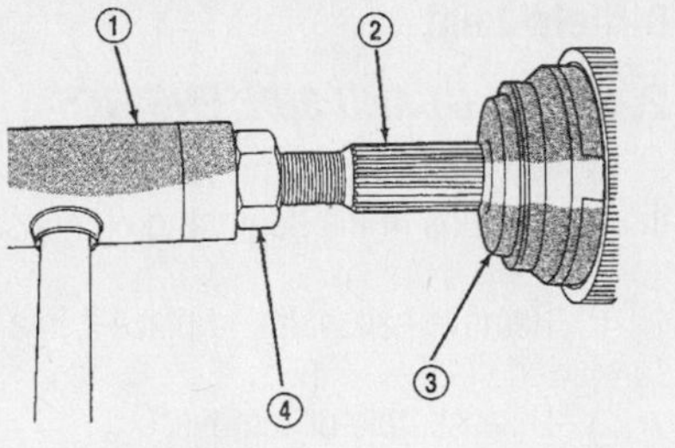

1 - SOFT FACED HAMMER
2 - STUB AXLE
3 - OUTER C/V JOINT
4 - NUT

9306EG10

Driving the Rzeppa joint onto the shaft

- Boot and slide it onto the shaft
- Circlip, if removed

3. Position the boot so that the raised bead on the inside the boot seal is in the shaft groove.

4. Install or assemble the following:
- Halfshaft hub nut onto the Rzeppa joint threaded shaft so it is flush with the end
- Rzeppa joint and align the shaft splines and tap it onto the shaft with a soft-faced hammer so it locks on the circlip

5. Distribute ½ of the grease in the service package inside the Rzeppa joint housing and the other ½ inside the boot.

6. Install or connect the following:
- New small boot clamp and position it evenly on the sealing boot

7. Using the Crimper Tool C-4975-A, place the tool over the clamp bridge, tighten the tool nut until the jaws are completely closed (face-to-face).

⁂ WARNING

The seal must not be dimpled, stretched or out of shape. If necessary, equalize the seal pressure and shape it by hand.

8. Position the boot onto the Rzeppa housing retaining groove and install the retaining clamp evenly on the boot.

9. Using the Crimper Tool C-4975-A, place the tool over the clamp bridge, tighten the tool nut until the jaws are completely closed (face-to-face).

⁂ WARNING

The seal must not be dimpled, stretched or out of shape. If necessary, equalize the seal pressure and shape it by hand.

10. Install the halfshaft into the vehicle.

Birfield Joint

2.4L (VIN G) AND 3.0L ENGINES

1. Before servicing the vehicle, refer to the precautions in the beginning of this section.
2. Remove halfshaft and place it in a soft jawed vise.
3. Disassemble or remove:
   - Tri-pot joint
   - Dynamic damper
   - Birfield joint boot bands
   - Birfield joint boot

To install:

4. Assemble or install:
   - Birfield joint boot
   - Birfield joint boot small band
5. Place the halfshaft in a soft jawed vise so that the birfield joint is standing vertically.
6. Position the boots small diameter so that only 1 groove is exposed on the shaft.
7. Adjust the Band Crimper Tool MB991561 so that the jaw opening is 0.114 in. (2.9mm).
8. Position the small band so there is clearance between the boot and the bands mating edge.
9. Using a Band Crimper Tool MB991561, crimp the small band until the internal measurement of the crimp is 0.094–0.110 in. (2.4–2.8mm).

⁂ WARNING

If the crimp dimension is not correct, remove the band install a new one.

10. Using 5.47 oz. (155g) specified amount of grease in the repair kit, pack the birfield boot.
11. Position the boot so 0.004–0.061 in. (0.1–1.55mm) of clearance exists between the boots large diameter end and the birfield joint housing shoulder.
12. Adjust the Band Crimper Tool MB991561 so that the jaw opening is 0.126 in. (3.2mm).
13. Install and position the large boot band so that it rests against the boot(s) projection and a gap exists between the clamp and the boot.
14. Using a Band Crimper Tool MB991561, crimp the small band until the internal measurement of the crimp is 0.094–0.110 in. (2.4–2.8mm).

⁂ WARNING

If the crimp dimension is not correct, remove the band install a new one.

15. If equipped with a 2.0L (DOHC) engine, install the dynamic damper by performing the following procedure:
   a. Slide the damper onto the halfshaft with new bands.
   b. Position the damper so the distance from the front of the birfield joint to the front edge of the damper assembly is 14.60–14.84 in. (371–377mm) for the right halfshaft or 7.52–7.76 in. (191–197mm) for the left halfshaft.
   c. Tighten and secure the damper bands.
16. Install the tri-pot joint.

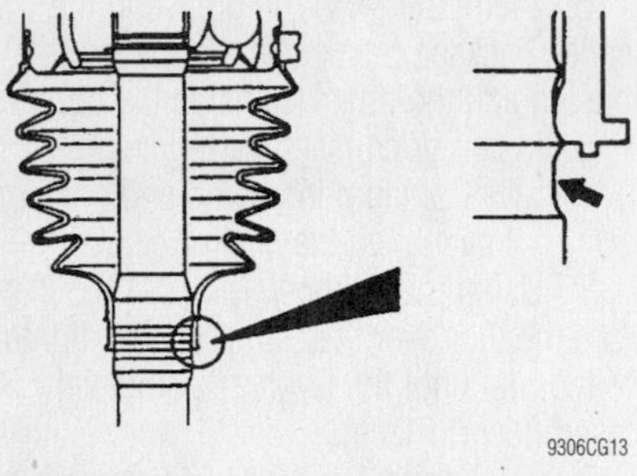

Positioning the birfield joint boot's small diameter

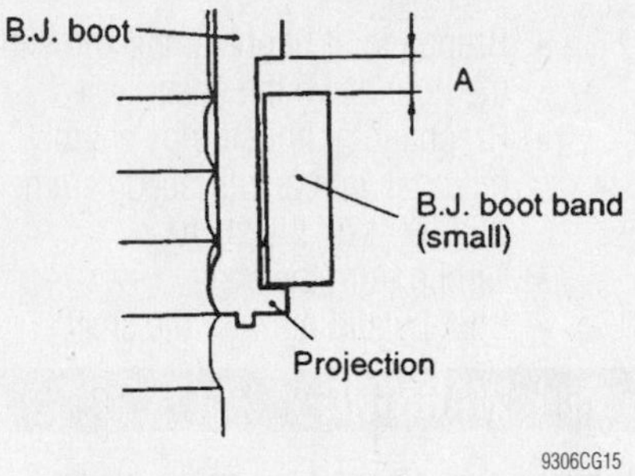

Positioning the birfield joint boot's small band

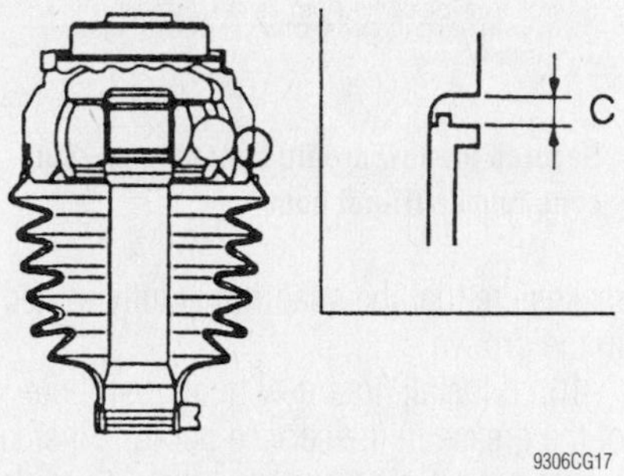

Positioning the birfield joint boot's large diameter

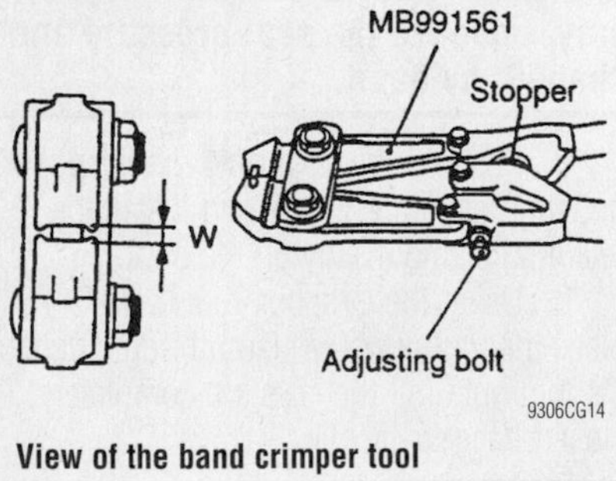

View of the band crimper tool

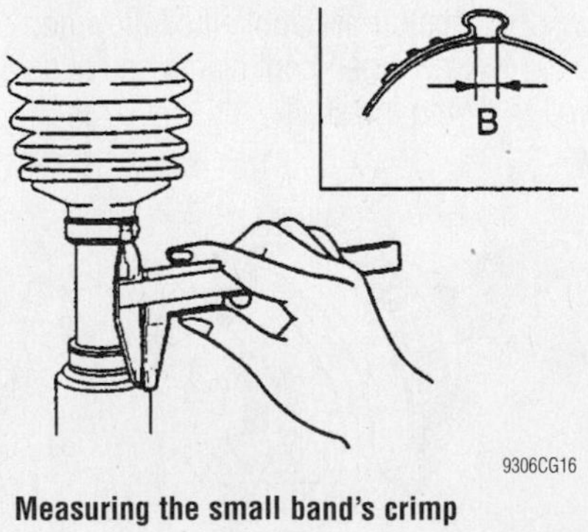

Measuring the small band's crimp

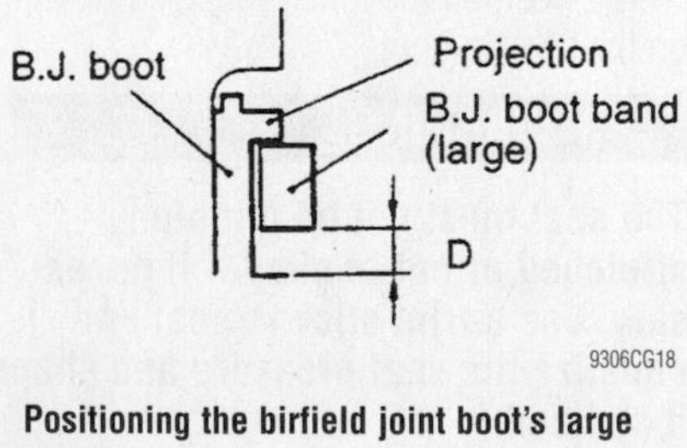

Positioning the birfield joint boot's large band

STEERING AND SUSPENSION

Air Bag

⁂ CAUTION

Some vehicles are equipped with an air bag system, also known as the Supplemental Inflatable Restraint (SIR) or Supplemental Restraint System (SRS). The system must be disabled before performing service on or around system components, steering column, instrument panel components, wiring and sensors. Failure to follow safety and disabling procedures could result in accidental air bag deployment, possible personal injury and unnecessary system repairs.

PRECAUTIONS

Several precautions must be observed when handling the inflator module to avoid accidental deployment and possible personal injury.

➡**Before servicing the vehicle, also refer to the precautions in the beginning of this section.**

- Never carry the inflator module by the wires or connector on the underside of the module.
- When carrying a live inflator module, hold securely with both hands, and ensure that the bag and trim cover are pointed away.
- Place the inflator module on a bench or other surface with the bag and trim cover facing up.
- With the inflator module on the bench, never place anything on or close to the module which may be thrown in the event of accidental deployment.

DISARMING

This air bag system is a sensitive, complex, electromechanical unit. Proper SRS (also called Supplemental Inflatable Restraint, or SIR, or air bag system) disarming can be obtained by disconnecting and isolating the negative battery cable (wrapping the battery cable end with electrical tape is a good method for isolating it). Failure to disconnect the battery could result in accidental air bag deployment and possible personal injury. Before beginning service work, allow the system capacitor 2 minutes to discharge after disconnecting and isolating the negative battery cable.

Power Rack and Pinion Steering Gear

REMOVAL & INSTALLATION

2.0L, 2.4L (VIN X), 2.5L and 2.7L Engines

1. Before servicing the vehicle, refer to the precautions in the beginning of this section.
2. Remove or disconnect the following:
   - Negative battery cable from the left shock tower

➡**The ground cable is equipped with an insulator grommet which should be placed on the stud to prevent the negative battery cable from accidentally grounding.**

   - Intermediate shaft coupler pin bolt retaining pin
   - Intermediate shaft coupler pin bolt
   - Front wheels
   - Tie rod ends by holding the tie rod end stud with a $11/32$ in. socket and loosen the retaining nut
   - Tie rod end from the steering knuckle

➡**Before removing the front suspension crossmember from the vehicle scribe the front suspension crossmember and the vehicle body. This must be done to retain the proper alignment. The caster and camber are not adjustable.**

3. Scribe a line on the body and on the crossmember on all 4 sides.
4. Remove or disconnect the following:
   - Stabilizer
   - 3 anti-lock brake controller-to-crossmember bolts and secure it to the chassis, if equipped with Anti-lock Brake System (ABS)
   - Strut clevis from the lower control arm
   - Both engine support bracket-to-crossmember bolts
   - Engine support bracket-to-transaxle mounting bracket bolt
5. Place a lifting device under the front suspension crossmember.
6. Remove or disconnect the following:
   - 8 crossmember-to-chassis bolts
7. Lower the lifting device enough to gain access to the steering rack.
8. Remove or disconnect the following:
   - Power steering lines and drain the fluid
   - Power steering pressure switch electrical connector
   - Solenoid control module electrical connector, if equipped with speed proportional steering
   - Both steering rack isolator bolts
   - Both steering rack saddle bracket bolts
   - Steering rack

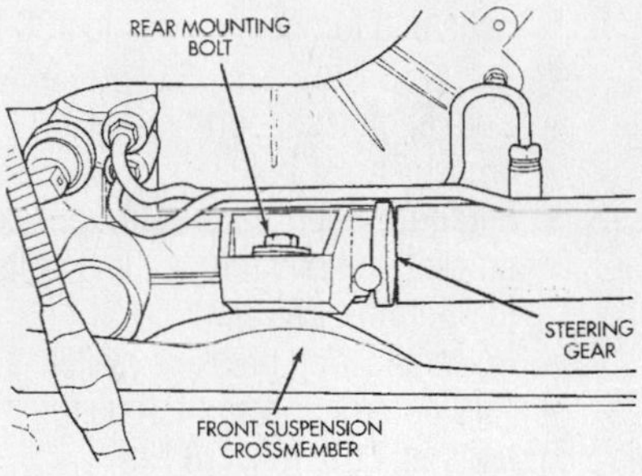

7922FG55

Steering gear rear mounting bolt location

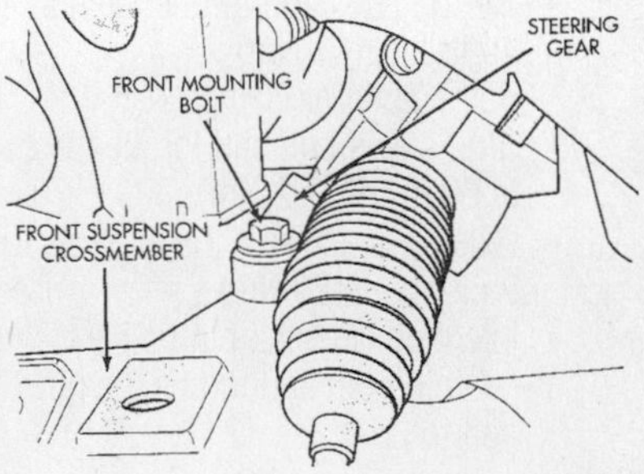

7922FG57

Steering gear front mounting bolt location

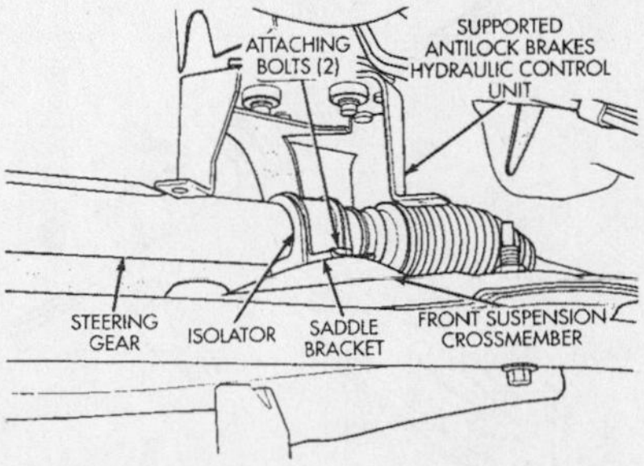

7922FG56

Check the condition of the isolator—if worn or oil soaked, replace

To install:

9. Install or connect the following:
   - Steering rack
   - Isolator and saddle bracket and torque the bolts to 50 ft. lbs. (68 Nm)
   - Power steering pressure and return lines and torque the lines to 23 ft. lbs. (31 Nm)
   - Crossmember, install the rear bolts first and torque the bolts to 20 inch lbs. (2 Nm)
10. Using a soft faced hammer tap the crossmember into position.

➡**Be sure to align the scribed marks on the crossmember.**

11. Install or connect the following:
    - Crossmember, starting with the rear bolts and torque the bolts to 120 ft. lbs. (163 Nm)
    - Both engine support bracket-to-crossmember bolts
    - Engine support bracket-to-transaxle bracket bolt and torque the 3 bolts to 55 ft. lbs. (75 Nm)
    - Power steering pressure switch
    - ABS control unit and torque the bolts to 21 ft. lbs. (28 Nm)
    - Heat shield on the tie rod ends
    - Shock clevis to the lower control arm
    - Tie rod ends and torque the nuts to 45 ft. lbs. (61 Nm)
    - Both stabilizer clamps
    - Strut clevis bolt and torque the bolt to 68 ft. lbs. (92 Nm)
    - Wheels and torque the lug nuts to 95 ft. lbs. (129 Nm)
    - Intermediate shaft pin bolt and retaining pin and torque the pin bolt to 20 ft. lbs. (27 Nm)
    - Negative battery cable
12. Refill the power steering system.
13. Start the engine and allow it to run for a few minutes.
14. Shut **OFF** the engine and check the power steering fluid.
15. Add power steering fluid if necessary.
16. Raise the front wheels off the ground.
17. Start the engine and turn the wheel from stop-to-stop to bleed any air from the system.
18. Check the fluid level and add, if necessary.
19. Check and adjust the alignment.

2.4L (VIN G) and 3.0L Engines

1. Before servicing the vehicle, refer to the precautions in the beginning of this section.
2. Drain the power steering fluid.
3. Remove or disconnect the following:
   - Negative battery cable from the left shock tower
   - Front exhaust pipe
   - Center member
   - Sway bar, 2.4L only
   - Steering shaft assembly
   - Stay, 2.4L only
   - Tie rod end from the steering knuckle
   - Return hose and pressure tube connections
   - Cylinder clamp
   - Steering gear from the vehicle

To install:

4. Install or connect the following:
   - Steering gear to the vehicle
   - Cylinder clamp and torque the bolts to 51 ft. lbs. (69 Nm)
   - Pressure tube and return hoses and torque to 11 ft. lbs. (15 Nm)
   - Tie rod ends to the steering knuckle and torque the nut to 21 ft. lbs. (29 Nm)
   - Stay and torque the bolts to 55 ft. lbs. (74 Nm), 2.4L only
   - Steering shaft assembly and torque the bolt to 13 ft. lbs. (18 Nm)
   - Sway bar, 2.4L Only
   - Center member
   - Front exhaust pipe, 2.4L only
   - Negative battery cable
5. Fill and bleed the power steering system.
6. Start the vehicle, check for leaks and repair if necessary.
7. Perform a front wheel alignment.

Strut

REMOVAL & INSTALLATION

Front

2.0L, 2.4L (VIN X), 2.5L AND 2.7L ENGINES

1. Before servicing the vehicle, refer to the precautions in the beginning of this section.
2. Remove or disconnect the following:
   - Front wheel
   - Steering knuckle
   - Strut-to-shock clevis pin bolt
   - Clevis-to-lower control arm through-bolt
   - Clevis from the strut by tapping it with a brass drift
   - 4 strut-to-shock tower bolts
   - Strut/upper control arm mounting bracket as an assembly

To install:

3. Install or connect the following:
   - Strut into the shock tower
   - 4 upper strut mounting bolts and

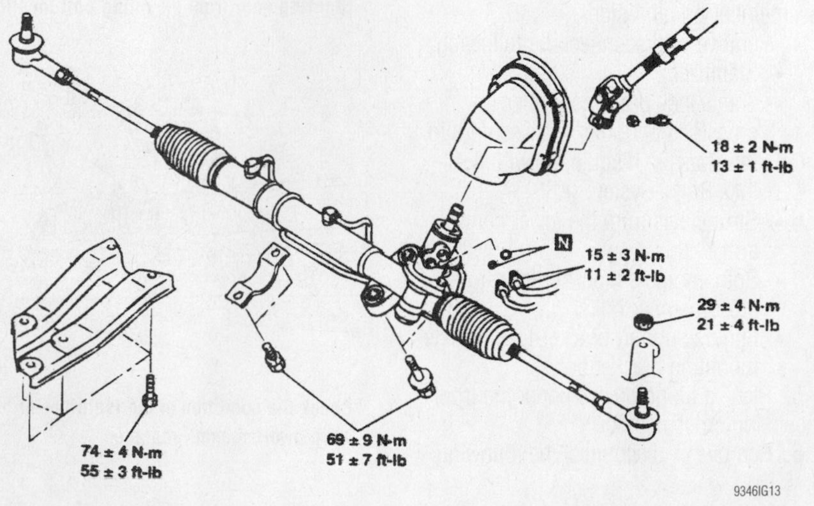

Exploded view of the steering gear assembly—2.4L (VIN G)

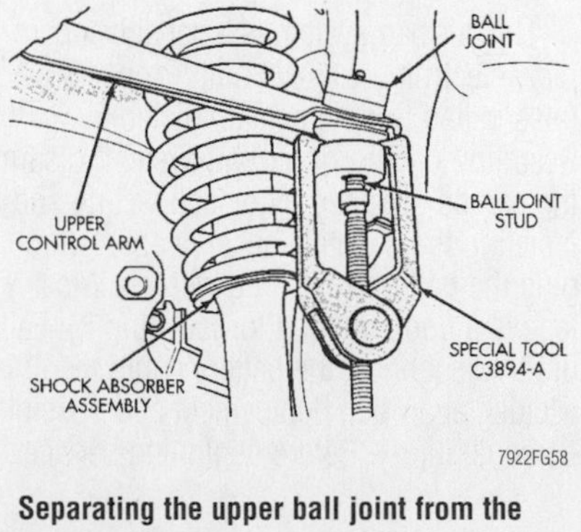

Separating the upper ball joint from the steering knuckle

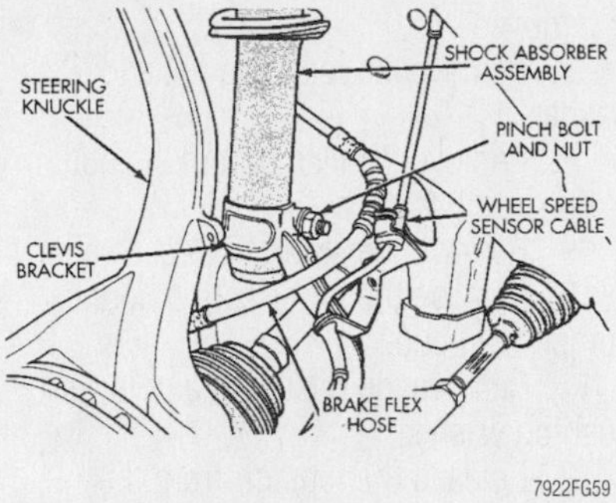

View of the front strut mount and related components

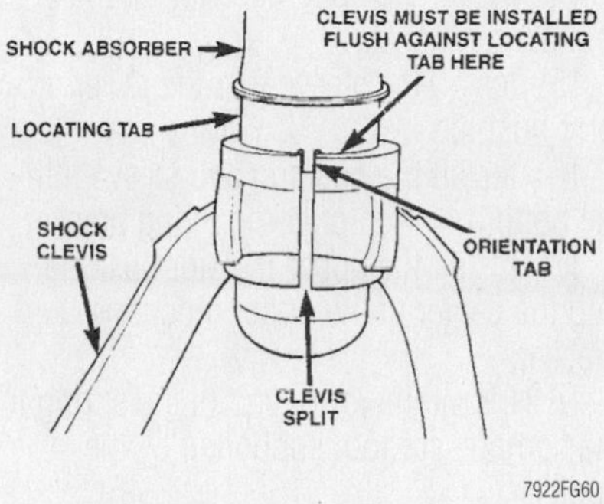

Be sure the orientation tab is situated into the clevis split

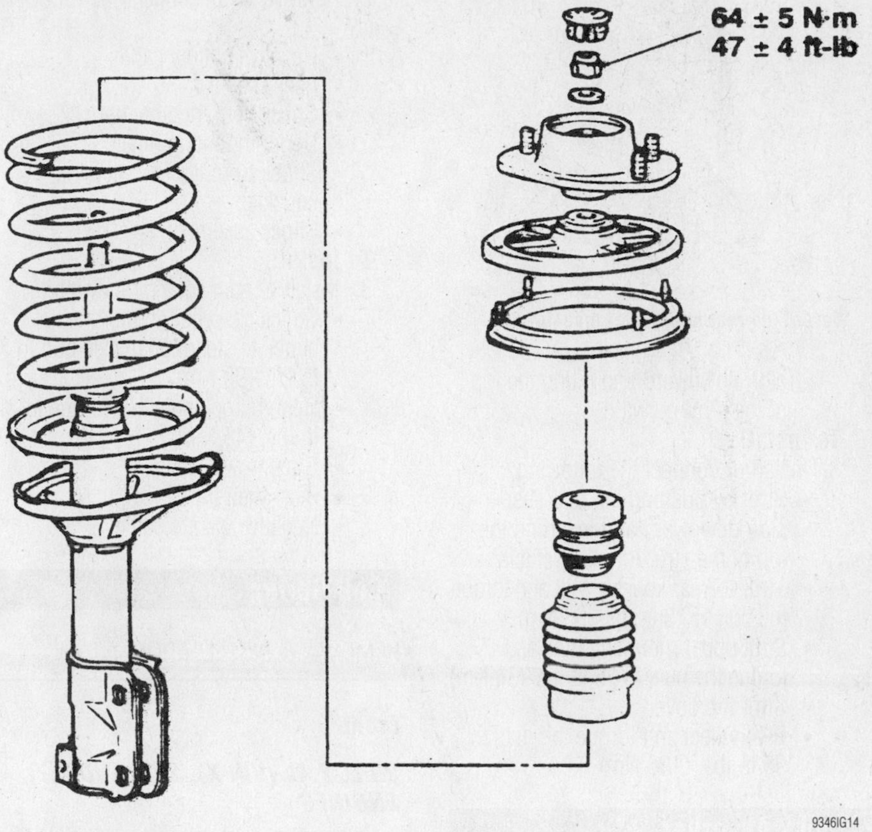

Exploded view of the front strut assembly—2.4L (VIN G) and 3.0L

torque the bolts to 23 ft. lbs. (31 Nm)
- Clevis onto the strut, using a brass drift until the clevis is fully seated against the locating tab
- Clevis pin bolt
- Clevis onto the lower control arm
- Clevis through-bolt
- Steering knuckle and torque the clevis-to-strut pin bolt to 65 ft. lbs. (88 Nm)

4. Lower the vehicle to support the lower control arm.
5. Torque the clevis-to-lower control arm mounting bolt 40 ft. lbs. (54 Nm)
6. Install the front wheel

2.4L (VIN G) AND 3.0L ENGINES

1. Before servicing the vehicle, refer to the precautions in the beginning of this section.
2. Remove or disconnect the following:
   - Negative battery cable
   - Front wheel
   - Brake hose clamp
   - Front speed sensor harness clamp, if equipped
   - Sway bar link
   - Lower control arm
   - Strut from the steering knuckle
   - Upper mounting bolts
   - Strut assembly

To install:

3. Install or connect the following:
   - Strut assembly and torque the upper bolts to 33 ft. lbs. (44 Nm)
   - Strut assembly to the lower control arm and steering knuckle. Torque the bolts to 221 ft. lbs. (300 Nm)
   - Sway bar link and torque the nut to 33 ft. lbs. (44 Nm)
   - Front speed sensor harness clamp, if equipped
   - Brake hose clamp
   - Front wheel
   - Negative battery cable
4. Check and adjust the front end alignment, if needed.

Rear

2.0L, 2.4L (VIN X), 2.5L AND 2.7L ENGINES

1. Before servicing the vehicle, refer to the precautions in the beginning of this section.
2. Remove or disconnect the following:
   - Carpet, pull it back from the rear strut tower
   - Plastic cover from the top of the strut tower
   - Both strut assembly-to-chassis nuts
   - Rear wheel
   - Splash shield from the upper strut mount
   - Strut-to-rear knuckle bolt
   - Strut by pushing the rear suspen-

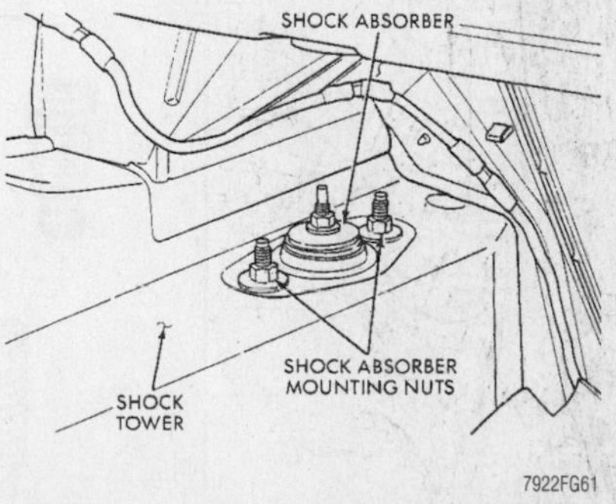

Access the rear strut upper mounting from inside the trunk

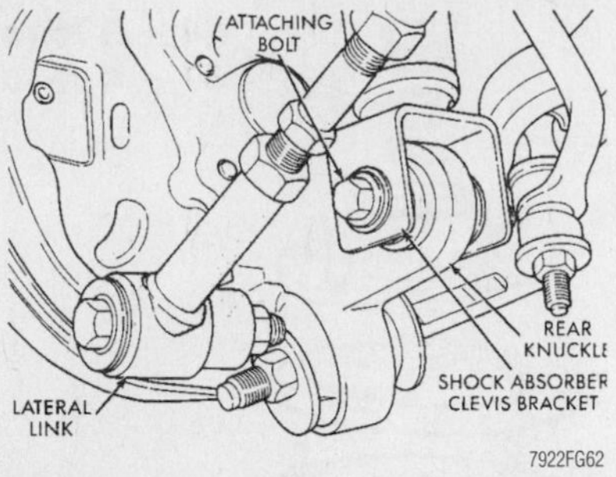

View of the rear strut lower mounting

sion downward and tilting the top of the strut outward

To install:

3. Install or connect the following:
   - Strut by pushing the rear suspension downward and inserting the top of the strut into the vehicle
   - Strut-to-rear knuckle bolt and torque the bolt to 70 ft. lbs. (95 Nm)
   - Strut upper mounting nuts and torque the nuts to 25 ft. lbs. (34 Nm)
   - Strut top cover
   - Rear wheel and torque the nuts to 95 ft. lbs. (125 Nm)

Shock Absorber

REMOVAL & INSTALLATION

Rear

2.4L (VIN G) AND 3.0L ENGINES

1. Before servicing the vehicle, refer to the precautions in the beginning of this section.

2. Remove or disconnect the following:
   - Negative battery cable
   - Rear shelf trim
   - Cover from the upper shock tower
   - Upper mounting nuts
   - Shock from the rear steering knuckle
   - Shock assembly

To install:

3. Install or connect the following:
   - Shock absorber and torque the shock to steering knuckle bolt to 73 ft. lbs. (98 Nm)
   - Upper shock nuts and torque to 32 ft. lbs. (45 Nm)
   - Upper tower cover
   - Rear shelf trim
   - Negative battery cable

Coil Spring

REMOVAL & INSTALLATION

Front

2.0L, 2.4L (VIN X), 2.5L AND 2.7L ENGINES

1. Before servicing the vehicle, refer to the precautions in the beginning of this section.
2. Disassemble as follows:
3. Remove the strut/shock assembly and place it in a suitable compressor tool
4. Set the lower hooks and install the clamp on the lower end of the spring.
5. Rotate the spring so the ball joint sits directly below the front upper hook.

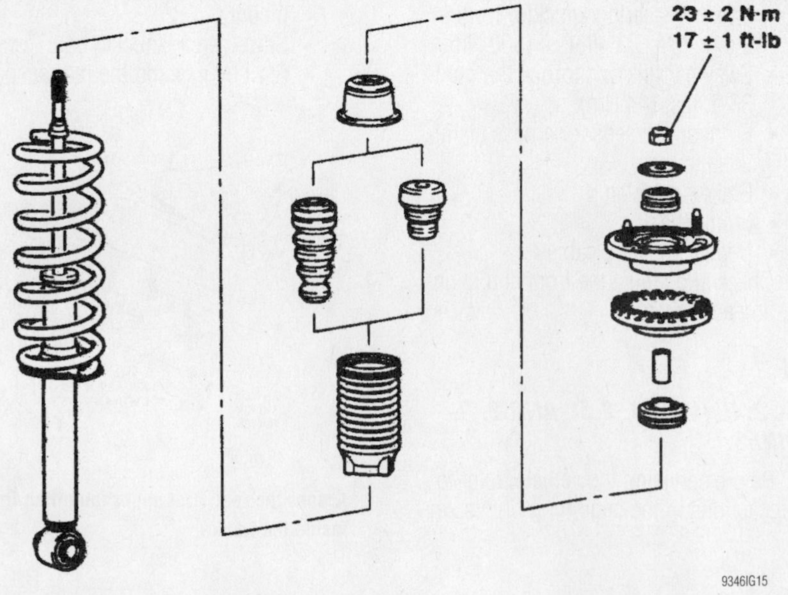

Exploded view of the shock absorber

6. Compress the spring until all tension is removed.
7. Remove the retaining nut and washer.
8. Remove the clamp from the bottom of the spring.
9. Remove the strut/shock assembly.
10. Remove the lower spring, jounce bumper and cup.
11. Remove the dust boot and lower bushing washer.
12. Release the tension from the spring.
13. Remove the spring from the compressor tool.

To assemble:

14. Install the upper isolator on the mounting bracket.
15. Install the sleeve into the lower isolator bushing.
16. Install the bushing and sleeve into the bottom of the upper mounting bracket.
17. Install the upper isolator bushing into the center of the upper mounting bracket.
18. Install the lower end of the spring in the compressor tool supported by the hooks.
19. Install the upper mounting bracket on top of the coil spring matching the spring to the isolator on the upper mounting bracket.
20. Install the control arm ball joint directly below the front upper hook.
21. Compress the spring.
22. Install the lower spring isolator on the spring seat.
23. Install the jounce bumper with the pointed end downward.
24. Install the collar with the undercut side facing down.
25. Install the dust shield and cup
26. Install the lower bushing retainer washer
27. Install the strut/shock assembly through the coil spring.
28. When properly aligned, install the upper mounting nut and torque the nut to 40 ft. lbs. (55 Nm).
29. Slowly release the tension from the compressor tool.
30. Install the strut/shock assembly in the vehicle.

2.4L (VIN G) AND 3.0L ENGINES

1. Before servicing the vehicle, refer to the precautions in the beginning of this section.
2. Disassemble as follows:
3. Remove the strut/shock assembly and place it in a suitable compressor tool.

4. Install special Tool MB991176 to secure the strut and remove the dust cover and jam nut.
5. Remove the strut insulator.
6. Remove the upper spring seat and pad.
7. Rempve the bump rubber and dust cover.
8. Remove the coil spring.
9. Remove the lower spring pad.

To assemble:

10. Install the lower spring pad and coil spring.
11. Install the dust cover and bump rubber.
12. Install the upper spring pad and seat.
13. Carefully compress the spring and install the insulator and jam nut.
14. When properly aligned, torque the jam nut to 47 ft. lbs. (64 Nm).
15. Install the dust cover and remove the compressor tools.
16. Install the strut/shock to the vehicle.

Rear

1. Before servicing the vehicle, refer to the precautions in the beginning of this section.
2. Disassemble as follows:
3. Remove the rear shock from the vehicle.
4. Install the assembly in a spring compressor tool.
5. Remove the jam nut, washer and upper bushing.
6. Remove the upper bracket assembly and spring pad.
7. Remove the collar and bushing.
8. Remove the cup, bump rubber and dust cover.
9. Remove the coil spring

To assemble:

10. Install the coil spring to the shock absorber using compressor Tools MB991237 and MB991239.
11. Install the dust cover and bump rubber.
12. Install the cup and bushing
13. Install the collar, supper spring pad and bracket assembly.
14. Install the upper bushing and washer.
15. When properly aligned, install the jam nut and torque to 17 ft. lbs. (23 Nm).
16. Remove the compressor tools and install the shock absorber.

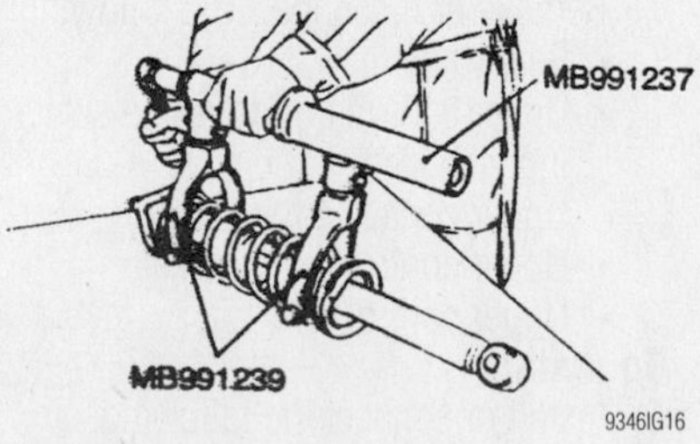

Compress the coil spring as shown—2.4L (VIN G) and 3.0L

Upper Ball Joint

REMOVAL & INSTALLATION

The upper ball joint is an integrated part of the upper control arm assembly, and cannot be serviced separately. A worn or damaged ball joint requires replacement of upper control arm assembly.

Lower Ball Joint

REMOVAL & INSTALLATION

On all vehicles, the ball joint cannot be serviced separately. If the ball joint is defective it will require replacement of the lower control arm.

Upper Control Arm

REMOVAL & INSTALLATION

Front

2.0L, 2.4L (VIN X), 2.5L AND 2.7L ENGINES

1. Before servicing the vehicle, refer to the precautions in the beginning of this section.
2. Remove or disconnect the following:
   - Front wheel
   - Ball joint from the steering knuckle using the joint separation Tool MB991113
   - Strut assembly
   - Upper control arm mounting bracket from the strut assembly
   - Upper control arm shaft-to-bracket nuts
   - Upper control arm, from the bracket using the joint separation tool
   - Upper control arm assembly

To install:

3. Install or connect the following:
   - Upper control arm assembly
   - Upper control arm to the bracket and torque the nuts to 62 ft. lbs. (86 Nm)
   - Upper control arm mounting bracket to the strut assembly
   - Strut assembly

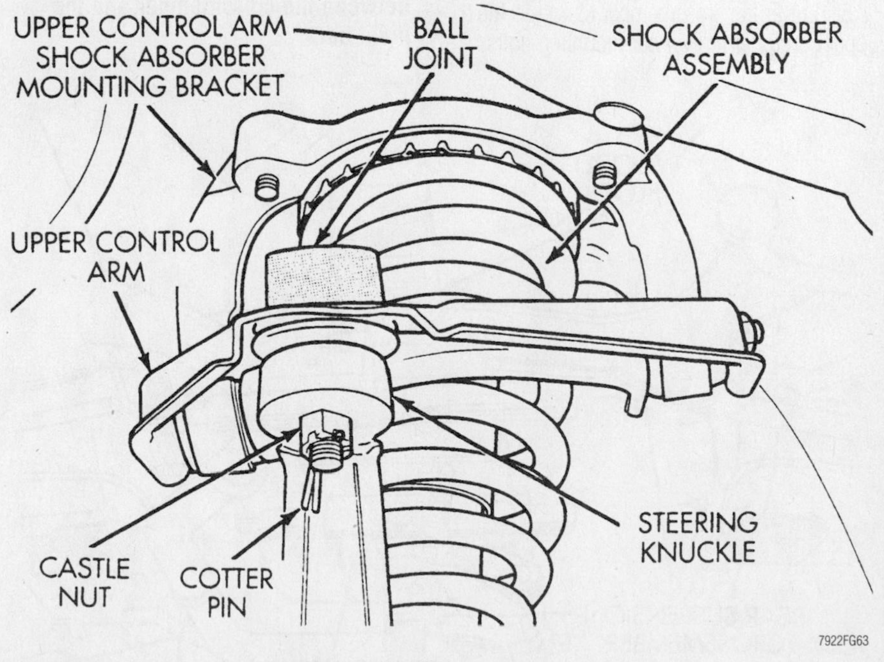

Front upper control arm component identification

- Ball joint to the steering knuckle and torque the locking nut to 20 ft. lbs. (28 Nm)
- Front wheel

4. Check and/or adjust the wheel alignment, if necessary.

Rear

2.0L, 2.4L (VIN X), 2.5L AND 2.7L ENGINES

1. Before servicing the vehicle, refer to the precautions in the beginning of this section.
2. Remove or disconnect the following:
   - Both rear wheels
   - Both struts from the knuckles
   - Muffler hanger from the frame rail
   - Rear exhaust pipe hanger from the rear crossmember and allow it to hang
   - Upper control arm ball joint stud from the knuckle
3. Support the center of the rear crossmember with a jack and a block of wood.
4. Remove or disconnect the following:
   - Speed sensor wiring from the upper control arms, if equipped with an Anti-lock Brake System (ABS)
   - 4 crossmember-to-frame bolts

⁂ WARNING

Do not damage the rear brake hoses while lowering the crossmember.

5. Lower the crossmember to access the 2 upper control arm pivot bar mounting bolts.

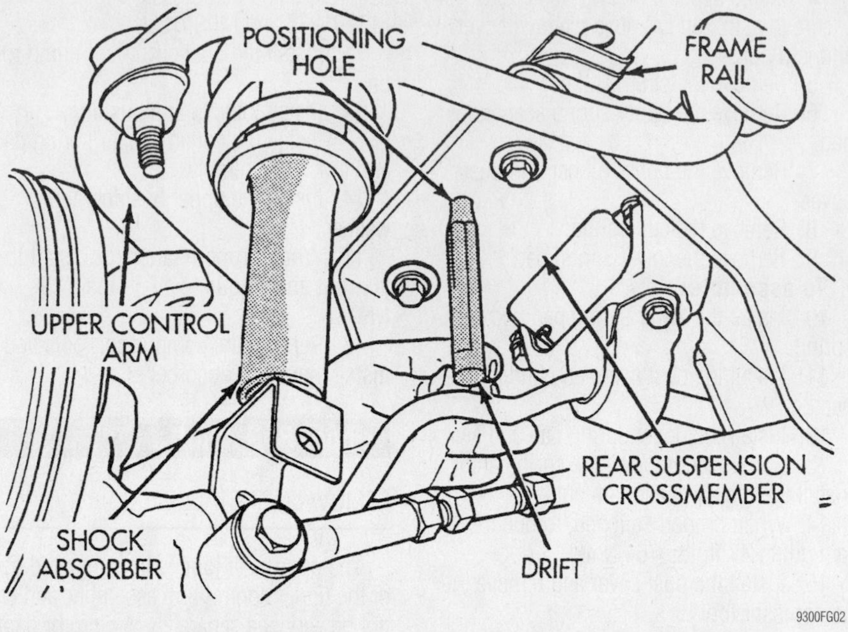

Use a drift to align the rear crossmember to the frame

6. Remove or disconnect the following:
   - Both upper control arm-to-crossmember bolts
   - Upper control arm

To install:

7. Install or connect the following:
   - Upper control arm to the crossmember and torque the bolts to 80 ft. lbs. (108 Nm)

➡Be sure to place the flat washers between the crossmember and the pivot bar.

   - Crossmember using a drift to align the holes and tighten the bolts to 80 ft. lbs. (108 Nm)
8. Remove the jack
9. Install or connect the following:
   - Speed sensor wiring to the upper control arms, if equipped with ABS
   - Upper control arm ball joint stud to the knuckle and torque the ball joint stud nut to 63 ft. lbs. (85 Nm)
   - Rear exhaust pipe hanger to the rear crossmember
   - Muffler hanger to the frame rail
   - Both struts to the knuckles and torque the lower strut bolt to 70 ft. lbs. (95 Nm)
   - Both rear wheels

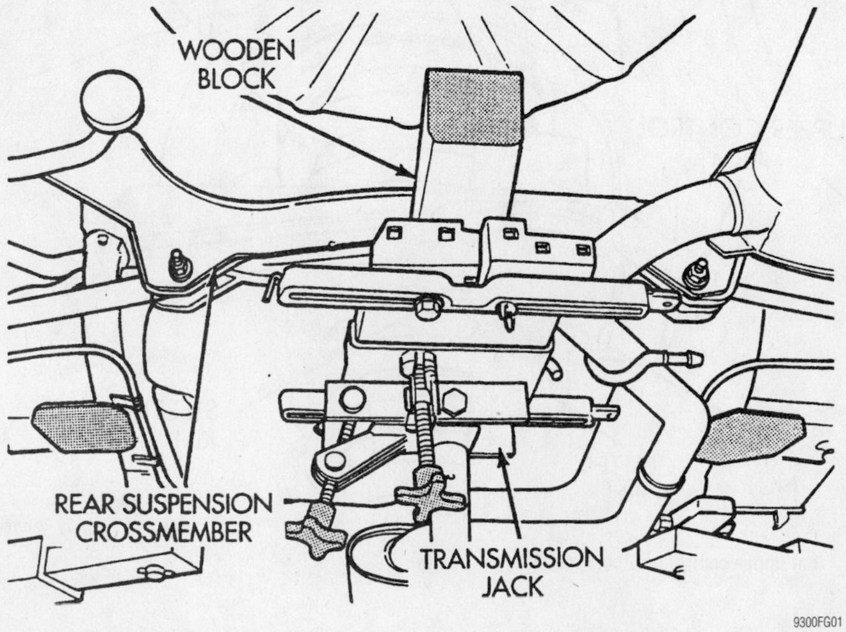

A jack with a block of wood should be used to support the rear crossmember

2.4L (VIN G) AND 3.0L ENGINES

1. Before servicing the vehicle, refer to the precautions in the beginning of this section.
2. Remove or disconnect the following:
   - Rear wheel
   - Upper control arm from the steering knuckle
   - Upper control arm assembly
   - Upper control arm bracket
   - Upper control arm

To install:

3. Install or connect the following:
   - Upper control arm and brackets and torque the bolts to 42 ft. lbs. (57 Nm)
   - Upper control arm assembly and torque the bolts to 29 ft. lbs. (39 Nm)

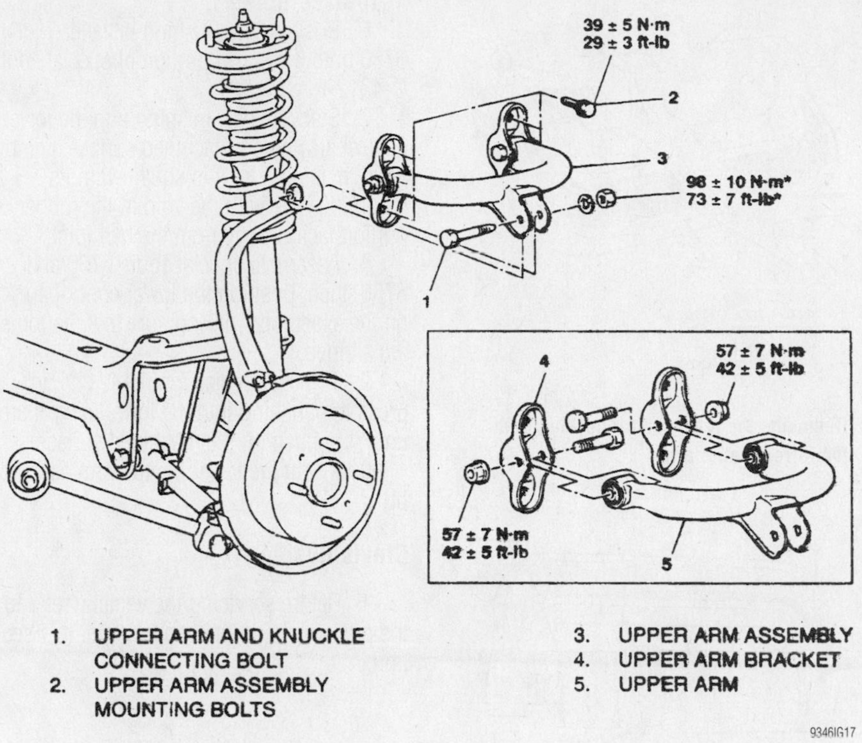

Exploded view of the upper control arm assembly

- Upper control arm to the steering knuckle and torque the bolt 73 ft. lbs. (98 Nm)
- Rear wheel

Lower Control Arm

REMOVAL & INSTALLATION

Front

2.0L, 2.4L (VIN X), 2.5L AND 2.7L ENGINES

1. Before servicing the vehicle, refer to the precautions in the beginning of this section.
2. Remove or disconnect the following:
   - Front wheels
   - Heat shield, if equipped with 15 inch wheels
   - Ball joint clamp nut
   - Sway bar-to-control arm bolts
   - Strut clevis from the lower control arm
   - Sway bar-to-crossmember bolts, loosen them and rotate the sway bar away from the control arm
   - Lower control arm from the steering knuckle using a prybar
   - Both lower control arm bolts
   - Control arm

To install:

3. Install or connect the following:
   - Lower control arm and torque the bolts to 120 ft. lbs. (163 Nm)
   - Control arm to the steering knuckle and torque the nut to 70 ft. lbs. (95 Nm)
4. Rotate the sway bar up to the control arms.
   - Strut clevis
   - Sway bar and torque the bolts to 21 ft. lbs. (28 Nm)
   - Rear wheels

CONTROL ARM BUSHING REPLACEMENT

Front Isolator Bushing

1. Before servicing the vehicle, refer to the precautions in the beginning of this section.
2. Remove the lower control arm assembly
3. Install the Bushing Remover Tool 6602-5 and Bushing Receiver Tool MB-990799 on Special Tool C-4212-F.
4. Position the lower control arm on the assembled removal tools.

➡Be sure the Bushing Receiver Tool MB-990799 is square on the lower control arm and Bushing Remover Tool 6602-5 is positioned correctly on the isolator bushing.

5. Tighten Special Tool C-4212-F to press the bushing from the lower control arm.

To assemble:

6. Position the Bushing Installer Tool

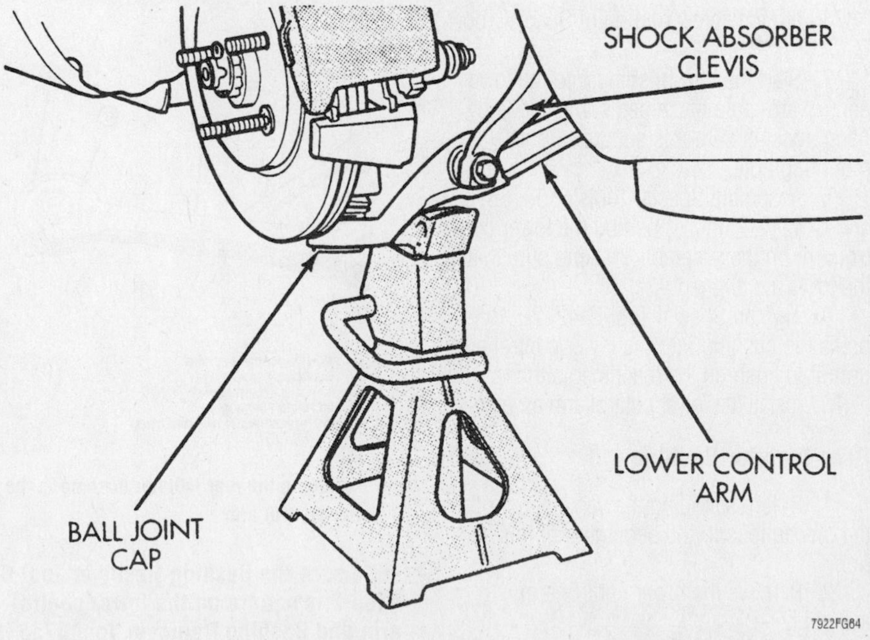

Pre-load the suspension before final tightening the control arm to the steering knuckle

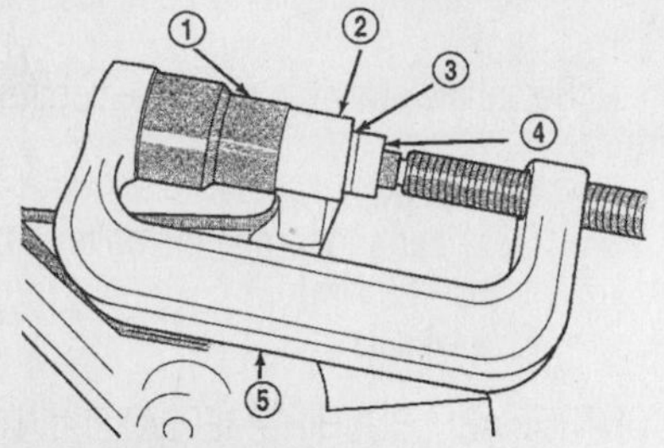

1 – SPECIAL TOOL MB-990799
2 – LOWER CONTROL ARM
3 – FRONT ISOLATOR BUSHING
4 – SPECIAL TOOL 6602-5
5 – SPECIAL TOOL C-4212-F

9306FG09

Removing the front isolator bushing from the lower control arm

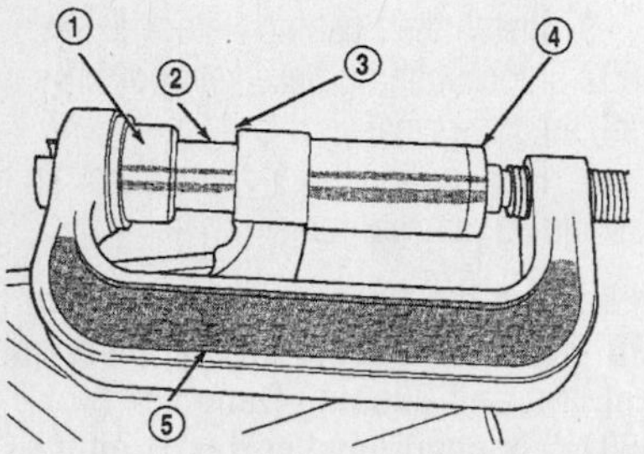

1 – SPECIAL TOOL 6876
2 – ISOLATOR BUSHING
3 – MACHINED SURFACE SIDE OF LOWER CONTROL ARM
4 – SPECIAL TOOL 6758
5 – SPECIAL TOOL C-4212-F

9306FG10

Installing the front isolator bushing to the lower control arm

6876 onto the screw portion of Special Tool C-4212-F.

7. Start the new bushing into the lower control arm hole machined surface side by hand, making sure it is square with its mounting hole.

8. Assemble Special Tools 6758, 6876 and C-4212-F; then, position the lower control arm on the assembly, making sure that the tools are aligned.

9. Tighten Special Tool C-4212-F to press the bushing into the lower control arm until it is flush on the machined surface.

10. Install the lower control arm assembly.

Rear Isolator Bushing

1. Before servicing the vehicle, refer to the precautions in the beginning of this section.

2. Remove the lower control arm assembly

3. Install the Bushing Remover Tool 6756 and Bushing Receiver Tool C-4366-2 on Special Tool C-4212-F.

4. Position the lower control arm on the assembled removal tools.

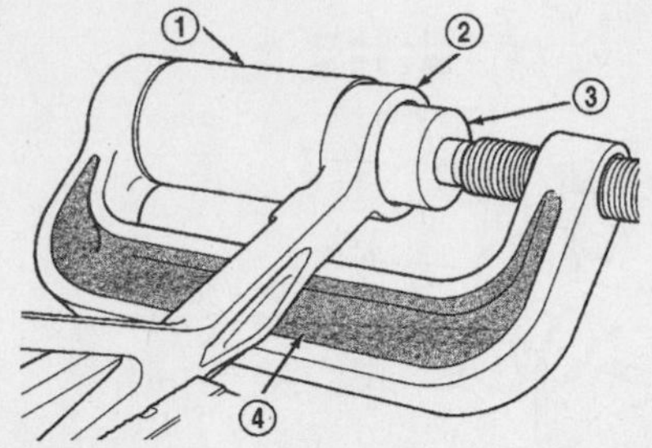

1 – SPECIAL TOOL C-4366-2
2 – LOWER CONTROL ARM
3 – SPECIAL TOOL 6756
4 – SPECIAL TOOL C-4212-F

9306FG11

Removing the rear isolator bushing from the lower control arm

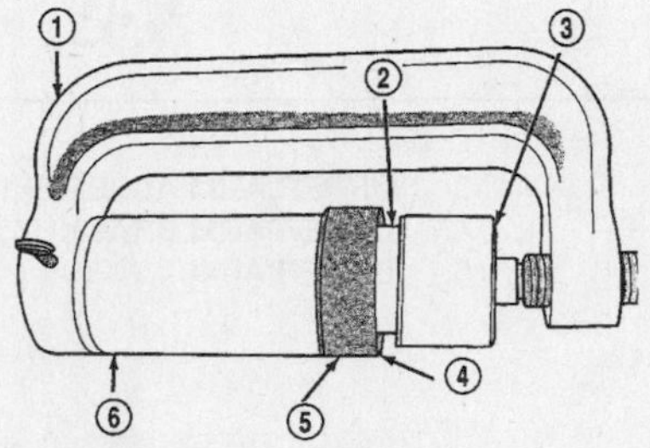

1 – SPECIAL TOOL C-4212-F
2 – REAR BUSHING
3 – SPECIAL TOOL 6760
4 – MACHINED SURFACE ON LOWER CONTROL ARM
5 – LOWER CONTROL ARM
6 – SPECIAL TOOL 6756

9306FG13

Installing the rear isolator bushing to the lower control arm

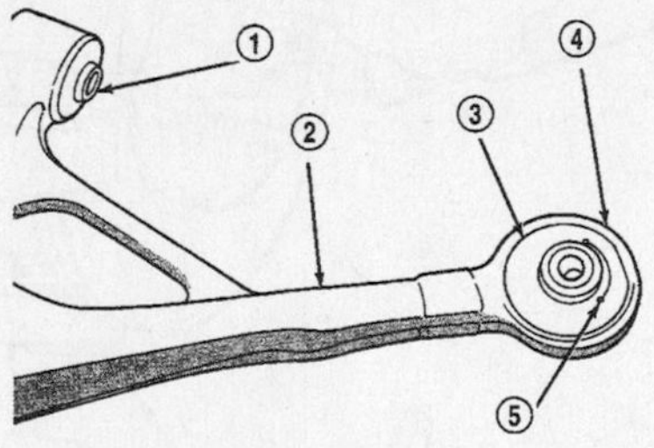

1 – FRONT ISOLATOR BUSHING
2 – LOWER CONTROL ARM
3 – REAR ISOLATOR BUSHING
4 – MACHINED SURFACE
5 – VOID IN BUSHING IN THIS DIRECTION

9306FG12

Positioning the rear isolator bushing to the lower control arm

➡Be sure the Bushing Receiver Tool C-4366-2 is square on the lower control arm and Bushing Remover Tool 6756 is positioned correctly on the isolator bushing.

5. Tighten Special Tool C-4212-F to press the bushing from the lower control arm.

To assemble:

6. Position the Bushing Installer Tool 6760 onto the screw portion of special Tool C-4212-F.

7. Start the new bushing into the lower control arm hole's machined surface side by hand, making sure it is square with its mounting hole with the void in the rubber portion facing away from the ball joint.

8. Assemble special Tools 6760 and 6756; then, position the lower control arm on the assembly, making sure that the tools are aligned.

9. Tighten special Tool C-4212-F to press the bushing into the lower control arm until it is flush on the machined surface.

10. Install the lower control arm assembly.

Clevis Bushing

1. Before servicing the vehicle, refer to the precautions in the beginning of this section.

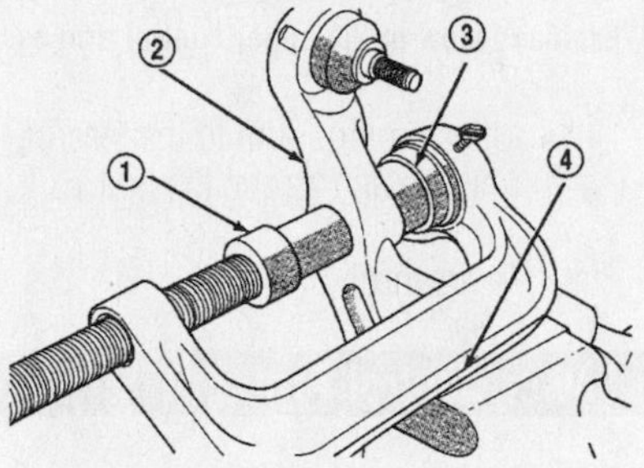

1 – SPECIAL TOOL 6877
2 – LOWER CONTROL ARM
3 – SPECIAL TOOL 6876
4 – SPECIAL TOOL C-4212-F

9306FG14

Removing the clevis bushing from the lower control arm

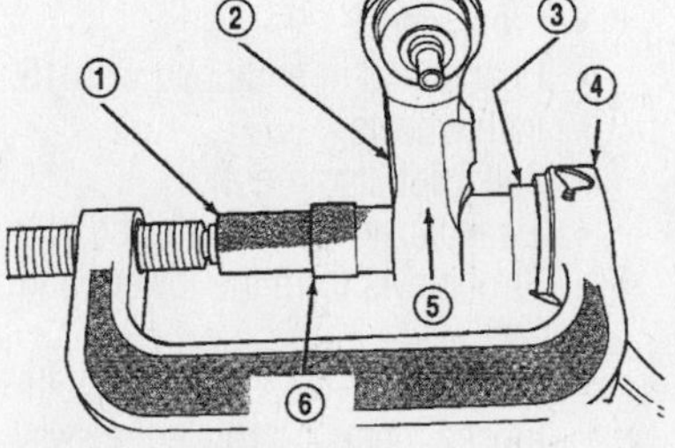

1 – SPECIAL TOOL 6877
2 – MACHINED SURFACE SIDE OF LOWER CONTROL ARM
3 – SPECIAL TOOL 6876
4 – SPECIAL TOOL C-4212-F
5 – LOWER CONTROL ARM
6 – CLEVIS BUSHING

9306FG15

Installing the clevis bushing to the lower control arm

2. Remove the lower control arm assembly
3. Install the Bushing Remover Tool 6877 and Bushing Receiver Tool 6876 on special Tool C-4212-F.
4. Position the lower control arm on the assembled removal tools.

➡Be sure the Bushing Receiver Tool 6876 is square on the lower control arm and Bushing Remover Tool 6877 is positioned correctly on the clevis bushing.

5. Tighten special Tool C-4212-F to press the bushing from the lower control arm.

To assemble:

6. Position the Bushing Installer Tool 6877 onto the screw portion of special Tool C-4212-F.
7. Start the new bushing into the lower control arm hole's machined surface side by hand, making sure it is square with its mounting hole with the void in the rubber portion facing away from the ball joint.
8. Assemble special Tools 6876 and 6877; then, position the lower control arm on the assembly, making sure that the tools are aligned.
9. Tighten special Tool C-4212-F to press the bushing into the lower control arm until it is flush on the machined surface.
10. Install the lower control arm assembly.

2.4L (VIN G) AND 3.0L ENGINES

1. Before servicing the vehicle, refer to the precautions in the beginning of this section.
2. Remove or disconnect the following:
   - Front wheel
   - Lower control arm from the steering knuckle
   - Lower control arm mounting bolt and clamp
   - Lower control arm
3. If replacing the bushing, perform the following:
   a. Apply soapy water between the shaft and old bushing and pry the old bushing out.

To install:

4. Apply soapy water to the shaft and new bushing and install the new bushing, if removed.
5. Install or connect the following:
   - Lower control arm
   - Lower control arm clamp and torque the bolts to 60 ft. lbs. (81 Nm)
   - Lower control arm to the steering knuckle and torque the bolts to 80 ft. lbs. (108 Nm)
   - Front wheel
6. Check and adjust the front toe, if needed.

Rear

2.4L (VIN G) AND 3.0L ENGINES

1. Before servicing the vehicle, refer to the precautions in the beginning of this section.
2. Remove or disconnect the following:
   - Rear wheel
   - Stabilizer link
   - Wheel Speed Sensor (WSS), if equipped
   - Lower control arm assembly to steering knuckle
   - Lower control arm

To install:

3. Install or connect the following:
   - Lower control arm and torque the bolt to 55 ft. lbs. (74 Nm)
   - Lower control arm to the steering knuckle and torque the bolt to 80 ft. lbs. (108 Nm)
   - WSS, if equipped
   - Stabilizer link and torque the bolt to 29 ft. lbs. (39 Nm)
   - Rear wheel

Wheel Bearings

ADJUSTMENT

Front

The front hub wheel bearing is designed for the life of the vehicle and requires no type of adjustment or periodic maintenance. The bearing is a sealed unit with the wheel hub and can only be removed and/or replaced as one unit.

Rear

The rear hub and wheel bearing assembly is designed for the life of the vehicle and requires no type of adjustment or periodic maintenance. The bearing is a sealed unit with the wheel hub and can only be removed and/or replaced as one unit.

The following procedure may be used for evaluation of bearing condition:

1. Raise and safely support the vehicle.
2. Remove the rear wheels and brake drums.
3. Turn the hub flange carefully. Excessive roughness, lateral play or resistance to rotation may indicate dirt intrusion or bearing failure.
4. If the rear wheel bearings exhibit the conditions during inspection, the hub and bearing assembly should be replaced.
5. Damaged bearing seals and resulting excessive grease loss may also require bearing replacement. Moderate grease loss from the bearing is considered normal and should not require replacement of the hub and bearing assembly.

REMOVAL & INSTALLATION

Front

The front wheel bearing used on this vehicle is a bolt-in type wheel bearing.

The wheel bearing is serviced separately from the front steering knuckle and front hub assembly. Retention of the front wheel bearing into the steering knuckle is by means of 3 bolts installed from the rear of the steering knuckle. The 3 bolts attach the hub/bearing to the front surface of the steering knuckle. Removal and installation of the hub/bearing assembly from the steering knuckle must be done with the steering knuckle removed from the vehicle.

The face of the outer CV-joint has a metal bearing shield pressed on it. This design deters direct water splash on the bearing seal while allowing any water that gets in to run out the bottom of the steering knuckle. It is important to thoroughly clean the outer CV-joint and the wheel bearing area in the steering knuckle before it is assembled after servicing the front wheel bearing or driveshaft.

At no time when servicing this vehicle, can a sheet metal screw, bolt or other metal fastener be installed in the shock tower to take the place of an original plastic clip. Also, NO holes can be drilled into the front shock tower for the installation of any metal fasteners into the shock tower. Because of the minimum clearance in this area installation of metal fasteners could damage the coil spring's protective coating and lead to corrosion failure of the spring. If a plastic clip is missing, lost or broken during servicing a vehicle, replace only with the equivalent part listed in the Mopar parts catalog.

1. Before servicing the vehicle, refer to the precautions in the beginning of this section.

2. Remove or disconnect the following:
   - Front wheel
   - Steering knuckle assembly
   - 3 hub/bearing assembly-to-steering knuckle bolts
   - Hub/bearing assembly from the steering knuckle. If necessary, tap the bearing out, using a soft-faced hammer

➡The wheel bearing is transferable to a replacement steering knuckle if the bearing is found in serviceable condition.

To install:

3. Clean all parts well. Thoroughly, clean all the hub/bearing assembly mounting surfaces on the steering knuckle.
4. Install or connect the following:
   - Hub/bearing assembly and torque the bolts to 80 ft. lbs. (110 Nm)
   - Steering knuckle
   - Front wheel

Rear

All vehicles are equipped with permanently lubricated and sealed for life rear wheel bearings. There is no periodic lubrication or maintenance recommended for these units.

To evaluate the condition of the rear wheel bearings, remove the wheel and brake drum or rotor and rotate the flanged outer ring of the hub. Excessive roughness or resistance to rotation may indicate dirt intrusion or wheel bearing failure. If the rear wheel bearings exhibit these conditions during inspection, the hub and bearing assembly should be replaced. Damaged bearing seals and resulting excessive grease loss may also require bearing replacement. Moderate grease loss from the bearing is considered normal and should not require replacement of the hub and bearing assembly. If service requires removal for inspection or replacement of the rear wheel bearing and hub assembly, use the following procedure.

1. Before servicing the vehicle, refer to the precautions in the beginning of this section.
2. Remove or disconnect the following:
   - Rear wheel
   - Brake drum, if equipped with drum brakes
   - Brake caliper and rotor, if equipped with disc brakes
   - Hub dust cap
   - Hub nut and discard it
   - Hub/bearing assembly by pulling straight off the spindle

To install:

3. Install or connect the following:
   - New bearing on the spindle
   - Hub using a new nut and torque the nut to 185 ft. lbs. (250 Nm)
   - Dust cap
   - Brake drum or rotor and caliper
   - Rear wheel and torque the lug nuts to 95 ft. lbs. (129 Nm)

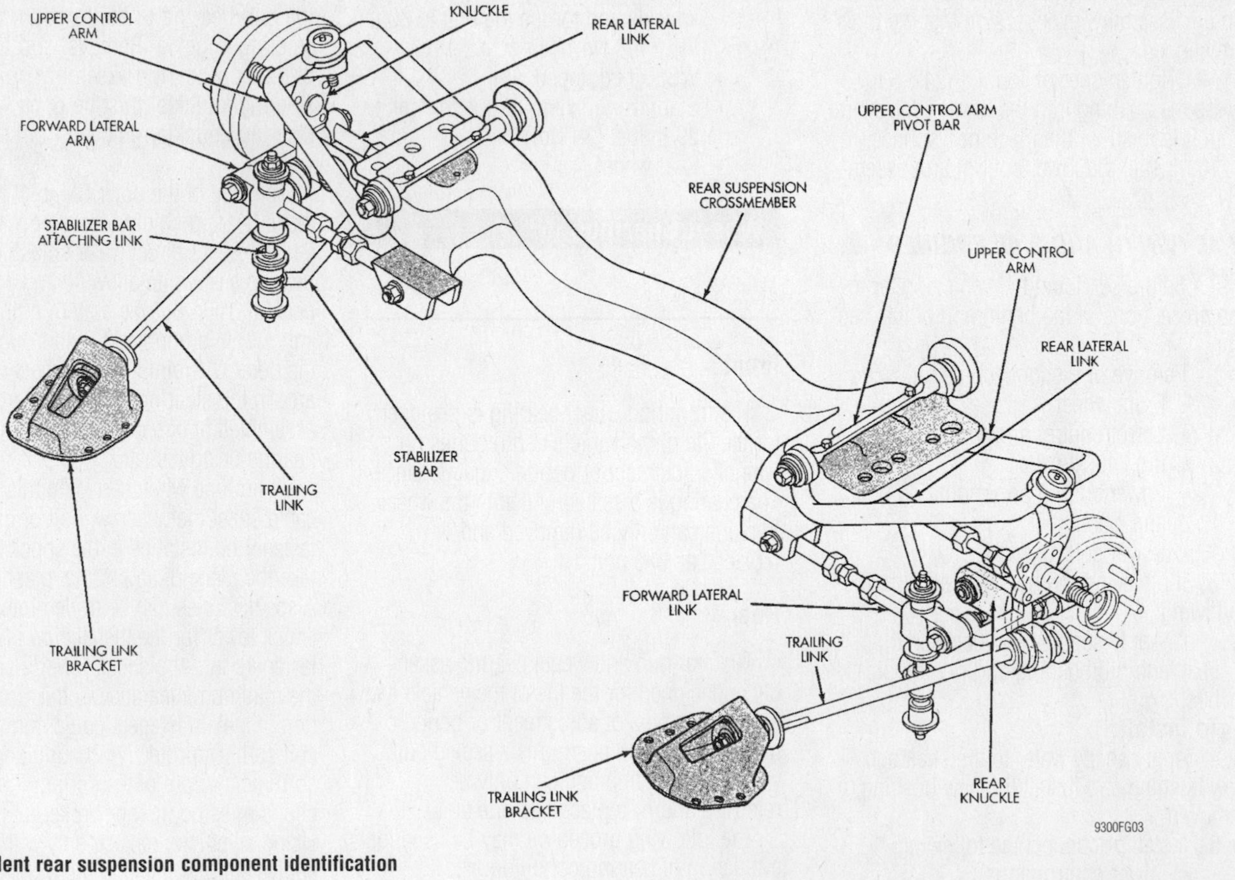

Independent rear suspension component identification

CHRYSLER CORP.

10

2001–02

PT Cruiser

PRECAUTIONS

Before servicing any vehicle, please be sure to read all of the following precautions, which deal with personal safety, prevention of component damage, and important points to take into consideration when servicing a motor vehicle:

- Never open, service or drain the radiator or cooling system when the engine is hot; serious burns can occur from the steam and hot coolant.
- Observe all applicable safety precautions when working around fuel. Whenever servicing the fuel system, always work in a well-ventilated area. Do not allow fuel spray or vapors to come in contact with a spark, open flame, or excessive heat (a hot drop light, for example). Keep a dry chemical fire extinguisher near the work area. Always keep fuel in a container specifically designed for fuel storage; also, always properly seal fuel containers to avoid the possibility of fire or explosion. Refer to the additional fuel system precautions later in this section.
- Fuel injection systems often remain pressurized, even after the engine has been turned **OFF**. The fuel system pressure must be relieved before disconnecting any fuel lines. Failure to do so may result in fire and/or personal injury.
- Brake fluid often contains polyglycol ethers and polyglycols. Avoid contact with the eyes and wash your hands thoroughly after handling brake fluid. If you do get brake fluid in your eyes, flush your eyes with clean, running water for 15 minutes. If eye irritation persists, or if you have taken brake fluid internally, IMMEDIATELY seek medical assistance.
- The EPA warns that prolonged contact with used engine oil may cause a number of skin disorders, including cancer! You should make every effort to minimize your exposure to used engine oil. Protective gloves should be worn when changing oil. Wash your hands and any other exposed skin areas as soon as possible after exposure to used engine oil. Soap and water, or waterless hand cleaner should be used.
- All new vehicles are now equipped with an air bag system, often referred to as a Supplemental Restraint System (SRS) or Supplemental Inflatable Restraint (SIR) system. The system must be disabled before performing service on or around system components, steering column, instrument panel components, wiring and sensors. Failure to follow safety and disabling procedures could result in accidental air bag deployment, possible personal injury and unnecessary system repairs.
- Always wear safety goggles when working with, or around, the air bag system. When carrying a non-deployed air bag, be sure the bag and trim cover are pointed away from your body. When placing a non-deployed air bag on a work surface, always face the bag and trim cover upward, away from the surface. This will reduce the motion of the module if it is accidentally deployed. Refer to the additional air bag system precautions later in this section.
- Clean, high quality brake fluid from a sealed container is essential to the safe and proper operation of the brake system. You should always buy the correct type of brake fluid for your vehicle. If the brake fluid becomes contaminated, completely flush the system with new fluid. Never reuse any brake fluid. Any brake fluid that is removed from the system should be discarded. Also, do not allow any brake fluid to come in contact with a painted surface; it will damage the paint.
- Never operate the engine without the proper amount and type of engine oil; doing so WILL result in severe engine damage.
- Timing belt maintenance is extremely important! Many models utilize an interference-type, non-freewheeling engine. If the timing belt breaks, the valves in the cylinder head may strike the pistons, causing potentially serious (also time-consuming and expensive) engine damage. Refer to the maintenance interval charts in the front of this manual for the recommended replacement interval for the timing belt, and to the timing belt section for belt replacement and inspection.
- Disconnecting the negative battery cable on some vehicles may interfere with the functions of the on-board computer system(s) and may require the computer to undergo a relearning process once the negative battery cable is reconnected.
- When servicing drum brakes, only disassemble and assemble one side at a time, leaving the remaining side intact for reference.
- Only an MVAC-trained, EPA-certified automotive technician should service the air conditioning system or its components.

ENGINE REPAIR

Distributor

The PT Cruiser uses a Direct Ignition System (DIS) and basic ignition timing is not adjustable.

Alternator

REMOVAL

Remove or disconnect the following:

- Negative battery cable
- Air cleaner lid
- Inlet Air Temperature (IAT) sensor and make-up hose
- Upper alternator adjustment lock nut, loosen only
- Right front wheel
- Splash shield
- Support bracket
- Lower pivot bolt, loosen only
- Drive belt T-bolt, loosen only
- Alternator wiring connectors
- Alternator belt
- Axle retaining nut
- Lower control arm from the steering knuckle
- Axle shaft
- Lower mounting bolt from the upper adjustment bracket
- Alternator

INSTALLATION

Install or connect the following:

- Alternator
- Alternator drive belt
- Axle shaft
- Lower control arm to the steering knuckle
- Lower ball joint nut and torque the nut to 70 ft. lbs. (95 Nm)
- Axle retaining nut and torque the nut to 120 ft. lbs. (163 Nm)
- Support bracket
- Alternator electrical connectors and torque the B+ terminal nut to 100 inch lbs. (11 Nm)
- Splash shield and right front wheel
- Right front wheel
- Upper pivot nut and torque to 40 ft. lbs. (54 Nm)
- Negative battery cable
- Air cleaner lid, IAT sensor and make up hose

1. Check the transmission fluid level and top off, if necessary.

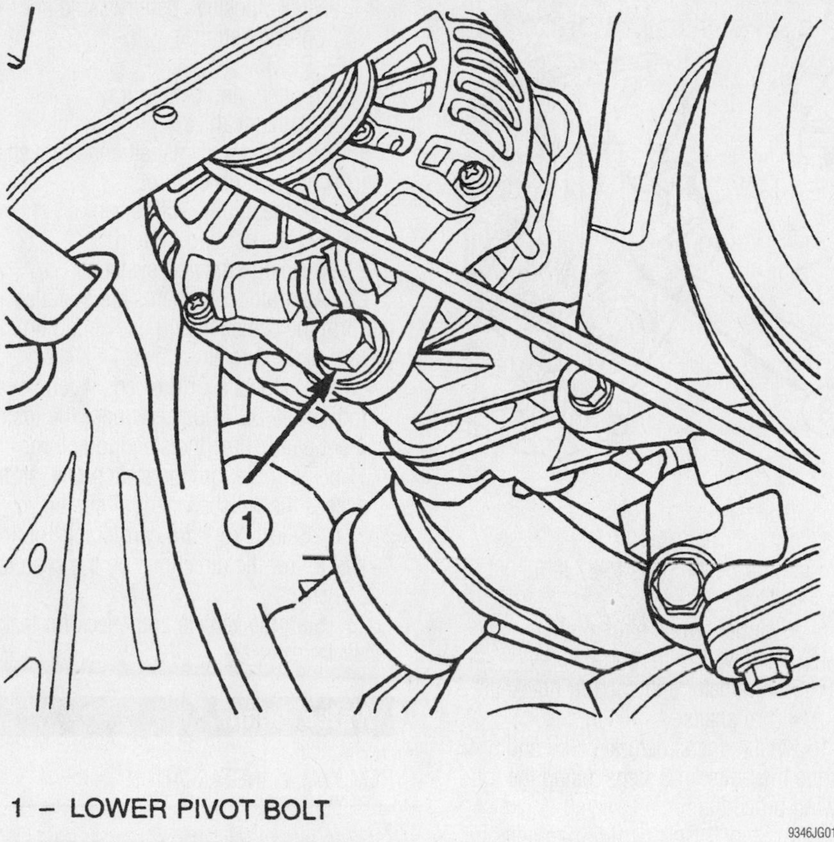

Remove the lower pivot bolt from the alternator

Ignition Timing

ADJUSTMENT

Ignition timing is controlled by the Powertrain Control Module (PCM). No adjustment is necessary or possible.

Engine Assembly

REMOVAL & INSTALLATION

➡After all components are installed on the engine, a DRB scan tool is necessary to perform the camshaft and crankshaft timing relearn procedure.

1. Before servicing the vehicle, refer to the precautions in the beginning of this section.
2. Properly recover the air conditioning system refrigerant.
3. Properly relieve the fuel system pressure.
4. Drain the cooling system.
5. Drain the engine oil.
6. Remove or disconnect the following:

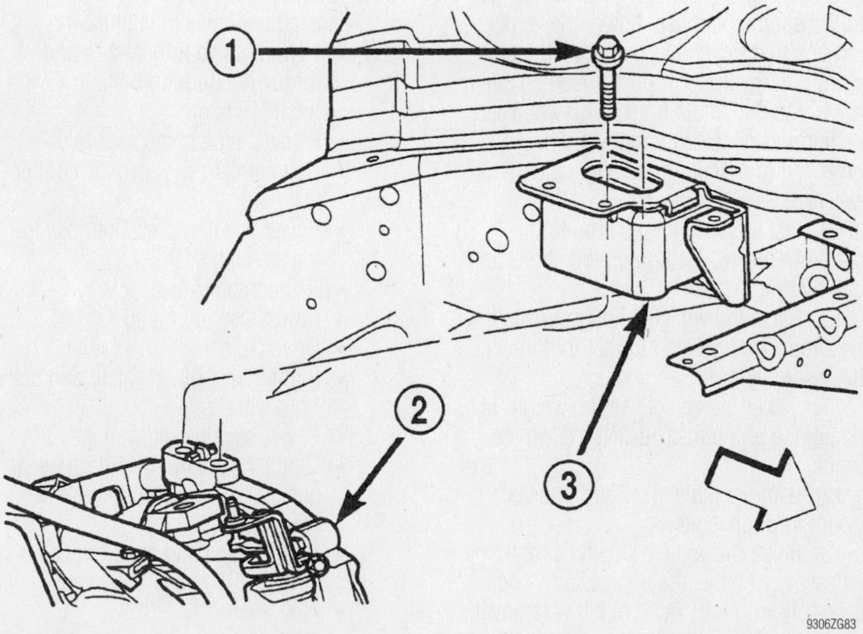

Front engine mount location and bolt identification

- Battery and battery tray
- Throttle and cruise control cables
- Powertrain Control Module (PCM) wiring harness
- Positive cable from the Power Distribution Center (PDC) and ground wire
- Ground wire from the body to the engine
- Brake booster vacuum hose
- Proportional purge hoses from the intake manifold
- Coolant recovery hose
- Heater hoses
- Upper radiator support crossmember
- Upper and lower radiator hoses
- A/C line from the condenser
- A/C lines from the junction at the upper torque strut
- Cooler lines, if equipped
- Fan cooling module assembly
- Transmission shift linkage and electrical connectors
- Clutch hydraulic lines, if equipped
- Both front wheels and the right inner splash shield
- Halfshafts
- Accessory drive belts
- Alternator and support brackets

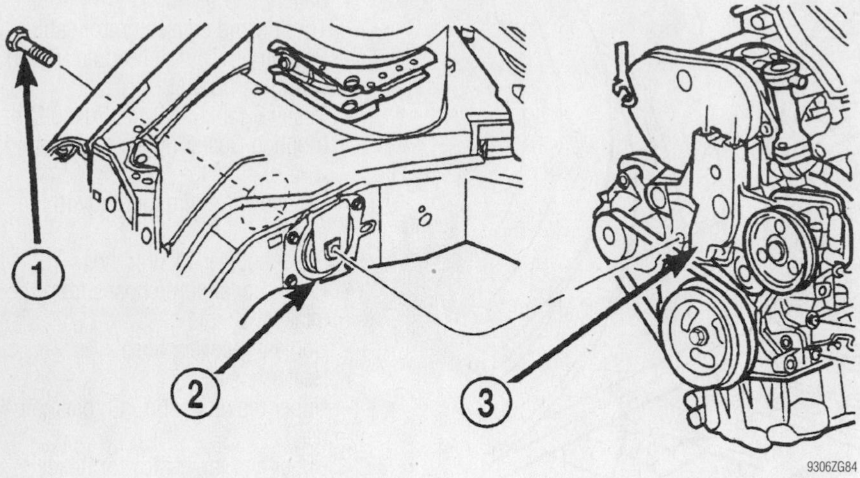

Exploded view of the left mount through bolt

- Downstream Oxygen (O_2S) Sensor
- Exhaust system from the manifold
- Power steering pressure hose
- Lower engine torque strut
- Structural collar
- Torque converter bolts
- A/C compressor
- Power steering return line
- Power steering pump

7. Raise the vehicle enough to allow an engine dolly and cradle to be placed under the engine.

8. Loosen the engine support posts in order to allow movement for positioning onto the engine locating holes and flange on the engine bedplate. Lower the vehicle and position the cradle until the engine is resting on the support posts. Tighten the mounts to the cradle frame. This will keep the support posts from moving when removing or installing the engine and transaxle.

9. Install safety straps around the engine to the cradle; tighten the straps and lock them into position.

10. Raise the vehicle enough to see if the straps are tight enough to hold the cradle assembly to the engine.

11. Lower the vehicle so the weight of the engine and transaxle ONLY is on the cradle.

12. Remove the engine and transaxle mount through-bolts.

13. Raise the vehicle slowly, it might be necessary to move the engine/transaxle assembly with the cradle to allow removal around the body flanges.

To install:

14. Install or connect the following:
    - Engine/transaxle assembly by lowering the vehicle over the assembly
    - Engine and transaxle mounts and torque the bolts to 87 ft. lbs. (118 Nm)
    - Upper torque strut

15. Remove the support fixtures.
    - Alternator and support brackets
    - Halfshafts

16. Install the structural collar and torque the bolts, in 3 steps, using the following procedure:
    a. Step 1: Collar-to-oil pan bolts to 30 inch lbs. (3 Nm).
    b. Step 2: Collar-to-transmission bolts to 80 ft. lbs. (108 Nm).
    c. Step 3: Collar-to-oil pan bolts to 40 ft. lbs. (54 Nm).

17. Install or connect the following:
    - Lower engine torque strut and torque the through bolts to 87 ft. lbs. (118 Nm)
    - Exhaust pipe to the manifold
    - Downstream O_2S sensor connector
    - Power steering pressure hose to the steering gear
    - A/C compressor
    - Power steering pump
    - Drive belts
    - Right inner splash shield and both front wheels
    - Power steering return hose
    - Clutch hydraulic line, electrical connectors and shift linkage, if equipped
    - Shift linkage and cooler lines, if equipped
    - Fuel lines
    - Heater hoses
    - Engine wiring harness and ground strap
    - Lower and upper radiator hoses
    - Fan module assembly
    - Throttle and speed control cables
    - PDC, positive battery cable and ground strap
    - PCM
    - Battery and battery tray
    - Battery cables

18. Perform the camshaft and crankshaft synchronization procedure.

19. Fill the engine with clean oil.

20. Fill the cooling system.

21. Recharge the A/C system.

22. After all components are installed, perform the camshaft and crankshaft timing relearn procedure as follows:
    a. Connect a DRB or equivalent, scan tool to the DLC (located under the instrument panel, near the steering column).
    b. Turn the ignition switch **ON**, and access the "miscellaneous" screen.
    c. Select "re-learn cam/crank" option and follow the directions on the scan tool screen.

23. Start the vehicle and check for leaks, repair if necessary.

Water Pump

REMOVAL & INSTALLATION

1. Before servicing the vehicle, refer to the precautions in the beginning of this section.

2. Drain the cooling system.

3. Remove or disconnect the following:
    - Negative battery cable
    - Timing belt
    - Camshaft sprockets
    - Rear timing belt cover
    - Water pump and discard the O-ring

To install:

4. Install or connect the following:
    - New O-ring in the water pump groove

⁂ WARNING

Before proceeding, be sure the O-ring gasket is properly seated in the water pump groove before tightening the screws. An improperly installed O-ring could cause a coolant leak.

- Water pump and torque the bolts to 105 inch lbs. (12 Nm)

➡Rotate the pump by hand to check for freedom of movement.

- Rear timing belt cover
- Camshaft sprockets
- Timing belt
- Negative battery cable

5. Fill the cooling system.

6. Start the vehicle and check for leaks, repair if necessary.

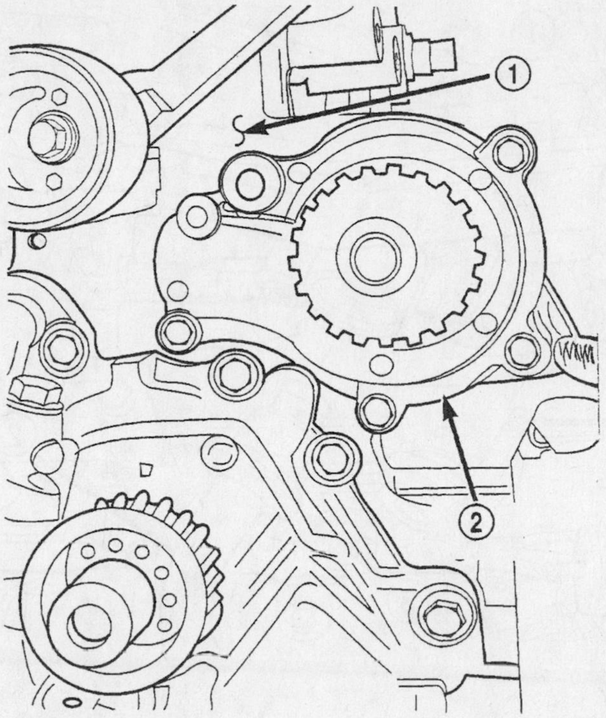

1 – CYLINDER BLOCK
2 – WATER PUMP

9306ZG85

Exploded view of the water pump

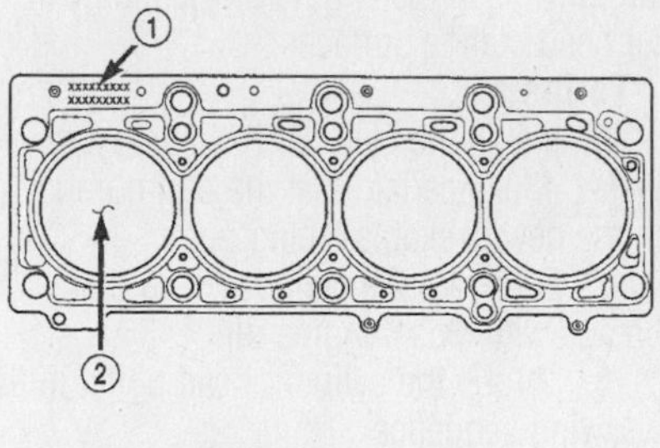

1 – PART NUMBER FACES UP
2 – NO. 1 CYLINDER

9346JG04

Install the new gasket with the part number facing upwards

Cylinder Head

REMOVAL & INSTALLATION

➡After all components are installed on the engine, a DRB scan tool is necessary to perform the camshaft and crankshaft timing relearn procedure.

1. Before servicing the vehicle, refer to the precautions in the beginning of this section.
2. Properly relieve the fuel system pressure.
3. Drain the cooling system.
4. Remove or disconnect the following:
   - Negative battery cable
   - Air cleaner inlet duct and air cleaner
   - Inlet Air Temperature (IAT) sensor and make-up air hose
   - Upper intake manifold
   - Fuel supply line from the fuel rail
   - Heater tube support bracket
   - Upper radiator hose
   - Heater supply hoses
   - Accessory drive belt
   - Exhaust pipe from the manifold
   - Power steering pump, move it aside. DO NOT disconnect the fluid lines.
   - Ignition coil pack wiring connector
   - Ignition coil pack and bracket
   - Cam sensor electrical connector
   - Timing belt
   - Timing belt idler pulley
   - Camshaft sprocket
   - Power steering pump reservoir and bracket
   - Cylinder head cover
   - Camshaft and cam followers
   - Cylinder head bolts, working from the center outward
   - Cylinder head

➡The cylinder head bolts must be inspected before they can be reused. If the threads of bolts are stretched, they must be replaced. Check for thread stretching by holding a scale or other straightedge against the threads. If all the threads do not contact the scale, the bolts must be replaced.

⁂ WARNING

Use only a plastic scraper to clean the mating surfaces. NEVER use metal, as this may gouge the surfaces and cause leaks.

5. Cover the combustion chambers, then use a plastic scraper to thoroughly and

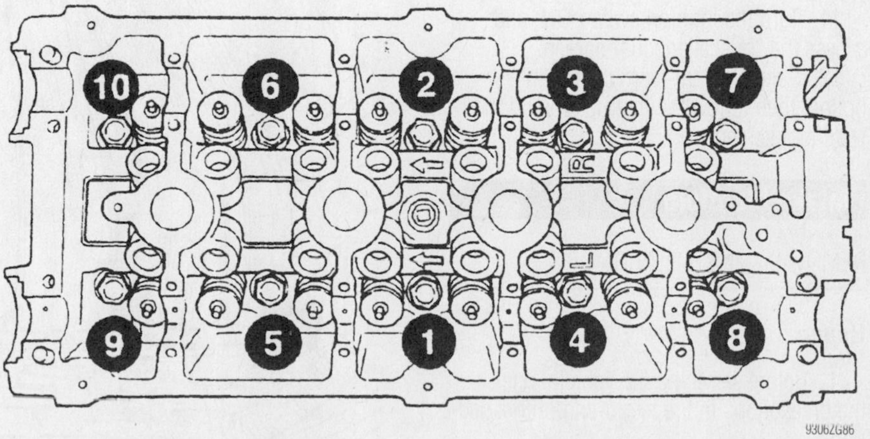

93062G86

Cylinder head bolt torque sequence

carefully clean the engine block and cylinder head mating surfaces.

To install:

6. Install the cylinder head with a new gasket. Make certain that the part number on the new gasket is facing up.
7. Lubricate the cylinder head bolt threads with clean engine oil.
8. Torque the cylinder head bolts, in the following sequence:
   a. Step 1: 25 ft. lbs. (34 Nm).
   b. Step 2: 50 ft. lbs. (68 Nm).
   c. Step 3: 50 ft. lbs. (68 Nm).
   d. Step 4: An additional ¼ turn.
9. Install or connect the following:
   - Camshaft and cam follower assemblies
   - Cylinder head cover
   - Rear timing belt cover and pulley
   - Camshaft sprockets
   - Timing belt
   - Cam sensor wiring connector
   - Ignition coil and spark plug wires
   - Power steering pump reservoir and bracket
   - Exhaust pipe to the manifold
   - Accessory drive belts
   - Lower intake manifold
   - Upper radiator and heater supply hose
   - Heater support bracket
   - Dipstick tube fastener to the intake manifold
   - Power brake vacuum hose to the intake manifold
   - Fuel supply line to the fuel rail
   - Vacuum lines and electrical wiring
   - Upper intake manifold
   - Air cleaner inlet duct and air cleaner
   - IAT sensor and make up hose
   - Negative battery cable
10. Fill the cooling system.
11. Turn the ignition switch **ON**, and access the "miscellaneous" screen.
12. Select the "re-learn cam/crank" option, then follow the instructions on the scan tool screen.

Intake Manifold

REMOVAL & INSTALLATION

Upper

1. Before servicing the vehicle, refer to the precautions in the beginning of this section.
2. Properly relieve the fuel system pressure.
3. Remove or disconnect the following:
   - Negative battery cable
   - Inlet Air Temperature (IAT) sensor and air hose
   - Engine cover
   - Throttle and cruise control cables from the throttle lever bracket
   - Manifold Absolute Pressure (MAP) sensor
   - Idle Air Control (IAC) motor electrical connector
   - Throttle Position (TPS) sensor wiring connector

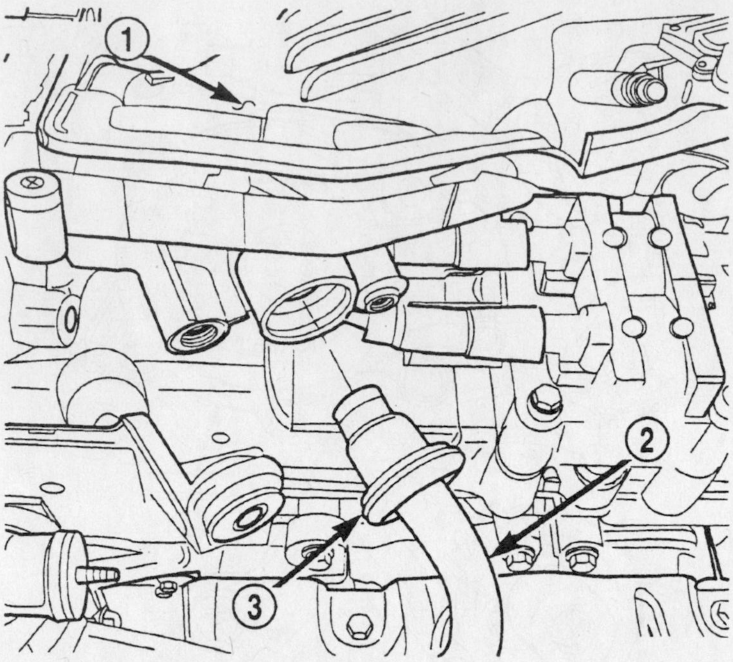

1 – INTAKE MANIFOLD
2 – EGR TUBE
3 – SEAL

9346JG05

Remove the EGR tube from the upper intake manifold

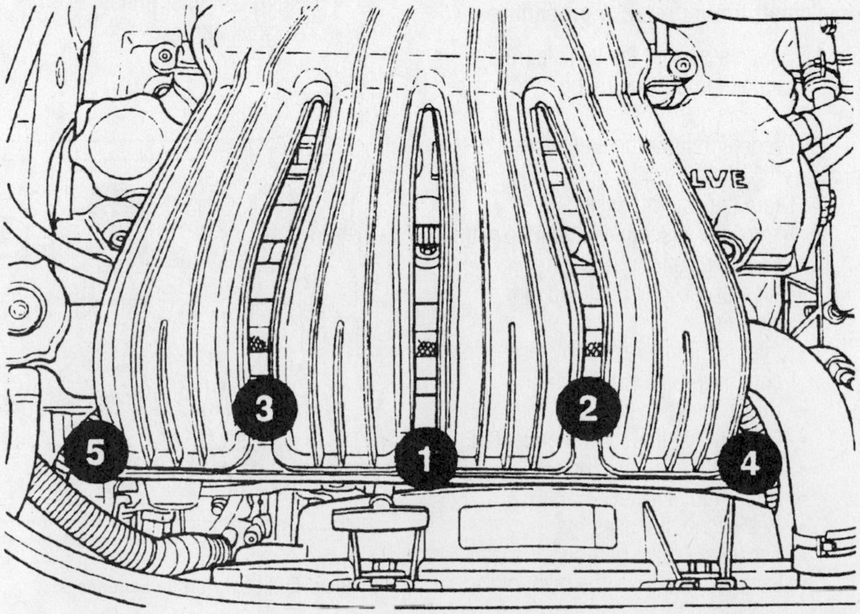

9306ZG87

Intake manifold bolt torque sequence

- Proportional purge hoses
- Brake booster vacuum hose
- Positive Crankcase Ventilation (PCV) hose from the intake manifold
- Throttle body support bracket bolt
- Exhaust Gas Recirculation (EGR) tube from the upper intake manifold
- Upper intake manifold

4. Clean the mating surfaces.

To install:

5. Install or connect the following:
   - New gaskets and seals
   - Intake manifold on the EGR tube then on the lower intake manifold. Torque the bolts, in sequence, to 105 inch lbs. (12 Nm).
   - Throttle body support bracket and torque the bolt to 28 ft. lbs. (20 Nm)
   - EGR retainer plate and torque the smaller bolt to 95 inch lbs. (11 Nm) and the large bolt to 28 ft. lbs. (21 Nm)
   - PCV hose to the intake manifold
   - MAP sensor electrical connector
   - Proportional purge hoses
   - Brake booster hose
   - IAC motor and TPS connectors
   - Throttle and speed control cables
   - Air cleaner assembly
   - Engine cover
   - IAT sensor
   - Negative battery cable

Lower

1. Before servicing the vehicle, refer to the precautions in the beginning of this section.
2. Properly relieve the fuel system pressure.
3. Drain the coolant system.
4. Remove or disconnect the following:
   - Negative battery cable
   - Inlet Air Temperature (IAT) sensor and make-up hose
   - Air cleaner
   - Upper intake manifold
   - Upper radiator hose and coolant outlet connector
   - Fuel supply line quick-connect from the fuel rail
   - Fuel injector wiring harness
   - Oil dipstick tube from the lower intake manifold
   - Intake manifold and discard the gaskets and seals
5. Thoroughly clean the gasket mating surfaces.

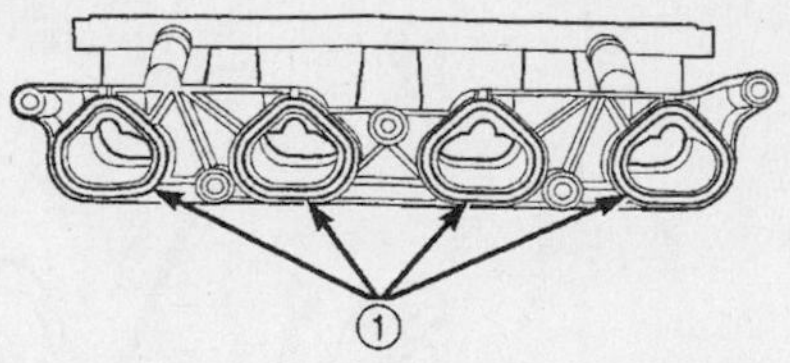

Install new seals on the lower intake manifold

To install:

6. Install or connect the following:
   - New gaskets and seals
   - Intake manifold and torque the bolts in sequence to 105 inch lbs. (12 Nm)
   - Intake manifold-to-lower support bracket bolts and torque to 17 ft. lbs. (23 Nm)
   - Fuel injector wiring harness
   - Fuel supply line quick-connect to the fuel tube assembly
   - Oil dipstick tube to the lower intake manifold
   - Upper radiator hose
   - Air cleaner assembly
   - IAT sensor
   - Negative battery cable
7. Fill the coolant system.
8. Pressurize the fuel system.
9. Start the vehicle and check for leaks, repair if necessary.

Exhaust Manifold

REMOVAL & INSTALLATION

1. Before servicing the vehicle, refer to the precautions in the beginning of this section.
2. Remove or disconnect the following:
   - Negative battery cable
   - Air cleaner assembly and bracket
   - Throttle and speed control cables
   - Manifold Absolute Pressure (MAP) sensor electrical connector
   - Power steering reservoir, move it aside. DO NOT disconnect the fluid lines.
   - Coolant recovery bottle
   - Exhaust manifold upper heat shield
   - Exhaust pipe from the manifold
   - Engine wiring heat shield
   - Manifold support bracket
   - Lower exhaust manifold heat shield
   - Upstream Heated Oxygen (HO_2S) sensor connector
   - Exhaust manifold and discard the gasket

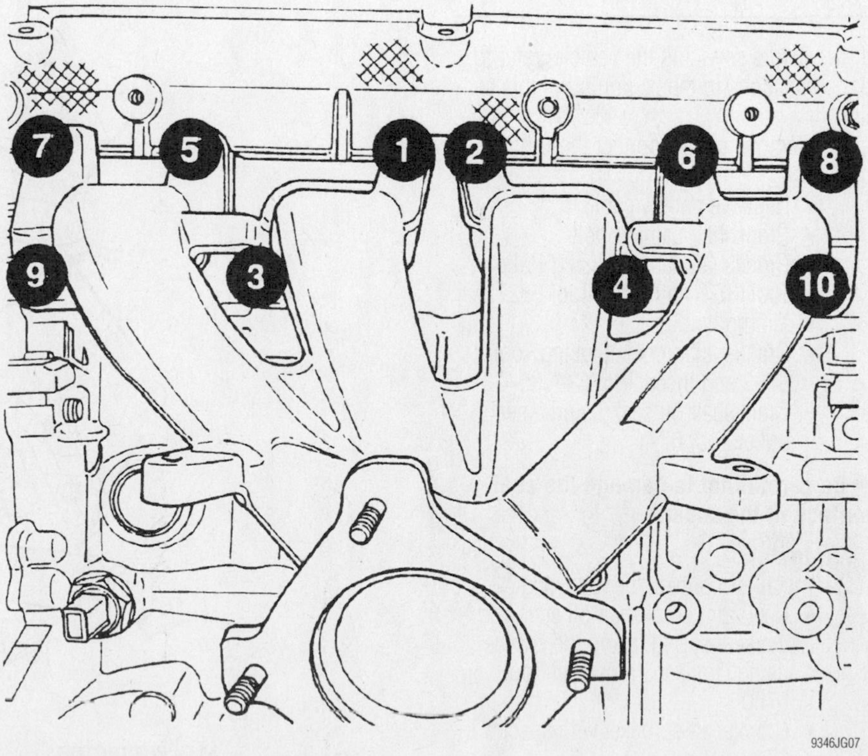

Torque the exhaust manifold bolts in sequence

3. Thoroughly clean the mating surfaces.

To install:

4. Install or connect the following:
   - New gasket
   - Exhaust manifold and torque the bolts, in sequence, to 17 ft. lbs. (23 Nm)
   - Alternator bracket bolt, if loosened
   - Exhaust manifold heat shields and torque the bolts to 105 inch lbs. (12 Nm)
   - Exhaust manifold support bracket
   - Engine wiring heat shield
   - Upstream HO_2sensor wiring connector
   - Exhaust pipe to the manifold and torque the bolts to 21 ft. lbs. (28 Nm)
   - Coolant recovery bottle
   - Power steering pump reservoir
   - MAP sensor connector
   - Throttle and speed control cables
   - Air cleaner bracket and assembly
   - Negative battery cable
5. Start the vehicle and check for leaks, repair if necessary.

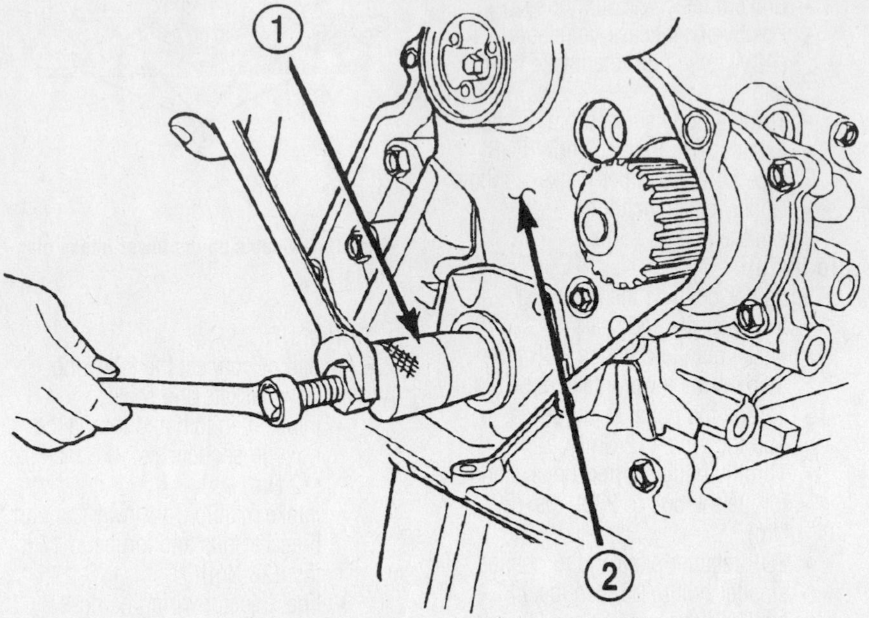

Removing the front crankshaft oil seal

Front Crankshaft Seal

REMOVAL & INSTALLATION

1. Before servicing the vehicle, refer to the precautions in the beginning of this section.
2. Remove or disconnect the following:
   - Negative battery cable
   - Crankshaft damper bolt
   - Crankshaft damper, using Puller Tool 1026 and Insert Tool 6827-A
   - Timing belt
   - Crankshaft sprocket, using Tool 6793 and Insert Tool C-4685-C2
   - Crankshaft oil seal, using a seal puller Tool 6771

➡Be careful not to damage the seal surface of the cover.

To install:

3. Install or connect the following:
   - New front crankshaft oil seal with the seal spring facing the engine, using Crankshaft Installer Tool 6780
   - Crankshaft sprocket with special T 6792

➡Make sure the word FRONT on the crankshaft sprocket is facing outward.

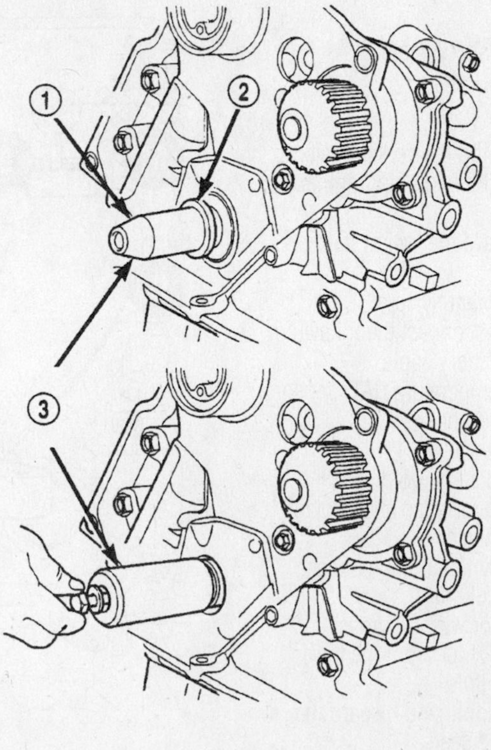

Installing a new seal using seal installer 6780-1; proceed with caution if using substitute tools

- Timing belt
- Crankshaft damper using thrust bearing washer and bolt from Installer Tool 6792. Torque the damper bolt to 105 ft. lbs. (142 Nm)
- Negative battery cable

4. Start the vehicle and check for leaks, repair if necessary.

Camshaft and Lifters

REMOVAL & INSTALLATION

➡After all components are installed on the engine, a DRB scan tool is necessary to perform the camshaft and crankshaft timing relearn procedure.

1. Before servicing the vehicle, refer to the precautions in the beginning of this section.
2. Relieve the fuel system pressure.
3. Remove or disconnect the following:
   - Negative battery cable
   - Cylinder head cover
   - Camshaft Position (CMP) sensor and target magnet
   - Timing belt
   - Camshaft sprocket
   - Rear timing belt cover
   - Loosen the camshaft bearing caps in sequence from the rear of the cylinder head
   - Remove the camshafts

To install:

Make certain that the pistons are **NOT** at Top Dead Center (TDC) before installing the camshafts.

4. Install the camshafts and cam followers, lubricate the bearing journals thoroughly.
5. Install the left and right camshaft bearing caps, No's 2–5 and right side No. 6. Torque these fasteners to 105 inch lbs. (12 Nm). Apply Mopar® gasket maker to No.1 and left side No. 6. Torque these fasteners to 18 ft. lbs. (24 Nm).
6. Install or connect the following:
   - Camshaft seals
   - Rear timing belt cover
   - Camshaft sprockets and torque the bolt to 85 ft. lbs. (115 Nm)
   - Timing belt
   - Target magnet
   - CMP sensor and torque the screws to 85 inch lbs. (9.6 Nm)
   - Cylinder head cover
   - Negative battery cable

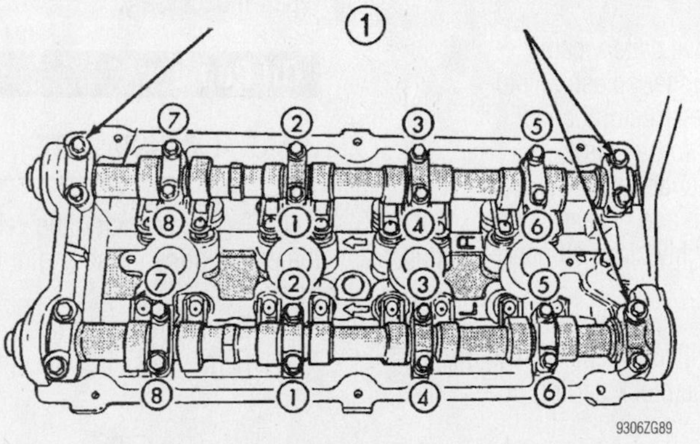

Remove the camshaft bearing caps in sequence

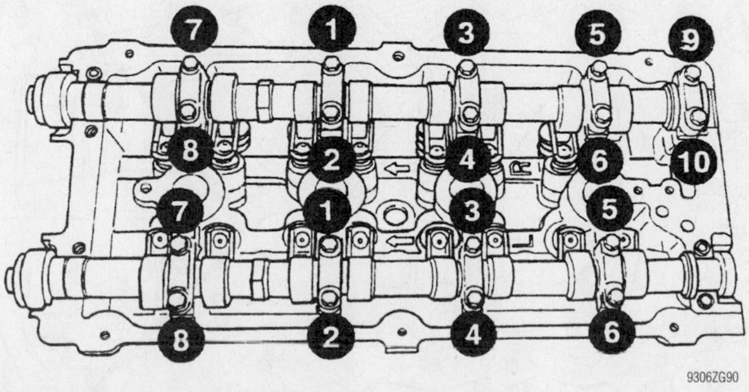

Camshaft bearing cap tightening sequence

➡An oil and filter change are recommended.

7. Use a DRB scan tool to perform the camshaft and crankshaft timing relearn procedure, as follows:
   a. Connect the scan tool to the DLC (located under the instrument panel, near the steering column).
   b. Turn the ignition switch **ON** and access the "miscellaneous" screen.
   c. Select the "re-learn cam/crank" option, then follow the instructions on the scan tool screen.
8. Start the engine and check for leaks. Run the engine with the radiator cap off so as the engine warms and the thermostat opens, coolant can be added to the radiator. Test drive vehicle to check for proper operation.

Valve Lash

ADJUSTMENT

The engines in these vehicles do not require periodic valve lash adjustment.

Starter Motor

REMOVAL & INSTALLATION

1. Before servicing the vehicle, refer to the precautions in the beginning of this section.
2. Remove or disconnect the following:
   - Negative battery cable
   - Air cleaner box cover
   - Engine structural collar
   - Starter electrical connectors
   - Starter

To install:

3. Install the starter and torque the bolts to 40 ft. lbs. (54 Nm). Attach the starter electrical connectors.
4. Install the engine structural collar. Torque the bolts in the following sequence:
   a. Step 1: Collar-to-oil pan bolts to 30 inch lbs. (3 Nm).
   b. Step 2: Collar-to-transmission bolts to 80 ft. lbs. (108 Nm).

Heater Core replacement is covered in Section 2 of this manual

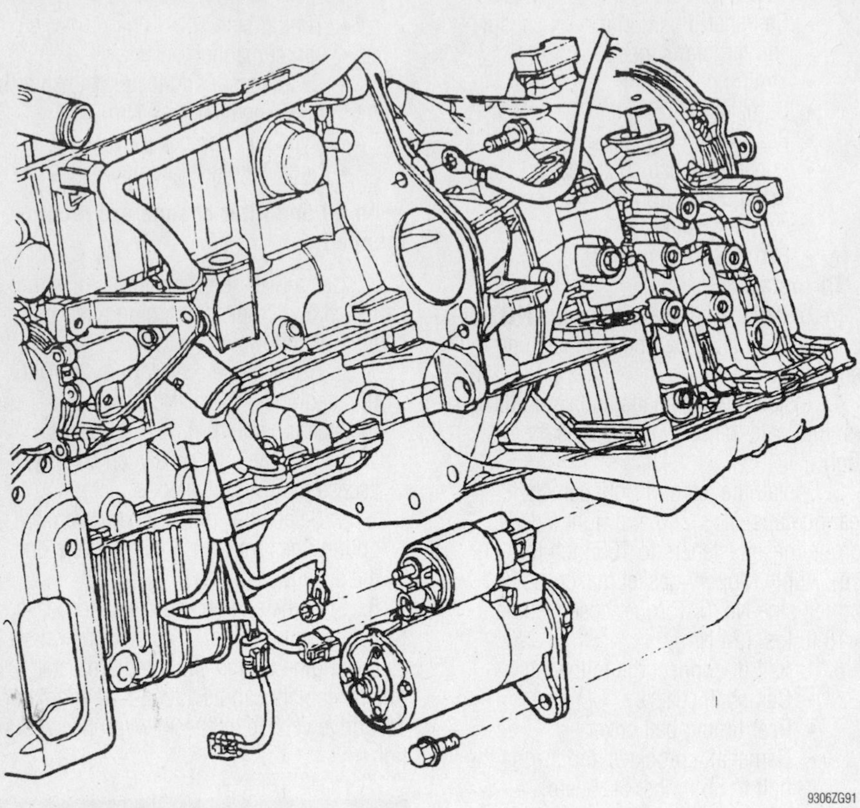

Removal of the starter motor mounting

c. Step 3: Collar-to-oil pan bolts to 40 ft. lbs. (54 Nm).

5. Install the air cleaner box cover and connect the negative battery cable.

Oil Pan

REMOVAL & INSTALLATION

1. Before servicing the vehicle, refer to the precautions in the beginning of this section.
2. Drain the engine oil and remove the oil filter.
3. Support the powertrain assembly.
4. Remove or disconnect the following:
   - Negative battery cable
   - Right inner splash shield
   - Engine structural collar
   - Lower torque strut
   - Oil filter adapter
   - Oil pan and gasket
5. Thoroughly clean all gasket mating surfaces.

To install:

6. Apply silicone sealer to the oil pump-to-engine block parting line.
7. Install or connect the following:
   - New gasket on the oil pan
   - Oil pan and torque the bolts to 105 inch lbs. (12 Nm)
   - Oil filter and adapter and torque the screws to 105 inch lbs. (12 Nm)

**** WARNING**

Follow the proper tightening sequence for the structural collar or damage to the collar or oil pan may occur

8. Install the structural collar and torque the bolts as follows:
   a. Collar-to-oil pan bolts: 30 inch lbs. (3 Nm).
   b. Collar-to-transaxle bolts: 80 ft. lbs. (108 Nm).
   c. Collar-to-oil pan bolts: 40 ft. lbs. (54 Nm), final torque.
9. Install or connect the following:
   - Lower torque strut
   - Right inner splash shield
   - Negative battery cable
10. Fill the engine with clean oil and a new filter.
11. Start the vehicle and check for leaks, repair if necessary.

Oil Pump

REMOVAL & INSTALLATION

1. Before servicing the vehicle, refer to the precautions in the beginning of this section.
2. Drain the engine oil.
3. Remove or disconnect the following:

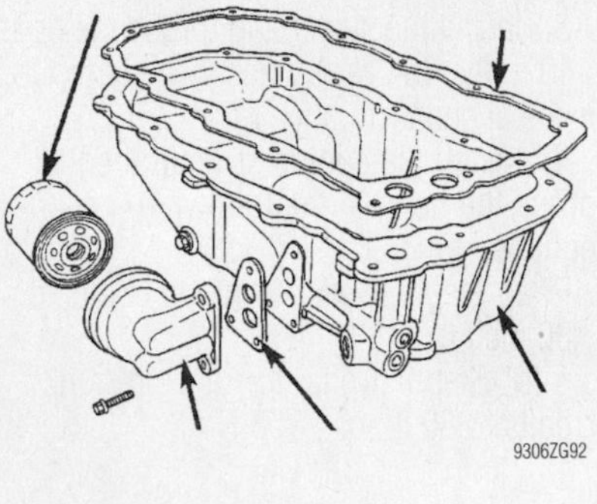

Remove the oil filter adapter

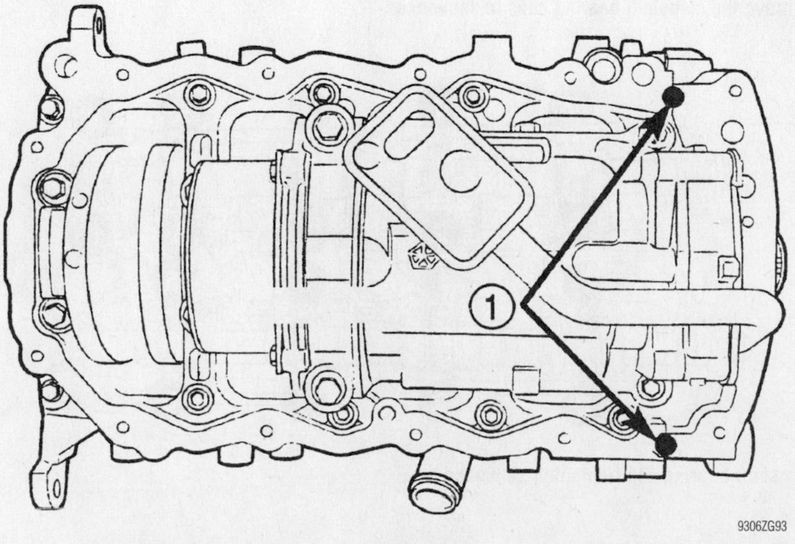

Silicone sealer application locations

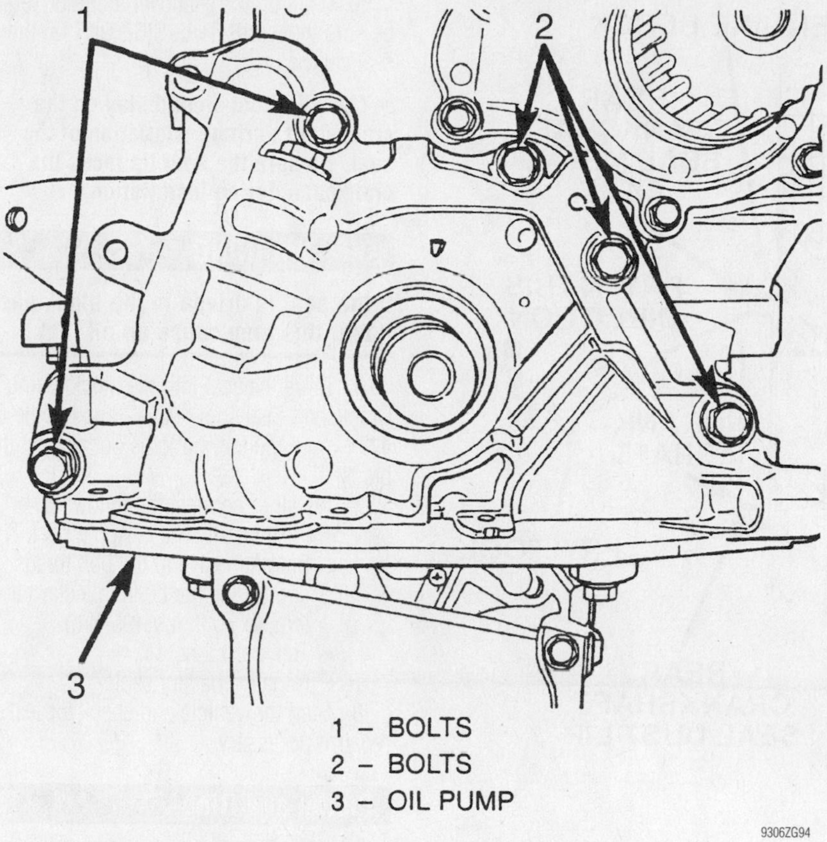

Exploded view of the oil pump mounting bolts

- Negative battery cable
- Timing belt and rear cover
- Oil pan
- Crankshaft sprocket, using Tool 6795 and Insert Tool C-4685-C2
- Crankshaft key
- Oil pickup tube
- Oil pump

To install:

4. Wash all parts in a solvent; then, inspect carefully for damage or wear, as follows:

a. Inspect the mating surface of the oil pump should be smooth. Replace the pump cover, if scratched or grooved.

b. Apply Mopar® gasket maker to the oil pump.

c. Install the O-ring into the oil pump body discharge passage.

5. Prime the oil pump before installation by filling the rotor cavity with engine oil.

6. Install or connect the following:

- Oil pump, align the rotor flats with the crankshaft flats and torque the bolts to 21 ft. lbs. (28 Nm).

WARNING

The front crankshaft seal MUST be out of the pump to align or damage may result.

- New front crankshaft seal, using Seal Driver Tool 6780
- Crankshaft key
- Crankshaft sprocket, using a Crankshaft Sprocket Installer Tool 6792
- Oil pump pickup tube
- Oil pan
- Rear timing belt cover
- Timing belt
- Negative battery cable

7. Fill the engine with clean oil.

8. Start the engine and check for leaks; repair if necessary.

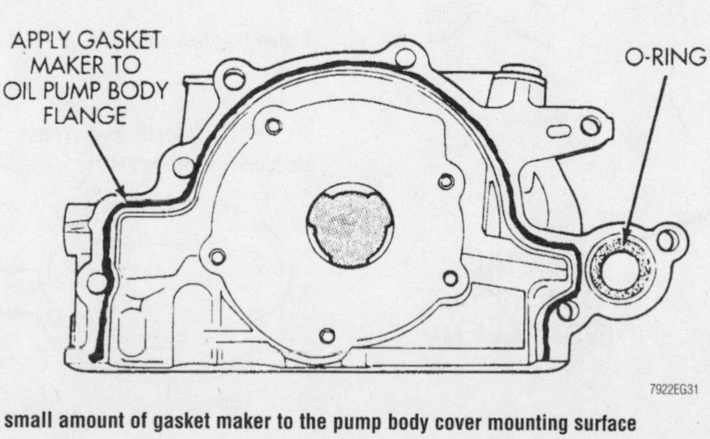

Apply a small amount of gasket maker to the pump body cover mounting surface

Rear Main Seal

REMOVAL & INSTALLATION

1. Before servicing the vehicle, refer to the precautions in the beginning of this section.

2. Remove or disconnect the following:

- Transmission
- Flexplate/flywheel
- Rear main seal

3. Insert a seal remover between the dust lip and the metal case of the crankshaft seal. Angle the tool through the dust lip against the metal case of the seal. Pry out the seal.

WARNING

DO NOT let the prytool contact the crankshaft seal surface. Contact of the tool blade against the crankshaft edge (chamfer) is permitted.

To install:

WARNING

If the crankshaft edge (chamfer) has any burrs or scratches on the, clean it up with 400 grit sand paper to prevent seal damage during installation of the new seal.

➡No lubrication is necessary when installing the seal.

4. Place Crankcase Seal Pilot Tool 6926-1 on the crankshaft; this is a pilot tool with a magnetic base

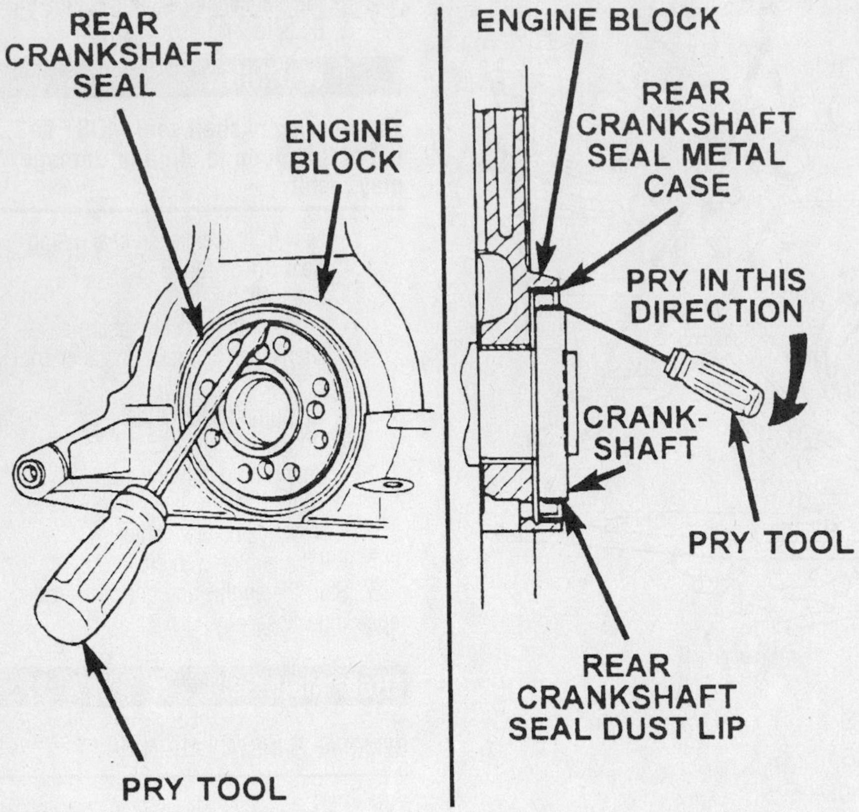

7922EG32

When prying the seal out, be sure to use the prytool at the proper angle

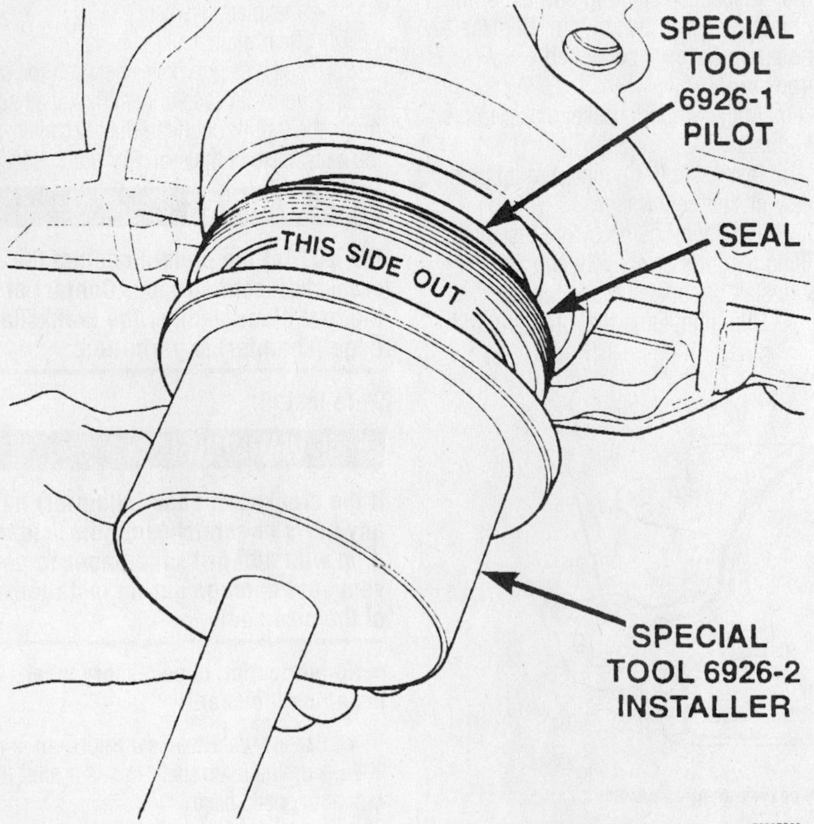

7922EG33

Place a proper size pilot tool with a magnetic base on the crankshaft

5. Position the seal over the pilot Tool; be sure the words THIS SIDE OUT on the seal can be read.

➡The pilot tool should stay on the crankshaft during installation of the seal. Be sure the seal lip faces the crankcase during installation.

⁂ WARNING

If the seal is driven in the block past flush, this may cause an oil leak.

6. Drive the seal into the block, using Crankshaft Seal Tool 6926-2 and handle C-4171, until the tool bottoms out against the block.

7. Install or connect the following:
   - Flexplate/flywheel. Apply Lock & Seal Adhesive to the bolt treads and torque the bolts in a star pattern, to 70 ft. lbs. (95 Nm).
   - Transmission
   - Negative battery cable

8. Start the vehicle and check for leaks, repair if necessary.

Piston and Ring

POSITIONING

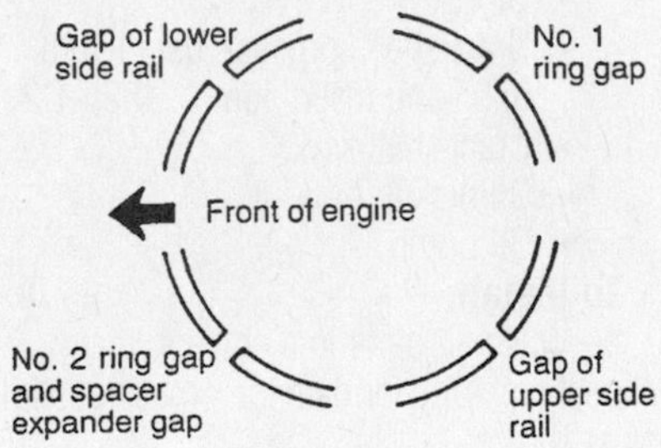

7922AG05

Piston ring end-gap spacing—2.4L engine

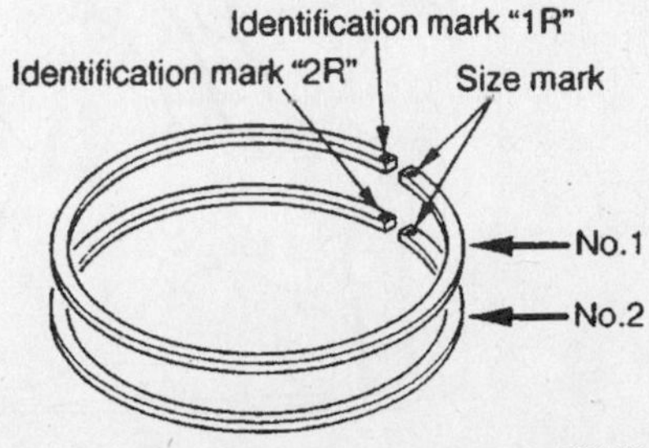

7922AG02

Common Chrysler piston ring identification mark locations

FUEL SYSTEM

Fuel System Service Precautions

Safety is an important factor when servicing the fuel system. Failure to conduct maintenance and repairs in a safe manner may result in serious personal injury. Maintenance and testing of the vehicle's fuel system components can be accomplished safely and effectively by adhering to the following rules and guidelines:

- To avoid the possibility of fire and personal injury, always disconnect the negative battery cable unless the repair or test procedure requires that battery voltage be applied.
- Always relieve the fuel system pressure prior to disconnecting any fuel system component (injector, fuel rail, pressure regulator, etc.), fitting or fuel line connection. Exercise extreme caution whenever relieving fuel system pressure, to avoid exposing skin, face and eyes to fuel spray. Please be advised that fuel under pressure may penetrate the skin or any part of the body that it contacts.
- Always place a shop towel or cloth around the fitting or connection prior to loosening to absorb any excess fuel due to spillage. Ensure that all fuel spillage is quickly removed from engine surfaces. Ensure that all fuel soaked cloths or towels are deposited into a suitable waste container.
- Always keep a dry chemical (Class B) fire extinguisher near the work area.
- Do not allow fuel spray or fuel vapors to come into contact with a spark or open flame.
- Always use a back-up wrench when loosening and tightening fuel line connection fittings. This will prevent unnecessary stress and torsion to fuel line piping.
- Always replace worn fuel fitting O-rings. Do not substitute fuel hose where fuel pipe is installed.

Fuel System Pressure

RELIEVING

⁂ CAUTION

Relieve the fuel system pressure before servicing any components of the fuel system. Service vehicles in well ventilated areas and avoid ignition sources. NEVER smoke while servicing the vehicle!

1. Before servicing the vehicle, refer to the precautions in the beginning of this section.
2. Remove the negative battery cable.
3. Remove the fuel pump relay from the Power Distribution Center (PDC).
4. Start and run the engine until it stalls.
5. Turn the ignition key to the OFF position.

Fuel Filter

The fuel filter is part of the fuel pump module located in the fuel tank. Refer to the fuel pump module procedure for the fuel filter.

Fuel Pump

The fuel pump is integral with the pump module, which also contains the fuel reservoir, level sensor, inlet strainer and fuel pressure regulator. The inlet strainer, fuel pressure regulator and level sensor are the only serviceable items. If the fuel pump requires service, replace the entire fuel pump module.

REMOVAL & INSTALLATION

1. Before servicing the vehicle, refer to the precautions in the beginning of this section.
2. Properly relieve the fuel system pressure.
3. Remove or disconnect the following:
   - Negative battery cable
   - Air cleaner lid
   - Inlet Air Temperature (IAT) sensor and make up air hose
   - Fuel tank
   - Fuel pump module and seal from the tank
   - Locknut to release the fuel pump module, using a Ring Spanner Tool No. 6856
   - Fuel filter from the fuel pump module
   - Fuel pump and seal from the fuel tank

To install:

4. Install or connect the following:
   - Fuel filter to the fuel pump module
   - Fuel filter lines to the fuel pump module
   - Fuel pump module in the tank
   - Locknut while holding the fuel pump in position. Using Tool 6856, torque the nut to 56 ft. lbs. (75 Nm)
   - Fuel tank
   - IAT sensor and make-up hose
   - Air cleaner lid
   - Negative battery cable
5. Fill the fuel tank.
6. Start the vehicle and check for leaks, repair if necessary.

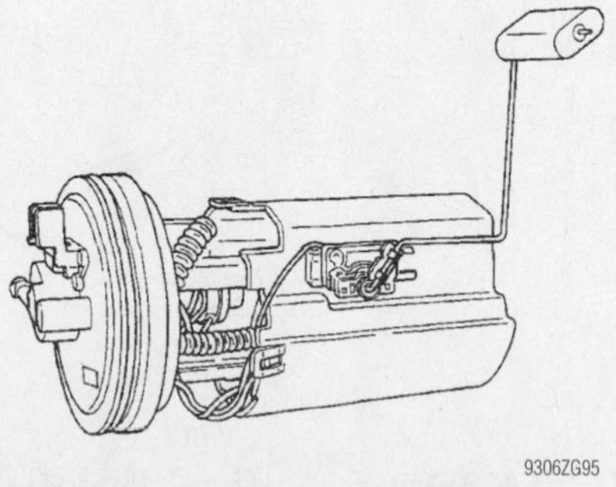

Exploded view of the fuel pump module

Fuel Injector

REMOVAL & INSTALLATION

1. Before servicing the vehicle, refer to the precautions in the beginning of this section.
2. Release the fuel system pressure.
3. Remove or disconnect the following:
   - Negative battery cable
   - Remove the air cleaner lid
   - Inlet Air Temperature (IAT) sensor and the make-up hose
   - Engine cover or throttle control shield, if equipped
   - Fuel supply tube from fuel rail
   - Intake manifold

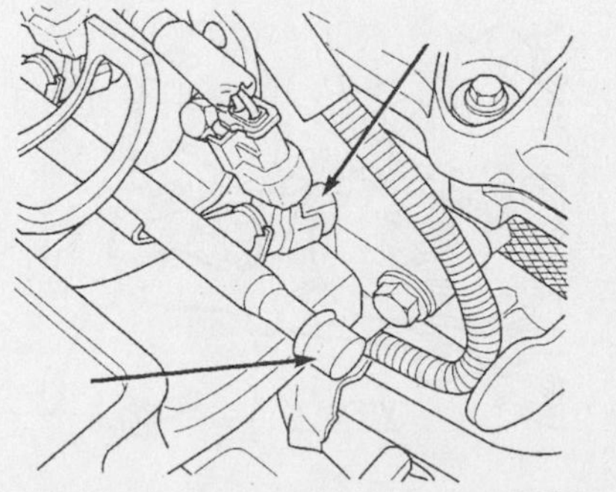

Remove the fuel rail and injectors as an assembly

- Fuel injector electrical connectors
- Fuel rail with the fuel injectors
- Fuel injector(s) and discard the O-rings

To install:

➡Lubricate the O-rings with clean engine oil.

4. Install or connect the following:
   - Fuel injector(s) to the fuel rail using new O-rings
   - Fuel injector nozzles into the intake manifold and torque the fuel rail bolts to 8 ft. lbs. (12 Nm)
   - Fuel injector electrical connectors
   - Fuel supply tube to fuel rail
   - Intake manifold
   - Engine cover or throttle control shield, if equipped
   - IAT sensor and make-up hose
   - Air cleaner lid
   - Negative battery cable
5. Start the vehicle and check for leaks, repair if necessary.

DRIVE TRAIN

Transaxle Assembly

REMOVAL & INSTALLATION

Manual

1. Before servicing the vehicle, refer to the precautions in the beginning of this section.
2. Drain the transaxle fluid.
3. Remove or disconnect the following:
   - Battery cables
   - Battery and tray
   - Air cleaner
   - Back-up lamp switch wiring from the transaxle
   - Shift selector and crossover cable and move them out of the way
   - Vehicle Speed Sensor (VSS) wire
   - Clutch master cylinder hydraulic tube from the slave cylinder
   - Both halfshafts
   - Bell housing dust cover
   - Power steering hose from the structural collar
   - Left side engine-to-transaxle lateral bending brace and structural collar
   - Right side engine-to-transaxle lateral bending brace
   - Starter
   - Driveplate-to-clutch module bolts and support the engine at the oil pan
   - Transaxle upper mount bolts and lower the powertrain assembly onto the support
   - Transaxle-to-engine mounting bolts
   - Transaxle
   - Clutch module from the input shaft
   - Slave cylinder
   - Upper transaxle mount

To install:

4. Install or connect the following:
   - Clutch module to the transaxle input shaft and place the transaxle into position
   - Transaxle-to-engine mounting bolts and torque them to 80 ft. lbs. (108 Nm)
   - Upper transaxle mount and torque the bolts to 45 ft. lbs. (62 Nm)
   - Driveplate-to-clutch module bolts and torque to 65 ft. lbs. (88 Nm)
   - Starter
   - Starter electrical connectors
   - Bell housing dust cover

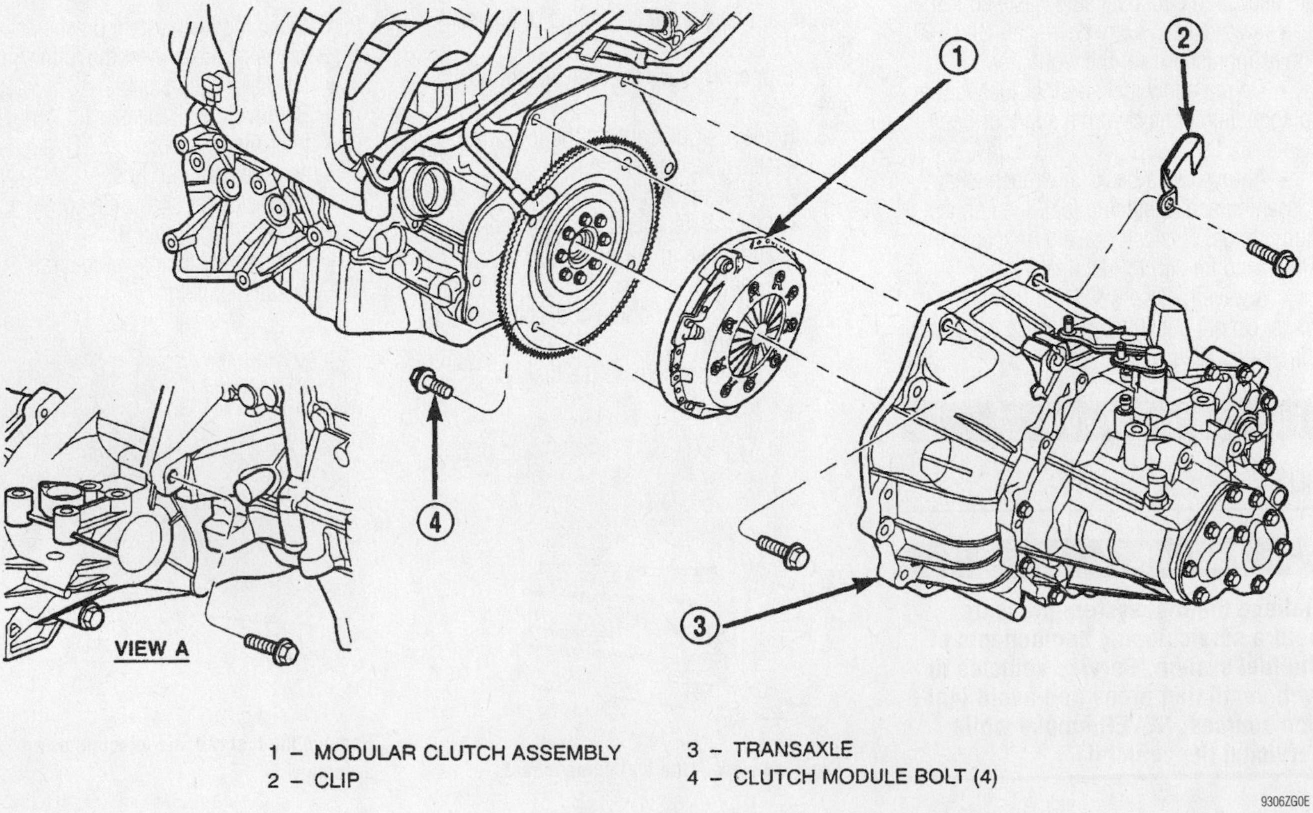

Remove the clutch module from the transaxle assembly

- Left side engine-to-transaxle bending brace and structural brace. Torque the bolts to 60 ft. lbs. (81 Nm).
- Right lateral bending brace and torque the bolts to 60 ft. lbs. (81 Nm)
- Power steering hose to the structural collar
- Both halfshafts
- Clutch master cylinder tube to the slave cylinder
- VSS electrical connector
- Shift crossover and selector cables to the shift lever
- Cables to the bracket and install a new retainer clip
- Back-up lamp switch connector
- Battery tray
- Battery and cables
- Air cleaner assembly

5. Fill the transaxle fluid to the proper level.
6. Be sure the vehicle's back-up lights and speedometer are functioning properly.
7. Start the vehicle and check for leaks, repair if necessary.

Automatic

1. Before servicing the vehicle, refer to the precautions in the beginning of this section.
2. Drain the transaxle fluid.
3. Remove or disconnect the following:
   - Battery cables
   - Air cleaner assembly
   - Battery and tray
   - Upper starter-to-transaxle bell housing bolt
   - Transaxle dipstick and tube
   - Gearshift cable end from the transaxle shift lever
   - Gearshift cable bracket bolt from the transaxle
   - Transaxle oil cooler lines. Plug the lines to prevent contamination.
   - Input and output speed sensor electrical connectors
   - Transaxle range sensor connector
   - Solenoid/pressure switch assembly connector
   - Both front wheels
   - Left front splash shield
   - Both halfshafts
   - Power steering hose from the structural collar
   - Left and right side engine-to-transaxle lateral brace and structural collar
   - Starter motor electrical connectors
   - Starter
   - Gearshift cable bracket
   - Driveplate-to-torque converter bolts and support the powertrain assembly
   - Transaxle upper mount to bracket bolts and lower the powertrain assembly
   - Transaxle

To install:

4. Install or connect the following:
   - Transaxle and torque the bolts to 80 ft. lbs. (105 Nm) while supporting the transaxle with a jack
   - Mount-to-transaxle bracket. Torque the bolts to 50 ft. lbs. (68 Nm) and remove the jack.
   - Driveplate-to-torque converter bolts and torque the bolts to 65 ft. lbs. (88 Nm)
   - Starter and hand tighten the bolts
   - Dipstick tube and secure the bracket to the transaxle. Torque the upper starter bolt to 40 ft. lbs. (54 Nm).
   - Cable bracket to the bell housing and torque the bolt to 45 ft. lbs. (61 Nm)
   - Starter lower bolt and torque the bolt to 40 ft. lbs. (54 Nm)
   - Starter electrical connections
   - Bell housing dust cover

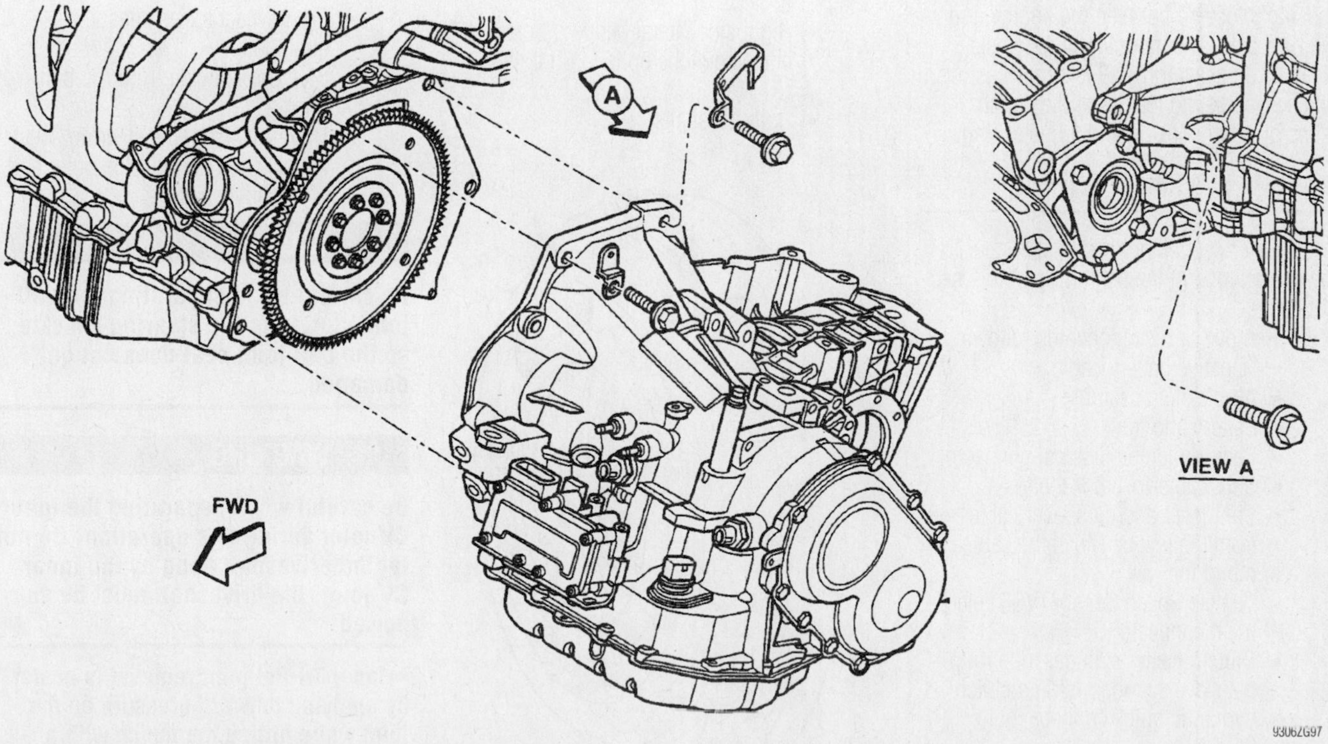

Exploded view of automatic transaxle removal

- Lower dust shield screws and torque the bell housing cover bolts to 108 inch lbs. (12 Nm)
- Right side lateral bending brace and torque the bolts to 60 ft. lbs. (81 Nm)
- Both halfshafts
- Left side splash shield and both wheels
- Cooler lines to the transaxle and secure with constant tension clamps
- Solenoid/pressure switch assembly
- Transmission range sensor connector
- Input/output sensor connectors
- Gearshift cable to the bracket and connect it to the manual valve lever
- Battery tray and battery
- Air cleaner assembly
- VSS wiring
- Battery cables

5. Fill the transaxle fluid to the proper level.
6. Be sure the vehicle's back-up lights and speedometer are working properly.

Clutch

ADJUSTMENT

This vehicle utilizes a modular clutch assembly located between the engine and the transaxle. The modular clutch is serviced as an assembly and is self-adjusting. The self-adjusting feature of the clutch relies on a sensor ring and adjuster ring.

REMOVAL & INSTALLATION

1. Before servicing the vehicle, refer to the precautions in the beginning of this section.
2. Remove or disconnect the following:
   - Negative battery cable
   - Air cleaner assembly
   - Battery and tray
   - Back-up lamp electrical connector
   - Shift cable-to-bracket clips
   - Shift lever and crossover cable from the levers. Move the cables out of the way.
   - Vehicle Speed Sensor (VSS) electrical connector
   - Clutch master cylinder tube from the slave cylinder using a clutch hydraulic quick connect Tool 6638A
   - Both halfshafts
   - Power steering hose from the structural collar
   - Left side lateral bending brace and structural collar
   - Bell housing dust cover
   - Right side lateral bending brace
   - Starter
   - Driveplate-to-clutch module bolts and support the engine at the oil pan
   - Transaxle upper mount bolts and lower the engine/transaxle assembly
   - Modular clutch from the input shaft

To install:

3. Install or connect the following:
   - Clutch module to the input shaft
   - Transaxle-to-engine mount bolts and torque them to 80 ft. lbs. (108 Nm)
   - Upper mount bolts and torque them to 65 ft. lbs. (88 Nm)
   - Driveplate-to-clutch module bolts and torque them to 65 ft. lbs. (88 Nm)
   - Starter and torque the bolts to 40 ft. lbs. (54 Nm). Make certain that the ground cable is fastened to the upper bolt.
   - Starter electrical connectors
   - Bell housing dust cover
   - Left side lateral bending brace and structural collar
   - Power steering hose to the structural collar
   - Right side lateral bending brace and torque the bolts to 60 ft. lbs. (81 Nm)
   - Both halfshafts

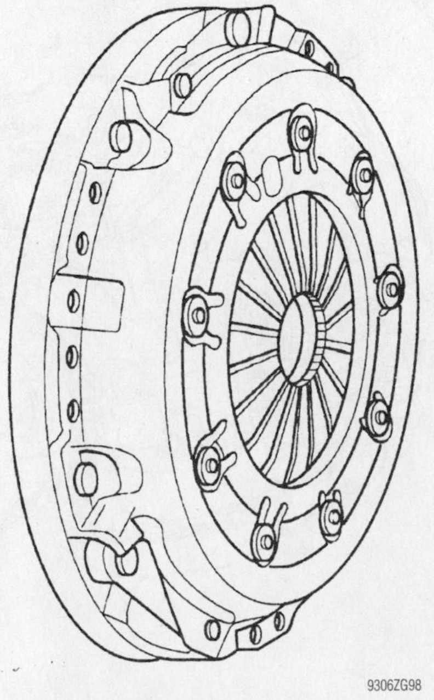

Exploded view of the modular clutch

   - Clutch master cylinder tube to the slave cylinder
   - VSS electrical connector
   - Back-up lamp switch electrical connector
   - Air cleaner assembly
   - Battery and tray
   - Battery cables
4. Fill the transmission to the proper level.
5. Road test the vehicle and check for proper clutch operation.
6. Check the fluid level and adjust if needed.

Halfshafts

REMOVAL & INSTALLATION

1. Before servicing the vehicle, refer to the precautions in the beginning of this section.
2. Place the transaxle in the **P** position, for automatic and neutral for manual.
3. Remove or disconnect the following:
   - Negative battery cable
   - Front wheel
   - Cotter pin, locknut and spring washer from the end of the outer Constant Velocity (CV) joint stub axle
   - Driveshaft to hub and bearing nut
   - Front wheel speed sensor, if equipped
   - Steering knuckle from the ball joint
   - Driveshaft from the steering knuckle and support the outer end of the driveshaft

⁂ WARNING

Be careful when separating the ball joint stud from the steering knuckle, so the ball joint seal does not get damaged.

⁂ WARNING

Be careful when separating the inner CV-joint during this operation. Do not let the driveshaft hang by the inner CV-joint, the driveshaft must be supported.

➡Inner Tri-Pot joint removal is easier by applying outward pressure on the joint while hitting the punch with a hammer.

4. Inner Tri-Pot joints from the transmission side gears using a punch to dis-

lodge the inner Tri-Pot joint retaining ring from the transmission side gear. If removing the right side inner Tri-Pot joint, position the punch against the inner Tri-Pot joint. Hit the punch sharply with a hammer to dislodge the right inner joint from the side gear. If removing the left side inner Tri-Pot joint, position the punch in the groove of the inner Tri-Pot joint. Hit the punch sharply with a hammer to dislodge the left inner Tri-Pot joint from the side gear.

5. Hold the inner Tri-Pot joint and interconnecting shaft of the driveshaft assembly. Remove the inner Tri-Pot joint from the transaxle by pulling it straight out of the transaxle side gear and transmission oil seal. When removing the Tri-Pot joint, do not let the spline or snapring drag across the sealing lip of the transaxle-to-Tri-Pot joint oil seal.

**** WARNING**

The driveshaft, when installed, acts as a bolt which secures the front hub and bearing assembly. If the vehicle is to be supported or moved on its wheels with a driveshaft removed, install a proper-sized bolt and nut through the front hub. Tighten the bolt and nut to 135 ft. lbs. (183 Nm). This will ensure that the hub bearing cannot loosen.

To install:

6. Thoroughly clean the spline and oil seal sealing surface on the Tri-Pot joint. Lightly lubricate the oil seal sealing surface on the Tri-Pot joint with fresh, clean transmission fluid.

7. Holding the driveshaft assembly by the Tri-Pot joint and interconnecting shaft, install the Tri-Pot joint into the transaxle side gear as far as possible by hand

8. Align the Tri-Pot joint with the transmission side gears, grasp the driveshaft interconnecting shaft and push the Tri-Pot joint into the transaxle side gear until fully seated. Be sure the snapring is fully engaged with the side gear by trying to remove the Tri-Pot joint from the transaxle by hand. If the snapring is fully seated with the side gear, the Tri-Pot joint will not be removable by hand.

9. Install or connect the following:
- Driveshaft back into the front hub
- Steering knuckle into the ball joint stud
- New steering knuckle-to-ball joint stud bolt and nut. Torque the nut and bolt to 70 ft. lbs. (95 Nm).
- Washer and hub nut to the stub axle and torque the nut to 180 ft. lbs. (244 Nm)
- Spring washer, locknut and cotter pin
- VSS, if equipped
- Front wheel
- Negative battery cable

10. Check the transaxle fluid and adjust if needed.

CV-Joints

OVERHAUL

Tri-Pot (Inner) Joint

1. Before servicing the vehicle, refer to the precautions in the beginning of this section.
2. Remove the negative battery cable.
3. Remove the halfshaft.
4. Remove the tri-pot joint boot clamps and slide the boot down the shaft.

**** WARNING**

When removing the spider joint, hold the rollers in place on the trunions to keep the rollers and needle bearings in place.

5. Slide the interconnecting shaft and spider assembly from the tri-pot housing.
6. Remove the snapring from the shaft.
7. Remove the spider assembly.

➡If necessary, tap the spider assembly from the shaft with a brass drift.

**** WARNING**

When removing the spider assembly, do not hit the outer bearings.

8. Remove the boot by sliding it off the shaft.

**** WARNING**

If any parts show excessive wear, replace the halfshaft assembly; the component parts are not serviceable.

To install:

**** WARNING**

The Tri-pot sealing boots are made of 2 different types of material, silicon rubber (high temperature) which is soft and pliable or hytrel plastic (standard temperature) which is stiff and rigid. Be sure to replace the boot made of the correct material.

9. Install a new small boot clamp and slide it on the shaft.
10. Install the boot and slide it on the shaft.
11. Position the boot so that the raised

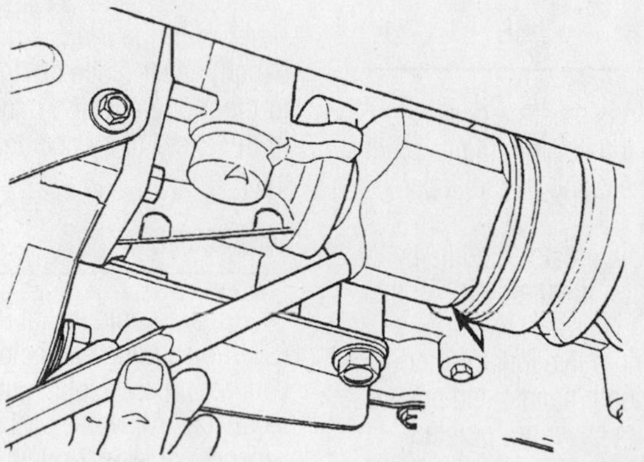

9306ZG99

Remove the Tri-Pot joint from the transaxle

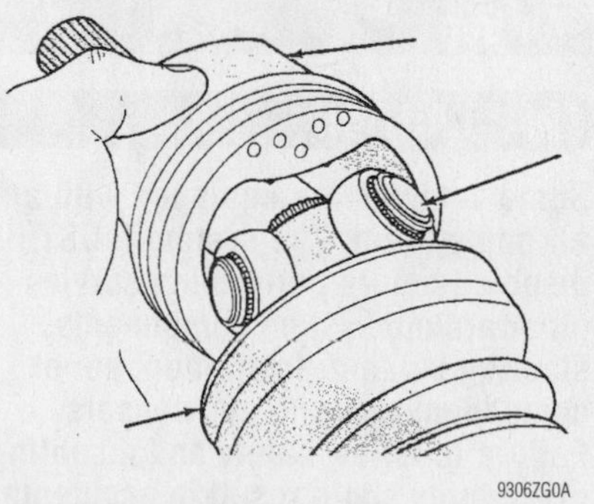

9306ZG0A

Install the Tri-Pot housing on to the spider assembly

bead on the inside the boot seal is in the shaft groove.

12. Install the spider assembly, face the chamfered side toward the shaft.
13. Install the snapring making sure it is fully seated in the groove.
14. Install the spider/shaft assembly into the tri-pot housing.
15. Install a new inner boot clamp and position it evenly on the sealing boot.
16. Using a trim stick, adjust the boot length to 115mm (hytrel plastic) or 115mm (silicone rubber).
17. If installing a high profile boot clamp, perform the following procedure:
    a. Using the Crimper Tool C-4975-A, place the tool over the clamp bridge, tighten the tool nut until the jaws are completely closed (face-to-face).

⁂ WARNING

The seal must not be dimpled, stretched or out of shape. If necessary, equalize the seal pressure and shape it by hand.

    b. Position the boot onto the tri-pot housing retaining groove and install the retaining clamp evenly on the boot.
    c. Using the Crimper Tool C-4975-A, place the tool over the clamp bridge, tighten the tool nut until the jaws are completely closed (face-to-face).
18. If installing a low profile latching type boot clamp, position Snap-On® Clamp Locking Tool YA3050 prongs in the clamp holes and squeeze the tool until the upper clamp band is latched behind the 2 tabs on the lower clamp band.
19. Install the halfshaft.
20. Connect the negative battery cable.

Outer Joint

1. Before servicing the vehicle, refer to the precautions in the beginning of this section.
2. Remove the halfshaft.
3. Remove the clamps from the CV-joint boot and discard.
4. Remove the boot from the CV-joint housing and slide it down the interconnecting shaft.
5. Remove the outer CV-joint from the interconnecting shaft by sharply hitting it with a soft-faced hammer to drive it off the shaft.
6. Remove the circlip from the shaft.
7. Remove the CV-joint by sliding it off the shaft.

⁂ WARNING

If any parts show excessive wear, replace the halfshaft assembly; the component parts are not serviceable.

To install:

8. Install a new small boot clamp and slide it onto the shaft.
9. Install the boot and slide it onto the shaft.
10. Install the circlip.
11. Position the boot so that the raised bead on the inside the boot seal is in the shaft groove.
12. Install the halfshaft hub nut onto the joint threaded shaft so it is flush with the end.
13. Align the shaft splines and tap it onto the shaft with a soft-faced hammer so it locks on the circlip.
14. Distribute ½ of the grease in the service package inside the joint housing and the other ½ inside the boot.

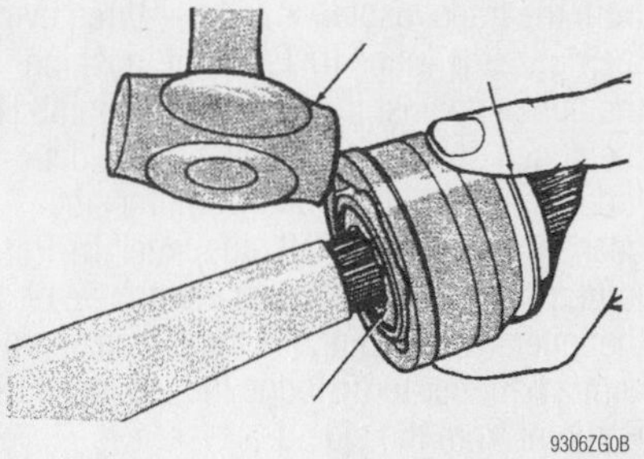

Remove the outer C/V joint from the interconnecting shaft

15. Install a new small boot clamp and position it evenly on the sealing boot.

⁂ CAUTION

Clamp the boot to the shaft using the Crimper Tool C-4975-A, place the tool over the clamp bridge, tighten the tool nut until the jaws are completely closed (face-to-face).

⁂ WARNING

The seal must not be dimpled, stretched or out of shape. If necessary, equalize the seal pressure and shape it by hand.

16. Position the boot onto the retaining groove and install the retaining clamp evenly on the boot.
17. Clamp the boot to the outer CV-joint housing using the Crimper Tool C-4975-A, place the tool over the clamp bridge, tighten the tool nut until the jaws are completely closed (face-to-face).
18. Install the halfshaft

STEERING AND SUSPENSION

Air Bag

⁂ CAUTION

Some vehicles are equipped with an air bag system. The system MUST BE disabled before performing service on or around system components, steering column, instrument panel components, wiring and sensors. Failure to follow safety and disabling procedures could result in accidental air bag deployment, possible personal injury and unnecessary system repairs.

PRECAUTIONS

Several precautions must be observed when handling the inflator module to avoid accidental deployment and possible personal injury:

1. Never carry the inflator module by the wires or connector on the underside of the module.
2. When carrying a live inflator module, hold securely with both hands, and ensure that the bag and trim cover are pointed away.
3. Place the inflator module on a bench or other surface with the bag and trim cover facing up.
4. With the inflator module on the bench, never place anything on or close to the module which may be thrown in the event of an accidental deployment.

DISARMING

Proper Supplemental Restraint System (SRS) disarming can be obtained by disconnecting and isolating the negative battery cable. Allow the air bag system capacitor at least 2 minutes to discharge before removing any air bag system components.

Rack and Pinion Steering Gear

REMOVAL & INSTALLATION

1. Before servicing the vehicle, refer to the precautions in the beginning of this section.
2. Place the steering wheel in the straight-ahead position. Lock the steering wheel in place, using a steering wheel holder.

➡**Locking the steering wheel keeps the clockspring in alignment position.**

3. Remove or disconnect the following:
   - Silencer pad from below the knee blocker panel
   - Knee blocker
   - Steering column coupling retainer pin, pinch bolt nut/bolt and separate the couplings
   - Both front wheels
   - Outer tie rod-to-steering knuckle nuts
   - Outer tie rod ends from the steering knuckle using remover Tool MB991113
   - Tie rod heat shield
   - Power steering fluid pressure switch wiring connector by releasing the locking tab
   - Power steering fluid pressure hose from the steering gear
   - Power steering fluid return hose from the steering gear, if not equipped with a power steering fluid cooler
   - Power steering fluid cooler hose from the steering gear, if equipped with a power steering fluid cooler
   - Power steering fluid return hose from the routing clip C-clamps, if not equipped with a power steering fluid cooler
   - Power steering fluid pressure hose from the steering gear's routing clips
   - Power steering cooler hose from the steering gear's right routing clip, if equipped
   - Both power steering cooler screws from the front suspension crossmember, if equipped, and move the cooler aside
   - Drive belt splash shield
   - Engine torque strut-to-front suspension crossmember bolt from the right forward corner of the crossmember
4. Matchmark the front suspension crossmember-to-chassis location.

✻✻ WARNING

If the front suspension crossmember-to-chassis location is not matchmarked, the front wheel alignment setting will be lost.

5. Place a transmission jack under the front crossmember and support it.
6. Remove the front suspension crossmember-to-frame rail bolts, one located at each side.
7. Loosen both rear suspension crossmember-to-frame rail bolts, one located at each side, until they release from the threaded tapping plates in the bolt.

✻✻ WARNING

Do not completely remove the rear bolts for they are designed to disengage from the body threads and will stay within the lower control arm rear isolator bushing.

➡**The threaded tapping plates allow the lower control arm to stay in place on the crossmember.**

8. Using the transmission jack, lower the front suspension crossmember enough to allow the power steering gear to be removed form the rear of the crossmember. Use the jack to support the crossmember's weight.
9. Remove or disconnect the following:
   - Lower steering column coupling-to-power steering gear pinion shaft's roll pin, using a roll pin punch
   - Lower steering column coupling from the power steering column pinion shaft
   - Pinion shaft dash cover seal from the tabs cast into the power steering gear housing
   - Power steering gear from the front suspension crossmember

To install:

10. Install or connect the following:
    - Power steering gear onto the front suspension crossmember and torque the bolts to 45 ft. lbs. (61 Nm)
    - Pinion shaft dash cover seal over the shaft and onto the power steering gear housing. Align the seal holes with the tabs cast into the power steering gear housing.
    - Lower steering column coupling by

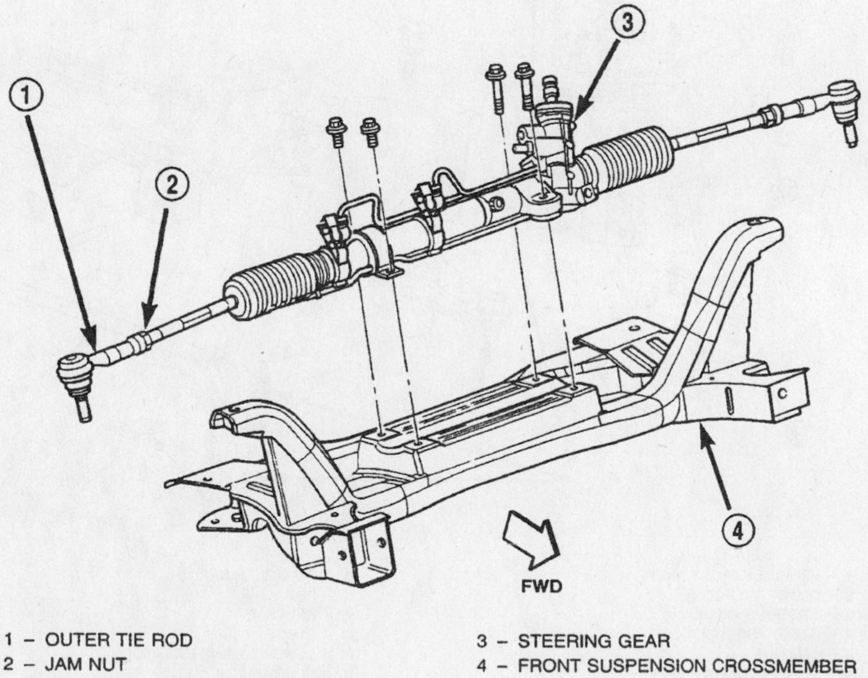

1 – OUTER TIE ROD
2 – JAM NUT
3 – STEERING GEAR
4 – FRONT SUSPENSION CROSSMEMBER

View of the power steering gear and crossmember

aligning the coupling and steering gear pinion shaft flats

- Lower steering column coupling-to-pinion shaft's roll pin until it is centered. Center the power steering gear rack's travel.
- Front suspension crossmember/power steering gear assembly by raising it with the jack until is aligns with its matchmarks
- Lower steering column coupling, guide it through the dash panel hole as it is raised
- Both rear crossmember-to-tapping plate bolts
- Both front crossmember-to-frame rail bolts and torque the 4 bolts to 20 inch lbs. (2 Nm)

⁂ WARNING

Be sure to align the front suspension crossmember-to-chassis matchmarks; otherwise, the front wheel alignment setting will be lost.

11. Once aligned, torque both rear crossmember-to-rear lower control arm bolts to 185 ft. lbs. (250 Nm) and both front crossmember bolts to 113 ft. lbs. (153 Nm).
12. Install the engine torque strut to the right forward corner of the front suspension crossmember.
13. Adjust the engine torque strut by performing the following procedure:
    a. Loosen the upper torque strut at the shock tower bracket.
    b. Position a floor jack on the forward edge of the bell housing to prevent the least amount of upward lifting of the engine.
    c. Slowly, lift the assembly, allowing the engine to rotate rearward so the distance between center of the engine mount bracket's rearmost attaching the and the center of the shock tower bracket's washer hose clip hole is 4.70 in. (119mm).
    d. Torque the upper and lower torque strut bolts to 87 ft. lbs. (118 Nm).
    e. Remove the floor jack.
14. Install or connect the following:
    - Drive belt splash shield
    - Power steering hose-to-power steering gear using a new O-ring lubricated with power steering oil, if not equipped with a power steering cooler
    - Power steering fluid cooler line-to-power steering gear, if equipped with a power steering cooler
    - Power steering fluid return hose to the routing clip C-clamps
    - Tie rod ends to the steering knuckle. Torque the nut, using a crowsfoot wrench, to 40 ft. lbs. (55 Nm), while holding the tie rod stationary.
    - Both front wheels
    - Dash-to-lower coupling seal over the lower coupling's plastic collar

➡Verify that the seal's lip shows grease at the coupling's plastic collar contact.

- Steering column lower coupling-to-steering column upper coupling pinch bolt and torque the nut to 21 ft. lbs. (28 Nm)
- Pinch bolt retainer pin
- Knee blocker and silencer pad

15. Remove the steering wheel holder.
16. Fill and bleed the power steering system.
17. Start the vehicle and check for leaks, repair if necessary.
18. Check and/or adjust the front toe setting.

Strut

REMOVAL & INSTALLATION

Front

1. Before servicing the vehicle, refer to the precautions in the beginning of this section.

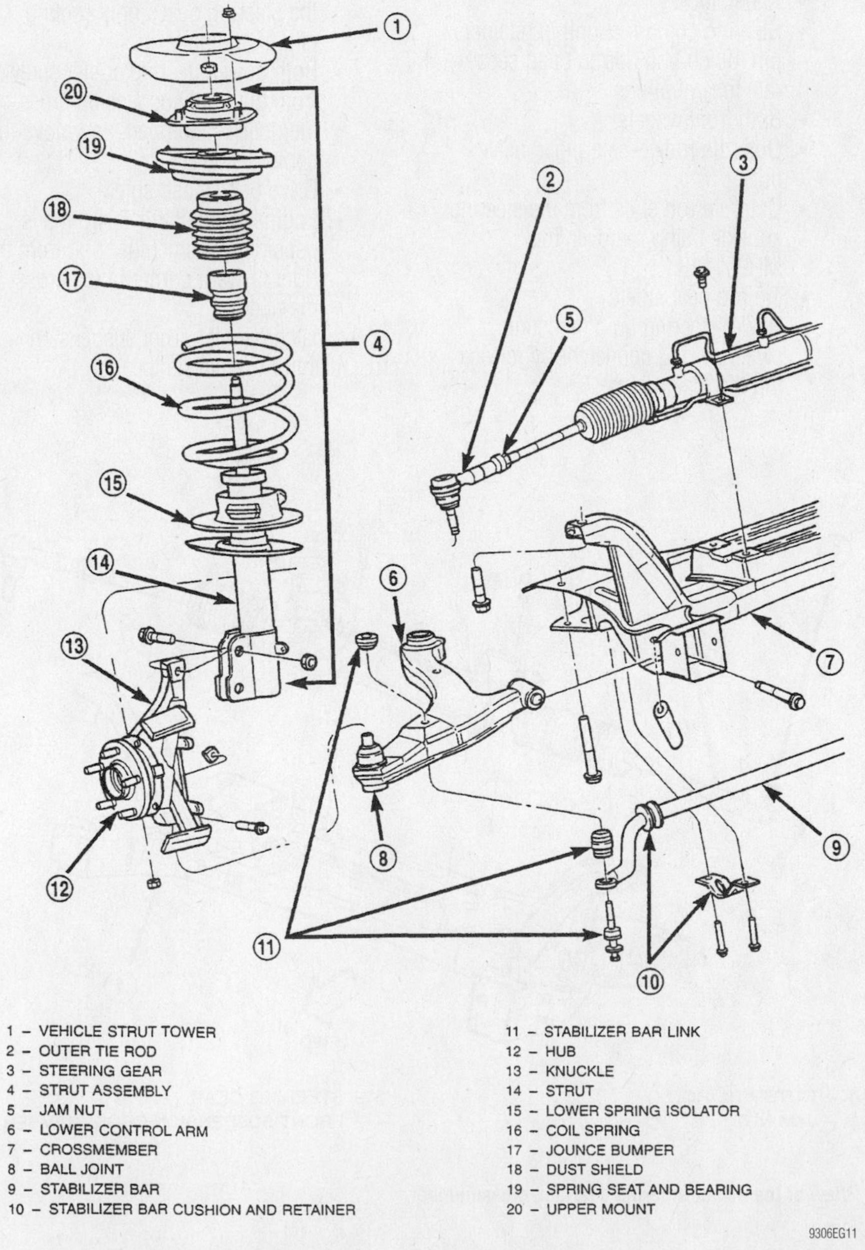

Exploded view of the front suspension

2. Install or connect the following:
   - Negative battery cable
   - Front wheels
   - Mark each one right or left, as applicable, if both struts are being removed
   - Ground strap from the rear of the strut
   - Anti-lock Brake System (ABS) wheel speed sensor from the strut, if equipped

⁂ WARNING

The steering knuckle-to-strut assembly attaching bolts are serrated and must not be turned during removal.

   - Steering knuckle nuts while holding the bolts stationary
   - 3 upper strut mount-to-strut tower nuts
   - Strut assembly

To install:

3. Install or connect the following:
   - Strut assembly into the strut tower by aligning the 3 upper strut mount studs with the shock tower holes. Torque the 3 upper strut mount nut/washer assemblies to 25 ft. lbs. (34 Nm)

⁂ WARNING

The steering knuckle-to-strut assembly attaching bolts are serrated and must not be turned during installation.

   - Steering knuckle nuts while holding the bolts stationary
   - Steering knuckle arm and position it into the strut assembly by aligning the strut assembly-to-steering knuckle holes
   - Both strut-to-steering knuckle bolts. Torque both bolts to 40 ft. lbs. (53 Nm), plus an additional 90 degrees after the specified torque is met.

➡The bolts should be installed with the nuts facing the front of the vehicle.

   - ABS wheel sensor to the rear of the strut and torque the screw to 120 inch lbs. (13 Nm)
   - Ground strap to the rear of the strut and torque the screw to 120 inch lbs. (13 Nm)
   - Front wheel
   - Negative battery cable

Shock Absorber

REMOVAL & INSTALLATION

Rear

1. Before servicing the vehicle, refer to the precautions in the beginning of this section.
2. Remove the rear wheel and position a transmission jack under the center of the axle. Raise the jack enough to support the axle.
3. Remove or disconnect the following:
   - Shock absorber lower mounting bolt from the axle
   - Upper mounting bolt
   - Shock absorber

To install:

4. Install or connect the following:
   - Shock absorber eye to the body bracket. Hand tighten the upper mounting bolt.
   - Lower the jack and install the lower mounting bolt through the axle flange and shock absorber. Torque the bolt to 50 ft. lbs. (68 Nm). Torque the upper mounting bolt to 73 ft. lbs. (99 Nm).
   - Rear wheel
   - Negative battery cable

Coil Spring

REMOVAL & INSTALLATION

Front

1. Before servicing the vehicle, refer to the precautions in the beginning of this section.
2. Remove the wheel.
3. Remove the strut assembly and place it in a Strut Spring Compressor Tool W-7200.
4. Set the lower hooks then the upper hooks. Position the clevis bracket straight outward away from the compressor tool
5. Install a clamp on the lower end of the coil spring to secure the strut when the nut is removed.

➡Do not remove the strut shaft nut until the coil spring is compressed. The coil spring is under pressure and must be compressed before the shaft nut is removed.

6. Compress the coil spring until all tension is removed from the upper mount.
7. Install a Strut Nut Socket Tool 6864 once the spring is compressed.
8. Install a socket on the hex end of the strut shaft and remove the nut.
9. Remove or disconnect the following:
   - Upper mount from the strut shaft

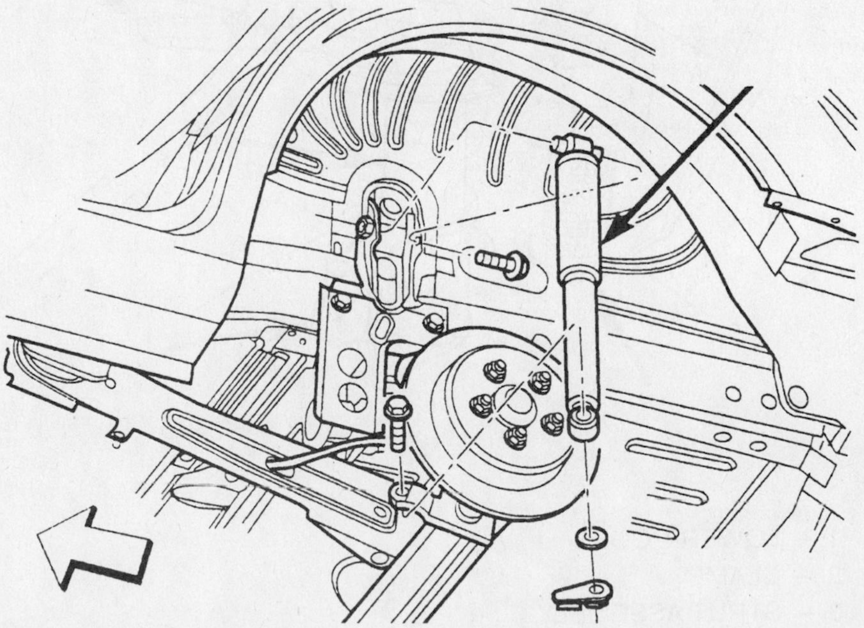

Shock absorber mounting bolts

- Upper spring seat, bearing and upper isolator as an assembly
- Dust shield and jounce bumper
- Clamp from the bottom of the coil spring
- Strut through the bottom of the coil
- Release the tension from the coil spring by backing off the compressor drive completely
- Coil spring

10. Inspect the coil spring for any signs of damage

To install:

11. Install or connect the following:

- Coil spring in the compressor. Rotate the spring so that the end of the top coil is directly in the front.
- Slowly compress the coil until enough room is available to install the strut
- Lower spring isolator on the lower spring seat
- Strut through the bottom of the of coil spring until the lower spring seat contacts the lower end of the coil spring. Rotate the strut until the clevis bracket is positioned straight outward away from the compressor.
- Clamp on the lower end of the coil spring and strut
- Jounce bumper on the strut shaft with the smaller end pointing downward
- Dust shield until the bottom of the shield snaps on to the retainer
- Upper spring isolator
- Upper spring seat and bearing on top of the coil spring. Position the notch formed into the edge of the upper seat straight out away from the compressor.
- Strut upper mount over the strut shaft and onto the top of the upper spring seat and bearing. Position the mount so that the third mounting stud is inward toward the compressor.
- Retaining nut on the strut shaft, loosely

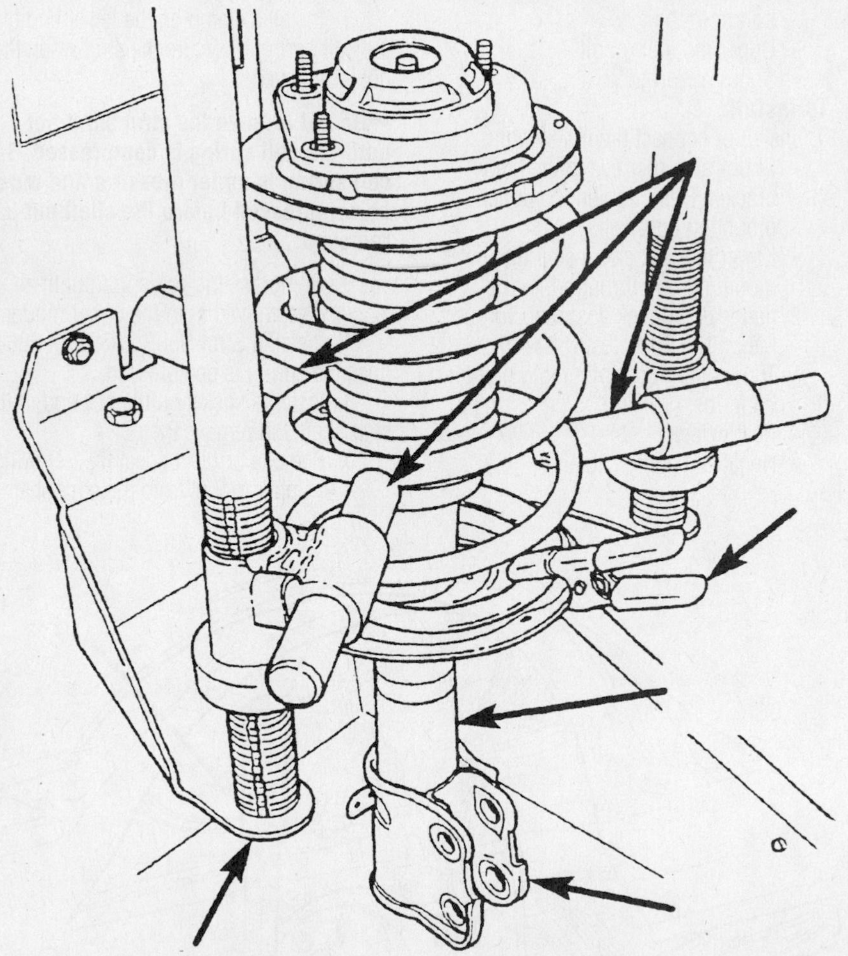

1 – LOWER HOOKS
2 – CLAMP
3 – STRUT ASSEMBLY
4 – CLEVIS BRACKET
5 – SPRING COMPRESSOR

9306ZG0D

Coil spring mounted in the coil spring compressor tool

12. Install the strut nut socket on the strut shaft retaining nut. Install a socket on the hex end of the shaft. Secure the strut shaft and torque the nut to 55 ft. lbs. (75 Nm).
13. Slowly release the tension from the coil spring by backing off the compressor completely.
14. Remove the clamp from the bottom of the coil spring and strut. Push back the spring compressor upper and lower hooks and remove the strut from the compressor.
15. Install the strut assembly to the vehicle.

Rear

1. Before servicing the vehicle, refer to the precautions in the beginning of this section.
2. Remove or disconnect the following:
   - Both rear wheels
   - Watts link bell crank from the center of the axle
   - Sway bar cushion retainers
   - Sway bar from the rear axle and place a jack under the rear axle
   - Shock absorber and lower the jack
   - Coil springs and rubber isolators

To install:

3. Install or connect the following:
   - Rubber isolator on each end of the coil spring and wrap the finger around the coil
   - Coil springs on top of the axle spring perches and make certain that the upper coils end near the

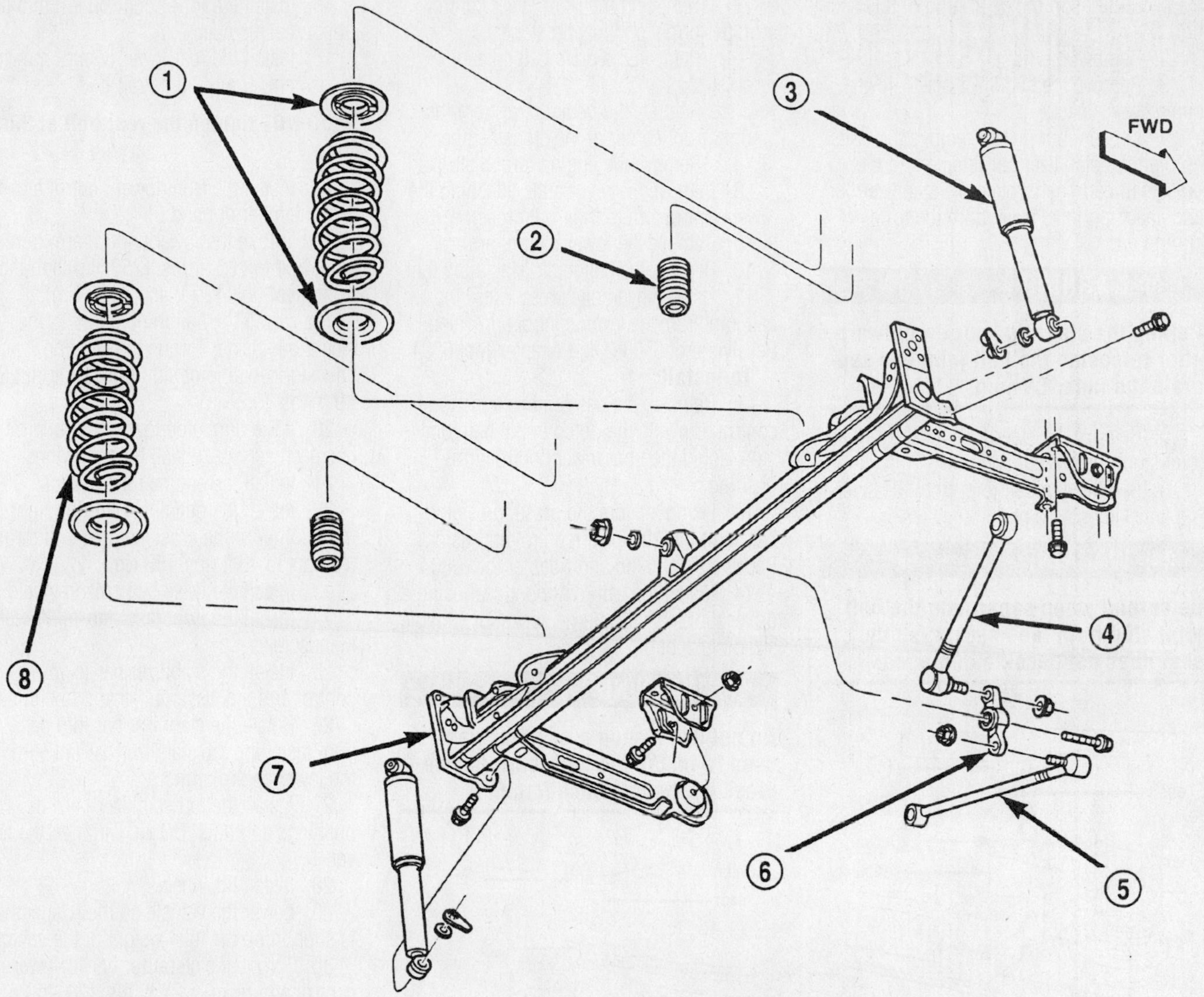

1 – ISOLATORS
2 – JOUNCE BUMPER
3 – SHOCK ABSORBER
4 – WATTS LINK (UPPER)
5 – WATTS LINK (LOWER)
6 – BELL CRANK
7 – AXLE
8 – COIL SPRING

Exploded view of the rear suspension

outboard sides of the vehicle and not at 180 degrees of that location

- Coil springs into the spring mounting brackets
- Washer and nut on the lower mounting bolts and torque the bolts to 50 ft. lbs. (68 Nm)
- Lower end of the sway bar retainers in the slots at the back of the axle
- Mounting bolt through the cushion retainer and torque to 40 ft. lbs. (54 Nm)
- Watts link bell crank to the center of the axle and torque the bolts to 90 ft. lbs. (122 Nm)
- Rear wheels

Lower Ball Joint

REMOVAL & INSTALLATION

The front suspension ball joints operate with no free-play. The ball joints are replaceable ONLY as an assembly. Do not attempt any type of repair on the ball joint assembly. The ball joint is a press fit into the lower control arm with the joint stud retained in the steering knuckle by the clamp bolt. To check the ball joint, with the weight of the vehicle resting on the road wheels, grasp the grease fitting and without using any tools, attempt to move the grease fitting. If the ball joint is worn the grease fitting will move easily. If movement is noted, replacement of the ball joint is recommended.

1. Before servicing the vehicle, refer to

the precautions in the beginning of this section.

2. Remove the wheel.

3. Remove the stabilizer bar-to-lower control arm links.

4. Loosen, but do not remove the bolts holding the stabilizer bar retainers to the crossmember. Then, rotate the stabilizer bar and attaching links away from the lower control arms.

✲✲ WARNING

Pulling the steering knuckle outward after releasing the ball joint can separate the inner CV-joint.

5. Remove the steering knuckle-to-ball joint stud's pinch bolt and nut.

6. Remove the ball joint from the steering knuckle using a prybar.

✲✲ WARNING

Be careful when separating the ball joint stud from the knuckle, so the seal does not become damaged.

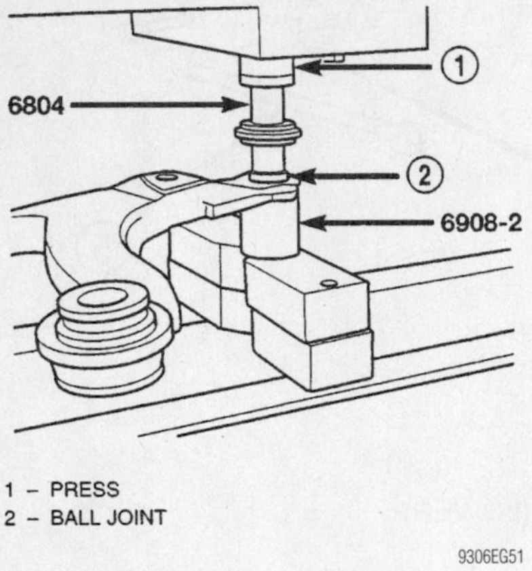

Removing the ball joint from the control arm

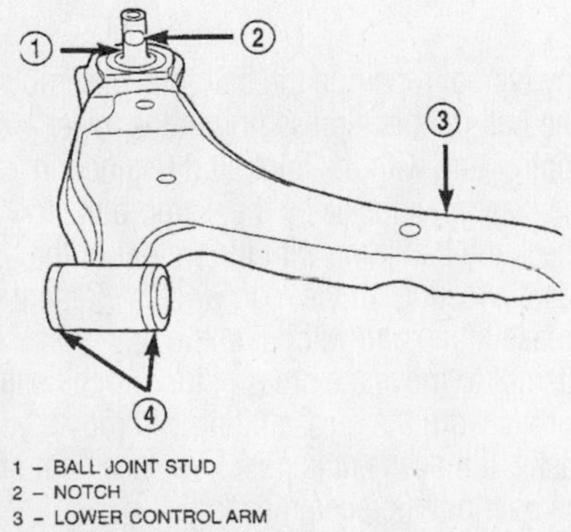

Aligning the ball joint stud notch to the control arm

7. If removing the right lower control arm, perform the following steps:

a. Remove the drive belt splash shield.

b. Remove the pencil strut from the right front corner of the crossmember

c. Remove the engine torque strut.

8. Remove the pivot bolts attaching the lower control arm to the front crossmember.

9. Remove the lower control arm.

10. Remove the ball joint using a pry tool.

11. Using a hydraulic press, press the ball joint from the lower control arm using Receiver tool 6908-2 and Adapter Tool 6804.

To install:

12. Reinstall the ball joint into the lower control arm with the notch in the ball joint stud facing the front lower control arm bushing.

13. Using a hydraulic press, press the ball joint into the lower control arm using Receiver Tool 6758 and Adapter tool 6804.

14. Install the ball joint boot seal using a driver tool such as a large socket or suitable sized piece of pipe.

✲✲ WARNING

Do not use a shop press that was used to install the ball joint, for the press exerts too much force.

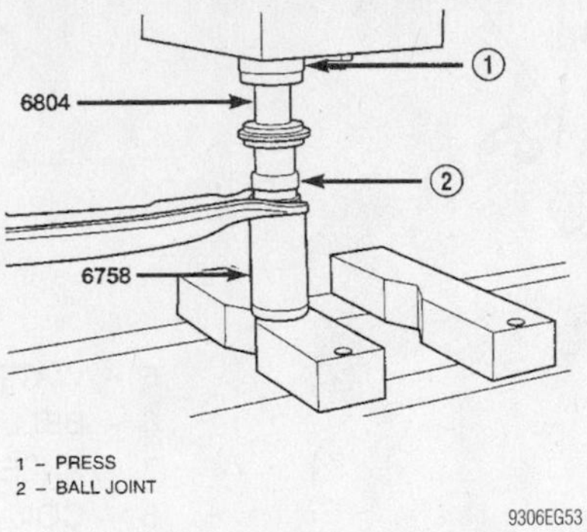

Installing the ball joint to the control arm

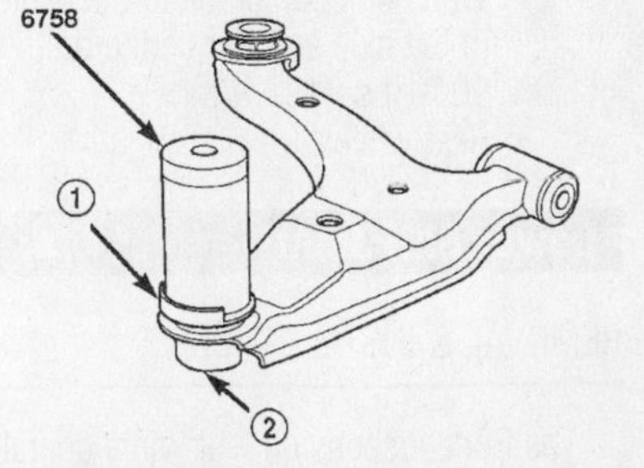

Installing the ball joint boot seal

15. Install the lower control arm into the front crossmember.

16. Install the rear lower control arm-to-crossmember and frame rail bolt.

➡DO NOT tighten the rear bolt at this time.

17. Install the front lower control arm-to-crossmember nut and bolt.

18. Torque the lower control arm to rear pivot bolt to 185 ft. lbs. (250 Nm) and the front pivot bolt 120 ft. lbs. (163 Nm).

19. Install the ball joint stud into the steering knuckle. Torque the steering knuckle-to-ball joint stud pinch bolt/nut to 70 ft. lbs. (95 Nm).

20. If the right side lower control arm has been service, install the following:

21. Install the engine torque strut.

22. Install the pencil strut to the right front corner of the crossmember and torque the nuts to 43 ft. lbs. (58 Nm).

23. Install the drive belt splash shield.

24. Install the front fascia-to-reinforcement screws.

25. Install the stabilizer bar-to-lower control arm link assemblies and bushings.

26. Rotate the stabilizer bar into position, installing the stabilizer bar links into the lower control arms.

27. Install the top stabilizer bar link bushings and nuts. DO NOT tighten the link yet.

28. Install the wheel.

29. Lower the vehicle so the suspension is supporting the total weight of the vehicle.

30. Torque the stabilizer bar-to-lower control arm links to 21 ft. lbs. (28 Nm).

31. Torque the stabilizer bar bushing retainer-to-crossmember bolts to 21 ft. lbs. (28 Nm).

32. Check and/or adjust the toe, as necessary.

Wheel Bearings

ADJUSTMENT

The PT Cruiser is equipped with a sealed hub and bearing assemblies. The hub and bearing assembly is non-serviceable. If the assembly is damaged, the complete unit must be replaced.

REMOVAL & INSTALLATION

Front

1. Before servicing the vehicle, refer to the precautions in the beginning of this section.

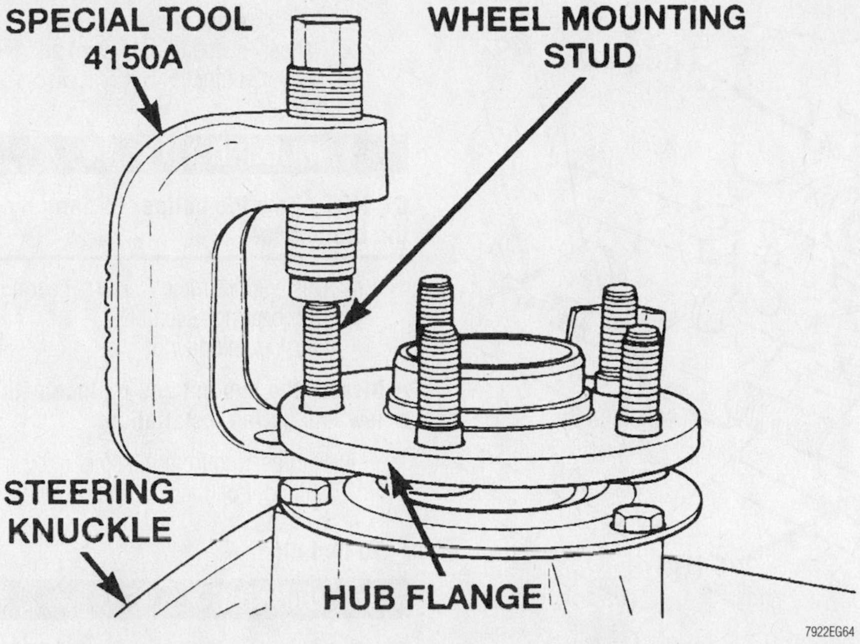

Use a proper C-clamp and adapter tool to press out one of the lug studs

2. Remove the steering knuckle and hub and bearing assembly.

3. Remove the wheel lug stud from the hub flange, using a C-clamp and Adapter Tool 4150A.

4. Rotate the hub to align the removed lug stud with the notch in the bearing retainer plate.

5. Rotate the hub so the stud hole is facing away from the brake caliper's lower rail on the steering knuckle.

6. Install ½ of a Bearing Splitter Tool 1130, between the hub and the bearing retainer plate. The threaded hole in this ½ is to be aligned with the caliper rail on the steering knuckle.

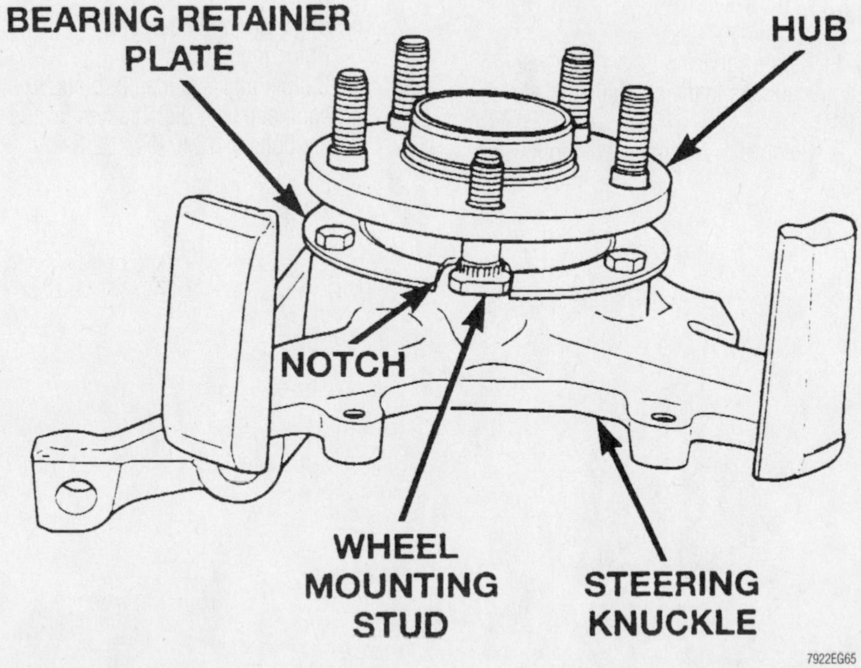

Rotate the hub in order to remove the lug stud

7. Install the remaining pieces of the bearing splitter on the steering knuckle. Hand-tighten the nuts to hold the splitter in place on the knuckle.

8. When the bearing splitter is installed, be sure the 3 bolts attaching the bearing retainer plate to the knuckle are contacting the bearing splitter. The bearing retainer plate should not support the knuckle or contact the splitter.

9. Place the steering knuckle in a hydraulic press, supported by the bearing splitter.

10. Position a driver on the small end of the hub. Using the press, remove the hub from the wheel bearing. The outer bearing race will come out of the wheel bearing when the hub is pressed out of the bearing.

11. Remove the bearing splitter tool from the knuckle.

12. Remove the 3 bolts mounting the bearing retainer plate to the steering knuckle.

13. Place the knuckle in a press supported by the press block. The blocks must not obstruct the bore in the steering knuckle so the wheel bearing can be pressed out of the knuckle. Place a driver on the outer race of the wheel bearing, then press the bearing out of the knuckle.

14. Install the bearing splitter on the hub. The splitter is to be installed on the hub so it is between the flange of the hub and the bearing race on the hub. Place the hub, bearing race and splitter in a press. Use a driver to press the hub out of the bearing race.

To install:

15. Use clean, dry cloth to wipe and grease or dirt from the bore of the steering knuckle.

16. Install a new wheel bearing into the bore of the steering knuckle. Be sure the bearing is placed squarely into the bore. Place the knuckle in a press with a Receiver Tool, C-4698-2 supporting the steering knuckle. Place a driver tool on the outer race of the wheel bearing. Press the wheel bearing into the steering knuckle until it is fully bottomed in the bore of the steering knuckle.

➡Only the original or original equipment replacement bolts should be used to mounting the bearing retainer to the knuckle. If a bolt requires replacement when installing the bearing retainer plate, be sure to get the proper type of replacement.

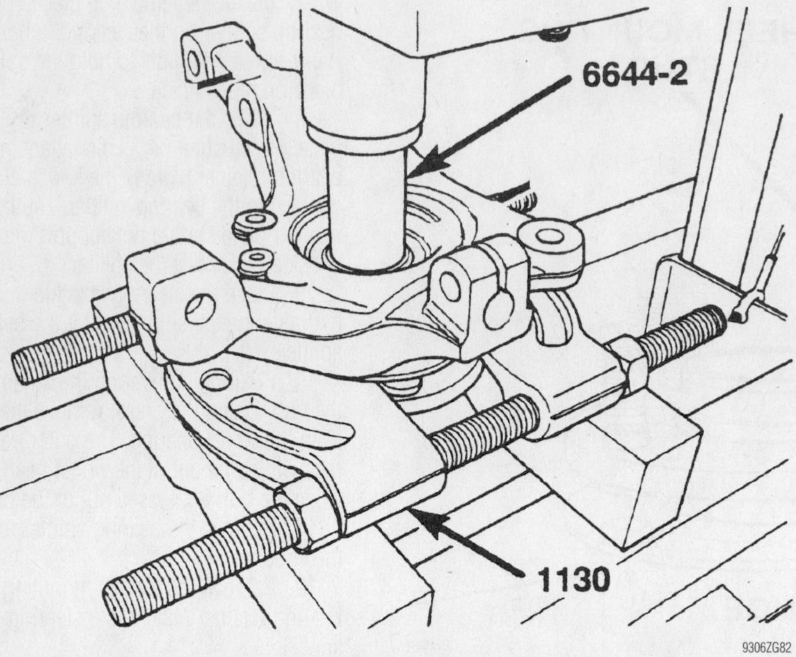

Properly support the steering knuckle for hub and bearing removal

17. Install the bearing retainer plate on the steering knuckle. Install the 3 bearing retainer mounting bolts. Tighten the bolts to 21 ft. lbs. (28 Nm).

18. Install the wheel lug stud into the hub flange.

19. Place the hub with the lug stud installed, in a press supported by an Adapter Tool C-4698-1 and press the wheel lug stud into the hub flange until it is fully seated against the back side on the hub flange.

20. Place the steering knuckle with the wheel bearing installed, in a press with special Receiver Tool MB-990799 supporting the inner race of the wheel bearing. Place the hub in the wheel bearing, making sure it is square with the bearing. Press the hub into the wheel bearing until it is fully bottomed in the wheel bearing.

21. Install the steering knuckle and the wheel.

22. Check and/or adjust the front alignment.

Rear

1. Before servicing the vehicle, refer to the precautions in the beginning of this section.

2. Remove or disconnect the following:

- Wheel
- Rear brake drum, if equipped
- Caliper (suspend on a wire) and the rotor, if equipped with rear disc brakes

✻✻ WARNING

DO NOT allow the caliper to hang by the brake hose.

- Dust cap from the rear hub/bearing
- Hub/bearing assembly-to-knuckle/spindle nut

➡Discard the hub nut and replace with a new one during installation.

- Hub/bearing from the spindle by pulling it off the end of the spindle by hand

To install:

✻✻ WARNING

The hub/bearing nut must be tightened to, but NOT over, its specified torque value. The proper specification is crucial to the life of the hub bearing.

3. Position the hub/bearing assembly on the rear spindle/knuckle. Install a NEW hub nut and tighten to 160 ft. lbs. (217 Nm)

4. Install or connect the following:

- Dust cap and seat it using a soft face hammer to carefully tap it into place
- Brake drum, if equipped
- Rotor, if equipped
- Caliper and 2 guide pin bolts, if equipped with disc brakes. Torque the bolts to 16 ft. lbs. (22 Nm).

FORD MOTOR CO.

11

1998–01

Ford-Contour • **Mercury**-Mystique • 1999–01 Cougar

PRECAUTIONS

Before servicing any vehicle, please be sure to read all of the following precautions, which deal with personal safety, prevention of component damage, and important points to take into consideration when servicing a motor vehicle:

- Never open, service or drain the radiator or cooling system when the engine is hot; serious burns can occur from the steam and hot coolant.
- Observe all applicable safety precautions when working around fuel. Whenever servicing the fuel system, always work in a well-ventilated area. Do not allow fuel spray or vapors to come in contact with a spark, open flame, or excessive heat (a hot drop light, for example). Keep a dry chemical fire extinguisher near the work area. Always keep fuel in a container specifically designed for fuel storage; also, always properly seal fuel containers to avoid the possibility of fire or explosion. Refer to the additional fuel system precautions later in this section.
- Fuel injection systems often remain pressurized, even after the engine has been turned **OFF**. The fuel system pressure must be relieved before disconnecting any fuel lines. Failure to do so may result in fire and/or personal injury.
- Brake fluid often contains polyglycol ethers and polyglycols. Avoid contact with the eyes and wash your hands thoroughly after handling brake fluid. If you do get brake fluid in your eyes, flush your eyes with clean, running water for 15 minutes. If eye irritation persists, or if you have taken brake fluid internally, IMMEDIATELY seek medical assistance.
- The EPA warns that prolonged contact with used engine oil may cause a number of skin disorders, including cancer. You should make every effort to minimize your exposure to used engine oil. Protective gloves should be worn when changing oil. Wash your hands and any other exposed skin areas as soon as possible after exposure to used engine oil. Soap and water, or waterless hand cleaner should be used.
- All new vehicles are now equipped with an air bag system, often referred to as a Supplemental Restraint System (SRS) or Supplemental Inflatable Restraint (SIR) system. The system must be disabled before performing service on or around system components, steering column, instrument panel components, wiring and sensors. Failure to follow safety and disabling procedures could result in accidental air bag deployment, possible personal injury and unnecessary system repairs.
- Always wear safety goggles when working with, or around, the air bag system. When carrying a non-deployed air bag, be sure the bag and trim cover are pointed away from your body. When placing a non-deployed air bag on a work surface, always face the bag and trim cover upward, away from the surface. This will reduce the motion of the module if it is accidentally deployed. Refer to the additional air bag system precautions later in this section.
- Clean, high quality brake fluid from a sealed container is essential to the safe and proper operation of the brake system. You should always buy the correct type of brake fluid for your vehicle. If the brake fluid becomes contaminated, completely flush the system with new fluid. Never reuse any brake fluid. Any brake fluid that is removed from the system should be discarded. Also, do not allow any brake fluid to come in contact with a painted surface; it will damage the paint.
- Never operate the engine without the proper amount and type of engine oil; doing so WILL result in severe engine damage.
- Timing belt maintenance is extremely important. Many models utilize an interference-type, non-freewheeling engine. If the timing belt breaks, the valves in the cylinder head may strike the pistons, causing potentially serious (also time-consuming and expensive) engine damage. Refer to the maintenance interval charts in the front of this manual for the recommended replacement interval for the timing belt, and to the timing belt section for belt replacement and inspection.
- Disconnecting the negative battery cable on some vehicles may interfere with the functions of the on-board computer system(s) and may require the computer to undergo a relearning process once the negative battery cable is reconnected.
- When servicing drum brakes, only disassemble and assemble one side at a time, leaving the remaining side intact for reference.
- Only an MVAC-trained, EPA-certified automotive technician should service the air conditioning system or its components.

ENGINE REPAIR

➡Disconnecting the negative battery cable on some vehicles may interfere with the operation of the on-board computer system. The computer may undergo a relearning process once the negative battery cable is reconnected.

Alternator

REMOVAL

2.0L Engine

1. Before servicing the vehicle, refer to the precautions in the beginning of this section.
2. Remove or disconnect the following:
   - Negative battery cable
   - Air intake resonator
   - Alternator electrical connections
   - Ground cable
   - Right front wheel
   - Right splash shield
   - Radiator splash shield
   - Accessory drive belt
   - Alternator. Remove the upper mounting bolt last.

➡There is a limited amount of space in which to remove the alternator. Do not damage any hoses or cables.

2.5L Engine

1998 MODELS

1. Before servicing the vehicle, refer to the precautions in the beginning of this section.
2. Remove or disconnect the following:
   - Negative battery cable
   - Accessory drive belt
   - Right halfshaft
   - Exhaust system Y pipe
   - Right splash shield
   - Alternator wiring connectors
   - Alternator rear support bracket
   - Alternator

1999–01 MODELS

1. Before servicing the vehicle, refer to the precautions in the beginning of this section.
2. Remove or disconnect the following:
   - Negative battery cable
   - Right front wheel
   - Right splash shield

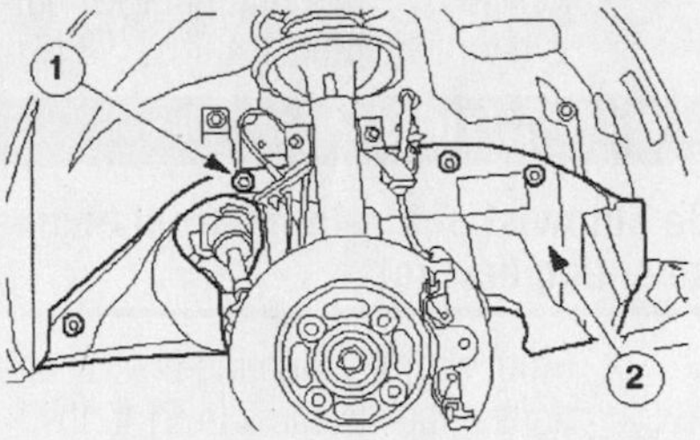

Remove the splash shield retainers (1) and the splash shield (2)—2.5L engines

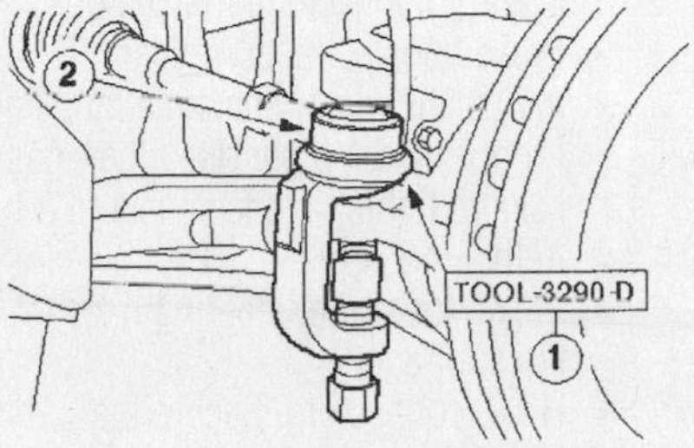

Separate the outer tie rod end from the steering knuckle—2.5L engines

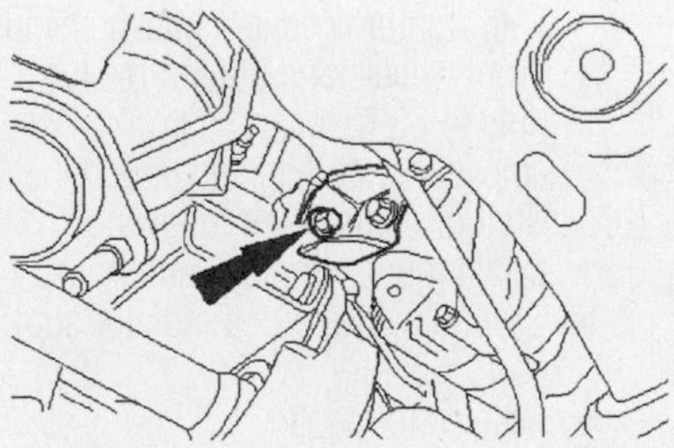

Alternator rear support bracket—2.5L engines

- Accessory drive belt
- Right outer tie rod end
- Alternator wiring connectors
- Alternator rear support bracket
- Alternator

INSTALLATION

2.0L Engine

Install or connect the following:

- Alternator and tighten the mounting bolts to 33 ft. lbs. (45 Nm)
- Accessory drive belt
- Radiator splash shield
- Right splash shield
- Right front wheel and tighten the lug nuts to 63 ft. lbs. (86 Nm)
- Ground cable
- Alternator electrical connections
- Air intake resonator
- Negative battery cable

2.5L Engine

1998 MODELS

Install or connect the following:

- Alternator and tighten the mounting bolts to 33 ft. lbs. (45 Nm)
- Alternator rear support bracket and tighten the bracket bolts to 18 ft. lbs. (25 Nm)
- Alternator wiring connectors
- Right splash shield
- Exhaust system Y pipe
- Right halfshaft
- Accessory drive belt
- Negative battery cable

1999–01 MODELS

Install or connect the following:

- Alternator and tighten the mounting bolts to 33 ft. lbs. (45 Nm)
- Alternator rear support bracket and tighten the bracket bolts to 18 ft. lbs. (25 Nm)
- Alternator wiring connectors
- Right outer tie rod end and tighten the nut to 21 ft. lbs. (28 Nm)
- Accessory drive belt
- Right splash shield
- Right front wheel and tighten the lug nuts to 63 ft. lbs. (86 Nm)
- Negative battery cable

Ignition Timing

ADJUSTMENT

The ignition timing is set at 10 degrees Before Top Dead Center (BTDC) and is not adjustable.

Engine Assembly

REMOVAL & INSTALLATION

2.0L Engine

➡The engine and transaxle are removed as an assembly.

1. Before servicing the vehicle, refer to the precautions in the beginning of this section.
2. Drain the engine oil and coolant.
3. Drain the transaxle fluid.
4. Attach an engine support fixture to the engine lifting eyes.
5. Secure the A/C condenser to the radiator support with safety wire.
6. Loosen the suspension strut lock nuts five turns.
7. Remove or disconnect the following:
   - Battery
   - Mass Air Flow (MAF) sensor connector
   - Intake Air Temperature (IAT) sensor connector
   - Air cleaner outlet tube
   - Air cleaner
   - Exhaust Gas Recirculation (EGR) vacuum supply hose, if equipped
   - EGR pressure sensor wiring, if equipped
   - EGR exhaust tube, if equipped
   - Engine wiring harness connector
   - Accelerator cable
   - Upper drive belt cover
   - Front wheels
   - Left, right and center splash shields
   - Accessory drive belt
   - Stabilizer bar links
   - Outer tie rod ends
   - Lower ball joints
   - Lower radiator retainers
   - Both Heated Oxygen (HO_2S) sensor connectors
   - Power steering hoses
   - Brake booster vacuum line at intake manifold
   - Steering gear and heat shield
   - Right support insulator and bracket
   - Front exhaust pipe
   - Left support insulator
   - A/C accumulator attachment bolts
   - 4 subframe bolts and the subframe
8. If equipped with a manual transaxle, place the shift lever in neutral and lock it in place with special tool Gear Lever Aligner T97P-7025-A.
   - Clutch hydraulic line
   - Shift cables
   - Reverse lamp switch connector
   - Vehicle Speed (VSS) sensor connector
9. If equipped with an automatic transaxle, remove or disconnect the following:
   - Torque converter
   - Transaxle control wiring connector
   - Range switch connector

- Ground strap
- Shift cable
- VSS and bracket
- Transaxle cooler lines and fluid tube
- Turbine Shaft Speed (TSS) sensor connector

10. For all vehicles, remove or disconnect the following:
    - Axle halfshafts
    - Upper and lower radiator hoses
    - 2 ground cables from radiator support
    - Coolant hose bracket and power steering hose
    - Coolant hoses at the recovery tank
    - Radiator and cooling fan
    - A/C compressor
    - Power steering pump
    - Heater hoses and bracket
    - Powertrain Control Module (PCM) connector
    - Vacuum hoses at the intake manifold
    - Battery cables at the battery tray
    - Fuel supply and return lines
11. Support the engine and transaxle assembly from below.
12. Remove or disconnect the following:
    - Top support fixture
    - Coolant recovery tank
    - Front and rear engine mount brackets
13. Raise the vehicle away from the engine and transaxle assembly
14. Support the engine with a shop hoist.
15. Support the transaxle with a transmission jack.
16. Remove or disconnect the following:
    - Starter
    - Ground cable
    - Transaxle flange bolts
    - Engine from the transaxle

To install:

➡Replace all snaprings, split pins and self-locking nuts.

17. Install or connect the following:
    - Engine to the transaxle and tighten the flange bolts to 35 ft. lbs. (48 Nm)
    - Ground cable
    - Starter motor and tighten the bolts to 35 ft. lbs. (48 Nm)
18. Lower the vehicle onto the engine and transaxle assembly.
19. Install or connect the following:
    - Front and rear engine mount brackets and tighten the nuts and bolts finger tight
    - Top support fixture
    - Fuel supply and return lines
    - Battery cables to the battery tray. Secure with cable ties.
    - Vacuum hoses to the intake manifold
    - PCM connector
    - Heater hoses and bracket
    - Power steering pump and tighten the bolts to 35 ft. lbs. (47 Nm)
    - A/C compressor and tighten the bolts to 18 ft. lbs. (25 Nm)
    - Radiator and cooling fan
    - Power steering hose and coolant hose bracket
    - 2 ground cables to the radiator support
    - Upper and lower radiator hoses
    - Axle halfshafts
20. If equipped with a manual transaxle, install or connect the following:
    - Clutch hydraulic line
    - Shift cables
    - VSS connector
    - Reverse lamp switch connector
21. Remove the Gear Lever Aligner.
22. If equipped with an automatic transaxle, install or connect the following:
    - TSS connector
    - Torque converter and tighten the nuts to 27 ft. lbs. (36 Nm)
    - Transaxle cooler lines and fluid tube
    - VSS and bracket
    - Shift cable
    - Ground strap
    - Range switch connector
    - Transaxle control wiring connector
    - Powertrain Alignment Gauge T94P-6000-AH to the subframe
    - Lower radiator retainers to the subframe
    - Subframe and attach the front bolts. Do not tighten the bolts at this time.
    - Steering gear to the subframe and tighten the bolts to 96 ft. lbs. (130 Nm)
    - Steering gear heat shield
    - Power steering line and coolant line brackets
    - Subframe Alignment Pin Set T94P-2100-AH
    - Subframe bolts and tighten the bolts in a diagonal pattern to 96 ft. lbs. (130 Nm)
    - Center bolt to the Powertrain Alignment Tool and tighten it to 22 ft. lbs. (30 Nm)
    - Right engine support insulator
23. Tighten the support insulator center bolt to 88 ft. lbs. (120 Nm) and the restrictor mounting bolts to 35 ft. lbs. (48 Nm).
    - Rear engine mounting bracket nut and tighten it to 61 ft. lbs. (83 Nm)

⁂ CAUTION

Do not twist or strain the front engine mounting bracket.

- Front engine mounting bracket and tighten the fasteners to 61 ft. lbs. (83 Nm)

24. Remove the Powertrain Alignment Tool.
25. Install the left engine support insulator. Tighten the center bolt to 88 ft. lbs. (120 Nm) and the restrictor mounting bolts to 35 ft. lbs. (48 Nm).
26. Install or connect the following:
    - Brake booster vacuum hose
    - A/C accumulator attachment bolts
    - Front exhaust pipe. Use a new gasket and tighten the fasteners to 33 ft. lbs. (45 Nm)
    - Attach the power steering lines to the subframe
    - HO_2S connectors. Secure the wiring harness with cable ties
    - Lower ball joints and tighten the nuts to 61 ft. lbs. (83 Nm)
    - Outer tie rod ends and tighten the nuts to 19 ft. lbs. (26 Nm)
    - Stabilizer bar links and tighten the fasteners to 35 ft. lbs. (47 Nm)
    - Left, right and center splash shields
    - Front wheels and tighten the lug nuts to 63 ft. lbs. (86 Nm)
27. Remove the top engine support.
28. Install or connect the following:
    - Coolant recovery tank and hoses
    - Ground cables at the radiator support
    - Accelerator cable
    - Accessory drive belt
    - Upper drive belt cover and tighten the bolts to 18 ft. lbs. (24 Nm)
    - EGR exhaust tube, if equipped
    - EGR pressure sensor connector, if equipped
    - EGR vacuum supply, if equipped
    - CKP sensor connector
    - ECT sensor connector
    - Main wiring harness connector
    - Air cleaner and air outlet tube
    - IAT sensor connector
    - MAF sensor connector
    - Battery
29. Tighten the suspension strut locknuts to 34 ft. lbs. (46 Nm).
30. Fill the engine crankcase and cooling system.
31. Fill the transaxle.
32. Start the engine and check for leaks and proper operation.

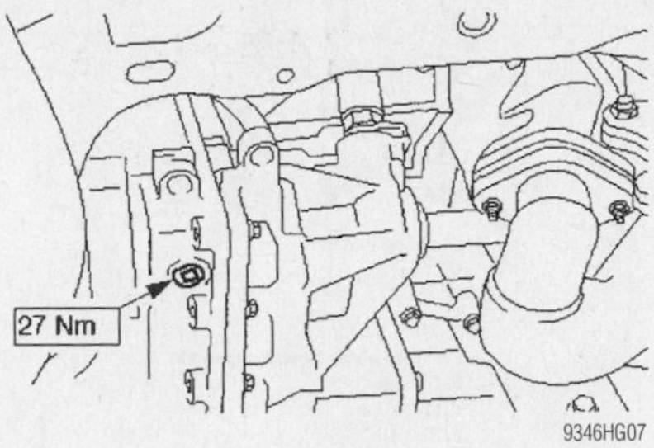

Automatic transaxle drain plug—all models

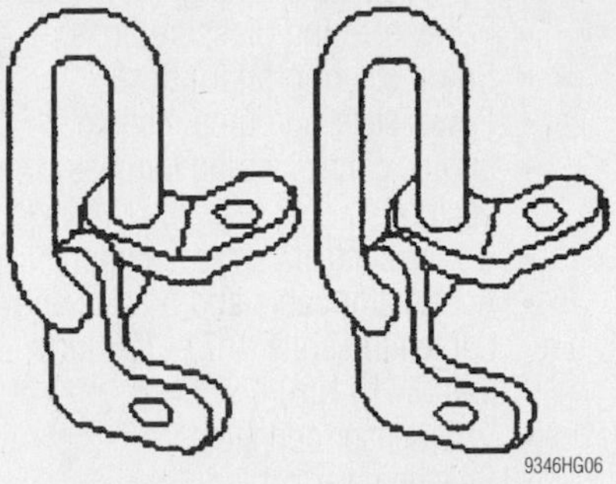

Engine lifting brackets 303-050—2.5L engines

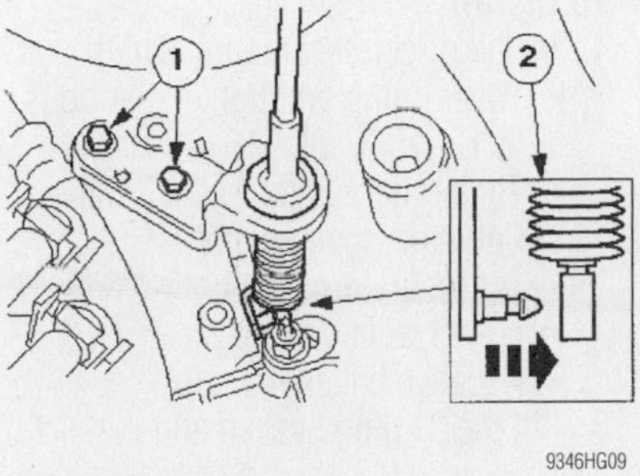

Shift cable bracket (1) and cable (2) removal—2.5L engine

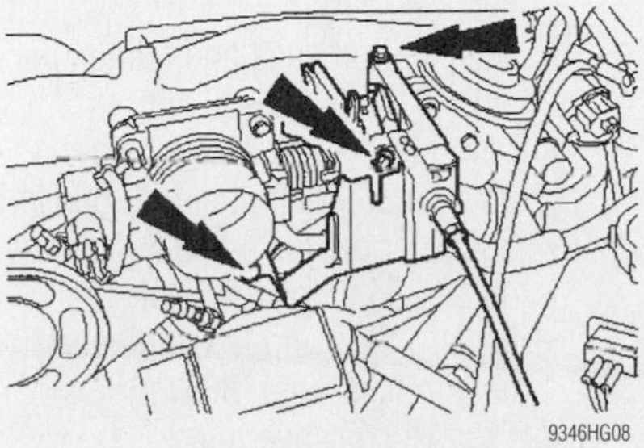

Throttle cable bracket fasteners—2.5L engines

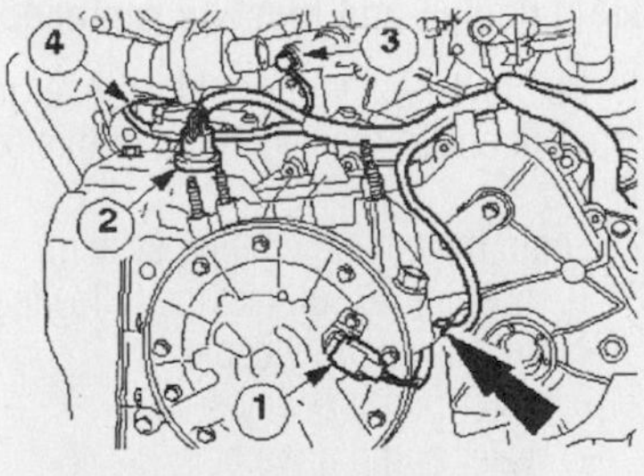

1 Turbine shaft speed (TSS) sensor
2 Transmission control
3 Temperature gauge sender unit
4 Transmission range (TR) sensor

9346HG10

Automatic transaxle harness connections—all models

➡Whenever the vehicle subframe is removed or lowered, the wheel alignment should be checked.

2.5L Engine

➡The engine and transaxle are removed as an assembly.

1. Before servicing the vehicle, refer to the precautions in the beginning of this section.
2. Drain the engine cooling system and engine oil.
3. If equipped, drain the automatic transaxle fluid.
4. Recover the A/C refrigerant.
5. Secure the radiator and fan shroud assembly to the radiator support.
6. Remove or disconnect the following:
   - Battery cables
   - Water pump pulley shield
   - Steering shaft joint at the cowl inside the vehicle
   - Air cleaner
   - Catalytic converter
   - Exhaust crossover pipe
   - Front wheels
   - Stabilizer bar links
   - Outer tie rod ends
   - Lower ball joints
   - Axle halfshafts from the hubs
   - A/C accumulator mounting screws
   - Speedometer cable
   - Vehicle Speed (VSS) sensor connector
   - Accelerator cable and bracket
   - Cruise control cable, if equipped
   - Engine control wiring harness connectors
   - Power steering pump reservoir
   - Power steering return hose
   - Power steering pressure hose
   - Powertrain Control Module (PCM) connector and ground strap
   - Coolant recovery tank
7. If equipped with an automatic transaxle, remove or disconnect the following:
   - Shift cable and bracket
   - Transaxle range switch connector
   - Transaxle oil cooler lines
   - Torque converter
8. If equipped with a manual transaxle, remove or disconnect the following:
   - Shift rod and stabilizer bar
   - Clutch slave cylinder fluid pipe
9. For all models, remove or disconnect the following:
   - Power brake booster vacuum pipe
   - Radiator hoses
   - Heater hoses
   - Block heater wiring, if equipped
   - A/C suction and discharge hoses
   - Lower radiator supports
   - A/C compressor harness connector
   - Bumper cover braces
   - Front engine support insulator and bracket
   - Transaxle support insulator
10. Support the powertrain from below.
    - Subframe bolts
    - Powertrain from the vehicle
11. Attach an engine hoist to the engine lifting eyes.
12. Remove the left and right engine support insulators and lift the powertrain away from the subframe.
13. Support the transaxle from below.
14. Remove or disconnect the following:
    - Axle halfshafts
    - Starter, wiring harness and ground cable
    - Engine from the transaxle

To install:

➡Replace all snaprings, split pins and self-locking nuts.

15. Install or connect the following:
    - Engine to the transaxle and tighten the transaxle flange bolts to 25–34 ft. lbs. (34–46 Nm)
    - Axle halfshafts to the transaxle
    - Starter and wiring harness connectors
    - Ground cable
    - Powertrain Alignment Gauge T94P-6000-AH to the subframe and position the powertrain and tighten the center bolt to 20 ft. lbs. (27 Nm)
    - Right engine support insulator. Leave the bolts finger-tight.
    - Subframe Alignment Pin Set 94P-

2100-AH and raise the subframe to the vehicle and tighten the subframe bolts to 92–100 ft. lbs. (125–135 Nm)

16. Remove the alignment pins.
   - Front engine support insulator and bracket and tighten the nuts to 84 inch lbs. (10 Nm)

➡The left and right support insulators must be aligned in the middle of the support insulator brackets.

   - Right engine support insulator-to-subframe bolts to 30–41 ft. lbs. (41–55 Nm)
   - Front engine support bracket nuts to 52–70 ft. lbs. (70–95 Nm)
   - Engine and transmission support insulator nuts to 30–41 ft. lbs. (41–55 Nm) for automatic transaxles or 52–70 ft. lbs. (70–95 Nm) for manual transaxles
   - Right front engine support insulator through-bolt to 75–102 ft. lbs. (103–137 Nm)

17. Remove the powertrain alignment gauge.
   - Left front engine support insulator to the subframe. Tighten the retaining bolts to 30–41 ft. lbs. (41–55 Nm) and the through-bolt to 75–102 ft. lbs. (103–137 Nm).
   - A/C compressor wiring harness
   - Lower radiator supports and tighten the retaining bolts to 71–97 inch lbs. (8–11 Nm)
   - A/C suction and discharge hoses
   - Block heater wiring, if equipped
   - Radiator hoses
   - Heater hoses
   - Power brake booster vacuum pipe

18. If equipped with an automatic transaxle, install or connect the following:
   - Transaxle oil cooler lines
   - Transaxle range switch connector
   - Shift cable and bracket and tighten the retaining bolts to 15–19 ft. lbs. (20–25 Nm)
   - Torque converter and tighten the retaining nuts in an alternating sequence to 54–64 ft. lbs. (73–87 Nm)

19. If equipped with a manual transaxle, install or connect the following:
   - Clutch slave cylinder fluid pipe
   - Shift rod and stabilizer bar. Tighten the shift rod bolt to 17 ft. lbs. (23 Nm) and the stabilizer nut to 41 ft. lbs. (55 Nm).

20. For all models, install or connect the following:
   - Coolant recovery tank
   - PCM connector and ground strap
   - Power steering pressure hose
   - Power steering return hose
   - Power steering pump reservoir
   - Engine control wiring harness connectors
   - Cruise control cable, if equipped
   - Accelerator cable and bracket and tighten the bolts to 71–106 inch lbs. (8–12 Nm)
   - VSS sensor connector
   - Speedometer cable
   - A/C accumulator mounting screws
   - Axle halfshafts to the front hubs. Use new nuts and tighten to 246 ft. lbs. (340 Nm).
   - Outer tie rod ends and tighten the nuts to 21 ft. lbs. (28 Nm)
   - Lower ball joints and tighten the bolts to 37–43 ft. lbs. (50–58 Nm)
   - Stabilizer bar links and tighten the nuts to 35 ft. lbs. (48 Nm)
   - Front wheels and tighten the lug nuts to 63 ft. lbs. (86 Nm)
   - Exhaust crossover pipe
   - Catalytic converter
   - Air cleaner
   - Steering shaft joint and tighten the bolt to 18 ft. lbs. (24 Nm)
   - Water pump pulley shield
   - Battery cables

21. Fill the crankcase and cooling system.

22. If equipped, fill the automatic transaxle.

23. Check all fluid levels.

24. Evacuate and recharge the air conditioning system.

25. Start the engine and check for leaks and proper operation.

➡Whenever the vehicle subframe is removed or lowered, the wheel alignment should be checked.

Water Pump

REMOVAL & INSTALLATION

2.0L Engine

1. Before servicing the vehicle, refer to the precautions in the beginning of this section.
2. Drain the cooling system.
3. Remove or disconnect the following:
   - Negative battery cable
   - Lower radiator hose
   - Accessory drive belt
   - Timing belt. Refer to the Timing Belt unit repair section.
   - Water pump

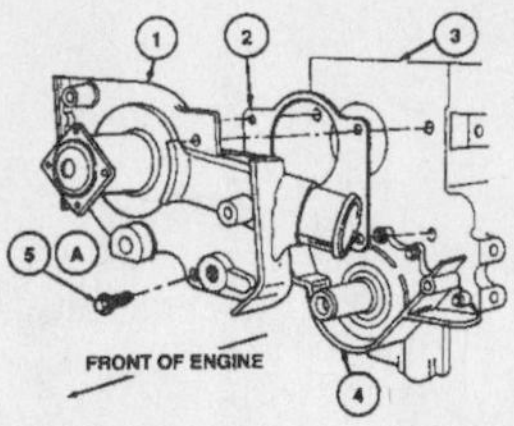

Exploded view of the water pump mounting—2.0L engine

To install:

4. Install or connect the following:
   - Water pump and tighten the bolts to 12–15 ft. lbs. (16–20 Nm)
   - Timing belt. Refer to the Timing Belt unit repair section.
   - Accessory drive belt
   - Lower radiator hose
   - Negative battery cable
5. Fill the cooling system and inspect for leaks

2.5L Engine

➡The 3 water pump bolts are a torque-to-yield design and must be replaced.

1. Before servicing the vehicle, refer to the precautions in the beginning of this section.
2. Drain the engine cooling system.
3. Remove or disconnect the following:
   - Negative battery cable
   - Water pump pulley shield
   - Water pump drive belt
   - Coolant hoses
   - Water pump and pump housing
4. Separate the water pump from the pump housing.

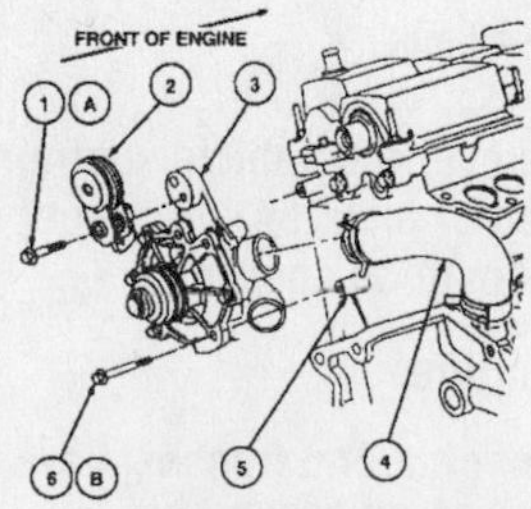

Exploded view of the water pump mounting—2.5L engine

To install:

5. Install or connect the following:
   - Water pump to the pump housing and tighten the fasteners to 16–18 ft. lbs. (22–25 Nm)
   - Water pump and pump housing to the engine. Use new bolts and tighten them to 11–13 ft. lbs. (15–18 Nm), plus 85–95 degrees.
   - Coolant hoses
   - Water pump drive belt
   - Water pump pulley shield
   - Negative battery cable
6. Fill the cooling system and check for leaks.

Cylinder Head

REMOVAL & INSTALLATION

2.0L Engine

➡**The cylinder head bolts are a torque-to-yield design and must be replaced.**

1. Before servicing the vehicle, refer to the precautions in the beginning of this section.
2. Drain the cooling system.
3. Remove or disconnect the following:
   - Negative battery cable
   - Accessory drive belt
   - Power steering pump pulley cover
   - Power steering pump
   - Power steering pump support bracket
   - Alternator bracket
   - Valve cover
   - Intake manifold
   - Exhaust manifold
   - Front cover
   - Timing belt. Refer to the Timing Belt unit repair section.
   - Camshafts and valve tappets

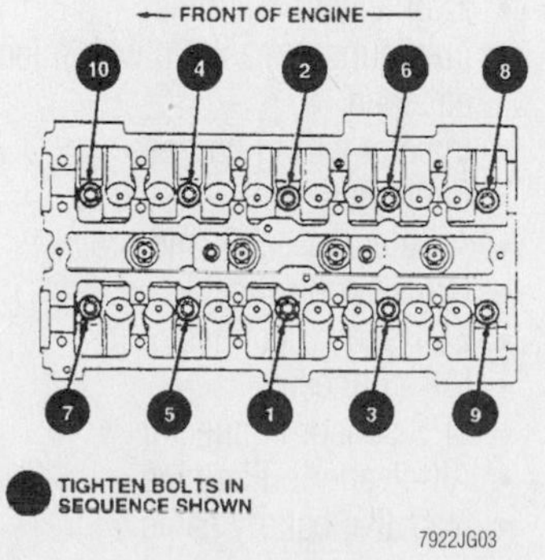

Cylinder head torque sequence—2.0L engine

   - Right lifting eye
   - Timing belt tensioner pulley
   - Thermostat housing
   - Ignition coil and bracket
   - Spark plugs
   - Cylinder head bolts in the reverse of the installation order
   - Cylinder head

To install:

4. Use a new gasket and install the cylinder head.
5. Use new cylinder head bolts and oil the threads.
6. Tighten the cylinder head bolts in sequence as follows:
   a. Step 1: 15–22 ft. lbs. (20–30 Nm)
   b. Step 2: 30–37 ft. lbs. (40–50 Nm)
   c. Step 3: Plus 90–120 degrees
7. Install or connect the following:
   - Ignition coil and bracket
   - Thermostat housing
   - Timing belt tensioner pulley
   - Right lifting eye
   - Camshafts and valve tappets
   - Timing belt. Refer to the Timing Belt unit repair section.
   - Front cover
   - Exhaust manifold
   - Intake manifold
   - Valve cover
   - Alternator bracket
   - Power steering pump support bracket
   - Power steering pump
   - Accessory drive belt
   - Power steering pump pulley cover
   - Spark plugs
   - Negative battery cable
8. Fill the cooling system and check for leaks.
9. Start the engine and check for proper operation.

2.5L Engine

➡**The cylinder head bolts are a torque-to-yield design and must be replaced.**

1. Before servicing the vehicle, refer to the precautions in the beginning of this section.
2. Drain the crankcase and cooling system.
3. Remove or disconnect the following:
   - Negative battery cable
   - Air cleaner housing
   - Accessory drive belts
   - Upper and lower intake manifolds
   - Valve covers
   - Oil pan
   - Alternator and bracket

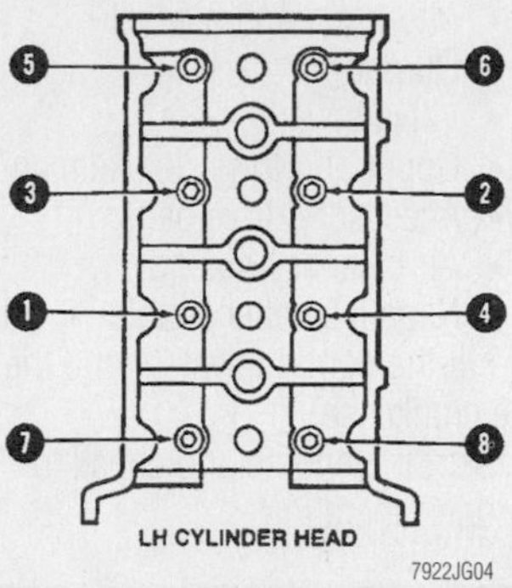

Cylinder head torque sequence—2.5L engine—left side shown

   - Left and right Heated Oxygen (HO_2S) sensors
   - Catalytic converter and exhaust crossover tube
   - Water pump
   - Front cover
   - Timing chains
   - Camshafts and valve tappets
   - Exhaust Gas Recirculation (EGR) pressure sensor
   - EGR transducer
   - EGR tube
   - Coolant bypass
   - Oil dipstick tube
   - Cylinder head bolts in the reverse of the installation order
   - Cylinder heads

To install:

4. Use new gaskets and install the cylinder heads.
5. Use new cylinder head bolts and oil the threads.
6. Tighten the cylinder head bolts in sequence as follows:
   a. Step 1: 27–32 ft. lbs. (37–43 Nm)
   b. Step 2: Plus 85–95 degrees
   c. Step 3: Loosen the bolts, in sequence, 1 full turn
   d. Step 4: 27–32 ft. lbs. (37–43 Nm)
   e. Step 5: Plus 85–95 degrees
   f. Step 6: Plus 85–95 degrees
7. Install or connect the following:
   - Oil dipstick tube
   - Coolant bypass
   - EGR tube
   - EGR transducer
   - EGR pressure sensor
   - Camshafts and valve tappets
   - Timing chains
   - Front cover
   - Water pump
   - Catalytic converter and exhaust crossover tube
   - Left and right HO_2S sensors

- Alternator and bracket
- Oil pan
- Valve covers
- Upper and lower intake manifolds
- Accessory drive belts
- Air cleaner housing
- Negative battery cable

8. Fill the cooling system and the engine crankcase.
9. Start the engine and check for leaks and proper operation.

Rocker Arms

REMOVAL & INSTALLATION

The 2.0L engine is not equipped with rocker arms, the camshaft directly actuates the valves through a hydraulic lifter.

2.5L Engine

1. Before servicing the vehicle, refer to the precautions in the beginning of this section.
2. Remove or disconnect the following:
   - Battery
   - Upper manifold
   - Intake Manifold Runner Control (IMRC) actuator
   - Spark plug wires and bracket
   - Coolant reservoir breather hose bracket and wiring
   - Fuel injector wiring harness
   - Ignition coil
   - Wiring harness brackets
   - Valve covers. Remove the bolts in reverse sequence.
   - Spark plugs.
3. Rotate the crankshaft so that the piston on the cylinder to be serviced is at bottom dead center with the valves closed.
4. Install an adapter in the spark plug hole and connect a compressed air supply at 102–144 psi.
5. Install special tool Valve Spring Compressor T94P-6565-BH.

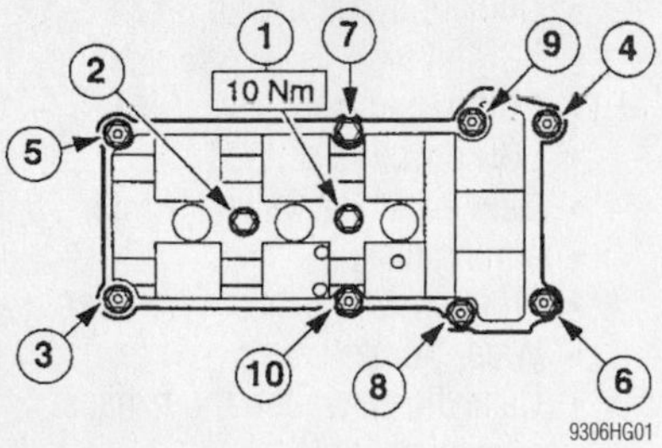

Valve cover bolt tightening sequence—2.5L engine

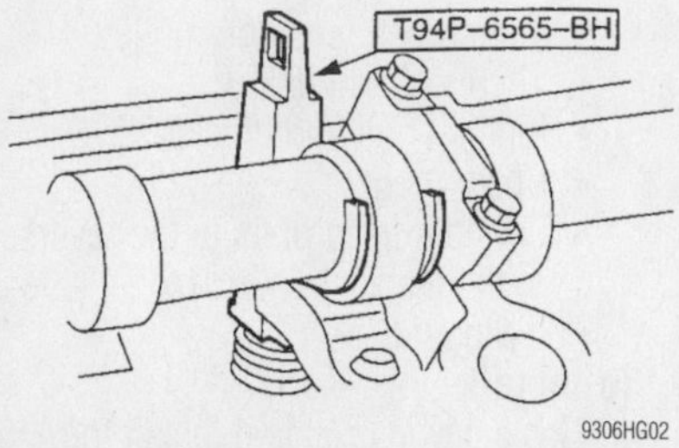

Valve spring compressor—2.5L engine

6. Compress the valve spring and remove the roller finger follower. Repeat for each follower to be removed.

➡If the followers are to be reused, ensure that they are installed in the same position that they were removed from.

To install:

7. Compress the valve spring and install the roller finger follower. Repeat for each follower to be installed.
8. Install or connect the following:
   - Spark plugs
   - Valve covers and tighten the bolts in sequence to 88 inch lbs. (10 Nm)
   - Wiring harness brackets
   - Ignition coil
   - Fuel injector wiring harness
   - Coolant reservoir breather hose bracket and wiring
   - Spark plug wires and bracket
   - IMRC actuator
   - Upper manifold
   - Battery
9. Run the engine and check for leaks and proper operation.

Intake Manifold

REMOVAL & INSTALLATION

2.0L Engine

1. Before servicing the vehicle, refer to the precautions in the beginning of this section.
2. Remove or disconnect the following:
   - Negative battery cable
   - Air cleaner outlet pipe
   - Cruise control cable
   - Accelerator cable and bracket
   - Throttle Position (TPS) sensor connector
   - Idle Air Control (IAC) connector
   - Engine Coolant Temperature (ECT) sensor connector
   - Fuel lines

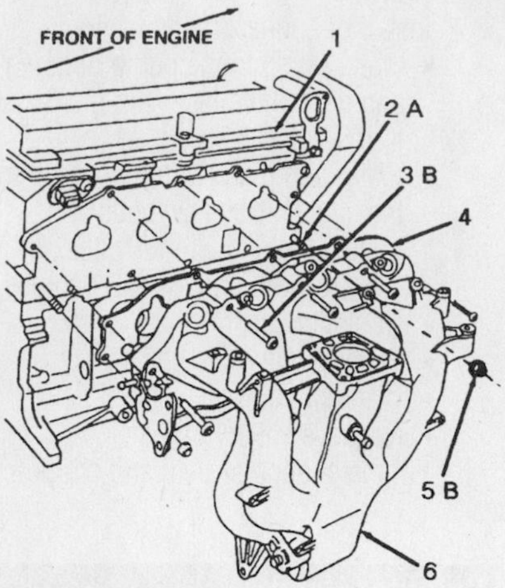

1. Cylinder head
2. Stud (2 req'd)
3. Bolt (8 req'd)
4. Intake manifold gasket
5. Nut (2 req'd)
6. Intake manifold
A. Tighten to 0-10 Nm (0-89 lb. in.)
B. Tighten to 16-20 Nm (12-15 lb. ft.)

7922JG06

Exploded view of the intake manifold mounting—2.0L engine

   - Injector wiring harness
   - Fuel supply manifold with injectors attached
   - Engine wiring harness
   - Vacuum hoses
   - Brake booster vacuum pipe
   - Accessory drive belt
   - Alternator
   - Intake manifold

To install:

3. Install or connect the following:
   - Intake manifold with a new gasket and tighten the fasteners to 13 ft. lbs. (18 Nm)
   - Alternator
   - Accessory drive belt
   - Brake booster vacuum pipe
   - Vacuum hoses
   - Engine wiring harness
   - Fuel supply manifold with injectors attached
   - Injector wiring harness
   - Fuel lines
   - Accelerator cable and bracket
   - Cruise control cable
   - ECT sensor connector
   - IAC connector
   - TPS sensor connector
   - Air cleaner outlet pipe
   - Negative battery cable

2.5L Engine

1. Before servicing the vehicle, refer to the precautions in the beginning of this section.

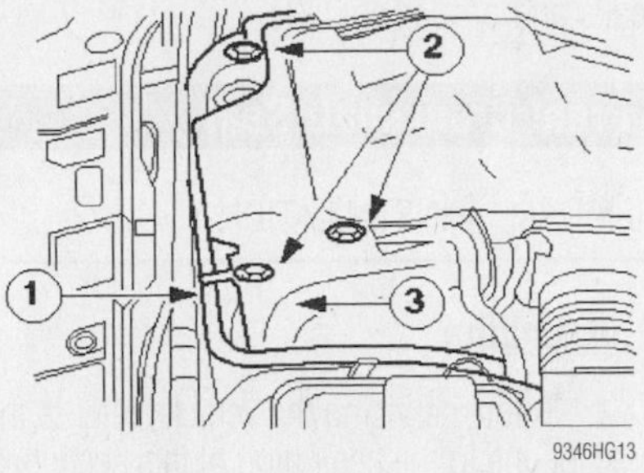

Coolant line (1), Cover bolts (2) and Water Pump Cover (3)—2.5L engine

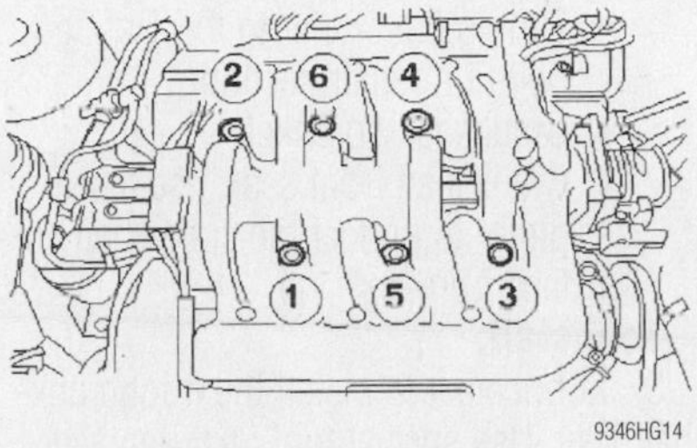

Upper intake manifold loosening sequence—2.5L engine

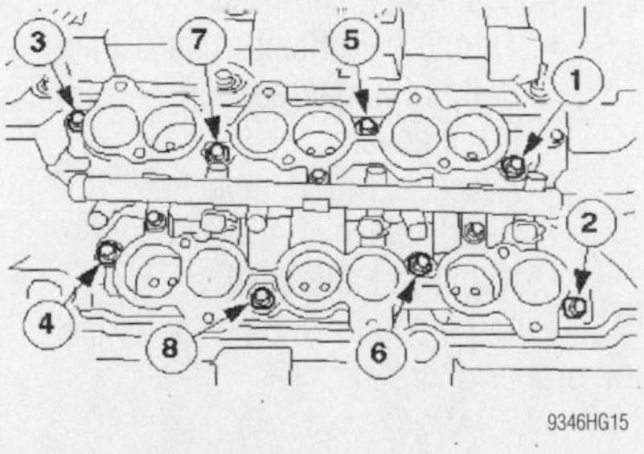

Lower intake manifold loosening sequence—2.5L engine

2. Remove or disconnect the following:
   - Negative battery cable
   - Water pump pulley shield
   - Brake booster vacuum pipe
   - Emission vacuum control
   - Cruise control cable
   - Accelerator cable and bracket
   - Throttle Position (TPS) sensor connector
   - Idle Air Control (IAC) valve connector
   - Exhaust Gas Recirculation (EGR) vacuum regulator connector and vacuum hoses
   - PCV valve vacuum hose
   - EGR tube
   - Intake Manifold Runner Control (IMRC)

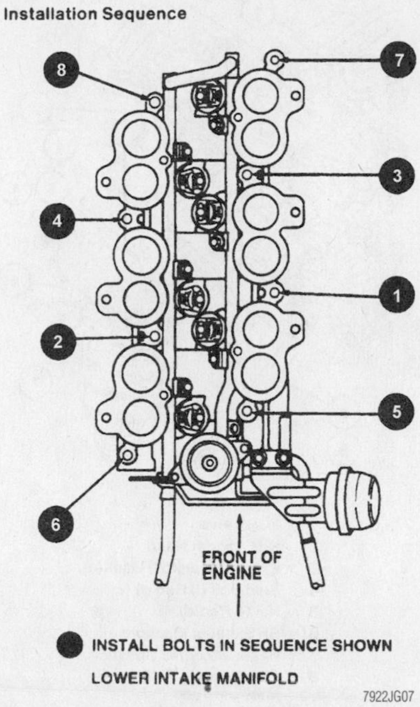

Lower intake manifold torque sequence—2.5L engine

   - Upper intake manifold. Remove the bolts in the sequence shown.
   - Fuel injector wiring harness
   - Fuel lines
   - Lower intake manifold. Remove the bolts in the sequence shown.

To install:

3. Install or connect the following:
   - Lower intake manifold. Use new gaskets and tighten the bolts to 71–106 inch lbs. (8–12 Nm).
   - Fuel lines
   - Fuel injector wiring harness
   - Upper intake manifold. Use new gaskets and tighten the bolts to 71–106 inch lbs. (8–12 Nm).
   - IMRC
   - EGR tube
   - PCV valve vacuum hose
   - EGR vacuum regulator connector and vacuum hoses
   - IAC valve connector
   - TPS sensor connector
   - Accelerator cable and bracket
   - Cruise control cable
   - Emission vacuum control
   - Brake booster vacuum pipe
   - Water pump pulley shield
   - Negative battery cable
4. Run the engine and check for leaks

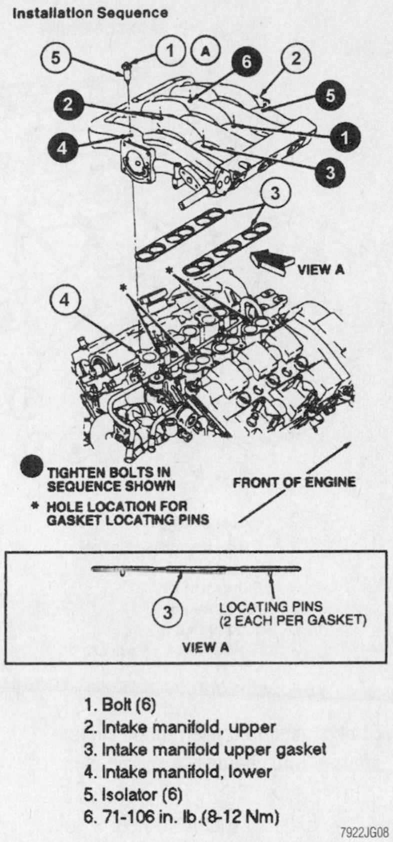

Upper intake manifold torque sequence—2.5L engine

Exhaust Manifold

REMOVAL & INSTALLATION

2.0L Engine

1. Before servicing the vehicle, refer to the precautions in the beginning of this section.
2. Remove or disconnect the following:
   - Negative battery cable
   - Air intake resonators
   - Oil dipstick tube
   - Exhaust manifold heat shield
   - Heated Oxygen (HO_2S) sensor
   - Exhaust Gas Recirculation (EGR) tube and bracket, if equipped
   - Catalytic converter
   - 9 manifold retaining nuts and the exhaust manifold

To install:

3. Install or connect the following:
   - New gasket
   - Exhaust manifold and tighten the nuts to 12 ft. lbs. (16 Nm)
   - Catalytic converter
   - EGR tube and bracket, if equipped
   - HO_2S sensor

Heater Core replacement is covered in Section 2 of this manual

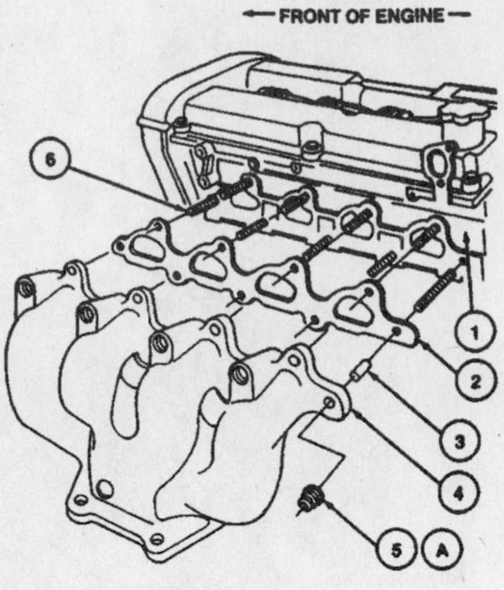

1 Cylinder Head
2 Exhaust Manifold Gasket
3 Spacer
4 Exhaust Manifold
5 Nut (9 Req'd)
6 Stud (9 Req'd)
A Tighten to 14-17 N·m (13-16 Lb-Ft)

7922JG09

Exhaust manifold and related components—2.0L engine

- Exhaust manifold heat shield
- Oil dipstick tube
- Air intake resonators
- Negative battery cable

4. Run the engine and check for leaks and proper operation

2.5L Engine

RIGHT SIDE

1. Before servicing the vehicle, refer to the precautions in the beginning of this section.
2. Remove or disconnect the following:
   - Negative battery cable
   - Alternator and bracket
   - Heated Oxygen (HO_2S) sensor
   - Catalytic converter
   - Halfshaft bearing support bracket
   - Exhaust Gas Recirculation (EGR) tube
   - Exhaust manifold

To install:

3. Install or connect the following:
   - New gasket
   - Exhaust manifold and tighten the fasteners to 15 ft. lbs. (20 Nm)
   - EGR tube
   - Halfshaft bearing support bracket
   - Catalytic converter
   - HO_2S sensor
   - Alternator and bracket
   - Negative battery cable
4. Run the engine and check for exhaust leaks and proper operation.

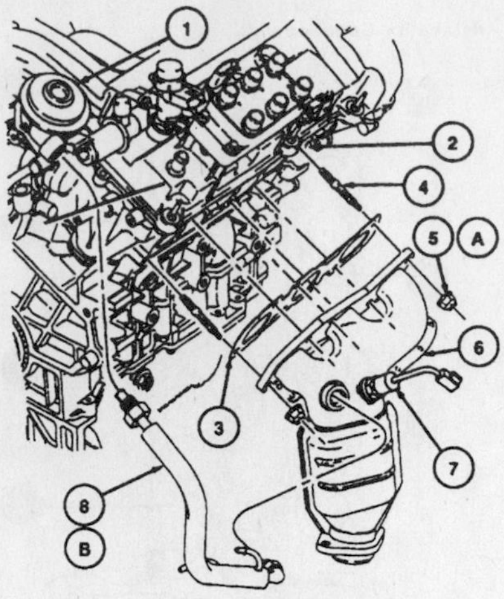

1 EGR Valve
2 RH Cylinder Head
3 Exhaust Manifold Gasket
4 Stud Bolt (6 Req'd)
5 Nut (6 Req'd)
6 RH Exhaust Manifold
7 Heated Oxygen Sensor
8 EGR Valve to Exhaust Manifold Tube
A Tighten to 18-22 N·m (13-16 Lb-Ft)
B Tighten to 35-45 N·m (26-33 Lb-Ft)

7922JG10

Exploded view of the right-side exhaust manifold mounting—2.5L engine

LEFT SIDE

1. Before servicing the vehicle, refer to the precautions in the beginning of this section.
2. Remove or disconnect the following:
   - Negative battery cable
   - Heated Oxygen (HO_2S) sensor
   - Exhaust crossover tube
   - Radiator hose bracket
   - 6 manifold retaining nuts and the exhaust manifold

To install:

3. Install or connect the following:
   - New gasket
   - Exhaust manifold and tighten the nuts to 15 ft. lbs. (20 Nm)
   - Radiator hose bracket
   - Exhaust crossover tube
   - HO_2S sensor
   - Negative battery cable

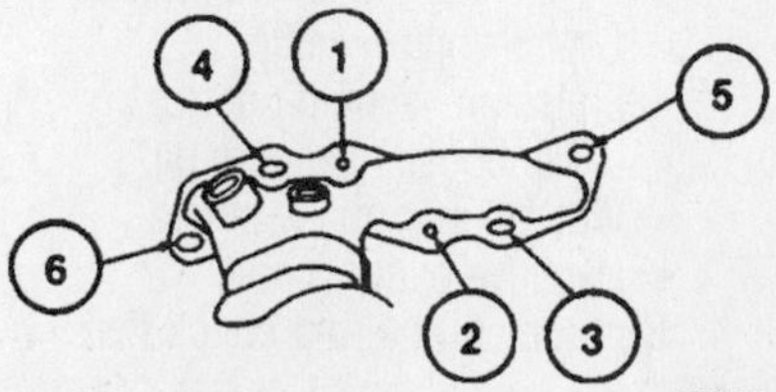

7922JG11

Exhaust manifold mounting bolt tightening sequence—2.5L engine

4. Run the engine and check for exhaust leaks and proper operation.

Front Crankshaft Seal

REMOVAL & INSTALLATION

2.0L Engine

1. Before servicing the vehicle, refer to the precautions in the beginning of this section.
2. Remove or disconnect the following:
   - Negative battery cable
   - Right splash shield
   - Accessory drive belt
   - Front cover
   - Timing belt. Refer to the Timing Belt unit repair section.
   - Crankshaft sprocket
   - Crankshaft front seal. Use a seal puller to protect the crankshaft surface from damage.

To install:

3. Lubricate and install the front crankshaft seal. Use special tool Seal Replacer T81P-6700-A and the crankshaft pulley bolt to press the seal into the oil pump housing.
4. Install or connect the following:
   - Crankshaft sprocket
   - Timing belt. Refer to the Timing Belt unit repair section.
   - Front cover
   - Accessory drive belt
   - Right splash shield
   - Negative battery cable
5. Run the engine and check for leaks and proper operation.

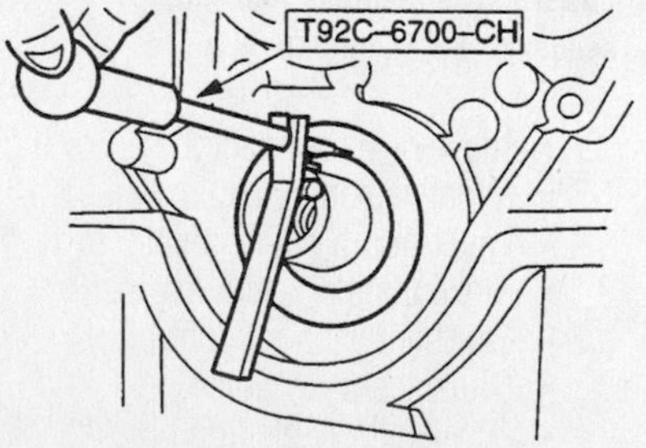

7922JG12

Removing the crankshaft front seal with Ford seal remover T92C-6700-CH—2.0L engine

Camshaft and Valve Lifters

REMOVAL & INSTALLATION

2.0L Engine

1. Before servicing the vehicle, refer to the precautions in the beginning of this section.

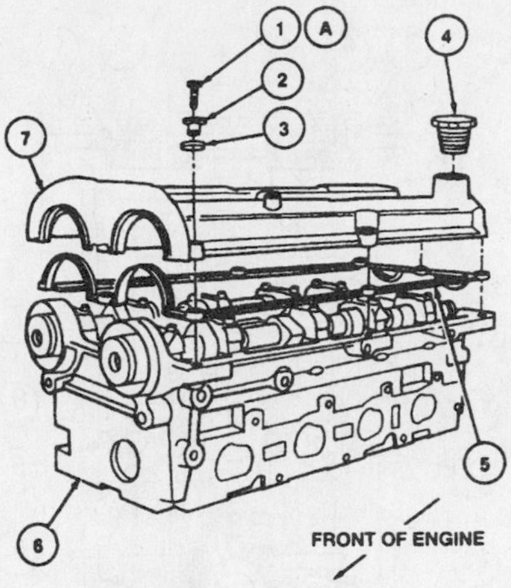
FRONT OF ENGINE

1. Bolt (10)
2. Spacer (10)
3. O-ring (10)
4. Oil filler cap
5. Valve cover gasket
6. Cylinder head
7. Valve cover
A. 53-71 in. lb.(6-8 Nm)

7922JG13

Valve cover and related components—2.0L engine

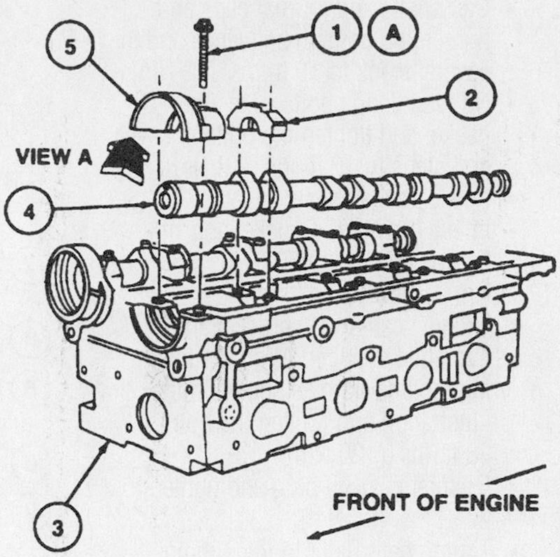

1. Bolt(20)
2. Camshaft journal cap(8)
3. Cylinder head
4. Camshaft
5. Camshaft journal thrust cap(2)
A. 13-15 ft. lb.(17-21 Nm)

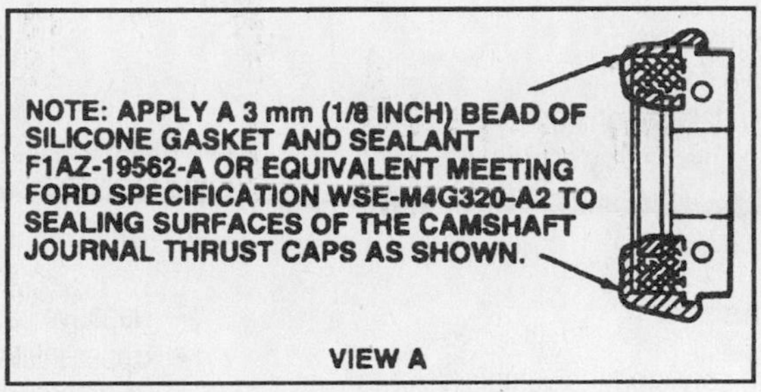
NOTE: APPLY A 3 mm (1/8 INCH) BEAD OF SILICONE GASKET AND SEALANT F1AZ-19562-A OR EQUIVALENT MEETING FORD SPECIFICATION WSE-M4G320-A2 TO SEALING SURFACES OF THE CAMSHAFT JOURNAL THRUST CAPS AS SHOWN.
VIEW A

7922JG14

During assembly, apply sealant to the camshaft journal thrust cap as shown—2.0L engine

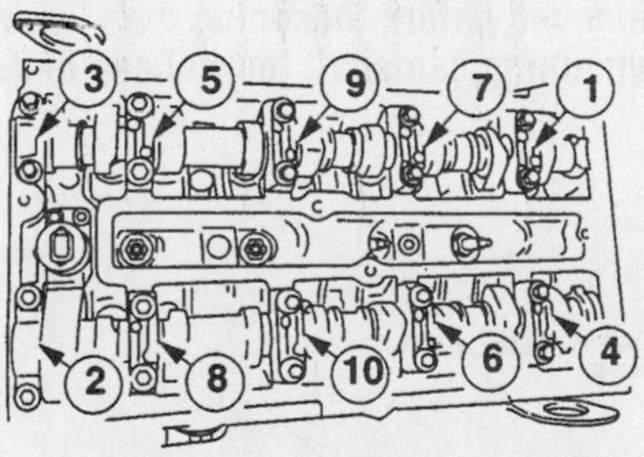

7922JG15

Remove the bearing cap bolts in pairs using the sequence shown—2.0L engine

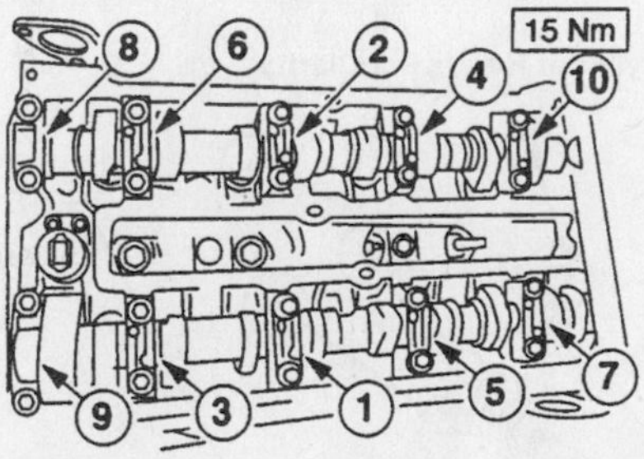
15 Nm

7922JG16

Camshaft journal cap retaining bolt tightening sequence—2.0L engine

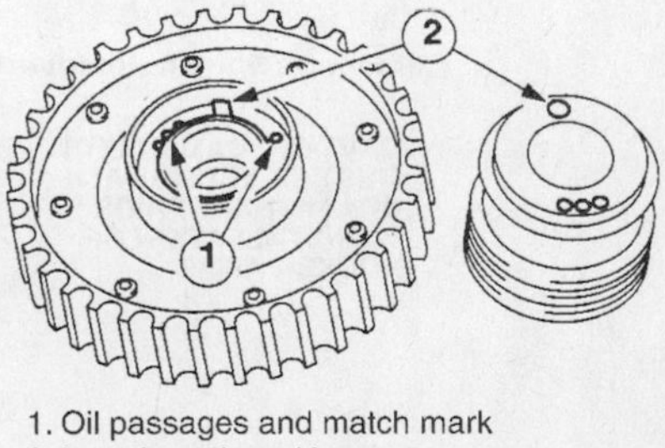

1. Oil passages and match mark
2. Locating tab and bore

9306HG03

Alignment of the exhaust camshaft sprocket and the VCT hydraulic cylinder—2.0L engines

2. Remove or disconnect the following:
   - Negative battery cable
   - Air intake resonators
   - Valve cover
   - Timing cover
   - Timing belt. Refer to the Timing Belt unit repair section.
   - Intake camshaft sprocket
   - Exhaust gear oil plug
   - Exhaust camshaft sprocket and Variable Camshaft Timing (VCT) hydraulic cylinder
   - Oil feed flange bolts

➡Mark the camshaft journal caps to the cylinder head for installation in the same position.

3. Loosen all of the camshaft journal cap bolts in pairs and in sequence 1 turn at a time.

➡Remove the camshaft journal thrust caps last.

4. Remove or disconnect the following:
   - All of the camshaft journal caps
   - Intake and exhaust camshafts and the camshaft front seals from the cylinder head

➡If the valve lifters are to be reused, mark their locations to ensure that they will be installed into their correct positions.

   - Valve lifters from the cylinder head

To install:

➡Before installing the camshafts, the crankshaft must be positioned so that No. 1 cylinder is at TDC on its compression stroke.

5. Install or connect the following:
   - Valve lifters in their original positions

- Camshafts and journal caps and tighten the bolts in sequence and in several steps to 10 ft. lbs. (13 Nm)
- Oil feed flange bolts. Use a new gasket and tighten the bolts in several steps to 10 ft. lbs. (13 Nm).
- New camshaft seals
- Intake camshaft sprocket and tighten the bolt to 50 ft. lbs. (68 Nm)
- Exhaust camshaft sprocket and VCT hydraulic cylinder. Align the locating tab and bore as shown in the illustration and tighten the bolt to 88 ft. lbs. (120 Nm).
- Exhaust gear oil plug and tighten to 27 ft. lbs. (37 Nm)
- Timing belt. Refer to the Timing Belt unit repair section.
- Timing cover
- Valve cover
- Air intake resonators
- Negative battery cable

6. Run the engine and check for leaks and proper operation.

2.5L Engine

1. Before servicing the vehicle, refer to the precautions in the beginning of this section.
2. Remove or disconnect the following:

- Negative battery cable
- Upper intake manifold
- Valve covers
- Accessory drive belts
- Front cover and timing chains

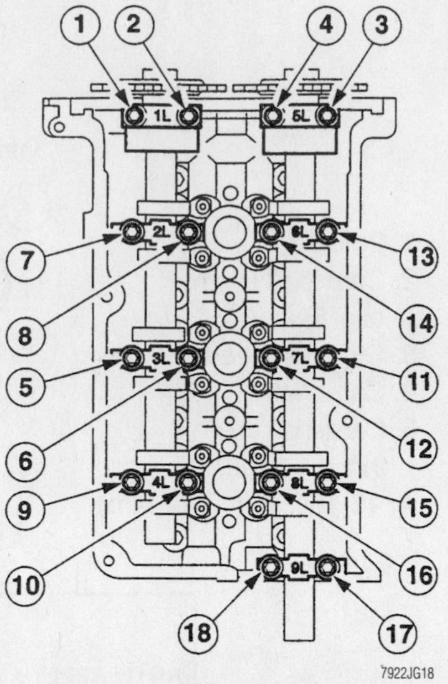
1L
2L
3L
4L
5L
6L
7L
8L
9L
7922JG18

Left (front) cylinder head camshaft journal cap retaining bolt loosening sequence—2.5L engine

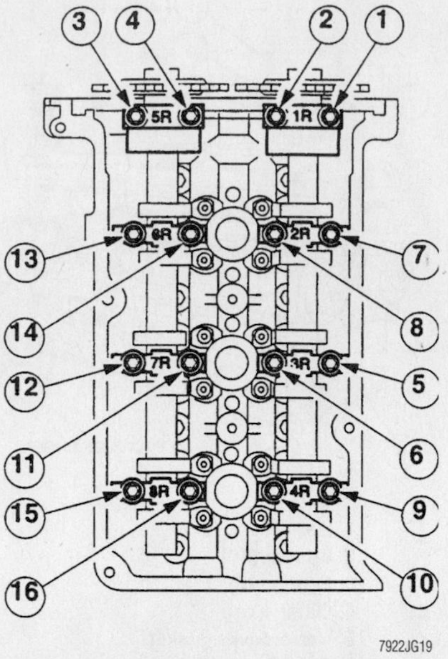
1R
2R
3R
4R
5R
6R
7R
8R
7922JG19

Right (rear) cylinder head camshaft journal cap retaining bolt loosening sequence—2.5L engine

⁂ WARNING

The camshaft thrust caps must be removed before loosening the remaining camshaft journal cap bolts

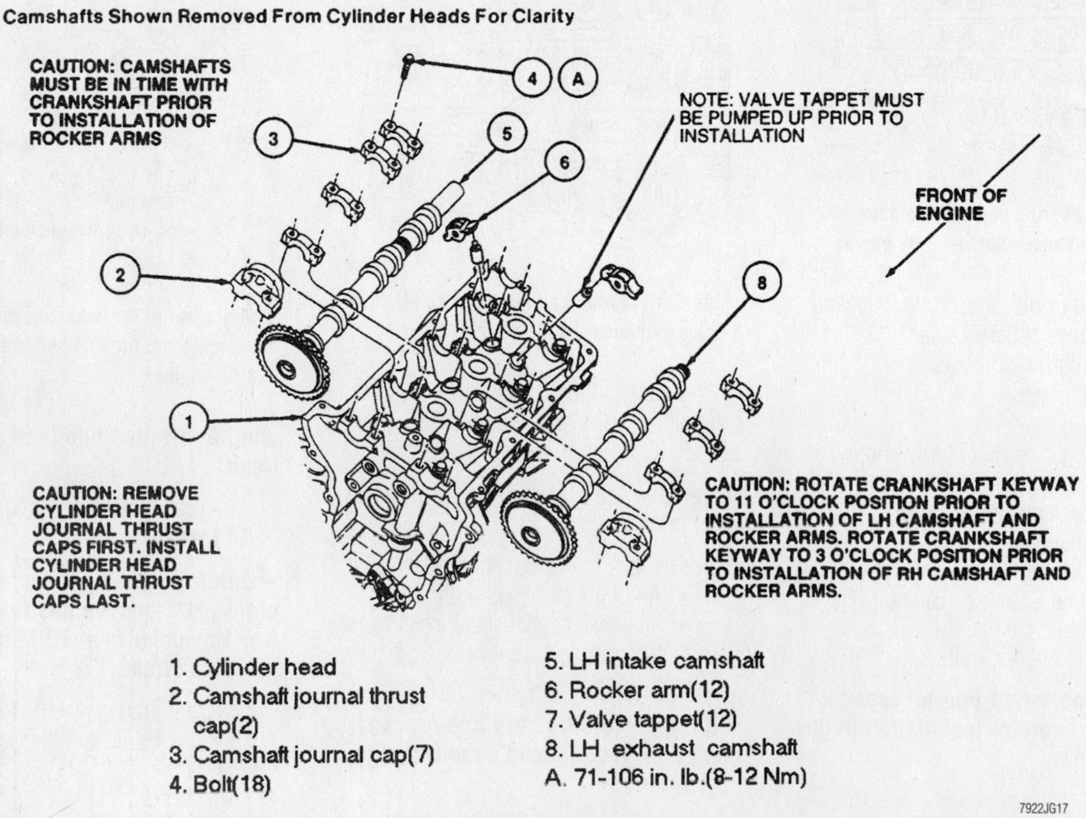
Camshafts Shown Removed From Cylinder Heads For Clarity
CAUTION: CAMSHAFTS MUST BE IN TIME WITH CRANKSHAFT PRIOR TO INSTALLATION OF ROCKER ARMS
NOTE: VALVE TAPPET MUST BE PUMPED UP PRIOR TO INSTALLATION
FRONT OF ENGINE
CAUTION: REMOVE CYLINDER HEAD JOURNAL THRUST CAPS FIRST. INSTALL CYLINDER HEAD JOURNAL THRUST CAPS LAST.
CAUTION: ROTATE CRANKSHAFT KEYWAY TO 11 O'CLOCK POSITION PRIOR TO INSTALLATION OF LH CAMSHAFT AND ROCKER ARMS. ROTATE CRANKSHAFT KEYWAY TO 3 O'CLOCK POSITION PRIOR TO INSTALLATION OF RH CAMSHAFT AND ROCKER ARMS.
1. Cylinder head
2. Camshaft journal thrust cap(2)
3. Camshaft journal cap(7)
4. Bolt(18)
5. LH intake camshaft
6. Rocker arm(12)
7. Valve tappet(12)
8. LH exhaust camshaft
A. 71-106 in. lb.(8-12 Nm)
7922JG17

Exploded view of camshaft mounting—2.5L engine

to ensure that the thrust caps are not damaged.

- Camshaft thrust caps
- Camshaft journal caps. Loosen the bolts in sequence and in several passes to allow the camshaft to raise off the cylinder head evenly.

⁂ WARNING

The camshaft journal caps and cylinder heads are numbered to ensure that they are assembled in their original positions. Keep the camshaft journal caps from each cylinder head together; do not mix them with caps from another cylinder head. Failure to do so may result in engine damage.

- Camshaft journal caps with the retaining bolts installed
- Camshafts from the cylinder head.

➡If the rocker arms and hydraulic lifters are to be reused, they must be installed in their original positions.

- Rocker arms.
- Hydraulic lifters

To install:

⁂ WARNING

The crankshaft keyway must be at the 11 o'clock position before reassembly. Failure to do so may lead to engine damage.

3. Rotate the crankshaft so that the keyway is at the 11 o'clock position for installation of the camshafts.
4. Install or connect the following:
   - Hydraulic lifters
   - Rocker arms.
   - Camshafts. Align the sprocket timing marks.

➡Do not install the camshaft journal thrust caps until the rocker arms and timing chains have been installed and the camshaft journal caps are secured into position.

   - All camshaft journal caps except the thrust caps.
   - Timing chains. Tighten the camshaft journal cap bolts in reverse of the loosening order and in several steps to 71–106 inch lbs. (8–12 Nm).
   - Thrust caps and tighten the bolts to 71–106 inch lbs. (8–12 Nm)
   - Front cover
   - Accessory drive belts
   - Valve covers
   - Upper intake manifold. Use new gaskets and tighten the bolts to 71–106 inch lbs. (8–12 Nm).
   - Negative battery cable
5. Run the engine and check for leaks and proper engine operation.

Valve Lash

ADJUSTMENT

2.0L Engine

1. Before servicing the vehicle, refer to the precautions in the beginning of this section.
2. Remove the valve cover.
3. Rotate the crankshaft so that the valve is closed and measure the clearance between the lifter shim and the camshaft base circle. Intake valve clearance should be 0.004–0.007 inches. Exhaust valve clearance should be 0.010–0.013 inches.
4. Measure each valve and note the clearance.
5. If adjustment is necessary, remove the camshafts and replace the shims. To obtain the correct shim size, measure the current shim and add the measured clearance. Subtract 0.006 inches for intake valves, and 0.012 inches for exhaust valves.
6. Install the camshafts and valve cover.

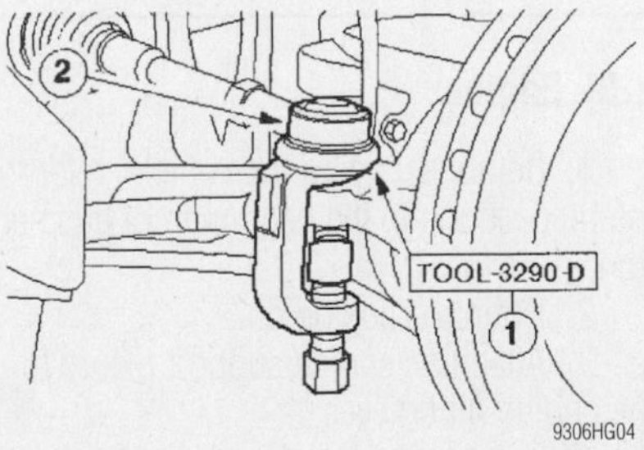

Checking valve clearance—2.0L engine

2.5L Engine

The lash adjusters are hydraulic and are not adjustable.

Starter Motor

REMOVAL & INSTALLATION

2.0L Engine

1. Before servicing the vehicle, refer to the precautions in the beginning of this section.
2. Remove or disconnect the following:
   - Negative battery cable
   - Air cleaner
   - Starter electrical connectors
   - Starter motor

To install:

3. Install or connect the following:
   - Starter motor and tighten the bolts to 18 ft. lbs. (25 Nm)
   - Starter electrical connectors
   - Air cleaner
   - Negative battery cable

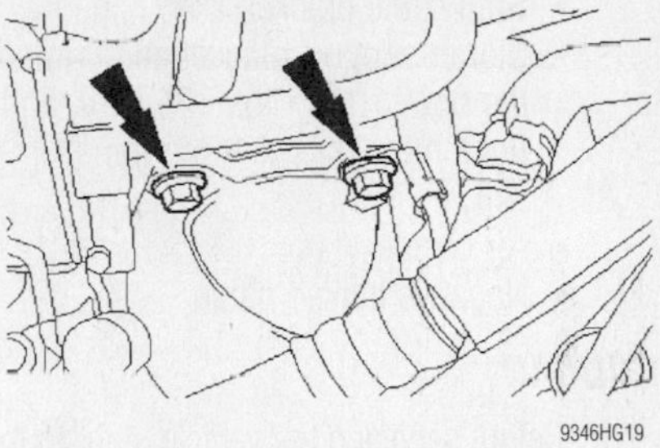

Starter motor upper bolts—2.0L engine

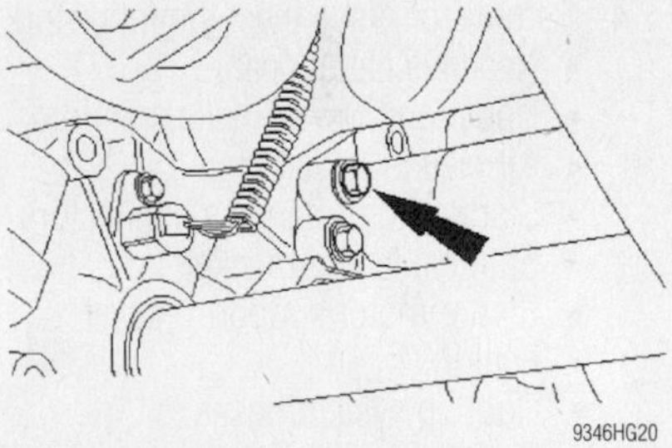

Starter motor lower bolt—2.0L engine

2.5L Engine

CONTOUR AND MYSTIQUE

1. Before servicing the vehicle, refer to the precautions in the beginning of this section.
2. Remove or disconnect the following:
   - Negative battery cable
   - Air cleaner
   - Starter support bracket
   - Shift cable bracket, if equipped

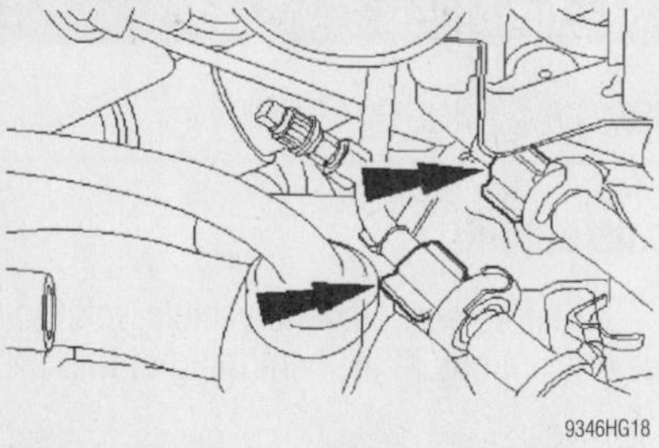

Remove the fuel lines from the support bracket—Contour and Mystique with 2.5L engine

- Fuel lines
- Starter electrical connectors
- Starter motor

To install:

3. Install or connect the following:
   - Starter motor and tighten the bolts to 18 ft. lbs. (25 Nm)
   - Starter electrical connectors
   - Fuel lines
   - Shift cable bracket, if equipped
   - Starter support bracket and tighten the bolt to 18 ft. lbs. (25 Nm), and tighten the nuts to 13 ft. lbs. (17 Nm)
   - Air cleaner
   - Negative battery cable

COUGAR

1. Before servicing the vehicle, refer to the precautions in the beginning of this section.
2. Drain the cooling system.
3. Relieve the fuel system pressure.
4. Remove or disconnect the following:
   - Negative battery cable
   - Upper and lower intake manifold
   - Air cleaner bracket
   - Starter motor harness connectors
   - Shift cable and bracket
   - Starter motor support bracket
   - Fuel lines
   - Cooling system hoses
   - Engine Coolant Temperature (ECT) sensor connector
   - Positive Crankcase Ventilation (PCV) pipe and hose

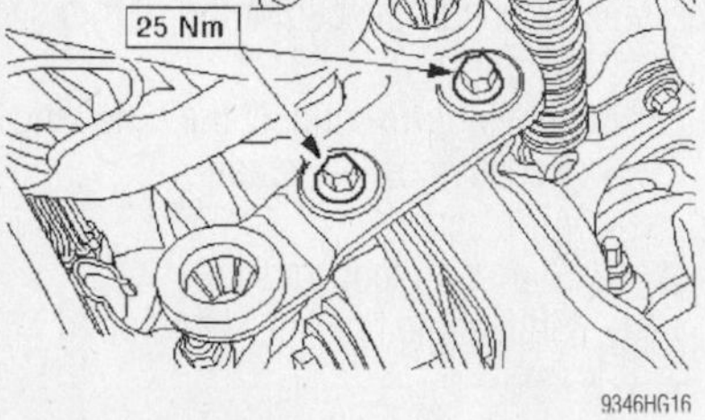

Air cleaner bracket—2.5L engine

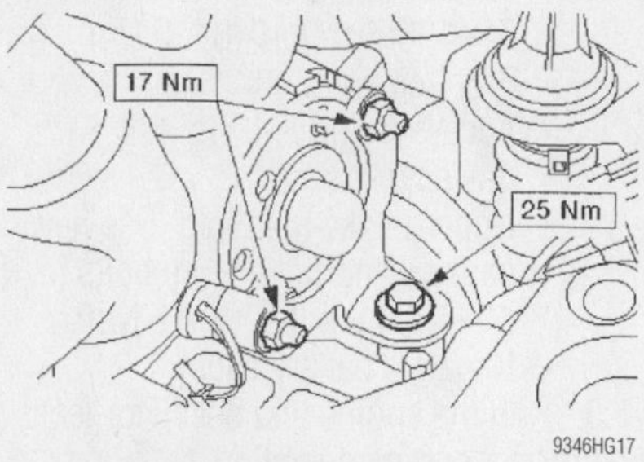

Starter motor support bracket—2.5L engine

- Water crossover pipe
- Starter motor

To install:

5. Install or connect the following:
   - Starter motor and tighten the bolts to 18 ft. lbs. (25 Nm)
   - Water crossover pipe and tighten the bolts to 88 inch lbs. (10 Nm)
   - PCV pipe and hose
   - ECT sensor connector
   - Cooling system hoses
   - Fuel lines
   - Starter motor support bracket. Tighten the nuts to 13 ft. lbs. (17 Nm) and the bolt to 18 ft. lbs. (25 Nm).
   - Shift cable and bracket
   - Starter motor harness connectors
   - Air cleaner bracket and tighten the bolts to 18 ft. lbs. (25 Nm)
   - Upper and lower intake manifold
   - Negative battery cable
6. Fill the cooling system.
7. Start the engine and check for leaks.

Oil Pan

REMOVAL & INSTALLATION

2.0L Engine

1. Before servicing the vehicle, refer to the precautions in the beginning of this section.
2. Drain the engine oil.
3. Attach an engine support fixture to the engine lifting eyes.
4. Remove or disconnect the following:
   - Negative battery cable
   - Catalytic converter
   - Oil lever sensor connector
   - Heater coolant pipe
   - Lower engine rear plate
   - Left and right engine support insulator center bolts
   - Front engine support bracket

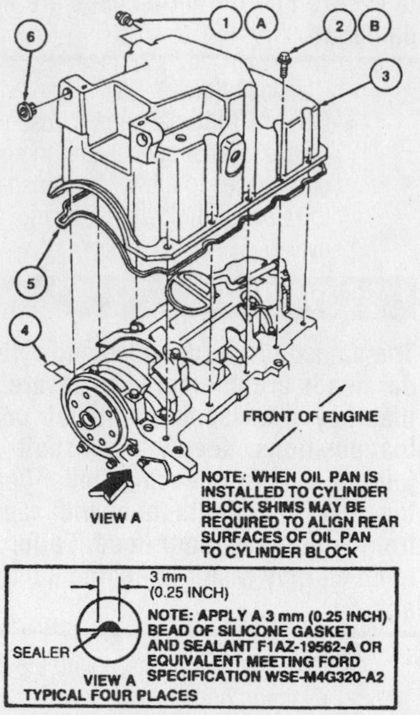

1 Oil pan drain plug
2 Bolt (10 req'd)
3 Oil pan
4 Cylinder block
5 Oil pan gasket
6 Oil pan spacer (as req'd)
A Tighten to 21-28 Nm (15-21 lb-ft)
B Tighten to 20-24 Nm (15-18 lb-ft)

7922JG20

Exploded view of the oil pan mounting—2.0L engine

➡Mark the location of the upper front engine support before removing it from the front engine support bracket.

5. Raise the engine to allow room for removal of the engine oil pan.
6. Remove the oil pan.

To install:

➡Apply a bead of silicone gasket sealer to the oil pump parting lines and at the crankshaft rear main seal retainer on the cylinder block.

7. Install or connect the following:
   - Oil pan. Use a new gasket and tighten the bolts in several passes to 15–18 ft. lbs. (20–24 Nm), working from the center of the block towards the ends. Tighten the transaxle case bolts to 25–34 ft. lbs. (34–46 Nm).
   - Lower engine rear plate
8. Lower the engine and install or connect:
   - Front engine support insulator
   - Left and right engine support insulator center bolts
   - Heater coolant pipe
   - Oil lever sensor connector

- Catalytic converter
- Negative battery cable

9. Fill the crankcase with the proper amount of engine oil.
10. Run the engine and check for leaks and proper operation.

2.5L Engine

1. Before servicing the vehicle, refer to the precautions in the beginning of this section.
2. Drain the engine oil.
3. Attach an engine support fixture to the engine lifting eyes.
4. Remove or disconnect the following:
   - Negative battery cable
   - Water pump pulley shield
   - Exhaust crossover pipe and bracket
   - Exhaust heat shields
   - Lower engine rear plate
   - Left and right engine support insulator center bolts
   - Front engine support bracket

➡Mark the location of the upper front engine support before removing it from the front engine support bracket.

5. Raise the engine to allow room for removal of the engine oil pan.
6. Remove the oil pan.

To install:

7. Apply a bead of silicone sealer to the gasket area where the pan meets the parting lines of the lower cylinder block and the front engine cover.
8. Install or connect the following:
   - Oil pan. Use a new gasket and tighten the pan bolts in several passes to 15–22 ft. lbs. (20–30 Nm). Tighten the transaxle case bolts to 25–34 ft. lbs. (34–46 Nm).
   - Lower engine rear plate
9. Lower the engine and install or connect:
   - Front engine support insulator
   - Left and right engine support insulator center bolts
   - Exhaust heat shields
   - Exhaust crossover pipe and bracket
   - Water pump pulley shield
   - Negative battery cable
10. Fill the crankcase with the proper amount of engine oil.
11. Run the engine and check for leaks and proper operation.

Oil Pump

REMOVAL & INSTALLATION

2.0L Engine

1. Before servicing the vehicle, refer to the precautions in the beginning of this section.
2. Drain the engine oil.
3. Remove or disconnect the following:
   - Negative battery cable
   - Accessory drive belt
   - Timing belt cover and timing belt. Refer to the Timing Belt unit repair section.
   - Crankshaft sprocket
   - Catalytic converter
   - Oil pan
   - Oil pump pickup tube
   - Oil filter
   - Oil pump

To install:

➡Clearance between the cylinder block oil pan sealing surface to the oil pump oil pan sealing surface should not exceed 0.012–0.031 in. (0.3–0.8mm).

4. Install or connect the following:
   - Oil pump. Use a straight-edge to align the oil pump oil pan sealing surface with the cylinder block oil pan sealing surface and tighten the oil pump retaining bolts to 71–102 inch lbs. (8–11.5 Nm).
   - Oil filter
   - Oil pump pickup tube. Use a new

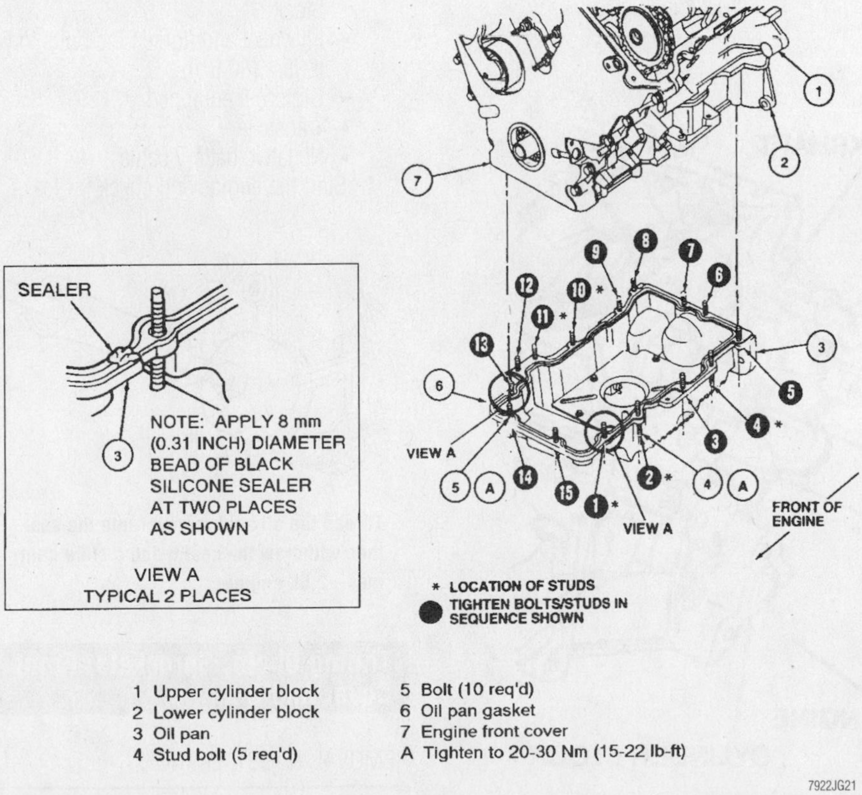

Exploded view of the oil pan mounting, showing the mounting bolt and stud tightening sequence—2.5L engine

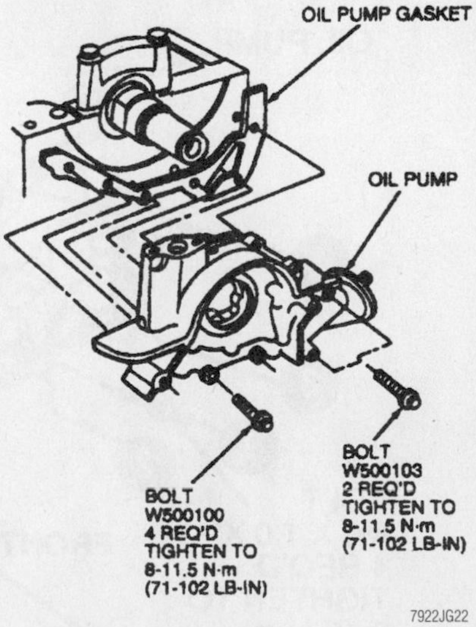

Exploded view of the oil pump mounting—2.0L engine

For Accessory Drive Belt illustrations, see Section 1 of this manual

gasket and tighten the retaining bolts to 71–97 inch lbs. (8–11 Nm). Use a new self-locking nut and tighten it to 13–15 ft. lbs. (17–21 Nm).
- Oil pan
- Catalytic converter
- Crankshaft sprocket
- Timing belt and timing belt cover. Refer to the Timing Belt unit repair section.
- Accessory drive belt
- Negative battery cable

5. Fill the crankcase with the proper amount of engine oil.
6. Run the engine and check for leaks and proper operation.

2.5L Engine

1. Before servicing the vehicle, refer to the precautions in the beginning of this section.
2. Drain the engine oil.
3. Remove or disconnect the following:
   - Negative battery cable
   - Oil pan
   - Timing cover and timing chains
   - Crankshaft sprockets
   - Oil pump pickup tube
   - Oil pump

To install:

4. Install or connect the following:
   - Oil pump and tighten the bolts to 71–106 inch lbs. (8–12 Nm)
   - Oil pump pickup tube. Use a new O-ring and tighten the retaining bolts to 71–97 inch lbs. (8–11 Nm). Use a new self-locking nut and tighten it to 15–22 ft. lbs. (20–30 Nm).
   - Crankshaft sprockets
   - Timing cover and timing chains
   - Oil pan
   - Negative battery cable
5. Fill the crankcase with the proper amount of engine oil.
6. Run the engine and check for leaks and proper operation.

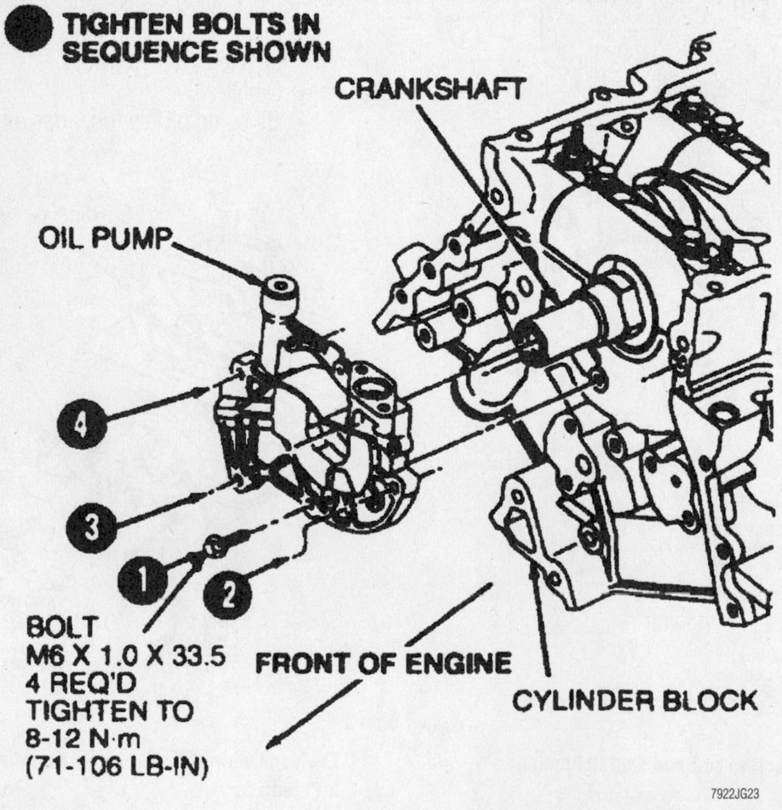

Exploded view of the oil pump mounting, showing the retaining bolt tightening sequence—2.5L engine

Rear Main Seal

REMOVAL & INSTALLATION

2.0L Engine

1. Before servicing the vehicle, refer to the precautions in the beginning of this section.
2. Remove or disconnect the following:
   - Negative battery cable
   - Transaxle
   - Clutch, if equipped
   - Flywheel
   - Oil seal

To install:

3. Install or connect the following:
   - Oil seal. Seat the seal flush with the rear of the crankshaft oil seal retainer.
   - Flywheel and tighten the bolts to 82 ft. lbs. (112 Nm)
   - Clutch, if equipped
   - Transaxle
   - Negative battery cable
4. Start the engine and check for leaks.

2.5L Engine

1. Before servicing the vehicle, refer to the precautions in the beginning of this section.
2. Remove or disconnect the following:
   - Negative battery cable
   - Transaxle
   - Clutch, if equipped
   - Flywheel
   - Oil seal. Use special tool Seal Remover T95P-6701-EH and a slide hammer to remove the seal.

To install:

3. Install or connect the following:
   - Oil seal. Press the seal in evenly until it is flush with the cylinder block.
   - Flywheel and tighten the bolts to 59 ft. lbs. (80 Nm)
   - Clutch, if equipped
   - Transaxle
   - Negative battery cable
4. Start the engine and check for leaks.

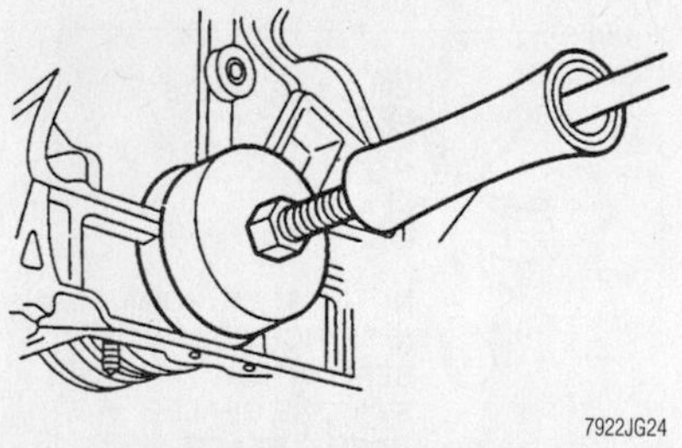

Thread the oil seal remover into the seal, then withdraw the seal using a slide hammer—2.5L engine

Timing Chain, Sprockets, Front Cover and Seal

REMOVAL & INSTALLATION

2.5L Engine

1. Before servicing the vehicle, refer to the precautions in the beginning of this section.

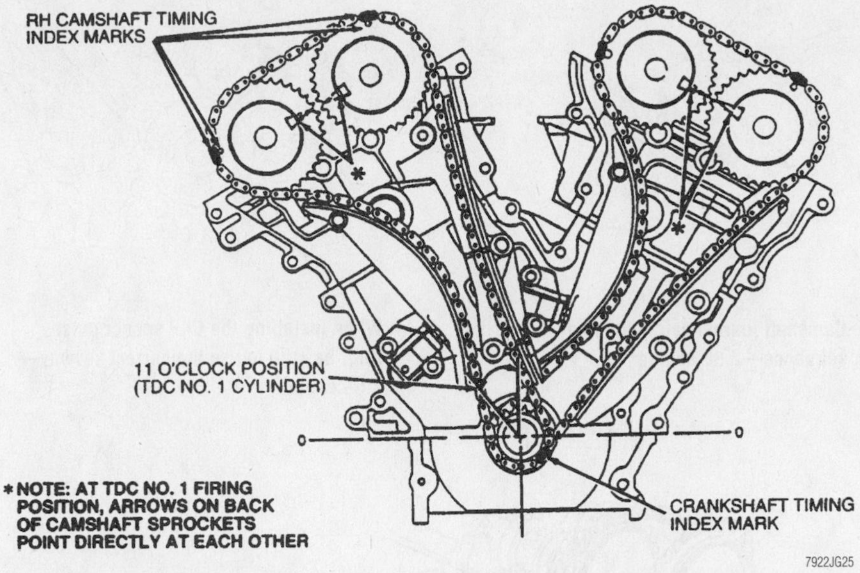

View of the timing chains and gears, showing the right timing chain alignment marks properly positioned—2.5L engine

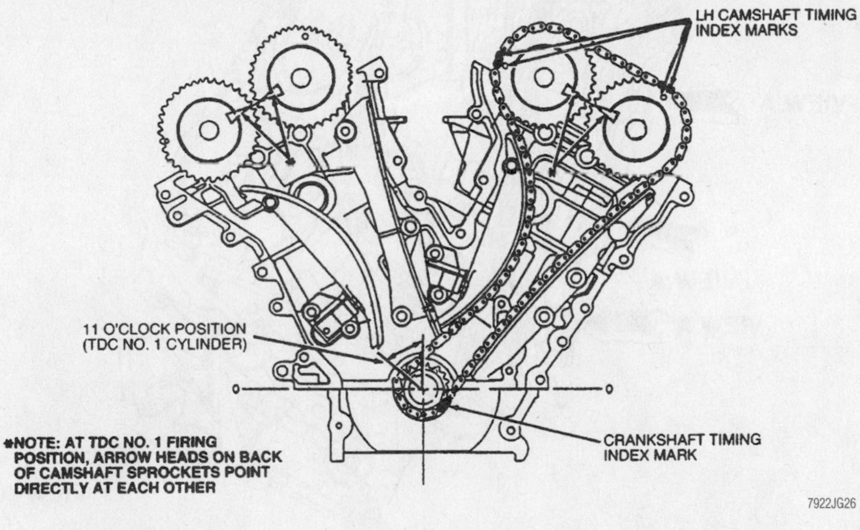

Left timing chain alignment mark positioning for servicing the chain—2.5L engine

2. Attach an engine support fixture to the engine lifting eyes.
3. Drain the engine oil.
4. Remove or disconnect the following:
   - Negative battery cable
   - Upper intake manifold
   - Valve covers
   - Low coolant level sensor connector
   - Coolant recovery reservoir

➡Mark the position of the upper front engine support bracket before removing.

   - Front engine support insulator and bracket
   - Engine wiring harness connectors
   - Accessory drive belt
   - Power steering pump and bracket
   - Right splash shield
   - Alternator
   - Crankshaft pulley
   - Crankshaft Position (CKP) sensor connector
   - Oil pan
   - A/C compressor
   - A/C hose bracket
   - Front cover. Remove the bolts in reverse of the installation sequence.
   - CKP sensor pulse ring

5. Rotate the crankshaft so that the keyway is at the 11 o'clock position to locate the crankshaft at TDC for No. 1 cylinder.
6. Verify that the alignment arrows on the camshafts are aligned. If not, rotate the crankshaft 1 complete revolution and recheck.
7. Rotate the crankshaft so that the keyway is at the 3 o'clock position. This positions the right cylinder head camshafts to the neutral position.
8. Remove the right cylinder head timing chain tensioner retaining bolts and the timing chain tensioner.

WARNING

The camshaft thrust caps must be removed before loosening the remaining camshaft journal cap bolts to ensure that the thrust caps are not damaged.

➡The camshaft journal caps and cylinder heads are numbered to ensure that they are assembled in their original positions.

9. Remove or disconnect the following:
   - Camshaft thrust caps
   - Camshaft journal caps. Loosen the bolts in sequence and in several passes to allow the camshaft to be raised from the cylinder head evenly.
   - Rocker arms. Keep the rocker arms in order for installation.
   - Right timing chain tensioner arm
   - Right timing chain and crankshaft sprocket
10. Rotate the crankshaft 2 revolutions and locate the crankshaft keyway at the 11 o'clock position. This will position the left cylinder head camshafts to their neutral position.
11. Verify that the alignment arrows on the camshafts are aligned.
12. Remove the left cylinder head timing chain tensioner retaining bolts and the timing chain tensioner.

➡The camshaft thrust caps must be removed before loosening the remaining camshaft journal cap bolts to ensure that the thrust caps are not damaged.

➡The camshaft journal caps and cylinder heads are numbered to ensure that they are assembled in their original positions.

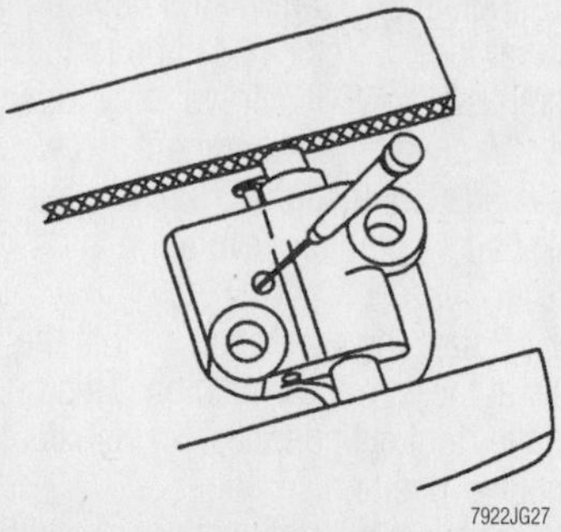

Using a thin prytool, release and hold the timing chain tensioner ratchet/pawl mechanism—2.5L engine

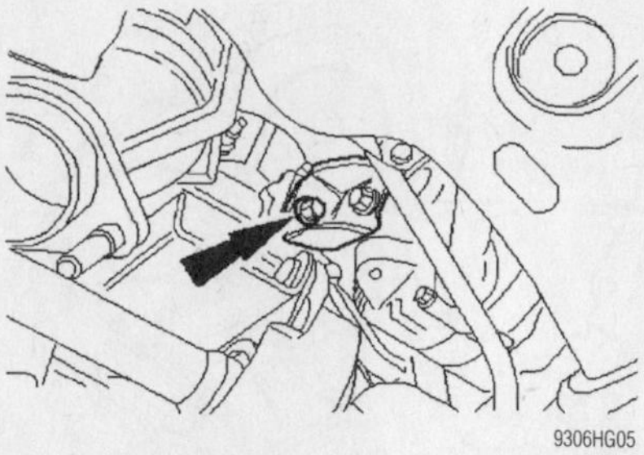

Camshaft journal cap tightening sequence—2.5L engine

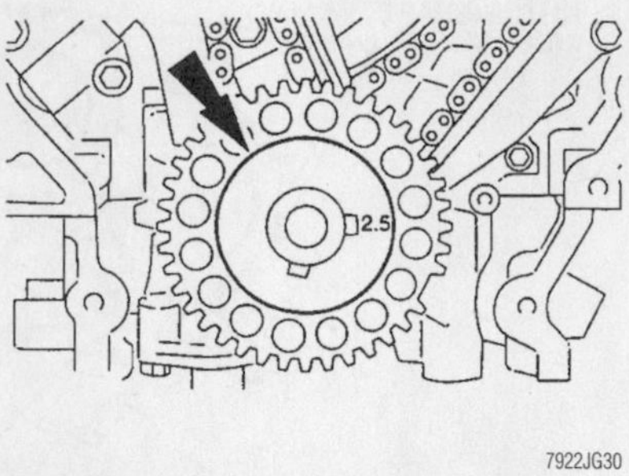

When installing the CKP sensor pulse ring, be sure to use the correct keyway—2.5L engine

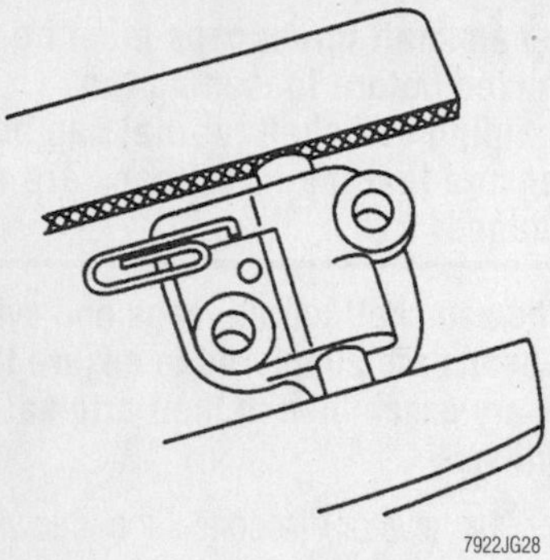

Retain the piston with a 1.5mm wire or paperclip—2.5L engine

13. Remove or disconnect the following:
    - Camshaft thrust caps
    - Camshaft journal caps. Loosen the bolts in sequence and in several passes to allow the camshaft to be raised from the cylinder head evenly.
    - Rocker arms. Keep the rocker arms in order for installation.
    - Left timing chain tensioner arm
    - Left timing chain and crankshaft sprocket

To install:

14. Prepare the timing chain tensioners for installation as follows:

a. Place the left chain tensioner in a vise.

b. Using a small prytool, release and hold the timing chain tensioner ratchet/pawl mechanism through the access hole in the timing chain tensioner.

c. Slowly compress the tensioner.

d. Lock the piston with a 1.5mm wire or paperclip.

e. Repeat for the right chain tensioner.

➡Be sure that the crankshaft keyway is still at the 11 o'clock position.

15. Install or connect the following:
    - Left timing chain and crankshaft

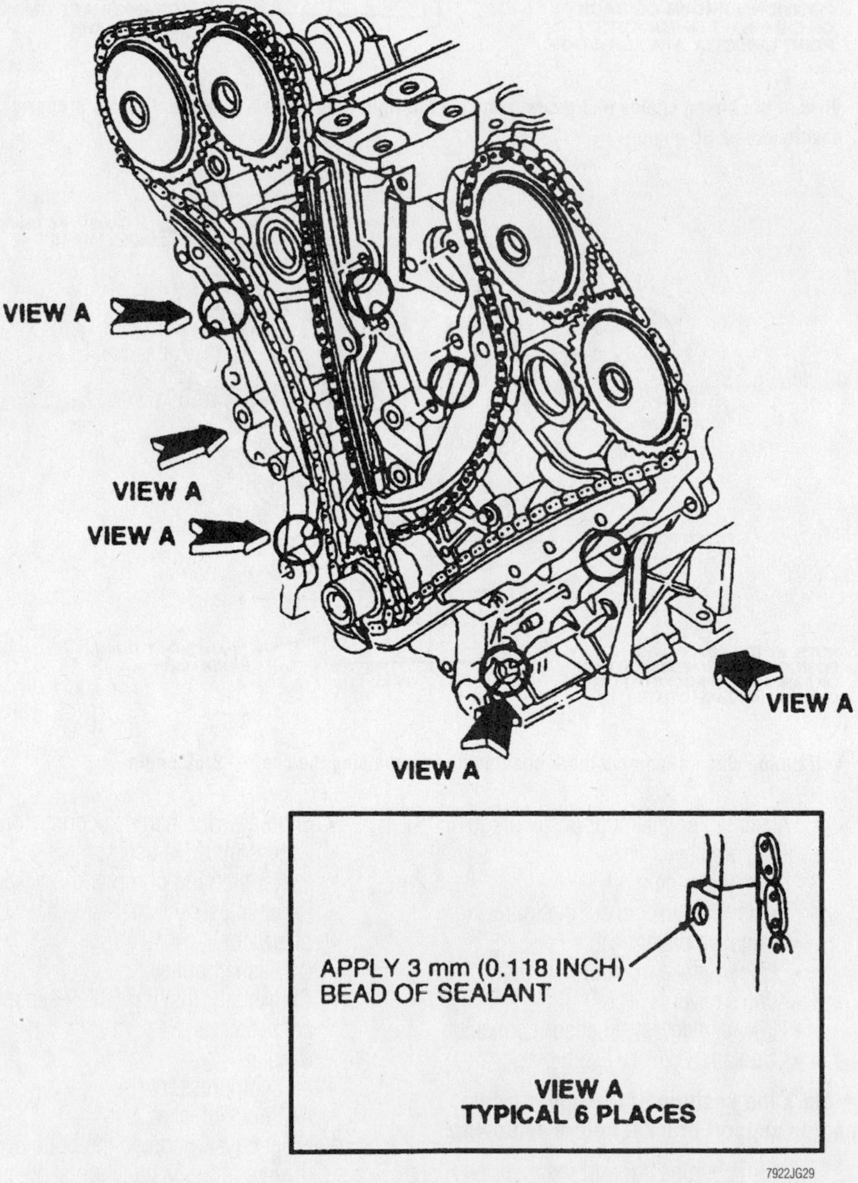

To prevent oil leakage, apply sealant to the places indicated—2.5L engine

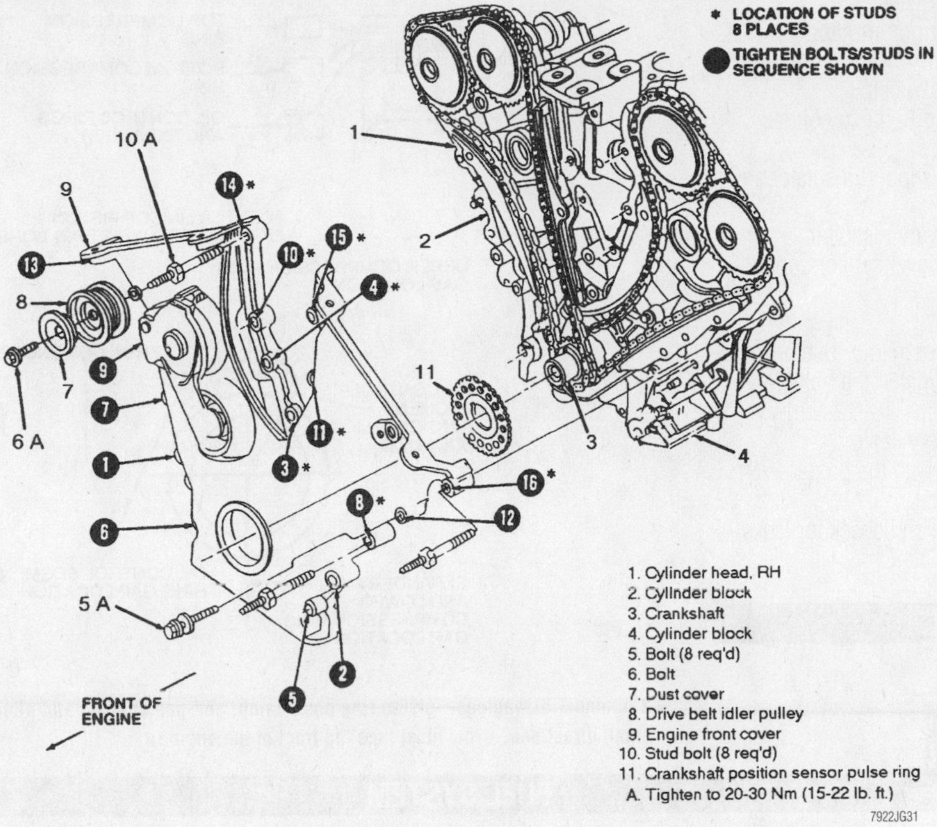

Exploded view of the front cover mounting, showing the retaining bolt and nut tightening sequence—2.5L engine

sprocket. Align the colored links with the index marks on the camshaft and crankshaft sprockets.

- Left timing chain tensioner arm
- Left timing chain tensioner and tighten the retaining bolts to 15–22 ft. lbs. (20–30 Nm)
- Right timing chain and crankshaft sprocket. Align the colored links with the index marks on the camshaft and crankshaft sprockets.
- Right timing chain tensioner arm
- Right timing chain tensioner and tighten the retaining bolts to 15–22 ft. lbs. (20–30 Nm)

➡The crankshaft keyway must be in the 11 o'clock position to install the left cylinder head rocker arms.

➡Do not install the camshaft journal thrust caps until the other journal caps have been installed and tightened.

16. Install the left cylinder head rocker arms in their original positions.
17. Tighten the left camshaft journal caps in the order shown and in several passes to 88 inch lbs. (10 Nm).
18. Install the left camshaft journal thrust caps and tighten the bolts to 88 inch lbs. (10 Nm).
19. Remove the retaining wire from the left timing chain tensioner.

➡The crankshaft keyway must be in the 3 o'clock position to install the right cylinder head rocker arms.

20. Rotate the crankshaft so that the keyway is in the 3 o'clock position.
21. Install the right cylinder head rocker arms in their original positions.

➡Do not install the camshaft journal thrust caps until the other journal caps have been installed and tightened.

22. Tighten the right camshaft journal caps in the order shown and in several passes to 88 inch lbs. (10 Nm).
23. Install the right camshaft journal thrust caps and tighten the bolts to 88 inch lbs. (10 Nm).
24. Remove the retaining wire from the right timing chain tensioner.
25. Install the CKP sensor pulse ring. Use the keyway for the 2.5L engine as shown.
26. Replace the crankshaft seal in the front cover with a new one. Apply clean engine oil to the seal lip.
27. Apply silicone sealer to the 6 critical areas shown in View **A**.
28. Place new front cover gaskets onto the dowel pins on the cylinder block and heads.
29. Place the front cover into position.
30. Install the 6 front cover retaining bolts and stud bolts where the silicone sealer was applied.
31. Tighten the bolts and stud bolts until the front cover contacts the cylinder block and heads, then turn the bolts and stud bolts an additional ¼ turn.
32. Install the remaining front cover retaining bolts and stud bolts.
33. Tighten all of the front cover retaining bolts and stud bolts in sequence to 15–22 ft. lbs. (20–30 Nm).
34. Install or connect the following:
    - A/C hose bracket
    - A/C compressor
    - Oil pan
    - CKP sensor connector
    - Crankshaft pulley
    - Alternator

- Right splash shield
- Power steering pump and bracket
- Accessory drive belt
- Engine wiring harness connectors
- Front engine support insulator and bracket
- Coolant recovery reservoir
- Low coolant level sensor connector
- Valve covers
- Upper intake manifold. Use new gaskets and tighten the bolts to 71–106 inch lbs. (8–12 Nm).
- Negative battery cable

35. Fill the engine with the proper amount and grade of oil.

36. Run the engine and check for leaks and proper operation.

Piston and Ring

POSITIONING

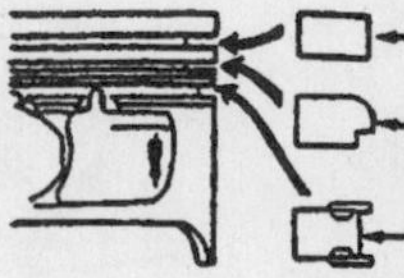

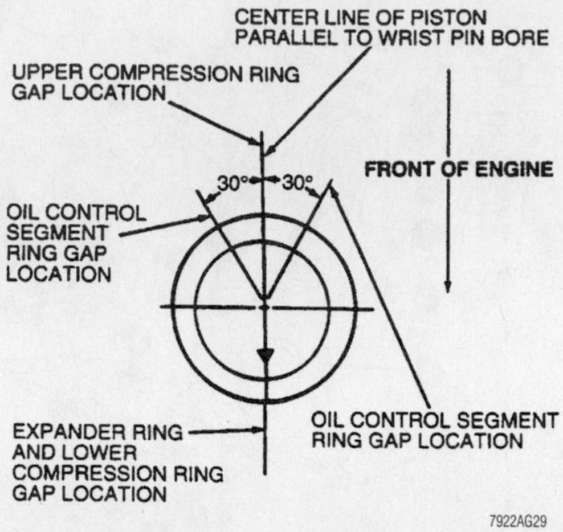

2.0L and 2.5L engines—piston ring positioning, end-gap spacing and piston positioning. The small directional arrow must face the front of the engine.

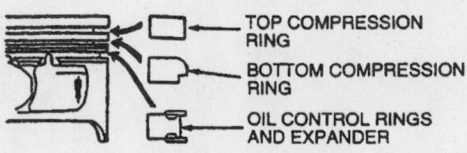

FUEL SYSTEM

Fuel System Service Precautions

Safety is the most important factor when performing not only fuel system maintenance but any type of maintenance. Failure to conduct maintenance and repairs in a safe manner may result in serious personal injury or death. Maintenance and testing of the vehicle's fuel system components can be accomplished safely and effectively by adhering to the following rules and guidelines.

- To avoid the possibility of fire and personal injury, always disconnect the negative battery cable unless the repair or test procedure requires that battery voltage be applied.
- Always relieve the fuel system pressure prior to disconnecting any fuel system component (injector, fuel rail, pressure regulator, etc.), fitting or fuel line connection. Exercise extreme caution whenever relieving fuel system pressure, to avoid exposing skin, face and eyes to fuel spray. Please be advised that fuel under pressure may penetrate the skin or any part of the body that it contacts.
- Always place a shop towel or cloth around the fitting or connection prior to loosening to absorb any excess fuel due to spillage. Ensure that all fuel spillage (should it occur) is quickly removed from engine surfaces. Ensure that all fuel soaked cloths or towels are deposited into a suitable waste container.
- Always keep a dry chemical (Class B) fire extinguisher near the work area.
- Do not allow fuel spray or fuel vapors to come into contact with a spark or open flame.
- Always use a back-up wrench when loosening and tightening fuel line connection fittings. This will prevent unnecessary stress and torsion to fuel line piping.
- Always replace worn fuel fitting O-rings with new. Do not substitute fuel hose or equivalent, where fuel pipe is installed.

Before servicing the vehicle, make sure to refer to the precautions in the beginning of this section as well.

Fuel System Pressure

RELIEVING

1. Before servicing the vehicle, refer to the precautions in the beginning of this section.
2. Disconnect the negative battery cable.
3. Remove the engine air cleaner assembly.
4. Loosen the fuel tank filler cap to relieve pressure in the fuel tank.
5. Connect a fuel pressure gauge to the fuel pressure relief valve located on the fuel rail.
6. Open the manual valve on the fuel pressure gauge and drain the fuel through the drain tube into a suitable container.
7. Remove the fuel pressure gauge.
8. When service on the vehicle is complete, install the engine air cleaner assembly, tighten the fuel tank filler cap and connect the negative battery cable.

Fuel Filter

REMOVAL & INSTALLATION

1. Before servicing the vehicle, refer to the precautions in the beginning of this section.
2. Relieve the fuel system pressure.

➡The fuel filter is located underneath the vehicle, near the fuel tank.

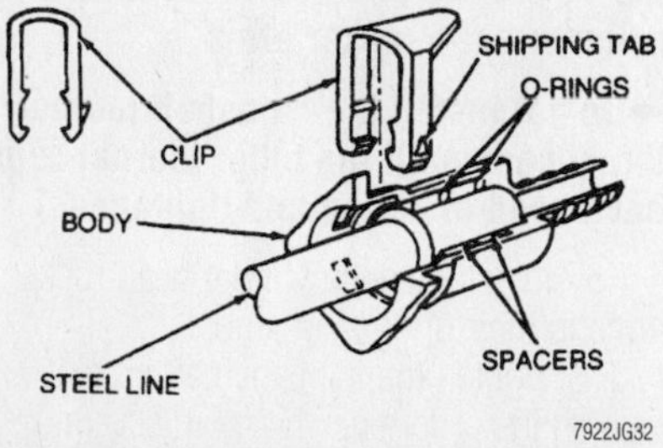

Once the retainer clip has been removed from the connector, the fuel line can be removed from the filter

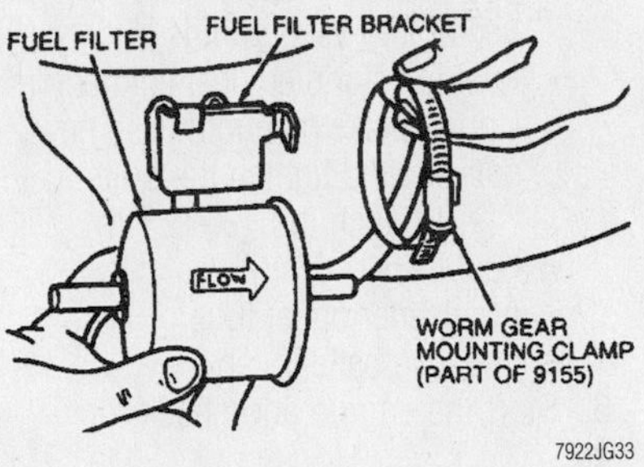

Be sure to install the fuel filter with the arrow pointing in the direction of the fuel flow

3. Remove or disconnect the following:
   - Retainer clips at both ends of the fuel filter
   - Fuel lines from the fuel filter
   - Fuel filter

To install:

4. Install or connect the following:
   - New fuel filter with the arrow pointed in the direction of flow
   - New retainer clips onto the fuel line fittings before placing the lines onto the fuel filter ends
   - Fuel lines onto the fuel filter until an audible click is heard. Pull on the fitting to verify a good connection.
5. Start the engine and check for fuel leaks and proper operation.

Fuel Pump

REMOVAL & INSTALLATION

1. Before servicing the vehicle, refer to the precautions in the beginning of this section.
2. Disconnect the negative battery cable.

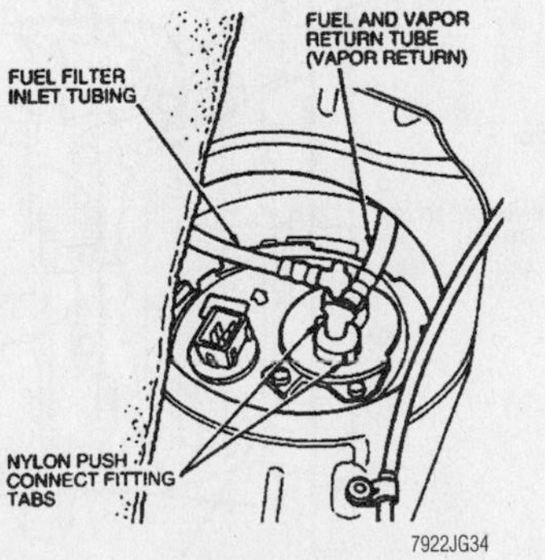

View of fuel pump fittings through the floor pan

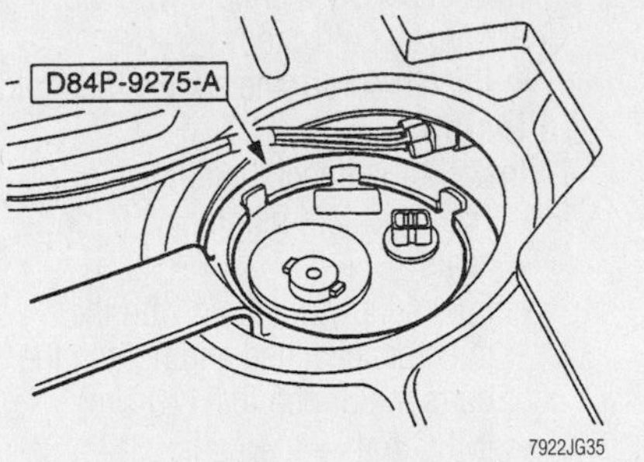

Remove the locking ring from the fuel pump sender with special tool Fuel Tank Sender Wrench D84P-9275-A

3. Relieve the fuel pressure.
4. Remove or disconnect the following:
   - Rear seat cushion
   - Plastic access panel
   - Fuel pump wiring harness connector
   - Fuel lines
   - Fuel pump module lock ring. Use special tool Fuel Tank Sender Wrench D84P-9275-A.
   - Fuel pump module

To install:

5. Install or connect the following:
   - Fuel pump module with a new O-ring seal
   - Fuel pump module lock ring using special tool Fuel Tank Sender Wrench D84P-9275-A
   - Fuel lines
   - Fuel pump wiring harness connector
   - Plastic access panel
   - Rear seat cushion
   - Negative battery cable
6. Start the engine and check for leaks and proper operation.

Fuel Injector

REMOVAL & INSTALLATION

2.0L Engine

1. Before servicing the vehicle, refer to the precautions in the beginning of this section.
2. Relieve the fuel system pressure.
3. Remove or disconnect the following:
   - Negative battery cable
   - Fuel lines and retaining clip
   - Pressure regulator vacuum hose
   - Accelerator cable
   - Fuel injector electrical connectors
   - Fuel supply manifold with the injectors attached
   - Injectors from the supply manifold

To install:

4. Install or connect the following:
   - Fuel injectors using new O-ring seals
   - Fuel supply manifold with the injectors attached and tighten the bolts to 88 inch lbs. (10 Nm)
   - Fuel injector electrical connectors
   - Accelerator cable
   - Pressure regulator vacuum hose
   - Fuel lines and retaining clip
   - Negative battery cable

2.5L Engine

EXCEPT RETURNLESS FUEL DELIVERY SYSTEM

1. Before servicing the vehicle, refer to the precautions in the beginning of this section.
2. Relieve fuel system pressure.
3. Remove or disconnect the following:
   - Negative battery cable
   - Air cleaner outlet tube
   - Fuel lines
   - Upper intake manifold
   - Fuel injector electrical connectors
   - Fuel pressure regulator vacuum line
   - Intake Manifold Runner Control (IMRC) rod
   - Fuel supply manifold with the injectors attached
   - Injectors from the supply manifold

To install:

4. Install or connect the following:
   - Fuel injectors using new O-ring seals
   - Fuel supply manifold with the injectors attached and tighten the bolts to 88 inch lbs. (10 Nm)
   - IMRC rod
   - Fuel pressure regulator vacuum line
   - Fuel injector electrical connectors
   - Upper intake manifold. Use new gaskets and tighten the bolts to 71–106 inch lbs. (8–12 Nm).
   - Fuel lines
   - Air cleaner outlet tube
   - Negative battery cable
5. Start the engine and check for leaks.

RETURNLESS FUEL DELIVERY SYSTEM

1. Before servicing the vehicle, refer to the precautions in the beginning of this section.

2. Relieve fuel system pressure.
3. Remove or disconnect the following:
   - Negative battery cable
   - Air cleaner outlet tube
   - Fuel line
   - Upper intake manifold
   - Fuel injector electrical connectors
   - Fuel pressure sensor electrical connector and vacuum line
   - Intake Manifold Runner Control (IMRC) rod
   - Fuel supply manifold with the injectors attached
   - Injectors from the supply manifold

To install:

4. Install or connect the following:
   - Fuel injectors using new O-ring seals
   - Fuel supply manifold with the injectors attached and tighten the bolts to 88 inch lbs. (10 Nm)
   - IMRC rod
   - Fuel pressure sensor electrical connector and vacuum line
   - Fuel injector electrical connectors
   - Upper intake manifold. Use new gaskets and tighten the bolts to 71–106 inch lbs. (8–12 Nm).
   - Fuel line
   - Air cleaner outlet tube
   - Negative battery cable
5. Start the engine and check for leaks.

DRIVE TRAIN

Transaxle Assembly

REMOVAL & INSTALLATION

Automatic

1. Before servicing the vehicle, refer to the precautions in the beginning of this section.
2. Secure the radiator and fan shroud to the radiator support.
3. Loosen the left and right upper strut mounting nuts 5 turns.
4. Attach an engine support fixture to the engine lifting eyes.
5. Drain the transaxle fluid.
6. Remove or disconnect the following:
   - Battery
   - Air cleaner assembly and mounting bracket
   - Shift cable and bracket
   - Transaxle control wiring harness connector
   - Transaxle range sensor
   - Rear transaxle support insulator
   - Oil dipstick tube and exhaust manifold heat shield (2.0L engine)
   - Water pump pulley shield (2.5L engine)
   - Steering column pinch bolt
   - Front wheels
   - Left, right and center splash shields
   - Lower ball joints
   - Outer tie rod ends
   - Stabilizer bar links
   - Left and right engine support insulators
   - Power steering oil cooler hoses
   - Bumper cover braces
   - Radiator air deflector
   - Lower radiator supports
   - Exhaust system
   - A/C accumulator mounting bolts
   - Transaxle cooler lines and bracket
   - Power steering hoses. Lower the subframe for access.
   - Subframe
   - Turbine Speed (TSS) sensor connector
   - Halfshafts and intermediate shaft
   - Vehicle Speed (VSS) sensor connector
   - Torque converter nuts
   - Starter
   - Front engine support insulator bracket
7. Lower the powertrain until the transaxle is level with the left frame member.
8. Support the transaxle on a transmission jack.
9. Remove or disconnect the following:
   - Transaxle flange bolts
   - Transaxle from the engine
   - Transaxle

➡**Use care when removing the transaxle to prevent the torque converter from falling out.**

To install:

➡**Replace all snaprings, split pins and self-locking nuts.**

10. Install or connect the following:
    - Transaxle and tighten the flange bolts to 41–50 ft. lbs. (55–68 Nm)
    - Torque converter and tighten the nuts to 23–39 ft. lbs. (31–53 Nm)
    - TSS connector
    - Subframe. Attach the power steering hoses before installing the subframe bolts.
    - Special tool Subframe Alignment Pin Set T95P-2100-AH and the subframe bolts. Tighten the subframe bolts to 81–110 ft. lbs. (110–150 Nm). Remove the alignment pins.

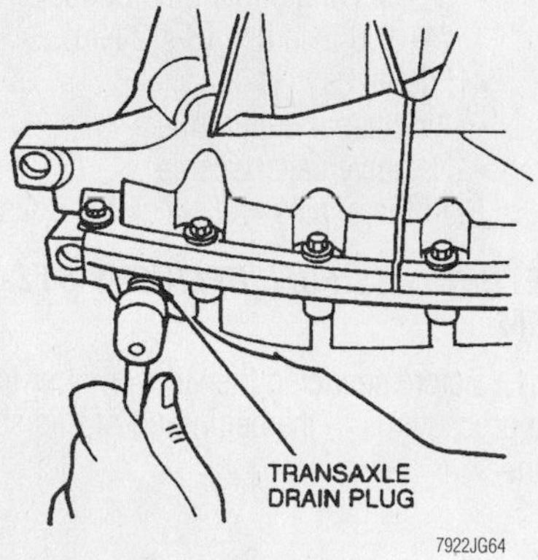

Transaxle drain plug location

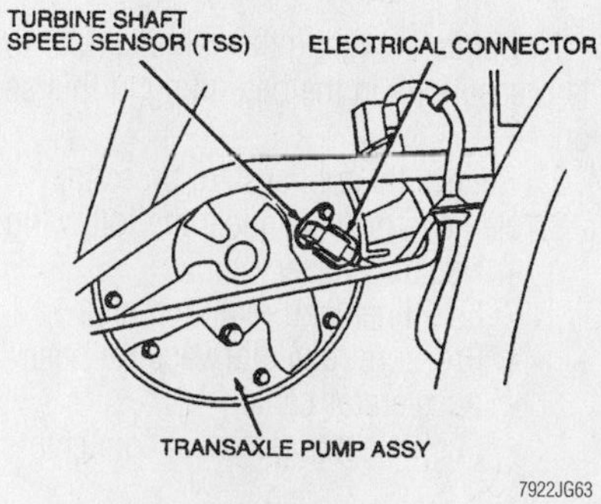

TSS sensor location

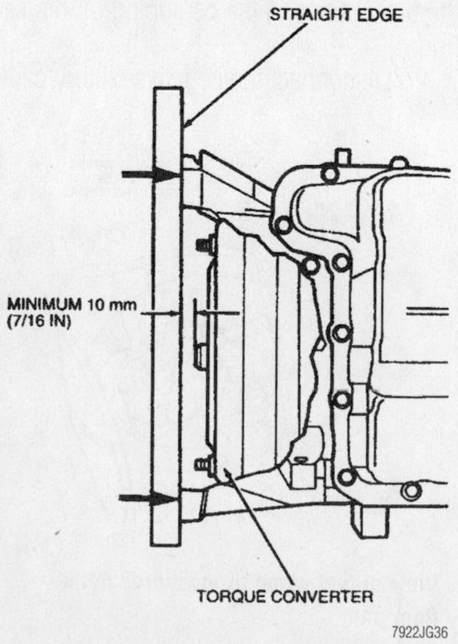

Be sure that the torque converter is properly seated in the transaxle

- A/C accumulator bolts
- Power steering oil cooler lines
- Special tool Powertrain Alignment Gauge T94P-6000-AH in place of the left engine support insulator. Tighten the 2 retaining bolts to 20 ft. lbs. (27 Nm) and snug the through-bolt.

➡**The left and right support insulators must be aligned in the middle of the support insulator brackets.**

- Right engine support insulator. Tighten the 2 subframe retaining bolts to 30–41 ft. lbs. (41–55 Nm) and the through-bolt to 75–102 ft. lbs. (103–137 Nm).
- Rear engine support insulator and tighten the nuts to 61 ft. lbs. (83 Nm)
- Front engine support insulator and tighten the nuts to 61 ft. lbs. (83 Nm)

11. Remove the Powertrain Alignment Tool and install the left engine support insulator. Tighten the center bolt to 88 ft. lbs. (120 Nm), and the mounting bolts to 35 ft. lbs. (48 Nm).
   - Transaxle oil cooler lines
   - Starter and tighten the bolts to 43–58 ft. lbs. (59–79 Nm) for the 2.5L engine or 15–20 ft. lbs. (20–27 Nm) for the 2.0L engine
   - VSS sensor connector
   - Halfshafts and intermediate shaft
   - Exhaust system
   - Lower radiator supports
   - Radiator air deflector
   - Bumper cover braces
   - Power steering oil cooler hoses
   - Left and right engine support insulators
   - Stabilizer bar links and tighten the nuts to 35–48 ft. lbs. (47–65 Nm)
   - Outer tie rod ends and tighten the nuts to 23–35 ft. lbs. (31–47 Nm)
   - Lower ball joints and tighten the pinch bolt to 61 ft. lbs. (83 Nm)
   - Left, right and center splash shields
   - Front wheels and tighten the lug nuts to 63 ft. lbs. (86 Nm)
   - Steering column pinch bolt and tighten the bolt to 15–20 ft. lbs. (20–27 Nm)
   - Water pump pulley shield (2.5L engine)
   - Oil dipstick tube and exhaust manifold heat shield (2.0L engine)
   - Transaxle range sensor
   - Transaxle control wiring harness connector
   - Shift cable and bracket
   - Air cleaner assembly and mounting bracket
   - Battery

12. Tighten the upper strut mount nuts to 34 ft. lbs. (46 Nm).
13. Fill the transaxle.
14. Check for leaks and proper operation.

➡**Whenever the vehicle subframe is removed or lowered, the wheel alignment should be checked.**

Manual

1. Before servicing the vehicle, refer to the precautions in the beginning of this section.
2. Attach an engine support fixture to the engine lifting eyes.
3. Secure the radiator and fan shroud to the radiator support.
4. Loosen the front strut upper mount nuts 5 turns.
5. Remove or disconnect the following:
   - Battery
   - Steering column pinch bolt
   - Air cleaner assembly and bracket
   - Front and rear engine support insulators
   - Reverse lamp switch connector
   - Ground strap
   - Left, right and center splash shields
   - Clutch slave cylinder hydraulic line
   - Accessory drive belt cover
   - Power steering pulley shield (2.5L engine)
   - Exhaust crossover pipe (2.5L engine)
   - Catalytic converter
   - Front wheels
   - Vehicle Speed Sensor (VSS)
   - Radiator air deflector
   - Shift rod and stabilizer bar
   - A/C accumulator mounting bolts.
   - Halfshafts and intermediate shaft
   - Left and right engine support insulators
   - Lower ball joints
   - Outer tie rod ends
   - Stabilizer bar links
   - Power steering oil cooler lines
   - Lower radiator supports
   - Bumper cover braces
   - Power steering hoses. Lower the subframe for access.
   - Subframe
   - Starter
6. Lower the powertrain until the transaxle is level with the left frame member.
7. Support the transaxle on a transmission jack.
8. Remove or disconnect the following:
   - Transaxle flange bolts
   - Transaxle from the engine

To install:

➡**Replace all snaprings, split pins and self-locking nuts.**

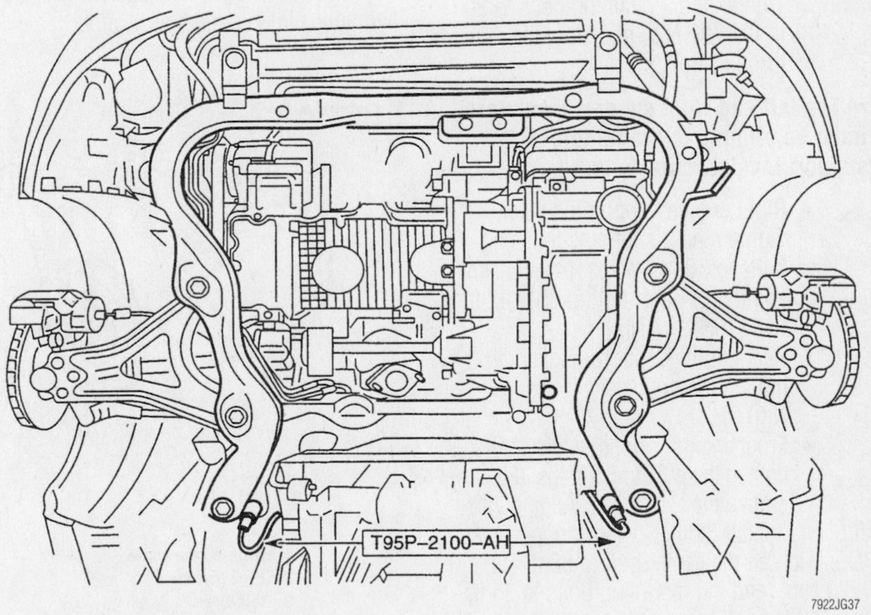

To properly align the subframe, install Subframe Alignment Pin Set T95P-2100-AH—manual and automatic transaxles

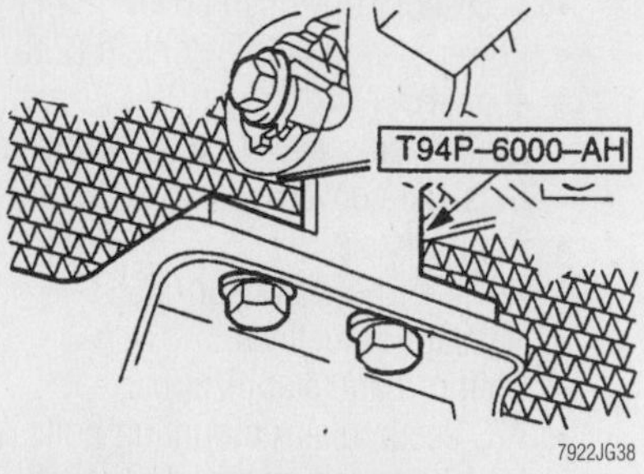

The Powertrain Alignment Gauge (T94P-6000-AH) tool must be installed in the correct position to ensure proper engine/transaxle orientation

9. Install or connect the following:
   - Transaxle to the engine and tighten the flange bolts to 28–38 ft. lbs. (38–51 Nm)
   - Starter and tighten the bolts to 35 ft. lbs. (48 Nm)
   - Subframe. Attach the power steering hoses before installing the subframe bolts.
   - Special tool Subframe Alignment Pin Set T95P-2100-AH and the subframe bolts. Tighten the subframe bolts to 81–110 ft. lbs. (110–150 Nm). Remove the alignment pins.
   - A/C accumulator bolts
   - Power steering oil cooler lines
   - Special tool Powertrain Alignment Gauge T94P-6000-AH in place of the left engine support insulator. Tighten the 2 retaining bolts to 20 ft. lbs. (27 Nm) and snug the through-bolt.

➡The left and right support insulators must be aligned in the middle of the support insulator brackets.

   - Right engine support insulator. Tighten the 2 subframe retaining bolts to 30–41 ft. lbs. (41–55 Nm) and the through-bolt to 75–102 ft. lbs. (103–137 Nm).
   - Rear engine support insulator and tighten the nuts to 61 ft. lbs. (83 Nm)
   - Front engine support insulator and tighten the nuts to 61 ft. lbs. (83 Nm)

10. Remove the Powertrain Alignment Tool and install the left engine support insulator. Tighten the center bolt to 88 ft. lbs. (120 Nm), and the mounting bolts to 35 ft. lbs. (48 Nm).
   - Bumper cover braces
   - Lower radiator supports and tighten the bolts to 71–97 inch lbs. (8–11 Nm)
   - Power steering oil cooler lines
   - Stabilizer bar links and tighten the fasteners to 35–48 ft. lbs. (47–65 Nm)
   - Outer tie rod ends and tighten the nuts to 23–35 ft. lbs. (31–47 Nm)
   - Lower ball joints and tighten the pinch bolt to 61 ft. lbs. (83 Nm)
   - Halfshafts and intermediate shaft
   - A/C accumulator mounting bolts
   - Shift rod and stabilizer bar. Tighten the shift rod bolt to 14–18 ft. lbs. (19–25 Nm) and the stabilizer bar nut to 28–38 ft. lbs. (38–51 Nm).
   - Radiator air deflector
   - VSS sensor
   - Front wheels and tighten the lug nuts to 63 ft. lbs. (86 Nm)
   - Catalytic converter
   - Exhaust crossover pipe (2.5L engine)
   - Power steering pulley shield (2.5L engine)
   - Accessory drive belt cover
   - Clutch slave cylinder hydraulic line
   - Left, right and center splash shields
   - Ground strap
   - Reverse lamp switch connector
   - Air cleaner assembly and bracket
   - Steering column pinch bolt and tighten the bolt to 15–20 ft. lbs. (20–27 Nm)
   - Battery

11. Tighten the strut mounting nuts to 34 ft. lbs. (46 Nm).

12. Adjust the shift linkage and bleed the hydraulic clutch system as required.

13. Road test the vehicle and check for proper operation.

➡Whenever the vehicle subframe is removed or lowered, the wheel alignment should be checked.

Clutch

ADJUSTMENTS

These vehicles are equipped with a hydraulic clutch system. No adjustment is necessary.

REMOVAL & INSTALLATION

1. Before servicing the vehicle, refer to the precautions in the beginning of this section.
2. Remove or disconnect the following:
   - Transaxle
   - Clutch pressure plate
   - Clutch disk
   - Flywheel

To install:

3. Install or connect the following:
   - Flywheel and tighten the bolts to 82 ft. lbs. (112 Nm) for 2.0L engines or to 59 ft. lbs. (80 Nm) for 2.5L engines
   - Clutch disk and pressure plate and tighten the pressure plate bolts evenly in several passes to 21 ft. lbs. (29 Nm)
   - Transaxle
4. Bleed the hydraulic clutch system, if required.

Clutch System

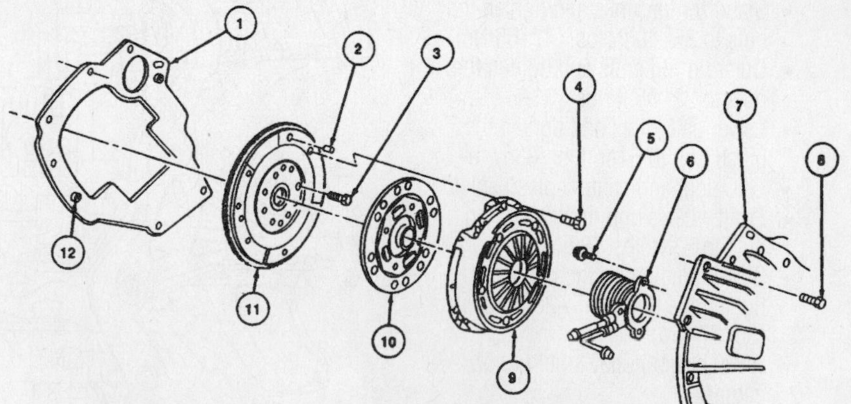

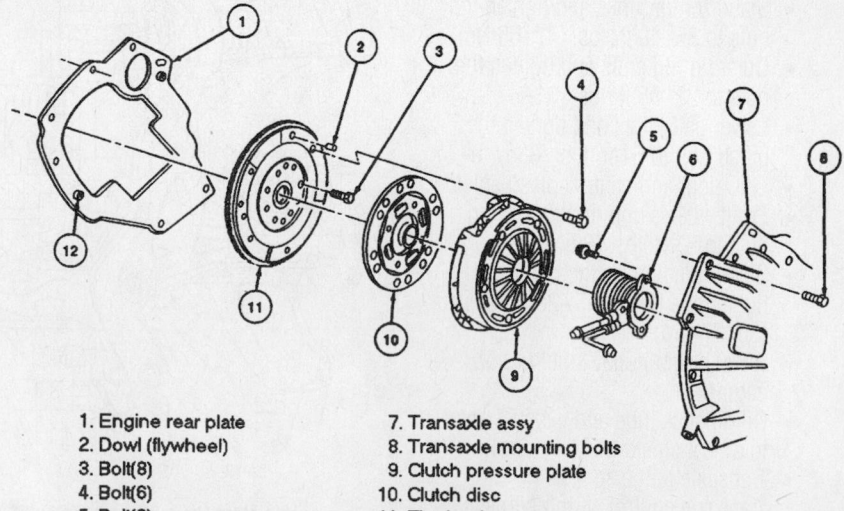

1. Engine rear plate
2. Dowl (flywheel)
3. Bolt(8)
4. Bolt(6)
5. Bolt(3)
6. Clutch slave cylinder
7. Transaxle assy
8. Transaxle mounting bolts
9. Clutch pressure plate
10. Clutch disc
11. Flywheel
12. Dowl bushing (engine plate)

7922JG39

Exploded view of the clutch disc, pressure plate and related component mounting

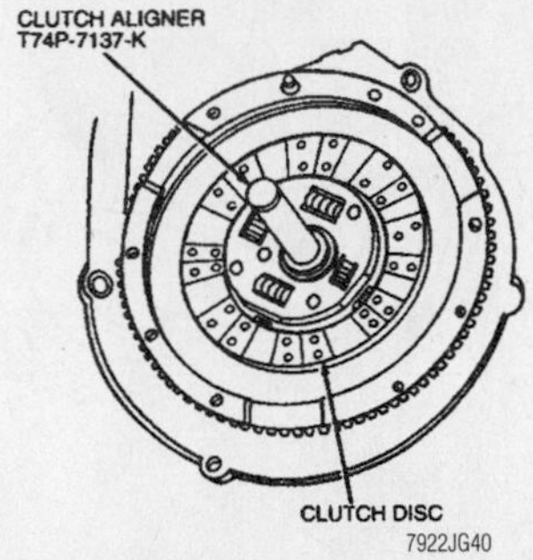

Insert an alignment tool through the clutch disc to ensure that it is centered after the pressure plate is installed

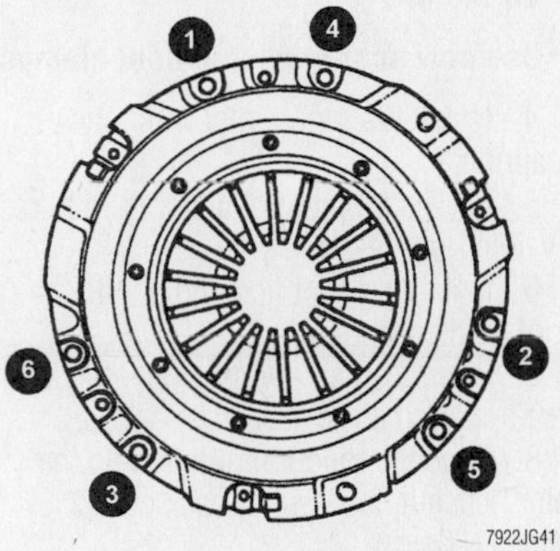

Pressure plate tightening sequence

5. Check the clutch system for proper operation.

Hydraulic Clutch System

BLEEDING

1. Before servicing the vehicle, refer to the precautions in the beginning of this section.
2. Remove the negative battery cable, air cleaner outlet tube and the Mass Airflow (MAF) sensor.

➡**The brake master cylinder fluid reservoir is also the reservoir for the hydraulic clutch master cylinder.**

3. Remove the rubber inspection cover from the bell housing.
4. Connect a hose to the bleeder valve fitting on the clutch slave cylinder. Submerge the other end of the hose into a container of clean brake fluid.
5. Open the bleeder valve and have an assistant depress the clutch pedal.
6. Close the bleeder before releasing the clutch pedal.
7. Repeat the procedure until no more air bubbles are seen.
8. Install the rubber inspection cover to the bell housing.
9. Top off the brake master cylinder fluid reservoir and install the diaphragm and cap securely.
10. Install the MAF sensor and air cleaner outlet tube.
11. Connect the negative battery cable.
12. Check the clutch for proper operation.

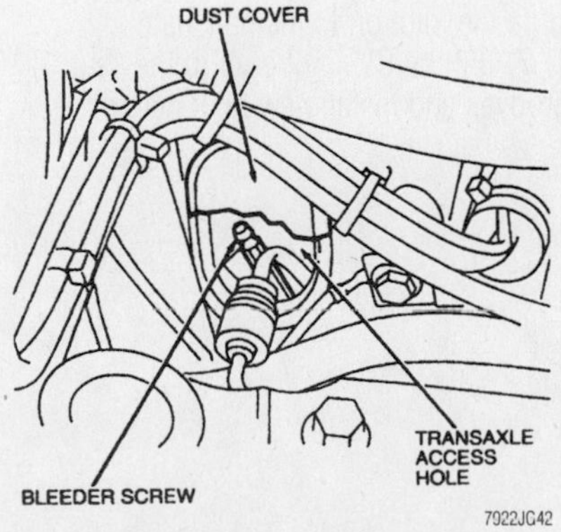

Clutch slave cylinder bleeder valve

Halfshaft

REMOVAL & INSTALLATION

Left Axle Halfshaft

➡**Replace all snaprings, split pins and self-locking nuts.**

1. Before servicing the vehicle, refer to the precautions in the beginning of this section.
2. Loosen the strut upper mount nut 5 turns.
3. Remove or disconnect the following:
   - Wheel
   - Splash shield
   - Lower ball joint

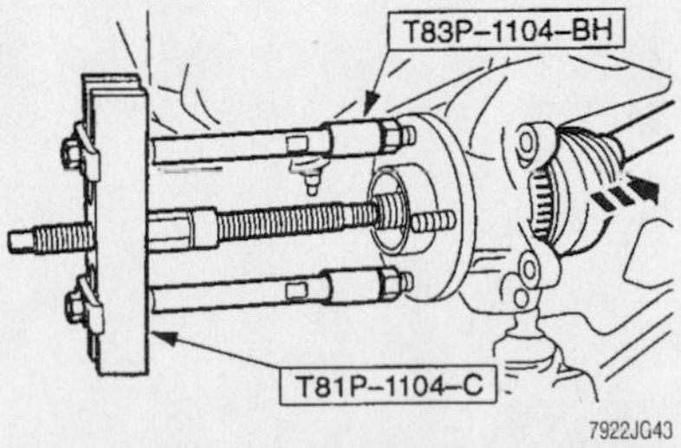

Front Hub Remover/Replacer T81P-1104-C

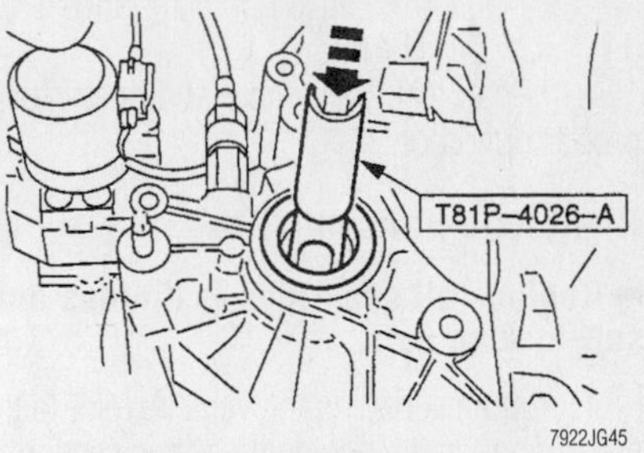

If the intermediate shaft has been removed, install Differential Rotator T81P-4026-A before removing the left halfshaft

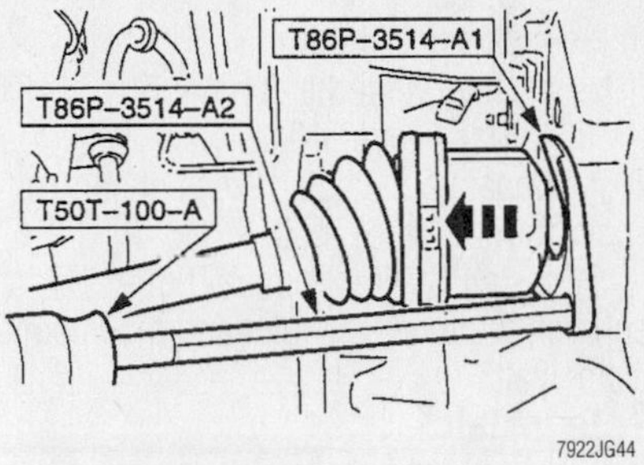

Use a slide hammer to pull the inner CV-joint from the transaxle

   - Outer tie rod end
   - Stabilizer bar link
4. Separate the outer CV-joint and halfshaft from the wheel hub.
5. Install CV-joint puller T86P-3514-A1 between the inner CV-joint and the transaxle case.

➡**If the intermediate shaft has already been removed, install Differential Rotator T81P-4026-A into the right side of the differential before removing the left shaft to maintain alignment within the differential.**

6. Remove the halfshaft.

To install:

7. Install the halfshaft. A non-metallic mallet may be used to aid in seating the inner CV-joint into the differential side gear. Insert the stub shaft into the wheel hub and tighten the hub nut to 246 ft. lbs. (340 Nm).
8. Install or connect the following:
   - Lower ball joint and tighten the pinch bolt to 61 ft. lbs. (83 Nm)
   - Outer tie rod end and tighten the nut to 20 ft. lbs. (28 Nm)
   - Stabilizer bar link and tighten the nut to 14–23 ft. lbs. (20–32 Nm)
   - Splash shield
   - Wheel and tighten the lug nuts to 63 ft. lbs. (86 Nm)

9. Tighten the strut upper mount nut to 34 ft. lbs. (46 Nm).
10. Road test the vehicle and check for proper operation.

Right Axle Halfshaft

➡Replace all snaprings, split pins and self-locking nuts.

1. Before servicing the vehicle, refer to the precautions in the beginning of this section.
2. Loosen the strut upper mount nut 5 turns.
3. Remove or disconnect the following:
   - Wheel
   - Splash shield
   - Lower ball joint
   - Outer tie rod end
   - Stabilizer bar link
4. Separate the outer CV-joint and halfshaft from the wheel hub.
5. Install CV-joint puller T86P-3514-A1 between the inner CV-joint and the intermediate shaft.

To install:

6. Install the halfshaft. A non-metallic mallet may be used to aid in seating the inner CV-joint into the differential side gear. Insert the stub shaft into the wheel hub and tighten the hub nut to 246 ft. lbs. (340 Nm).
7. Install or connect the following:
   - Lower ball joint and tighten the pinch bolt to 61 ft. lbs. (83 Nm)
   - Outer tie rod end and tighten the nut to 20 ft. lbs. (28 Nm)
   - Stabilizer bar link and tighten the nut to 14–23 ft. lbs. (20–32 Nm)
   - Splash shield
   - Wheel and tighten the lug nuts to 63 ft. lbs. (86 Nm)
8. Tighten the strut upper mount nut to 34 ft. lbs. (46 Nm).
9. Road test the vehicle and check for proper operation.

Intermediate Shaft

1. Before servicing the vehicle, refer to the precautions in the beginning of this section.
2. Remove the right halfshaft.

➡If the left halfshaft has already been removed, install Differential Rotator T81P-4026-A into the left side of the differential before removing the intermediate shaft to maintain alignment within the differential.

3. Remove the intermediate shaft and support bearing.

To install:

4. Install the intermediate shaft and tighten the support bearing nuts to 17–22 ft. lbs. (24–30 Nm).
5. Install the right halfshaft.

CV-Joints

REMOVAL & REPLACEMENT

Inner Tripod Joint

1. Before servicing the vehicle, refer to the precautions in the beginning of this section.
2. Place the halfshaft in a vise.
3. Remove or disconnect the following:
   - Tripod joint boot
   - Tripod joint housing
   - Tripod assembly snapring
   - Tripod assembly from the halfshaft with Bearing Puller D79L-4621-A and Remover/Installer T81P-1104-C

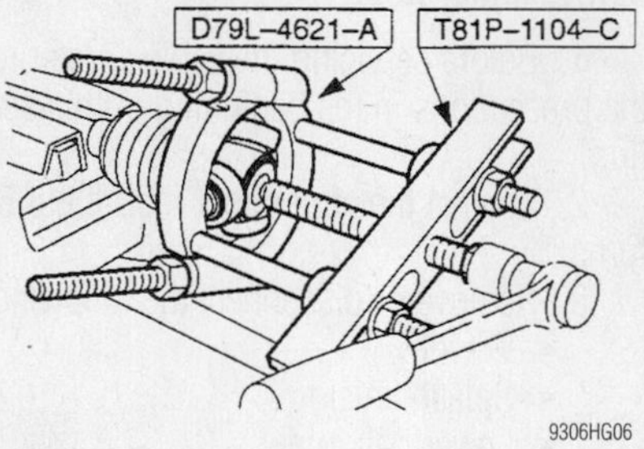

Removing the tripod assembly

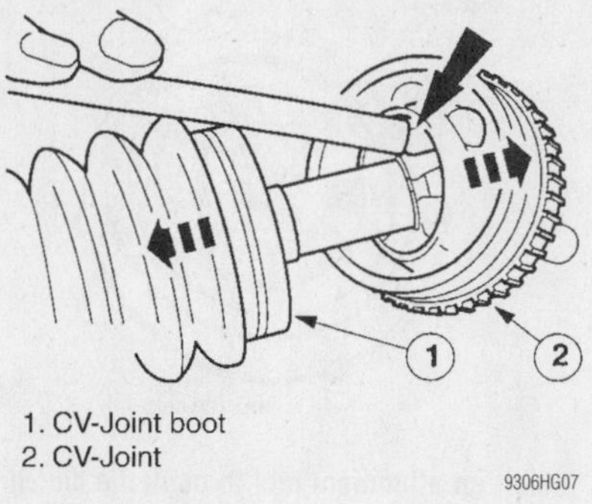

Removing the outer CV-joint

To install:

➡Use new snaprings and boot clamps.

4. Install the tripod joint with a new snapring.
5. Fill the tripod joint housing with fresh CV-joint grease.
6. Install the joint housing and fit the boot to the seat groove.
7. Insert a small prytool under the boot seat to permit air to escape.
8. Slide the tripod joint fully into the joint housing, then pull it back out ¾ inches (20mm).
9. Remove the prytool and install the boot clamps.

Outer CV-Joint

1. Before servicing the vehicle, refer to the precautions in the beginning of this section.
2. Place the halfshaft in a vise.
3. Remove the CV-joint boot clamps and slide the boot away from the joint.
4. Drive the CV-joint off the halfshaft with a brass drift and a hammer.

To install:

5. Replace the snapring.
6. Fill the CV-joint with fresh grease and slide the joint on to the halfshaft.
7. Fit the CV-joint boot to the seat grooves and install new boot clamps.

STEERING AND SUSPENSION

Air Bag

⁂ CAUTION

Some vehicles are equipped with an air bag system. The system must be disarmed before performing service on, or around, system components, the steering column, instrument panel components, wiring and sensors. Failure to follow the safety precautions and the disarming procedure could result in accidental air bag deployment, possible injury and unnecessary system repairs.

PRECAUTIONS

Several precautions must be observed when handling the inflator module to avoid accidental deployment and possible personal injury.

- Never carry the inflator module by the wires or connector on the underside of the module
- When carrying a live inflator module, hold securely with both hands, and ensure that the bag and trim cover are pointed away
- Place the inflator module on a bench or other surface with the bag and trim cover facing up
- With the inflator module on the bench, never place anything on or close to the module which may be thrown in the event of an accidental deployment

Before servicing the vehicle, also be sure to refer to the precautions in the beginning of this section as well

DISARMING

1. Before servicing the vehicle, refer to the precautions in the beginning of this section.
2. Position the vehicle with the front wheels in a straight-ahead position.
3. Disconnect the battery cables, negative cable first.
4. Wait at least 1 minute for the air bag back-up power supply to drain before continuing.
5. Proceed with the repair.

ARMING

1. After repairs are completed, connect the battery cables.
2. Check the functioning of the air bag system by turning the ignition key to the **RUN** position and visually monitoring the air bag indicator lamp in the instrument cluster. The indicator lamp should illuminate for approximately 6 seconds, then turn **OFF**. If the indicator lamp does not illuminate, stays on, or flashes at any time, a fault has been detected by the air bag diagnostic monitor.

Power Rack and Pinion Steering Gear

REMOVAL & INSTALLATION

1. Before servicing the vehicle, refer to the precautions in the beginning of this section.
2. Attach an engine support fixture to the engine lifting eyes.
3. Secure the radiator and fan shroud to the radiator support.
4. Remove or disconnect the following:
   - Negative battery cable
   - Steering gear flexible coupling
   - Left, right and center splash shields
   - Catalytic converter
   - Front wheels
   - Lower ball joints
   - Outer tie rod ends
   - Stabilizer bar links
   - Left and right engine support insulators
   - Power steering oil cooler lines
   - A/C accumulator bolts
   - Lower radiator supports
   - Bumper cover braces
   - Power steering hoses. Lower the subframe for access.
   - Subframe with steering gear attached
   - Steering gear cover plate
   - Steering gear

To install:

5. Install or connect the following:
   - Steering gear and tighten the bolts to 101 ft. lbs. (137 Nm)
   - Steering gear cover plate and tighten the bolts to 37 ft. lbs. (50 Nm)
   - Subframe. Use new seals and attach the power steering hoses before bolting the subframe in place.
   - Special tool Subframe Alignment Pin Set T95P-2100-AH and the subframe bolts. Tighten the subframe bolts to 81–110 ft. lbs. (110–150 Nm). Remove the alignment pins.
   - Special tool Powertrain Alignment Gauge T94P-6000-AH in place of the left engine support insulator. Tighten the 2 retaining bolts to 20 ft. lbs. (27 Nm) and snug the through-bolt.

➡The left and right support insulators must be aligned in the middle of the support insulator brackets.

   - Right engine support insulator. Tighten the 2 subframe retaining bolts to 30–41 ft. lbs. (41–55 Nm) and the through-bolt to 75–102 ft. lbs. (103–137 Nm).

6. Remove the Powertrain Alignment Tool and install the left engine support insulator and tighten the center bolt to 88 ft. lbs. (120 Nm), and the mounting bolts to 35 ft. lbs. (48 Nm).
   - Bumper cover braces
   - Lower radiator supports

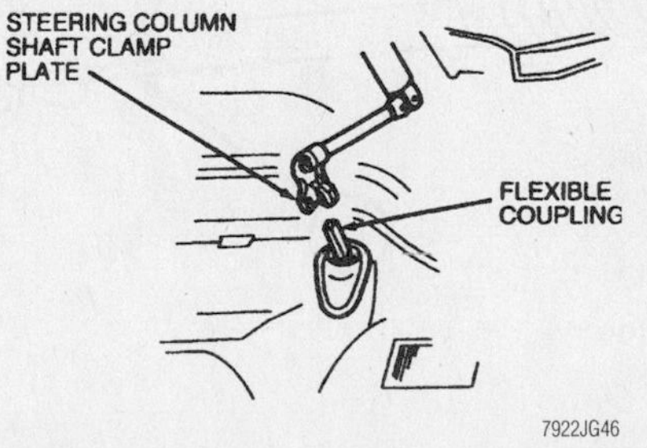

Disconnect the steering column shaft from the flexible coupling

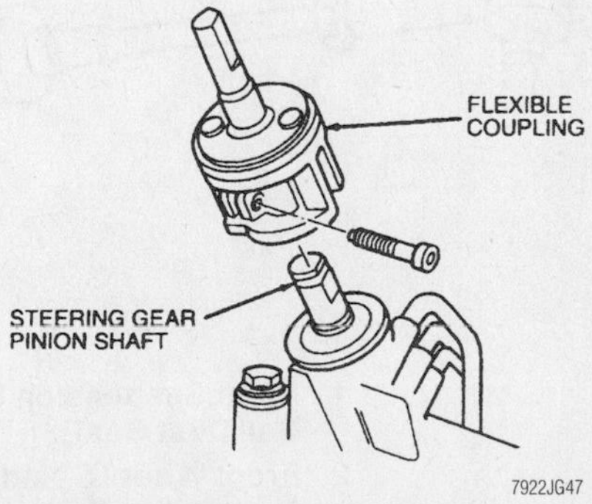

Disengage the flexible coupling from the steering gear pinion shaft

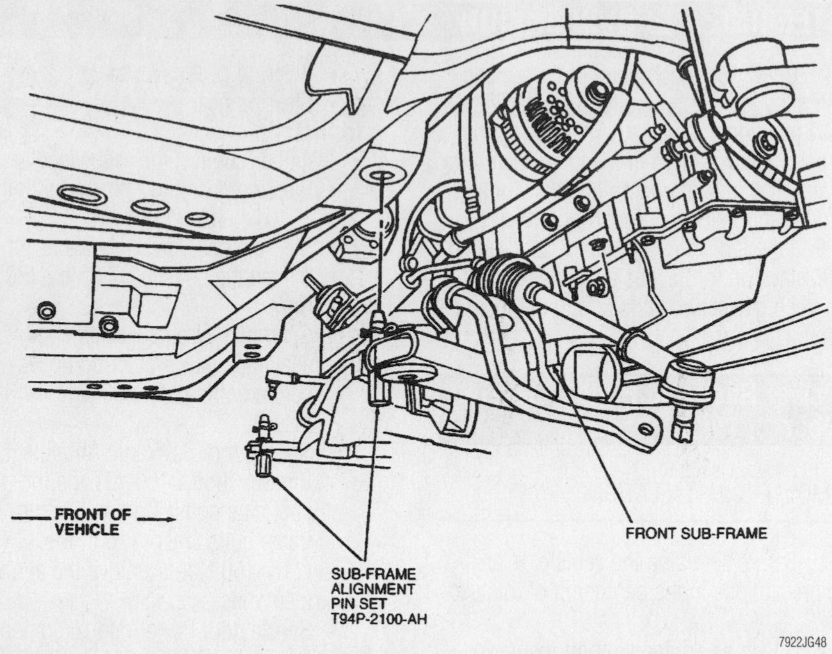

Install Subframe Alignment Pin Set T94P-2100-AH into the subframe to ensure correct positioning

- A/C accumulator bolts
- Power steering oil cooler lines
- Stabilizer bar links
- Outer tie rod ends and tighten the nuts to 23–35 ft. lbs. (31–47 Nm)
- Lower ball joints and tighten the pinch bolts to 61 ft. lbs. (83 Nm)
- Front wheels and tighten the lug nuts to 63 ft. lbs. (86 Nm)
- Catalytic converter
- Left, right and center splash shields
- Steering gear flexible coupling. Tighten the upper bolt to 21 ft. lbs. (28 Nm), and the lower bolt to 18 ft. lbs. (24 Nm).
- Negative battery cable

7. Fill the power steering system.
8. Run the engine and check for leaks and proper operation.

➡Whenever the vehicle subframe is removed or lowered, the wheel alignment should be checked.

Strut and Spring

REMOVAL & INSTALLATION

Front

1. Before servicing the vehicle, refer to the precautions in the beginning of this section.
2. Remove or disconnect the following:
   - Negative battery cable
   - Front wheels
   - Upper strut mount nut
   - Stabilizer bar link
   - Brake hose bracket
   - Lower ball joint
   - Wheel speed sensor
   - Brake caliper and caliper support
   - Brake rotor
   - Steering knuckle pinch bolt
3. Work the strut out of the steering knuckle and lower the strut out of the strut tower.
4. Compress the spring and remove the upper strut mount.
5. Remove the spring from the strut.

To install:

6. Transfer parts as necessary and install the spring onto the strut.
7. Install or connect the following:
   - Upper strut mount and tighten the nut to 44 ft. lbs. (59 Nm)
   - Strut. Guide the strut into the strut tower and attach the steering knuckle. Tighten the upper mounting nut to 34 ft. lbs. (46 Nm), and the knuckle pinch bolt to 40 ft. lbs. (54 Nm).
   - Brake rotor

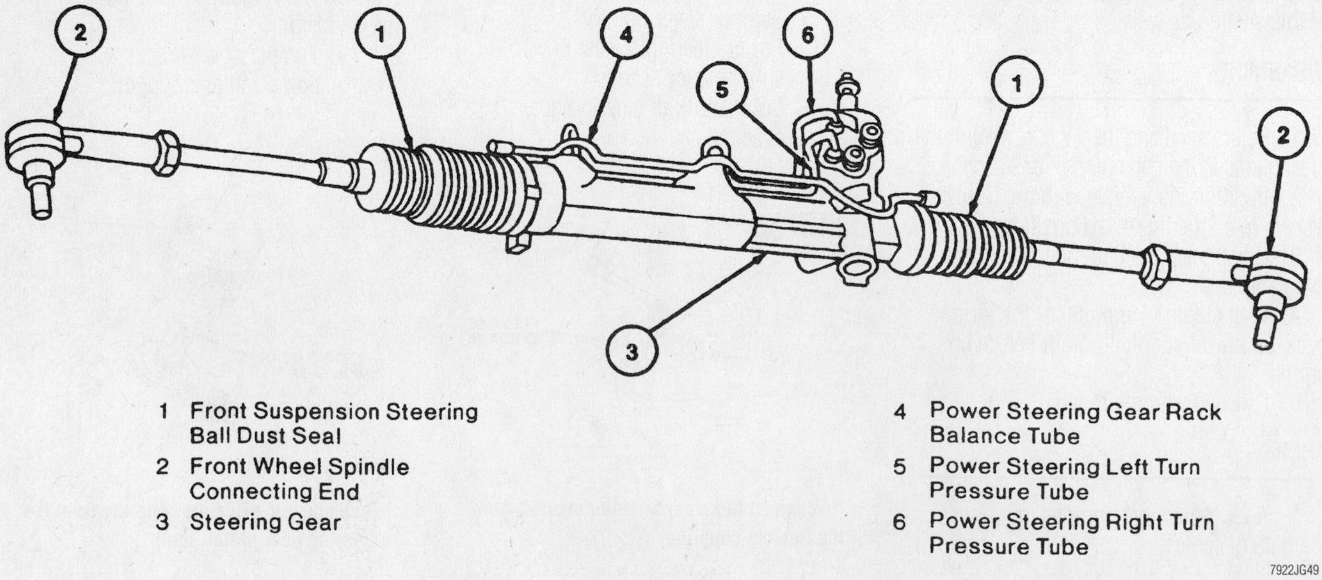

Power rack and pinion steering gear assembly component identification

- Brake caliper support and tighten the bolts to 88 ft. lbs. (120 Nm)
- Wheel speed sensor
- Lower ball joint and tighten the pinch bolt to 61 ft. lbs. (83 Nm)

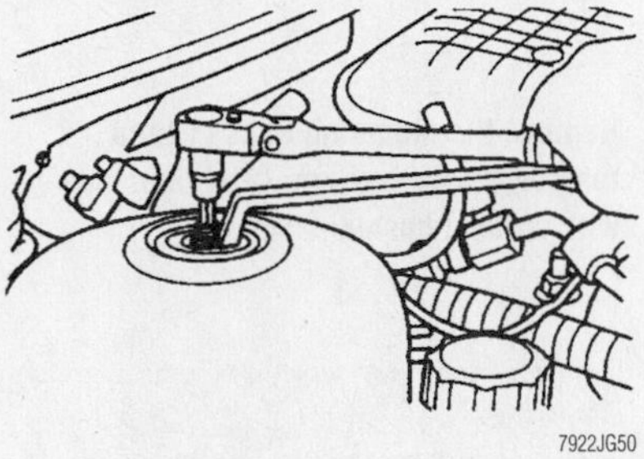

Hold the strut piston with an Allen wrench while removing the retaining nut

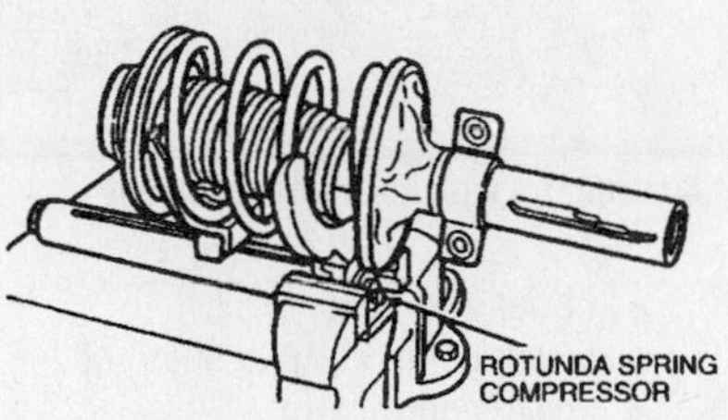

Compress the coil spring assembly before removing the retaining nut

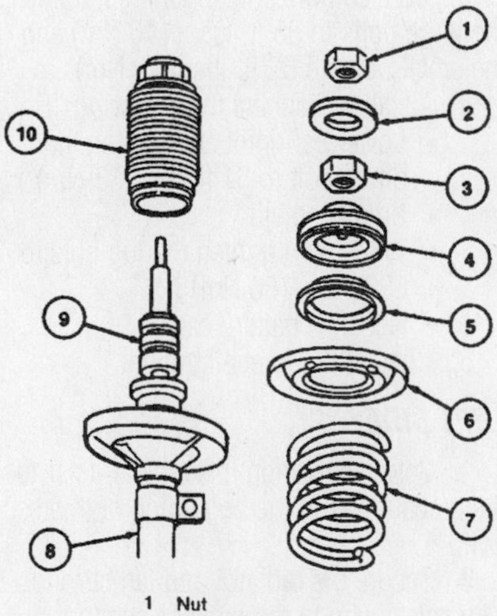

Exploded view of the strut assembly

- Brake hose bracket
- Stabilizer bar link and tighten the nuts to 35–48 ft. lbs. (47–65 Nm)
- Front wheels and tighten the lug nuts to 63 ft. lbs. (86 Nm)
- Negative battery cable

Rear

1. Before servicing the vehicle, refer to the precautions in the beginning of this section.
2. Remove or disconnect the following:
   - Negative battery cable
   - Rear wheels
   - Wheel speed sensor
   - Rear brake hose
   - Stabilizer bar link
   - Tie rod

➡The front and rear control arms must be supported before the removal of the strut attachments.

   - Spindle pinch bolt
3. Separate the spindle from the strut by tapping down on the wheel spindle.
4. Compress the coil spring.
5. Remove the 2 top retaining bolts and remove the strut assembly.
6. Compress the coil spring enough to relieve the tension on the spring seat.
7. Remove the top mount nut, rear strut bracket, bushing and spring seat.
8. Remove the coil spring from the strut.

To install:

9. Transfer parts as necessary and install the spring to the strut.
10. Install the spring seat, bushing, rear strut bracket and the top mount nut. Tighten the top mount nut to 30–43 ft. lbs. (41–58 Nm).
11. With the coil spring compressed, install the strut assembly into position.
12. Install or connect the following:
   - Strut bracket mounting bolts and tighten the bolts to 17–22 ft. lbs. (23–30 Nm)
   - Spindle pinch bolt. Tighten the bolt to 52–72 ft. lbs. (70–98 Nm). Remove the spring compressor.
   - Tie rod and tighten the bolt to 75–102 ft. lbs. (102–138 Nm)
   - Stabilizer bar link
   - Rear brake hose

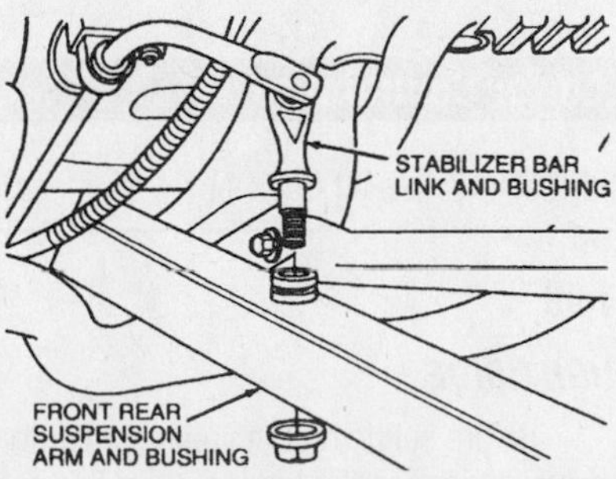

For rear strut removal, disconnect the sway bar link from the control arm

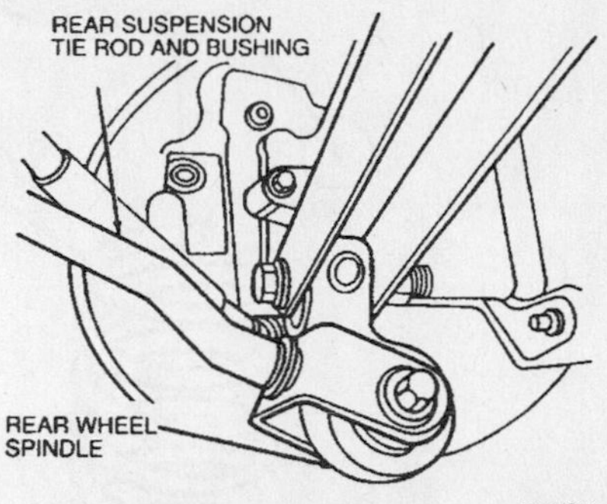

. . . and separate the tie rod from the rear wheel spindle

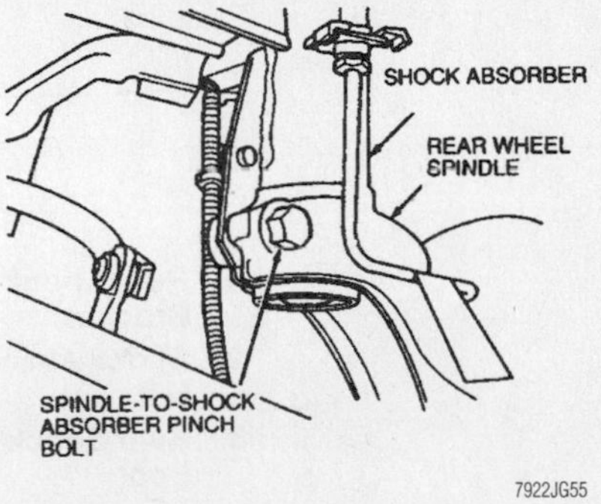

Remove the rear strut pinch bolt to release the strut assembly from the spin-

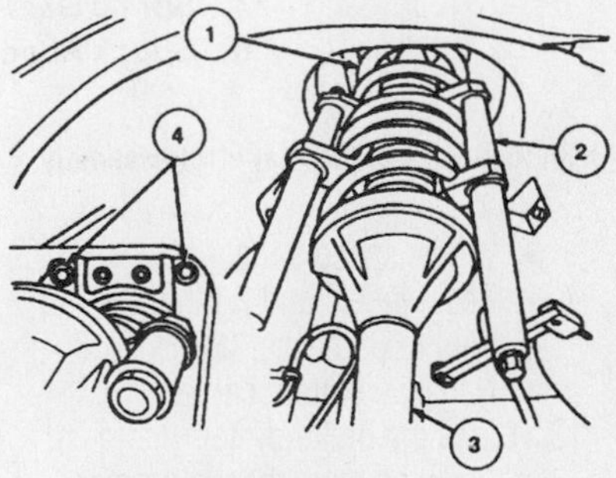

Compress the coil spring, then unthread the 2 top retaining bolts and remove the strut from the vehicle

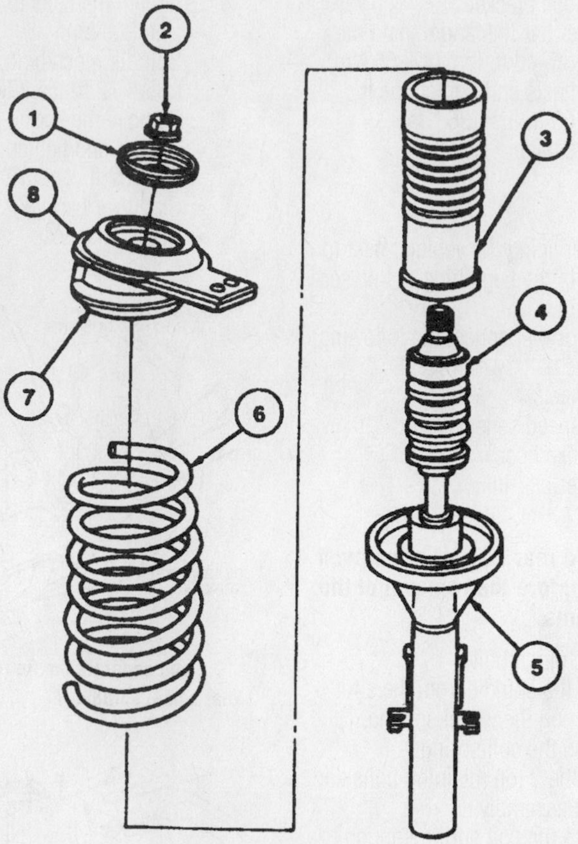

Exploded view of the rear strut assembly

- Wheel speed sensor
- Rear wheels and tighten the lug nuts to 63 ft. lbs. (86 Nm)
- Negative battery cable

13. Bleed the brake system.
14. Check the rear wheel alignment.

Lower Ball Joint

REMOVAL & INSTALLATION

The lower ball joint is serviced with the lower control arm as an assembly.

Lower Control Arm

REMOVAL & INSTALLATION

Front

RIGHT SIDE

1. Before servicing the vehicle, refer to the precautions in the beginning of this section.
2. Remove or disconnect the following:
   - Negative battery cable
   - Wheel
   - Splash shield

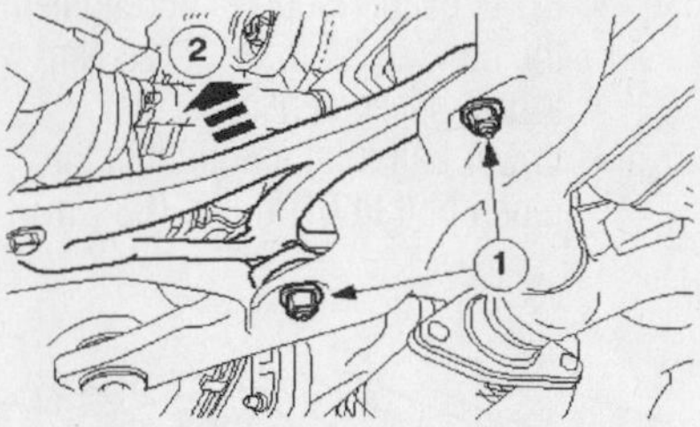

Remove the mounting bolts (1) and remove the control arm (2)—right side with vertical bushings shown

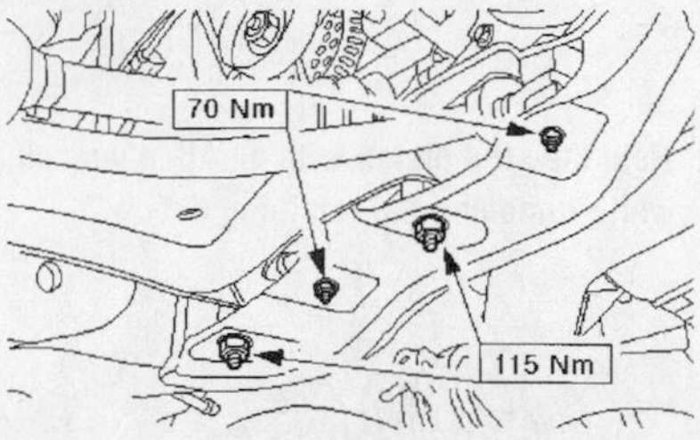

Control arm with horizontal bushings

   - Lower ball joint
   - Lower control arm bushing bolts
   - Lower control arm

To install:

3. Install the lower control arm. If equipped with vertical bushings, tighten the mounting bolts to 85 ft. lbs. (115 Nm). If equipped with horizontal bushings, tighten the large bolts to 85 ft. lbs. (115 Nm) and the small bolts to 52 ft. lbs. 970 Nm).
4. Install or connect the following:
   - Lower ball joint and tighten the pinch bolt to 61 ft. lbs. (83 Nm)
   - Splash shield
   - Wheel and tighten the lug nuts to 63 ft. lbs. (86 Nm)
   - Negative battery cable
5. Check the wheel alignment.

LEFT SIDE

1. Before servicing the vehicle, refer to the precautions in the beginning of this section.
2. Secure the radiator and fan shroud with safety wire to the radiator support.
3. Attach an engine support fixture to the engine lifting eyes.
4. Remove or disconnect the following:
   - Negative battery cable
   - Steering column pinch bolt
   - Left, right and center splash shields
   - Exhaust crossover pipe (2.5L engine)
   - Catalytic converter
   - Front wheel
   - Vehicle Speed Sensor (VSS)

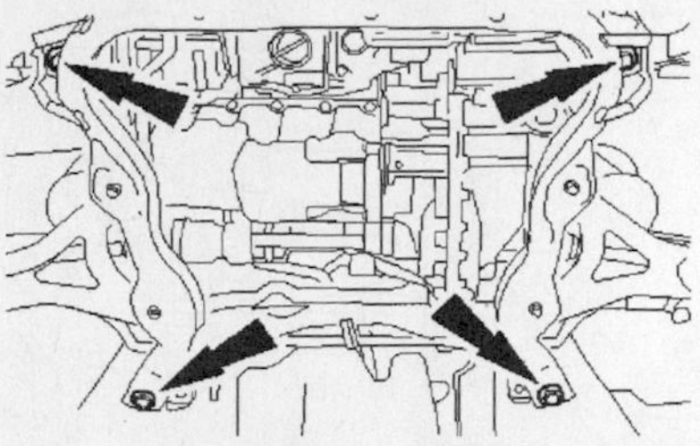

Remove these bolts to lower the subframe

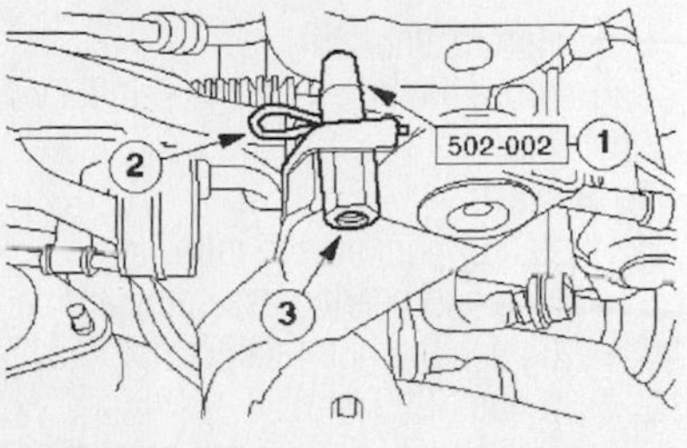

Subframe alignment guide pin (1), locking pin (2) and guide pin sleeve (3)

- Radiator air deflector
- A/C accumulator mounting bolts
- Left and right engine support insulators
- Lower ball joint
- Power steering oil cooler lines
- Lower radiator supports
- Bumper cover braces

5. Unbolt and lower the subframe to access the left control arm bushing bolts.
6. Remove the control arm.

To install:

➡Replace all snaprings, split pins and self-locking nuts.

7. Install or connect the following:
   - Left control arm and tighten the mounting bolts and nuts to 96 ft. lbs. (130 Nm)
   - Subframe
   - Special tool Subframe Alignment Pin Set T95P-2100-AH and the subframe bolts. Tighten the subframe bolts to 81–110 ft. lbs. (110–150 Nm). Remove the alignment pins.
   - A/C accumulator bolts
   - Power steering oil cooler lines
   - Special tool Powertrain Alignment Gauge T94P-6000-AH in place of the left engine support insulator. Tighten the 2 retaining bolts to 20 ft. lbs. (27 Nm) and snug the through-bolt.

➡The left and right support insulators must be aligned in the middle of the support insulator brackets.

   - Right engine support insulator. Tighten the 2 subframe retaining bolts to 30–41 ft. lbs. (41–55 Nm) and the through-bolt to 75–102 ft. lbs. (103–137 Nm).

8. Remove the Powertrain Alignment Tool and install the left engine support insulator. Tighten the center bolt to 88 ft. lbs. (120 Nm), and the mounting bolts to 35 ft. lbs. (48 Nm).
   - Bumper cover braces
   - Lower radiator supports and tighten the bolts to 71–97 inch lbs. (8–11 Nm)
   - Power steering oil cooler lines
   - Lower ball joint and tighten the pinch bolt to 61 ft. lbs. (83 Nm)
   - A/C accumulator mounting bolts
   - Radiator air deflector
   - VSS sensor
   - Front wheel and tighten the lug nuts to 63 ft. lbs. (86 Nm)
   - Catalytic converter
   - Exhaust crossover pipe (2.5L engine)
   - Left, right and center splash shields
   - Steering column pinch bolt and tighten the bolt to 15–20 ft. lbs. (20–27 Nm)
   - Battery
9. Road test the vehicle and check for proper operation.

➡Whenever the vehicle subframe is removed or lowered, the wheel alignment should be checked.

Rear

1. Before servicing the vehicle, refer to the precautions in the beginning of this section.
2. Remove or disconnect the following:
   - Rear wheels
   - Rear control arms
   - Stabilizer bar link
   - Exhaust system hangers
   - Wheel spindle tie bar
3. Support the rear crossmember and remove the 4 attaching bolts.
4. Lower the rear crossmember to access the front control arm bolts.
5. Remove the front control arm.

To install:

➡The final tightening of the rear suspension components is carried out with the full weight of the vehicle on its wheels.

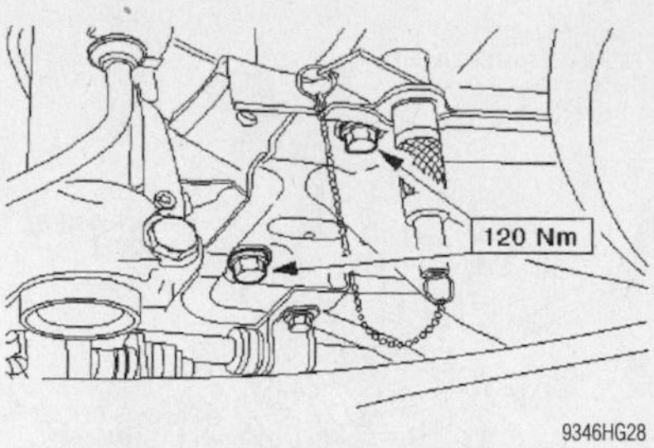

Tighten the rear crossmember bolts with the alignment pins installed

6. Install or connect the following:
   - Front control arm and tighten the bolts to 62 ft. lbs. (84 Nm)
   - Rear crossmember. Use special tools Subframe Alignment Pins T94P-2100-AH, and tighten the crossmember bolts to 88 ft. lbs. (120 Nm).
   - Wheel spindle tie bar and tighten the bolt to 88 ft. lbs. (120 Nm)
   - Exhaust system hangers
   - Stabilizer bar link and tighten the nuts to 26 ft. lbs. (35 Nm)
   - Rear control arms and tighten the bolts to 62 ft. lbs. (84 Nm)
   - Rear wheels and tighten the lug nuts to 63 ft. lbs. (86 Nm)
7. Complete the final tightening of the rear suspension components with the full weight of the vehicle resting on its wheels.

CONTROL ARM BUSHING REPLACEMENT

The control arm bushings are serviced with the control arm as an assembly

Wheel Bearings

ADJUSTMENT

Front and Rear

The wheel bearings are not adjustable. If the bearings make noise or become loose, they must be replaced.

REMOVAL & INSTALLATION

Front

1. Before servicing the vehicle, refer to the precautions in the beginning of this section.
2. Remove or disconnect the following:
   - Negative battery cable
   - Front wheel

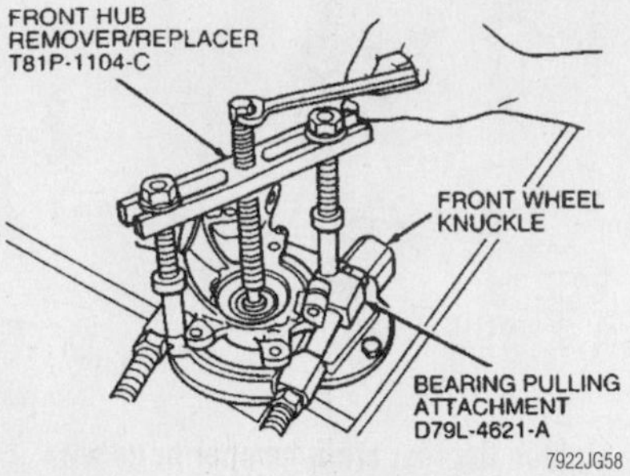

Use gear or bearing pulling tools to remove the hub from the steering knuckle

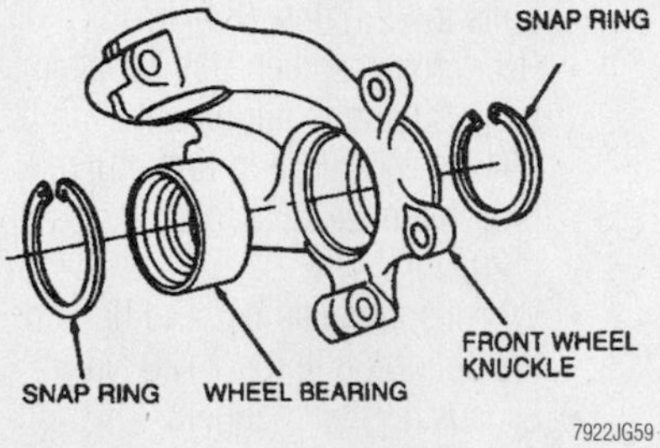

The wheel bearing is retained in the knuckle by 2 snaprings

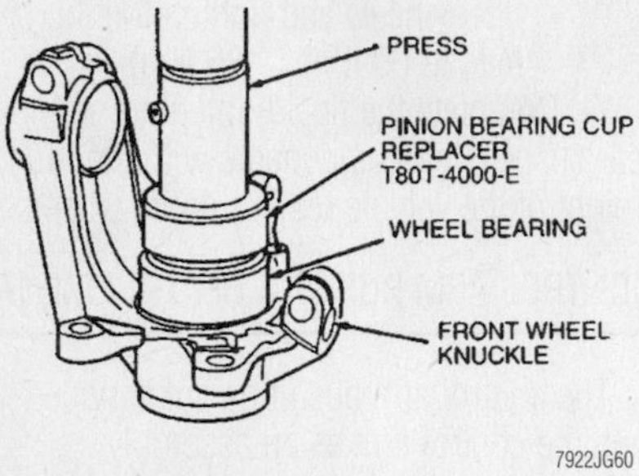

Using a press, install the new wheel bearing in the knuckle

- Brake caliper support and rotor
- Wheel speed sensor
- Outer tie rod end
- Wheel hub retaining nut
- Lower ball joint
- Steering knuckle pinch bolt
- Halfshaft from the wheel hub
- Steering knuckle from the vehicle

3. Install Front Hub Remover/Replacer T81P-1104-C or equivalent with the appropriate adapters and separate the hub from the steering knuckle.
   - Inner and outer snaprings securing the wheel bearing
   - Wheel bearing from the steering knuckle. Drive or press the old wheel bearing out as required.

To install:

➡Replace all snaprings, split pins and self-locking nuts.

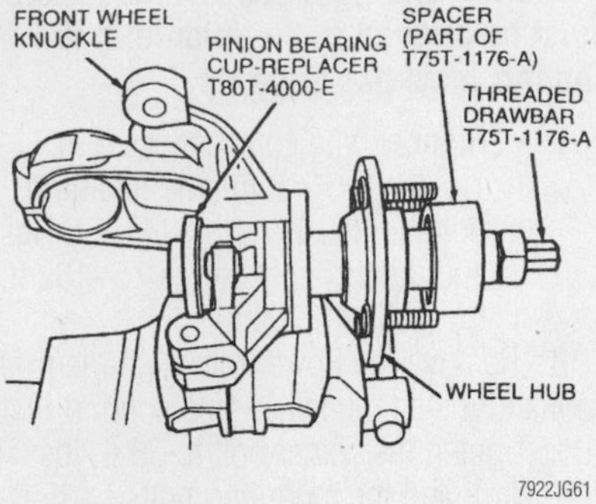

Use the special tools shown or a press to install the hub in the knuckle assembly

4. Install the outer snapring into the steering knuckle.
5. Install the wheel bearing using a hydraulic press with Pinion Bearing Cup Replacer T80T-4000-E.
6. Install or connect the following:
   - Inner snapring in the steering knuckle
   - Hub to the steering knuckle using Threaded Drawbar T75T-1176-A or equivalent
   - Steering knuckle. Tighten the strut pinch bolt to 40 ft. lbs. (54 Nm), and the ball joint pinch bolt to 61 ft. lbs. (83 Nm).
   - Wheel speed sensor
   - Wheel hub retaining nut and tighten the hub nut to 246 ft. lbs. (340 Nm)
   - Outer tie rod end and tighten the nut to 18–22 ft. lbs. (24–30 Nm)
   - Wheel speed sensor
   - Brake caliper support and rotor and tighten the bolts to 88 ft. lbs. (120 Nm)
   - Front wheel and tighten the lug nuts to 63 ft. lbs. (86 Nm)
   - Negative battery cable
7. Road test the vehicle and check for proper operation.

Rear

➡The wheel bearings are contained within the wheel hub and must be replaced as an assembly.

WITH REAR DISC BRAKES

1. Before servicing the vehicle, refer to the precautions in the beginning of this section.
2. Remove or disconnect the following:
   - Negative battery cable
   - Rear wheel
   - Wheel speed sensor
   - Brake caliper bracket and rotor

➡Do not use an impact gun to remove the hub retainer nut.

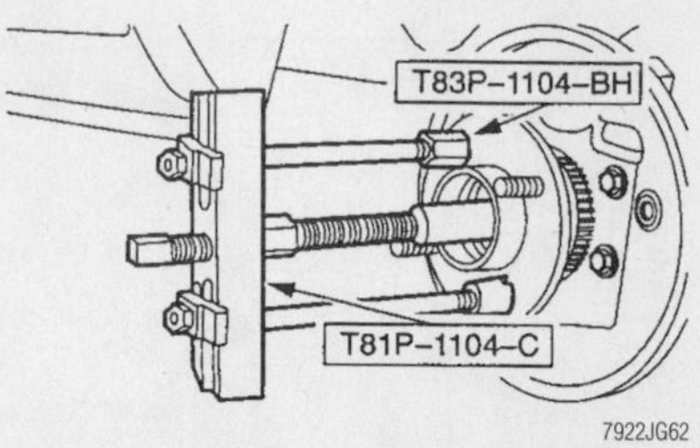

After removing the spindle nut, use the special tools or a puller to remove the rear hub/bearing assembly—all models

- Hub retainer nut

3. Remove the hub from the spindle with a hub puller.

To install:

4. Install or connect the following:
   - Hub and bearing assembly and tighten the hub retainer nut to 213 ft. lbs. (290 Nm)
   - Brake rotor and caliper bracket and tighten the bracket bolts to 88 ft. lbs. (120 Nm)
   - Wheel speed sensor
   - Rear wheel and tighten the lug nuts to 63 ft. lbs. (86 Nm)
   - Negative battery cable
5. Road test the vehicle and check for proper operation.

WITH REAR DRUM BRAKES

➡The wheel bearings are contained within the wheel hub and must be replaced as an assembly.

1. Before servicing the vehicle, refer to the precautions in the beginning of this section.
2. Remove or disconnect the following:
   - Negative battery cable
   - Rear wheel
   - Brake drum
   - Wheel speed sensor

➡Do not use an impact gun to remove the hub retainer nut.

- Wheel hub retainer nut

3. Remove the hub from the spindle with a hub puller.

To install:

4. Install or connect the following:
   - Hub and bearing assembly and tighten the hub retainer nut to 213 ft. lbs. (290 Nm)
   - Brake drum
   - Wheel speed sensor
   - Rear wheel and tighten the lug nuts to 63 ft. lbs. (86 Nm)
   - Negative battery cable
5. Road test the vehicle and check for proper operation.

FORD MOTOR CO.

12

1998–01

Ford-Taurus • **Mercury**-Sable

PRECAUTIONS

Before servicing any vehicle, please be sure to read all of the following precautions, which deal with personal safety, prevention of component damage, and important points to take into consideration when servicing a motor vehicle:

- Never open, service or drain the radiator or cooling system when the engine is hot; serious burns can occur from the steam and hot coolant.
- Observe all applicable safety precautions when working around fuel. Whenever servicing the fuel system, always work in a well-ventilated area. Do not allow fuel spray or vapors to come in contact with a spark, open flame, or excessive heat (a hot drop light, for example). Keep a dry chemical fire extinguisher near the work area. Always keep fuel in a container specifically designed for fuel storage; also, always properly seal fuel containers to avoid the possibility of fire or explosion. Refer to the additional fuel system precautions later in this section.
- Fuel injection systems often remain pressurized, even after the engine has been turned **OFF**. The fuel system pressure must be relieved before disconnecting any fuel lines. Failure to do so may result in fire and/or personal injury.
- Brake fluid often contains polyglycol ethers and polyglycols. Avoid contact with the eyes and wash your hands thoroughly after handling brake fluid. If you do get brake fluid in your eyes, flush your eyes with clean, running water for 15 minutes. If eye irritation persists, or if you have taken brake fluid internally, IMMEDIATELY seek medical assistance.
- The EPA warns that prolonged contact with used engine oil may cause a number of skin disorders, including cancer. You should make every effort to minimize your exposure to used engine oil. Protective gloves should be worn when changing oil. Wash your hands and any other exposed skin areas as soon as possible after exposure to used engine oil. Soap and water, or waterless hand cleaner should be used.
- All new vehicles are now equipped with an air bag system, often referred to as a Supplemental Restraint System (SRS) or Supplemental Inflatable Restraint (SIR) system. The system must be disabled before performing service on or around system components, steering column, instrument panel components, wiring and sensors. Failure to follow safety and disabling procedures could result in accidental air bag deployment, possible personal injury and unnecessary system repairs.
- Always wear safety goggles when working with, or around, the air bag system. When carrying a non-deployed air bag, be sure the bag and trim cover are pointed away from your body. When placing a non-deployed air bag on a work surface, always face the bag and trim cover upward, away from the surface. This will reduce the motion of the module if it is accidentally deployed. Refer to the additional air bag system precautions later in this section.
- Clean, high quality brake fluid from a sealed container is essential to the safe and proper operation of the brake system. You should always buy the correct type of brake fluid for your vehicle. If the brake fluid becomes contaminated, completely flush the system with new fluid. Never reuse any brake fluid. Any brake fluid that is removed from the system should be discarded. Also, do not allow any brake fluid to come in contact with a painted surface; it will damage the paint.
- Never operate the engine without the proper amount and type of engine oil; doing so will result in severe engine damage.
- Timing belt maintenance is extremely important. Many models utilize an interference-type, non-freewheeling engine. If the timing belt breaks, the valves in the cylinder head may strike the pistons, causing potentially serious (also time-consuming and expensive) engine damage. Refer to the maintenance interval charts in the front of this manual for the recommended replacement interval for the timing belt, and to the timing belt section for belt replacement and inspection.
- Disconnecting the negative battery cable on some vehicles may interfere with the functions of the on-board computer system(s) and may require the computer to undergo a relearning process once the negative battery cable is reconnected.
- When servicing drum brakes, only disassemble and assemble one side at a time, leaving the remaining side intact for reference.

ENGINE REPAIR

➡Disconnecting the negative battery cable on some vehicles may interfere with the functions of the on board computer system. The computer may undergo a relearning process once the negative battery cable is reconnected.

Alternator

REMOVAL

3.0L (VIN U) Engine

1. Before servicing the vehicle, refer to the precautions at the beginning of this section.
2. Remove or disconnect the following:
   - Negative battery cable
   - Accessory drive belt
   - Alternator wiring connectors
   - Alternator brace
   - Alternator

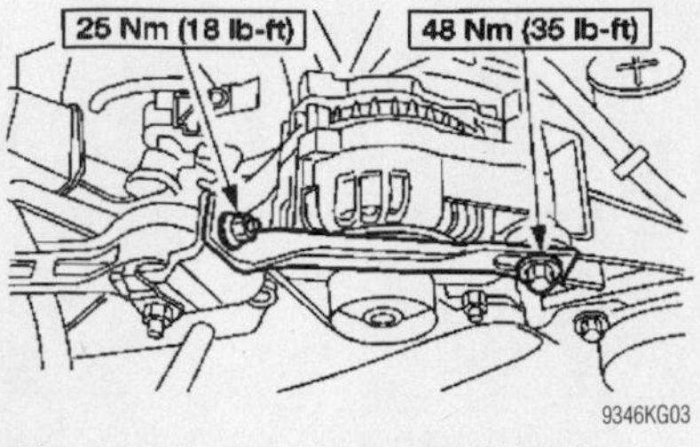

Alternator bracket–3.0L (VIN U) engine

3.0L (VIN S) Engine

1998–99 MODELS

1. Before servicing the vehicle, refer to the precautions at the beginning of this section.
2. Remove or disconnect the following:
   - Negative battery cable
   - Accessory drive belt
   - Power steering line and heater hose bracket
   - Upper alternator bolts
   - Right front wheel
   - Right splash shield
   - Alternator splash shield
   - Alternator electrical connectors
   - Lower alternator bolt

3. Lower the right side of the subframe for clearance and remove the alternator.

2000–01 MODELS

1. Before servicing the vehicle, refer to the precautions in the beginning of this section.
2. Remove or disconnect the following:
   - Negative battery cable
   - Right front wheel
   - Right fender splash shield
   - Accessory drive belt

➡The crankshaft pulley retainer has left-hand threads.

- Crankshaft pulley
- Alternator harness connectors
- Alternator

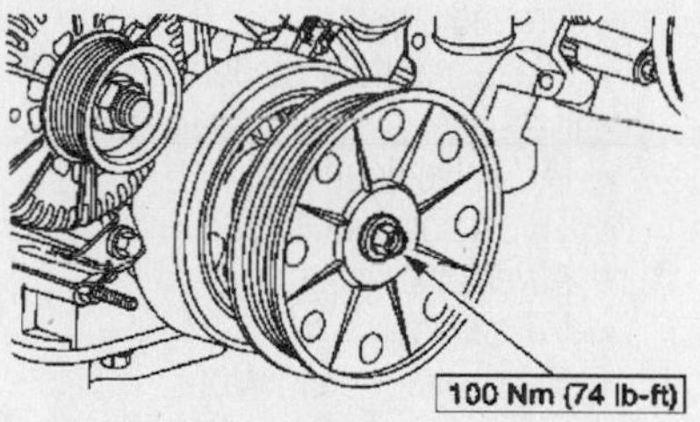

9346KG04

Remove the crankshaft pulley to provide clearance—2000–01 3.0L (VIN S) engine

3.4L (VIN N) Engine

1. Before servicing the vehicle, refer to the precautions at the beginning of this section.
2. Remove or disconnect the following:
   - Negative battery cable
   - Right cowl vent screen
   - Right front wheel
   - Right outer tie rod end
   - Accessory drive belt
   - Alternator electrical connectors
   - Engine control sensor wiring
   - Alternator

INSTALLATION

3.0L (VIN U) Engine

➡Do not tighten the alternator pivot and bracket bolts until the accessory drive belt is installed.

1. Install or connect the following:
   - Alternator
   - Alternator brace. Tighten the bolts to 76–97 inch lbs. (8–11 Nm), and the nut to 15–22 ft. lbs. (20–30 Nm).
   - Alternator wiring connectors
   - Accessory drive belt
2. Tighten the alternator pivot bolt to 30–40 ft. lbs. (40–55 Nm), and the bracket bolt to 15–22 ft. lbs. (20–30 Nm).
3. Install the negative battery cable. Start the engine and check for proper operation.

3.0L (VIN S) Engine

1998–99 MODELS

1. Install or connect the following:
   - Alternator and tighten the bolts to 18 ft. lbs. (25 Nm)
   - Right subframe bolts and tighten to 66 ft. lbs. (90 Nm)
   - Alternator electrical connectors
   - Alternator splash shield
   - Right splash shield
   - Right front wheel
   - Power steering line and heater hose bracket
   - Accessory drive belt
   - Negative battery cable
2. Start the engine and check for proper operation.

2000–01 MODELS

1. Install or connect the following:
   - Alternator and tighten the fasteners to 18 ft. lbs. (25 Nm)
   - Alternator harness connectors

➡The crankshaft pulley retainer has left-hand threads.

- Crankshaft pulley and tighten to 74 ft. lbs. (100 Nm)
- Accessory drive belt
- Right fender splash shield
- Right front wheel
- Negative battery cable

2. Start the engine and check for proper operation.

3.4L (VIN N) Engine

1. Install or connect the following:
   - Alternator and tighten the bolts to 15–22 ft. lbs. (20–30 Nm)
   - Engine control sensor wiring
   - Alternator electrical connectors
   - Accessory drive belt
   - Right outer tie rod end
   - Right front wheel
   - Right cowl vent screen
   - Negative battery cable
2. Start the engine and check for proper operation.

Ignition Timing

ADJUSTMENT

The base ignition timing is set at 10 degrees Before Top Dead Center (BTDC) and is not adjustable.

Engine Assembly

REMOVAL & INSTALLATION

1. Before servicing the vehicle, refer to the precautions at the beginning of this section.
2. Disconnect the negative battery cable.
3. Drain the engine oil.
4. Drain the cooling system.
5. Recover the A/C refrigerant.
6. On the 3.0L (VIN S) engine, remove or disconnect the following:
   - Windshield wipers
   - Cowl extension
   - Emission vacuum control connector at the right side of the dash panel
7. On the 3.4L (VIN N) engine, remove or disconnect the following:
   - Battery and tray
   - Engine appearance cover
   - Cowl top extension
   - Windshield wiper module
8. For all vehicles, remove or disconnect the following:
   - Hood
   - Steering column pinch bolt
   - Mass Airflow (MAF) sensor connector
   - Intake Air Temperature (IAT) sensor
   - Air cleaner assembly
   - Idle Air Control (IAC) valve connector
   - Throttle Position (TP) sensor connector
   - Fuel lines
   - Intake manifold vacuum hoses
   - Ground straps
   - Powertrain Control Module (PCM) connector
   - Engine control sensor wiring harness connectors from the bracket located at the top of the transaxle
   - Evaporative emission canister purge valve connector
   - Crankcase vent hose
   - Accelerator cable
   - Cruise control actuator

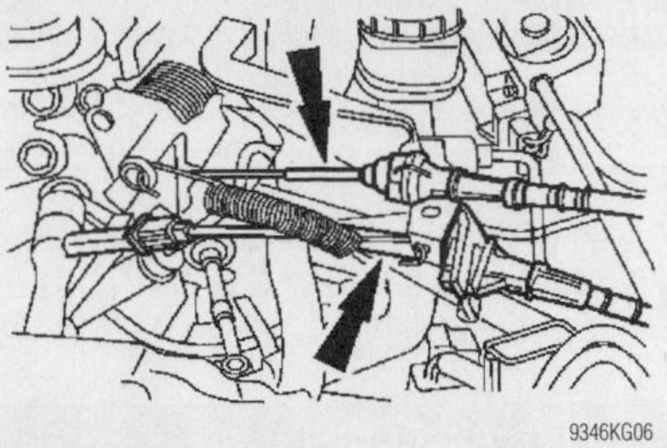

Throttle and cruise control cables—3.0L (VIN U) engine shown

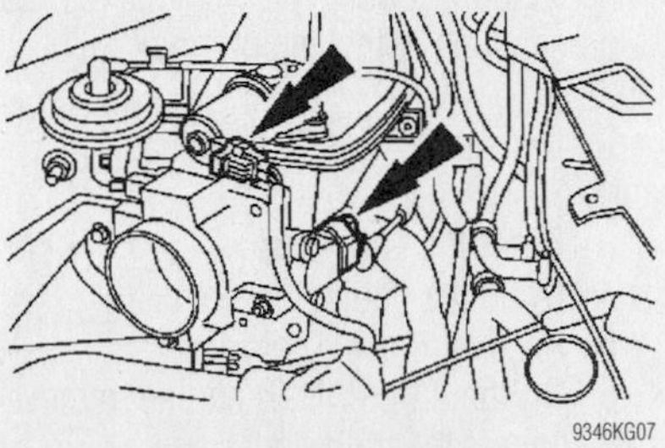

Throttle Position (TP) sensor and Intake Air Control (IAC) valve connectors—3.0L (VIN U) engine shown

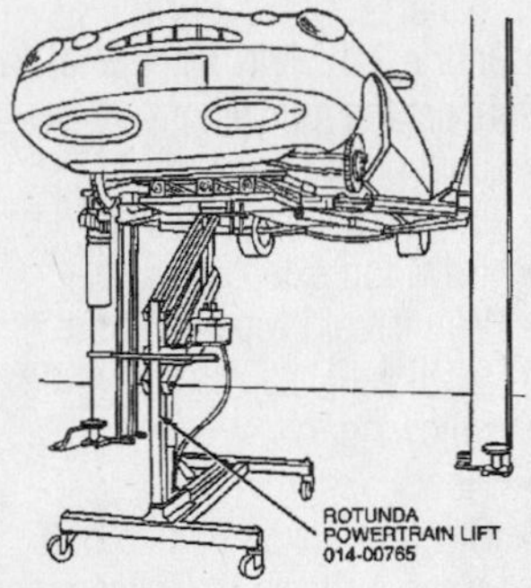

Support the powertrain when removing the subframe bolts

- Shift cable and lever
- Secondary air injection pump relay connector
- Main emission vacuum control connector near the fan shroud
- Radiator hoses
- Heater hoses
- Transaxle oil cooler lines
- Power steering return hose
- Alternator wiring harness connectors
- A/C compressor lines
- A/C pressure cutoff switch, if equipped
- Front wheels
- Stabilizer bar links
- Lower ball joints
- Outer tie rod ends
- Halfshafts
- Radiator splash shield
- Heated Oxygen (HO_2S) sensors
- Dual converter Y-pipe
- Power steering pressure switch connector
- Power steering pressure hose

9. Support the powertrain from below and remove the subframe bolts.

10. Raise the vehicle away from the powertrain.

11. Attach an engine hoist to the powertrain.

12. Remove the left, right, and rear powertrain support insulators and lift the powertrain away from the subframe.

13. Support the transaxle from below.

14. Remove or disconnect the following:
    - Starter motor
    - Torque converter
    - Transaxle flange bolts

15. Separate the engine from the transaxle.

To install:

16. Install the engine to the transaxle. For 1999–01 3.0L (VIN S) engines, tighten the flange bolts to 25–33 ft. lbs. (33–46 Nm). For all other engines, tighten the flange bolts to 30–44 ft. lbs. (40–60 Nm). Tighten the torque converter nuts to 20–34 ft. lbs. (27–46 Nm).

17. Install or connect the following:
    - Starter motor and tighten the fasteners to 21 ft. lbs. (29 Nm)
    - Powertrain support insulators. Tighten the bracket bolts to 65 ft. lbs. (88 Nm) and the subframe bolts to 90 ft. lbs. (122 Nm).

18. Lower the vehicle on to the powertrain assembly. Use 2 pieces of ¾ inch outside diameter pipe in the alignment holes behind the front subframe mounts to align the subframe to the body. Tighten the subframe mounting bolts to 57–76 ft. lbs. (77–103 Nm).

19. Install or connect the following:
    - Power steering pressure hose
    - Power steering pressure switch connector
    - Dual converter Y-pipe
    - HO_2S sensors
    - Radiator splash shield
    - Halfshafts and tighten the hub retainer nuts to 170–202 ft. lbs. (230–275 Nm)
    - Outer tie rod ends and tighten the nuts to 35–46 ft. lbs. (47–63 Nm)
    - Lower ball joints and tighten the nuts to 50–68 ft. lbs. (68–92 Nm)
    - Stabilizer bar links and tighten the nuts to 30–40 ft. lbs. (40–55 Nm)
    - Front wheels
    - A/C pressure cutoff switch, if equipped
    - A/C compressor lines
    - Alternator wiring harness connectors
    - Power steering return hose
    - Transaxle oil cooler lines
    - Heater hoses
    - Radiator hoses
    - Main emission vacuum control connector near the fan shroud
    - Secondary air injection pump relay connector
    - Shift cable and lever
    - Cruise control actuator
    - Accelerator cable
    - Crankcase vent hose
    - Evaporative emission canister purge valve connector
    - Engine control sensor wiring harness connectors from the bracket located at the top of the transaxle
    - PCM connector
    - Ground straps
    - Intake manifold vacuum hoses
    - Fuel lines
    - IAC valve connector
    - TP sensor connector
    - Air cleaner assembly
    - IAT sensor
    - MAF sensor connector
    - Steering column pinch bolt
    - Hood

20. On the 3.0L (VIN S) engine, install or connect the following:
    - Windshield wipers
    - Cowl extension
    - Emission vacuum control connector at the right side of the dash panel

21. On the 3.4L (VIN N) engine, install or connect the following:
    - Battery and tray
    - Engine appearance cover
    - Cowl top extension
    - Windshield wiper module

22. Fill the cooling system, fill the crankcase and recharge the A/C system.

23. Connect the negative battery cable.

24. Start the engine. Check for leaks and proper operation.

➡Whenever the vehicle's subframe is removed or lowered, the wheel alignment should be checked.

Water Pump

REMOVAL & INSTALLATION

3.0L (VIN U) Engine

1. Before servicing the vehicle, refer to the precautions at the beginning of this section.
2. Drain the cooling system.

3. Remove or disconnect the following:
   - Negative battery cable
   - Accessory drive belt and tensioner
   - Water pump pulley
   - Heater hose
   - Engine control sensor wiring
   - Water pump

To install:

➡The water pump bolts are of different lengths and must be installed in the correct locations.

4. Install the water pump. Tighten bolts 1, 2, 3, 4, 5, 6, 7, 8, 9 and 15 to 15–22 ft. lbs. (20–30 Nm). Tighten bolts 10, 11, 12, 13 and 14 to 72–106 inch lbs. (8–12 Nm). Refer to the accompanying illustration for the bolt locations.
5. Install or connect the following:
   - Engine control sensor wiring
   - Heater hose
   - Water pump pulley and tighten the bolts to 15–22 ft. lbs. (20–30 Nm)
   - Accessory drive belt and tensioner
   - Negative battery cable
6. Fill the cooling system and check for leaks.

3.0L (VIN S) Engine

1. Before servicing the vehicle, refer to the precautions at the beginning of this section.
2. Drain the cooling system.
3. Remove or disconnect the following:
   - Splash shield
   - Lower heater hose

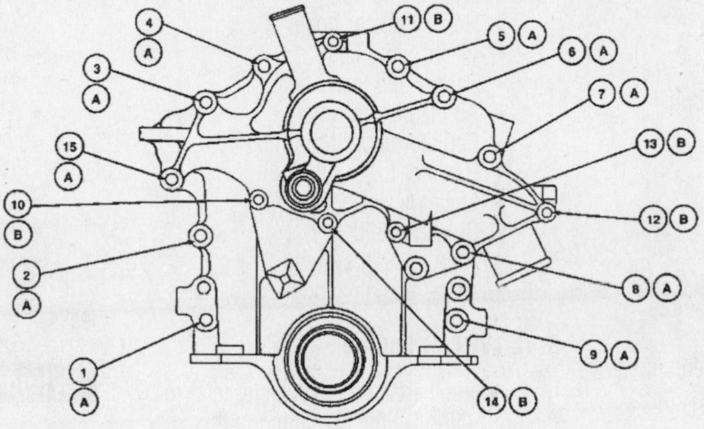

	Fasteners		Torque Specifications	
Fastener And Hole No.	Size	Fastener Application	Nm	Lb-Ft
1A	M8 x 1.25 x 43.5	F/C TO BLOCK	20-30	15-22
2A	M8 x 1.25 x 73	W/P & F/C TO BLOCK	20-30	15-22
3A	M8 x 1.25 x 104.3	W/P & F/C TO BLOCK	20-30	15-22
4A	M8 x 1.25 x 71.3	F/C TO BLOCK	20-30	15-22
5A	M8 x 1.25 x 71.3	W/P & F/C TO BLOCK	20-30	15-22
6A	M8 x 1.25 x 71.3	W/P & F/C TO BLOCK	20-30	15-22
7A	M8 x 1.25 x 104.3	W/P & F/C TO BLOCK	20-30	15-22
8A	M8 x 1.25 x 104.3	W/P & F/C TO BLOCK	20-30	15-22
9A	M8 x 1.25 x 52	F/C TO BLOCK	20-30	15-22
10B	M6 x 1 x 28.5	W/P TO F/C	8-12	71-106 (lb-in)
11B	M6 x 1 x 28.5	W/P TO F/C	8-12	71-106 (lb-in)
12B	M6 x 1 x 28.5	W/P TO F/C	8-12	71-106 (lb-in)
13B	M6 x 1 x 28.5	W/P TO F/C	8-12	71-106 (lb-in)
14B	M6 x 1 x 28.5	W/P TO F/C	8-12	71-106 (lb-in)
15A	M8 x 1.25 x 71.3	W/P & F/C TO BLOCK	20-30	15-22

W/P–Water Pump
F/C–Engine Front Cover

9300KG01

Exploded view of the water pump and timing front cover bolt locations—3.0L (VIN U) engine

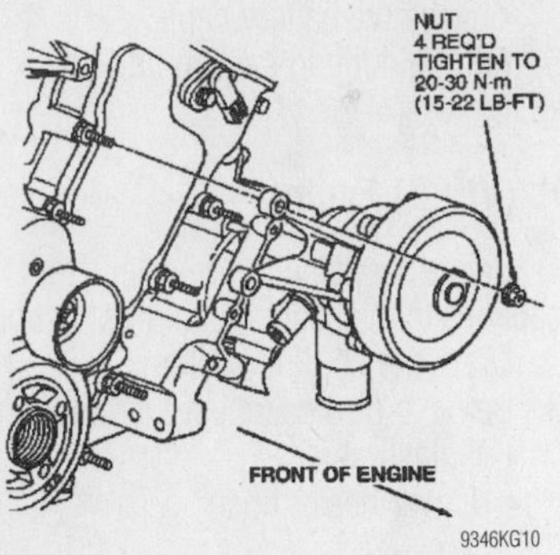

Water pump mounting—3.0L (VIN S) engine

- Radiator lower tube bolt
- Radiator upper front tube bolt
- Air cleaner assembly
- Battery and tray
- Water pump belt
- Upper radiator hose
- Engine vent hose
- Transaxle 10 pin connector
- Radiator bypass hose assembly
- Thermostat housing
- A/C compressor brace, if equipped
- Water pump mounting nuts
- Connector hose
- Water pump

To install:

4. Install or connect the following:
   - Water pump
   - Connector hose
   - Water pump mounting nuts and tighten them to 15–22 ft. lbs. (20–30 Nm)
   - A/C compressor brace, if equipped, and tighten the nuts to 15–22 ft. lbs. (20–30 Nm)
   - Thermostat housing
   - Radiator bypass hose assembly
   - Transaxle 10 pin connector
   - Engine vent hose
   - Upper radiator hose
   - Water pump belt
   - Battery and tray
   - Air cleaner assembly
   - Radiator upper front tube bolt
   - Radiator lower tube bolt
   - Lower heater hose
   - Splash shield
5. Fill the cooling system.
6. Start the engine and check for leaks.

3.4L (VIN N) Engine

1. Before servicing the vehicle, refer to the precautions at the beginning of this section.
2. Drain the cooling system.
3. Remove or disconnect the following:
   - Engine appearance cover
   - Battery and tray
   - Water pump drive belt
   - Coolant hoses
   - Thermostat
   - Water pump

To install:

4. If replacing the water pump, transfer the idler pulley and tighten the bolt to 89–141 inch lbs. (10–16 Nm).
5. Install or connect the following:
   - Water pump. Tighten the 2 water inlet bolts to 71–106 inch lbs. (8–12 Nm) and the 2 remaining bolts to 14–20 ft. lbs. (18–28 Nm).
   - Thermostat and tighten the bolts to 71–106 inch lbs. (8–12 Nm)
   - Coolant hoses
   - Water pump drive belt
   - Battery and tray
   - Engine appearance cover
6. Fill the cooling system and check for leaks.

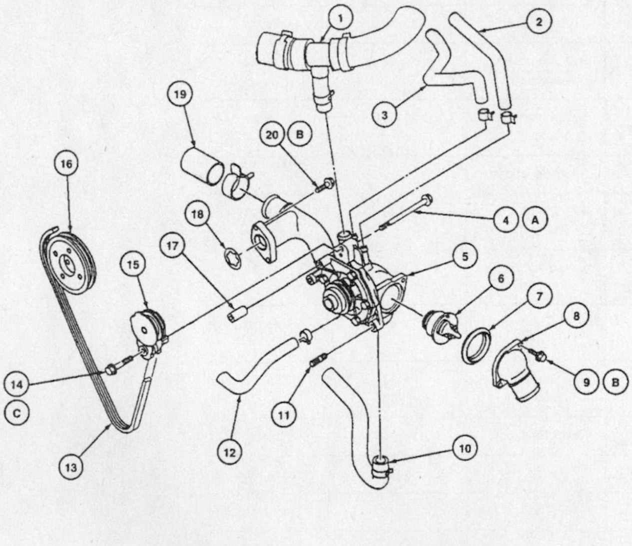

1 Water Outlet Hose
2 Throttle Body to Water Pump Return Hose
3 Water Pump to Throttle Body Supply Hose
4 Bolt (2 Req'd)
5 Water Pump
6 Water Thermostat
7 Oil Cooler Return Tube Gasket
8 Water Hose Connection
9 Bolt (2 Req'd)
10 Heater Core to Water Pump Return Hose
11 Stud
12 Oil Cooler to Water Pump Return Hose
13 Drive Belt
14 Bolt
15 Belt Idler Pulley
16 Water Pump Drive Pulley
17 Collar
18 O-Ring
19 Water Inlet Hose
20 Bolt
A Tighten to 18-28 N·m (14-20 Lb-Ft)
B Tighten to 10-16 N·m (89-141 Lb-In)
C Tighten to 8-12 N·m (71-106 Lb-In)

7922KG03

Exploded view of the water pump and related components—3.4L (VIN N) engine

Cylinder Head

REMOVAL & INSTALLATION

3.0L (VIN U) Engine

1. Before servicing the vehicle, refer to the precautions at the beginning of this section.
2. Drain the cooling system.
3. Set the No. 1 cylinder to Top Dead Center (TDC) of the compression stroke.
4. Remove or disconnect the following:
   - Negative battery cable
   - Air cleaner outlet hose
   - Vacuum hoses from the intake manifold
   - Exhaust Gas Recirculation (EGR) valve
   - EGR transducer
   - EGR vacuum regulator solenoid
   - Intake Air Temperature (IAT) sensor connector
   - Throttle Position (TP) sensor connector
   - Intake Air Control (IAC) valve connector
   - Camshaft Position (CMP) sensor and housing
   - Engine Coolant Temperature (ECT) sensor connector
   - Accelerator cable
   - Cruise control cable
   - Fuel lines
   - Upper intake manifold

- Fuel injector wiring harness
- Coolant hoses
- Ignition coil and bracket
- Spark plug wires
- Accessory drive belt and tensioner
- Alternator
- Power steering pump
- Oil dipstick tube
- Heater supply tube brackets
- Valve covers
- Rocker arms and pushrods. Keep the rocker arms and pushrods in order for installation.
- Lower intake manifold
- Spark plugs
- Exhaust manifolds
- Cylinder heads

To install:

➡The cylinder head bolts are a torque-to-yield design and cannot be reused.

5. Install the cylinder heads with new gaskets.
6. On 1998 models, tighten the bolts in sequence as follows:
   a. Step 1: 53–66 ft. lbs. (70–90 Nm).
   b. Step 2: Loosen the bolts 1 complete turn.
   c. Step 3: Tighten the bolts to 34–40 ft. lbs. (45–55 Nm).
   d. Step 4: Tighten the bolts to 63–73 ft. lbs. (85–99 Nm).

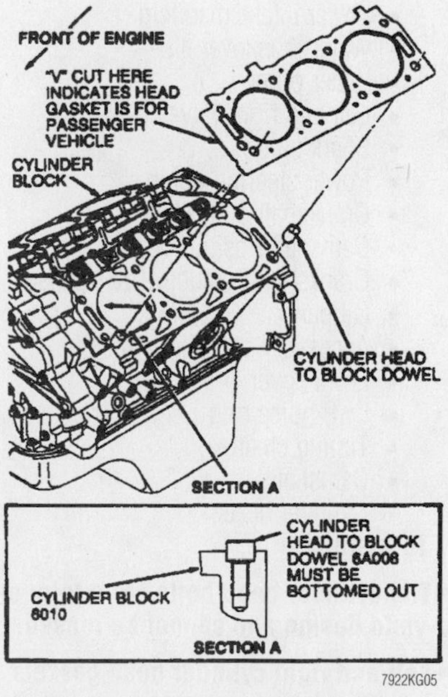

Install the cylinder head gasket with the "UP" designation on gasket facing the cylinder head—3.0L (VIN U) engines

7. On 1999–01 models, tighten the bolts in sequence as follows:
   a. Step 1: 36–39 ft. lbs. (47–53 Nm).
   b. Step 2: Loosen the bolts 1 complete turn.
   c. Step 3: Tighten the bolts to 20–24 ft. lbs. (27–33 Nm).
   d. Step 4: Tighten the bolts 85–95 degrees.
   e. Step 5: Tighten the bolts 85–95 degrees.
8. Install or connect the following:
   - Exhaust manifolds
   - Spark plugs
   - Lower intake manifold
   - Rocker arms and pushrods in their original locations. Tighten the rocker arm bolts to 24 ft. lbs. (32 Nm). Tighten each rocker arm bolt with the valve closed and the lifter on the camshaft lobe base circle. Rotate the crankshaft as necessary.
   - Valve covers
   - Heater supply tube brackets
   - Oil dipstick tube and tighten the nut to 13 ft. lbs. (18 Nm)
   - Power steering pump
   - Alternator
   - Accessory drive belt and tensioner
   - Spark plug wires
   - Ignition coil and bracket. Tighten the fasteners to 30–39 ft. lbs. (40–55 Nm).
   - Coolant hoses
   - Fuel injector wiring harness
   - Upper intake manifold

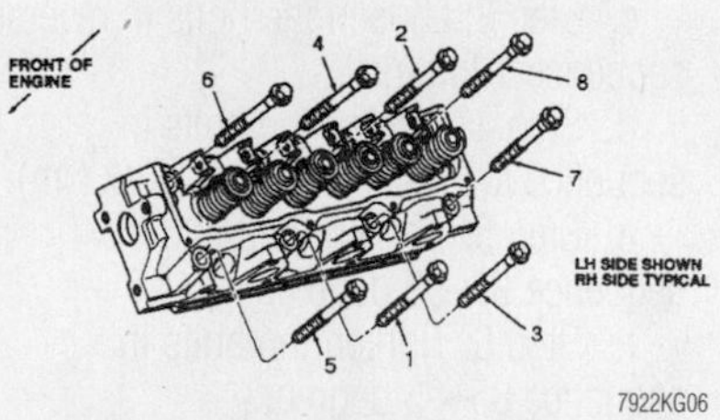

Cylinder head torque sequence—3.0L (VIN U) engine

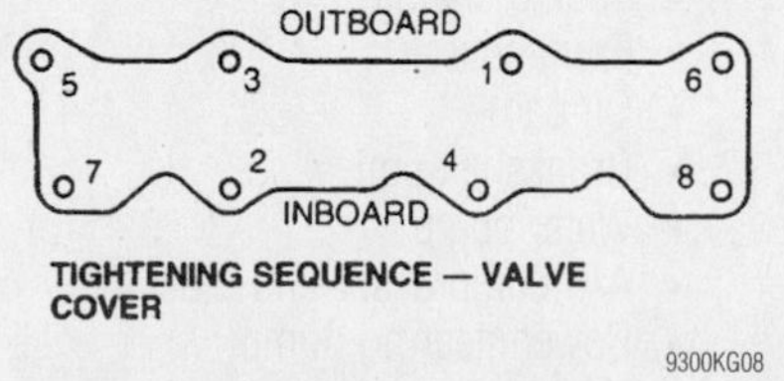

Valve cover tightening sequence—3.0L (VIN U) engine

   - Fuel lines
   - Cruise control cable
   - Accelerator cable
   - ECT sensor connector
   - CMP sensor and housing
   - IAC valve connector
   - TP sensor connector
   - IAT sensor connector
   - EGR vacuum regulator solenoid
   - EGR transducer
   - EGR valve
   - Vacuum hoses from the intake manifold
   - Air cleaner outlet hose
   - Negative battery cable
9. Fill the cooling system and check for leaks.

3.0L (VIN S) Engine

1. Before servicing the vehicle, refer to the precautions at the beginning of this section.
2. Remove the engine from the vehicle and mount it on an engine stand.
3. Remove or disconnect the following:
   - Upper intake manifold
   - Lower intake manifold
   - Exhaust manifolds
   - Valve covers
   - Radiator hose tube and bracket
   - Accessory drive belt
   - Crankcase ventilation tube
   - Heater hose bypass tube
   - Power steering pump
   - A/C compressor and bracket
   - Water pump

➡The crankshaft accessory drive pulley shaft has left-hand threads. Rotate the pulley shaft clockwise to remove.

   - Crankshaft pulley
   - Oil pan
   - Front cover

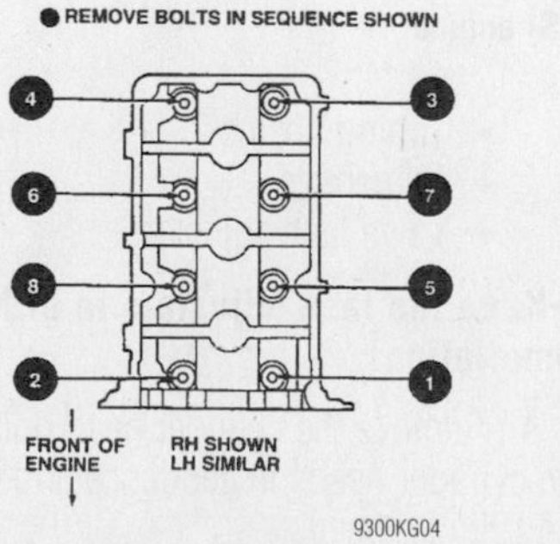

Cylinder head bolt removal sequence–right side head shown, left side similar—3.0L (VIN S) engine

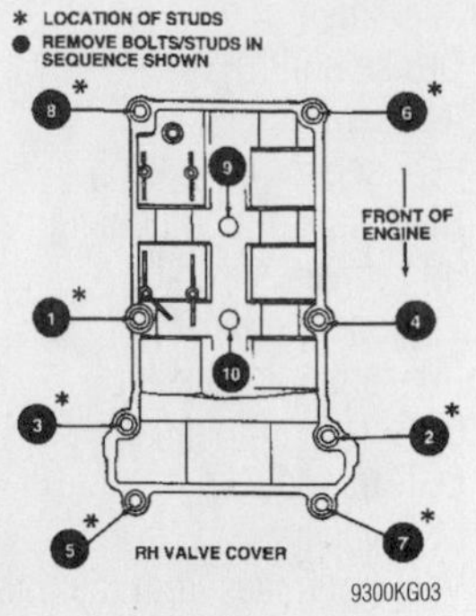

Right side valve cover bolt removal sequence—3.0L (VIN S) engine

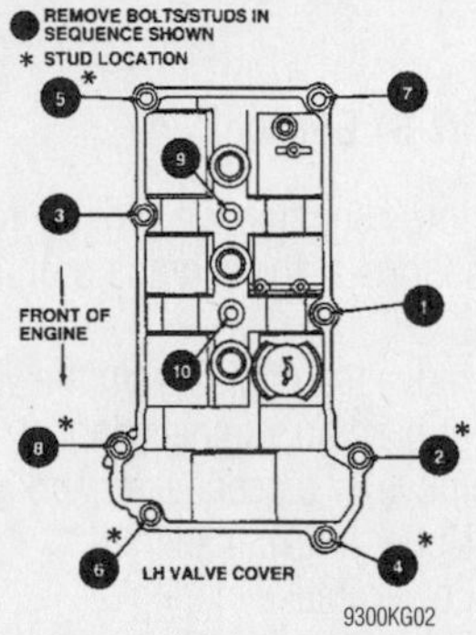

Left side valve cover bolt removal sequence—3.0L (VIN S) engine

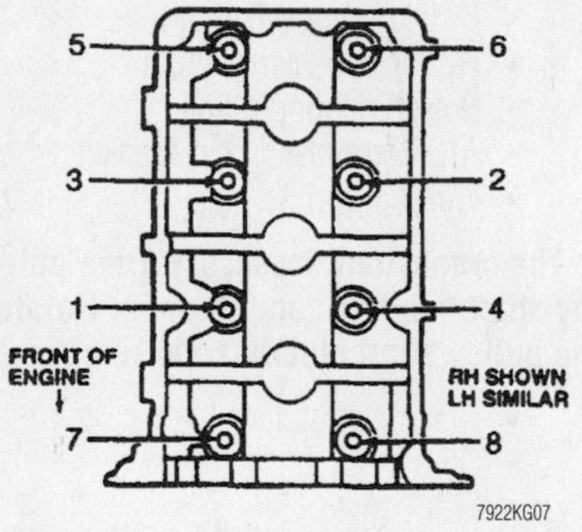

Cylinder head torque sequence—3.0L (VIN S) engine

- Timing chains
- Camshafts
- Valve lash adjusters

➡Keep the lash adjusters in order for installation.

4. Remove the cylinder head bolts from the cylinder heads in sequence and remove the cylinder heads.

To install:

➡The cylinder head bolts are a torque-to-yield design and cannot be reused.

➡Left and right cylinder head gaskets are not interchangeable.

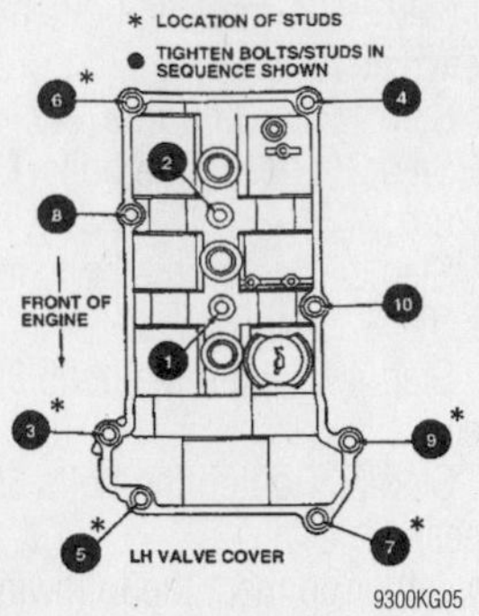

Left side valve cover bolt tightening sequence—3.0L (VIN S) engine

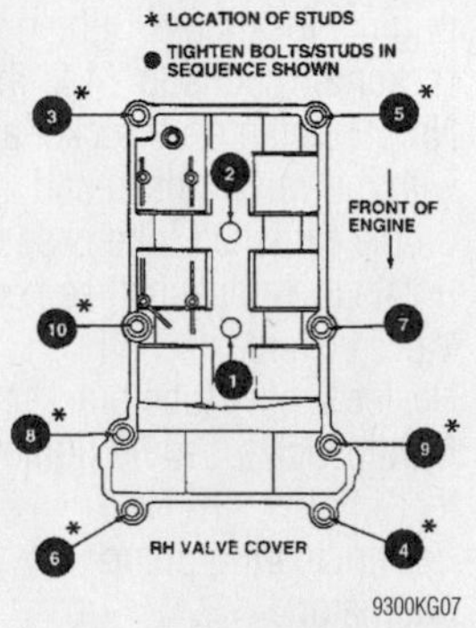

Right side valve cover bolt tightening sequence—3.0L (VIN S) engine

5. Install new gaskets and install the cylinder heads. Use new bolts and tighten as follows:
   a. Step 1: Tighten the bolts in sequence to 28–31 ft. lbs. (37–43 Nm).
   b. Step 2: Plus 85–95 degrees.
   c. Step 3: Loosen the bolts in reverse sequence 1 full turn.
   d. Step 4: Tighten the bolts in sequence to 28–31 ft. lbs. (37–43 Nm).
   e. Step 5: Tighten the bolts in sequence 85–95 degrees.
   f. Step 6: Tighten the bolts in sequence 85–95 degrees.
6. Install or connect the following:
   - Valve lash adjusters in their original locations
   - Camshafts
   - Timing chains
   - Front cover
   - Oil pan
   - Crankshaft pulley
   - Water pump
   - A/C compressor and bracket
   - Power steering pump
   - Heater hose bypass tube
   - Crankcase ventilation tube
   - Accessory drive belt
   - Radiator hose tube and bracket
   - Valve covers
   - Exhaust manifolds
   - Lower intake manifold
   - Upper intake manifold
   - Engine assembly into the vehicle.
7. Run the engine and check for leaks.

3.4L (VIN N) Engine

1. Before servicing the vehicle, refer to the precautions at the beginning of this section.
2. Remove the engine from the vehicle and mount it on an engine stand.
3. Set the crankshaft at Top Dead Center of the compression stroke for the No. 1 cylinder.
4. Remove or disconnect the following:
   - Accessory drive belts
   - A/C compressor
   - Alternator
   - Water pump drive pulley
   - Oil dipstick tube
   - Intake vacuum lines
   - Exhaust Gas Recirculation (EGR) tube
   - Throttle body
   - Surge tank and supports
   - Left and right intake manifolds
   - Engine control sensor wiring harness and brackets
   - Secondary air injection valve, tubes and brackets
   - Exhaust manifolds
   - Intake Manifold Runner Control (IMRC) cable and bracket
   - Lower intake manifold
   - Water crossover pipe
   - Water pump
   - Left and right valve covers
   - Spark plugs
   - Power steering pump
   - Crankshaft pulley
   - Camshaft Position (CMP) sensor
   - Crankshaft Position (CKP) sensor
   - Oil pan
   - Accessory belt idler pulleys
   - Front cover
   - CKP pulse ring
   - Timing chain
   - Camshafts
   - Cylinder heads

To install:

➡The cylinder head bolts are a torque-to-yield design and cannot be reused.

➡Left and right cylinder head gaskets are not interchangeable.

5. Install new gaskets and install the cylinder heads. Use new bolts and tighten in sequence as follows:
   a. Step 1: 20–23 ft. lbs. (27–32 Nm).
   b. Step 2: Plus 85–95 degrees.

Removal Sequence

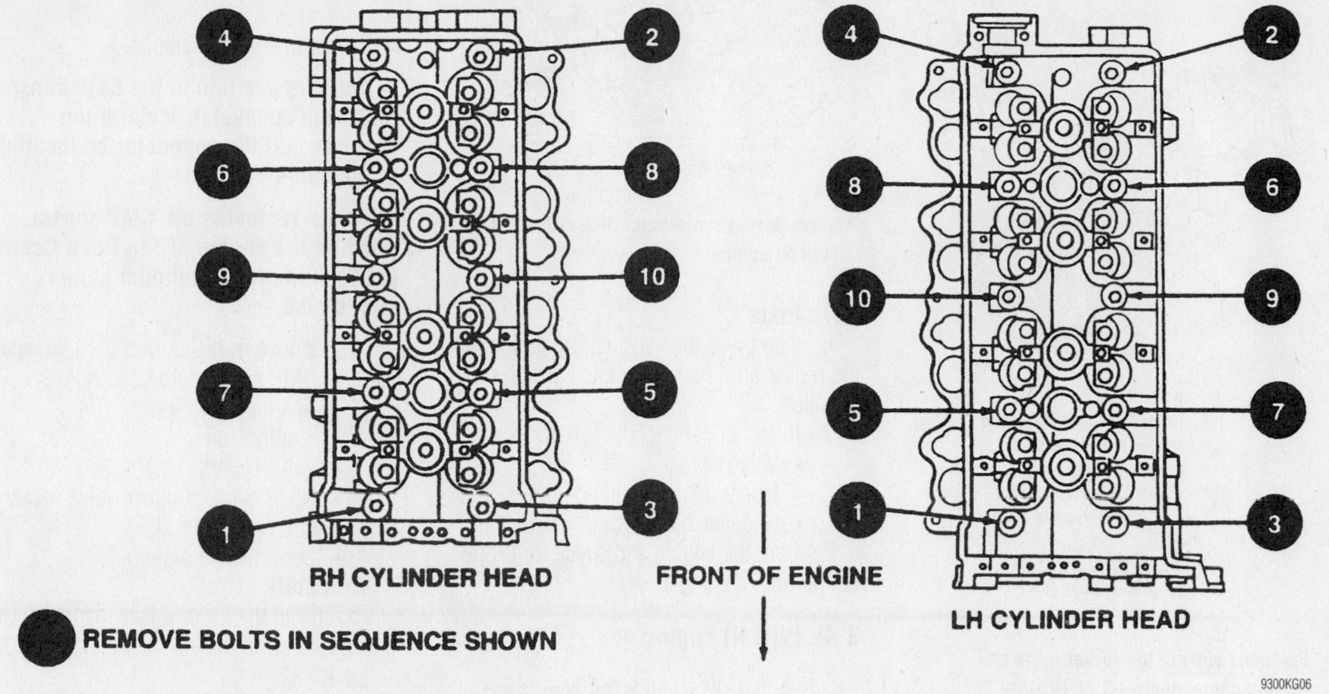

Cylinder head bolt removal sequence—3.4L (VIN N) engine

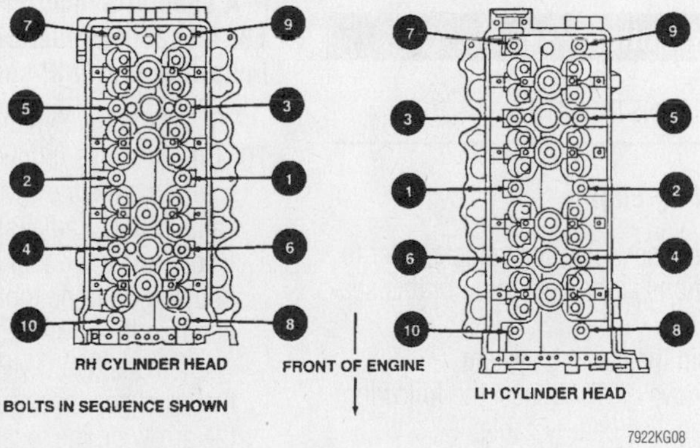

Cylinder head torque sequence—3.4L (VIN N) engine

6. Install or connect the following:
   - Camshafts
   - Timing chain
   - CKP pulse ring
   - Front cover
   - Accessory belt idler pulleys
   - Oil pan
   - CKP sensor
   - CMP sensor
   - Crankshaft pulley
   - Power steering pump
   - Spark plugs
   - Left and right valve covers
   - Water pump
   - Water crossover pipe
   - Lower intake manifold
   - IMRC cable and bracket
   - Exhaust manifolds
   - Secondary air injection valve, tubes, and brackets
   - Engine control sensor wiring harness and brackets
   - Left and right intake manifolds
   - Surge tank and supports
   - Throttle body
   - EGR tube
   - Intake vacuum lines
   - Oil dipstick tube
   - Water pump drive pulley
   - Alternator
   - A/C compressor
   - Accessory drive belts
   - Engine assembly into the vehicle.
7. Run the engine and check for leaks.

Rocker Arms/Shafts

REMOVAL & INSTALLATION

3.0L (VIN U) Engine

1. Before servicing the vehicle, refer to the precautions at the beginning of this section.
2. Remove or disconnect the following:
   - Negative battery cable
   - Upper intake manifold
   - Valve covers
   - Rocker arms

➡Keep the rocker arms in order for installation.

To install:

➡The rocker arm bolts are tightened with the valves closed. Rotate the crankshaft as necessary to position the lifter on the base circle of the camshaft lobe before tightening the corresponding rocker arm bolt.

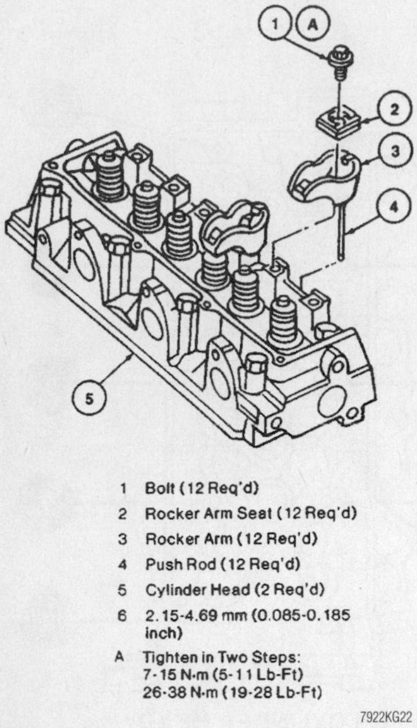

Exploded view of the rocker arms and related components—3.0L (VIN U)

3. Install the rocker arms. The rocker arm bolts are tightened in two steps as follows:
   a. Step 1: 96 inch lbs. (11 Nm).
   b. Step 2: 24 ft. lbs. (32 Nm).
4. Install or connect the following:
   - Valve covers
   - Upper intake manifold
   - Negative battery cable
5. Start the engine and check for proper operation.

3.0L (VIN S) Engine

1. Before servicing the vehicle, refer to the precautions at the beginning of this section.
2. Remove or disconnect the following:
   - Negative battery cable
   - Upper intake manifold
   - Valve covers
3. Rotate the crankshaft so that the camshaft lobe on the valve to be serviced is pointing directly away from the valve.
4. Install special tool Valve Spring Compressor 303-473.
5. Compress the valve spring and remove the rocker arm. Repeat for each arm to be removed.

➡If the rocker arms are to be reused, ensure that they are installed in the same position that they were removed from.

Valve spring compressor 303-473—3.0L (VIN S) engine

To install:

6. Compress the valve spring and install the rocker arm. Repeat for each arm to be installed.
7. Install or connect the following:
   - Valve covers
   - Upper intake manifold
   - Negative battery cable
8. Start the engine and check for proper operation.

3.4L (VIN N) Engine

The 3.4L (VIN N) engine does not have rocker arms. The camshaft lobes work directly on the valves.

Intake Manifold

REMOVAL & INSTALLATION

3.0L (VIN U) Engine

1. Before servicing the vehicle, refer to the precautions at the beginning of this section.
2. Drain the cooling system.
3. Remove or disconnect the following:
   - Negative battery cable
   - Air cleaner and outlet tube
   - Cruise control cable
   - Accelerator cable and bracket
   - Fuel lines
   - Vacuum lines
   - Exhaust Gas Recirculation (EGR) valve
   - EGR transducer
   - EGR vacuum solenoid
   - Throttle Position (TP) sensor connector
   - Intake Air Control (IAC) valve connector
   - Camshaft Position (CMP) sensor connector
   - Engine Coolant Temperature (ECT) sensor connector
   - Coolant temperature indicator sender connector
   - Coolant hoses
   - Upper alternator bracket
   - Engine control sensor wiring harness and bracket
   - Upper intake manifold

➡Note the position of the CMP sensor electrical connector. Installation requires that the connector be located in the same position.

➡Before removing the CMP sensor, position the engine at Top Dead Center (TDC) of the No. 1 cylinder compression stroke.

4. Remove or disconnect the following:
   - CMP sensor and housing
   - Spark plug wires
   - Ignition coil
   - Valve covers
   - No. 3 cylinder intake valve rocker arm and pushrod
   - Lower intake manifold

To install:

5. Install the lower intake manifold. Use new gaskets and tighten the bolts in sequence as follows:
   a. Step 1: 15–22 ft. lbs. (20–30 Nm).
   b. Step 2: 19–24 ft. lbs. (26–32 Nm).

➡A special Synchro Positioning tool T95T-12200-A must be used when installing the CMP sensor housing.

6. Attach the Synchro Position tool T95T-12200-A as follows:
   a. Engage the CMP sensor housing vane into the radial slot of the tool.
   b. Rotate the tool on the CMP sensor housing until the tool boss engages the notch in the CMP sensor housing.
   c. Install the CMP sensor housing so the drive gear engagement occurs when the arrow on the locator tool is pointed about 75 degrees counterclockwise from the rear face of the cylinder block. This step will locate the CMP sensor electrical connector in the same position as was noted on removal.
   d. Install the hold-down clamp and tighten the bolt to 14–22 ft. lbs. (19–30 Nm).
7. Remove the Synchro Position tool.
8. Install or connect the following:
   - CMP sensor
   - No. 3 cylinder intake valve rocker arm and pushrod
   - Valve covers
   - Ignition coil and tighten the bolts to 30–40 ft. lbs. (40–55 Nm)
   - Spark plug wires
   - Upper intake manifold and tighten the bolts to 15–22 ft. lbs. (20–30 Nm)

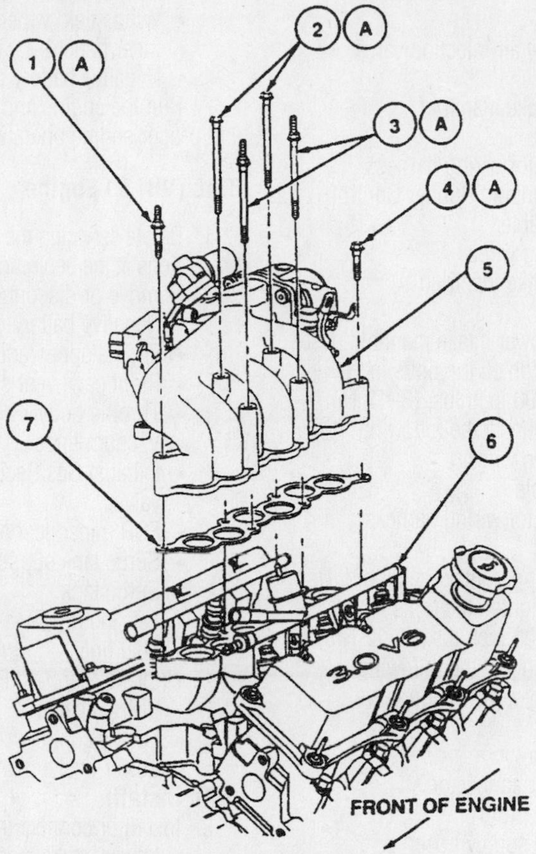

Item	Description
1	Stud Bolt
2	Bolt (2 Req'd)
3	Stud Bolt (2 Req'd)
4	Bolt
5	Upper Intake Manifold
6	Lower Intake Manifold
7	Intake Manifold Upper Gasket
A	Tighten to 20-30 N·m (15-22 Lb-Ft)

9300KG09

Exploded view of the upper intake manifold mounting—3.0L (VIN U) engine

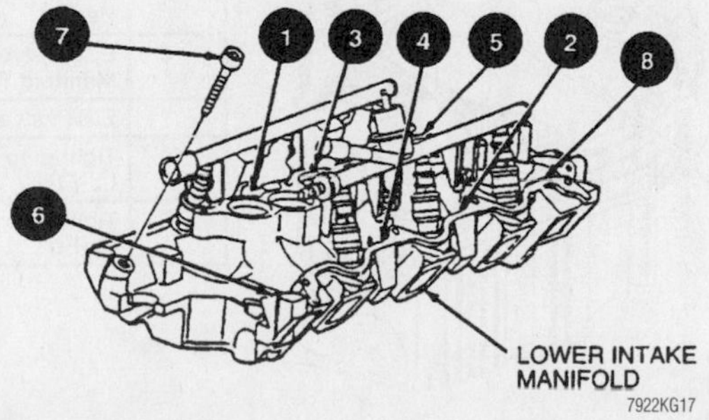

Intake manifold torque sequence—3.0L (VIN U) engine

- Engine control sensor wiring harness and bracket
- Upper alternator bracket
- Coolant hoses
- Coolant temperature indicator sender connector
- ECT sensor connector
- CMP sensor connector
- IAC valve connector
- TP sensor connector
- EGR vacuum solenoid
- EGR transducer
- EGR valve
- Vacuum lines
- Fuel lines
- Accelerator cable and bracket
- Cruise control cable
- Air cleaner and outlet tube
- Negative battery cable

9. Fill the cooling system.
10. Start the engine and check for leaks and proper operation.

3.0L (VIN S) Engine

1. Before servicing the vehicle, refer to the precautions at the beginning of this section.
2. Disconnect or remove:
   - Negative battery cable
   - Windshield wipers and cowl top inner panels
   - Cruise control cable
   - Accelerator cable and bracket
   - Vacuum hoses

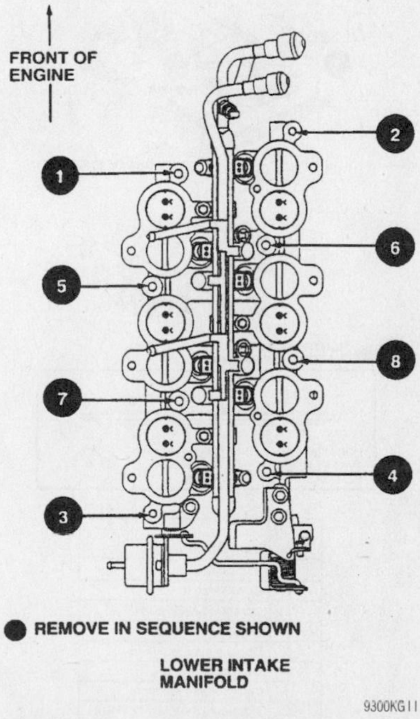

Lower intake manifold bolt removal

- Intake Air Control (IAC) valve supply hose
- IAC connector
- Throttle Position (TP) sensor connector

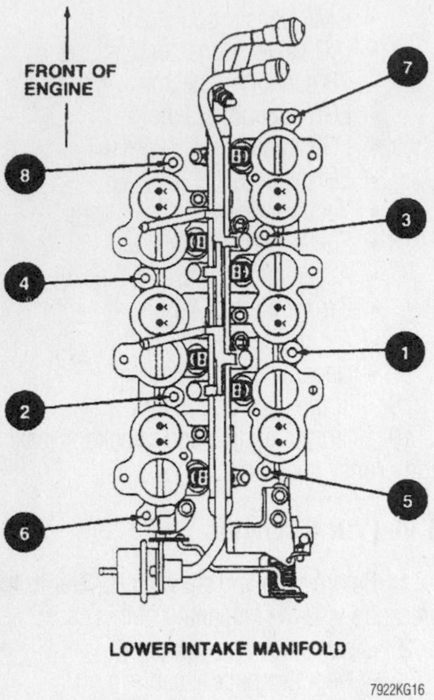

Lower intake manifold torque sequence—3.0L (VIN S) engine

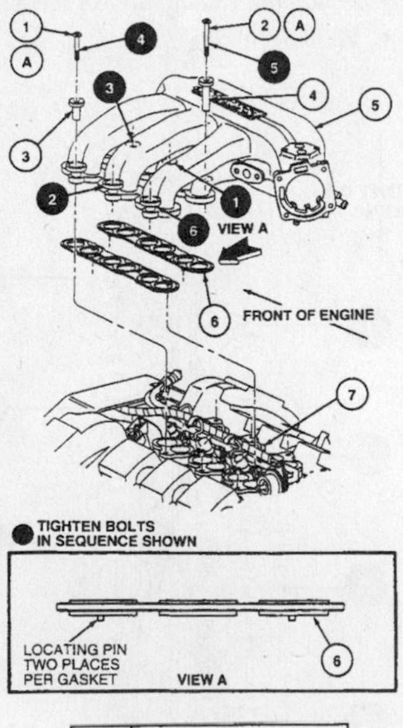

Item	Description
1	Bolt (3 Req'd)
2	Bolt (3 Req'd)
3	Isolator (3 Req'd)
4	Isolator (3 Req'd)
5	Upper Intake Manifold
6	Gasket (2 Req'd)
7	Lower Intake Manifold
A	Tighten to 8-12 N·m (71-106 Lb-in)

9300KG10

Upper intake manifold torque sequence—3.0L (VIN S) engine

- Exhaust Gas Recirculation (EGR) valve
- Secondary air injection valve bracket
- Upper intake manifold
- Fuel lines
- Fuel injector wiring harness
- Intake Manifold Runner Control (IMRC) cable
- Spark plug wires
- Lower intake manifold

To install:

3. Install the lower intake manifold. Use new gaskets and tighten the bolts in sequence to 71–106 inch lbs. (8–12 Nm).
4. Install or connect the following:
   - Spark plug wires
   - IMRC cable
   - Fuel injector wiring harness
   - Fuel lines
5. Install the upper intake manifold. Use new gaskets and tighten the bolts in sequence to 71–106 inch lbs. (8–12 Nm).
6. Install or connect the following:
   - Secondary air injection valve bracket
   - EGR valve
   - TP sensor connector
   - IAC connector
   - IAC valve supply hose
   - Vacuum hoses
   - Accelerator cable and bracket
   - Cruise control cable
   - Windshield wipers and cowl top inner panels
   - Negative battery cable
7. Run the engine and check for leaks and proper engine operation.

3.4L (VIN N) Engine

1. Before servicing the vehicle, refer to the precautions at the beginning of this section.
2. Remove or disconnect the following:
   - Negative battery cable
   - Engine appearance cover
   - Right cowl vent screen
   - Throttle body
   - Vacuum lines
   - Exhaust Gas Recirculation (EGR) valve
   - EGR transducer and bracket
   - Surge tank supports
   - Surge tank
   - Left and right intake manifold runners
   - Fuel lines
   - Intake Manifold Runner Control (IMRC) cable
   - Engine control sensor wiring harness
   - Lower intake manifold

To install:

3. Install or connect the following:
   - Lower intake manifold. Use new gaskets and tighten the bolts to 14–20 ft. lbs. (18–28 Nm).

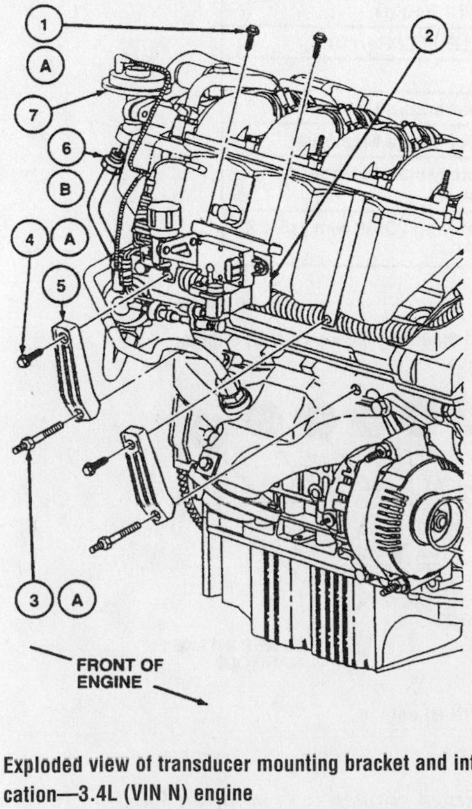

Item	Description
1	Bolt (2 Req'd)
2	Transducer Mounting Bracket
3	Stud Bolt (2 Req'd)
4	Bolt (2 Req'd)
5	Intake Manifold Support (2 Req'd)
6	EGR Valve to Exhaust Manifold Tube
7	EGR Valve
A	Tighten to 18-28 N·m (14-20 Lb-Ft)
B	Tighten to 25-35 N·m (19-25 Lb-Ft)

9300KG12

Exploded view of transducer mounting bracket and intake manifold support with torque specification—3.4L (VIN N) engine

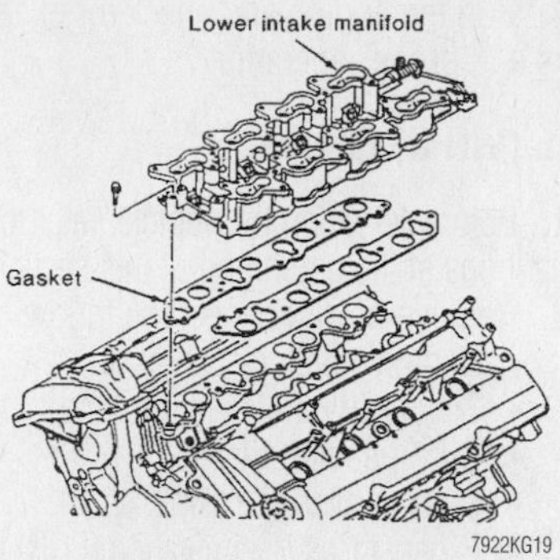

Exploded view of the lower intake manifold mounting—3.4L (VIN N) engine

- Engine control sensor wiring harness
- IMRC cable
- Fuel lines
- Left and right intake manifold runners. Tighten the bolts to 14–20 ft. lbs. (18–28 Nm).
- Surge tank
- Surge tank supports and tighten the fasteners to 14–20 ft. lbs. (18–28 Nm)
- EGR transducer and bracket
- EGR valve
- Vacuum lines
- Throttle body
- Right cowl vent screen
- Engine appearance cover
- Negative battery cable

4. Run the engine and check for leaks and proper engine operation.

Exhaust Manifold

REMOVAL & INSTALLATION

3.0L (VIN U) Engine

RIGHT SIDE

1. Before servicing the vehicle, refer to the precautions at the beginning of this section.
2. Remove or disconnect the following:
    - Negative battery cable
    - Cowl vent screen and extension
    - Heated Oxygen (HO_2S) sensor connector
    - Exhaust Gas Recirculation (EGR) tube
    - Exhaust manifold heat shield
    - Catalytic converter
    - Exhaust manifold

To install:

3. Install the exhaust manifold with a new gasket. Tighten the bolts in sequence as follows:
    a. Step 1: 89 inch lbs. (10 Nm)
    b. Step 2: 16 ft. lbs. (22 Nm)
4. Install or connect the following:
    - Catalytic converter and tighten the bolts to 30 ft. lbs. (40 Nm)
    - Exhaust manifold heat shield and tighten the bolts to 89 inch lbs. (10 Nm)

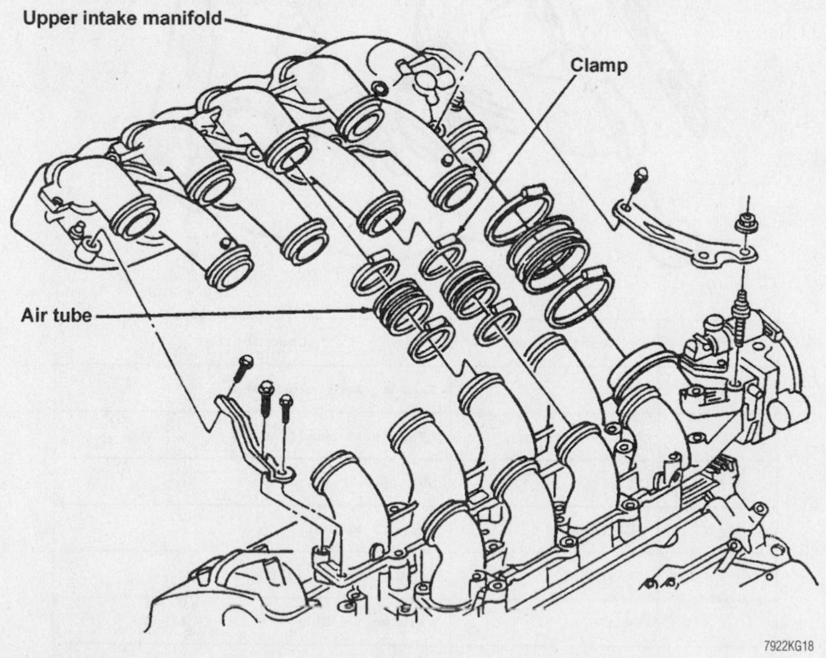

Exploded view of the upper intake manifold mounting—3.4L (VIN N) engine

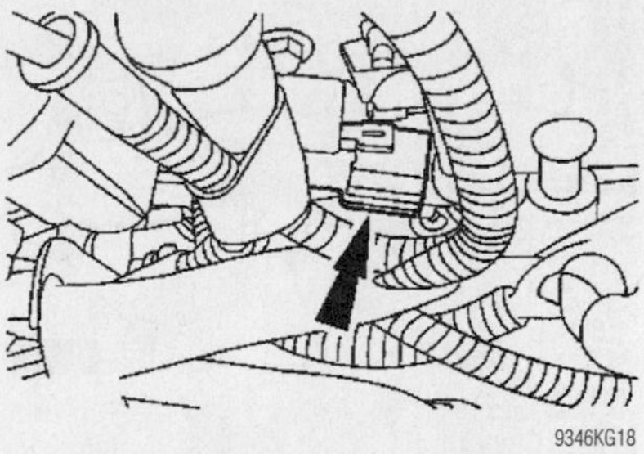

Right side Heated Oxygen (HO_2S) sensor connector—3.0L (VIN U) engine

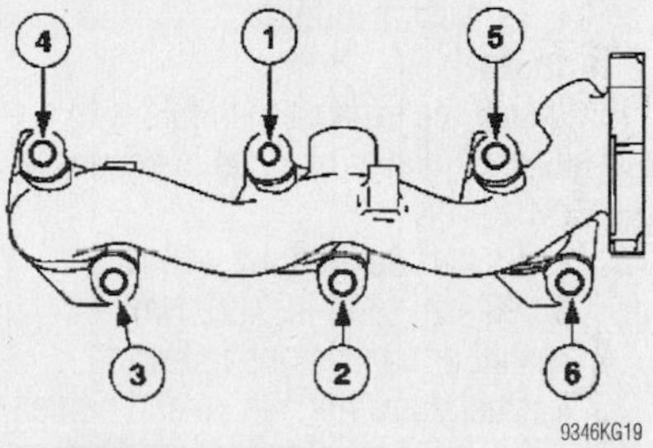

Right exhaust manifold torque sequence—3.0L (VIN U) engine

- EGR tube
- HO_2S sensor connector
- Cowl vent screen and extension
- Negative battery cable

5. Start the engine and check for leaks.

LEFT SIDE

1. Before servicing the vehicle, refer to the precautions in the beginning of this section.
2. Remove or disconnect the following:
    - Negative battery cable
    - Heated Oxygen (HO_2S) sensor connector
    - Oil dipstick tube
    - Power steering pressure line
    - Secondary air injection tube, if equipped

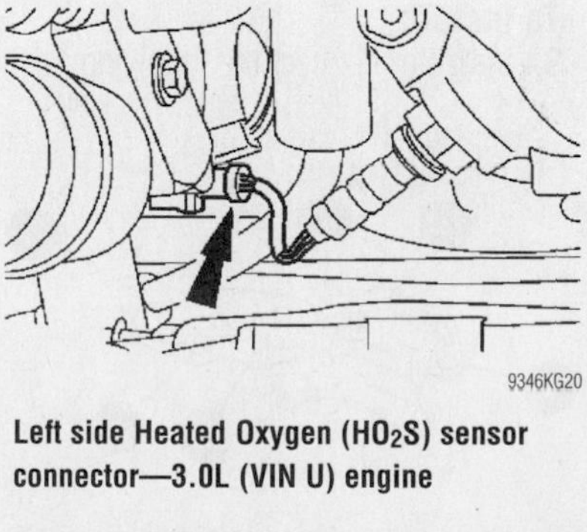

Left side Heated Oxygen (HO_2S) sensor connector—3.0L (VIN U) engine

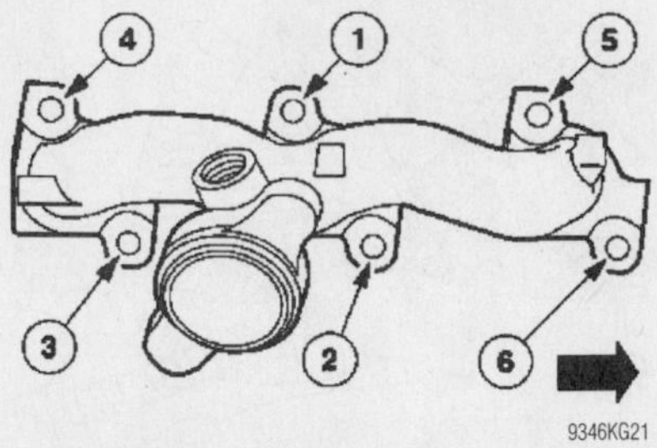

Left exhaust manifold torque sequence—3.0L (VIN U) engine

- Dual converter Y-pipe
- Exhaust manifold

To install:

3. Install the exhaust manifold with a new gasket. Tighten the bolts in sequence as follows:
   a. Step 1: 89 inch lbs. (10 Nm)
   b. Step 2: 16 ft. lbs. (22 Nm)
4. Install or connect the following:
   - Dual converter Y-pipe and tighten the bolts to 30 ft. lbs. (40 Nm)
   - Secondary air injection tube, if equipped
   - Power steering pressure line
   - Oil dipstick tube
   - HO_2S sensor connector
   - Negative battery cable
5. Start the engine and check for leaks.

3.0L (VIN S) Engine

1. Before servicing the vehicle, refer to the precautions at the beginning of this section.
2. Remove or disconnect the following:
   - Negative battery cable
   - Upper intake manifold
   - Right exhaust manifold heat shield
   - Ignition coil
   - Exhaust Gas Recirculation (EGR) tube
   - Heated Oxygen (HO_2S) sensors
   - Secondary air injection tube
   - Dual converter Y-pipe
   - Exhaust manifolds

To install:

3. Install or connect the following:

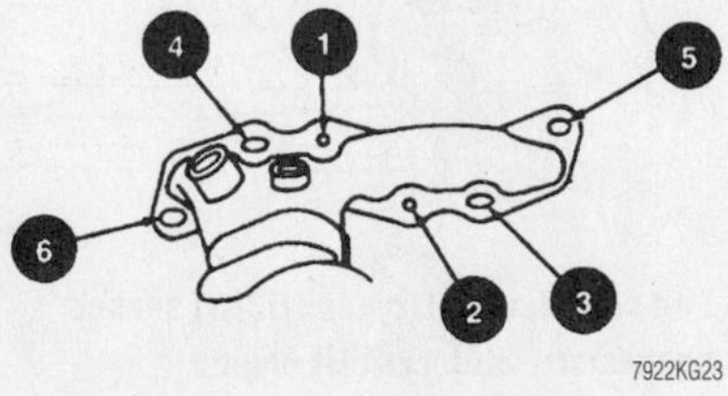

Right side exhaust manifold mounting bolt tightening sequence—3.0L (VIN S) engine

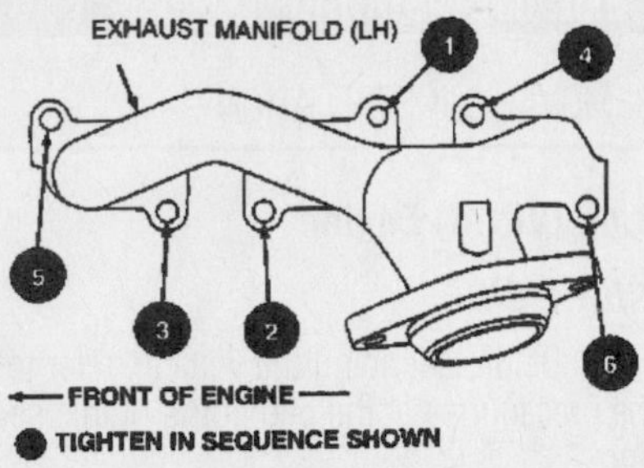

Left side exhaust manifold mounting bolt tightening sequence—3.0L (VIN S) engine

- Exhaust manifolds. Use new gaskets and tighten the bolts in sequence to 13–16 ft. lbs. (18–22 Nm).
- Dual converter Y-pipe
- Secondary air injection tube
- HO_2S sensors
- EGR tube
- Ignition coil
- Right exhaust manifold heat shield
- Upper intake manifold
- Negative battery cable

4. Run the engine and check for exhaust leaks and proper operation.

3.4L (VIN N) Engine

1. Before servicing the vehicle, refer to the precautions at the beginning of this section.
2. Remove or disconnect the following:
   - Negative battery cable
   - Secondary air injection tube
   - Oil dipstick tube
   - Dual converter Y-pipe
   - Exhaust Gas Recirculation (EGR) tube
   - Exhaust manifold heat shields
   - Exhaust manifolds

To install:

3. Install the exhaust manifolds. Use new gaskets and tighten the nuts to 30–44 ft. lbs. (40–60 Nm).
4. Install or connect the following:
   - Exhaust manifold heat shields
   - EGR tube
   - Dual converter Y-pipe
   - Oil dipstick tube
   - Secondary air injection tube
   - Negative battery cable
5. Run the engine and check for exhaust leaks and proper operation.

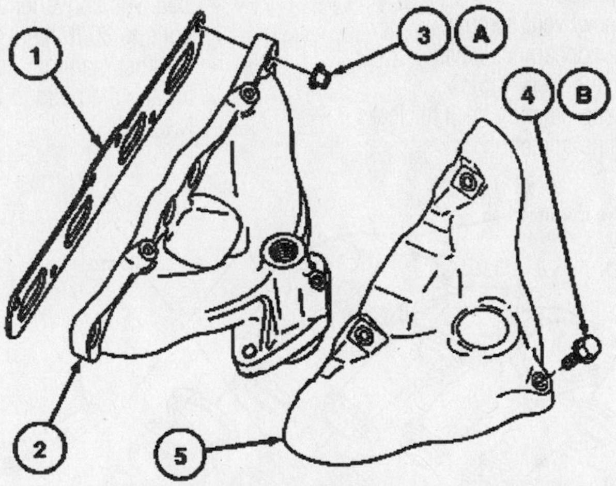

Item	Part Number	Description
1	9448	Exhaust Manifold Gasket
2	9431	LH Exhaust Manifold
3	A9921	Nut (6 Req'd)
4	90105-08512	Bolt (3 Req'd)
5	9A462	Exhaust Manifold Shield
A	--	Tighten to 40-60 Nm (30-44 Lb-Ft)
B	--	Tighten to 16-23 Nm (12-16 Lb-Ft)

9346KG22

Left exhaust manifold and related components—3.4L (VIN N) engine

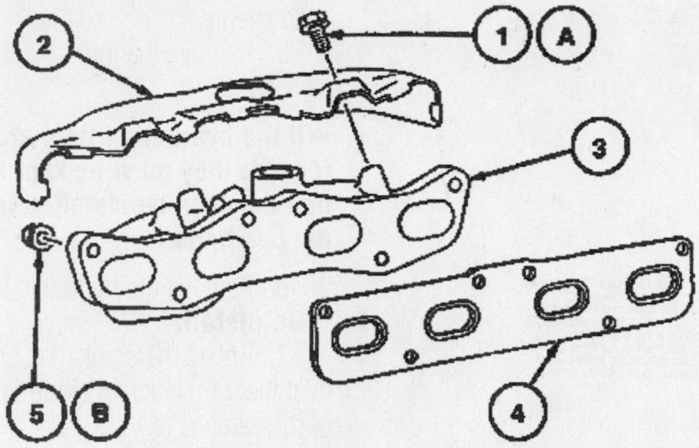

Item	Part Number	Description
1	90105-08512	Bolt (3 Req'd)
2	9A462	Exhaust Manifold Shield
3	9430	Exhaust Manifold
4	9448	Exhaust Manifold Gasket
5	A9221	Nut (6 Req'd)
A	--	Tighten to 16-23 Nm (12-16 Lb-Ft)
B	--	Tighten to 40-60 Nm (30-44 Lb-Ft)

9346KG23

Right exhaust manifold and related components—3.4L (VIN N) engine

Camshaft and Valve Lifters

REMOVAL & INSTALLATION

3.0L (VIN U) Engine

➡If the rocker arms, pushrods or lifters are to be reused, they must be installed in the same positions they were removed from.

1. Before servicing the vehicle, refer to the precautions at the beginning of this section.
2. Remove the engine from the vehicle and mount it on an engine stand.
3. Remove or disconnect the following:
   - Valve covers
   - Rocker arms and pushrods
   - Accessory drive belt and tensioner
   - Alternator and brackets
   - Intake manifold
   - Valve lifter guide plate
   - Valve lifters
   - Crankshaft damper
   - Oil pan
   - Front cover
   - Timing chain and gears
   - Camshaft thrust plate
   - Camshaft

To install:

➡If replacing the camshaft, the valve lifters must also be replaced.

4. Install or connect the following:
   - Camshaft
   - Camshaft thrust plate and tighten the bolts to 84 inch lbs. (10 Nm)
   - Timing chain and gears. Align the timing marks on the sprockets.
   - Front cover
   - Oil pan
   - Crankshaft damper
   - Valve lifters
   - Valve lifter guide plate
   - Intake manifold
   - Alternator and brackets
   - Accessory drive belt and tensioner
   - Rocker arms and pushrods
   - Valve covers
5. Install the engine assembly into the vehicle.
6. Start the engine and check for leaks and proper engine operation.

3.0L (VIN S) Engine

1. Before servicing the vehicle, refer to the precautions at the beginning of this section.

7922KG30

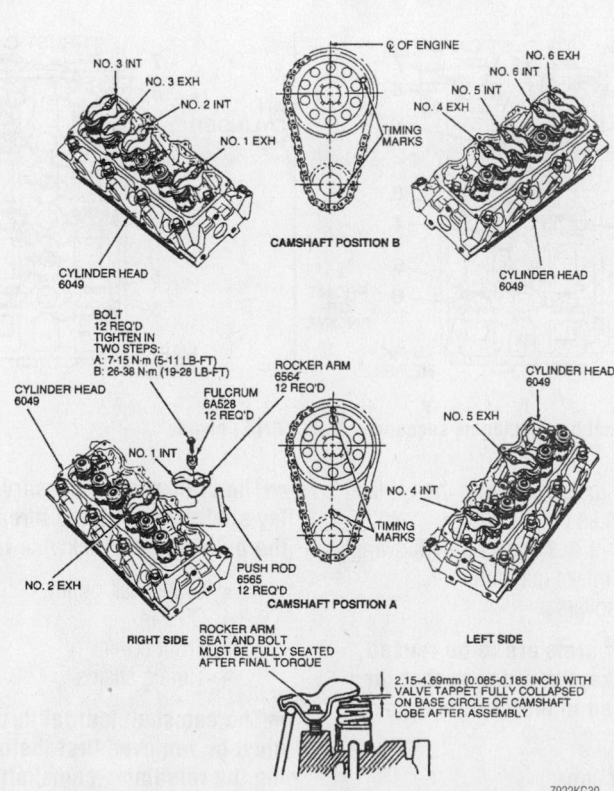

Tighten the rocker arm nuts according to the position of the camshaft—3.0L (VIN U) engine

For Accessory Drive Belt illustrations, see Section 1 of this manual

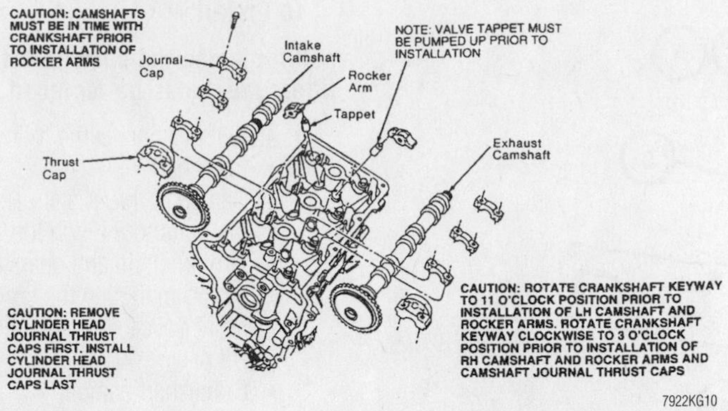
CAUTION: CAMSHAFTS MUST BE IN TIME WITH CRANKSHAFT PRIOR TO INSTALLATION OF ROCKER ARMS
Journal Cap
Intake Camshaft
NOTE: VALVE TAPPET MUST BE PUMPED UP PRIOR TO INSTALLATION
Rocker Arm
Tappet
Exhaust Camshaft
Thrust Cap
CAUTION: REMOVE CYLINDER HEAD JOURNAL THRUST CAPS FIRST. INSTALL CYLINDER HEAD JOURNAL THRUST CAPS LAST
CAUTION: ROTATE CRANKSHAFT KEYWAY TO 11 O'CLOCK POSITION PRIOR TO INSTALLATION OF LH CAMSHAFT AND ROCKER ARMS. ROTATE CRANKSHAFT KEYWAY CLOCKWISE TO 3 O'CLOCK POSITION PRIOR TO INSTALLATION OF RH CAMSHAFT AND ROCKER ARMS AND CAMSHAFT JOURNAL THRUST CAPS
7922KG10

Exploded view of camshaft mounting—3.0L (VIN S) engine

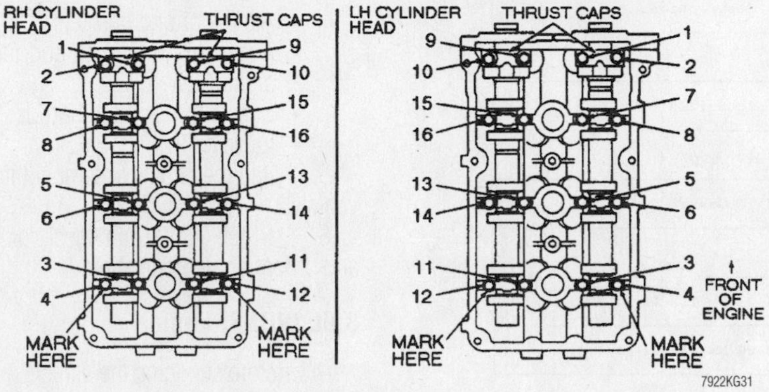
RH CYLINDER HEAD
THRUST CAPS
LH CYLINDER HEAD
THRUST CAPS
FRONT OF ENGINE
MARK HERE
MARK HERE
MARK HERE
MARK HERE
7922KG31

Camshaft journal bolt removal sequence—3.0L (VIN S) engine

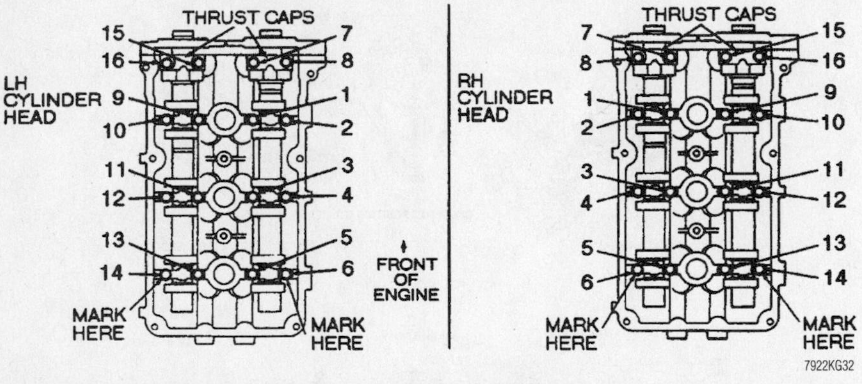
THRUST CAPS
LH CYLINDER HEAD
FRONT OF ENGINE
RH CYLINDER HEAD
THRUST CAPS
MARK HERE
MARK HERE
MARK HERE
MARK HERE
7922KG32

Camshaft journal bolt tightening sequence—3.0L (VIN S) engine

2. Remove the engine from the vehicle and mount it on an engine stand.
3. Remove or disconnect the following:
   - Upper intake manifold
   - Valve covers

➡If the rocker arms are to be reused, they must be kept in order so that they can be installed in their original positions.

   - Rocker arms
   - Accessory drive belt
   - Power steering pump
   - Water pump
   - A/C compressor and bracket

➡The crankshaft accessory drive pulley shaft has left-hand threads. Rotate the pulley shaft clockwise to remove.

   - Crankshaft pulley
   - Oil pan
   - Front cover
   - Timing chains

➡The camshaft journal thrust caps must be removed first, before loosening the remaining camshaft journal cap bolts, to ensure that the camshaft journal thrust caps are not damaged.

4. Loosen the camshaft journal cap bolts in sequence and in several passes to allow the camshaft to raise off the cylinder head evenly.
5. Remove the journal caps and the camshafts.

➡If the hydraulic lifters are to be reused, they must be kept in order so that they can be installed in their original positions.

6. Remove the hydraulic lifters.

To install:

7. Rotate the crankshaft so the keyway is at the 11 o'clock position for installation of the camshafts.
8. Install or connect the following:
   - Hydraulic lifters in their original positions
   - Camshafts with the timing marks (marked RFF) on the back of the camshaft sprockets aligned

➡The camshaft journal caps and cylinder heads are numbered to ensure that they are assembled in their original positions.

➡Do not install the camshaft journal thrust caps until the rocker arms and timing chains have been installed and the camshaft journal caps are tightened into position.

   - Camshaft journal caps in their original positions
   - Timing chains

9. Tighten the camshaft journal cap bolts, in sequence, to 71–106 inch lbs. (8–12 Nm).
10. Install the thrust caps and tighten the bolts to 71–106 inch lbs. (8–12 Nm).
11. Install or connect the following:
   - Front cover
   - Oil pan
   - Crankshaft pulley
   - A/C compressor and bracket
   - Water pump
   - Power steering pump
   - Accessory drive belt
   - Rocker arms in their original positions
   - Valve covers
   - Upper intake manifold
   - Engine assembly into the vehicle.
12. Run the engine and check for leaks and proper operation.

3.4L (VIN N) Engine

1. Before servicing the vehicle, refer to the precautions at the beginning of this section.
2. Remove the engine from the vehicle and mount it on an engine stand.
3. Set the crankshaft at Top Dead Cen-

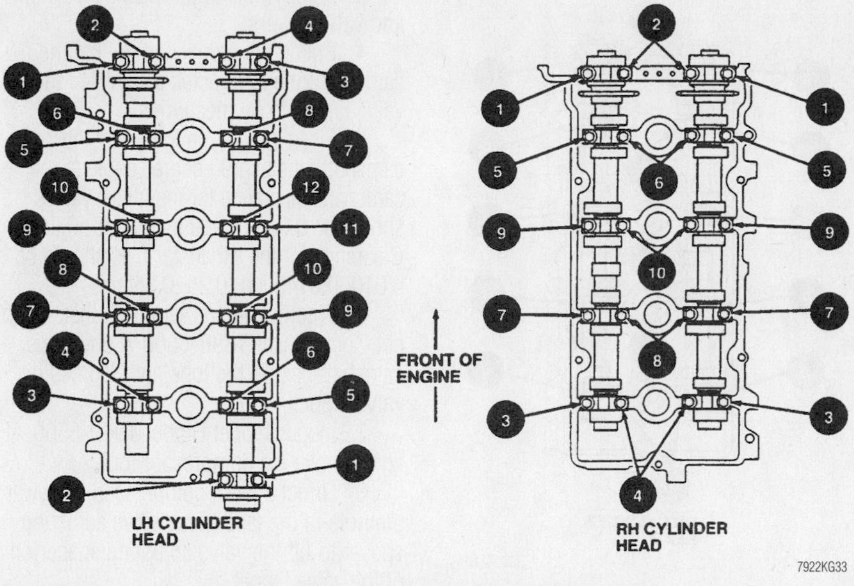

Camshaft journal cap bolt removal sequence—3.4L (VIN N) engine

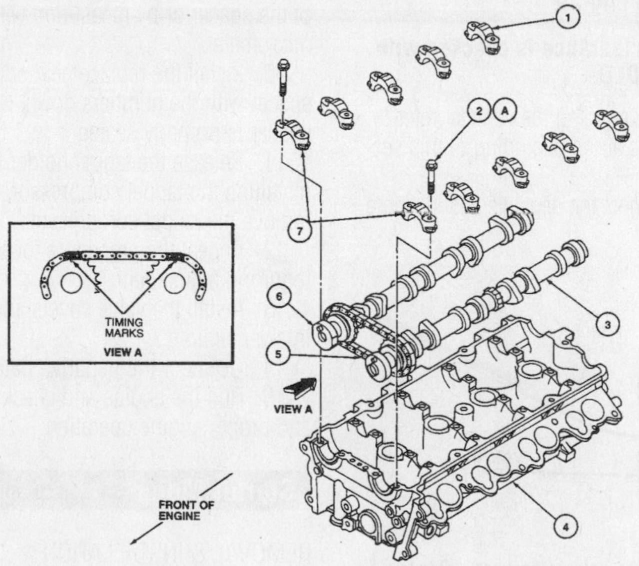

Item	Part Number	Description
1	—	Camshaft Journal Cap (Part of 6049) (8 Req'd)
2	6L293	Hex Bolt with Washer (20 Req'd)
3	6250	Camshaft , RH Intake
4	6010	Cylinder Block , RH
5	6268	Timing Chain / Belt

Item	Part Number	Description
6	6250	Camshaft , RH Exhaust
7	—	Camshaft Journal Thrust Cap (Part of 6049) (2 Req'd)
A	—	Tighten in sequence in two steps: 7-12 N·m (62-106 Lb-In) and 16-21 N·m (12-15 Lb-Ft)

9306KG01

Camshaft installation—3.4L (VIN N) engine—Right head shown, left head similar

ter (TDC) of the compression stroke for the No. 1 cylinder.

4. Remove or disconnect the following:

- Accessory drive belts
- A/C compressor
- Alternator
- Water pump drive pulley
- Intake vacuum lines
- Exhaust Gas Recirculation (EGR) tube
- Throttle body
- Surge tank and supports
- Left and right intake manifolds
- Engine control sensor wiring harness and brackets
- Intake Manifold Runner Control (IMRC) cable and bracket
- Left and right valve covers
- Power steering pump
- Crankshaft pulley
- Camshaft Position (CMP) sensor
- Crankshaft Position (CKP) sensor
- Oil pan
- Accessory belt idler pulleys
- Front cover
- CKP pulse ring
- Timing chain
- Left and right camshaft chain tensioners

➡The camshaft journal thrust caps must be removed before loosening the remaining camshaft journal cap bolts to ensure that the camshaft journal thrust caps are not damaged.

5. Loosen the camshaft journal cap bolts in sequence, in several passes, to allow the camshaft to raise off the cylinder head evenly.
6. Remove the journal caps and the camshafts.

➡The camshaft journal caps and cylinder heads are numbered to ensure that they are assembled in their original positions.

➡If the hydraulic lifters are to be reused, they must be kept in order so that they can be installed in their original positions.

7. Remove the hydraulic lifters.

To install:

8. Rotate the crankshaft so the keyway is aligned with the mark on the oil pump at the 11 o'clock position for installation of the camshafts.
9. Install the hydraulic lifters in their original positions.
10. Install the camshafts. Match the camshaft and chain timing marks as shown, and install the camshafts with the timing marks facing up.
11. Apply a 0.08–0.11 inch (2–3mm) bead of silicone gasket and sealant to the left side cylinder head intake camshaft journal thrust cap.
12. Install the camshaft journal caps in their original positions.
13. Tighten the camshaft journal cap bolts in sequence as follows:
    a. Step 1: 62–106 inch lbs. (7–12 Nm).
    b. Step 2: 12–15 ft. lbs. (16–21 Nm).
14. Install a new camshaft seal with a

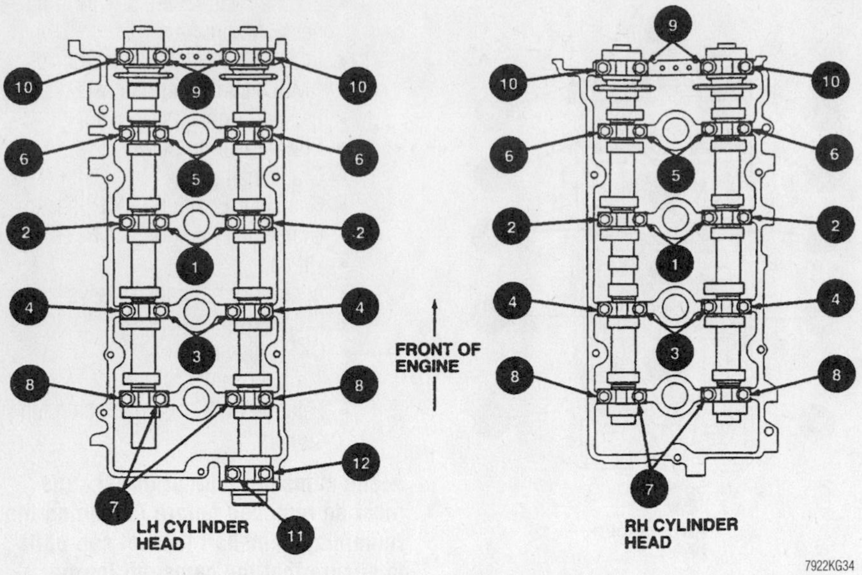

Camshaft journal cap bolt tightening sequence—3.4L (VIN N) engine

Cam Seal Expander T89P-6256-B and Cam Seal Replacer T89-6256-A.

15. Install or connect the following:
   - Left and right camshaft chain tensioners
   - Timing chain
   - CKP pulse ring
   - Front cover
   - Accessory belt idler pulleys
   - Oil pan
   - CKP sensor
   - CMP sensor
   - Crankshaft pulley
   - Power steering pump
   - Left and right valve covers
   - IMRC cable and bracket
   - Engine control sensor wiring harness and brackets
   - Left and right intake manifolds
   - Surge tank and supports
   - Throttle body
   - EGR tube
   - Intake vacuum lines
   - Water pump drive pulley
   - Alternator
   - A/C compressor
   - Accessory drive belts
   - Engine assembly into the vehicle.
16. Run the engine and check for leaks and proper operation.

Valve Lash

ADJUSTMENT

3.0L (VIN U and S) Engines

The lash adjusters (valve tappets), are hydraulic and are not adjustable.

3.4L (VIN N) engine

➡The valve clearance is checked with the engine COLD.

1. Before servicing the vehicle, refer to the precautions at the beginning of this section.
2. Disconnect the negative battery cable.

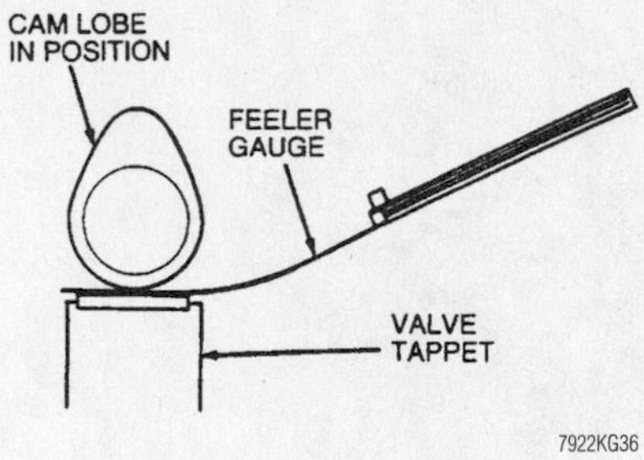

To measure the valve clearance, slide the feeler gauge between the base circle of the cam and the tappet—3.4L (VIN N) engine

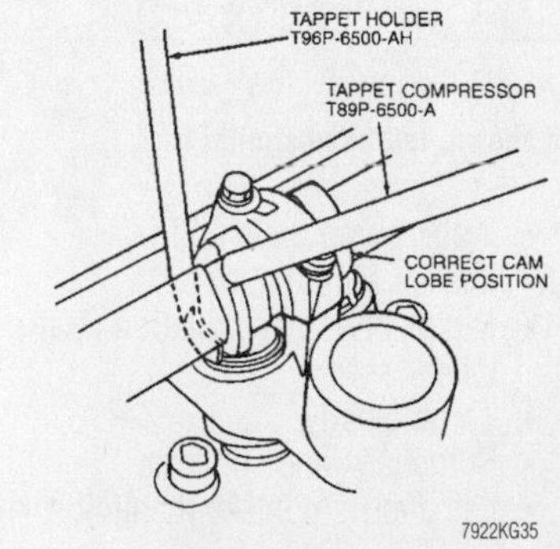

Correct positioning of the tappet compressor and holder tools—3.4L (VIN N) engine

3. Remove the upper intake manifold and valve covers.
4. Rotate the crankshaft so that the camshaft lobe is directed away from the valve tappet to be measured.
5. Insert a feeler gauge under the camshaft lobe at a 90 degree angle to the camshaft. Clearance for the intake valves should be 0.006–0.010 inch (0.15–0.25mm). Clearance for the exhaust valves should be 0.010–0.014 inch (0.25–0.35mm).
6. If adjustment is required, install Tappet Compressor T89P-6500-A under the camshaft next to the lobe and depress the valve tappet.
7. Install Tappet Holder T96P-6500-AH and remove the compressor tool.
8. Direct a jet of compressed air toward the hole in the face of the valve adjusting spacer to lift the valve adjusting spacer off of the valve tappet.
9. Determine the size of the adjusting spacer by the numbers on the bottom face of the spacer or by measuring with a micrometer.
10. Install the replacement adjusting spacer with the numbers down. Be sure the spacer is properly seated.
11. Release the tappet holder by installing the tappet compressor, then remove the tappet compressor.
12. Repeat the procedure for each valve requiring adjustment.
13. Install the valve covers and the upper intake manifold.
14. Connect the negative battery cable.
15. Run the engine and check for leaks and proper engine operation.

Starter Motor

REMOVAL & INSTALLATION

1. Before servicing the vehicle, refer to the precautions at the beginning of this section.
2. Remove or disconnect the following:
   - Negative battery cable
   - Splash shield
   - Starter electrical connectors
   - Starter

To install:

3. Install or connect the following:
   - Starter and tighten the bolts to 16–21 ft. lbs. (21–29 Nm)
   - Starter electrical connectors and tighten the battery cable nut to 80–123 inch lbs. (9–14 Nm)
   - Splash shield
   - Negative battery cable

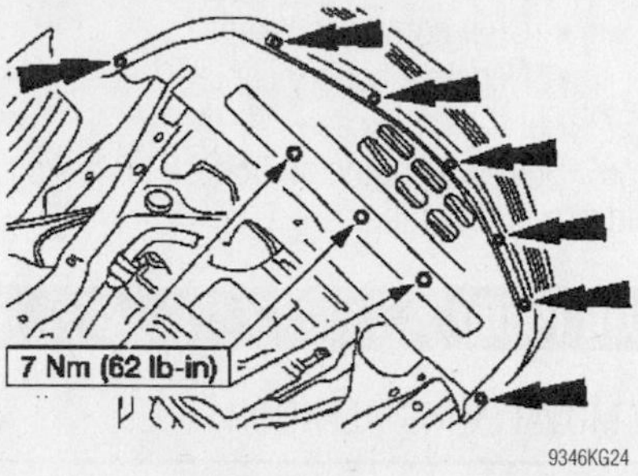

Remove the front splash shield for access

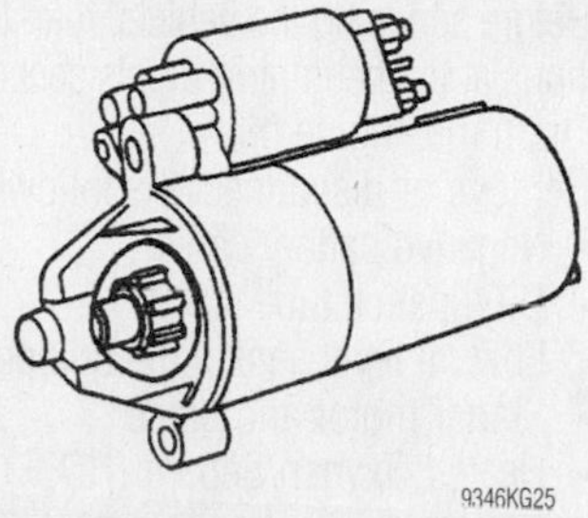

Starter motor—3.0L (VIN U) engine

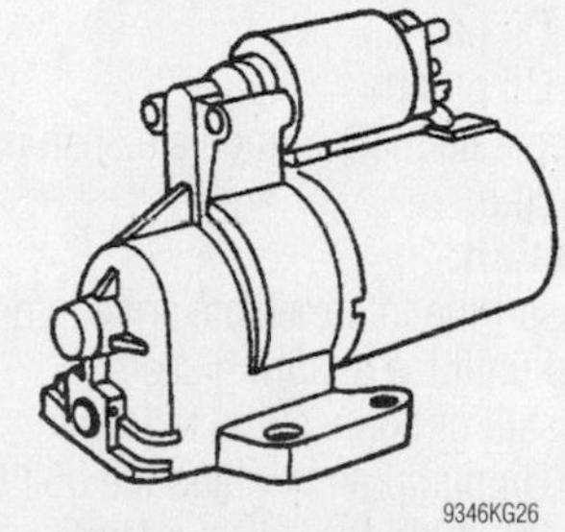

Starter motor—3.0L (VIN S) and 3.4L (VIN N) engines

Oil Pan

REMOVAL & INSTALLATION

3.0L (VIN U) Engine

1998–99 ENGINES

1. Before servicing the vehicle, refer to the precautions at the beginning of this section.
2. Drain the engine oil.
3. Remove or disconnect the following:
   - Negative battery cable
   - Oil dipstick tube
   - Low oil level sensor, if equipped
   - Starter motor and brace
   - Heated Oxygen (HO_2S) sensor connectors
   - Dual converter Y-pipe
   - Engine rear plate
   - Oil pan

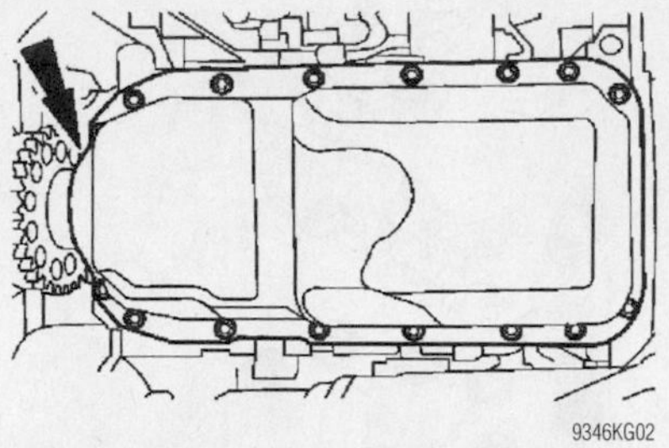

Oil pan—3.0L (VIN U) engine

To install:

4. Install the oil pan with new gaskets. Tighten the bolts as follows:
   a. Step 1: Tighten all bolts to 10 ft. lbs. (14 Nm).
   b. Step 2: Loosen all bolts.
   c. Step 3: Tighten all bolts to 10 ft. lbs. (14 Nm).
5. Install or connect the following:
   - Engine rear plate
   - Dual converter Y-pipe
   - HO_2S sensor connectors
   - Starter motor and brace
   - Low oil level sensor, if equipped
   - Oil dipstick tube
   - Negative battery cable
6. Fill the crankcase.
7. Start the engine and check for leaks.

2000–01 ENGINES

1. Before servicing the vehicle, refer to the precautions at the beginning of this section.
2. Drain the engine oil.
3. Remove or disconnect the following:
   - Negative battery cable
   - Oil dipstick tube
   - Low oil level sensor, if equipped
   - Starter motor and brace
   - Heated Oxygen (HO_2S) sensor connectors
   - Dual converter Y-pipe
   - Engine rear plate
   - Oil pan

To install:

4. Install the oil pan with new gaskets. Tighten the bolts as follows:
   a. Step 1: Tighten the 4 corner bolts to 108 inch lbs. (12 Nm).
   b. Step 2: Tighten the remaining bolts to 108 inch lbs. (12 Nm).
   c. Step 3: Retighten all bolts to 108 inch lbs. (12 Nm).
5. Install or connect the following:
   - Engine rear plate
   - Dual converter Y-pipe
   - HO_2S sensor connectors
   - Starter motor and brace
   - Low oil level sensor, if equipped
   - Oil dipstick tube
   - Negative battery cable
6. Fill the crankcase.
7. Start the engine and check for leaks.

3.0L (VIN S) Engine

1. Before servicing the vehicle, refer to the precautions at the beginning of this section.

Item	Description
1	Upper Cylinder Block
2	Lower Cylinder Block (Part of 6010)
3	Oil Pan
4	Stud Bolt (5 Req'd)
5	Bolt (10 Req'd)
6	Oil Pan Gasket
7	Engine Front Cover

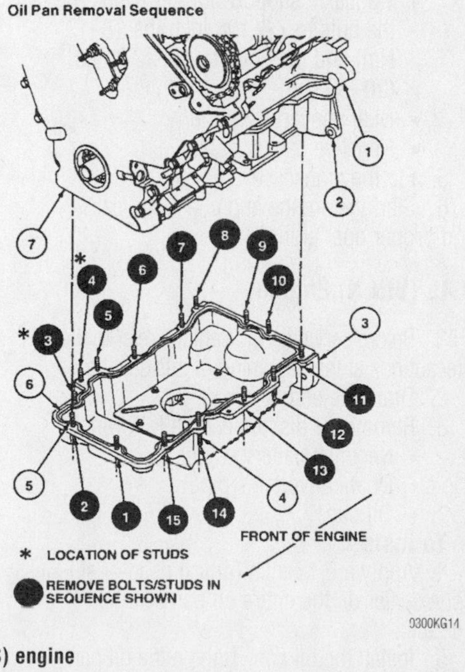

Oil pan bolt removal sequence—3.0L (VIN S) engine

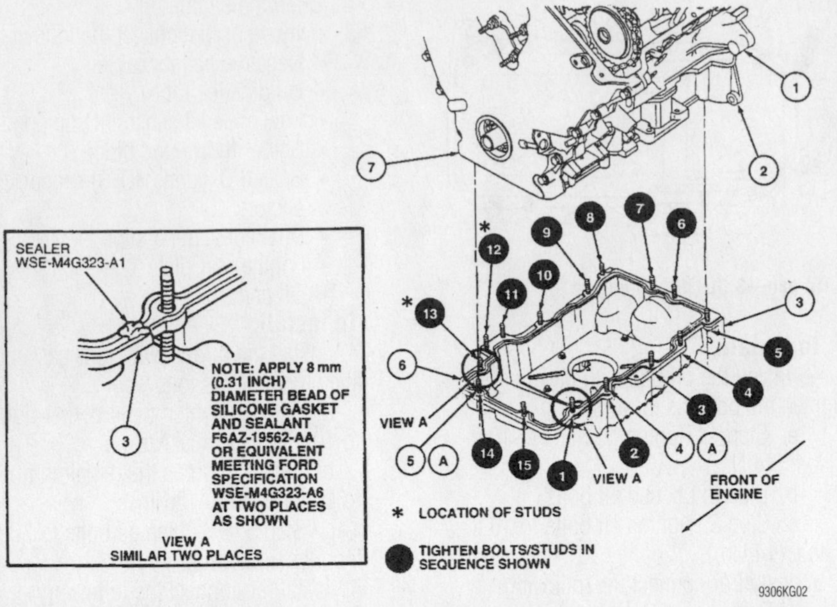

Oil pan bolt tightening sequence—3.0L (VIN S) engine

2. Drain the engine oil.
3. Remove or disconnect the following:
   - Negative battery cable
   - Dual converter Y-pipe
   - Transaxle support bracket
   - Oil pan by following the bolt removal sequence shown

To install:

4. Install or connect the following:
   - Oil pan. Use a new gasket and tighten the bolts in sequence to 15–22 ft. lbs. (20–30 Nm).
   - Transaxle support bracket. Tighten the nuts to 71–106 inch lbs. (8–12 Nm) and the bolts to 15–22 ft. lbs. (20–30 Nm).
   - Dual converter Y-pipe
   - Negative battery cable
5. Fill the crankcase.
6. Start the engine and check for leaks and proper operation.

3.4L (VIN N) Engine

1. Before servicing the vehicle, refer to the precautions at the beginning of this section.
2. Drain the engine oil.
3. Remove or disconnect the following:
   - Negative battery cable
   - Dual converter Y-pipe
   - Oil pan

To install:

4. Apply a 0.16 inch (4mm) bead of silicone sealer on the entire oil pan sealing surface.
5. Install the oil pan. Tighten the oil pan bolts and studs in sequence to 15–22 ft. lbs. (20–30 Nm). Tighten the oil pan-to-transaxle case bolts to 25–34 ft. lbs. (34–46 Nm).

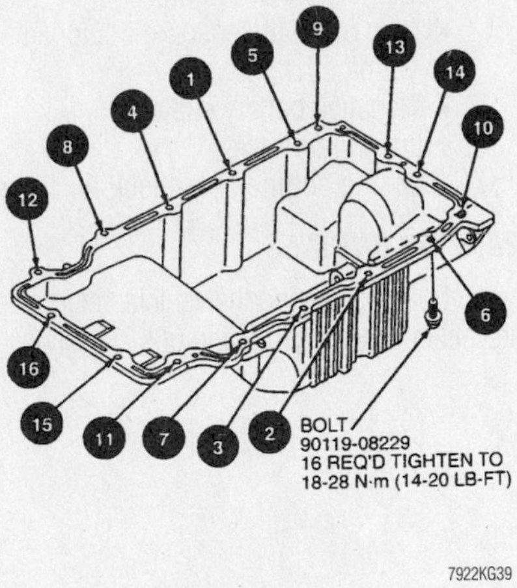

Oil pan bolt tightening sequence—3.4L (VIN N) engine

6. Install or connect the following:
   - Dual converter Y-pipe
   - Negative battery cable
7. Fill the crankcase.
8. Start the engine and check for leaks and proper operation.

Oil Pump

REMOVAL & INSTALLATION

3.0L (VIN U) Engine

1. Before servicing the vehicle, refer to the precautions at the beginning of this section.
2. Drain the engine oil.
3. Remove or disconnect the following:
   - Negative battery cable
   - Oil dipstick tube
   - Low oil level sensor, if equipped
   - Starter motor and brace
   - Heated Oxygen Sensor (HO_2S) connectors
   - Dual converter Y-pipe
   - Engine rear plate
   - Oil pan
   - Oil pump
4. Separate the intermediate shaft from the oil pump.

To install:

5. Insert the intermediate shaft into the oil pump until the snapring seats.
6. Install or connect the following:
   - Oil pump and tighten the bolt to 30–40 ft. lbs. (40–55 Nm)
   - Oil pan
   - Engine rear plate
   - Dual converter Y-pipe
   - HO_2S sensor connectors
   - Starter motor and brace
   - Low oil level sensor, if equipped
   - Oil dipstick tube
   - Negative battery cable

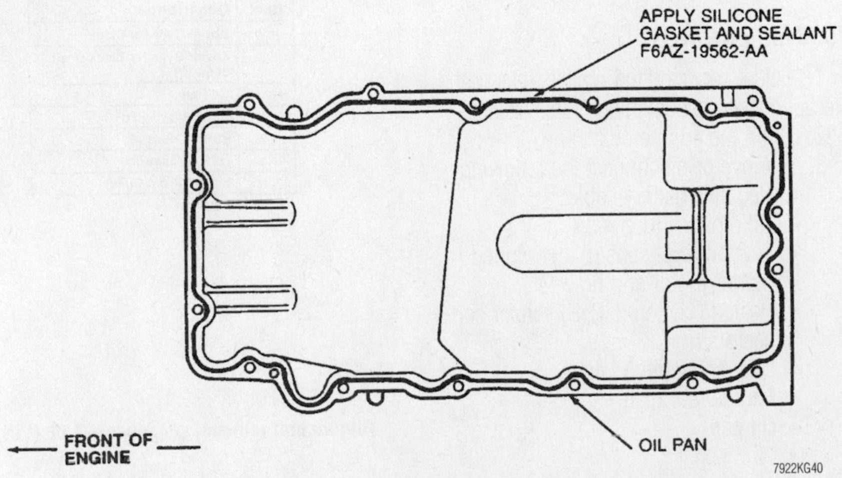

Apply silicone sealant to the entire sealing surface of the oil pan—3.4L (VIN N) engine

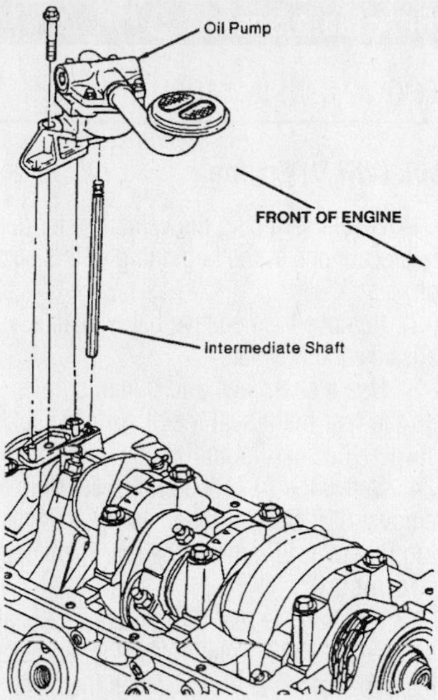

Exploded view of the oil pump mounting—3.0L (VIN U) engine

7. Fill the crankcase.
8. Start the engine and check for leaks and proper operation.

3.0L (VIN S) Engine

1. Before servicing the vehicle, refer to the precautions at the beginning of this section.
2. Remove the engine from the vehicle and mount it on an engine stand.
3. Remove or disconnect the following:
   - Upper intake manifold
   - Valve covers
   - Accessory drive belt
   - Power steering pump

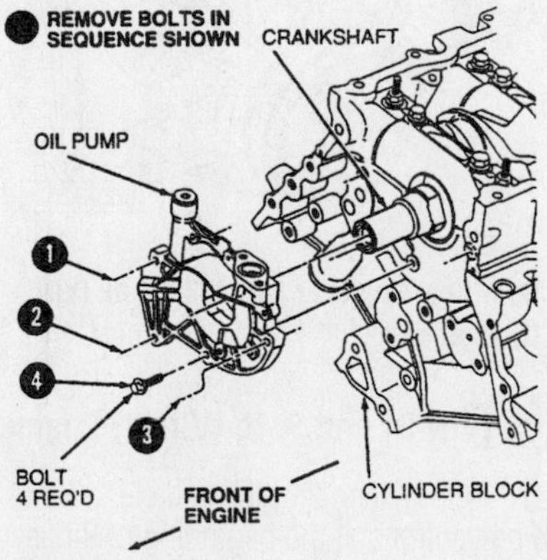

Oil pump bolt removal sequence—3.0L (VIN S) engine—3.4L (VIN N) engine is similar

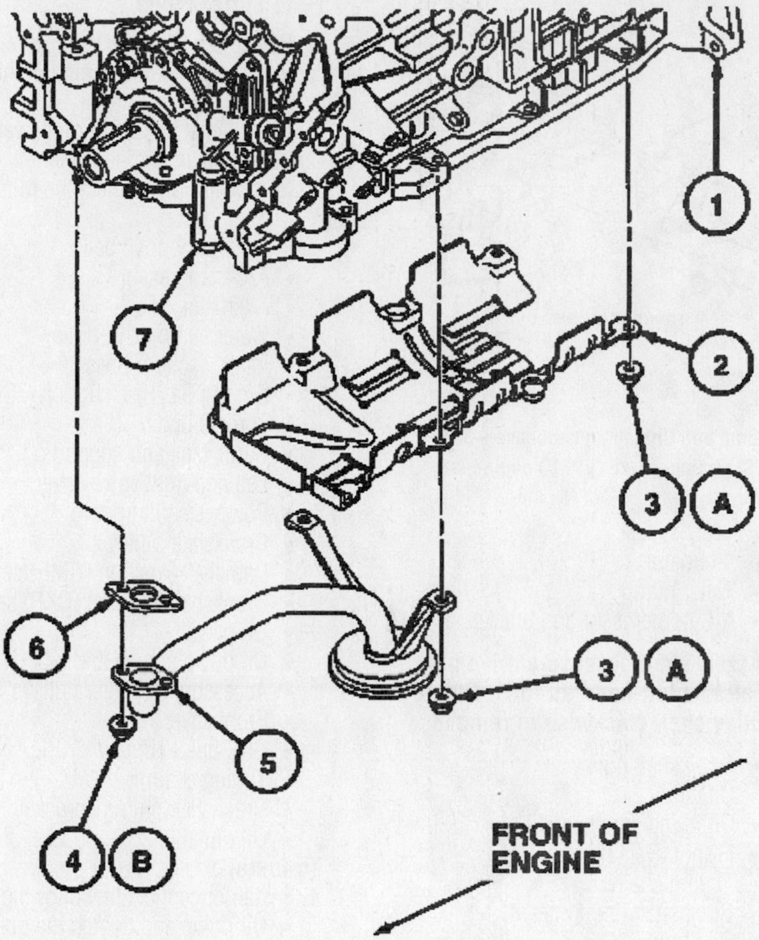

Item	Description
1	Cylinder Block
2	Oil Pan Baffle
3	Nut (5 Req'd)
4	Nut (2 Req'd)
5	Oil Pump Screen Cover and Tube
6	Oil Pump Inlet Tube Gasket
7	Oil Pump
A	Tighten to 14-25 Nm (11-18 Lb-Ft)
B	Tighten to 8-14 Nm (71-123 Lb-In)

9300KG16

Exploded view of the oil pump pickup tube and screen mounting—3.4L (VIN N) engine—3.0L (VIN S) engine is similar

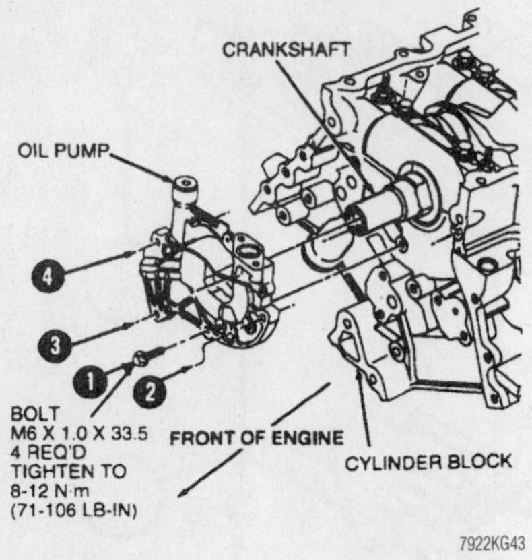

Oil pump bolt tightening sequence—3.0L (VIN S) engine—3.4L (VIN N) engine is similar

- Alternator
- Water pump
- A/C compressor and bracket

➡The crankshaft accessory drive pulley shaft has left-hand threads. Rotate the pulley shaft clockwise to remove.

- Crankshaft pulley
- Oil pan
- Oil pump screen and tube
- Front cover
- Timing chains
- Crankshaft timing gears
- Oil pump

To install:

4. Install or connect the following:
   - Oil pump and tighten the bolts in sequence to 71–106 inch lbs. (8–12 Nm)
   - Crankshaft timing gears
   - Timing chains
   - Front cover
   - Oil pump screen and tube. Tighten the bolts to 71–106 inch lbs. (8–12 Nm), and the nut to 15–22 ft. lbs. (20–30 Nm).
   - Oil pan
   - Crankshaft pulley

➡The air conditioning mounting bracket bolts are torque-to-yield bolts and must be replaced when removed.

   - A/C compressor and bracket
   - Water pump
   - Alternator
   - Power steering pump
   - Accessory drive belt
   - Valve covers
   - Upper intake manifold
   - Engine in the vehicle.
5. Fill the crankcase.
6. Start the engine and check for leaks and proper operation.

3.4L (VIN N) Engine

1. Before servicing the vehicle, refer to the precautions at the beginning of this section.
2. Remove the engine from the vehicle and mount it on an engine stand.
3. Remove or disconnect the following:
   - Accessory drive belts
   - A/C compressor
   - Alternator
   - Water pump drive pulley
   - Intake vacuum lines
   - Exhaust Gas Recirculation (EGR) tube
   - Throttle body
   - Surge tank and supports
   - Left and right valve covers
   - Power steering pump
   - Crankshaft pulley
   - Camshaft Position (CMP) sensor
   - Crankshaft Position (CKP) sensor
   - Oil pan
   - Oil pump screen and tube
   - Accessory belt idler pulleys
   - Front cover
   - CKP pulse ring
   - Timing chain
   - Crankshaft timing sprocket
   - Oil pump

To install:

4. Install or connect the following:
   - Oil pump and tighten the bolts in sequence to 80–115 inch lbs. (9–13 Nm)
   - Crankshaft timing sprocket
   - Timing chain
   - CKP pulse ring
   - Front cover
   - Accessory belt idler pulleys
   - Oil pump screen and tube. Tighten the tube nuts to 71–123 inch lbs. (8–14 Nm) and the bracket nuts to 11–18 ft. lbs. (14–25 Nm).
   - Oil pan
   - CKP sensor
   - CMP sensor
   - Crankshaft pulley
   - Power steering pump
   - Left and right valve covers
   - Surge tank and supports
   - Throttle body
   - EGR tube
   - Intake vacuum lines
   - Water pump drive pulley
   - Alternator
   - A/C compressor
   - Accessory drive belts
   - Engine into the vehicle
5. Fill the crankcase.
6. Start the engine and check for leaks and proper operation.

Rear Main Seal

REMOVAL & INSTALLATION

3.0L (VIN U) Engine

1. Before servicing the vehicle, refer to the precautions at the beginning of this section.
2. Remove the negative battery cable, transaxle and flexplate.
3. Use a sharp awl and punch a hole into the rear main seal metal surface between the seal lip and the cylinder block.
4. Screw the threaded end of Jet Plug Remover 310-005 into the seal. Use the Jet Plug Remover to remove the rear main seal.

To install:

5. Position the rear oil seal on Rear Seal Replacer 303-323. Position the tool and seal on the rear of the engine. Tighten the bolts alternately to seat the rear main oil seal.
6. Install the flexplate and tighten the bolts to 54–64 ft. lbs. (73–87 Nm).
7. Install the transaxle and the negative battery cable.
8. Check all fluid levels and fill as needed.
9. Start the engine and check for leaks.

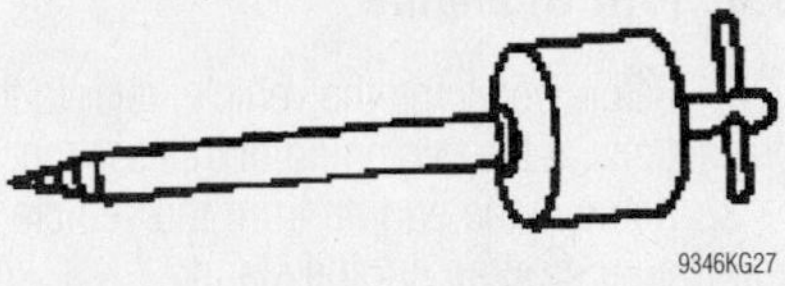

Jet Plug Remover 310-005—3.0L (VIN U) engine

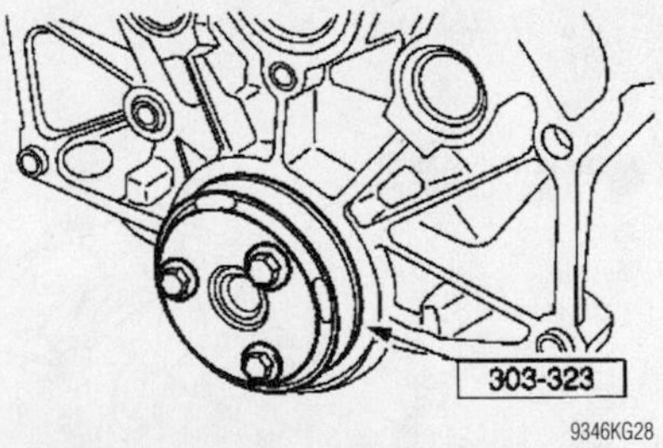

Rear Seal Replacer 303-323—3.0L (VIN U) engine

3.0L (VIN S) and 3.4L (VIN N) Engines

1. Before servicing the vehicle, refer to the precautions at the beginning of this section.
2. Attach an engine support fixture to the engine lifting eyes.
3. Remove or disconnect the following:
   - Negative battery cable
   - Transaxle
   - Flexplate

4. Use Seal Remover T95P-6701-EH and a slide hammer to remove the rear crankshaft seal.

To install:

5. Install or connect the following:
   - Rear main seal flush with the cylinder block surface. Use Rear Main Seal Replacer T82L-6701-A and Adapter Bolts T91P-6701-A.
   - Flexplate and tighten the bolts to 54–64 ft. lbs. (73–87 Nm)
   - Transaxle
   - Negative battery cable
6. Run the engine and check for leaks.

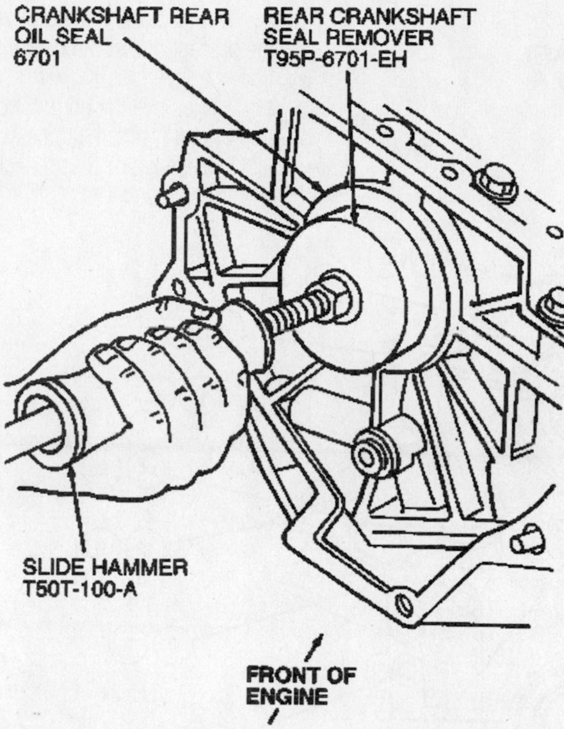

Rear Crankshaft Seal Remover T95P-6701-EH—3.0L (VIN S) and 3.4L (VIN N) engines

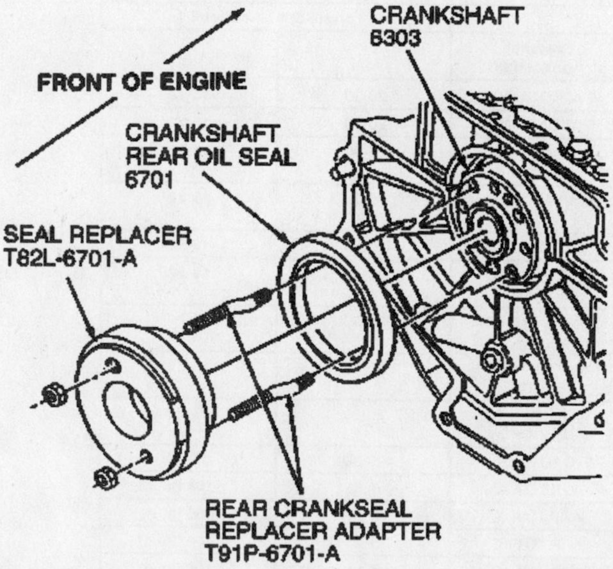

Rear Seal Replacer T82L-6701-A and Adapters T91P-6701-A—3.0L (VIN S) and 3.4L (VIN N) engines

Timing Chain, Sprockets and Front Cover and Seal

REMOVAL & INSTALLATION

3.0L (VIN U) Engine

1. Before servicing the vehicle, refer to the precautions at the beginning of this section.
2. Drain the engine oil and the cooling system.
3. Remove or disconnect the following:
   - Negative battery cable
   - Accessory drive belt
   - Accessory drive belt tensioner and idler pulley
   - Water pump pulley
   - Crankshaft pulley
   - Crankshaft Position (CKP) sensor
   - Coolant hoses
   - Dual converter Y-pipe
   - Oil pan
   - Front crankshaft seal
   - Front cover
4. Rotate the crankshaft until the No. 1 piston is at Top Dead Center (TDC) and the timing sprocket marks are aligned.
5. Remove the timing chain and sprockets.

To install:

➡The camshaft bolt has a drilled oil passage in it for timing chain lubrication. Prior to installation, clean the passage and be sure it is clear. Never replace the camshaft bolt with a standard bolt.

6. Install the timing chain and sprockets with the timing marks aligned. Tighten the camshaft sprocket bolt to 46 ft. lbs. (63 Nm).
7. Use a new gasket and install the front cover. Apply sealant to bolts 1, 2 and 3. Tighten bolts 1–10 to 19 ft. lbs. (25 Nm) and bolts 11–15 to 84 inch lbs. (10 Nm).
8. Install or connect the following:
   - Front crankshaft seal
   - Oil pan
   - Dual converter Y-pipe
   - Coolant hoses
   - CKP sensor
   - Crankshaft pulley and tighten the damper bolt to 107 ft. lbs. (145 Nm)
   - Water pump pulley and tighten the bolts to 16 ft. lbs. (21 Nm)
   - Accessory drive belt tensioner and idler pulley

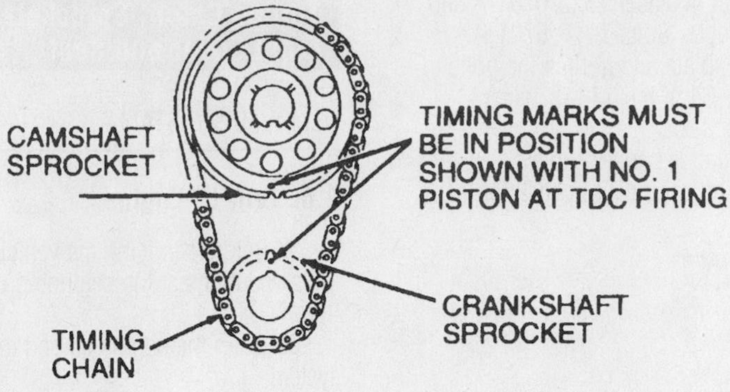

Timing chain alignment marks—3.0L (VIN U) engine

- Accessory drive belt
- Negative battery cable

9. Fill the cooling system.
10. Fill the crankcase.
11. Start the engine and check for leaks and proper operation.

3.0L (VIN S) Engine

1. Before servicing the vehicle, refer to the precautions at the beginning of this section.
2. Remove the engine from the vehicle and mount it on an engine stand.
3. Remove or disconnect the following:

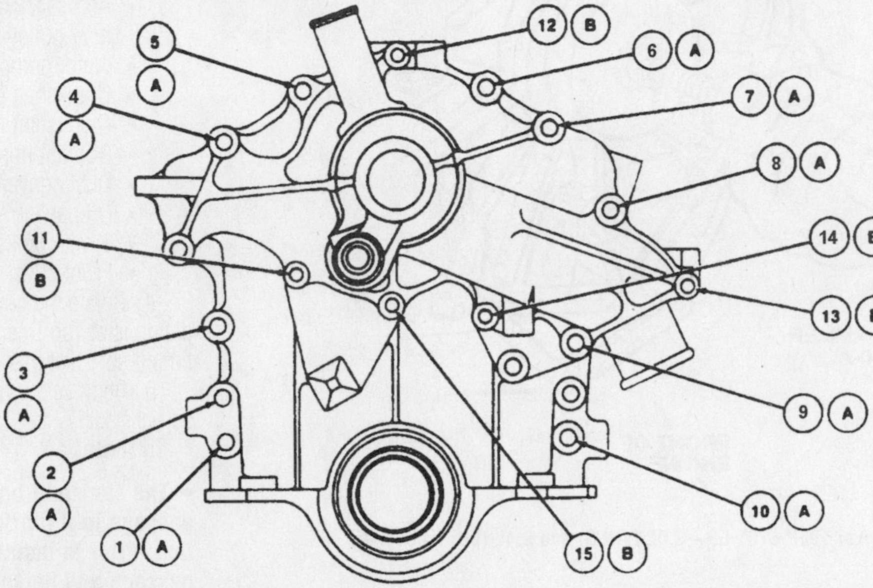

Fastener And Hole No.	Fasteners		Torque Specifications	
	Size	Fastener Application	N·m	LB-FT
1A	M8 x 1.25 x 43.5	F/C TO BLOCK	20-30	15-22
2A	M8 x 1.25 x 43.5	F/C TO BLOCK	20-30	15-22
3A	M8 x 1.25 x 73	W/P & F/C TO BLOCK	20-30	15-22
4A	M8 x 1.25 x 104 3	W/P & F/C TO BLOCK	20-30	15-22
5A	M8 x 1.25 x 73	F/C TO BLOCK	20-30	15-22
6A	M8 x 1 25 x 73	W/P & F/C TO BLOCK	20-30	15-22
7A	M8 x 1.25 x 73	W/P & F/C TO BLOCK	20-30	15-22
8A	M8 x 1 25 x 104.3	W/P & F/C TO BLOCK	20-30	15-22
9A	M8 x 1.25 x 104.3	W/P & F/C TO BLOCK	20-30	15-22
10A	M8 x 1.25 x 52	F/C TO BLOCK	20-30	15-22
11B	M6 x 1 x 28.5	W/P TO F/C	8-12	71-106 (lb-in)
12B	M6 x 1 x 28.5	W/P TO F/C	8-12	71-106 (lb-in)
13B	M6 x 1 x 28.5	W/P TO F/C	8-12	71-106 (lb-in)
14B	M6 x 1 x 28.5	W/P TO F/C	8-12	71-106 (lb-in)
15B	M6 x 1 x 28.5	W/P TO F/C	8-12	71-106 (lb-in)

W/P—Water Pump
F/C—Engine Front Cover

7922KG48

Timing chain front cover bolt location and identification—3.0L (VIN U) engine

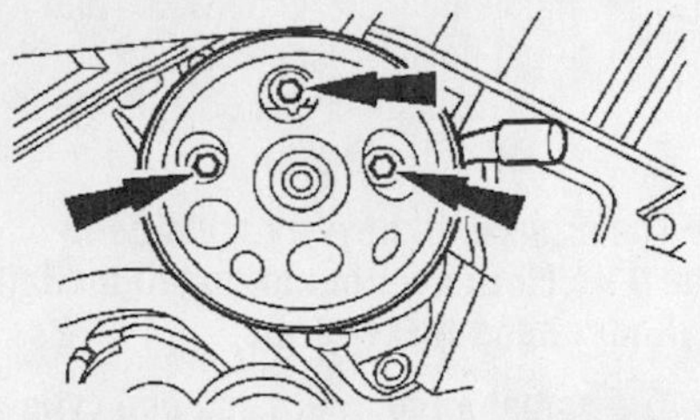

9346KG31

Access the power steering pump bolts through the holes in the pulley—3.0L (VIN S) engine

- Upper intake manifold
- Valve covers
- Accessory drive belt
- Power steering pump
- Alternator
- Water pump
- A/C compressor and bracket

➡The crankshaft accessory drive pulley shaft has left-hand threads. Rotate the pulley shaft clockwise to remove.

- Crankshaft pulley
- Crankshaft damper
- Oil pan
- Oil pump screen and tube
- Crankshaft Position (CKP) sensor connector
- Camshaft Position (CMP) sensor connector
- Front cover
- CKP sensor pulse ring. Note the keyway alignment for installation.

➡The Crankshaft Position Sensor (CKP) pulse ring is used on several different engines. Note the keyway alignment for installation.

4. Rotate the crankshaft so that the keyway is at the 11 o'clock position to locate the crankshaft at Top Dead Center (TDC) for No. 1 cylinder.
5. Verify that the alignment arrows on the camshafts are aligned. If not, rotate the crankshaft 1 complete revolution and recheck.
6. Rotate the crankshaft so that the keyway is at the 3 o'clock position. This positions the right cylinder head camshafts to the neutral position.
7. Remove or disconnect the following:
   - Right timing chain tensioner

➡The camshaft thrust caps must be removed before loosening the remaining camshaft journal cap bolts to ensure that the thrust caps are not damaged.

   - Camshaft thrust caps
   - Camshaft journal caps. Loosen the bolts in sequence and in several passes to allow the camshaft to be raised from the cylinder head evenly.
   - Rocker arms. Keep the rocker arms in order for installation.
   - Right timing chain tensioner arm
   - Right timing chain and crankshaft sprocket

8. Rotate the crankshaft 2 revolutions and locate the crankshaft keyway at the 11 o'clock position. This will position the left cylinder head camshafts to their neutral position.
9. Verify that the alignment arrows on the camshafts are aligned.
10. Remove the left cylinder head timing chain tensioner retaining bolts and the timing chain tensioner.

➡The camshaft thrust caps must be removed before loosening the remain-

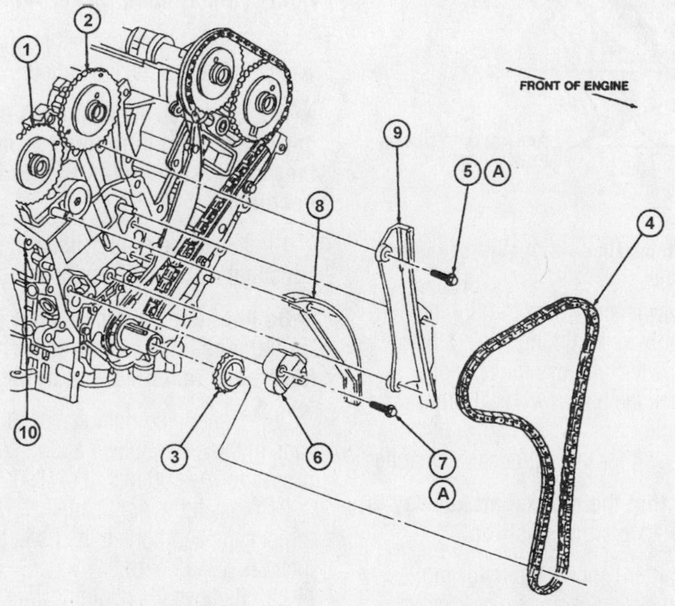

1 RH Exhaust Camshaft
2 RH Intake Camshaft
3 RH Timing Chain Crankshaft Sprocket
4 RH Timing Chain
5 Bolt (2 Req'd)
6 Timing Chain Tensioner
7 Bolt (2 Req'd)
8 Timing Chain Tensioner Arm
9 Timing Chain Guide
10 RH Cylinder Head
A Tighten to 20-30 N·m (15-22 Lb-Ft)

7922KG49

Exploded view of the right cylinder head timing chain and related components—3.0L (VIN S) engine—left side similar

7922JG27

Using a thin prytool, release and hold the timing chain tensioner ratchet/pawl mechanism—3.0L (VIN S) engine

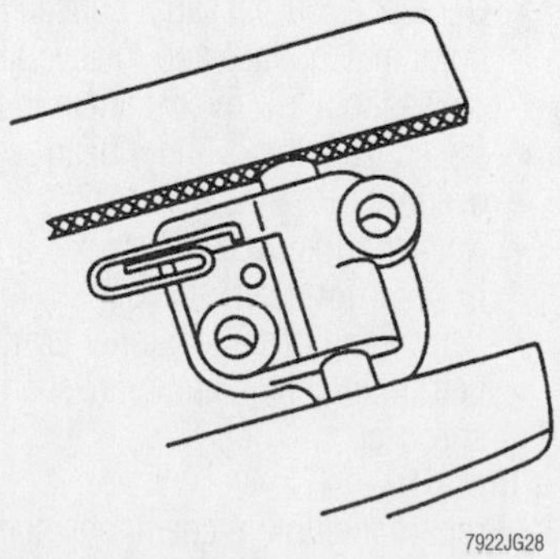

7922JG28

Retain the piston with a 1.5mm wire or paperclip—3.0L (VIN S) engine

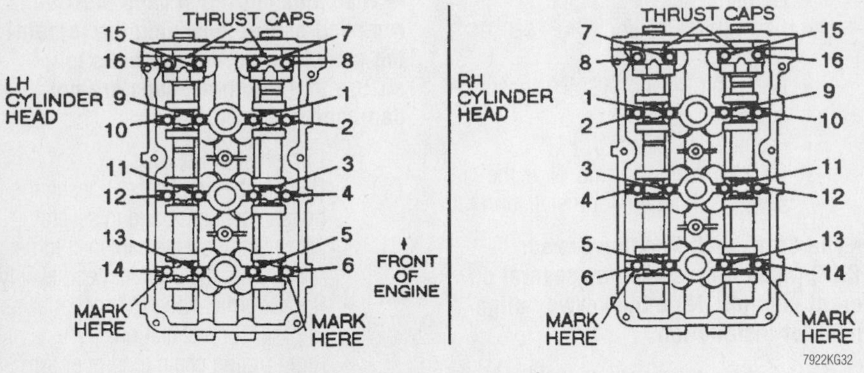

Camshaft journal bolt tightening sequence—3.0L (VIN S) engine

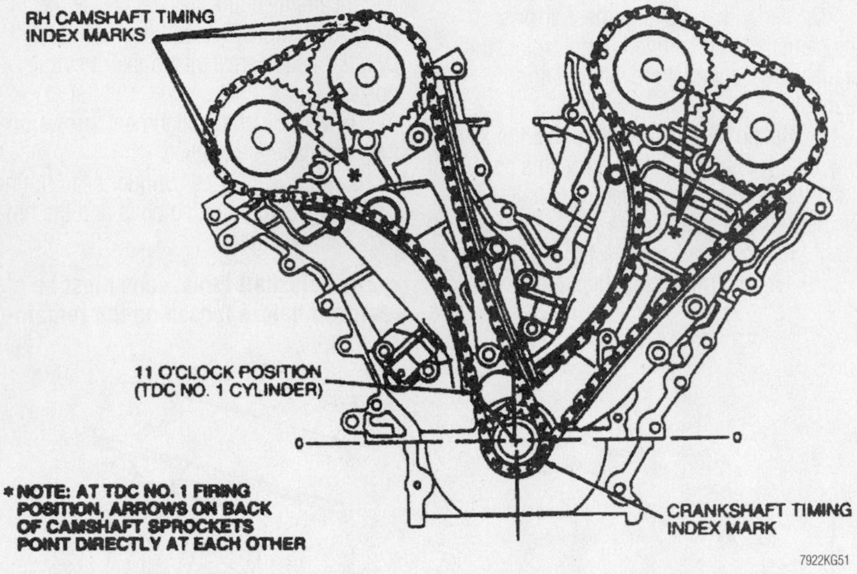

Be sure the timing marks are as shown after the chain has been installed—3.0L (VIN S) engine

ing camshaft journal cap bolts to ensure that the thrust caps are not damaged.

11. Remove or disconnect the following:
    - Camshaft thrust caps
    - Camshaft journal caps. Loosen the bolts in sequence and in several passes to allow the camshaft to be raised from the cylinder head evenly.
    - Rocker arms. Keep the rocker arms in order for installation.
    - Left timing chain tensioner arm
    - Left timing chain and crankshaft sprocket

To install:

12. Prepare the timing chain tensioners for installation as follows:
    a. Place the left chain tensioner in a vise.
    b. Using a small prytool, release and hold the timing chain tensioner ratchet/pawl mechanism through the access hole in the timing chain tensioner.
    c. Slowly compress the tensioner.
    d. Lock the piston with a 1.5mm wire or paperclip.
    e. Repeat for the right chain tensioner.

➡Be sure that the crankshaft keyway is still at the 11 o'clock position.

13. Install or connect the following:
    - Left timing chain and crankshaft sprocket. Align the colored links with the index marks on the camshaft and crankshaft sprockets.
    - Left timing chain tensioner arm
    - Left timing chain tensioner and tighten the retaining bolts to 15–22 ft. lbs. (20–30 Nm)
    - Right timing chain and crankshaft sprocket. Align the colored links with the index marks on the camshaft and crankshaft sprockets.
    - Right timing chain tensioner arm
    - Right timing chain tensioner and tighten the retaining bolts to 15–22 ft. lbs. (20–30 Nm)

➡The crankshaft keyway must be in the 11 o'clock position to install the left cylinder head rocker arms.

➡The camshaft journal caps and cylinder heads are numbered to ensure that they are assembled in their original positions.

14. Install the left cylinder head rocker arms in their original positions.

➡Do not install the camshaft journal thrust caps until the other journal caps have been installed and tightened.

15. Tighten the left camshaft journal caps in the order shown and in several passes to 71–106 inch lbs. (8–12 Nm).
16. Install the left camshaft journal thrust caps and tighten the bolts to 71–106 inch lbs. (8–12 Nm).
17. Remove the retaining wire from the left timing chain tensioner.

➡The crankshaft keyway must be in the 3 o'clock position to install the right cylinder head rocker arms.

18. Rotate the crankshaft so that the keyway is in the 3 o'clock position.

➡The camshaft journal caps and cylinder heads are numbered to ensure that they are assembled in their original positions.

19. Install the right cylinder head rocker arms in their original positions.

➡Do not install the camshaft journal thrust caps until the other journal caps have been installed and tightened.

20. Tighten the right camshaft journal caps in the order shown and in several passes to 71–106 inch lbs. (8–12 Nm).
21. Install the right camshaft journal thrust caps and tighten the bolts to 71–106 inch lbs. (8–12 Nm).
22. Remove the retaining wire from the right timing chain tensioner.

➡The Crankshaft Position Sensor (CKP) pulse ring is used on several different engines. The keyway position for the 3.0L (VIN S) engine will be marked 30, 30RFF or with an ORANGE stripe.

23. Install the CKP sensor pulse ring. Align the keyway with the proper slot.
24. Replace the crankshaft seal in the front cover with a new one. Apply clean engine oil to the seal lip.

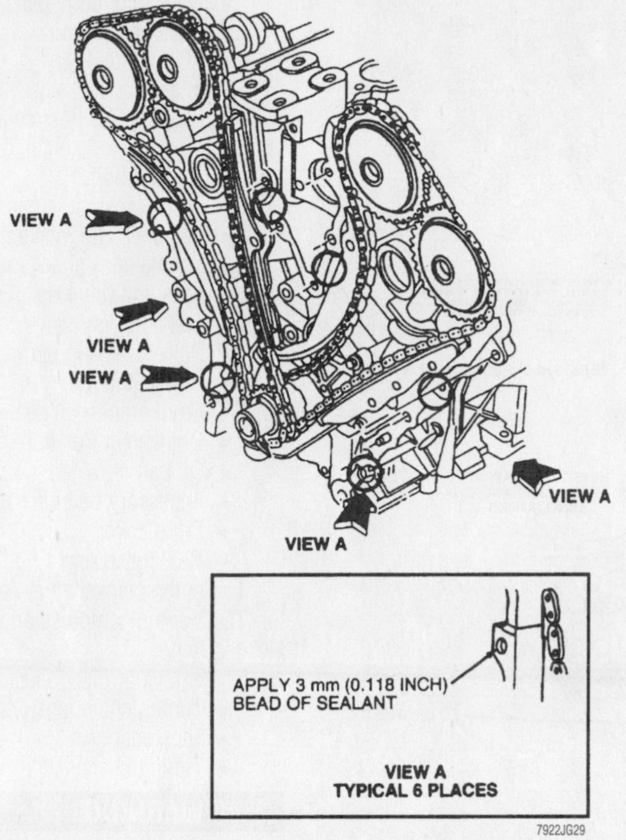

To prevent oil leakage, apply sealant to the places indicated—3.0L (VIN S) engine

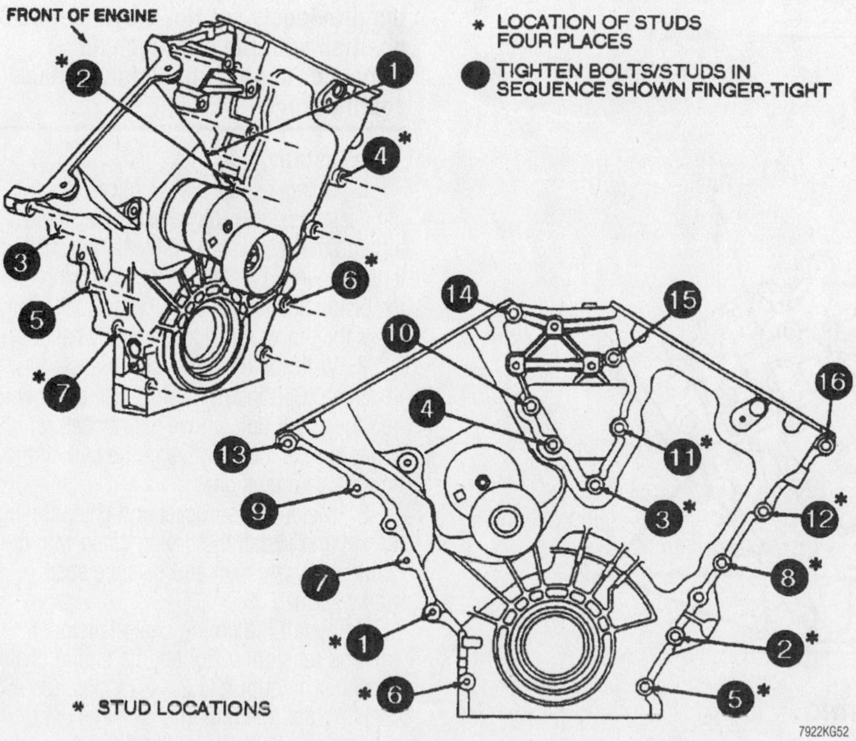

Front cover bolt torque sequence—3.0L (VIN S) engine

25. Apply silicone sealer to the 6 critical areas shown in View **A**, to the cylinder block to prevent oil seepage.
26. Place new front cover gaskets onto the dowel pins on the cylinder block and heads.
27. Place the front cover into position.
28. Install the 6 front cover retaining bolts and stud bolts where the silicone sealer was applied.
29. Tighten the bolts and stud bolts until the front cover contacts the cylinder block and heads, then turn the bolts and stud bolts an additional 1/4 turn.
30. Install the remaining front cover retaining bolts and stud bolts.
31. Tighten all of the front cover retaining bolts and stud bolts in sequence to 15–22 ft. lbs. (20–30 Nm).

➡The air conditioning mounting bracket bolts are torque-to-yield bolts and must be replaced.

32. Install or connect the following:
    - CMP sensor connector
    - CKP sensor connector
    - Oil pump screen and tube
    - Oil pan
33. Install the crankshaft damper and tighten the bolt as follows:
    a. Step 1: Tighten the bolt to 78–99 ft. lbs. (105–135 Nm).
    b. Step 2: Loosen the bolt one full turn.
    c. Step 3: Tighten the bolt to 35–39 ft. lbs. (47–53 Nm).
    d. Step 4: Tighten the bolt 85–95 degrees.
34. Install or connect the following:
    - Crankshaft pulley and tighten counterclockwise to 70–77 ft. lbs. (95–105 Nm)
    - A/C compressor and bracket
    - Water pump
    - Alternator
    - Power steering pump and tighten the bolts to 18 ft. lbs. (25 Nm)
    - Accessory drive belt
    - Valve covers
    - Upper intake manifold
    - Engine assembly into the vehicle
35. Run the engine and check for leaks and proper operation.

3.4L (VIN N) Engine

1. Before servicing the vehicle, refer to the precautions at the beginning of this section.
2. Remove the engine from the vehicle and mount it on an engine stand.
3. Remove or disconnect the following:

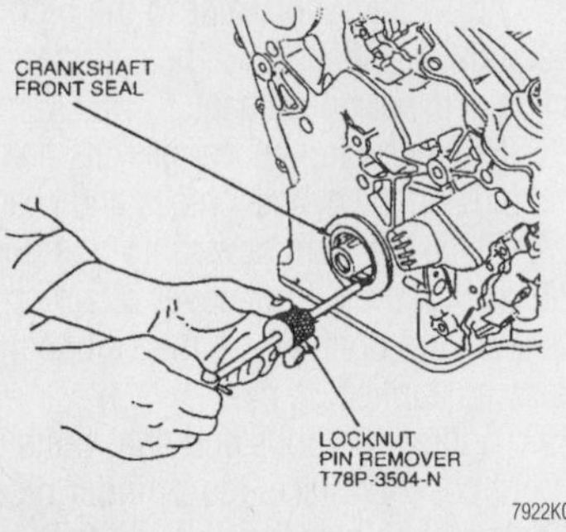

Use the special tool or a slide hammer to remove the crankshaft front oil seal—3.4L (VIN N) engine

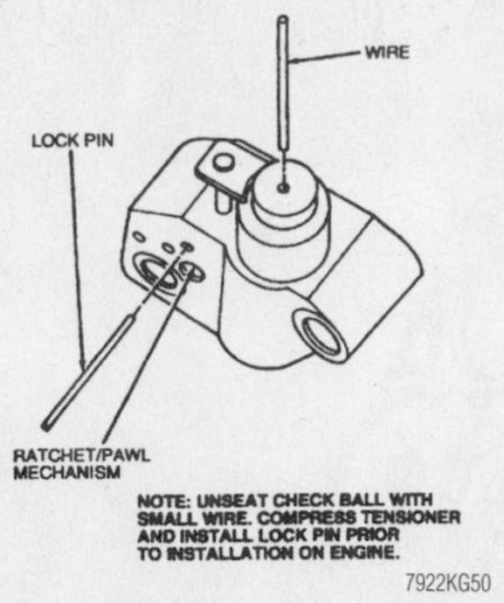

Timing chain tensioner—3.4L (VIN N) engine

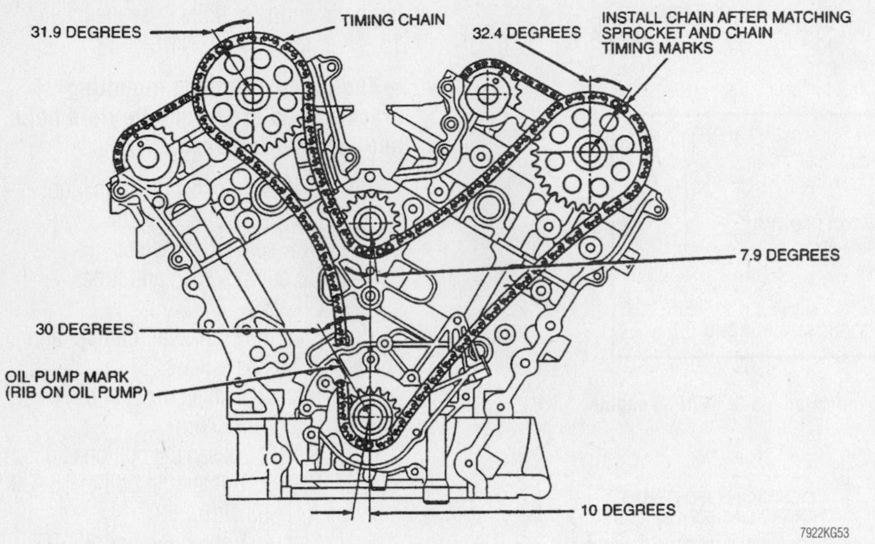

Timing chain alignment marks—3.4L (VIN N) engine

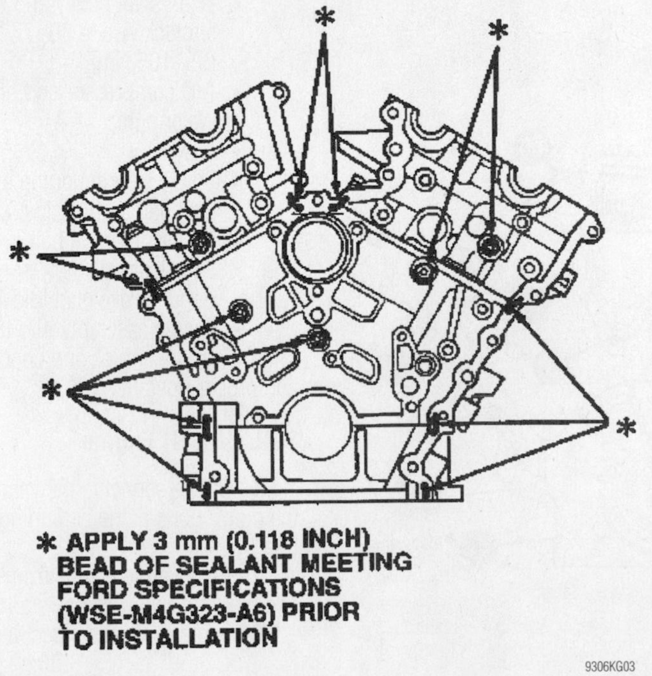

* APPLY 3 mm (0.118 INCH) BEAD OF SEALANT MEETING FORD SPECIFICATIONS (WSE-M4G323-A6) PRIOR TO INSTALLATION
9306KG03

Apply sealant to the cylinder block and heads—3.4L (VIN N) engine

- Accessory drive belts
- A/C compressor
- Alternator
- Intake vacuum lines
- Exhaust Gas Recirculation (EGR) tube
- Throttle body
- Surge tank and supports
- Left and right intake manifolds
- Engine control sensor wiring harness and brackets
- Left and right valve covers
- Power steering pump
- Crankshaft pulley
- Front crankshaft seal
- Crankshaft Position (CKP) sensor
- Oil pan
- Accessory belt idler pulleys
- Front cover
- CKP pulse ring

4. Set the crankshaft at Top Dead Center (TDC) of the compression stroke for the No. 1 cylinder.
5. Remove or disconnect the following:
   - Timing chain tensioner
   - Tensioner arm
   - Timing chain

⁂ WARNING

Do not rotate the crankshaft more than 45 degrees counterclockwise or more than 90 degrees clockwise from the previously set No. 1 cylinder TDC position with the timing chain removed, or valve to piston contact could occur.

To install:

6. Using a small prybar, release the timing chain tensioner ratchet/pawl mechanism through the access hole in the timing chain tensioner. Insert a small wire into the top of the piston and unseat the oil check ball. Compress the timing chain tensioner by hand.
7. With the tensioner compressed, install a 0.060 inch (1.5mm) drill bit or wire into the small hole above the ratchet, engaging the lock groove in the rack of the timing chain tensioner.
8. Match the sprocket and chain timing marks, and install the timing chain with the crankshaft, camshaft and balance shaft marks aligned.
9. Install the timing chain tensioner arm and tensioner. Tighten the timing chain tensioner pivot bolt to 25–39 ft. lbs. (34–53 Nm). Tighten the timing chain tensioner bolts to 14–20 ft. lbs. (18–27 Nm).
10. Remove the locking wire from the timing chain tensioner.
11. Apply silicone sealer to the 13 critical areas of the cylinder block as shown.

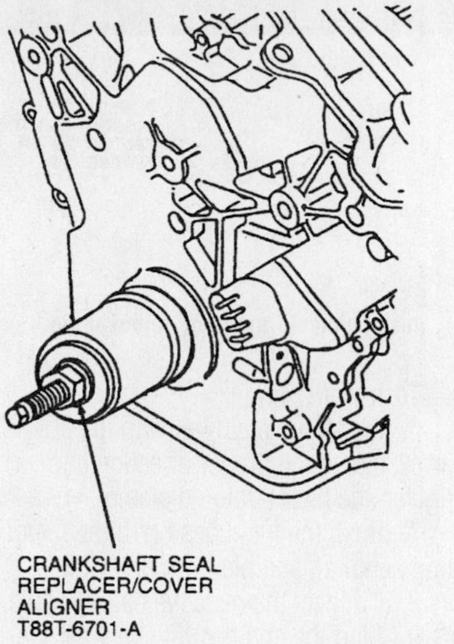

Install the new crankshaft front oil seal using a suitable seal driver or the special tool—3.4L (VIN N) engine

12. Install or connect the following:
- CKP pulse ring
- Front cover and tighten the bolts to 14–20 ft. lbs. (18–28 Nm)
- Accessory belt idler pulleys
- Oil pan
- CKP sensor
- Front crankshaft seal

13. Install the crankshaft pulley and tighten the bolt as follows:

a. Step 1: Tighten the bolt to 78–99 ft. lbs. (105–135 Nm).

b. Step 2: Loosen the bolt one full turn.

c. Step 3: Tighten the bolt to 35–39 ft. lbs. (47–53 Nm).

d. Step 4: Tighten the bolt 85–95 degrees.

14. Install or connect the following:
- Power steering pump
- Left and right valve covers
- Engine control sensor wiring harness and brackets
- Left and right intake manifolds
- Surge tank and supports
- Throttle body
- EGR tube
- Intake vacuum lines
- Alternator
- A/C compressor
- Accessory drive belts
- Engine assembly into the vehicle.

15. Run the engine and check for leaks and proper operation.

Piston and Ring

POSITIONING

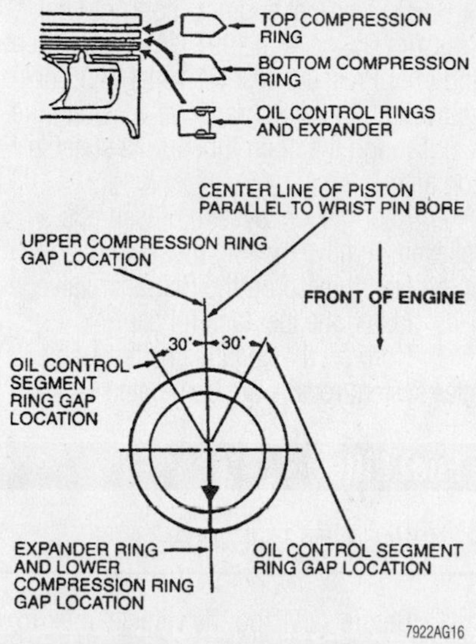

Ring end-gap spacing—3.0L (VIN S and U) engines

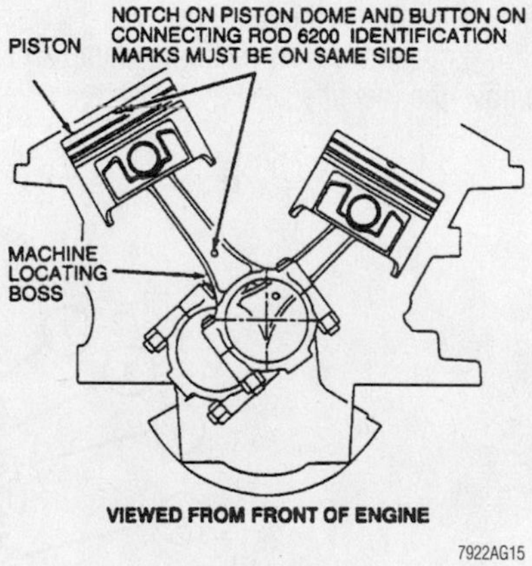

Piston and connecting rod positioning—3.0L (VIN S and U) engines

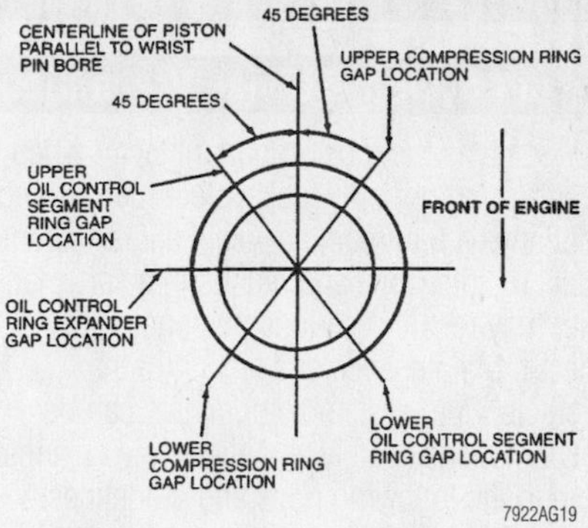

Ring end-gap positioning—3.4L (VIN N) engine

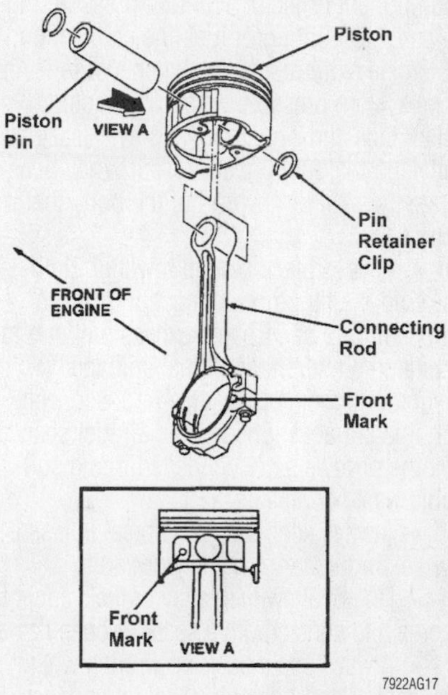

Piston and connecting rod positioning—3.4L (VIN N) engine

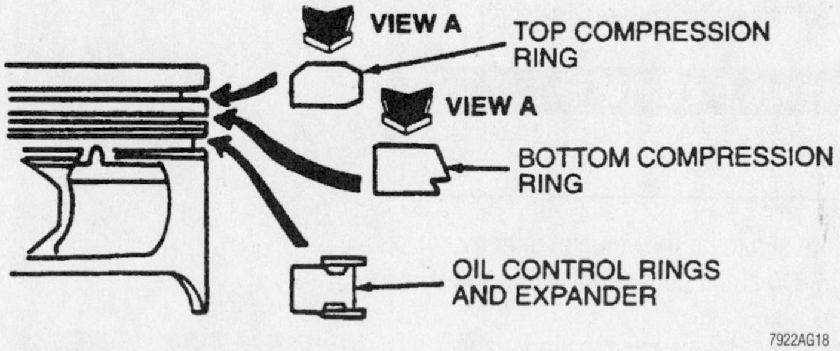

Piston ring positioning—3.4L (VIN N) engine

FUEL SYSTEM

Fuel System Service Precautions

Safety is the most important factor when performing not only fuel system maintenance but any type of maintenance. Failure to conduct maintenance and repairs in a safe manner may result in serious personal injury or death. Maintenance and testing of the vehicle's fuel system components can be accomplished safely and effectively by adhering to the following rules and guidelines.

- To avoid the possibility of fire and personal injury, always disconnect the negative battery cable unless the repair or test procedure requires that battery voltage be applied.
- Always relieve the fuel system pressure prior to disconnecting any fuel system component (injector, fuel rail, pressure regulator, etc.), fitting or fuel line connection. Exercise extreme caution whenever relieving fuel system pressure, to avoid exposing skin, face and eyes to fuel spray. Please be advised that fuel under pressure may penetrate the skin or any part of the body that it contacts.
- Always place a shop towel or cloth around the fitting or connection prior to loosening to absorb any excess fuel due to spillage. Ensure that all fuel spillage (should it occur) is quickly removed from engine surfaces. Ensure that all fuel soaked cloths or towels are deposited into a suitable waste container.
- Always keep a dry chemical (Class B) fire extinguisher near the work area.
- Do not allow fuel spray or fuel vapors to come into contact with a spark or open flame.
- Always use a back-up wrench when loosening and tightening fuel line connection fittings. This will prevent unnecessary stress and torsion to fuel line piping. Always follow the proper torque specifications.
- Always replace worn fuel fitting O-rings with new. Do not substitute fuel hose or equivalent, where fuel pipe is installed.

Fuel System Pressure

RELIEVING

1. Before servicing the vehicle, refer to the precautions at the beginning of this section.
2. Disconnect the negative battery cable.
3. Remove the fuel tank fill cap to relieve the pressure in the fuel tank.
4. Remove the cap from the Schrader valve located on the fuel supply manifold.
5. On gasoline engines, attach Fuel Pressure Gauge T80L-9974-A or equivalent, to the valve and drain the fuel through the drain tube into a suitable container.
6. On flex-fuel engines, connect Fuel Pressure Gauge T80L-9974-A or equivalent and Fuel Pressure Test Kit 134-R0035 or equivalent, to the Schrader valve. Drain the fuel through the drain tube into a suitable container.
7. After the fuel system pressure is relieved, remove the fuel pressure gauge and install the cap on the Schrader valve.
8. Install the fuel tank fill cap.
9. Connect the negative battery cable after system repairs are completed.

Fuel Filter

REMOVAL & INSTALLATION

1. Before servicing the vehicle, refer to the precautions at the beginning of this section.
2. Disconnect the negative battery cable.
3. Relieve the fuel system pressure.
4. Disconnect the fuel lines.
5. Loosen the filter retaining clamp and remove the fuel filter.

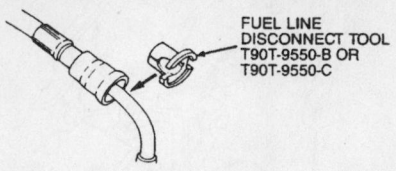

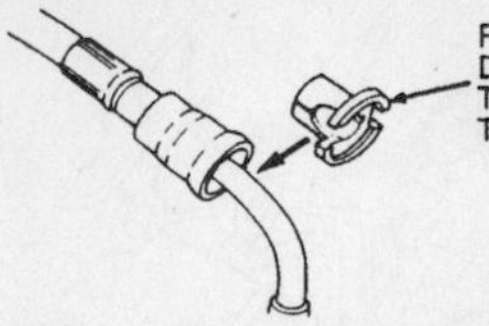

Push connect fitting and removal tool

To install:

6. Install the fuel filter with the flow arrow facing the proper direction and tighten the filter retaining clamp.
7. Push the fuel lines on to the filter fittings until an audible click is heard.
8. Connect the negative battery cable.
9. Start the engine and check for fuel leaks and proper operation.

Fuel Pump

REMOVAL & INSTALLATION

1. Before servicing the vehicle, refer to the precautions at the beginning of this section.
2. Relieve the fuel system pressure.

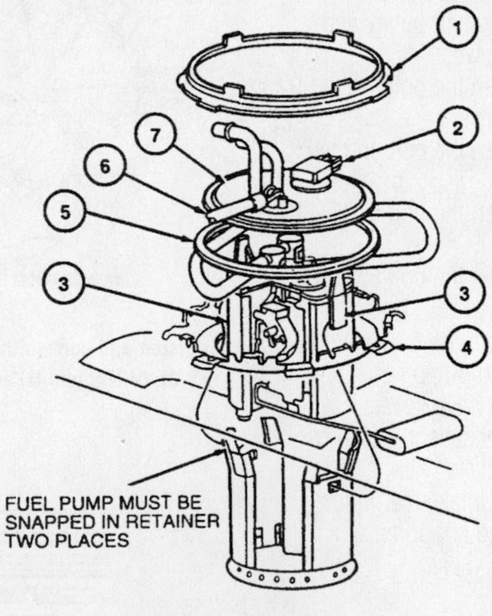

Exploded view of the fuel pump module mounting

3. Drain the fuel tank.
4. Remove or disconnect the following:
   - Negative battery cable
   - Fuel lines
   - Fuel pump module electrical connector
   - Fuel tank
   - Fuel pump module

To install:

5. Install the fuel pump module carefully to ensure the filter and hoses and float rod are not damaged. Use a new O-ring seal.
6. Align the fuel pump module and the fuel tank retainer and push the fuel pump module into the fuel tank retainer. When the fuel pump module is properly engaged, a definite click will be heard engaging 2 locking tabs on the outside of the fuel pump.
7. Install or connect the following:
   - Fuel pump module locking ring
   - Fuel tank
   - Fuel pump module electrical connector
   - Fuel lines
   - Negative battery cable
8. Add a minimum of 10 gallons of clean fuel to the tank.
9. Start the engine and check for leaks.

Fuel Injector

REMOVAL & INSTALLATION

3.0L (VIN U) Engine

1. Before servicing the vehicle, refer to the precautions in the beginning of this section.
2. Relieve fuel system pressure.
3. Remove or disconnect the following:
   - Negative battery cable
   - Air cleaner outlet tube
   - Fuel lines
   - Upper intake manifold
   - Fuel injector electrical connectors
   - Fuel pressure regulator vacuum line
   - Fuel supply manifold with the injectors attached
   - Injectors from the supply manifold

To install:

4. Install or connect the following:
   - Fuel injectors. Use new O-ring seals.

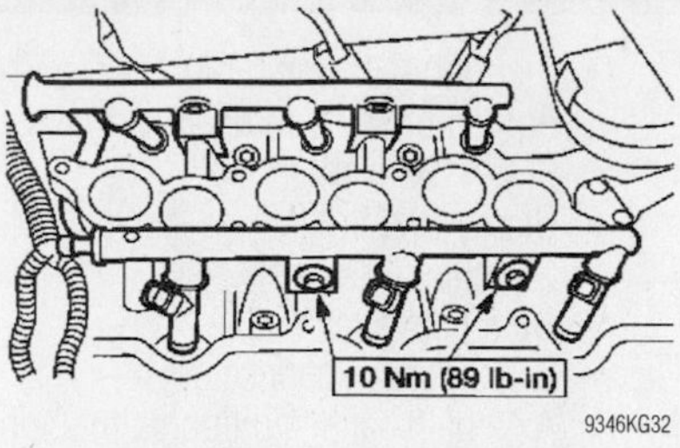

Fuel supply manifold—3.0L (VIN U) engine

   - Fuel supply manifold with the injectors attached and tighten the bolts to 89 inch lbs. (10 Nm)
   - Fuel pressure regulator vacuum line
   - Fuel injector electrical connectors
   - Upper intake manifold
   - Fuel lines
   - Air cleaner outlet tube
   - Negative battery cable
5. Start the engine and check for leaks.

3.0L (VIN S) Engine

1. Before servicing the vehicle, refer to the precautions in the beginning of this section.
2. Relieve fuel system pressure.
3. Remove or disconnect the following:
   - Negative battery cable
   - Air cleaner outlet tube
   - Fuel lines
   - Upper intake manifold
   - Fuel injector electrical connectors
   - Fuel pressure regulator vacuum line
   - Fuel supply manifold
   - Injectors from the lower intake manifold

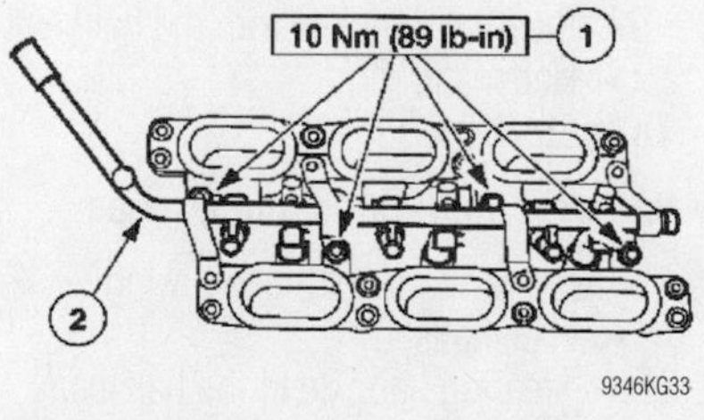

Retaining bolts (1) and Fuel supply Manifold (2)—3.0L (VIN S) engine

To install:

4. Install or connect the following:
   - Fuel injectors. Use new O-ring seals.
   - Fuel supply manifold and tighten the bolts 89 inch lbs. (10 Nm)
   - Fuel pressure regulator vacuum line
   - Fuel injector electrical connectors
   - Upper intake manifold
   - Fuel lines
   - Air cleaner outlet tube
   - Negative battery cable
5. Start the engine and check for leaks.

3.4L (VIN N) Engine

1. Before servicing the vehicle, refer to the precautions in the beginning of this section.
2. Relieve fuel system pressure.
3. Remove or disconnect the following:
   - Negative battery cable
   - Air cleaner outlet tube
   - Fuel lines
   - Throttle body
   - Surge tank
   - Left and right intake manifolds
   - Fuel injector electrical connectors
   - Fuel pressure regulator vacuum line
   - Fuel supply manifold with the injectors attached
   - Injectors from the supply manifold

To install:

4. Install or connect the following:
   - Fuel injectors. Use new O-ring seals.
   - Fuel supply manifold with the injectors attached and tighten the bolts to 11–16 ft. lbs. (15–23 Nm)
   - Fuel pressure regulator vacuum line
   - Fuel injector electrical connectors
   - Left and right intake manifolds
   - Surge tank
   - Throttle body
   - Fuel lines
   - Air cleaner outlet tube
   - Negative battery cable
5. Start the engine and check for leaks.

DRIVE TRAIN

Transaxle

REMOVAL & INSTALLATION

1. Before servicing the vehicle, refer to the precautions at the beginning of this section.
2. Attach a powertrain support to the engine lifting eyes.
3. Remove or disconnect the following:
   - Battery and tray
   - Air cleaner assembly
   - Transaxle electrical connectors
   - Shift cable
   - Transaxle cooler lines
   - Front wheels
   - Stabilizer bar links
   - Lower ball joints
   - Axle halfshafts
   - Heated Oxygen Sensor (HO_2S) connectors
   - Dual converter Y-pipe
   - Starter
   - Steering rack and pinion gear. Support the gear with safety wire.
   - Powertrain support insulators
   - Subframe
   - Torque converter nuts
4. Support the transaxle with a transmission jack.
5. Remove the transaxle flange bolts and remove the transaxle.

To install:

6. Install the transaxle. For 1999–01 3.0L (VIN S) engines, tighten the flange bolts to 25–33 ft. lbs. (33–46 Nm). For all other engines, tighten the flange bolts to 30–44 ft. lbs. (40–60 Nm).
7. Tighten the torque converter nuts to 20–34 ft. lbs. (27–46 Nm).
8. Install the subframe. Use 2 pieces of ¾ inch outside diameter pipe in the alignment holes behind the front subframe mounts to align the subframe to the body. Tighten the subframe mounting bolts to 57–76 ft. lbs. (77–103 Nm).
9. Install or connect the following:
   - Powertrain support insulators. Tighten the bracket bolts to 65 ft. lbs. (88 Nm) and the subframe bolts to 90 ft. lbs. (122 Nm).
   - Steering rack and pinion gear and tighten the nuts to 84–113 ft. lbs. (113–133 Nm)
   - Starter and tighten the fasteners to 21 ft. lbs. (29 Nm)
   - Dual converter Y-pipe
   - HO_2S sensor connectors
   - Axle halfshafts and tighten the hub retainer nuts to 170–202 ft. lbs. (230–275 Nm)
   - Lower ball joints and tighten the nuts to 51–67 ft. lbs. (68–92 Nm)
   - Stabilizer bar links and tighten the nuts to 35–46 ft. lbs. (47–63 Nm)
   - Front wheels
   - Transaxle cooler lines
   - Shift cable
   - Transaxle electrical connectors
   - Air cleaner assembly
   - Battery and tray
10. Start the engine. Check for leaks and proper operation.

➡**Whenever the vehicle subframe is removed or lowered, the wheel alignment should be checked.**

Halfshaft

REMOVAL & INSTALLATION

1. Before servicing the vehicle, refer to the precautions at the beginning of this section.
2. Remove or disconnect the following:
   - Front wheels
   - Lower ball joints
   - Outer tie rod ends
   - Stabilizer bar links
   - Height sensors, if equipped with air suspension
   - Wheel speed sensors
   - Steering knuckle from the halfshaft
   - Halfshafts

To install:

➡**Use new nuts, bolts and circlips.**

3. Install or connect the following:
   - Halfshafts
   - Steering knuckle to the halfshaft
   - Hub retainer nut and tighten to 170–202 ft. lbs. (230–275 Nm)
   - Wheel speed sensors
   - Height sensors, if equipped with air suspension
   - Stabilizer bar links and tighten the nuts to 57–75 ft. lbs. (77–103 Nm)
   - Outer tie rod ends and tighten the nuts to 35–46 ft. lbs. (47–63 Nm)
   - Lower ball joints and tighten the fasteners to 50–68 ft. lbs. (68–92 Nm)
   - Front wheels

CV-Joint

REMOVAL & REPLACEMENT

Inner Tripod Joint

The inner CV-joint is serviced with the axle shaft as an assembly. The inner CV-joint boot can be serviced by removing the outer CV-joint.

Outer CV-Joint

1. Before servicing the vehicle, refer to the precautions in the beginning of this section.
2. Place the halfshaft in a vise.
3. Remove the CV-joint boot clamps and slide the boot away from the joint.
4. Drive the CV-joint off the halfshaft with a brass drift and a hammer.

To install:

5. Replace the snapring.
6. Fill the CV-joint with fresh grease and slide the joint on to the halfshaft.
7. Use new clamps and install the CV-joint boot.

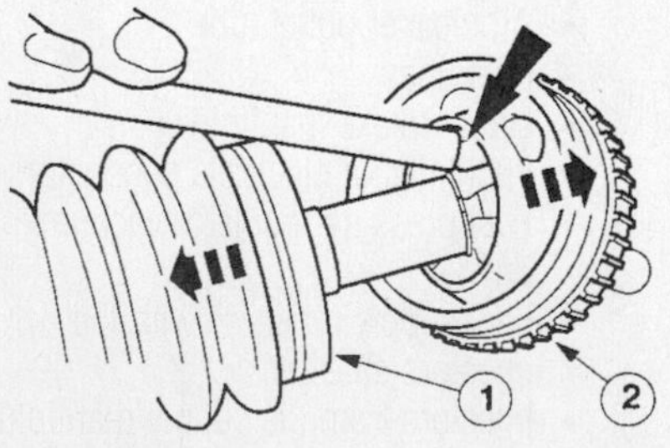

Removing the outer CV-joint

STEERING AND SUSPENSION

Air Bag

⁂ CAUTION

Some vehicles are equipped with an air bag system. The system must be disarmed before performing service on, or around, system components, the steering column, instrument panel components, wiring and sensors. Failure to follow the safety precautions and the disarming procedure could result in accidental air bag deployment, possible injury and unnecessary system repairs.

PRECAUTIONS

Several precautions must be observed when handling the inflator module to avoid accidental deployment and possible personal injury.

- Never carry the inflator module by the wires or connector on the underside of the module.
- When carrying a live inflator module, hold securely with both hands, and ensure that the bag and trim cover are pointed away.
- Place the inflator module on a bench or other surface with the bag and trim cover facing up.
- With the inflator module on the bench, never place anything on or close to the module which may be thrown in the event of an accidental deployment.

DISARMING

1. Before servicing the vehicle, refer to the precautions in the beginning of this section.
2. Position the vehicle with the front wheels in a straight-ahead position.
3. Disconnect the negative battery cable.
4. Disconnect the positive battery cable.
5. Wait at least 1 minute for the air bag back-up power supply to drain before continuing.
6. Proceed with the repair.

ARMING

1. After service is completed, connect the battery cables, negative cable last.
2. Check the functioning of the air bag system by turning the ignition key to the **RUN** position and visually monitoring the air bag indicator lamp in the instrument cluster. The indicator lamp should illuminate for approximately 6 seconds, then turn **OFF**. If the indicator lamp does not illuminate, stays ON, or flashes at any time, a fault has been detected by the air bag diagnostic monitor.

Power Rack and Pinion Steering Gear

REMOVAL & INSTALLATION

1. Before servicing the vehicle, refer to the precautions at the beginning of this section.
2. Remove or disconnect the following:
   - Negative battery cable
   - Intermediate steering shaft
   - Front wheels
   - Dual converter Y-pipe
   - Outer tie rod ends
   - Left stabilizer link
   - Heat shield
   - Power steering hose bracket
   - Auxiliary actuator connector, if equipped
   - Power steering pressure switch connector
   - Rear subframe bolts and lower the rear of the subframe about 4 inches
   - Steering gear mounting nuts, then move the steering gear to the left to access the power steering hoses
   - Power steering hoses
   - Steering gear through the left wheel opening

To install:

3. Install or connect the following:
   - Steering gear. Use new seals on the hydraulic fittings. Tighten the mounting nuts to 85–100 ft. lbs. (115–135 Nm).
   - Rear subframe bolts and tighten them to 57–76 ft. lbs. (77–103 Nm)
   - Power steering pressure switch connector
   - Auxiliary actuator connector, if equipped
   - Power steering hose bracket

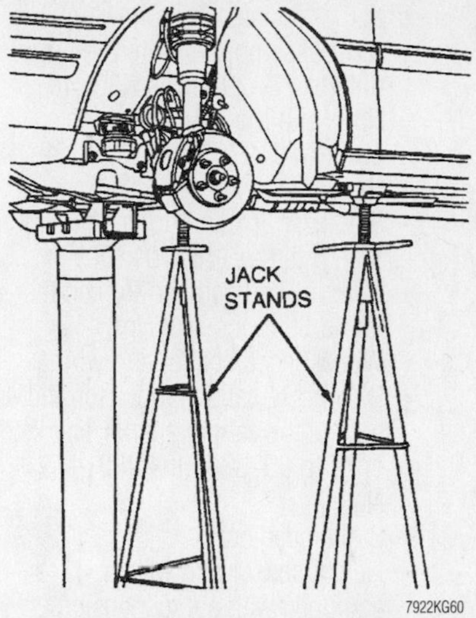

Support the rear of the subframe with 2 tall jackstands while removing the steering gear

   - Heat shield
   - Left stabilizer link
   - Outer tie rod ends and tighten the nuts to 35 ft. lbs. (48 Nm)
   - Dual converter Y-pipe
   - Front wheels
   - Intermediate steering shaft
   - Negative battery cable
4. Fill and bleed the power steering system. Check the system for leaks and proper operation.
5. Check the alignment and adjust as necessary.

Strut

REMOVAL & INSTALLATION

Front

1. Before servicing the vehicle, refer to the precautions at the beginning of this section.
2. Remove or disconnect the following:
   - Negative battery cable
   - Front wheels
   - Height sensor and wiring, if equipped with air suspension
   - Brake hose bracket
   - Disc brake caliper and rotor
   - Wheel speed sensor and wiring
   - Outer tie rod end
   - Stabilizer link

- Steering knuckle pinch bolt
- Strut assembly

To install:

3. Install or connect the following:
   - Strut assembly. Tighten the upper nuts to 22–29 ft. lbs. (30–40 Nm), and the knuckle pinch bolt to 73–97 ft. lbs. (98–132 Nm).
   - Stabilizer link and tighten the nut to 55–75 ft. lbs. (75–101 Nm)
   - Outer tie rod end and tighten the nut to 35 ft. lbs. (48 Nm)
   - Wheel speed sensor and wiring
   - Disc brake caliper and rotor and tighten the caliper anchor bracket bolts to 65–87 ft. lbs. (88–118 Nm)
   - Brake hose bracket
   - Height sensor and wiring, if equipped with air suspension
   - Front wheels
   - Negative battery cable
4. Road test the vehicle and check for proper operation.

Rear

1. Before servicing the vehicle, refer to the precautions at the beginning of this section.
2. Remove or disconnect the following:
   - Rear package tray trim panel
   - Rear wheels
   - Brake load sensor
   - Brake hose
   - Stabilizer bar bracket and link
   - Tension strut
   - Spindle pinch bolt
   - Strut assembly

To install:

➡Use new mounting nuts and bolts.

3. Install or connect the following:
   - Strut assembly. Use a new pinch bolt and tighten to 50–67 ft. lbs. (68–92 Nm). Tighten the 3 upper mounting nuts to 19–25 ft. lbs. (25–34 Nm).
   - Tension strut and tighten the nut to 35–46 ft. lbs. (68–92 Nm)
   - Stabilizer bar link and tighten the nut to 60–81 inch lbs. (7–9 Nm)
   - Stabilizer bar bracket and tighten the bolts to 25 –33 ft. lbs. (34–46 Nm)
   - Brake hose
   - Brake load sensor
   - Rear wheels
   - Rear package tray trim panel

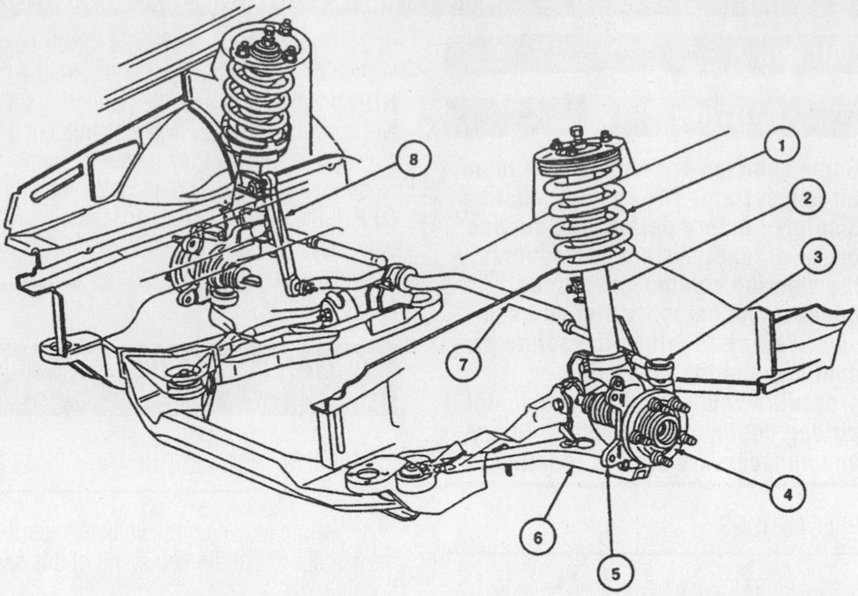

Item	Description
1	Front Coil Spring
2	Front Shock Absorber
3	Tie Rod End
4	Wheel Hub

Item	Description
5	Front Wheel Knuckle
6	Front Suspension Lower Arm
7	Front Stabilizer Bar
8	Stabilizer Bar Link

7922KG61

Front suspension component identification

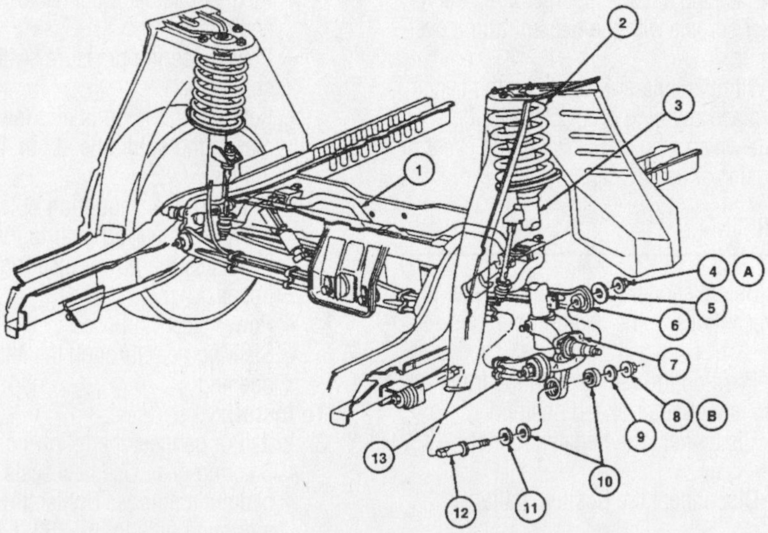

Item	Description
1	Rear Stabilizer Bar
2	Rear Spring
3	Shock Absorber
4	Nut
5	Washer
6	Lower Suspension Arm (Rear)
7	Rear Wheel Spindle
8	Nut (4 Req'd)
9	Washer (2 Req'd)

Item	Description
10	Rear Suspension Tie Rod Bushing (4 Req'd)
11	Washer (2 Req'd)
12	Rear Suspension Tension Strut and Bushing (2 Req'd)
13	Rear Suspension Lower Arm (Front)
A	Tighten to 68-92 N·m (50-67 Lb-Ft)
B	Tighten to 46.7-63.3 N·m (35-46 Lb-Ft)

7922KG62

Rear suspension component identification—sedan models

Shock Absorber

REMOVAL & INSTALLATION

Wagons

1. Before servicing the vehicle, refer to the precautions at the beginning of this section.
2. Remove the rear wheels and support the rear control arms on jackstands.
3. Remove or disconnect the following:
   - Rear compartment access panels
   - Upper shock mounting nuts and insulators
   - Lower shock mounting bolts
   - Shock absorbers

To install:

➡Use new mounting fasteners and insulators.

4. Install or connect the following:
   - Shock absorbers. Tighten the lower bolt to 50–68 ft. lbs. (68–92 Nm), and the upper nuts to19–25 ft. lbs. (26–34 Nm).
   - Rear compartment access panels
   - Rear wheels

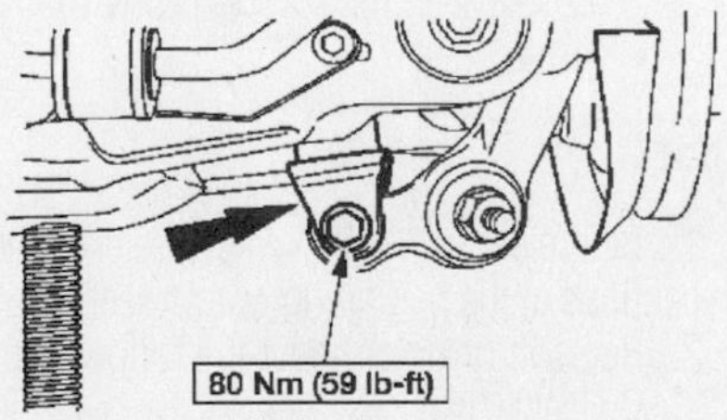

9346KG34

Shock absorber lower mounting bolt—Wagon

Coil Spring

REMOVAL & INSTALLATION

Struts

1. Before servicing the vehicle, refer to the precautions at the beginning of this section.
2. Remove the strut from the vehicle.
3. Compress the coil spring using a suitable spring compressor until the spring comes away from the seat.
4. Remove the large center nut and slowly release the spring compressor.

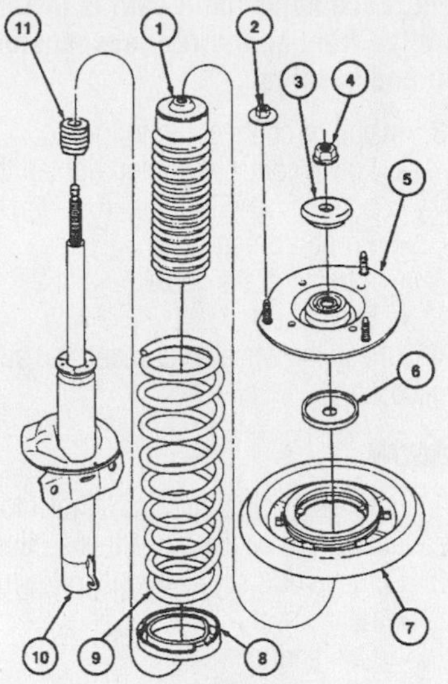

Item	Description
1	Dust Boot (Part of 18124)
2	Nut (3 Req'd)
3	Washer
4	Nut
5	Front Shock Absorber Mounting Bracket
6	Washer
7	Front Suspension Bearing and Seal
8	Front Spring Insulator (Part of 18124)
9	Front Coil Spring
10	Front Shock Absorber
11	Jounce Bumper (Part of 18124)

7922KG63

Exploded view of the front strut and coil spring assembly

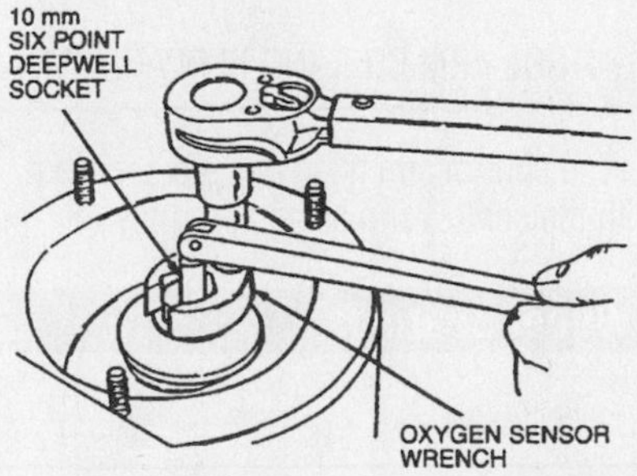

7922KG64

Hold the strut rod while loosening or tightening the nut

To install:

5. Compress the spring and install it on the strut.
6. Install the lower washer and mounting bracket.
7. Install the upper washer and a new nut. Tighten the nut to 39–53 ft. lbs. (53–72 Nm).
8. Install the strut assembly in the vehicle.

Wagons with rear shock absorbers

1. Before servicing the vehicle, refer to the precautions at the beginning of this section.
2. Remove or disconnect the following:
   - Rear shock absorber
   - Stabilizer bar link and bracket
   - Brake hose bracket
   - Upper ball joint
3. Install spring keepers on the coil springs.
4. Slowly lower the lower control arm until the tension is relaxed on the coil spring. Remove the coil spring and the upper and lower spring insulators.

To install:

➡Use new mounting nuts and bolts.

5. Install or connect the following:
   - Coil spring with upper and lower spring insulators
   - Upper ball joint and tighten the nut to 50–68 ft. lbs. (68–92 Nm)
   - Brake hose bracket
   - Stabilizer bar link and bracket
   - Rear shock absorber

Lower Ball Joints

REMOVAL & INSTALLATION

The lower ball joint is an integral part of the steering knuckle. If the lower ball joint is found to be defective, the entire steering knuckle must be replaced.

Upper Control Arm

REMOVAL AND INSTALLATION

Rear

WAGON ONLY

1. Before servicing the vehicle, refer to the precautions at the beginning of this section.
2. Support the lower control arm on a jackstand.
3. Remove or disconnect the following:
   - Rear wheel
   - Brake hose bracket
   - Upper ball joint
   - Upper control arm

To install:

➡Use new mounting nuts and bolts.

4. Install or connect the following:
   - Upper control arm. Tighten the ball joint nut to 50–67 ft. lbs. (68–92 Nm), then tighten the control arm

mounting bolts to 73–97 ft. lbs. (98–132 Nm).
- Brake hose bracket
- Rear wheel

5. Check the wheel alignment and adjust as necessary.

CONTROL ARM BUSHING REPLACEMENT

The control arm bushings are serviced with the control arm as an assembly.

Lower Control Arm

REMOVAL & INSTALLATION

Front

1. Before servicing the vehicle, refer to the precautions at the beginning of this section.
2. Remove or disconnect the following:
   - Front wheel
   - Wheel speed sensor wiring harness
   - Lower ball joint
   - Lower control arm

To install:

3. Install or connect the following:
   - Lower control arm. Tighten the front bolt to 57–75 ft. lbs. (77–103 Nm), and the rear bolt to 72–97 ft. lbs. (98–132 Nm).
   - Lower ball joint. Use a new nut and tighten to 50–67 ft. lbs. (68–92 Nm).
   - Wheel speed sensor wiring harness
   - Front wheel
4. Check the wheel alignment and adjust as necessary.

Rear

SEDAN

1. Before servicing the vehicle, refer to the precautions at the beginning of this section.
2. Remove or disconnect the following:
   - Rear wheel
   - Parking brake cable
   - Proportioning valve
   - Lower control arm

To install:

➡Use new bolts and nuts.

➡The rear suspension lower control arms are marked BOTTOM on the lower edge. The flange edge of the right side rear suspension arm and bushing stamping must face the front of the vehicle. The other three must face the rear of the vehicle.

➡The rear suspension arms have two adjustment cams that fit inside the bushings at the arm-to-body attachment. Each adjustment cam is installed from the front on the rear suspension arm and bushing.

3. Install or connect the following:
   - Lower control arm and tighten the bolts to 50–67 ft. lbs. (68–92 Nm)
   - Proportioning valve
   - Parking brake cable
   - Rear wheel
4. Check the wheel alignment and adjust as necessary.

WAGON

1. Before servicing the vehicle, refer to the precautions at the beginning of this section.
2. Remove or disconnect the following:
   - Rear wheel
   - Coil spring
   - Wheel spindle
   - Lower control arm

To install:

➡Use new nuts and bolts.

➡Tighten the control arm fasteners with the vehicle weight resting on the wheels.

3. Install or connect the following:
   - Lower control arm
   - Wheel spindle
   - Coil spring
   - Rear wheel
4. Tighten the control arm-to-body bolt to 40–52 ft. lbs. (54–71 Nm), and the arm-to-spindle bolt to 50–67 ft. lbs. (68–92 Nm).
5. Check the wheel alignment and adjust as necessary.

CONTROL ARM BUSHING REPLACEMENT

The control arm bushings are serviced with the control arm as an assembly.

Wheel Bearings

ADJUSTMENT

There is no adjustment for the front or rear wheel bearings due to the nature of their design. These bearings are permanently lubricated and require no periodic maintenance.

REMOVAL & INSTALLATION

Front

1. Before servicing the vehicle, refer to the precautions at the beginning of this section.
2. Remove or disconnect the following:
   - Front wheel
   - Hub retainer nut
   - Brake caliper and rotor
   - Outer tie rod end
   - Stabilizer bar link
   - Wheel speed sensor
   - Lower ball joint
   - Steering knuckle
   - Hub and bearing assembly

To install:

➡Use new nuts, bolts, and split pins.

➡The knuckle must be clean enough to allow the wheel hub to be completely seated by hand. Do not press or draw the wheel hub into place.

3. Install or connect the following:
   - Hub and bearing assembly and tighten the bolts to 61–78 ft. lbs. (83–107 Nm)
   - Steering knuckle and tighten the pinch bolt to 72–97 ft. lbs. (98–132 Nm)
   - Lower ball joint and tighten the nut to 50–67 ft. lbs. (68–92 Nm)
   - Wheel speed sensor
   - Stabilizer bar link and tighten the nut to 57–75 ft. lbs. (77–103 Nm)
   - Outer tie rod end and tighten the nut to 35–46 ft. lbs. (47–63 Nm)
   - Brake caliper and rotor and tighten the caliper anchor bracket bolts to 65–87 ft. lbs. (88–118 Nm)
   - Hub retainer nut and tighten to 170–202 ft. lbs. (230–275 Nm)
   - Front wheel

Rear

1. Before servicing the vehicle, refer to the precautions at the beginning of this section.
2. Remove or disconnect the following:
   - Rear wheel
   - Brake hose bracket
   - Brake caliper and rotor, if equipped with rear disc brakes
   - Brake drum, if equipped with rear drum brakes
   - Hub and bearing assembly grease cap and retaining nut
   - Hub and bearing assembly

To install:

➡Use new retaining nuts and grease caps

3. Install or connect the following:
   - Hub and bearing assembly and tighten the retaining nut to 188–254 ft. lbs. (255–345 Nm)
   - Grease cap
   - Brake caliper and rotor, if equipped with rear disc brakes
   - Brake drum, if equipped with rear drum brakes
   - Brake hose bracket
   - Rear wheel

FORD MOTOR CO.

13

1998–01

Lincoln-Continental

PRECAUTIONS

Before servicing any vehicle, please be sure to read all of the following precautions, which deal with personal safety, prevention of component damage, and important points to take into consideration when servicing a motor vehicle:

- Never open, service or drain the radiator or cooling system when the engine is hot; serious burns can occur from the steam and hot coolant.
- Observe all applicable safety precautions when working around fuel. Whenever servicing the fuel system, always work in a well-ventilated area. Do not allow fuel spray or vapors to come in contact with a spark, open flame, or excessive heat (a hot drop light, for example). Keep a dry chemical fire extinguisher near the work area. Always keep fuel in a container specifically designed for fuel storage; also, always properly seal fuel containers to avoid the possibility of fire or explosion. Refer to the additional fuel system precautions later in this section.
- Fuel injection systems often remain pressurized, even after the engine has been turned **OFF**. The fuel system pressure must be relieved before disconnecting any fuel lines. Failure to do so may result in fire and/or personal injury.
- Brake fluid often contains polyglycol ethers and polyglycols. Avoid contact with the eyes and wash your hands thoroughly after handling brake fluid. If you do get brake fluid in your eyes, flush your eyes with clean, running water for 15 minutes. If eye irritation persists, or if you have taken brake fluid internally, IMMEDIATELY seek medical assistance.
- The EPA warns that prolonged contact with used engine oil may cause a number of skin disorders, including cancer. You should make every effort to minimize your exposure to used engine oil. Protective gloves should be worn when changing oil. Wash your hands and any other exposed skin areas as soon as possible after exposure to used engine oil. Soap and water, or waterless hand cleaner should be used.
- All new vehicles are now equipped with an air bag system, often referred to as a Supplemental Restraint System (SRS) or Supplemental Inflatable Restraint (SIR) system. The system must be disabled before performing service on or around system components, steering column, instrument panel components, wiring and sensors. Failure to follow safety and disabling procedures could result in accidental air bag deployment, possible personal injury and unnecessary system repairs.
- Always wear safety goggles when working with, or around, the air bag system. When carrying a non-deployed air bag, be sure the bag and trim cover are pointed away from your body. When placing a non-deployed air bag on a work surface, always face the bag and trim cover upward, away from the surface. This will reduce the motion of the module if it is accidentally deployed. Refer to the additional air bag system precautions later in this section.
- Clean, high quality brake fluid from a sealed container is essential to the safe and proper operation of the brake system. You should always buy the correct type of brake fluid for your vehicle. If the brake fluid becomes contaminated, completely flush the system with new fluid. Never reuse any brake fluid. Any brake fluid that is removed from the system should be discarded. Also, do not allow any brake fluid to come in contact with a painted surface; it will damage the paint.
- Never operate the engine without the proper amount and type of engine oil; doing so WILL result in severe engine damage.
- Timing belt maintenance is extremely important. Many models utilize an interference-type, non-freewheeling engine. If the timing belt breaks, the valves in the cylinder head may strike the pistons, causing potentially serious (also time-consuming and expensive) engine damage. Refer to the maintenance interval charts in the front of this manual for the recommended replacement interval for the timing belt, and to the timing belt section for belt replacement and inspection.
- Disconnecting the negative battery cable on some vehicles may interfere with the functions of the on-board computer system(s) and may require the computer to undergo a relearning process once the negative battery cable is reconnected.
- When servicing drum brakes, only disassemble and assemble one side at a time, leaving the remaining side intact for reference.

ENGINE REPAIR

➡Disconnecting the negative battery cable on some vehicles may interfere with the functions of the on board computer system. The computer may undergo a relearning process once the negative battery cable is reconnected.

Alternator

REMOVAL

1. Before servicing the vehicle, refer to the precautions in the beginning of this section.
2. Remove or disconnect the following:
   - Negative battery cable
   - Coolant recovery reservoir
   - Accessory drive belt
   - Alternator bracket
   - Alternator

INSTALLATION

1. Install or connect the following:
   - Alternator. Tighten the bolts to 15–22 ft. lbs. (20–30 Nm).
   - Alternator bracket. Tighten the nuts to 71–106 inch lbs. (8–12 Nm).
   - Accessory drive belt
   - Coolant recovery reservoir
   - Negative battery cable

Ignition Timing

ADJUSTMENT

The ignition timing is controlled by the Powertrain Control Module (PCM) and is not adjustable.

Engine Assembly

REMOVAL & INSTALLATION

➡Disable the air suspension before raising the vehicle. The switch is located on the left side of the luggage compartment.

1. Before servicing the vehicle, refer to the precautions in the beginning of this section.
2. Turn the air suspension switch to the **OFF** position.
3. Recover the A/C refrigerant.
4. Drain the cooling system.
5. Remove or disconnect the following:
   - Negative battery cable
   - Hood
   - Steering column intermediate shaft

- Engine appearance cover
- Intake Air Temperature (IAT) sensor connector
- Air cleaner outlet tube
- Upper motor mount
- Fuel line
- Dash panel ground straps
- Powertrain Control Module (PCM) connector
- Mass Air Flow (MAF) sensor connector
- Traction control motor connector
- Throttle Position (TP) sensor connector
- Cruise control cable
- Accelerator cable and bracket
- Shift selector cable
- Engine control wiring harness (3 connectors) and bracket
- Intake manifold vacuum lines
- Transaxle cooler lines
- Power steering return hose at the fluid reservoir
- Alternator wiring harness
- Radiator splash shield
- Heater hoses
- Radiator hoses
- Front wheels
- Suspension height sensor links
- Lower ball joints
- Stabilizer bar links
- Outer tie rod ends
- Halfshafts
- Heated Oxygen (HO_2S) sensor connectors
- Dual converter Y-pipe
- Block heater connector
- Power steering cooler lines
- A/C compressor lines
- Starter motor
- Subframe brackets
- Torque converter shield and torque converter

6. Support the powertrain from below and remove the subframe bolts.
7. Raise the vehicle away from the powertrain.
8. Attach an engine hoist to the powertrain.
9. Remove or disconnect the following:
   - Power steering sensor connector
   - Power steering return line from the steering gear
   - Turbine shaft speed sensor connector
   - Transaxle range sensor connector
   - Transaxle control harness connector
   - Left and right engine support insulators
   - Right engine support insulator bracket
10. Remove the transaxle flange bolts and separate the engine from the transaxle.

To install:

➡When installing suspension components and halfshafts, use new nuts, bolts, circlips and split pins.

11. Install or connect the following:
    - Transaxle. Tighten the flange bolts to 25–34 ft. lbs. (34–46 Nm).
    - Right engine support insulator bracket. Tighten the bolts to 16–21 ft. lbs. (22–29 Nm).
    - Left and right engine support insulators. Tighten the through-bolts to 64–88 ft. lbs. (87–119 Nm).
    - Transaxle control harness connector
    - Transaxle range sensor connector
    - Turbine shaft speed sensor connector
    - Power steering return line from the steering gear
    - Power steering sensor connector
12. Lower the vehicle on to the powertrain assembly. Use 2 pieces of ¾ inch outside diameter pipe in the alignment holes behind the front subframe mounts to align the subframe to the body. Tighten the subframe mounting bolts to 57–76 ft. lbs. (77–103 Nm).
13. Install or connect the following:
    - Torque converter shield and torque converter. Tighten the torque converter nuts to 20–34 ft. lbs. (27–46 Nm).
    - Subframe brackets
    - Starter motor
    - A/C compressor lines
    - Power steering cooler lines
    - Block heater connector
    - Dual converter Y-pipe
    - HO_2S sensor connectors
    - Halfshafts. Tighten the hub retainer nuts to 170–202 ft. lbs. (230–275 Nm).
    - Outer tie rod ends. Tighten the nuts to 35–46 ft. lbs. (47–63 Nm).
    - Stabilizer bar links. Tighten the nuts to 30–40 ft. lbs. (40–55 Nm).
    - Lower ball joints. Tighten the nuts to 50–68 ft. lbs. (68–92 Nm).
    - Suspension height sensor links
    - Front wheels
    - Radiator hoses
    - Heater hoses
    - Radiator splash shield
    - Alternator wiring harness
    - Power steering return hose at the fluid reservoir
    - Transaxle cooler lines
    - Intake manifold vacuum lines
    - Engine control wiring harness (3 connectors) and bracket
    - Shift selector cable
    - Accelerator cable and bracket
    - Cruise control cable
    - TP sensor connector
    - Traction control motor connector
    - MAF sensor connector
    - PCM connector
    - Dash panel ground straps
    - Fuel line
    - Upper motor mount
    - Air cleaner outlet tube
    - IAT sensor connector
    - Engine appearance cover
    - Steering column intermediate shaft
    - Hood
    - Negative battery cable
14. Fill the cooling system.
15. Check and adjust fluid levels as necessary.
16. Recharge the A/C system.
17. Turn the air suspension switch to the **ON** position.
18. Run the engine and check for leaks.

➡Whenever the subframe is removed or lowered, the wheel alignment should be checked.

Water Pump

REMOVAL & INSTALLATION

1. Before servicing the vehicle, refer to the precautions in the beginning of this section.
2. Drain the cooling system.
3. Remove or disconnect the following:
   - Negative battery cable
   - Coolant reservoir
   - Accessory drive belt
   - Alternator
   - Water pump pulley
   - Water pump

To install:

4. Install or connect the following:
   - Water pump. Use a new O-ring seal and tighten the bolts to 15–22 ft. lbs. (20–30 Nm).
   - Water pump pulley. Tighten the bolts to 15–22 ft. lbs. (20–30 Nm).
   - Alternator

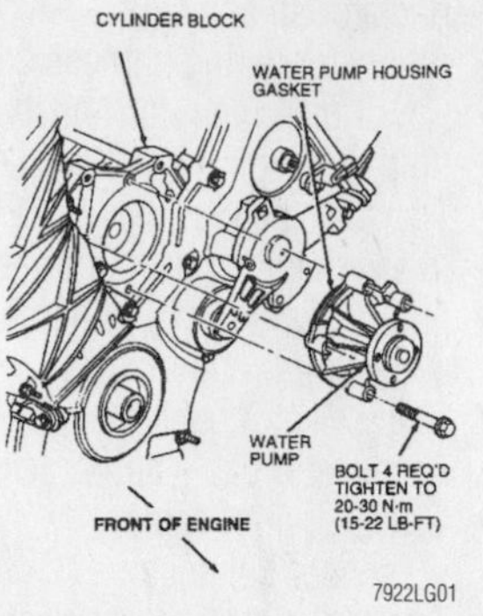

Exploded view of the water pump mounting

- Accessory drive belt
- Coolant reservoir
- Negative battery cable

5. Fill the cooling system.
6. Start the engine and check for leaks.

Cylinder Head

REMOVAL & INSTALLATION

1. Before servicing the vehicle, refer to the precautions in the beginning of this section.
2. Remove the engine from the vehicle and mount it on a suitable workstand.
3. Remove or disconnect the following:
   - Accessory drive belt and tensioner
   - Idler pulley
   - Engine Coolant Temperature (ECT) sensor connector
   - Water bypass tube
   - Alternator
   - Water pump
   - Power steering pump
   - Crankshaft pulley
   - Camshaft Position (CMP) sensor
   - Ignition coils
   - Fuel pressure sensor connector and vacuum line
   - Valve covers
   - Rocker arms
   - Exhaust Gas Recirculation (EGR) vacuum regulator valve
   - EGR tube
   - Injector wiring connectors
   - Intake manifold
   - Left and right exhaust manifolds
   - Front cover
   - Crankshaft Position (CKP) sensor pulse wheel
   - Timing chains
   - Cylinder heads

To install:

➡**The cylinder head bolts are a torque-to-yield design and cannot be reused.**

4. Use new gaskets and install the cylinder heads. Tighten the cylinder head bolts in sequence as follows:
   a. Step 1: 28–31 ft. lbs. (37–43 Nm).
   b. Step 2: Tighten the bolts 85–95 degrees.
   c. Step 3: Loosen all bolts 1 full turn.
   d. Step 4: Tighten all bolts to 28–31 ft. lbs. (37–43 Nm).
   e. Step 5: Tighten the bolts 85–95 degrees.
   f. Step 6: Tighten the bolts 85–95 degrees.
5. Install or connect the following:
   - Timing chains
   - CKP sensor pulse wheel
   - Front cover
   - Left and right exhaust manifolds
   - Intake manifold
   - Injector wiring connectors
   - EGR tube
   - EGR vacuum regulator valve
   - Rocker arms
   - Valve covers
   - Fuel pressure sensor connector and vacuum line
   - Ignition coils
   - CMP sensor
   - Crankshaft pulley
   - Water pump
   - Power steering pump
   - Alternator
   - Water bypass tube
   - ECT sensor connector
   - Idler pulley
   - Accessory drive belt and tensioner

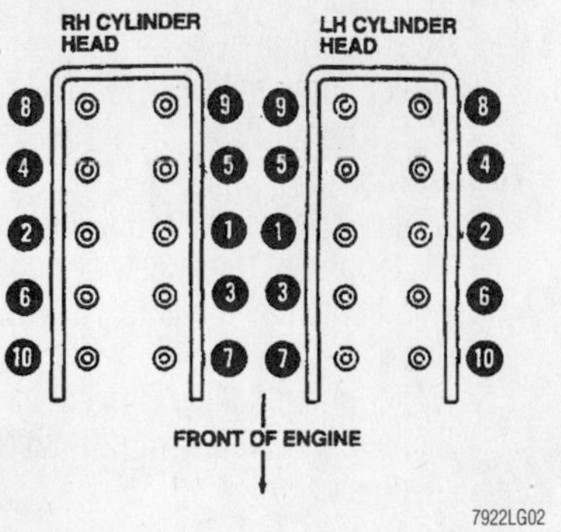

Cylinder head torque sequence—4.6L engine

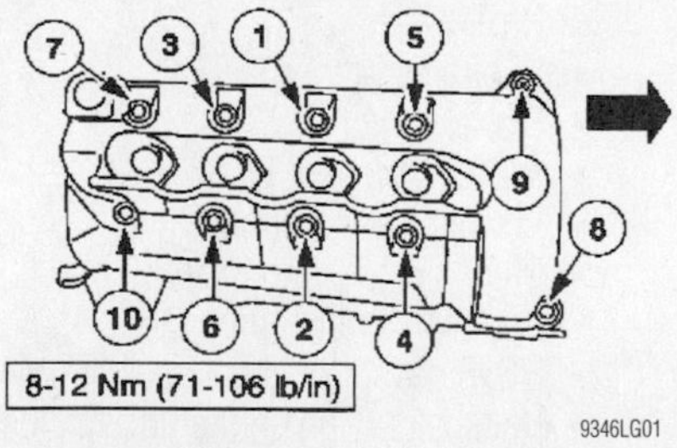

Right valve cover torque sequence

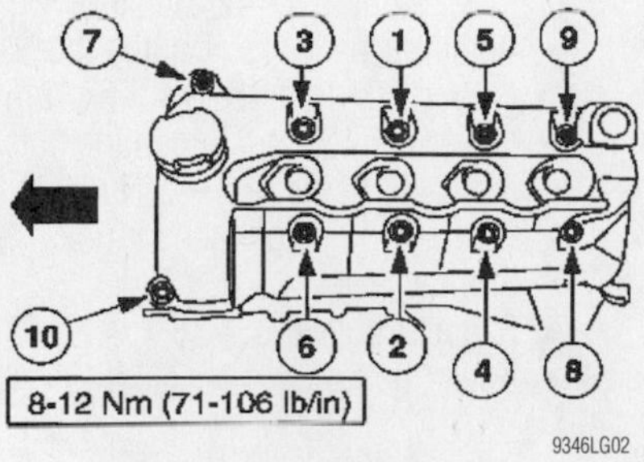

Left valve cover torque sequence

Rocker Arms

REMOVAL & INSTALLATION

1. Before servicing the vehicle, refer to the precautions at the beginning of this section.
2. Remove or disconnect the following:
   - Negative battery cable
   - Ignition coils
   - Valve covers
3. Rotate the crankshaft so that the piston on the cylinder to be serviced is at bottom dead center with the valves closed.
4. Install special tool Valve Spring Compressor T91P-6565-A for exhaust valves, and Valve Spring Compressor T93P-6565-A for intake valves.
5. Compress the valve spring and remove the rocker arm. Repeat for each arm to be removed.

➡**If the rocker arms are to be reused, ensure that they are installed in the same position that they were removed from.**

To install:

6. Compress the valve spring and install the rocker arm. Repeat for each arm to be installed.

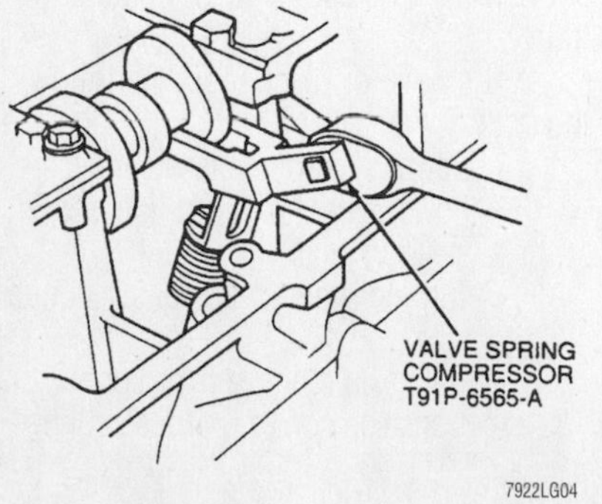

Rocker arm service tool—exhaust valve tool shown

7. Install or connect the following:
   - Valve covers
   - Ignition coils
   - Negative battery cable
8. Start the engine and check for proper operation.

Intake Manifold

REMOVAL & INSTALLATION

1. Before servicing the vehicle, refer to the precautions in the beginning of this section.
2. Drain the cooling system.
3. Remove or disconnect the following:
   - Negative battery cable
   - Engine appearance cover
   - Air cleaner outlet tube
   - Fuel lines
   - Fuel pressure sensor vacuum and wiring connectors
   - Upper radiator hose
   - Heater hose
   - Engine Coolant Temperature (ECT) sensor connector
   - Water bypass tube
   - Cruise control cable
   - Accelerator cable and bracket
   - Crankcase vent hose
   - Chassis vacuum supply line
   - Secondary air injection vacuum lines
   - Exhaust Gas Recirculation (EGR) vacuum regulator
   - EGR valve
   - Intake Manifold Runner Control (IMRC) cables and actuator
   - Fuel injector connectors
   - Throttle Position (TP) sensor connector
   - Idle Air Control (IAC) connector
   - Fuel temperature sensor connector
   - Fuel injection supply manifold
   - Intake manifold. Loosen the bolts in three steps in the sequence shown.
   - IMRC housings

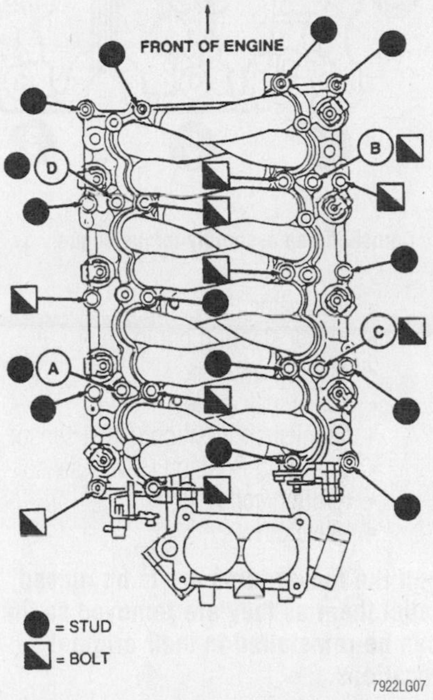

IMRC torque sequence—4.6L engine

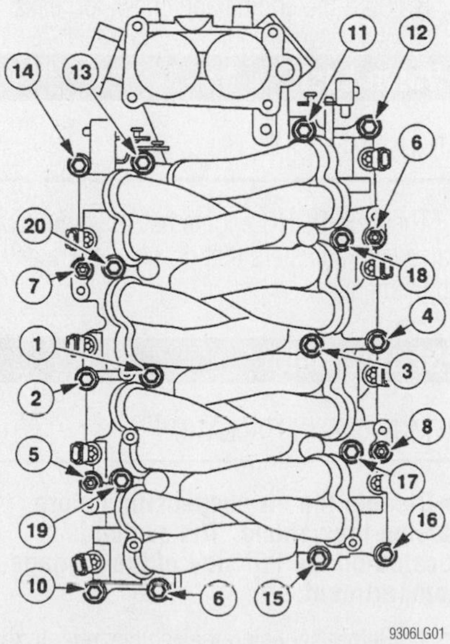

Intake manifold bolt removal sequence

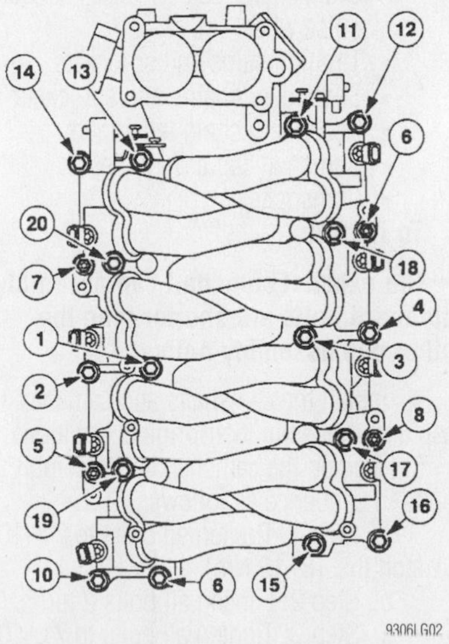

Intake manifold torque sequence—4.6L engine

To install:

4. Install or connect the following:
   - IMRC housings to the intake manifold
   - Fuel supply manifold and tighten the bolts to 71–106 inch lbs. (8–12 Nm)

➡Install the long bolts and studs in the outer holes. Install the short bolts and studs in the inner holes.

5. Install the intake manifold assembly. Tighten the bolts in sequence as follows:
   a. Step 1: Tighten bolts 5, 7, 9, and 11 to 9–11 ft. lbs. (12–15 Nm).
   b. Step 2: Tighten all other bolts to 13–16 ft. lbs. (18–22 Nm).
   c. Step 3: Tighten all bolts in sequence 85–95 degrees.
6. Tighten the IMRC housing bolts as follows:
   a. Step 1: 71–89 inch lbs. (8–10 Nm).
   b. Step 2: Plus 85–95 degrees.
7. Install or connect the following:
   - Fuel temperature sensor connector
   - IAC connector
   - TP sensor connector
   - Fuel injector connectors
   - IMRC cables and actuator
   - EGR valve
   - EGR vacuum regulator
   - Secondary air injection vacuum lines
   - Chassis vacuum supply line
   - Crankcase vent hose
   - Accelerator cable and bracket
   - Cruise control cable
   - Water bypass tube
   - ECT sensor connector
   - Heater hose
   - Upper radiator hose
   - Fuel pressure sensor vacuum and wiring connectors
   - Fuel lines
   - Air cleaner outlet tube
   - Engine appearance cover
   - Negative battery cable
8. Fill the cooling system.
9. Start the engine and check for leaks.

Exhaust Manifold

REMOVAL & INSTALLATION

➡Disable the air suspension before raising the vehicle. The switch is located on the left side of the luggage compartment.

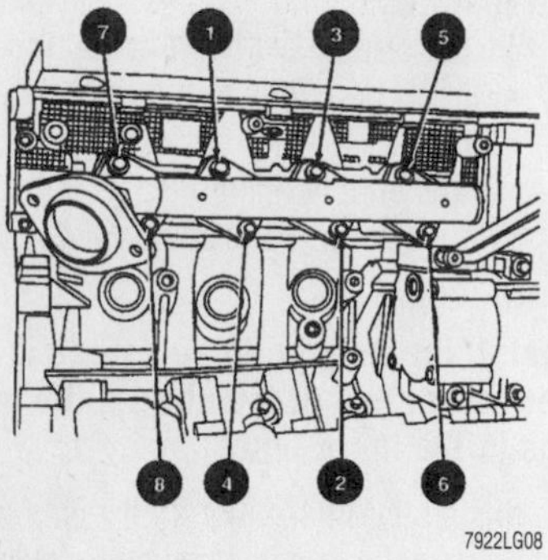

Exhaust manifold torque sequence—right side shown, left side similar

1. Before servicing the vehicle, refer to the precautions in the beginning of this section.
2. Turn the air suspension switch to the **OFF** position.
3. Remove or disconnect the following:
   - Negative battery cable
   - Radiator splash shield
   - Heated Oxygen (HO_2S) sensor connectors
   - Dual converter Y-pipe
   - Exhaust Gas Recirculation (EGR) tube
   - Secondary air injection tubes
   - Exhaust manifolds

To install:

4. Install or connect the following:
   - Exhaust manifolds. Tighten the nuts in sequence to 13–16 ft. lbs. (18–22 Nm).
   - Secondary air injection tubes. Tighten the nuts to 25–34 ft. lbs. (34–46 Nm).
   - EGR tube. Tighten the nut to 30–33 ft. lbs. (40–45 Nm).
   - Dual converter Y-pipe
   - HO_2S sensor connectors
   - Radiator splash shield
   - Negative battery cable
5. Turn the air suspension switch to the **ON** position.
6. Start the engine and check for leaks.

Camshaft and Valve Lifters

REMOVAL & INSTALLATION

1. Before servicing the vehicle, refer to the precautions in the beginning of this section.
2. Remove the engine from the vehicle and mount it on a suitable workstand.
3. Remove or disconnect the following:
   - Accessory drive belt and tensioner
   - Idler pulley
   - Alternator
   - Power steering pump
   - Water pump
   - Crankshaft pulley
   - Crankshaft Position (CKP) sensor
   - Camshaft Position (CMP) sensor
   - Ignition coils
   - Valve covers

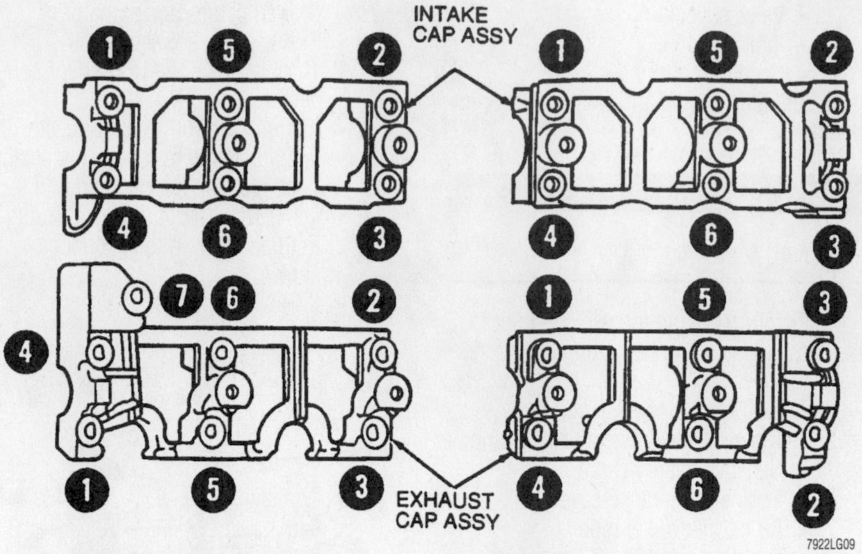

Camshaft cap assembly torque sequence

➡If the rocker arms are to be reused, label them as they are removed so they can be reinstalled in their original locations.

   - Rocker arms
   - Front cover
   - Crankshaft Position (CKP) sensor pulse wheel
   - Timing chains and sprockets
   - Secondary chains and sprockets
   - Secondary chain tensioners
   - Camshaft cap assemblies
   - Camshafts

To install:

➡The exhaust camshaft cap assembly outboard bolts are shorter than the other cap assembly bolts.

4. Install the camshafts and camshaft cap assemblies in their original positions.
5. Tighten the camshaft cap assembly bolts in sequence as follows:
   a. Step 1: Tighten all bolts to 71–106 inch lbs. (8–12 Nm)
   b. Step 2: Loosen all bolts 2 turns
   c. Step 3: Tighten all bolts to 71–106 inch lbs. (8–12 Nm)
6. Install or connect the following:
   - Secondary chain tensioners
   - Secondary chains and sprockets
   - Timing chains and sprockets
   - CKP sensor pulse wheel
   - Front cover
   - Rocker arms in the original locations
   - Valve covers
   - Ignition coils
   - CMP sensor
   - CKP sensor
   - Crankshaft pulley
   - Water pump
   - Power steering pump
   - Alternator
   - Idler pulley
   - Accessory drive belt and tensioner
7. Remove the engine from the workstand and install in the vehicle.
8. Start the engine and check for leaks.

Valve Lash

ADJUSTMENT

The 4.6L DOHC engine uses hydraulic valve lash adjusters that do not require any adjustment.

Starter Motor

REMOVAL & INSTALLATION

➡Disable the air suspension before raising the vehicle. The switch is located on the left side of the luggage compartment.

1. Before servicing the vehicle, refer to the precautions in the beginning of this section.
2. Turn the air suspension switch to the **OFF** position.

3. Remove or disconnect the following:
   - Negative battery cable
   - Radiator splash shield
   - Starter solenoid connections
   - Starter

To install:

4. Install or connect the following:
   - Starter. Tighten the bolt and stud to 15–20 ft. lbs. (20–27 Nm).
   - Starter solenoid connections
   - Radiator splash shield
   - Negative battery cable
5. Turn the air suspension switch to the **ON** position.

Oil Pan

REMOVAL & INSTALLATION

➡Disable the air suspension before raising the vehicle. The switch is located on the left side of the luggage compartment.

1. Before servicing the vehicle, refer to the precautions in the beginning of this section.
2. Turn the air suspension switch to the **OFF** position.
3. Remove or disconnect the following:
   - Negative battery cable
   - Oil level dipstick
   - Dual converter Y-pipe
   - Oil level sensor connector
   - Power steering pressure hose bracket
   - Oil pan

To install:

4. Apply a bead of silicone sealer to the oil pan flange. Also apply a bead of sealer to the front cover/cylinder block joint and fill the grooves on both sides of the rear main seal cap.
5. Use a new gasket and install the oil pan. Tighten the bolts in sequence as follows:
   a. Step 1: 14 ft. lbs. (20 Nm)
   b. Step 2: Plus 60 degrees
6. Install or connect the following:
   - Power steering pressure hose bracket
   - Oil level sensor connector
   - Dual converter Y-pipe
   - Oil level dipstick
   - Negative battery cable
7. Fill the crankcase with the correct type and quantity of engine oil.
8. Turn the air suspension switch to the **ON** position.
9. Start the engine and check for leaks.

Oil Pump

REMOVAL & INSTALLATION

1. Before servicing the vehicle, refer to the precautions in the beginning of this section.
2. Remove the engine from the vehicle and mount it on a suitable workstand.
3. Remove or disconnect the following:
   - Accessory drive belt and tensioner
   - Idler pulley
   - Alternator
   - Power steering pump
   - Water pump
   - Crankshaft pulley
   - Camshaft Position (CMP) sensor
   - Ignition coils
   - Valve covers
   - Rocker arms
   - Front cover
   - Crankshaft Position (CKP) sensor pulse wheel
   - Timing chains and sprockets
   - Oil pan
   - Oil pump pickup screen and tube
   - Oil pump

To install:

4. Install or connect the following:
   - Oil pump. Tighten the bolts in sequence to 71–106 inch lbs. (8–12 Nm).
   - Oil pump pickup screen and tube. Use a new O-ring seal and tighten the mounting bolts to 71–106 inch lbs. (8–12 Nm). Tighten the bracket bolt to 15–22 ft. lbs. (20–30 Nm).
   - Oil pan
   - Timing chains and sprockets
   - CKP sensor pulse wheel
   - Front cover
   - Rocker arms
   - Valve covers
   - Ignition coils
   - CMP sensor
   - Crankshaft pulley
   - Water pump
   - Power steering pump
   - Alternator
   - Idler pulley
   - Accessory drive belt and tensioner
5. Remove the engine from the workstand and install in the vehicle.
6. Restore all fluid levels.
7. Run the engine and check for leaks and proper operation.

Rear Main Seal

REMOVAL & INSTALLATION

1. Before servicing the vehicle, refer to the precautions in the beginning of this section.
2. Remove or disconnect the following:
   - Transaxle
   - Flywheel
   - Oil seal retainer
   - Oil seal

To install:

3. Apply a 0.060 inch (1.5mm) continuous bead of silicone gasket sealer to the cylinder block.
4. Install or connect the following:
   - Oil seal retainer. Tighten the bolts in sequence to 71–106 inch lbs. (8–12 Nm).
   - Oil seal. Use special tool Seal Installer T82L-6701-A and adapter T91P-6701-A to press the seal into place.

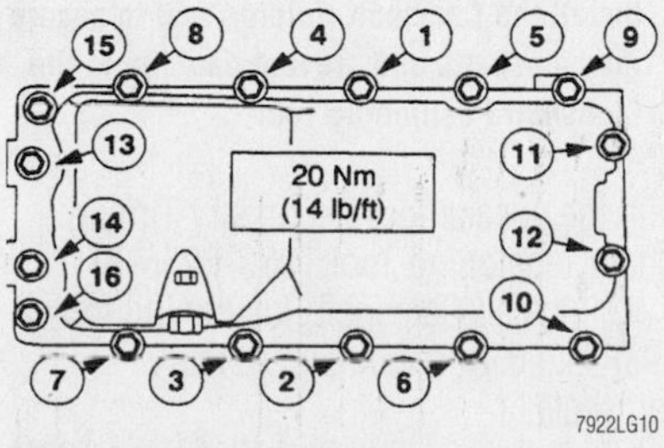

Oil pan bolt torque sequence

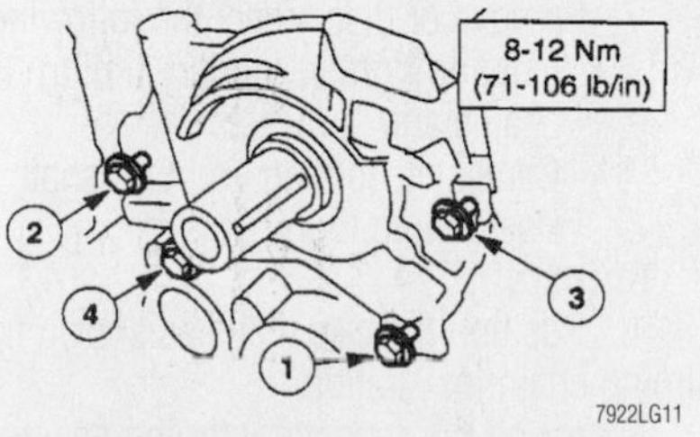

Oil pump mounting bolt sequence

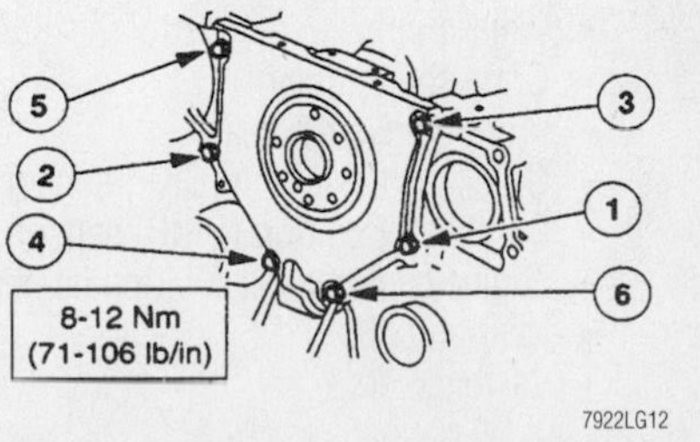

Oil seal retainer torque sequence

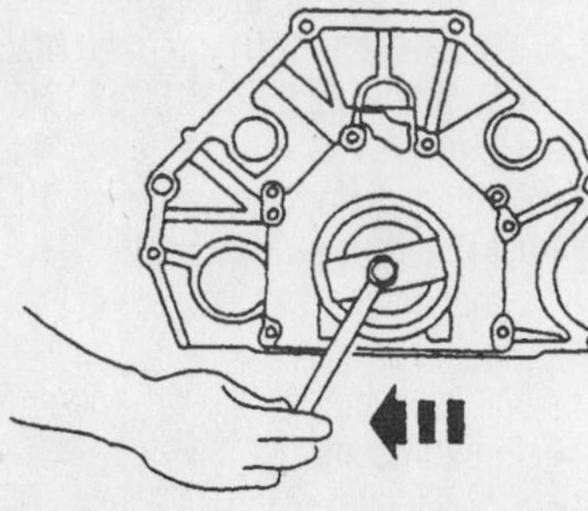
7922LG13

Use Seal Installer T82L-6701-A and adapter T91P-6701-A to press the seal into the retainer

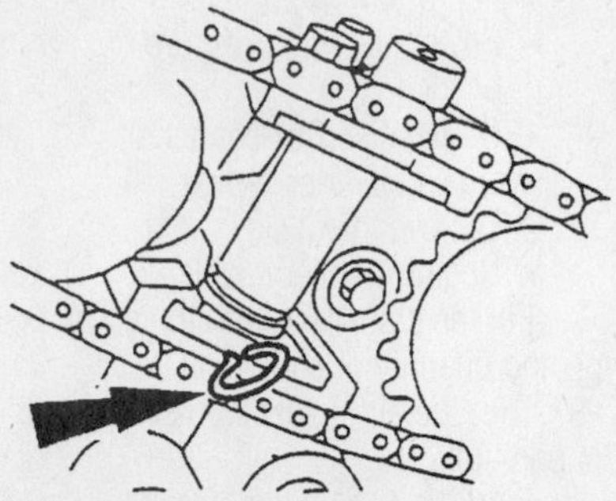
9306LG03

Secondary timing chain tensioner and locking pin

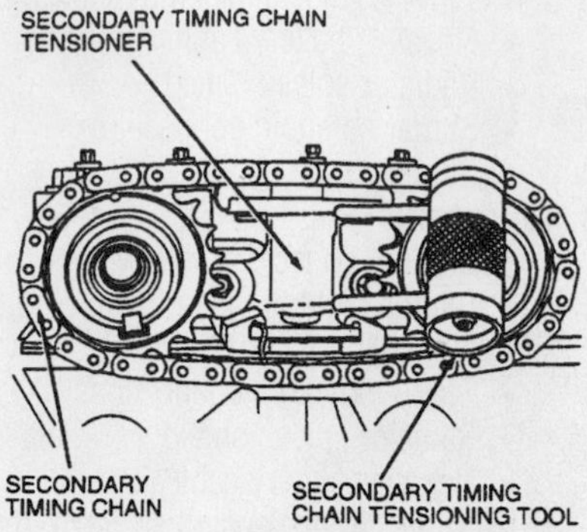

7922LG17

Secondary timing chain tensioning tool

- Flywheel. Tighten the bolts to 54–64 ft. lbs. (73–87 Nm).
- Transaxle

5. Check and adjust all fluid levels as necessary.
6. Start the engine and check for leaks.

Timing Chain, Sprockets, Front Cover and Seal

REMOVAL & INSTALLATION

※ WARNING

This is an interference engine. When the timing chains are removed and the cylinder heads are installed, the crankshaft and/or camshafts must not be rotated unless as directed in this procedure. Failure to follow these instructions will result in valve and/or piston damage.

1. Before servicing the vehicle, refer to the precautions in the beginning of this section.
2. Remove the engine from the vehicle and mount it on a suitable workstand.
3. Remove or disconnect the following:
   - Accessory drive belt and tensioner
   - Idler pulley
   - Alternator
   - Power steering pump
   - Water pump
   - Crankshaft pulley
   - Front crankshaft seal
   - Crankshaft Position (CKP) sensor
   - Camshaft Position (CMP) sensor
   - Engine control sensor wiring harness
   - Ignition coils
   - Valve covers

➡If the rocker arms are to be reused, label them as they are removed so they can be reinstalled in their original locations.

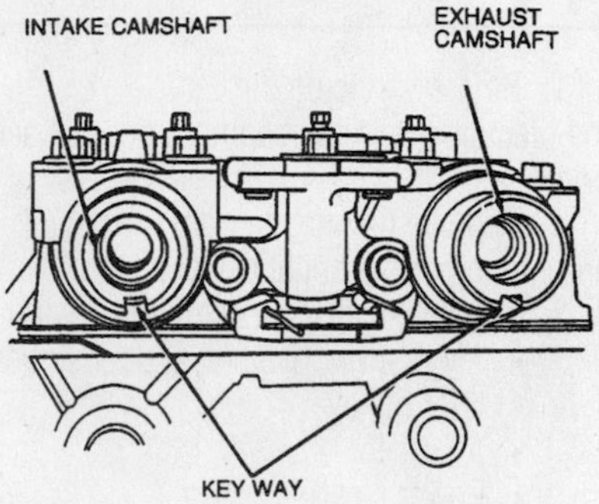

7922LG14

Position the camshafts for timing chain installation

   - Rocker arms
   - Front cover

4. Rotate the crankshaft to place the piston for No. 1 cylinder at Top Dead Center (TDC) on its compression stroke.
5. Remove or disconnect the following:
   - Right primary timing chain tensioner
   - Right primary timing chain tensioner arm and chain guide
   - Right primary timing chain and sprockets
   - Left primary timing chain tensioner
   - Left primary timing chain tensioner arm and chain guide
   - Left primary timing chain and sprockets
6. Compress the secondary chain tensioners and install locking pins.
7. Remove or disconnect the following:
   - Left and right secondary timing chains and sprockets
   - Camshaft Holding and Camshaft Positioning tools

To install:

8. Position the camshafts as shown for timing chain installation.
9. Install the secondary timing chains and sprockets. Do not tighten the sprocket bolts at this time.
10. Tension the secondary timing chains

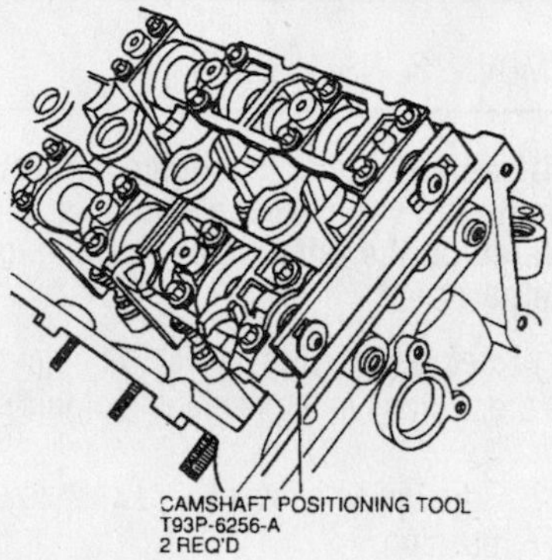

9306LG04

Install the Camshaft Positioning tool

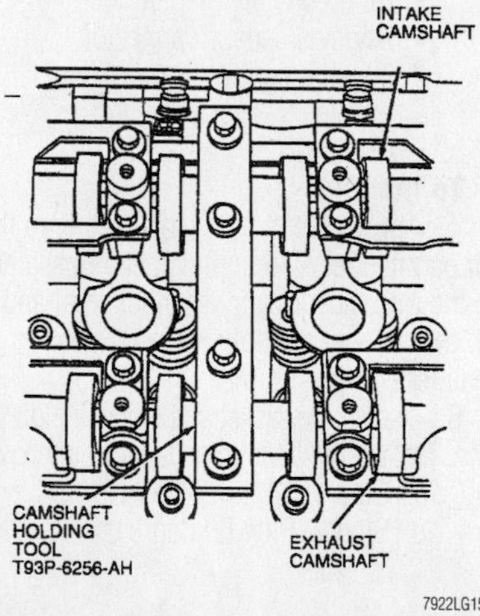

7922LG15

Install the Camshaft Holding tool to secure the camshafts and prevent damage to the Camshaft Positioning Tool

with the special tool Secondary Timing Chain Tensioning Tool T93P-6256-BH.

11. Install Camshaft Positioning tool T93P-6256-A in the rear D-slots of the camshaft.
12. Install Camshaft Holding tool T93P-6256-AH onto the camshafts to keep the camshafts from rotating and to prevent damaging the camshaft positioning tool.

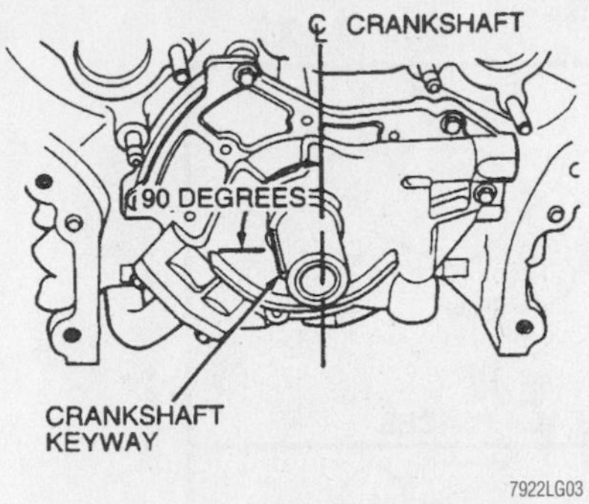

Turn the crankshaft to the 9 o'clock position for timing chain installation

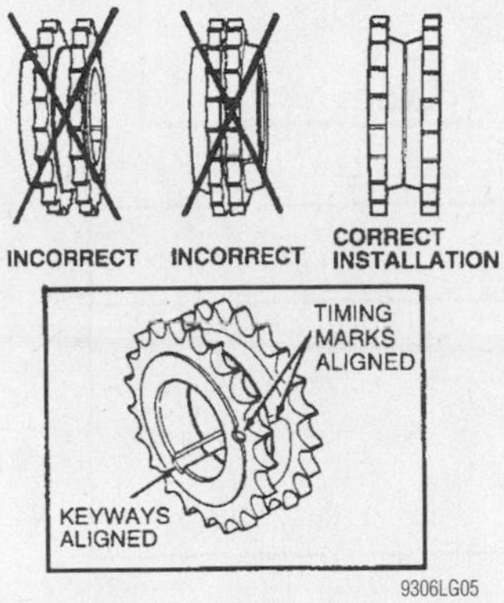

Crankshaft timing sprocket installation

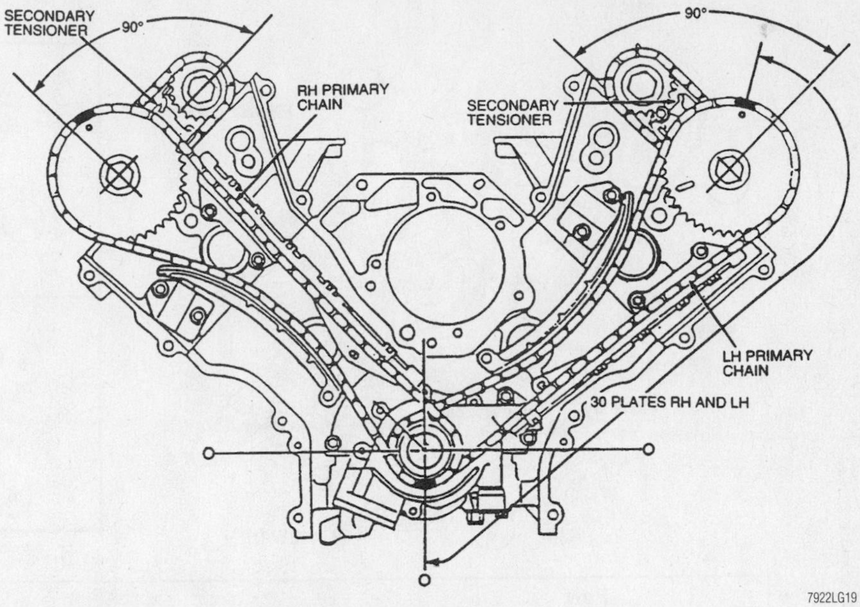

Correct alignment of the primary timing chains and sprockets

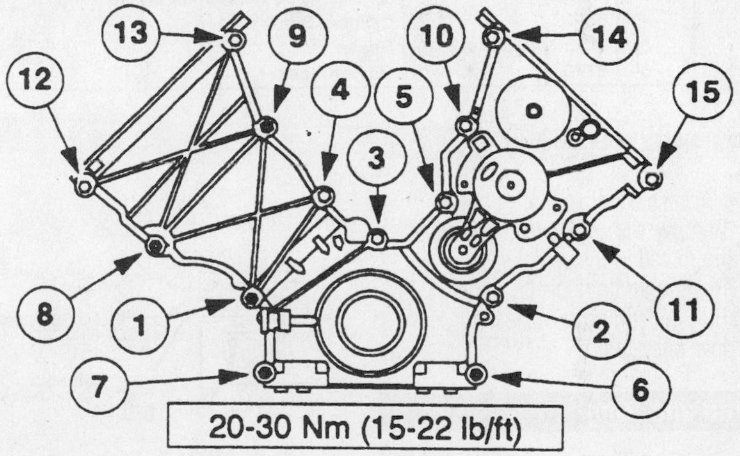

Front cover torque sequence

13. Tighten the secondary timing chain sprocket bolts to 81–95 ft. lbs. (110–130 Nm).
14. Remove the secondary timing chain tensioner locking pins and remove the tensioner tool.
15. Position the crankshaft so that the keyway is at 9 o'clock as shown.
16. Compress the primary timing chain tensioners and install locking pins.
17. Install the crankshaft timing sprockets arranged as shown.
18. Align the copper colored primary timing chain links with the sprocket timing marks as shown.
19. Install or connect the following:
    - Left primary timing chain and sprockets
    - Left primary timing chain tensioner arm and chain guide
    - Left primary timing chain tensioner
    - Right primary timing chain and sprockets
    - Right primary timing chain tensioner arm and chain guide
    - Right primary timing chain tensioner
20. Tighten the camshaft sprocket bolts to 81–95 ft. lbs. (110–130 Nm).
21. Remove the primary timing chain tensioner locking pins.
22. Remove the Camshaft Holding and Camshaft Positioning tools.
23. Install or connect the following:
    - Front cover. Tighten the bolts in sequence to 15–22 ft. lbs. (20–30 Nm).
    - Rocker arms in their original positions
    - Valve covers
    - Ignition coils
    - CMP sensor
    - CKP sensor
    - Engine control sensor wiring harness
    - Front crankshaft seal
24. Install the crankshaft pulley and tighten the bolt as follows:
    a. Step 1: 89 ft. lbs. (120 Nm)
    b. Step 2: Loosen the bolt
    c. Step 3: 39 ft. lbs. (53 Nm)
    d. Step 4: Plus 90 degrees
25. Install or connect the following:
    - Water pump
    - Power steering pump
    - Alternator
    - Idler pulley and tensioner. Tighten the bolts to 15–22 ft. lbs. (20–30 Nm).

Heater Core replacement is covered in Section 2 of this manual

Sealer Location

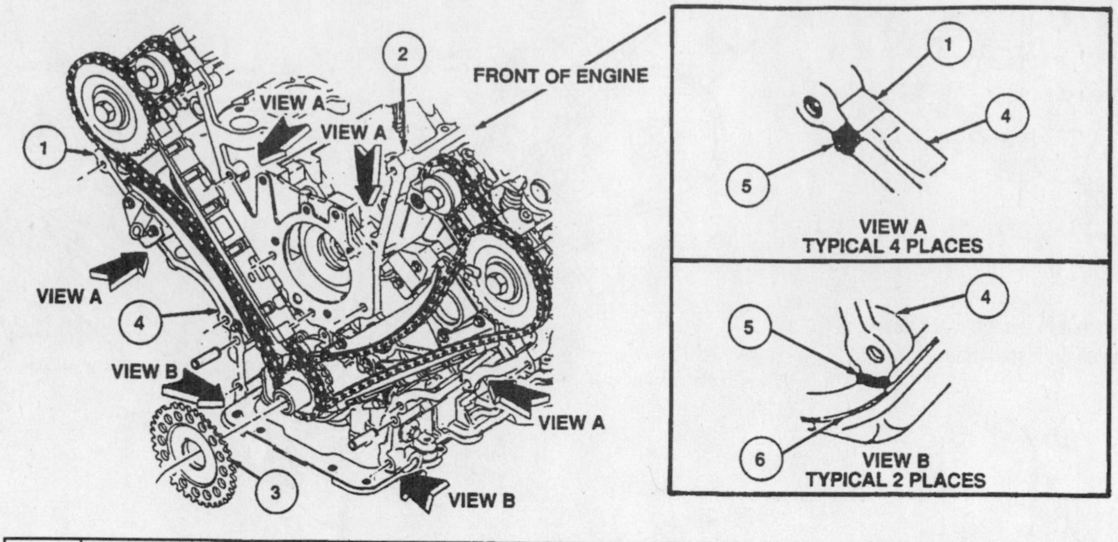

Item	Part Number	Description
1	6049	RH Cylinder Head
2	6049	LH Cylinder Head
3	12A227	Ignition Pulse Crankshaft Sensor Ring
4	6010	Cylinder Block
5	WSE-M4G320-A2	Sealer
6	6710	Oil Pan Gasket

9306LG06

Apply sealer to these locations

- Accessory drive belt

26. Remove the engine from the workstand and install in the vehicle.
27. Restore all fluid levels.
28. Run the engine and check for leaks and proper operation.

Piston and Ring

POSITIONING

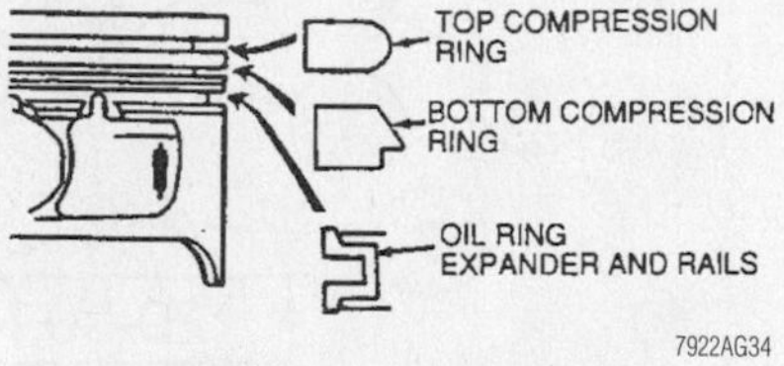

Piston ring positioning—4.6L VIN V engine

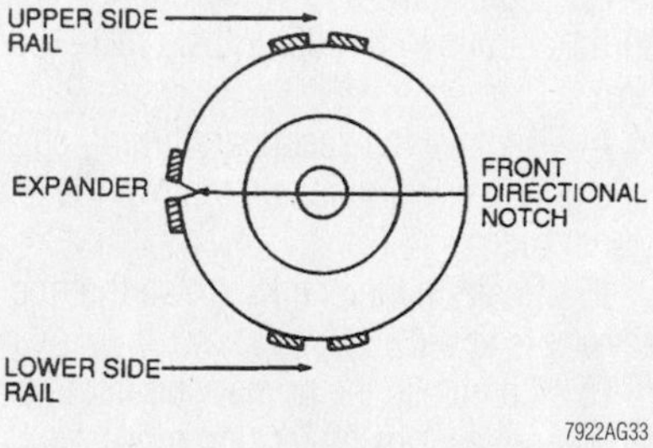

Piston positioning and ring end-gap spacing—4.6L VIN V engine

FUEL SYSTEM

Fuel System Service Precautions

Safety is the most important factor when performing not only fuel system maintenance but any type of maintenance. Failure to conduct maintenance and repairs in a safe manner may result in serious personal injury or death. Maintenance and testing of the vehicle's fuel system components can be accomplished safely and effectively by adhering to the following rules and guidelines.

- To avoid the possibility of fire and personal injury, always disconnect the negative battery cable unless the repair or test procedure requires that battery voltage be applied.
- Always relieve the fuel system pressure prior to disconnecting any fuel system component (injector, fuel rail, pressure regulator, etc.), fitting or fuel line connection. Exercise extreme caution whenever relieving fuel system pressure, to avoid exposing skin, face and eyes to fuel spray. Please be advised that fuel under pressure may penetrate the skin or any part of the body that it contacts.
- Always place a shop towel or cloth around the fitting or connection prior to loosening to absorb any excess fuel due to spillage. Ensure that all fuel spillage (should it occur) is quickly removed from engine surfaces. Ensure that all fuel soaked cloths or towels are deposited into a suitable waste container.
- Always keep a dry chemical (Class B) fire extinguisher near the work area.
- Do not allow fuel spray or fuel vapors to come into contact with a spark or open flame.
- Always use a back-up wrench when loosening and tightening fuel line connection fittings. This will prevent unnecessary stress and torsion to fuel line piping. Always follow the proper torque specifications.

• Always replace worn fuel fitting O-rings with new. Do not substitute fuel hose or equivalent, where fuel pipe is installed.

Fuel System Pressure

RELIEVING

1. Before servicing the vehicle, refer to the precautions in the beginning of this section.
2. Disconnect the negative battery cable.
3. Remove the fuel tank fill cap to relieve the pressure in the fuel tank.
4. Remove the cap from the Schrader valve located on the fuel supply manifold.
5. Attach Fuel Pressure Gauge T80L-9974-A to the valve and drain the fuel through the drain tube into a suitable container.
6. After the fuel system pressure is relieved, remove the fuel pressure gauge and install the cap on the Schrader valve.
7. Install the fuel tank fill cap.
8. Connect the negative battery cable only after system repairs are completed.

Fuel Filter

REMOVAL & INSTALLATION

➡Disable the air suspension before raising the vehicle. The switch is located on the left side of the luggage compartment.

1. Before servicing the vehicle, refer to the precautions in the beginning of this section.
2. Turn the air suspension service switch **OFF**.
3. Disconnect the negative battery cable.
4. Relieve the fuel system pressure.
5. Disconnect the fuel lines.

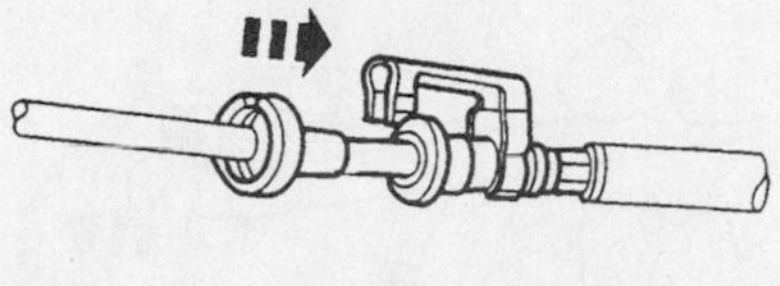

Slide the special tool into the fitting to disengage it

6. Loosen the filter retaining clamp and remove the fuel filter.

To install:

7. Install the fuel filter with the flow arrow facing the proper direction and tighten the filter retaining clamp.
8. Push the fuel lines on to the filter fittings until an audible click is heard.
9. Connect the negative battery cable.
10. Start the engine and check for fuel leaks and proper operation.
11. Turn the air suspension switch to the **ON** position.

Fuel Pump

REMOVAL & INSTALLATION

➡Disable the air suspension before raising the vehicle. The switch is located on the left side of the luggage compartment.

1. Before servicing the vehicle, refer to the precautions in the beginning of this section.
2. Turn the air suspension switch to the **OFF** position.
3. Relieve the fuel system pressure.
4. Drain the fuel tank.
5. Remove or disconnect the following:
   - Fuel fill and vent hoses
   - Fuel supply line
   - Fuel pump module wiring connector
   - Fuel tank
   - Fuel pump module locking ring
6. Pull the fuel pump module up and out of the fuel tank until the locking tabs for the fuel pump module are accessible.

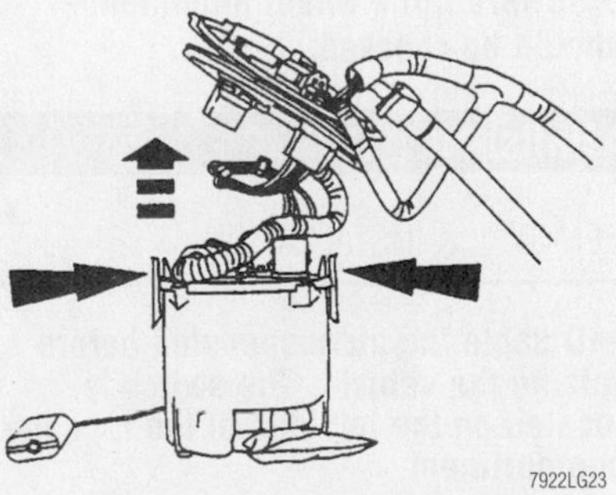

To remove the fuel pump, reach into the tank and press the locking tabs inward

Squeeze both locking tabs together and remove the fuel pump module from the fuel tank.

To install:

7. Install or connect the following:
   - Fuel pump module. Use a new O-ring seal and push the module into the tank so that the locking tabs engage.
   - Fuel pump module lock ring
   - Fuel tank. Tighten the tank strap bolts to 22–30 ft. lbs. (30–40 Nm).
   - Fuel pump module wiring connector
   - Fuel supply line
   - Fuel fill and vent hoses
8. Add 10 gallons of clean fuel to the tank.
9. Turn the air suspension switch to the **ON** position.
10. Start the engine and check for leaks.

Fuel Injector

REMOVAL & INSTALLATION

1. Before servicing the vehicle, refer to the precautions in the beginning of this section.
2. Relieve the fuel system pressure.
3. Remove or disconnect the following:
   - Negative battery cable
   - Fuel pressure sensor vacuum line and wiring connector
   - Fuel injector wiring connectors
   - Fuel supply line
   - Fuel supply manifold with injectors attached
4. Remove the clips and the injectors from the fuel supply manifold.

To install:

5. Install the injectors with new O-ring seals.
6. Install the fuel supply manifold and tighten the bolts as follows:
   a. Step 1: 15 ft. lbs. (20 Nm)
   b. Step 2: Plus 85–95 degrees
7. Install or connect the following:
   - Fuel supply line
   - Fuel injector wiring connectors
   - Fuel pressure sensor vacuum line and wiring connector
   - Negative battery cable
8. Start the engine and check for leaks.

DRIVE TRAIN

Transaxle Assembly

REMOVAL & INSTALLATION

➡Disable the air suspension before raising the vehicle. The switch is located on the left side of the luggage compartment.

1. Before servicing the vehicle, refer to the precautions in the beginning of this section.
2. Turn the air suspension switch to the **OFF** position.
3. Attach a powertrain support fixture to the engine lifting eyes.
4. Drain the transaxle fluid.
5. Remove or disconnect the following:
   - Battery and battery tray
   - Intake Air Temperature (IAT) sensor connector
   - Mass Air Flow (MAF) sensor connector
   - Air cleaner assembly
   - Transaxle control harness connector
   - Transaxle range sensor
   - Turbine shaft speed sensor connector
   - Shift cable
   - Transaxle oil cooler lines
   - Front wheels
   - Suspension height sensors
   - Wheel speed sensors
   - Outer tie rod ends
   - Lower ball joints
   - Stabilizer bar links
   - Halfshafts
   - Heated Oxygen (HO_2S) sensor connectors
   - Dual converter Y-pipe
   - Starter
   - Left, right and rear engine support insulators
   - Subframe
   - Transaxle housing cover
   - Torque converter
   - Transaxle flange bolts
6. Separate the transaxle from the engine and lower it from the vehicle.

To install:

➡When installing suspension components and halfshafts, use new nuts, bolts, circlips and split pins.

7. Install or connect the following:
   - Transaxle. Tighten the flange bolts to 25–34 ft. lbs. (34–46 Nm).
   - Torque converter. Tighten the nuts to 20–34 ft. lbs. (27–46 Nm).
   - Transaxle housing cover
8. Install the subframe. Use 2 pieces of ¾ inch outside diameter pipe in the alignment holes behind the front subframe mounts to align the subframe to the body. Tighten the subframe mounting bolts to 57–76 ft. lbs. (77–103 Nm).
9. Install or connect the following:
   - Left, right and rear engine support insulators. Tighten the through-bolts to 64–88 ft. lbs. (87–119 Nm).
   - Starter. Tighten the bolt and stud to 15–21 ft. lbs. (21–29 Nm).
   - Dual converter Y-pipe
   - HO_2S sensor connectors
   - Halfshafts
   - Stabilizer bar links. Tighten the nuts to 30–40 ft. lbs. (40–55 Nm).
   - Lower ball joints. Tighten the pinch bolts to 50–68 ft. lbs. (68–92 Nm).
   - Outer tie rod ends. Tighten the nuts to 35–46 ft. lbs. (47–63 Nm).
   - Wheel speed sensors
   - Suspension height sensors
   - Front wheels
   - Transaxle oil cooler lines
   - Shift cable
   - Turbine shaft speed sensor connector
   - Transaxle range sensor
   - Transaxle control harness connector
   - Air cleaner assembly
   - MAF sensor connector
   - IAT sensor connector
   - Battery and battery tray
10. Fill the transaxle.
11. Check for leaks and proper operation.

➡Whenever the subframe is removed or lowered, the wheel alignment should be checked.

Halfshaft

REMOVAL & INSTALLATION

➡Disable the air suspension before raising the vehicle. The switch is located on the left side of the luggage compartment.

1. Before servicing the vehicle, refer to the precautions in the beginning of this section.
2. Turn the air suspension switch to the **OFF** position.
3. Remove or disconnect the following:
   - Negative battery cable
   - Front wheels
   - Suspension height sensors
   - Wheel speed sensors
   - Outer tie rod ends
   - Lower ball joints
   - Stabilizer bar links
4. Separate the halfshafts from the wheel hubs and the transaxle and remove the halfshafts.

To install:

➡When installing suspension components and halfshafts, use new nuts, bolts, circlips and split pins.

5. Install or connect the following:
   - Halfshafts. Tighten the hub retainer nuts to 180–200 ft. lbs. (245–270 Nm).
   - Stabilizer bar links. Tighten the nuts to 30–40 ft. lbs. (40–55 Nm).
   - Lower ball joints. Tighten the pinch bolts to 50–68 ft. lbs. (68–92 Nm).
   - Outer tie rod ends. Tighten the nuts to 35–46 ft. lbs. (47–63 Nm).
   - Wheel speed sensors
   - Suspension height sensors
   - Front wheels
6. Turn the air suspension switch to the **ON** position.
7. Road test the vehicle and check for proper operation.

CV-Joint

REMOVAL AND REPLACEMENT

Inner Tripod Joint

The inner CV-joint is serviced with the halfshaft as an assembly. The inner CV-joint boot can be serviced by removing the outer CV-joint.

Outer CV-Joint

1. Before servicing the vehicle, refer to the precautions in the beginning of this section.

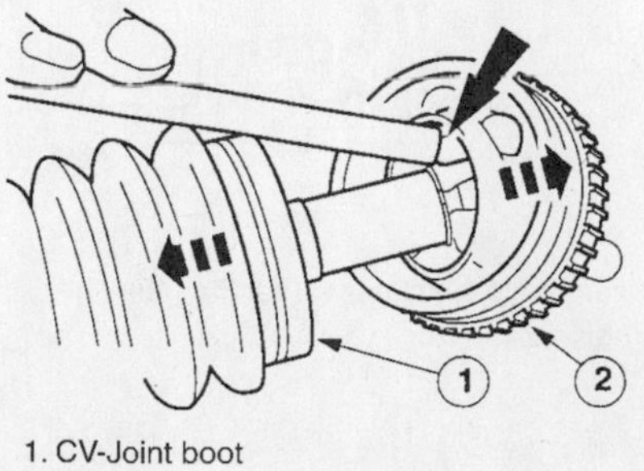

1. CV-Joint boot
2. CV-Joint

9306HG07

Removing the outer CV-joint

2. Place the halfshaft in a vise.
3. Remove the CV-joint boot clamps and slide the boot away from the joint.
4. Drive the CV-joint off the halfshaft with a brass drift and a hammer.

To install:

5. Replace the snapring.
6. Fill the CV-joint with fresh grease and slide the joint onto the halfshaft.
7. Use new clamps and install the CV-joint boot.

STEERING AND SUSPENSION

Air Bag

CAUTION

Some vehicles are equipped with an air bag system. The system must be disarmed before performing service on, or around, system components, the steering column, instrument panel components, wiring and sensors. Failure to follow the safety precautions and the disarming procedure could result in accidental air bag deployment, possible injury and unnecessary system repairs.

PRECAUTIONS

Several precautions must be observed when handling the inflator module to avoid accidental deployment and possible personal injury.

- Never carry the inflator module by the wires or connector on the underside of the module.
- When carrying a live inflator module, hold securely with both hands, and ensure that the bag and trim cover are pointed away.
- Place the inflator module on a bench or other surface with the bag and trim cover facing up.
- With the inflator module on the bench, never place anything on or close to the module that may be thrown in the event of an accidental deployment.

DISARMING

1. Before servicing the vehicle, refer to the precautions in the beginning of this section.
2. Disconnect both battery cables from the battery, negative cable first.
3. Wait 1 minute before proceeding with the service procedure. This is the time required for the back-up power supply in the air bag diagnostic monitor to deplete its stored energy.
4. After service is completed, reconnect the battery cables, negative cable last.
5. Turn the ignition switch to the **RUN** position. The air bag indicator should light continuously for approximately 6 seconds, then turn **OFF**. If the indicator fails to light, flashes or remains lit continuously, there is a fault in the air bag system.

Power Rack and Pinion Steering Gear

REMOVAL & INSTALLATION

CAUTION

Do not rotate the steering wheel when the intermediate shaft is disconnected. Damage to the air bag sliding contact will result.

➡Disable the air suspension before raising the vehicle. The switch is located on the left side of the luggage compartment.

1. Before servicing the vehicle, refer to the precautions in the beginning of this section.
2. Turn the air suspension switch to the **OFF** position.
3. Remove or disconnect the following:
   - Negative battery cable
   - Steering column intermediate shaft
   - Powertrain Control Module (PCM)

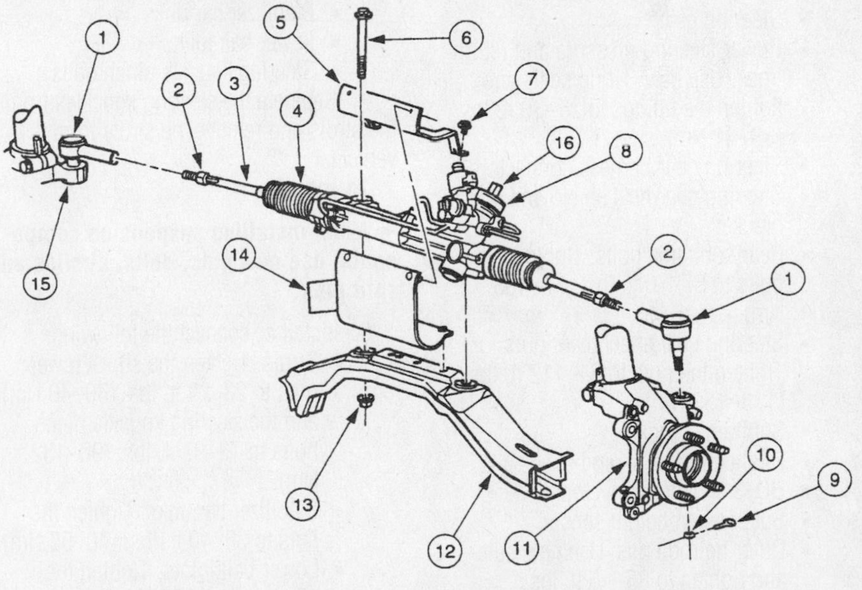

1 Tie Rod End
2 Nut
3 Front Wheel Spindle Tie Rod (2 Req'd)
4 Front Suspension Steering Ball Stud Dust Seal
5 Power Steering Hose Bracket
6 Bolt (2 Req'd)
7 Screw (2 Req'd)
8 Steering Gear
9 Screw (2 Req'd)
10 Nut (2 Req'd)
11 Front Wheel Knuckle (LH)
12 Crossmember
13 Nut (2 Req'd)
14 Steering Shaft U-Joint Shield
15 Front Wheel Knuckle (RH)
16 Power Steering Gear Input Shaft and Control

7922LG26

Exploded view of the power rack and pinion steering gear mounting

- harness connector and ground straps
- Upper motor mount
- Front wheels
- Outer tie rod ends
- Suspension height sensor
- Heated Oxygen (HO_2S) sensor connectors
- Dual converter Y-pipe
- Subframe brackets
- Steering gear attachment nuts
- Rear subframe bolts. Lower the rear of the subframe about 4 inches for access.
- Steering gear heat shield and bracket
- Steering gear harness connectors

4. Rotate the steering gear to clear the mounting bolts from the subframe, and pull the gear to the left to access the fluid lines.
   - Power steering pressure and return lines
   - Steering gear through the left fender opening

To install:

5. Install or connect the following:
   - Steering gear
   - Power steering pressure and return lines. Use new O-ring seals and tighten the fittings to 25–30 ft. lbs. (33–41 Nm).
   - Steering gear harness connectors
   - Steering gear heat shield and bracket
   - Rear subframe bolts. Tighten the bolts to 57–76 ft. lbs. (77–103 Nm).
   - Steering gear attachment nuts. Tighten the nuts to 84–112 ft. lbs. (113–153 Nm).
   - Subframe brackets
   - Dual converter Y-pipe
   - HO_2S sensor connectors
   - Suspension height sensor
   - Outer tie rod ends. Use new nuts and tighten to 35–46 ft. lbs. (47–63 Nm).
   - Front wheels
   - Upper motor mount
   - PCM harness connector and ground straps
   - Steering column intermediate shaft. Tighten the lower bolt to 30–38 ft. lbs. (41–51 Nm), and the upper bolt to 24–30 ft. lbs. (33–41 Nm).
   - Negative battery cable
6. Fill the power steering system.
7. Turn the air suspension switch to the **ON** position.
8. Check the system for leaks and proper operation. Adjust the toe setting as necessary.

Strut

REMOVAL & INSTALLATION

Front

➡Disable the air suspension before raising the vehicle. The switch is located on the left side of the luggage compartment.

1. Before servicing the vehicle, refer to the precautions in the beginning of this section.
2. Turn the air suspension switch to the **OFF** position.
3. Remove or disconnect the following:
   - Negative battery cable
   - Front wheels
   - Left and right splash shields
   - Wheel speed sensors
   - Strut actuator connectors, if equipped with semi-active suspension
   - Suspension height sensors
   - Brake line brackets
   - Stabilizer bar links
   - Lower ball joints
   - Steering knuckle pinch bolts
4. Separate the steering knuckles from the struts and remove the struts from the vehicle.

To install:

➡When installing suspension components, use new nuts, bolts, circlips and split pins.

5. Install or connect the following:
   - Struts. Tighten the shock tower nuts to 23–29 ft. lbs. (30–40 Nm), and the steering knuckle pinch bolts to 73–97 ft. lbs. (98–132 Nm).
   - Stabilizer bar links. Tighten the nuts to 30–40 ft. lbs. (40–55 Nm).
   - Lower ball joints. Tighten the pinch bolts to 50–68 ft. lbs. (68–92 Nm).
   - Brake line brackets. Tighten the screws to 11 ft. lbs. (15 Nm).
   - Suspension height sensors
   - Strut actuator connectors, if equipped with semi-active suspension
   - Wheel speed sensors
   - Left and right splash shields
   - Front wheels
   - Negative battery cable
6. Turn the air suspension switch to the **ON** position.
7. Road test the vehicle and check for proper operation.

Shock Absorber

REMOVAL & INSTALLATION

Rear

➡Disable the air suspension before raising the vehicle. The switch is located on the left side of the luggage compartment.

1. Before servicing the vehicle, refer to the precautions in the beginning of this section.
2. Turn the air suspension switch to the **OFF** position.
3. Remove or disconnect the following:
   - Negative battery cable
   - Luggage compartment side trim panels
   - Rear wheels
   - Shock absorber actuator harness connectors, if equipped with semi-active suspension
   - Shock absorber assemblies

To install:

4. If replacing the shock absorbers, transfer the mass dampers and mounting brackets.
5. Turn the shock absorber upside down. Compress and expand the piston several times to remove the air from the working cylinder.
6. Install or connect the following:
   - Shock absorber assemblies. Tighten the upper bolts to 25–34 ft. lbs. (34–46 Nm), and the lower bolts to 58–68 ft. lbs. (68–92 Nm).
   - Shock absorber actuator harness connectors, if equipped with semi-active suspension
   - Rear wheels
   - Luggage compartment side trim panels
   - Negative battery cable
7. Turn the air suspension system **ON**.

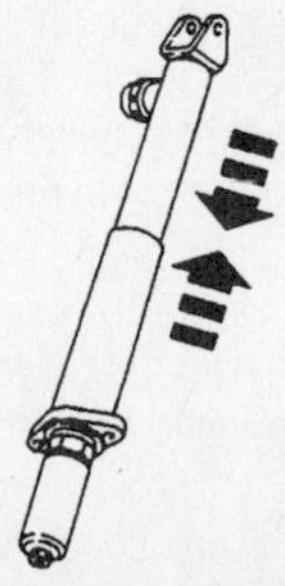

To bleed the air from the shock, turn it upside down, then compress and expand it several times

Coil Spring

REMOVAL & INSTALLATION

Front

1. Before servicing the vehicle, refer to the precautions in the beginning of this section.
2. Remove the strut assembly from the vehicle.
3. Install a spring compressor and retract the spring until the upper mount rotates freely.
4. Remove or disconnect the following:
   - Flange nut
   - Strut mounting bracket
   - Strut bearing plate
   - Coil spring

To install:

5. Install or connect the following:
   - Coil spring
   - Strut bearing plate
   - Strut mounting bracket
   - Flange nut. Tighten the nut to 63–70 ft. lbs. (85–95 Nm).
6. Remove the spring compressor and ensure that the spring seats correctly in the insulators.
7. Install the strut assembly to the vehicle.

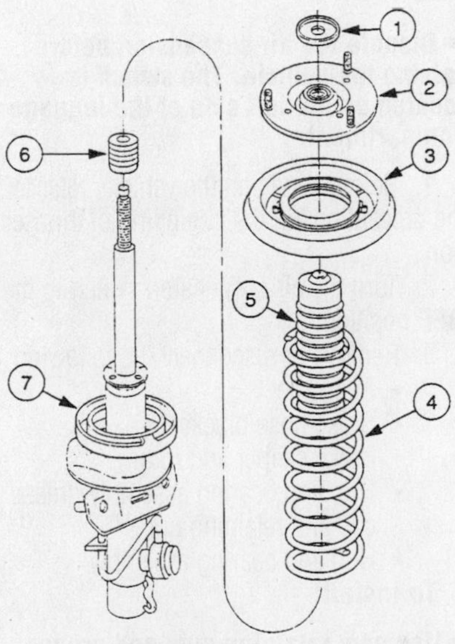

Exploded view of the front strut and coil spring assembly

Air Spring

REMOVAL & INSTALLATION

Rear

⁂ CAUTION

Do not attempt to service the rear suspension without first deflating the air springs.

➡Servicing the air suspension requires a scan tool with bi-directional capabilities to deflate the air springs.

1. Properly deflate the air springs.
2. Turn the air suspension switch to the **OFF** position. The switch located on the right kick panel.
3. Remove the rear wheels.
4. Disconnect the air spring solenoid wiring connector and air line.
5. Depress the locking tabs and remove the air spring from the vehicle.

To install:

6. Install or connect the following:
   - Air spring. Press the plastic tab through the suspension until it clicks into place.
   - Solenoid air line and electrical connectors
   - Rear wheels
7. Turn the air suspension switch to the **ON** position and check for proper operation.

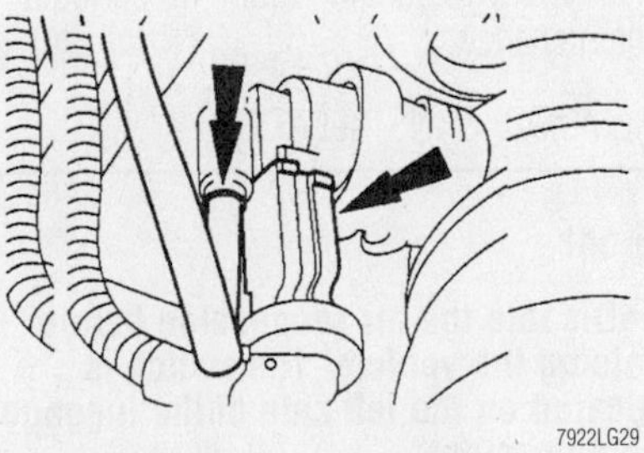

Air spring solenoid connectors

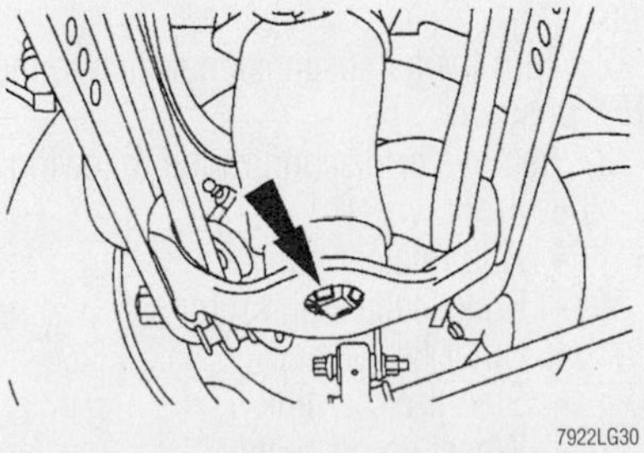

Depress the plastic locking tabs to disengage the air spring from the suspension

Lower Ball Joints

REMOVAL & INSTALLATION

The lower ball joint is an integral part of the steering knuckle. If the lower ball joint is found to be defective, the entire steering knuckle must be replaced.

Upper Control Arm

REMOVAL & INSTALLATION

Rear

⁂ CAUTION

Do not attempt to service the rear suspension without first deflating the air springs.

➡Servicing the air suspension requires a scan tool with bi-directional capabilities to deflate the air springs.

1. Properly deflate the air springs.
2. Turn the air suspension switch to the **OFF** position. The switch located on the right kick panel.
3. Support the lower control arm on a jackstand.
4. Remove or disconnect the following:
   - Rear wheel
   - Brake hose bracket
   - Upper ball joint
   - Upper control arm

To install:

➡Use new mounting nuts and bolts.

5. Install the upper control arm. Tighten the ball joint nut to 50–67 ft. lbs. (68–92 Nm), then tighten the control arm mounting bolts to 73–97 ft. lbs. (98–132 Nm).
6. Install or connect the following:
   - Brake hose bracket
   - Rear wheel
7. Turn the air suspension switch to the **ON** position and check for proper operation.
8. Check the wheel alignment and adjust as necessary.

CONTROL ARM BUSHING REPLACEMENT

The control arm bushings are serviced with the control arm as an assembly.

Lower Control Arm

REMOVAL & INSTALLATION

Front

➡**Disable the air suspension before raising the vehicle. The switch is located on the left side of the luggage compartment.**

1. Before servicing the vehicle, refer to the precautions in the beginning of this section.
2. Turn the air suspension switch to the **OFF** position.
3. Remove or disconnect the following:
   - Front wheel
   - Wheel speed sensor wiring harness
   - Height sensor
   - Lower ball joint
   - Lower control arm strut
   - Lower control arm

To install:

4. Install or connect the following:
   - Lower control arm. Tighten the pivot bolt to 73–98 ft. lbs. (98–132 Nm).
   - Lower control arm strut. Tighten the nut to 73–98 ft. lbs. (98–132 Nm).
   - Lower ball joint. Use a new nut and tighten to 50–67 ft. lbs. (68–92 Nm).
   - Wheel speed sensor wiring harness
   - Height sensor
   - Front wheel
5. Turn the air suspension switch to the **ON** position.
6. Check the wheel alignment and adjust as necessary.

Rear

⁂ CAUTION

Do not attempt to service the rear suspension without first deflating the air springs.

➡**Disable the air suspension before raising the vehicle. The switch is located on the left side of the luggage compartment.**

1. Before servicing the vehicle, refer to the precautions in the beginning of this section.
2. Turn the air suspension switch to the **OFF** position.
3. Remove or disconnect the following:
   - Rear wheel
   - Shock absorber actuator wiring, if equipped with semi-active suspension
   - Air spring
   - Stabilizer bar link
   - Lower control arm

To install:

4. Install or connect the following:
   - Lower control arm. Tighten the bolts to 50–68 ft. lbs. (68–92 Nm).
   - Stabilizer bar link. Tighten the nut to 25–34 ft. lbs. (34–46 Nm).
   - Air spring
   - Shock absorber actuator wiring, if equipped with semi-active suspension
   - Rear wheel
5. Turn the air suspension switch to the **ON** position.

CONTROL ARM BUSHING REPLACEMENT

The control arm bushings are serviced with the control arm as an assembly.

Wheel Bearings

ADJUSTMENT

There is no adjustment for the front or rear wheel bearings due to the nature of their design. These bearings are permanently lubricated and require no periodic maintenance.

REMOVAL & INSTALLATION

Front

➡**Disable the air suspension before raising the vehicle. The switch is located on the left side of the luggage compartment.**

1. Before servicing the vehicle, refer to the precautions in the beginning of this section.
2. Turn the air suspension switch to the **OFF** position.
3. Remove or disconnect the following:
   - Front wheel
   - Hub retainer nut
   - Brake caliper and rotor
   - Outer tie rod end
   - Stabilizer bar link
   - Wheel speed sensor
   - Lower ball joint
   - Steering knuckle
4. Unbolt and remove the hub and bearing assembly.

To install:

➡**Use new nuts, bolts and split pins.**

➡**The knuckle must be clean enough to allow the wheel hub to be completely seated by hand. Do not press or draw the wheel hub into place.**

5. Install or connect the following:
   - Hub and bearing assembly. Tighten the bolts to 61–78 ft. lbs. (83–107 Nm).
   - Steering knuckle. Tighten the pinch bolt to 72–97 ft. lbs. (98–132 Nm).
   - Lower ball joint. Tighten the nut to 50–67 ft. lbs. (68–92 Nm).
   - Wheel speed sensor
   - Stabilizer bar link. Tighten the nut to 57–75 ft. lbs. (77–103 Nm).
   - Outer tie rod end. Tighten the nut to 35–46 ft. lbs. (47–63 Nm).
   - Brake caliper and rotor. Tighten the caliper anchor bracket bolts to 65–87 ft. lbs. (88–118 Nm).
   - Hub retainer nut. Tighten the nut to 170–202 ft. lbs. (230–275 Nm).
   - Front wheel
6. Turn the air suspension switch to the **ON** position.

Rear

➡**Disable the air suspension before raising the vehicle. The switch is located on the left side of the luggage compartment.**

1. Before servicing the vehicle, refer to the precautions in the beginning of this section.
2. Turn the air suspension switch to the **OFF** position.
3. Remove or disconnect the following:
   - Rear wheel
   - Brake hose bracket
   - Brake caliper and rotor
   - Hub and bearing assembly grease cap and retaining nut
   - Hub and bearing assembly

To install:

➡**Use new retaining nuts and grease caps**

4. Install or connect the following:
   - Hub and bearing assembly. Tighten the retaining nut to 188–254 ft. lbs. (255–345 Nm).
   - Grease cap
   - Brake caliper and rotor
   - Brake hose bracket
   - Rear wheel
5. Turn the air suspension switch to the **ON** position.

FORD MOTOR CO.

14

1998–01

Ford-Escort • Escort ZX2 • **Mercury**-Tracer

PRECAUTIONS

Before servicing any vehicle, please be sure to read all of the following precautions, which deal with personal safety, prevention of component damage, and important points to take into consideration when servicing a motor vehicle:

- Never open, service or drain the radiator or cooling system when the engine is hot; serious burns can occur from the steam and hot coolant.
- Observe all applicable safety precautions when working around fuel. Whenever servicing the fuel system, always work in a well-ventilated area. Do not allow fuel spray or vapors to come in contact with a spark, open flame, or excessive heat (a hot drop light, for example). Keep a dry chemical fire extinguisher near the work area. Always keep fuel in a container specifically designed for fuel storage; also, always properly seal fuel containers to avoid the possibility of fire or explosion. Refer to the additional fuel system precautions later in this section.
- Fuel injection systems often remain pressurized, even after the engine has been turned **OFF**. The fuel system pressure must be relieved before disconnecting any fuel lines. Failure to do so may result in fire and/or personal injury.
- Brake fluid often contains polyglycol ethers and polyglycols. Avoid contact with the eyes and wash your hands thoroughly after handling brake fluid. If you do get brake fluid in your eyes, flush your eyes with clean, running water for 15 minutes. If eye irritation persists, or if you have taken brake fluid internally, IMMEDIATELY seek medical assistance.
- The EPA warns that prolonged contact with used engine oil may cause a number of skin disorders, including cancer! You should make every effort to minimize your exposure to used engine oil. Protective gloves should be worn when changing oil. Wash your hands and any other exposed skin areas as soon as possible after exposure to used engine oil. Soap and water, or waterless hand cleaner should be used.
- All new vehicles are now equipped with an air bag system, often referred to as a Supplemental Restraint System (SRS) or Supplemental Inflatable Restraint (SIR) system. The system must be disabled before performing service on or around system components, steering column, instrument panel components, wiring and sensors. Failure to follow safety and disabling procedures could result in accidental air bag deployment, possible personal injury and unnecessary system repairs.
- Always wear safety goggles when working with, or around, the air bag system. When carrying a non-deployed air bag, be sure the bag and trim cover are pointed away from your body. When placing a non-deployed air bag on a work surface, always face the bag and trim cover upward, away from the surface. This will reduce the motion of the module if it is accidentally deployed. Refer to the additional air bag system precautions later in this section.
- Clean, high quality brake fluid from a sealed container is essential to the safe and proper operation of the brake system. You should always buy the correct type of brake fluid for your vehicle. If the brake fluid becomes contaminated, completely flush the system with new fluid. Never reuse any brake fluid. Any brake fluid that is removed from the system should be discarded. Also, do not allow any brake fluid to come in contact with a painted surface; it will damage the paint.
- Never operate the engine without the proper amount and type of engine oil; doing so WILL result in severe engine damage.
- Timing belt maintenance is extremely important! Many models utilize an interference-type, non-freewheeling engine. If the timing belt breaks, the valves in the cylinder head may strike the pistons, causing potentially serious (also time-consuming and expensive) engine damage. Refer to the maintenance interval charts in the front of this manual for the recommended replacement interval for the timing belt, and to the timing belt section for belt replacement and inspection.
- Disconnecting the negative battery cable on some vehicles may interfere with the functions of the on-board computer system(s) and may require the computer to undergo a relearning process once the negative battery cable is reconnected.
- When servicing drum brakes, only disassemble and assemble one side at a time, leaving the remaining side intact for reference.
- Only an MVAC-trained, EPA-certified automotive technician should service the air conditioning system or its components.

ENGINE REPAIR

➡Disconnecting the negative battery cable on some vehicles may interfere with the operation of the on-board computer system. The computer may undergo a relearning process once the negative battery cable is reconnected.

Alternator

REMOVAL

Sedan and Wagon

1. Before servicing the vehicle, refer to the precautions in the beginning of this section.
2. Remove or disconnect the following:
   - Negative battery cable
   - Drive belt
   - Power steering hose bracket from the alternator bracket
   - Alternator mounting bolts
   - Alternator electrical connectors
   - Battery positive cable
   - Alternator from the vehicle

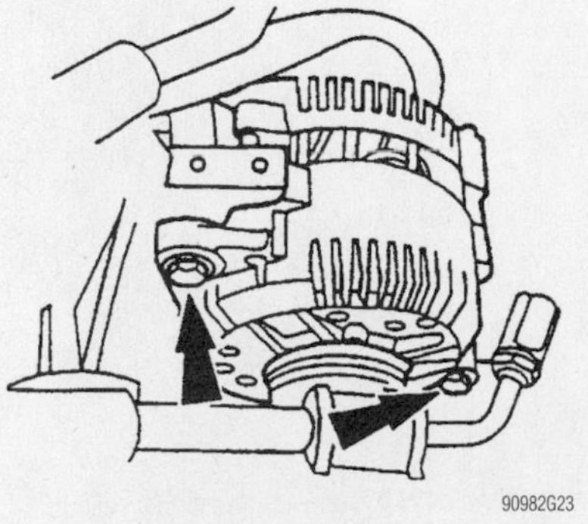

Location of the alternator upper and lower mounting bolts—sedan and wagon models

Coupe

1. Before servicing the vehicle, refer to the precautions in the beginning of this section.
2. Remove or disconnect the following:
   - Negative battery cable
   - Drive belt
   - Coolant tank reservoir
   - Alternator electrical connectors
   - Alternator mounting bolts
   - Battery positive cablc
   - Alternator from the vehicle

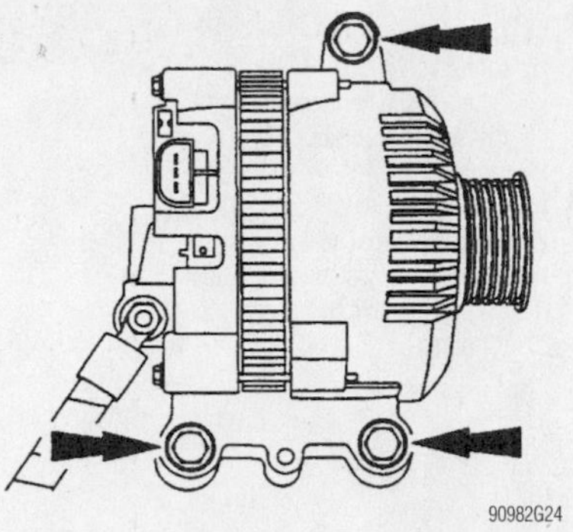

Location of the alternator upper and lower mounting bolts—coupe model

INSTALLATION

Sedan and Wagon

1. Install or connect the following:
   - Battery positive cable to the alternator. Tighten the nut to 7–9 ft. lbs. (9–12 Nm).
   - Alternator electrical connections
   - Alternator in position
   - Alternator mounting bolts. Tighten the upper mounting bolt to 29–40 ft. lbs. (40–55 Nm) and the lower mounting bolt to 15–22 ft. lbs. (20–30 Nm).
   - Power steering hose bracket to the alternator bracket
   - Drive belt
   - Negative battery cable

Coupe

1. Install or connect the following:
   - Battery positive cable to the alternator. Tighten the nut to 5–7 ft. lbs. (7–9 Nm).
   - Alternator in position
   - Alternator mounting bolts and tighten to 29–40 ft. lbs. (40–55 Nm)
   - Alternator electrical connections
   - Coolant tank reservoir
   - Drive belt
   - Negative battery cable

Ignition Timing

ADJUSTMENT

The Powertrain Control Module (PCM) controls ignition timing on these models. No adjustment is necessary or possible.

Engine Assembly

REMOVAL & INSTALLATION

1. Before servicing the vehicle, refer to the precautions in the beginning of this section.
2. Drain the cooling system.
3. Relieve the fuel system pressure.
4. Drain the engine oil.
5. Recover the A/C refrigerant, if equipped.
6. Remove or disconnect the following:
   - Battery and tray
   - Hood
   - Air cleaner outlet tube
   - Fuel pressure regulator vacuum hose
   - Intake manifold vacuum hoses
   - Exhaust Gas Recirculation (EGR) vacuum line
   - Positive Crankcase Ventilation (PCV) valve vacuum line
   - Vacuum tree
   - Throttle cable
   - Cruise control cable, if equipped
   - Shift cable
   - Constant Control Relay Module (CCRM)
   - EGR backpressure transducer
   - Transaxle range sensor connector, if equipped
   - Transaxle solenoid connector, if equipped
   - Turbine speed sensor connector, if equipped
   - Vehicle Speed (VSS) sensor
   - Heated Oxygen (HO_2S) sensor connectors
   - Ground strap
   - Engine cooling fan connector
   - Power steering hose brackets
   - Fuel line
   - Exhaust manifold heat shield
   - Starter bracket
   - Clutch cylinder fluid line, if equipped
   - Transaxle cooler lines, if equipped
   - A/C line bracket, if equipped
   - A/C condenser tube, if equipped
   - A/C accumulator tube, if equipped
   - Accessory drive belt and tensioner
   - Accessory drive belt idler pulley, if equipped
   - Power steering hoses
   - Front wheels
   - Left and right splash shields
   - Front subframe crossmember
   - Catalytic converter
   - Axle halfshafts
   - Starter motor
   - Oil pressure switch connector
   - A/C compressor, if equipped
   - Radiator
   - Heater hoses
   - Alternator harness connectors
7. Attach a support fixture to the engine.
8. Remove or disconnect the following:
   - Left front engine mount
   - Transaxle mount
   - Transaxle support crossmember
   - Upper transaxle mount
   - Right engine mount
9. Attach a hoist to the engine and remove the support fixture.
10. Raise the engine and transaxle from the vehicle.

To install:

11. Installation is the reverse of removal, but please note the following important steps.
12. On 2.0L SOHC engines pay special

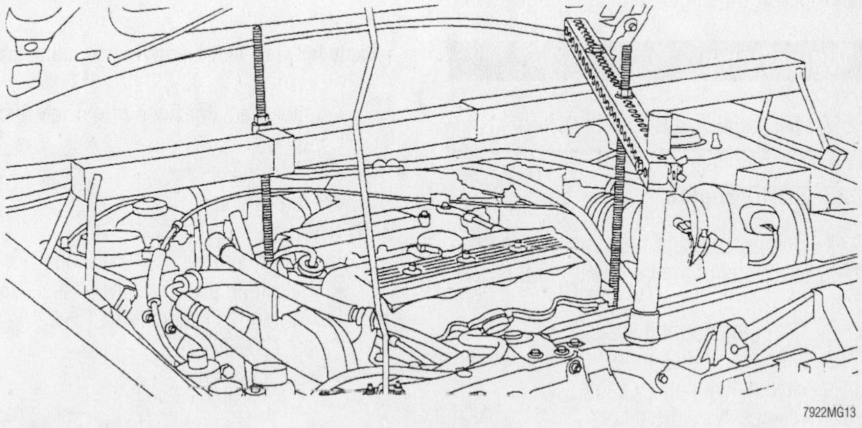

Properly support the engine as shown—2.0L SOHC engines

attention to the following torque specifications:

- Right-hand insulator through-bolt: 50–69 ft. lbs. (67–93 Nm)
- Engine mount bolts: 50–69 ft. lbs. (67–93 Nm)
- Transaxle support crossmember inner retainers: 28–38 ft. lbs. (38–51 Nm)
- Transaxle support crossmember outer retainers: 48–65 ft. lbs. (64–89 Nm)
- Air conditioning compressor mounting bolts: 15–22 ft. lbs.)
- Muffler-to-catalytic converter: 30–40 ft. lbs. (40–55 Nm)
- Crossmember bolts: 69–97 ft. lbs. (94–131 Nm)
- Power steering pressure hose connection: 15–18 ft. lbs. (20–25 Nm)
- Air conditioning line bracket bolt: 15–18 ft. lbs. (20–25 Nm)
- Exhaust manifold-to-catalytic converter nuts: 26–34 ft. lbs. (34–47 Nm)

13. On 2.0L DOHC engines pay special attention to the following torque specifications:

- Front isolator nuts and bolts: 50–68 ft. lbs. (67–93 Nm)
- Front engine support isolator through-bolt: 50–68 ft. lbs. (67–93 Nm)
- Front roll restrictor bolts: 48–65 ft. lbs. (64–89 Nm) and the nuts to 50–69 ft. lbs. (67–93 Nm)
- Rear roll restrictor nuts: 28–38 ft. lbs. (38–51 Nm)
- Torque converter nuts: 27 ft. lbs. (37 Nm)

14. Fill the crankcase to the correct level.
15. Fill the cooling system.
16. Recharge the A/C system, if equipped.
17. Start the engine and check for leaks.

1. Water pump housing gasket
2. Water pump
3. Timing belt tensioner
4. Front engine support mounting bracket
5. Oil pressure sensor
6. Knock sensor
7. Engine block
8. Oil pan
9. Oil filter
10. Crankshaft Position (CKP) sensor
11. Camshaft sprocket
12. Crankshaft pulley
13. Outer engine front cover
14. Crankshaft sprocket
15. Timing belt
16. Timing belt guide
17. Crankshaft front seal
18. Oil pump screen cover and tube
19. Oil pump inlet tube gasket
20. Oil pump

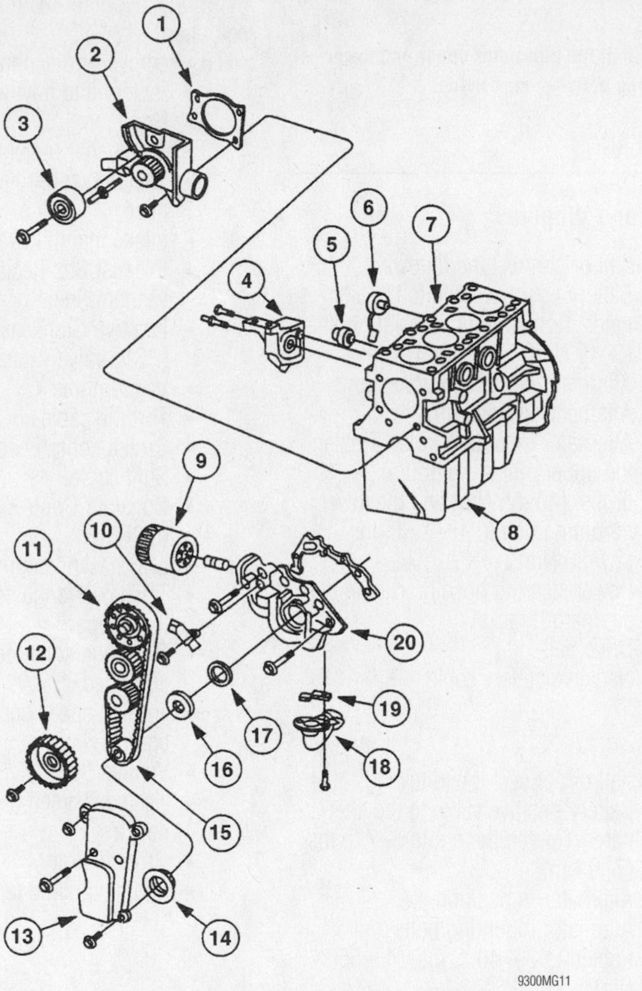

Exploded view of the water pump, oil pump and timing belt mounting—2.0L SOHC SPI engine

Water Pump

REMOVAL & INSTALLATION

2.0L SOHC Engine

1. Before servicing the vehicle, refer to the precautions in the beginning of this section.
2. Drain the cooling system.
3. Remove or disconnect the following:
   - Negative battery cable
   - Accessory drive belt
   - Front cover
   - Timing belt and tensioner. Refer to the Timing Belt unit repair section.
   - Lower radiator hose from the water pump
   - Heater hose
   - Water pump

To install:

4. Install or connect the following:
   - Water pump with a new gasket. Tighten the fasteners to 15–22 ft. lbs. (20–30 Nm).
   - Heater hose
   - Lower radiator hose
   - Timing belt tensioner. Tighten the bolt to 15–22 ft. lbs. (20–30 Nm)
   - Timing belt
   - Front cover

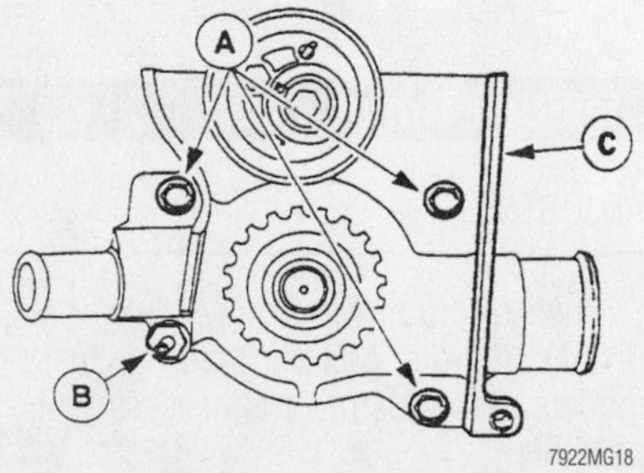

Location of the water pump (C) mounting bolts (A) and stud (B)—2.0L SOHC engine

- Accessory drive belt
- Negative battery cable
5. Fill the cooling system.
6. Start the engine and check for leaks.

2.0L DOHC Engine

1. Before servicing the vehicle, refer to the precautions in the beginning of this section.
2. Drain the cooling system.
3. Remove or disconnect the following:
   - Negative battery cable
   - Splash shield
   - Drive belt
   - Air conditioning compressor
   - Water pump pulley
   - Water pump

To install:

4. Install or connect the following:
   - Water pump. Tighten the bolts to 17 ft. lbs. (24 Nm).
   - Water pump pulley. Tighten the bolts to 17 ft. lbs. (24 Nm).
   - Air conditioning compressor
   - Drive belt

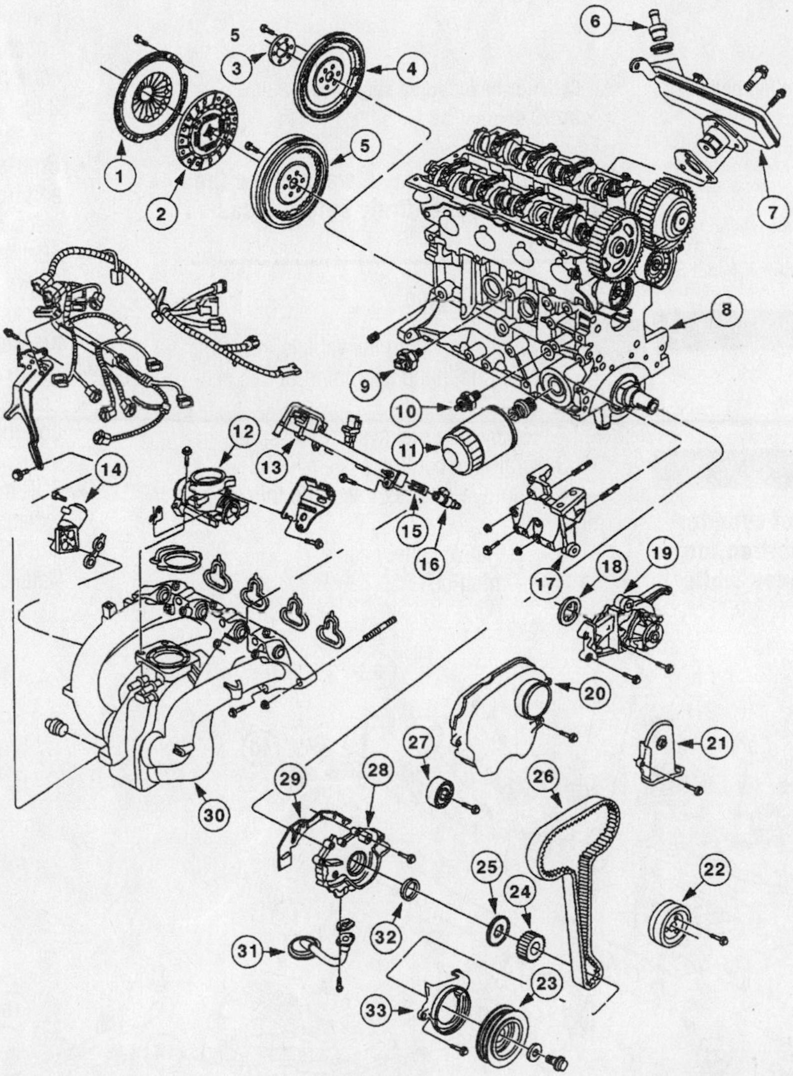

1. Clutch pressure plate
2. Clutch disc
3. Support
4. Flywheel (automatic trans.)
5. Flywheel (manual trans.)
6. PVC valve
7. Oil separator
8. Engine block
9. Knock sensor
10. Oil pressure sensor
11. Oil filter
12. Throttle body
13. Fuel injection supply manifold
14. Idle Air Control (IAC) valve
15. Retainer clip
16. Fuel injector
17. Bracket
18. Water pump housing gasket
19. Water pump
20. Upper timing belt cover
21. Center timing belt cover
22. Water pump pulley
23. Crankshaft pulley
24. Crankshaft sprocket
25. Timing belt guide
26. Timing belt
27. Timing belt idler pulley
28. Oil pump
29. Oil pump housing gasket
30. Intake manifold
31. Oil pump pick-up
32. Crankshaft front seal
33. Lower timing belt cover

9300MG10

Exploded view of peripheral engine component mounting, including water pump, oil pump and intake manifold—2.0L DOHC Zetec engine

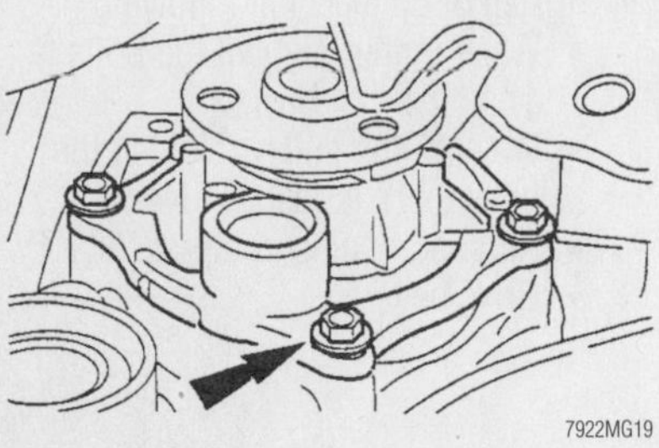

Remove the water pump mounting bolts—2.0L Zetec engine

- Splash shield
- Negative battery cable

5. Fill the cooling system.
6. Start the engine and check for leaks.

Cylinder Head

REMOVAL & INSTALLATION

⁂ WARNING

To reduce the possibility of cylinder head warpage and/or distortion, do not remove the cylinder head while the engine is warm. Always allow the engine to cool entirely before disassembly.

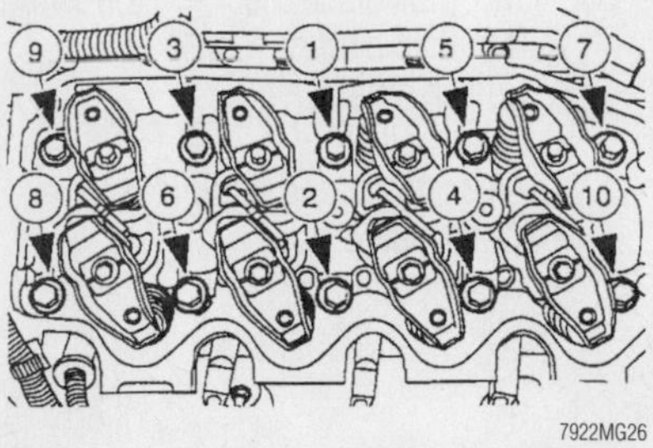

Cylinder head torque sequence—2.0L SOHC engine

2.0L DOHC Engine

1. Before servicing the vehicle, refer to the precautions in the beginning of this section.
2. Relieve the fuel system pressure.
3. Drain the cooling system.
4. Remove or disconnect the following:
   - Negative battery cable
   - Timing belt
   - Engine air cleaner intake tube
   - Exhaust Gas Recirculation (EGR) valve vacuum hose
   - Positive Crankcase Ventilation (PCV) valve vacuum hose
   - Throttle body vacuum hoses
   - Fuel pressure regulator vacuum hose
   - Intake manifold vacuum hoses
   - Accelerator cable, throttle control lever and if equipped, the speed control cable
   - Speed control cable bracket bolt, if equipped
   - 2 fuel charging wiring electrical connections
   - Crankshaft Position (CKP) sensor and Heated Oxygen (HO_2S) sensor electrical connections
   - Fuel supply line
   - Power steering pressure hose bracket bolts, then position the bracket and hose aside
   - Alternator
   - Engine oil dipstick tube bracket bolt that attaches the tube to the intake manifold
   - EGR manifold tube
   - Exhaust manifold heat shield nuts and the heat shield
   - Catalytic converter-to-exhaust man-

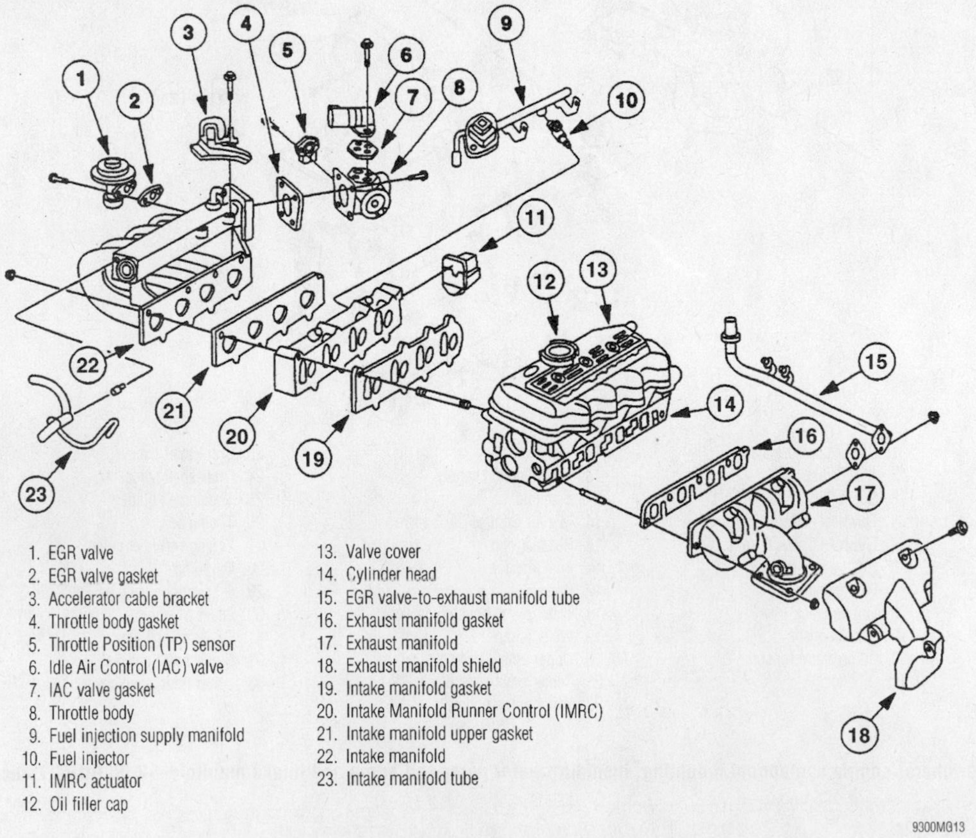

1. EGR valve
2. EGR valve gasket
3. Accelerator cable bracket
4. Throttle body gasket
5. Throttle Position (TP) sensor
6. Idle Air Control (IAC) valve
7. IAC valve gasket
8. Throttle body
9. Fuel injection supply manifold
10. Fuel injector
11. IMRC actuator
12. Oil filler cap
13. Valve cover
14. Cylinder head
15. EGR valve-to-exhaust manifold tube
16. Exhaust manifold gasket
17. Exhaust manifold
18. Exhaust manifold shield
19. Intake manifold gasket
20. Intake Manifold Runner Control (IMRC)
21. Intake manifold upper gasket
22. Intake manifold
23. Intake manifold tube

Exploded view of the cylinder head, exhaust manifold and intake manifold mounting—2.0L SOHC SPI engine

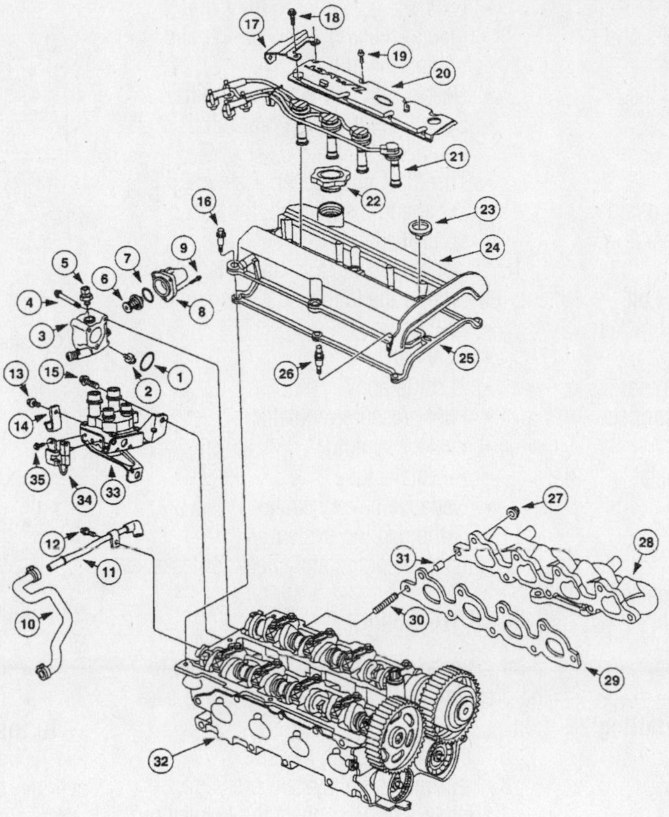

Item	Description
1	Water Thermostat Housing Oil Ring
2	Water Temperature Indicator Sender Unit
3	Water Thermostat Housing
4	Bolt (3 Req'd)
5	Engine Coolant Temperature Sensor
6	Water Thermostat
7	Water Thermostat Housing O-Ring
8	Water Outlet Connection
9	Bolt (3 Req'd)
10	Oil Separator Hose
11	Crankcase Ventilation Tube
12	Bolt
13	Bolt
14	Support Bracket
15	Bolt (3 Req'd)
16	Bolt (9 Req'd)

Item	Description
17	Bracket
18	Bolt (2 Req'd)
19	Bolt (6 Req'd)
20	Engine Appearance Cover
21	Ignition Wire (4 Req'd)
22	Oil Filler Cap
23	Grommet
24	Valve Cover
25	Valve Cover Gasket
26	Stud Bolt
27	Nut (9 Req'd)
28	Exhaust Manifold
29	Exhaust Manifold Gasket
30	Stud (9 Req'd)
31	Spacer
32	Cylinder Head
33	Ignition Coil Bracket
34	Radio Ignition Interference Capacitor
35	Bolt

9306MG77

Exploded view of the upper engine components—2.0L DOHC engine

Timing belt service is covered in Section 3 of this manual

ifold nuts and separate the components

- Right-hand splash shield bolts and the shield
- Engine oil dipstick tube from the cylinder block
- Air conditioning compressor electrical connection
- Air conditioning compressor retainers and set the compressor aside with the lines still attached
- Air conditioning line bracket bolt
- Front engine accessory drive bracket
- Valve cover
- Upper radiator hose from the thermostat housing water hose connection
- Heater hose from the thermostat housing
- Cylinder head bolts
- Cylinder head and gasket

5. Discard the old bolts and the gasket.

To install:

6. Install a new head gasket on the cylinder block.

➡Always use new bolts when installing the cylinder head.

7. Lubricate the new cylinder bolts when engine oil.
8. Install the cylinder head and tighten the bolts in the sequence illustrated in the following order:
   a. Step 1: 30–44 ft. lbs. (40–60 Nm)
   b. Step 2: Back all the bolts off ½ a turn
   c. Step 3: 30–44 ft. lbs. (40–60 Nm)
   d. Step 4: Plus 90 degrees
   e. Step 5: Plus 90 degrees
9. Install or connect the following:
   - Front engine accessory drive bracket. Tighten the fasteners to 30–40 ft. lbs. (40–55 Nm).
   - Air conditioning line bracket and tighten the bolt to 15–18 ft. lbs. (20–25 Nm)
   - Valve cover
   - Air conditioning compressor and tighten the retaining bolts
   - Right-hand splash shield and tighten its retainers
   - Engine oil dipstick tube
   - Heater and upper radiator hoses
   - Timing belt and alternator
   - Power steering pressure hose
   - CKP sensor and 2 main fuel charging electrical connections
   - Engine oil dipstick tube bolt
   - EGR valve manifold tube connection
   - EGR valve manifold tube to the exhaust manifold
   - Fuel supply line to the fuel rail
   - Catalytic converter to the exhaust manifold using a new gasket and tighten the nuts
   - Heat shield and tighten the nuts
   - Speed control cable bracket, if equipped
   - Throttle control lever, accelerator cable and if equipped, the speed control cable
10. Connect the vacuum hoses tagged and removed to the following components:
    - EGR valve
    - PCV valve
    - Throttle body
    - Fuel pressure regulator
    - Intake manifold
    - Air intake tube
    - Negative battery cable
11. Fill the cooling system.
12. Start the engine and check for leaks.

2.0L DOHC Engine

1. Before servicing the vehicle, refer to the precautions in the beginning of this section.
2. Drain the cooling system.
3. Relieve the fuel system pressure.
4. Remove or disconnect the following:
   - Negative battery cable
   - Air cleaner assembly outlet tube
   - Timing belt
5. Tag and disconnect the vacuum hoses from the following components:
   - Positive Crankcase Ventilation Valve (PCV) valve
   - Throttle body
   - Fuel pressure sensor
   - Intake manifold
6. Remove or disconnect the following:
   - Speed control and accelerator cables from the control lever
   - Fuel charging electrical connectors at the main engine connector
   - Crankshaft Position (CKP) sensor and Heated Oxygen (HO_2S) sensor electrical connections
   - Fuel line
   - Power steering pump and bracket, then set them aside with the lines still attached
   - Alternator
   - Engine oil dipstick tube
   - Splash shield bolts and the shield
   - Air conditioning compressor electrical connection
   - Air conditioning compressor retainers and set the compressor aside with the lines still attached
   - Spark plug wircs and the plugs
   - Valve cover

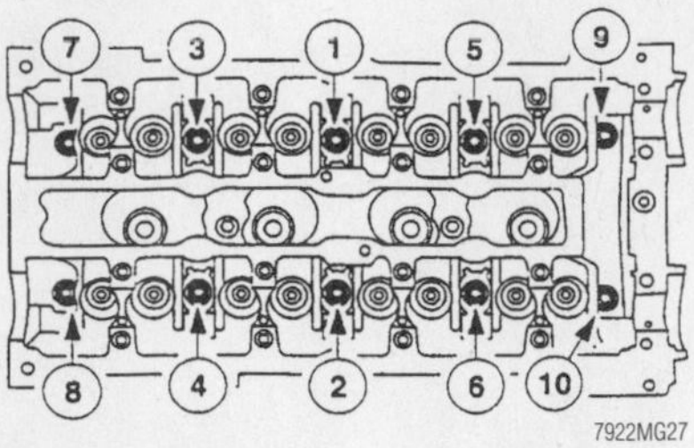

Cylinder head torque sequence—2.0L DOHC engine

   - Upper radiator hose from the thermostat housing water hose connection
   - Heater hose from the thermostat housing
   - Camshafts
   - Ignition coil
   - Thermostat housing
   - Cylinder head bolts
   - Cylinder head and the gasket

To install:

7. Install a new head gasket on the cylinder block.

➡Always use new bolts when installing the cylinder head.

8. Lubricate the new cylinder bolts when engine oil.
9. Install the cylinder head and tighten the bolts in the sequence illustrated in the following steps:
   a. Step 1: 12–18 ft. lbs. (15–25 Nm)
   b. Step 2: 26–33 ft. lbs. (35–45 Nm)
   c. Step 3: Plus 90 degrees
10. Install or connect the following:
    - Thermostat housing
    - Ignition coil and the camshafts
    - Heater and upper radiator hoses
    - Valve cover
    - Spark plugs
    - Spark plug wires
    - Air conditioning compressor and tighten the retaining bolts
    - Air conditioning compressor electrical connection
    - Splash shield and tighten its retainers
    - Engine oil dipstick tube
    - Alternator and the power steering reservoir and bracket
    - Fuel line
    - CKP sensor, HO_2sensor, and 2 main fuel charging electrical connections
    - Accelerator cable and if equipped,

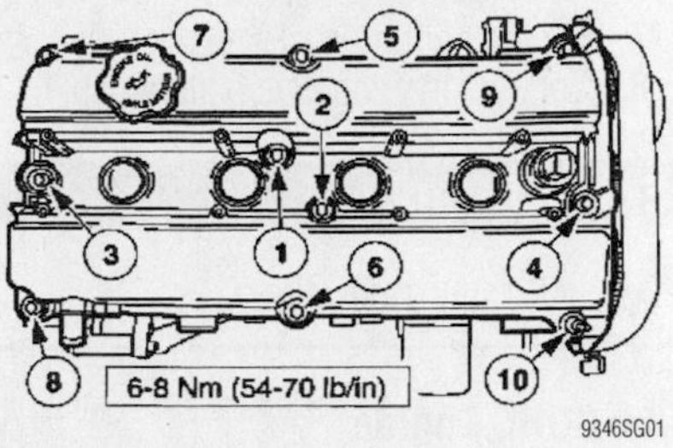

Valve cover torque sequence—DOHC Engine

the speed control cable to the control lever

11. Connect the vacuum hoses tagged and removed to the following components:
    - PCV valve
    - Throttle body
    - Fuel pressure sensor
    - Intake manifold
12. Install or connect the following:
    - Timing belt
    - Air cleaner outlet tube
    - Negative battery cable
13. Fill the cooling system.
14. Start the engine and check for leaks.

Rocker Arms

REMOVAL & INSTALLATION

2.0L SOHC Engine

1. Before servicing the vehicle, refer to the precautions in the beginning of this section.
2. Remove or disconnect the following:
    - Negative battery cable
    - Valve cover

➡Note the location of the valvetrain components for assembly.

- Rocker arm bolts
- Rocker arms and seats

To install:

3. Install or connect the following:
    - Seats and rocker arms in their original positions
    - Rocker arm bolts and tighten to 17–22 ft. lbs. (23–30 Nm)
    - Valve cover
    - Negative battery cable

2.0L DOHC Engine

This engine is not equipped with rocker arms. The camshafts act directly on the valves.

Intake Manifold

REMOVAL & INSTALLATION

2.0L SOHC Engine

1. Before servicing the vehicle, refer to the precautions in the beginning of this section.
2. Drain the cooling system.
3. Remove or disconnect the following:
    - Negative battery cable
    - Air cleaner outlet tube
4. Tag and disconnect the vacuum hoses from the following components:
    - Exhaust gas Recirculation (EGR) valve
    - Positive Crankcase Ventilation (PCV) valve
    - Throttle body
    - Fuel pressure regulator
    - Intake manifold
5. Remove or disconnect the following:
    - Accelerator cable, throttle control lever and if equipped, the speed control cable
    - Speed control cable bracket bolt, if equipped
    - Idle Air Control (IAC) valve and Throttle Position (TP) sensor electrical connections
    - EGR manifold tube located below the EGR valve
    - Engine oil dipstick tube
    - Intake manifold lower nuts
    - Intake manifold upper nuts
    - Manifold
    - Manifold gasket

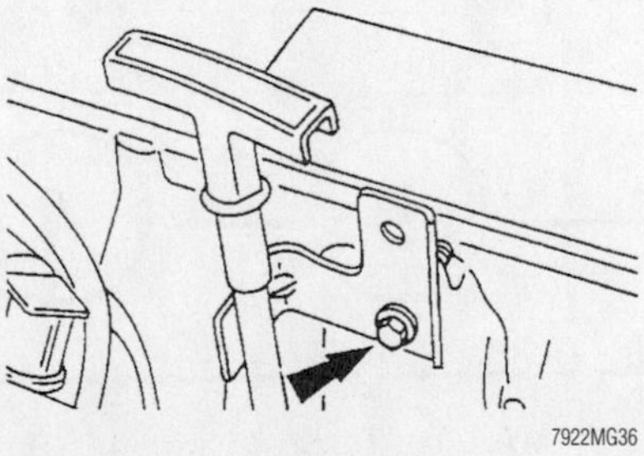

Remove the engine oil dipstick tube bracket bolt that attaches the tube to the intake manifold—2.0L SOHC engine

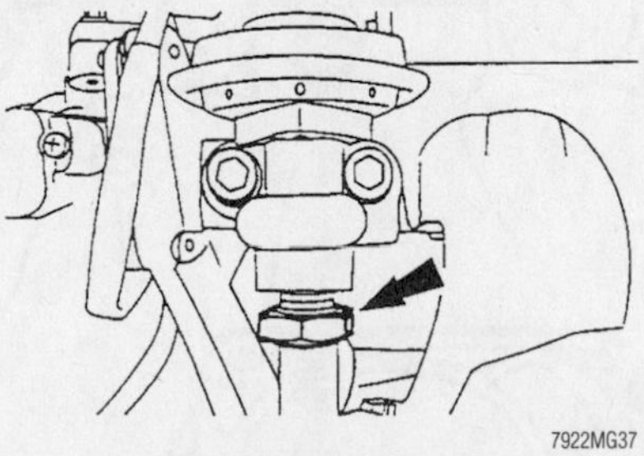

Disconnect the EGR manifold tube located below the EGR valve—2.0L SOHC engine

To install:

6. Clean and oil the intake manifold mounting studs.
7. Install or connect the following:
    - Intake manifold and hand-tighten the upper nuts
    - Lower manifold nuts and tighten them to 15–22 ft. lbs. (20–30 Nm)
    - Engine oil dipstick tube to the block
    - Manifold upper nuts and tighten to 15–22 ft. lbs. (20–30 Nm)
    - EGR manifold tube and tighten it to 15–20 ft. lbs. (20–28 Nm)
    - Engine oil dipstick tube bolt and tighten to 71–97 inch lbs. (8–11 Nm)
    - IAC valve and TP sensor electrical connections
    - Speed control cable bracket bolt, if equipped and tighten to 71–88 inch lbs. (8–10 Nm)
    - Throttle control lever, accelerator cable and if equipped, the speed control cable
8. Connect the vacuum hoses tagged and removed to the following components:
    - EGR valve
    - PCV valve
    - Throttle body
    - Fuel pressure regulator.
    - Intake manifold
9. Install the air cleaner outlet tube.
10. Fill the cooling system.
11. Connect the negative battery cable.
12. Start the engine and check for coolant leaks.

2.0L DOHC Engine

1. Before servicing the vehicle, refer to the precautions in the beginning of this section.
2. Drain the cooling system.
3. Relieve the fuel system pressure.
4. Remove or disconnect the following:
    - Negative battery cable
    - Air cleaner outlet tube

- Throttle Position (TP) sensor electrical connection
- Bolt that secures the pipe located by the crankshaft pulley
- Heater hoses from the core
- Main engine control sensor wiring
- Connectors from the mounting bracket
- Vacuum hoses from the intake manifold by squeezing the tabs, twisting the hoses and pulling them away from the manifold
- Crankcase ventilation hose from the valve cover
- Drive belt
- Alternator mounting bolts and move the alternator aside
- Fuel lines
- Intake manifold nuts and bolts in the sequence illustrated
- Intake manifold and gasket

To install:

5. Clean all dirt and gasket residue from the intake manifold mating surfaces.
6. Install or connect the following:
   - New gasket and place the manifold in position
   - Manifold bolts and nuts. Tighten in the sequence illustrated to 10–12 ft. lbs. (14–17 Nm).

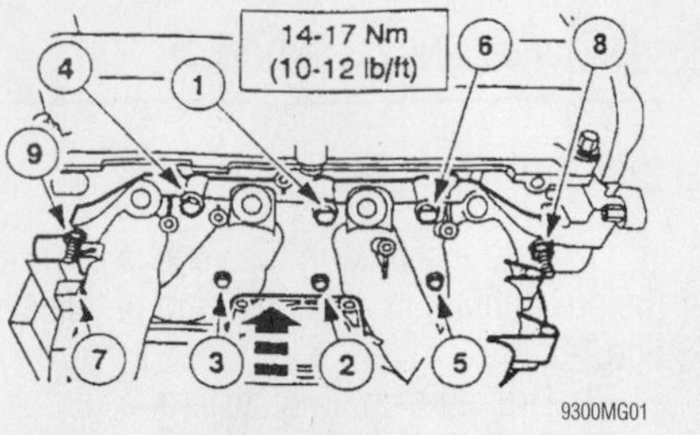

Intake manifold torque sequence—2.0L DOHC engine

   - Fuel lines
   - Alternator
   - Drive belt
   - Crankcase ventilation hose to the valve cover
   - Vacuum hoses to the manifold making sure the tabs are firmly engaged
   - Connectors in the mounting bracket
   - Main engine control sensor wiring
   - Heater hoses to the heater core
   - Bolt that secures the pipe located by the crankshaft pulley. Tighten the bolt to 71–97 inch lbs. (8–11 Nm).
   - TP sensor electrical connection
   - Air cleaner outlet tube
7. Fill the cooling system.
8. Connect the negative battery cable.

Exhaust Manifold

REMOVAL & INSTALLATION

2.0L SOHC Engine

1. Before servicing the vehicle, refer to the precautions in the beginning of this section.
2. Remove or disconnect the following:
   - Negative battery cable
   - Drive belt
   - Heated Oxygen (HO_2S) sensor electrical connection which is mounted on the cooling fan shroud
   - 5 exhaust manifold heat shield nuts and the shield
   - 2 Exhaust Gas Recirculation (EGR) tube-to-exhaust manifold nuts studs
   - Power steering pressure hose bracket bolts, then position the bracket and hose aside
   - Alternator lower bolt, loosen only
   - Alternator upper bolt and pivot the alternator forward

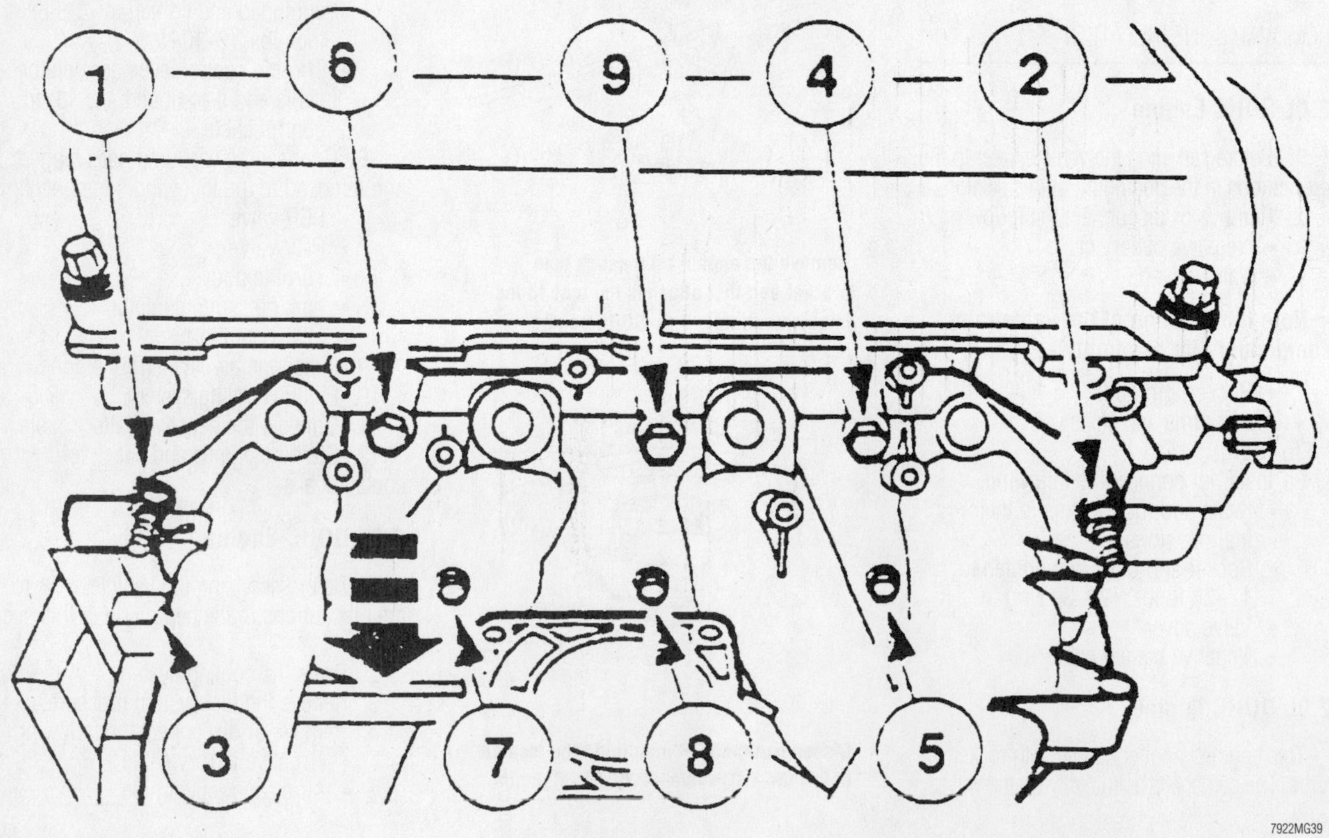

Remove the intake manifold bolts and nuts in this sequence—2.0L Zetec engine

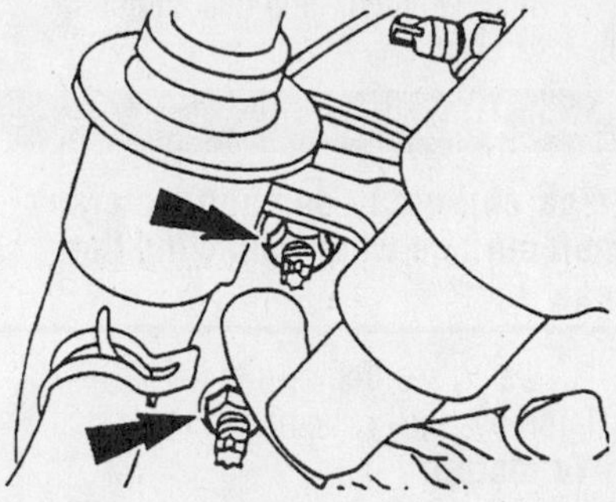

7922MG43

Remove the 2 EGR tube-to-exhaust manifold stud nuts—2.0L SOHC engine

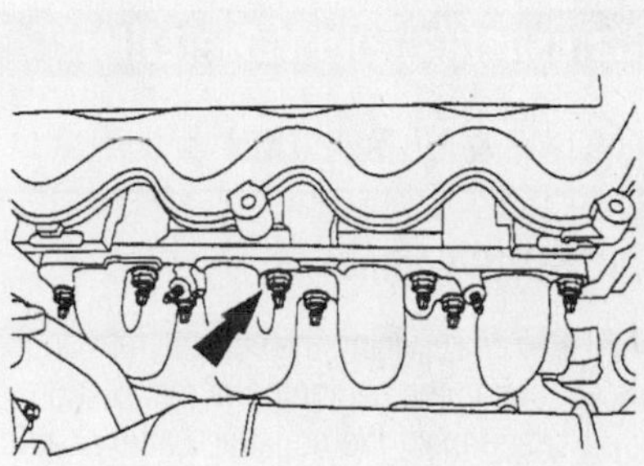

7922MG44

Remove the exhaust manifold retaining nuts—2.0L SOHC engine

- 4 exhaust manifold-to-catalytic converter nuts
- 8 exhaust manifold nuts
- Manifold and gasket

To install:

3. Install or connect the following:
   - New gasket and the exhaust manifold
   - Manifold retaining nuts and tighten to 15–18 ft. lbs. (20–24 Nm)
   - Exhaust manifold-to-catalytic converter nuts and tighten to 26–34 ft. lbs. (34–47 Nm)
   - Alternator and tighten the upper and lower mounting bolts
   - Power steering pressure hose and bracket, then tighten the retaining bolts
   - 2 EGR tube-to-exhaust manifold nuts studs and tighten to 44.6–62 inch lbs. (5–7 Nm)
   - Exhaust manifold heat shield and tighten the nuts to 45–61 inch lbs. (5–7 Nm)
   - HO_2S sensor electrical connection
   - Drive belt
   - Negative battery cable

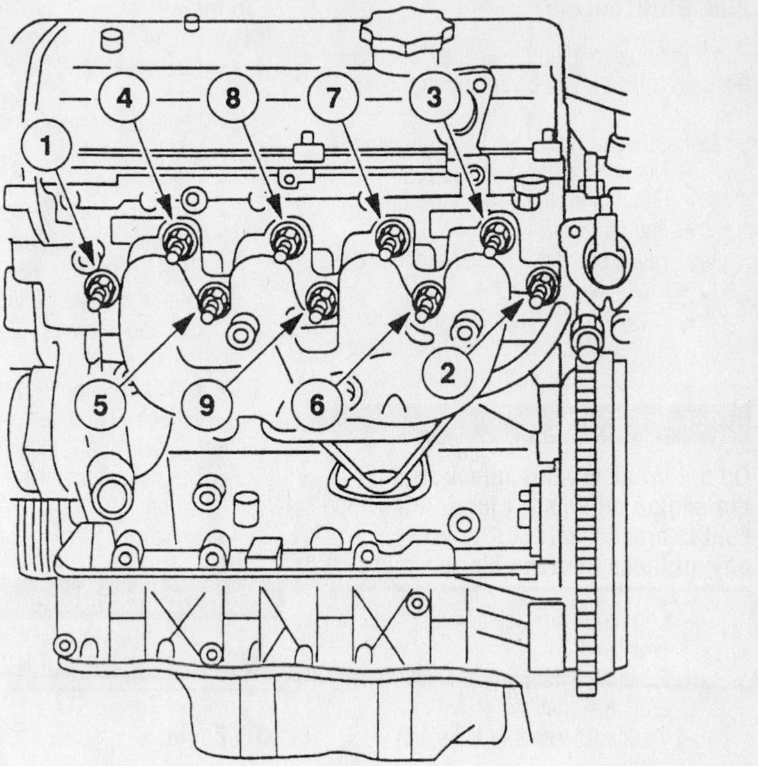

7922MG45

Remove the exhaust manifold nuts and bolts in the sequence illustrated—2.0L Zetec engine

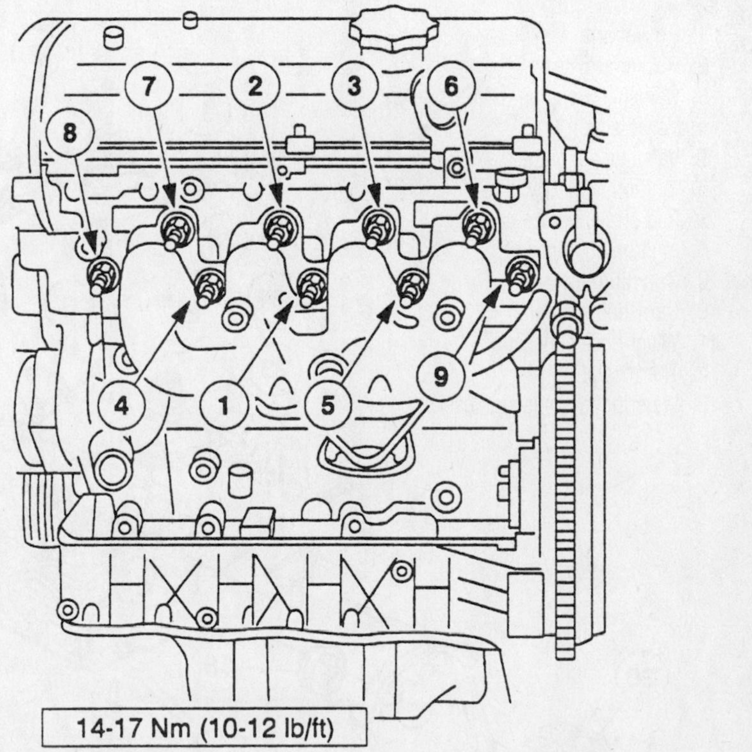

7922MG46

Tighten the exhaust manifold nuts and bolts in the sequence illustrated to the proper specification—2.0L Zetec engine

Brake service is covered in Section 4 of this manual

2.0L DOHC Engine

1. Before servicing the vehicle, refer to the precautions in the beginning of this section.
2. Remove or disconnect the following:
   - Negative battery cable
   - Fan motor electrical connection
   - Fan shroud
   - Heated Oxygen (HO_2S) sensor electrical connection
   - Catalytic converter-to-exhaust manifold bolt and nut, then disconnect the converter from the manifold

⁂ WARNING

Do not break the threadlock seal on the engine oil dipstick tube. If the seal is broken, relock it to prevent any oil leaks from the block.

   - Engine oil dipstick tube bracket bolt
   - Exhaust manifold heat shield bolts and nuts, then the shield
   - 7 exhaust manifold bolts and 2 studs
   - Manifold and gasket

To install:

3. Install or connect the following:
   - New gasket and the exhaust manifold
   - Manifold retaining bolts and nuts and tighten to 13–16 ft. lbs. (14–17 Nm)
   - Exhaust manifold heat shield and retainers. Tighten the retainers to 71–101 inch lbs. (8–11 Nm).
   - Engine oil dipstick tube bracket bolt and tighten to 71–101 inch lbs. (8–11 Nm)
   - Exhaust manifold-to-catalytic converter bolt and nut
   - HO_2sensor electrical connection
   - Fan shroud and the fan motor wiring
   - Negative battery cable

Front Crankshaft Seal

REMOVAL & INSTALLATION

2.0L Engines

1. Before servicing the vehicle, refer to the precautions in the beginning of this section.
2. Remove the timing belt and crankshaft sprocket.

⁂ WARNING

Be careful not to damage the crankshaft surface when removing the seal.

3. Using seal remover tool T92C-6700-CH, remove the crankshaft front oil seal.

To install:

4. Use seal replacer tool T81P-6700-A, install the new seal.
5. Install the crankshaft sprocket and the timing belt.

Camshaft(s) and Valve Lifters

REMOVAL & INSTALLATION

2.0L SOHC Engine

1. Before servicing the vehicle, refer to the precautions in the beginning of this section.
2. Disconnect the negative battery cable.
3. Remove the air cleaner and valve cover.

1. Rocker arm
2. Rocker arm seats
3. Camshaft Position (CMP) sensor
4. Valve tappet
5. Valve tappet guide plate
6. Ignition coil
7. Radio interference capacitor
8. Ignition coil bracket
9. Cup plug
10. Camshaft
11. Water hose connection
12. Thermostat
13. Water hose connection gasket

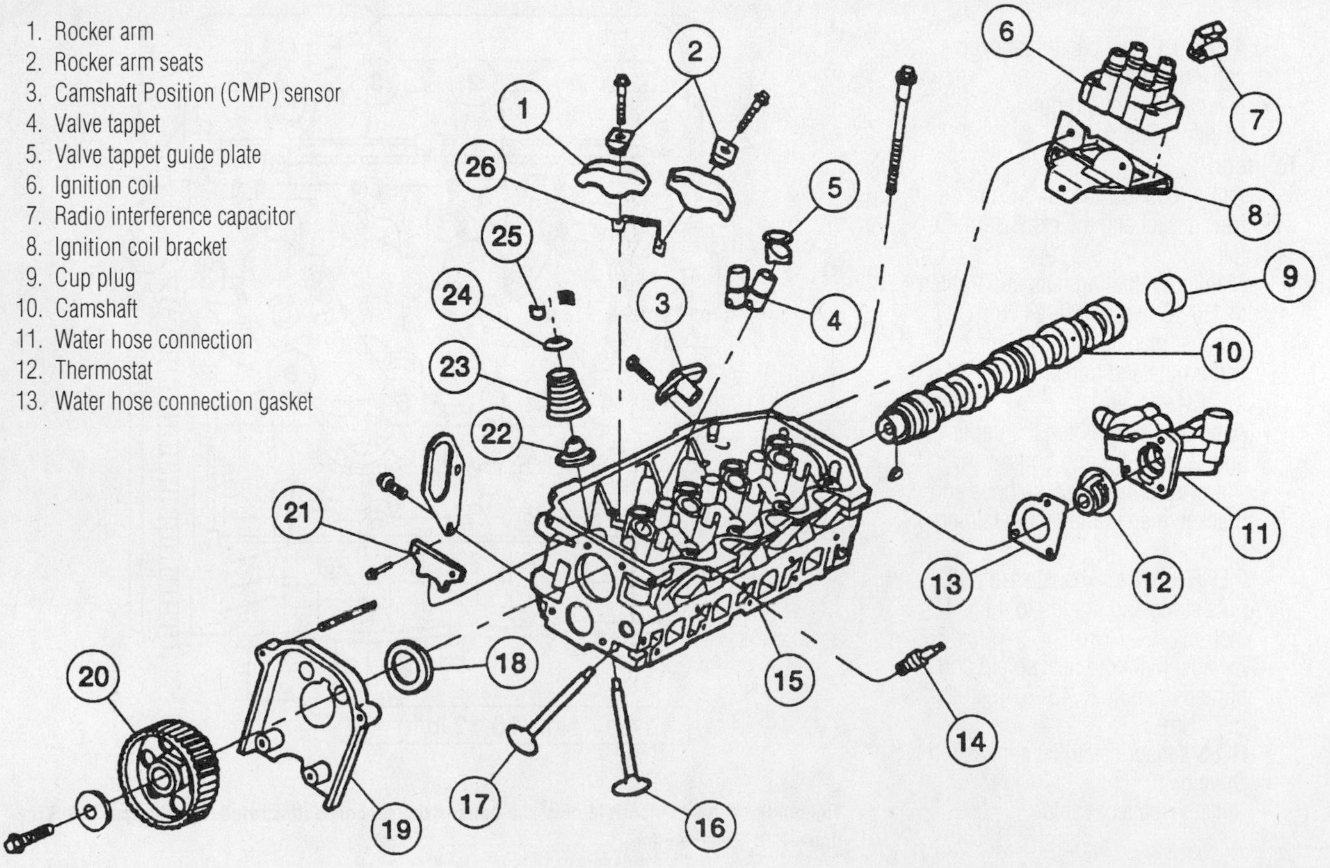

Exploded view of the cylinder head—2.0L SOHC SPI engine

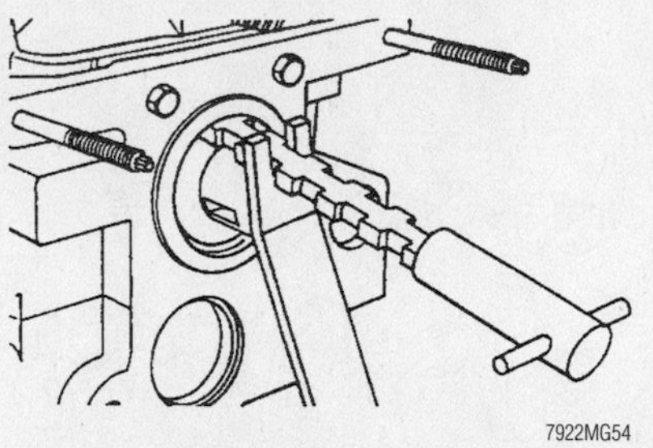

Removing the camshaft front seal—2.0L SOHC engine

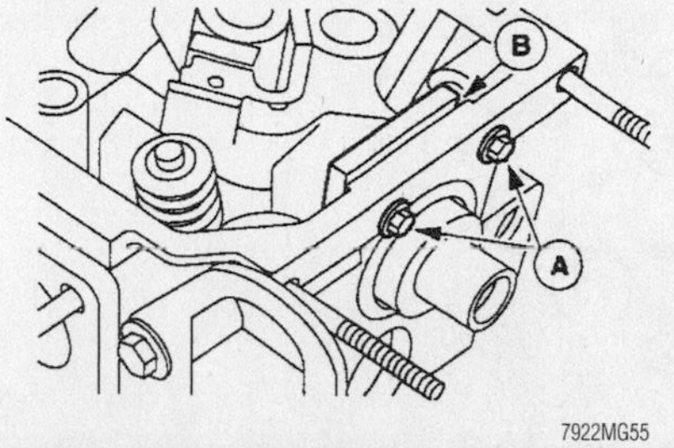

Remove the camshaft thrust plate retaining bolts (A) and the plate (B)—2.0L SOHC engine

4. Remove the camshaft front seal as follows:
   a. Align the timing marks and remove the timing belt.
   b. Use cam sprocket holding/removing tool T74P-6256-B, and remove the camshaft sprocket.
   c. Unfasten the timing belt tensioner bolt and remove the tensioner.
   d. Remove the inner engine front cover.
   e. Use seal removal tool T92C-6700-CH, to remove the camshaft front seal.
5. Remove or disconnect the following:
   - Ignition coil and bracket
   - Rocker arms and valve tappets
   - Camshaft thrust plate bolts and the plate
   - Cup plug from the rear of the cylinder head and discard
   - Camshaft from the rear of the cylinder head

To install:

➡Liberally coat the cam bore in the cylinder with clean 5W30 motor oil.

6. Install or connect the following:
   - Camshaft through the rear of the cylinder head
   - Camshaft thrust plate and the bolts. Tighten the bolts to 71–115 inch lbs. (8–13 Nm).

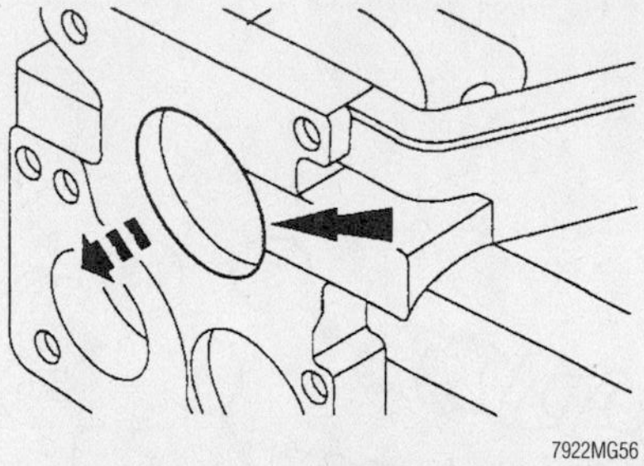

Remove and discard the cup plug located at the rear of the cylinder head—2.0L engine

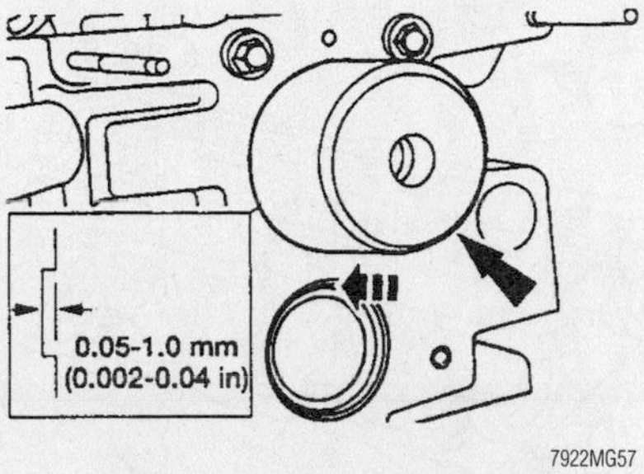

Use seal Replacer T81P-6292-At, to install the camshaft front seal 0.002–0.04 in. (0.05–1.0mm) below flush with the cylinder head front face—2.0L SOHC engine

   - New cup plug and the tappets
   - Rocker arms
   - Ignition coil bracket and coil
7. Install the camshaft front seal as follows:
   a. Apply a thin film of 5W30 motor oil to the lip of the seal.

➡The seal depth should be 0.002–0.04 in. (0.05–1.0mm) below flush with the cylinder head front face.

   b. Use seal replacer T81P-6292-A, to install the camshaft front seal.
   c. Install the inner engine front cover.
   d. Install the timing belt tensioner. Tighten the tensioner bolt to15–22 ft. lbs. (20–30 Nm).
   e. Install the camshaft sprocket and the timing belt
8. Install the valve cover and the air cleaner assembly.
9. Connect the negative battery cable.

2.0L DOHC Engine

1. Before servicing the vehicle, refer to the precautions in the beginning of this section.
2. Remove or disconnect the following:
   - Timing belt
   - Valve cover and camshaft sprockets

➡It may be necessary to rotate the oil control solenoid flange 90 degrees prior to removal.

   - Oil control solenoid flange bolts and the flange

➡Mark the camshaft journal caps with a number to identify their location, as they must be replaced in their original position.

   - Camshaft journal caps in several passes in the sequence illustrated, then remove the bolts and caps
   - Camshafts from the cylinder head
   - Oil control sensor and bushing
3. Inspect the camshafts for damage and wear.

To install:

4. Be sure the valve clearance is correct.
5. Install the oil control solenoid bushing and flange on the exhaust camshaft.
6. Coat the surface of the front camshaft journal cap with gasket maker E2AZ-19562-B.
7. Place the camshatts in position and lubricate the bearing surfaces with engine assembly lubricant D9AZ-19579-D.
8. Install new camshaft front oil seals.
9. Apply a thin coat of silicone gasket and sealant F6AZ-19562-AA to the sealing surface of the of the front camshaft journal bearing cap.
10. Install or connect the following:
    - Caps and tighten the bolts in several 2-turn passes in the sequence illustrated to 10–12 ft. lbs. (13–17 Nm)

WARNING

The oil control O-ring seals must not fall out of position in the oil control solenoid flange or poor engine performance may occur.

11. Inspect the oil control solenoid flange O-rings for damage or wear and replace as necessary.
    - Oil control solenoid flange and tighten the bolts to 84–92 inch lbs. (9.5–10.5 Nm) in the sequence illustrated
12. Rotate the camshafts a full turn and check for binding.

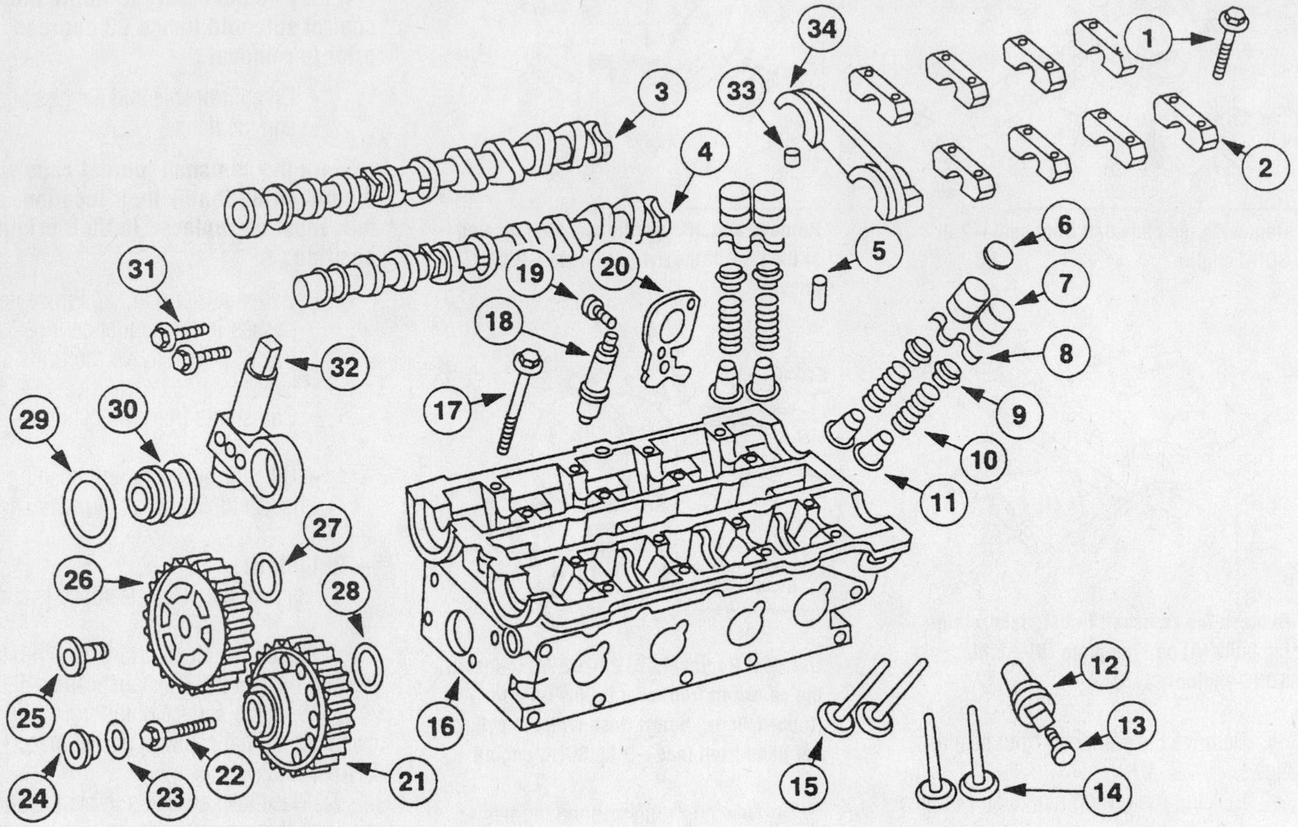

1	Bolt—Camshaft Bearing Cap
2	Camshaft Bearing Cap
3	Camshaft Intake
4	Camshaft Exhaust
5	Plug
6	Adjusting Shim—Valve Clearance
7	Valve Tappets
8	Valve Spring Retainer Key
9	Valve Spring Retainers
10	Valve Spring (Color Coding: Exhaust = Blue, Intake = Red)
11	Valve Stem Seal
12	Camshaft Position Sensor
13	CMP Sensor Bolt
14	Intake Valves
15	Exhaust Valves
16	Cylinder Head
17	Cylinder Head Bolts

18	Spark Plug
19	Engine Lifting Eye Bolt
20	Engine Lifting Eye
21	Camshaft Sprocket —Exhaust
22	Exhaust Camshaft Sprocket Bolt
23	Blanking Plug O-Ring
24	Blanking Plug
25	Bolt—Intake Camshaft
26	Camshaft Sprocket Intake
27	Camshaft Front Seal
28	O-Ring
29	Camshaft Front Seal
30	Oil Feed Ring
31	Oil Feed Flange Bolts
32	Oil Feed Flange
33	Guide Sleeve—Front Camshaft Bearing Cap
34	Fifth Camshaft Bearing Cap

9300MG06

Exploded view of the cylinder head, showing camshaft mounting—2.0L DOHC Zetec engine

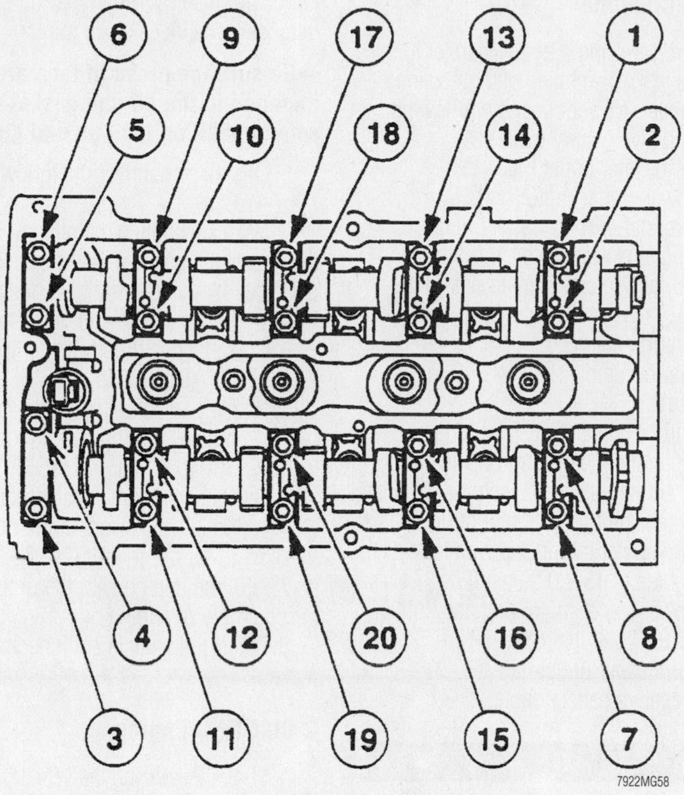

Loosen the camshaft journal caps in several passes in this sequence—2.0L Zetec engine

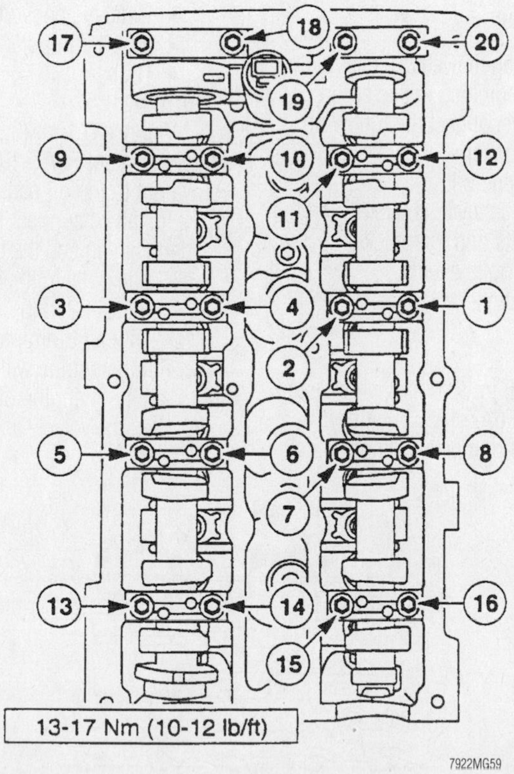

Tighten the camshaft journal caps in the sequence illustrated to the proper specification—2.0L Zetec engine

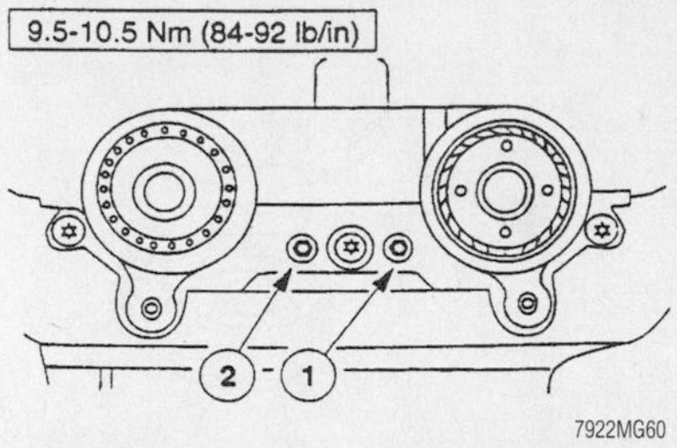

Tighten the oil control solenoid flange bolts in this sequence—2.0L Zetec engine

- Camshaft sprockets and timing belt
- Valve cover

Valve Lash

ADJUSTMENT

2.0L DOHC Zetec Engine

1. Before servicing the vehicle, refer to the precautions in the beginning of this section.
2. Remove the valve cover.
3. Remove the timing belt.

➡Measure each valve's clearance at the base circle before removing the camshafts. The shims are not serviceable with the camshafts in place.

4. Measure the valve clearances, then remove the camshafts and shims.

The correct shims allow the following valve clearances:

- Intake valve clearance: 0.0043–0.0071 in. (0.11–0.18mm)
- Exhaust valve clearance: 0.0106–0.0134 in. (0.27–0.34mm)

➡A midrange clearance is the most desirable.

5. Select the shims using the following formula: shim thickness = measured clear-

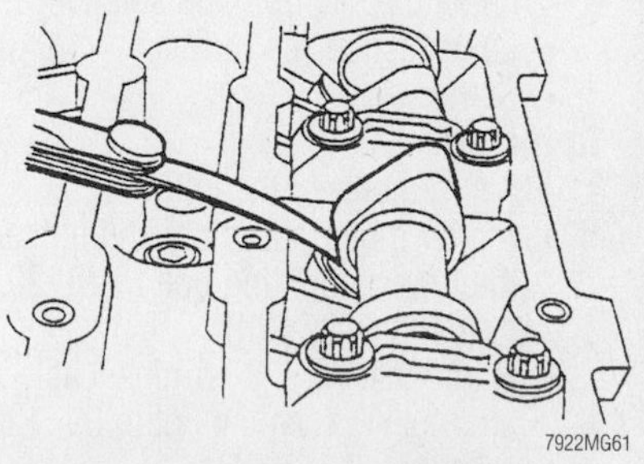

Measure each valve's clearance at the base circle before removing the camshafts—2.0L Zetec DOHC engine

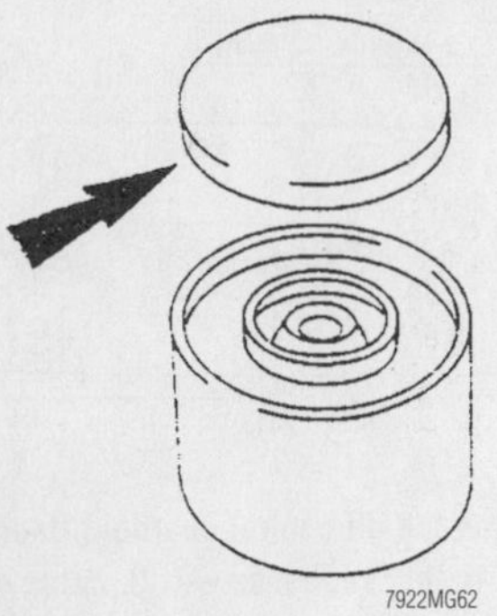

7922MG62

Example of a valve lifter shim (arrow)—2.0L Zetec DOHC engine

ance plus the base shim thickness minus the most desirable thickness.

6. Select the correct shims and mark their installation location.
7. Replace the shims and install the camshafts.
8. Check the new valve clearances.
9. Install the timing belt and covers.
10. Install the valve cover.
11. Start the vehicle and check for proper operation.

Starter Motor

REMOVAL & INSTALLATION

2.0L SOHC Engine

1. Before servicing the vehicle, refer to the precautions in the beginning of this section.
2. Remove or disconnect the following:
   - Negative battery cable
   - Air cleaner outlet tube
   - Two top starter motor bolts

➡A protective cap is covering the B terminal and must be replaced after servicing the starter motor or solenoid.

   - **S** terminal wire and the **B** terminal nut and cable from the solenoid
   - Lower starter motor bolt
   - Starter

To install:

3. Install or connect the following:
   - Starter motor assembly and tighten the lower mounting bolt to 18–20 ft. lbs. (25–27 Nm)
   - **S** terminal wire, **B** terminal cable and nut. Tighten the nut to 80–120 inch lbs. (9–13.5 Nm).
   - Top starter motor bolts. Tighten the bolts to 18–20 ft. lbs. (25–27 Nm).
   - Air cleaner outlet tube
   - Negative battery cable

2.0L DOHC Engine

1. Before servicing the vehicle, refer to the precautions in the beginning of this section.
2. Remove or disconnect the following:
   - Negative battery cable
   - Air cleaner outlet tube
   - Lower starter motor bolt
   - Two starter motor upper bolts
3. Slide the starter motor from its mounting until you can gain access to the wiring.
   - Integral connector
   - Starter motor

To install:

4. Install or connect the following:
   - Integral connector and tighten the retaining nuts to 61 inch lbs. (7 Nm)
   - Starter motor to its mounting
   - Two starter motor upper bolts and tighten to 15–20 ft. lbs. (20–27 Nm)
   - Lower starter motor bolt and tighten to 15–20 ft. lbs. (20–27 Nm)
   - Air cleaner outlet tube
   - Negative battery cable

Oil Pan

REMOVAL & INSTALLATION

2.0L SOHC Engine

1. Before servicing the vehicle, refer to the precautions in the beginning of this section.
2. Remove or disconnect the following:
   - Catalytic converter
3. Drain the engine oil.
   - Oil pan-to-catalytic converter bracket bolts and the bracket
   - 10 oil pan bolts evenly
   - Oil pan and gasket. Discard the old gasket.

To install:

4. Apply a 0.125 in. (3.0mm) wide bead of silicone sealant F6AZ-19562-AA or its equivalent at the oil pump-to-cylinder block joints and the crankshaft rear oil seal retainer-to-cylinder block joints.

➡Be sure the press fit tabs are fully engaged in the oil pan gasket channel when installing the pan and gasket.

5. Install or connect the following:
   - Gasket
6. When installing the oil pan, be sure the rear of the pan is aligned with the cylinder block by using a straight edge as a guide.
   - Oil pan bolts and tighten in the sequence illustrated to 15–22 ft. lbs. (20–30 Nm)
   - Oil pan-to-catalytic converter bracket and tighten the bolts to 30–40 ft. lbs. (40–55 Nm)
   - Catalytic converter and oil pan drain plug. Tighten the drain plug to 15–22 ft. lbs. (20–30 Nm).
7. Fill the engine with the proper type and quantity of engine oil.
8. Start the vehicle and check for oil leaks.

2.0L DOHC Engine

1. Before servicing the vehicle, refer to the precautions in the beginning of this section.
2. Drain the engine oil.
3. Remove or disconnect the following:
   - Catalytic converter
   - 17 oil pan bolts evenly
   - Oil pan

To install:

4. Apply a 0.1 in. (3.0mm) wide bead of silicone sealant F6AZ-19562-AA or its equivalent to the oil pan.
5. Install or connect the following:
   - Oil pan and the bolts. Tighten the oil pan bolts in the sequence illustrated to 15–22 ft. lbs. (20–30 Nm).
   - Oil pan drain plug
6. Fill the engine with the proper type and quantity of engine oil.
7. Start the vehicle and check for oil leaks.

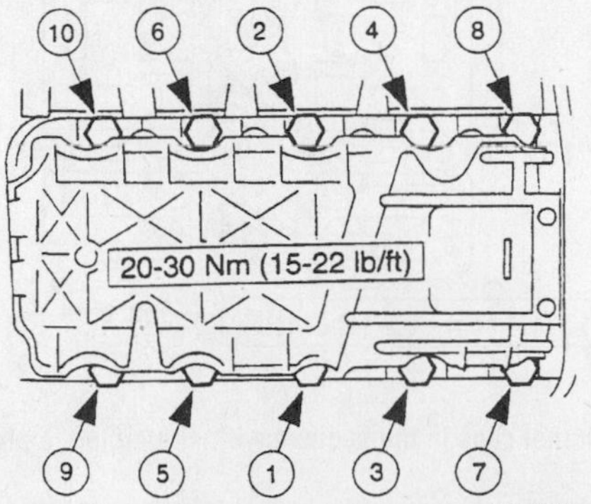

7922MG68

Tighten the oil pan bolts in this sequence to the proper specification—2.0L SOHC engine

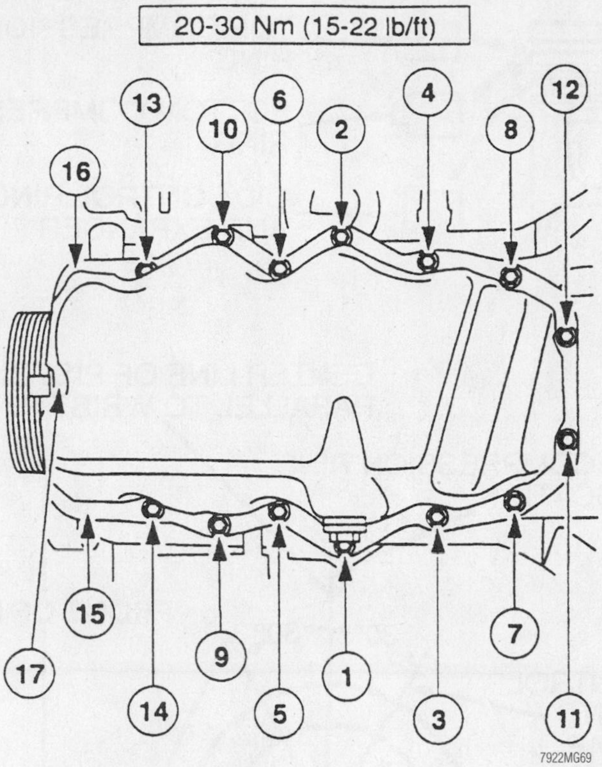

Tighten the oil pan bolts in the proper sequence to the correct torque specification—2.0L Zetec

Oil Pump

REMOVAL & INSTALLATION

2.0L SOHC Engine

1. Before servicing the vehicle, refer to the precautions in the beginning of this section.
2. Remove or disconnect the following:
   - Timing belt
   - Oil pan
   - Crankshaft Position (CKP) sensor electrical connection
   - Oil pump screen cover and tube bolts
   - Cover and tube
   - 6 oil pump retaining bolts
   - Oil pump and gasket. Discard the gasket.

To install:

3. Lubricate the crankshaft front oil seal lip with clean engine oil.
4. Install or connect the following:
   - Pump gasket and the pump
   - Oil pump and tighten the bolts to 10–12 ft. lbs. (13–16 Nm)
   - Oil pump screen cover and tube. Tighten the retaining bolts to 71–97 inch lbs. (8–11 Nm).
   - CKP sensor electrical connection
   - Oil pan
   - Timing belt
5. Fill the engine with the proper type and quantity of engine oil.
6. Start the vehicle and check for oil leaks.

2.0L DOHC Engine

1. Before servicing the vehicle, refer to the precautions in the beginning of this section.
2. Remove or disconnect the following:
   - Timing covers and belt
   - Crankshaft pulley, sprocket and the timing chain guide
   - Oil pan
   - Oil pump cover and screen bolts
   - Cover and screen
   - Lower cylinder block shims, if necessary
   - Lower cylinder block, the gasket and the crankshaft front oil seal, if necessary
   - Oil pump retaining bolts
   - Oil pump and gasket

To install:

➡The clearance between the lower cylinder block sealing surfaces on the oil pump and the cylinder block cannot exceed 0.012–0.031 in. (0.3–0.8mm).

3. Install or connect the following:
   - New oil pump gasket and the pump
   - Oil pump bolts and tighten to 88–97 inch lbs. (10–11 Nm)
   - Crankshaft front oil seal, if removed
   - Lower cylinder block and gasket, if removed. Tighten the lower cylinder

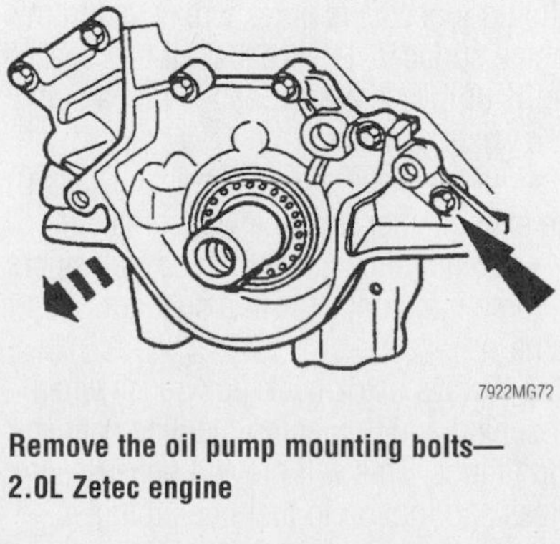

Remove the oil pump mounting bolts—2.0L Zetec engine

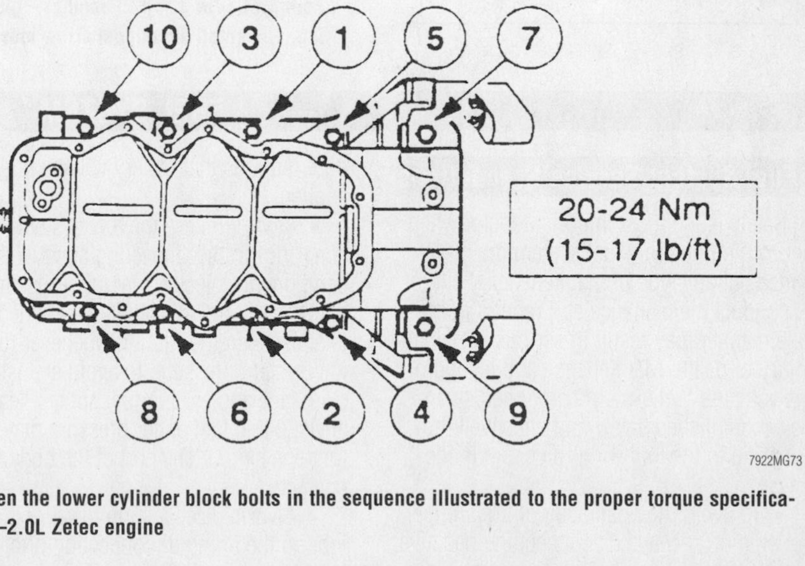

Tighten the lower cylinder block bolts in the sequence illustrated to the proper torque specification—2.0L Zetec engine block bolts to 15–17 ft. lbs. (20–24 Nm) in the sequence illustrated.

- Oil pump screen cover and tube. Tighten the retaining bolts to 71–97 inch lbs. (8–11 Nm).
- Oil pan, timing chain guide, crankshaft sprocket and the pulley
- Timing belt and covers

4. Fill the engine with the proper type and quantity of engine oil.
5. Start the vehicle and check for oil leaks.

Rear Main Seal

REMOVAL & INSTALLATION

2.0L Engine

1. Before servicing the vehicle, refer to the precautions in the beginning of this section.
2. Remove the transaxle and flywheel.

**** WARNING**

Be careful not to damage the crankshaft surface when removing the seal.

3. Use seal replacer tool T92C-6700-Ch, to remove the old seal.

To install:

4. Coat the lip of the new seal with clean 5W30 engine oil.
5. Install the new seal using crankshaft rear seal pilot tool T88P-6701-B2 and Rear seal replacer tool T88P-6701-B1.
6. Install the flywheel and the transaxle.

Piston and Ring

POSITIONING

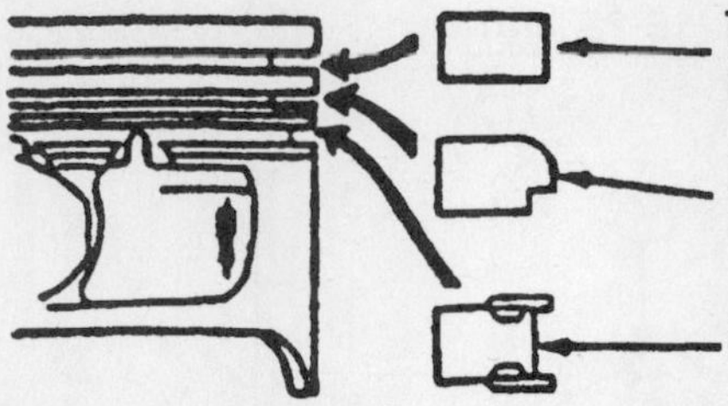

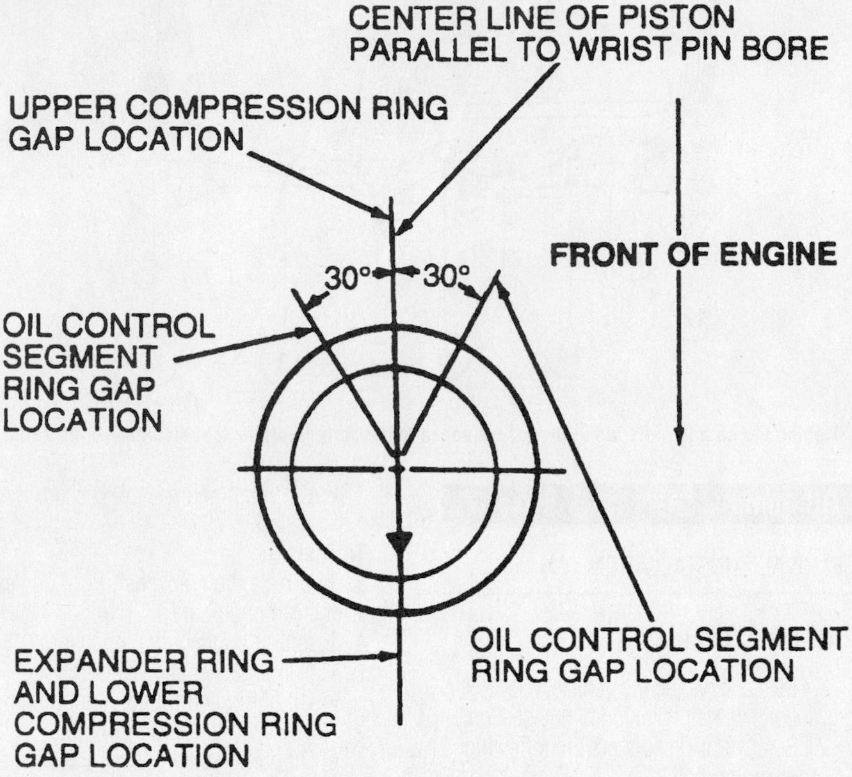

7922AG29

Ford 2.0L (VIN 3 and P) engines—piston ring positioning, end-gap spacing and piston positioning. The small directional arrow must face the front of the engine.

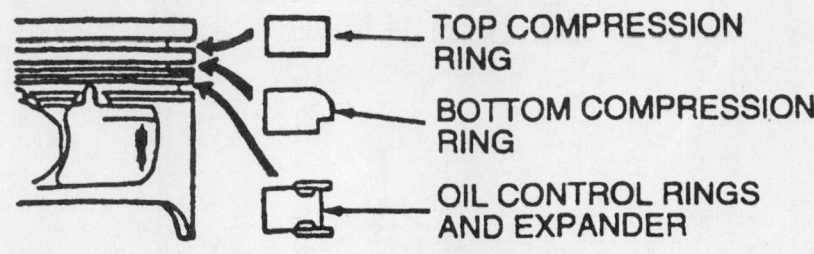

FUEL SYSTEM

Fuel System Service Precautions

Safety is the most important factor when performing not only fuel system maintenance but any type of maintenance. Failure to conduct maintenance and repairs in a safe manner may result in serious personal injury or death. Maintenance and testing of the vehicle's fuel system components can be accomplished safely and effectively by adhering to the following rules and guidelines.

- To avoid the possibility of fire and personal injury, always disconnect the negative battery cable unless the repair or test procedure requires that battery voltage be applied.
- Always relieve the fuel system pressure prior to disconnecting any fuel system component (injector, fuel rail, pressure regulator, etc.), fitting or fuel line connection. Exercise extreme caution whenever relieving fuel system pressure, to avoid exposing skin, face and eyes to fuel spray. Please be advised that fuel under pressure may penetrate the skin or any part of the body that it contacts.
- Always place a shop towel or cloth around the fitting or connection prior to loosening to absorb any excess fuel due to spillage. Ensure that all fuel spillage (should it occur) is quickly removed from engine surfaces. Ensure that all fuel soaked cloths or towels are deposited into a suitable waste container.
- Always keep a dry chemical (Class B) fire extinguisher near the work area.
- Do not allow fuel spray or fuel vapors to come into contact with a spark or open flame.
- Always use a back-up wrench when loosening and tightening fuel line connection fittings. This will prevent unnecessary stress and torsion to fuel line piping.
- Always replace worn fuel fitting O-

rings with new. Do not substitute fuel hose or equivalent, where fuel pipe is installed.

Fuel System Pressure

RELIEVING

1. Before servicing the vehicle, refer to the precautions in the beginning of this section.
2. Remove the Schrader valve cap at the end of the fuel rail and attach a fuel pressure gauge.
3. Open the manual relief valve on the fuel pressure gauge slowly to relive the fuel pressure

Fuel Filter

REMOVAL & INSTALLATION

The inline fuel filter is located in the engine compartment between the fuel tank and the fuel rail.

1. Before servicing the vehicle, refer to the precautions in the beginning of this section.
2. Relieve the fuel system pressure.
3. Disconnect the negative battery cable.
4. Position a suitable container below the fuel filter to collect any excess fuel that may leak from the filter and lines.
5. Remove or disconnect the following:
   - Fuel filter mounting clamp, loosen only
   - Filter from the mounting bracket
   - Retaining clips from the upper fuel filter hose
   - Upper hose from the fuel filter and drain any excess fuel into the container. Plug the hose.
   - Retaining clip from the fuel filter lower hose
   - Lower hose from the fuel filter and drain any excess fuel into the container. Plug the hose.
   - Fuel filter

To install:

6. Remove the plugs from the lower and upper hoses.
7. Install or connect the following:
   - Lower hose to the filter
   - Hose retaining clip
   - Upper hose to the filter
   - Hose retaining clip
   - Fuel filter and tighten the filter mounting clamp
   - negative battery cable
8. Start the engine and check for leaks.

Fuel Pump

The fuel pump is located inside the fuel tank.

REMOVAL & INSTALLATION

This procedure will require a new fuel pump gasket for pump installation, so be sure to have one before starting.

1. Before servicing the vehicle, refer to the precautions in the beginning of this section.
2. Relieve the fuel system pressure.
3. Remove or disconnect the following:
   - Negative battery cable
   - Rear seat cushion
   - Electrical connectors from the fuel pump
   - 4 pump access cover screws and the cover
   - Pump electrical connector located under the cover, if equipped
   - Fuel line clip(s)
   - Fuel line(s) from the pump
   - Pump locking retaining ring., using a fuel pump locking ring removal tool or a brass drift and a hammer
   - Fuel pump and the gasket from the tank

To install:

4. Install or connect the following:
   - New gasket and place the pump into position in the tank
   - Pump locking retaining ring
   - Fuel line(s) to the pump
   - Line retaining clip(s)
   - Fuel pump electrical connection(s), if equipped
   - Access cover and tighten the retainers
   - Pump electrical connector(s)
   - Rear seat cushion
   - Negative battery cable
5. Start the vehicle and check for proper operation.

Fuel Injector

REMOVAL & INSTALLATION

⁂ CAUTION

Fuel injection systems remain under pressure, even after the engine has been turned OFF. The fuel system pressure must be relieved before disconnecting any fuel lines. Failure to do so may result in fire and/or personal injury.

1. Before servicing the vehicle, refer to the precautions in the beginning of this section.
2. Relieve the fuel system pressure.
3. Remove or disconnect the following:
   - Negative battery cable

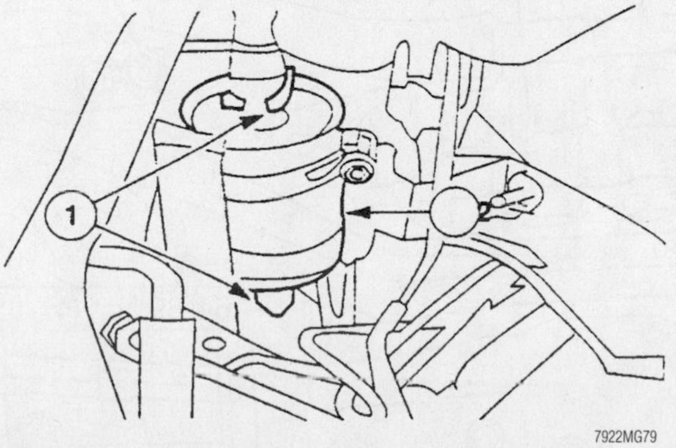

Detach the upper and lower hose clips before disconnecting hoses (1) from the fuel filter

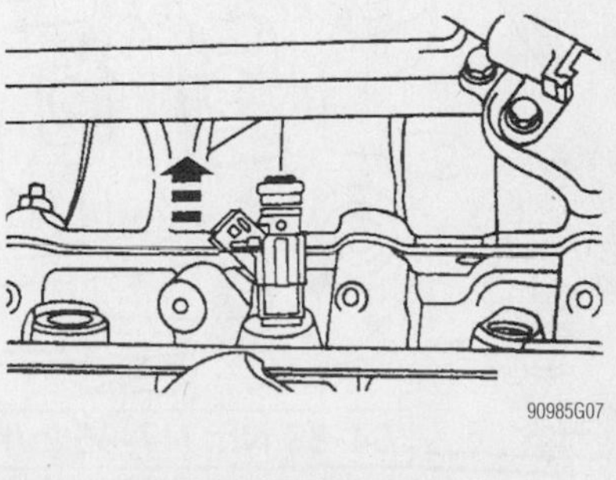

Grasp the fuel injector's body and pull upward, while gently rocking the fuel injector from side to side

- Fuel injection fuel rail
- Injector retaining clips, on DOHC engines

4. Grasp the fuel injector body and pull up while gently rocking the fuel injector from side to side.

5. Once removed, inspect the fuel injector cap and body for signs of deterioration. Replace as required.

6. Remove the O-rings and discard. If an O-ring or end cap is missing, look in the intake manifold for the missing part.

To install:

7. Install or connect the following:
   - New O-rings onto each injector and apply a small amount of clean engine oil to the O-rings
   - Injectors using a slight twisting downward motion
   - Injector retaining clips, on DOHC engines
   - Fuel injection fuel rail
   - Negative battery cable

8. Run the engine at idle for 2 minutes, then turn the engine **OFF** and check for fuel leaks and proper operation.

DRIVE TRAIN

Transaxle Assembly

REMOVAL & INSTALLATION

Manual

1. Before servicing the vehicle, refer to the precautions in the beginning of this section.

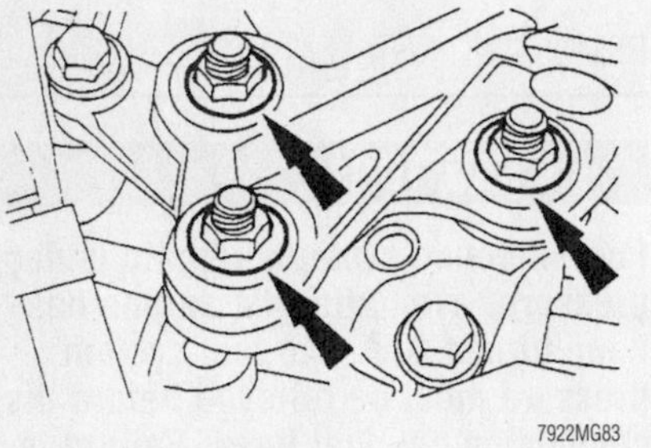

Location of the left-hand engine support bracket nuts—2.0L engines

2. Remove or disconnect the following:
   - Both battery cables, negative cable first
   - Battery and tray
   - Engine air cleaner outlet tube
   - Constant Control Relay Module (CCRM) module and bracket
   - Slave cylinder line from the slave cylinder hose and plug the hose

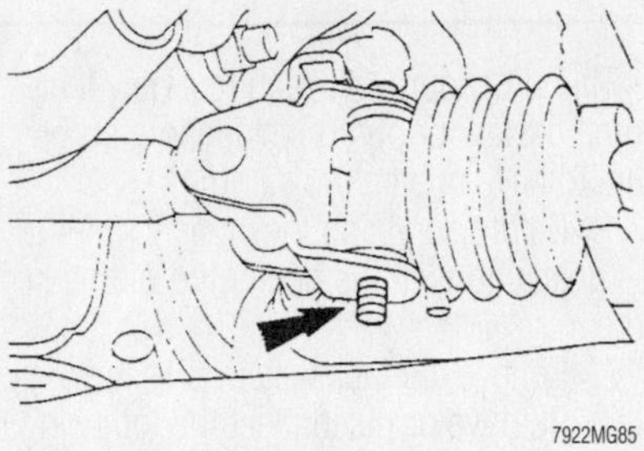

Loosen the transaxle gearshift bolt and remove the gearshift rod and clevis pin from the input shift shaft

   - Slave cylinder line retaining clip
   - Slave cylinder line from the bracket
   - Heated Oxygen (HO_2S) sensor electrical connection
   - Back-up light electrical connection
   - Electrical connector bracket bolt and the bracket
   - Vehicle Speed Sensor (VSS) electrical connection
   - Halfshafts

3. Install Engine Support Bar D88–6000A, and attach it to the engine lifting eyes with suitable chains or cables.
   - Insulator nuts from the left-hand engine support bracket
   - Front and rear upper transaxle-to-engine bolts

4. Drain the transaxle fluid and install the drain plug.
   - Air conditioning line from the retainer located on the engine support crossmember
   - Engine support crossmember bolts and nuts

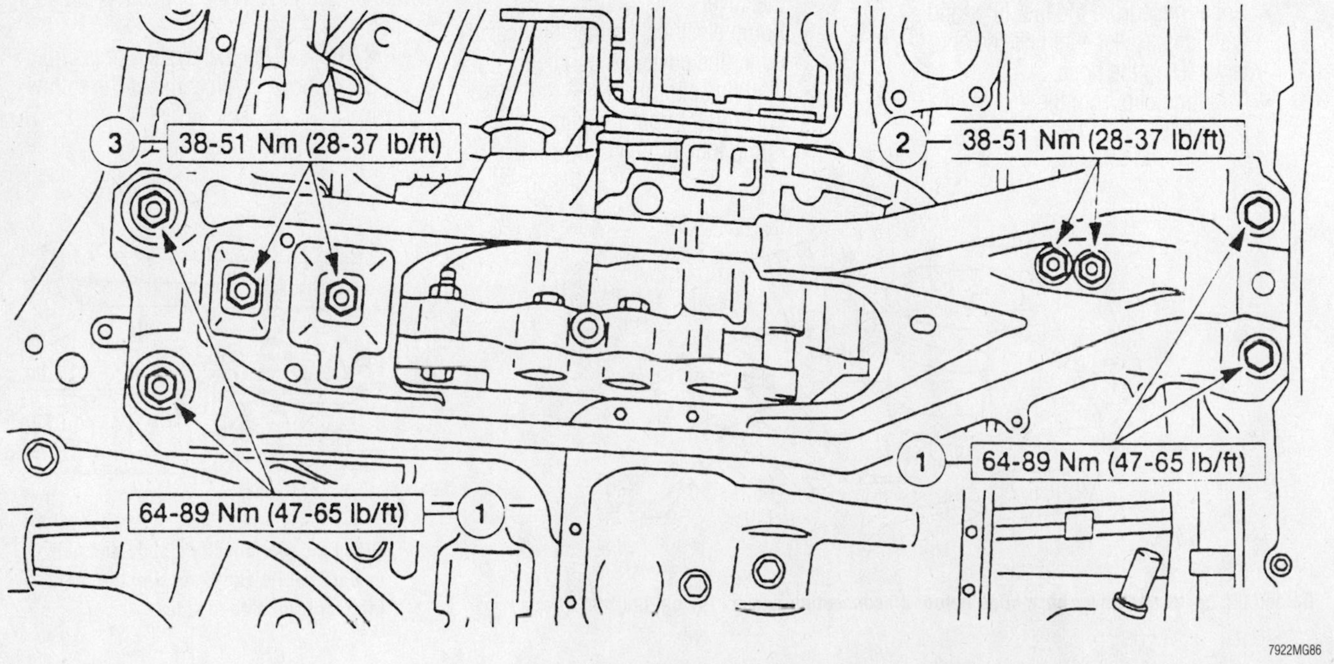

Install the engine support crossmember and tighten the retainers to specification

- Crossmember
- Gearshift stabilizer bar-to-transaxle nut
- Stabilizer bar and support from the transaxle
- Transaxle gearshift rod nut
- Transaxle gearshift bolt
- Gearshift rod and clevis pin from the input shift shaft
- Starter motor
- Lower slave cylinder tube and the slave cylinder
- Lower transaxle-to-engine bolts

5. Position and secure a suitable transaxle jack under the transaxle.
   - Catalytic converter
   - Middle transaxle-to-engine bolts
   - Transaxle from the vehicle

To install:

6. Apply a thin coating of suitable grease to the splines of the input shaft.
7. Place the transaxle onto a suitable transaxle jack. Be sure the transaxle is secure.
8. Install or connect the following:
   - Transaxle
   - Middle transaxle-to-engine bolts and tighten them to 23–38 ft. lbs. (38–51 Nm)
   - Lower engine-to-transaxle bolts and tighten them to 23–38 ft. lbs. (38–51 Nm)
   - clutch slave cylinder and tighten the nuts to 12–17 ft. lbs. (16–23 Nm)
   - Lower slave cylinder tube and tighten the fitting to 10–16 ft. lbs. (13–21 Nm)
   - Starter motor
   - Gearshift rod and clevis into position on the input shift shaft
   - Gearshift bolt and gearshift rod nut and tighten to 12–17 ft. lbs. (16–23 Nm)
   - Gearshift stabilizer bar and tighten the nut to 23–34 ft. lbs. (31–46 Nm)
   - Halfshafts and the catalytic converter
   - Engine support crossmember. Tighten the crossmember bolts and nuts to 47–65 ft. lbs. (64–89 Nm) and the insulator nuts to 28–37 ft. lbs. (38–51 Nm).
   - Front and rear upper transaxle-to-engine bolt and tighten it to 28–38 ft. lbs. (38–51 Nm)
   - Left-hand engine support insulator nuts and tighten them to 50–68 ft. lbs. (67–93 Nm)
9. Remove the engine support bar.
   - Electrical connector support bracket
   - All the electrical connectors
   - Slave cylinder line to the slave cylinder hose and attach the retaining clip. Tighten the fitting to 10–16 ft. lbs. (13–21 Nm).
10. Bleed the clutch hydraulic system.
11. Add the proper type and amount of fluid to the transaxle.
12. Install or connect the following:
    - CCRM and bracket assembly, if removed
    - CCRM retainers and attach the electrical connection
    - Engine air cleaner tube
    - Battery tray and the battery
    - Battery cables, negative cable last
13. Check for fluid leaks and proper clutch operation.
14. Road test the vehicle and check for proper transaxle operation.

Automatic

1. Before servicing the vehicle, refer to the precautions in the beginning of this section.
2. Remove or disconnect the following:
   - Battery cables, negative cable first
   - Battery and tray
   - Engine air cleaner assembly
   - Computer Control Relay Module (CCRM) electrical connections, relay retainers, the relay and bracket from the engine
   - Shift control cable and bracket retaining nut from the manual shift lever
   - Shift cable and bracket clip, then set the bracket and cable assembly aside
   - All electrical connections from the transaxle
   - Starter motor
   - Throttle valve actuating cable bolts and disconnect the cable from the throttle cam, if equipped
3. Install Engine Support D88L-6000-A, to the engine. The engine must be properly supported for transaxle removal.
4. Remove or disconnect the following:
   - Transaxle cooler lines at the transaxle
   - Left-hand engine mount
   - Upper transaxle housing bolts
   - Left-hand splash shields, if equipped
   - Transaxle plug and drain the fluid
   - Halfshafts. Install 2 transaxle plugs T88C-7025-AH, into the differential side gears

✱✱ WARNING

Failure to install the transaxle plugs may cause the differential side gears to become improperly positioned. If the gears become misaligned, the differential will have to be removed from the transaxle to align them.

   - Engine support crossmember and the catalytic converter
   - Air conditioning line from the retainer located on the engine support crossmember
   - Transaxle support crossmember bolts and nuts, the left-hand engine isolator nuts, then the crossmember
   - Transaxle housing cover bolts and the cover
   - 4 flywheel-to-torque converter nuts
5. Position a suitable transaxle jack under the transaxle. Secure the transaxle to the jack.
   - Remaining transaxle-to-engine bolts
6. Slowly lower the transaxle out of the vehicle.
7. Inspect all components including mounts and brackets.

To install:

➡Prior to installing the transaxle, lubricate the torque converter pilot hub with multi-purpose grease.

8. Align the torque converter studs to the flywheel.
9. Secure the transaxle on the transaxle jack.
10. Install or connect the following:
    - Transaxle. Tighten the bolts to 40–58 ft. lbs. (55–80 Nm).
    - Flywheel-to-converter nuts and tighten them to 26–36 ft. lbs. (35–49 Nm)
    - Transaxle housing cover and tighten the bolts to 40–58 ft. lbs. (55–80 Nm)
    - Engine support crossmember. Tighten the crossmember bolts and nuts to 47–65 ft. lbs. (64–89 Nm) and the insulator nuts to 28–37 ft. lbs. (38–51 Nm).
    - Front and rear upper transaxle-to-engine bolt and tighten it to 28–38 ft. lbs. (38–51 Nm)

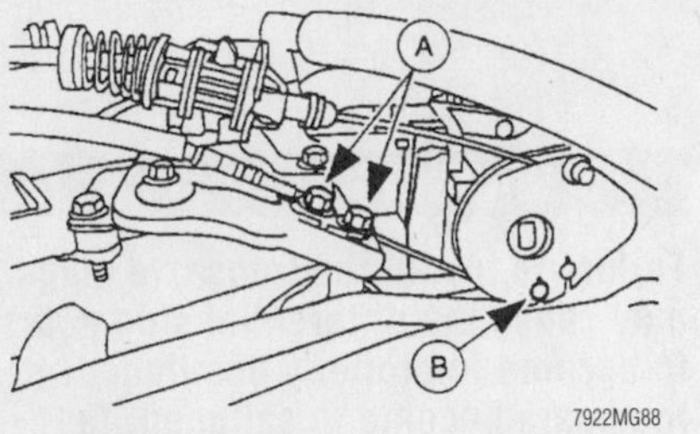

Connect the throttle valve cable at the throttle cam (B) and tighten the cable retaining bolts (A)

11. Install or connect the following:
    - Halfshafts
    - Crossmember to the transaxle mounts and the chassis. Tighten the crossmember-to-transaxle mount nuts to 27–38 ft. lbs. (37–52 Nm). Tighten the crossmember-to-chassis nuts and bolts to 47–66 ft. lbs. (64–89 Nm).
    - Lower transaxle-to-engine oil pan bolts and tighten to 27–38 ft. lbs. (37–52 Nm)
    - Engine/transaxle splash shields
    - Starter motor
    - All the electrical connections
    - Upper transaxle-to-engine bolts and tighten to 40–58 ft. lbs. (55–80 Nm)
    - Left-hand engine mount and tighten the nuts to 50–68 ft. lbs. (67–93 Nm)
    - Throttle valve cable at the throttle cam, if equipped
12. Remove the engine support.
    - Shift cable and bracket to the manual shift lever, the shift cable and bracket clip. Tighten the nut to 12–16 ft. lbs. (16–22 Nm).
    - Engine air cleaner assembly
    - Battery tray and battery
    - Both battery cables, negative cable last

Align the marks on the MLP switch and the manual control lever as shown

13. Add the proper type and quantity of transaxle fluid.
14. Check the transaxle for leaks and for proper operation.
15. Check the MLP switch for proper adjustment, as follows:
    a. Shift the transaxle into NEUTRAL, then align the marks on the transmission range switch and the manual control lever.
    b. Install and finger-tighten the switch retaining bolts.
    c. Attach an ohmmeter between terminals **A** and **B** on the switch as shown in the accompanying illustration.
    d. Adjust the switch by rotating the switch housing on the manual control lever until there is no continuity between the terminals.
    e. Hold the switch in place, then tighten its retaining bolts to 70–95 inch lbs. (8–11 Nm).
    f. Remove the ohmmeter and attach the switch electrical connection.
    g. Place the manual control shift outer lever in position and tighten its retaining nut to 33–47 ft. lbs. (44–64 Nm).
16. Road test the vehicle and check for proper operation.

Clutch

ADJUSTMENT

Pedal Free-Play

1. Before servicing the vehicle, refer to the precautions in the beginning of this section.
2. Depress the clutch pedal until resistance can be felt, and measure the distance between the upper clutch height and where the resistance is felt. The free-play should be 0.0–0.40 in. (5.0–13.9mm).
3. If an adjustment is necessary, turn the locknut and equalizer bar-to-clutch release lever rod.
4. After adjustment, measure the disengagement height from the upper surface of the clutch pedal pad to the carpet. The distance should be 2.3 in. (59mm).

Pedal Height

1. Measure the distance from the upper surface of the pedal pad to the carpet. The measurement should be 8.35–8.54 in. (212–217mm).
2. If an adjustment is necessary, turn the locknut and the Clutch Pedal Position

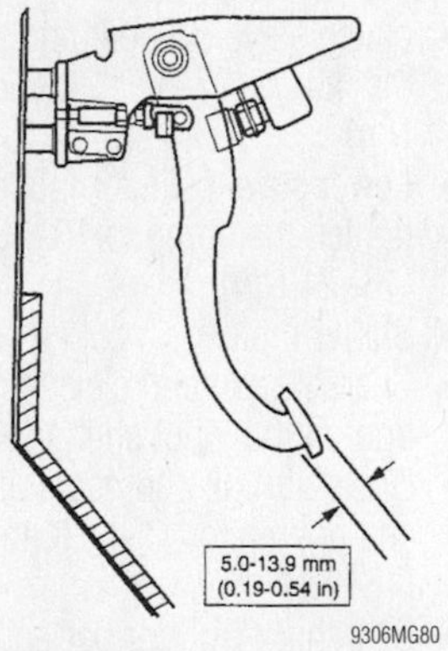

Clutch pedal free play measurement

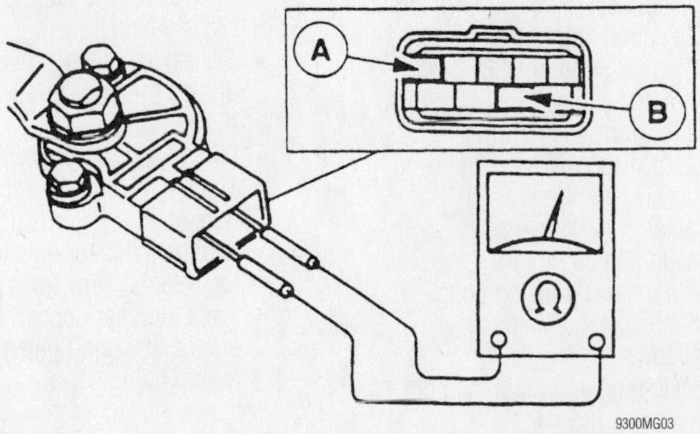

Attach an ohmmeter to terminals A and B on the transmission range switch

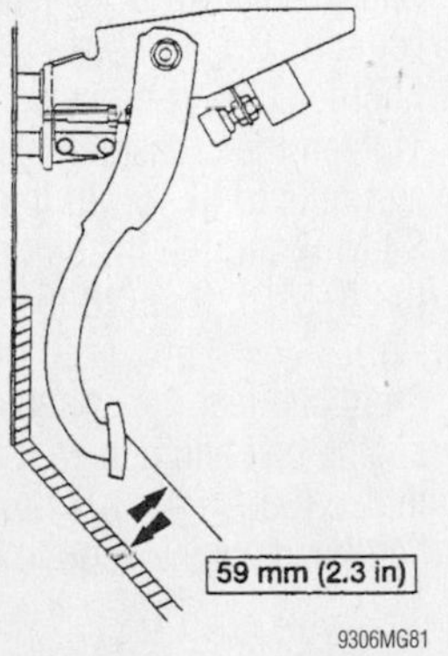

Clutch pedal free play adjustment

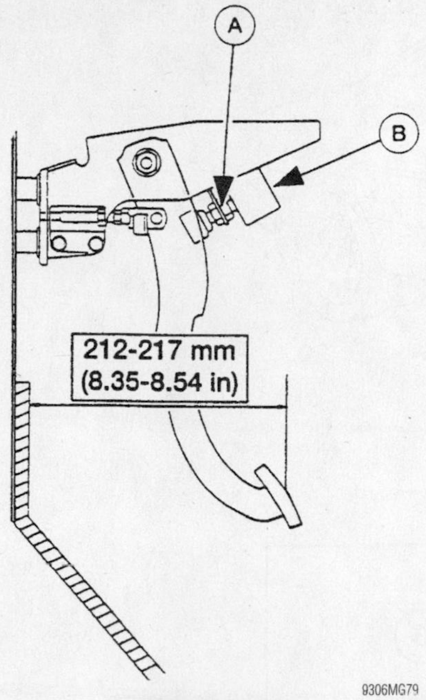

Clutch pedal height adjustment

(CPP) switch until the pedal height is correct.

REMOVAL & INSTALLATION

1. Before servicing the vehicle, refer to the precautions in the beginning of this section.
2. Disconnect the negative battery cable.
3. Raise and safely support the vehicle.
4. Remove the transaxle assembly.
5. If the clutch assembly is to be reused, matchmark the pressure plate and the flywheel so they can be assembled in the same position.
6. Install a flywheel holding tool in a transaxle mounting hole on the engine and engage the tooth of the holding tool into the flywheel ring gear.
7. Install a clutch alignment tool.
8. Loosen the pressure plate-to-flywheel retaining bolts 1 turn at a time, in a crisscross pattern, until the spring tension is relieved, to prevent pressure plate damage.

WARNING

Do not use any cleaners with a petroleum base and do not immerse the clutch pressure plate in solvent.

9. Clean the clutch pressure plate with a suitable commercial alcohol base solvent.
10. Inspect the pressure plate surface for burns, scores, flatness or ridges. Reface or replace the pressure plate.
11. Inspect the pressure plate diaphragm fingers for wear. Replace the pressure plate if necessary.
12. Using a slide caliper, measure the depth of the rivet heads. If the rivet head is within 0.012 in. (0.3mm) from the clutch surface, replace the clutch.

➡Use emery cloth to remove minor imperfections from the clutch disc lining surface.

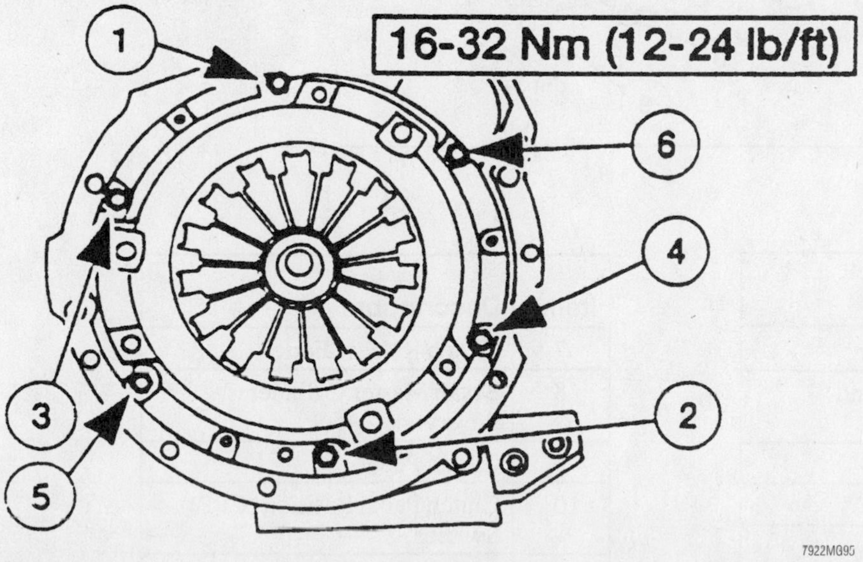

Tighten the pressure plate-to-flywheel retaining bolts in the sequence illustrated to specification

13. Inspect the clutch disc for the following:
    - Oil or grease saturation
    - Worn or loose facings
    - Warpage or loose rivets at the hub
    - Loose or broken torsion dampening springs
    - Wear or rust on the splines.
14. If the clutch disc shows any of these conditions, it should be replaced.
15. Use a dial indicator mounted on a metal base to measure the clutch disc run-out. If the run-out exceeds 0.0276 in. (0.700mm), replace the disc.
16. Inspect the clutch release for distortion, cracks, excessive release bearing surface wear or damaged tines and replace as necessary.

To install:

17. Clean the pressure plate and flywheel surfaces thoroughly. Position the clutch disc and pressure plate into the installed position and support them with a clutch aligning tool. If the clutch assembly is being reused, align the matchmarks that were made during the removal procedure.
18. Install the pressure plate-to-flywheel retaining bolts. Tighten the bolts in the correct sequence to 12–24 ft. lbs. (16–32 Nm). Remove the alignment tool.
19. Install the transaxle assembly.
20. Lower the vehicle.
21. Bleed the hydraulic clutch system.
22. Adjust the clutch pedal free-play.
23. Connect the negative battery cable.
24. Road test the vehicle and check the clutch for proper operation.

Hydraulic Clutch System

BLEEDING

1. Before servicing the vehicle, refer to the precautions in the beginning of this section.
2. Check that the brake master cylinder is at least ¾ full during the entire bleeding process.
3. Remove the bleeder screw cap from the clutch slave cylinder and attach a hose to the bleeder screw. Place the other end of the hose into a container to catch the fluid.
4. Have an assistant slowly pump the clutch pedal several times, then hold the clutch pedal down.
5. Loosen the bleeder screw to release the fluid and air. Tighten the bleeder screw.
6. Repeat the bleeding procedure until no more air bubbles are seen in the fluid.

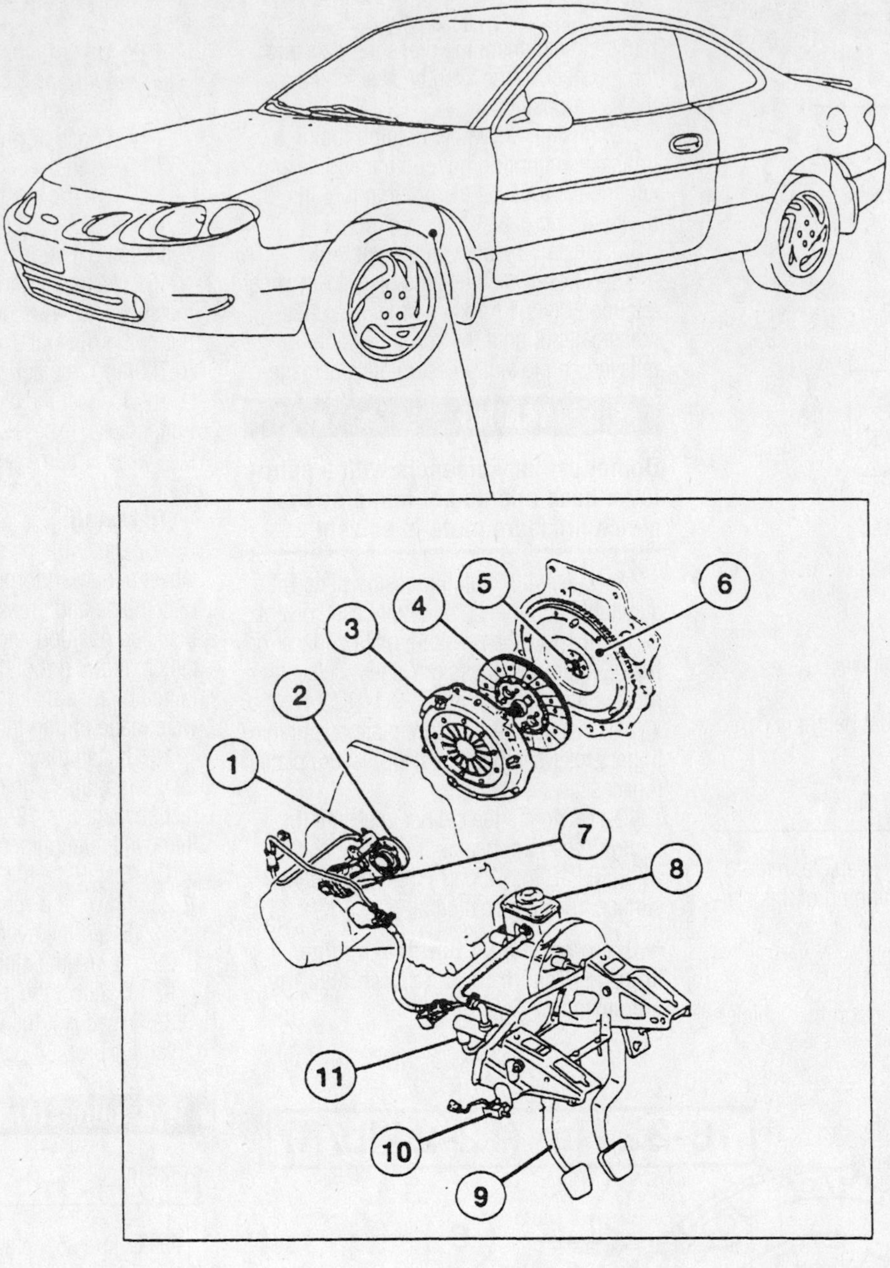

Item	Description
1	Clutch Slave Cylinder
2	Clutch Release Hub and Bearing
3	Clutch Pressure Plate
4	Clutch Disc
5	Pilot Bearing
6	Flywheel

Item	Description
7	Clutch Release Fork
8	Brake Master Cylinder Reservoir
9	Clutch Pedal
10	Clutch Pedal Position (CPP) Switch
11	Clutch Master Cylinder

9306MG78

Exploded view of the clutch system components

7. Tighten the bleeder screw to 52–78 inch lbs. (6–9 Nm).
8. Top off the brake master cylinder to the full line.
9. Check for proper clutch system operation.

Halfshafts

REMOVAL & INSTALLATION

1. Before servicing the vehicle, refer to the precautions in the beginning of this section.
2. Disconnect the negative battery cable.
3. With the vehicle sitting on the ground, carefully raise the staked portion of the halfshaft retaining nut using a suitable small chisel. Loosen the nut.
4. Remove or disconnect the following:
   - Wheel and tire assembly
   - Halfshaft retaining nut and discard
   - Cotter pin and nut from the tie rod end, then separate the tie rod end from the steering knuckle using a suitable removal tool. Discard the cotter pin.
5. Remove or disconnect the following components to remove the stabilizer bar link:
   - Stabilizer bar end nut (1) and bolt (2)
   - Stabilizer bar end retainer (3)
   - End bushing (4) above the stabilizer bar
   - End bushing (5) below the stabilizer bar
   - Stabilizer bar spacer (6)
   - Stabilizer bar end bushings from above (7) and below (8) the subframe
   - Lower stabilizer bar end retainer (9)
   - Stabilizer bar

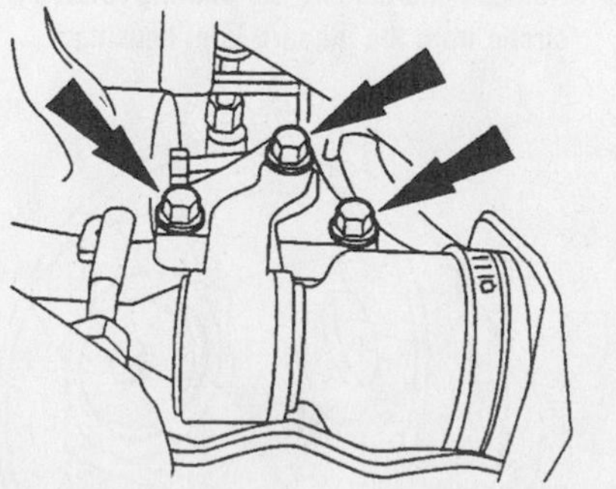
7922MG97

On models with a manual transaxle, unfasten the center support bearing bolts

6. Remove the ball joint bolt and nut. Carefully pry down on the lower control arm to separate the ball joint stud from the steering knuckle.
7. Pull outward on the steering knuckle/brake assembly. Carefully pull the halfshaft from the hub and position it aside.

➡Removal of the left side halfshaft requires removal of the crossmember to allow access with a prybar.

8. Support the transaxle with a suitable transaxle jack.
9. Unfasten the 4 transaxle crossmember retainers and remove the crossmember.
10. If removing the right side halfshaft, remove the right-hand shield and splash shield.
11. Position a drain pan under the transaxle.
12. Remove the left-hand halfshaft as follows:
    a. Insert a prybar between the halfshaft and the transaxle case. Gently pry outward to release the halfshaft from the differential side gear. Be careful not to damage the transaxle case, oil seal, CV-joint or CV-joint boot.
    b. Remove the halfshaft.
13. Remove the right-hand halfshaft as follows:
    a. On models with a manual transaxle, unfasten the center support bearing bolts.
    b. Lower the halfshaft and remove it from the transaxle.
    c. Separate the halfshaft from the center support bearing and remove the halfshaft from the vehicle.
    d. Inspect the center support bearing for damage and replace as necessary.

➡Install suitable plugs after removing the halfshafts to prevent the differential side gears from moving out of place. Should the gears become misaligned, the differential will have to be removed from the transaxle to align the gears.

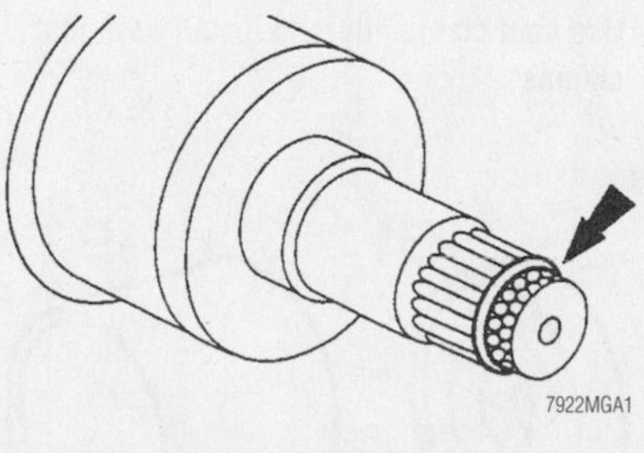
7922MGA1

During assembly, be sure to install a new circlip on the inner CV-joint spline—2.0L engine

To install:

➡Use new locknuts, split pins and circlips for assembly.

14. Position a new circlip on the inner CV-joint spline so the circlip gap is at the top. Lubricate the splines lightly with a suitable grease.
15. Remove the plugs that were installed in the differential side gears.
16. Position the halfshaft so the CV-joint splines are aligned with the differential side gear splines. Push the halfshaft into the differential.

➡When seated properly, the circlip can be felt snapping into the differential side gear groove.

17. Install the right-hand halfshaft on models with a manual transaxle as follows:
    a. Position the halfshaft and joint so that the splines line up with the splines in the halfshaft, and push the halfshaft, joint and halfshaft together with the center support bearing.
    b. Install the halfshaft in the transaxle.
18. Place the center support bearing into position and tighten it retaining bolts to 32–46 ft. lbs. (46–62 Nm).
    a. Install the right-hand shield and splash shield.
19. Pull outward on the steering knuckle/brake assembly and insert the halfshaft into the hub.
20. Pry downward on the lower control arm and position the lower ball joint stud in the steering knuckle.
21. Install the crossmember and the crossmember-to-frame bolts. Tighten the bolts to 69–93 ft. lbs. (94–126 Nm).
22. Remove the transaxle jack.
23. Install or connect the following:
    - Steering knuckle and ball joint nut and bolt. Tighten the nut and bolt to 32–43 ft. lbs. (43–59 Nm).
    - Stabilizer bar link in reverse order of removal. Tighten the bar end nut until the protruding bar end bolt length is 0.67–75 in. (17–19mm).
    - Tie rod end to the steering knuckle
    - Tie rod end nut and tighten to 32–41 ft. lbs. (43–56 Nm). Install a new cotter pin.
    - Wheel and tire assembly
    - New halfshaft retaining nut and

tighten to 174–235 ft. lbs. (235–319 Nm). Stake the halfshaft retaining nut using a suitable chisel with a rounded cutting edge.

24. Check and refill the transaxle with the proper type and quantity of fluid.
25. Connect the negative battery cable.
26. Road test the vehicle and check for proper operation.

CV-Joints

OVERHAUL

Outboard Side

1. Before servicing the vehicle, refer to the precautions in the beginning of this section.
2. Remove or disconnect the following:
   - Halfshaft from the vehicle. Support the assembly in a vise with soft jaws.
   - Front brake anti-lock sensor indicator off the halfshaft using a punch to drive it off
   - Two halfshaft boot clamp and discard
3. Slide the boot back out of the way to access the outboard joint.
4. Matchmark the joint-to-halfshaft for reassembly.
5. Use a soft faced mallet to separate the outboard joint by gently tapping it off the halfshaft.
6. Remove or disconnect the following:
   - Halfshaft bearing retaining circlip and discard
   - snapring from the outboard side of the halfshaft

➡Do not remove the tape until after installation of the boot.

7. Wrap the outboard halfshaft splines with tape and then slide the boot from the shaft.
8. Clean and inspect the outboard bearings for damage, grit in the grease, pitting or cracks and replace as necessary.

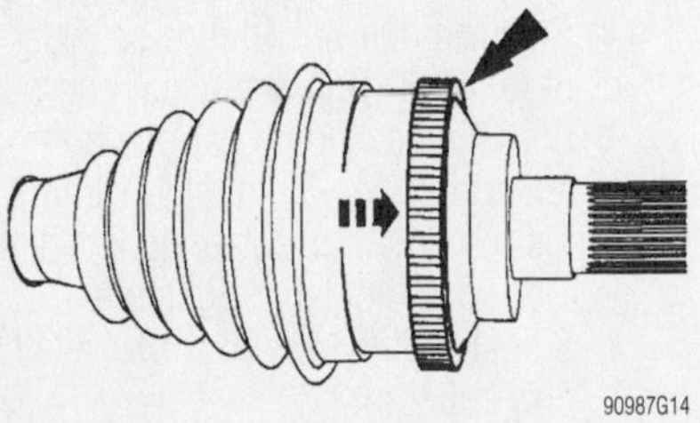

90987G14

If equipped with ABS, use a punch to drive the front brake anti-lock sensor indicator off the halfshaft

To install:

9. Lubricate the joint bearings with constant velocity grease E43Z-19590-A.
10. Install or connect the following:
    - Boot and remove the tape
    - Snapring on the outboard side of the halfshaft
    - New bearing retaining circlip
11. Use a soft faced mallet to install the outboard joint by gently tapping it onto the halfshaft.
12. Remove any excess grease from the mating surfaces and slide the boot forward onto the joint.
13. Insert a suitable cloth covered tool between the boot and the outer bearing race to allow trapped air to escape from the boot.
14. Use boot clamp pliers to install two new boot clamps.
15. If equipped with anti-lock brakes, use sensing ring replacer tool T94P-20202-B, to install the front brake anti-lock sensor indicator onto the halfshaft.
16. Install the halfshaft.

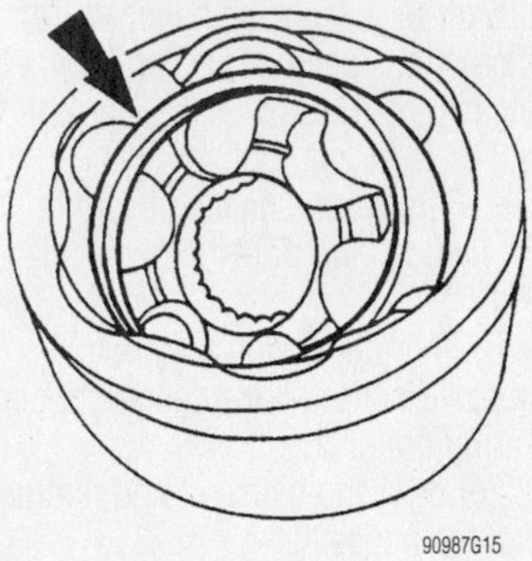

90987G15

Clean and inspect the outboard bearings for damage

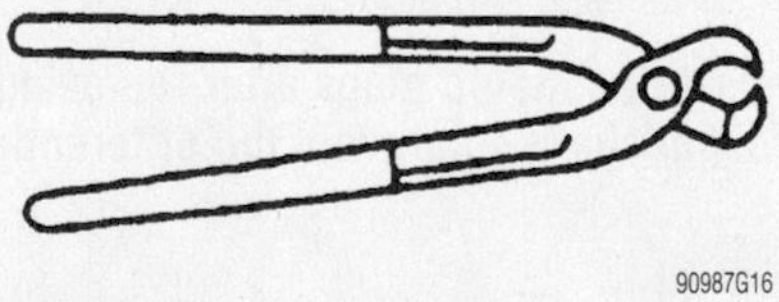

90987G16

Use boot clamp pliers to install new boot clamps

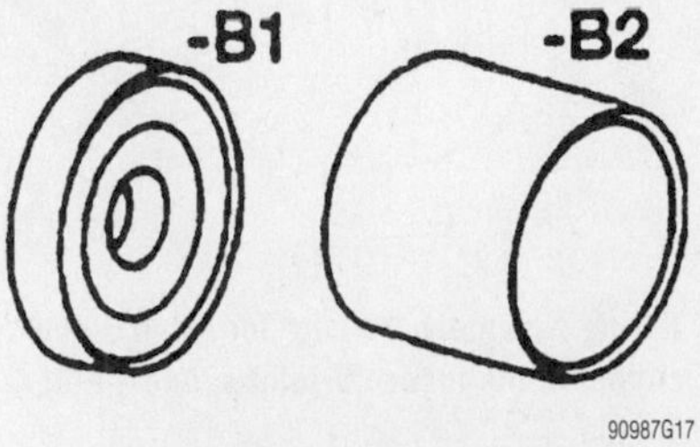

90987G17

Use sensing ring replacer tool T94P-20202-B to install the front brake anti-lock sensor indicator onto the halfshaft

INBOARD SIDE

1. Before servicing the vehicle, refer to the precautions in the beginning of this section.
2. Remove or disconnect the following:
   - Halfshaft from the vehicle. Support the assembly in a vise with soft jaws.

➡The right-hand side halfshaft on models equipped with a manual transaxle do not have an inboard circlip.

   - Bearing retaining circlip from the inboard joint housing and discard
   - Joint boot clamps and discard
   - Joint boot from the joint housing
   - Joint housing from the tripod bearing and halfshaft
   - Tripod bearing snapring
3. Matchmark the tripod bearing-to-halfshaft for reassembly.
4. Remove the tripod bearing from the halfshaft.

➡Do not remove the tape until after installation of the boot.

5. Wrap the halfshaft splines with tape and then slide the boot from the shaft.

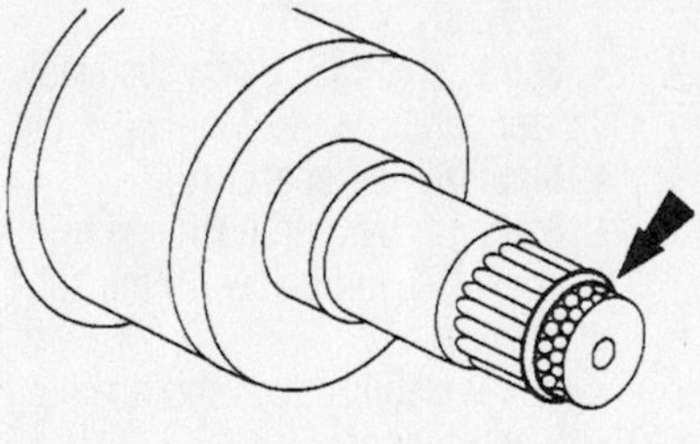

90987G18

Remove and discard the bearing retaining circlip from the inboard joint housing

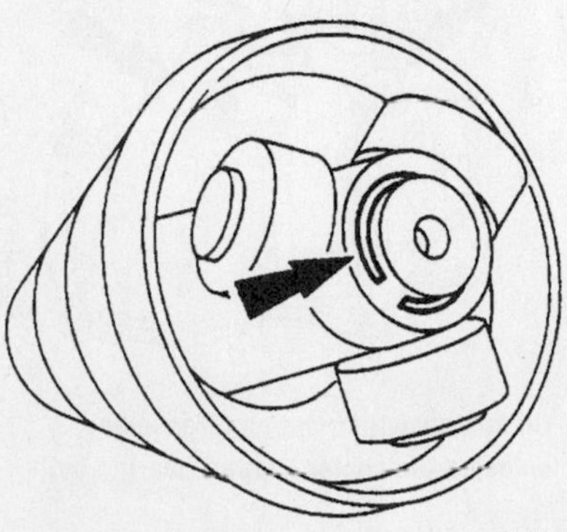

90987G19

Location of the tripod bearing snapring

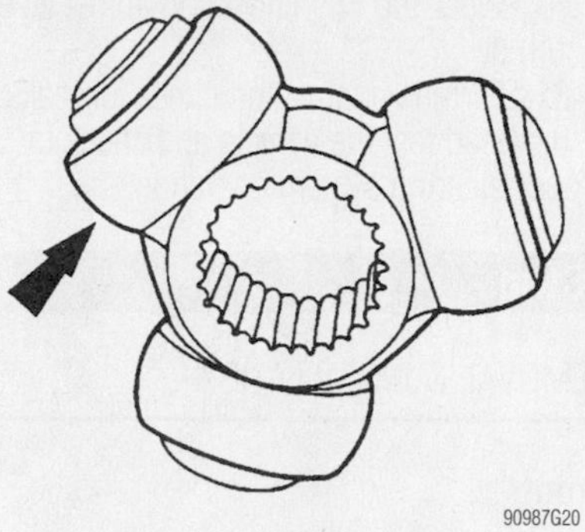

Clean and inspect the tripod bearing assembly . . .

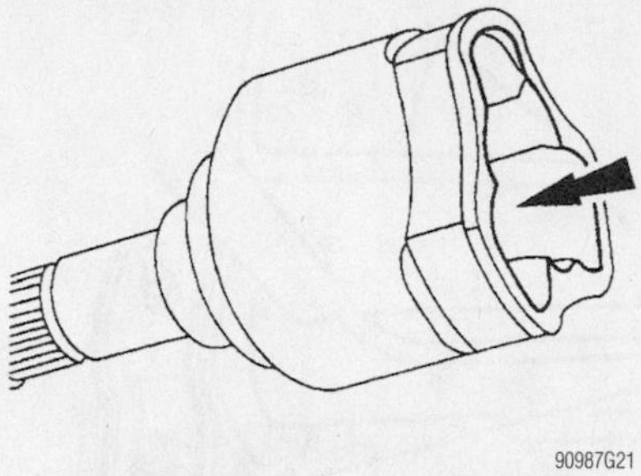

. . . and outboard joint housing for damage, grit in the grease, pitting or cracks, and replace as necessary

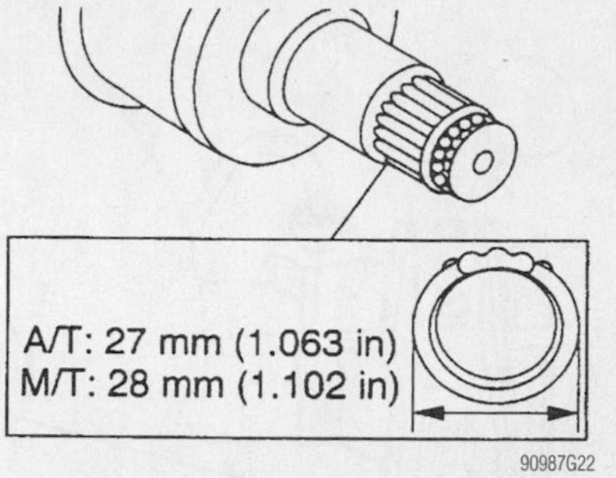

Measure the relaxed state of the halfshaft bearing circlip installed on the shaft, and make sure it is within specification

6. Clean and inspect the tripod bearing assembly and outboard joint housing for damage, grit in the grease, pitting or cracks and replace as necessary.

To install:

7. Lubricate the tripod bearing and inboard joint housing with constant velocity grease E43Z-19590-A.
8. Install or connect the following:
   - Boot and remove the tape
   - Tripod bearing onto the halfshaft
   - Tripod bearing snapring
   - Inboard joint housing onto the tripod bearing
   - Boot onto the inboard joint housing
9. Insert a suitable cloth covered tool between the boot and the bearing to allow trapped air to escape from the boot.
10. Use boot clamp pliers to install two new boot clamps.

➡Measure the relaxed state of the halfshaft bearing circlip installed on the shaft. Refer to the accompanying illustration. The wrong size clip will not retain the halfshaft properly.

11. Install a new halfshaft bearing retainer circlip of the correct size onto the outboard halfshaft joint.
12. Install the halfshaft.

STEERING AND SUSPENSION

Air Bag

PRECAUTIONS

Several precautions must be observed when handling the inflator module to avoid accidental deployment and possible personal injury.

- Never carry the inflator module by the wires or connector on the underside of the module.
- When carrying a live inflator module, hold securely with both hands, and ensure that the bag and trim cover are pointed away.
- Place the inflator module on a bench or other surface with the bag and trim cover facing up.
- With the inflator module on the bench, never place anything on or close to the module that may be thrown in the event of an accidental deployment.

DISARMING

1. Before servicing the vehicle, refer to the precautions in the beginning of this section.
2. Disconnect both battery cables, negative cable first.
3. Wait at least 1 minute. This allows time for the back-up power supply to deplete its stored energy.
4. Remove the driver's side air bag module, then the passenger side if required.
5. Use caution when carrying live air bags. Always place the air bag with the cover up.
6. If the battery needs to be reconnected while one or both of the air bags are removed from the system, install Air Bag Simulator 105–00010, to the drivers side and/or passenger side air bag harness connectors as required. Before removing either air bag simulator, disconnect both battery cables and wait at least 1 minute before continuing.

ARMING

1. Once service is completed and the air bag modules are back in place, connect the negative battery cable and prove out the air bag system by turning the ignition key to the **RUN** position and visually monitoring the air bag indicator lamp in the instrument cluster. The indicator lamp should illuminate for approximately 6 seconds, then turn **OFF**. If the indicator lamp does not illuminate, stays **ON**, or flashes at any time, a fault has been detected by the air bag diagnostic monitor requiring immediate attention.

Rack and Pinion Steering Gear

REMOVAL & INSTALLATION

1. Before servicing the vehicle, refer to the precautions in the beginning of this section.
2. Turn the key to the **ACC** position.
3. Remove or disconnect the following:
   - Steering column tube boot nuts at the base of the column and the tube boots, from inside the passenger compartment
   - Steering column input shaft cou-

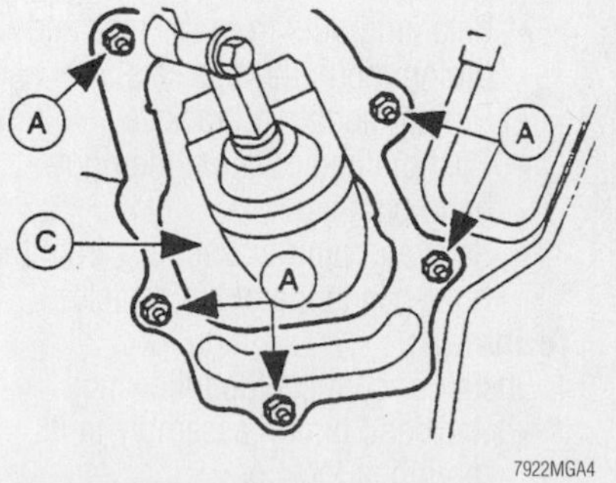

Unfasten the steering column tube boot nuts at the base of the column and remove the tube boots

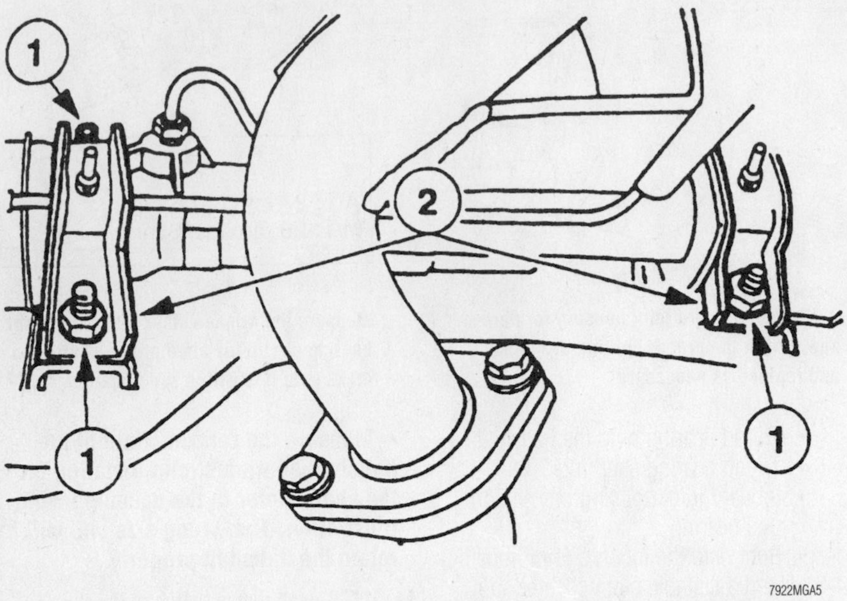

Unfasten the retaining nuts from the rack and pinion mounting brackets and remove the brackets

pling-to-steering gear input shaft pinch bolt
- Front wheel and tire assemblies
- Separate the tie rod ends from the steering knuckles and discard the cotter pins
- Right-hand lower splash shield
- Crossmember
- Pressure and return lines from the rack and pinion assembly and plug the lines
- Strap holding the hoses to the steering gear and discard
- Gearshift rod and clevis from the transaxle, if equipped with a manual transaxle
- Extension bar nut and disconnect the gearshift lever stabilizer bar and support from the transaxle, if equipped with a manual transaxle
- Retaining nuts from the rack and pinion mounting brackets
- Pinion mounting brackets
- Pushpin and position the right-hand boot aside
- Rack and pinion assembly from the right-hand side of the vehicle

To install:

4. Install or connect the following:
   - Rack and pinion assembly in its mounting location
5. Align the steering column input shaft coupling and the steering gear input shaft.
   - Rack and pinion mounting brackcts. Tighten the retaining nuts to 28–38 ft. lbs. (37–57 Nm).
   - Gearshift stabilizer bar, support and the gearshift rod and clevis, if equipped with a manual transaxle

➡Install new Teflon seals on the power steering pressure hose fitting and be sure the threads are clean before connecting the hose.

   - Plugs and connect the pressure and return lines to the rack and pinion assembly. Tighten the pressure hose to 21–25 ft. lbs. (28–33 Nm) and the return line fitting to 20–25 ft. lbs. (27.3–33.9 Nm).
   - New strap to hold the power steering lines to the rack and pinion housing
   - Crossmember and tighten the bolts to 69–97 ft. lbs. (94–131 Nm)
   - Right-hand boot shield into position
   - Shield pushpin
   - Right-hand splash shield
   - Tie rod ends to the steering knuckles and the castellated nuts. Tighten to specification.
   - New cotter pins
   - Wheel and tire assemblies
   - Steering column input shaft coupling-to-steering gear input shaft pinch bolt and tighten the bolt to 30–36 ft. lbs. (40–50 Nm)
   - Steering column tube boots and the 5 retainers. Tighten the retainers to 18–52 inch lbs. (2–5.9 Nm).
   - Negative battery cable
6. Fill and bleed the power steering system.
7. Check the alignment and adjust as required.
8. Start the engine and check for leaks.
9. Road test the vehicle and check for proper steering system operation.

Strut

REMOVAL & INSTALLATION

Front

1. Before servicing the vehicle, refer to the precautions in the beginning of this section.
2. Remove or disconnect the following:
   - Negative battery cable
   - Front wheel and tire assembly
   - Clip securing the brake hose to the strut (spring and shock) assembly.
   - Anti-lock brake harness cable and clip, if equipped with anti-lock brakes
   - 2 nuts and 2 bolts securing the strut assembly to the steering knuckle
   - 4 upper strut retaining nuts
   - Strut assembly from the vehicle

⁂ CAUTION

Never remove the strut piston rod nut unless the coil spring is compressed. Always wear safety glasses when using a spring compressor.

3. Inspect all components and replace as needed.

To install:

4. Install or connect the following:
   - Strut assembly into the wheel housing. Be sure the direction indicator on the upper mounting bracket faces inboard.
   - Upper mounting bracket to the strut tower with the retaining nuts. Tighten the nuts to 22–30 ft. lbs. (29–40 Nm).
   - Strut assembly to the steering knuckle and the retaining bolts and nuts. Tighten to 69–93 ft. lbs. (93–127 Nm).
   - Brake hose on the strut assembly and secure it with the brake hose clip
   - Anti-lock harness cable and clip, if equipped with anti-lock brakes
   - Front wheel and tire assembly
   - Negative battery cable
5. Check the front wheel alignment.
6. Road test the vehicle and check for proper operation.

1	Front Shock Absorber Upper Mounting Bracket
2	Upper Spring Seat (Part of 18198)
3	Front Brake Hose
4	Disc Brake Caliper
5	Front Disc Brake Rotor
6	Front Wheel Knuckle
7	Tie Rod End
8	Front Stabilizer Bar

9	Front Suspension Lower Arm Mounting Bolt Bushing (Rear)
10	Front Wheel Spindle Tie Rod
11	Front Suspension Lower Arm
12	Front Suspension Lower Arm Mounting Bolt Bushing (Front)
13	Front Shock Absorber
14	Front Coil Spring

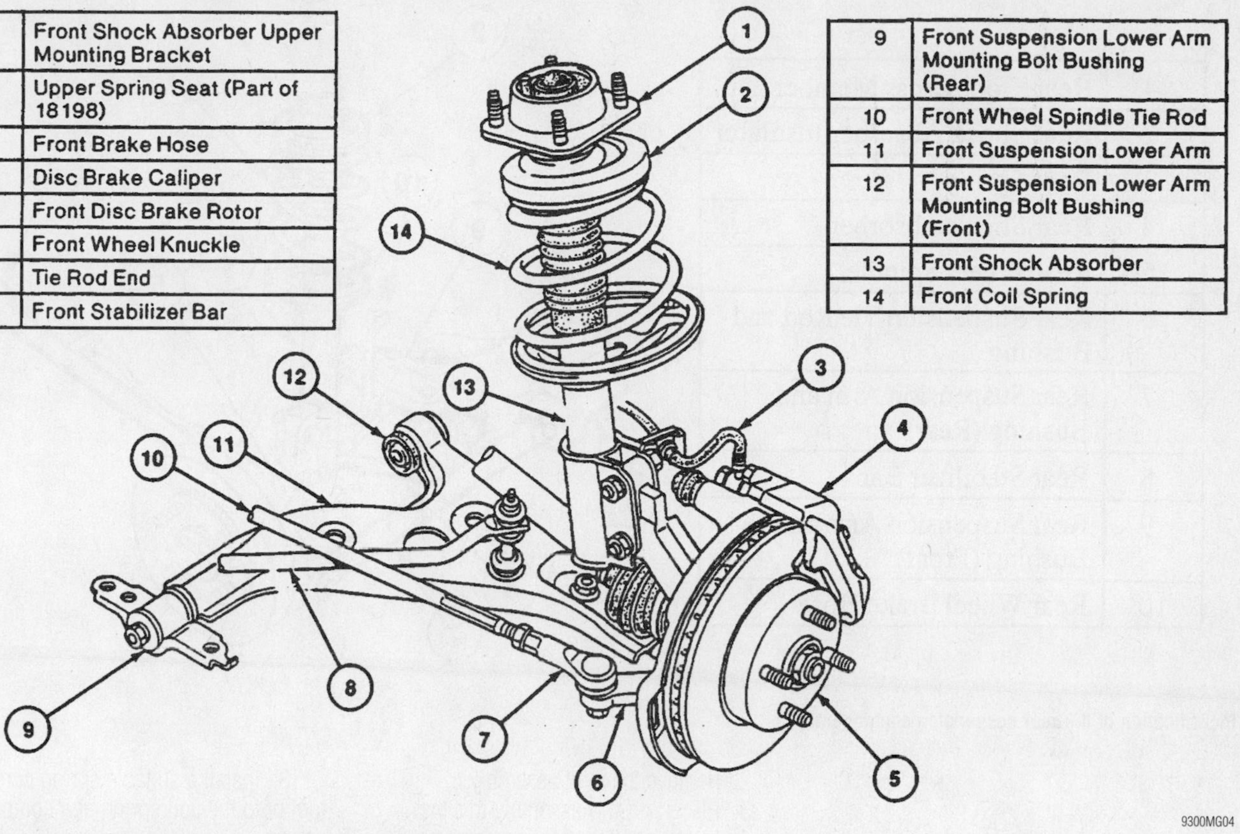

Identification of the front suspension components

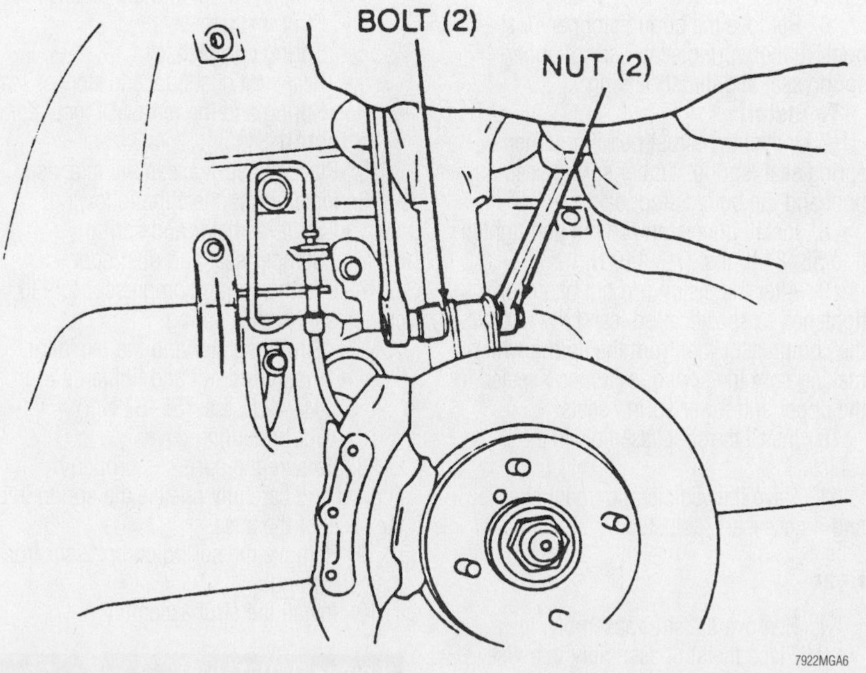

Remove the 2 nuts and 2 bolts securing the strut assembly to the steering knuckle

Rear

1. Before servicing the vehicle, refer to the precautions in the beginning of this section.
2. Remove or disconnect the following:
   - Negative battery cable
   - Package tray trim panel, on sedan and coupe models
   - Quarter trim pane, on wagon models
   - 2 upper strut retaining nuts
   - Wheel and tire assembly
   - Clip securing the brake hose to the rear strut assembly
   - ABS sensor bolt, if equipped
   - Nuts and bolts securing the rear strut assembly to the wheel spindle
   - strut assembly from the vehicle

To install:

3. Install or connect the following:
   - Strut assembly into the vehicle wheel housing
   - Nuts and bolts securing the strut assembly to the rear wheel spindle assembly. Tighten the lower strut bolts and nuts to 76–100 ft. lbs. (103–136 Nm).

1	Rear Floor Cross Member
2	Rear Shock Absorber Insulator
3	Rear Spring
4	Rear Shock Absorber
5	Rear Wheel Spindle
6	Rear Suspension Tie Rod and Bushing
7	Rear Suspension Arm and Bushing (Rear)
8	Rear Stabilizer Bar
9	Rear Suspension Arm and Bushing (Front)
10	Rear Wheel Brake Hose

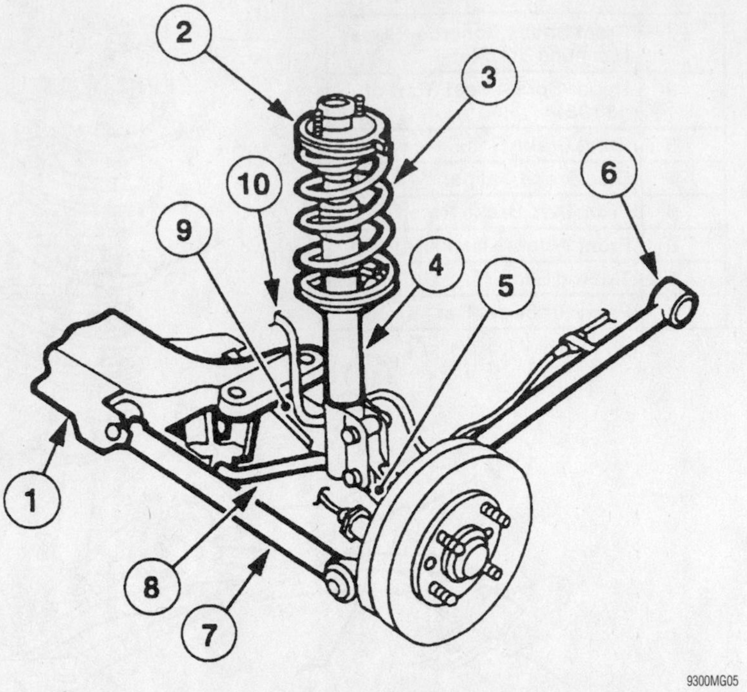

Identification of the rear suspension components

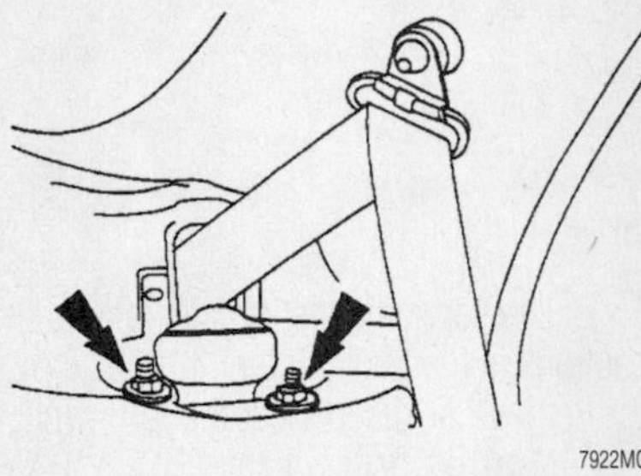

Location of the 2 rear strut upper retaining nuts (arrows) inside the vehicle

- ABS sensor bolt, if equipped
- Clip securing the flexible brake hose to the rear strut assembly
- Wheel and tire assembly
- 2 upper strut retaining nuts and tighten to 34–46 ft. lbs. (47–62 Nm)
- Trim panel
- Negative battery cable

4. Check the rear wheel alignment.
5. Road test the vehicle and check for proper operation.

Coil Spring

REMOVAL & INSTALLATION

Front

1. Before servicing the vehicle, refer to the precautions in the beginning of this section.
2. Remove the strut assembly.
3. Place the strut assembly in a vise.
4. Install a suitable spring compressor tool onto the coil spring and compress the spring.
5. Unfasten the piston rod nut and remove the upper mounting bracket.
6. Remove the bound stopper, dust boot, rubber spring seat, spring, upper spring seat and thrust bearing.

To install:

7. Install the thrust bearing, upper spring seat, spring, rubber spring seat, dust boot and the bound stopper.
8. Install the piston rod nut and tighten it to 58–81 ft. lbs. (79–110 Nm).
9. After the piston rod nut has been tightened to specification, carefully remove the compressor tool from the spring while making sure the spring is properly seated in the upper and lower spring seats.
10. Install the strut assembly in the vehicle.
11. Have the vehicle alignment checked and if necessary, adjusted.

Rear

1. Remove the strut assembly.
2. Place the strut assembly in a vise.

CAUTION

Always take the necessary precautions when using a spring compressor

3. Install a suitable spring compressor tool onto the coil spring and compress the spring.
4. Remove or disconnect the following:
   - Rear strut assembly top mounting cover
   - Piston rod nut and the retainer
   - Strut insulator
   - Spring compressor
   - Rear strut dust boot and stopper seat
   - Spring and the rear strut insulator

To install:

5. Place the strut assembly in a vise.
6. Install or connect the following:
   - Strut insulator and spring
   - Stopper seat and dust boot
7. Use the spring compressor tool to compress the strut spring.
   - Strut insulator and the retainer
   - Piston rod nut and tighten it to 41–49 ft. lbs. (55–67 Nm)
   - Top mounting cover
8. Be sure the spring is properly aligned and carefully release the spring into the seats of the strut.
9. Remove the spring compressor from the coil spring.
10. Install the strut assembly.

Lower Ball Joint

REMOVAL & INSTALLATION

1. Before servicing the vehicle, refer to the precautions in the beginning of this section.

2. Remove or disconnect the following:
   - Wheel assembly
   - Nut and bolt attaching the lower ball joint to the steering knuckle
   - Ball joint from the steering knuckle
   - Lower control arm ball joint nuts and bolt
   - Ball joint

To install:

3. Install or connect the following:
   - Ball joint into position on the lower control arm
   - Ball joint-to-lower control arm retainers. Tighten the nuts and bolt to 69–86 ft. lbs. (93–117 Nm).
4. Apply Loctite® 290 thread locking compound to the ball joint nut and threads.
   - Ball joint to the steering knuckle
   - Ball joint nut and bolt. Tighten the nut and bolt to 32–43 ft. lbs. (43–59 Nm).
   - Wheel assembly
   - Check and adjust the wheel alignment as necessary

Control Arm

REMOVAL & INSTALLATION

Front

1. Before servicing the vehicle, refer to the precautions in the beginning of this section.
2. Remove or disconnect the following:
   - Front wheel and tire assembly
   - Stabilizer bar link nuts, retainers, bushings, bolts and sleeves
   - Pinch bolt and nut securing the ball joint to the steering knuckle
   - Lower ball joint from the steering knuckle by prying the lower control arm down with a prybar
   - Front lower control arm pivot bolt
   - Three lower control arm retaining bolts and the lower control arm

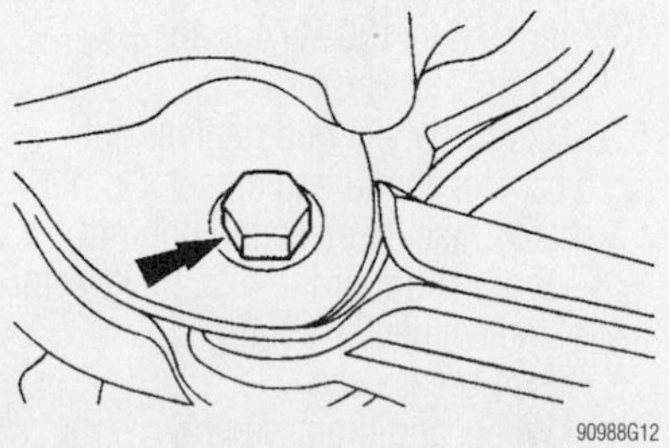

Unfasten of the lower control arm pivot bolt (arrow)

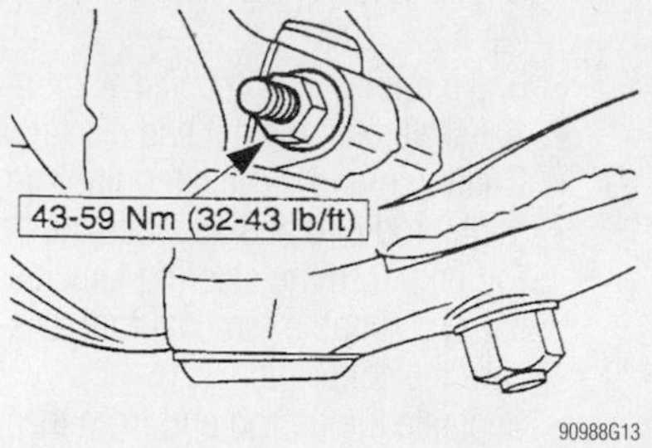

Engage the lower ball joint to the steering knuckle and tighten the pinch bolt and nut to specification

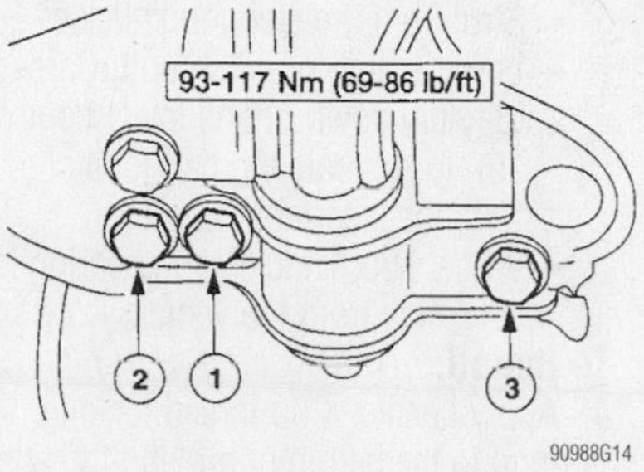

Tighten the lower control arm retaining bolts in the sequence shown to the proper specification

3. Inspect the lower control arm, lower control arm bushings and the lower ball joint. The ball joint and bushings can be replaced individually.

To install:

4. Install or connect the following:
   - Lower control arm into position at the bushings
   - Lower ball joint to the steering knuckle and tighten the pinch bolt and nut to 32–43 ft. lbs. (43–59 Nm)
5. Use a transmission jack to raise the lower control arm so that it is parallel with the ground.
   - Control arm retaining bolts and tighten them to 69–86 ft. lbs. (93–117 Nm) in the order shown in the accompanying illustration
   - Control arm pivot bolt and tighten to 69–86 ft. lbs. (93–117 Nm)
   - Stabilizer bolts, washers, bushings, sleeves and nuts. Tighten the stabilizer nuts so 0.67–0.75 inches (17–19mm) of thread is exposed at the end of the bolt.
   - Wheel and tire assembly
6. Check the front wheel alignment.
7. Road test the vehicle and check for proper operation.

CONTROL ARM BUSHING REPLACEMENT

1. Before servicing the vehicle, refer to the precautions in the beginning of this section.
2. Remove or disconnect the following:
   - Control arm
   - Nut and control arm mounting bushing (rear) and washer
   - Front bushing from the arm using a suitable control arm bushing tool, C-frame and clamp assembly
   - Front bushing into the arm using a suitable control arm bushing tool, C-frame and clamp assembly
   - Washer, control arm mounting bushing (rear) and nut. Tighten the nut to 69 ft. lbs. (93 Nm).
   - Control arm

Rear

1. Before servicing the vehicle, refer to the precautions in the beginning of this section.
2. Position a floor jackstand beneath the rear suspension crossmember.
3. Remove or disconnect the following:
   - Wheel and tire assembly
   - Stabilizer nuts, washers, bushings, sleeves and bolts
   - Bolts securing the stabilizer bar brackets and grommets to the rear suspension crossmember
   - Stabilizer bar
   - Bolts securing the rear suspension crossmember to the vehicle frame
4. Lower the floor jackstand to allow the rear suspension crossmember to be lowered from the vehicle frame.
   - Control arm and bushing nut
   - Control arm and bushing bolt
   - Control arm

To install:

5. Install or connect the following:
   - Control arm into position
   - Retaining bolt and tighten it to 50–70 ft. lbs. (68–95 Nm)

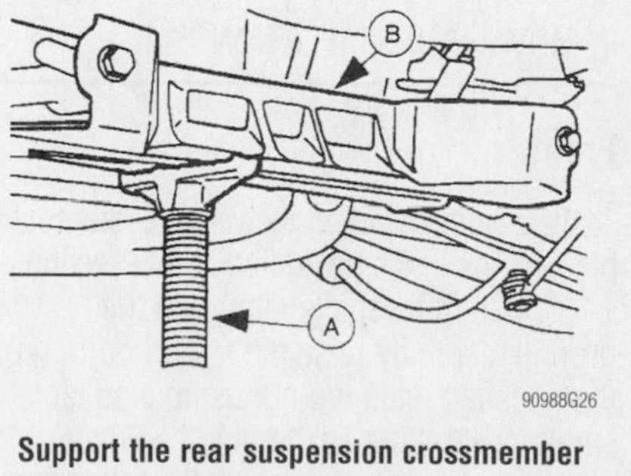

Support the rear suspension crossmember (B) with a floor jackstand (A)

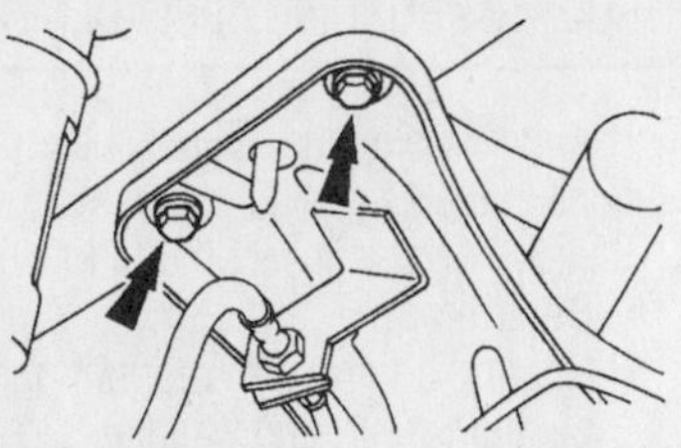

90988G27

Unfasten the rear suspension crossmember bolts, then lower the floor jackstand and crossmember

- Control arm and bushing retaining nut, then tighten the nut to 64–86 ft. lbs. (87–116 Nm)

6. Raise the floor jack and place the crossmember into position.
   - Crossmember bolts and tighten them to 34–46 ft. lbs. (47–62 Nm)
   - Grommets onto the stabilizer bar and align the grommets to the positions painted on the bar
   - Stabilizer bar to the rear suspension crossmember and secure it in place with the straps and bolts. Tighten the bolts to 32–43 ft. lbs. (43–59 Nm).
   - Stabilizer bolts, washers, grommets, sleeves and nuts. Tighten the stabilizer nuts so 0.64–0.71 in. (16.2–18.2mm) of thread is exposed at the end of the bolt.
   - Wheel and tire assembly
7. Check the wheel alignment.

Wheel Bearings

ADJUSTMENT

The bearings on the front and rear wheels are a one piece cartridge design and cannot be adjusted. Wheel bearing play can be checked with a dial indicator. If wheel bearing play exceeds 0.002 in. (0.05mm) check the wheel hub retainer nut for proper torque. If the torque is correct, replacement of the wheel bearing is required.

REMOVAL & INSTALLATION

Front

1. Before servicing the vehicle, refer to the precautions in the beginning of this section.
2. With the vehicle sitting on the ground, carefully raise the staked portion of the halfshaft retaining nut using a suitable small chisel. Loosen the nut.
3. Remove or disconnect the following:
   - Wheel and tire assembly
   - Brake caliper and secure it out of the way with a piece of mechanic's wire. Do not let the caliper hang on the hose.
   - Brake rotor
   - Halfshaft retaining nut and discard it
   - Cotter pin and castellated nut from the tie rod end and separate the tie rod end from the steering knuckle using a suitable removal tool. Discard the cotter pin.
   - Separate the tie rod end from the wheel knuckle
   - ABS sensor bolt and the sensor, if equipped
   - Strut mounting nuts and the studs which attach the strut assembly to the steering knuckle
   - Strut from the steering knuckle
   - Lower ball joint pinch bolt. Carefully pry down on the lower control arm to separate the ball joint stud from the steering knuckle.
   - Wheel hub, knuckle and bearing assembly from the vehicle

To install:

4. Apply Loctite® 290 thread locking compound to the ball joint nut and threads.
5. Install or connect the following:
   - Wheel hub, knuckle and bearing assembly onto the ball joint and tighten the pinch bolt and nut to 32–43 ft. lbs. (43–59 Nm)
   - Tie rod end and tighten the nut to 25–33 ft. lbs. (34–46 Nm). Install a new cotter pin.
   - ABS sensor and tighten the bolt, if equipped
   - Knuckle to the strut assembly and tighten the strut mounting nuts to 69–93 ft. lbs. (93–127 Nm)
   - New halfshaft retaining nut and tighten to 174–235 ft. lbs. (235–319 Nm). Stake the retaining nut using a suitable chisel with the cutting edge rounded off.
   - Brake rotor and caliper
   - Wheel and tire assembly
6. Check the front wheel alignment.
7. Road test the vehicle and check for proper operation.

Rear

➡The wheel bearings are a cartridge design and are not serviceable. If bearing replacement is required, the bearings and hub must be replaced as an assembly.

1. Before servicing the vehicle, refer to the precautions in the beginning of this section.
2. Remove or disconnect the following:
   - Wheel and tire assembly
   - Hub grease cap
   - Brake drum or disc brake caliper and rotor, as necessary
   - Wheel hub retainer nut securing the hub to the spindle, unstake the nut prior to loosening
   - Hub and bearing assembly. Discard the hub retainer nut.
   - Disc brake shield retaining bolts and the shield, if equipped with disc brakes
   - Backing plate, if equipped with drum brakes
   - Strut-to-spindle retaining bolts and nuts
   - Trailing arm bolt securing the trailing arm to the spindle
   - Stabilizer bar link nuts, retainers, bushings, sleeves and bolts
   - Control arm nut and bolt securing both control arms to the wheel spindle
   - Spindle from the vehicle
3. Inspect all components. If the wheel spindle is damaged, replace it. Wheel bearings are sealed and must be replaced if damaged with the wheel hub.

To install:

4. Install or connect the following:
   - Wheel spindle in position to the strut
   - Spindle retaining bolts and nuts. Tighten to 69–93 ft. lbs. (93–127 Nm).
   - Trailing arm to the spindle and tighten retaining bolt to 69–93 ft. lbs. (93–127 Nm)
   - Both control arms in position to the spindle and tighten the retaining bolt and nut to 63–86 ft. lbs. (85–117 Nm)
   - Stabilizer bar link nuts, retainers, bushings, sleeves and bolts
   - Brake backing plate, if equipped with drum brakes
   - Brake shield and the disc brake shield retaining bolts, if equipped with disc brakes
   - Wheel hub and bearing assembly to the wheel spindle
   - New wheel hub retainer nut and tighten to 130–174 ft. lbs. (177–235 Nm)
5. Stake the wheel hub retainer nut using a cape or round end chisel. Do not use a sharp chisel to stake the hub nut.
   - Brake drum or the disc brake rotor and caliper, as equipped
   - Hub grease cap
   - Wheel and tire assembly
6. Pump the brake pedal several times to position the brake lining before attempting to move the vehicle.
7. Road test the vehicle and check for proper operation.

FORD MOTOR CO.

15

2000–01
Ford-Focus

PRECAUTIONS

Before servicing any vehicle, please be sure to read all of the following precautions, which deal with personal safety, prevention of component damage, and important points to take into consideration when servicing a motor vehicle:

- Never open, service or drain the radiator or cooling system when the engine is hot; serious burns can occur from the steam and hot coolant.
- Observe all applicable safety precautions when working around fuel. Whenever servicing the fuel system, always work in a well-ventilated area. Do not allow fuel spray or vapors to come in contact with a spark, open flame, or excessive heat (a hot drop light, for example). Keep a dry chemical fire extinguisher near the work area. Always keep fuel in a container specifically designed for fuel storage; also, always properly seal fuel containers to avoid the possibility of fire or explosion. Refer to the additional fuel system precautions later in this section.
- Fuel injection systems often remain pressurized, even after the engine has been turned **OFF**. The fuel system pressure must be relieved before disconnecting any fuel lines. Failure to do so may result in fire and/or personal injury.
- Brake fluid often contains polyglycol ethers and polyglycols. Avoid contact with the eyes and wash your hands thoroughly after handling brake fluid. If you do get brake fluid in your eyes, flush your eyes with clean, running water for 15 minutes. If eye irritation persists, or if you have taken brake fluid internally, IMMEDIATELY seek medical assistance.
- The EPA warns that prolonged contact with used engine oil may cause a number of skin disorders, including cancer. You should make every effort to minimize your exposure to used engine oil. Protective gloves should be worn when changing oil. Wash your hands and any other exposed skin areas as soon as possible after exposure to used engine oil. Soap and water, or waterless hand cleaner should be used.
- All new vehicles are now equipped with an air bag system, often referred to as a Supplemental Restraint System (SRS) or Supplemental Inflatable Restraint (SIR) system. The system must be disabled before performing service on or around system components, steering column, instrument panel components, wiring and sensors. Failure to follow safety and disabling procedures could result in accidental air bag deployment, possible personal injury and unnecessary system repairs.
- Always wear safety goggles when working with, or around, the air bag system. When carrying a non-deployed air bag, be sure the bag and trim cover are pointed away from your body. When placing a non-deployed air bag on a work surface, always face the bag and trim cover upward, away from the surface. This will reduce the motion of the module if it is accidentally deployed. Refer to the additional air bag system precautions later in this section.
- Clean, high quality brake fluid from a sealed container is essential to the safe and proper operation of the brake system. You should always buy the correct type of brake fluid for your vehicle. If the brake fluid becomes contaminated, completely flush the system with new fluid. Never reuse any brake fluid. Any brake fluid that is removed from the system should be discarded. Also, do not allow any brake fluid to come in contact with a painted surface; it will damage the paint.
- Never operate the engine without the proper amount and type of engine oil; doing so WILL result in severe engine damage.
- Timing belt maintenance is extremely important. Many models utilize an interference-type, non-freewheeling engine. If the timing belt breaks, the valves in the cylinder head may strike the pistons, causing potentially serious (also time-consuming and expensive) engine damage. Refer to the maintenance interval charts in the front of this manual for the recommended replacement interval for the timing belt, and to the timing belt section for belt replacement and inspection.
- Disconnecting the negative battery cable on some vehicles may interfere with the functions of the on-board computer system(s) and may require the computer to undergo a relearning process once the negative battery cable is reconnected.
- When servicing drum brakes, only disassemble and assemble one side at a time, leaving the remaining side intact for reference.

ENGINE REPAIR

➡Disconnecting the negative battery cable on some vehicles may interfere with the functions of the on board computer system. The computer may undergo a relearning process once the negative battery cable is reconnected.

Alternator

REMOVAL

SOHC Engine

1. Before servicing the vehicle, refer to the precautions in the beginning of this section.
2. Remove or disconnect the following:
   - Negative battery cable
   - Accessory drive belt
   - Power steering pipe brackets
   - Exhaust manifold heat shield
   - Alternator harness connectors
   - Alternator

DOHC Engine

1. Before servicing the vehicle, refer to the precautions in the beginning of this section.
2. Remove or disconnect the following:
   - Negative battery cable
   - Accessory drive belt
   - Alternator harness connectors
   - Coolant expansion tank
   - Power steering reservoir
   - Engine wiring harness bracket
   - Ground cable
   - Evaporative Emissions (EVAP) canister purge valve
   - Alternator

INSTALLATION

SOHC Engine

Install or connect the following:

- Alternator. Tighten the bolts to 35 ft. lbs. (45 Nm).
- Alternator harness connectors
- Exhaust manifold heat shield
- Power steering pipe brackets
- Accessory drive belt
- Negative battery cable

DOHC Engine

Install or connect the following:

- Alternator. Tighten the bolts to 18 ft. lbs. (25 Nm).
- EVAP canister purge valve
- Ground cable

- Engine wiring harness bracket
- Power steering reservoir
- Coolant expansion tank
- Alternator harness connectors
- Accessory drive belt
- Negative battery cable

Ignition Timing

ADJUSTMENT

The engines are equipped with a Distributorless Ignition System (DIS). No adjustment is necessary.

Engine Assembly

REMOVAL & INSTALLATION

Manual Transaxle

1. Before servicing the vehicle, refer to the precautions in the beginning of this section.
2. Drain the cooling system.
3. Relieve the fuel system pressure.
4. Loosen the front strut center nuts 5 turns.
5. Remove or disconnect the following:
   - Battery, tray and cables
   - Mass Air Flow (MAF) sensor connector
   - Air cleaner housing and tubes
   - Accelerator cable
   - Cruise control cable, if equipped
   - Ignition coil connectors
   - Power steering pump pressure switch connector
   - Heated Oxygen (HO_2S) sensor connectors
   - Fuel injector harness connectors
   - Powertrain Control Module (PCM) connectors
   - Vehicle Speed (VSS) sensor connector
   - Reverse lamp switch connector
   - Clutch slave cylinder fluid pipe
   - Evaporative Emissions (EVAP) canister vacuum lines
   - Power brake booster vacuum line
   - Delta Pressure Feedback Electronic (DPFE) system sensor vacuum line
   - Exhaust Gas Recirculation (EGR) vacuum line
   - Intake manifold vacuum line
   - Fuel line
   - Coolant bypass hoses
   - Radiator hoses
   - Coolant expansion tank
   - Accessory drive belt
   - Power steering pump and reservoir
   - Power steering high pressure pipe
   - Radiator cooling fan
   - Catalytic converter
   - Exhaust flex pipe
   - Drive belt cover
   - Shift cable cover
   - Shift cables
   - A/C compressor, if equipped
   - Right engine mount
   - Lower ball joints
   - Axle halfshafts
6. Support the powertrain from below
7. Remove the front and rear engine mounts.
8. Raise the vehicle away from the powertrain.

To install:

9. Lower the vehicle over the powertrain.
10. Install or connect the following:
    - Front engine mount. Tighten the nuts to 59 ft. lbs. (80 Nm) and the bolts to 35 ft. lbs. (48 Nm).
    - Rear engine mount. Tighten the outer nuts to 35 ft. lbs. (48 Nm) and the center nut to 98 ft. lbs. (133 Nm).
    - Axle halfshafts
    - Lower ball joints. Tighten the pinch bolts to 37 ft. lbs. (50 Nm).
    - Right engine mount. Tighten the bolts to 35 ft. lbs. (48 Nm).
    - A/C compressor, if equipped
    - Shift cables
    - Shift cable cover
    - Drive belt cover
    - Exhaust flex pipe
    - Catalytic converter
    - Radiator cooling fan
    - Power steering high pressure pipe
    - Power steering pump and reservoir
    - Accessory drive belt
    - Coolant expansion tank
    - Radiator hoses
    - Coolant bypass hoses
    - Fuel line
    - Intake manifold vacuum line
    - EGR vacuum line
    - DPFE system sensor vacuum line
    - Power brake booster vacuum line
    - EVAP canister vacuum lines
    - Clutch slave cylinder fluid pipe
    - Reverse lamp switch connector
    - VSS sensor connector
    - PCM connectors
    - Fuel injector harness connectors
    - HO_2S sensor connectors
    - Power steering pump pressure switch connector
    - Ignition coil connectors
    - Cruise control cable, if equipped
    - Accelerator cable
    - Air cleaner housing and tubes
    - MAF sensor connector
    - Battery, tray and cables
11. Tighten the strut center nuts to 35 ft. lbs. (48 Nm).
12. Fill the cooling system.
13. Start the engine and check for leaks.

Automatic Transaxle

SOHC ENGINE

1. Before servicing the vehicle, refer to the precautions in the beginning of this section.
2. Drain the cooling system.
3. Relieve the fuel system pressure.
4. Loosen the front strut center nuts 5 turns.
5. Remove or disconnect the following:
   - Battery, tray and cables
   - Mass Air Flow (MAF) sensor connector
   - Air cleaner housing and tubes
   - Accelerator cable
   - Cruise control cable, if equipped
   - Ignition coil connectors
   - Power steering pump pressure switch connector
   - Heated Oxygen (HO_2S) sensor connectors
   - Fuel injection wiring harness connector
   - Powertrain Control Module (PCM) connectors
   - Radiator cooling fan
   - Evaporative Emissions (EVAP) canister vacuum lines
   - Power brake booster vacuum line
   - Delta Pressure Feedback Electronic (DPFE) system sensor vacuum line
   - Exhaust Gas Recirculation (EGR) vacuum line
   - Intake manifold vacuum line
   - Fuel line
   - Coolant bypass hoses
   - Radiator hoses
   - Coolant expansion tank
   - Accessory drive belt
   - Power steering pump and reservoir
   - Power steering high pressure pipe
   - Alternator and bracket
   - Drive belt cover
   - A/C compressor, if equipped
   - Lower ball joints
   - Exhaust front pipe

- Torque converter cover
- Torque converter
- Transaxle fluid cooler lines
- Axle halfshafts
- Intermediate shaft bracket
- Starter motor
- Crankshaft pulley
- Right engine mount
- Transaxle flange bolts. Support the transaxle
- Engine front mount
- Crankshaft Position (CKP) sensor
- Engine

To install:

6. Install or connect the following:
   - Engine
   - CKP sensor
   - Engine front mount. Tighten the nuts to 59 ft. lbs. (80 Nm) and the bolts to 35 ft. lbs. (48 Nm).
   - Transaxle flange bolts. Tighten the large bolts to 35 ft. lbs. (48 Nm) and the small bolts to 16 ft. lbs. (22 Nm).
   - Right engine mount. Tighten the bolts to 35 ft. lbs. (48 Nm).
   - Crankshaft pulley. Tighten the bolt to 88 ft. lbs. (120 Nm).
   - Starter motor
   - Intermediate shaft bracket. Tighten the bolts to 35 ft. lbs. (48 Nm).
   - Axle halfshafts
   - Transaxle fluid cooler lines
   - Torque converter. Tighten the nuts to 27 ft. lbs. (37 Nm).
   - Torque converter cover
   - Exhaust front pipe
   - Lower ball joints. Tighten the pinch bolts to 37 ft. lbs. (50 Nm).
   - A/C compressor, if equipped
   - Drive belt cover
   - Alternator and bracket
   - Power steering high pressure pipe
   - Power steering pump and reservoir
   - Accessory drive belt
   - Coolant expansion tank
   - Radiator hoses
   - Coolant bypass hoses
   - Fuel line
   - Intake manifold vacuum line
   - EGR vacuum line
   - DPFE system sensor vacuum line
   - Power brake booster vacuum line
   - EVAP canister vacuum lines
   - Radiator cooling fan
   - PCM connectors
   - HO_2S sensor connectors
   - Fuel injection wiring harness connector
   - Power steering pump pressure switch connector
   - Ignition coil connectors
   - Cruise control cable, if equipped
   - Accelerator cable
   - Air cleaner housing and tubes
   - MAF sensor connector
   - Battery, tray and cables
7. Tighten the strut center nuts to 35 ft. lbs. (48 Nm).
8. Fill the cooling system.
9. Start the engine and check for leaks.

DOHC ENGINE

1. Before servicing the vehicle, refer to the precautions in the beginning of this section.
2. Drain the cooling system.
3. Relieve the fuel system pressure.
4. Loosen the front strut center nuts 5 turns.
5. Remove or disconnect the following:
   - Battery, tray and cables
   - Mass Air Flow (MAF) sensor connector
   - Positive Crankcase Ventilation (PCV) valve and hose
   - Air cleaner assembly and tubes
   - Accelerator cable
   - Cruise control cable, if equipped
   - Ignition coil connectors
   - Heated Oxygen (HO_2S) sensor connectors
   - Fuel injector harness connectors
   - Powertrain Control Module (PCM) connectors
   - Radiator cooling fan
   - Evaporative Emissions (EVAP) canister vacuum lines
   - Power brake booster vacuum line
   - Delta Pressure Feedback Electronic (DPFE) system sensor vacuum line
   - Exhaust Gas Recirculation (EGR) vacuum line
   - Intake manifold vacuum line
   - Fuel line
   - Coolant bypass hoses
   - Radiator hoses
   - Coolant expansion tank
   - Shift select cable and bracket
   - Transaxle fluid cooler lines
   - Drive belt cover
   - Power steering pressure switch connector
   - Accessory drive belt
   - A/C compressor, if equipped
   - Power steering pump
   - Power steering pressure pipe
   - Exhaust flex pipe
   - Right engine mount
   - Lower ball joints
   - Axle halfshafts and intermediate shaft
6. Support the powertrain from below
7. Remove the front and rear engine mounts.
8. Raise the vehicle away from the powertrain.

To install:

9. Lower the vehicle over the powertrain.
10. Install or connect the following:
    - Front engine mount. Tighten the nuts to 59 ft. lbs. (80 Nm) and the bolts to 35 ft. lbs. (48 Nm).
    - Rear engine mount. Tighten the outer nuts to 35 ft. lbs. (48 Nm) and the center nut to 98 ft. lbs. (133 Nm).
    - Axle halfshafts and intermediate shaft
    - Lower ball joints. Tighten the pinch bolts to 37 ft. lbs. (50 Nm).
    - Right engine mount. Tighten the bolts to 35 ft. lbs. (48 Nm).
    - Exhaust flex pipe
    - Power steering pressure pipe
    - Power steering pump
    - A/C compressor, if equipped
    - Accessory drive belt
    - Power steering pressure switch connector
    - Drive belt cover
    - Transaxle fluid cooler lines
    - Shift select cable and bracket
    - Coolant expansion tank
    - Radiator hoses
    - Coolant bypass hoses
    - Fuel line
    - Intake manifold vacuum line
    - EGR vacuum line
    - DPFE system sensor vacuum line
    - Power brake booster vacuum line
    - EVAP canister vacuum lines
    - Radiator cooling fan
    - PCM connectors
    - Fuel injector harness connectors
    - HO_2S sensor connectors
    - Ignition coil connectors
    - Cruise control cable, if equipped
    - Accelerator cable
    - Air cleaner assembly and tubes
    - PCV valve and hose
    - MAF sensor connector
    - Battery, tray and cables
11. Tighten the strut center nuts to 35 ft. lbs. (48 Nm).
12. Fill the cooling system.
13. Start the engine and check for leaks.

Water Pump

REMOVAL & INSTALLATION

SOHC Engine

1. Before servicing the vehicle, refer to the precautions in the beginning of this section.

2. Drain the cooling system.
3. Remove or disconnect the following:
   - Negative battery cable
   - Accessory drive belt
   - Front cover
   - Timing belt. Refer to the Timing Belt unit repair section.
   - Timing belt tensioner
   - Water pump

To install:

4. Install or connect the following:
   - Water pump. Tighten the bolts to 18 ft. lbs. (25 Nm).
   - Timing belt tensioner
   - Timing belt
   - Accessory drive belt
   - Negative battery cable
5. Fill the cooling system.
6. Start the engine and check for leaks.

DOHC Engine

1. Before servicing the vehicle, refer to the precautions in the beginning of this section.
2. Drain the cooling system.
3. Remove or disconnect the following:
   - Negative battery cable
   - Accessory drive belt
   - Water pump pulley
   - Front cover
   - Timing belt. Refer to the Timing Belt unit repair section.
   - Timing belt tensioner
   - Water pump

To install:

4. Install or connect the following:
   - Water pump. Tighten the bolts to 13 ft. lbs. (18 Nm).
   - Timing belt tensioner
   - Timing belt
   - Front cover
   - Water pump pulley
   - Accessory drive belt
   - Negative battery cable
5. Fill the cooling system.
6. Start the engine and check for leaks.

Cylinder Head

REMOVAL & INSTALLATION

SOHC Engine

1. Before servicing the vehicle, refer to the precautions in the beginning of this section.
2. Disconnect the negative battery cable.
3. Drain the cooling system.
4. Relieve the fuel system pressure.
5. Remove or disconnect the following:
   - Timing belt. Refer to the Timing Belt unit repair section.
   - Engine air cleaner intake tube
6. Tag and disconnect the vacuum hoses from the following components:
   - Exhaust Gas Recirculation (EGR) valve
   - Positive Crankcase Ventilation (PCV) valve
   - Throttle body
   - Fuel pressure regulator
   - Intake manifold
7. Remove or disconnect the following:
   - Accelerator cable, throttle control lever and if equipped, the speed control cable
   - Speed control cable bracket bolt, if equipped
   - 2 fuel charging wiring electrical connections
   - Crankshaft Position (CKP) sensor and Heated Oxygen (HO_2S) sensor electrical connections
   - Fuel supply line
   - Power steering pressure hose bracket bolts, then position the bracket and hose aside
   - Alternator lower bolt, loosen only
   - Alternator upper bolt and pivot the alternator forward
   - Engine oil dipstick tube bracket bolt
   - Exhaust Gas Recirculation (EGR) tube
   - Exhaust manifold heat shield
   - Catalytic converter
   - Right-hand splash shield bolts and the shield
   - Engine oil dipstick tube from the cylinder block
   - Air conditioning compressor
   - Engine accessory drive bracket
   - Valve cover

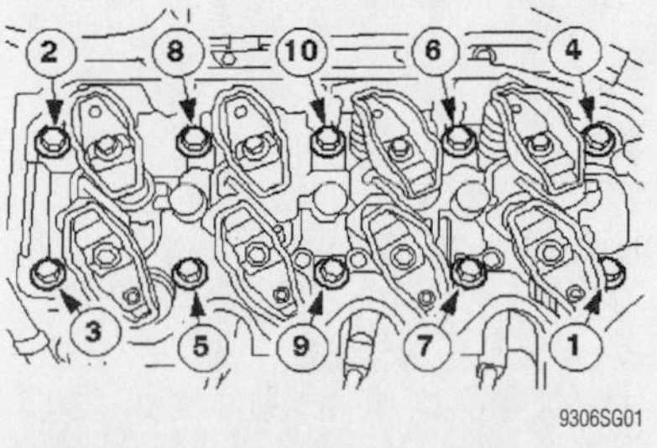

Cylinder head loosening sequence—SOHC engine

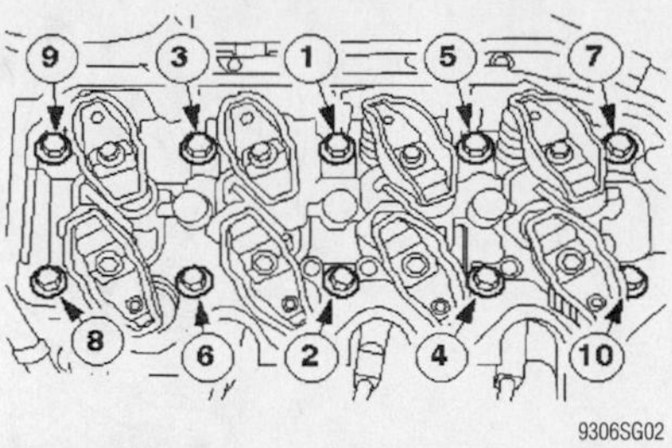

Cylinder head torque sequence—2.0L SOHC engine

   - Upper radiator hose
   - Heater hose
   - Cylinder head bolts. Loosen the bolts in sequence.
   - Cylinder head and gasket

To install:

➡Use new cylinder head bolts for assembly.

8. Install the cylinder head and tighten the bolts in sequence as follows:
   a. Step 1: 37 ft. lbs. (50 Nm)
   b. Step 2: Loosen all bolts ½ turn
   c. Step 3: 37 ft. lbs. (50 Nm)
   d. Step 4: Plus 90 degrees
   e. Step 6: Plus 90 degrees
9. Install or connect the following:
   - Accessory drive bracket. Tighten the fasteners to 30–40 ft. lbs. (40–55 Nm)
   - Air conditioning line bracket and tighten the bolt to 15–18 ft. lbs. (20–25 Nm)
   - Valve cover
   - Air conditioning compressor
   - Right-hand splash shield
   - Engine oil dipstick tube
   - Heater and upper radiator hoses
   - Timing belt and alternator
   - Power steering pressure hose
   - CKP sensor and 2 main fuel charging electrical connections
   - Engine oil dipstick tube bolt
   - EGR valve tube
   - Fuel supply line
   - Catalytic converter
   - Heat shield
   - Speed control cable bracket, if equipped
   - Throttle control lever, accelerator cable and if equipped, the speed control cable
10. Connect the vacuum hoses to the following components:
   - EGR valve
   - PCV valve

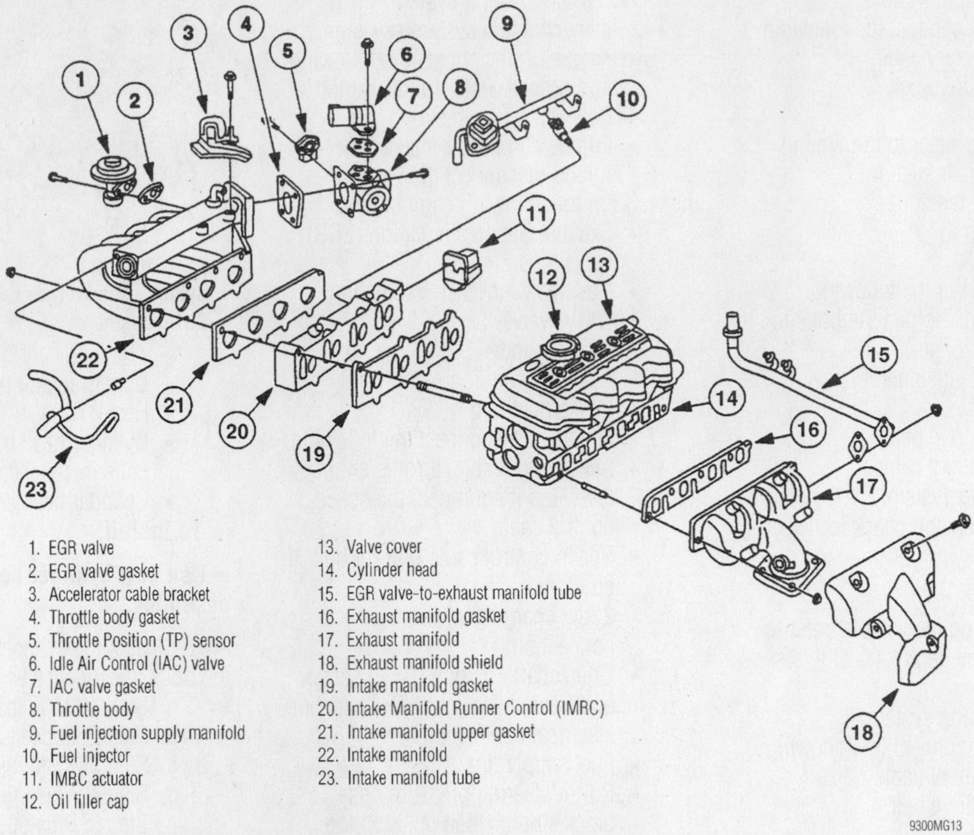

Exploded view of the cylinder head, exhaust manifold and intake manifold mounting—2.0L SOHC engine

- Throttle body
- Fuel pressure regulator
- Intake manifold
- Air cleaner outlet tube
- Negative battery cable

11. Fill the cooling system.
12. Start the engine and check for leaks.

DOHC Engine

1. Before servicing the vehicle, refer to the precautions in the beginning of this section.
2. Disconnect the negative battery cable.
3. Drain the cooling system.
4. Relieve the fuel system pressure.
5. Remove or disconnect the following:
   - Air cleaner assembly outlet tube
   - Timing belt. Refer to the Timing Belt unit repair section.
6. Tag and disconnect the vacuum hoses from the following components:
   - Positive Crankcase Ventilation (PCV) valve
   - Throttle body
   - Fuel pressure sensor
   - Intake manifold
7. Remove or disconnect the following:
   - Speed control and accelerator cables from the control lever
   - Fuel charging electrical connectors at the main engine connector
   - Crankshaft Position (CKP) sensor and Heated Oxygen (HO_2S) sensor electrical connections
   - Fuel line
   - Power steering pump and bracket
   - Alternator
   - Engine oil dipstick tube
   - Splash shield
   - Air conditioning compressor
   - Spark plug wires
   - Valve cover
   - Upper radiator hose
   - Heater hose
   - Camshafts
   - Ignition coil
   - Thermostat housing
   - Cylinder head bolts. Mark each bolt with a punch and loosen them in the sequence shown.
   - Cylinder head and the gasket

To install:

➡The cylinder head bolts may be reused twice. Replace bolts that have two punch marks.

8. Lubricate the cylinder bolts when engine oil.
9. Install the cylinder head and tighten the bolts in sequence as follows:
   a. Step 1: 15 ft. lbs. (20 Nm)
   b. Step 2: 30 ft. lbs. (40 Nm)
   c. Step 3: Plus 90 degrees

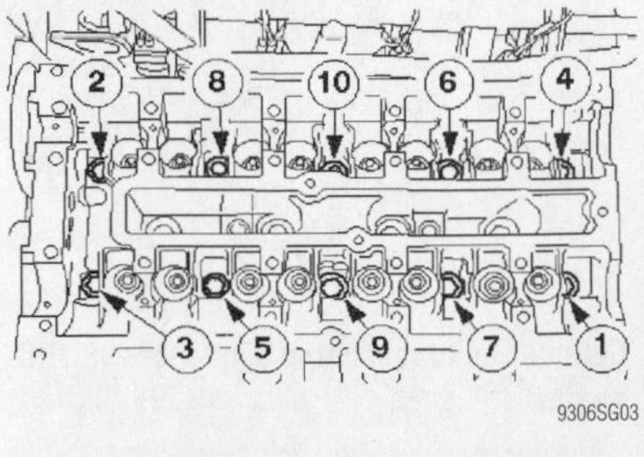

Cylinder head bolt loosening sequence—DOHC engine

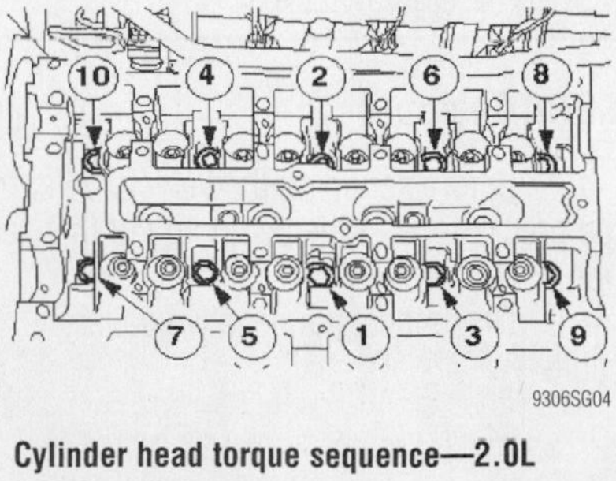

Cylinder head torque sequence—2.0L DOHC engine

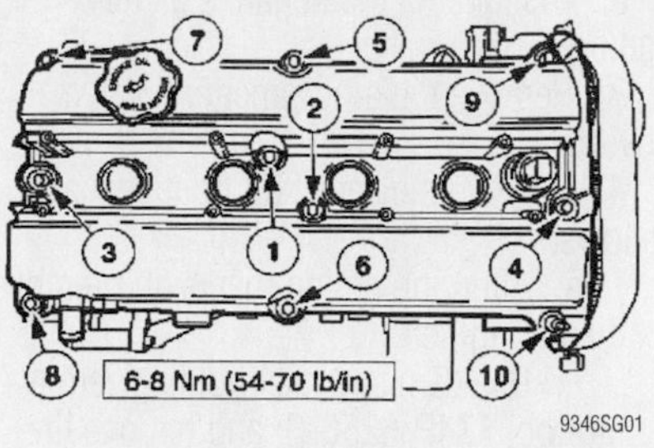

Valve cover torque sequence—DOHC Engine

10. Install or connect the following:
    - Thermostat housing
    - Ignition coil and the camshafts
    - Heater and upper radiator hoses
    - Valve cover
    - Spark plugs
    - Spark plug wires
    - Air conditioning compressor
    - Splash shield
    - Engine oil dipstick tube
    - Alternator and the power steering reservoir and bracket
    - Fuel line
    - CKP sensor, HO_2sensor and 2 main fuel charging electrical connections
    - Accelerator cable and if equipped, the speed control cable to the control lever
11. Connect the vacuum hoses to the following components:
    - PCV valve
    - Throttle body
    - Fuel pressure sensor
    - Intake manifold
12. Install or connect the following:
    - Timing belt
    - Air cleaner outlet tube
    - Negative battery cable
13. Fill the cooling system.
14. Start the engine and check for leaks.

Rocker Arms/Shafts

REMOVAL & INSTALLATION

SOHC Engine

1. Before servicing the vehicle, refer to the precautions in the beginning of this section.
2. Remove or disconnect the following:
    - Negative battery cable
    - Valve cover

➡Keep all valve train components in order for assembly.

- Rocker arm bolts
- Rocker arms and seats

To install:

3. Install or connect the following:
    - Seats and rocker arms in their original positions
    - Rocker arm bolts. Tighten to 17–22 ft. lbs. (23–30 Nm).
    - Valve cover
    - Negative battery cable

DOHC Engine

The DOHC engine is not equipped with rocker arms. The camshafts act directly on the valves.

Intake Manifold

REMOVAL & INSTALLATION

SOHC Engine

1. Before servicing the vehicle, refer to the precautions in the beginning of this section.
2. Drain the cooling system.
3. Relieve the fuel system pressure.
4. Remove or disconnect the following:
    - Negative battery cable
    - Air cleaner outlet tube
5. Tag and disconnect the vacuum hoses from the following components:
    - Exhaust gas Recirculation (EGR) valve
    - Positive Crankcase Ventilation (PCV) valve
    - Throttle body
    - Fuel pressure regulator
    - Intake manifold
6. Remove or disconnect the following:
    - Accelerator cable, throttle control lever and if equipped, the speed control cable
    - Speed control cable bracket bolt, if equipped
    - Idle Air Control (IAC) valve and Throttle Position (TP) sensor electrical connections
    - Engine oil dipstick tube bracket bolt that attaches the tube to the intake manifold
    - EGR manifold tube located below the EGR valve
    - Engine oil dipstick tube from the block
    - Intake manifold lower nuts
    - Intake manifold upper nuts
    - Intake manifold

To install:

7. Clean and oil the intake manifold mounting studs.
8. Install or connect the following:
    - Intake manifold with a new gasket. Tighten the nuts to 15–22 ft. lbs. (20–30 Nm).
    - Engine oil dipstick tube
    - EGR manifold tube. Tighten it to 15–20 ft. lbs. (20–28 Nm).
    - IAC valve and TP sensor electrical connections
    - Speed control cable bracket bolt, if equipped. Tighten to 71–88 inch lbs. (8–10 Nm).
    - Throttle control lever, accelerator cable and if equipped, the speed control cable
9. Connect the vacuum hoses to the following components:
    - EGR valve
    - PCV valve
    - Throttle body
    - Fuel pressure regulator
    - Intake manifold
    - Air cleaner outlet tube
    - Negative battery cable
10. Fill the cooling system.
11. Start the engine and check for coolant leaks.

DOHC Engine

1. Before servicing the vehicle, refer to the precautions in the beginning of this section.
2. Drain the cooling system.
3. Relieve the fuel system pressure.
4. Remove or disconnect the following:
    - Negative battery cable
    - Air cleaner outlet tube
    - Throttle Position (TP) sensor connector
    - Bolt that secures the pipe located by the crankshaft pulley
    - Heater hoses
    - Main engine control sensor wiring
    - Connectors from the mounting bracket
    - Intake manifold vacuum lines
    - Positive Crankcase Ventilation (PCV) hose
    - Drive belt
    - Alternator
    - Fuel line
    - Intake manifold

To install:

5. Install or connect the following:
    - Intake manifold. Tighten the fasteners to 13 ft. lbs. (18 Nm).

- Fuel line
- Alternator
- Drive belt
- PCV hose
- Intake manifold vacuum lines
- Connectors in the mounting bracket
- Main engine control sensor wiring
- Heater hoses
- Bolt that secures the pipe located by the crankshaft pulley. Tighten the bolt to 71–97 inch lbs. (8–11 Nm).
- TP sensor electrical connection
- Air cleaner outlet tube
- Negative battery cable

6. Fill the cooling system.
7. Start the engine and check for leaks.

Exhaust Manifold

REMOVAL & INSTALLATION

SOHC Engine

1. Before servicing the vehicle, refer to the precautions in the beginning of this section.
2. Remove or disconnect the following:
   - Negative battery cable
   - Drive belt
   - Heated Oxygen (HO_2S) sensor electrical connection
   - Exhaust manifold heat shield
   - Exhaust Gas Recirculation (EGR) tube
   - Power steering pressure hose bracket
   - Alternator
   - Catalytic converter
   - Exhaust manifold

To install:

3. Install or connect the following:
   - Exhaust manifold. Tighten the nuts to 15–18 ft. lbs. (20–24 Nm).
   - Catalytic converter. Tighten the nuts to 26–34 ft. lbs. (34–48 Nm).
   - Alternator
   - Power steering pressure hose bracket
   - EGR tube. Tighten the fasteners to 45–62 inch lbs. (5–7 Nm).
   - Exhaust manifold heat shield. Tighten the nuts to 45–62 inch lbs. (5–7 Nm).
   - HO_2S sensor electrical connection
   - Drive belt
   - Negative battery cable
4. Start the engine and check for leaks.

DOHC Engine

1. Before servicing the vehicle, refer to the precautions in the beginning of this section.
2. Remove or disconnect the following:
   - Negative battery cable
   - Fan motor electrical connection
   - Fan shroud
   - Heated Oxygen (HO_2S) sensor electrical connection
   - Catalytic converter
   - Engine oil dipstick tube bracket bolt
   - Exhaust manifold heat shield
   - Exhaust manifold

To install:

3. Install or connect the following:
   - Exhaust manifold. Tighten the fasteners to 13 ft. lbs. (18 Nm).
   - Exhaust manifold heat shield. Tighten the retainers to 71–101 inch lbs. (8–11 Nm).
   - Engine oil dipstick tube bracket. Tighten the bolt to 71–101 inch lbs. (8–11 Nm).
   - Catalytic converter
   - HO_2sensor electrical connection
   - Fan shroud and the fan motor wiring
   - Negative battery cable
4. Start the engine and check for leaks.

Front Crankshaft Seal

REMOVAL & INSTALLATION

1. Before servicing the vehicle, refer to the precautions in the beginning of this section.
2. Remove the timing belt and crankshaft sprocket.

WARNING

Be careful not to damage the crankshaft surface when removing the seal.

3. Using seal remover tool T92C-6700-CH, remove the crankshaft front oil seal.

To install:

4. Use seal replacer tool T81P-6700-A and install the new seal.
5. Install the crankshaft sprocket and the timing belt.

Camshaft and Valve Lifters

REMOVAL & INSTALLATION

SOHC Engine

1. Before servicing the vehicle, refer to the precautions in the beginning of this section.
2. Disconnect the negative battery cable.
3. Remove the air cleaner and valve cover.
4. Remove the camshaft front seal as follows:
   a. Align the timing marks and remove the timing belt.
   b. Use cam sprocket holding/removing tool T74P-6256-B, and remove the camshaft sprocket.
   c. Unfasten the timing belt tensioner bolt and remove the tensioner.
   d. Remove the inner engine front cover.
   e. Use seal removal tool T92C-6700-CH, to remove the camshaft front seal.
5. Remove or disconnect the following:
   - Ignition coil and bracket
   - Rocker arms and valve tappets
   - Camshaft thrust plate bolts and the plate
   - Cup plug from the rear of the cylinder head and discard
   - Camshaft from the rear of the cylinder head

To install:

➡Liberally coat the cam bore in the cylinder with clean 5W-30 motor oil.

6. Install or connect the following:
   - Camshaft through the rear of the cylinder head
   - Camshaft thrust plate and the bolts. Tighten the bolts to 71–115 inch lbs. (8–13 Nm).
   - New cup plug and the tappets
   - Rocker arms
   - Ignition coil bracket and coil
7. Install the camshaft front seal as follows:
   a. Apply a thin film of 5W-30 motor oil to the lip of the seal.
   b. Use seal replacer T81P-6292-A, to install the camshaft front seal.

➡The seal depth should be 0.002–0.04 in. (0.05–1.0mm) below flush with the cylinder head front face.

   c. Install the inner engine front cover.
   d. Install the timing belt tensioner. Tighten the tensioner bolt to 15–22 ft. lbs. (20–30 Nm).
   e. Install the camshaft sprocket and the timing belt
8. Install the valve cover and the air cleaner assembly.
9. Connect the negative battery cable.

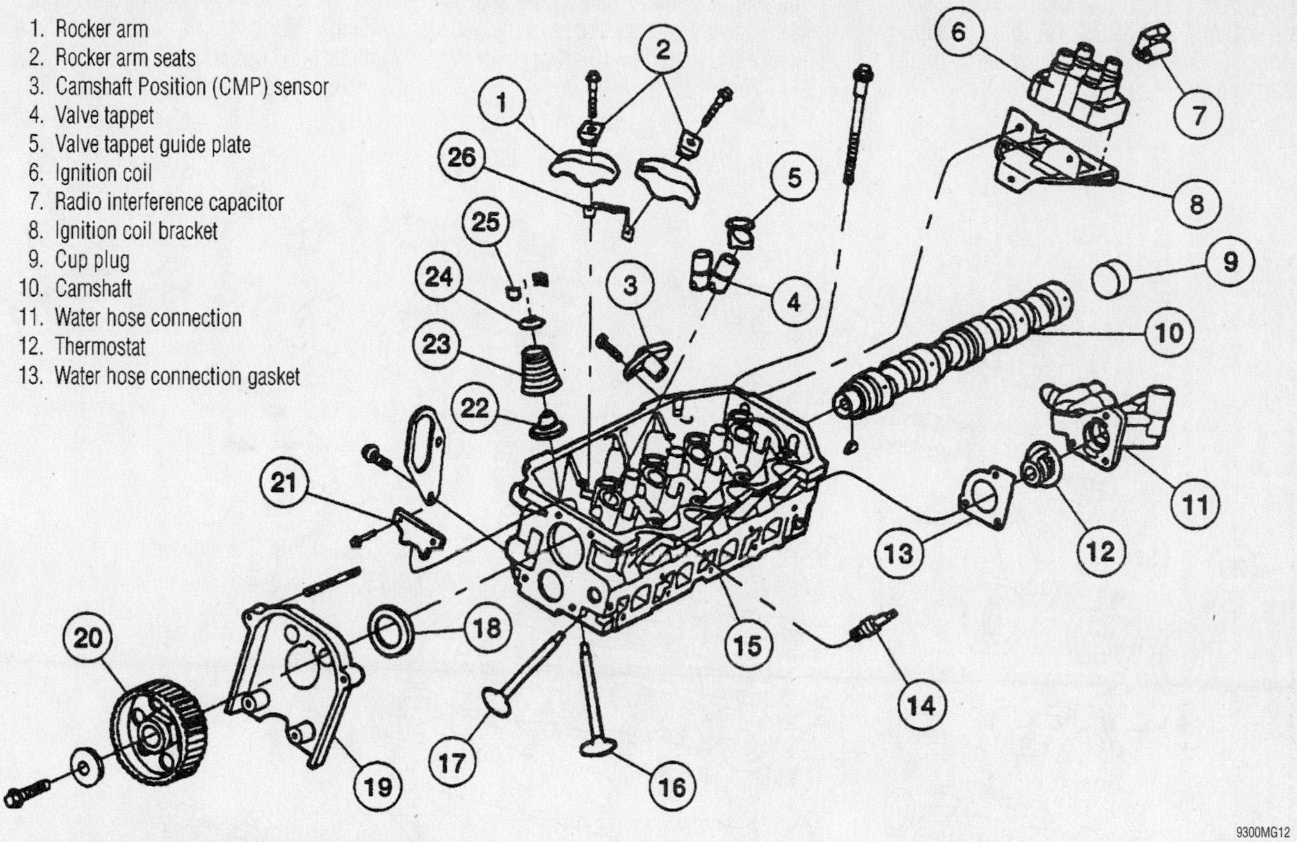

Exploded view of the cylinder head—SOHC engine

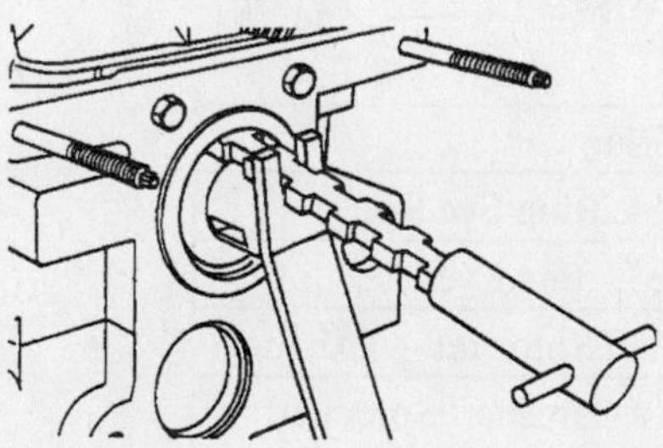

Removing the camshaft front seal—SOHC engine

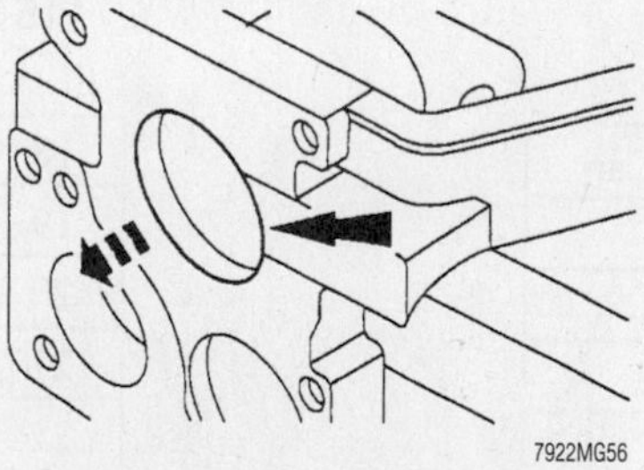

Remove and discard the cup plug located at the rear of the cylinder head—SOHC engine

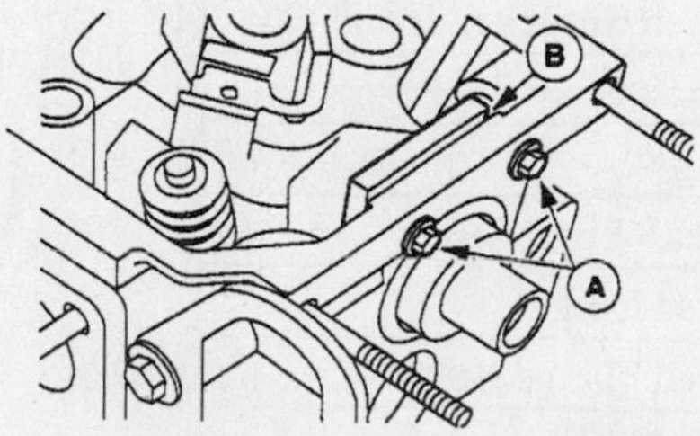

Remove the camshaft thrust plate retaining bolts (A) and the plate (B)—SOHC engine

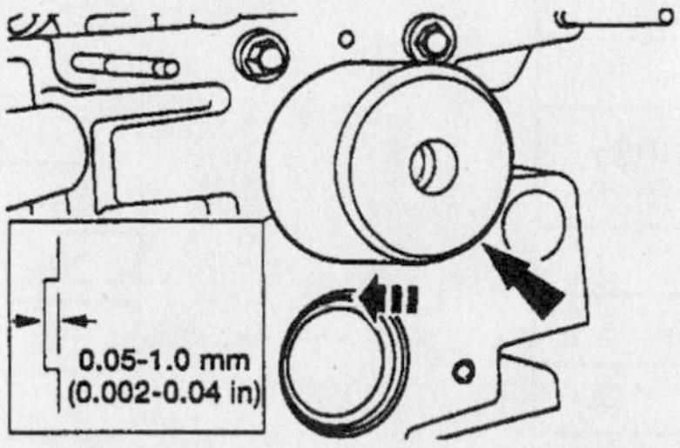

Use seal Replacer T81P-6292-A, to install the camshaft front seal 0.002–0.04 in. (0.05–1.0mm) below flush with the cylinder head front face—2.0L SOHC engine

DOHC Engine

1. Before servicing the vehicle, refer to the precautions in the beginning of this section.
2. Remove or disconnect the following:
   - Timing belt. Refer to the Timing Belt unit repair section.
   - Valve cover and camshaft sprockets
   - Oil control solenoid flange

➡Mark the camshaft journal caps with a number to identify their location.

   - Camshaft journal caps. Loosen the bolts in sequence and in several passes.
   - Camshafts
   - Oil control sensor and bushing

To install:

3. Install the oil control solenoid bushing and flange on the exhaust camshaft.
4. Coat the surface of the front camshaft journal cap with gasket maker E2AZ-19562-B.
5. Place the camshafts in position and lubricate the bearing surfaces with engine assembly lubricant D9AZ-19579-D.
6. Install new camshaft front oil seals.

7. Apply a thin coat of silicone gasket and sealant F6AZ-19562-AA to the sealing surface of the of the front camshaft journal bearing cap.

8. Install the caps and tighten the bolts in several 2-turn passes in the sequence illustrated to 10–12 ft. lbs. (13–17 Nm).

9. Inspect the oil control solenoid flange O-rings for damage or wear and replace, as necessary.

10. Install the oil control solenoid flange

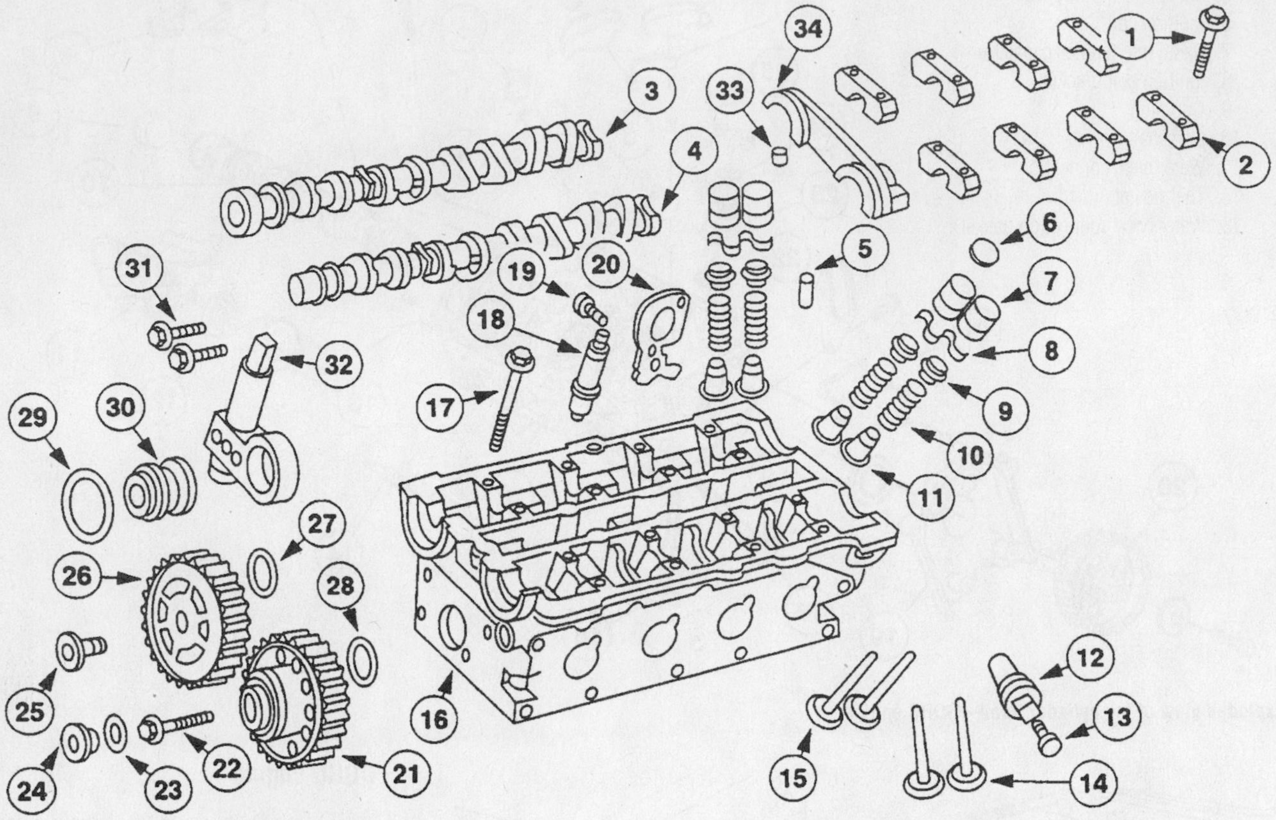

1	Bolt—Camshaft Bearing Cap
2	Camshaft Bearing Cap
3	Camshaft Intake
4	Camshaft Exhaust
5	Plug
6	Adjusting Shim—Valve Clearance
7	Valve Tappets
8	Valve Spring Retainer Key
9	Valve Spring Retainers
10	Valve Spring (Color Coding: Exhaust = Blue, Intake = Red)
11	Valve Stem Seal
12	Camshaft Position Sensor
13	CMP Sensor Bolt
14	Intake Valves
15	Exhaust Valves
16	Cylinder Head
17	Cylinder Head Bolts
18	Spark Plug
19	Engine Lifting Eye Bolt
20	Engine Lifting Eye
21	Camshaft Sprocket —Exhaust
22	Exhaust Camshaft Sprocket Bolt
23	Blanking Plug O-Ring
24	Blanking Plug
25	Bolt—Intake Camshaft
26	Camshaft Sprocket Intake
27	Camshaft Front Seal
28	O-Ring
29	Camshaft Front Seal
30	Oil Feed Ring
31	Oil Feed Flange Bolts
32	Oil Feed Flange
33	Guide Sleeve—Front Camshaft Bearing Cap
34	Fifth Camshaft Bearing Cap

9300MG06

Exploded view of the cylinder head and camshaft mounting—DOHC engine

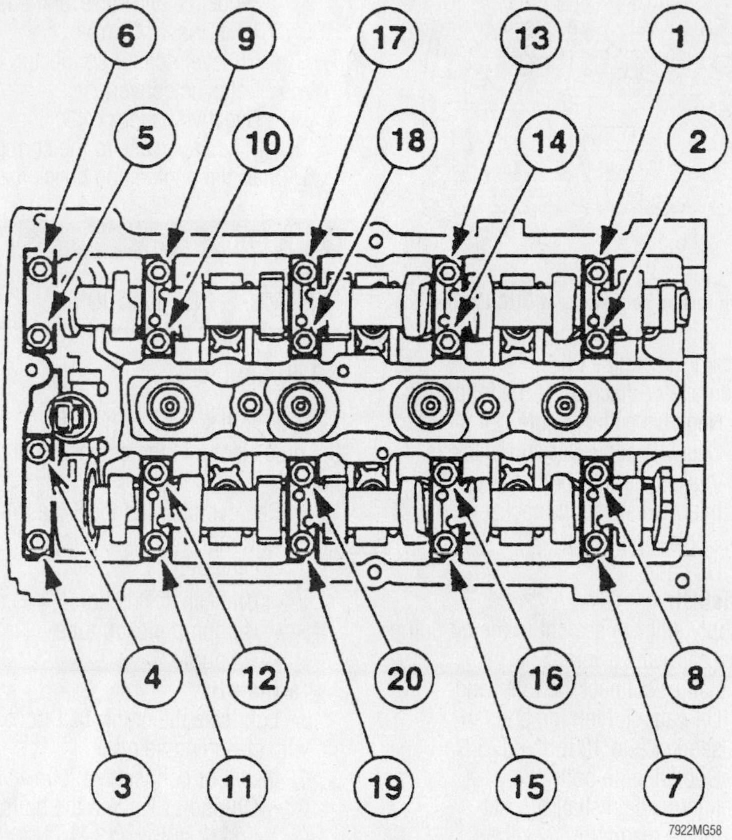

Camshaft journal cap loosening sequence—DOHC engine

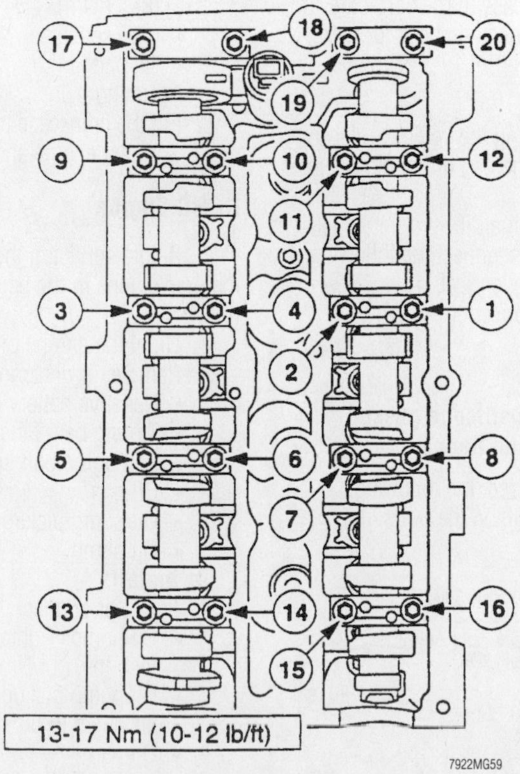

Camshaft journal cap torque sequence—DOHC engine

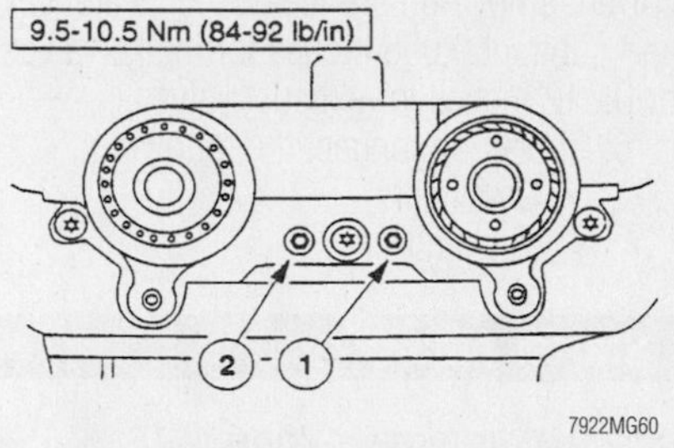

Tighten the oil control solenoid flange bolts in this sequence—DOHC engine

and tighten the bolts to 84–92 inch lbs. (9.5–10.5 Nm) in the sequence.

11. Rotate the camshafts a full turn and check for binding.
12. Install or connect the following:
    - Camshaft sprockets and timing belt
    - Valve cover

Valve Lash

ADJUSTMENT

SOHC Engine

The lash adjusters are hydraulic and are not adjustable.

DOHC Engine

1. Before servicing the vehicle, refer to the precautions in the beginning of this section.
2. Remove the valve cover.
3. Rotate the crankshaft so that the valve is closed and measure the clearance between the lifter shim and the camshaft base circle. Intake valve clearance should be 0.004–0.007 inches. Exhaust valve clearance should be 0.010–0.013 inches.
4. Measure each valve and note the clearance.
5. If adjustment is necessary, remove the camshafts and replace the shims. To obtain the correct shim size, measure the

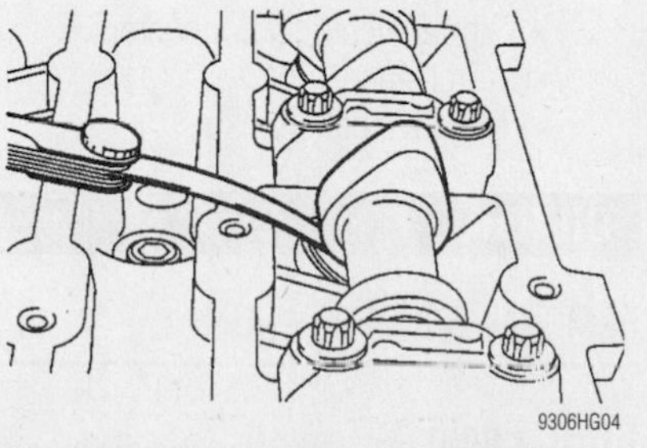

Checking valve clearance—2.0L engine

current shim, add the measured clearance, and subtract 0.006 inches for intake valves or 0.012 inches for exhaust valves.

6. Install or connect the following:
   - Camshafts
   - Valve cover

Starter Motor

REMOVAL & INSTALLATION

SOHC Engine

1. Before servicing the vehicle, refer to the precautions in the beginning of this section.
2. Remove or disconnect the following:
   - Negative battery cable
   - Air cleaner outlet tube
   - Two top starter motor bolts
   - **S** terminal wire and the **B** terminal nut and cable from the solenoid
   - Lower starter motor bolt
   - Starter

To install:

3. Install or connect the following:
   - Starter motor assembly and tighten the lower mounting bolt to 18–20 ft. lbs. (25–27 Nm)
   - **S** terminal wire, **B** terminal cable and nut. Tighten the nut to 80–120 inch lbs. (9–13.5 Nm).
   - Top starter motor bolts. Tighten the bolts to 18–20 ft. lbs. (25–27 Nm).
   - Air cleaner outlet tube
   - Negative battery cable

DOHC Engine

1. Before servicing the vehicle, refer to the precautions in the beginning of this section.
2. Remove or disconnect the following:
   - Negative battery cable
   - Air cleaner
   - Starter electrical connectors
   - Starter

To install:

3. Install or connect the following:
   - Starter. Tighten the bolts to 26 ft. lbs. (35 Nm).
   - Starter electrical connectors
   - Air cleaner
   - Negative battery cable

Oil Pan

REMOVAL & INSTALLATION

SOHC Engine

1. Before servicing the vehicle, refer to the precautions in the beginning of this section.

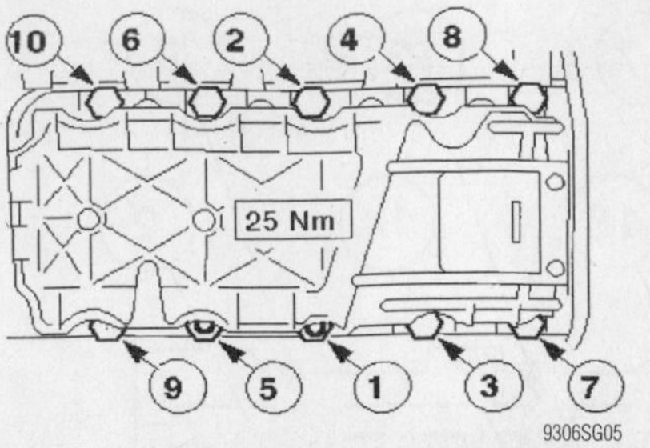

Oil pan torque sequence—SOHC engine

2. Drain the engine oil.
3. Remove or disconnect the following:
   - Negative battery cable
   - Catalytic converter and brackets
   - Intake manifold bracket
   - Intermediate shaft bracket
   - Coolant pipe bolt
   - Oil pan

To install:

4. Apply silicone sealant to the oil pump and rear seal retainer joints.
5. Install or connect the following:
   - Oil pan. Tighten the bolts in sequence to 18 ft. lbs. (25 Nm).
   - Coolant pipe bolt
   - Intermediate shaft bracket
   - Intake manifold bracket
   - Catalytic converter and brackets
   - Negative battery cable
6. Fill the crankcase to the correct level.
7. Start the engine and check for leaks.

DOHC Engine

1. Before servicing the vehicle, refer to the precautions in the beginning of this section.
2. Drain the engine oil.
3. Remove or disconnect the following:
   - Negative battery cable
   - Catalytic converter
   - Oil pan

To install:

➡Apply a bead of silicone gasket sealer to the oil pan flange.

4. Install or connect the following:
   - Oil pan. Tighten the bolts in sequence and in several passes to 18 ft. lbs. (24 Nm).

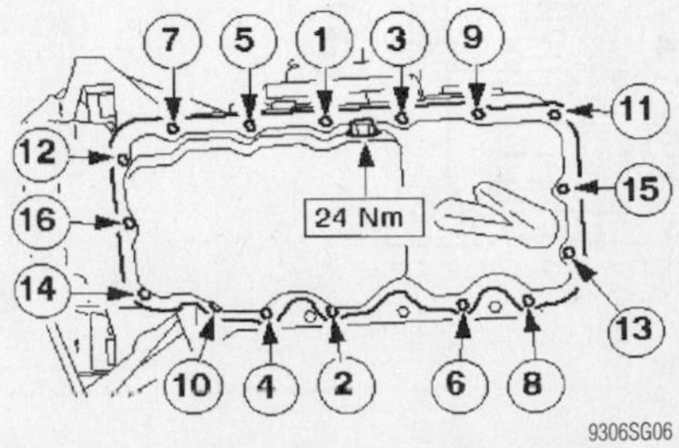

Oil pan torque sequence—DOHC engine

   - Oil lever sensor connector
   - Catalytic converter
   - Negative battery cable
5. Fill the crankcase to the correct level.
6. Start the engine and check for leaks.

Oil Pump

REMOVAL & INSTALLATION

SOHC Engine

1. Before servicing the vehicle, refer to the precautions in the beginning of this section.
2. Remove or disconnect the following:
   - Timing belt
   - Oil pan
   - Crankshaft Position (CKP) sensor
   - Oil pump pickup tube
   - Oil pump

To install:

3. Lubricate the crankshaft front oil seal lip with clean engine oil.
4. Install or connect the following:
   - Oil pump. Tighten the bolts to 10–12 ft. lbs. (13–16 Nm).
   - Oil pump pickup tube. Tighten the retaining bolts to 71–97 inch lbs. (8–11 Nm).
   - CKP sensor
   - Oil pan
   - Timing belt
5. Fill the crankcase to the correct level.
6. Start the engine and check for leaks.

DOHC Engine

1. Before servicing the vehicle, refer to the precautions in the beginning of this section.
2. Drain the engine oil.
3. Remove or disconnect the following:
   - Negative battery cable
   - Timing belt. Refer to the Timing Belt unit repair section.
   - Oil pan
   - Oil pump pickup tube
   - Oil pump

To install:

4. Install or connect the following:
   - Oil pump. Tighten the bolts to 97 inch lbs. (11 Nm)
   - Oil pump pickup tube. Tighten the fasteners to 88 inch lbs. (10 Nm).
   - Oil pan
   - Timing belt
   - Negative battery cable
5. Fill the crankcase to the correct level.
6. Start the engine and check for leaks.

Rear Main Seal

REMOVAL & INSTALLATION

1. Before servicing the vehicle, refer to the precautions in the beginning of this section.
2. Remove or disconnect the following:
   - Negative battery cable
   - Transaxle
   - Clutch, if equipped
   - Flexplate/flywheel
   - Oil seal

To install:

3. Install or connect the following:
   - Oil seal. Seat the seal flush with the rear of the crankshaft oil seal retainer.
   - Flexplate/flywheel. Tighten the bolts to 82 ft. lbs. (112 Nm).
   - Clutch, if equipped
   - Transaxle
   - Negative battery cable
4. Start the engine and check for leaks.

Piston and Ring

POSITIONING

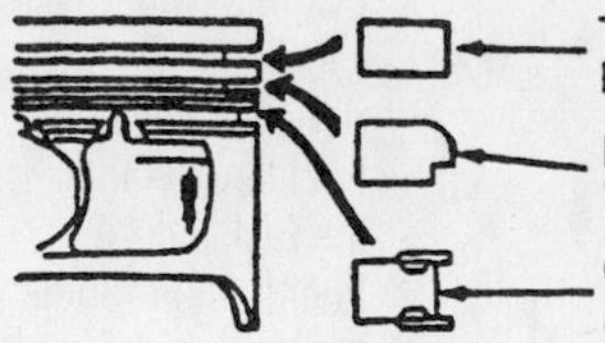

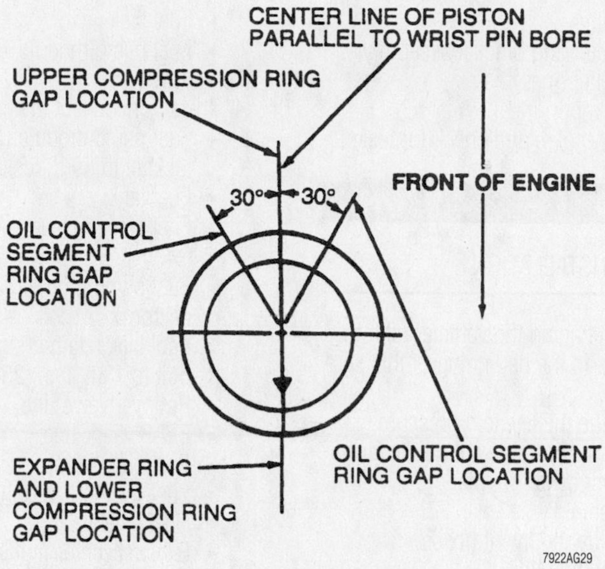

Ford 2.0L (VIN 3 and P) engines—piston ring positioning, end-gap spacing and piston positioning. The small directional arrow must face the front of the engine.

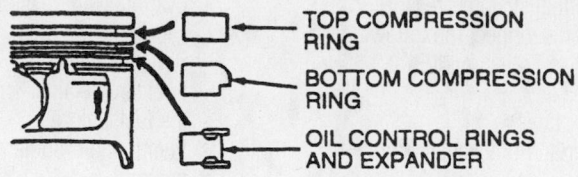

FUEL SYSTEM

Fuel System Service Precautions

Safety is the most important factor when performing not only fuel system maintenance but any type of maintenance. Failure to conduct maintenance and repairs in a safe manner may result in serious personal injury or death. Maintenance and testing of the vehicle's fuel system components can be accomplished safely and effectively by adhering to the following rules and guidelines.

- To avoid the possibility of fire and personal injury, always disconnect the negative battery cable unless the repair or test procedure requires that battery voltage be applied.
- Always relieve the fuel system pressure prior to disconnecting any fuel system component (injector, fuel rail, pressure regulator, etc.), fitting or fuel line connection. Exercise extreme caution whenever relieving fuel system pressure, to avoid exposing skin, face and eyes to fuel spray. Please be advised that fuel under pressure may penetrate the skin or any part of the body that it contacts.
- Always place a shop towel or cloth around the fitting or connection prior to loosening to absorb any excess fuel due to spillage. Ensure that all fuel spillage (should it occur) is quickly removed from engine surfaces. Ensure that all fuel soaked cloths or towels are deposited into a suitable waste container.
- Always keep a dry chemical (Class B) fire extinguisher near the work area.
- Do not allow fuel spray or fuel vapors to come into contact with a spark or open flame.
- Always use a backup wrench when loosening and tightening fuel line connection fittings. This will prevent unnecessary stress and torsion to fuel line piping.
- Always replace worn fuel fitting O-rings with new. Do not substitute fuel hose or equivalent, where fuel pipe is installed.

Before servicing the vehicle, make sure to refer to the precautions in the beginning of this section as well.

Fuel System Pressure

RELIEVING

1. Before servicing the vehicle, refer to the precautions in the beginning of this section.
2. Remove the fuel pump fuse from the battery junction box.
3. Start the engine and allow it to idle until it stalls.
4. Crank the engine for 5 seconds to ensure that the fuel pressure has been relieved.
5. Disconnect the negative battery cable.
6. When repairs are completed, connect the negative battery cable and replace the fuel pump fuse.

Fuel Filter

REMOVAL & INSTALLATION

1. Before servicing the vehicle, refer to the precautions in the beginning of this section.

2. Relieve the fuel system pressure.
3. Remove or disconnect the following:
   - Evaporative Emissions (EVAP) vapor line
   - Fuel filter lines
   - Fuel filter and bracket assembly
   - Fuel filter

To install:

4. Install or connect the following:
   - Fuel filter
   - Fuel filter and bracket assembly
   - Fuel filter lines
   - EVAP vapor line
5. Start the engine and check for leaks.

Fuel Pump

REMOVAL & INSTALLATION

1. Before servicing the vehicle, refer to the precautions in the beginning of this section.
2. Relieve the fuel system pressure.
3. Remove or disconnect the following:
   - Exhaust center muffler
   - Heat shield
   - Fuel filler and vent hoses
   - Fuel supply line
   - Evaporative Emissions (EVAP) vapor line
   - Rollover valve line
   - Fuel tank retainer strap. Support the fuel tank.
   - Fuel pump module harness connector
   - Fuel tank pressure sensor connector
   - Fuel tank
   - Fuel pump module

To install:

4. Install or connect the following:
   - Fuel pump module. Tighten the locknut to 59 ft. lbs. (80 Nm).
   - Fuel tank
   - Fuel tank pressure sensor connector
   - Fuel pump module harness connector
   - Fuel tank retainer strap. Tighten the bolt to 18 ft. lbs. (25 Nm).
   - Rollover valve line
   - EVAP vapor line
   - Fuel supply line
   - Fuel filler and vent hoses
   - Heat shield
   - Exhaust center muffler
5. Start the engine and check for leaks.

Fuel Injector

REMOVAL & INSTALLATION

1. Before servicing the vehicle, refer to the precautions in the beginning of this section.
2. Relieve fuel system pressure.
3. Remove or disconnect the following:
   - Negative battery cable
   - Fuel line and retaining clip
   - Pressure regulator vacuum hose
   - Accelerator cable
   - Fuel injector electrical connectors
   - Fuel supply manifold with the injectors attached
   - Injectors from the supply manifold

To install:

4. Install or connect the following:
   - Fuel injectors using new O-ring seals
   - Fuel supply manifold with the injectors attached. Tighten the bolts to 88 inch lbs. (10 Nm).
   - Fuel injector electrical connectors
   - Accelerator cable
   - Pressure regulator vacuum hose
   - Fuel line and retaining clip
   - Negative battery cable

DRIVE TRAIN

Transaxle Assembly

REMOVAL & INSTALLATION

Manual

1. Before servicing the vehicle, refer to the precautions in the beginning of this section.
2. Attach a support fixture to the engine lifting eyes.
3. Loosen the front strut center nuts 5 turns.
4. Remove or disconnect the following:
   - Battery and tray
   - Mass Air Flow (MAF) sensor connector
   - Air cleaner assembly
   - Air intake pipe
   - Clutch slave cylinder fluid line
   - Shift cables
   - Exhaust flex pipe
   - Engine front mount
   - Lower ball joints
   - Axle halfshafts and intermediate shaft
   - Drive belt cover
   - Reverse light switch connector
   - Vehicle Speed (VSS) sensor connector
   - Rear engine mount and bracket
   - Starter motor
   - Transaxle flange bolts. Support the transaxle.
   - Transaxle

To install:

5. Install or connect the following:
   - Transaxle. Tighten the flange bolts to 35 ft. lbs. (48 Nm).
   - Starter motor
   - Rear engine mount bracket. Tighten the fastener to 59 ft. lbs. (80 Nm).
   - Rear engine mount. Tighten the outer nuts to 35 ft. lbs. (48 Nm) and the center nut to 98 ft. lbs. (133 Nm).
   - VSS sensor connector
   - Reverse light switch connector
   - Drive belt cover
   - Axle halfshafts and intermediate shaft
   - Lower ball joints. Tighten the pinch bolts to 37 ft. lbs. (50 Nm).
   - Right engine mount. Tighten the bolts to 35 ft. lbs. (48 Nm).
   - Exhaust flex pipe
   - Shift cables
   - Clutch slave cylinder fluid line
   - Air intake pipe
   - Air cleaner assembly
   - MAF sensor connector
   - Battery and tray
6. Tighten the strut center nuts to 35 ft. lbs. (48 Nm).

Automatic

1. Before servicing the vehicle, refer to the precautions in the beginning of this section.
2. Attach a support fixture to the engine lifting eyes.
3. Loosen the front strut center nuts 5 turns.
4. Remove or disconnect the following:
   - Battery and tray
   - Mass Air Flow (MAF) sensor connector
   - Positive Crankcase Ventilation (PCV) hose
   - Air intake pipe and resonator
   - Turbine Shaft Speed (TSS) sensor connector
   - Transmission range sensor connector
   - Transaxle harness connector
   - Starter motor
   - Shift cable and bracket

- Front wheels
- Exhaust front pipe
- Right engine support
- Outer tie rod ends
- Lower ball joints
- Axle halfshafts and intermediate shaft
- Output Shaft Speed (OSS) sensor connector
- Transaxle oil cooler lines
- Torque converter
- Transaxle oil dipstick tube
- Rear engine mount and bracket

➡The transaxle flange bolts vary in length. Note their locations for installation.

- Transaxle flange bolts. Support the transaxle.
- Transaxle

To install:

5. Install or connect the following:
   - Transaxle. Tighten the flange bolts to 35 ft. lbs. (48 Nm).
   - Rear engine mount bracket. Tighten the fastener to 59 ft. lbs. (80 Nm).
   - Rear engine mount. Tighten the outer nuts to 35 ft. lbs. (48 Nm) and the center nut to 98 ft. lbs. (133 Nm).
   - Transaxle oil dipstick tube
   - Torque converter. Tighten the nuts to 27 ft. lbs. (37 Nm).
   - Transaxle oil cooler lines
   - OSS sensor connector
   - Axle halfshafts and intermediate shaft
   - Lower ball joints. Tighten the pinch bolts to 37 ft. lbs. (50 Nm).
   - Outer tie rod ends. Tighten the nuts to 35 ft. lbs. (48 Nm).
   - Right engine support. Tighten the bolts to 35 ft. lbs. (48 Nm).
   - Exhaust front pipe. Tighten the fasteners to 35 ft. lbs. (48 Nm).
   - Front wheels
   - Shift cable and bracket
   - Starter motor
   - Transaxle harness connector
   - Transmission range sensor connector
   - TSS sensor connector
   - Air intake pipe and resonator
   - PCV hose
   - MAF sensor connector
   - Battery and tray
6. Tighten the strut center nuts to 35 ft. lbs. (48 Nm).

Clutch

REMOVAL & INSTALLATION

1. Before servicing the vehicle, refer to the precautions in the beginning of this section.
2. Remove or disconnect the following:
   - Negative battery cable
   - Transaxle
   - Pressure plate. Loosen the bolts evenly in ½ turn steps.
   - Clutch disc

To install:

3. Install or connect the following:
   - Clutch disc and pressure plate. Tighten the pressure plate bolts evenly in ½ turns to 21 ft. lbs. (29 Nm).
   - Transaxle
   - Negative battery cable

Hydraulic Clutch System

BLEEDING

1. Before servicing the vehicle, refer to the precautions in the beginning of this section.
2. Remove or disconnect the following:
   - Air cleaner assembly
   - Mass Air Flow (MAF) sensor connector
   - Air intake pipe
3. Remove fluid from the brake fluid reservoir until the level reaches the **MIN** mark.
4. Fill the reservoir of Special Tool 416-D002 with DOT 4 brake fluid.
5. Attach the tool to the slave cylinder bleed nipple.
6. Open the bleed nipple and pump 80 ml brake fluid into the clutch control system.
7. Tighten the bleed nipple to 88 inch lbs. (10 Nm) and remove the tool.
8. Have an assistant depress the clutch pedal 4–5 times and hold the pedal at full travel.
9. Close the bleeder before releasing the clutch pedal.
10. Repeat the procedure until no more air bubbles are seen.
11. Check the clutch for proper operation.
12. Check the brake fluid reservoir level and fill with DOT 4 brake fluid, as necessary.
13. Install or connect the following:
    - Air intake pipe
    - MAF sensor connector
    - Air cleaner assembly

Halfshaft

REMOVAL & INSTALLATION

Left

➡The IB5 manual transmission is used with the SOHC engine and the MTX75 manual transmission is used with the DOHC engine.

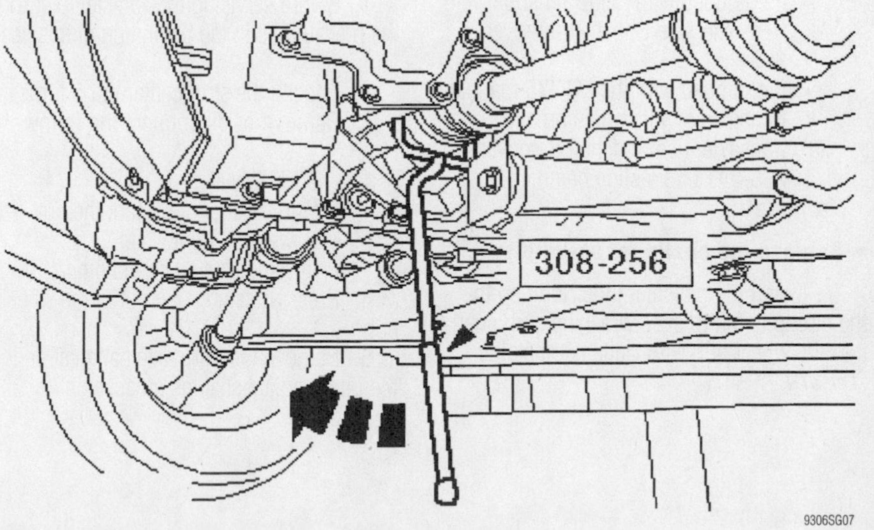

Left halfshaft removal—IB5 manual transaxle

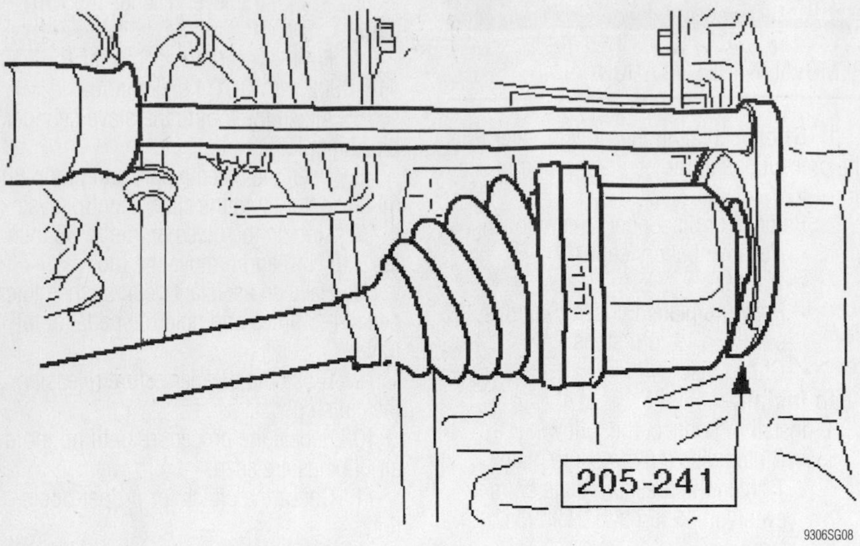

Left halfshaft removal—MTX75 manual transaxle and automatic transaxle

➡**The hub nut may be reused 4 times. Mark the nut at removal and only use hub nuts with 3 or fewer marks for assembly.**

1. Before servicing the vehicle, refer to the precautions in the beginning of this section.
2. Loosen the strut center nut 5 turns.
3. Remove or disconnect the following:
   - Front wheel
   - Hub retainer nut. Mark the nut.
   - Lower ball joint
4. Press the stub shaft out of the wheel hub.
5. Separate the inner CV-joint from the transaxle as follows:
   - If equipped with the IB5 manual transaxle, use Halfshaft remover 308-256
   - If equipped with the MTX75 manual transaxle or with an automatic transaxle, use Halfshaft Remover 205-241 and a slide hammer

To install:

➡**Replace the circlip for assembly.**

6. Install the halfshaft inner joint so that the circlip is felt to seat. Draw the stub shaft into the wheel hub with Halfshaft Installer 205-379.
7. Install or connect the following:
   - Lower ball joint. Tighten the pinch bolt to 37 ft. lbs. (50 Nm).
   - Hub retainer nut. Tighten the nut to 232 ft. lbs. (316 Nm).
   - Front wheel
8. Check the transaxle fluid and add, as necessary.
9. Tighten the strut center nuts to 35 ft. lbs. (48 Nm).

Right

➡**The hub nut may be reused 4 times. Mark the nut at removal and only use hub nuts with 3 or fewer marks for assembly.**

1. Before servicing the vehicle, refer to the precautions in the beginning of this section.
2. Loosen the strut center nut 5 turns.
3. Remove or disconnect the following:
   - Front wheel
   - Hub retainer nut. Mark the nut.
   - Lower ball joint
   - Intermediate shaft retaining clip
4. Press the stub shaft out of the wheel hub.
5. Remove the right axle halfshaft and the intermediate shaft as an assembly.

To install:

➡**Use a new retaining clip for assembly.**

6. Install or connect the following:
   - Axle halfshaft and intermediate shaft assembly. Tighten the nuts to 18 ft. lbs. (25 Nm).
   - Lower ball joint. Tighten the pinch bolt to 37 ft. lbs. (50 Nm).
   - Hub retainer nut. Tighten the nut to 232 ft. lbs. (316 Nm).
   - Front wheel
7. Check the transaxle fluid and add, as necessary.
8. Tighten the strut center nuts to 35 ft. lbs. (48 Nm).

CV-Joints

OVERHAUL

Inner Tripod Joint

1. Before servicing the vehicle, refer to the precautions in the beginning of this section.
2. Remove the axle halfshaft from the vehicle and place it in a vise.
3. Remove or disconnect the following:
   - Tripod joint boot clamps
   - Tripod joint boot
   - Tripod joint housing
   - Snapring
   - Tripod joint

To install:

➡**Use new snaprings and boot clamps for assembly.**

4. Install or connect the following:
   - Tripod joint
   - Snapring
   - Tripod joint housing. Fill the joint housing with grease.
   - Tripod joint boot
   - Tripod joint boot clamps
5. Install the halfshaft.

Outer CV-Joint

The outer CV-joint is serviced with the halfshaft as an assembly. The outer CV-joint boot can be serviced by removing the inner tripod joint.

STEERING AND SUSPENSION

Air Bag

⁂ CAUTION

These vehicles are equipped with an air bag system. The system must be disarmed before performing service on, or around, system components, the steering column, instrument panel components, wiring and sensors. Failure to follow the safety precautions and the disarming procedure could result in accidental air bag deployment, possible injury and unnecessary system repairs.

PRECAUTIONS

Several precautions must be observed when handling the inflator module to avoid accidental deployment and possible personal injury.

- Never carry the inflator module by the wires or connector on the underside of the module
- When carrying a live inflator module, hold securely with both hands, and ensure that the bag and trim cover are pointed away
- Place the inflator module on a bench or other surface with the bag and trim cover facing up
- With the inflator module on the bench, never place anything on or close to the module that may be thrown in the event of an accidental deployment

Before servicing the vehicle, also be sure to refer to the precautions in the beginning of this section as well

DISARMING

1. Before servicing the vehicle, refer to the precautions in the beginning of this section.
2. Position the vehicle with the front wheels in a straight-ahead position.
3. Disconnect the negative battery cable.
4. Disconnect the positive battery cable.
5. Wait at least 1 minute for the air bag backup power supply to drain before continuing.
6. Proceed with the repair.
7. Once repairs are completed, connect the battery cables, negative cable last.
8. Check the functioning of the air bag system by turning the ignition key to the **RUN** position and visually monitoring the air bag indicator lamp in the instrument cluster. The indicator lamp should illuminate for approximately 6 seconds, then turn **OFF**. If the indicator lamp does not illuminate, stays on, or flashes at any time, a fault has been detected by the air bag diagnostic monitor.

Power Rack & Pinion Steering Gear

REMOVAL & INSTALLATION

1. Before servicing the vehicle, refer to the precautions in the beginning of this section.
2. Center the steering wheel and turn the ignition to the **LOCK** position.
3. Remove or disconnect the following:
   - Negative battery cable
   - Lower instrument panel cover
   - Steering shaft pinch bolt
   - Front wheels
   - Outer tie rod ends
   - Stabilizer bar links
   - Power steering fluid cooler hose
   - Right engine mount
   - Steering gear heat shield
   - Power steering hose support clamp
   - Power steering hoses
   - 6 subframe bolts. Support the subframe.
   - Floor seal
   - Steering gear pinion extension
   - Steering gear

To install:

4. Install or connect the following:
   - Steering gear. Tighten the bolts to 59 ft. lbs. (80 Nm).
   - Steering gear pinion extension. Tighten the pinch bolt to 26 ft. lbs. (35 Nm).
   - Floor seal
5. Install Subframe Alignment Pins 502-002 and raise the subframe into position. Tighten the 4 rear bolts to 147 ft. lbs. (200 Nm) and the other bolts to 85 ft. lbs. (115 Nm).
6. Install or connect the following:
   - Power steering hoses
   - Power steering hose support clamp
   - Steering gear heat shield
   - Right engine mount. Tighten the bolt to 37 ft. lbs. (50 Nm).
   - Power steering fluid cooler hose
   - Stabilizer bar links. Tighten the nuts to 37 ft. lbs. (50 Nm).
   - Outer tie rod ends. Tighten the nuts to 35 ft. lbs. (48 Nm).
   - Front wheels
   - Steering shaft pinch bolt. Tighten the bolt to 21 ft. lbs. (28 Nm).
   - Lower instrument panel cover
   - Negative battery cable
7. Check the wheel alignment and adjust as necessary.

Strut

REMOVAL & INSTALLATION

1. Before servicing the vehicle, refer to the precautions in the beginning of this section.
2. Remove or disconnect the following:
   - Front wheel
   - Brake hose bracket
   - Stabilizer bar link
   - Wheel speed sensor, if equipped
   - Brake caliper and rotor
   - Outer tie rod end
   - Lower ball joint
   - Steering knuckle pinch bolt. Separate the knuckle from the strut.
   - Upper strut mount nuts
   - Strut assembly

To install:

3. Install or connect the following:
   - Strut assembly. Tighten the upper mount nuts to 18 ft. lbs. (25 Nm).
   - Steering knuckle. Tighten the pinch bolt to 66 ft. lbs. (90 Nm).
   - Lower ball joint. Tighten the pinch bolt to 37 ft. lbs. (50 Nm).
   - Outer tie rod end. Tighten the nut to 35 ft. lbs. (48 Nm).
   - Brake caliper and rotor
   - Wheel speed sensor, if equipped
   - Stabilizer bar link. Tighten the nut to 37 ft. lbs. (50 Nm).
   - Brake hose bracket
   - Front wheel
4. Check the wheel alignment and adjust as necessary.

Shock Absorber

REMOVAL & INSTALLATION

3 Door Model

1. Before servicing the vehicle, refer to the precautions in the beginning of this section.

2. Remove or disconnect the following:
   - Luggage compartment interior trim panel
   - Upper shock absorber mounting nut
   - Lower shock absorber mounting bolt
   - Shock absorber

To install:

➡Tighten the shock absorber mounting fasteners with the suspension at curb height and the vehicle weight supported by the wheels.

3. Install or connect the following:
   - Shock absorber. Guide the rod into the locating hole.
   - Lower shock absorber mounting bolt. Tighten the bolt to 85 ft. lbs. (115 Nm).
   - Upper shock absorber mounting nut. Tighten the nut to 13 ft. lbs. (18 Nm).
   - Luggage compartment interior trim panel

Wagon

1. Before servicing the vehicle, refer to the precautions in the beginning of this section.
2. Remove the upper and lower mounting bolts and remove the shock absorber.

To install:

➡Tighten the shock absorber mounting fasteners with the suspension at curb height and the vehicle weight supported by the wheels.

3. Install the shock absorber and tighten the mounting bolts to 85 ft. lbs. (115 Nm).

Coil spring

REMOVAL & INSTALLATION

Front

1. Before servicing the vehicle, refer to the precautions at the beginning of this section.
2. Remove the strut from the vehicle.
3. Compress the coil spring using a suitable spring compressor until the spring comes away from the seat.
4. Remove the large center nut and slowly release the spring compressor.

To install:

5. Compress the spring and install it on the strut.

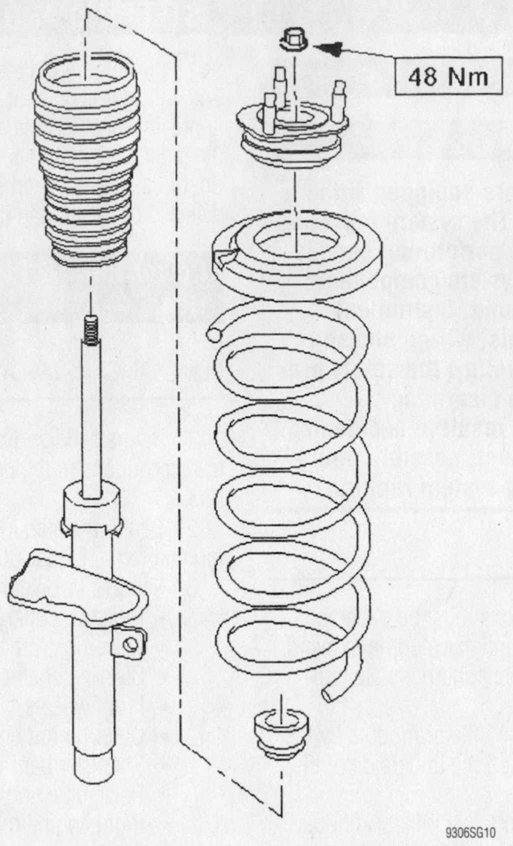

Front strut assembly exploded view

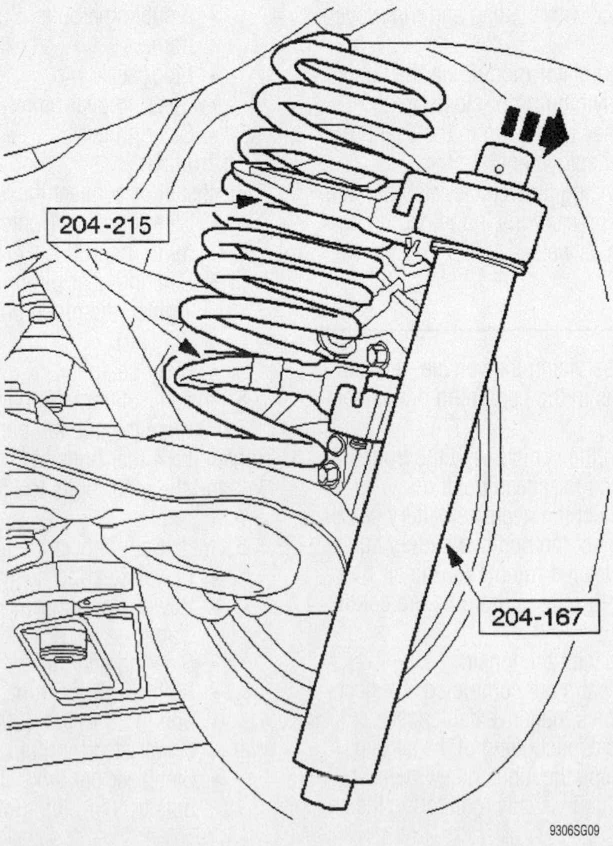

Coil spring compressor and adapters

6. Install the upper strut mount. Tighten the nut to 35 ft. lbs. (48 Nm).
7. Install the strut assembly in the vehicle.

Rear

1. Before servicing the vehicle, refer to the precautions in the beginning of this section.
2. Raise and support the vehicle.
3. Install spring compressor 204-167 with Adapters 204-215.
4. Compress the coil spring and remove it.

To install:

5. Install the coil spring and remove the compressor.

Lower Ball Joint

REMOVAL & INSTALLATION

The lower ball joint is replaced with the lower control arm as an assembly.

Lower Control Arm

REMOVAL & INSTALLATION

1. Before servicing the vehicle, refer to the precautions in the beginning of this section.
2. Remove or disconnect the following:
   - Front wheel
   - Lower ball joint
   - Rear bracket bolts
   - Front bolt
   - Control arm

To install:

➡Use new nuts, bolts and ball bearing washers for assembly.

3. Install the control arm. Tighten the fasteners in sequence as follows:
   a. Step 1: Tighten nut No. 1 to 74 ft. lbs. (100 Nm) plus 60 degrees
   b. Step 2: Tighten nut No. 2 to 88 ft. lbs. (120 Nm)
   c. Step 3: Tighten bolt No. 3 to 88 ft. lbs. (120 Nm) plus 90 degrees
   d. Step 4: Check that bolt No. 3 is tightened to 125–169 ft. lbs. (170–230 Nm)
4. Install or connect the following:
   - Lower ball joint. Tighten the pinch bolt to 37 ft. lbs. (50 Nm).
   - Front wheel
5. Check the wheel alignment and adjust as necessary.

CONTROL ARM BUSHING REPLACEMENT

The lower control arm bushings are replaced with the lower control arm as an assembly.

Wheel Bearings

ADJUSTMENT

The bearings on the front and rear wheels are a one piece cartridge design and cannot be adjusted. If wheel bearing play is excessive, check the wheel hub retainer nut for proper torque. If the torque is correct, replacement of the wheel bearing is required.

REMOVAL & REPLACEMENT

Front

➡The hub nut may be reused 4 times. Mark the nut at removal and only use hub nuts with 3 or fewer marks for assembly.

1. Before servicing the vehicle, refer to the precautions in the beginning of this section.
2. Loosen the strut center nut 5 turns.
3. Remove or disconnect the following:
   - Front wheel
   - Wheel speed sensor, if equipped
   - Hub retainer nut. Mark the nut.
   - Brake caliper and rotor
   - Outer tie rod end
   - Lower ball joint
   - Steering knuckle pinch bolt
4. Press the stub shaft out of the wheel hub.
5. Press the hub out of the wheel bearing.
6. Remove the snapring.
7. Press the bearing out of the hub.

To install:

8. Press the bearing into the hub.
9. Install the snapring.
10. Press the hub into the wheel bearing.
11. Draw the stub shaft into the wheel hub with Halfshaft installer 205-379.
12. Install or connect the following:
    - Steering knuckle. Tighten the pinch bolt to 66 ft. lbs. (90 Nm).
    - Lower ball joint. Tighten the pinch bolt to 37 ft. lbs. (50 Nm).
    - Outer tie rod end. Tighten the nut to 35 ft. lbs. (48 Nm).
    - Brake caliper and rotor
    - Hub retainer nut. Tighten the nut to 232 ft. lbs. (316 Nm).
    - Wheel speed sensor, if equipped
    - Front wheel
13. Tighten the strut center nuts to 35 ft. lbs. (48 Nm).
14. Check the wheel alignment and adjust as necessary.

Rear

1. Before servicing the vehicle, refer to the precautions in the beginning of this section.

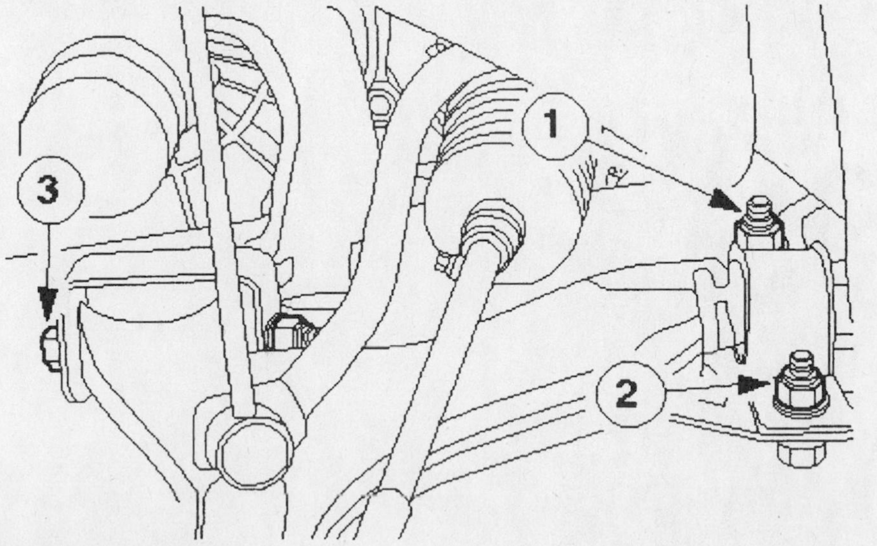

Control arm torque sequence

2. Remove or disconnect the following:
   - Rear wheel
   - Dust cap
   - Hub nut
   - Brake drum
   - Wheel speed sensor tone ring, if equipped
   - Snapring
3. Remove the wheel bearing with a press.

To install:

➡Use a new wheel speed sensor tone ring for assembly.

4. Press the wheel bearing into the brake drum hub.
5. Install or connect the following:
   - Snapring
   - Wheel speed sensor tone ring, if equipped
   - Brake drum
   - Hub nut. Tighten the hub nut to 173 ft. lbs. (235 Nm).
   - Dust cap
   - Rear wheel

FORD MOTOR CO. 16

2000–01
Lincoln LS

PRECAUTIONS

Before servicing any vehicle, please be sure to read all of the following precautions, which deal with personal safety, prevention of component damage, and important points to take into consideration when servicing a motor vehicle:

- Never open, service or drain the radiator or cooling system when the engine is hot; serious burns can occur from the steam and hot coolant.
- Observe all applicable safety precautions when working around fuel. Whenever servicing the fuel system, always work in a well-ventilated area. Do not allow fuel spray or vapors to come in contact with a spark, open flame, or excessive heat (a hot drop light, for example). Keep a dry chemical fire extinguisher near the work area. Always keep fuel in a container specifically designed for fuel storage; also, always properly seal fuel containers to avoid the possibility of fire or explosion. Refer to the additional fuel system precautions later in this section.
- Fuel injection systems often remain pressurized, even after the engine has been turned **OFF**. The fuel system pressure must be relieved before disconnecting any fuel lines. Failure to do so may result in fire and/or personal injury.
- Brake fluid often contains polyglycol ethers and polyglycols. Avoid contact with the eyes and wash your hands thoroughly after handling brake fluid. If you do get brake fluid in your eyes, flush your eyes with clean, running water for 15 minutes. If eye irritation persists, or if you have taken brake fluid internally, IMMEDIATELY seek medical assistance.
- The EPA warns that prolonged contact with used engine oil may cause a number of skin disorders, including cancer. You should make every effort to minimize your exposure to used engine oil. Protective gloves should be worn when changing oil. Wash your hands and any other exposed skin areas as soon as possible after exposure to used engine oil. Soap and water, or waterless hand cleaner should be used.
- All new vehicles are now equipped with an air bag system, often referred to as a Supplemental Restraint System (SRS) or Supplemental Inflatable Restraint (SIR) system. The system must be disabled before performing service on or around system components, steering column, instrument panel components, wiring and sensors. Failure to follow safety and disabling procedures could result in accidental air bag deployment, possible personal injury and unnecessary system repairs.
- Always wear safety goggles when working with, or around, the air bag system. When carrying a non-deployed air bag, be sure the bag and trim cover are pointed away from your body. When placing a non-deployed air bag on a work surface, always face the bag and trim cover upward, away from the surface. This will reduce the motion of the module if it is accidentally deployed. Refer to the additional air bag system precautions later in this section.
- Clean, high quality brake fluid from a sealed container is essential to the safe and proper operation of the brake system. You should always buy the correct type of brake fluid for your vehicle. If the brake fluid becomes contaminated, completely flush the system with new fluid. Never reuse any brake fluid. Any brake fluid that is removed from the system should be discarded. Also, do not allow any brake fluid to come in contact with a painted surface; it will damage the paint.
- Never operate the engine without the proper amount and type of engine oil; doing so WILL result in severe engine damage.
- Timing belt maintenance is extremely important. Many models utilize an interference-type, non-freewheeling engine. If the timing belt breaks, the valves in the cylinder head may strike the pistons, causing potentially serious (also time-consuming and expensive) engine damage. Refer to the maintenance interval charts in the front of this manual for the recommended replacement interval for the timing belt, and to the timing belt section for belt replacement and inspection.
- Disconnecting the negative battery cable on some vehicles may interfere with the functions of the on-board computer system(s) and may require the computer to undergo a relearning process once the negative battery cable is reconnected.
- When servicing drum brakes, only disassemble and assemble one side at a time, leaving the remaining side intact for reference.

ENGINE REPAIR

➡Disconnecting the negative battery cable on some vehicles may interfere with the functions of the on board computer system. The computer may undergo a relearning process once the negative battery cable is reconnected.

Alternator

REMOVAL

3.0L Engine

1. Before servicing the vehicle, refer to the precautions in the beginning of this section.
2. Remove or disconnect the following:
   - Negative battery cable
   - Accessory drive belt
   - Lower splash shield
   - Alternator mounting bolts
   - Alternator harness connectors
   - Alternator

3.9L Engine

1. Before servicing the vehicle, refer to the precautions in the beginning of this section.
2. Remove or disconnect the following:
   - Negative battery cable
   - Air intake tube
   - Accessory drive belt
   - Lower splash shield
   - Alternator mounting bolts
   - Alternator harness connectors
   - Alternator

INSTALLATION

3.0L Engine

Install or connect the following:

- Alternator
- Alternator harness connectors. Tighten the battery cable terminal nut to 70 inch lbs. (8 Nm).
- Alternator mounting bolts. Tighten the bolts to 35 ft. lbs. (48 Nm).
- Lower splash shield
- Accessory drive belt
- Negative battery cable

3.9L Engine

1. Install the alternator harness connectors, then install the alternator. Tighten the bolts in sequence as follows:

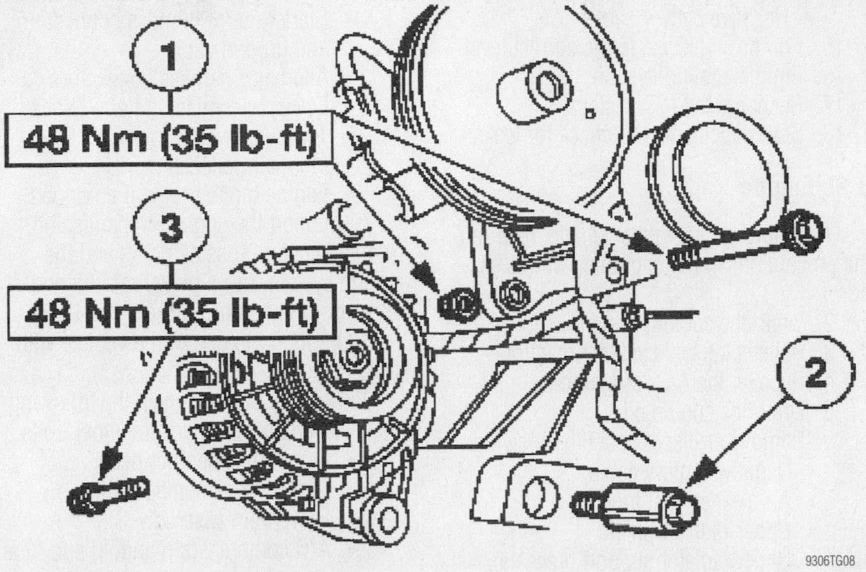

Alternator torque sequence—3.9L engine

a. Step 1: Tighten bolt No. 1 to 35 ft. lbs. (48 Nm)

b. Step 2: Tighten bolt No. 2 to 15 ft. lbs. (20 Nm) plus 90 degrees

c. Step 3: Tighten bolt No. 3 to 35 ft. lbs. (48 Nm)

2. Install or connect the following:
   - Lower splash shield
   - Accessory drive belt
   - Air intake tube
   - Negative battery cable

Ignition Timing

ADJUSTMENT

This vehicle is equipped with a Distributorless Ignition System (DIS). The ignition timing is not adjustable. It is controlled by the PCM.

Engine Assembly

REMOVAL & INSTALLATION

3.0L Engine

1. Before servicing the vehicle, refer to the precautions in the beginning of this section.
2. Drain the cooling system.
3. Relieve the fuel system pressure.
4. Recover the A/C refrigerant.
5. Drain the engine oil.
6. Remove or disconnect the following:
   - Negative battery cable
   - Air cleaner housing and outlet tube
   - Engine appearance cover
   - Upper radiator shield
   - Upper radiator support brackets
   - A/C pressure switch connector
   - Power steering reservoir
   - Fuel line
   - Vapor Management Valve (VMV) cover
   - VMV vacuum hose
   - Cowl leaf screens
   - Chassis vacuum lines
   - Cross vehicle support bar
   - Fresh air intake housing
   - Intake manifold rear main vacuum hose
   - Accelerator cable and cruise control cable
   - Cable bracket
   - Ground strap
   - Main engine wiring harness connector
   - Main transmission wiring harness connector
   - 2 fuel charging harness connectors
   - A/C line mounting bracket
   - Hydraulic cooling fan reservoir
   - Left, right and center splash shields
   - A/C compressor manifold and tube assembly
   - Coolant hoses
   - Exhaust front pipe and heat shields
   - Driveshaft
   - Shift cable and bracket
   - Front wheels
   - Front wheel speed sensor connectors
   - Front brake calipers
   - Lower stabilizer bar links
   - Upper ball joints
   - Lower strut mount bolts
   - Starter motor wiring harness connectors
   - Power Steering Pressure (PSP) switch connector
   - Steering shaft pinch bolt
   - Torque converter, if equipped
7. Support the engine, transmission, front and center crossmembers and the cooling system with a powertrain lift and transmission support bracket.
8. Support the rear of the vehicle with safety stands.
9. Remove or disconnect the following:
   - Transmission crossmember bolts
   - Subframe bolts
   - Crossmember bolts
   - Powertrain assembly
10. Attach a hoist to the engine.
11. Remove or disconnect the following:
    - Wire harness retainers
    - Starter motor
    - Heated Oxygen (HO_2S) sensor bracket
    - Motor mount nuts
    - Accessory drive belt
    - Power steering pump
    - Hydraulic cooling fan pump
    - Upper radiator hose
    - Transmission oil cooler lines, if equipped
12. Lift the engine and transmission out of the subframe.
13. Remove or disconnect the following:
    - Oil cooler hoses
    - Transmission flange bolts
    - Transmission from the engine

To install:

14. Install or connect the following:
    - Transmission to the engine. Tighten the flange bolts to 35 ft. lbs. (48 Nm).
    - Powertrain on the subframe. Tighten the motor mount nuts to 46 ft. lbs. (63 Nm).
    - Oil cooler hoses
    - Transmission oil cooler lines, if equipped
    - Upper radiator hose
    - Hydraulic cooling fan pump. Tighten the bolts to 18 ft. lbs. (25 Nm).
    - Power steering pump. Tighten the bolts to 18 ft. lbs. (25 Nm).
    - Accessory drive belt
    - HO_2S sensor bracket. Tighten the nut to 89 inch lbs. (10 Nm).

- Starter motor. Tighten the bolts to 18 ft. lbs. (25 Nm).
- Wire harness retainers
- Powertrain assembly. Tighten the crossmember bolts to 76 ft. lbs. (103 Nm).
- Torque converter, if equipped.
- Steering shaft pinch bolt. Tighten the bolts to 18 ft. lbs. (25 Nm).
- PSP switch connector
- Starter motor wiring harness connectors
- Lower strut mount bolts. Tighten the bolts to 129 ft. lbs. (175 Nm).
- Upper ball joints. Tighten the nuts to 66 ft. lbs. (90 Nm).
- Lower stabilizer bar links. Tighten the nuts to 41 ft. lbs. (55 Nm).
- Front brake calipers
- Front wheel speed sensor connectors
- Front wheels
- Shift cable and bracket
- Driveshaft
- Exhaust front pipe and heat shields
- Coolant hoses
- A/C compressor manifold and tube assembly. Tighten the bolt to 15 ft. lbs. (21 Nm).
- Left, right and center splash shields
- Hydraulic cooling fan reservoir
- A/C line mounting bracket
- 2 fuel charging harness connectors
- Main transmission wiring harness connector. Tighten the bolt to 89 inch lbs. (10 Nm).
- Main engine wiring harness connector. Tighten the bolt to 89 inch lbs. (10 Nm).
- Ground strap. Tighten the bolt to 89 inch lbs. (10 Nm).
- Cable bracket. Tighten the bolts to 89 inch lbs. (10 Nm).
- Accelerator cable and cruise control cable
- Intake manifold rear main vacuum hose
- Fresh air intake housing
- Cross vehicle support bar. Tighten the bolts to 15 ft. lbs. (20 Nm).
- Chassis vacuum lines
- Cowl leaf screens
- VMV vacuum hose
- VMV cover
- Fuel line
- Power steering reservoir
- A/C pressure switch connector
- Upper radiator support brackets. Tighten the bolts to 89 inch lbs. (10 Nm).
- Upper radiator shield
- Engine appearance cover
- Air cleaner housing and outlet tube
- Negative battery cable

15. Fill the crankcase to the correct level.
16. Fill the cooling system.
17. Recharge the A/C system.
18. Start the engine and check for leaks.

3.9L Engine

1. Before servicing the vehicle, refer to the precautions in the beginning of this section.
2. Drain the cooling system.
3. Relieve the fuel system pressure.
4. Recover the A/C refrigerant.
5. Drain the engine oil.
6. Remove or disconnect the following:
   - Negative battery cable
   - Air cleaner inlet tube
   - Upper radiator shield
   - Upper radiator support brackets
   - A/C pressure switch connector
   - Power steering return line clip
   - Power steering reservoir
   - Vapor Management Valve (VMV) cover
   - VMV vacuum hose and canister purge hose
   - Main vacuum supply hose
   - Cowl vent screens
   - Cross vehicle support bar
   - Degas bottle hose
   - Accelerator cable
   - Cruise control cable
   - Ground strap
   - Fresh air filter and housing
   - Powertrain harness connectors at right strut tower
   - Fresh air filter panel
   - Main engine wiring harness connector
   - Main transmission wiring harness connector
   - Heater hoses at the water control valve. Note the locations for assembly.
   - Hydraulic cooling fan reservoir
   - Water control valve harness connector
   - Front wheels
   - Inner splash shields
   - Wheel speed sensor connectors and harness clips
   - Brake calipers
   - Lower stabilizer bar links
   - Upper ball joints
   - Lower strut mount bolts
   - Left, right and center splash shields
   - A/C suction and discharge lines
   - Shift cable and bracket
   - Power steering line frame rail clip
   - Rack and pinion harness connectors
   - Steering shaft bolt and coupling
   - Starter motor harness connectors and ground cable
   - Alternator harness connectors
   - Lower transmission flange bolts
   - Torque converter
   - Inner air deflector
   - Engine block heater, if equipped
7. Support the engine, transmission, front and center crossmembers and the cooling system with a powertrain lift and transmission support bracket.
8. Support the rear of the vehicle with safety stands.
9. Remove or disconnect the following:
   - Transmission crossmember bolts
   - Front crossmember bolts
   - Center crossmember bolts
   - Powertrain assembly
   - A/C compressor manifold and tube assembly
   - Power steering pump return hose
   - Hydraulic cooling fan return hose
   - Lower radiator hose
   - Upper radiator hoses
   - Knock Sensor (KS) connector
   - Heater hose
   - Transmission cooler lines and bracket
   - Power steering pressure line and bracket
   - Hydraulic cooling fan pressure line and bracket
10. Attach a hoist to the engine.
11. Remove the motor mount nuts and lift the powertrain out of the subframe.
12. Remove or disconnect the following:
    - Wiring harness retainers
    - Upper transmission flange bolts
    - Transmission from the engine

To install:

13. Install or connect the following:
    - Transmission to the engine. Tighten the upper flange bolts to 35 ft. lbs. (48 Nm).
    - Wiring harness retainers. Tighten the nuts to 89 inch lbs. (10 Nm).
    - Powertrain to the subframe. Tighten the mount nuts to 30 ft. lbs. (40 Nm).
    - Hydraulic cooling fan pressure line and bracket
    - Power steering pressure line and bracket
    - Transmission cooler lines and bracket
    - Heater hose
    - Knock sensor connector
    - Upper radiator hoses
    - Lower radiator hose
    - Hydraulic cooling fan return hose

- Power steering pump return hose
- A/C compressor manifold and tube assembly. Tighten the bolt to 15 ft. lbs. (21 Nm).
- Powertrain assembly. Tighten the front and center crossmember bolts to 76 ft. lbs. (103 Nm) and the transmission crossmember bolts to 30 ft. lbs. (40 Nm).
- Engine block heater, if equipped
- Inner air deflector
- Torque converter. Tighten the nuts to 28 ft. lbs. (38 Nm).
- Lower transmission flange bolts. Tighten the bolts to 35 ft. lbs. (47 Nm).
- Alternator harness connectors
- Starter motor harness connectors and ground cable
- Steering shaft bolt and coupling. Tighten the coupling pinch bolt to 26 ft. lbs. (35 Nm) and the shaft bolt to 22 ft. lbs. (30 Nm).
- Rack and pinion harness connectors
- Power steering line frame rail clip
- Shift cable and bracket
- A/C suction and discharge lines
- Left, right and center splash shields
- Lower strut mount bolts. Tighten the bolts to 129 ft. lbs. (175 Nm).
- Upper ball joints. Tighten the nuts to 66 ft. lbs. (90 Nm).
- Lower stabilizer bar links. Tighten the nuts to 41 ft. lbs. (55 Nm).
- Brake calipers
- Wheel speed sensor connectors and harness clips
- Inner splash shields
- Front wheels
- Water control valve harness connector
- Hydraulic cooling fan reservoir
- Heater hoses at the water control valve
- Main transmission wiring harness connector
- Main engine wiring harness connector
- Fresh air filter panel
- Powertrain harness connectors at right strut tower
- Fresh air filter and housing
- Ground strap
- Cruise control cable
- Accelerator cable
- Degas bottle hose
- Cross vehicle support bar. Tighten the bolts to 15 ft. lbs. (20 Nm).
- Cowl vent screens
- Main vacuum supply hose
- VMV vacuum hose and canister purge hose
- VMV cover
- Power steering reservoir
- Power steering return line clip
- A/C pressure switch connector
- Upper radiator support brackets
- Upper radiator shield
- Air cleaner inlet tube
- Negative battery cable

14. Fill the crankcase to the correct level.
15. Fill the cooling system.
16. Recharge the A/C system.
17. Start the engine and check for leaks.

Water Pump

REMOVAL & INSTALLATION

3.0L Engine

1. Before servicing the vehicle, refer to the precautions in the beginning of this section.
2. Drain the cooling system.
3. Remove or disconnect the following:
   - Negative battery cable
   - Air cleaner outlet tube
   - Engine vent hose
   - Upper radiator hose
   - Heater supply hose
   - Water pump hose
   - Lower radiator hose
   - Water crossover assembly
   - Water inlet hose
   - Accessory drive belt and idler pulley
   - Bracket assembly
   - Water pump

To install:

4. Install or connect the following:
   - Water pump. Tighten the bolts to 18 ft. lbs. (25 Nm).
   - Bracket assembly. Tighten the fasteners to 89 inch lbs. (10 Nm).
   - Accessory drive belt and idler pulley
   - Water inlet hose
   - Water crossover assembly
   - Lower radiator hose
   - Water pump hose
   - Heater supply hose
   - Upper radiator hose
   - Engine vent hose
   - Air cleaner outlet tube
   - Negative battery cable
5. Fill the cooling system.
6. Start the engine and check for leaks.

3.9L Engine

1. Before servicing the vehicle, refer to the precautions in the beginning of this section.
2. Drain the cooling system.
3. Remove or disconnect the following:
   - Accessory drive belt
   - Water pump pulley
   - Water pump

To install:

4. Install or connect the following:
   - Water pump. Tighten the bolts to 71 inch lbs. (8 Nm) plus 90 degrees.
   - Water pump pulley. Tighten the bolts to 89 inch lbs. (10 Nm) plus 45 degrees.
   - Accessory drive belt
5. Fill the cooling system.
6. Start the engine and check for leaks.

Cylinder Head

REMOVAL & INSTALLATION

3.0L Engine

1. Before servicing the vehicle, refer to the precautions in the beginning of this section.
2. Drain the cooling system.
3. Relieve the fuel system pressure.
4. Drain the engine oil.
5. Remove or disconnect the following:
   - Negative battery cable
   - Upper and lower intake manifolds
   - Valve covers
   - Accessory drive belts
6. Install a support fixture to the engine lifting eyes.
7. Remove or disconnect the following:
   - Motor mount nuts
   - Subframe bolts
   - Oil pan
   - Front cover
   - Timing chains
   - Camshafts
   - Exhaust manifolds
   - Ground strap
   - Positive Crankcase Ventilation (PCV) tube
   - Ignition noise suppressor
   - Coolant outlet tube
   - Engine oil dipstick tube
8. Install the subframe bolts and remove the engine support fixture.
9. Remove the cylinder heads.

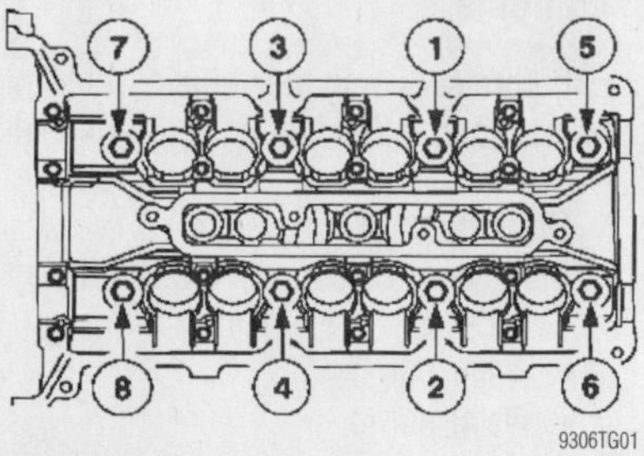

Cylinder head torque sequence—3.0L engine

To install:

10. Install the cylinder heads. Tighten the bolts in sequence as follows:
    a. Step 1: 22 ft. lbs. (30 Nm)
    b. Step 2: Plus 90 degrees
    c. Step 3: Loosen all bolts one full turn
    d. Step 4: 22 ft. lbs. (30 Nm)
    e. Step 5: Plus 90 degrees
    f. Step 6: Plus 90 degrees
11. Install the engine support fixture and remove the subframe bolts.
12. Install or connect the following:
    - Engine oil dipstick tube
    - Coolant outlet tube
    - Ignition noise suppressor
    - PCV tube
    - Ground strap
    - Exhaust manifolds
    - Camshafts
    - Timing chains

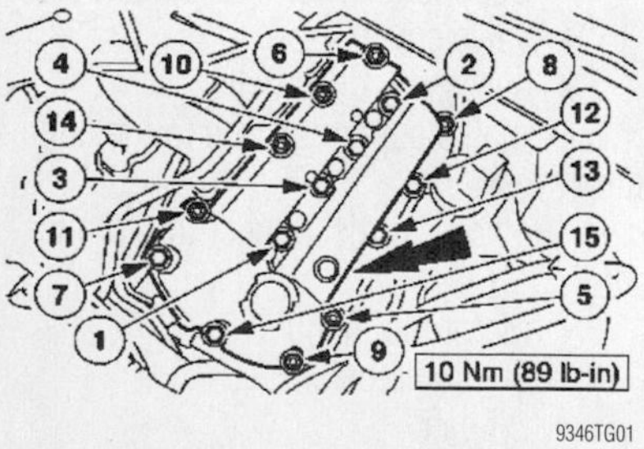

Left valve cover torque sequence—3.0L Engine

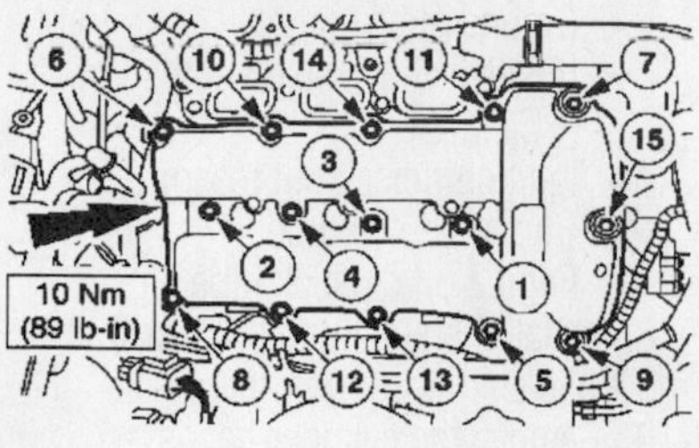

Right valve cover torque sequence—3.0L Engine

    - Front cover
    - Oil pan
    - Subframe bolts. Tighten the bolts to 76 ft. lbs. (103 Nm).
    - Motor mount nuts. Tighten the nuts to 46 ft. lbs. (63 Nm).
    - Accessory drive belts
    - Valve covers
    - Upper and lower intake manifolds
    - Negative battery cable
13. Fill the crankcase to the correct level.
14. Fill the cooling system.
15. Start the engine and check for leaks.

3.9L Engine

1. Before servicing the vehicle, refer to the precautions in the beginning of this section.
2. Drain the cooling system.
3. Relieve the fuel system pressure.
4. Remove or disconnect the following:
    - Negative battery cable

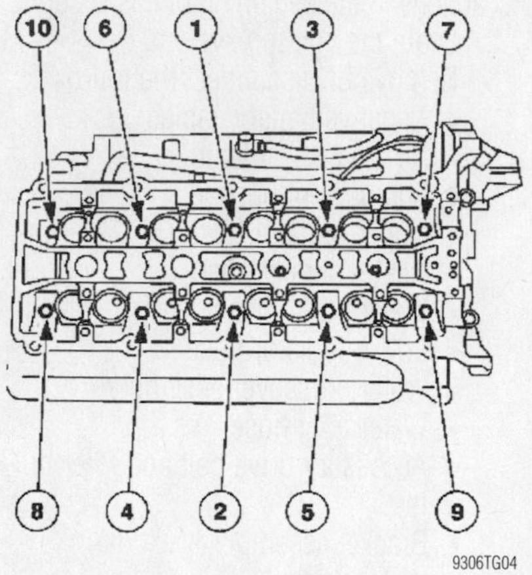

Right cylinder head torque sequence—3.9L engine

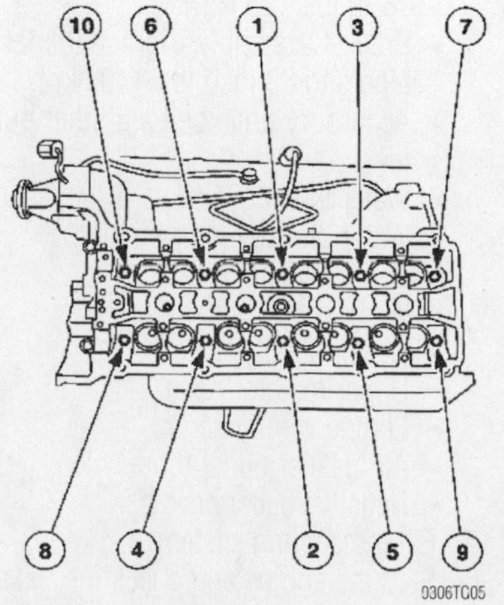

Left cylinder head torque sequence—3.9L engine

    - Engine appearance cover
    - Intake manifold
    - Valve covers
    - Accessory drive belts
    - Front cover
    - Timing chains
    - Camshafts
    - Water outlet pipe
    - Cylinder head temperature sensor connector
    - Exhaust front pipes
    - Exhaust Gas Recirculation (EGR) tube
    - Bolts and stud bolts at the rear of the cylinder heads
    - Cylinder heads. Loosen the bolts in reverse of the tightening sequence.

To install:

5. Install the cylinder heads. Tighten the bolts in sequence as follows:
    a. Step 1: M10 bolts to 15 ft. lbs. (20 Nm)
    b. Step 2: M10 bolts to 26 ft. lbs. (35 Nm)

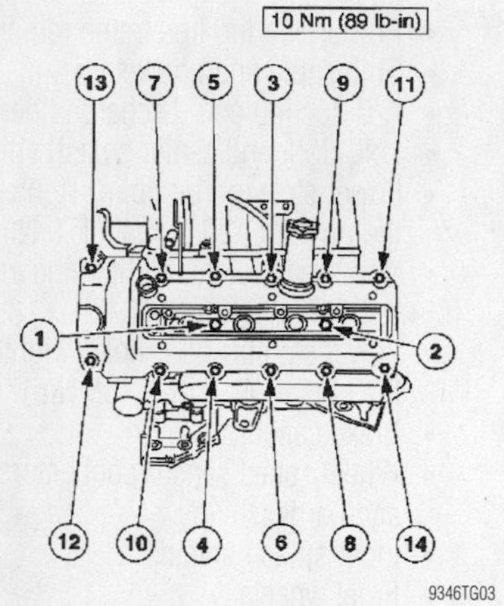

Left valve cover torque sequence—3.9L Engine

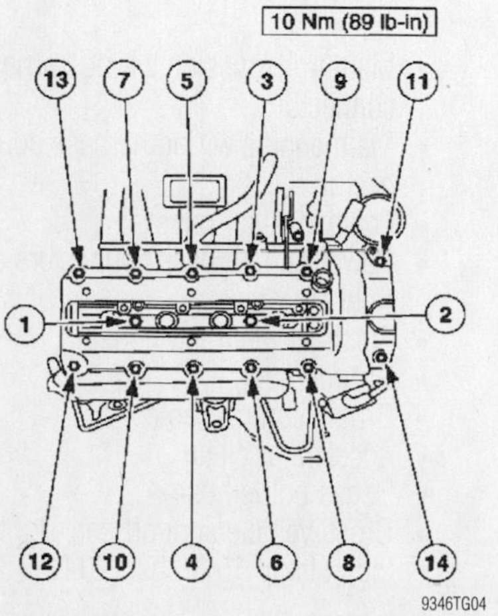

Right valve cover torque sequence—3.9L Engine

c. Step 3: M10 bolts to 33 ft. lbs. (45 Nm)
d. Step 4: M10 bolts plus 90 degrees
e. Step 5: M10 bolts plus 90 degrees
f. Step 6: M8 bolts to 15 ft. lbs. (20 Nm)
g. Step 7: M8 bolts plus 90 degrees

6. Install or connect the following:
   - Bolts and stud bolts at the rear of the cylinder heads. Tighten the bolts to 37 ft. lbs. (50 Nm).
   - EGR tube
   - Exhaust front pipes
   - Cylinder head temperature sensor connector
   - Water outlet pipe
   - Camshafts
   - Timing chains
   - Front cover
   - Accessory drive belts
   - Valve covers
   - Intake manifold
   - Engine appearance cover
   - Negative battery cable
7. Fill the cooling system.
8. Start the engine and check for leaks.

Rocker Arms/Shafts

REMOVAL & INSTALLATION

The vehicles covered in this section are not equipped with rocker arms/shafts. The camshaft directly actuates the valves.

Intake Manifold

REMOVAL & INSTALLATION

3.0L Engine

1. Before servicing the vehicle, refer to the precautions in the beginning of this section.
2. Drain the cooling system.
3. Relieve the fuel system pressure.
4. Remove or disconnect the following:
   - Negative battery cable
   - Engine appearance cover
   - Air cleaner outlet tube
   - Throttle Position (TP) sensor connector
   - Idle Air Control (IAC) valve connector
   - Accelerator cable
   - Cruise control cable
   - Cable bracket
   - Throttle body coolant bypass hose
   - Positive Crankcase Ventilation (PCV) hose
   - Evaporative Emissions (EVAP) canister purge hose
   - Exhaust Gas Recirculation (EGR) vacuum line
   - EGR tube
   - Cowl vent screen
   - Cross vehicle support bar
   - Chassis vacuum supply hose
   - EGR pressure transducer
   - Fuel pressure sensor shield
   - Upper intake manifold vacuum hose
   - Intake Manifold Tuning Valve (IMTV) connector
   - Exhaust Vacuum Regulator (EVR) valve connector and vacuum line
   - Upper intake manifold support brackets
   - Upper intake manifold
   - Fuel line and bracket
   - Fuel pressure sensor vacuum line
   - Fuel injector harness connectors
   - PCV tube
   - Lower intake manifold assembly

To install:

5. Install or connect the following:
   - Lower intake manifold assembly. Tighten the bolts in sequence to 89 inch lbs. (10 Nm).
   - PCV tube
   - Fuel injector harness connectors

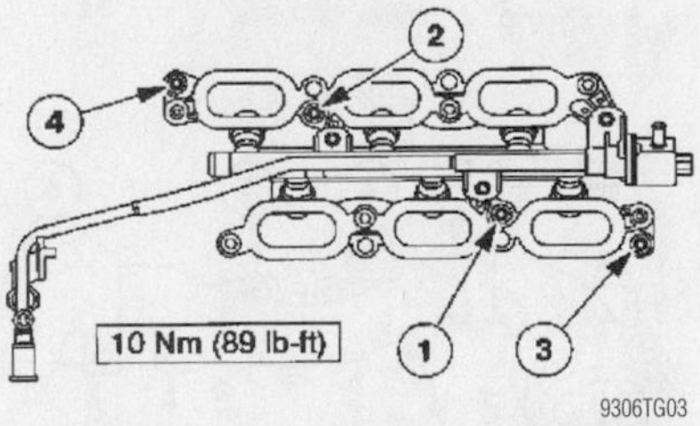

Lower intake manifold torque sequence—3.0L engine

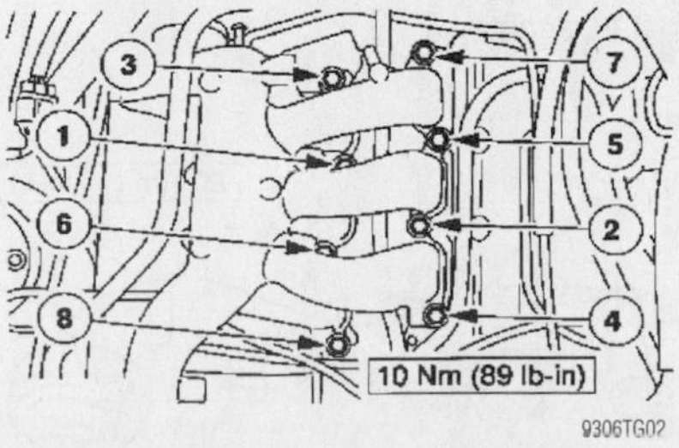

Upper intake manifold torque sequence—3.0L engine

   - Fuel pressure sensor vacuum line
   - Fuel line and bracket. Tighten the bolt to 89 inch lbs. (10 Nm).
   - Upper intake manifold. Tighten the bolts in sequence to 89 inch lbs. (10 Nm).
   - Upper intake manifold support brackets. Tighten the bolts to 89 inch lbs. (10 Nm).
   - EVR valve connector and vacuum line
   - IMTV connector
   - Upper intake manifold vacuum hose
   - Fuel pressure sensor shield
   - EGR pressure transducer
   - Chassis vacuum supply hose
   - Cross vehicle support bar. Tighten the bolts to 15 ft. lbs. (20 Nm).
   - Cowl vent screen
   - EGR tube
   - EGR vacuum line
   - EVAP canister purge hose
   - PCV hose
   - Throttle body coolant bypass hose
   - Cable bracket
   - Cruise control cable
   - Accelerator cable
   - IAC valve connector
   - TP sensor connector
   - Air cleaner outlet tube
   - Engine appearance cover
   - Negative battery cable
6. Fill the cooling system.
7. Start the engine and check for leaks.

3.9L Engine

1. Before servicing the vehicle, refer to the precautions in the beginning of this section.
2. Drain the cooling system.
3. Relieve the fuel system pressure.
4. Remove or disconnect the following:
   - Negative battery cable
   - Air cleaner outlet tube
   - Cowl vent screen
   - Cross vehicle support bar
   - Accelerator cable
   - Cruise control cable
   - Intake manifold vacuum hoses
   - Exhaust Gas Recirculation (EGR) valve
   - Camshaft Position (CMP) sensor connector
   - Evaporative Emissions (EVAP) canister purge valve line
   - Fuel pressure sensor connector and vacuum line
   - Fuel line

- Knock Sensor (KS) connector
- Cylinder head temperature sensor connector
- Sensor connector bracket
- Wiring harness and hose bracket
- Left bank fuel injector connectors
- Idle Air Control (IAC) valve connector
- Throttle Position (TP) sensor connector
- Positive Crankcase Ventilation (PCV) tube
- Throttle body coolant hoses
- Right bank fuel injector connectors
- Delta Pressure Feedback Electronic (DPFE) system sensor
- Intake manifold. Loosen the bolts in reverse of the tightening sequence.

To install:

5. Install or connect the following:
   - Intake manifold. Tighten the bolts in sequence to 18 ft. lbs. (25 Nm).
   - DPFE system sensor
   - Right bank fuel injector connectors
   - Throttle body coolant hoses
   - PCV tube
   - TP sensor connector
   - IAC valve connector
   - Left bank fuel injector connectors
   - Wiring harness and hose bracket
   - Sensor connector bracket
   - Cylinder head temperature sensor connector
   - KS sensor connector
   - Fuel line
   - Fuel pressure sensor connector and vacuum line
   - EVAP canister purge valve line
   - CMP sensor connector
   - EGR valve

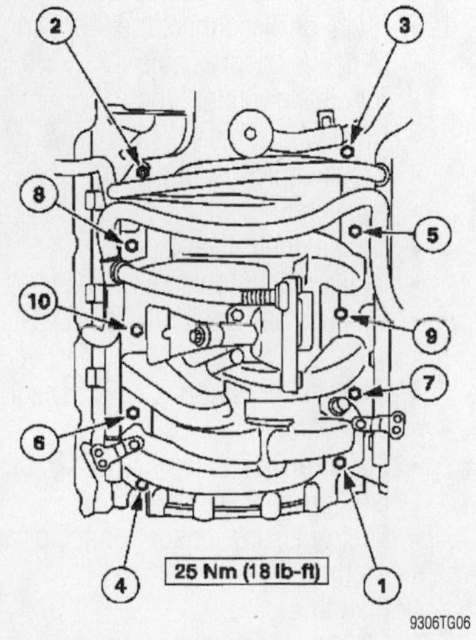

Intake manifold torque sequence—3.9L engine

   - Intake manifold vacuum hoses
   - Cruise control cable
   - Accelerator cable
   - Cross vehicle support bar. Tighten the bolts to 15 ft. lbs. (20 Nm).
   - Cowl vent screen
   - Air cleaner outlet tube
   - Negative battery cable
6. Fill the cooling system.
7. Start the engine and check for leaks.

Exhaust Manifolds

REMOVAL & INSTALLATION

3.0L Engine

1. Before servicing the vehicle, refer to the precautions in the beginning of this section.
2. Remove or disconnect the following:
   - Negative battery cable
   - Heat shields
   - Lower splash shields
   - Exhaust front pipes
   - Secondary air tubes
   - Exhaust Gas Recirculation (EGR) tube
   - Exhaust manifolds

To install:

3. Install the exhaust manifolds. Tighten the fasteners in sequence as follows:
   a. Step 1: 15 ft. lbs. (20 Nm)
   b. Step 2: 15 ft. lbs. (20 Nm)
4. Install or connect the following:
   - Exhaust Gas Recirculation (EGR) tube

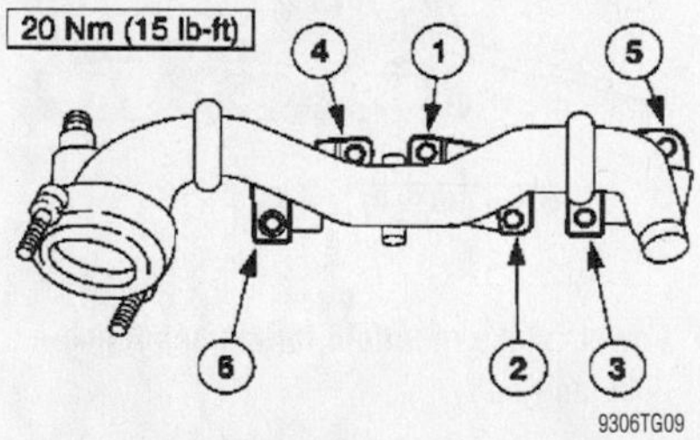

Right exhaust manifold torque sequence—3.0L engine

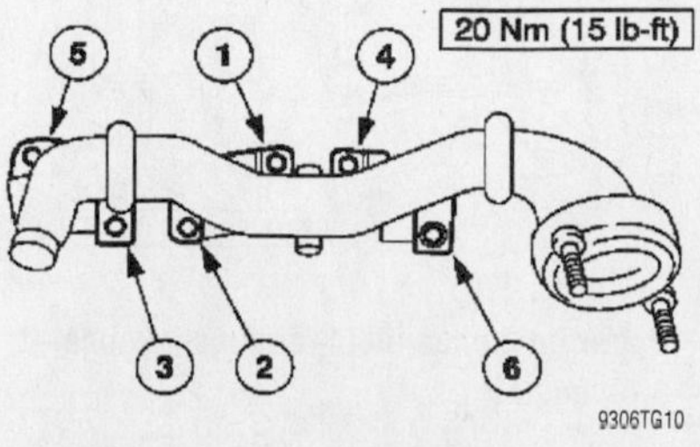

Left exhaust manifold torque sequence—3.0L engine

   - Secondary air tubes
   - Exhaust front pipes
   - Lower splash shields
   - Heat shields
   - Negative battery cable
5. Start the engine and check for leaks.

3.9L Engine

1. Before servicing the vehicle, refer to the precautions in the beginning of this section.
2. Remove or disconnect the following:
   - Negative battery cable
   - Power steering pump reservoir
   - Oil dipstick tube
   - Exhaust front pipes
   - Exhaust Gas Recirculation (EGR) tube
   - Exhaust manifolds

To install:

3. Install or connect the following:
   - Exhaust manifolds. Tighten the bolts to 18 ft. lbs. (25 Nm).
   - EGR tube
   - Exhaust front pipes
   - Oil dipstick tube
   - Power steering pump reservoir
   - Negative battery cable
4. Start the engine and check for leaks.

Camshaft and Valve Lifters

REMOVAL & INSTALLATION

3.0L Engine

1. Before servicing the vehicle, refer to the precautions in the beginning of this section.
2. Remove or disconnect the following:
   - Negative battery cable
   - Valve covers
   - Front cover
   - Timing chains

WARNING

The camshaft journal thrust caps must be removed before loosening the remaining camshaft journal cap bolts to ensure that the camshaft journal thrust caps are not damaged.

➡Keep all valvetrain components in order for assembly.

   - Camshaft journal thrust caps
   - Remaining camshaft journal caps
   - Camshafts
   - Valve tappets and shims

To install:

3. Install or connect the following:

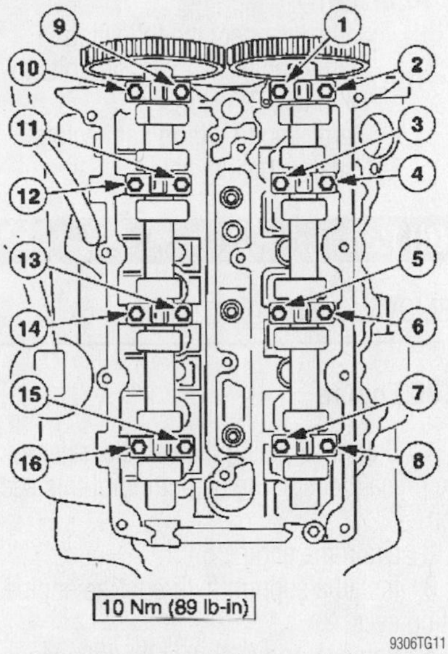

Camshaft journal cap torque sequence—3.0L engine

- Valve tappets and shims in their original locations
- Camshafts
- Camshaft journal caps in their original positions. Install the thrust journal caps last. Tighten the bolts in sequence to 89 inch lbs. (10 Nm).
- Timing chains
- Front cover
- Valve covers
- Negative battery cable

4. Start the engine and check for leaks.

3.9L Engine

1. Before servicing the vehicle, refer to the precautions in the beginning of this section.
2. Remove or disconnect the following:
   - Negative battery cable
   - Valve covers
   - Front cover
   - Timing chains

➡Keep all valvetrain components in order for assembly.

- Camshaft journal bearing caps
- Camshafts
- Valve tappets and shims

To install:

3. Install or connect the following:
   - Valve tappets and shims
   - Camshafts
4. Install the camshaft journal bearing caps in their original positions. Tighten the bolts in sequence as follows:
   a. Step 1: Finger tight

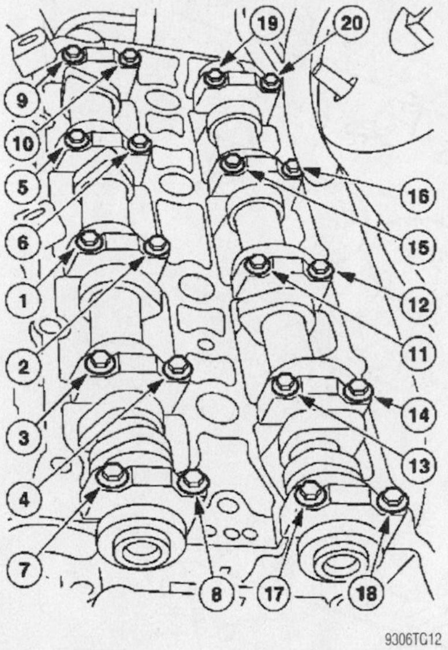

Camshaft journal bearing cap torque sequence—3.9L engine

   b. Step 2: 53 inch lbs. (6 Nm)
   c. Step 3: Plus 90 degrees
5. Install or connect the following:
   - Timing chains
   - Front cover
   - Valve covers
   - Negative battery cable
6. Start the engine and check for leaks.

Valve Lash

ADJUSTMENT

3.0L Engine

1. Before servicing the vehicle, refer to the precautions in the beginning of this section.
2. Remove or disconnect the following:
   - Negative battery cable
   - Engine appearance covers
   - Ignition coils
   - Valve covers
3. Measure the valve clearance while the camshaft lobe is pointed away from the valve shim. Rotate the crankshaft as necessary for each valve to be measured.

➡Keep all valvetrain components in order for assembly.

4. Remove the camshaft thrust cap and the rear camshaft bearing journal cap from the camshaft that requires shim adjustment.
5. Install Service Tool set 303-659 in place of the bearing journal caps, with the taller tool in place of the rear bearing cap.

Camshaft lift tools 303-659—3.0L engine

6. Remove the center camshaft bearing journal caps.
7. Remove the valve shims with compressed air and replace as required to achieve the correct adjustment.
8. Valve clearance should be 0.007–0.009 in. (0.18–0.23mm) for intake valves or 0.013–0.015 in. (0.33–0.38mm) for exhaust valves.
9. Install the center bearing journal caps.
10. Remove the service tools.
11. Install or connect the following:
    - Rear bearing journal cap and the thrust journal cap. Tighten the bearing journal caps in sequence to 89 inch lbs. (10 Nm).
    - Valve covers
    - Ignition coils
    - Engine appearance covers
    - Negative battery cable

3.9L Engine

1. Before servicing the vehicle, refer to the precautions in the beginning of this section.
2. Remove or disconnect the following:
   - Negative battery cable
   - Engine appearance covers
   - Ignition coils

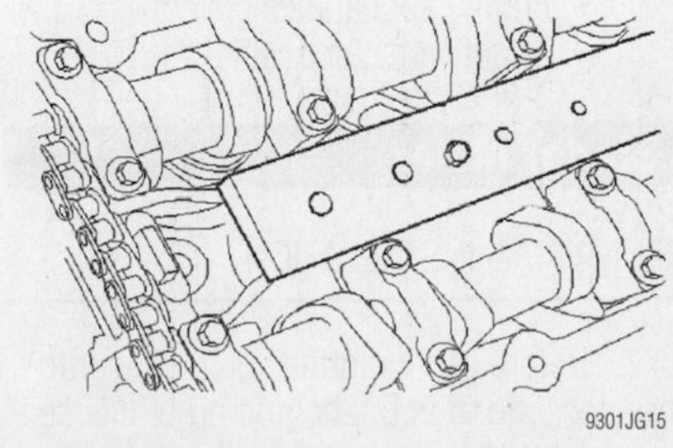

Valve adjustment tool base plate—3.9L engine

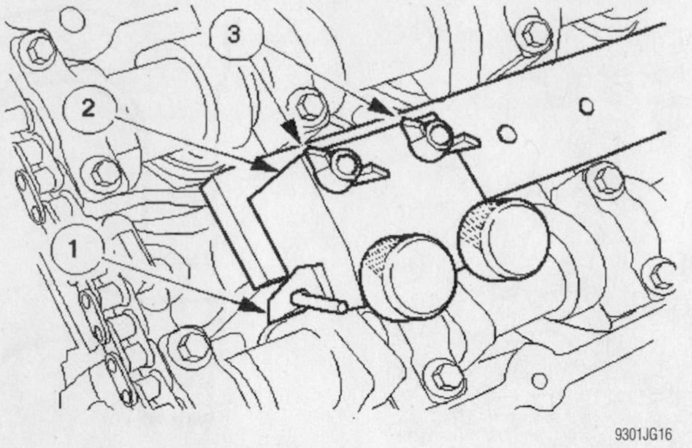

Valve adjustment tool attachment—3.9L engine

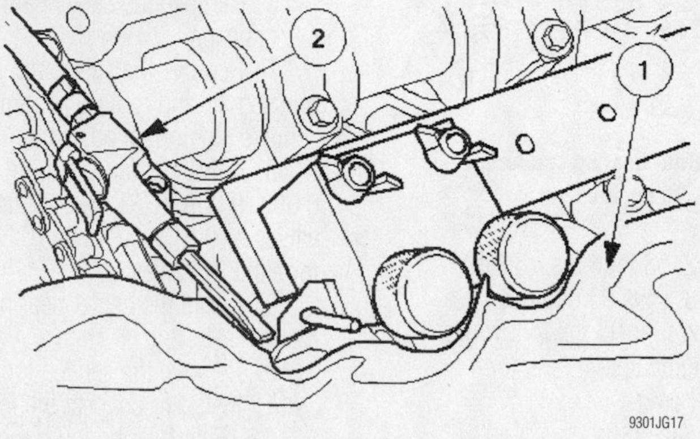

Remove the shims with compressed air—3.9L engine

- Valve covers

3. Measure the valve clearance while the camshaft lobe is pointed away from the valve shim. Rotate the crankshaft as necessary for each valve to be measured.

4. Valve clearance should be 0.007–0.009 in. (0.18–0.23mm) for intake valves or 0.009–0.011 in. (0.23–0.28mm) for exhaust valves.

5. If adjustment is necessary, compress the valves with the special tools and remove the shim with compressed air. Repeat for each valve to be adjusted.

6. Install or connect the following:
   - Valve covers
   - Ignition coils
   - Engine appearance covers
   - Negative battery cable

Starter Motor

REMOVAL & INSTALLATION

1. Before servicing the vehicle, refer to the precautions in the beginning of this section.

2. Remove or disconnect the following:
   - Negative battery cable
   - Starter motor wiring connectors
   - Starter motor

To install:

3. Install or connect the following:
   - Starter motor. Tighten the bolts to 18 ft. lbs. (25 Nm).
   - Starter motor wiring connectors
   - Negative battery cable

Oil Pan

REMOVAL & INSTALLATION

3.0L Engine

1. Before servicing the vehicle, refer to the precautions in the beginning of this section.

2. Drain the engine oil.

3. Install a support fixture to the engine lifting eyes.

4. Remove or disconnect the following:
   - Negative battery cable
   - Air cleaner outlet tube
   - Accessory drive belt
   - Alternator
   - Center and right splash shields
   - A/C compressor
   - Electronic Thermactor Air (ETA) bracket, if equipped
   - Rack and pinion steering gear
   - Lower control arm through bolts

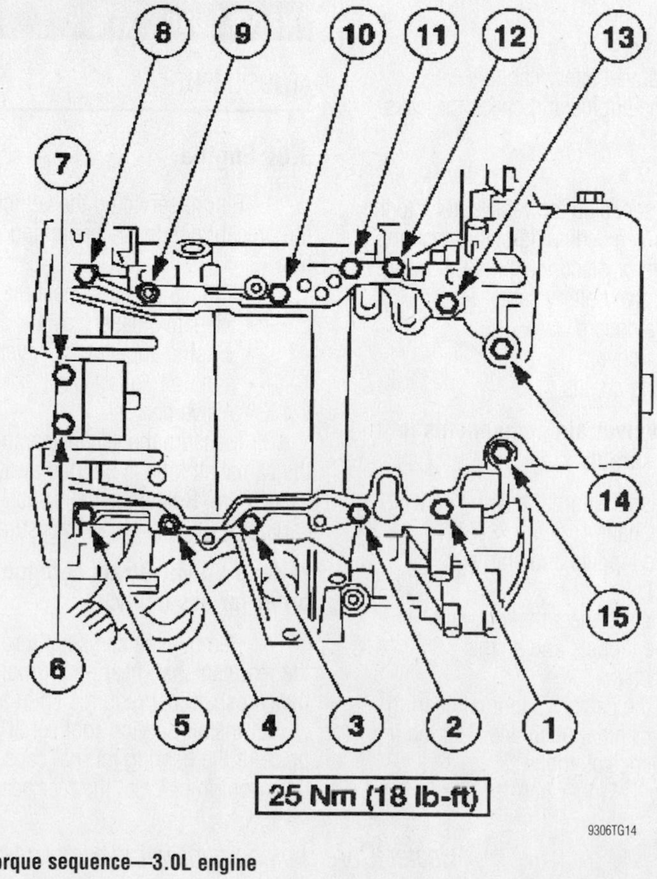

Oil pan torque sequence—3.0L engine

- Transmission cooler line bracket
- Engine mount nuts
- Left and right subframe bolts. Pry the subframe down for clearance.
- Wiring harness bracket
- Oil pan

To install:

5. Install or connect the following:
   - Oil pan. Tighten the pan bolts in sequence to 18 ft. lbs. (25 Nm) and the transmission bolts to 35 ft. lbs. (47 Nm).
   - Wiring harness bracket
   - Left and right subframe bolts. Tighten the bolts to 76 ft. lbs. (103 Nm).
   - Engine mount nuts. Tighten the nuts to 46 ft. lbs. (63 Nm).
   - Transmission cooler line bracket
   - Lower control arm through bolts. Tighten the bolts to 129 ft. lbs. (175 Nm).
   - Rack and pinion steering gear. Tighten the nuts to 76 ft. lbs. (103 Nm).
   - ETA bracket, if equipped
   - A/C compressor
   - Center and right splash shields
   - Alternator
   - Accessory drive belt
   - Air cleaner outlet tube
   - Negative battery cable
6. Fill the crankcase to the correct level.
7. Start the engine and check for leaks.

3.9L Engine

1. Before servicing the vehicle, refer to the precautions in the beginning of this section.
2. Drain the engine oil.
3. Remove the oil pan.

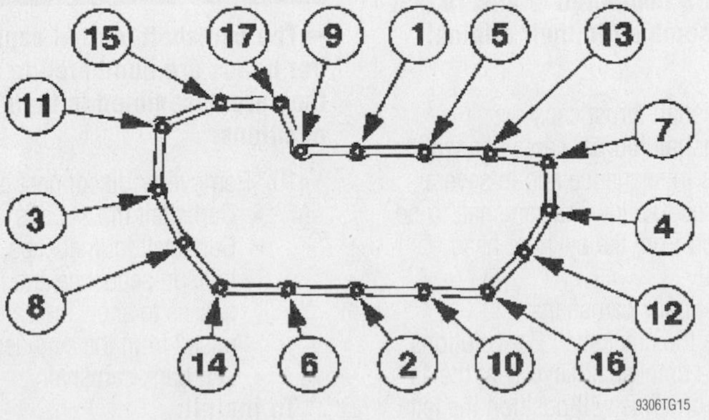

Oil pan torque sequence—3.9L engine

To install:

4. Install the oil pan. Tighten the bolts in sequence as follows:
   a. Step 1: 44 inch lbs. (5 Nm)
   b. Step 2: 108 inch lbs. (12 Nm)
5. Fill the crankcase to the correct level.
6. Start the engine and check for leaks.

Oil Pump

REMOVAL & INSTALLATION

3.0L Engine

1. Before servicing the vehicle, refer to the precautions at the beginning of this section.
2. Remove or disconnect the following:
   - Upper intake manifold
   - Valve covers
   - Accessory drive belt
   - Power steering pump
   - Alternator
   - Water pump
   - A/C compressor and bracket
   - Crankshaft pulley
   - Oil pan
   - Oil pump screen and tube
   - Front cover
   - Timing chains
   - Crankshaft timing gears
   - Oil pump

To install:

3. Install or connect the following:
   - Oil pump. Tighten the bolts to 89 inch lbs. (10 Nm).
   - Crankshaft timing gears
   - Timing chains
   - Front cover
   - Oil pump screen and tube. Tighten the bolts to 89 inch lbs. (10 Nm).
   - Oil pan
   - Crankshaft pulley
   - A/C compressor and bracket
   - Water pump
   - Alternator
   - Power steering pump
   - Accessory drive belt
   - Valve covers
   - Upper intake manifold
4. Fill the crankcase.
5. Start the engine and check for leaks.

3.9L Engine

1. Before servicing the vehicle, refer to the precautions in the beginning of this section.
2. Remove or disconnect the following:
   - Crankshaft pulley
   - Front cover
   - Primary timing chains
   - Oil pump mounting bolts
   - Oil pump

To install:

3. Install or connect the following:
   - New gasket
   - Oil pump. Tighten the bolts to 53 inch lbs. (6 Nm) plus 90 degrees.
   - Primary timing chains
   - Front cover
   - Crankshaft pulley

Rear Main Seal

REMOVAL & INSTALLATION

3.0L Engine

1. Before servicing the vehicle, refer to the precautions at the beginning of this section.
2. Attach an engine support fixture to the engine lifting eyes.
3. Remove or disconnect the following:
   - Negative battery cable
   - Transmission
   - Clutch, if equipped
   - Flywheel
   - Rear crankshaft seal

To install:

4. Install or connect the following:
   - Rear main seal flush with the cylinder block surface
   - Flywheel. Tighten the bolts to 59 ft. lbs. (80 Nm).
   - Clutch, if equipped
   - Transmission
   - Negative battery cable
5. Start the engine and check for leaks.

3.9L Engine

1. Before servicing the vehicle, refer to the precautions at the beginning of this section.

2. Attach an engine support fixture to the engine lifting eyes.
3. Remove or disconnect the following:
   - Negative battery cable
   - Transmission
   - Flywheel
   - Rear crankshaft seal

To install:

4. Install or connect the following:
   - Rear main seal flush with the cylinder block surface
   - Flywheel
5. Tighten the flywheel bolts in a crossing pattern as follows:
   a. Step 1: 11 ft. lbs. (15 Nm)
   b. Step 2: 81 ft. lbs. (110 Nm)
6. Install or connect the following:
   - Transmission
   - Negative battery cable
7. Start the engine and check for leaks.

Timing Chain, Sprockets, Front Cover and Seal

REMOVAL & INSTALLATION

3.0L Engine

1. Before servicing the vehicle, refer to the precautions at the beginning of this section.
2. Remove or disconnect the following:
   - Upper intake manifold
   - Valve covers
   - Accessory drive belt
   - Power steering pump
   - Alternator
   - Water pump
   - A/C compressor and bracket
   - Crankshaft pulley
   - Oil pan
   - Oil pump screen and tube
   - Crankshaft Position (CKP) sensor connector
   - Camshaft Position (CMP) sensor connector
   - Front cover
   - CKP sensor pulse ring
3. Rotate the crankshaft so that the keyway is at the 11 o'clock position to locate the crankshaft at Top Dead Center (TDC) for No. 1 cylinder.
4. Verify that the alignment arrows on the camshafts are aligned. If not, rotate the crankshaft 1 complete revolution and recheck.
5. Rotate the crankshaft so that the keyway is at the 3 o'clock position. This positions the right cylinder head camshafts to the neutral position.
6. Remove or disconnect the following:

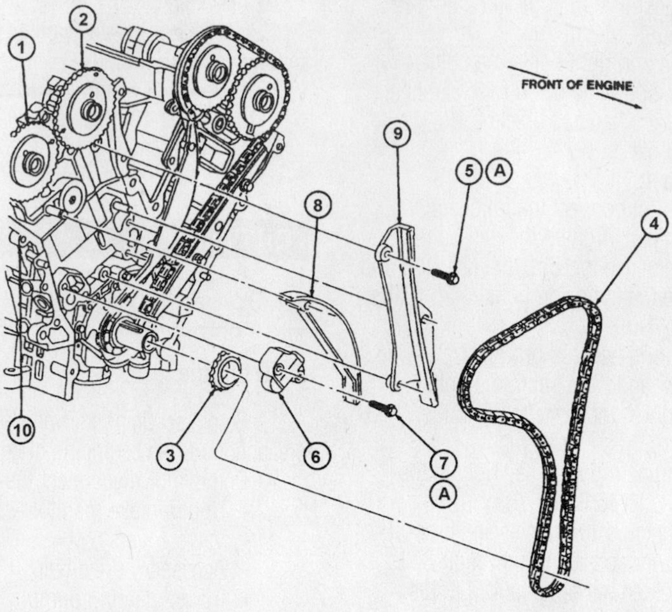

1 RH Exhaust Camshaft
2 RH Intake Camshaft
3 RH Timing Chain Crankshaft Sprocket
4 RH Timing Chain
5 Bolt (2 Req'd)
6 Timing Chain Tensioner
7 Bolt (2 Req'd)
8 Timing Chain Tensioner Arm
9 Timing Chain Guide
10 RH Cylinder Head
A Tighten to 20-30 N·m (15-22 Lb-Ft)

7922KG49

Exploded view of the right cylinder head timing chain and related components—3.0L engine—left side similar

   - Right timing chain tensioner
   - Right timing chain tensioner arm
   - Right timing chain and crankshaft sprocket

WARNING

The camshaft thrust caps must be removed before loosening the remaining camshaft journal cap bolts to ensure that the thrust caps are not damaged.

➡The camshaft journal caps and cylinder heads are numbered to ensure that they are assembled in their original positions.

   - Camshaft thrust caps
   - Camshaft journal caps. Loosen the bolts in sequence and in several passes to allow the camshaft to be raised from the cylinder head evenly.
   - Right bank camshafts
7. Rotate the crankshaft 2 revolutions and locate the crankshaft keyway at the 11 o'clock position. This will position the left cylinder head camshafts to their neutral position.
8. Verify that the alignment arrows on the camshafts are aligned.
9. Remove or disconnect the following:
   - Left cylinder head timing chain tensioner
   - Left timing chain tensioner arm
   - Left timing chain and crankshaft sprocket

WARNING

The camshaft thrust caps must be removed before loosening the remaining camshaft journal cap bolts to ensure that the thrust caps are not damaged.

➡The camshaft journal caps and cylinder heads are numbered to ensure that they are assembled in their original positions.

10. Remove or disconnect the following:
   - Camshaft thrust caps
   - Camshaft journal caps. Loosen the bolts in sequence and in several passes to allow the camshaft to be raised from the cylinder head evenly.
   - Left bank camshafts

To install:

11. Prepare the timing chain tensioners for installation as follows:
   a. Place the left chain tensioner in a vise.

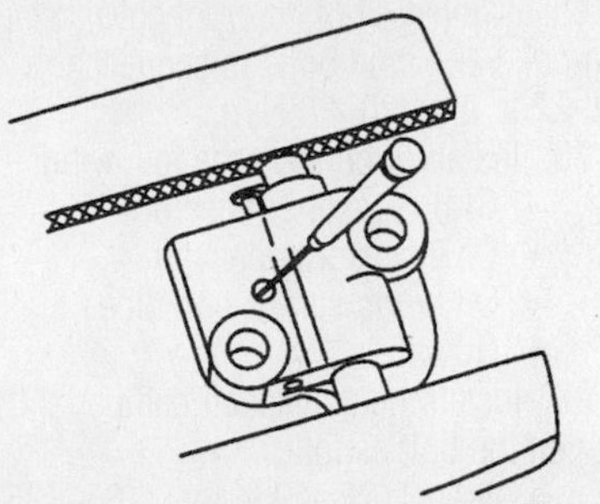
7922JG27

Using a thin prytool, release and hold the timing chain tensioner ratchet/pawl mechanism—3.0L engine

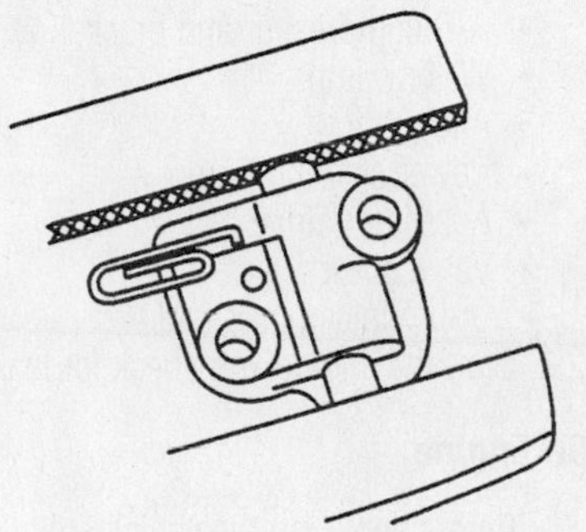
7922JG28

Retain the piston with a 1.5mm wire or paperclip—3.0L engine

b. Using a small prytool, release and hold the timing chain tensioner ratchet/pawl mechanism through the access hole in the timing chain tensioner.
c. Slowly compress the tensioner.
d. Lock the piston with a 1.5mm wire or paperclip.
e. Repeat for the right chain tensioner.

➡Be sure that the crankshaft keyway is still at the 11 o'clock position.

12. Install or connect the following:
   - Left bank camshafts
   - Camshaft journal caps
   - Camshaft thrust caps
   - Left timing chain and crankshaft sprocket. Align the colored links with the index marks on the camshaft and crankshaft sprockets.
   - Left timing chain tensioner arm
   - Left timing chain tensioner. Tighten the retaining bolts to 15–22 ft. lbs. (20–30 Nm).
13. Remove the retaining wire from the left timing chain tensioner.
14. Rotate the crankshaft so that the keyway is in the 3 o'clock position.
15. Install or connect the following:
   - Right bank camshafts
   - Camshaft journal caps
   - Camshaft thrust caps
   - Right timing chain and crankshaft sprocket. Align the colored links with the index marks on the camshaft and crankshaft sprockets.
   - Right timing chain tensioner arm
   - Right timing chain tensioner. Tighten the retaining bolts to 15–22 ft. lbs. (20–30 Nm).
16. Remove the retaining wire from the right timing chain tensioner.
17. Install the CKP sensor pulse ring.
18. Replace the crankshaft seal in the front cover with a new one. Apply clean engine oil to the seal lip.
19. Apply silicone sealer to the 6 critical areas shown in View **A**, to the cylinder block.
20. Place new front cover gaskets onto the dowel pins on the cylinder block and heads.
21. Place the front cover into position.
22. Install the 6 front cover retaining bolts and stud bolts where the silicone sealer was applied.
23. Tighten the bolts and stud bolts until the front cover contacts the cylinder block and heads an, then turn the bolts and stud bolts an additional ¼ turn.
24. Install the remaining front cover retaining bolts and stud bolts.

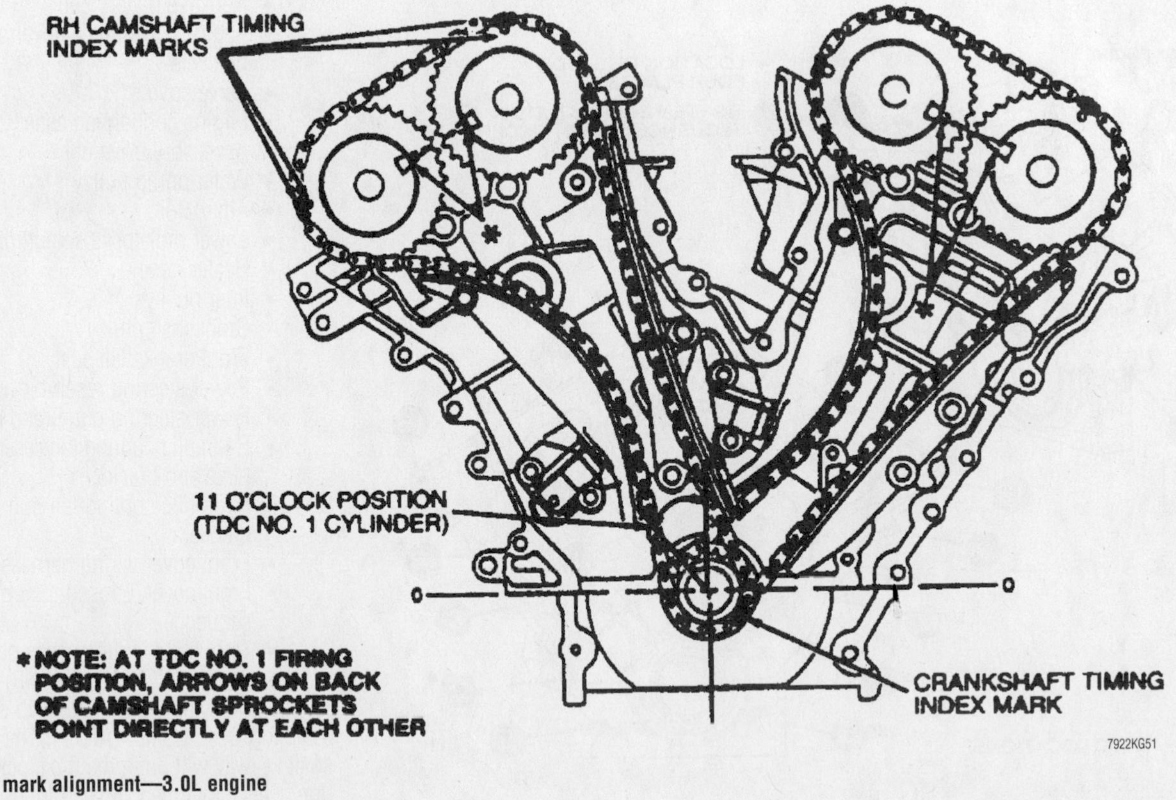

Timing mark alignment—3.0L engine

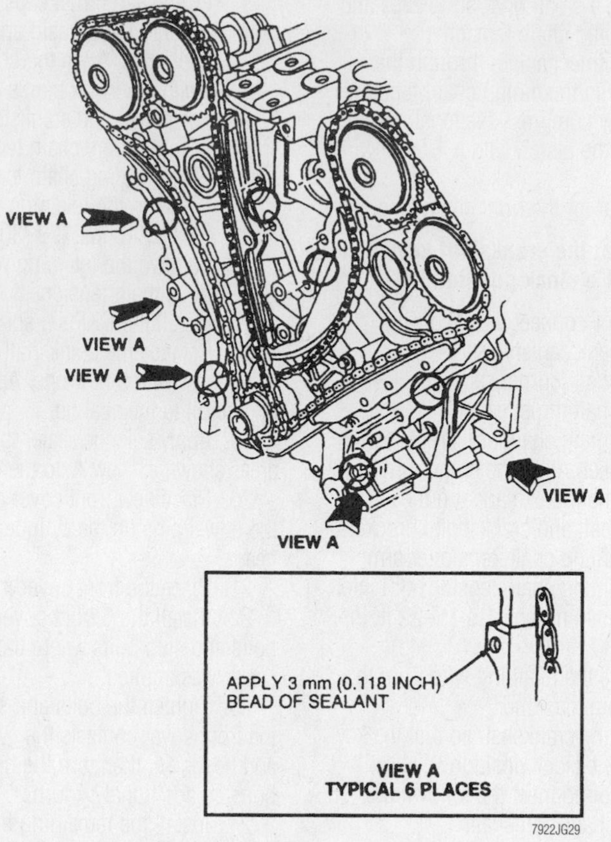

Apply sealant to the places indicated—3.0L engine

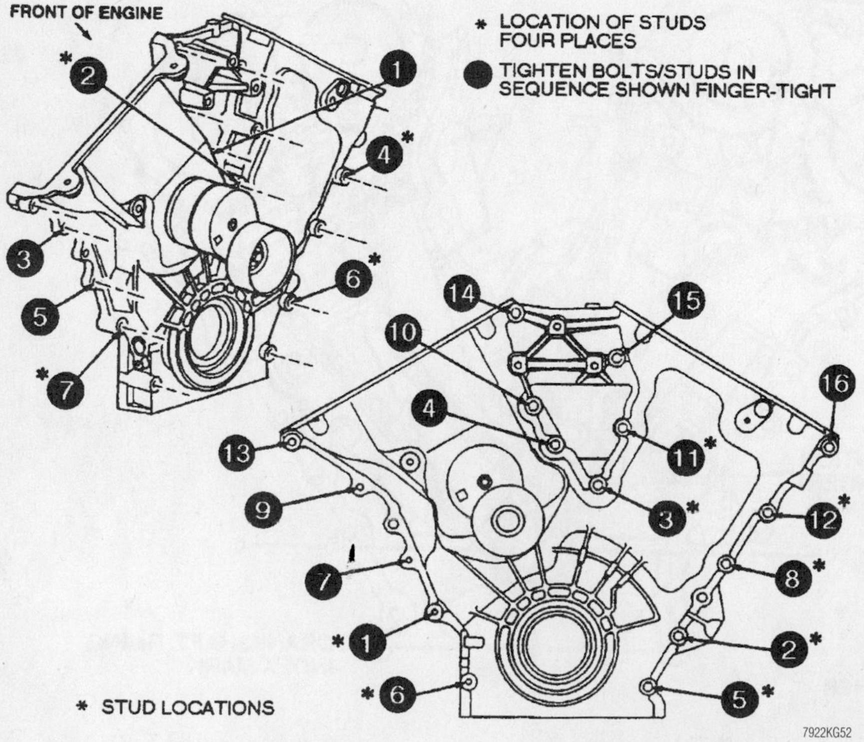

Front cover torque sequence—3.0L engine

25. Tighten all of the front cover retaining bolts and stud bolts in sequence to 15–22 ft. lbs. (20–30 Nm).
26. Install or connect the following:
    - CMP sensor connector
    - CKP sensor connector
    - Oil pump screen and tube
    - Oil pan
27. Install the crankshaft damper and tighten the bolt as follows:
    a. Step 1: 78–99 ft. lbs. (105–135 Nm).
    b. Step 2: Loosen one full turn.
    c. Step 3: 35–39 ft. lbs. (47–53 Nm).
    d. Step 4: Plus an 85–95 degrees.
28. Install or connect the following:
    - A/C compressor and bracket
    - Water pump
    - Alternator
    - Power steering pump
    - Accessory drive belt
    - Valve covers
    - Upper intake manifold
29. Start the engine and check for leaks.

3.9L Engine

1. Before servicing the vehicle, refer to the precautions in the beginning of this section.
2. Drain the cooling system.
3. Remove or disconnect the following:
    - Negative battery cable
    - Engine appearance cover and brackets
    - Valve covers
    - Engine cooling fan assembly
    - Accessory drive belt
    - Water pump pulley
    - Alternator
    - Lower radiator hose and pipe
    - Heater hose
    - Idler pulleys
    - Crankshaft pulley
    - Front crankshaft seal
    - Power steering reservoir hose
    - Power steering pump and bracket
    - Hydraulic cooling fan reservoir hose and bracket
    - Hydraulic cooling fan pump and bracket
    - Front cover wiring harness clips
    - Front cover. Loosen the bolts in sequence.
    - Torque converter access panel
    - Crankshaft Position (CKP) sensor
4. Rotate the crankshaft to 45 degrees After Top Dead Center (ATDC). The crankshaft keyway will be in the 6 o'clock position. Check that the camshaft lobes are facing upwards. If not, rotate the crankshaft 1 full turn.

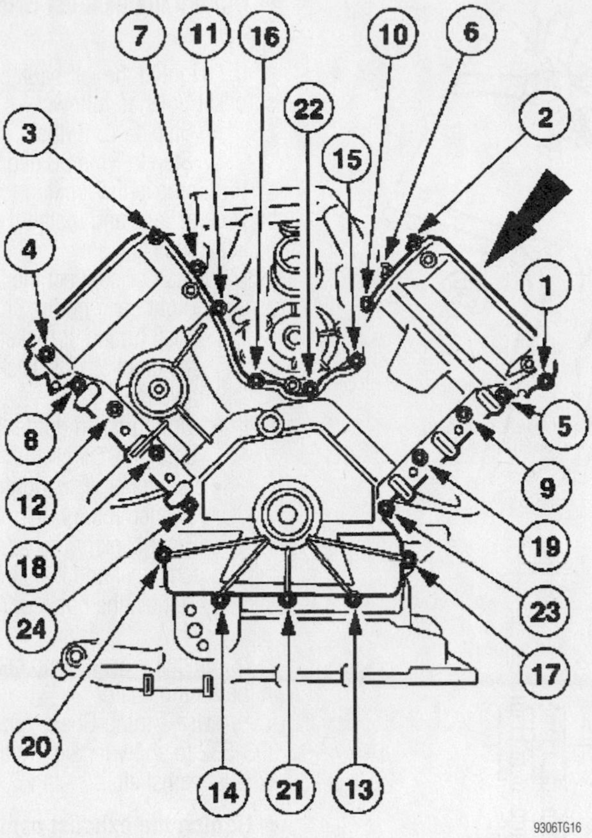

Front cover loosening sequence—3.9L engine

5. Install Crankshaft Holding Tool 303-645 in place of the CKP sensor.
6. Install Camshaft Locking Tool 303-530 to the right bank camshafts.
7. Loosen the exhaust and intake camshaft sprocket bolts and slide the sprockets forward on the bolts.
8. Remove or disconnect the following:
- Right bank primary timing chain tensioner and blanking plate
- Tensioner arm
- Timing chain guide
- Right bank timing chain and crankshaft timing sprocket

9. Remove the locking tool from the right bank camshafts and install it on the left bank camshafts.
10. Remove or disconnect the following:
- Left bank primary timing chain tensioner and blanking plate
- Tensioner arm
- Timing chain guide
- Left bank timing chain and crankshaft timing sprocket
- Intake and exhaust camshaft sprocket bolts
- Intake and exhaust camshaft sprockets and secondary timing chain assemblies
- Secondary timing chain tensioners

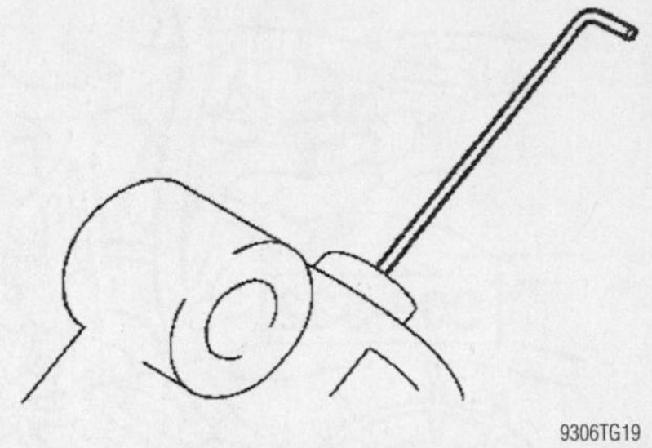

Timing chain tensioner preparation—3.9L engine

To install:

11. Prepare the primary and secondary timing chain tensioners for installation as follows:
    a. Step 1: Use a thin wire to unseat the check valve
    b. Step 2: Compress the tensioner piston fully into the bore by hand
    c. Step 3: Remove the wire
12. Install or connect the following:
- Secondary timing chain tensioners. Tighten the bolts to 97 inch lbs. (11 Nm).
- Intake and exhaust camshaft sprockets and secondary timing chain assemblies
- Left bank timing chain and crankshaft timing sprocket
- Timing chain guide. Tighten the bolts to 97 inch lbs. (11 Nm).
- Tensioner arm. Tighten the bolts to 97 inch lbs. (11 Nm).
- Left bank primary timing chain tensioner and blanking plate. Tighten the bolts to 97 inch lbs. (11 Nm).

13. Install a tie strap to take up the slack in the timing chain.
14. Use Timing Chain Tensioning Tool 303-532 to apply tension to the left bank exhaust camshaft.

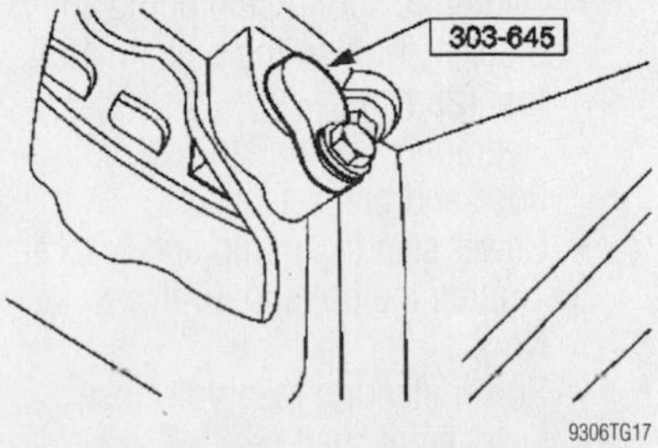

Crankshaft Holding Tool—3.9L engine

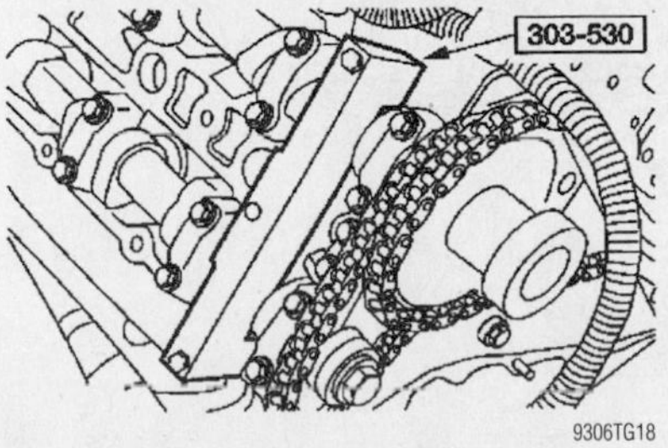

Camshaft Locking Tool—3.9L engine

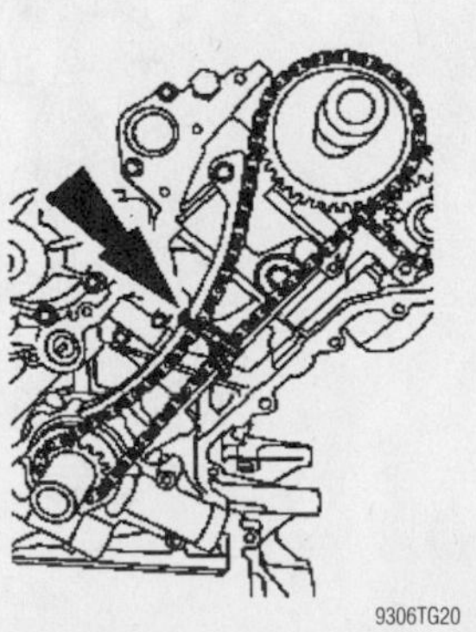

Use a tie strap to remove slack from the primary timing chain—3.9L engine

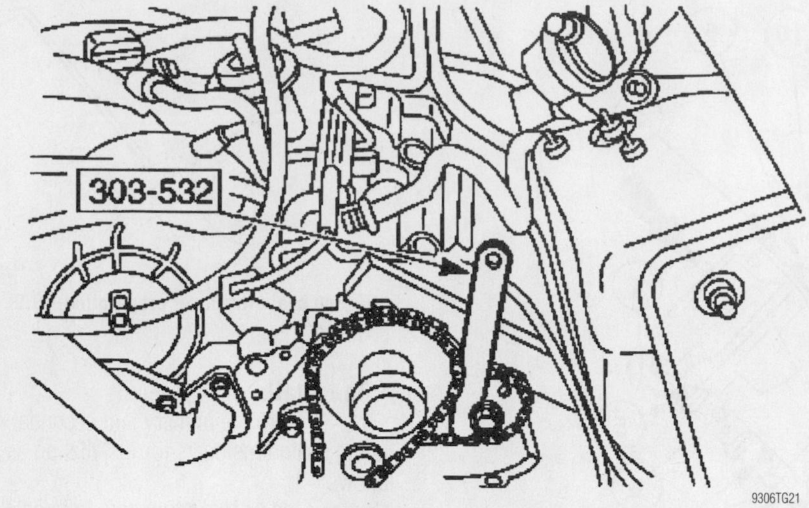

Timing Chain Tensioning Tool—3.9L engine

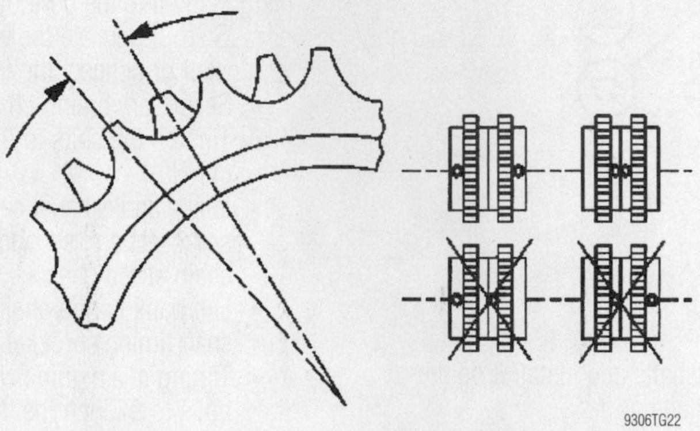

Crankshaft sprocket alignment—3.9L engine

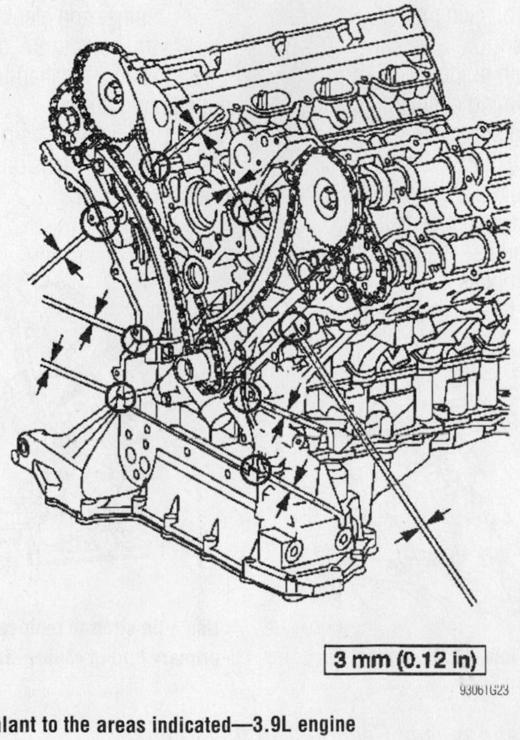

Apply silicone sealant to the areas indicated—3.9L engine

➡Tighten the exhaust camshaft bolt first.

15. Tighten the left bank camshaft sprocket bolts as follows:
    a. Step 1: 15 ft. lbs. (20 Nm)
    b. Step 2: Plus 90 degrees
16. Remove the locking tool from the left bank camshafts and install it on the right bank camshafts.
17. Install or connect the following:
    - Right bank timing chain and crankshaft timing sprocket. Ensure that the crankshaft sprocket is installed as shown.
    - Timing chain guide. Tighten the bolts to 97 inch lbs. (11 Nm).
    - Tensioner arm. Tighten the bolts to 97 inch lbs. (11 Nm).
    - Right bank primary timing chain tensioner and blanking plate. Tighten the bolts to 97 inch lbs. (11 Nm).
18. Install a tie strap to take up the slack in the timing chain.
19. Use Timing Chain Tensioning Tool 303-532 to apply tension to the right bank exhaust camshaft.

➡Tighten the exhaust camshaft bolt first.

20. Tighten the right bank camshaft sprocket bolts as follows:
    a. Step 1: 15 ft. lbs. (20 Nm)
    b. Step 2: Plus 90 degrees
21. Remove the camshaft locking tool.
22. Apply silicone sealant to the areas indicated and install the front cover.
23. Tighten the front cover bolts in sequence as follows:
    a. Step 1: 44 inch lbs. (5 Nm)
    b. Step 2: 89 inch lbs. (10 Nm)
24. Remove the crankshaft locking tool.
25. Install or connect the following:
    - CKP sensor
    - Torque converter access panel
    - Front cover wiring harness clips
    - Hydraulic cooling fan pump and bracket. Tighten the bolts to 18 ft. lbs. (25 Nm).
    - Hydraulic cooling fan reservoir hose and bracket
    - Power steering pump and bracket. Tighten the bolts to 18 ft. lbs. (25 Nm).
    - Power steering reservoir hose
    - Front crankshaft seal
26. Install the crankshaft pulley and tighten the bolt as follows:
    a. Step 1: 59 ft. lbs. (80 Nm)

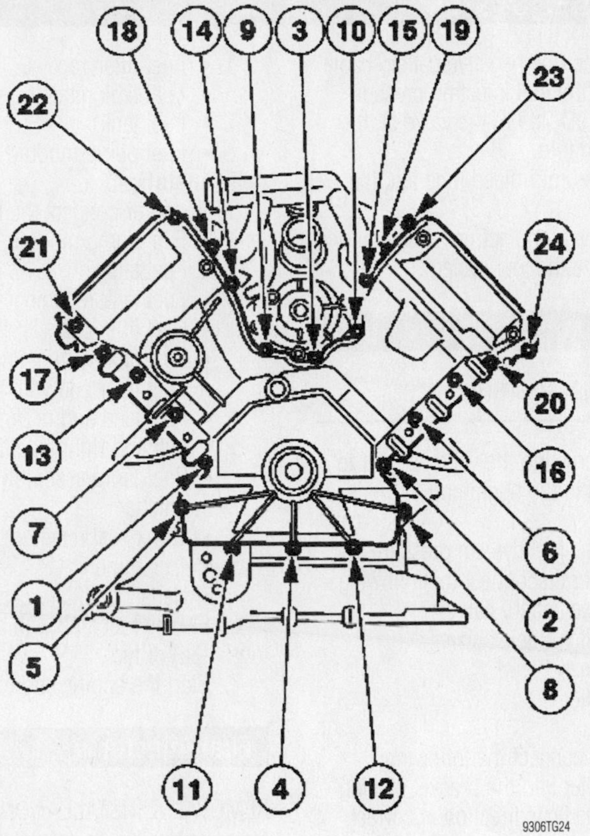

Front cover torque sequence—3.9L engine

b. Step 2: Loosen the bolt 2 complete turns
c. Step 3: 37 ft. lbs. (50 Nm)
d. Step 4: Plus 90 degrees

27. Install or connect the following:
    - Idler pulleys. Tighten the bolts to 18 ft. lbs. (25 Nm).
    - Heater hose
    - Lower radiator hose and pipe
    - Alternator
    - Water pump pulley. Tighten the bolts to 89 inch lbs. (10 Nm) plus 45 degrees.
    - Accessory drive belt
    - Engine cooling fan assembly
    - Valve covers
    - Engine appearance cover and brackets
    - Negative battery cable
28. Fill the cooling system.
29. Start the engine and check for leaks.

Piston and Ring

POSITIONING

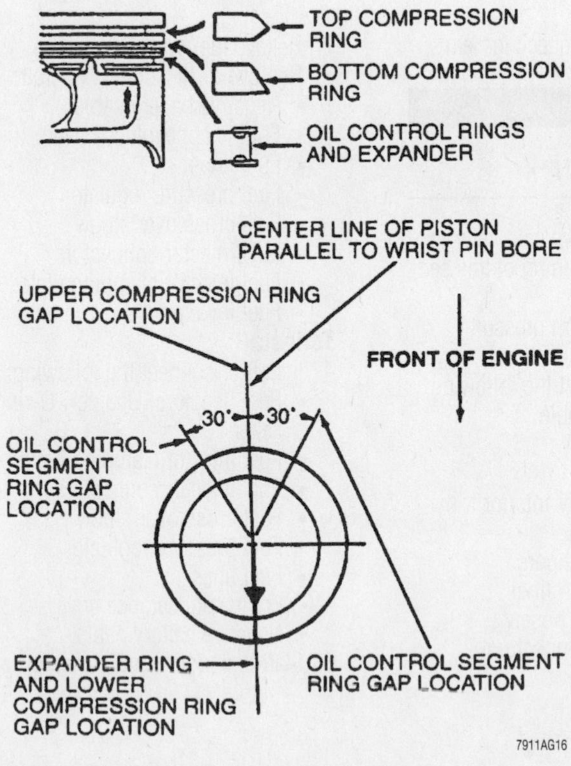

Piston ring end-gap spacing—3.0L (VIN S) engine

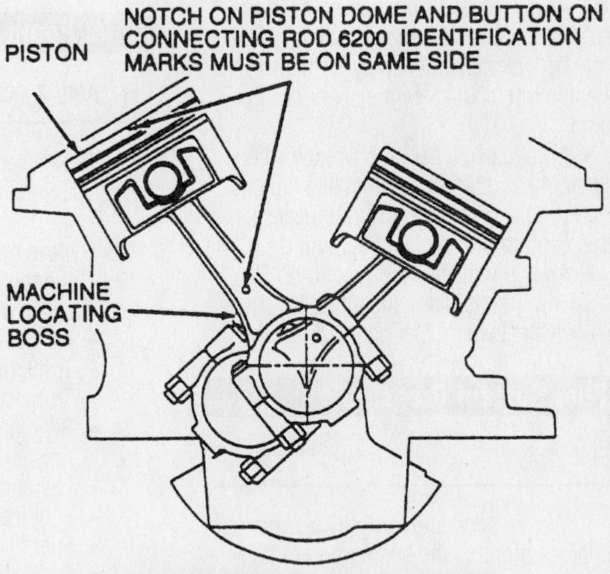

Piston ring end-gap spacing—3.0L (VIN S) engine

FUEL SYSTEM

Fuel System Service Precautions

Safety is the most important factor when performing not only fuel system maintenance but any type of maintenance. Failure to conduct maintenance and repairs in a safe manner may result in serious personal injury or death. Maintenance and testing of the vehicle's fuel system components can be accomplished safely and effectively by adhering to the following rules and guidelines.

- To avoid the possibility of fire and personal injury, always disconnect the negative battery cable unless the repair or test procedure requires that battery voltage be applied.
- Always relieve the fuel system pressure prior to disconnecting any fuel system component (injector, fuel rail, pressure regulator, etc.), fitting or fuel line connection. Exercise extreme caution whenever relieving fuel system pressure, to avoid exposing skin, face and eyes to fuel spray. Please be advised that fuel under pressure may penetrate the skin or any part of the body that it contacts.
- Always place a shop towel or cloth around the fitting or connection prior to loosening to absorb any excess fuel due to spillage. Ensure that all fuel spillage (should it occur) is quickly removed from engine surfaces. Ensure that all fuel soaked cloths or towels are deposited into a suitable waste container.
- Always keep a dry chemical (Class B) fire extinguisher near the work area.
- Do not allow fuel spray or fuel vapors to come into contact with a spark or open flame.
- Always use a back-up wrench when loosening and tightening fuel line connection fittings. This will prevent unnecessary stress and torsion to fuel line piping.
- Always replace worn fuel fitting O-rings with new. Do not substitute fuel hose or equivalent where fuel pipe is installed.

Fuel System Pressure

RELIEVING

1. Before servicing the vehicle, refer to the precautions in the beginning of this section.
2. Disconnect the negative battery cable.
3. Connect the fuel injection pressure test equipment JD 209 to the valve on the fuel supply manifold.
4. Insert the drain/bleed tube into the fuel container.
5. Follow the manufacturer's instructions and depressurize the fuel system.

Fuel Filter

REMOVAL & INSTALLATION

1. Before servicing the vehicle, refer to the precautions in the beginning of this section.
2. Relieve the fuel system pressure.
3. Remove or disconnect the following:
   - Negative battery cable
   - Fuel filter bracket cover
   - Fuel lines
   - Fuel filter

To install:

4. Install or connect the following:
   - Fuel filter into the bracket making sure the flow direction is correct. Tighten the clamp to 15–25 inch lbs. (2–3 Nm).
   - Fuel lines. Tighten the fittings to 22 ft. lbs. (30 Nm).
5. Start the engine and check for leaks.

Fuel Pump

REMOVAL & INSTALLATION

1. Before servicing the vehicle, refer to the precautions in the beginning of this section.
2. Relieve the fuel system pressure.
3. Drain the fuel tank.
4. Remove or disconnect the following:
   - Negative battery cable
   - Trunk liner
   - Trunk seal retainer
   - Rear lamp assembly interior trim finisher
   - Left and right side liners
   - Fuel feed and return lines
   - Fuel filler and vent hoses
   - Fuel tank wiring connectors
   - Fuel filler cap
   - Fuel tank retaining straps
   - Fuel tank
   - Fuel pump module

To install:

5. Install or connect the following:
   - Fuel pump module
   - Fuel tank
   - Fuel tank retaining straps
   - Fuel filler cap
   - Fuel tank wiring connectors
   - Fuel filler and vent hoses
   - Fuel feed and return lines
   - Left and right side liners
   - Rear lamp assembly interior trim finisher
   - Trunk seal retainer
   - Trunk liner
   - Negative battery cable
6. Fill the fuel tank with at least 10 gallons (38L) of fuel.
7. Start the engine and check for leaks.

Fuel Injector

REMOVAL & INSTALLATION

1. Before servicing the vehicle, refer to the precautions in the beginning of this section.
2. Relieve fuel system pressure.
3. Remove or disconnect the following:
   - Negative battery cable
   - Engine appearance covers
   - Fuel lines
   - Fuel pressure regulator
   - Fuel cross over elbow
   - Fuel injector connectors
   - Fuel injector clamping plates
   - Fuel injectors

To install:

4. Install or connect the following:
   - Fuel injectors. Use new O-ring seals.
   - Fuel injector clamping plates
   - Fuel injector connectors
   - Fuel cross over elbow
   - Fuel pressure regulator
   - Fuel lines
   - Engine appearance covers
   - Negative battery cable
5. Start the engine and check for leaks.

DRIVE TRAIN

Transmission Assembly

REMOVAL & INSTALLATION

Automatic

1. Before servicing the vehicle, refer to the precautions in the beginning of this section.
2. Install a support fixture to the engine lifting eyes.
3. Drain the transmission fluid.
4. Remove or disconnect the following:
   - Negative battery cable
   - Engine appearance covers
   - Mass Air Flow (MAF) meter
   - Air intake assembly
   - Coolant recovery tank
   - Exhaust front pipes
   - Driveshaft
   - Shift selector cable
   - Transmission electrical connectors
   - Transmission oil cooler lines
   - Torque converter
   - Transmission mount and bracket
   - Transmission flange bolts
   - Transmission

To install:

5. Install or connect the following:
   - Transmission to the engine. Tighten the flange bolts to 32–42 ft. lbs. (43–57 Nm).
   - Transmission mount and bracket. Tighten the mount bolts to 22–30 ft. lbs. (30–40 Nm) and the bracket bolts to 16–21 ft. lbs. (22–28 Nm).
   - Torque converter. Tighten the bolts to 32–42 ft. lbs. (43–57 Nm).
   - Transmission oil cooler lines
   - Transmission electrical connectors
   - Shift selector cable
   - Driveshaft. Tighten the bolts to 55–65 ft. lbs. (75–88 Nm).
   - Exhaust front pipes
   - Coolant recovery tank
   - Air intake assembly
   - MAF meter
   - Engine appearance covers
   - Negative battery cable
6. Fill the transmission to the correct level with the proper fluid. Do not over-fill.
7. Start the engine and check for leaks.

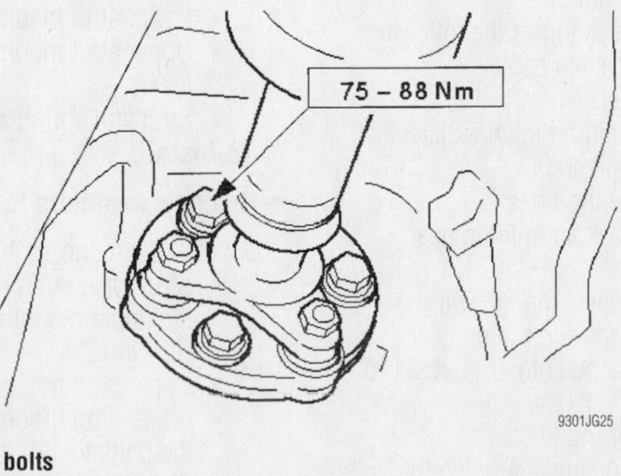

Driveshaft flange bolts

Manual

1. Before servicing the vehicle, refer to the precautions in the beginning of this section.
2. Remove or disconnect the following:
   - Negative battery cable
   - Shift linkage
   - Exhaust center section

➡The driveshaft flange may have additional nuts added for balance. These nuts must be replaced in the same position to maintain the driveshaft balance.

   - Driveshaft. Matchmark the flange and note the location of the fasteners for assembly.
   - Heated Oxygen (HO_2S) sensor connectors and harness
   - Vehicle Speed (VSS) sensor connector
   - Reverse light switch connector and harness
   - Transmission mount and cross-member. Support the transmission.
   - Clutch slave cylinder fluid line
   - Starter motor
   - Transmission flange bolts
   - Transmission

To install:

3. Install or connect the following:
   - Transmission. Tighten the flange bolts to 35 ft. lbs. (47 Nm).
   - Starter motor
   - Clutch slave cylinder fluid line
   - Transmission mount and cross-member. Tighten the bolts to 41 ft. lbs. (55 Nm) and the nut to 30 ft. lbs. (40 Nm).
   - Reverse light switch connector and harness
   - VSS sensor connector
   - HO_2S sensor connectors and harness
   - Driveshaft. Align the matchmarks and tighten the retainers to 63 ft. lbs. (85 Nm). Tighten any balance nuts to 18 ft. lbs. (24 Nm).
   - Exhaust center section
   - Shift linkage
   - Negative battery cable

Clutch

ADJUSTMENTS

Because the clutch system is hydraulic, the clutch pedal free-play is self-adjusting and requires no additional maintenance.

REMOVAL & INSTALLATION

1. Before servicing the vehicle, refer to the precautions in the beginning of this section.
2. Remove or disconnect the following:
   - Transaxle
   - Clutch pressure plate
   - Clutch disk

To install:

3. Install or connect the following:
   - Clutch disk and pressure plate. Tighten the pressure plate bolts evenly in several passes to 17 ft. lbs. (23 Nm).
   - Transmission
4. Bleed the hydraulic clutch system, if required.
5. Check the clutch system for proper operation.

Hydraulic Clutch System

BLEEDING

1. Before servicing the vehicle, refer to the precautions in the beginning of this section.

2. Remove the inspection cover.
3. Connect a hose to the bleeder valve fitting on the clutch slave cylinder. Submerge the other end of the hose into a container of clean brake fluid.
4. Open the bleeder valve and have an assistant depress the clutch pedal.
5. Close the bleeder before releasing the clutch pedal.
6. Repeat the procedure until no more air bubbles are seen.
7. Install the inspection cover to the bell housing.
8. Top off the clutch master cylinder fluid reservoir and install the diaphragm and cap.

Halfshaft

REMOVAL & INSTALLATION

1. Before servicing the vehicle, refer to the precautions in the beginning of this section.
2. Remove or disconnect the following:
   - Negative battery cable
   - Rear wheel
   - Brake caliper
   - Wheel speed sensor
   - Hub carrier pivot bolt
   - Hub retaining nut
3. Press the stub shaft out of the hub and pry the inner joint out of the differential.

To install:

4. Install or connect the following:
   - Halfshaft inner joint to the differential
   - Halfshaft in the wheel hub by applying Loctite® 270 thread locking compound to the splines
   - Hub carrier pivot bolt by aligning the bolt head matchmarks. Tighten it to 66–81 ft. lbs. (90–110 Nm).
   - Hub retaining nut. Tighten the new nut to 221 ft. lbs. (300 Nm).
   - Wheel speed sensor
   - Brake caliper
   - Rear wheel
   - Negative battery cable
5. Check the wheel alignment and adjust as necessary.

➡**If the hub is removed for any reason, a new bearing assembly must be installed. Never attempt to re-use a bearing.**

CV-Joint

OVERHAUL

The CV-joints are serviced with the axle halfshaft as an assembly.

STEERING AND SUSPENSION

Air Bag

⁂ CAUTION

Some vehicles are equipped with an air bag system. The system must be disarmed before performing service on, or around, system components, the steering column, instrument panel components, wiring and sensors. Failure to follow the safety precautions and the disarming procedure could result in accidental air bag deployment, possible injury and unnecessary system repairs.

PRECAUTIONS

Several precautions must be observed when handling the inflator module to avoid accidental deployment and possible personal injury.

- Never carry the inflator module by the wires or connector on the underside of the module.
- When carrying a live inflator module, hold securely with both hands, and ensure that the bag and trim cover are pointed away.
- Place the inflator module on a bench or other surface with the bag and trim cover facing up.
- With the inflator module on the bench, never place anything on or close to the module that may be thrown in the event of an accidental deployment.

DISARMING

Proper SRS disarming can be obtained by disconnecting and isolating the negative battery cable. Allow the air bag system capacitor at least 2 minutes to discharge before removing any air bag system components.

Power Rack and Pinion Steering Gear

REMOVAL & INSTALLATION

1. Before servicing the vehicle, refer to the precautions in the beginning of this section.
2. Lock the steering wheel in the straight-ahead position.
3. Remove or disconnect the following:
   - Negative battery cable
   - Front wheels
   - Steering column intermediate shaft
   - Outer tie rod ends
   - Power steering lines
   - Steering rack and pinion gear

To install:

4. Install or connect the following:
   - Steering rack and pinion gear. Tighten the bolts to 76 ft. lbs. (103 Nm).
   - Power steering lines
   - Outer tie rod ends. Tighten the nuts to 52–63 ft. lbs. (71–85 Nm).
   - Steering column intermediate shaft
   - Front wheels
   - Negative battery cable
5. Fill the power steering fluid reservoir.
6. Start the engine and check for leaks.
7. Check the wheel alignment and adjust, as necessary.

Strut

REMOVAL & INSTALLATION

Front

1. Before servicing the vehicle, refer to the precautions in the beginning of this section.
2. Remove or disconnect the following:
   - Front wheel
   - Stabilizer bar link
   - Lower strut mounting bolt
   - Upper strut mount cover and fasteners
   - Strut and spring assembly

To install:

➡**Use new fasteners for assembly.**

3. Install or connect the following:
   - Strut and spring assembly. Tighten the upper mount nuts to 21 ft. lbs. (28 Nm).
   - Upper strut mount cover
   - Lower strut mounting bolt. Tighten the bolts to 129 ft. lbs. (175 Nm).
   - Stabilizer bar link. Tighten the nut to 41 ft. lbs. (55 Nm).
   - Front wheel

Rear

1. Before servicing the vehicle, refer to the precautions in the beginning of this section.
2. Remove or disconnect the following:
   - Trunk trim covers
   - Upper strut mount nuts
   - Lower strut mount bolt
   - Strut and spring assembly

To install:

➡Use new fasteners for assembly.

3. Install or connect the following:
   - Strut and spring assembly. Tighten the lower bolt to 98 ft. lbs. (133 Nm) and the upper nuts to 21 ft. lbs. (28 Nm).
   - Trunk trim covers

Coil Spring

REMOVAL & INSTALLATION

1. Before servicing the vehicle, refer to the precautions in the beginning of this section.
2. Remove the strut assembly from the vehicle.
3. Compress the coil spring and remove the piston rod nut.
4. Remove or disconnect the following:
   - Upper strut mount
   - Spring upper seat
   - Coil spring

To install:

5. Install or connect the following:
   - Coil spring
   - Spring upper seat
   - Upper strut mount. Tighten the piston rod nut to 37 ft. lbs. (50 Nm).
6. Remove the spring compressor and install the strut assembly to the vehicle.

Upper Ball Joint

REMOVAL & INSTALLATION

The upper ball joint is serviced with the upper control arm as an assembly.

Lower Ball Joint

REMOVAL & INSTALLATION

The lower ball joint is serviced with the lower control arm as an assembly.

Upper Control Arm

REMOVAL & INSTALLATION

1. Before servicing the vehicle, refer to the precautions in the beginning of this section.
2. Remove or disconnect the following:
   - Front wheel
   - Strut and spring assembly
   - Upper ball joint and tapered washer
   - Inner control arm fasteners
   - Upper control arm

To install:

➡Use new fasteners for assembly.

3. Install or connect the following:
   - Upper control arm. Tighten the inner fasteners to 35 ft. lbs. (48 Nm).
   - Upper ball joint and tapered washer. Tighten the nut to 66 ft. lbs. (90 Nm).
   - Strut and spring assembly
   - Front wheel

CONTROL ARM BUSHING REPLACEMENT

The control arm bushings are serviced with the control arm as an assembly.

Lower Control Arm

REMOVAL & INSTALLATION

1. Before servicing the vehicle, refer to the precautions in the beginning of this section.
2. Remove or disconnect the following:
   - Front wheel
   - Splash shield
   - Stabilizer bar link
   - Lower strut mounting bolt
   - Lower ball joint
   - Rack and pinion steering gear
   - Inner control arm mounting bolts
   - Lower control arm

To install:

➡Use new fasteners for assembly.

3. Install or connect the following:
   - Lower control arm. Tighten the inner mounting bolts to 129 ft. lbs. (175 Nm).
   - Rack and pinion steering gear
   - Lower ball joint. Tighten the nut to 111 ft. lbs. (150 Nm).
   - Lower strut mounting bolt. Tighten the bolt to 129 ft. lbs. (175 Nm).
   - Stabilizer bar link. Tighten the nut to 41 ft. lbs. (55 Nm).
   - Splash shield
   - Front wheel
4. Check the wheel alignment and adjust as necessary.

CONTROL ARM BUSHING REPLACEMENT

The control arm bushings are serviced with the control arm as an assembly.

Wheel Bearings

ADJUSTMENT

The wheel bearings are not adjustable.

REMOVAL & REPLACEMENT

Front

1. Before servicing the vehicle, refer to the precautions in the beginning of this section.
2. Remove or disconnect the following:
   - Front wheel
   - Brake caliper and rotor
   - Wheel speed sensor connector
   - Hub and bearing assembly

➡The hub and bearing assembly is not pressed into the knuckle. Do not use a slide hammer or press to remove the hub and bearing assembly. Damage to the hub and bearing assembly may result.

To install:

➡Do not remove the wheel speed sensor from the hub and bearing assembly unless it is being replaced. If installing a new hub and bearing assembly, a new wheel speed sensor must be installed.

➡Use new fasteners for assembly.

3. Install or connect the following:
   - Hub and bearing assembly. Tighten the bolts to 66 ft. lbs. (90 Nm).
   - Wheel speed sensor connector
   - Brake caliper and rotor
   - Front wheel

Rear

1. Before servicing the vehicle, refer to the precautions in the beginning of this section.

2. Remove or disconnect the following:
   - Rear wheel
   - Wheel speed sensor
   - Brake caliper and rotor
   - Hub, bearing and knuckle assembly
   - Disc brake dust shield
3. Press the hub from the knuckle and bearing assembly.
4. Remove the snapring and press the bearing assembly out of the knuckle.

To install:

5. Press the bearing assembly into the knuckle.
6. Install the snapring and press the hub into the knuckle and bearing assembly.
7. Install or connect the following:
   - Disc brake dust shield. Use aluminum rivets.
   - Hub, bearing and knuckle assembly
   - Brake caliper and rotor
   - Wheel speed sensor
   - Rear wheel
8. Check the wheel alignment and adjust as necessary.

FORD MOTOR CO.

1998–01
Mustang

17

PRECAUTIONS

Before servicing any vehicle, please be sure to read all of the following precautions, which deal with personal safety, prevention of component damage, and important points to take into consideration when servicing a motor vehicle:

- Never open, service or drain the radiator or cooling system when the engine is hot; serious burns can occur from the steam and hot coolant.
- Observe all applicable safety precautions when working around fuel. Whenever servicing the fuel system, always work in a well-ventilated area. Do not allow fuel spray or vapors to come in contact with a spark, open flame, or excessive heat (a hot drop light, for example). Keep a dry chemical fire extinguisher near the work area. Always keep fuel in a container specifically designed for fuel storage; also, always properly seal fuel containers to avoid the possibility of fire or explosion. Refer to the additional fuel system precautions later in this section.
- Fuel injection systems often remain pressurized, even after the engine has been turned **OFF**. The fuel system pressure must be relieved before disconnecting any fuel lines. Failure to do so may result in fire and/or personal injury.
- Brake fluid often contains polyglycol ethers and polyglycols. Avoid contact with the eyes and wash your hands thoroughly after handling brake fluid. If you do get brake fluid in your eyes, flush your eyes with clean, running water for 15 minutes. If eye irritation persists, or if you have taken brake fluid internally, IMMEDIATELY seek medical assistance.
- The EPA warns that prolonged contact with used engine oil may cause a number of skin disorders, including cancer. You should make every effort to minimize your exposure to used engine oil. Protective gloves should be worn when changing oil. Wash your hands and any other exposed skin areas as soon as possible after exposure to used engine oil. Soap and water, or waterless hand cleaner should be used.
- All new vehicles are now equipped with an air bag system, often referred to as a Supplemental Restraint System (SRS) or Supplemental Inflatable Restraint (SIR) system. The system must be disabled before performing service on or around system components, steering column, instrument panel components, wiring and sensors. Failure to follow safety and disabling procedures could result in accidental air bag deployment, possible personal injury and unnecessary system repairs.
- Always wear safety goggles when working with, or around, the air bag system. When carrying a non-deployed air bag, be sure the bag and trim cover are pointed away from your body. When placing a non-deployed air bag on a work surface, always face the bag and trim cover upward, away from the surface. This will reduce the motion of the module if it is accidentally deployed. Refer to the additional air bag system precautions later in this section.
- Clean, high quality brake fluid from a sealed container is essential to the safe and proper operation of the brake system. You should always buy the correct type of brake fluid for your vehicle. If the brake fluid becomes contaminated, completely flush the system with new fluid. Never reuse any brake fluid. Any brake fluid that is removed from the system should be discarded. Also, do not allow any brake fluid to come in contact with a painted surface; it will damage the paint.
- Never operate the engine without the proper amount and type of engine oil; doing so WILL result in severe engine damage.
- Timing belt maintenance is extremely important. Many models utilize an interference-type, non-freewheeling engine. If the timing belt breaks, the valves in the cylinder head may strike the pistons, causing potentially serious (also time-consuming and expensive) engine damage. Refer to the maintenance interval charts in the front of this manual for the recommended replacement interval for the timing belt, and to the timing belt section for belt replacement and inspection.
- Disconnecting the negative battery cable on some vehicles may interfere with the functions of the on-board computer system(s) and may require the computer to undergo a relearning process once the negative battery cable is reconnected.
- When servicing drum brakes, only disassemble and assemble one side at a time, leaving the remaining side intact for reference.
- Only an MVAC-trained, EPA-certified automotive technician should service the air conditioning system or its components.

ENGINE REPAIR

➡Disconnecting the negative battery cable on some vehicles may interfere with the functions of the on board computer system. The computer may undergo a relearning process once the negative battery cable is reconnected.

Distributor

The 3.8L and 4.6L engines use Distributorless Ignition Systems (DIS).

Alternator

REMOVAL

3.8L Engine

1. Before servicing the vehicle, refer to the precautions in the beginning of this section.
2. Remove or disconnect the following:
   - Negative battery cable
   - Accessory drive belt
   - Alternator electrical connectors
   - Alternator bolts
   - Alternator

4.6L Engines

1. Before servicing the vehicle, refer to the precautions in the beginning of this section.
2. Remove or disconnect the following:
   - Negative battery cable
   - Accessory drive belt
   - Upper alternator bracket
   - Alternator electrical connectors
   - Alternator bolts
   - Alternator

INSTALLATION

3.8L Engine

Install or connect the following:

- Alternator. Tighten the lower bolt to 35 ft. lbs. (47 Nm) and the upper bolt to 18 ft. lbs. (25 Nm).
- Alternator electrical connectors
- Accessory drive belt
- Negative battery cable

4.6L Engines

Install or connect the following:

- Alternator. Tighten the mounting bolts to 18 ft. lbs. (25 Nm).
- Alternator electrical connectors
- Upper alternator bracket. Tighten the bolts to 89 inch lbs. (10 Nm).

- Accessory drive belt
- Negative battery cable

Ignition Timing

ADJUSTMENT

The ignition timing is controlled by the Powertrain Control Module (PCM). No adjustment is necessary.

Engine Assembly

REMOVAL & INSTALLATION

3.8L Engine

1. Before servicing the vehicle, refer to the precautions in the beginning of this section.
2. Discharge the air conditioning system.
3. Drain the engine cooling system.
4. Drain the engine oil and remove the oil filter.
5. Remove or disconnect the following:
   - Negative battery cable
   - Hood
   - Cooling fan
   - Radiator and hoses
   - Heater hoses
   - Intake Air Temperature (IAT) sensor connector
   - Air intake duct
   - Accessory drive belt
   - Alternator wiring connectors
   - Coolant recovery tank
   - Power steering pump and bracket
   - Power steering pressure switch connector
   - A/C compressor and bracket
   - Cruise control cable
   - Accelerator cable and bracket
   - Fuel supply line
   - Fuel return line, if equipped
   - Fuel charging wiring harness connectors
   - Main vacuum source hose
   - Evaporative emissions hose
   - Heated Oxygen (HO_2S) sensor connectors
   - Dual converter Y-pipe
   - Starter
   - Starter wiring harness retainers
   - Transmission flange bolts
   - Torque converter, automatic transmission models only
   - Transmission oil cooler lines, automatic transmission models only
   - Left and right engine support insulators
6. Attach an engine hoist and remove the engine from the vehicle.

To install:

7. Lower the engine into the vehicle.
8. Install or connect the following:
   - Left and right engine support insulators. Tighten the through bolts to 35–50 ft. lbs. (47–68 Nm).
   - Transmission flange bolts. Tighten the bolts to 25–33 ft. lbs. (34–46 Nm).
   - Torque converter, automatic transmission models only. Tighten the nuts to 20–33 ft. lbs. (27–46 Nm).
   - Transmission oil cooler lines, automatic transmission models only
   - Starter wiring harness retainers
   - Starter
   - Dual converter Y-pipe
   - HO_2S sensor connectors
   - Evaporative emissions hose
   - Main vacuum source hose
   - Fuel charging wiring harness connectors
   - Fuel supply line
   - Fuel return line, if equipped
   - Accelerator cable and bracket
   - Cruise control cable
   - A/C compressor and bracket
   - Power steering pressure switch connector
   - Power steering pump and bracket
   - Coolant recovery tank
   - Alternator wiring connectors
   - Accessory drive belt
   - Air intake duct
   - IAT sensor connector
   - Heater hoses
   - Radiator and hoses
   - Cooling fan
   - Hood
   - Oil and oil filter
   - Negative battery cable
9. Fill the cooling system.
10. Recharge the air conditioning.
11. Start the engine and check for leaks.

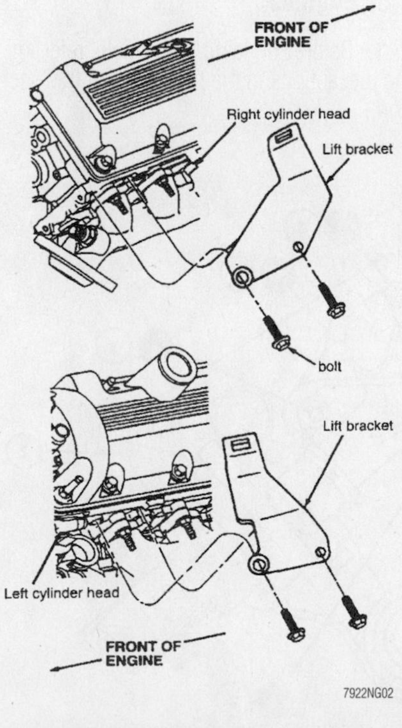

Install the engine lifting brackets on both sides of the engine—4.6L SOHC engine shown, other engines similar

4.6L Engines

1. Before servicing the vehicle, refer to the precautions in the beginning of this section.
2. Discharge the air conditioning system.
3. Drain the engine cooling system.
4. Drain the engine oil and remove the oil filter.
5. Remove the transmission. Refer to the transmission procedure in this section.
6. Remove or disconnect the following:
   - Negative battery cable
   - Hood
   - Engine compartment brace
   - Cooling fan
   - Radiator hoses
   - Heater hoses
   - Intake Air Temperature (IAT) sensor connector
   - Air intake duct
   - Accessory drive belt
   - Alternator wiring connectors
   - Coolant recovery tank
   - A/C compressor suction and discharge lines
   - Cruise control cable
   - Accelerator cable
   - Fuel supply line
   - Fuel return line, if equipped
   - Fuel charging wiring harness connectors
   - Main vacuum source hose

- Evaporative emissions hose
- Canister purge valve hose
- Secondary air injection switching valve, on DOHC engines
- Power steering reservoir
- Power distribution box
- Heated Oxygen (HO_2S) sensor connectors
- Dual converter H-pipe
- Power steering fluid cooler hoses
- Power steering pressure and return hoses
- Starter
- Left and right engine support insulators

7. Attach an engine hoist and remove the engine from the vehicle.

To install:

8. Lower the engine into the vehicle.
9. Install or connect the following:
   - Left and right engine support insulators. Tighten the through bolts to 15–22 ft. lbs. (20–30 Nm).
   - Starter
   - Power steering pressure and return hoses
   - Power steering fluid cooler hoses
   - Dual converter H-pipe
   - HO_2S sensor connectors
   - Power distribution box
   - Power steering reservoir
   - Secondary air injection switching valve, on DOHC engines
   - Canister purge valve hose
   - Evaporative emissions hose
   - Main vacuum source hose
   - Fuel charging wiring harness connectors
   - Fuel supply line
   - Fuel return line, if equipped
   - Accelerator cable
   - Cruise control cable
   - A/C compressor suction and discharge lines
   - Coolant recovery tank
   - Alternator wiring connectors
   - Accessory drive belt
   - Air intake duct
   - IAT sensor connector
   - Heater hoses
   - Radiator hoses
   - Cooling fan
   - Engine compartment brace
   - Hood
   - Transmission
   - Oil and oil filter
   - Negative battery cable
10. Fill the cooling system.
11. Recharge the air conditioning.
12. Start the engine and check for leaks.

Water Pump

REMOVAL & INSTALLATION

3.8L Engine

1. Before servicing the vehicle, refer to the precautions in the beginning of this section.
2. Drain the engine cooling system.
3. Remove or disconnect the following:
   - Electric cooling fan
   - Accessory drive belt
   - Water pump pulley
   - Ignition coil and bracket

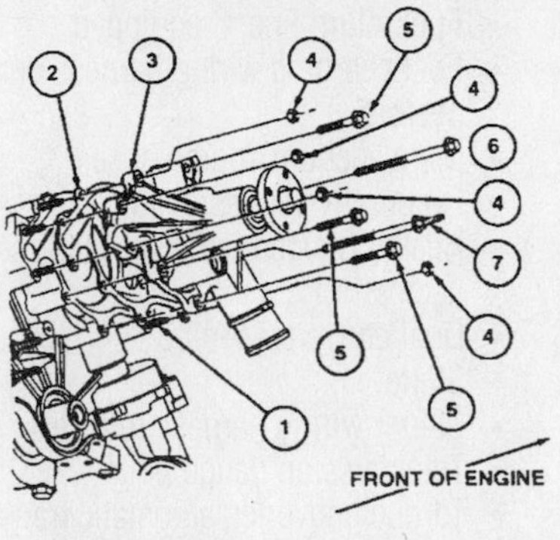

1. Mounting stud
2. Water pump housing gasket
3. Water pump
4. Mounting nuts
5. Short mounting bolts
6. Long mounting bolt
7. Mounting stud bolt

7922NG03

Exploded view of the water pump mounting on the 3.8L engine—during removal, note the original positions of the different length bolts for reassembly

- Power steering pump pulley
- Power steering pump brace
- Heater water outlet tube
- Lower radiator hose
- Water pump

To install:

➡The threads of the No. 1 water pump retaining bolt must be coated with a Teflon® sealant prior to installation.

4. Install or connect the following:
   - Water pump. Use a new gasket and tighten the bolts and studs to 15–22 ft. lbs. (20–30 Nm). Tighten the nuts to 71–106 inch lbs. (8–12 Nm).
   - Lower radiator hose
   - Heater water outlet tube. Use a new O-ring seal and tighten the bolts to 71–106 inch lbs. (8–12 Nm).
   - Power steering pump brace
   - Power steering pump pulley
   - Ignition coil and bracket
   - Water pump pulley. Tighten the bolts to 15–21 ft. lbs. (21–29 Nm).
   - Accessory drive belt
   - Electric cooling fan
5. Fill the cooling system.
6. Start the engine and check for coolant leaks.

4.6L Engines

1. Before servicing the vehicle, refer to the precautions in the beginning of this section.

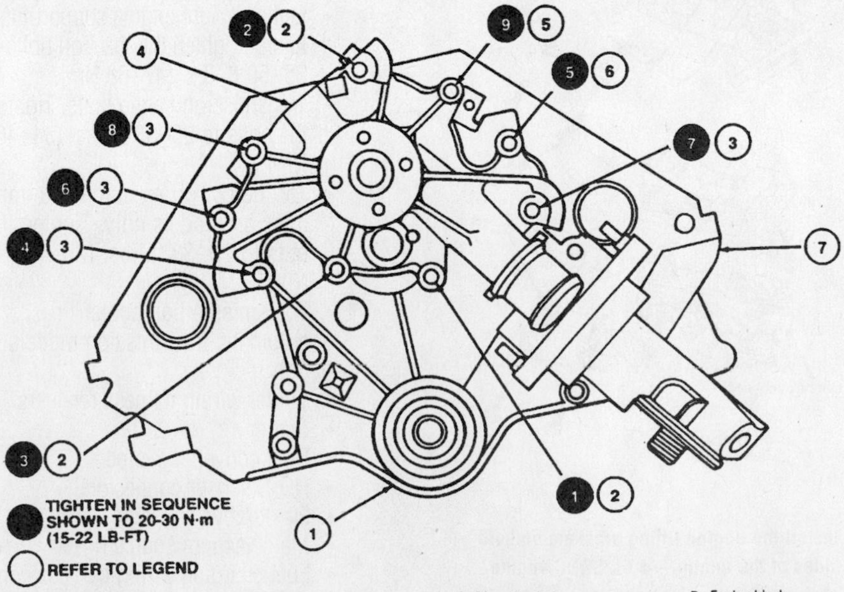

1. Engine front cover
2. Mounting bolts
3. Mounting nuts
4. Water pump
5. Mounting bolt
6. Mounting stud bolt
7. Engine block

7922NG04

Water pump torque sequence—3.8L engine

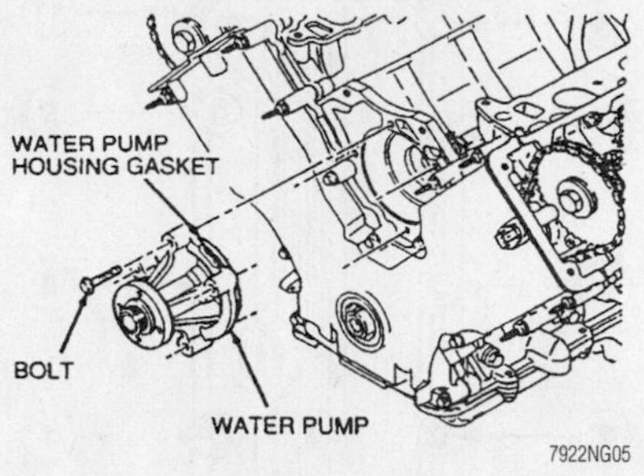

Exploded view of the water pump mounting—4.6L engines

2. Drain the engine cooling system.
3. Remove or disconnect the following:
   - Electric cooling fan
   - Accessory drive bolt
   - Water pump pulley
   - Water pump

To install:

4. Install or connect the following:
   - Water pump. Use a new O-ring seal and tighten the bolts in a crossing pattern to 15–22 ft. lbs. (20–30 Nm).
   - Water pump pulley. Tighten the bolts to 15–21 ft. lbs. (21–29 Nm).
   - Accessory drive belt
   - Electric cooling fan
5. Fill the cooling system.
6. Start the engine and check for coolant leaks.

Cylinder Head

REMOVAL & INSTALLATION

3.8L Engine

1. Before servicing the vehicle, refer to the precautions in the beginning of this section.
2. Drain the cooling system.
3. Remove or disconnect the following:
   - Negative battery cable
   - Intake Air Temperature (IAT) sensor connector
   - Air intake tube
   - Accessory drive belt
   - Alternator and bracket
   - Power steering pump and bracket
   - A/C compressor and bracket
   - Positive Crankcase Ventilation (PCV) valve
   - Upper intake manifold
   - Spark plug wires
   - Valve covers
   - Fuel supply line
   - Fuel return line, if equipped
   - Fuel injector wiring connectors
   - Fuel injection supply manifold
   - Lower intake manifold
   - Exhaust manifolds

➡Keep rocker arms and pushrods in order for installation.

   - Rocker arms
   - Pushrods
   - Cylinder heads

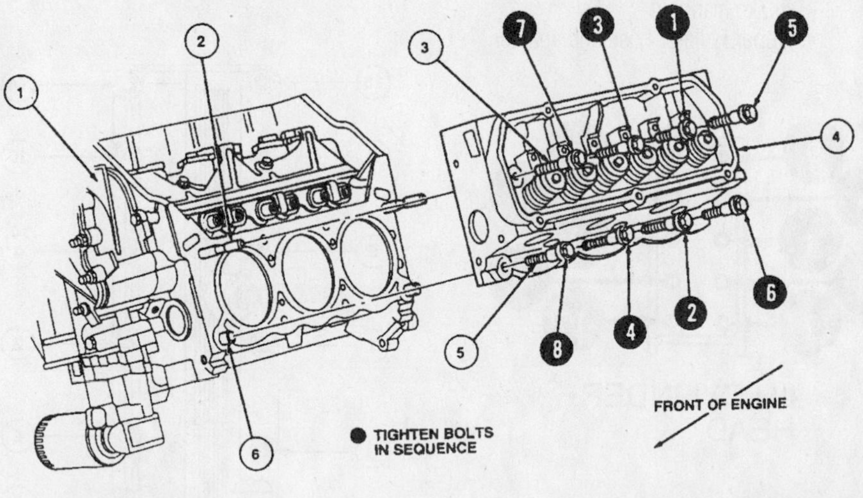

1. Engine block
2. Locating pin
3. Long cylinder head mounting bolts
4. Cylinder head
5. Short cylinder head mounting bolts
6. Alignment dowel

7922NG07

Cylinder head torque sequence—3.8L engine

To install:

➡The cylinder head bolts are a torque-to-yield design and cannot be reused.

➡Left and right cylinder head gaskets are not interchangeable.

4. Install the new cylinder heads. Lubricate the new cylinder head bolt threads with clean oil and tighten them in sequence as follows:
   a. Step 1: 15 ft. lbs. (20 Nm).
   b. Step 2: 29 ft. lbs. (40 Nm).
   c. Step 3: 36 ft. lbs. (50 Nm).

➡Loosen and tighten each bolt individually for the remainder of the procedure. Do not loosen or tighten all the bolts at the same time.

   d. Step 4: Loosen each bolt, one at a time, 2–3 turns. Tighten the long cylinder head bolts to 29–37 ft. lbs. (40–50 Nm) plus 180 degrees. Tighten the short cylinder head bolts to 15–22 ft. lbs. (20 –30 Nm) plus 180 degrees.
   e. Step 5: Repeat step 4 for the next bolt in the tightening sequence.
5. Install or connect the following:
   - Pushrods and rocker arms in their original locations
   - Exhaust manifolds
   - Lower intake manifold
   - Fuel injection supply manifold
   - Fuel injector wiring connectors
   - Fuel supply line
   - Fuel return line, if equipped
   - Valve covers
   - Spark plug wires
   - Upper intake manifold
   - PCV valve
   - A/C compressor and bracket
   - Power steering pump and bracket
   - Alternator and bracket
   - Accessory drive belt
   - Air intake tube
   - IAT sensor connector
   - Negative battery cable
6. Fill the cooling system.
7. Start the engine and check for leaks.

4.6L SOHC Engine

1. Before servicing the vehicle, refer to the precautions in the beginning of this section.
2. Drain the cooling system.
3. Remove or disconnect the following:
   - Negative battery cable
   - Electric cooling fan
   - Fuel supply line
   - Fuel return line, if equipped

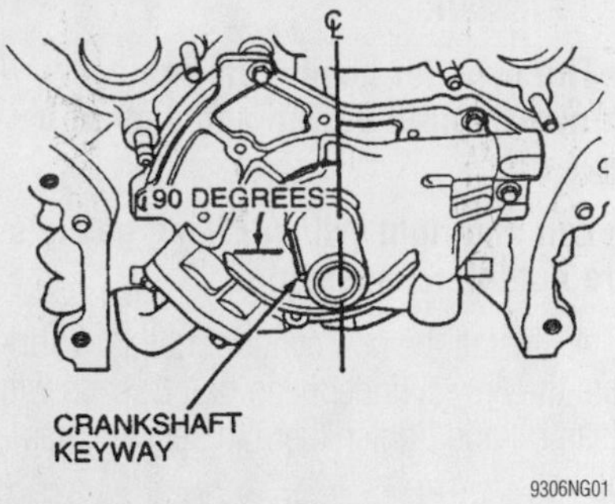

Place the crankshaft keyway at 9 o'clock for cylinder head installation—4.6L SOHC engine

- Air intake tube
- Windshield wiper governor
- Accessory drive belt
- Spark plug wires, if equipped
- Ignition coils
- Camshaft Position (CMP) sensor connector
- Power steering reservoir
- Alternator and bracket
- Water pump pulley
- Mass Air Flow (MAF) sensor connector
- Fuel injector wiring connectors
- Crankshaft Position (CKP) sensor connector
- A/C compressor clutch connector
- Cruise control cable
- Accelerator cable and bracket
- Canister purge valve connector
- Power steering pump
- Oil pan
- Crankshaft pulley
- Oil pressure sensor connector
- Exhaust Gas Recirculation (EGR) tube
- Heated Oxygen (HO_2S) sensor connectors
- Dual converter H-pipe
- Starter motor wiring harness retainer
- Valve covers
- Throttle body vacuum adapter
- Heater hoses
- Upper radiator hose and adapter
- Intake manifold
- Front cover
- Timing chains
- Heater water tube
- Cylinder heads

To install:

➡The cylinder head bolts are a torque-to-yield design and cannot be reused.

4. Rotate the crankshaft so that the keyway is at the 9 o'clock position.
5. Rotate each camshaft to a stable position where the valves do not extend below the cylinder head face.
6. Install the new cylinder heads. Lubricate the new cylinder head bolt threads with clean oil and tighten them in sequence as follows:
   a. Step 1: 28–31 ft. lbs. (37–43 Nm).
   b. Step 2: Plus 85–95 degrees.
   c. Step 3: Loosen all bolts one full turn.
   d. Step 4: 28–31 ft. lbs. (37–43 Nm).
   e. Step 5: Plus 85–95 degrees.
   f. Step 6: Plus 85–95 degrees.
7. Install or connect the following:
   - Heater water tube
   - Timing chains
   - Front cover
   - Intake manifold
   - Upper radiator hose and adapter

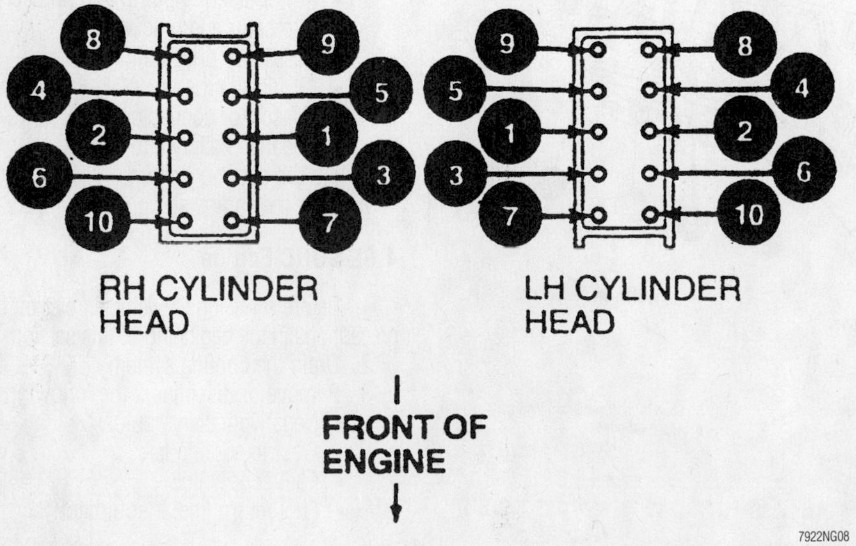

Cylinder head torque sequence—4.6L SOHC engine

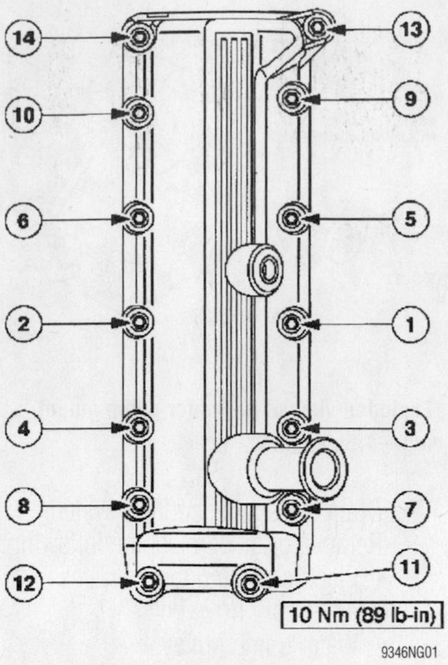

Right valve cover torque sequence—4.6L SOHC Engine

- Heater hoses
- Throttle body vacuum adapter
- Valve covers
- Starter motor wiring harness retainer
- Dual converter H-pipe
- HO_2S sensor connectors
- EGR tube
- Oil pressure sensor connector
- Crankshaft pulley
- Oil pan
- Power steering pump
- Canister purge valve connector
- Accelerator cable and bracket

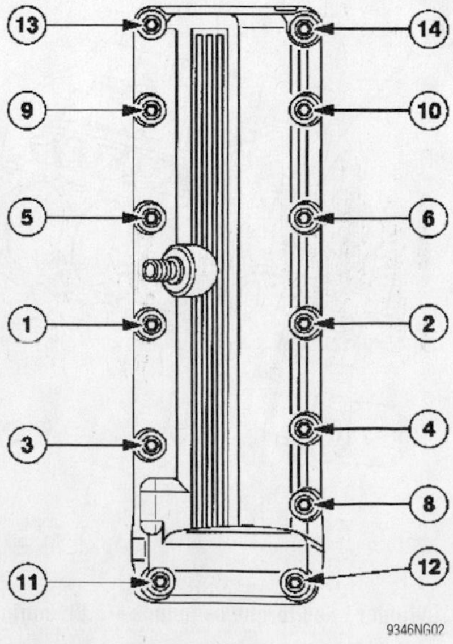

Left valve cover torque sequence—4.6 SOHC Engine

- Cruise control cable
- A/C compressor clutch connector
- CKP sensor connector
- Fuel injector wiring connectors
- MAF sensor connector
- Water pump pulley
- Alternator and bracket
- Power steering reservoir
- CMP sensor connector
- Ignition coils
- Spark plug wires, if equipped
- Accessory drive belt
- Windshield wiper governor
- Air intake tube
- Fuel supply line
- Fuel return line, if equipped
- Electric cooling fan
- Negative battery cable

8. Fill the cooling system.
9. Start the engine and check for leaks.

4.6L DOHC Engine

1. Before servicing the vehicle, refer to the precautions in the beginning of this section.
2. Remove the engine from the vehicle and mount it on a suitable workstand.
3. Remove or disconnect the following:
   - Accessory drive belt and tensioner
   - Idler pulley
   - Engine Coolant Temperature (ECT) sensor connector
   - Water bypass tube
   - Alternator
   - Water pump
   - Power steering pump
   - Crankshaft pulley
   - Camshaft Position (CMP) sensor
   - Spark plug wires, if equipped
   - Ignition coils
   - Fuel pressure sensor connector and vacuum line
   - Valve covers
   - Rocker arms
   - Exhaust Gas Recirculation (EGR) vacuum regulator valve
   - EGR tube
   - Injector wiring connectors
   - Intake manifold
   - Left and right exhaust manifolds
   - Front cover
   - Crankshaft Position (CKP) sensor pulse wheel
   - Timing chains
   - Cylinder heads

To install:

The cylinder head bolts are a torque-to-yield design and cannot be reused.

4. Use new gaskets and install the cylinder heads.
5. For 1998 engines, tighten the new cylinder head bolts in sequence as follows:
   a. Step 1: 27–32 ft. lbs. (37–43 Nm).
   b. Step 2: Tighten the bolts 85–95 degrees.
   c. Step 3: Tighten the bolts 85–95 degrees.
6. For 1999–01 engines, tighten the cylinder head bolts in sequence as follows:
   a. Step 1: 28–31 ft. lbs. (37–43 Nm).
   b. Step 2: Tighten the bolts 85–95 degrees.
   c. Step 3: Loosen all bolts 1 full turn.
   d. Step 4: Tighten all bolts to 28–31 ft. lbs. (37–43 Nm).
   e. Step 5: Tighten the bolts 85–95 degrees.
   f. Step 6: Tighten the bolts 85–95 degrees.
7. Install or connect the following:
   - Timing chains
   - CKP sensor pulse wheel
   - Front cover
   - Left and right exhaust manifolds
   - Intake manifold
   - Injector wiring connectors

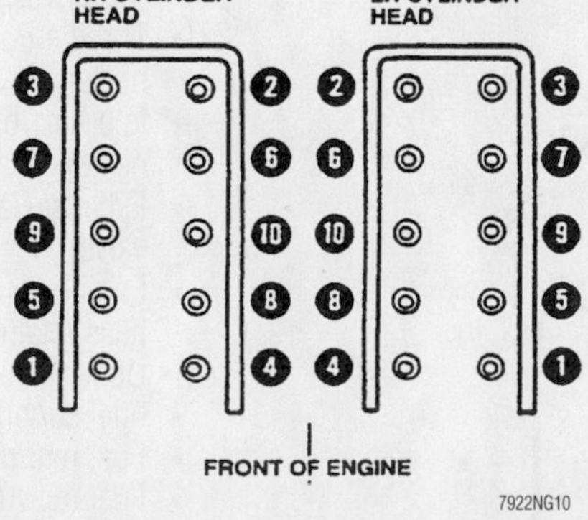

Cylinder head loosening sequence—4.6L DOHC engine

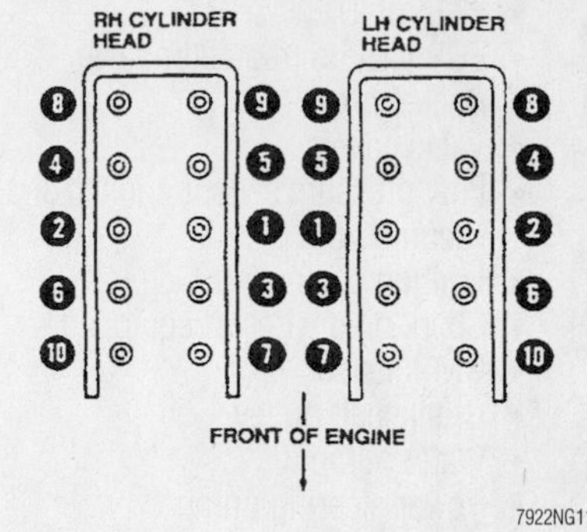

Cylinder head torque sequence—4.6L DOHC engine

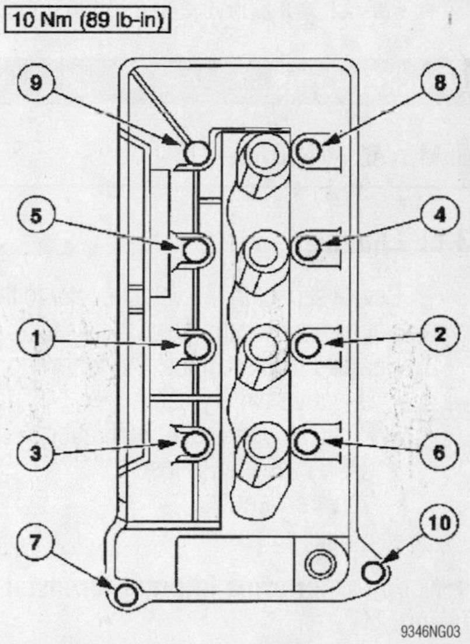

Right valve cover torque sequence—4.6L DOHC Engine

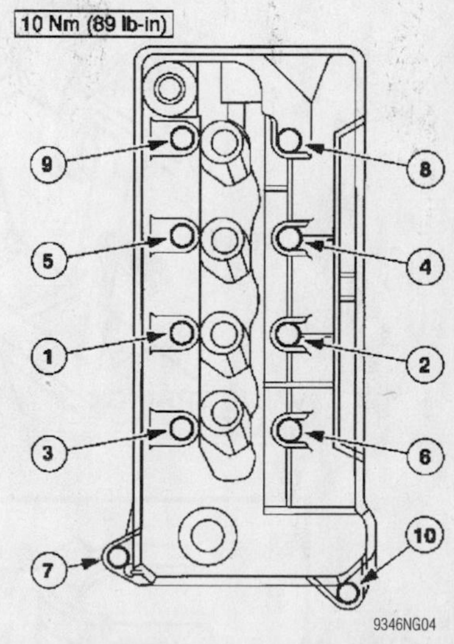

Left valve cover torque sequence—4.6L DOHC Engine

- EGR tube
- EGR vacuum regulator valve
- Rocker arms
- Valve covers
- Fuel pressure sensor connector and vacuum line
- Ignition coils
- Spark plug wires, if equipped
- CMP sensor
- Crankshaft pulley
- Water pump
- Power steering pump
- Alternator
- Water bypass tube
- ECT sensor connector
- Idler pulley
- Accessory drive belt and tensioner

Rocker Arms

REMOVAL & INSTALLATION

3.8L Engine

1. Before servicing the vehicle, refer to the precautions in the beginning of this section.
2. Remove or disconnect the following:
   - Negative battery cable
   - Positive Crankcase Ventilation (PCV) valve and hose
   - Valve covers
   - Rocker arms

➡Keep rocker arms in order for installation.

To install:

➡The rocker arm bolts are tightened with the valves closed. Rotate the crankshaft as necessary to position the lifter on the base circle of the camshaft lobe before tightening the corresponding rocker arm bolt.

3. Install the rocker arms. The rocker arm bolts are tightened in two steps as follows:
   a. Step 1: 44 inch lbs. (5 Nm).
   b. Step 2: 23–29 ft. lbs. (30–40 Nm).
4. Install or connect the following:
   - Valve covers
   - PCV valve and hose
   - Negative battery cable
5. Start the engine and check for proper operation.

4.6L Engines

1. Before servicing the vehicle, refer to the precautions at the beginning of this section.
2. Remove or disconnect the following:
   - Negative battery cable
   - Spark plug wires, if equipped
   - Ignition coils, on 1999–01 engines
   - Valve covers

➡The 4.6L DOHC engine requires special tool Valve Spring Compressor T91P-6565-A for exhaust valves, and Valve Spring Compressor T93P-6565-A for intake valves. The 4.6L SOHC engine requires special tool Valve Spring Compressor T91P-6565-A and Valve Spring Spacer T91P-6565-AH.

3. Rotate the crankshaft so that the piston on the cylinder to be serviced is at bottom dead center with the valves closed.
4. Compress the valve spring and remove the rocker arm. Repeat for each arm to be removed.

➡If the rocker arms are to be reused, ensure that they are installed in the same position that they were removed from.

To install:

5. Compress the valve spring and install the rocker arm. Repeat for each arm to be installed.
6. Install or connect the following:
   - Valve covers
   - Spark plug wires, if equipped
   - Ignition coils, on 1999–01 engines
   - Negative battery cable
7. Start the engine and check for proper operation.

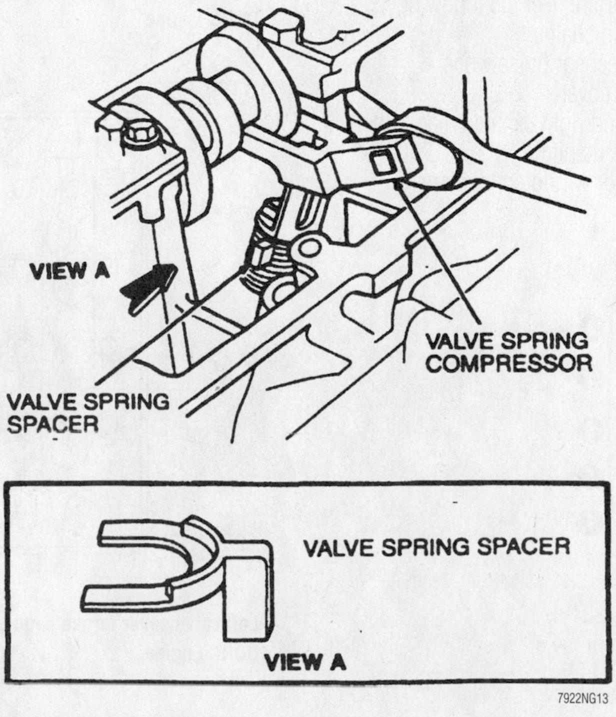

Valve spring compression tool and spacer—4.6L SOHC engine

Intake Manifold

REMOVAL & INSTALLATION

3.8L Engine

1. Before servicing the vehicle, refer to the precautions in the beginning of this section.
2. Drain the cooling system.
3. Remove or disconnect the following:
   - Negative battery cable
   - Intake Air Temperature (IAT) sensor
   - Air intake tube
   - Cruise control cable
   - Accelerator cable and bracket
   - Spark plug wires and ignition coil assembly
   - Upper intake vacuum lines
   - Positive Crankcase Ventilation (PCV) valve and hose
   - Throttle Position (TP) sensor connector
   - Idle Air Control (IAC) valve connector
   - Exhaust Gas Recirculation (EGR) valve
   - Engine control sensor wiring harness retainer bracket
   - Upper intake manifold
   - Fuel supply line
   - Fuel return line, if equipped
   - Fuel injector connectors
   - Engine Coolant Temperature (ECT) sensor connector
   - Fuel supply manifold and injectors
   - Upper radiator hose

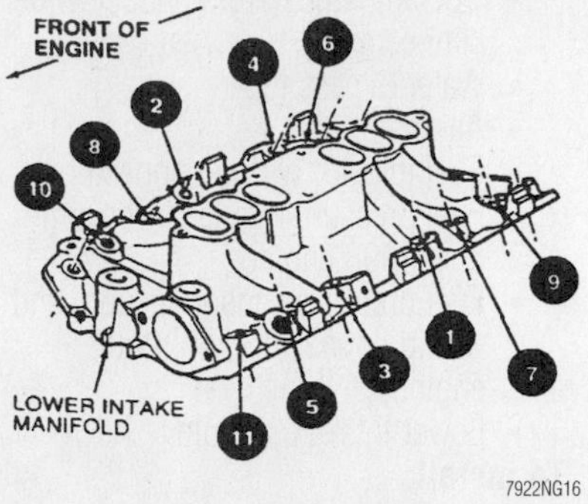

Lower intake manifold torque sequence—3.8L engine

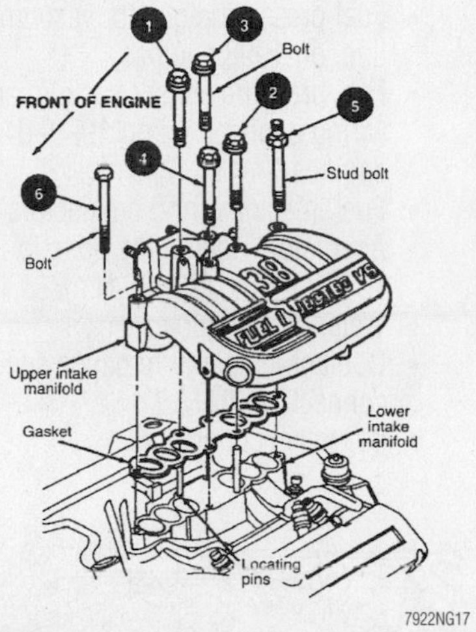

Upper intake manifold torque sequence—3.8L engine

- Heater hose
- Lower intake manifold

To install:

4. Install the lower intake manifold. Use new gaskets and tighten the bolts in sequence as follows:
   a. Step 1: 45 inch lbs. (5 Nm).
   b. Step 2: 71–106 inch lbs. (8–12 Nm).
5. Install or connect the following:
   - Heater hose
   - Upper radiator hose
   - Fuel supply manifold and injectors
   - ECT sensor connector
   - Fuel injector connectors
   - Fuel supply line
   - Fuel return line, if equipped
6. Install the upper intake manifold. Use a new gasket and tighten the bolts as follows:
   a. Step 1: 88 inch lbs. (10 Nm).
   b. Step 2: 15 ft. lbs. (20 Nm).
   c. Step 3: 24 ft. lbs. (32 Nm).
7. Install or connect the following:
   - Engine control sensor wiring harness retainer bracket
   - EGR valve
   - IAC valve connector
   - TP sensor connector
   - PCV valve and hose
   - Upper intake vacuum lines
   - Spark plug wires and ignition coil assembly
   - Accelerator cable and bracket
   - Cruise control cable
   - Air intake tube
   - IAT sensor connector
   - Negative battery cable
8. Fill the cooling system.
9. Start the engine and check for leaks.

4.6L SOHC Engine

1. Before servicing the vehicle, refer to the precautions in the beginning of this section.
2. Drain the cooling system.
3. Remove or disconnect the following:
   - Negative battery cable
   - Fuel supply line
   - Fuel return line, if equipped
   - Intake Air Temperature (IAT) sensor connector
   - Air intake tube
   - Accessory drive belt
   - Spark plug wires, if equipped
   - Ignition coils, on 1998 engines
   - Camshaft Position (CMP) sensor connector
   - Alternator and bracket
   - Fuel injector wiring connectors
   - Oil pressure sensor connector
   - Exhaust Gas Recirculation (EGR) tube
   - Cruise control cable
   - Accelerator cable and bracket
   - Throttle body vacuum adapter hose
   - Heater hose
   - Upper radiator hose adapter
   - Intake manifold

To install:

4. Install or connect the following:
   - Intake manifold. Use new gaskets and tighten the bolts to 15–22 ft. lbs. (20–30 Nm).
   - Upper radiator hose adapter. Use a new O-ring and tighten the bolts to 15–22 ft. lbs. (20–30 Nm).
   - Heater hose
   - Throttle body vacuum adapter hose

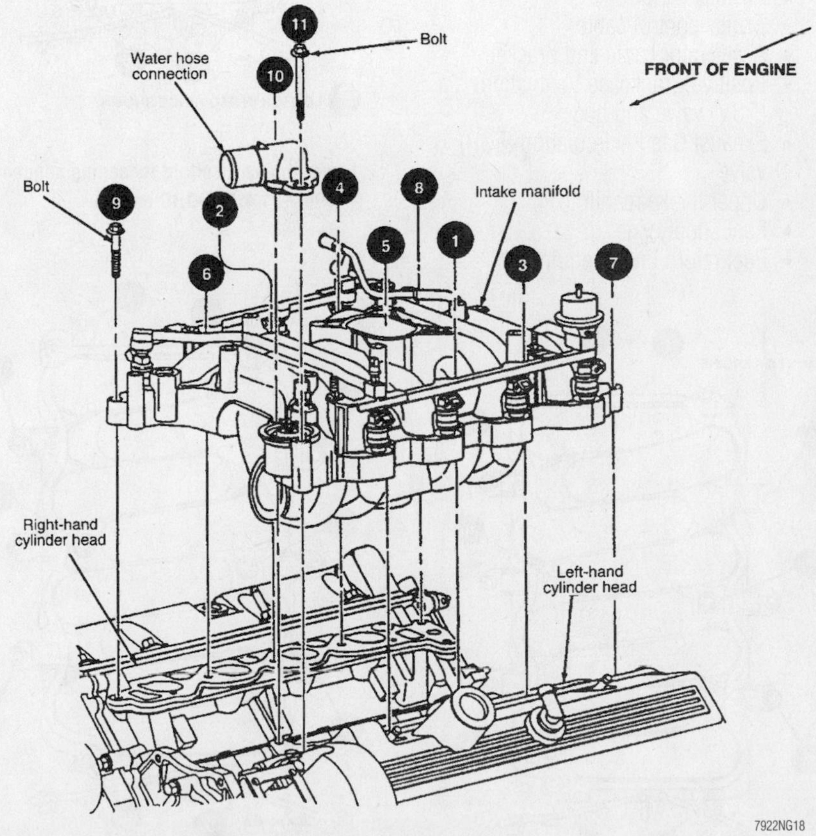

Intake manifold torque sequence—4.6L SOHC engine

Heater Core replacement is covered in Section 2 of this manual

- Accelerator cable and bracket
- Cruise control cable
- EGR tube. Tighten the nut to 26–33 ft. lbs. (35–45 Nm).
- Oil pressure sensor connector
- Fuel injector wiring connectors
- Alternator and bracket
- CMP sensor connector
- Ignition coils, on 1998 engines
- Spark plug wires, if equipped
- Accessory drive belt
- Air intake tube
- IAT sensor connector
- Fuel supply line
- Fuel return line, if equipped
- Negative battery cable

5. Fill the engine cooling system.
6. Start the engine and check for leaks.

4.6L DOHC Engine

1. Before servicing the vehicle, refer to the precautions in the beginning of this section.
2. Drain the cooling system.
3. Remove or disconnect the following:
   - Negative battery cable
   - Engine compartment brace
   - Air intake tube
   - Cruise control cable
   - Accelerator cable and bracket
   - Positive Crankcase Ventilation (PCV) valve and hose
   - Exhaust Gas Recirculation (EGR) valve
   - Upper intake manifold
   - Fuel supply line
   - Fuel return line, if equipped
   - Accessory drive belt
   - Spark plug wires, if equipped
   - Alternator bracket
   - Upper radiator hose
   - Throttle Position (TP) sensor connector
   - Engine Coolant Temperature (ECT) sensor connector
   - Idle Air Control (IAC) valve connector
   - Coolant temperature gauge sender connector
   - Water bypass tube
   - Alternator
   - Fuel injector wiring connectors
   - Fuel pressure regulator vacuum line, 1998 engines
   - Fuel pressure sensor vacuum and wiring connectors, 1999–01 engines
   - Lower intake manifold

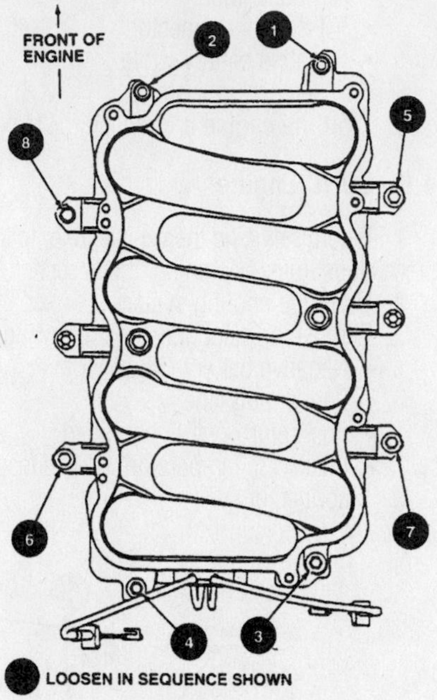

9300NG01

Lower intake manifold loosening sequence —1999–01 4.6L DOHC engine

To install:

4. Install or connect the following:
   - Lower intake manifold. Tighten the bolts in sequence to 89 inch lbs. (10 Nm).
   - Fuel pressure regulator vacuum line, on 1998 engines
   - Fuel pressure sensor vacuum and wiring connectors, on 1999–01 engines
   - Fuel injector wiring connectors
   - Alternator. Tighten the bolts to 15–22 ft. lbs. (20–30 Nm).
   - Water bypass tube
   - Coolant temperature gauge sender connector
   - IAC valve connector

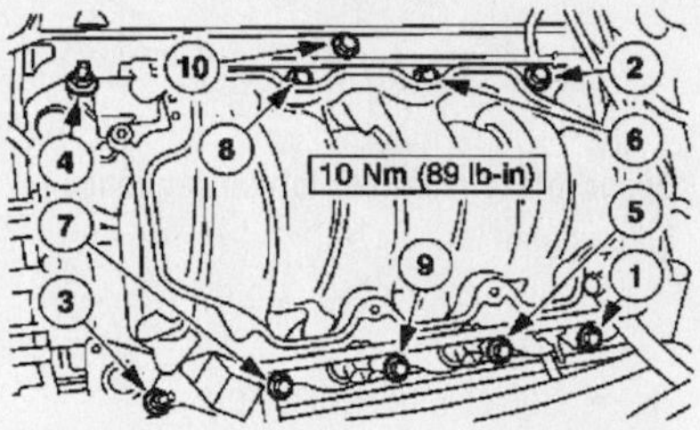

9306NG02

Lower intake manifold torque sequence—1999–01 4.6L DOHC engines

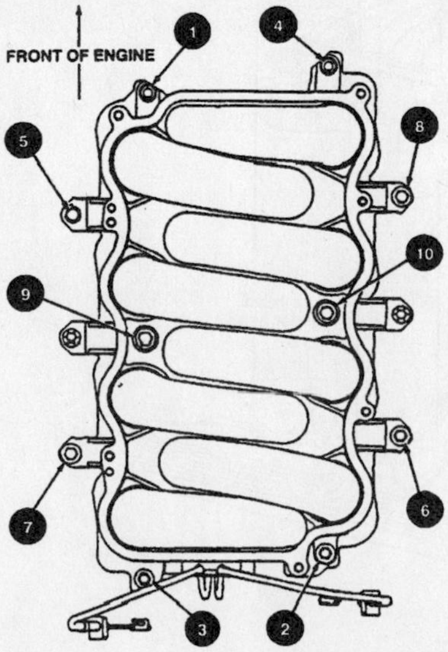

7922NG19

Lower intake manifold loosening sequence—1998 4.6L DOHC engines

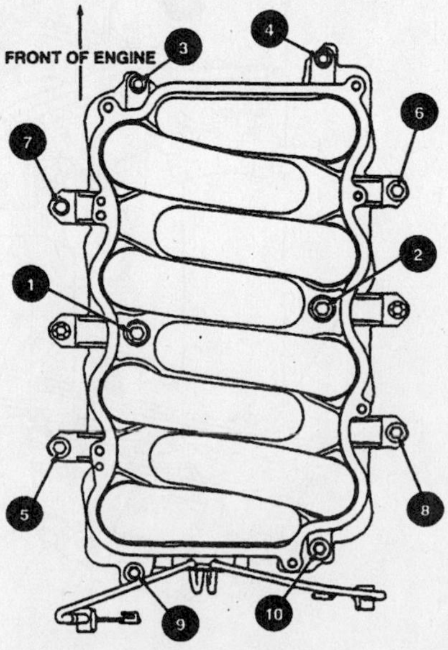

7922NG20

Lower intake manifold torque sequence—1998 4.6L DOHC engine

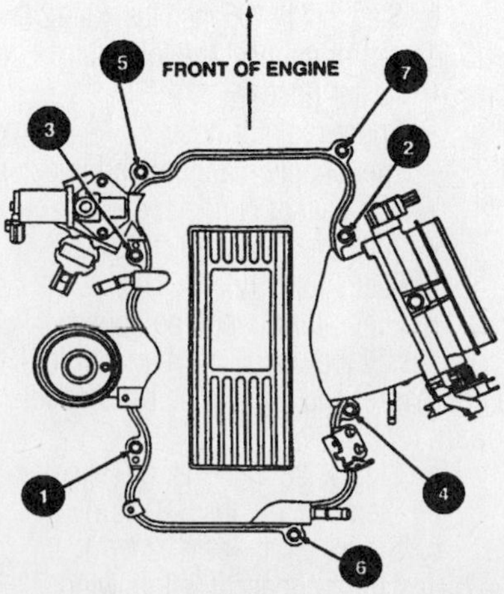

7922NG21

Upper intake manifold torque sequence—4.6L DOHC engines

- ECT sensor connector
- TP sensor connector
- Upper radiator hose
- Alternator bracket. Tighten the bolts to 71–106 inch lbs. (8–12 Nm).
- Spark plug wires, if equipped
- Accessory drive belt
- Fuel supply line
- Fuel return line, if equipped
- Upper intake manifold. Tighten the bolts to 71–106 inch lbs. (8–12 Nm).
- EGR valve
- PCV valve and hose
- Accelerator cable and bracket
- Cruise control cable
- Air intake tube
- Engine compartment brace
- Negative battery cable

5. Fill the engine cooling system.
6. Start the engine and check for leaks.

Exhaust Manifold

REMOVAL & INSTALLATION

3.8L Engine

1. Before servicing the vehicle, refer to the precautions in the beginning of this section.
2. Remove or disconnect the following:
   - Negative battery cable
   - Spark plug wires
   - Secondary air injection diverter valve, if equipped
   - Secondary air injection tubes, if equipped
   - Exhaust Gas Recirculation (EGR) tube
   - Oil dipstick tube
   - Heated Oxygen (HO_2S) sensors
   - Dual converter Y-pipe
   - Exhaust manifolds. Note the locations of the studs and bolts for reassembly.

To install:

3. Install or connect the following:
   - Exhaust manifolds. Tighten the bolts in sequence to 22–26 ft. lbs. (30–36 Nm).
   - Dual converter Y-pipe
   - HO_2S sensors. Tighten the sensors to 28–33 ft. lbs. (37–45 Nm).
   - Oil dipstick tube
   - EGR tube

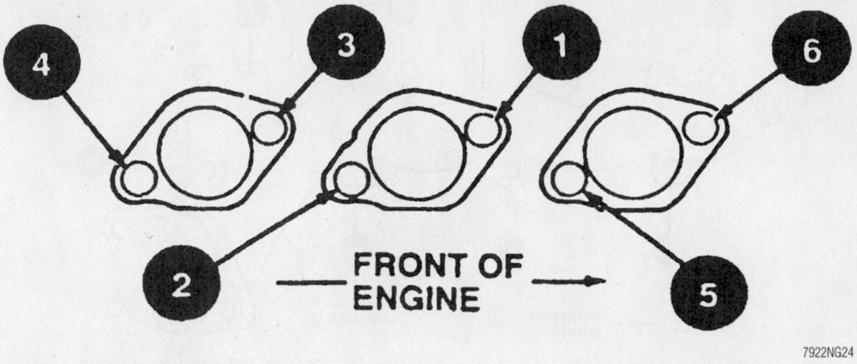

Exhaust manifold torque sequence—3.8L engine

   - Secondary air injection tubes, if equipped
   - Secondary air injection diverter valve, if equipped
   - Spark plug wires
   - Negative battery cable

4.6L Engines

1. Before servicing the vehicle, refer to the precautions in the beginning of this section.
2. Remove or disconnect the following:

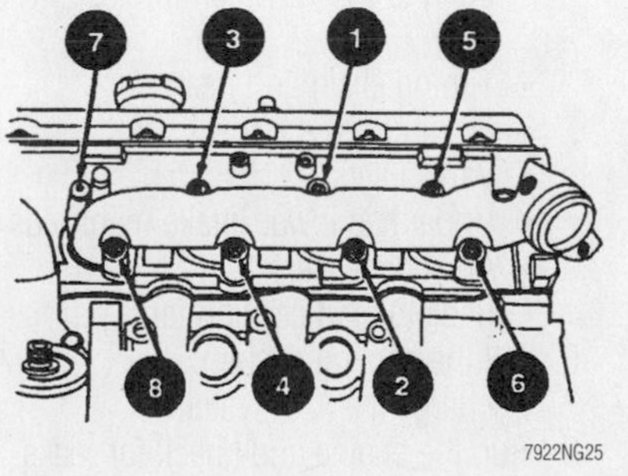

Exhaust manifold torque sequence—4.6L SOHC engine

   - Negative battery cable
   - Engine compartment brace
   - Dual converter H-pipe
   - Starter
   - Steering column intermediate shaft
   - Secondary air injection tubes, if equipped
   - Exhaust Gas Recirculation (EGR) tube
   - Oil filter
   - Left and right motor mounts

3. Raise the engine about 1.5 inches for clearance and remove the exhaust manifolds.

To install:

4. Install or connect the following:
   - Exhaust manifolds. Use new gaskets and tighten the nuts in sequence to 15 ft. lbs. (20 Nm).
   - Left and right motor mounts
   - Oil filter
   - EGR tube
   - Secondary air injection tubes, if equipped
   - Steering column intermediate shaft
   - Starter

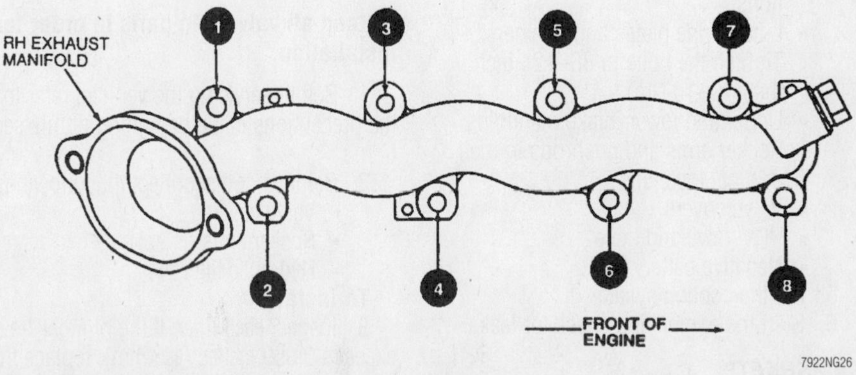

Left exhaust manifold torque sequence—4.6L DOHC engine

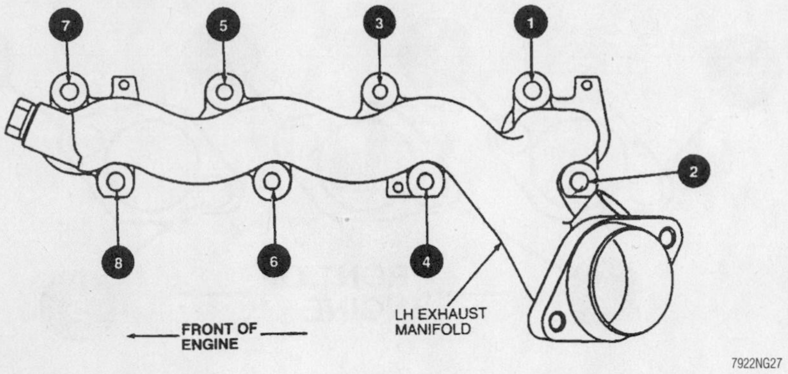

Right exhaust manifold torque sequence—4.6L DOHC engine

- Dual converter H-pipe
- Engine compartment brace
- Negative battery cable

Camshaft and Valve Lifters

REMOVAL & INSTALLATION

3.8L Engine

VALVE LIFTERS

➡Keep all valvetrain parts in order for installation.

1. Before servicing the vehicle, refer to the precautions in the beginning of this section.
2. Drain the cooling system
3. Remove or disconnect the following:
   - Negative battery cable
   - Positive Crankcase Ventilation (PCV) valve and hose
   - Valve covers
   - Rocker arms and pushrods
   - Upper and lower intake manifolds
   - Lifter guide plates and retainers
   - Valve lifters

To install:

4. Install or connect the following:
   - Valve lifters in their original locations
   - Lifter guide plates and retainers. Tighten the bolts to 88–124 inch lbs. (10–14 Nm).
   - Upper and lower intake manifolds
   - Rocker arms and pushrods in their original locations
   - Valve covers
   - PCV valve and hose
   - Negative battery cable
5. Fill the cooling system.
6. Start the engine and check for leaks.

CAMSHAFT

1. Before servicing the vehicle, refer to the precautions in the beginning of this section.
2. Drain the cooling system.
3. Recover the A/C refrigerant.
4. Remove or disconnect the following:
   - Radiator and cooling fan
   - A/C condenser
   - Upper and lower intake manifolds
   - Valve lifters
   - Front cover
   - Timing chain and gears
   - Camshaft thrust plate
   - Camshaft

To install:

5. Install or connect the following:
   - Camshaft
   - Camshaft thrust plate. Tighten the bolts to 71–124 inch lbs. (8–14 Nm).
   - Timing chain and gears
   - Front cover
   - Valve lifters
   - Upper and lower intake manifolds
   - A/C condenser
   - Radiator and cooling fan
6. Fill the cooling system.
7. Recharge the A/C system.
8. Run the engine and check for leaks.

4.6L Engines

VALVE LIFTERS

➡Keep all valvetrain parts in order for installation.

1. Before servicing the vehicle, refer to the precautions in the beginning of this section.
2. Remove or disconnect the following:
   - Valve covers
   - Rocker arms
   - Hydraulic lifters

To install:

3. Inspect each lifter. If the plunger travel exceeds 0.059 inches (1.5 mm), replace the lifter.
4. Install or connect the following:
   - Hydraulic lifters in their original positions
   - Rocker arms
   - Valve covers

CAMSHAFTS

1. Before servicing the vehicle, refer to the precautions in the beginning of this section.
2. Remove or disconnect the following:
   - Valve covers
   - Rocker arms
   - Oil pan (SOHC engine)
   - Front cover
   - Timing chains and sprockets
   - Camshaft bearing caps
   - Camshafts

To install:

➡On 4.6L DOHC engines, the outboard exhaust camshaft bearing cap bolts are shorter than the other bearing cap bolts.

3. Install the camshafts. Tighten the bearing cap bolts in sequence as follows:
   a. Step 1: 71–106 inch lbs. (8–12 Nm).
   b. Step 2: Loosen all bolts 2 turns.
   c. Step 3: 71–106 inch lbs. (8–12 Nm).
4. Install or connect the following:
   - Timing chains and sprockets

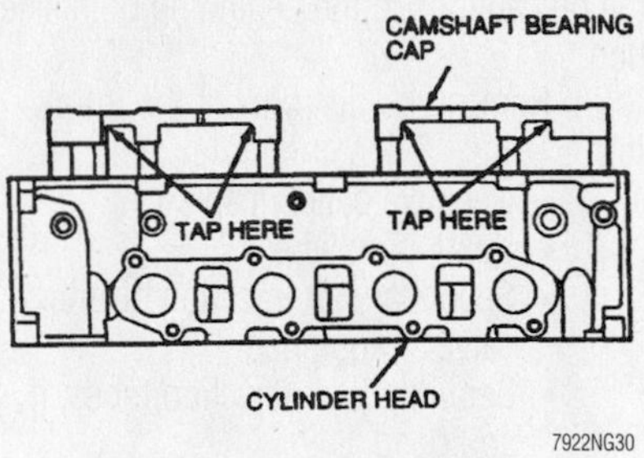

To remove the camshaft bearing caps, tap the caps with a rubber or leather mallet where shown—4.6L engines

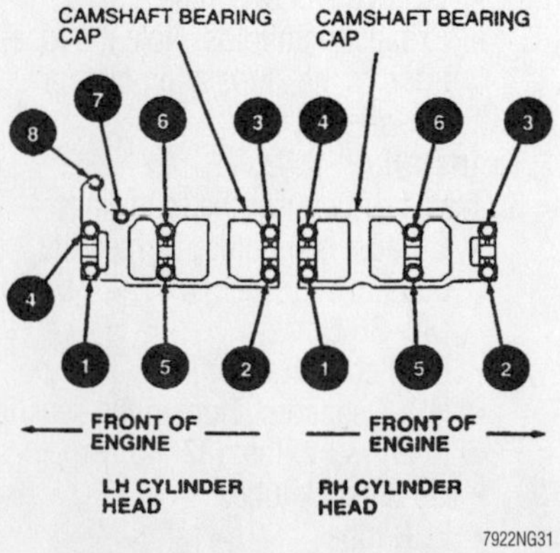

Camshaft bearing cap torque sequence—4.6L SOHC engine

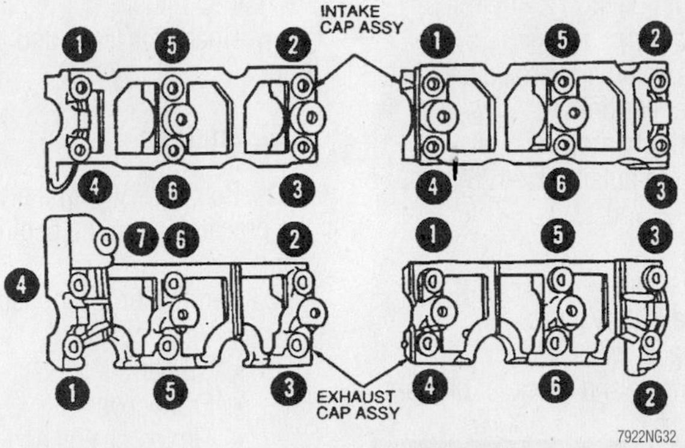

Camshaft bearing cap torque sequence—4.6L DOHC engine

- Front cover
- Oil pan (SOHC engine)
- Rocker arms
- Valve covers

5. Start the engine and check for leaks.

Valve Lash

ADJUSTMENT

The 3.8L and 4.6L engines are equipped with hydraulic lash adjusters. Valve clearance is not adjustable.

Starter Motor

REMOVAL & INSTALLATION

3.8L Engine

1. Before servicing the vehicle, refer to the precautions in the beginning of this section.
2. Remove or disconnect the following:
   - Negative battery cable
   - Starter electrical connections
   - Ground cable
   - Starter bolts
   - Starter

To install:

3. Install or connect the following:
   - Starter. Tighten the bolts to 17 ft. lbs. (23 Nm).
   - Ground cable. Tighten the bolt to 17 ft. lbs. (23 Nm).
   - Starter electrical connections
   - Negative battery cable

4.6L Engines

1. Before servicing the vehicle, refer to the precautions in the beginning of this section.
2. Remove or disconnect the following:
   - Negative battery cable
   - Heated Oxygen (HO_2S) sensor connector and bracket
   - Starter electrical connections
   - Starter bolts
   - Starter

To install:

3. Install or connect the following:
   - Starter. Tighten the bolts to 17 ft. lbs. (23 Nm).
   - Starter electrical connections
   - HO_2S sensor connector and bracket
   - Negative battery cable

Oil Pan

REMOVAL & INSTALLATION

3.8L Engine

1. Before servicing the vehicle, refer to the precautions in the beginning of this section.
2. Drain the engine oil.
3. Install engine lifting brackets and attach an engine support fixture.

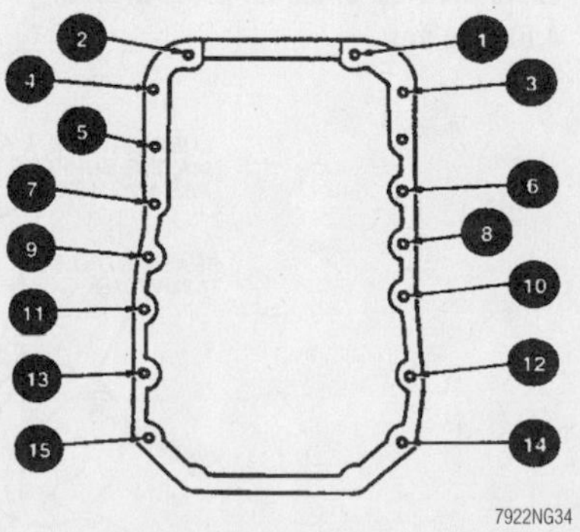

Oil pan torque sequence—3.8L engine

4. Remove or disconnect the following:
   - Negative battery cable
   - Oil filter
   - Left and right motor mounts
   - Starter motor
   - Steering column intermediate shaft
   - Subframe, lower it for clearance
   - Oil pan

To install:

5. Apply a bead of silicone sealant to the oil pan mounting flange. Install the oil pan and tighten the bolts as follows:
   a. Step 1: 44 inch lbs. (5 Nm).
   b. Step 2: 88 inch lbs. (10 Nm).
   c. Step 3: Tighten the bell housing bolts to 33 ft. lbs. (45 Nm).
6. Raise the subframe and tighten the upper bolts to 85 ft. lbs. (115 Nm). Tighten the lower bolts to 68 ft. lbs. (90 Nm).
7. Install or connect the following:
   - Steering column intermediate shaft
   - Starter motor
   - Left and right motor mounts
   - Oil filter
   - Negative battery cable
8. Fill the engine with oil.
9. Run the engine and check for leaks.

4.6L SOHC Engine

1. Before servicing the vehicle, refer to the precautions in the beginning of this section.
2. Drain the engine oil.
3. Remove or disconnect the following:
   - Negative battery cable
   - Engine compartment brace
   - Front subframe crossmember brace
   - Left and right motor mounts
   - Transmission housing cover
4. Raise the engine about 4 inches and support with wood blocks under each motor mount.
5. Remove the oil pan.

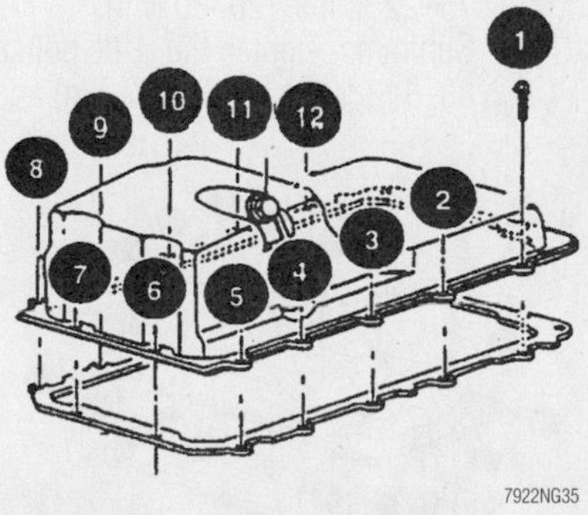

Oil pan torque sequence—4.6L SOHC engine

To install:

6. Install the oil pan. Use a new gasket and tighten the bolts in sequence as follows:
   a. Step 1: 15 ft. lbs. (20 Nm).
   b. Step 2: Plus 60 degrees.
7. Remove the wood blocks and lower the engine into position.
8. Install or connect the following:
   - Transmission housing cover
   - Left and right motor mounts. Tighten the nuts to 95–126 ft. lbs. (128–172 Nm).
   - Front subframe crossmember brace. Tighten the bolts to 30–40 ft. lbs. (40–55 Nm).
   - Engine compartment brace
   - Negative battery cable
9. Fill the engine with oil.
10. Start the engine and check for leaks.

4.6L DOHC Engine

1. Before servicing the vehicle, refer to the precautions in the beginning of this section.
2. Drain the engine oil.
3. Install engine lifting brackets and attach an engine support fixture.
4. Remove or disconnect the following:
   - Negative battery cable
   - Transmission
   - Oil filter
   - Left and right motor mounts
   - Stabilizer bar links
   - Steering column intermediate shaft
   - Subframe, lower it for clearance
   - Oil pan

To install:

5. Apply silicone sealant to the joints where the front cover and the rear seal retainer meet the cylinder block.
6. Install or connect the following:
   - Oil pan. Use a new gasket and tighten the bolts in sequence to 15–22 ft. lbs. (20–30 Nm).
   - Subframe. Tighten the large bolts to 83–113 ft. lbs. (113–153 Nm) and the small bolts to 72–97 ft. lbs. (98–132 Nm).
   - Steering column intermediate shaft
   - Stabilizer bar links
   - Left and right motor mounts. Tighten the nuts to 95–126 ft. lbs. (128–172 Nm).
   - Oil filter
   - Transmission
   - Negative battery cable
7. Fill the engine with oil.
8. Start the engine and check for oil leaks.

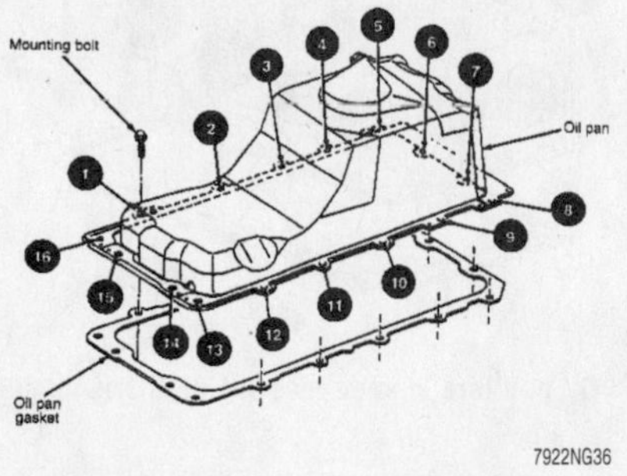

Oil pan torque sequence—4.6L DOHC engine

Oil Pump

REMOVAL & INSTALLATION

3.8L Engine

1. Before servicing the vehicle, refer to the precautions in the beginning of this section.
2. Remove or disconnect the following:
   - Oil filter
   - Oil pump body and gears

To install:

3. Install or connect the following:
   - Oil pump. Use a new O-ring seal. Tighten the large bolts to 18 ft. lbs. (25 Nm) and the small bolts to 88 inch lbs. (10 Nm).

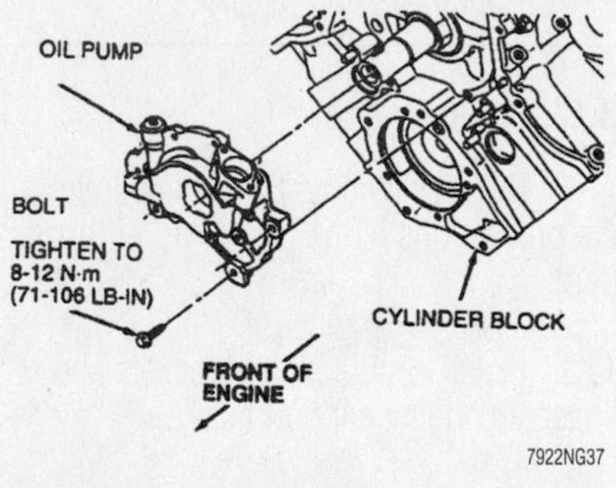

Exploded view of the oil pump mounting—4.6L engines

- Oil filter

4. Check for leaks and proper operation.

4.6L Engines

1. Before servicing the vehicle, refer to the precautions in the beginning of this section.
2. Remove or disconnect the following:
   - Valve covers
   - Oil pan
   - Front cover
   - Timing chains and sprockets
   - Oil pump

To install:

3. Install or connect the following:
   - Oil pump. Tighten the bolts to 71–106 inch lbs. (8–12 Nm).
   - Timing chains and sprockets
   - Front cover
   - Oil pan
   - Valve covers
4. Check for leaks and proper operation.

Rear Main Seal

REMOVAL & INSTALLATION

3.8L Engine

1. Before servicing the vehicle, refer to the precautions in the beginning of this section.
2. Remove or disconnect the following:
   - Negative battery cable
   - Transmission
   - Clutch pressure plate and disc, if equipped
   - Flywheel
   - Engine rear plate
   - Rear main seal

To install:

3. Install or connect the following:
   - Rear main seal
   - Engine rear plate

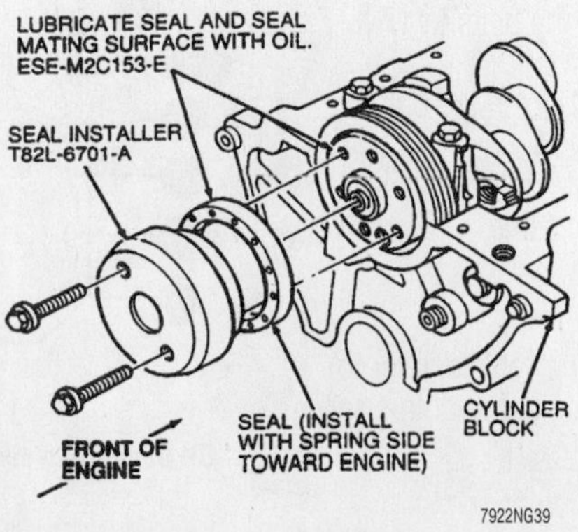

Installing the rear main seal—3.8L engine

- Flywheel. Tighten the bolts to 54–64 ft. lbs. (73–87 Nm).
- Clutch pressure plate and disc, if equipped
- Transmission
- Negative battery cable

4.6L Engines

1. Before servicing the vehicle, refer to the precautions in the beginning of this section.
2. Remove or disconnect the following:
   - Negative battery cable
   - Transmission
   - Clutch pressure plate and disc, if equipped
   - Flywheel
   - Oil slinger. Use Rear Crankshaft Oil Slinger Remover T95P-6701-AH.
   - Rear oil seal retainer
   - Rear oil seal

To install:

3. Install the rear oil seal retainer. Apply silicone sealant as shown. Tighten the bolts in sequence to 71–106 inch lbs. (8–12 Nm).
4. Install or connect the following:
   - Rear oil seal. Use Rear Crankshaft Seal Replacer T95P-6701-BH.
   - Oil slinger. Use Rear Crankshaft Oil Slinger Replacer T95P-6701-CH.
   - Flywheel. Tighten the bolts in a crossing pattern to 54–64 ft. lbs. (73–87 Nm).
   - Clutch pressure plate and disc, if equipped
   - Transmission
   - Negative battery cable
5. Run the engine and check for leaks.

Timing Chain, Sprockets, Front Cover and Seal

REMOVAL & INSTALLATION

3.8L Engine

1. Before servicing the vehicle, refer to the precautions in the beginning of this section.
2. Drain the cooling system and the engine oil.
3. Rotate the crankshaft so that the No. 1 cylinder is at Top Dead Center (TDC) of the compression stroke.
4. Remove or disconnect the following:
   - Negative battery cable
   - Intake Air Temperature (IAT) sensor connector
   - Air cleaner and air intake tube
   - Cooling fan and shroud
   - Accessory drive belt
   - Power steering pump and bracket
   - A/C compressor front bracket, if equipped
   - Oil filter
   - Radiator hoses
   - Heater water outlet tube
   - Crankshaft pulley and damper
   - Front crankshaft seal

➡Note the position of the Camshaft Position (CMP) sensor connector. The installation procedure requires that the connector be located in the same position.

   - Camshaft Position (CMP) sensor housing
   - Crankshaft Position (CKP) sensor
   - Oil pan
   - Front cover
   - Distributor drive gear
   - Timing chain and sprockets

To install:

5. Compress the timing chain vibration damper and install a retaining pin.
6. Install or connect the following:
   - Timing chain and sprockets with the timing marks aligned as shown
   - Distributor drive gear. Tighten the bolt to 30–36 ft. lbs. (40–50 Nm) and remove the vibration damper retaining pin.

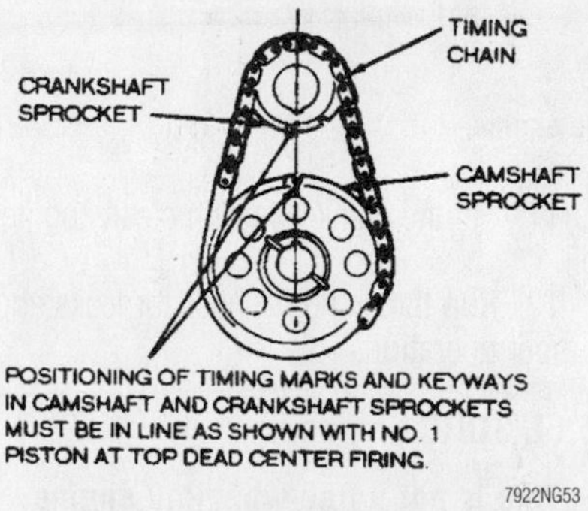

Camshaft timing marks—3.8L engine

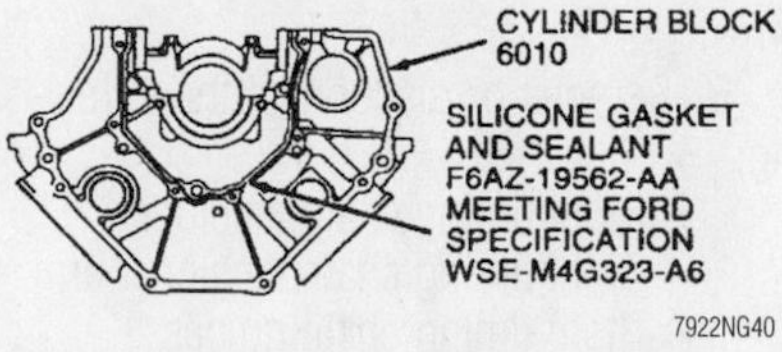

Apply silicone sealant to the engine block when installing the oil seal retainer—4.6L engines

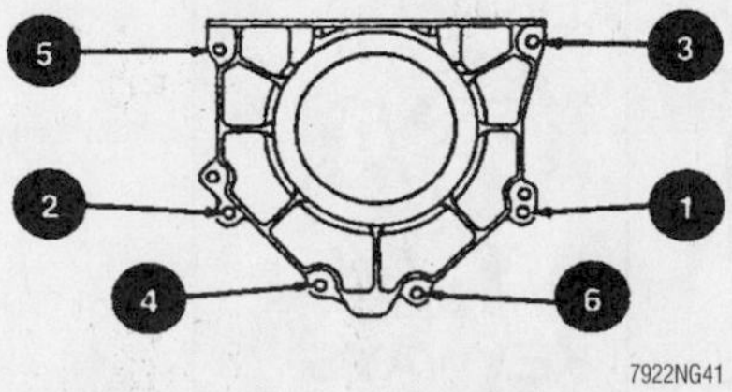

Oil seal retainer torque sequence—4.6L engines

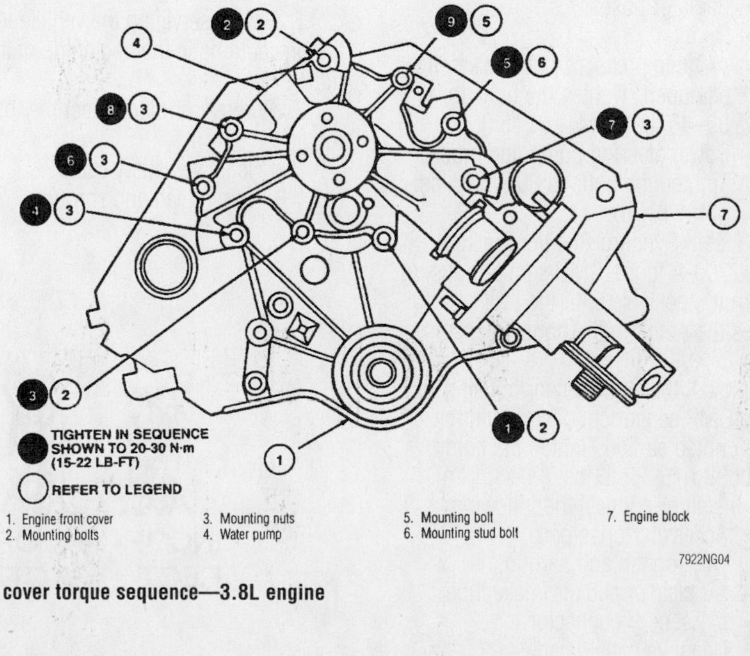

Front cover torque sequence—3.8L engine

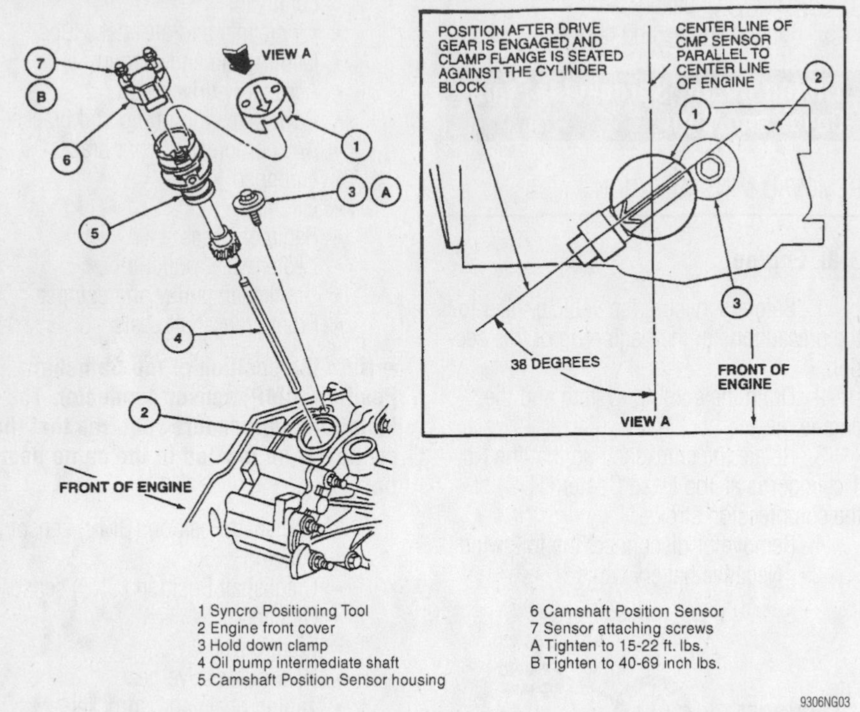

Camshaft Position Sensor housing installation—3.8L engine

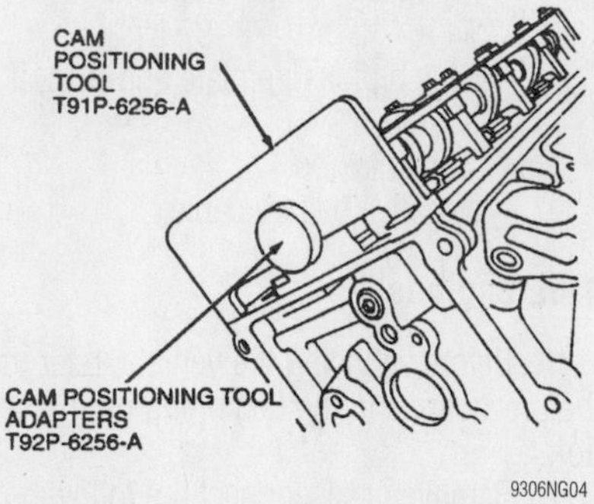

Camshaft Positioning Tool and Adapter—4.6L SOHC engine

- Front cover. Tighten the bolts in sequence to 15–22 ft. lbs. (20–30 Nm).
- Oil pan
- CKP sensor
- Front crankshaft seal
- Crankshaft pulley and damper. Tighten the bolt to 103–132 ft. lbs. (140–180 Nm).
- Heater water outlet tube
- Radiator hoses
- Oil filter
- A/C compressor front bracket, if equipped. Tighten the bolts to 30–45 ft. lbs. (41–61 Nm).
- Power steering pump and bracket. Tighten the bolts to 30–45 ft. lbs. (41–61 Nm).

7. Install Synchro Positioning Tool T96T-12200-A to the CMP sensor housing and turn it clockwise until the tool boss engages the notch in the housing assembly.

8. Install the CMP sensor housing so that the CMP sensor connector is in the position noted earlier. Tighten the hold down bolt to 15–22 ft. lbs. (20–30 Nm).

9. Install or connect the following:

- Accessory drive belt
- Cooling fan and shroud
- Air cleaner and air intake tube
- IAT sensor connector
- Negative battery cable

10. Fill the cooling system.

11. Fill the crankcase with clean engine oil.

12. Run the engine. Check for leaks and proper operation.

4.6L SOHC Engine

➡This is not a free wheeling engine. Do not rotate the crankshaft or camshafts with the timing chains removed.

1. Before servicing the vehicle, refer to the precautions in the beginning of this section.

2. Remove or disconnect the following:

- Negative battery cable
- Cooling fan and shroud
- Accessory drive belt
- Water pump pulley
- Power steering pump and reservoir
- Oil pan
- Crankshaft pulley
- Front crankshaft seal
- Valve covers
- Ignition coils and brackets, on 1998 engines
- Idler pulley
- Camshaft Position (CMP) sensor connector
- Crankshaft Position (CKP) sensor connector
- Front cover
- CKP sensor pulse wheel

3. Rotate the crankshaft so that the No. 1 cylinder is at Top Dead Center (TDC) of the compression stroke.

4. Install Camshaft Positioning Tool Adapters T92P-6256-A and Camshaft Positioning Tool T91P-6256-A on the flats of the camshafts.

5. Remove or disconnect the following:

- Right timing chain tensioner
- Right timing chain tensioner arm
- Right timing chain guides
- Right timing chain and sprockets

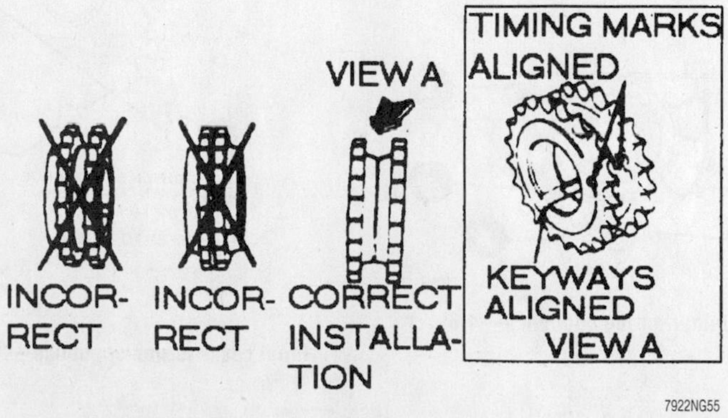

Crankshaft sprocket positioning—4.6L SOHC engine

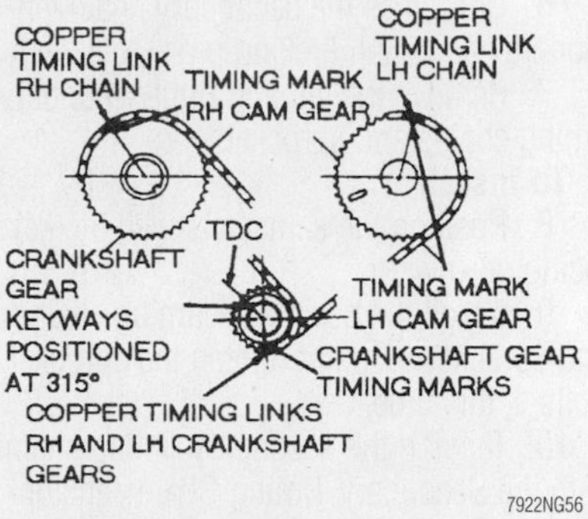

Timing chain and sprocket alignment—4.6L SOHC engine

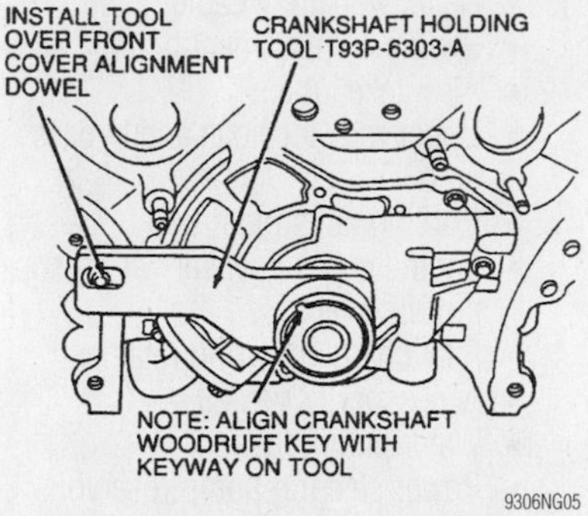

Crankshaft Holding Tool—4.6L SOHC engine

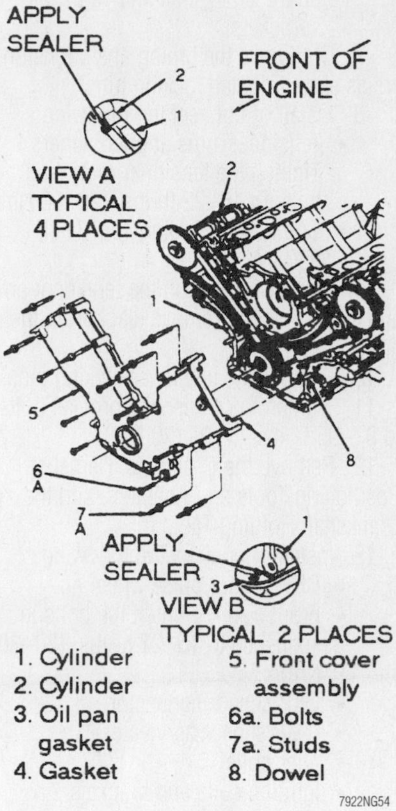

Apply sealer when installing the front cover—4.6L SOHC engine

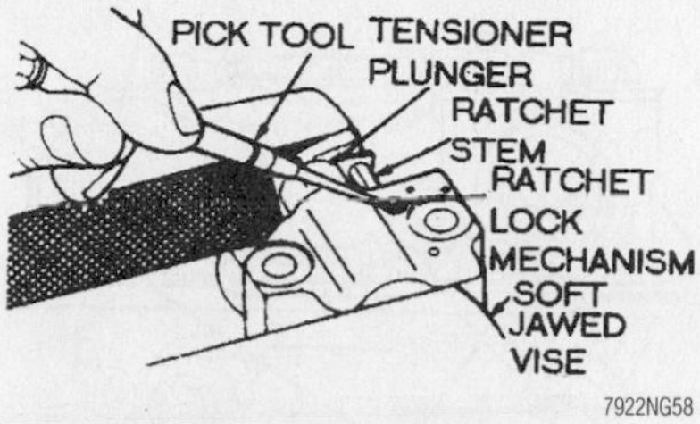

Timing chain tensioner bleeding procedure—4.6L SOHC engine

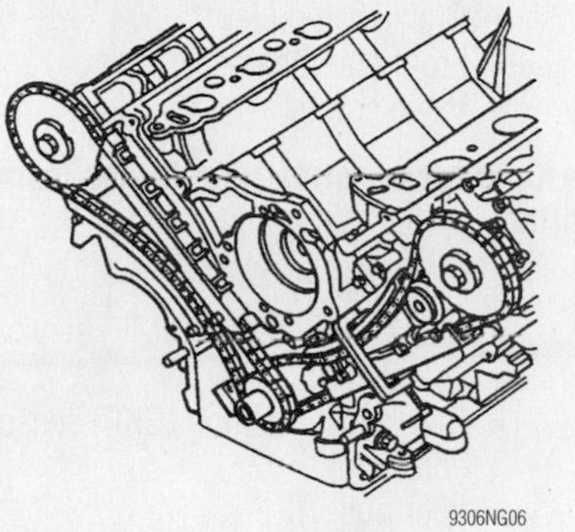

Remove slack from the timing chains with a C-clamp—4.6L SOHC engine

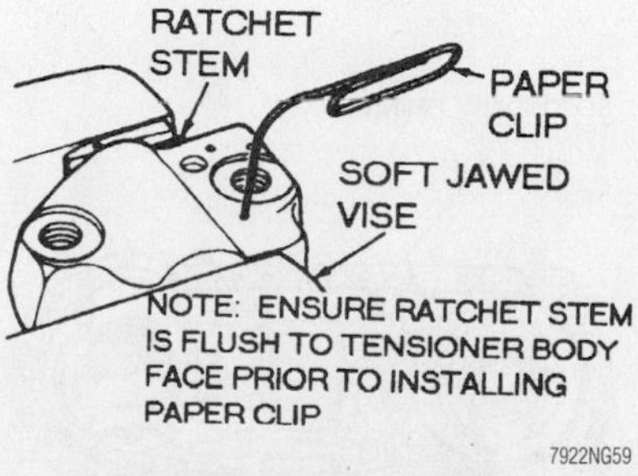

Timing chain tensioner locking procedure—4.6L SOHC engine

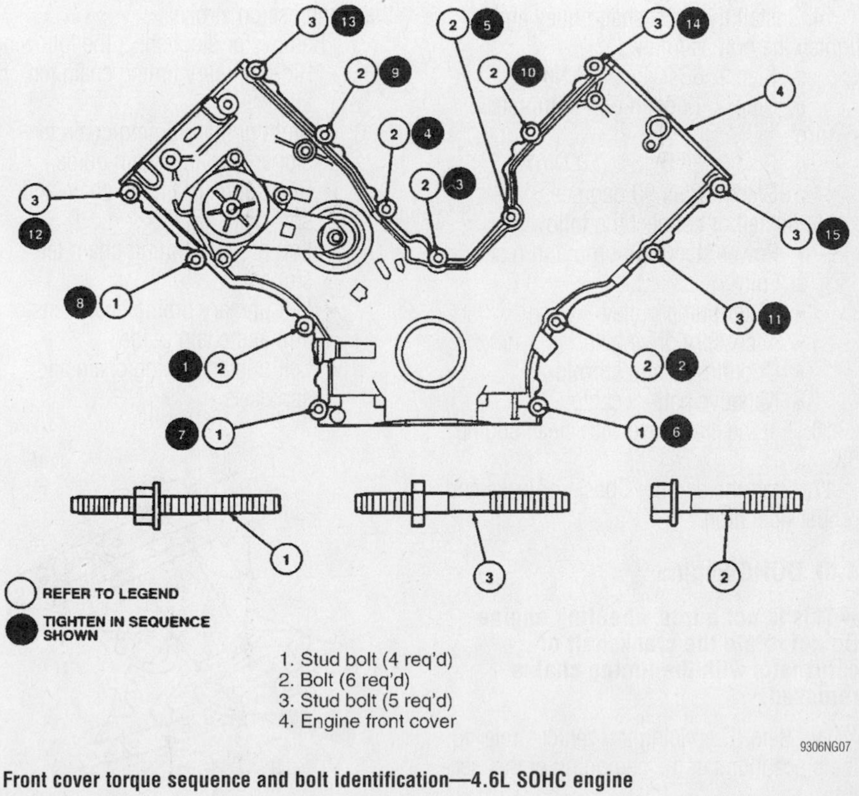

Front cover torque sequence and bolt identification—4.6L SOHC engine

- Left timing chain tensioner
- Left timing chain tensioner arm
- Left timing chain guides
- Left timing chain and sprockets

To install:

6. Install or connect the following:
   - Timing chain guides. Tighten the bolts to 71–106 inch lbs. (8–12 Nm).
   - Camshaft sprockets. Do not tighten the bolts at this time.
   - Crankshaft sprockets
   - Left timing chain with the copper links aligned with the timing marks on the crankshaft and camshaft sprockets
   - Right timing chain with the copper links aligned with the timing marks

on the crankshaft and camshaft sprockets.

7. Compress the timing chain tensioners as shown. Install locking pins.

8. Install or connect the following:

- Tensioner arms and tensioners. Tighten the tensioner mounting bolts to 15–22 ft. lbs. (20–30 Nm).
- Crankshaft Holding Tool T93P-6303-A

9. Use a C-clamp on the tensioner arm and chain guide to remove slack from the timing chain.

10. Remove the tensioner locking pins.

11. Tighten the camshaft sprocket bolts to 81–95 ft. lbs. (110–130 Nm).

12. Remove the C-clamps, Camshaft Positioning Tools and Adapters, and the Crankshaft Holding Tool.

13. Install or connect the following:

- CKP sensor pulse wheel
- Front cover. Tighten the bolts in sequence to 15–22 ft. lbs. (20–30 Nm).
- CKP sensor connector
- CMP sensor connector
- Idler pulley
- Ignition coils and brackets, on 1998 engines
- Valve covers
- Front crankshaft seal
- Oil pan

14. Install the crankshaft pulley and tighten the bolt as follows:

a. Step 1: 66 ft. lbs. (90 Nm).

b. Step 2: Loosen one complete turn.

c. Step 3: 36 ft. lbs. (50 Nm).

d. Step 4: Plus 90 degrees.

15. Install or connect the following:

- Power steering pump and reservoir
- Water pump pulley
- Accessory drive belt
- Cooling fan and shroud
- Negative battery cable

16. Fill the crankcase with clean engine oil.

17. Start the engine. Check for leaks and proper operation.

4.6L DOHC Engine

➡This is not a free wheeling engine. Do not rotate the crankshaft or camshafts with the timing chains removed.

1. Before servicing the vehicle, refer to the precautions in the beginning of this section.

2. Drain the cooling system.

3. Remove or disconnect the following:

- Negative battery cable
- Engine compartment brace
- Air intake tube
- Upper radiator hose and bypass tube
- Cooling fan and shroud
- Engine control sensor wiring support bracket
- Fuel charging wiring retainer
- Accessory drive belt
- Water pump pulley
- Power steering pump reservoir
- Ignition coils
- Ignition coil brackets, on 1998 engines
- Power steering pump
- Crankshaft pulley
- Front crankshaft seal
- Valve covers

➡Keep rocker arms in order for installation.

- Rocker arms
- Camshaft Position (CMP) sensor connector
- Crankshaft Position (CKP) sensor connector
- Idler pulley
- Front cover
- CMP sensor pulse wheel

4. Rotate the crankshaft so that the No. 1 cylinder is at Top Dead Center (TDC) of the compression stroke.

5. Remove or disconnect the following:

- Right primary timing chain tensioner
- Right primary timing chain tensioner arm and chain guide
- Right primary timing chain and sprockets
- Left primary timing chain tensioner
- Left primary timing chain tensioner arm and chain guide
- Left primary timing chain and sprockets

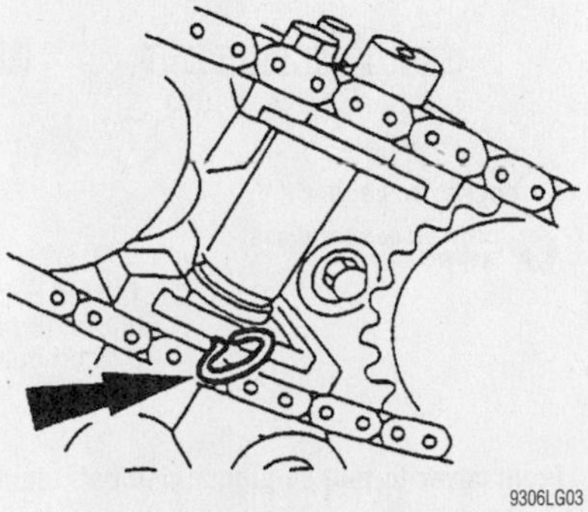

Secondary timing chain tensioner and locking pin—4.6L DOHC engines

6. Compress the secondary chain tensioners and install locking pins.

7. Remove the left and right secondary timing chains and sprockets.

To install:

8. Position the camshafts with the keys facing downward.

9. Install the secondary timing chains and sprockets. Do not tighten the sprocket bolts at this time.

10. Tension the secondary timing chains with the Secondary Timing Chain Tensioning tool T93P-6256-BH.

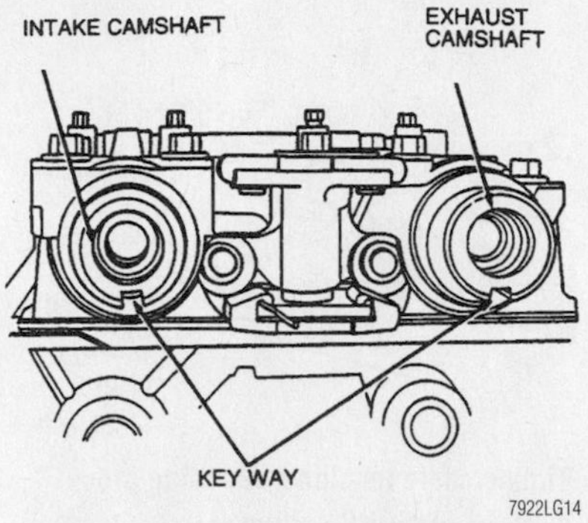

Position the camshafts for timing chain installation—4.6L DOHC engines

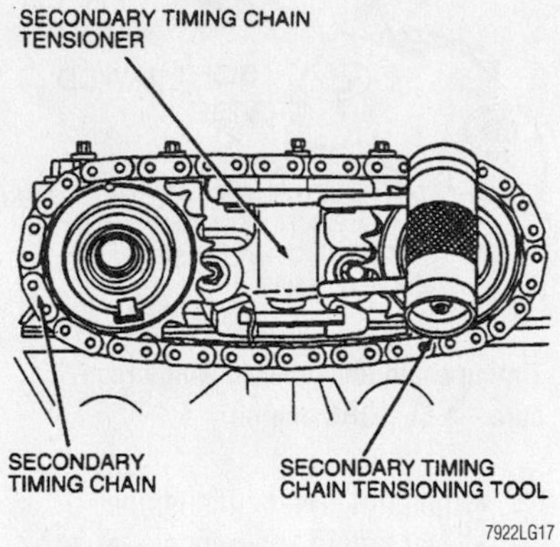

Secondary timing chain tensioning tool—4.6L DOHC engines

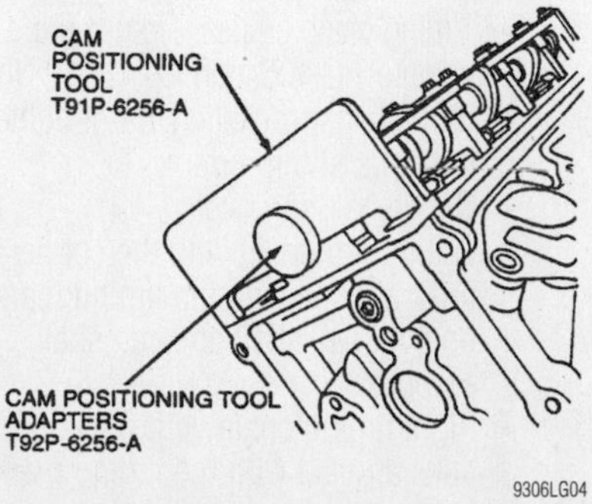

Install the camshaft positioning tool—4.6L DOHC engines

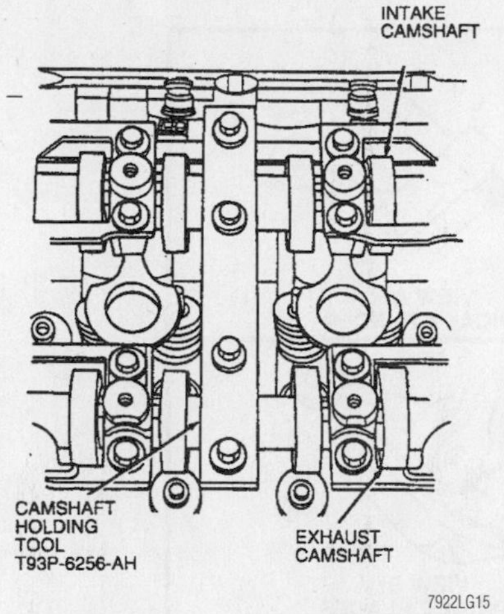

Install the Camshaft Holding tool to secure the camshafts and prevent damage to the camshaft positioning tool—4.6L DOHC engines

11. Install Camshaft Positioning tool T93P-6256-A in the rear D-slots of the camshaft.
12. Install Camshaft Holding tool T93P-6256-AH onto the camshafts to keep the camshafts from rotating and to prevent damaging the camshaft positioning tool.
13. Tighten the secondary timing chain sprocket bolts to 81–95 ft. lbs. (110–130 Nm).
14. Remove the secondary timing chain tensioner locking pins and remove the tensioner tool.

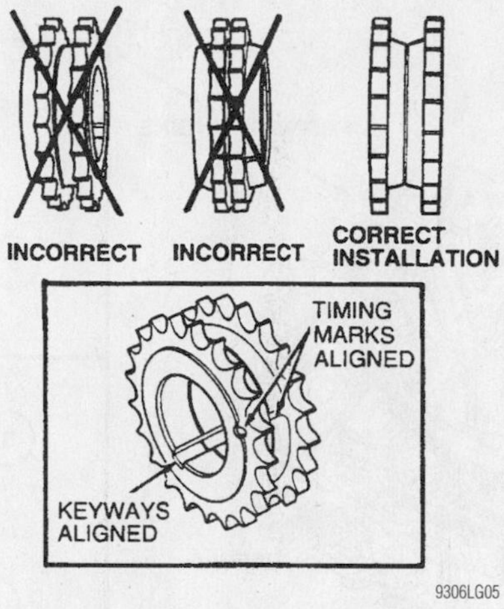

Crankshaft timing sprocket installation—4.6L DOHC engines

15. Compress the primary timing chain tensioners and install locking pins.
16. Install the crankshaft timing sprockets by aligning the timing marks.
17. Align the copper colored primary timing chain links with the sprocket timing marks as shown.
18. Install or connect the following:
    - Left primary timing chain and sprockets
    - Left primary timing chain tensioner arm and chain guide
    - Left primary timing chain tensioner
    - Right primary timing chain and sprockets

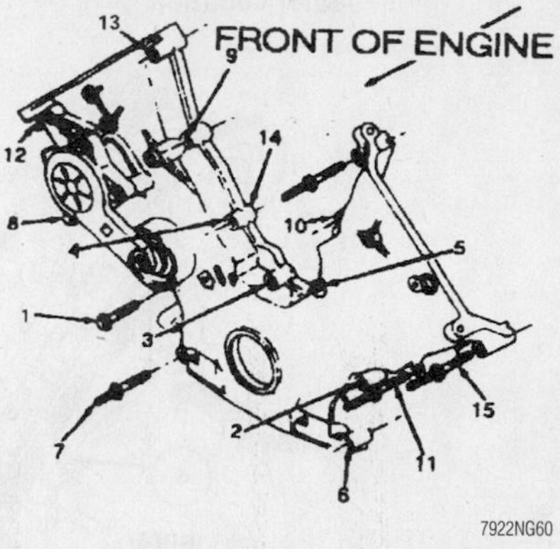

Front cover torque sequence—4.6L DOHC engines

    - Right primary timing chain tensioner arm and chain guide
    - Right primary timing chain tensioner
19. Tighten the camshaft sprocket bolts to 81–95 ft. lbs. (110–130 Nm).
20. Remove the primary timing chain tensioner locking pins.
21. Remove the Camshaft Holding and Camshaft Positioning tools.
22. Install the CMP pulse wheel
23. Install the front cover. Tighten the bolts in sequence as follows:
    a. Step 1: 14 ft. lbs. (20 Nm).
    b. Step 2: Plus 60 degrees.
24. Install or connect the following:
    - Idler pulley
    - Rocker arms in their original positions
    - Valve covers
    - Front crankshaft seal
    - CKP sensor connector
    - CMP sensor connector
    - Crankshaft pulley
    - Power steering pump
    - Ignition coil brackets, on 1998 engines
    - Ignition coils
    - Power steering pump reservoir
    - Water pump pulley
    - Accessory drive belt
    - Fuel charging wiring retainer
    - Engine control sensor wiring support bracket
    - Cooling fan and shroud
    - Upper radiator hose and bypass tube
    - Air intake tube
    - Engine compartment brace
    - Negative battery cable
25. Fill the cooling system.
26. Start the engine. Check for leaks and proper operation.

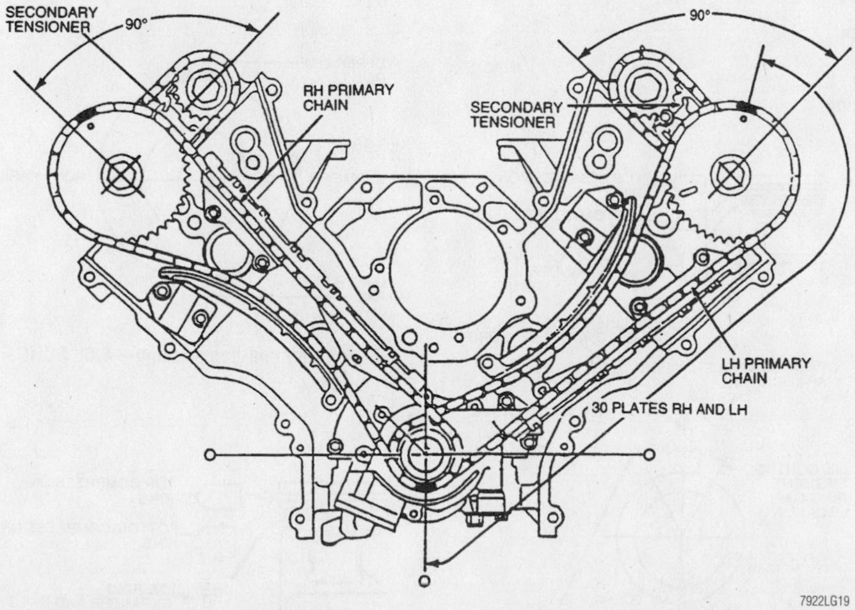

Correct alignment of the primary timing chains and sprockets—4.6L DOHC engines

Sealer Location

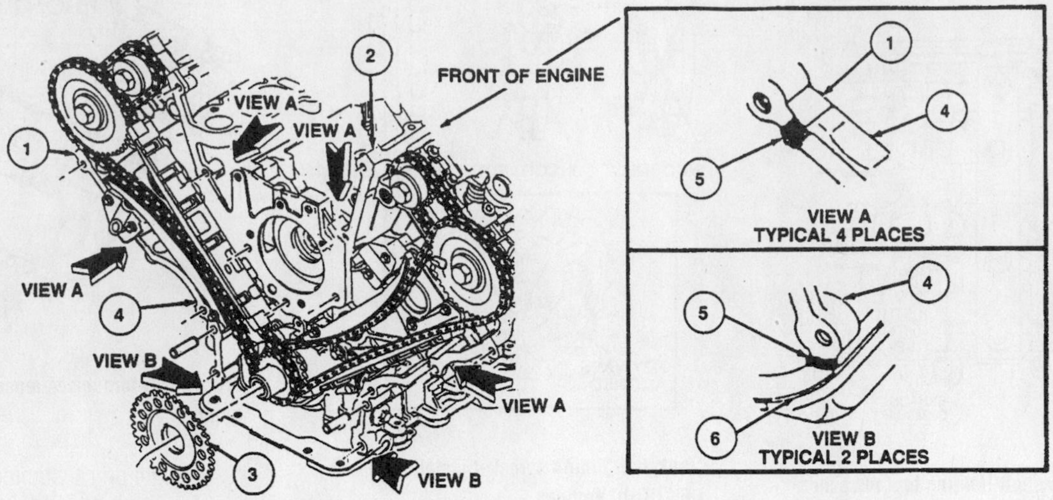

1 RH Cylinder Head
2 LH Cylinder Head
3 Ignition Pulse Crankshaft Sensor Ring
4 Cylinder Block
5 Sealer
6 Oil Pan Gasket

Apply sealer to these locations—4.6L DOHC engines

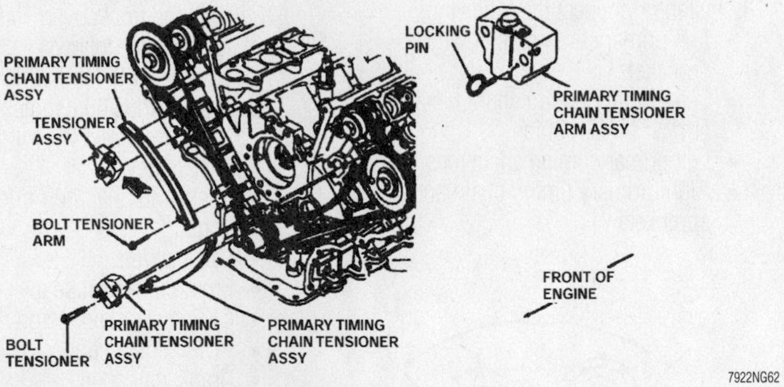

Timing chain tensioner installation—4.6L DOHC engine

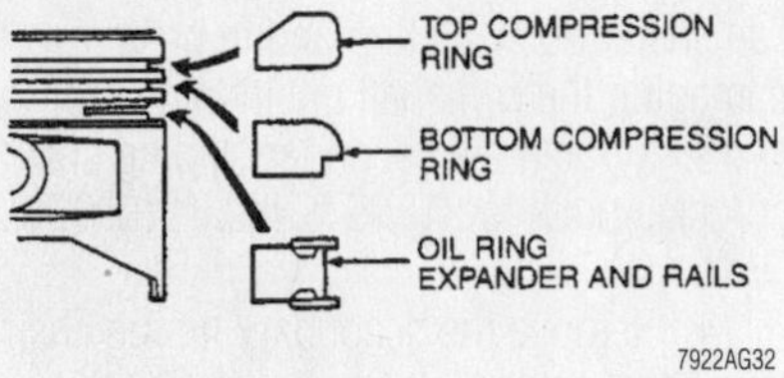

Piston ring positioning—4.6L SOHC engine

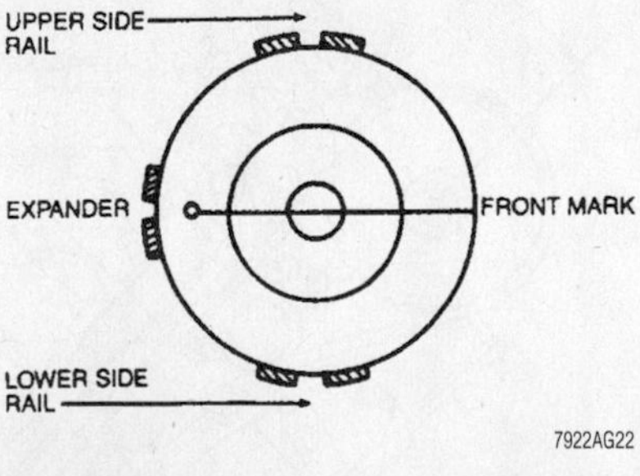

Piston ring end-gap spacing—4.6L SOHC engine

Piston and Ring

POSITIONING

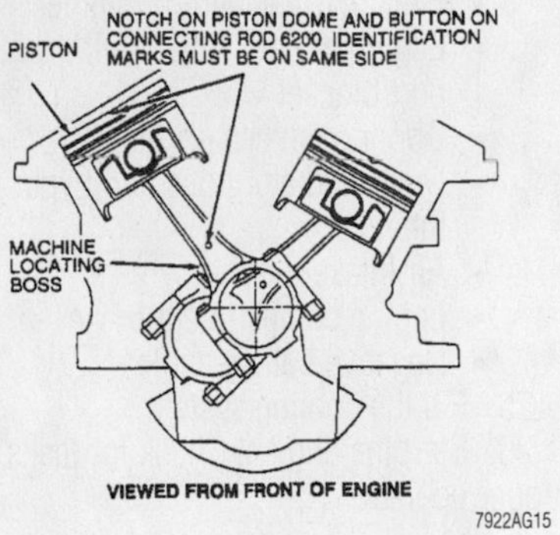

Piston and connecting rod positioning—3.8L engine

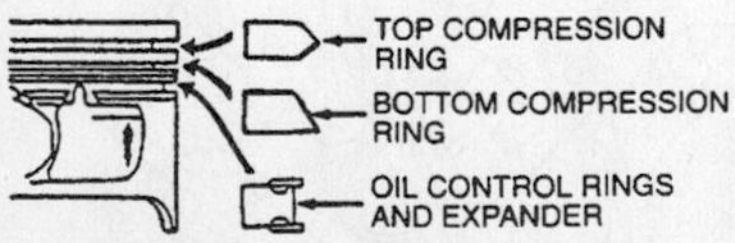

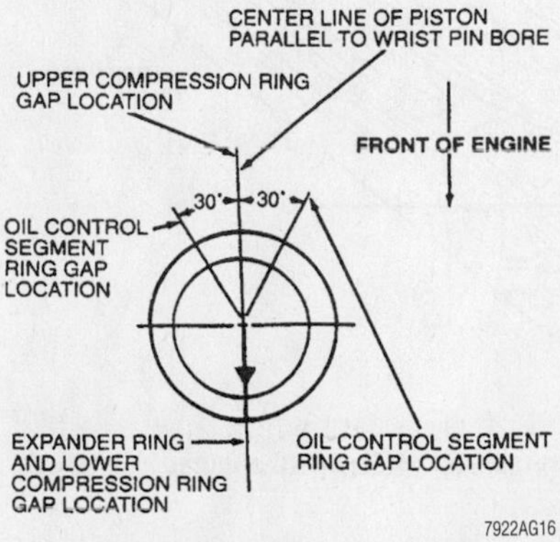

Piston ring end-gap spacing—3.8L engine

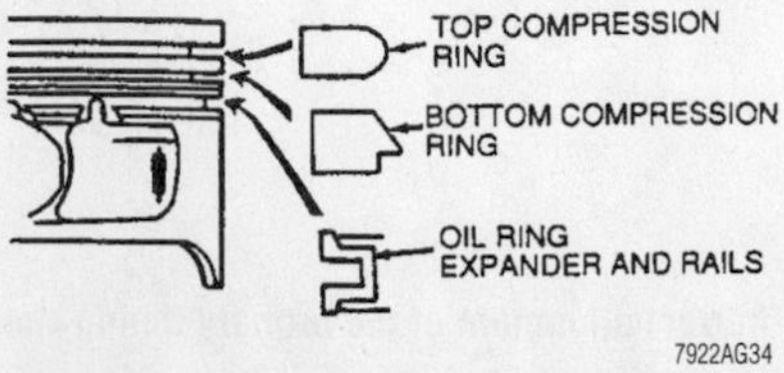

Piston ring positioning—4.6L DOHC engine

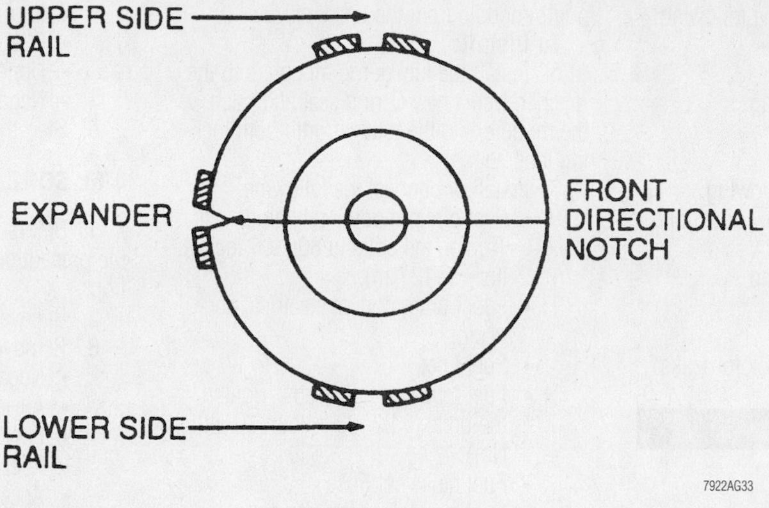

Piston ring end-gap spacing—4.6L DOHC engine

FUEL SYSTEM

Fuel System Service Precautions

Safety is the most important factor when performing not only fuel system maintenance, but any type of maintenance. Failure to conduct maintenance and repairs in a safe manner may result in serious personal injury or death. Work on a vehicle's fuel system components can be accomplished safely and effectively by adhering to the following rules and guidelines.

- To avoid the possibility of fire and personal injury, always disconnect the negative battery cable unless the repair or test procedure requires that battery voltage be applied.
- Always relieve the fuel system pressure prior to disconnecting any fuel system component (injector, fuel rail, pressure regulator, etc.) fitting or fuel line connection. Exercise extreme caution whenever relieving fuel system pressure to avoid exposing skin, face and eyes to fuel spray. Please be advised that fuel under pressure may penetrate the skin or any part of the body that it contacts.
- Always place a shop towel or cloth around the fitting or connection prior to loosening to absorb any excess fuel due to spillage. Ensure that all fuel spillage is quickly removed from engine surfaces. Ensure that all fuel-soaked cloths or towels are deposited into a flame-proof waste container with a lid.
- Always keep a dry chemical (Class B) fire extinguisher near the work area.
- Do not allow fuel spray or fuel vapors to come into contact with a spark or open flame.
- Always use a second wrench when loosening or tightening fuel line connection fittings. This will prevent unnecessary stress and torsion on fuel piping. Always follow the proper torque specifications.
- Always replace worn fuel fitting O-rings with new ones. Do not substitute fuel hose where rigid pipe is installed.

Fuel System Pressure

RELIEVING

All Sequential Fuel Injection (SFI) engines are equipped with a pressure relief valve located on the fuel supply manifold. Remove the fuel tank cap and attach fuel pressure gauge T80L-9974-B, or equivalent, to the valve on the fuel rail. Place the tube from the tool into a small container and open the valve on the tool to release the fuel pressure. Be sure to drain the fuel into a suitable container and to avoid gasoline spillage. If a pressure gauge is not available, disconnect the vacuum hose from the fuel pressure regulator and attach a hand-held vacuum pump. Apply about 25 in. Hg (84 kPa) of vacuum to the regulator to vent the fuel system pressure into the fuel tank through the fuel return hose.

➡This procedure will remove the fuel pressure from the lines, but not the fuel. Take precautions to avoid the risk of fire and use clean rags to soak up any spilled fuel when the lines are disconnected.

Fuel Filter

REMOVAL & INSTALLATION

➡Always replace fuel line fitting plastic clips.

1. Before servicing the vehicle, refer to the precautions in the beginning of this section.

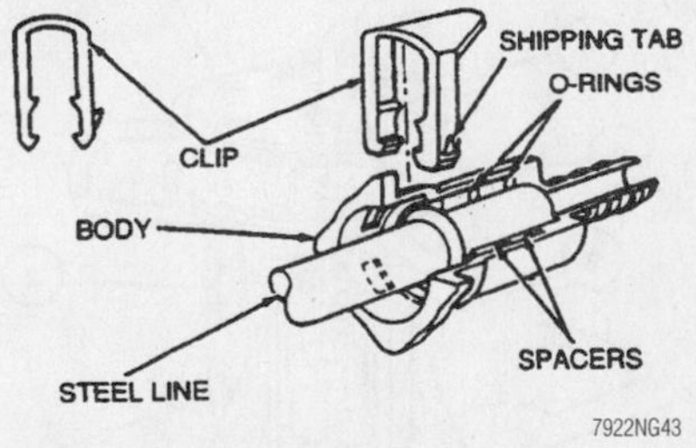

Hairpin clip fuel fitting

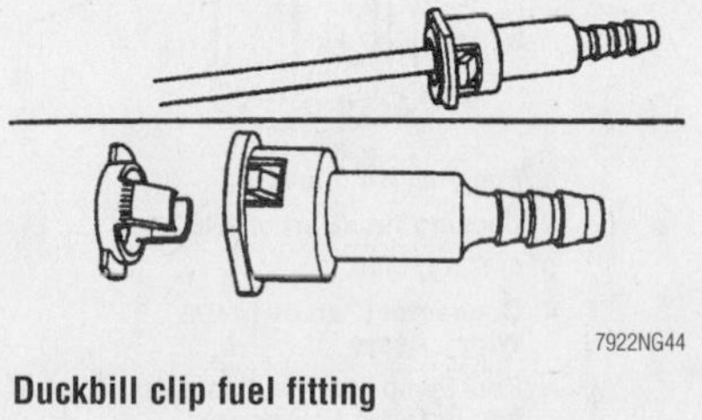

Duckbill clip fuel fitting

2. Relieve fuel system pressure
3. Remove or disconnect the following:
   - Negative battery cable
   - Fuel line fittings
   - Fuel filter bracket clamp
   - Fuel filter

To install:

4. Install or connect the following:
   - Fuel filter. Note the flow direction arrow.
   - Fuel filter bracket clamp
   - Fuel line fittings
   - Negative battery cable
5. Start the engine and check for leaks.

Fuel Pump

REMOVAL & INSTALLATION

1. Before servicing the vehicle, refer to the precautions in the beginning of this section.
2. Relieve fuel system pressure.
3. Drain the fuel tank.
4. Remove or disconnect the following:
   - Negative battery cable
   - Fuel tank filler pipe retainer
   - Fuel tank vent hose
   - Fuel tank support straps
   - Fuel lines
   - Fuel pump module electrical connector
   - Fuel pump module retaining bolts
5. Raise the fuel pump module until the locking tabs are accessible. Squeeze the locking tabs together and remove the fuel pump module from the fuel tank.

To install:

6. Install the fuel pump module into the retainer. Use a new O-ring seal and push the module into the retainer until both locking tabs engage.
7. Install or connect the following:
   - Fuel pump module retaining bolts. Tighten the bolts to 80–106 inch lbs. (9–12 Nm).
   - Fuel pump module electrical connector
   - Fuel lines
   - Fuel tank support straps. Tighten the bolts to 22–29 ft. lbs. (30–40 Nm).
   - Fuel tank vent hose
   - Fuel tank filler pipe retainer
   - Negative battery cable
8. Add fuel (10 gallons minimum) to the tank.
9. Start the engine and check for leaks.

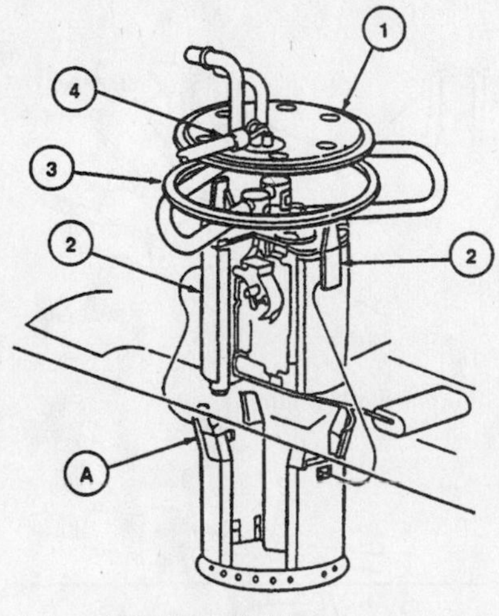

1 Fuel Pump Module
2 Locking Tab (Part of 9H307)
3 O-Ring Seal
4 Connector (Part of 14405 Wiring Assy)
A Fuel Pump Must Be Snapped Into Retainer (2 Places)

9306NG08

Fuel pump module assembly

Fuel Injector

REMOVAL & INSTALLATION

3.8L Engine

1. Before servicing the vehicle, refer to the precautions in the beginning of this section.
2. Relieve fuel system pressure.
3. Remove or disconnect the following:
   - Negative battery cable
   - Upper intake manifold
   - Fuel supply line
   - Fuel return line, if equipped
   - Fuel injector connectors
   - Fuel pressure regulator vacuum hose, on 1998 engines
   - Fuel pressure sensor vacuum hose and electrical connector, on 1999–01 engines
   - Fuel supply manifold with injectors attached
   - Fuel injectors from the supply manifold

To install:

4. Install or connect the following:
   - New O-rings
   - Fuel injectors to the supply manifold
   - Fuel supply manifold with injectors. Tighten the bolts to 80 inch lbs. (9 Nm).
   - Fuel pressure regulator vacuum hose, on 1998 engines
   - Fuel pressure sensor vacuum hose and electrical connector, on 1999–01 engines
   - Fuel injector connectors
   - Fuel supply line
   - Fuel return line, if equipped
   - Upper intake manifold
   - Negative battery cable
5. Start the engine and check for leaks.

4.6L SOHC Engine

1. Before servicing the vehicle, refer to the precautions in the beginning of this section.
2. Relieve fuel system pressure.
3. Remove or disconnect the following:
   - Negative battery cable
   - Air intake tube
   - Fuel supply line
   - Fuel return line, if equipped
   - Accelerator cable
   - Cruise control cable
   - Throttle body
   - Idle Air Control (IAC) valve connector and hose
   - Positive Crankcase Ventilation (PCV) hose
   - Main chassis vacuum supply hose
   - Fuel injector connectors
   - Fuel pressure regulator vacuum hose, on 1998 engines
   - Fuel pressure sensor vacuum hose and electrical connector, on 1999–01 engines
   - Exhaust Vacuum Regulator (EVR) solenoid vacuum lines
   - Exhaust Gas Recirculation (EGR) pressure transducer bracket
   - EGR tube and vacuum line
   - Fuel supply manifold with injectors attached
   - Fuel injectors from the supply manifold

To install:

4. Install or connect the following:
   - New O-rings
   - Fuel injectors to the supply manifold
   - Fuel supply manifold with injectors attached. Tighten the bolts to 80 inch lbs. (9 Nm).
   - EGR tube and vacuum line
   - EGR pressure transducer bracket
   - EVR solenoid vacuum lines
   - Fuel pressure regulator vacuum hose, on 1998 engines
   - Fuel pressure sensor vacuum hose and electrical connector, on 1999–01 engines
   - Fuel injector connectors
   - Main chassis vacuum supply hose
   - PCV hose
   - IAC valve connector and hose

- Throttle body
- Cruise control cable
- Accelerator cable
- Fuel supply line
- Fuel return line, if equipped
- Air intake tube
- Negative battery cable

5. Start the engine and check for leaks.

4.6L DOHC Engine

1. Before servicing the vehicle, refer to the precautions in the beginning of this section.
2. Relieve fuel system pressure.
3. Remove or disconnect the following:
   - Negative battery cable
   - Upper intake manifold
   - Fuel supply line
   - Fuel return line, if equipped
   - Fuel injector connectors
   - Fuel pressure regulator vacuum hose, on 1998 engines
   - Fuel pressure sensor vacuum hose and electrical connector, on 1999–01 engines
   - Fuel supply manifold with injectors attached
   - Fuel injectors from the supply manifold

To install:

4. Install or connect the following:
   - New O-rings
   - Fuel injectors to the supply manifold
   - Fuel supply manifold with injectors attached. Tighten the bolts to 80 inch lbs. (9 Nm).
   - Fuel pressure regulator vacuum hose, on 1998 engines
   - Fuel pressure sensor vacuum hose and electrical connector, on 1999–01 engines
   - Fuel injector connectors
   - Fuel supply line
   - Fuel return line, if equipped
   - Upper intake manifold
   - Negative battery cable
5. Start the engine and check for leaks.

DRIVE TRAIN

Transmission

REMOVAL & INSTALLATION

Manual

➡The clutch pedal must be held in the uppermost position during clutch release cable removal and installation. Failure to properly position and support the clutch pedal can result in damage to the self-adjusting mechanism.

1. Before servicing the vehicle, refer to the precautions in the beginning of this section.
2. Lift the clutch pedal and secure in place with safety wire.
3. Remove or disconnect the following:
   - Negative battery cable
   - Shift lever and boot
   - Driveshaft
   - Clutch cable
   - Heated Oxygen (HO_2S) sensor connectors
   - Dual converter "Y" or "H" pipe
   - Backup lamp switch connector
   - Vehicle Speed (VSS) sensor connector
   - Flywheel cover
   - Starter motor
4. Support the transmission with a jack and remove the rear transmission support crossmember.
5. Lower the transmission jack and remove the transmission flange bolts.
6. Slide the transmission input shaft out of the clutch and lower the transmission away from the vehicle.

To install:

7. Install or connect the following:
   - Transmission. Tighten the flange bolts to 55 ft. lbs. (75 Nm).
   - Transmission crossmember. Tighten the transmission mount bolts to 43 ft. lbs. (58 Nm) and the crossmember bolts to 30 ft. lbs. (41 Nm).
   - Starter motor. Tighten the bolts to 17 ft. lbs. (23 Nm).
   - Flywheel cover. Tighten the bolts to 20 ft. lbs. (27 Nm).
   - VSS sensor connector
   - Backup lamp switch connector
   - Dual converter "Y" or "H" pipe
   - HO_2S sensor connectors
   - Clutch cable
   - Driveshaft
   - Shift lever and boot
   - Negative battery cable
8. Remove the safety wire from the clutch pedal and check for proper operation.

Automatic

1. Before servicing the vehicle, refer to the precautions in the beginning of this section.
2. Drain the transmission fluid.
3. Remove the torque converter access cover and drain the torque converter.
4. Remove or disconnect the following:
   - Negative battery cable
   - Heated Oxygen (HO_2S) sensor connectors
   - Dual converter "Y" or "H" pipe
   - Torque converter
   - Driveshaft
   - Shift cable
   - Transmission wiring connectors
   - Transmission fluid cooler lines
   - Starter motor
5. Support the transmission with a jack and remove the rear transmission support crossmember.
6. Lower the transmission jack and remove the transmission flange bolts.
7. Lower the transmission away from the vehicle.

To install:

8. Install or connect the following:
   - Transmission. Tighten the flange bolts to 41–50 ft. lbs. (55–68 Nm).
   - Transmission crossmember. Tighten the transmission mount bolts to 72 ft. lbs. (98 Nm) and the crossmember bolts to 41 ft. lbs. (55 Nm).
   - Starter motor. Tighten the bolts to 18 ft. lbs. (25 Nm).
   - Transmission fluid cooler lines
   - Transmission wiring connectors
   - Shift cable
   - Driveshaft
   - Torque converter. Tighten the nuts to 20–33 ft. lbs. (27–46 Nm).
   - Torque converter access cover. Tighten the bolts to 12–16 ft. lbs. (16–22 Nm).
   - Dual converter "Y" or "H" pipe
   - HO_2S sensor connectors
   - Negative battery cable
9. Fill the transmission with f
10. Start the engine. Check proper operation.

Clutch

ADJUSTMENTS

The clutch is equipped with a self-adjusting mechanism. Pull the clutch pedal up to activate the adjuster.

REMOVAL & INSTALLATION

➡The clutch pedal must be held in the uppermost position during clutch release cable removal and installation. Failure to properly position and support the clutch pedal can result in damage to the self-adjusting mechanism.

1. Before servicing the vehicle, refer to the precautions in the beginning of this section.
2. Lift the clutch pedal and secure in place with safety wire.
3. Remove or disconnect the following:
   - Transmission

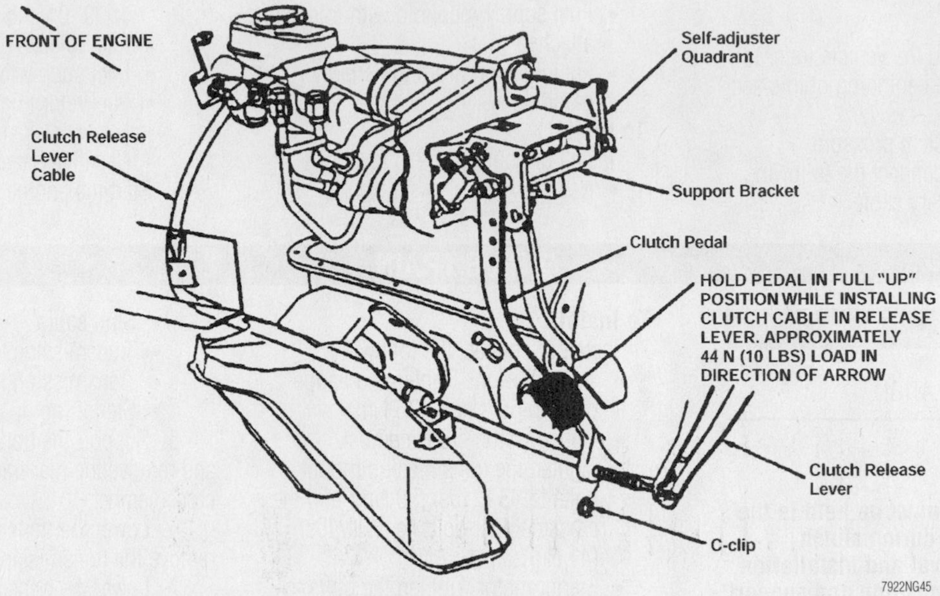

Clutch self-adjusting system component identification—4.6L engine—3.8L engine is similar

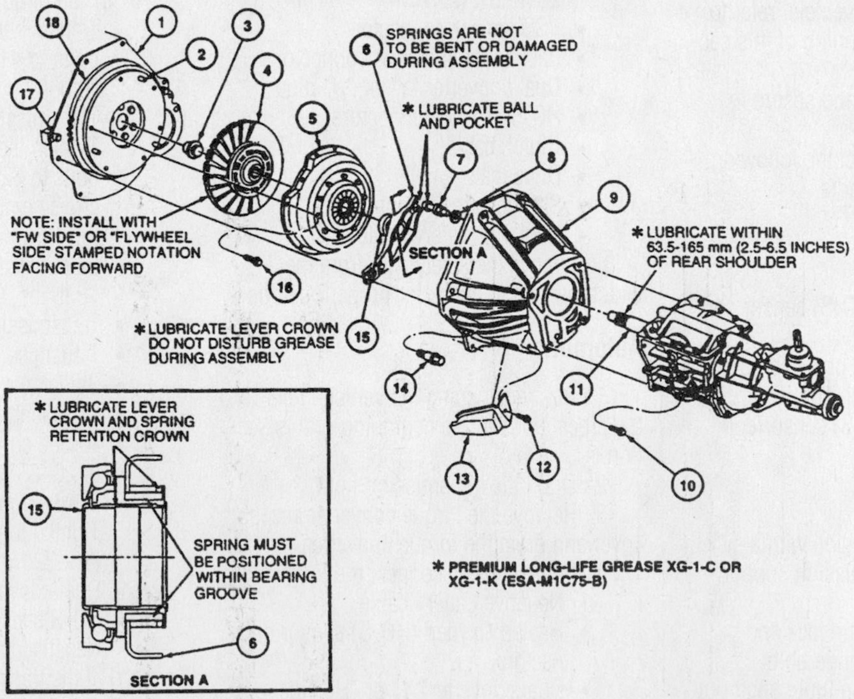

1. Rear face of engine block
2. Flywheel-to-clutch cover alignment dowel
3. Pilot bearing
4. Clutch disc
5. Clutch pleasure plate
6. Clutch release lever
7. Clutch release lever stud
8. Lockwasher
9. Flywheel housing
10. Bolt
11. Input shaft
12. Bolt
13. Clutch release lever dust shield
14. Bolt
15. Clutch release bearing
16. Bolt
17. Flywheel housing-to-engine block dowel

7922NG46

ded view of the clutch system components

- Pressure plate bolts, loosen them evenly in several passes to avoid distortion of the pressure plate.
- Pressure plate and clutch disk

To install:

4. Install or connect the following:
    - Clutch disk and pressure plate. Tighten the pressure plate bolts evenly in several passes to 20–28 ft. lbs. (27–39 Nm) for 3.8L engines or to 19–24 ft. lbs. (25–33 Nm) for 4.6L engines.
    - Transmission

5. Remove the clutch pedal safety wire and check for proper operation.

Halfshaft

REMOVAL & INSTALLATION

SVT Cobra

1. Before servicing the vehicle, refer to the precautions in the beginning of this section.
2. Before raising the vehicle, matchmark the rear shock absorber to indicate curb height.
3. Remove or disconnect the following:
    - Rear wheel
    - Hub retainer nut
    - Disc brake caliper and rotor
    - Wheel speed sensor
    - Tie rod link

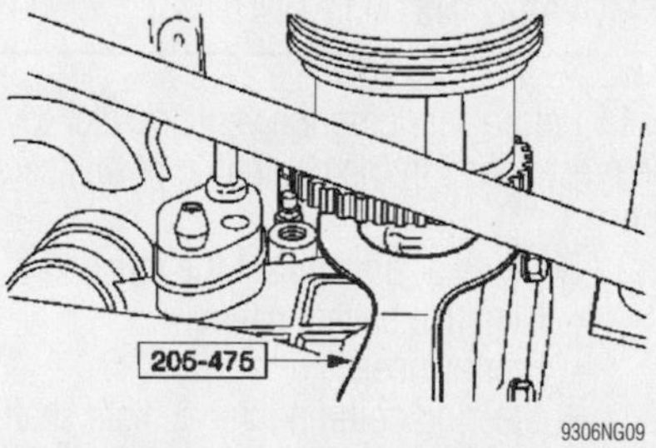

9306NG09

Halfshaft Removal Tool—SVT Cobra

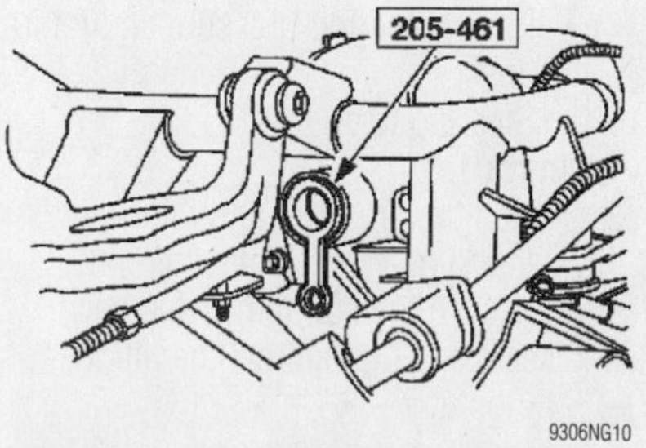

9306NG10

Differential Seal Protector—SVT Cobra

4. Place a support under the lower control arm
5. Remove or disconnect the following:
    - Lower shock absorber mounting bolt
    - Upper control arm from the knuckle
6. Press the outboard CV-joint from the hub. Lower the knuckle until the stub shaft clears the hub.
7. Use Halfshaft Removal Tool 205-475 to remove the halfshaft from the differential.

To install:

➡Use new nuts, bolts, circlips, and split pins for assembly.

8. Install Differential Seal Protector 205-461.
9. Install the axle inner stub shaft into the differential until the splines are past the seal.
10. Remove the Differential Seal Protector.
11. Install or connect the following:
    - Inner stub shaft into the differential until the circlip seats
    - Outer stub shaft into the knuckle
    - Upper control arm to the knuckle and install the lower shock absorber bolt
    - Lower control arm to align the matchmark on the shock absorber
    - Lower shock absorber mounting bolt. Torque it to 98 ft. lbs. (133 Nm).
    - Upper control arm bolt. Torque it to 66 ft. lbs. (90 Nm).
12. Remove the lower control arm support.
13. Install or connect the following:
    - Tie rod link. Tighten the nut to 35 ft. lbs. (47 Nm).
    - Wheel speed sensor
    - Disc brake rotor and caliper. Tighten the caliper support bolts to 76 ft. lbs. (103 Nm).

➡The hub retainer nut must be tightened with the brakes applied and the wheels off the ground to ensure correct bearing seating.

- Hub nut. Tighten it to 240 ft. lbs. (325 Nm).
- Wheel

14. Check the rear wheel alignment and adjust as necessary.

CV-Joints

REMOVAL & REPLACEMENT

Inner CV-Joint

1. Before servicing the vehicle, refer to the precautions in the beginning of this section.
2. Remove or disconnect the following:
    - Halfshaft
    - Inner CV-joint boot clamps
    - CV-joint boot, slide it away from the joint
    - Snapring, release it
    - Inner CV-joint from the halfshaft.

To install:

3. CV-joint, fill it with fresh grease and slide it onto the halfshaft.
4. Install or connect the following:
    - New snapring
    - New CV-joint clamps
    - CV-joint boot
    - Halfshaft

Outer CV-Joint

The outer CV-joint is serviced with the halfshaft as an assembly. The outer CV-joint boot can be serviced by removing the inner CV-joint.

Axle Shaft, Bearing and Seal

REMOVAL & INSTALLATION

1. Before servicing the vehicle, refer to the precautions in the beginning of this section.
2. Remove or disconnect the following:
    - Rear wheel
    - Disc brake caliper and rotor
    - Wheel speed sensor
    - Axle housing cover
    - Differential pinion shaft
    - Axle retaining U-washer
    - Axle shaft
    - Bearing and seal, using a slide hammer

To install:

3. Install or connect the following:
    - Bearing, so that it is fully seated in the axle tube
    - Axle seal
    - Axle shaft
    - Axle retaining U-washer
    - Differential pinion shaft. Tighten the

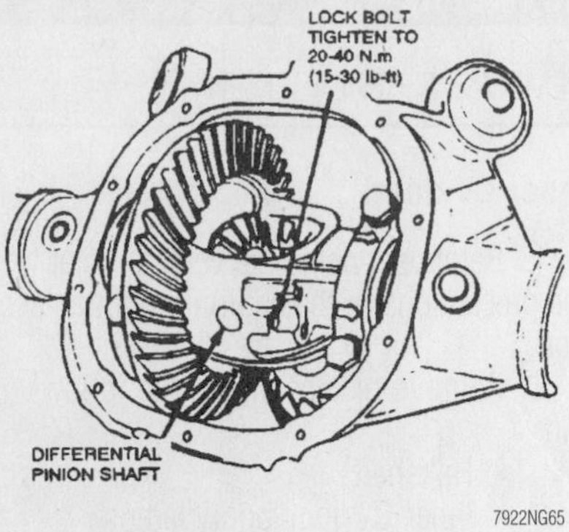

Differential pinion shaft and lockbolt

lockbolt to 15–30 ft. lbs. (20–41 Nm).
- Axle housing cover. Tighten the cover bolts to 18–28 ft. lbs. (24–38 Nm).
- Wheel speed sensor
- Disc brake rotor and caliper. Tighten the caliper mounting bolts to 65–87 ft. lbs. (87–119 Nm).
- Rear wheel

4. Fill the differential with gear lubricant. Tighten the filler plug to 15–30 ft. lbs. (20–41 Nm).

Pinion Seal

REMOVAL & INSTALLATION

1. Before servicing the vehicle, refer to the precautions in the beginning of this section.
2. Remove or disconnect the following:
   - Driveshaft
   - Rear wheels
   - Rear brake calipers

➡The rear brake calipers must be removed so that there is no additional drag when measuring pinion bearing preload.

3. Use an inch lb. torque wrench and measure the amount of torque required to maintain pinion rotation through several revolutions.
4. Remove the pinion flange and remove the seal.

To install:

5. Install or connect the following:
   - Pinion seal and flange
   - New pinion flange nut
6. Rotate the pinion flange occasionally while tightening the flange nut to make sure the pinion bearings seat correctly.
7. Take frequent bearing preload torque readings.
8. If the preload recorded prior to disassembly is **lower** than the specification for used bearings, then tighten the pinion flange nut to specification. If the preload recorder prior to disassembly is **higher** than the specification for used bearings, then tighten the pinion flange nut to the original reading as recorded.
9. The pinion bearing preload specifications are as follows:
   a. Used bearings: 8–14 inch lbs. (0.9–1.6 Nm).
   b. New bearings: 16–29 inch lbs. (1.8–3.2 Nm).

⁂ CAUTION

Never loosen the pinion nut to reduce bearing preload. If it is necessary to reduce bearing preload, install a new collapsible spacer and pinion nut.

10. Install or connect the following:
    - Driveshaft
    - Brake calipers
    - Rear wheels
11. Fill the differential with gear lubricant and check for leaks.

STEERING AND SUSPENSION

Air Bag

⁂ CAUTION

Some vehicles are equipped with an air bag system. The system must be disarmed before performing service on, or around, system components, the steering column, instrument panel components, wiring and sensors. Failure to follow the safety precautions and the disarming procedure could result in accidental air bag deployment, possible injury and unnecessary system repairs.

PRECAUTIONS

Several precautions must be observed when handling the inflator module to avoid accidental deployment and possible personal injury.

- Never carry the inflator module by the wires or connector on the underside of the module.
- When carrying a live inflator module, hold securely with both hands, and ensure that the bag and trim cover are pointed away.
- Place the inflator module on a bench or other surface with the bag and trim cover facing up.
- With the inflator module on the bench, never place anything on or close to the module that may be thrown in the event of an accidental deployment.

DISARMING

1. Before servicing the vehicle, refer to the precautions in the beginning of this section.
2. Disconnect both battery cables from the battery, negative cable first.
3. Wait 1 minute before proceeding with the service procedure. This is the time required for the back-up power supply in the air bag diagnostic monitor to deplete its stored energy.
4. After service is completed, reconnect the battery cables, negative cable last.
5. Turn the ignition switch to the **RUN** position. The air bag indicator should light continuously for approximately 6 seconds, then turn **OFF**. If the indicator fails to light, flashes or remains lit continuously, there is a fault in the air bag system.

Power Rack and Pinion Steering Gear

REMOVAL & INSTALLATION

1. Before servicing the vehicle, refer to the precautions in the beginning of this section.
2. Remove or disconnect the following:
   - Negative battery cable
   - Front wheels
   - Steering column intermediate shaft coupling
   - Outer tie rod ends
   - Steering gear retaining bolts
   - Power steering pressure and return lines
   - Steering gear

To install:

3. Attach the pressure and return lines to the steering gear before attaching the steering gear to the subframe. Use new plastic seals and tighten the line fittings to 20–25 ft. lbs. (27–34 Nm).

➡The power steering fluid lines are designed to swivel when properly tightened. Do not overtighten the fittings.

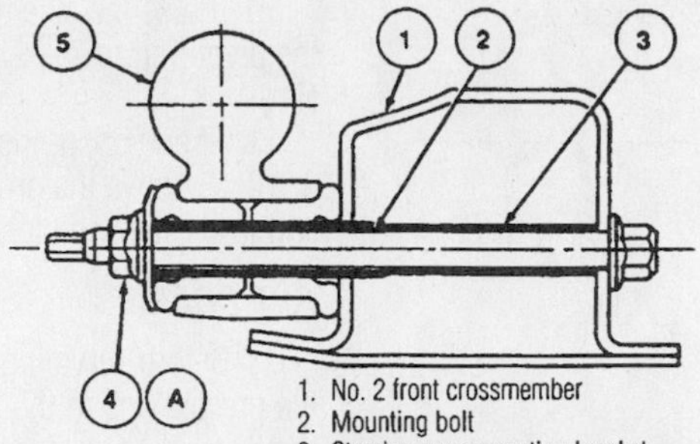

Power rack and pinion steering gear mounting on the No. 2 crossmember

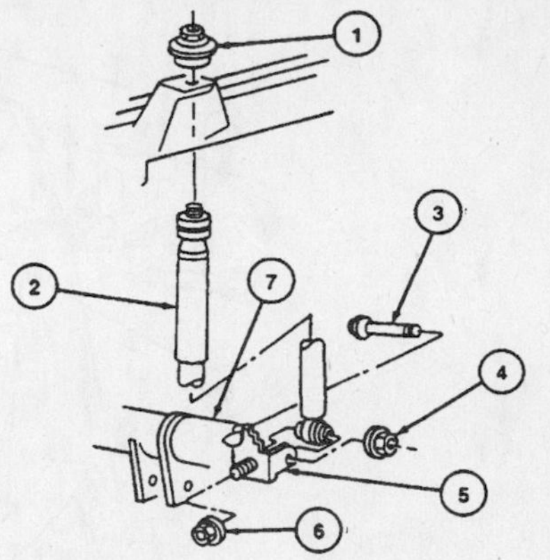

Exploded view of the rear shock absorber mounting

4. Install or connect the following:
   - Steering gear. Tighten the bolts to 31–39 ft. lbs. (41–54 Nm).
   - Outer tie rod ends. Tighten the nuts to 36–46 ft. lbs. (48–63 Nm).
   - Steering column intermediate shaft coupling. Tighten the pinch bolt to 21–29 ft. lbs. (28–40 Nm).
   - Front wheels
   - Negative battery cable
5. Fill the power steering system with the proper type and quantity of fluid.
6. Check the front end alignment and adjust as necessary.

Strut

REMOVAL & INSTALLATION

Front

1. Before servicing the vehicle, refer to the precautions in the beginning of this section.
2. Support the front of the vehicle on jackstands placed under the control arms.
3. Remove or disconnect the following:
   - Front wheel
   - Disc brake caliper
   - Wheel speed sensor and bracket
   - Upper strut retaining fasteners
   - Wheel spindle attachment bolts
4. Compress the strut assembly and remove it from the vehicle.

To install:

5. If replacing the strut, transfer the upper mounting bracket.
6. Install or connect the following:
   - Strut assembly. Tighten the spindle bolts to 141–191 ft. lbs. (191–259 Nm).
   - Upper strut retaining fasteners. Tighten to 25–34 ft. lbs. (34–46 Nm).

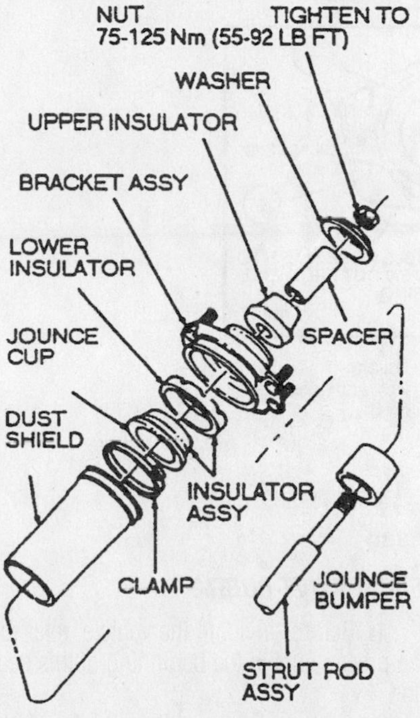

Exploded view of the front strut upper mounting

   - Wheel speed sensor and bracket
   - Disc brake caliper. Tighten the caliper mounting bolts to 96 ft. lbs. (130 Nm).
   - Front wheel
7. Remove the jackstands and check the alignment. Adjust as necessary.

Shock Absorber

REMOVAL & INSTALLATION

Rear

1. Before servicing the vehicle, refer to the precautions in the beginning of this section.
2. Remove or disconnect the following:
   - Rear compartment trim panels
   - Upper shock absorber retaining nut
   - Lower shock absorber bolt
   - Shock absorber

To install:

3. Prime the new shock absorber as follows:
   a. Step 1: With the shock absorber right side up, extend it fully.
   b. Step 2: Turn the shock absorber upside down and fully compress it.
   c. Step 3: Repeat for 3 cycles.
4. Install the shock absorber and tighten the lower bolt as follows:
   a. SVT Cobra: 98 ft. lbs. (133 Nm).
   b. All others: 57–75 ft. lbs. (76–103 Nm).
5. Install or connect the following:
   - Upper shock absorber nut. Tighten it to 25–33 ft. lbs. (34–46 Nm).
   - Rear compartment trim panels

Coil Spring

REMOVAL & INSTALLATION

Front

1. Before servicing the vehicle, refer to the precautions in the beginning of this section.
2. Install an internal spring compressor to the spring to be serviced. Tighten the compressor to relieve spring pressure on the lower control arm.
3. Remove or disconnect the following:

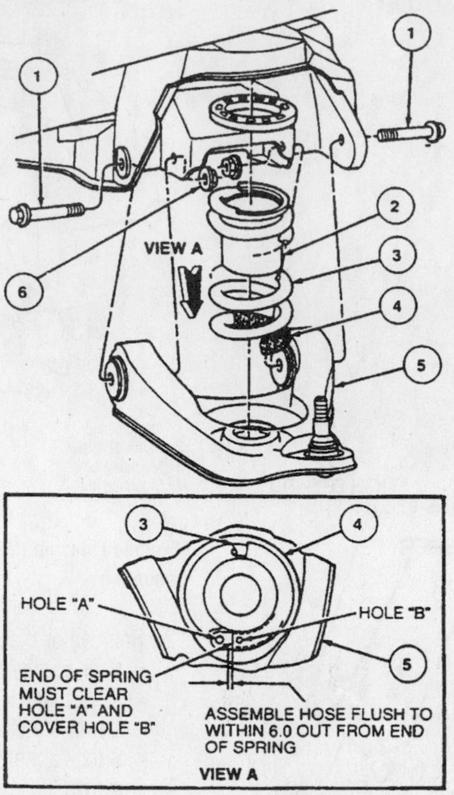

Exploded view of the front coil spring mounting

- Front wheel
- Disc brake caliper
- Outer tie rod end
- Stabilizer bar link
- Power steering gear
- Lower control arm mounting bolts
- Coil spring

To install:

4. If replacing the coil spring, transfer the spring compressor to the new spring.
5. Install or connect the following:
   - Coil spring
   - Lower control arm mounting bolts. Use a jack to raise the control arm to a normal position and tighten the bolts to 141–191 ft. lbs. (191–259 Nm).
   - Power steering gear
   - Stabilizer bar link. Tighten the nut to 11–16 ft. lbs. (16–22 Nm).
   - Outer tie rod end. Tighten the nut to 36–46 ft. lbs. (48–63 Nm).
   - Disc brake caliper. Tighten the caliper mounting bolts to 96 ft. lbs. (130 Nm).
   - Front wheel
6. Position the coil spring and remove the spring compressor.

Rear

EXCEPT SVT COBRA

1. Before servicing the vehicle, refer to the precautions in the beginning of this section.
2. Raise and support the vehicle safely under the frame. Support the body at the rear body crossmember.
3. If equipped, remove the stabilizer bar.
4. Support the axle with a jack.
5. Place another jack under the lower arm axle pivot bolt. Remove and discard the bolt and nut. Lower the jack slowly until the coil spring load is relieved.
6. Remove the coil spring and insulator from the vehicle.

To install:

7. Place the upper spring insulator on top of the spring. Place the lower spring insulator on the lower arm.
8. Position the coil spring on the lower arm spring seat with the pigtail on the lower arm at the rear of the vehicle and pointing toward the left side of the vehicle.
9. Slowly raise the jack until the arm is in position. Insert a new rear pivot bolt and nut.
10. Raise the axle to curb height. Tighten the pivot bolt to 71–97 ft. lbs. (97–132 Nm).
11. If equipped, install the stabilizer bar.
12. Remove the crossmember supports and lower the vehicle.

SVT COBRA

1. Before servicing the vehicle, refer to the precautions in the beginning of this section.
2. Support the rear subframe with a jack.
3. Remove or disconnect the following:
   - Rear wheels
   - Mufflers
   - Driveshaft
   - Parking brake cables and brackets
   - Brake fluid lines
   - Wheel speed sensors
   - Tie rod links
   - Shock absorbers
   - Lower control arms
4. Loosen the front subframe bolts and remove the rear subframe bolts.
5. Lower the rear subframe and remove the coil springs

To install:

➡Use new nuts, bolts and split pins for assembly.

6. Install the coil springs and raise the rear subframe. Install new subframe bolts and tighten them to 76 ft. lbs. (103 Nm).
7. Install or connect the following:
   - Lower control arms
   - Shock absorbers. Tighten the bolts to 98 ft. lbs. (133 Nm).
   - Tie rod links
   - Wheel speed sensors
   - Brake fluid lines
   - Parking brake cables and brackets
   - Driveshaft
   - Mufflers
   - Rear wheels
8. Check the wheel alignment and adjust as necessary.

Lower Ball Joint

REMOVAL & INSTALLATION

1. Before servicing the vehicle, refer to the precautions in the beginning of this section.
2. Remove or disconnect the following:
   - Front wheel
   - Disc brake caliper and rotor
   - Wheel speed sensor
   - Outer tie rod end
3. Support the lower control arm with a jackstand and remove the spindle.

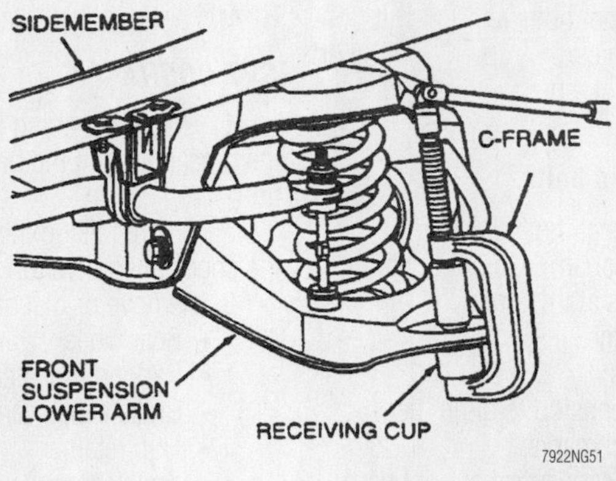

Use the C-clamp, cup and adapters to press the ball joint out of the lower control arm

4. Press the lower ball joint out of the control arm.

To install:

5. Press the lower ball joint into the control arm so that the joint is fully seated in the control arm.
6. Install or connect the following:
   - Spindle. Tighten the ball joint nut to 109–149 ft. lbs. (148–202 Nm). Tighten the strut bolts to 141–191 ft. lbs. (190–259 Nm).
   - Outer tie rod end. Tighten the nut to 36–46 ft. lbs. (48–63 Nm).
   - Disc brake rotor and caliper. Tighten the caliper mounting bolts to 96 ft. lbs. (130 Nm).
   - Wheel speed sensor
   - Front wheel

Upper Control Arm

REMOVAL & INSTALLATION

Rear

EXCEPT SVT COBRA

1. Before servicing the vehicle, refer to the precautions in the beginning of this section.
2. Before raising the vehicle, matchmark the shock absorbers to indicate curb height.
3. Support the axle housing with a jack.
4. Remove the upper control arms.

To install:

➡Use new nuts and bolts.

5. Install the upper control arms.
6. Align the matchmarks on the shock absorbers and tighten the control arm bolts as follows:
   a. Step 1: Tighten the mounting bracket bolt to 76 ft. lbs. (103 Nm).
   b. Step 2: Tighten the axle housing bolt to 66 ft. lbs. (90 Nm).

SVT COBRA

1. Before servicing the vehicle, refer to the precautions in the beginning of this section.
2. Before raising the vehicle, matchmark the shock absorbers to indicate curb height.
3. Remove the coil springs.
4. Remove the subframe front bolts and lower the subframe from the vehicle.
5. Matchmark the upper control arm cam bolt to the knuckle.
6. Remove the upper control arm.

To install:

➡Use new nuts and bolts.

7. Transfer the cam bolt matchmark to the new cam bolt.
8. Install the upper control arm. Do not tighten the fasteners at this time.
9. Install the coil springs.
10. Raise the suspension to align the shock absorber matchmarks and tighten the upper control arm inner bolts to 66 ft. lbs. (90 Nm).
11. Align the cam bolt matchmark and tighten the nut to 66 ft. lbs. (90 Nm).
12. Check the wheel alignment and adjust as necessary.

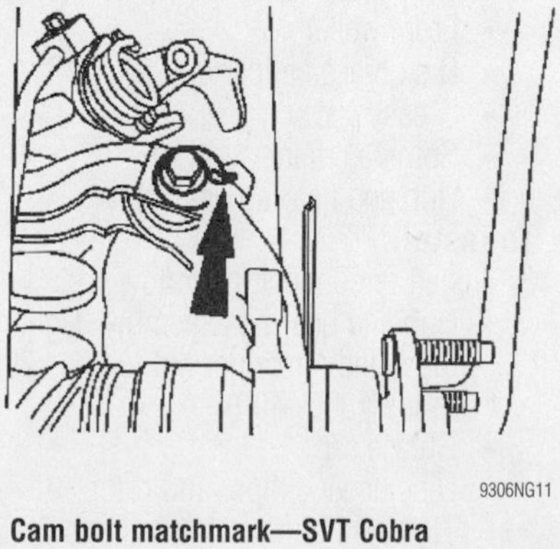

Cam bolt matchmark—SVT Cobra

CONTROL ARM BUSHING REPLACEMENT

Rear

EXCEPT SVT COBRA

The inboard control arm bushing is serviced with the control arm as an assembly.

1. Before servicing the vehicle, refer to the precautions in the beginning of this section.
2. Remove the upper control arm.
3. Press the bushing out of the axle housing.
4. Press a new bushing into the axle housing and install the upper control arm.

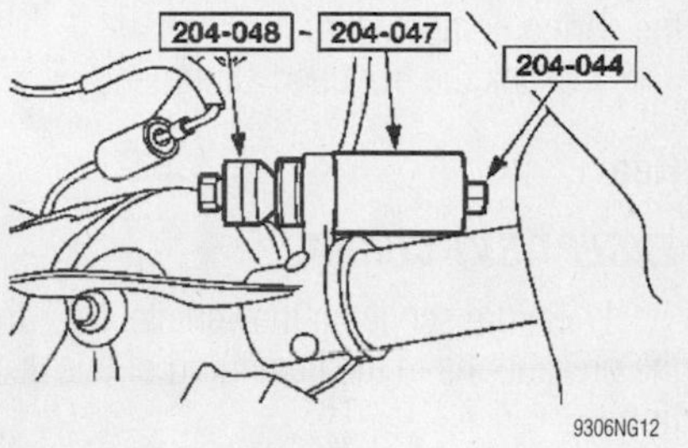

Axle bushing removal

SVT COBRA

The upper control arm bushing are serviced with the control arm as an assembly.

Lower Control Arm

REMOVAL & INSTALLATION

Front

1. Before servicing the vehicle, refer to the precautions in the beginning of this section.
2. Install an internal spring compressor to the coil spring. Tighten the compressor to relieve spring pressure on the lower control arm.
3. Remove or disconnect the following:
   - Front wheel
   - Disc brake caliper
   - Outer tie rod end
   - Stabilizer bar link
   - Power steering gear
   - Lower control arm mounting bolts
   - Coil spring
   - Lower ball joint
   - Lower control arm

To install:

4. Connect the lower ball joint to the spindle and install the nut hand tight.
5. Install or connect the following:
   - Coil spring

- Lower control arm mounting bolts. Use a jack to raise the control arm to curb height and tighten the bolts to 141–191 ft. lbs. (191–259 Nm). Tighten the ball joint nut to 109–149 ft. lbs. (148–202 Nm).
- Power steering gear
- Stabilizer bar link. Tighten the nut to 11–16 ft. lbs. (16–22 Nm).
- Outer tie rod end. Tighten the nut to 36–46 ft. lbs. (48–63 Nm).
- Disc brake caliper. Tighten the caliper mounting bolts to 96 ft. lbs. (130 Nm).
- Front wheel

6. Position the coil spring and remove the spring compressor.
7. Check the front end alignment.

Rear

EXCEPT SVT COBRA

1. Before servicing the vehicle, refer to the precautions in the beginning of this section.
2. Before raising the vehicle, matchmark the shock absorbers to indicate curb height.
3. Raise and support the vehicle safely under the frame. Support the body at the rear body crossmember.
4. If equipped, remove the stabilizer bar.
5. Support the axle with a jack.
6. Place another jack under the lower arm axle pivot bolt. Remove and discard the bolt and nut. Lower the jack slowly until the coil spring load is relieved.
7. Remove or disconnect the following:
   - Coil spring
   - Lower control arm

To install:

➡Use new nuts and bolts.

8. Install or connect the following:
   - Lower control arm to the axle housing
   - Coil spring
9. Raise the control arm and install the pivot bolt.
10. Raise the axle to align the matchmarks on the shock absorbers and tighten the lower control arm bolts to 71–97 ft. lbs. (97–132 Nm).
11. install the stabilizer bar, if equipped.

SVT COBRA

1. Before servicing the vehicle, refer to the precautions in the beginning of this section.
2. Before raising the vehicle, matchmark the shock absorbers to indicate curb height.
3. Remove or disconnect the following:
   - Coil springs. Refer to the coil spring procedure in this section.
   - Subframe front bolts and lower it from the vehicle
   - Lower control arm

To install:

➡Use new nuts and bolts.

4. Install or connect the following:
   - Lower control arm. Do not tighten the fasteners at this time.
   - Rear subframe
   - Coil springs
5. Raise the suspension to align the shock absorber matchmarks.
6. Tighten the lower control arm inboard bolts to 184 ft. lbs. (250 Nm) and the knuckle bolt to 85 ft. lbs. (115 Nm).
7. Check the wheel alignment and adjust as necessary.

CONTROL ARM BUSHING REPLACEMENT

Front and Rear

ALL MODELS

The lower control arm bushings are serviced with the lower control arm as an assembly.

Wheel Bearings

ADJUSTMENT

The front wheel bearings are an integral part of the hub assembly. They require no periodic maintenance or adjustment. If the bearings are found to be defective, they must be replaced along with the hub assembly.

The rear wheel bearings are not adjustable.

REMOVAL & INSTALLATION

Front

1. Before servicing the vehicle, refer to the precautions in the beginning of this section.
2. Remove or disconnect the following:
   - Front wheel
   - Disc brake caliper and rotor
   - Grease cap
   - Spindle retainer nut
   - Hub and bearing assembly

To install:

3. Install or connect the following:
   - Hub and bearing assembly. Tighten the spindle retainer nut to 221–295 ft. lbs. (300–400).
   - Grease cap
   - Disc brake caliper and rotor. Tighten the mounting bolts to 96 ft. lbs. (130 Nm).
   - Front wheel

Rear

SVT COBRA

1. Before servicing the vehicle, refer to the precautions in the beginning of this section.
2. Before raising the vehicle, matchmark the shock absorbers to indicate curb height.
3. Remove or disconnect the following:
   - Rear wheel
   - Parking brake cable
   - Brake caliper and disc
   - Hub retainer
   - Shock absorber
   - Tie rod link
4. Matchmark the upper control arm cam bolt to the knuckle.
5. Remove or disconnect the following:
   - Knuckle from the vehicle
   - Dust shield and press the hub out of the bearing
   - Snapring and press the bearing out of the knuckle

To install:

➡Use new nuts, bolts, snaprings, and split pins for assembly.

6. Install or connect the following:
   - Bearing so that it is fully seated in the knuckle bore
   - Snapring
7. Support the bearing inner race and press the hub into the bearing.
8. Install the dust shield and tighten the bolts to 88 inch lbs. (10 Nm).
9. Transfer the cam bolt matchmark to a new cam bolt.
10. Install or connect the following:
    - Knuckle. Do not tighten the fasteners at this time.
    - Tie rod link. Tighten the nut to 35 ft. lbs. (47 Nm).
    - Shock absorber. Tighten the bolt to 98 ft. lbs. (133 Nm).
    - Hub retainer. Do not tighten the nut at this time.
11. Raise the suspension to align the shock absorber matchmarks and tighten the lower control arm bolt to 85 ft. lbs. (115 Nm).
12. Align the cam bolt matchmarks and tighten the nut to 66 ft. lbs. (90 Nm).
13. Install the brake disc and caliper. Install the parking brake cable.

➡The hub retainer nut must be tightened with the brakes applied and the wheels off the ground to ensure correct bearing seating.

14. Tighten the hub retainer to 184 ft. lbs. (250 Nm) and install the wheel.
15. Check the wheel alignment and adjust as necessary.

FORD MOTOR CO.

18

1998

Mark VIII

PRECAUTIONS

Before servicing any vehicle, please be sure to read all of the following precautions, which deal with personal safety, prevention of component damage, and important points to take into consideration when servicing a motor vehicle:

- Never open, service or drain the radiator or cooling system when the engine is hot; serious burns can occur from the steam and hot coolant.
- Observe all applicable safety precautions when working around fuel. Whenever servicing the fuel system, always work in a well-ventilated area. Do not allow fuel spray or vapors to come in contact with a spark, open flame or excessive heat (a hot drop light, for example). Keep a dry chemical fire extinguisher near the work area. Always keep fuel in a container specifically designed for fuel storage; also, always properly seal fuel containers to avoid the possibility of fire or explosion. Refer to the additional fuel system precautions later in this section.
- Fuel injection systems often remain pressurized, even after the engine has been turned **OFF**. The fuel system pressure must be relieved before disconnecting any fuel lines. Failure to do so may result in fire and/or personal injury.
- Brake fluid often contains polyglycol ethers and polyglycols. Avoid contact with the eyes and wash your hands thoroughly after handling brake fluid. If you do get brake fluid in your eyes, flush your eyes with clean, running water for 15 minutes. If eye irritation persists, or if you have taken brake fluid internally, IMMEDIATELY seek medical assistance.
- The EPA warns that prolonged contact with used engine oil may cause a number of skin disorders, including cancer. You should make every effort to minimize your exposure to used engine oil. Protective gloves should be worn when changing oil. Wash your hands and any other exposed skin areas as soon as possible after exposure to used engine oil. Soap and water, or waterless hand cleaner should be used.
- All new vehicles are now equipped with an air bag system, often referred to as a Supplemental Restraint System (SRS) or Supplemental Inflatable Restraint (SIR) system. The system must be disabled before performing service on or around system components, steering column, instrument panel components, wiring and sensors. Failure to follow safety and disabling procedures could result in accidental air bag deployment, possible personal injury and unnecessary system repairs.
- Always wear safety goggles when working with, or around, the air bag system. When carrying a non-deployed air bag, be sure the bag and trim cover are pointed away from your body. When placing a non-deployed air bag on a work surface, always face the bag and trim cover upward, away from the surface. This will reduce the motion of the module if it is accidentally deployed. Refer to the additional air bag system precautions later in this section.
- Clean, high quality brake fluid from a sealed container is essential to the safe and proper operation of the brake system. You should always buy the correct type of brake fluid for your vehicle. If the brake fluid becomes contaminated, completely flush the system with new fluid. Never reuse any brake fluid. Any brake fluid that is removed from the system should be discarded. Also, do not allow any brake fluid to come in contact with a painted surface; it will damage the paint.
- Never operate the engine without the proper amount and type of engine oil; doing so WILL result in severe engine damage.
- Timing belt maintenance is extremely important. Many models utilize an interference-type, non-freewheeling engine. If the timing belt breaks, the valves in the cylinder head may strike the pistons, causing potentially serious (also time-consuming and expensive) engine damage. Refer to the maintenance interval charts in the front of this manual for the recommended replacement interval for the timing belt, and to the timing belt section for belt replacement and inspection.
- Disconnecting the negative battery cable on some vehicles may interfere with the functions of the on-board computer system(s) and may require the computer to undergo a relearning process once the negative battery cable is reconnected.
- When servicing drum brakes, only disassemble and assemble one side at a time, leaving the remaining side intact for reference.

ENGINE REPAIR

Alternator

REMOVAL

1. Before servicing the vehicle, refer to the precautions in the beginning of this section.
2. Remove or disconnect the following:
   - Negative battery cable
   - Alternator cover
   - Pushpins and wiring harness
   - Alternator mounting bracket
   - Alternator harness connectors
   - Accessory drive belt
   - Alternator

INSTALLATION

Install or connect the following:

- Alternator. Tighten the bolts to 15–22 ft. lbs. (20–30 Nm).
- Accessory drive belt
- Alternator harness connectors
- Alternator mounting bracket. Tighten the bolts to 71–106 inch lbs. (8–12 Nm).
- Pushpins and wiring harness
- Alternator cover
- Negative battery cable

Ignition Timing

ADJUSTMENT

The 4.6L engine is equipped with a Distributorless Ignition System (DIS). No adjustments are necessary.

Engine Assembly

REMOVAL & INSTALLATION

1. Before servicing the vehicle, refer to the precautions in the beginning of this section.

2. The air suspension switch, located on the left-hand side of the luggage compartment, must be turned to the **OFF** position before raising the vehicle.
3. Drain the cooling system.
4. Relieve the fuel system pressure.
5. Recover the A/C refrigerant.
6. Position the vehicle on a suitable lift, then discharge the air suspension system.
7. Remove or disconnect the following:
   - Battery
   - Air cleaner outlet tube
   - Windshield wiper module
   - Secondary air injection inlet tube
   - Cooling fan and shroud
   - 42 pin wiring connector at the rear of the left valve cover
   - Power distribution box
   - Alternator wires at the power distribution box and dash panel
   - Headlamp dash panel junction wire
   - Evaporative Emissions (EVAP) hose at the upper intake manifold
   - Chassis vacuum supply hose
   - Heater hoses
   - Alternator mounting bracket
   - Upper radiator hose
   - Bypass hose
   - Coolant overflow hose
   - Power steering pressure and return lines
   - Transmission fluid cooler lines
   - Lower radiator hose
   - A/C compressor lines
   - Front wheels
   - Left and right air suspension height sensor connectors
   - Sway bar links
   - Outer tie rod ends
   - Lower ball joints
   - Lower spring and shock assembly bolts
   - Dual converter Y pipe
   - Starter motor
   - Steering intermediate shaft
   - Right fender apron ground strap
   - Differential mounting nuts. Move the differential to provide clearance for driveshaft removal.
   - Driveshaft. Matchmark the driveshaft flange to the axle yoke.
   - Left front shock absorber strut cover cup
   - Wiring connectors and vacuum lines near the left fender wall
   - Brake booster vacuum hose
   - Fuel lines
   - Shift cable and bracket
   - Transmission mount crossmember. Support the engine and transmission.
   - Front subframe bolts
8. Lower the engine and transmission from the vehicle.
9. Separate the engine from the transmission and mount it on a workstand.
10. Remove the subframe from the engine.

To install:

11. Install the subframe to the engine. Tighten the motor mount fasteners to 15–22 ft. lbs. (20–30 Nm).
12. Install the engine to the transmission. Tighten the bolts to 30–44 ft. lbs. (40–60 Nm).
13. Raise the engine and transmission into position. Tighten the subframe bolts to 57–77 ft. lbs. (77–104 Nm).
14. Install or connect the following:
   - Transmission mount crossmember. Tighten the bolts to 47–63 ft. lbs. (35–46 Nm) and the nuts to 65–87 ft. lbs. (88–118 Nm).
   - Shift cable and bracket
   - Fuel lines
   - Brake booster vacuum hose
   - Wiring connectors and vacuum lines near the left fender wall
   - Left front shock absorber strut cover cup
   - Driveshaft. Align the matchmarks and tighten the bolts to 57–75 ft. lbs. (88–119 Nm).
   - Differential mounting nuts. Tighten the front nuts to 57–75 ft. lbs. (88–119 Nm) and the rear nuts to 45–59 ft. lbs. (60–80 Nm).
   - Right fender apron ground strap
   - Steering intermediate shaft. Tighten the pinch bolt to 21–29 ft. lbs. (28–40 Nm).
   - Starter motor
   - Dual converter Y pipe
   - Lower spring and shock assembly bolts. Do not tighten the bolts at this time.
   - Lower ball joints. Tighten the nuts to 83–113 ft. lbs. (113–153 Nm).
   - Outer tie rod ends. Tighten the nuts to 39–53 ft. lbs. (53–72 Nm).
   - Sway bar links. Tighten the nuts to 30–39 ft. lbs. (40–54 Nm).
   - Left and right air suspension height sensor connectors
   - Front wheels
   - A/C compressor lines
   - Lower radiator hose
   - Transmission fluid cooler lines
   - Power steering pressure and return lines
   - Coolant overflow hose
   - Bypass hose
   - Upper radiator hose
   - Alternator mounting bracket
   - Heater hoses
   - Chassis vacuum supply hose
   - EVAP hose at the upper intake manifold
   - Headlamp dash panel junction wire
   - Alternator wires at the power distribution box and dash panel
   - Power distribution box
   - 42 pin wiring connector at the rear of the left valve cover
   - Cooling fan and shroud
   - Secondary air injection inlet tube
   - Windshield wiper module
   - Air cleaner outlet tube
   - Battery
15. Turn the air suspension switch to the **ON** position.
16. With the weight of the vehicle supported by the wheels, tighten the lower spring and shock assembly bolts to 126–169 ft. lbs. (170–230 Nm).
17. Fill the cooling system.
18. Recharge the A/C system.
19. Start the engine and check for leaks.

Water Pump

REMOVAL & INSTALLATION

1. Before servicing the vehicle, refer to the precautions in the beginning of this section.
2. Drain the cooling system.
3. Remove or disconnect the following:
   - Negative battery cable
   - Accessory drive belt
   - Water pump pulley
   - Water pump

To install:

4. Clean the sealing surfaces of the water pump and block.
5. Install or connect the following:
   - Water pump. Use a new O-ring seal and tighten the bolts to 15–22 ft. lbs. (20–30 Nm).
   - Water pump pulley. Tighten the bolts to 15–22 ft. lbs. (20–30 Nm).
   - Accessory drive belt
   - Negative battery cable
6. Fill the cooling system.
7. Check the front end alignment and adjust as necessary.

8. Operate the engine to normal operating temperatures and check for leaks.

Cylinder Head

REMOVAL & INSTALLATION

1. Before servicing the vehicle, refer to the precautions in the beginning of this section.
2. Remove the engine from the vehicle and mount it on a suitable workstand.
3. Remove or disconnect the following:
   - Fuel injectors
   - Intake manifold
   - Front cover
   - Timing chains
   - Valve covers
   - Exhaust manifolds
   - Cylinder heads

To install:

➡Use new cylinder head bolts for assembly.

4. Install the cylinder heads with new gaskets and bolts. Tighten the bolts in sequence as follows:
   a. Step 1: 28–31 ft. lbs. (37–43 Nm)
   b. Step 2: Plus 85–95 degrees

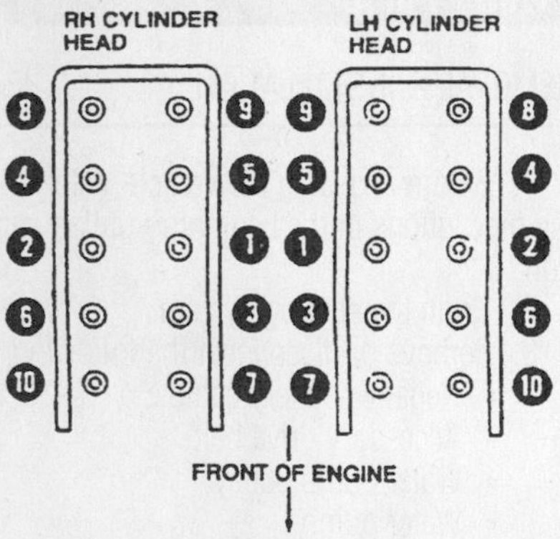

7922NG11

Cylinder head torque sequence—4.6L DOHC engine

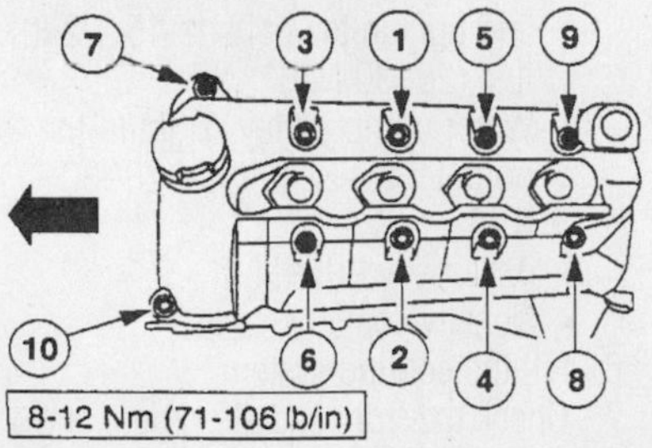

9346GG08

Left valve cover torque sequence—4.6L (VIN V) Engine

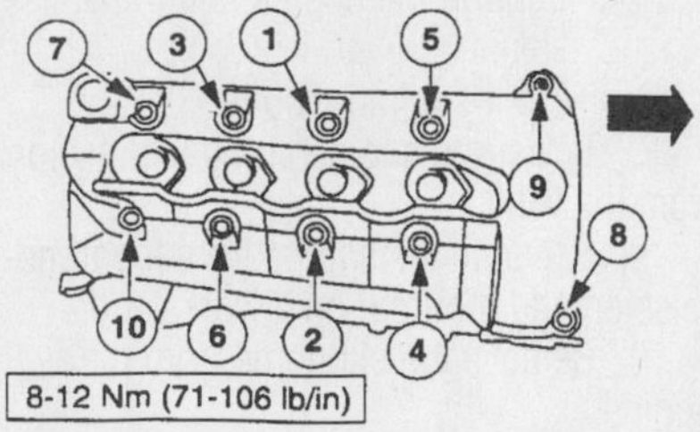

9346GG09

Right valve cover torque sequence—4.6L (VIN V) Engine

   c. Step 3: Loosen all bolts one full turn
   d. Step 4: 28–31 ft. lbs. (37–43 Nm)
   e. Step 5: Plus 85–95 degrees
   f. Step 6: Plus 85–95 degrees
5. Install or connect the following:
   - Exhaust manifolds
   - Valve covers
   - Timing chains
   - Front cover
   - Intake manifold
   - Fuel injectors
6. Install the engine into the vehicle.

Rocker Arms

REMOVAL & INSTALLATION

1. Before servicing the vehicle, refer to the precautions at the beginning of this section.
2. Remove or disconnect the following:
   - Negative battery cable
   - Ignition coils
   - Valve covers
3. Rotate the crankshaft so that the piston on the cylinder to be serviced is at bottom dead center with the valves closed.
4. Install special tool Valve Spring Compressor T97P-6565-AH for exhaust valves, and Valve Spring Compressor T93P-6565-A for intake valves.
5. Compress the valve spring and remove the rocker arm. Repeat for each arm to be removed.

➡If the rocker arms are to be reused, ensure that they are installed in the same position that they were removed from.

To install:

6. Compress the valve spring and install the rocker arm. Repeat for each arm to be installed.
7. Install or connect the following:
   - Valve covers
   - Ignition coils
   - Negative battery cable

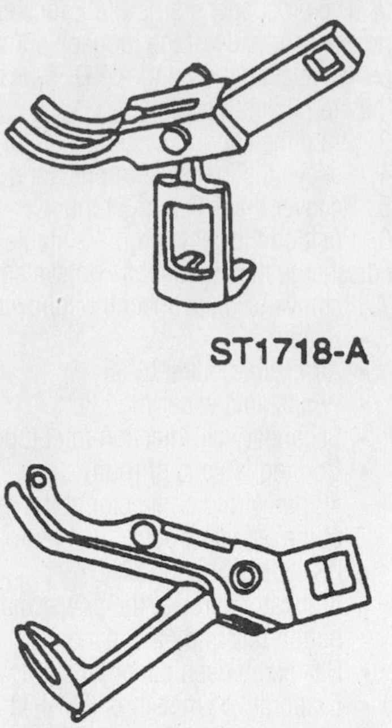

9346GG01

Rocker arm service tools—4.6L (VIN V) Engine

8. Start the engine and check for proper operation.

Intake Manifold

REMOVAL & INSTALLATION

1. Before servicing the vehicle, refer to the precautions in the beginning of this section.
2. Drain the cooling system.
3. Relieve the fuel system pressure.
4. Remove or disconnect the following:
   - Negative battery cable

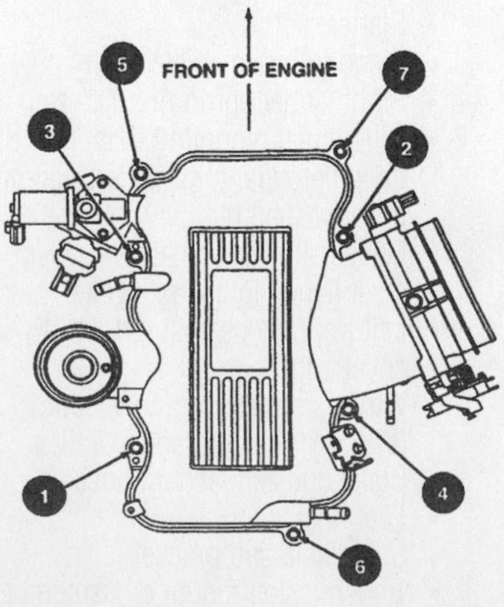

7922NG21

Upper intake manifold torque sequence—4.6L DOHC engine

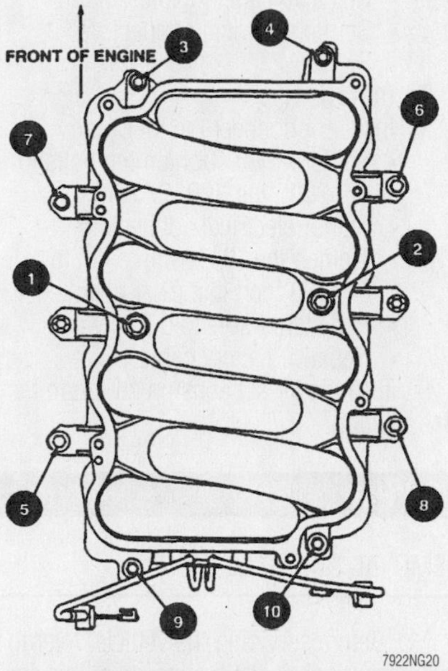

Lower intake manifold torque sequence—4.6L DOHC engine

- Coolant bypass tube
- Alternator
- Throttle cable
- Cruise control cable
- Evaporative Emissions (EVAP) hose
- Throttle Position (TP) sensor connector
- Positive Crankcase Ventilation (PCV) hose
- Idle Air Control (IAC) valve connector
- Exhaust Gas Recirculation (EGR) tube
- EGR vacuum hose
- Upper intake manifold
- Fuel injection supply manifold
- Lower intake manifold

To install:

5. Install or connect the following:
   - Lower intake manifold. Tighten the bolts in sequence to 15–22 ft. lbs. (20–30 Nm).
   - Fuel injection supply manifold
   - Upper intake manifold. Tighten the bolts to 15–22 ft. lbs. (20–30 Nm).
   - EGR vacuum hose
   - EGR tube
   - IAC valve connector
   - PCV hose
   - TP sensor connector
   - EVAP hose
   - Cruise control cable
   - Throttle cable
   - Alternator
   - Coolant bypass tube
   - Negative battery cable
6. Fill the cooling system.
7. Start the engine and check for leaks.

Exhaust Manifold

REMOVAL & INSTALLATION

1. Before servicing the vehicle, refer to the precautions in the beginning of this section.
2. The air suspension switch, located on the left-hand side of the luggage compartment, must be turned to the **OFF** position before raising the vehicle.
3. Remove or disconnect the following:
   - Negative battery cable
   - Heated Oxygen (HO_2S) sensor connectors

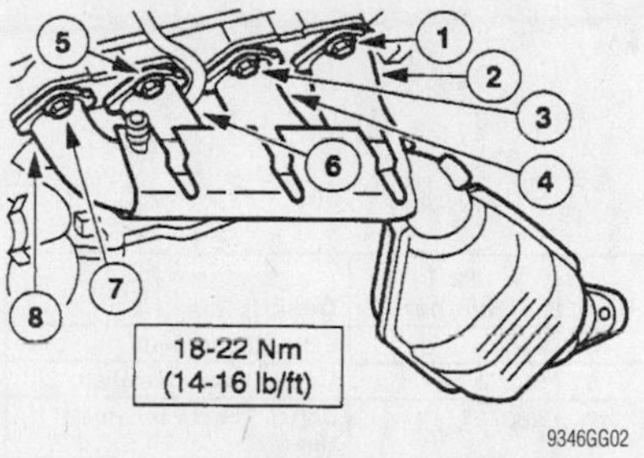

Left exhaust manifold torque sequence—4.6L (VIN V) Engine

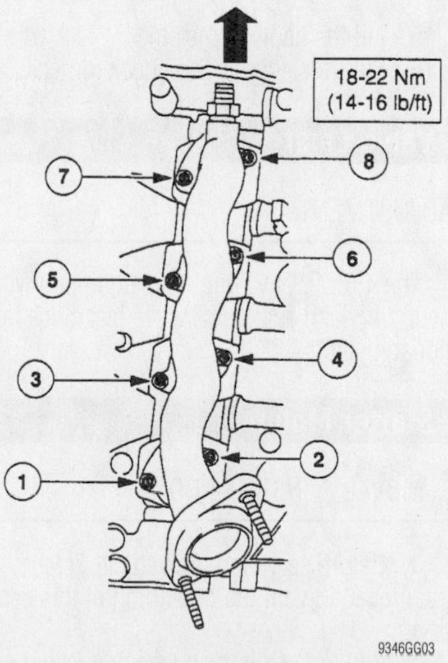

Right exhaust manifold torque sequence—4.6L (VIN V) Engine

   - Dual converter Y-pipe
   - Secondary air injection tubes
   - Exhaust Gas Recirculation (EGR) tube
   - Exhaust manifolds and gaskets

To install:

4. Install or connect the following:
   - Exhaust manifolds with new gaskets. Tighten the nuts in sequence to 14–16 ft. lbs. (18–22 Nm).
   - EGR tube
   - Secondary air injection tubes
   - Dual converter Y-pipe
   - Heated Oxygen (HO_2S) sensor connectors
   - Negative battery cable
5. Turn the air suspension switch to the **ON** position.
6. Start the engine and check for leaks.

Camshaft and Valve Lifters

REMOVAL & INSTALLATION

1. Before servicing the vehicle, refer to the precautions in the beginning of this section.
2. Drain the cooling system.
3. Remove or disconnect the following:
   - Negative battery cable
   - Windshield wiper module
   - Engine front cover
   - Timing chains and sprockets
   - Valve covers
   - Rocker arms
   - Hydraulic lifters
   - Camshaft bearing cap assembly
   - Camshafts

➡Keep all valvetrain components in order for assembly in their original locations.

To install:

➡The outboard exhaust camshaft bearing cap bolts are shorter than the other bearing cap bolts.

4. Install or connect the following:
   - Camshafts
   - Camshaft bearing cap assembly. Tighten the bolts in sequence to 71–106 inch lbs. (8–12 Nm).
   - Hydraulic lifters
   - Rocker arms
   - Valve covers
   - Timing chains and sprockets
   - Engine front cover
   - Windshield wiper module
   - Negative battery cable

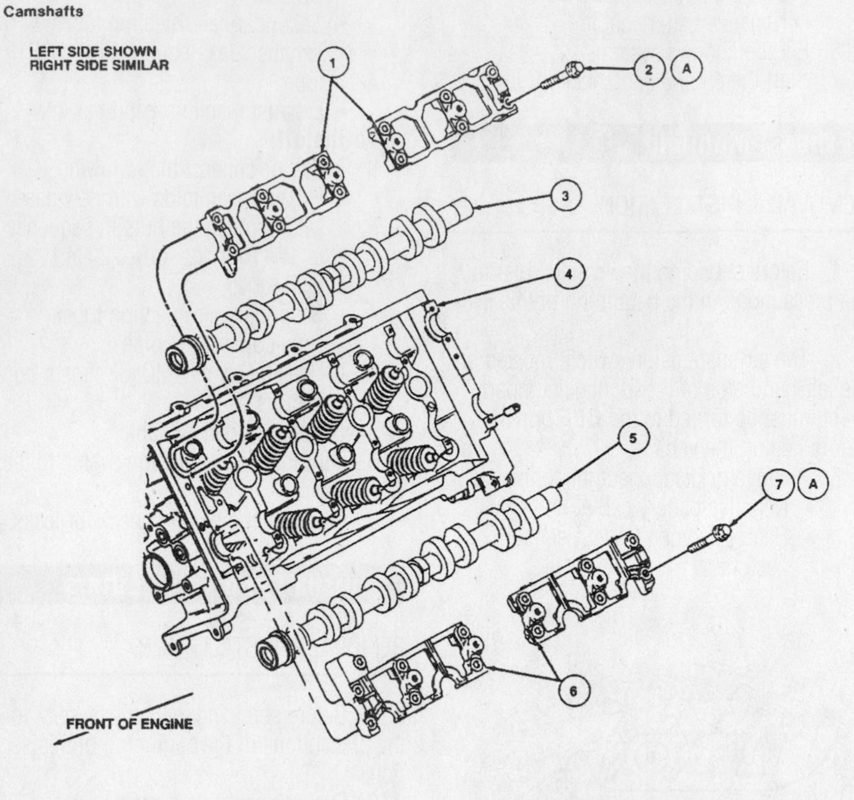

Item	Part Number	Description
1	6B277	Cap, Intake Camshaft
2	N806070	Bolt (19 Req'd Per Cylinder Head)
3	6250	Intake Camshaft
4	6049	LH Cylinder Head

Item	Part Number	Description
5	6250	Exhaust Camshaft
6	6B278	Cap, Exhaust Camshaft
7	N80757	Bolt (6 Req'd Per Cylinder Head)
A	—	Tighten to 8-12 N-m (71-106 Lb-In)

9346GG05

Camshaft mounting exploded view—4.6L (VIN V) Engine

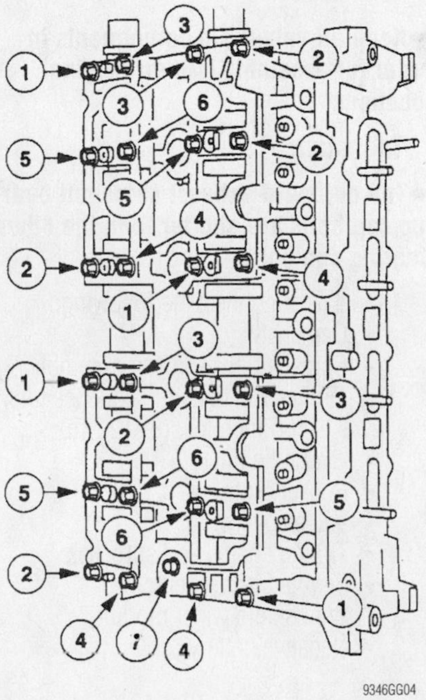

Camshaft bearing cap torque sequence—

5. Fill the cooling system.
6. Start the engine and check for leaks.

Valve Lash

ADJUSTMENT

The 4.6L (VIN V) engine is equipped with hydraulic lash adjusters. Valve clearance is not adjustable.

Starter Motor

REMOVAL & INSTALLATION

1. Before servicing the vehicle, refer to the precautions in the beginning of this section.
2. The air suspension switch, located on the left-hand side of the luggage compartment, must be turned to the **OFF** position before raising the vehicle.
3. Remove or disconnect the following:
   - Negative battery cable
   - Starter electrical connectors
   - Starter

To install:

4. Install or connect the following:
   - Starter motor. Tighten the bolts to 15–20 ft. lbs. (20–27 Nm).
   - Starter electrical connectors. Tighten the "B" terminal nut to 60–123 inch lbs. (9–14 Nm).
   - Red starter solenoid safety cap
   - Negative battery cable
5. Turn the air suspension switch to the **ON** position.

Oil Pan

REMOVAL & INSTALLATION

1. Before servicing the vehicle, refer to the precautions in the beginning of this section.
2. The air suspension switch, located on the left-hand side of the luggage compartment, must be turned to the **OFF** position before raising the vehicle.
3. Install Engine Lifting Set 303-DS086 and Three-Bar Engine Support 303-D063 to support the engine.
4. Drain the engine oil.
5. Remove or disconnect the following:
   - Negative battery cable
   - Front wheels
   - Left and right height sensor connectors
   - Disc brake calipers
   - Upper ball joints
   - Steering column intermediate shaft
   - Power steering pressure and return lines
   - Left and right motor mount through-bolts
   - Lower spring and shock assembly bolts.

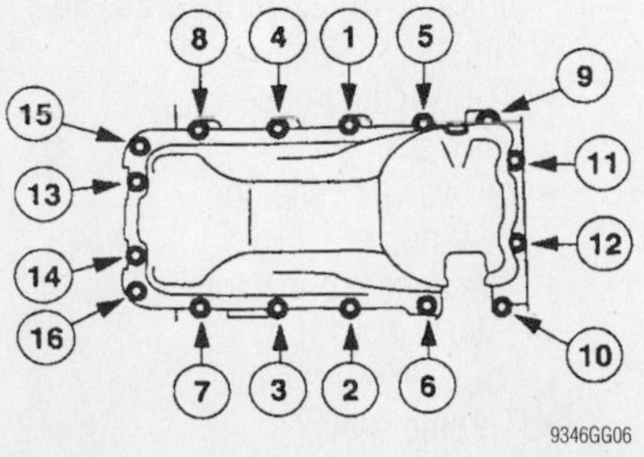

Oil pan torque sequence—4.6L (VIN V) engine

- Front subframe bolts. Lower and support the subframe.
- Oil level sensor connector
- Oil pan and gasket

To install:

6. Apply RTV silicone sealer where the front cover and rear oil seal retainer meet the cylinder block.

7. Install the oil pan with a new gasket and tighten the bolts in sequence as follows:

a. Step 1: 14 ft. lbs. (20 Nm)
b. Step 2: Plus 60 degrees

8. Install or connect the following:

- Oil level sensor connector
- Front subframe bolts. Tighten the bolts to 57–77 ft. lbs. (77–104 Nm).
- Lower spring and shock assembly bolts. Do not tighten the bolts at this time.
- Left and right motor mount through-bolts
- Power steering pressure and return lines
- Steering column intermediate shaft
- Upper ball joints
- Disc brake calipers
- Left and right height sensor connectors
- Front wheels
- Negative battery cable

9. With the weight of the vehicle supported by the wheels, tighten the lower spring and shock assembly bolts to 126–169 ft. lbs. (170–230 Nm).

10. Fill the crankcase to the correct level.

11. Turn the air suspension switch to the **ON** position.

12. Start the engine and check for leaks.

13. Check the alignment and adjust as necessary.

Oil Pump

REMOVAL & INSTALLATION

1. Before servicing the vehicle, refer to the precautions in the beginning of this section.

2. The air suspension switch, located on the left-hand side of the luggage compartment, must be turned to the **OFF** position before raising the vehicle.

3. Remove or disconnect the following:

- Negative battery cable
- Cylinder head covers
- Engine front cover
- Engine oil pan

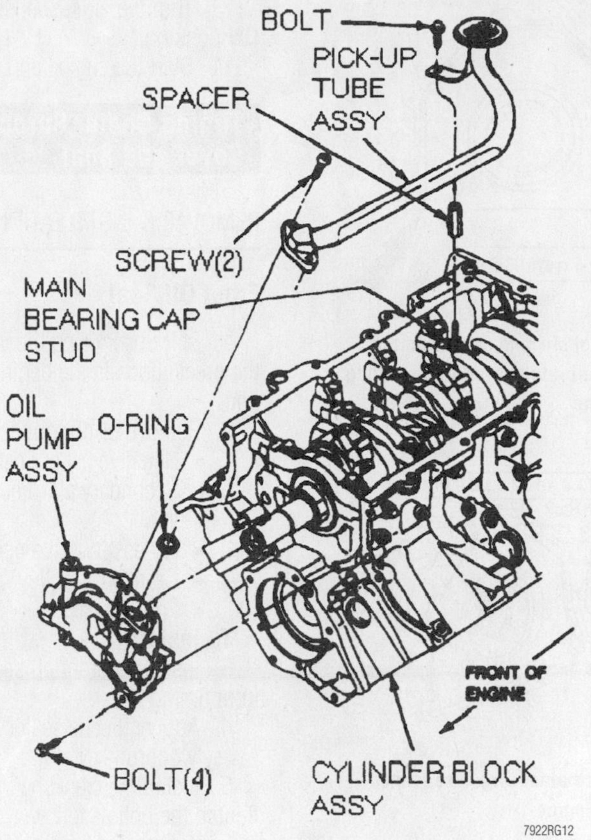

The oil pump is mounted on the crankshaft at the front of the engine

- Timing chains
- 2 bolts retaining the oil pick-up tube to the oil pump and the bolt attaching the oil pick-up tube to the main bearing stud spacer
- Pick-up tube
- 4 bolts retaining the oil pump to the cylinder block
- Oil pump

To install:

4. Rotate the inner rotor of the oil pump to align with the flats on the crankshaft and install the oil pump flush with the cylinder block. Install the 4 retaining bolts and tighten to 72–106 inch lbs. (8–12 Nm).

5. Clean the oil pick-up tube and replace the O-ring.

6. Place the pick-up tube on the oil pump and hand start the 2 retaining bolts. Install the bolt retaining the pick-up tube to the main bearing stud spacer hand tight. Tighten the pick-up tube-to-oil pump bolts to 72–106 inch lbs. (8–12 Nm). Tighten the pick-up tube to main bearing stud spacer bolt to 15–22 ft. lbs. (20–30 Nm).

7. Install or connect the following:

- New engine oil filter
- Timing chains
- Engine oil pan
- Engine front cover
- Cylinder head covers
- Negative battery cable

8. Fill the crankcase.

9. Turn the air suspension switch to the **ON** position.

10. Start the engine and check for leaks and proper engine oil pressure.

11. Road test the vehicle and check for proper engine operation.

Rear Main Seal

REMOVAL & INSTALLATION

1. Before servicing the vehicle, refer to the precautions in the beginning of this section.

2. The air suspension switch, located on the left-hand side of the luggage compartment, must be turned to the **OFF** position before raising the vehicle.

3. Remove or disconnect the following:

- Transmission
- Flywheel

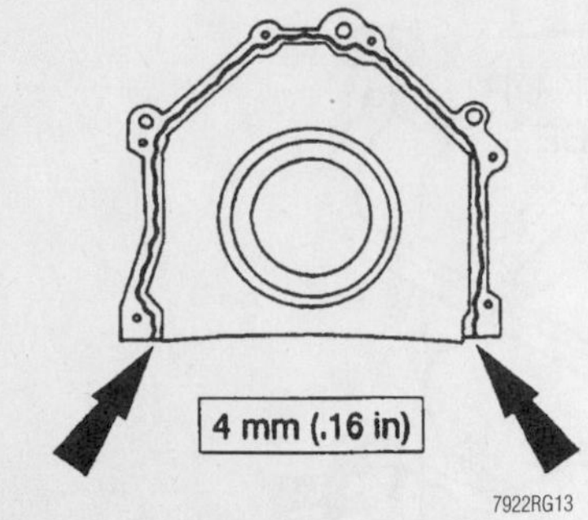

Apply a bead of silicone sealant to the back of the seal retainer before installing it on the engine

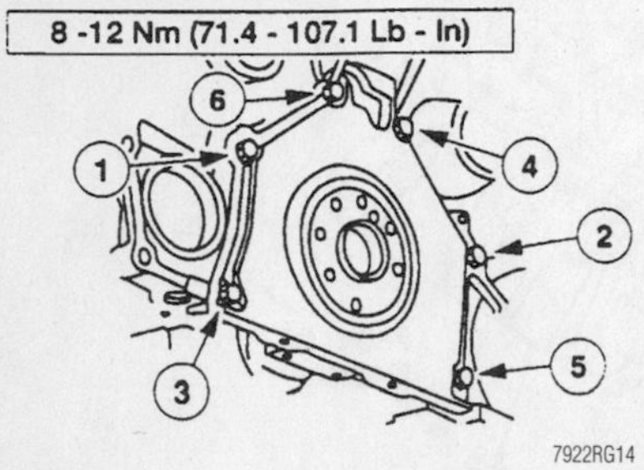

Rear oil seal retainer torque sequence—4.6L (VIN V) Engine

4. With a sharp awl, carefully punch a small hole in the metal portion of the seal.

5. Remove the seal using a slide hammer with a sheet metal screw attached.

➡If the oil leak is coming from around the seal retainer, the retainer must also be removed and resealed.

To install:

6. If the seal retainer was removed, carefully clean the sealant from the retainer and engine block using a plastic scraper. Remove any oil or grease residue from the sealing surfaces with a solvent.

7. Apply silicone sealant to the back of the retainer and immediately install it on the engine block. Tighten the bolts in sequence to 71–106 inch lbs. (8–12 Nm).

8. Lubricate the seal and the crankshaft with clean engine oil.

9. Install the seal with the spring side toward the engine.

10. Remove the installation tool.

11. Install or connect the following:
- Flywheel. Tighten the bolts, in a crisscross pattern, to 54–64 ft. lbs. (73–87 Nm).
- Transmission
- Negative battery cable

12. Check the engine oil level.

13. Turn the air suspension switch to the **ON** position.

14. Start the engine and check for leaks.

Timing Chain, Sprockets, Front Cover and Seal

REMOVAL & INSTALLATION

Front Oil Seal

1. Before servicing the vehicle, refer to the precautions in the beginning of this section.

2. Remove or disconnect the following:
- Negative battery cable
- Secondary air injection diverter valve
- Accessory drive belt
- Crankshaft pulley
- Front oil seal

To install:

3. Install the front oil seal flush with the front timing cover.

4. Apply silicone sealer to the crankshaft pulley woodruff key slot.

5. Install the crankshaft pulley and tighten the bolt as follows:
   a. Step 1: 66 ft. lbs. (90 Nm)
   b. Step 2: Loosen the bolt one full turn
   c. Step 3: 35–39 ft. lbs. (47–53 Nm)
   d. Step 4: Plus 85–90 degrees

6. Install or connect the following:
- Accessory drive belt
- Secondary air injection diverter valve. Tighten the nuts to 71–106 inch lbs. (8–12 Nm).
- Negative battery cable

Front Cover, Timing Chains and Sprockets

1. Before servicing the vehicle, refer to the precautions in the beginning of this section.

2. The air suspension switch, located on the left-hand side of the luggage compartment, must be turned to the **OFF** position before raising the vehicle.

3. Drain the cooling system.

4. Drain the engine oil.

5. Remove or disconnect the following:
- Negative battery cable
- Water outlet hose
- Water bypass hose
- Cooling fan assembly
- Accessory drive belt
- Water pump pulley
- Secondary air injection valve
- Power steering pump harness connector
- Starter relay cable and bracket
- Camshaft Position (CMP) sensor connector
- Power steering pump and reservoir
- Crankshaft pulley
- Valve covers
- Rocker arms
- Accessory drive belt tensioner and idler pulley
- Crankshaft Position (CKP) sensor connector
- Front cover
- CKP sensor pulse wheel

6. Rotate the crankshaft so that the No. 1 cylinder is at Top Dead Center (TDC) of the compression stroke.

7. Remove or disconnect the following:
- Right primary timing chain tensioner
- Right primary timing chain tensioner arm and chain guide
- Right primary timing chain and sprockets
- Left primary timing chain tensioner
- Left primary timing chain tensioner arm and chain guide
- Left primary timing chain and sprockets

8. Compress the secondary chain tensioners and install locking pins.

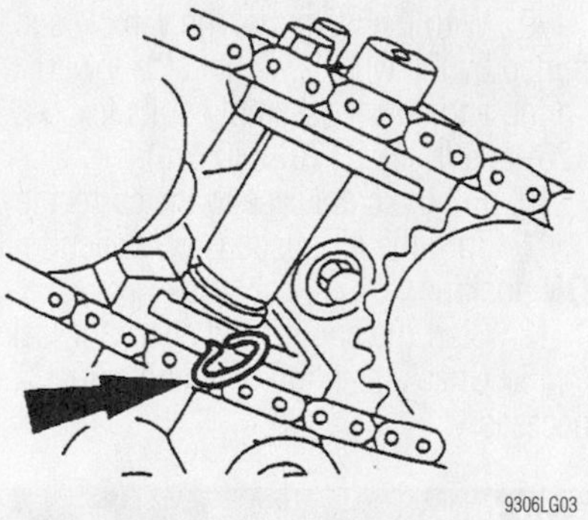

Secondary timing chain tensioner and locking pin—4.6L (VIN V) Engine

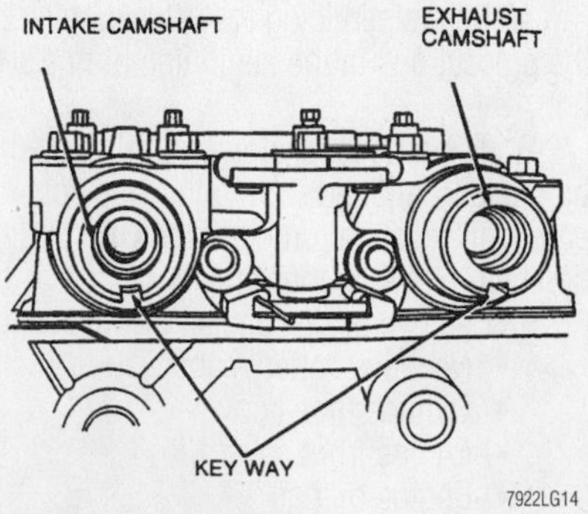

Position the camshafts for timing chain installation—4.6L (VIN V) Engine

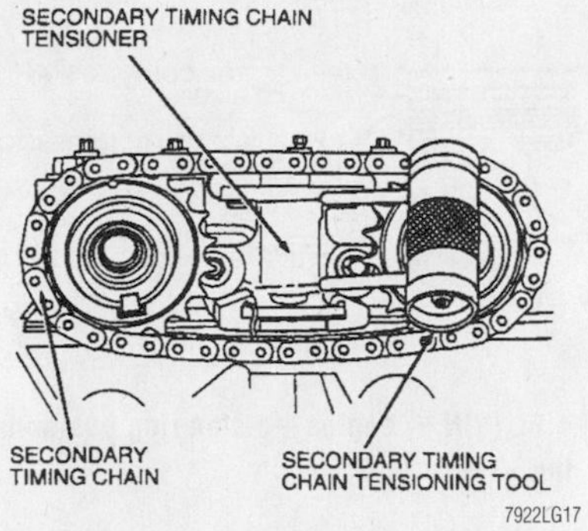

Secondary timing chain tensioning tool—4.6L (VIN V) Engine

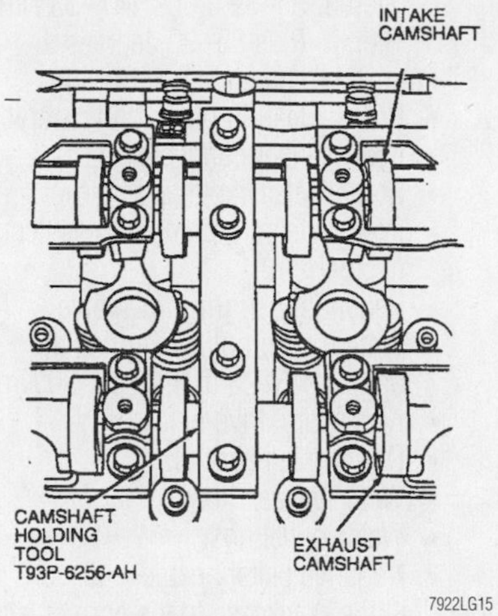

Install the Camshaft Holding tool to secure the camshafts and prevent damage to the camshaft positioning tool—4.6L (VIN V) Engine

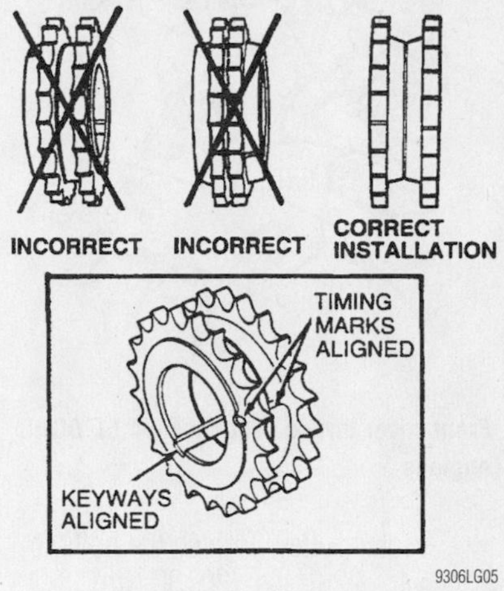

Crankshaft timing sprocket installation—4.6L (VIN V) Engine

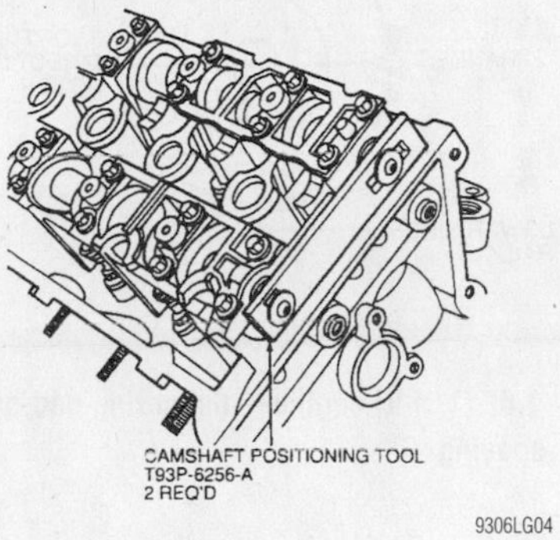

Install the camshaft positioning tool—4.6L (VIN V) Engine

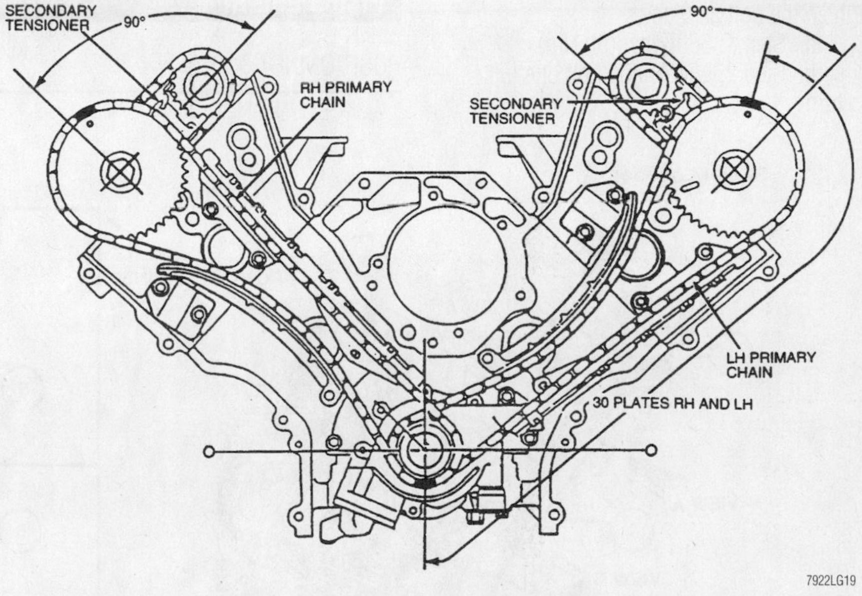

Correct alignment of the primary timing chains and sprockets—4.6L (VIN V) Engine

9. Remove the left and right secondary timing chains and sprockets.

To install:

10. Position the camshafts with the keys facing downward.

11. Install the secondary timing chains and sprockets. Do not tighten the sprocket bolts at this time.

12. Tension the secondary timing chains with the Secondary Timing Chain Tensioning tool T93P-6256-BH.

13. Install Camshaft Positioning tool T93P-6256-A in the rear D-slots of the camshaft.

14. Install Camshaft Holding tool T93P-6256-AH onto the camshafts to keep the camshafts from rotating and to prevent damaging the camshaft positioning tool.

15. Tighten the secondary timing chain sprocket bolts to 81–95 ft. lbs. (110–130 Nm).

16. Remove the secondary timing chain tensioner locking pins and remove the tensioner tool.

17. Compress the primary timing chain tensioners and install locking pins.

18. Install the crankshaft timing sprockets.

19. Align the copper colored primary timing chain links with the sprocket timing marks as shown.

20. Install or connect the following:
- Left primary timing chain and sprockets
- Left primary timing chain tensioner arm and chain guide
- Left primary timing chain tensioner
- Right primary timing chain and sprockets
- Right primary timing chain tensioner arm and chain guide
- Right primary timing chain tensioner

21. Tighten the camshaft sprocket bolts to 81–95 ft. lbs. (110–130 Nm).

22. Remove the primary timing chain tensioner locking pins.

23. Remove the Camshaft Holding and Camshaft Positioning tools.

24. Install or connect the following:
- CKP sensor pulse wheel
- Front cover. Apply silicone sealant as shown. Tighten the bolts in sequence to 15–22 ft. lbs. (20–30 Nm).
- CKP sensor connector
- Accessory drive belt tensioner and

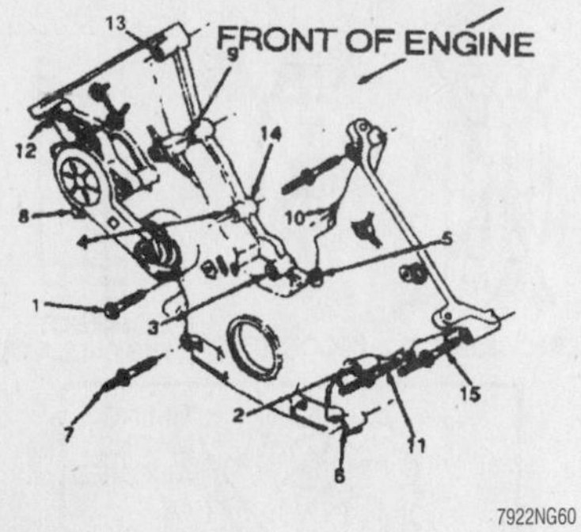

Front cover torque sequence—4.6L DOHC engines

idler pulley. Tighten the bolts to 15–22 ft. lbs. (20–30 Nm).

- Rocker arms
- Valve covers

25. Apply silicone sealer to the crankshaft pulley woodruff key slot.

26. Install the crankshaft pulley and tighten the bolt as follows:

a. Step 1: 66 ft. lbs. (90 Nm)

b. Step 2: Loosen the bolt one full turn

c. Step 3: 35–39 ft. lbs. (47–53 Nm)

d. Step 4: Plus 85–90 degrees

27. Install or connect the following:

- Power steering pump and reservoir
- CMP sensor connector
- Starter relay cable and bracket
- Power steering pump harness connector
- Secondary air injection valve
- Water pump pulley. Tighten the bolts to 15–22 ft. lbs. (20–30 Nm).
- Accessory drive belt
- Cooling fan assembly
- Water bypass hose
- Water outlet hose
- Negative battery cable

28. Fill the crankcase to the correct level.

29. Fill the cooling system.

30. Turn the air suspension switch to the **ON** position.

31. Start the engine and check for leaks.

Piston and Ring

POSITIONING

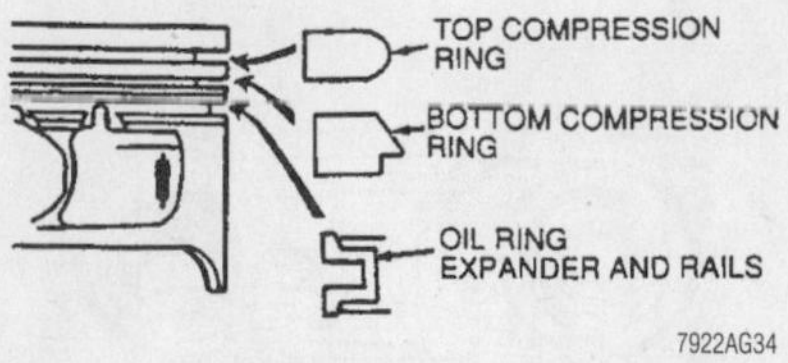

4.6L (VIN V) engine—piston ring positioning

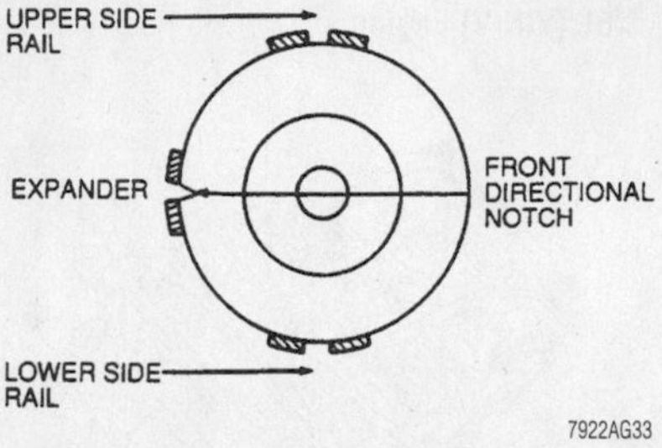

4.6L (VIN V) engine—piston ring end-gap spacing

Sealer Location

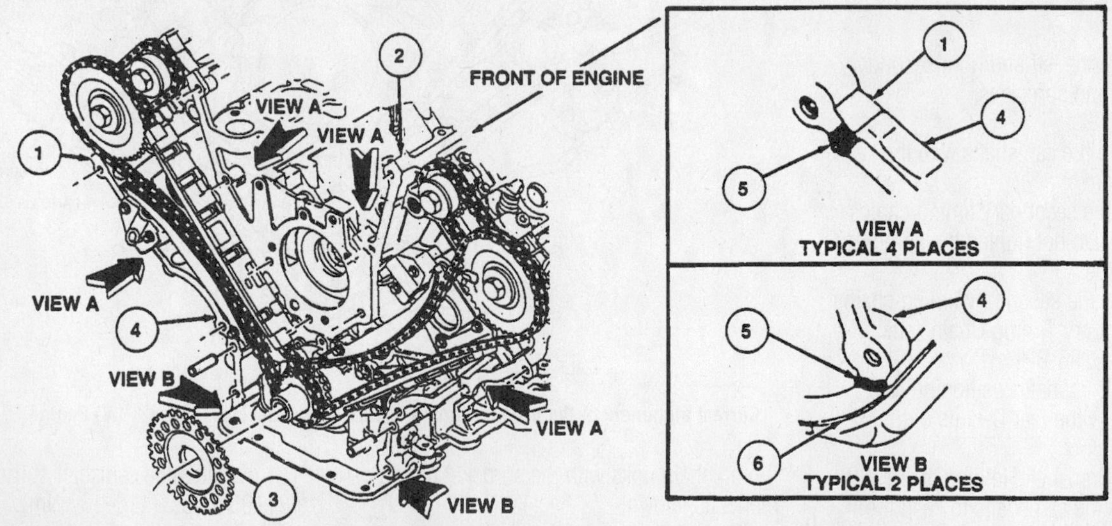

Item	Part Number	Description
1	6049	RH Cylinder Head
2	6049	LH Cylinder Head
3	12A227	Ignition Pulse Crankshaft Sensor Ring
4	6010	Cylinder Block
5	WSE-M4G320-A2	Sealer
6	6710	Oil Pan Gasket

9306LG06

Apply sealer to these locations—4.6L DOHC engines

FUEL SYSTEM

Fuel System Service Precautions

Safety is the most important factor when performing not only fuel system maintenance but any type of maintenance. Failure to conduct maintenance and repairs in a safe manner may result in serious personal injury or death. Maintenance and testing of the vehicle's fuel system components can be accomplished safely and effectively by adhering to the following rules and guidelines.

- To avoid the possibility of fire and personal injury, always disconnect the negative battery cable unless the repair or test procedure requires that battery voltage be applied.
- Always relieve the fuel system pressure prior to disconnecting any fuel system component (injector, fuel rail, pressure regulator, etc.), fitting or fuel line connection. Exercise extreme caution whenever relieving fuel system pressure, to avoid exposing skin, face and eyes to fuel spray. Please be advised that fuel under pressure may penetrate the skin or any part of the body that it contacts.
- Always place a shop towel or cloth around the fitting or connection prior to loosening to absorb any excess fuel due to spillage. Ensure that all fuel spillage (should it occur) is quickly removed from engine surfaces. Ensure that all fuel soaked cloths or towels are deposited into a waste container.
- Always keep a dry chemical (Class B) fire extinguisher near the work area.
- Do not allow fuel spray or fuel vapors to come into contact with a spark or open flame.
- Always use a back-up wrench when loosening and tightening fuel line connection fittings. This will prevent unnecessary stress and torsion to fuel line piping. Always follow the proper torque specifications.
- Always replace worn fuel fitting O-rings with new. Do not substitute fuel hose or equivalent, where fuel pipe is installed.

Fuel System Pressure

RELIEVING

Fuel supply lines on all fuel injected engines will remain pressurized for some period of time after the engine is shut **OFF**. This pressure must be relieved before servicing the fuel system. Pressure is relieved through the fuel pressure relief valve, located on the fuel rail.

To relieve the fuel system pressure, first remove the fuel tank cap to relieve pressure in the tank, then remove the cap on the fuel pressure relief valve. Attach a fuel pressure gauge and drain the system through the drain tube into a container. Remove the fuel pressure gauge and replace the cap on the relief valve.

Fuel Filter

REMOVAL & INSTALLATION

1. Before servicing the vehicle, refer to the precautions in the beginning of this section.
2. Relieve the fuel system pressure.
3. The air suspension switch, located on the left-hand side of the luggage compartment, must be turned to the **OFF** position before raising the vehicle.
4. Remove or disconnect the following:
   - Negative battery cable
   - Front fender splash shield
   - Fuel filter spring-lock fittings
   - Fuel filter

To install:

5. Install or connect the following:
   - Fuel filter
   - Fuel filter spring-lock fittings
   - Front fender splash shield
   - Negative battery cable
6. Turn the air suspension switch to the **ON** position.
7. Start the engine and check for leaks.

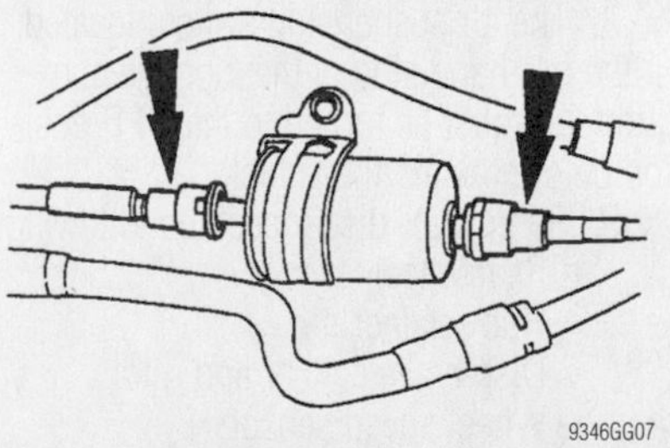

Fuel filter mounting—4.6L (VIN V) Engine

Fuel Pump

REMOVAL & INSTALLATION

1. Before servicing the vehicle, refer to the precautions in the beginning of this section.
2. Relieve the fuel system pressure.
3. The air suspension switch, located on the left-hand side of the luggage compartment, must be turned to the **OFF** position before raising the vehicle.
4. Drain the fuel tank.
5. Remove or disconnect the following:
   - Negative battery cable
   - Exhaust system
   - Fuel tank filler pipe
   - Control valve signal jumper
   - Control valve vapor jumper
   - Fuel tank wiring harness connector
   - Fuel tank support straps. Support the fuel tank.
   - Fuel lines. Lower the tank for access.
   - Fuel pressure sensor connector
   - Fuel pump top flange plate bolts
6. Lift the top flange plate so that the fuel pump locking taps are accessible.
7. Squeeze the tabs together and remove the fuel pump.

To install:

8. Install or connect the following:
   - Fuel pump with a new mounting gasket
   - Fuel pump top flange. Tighten the bolts to 71–88 inch lbs. (8–10 Nm).
   - Fuel pressure sensor connector
   - Fuel lines. Lower the tank for access.
   - Fuel tank support straps. Tighten the bolts to 23–29 ft. lbs. (30–40 Nm).
   - Fuel tank wiring harness connector
   - Control valve vapor jumper
   - Control valve signal jumper
   - Fuel tank filler pipe
   - Exhaust system
   - Negative battery cable
9. Turn the air suspension switch to the **ON** position.
10. Fill the fuel tank.
11. Start the engine and check for leaks.

Fuel Injector

REMOVAL & INSTALLATION

1. Before servicing the vehicle, refer to the precautions in the beginning of this section.
2. Relieve the fuel system pressure.
3. Remove or disconnect the following:
   - Negative battery cable

- Air cleaner outlet tube
- Upper intake manifold
- Fuel pressure regulator vacuum line
- Fuel pulse damper vacuum line
- Brake booster vacuum line
- Fuel lines
- Fuel injector harness connectors
- Injector supply manifold
- Fuel injectors

To install:

4. Install or connect the following:
   - Fuel injectors with new O-ring seals
   - Injector supply manifold. Tighten the bolts to 71–106 inch lbs. (8–12 Nm).
   - Fuel injector harness connectors
   - Fuel lines
   - Brake booster vacuum line
   - Fuel pulse damper vacuum line
   - Fuel pressure regulator vacuum line
   - Upper intake manifold
   - Air cleaner outlet tube
   - Negative battery cable
5. Start the engine and check for leaks.

DRIVE TRAIN

Transmission Assembly

REMOVAL & INSTALLATION

1. Before servicing the vehicle, refer to the precautions in the beginning of this section.
2. The air suspension switch, located on the left-hand side of the luggage compartment, must be turned to the **OFF** position before raising the vehicle.
3. Drain the transmission.
4. Remove or disconnect the following:
   - Negative battery cable
   - Transmission wiring harness connector
   - Differential mounting nuts. Move the differential to provide clearance for driveshaft removal.
   - Driveshaft. Matchmark the driveshaft flange to the axle yoke.
   - Flywheel access cover
   - Torque converter
   - Starter motor
   - Transmission shift linkage
   - Transmission fluid cooler lines
   - Heated Oxygen (HO_2S) sensor connectors
   - Dual converter Y-pipe
   - Transmission mount crossmember. Support the transmission.
   - Transmission dipstick tube
   - Transmission flange bolts
   - Transmission

To install:

5. Install or connect the following:
   - Transmission. Tighten the flange bolts to 30–40 ft. lbs. (40–55 Nm).
   - Transmission dipstick tube
   - Transmission mount crossmember. Tighten the bolts to 47–63 ft. lbs. (35–46 Nm) and the nuts to 65–87 ft. lbs. (88–118 Nm).
   - Dual converter Y-pipe
   - HO_2S sensor connectors
   - Transmission fluid cooler lines
   - Transmission shift linkage
   - Starter motor
   - Torque converter. Tighten the nuts to 20–34 ft. lbs. (27–46 Nm).
   - Flywheel access cover
   - Driveshaft. Align the matchmarks and tighten the bolts to 57–75 ft. lbs. (88–119 Nm).
   - Differential mounting nuts. Tighten the front nuts to 57–75 ft. lbs. (88–119 Nm) and the rear nuts to 45–59 ft. lbs. (60–80 Nm).
   - Transmission wiring harness connector
   - Negative battery cable
6. Turn the air suspension switch to the **ON** position.
7. Fill the transmission.
8. Start the engine and check the transmission for leakage.
9. Road test the vehicle and check for proper transmission operation.

Halfshaft

REMOVAL & INSTALLATION

1. Before servicing the vehicle, refer to the precautions in the beginning of this section.
2. The air suspension switch, located on the left-hand side of the luggage compartment, must be turned to the **OFF** position before raising the vehicle.
3. Remove or disconnect the following:
   - Rear wheel
   - Hub retainer nut
   - Disc brake caliper and rotor
   - Wheel speed sensor
4. Place a support under the lower control arm
5. Remove or disconnect the following:
   - Upper control arm from the knuckle
   - Lower control arm bolts
6. Press the outboard CV-joint from the hub and remove the knuckle.
7. Remove the halfshaft from the differential.

To install:

➡Use new nuts, bolts, circlips, and split pins for assembly.

8. Install or connect the following:
   - Inner stub shaft into the differential until the circlip seats
   - Outer stub shaft into the knuckle
   - Upper control arm to the knuckle. Tighten the bolt to 117–142 ft. lbs. (158–193 Nm).
   - Lower control arm bolts. Tighten the bolts to 83–113 ft. lbs. (113–153 Nm).
   - Wheel speed sensor
   - Disc brake rotor and caliper. Tighten the caliper support bolts to 73–97 ft. lbs. (98–132 Nm).
   - Hub nut. Tighten it to 188–254 ft. lbs. (255–345 Nm).
   - Wheel
9. Turn the air suspension switch to the **ON** position.
10. Check the rear wheel alignment and adjust as necessary.

CV-Joints

OVERHAUL

Inner Tripod Joint

1. Before servicing the vehicle, refer to the precautions in the beginning of this section.
2. Remove or disconnect the following:
3. Remove the axle halfshaft from the vehicle.
4. Remove or disconnect the following:
   - Tripod joint boot clamps
   - Tripod joint housing
   - Tripod joint retaining ring
   - Tripod joint circlip
   - Tripod joint
   - Tripod joint boot

To install:

➡Use new circlips and boot clamps for assembly.

5. Install or connect the following:
   - Tripod joint boot
   - Tripod joint
   - Tripod joint circlip
   - Tripod joint retaining ring

- Tripod joint housing
- Tripod joint boot clamps

Outer CV-Joint

The outer CV-joint is serviced with the halfshaft as an assembly. The outer CV-joint boot can be serviced by removing the inner CV-joint.

Pinion Seal

REMOVAL & INSTALLATION

1. Before servicing the vehicle, refer to the precautions in the beginning of this section.
2. The air suspension switch, located on the left-hand side of the luggage compartment, must be turned to the **OFF** position before raising the vehicle.
3. Remove or disconnect the following:
   - Rear wheels
   - Rear brake calipers

➡The rear brake calipers must be removed so that there is no additional drag when measuring pinion bearing preload.

   - Differential front retaining nuts. Support the differential.
   - Rear differential upper bolt
   - Driveshaft
4. Use an inch lb. torque wrench and measure the amount of torque required to maintain pinion rotation through several revolutions.
5. Remove the pinion flange and remove the seal.

To install:

6. Install or connect the following:
   - Pinion seal and flange
   - New pinion flange nut
7. Rotate the pinion flange occasionally while tightening the flange nut to make sure the pinion bearings seat correctly.
8. Take frequent bearing preload torque readings.
9. If the preload recorded prior to disassembly is **lower** than the specification for used bearings, then tighten the pinion flange nut to specification. If the preload recorder prior to disassembly is **higher** than the specification for used bearings, then tighten the pinion flange nut to the original reading as recorded.
10. The pinion bearing preload specifications are as follows:
    a. Used bearings: 8–14 inch lbs. (0.9–1.6 Nm).
    b. New bearings: 16–29 inch lbs. (1.8–3.2 Nm).

⁂ CAUTION

Never loosen the pinion nut to reduce bearing preload. If it is necessary to reduce bearing preload, install a new collapsible spacer and pinion nut.

11. Install or connect the following:
    - Differential front retaining nuts. Tighten the nuts to 68–100 ft. lbs. (92–136 Nm).
    - Rear differential upper bolt. Tighten the bolts to 80–100 ft. lbs. (108–136 Nm).
    - Driveshaft
    - Brake calipers
    - Rear wheels
12. Fill the differential with gear lubricant and check for leaks.
13. Turn the air suspension switch to the **ON** position.

STEERING AND SUSPENSION

Air Bag

⁂ CAUTION

Some vehicles are equipped with an air bag system. The system must be disarmed before performing service on, or around, system components, the steering column, instrument panel components, wiring and sensors. Failure to follow the safety precautions and the disarming procedure could result in accidental air bag deployment, possible injury and unnecessary system repairs.

PRECAUTIONS

Several precautions must be observed when handling the inflator module to avoid accidental deployment and possible personal injury.

- Never carry the inflator module by the wires or connector on the underside of the module.
- When carrying a live inflator module, hold securely with both hands, and ensure that the bag and trim cover are pointed away.
- Place the inflator module on a bench or other surface with the bag and trim cover facing up.
- With the inflator module on the bench, never place anything on or close to the module that may be thrown in the event of an accidental deployment.

DISARMING

1. Before servicing the vehicle, refer to the precautions in the beginning of this section.
2. Disconnect both battery cables from the battery, negative cable first.
3. Wait 1 minute before proceeding with the service procedure. This is the time required for the back-up power supply in the air bag diagnostic monitor to deplete its stored energy.
4. After service is completed, reconnect the battery cables, negative cable last.
5. Turn the ignition switch to the **RUN** position. The air bag indicator should light continuously for approximately 6 seconds, then turn **OFF**. If the indicator fails to light, flashes or remains lit continuously, there is a fault in the air bag system.

Power Rack and Pinion Steering Gear

REMOVAL & INSTALLATION

1. Before servicing the vehicle, refer to the precautions in the beginning of this section.
2. The air suspension switch, located on the left-hand side of the luggage compartment, must be turned to the **OFF** position before raising the vehicle.
3. Remove or disconnect the following:
   - Front wheels
   - Outer tie rod ends
   - Engine oil filter
   - Power steering fluid pressure and return lines
   - Steering shaft pinch bolt
   - Steering gear retaining nuts
   - Steering gear

To install:

4. Install or connect the following:
   - Steering gear. Tighten the retaining

nuts to 100–143 ft. lbs. (135–195 Nm).
- Steering shaft pinch bolt. Tighten the bolt to 30–42 ft. lbs. (41–57 Nm).
- Power steering fluid pressure and return lines
- Engine oil filter
- Outer tie rod ends. Tighten the retaining nuts to 40–53 ft. lbs. (53–72 Nm).
- Front wheels

5. Fill the power steering fluid reservoir.
6. Check the engine oil level.
7. Turn the air suspension switch to the **ON** position.
8. Align the vehicle.

Strut

REMOVAL & INSTALLATION

Front

CAUTION

Do not attempt to service the suspension without first deflating the air springs.

➡Servicing the air suspension requires a scan tool with bi-directional capabilities to deflate the air springs.

1. Before servicing the vehicle, refer to the precautions in the beginning of this section.
2. The air suspension switch, located on the left-hand side of the luggage compartment, must be turned to the **OFF** position before raising the vehicle.
3. Deflate the air springs.
4. Remove or disconnect the following:
   - Front wheel
   - Suspension height sensor
   - Air spring solenoid valve
   - Front strut cover cup
   - Upper strut mount nuts
   - Lower strut mount bolt
5. Remove the strut and spring assembly from the vehicle.

To install:

6. Install or connect the following:
   - Strut and spring assembly. Tighten the upper mount nuts to 17–23 ft. lbs. (23–32 Nm) and the lower mounting bolt to 126–169 ft. lbs. (170–230 Nm).
   - Front strut cover cup
   - Air spring solenoid valve
   - Suspension height sensor
   - Front wheel
7. Turn the air suspension switch to the **ON** position.

Shock Absorber

REMOVAL & INSTALLATION

Rear

➡The weight of the vehicle must be supported by the suspension during this procedure.

1. Before servicing the vehicle, refer to the precautions in the beginning of this section.
2. The air suspension switch, located on the left-hand side of the luggage compartment, must be turned to the **OFF** position before raising the vehicle.
3. Remove or disconnect the follow-ing:
   - Rear compartment trim panels
   - Upper shock absorber retaining nuts
   - Lower shock absorber bolt
   - Shock absorber

To install:

4. Install or connect the following:
   - Shock absorber
   - Upper shock absorber retaining nuts. Tighten the nuts to 17–23 ft. lbs. (23–31 Nm).
   - Lower shock absorber bolt. Tighten the bolt to 83–113 ft. lbs. (113–153 Nm).
   - Rear compartment trim panels
5. Turn the air suspension switch to the **ON** position.

Air Spring

REMOVAL & INSTALLATION

Front

The front air spring is serviced with the front strut as an assembly.

Rear

CAUTION

Do not attempt to service the suspension without first deflating the air springs.

➡Servicing the air suspension requires a scan tool with bi-directional capabilities to deflate the air springs.

1. Before servicing the vehicle, refer to the precautions in the beginning of this section.
2. The air suspension switch, located on the left-hand side of the luggage compartment, must be turned to the **OFF** position before raising the vehicle.
3. Deflate the air springs.
4. Remove or disconnect the following:
   - Rear wheel
   - Air spring solenoid valve harness connector
   - Air spring solenoid valve air line
   - Air spring solenoid valve
5. Depress the 4 plastic locking fingers at the bottom of the air spring piston and detach the air spring from the lower control arm.
6. Depress the metal locking tab at the top of the air spring and rotate to disengage the air spring cap from the body bracket.
7. Remove the air spring.

To install:

8. Install the air spring solenoid valve.
9. Install the air spring cap to the body bracket and rotate to lock in place.
10. Insert the 4 plastic locking fingers to the lower control arm.
11. Install or connect the following:
    - Air spring solenoid valve air line
    - Air spring solenoid valve harness connector
    - Rear wheel
12. Turn the air suspension switch to the **ON** position.

Upper Ball Joint

REMOVAL & INSTALLATION

The upper ball joint is serviced with the upper control arm as an assembly.

Lower Ball Joint

REMOVAL & INSTALLATION

1. Before servicing the vehicle, refer to the precautions in the beginning of this section.
2. Remove the lower control arm from the vehicle and place it in a vise.
3. Press the lower ball joint out of the control arm.

To install:

4. Press lower ball joint into the control arm so that it is fully seated to the control arm.
5. Install the control arm to the vehicle.
6. Align the vehicle.

Upper Control Arm

REMOVAL & INSTALLATION

Front

⁂ CAUTION

Do not attempt to service the suspension without first deflating the air springs.

➡Servicing the air suspension requires a scan tool with bi-directional capabilities to deflate the air springs.

1. Before servicing the vehicle, refer to the precautions in the beginning of this section.
2. The air suspension switch, located on the left-hand side of the luggage compartment, must be turned to the **OFF** position before raising the vehicle.
3. Deflate the air springs.
4. Remove or disconnect the follow-ing:
   - Front wheel
   - Air suspension height sensor
   - Upper ball joint
   - Upper control arm bushing joint bolts
   - Upper control arm

To install:

➡Use new control arm bushing joint and ball joint pinch bolts for assembly.

➡Tighten the bushing bolts with the suspension at curb height.

5. Install or connect the following:
   - Upper control arm
   - Upper control arm bushing joint bolts. Tighten the bolts to 56–88 ft. lbs. (75–120 Nm).
   - Upper ball joint. Tighten the pinch bolt to 50–68 ft. lbs. (68–92 Nm).
   - Air suspension height sensor
   - Front wheel
6. Turn the air suspension switch to the **ON** position.

Rear

⁂ CAUTION

Do not attempt to service the suspension without first deflating the air springs.

➡Servicing the air suspension requires a scan tool with bi-directional capabilities to deflate the air springs.

1. Before servicing the vehicle, refer to the precautions in the beginning of this section.
2. The air suspension switch, located on the left-hand side of the luggage compartment, must be turned to the **OFF** position before raising the vehicle.
3. Deflate the air springs.
4. Remove or disconnect the following:
   - Rear wheel
   - Upper control arm pivot bolts
   - Upper control arm

To install:

➡Tighten the bushing bolts with the suspension at curb height.

5. Install or connect the following:
   - Upper control arm
   - Upper control arm pivot bolts. Tighten the inner bolt to 50–68 ft. lbs. (68–92 Nm) and the outer bolt to 117–142 ft. lbs. (158–193 Nm).
   - Rear wheel
6. Check the wheel alignment and adjust as necessary.

CONTROL ARM BUSHING REPLACEMENT

Front

1. Before servicing the vehicle, refer to the precautions in the beginning of this section.
2. Remove the control arm from the vehicle.
3. Remove the control arm bushings with a hydraulic press.

To install:

4. Press the bushings into the control arm.
5. Install the control arm to the vehicle.
6. Check the wheel alignment and adjust as necessary.

Rear

The control arm bushings are serviced with the control arm as an assembly.

Lower Control Arm

REMOVAL & INSTALLATION

Front

⁂ CAUTION

Do not attempt to service the suspension without first deflating the air springs.

➡Servicing the air suspension requires a scan tool with bi-directional capabilities to deflate the air springs.

1. Before servicing the vehicle, refer to the precautions in the beginning of this section.
2. The air suspension switch, located on the left-hand side of the luggage compartment, must be turned to the **OFF** position before raising the vehicle.
3. Deflate the air springs.
4. Remove or disconnect the following:
   - Front wheel
   - Lower ball joint
   - Lower arm strut
   - Strut and spring assembly lower bolt
   - Pivot bolt
   - Lower control arm

To install:

➡Tighten the bushing bolts with the suspension at curb height.

5. Install or connect the following:
   - Lower control arm
   - Pivot bolt. Tighten the bolt to 82–114 ft. lbs. (110–155 Nm).
   - Strut and spring assembly lower bolt. 118–162 ft. lbs. (160–220 Nm).
   - Lower arm strut. Tighten the nut to 89–118 ft. lbs. (120–160 Nm).
   - Lower ball joint. Tighten the nut to 83–113 ft. lbs. (113–153 Nm).
   - Front wheel
6. Turn the air suspension switch to the **ON** position.
7. Check the wheel alignment and adjust as necessary.

Rear

CAUTION

Do not attempt to service the suspension without first deflating the air springs.

➡Servicing the air suspension requires a scan tool with bi-directional capabilities to deflate the air springs.

1. Before servicing the vehicle, refer to the precautions in the beginning of this section.
2. The air suspension switch, located on the left-hand side of the luggage compartment, must be turned to the **OFF** position before raising the vehicle.
3. Deflate the air springs.
4. Remove or disconnect the following:
   - Rear wheel
   - Brake caliper and rotor
   - Rear wheel knuckle and hub
   - Air spring
   - Parking brake cable
   - Sway bar link
   - Lower shock absorber bolt
   - Control arm inboard pivot bolts
   - Lower control arm

To install:

➡Use a new axle hub nut for assembly.

➡Tighten the bushing bolts with the suspension at curb height.

5. Install or connect the following:
   - Lower control arm
   - Control arm inboard pivot bolts. Tighten the front bolt to 166–203 ft. lbs. (225–275 Nm) and the rear bolt to 141–191 ft. lbs. (191–259 Nm).
   - Lower shock absorber bolt. Tighten the bolt to 83–113 ft. lbs. (113–153 Nm).
   - Sway bar link. Tighten the nuts to 96–120 inch lbs. (10–14 Nm).
   - Parking brake cable
   - Air spring
   - Rear wheel knuckle and hub. Tighten the lower bolts to 94–131 ft. lbs. (128–178 Nm) and the upper bolt to 117–142 ft. lbs. (158–193 Nm).
   - Axle hub nut. Tighten the hub nut to 188–254 ft. lbs. (255–345 Nm).
   - Brake caliper and rotor
   - Rear wheel
6. Turn the air suspension switch to the **ON** position.
7. Check the wheel alignment and adjust as necessary.

CONTROL ARM BUSHING REPLACEMENT

Front

1. Before servicing the vehicle, refer to the precautions in the beginning of this section.
2. Remove the control arm from the vehicle.
3. Remove the control arm bushings with a hydraulic press.

To install:

4. Press the bushings into the control arm.
5. Install the control arm to the vehicle.
6. Check the wheel alignment and adjust as necessary.

Rear

The control arm bushings are serviced with the control arm as an assembly.

Wheel Bearings

ADJUSTMENT

The wheel bearings are not adjustable. They are serviced by replacement only.

REMOVAL & INSTALLATION

Front

1. Before servicing the vehicle, refer to the precautions in the beginning of this section.
2. Remove or disconnect the following:
   - Front wheel
   - Disc brake caliper and rotor
   - Grease cap
   - Spindle retainer nut
   - Hub and bearing assembly

To install:

3. Install or connect the following:
   - Hub and bearing assembly. Tighten the spindle retainer nut to 188–254 ft. lbs. (255–345).
   - Grease cap
   - Disc brake caliper and rotor. Tighten the mounting bolts to 87 ft. lbs. (119 Nm).
   - Front wheel

Rear

CAUTION

Do not attempt to service the suspension without first deflating the air springs.

➡Servicing the air suspension requires a scan tool with bi-directional capabilities to deflate the air springs.

1. Before servicing the vehicle, refer to the precautions in the beginning of this section.
2. The air suspension switch, located on the left-hand side of the luggage compartment, must be turned to the **OFF** position before raising the vehicle.
3. Deflate the air springs.
4. Remove or disconnect the following:
   - Rear wheel
   - Axle hub nut
   - Brake caliper and support
   - Upper control arm pivot bolt
   - Lower control arm pivot bolts
   - Rear wheel knuckle and hub
5. Press the hub out of the wheel bearing.
6. Remove the wheel bearing snapring.
7. Press the bearing out of the knuckle.

To install:

➡Use a new axle hub nut for assembly.

➡Tighten the bushing bolts with the suspension at curb height.

8. Press the bearing into the knuckle until it bottoms out in the bore.
9. Install the snapring.
10. Press the hub into the bearing.
11. Install or connect the following:
    - Rear wheel knuckle and hub. Tighten the lower bolts to 94–131 ft. lbs. (128–178 Nm) and the upper bolt to 117–142 ft. lbs. (158–193 Nm).
    - Axle hub nut. Tighten the hub nut to 188–254 ft. lbs. (255–345 Nm).
    - Brake caliper and rotor
    - Rear wheel
12. Turn the air suspension switch to the **ON** position.
13. Check the wheel alignment and adjust as necessary.

FORD MOTOR CO.

19

1998–01

Ford-Crown Victoria • **Lincoln-**Town Car • **Mercury-**Grand Marquis

PRECAUTIONS

Before servicing any vehicle, please be sure to read all of the following precautions, which deal with personal safety, prevention of component damage, and important points to take into consideration when servicing a motor vehicle:

- Never open, service or drain the radiator or cooling system when the engine is hot; serious burns can occur from the steam and hot coolant.
- Observe all applicable safety precautions when working around fuel. Whenever servicing the fuel system, always work in a well-ventilated area. Do not allow fuel spray or vapors to come in contact with a spark, open flame, or excessive heat (a hot drop light, for example). Keep a dry chemical fire extinguisher near the work area. Always keep fuel in a container specifically designed for fuel storage; also, always properly seal fuel containers to avoid the possibility of fire or explosion. Refer to the additional fuel system precautions later in this section.
- Fuel injection systems often remain pressurized, even after the engine has been turned **OFF**. The fuel system pressure must be relieved before disconnecting any fuel lines. Failure to do so may result in fire and/or personal injury.
- Brake fluid often contains polyglycol ethers and polyglycols. Avoid contact with the eyes and wash your hands thoroughly after handling brake fluid. If you do get brake fluid in your eyes, flush your eyes with clean, running water for 15 minutes. If eye irritation persists, or if you have taken brake fluid internally, IMMEDIATELY seek medical assistance.
- The EPA warns that prolonged contact with used engine oil may cause a number of skin disorders, including cancer. You should make every effort to minimize your exposure to used engine oil. Protective gloves should be worn when changing oil. Wash your hands and any other exposed skin areas as soon as possible after exposure to used engine oil. Soap and water, or waterless hand cleaner should be used.
- All new vehicles are now equipped with an air bag system, often referred to as a Supplemental Restraint System (SRS) or Supplemental Inflatable Restraint (SIR) system. The system must be disabled before performing service on or around system components, steering column, instrument panel components, wiring and sensors. Failure to follow safety and disabling procedures could result in accidental air bag deployment, possible personal injury and unnecessary system repairs.
- Always wear safety goggles when working with, or around, the air bag system. When carrying a non-deployed air bag, be sure the bag and trim cover are pointed away from your body. When placing a non-deployed air bag on a work surface, always face the bag and trim cover upward, away from the surface. This will reduce the motion of the module if it is accidentally deployed. Refer to the additional air bag system precautions later in this section.
- Clean, high quality brake fluid from a sealed container is essential to the safe and proper operation of the brake system. You should always buy the correct type of brake fluid for your vehicle. If the brake fluid becomes contaminated, completely flush the system with new fluid. Never reuse any brake fluid. Any brake fluid that is removed from the system should be discarded. Also, do not allow any brake fluid to come in contact with a painted surface; it will damage the paint.
- Never operate the engine without the proper amount and type of engine oil; doing so WILL result in severe engine damage.
- Timing belt maintenance is extremely important. Many models utilize an interference-type, non-freewheeling engine. If the timing belt breaks, the valves in the cylinder head may strike the pistons, causing potentially serious (also time-consuming and expensive) engine damage. Refer to the maintenance interval charts in the front of this manual for the recommended replacement interval for the timing belt, and to the timing belt section for belt replacement and inspection.
- Disconnecting the negative battery cable on some vehicles may interfere with the functions of the on-board computer system(s) and may require the computer to undergo a relearning process once the negative battery cable is reconnected.
- When servicing drum brakes, only disassemble and assemble one side at a time, leaving the remaining side intact for reference.

ENGINE REPAIR

Alternator

REMOVAL

1. Before servicing the vehicle, refer to the precautions in the beginning of this section.
2. Remove or disconnect the following:
   - Negative battery cable
   - Engine cover
   - Pushpins and wiring harness
   - Mounting bolts
   - Alternator bracket
   - Accessory drive belt
   - Alternator

INSTALLATION

Install or connect the following:

- Alternator. Tighten the mounting bolts to 15–22 ft. lbs. (20–30 Nm).
- Alternator bracket. Tighten the bolts to 71–106 inch lbs. (8–12 Nm).
- Accessory drive belt
- Pushpins and wiring harness
- Engine cover
- Negative battery cable

ADJUSTMENT

The 4.6L engine is equipped with a Distributorless Ignition System (DIS). No adjustments are necessary.

Engine Assembly

REMOVAL & INSTALLATION

1. Before servicing the vehicle, refer to the precautions in the beginning of this section.

2. Drain the engine cooling system.
3. Recover the refrigerant from the air conditioning system.
4. Relieve the fuel system pressure.
5. Drain the engine oil.
6. Remove or disconnect the following:
- Battery
- Hood
- Engine cooling fan, shroud and radiator
- Windshield wiper governor and support bracket
- Engine air cleaner outlet tube
- Engine/transmission harness connector from the retaining bracket
- Accelerator and cruise control cables at the throttle body
- Electrical connector and vacuum hose from the evaporative emission canister purge valve
- Positive battery cable from the power distribution box and harness
- Vacuum supply hose from the throttle body adapter vacuum port
- Both heater hoses
- Alternator harness from the front fender apron and the power distribution box
- Air conditioning hoses from the air conditioning compressor
- Power steering control valve harness connector
- Body ground strap from the dash panel
- Exhaust system from the exhaust manifolds and support with wire hung from the crossmember
- Retaining nut from the transmission line bracket
- 3 bolts and 1 stud retaining the engine to the transmission knee braces
- Starter motor
- 4 bolts retaining the power steering pump to the cylinder block and position aside

7. Transmission housing cover from the cylinder block to access the torque converter nuts.

Rotate the crankshaft until each of the 4 nuts is accessible and remove the nuts

8. Remove or disconnect the following:
- 6 transmission-to-engine retaining bolts
- Engine support insulator (mount) through-bolts

9. Support the transmission with a floor jack and a block of wood.
- Bolt retaining the right-hand front engine support insulator to the front engine mount insulator support bracket

10. Install engine lifting bracket to the front of the left-hand cylinder head and to the rear of the right-hand cylinder head. Connect engine lifting equipment to the lifting brackets

11. Raise the engine slightly using a floor crane and carefully separate the engine from the transmission. Do not let the torque converter fall out of the transmission.

12. Carefully lift the engine out of the engine compartment and position on a workstand. Remove the engine lifting equipment.

To install:

13. Engine lifting brackets. Support the engine using a floor crane installed to the lifting equipment and remove the engine from the workstand

14. Lower the engine into the engine compartment. Start the converter pilot into the flywheel and align the paint marks on the flywheel and torque converter. Be sure the studs on the torque converter align with the holes in the flywheel.

15. Fully engage the engine to the transmission and lower onto front engine support insulators.

16. Install or connect the following:
- Engine lifting equipment and brackets
- Bolt retaining the right-hand front engine support insulator to the front engine mount insulator support bracket
- 6 engine-to-transmission retaining bolts and tighten to 30–44 ft. lbs. (40–60 Nm)
- Front engine support insulator through-bolts and tighten to 15–22 ft. lbs. (20–30 Nm)
- 4 torque converter retaining nuts and tighten to 22–25 ft. lbs. (20–30 Nm)
- Transmission housing cover to the cylinder block
- Power steering pump on the cylinder block and the 4 retaining nuts. Tighten to 15–22 ft. lbs. (20–30 Nm).
- Starter motor
- Engine-to-transmission brace and the 3 bolts and 1 stud. Tighten the bolts and stud to 18–31 ft. lbs. (25–43 Nm).
- Transmission line bracket to the brace stud and 1 retaining nut. Tighten to 15–22 ft. lbs. (20–30 Nm).
- Exhaust system to the exhaust manifolds. Tighten the 4 nuts to 20–30 ft. lbs. (27–41 Nm). Be sure the exhaust system clears the No. 3 crossmember. Adjust as necessary.
- Power steering valve harness connector
- Ground strap to the dash panel
- Air conditioning lines to the air conditioning compressor
- Alternator harness at the front fender apron and the power distribution box
- Both heater hoses
- Vacuum supply hose to the throttle body adapter vacuum port
- Positive battery cable to the power distribution box and harness
- Electrical connector and vacuum hose to the evaporative emission canister purge valve
- Accelerator and cruise control cables at the throttle body
- Engine/transmission harness connector to the retaining bracket on the power brake booster
- Windshield wiper governor and support bracket
- Fuel supply and return lines
- Radiator, cooling fan and shroud
- Engine air cleaner outlet tube
- Hood
- Battery

17. Fill the crankcase to the correct level.
18. Fill the cooling system.
19. Start the engine and allow it to reach normal operating temperature.
20. Check for leaks and proper fluid levels.
21. Evacuate and recharge the air conditioning system.
22. Road test the vehicle and check the engine and transmission for proper operation.

Water Pump

REMOVAL & INSTALLATION

1. Before servicing the vehicle, refer to the precautions in the beginning of this section.
2. Drain the cooling system.
3. Remove or disconnect the following:
- Negative battery cable
- Cooling fan and shroud
- Accessory drive belt

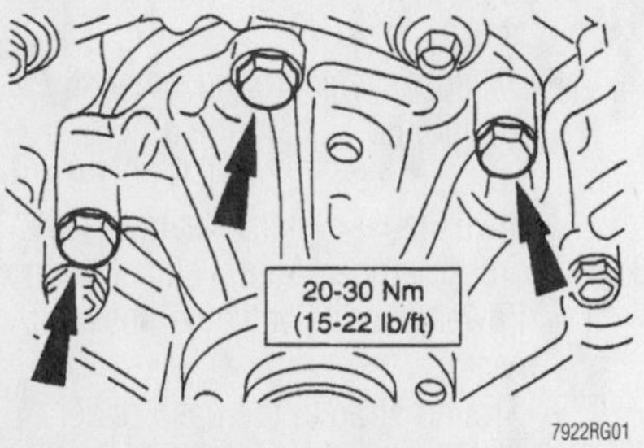

Be sure to tighten the water pump mounting bolts to the specification

- 4 water pump pulley-to-water pump bolts
- Pulley
- 4 water pump-to-engine bolts
- Water pump

To install:

4. Clean the sealing surfaces of the water pump and block.
5. Install or connect the following:
   - New O-ring, lubricate it with clean antifreeze prior to installation
   - Water pump. Tighten the bolts to 15–22 ft. lbs. (20–30 Nm).
   - Water pump pulley. Tighten the bolts to 15–22 ft. lbs. (20–30 Nm).
   - Accessory drive belt
6. Fill the cooling system.
7. Operate the engine to normal operating temperatures and check for leaks.

Cylinder Head

REMOVAL & INSTALLATION

➡The cylinder head bolts are a torque-to-yield design and cannot be reused. Always use new cylinder head bolts for assembly.

1. Before servicing the vehicle, refer to the precautions in the beginning of this section.
2. If equipped with air suspension, the air suspension switch, located on the right-hand side of the luggage compartment, must be turned to the **OFF** position before raising the vehicle.
3. Drain the engine cooling system.
4. Properly relieve the fuel system pressure.
5. Remove or disconnect the following:
   - Negative battery cable
   - Cooling fan and shroud assembly
   - Engine air cleaner outlet tube
   - Windshield wiper governor (module)
   - Accessory drive belt
   - Ignition wires from the spark plugs
   - Ignition wire brackets from the cylinder head cover studs
   - 2 ignition wire tray-to-ignition coil brackets bolts
   - Bolt retaining the air conditioning pressure line to the right-hand ignition coil bracket
   - Wiring to both ignition coils and the Camshaft Position (CMP) sensor
   - Ignition coil brackets-to-engine front cover nuts. Slide the ignition coil brackets and ignition wire assemblies off the mounting studs and from the vehicle
   - Water pump pulley
   - Alternator wiring harness from the junction block, fender apron and alternator
   - Alternator
   - Positive battery cable at the power distribution box
   - Retaining bolt from the positive battery cable bracket located on the side of the right-hand cylinder head
   - Vent hose from the canister purge solenoid and position the positive battery cable aside
   - Positive Crankcase Ventilation (PCV) valve from the cylinder head cover
   - Engine/transmission harness connector from the retaining bracket on the power brake booster
   - Crankshaft Position (CKP) sensor, air conditioning compressor clutch and canister purge solenoid electrical connectors
6. Remove the bolts retaining the power steering pump to the cylinder block and engine front cover. The front lower bolt on the power steering pump will not come all the way out. Wire the power steering pump aside.
7. Remove or disconnect the following:
   - Engine oil pan and oil pan gasket
   - Crankshaft pulley retaining bolt
   - Pulley
   - Power steering control valve actuator and oil pressure sensor wiring connectors and position aside.
   - Exhaust Gas Recirculation (EGR) tube from the right-hand exhaust manifold
   - Exhaust pipes from the exhaust manifolds. Lower the exhaust pipes and hang with wire from the crossmember.
   - Bolts retaining the starter wiring harness to the rear of the right-hand cylinder head
   - Cylinder head covers to the cylinder heads
   - Accelerator and cruise control cables
   - Accelerator cable bracket from the intake manifold and position aside
   - Vacuum hose from the throttle body elbow vacuum port
   - Heated Oxygen (HO_2S) sensors and the heater water hose
   - 2 bolts retaining the thermostat housing to the intake manifold and position the upper hose and thermostat housing aside

➡The 2 thermostat housing bolts also retain the intake manifold.

   - 9 intake manifold-to-cylinder heads bolts
   - Intake manifold and gaskets
   - 7 stud bolts and the 4 bolts attaching the engine front cover to the engine
   - Front cover
   - Both timing chains
   - 10 left-hand cylinder head-to-cylinder block bolts
8. Remove the cylinder head. The lower rear cylinder head bolt must stay in the cylinder head until the cylinder head is removed due to lack of clearance for removal in the vehicle. Use a rubber band to secure the cylinder head bolt in the cylinder head during removal and installation of the cylinder head and to prevent the bolt from damaging the cylinder block or head gasket

➡The lower rear cylinder head bolt cannot be removed due to interference with the power brake booster. Use a rubber band to hold the bolt away from the cylinder block.

9. Remove or disconnect the following:
   - Ground strap, 1 stud and 1 bolt retaining the heater return line to the right-hand cylinder head
   - 10 right-hand cylinder head-to-cylinder block bolts
10. Remove the cylinder head. The lower

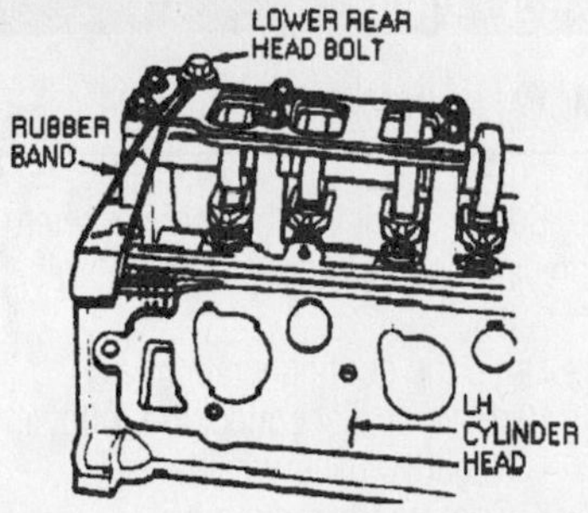

Use a rubber band to support the rear cylinder head bolt to ease removal of the head

rear cylinder head bolt must stay in the cylinder head until the cylinder head is removed due to lack of clearance for removal in the vehicle. Use a rubber band to secure the cylinder head bolt in the cylinder head during removal and installation of the cylinder head and to prevent the bolt from damaging the cylinder block or head gasket.

➡The lower rear cylinder head bolt cannot be removed due to interference with the evaporator housing. Use a rubber band to hold the bolt away from the cylinder block.

11. Clean all gaskets mating surfaces. Check the cylinder heads and cylinder block for flatness. Check the cylinder heads for scratches near the coolant passages and combustion chambers that could provide leak paths.

To install:

12. Rotate the crankshaft counterclockwise 45 degrees. The crankshaft keyway should be at the 9 o'clock position viewed from the front of the engine. This ensures that all pistons are below the top of the engine block deck face.
13. Rotate the camshaft to a stable position where the valves do not extend below the head face.
14. Install or connect the following:
    - New head gaskets on the cylinder block
    - New bolts in the lower rear bolt holes on both cylinder heads and retain with rubber bands as explained during the removal procedure
15. Position the cylinder heads on the cylinder block dowels, being careful not to score the surface of the head face. Apply clean oil to the new cylinder head bolts, remove the rubber bands from the lower rear bolts and install all bolts hand-tight.
16. Tighten the new cylinder head bolts, in sequence, as follows:
    a. Step 1: 28–31 ft. lbs. (37–43 Nm).

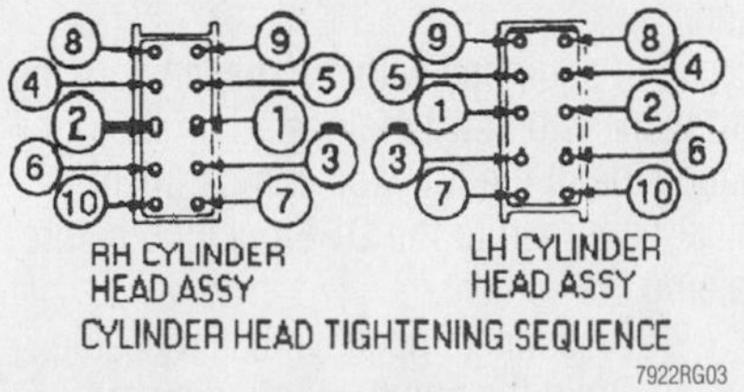

Cylinder head torque sequence—4.6L SOHC engine

    b. Step 2: plus 85–95 degrees.
    c. Step 3: loosen all bolts at least 1 full turn.
    d. Step 4: 27–32 ft. lbs. (37–43 Nm).
    e. Step 5: plus 85–95 degrees.
    f. Step 6: again, plus 85–95 degrees.
17. Position the heater return hose and install the 2 retaining bolts.
18. Rotate the camshafts using the flats matched at the center of the camshaft until both are in time. Install Camshaft Positioning Tools T91P-6256-A, on the flats of the camshafts to keep them from rotating.
19. Rotate the crankshaft clockwise 45 degrees to position the crankshaft at Top Dead Center (TDC) for the No. 1 cylinder.

➡The crankshaft must only be rotated in the clockwise direction and only as far as TDC.

20. Install or connect the following:
    - Both timing chains
    - New engine front cover seal and gasket. Apply silicone sealer to the lower corners of the cover where it meets the junction of the engine oil pan and cylinder block and to the points where the cover contacts the junction of the cylinder block and the cylinder heads.
    - Engine front cover and the bolts. Tighten to 15–22 ft. lbs. (20–30 Nm).
    - New intake manifold gaskets on the cylinder heads. Be sure the alignment tabs on the gaskets are aligned with the holes in the cylinder heads.
    - Intake manifold on the cylinder heads and the retaining bolts. Tighten the bolts in sequence, to 15–22 ft. lbs. (20–30 Nm).
    - Thermostat, O-ring, thermostat housing and upper hose. Tighten the 2 retaining bolts to 15–22 ft. lbs. (20–30 Nm).
    - Heater water hose and both HO_2S sensors
    - Vacuum hose to the throttle body adapter vacuum port
    - Accelerator cable bracket on the intake manifold
    - Accelerator and cruise control cables to the throttle body
21. Apply silicone sealer to both places where the engine front cover meets the cylinder heads.
    - Cylinder head covers with new gasket on the cylinder heads. Tighten the bolts and stud bolts to 71–106 inch lbs. (8–12 Nm).
    - Starter motor wiring harness to the right-hand cylinder head and tighten the retaining bolt
    - Exhaust pipes to the exhaust manifolds. Tighten the 4 nuts to 20–30 ft. lbs. (27–41 Nm).

➡Be sure the exhaust system clears the No. 3 crossmember. Adjust as necessary.

    - EGR tube to the right-hand exhaust manifold and tighten the line nut to 26–33 ft. lbs. (35–45 Nm).
    - Power steering control valve actuator and oil pressure sensor electrical connectors
22. Apply a small amount of silicone sealer in the rear of the keyway on the crankshaft pulley.
    - Pulley on the crankshaft, making sure the crankshaft key and keyway are aligned.
23. Install the crankshaft pulley and tighten the bolt as follows:
    a. Step 1: Tighten to 66 ft. lbs. (90 Nm).
    b. Step 2: Loosen one complete turn.
    c. Step 3: Tighten to 35–39 ft. lbs. (47–53 Nm).
    d. Step 4: Tighten an additional 85–95 degrees.
24. Install or connect the following:
    - Engine oil pan and a new gasket
    - Power steering pump in position on the cylinder block
    - 4 retaining bolts. Tighten the bolts to 15–22 ft. lbs. (20–30 Nm).
    - Air conditioning compressor, CKP sensor and canister purge solenoid electrical connectors
    - Engine/transmission harness connector on the power brake booster
    - PCV valve in the right-hand cylinder head cover and connect the canister purge solenoid vent hose
    - Positive battery cable harness on the right-hand cylinder head
    - Bolt retaining the cable bracket to the cylinder head
    - Positive battery cable at the power distribution box and battery
    - Alternator and the 2 retaining bolts. Tighten the bolts to 15–22 ft. lbs. (20–30 Nm).
    - 2 bolts retaining the alternator brace to the intake manifold. Tighten to 72–96 inch lbs. (8–12 Nm).

- Water pump pulley. Tighten the bolts to 15–22 ft. lbs. (20–30 Nm).
- Ignition coil brackets and ignition wire assemblies onto the mounting studs
- 7 nuts retaining the ignition coil brackets to the engine front cover and tighten to 15–22 ft. lbs. (20–30 Nm)
- 2 bolts retaining the ignition wire tray to the ignition coil bracket and tighten to 71–106 inch lbs. (8–12 Nm)
- Ignition coil and CMP sensor harness connectors
- Air conditioning pressure line on the right-hand ignition coil bracket and tighten the retaining bolt
- Ignition wires to the spark plugs and the bracket onto the cylinder head cover studs
- Accessory drive belt and the windshield wiper governor
- Fuel supply and return lines
- Cooling fan and shroud
- Engine air cleaner outlet tube
- Negative battery cable

25. Fill the cooling system.
26. If equipped with air suspension, turn the air suspension switch to the **ON** position.
27. Refill the engine with the correct amount of oil and replace the filter.
28. Start the engine and bring to normal operating temperature while checking for leaks.
29. Road test the vehicle and check for proper engine operation.

Rocker Arms

REMOVAL & INSTALLATION

1. Before servicing the vehicle, refer to the precautions in the beginning of this section.
2. Relieve the fuel system pressure.
3. Disconnect the negative battery cable.
4. Remove the right camshaft cover by removing or disconnecting the following:
   - Positive battery cable at the battery and at the power distribution box
   - Retaining bolt from the positive battery cable bracket located on the side of the right cylinder head
   - Crankshaft Position (CKP) sensor, air conditioning compressor clutch and canister purge solenoid connectors. Position the harness aside.
   - Vent hose from the purge solenoid and position the positive battery cable aside
   - Ignition wires from the spark plugs
   - Ignition wire brackets from the camshaft cover studs and position the wires aside
   - PCV valve from the camshaft cover grommet and position aside
   - Bolts and stud bolts and remove the camshaft cover.
5. Remove the left camshaft cover by removing or disconnecting the following:
   - Air inlet tube
   - Fuel lines
   - PSP switch and oil pressure sending unit and position the harness aside
   - 42-pin engine harness connector from the retaining bracket on the brake vacuum booster and position aside
   - Windshield wiper module
   - Ignition wires from the spark plugs
   - Ignition wire brackets from the studs and position the wires aside
   - Camshaft cover
6. Position the piston of the cylinder being serviced at the bottom of its stroke and position the camshaft lobe on the base circle.
7. Install valve spring spacer tool T91P-6565-AH between the spring coils to prevent valve seal damage.
8. Install a valve spring compressor under the camshaft and on top of the valve spring retainer.
9. Compress the valve spring and remove the roller follower. Remove the valve spring compressor and spacer.

To install:

10. Apply engine oil to the valve stem and tip and roller follower contact surfaces.
11. Install valve spring spacer tool T91P-6565-AH between the spring coils. Compress the valve spring, and install the roller follower.

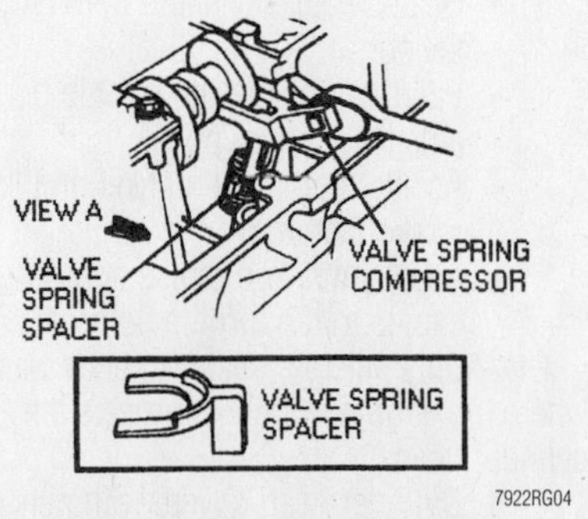

After installing the spring spacer, compress the valve spring and remove the rocker arm

➡**The piston must be at the bottom of its stroke and the camshaft at the base circle.**

12. Remove the valve spring compressor and spacer.
13. Clean the sealing surfaces of the camshaft covers and cylinder heads. Apply silicone sealer to the places where the front cover meets the cylinder head.
14. Position new gaskets onto the camshaft covers and install the covers. Install the bolts and stud bolts and tighten to 72–106 inch lbs. (8–12 Nm).
15. When installing the right camshaft cover, install or connect the following:
   - PCV into the camshaft cover grommet
   - Ignition wire brackets on the studs
   - Wires to the spark plugs
   - Canister purge solenoid, air conditioning compressor clutch and CKP sensor
   - Positive battery cable harness on the right cylinder head
   - Bolt retaining the cable bracket to the cylinder head
   - Positive battery cable at the power distribution box and the battery
16. When installing the left camshaft cover, install or connect the following:
   - Ignition wire brackets on the studs
   - Wires to the spark plugs
   - Windshield wiper module
   - 42-pin and transmission harness connectors
   - Retaining bracket
   - PSP switch and oil pressure sending unit harness.
   - Fuel lines
   - Negative battery cable
17. Start the engine and check for leaks.

Intake Manifold

REMOVAL & INSTALLATION

1. Before servicing the vehicle, refer to the precautions in the beginning of this section.
2. If equipped with air suspension, the air suspension switch, located on the right-hand side of the luggage compartment, must be turned to the **OFF** position before raising the vehicle.
3. Disconnect negative battery cable.
4. Drain the engine cooling system.
5. Properly relieve the fuel system pressure.
6. Remove or disconnect the following:
   - Fuel supply and return lines

- Windshield wiper governor (module)
- Engine air cleaner outlet tube
- Accessory drive belt
- Ignition wires from the spark plugs
- Ignition wire brackets from the cylinder head cover studs
- Ignition coils and the Camshaft Position (CMP) sensor
- Ignition wires from both ignition coils
- 2 bolts retaining the ignition wire bracket to the ignition coil brackets
- Ignition wire assembly
- Alternator wiring harness from the junction block at the fender apron and alternator
- Bolts retaining the alternator brace to the intake manifold and the alternator to the cylinder block
- Alternator
- Oil pressure sensor and power steering control valve actuator wiring and position the wiring harness aside
- Exhaust Gas Recirculation (EGR) valve-to-exhaust manifold tube from the right-hand exhaust manifold
- Engine/transmission harness connector from the retaining bracket on the power brake booster
- Air conditioning compressor clutch, Crankshaft position (CKP) sensor and the canister purge solenoid wiring connectors
- Positive Crankcase Ventilation (PCV) valve from the cylinder head cover
- Canister purge vent hose from the PCV valve
- Accelerator and cruise control cables from the throttle body
- Accelerator cable bracket from the intake manifold and position aside
- Vacuum hose from the throttle body adapter port
- Heated Oxygen (HO_2S) sensor and the heater water hose
- 2 bolts retaining the thermostat housing to the intake manifold and position the upper hose and thermostat housing aside

➡The 2 thermostat housing bolts are also used to retain the intake manifold.

- 9 bolts retaining the intake manifold to the cylinder heads
- Intake manifold and gaskets

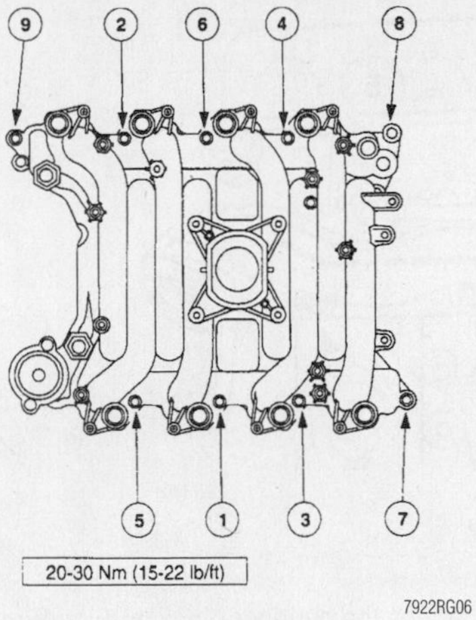

Intake manifold torque sequence—4.6L SOHC engine

7. If replacing the intake manifold, swap over the necessary parts.

To install:

8. Clean all gaskets mating surfaces.
9. Position new intake manifold gaskets on the cylinder heads. Be sure the alignment tabs on the gaskets are aligned with the holes in the cylinder heads.
10. Install the intake manifold and the 9 retaining bolts. Hand-tighten the right-rear bolt (viewed from the front of the engine) before final tightening, then tighten the bolts, in sequence, to 15–22 ft. lbs. (20–30 Nm).
11. Inspect and if necessary, replace the O-ring seal on the thermostat housing. Position the housing and upper hose and install the 2 retaining bolts. Tighten to 15–22 ft. lbs. (20–30 Nm).
12. Install or connect the following:
    - Heater water hose
    - HO_2S sensor
    - Vacuum hose to the throttle body adapter vacuum port
    - Accelerator cable bracket on the intake manifold
    - Accelerator and cruise control cables to the throttle body
    - PCV valve in the cylinder head cover
    - Canister purge solenoid vent hose
    - Air conditioning compressor clutch, CKP sensor and canister purge solenoid wiring connectors
    - Engine/transmission harness connector the retaining bracket on the power brake booster
    - EGR valve-to-exhaust manifold tube to the right-hand exhaust manifold. Tighten the tube nut to 26–33 ft. lbs. (35–45 Nm).
    - Power steering control valve actuator
    - Oil pressure sensor wiring connectors
    - Alternator. Tighten the bolts to 15–22 ft. lbs. (20–30 Nm).
    - 2 bolts retaining the alternator brace to the intake manifold and tighten to 71–106 inch lbs. (8–12 Nm)
    - Alternator wiring harness to the alternator, right-hand fender apron and junction block
    - Ignition wire assembly on the engine
    - 2 bolts retaining the ignition wire bracket to the ignition coil brackets. Tighten the bolts to 71–106 inch lbs. (8–12 Nm).
    - Ignition wires to the ignition coils
    - Ignition wires to the spark plugs
    - Ignition wire brackets on the cylinder head cover studs
    - Wiring connectors to both ignition coils and the CMP sensor
    - Accessory drive belt
    - Air cleaner outlet tube
    - Windshield wiper governor
    - Fuel supply and return lines
    - Negative battery cable
13. Fill the engine cooling system.
14. If equipped with air suspension, turn the air suspension switch to the **ON** position.
15. Start the engine and check for leaks.
16. Road test the vehicle and check for proper operation.

Exhaust Manifold

REMOVAL & INSTALLATION

1. Before servicing the vehicle, refer to the precautions in the beginning of this section.
2. Drain the engine cooling system.
3. Relieve the fuel system pressure.
4. Discharge the air conditioning system
5. Remove or disconnect the following:
    - Battery cables
    - Engine air inlet tube
    - Cooling fan and shroud assembly
    - Fuel supply and return lines
    - Upper radiator hose
    - Windshield wiper governor and support bracket
    - Compressor outlet hose at the

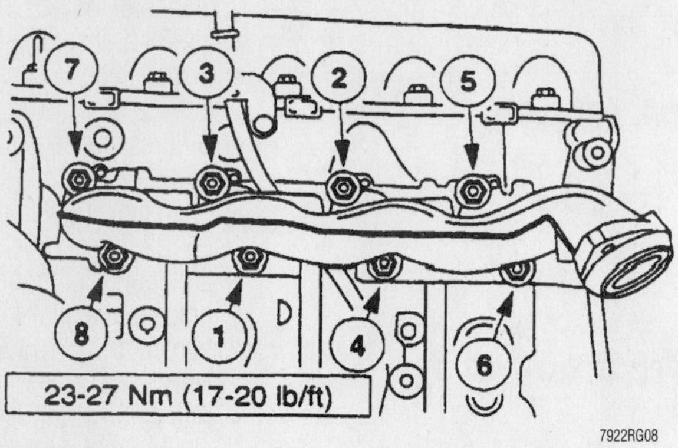

Left exhaust manifold torque sequence

compressor and the hose assembly-to-right-hand ignition coil bracket bolt. Plug both openings.

- Engine/transmission harness connector from the retaining bracket on the power brake booster
- Heater hose
- Ground strap-to-right-hand cylinder head nut
- Upper stud and lower bolt retaining the heater hose to the right cylinder head and position aside
- Heater blower motor switch resistor
- Bolt retaining the right-hand front engine support insulator to the sub-frame
- Both Heated Oxygen (HO_2S) sensors
- Engine support insulator through-bolts
- Exhaust Gas Recirculation (EGR) valve-to-exhaust manifold tube nut from the right-hand exhaust manifold.
- Catalytic converter pipes from both exhaust manifolds. Lower the exhaust system and hang it from the crossmember with wire.
- Left-hand exhaust manifold
- Front engine support insulator from the cylinder block and the 8 nuts retaining the exhaust manifold
- Left-hand exhaust manifold and the 2 manifold gaskets

6. Position an adjustable jackstand and a block of wood under the engine oil pan, rearward of the oil drain hole. Raise the engine approximately 4 inches (100mm).
7. Install or connect the following:
   - 8 exhaust manifold retaining nuts and right-hand exhaust manifold
   - Manifold and gasket

To install:

8. If the exhaust manifolds are being replaced, transfer the heated O_2sensors and tighten to 27–33 ft. lbs. (37–45 Nm). On the right-hand exhaust manifold, transfer the EGR tube connector and tighten to 33–48 ft. lbs. (45–65 Nm).
9. Clean the mating surfaces of the exhaust manifolds and cylinder heads.
10. Install or connect the following:
   - Exhaust manifolds, using new gaskets. Tighten the bolts in sequence to 15–22 ft. lbs. (20–30 Nm).
   - EGR valve and tube assembly to the exhaust manifold. Tighten the line nut to 26–33 ft. lbs. (35–45 Nm).
   - Left-hand front engine support insulator to the cylinder block and tighten the bolts to 15–22 ft. lbs. (20–30 Nm)
11. Lower the engine onto the front engine support insulator and remove the jack.
12. Install or connect the following:
   - Left-hand and right-hand engine support insulator through-bolts and tighten to 15–22 ft. lbs. (20–30 Nm).
   - Catalytic converter pipes to both exhaust manifolds. Tighten he nuts to 20–30 ft. lbs. (27–41 Nm).

➡**Be sure the exhaust system clears the No. 3 crossmember. Adjust as necessary.**

   - Both HO_2S sensors
   - Bolt retaining the right-hand front engine support insulator to the sub-frame. Tighten to 15–22 ft. lbs. (20–30 Nm)
   - Heater blower motor switch resistor using the 2 retaining screws
   - Heater hose in position
   - Upper stud and lower bolt and tighten to 15–22 ft. lbs. (20–30 Nm)
   - Ground strap onto the stud and tighten the nut to 15–22 ft. lbs. (20–30 Nm)
   - Heater hose
   - Engine/transmission harness connector
   - Retaining bracket on the power brake booster
   - Air conditioning compressor outlet hose to the compressor
   - Bolt retaining the hose assembly to the right-hand ignition coil bracket
   - Upper radiator hose
   - Fuel supply and return lines
   - Windshield wiper governor and retaining bracket
   - Engine cooling fan blade and fan shroud
   - Engine air inlet tube
   - Battery cables
13. Fill the cooling system.
14. Start the engine and check for leaks.
15. Properly evacuate and charge the air conditioning system.
16. Road test the vehicle and check for proper operation.

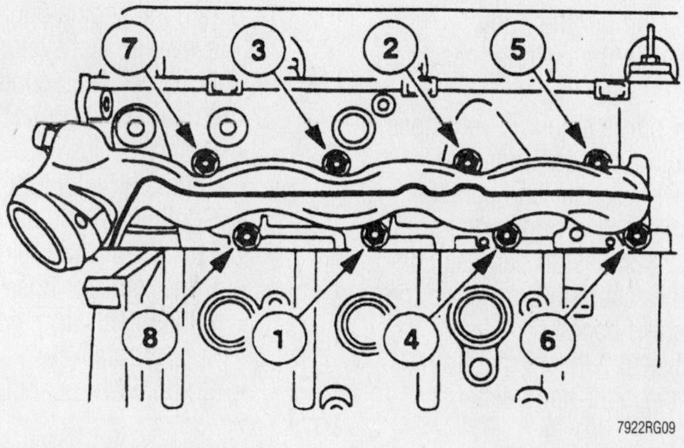

Right exhaust manifold torque sequence

Camshaft and Valve Lifters

REMOVAL & INSTALLATION

1. Before servicing the vehicle, refer to the precautions in the beginning of this section.

2. If equipped with air suspension, the air suspension switch, must be turned to the **OFF** position before raising the vehicle.

3. Properly relieve the fuel system pressure.

4. Drain the engine oil

5. Remove or disconnect the following:

- Negative battery cable
- Fan blade and fan shroud assembly
- Fuel supply and return lines from the fuel injection supply manifold
- Windshield wiper governor (module) assembly from the vehicle
- Engine air cleaner outlet tube
- Accessory drive belt
- Ignition wires from the spark plugs
- Ignition wire brackets from the cylinder head cover studs
- 2 bolts retaining the ignition wire separator to the ignition coil brackets and the bolt retaining the air conditioning pressure line to the right-hand ignition coil bracket.
- Connectors from both ignition coils and the Crankshaft Position (CMP) sensor
- Ignition coils with brackets attached
- Electrical connector from the alternator and at the power distribution box
- Water pump pulley
- Positive battery cable at the power distribution box
- Bolt from the positive battery cable bracket located on the right-hand cylinder head
- Fuel vapor hose from the EVAP canister purge valve and position the positive battery cable aside
- Positive Crankcase Ventilation (PCV) valve from the cylinder head cover and position aside
- Engine/transmission harness connector from the bracket on the power brake booster
- Crankshaft Position (CKP) sensor and air conditioning clutch harness connectors
- Bolts retaining the power steering pump to the cylinder block and wire the pump aside

➡The front lower bolt on the power steering pump will not come all the way out.

- Oil pan
- Crankshaft pulley bolt and washer and crankshaft pulley
- Engine oil filter
- Power steering control valve actuator and oil pressure sensor
- Oil filter adapter
- Cylinder head covers
- Engine front cover
- Timing chains

6. Rotate the crankshaft counterclockwise no more than 45 degrees from Top Dead Center (TDC) to ensure that all pistons are below the top of the engine block deck face.

7. Install a valve spring compressor under the camshaft and on top of one of the valve spring retainers.

8. Install Valve Spring Spacer T91P-6565-AH between the valve spring coils. Be sure that the valve being compressed is on its base circle. Compress the valve spring and remove the rocker arm. Repeat the procedure until all rocker arms are removed.

9. If required, pull the lash adjusters out of their bores in the cylinder head. Note their locations, they must be installed in the same bore they were removed from.

➡Do not mix the camshaft bearing caps. Note the camshaft bearing cap locations for installation.

10. To remove each camshaft, unfasten the 14 bolts retaining the camshaft bearing caps (cluster assemblies) to the cylinder head. Tap upward on the camshaft bearing caps at points near the upper bearing halves and gradually lift the camshaft bearing cap clusters from the cylinder head.

11. Repeat the removal procedure for the opposite cylinder head.

12. Remove the camshaft straight upward to avoid bearing damage.

13. Clean and inspect the camshafts and related components for unusual wear or damage.

To install:

14. Clean and inspect the cylinder head covers, engine front cover and cylinder head sealing surfaces.

15. Apply clean engine oil to the camshaft journals and lobes. Position the camshafts on the cylinder heads.

16. Install and seat the camshaft bearing cap cluster assemblies. Install and hand start the retaining bolts. Tighten the camshaft cluster retaining bolts in sequence to 71–106 inch lbs. (8–12 Nm). Be sure to tighten each camshaft bearing cap cluster individually.

➡Each camshaft bearing cap cluster assembly is tightened individually.

17. Loosen the camshaft bearing cap cluster retaining bolts approximately 2 turns or until the heads of the bolts are free. Tighten all bolts, again in sequence, to 71–106 inch lbs. (8–12 Nm).

18. Repeat the camshaft bearing cap installation for the opposite cylinder head.

➡The camshafts should turn freely but with a slight drag.

19. Check camshaft end-play as follows:

a. Step 1: Install a dial indicator on the front of the engine. Position it so the indicator foot is resting on the camshaft sprocket bolt or the front of the camshaft.

b. Step 2: Push the camshaft toward the rear of the engine and zero the dial indicator.

c. Step 3: Pull the camshaft forward and release it. Specified end-play is 0.0901–0.006 inch (0.025–0.190mm).

d. Step 4: If end-play is too tight, check for binding or foreign material in the camshaft thrust bearing. If end-play is excessive, check for worn camshaft thrust plate and replace the cylinder head, as required.

e. Step 5: Remove the dial indicator.

20. If removed, install the lash adjusters in their original positions.

21. If necessary, install Camshaft Positioning Tools T92P-6256-A on the flats of the camshafts and install the spacers and camshaft sprockets. Install the bolts and washers and tighten to 81–95 ft. lbs. (110–130 Nm).

22. Install a valve spring compressor under the camshaft and on top of the valve spring retainer.

23. Install Valve Spring Spacer T91P-6565-AH between the valve spring coils. Be sure that the valve being compressed is on its base circle. Compress the valve spring and install the rocker arm. Repeat the procedure until all rocker arms are installed.

24. Rotate the crankshaft clockwise 45 degrees to position the crankshaft at Top Dead Center (TDC).

➡The crankshaft must only be rotated in the clockwise direction and only as far as TDC.

25. Install or connect the following:

- Timing chains
- Engine front cover
- Cylinder head covers. Tighten the cylinder head cover bolts to 71–106 inch lbs. (8–12 Nm).

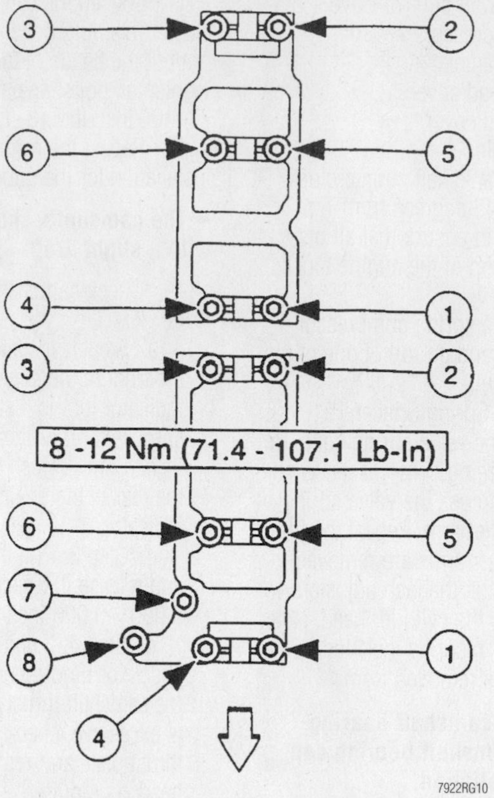

Tighten the camshaft bearing cap cluster bolts in the sequence to avoid damage to the camshaft or bearings

- Power steering control valve actuator connector
- Oil pressure sensor harness connector

26. Apply silicone sealer to the crankshaft keyway

27. Install the crankshaft pulley and tighten the bolt as follows:
    a. Step 1: Tighten to 66 ft. lbs. (90 Nm).
    b. Step 2: Loosen one complete turn.
    c. Step 3: Tighten to 35–39 ft. lbs. (47–53 Nm).
    d. Step 4: Tighten an additional 85–95 degrees.

28. Install or connect the following:
    - Engine oil pan
    - Power steering pump on the engine and the 4 retaining bolts. Tighten the bolts to 15–22 ft. lbs. (20–30 Nm).
    - Air conditioning clutch and CKP sensor
    - Evaporative emission canister purge valve harness connector
    - Engine/transmission harness connectors on the power brake booster retaining bracket
    - PCV valve to the right-hand cylinder head cover
    - Positive battery cable harness on the right-hand cylinder head
    - Bolt retaining the battery cable bracket to the cylinder head
    - Evaporative emission hose to the canister purge valve
    - Positive battery cable at the power distribution box
    - Water pump pulley and tighten the bolts to 15–22 ft. lbs. (20–30 Nm)
    - Ignition coil brackets and ignition wires to the engine front cover. Tighten the retaining nuts to 15–22 ft. lbs. (20–30 Nm).
    - Harness connectors to the ignition coils and the CMP sensor
    - Air conditioning pressure line on the right-hand ignition coil bracket and the retaining bolt
    - Ignition wires to the spark plugs and the brackets onto the cylinder head cover studs
    - Accessory drive belt
    - Windshield wiper governor
    - Fuel supply and return lines
    - Fan and shroud assembly
    - Negative battery cable

29. Fill the engine cooling system.
30. Fill the crankcase.
31. If equipped with air suspension, turn the air suspension switch to the **ON** position.
32. Start the engine and check for leaks.
33. Road test the vehicle and check for proper engine operation.

Valve Lash

ADJUSTMENT

The valve lash is not adjustable. If the collapsed lash adjuster clearance is incorrect, check the camshaft, roller follower and valve for wear or damage.

1. Before servicing the vehicle, refer to the precautions in the beginning of this section.
2. Disconnect the negative battery cable.
3. Remove the camshaft covers.
4. Rotate the crankshaft until the camshaft base circle is contacting the roller follower.
5. Use a suitable tool to bleed down the lash adjuster. Slowly compress the lash adjuster until the plunger is bottomed.
6. Use a feeler gauge to check the clearance between the camshaft and the roller follower. The clearance should be 0.018–0.033 inch (0.45–0.85mm).

Starter Motor

REMOVAL & INSTALLATION

1. Before servicing the vehicle, refer to the precautions in the beginning of this section.
2. Remove or disconnect the following:
    - Negative battery cable
    - Red solenoid safety cap
    - Wires from solenoid
    - 2 upper bolts
    - 1 lower bolt and starter

To install:

3. Install or connect the following:
    - Starter and lower mounting bolt
    - 2 upper mounting bolts. Tighten all 3 bolts to 15–20 ft. lbs. (20–27 Nm).
    - Wires from solenoid and tighten the nut to 40–50 inch lbs.
    - Red solenoid safety cap
    - Negative battery cable

Oil Pan

REMOVAL & INSTALLATION

1. Before servicing the vehicle, refer to the precautions in the beginning of this section.

2. Drain the engine cooling system

3. Properly discharge the air conditioning system.

4. Relieve the fuel system pressure.

5. Remove or disconnect the following:

- Negative battery cable
- Engine air cleaner outlet tube
- Fuel supply and return lines at the fuel injection supply manifold
- Cooling fan and fan shroud
- Upper radiator hose
- Wiper governor and support bracket
- A/C compressor outlet hose
- Engine/transmission electrical harness connector from the retaining bracket on the power brake booster
- Heater water hose
- Nut retaining the ground strap to the right-hand cylinder head
- Upper stud and loosen the lower bolt retaining the heater outlet hose to the right-hand cylinder head and position aside
- Heater blower motor switch resistor

6. Drain the engine oil and reinstall the oil pan drain plug with a new gasket. Tighten the plug to 10–12 ft. lbs. (13–16 Nm).

7. Remove or disconnect the following:

- Bolt retaining the right-hand engine support insulator to the lower front sub-frame
- Bolts retaining the left-hand and right-hand front engine support insulators to the engine mount supports
- Catalytic converter pipes from both exhaust manifolds. Lower the exhaust system and support it with wire from the transmission crossmember.

8. Position a jack and a block of wood under the oil pan, rearward of the oil drain hole. Raise the engine approximately 4 inches (100mm) and insert 2 wood blocks approximately 2½–2¾ inch (60–70mm) thick under each front engine support insulator. Lower the engine onto the wood blocks and remove the jack.

9. Remove the oil pan.

10. If necessary, remove the 2 bolts retaining the oil pick-up tube to the oil pump and remove the bolt retaining the pick-up tube to the main bearing stud spacer. Remove the pick-up tube.

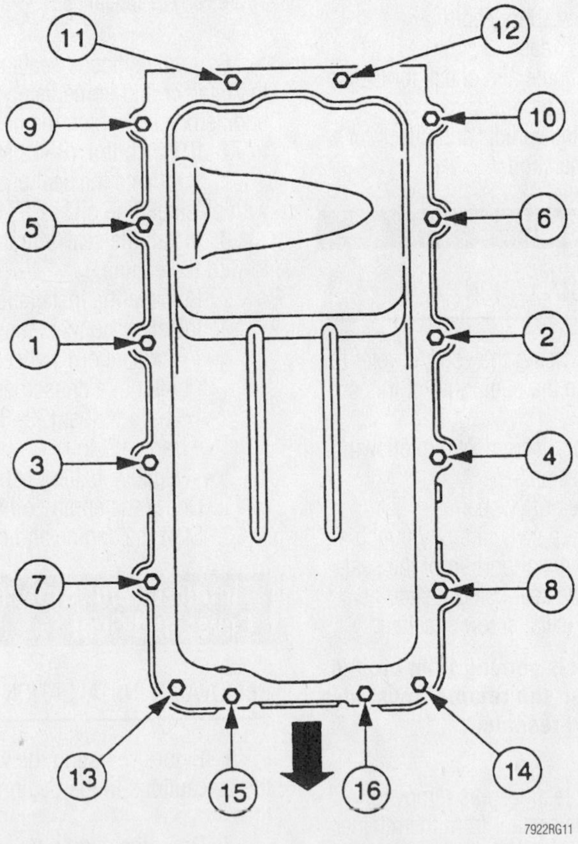

To prevent oil leaks, tighten the oil pan bolts in the sequence shown

To install:

11. Clean the engine oil pan and inspect for damage. Clean the sealing surfaces of the front cover and cylinder block. Clean and inspect the oil pick-up tube and replace the O-ring.

12. If removed, position the oil pick-up tube on the oil pump and hand start the 2 retaining bolts. Install the bolt retaining the pick-up tube on the main bearing stud spacer, hand tight.

13. Tighten the pick-up tube-to-oil pump bolts to 72–108 inch lbs. (8–12 Nm), then tighten the pick-up tube-to-main bearing stud spacer bolt to 15–22 ft. lbs. (20–30 Nm).

14. Position a new gasket on the oil pan. Apply silicone sealer to where the front cover meets the cylinder block and the crankshaft rear oil seal and retainer meets the cylinder block. Position the oil pan to the engine and install the retaining bolts. Tighten the bolts in sequence, to 14 ft. lbs. (20 Nm), then rotate the oil pan retaining bolts, in sequence an additional 60 degrees within 4 minutes of applying the silicone sealer.

15. Position the jack and wood block under the engine oil pan, rearward of the oil drain hole, and raise the engine enough to remove the wood blocks. Lower the engine and remove the jack.

16. Install or connect the following:

- Left-hand and right-hand engine support insulator through-bolts and tighten to 15–22 ft. lbs. (20–30 Nm)
- Bolt retaining the right-hand engine support insulator to the lower front sub-frame. Tighten the bolt to 15–22 ft. lbs. (20–30 Nm).
- Exhaust system to the exhaust manifolds and tighten the 4 retaining nuts to 20–30 ft. lbs. (27–41 Nm). Be sure the exhaust system clears the crossmember. Adjust as necessary.
- New engine oil filter
- Heater blower motor switch resistor using the 2 retaining screws
- Heater water hose
- Upper stud and tighten the upper and lower bolts to 15–22 ft. lbs. (20–30 Nm)
- Ground strap on the stud and tighten to 15–22 ft. lbs. (20–30 Nm)
- Heater water hose
- Throttle valve cable, if equipped

- Engine/transmission electrical harness connector
- Harness connector on the power brake booster bracket
- Air conditioning compressor outlet hose to the compressor
- Bolt retaining the hose to the right-hand ignition coil bracket
- Upper radiator hose
- Fuel supply and return lines
- Wiper governor and retaining bracket
- Engine cooling fan and fan shroud
- Engine air cleaner outlet tube
- Negative battery cable

17. Fill the cooling system.
18. Fill the engine crankcase with engine oil.
19. Start the engine and check for leaks.
20. Properly evacuate and recharge the air conditioning system.
21. Road test the vehicle and check for proper engine operation.

Oil Pump

REMOVAL & INSTALLATION

1. Before servicing the vehicle, refer to the precautions in the beginning of this section.
2. Remove or disconnect the following:
   - Negative battery cable
   - Cylinder head covers
   - Engine front cover
   - Engine oil pan
   - Timing chains
   - 2 bolts retaining the oil pick-up tube to the oil pump and the bolt attaching the oil pick-up tube to the main bearing stud spacer
   - Pick-up tube
   - 4 bolts retaining the oil pump to the cylinder block
   - Oil pump

To install:

3. Rotate the inner rotor of the oil pump to align with the flats on the crankshaft and install the oil pump flush with the cylinder block. Install the 4 retaining bolts and tighten to 72–106 inch lbs. (8–12 Nm).
4. Clean the oil pick-up tube and replace the O-ring.
5. Place the pick-up tube on the oil pump and hand start the 2 retaining bolts. Install the bolt retaining the pick-up tube to the main bearing stud spacer hand tight. Tighten the pick-up tube-to-oil pump bolts to 72–106 inch lbs. (8–12 Nm). Tighten the pick-up tube to main bearing stud spacer bolt to 15–22 ft. lbs. (20–30 Nm).

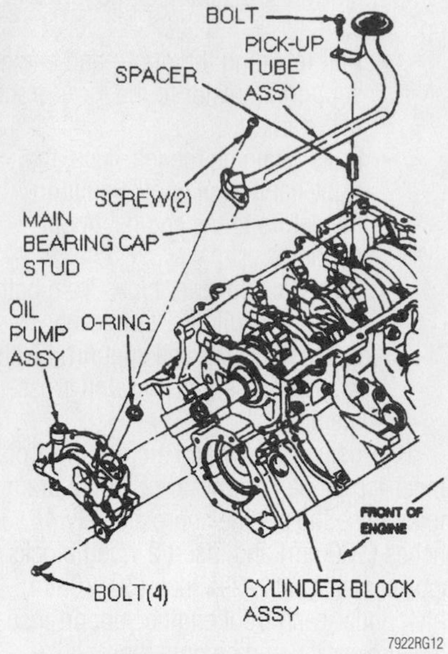

The oil pump is mounted on the crankshaft at the front of the engine

6. Install or connect the following:
   - New engine oil filter
   - Timing chains
   - Engine oil pan
   - Engine front cover
   - Cylinder head covers
   - Negative battery cable
7. Fill the crankcase.
8. Start the engine and check for leaks and proper engine oil pressure.
9. Road test the vehicle and check for proper engine operation.

Rear Main Seal

REMOVAL & INSTALLATION

1. Before servicing the vehicle, refer to the precautions in the beginning of this section.
2. Remove or disconnect the following:
   - Transmission
   - Flexplate or flywheel
3. With a sharp awl, carefully punch a small hole in the metal portion of the seal.
4. Remove the seal using a slide hammer with a sheet metal screw attached.

➡If the oil leak is coming from around the seal retainer, the retainer must also be removed and resealed.

To install:

5. If the seal retainer was removed, carefully clean the sealant from the retainer and engine block using a plastic scraper. Remove any oil or grease residue from the sealing surfaces with a solvent.

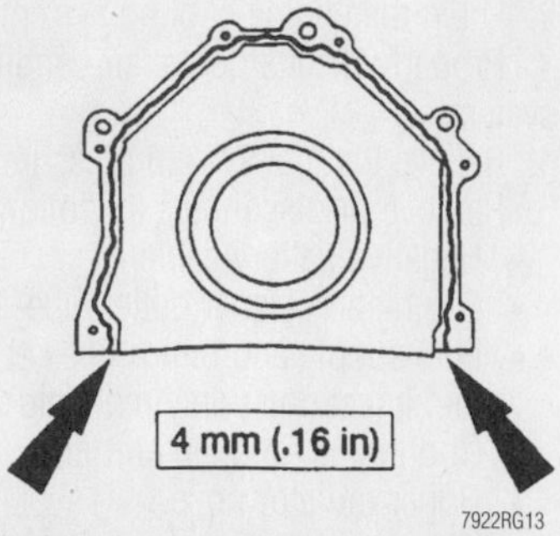

Apply a continuous bead of silicone sealant to the back of the seal retainer before installing it on the engine

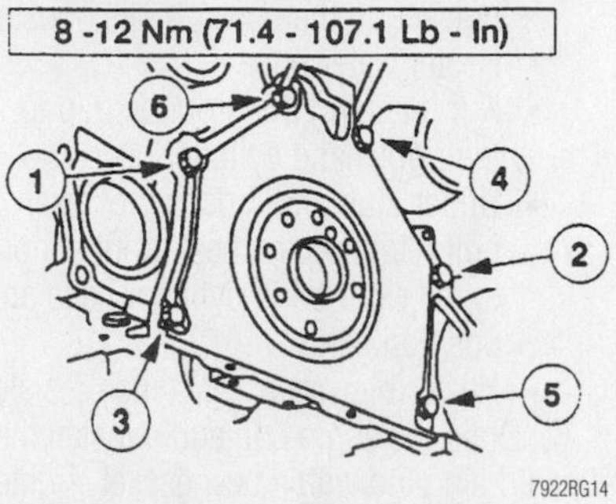

To avoid leakage, be sure to tighten the crankshaft rear oil seal retainer bolts in the correct sequence

6. Apply silicone sealant to the back of the retainer and immediately install it on the engine block. Tighten the bolts in sequence to 71–107 inch lbs. (8–12 Nm).
7. Lubricate the seal and the crankshaft with clean engine oil.
8. Install the seal with the spring side toward the engine.
9. Remove the installation tool.
10. Install or connect the following:
    - Flexplate or flywheel. Tighten the bolts, in a crisscross pattern, to 54–64 ft. lbs. (73–87 Nm).
    - Transmission
    - Negative battery cable
11. Check the engine oil level.
12. Start the engine and check for leaks.

Timing Chain, Sprockets, Front Cover and Seal

REMOVAL & INSTALLATION

1. Before servicing the vehicle, refer to the precautions in the beginning of this section.
2. Drain the engine oil.
3. Remove or disconnect the following:
   - Negative battery cable
   - Cooling fan and shroud

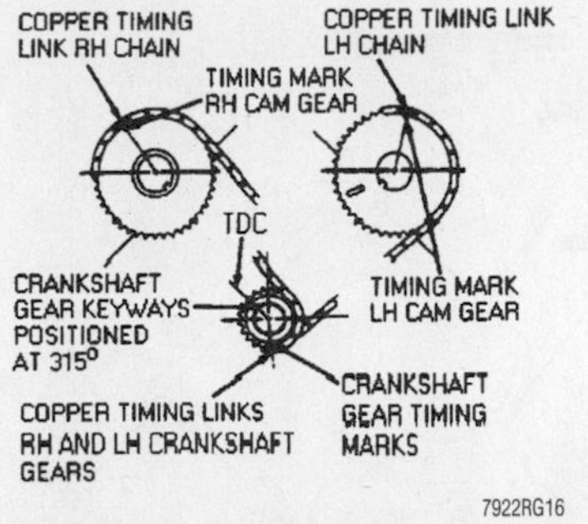

Be sure that the timing marks are aligned when the No. 1 piston is at TDC on compression

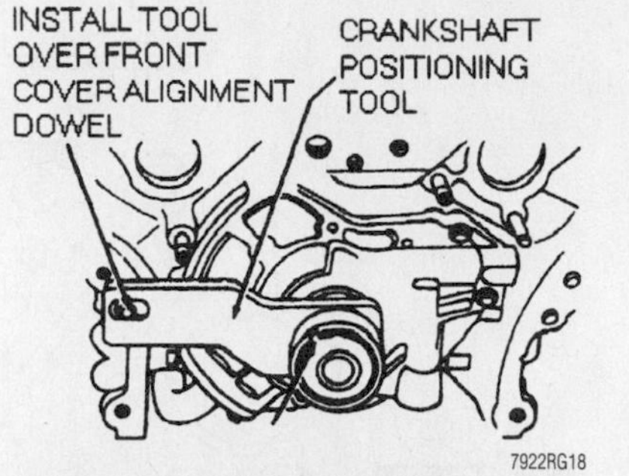

Install the crankshaft positioning tool to be sure the crankshaft does not turn while installing the timing chains

- Accessory drive belt
- Water pump pulley
- Power steering pump
- Oil pan
- Crankshaft pulley retaining bolt and washer
- Crankshaft pulley
- Bolt retaining the air conditioning pressure line to the right-hand ignition coil bracket
- Cylinder head covers
- Wiring at both ignition coils and the Crankshaft Position (CMP) sensor
- 3 bolts retaining the right-hand ignition coil bracket to the engine front cover. Position the power steering hose aside.
- 3 nuts retaining the left-hand ignition coil bracket to the engine front cover. Slide both ignition coil brackets and ignition wires off the mounting studs and lay the assembly on top of the engine.
- Bolts retaining the drive belt idler pulley and the pulley
- Wiring to the Crankshaft Position (CKP) sensor and the sensor

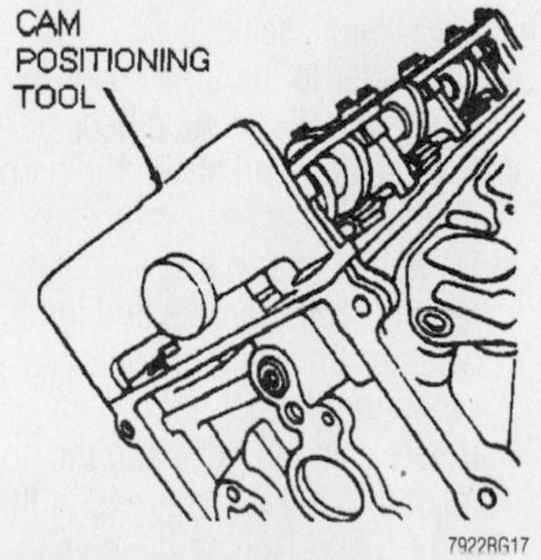

Use the special tool to maintain camshaft position while installing the timing chains

4. If equipped, remove the retainers for the oil cooler from the engine front cover retaining stud bolts and position the oil cooler aside.
   - 9 stud bolts and the 6 standard engine front cover bolts and the cover
   - Crankshaft oil seal from the cover using a seal driver
   - CKP sensor pulse wheel
5. Rotate the engine to set the piston for No. 1 to Top Dead Center (TDC) on its compression stroke.
6. Install Camshaft Positioning Adapters T92P-6256-A on the flats of both camshafts. This will prevent accidental rotation of the camshafts.
7. Remove or disconnect the following:
   - 2 bolts retaining the tensioner to the right-hand cylinder head and the tensioner
   - Right-hand timing chain tensioner arm
   - 2 bolts retaining the right-hand timing chain guide to the cylinder head and remove the timing chain guide.
   - Right-hand timing chain from the camshaft and crankshaft sprockets
   - Right-hand camshaft sprocket retaining bolt, washer, sprocket and spacer, if necessary
   - 2 bolts retaining the timing chain tensioner to the left-hand cylinder head
   - Timing chain tensioner
   - Left-hand timing chain tensioner arm
   - 2 bolts retaining the timing chain guide to the left-hand cylinder head
   - Timing chain guide
   - Left-hand timing chain from the camshaft and crankshaft sprockets
   - Left-hand camshaft sprocket retaining bolt, washer, sprocket and spacer, if necessary
8. If necessary, note the position of the crankshaft sprockets and remove the crankshaft sprockets by sliding them off the front of the crankshaft.
9. Inspect the plastic running face on the tensioner arms and chain guides. If worn or damaged, inspect the engine oil pan for contamination and thoroughly clean the oil pan. Replace the oil pick-up tube.

To install:

10. Examine the timing chains, looking for the copper links. If the copper links are not visible, lay the chain on a flat surface and pull the chain taught until the opposite sides of the chain contact one another. Mark the links at each end of the chain and use these marks in place of the copper links.
11. Be sure Camshaft Positioning Adapters T92P-6256-A are installed on the flats of the camshafts to prevent them from rotating.
12. Install or connect the following:
    - Left-hand and right-hand timing chain guides and retaining bolts. Tighten the retaining bolts to 71–106 inch lbs. (8–12 Nm).
    - Left-hand and right-hand camshaft spacers and sprockets, (if removed) on the camshafts, the washers and retaining bolts but do not tighten at this time.
    - Left-hand crankshaft sprocket with the tapered part of the sprocket facing away from the engine block
    - Left-hand timing chain on the camshaft and crankshaft sprockets. Be sure the copper links of the timing chain line up with the timing marks on both sprockets.
    - Right-hand crankshaft sprocket

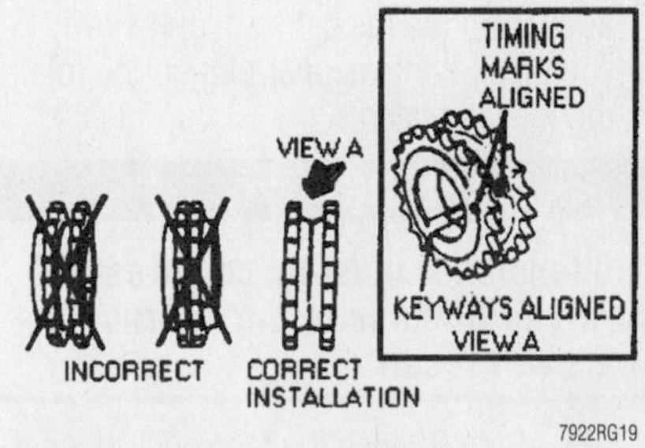

Install the crankshaft sprockets with the tapered sides facing each other

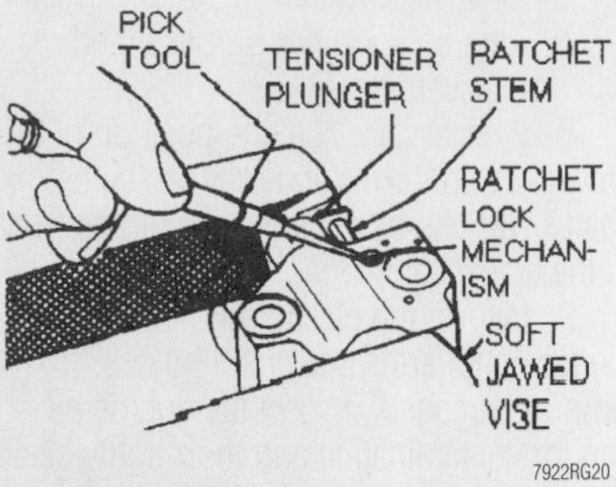

Slowly compress the timing chain tensioner while holding the ratchet lock away from the stem with a suitable tool

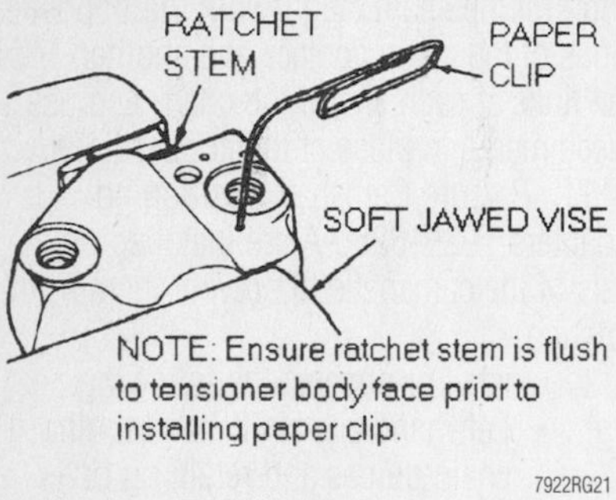

Install a paper clip or wire into the tensioner to hold the plunger in during assembly

with the tapered part of the sprocket facing the left-hand crankshaft sprocket, if removed
- Right-hand timing chain on the camshaft and crankshaft sprockets. Be sure the copper links of the timing chain line up with the timing marks on both sprockets.

13. It is necessary to bleed the timing chain tensioners before installation. Proceed as follows:

a. Step 1: position the timing chain tensioner in a soft-jawed vise.

b. Step 2: using a small pick or similar tool, hold the ratchet lock mechanism away from the ratchet stem and slowly compress the tensioner plunger by rotating the vise handle.

⁂ WARNING

The tensioner must be compressed slowly or damage to the internal seals will result.

c. Step 3: once the tensioner plunger bottoms in the tensioner bore, continue to hold the ratchet lock mechanism and push down on the ratchet stem until flush with the tensioner face.

d. Step 4: while holding the ratchet stem flush with the tensioner face, release the ratchet lock mechanism and install a paper clip or similar tool in the tensioner body to lock the tensioner in the collapsed position.

e. Step 5: the paper clip must not be removed until the timing chain, tensioner, tensioner arm and timing chain guide are completely installed on the engine.

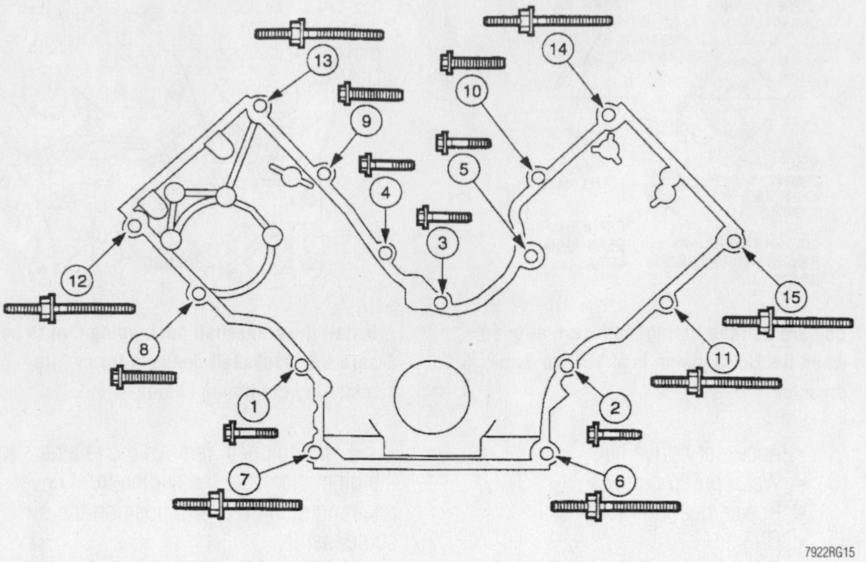

Timing chain front cover bolt tightening sequence

14. Install the right-hand and left-hand timing chain tensioners and 2 bolts on each. Tighten the bolts to 15–22 ft. lbs. (20–30 Nm).

15. Crankshaft Positioning Tool T93P-6265-A over the crankshaft and the engine front cover alignment dowel to position the crankshaft.

16. Lubricate the timing chain tensioner arm contact surfaces with clean engine oil and install the right-hand and left-hand tensioner arms on their dowel pins.

17. Position a C-clamp around the timing chain tensioner arm and timing chain guide to remove all slack from the timing chain. Use care not to bend the timing chain guide.

18. Remove the locking pins or paper clips from the timing chain tensioners and be sure that all timing marks are aligned.

19. Using Camshaft Positioning Adapters T92P-6265-A to align and hold the camshafts, tighten the camshaft sprocket retaining bolts to 81–95 ft. lbs. (110–130 Nm).

20. Position a dial indicator in the No. 1 cylinder spark plug hole to measure intake valve lift. The intake valve should be at maximum lift when the crankshaft is at 114 degrees after TDC. If the intake valve lift is not at maximum lift, loosen the camshaft sprocket bolt and repeat the steps detailing the installation of the timing chain tensioners to the tightening of the camshaft sprockets.

21. Remove the camshaft and crankshaft positioning tools.

22. Install a new crankshaft seal in the front cover. Apply engine oil to the lip of the seal.

23. Thoroughly clean the sealing surfaces of the front cover, cylinder block and oil pan. Apply silicone sealer to the points where the cylinder head meets the cylinder block.

24. Install or connect the following:
- Front cover in position using new gaskets
- Retaining bolts and studs in their proper locations. Tighten in sequence to 15–22 ft. lbs. (20–30 Nm) within 4 minutes of applying the silicone sealer.
- Oil cooler to the front cover retaining stud bolts, if equipped
- CKP sensor and attach the harness connector
- Drive belt idler pulley
- Ignition coil brackets and ignition wires as an assembly onto the mounting studs
- Power steering hose and the nuts retaining the coil brackets to the front cover. Tighten the nuts to 15–22 ft. lbs. (20–30 Nm).
- Wiring to both ignition coils and thc CMP sensor
- Cylinder head covers

- Air conditioning pressure line on the right-hand ignition coil bracket and tighten the retaining bolt

25. Apply a small amount of silicone sealer in the rear of the keyway in the crankshaft pulley.
26. Install or connect the following:
    - Pulley on the crankshaft
    - Crankshaft pulley bolt and washer and tighten to 114–121 ft. lbs. (155–165 Nm)
    - Oil pan
    - Power steering pump on the engine. Tighten the bolts to 15–22 ft. lbs. (20–30 Nm).
    - Water pump pulley. Tighten the bolts to 15–22 ft. lbs. (20–30 Nm).
    - Accessory drive belt
    - Engine cooling fan and shroud
    - Negative battery cable
27. Fill the engine.
28. Start the engine and check for leaks.
29. Road test the vehicle and check for proper engine operation.

Piston and Ring

POSITIONING

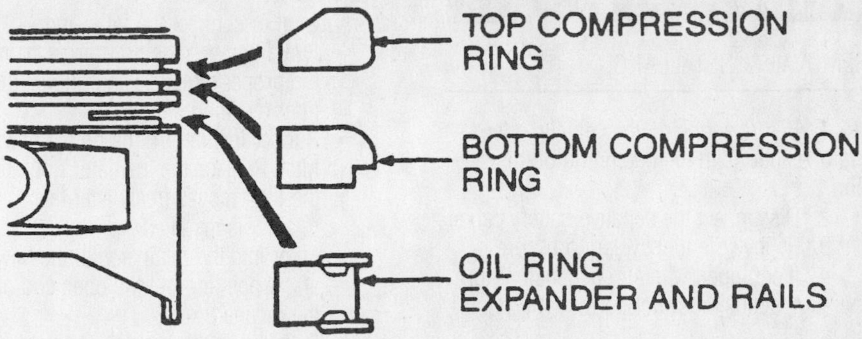

Ford 4.6L (VIN W and X) engines—piston ring positioning

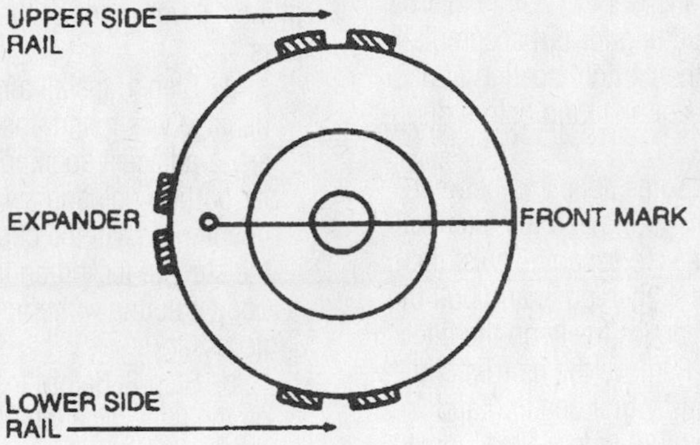

Ford 4.6L (VIN W and X) engines—piston ring end-gap spacing and piston positioning

FUEL SYSTEM

Fuel System Service Precautions

Safety is the most important factor when performing not only fuel system maintenance but any type of maintenance. Failure to conduct maintenance and repairs in a safe manner may result in serious personal injury or death. Maintenance and testing of the vehicle's fuel system components can be accomplished safely and effectively by adhering to the following rules and guidelines.

- To avoid the possibility of fire and personal injury, always disconnect the negative battery cable unless the repair or test procedure requires that battery voltage be applied.
- Always relieve the fuel system pressure prior to disconnecting any fuel system component (injector, fuel rail, pressure regulator, etc.), fitting or fuel line connection. Exercise extreme caution whenever relieving fuel system pressure, to avoid exposing skin, face and eyes to fuel spray. Please be advised that fuel under pressure may penetrate the skin or any part of the body that it contacts.
- Always place a shop towel or cloth around the fitting or connection prior to loosening to absorb any excess fuel due to spillage. Ensure that all fuel spillage (should it occur) is quickly removed from engine surfaces. Ensure that all fuel soaked cloths or towels are deposited into a waste container.
- Always keep a dry chemical (Class B) fire extinguisher near the work area.
- Do not allow fuel spray or fuel vapors to come into contact with a spark or open flame.
- Always use a back-up wrench when loosening and tightening fuel line connection fittings. This will prevent unnecessary stress and torsion to fuel line piping. Always follow the proper torque specifications.
- Always replace worn fuel fitting O-rings with new. Do not substitute fuel hose or equivalent, where fuel pipe is installed.

Fuel System Pressure

RELIEVING

Fuel supply lines on all fuel injected engines will remain pressurized for some period of time after the engine is shut **OFF**. This pressure must be relieved before servicing the fuel system. Pressure is relieved through the fuel pressure relief valve, located on the fuel rail.

To relieve the fuel system pressure, first remove the fuel tank cap to relieve pressure in the tank, then remove the cap on the fuel pressure relief valve. Attach a fuel pressure gauge and drain the system through the drain tube into a container. Remove the fuel pressure gauge and replace the cap on the relief valve.

For Accessory Drive Belt illustrations, see Section 1 of this manual

Fuel Filter

REMOVAL & INSTALLATION

1. Before servicing the vehicle, refer to the precautions in the beginning of this section.
2. Disconnect the negative battery cable.
3. Relieve the fuel system pressure.
4. If equipped with air suspension, turn the air suspension switch to the **OFF** position.
5. Remove the hairpin clip push connect fittings from both ends of the fuel filter as follows:
   a. Step 1: Inspect the visible internal portion of the fitting for dirt accumulation. If more than a light coating of dust is present, clean the fitting before disassembly.
   b. Step 2: Some adhesion between the seals in the fitting and the filter will occur with time. To separate, twist the fitting on the filter, then push and pull the fitting until it moves freely on the filter.
   c. Step 3: Remove the hairpin clip from the fitting by first bending and breaking the shipping tab. Next, spread the 2 clip legs by hand about ⅛ inch each, to disengage the body and push the legs into the fitting. Lightly pull the triangular end of the clip and work it clear of the filter and fitting.
   d. Step 4: Grasp the fitting and pull in an axial direction to remove the fitting from the filter. Be careful on 90 degree elbow connectors, as excessive side loading could break the connector body.
   e. Step 5: After disassembly, inspect the inside of the fitting for any internal parts such as O-rings and spacers that may have been dislodged from the fitting. Replace any damaged connector.
6. Remove the filter retaining clamp and remove the fuel filter. Note the direction of the flow arrow on the filter, so the replacement filter can be reinstalled in the same position.

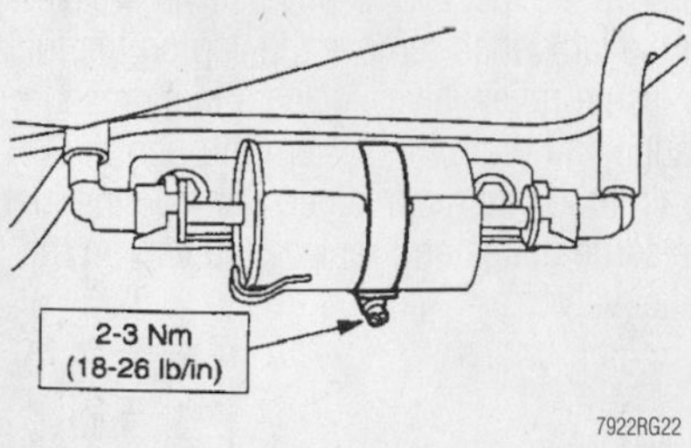

The fuel filter is located near the center of the vehicle on the frame rail

To install:

7. Install or connect the following:
   - Fuel filter with the flow arrow facing the proper direction and tighten the filter retaining clamp
   - Rubber insulator rings on the new filter. Replace the insulator rings if the filter moves freely after the retainer is installed.
   - Filter into the retainer with the flow arrow pointing out the open end of the retainer
   - Retainer on the bracket and tighten the mounting bolts to 27–44 inch lbs. (3–5 Nm)
8. Install the hairpin clip push connect fittings at both ends of the fuel filter as follows:
   a. Step 1: Install a new connector if damage was found. Insert a new clip into any 2 adjacent openings with the triangular portion pointing away from the fitting opening. Install the clip until the legs of the clip are locked on the outside of the body. Piloting with an index finger is necessary.
   b. Step 2: Before installing the fitting on the filter, wipe the filter end with a clean cloth. Inspect the inside of the fitting to be sure it is free of dirt and/or obstructions.
   c. Step 3: Apply a light coating of engine oil to the filter end. Align the fitting and filter axially and push the fitting onto the filter end. When the fitting is engaged, a definite click will be heard. Pull on the fitting to be sure it is fully engaged.
9. If equipped with air suspension, turn the air suspension switch to the **ON** position.
10. Reconnect the negative battery cable.
11. Start the engine and check for fuel leaks and proper operation.

Fuel Pump

REMOVAL & INSTALLATION

1. Before servicing the vehicle, refer to the precautions in the beginning of this section.
2. Disconnect the negative battery cable.
3. Relieve the fuel system pressure.
4. Install a hose into the fuel filler pipe and drain or siphon the fuel into a storage tank designed for fuel storage.
5. Remove any dirt that has accumulated around the fuel pump and fuel lines to prevent the entry of contaminants into the tank during fuel pump removal and installation.
6. Remove or disconnect the following:
   - Fuel supply and return line fittings at the fuel pump using fuel line disconnect tools
   - Fuel pump module electrical connector
   - 6 retaining bolts around the perimeter of the fuel pump module
   - Fuel pump module and seal from the fuel tank

To install:

7. Clean the fuel pump module mounting flange and fuel tank mounting surface.
8. Install or connect the following:
   - New seal and the fuel pump module using care not to damage the inlet filter and fuel sending unit float arm
   - 6 retaining bolts and tighten to 80–107 inch lbs. (9–12 Nm)
   - Fuel pump module electrical connector
   - Fuel supply and return lines to the fuel pump module. Pull on the fuel line fittings to verify engagement.
9. Minimum of 10 gallons (38L) of clean fuel to the fuel tank and check for leaks.
10. Install a Fuel pressure gauge to the Schrader valve on the fuel injection supply manifold.
11. Connect the negative battery cable.
12. Cycle the ignition switch from the **OFF** to **ON** position 5–10 times for 3 second intervals or until the fuel pressure gauge shows at least 35 psi (241 kPa).
13. Check for fuel leaks.
14. Remove the fuel pressure gauge.
15. Start the engine and recheck for fuel leaks.
16. Road test the vehicle and check for proper operation.

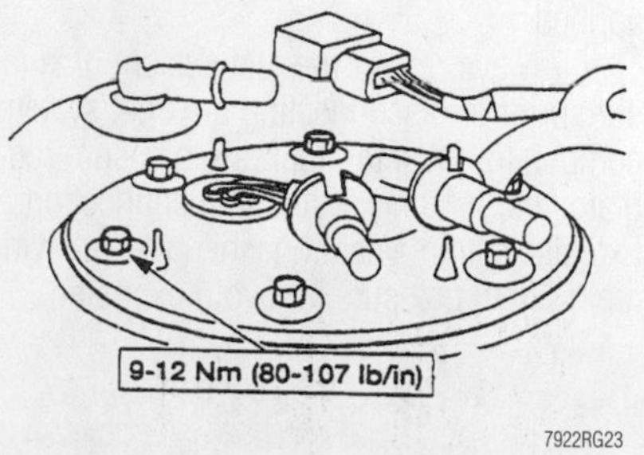

Tighten the fuel pump mounting bolts to 80–107 inch lbs. (9–12 Nm)

Fuel Injector

REMOVAL & INSTALLATION

1. Before servicing the vehicle, refer to the precautions in the beginning of this section.
2. Relieve the fuel system pressure.
3. Remove or disconnect the following:
   - Injector supply manifold
   - Wiring
   - Injector by pulling it up and gently rocking it side to side
   - O-rings and discard

To install:

4. Lubricate new O-rings with clean engine oil.
5. Install or connect the following:
   - O-rings
   - Fuel injector using a light, twisting and pushing motion
   - Injector supply manifold
   - Wiring

DRIVE TRAIN

Transmission Assembly

REMOVAL & INSTALLATION

1. Before servicing the vehicle, refer to the precautions in the beginning of this section.
2. Drain the transmission.
3. If equipped with air suspension, the air suspension switch, located on the right-hand side of the luggage compartment, must be turned to the **OFF** position before raising the vehicle.
4. Remove or disconnect the following:
   - Negative battery cable
   - Exhaust system as necessary for transmission removal
   - Converter bottom access cover and adapter plate bolts
   - Torque converter drain plug, to allow the converter to drain into a container, if equipped. After the converter has drained, reinstall the drain plug and tighten.
   - 4 torque converter-to-flywheel retaining nuts
   - Driveshaft (mark for installation), plug the transmission extension housing to prevent fluid leakage
   - Vehicle Speed Sensor (VSS) or if equipped, the speedometer cable from the transmission extension housing
   - Shift cable from the transmission manual control lever the throttle valve cable from the transmission throttle valve lever, if equipped
   - Transmission wiring harness connectors.
   - Starter motor retaining bolts and place the starter motor aside
5. Position a transmission jack under the transmission and raise it enough to allow crossmember removal.
6. Remove or disconnect the following:
   - Engine rear support-to-crossmember bolts and the crossmember-to-frame side support retaining bolts
   - Crossmember and transmission support insulator
7. Lower the transmission jack and allow the transmission to hang.
8. Place a jack to the front of the engine and raise the engine enough to gain access to the 2 upper transmission-to-cylinder block retaining bolts. Do not remove the bolts at this time.
9. Remove or disconnect the following:
   - Transmission cooler lines at the transmission. Plug all openings to keep dirt out.
   - Lower transmission-to-cylinder block retaining bolts
   - Transmission fluid fill tube and plug the opening in the transmission
10. Secure the transmission to the transmission jack with a safety strap or chain.
11. Remove the 2 upper transmission-to-cylinder block bolts.
12. Carefully move the transmission rearward to disengage the bell housing from the dowel pins and the torque converter studs from the flywheel.
13. Remove or disconnect the following:
    - Transmission
    - Torque converter to prevent the converter from dropping out of the transmission causing possible damage or personal injury

To install:

14. Remove the safety stand and block of wood supporting the rear of the engine, if installed.
15. Install or connect the following:
    - Torque converter drain plug to 21–23 ft. lbs. (28–30 Nm), if equipped
    - Torque converter on the transmission and rotate into position to be sure the drive flats are fully engaged in the pump gear. When fully seated, the center of the torque converter should be about 7/16–9/16 inch (10.2–14.4mm) below the transmission mounting surface
16. Mount the transmission on a transmission jack and secure with a safely strap or chain. Raise the transmission and align with the cylinder block dowel pins.
17. Rotate the converter until the studs and drain plug are in alignment with the holes in the flywheel. Align the orange balancing marks on the converter stud and flywheel bolt hole, if balancing marks are present.
18. Slide the transmission assembly forward into position, being careful not to damage the flywheel and converter pilot.
19. Install or connect the following:
    - 2 transmission housing-to-cylinder block bolts at the engine dowel pin locations. Tighten the bolts to 41–50 ft. lbs. (55–68 Nm).
    - Transmission housing-to-cylinder block bolts. Tighten the bolts to 41–50 ft. lbs. (55–68 Nm).
20. Remove the safety strap or chain from around the transmission.
21. Install or connect the following:
    - Transmission fluid fill tube. Tighten the bolt to 28–38 ft. lbs. (38–51 Nm).
    - Oil cooler lines to the transmission case. Tighten the cooler line fittings to 15–19 ft. lbs. (20–26 Nm).

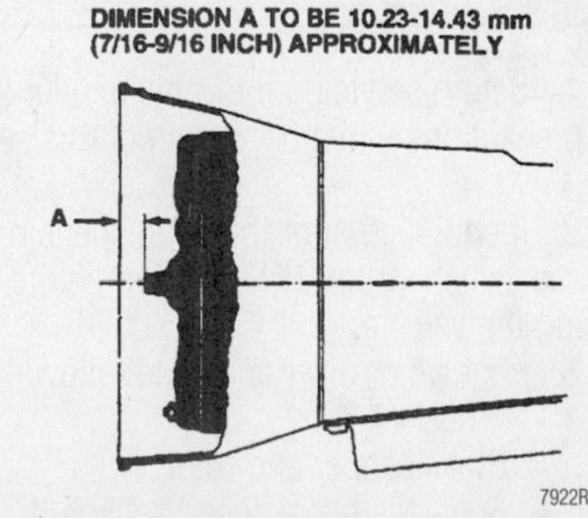

To prevent transmission damage, be sure that the torque converter is fully seated in the front pump of the transmission

22. Remove the jack supporting the front of the engine.

23. Install or connect the following:
- Crossmember using the proper jack
- Crossmember and transmission support insulators in position
- Engine rear support-to-crossmember retaining bolts and the cross-member-to-frame side support retaining bolts

24. Remove the transmission jack.

25. Install or connect the following:
- Transmission wiring harness connectors
- Starter motor and wiring
- 4 torque converter-to-flywheel retaining nuts. Tighten to 20–33 ft. lbs. (27–46 Nm).
- Torque converter access cover and cover plate bolts. Tighten the bolts to 12–16 ft. lbs. (16–22 Nm).
- Exhaust system
- VSS and the wiring, or if equipped, the speedometer cable to the transmission extension housing
- Driveshaft, aligning the marks that were made during removal
- Shift cable to the transmission manual control lever
- Throttle valve cable to the transmission throttle valve lever, if equipped

26. If equipped with air suspension, turn the air suspension switch to the **ON** position.

27. Fill the transmission.

28. Start the engine and check the transmission for leakage.

29. Road test the vehicle and check for proper transmission operation.

Axle Shaft, Bearing and Seal

REMOVAL & INSTALLATION

1. Before servicing the vehicle, refer to the precautions in the beginning of this section.

2. If equipped, turn the air suspension service switch to the **OFF** position before raising the vehicle.

3. Remove or disconnect the following
- Wheel
- Brake caliper and rotor
- Anti-lock brake speed sensor, if equipped

4. Clean all dirt from the area of the axle housing cover.

5. Place a drain pan under the axle housing.

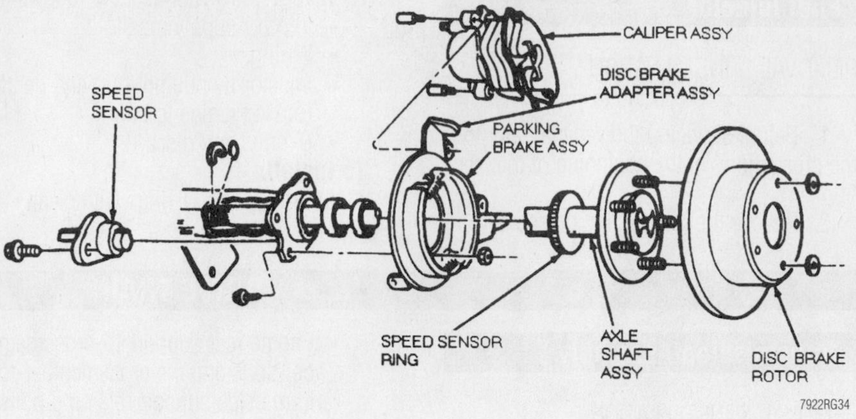

Exploded view of the rear axle shaft assembly

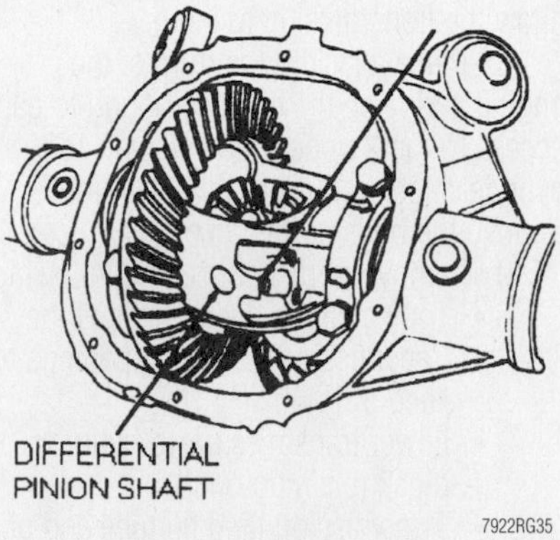

Removal of differential pinion shaft

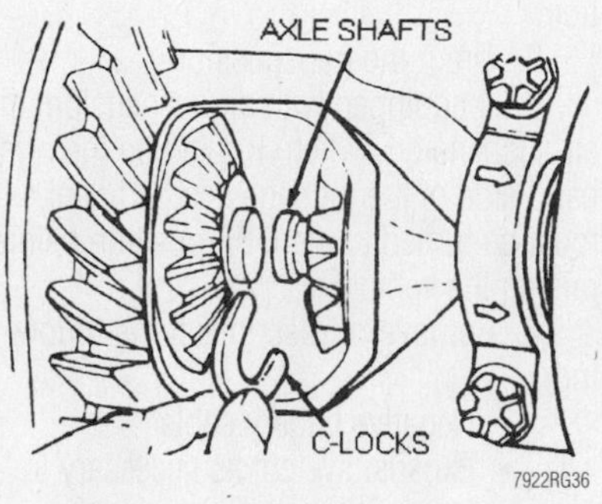

Removing axle shaft C-lock clips

6. Remove or disconnect the following:
- Axle housing cover retaining bolts and the cover, draining the axle lubricant from the housing
- Differential pinion shaft lockbolt and the differential pinion shaft

7. Push the flanged end of the axle shaft being removed toward the center of the vehicle

8. Remove or disconnect the following:
- C-lock from the button end of the axle shaft
- Axle shaft from the housing, being careful not to damage the oil seal and anti-lock brake sensor ring, if equipped.

9. Insert an axle bearing remover in the axle housing bore and position it behind the wheel bearing so the tangs on the tool engage the bearing outer race.

10. Remove the wheel bearing and seal as an assembly using an impact slide hammer attached to the bearing remover tool

To install:

11. Lubricate the new wheel bearing with rear axle lubricant.

12. Install the wheel bearing into the axle housing bore using a bearing replacer.

13. Lubricate the lips of a new wheel bearing oil seal with wheel bearing grease.

14. Install or connect the following:
- New wheel bearing seal using a seal replacer
- Axle shaft into the axle housing without damaging the bearing/seal assembly or anti-lock brake sensor ring, if equipped. Start the splines into the side gear and push firmly until the button end of the axle shaft can be seen in the differential case.
- C-lock on the button end of the axle shaft splines, then push the shaft outboard until the shaft splines engage and the C-lock seats in the counterbore of the differential side gear.
- Differential pinion shaft through the case and pinion gears, aligning the hole in the shaft with the lockbolt hole
- Apply a thread locking compound to the lockbolt threads and place in the case and pinion shaft. Tighten to 15–30 ft. lbs. (20–41 Nm).

15. Cover the inside of the differential case with a shop rag and clean the sealing surface of the axle housing and the axle housing cover. Remove the shop rag.

16. Apply a ⅛–³⁄₁₆ inch (3.18–4.76mm) wide bead of silicone sealer to the cover.

17. Install the axle housing and bolts and tighten in a crisscross pattern. Final torque the cover retaining bolts to 28–38 ft. lbs. (38–52 Nm).

18. Add the appropriate rear axle lubricant to the axle housing to a level ¼–⁹⁄₁₆ inch (6–14mm) below the bottom of the fill hole. If equipped with a limited slip differential, add 4 oz. (118.3 ml) of the appropriate friction modifier.

19. Install or connect the following:
- Axle housing fill plug and tighten to 15–30 ft. lbs. (20–41 Nm)
- Anti-lock brake speed sensor, if equipped. Tighten the retaining bolt to 40–60 inch lbs. (4.5–6.8 Nm).
- Brake calipers and rotors
- Wheel

20. If equipped with air suspension, turn the air suspension switch to the **ON** position.

21. Road test the vehicle and check for proper operation.

Pinion Seal

REMOVAL & INSTALLATION

1. Before servicing the vehicle, refer to the precautions in the beginning of this section.

2. Remove or disconnect the following:
- Driveshaft
- Rear wheels
- Rear brake calipers

➡The rear brake calipers must be removed so that there is no additional drag when measuring pinion bearing preload.

3. Use an inch lb. torque wrench and measure the amount of torque required to maintain pinion rotation through several revolutions.

4. Remove the pinion flange and remove the seal.

To install:

5. Install or connect the following:
- Pinion seal and flange
- New pinion flange nut

6. Rotate the pinion flange occasionally while tightening the flange nut to make sure the pinion bearings seat correctly.

7. Take frequent bearing preload torque readings.

8. If the preload recorded prior to disassembly is **lower** than the specification for used bearings, then tighten the pinion flange nut to specification. If the preload recorder prior to disassembly is **higher** than the specification for used bearings, then tighten the pinion flange nut to the original reading as recorded.

9. The pinion bearing preload specifications are as follows:

a. Used bearings: 8–14 inch lbs. (0.9–1.6 Nm).

b. New bearings: 16–29 inch lbs. (1.8–3.2 Nm).

✲✲ CAUTION

Never loosen the pinion nut to reduce bearing preload. If it is necessary to reduce bearing preload, install a new collapsible spacer and pinion nut.

10. Install or connect the following:
- Driveshaft
- Brake calipers
- Rear wheels

11. Fill the differential with gear lubricant and check for leaks.

STEERING AND SUSPENSION

Air Bag

✲✲ CAUTION

Some vehicles are equipped with an air bag system. The system must be disarmed before performing service on, or around, system components, the steering column, instrument panel components, wiring and sensors. Failure to follow the safety precautions and the disarming procedure could result in accidental air bag deployment, possible injury and unnecessary system repairs.

PRECAUTIONS

Several precautions must be observed when handling the inflator module to avoid accidental deployment and possible personal injury.

- Never carry the inflator module by the wires or connector on the underside of the module.
- When carrying a live inflator module, hold securely with both hands, and ensure that the bag and trim cover are pointed away.
- Place the inflator module on a bench or other surface with the bag and trim cover facing up.
- With the inflator module on the bench, never place anything on or close to the module which may be thrown in the event of an accidental deployment.

DISARMING

1. Before servicing the vehicle, refer to the precautions in the beginning of this section.

2. Position the vehicle with the front wheels in a straight-ahead position.

3. Disconnect both battery cables.

4. Wait at least 1 minute for the air bag back-up power supply to deplete its stored energy before continuing.

5. Proceed with the repair.

6. Once the repair is complete.

7. Reconnect both battery cables.

8. Prove out the air bag system by turning the ignition key to the **RUN** position and visually monitoring the air bag indicator lamp in the instrument cluster. The indicator lamp should illuminate for approximately 6 seconds, then turn **OFF**. If the indicator lamp does not illuminate, stays on, or flashes at any time, a fault has been detected by the air bag diagnostic monitor.

Power Steering Gear

REMOVAL & INSTALLATION

1. Before servicing the vehicle, refer to the precautions in the beginning of this section.

2. If equipped with air suspension, the air suspension switch must be turned to the **OFF** position before raising the vehicle.

3. Center the steering wheel and turn the key to the locked position.

4. Remove or disconnect the following:
- Negative battery cable
- Bolt and the intermediate shaft from the steering gear

5. On the Town Car, separate the 2 halves and remove the steering gear cover.

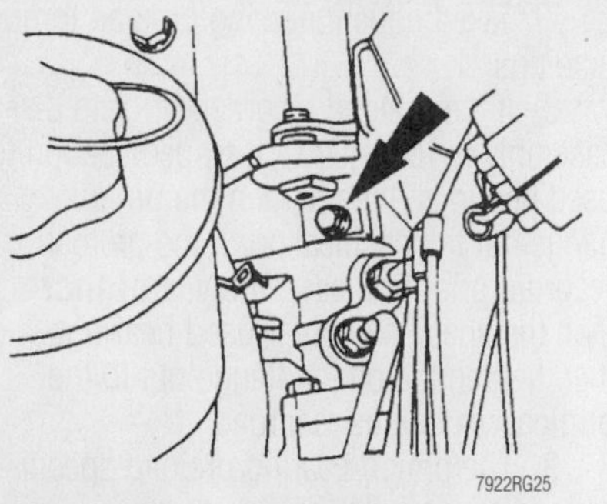

Remove the bolt and separate the intermediate shaft from the steering gear input shaft

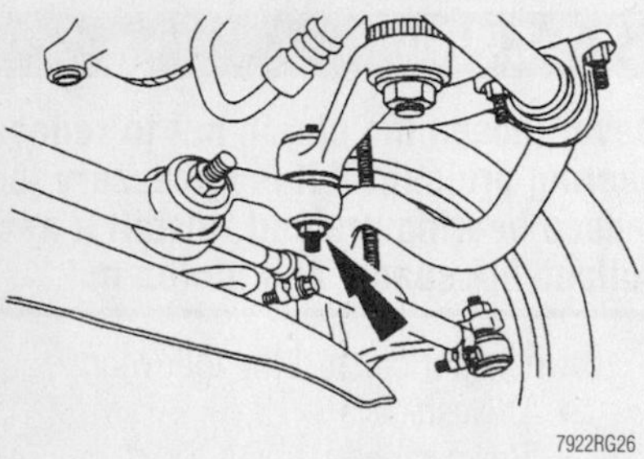

Remove the locknut and separate the Pitman arm from the center link using the appropriate puller

6. Tag the power steering pressure and return lines so they may be reassembled in their original positions.
7. Place a drain pan under the steering gear
8. Remove or disconnect the following:
   - Pressure and return lines. Plug the lines and ports in the gear to prevent the entry of dirt.
   - Pitman arm from the center link using a puller. It is not necessary to remove the Pitman arm from the steering gear.
9. Support the steering gear
10. Remove or disconnect the following:
    - Steering gear-to-frame rail retaining bolts
    - Steering gear

To install:

11. Install or connect the following:
    - Steering gear on the frame rail. Tighten the steering gear-to-frame retaining bolts to 50–67 ft. lbs. (66–90 Nm).
    - Pitman arm to the center link. Tighten the retaining nut to 52–60 ft. lbs. (70–81 Nm).
    - Power steering pressure and return lines to the steering gear and tighten the lines to 12–18 ft. lbs. (16–24 Nm)
    - Intermediate shaft to the steering gear. Tighten the bolt to 31–41 ft. lbs. (41–55 Nm).
    - Negative battery cable
12. If equipped with air suspension, turn the air suspension switch to the **ON** position.
13. Fill the reservoir with the correct power steering fluid and turn the steering wheel from stop-to-stop to distribute the fluid. Check the fluid level and add fluid, if necessary.
14. Start the engine and turn the steering wheel from left to right. Check for leaks.
15. On the Town Car, install the steering gear cover.

Shock Absorber

REMOVAL & INSTALLATION

Front

1. Before servicing the vehicle, refer to the precautions in the beginning of this section.
2. If equipped with air suspension, the air suspension switch, located on the right-hand side of the luggage compartment, must be turned to the **OFF** position before raising the vehicle.
3. Remove or disconnect the following:
   - Nut, washer and bushing from the upper end of the shock absorber
   - 2 bolts retaining the shock absorber to the lower control arm
   - Shock absorber

To install:

4. Prior to installation, prime the new shock absorber. Fully extend the shock absorber while in the right side up (installed) position. Turn the shock absorber upside down and fully compress it. Repeat the procedure at least 3 times to purge any air trapped in the shock absorber.
5. Install or connect the following:
   - New bushing and washer on the stud on the top of the new shock absorber and position the unit inside the front coil spring
   - 2 lower retaining bolts and tighten them to 10–12 ft lbs. (13–17 Nm).
   - New bushing and washer on the shock absorber top stud
   - New retaining nut. Tighten the retaining nut to 25–34 ft. lbs. (34–46 Nm).
6. If equipped with air suspension, turn the air suspension switch to the **ON** position.

Rear

1. Before servicing the vehicle, refer to the precautions in the beginning of this section.
2. If equipped with air suspension, turn the air suspension service switch **OFF**.
3. Be sure the ignition switch is in the **OFF** position.
4. Support the rear axle assembly with a jack.
5. Remove or disconnect the following:
   - Top retaining nut, washer and bushing
   - Bottom retaining nut and washer
   - Shock absorber

To install:

6. Install or connect the following:
   - Shock absorber so the upper stud enters the hole in the frame
   - Top bushing, washer and retaining nut. Tighten to 26–34 ft. lbs. (34–46 Nm).
7. Extend the shock absorber and place the lower stud through the hole in the bracket
8. Bottom retaining washer and nut. Tighten to 57–75 ft. lbs. (76–103 Nm).
9. Remove the jack from the axle assembly.
10. Turn the air suspension service switch to the **ON** position.

Coil Spring

REMOVAL & INSTALLATION

Front

1. Before servicing the vehicle, refer to the precautions in the beginning of this section.

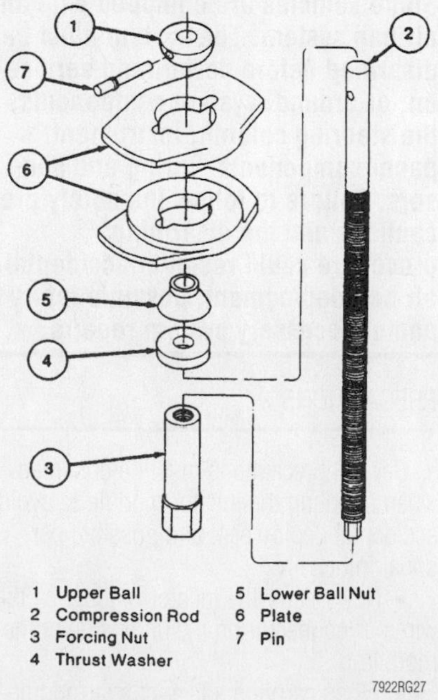

Exploded view of Spring Compressor D78P-5310-A

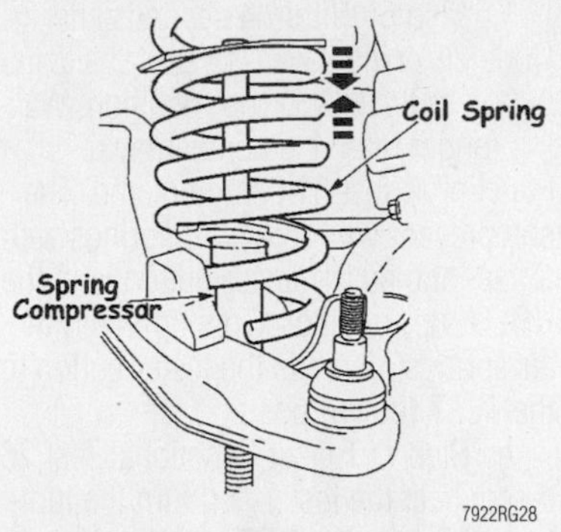

Compress the coil spring until it moves away from its seat

2. If equipped with air suspension, turn the air suspension service switch to the **OFF** position before raising the vehicle.

3. Remove or disconnect the following:
   - Wheel
   - Shock absorber
   - Center link from the Pitman arm

4. Using a spring compressor perform the following steps:
   a. Step 1: install 1 plate with the pivot ball seat facing downward into the coils of the spring. Rotate the plate so it is flush with the upper surface of the lower arm.
   b. Step 2: Install the other plate with the pivot ball seat facing upward into the coils of the spring. Insert the upper ball nut through the coils of the spring, so the nut rests in the upper plate.
   c. Step 3: Insert the compression rod into the opening in the lower arm, through the upper and lower plate and upper ball nut. Insert the securing pin through the upper ball nut and compression rod.
   d. Step 4: With the upper ball nut secured, turn the upper plate so it walks up the coil until it contacts the upper spring seat. Then, back off ½ turn.
   e. Step 5: Install the lower ball nut

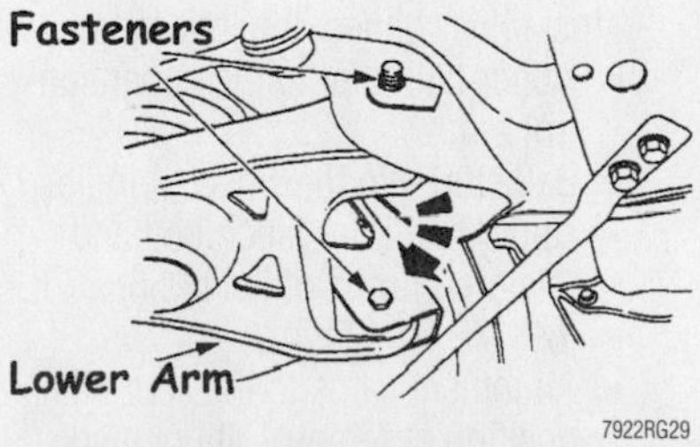

Remove the fasteners attaching the lower arm to the frame, then lower the arm and remove the spring with the compressor

and thrust washer on the compression rod and screw on the forcing nut. Tighten the forcing nut until the spring is compressed enough so it is free in its seat.

5. Remove or disconnect the following:
   - 2 lower control arm pivot bolts
   - Lower arm from the frame crossmember
   - Coil spring

6. If a new coil spring is to be installed, mark the position of the upper and lower plates on the spring with chalk. With an assistant, compress a new spring for installation and measure the compressed length and the amount of curvature of the old spring.

7. Loosen the forcing nut to relieve the spring tension and remove the tools from the spring.

To install:

8. Assemble the spring compressor and locate in the same position as marked during disassembly.

9. Before compressing the coil spring, be sure the upper ball nut securing the pin is inserted properly.

10. Compress the coil spring until the spring height reaches the dimension measured during disassembly.

11. Position the coil spring assembly into the lower arm and position the lower arm into the frame crossmember.

12. Install both the front and rear lower control arm pivot bolts through the frame and lower arm bushings. Tighten the bolts and nuts to 109–148 ft. lbs. (148–201 Nm).

13. Remove the spring compressor from the coil spring.

14. Install or connect the following:
   - Drag link to the Pitman arm. Tighten the retaining nut to 60 ft. lbs. (80 Nm).
   - Shock absorber inside the coil spring
   - Retaining bolts
   - Wheel. Tighten the lug nuts to 85–105 ft. lbs. (115–142 Nm).

15. Place a washer and retaining nut on the shock absorber top stud. Tighten the nut to 25–34 ft. lbs. (34–46 Nm).

16. If equipped with air suspension, turn the air suspension switch to the **ON** position.

17. Check the front end alignment.

Rear

1. Before servicing the vehicle, refer to the precautions in the beginning of this section.

2. Place a hoist under the rear axle housing and raise and safely support the vehicle.

3. Support the frame side rails with 2 jackstands.

4. Remove or disconnect the following:
   - Rear sway bar
   - Lower studs of both rear shock absorbers from the mounting brackets on the axle tube
   - Parking brake cable from the upper arm retainer before lowering the axle housing

5. Lower the axle housing until the coil springs are released. If the axle housing is supported by the hoist, lower the hoist allowing the rear of the vehicle to rest on the jackstands. If the vehicle's axle housing is supported by the jackstands, leave the hoist stationary and lower the jackstands or raise the hoist to release the tension on the coil springs.

6. Remove the coil springs and insulators.

To install:

7. Install or connect the following:
   - Coil spring in the upper and lower seats with an insulator between the upper end of the spring and frame seat
   - Axle housing and connect the lower studs of the shock absorbers to the mounting brackets
   - Parking cable into the upper arm retainer
   - Sway bar

8. Road test the vehicle and check for proper operation.

Air Spring

REMOVAL & INSTALLATION

CAUTION

Before servicing any air suspension component, disconnect power to the system by turning the air suspension service switch OFF or by disconnecting the negative battery cable. Do not remove an air spring under any circumstances when there is pressure in the air spring. Do not remove any components supporting an air spring without either exhausting the air or providing support for the air spring.

1. Before servicing the vehicle, refer to the precautions in the beginning of this section.

2. Turn the air suspension switch to the **OFF** position.

3. Raise and safely support the vehicle so the suspension is fully down with no load.

4. Remove or disconnect the following:

- Heat shield, as required
- Spring retainer clip
- Air spring solenoid valve electrical connector
- Air line
- Air spring solenoid retainer

5. Rotate the solenoid valve counterclockwise to the first stop.

6. Pull the solenoid valve straight out slowly to the second stop to bleed air from the system.

✻✻ CAUTION

Do not fully release the solenoid until the air is completely bled from the air spring or personal injury may result.

7. After the air is fully bled from the system, rotate the solenoid valve counterclockwise to the third stop and remove the solenoid valve from the solenoid housing. Remove the large O-ring from the solenoid housing.

8. Remove the air spring.

To install:

9. Check the solenoid valve O-rings for cuts or abrasions. Replace the O-rings as required. Lightly grease the O-ring area of the solenoid valve and the larger solenoid housing O-ring with silicone dielectric compound.

10. Insert the solenoid into the air spring end cap and rotate clockwise to the third stop, push in to the second stop, then rotate clockwise to the first stop.

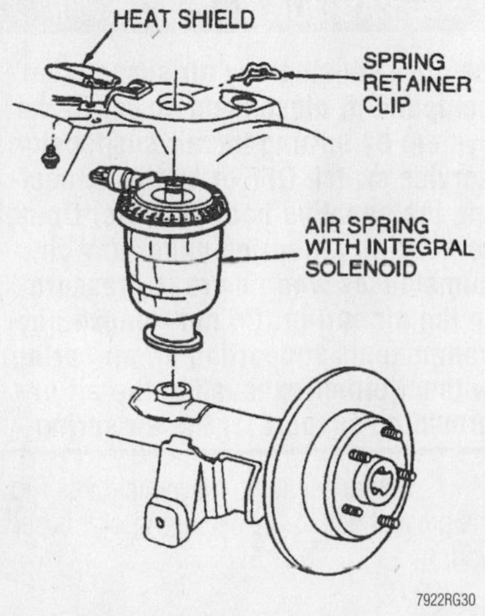

Exploded view of the air spring mounting

11. Install or connect the following:

- Air spring solenoid retainer. Inspect the wiring harness connector and ensure the rubber gasket is in place at the bottom of the connector cavity
- Air spring into the frame (upper) spring seat, taking care to keep the solenoid air and electrical connections clean and free of damage
- Push-on ring spring retainer clip to the knob of the spring cap from the top side of the frame spring seat
- Air line and electrical connector to the solenoid
- Heat shield to the frame spring seat, if removed
- Align the air spring piston-to-axle (lower) seat. Squeeze to increase pressure and push downward on the piston, snapping the piston to the axle seat at rebound and supported by the shock absorber.
- Negative battery cable.

✻✻ WARNING

The air springs may be damaged if the suspension is allowed to compress before the spring is inflated.

12. Refill the air spring as follows:

a. Step 1: Turn the air suspension switch to the **ON** position. The ignition switch must be **ON** and the engine running or a battery charger must be connected to the battery to reduce battery drain.

b. Step 2: Fold back or remove the right-hand luggage compartment trim panel and connect Super Star II Tester 007–0041-A to the air suspension DLC, which is located near the air suspension switch.

c. Step 3: Set the tester to EEC-IV/MCU mode. Also set the tester to FAST mode. Release the tester button to the HOLD (up) position and turn the tester **ON**.

d. Step 4: Depress the tester button to TEST (down) position. A Code 10 will be displayed. Within 2 minutes a Code 13 will be displayed. After Code 13 is displayed, release the tester button to the HOLD (up) position, wait 5 seconds and depress the tester button to TEST (down) position. Ignore any codes displayed.

e. Step 5: Release the tester button to the HOLD (up) position. Wait at least 20 seconds, then depress the tester button to TEST (down) position. Within 10 seconds, the codes will be displayed in the order shown.

f. Step 6: Within 4 seconds after Code 26 is displayed, release the tester button to the HOLD (up) position. Waiting longer than 4 seconds may result in Functional Test 31 being entered. The compressor will fill the air springs with air as long as the tester button is in the HOLD (up) position. To stop filling the air springs, depress the tester button to the TEST (down) position.

g. Step 7: To exit Functional Test 26, disconnect the tester and turn the ignition switch to the **OFF** position.

13. Install the luggage trim panel, if removed.

Upper Ball Joint

REMOVAL & INSTALLATION

1. Before servicing the vehicle, refer to the precautions in the beginning of this section.

2. If equipped with air suspension, the air suspension switch to the **OFF** position before raising the vehicle.

3. Place supports under both sides of the frame just behind the lower control arms.

4. Remove the wheel.

5. Place a floor jack under the lower control arm at the lower ball joint area. The floor jack will support the spring load on the lower control arm.

6. Remove the retaining nut and pinch bolt from the upper ball joint stud.

7. Mark the position of the alignment cams. When replacing the upper ball joint this will approximate the current alignment.

8. Remove or disconnect the following:

- 2 nuts retaining the upper ball joint to the upper control arm
- Upper ball joint from the upper control arm and spread the slot in the wheel spindle with a prybar to remove the ball joint stud from the wheel spindle.

To install:

9. Install or connect the following:

- Upper ball joint to the upper control arm
- Ball stud into the wheel spindle
- Upper ball joint pinch bolt and retaining nut. Tighten to 56–77 ft. lbs. (76–104 Nm).
- Alignment cams to the approximate position at removal. If not marked, install in the neutral positions.
- 2 nuts retaining the upper ball joint to the upper control arm. Hold the cams and tighten the nuts to 107–129 ft. lbs. (145–175 Nm).

- Wheel and tire assembly. Tighten the lug nuts in a star pattern to 85–105 ft. lbs. (115–142 Nm).

10. Remove the floor jack from under the lower control arm.
11. If equipped with air suspension, turn the air suspension switch to the **ON** position.
12. Check and adjust the front wheel alignment.

Lower Ball Joint

REMOVAL & INSTALLATION

1. Before servicing the vehicle, refer to the precautions in the beginning of this section.
2. If equipped with air suspension, the air suspension service switch to the **OFF** position before raising the vehicle.
3. Place supports under both sides of the frame behind the lower control arms.
4. Remove or disconnect the following:
   - Wheel
   - Wheel spindle
   - Ball joint boot seal and discard
   - Ball joint

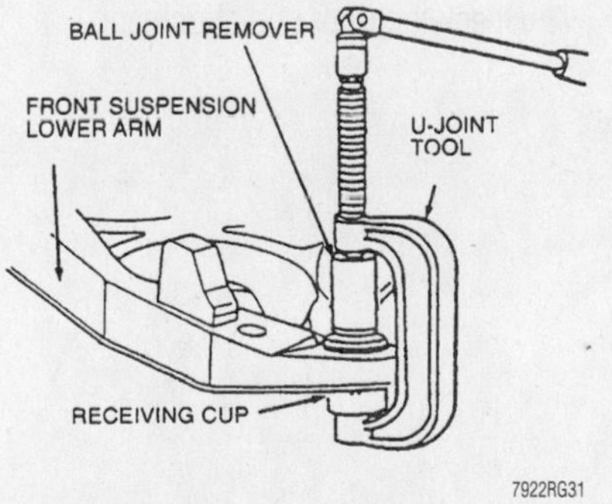

Use a ball joint press to remove the ball joint from the lower control arm

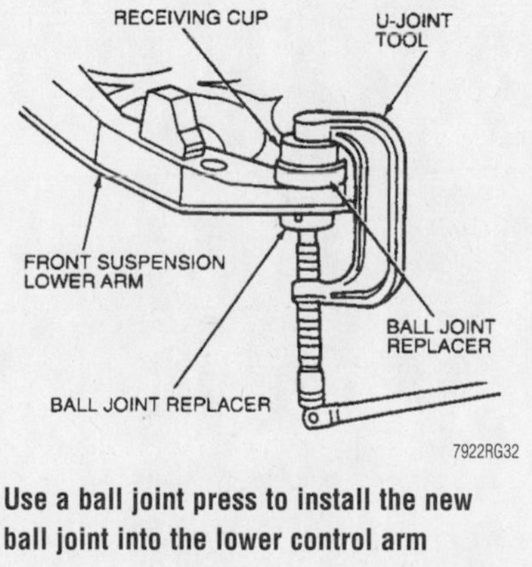

Use a ball joint press to install the new ball joint into the lower control arm

To install:

➡When installing a new ball joint, the protective cover should be left on to protect the ball joint seal during installation. It may be necessary to trim the cover so it can pass through the installation tool.

5. Install the ball joint.
6. Discard the protective cover and be sure the new ball joint is fully seated in the lower control arm. Ensure that the ball joint seal is not damaged.
7. Install or connect the following:
   - Wheel spindle
   - Wheel. Tighten the lug nuts to 85–105 ft. lbs. (115–142 Nm).
8. If equipped with air suspension, turn the air suspension service switch to the **ON** position.
9. Check the front end alignment.

Upper Control Arm

REMOVAL & INSTALLATION

1. Before servicing the vehicle, refer to the precautions in the beginning of this section.
2. If equipped with air suspension, the air suspension switch, located on the right-hand side of the luggage compartment, must be turned to the **OFF** position before raising the vehicle.
3. Remove or disconnect the following:
   - Front wheel
   - Upper ball joint. Support the lower control arm.
   - Upper control arm pivot bolts
   - Upper control arm

To install:

4. Install or connect the following:
   - Upper control arm. Tighten the pivot bolts to 110–148 ft. lbs. (148–201 Nm).
   - Upper ball joint. Tighten the pinch bolt to 56–76 ft. lbs. (76–104 Nm).
   - Front wheel
5. If equipped with air suspension, turn the air suspension service switch to the **ON** position.
6. Align the vehicle.

CONTROL ARM BUSHING REPLACEMENT

The control arm bushings are serviced with the control arm as an assembly.

Lower Control Arm

REMOVAL & INSTALLATION

1. Before servicing the vehicle, refer to the precautions in the beginning of this section.
2. If equipped with air suspension, the air suspension switch, located on the right-hand side of the luggage compartment, must be turned to the **OFF** position before raising the vehicle.
3. Remove or disconnect the following:
   - Front wheel
   - Wheel speed sensor
   - Brake caliper and rotor
   - Sway bar link
   - Shock absorber
   - Drag link
   - Lower ball joint
   - Coil spring
   - Lower control arm pivot bolts
   - Lower control arm

To install:

4. Install or connect the following:
   - Lower control arm. Tighten the pivot bolts to 110–148 ft. lbs. (148–201 Nm).
   - Coil spring
   - Lower ball joint. Tighten the nut to 110–148 ft. lbs. (148–201 Nm).
   - Drag link. Tighten the nut to 35–46 ft. lbs. (47–63 Nm).
   - Shock absorber
   - Sway bar link
   - Brake caliper and rotor
   - Wheel speed sensor
   - Front wheel
5. If equipped with air suspension, turn the air suspension service switch to the **ON** position.
6. Align the vehicle.

CONTROL ARM BUSHING REPLACEMENT

The control arm bushings are serviced with the control arm as an assembly.

Wheel Bearings

ADJUSTMENT

The front wheel bearings are of a hub unit design and are pre-greased, sealed and require no maintenance. The bearings are preset and cannot be adjusted.

REMOVAL & INSTALLATION

Front

1. Before servicing the vehicle, refer to the precautions in the beginning of this section.
2. If equipped, turn the air suspension service switch to the **OFF** position before raising the vehicle.
3. Remove or disconnect the following
   - Front wheel
   - Grease cap from the hub
   - Disc brake caliper. Suspend the caliper with a length of wire. Do not let it hang from the brake hose. Discard the disc brake caliper mounting bolts.
   - Disc brake rotor. If the factory installed push on nuts are installed, remove them first.
   - Wheel hub retainer nut and discard
   - Hub and bearing assembly

To install:

4. Install or connect the following:
   - Hub and bearing assembly
   - New wheel hub retainer nut and tighten to 189–254 ft. lbs. (255–345 Nm)
   - Disc brake rotor and push on nuts, if equipped
   - New grease cap seal
   - Disc brake caliper using the 2 new disc brake caliper mounting bolts. Tighten the bolts to 125–170 ft. lbs. (170–230 Nm).
   - Wheel. Tighten the lug nuts to 85–104 ft. lbs. (115–142 Nm).
5. If equipped with air suspension, turn the air suspension switch to the **ON** position.
6. Pump the brake pedal several times to position the brake pads prior to moving the vehicle.
7. Check the front end alignment.

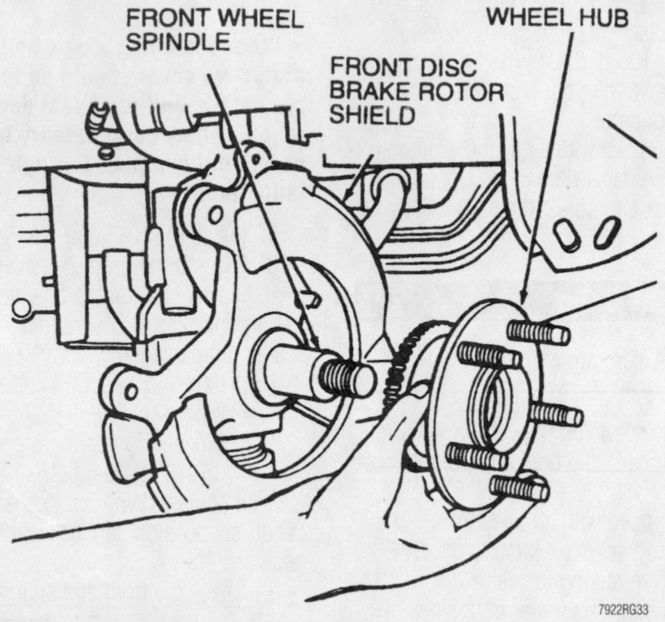

Front hub and bearing assembly

FORD MOTOR CO.

20

2002

Thunderbird

PRECAUTIONS

Before servicing any vehicle, please be sure to read all of the following precautions, which deal with personal safety, prevention of component damage, and important points to take into consideration when servicing a motor vehicle:

- Never open, service or drain the radiator or cooling system when the engine is hot; serious burns can occur from the steam and hot coolant.
- Observe all applicable safety precautions when working around fuel. Whenever servicing the fuel system, always work in a well-ventilated area. Do not allow fuel spray or vapors to come in contact with a spark, open flame, or excessive heat (a hot drop light, for example). Keep a dry chemical fire extinguisher near the work area. Always keep fuel in a container specifically designed for fuel storage; also, always properly seal fuel containers to avoid the possibility of fire or explosion. Refer to the additional fuel system precautions later in this section.
- Fuel injection systems often remain pressurized, even after the engine has been turned **OFF**. The fuel system pressure must be relieved before disconnecting any fuel lines. Failure to do so may result in fire and/or personal injury.
- Brake fluid often contains polyglycol ethers and polyglycols. Avoid contact with the eyes and wash your hands thoroughly after handling brake fluid. If you do get brake fluid in your eyes, flush your eyes with clean, running water for 15 minutes. If eye irritation persists, or if you have taken brake fluid internally, IMMEDIATELY seek medical assistance.
- The EPA warns that prolonged contact with used engine oil may cause a number of skin disorders, including cancer. You should make every effort to minimize your exposure to used engine oil. Protective gloves should be worn when changing oil. Wash your hands and any other exposed skin areas as soon as possible after exposure to used engine oil. Soap and water, or waterless hand cleaner should be used.
- All new vehicles are now equipped with an air bag system, often referred to as a Supplemental Restraint System (SRS) or Supplemental Inflatable Restraint (SIR) system. The system must be disabled before performing service on or around system components, steering column, instrument panel components, wiring and sensors. Failure to follow safety and disabling procedures could result in accidental air bag deployment, possible personal injury and unnecessary system repairs.
- Always wear safety goggles when working with, or around, the air bag system. When carrying a non-deployed air bag, be sure the bag and trim cover are pointed away from your body. When placing a non-deployed air bag on a work surface, always face the bag and trim cover upward, away from the surface. This will reduce the motion of the module if it is accidentally deployed. Refer to the additional air bag system precautions later in this section.
- Clean, high quality brake fluid from a sealed container is essential to the safe and proper operation of the brake system. You should always buy the correct type of brake fluid for your vehicle. If the brake fluid becomes contaminated, completely flush the system with new fluid. Never reuse any brake fluid. Any brake fluid that is removed from the system should be discarded. Also, do not allow any brake fluid to come in contact with a painted surface; it will damage the paint.
- Never operate the engine without the proper amount and type of engine oil; doing so WILL result in severe engine damage.
- Timing belt maintenance is extremely important. Many models utilize an interference-type, non-freewheeling engine. If the timing belt breaks, the valves in the cylinder head may strike the pistons, causing potentially serious (also time-consuming and expensive) engine damage. Refer to the maintenance interval charts in the front of this manual for the recommended replacement interval for the timing belt, and to the timing belt section for belt replacement and inspection.
- Disconnecting the negative battery cable on some vehicles may interfere with the functions of the on-board computer system(s) and may require the computer to undergo a relearning process once the negative battery cable is reconnected.
- When servicing drum brakes, only disassemble and assemble one side at a time, leaving the remaining side intact for reference.

ENGINE REPAIR

➡Disconnecting the negative battery cable on some vehicles may interfere with the functions of the on board computer system. The computer may undergo a relearning process once the negative battery cable is reconnected.

Alternator

REMOVAL

1. Before servicing the vehicle, refer to the precautions in the beginning of this section.
2. Remove or disconnect the following:
   - Negative battery cable
   - Air intake tube
   - Accessory drive belt
   - Lower splash shield
   - Alternator mounting bolts

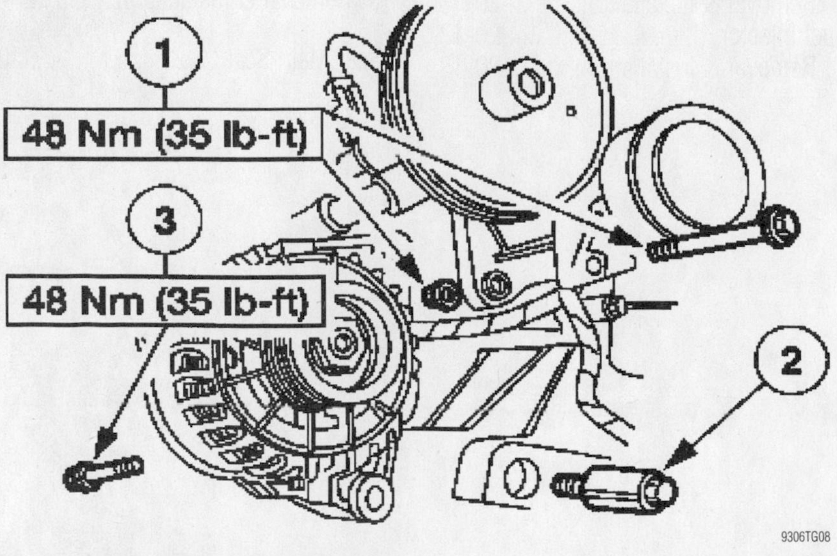

Alternator torque sequence—3.9L engine

- Alternator harness connectors
- Alternator

INSTALLATION

1. Install the alternator harness connectors, then install the alternator. Tighten the bolts in sequence as follows:
   a. Step 1: Tighten bolt No. 1 to 35 ft. lbs. (48 Nm)
   b. Step 2: Tighten bolt No. 2 to 15 ft. lbs. (20 Nm) plus 90 degrees
   c. Step 3: Tighten bolt No. 3 to 35 ft. lbs. (48 Nm)
2. Install or connect the following:
   - Lower splash shield
   - Accessory drive belt
   - Air intake tube
   - Negative battery cable

Ignition Timing

ADJUSTMENT

This vehicle is equipped with a Distributorless Ignition System (DIS). The ignition timing is not adjustable. It is controlled by the PCM.

Engine Assembly

REMOVAL & INSTALLATION

1. Before servicing the vehicle, refer to the precautions in the beginning of this section.
2. Drain the cooling system.
3. Relieve the fuel system pressure.
4. Recover the A/C refrigerant.
5. Drain the engine oil.
6. Remove or disconnect the following:
   - Negative battery cable
   - Air cleaner inlet tube
   - Upper radiator shield
   - Upper radiator support brackets
   - A/C pressure switch connector
   - Power steering return line clip
   - Power steering reservoir
   - Vapor Management Valve (VMV) cover
   - VMV vacuum hose and canister purge hose
   - Main vacuum supply hose
   - Cowl vent screens
   - Cross vehicle support bar
   - Degas bottle hose
   - Accelerator cable
   - Cruise control cable
   - Ground strap
   - Fresh air filter and housing
   - Powertrain harness connectors at right strut tower
   - Fresh air filter panel
   - Main engine wiring harness connector
   - Main transmission wiring harness connector
   - Heater hoses at the water control valve. Note the locations for assembly.
   - Hydraulic cooling fan reservoir
   - Water control valve harness connector
   - Front wheels
   - Inner splash shields
   - Wheel speed sensor connectors and harness clips
   - Brake calipers
   - Lower stabilizer bar links
   - Upper ball joints
   - Lower strut mount bolts
   - Left, right and center splash shields
   - A/C suction and discharge lines
   - Shift cable and bracket
   - Power steering line frame rail clip
   - Rack and pinion harness connectors
   - Steering shaft bolt and coupling
   - Starter motor harness connectors and ground cable
   - Alternator harness connectors
   - Lower transmission flange bolts
   - Torque converter
   - Inner air deflector
   - Engine block heater, if equipped
7. Support the engine, transmission, front and center crossmembers and the cooling system with a powertrain lift and transmission support bracket.
8. Support the rear of the vehicle with safety stands.
9. Remove or disconnect the following:
   - Transmission crossmember bolts
   - Front crossmember bolts
   - Center crossmember bolts
   - Powertrain assembly
   - A/C compressor manifold and tube assembly
   - Power steering pump return hose
   - Hydraulic cooling fan return hose
   - Lower radiator hose
   - Upper radiator hoses
   - Knock Sensor (KS) connector
   - Heater hose
   - Transmission cooler lines and bracket
   - Power steering pressure line and bracket
   - Hydraulic cooling fan pressure line and bracket
10. Attach a hoist to the engine.
11. Remove the motor mount nuts and lift the powertrain out of the subframe.
12. Remove or disconnect the following:
    - Wiring harness retainers
    - Upper transmission flange bolts
    - Transmission from the engine

To install:

13. Install or connect the following:
    - Transmission to the engine. Tighten the upper flange bolts to 35 ft. lbs. (48 Nm).
    - Wiring harness retainers. Tighten the nuts to 89 inch lbs. (10 Nm).
    - Powertrain to the subframe. Tighten the mount nuts to 30 ft. lbs. (40 Nm).
    - Hydraulic cooling fan pressure line and bracket
    - Power steering pressure line and bracket
    - Transmission cooler lines and bracket
    - Heater hose
    - Knock sensor connector
    - Upper radiator hoses
    - Lower radiator hose
    - Hydraulic cooling fan return hose
    - Power steering pump return hose
    - A/C compressor manifold and tube assembly. Tighten the bolt to 15 ft. lbs. (21 Nm).
    - Powertrain assembly. Tighten the front and center crossmember bolts to 76 ft. lbs. (103 Nm) and the transmission crossmember bolts to 30 ft. lbs. (40 Nm).
    - Engine block heater, if equipped
    - Inner air deflector
    - Torque converter. Tighten the nuts to 28 ft. lbs. (38 Nm).
    - Lower transmission flange bolts. Tighten the bolts to 35 ft. lbs. (47 Nm).
    - Alternator harness connectors
    - Starter motor harness connectors and ground cable
    - Steering shaft bolt and coupling. Tighten the coupling pinch bolt to 26 ft. lbs. (35 Nm) and the shaft bolt to 22 ft. lbs. (30 Nm).
    - Rack and pinion harness connectors
    - Power steering line frame rail clip
    - Shift cable and bracket
    - A/C suction and discharge lines
    - Left, right and center splash shields

- Lower strut mount bolts. Tighten the bolts to 129 ft. lbs. (175 Nm).
- Upper ball joints. Tighten the nuts to 66 ft. lbs. (90 Nm).
- Lower stabilizer bar links. Tighten the nuts to 41 ft. lbs. (55 Nm).
- Brake calipers
- Wheel speed sensor connectors and harness clips
- Inner splash shields
- Front wheels
- Water control valve harness connector
- Hydraulic cooling fan reservoir
- Heater hoses at the water control valve
- Main transmission wiring harness connector
- Main engine wiring harness connector
- Fresh air filter panel
- Powertrain harness connectors at right strut tower
- Fresh air filter and housing
- Ground strap
- Cruise control cable
- Accelerator cable
- Degas bottle hose
- Cross vehicle support bar. Tighten the bolts to 15 ft. lbs. (20 Nm).
- Cowl vent screens
- Main vacuum supply hose
- VMV vacuum hose and canister purge hose
- VMV cover
- Power steering reservoir
- Power steering return line clip
- A/C pressure switch connector
- Upper radiator support brackets
- Upper radiator shield
- Air cleaner inlet tube
- Negative battery cable

14. Fill the crankcase to the correct level.
15. Fill the cooling system.
16. Recharge the A/C system.
17. Start the engine and check for leaks.

Water Pump

REMOVAL & INSTALLATION

1. Before servicing the vehicle, refer to the precautions in the beginning of this section.
2. Drain the cooling system.
3. Remove or disconnect the following:
   - Accessory drive belt
   - Water pump pulley
   - Water pump

To install:

4. Install or connect the following:
   - Water pump. Tighten the bolts to 71 inch lbs. (8 Nm) plus 90 degrees.
   - Water pump pulley. Tighten the bolts to 89 inch lbs. (10 Nm) plus 45 degrees.
   - Accessory drive belt
5. Fill the cooling system.
6. Start the engine and check for leaks.

Cylinder Head

REMOVAL & INSTALLATION

1. Before servicing the vehicle, refer to the precautions in the beginning of this section.
2. Drain the cooling system.

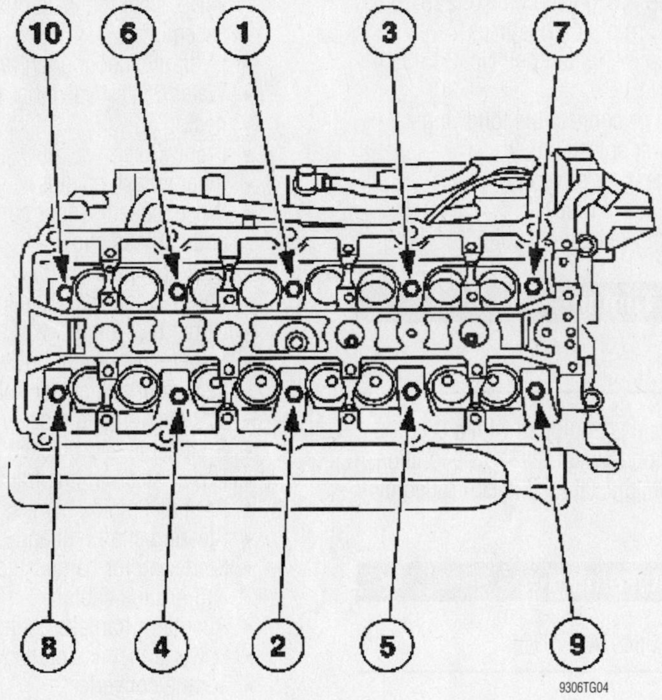

Right cylinder head torque sequence—3.9L engine

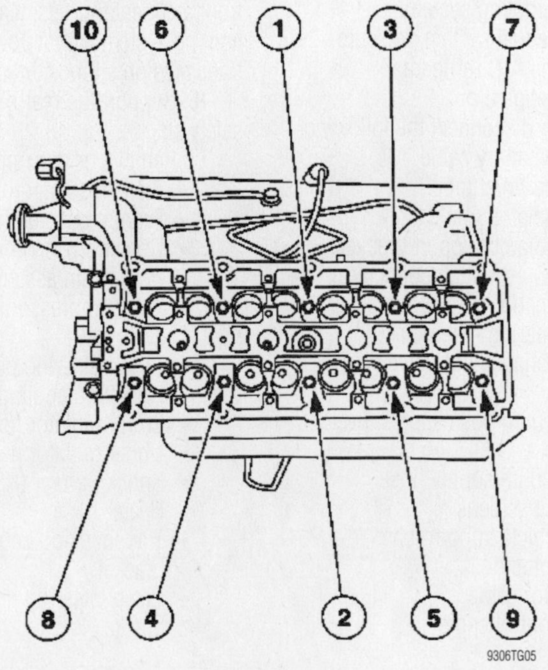

Left cylinder head torque sequence—3.9L engine

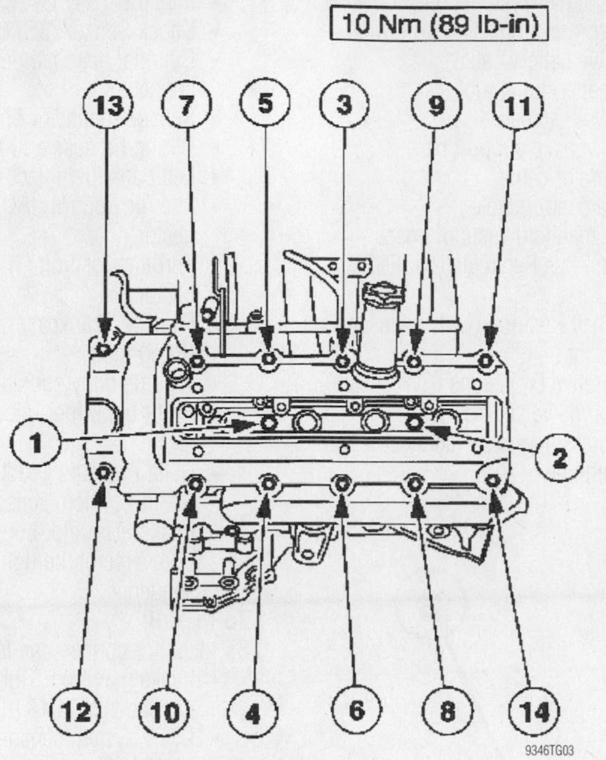

Left valve cover torque sequence—3.9L Engine

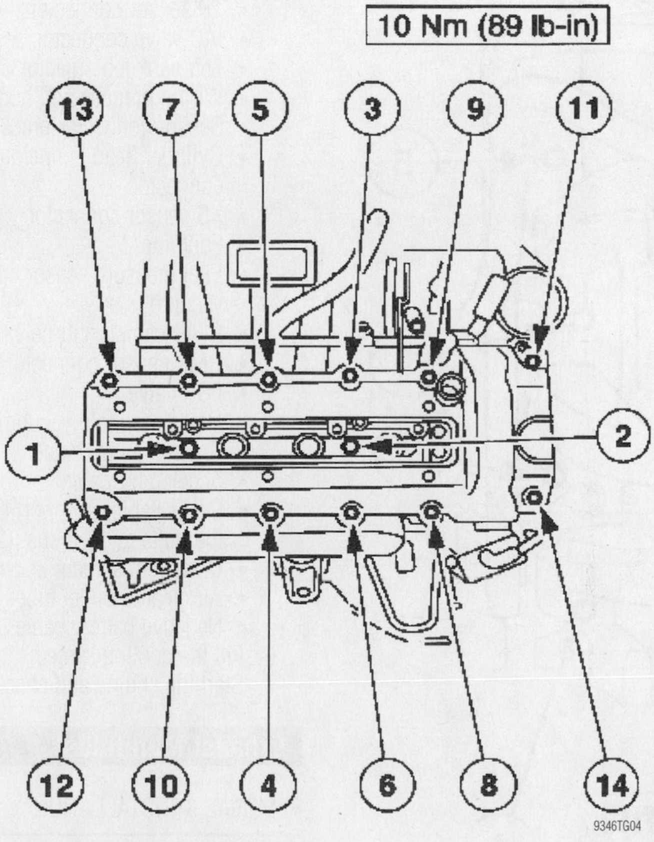

Right valve cover torque sequence—3.9L Engine

3. Relieve the fuel system pressure.
4. Remove or disconnect the following:
   - Negative battery cable
   - Engine appearance cover
   - Intake manifold
   - Valve covers
   - Accessory drive belts
   - Front cover
   - Timing chains
   - Camshafts
   - Water outlet pipe
   - Cylinder head temperature sensor connector
   - Exhaust front pipes
   - Exhaust Gas Recirculation (EGR) tube
   - Bolts and stud bolts at the rear of the cylinder heads
   - Cylinder heads. Loosen the bolts in reverse of the tightening sequence.

To install:

5. Install the cylinder heads. Tighten the bolts in sequence as follows:
   a. Step 1: M10 bolts to 15 ft. lbs. (20 Nm)
   b. Step 2: M10 bolts to 26 ft. lbs. (35 Nm)
   c. Step 3: M10 bolts to 33 ft. lbs. (45 Nm)
   d. Step 4: M10 bolts plus 90 degrees
   e. Step 5: M10 bolts plus 90 degrees
   f. Step 6: M8 bolts to 15 ft. lbs. (20 Nm)
   g. Step 7: M8 bolts plus 90 degrees
6. Install or connect the following:
   - Bolts and stud bolts at the rear of the cylinder heads. Tighten the bolts to 37 ft. lbs. (50 Nm).
   - EGR tube
   - Exhaust front pipes
   - Cylinder head temperature sensor connector
   - Water outlet pipe
   - Camshafts
   - Timing chains
   - Front cover
   - Accessory drive belts
   - Valve covers
   - Intake manifold
   - Engine appearance cover
   - Negative battery cable
7. Fill the cooling system.
8. Start the engine and check for leaks.

Rocker Arms/Shafts

REMOVAL & INSTALLATION

The vehicles covered in this section are not equipped with rocker arms/shafts. The camshaft directly actuates the valves.

Intake Manifold

REMOVAL & INSTALLATION

1. Before servicing the vehicle, refer to the precautions in the beginning of this section.
2. Drain the cooling system.
3. Relieve the fuel system pressure.
4. Remove or disconnect the following:
   - Negative battery cable
   - Air cleaner outlet tube
   - Cowl vent screen
   - Cross vehicle support bar
   - Accelerator cable
   - Cruise control cable
   - Intake manifold vacuum hoses
   - Exhaust Gas Recirculation (EGR) valve
   - Camshaft Position (CMP) sensor connector
   - Evaporative Emissions (EVAP) canister purge valve line
   - Fuel pressure sensor connector and vacuum line
   - Fuel line
   - Knock Sensor (KS) connector
   - Cylinder head temperature sensor connector
   - Sensor connector bracket
   - Wiring harness and hose bracket
   - Left bank fuel injector connectors
   - Idle Air Control (IAC) valve connector
   - Throttle Position (TP) sensor connector
   - Positive Crankcase Ventilation (PCV) tube
   - Throttle body coolant hoses
   - Right bank fuel injector connectors
   - Delta Pressure Feedback Electronic (DPFE) system sensor
   - Intake manifold. Loosen the bolts in reverse of the tightening sequence.

To install:

5. Install or connect the following:
   - Intake manifold. Tighten the bolts in sequence to 18 ft. lbs. (25 Nm).
   - DPFE system sensor
   - Right bank fuel injector connectors
   - Throttle body coolant hoses
   - PCV tube
   - TP sensor connector
   - IAC valve connector
   - Left bank fuel injector connectors
   - Wiring harness and hose bracket
   - Sensor connector bracket
   - Cylinder head temperature sensor connector
   - KS sensor connector
   - Fuel line
   - Fuel pressure sensor connector and vacuum line
   - EVAP canister purge valve line
   - CMP sensor connector
   - EGR valve
   - Intake manifold vacuum hoses
   - Cruise control cable
   - Accelerator cable
   - Cross vehicle support bar. Tighten the bolts to 15 ft. lbs. (20 Nm).
   - Cowl vent screen
   - Air cleaner outlet tube
   - Negative battery cable
6. Fill the cooling system.
7. Start the engine and check for leaks.

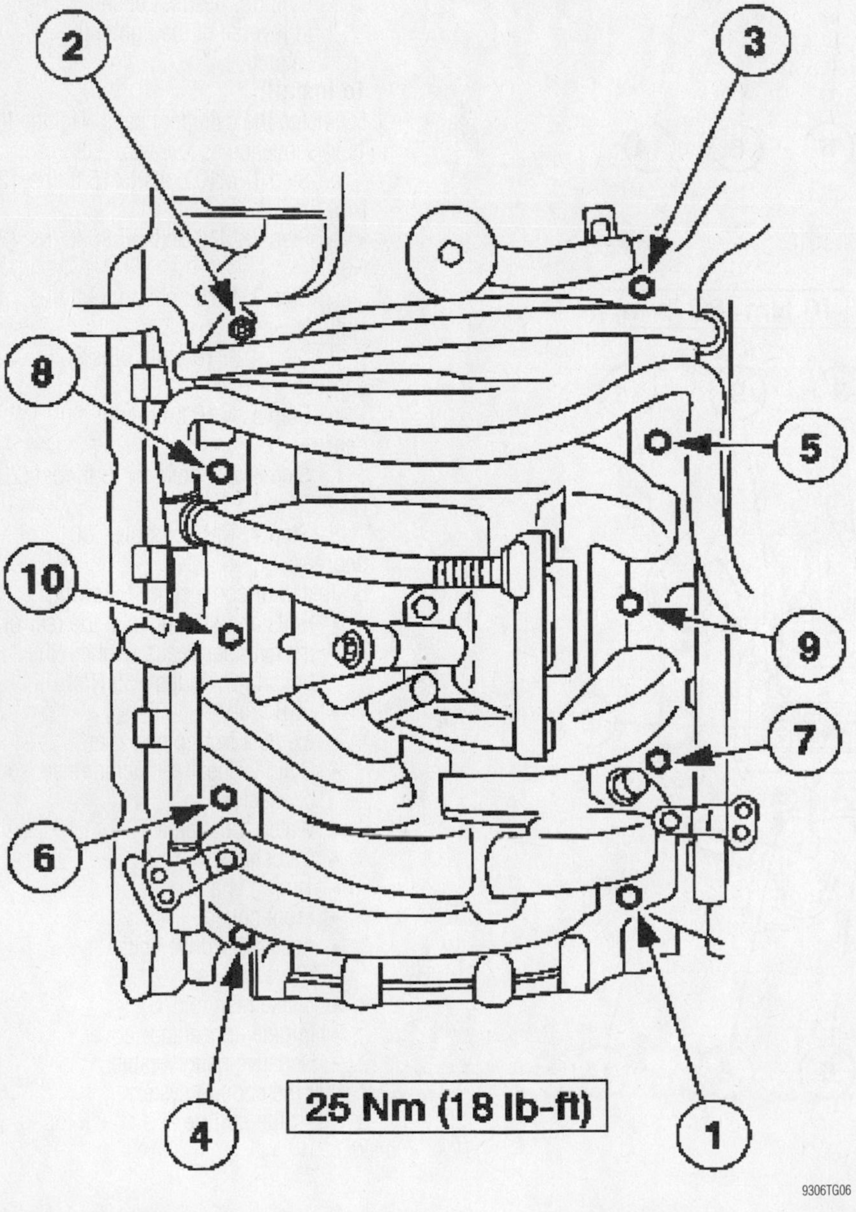

Intake manifold torque sequence—3.9L engine

Exhaust Manifolds

REMOVAL & INSTALLATION

1. Before servicing the vehicle, refer to the precautions in the beginning of this section.

2. Remove or disconnect the following:
   - Negative battery cable
   - Power steering pump reservoir
   - Oil dipstick tube
   - Exhaust front pipes
   - Exhaust Gas Recirculation (EGR) tube
   - Exhaust manifolds

To install:

3. Install or connect the following:
   - Exhaust manifolds. Tighten the bolts to 18 ft. lbs. (25 Nm).
   - EGR tube
   - Exhaust front pipes
   - Oil dipstick tube
   - Power steering pump reservoir
   - Negative battery cable
4. Start the engine and check for leaks.

Camshaft and Valve Lifters

REMOVAL & INSTALLATION

1. Before servicing the vehicle, refer to the precautions in the beginning of this section.
2. Remove or disconnect the following:
   - Negative battery cable
   - Valve covers
   - Front cover
   - Timing chains

➡**Keep all valve train components in order for assembly.**

   - Camshaft journal bearing caps
   - Camshafts
   - Valve tappets and shims

To install:

3. Install or connect the following:
   - Valve tappets and shims
   - Camshafts
4. Install the camshaft journal bearing caps in their original positions. Tighten the bolts in sequence as follows:
   a. Step 1: Finger tight
   b. Step 2: 53 inch lbs. (6 Nm)
   c. Step 3: Plus 90 degrees
5. Install or connect the following:
   - Timing chains
   - Front cover
   - Valve covers
   - Negative battery cable
6. Start the engine and check for leaks.

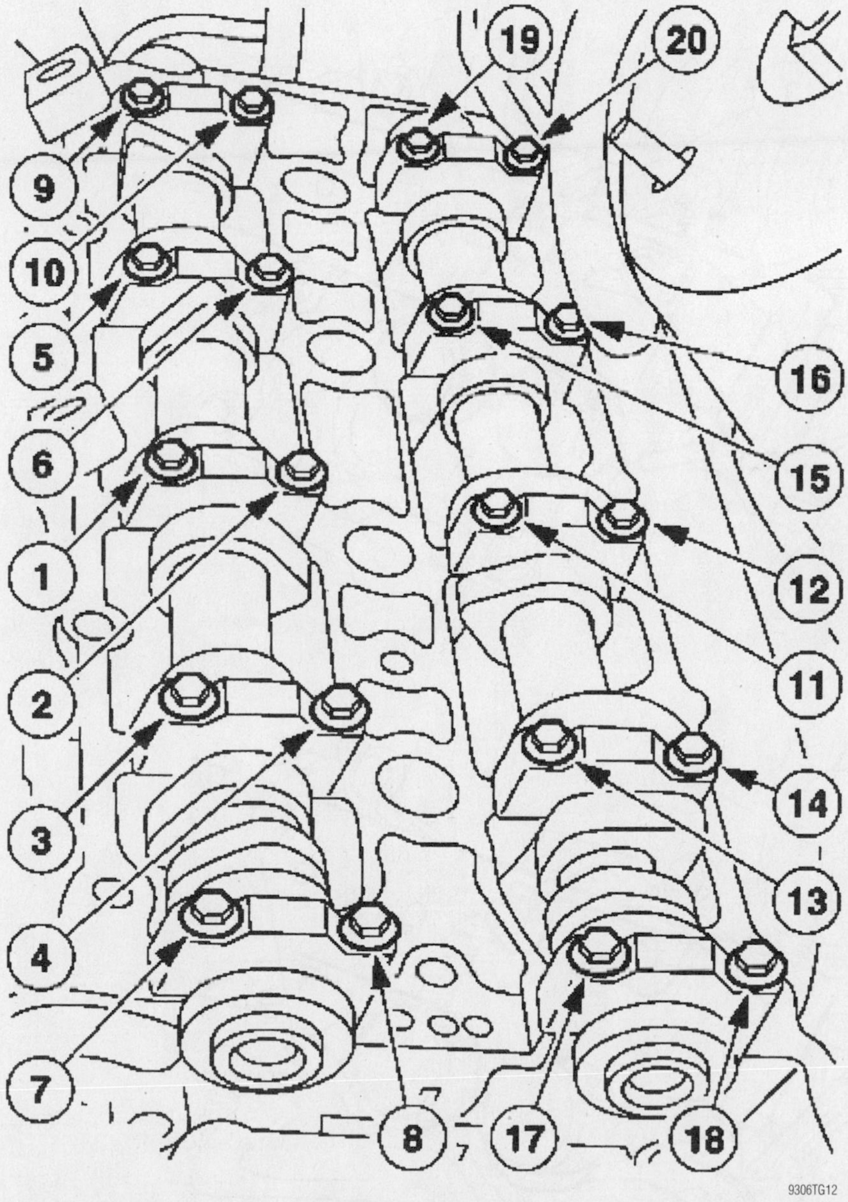

Camshaft journal bearing cap torque sequence—3.9L engine

Valve Lash

ADJUSTMENT

1. Before servicing the vehicle, refer to the precautions in the beginning of this section.
2. Remove or disconnect the following:
   - Negative battery cable
   - Engine appearance covers
   - Ignition coils
   - Valve covers
3. Measure the valve clearance while the camshaft lobe is pointed away from the valve shim. Rotate the crankshaft as necessary for each valve to be measured.
4. Valve clearance should be 0.007–0.009 in. (0.18–0.23mm) for intake valves or 0.009–0.011 in. (0.23–0.28mm) for exhaust valves.
5. If adjustment is necessary, compress the valves with the special tools and remove the shim with compressed air. Repeat for each valve to be adjusted.
6. Install or connect the following:
   - Valve covers
   - Ignition coils

- Engine appearance covers
- Negative battery cable

Starter Motor

REMOVAL & INSTALLATION

1. Before servicing the vehicle, refer to the precautions in the beginning of this section.
2. Remove or disconnect the following:
   - Negative battery cable
   - Starter motor wiring connectors
   - Starter motor

To install:

3. Install or connect the following:

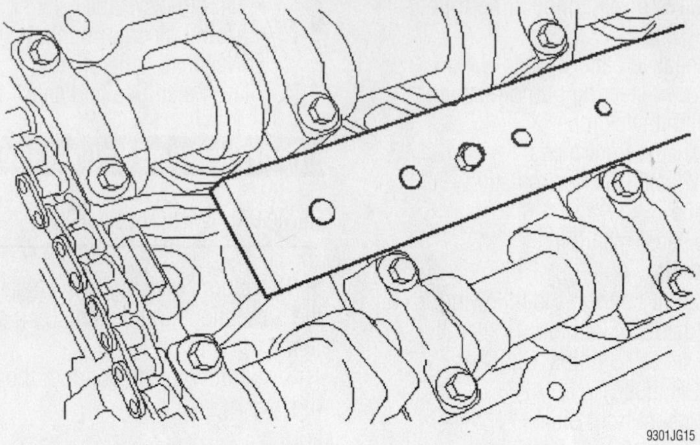

Valve adjustment tool base plate—3.9L engine

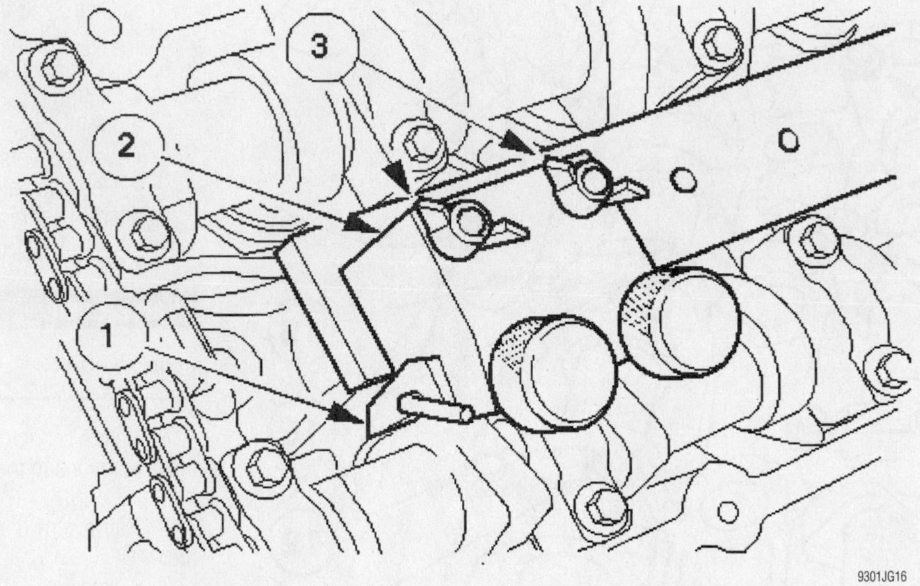

Valve adjustment tool attachment—3.9L engine

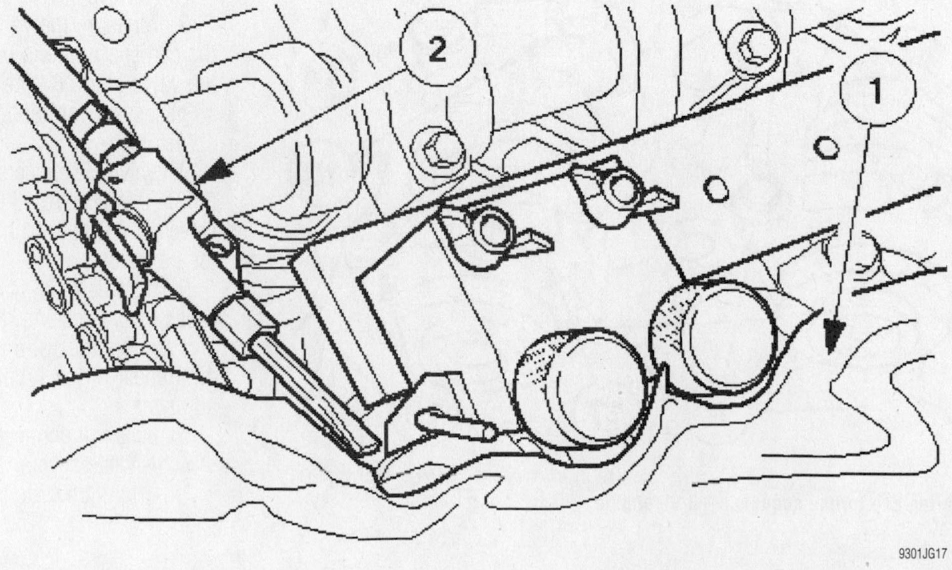

Remove the shims with compressed air—3.9L engine

- Starter motor. Tighten the bolts to 18 ft. lbs. (25 Nm).
- Starter motor wiring connectors
- Negative battery cable

Oil Pan

REMOVAL & INSTALLATION

1. Before servicing the vehicle, refer to the precautions in the beginning of this section.
2. Drain the engine oil.
3. Remove the oil pan.

To install:

4. Install the oil pan. Tighten the bolts in sequence as follows:
   a. Step 1: 44 inch lbs. (5 Nm)
   b. Step 2: 108 inch lbs. (12 Nm)
5. Fill the crankcase to the correct level.
6. Start the engine and check for leaks.

Oil Pump

REMOVAL & INSTALLATION

1. Before servicing the vehicle, refer to the precautions in the beginning of this section.
2. Remove or disconnect the following:
   - Crankshaft pulley
   - Front cover
   - Primary timing chains
   - Oil pump mounting bolts
   - Oil pump

To install:

3. Install or connect the following:
   - New gasket
   - Oil pump. Tighten the bolts to 53 inch lbs. (6 Nm) plus 90 degrees.
   - Primary timing chains
   - Front cover
   - Crankshaft pulley

Rear Main Seal

REMOVAL & INSTALLATION

1. Before servicing the vehicle, refer to the precautions at the beginning of this section.
2. Attach an engine support fixture to the engine lifting eyes.
3. Remove or disconnect the following:
   - Negative battery cable
   - Transmission
   - Flywheel
   - Rear crankshaft seal

To install:

4. Install or connect the following:
   - Rear main seal flush with the cylinder block surface
   - Flywheel
5. Tighten the flywheel bolts in a crossing pattern as follows:
   a. Step 1: 11 ft. lbs. (15 Nm)
   b. Step 2: 81 ft. lbs. (110 Nm)
6. Install or connect the following:
   - Transmission
   - Negative battery cable
7. Start the engine and check for leaks.

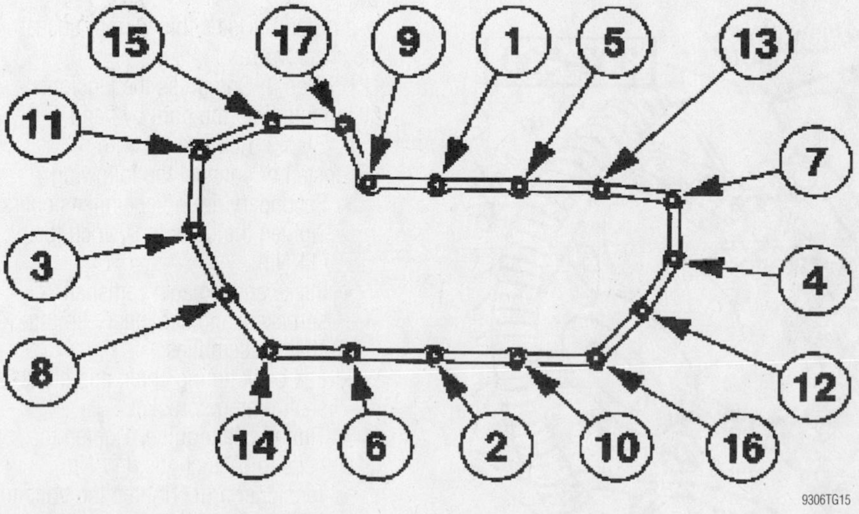

Oil pan torque sequence—3.9L engine

Timing Chain, Sprockets, Front Cover and Seal

REMOVAL & INSTALLATION

1. Before servicing the vehicle, refer to the precautions in the beginning of this section.
2. Drain the cooling system.
3. Remove or disconnect the following:
   - Negative battery cable
   - Engine appearance cover and brackets
   - Valve covers
   - Engine cooling fan assembly
   - Accessory drive belt
   - Water pump pulley
   - Alternator
   - Lower radiator hose and pipe
   - Heater hose
   - Idler pulleys
   - Crankshaft pulley
   - Front crankshaft seal
   - Power steering reservoir hose
   - Power steering pump and bracket
   - Hydraulic cooling fan reservoir hose and bracket
   - Hydraulic cooling fan pump and bracket
   - Front cover wiring harness clips
   - Front cover. Loosen the bolts in sequence.
   - Torque converter access panel
   - Crankshaft Position (CKP) sensor
4. Rotate the crankshaft to 45 degrees After Top Dead Center (ATDC). The crankshaft keyway will be in the 6 o'clock position. Check that the camshaft lobes are facing upwards. If not, rotate the crankshaft 1 full turn.
5. Install Crankshaft Holding Tool 303-645 in place of the CKP sensor.
6. Install Camshaft Locking Tool 303-530 to the right bank camshafts.
7. Loosen the exhaust and intake camshaft sprocket bolts and slide the sprockets forward on the bolts.
8. Remove or disconnect the following:
   - Right bank primary timing chain tensioner and blanking plate
   - Tensioner arm
   - Timing chain guide
   - Right bank timing chain and crankshaft timing sprocket
9. Remove the locking tool from the right bank camshafts and install it on the left bank camshafts.
10. Remove or disconnect the following:

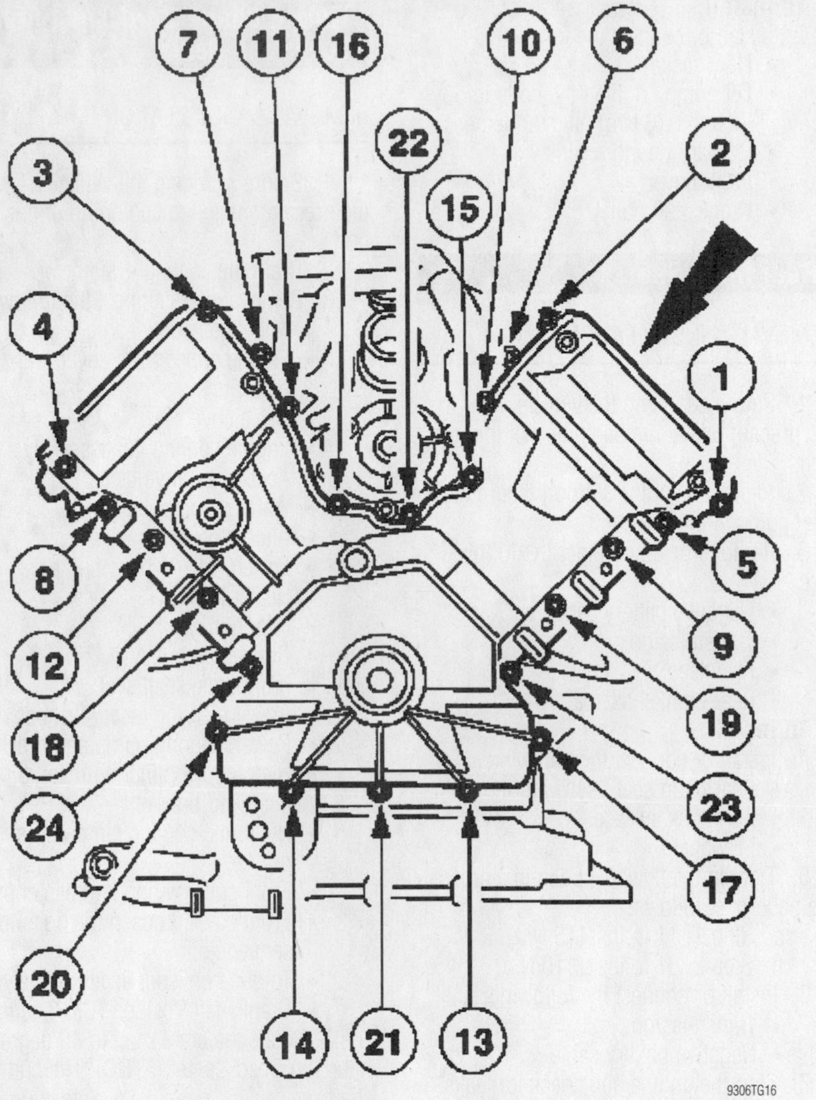

Front cover loosening sequence—3.9L engine

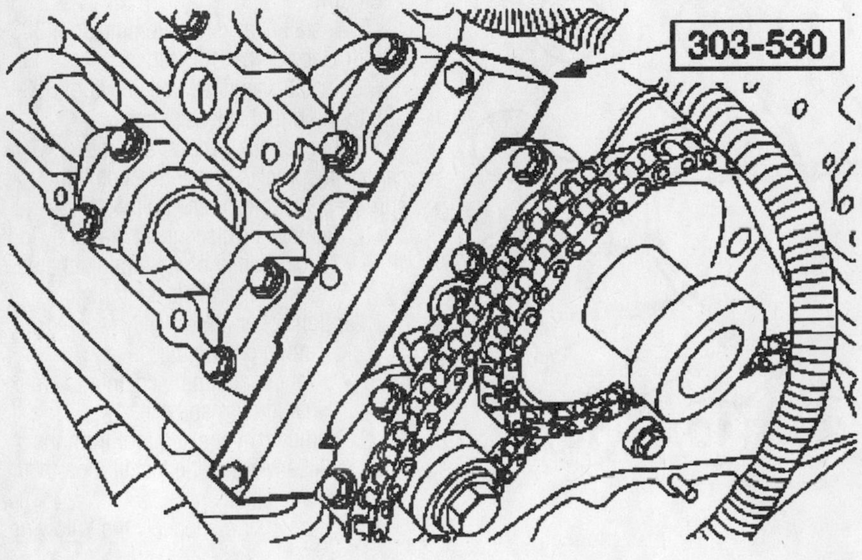

Camshaft Locking Tool—3.9L engine

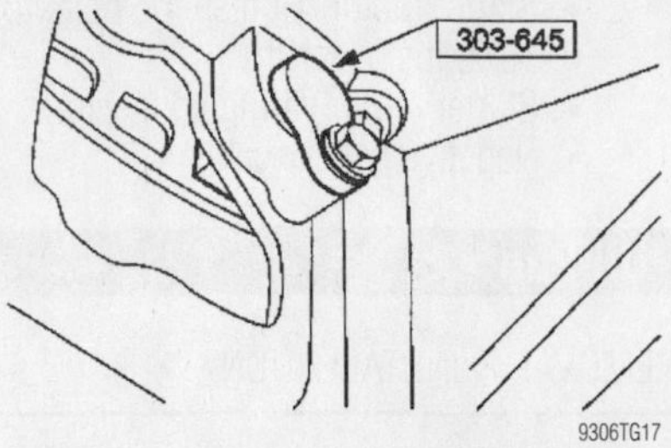

Crankshaft Holding Tool—3.9L engine

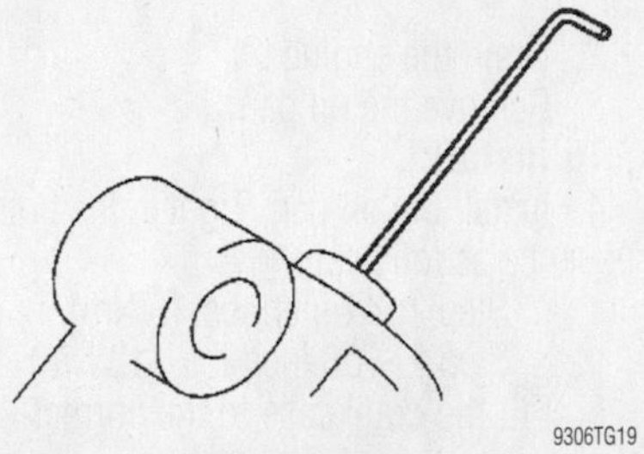

Timing chain tensioner preparation—3.9L engine

- Left bank primary timing chain tensioner and blanking plate
- Tensioner arm
- Timing chain guide
- Left bank timing chain and crankshaft timing sprocket
- Intake and exhaust camshaft sprocket bolts
- Intake and exhaust camshaft sprockets and secondary timing chain assemblies
- Secondary timing chain tensioners

To install:

11. Prepare the primary and secondary timing chain tensioners for installation as follows:
    a. Step 1: Use a thin wire to unseat the check valve
    b. Step 2: Compress the tensioner piston fully into the bore by hand
    c. Step 3: Remove the wire
12. Install or connect the following:
    - Secondary timing chain tensioners. Tighten the bolts to 97 inch lbs. (11 Nm).
    - Intake and exhaust camshaft sprockets and secondary timing chain assemblies
    - Left bank timing chain and crankshaft timing sprocket
    - Timing chain guide. Tighten the bolts to 97 inch lbs. (11 Nm).
    - Tensioner arm. Tighten the bolts to 97 inch lbs. (11 Nm).
    - Left bank primary timing chain tensioner and blanking plate. Tighten the bolts to 97 inch lbs. (11 Nm).

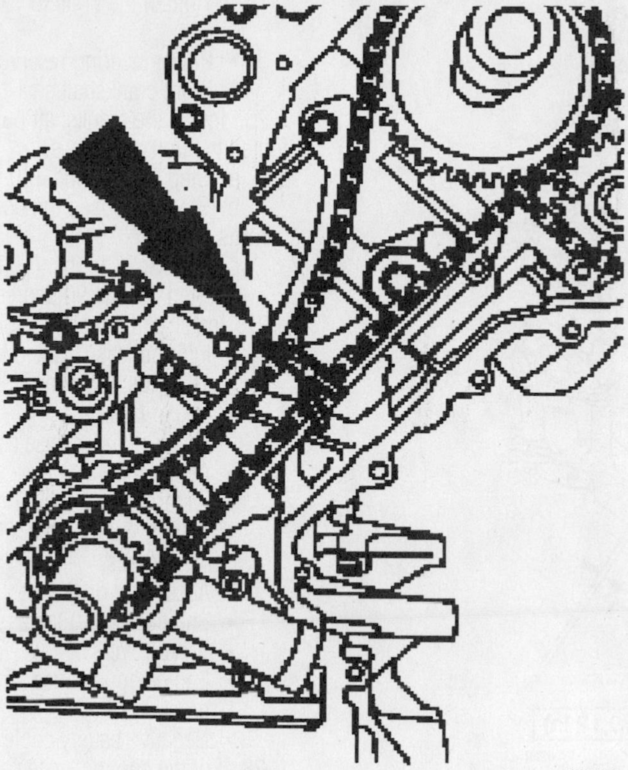

Use a tie strap to remove slack from the primary timing chain—3.9L engine

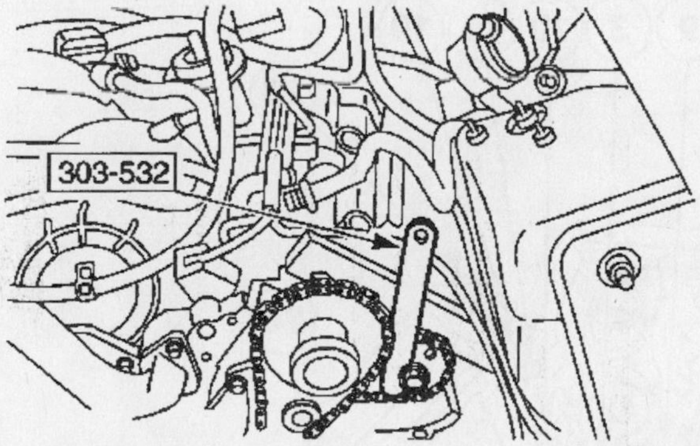

Timing Chain Tensioning Tool—3.9L engine

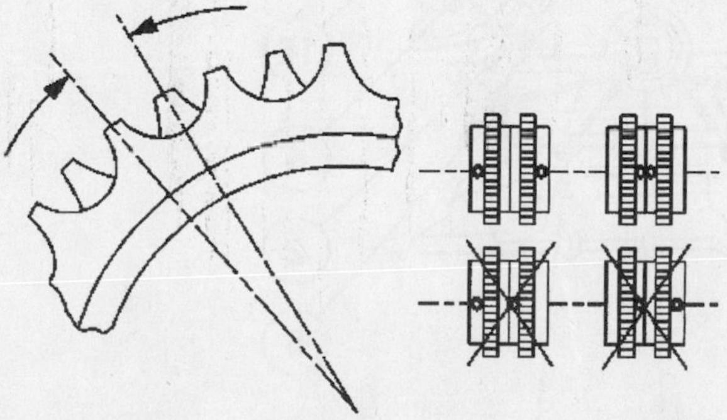

Crankshaft sprocket alignment—3.9L engine

13. Install a tie strap to take up the slack in the timing chain.
14. Use Timing Chain Tensioning Tool 303-532 to apply tension to the left bank exhaust camshaft.

➡**Tighten the exhaust camshaft bolt first.**

15. Tighten the left bank camshaft sprocket bolts as follows:
    a. Step 1: 15 ft. lbs. (20 Nm)
    b. Step 2: Plus 90 degrees
16. Remove the locking tool from the left bank camshafts and install it on the right bank camshafts.
17. Install or connect the following:
    - Right bank timing chain and crankshaft timing sprocket. Ensure that the crankshaft sprocket is installed as shown.
    - Timing chain guide. Tighten the bolts to 97 inch lbs. (11 Nm).
    - Tensioner arm. Tighten the bolts to 97 inch lbs. (11 Nm).
    - Right bank primary timing chain tensioner and blanking plate. Tighten the bolts to 97 inch lbs. (11 Nm).
18. Install a tie strap to take up the slack in the timing chain.
19. Use Timing Chain Tensioning Tool 303-532 to apply tension to the right bank exhaust camshaft.

➡**Tighten the exhaust camshaft bolt first.**

20. Tighten the right bank camshaft sprocket bolts as follows:
    a. Step 1: 15 ft. lbs. (20 Nm)
    b. Step 2: Plus 90 degrees
21. Remove the camshaft locking tool.
22. Apply silicone sealant to the areas indicated and install the front cover.
23. Tighten the front cover bolts in sequence as follows:
    a. Step 1: 44 inch lbs. (5 Nm)
    b. Step 2: 89 inch lbs. (10 Nm)
24. Remove the crankshaft locking tool.
25. Install or connect the following:
    - CKP sensor
    - Torque converter access panel
    - Front cover wiring harness clips
    - Hydraulic cooling fan pump and bracket. Tighten the bolts to 18 ft. lbs. (25 Nm).
    - Hydraulic cooling fan reservoir hose and bracket
    - Power steering pump and bracket.

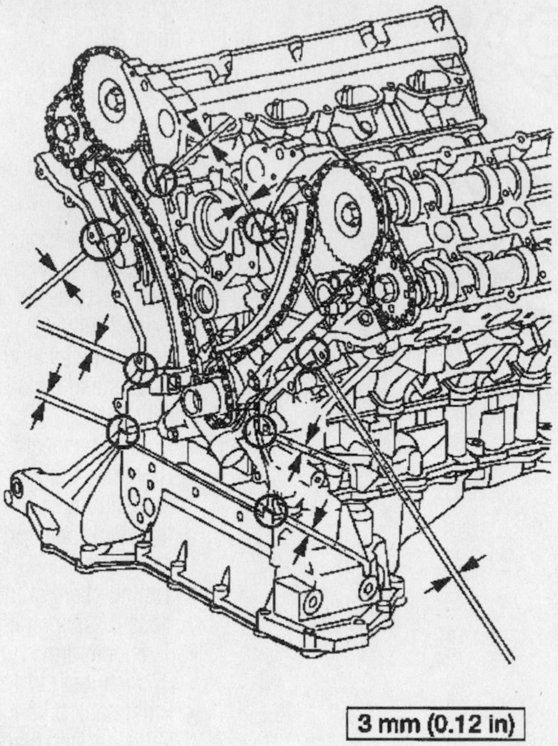

Apply silicone sealant to the areas indicated—3.9L engine

Tighten the bolts to 18 ft. lbs. (25 Nm).

- Power steering reservoir hose
- Front crankshaft seal

26. Install the crankshaft pulley and tighten the bolt as follows:
    a. Step 1: 59 ft. lbs. (80 Nm)
    b. Step 2: Loosen the bolt 2 complete turns
    c. Step 3: 37 ft. lbs. (50 Nm)
    d. Step 4: Plus 90 degrees
27. Install or connect the following:
    - Idler pulleys. Tighten the bolts to 18 ft. lbs. (25 Nm).
    - Heater hose
    - Lower radiator hose and pipe
    - Alternator
    - Water pump pulley. Tighten the bolts to 89 inch lbs. (10 Nm) plus 45 degrees.
    - Accessory drive belt
    - Engine cooling fan assembly
    - Valve covers
    - Engine appearance cover and brackets
    - Negative battery cable
28. Fill the cooling system.
29. Start the engine and check for leaks.

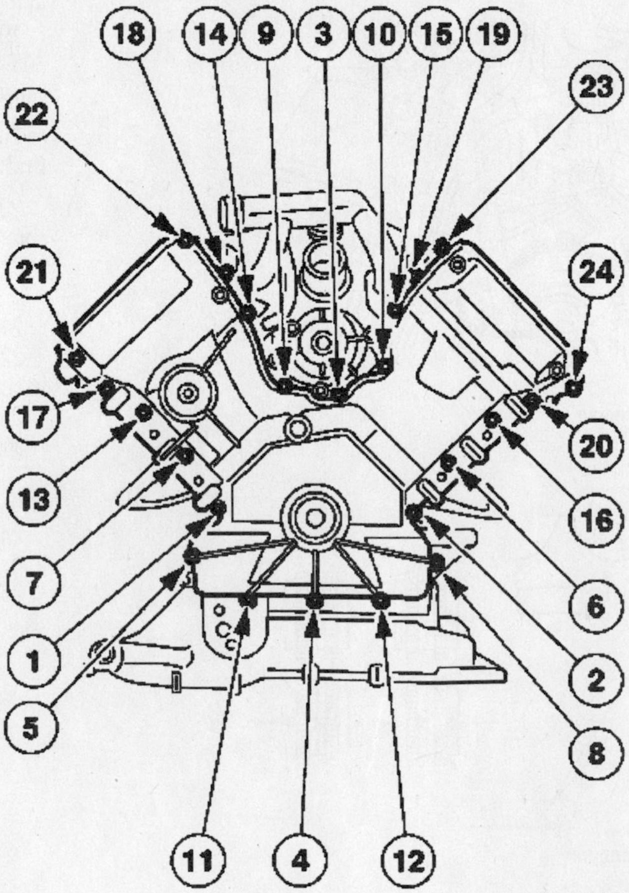

Front cover torque sequence—3.9L engine

FUEL SYSTEM

Fuel System Service Precautions

Safety is the most important factor when performing not only fuel system maintenance but also any type of maintenance. Failure to conduct maintenance and repairs in a safe manner may result in serious personal injury or death. Maintenance and testing of the vehicle's fuel system components can be accomplished safely and effectively by adhering to the following rules and guidelines.

- To avoid the possibility of fire and personal injury, always disconnect the negative battery cable unless the repair or test procedure requires that battery voltage be applied.
- Always relieve the fuel system pressure prior to disconnecting any fuel system component (injector, fuel rail, pressure regulator, etc.), fitting or fuel line connection. Exercise extreme caution whenever relieving fuel system pressure, to avoid exposing skin, face and eyes to fuel spray. Please be advised that fuel under pressure may penetrate the skin or any part of the body that it contacts.
- Always place a shop towel or cloth around the fitting or connection prior to loosening to absorb any excess fuel due to spillage. Ensure that all fuel spillage (should it occur) is quickly removed from engine surfaces. Ensure that all fuel soaked cloths or towels are deposited into a suitable waste container.
- Always keep a dry chemical (Class B) fire extinguisher near the work area.
- Do not allow fuel spray or fuel vapors to come into contact with a spark or open flame.
- Always use a back-up wrench when loosening and tightening fuel line connection fittings. This will prevent unnecessary stress and torsion to fuel line piping.
- Always replace worn fuel fitting O-rings with new. Do not substitute fuel hose or equivalent where fuel pipe is installed.

Fuel System Pressure

RELIEVING

1. Before servicing the vehicle, refer to the precautions in the beginning of this section.
2. Disconnect the negative battery cable.
3. Connect the fuel injection pressure test equipment JD 209 to the valve on the fuel supply manifold.
4. Insert the drain/bleed tube into the fuel container.
5. Follow the manufacturer's instructions and depressurize the fuel system.

Fuel Filter

REMOVAL & INSTALLATION

1. Before servicing the vehicle, refer to the precautions in the beginning of this section.
2. Relieve the fuel system pressure.
3. Remove or disconnect the following:
   - Negative battery cable
   - Fuel filter bracket cover
   - Fuel lines
   - Fuel filter

To install:

4. Install or connect the following:
   - Fuel filter into the bracket making sure the flow direction is correct. Tighten the clamp to 15–25 inch lbs. (2–3 Nm).
   - Fuel lines. Tighten the fittings to 22 ft. lbs. (30 Nm).
5. Start the engine and check for leaks.

Fuel Pump

REMOVAL & INSTALLATION

1. Before servicing the vehicle, refer to the precautions in the beginning of this section.
2. Relieve the fuel system pressure.
3. Drain the fuel tank.
4. Remove or disconnect the following:
   - Negative battery cable
   - Trunk liner
   - Trunk seal retainer
   - Rear lamp assembly interior trim finisher
   - Left and right side liners
   - Fuel feed and return lines
   - Fuel filler and vent hoses
   - Fuel tank wiring connectors
   - Fuel filler cap
   - Fuel tank retaining straps
   - Fuel tank
   - Fuel pump module

To install:

5. Install or connect the following:
   - Fuel pump module
   - Fuel tank
   - Fuel tank retaining straps
   - Fuel filler cap
   - Fuel tank wiring connectors
   - Fuel filler and vent hoses
   - Fuel feed and return lines
   - Left and right side liners
   - Rear lamp assembly interior trim finisher
   - Trunk seal retainer
   - Trunk liner
   - Negative battery cable
6. Fill the fuel tank with at least 10 gallons (38L) of fuel.
7. Start the engine and check for leaks.

Fuel Injector

REMOVAL & INSTALLATION

1. Before servicing the vehicle, refer to the precautions in the beginning of this section.
2. Relieve fuel system pressure.
3. Remove or disconnect the following:
   - Negative battery cable
   - Engine appearance covers
   - Fuel lines
   - Fuel pressure regulator
   - Fuel cross over elbow
   - Fuel injector connectors
   - Fuel injector clamping plates
   - Fuel injectors

To install:

4. Install or connect the following:
   - Fuel injectors. Use new O-ring seals.
   - Fuel injector clamping plates
   - Fuel injector connectors
   - Fuel cross over elbow
   - Fuel pressure regulator
   - Fuel lines
   - Engine appearance covers
   - Negative battery cable
5. Start the engine and check for leaks.

DRIVE TRAIN

Transmission Assembly

REMOVAL & INSTALLATION

1. Before servicing the vehicle, refer to the precautions in the beginning of this section.
2. Install a support fixture to the engine lifting eyes.
3. Drain the transmission fluid.
4. Remove or disconnect the following:
   - Negative battery cable
   - Engine appearance covers
   - Mass Air Flow (MAF) meter
   - Air intake assembly
   - Coolant recovery tank
   - Exhaust front pipes
   - Driveshaft
   - Shift selector cable
   - Transmission electrical connectors
   - Transmission oil cooler lines
   - Torque converter
   - Transmission mount and bracket
   - Transmission flange bolts
   - Transmission

To install:

5. Install or connect the following:
   - Transmission to the engine. Tighten the flange bolts to 32–42 ft. lbs. (43–57 Nm).
   - Transmission mount and bracket. Tighten the mount bolts to 22–30 ft. lbs. (30–40 Nm) and the bracket bolts to 16–21 ft. lbs. (22–28 Nm).
   - Torque converter. Tighten the bolts to 32–42 ft. lbs. (43–57 Nm).
   - Transmission oil cooler lines
   - Transmission electrical connectors
   - Shift selector cable
   - Driveshaft. Tighten the bolts to 55–65 ft. lbs. (75–88 Nm).
   - Exhaust front pipes
   - Coolant recovery tank
   - Air intake assembly
   - MAF meter
   - Engine appearance covers
   - Negative battery cable

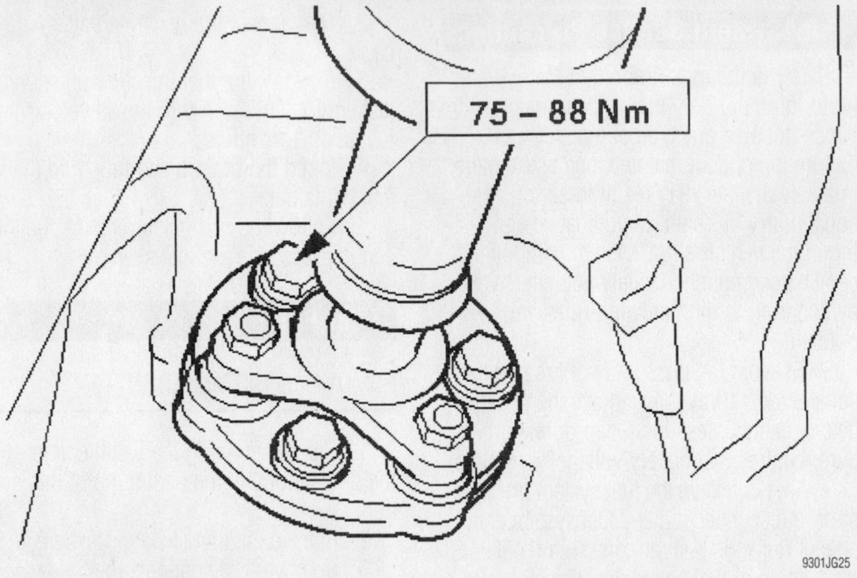

Driveshaft flange bolts

6. Fill the transmission to the correct level with the proper fluid. Do not over-fill.
7. Start the engine and check for leaks.

Halfshaft

REMOVAL & INSTALLATION

1. Before servicing the vehicle, refer to the precautions in the beginning of this section.
2. Remove or disconnect the following:
   - Negative battery cable
   - Rear wheel
   - Brake caliper
   - Wheel speed sensor
   - Hub carrier pivot bolt
   - Hub retaining nut
3. Press the stub shaft out of the hub and pry the inner joint out of the differential.

To install:

4. Install or connect the following:
   - Halfshaft inner joint to the differential
   - Halfshaft in the wheel hub by applying Loctite® 270 thread locking compound to the splines
   - Hub carrier pivot bolt by aligning the bolt head matchmarks. Tighten it to 66–81 ft. lbs. (90–110 Nm).
   - Hub retaining nut. Tighten the new nut to 221 ft. lbs. (300 Nm).
   - Wheel speed sensor
   - Brake caliper
   - Rear wheel
   - Negative battery cable
5. Check the wheel alignment and adjust as necessary.

➡If the hub is removed for any reason, a new bearing assembly must be installed. Never attempt to re-use a bearing.

CV-Joint

OVERHAUL

The CV-joints are serviced with the axle halfshaft as an assembly.

GENERAL MOTORS CORPORATION—C & H-BODIES

21

1998–01

Buick-Le Sabre • Park Ave. • **Oldsmobile-**Eighty Eight • LSS • Regency • **Pontiac-**Bonneville

PRECAUTIONS

Before servicing any vehicle, please be sure to read all of the following precautions, which deal with personal safety, prevention of component damage, and important points to take into consideration when servicing a motor vehicle:

- Never open, service or drain the radiator or cooling system when the engine is hot; serious burns can occur from the steam and hot coolant.
- Observe all applicable safety precautions when working around fuel. Whenever servicing the fuel system, always work in a well-ventilated area. Do not allow fuel spray or vapors to come in contact with a spark, open flame, or excessive heat (a hot drop light, for example). Keep a dry chemical fire extinguisher near the work area. Always keep fuel in a container specifically designed for fuel storage; also, always properly seal fuel containers to avoid the possibility of fire or explosion. Refer to the additional fuel system precautions later in this section.
- Fuel injection systems often remain pressurized, even after the engine has been turned **OFF**. The fuel system pressure must be relieved before disconnecting any fuel lines. Failure to do so may result in fire and/or personal injury.
- Brake fluid often contains polyglycol ethers and polyglycols. Avoid contact with the eyes and wash your hands thoroughly after handling brake fluid. If you do get brake fluid in your eyes, flush your eyes with clean, running water for 15 minutes. If eye irritation persists, or if you have taken brake fluid internally, IMMEDIATELY seek medical assistance.
- The EPA warns that prolonged contact with used engine oil may cause a number of skin disorders, including cancer! You should make every effort to minimize your exposure to used engine oil. Protective gloves should be worn when changing oil. Wash your hands and any other exposed skin areas as soon as possible after exposure to used engine oil. Soap and water, or waterless hand cleaner should be used.
- All new vehicles are now equipped with an air bag system, often referred to as a Supplemental Restraint System (SRS) or Supplemental Inflatable Restraint (SIR) system. The system must be disabled before performing service on or around system components, steering column, instrument panel components, wiring and sensors. Failure to follow safety and disabling procedures could result in accidental air bag deployment, possible personal injury and unnecessary system repairs.
- Always wear safety goggles when working with, or around, the air bag system. When carrying a non-deployed air bag, be sure the bag and trim cover are pointed away from your body. When placing a non-deployed air bag on a work surface, always face the bag and trim cover upward, away from the surface. This will reduce the motion of the module if it is accidentally deployed. Refer to the additional air bag system precautions later in this section.
- Clean, high quality brake fluid from a sealed container is essential to the safe and proper operation of the brake system. You should always buy the correct type of brake fluid for your vehicle. If the brake fluid becomes contaminated, completely flush the system with new fluid. Never reuse any brake fluid. Any brake fluid that is removed from the system should be discarded. Also, do not allow any brake fluid to come in contact with a painted surface; it will damage the paint.
- Never operate the engine without the proper amount and type of engine oil; doing so WILL result in severe engine damage.
- Timing belt maintenance is extremely important! Many models utilize an interference-type, non-freewheeling engine. If the timing belt breaks, the valves in the cylinder head may strike the pistons, causing potentially serious (also time-consuming and expensive) engine damage. Refer to the maintenance interval charts in the front of this manual for the recommended replacement interval for the timing belt, and to the timing belt section for belt replacement and inspection.
- Disconnecting the negative battery cable on some vehicles may interfere with the functions of the on-board computer system(s) and may require the computer to undergo a relearning process once the negative battery cable is reconnected.
- When servicing drum brakes, only disassemble and assemble one side at a time, leaving the remaining side intact for reference.
- Only an MVAC-trained, EPA-certified automotive technician should service the air conditioning system or its components.

ENGINE REPAIR

➡Disconnecting the negative battery cable on some vehicles may interfere with the operation of the onboard computer system. The computer may undergo a relearning process once the negative battery cable is reconnected.

Alternator

REMOVAL

1998–99 Models

1. Before servicing the vehicle, refer to the precautions in the beginning of this section.
2. Remove or disconnect the following:
   - Negative battery cable
   - Supercharger belt, if equipped
   - Accessory drive belt
   - Engine cover
   - Electrical connections
   - Support brackets
   - Alternator

2000–01 Models

1. Before servicing the vehicle, refer to the precautions in the beginning of this section.
2. Remove or disconnect the following:
   - Negative battery cable
   - Accessory drive belt
   - Fuel injector sight shield
   - Alternator brace
   - Electrical connections
   - Alternator bolts and the alternator

INSTALLATION

1998–99 Models

1. Install or connect the following:
   - Alternator and torque the bolts to 37 ft. lbs. (50 Nm)
   - Electrical connections and torque the nut to 15 ft. lbs. (20 Nm)
   - Support brackets. Torque the nut to 37 ft. lbs. (50 Nm) and the bolt to 18 ft. lbs. (25 Nm).
   - Engine cover
   - Accessory drive belt
   - Supercharger belt, if equipped
   - Negative battery cable

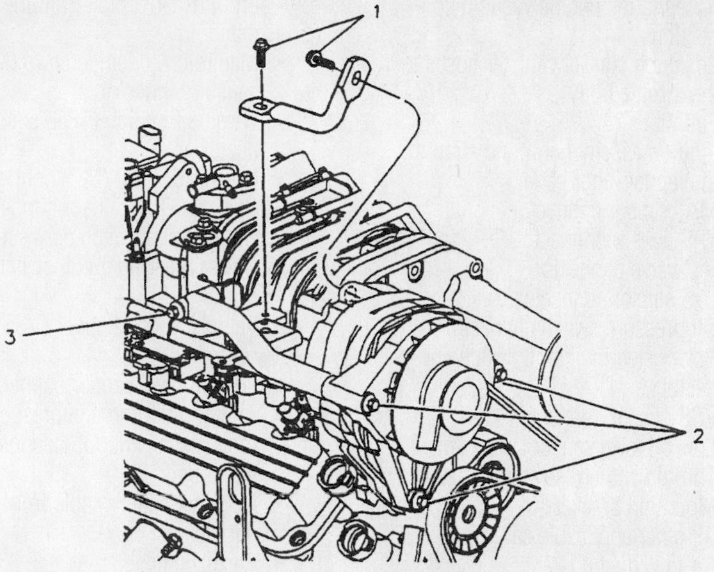

Exploded view of the alternator mounting—1998–99 models

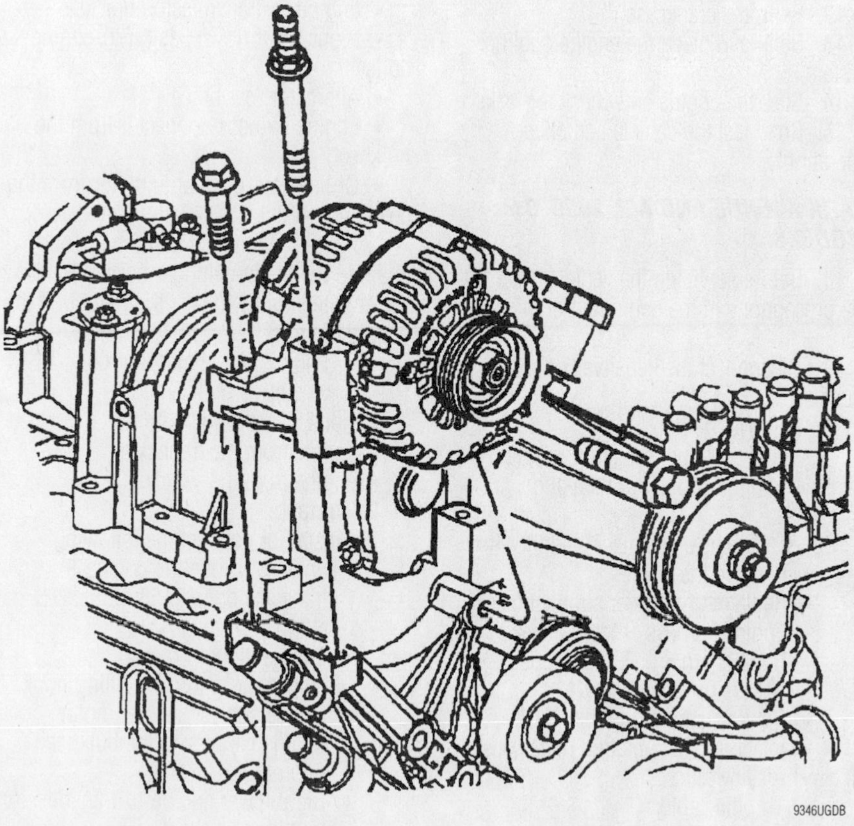

Exploded view of the alternator mounting—2000–01 supercharged models

2000–01 Models

1. Install or connect the following:
   - Alternator and torque the bolts to 37 ft. lbs. (50 Nm)
   - Electrical connections and torque the nut to 15 ft. lbs. (20 Nm)
   - Alternator brace. Torque the nut to 37 ft. lbs. (50 Nm) and the bolt to 22 ft. lbs. (30 Nm).
   - Fuel injector sight shield
   - Accessory drive belt
   - Negative battery cable

Ignition Timing

The 3.8L (VIN K and 1) engines utilize a Distributorless Ignition System (DIS). This system uses 3 twin tower coils which fire 2 spark plugs simultaneously.

ADJUSTMENT

The ignition timing is not adjustable, and is set according to engine demand electronically. The Powertrain Control Module (PCM) controls the ignition timing for all driving conditions.

Engine Assembly

REMOVAL & INSTALLATION

1998–99 Models

EXCEPT PARK AVENUE AND ALL 2000–01 MODELS

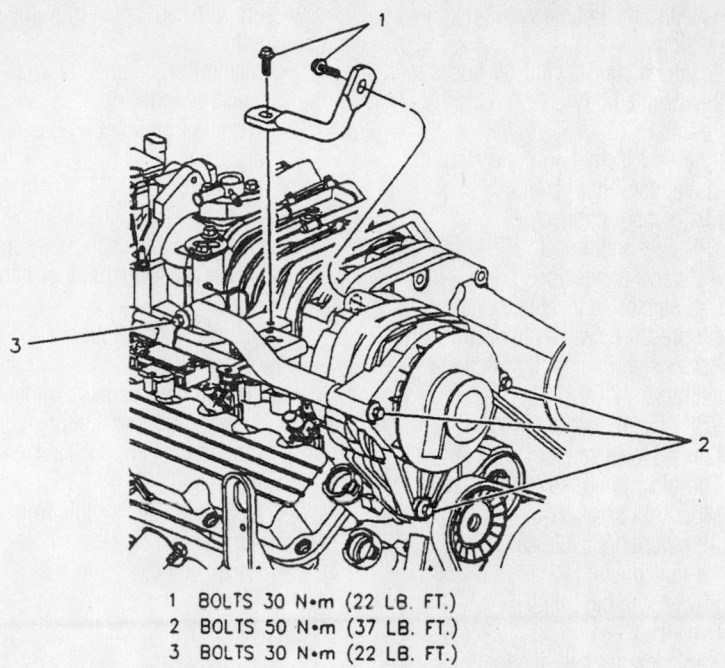

1. Before servicing the vehicle, refer to the precautions in the beginning of this section.
2. Disconnect the negative battery cable.
3. Remove the hood.
4. Relieve the fuel system pressure.
5. Drain the coolant system and crankcase.
6. Remove or disconnect the following:
   - Radiator and heater supply hoses
   - Negative battery cable from the engine block
   - Engine harness connector at the bulkhead
   - Accessory drive belt
   - Supercharger belt, if equipped
   - Power steering pump
   - Air flow duct
   - Throttle cable
   - Cruise control cable, if equipped
   - Mounting bracket

- All other cables from the engine
- Manifold Air Temperature (MAT) sensor connector
- Throttle Position (TPS) sensor connector
- Idle Air Control (IAC) connector
- Oxygen (O_2S) sensor connector
- Oil pressure switch connector
- Power Steering cutoff switch connector
- Vehicle Speed Sensor (VSS) connector
- Low oil level sensor connector
- Ignition assembly ground strap from the fender inner panel
- Fuel lines from the fuel rail and the pressure regulator
- Emission control canister hoses from the throttle body
- Brake booster and heater control hoses from the engine vacuum connections
- Front exhaust pipe from the right manifold
- Air conditioning compressor and move aside without disconnecting the lines
- Engine-to-transaxle bracket
- Flexplate cover
- Starter
- Torque converter bolts

➡Matchmark the flexplate-to-torque converter relationship for reassembly.

7. Attach a suitable lifting hook and chain to the engine lifting brackets. Raise the engine slightly to take the weight off the engine mounts.
8. Support the transaxle.
9. Remove or disconnect the following:
   - Drive belt
   - Engine torque axis engine mount
   - Transaxle-to-engine bolts
   - Engine from the transaxle
   - Engine from the vehicle

To install:

10. Install or connect the following:
    - Engine. Torque the engine-to-transaxle bolts to 55 ft. lbs. (75 Nm).
    - Torque axis engine mount bolts. Torque them to 52 ft. lbs. (87 Nm).
11. Align the torque converter-to-flexplate matchmark.
    - Torque converter bolts and torque them to 46 ft. lbs. (62 Nm)
    - Starter
    - Flexplate cover
    - Engine-to-transaxle bracket
    - Air conditioning compressor
    - Exhaust pipe to the manifold
    - Brake booster and heater control hoses to the engine vacuum connections
    - Emission control canister hoses to the throttle body
    - Fuel lines
    - Ignition assembly ground strap to the fender inner panel
    - MAT sensor connector
    - TPS sensor connector
    - IAC valve connector
    - O_2S sensor connector
    - Oil pressure switch connector
    - Power steering cutoff switch connector
    - VSS sensor connector
    - Low oil level sensor connector
    - Throttle and cruise control cables
    - Mounting bracket
    - All other applicable cables
    - Air flow duct
    - Power steering pump
    - Drive belt
    - Supercharger belt, if equipped
    - Engine harness at the bulkhead
    - Negative battery cable to the engine
    - Radiator and heater supply hoses
    - Hood
    - Negative battery cable
12. Refill the crankcase.
13. Refill and bleed the engine cooling system.
14. Start the engine and check for leaks.
15. Road test the vehicle and check operation.

PARK AVENUE AND ALL 2000–01 MODELS

1. Before servicing the vehicle, refer to the precautions in the beginning of this section.
2. Disconnect the negative battery cable.
3. Remove the hood.
4. Relieve the fuel system pressure.
5. Drain the coolant system and crankcase.
6. Remove or disconnect the following:
   - Remove the hood
   - Radiator and heater supply hoses
   - Engine harness at the bulkhead
   - Supercharger belt, if equipped
   - Accessory drive belt
   - Power steering pump
   - Intake Air Temperature (IAT) sensor
   - Air flow duct
   - Throttle cable
   - Cruise control cable, if equipped
   - Battery wire from the alternator
   - Fuel lines from the fuel rail and the pressure regulator
   - Brake booster from the engine vacuum connections
   - Front exhaust pipe from the right manifold
   - Manifold Air Temperature (MAT) sensor connector
   - Throttle Position (TPS) sensor connector
   - Idle Air Control (IAC) connector
   - Oxygen (O_2S) sensor connector
   - Oil pressure switch connector
   - Power Steering cutoff switch connector
   - Vehicle Speed Sensor (VSS) connector
   - Low oil level sensor connector
   - Air conditioning compressor and move aside without disconnecting the lines
   - Negative battery cable from the engine
   - Flexplate cover
   - Starter
   - Torque converter bolts

➡Matchmark the flexplate-to-torque converter relationship for reassembly.

- Right front tire and wheel
- Right engine splash shield
- Crankshaft balancer
- Engine-to-transmission brackets
- Right lower transmission-to-engine bolts
- Alternator rear brace
- Engine harness connector from the bulkhead
- Cruise control stepper motor from the cowl and set aside

7. Attach a suitable lifting hook and chain to the engine lifting brackets. Raise the engine slightly to take the weight off the engine mounts.
   - Engine mount and bracket
   - Drive belt
   - Transaxle-to-engine bolts
   - Engine from the transaxle
   - Engine from the vehicle

To install:

8. Install or connect the following:
   - Engine. Torque the engine-to-transaxle bolts to 55 ft. lbs. (75 Nm).
   - Engine mount bracket
9. Remove the chain and lifting hook.
   - Cruise control stepper motor
   - Engine harness to the bulkhead
   - Alternator rear brace
   - Right lower engine-to-transmission bolt
   - Engine-to-transmission bracket
   - Crankshaft balancer. Tighten the balancer bolt to 111 ft. lbs. (150 Nm).
   - Right engine splash shield

- Right front tire and wheel. Tighten lug nuts to 100 ft. lbs. (140 Nm).

10. Align the torque converter-to-flexplate matchmark.
    - Torque converter bolts. Torque them to 46 ft. lbs. (62 Nm).
    - Starter
    - Flexplate cover
    - Negative battery cable to the engine
    - Air conditioning compressor
    - MAT sensor connector
    - TPS sensor connector
    - IAC valve connector
    - O_2S sensor connector
    - Oil pressure switch connector
    - Power Steering cutoff switch connector
    - VSS sensor connector
    - Low oil level sensor connector
    - Front engine splash shield
    - Exhaust pipe to the manifold
    - Brake booster hoses to the engine vacuum connections
    - Fuel lines
    - Battery wire to the alternator
    - Throttle and cruise control cables
    - Air flow duct
    - IAT sensor
    - Power steering pump
    - Drive belt
    - Radiator and heater supply hoses
    - Hood
    - Negative battery cable
11. Refill the crankcase.
12. Refill and bleed the engine cooling system.
13. Start the engine and check for leaks.
14. Road test the vehicle and check operation.

Water Pump

REMOVAL & INSTALLATION

3.8L (VIN K) Engine

1. Before servicing the vehicle, refer to the precautions in the beginning of this section.
2. Drain the cooling system.
3. Remove or disconnect the following:
    - Negative battery cable
    - Supercharger belt, if equipped
    - Accessory drive belt
    - Coolant hoses from the water pump
    - Water pump pulley bolts

➡The long bolt can be removed by aligning the bolt head up with the hole in the frame rail.

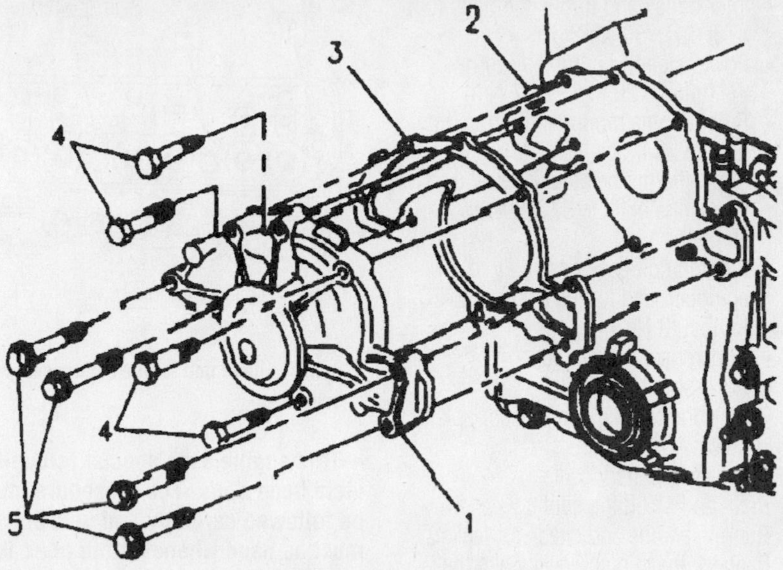

Exploded view of the water pump—3.8L (VIN K and 1) engines

- Pulley
- Water pump bolts
- Water pump

To install:

4. Apply a thin bead of sealer around the outside edge of the water pump.
5. Install or connect the following:
    - Water pump with new gasket. Torque the water pump short bolts to 11 ft. lbs. (15 Nm) and the long bolts to 22 ft. lbs. (30 Nm) plus an additional 80 degree turn.
    - Water pump pulley. Torque the bolts to 115 inch lbs. (13 Nm).
    - Coolant hoses to the water pump
    - Drive belt
    - Supercharger belt, if equipped
6. Refill and bleed the cooling system.
7. Run the engine and check for leaks.
8. Recheck the coolant level when the engine has cooled.

3.8L (VIN 1) Engine

1. Before servicing the vehicle, refer to the precautions in the beginning of this section.
2. Drain the cooling system.
3. Support the engine using an engine support fixture.
4. Remove or disconnect the following:
    - Negative battery cable
    - Air conditioning compressor splash shield
    - Supercharger and accessory drive belts
    - Coil pack, if equipped
    - Supercharger belt tensioner
    - Front engine mount
    - Power steering pump
    - Engine mount bracket
    - Idler pulley
    - Water pump pulley
    - Water pump

To install:

5. Apply a thin bead of sealer around the outside edge of the water pump.
6. Install or connect the following:
    - Water pump with a new gasket. Torque the short bolts to 11 ft. lbs. (15 Nm) and the long bolts to 22 ft. lbs. (30 Nm).
    - Water pump pulley and torque the bolts to 115 inch lbs. (13 Nm)
    - Engine mount bracket. Torque the nut closest to the a/c compressor to 33 ft. lbs. (45 Nm), and the others to 58 ft. lbs. (78 Nm).

- Idler pulley and torque the bolts to 37 ft. lbs. (50 Nm)
- Power steering pump and torque the bolts to 20 ft. lbs. (27 Nm)
- Front engine mount and torque the nuts to 58 ft. lbs. (78 Nm)
- Supercharger belt tensioner and torque the bolts to 37 ft. lbs. (50 Nm)
- Ignition coil pack assembly, if equipped and torque the nuts to 70 in. lbs. (8 Nm)
- Supercharger drive belt
- Accessory drive belt
- Air conditioning compressor splash shield
- Negative battery cable

7. Refill and bleed the cooling system.
8. Run the engine and check for leaks.
9. Recheck the coolant level when the engine has cooled.

Cylinder Head

REMOVAL & INSTALLATION

1. Before servicing the vehicle, refer to the precautions in the beginning of this section.
2. Disconnect the negative battery cable.
3. Relieve the fuel system pressure.
4. Drain the cooling system.
5. Remove or disconnect the following:
   - Intake manifold
   - Exhaust manifold
   - Valve covers
   - Ignition wires and ignition coil/module assembly
   - Alternator front mounting bracket and alternator
   - Air conditioning bracket-to-cylinder head bolt
   - Power steering pump
   - Accessory drive belt tensioner
   - Supercharger belt tensioner, if equipped
   - Fuel pipe heat shield
   - Rocker arm assemblies, note their original position
   - Pushrods and guide plate
   - Cylinder head bolts
   - Cylinder head

To install:

6. Place the new cylinder head gasket on the engine block dowels with the note **THIS SIDE UP** facing the cylinder head and the arrow facing the front of the engine. Position the cylinder head on the engine block.

➡The head gasket is identified by either a L or a R stamped on it next to the arrow.

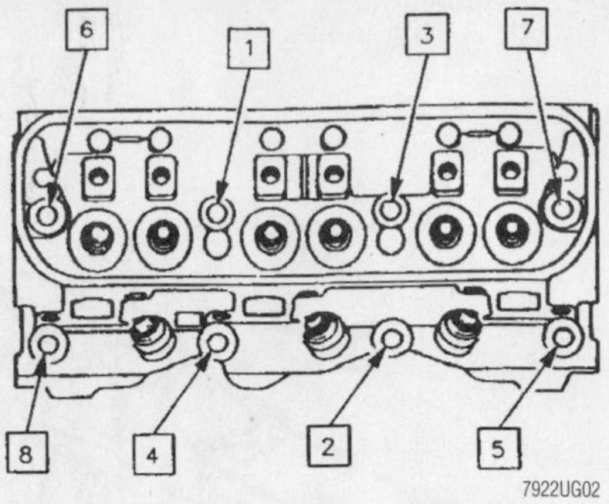

Cylinder head bolt torque sequence—3.8L engine

➡This engine uses special torque-to-yield head bolts. The procedure must be followed carefully and new bolts must be used whenever the head is removed. Total bolt torque should not exceed 60 ft. lbs. (81 Nm).

7. Install new cylinder head bolts and torque them in sequence as follows:
   a. Step 1: 37 ft. lbs. (50 Nm).
   b. Step 2: Plus 130 degrees.
   c. Step 3: Rotate the center 4 bolts an additional 30 degrees.
8. Install or connect the following:
   - Pushrods and guide plate
   - Rocker arm assemblies into their original location

➡Apply a thread lock compound to the rocker arm pedestal bolts before assembly.

   - Valve covers
   - Fuel pipe heat shield
   - Accessory drive belt tensioner
   - Supercharger belt tensioner (VIN 1 engine)
   - Power steering pump
   - Air conditioning compressor bracket bolt. Torque it to 52 ft. lbs. (70 Nm).
   - Alternator front mounting bracket, and alternator
   - Ignition coil/module assembly and spark plug wires
   - Exhaust manifold. Torque the bolts to 22 ft. lbs. (30 Nm).
   - Intake manifold
   - Negative battery cable
9. Refill and bleed the cooling system.
10. Start the engine and check for leaks and proper operation.

Rocker Arms

REMOVAL & INSTALLATION

➡When removing valvetrain components, it is very important that they are marked for installation reference, so that they can be reinstalled in their original location.

1. Before servicing the vehicle, refer to the precautions in the beginning of this section.
2. Remove or disconnect the following:
   - Negative battery cable
   - Spark plug wires
   - Rocker arm cover(s)
   - Rocker arm pedestal bolts and assemblies
   - Pushrods

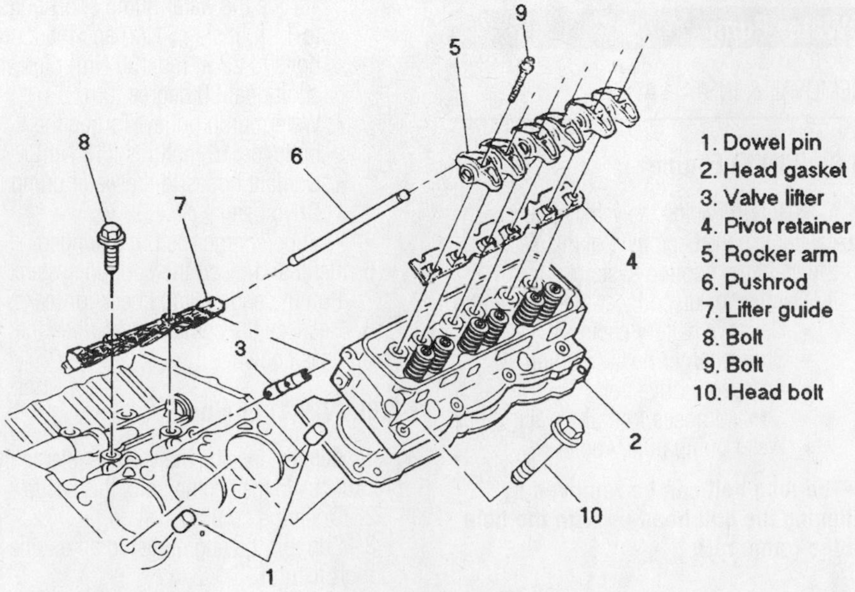

Exploded view of the rocker arms and related components mounting—3.8L engines

To install:

3. Lubricate the pushrod tips and put them in their proper locations.
4. Lubricate the rocker arms and pedestals and install them. Be sure the pushrod tips are properly seated in the rocker arms.
5. Apply thread locking compound to the rocker arm pedestal bolt threads. Torque the bolts to 11 ft. lbs. (15 Nm) plus an additional 90 degree turn.
6. Apply suitable thread locking compound to the rocker arm cover bolts.
7. Install or connect the following:
   - Rocker arm cover using a new gasket. Torque the bolts to 89 inch lbs. (10 Nm).
   - Spark plug wires
   - Accessory drive belt
   - Supercharger belt, if equipped
   - Negative battery cable
8. Run the engine and check for leaks and proper engine operation.

Supercharger

REMOVAL & INSTALLATION

➡A small amount of oil seepage through the front seal is normal. This seepage is caused by minute traces of oil escaping around the seal due to

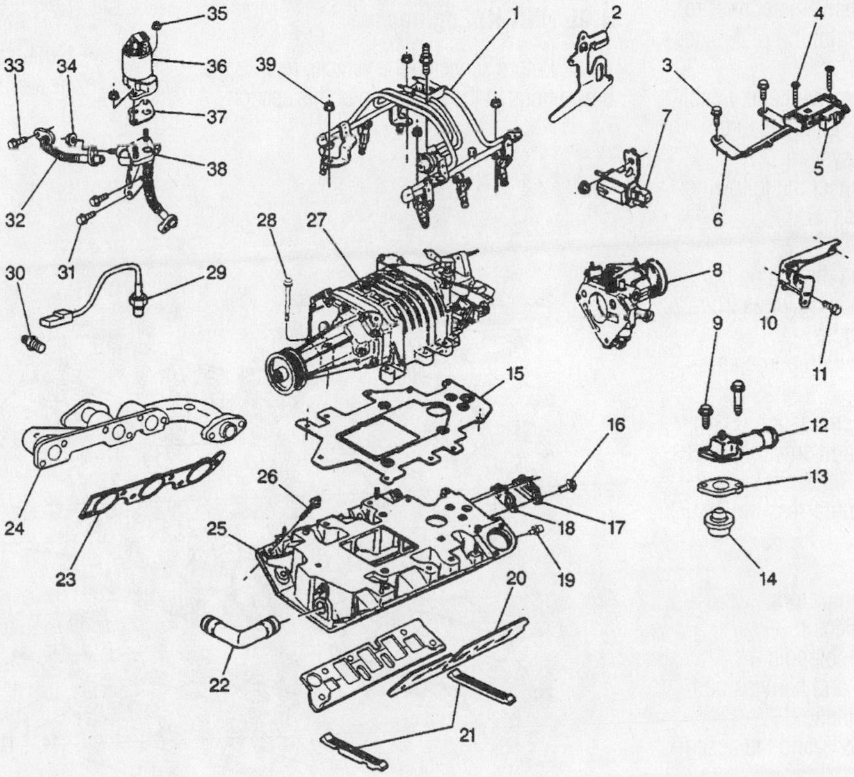

1 Fuel Injection Rail
2 Fuel Injector Sight Shield Bracket
3 MAP Sensor Bracket Bolt
4 MAP Sensor Bolt
5 MAP Sensor
6 MAP Sensor Bracket
7 Bypass Valve
8 Throttle Body
9 Water Outlet Bolt
10 Accelerator Cable Control Bracket
11 Accelerator Cable Control Bracket Bolt
12 Water Outlet
13 Water Outlet Gasket
14 Thermostat
15 Supercharger Gasket
16 Engine Coolant Manifold Bolt
17 Engine Coolant Manifold
18 Engine Coolant Manifold Gasket
19 Coolant Temperature Sensor
20 Lower Intake Manifold Gasket
21 Lower Intake Manifold Seal
22 Heater Inlet Pipe With Seal
23 Exhaust Manifold Gasket
24 Exhaust Manifold (Right)
25 Lower Intake Manifold
26 Lower Intake Manifold Bolt
27 Supercharger
28 Supercharger Bolt
29 Heated Oxygen Sensor
30 Exhaust Manifold Bolt/Stud (Right)
31 EGR Valve Adapter Bolt
32 EGR Valve Outlet Pipe
33 EGR Valve Outlet Pipe Bolt
34 EGR Valve Outlet Pipe Nut
35 EGR Valve Nut
36 EGR Valve
37 EGR Valve Gasket
38 EGR Valve Adapter
39 Fuel Injection Rail Nut

9300UG01

Exploded view of the supercharger and related components—3.8L (VIN 1) engine

normal pressure build up in the oil cavity within the supercharger. A build up of dust can stick to the thin oil film, which causes the oil seepage to appear worse than it really is. The supercharger should not be replaced for this seepage. However, if supercharger oil is visually dripping from the supercharger front seal, the supercharger will need to be replaced. The supercharger oil level should be checked every 30,000 miles or every 36 months.

1. Before servicing the vehicle, refer to the precautions in the beginning of this section.
2. Disconnect the negative battery cable.
3. Relieve the fuel system pressure.
4. Drain the cooling system.
5. Remove or disconnect the following:
   - Supercharger belt
   - Engine cover
   - Air duct from the throttle body
   - Right side spark plug wires from the ignition module
   - Alternator brace with purge solenoid
   - Exhaust Gas Recirculation (EGR) wiring harness and shield, if necessary
   - Manifold Absolute Pressure (MAP) sensor bracket
   - Fuel lines
   - Fuel injector connectors
   - Fuel rail with injectors
   - Booster bypass solenoid
   - Regulator valve and harness bolt
   - Boost control solenoid
   - Throttle body bolts and the assembly
   - Supercharger bolts and the assembly

To install:

6. Install or connect the following:
   - Supercharger with new gasket and torque the bolts, gradually and evenly, to 17 ft. lbs. (23 Nm).
   - MAP sensor bracket
   - Throttle body and torque the nuts to 89 inch lbs. (10 Nm)
   - Boost control solenoid and torque the nut to 72 inch lbs. (8 Nm)
   - Regulator valve and harness bolt
   - Booster bypass solenoid
   - Fuel rail with injectors
   - Fuel lines
   - Electrical connectors to the fuel injectors
   - Alternator brace with purge solenoid
   - EGR wiring harness and shield
   - Right side spark plug wires
   - Supercharger drive belt
   - Air duct
   - Engine cover
   - Negative battery cable
7. Refill and bleed the cooling system.
8. Run the engine and check for leaks and proper engine operation.

Intake Manifold

REMOVAL & INSTALLATION

3.8L (VIN K) Engine

1. Before servicing the vehicle, refer to the precautions in the beginning of this section.
2. Disconnect the negative battery cable.
3. Drain the cooling system.
4. Relieve the fuel system pressure.
5. Remove or disconnect the following:
   - Fuel injector sight shield
   - Air inlet duct
   - Spark plug wires from the right side
   - Vacuum lines from the intake manifold
   - Fuel lines
   - Fuel injector electrical connectors
   - Fuel regulator vacuum line
   - Fuel rail from the intake manifold
   - Exhaust Gas Recirculation (EGR) heat shield

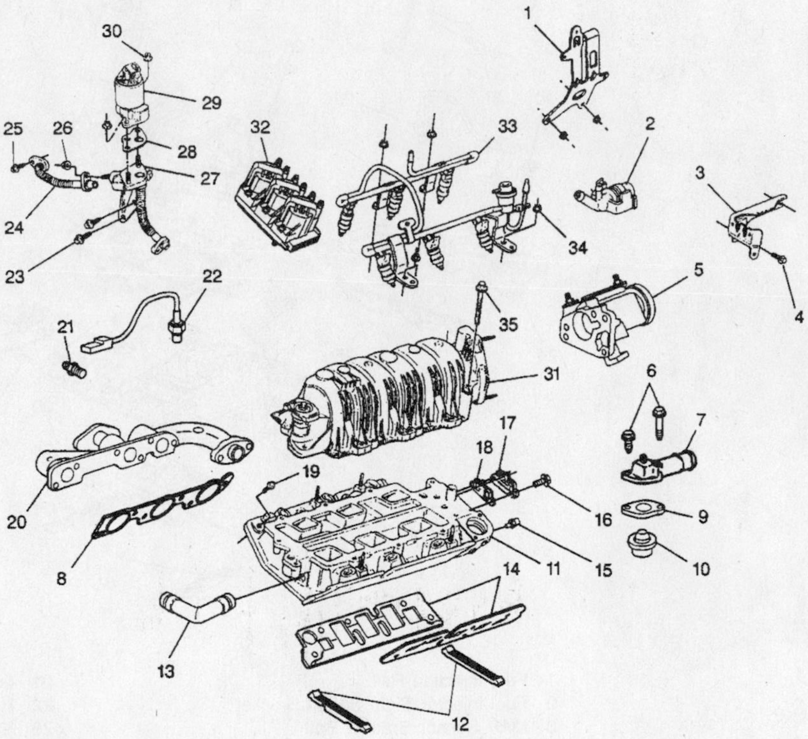

1 Fuel Injector Sight Shield Bracket
2 Vacuum Source Manifold
3 Accelerator Cable Control Bracket
4 Throttle Body Support Bolt
5 Throttle Body
6 Water Outlet Bolt
7 Water Outlet
8 Exhaust Manifold Gasket
9 Water Outlet Gasket
10 Thermostat
11 Lower Intake Manifold
12 Intake Manifold Seal
13 Heater Water Inlet Pipe
14 Lower Intake Manifold Gasket
15 Coolant Temperature Sensor
16 Engine Coolant Manifold Bolt
17 Engine Coolant Manifold
18 Engine Coolant Manifold Gasket
19 Lower Intake Manifold Bolt
20 Exhaust Manifold (Right)
21 Exhaust Manifold Bolt/Stud
22 Exhaust Oxygen Sensor
23 EGR Valve Adapter Bolt
24 EGR Valve Outlet Pipe
25 EGR Valve Outlet Pipe Bolt
26 EGR Valve Outlet Pipe Nut
27 EGR Valve Adapter
28 EGR Valve Gasket
29 EGR Valve
30 EGR Valve Nut
31 Upper Intake Manifold
32 ICM
33 Fuel Injection Rail
34 Fuel Injector Rail Nut
35 Upper Intake Manifold Bolt

9300UG02

Exploded view of the intake manifold and related components—3.8L (VIN K) engine

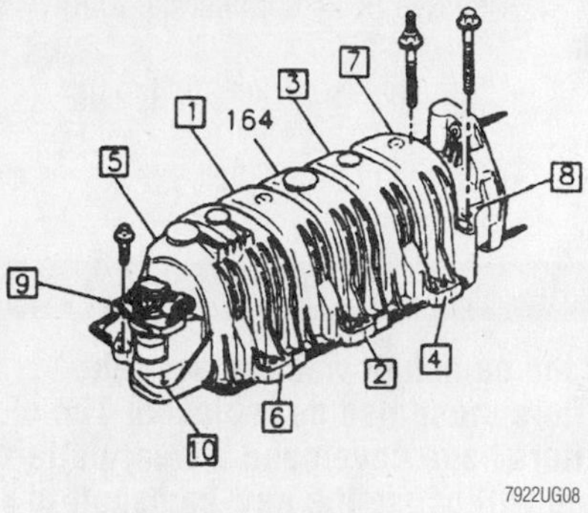

Upper intake manifold torque sequence—3.8L (VIN K) engine

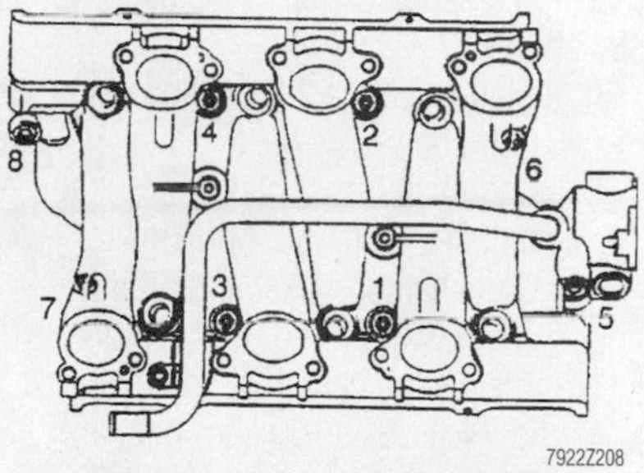

Lower intake manifold torque sequence—3.8L (VIN K) engine

- Throttle cable bracket from the cylinder head mounting bracket and the throttle body cables
- Throttle body support bracket
- Upper intake plenum and gasket
- Upper radiator hose
- Serpentine belt
- Alternator
- Drive belt tensioner assembly
- EGR valve outlet pipe
- Lower intake manifold

To install:

6. Install or connect the following:
   - Intake manifold using new manifold gaskets. Torque the bolts in sequence to 11 ft. lbs. (15 Nm); then, re-torque to 11 ft. lbs. (15 Nm).
   - EGR valve outlet pipe
   - Drive belt tensioner assembly. Torque the tensioner bolts to 37 ft. lbs. (50 Nm).
   - Alternator
   - Drive belt
   - Upper radiator hose to the thermostat housing
   - Upper intake plenum. Torque the intake plenum bolts to 11 ft. lbs. (15 Nm).
   - Throttle body support bracket
   - Throttle cable bracket to the cylinder head mounting bracket and the cables to the throttle body lever
   - EGR heat shield
   - Fuel rail. Torque the fuel rail bolts to 88 inch. lbs. (10 Nm).
   - Fuel lines
   - Fuel regulator vacuum line
   - Fuel injector electrical connectors
   - Vacuum lines to the intake manifold
   - Spark plug wires
   - Fuel injector sight shield and air inlet duct
   - Negative battery cable

7. Refill and bleed the cooling system.
8. Run the engine and check for leaks and proper engine operation.

3.8L (VIN 1) Engine

1. Before servicing the vehicle, refer to the precautions in the beginning of this section.
2. Relieve the fuel system pressure.
3. Drain the cooling system.
4. Remove or disconnect the following:
   - Negative battery cable
   - Supercharger
   - Thermostat housing
   - Exhaust Gas Recirculation (EGR) tube at the intake manifold
   - Engine Control Temperature (ECT) sensor
   - Intake manifold

To install:

5. Install or connect the following:
   - Intake manifold with new gaskets. Torque the bolts, working from the center out, to 11 ft. lbs. (15 Nm).
   - ECT sensor connector
   - EGR tube to the intake manifold
   - Thermostat housing
   - Supercharger
   - Negative battery cable
6. Refill and bleed the cooling system.
7. Run the engine and check for leaks and proper engine operation.

Exhaust Manifold

REMOVAL & INSTALLATION

Left Side (Front) Manifold

1. Before servicing the vehicle, refer to the precautions in the beginning of this section.
2. Remove or disconnect the following:

Exploded view of the left exhaust manifold mounting

   - Negative battery cable
   - 2 bolts attaching the left exhaust manifold to the crossover pipe
   - Spark plug wires
   - Engine oil dipstick and tube
   - Left side lift bracket, if necessary
   - Exhaust manifold

To install:

3. Install or connect the following:
   - Exhaust manifold with a new gasket. Torque the studs and bolts gradually and evenly to 22 ft. lbs. (30 Nm).
   - Left side lift bracket, if removed
   - Engine oil dipstick and tube
   - Spark plug wires
   - 2 bolts attaching the left exhaust manifold to the crossover pipe. Torque the bolts to 13 ft. lbs. (18 Nm).
   - Negative battery cable
4. Run the engine and check for exhaust leaks.

Right Side (Rear) Manifold

1998—99 MODELS

1. Before servicing the vehicle, refer to the precautions in the beginning of this section.
2. Remove or disconnect the following:
   - Negative battery cable
   - Spark plug wires
   - Transaxle fluid dipstick and tube
   - Oxygen (O_2S) sensor electrical connector
   - 2 bolts attaching the right exhaust manifold to the crossover pipe
   - Front exhaust pipe
   - Engine lift bracket
   - Exhaust manifold

To install:

3. Install or connect the following:
   - Manifold to the cylinder head and crossover pipe using new gaskets

- Manifold mounting studs. Torque the studs and bolts to 22 ft. lbs. (30 Nm), beginning at the center and working outwards.
- Engine lift bracket
- Exhaust pipe to the manifold
- Front exhaust pipe-to-manifold nuts. Torque the nuts to 18 ft. lbs. (25 Nm).
- Both exhaust manifold-to-crossover pipe bolts. Torque the manifold-to-crossover pipe bolts to 15 ft. lbs. (20 Nm).
- O_2S electrical connector
- Transaxle fluid dipstick and tube
- Spark plug wires to the spark plugs
- Negative battery cable

4. Run the engine and check for exhaust leaks.

2000–01 MODELS

1. Before servicing the vehicle, refer to the precautions in the beginning of this section.
2. Remove or disconnect the following:
   - Negative battery cable
   - Fuel injector sight shield
   - Air cleaner assembly
   - Brake booster heat shield
   - Crossover pipe
   - Engine harness from the right hand engine lift hook bracket
   - Spark plug wires
   - Transaxle fluid dipstick and tube
   - Oxygen (O_2S) sensor
   - Exhaust Gas Recirculation (EGR) feed pipe bolt from the manifold
   - Transmission oil level tube and seal
   - Exhaust manifold flange nuts
   - Front exhaust pipe
   - Engine lift bracket
   - Exhaust manifold

To install:

3. Install or connect the following:
   - Manifold to the cylinder head and crossover pipe using new gaskets
   - Manifold mounting studs. Torque the studs and bolts to 22 ft. lbs. (30 Nm), beginning at the center and working outwards.
   - Engine lift bracket
   - Front exhaust pipe
   - Front exhaust pipe-to-manifold nuts. Torque the nuts to 18 ft. lbs. (25 Nm).
   - Transmission dipstick tube seal and the tube
   - EGR feed pipe to the manifold
   - O_2S sensor
   - Spark plug wires to the spark plugs
   - Engine harness to the right hand engine lift hook bracket
   - Crossover pipe
   - Brake booster heat shield
   - Air cleaner assembly
   - Fuel injector sight shield
   - Negative battery cable
4. Run the engine and check for exhaust leaks.

Camshaft and Valve Lifters

REMOVAL & INSTALLATION

1. Before servicing the vehicle, refer to the precautions in the beginning of this section.
2. Relieve the fuel system pressure.
3. Remove the engine and mount it on an engine stand.
4. Remove or disconnect the following:
   - Negative battery cable
   - Supercharger, if equipped
   - Intake manifold
   - Rocker arm covers
   - Rocker arm assemblies
   - Pushrods
   - Lifters and guides

➡A magnet may be helpful when pulling the lifters out of their bores. Identify all parts as they are removed, so they can be reinstalled in their original locations.

   - Crankshaft balancer
   - Timing chain front cover
5. Set the engine to Top Dead Center (TDC) No. 1 cylinder (firing position) to align the timing marks, before disassembling the timing chain and sprockets.

⁂ WARNING

Align the timing marks of the camshaft and crankshaft sprockets to avoid burring the camshaft journals by the crankshaft.

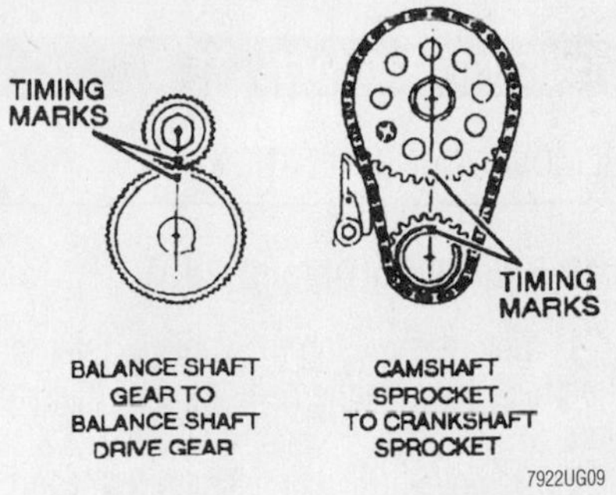

The timing marks should face each other when the chain and gears are installed properly

6. Remove or disconnect the following:
   - Camshaft sprocket and timing chain
   - Camshaft thrust plate
   - Camshaft

⁂ WARNING

If the camshaft was replaced the lifters must also be replaced. The old lifters have developed a wear pattern and will cause the new camshaft to wear prematurely.

To install:

7. Coat the camshaft lobes and bearings with camshaft break-in prelube prior to installation.
8. Install or connect the following:
   - Camshaft
   - Camshaft thrust plate. Torque the bolts to 10 ft. lbs. (14 Nm).
   - Camshaft sprocket and timing chain with timing marks aligned. Torque the camshaft sprocket bolt to 74 ft. lbs. (100 Nm) plus an additional 90 degree (¼) turn.
   - Timing chain front cover
   - Crankshaft balancer. Torque the mounting bolt to 111 ft. lbs. (150 Nm). plus an additional 76 degree turn.
9. Coat the valve lifters with camshaft break-in prelube.
10. Install or connect the following:
    - Valve lifters
    - Lifter guides and lifter guide retainer. Torque the retainer mounting bolts to 22 ft. lbs. (30 Nm).
    - Pushrods and rocker arms. Torque the rocker arm bolts to 11 ft. lbs. (15 Nm) plus an additional 90 degree turn.
    - Rocker arm covers
    - Intake manifold
    - Supercharger, if equipped
    - Engine
    - Negative battery cable
11. Verify that all fluid levels are full and correct.
12. Start the engine and check for leaks. Check engine operation.

Valve Lash

ADJUSTMENT

The valve clearance cannot be adjusted on these engines. The engine is equipped with hydraulic lifters, and adjustment is not necessary.

Starter Motor

REMOVAL & INSTALLATION

1. Before servicing the vehicle, refer to the precautions in the beginning of this section.
2. Remove or disconnect the following:
   - Negative battery cable
   - Splash shield
   - Electrical connectors
   - Flexplate inspection cover
   - Transmission cooler line clip from the transmission, if necessary
   - Starter motor wiring
   - Starter motor bolts
   - Starter

To install:

3. Install or connect the following:
   - Starter and torque the bolts to 32 ft. lbs. (43 Nm)
   - Wiring and torque the "B" terminal nut to 12 ft, lbs. (16 Nm) and the "S" terminal nut to 22 inch lbs. (3 Nm).
   - Flexplate inspection cover and torque the bolts to 62 inch lbs. (7 Nm)
   - Splash shield
   - Negative battery cable

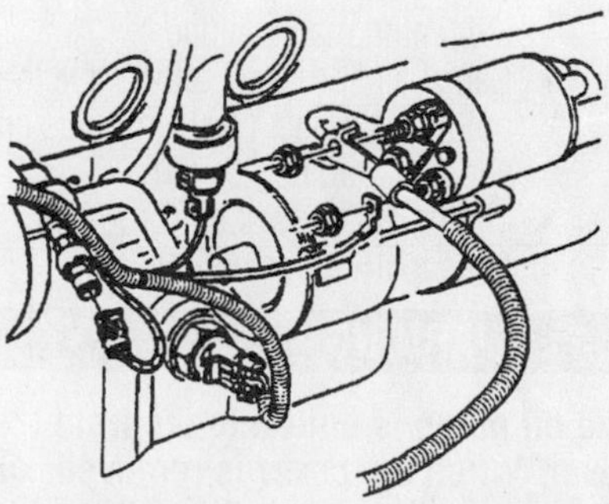
9306UG02

Starter in place with wiring

Oil Pan

REMOVAL & INSTALLATION

WARNING

The oil level sensor, located in the oil pan, must be removed prior to removal of the oil pan. If the oil pan is removed first, damage to the oil level sensor may occur.

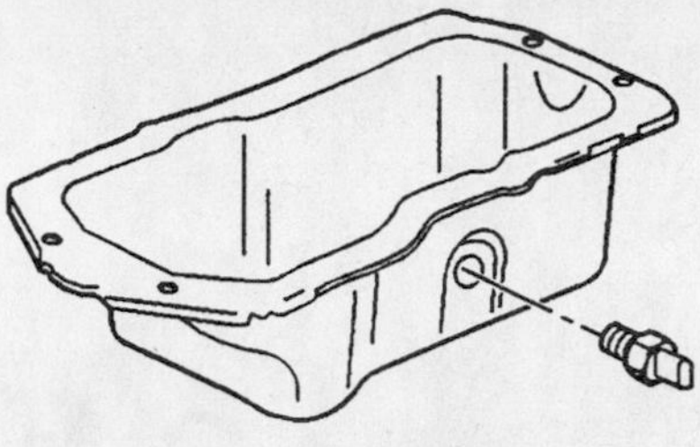
7922UG10

If equipped, be sure to remove the oil level sensor before removing the pan

1. Before servicing the vehicle, refer to the precautions in the beginning of this section.
2. Drain oil into an approved container.
3. Remove or disconnect the following:
   - Negative battery cable
   - Flexplate cover
   - Oil level sensor
   - Oil filter
   - Torque axis mount bracket bolts, if necessary
   - Oil pan bolts
   - Oil pan
   - Oil pan gasket
4. Clean the oil pan and cylinder block mating surfaces.

To install:

5. Install or connect the following:
   - Oil pan with a new gasket and torque the bolts to 125 inch lbs. (14 Nm)
   - Torque axis mount bracket bolts, if removed
   - Oil filter
   - Flexplate cover
   - Oil level sensor
   - Oil drain plug and torque the plug to 30 ft. lbs. (40 Nm)
   - Negative battery cable
6. Refill the crankcase.
7. Run the engine and check for leaks.

Oil Pump

REMOVAL & INSTALLATION

1. Before servicing the vehicle, refer to the precautions in the beginning of this section.
2. Support the engine using an engine support fixture.
3. Remove or disconnect the following:
   - Negative battery cable
   - Engine drive belts and tensioner assembly
   - Drive belt idler pulley and bracket

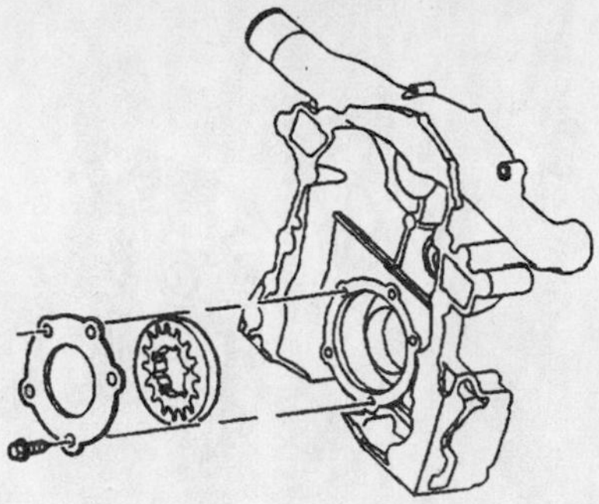
7922UG11

The oil pump is located inside the front engine cover—3.8L (VIN K and 1) engines

4. Remove or disconnect the following:
   - Torque axis mount bracket, if necessary
   - Engine front cover assembly
   - Oil filter adapter with pressure regulator valve and spring
   - Oil pump cover
   - Inner and outer pump gears

To install:

5. Lubricate the oil pump gears with petroleum jelly.
6. Install the gears into the oil pump housing.
7. Pack the gear cavity with petroleum jelly after the gears have been installed in the housing.
8. Install or connect the following:
   - Oil pump cover. Torque the screws to 97 inch lbs. (11 Nm).
   - Oil filter adapter with new gasket, pressure regulator valve and spring. Torque the bolts to 22 ft. lbs. (30 Nm) on 1998–99 models and 11 ft. lbs. (15 Nm) on 2000–01 models.
   - Front cover assembly
   - Tensioner assembly
   - Drive belt idler pulley and bracket, if removed
   - Drive belts
   - Torque axis mount bracket
   - Negative battery cable
9. Remove the engine support fixture.
10. Verify the correct engine oil level.
11. Start the vehicle and verify no leaks and proper oil pressure.

Rear Main Seal

REMOVAL & INSTALLATION

1. Before servicing the vehicle, refer to the precautions in the beginning of this section.

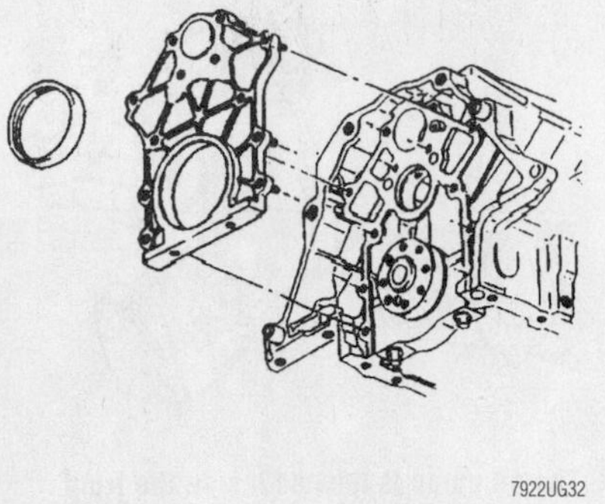

Rear main oil seal and rear cover

2. Remove or disconnect the following:
   - Transaxle assembly
   - Flexplate from the crankshaft
   - Rear main seal from engine block by inserting a small flat-bladed prytool through the dust lip at an angle, then pry out the crankshaft rear oil seal. Repeat as necessary around the seal until it is removed.

✻✻ WARNING

Do not damage or scratch the sealing surface of the crankshaft or the seal bore.

To install:

3. Lubricate new rear main with clean engine oil prior to installation.
4. Slide the oil seal on the mandrel of seal installer tool J-38196 until the back of the seal is seated squarely against the collar of the tool.
5. Attach the seal installer to the rear of the crankshaft with the 2 mounting bolts, then turn the T-handle until the oil seal is fully seated into the rear of the engine.
6. Loosen the T-handle of the tool completely.
7. Remove both bolts and the tool.
8. Install or connect the following:
   - Flexplate. Torque the bolts to 11 ft. lbs. (15 Nm), plus an additional 50 degrees.
   - Transaxle

Timing Chain, Sprockets, Front Cover and Seal

REMOVAL & INSTALLATION

1. Before servicing the vehicle, refer to the precautions in the beginning of this section.
2. Drain the cooling system.
3. Support the engine.

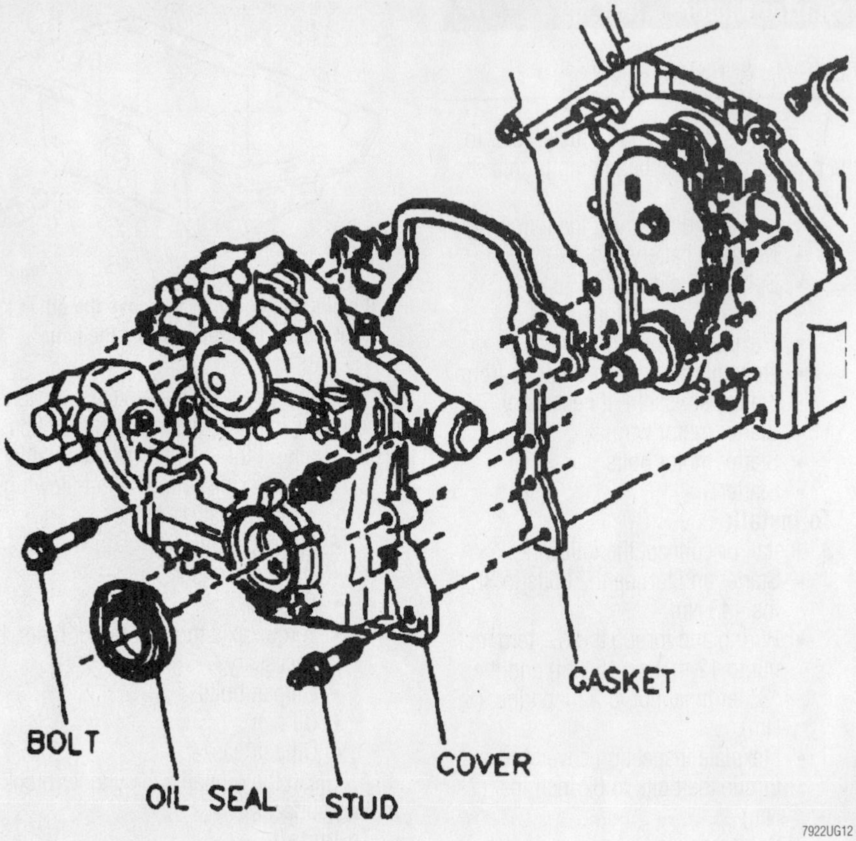

Timing chain front cover—3.8L (VIN K and 1) engines

4. Remove or disconnect the following:
   - Negative battery cable
   - Torque axis mount and bracket
   - Drive belt
   - Supercharger belt, if equipped
   - Drive belt idler pulley and bracket (VIN 1 engine)
   - Drive belt tensioner (for VIN K engine)
   - Crankshaft balancer
   - Crankshaft Position (CKP) sensor shield and the CKP sensor
   - Oil pan-to-front cover bolts
   - Timing chain front cover
5. Align the timing marks on the camshaft and crankshaft sprockets so they are as close together as possible.
   - Timing chain damper
   - Camshaft sprocket bolt, the camshaft sprocket and timing chain
   - Crankshaft sprocket

✻✻ WARNING

Do not rotate the camshaft or crankshaft while the timing chain and sprockets are removed.

To install:

6. Install or connect the following:
   - Timing chain and sprockets with the timing marks aligned
   - Camshaft sprocket bolt. Torque the bolt to 74 ft. lbs. (100 Nm) plus an additional 90 degree turn.
   - Timing chain damper. Torque the bolts to 16 ft. lbs. (22 Nm).

✻✻ WARNING

The oil pump is built into the front cover. When the cover is removed, oil drains from the pump. Since the pump "loses its prime" it may not establish oil pressure as soon as the engine starts. Therefore, it is important to remove the oil pump cover from the back of the timing chain front cover and pack the space around the oil pump gears completely full of petroleum jelly. If this is not done, the oil pump may not pump engine oil when the engine is started, resulting in severe engine damage.

7. Remove the screws and the oil pump cover from the back of the timing chain front cover. Pack the space around the oil pump gears completely full of petroleum jelly. There must be no air space left inside the pump.

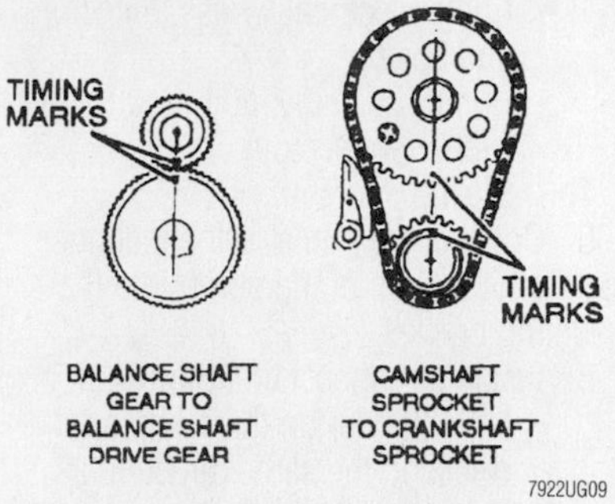

Timing chain sprocket and balance shaft gear alignment—3.8L (VIN K and 1) engines

8. Install or connect the following:
   - Pump cover with new gaskets. Torque the screws to 97 inch lbs. (11 Nm).
   - Timing chain front cover. Torque the front cover-to-engine bolts to 11 ft. lbs. (15 Nm) plus an additional 40 degrees.
   - Oil pan-to-front cover bolts. Torque the bolts to 125 inch lbs. (14 Nm).
   - CKP sensor. Torque the bolts to 14–28 ft. lbs. (20–40 Nm).
   - CKP sensor shield
   - Crankshaft balancer. Torque the bolt to 111 ft. lbs. (150 Nm) plus an additional 76 degree turn.
   - Drive belt tensioner assembly (VIN K engine)
   - Drive belt idler pulley (VIN 1 engine)
   - Right inner fender access panel and the right front wheel
   - Drive belt(s)
   - Engine mount
   - Coolant hoses
   - Negative battery cable
9. Remove the engine support fixture.
10. Refill and bleed the cooling system.
11. Start the vehicle and check for leaks and proper engine operation.

Piston and Ring

POSITIONING

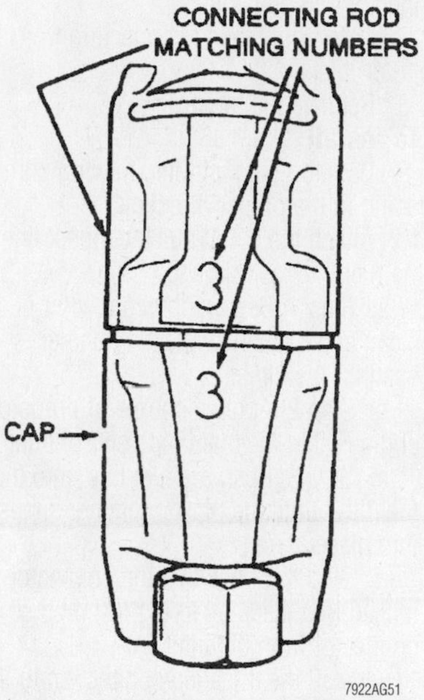

Engine connecting rod and cap installation. Be sure to matchmark the cap and rod prior to disassembly, as shown.

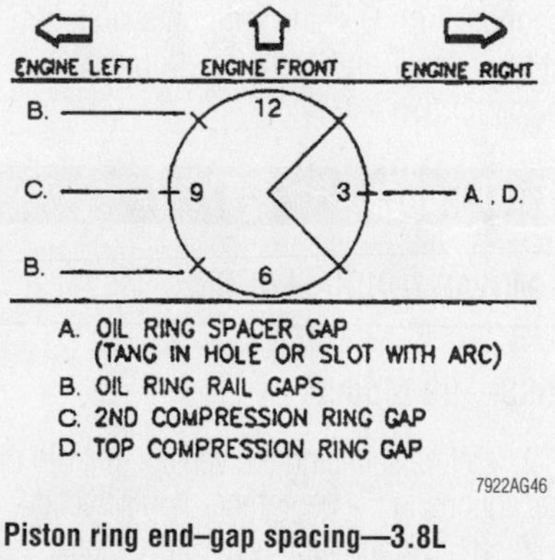

Piston ring end-gap spacing—3.8L engines

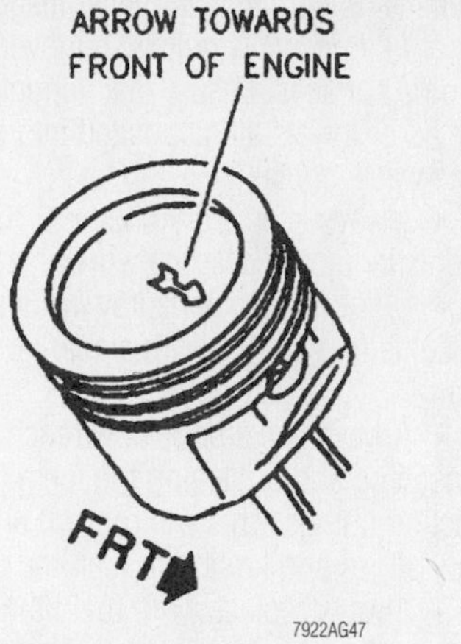

Piston positioning. Often the arrow is replaced by a notch, which also must face toward the front of the engine—3.8L engines

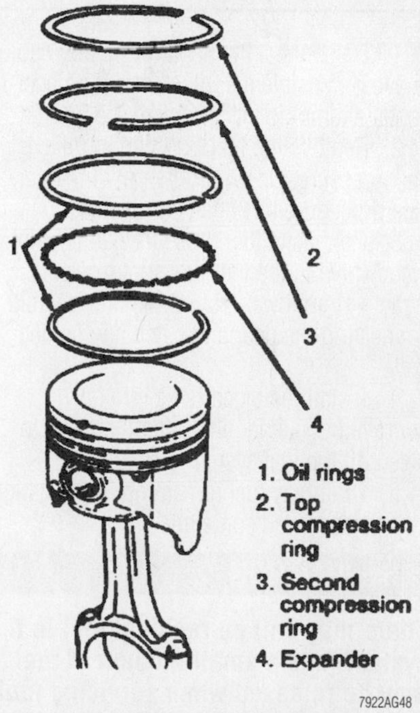

Piston ring positioning—3.8L engines

FUEL SYSTEM

Fuel System Service Precautions

Safety is the most important factor when performing not only fuel system maintenance but any type of maintenance. Failure to conduct maintenance and repairs in a safe manner may result in serious personal injury or death. Maintenance and testing of the vehicle's fuel system components can be accomplished safely and effectively by adhering to the following rules and guidelines.

- To avoid the possibility of fire and personal injury, always disconnect the negative battery cable unless the repair or test procedure requires that battery voltage be applied.
- Always relieve the fuel system pressure prior to disconnecting any fuel system component (injector, fuel rail, pressure regulator, etc.), fitting or fuel line connection. Exercise extreme caution whenever relieving fuel system pressure, to avoid exposing skin, face and eyes to fuel spray. Please be advised that fuel under pressure may penetrate the skin or any part of the body that it contacts.
- Always place a shop towel or cloth around the fitting or connection prior to loosening to absorb any excess fuel due to

spillage. Ensure that all fuel spillage (should it occur) is quickly removed from engine surfaces. Ensure that all fuel soaked cloths or towels are deposited into a suitable waste container.

- Always keep a dry chemical (Class B) fire extinguisher near the work area.
- Do not allow fuel spray or fuel vapors to come into contact with a spark or open flame.
- Always use a back-up wrench when loosening and tightening the fuel line connection fittings. This will prevent unnecessary stress and torsion to fuel line piping.
- Always replace worn fuel fitting O-rings with new. Do not substitute fuel hose or equivalent, where fuel pipe is installed.

Fuel System Pressure

RELIEVING

1. Disconnect the negative battery cable to avoid possible fuel discharge if an accidental attempt is made to start the engine.
2. Remove the fuel tank cap to relieve tank pressure. Do not tighten until the service procedure has been completed.
3. Connect a fuel pressure gauge with bleed valve to the fuel pressure test port. Wrap a shop towel around the fitting while connecting the gauge to catch any spilled fuel.
4. Install the bleed hose into an approved container and open the valve to bleed off the fuel system pressure.
5. Drain any fuel remaining in the gauge into an approved container.

⁂ CAUTION

There may still be residual fuel in the system, and a small amount of fuel may be released when servicing fuel lines or connections. In order to reduce the chance of personal injury, cover the fuel line fittings with a shop towel before disconnecting to catch any fuel that may leak out.

Fuel Filter

REMOVAL & INSTALLATION

1. Before servicing the vehicle, refer to the precautions in the beginning of this section.
2. Disconnect the negative battery cable.
3. Relieve the fuel system pressure.
4. Twist the quick connector ¼ turn in each direction to loosen any dirt that may have accumulated in the connector.
5. Use compressed air to remove any dirt in the connector.
6. Squeeze the plastic tabs of the male connector and pull apart.
7. Remove threaded connection from the filter inlet.
8. Remove the filter.

To install:

9. Position the fuel filter, making sure it is facing in the proper direction.
10. Attach the outlet quick connect line to the fuel filter as follows:
    a. Step 1: Be sure the connector is clean and that a new plastic retainer is used on the filter.
    b. Step 2: Apply a couple of drops of engine oil to the male pipe end of filter.
    c. Step 3: Push the fuel line onto the fuel filter until the plastic retainer snaps into place.
    d. Step 4: Check that the connector is locked into place by trying to pull the connector from the filter.
11. Install the threaded connection to the inlet side of filter and tighten to 22 ft. lbs. (30 Nm).
12. Connect the negative battery cable.
13. Pressurize the fuel system by turning the ignition switch to the **ON** position for 2 seconds. Turn OFF the ignition switch for 10 seconds. Turn ON the ignition switch. Check for fuel leaks.

Fuel Pump

REMOVAL & INSTALLATION

1998—99 Models

1. Before servicing the vehicle, refer to the precautions in the beginning of this section.
2. Relieve the fuel system pressure.
3. Drain the fuel tank.
4. Remove or disconnect the following:
   - Negative battery cable
   - Fuel tank filler pipe
   - Evaporative Emissions (EVAP) pipe
   - Quick connect fuel lines from the fuel tank
   - Rear rubber exhaust hangers
   - Fuel tank strap bolts, support the fuel tank
   - Fuel tank.
   - Lockring and the fuel sender assembly from the tank

➡Note the direction the strainer is pointing and remove the strainer from the pump by pulling it down and twisting.

   - Pump electrical wires connectors and hoses
   - Pump assembly, pull it out of the rubber connectors

To install:

5. Connect the pump to the fuel hose and tilt the bottom of the pump into the mounting bracket.
6. Install or connect the following:
   - New strainer on the pump so it points in the same direction as noted during removal
   - Electrical connectors and fuel lines to the pump
   - New O-ring on top of the fuel tank
   - Fuel sender assembly into the tank
   - Lockring
   - Fuel tank
   - Fuel sender electrical connector
   - Fuel tank straps. Torque the bolts to 25 ft. lbs. (34 Nm) on all models except Park Avenue and 35 ft. lbs. (48 Nm) on Park Avenue models.
7. Remove the support.
   - Rubber exhaust hangers to rear exhaust
   - Quick connect fuel lines onto the fuel tank
   - Fuel tank EVAP pipe to the tank.

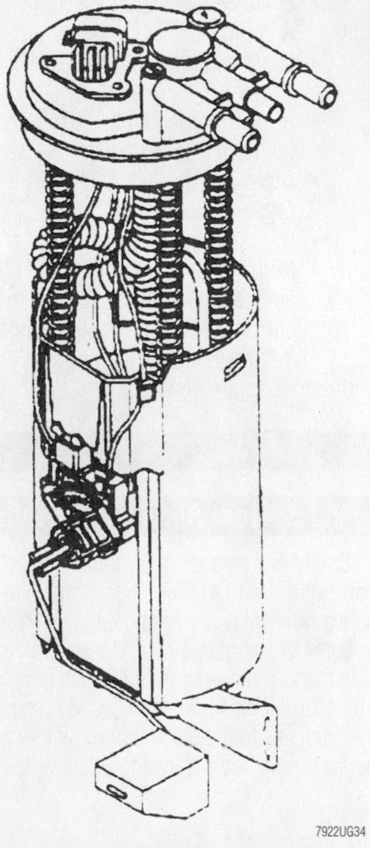

Fuel pump and level sender assembly

Torque the hose clamp to 25 inch lbs. (2.8 Nm).
- Fuel tank filler pipe to the tank. Torque the hose clamp to 25 inch lbs. (2.8 Nm).
- Negative battery cable

8. Refill the fuel tank.
9. Run the engine and check for leaks and proper engine operation.

Park Avenue And All 2000–01 Models

1. Before servicing the vehicle, refer to the precautions in the beginning of this section.
2. Relieve the fuel system pressure.
3. Drain the fuel tank.
4. Remove or disconnect the following:
   - Negative battery cable
   - Spare tire and jack
   - Trunk lining
   - Fuel sender access panel
   - Sender and quick connect fittings from the sender
   - Electrical connector from the sender and position harness and hoses aside

CAUTION

When removing the fuel sender from the tank, the reservoir bucket is full of fuel. Use caution in containing the fuel.

- Sender retaining ring
- Sender and take note of its position
- Fuel sender O-ring and discard it

➡Note the direction the strainer is pointing.

- Strainer from the pump by pulling it down and twisting
- Pump electrical wires and hoses
- Pump assembly out of the rubber connectors

To install:

5. Transfer any insulators and grommets from the old pump to the new one.
6. Connect the pump to the fuel hose and tilt the bottom of the pump into the mounting bracket.
7. Install or connect the following:
   - New strainer on the pump so it points in the same direction as noted during removal
   - Electrical connectors and fuel lines to the pump
   - New O-ring on top of the fuel tank
   - Fuel sender assembly into the tank

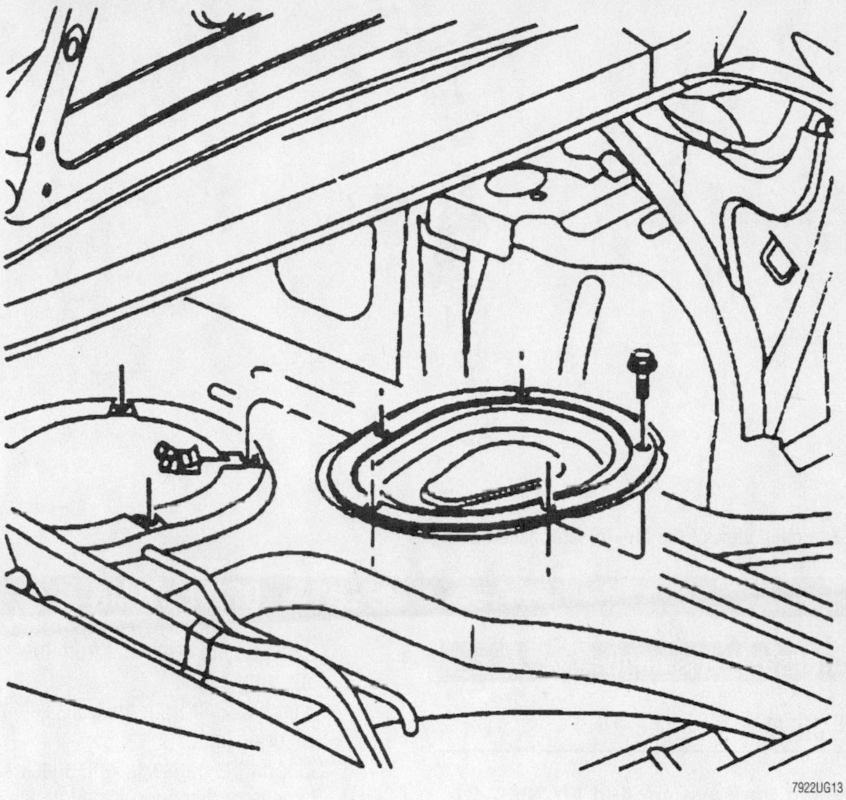

The fuel pump service cover is located in the luggage compartment under the spare tire—Park Avenue shown others similar

   - Lockring
   - Fuel line quick connectors
   - Sender electrical connector
   - Fuel sender access cover
   - Trunk liner
   - Spare tire and jack

8. Refill with fuel and check for leaks.

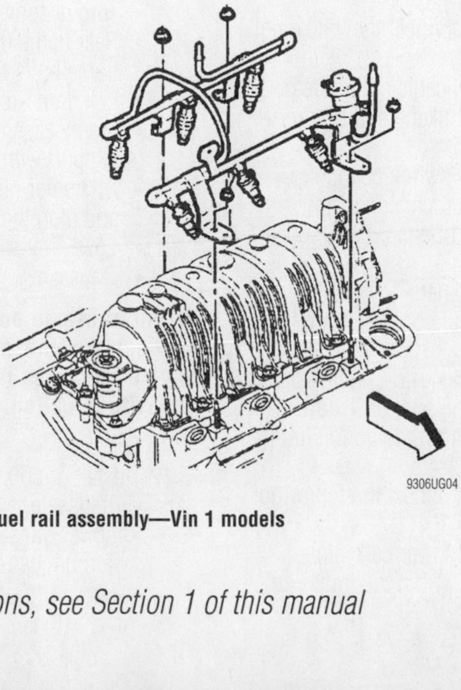

Exploded view of the fuel rail assembly—Vin 1 models

Fuel Injectors

REMOVAL & INSTALLATION

1. Before servicing the vehicle, refer to the precautions in the beginning of this section.
2. Relieve the fuel system pressure.

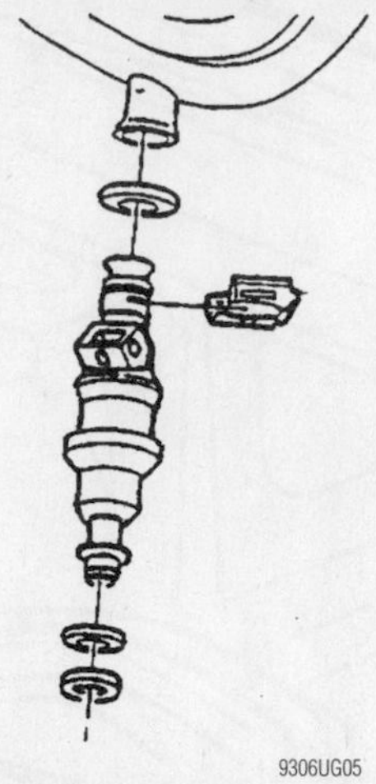

Exploded view of the fuel injector assembly

3. Remove or disconnect the following:
   - Negative battery cable
   - Fuel lines
   - Fuel rail
   - Injector retaining clips
   - Fuel injector

To install:

4. Install or connect the following:
   - Fuel injector with new O-rings, coat the o-rings with clean engine oil prior to installation
   - Injector retaining clips
   - Fuel rail and torque the bolts to 7 ft. lbs. (10 Nm). On Vin 1 models, tighten the fuel rail hold down stud to 18 ft. lbs. (25 Nm).
   - Fuel lines
   - Negative battery cable

DRIVE TRAIN

Transaxle Assembly

REMOVAL & INSTALLATION

Except Park Avenue And All 2000–01 Models

1. Before servicing the vehicle, refer to the precautions in the beginning of this section.
2. Remove or disconnect the following:
   - Negative battery cable
3. Remove the strut tower crossbrace by completing the following steps:
   a. Step 1: Loosen the through-bolts
   b. Step 2: Remove inboard strut nuts
   c. Step 3: Remove crossbrace
   d. Step 4: Reinstall the inboard strut retaining nuts
4. Remove or disconnect the following:
   - Air intake duct
   - Cruise control cable at throttle body
   - Shift control linkage and bracket at the transaxle
   - Neutral safety switch connector
   - Transaxle electrical connector
   - Vehicle Speed Sensor (VSS) connector
   - Fuel pipe retainers
   - Vacuum modulator hose from the modulator
   - 3 upper transaxle-to-engine bolts
5. Install an engine support fixture.
6. Turn the steering wheel to the full left position.
7. Remove or disconnect the following:
   - Front wheels
   - Right and left lower ball joints
   - Right halfshaft from the transaxle only; do not remove it from the steering knuckle
   - Left halfshaft from the transaxle and steering knuckle
8. Support the transaxle with a jack.
9. Remove or disconnect the following:
   - Left front transaxle mount
   - Torque strut bracket from the transaxle
   - Left rear transaxle mount-to-transaxle bolts
   - Transaxle brace from the engine bracket
   - Stabilizer shaft link-to-control arm bolt
   - Flexplate cover
10. Matchmark the flexplate to the torque converter.
11. Remove or disconnect the following:
   - Torque converter bolts
   - Rear frame member to the front frame bolts
   - Left frame-to-body bolts
   - Frame assembly, swing it aside and support with a jack stand
   - Oil cooler lines from the transaxle and plug the lines
   - Lower transaxle-to-engine bolts
   - Transaxle

➡One transaxle bolt is located between the transaxle case and the engine block and is installed in the opposite direction.

To install:

12. Install or connect the following:
   - Transaxle, support it with a jack
   - Lower transaxle-to-engine bolts. Torque the bolts to 55 ft. lbs. (76 Nm).
   - Oil cooler lines at the transaxle
   - Frame assembly. Torque the frame-to-body bolts to 83 ft. lbs. (112 Nm).
   - Front frame mount-to-right frame member and the left frame-to-body bolts. Torque the bolts to 83 ft. lbs. (112 Nm).
   - Torque converter bolts, by aligning the matchmarks. Torque the bolts to 46 ft. lbs. (62 Nm).
   - Oil cooler lines to the transaxle
   - Torque converter cover
   - Stabilizer shaft link-to-control arm bolt. Torque the bolt to 13 ft. lbs. (17 Nm).
   - Transaxle brace to the engine bracket. Torque the bolts to 70 ft. lbs. (95 Nm).
   - Left rear transaxle mount-to-transaxle bolts
   - Torque strut bracket to the transaxle
   - Left front transaxle mount
   - Halfshafts
   - Lower ball joints
   - Front wheels
13. Turn the steering wheel to the straight-ahead position.
14. Remove the support fixture from the engine.
15. Install or connect the following:
   - 3 top transaxle-to-engine bolts. Torque the bolts to 55 ft. lbs. (76 Nm).
   - Vacuum modulator hose to the modulator
   - Fuel pipe retainers
   - Transaxle park/neutral position switch
   - Back-up light switch

- Transaxle electrical connector
- Vehicle Speed Sensor (VSS)
- Shift control linkage and mounting bracket at the transaxle
- Cruise control cable at throttle body
- Air intake duct
- Inboard strut nuts
- Crossbrace assembly, with through-bolts and the inboard strut nuts. Torque the fasteners to 18 ft. lbs. (24 Nm).
- Negative battery cable

16. Check the transaxle fluid level.
17. Start the engine and check for leaks.
18. Road test the vehicle.

Park Avenue And All 2000–01 Models

1. Before servicing the vehicle, refer to the precautions in the beginning of this section.
2. Remove or disconnect the following:
   - Negative battery cable
   - Air cleaner cover
   - Transaxle electrical connector
   - Shift control cable bracket with cable attached
   - Top transaxle case bolts
   - Top bolt from rear transaxle mount
   - Left side steering rack mount bolts
   - Oxygen (O_2S) sensor connector
3. Install an engine support fixture.
   - Both front tires
   - Both splash shields
   - Power steering line bracket
   - Both transaxle mounts
   - Right steering rack mount bolts
   - Transaxle cooler lines
   - Sway bar mounts
   - Lower ball joints
   - Front subframe bolts and lower subframe
   - Both drive axles
   - Torque converter cover
   - Flexplate bolts
   - Input Speed (ISS) sensor connector
   - Engine to transaxle bracket
   - Transaxle fluid filler tube
   - Rear engine mount
   - Remaining transaxle to engine bolts
   - Transaxle from vehicle

To install

4. Install or connect the following:
   - Transaxle to vehicle
   - Transaxle to engine bolts. Torque the bolts to 55 ft. lbs. (75 Nm).
   - Rear engine mount
   - Transaxle fluid filler tube. Torque the filler tube bolt to 15 ft. lbs. (20 Nm).
   - Engine to transaxle bracket. Torque the bolts to 44 ft. lbs. (60 Nm).
   - ISS electrical connectors
   - Flexplate bolts. Torque the bolts to 46 ft. lbs. (62 Nm).
   - Torque converter cover

➡Use care when installing the right side drive axle into the transaxle case. The splined shaft of the axle can easily damage the seal.

   - Left and right drive axles to transaxle
   - Engine frame bolts. Torque the bolts to 83 ft. lbs. (112 Nm).
   - Lower ball joints to steering knuckles. Torque the nuts to 88 in. lbs. (10 Nm).
   - Sway bar mounts to the lower control arms
   - Transaxle cooler lines to the frame
   - Right steering rack mount bolts to engine frame. Torque the bolts to 45 ft. lbs. (65 Nm).
   - Right and left transaxle mount to frame
   - Power steering line bracket to frame
   - Right and left splash shields
   - Both front tires. Torque the wheel nuts to 100 ft. lbs. (140 Nm).
   - O_2S sensor electrical connector
   - Left side steering rack mount bolts. Torque the bolts to 48 ft. lbs. (65 Nm).
   - Top bolt of the rear transaxle mount. Torque the bolt to 42 ft. lbs. (58 Nm).
   - Shift control cable and bracket
   - Transaxle electrical connector
   - Air cleaner cover
   - Negative battery cable
5. Check and adjust transaxle fluid level.
6. Road test vehicle and check for transaxle leaks.

Halfshaft

REMOVAL & INSTALLATION

1998–99 Models

⁂ WARNING

Use care when removing the halfshaft to prevent the inner CV-joint from becoming over-extended. Overextension of the joint could result in separation of internal components and possible joint failure.

1. Before servicing the vehicle, refer to the precautions in the beginning of this section.
2. Install a boot protector on the outer CV-joint boot.
3. Remove or disconnect the following:
   - Front wheel
4. Loosen the stabilizer shaft link assembly bolt (to accommodate ball joint separation).
5. Remove the ball joint cotter pin and nut. Loosen the joint.

➡The grease fitting may have to be removed from the ball joint for tool access.

6. Separate the lower control arm from the joint.
7. Remove or disconnect the following:
   - Hub nut
   - Halfshaft from the hub
8. Move the strut and knuckle rearward.
9. Remove the halfshaft from the transaxle.

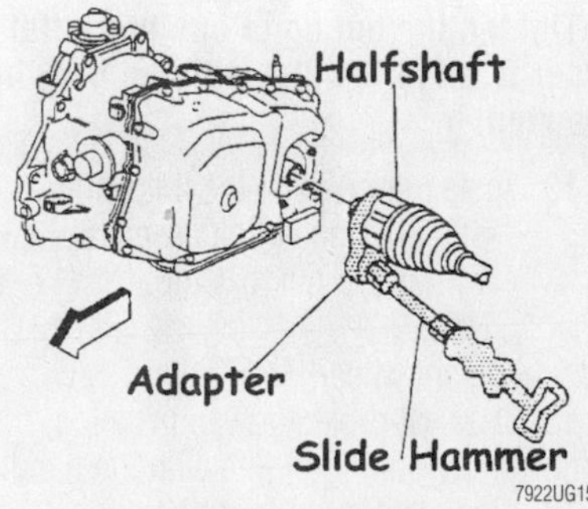

Use a slide hammer with the special adapter J-3308, to remove the halfshaft from the transaxle—1998–99 models

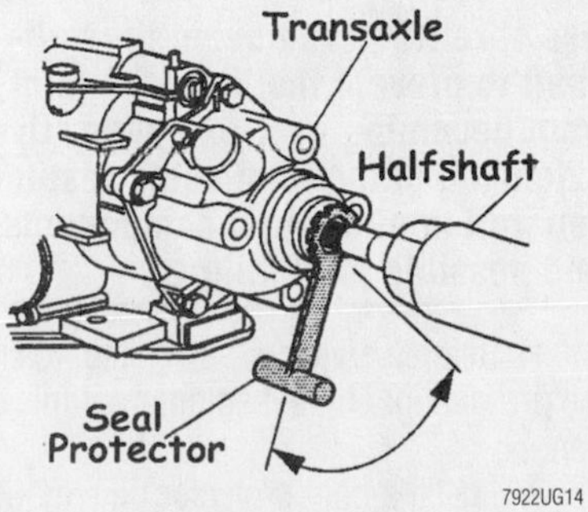

Insert the seal protector in the transaxle to prevent seal damage while the halfshaft is being installed—1998–99 models

WARNING

If equipped with anti-lock brakes, care must be used to prevent damage to the toothed sensor ring on the halfshaft and the wheel speed sensor on the steering knuckle.

To install:

➡If installing the right halfshaft, install a seal protector, so that it can be pulled out after the halfshaft is installed.

10. Install the halfshaft into the transaxle by placing a drift pin or punch into the groove on the joint housing and tapping lightly until seated. Verify that the halfshaft is seated by grasping the inner joint housing and pulling. DO NOT pull on the halfshaft.
11. Install or connect the following:
    - Halfshaft into the hub/bearing assembly with new hub nut and tighten to 107 ft. lbs. (145 Nm)
    - Ball joint into the steering knuckle. Torque the nut to 88 inch lbs. (10 Nm), plus an additional 120 degree turn during which a torque of 41 ft. lbs. (55 Nm) must be obtained.

➡Tighten the nut up to one more flat in order to align the slot with the hole in the stud.

12. Install or connect the following:
    - Stabilizer shaft link assembly. Torque the nut to 14 ft. lbs. (17 Nm).
    - Front wheel
13. If a seal protector was installed, remove it by pulling in line with the handle.
14. Road test for proper operation.

2000–01 Models

WARNING

Use care when removing the half-shaft to prevent the inner CV-joint from becoming over-extended. Over-extension of the joint could result in separation of internal components and possible joint failure.

1. Before servicing the vehicle, refer to the precautions in the beginning of this section.
2. Install a boot protector on the outer CV-joint boot.
3. Remove or disconnect the following:
    - Front wheel
4. Loosen the stabilizer shaft link assembly bolt (to accommodate ball joint separation).

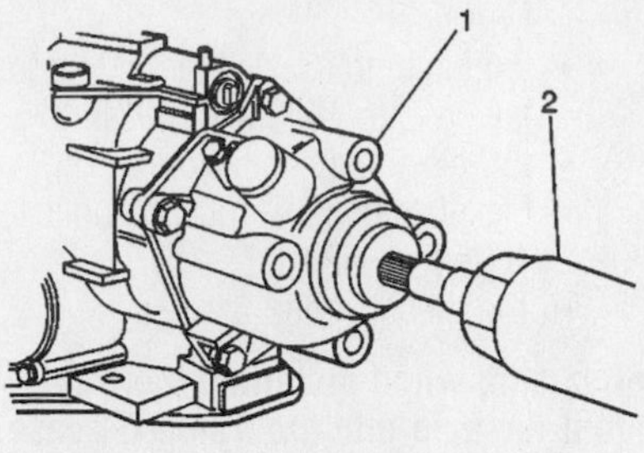

9346UG07

Remove the halfshaft from the transaxle using tool J 42128 Axle Shaft Remover and tool J 2619-01 Slide Hammer

5. Remove the ball joint cotter pin and nut. Loosen the joint.

➡The grease fitting may have to be removed from the ball joint for tool access.

6. Separate the lower control arm from the joint.
7. Remove or disconnect the following:
    - Hub nut
    - Halfshaft from the hub using tool J 28733
8. Move the strut and knuckle rearward.
9. Remove the halfshaft from the transaxle using tool J 42128 Axle Shaft Remover and tool J 2619-01 Slide Hammer.

WARNING

If equipped with anti-lock brakes, care must be used to prevent damage to the toothed sensor ring on the halfshaft and the wheel speed sensor on the steering knuckle.

To install:

➡If installing the right halfshaft, install a seal protector, so that it can be pulled out after the halfshaft is installed.

10. Install the halfshaft into the transaxle by placing a drift pin or punch into the groove on the joint housing and tapping lightly until seated. Verify that the halfshaft is seated by grasping the inner joint housing and pulling. DO NOT pull on the half-shaft.
11. Install or connect the following:
    - Halfshaft into the hub/bearing assembly with new hub nut and tighten to 118 ft. lbs. (160 Nm)
    - Ball joint into the steering knuckle. Torque the nut to 88 inch lbs. (10 Nm), plus an additional 120 degree turn during which a torque of 41 ft. lbs. (55 Nm) must be obtained.

➡Tighten the nut up to one more flat in order to align the slot with the hole in the stud.

12. Install or connect the following:
    - Stabilizer shaft link assembly. Torque the nut to 14 ft. lbs. (17 Nm).
    - Front wheel
13. If a seal protector was installed, remove it by pulling in line with the handle.
14. Road test for proper operation.

CV-Joints

OVERHAUL

Inner (Tripod) Joint

1. Before servicing the vehicle, refer to the precautions in the beginning of this section.
2. Raise and safely support the vehicle.
3. Remove or disconnect the following:
    - Front wheel
    - Halfshaft and place it in a vise
    - Small CV-joint boot clamp, cut and discard it
    - Large CV-joint boot clamp, cut and discard it
    - CV-joint boot by sliding it away from the tripod joint
    - Tripod housing from the tripod spider
    - Inboard spacer ring and slide it rearward on the shaft
    - Outboard retaining ring
    - Tripod joint spider assembly
    - Inboard spacer ring and discard it
    - Tripod joint spider assembly by tapping it from the halfshaft with a brass drift
    - Tripod spider retaining ring and discard it
    - Trilobal tripod bushing from the housing
    - CV-joint boot
4. Thoroughly clean and inspect all parts.

To install:

5. Install or connect the following:
    - Small boot clamp
    - CV-joint boot
    - New inboard spacer ring. Slide it rearward on the shaft past the 2nd groove
    - Tripod joint spider assembly onto the shaft until it passes the 2nd groove
6. Assemble the tripod spider assembly onto the halfshaft as follows:
    a. Position the tripod spider assembly onto the shop press plate.

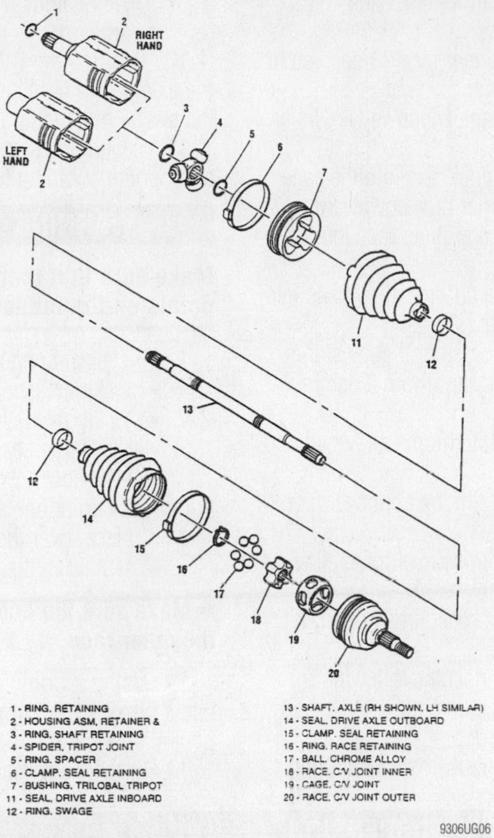

Exploded view of the halfshaft assembly

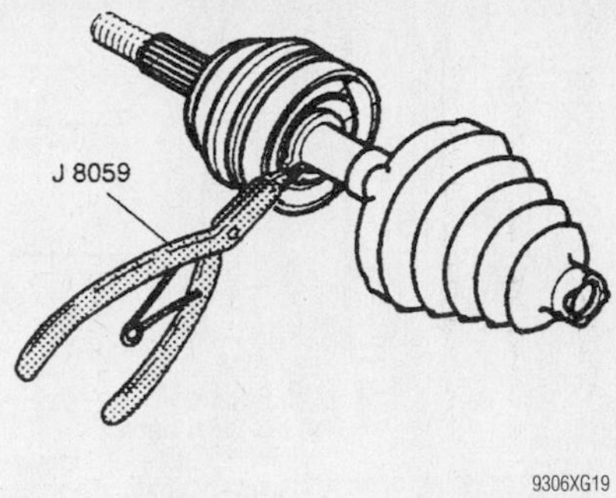

Disconnecting the outer CV-joint from the axle shaft

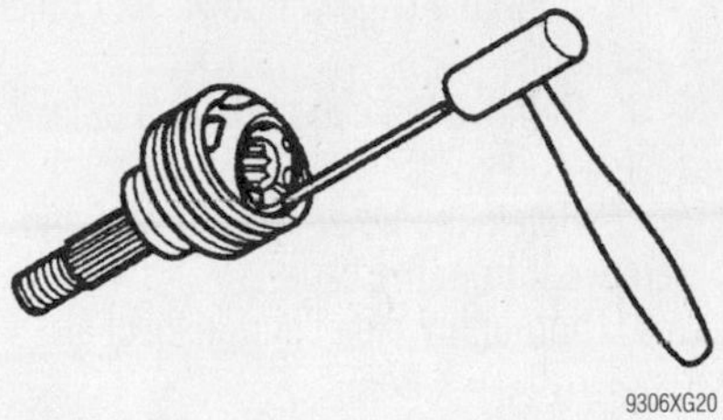

Tilting the cage—Outer CV-joint

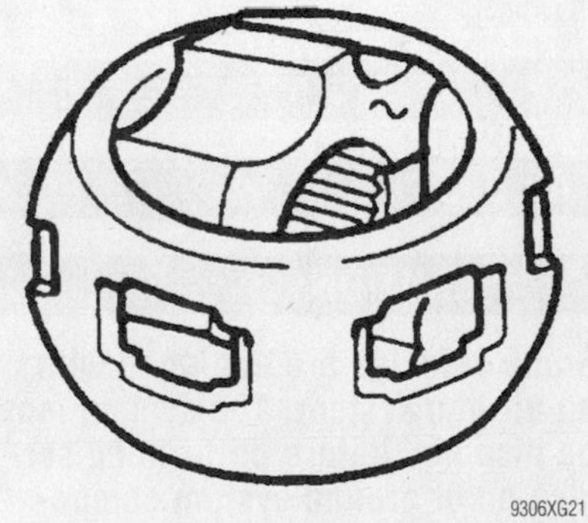

View the cage and inner race—Outer CV-joint

b. Position the halfshaft onto the tripod spider assembly, in the shop press.

c. Press the halfshaft into the tripod spider assembly until the spider assembly passes the 2nd groove.

⁂ WARNING

When assembling the tripod assembly onto the halfshaft, do not exceed 4,000 lbs. pressure.

7. Remove the halfshaft from the shop press and place it in vise.
8. Install or connect the following:
   - New outboard retaining ring into the axle shaft groove
   - Tripod joint spider assembly, slide it against the outboard retaining ring using a brass drift
   - Inboard spacer ring, seat it in the groove
9. Use ½ of the grease supplied in the kit into the boot and the other ½ into the tripod housing.
   - Trilobal tripod bushing flush with the tripod housing face
   - New large seal clamp onto the CV-joint boot
   - Tripod housing, slide it over the tripod joint spider assembly
   - CV-joint boot/clamp, slide it into place, over the trilobal tripod bushing with the seal lip in the groove

➡Make sure the boot lies flat against the trilobal bushing.

10. Using the crimp tool, a torque wrench and a breaker bar, crimp the small CV-joint boot clamp to 100 ft. lbs. (136 Nm).
11. Using the crimp tool, latch the large CV-joint boot clamp.
12. Install the halfshaft and the front wheel.

Outer CV-Joint

1. Before servicing the vehicle, refer to the precautions in the beginning of this section.
2. Remove or disconnect the following:
   - Front wheel
   - Halfshaft
   - Swage ring using a hand grinder
   - Large boot clamp
   - CV-joint boot, slide it away from the CV-joint
   - CV-joint assembly by spreading the inner race-to-axle shaft retaining ring ears using Snapring Pliers
   - CV-joint boot from the axle shaft
3. Disassemble the chrome alloy balls from the CV-joint cage as follows:

   a. Position a brass drift against the CV-joint cage and tap it with a hammer to tilt the cage.

   b. Chrome alloy ball from the cage.

   c. Tilt the cage in the opposite direction.

   d. Remove the opposite chrome alloy ball.

   e. Repeat the procedure until all 6 balls are removed.
4. Disassemble the CV-joint cage and inner race as follows:

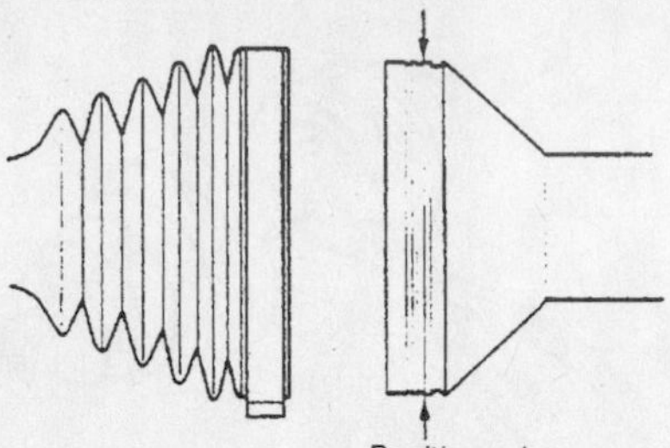

Positioning the boot—Outer CV-joint

a. Pivot the cage and race 90 degrees to the center line of the outer race.

b. Align the cage windows with outer race lands.

c. Remove the cage from the outer race.

d. Rotate the inner race upward and remove it from the cage.

5. Thoroughly clean and inspect all parts.

To install:

6. Lubricate the parts with a light coat of grease.

7. Assemble the CV-joint cage and inner race, as follows:

a. Rotate the inner race 90 degrees to the cage centerline.

b. Align the cage windows with inner race lands.

c. Insert the inner race into the cage by rotating the inner race downward.

d. Insert the cage/inner race into the outer race.

8. Assemble the chrome alloy balls into the CV-joint cage, as follows:

a. Position a brass drift against the CV-joint cage and tap it with a hammer to tilt the cage.

b. Insert the 1st chrome alloy ball into the cage.

c. Tilt the cage in the opposite direction.

d. Insert the opposite chrome alloy ball.

e. Repeat the procedure until all 6 balls are inserted.

9. Install or connect the following:

- Swage ring clamp
- CV-joint boot
- CV-joint onto the axle shaft until the retaining ring seats into the groove

10. Position the CV-joint boot seal into the axle shaft's joint seal groove and align the swage ring clamp on the boot.

11. Secure the swage ring clamp using appropriate crimping tool.

WARNING

Make sure that there are no pinch points on the inboard seal.

12. Install or connect the following:

- ½ kit grease into the CV-joint boot
- ½ kit grease into the CV-joint
- New large seal clamp onto the CV-joint boot
- CV-joint boot/clamp, slide it into place, over the outer race with the seal lip in the groove

➡Make sure the boot lies flat against the outer race.

13. Using a Crimp tool, a torque wrench and a breaker bar, crimp the large CV-joint boot clamp to 130 ft. lbs. (176 Nm).

14. Install the halfshaft and the front wheel.

STEERING AND SUSPENSION

Air Bag

CAUTION

Some vehicles are equipped with an air bag system. The system must be disabled before performing service on or around system components, the steering column, instrument panel components, wiring and sensors. Failure to follow safety precautions and the disarming procedures could result in accidental air bag deployment, possible personal injury and unnecessary system repairs.

PRECAUTIONS

Several precautions must be observed when handling the inflator module to avoid accidental deployment and possible personal injury.

- Never carry the inflator module by the wires or connector on the underside of the module.
- When carrying a live inflator module, hold it securely with both hands, and ensure that the bag and trim cover are pointed away.
- Place the inflator module on a bench or other surface with the bag and trim cover facing up.
- With the inflator module on the bench, never place anything on or close to the module which may be thrown in the event of an accidental deployment.

DISARMING

CAUTION

The Supplemental Inflatable Restraint system (SIR) must be disarmed before performing service around the air bag or SIR wiring. Failure to do so may cause accidental deployment of the air bag, resulting in unnecessary SIR repairs and/or personal injury.

1. Turn the steering wheel so the front wheels are in the straight-ahead position.

2. Turn the ignition switch to the **LOCK** position.

3. Remove or disconnect the following:

- Negative battery cable
- Air bag fuse from the fuse panel

➡The position of the fuse on the panel varies according to model and year. Consult the vehicle owner's manual for fuse location.

- Left-hand sound insulator trim panel under the instrument panel
- Connector retainer clip and the yellow 2-way connector at the base of the steering column
- Connector retainer and detach the passenger side yellow 2-way connector. Located behind the right side sound insulator.

ARMING

After the necessary repairs have been made, re-enable the air bag system as follows:

1. Turn the steering wheel so the front wheels are in the straight-ahead position.

2. Turn the ignition switch to the **LOCK** position.

3. Disconnect the negative battery cable.

4. Install or connect the following:

- Yellow 2-way connector at the base of the steering column and the connector retainer
- Left-hand sound insulator
- Yellow 2-way connector on the right side and the connector retainer
- Sound insulator and/or glove box
- Air bag fuse
- Negative battery cable

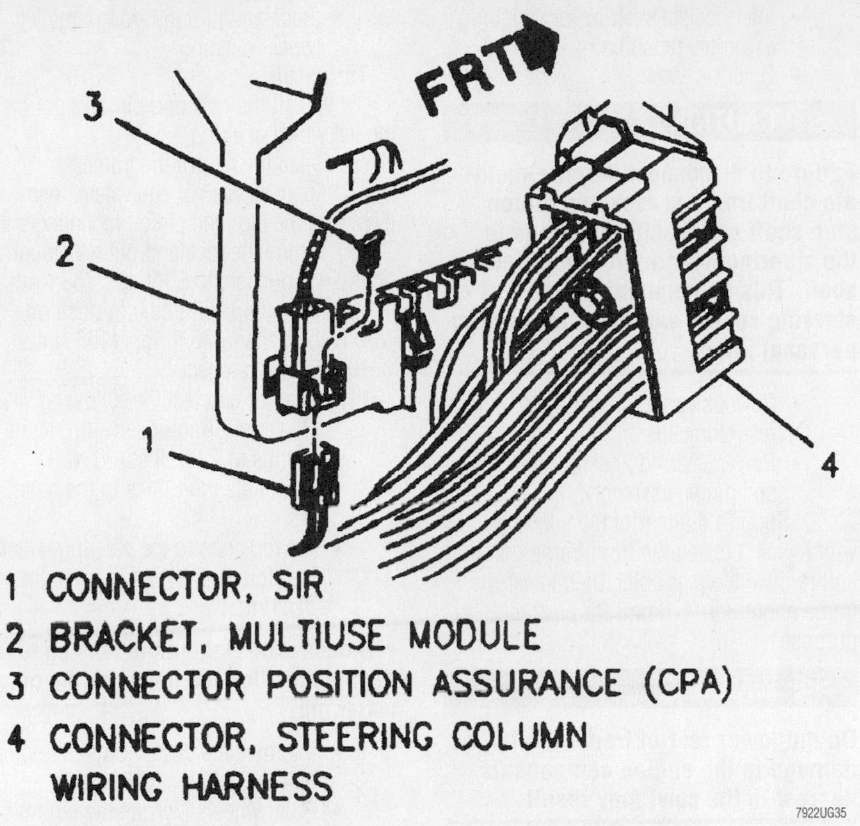

Drivers side air bag connector

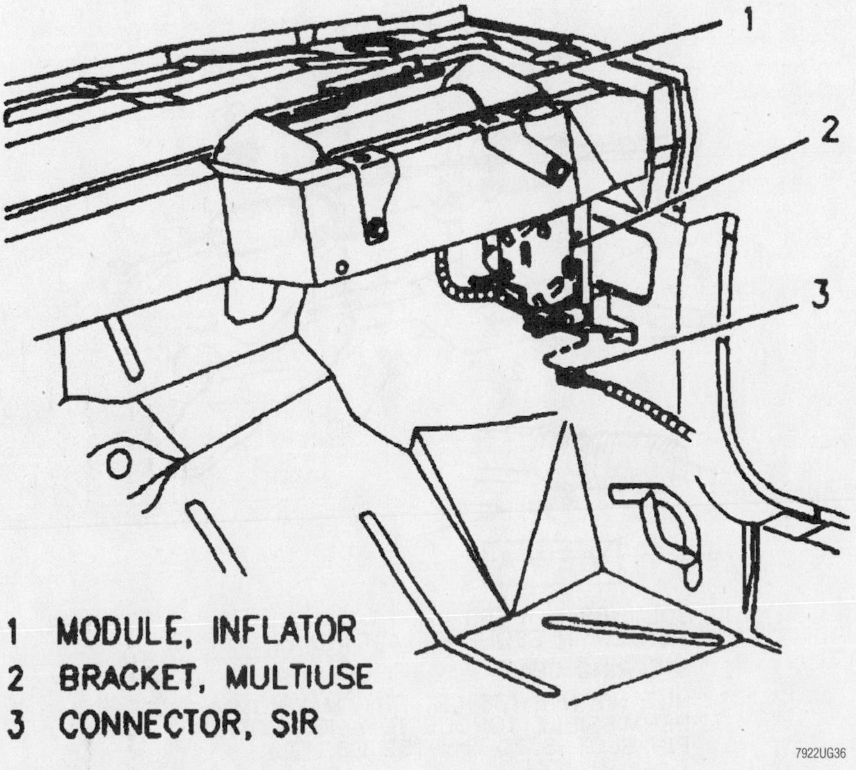

Passenger's side air bag connector

5. Turn the ignition switch to the **RUN** position. Verify that the INFLATABLE RESTRAINT indicator lamp flashes 7–9 times, then remains OFF. If the lamp does not function as specified, there is a malfunction in the air bag system.

Power Rack and Pinion Steering Gear

REMOVAL & INSTALLATION

1998–99 Park Avenue

⁂ WARNING

The wheels of the vehicle must be straight-ahead and the steering column in the LOCK position before disconnecting the steering column or intermediate shaft from the steering gear. Failure to do so will cause the Supplemental Inflatable Restraint (SIR) coil assembly in the steering column to become off-center, which will cause damage to the coil assembly.

1. Before servicing the vehicle, refer to the precautions at the beginning of this section.
2. Disconnect negative battery cable.
3. Remove or disconnect the following:
   - Both front tires
   - Exhaust system
   - Wheel house splash shield
   - Intermediate shaft lower connection
   - Tie rod ends from steering knuckles
   - Power steering heat shield
   - Electrical connector
   - Power steering gear outlet and inlet hoses from steering gear
   - Steering gear mounting bolts
4. Support the rear the of frame.
5. Remove the rear frame bolts, and lower the rear the of frame to allow steering gear removal.
6. Remove the steering gear through the right wheel opening.

To install

7. Install or connect the following:
   - Steering gear through right wheel opening
8. Raise the rear the of frame and torque the frame bolts to 142 ft. lbs. (192 Nm).
9. Install three steering gear mounting bolts. Tighten bolts in the sequence shown to 48 ft. lbs. (65 Nm).

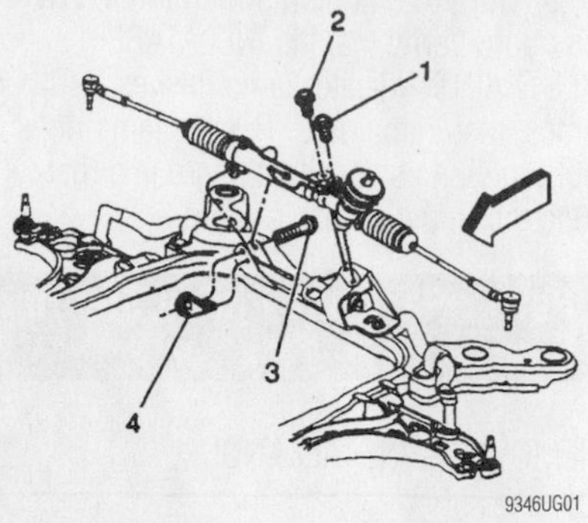

Torque sequence for steering gear mounting bolts—Park Avenue models

10. Install or connect the following:
    - Power steering gear outlet and inlet hoses. Torque the hose connections to 20 ft. lbs. (27 Nm).
    - Electrical connector
    - Steering gear heat shield
    - Tie rod ends to steering knuckles. Torque the tie rod end nuts to 35 ft. lbs. (47 Nm).
    - Cotter pin. Tighten the nut up to 52 ft. lbs. (70 Nm) max to align cotter pin slot.
    - Intermediate shaft lower connection. Torque the bolt to 35 ft. lbs. (47 Nm).
    - Exhaust system. Torque the bolts to 18 ft. lbs. (25 Nm).
    - Wheelhouse splash shield
    - Tires. Torque the lug nuts to 100 ft. lbs. (140 Nm).
11. Fill and bleed power steering system.
12. Inspect the system for leaks.

1998–99 Models—Except Park Avenue

✲✲ WARNING

The wheels of the vehicle must be straight-ahead and the steering column in the LOCK position before disconnecting the steering column or intermediate shaft from the steering gear. Failure to do so will cause the Supplemental Inflatable Restraint (SIR) coil assembly in the steering column to become off-center, which will cause damage to the coil assembly.

1. Before servicing the vehicle, refer to the precautions in the beginning of this section.
2. Remove or disconnect the following:
    - Negative battery cable
    - Both front wheels
    - Intermediate shaft-to-rack/pinion assembly pinch bolt
    - Outer tie rods

✲✲ CAUTION

Failure to disconnect the intermediate shaft from the rack and pinion stub shaft can result in damage to the steering gear and/or intermediate shaft. This damage can cause loss of steering control which could result in personal injury.

- Power steering line retainers and retaining clips
- Power steering lines from the rack and pinion assembly

3. Support the rear of the subframe with a jack. Loosen the front frame bolts and remove the rear bolts, then lower the frame about 3 in. (76mm) for clearance purposes.

✲✲ WARNING

Do not lower rear of frame too far, as damage to the engine components nearest to the cowl may result.

4. Remove or disconnect the following:
    - Rack and pinion mounting bolts
    - Rack and pinion through the left wheel opening

To install:

5. Install the rack and pinion through the left wheel opening.
6. Raise the rear of the frame.
7. Apply Loctite® or equivalent, to the threads of the rack and pinion mounting bolts.
8. Torque the rack and pinion mounting bolts, in sequence, to 50 ft. lbs. (68 Nm).
9. Raise the subframe into position. Torque the bolts to 76 ft. lbs. (103 Nm).
10. Remove the jack.
11. Install or connect the following:
    - Power steering hoses. Torque the fittings to 22 ft. lbs. (30 Nm).
    - Power steering lines to the retainers
    - Tie rod ends to the steering knuckles. Torque the nuts to 35 ft. lbs. (47 Nm).

➡If necessary tighten the nuts up to 60 degrees to align the holes for the new cotter pin.

- Intermediate shaft. Torque the pinch bolt to 33 ft. lbs. (45 Nm).
- Front wheels. Torque the lug nuts to 100 ft. lbs. (140 Nm).
- Negative battery cable

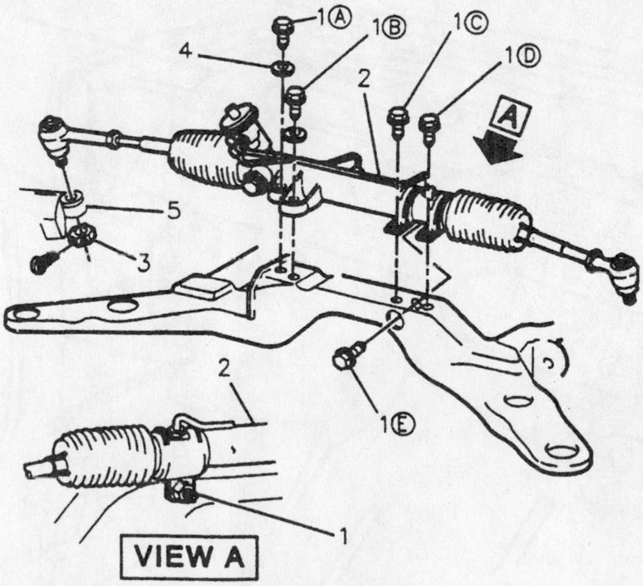

1 BOLT; 68 N•m (50 LB. FT.). TIGHTEN IN SEQUENCE A THRU E.
2 STEERING GEAR
3 NUT; 47 N•m (35 LB. FT.). MAXIMUM PERMISSIBLE TORQUE TO ALIGN COTTER PIN SLOT IS 70 N•m (52 LB. FT.).
4 WASHER
5 STEERING KNUCKLE

7922UG16

Torque the steering gear mounting bolts in the sequence shown

12. Refill the reservoir with fluid and bleed the air from the system.
13. Start the engine, check for leaks and proper steering operation.
14. Check the wheel alignment.

2000–01 Models

1. Before servicing the vehicle, refer to the precautions at the beginning of this section.

CAUTION

Make sure the wheels of the vehicle are straight ahead and the steering column in the LOCK position before disconnecting the steering column or intermediate shaft from the steering gear. Failure to do so will cause the coil assembly in the steering column to become uncentered which will cause damage to the coil assembly.

2. Lock the steering column by installing the Steering Column Anti Rotation Pin tool J 42640 into the underside of the steering column.
3. Disconnect negative battery cable.
4. Remove or disconnect the following:
   - Intermediate shaft lower coupling
   - Pressure and return pipes from the rack and pinion gear
   - Electrical connector, if equipped
   - Stabilizer shaft links at lower control arms
5. Rotate the stabilizer shaft to access steering gear bolts.
   - Rack and pinion attaching bolts
   - Rack and pinion assembly

To install:

6. Install or connect the following:
   - Rack and pinion assembly and tighten the bolts to 70 ft. lbs. (95 Nm)
   - Electrical connector, if equipped

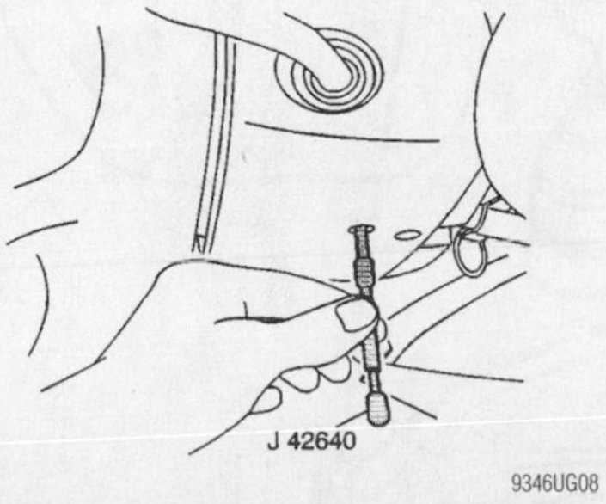

9346UG08

Lock the steering column by installing the Steering Column Anti Rotation Pin tool J 42640 into the underside of the steering column

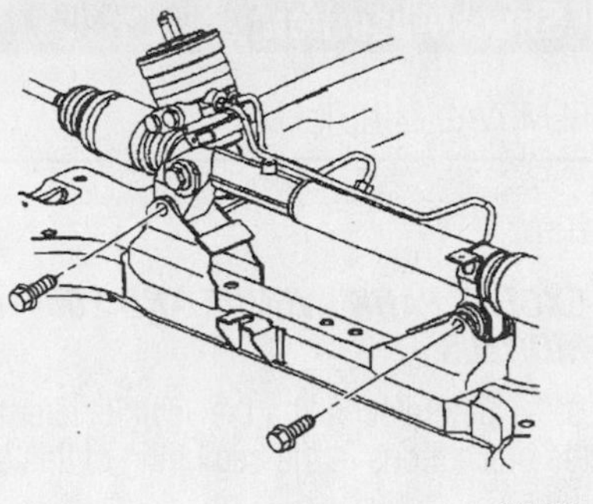

9346UG09

Steering gear mounting bolts—2000–01 models

   - Pressure and return lines to the steering gear and tighten the fittings to 22 ft. lbs. (30 Nm)
   - Heat shield and bolts. Tighten the bolts to 7 ft6. Lbs. (10 Nm)
7. Align stabilizer shaft.
   - Links to lower control arms
   - Intermediate shaft to the steering gear and tighten the pinch bolt to 33 ft. lbs. (45 Nm)
8. Remove the steering column anti rotation pin from the steering column.
9. Fill and bleed the power steering system.

Strut

REMOVAL & INSTALLATION

Front

WARNING

The steering knuckle must be retained after the strut-to-steering knuckle bolts have been removed. Failure to observe this may cause ball joint and/or halfshaft damage.

1. Before servicing the vehicle, refer to the precautions in the beginning of this section.

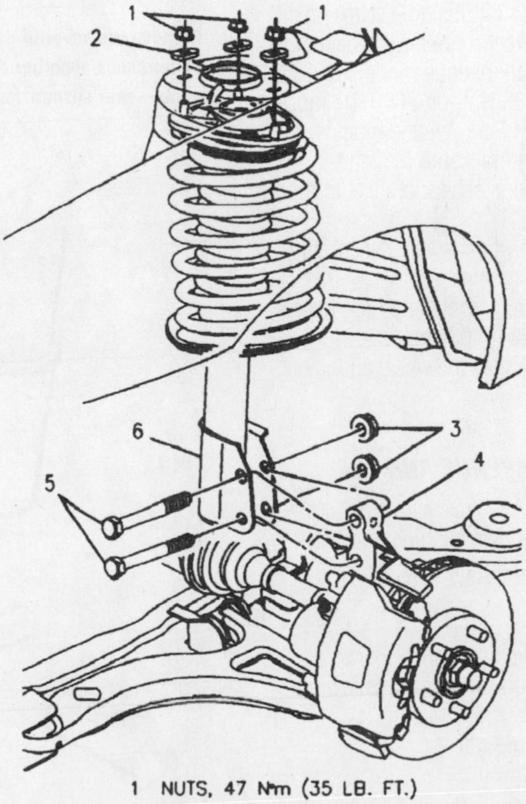

7922UG17

The strut assembly is mounted between the steering knuckle and the body—front strut shown

2. Matchmark the strut-to-steering knuckle location.
3. If equipped with electronic ride control, detach the electrical connection.
4. Remove or disconnect the following:
- 3 upper strut mount nuts and allow the control arms to hang free
- Anti-lock Brakes System (ABS) front wheel speed sensor
- Wheel speed sensor bracket from the strut
- Brake line bracket from the strut
- Strut-to-steering knuckle bolts and the strut

To install:

5. Install or connect the following:
- Strut. Torque the 3 nuts to 18 ft. lbs. (24 Nm) on all models except 1998–99 models and all Park Avenue models or 30–35 ft. lbs. (40–47 Nm) on 2000–01 models and all Park Avenue models.
- Electronic ride control electrical connector, if removed
- Strut-to-knuckle bolts. Torque the bolts to 140 ft. lbs. (190 Nm) on all models except 1998–99 models and all Park Avenue models or 135 ft. lbs. (185 Nm) on 2000–01 models and all Park Avenue models.
- Brake line bracket to the strut
- Wheel speed sensor bracket to the strut
- ABS front wheel speed sensor connector
- Front wheel. Torque the lug nuts to 100 ft. lbs. (140 Nm).

6. Check and adjust the wheel alignment.

Rear

EXCEPT PARK AVENUE AND ALL 2000–01 MODELS

1. Remove rear seat cushion and seatback to gain access to the strut tower mounting nuts.
2. Remove tire.
3. Support the control arm.
4. Remove or disconnect the following:
- Strut to knuckle bolts
- Upper strut nuts
- Strut from vehicle

To install:

5. Install or connect the following:
- Strut
- Upper strut nuts. Torque the nuts to 35 ft. lbs. (47 Nm).
- Strut to knuckle bolts. Torque the bolts to 140 ft. lbs. (190 Nm).
- Rear seatback and cushion
- Tire. Torque the lug nuts to 100 ft. lbs. (140 Nm).

Shock Absorber

REMOVAL & INSTALLATION

Rear

EXCEPT PARK AVENUE AND 2000–01 MODELS

1. Before servicing the vehicle, refer to the precautions in the beginning of this section.
2. Remove the rear seat cushion and seatback to gain access to the upper shock mounting nuts,.
3. Using a floor jack and a block of wood, support the lower control arm.
4. Remove or disconnect the following:
- Rear wheel
- 2 upper shock absorber mounting nuts
- Electronic Leveling Control (ELC), if equipped
- Lower shock absorber bracket-to-knuckle bolts
- Shock absorber

To install:

5. Install or connect the following:
- Shock absorber and loosely thread the upper and lower shock absorber mounting nuts
- Upper shock absorber mounting nuts and torque to 35 ft. lbs. (47 Nm).
- Strut bracket-to-knuckle bolts and torque to 140 ft. lbs. (190 Nm).
- ELC, if equipped

6. For ELC equipped vehicles, lightly pressurize the ELC system by momentarily grounding terminal "B" at the height sensor connector before lowering the vehicle.
- Rear wheel. Torque the lug nuts to 100 ft. lbs. (140 Nm).

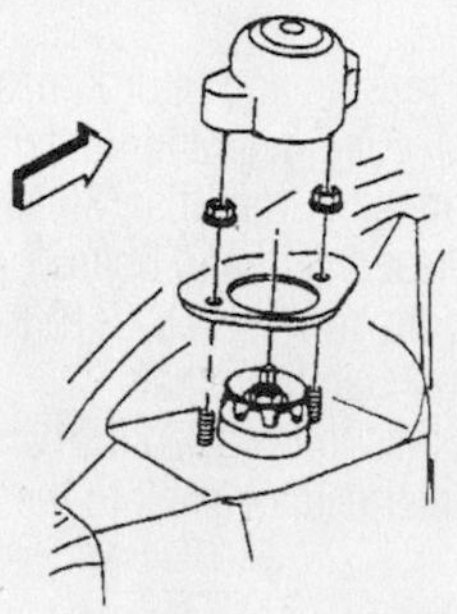

7922UG21

Remove the cover to gain access to the upper shock absorber mounting components—rear suspension

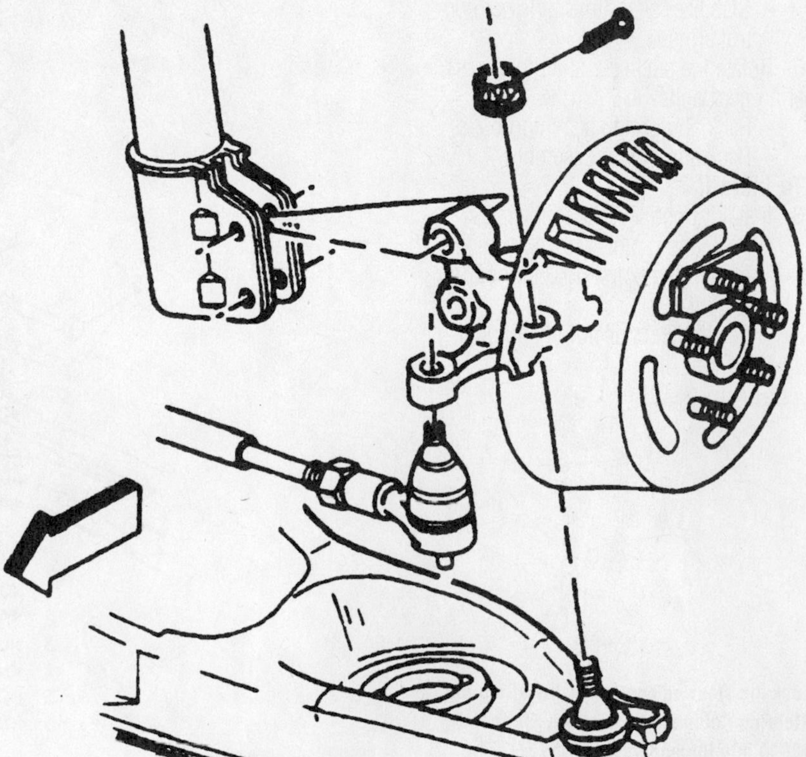

7922UG22

Exploded view of the lower shock absorber mounting to the knuckle assembly—rear suspension

7. Check the wheel alignment and adjust if necessary.

2000–01 MODELS AND ALL PARK AVENUE MODELS

1. Before servicing the vehicle, refer to the precautions in the beginning of this section.
2. Remove the tire and wheel assembly.
3. Support the control arm with a jack stand.
4. Remove or disconnect the following:
   - Electronic Ride Control (ELC) air tube from the shock
   - Two bolts securing the shock to the control arm
   - Trunk trim to gain access to the shock upper mounting nuts.
   - Cover, the two nuts, and the reinforcement from the top of the shock
   - Shock from the vehicle

To install:

5. Install or connect the following:
   - Shock, reinforcement, and the two nuts. Tighten the mounting nuts to 15 ft. lbs. (20 Nm).
   - Shock cover
   - Trunk trim
   - Shock-to-control arm bolts and tighten the bolts to 18ft. lbs. (24 Nm)
   - ELC air tube to the shock.
   - Tire and wheel assembly

Coil Spring

REMOVAL & INSTALLATION

CAUTION

The coil springs are under a considerable amount of tension. Be very careful when removing or installing them; they can exert enough force to cause very serious injury.

Front

1. Remove the strut from the vehicle.
2. Disassemble the strut as follows:
   a. Step 1: Place the strut assembly into compressor tool, to compress the coil spring.
   b. Step 2: Compress the spring slightly.
   c. Step 3: Hold the strut shaft from turning using a No. 50 Torx® socket and remove the 24mm nut on the top end of the strut.

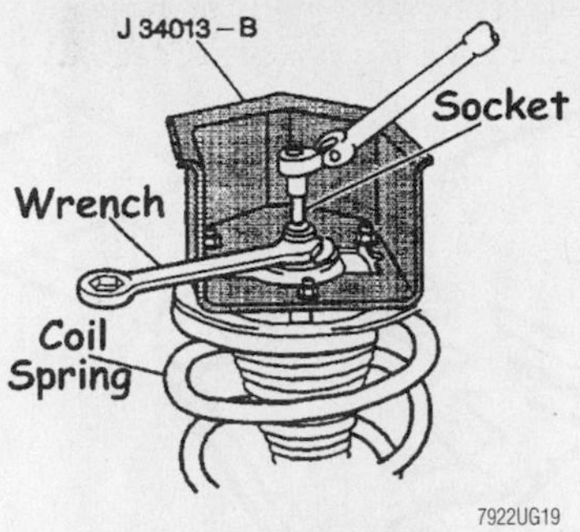

Use a Torx® socket to keep the piston rod from turning while removing the upper nut—front strut shown

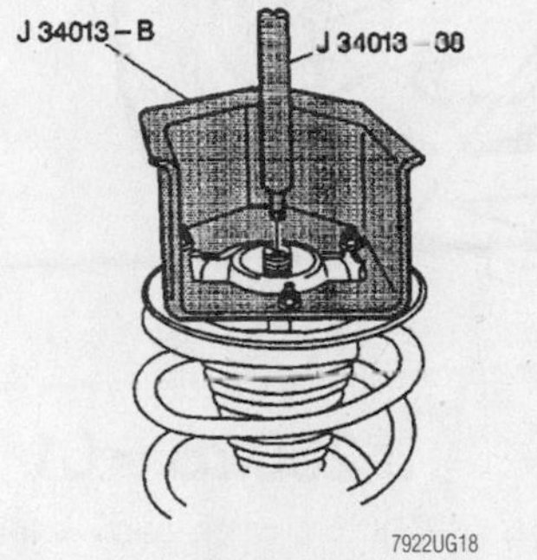

Install Rod J 34013-38 to help guide the strut shaft from the upper mount assembly—front strut shown

   d. Step 4: Install Rod tool J 34013-38 to help guide the strut shaft from the upper mount assembly.
   e. Step 5: Loosen the spring compressor tool until the coil spring and mount can be removed as an assembly. Remove the lower spring insulator, if equipped.

To install:

3. Assemble the strut as follows:
   a. Step 1: Place the strut in compressor tool.
   b. Step 2: Install the coil spring over the strut.
   c. Step 3: Compress the coil spring while guiding strut shaft through the top of the strut assembly.
   d. Step 4: Install the top strut nut. Torque the nut to 55 ft. lbs. (75 Nm).
   e. Step 5: Remove the strut from the compressor.

Rear

EXCEPT 2000–01 MODELS AND ALL PARK AVENUE MODELS

1. Before servicing the vehicle, refer to the precautions in the beginning of this section.
2. Remove or disconnect the following:

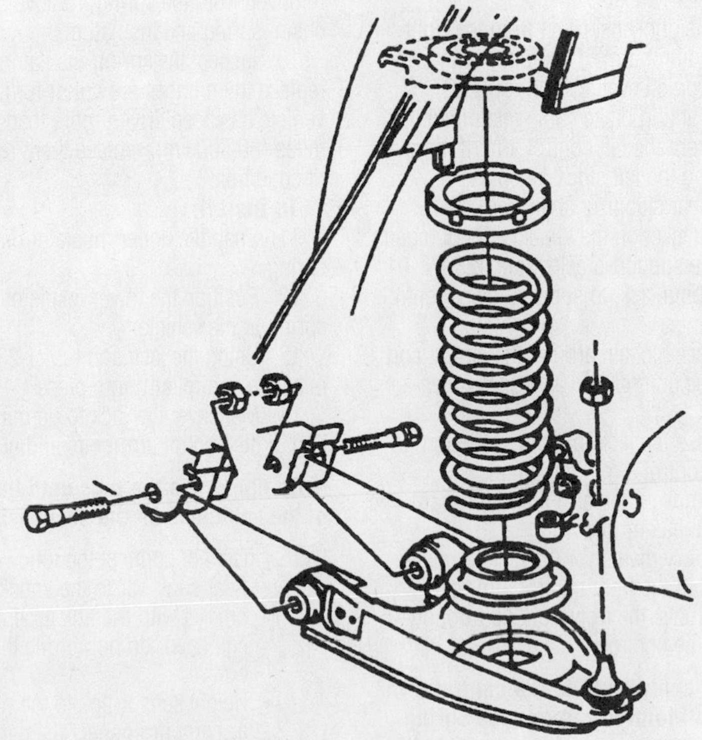

Exploded view of the rear coil spring mounting—H body

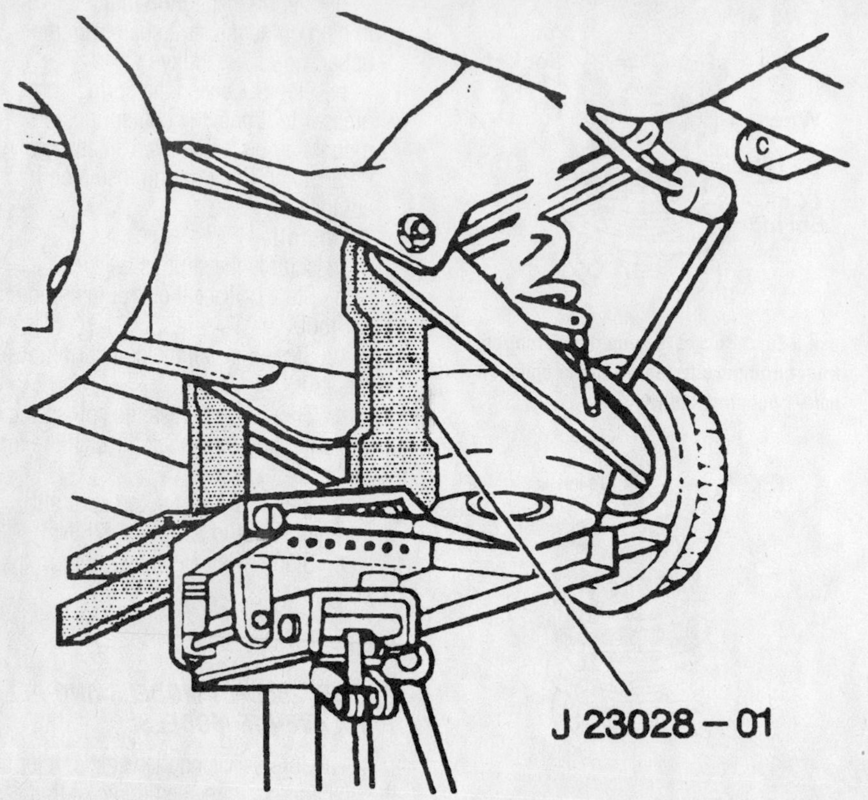

Use support bracket tool J-23028-01 mounted on a jack to support the rear lower control arm—H body

- Rear wheel
- Height sensor link from the right control arm, if equipped with Electronic Leveling Control (ELC)
- Parking brake cable retaining clip from the left control arm, if removing the left side coil spring
- Rear stabilizer shaft from the bracket on the knuckle, if equipped

3. Use support bracket tool J23028-01, mounted on a jack, to support lower control arm bushings.
4. Place a chain around the spring and through the control arm as a safety measure.
5. Raise the jack to remove tension from the control arm pivot bolts.
6. Remove the rear nut and through-bolt from the control arm.
7. Slowly maneuver the jack to relieve tension from the front control arm bolt.
8. Remove the front nut and through-bolt from the control arm.

➡Do not apply force to the control arm and/or ball joint to remove the spring. Proper maneuvering of the spring will allow for easy removal.

9. Lower the jack to pivot the control arm downward. When all compression is removed from the spring, remove the safety chain, spring and insulators.
10. Inspect the spring insulators and replace them if they are cut or torn. If the vehicle has been driven more than 50,000 miles (80,500 km), replace them regardless of condition.

To install:

11. Snap the upper insulator onto the spring.
12. Position the lower insulator and spring in the vehicle.
13. Using the jack and tool J-23028-01, raise the control arm into place.
14. Maneuver the jack to permit installation of the control arm bolts and nuts.

➡Do not torque the nuts until the weight of the vehicle is on the suspension.

15. Install or connect the following:
    - Rear sway bar to the knuckle bracket with the link assembly if equipped, do not torque the link bolt
    - Height sensor link to the right control arm or connect the parking brake cable retaining clip to the left control arm, as required
    - Rear wheel. Torque the lug nuts to 100 ft. lbs. (140 Nm).
16. With the vehicle resting on its wheels, torque the control arm through-bolt nuts to 85 ft. lbs. (115 Nm).
17. Torque the sway bar link bolt to 13 ft. lbs. (17 Nm).
18. Check wheel alignment and adjust if necessary.
19. Road test the vehicle for proper operation.

2000–01 MODELS AND ALL PARK AVENUE MODELS

1. Before servicing the vehicle, refer to the precautions in the beginning of this section.
2. Remove or disconnect the following:
    - Tire
    - Electronic Leveling Control (ELC) air tube
3. Support the control arm.
    - Two lower shock bolts
    - Trunk trim
    - Cover, 2 nuts and the reinforcement from the top of the shock
    - Cotter pin and hex nut on control arm
4. Separate the adjustment link from the knuckle.
5. Slowly lower the control arm.
6. Remove the spring with the lower insulator.

To install

7. Install or connect the following:
    - Coil spring with insulator
    - Lower shock bolts. Torque the bolts to 18 ft. lbs. (24 Nm).
    - Adjustment link. Torque the nut to 88 inch lbs. (10 Nm) on 1998–99 Park Avenue models or 36 ft. lbs. (50 Nm) on all 2000–01 models including Park Avenue. Tighten an additional turn to align the cotter pin.
    - ELC air tube
    - Tire. Torque the lug nuts to 100 ft. lbs. (140 Nm).

Lower Ball Joint

REMOVAL & INSTALLATION

Front

EXCEPT ALL 2000–01 AND PARK AVENUE MODELS

1. Before servicing the vehicle, refer to the precautions in the beginning of this section.
2. Remove or disconnect the following:
    - Front wheel
    - Cotter pin from the lower ball joint and loosen the nut

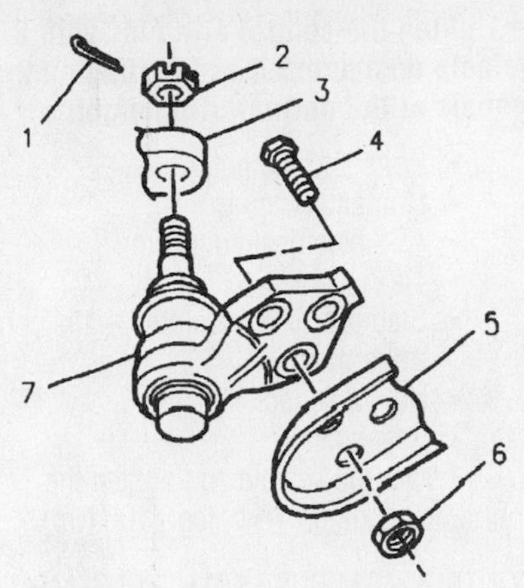

1 PIN
2 NUT, BALL JOINT TO KNUCKLE; TIGHTEN TO 10 N•m (88 LB. IN.) THEN TIGHTEN 2 FLATS TO 55 N•m (41 LB. FT.), MIN.
3 KNUCKLE
4 BALL JOINT MOUNTING BOLTS MUST FACE DOWN
5 CONTROL ARM
6 BALL JOINT MOUNTING NUTS 68 N•m (50 LB. FT.)
7 SERVICE BALL JOINT

7922UG25

The replacement ball joint should be attached to the control arm using 3 bolts and nuts—except Park Avenue

- Ball joint from the steering knuckle
- Stabilizer shaft link assembly

3. Drill out the 3 rivets retaining the ball joint.
4. Remove the ball joint.

To install:

5. Install or connect the following:
   - Lower ball joint to the steering knuckle
   - Ball joint to the lower control arm, making sure the bolts face downward. Torque the nuts to 50 ft. lbs. (68 Nm).
   - Stabilizer shaft link assembly. Torque the nut to 13 ft. lbs. (17 Nm).
   - Lower ball joint nut. Torque the nut to 88 inch lbs. (10 Nm) plus an additional 120 degree turn, during which a torque of 41 ft. lbs. (55 Nm) must be obtained.
   - New cotter pin to the lower ball joint

➡Tighten the nut up to one more flat in order to align the slot wlth the hole in the stud.

- Front wheel. Torque the lug nuts to 100 ft. lbs. (140 Nm).

6. Check and adjust the wheel alignment.

2000–01 AND ALL PARK AVENUE MODELS

The ball joint is an integral part of the lower control arm and if found to be defective the control arm should be replaced.

Rear

1. Before servicing the vehicle, refer to the precautions in the beginning of this section.
2. Remove or disconnect the following:
   - Rear wheel
   - Height sensor link from the right control arm
   - Parking brake cable retaining clip from the left control arm
   - Adjustment link from the knuckle
3. Support the control arm with a jack.
4. Remove the ball joint stud cotter pin and nut.
5. Reinstall the nut on the stud with the flat side of the nut facing up; do not torque the nut.
6. Remove or disconnect the following:
   - Stud from the knuckle
   - Slotted hex nut from the ball joint stud
7. Press the ball joint out of the control arm

To install:

8. Install or connect the following:
   - New ball joint, press it into the control arm
   - Ball joint stud into the knuckle. Torque the nut to 88 inch lbs. (10 Nm), plus an additional 4 flats. The minimum torque on the nut should be 40 ft. lbs. (55 Nm).
   - New cotter pin

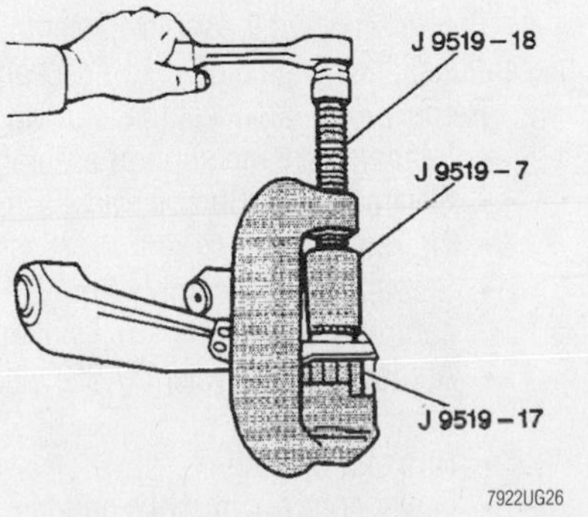

Removing the ball joint from the lower control arm—except Park Avenue

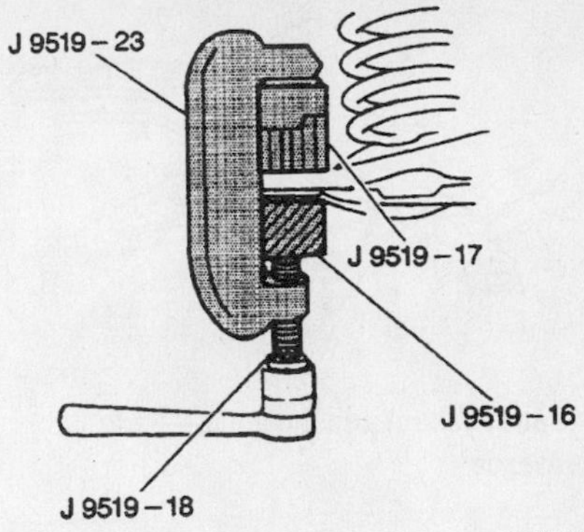

Installing the ball joint into the lower control arm—except Park Avenue

➡Tighten the nut up to one more flat in order to align the slot with the hole in the stud.

- Adjustment link to the knuckle. Torque the slotted nut to 33 ft. lbs. (45 Nm).
- New cotter pin

➡Tighten the nut up to one more flat in order to align the slot with the hole in the stud.

- Height sensor link to the right control arm
- Parking brake cable retainer to the left control arm
- Rear wheel

Lower Control Arm

REMOVAL & INSTALLATION

Front

1. Before servicing the vehicle, refer to the precautions in the beginning of this section.
2. Allow the control arms to hang free.
3. Remove or disconnect the following:

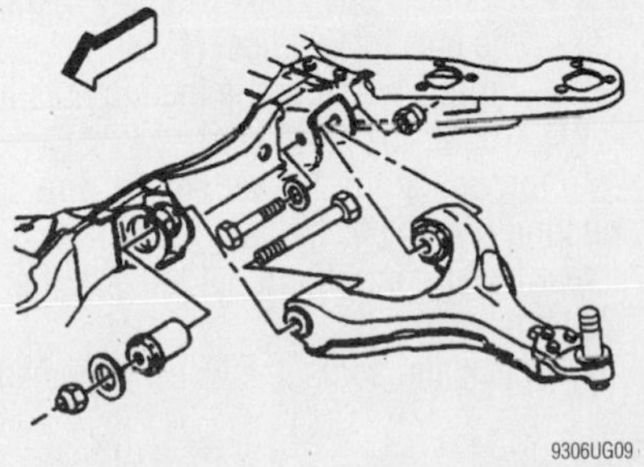

Lower control arm assembly—except Park Avenue

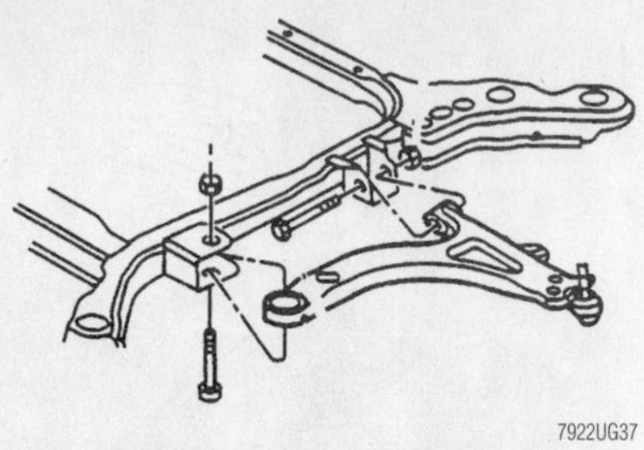

Lower control arm mounting—Park Avenue

- Front wheel
- Stabilizer shaft link assembly from the control arm
- Cotter pin from the lower ball joint and remove the nut
- Ball joint from the steering knuckle
- Lower control arm from the frame

To install:

4. Install or connect the following:
   - Lower control arm to the frame

➡Do not tighten the lower control arm nuts at this time. The weight of the vehicle must be supported by the control arms, since the vehicle design trim heights are obtained before tightening the lower control arm mounting nuts.

- Ball joint stud to the steering knuckle. Torque the lower ball joint nut to 88 inch lbs. (10 Nm), then an additional 120 degree turn, during which a torque of 41 ft. lbs. (55 Nm) must be obtained.
- New cotter pin

➡Align the slot in the nut with the hole in the lower ball joint stud. Torque the ball joint nut up to one more flat. Never loosen the nut to align the slot.

- Control arm rear mounting bolts to 117 ft. lbs. (158 Nm), then the front mounting nut to 117 ft. lbs. (158 Nm)
- Stabilizer shaft link assembly. Torque the nuts to 13 ft. lbs. (17 Nm).
- Front wheel. Torque the lug nuts to 100 ft. lbs. (140 Nm).

5. Torque the front lower control arm mounting nut to 140 ft. lbs. (190 Nm) and the rear lower control arm nut to 91 ft. lbs. (123 Nm).
6. Check and adjust the wheel alignment.

Rear

EXCEPT 2000–01 AND ALL PARK AVENUE MODELS

1. Before servicing the vehicle, refer to the precautions in the beginning of this section.
2. Remove or disconnect the following:
   - Adjustment link retaining nut
   - Adjustment link from the control arm
   - Spring
   - Cotter pin and hex nut from the ball joint. Turn the nut over and install the nut with the flat portion facing upward. Do not tighten the nut.
   - Separate the knuckle from the ball joint stud and remove the nut
   - Control arm

To install:

3. Install or connect the following:
   - Lower control arm ball joint stud in the knuckle boss
   - New hex nut and tighten to 88 inch lbs. (10 Nm). Tighten the nut an additional four flats; torque should be 40 ft. lbs. (55 Nm) minimum. It may be necessary to partially load the joint to keep the ball stud from rotating while the nut is being tightened. Tighten the nut up to one more flat to align the slot with the cotter pin hole.
   - Cotter pin
   - Spring
   - Adjustment link to the control arm and tighten the link retaining nut to 63 ft. Lbs. (85 Nm)

➡Tighten the control arm nuts with the vehicle on its wheels at its normal ride height. Failure to follow this procedure may adversely affect the ride and handling of the vehicle.

4. Tighten the control arm nuts to 85 ft. lbs. (115 Nm). Optional torque: Tighten the control arm bolts to 143 ft. lbs. (182 Nm)
5. Tighten the stabilizer shaft link bolt to 13 ft. lbs. (17 Nm).

2000–01 AND ALL PARK AVENUE MODELS

1. Before servicing the vehicle, refer to the precautions in the beginning of this section.
2. Remove or disconnect the following:
   - Rear suspension support assembly
   - Electronic Ride Control height sensor
   - Tie rod
   - Stabilizer link assembly from the control arm
   - Antilock Brake System (ABS) electrical connector
   - Hub and bearing
   - Lower control arm bolts and the control arm

To install:

3. Install or connect the following:
   - Lower control arm to the frame.

➡Tighten the control arm nuts with the vehicle unsupported and resting on the wheels at the normal trim height.

- Bolts and the nuts
- Hub and bearing
- ABS electrical connector
- Tie rod
- Stabilizer link assembly to the control arm
- Height sensor
- Rear wheel

4. Lower the vehicle and tighten the control arm nuts to 78 ft. lbs. (106 Nm).

CONTROL ARM BUSHING REPLACEMENT

Except Park Avenue

FRONT

1. Before servicing the vehicle, refer to the precautions in the beginning of this section.
2. Remove or disconnect the following:
   - Front wheel
   - Lower control arm
3. Press the control arm bushing out of the control arm.

To install:

4. Insert the new bushing into control arm.
5. Press the new bushing into the control arm.
6. Install the lower control arm.
7. Install the front wheel. Torque the lug nuts to 100 ft. lbs. (140 Nm).

REAR—1998–99 MODELS

1. Before servicing the vehicle, refer to the precautions in the beginning of this section.

➡There are two different sized bushings used in the control arm. These

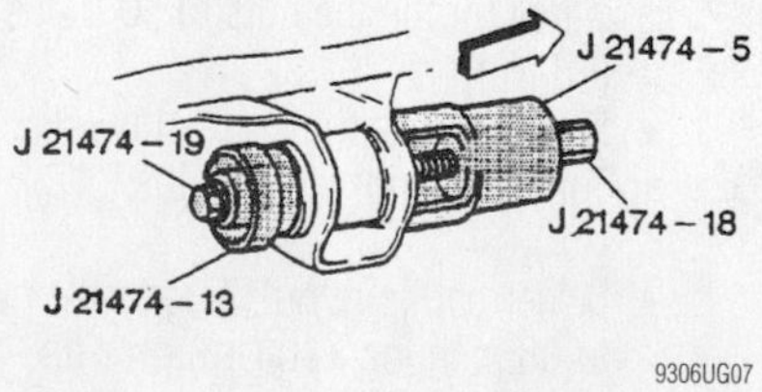

Removing the bushing from the control arm

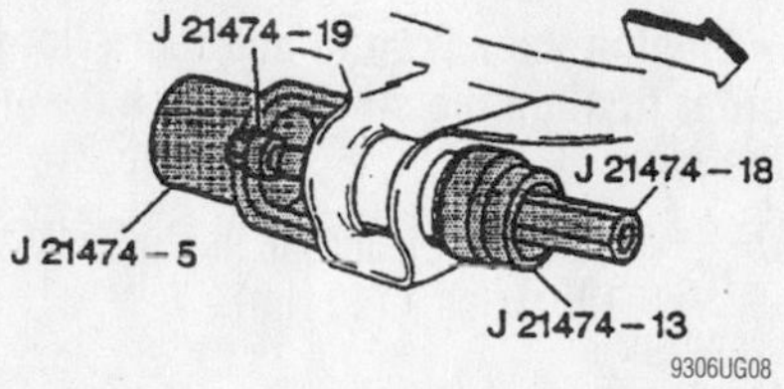

Installing the bushing into the control arm

bushings are not interchangeable, and require different combinations of special tools for removal and installation.

2. Remove the control arm.
3. Place the control arm upside-down on the bench and install the tools shown in the accompanying illustration.
4. Installl spacer J 22222-5 on bushing 1 or on bushing 2 use J 21474-12 (refer to the accompanying illustration).
5. Position the receiver tube and the J 29376-7 on the outer side of the control arm. Be sure the receiver tube does not contact the bushing flange.
6. Coat the threaded area of the long bolt with the high pressure lubricant J 23444-A and install the bolt through the receiver, the cap, and the bushing.
7. Install the remover on the bolt at the inner side of the control arm. Position the remover so the small diameter end contacts the bushing. On bushing 1 use J 22222-2 or bushing 2 use J 28685.
8. Place the bearing on the bolt and install the J 21474-18 long nut. The bearing must be installed between the nut and the remover.

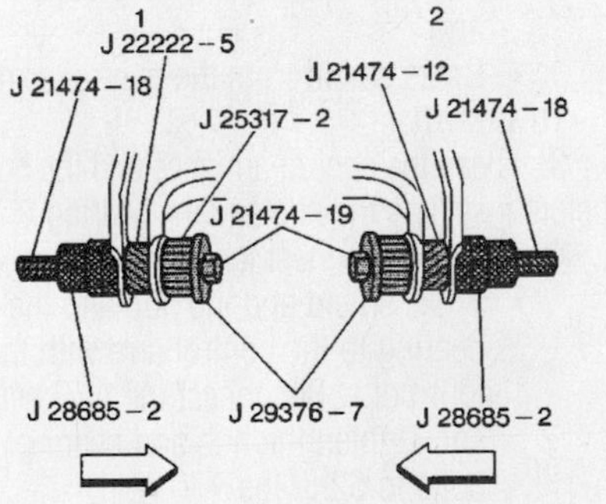

Removing the bushing from the rear control arm—1998-99 models

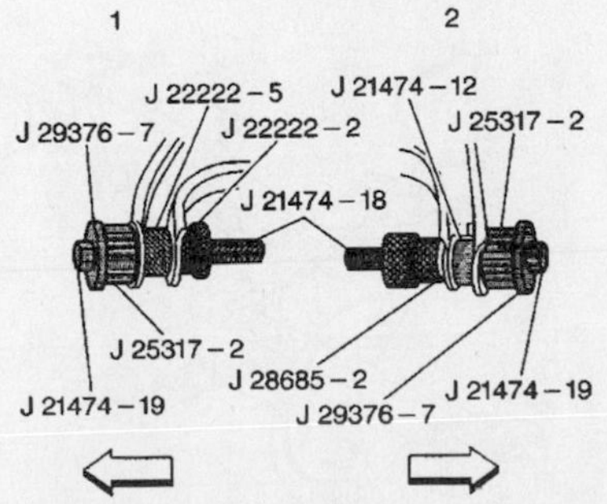

Installing the bushing into the rear control arm—1998-99 models

Draw the bushing out of the control arm by tightening the nut.

To install:

9. With the control arm upside-down on the bench, position the new bushing in the control arm.

➡When installing new bushings, the flange end of the bushing must face outward when installed in the control arm.

10. Install the appropriate tools: as follows:
    a. Install the spacer. On bushing 1 use tool J 22222-5, or on bushing 2 use tool J 21474-12.
11. Position the receiver tube and the J 29376-7 cap on the inner side of the control arm. Center the receiver tube over the hole.
12. Coat the threaded area of the long bolt with the high pressure lubricant and install the bolt through the receiver, the cap, and the bushing.
13. Position installer J 28685 on the bolt at the outer side of the control arm. Position the installer with the large diameter contacting the bushing flange.
14. Place the bearing on the bolt and install J 21474-18 long nut. The bearing must be positioned between the nut and the installer.
15. Draw the bushing into the control arm by tightening the nut. Tighten the nut until the bushing flange seats firmly against the control arm.
16. Reinstall the control arm in the vehicle and inspect the rear wheel alignment.

REAR—2000-01 MODELS

1. Before servicing the vehicle, refer to the precautions in the beginning of this section.
2. Remove the control arm.
3. Place tool J 21474-27 with the washer through tool J 41014-2 over the bushing against the control arm.
4. Lubricate the bolt threads with high pressure lubricant.
5. Install tool J 41014-1 (with small end facing the bushing), the thrust bearing and tool J 21474-4 onto tool J 21474-27.
6. Tighten the nut until the bushing is driven out of the control arm.
7. Remove the bushing tools.

To install:

8. Start the bushing into the control arm.

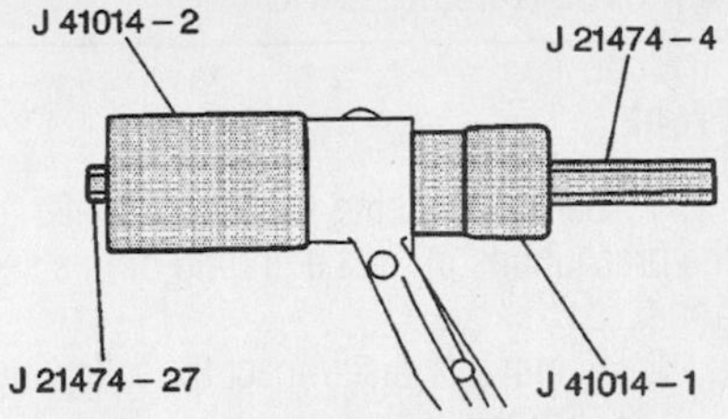

Removing the bushing from the rear control arm—2000-01 models

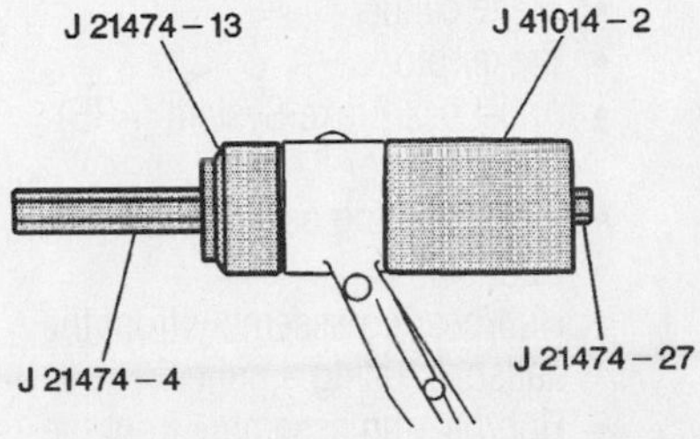

Installing the bushing into the rear control arm—2000-01 models

9. Make sure that the flat on the bushing is vertical and facing rearward. Place tool J 21474-27 with the washer through tool J41014-1 over the bushing against the control arm.
10. Lubricate the bolt threads with high pressure lubricant.
11. Install tool J 21474-13 (making sure the large end is facing the bushing), the thrust bearing and tool J 21474-4 onto tool J 21474-27.
12. Tighten the bolt until the bushing is fully seated into the control arm.
13. Remove the bushing tools.
14. Install the control arm.

Park Avenue

The lower control arm is replaced as a unit. The bushing can not be removed.

Wheel Bearings

ADJUSTMENT

The wheel bearings are not adjustable. If a wheel bearing is out of specifications, it must be replaced. Using a dial indicator, check for looseness. If play exceeds 0.005 inch (0.127mm), the bearing wear is excessive and the hub/bearing should be replaced.

REMOVAL & INSTALLATION

Front

1. Before servicing the vehicle, refer to the precautions in the beginning of this section.
2. Remove or disconnect the following:
   - Front wheel

➡Insert a drift punch through the caliper and into the rotor cooling fins to prevent the rotor from turning.

- Halfshaft nut and washer
- Brake caliper
- Brake rotor
- Anti-Lock Brake System (ABS) speed sensor
- 3 hub/bearing assembly bolts
- Dust shield
- Hub/bearing assembly from the halfshaft, using a puller
- Hub/bearing assembly from the steering knuckle

To install:

3. Install the hub/bearing assembly over the halfshaft splines. Be sure the splines engage smoothly.

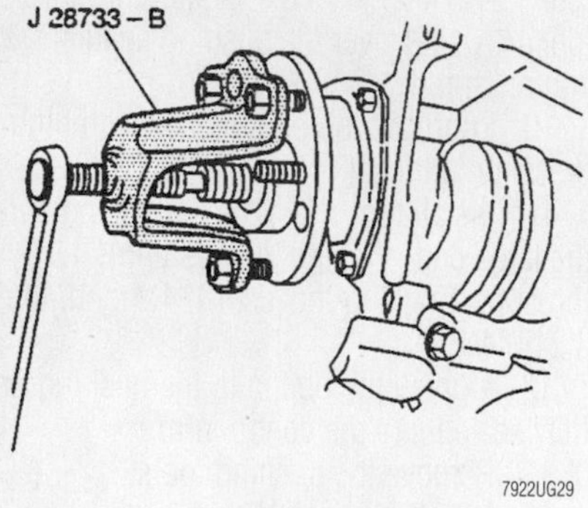

Use a puller such as J 28733-B to press the halfshaft from the hub assembly

4. Apply a light coating of grease to the steering knuckle bore.
5. Slide the hub assembly onto the halfshaft as far as possible. If the hub will not bottom out on the halfshaft, install the hub mounting bolts and use the halfshaft nut to draw the hub onto the halfshaft.
6. Once the hub is flush with the steering knuckle, remove the mounting bolts and install the dust shield.
7. Install or connect the following:
8. Install the mounting bolts. Torque the bolts to 70 ft. lbs. (95 Nm).
9. Place the transaxle in **N**.
10. Install or connect the following:
    - ABS front wheel speed sensor connector, and clip to the dust shield
    - Brake rotor
    - Caliper. Torque the bolts to 38 ft. lbs. (51 Nm).
11. Torque the halfshaft nut to 107 ft. lbs. (145 Nm) on 1998–99 models or 118 ft. lbs. (160 Nm) on 2000–01 models.
12. Install the front wheels. Torque the lug nuts to 100 ft. lbs. (140 Nm).
13. Road test the vehicle.

Rear

1998–99 MODELS

1. Before servicing the vehicle, refer to the precautions in the beginning of this section.
2. Remove or disconnect the following:
   - Rear wheel
   - Brake drum
   - Anti-Lock Brake Systems (ABS) sensor wire, if equipped

**** WARNING**

The hub assembly mounting bolts also secure the backing plate assembly. Once the bolts are removed, the backing plate must be supported with wire or other means. Do not allow the brake line or ABS electrical wire to support the brake assembly.

- 4 hub/bearing assembly bolts and the hub assembly

To install:

3. Install or connect the following:
   - Hub/bearing assembly onto the rear knuckle. Torque the bolts to 52 ft. lbs. (70 Nm).
   - ABS sensor connector, if equipped
   - Brake drum
   - Rear wheel. Torque the nuts to 100 ft. lbs. (140 Nm).
4. Road test for proper operation.

2000–01 MODELS

1. Before servicing the vehicle, refer to the precautions in the beginning of this section.
2. Remove or disconnect the following:
   - Wheel and the tire
   - Brake caliper
   - Brake rotor
   - ABS sensor wire connector.
   - Four bolts from the control arm
   - Hub and bearing from the control arm
   - Brake shield from the control arm

To install:

3. Clean the control arm face and the bore before installing the hub and the bearing.
4. Install or connect the following:
   - Brake shield and the hub and the bearing to the control arm with the four bolts. Reconnect the ABS sensor. Tighten the hub and bearing bolts to 52 ft. lbs. (70 Nm).
   - Brake rotor
   - Brake caliper
   - Wheel and the tire

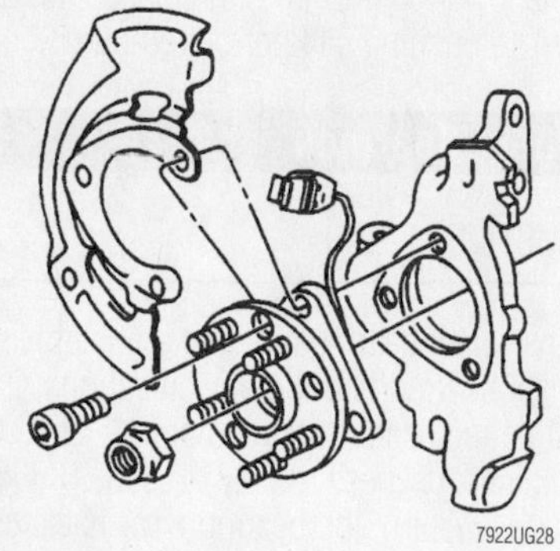

Exploded view of the front hub and wheel bearing assembly

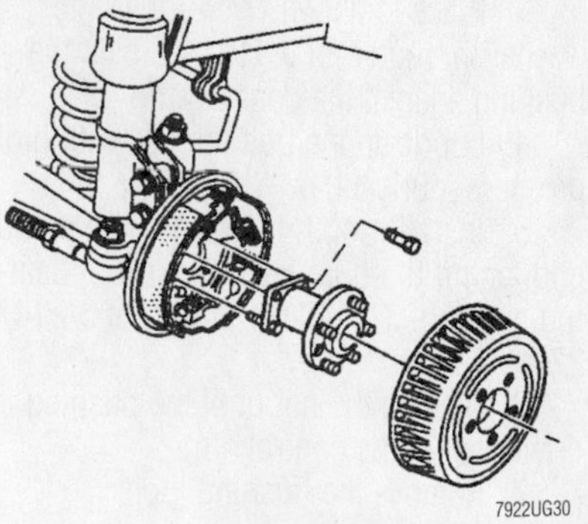

Exploded view of the rear hub/wheel bearing assembly—1998–99 models

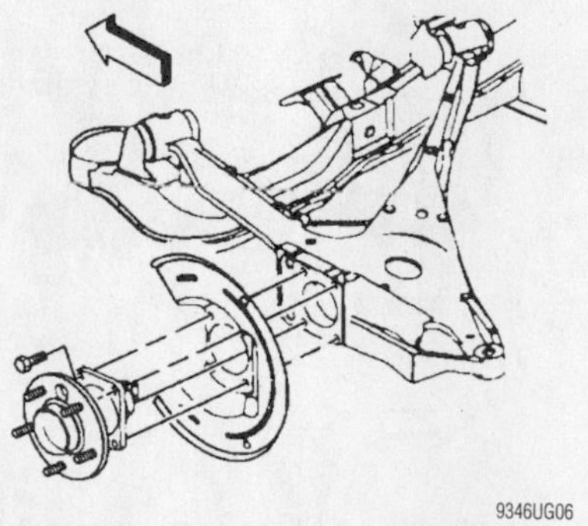

Exploded view of the rear hub/wheel bearing assembly—2000–01 models

GENERAL MOTORS CORPORATION–E & K- BODIES

22

1998–01

Cadillac-Deville • Eldorado • Seville

PRECAUTIONS

Before servicing any vehicle, please be sure to read all of the following precautions, which deal with personal safety, prevention of component damage, and important points to take into consideration when servicing a motor vehicle:

- Never open, service or drain the radiator or cooling system when the engine is hot; serious burns can occur from the steam and hot coolant.
- Observe all applicable safety precautions when working around fuel. Whenever servicing the fuel system, always work in a well-ventilated area. Do not allow fuel spray or vapors to come in contact with a spark, open flame or excessive heat (a hot drop light, for example). Keep a dry chemical fire extinguisher near the work area. Always keep fuel in a container specifically designed for fuel storage; also, always properly seal fuel containers to avoid the possibility of fire or explosion. Refer to the additional fuel system precautions later in this section.
- Fuel injection systems often remain pressurized, even after the engine has been turned **OFF**. The fuel system pressure must be relieved before disconnecting any fuel lines. Failure to do so may result in fire and/or personal injury.
- Brake fluid often contains polyglycol ethers and polyglycols. Avoid contact with the eyes and wash your hands thoroughly after handling brake fluid. If you do get brake fluid in your eyes, flush your eyes with clean, running water for 15 minutes. If eye irritation persists, or if you have taken brake fluid internally, IMMEDIATELY seek medical assistance.
- The EPA warns that prolonged contact with used engine oil may cause a number of skin disorders, including cancer! You should make every effort to minimize your exposure to used engine oil. Protective gloves should be worn when changing oil. Wash your hands and any other exposed skin areas as soon as possible after exposure to used engine oil. Soap and water, or waterless hand cleaner should be used.
- All new vehicles are now equipped with an air bag system, often referred to as a Supplemental Restraint System (SRS) or Supplemental Inflatable Restraint (SIR) system. The system must be disabled before performing service on or around system components, steering column, instrument panel components, wiring and sensors. Failure to follow safety and disabling procedures could result in accidental air bag deployment, possible personal injury and unnecessary system repairs.
- Always wear safety goggles when working with, or around, the air bag system. When carrying a non-deployed air bag, be sure the bag and trim cover are pointed away from your body. When placing a non-deployed air bag on a work surface, always face the bag and trim cover upward, away from the surface. This will reduce the motion of the module if it is accidentally deployed. Refer to the additional air bag system precautions later in this section.
- Clean, high quality brake fluid from a sealed container is essential to the safe and proper operation of the brake system. You should always buy the correct type of brake fluid for your vehicle. If the brake fluid becomes contaminated, completely flush the system with new fluid. Never reuse any brake fluid. Any brake fluid that is removed from the system should be discarded. Also, do not allow any brake fluid to come in contact with a painted surface; it will damage the paint.
- Never operate the engine without the proper amount and type of engine oil; doing so WILL result in severe engine damage.
- Timing belt maintenance is extremely important! Many models utilize an interference-type, non-freewheeling engine. If the timing belt breaks, the valves in the cylinder head may strike the pistons, causing potentially serious (also time-consuming and expensive) engine damage. Refer to the maintenance interval charts in the front of this manual for the recommended replacement interval for the timing belt, and to the timing belt section for belt replacement and inspection.
- Disconnecting the negative battery cable on some vehicles may interfere with the functions of the on-board computer system(s) and may require the computer to undergo a relearning process once the negative battery cable is reconnected.
- When servicing drum brakes, only disassemble and assemble one side at a time, leaving the remaining side intact for reference.
- Only an MVAC-trained, EPA-certified automotive technician should service the air conditioning system or its components.

ENGINE REPAIR

Alternator

REMOVAL

Eldorado and 1998–99 Deville

1. Before servicing the vehicle, refer to the precautions in the beginning of this section.
2. Drain the cooling system.
3. Remove or disconnect the following:
   - Negative battery cable
   - Accessory drive belt
   - Alternator upper mounting bolt
   - Engine splash shield
   - Radiator support access panel
   - Rear alternator bracket from the engine
   - Alternator mounting bolts
   - Duct on the back of the alternator
   - Electrical connectors
   - Front alternator-to-A/C bracket
   - Alternator

1998 Seville and 2000–01 Deville

4. Before servicing the vehicle, refer to the precautions in the beginning of this section.
5. Drain the cooling system.
6. Remove or disconnect the following:
   - Negative battery cable
   - Accessory drive belt
   - Radiator
   - Air conditioning condenser
   - Alternator cooler outlet hose
   - Alternator cooler inlet hose
   - Electrical connectors
   - Alternator mounting bolts
   - Alternator

1999–01 Seville

1. Before servicing the vehicle, refer to the precautions in the beginning of this section.
2. Drain the cooling system.
3. Remove or disconnect the following:
   - Negative battery cable
   - Accessory drive belt
   - Cooling fans
   - Alternator cooler outlet hose
   - Alternator cooler inlet hose
   - Electrical connectors
   - Alternator mounting bolts
   - Alternator

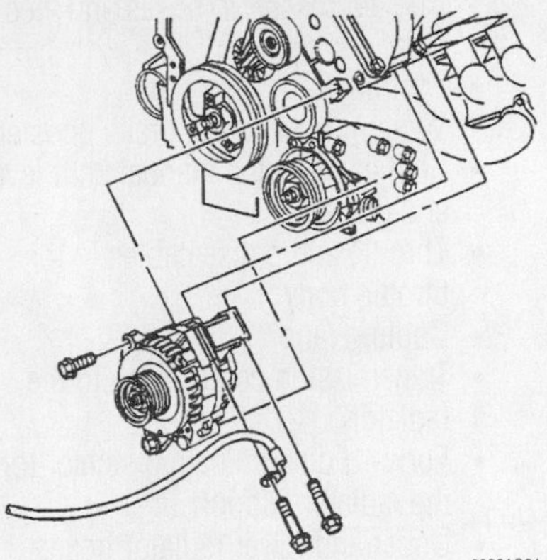

9306AG01

Alternator mounting and wires

INSTALLATION

Eldorado and 1998–99 Deville

1. Install or connect the following:
   - Alternator
   - Front alternator-to-A/C bracket. Torque the bolt to 35 ft. lbs. (47 Nm).
   - Electrical connectors. Torque the battery connection nut to 115 inch lbs. (13 Nm).
   - Duct on the back of the alternator
   - Alternator mounting bolts. Torque the bolts to 35 ft. lbs. (47 Nm).
   - Upper front bolt. Torque the bolt to 35 ft. lbs. (47 Nm).
   - Rear bracket to the engine. Torque the bolt to 35 ft. lbs. (47 Nm).
   - Radiator support access panel
   - Engine splash shield
   - Accessory drive belt
   - Negative battery cable
2. Refill the cooling system.

1998 Seville and 2000–01 Deville

1. Install or connect the following:
   - Alternator
   - Alternator mounting bolts. Torque the bolts to 37 ft. lbs. (50 Nm).
   - Electrical connectors. Torque the battery connection nut to 115 inch lbs. (13 Nm).
   - Alternator cooler inlet hose
   - Alternator cooler outlet hose. Torque the bolt to 80 inch lbs. (9 Nm).
   - Radiator
   - Air conditioning condenser
   - Accessory drive belt
   - Negative battery cable
2. Refill the cooling system.

1999–01 Seville

1. Install or connect the following:
   - Alternator. Torque the bolt to 37 ft. lbs. (50 Nm).
   - Electrical connectors. Torque the battery connection nut to 115 inch lbs. (13 Nm).
   - Alternator cooler inlet hose
   - Alternator cooler outlet hose. Torque the bolt to 80 inch lbs. (9 Nm).
   - Cooling fans
   - Accessory drive belt
   - Negative battery cable
2. Refill the cooling system.

Ignition Timing

ADJUSTMENT

The 4.6L Northstar engine is equipped with a Distributorless Ignition System (DIS). The ignition timing is controlled by the Powertrain Control Module (PCM). Therefore no adjustments are necessary.

Engine Assembly

REMOVAL & INSTALLATION

1998–99 Models

1. Before servicing the vehicle, refer to the precautions in the beginning of this section.
2. Disable the Air bag system.
3. Recover the air-conditioning system.
4. Drain the cooling system.
5. Recover the refrigerant from the air conditioning system.
6. Relieve the fuel pressure.
7. Drain the crankcase.
8. Remove or disconnect the following:
   - Tower-to-tower brace
   - Air intake duct
   - Air cleaner assembly
   - Radiator hose at the crossover
   - Engine cover
   - Radiator upper cover
   - Upper radiator hoses from the engine
   - Forward discriminating sensor from the radiator support and set it aside
   - Cooling fans
   - Transmission cooler lines from the radiator
   - Cruise control servo connections
   - Throttle and cruise control cables from the throttle body
   - Shift cable from the manual shift lever and the bracket
   - Vacuum hose from the brake booster
   - Fuel lines
   - Coolant reservoir
   - Coolant hoses from the pipes at the front of the engine
   - Negative battery cable from the left cylinder head
   - Positive battery cable from the battery and body retainer
   - Powertrain Control Module (PCM) case and the PCM
   - Right sound insulator
   - Engine wire harness electrical connector and the vacuum line under the instrument panel
   - Wire harness from the pass through at the cowl and pull the harness through the cowl
   - 3 electrical connectors from the A/C line near the engine cowl
   - Vacuum hose from the **T"** fitting near the engine harness pass through
   - Right and left front wheel speed sensor electrical connectors
   - Manifold Absolute Pressure (MAP) sensor and the control motor near the master cylinder
   - Engine harness retainer from the brake booster
   - 2 hoses from the power steering cooler
   - Engine harness from the hood fuse panel
   - 3 relays from the lower radiator and the electrical connector from the Antilock Brake System (ABS) modulator
   - Tires and wheels
   - Drive axles from the steering knuckle
   - Post Oxygen (O_2S) sensor heat shield and wire harness connector from the body
   - Intermediate pipe hangers from the intermediate pipe
   - Catalytic converter from the pipe
   - Steering coupling from the steering gear
   - Fuel line bundle from the transmission
   - Battery wire from the generator
   - A/C manifold from the A/C compressor

- ABS modulator from the frame and support

9. Position the powertrain support table J39580 under the frame and lower the vehicle onto the table.
10. Remove the 6 frame-to-body bolts.
11. Raise the body off the powertrain.
12. Remove or disconnect the following:
    - Coolant crossover pipe from the engine
    - Park/Neutral switch from the transmission
    - Engine wire harness
    - Exhaust **Y** pipe
    - Accessory drive belt
    - Power steering pump and hoses
    - Alternator and the bracket
    - A/C compressor and bracket
    - Idler pulley
    - Right engine-to-transmission brace
    - Lower engine-to-the transmission brace
    - Left engine-to-transmission brace
13. Attach the engine lift to the engine.
    - Left engine mount and bracket
    - Left exhaust manifold
    - Flywheel inspection cover
    - Torque converter from the flywheel
    - Right engine mount bracket from the engine
    - Transmission-to-engine bolts.
    - Engine from the transmission and frame
    - Exhaust Gas Recirculation (EGR) cross under pipe
    - Right exhaust manifold
    - Engine

To install:

14. Install or connect the following:
    - Engine-to-transmission and frame assembly
    - 4 transmission-to-engine bolts and tighten to 55 ft. lbs. (75 Nm)
    - Right engine mount bracket to the engine
    - Torque converter to the flywheel and tighten the bolts to 35 ft. lbs. (47 Nm)
    - Flywheel inspection cover
    - Left exhaust manifold
    - Left engine mount and bracket
15. Remove the engine lift from the engine.
    - Left engine-to-transmission brace
    - Lower engine-to-transmission brace
    - Right side engine-to-transmission brace
    - Right side coolant pipes to the engine
    - Power steering lines to the pump and the pump to the engine
    - Idler pulley and tighten the bolts to 37 ft. lbs. (50 Nm)
    - A/C compressor and the brackets
    - Alternator and bracket
    - Power steering pump pulley
    - Accessory drive belt
    - Exhaust manifold rear pipe
    - Engine wire harness
    - Park/Neutral switch and bracket
    - Coolant crossover pipe
16. Position the powertrain assembly under the body and lower the body onto the powertrain assembly.
    - Frame-to-body bolts and tighten to 74 ft. lbs. (100 Nm)
    - ABS modulator
    - A/C manifold to the A/C compressor
    - Battery wire to the alternator
    - Fuel line bundle to the transmission
    - Steering coupling to the steering gear
    - Catalytic converter
    - Intermediate pipe hangers
    - Post converter O_2sensor electrical connection and heat shield
    - Drive axle to the hubs
    - Tie rods
    - Height sensor links
    - Stabilizer shaft links
    - Tire and the wheels
17. Lower the vehicle.
    - 3 relays to the radiator support
    - 2 hoses to the power steering cooler
    - Engine harness to under the hood fuse panel
    - Engine harness retainer to the brake booster
    - Engine harness left front wheel speed sensor, cruise control motor and MAP sensor electrical connections
    - Right front wheel speed sensor connection
    - Vacuum hose to the **T** fitting
    - 3 electrical connections to the A/C line near the cowl
18. Push the engine wire harness through the cowl and attach pass through to the cowl.
    - 2 engine wire harness electrical connections and the vacuum line
    - Right sound insulator
    - PCM case, the PCM and the electrical connection
    - Positive battery cable to the battery and body retainer
    - Negative battery cable to the left cylinder head
    - Coolant hoses to the pipes at the front of the engine
    - Coolant reservoir, hoses and electrical connector
    - Fuel lines
    - Vacuum hose to the brake booster
    - Shift cable to the manual shift lever and bracket
    - Throttle and cruise cables to the throttle body
    - Cooling fans
    - Transmission cooler lines to the radiator
    - Forward discriminating sensor to the radiator support
    - Upper and lower radiator hoses
    - Engine mount struts and adjust to zero preload
    - Radiator upper cover
    - Engine cover
    - Air cleaner assembly
    - Aiir intake duct
19. Fill cooling system.
20. Recharge the A/C system.
21. Enable the air bag system
22. Fill the cooling system.
23. Fill the crankcase.
24. Bleed the power steering system.

2000–01 Models

1. Before servicing the vehicle, refer to the precautions in the beginning of this section.
2. Disable the air bag system.
3. Recover the air-conditioning system.
4. Drain the cooling system.
5. Recover the refrigerant from the air conditioning system.
6. Relieve the fuel pressure.
7. Drain the crankcase.
8. Remove or disconnect the following:
    - Upper filler panel
    - Connectors from the Powertrain Control Module (PCM)
    - Air cleaner assembly
    - 2 nuts from the intake manifold sight shield
    - Sight shield from the engine
    - Inlet radiator hose from the engine and the outlet radiator hose from the thermostat housing
    - Upper transaxle oil cooler line retaining bolt from the fan shroud
    - Plastic cap off the upper transaxle oil cooler line fitting
    - Internal spring clip from the transaxle oil cooler fitting
    - Upper transaxle oil cooler line from the radiator
    - Lower transaxle oil cooler line from the radiator
    - Surge tank inlet pipe from the engine

CAUTION

In order to avoid possible injury or vehicle damage, always replace the accelerator cable with a NEW cable whenever you remove the engine from the vehicle. Also to avoid cruise control cable damage, position the cable out of the way when removing or installing the engine. Never pry or lean against the cruise control cable or kink the cable. If the cable is damaged it must be replaced.

- Accelerator cable from the throttle body
- Cruise control cable from the throttle body
- Heater hoses from the engine
- 2 brake pipes from the master cylinder and plug the lines to prevent system contamination
- Bracket and shift cable from the manual shift lever and set aside
- Vacuum hose from the brake booster
- Evaporative Emission (EVAP) hose from the EVAP purge valve
- Fuel inlet and return fittings from the fuel rail
- Engine ground cable from the body frame rail
- Main engine harness
- Electrical harness from the underhood fuse block
- Right and left strut tower bolts
- Wheels
- Real time dampening sensor links from the lower control arms
- Both wheel speed sensor connectors
- Both road sensing suspension electrical connectors
- 2 brake pipes from both front frame brackets
- 2 rear brake pipes at the rear of the engine frame
- Air deflector

➡The engine oil cooler quick connect fittings must be replaced whenever they are removed from the engine oil filter adapter.

- Engine oil cooler fittings from the engine oil filter adapter, with the oil lines still attached, and position aside
- Dust cover from the quick connect joint
- Internal spring clip from the engine oil cooler fittings
- Engine oil cooler lines from the cooler fittings
- A/C discharge and suction hoses from the compressor

➡Make sure the wheels of the vehicle are in the straight ahead position and the steering column in the Lock position before disconnecting the steering column or intermediate shaft from the steering gear. Failure to do so will cause the coil assembly in the steering column to become uncentered which will cause damage to the coil assembly.

9. Lock the steering column by removing the ignition key from the lock cylinder.
10. Remove or disconnect the following:
    - Intermediate shaft pinch bolt
    - Intermediate shaft from the steering gear
    - Wheel speed and brake wear sensors at the right and left strut towers
    - Post Heated Oxygen (HO_2S) sensor at the sensor pigtail
    - Exhaust front pipe
    - Brace between the engine oil pan and the transaxle case
    - Torque converter cover
    - Torque converter-to-flywheel bolts
11. Position a powertrain support dolly under the engine assembly. If a powertrain support dolly is unavailable, support the powertrain with four suitable jackstands.
    - Nut attaching the right engine mount to the engine mount bracket
    - Nut attaching the left transaxle mount to the transaxle mount bracket

CAUTION

To avoid any vehicle damage, serious personal injury or death when major components are removed from the vehicle and the vehicle is supported by a hoist, support the vehicle with jack stands at the opposite end from which the components are being removed.

12. Secure the front hoist pads to the vehicle.
13. Remove the 6 frame mount bolts (3), raise the vehicle slowly and attach an engine lift to the engine.
    - Heater pipes
    - Nut attaching the coil cassette ground wire to the cylinder head
    - Engine wiring harness from the engine
    - Bolt attaching the crossover pipe to the cylinder head
    - Power steering pump
    - Bolt attaching the rear transaxle brace to the transaxle
    - Nuts attaching the rear transaxle brace to the cylinder head
    - Rear transaxle brace
    - Bolt attaching the front transaxle brace to the cylinder head
    - Bolts attaching the transaxle brace to the transaxle
    - Bolts securing the transaxle brace to the engine
    - Nut attaching the front engine mount to the engine frame
    - Bolts attaching the engine to the transaxle
    - Engine from the frame and transaxle assembly

To install:

14. Attach the engine to the frame and transaxle assembly..
15. Install or connect the following:
    - Bolts attaching the engine to the transaxle and tighten to 55 ft. lbs. (75 Nm)
    - Nut attaching the front engine mount to the engine frame and tighten to 52 ft. Lbs. (70 Nm)
    - 4 bolts attaching the transaxle brace to the engine and tighten to 37 ft. lbs. (50 Nm)
    - Bolt attaching the front transaxle brace to the cylinder head and tighten the bolts to 37 ft. lbs. (50 Nm)
    - Rear transaxle brace and tighten the bolt to 37 ft. lbs. (50 Nm) and the nuts to 37 ft. lbs. (50 Nm)
    - Bolt attaching the crossover pipe to the cylinder head and tighten the bolt to 18 ft. lbs. (25 Nm)
    - Engine wiring harness to the engine
    - Nut attaching the coil cassette ground wire to the cylinder head and tighten to 13 ft. lbs. (17 Nm)
    - Heater pipes
    - Power steering pump
16. Remove the engine lift from the engine. Position the engine/transaxle assembly under the vehicle. Carefully lower the vehicle over the engine/transaxle

assembly. Place dowel pins in the alignment holes, align the engine frame with the vehicle.

- 6 frame mounting bolts and tighten to 141 ft. lbs. (191 Nm)
- Nut attaching the left transaxle mount to the transaxle mount bracket and tighten to 59 ft. lbs. (80 Nm)
- Nut securing the right engine mount to the engine mount bracket and tighten to 59 ft. lbs. (80 Nm)

17. Remove the powertrain support dolly from under the engine assembly.

- Torque converter to flywheel bolts and tighten to 44 ft. lbs. (60 Nm)
- Torque converter cover
- Oil pan to transaxle brace bolts and tighten and tighten to 37 ft. lbs. (50 Nm)
- Exhaust front pipe
- Post HO_2S sensor pigtail
- Wheel speed and brake wear sensors at the strut towers
- Intermediate shaft to the steering gear
- Pinch bolt and tighten to and tighten to 33 ft. lbs. (45 Nm)
- A/C suction and discharge hoses to the compressor and tighten the nuts to 15 ft. lbs. (20 Nm)

➡The engine oil cooler quick connect fittings must be replaced whenever they are removed from the engine oil filter adapter.

- Engine oil cooler quick connect fittings to the engine oil filter adapter and tighten to 13 ft. lbs. (18 Nm)

18. Push the engine oil cooler lines fully into the oil cooler quick connect fittings, until a click is heard. Slide the dust cover over the quick connect joint.

- Air deflector
- Dampening sensor links
- Road sensing suspension connectors
- Wheel speed sensor connectors
- Brake pipes to both front frame brackets and tighten to 11 ft. lbs. (15 Nm)
- Rear brake pipes at the rear of the engine frame and tighten to 11 ft. lbs. (15 Nm)
- Struts to the strut towers
- Wheels
- Strut tower bolts and washers and tighten to 30 ft. lbs. (40 Nm)
- Engine ground cable
- Main engine harness
- Electrical harness to the underhood fuse block
- Shift cable to the manual shift lever and bracket
- Vacuum hose to the brake booster
- EVAP hose to the EVAP purge valve
- Fuel lines
- Heater hoses
- Brake pipes to the master cylinder and tighten the brake pipes to 11 ft. lbs. (15 Nm)
- Surge tank inlet pipe
- NEW accelerator control cable
- Cruise control cable to the throttle body
- Lower transaxle oil cooler line to the radiator and tighten to 26 ft. lbs. (35 Nm)
- Spring clip to the fitting. Push the upper transaxle oil cooler line into the radiator quick connect fitting, until a "click" is heard. Tug gently on the cooler line to ensure proper retention. Slide the plastic cap (1) over the quick connect joint.
- Upper transaxle oil cooler line retaining bolt to the fan shroud and tighten to 53 inch lbs. (6 Nm)
- Inlet and outlet radiator hoses
- Intake manifold sight shield and tighten the nuts to 27 inch lbs. (3 Nm)
- Air cleaner assembly
- PCM connections
- Upper filler panel

19. Fill the cooling system
20. Recharge the A/C Refrigerant System
21. Connect the negative battery cable.
22. Fill the crankcase.
23. Perform the Crankcase Position (CKP) system variation learn procedure.
24. Bleed the brake system
25. Align the vehicle.

Water Pump

REMOVAL & INSTALLATION

1. Before servicing the vehicle, refer to the precautions in the beginning of this section.
2. Drain the coolant.
3. Remove or disconnect the following:
   - Negative battery cable
   - Upper fill panel, if equipped
   - AIR injection control valve, if equipped
   - Air cleaner
   - Oil level indicator tube nut, if necessary
   - Water pump belt cover
   - Water pump drive belt
   - Lower radiator hose

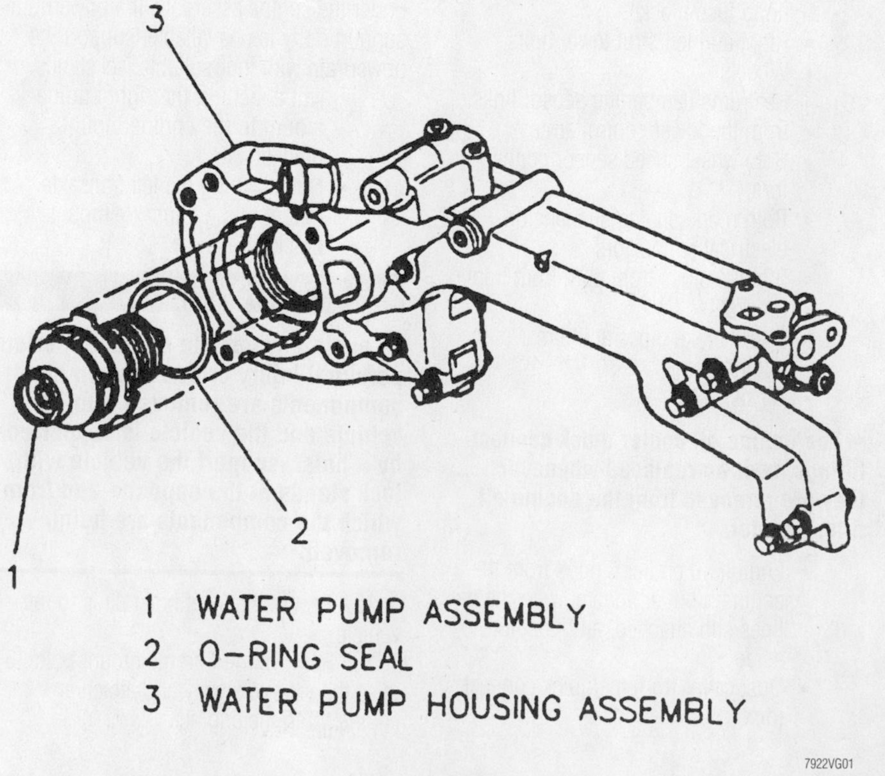

To ensure proper operation, be sure to install a new O-ring

- Thermostat bypass hose from the coolant inlet housing
- Water pump cover

4. Remove the water pump by rotating it clockwise using special Tool J 38816-1A (water pump remover/installer).

To install:

5. Install the Water pump with a new O-ring, by turning it counterclockwise using the J 38816-1A, until it stops. Torque the water pump to 73 ft. lbs. (100 Nm).
6. Install or connect the following:
   - Water pump cover. Torque the bolts to 89 inch lbs. (10 Nm).
   - Lower radiator hose
   - Thermostat bypass hose to the coolant inlet housing
   - Water pump drive belt
   - Water pump belt cover
   - Oil level indicator tube nut, if necessary
   - Air cleaner
   - AIR injection control valve, if equipped
   - Upper fill panel, if equipped
   - Negative battery cable
7. Refill and bleed the cooling system.
8. Run the engine and check for leaks.

Cylinder Head

REMOVAL & INSTALLATION

1998–99 Models

➡The manufacturer recommends that the entire powertrain be removed from the vehicle before removing the cylinder heads.

1. Before servicing the vehicle, refer to the precautions in the beginning of this section.
2. Drain the cooling system.
3. Properly relieve the fuel system pressure.
4. Remove or disconnect the following:
   - Negative battery cable
   - Powertrain assembly from the vehicle
   - Intake manifold
   - Camshaft covers
   - Crankshaft balancer
   - Timing chain front cover
   - Oil pump

✲✲ WARNING

Align all timing marks before performing the next step.

- Timing chain tensioner
- Camshaft sprockets

➡The timing chain remains in the chain case.

- Timing chain guides. Access for the retaining screw is through the plugs at the front of the cylinder head.
- Water crossover
- Exhaust manifolds
- Cylinder head bolts by reversing the torque sequence
- Cylinder head

✲✲ WARNING

With the camshafts remaining in the cylinder head some valves will be open at all times. Do not rest the cylinder head on a flat surface with the cylinder face down, or valve damage will result.

✲✲ WARNING

Be careful when cleaning aluminum gasket surfaces to prevent damage to the sealing surfaces.

5. Check the cylinder head for warpage; it must be less than 0.002 inches (0.05mm). If the cylinder head was resurfaced, the dimension between the combustion chamber gauge pad and the deck surface must be at least 10.5mm.

To install:

6. Install the cylinder head with a new gasket. Lubricate the bolts with engine oil prior to installation.
7. Torque the M11 bolts in sequence as follows:
   a. Step 1: 22 ft. lbs. (30 Nm), plus an additional 60 degrees.
   b. Step 2: Turn an additional 60 degrees.
   c. Step 3: Turn an additional 60 degrees (total 180 degrees).

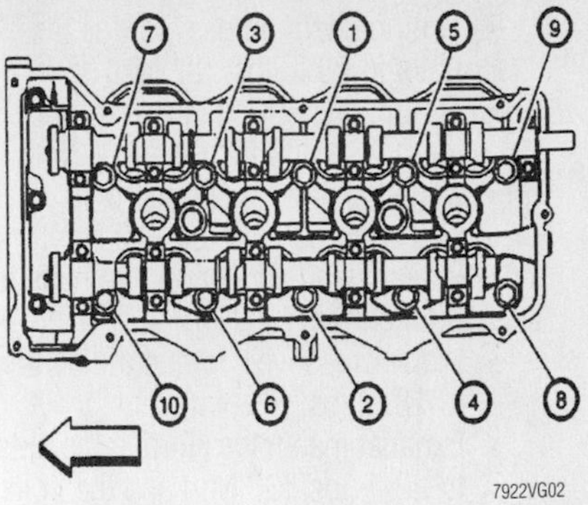

Left cylinder head bolt torque sequence—1998–99 4.6L engine

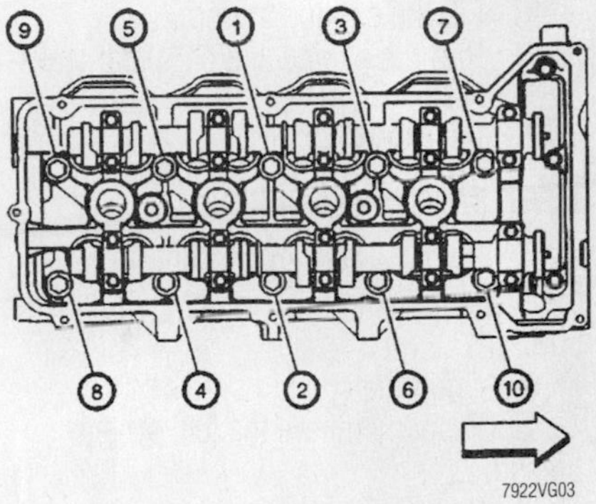

Right cylinder head bolt torque sequence—1998–99 4.6L engine

8. Torque the M6 bolts to 106 inch lbs. (12 Nm).
9. Install or connect the following:
   - Camshaft covers. Torque the screws to 89 inch lbs. (10 Nm).
   - Oil pump. Torque the bolts to 20 ft. lbs. (27 Nm).
   - Crankshaft balancer. Torque the bolt to 37 ft. lbs. (50 Nm) plus an additional 120 degrees.
   - Camshaft sprockets. Torque the bolts to 90 ft. lbs. (120 Nm).
   - Timing chain tensioner. Torque the bolts to 20 ft. lbs. (27 Nm).

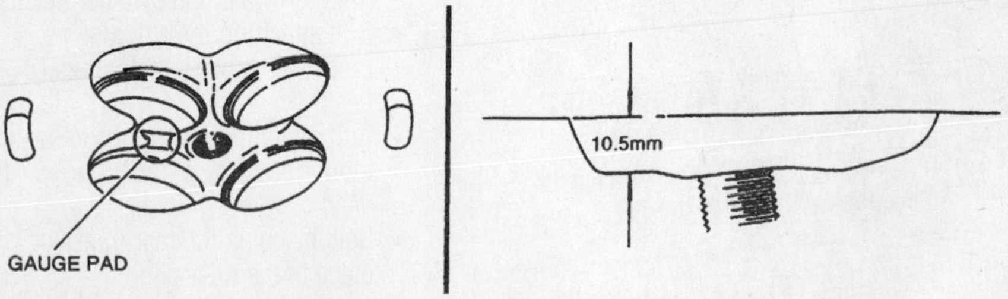

Minimum head resurface dimension

Timing belt service is covered in Section 3 of this manual

- Timing chain guides. Torque the bolts to 20 ft. lbs. (27 Nm).
- Timing chain front cover. Torque the bolts to 89 inch lbs. (10 Nm).
- Camshaft covers. Torque the screws to 89 inch lbs. (10 Nm).
- Intake manifold. Torque the bolts to 89 inch lbs. (10 Nm).
- Water crossover. Torque the bolts to 18 ft. lbs. (25 Nm).
- Exhaust manifolds. Torque the nuts to 22 ft. lbs. (30 Nm) and the bolts to 18 ft. lbs. (25 Nm).
- Powertrain assembly to the vehicle
- Negative battery cable

10. Fill the cooling system.
11. Run the engine and check for leaks.

2000–01 Models

LEFT SIDE

1. Before servicing the vehicle, refer to the precautions in the beginning of this section.
2. Drain the cooling system.
3. Properly relieve the fuel system pressure.
4. Remove or disconnect the following:
   - Negative battery cable
   - Exhaust manifold
   - Alternator
   - Water crossover
   - Intake manifold
   - Valve cover
   - Timing cover
   - Secondary camshaft drive chain
   - Power steering return hose retaining bolt from the cylinder head
   - 3 M6 external drive bolts from the front portion of the cylinder head
   - 10 M11 internal drive cylinder head bolts
   - Cylinder head and gasket.
5. Clean the head mating surfaces.

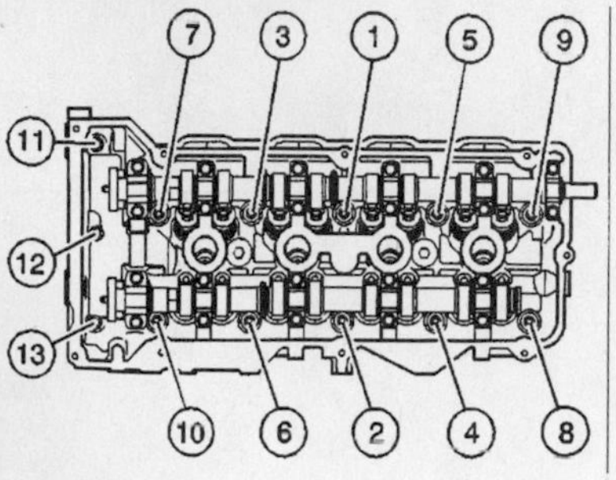

Left cylinder head bolt torque sequence—2000–01 4.6L engine

To install:

6. Install the cylinder head with a new gasket. Lubricate the bolts with engine oil prior to installation.
7. Torque the M11 bolts in sequence as follows:
   a. Step 1: 30 ft. lbs. (40 Nm), plus an additional 70 degrees
   b. Step 2: Turn an additional 60 degrees
   c. Step 3: Turn an additional 60 degrees (total 190 degrees)
8. Torque the M6 bolts to 106 inch lbs. (12 Nm)
9. Install or connect the following:
   - Power steering return hose retaining bolt to the cylinder head and tighten to 37 ft. lbs. (50 Nm)
   - Secondary camshaft drive chain
   - Timing cover
   - Valve cover
   - Intake manifold
   - Water crossover
   - Exhaust manifold
   - Negative battery cable
10. Change the oil and filter.
11. Fill the cooling system.

RIGHT SIDE

1. Before servicing the vehicle, refer to the precautions in the beginning of this section.
2. Drain the cooling system.
3. Properly relieve the fuel system pressure.
4. Remove or disconnect the following:
   - Negative battery cable
   - Exhaust manifold
   - Water crossover
   - Intake manifold
   - Valve cover
   - Timing cover
   - Secondary camshaft drive chain
   - Electrical connector from the Engine Coolant Temperature (ECT) sensor
   - Nut attaching the coil cassette ground wire to the cylinder head
   - Bolt attaching the exhaust crossover pipe to the cylinder head
   - Bolt attaching the right transaxle mount bracket to the cylinder head
   - Bolt attaching the rear transaxle brace to the transaxle
   - Nuts attaching the rear transaxle brace to the cylinder head
   - Rear transaxle brace
   - 3 M6 external drive bolts from the front portion of the cylinder head
   - 10 M11 internal drive cylinder head bolts
   - Cylinder head and gasket
5. Clean the head mating surfaces.

To install:

6. Install the cylinder head with a new gasket. Lubricate the bolts with engine oil prior to installation.
7. Torque the M11 bolts in sequence as follows:
   a. Step 1: 30 ft. lbs. (40 Nm), plus an additional 70 degrees.
   b. Step 2: Turn an additional 60 degrees.
   c. Step 3: Turn an additional 60 degrees (total 190 degrees).
8. Torque the M6 bolts to 106 inch lbs. (12 Nm).
9. Install or connect the following:
   - Rear transaxle brace over the studs located at the rear of the right cylinder head
   - Nuts attaching the rear transaxle brace to the cylinder head, loosely
   - Bolts attaching the rear transaxle braces to the transaxle and tighten to 37 ft. lbs. (50 Nm)
   - Nuts attaching the rear transaxle brace to the cylinder head and tighten 37 ft. lbs. (50 Nm)
   - Electrical connector to the ECT sensor
   - Bolt attaching the exhaust crossover pipe to the cylinder head and tighten to 18 ft. lbs. (25 Nm)
   - Nut attaching the coil cassette ground wire to the cylinder head and tighten to 13 ft. lbs. (17 Nm)
   - Secondary camshaft drive chain
   - Timing cover
   - Valve cover
   - Intake manifold
   - Water crossover
   - Exhaust manifold
   - Negative battery cable

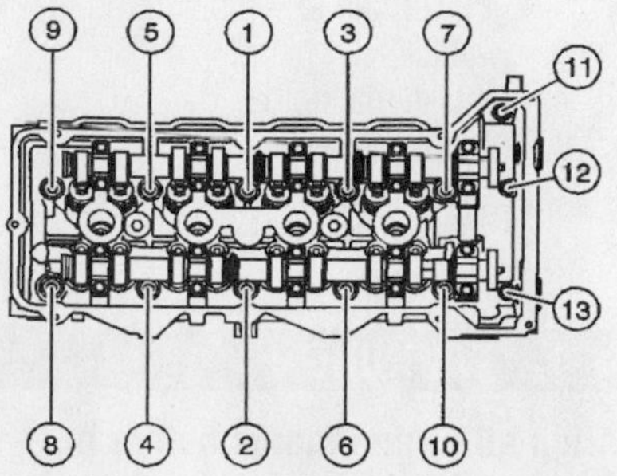

Right cylinder head bolt torque sequence—2000–01 4.6L engine

10. Change the oil and filter.
11. Fill the cooling system.

Intake Manifold

REMOVAL & INSTALLATION

1998–99 Models

1. Before servicing the vehicle, refer to the precautions in the beginning of this section.
2. Relieve the fuel system pressure.
3. Drain the cooling system.
4. Remove or disconnect the following:
   - Negative battery cable
   - Engine cover
   - Air inlet duct from the throttle body
   - Transaxle vent hose
   - Transaxle shift cable at the bracket
   - Throttle Position (TPS) sensor electrical connector
   - Idle Air Control (IAC) valve electrical connector
   - Throttle cable
   - Cruise control cable
   - Spark plug wires
   - Throttle body coolant hoses
   - Surge tank pipe
   - Exhaust Gas Recirculation (EGR) pipe
   - Crankcase ventilation pipe
   - Brake booster vacuum hose from the intake manifold
   - Fuel rail ground wire from the rear cylinder head
   - Quick-disconnect fuel rail fittings
   - Fuel rail bracket from the EGR valve
   - Positive Crankcase Ventilation (PCV) hose from the intake manifold
   - Fuel injector harness connector
   - Purge solenoid harness
   - Throttle body heater hose from the water crossover
   - Alternator coolant inlet hose from the heater outlet pipe
   - Throttle body heater outlet hose
   - Intake manifold

➡The intake manifold carrier gaskets are attached to the intake manifold through a snap-lock feature. When removing the intake manifold, the carrier gaskets will remain attached to the intake manifold. DO NOT replace the intake manifold gaskets after intake manifold removal. The gaskets are reusable. The gaskets should be

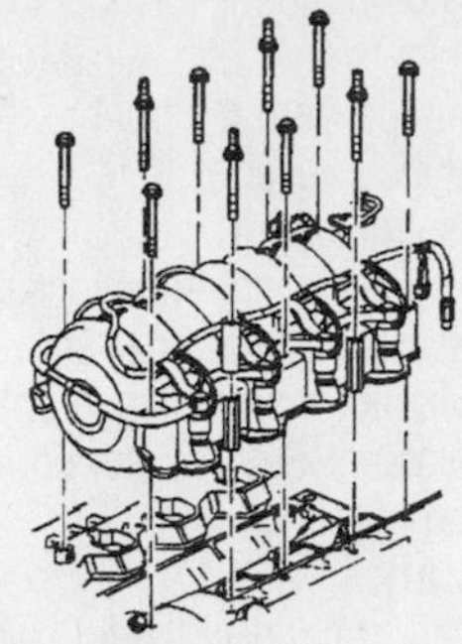

7922VG05

Exploded view of the intake manifold mounting—4.6L engine

replaced if the plastic housing or the rubber seals are damaged.

To install:

5. Intake manifold. Torque the bolts and studs to 89 inch lbs. (10 Nm). Start at the center of the manifold and work outward in a circular pattern. DO NOT torque the intake manifold bolts when the engine is HOT or at operating temperature.
6. Install or connect the following:
   - Fuel injector harness connector
   - PCV hose
   - Fuel bracket at the EGR valve
   - Fuel lines to the fuel rail
   - Fuel rail ground wire. Torque the bolt to 97 inch lbs. (11 Nm).
   - Brake booster vacuum hose
   - EGR pipe. Torque the bolt to 21 ft. lbs. (28 Nm).
   - PCV pipe at the throttle body spacer
   - Throttle body coolant hoses
   - Surge tank pipe
   - Alternator coolant hoses
   - Purge solenoid harness. Torque the retaining bolt to 106 inch lbs. (12 Nm).
   - Spark plug wires
   - Throttle cable
   - Cruise control cable
   - TPS sensor electrical connector
   - IAC valve electrical connector
   - Transaxle vent hose
   - Transaxle shift cable at the bracket
   - Air inlet duct
   - Negative battery cable
7. Turn the ignition switch to **RUN** to pressurize the fuel system and inspect for leaks.
8. Install the engine cover. Torque the cover nuts to 89 inch lbs. (10 Nm).
9. Fill and bleed the cooling system.
10. Road test the vehicle.

2000–01 Models

1. Before servicing the vehicle, refer to the precautions in the beginning of this section.
2. Relieve the fuel system pressure.
3. Drain the cooling system.
4. Remove or disconnect the following:
   - Negative battery cable
   - Intake manifold heat shield
   - Sight shield
   - Coil module connectors from the coil modules located on the valve covers
   - Positive Crankcase Ventilation (PCV) hose and valve from the valve cover
   - Fuel regulator vacuum tube
   - Vacuum tubes from the AIR solenoid
   - Fuel inlet and return lines
   - Fuel rail bracket retaining nut at the rear lift bracket
   - 2 pushnuts attaching the engine coolant heater wire and set it aside, if equipped
   - Alternator coolant pipe/surge tank pipe and position aside from the fuel rail studs
   - Fuel injector electrical connections
   - Fuel rail and injectors
   - Plenum duct clamp at the rear of the intake manifold, loosen
   - Intake manifold bolts and the manifold
5. Clean the manifold mating surfaces.

To install:

6. Grease the inside edge of the rubber plenum duct.
7. Position the intake manifold by performing the following sub-steps:
   a. Place the rear of the intake manifold into the plenum duct.
   b. Place the front of the intake manifold downward on to the cylinder heads.
8. Install the intake manifold bolts and torque to 89 inch lbs. (10 Nm). Start at the center of the manifold and work outward in a circular pattern. DO NOT torque the intake manifold bolts when the engine is HOT or at operating temperature.
9. Make sure the plenum duct is fully attached to the rear of the intake manifold and tighten the plenum duct clamp to 20 inch lbs. (2.25 Nm).
10. Install or connect the following:
   - Fuel rail and injectors and tighten the rail retainers to 35 inch lbs. (4 Nm)
   - Fuel injector electrical connections

- Alternator coolant pipe/surge tank pipe and position onto the fuel rail studs
- Engine coolant heater wire onto the studs and install the pushnuts, if equipped
- Fuel rail bracket retaining nut at the rear lift bracket and tighten to 35 inch lbs. (4 Nm)
- Fuel inlet and return lines
- Vacuum tubes to the AIR solenoid
- Fuel regulator vacuum tube
- Positive Crankcase Ventilation (PCV) valve and hose to the valve cover
- Coil module connectors to the coil modules located on the valve covers
- Sight shield
- Intake manifold heat shield
- Negative battery cable

11. Fill and bleed the cooling system.

Exhaust Manifold

REMOVAL & INSTALLATION

Left Side

1998–99 MODELS

1. Before servicing the vehicle, refer to the precautions in the beginning of this section.
2. Remove or disconnect the following:
   - Negative battery cable
   - Radiator cover panel
   - Air cleaner assembly
   - Left and right engine torque struts
   - Engine cooling fans
3. Support the engine.
   - Engine mount-to-engine cradle nuts
   - Engine mount bracket-to-crankcase bolts
   - Engine mount bracket-to-cylinder head bolts

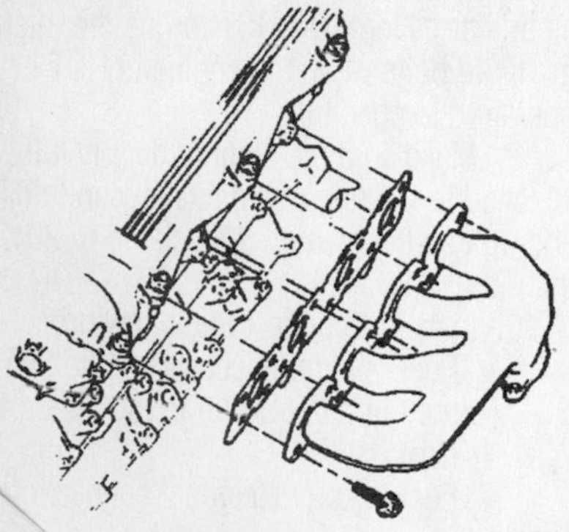

9300VG03

led view of the left exhaust manifold

   - Engine mount-to-mount bracket nuts
   - Y-pipe from the front of the catalytic converter
4. Raise the engine.
   - Engine mount and bracket
   - Rear alternator bracket
   - Manifold outlet flange bolts
   - Oxygen (O_2S) sensor connector
   - Exhaust manifold

To install:

5. Install or connect the following:
   - Exhaust manifold with a new gasket by inserting the outlet pipe partially into the exhaust crossover pipe. Torque all the bolts to 18 ft. lbs. (25 Nm).
   - O_2S sensor connector
   - Rear alternator bracket. Torque the crankcase bolts to 44 ft. lbs. (60 Nm), and the alternator bolts to 22 ft. lbs. (30 Nm).
   - Manifold outlet flange bolts. Torque the bolts to 25 ft. lbs. (30 Nm).
6. Lower the engine into position. Guide the engine mount studs into the cradle holes and loosely install the fasteners.
7. Install or connect the following:
   - Engine mount-to-engine cradle nuts. Torque the nuts to 22 ft. lbs. (30 Nm).
   - Engine mount bracket-to-crankcase bolts. Torque the bolts to 22 ft. lbs. (30 Nm).
   - Engine mount bracket-to-cylinder head bolts. Torque the bolts to 22 ft. lbs. (30 Nm).
   - Engine mount-to-mount bracket nuts. Torque the nuts to 22 ft. lbs. (30 Nm).
   - Converter-to-exhaust Y-pipe. Torque the bolts to 20 ft. lbs. (27 Nm).
8. Remove the engine support fixture.
   - Engine cooling fans
   - Air cleaner assembly
   - Left and right engine torque struts. Torque the bolts to 44 ft. lbs. (60 Nm).
   - Radiator cover panel
   - Negative battery cable
9. Run the engine and check for exhaust leaks.

2000–01 MODELS

1. Before servicing the vehicle, refer to the precautions in the beginning of this section.
2. Remove or disconnect the following:
   - Negative battery cable
   - 2 nuts attaching the AIR tube to the exhaust manifold
   - Engine mount and bracket
   - 2 bolts at the manifold outlet flange
   - Oxygen (O_2S) sensor, if necessary
   - Exhaust manifold bolts
   - Manifold and the gasket. Discard the gasket.

To install:

3. Install or connect the following:
   - Exhaust manifold by inserting the outlet pipe partially into the exhaust crossover pipe
   - Exhaust manifold with a new gasket. Torque the nuts to 25 ft. lbs. (30 Nm).
   - O_2S sensor. Torque the sensor to 30 ft. lbs. (40 Nm).
   - Engine mount and bracket
   - 2 nuts attaching the AIR tube to the exhaust manifold and tighten to 106 inch lbs. (12 Nm)
   - Negative battery cable

Right Side

1998–99 MODELS

1. Before servicing the vehicle, refer to the precautions in the beginning of this section.
2. Remove or disconnect the following:
   - Negative battery cable
   - Oxygen (O_2S) sensor and the harness clip
   - Y-pipe from the front of the catalytic converter
   - Suspension position sensor from the lower control arm on both sides
   - Intermediate shaft from the steering gear
3. Place a support below the rear crossmember of the engine cradle and remove the 4 cradle to body bolts.
4. Lower the rear of the engine cradle.
   - Y-pipe from the exhaust crossover and manifold
   - Exhaust manifold

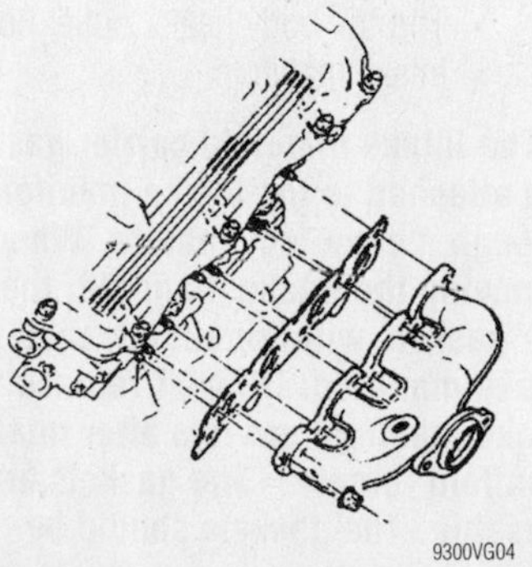

9300VG04

Exploded view of the right exhaust manifold

To install:

5. Install or connect the following:
   - O_2S sensor. Torque the sensor to 30 ft. lbs. (40 Nm).
   - Exhaust manifold with a new gasket. Torque the nuts to 25 ft. lbs. (30 Nm).
   - Y-pipe using 4 new bolts. Torque the M10 bolts to 35 ft. lbs. (50 Nm) and the M8 bolts to 25 ft. lbs. (30 Nm).
   - Engine cradle. Torque the bolts to 75 ft. lbs. (100 Nm).
   - Intermediate shaft to the steering gear. Torque the bolts to 35 ft. lbs. (50 Nm).
   - Y-pipe to the catalytic converter. Torque the bolts to 35 ft. lbs. (50 Nm).
   - Suspension position sensors to the lower control arms
   - O_2S sensor harness clip
   - Negative battery cable
6. Run the engine and check for exhaust leaks.

1998–99 MODELS

1. Before servicing the vehicle, refer to the precautions in the beginning of this section.
2. Remove or disconnect the following:
   - Negative battery cable
   - Oxygen (O_2S) sensor
   - 2 nuts attaching the AIR tube to the exhaust manifold
   - Exhaust front pipe
   - Real time dampening sensor links from the lower control arms
   - Intermediate shaft from the steering gear
3. Support the rear cross member of the engine cradle with a tall screw jack.
   - 4 rearward cradle-to-body bolts, then lower the rear of the engine cradle
   - Right side cylinder head to the transaxle brace
   - Exhaust manifold nuts
   - Manifold and the gasket

To install:

4. Install or connect the following:
   - Exhaust manifold with a new gasket. Torque the nuts to 25 ft. lbs. (30 Nm).
   - Right side cylinder head to the transaxle brace and tighten the bolt and nut to 35 ft. lbs. (47 Nm)
5. Raise the engine cradle into position.
   - 4 rearward cradle-to-body bolts and tighten to 141 ft. lbs. (191 Nm)
6. Remove the screw jack.
   - Itermediate shaft to the steering gear and tighten the pinch bolt to 35 ft. lbs. (47 Nm)
   - Real time dampening sensor links to the lower control arms
   - Exhaust front pipe
   - New gasket to the AIR tube
   - 2 nuts attaching the AIR tube to the exhaust manifold and tighten to 106 inch lbs. (12 Nm)
   - O_2S) sensor
   - Negative battery cable

Camshaft and Valve Lifters

REMOVAL & INSTALLATION

1998–99 Models

LEFT SIDE

1. Before servicing the vehicle, refer to the precautions in the beginning of this section.
2. Drain the coolant.
3. Disconnect the negative battery cable.
4. Remove the camshaft cover.
5. Secure the camshaft sprocket to the timing chain by installing tie-wraps through the camshaft sprocket holes. Use 4 tie-wraps per sprocket.

➡The sprocket/chain relationship must be maintained throughout this procedure or camshaft timing will be lost and require further engine disassembly to retime.

6. Working from behind the sprockets, install Camshaft Chain Holder J-38222 so that it is positioned between the chain tensioner and chain guide. Apply tension to the tool by tightening the tension adjusting screw.
7. Remove or disconnect the following:
   - Camshaft sprocket bolts
   - Sprockets off the camshafts

➡Alternately loosen the camshaft bearing cap screws a few turns at a time until all valve spring pressure has been released.

   - Camshaft bolts and caps
   - Camshaft
   - Valve lifters

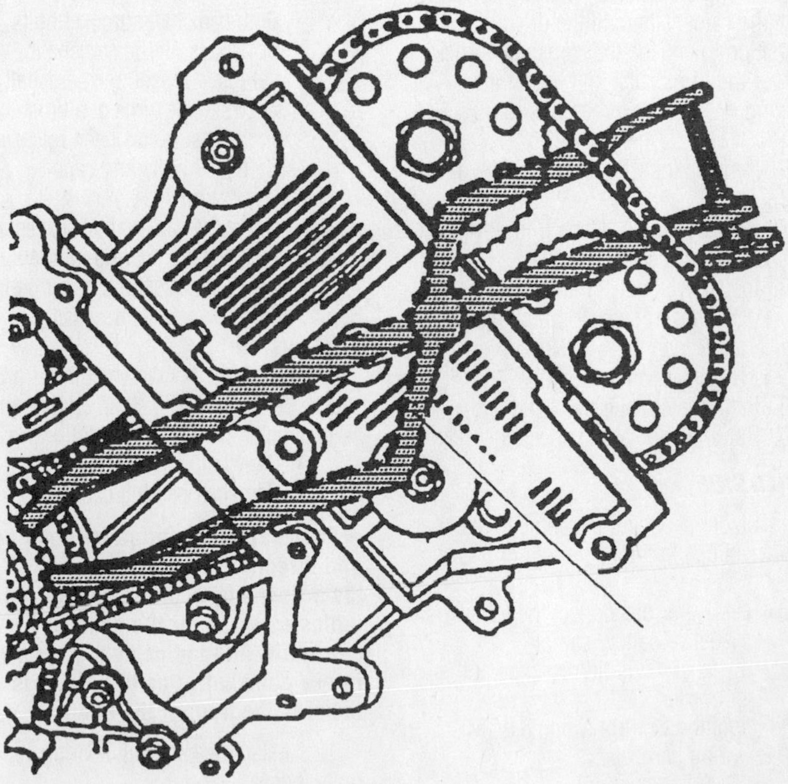

9300VG01

Use the Camshaft Chain Holder tool J-38822 to keep the chain in position while removing the camshafts

8. Inspect the lifters for wear and/or damage; replace as necessary.

9. Inspect the camshaft for excessive lobe wear. Check the bearing journals, making sure they are not scored or burned. Replace the camshaft as necessary.

To install:

10. Install the valve lifters in the same bore from which they were removed.

11. Lubricate the camshaft lobes and the camshaft journals with clean engine oil.

12. Install or connect the following:
- Camshaft
- Camshaft bearing caps

➡Each cap is identified for position and direction. The arrow points towards the front of the engine. AnE indicates a cap for the exhaust camshaft. AnI indicates a cap for the intake camshaft. Position No. 1 is towards the front of the engine.

13. Install the camshaft bearing cap bolts loosely.

14. Alternately, tighten the camshaft bearing cap bolts a few turns at a time against valve spring pressure until all the bolts are snug. Torque the bolts to 106 inch lbs. (12 Nm).

15. Using the hex cast into the camshaft, rotate the camshafts until the drive pins are in position to engage the camshaft sprockets over the camshafts, and install the retaining bolts. Torque the bolts to 90 ft. lbs. (120 Nm).

16. Remove the Chain Holder Tool J-38222.

17. Remove the tie-wraps from the camshaft sprockets.

18. Install or connect the following:
- Camshaft cover. Torque the screws to 89 inch lbs. (10 Nm).
- Negative battery cable

19. Refill the cooling system.

20. Run the engine and check for leaks.

RIGHT SIDE

1. Before servicing the vehicle, refer to the precautions in the beginning of this section.

2. Remove or disconnect the following:
- Negative battery cable
- Exhaust manifold rear pipe at the converter
- Ignition Control Module (ICM) wiring harnesses
- ICM
- Spark plug wires on the right bank
- Positive Crankcase Ventilation (PCV) valve
- Purge canister solenoid from the rear of the camshaft cover
- Wiring harness from the camshaft cover
- Camshaft cover screws

3. Support the front of the engine cradle.
- Mounting screws at the front of the cradle
- Right and left torque struts

4. Lower the engine cradle or raise the vehicle to provide clearance at the rear of the engine compartment.

5. Remove the camshaft cover.

➡The spark plug seals may be reused if undamaged.

6. Secure the camshaft sprocket to the timing chain by installing tie-wraps through the camshaft sprocket holes. Use 4 tie-wraps per sprocket.

➡The sprocket/chain relationship must be maintained throughout this procedure or camshaft timing will be lost and require further engine disassembly to retime.

7. Working from behind the sprockets, install Camshaft Chain Holder J-38222 so that it is positioned between the chain tensioner and chain guide. Apply tension to the tool by tightening the tension adjusting screw.
- Both camshaft sprocket bolts
- Sprockets off the camshafts

8. Alternately loosen the camshaft bearing cap screws a few turns at a time until all valve spring pressure has been released.
- Camshaft bolts and caps
- Camshaft

9. Inspect the camshaft for excessive lobe wear. Check the bearing journals, making sure they are not scored or burned. Replace the camshaft, as necessary.

To install:

10. Lubricate the camshaft lobes and camshaft journals with clean engine oil.

11. Install or connect the following:
- Camshaft
- Camshaft bearing caps

➡Each cap is identified for position and direction. The arrow points towards the front of the engine. An E indicates a cap for the exhaust camshaft. An I indicates a cap for the intake camshaft. Position No. 1 is towards the front of the engine.

12. Install the camshaft bearing cap bolts loosely.

13. Alternately, tighten the camshaft bearing cap bolts a few turns at a time against valve spring pressure until all the bolts are snug. Torque the bolts to 106 inch lbs. (12 Nm).

14. Using the hex cast into the camshaft, rotate the camshafts until the drive pins are in position to engage the camshaft sprockets over the camshafts and install the retaining bolts. Torque the bolts to 90 ft. lbs. (120 Nm).

15. Remove the Chain Holder Tool J-38222.

16. Remove the tie-wraps from the camshaft sprockets.

17. Install or connect the following:
- Spark plug and camshaft cover seals
- Camshaft cover. Torque the screws to 84 inch lbs. (10 Nm).
- Engine cradle. Torque the bolts to 75 ft. lbs. (100 Nm).

18. Install the right and left torque struts and torque the bolts to 45 ft. lbs. (60 Nm).

➡It is important during installation that the engine torque struts are not preloaded in their installed position. Adjustment is provided at the point the strut fastens to the core support bracket. Be sure this bolt is loose during assembly.

- Wiring harness to the camshaft cover
- Purge canister solenoid
- PCV valve
- ICM
- Spark plug wires on the right bank
- ICM wiring harnesses
- Negative battery cable
- Exhaust manifold rear pipe to the converter. Torque the bolts to 20 ft. lbs. (27 Nm).

19. Run the engine and check for leaks.

2000–01 Models

1. Before servicing the vehicle, refer to the precautions in the beginning of this section.

2. Remove both camshaft covers.

3. Secure the camshaft sprockets to the timing chain by installing tie wraps through the camshaft sprocket holes.

4. Working behind the sprockets, install camshaft Chain Holder J38822 so that it is positioned between the chain tensioner and guide. Apply tension to the tool by tightening the tension adjusting screw.

5. Remove or disconnect the following:
- Both camshaft sprocket bolts
- Both camshaft sprockets

6. Alternately loosen the camshaft bearing cap screws a few turns at a time until all valve spring pressure has been released.

7. Remove the bolts and bearing caps.

8. Remove the camshaft.

To install:

9. Lubricate the camshaft lobes and camshaft journals with clean engine oil.
10. Position the camshaft bearing caps in the cylinder head and loosely install the bolts.
11. Tighten the camshaft bearing cap bolts in the sequence shown, to 44 inch lbs. (5 Nm), plus 30 degrees.
12. Using the hex cast into the camshaft, position the camshafts to accept the camshaft sprockets.

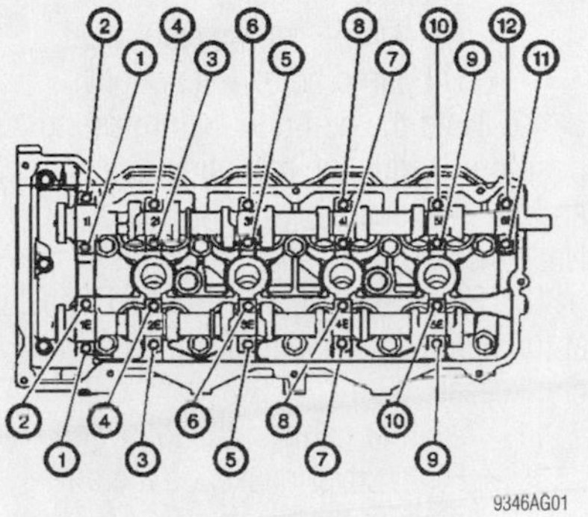

Torque sequence for the camshaft bearing cap bolts.

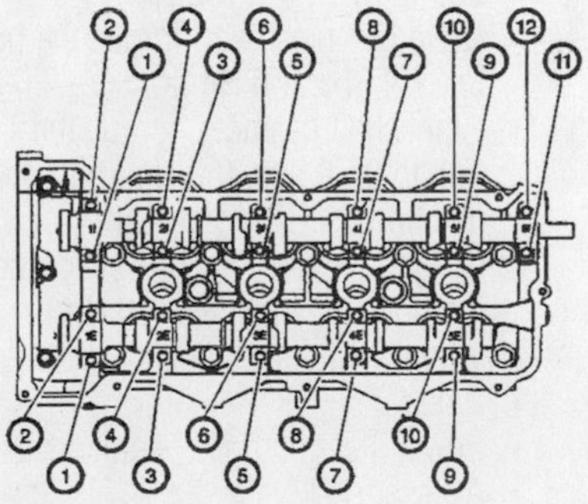

Left cylinder head camshaft bearing cap torque sequence—4.6L engine

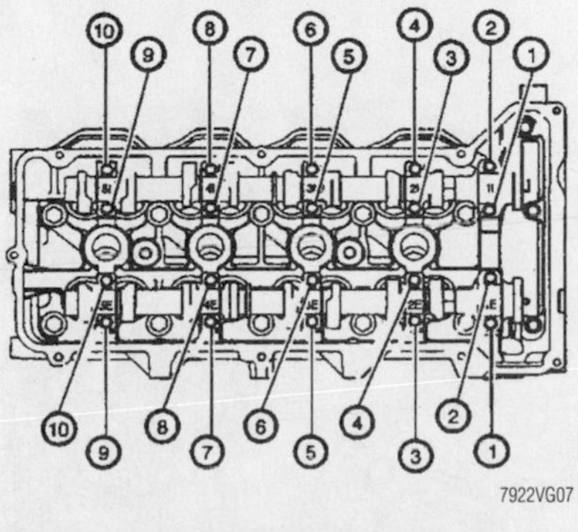

Right cylinder head camshaft bearing cap torque sequence—4.6L engine

13. Install the camshaft sprockets onto the camshafts. Torque the bolts to 89 ft. lbs. (120 Nm).
14. Remove the Chain Holder J38822 and the tie wraps.
15. Install the left and right camshaft covers.

Valve Lash

ADJUSTMENT

The valve clearance cannot be adjusted. The engines in this section are equipped with hydraulic lifters and no adjustment is possible.

Starter Motor

REMOVAL & INSTALLATION

1. Before servicing the vehicle, refer to the precautions in the beginning of this section.
2. Remove or disconnect the following:
   - Negative battery cable
   - Intake manifold
   - Starter electrical connectors
   - Starter motor

To install:

3. Install or connect the following:
   - "S" terminal wire. Torque the nut to 26 inch lbs. (3 Nm).
   - Starter electrical connectors. Torque the nut to 70 inch lbs. (8 Nm).
   - Starter motor
   - Mounting bolts. Torque the bolts to 22 ft. lbs. (30 Nm).
   - Intake manifold
   - Negative battery cable

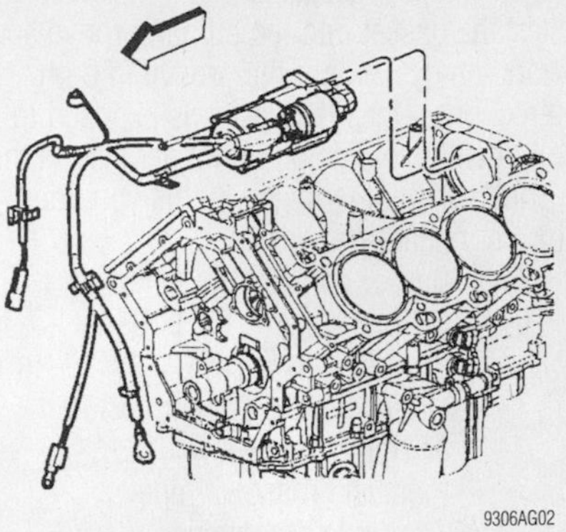

View of starter motor removal and wires

Oil Pan

REMOVAL & INSTALLATION

1. Before servicing the vehicle, refer to the precautions in the beginning of this section.
2. Drain the crankcase.
3. Remove or disconnect the following:
   - Negative battery cable
   - Oil level indicator harness connector, if equipped
   - Exhaust crossover pipe
   - Transaxle assembly from the vehicle
   - Oil pan bolts and the oil pan

To install:

➡The oil pan gasket is reusable unless it is damaged. Do not remove the gasket from the oil pan groove unless gasket replacement is required.

4. Thoroughly clean the inside of the oil pan and the cylinder block contact surface. If the oil pan gasket is being reused, be careful not to damage it. Do not expose the gasket to cleaning solvents.

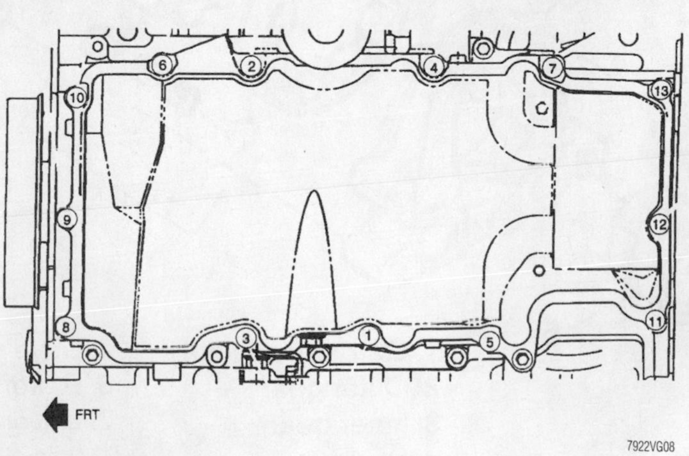

Oil pan bolt torque sequence

5. If a new gasket is being installed, start the gasket into the oil pan groove and work the gasket into the groove in both directions. Once the gasket is exposed to oil, it will expand and no longer stay in the groove without wrinkles. If this condition exists, replace the gasket.

6. Install or connect the following:
- Oil pan. Torque the bolts in sequence to 89 inch lbs. (10 Nm).
- Oil level indicator connector, if equipped
- Exhaust crossover pipe
- Transaxle assembly
- Oil pan drain plug. Torque it to 15 ft. lbs. (20 Nm).
- Negative battery cable

7. Refill the crankcase.
8. Run the engine and check for leaks.

Oil Pump

REMOVAL & INSTALLATION

1. Before servicing the vehicle, refer to the precautions in the beginning of this section.
2. Remove or disconnect the following:
- Negative battery cable
- Accessory drive belt
- Power steering hose retainer
- Right front wheel
- Wheel well splash shields
- Oil pan-to-transaxle brace
- Crankshaft balancer
- Drive belt tensioner
- Drive belt idler pulley
- Front cover and gasket

➡The front cover gasket is reusable as long as it is not damaged.

- Oil pump mounting bolts
- Oil pump
- Drive spacer from the pump housing
- Screws holding the pump housing halves together
- Inner (drive) and outer (driven) rotors from the housing
- Pressure relief valve

➡If any components show signs of excessive wear or damage, replace the pump assembly.

To install:

3. Install the inner and outer rotors to the pump cover in the same orientation as removed.
4. Install the pressure relief valve seat, spring and pilot in the pump housing.
5. Pack the pump housing halves with Amojell® or white petroleum grease to ensure pump priming.
6. Assemble the housing and cover over the locating dowel.
7. Insert a 9mm drill in the pump mounting hole on the opposite side to aid alignment of the housing and cover. Install the 2 screws and torque to 108 inch lbs. (12 Nm).
8. Install or connect the following:
- Oil pump drive spacer into the oil pump from the rear so the drive flat engages the pump rotor
- Oil pump over the crankshaft
- Oil pump bolts, finger tight

9. Hold the pump in its furthest up position. Torque the mounting bolts (1, 2 and 3), in sequence, to 89 inch lbs. (10 Nm); then an additional 35 degrees.
10. Place a small amount of RTV sealant at the split line of the upper and lower crankcases.
11. Install or connect the following:
- Front cover gasket on the dowel pins on the block
- Front cover. Torque the bolts to 89 inch lbs. (10 Nm).
- Drive belt idler pulley. Torque the bolt to 37 ft. lbs. (50 Nm).
- Drive belt tensioner. Torque the bolt to 37 ft. lbs. (50 Nm).
- Crankshaft balancer. Torque the bolt to 37 ft. lbs. (50 Nm) plus an additional 120 degree turn.
- Oil pan-to-transaxle brace. Torque the bolts to 37 ft. lbs. (50 Nm).
- Wheel well splash shield
- Wheel
- Power steering hose retainer
- Accessory drive belt
- Negative battery cable

12. Run the engine and check for proper engine oil pressure, and for possible leaks.

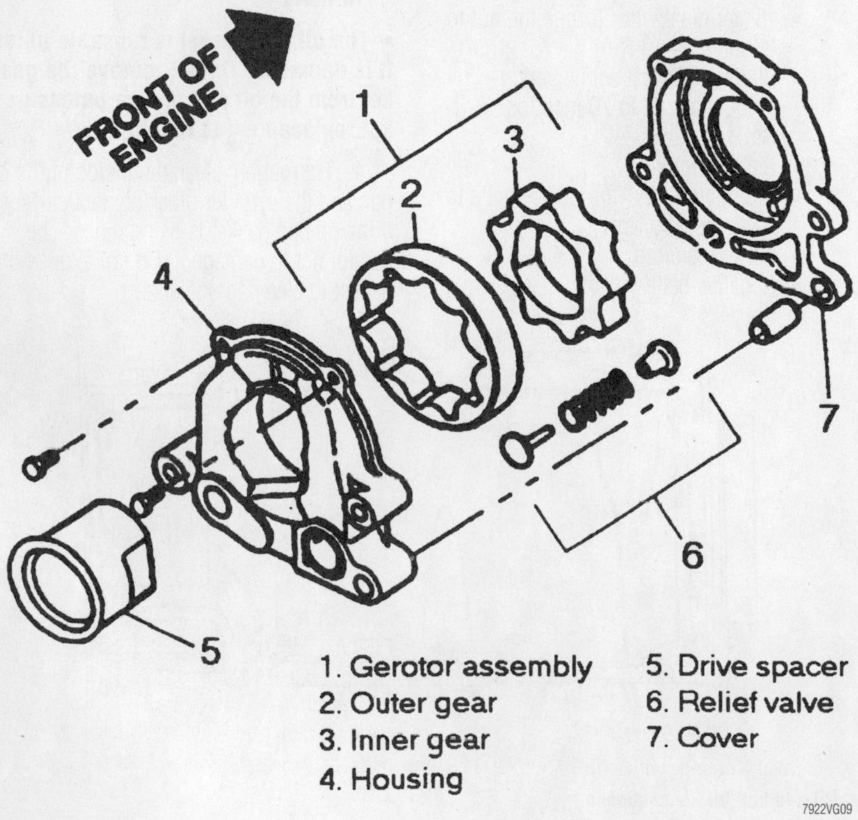

...xploded view of the oil pump

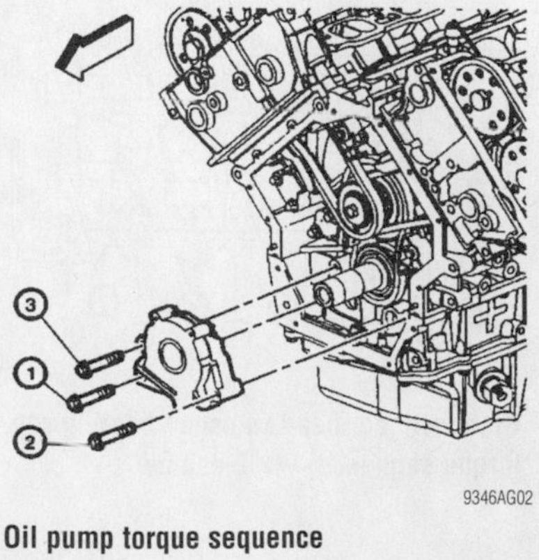

Oil pump torque sequence

Rear Main Seal

REMOVAL & INSTALLATION

1998–99 Models

1. Before servicing the vehicle, refer to the precautions in the beginning of this section.
2. Remove or disconnect the following:
   - Transaxle
   - Flexplate
   - Rear main seal using a suitable prytool

✲✲ WARNING

Use care not to damage the crankshaft seal surface with a pry tool.

To install:

3. Before installing, lubricate the seal bore and seal surface with clean engine oil.
4. Install the new seal using seal installer J 38817.

➡Check to see that the seal is squarely seated in the bore.

5. Install or connect the following:
   - Flexplate. Torque the bolts to 11 ft. lbs. (15 Nm) plus an additional 50 degree turn.
   - Transaxle
6. Start the engine and check for leaks.

2000–01 Models

1. Before servicing the vehicle, refer to the precautions in the beginning of this section.
2. Remove or disconnect the following:
   - Transaxle
   - Flexplate
3. Place oil seal removal Tool J 42841 on to the crankshaft.

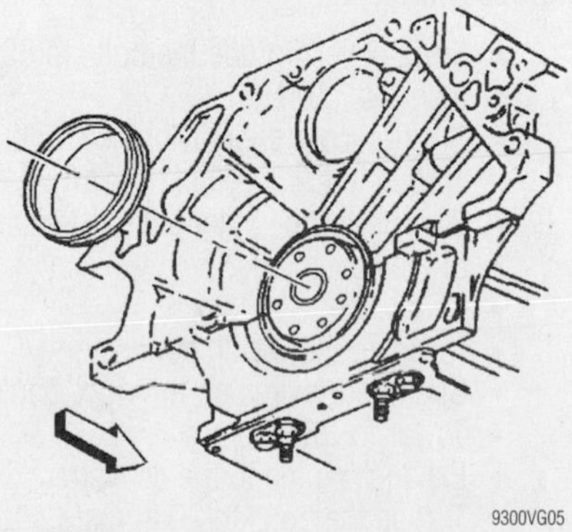

Exploded view of the rear main oil seal

4. Install the tool retaining bolts.
5. Using a drill with a socket adapter, install eight one-inch self-drilling crews into the seal using the guide holes in the removal tool. When drilling, make sure you reduce the drill speed when the screw begins threading into the seal.
6. With all eight removal screws installed, remove the removal tool retaining bolts.
7. Install the removal tool center screw.
8. Tighten the screw until the seal is removed.

To install:

9. Clean any debris from the crankshaft rear oil seal drain using wire or an unbound plastic tie-wrap.
10. Coat the outer diameter of the cylinder block crankshaft rear oil seal area with clean engine oil

➡DO NOT allow any engine oil on the area where the crankshaft rear oil seal is to be pressed onto the crankshaft. The green coating pre-applied to the inner diameter of the crankshaft rear oil seal must not be contaminated.

11. Wipe the outer diameter of the flywheel flange clean with a lint-free cloth.
12. Lubricate the outer rubber surface of the crankshaft rear oil seal with clean engine oil.
13. Loosen the center bolt of the removal tool until the center hub protrudes approximately one-half inch beyond the outer plate. It is not necessary to completely unthread the center bolt and separate the two pieces of the installation Tool J 42842. Place the installation tool on to the crankshaft.
14. Thread the three mounting bolts into the crankshaft flange.
15. Tighten the bolts until the installation tool is firmly mounted on the crankshaft.

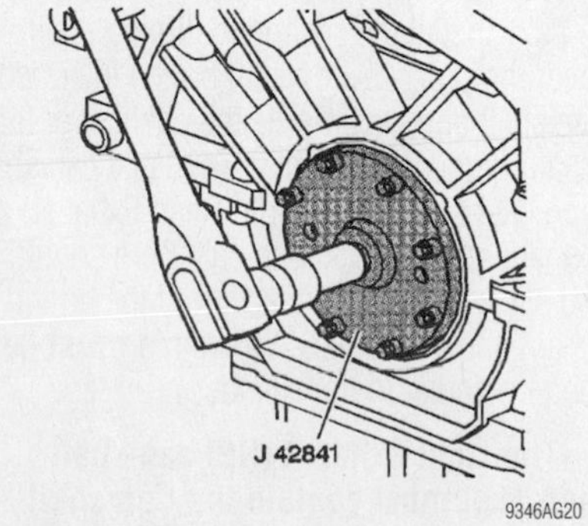

Removing the rear main seal—2000–01 models

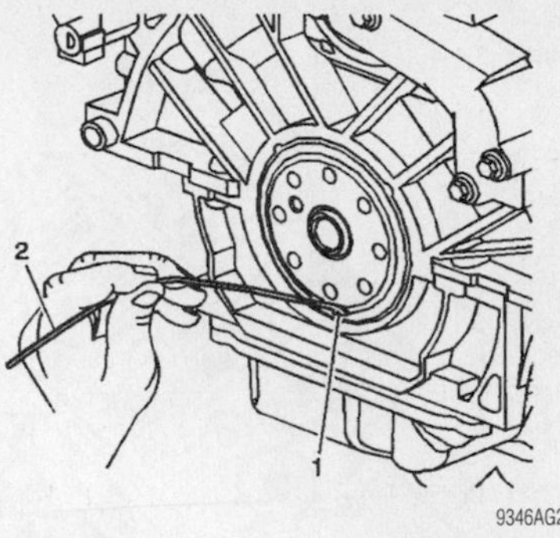

Clean any debris from the crankshaft rear oil seal drain using wire or an unbound plastic tie-wrap—2000–01 models

16. Install the crankshaft rear oil seal by tightening the center bolt until the installation tool bottoms against the crankcase.
17. Loosen the center bolt to release pressure on the crankcase.
18. Loosen the three mounting bolts.
19. Remove the installation tool from the crankshaft flange.
20. Inspect to ensure the installation depth is equal around the crankshaft rear oil seal's circumference. If the depth is not equal reinstall the tool and repeat the installation procedures.
21. Install the flexplate. Torque the bolts to 11 ft. lbs. (15 Nm) plus an additional 50 degree turn.
22. Install the transaxle.
23. Start the engine and check for leaks.

Timing Chain, Sprockets, Front Cover and Seal

REMOVAL & INSTALLATION

The left and right-side secondary timing chains can be removed with the engine in the vehicle. If the primary timing chain or intermediate shaft sprocket need to be replaced, the engine must be removed from the vehicle and supported on an engine stand.

➡Setting the camshaft timing is necessary whenever the camshaft drive system has been disturbed, meaning the relationship between any chain and sprocket has been lost. Correct timing exists when the crankshaft and intermediate shaft sprocket timing marks are in alignment and all 4 camshaft drive pins are perpendicular (90 degrees) to the cylinder head surface.

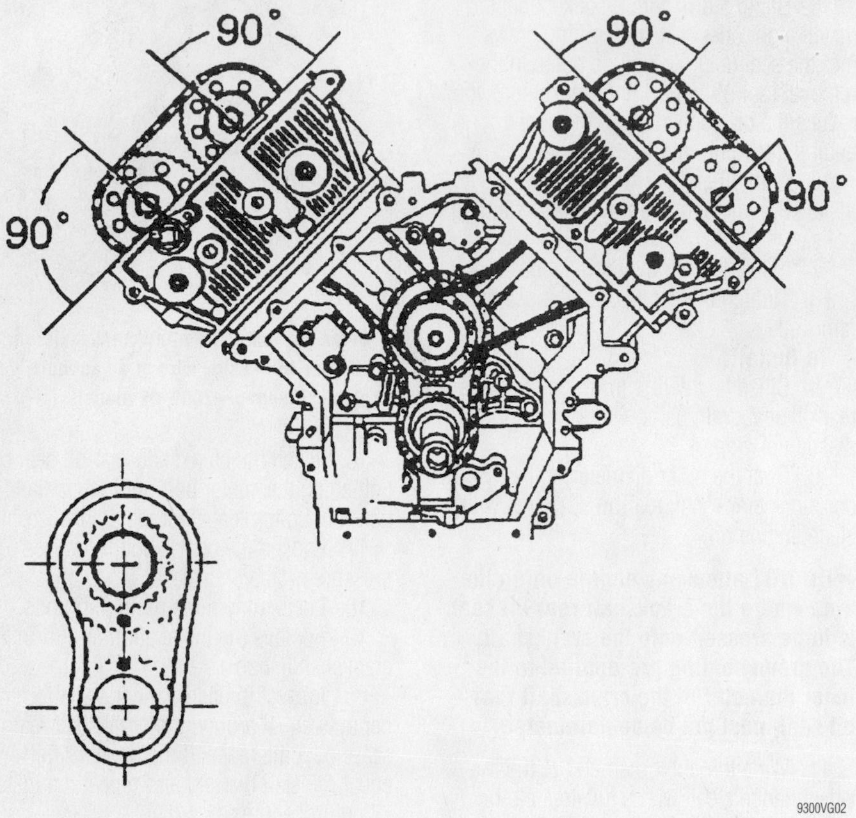

Correct timing chain alignment

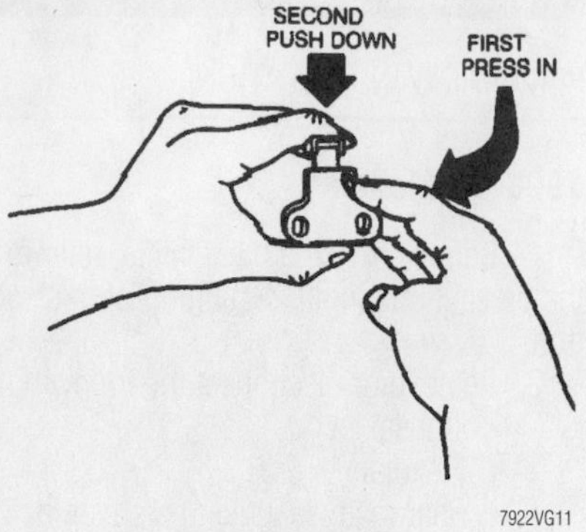

Rotating tensioner release lever

Right Side Secondary Chain

1. Before servicing the vehicle, refer to the precautions in the beginning of this section.
2. Remove or disconnect the following:
   - Negative battery cable
   - Exhaust Y-pipe at the converter
   - Tower-to-tower brace
   - Ignition Control Module (ICM) wiring connectors and mounting bolts
   - ICM
   - Spark plug wires on the right bank
   - Positive Crankcase Ventilation (PCV) valve
   - Canister purge solenoid from the rear of the cover
   - Front cover
   - Camshaft cover screws
   - Right and left torque struts
3. Safely support the front of the engine cradle and remove the 2 mounting bolts at the front of the cradle.
4. Lower the engine cradle or raise the vehicle to provide clearance at the rear of the engine compartment.
   - Camshaft cover

➡The camshaft cover gasket is reusable as long as it is not damaged.

   - Right side secondary chain tensioner
   - Right side chain guide. Access the upper chain guide mounting bolt through the hole in the cylinder head capped with the plastic plug.
   - Right side camshaft sprocket bolts and camshaft sprockets
   - Secondary drive chain

To install:

5. Install the secondary timing chain over the outer row of teeth on the intermediate shaft sprocket. Route the chain over the chain guide and install the exhaust camshaft sprocket so the **RE** (Right Head Exhaust) pin engages the sprocket notch. There should be no slack in the lower section of the timing chain and the camshaft drive pin **must** be perpendicular to the cylinder head face.

➡The right exhaust (RE) camshaft sprocket must contain the Camshaft Position (CMP) sensor pick-up.

6. Install the intake camshaft sprocket into the chain so the sprocket notch **RI** (Right Head Intake) engages the camshaft and the camshaft drive pin remains perpendicular to the cylinder head face. A hex is cast into the camshafts behind the lobes for cylinder No. 1, so an open end wrench may be used to provide minor repositioning of the camshafts.
7. Install or connect the following:
   - Both camshaft sprocket bolts loosely
   - Chain tensioner. Torque the mounting bolts to 20 ft. lbs. (27 Nm).
8. Torque the camshaft sprocket bolts to 90 ft. lbs. (120 Nm).
   - Spark plug and camshaft cover seals
   - Camshaft cover. Torque the screws to 84 inch lbs. (10 Nm).
   - Engine cradle. Torque both bolts to 75 ft. lbs. (100 Nm).
9. Install the right and left torque struts. Torque the retaining bolts to 45 ft. lbs. (60 Nm).

➡It is important during installation that the engine torque struts are not preloaded in their installed position. Adjustment is provided at the point the strut fastens to the core support bracket. Be sure this bolt is loose during assembly.

   - Front cover. Torque the bolts to 89 inch lbs. (10 Nm).
   - Strut-to-core support bracket bolt. Torque it to 45 ft. lbs. (60 Nm).
   - Canister purge solenoid to the rear of the cover
   - PCV valve
   - ICM and the wiring connectors
   - Spark plug wires on the right-bank
   - Tower-to-tower brace
   - Exhaust Y-pipe to the converter. Torque the bolts to 20 ft. lbs. (27 Nm).
   - Negative battery cable
10. Start the engine and check for leaks.

Left Side Secondary Chain

1. Before servicing the vehicle, refer to the precautions in the beginning of this section.
2. Drain the cooling system.
3. Remove or disconnect the following:
   - Negative battery cable
   - Accessory drive belt
   - Front cover bolts
   - Front cover and gasket
   - Power steering hose
   - Right front wheel
   - Splash shields from the wheel well
   - Flexplate cover
   - Crankshaft balancer bolt
4. Support the engine cradle.
   - 3 right-side engine cradle bolts
   - Road Sensing Suspension (RSS) sensor from the right control arm
5. Lower the cradle to gain access for the crankshaft balancer puller.
   - Crankshaft balancer
   - Drive belt tensioner
   - Drive belt idler pulley
   - Upper radiator hose at the water crossover
   - Spark plug wires
   - Right side cooling fan
   - Battery cable at the alternator
   - Battery cable harness at the camshaft cover
   - Positive Crankcase Ventilation (PCV) fresh air tube from the camshaft cover
   - Water pump pulley
   - Camshaft seal retainer screws and the seal
   - Battery cable retainer at the front of the camshaft cover
   - Camshaft cover by pivoting the entire cover around the water pump driveshaft. Continue moving the cover upward and pivoting so that the edge of the cover closely follows the left edge of the intake manifold cover.

➡The camshaft cover gasket is reusable as long as it is not damaged.

   - Right side secondary chain
   - Left side secondary chain tensioner
   - Left side chain guide. Access the upper chain guide mounting bolt through the hole in the cylinder head capped with the plastic plug.
   - Left side camshaft sprocket bolts and sprockets
   - Secondary drive chain

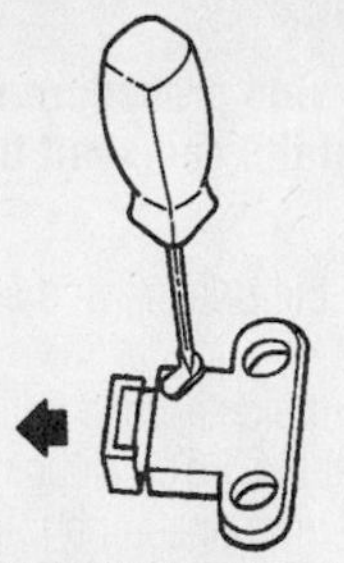

Locking the tensioner in the collapsed position

To install:

6. Route the secondary timing chain for the left side over the inner row of intermediate sprocket teeth.
7. Install or connect the following:
   - Secondary timing chain over the inner row of teeth on the intermediate shaft sprocket. Route the chain over the chain guide and install the exhaust camshaft sprocket so the **LE** (Left Head Exhaust) pin engages the sprocket notch. There should be no slack in the lower section of the timing chain and the camshaft drive pin **must** be perpendicular to the cylinder head face.
   - Intake camshaft sprocket into the chain so the sprocket notch **LI** (Left Head Intake) engages the camshaft and the camshaft drive pin remains perpendicular to the cylinder head face. A hex is cast into the camshafts behind the lobes for cylinder No. 2, so an open-end wrench may be used to provide minor repositioning of the camshafts.
   - Both camshaft sprocket bolts loosely
   - Chain tensioner. Torque the mounting bolts to 20 ft. lbs. (27 Nm).
8. Torque the camshaft sprocket bolts to 90 ft. lbs. (120 Nm).
   - Front cover gasket on the dowel pins on the block
   - Front cover. Torque the bolts to 89 inch lbs. (10 Nm). Apply a dab of RTV to the split line between the upper and lower crankcase assemblies.
   - Drive belt idler pulley. Torque the bolt to 35 ft. lbs. (47 Nm).
   - Drive belt tensioner. Torque the nut to 35 ft. lbs. (47 Nm).
   - Crankshaft balancer. Torque the bolt to 44 ft. lbs. (60 Nm) plus an additional 120 degree turn.
   - Engine cradle. Torque the bolts to 75 ft. lbs. (102 Nm).
   - RSS sensor
   - Wheel well splash shields
   - Flexplate cover
   - Spark plug and camshaft cover seals
   - Intake camshaft through the hole in the camshaft cover

WARNING

Use care to prevent the exposed section of the camshaft cover seal from being damaged by the edge of the cylinder head casting.

   - Camshaft cover
   - Camshaft cover screws. Torque the screws to 84 inch lbs. (10 Nm).
   - Battery cable retainer to the front of the camshaft cover
   - Battery cable at the alternator
   - Camshaft seal to the end of the intake camshaft. Seal the screw threads with sealer.
   - Water pump pulley
   - PCV fresh air tube to the camshaft cover
   - Right side fan
   - Spark plug wires
   - Upper radiator hose to the water crossover
9. Refill the cooling system.

Primary Chain/Intermediate Sprocket

1. Before servicing vehicle, refer to the precautions in the beginning of this section.
2. Remove or disconnect the following:
   - Engine
   - Accessory drive belt
   - Drive belt pulley
   - Dive belt tensioner
   - Front cover
   - Right and left camshaft covers

➡Align all timing marks prior to removal of the timing chains.

   - Timing chain tensioners
   - Camshaft sprocket bolts
   - Camshaft sprockets (4)
   - Right and left secondary chains
   - Intermediate shaft sprocket bolt and the sprocket
   - Primary timing sprockets and primary chain off the engine

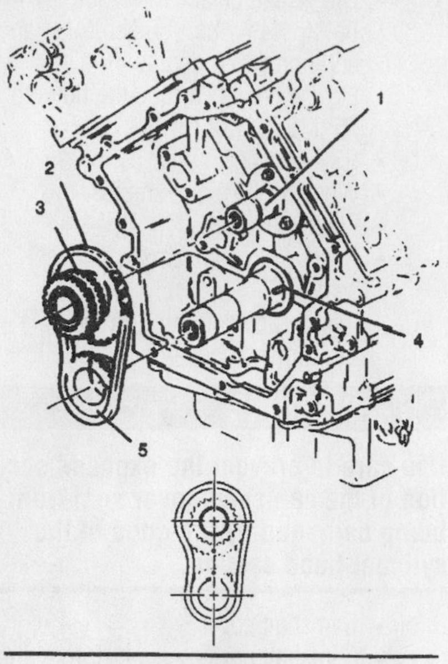

1 INTERMEDIATE SHAFT
2 PRIMARY CHAIN
3 INTERMEDIATE SHAFT SPROCKET
4 CRANKSHAFT SPROCKET KEY
5 SPROCKET

7922VG10

Primary drive chain components

To install:

➡The following procedure must be followed to set the camshaft timing on the vehicle.

3. Install the primary and secondary chain guides.
4. Rotate the crankshaft until the sprocket drive key is at the 1 o'clock position.
5. Install or connect the following:
   - Crankshaft sprocket and intermediate shaft sprocket in the primary timing chain so the timing marks are aligned
   - Primary timing chain assembly

➡The crankshaft sprocket keyway will have to slide over the key on the crankshaft. If it is necessary to turn the crankshaft sprocket, the intermediate shaft sprocket will also have to be turned so the timing mark remains aligned with the crankshaft sprocket.

   - Intermediate shaft sprocket bolt. Torque the bolt to 45 ft. lbs. (61 Nm).
   - Primary timing chain tensioner. Torque the tensioner mounting bolts to 20 ft. lbs. (27 Nm).
   - Right and left secondary chains with the timing marks aligned
   - Camshaft sprockets (4). Torque the bolts to 90 ft. lbs. (120 Nm).
   - Right and left camshaft covers. Torque the bolts to 89 inch lbs. (10 Nm).
   - Front engine cover. Torque the bolts to 89 inch lbs. (10 Nm).
   - Accessory drive belt
   - Drive belt pulley
   - Drive belt tensioner
   - Engine
6. Refill the cooling system.
7. Start the engine and check for leaks.

Piston and Ring

POSITIONING

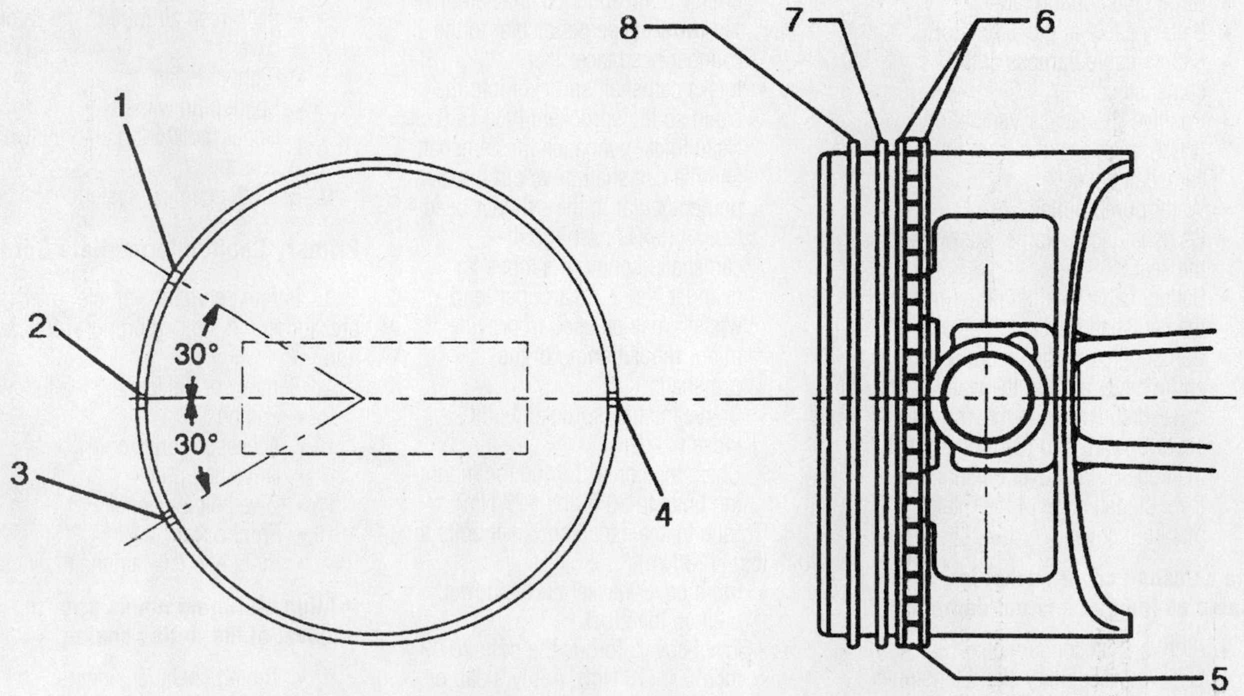

1. Oil ring segment gap
2. Upper compression ring gap
3. Oil ring segment gap
4. Expander & lower compression ring gaps
5. Expander ring
6. Oil segment rings
7. Lower compression ring
8. Upper compression ring

7922AG52

Piston ring and ring end-gap positioning—4.6L engine

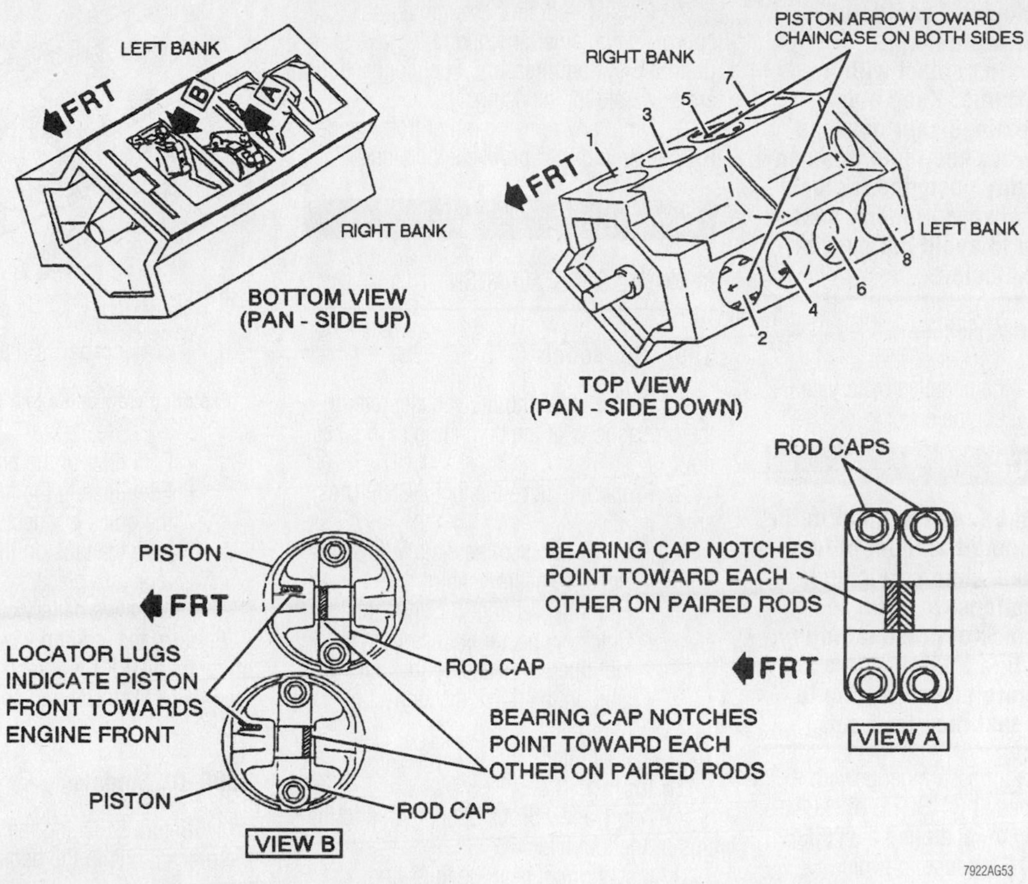

Piston and connecting rod assembly positioning—4.6L engine

FUEL SYSTEM

Fuel System Service Precautions

Safety is the most important factor when performing not only fuel system maintenance but also any type of maintenance. Failure to conduct maintenance and repairs in a safe manner may result in serious personal injury or death. Maintenance and testing of the vehicle's fuel system components can be accomplished safely and effectively by adhering to the following rules and guidelines.

- To avoid the possibility of fire and personal injury, always disconnect the negative battery cable unless the repair or test procedure requires that battery voltage be applied.
- Always relieve the fuel system pressure prior to disconnecting any fuel system component (injector, fuel rail, pressure regulator, etc.), fitting or fuel line connection. Exercise extreme caution whenever relieving fuel system pressure, to avoid exposing skin, face and eyes to fuel spray. Please be advised that fuel under pressure may penetrate the skin or any part of the body that it contacts.
- Always place a shop towel or cloth around the fitting or connection prior to loosening to absorb any excess fuel due to spillage. Ensure that all fuel spillage (should it occur) is quickly removed from engine surfaces. Ensure that all fuel soaked cloths or towels are deposited into a suitable waste container.
- Always keep a dry chemical (Class B) fire extinguisher near the work area.
- Do not allow fuel spray or fuel vapors to come into contact with a spark or open flame.
- Always use a back-up wrench when loosening. Torque the fuel line connection fittings. This will prevent unnecessary stress and torsion to fuel line piping.
- Always replace worn fuel fitting O-rings with new. Do not substitute fuel hose or equivalent, where fuel pipe is installed.

Fuel System Pressure

RELIEVING

✱✱ CAUTION

The fuel injection system remains under pressure, even when the engine has been turned OFF. The fuel system pressure must be relieved before disconnecting any fuel lines. Failure to do so may result in fire and/or personal injury.

1. Before servicing vehicle, refer to the precautions at the beginning of this section.
2. Loosen the fuel filler cap to relieve tank vapor pressure.

✱✱ CAUTION

Observe all applicable safety precautions when working around fuel. Whenever servicing the fuel system,

always work in a well-ventilated area. Do not allow fuel spray or vapors to come in contact with a spark or open flame. Keep a dry chemical fire extinguisher near the work area. Always keep fuel in a container specifically designed for fuel storage; also, always properly seal fuel containers to avoid the possibility of fire or explosion.

3. Be sure the ignition switch is in the **OFF** position.
4. Disconnect the negative battery cable.
5. Remove the engine cover.

✲✲ CAUTION

There may still be residual fuel in the system, and a small amount of fuel may be released when servicing fuel lines or connections. In order to reduce the chance of personal injury, cover the fuel line fittings with a shop towel before disconnecting to catch any fuel that may leak out.

6. Install a fuel pressure gauge with a drain hose attached, J-34730–1, or equivalent. Wrap a shop towel around the fitting while connecting the gauge to avoid spillage.
7. Install the drain hose into an approved container and open the valve to drain the system pressure. Fuel connections are now safe for servicing.
8. Drain any remaining fuel from inside the gauge into the approved container.

Fuel Filter

REMOVAL & INSTALLATION

1998–99 Models

1. Before servicing the vehicle, refer to the precautions in the beginning of this section.
2. Properly relieve the fuel system pressure.
3. Remove or disconnect the following:
   - Negative battery cable
   - Fuel filter retainer locking tabs
   - Quick connect fuel lines from the fuel filter by releasing the locking tabs on the quick disconnect fittings
   - Fuel filter

To install:

4. Apply a few drops of engine oil to the tips of the fuel filter.
5. Install or connect the following:

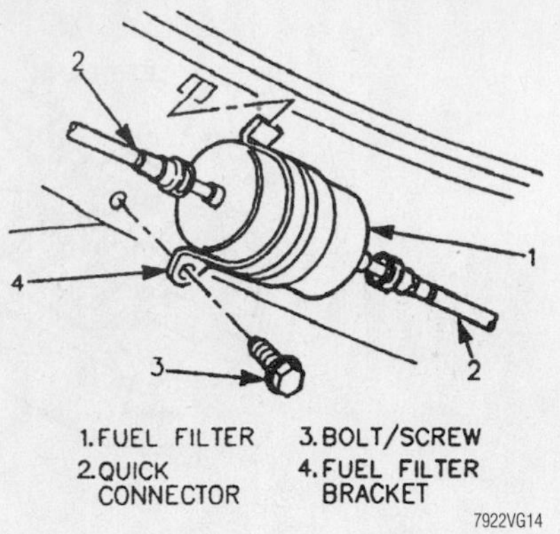

Exploded view of the fuel filter mounting

   - Fuel filter in the bracket
   - Fuel lines to the fuel filter and snap the quick-connect into place. Be sure the tabs on the quick-connects lock into place.
   - Negative battery cable
6. Turn the ignition key **ON** for 2 seconds, then **OFF** for 5 seconds. Again turn the ignition key **ON** and check for fuel leaks.

2000–01 Models

1. Before servicing the vehicle, refer to the precautions in the beginning of this section.

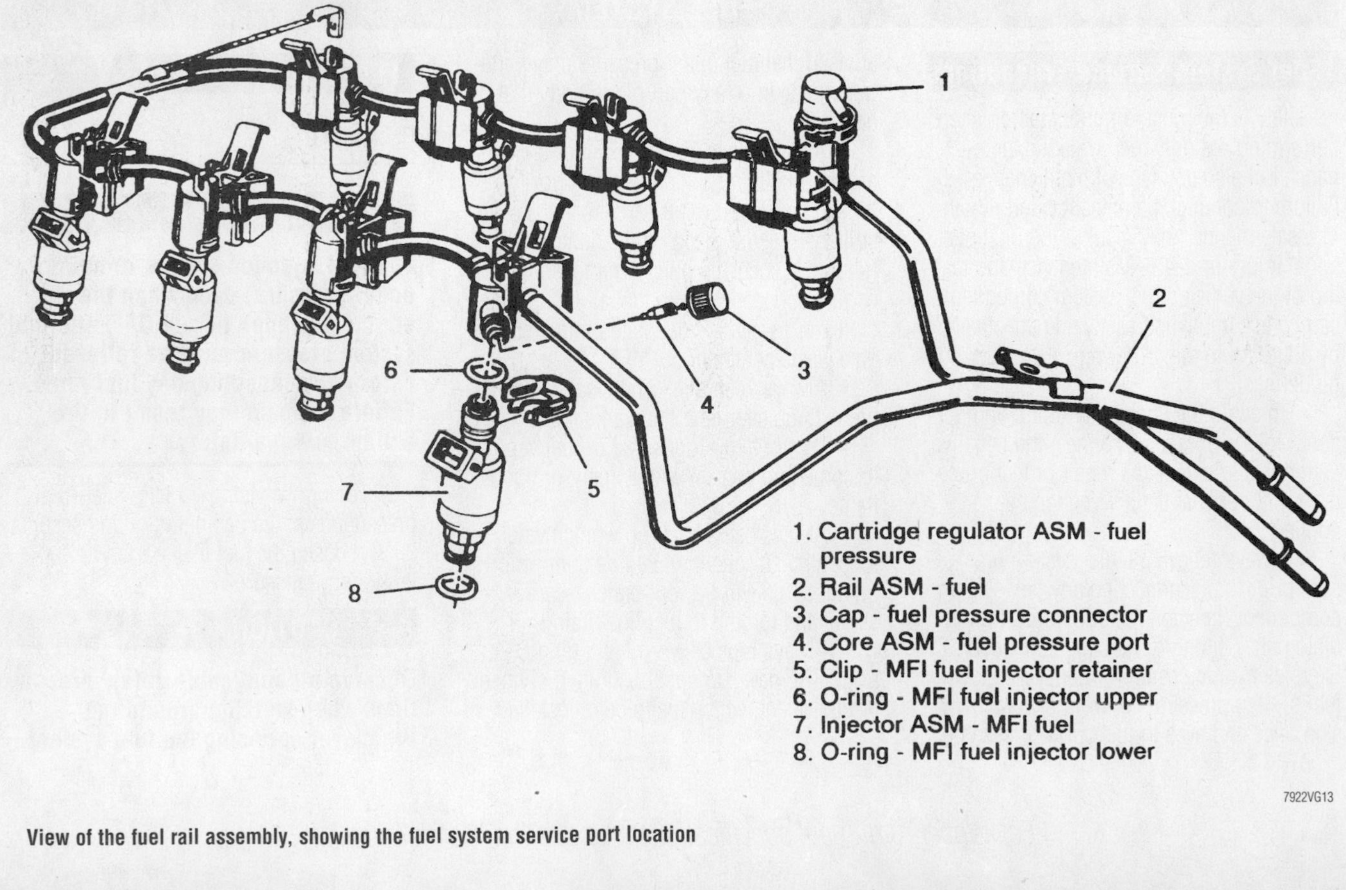

View of the fuel rail assembly, showing the fuel system service port location

2. Properly relieve the fuel system pressure.

3. Remove or disconnect the following:
- Negative battery cable
- Quick connect fuel line from the fuel filter inlet
- Threaded fitting at the outlet
- Fuel filter

To install:

4. Apply a few drops of engine oil to the tips of the fuel filter.

5. Install or connect the following:
- Fuel filter in the bracket
- Inlet line to the fuel filter and snap the quick-connect into place. Be sure the tabs on the quick-connects lock into place.
- Threaded fitting at the outlet
- Negative battery cable

6. Turn the ignition key **ON** for 2 seconds, then **OFF** for 5 seconds. Again turn the ignition key **ON** and check for fuel leaks.

Fuel Pump

REMOVAL & INSTALLATION

➡The modular fuel sender assembly must be disassembled in the exact order described.

1. Before servicing the vehicle, refer to the precautions in the beginning of this section.
2. Relieve the fuel system pressure.
3. Drain the fuel tank.
4. Clean the fuel tank in the area of the modular fuel sender assembly.
5. Remove or disconnect the following:
   - Fuel tank from the vehicle
   - Locking nut
   - Modular fuel sender assembly from the fuel tank

⁂ CAUTION

The modular fuel sender assembly may spring up from its position. When removing the assembly, be aware that the reservoir bucket is full of fuel. Tip the assembly slightly during removal to avoid damaging the float. Have a shop towel ready to absorb any leakage.

6. Slide the fuel sender seal downward, past the reservoir and carefully over the float arm assembly.
7. Disconnect the connector retainer clip from the wiring harness under the modular fuel sender assembly cover.
8. Locate the curved side of the modular unit's reservoir. Beginning at locking tab 1, squeeze the reservoir to release the first locking tab.
9. Moving clockwise to locking tab 2, apply gentle pressure to the guide rod to release the second locking tab.
10. At locking tab 3, gently twist and squeeze to release the reservoir from the retainer.
11. Remove or disconnect the following:
    - Cover and the retainer from the reservoir. Be careful not to damage the crossover tube. The unit will still be attached by the fuel pipe and the crossover tube.
    - External strainer by prying the strainer ferrule off the reservoir.

⁂ CAUTION

Excessive force may dislodge the jet pump.

- Rubber bumper pad. The fuel pump and the sleeve assembly are attached to the retainer when pulled from the reservoir. Depress the flex member on the pump sleeve and rotate the sleeve counterclockwise to remove the fuel pump from the retainer. Note the orientation of the pump to the retainer.

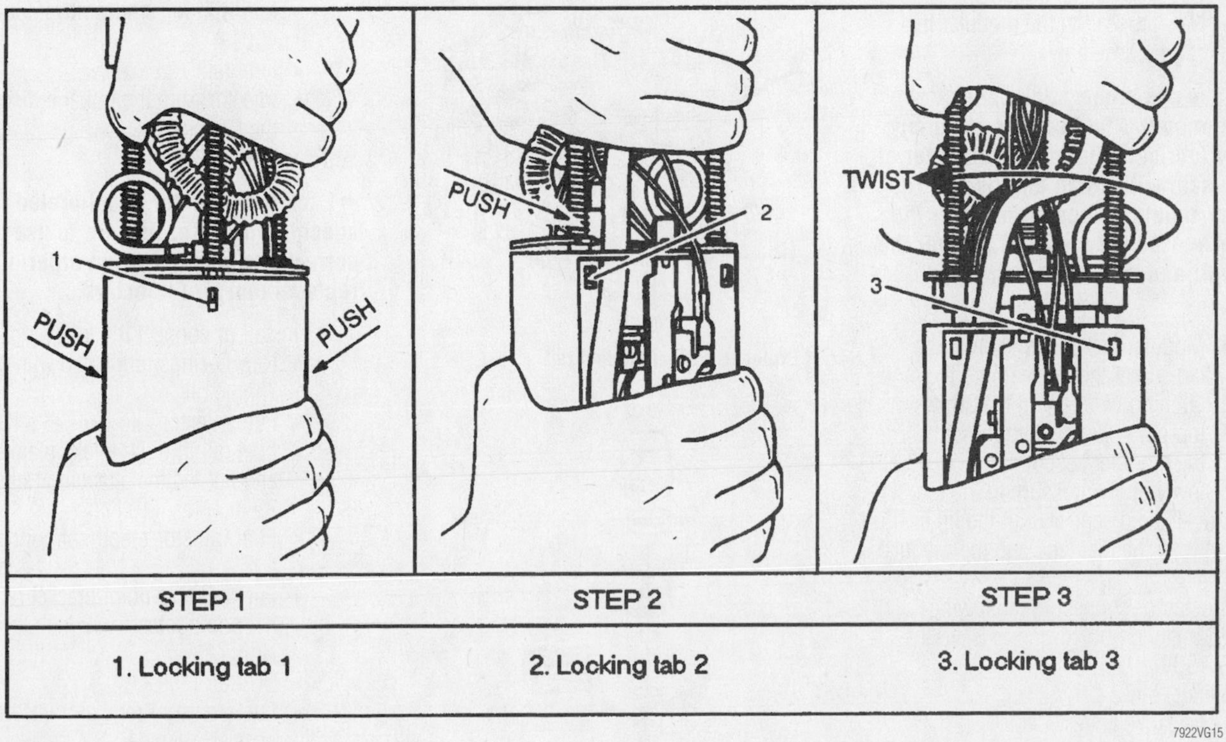

Modular fuel sender disassembly

12. Slide the lower connector assembly out of the retainer to remove the fuel pulse damper from the lower connector. Note the orientation of the seal (the modular unit is now held together by the crossover tube only). Discard the fuel pulse damper.

To install:

13. Install the fuel pulse damper.

➡Always use a new damper when installing a new fuel pump.

14. Slide the lower connector into the retainer.
15. Push the pump outlet tube into the fuel pulse damper and rotate the flex member back to its original position. Align the pump outlet tube into the retainer opening. All 3 sleeve tabs should protrude through the retainer before rotating. Rotate the pump clockwise until a click is heard. Be sure fit is snug before rotating. Place the fuel pump back into its reservoir. The crossover tube must be placed in its proper slot.
16. Install or connect the following:
    - New rubber bumper pad. Insert the drain tube into the proper retainer and bumper pad slots.
    - New strainer, being careful not to dislodge the jet pump
    - Fuel pump wire connector
    - Wiring harness connector
    - Connector retainer clip
    - New lip seal on the modular fuel sender assembly

➡Always use a new seal when servicing the modular fuel sender assembly. Lightly lubricate the inside diameter of the lip seal with clean engine oil. The lip seal should be positioned over the float arm assembly, moved up over the reservoir and half-way up the guide posts.

- Modular fuel assembly into the tank. Seat the lip seal into the tank opening by aligning the arrows on top of the fuel tank to the arrow on the modular assembly.

17. Slowly apply pressure to the top of the spring-loaded sender until the lip seal is flush between the fuel tank and the modular cover.
18. Install or connect the following:
    - Locking nut. Torque to 37 ft. lbs. (50 Nm).
    - Fuel tank
    - Negative battery cable
19. Refill the fuel tank.
20. Pressurize the fuel system and verify there are no fuel leaks.

Fuel Injectors

REMOVAL & INSTALLATION

1998–99 Models

1. Before servicing the vehicle, refer to the precautions in the beginning of this section.
2. Relieve the fuel system pressure.
3. Remove or disconnect the following:
    - Intake manifold cover
    - Fuel injector electrical connectors
    - Fuel rail from intake manifold
    - Fuel injectors
    - Upper and lower O-rings

To install:

4. Install or connect the following:
    - New O-rings lubricated with clean engine oil
    - Fuel injector into the fuel rail
    - Fuel rail into the intake manifold. Be sure all locking tabs engage fully.
    - Fuel injector electrical connectors
    - Negative battery cable
5. Inspect for fuel leaks as follows:
    a. Step 1: Turn ignition switch to the ON position for 2 seconds.
    b. Step 2: Turn ignition switch OFF for 10 seconds.
    c. Step 3: Turn ignition switch ON.
    d. Step 4: Check for leaks.
6. Install the intake manifold cover. Torque the nuts to 27 inch lbs. (3 Nm).

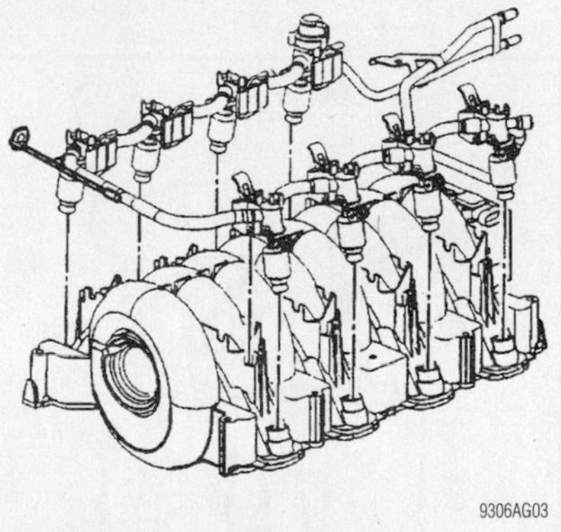

9306AG03

Exploded view of the fuel rail

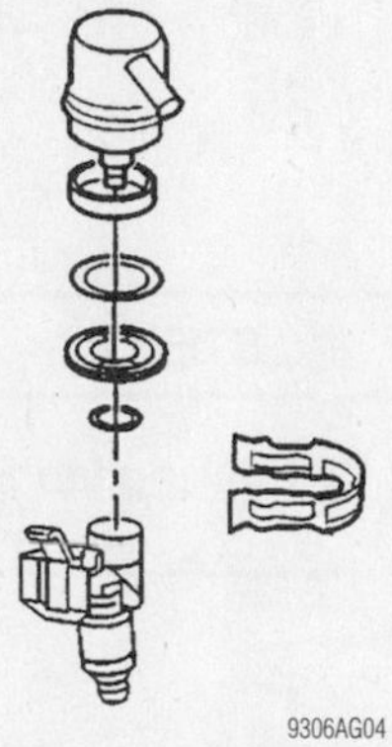

9306AG04

Exploded view of the fuel injector

2000–01 Models

1. Before servicing the vehicle, refer to the precautions in the beginning of this section.
2. Relieve the fuel system pressure.
3. Remove or disconnect the following:
    - Intake manifold cover
    - Intake Air Temperature (IAT) sensor electrical connector
    - Mass Airflow (MAF) sensor electrical connector
    - Air cleaner intake duct
    - Quick-connect fittings from the fuel rail
    - Fuel pressure regulator vacuum hose
    - Positive Crankcase Ventilation (PCV) air tube
    - Fuel rail end-point bracket retainer nut
    - Fuel injector electrical connectors
    - Fuel rail attaching bolts
    - Fuel injectors from the intake manifold with injector removal tool J 43013
    - Fuel rail
    - Injector from the rail and discard the O-rings

To install:

➡Each fuel injector is calibrated for a specific flow rate. Be sure to use the correct part number when ordering replacement fuel injectors.

4. Install or connect the following:
    - New O-rings lubricated with clean engine oil
    - Fuel injector into the fuel rail
    - Fuel rail into the intake manifold and tighten the attaching bolts to 89 inch lbs. (10 Nm)
    - Fuel injector electrical connectors
    - Fuel rail end-point bracket retainer nut and tighten to 89 inch lbs. (10 Nm)
    - PCV air tube
    - Fuel pressure regulator vacuum hose
    - Quick-connect fittings from the fuel rail
    - Air cleaner intake duct

- MAF sensor electrical connector
- IAT sensor electrical connector
- Negative battery cable

5. Inspect for fuel leaks as follows:

a. Step 1: Turn ignition switch to the ON position for 2 seconds.

b. Step 2: Turn ignition switch OFF for 10 seconds.

c. Step 3: Turn ignition switch ON.

d. Step 4: Check for leaks.

6. Install the intake manifold cover. Torque the nuts to 27 inch lbs. (3 Nm).

DRIVE TRAIN

Transaxle Assembly

REMOVAL & INSTALLATION

1998–99 Models

1. Before servicing the vehicle, refer to the precautions in the beginning of this section.
2. Remove or disconnect the following:
   - Negative battery cable
   - Headlight housing upper filler panel and diagonal brace
   - Air cleaner assembly
   - Shift control cable and bracket at the transaxle
   - Torque struts
   - Oil cooler lines at the cooler
   - Oil sending line at the transaxle
   - 2 upper transaxle-to-engine bolts
   - Power steering return hose at the auxiliary cooler
3. Support the engine to take the weight of the powertrain off the mounts and frame.
   - Front wheels
   - Splash shields from both front wheel wells
   - Both front suspension position sensors from the lower control arms
   - Both stabilizer links from the struts
   - Tie rod ends from the steering knuckles
   - Lower ball joints from the steering knuckles
   - Halfshafts
   - Fuel filter bracket
   - A/C splash shield
   - Power steering filter at the cradle
   - Air conditioning splash shield
   - Anti-lock Brake System (ABS) modulator from the bracket
   - Engine oil pan-to-transaxle bracket
   - Torque converter cover
   - Torque converter bolts
   - Powertrain mount nuts from the frame
4. Rotate the intermediate steering shaft until the steering gear stub shaft clamp bolt is accessible at the left wheel well. Remove the clamp bolt and disconnect the intermediate steering shaft from the steering gear.

✲✲ CAUTION

If the intermediate steering shaft is not disconnected from the steering gear stub shaft, damage to the steering gear and/or intermediate shaft may result. This damage can cause loss of steering control, which could result in personal injury.

✲✲ WARNING

Do not turn the steering wheel or move the position of the steering gear once the intermediate steering shaft is disconnected as this will off-center the air bag coil in the steering column. If the air bag coil becomes off-centered, it may be damaged during vehicle operation.

5. Disconnect the electrical harness and connector from the engine cradle.
6. Support the rear of the cradle with a jack, then remove the 4 rear cradle bolts.
7. Lower the jack a few inches to gain access to the power steering gear heat shield and return line titting.
8. Remove or disconnect the following:
   - Heat shield
   - Power steering return line
   - Power steering electrical connector
9. Raise the jack and reinstall 1 rear cradle bolt on each side finger-tight to support the cradle. Remove the jack.
10. Support the frame with a jack and remove the 6 frame mount bolts. Lower the frame and/or raise the vehicle with the steering gear attached.
    - Electrical connectors to the transaxle
    - Vehicle Speed Sensor (VSS) electrical connector
    - Transaxle harness from the transaxle clip

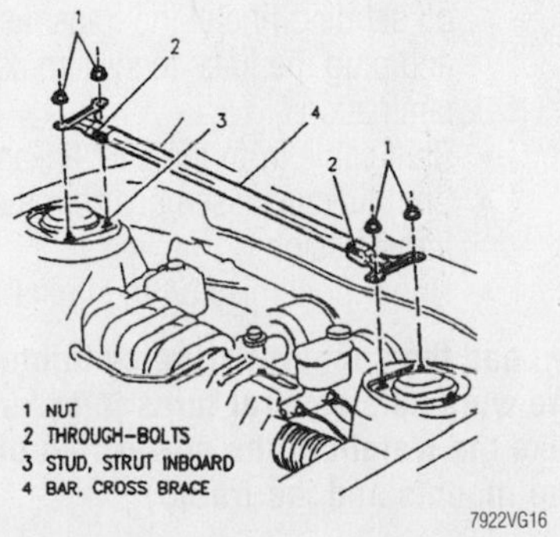

Exploded view of the crossbrace-to-strut towers mounting

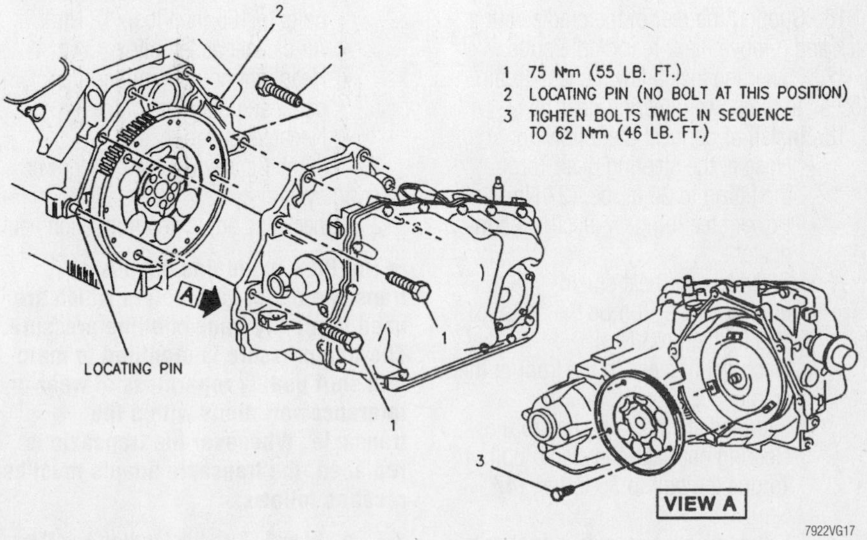

Exploded view of the engine-to-transaxle attachments

- Fuel line bundle from the transaxle
- Left and right transaxle mounts and brackets

11. Support the transaxle with a jack.
    - Engine-to-transaxle heat shield and bracket
    - Remaining transaxle-to-engine bolts
    - Manual shaft linkage
    - Neutral safety switch
    - VSS and oil return line
    - Transaxle

To install:

12. Install or connect the following:
    - Transaxle. Torque the transaxle to engine bolts to 55 ft. lbs. (75 Nm).
    - Oil return line and the VSS
    - Neutral safety switch. Torque the bolts to 106 inch lbs. (12 Nm).
    - Manual shaft linkage. Torque the manual shaft nut to 15 ft. lbs. (20 Nm).
    - Engine-to-transaxle bracket and heat shield. Torque the bolts to 35 ft. lbs. (47 Nm).
13. Remove the transaxle jack.
    - Right and left transaxle bracket and mount to the transaxle. Torque the nuts/bolts to 35 ft. lbs. (47 Nm).
    - Fuel line bundle to the transaxle
    - Electrical connectors to the transaxle and VSS
    - Transaxle harness to the transaxle clip
14. Raise the frame and/or lower the vehicle while locating the engine and transaxle mount studs into the frame.
15. Install the 2 front and 2 rear cradle bolts finger-tight to support the cradle, then remove the cradle support.
16. Support the rear of the cradle with a jack and remove the 2 rear cradle bolts.
17. Lower the jack a few inches to gain access to the power steering gear.
18. Install or connect the following:
    - Hose at the steering gear. Torque the fitting to 20 ft. lbs. (27 Nm).
    - Power steering gear electrical connector
    - Steering gear heat shield
    - Engine cradle. Torque the bolts to 120 ft. lbs. (163 Nm).
    - Electrical harness to the front of the cradle
    - Intermediate steering shaft to the steering gear and the clamp bolt. Torque the bolt to 35 ft. lbs. (47 Nm).
    - Left and right transaxle mount nuts and right engine mount nuts at the frame. Torque the nuts to 35 ft. lbs. (47 Nm).
    - Flexplate. Torque the bolts to 35 ft. lbs. (47 Nm).
    - Torque converter cover. Torque the bolts to 106 inch lbs. (12 Nm).
    - Engine oil pan-to-transaxle bracket. Torque the bolts to 35 ft. lbs. (47 Nm).
    - ABS modulator to the bracket and the air conditioning splash shield at the frame
    - Halfshafts. Torque the halfshaft nuts to 110 ft. lbs. (145 Nm).
    - Lower ball joints into the steering knuckles, use new cotter pins
    - Tie rod ends into the steering knuckles, use new cotter pins
    - Stabilizer links to the struts. Torque the nuts to 49 ft. lbs. (65 Nm).
    - Front suspension position sensors to the lower control arms
    - Power steering filter to the cradle
    - Splash shields in the wheel wells
    - Wheels. Torque the lug nuts to 100 ft. lbs. (140Nm).
    - Power steering hose at the auxiliary cooler
    - Remaining transaxle-to-engine bolts. Torque the bolts to 35 ft. lbs. (47 Nm).
19. Flush the transaxle oil cooler. The transaxle oil cooler and lines should be flushed before the oil cooler lines are connected to the transaxle.
    - Oil cooler lines to the transaxle. Torque the fittings to 16 ft. lbs. (22 Nm).
    - Torque struts
    - Neutral safety switch
    - Shift control cable and bracket to the transaxle. Torque the bracket bolts to 106 inch lbs. (12 Nm).
    - Air cleaner assembly
    - Headlight housing upper filler panel and diagonal brace
    - Negative battery cable
20. Fill the transaxle. Bleed the power steering system.
21. Check and adjust the front alignment.

➡The PCM maintains 3 types of transaxle adapt parameters which are used to modify transaxle line pressure. The line pressure is modified to maintain shift quality regardless of wear or tolerance variations within the transaxle. Whenever the transaxle is replaced, the transaxle adapts must be reset as follows:

a. Step 1: Turn the ignition key **ON**. Enter the self-diagnostic system.

b. Step 2: Select PCM override PS13 (TPS SENSOR LEARN).

c. Step 3: Press the WARMER button. The Driver Information Center (DIC) should display 09, indicating that the Garage Shift Adapt value has been reset.

d. Step 4: Select PCM override PS14 (TRAN ADAPT).

e. Step 5: Press the COOLER button. The DIC should display 90, indicating the Upshift Adapt (UA) value has been reset.

f. Step 6: Press the WARMER button. The DIC should display 09, indicating the Steady State Adapt (SSA) value has been reset.

➡The PCM maintains a value for transaxle oil life. This value indicates the percentage of oil life remaining and is calculated based on transaxle temperature and speed. When the vehicle is new, the transaxle oil life value is 100. As the vehicle operates, the percentage will decrease. Whenever the transaxle is replaced, the transaxle oil life indicator should be reset to 100 as follows:

g. Step 1: Turn the ignition key **ON**, but leave the engine OFF.

h. Step 2: Press and hold the OFF and REAR DEFOG buttons on the DIC until the message TRANSAXLE OIL LIFE RESET is displayed on the DIC.

2000–01 Models

1. Before servicing the vehicle, refer to the precautions in the beginning of this section.
2. Drain the cooling system.
3. Remove or disconnect the following:
    - Negative battery cable
    - Upper filler panel
    - Air cleaner upper plenum and intake tube
    - Surge tank outlet pipe
    - Oil cooler lines at the cooler and oil sending line at the transmission and plug the lines to system contamination
    - Shift cable from the shift linkage
    - Shift linkage assembly from the transmission
    - Upper transmission to engine bolts

➡Load the support fixture by tightening the wing nuts several turns in order to take the weight of the powertrain off of the mounts and the frame.

4. Install an engine support fixture. Raise the left side of the assembly 1 inch (25.4mm) above the resting position with the adjusting screws.

- Tires and wheels
- Lower slash shield
- Front suspension position sensors links from the lower control arms and position aside
- Stabilizer links from the lower control arms and let the stabilizer shaft sit loosely on the engine frame
- Stabilizer shaft insulator brackets from the engine frame
- Drive axle nuts
- Drive axles from the hubs
- Tie rod cotter pins and nuts
- Tie rods from the steering knuckles
- Lower ball joint cotter pins and nuts
- Ball joints from the steering knuckles
- Drive axles from the transmission
- Brake Pressure Modulator Valve (BPMV) and bracket from the engine frame and support the valve with a bungee cord
- Rear transmission mount bracket nut and bolts
- Rear transmission mount bolt and nuts from the left frame rail
- Left engine mount nut from the engine frame
- Right engine mount nuts from the engine frame
- Steering rack bolts from the engine frame
- Brake line and power steering line fasteners from the engine frame
- Post Oxygen (O_2S) sensor heat shield and sensor connector
- One center exhaust hanger to allow easy movement of the exhaust in the rearward direction

5. Support the engine frame.
    - 6 engine frame mount bolts and separate the tie rod ends and lower ball joints from the steering knuckles

6. Raise the vehicle slowly away from the engine frame.
    - Engine frame and support table from under the vehicle
    - Steering rack heat shield from the rack
    - Power steering lines from the steering rack
    - Right rear engine to transmission bracket upper bolt, nuts, and bracket
    - Right transmission mount bracket lower bolts and bracket from the transmission

7. It is necessary to support the steering rack in a position that allows access to the bracket bolts.
    - Steering rack from the vehicle
    - Right front engine-to-transmission brace bolt, nut, brace and heat shield
    - Right front engine-to-transmission bracket
    - Engine oil pan-to-transmission bracket
    - Torque converter cover brace and the cover
    - Flywheel-to-converter bolts
    - Transmission electrical connectors
    - Engine-to-transmission heat shield and bracket

8. Support the transmission with a transmission jack.
    - Engine-to-transmission bolts
    - Transmission

To install:

9. Install or connect the following:
    - Rear transmission mount bracket onto the stud before the transmission is all the way up into the vehicle
    - Rear transmission mount (loosely) to the left frame rail with the nuts and bolt

10. Raise the transmission into position. Guide the rear mount bracket and the rear mount together as the transmission is raised into the vehicle.

11. Align the transmission and position the transmission onto the dowels on the engine.
    - Engine-to-transmission bolts and tighten to 55 ft. lbs. (75 Nm), then remove the transmission jack
    - Right front engine-to-transmission heat shield and brace, then tighten the bolts to 55 ft. lbs. (75 Nm)
    - Right front engine-to-transmission bracket and tighten the bolts to 37 ft. lbs. (50 Nm)
    - Right transmission mount bracket and the lower bolts. Tighten the bolts to 81 inch lbs. (110 Nm)
    - Steering rack into position (loosely) onto the vehicle and insert the tie rod ends into the steering knuckles
    - Right rear engine-to-transmission bracket and tighten the bolt 81 inch lbs. (110 Nm) and the nuts to 37 ft. lbs. (50 Nm)
    - Power steering lines to the power steering rack and tighten to 20 ft. lbs. (27 Nm)
    - Power steering rack heat shield
    - Flywheel to converter bolts and tighten to 44 ft. lbs. (60 Nm)
    - Torque converter cover
    - All electrical connectors
    - Drive axles into the hubs and both drive axle nuts and tighten to 110 ft. lbs. (145 Nm)
    - Engine frame on to the engine support table
    - Stabilizer bar onto the engine frame in its approximate position

12. Lower the vehicle onto the engine frame while locating all the mounting points.
    - Engine frame mounting bolts, finger tight

13. Raise the vehicle away from the engine support table and tighten the engine frame bolts to 142 ft. lbs. (192 Nm).
    - Steering intermediate shaft to the steering gear and install the clamp bolt. Tighten the bolt to 37 ft. lbs. (50 Nm)
    - Steering gear bolts and tighten to 77 ft. lbs. (105 Nm)
    - Transmission mount nuts and the right engine mount nuts to the frame and tighten to 37 ft. lbs. (50 Nm)

➡When tightening the ball joint nut, a minimum torque of 37 ft. lbs. (50 Nm) must be reached. If the proper torque is not obtained, inspect for stripped threads. If threads are satisfactory, replace the ball joint and knuckle. If required, turn the nut up to an additional 60 degrees to allow for installation of the cotter pin.

- Lower ball joints into the steering knuckles
- Ball joint nuts and cotter pins. Tighten the ball joint nut to 84 inch lbs. (10 Nm), then tighten the nut an additional 120 degrees.

➡When tightening the tie rod end nut, a minimum torque of 33 ft. lbs. (45 Nm) must be reached. If the proper torque is not obtained, inspect for stripped threads. If the threads are satisfactory, replace the tie rod end and knuckle.

- Tie rod end nut to 74 ft. lbs. (100 Nm), then an additional 1/3 turn (or 2 flats)
- Brake and power steering line retainers to the engine frame

- Brake pressure modulator valve and bracket to the engine frame
- Sway bar insulators onto the sway bar
- Stabilizer links to the control arms and tighten the nuts to 49 inch lbs. (65 Nm)
- Front suspension position sensors to the lower control arms
- Lower close out panel
- Center exhaust hanger
- Post O_2S sensor and heat shield
- Front tire and wheel assemblies

14. Remove the Engine Support Fixture.
- Upper bell housing bolts
- Transmission cooler pipe fittings at the transmission and tighten to 16 ft. lbs. (22 Nm)
- Transmission oil sending pipe to the transmission
- Shift linkage assembly to the transmission
- Shift cable to the shift linkage
- Surge tank outlet hose
- Surge tank inlet hose
- Air cleaner assembly
- Negative battery cable

15. Bleed the power steering system.
16. Fill the transmission with fluid.
17. Fill the cooling system.
18. Check the front suspension alignment and as necessary.

➡The PCM maintains 3 types of transaxle adapt parameters which are used to modify transaxle line pressure. The line pressure is modified to maintain shift quality regardless of wear or tolerance variations within the transaxle. Whenever the transaxle is replaced, the transaxle adapts must be reset as follows:

a. Step 1: Turn the ignition key **ON**. Enter the self-diagnostic system.
b. Step 2: Select PCM override PS13 (TPS SENSOR LEARN).
c. Step 3: Press the WARMER button. The Driver Information Center (DIC) should display 09, indicating that the Garage Shift Adapt value has been reset.
d. Step 4: Select PCM override PS14 (TRAN ADAPT).
e. Step 5: Press the COOLER button. The DIC should display 90, indicating the Upshift Adapt (UA) value has been reset.
f. Step 6: Press the WARMER button. The DIC should display 09, indicating the Steady State Adapt (SSA) value has been reset.

➡The PCM maintains a value for transaxle oil life. This value indicates the percentage of oil life remaining and is calculated based on transaxle temperature and speed. When the vehicle is new, the transaxle oil life value is 100. As the vehicle operates, the percentage will decrease. Whenever the transaxle is replaced, the transaxle oil life indicator should be reset to 100 as follows:

g. Step 1: Turn the ignition key **ON**, but leave the engine OFF.
h. Step 2: Press and hold the OFF and REAR DEFOG buttons on the DIC until the message TRANSAXLE OIL LIFE RESET is displayed on the DIC.

Halfshaft

REMOVAL & INSTALLATION

⁂ WARNING

Use care when removing the halfshaft to prevent the inner CV-joint from becoming over-extended. Overextension of the joint could result in separation of internal components and possible joint failure.

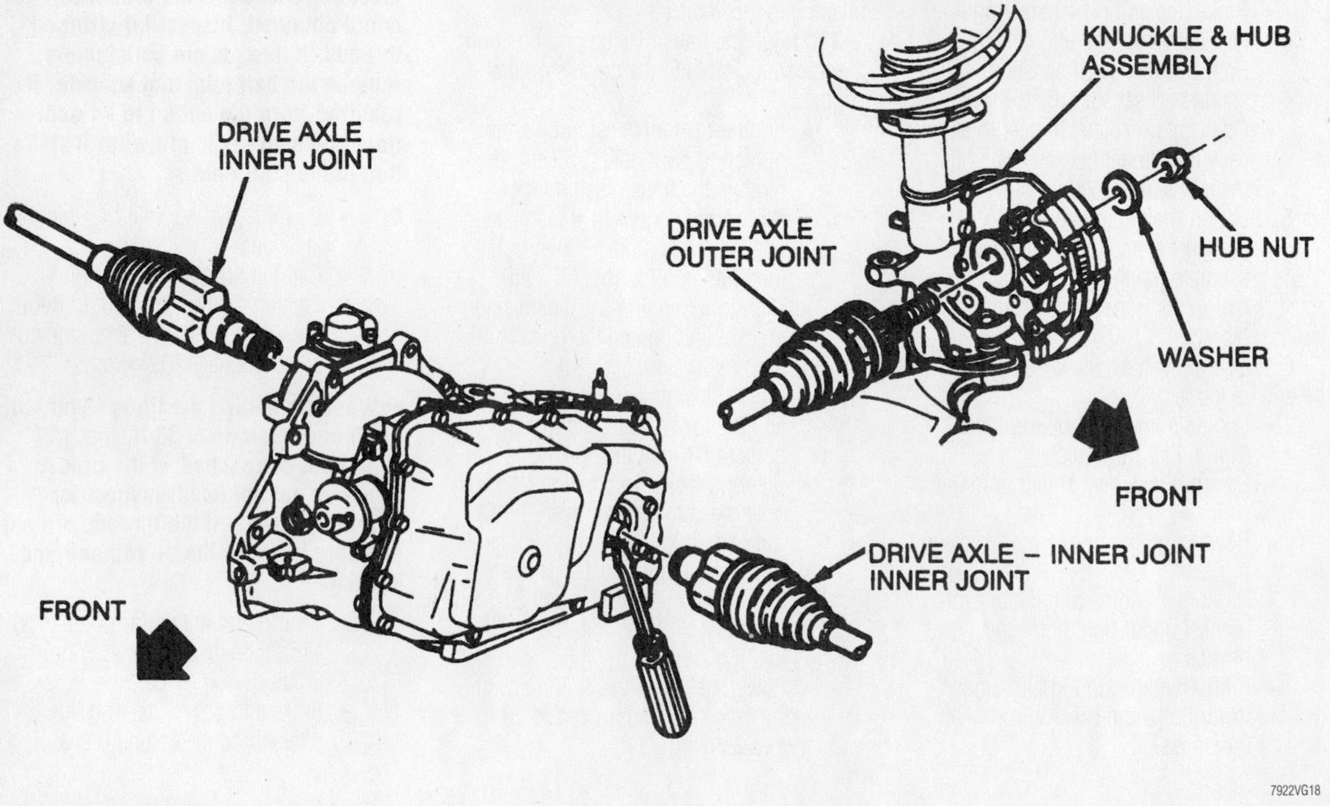

Removing the halfshaft from the transaxle

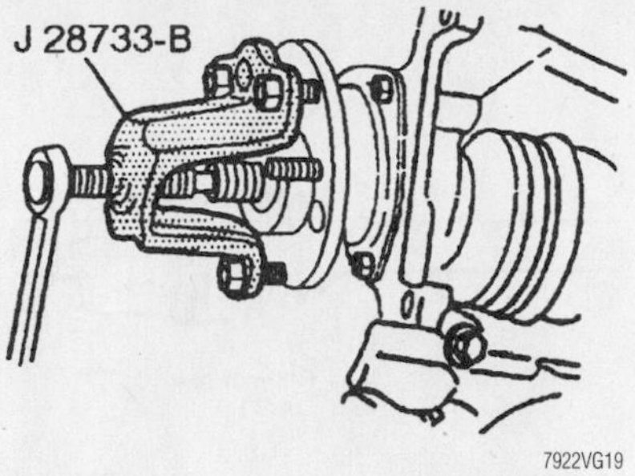

Removing the halfshaft from the hub

1. Before servicing the vehicle, refer to the precautions in the beginning of this section.
2. Remove the front wheel.
3. Install a boot protector on the outer CV-joint boot to protect it from damage.
4. Remove or disconnect the following:
   - Hub nut
   - Stabilizer link
   - Ball joint cotter pin and nut
   - Ball joint from the steering knuckle

➡Partially install the hub nut to protect the threads when removing the halfshaft from the hub.

5. Remove the halfshaft from the transaxle.

➡If equipped with anti-lock brakes, care must be used to prevent damage to the toothed sensor ring on the halfshaft and the wheel speed sensor on the steering knuckle.

To install:

6. Install or connect the following:
   - Halfshaft into the transaxle

➡To verify the halfshaft is properly seated, grasp the inner CV-joint housing and pull it outward. DO NOT pull on the halfshaft. If the CV-joint is properly seated, the halfshaft will not pull back out.

   - Halfshaft into the hub/bearing assembly
   - New hub nut loosely
   - Ball joint into the steering knuckle
   - Ball joint castle nut. Torque the nut to 88 inch lbs. plus an additional 150 degrees. A minimum of 41 ft. lbs. (51 Nm) of torque must be attained.

➡If necessary, the nut can be tightened up to 20 degrees additional. NEVER loosen the castle nut to install the cotter pin.

7. Torque the hub nut to 107 ft. lbs. (145 Nm) on 1998–99 models or 118 ft. lbs. (160 Nm) on 2000–01 models.
8. Install the stabilizer link. Torque the nut to 13 ft. lbs. (17 Nm).
9. If a halfshaft protector was used, remove it now.
10. Install the wheel. Torque the lug nuts to 100 ft. lbs. (140 Nm).
11. Road test and check vehicle operation.

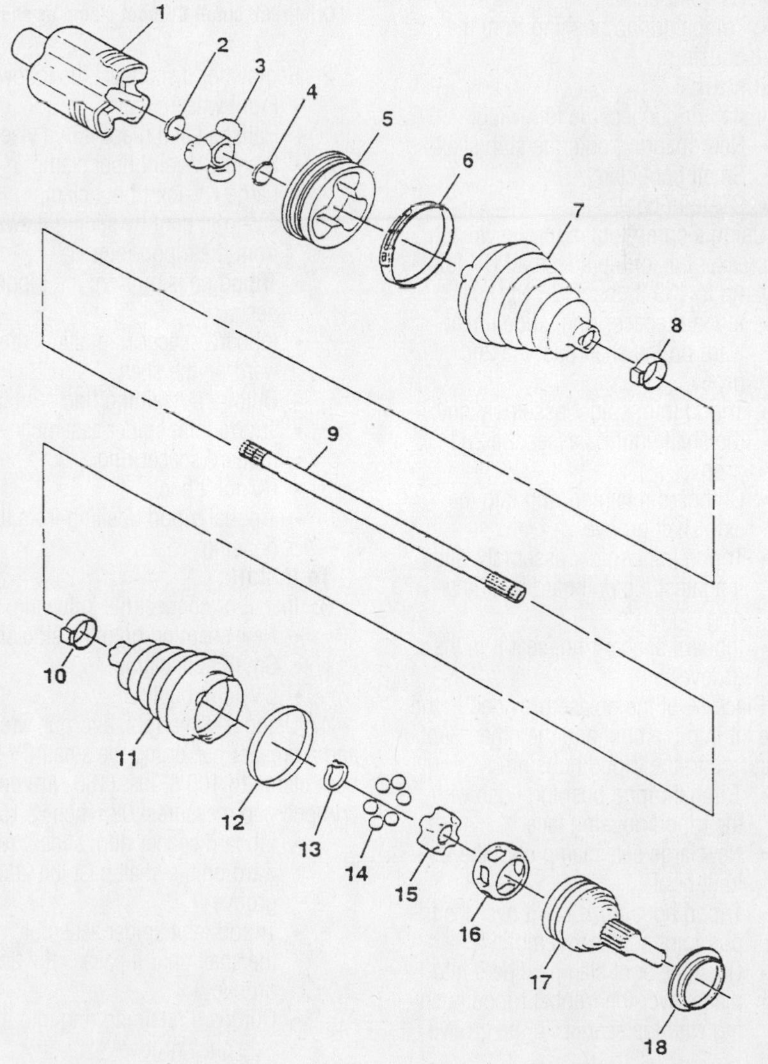

1. Assembly, Tripot Housing
2. Ring, Spacer
3. Spider Assembly, Tripot Joint
4. Ring, Spacer
5. Bushing, Tripot
6. Clamp, Seal Retaining
7. Seal, Tripot Joint
8. Clamp, Small Seal Retaining
9. Shaft, Axle
10. Clamp, Small Seal Retaining
11. Seal, CV Joint
12. Ring, Swage
13. Ring, Race Retaining
14. Ball
15. Race, CV Joint Inner
16. Cage, CV Joint
17. Race, CV Joint Outer
18. Ring, Deflector

9306AG08

Exploded view of CV-joints

CV-Joints

OVERHAUL

Inner (Tripod) Joint

1998–99 MODLES

1. Before servicing the vehicle, refer to the precautions in the beginning of this section.
2. Remove or disconnect the following:

- Front wheel
- Halfshaft and place it in a vise
- Small CV-joint boot clamp
- Large CV-joint boot clamp
- CV-joint boot by sliding it away from the tripod joint
- Tripod housing from the tripod spider
- Inboard spacer ring, slide it rearward on the shaft
- Outboard retaining ring
- Tripod joint spider assembly
- Inboard spacer ring
- CV-joint boot
- Trilobal tripod bushing from the housing

To install:

3. Install or connect the following:
   - New snapring onto the stub shaft
   - Small boot clamp
   - CV-joint boot
4. Using a crimp tool, a torque wrench and a breaker bar, crimp the small CV-joint boot clamp to 130 ft. lbs. (176 Nm).
   - Inboard spacer ring, slide it rearward on the shaft past the 2nd groove
   - Tripod joint spider assembly onto the shaft until it passes the 2nd groove
   - Outboard retaining ring into the axle shaft groove
   - Tripod joint spider assembly, slide it against the outboard retaining ring
   - Inboard spacer ring, seat it in the groove
5. Place ½ of the grease provided in the service kit into the boot and the other ½ of the grease into the tripod housing.
   - Trilobal tripod bushing flush with the tripod housing face
   - New large seal clamp onto the CV-joint boot
   - Tripod housing, slide it over the tripod joint spider assembly
   - CV-joint-boot clamp, slide it into place, over the trilobal tripod bushing with the seal lip in the groove

➡Make sure the boot lies flat against the trilobal bushing.

6. Using a crimp tool, latch the large CV-joint boot clamp.
7. Install the halfshaft and the front wheel. Torque the lug nuts to 100 ft. lbs. (140 Nm).

2000–01 MODELS

1. Before servicing the vehicle, refer to the precautions in the beginning of this section.

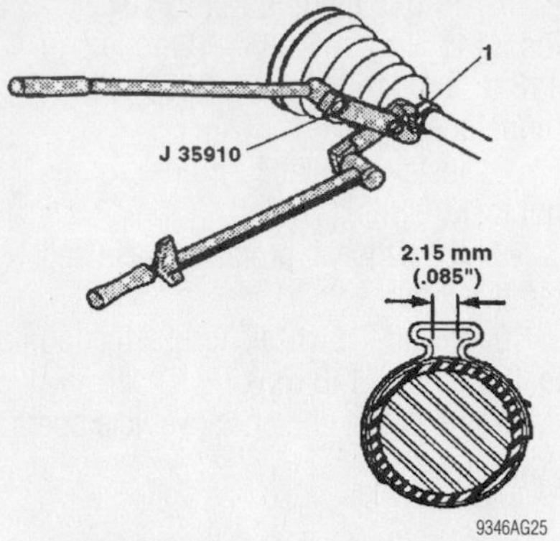

Crimp the small CV-boot clamp as shown

2. Remove or disconnect the following:
   - Front wheel
   - Halfshaft and place it in a vise
   - Small CV-joint boot clamp
   - Large CV-joint boot clamp
   - CV-joint boot by sliding it away from the tripod joint
   - Tripod housing from the tripod spider
   - Inboard spacer ring, slide it rearward on the shaft
   - Outboard retaining ring
   - Tripod joint spider assembly
   - Inboard spacer ring
   - CV-joint boot
   - Trilobal tripod bushing from the housing

To install:

3. Install or connect the following:
   - New snapring onto the stub shaft
   - Small boot clamp
   - CV-joint boot
4. Using a crimp tool, a torque wrench and a breaker bar, crimp the small CV-joint boot clamp to 100 ft. lbs. (136 Nm) until the crimped gap measures .085 inch (2.15mm).
   - Inboard spacer ring, slide it rearward on the shaft past the 2nd groove
   - Tripod joint spider assembly onto the shaft until it passes the 2nd groove
   - Outboard retaining ring into the axle shaft groove
   - Tripod joint spider assembly, slide it against the outboard retaining ring
   - Inboard spacer ring, seat it in the groove
5. Place ½ of the grease provided in the service kit into the boot and the other ½ of the grease into the tripod housing.
   - Trilobal tripod bushing flush with the tripod housing face
   - New large seal clamp onto the CV-joint boot

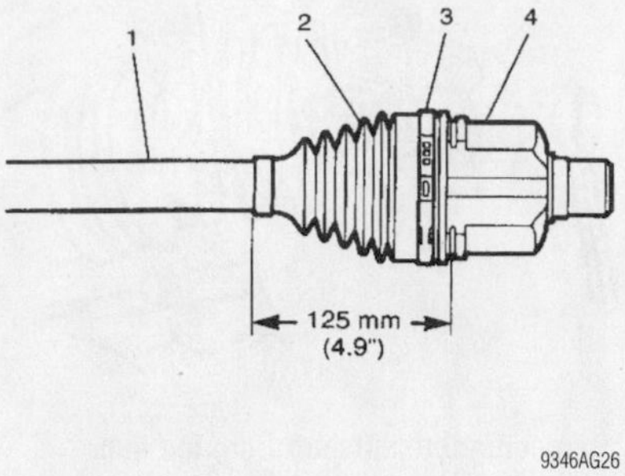

Position the boot as illustrated before clamping the large clamp

   - Tripod housing, slide it over the tripod joint spider assembly
   - CV-joint-boot clamp, slide it into place, over the trilobal tripod bushing with the seal lip in the groove

➡Make sure the boot lies flat against the trilobal bushing.

6. Make sure the boot is positioned as illustrated, then using a crimp tool, latch the large CV-joint boot clamp.
7. Install the halfshaft and the front wheel. Torque the lug nuts to 100 ft. lbs. (140 Nm).

Outer CV-Joint

1. Before servicing the vehicle, refer to the precautions in the beginning of this section.
2. Remove or disconnect the following:
   - Front wheel
   - Halfshaft
   - Swage ring using a hand grinder
   - Large boot clamp
   - CV-joint boot, slide it away from the CV-joint
   - CV-joint assembly by spreading the inner race-to-axle shaft retaining ring ears using Snapring Pliers
   - CV-joint boot from the axle shaft
3. Disassemble the chrome alloy balls from the CV-joint cage as follows:
   a. Position a brass drift against the CV-joint cage and tap it with a hammer to tilt the cage.
   b. Chrome alloy ball from the cage.
   c. Tilt the cage in the opposite direction.
   d. Remove the opposite chrome alloy ball.
   e. Repeat the procedure until all 6 balls are removed.
4. Disassemble the CV-joint cage and inner race as follows:
   a. Pivot the cage and race 90 degrees to the center line of the outer race.

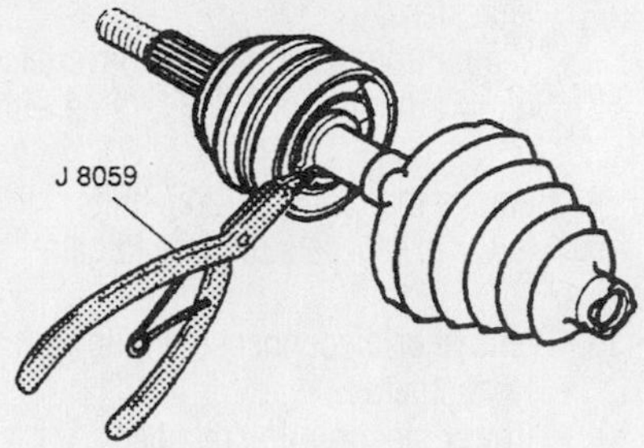

9306XG19

Disconnecting the outer CV-joint from the axle shaft

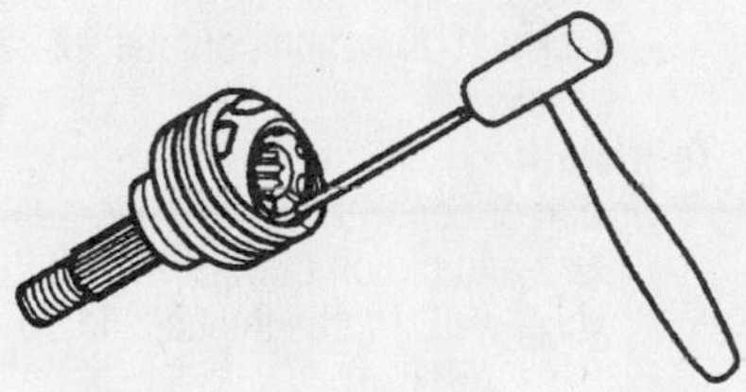

9306XG20

Tilting the cage—Outer CV-joint

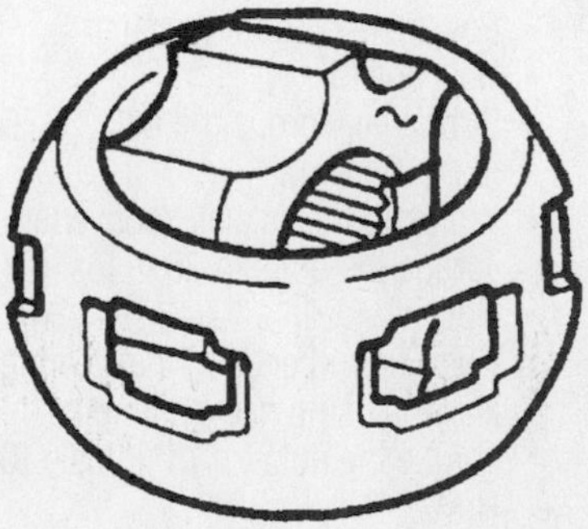

9306XG21

View the cage and inner race—Outer CV-joint

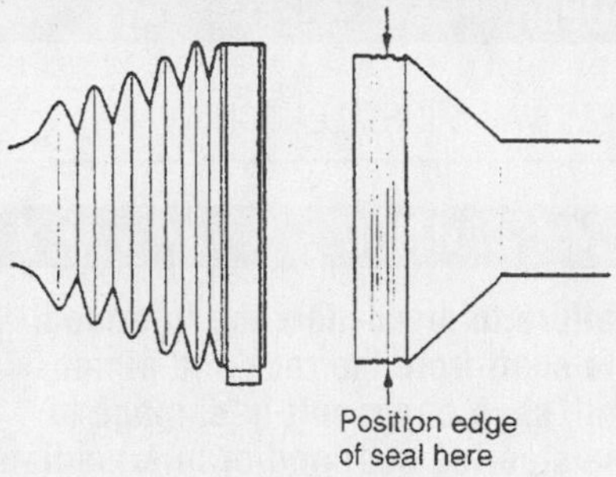

9306XG22

Positioning the boot—Outer CV-joint

b. Align the cage windows with outer race lands.

c. Remove the cage from the outer race.

d. Rotate the inner race upward and remove it from the cage.

5. Thoroughly clean and inspect all parts.

To install:

6. Lubricate the parts with a light coat of grease.

7. Assemble the CV-joint cage and inner race, as follows:

a. Rotate the inner race 90 degrees to the cage centerline.

b. Align the cage windows with inner race lands.

c. Insert the inner race into the cage by rotating the inner race downward.

d. Insert the cage/inner race into the outer race.

8. Assemble the chrome alloy balls into the CV-joint cage, as follows:

a. Position a brass drift against the CV-joint cage and tap it with a hammer to tilt the cage.

b. Insert the 1st chrome alloy ball into the cage.

c. Tilt the cage in the opposite direction.

d. Insert the opposite chrome alloy ball.

e. Repeat the procedure until all 6 balls are inserted.

9. Install or connect the following:
- Swage ring clamp
- CV-joint boot
- CV-joint onto the axle shaft until the retaining ring seats into the groove

10. Position the CV-joint boot seal into the axle shaft's joint seal groove and align the swage ring clamp on the boot.

11. Secure the swage ring clamp using appropriate crimping tool.

⁂ WARNING

Make sure that there are no pinch points on the inboard seal.

12. Install or connect the following:
- ½ kit grease into the CV-joint boot
- ½ kit grease into the CV-joint
- New large seal clamp onto the CV-joint boot
- CV-joint boot/clamp, slide it into place, over the outer race with the seal lip in the groove

➡Make sure the boot lies flat against the outer race.

13. Using a Crimp tool, a torque wrench and a breaker bar, crimp the large CV-joint boot clamp to 130 ft. lbs. (176 Nm).

14. Install the halfshaft and the front wheel.

STEERING AND SUSPENSION

Air Bag

⁂ CAUTION

Some vehicles are equipped with an air bag system. The system must be disabled before performing service on or around system components, steering column, instrument panel components, wiring and sensors. Failure to follow safety and disabling procedures could result in accidental air bag deployment, possible personal injury and unnecessary system repairs.

PRECAUTIONS

Several precautions must be observed when handling the inflator module to avoid accidental deployment and possible personal injury.

- Never carry the inflator module by the wires or connector on the underside of the module.
- When carrying a live inflator module, hold securely with both hands, and ensure that the bag and trim cover are pointed away.
- Place the inflator module on a bench or other surface with the bag and trim cover facing up.
- With the inflator module on the bench, never place anything on or close to the module, which may be thrown in the event of an accidental deployment.

DISARMING

⁂ CAUTION

The air bag system must be disarmed before performing service procedures around the air bag or wiring. Failure

Timing belt service is covered in Section 3 of this manual

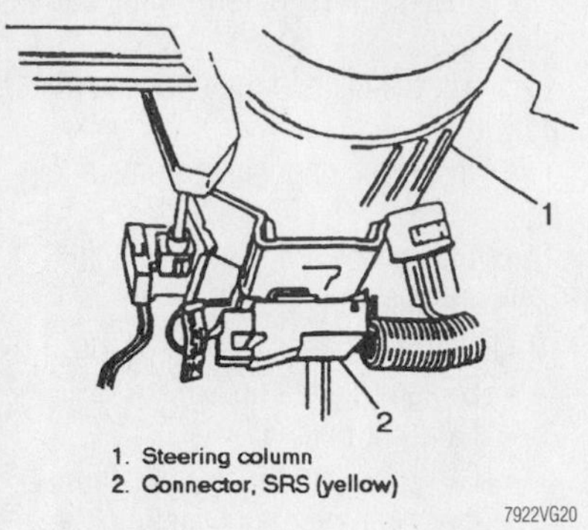

Detach the SRS yellow 2-way connector

to do so may cause accidental deployment, resulting in unnecessary repairs and/or personal injury.

1. Before servicing the vehicle, refer to the precautions in the beginning of this section.
2. Turn the steering wheel so that the vehicle's wheels are pointing straight-ahead.
3. Turn the ignition switch to the **LOCK** position and remove the key.
4. Remove or disconnect the following:
   - Negative battery cable
   - **AIR BAG** fuse from the fuse block
   - Left sound insulator
5. Disconnect the connector retainer clip from the yellow 2-way connector at the base of the steering column, and detach the harness.
6. Remove the connector retainer clip and detach the yellow 2-way connector from the passenger air bag lead located behind the instrument panel.

ARMING

After the applicable service is concluded, enable the air bag system as follows:

1. Turn the ignition switch to the **LOCK** position and remove the key.
2. Install or connect the following:
   - Yellow 2-way connector at the base of steering column and secure it with the connector retainer clip. Attach the yellow 2-way connector at the passenger air bag lead and secure it with the connector retainer clip.
   - Left sound insulator
   - **AIR BAG** fuse in the fuse block
3. Connect the negative battery cable.
4. Turn the ignition switch to the **RUN** position and verify that the **AIR BAG** warning lamp flashes 7 times, then turns **OFF**.

Power Rack and Pinion Steering Gear

REMOVAL & INSTALLATION

⁂ CAUTION

Failure to disconnect the intermediate shaft from the rack and pinion stub shaft can result in damage to the steering gear and/or intermediate shaft. This damage can cause loss of steering control, which could result in personal injury.

⁂ WARNING

The wheels of the vehicle must be straight-ahead and the steering column in the LOCK position before disconnecting the steering column or intermediate shaft from the steering gear. Failure to do so will cause the coil assembly in the steering column to become off-centered, which will cause damage to the coil assembly.

1998–01 Eldorado and 1998–99 Deville

1. Before servicing the vehicle, refer to the precautions in the beginning of this section.
2. Remove or disconnect the following:
   - Negative battery cable
   - Front wheels
   - Bolt holding the intermediate shaft to the steering shaft
   - Road Sensing Suspension (RSS) sensor from the lower control arm
   - Outer tie rod ends from the steering knuckles
   - Y-pipe from the catalytic converter
3. Support the rear of the subframe with a jack.
4. Remove the rear subframe bolts.
5. Slowly lower the subframe to gain access.
6. Remove or disconnect the following:
   - Heat shield
   - Power steering line retainer
   - Power steering pressure and return lines
   - Speed Sensitive Steering (SSS) solenoid valve connector
   - 5 power steering rack-to-subframe bolts
   - Rack and pinion unit out the left wheel well

To install:

7. Install or connect the following:
   - Rack and pinion through the left wheel well. Torque the bolts to 50 ft. lbs. (68 Nm).
   - SSS solenoid valve connector
   - Power steering pressure and return lines. Torque the fittings to 20 ft. lbs. (27 Nm).
   - Power steering line retainer
   - Heat shield
   - Subframe. Torque the bolts to 76 ft. lbs. (103 Nm).
   - Y-pipe to the catalytic converter. Torque the bolts to 20 ft. lbs. (27 Nm).
   - Outer tie rod ends to the steering knuckle using new cotter pins. Torque the nuts to 7.5 ft. lbs. (10 Nm).
   - RSS sensor to the lower control arm

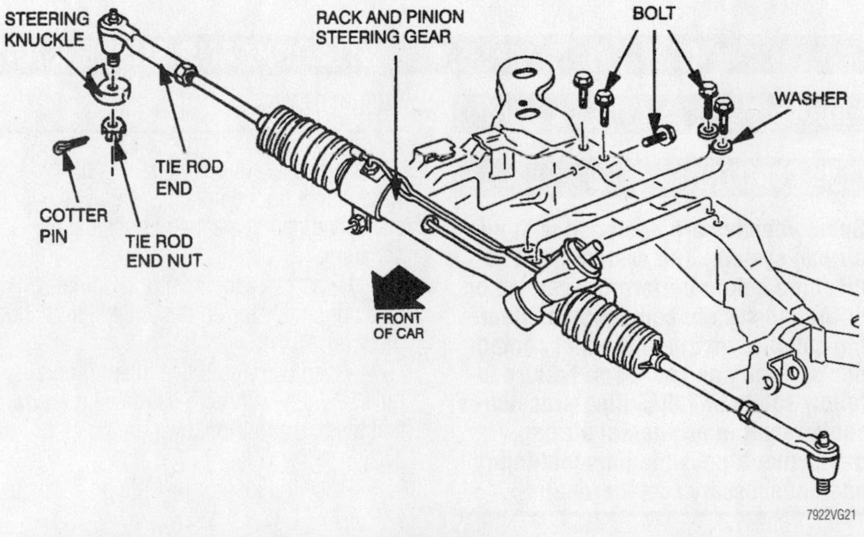

Exploded view of the power steering rack components

- Intermediate shaft to the steering shaft. Torque the bolt to 35 ft. lbs. (47 Nm).
- Front wheels. Torque the lug nuts to 100 ft. lbs. (140 Nm).
- Negative battery cable

8. Refill and bleed the power steering system.
9. Check the wheel alignment.
10. Road test the vehicle.

2000–01 Deville

1. Before servicing the vehicle, refer to the precautions in the beginning of this section.
2. Lock the steering column by placing the wheels in the straight ahead position and remove the keys.
3. Remove or disconnect the following:
   - Intermediate shaft pinch bolt and
   - Intermediate shaft lower coupling
   - Heat shield
   - Rear transaxle mount
   - Pressure and return lines from steering gear
   - Electrical connector
   - Steering gear attaching bolts
   - Steering gear assembly

To install

4. Install or connect the following:
   - Steering gear into vehicle. Torque the bolts to 89 ft. lbs. (120 Nm).
   - Electrical connector
   - Pressure and return lines. Torque the lines to 22 ft. lbs. (30 Nm).
   - Heat shield
   - Rear transaxle mount
   - Intermediate shaft to steering gear. Torque the pinch bolt to 33 ft. lbs. (45 Nm).
5. Fill and bleed the power steering system.

Seville

1. Before servicing the vehicle, refer to the precautions in the beginning of this section.
2. Lock the steering column by installing the Steering Column Anti Rotation Pin tool J 42640 into the underside of the steering column.
3. Remove or disconnect the following:
   - Intermediate shaft lower coupling
   - Heat shield
   - Right transaxle mount
   - Pressure and return lines from steering gear
   - Electrical connector

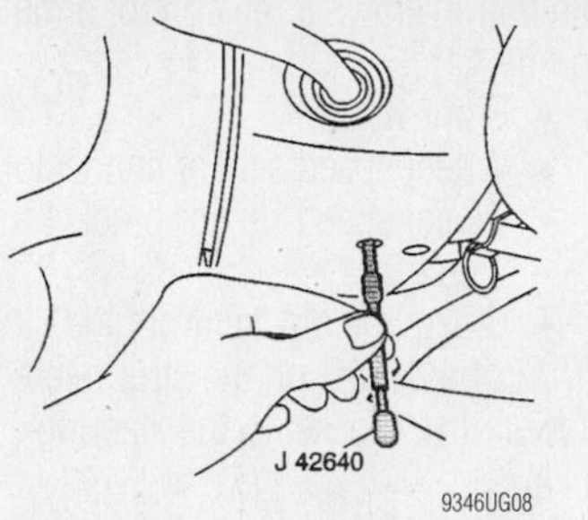

Lock the steering column by installing the Steering Column Anti Rotation Pin tool J 42640 into the underside of the steering column

   - Steering gear attaching bolts
   - Steering gear assembly

To install

4. Install or connect the following:
   - Steering gear into vehicle. Torque the bolts to 89 ft. lbs. (120 Nm).
   - Electrical connector
   - Pressure and return lines. Torque the lines to 22 ft. lbs. (30 Nm).
   - Heat shield
   - Right transaxle mount. Torque the nuts to 37 ft. lbs. (50 Nm).
   - Intermediate shaft to steering gear. Torque the pinch bolt to 33 ft. lbs. (45 Nm).
5. Remove the steering column anti rotation pin from the steering column.
6. Fill and bleed the power steering system.

Strut

REMOVAL & INSTALLATION

Front

1998–99 MODELS AND 1998–01 ELDORADO

✲✲ WARNING

When working near the halfshafts, use care to prevent the inner Tripod CV-joint from being overextended. If the joint is overextended, the internal joint components could separate, resulting in CV-joint failure.

1. Before servicing the vehicle, refer to the precautions in the beginning of this section.

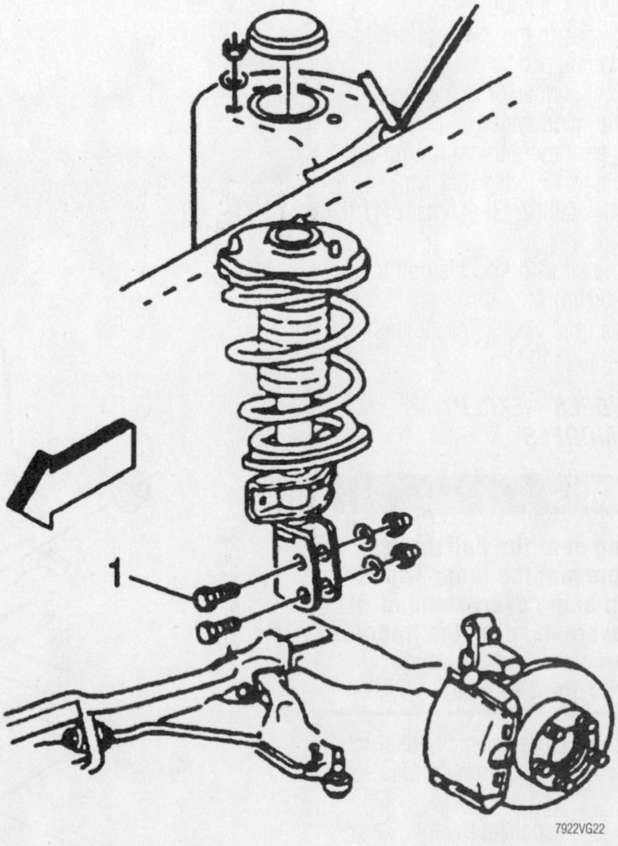

Exploded view of the strut mounting

Heater Core replacement is covered in Section 2 of this manual

2. Remove or disconnect the following:
   - Negative battery cable
   - Front wheel
   - Electrical connector from the top of the strut, if equipped
   - Upper strut mounting nuts
   - Road Sensing Suspension (RSS) position sensor from the lower control arm, if equipped
   - Anti-lock Brake System (ABS) wheel speed sensor, if equipped
   - Wheel speed sensor from the strut bracket, if equipped
   - Brake line bracket from the strut
   - Stabilizer link from the strut
3. Scribe a mark on the strut referencing the lower strut bracket to the steering knuckle.
4. Remove the strut.

To install:

5. Install or connect the following:
   - Strut assembly into the vehicle
   - Strut-to-knuckle bolts and nuts, but do not torque yet
   - Stabilizer link to the strut assembly, but do not torque the nuts yet
   - Brake line bracket to the strut
   - Speed sensor bracket on the strut, if equipped
   - ABS sensor, if equipped
   - RSS sensor to the lower control arm, if equipped
   - Electrical connector to the top of the strut, if equipped
   - Upper strut mounting nuts. Torque the nuts to 15 ft. lbs. (21 Nm).
6. Torque the stabilizer link nuts to 41 ft. lbs. (55 Nm).
7. Torque the strut-to-knuckle bolt to 140 ft. lbs. (190 Nm).
8. Install the front wheel. Torque the lug nuts to 100 ft. lbs. (140 Nm).

2000–01 MODELS—EXCEPT ELDORADO MODELS

WARNING

When working near the halfshafts, use care to prevent the inner Tripod CV-joint from being overextended. If the joint is overextended, the internal joint components could separate, resulting in CV-joint failure.

1. Before servicing the vehicle, refer to the precautions in the beginning of this section.
2. Remove or disconnect the following:
   - Negative battery cable
   - 3 strut mount bolts
3. Raise the vehicle and suitably support by the frame allowing the control arms to hang free.
   - Front wheel
   - Wheel speed sensor connector
   - Wheel speed sensor bracket from the strut
   - Brake line bracket from the strut
4. Scribe a mark on the strut referencing the lower strut bracket to the steering knuckle.
   - Strut-to-knuckle bolts
   - Strut from the vehicle

To install:

   - Strut
   - 3 upper strut mount bolts and washers. Tighten the bolts to 35 ft. lbs. (48 Nm) on 1998–99 Seville models or 49 ft. lbs. (66 Nm) on all models.
5. Align the scribe marks made on the steering knuckle with the strut.
   - Strut-toknuckle bolts and nuts and hand tighten
   - Brake line bracket to strut
   - Wheel speed sensor bracket to strut
   - Wheel speed sensor connector
6. Tighten the strut-to-knuckle bolts to 125 ft. lbs. (170 Nm) on 1998–99 Seville models or 131 ft. lbs. (177 Nm) on all other models.
7. Tighten the brake line and wheel speed sensor bracket to 13 ft. lbs. (17 Nm) on 1998–99 Seville models or 17 ft. lbs. (23 Nm) on all other models.
8. Install the front wheel.

Shock Absorber

REMOVAL & INSTALLATION

Rear

1998–99 DEVILLE AND 1998–01 ELDORADO

1. Before servicing the vehicle, refer to the precautions in the beginning of this section.
2. Remove or disconnect the following:
   - Rear wheel
   - Negative battery cable
   - Shock absorber electrical connector from the rear suspension support
3. Support the lower control arm with a jack to relieve the tension on the shock absorber.
   - Lower shock absorber mounting bolt and nut
   - Upper mounting nut, retainer and insulator

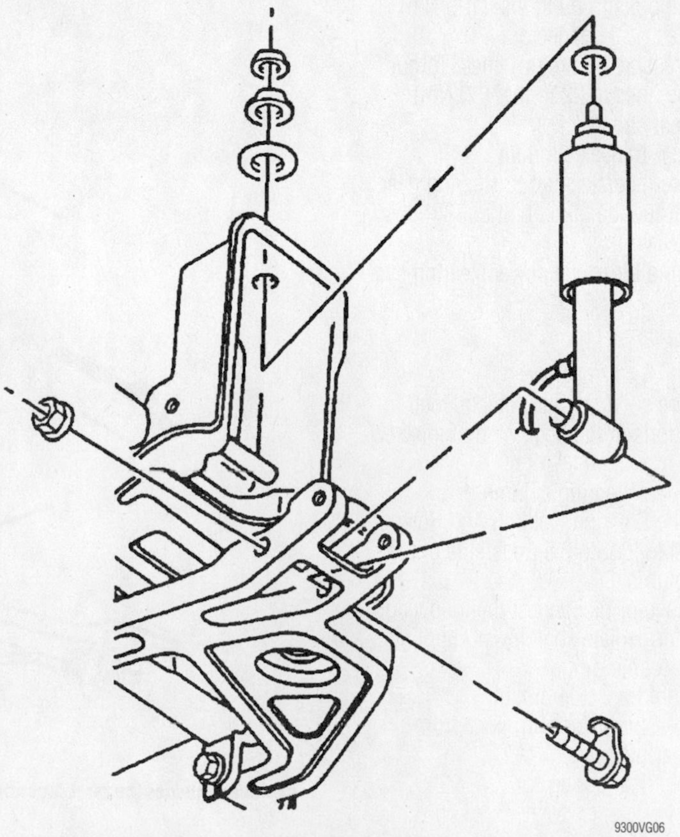

Exploded view of the rear shock absorber mounting

- Shock absorber, compress by hand and remove through the upper control arm

To install:

4. Position the top of the shock absorber with the insulator attached into the suspension support.
5. Install or connect the following:
   - Upper shock insulator, retainer and nut. Torque the nut to 55 ft. lbs. (74 Nm).
   - Shock absorber lower mounting nut and bolt. Torque the nut to 75 ft. lbs. (102 Nm).
   - Shock absorber electrical connector to the rear suspension support
   - Rear wheel. Torque the lug nuts to 100 ft. lbs. (140 Nm).
   - Negative battery cable

SEVILL AND 2000–01 DEVILLE

1. Before servicing the vehicle, refer to the precautions in the beginning of this section.
2. Remove the tire and wheel assembly.
3. Support the control arm with a jack stand.
4. Remove or disconnect the following:
   - Electronic Level Control (ELC) air tube from the shock
   - Two bolts securing the shock to the control arm
   - Trunk trim to gain access to the shock upper mounting nuts.
   - Cover, the two nuts, and the reinforcement from the top of the shock
   - Shock from the vehicle

To install:

5. Install or connect the following:
   - Shock, reinforcement, and the two nuts. Tighten the mounting nuts to 15 ft. lbs. (20 Nm).
   - Shock cover
   - Trunk trim
   - Shock-to-control arm bolts and tighten the bolts to 18ft. lbs. (24 Nm)
   - ELC air tube to the shock.
   - Tire and wheel assembly

Coil Spring

REMOVAL & INSTALLATION

Front

1. Before servicing the vehicle, refer to the precautions in the beginning of this section.
2. Remove the strut from the vehicle.
3. Disassemble the strut as follows:

a. Step 1: Place the strut assembly into compressor tool, to compress the coil spring.

b. Step 2: Compress the spring slightly.

c. Step 3: Hold the strut shaft from turning using a No. 50 Torx® socket and remove the 24mm nut on the top end of the strut.

S1>Step 4: Install Rod Tool J 34013-38 to help guide the strut shaft from the upper mount assembly.

d. Step 5: Loosen the spring compressor tool until the coil spring and mount can be removed as an assembly. Remove the lower spring insulator, if equipped.

To install:

4. Assemble the strut as follows:

a. Step 1: Place the strut in compressor tool.

b. Step 2: Install the coil spring over the strut.

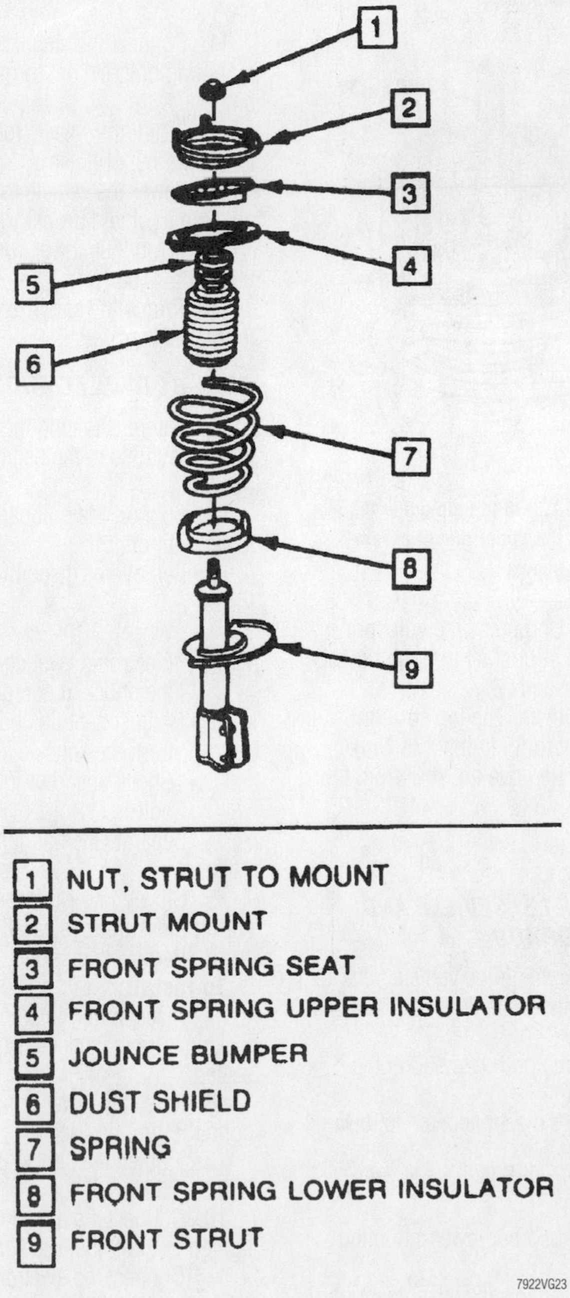

Disassembled view of strut

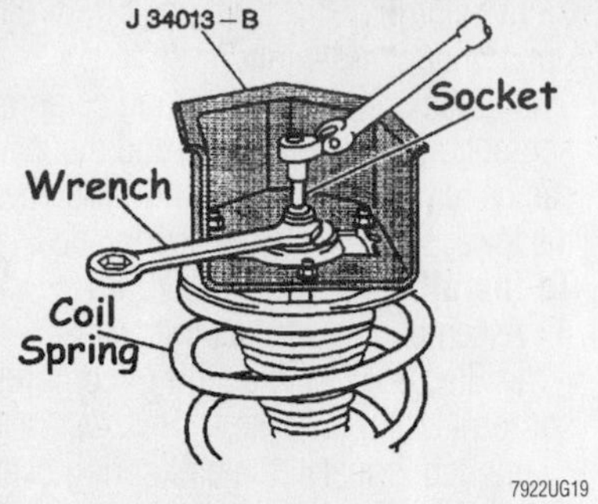

Use a Torx® socket to keep the piston rod from turning while removing the upper nut—front strut shown

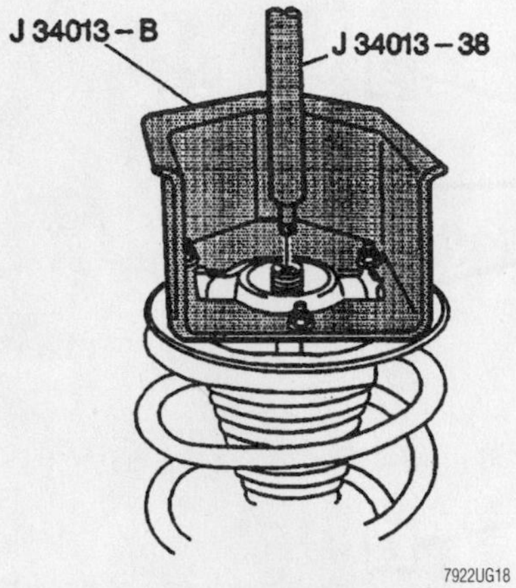

Install Rod J 34013-38 to help guide the strut shaft from the upper mount assembly—front strut shown

c. Step 3: Compress the coil spring while guiding strut shaft through the top of the strut assembly.

d. Step 4: Install the top strut nut. Torque the nut to 55 ft. lbs. (75 Nm).

e. Step 5: Remove the strut from the compressor.

Rear

1998–99 DEVILLE, SEVILLE AND 1998–01 ELDORADO

1. Before servicing the vehicle, refer to the precautions in the beginning of this section.
2. Support the outboard end of the lower control arm.
3. Remove or disconnect the following:
   - Wheel
   - Stabilizer link lower mounting bolt
   - Shock absorber lower mounting bolt
   - Height sensor link attachment from the lower control arm
4. Separate the ball joint from the control arm.
5. Slowly lower the control arm until spring pressure is released.
6. Remove the spring.

To install:

7. Place spring into lower control arm.
8. Raise the lower control arm until the spring is properly seated. Install the bolts finger tight, do not torque the bolts until proper trim height is obtained.
9. Insert the ball joint into the knuckle. Torque the nut to 129 ft. lbs. (175 Nm).
10. Install the shock absorber lower mounting bolt. Torque the bolt to 80 ft. lbs. (108 Nm).
11. Connect the stabilizer link lower mounting bolt. Do not torque the bolt at this time.
12. Install the wheel. Torque the lug nuts to 100 ft. lbs. (140 Nm).
13. Lower the vehicle to allow the suspension to obtain proper trim height.
14. Torque the lower control arm nuts to 80 ft. lbs. (108 Nm).
15. Torque the stabilizer link bolt to 38 ft. lbs. (52 Nm).

2000–01 DEVILLE AND SEVILLE

1. Before servicing the vehicle, refer to the precautions in the beginning of this section.
2. Support the outboard end of the lower control arm.
3. Remove or disconnect the following:
   - Wheel
   - Electronic level control tube from the shock, if equipped
   - Adjustment link bolt and separate it from the knuckle
   - Shock absorber lower mounting bolt
4. Separate the ball joint from the control arm.
5. Slowly lower the control arm until spring pressure is released.
6. Remove the spring.

To install:

7. Place spring into lower control arm making sure the insulator is properly seated.
8. Raise the lower control arm until the spring is properly seated.
9. Tighten the shock lower bolts to 18 ft. lbs. (25 Nm).
10. Connect the adjustment link bolt and tighten to 22 ft. lbs. (30 Nm)
11. Connect the electronic level control tube to the shock, if equipped.
12. Install the wheel.
13. Check the alignment.

Lower Ball Joint

REMOVAL & INSTALLATION

1998–99 Models

EXCEPT COMMERCIAL CHASSIS

1. Before servicing the vehicle, refer to the precautions in the beginning of this section.
2. Allow the front suspension to hang free.
3. Remove or disconnect the following:
   - Front wheel

CAUTION

Be careful when working in the area of the CV-boot. Damage to the boot could result in eventual joint failure.

   - Road Sensing Suspension (RSS) position sensor from the lower control arm, if equipped
   - Ball joint from the steering knuckle
   - Ball joint by drilling out the 3 ball joint-to-lower control arm rivets

To install:

4. Install the new ball joint to the lower control arm with 3 new bolts and nuts. The bolts must be installed from the bottom of the control arm. Torque the nuts to 50 ft. lbs. (68 Nm).
5. Install or connect the following:
   - Ball joint to the steering knuckle. Torque the nut to 84 inch lbs. (10 Nm) plus an additional 120 degree turn during which a minimum torque of 37 ft. lbs. (50 Nm) must be obtained.

➡If the minimum torque is not obtained, check for stripped threads. If the threads are okay, replace the ball joint and knuckle.

   - New cotter pin
   - RSS position sensor to the lower control arm if equipped
6. If used, remove the CV-Joint boot protector tool.
7. Install the front wheel. Torque the lug nuts to 100 ft. lbs. (140 Nm).

COMMERCIAL CHASSIS—FRONT

1. Before servicing the vehicle, refer to the precautions in the beginning of this section.

CAUTION

Be careful when working in the area of the CV-boot. Damage to the boot could result in eventual joint failure.

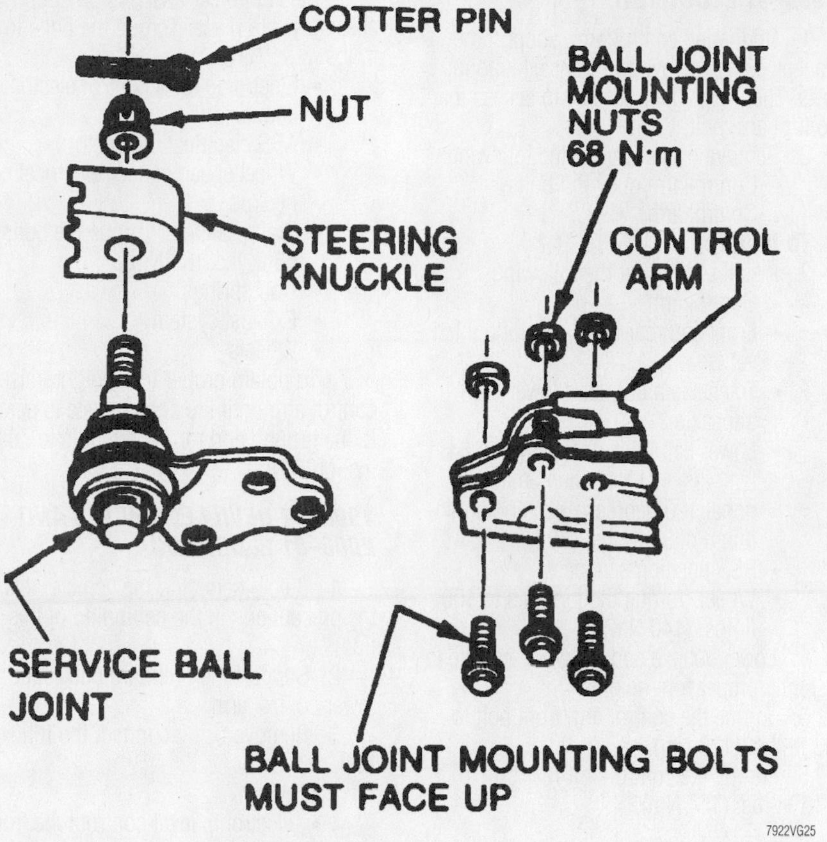

Exploded view of the lower ball joint mounting

2. Remove or disconnect the following:
   - Wheel
   - Ball joint cotter pin and nut
   - Ball joint from steering knuckle
3. Press the ball joint from the lower control arm.

To install:

4. Press the ball joint into the control arm.
5. Install or connect the following:
   - Hex nut. Torque the nut to 129 ft. lbs. (175 Nm).
   - New cotter pin

➡If the cotter pin cannot be installed because the hole in the stud does not align with a nut slot, torque the nut an additional 60 degrees to allow for installation. NEVER loosen the nut to provide for cotter pin installation.

6. Install the wheel and torque the lug nuts to 100 ft. lbs. (140 Nm).

COMMERCIAL CHASSIS—REAR

1. Before servicing the vehicle, refer to the precautions in the beginning of this section.
2. Remove the wheel.
3. Support the lower control arm.

➡The tension of the coil spring must be supported with a suitable jack stand.

4. Remove the cotter pin and ball stud nut.
5. Separate the ball joint from the knuckle.
6. Press the ball joint out of the control arm.

To install:

7. Press the ball joint into the control arm.
8. Install the ball joint into the knuckle.
9. Install the ball stud nut. Torque the nut to 129 ft. lbs. (175 Nm) and install a new cotter pin.
10. Remove the support from the lower control arm.
11. Install the wheel. Torque the lug nuts to 100 ft. lbs. (140 Nm).

2000–01 Models

The ball joint is an integral part of the lower control arm and if found to be defective the control arm should be replaced.

Lower Control Arm

REMOVAL & INSTALLATION

Front

1998–99 SEVILLE

1. Before servicing the vehicle, refer to the precautions in the beginning of this section.
2. Remove or disconnect the following:
   - Wheel
   - Road Sensing Suspension (RSS) sensor, if equipped
   - Stabilizer link bolt from the control arm
   - Ball joint from the steering knuckle
   - Mounting bolts
   - Lower control arm

To install:

3. Install or connect the following:
   - Lower control arm and hand-tighten the bolts
   - Stabilizer link bolt to the control arm and hand-tighten
   - Ball joint to lower control arm.
4. Torque the stabilizer link bolt to 13 ft. lbs. (17 Nm).
5. Torque the nut to 88 inch lbs. (10 Nm) plus an additional 120 degree turn till the torque is a minimum 41 ft. lbs. (55Nm).
   - New cotter pin
   - Wheel. Torque the lug nuts to 100 ft. lbs. (140 Nm).
6. Lower the vehicle and torque the control arm front nut to 116 ft. lbs. (157 Nm)) and the rear nut to 117 ft. lbs. (158 Nm).

1998–99 DEVILLE AND ELDORADO

1. Before servicing the vehicle, refer to the precautions in the beginning of this section.
2. Remove or disconnect the following:
   - Wheel
   - Road Sensing Suspension (RSS) sensor

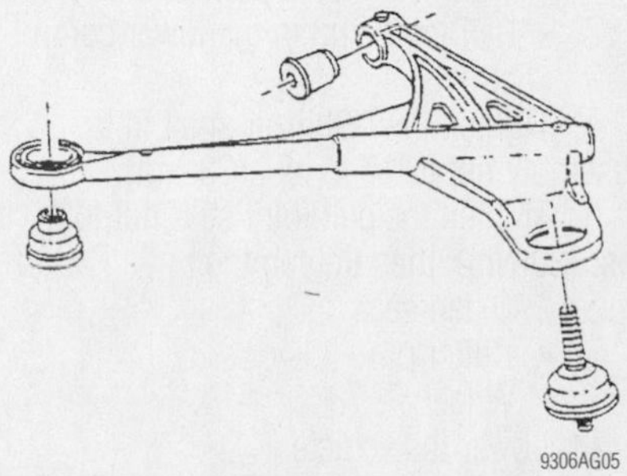

Exploded view of the lower control arm

- Lower ball joint from the steering knuckle

3. On the left control arm, support the transaxle and remove two transaxle mount nuts. Then raise the transaxle to access the control arm bolt.
4. Remove or disconnect the following:
   - Control arm nuts and bolts
   - Control arm

To install:

5. Install or connect the following:
   - Control arm
   - Front and rear bolts, do not tighten yet
   - Transaxle mount and lower the transaxle
   - Lower ball joint. Torque nut to 84 inch lbs. (10 Nm) then an additional 120 degrees, reaching a minimum of 37 ft. lbs. (50 Nm).
   - RSS position sensor
   - Wheel. Torque the lug nuts to 100 ft. lbs. (140 Nm).
6. Lower vehicle and allow the vehicle to assume proper trim heights.
7. Torque the control arm front bolt to 93 ft. lbs. (126 Nm).
8. Torque the control arm rear bolt to 116 ft. lbs. (157 Nm).

2000–01 DEVILLE AND SEVILLE

1. Before servicing the vehicle, refer to the precautions in the beginning of this section.
2. Remove or disconnect the following:
   - Wheel
   - Stabilizer link bolt from the control arm
   - Ball joint from the lower control arm
   - Control arm mounting nuts and bolts
   - Control arm

To install:

3. Install or connect the following:
   - Control arm
   - Front and rear bolts, do not tighten yet
   - Stabilizer link to control arm
   - Ball joint stud in the lower control arm
4. Tighten the stabilizer shaft link assembly nut to 17 ft. lbs. (23 Nm).
5. Tighten the ball joint stud nut to 22 ft. lbs. (30 Nm), then tighten the nut an additional 190 degrees.
   - Cotter pin
   - Wheel
6. Lower the vehicle.
7. Bounce the vehicle and tighten the front lower control arm nut to 120 ft. lbs. (162 Nm) and the rear nut to 108 ft. lbs. (146 Nm).

2000–01 ELDORADO

1. On the left control arm, support the transaxle and remove two transaxle mount nuts. Then raise the transaxle to access the control arm bolt.
2. Remove or disconnect the following:
   - Control arm nuts and bolts
   - Control arm

To install:

3. Install or connect the following:
   - Control arm
   - Front and rear bolts, do not tighten yet
   - Transaxle mount and lower the transaxle
   - Lower ball joint. Torque nut to 84 inch lbs. (10 Nm) then an additional 120 degrees, reaching a minimum of 37 ft. lbs. (50 Nm).
   - RSS position sensor
   - Wheel. Torque the lug nuts to 100 ft. lbs. (140 Nm).
4. Lower vehicle and allow the vehicle to assume proper trim heights.
5. Torque the control arm front bolt to 93 ft. lbs. (126 Nm).
6. Torque the control arm rear bolt to 116 ft. lbs. (157 Nm).

Rear

1998–99 DEVILLE AND SEVILLE

1. Before servicing the vehicle, refer to the precautions in the beginning of this section.
2. Remove or disconnect the following:
   - Wheels
   - Exhaust system
   - Coil springs
   - Brake calipers
   - Wheel speed sensor electrical connectors
   - Antilock Brake System (ABS) electrical connectors
   - Electronic level control electrical connectors
3. Support the rear suspension support assembly.
4. Remove the bolts and lower the support assembly from the vehicle.
   - Stabilizer link bolt
   - Hub and bearing
   - Lower control arm

To install

5. Install or connect the following:
   - Lower control arm. Do not torque the bolts at this time.
   - Hub and bearing. Torque the bolts to 52 ft. lbs. (70 Nm).
   - Stabilizer link. Torque the bolt to 11 ft. lbs. (15 Nm).
6. Raise the rear suspension support assembly into place. Torque the bolts to 141 ft. lbs. (191 Nm).
   - Electronic level control electrical connectors
   - ABS electrical connectors
   - Wheel speed sensor electrical connectors
   - Brake calipers. Torque the bolts to 63 ft. lbs. (85 Nm).
   - Coil springs
   - Exhaust system
   - Wheels
7. To obtain proper trim height, raise the control arm until the spindle face is parallel to the ground and torque the nuts to 78 ft. lbs. (106 Nm).

1998–99 DEVILLE, SEVILLE AND 2000–01 ELDORADO

1. Before servicing the vehicle, refer to the precautions in the beginning of this section.
2. Support the outboard end of the lower control arm.
3. Remove or disconnect the following:
   - Wheel
   - Electronic level control tube from the shock, if equipped
   - Adjustment link bolt and separate it from the knuckle
   - Shock absorber lower mounting bolt
4. Separate the ball joint from the control arm.
5. Slowly lower the control arm until spring pressure is released.
6. Remove the spring.
7. Remove the control arm bolt and the arm

To install:

8. Lower control arm and inboard control arm bolt and nut and tighten to 75 ft. lbs. (102 Nm).
9. Install the spring and insulators.
10. Raise the lower control arm until the spring is properly seated.
11. Install the stabilizer link lower attachment.
12. Remove the transmission jack. Place the jack under the outboard end of the lower control arm in order to bring the suspension to the suspension's design position.
13. The inner control arm nuts must be tightened in the design position in order to reduce wind up in the bushings as follows:
    a. Stabilizer link lower nut to 44 ft. lbs. (60 Nm).
    b. Tighten the shock absorber lower nut to 75 ft. lbs. (102 Nm).

c. Tighten the lower control arm inner nuts to 75 ft. lbs. (102 Nm).

14. Install the wheel.

2000–01 DEVILLE AND SEVILLE

1. Before servicing the vehicle, refer to the precautions in the beginning of this section.
2. Remove or disconnect the following:
   - Wheels
   - Exhaust system
   - Coil springs
   - Brake calipers
   - Wheel speed sensor electrical connectors
   - Antilock Brake System (ABS) electrical connectors
   - Electronic level control electrical connectors
3. Support the rear suspension support assembly.
4. Remove the bolts and lower the support assembly from the vehicle.
   - Stabilizer link bolt
   - Hub and bearing
   - Lower control arm

To install

5. Install or connect the following:
   - Lower control arm. Do not torque the bolts at this time.
   - Hub and bearing. Torque the bolts to 52 ft. lbs. (70 Nm).
   - Stabilizer link. Torque the bolt to 11 ft. lbs. (15 Nm).
6. Raise the rear suspension support assembly into place. Torque the bolts to 141 ft. lbs. (191 Nm).
   - Electronic level control electrical connectors
   - ABS electrical connectors
   - Wheel speed sensor electrical connectors
   - Brake calipers. Torque the bolts to 63 ft. lbs. (85 Nm).
   - Coil springs
   - Exhaust system
   - Wheels
7. To obtain proper trim height, raise the control arm until the spindle face is parallel to the ground and torque the nuts to 78 ft. lbs. (106 Nm).

CONTROL ARM BUSHING REPLACEMENT

Front

1. Before servicing the vehicle, refer to the precautions in the beginning of this section.
2. Remove or disconnect the following:
   - Wheel
   - Lower control arm
3. Press the bushing out of the lower control arm.

To install

4. Lubricate the outer case of the new bushing.
5. Press the new bushing into the control arm.
6. Install or connect the following:
   - Lower control arm
   - Wheel. Torque the lug nuts to 100 ft. lbs. (140 Nm).

Wheel Bearings

ADJUSTMENT

The wheel bearings are not adjustable. If a wheel bearing is out of specifications, it must be replaced. Using a dial indicator, check for looseness. If play exceeds 0.005 inches (0.127mm), the bearing wear is excessive and the hub and bearing should be replaced.

REMOVAL & INSTALLATION

Front

1. Before servicing the vehicle, refer to the precautions in the beginning of this section.
2. Remove or disconnect the following:
   - Front wheel
   - Halfshaft nut and washer
   - Caliper from the steering knuckle and support it on a wire
   - Brake rotor
   - Anti-lock Brake System (ABS) speed sensor electrical connector
   - 3 hub/bearing assembly to steering knuckle bolts
   - Dust shield
   - Hub and bearing assembly from the halfshaft
   - Hub and bearing assembly from the steering knuckle

To install:

3. Install the hub and bearing assembly over the halfshaft splines.

➡Be sure the splines engage smoothly.

4. Apply a light coating of grease to the steering knuckle bore.
5. Install the hub/bearing assembly onto the halfshaft as far as possible. If the hub will not bottom out on the halfshaft, install the hub bolts and use the hub nut to draw the hub onto the halfshaft.
6. Once the hub is flush with the steering knuckle, remove the mounting bolts and

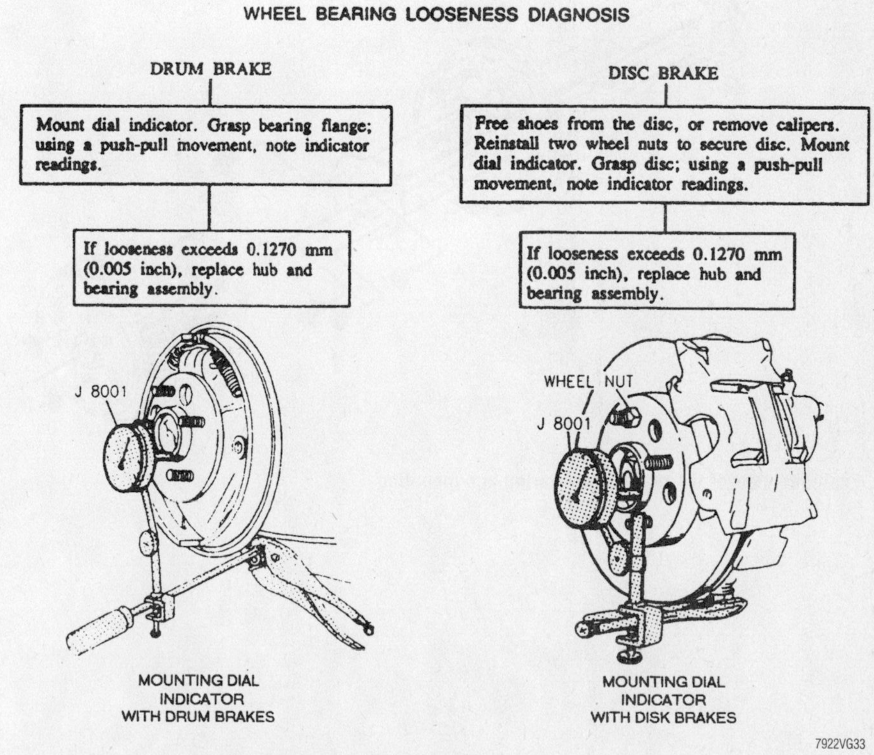

WHEEL BEARING LOOSENESS DIAGNOSIS
DRUM BRAKE
Mount dial indicator. Grasp bearing flange; using a push-pull movement, note indicator readings.
If looseness exceeds 0.1270 mm (0.005 inch), replace hub and bearing assembly.
J 8001
MOUNTING DIAL INDICATOR WITH DRUM BRAKES
DISC BRAKE
Free shoes from the disc, or remove calipers. Reinstall two wheel nuts to secure disc. Mount dial indicator. Grasp disc; using a push-pull movement, note indicator readings.
If looseness exceeds 0.1270 mm (0.005 inch), replace hub and bearing assembly.
WHEEL NUT
J 8001
MOUNTING DIAL INDICATOR WITH DISK BRAKES
7922VG33

Inspect the wheel bearings for play with a dial indicator

For Accessory Drive Belt illustrations, see Section 1 of this manual

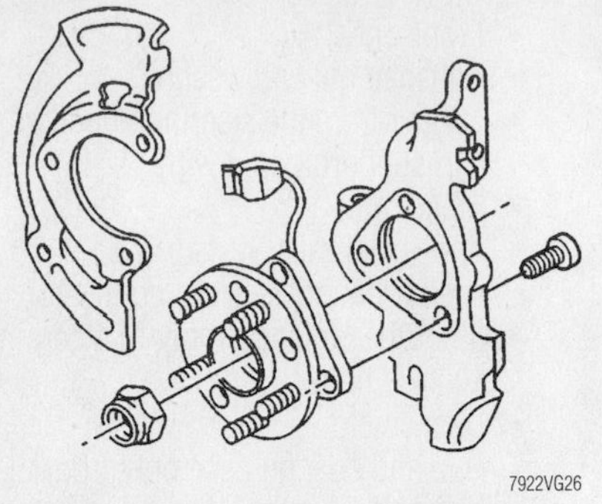

Exploded view of the front wheel bearing mounting

install the dust shield. Reinstall the mounting bolts and torque to 70 ft. lbs. (95 Nm) on 1998–99 models (except Seville) or 95 ft. lbs. (130 Nm) on all Seville and all 2000–01 models .

7. Install or connect the following:
   - ABS speed sensor electrical connector
   - Brake rotor
   - Caliper. Torque the bolts to 38 ft. lbs. (51 Nm).
   - Halfshaft nut. Torque it to 107 ft. lbs. (145 Nm).
   - Wheel. Torque the lug nuts to 100 ft. lbs. (140 Nm).
8. Road test the vehicle for proper operation.

Rear

EXCEPT COMMERCIAL CHASSIS

➡The wheel bearing and hub are serviced as an assembly. The individual components are not serviceable separately.

1. Before servicing the vehicle, refer to the precautions in the beginning of this section.
2. Remove or disconnect the following:
   - Rear wheel
   - Caliper bracket from the knuckle
   - Brake rotor
   - Hub and bearing assembly bolts
   - Hub and bearing assembly

To install:

3. Install or connect the following:
   - Hub and bearing assembly. Torque the bolts to 52 ft. lbs. (70 Nm).
   - Brake rotor
   - Caliper. Torque the new bolts to 83 ft. lbs. (113 Nm).
   - Wheel. Torque the lug nuts to 100 ft. lbs. (140 Nm).
4. Road test the vehicle.

COMMERCIAL CHASSIS

1. Before servicing the vehicle, refer to the precautions in the beginning of this section.
2. Remove or disconnect the following:
   - Wheel
   - Brake drum
   - Brake shoes and cable
   - Brake line
   - Wheel speed sensor
3. Support the lower control arm.
4. Separate the ball joint from the knuckle.
   - Upper knuckle bolt
   - Bearing retainer
   - Nut from backside of knuckle
5. Press the hub from the knuckle.
6. Remove the brake backing plate from the knuckle.
7. Press the bearing from the knuckle.

To install

8. Press the bearing into the knuckle.
9. Install the backing plate and bolts. Torque the bolts to 37 ft. lbs. (50 Nm).
10. Press the hub into the bearing.
11. Install the speed sensor ring and nut. Torque the nut to 145 ft. lbs. (200 Nm).
12. Install or connect the following:
    - Upper knuckle bolt and nut. Torque the nut to 80 ft. lbs. (108 Nm).
    - Ball joint to knuckle
    - Brake line
    - Wheel speed sensor
    - Brake shoes and cable
    - Brake drum
13. Bleed the brake system.
14. Install the wheel. Torque the lug nuts to 100 ft. lbs. (140 Nm).
15. Check the brake system for leaks.

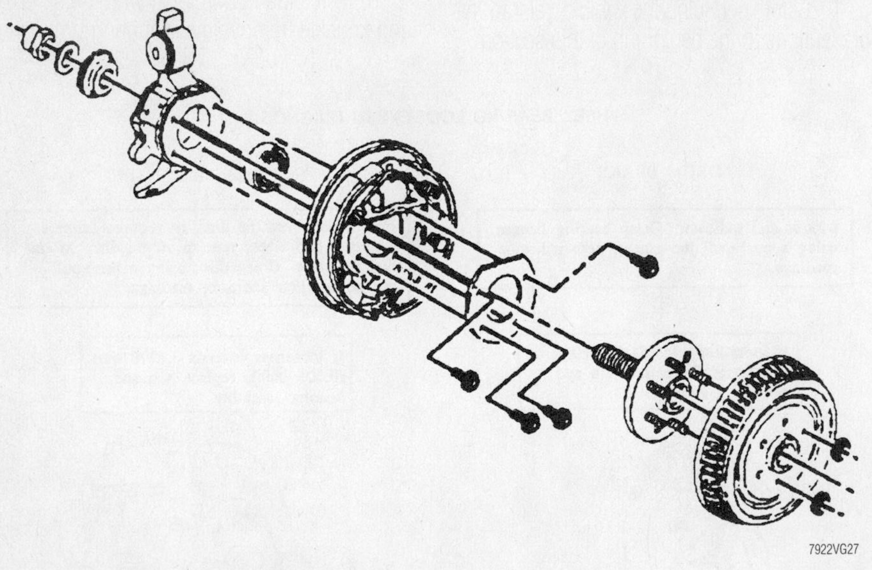

Exploded view of the rear wheel bearing hub mounting

GENERAL MOTORS CORPORATION—F-BODY

23

1998–01

Chevrolet-Camaro • Z28 • **Pontiac**-Firebird • Trans Am

PRECAUTIONS

Before servicing any vehicle, please be sure to read all of the following precautions, which deal with personal safety, prevention of component damage, and important points to take into consideration when servicing a motor vehicle:

- Never open, service or drain the radiator or cooling system when the engine is hot; serious burns can occur from the steam and hot coolant.
- Observe all applicable safety precautions when working around fuel. Whenever servicing the fuel system, always work in a well-ventilated area. Do not allow fuel spray or vapors to come in contact with a spark, open flame or excessive heat (a hot drop light, for example). Keep a dry chemical fire extinguisher near the work area. Always keep fuel in a container specifically designed for fuel storage; also, always properly seal fuel containers to avoid the possibility of fire or explosion. Refer to the additional fuel system precautions later in this section.
- Fuel injection systems often remain pressurized, even after the engine has been turned **OFF**. The fuel system pressure must be relieved before disconnecting any fuel lines. Failure to do so may result in fire and/or personal injury.
- Brake fluid often contains polyglycol ethers and polyglycols. Avoid contact with the eyes and wash your hands thoroughly after handling brake fluid. If you do get brake fluid in your eyes, flush your eyes with clean, running water for 15 minutes. If eye irritation persists, or if you have taken brake fluid internally, IMMEDIATELY seek medical assistance.
- The EPA warns that prolonged contact with used engine oil may cause a number of skin disorders, including cancer! You should make every effort to minimize your exposure to used engine oil. Protective gloves should be worn when changing oil. Wash your hands and any other exposed skin areas as soon as possible after exposure to used engine oil. Soap and water, or waterless hand cleaner should be used.
- All new vehicles are now equipped with an air bag system. The system must be disabled before performing service on or around system components, steering column, instrument panel components, wiring and sensors. Failure to follow safety and disabling procedures could result in accidental air bag deployment, possible personal injury and unnecessary system repairs.
- Always wear safety goggles when working with, or around, the air bag system. When carrying a non-deployed air bag, be sure the bag and trim cover are pointed away from your body. When placing a non-deployed air bag on a work surface, always face the bag and trim cover upward, away from the surface. This will reduce the motion of the module if it is accidentally deployed. Refer to the additional air bag system precautions later in this section.
- Clean, high quality brake fluid from a sealed container is essential to the safe and proper operation of the brake system. You should always buy the correct type of brake fluid for your vehicle. If the brake fluid becomes contaminated, completely flush the system with new fluid. Never reuse any brake fluid. Any brake fluid that is removed from the system should be discarded. Also, do not allow any brake fluid to come in contact with a painted surface; it will damage the paint.
- Never operate the engine without the proper amount and type of engine oil; doing so WILL result in severe engine damage.
- Timing belt maintenance is extremely important! Many models utilize an interference-type, non-freewheeling engine. If the timing belt breaks, the valves in the cylinder head may strike the pistons, causing potentially serious (also time-consuming and expensive) engine damage. Refer to the maintenance interval charts in the front of this manual for the recommended replacement interval for the timing belt, and to the timing belt section for belt replacement and inspection.
- Disconnecting the negative battery cable on some vehicles may interfere with the functions of the on-board computer system(s) and may require the computer to undergo a relearning process once the negative battery cable is reconnected.
- When servicing drum brakes, only disassemble and assemble one side at a time, leaving the remaining side intact for reference.
- Only an MVAC-trained, EPA-certified automotive technician should service the air conditioning system or its components.

ENGINE REPAIR

Alternator

REMOVAL

3.8L Engine

1. Before servicing the vehicle, refer to the precautions in the beginning of this section.
2. Remove or disconnect the following:
   - Negative battery cable
   - Accessory drive belt
   - Alternator electrical connectors
   - Evaporative Emissions (EVAP) canister purge solenoid
   - Rear alternator brace-to-alternator bolt
   - Both front alternator bolts
   - Alternator

5.7L Engine

1. Before servicing the vehicle, refer to the precautions in the beginning of this section.
2. Remove or disconnect the following:
   - Negative battery cable
   - Accessory drive belt
   - Alternator rear inner brace
   - Transmission oil cooler line clips
   - Alternator bolts
   - Alternator electrical connectors
   - Alternator

INSTALLATION

3.8L Engine

Install or connect the following:

- Alternator. Torque the upper (drive belt tensioner) bolt to 22 ft. lbs. (30 Nm) and the lower bolt to 37 ft. lbs. (50 Nm).

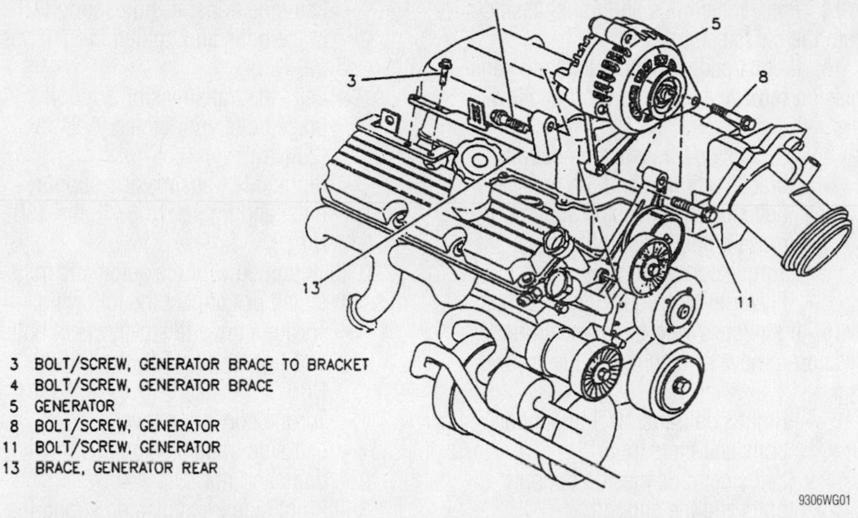

Exploded view of the alternator and related components—3.8L engine

- Rear alternator brace-to-alternator bolt. Torque the bolt to 22 ft. lbs. (30 Nm).
- EVAP canister purge solenoid
- Alternator electrical connectors. Torque the battery cable nut to 16 ft. lbs. (22 Nm).
- Accessory drive belt
- Negative battery cable

5.7L Engine

Install or connect the following:

- Alternator
- Alternator bolts. Torque the bolts to 37 ft. lbs. (50 Nm).
- Transmission oil cooler line clips
- Alternator rear inner brace. Torque the nut to 18 ft. lbs. (25 Nm).
- Alternator electrical connectors. Torque the battery cable nut to 16 ft. lbs. (22 Nm).
- Accessory drive belt
- Negative battery cable

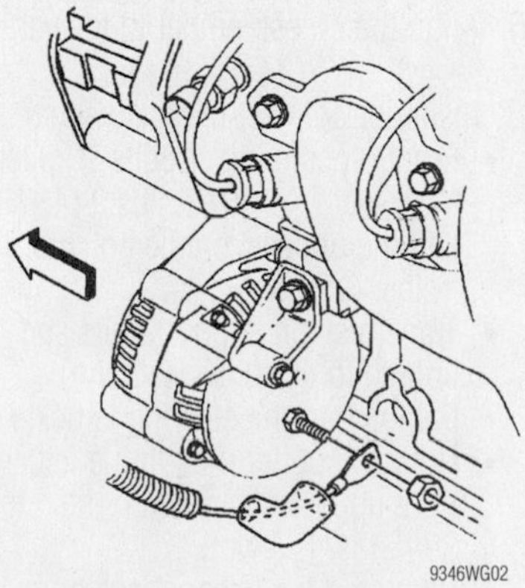

Exploded view of the alternator and related components—5.7L engine

Ignition Timing

ADJUSTMENT

On these vehicles, base timing is preset when the engine is manufactured. All timing changes are, then controlled directly by the Powertrain Control Module (PCM) based on information from the ignition and knock sensor systems. No adjustments are necessary or possible.

Engine Assembly

REMOVAL & INSTALLATION

➡The engine, transmission and suspension assembly is removed from the bottom of the vehicle. After the assembly is removed, separate the engine from the transmission and frame.

1. Before servicing the vehicle, refer to the precautions in the beginning of this section.
2. Remove both battery cables.
3. Relieve the fuel system pressure.
4. Recover the refrigerant from the AC system using approved equipment.
5. Drain the cooling system, crankcase and transmission fluid.
6. Remove or disconnect the following:
   - Intake air temperature (IAT) sensor electrical connector
   - Air intake duct
   - Front wheels
   - 3-way catalytic converter
   - Propeller shaft
   - Torque arm
   - Starter motor
   - Transmission range selector lever cable from the range selector lever at the transmission by unsnapping it, on models equipped with a automatic transmission
   - Retainer from the range selector lever cable
   - Range selector lever cable from the cable bracket
   - Clutch actuator cylinder line from the actuator cylinder using Tool J36221, on models equipped with a manual transmission
   - Left side front air deflector
   - Stabilizer bar bracket bolts and brackets
   - Steering gear coupling shield
   - Intermediate steering shaft bolt and shaft from the rack
   - Wiring harness ground bolt and radio frequency ground strap and cruise control ground lead from the front rail
   - Knock sensors electrical connectors
   - Wheel speed sensors electrical connectors
   - A/C compressor and condenser hose bolt. Discard the O-ring.
   - A/C compressor and condenser hose nut at the condenser. Discard the O-ring.
   - Front fuel pipe heat shield nuts
   - Front fuel pipe heat shield
   - Brake lines from the brake pipe clip
   - Engine wiring harness ground bolt
   - Engine ground
   - A/C compressor and condenser hose bolt at the accumulator. Discard the O-ring.
   - Heater hoses from the drive belt tensioner
   - Fuel feed pipe from the fuel rail
   - Fuel return pipe from the fuel rail
   - Evaporative Emissions (EVAP) pipe at the engine
   - Cruise control and accelerator control cables from the throttle body and bracket
   - Accelerator control cable bracket bolts and bracket
   - Inlet hose from the water outlet
   - Outlet hose from the water pump
   - Brake booster vacuum hose from the intake manifold fitting
   - Front two brake pipes from the brake modulator

- Secondary captured locks
- Forward lamp harness from the engine harness
- Engine harness vacuum tube from the bottom of the vacuum check valve
- Powertrain Control Module (PCM) connectors
- PCM
- Right side insulator panel
- Hinge pillar trim panel
- Engine wiring harness from the instrument panel wiring harness
- Engine wire harness through the front of the dash

7. Place the harness on top of the engine.
- Floor shift control, if equipped
- Transmission control, if equipped
- Right and left lower shock bolts from the lower control arms
- Cotter pins and nuts from the left and right upper ball joints
- Separate the upper control arms from the steering knuckles and support both steering knuckles

8. Position Engine Support Table J 39580 below the vehicle. Lower the vehicle and lay the engine wire harness on top of the engine.
- Transmission support bolts
- Engine and transmission from the vehicle

9. Secure the crossmember to the engine support table.

10. Position a hex-head socket on the belt tensioner pulley bolt.

11. Rotate the drive belt tensioner clockwise to relieve belt tension.

12. Remove or disconnect the following:
- Drive belt
- A/C compressor rear bolts
- A/C compressor rear bracket stud
- A/C compressor rear bracket
- A/C compressor bolts
- A/C compressor
- Drive belt tensioner bolts and the tensioner
- Power steering pump nuts and reposition the power steering pump to the crossmember
- Alternator electrical connector
- Positive cable nut from the alternator
- Alternator

13. Install a suitable lifting devise to the engine.
- Right engine mount bracket bolts and bracket
- Left engine mount bracket bolts, stud and the bracket

14. Raise the engine and transmission from the crossmember.

15. If equipped with the automatic transmission remove or disconnect the following:
- Right side transmission support brace bolts, nut and the brace
- Left side transmission support brace bolts, nut and the brace
- Torque converter cover
- Flywheel-to-torque converter bolts

16. If equipped with the manual transmission remove or disconnect the following:
- Right side transmission support bolts and the support
- Left side transmission support bolts and the support
- Electrical connectors from the transmission
- Transmission-to-engine bolts and nut, then separate the transmission from the engine, if equipped with an automatic transmission
- Flywheel housing to engine bolts and nuts, then separate the flywheel housing from the engine, if equipped with a manual transmission
- Pressure plate and clutch, if equipped with a manual transmission
- Flywheel bolts and flywheel, if equipped with a manual transmission

To install:

17. Install or connect the following:
- Flywheel bolts and flywheel, if equipped with a manual transmission. Tighten the flywheel bolts to 11ft. lbs. (15 Nm), then tighten an additional 50 degrees.
- Clutch and pressure plate, if equipped with a manual transmission. Tighten the pressure plate bolts to 15 ft. lbs. (20 Nm) and tighten an additional 45 degrees.

18. Align the flywheel housing to the engine.
- Flywheel housing-to-engine bolts and nuts, if equipped with a manual transmission and tighten to 70 ft. lbs. (95 Nm).
- Transmission-to-engine bolts and nut and tighten to 70 ft. lbs. (95 Nm), if equipped with an automatic transmission
- Electrical connectors to the transmission

19. If equipped with the manual transmission install or connect the following:
- Left side transmission support brace bolts and tighten to 37 ft. lbs. (50 Nm)
- Left side transmission support brace bolts and tighten to 21 ft. lbs. (28 Nm)
- Right side transmission support bolts and tighten to 37 ft. lbs. (50 Nm)

20. If equipped with the automatic transmission install or connect the following:
- Torque converter-to-flywheel bolts and tighten to 47 ft. lbs. (63 Nm)
- Torque converter cover
- Left side transmission support, bolts and nut
- Right side transmission support

21. Lower the engine and transmission to the crossmember.
- Left engine mount bracket, stud and bolts. Tighten the bolts to 74 ft. lbs. (100 Nm) and the stud 64 ft. lbs. (87 Nm).
- Right engine mount bracket and the bolts

22. Remove the engine lifting device.
- Alternator and wiring
- Power steering pump and tighten the nuts to 22 ft. lbs. (30 Nm)
- Drive belt tensioner and bolts, and tighten to 37 ft. lbs. (50 Nm)
- A/C compressor and tighten to 37 ft. lbs. (50 Nm)
- A/C compressor rear bracket. and tighten the stud to 37 ft. lbs. (50 Nm) and the bolts to 22 ft. lbs. (30 Nm)
- Drive belt

23. Remove the crossmember from the table security straps.

24. Lower the vehicle and install the engine and transmission to the vehicle.

25. Align the crossmember to the vehicle body.

26. Install or connect the following:
- Front crossmember bolts. Tighten the lower 2 to 107 ft. lbs. (145 Nm) and the upper 4 cradle bolts to 92 ft. lbs. (125 Nm).
- Transmission support bolts and tighten to 42 ft. lbs. (55 Nm)
- Ball stud to the steering knuckle
- Nuts and cotter pins to the left and right upper ball joints. Tighten to 39 ft. lbs. (53 Nm).
- Right and left lower shock bolts to the lower control arms
- Transmission control, if equipped
- Floor shift control, if equipped
- Engine wire harness through the front of dash

- Engine wiring harness to the instrument panel wiring harness
- Hinge pillar trim panel
- Right side insulator panel
- PCM and electrical connectors
- Engine harness vacuum tube
- Forward lamp harness
- Engine harness
- Secondary captured locks
- Front two brake pipes
- Brake booster vacuum hose
- Outlet hose to the water pump
- Inlet hose to the water outlet
- Accelerator control cable bracket and bolts. Tighten to 12 ft. lbs. (16 Nm).
- Accelerator control and cruise control cables to the throttle body and bracket
- EVAP pipe
- Fuel return pipe
- Fuel feed pipe
- Heater hoses
- New O-ring
- A/C compressor and condenser hose bolt at the accumulator and tighten to 36 ft. lbs. (48 Nm)
- Engine ground and bolt and tighten to 18 ft. lbs. (25 Nm)
- Brake lines
- Front fuel pipe heat shield and bolts
- A/C compressor and condenser hose nut at the condenser. Tighten to 12 ft. lbs. (16 Nm).
- Wheel speed sensors electrical connectors
- Knock sensors electrical connectors
- Radio frequency ground strap, cruise control ground lead and wiring harness ground bolt to the front rail
- Intermediate steering shaft and bolt to the rack. Tighten to 35 ft. lbs. (47 Nm).
- Steering gear coupling shield
- Stabilizer bar brackets and bolts Tighten to 41 ft. lbs. (55 Nm).
- Left side front air deflector
- Clutch actuator cylinder line to the actuator cylinder, if equipped
- Range selector lever cable, if equipped
- Starter motor
- Torque arm
- Propeller shaft
- 3-way catalytic converter
- Front wheels
- Air intake duct
- IAT sensor electrical connector
- Battery cables

27. Refill the engine oil, transmission fluid and coolant
28. Recharge the A/C system.
29. Bleed the brakes.
30. Bleed the clutch hydraulic system.
31. Connect the negative battery cable.

Water Pump

REMOVAL & INSTALLATION

3.8L Engine

1. Before servicing the vehicle, refer to the precautions in the beginning of this section.
2. Drain the cooling system.
3. Remove or disconnect the following:
   - Negative battery cable
   - Accessory drive belt
   - Radiator inlet hose from the water pump
   - Water pump pulley
   - Water pump

To install:

4. Install or connect the following:
   - Water pump using a new gasket. Torque the water pump bolts to 11 ft. lbs. (15 Nm), plus an additional 80 degree rotation.
   - Water pump pulley. Torque the bolts to 115 inch lbs. (13 Nm).
   - Accessory drive belt
   - Radiator inlet hose to the water pump
   - Negative battery cable
5. Refill the cooling system.
6. Start the engine and check for leaks.

5.7L Engines

1. Before servicing the vehicle, refer to the precautions in the beginning of this section.
2. Drain the cooling system by removing the block coolant drain plug and the knock sensor.
3. Remove or disconnect the following:
   - Negative battery cable
   - Electrical connector from the cooling fan
   - Intake Air Temperature (IAT) sensor electrical connector
   - Mass Airflow (MAF) sensor electrical connector
   - Air intake duct
   - Outlet hose from the water outlet
   - Inlet hose from the water pump
   - Air cleaner assembly
   - Inlet hose from the radiator
   - Both cooling fan assemblies
   - Drive belt
   - Drive belt tensioner pulley
   - Vent hose from the radiator
   - Water pump pulley
   - Water pump

To install:

4. Install or connect the following:
   - Drain plug
   - Water pump using a new gasket. Torque the bolts to 18 ft. lbs. (25 Nm) on 1998–99 models and 22 ft. lbs. (30 Nm)on 2000–01 models.
   - Water pump pulley. Torque the bolts to 18 ft. lbs. (25 Nm).
   - Vent hose to the radiator
   - Drive belt tensioner pulley. Torque the bolt to 37 ft. lbs. (50 Nm).
   - Drive belts
   - Cooling fans
   - Inlet hose to the radiator
   - Air cleaner assembly. Torque the

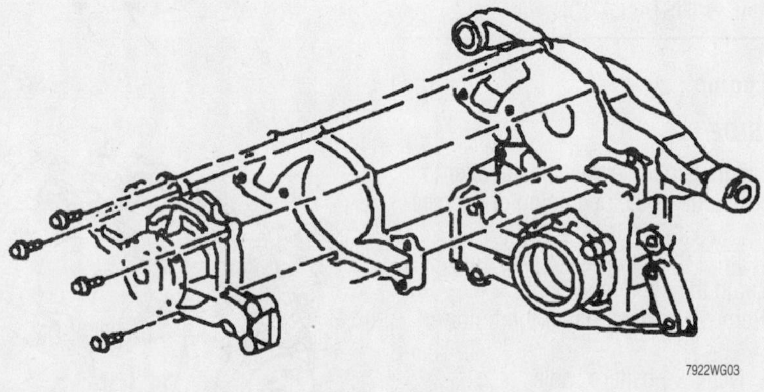

Exploded view of the water pump mounting—3.8L engine

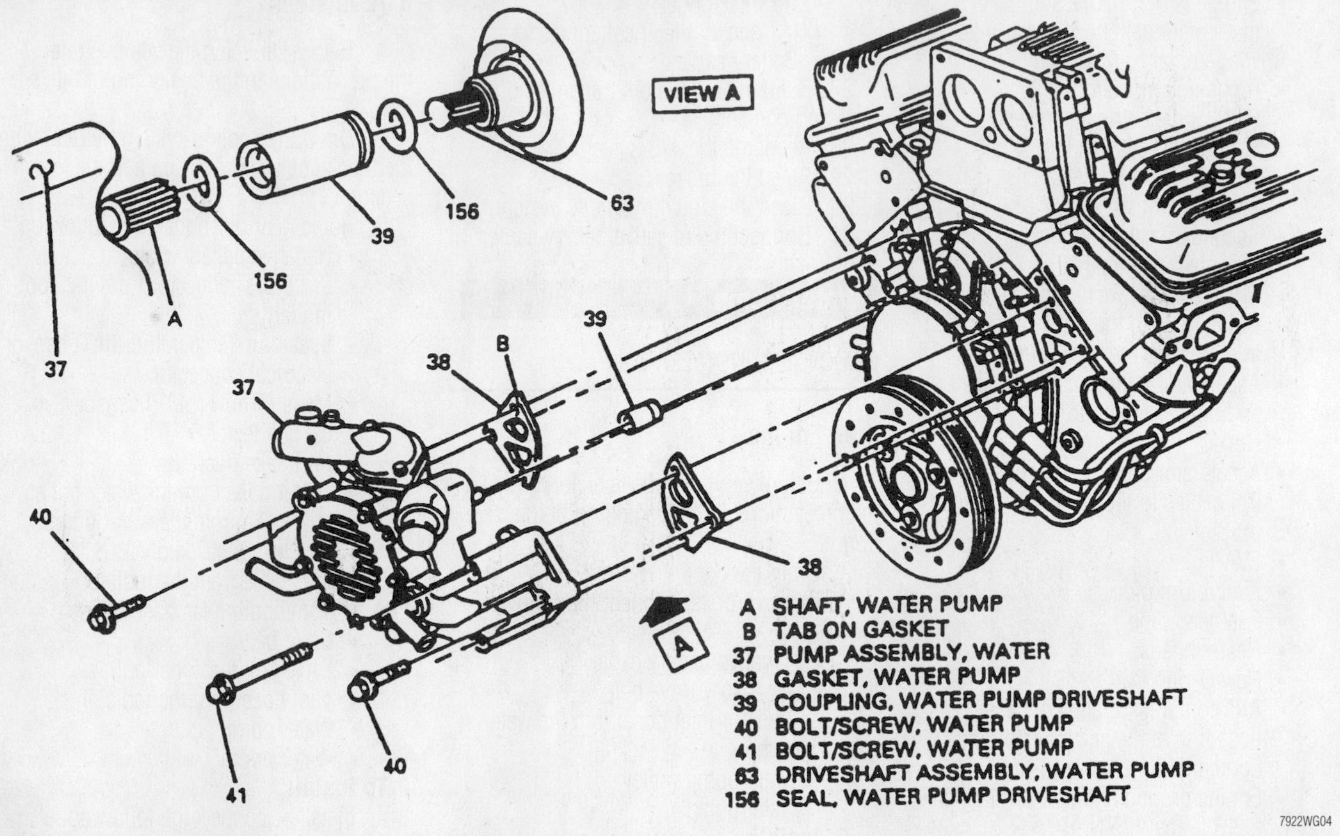

Exploded view of the water pump mounting and related components—5.7L engines

bolts to 106 inch lbs. (12 Nm).

- Inlet hose to the water pump
- Outlet hose to the water outlet
- Air intake duct
- IAT and MAF sensor electrical connectors
- Coolant fan electrical connectors
- Negative battery cable

5. Refill the cooling system.
6. Start the engine and check for leaks.

Cylinder Head

REMOVAL & INSTALLATION

3.8L Engine

LEFT SIDE

1. Before servicing the vehicle, refer to the precautions in the beginning of this section.
2. Relieve the fuel system pressure.
3. Drain the cooling system.
4. Remove or disconnect the following:
   - Negative battery cable
   - Fuel lines from the fuel rail
   - Upper and lower intake manifolds
   - Spark plugs
   - Left exhaust manifold
   - Rocker arm cover, rocker arms and push rods
   - Cylinder head

To install:

5. Install the cylinder head with a new gasket and secure with NEW head bolts.
6. Torque the head bolts in sequence as follows:
   a. Step 1: 37 ft. lbs. (50 Nm).
   b. Step 2: Plus 120 degree turn.
7. Install or connect the following:
   - Rocker arms and push rods
   - Rocker arm cover. Torque the bolts to 89 inch lbs. (10 Nm).
   - Left exhaust manifold

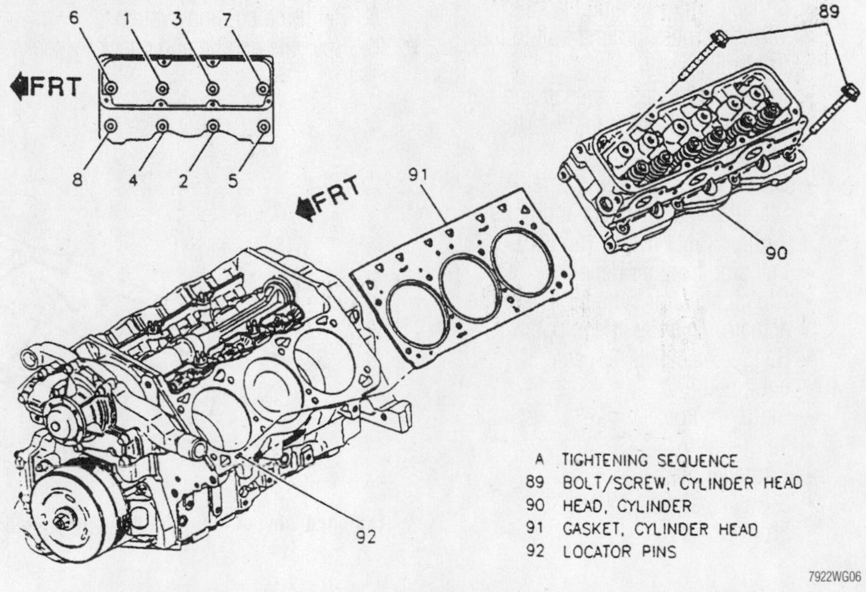

Cylinder head bolt torque sequence—3.8L (VIN K) engine

- Spark plugs and torque them to 20 ft. lbs. (27 Nm).
- Lower and upper intake manifolds
- Negative battery cable

8. Refill the cooling system.
9. Start the engine and check for leaks.

RIGHT SIDE

1. Before servicing the vehicle, refer to the precautions in the beginning of this section.
2. Relieve the fuel system pressure.
3. Drain the cooling system.
4. Remove or disconnect the following:
   - Negative battery cable
   - Fuel lines from the fuel rail
   - Upper and lower intake manifolds
   - Spark plugs
   - Right exhaust manifold
   - Rocker arm cover, rocker arms and push rods
   - Cylinder head

To install:

5. Install the cylinder head with a new gasket and secure with NEW head bolts. Torque the bolts in sequence as follows:
   a. Step 1: 37 ft. lbs. (50 Nm).
   b. Step 2: Plus a 120 degree turn
6. Install or connect the following:
   - Rear alternator brace. Torque the bolts to 18 ft. lbs. (25 Nm).
   - Drive belt tensioner. Torque the bolts to 37 ft. lbs. (50 Nm).
   - Rocker arms and push rods
   - Rocker arm cover. Torque the bolts to 89 inch lbs. (10 Nm).
   - Right exhaust manifold. Torque the bolts to 106 inch lbs. (12 Nm).
   - Spark plugs. Torque the plugs to 20 ft. lbs. (27 Nm).
   - Lower and upper intake manifolds
   - Negative battery cable
7. Refill the cooling system.
8. Start the engine and check for leaks.

5.7L Engines

LEFT SIDE

1. Before servicing the vehicle, refer to the precautions in the beginning of this section.
2. Drain the cooling system.
3. Remove or disconnect the following:
   - Negative battery cable
   - Coolant air bleed pipe
   - Power steering pump pulley and hoses
   - Power steering pump and bracket
   - Exhaust manifold

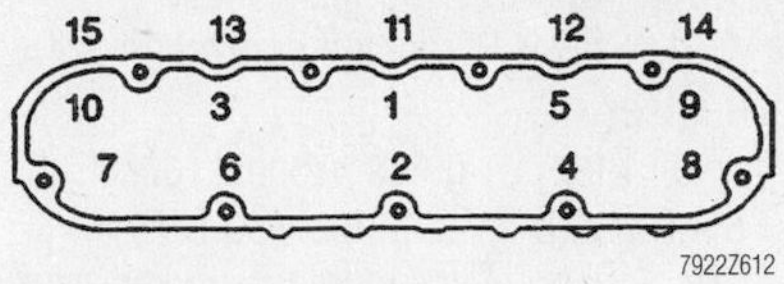

Cylinder head bolt tightening sequence—5.7L (VIN G) engine

   - Intake manifold
   - Rocker arm cover
   - Rocker arms, pedestals and push rods
   - Ground wires from the back of the head
   - Cylinder head

To install:

4. Install the cylinder head with a new gasket.

➡Before installing the head bolts, be sure to mark the short bolts so the correct tightening specifications are used.

5. Lubricate the M11 bolts with clean engine oil, then install them.
6. Apply a 0.2 inch (5mm) bead of threadlock GM P/N 12345382 to the threads of the M8 cylinder head bolts.
7. Torque the cylinder head bolts in sequence as follows:
   a. M11 bolts 1–10: 22 ft. lbs. (30 Nm).
   b. M11 bolts 1–10: an additional 90 degrees.
   c. M11 bolts 1–8: an additional 90 degrees.
   d. M11 bolts 9–10: an additional 50 degrees.
   e. M8 bolts 11–15: 22 ft. lbs. (30 Nm). Begin at the center and alternate side-to-side working outward.
8. Install or connect the following:
   - Ground wires to the back of the head. Torque the bolt to 37 ft. lbs. (50 Nm).
   - Rocker arms, pedestals and push rods
   - Exhaust manifold
   - Rocker arm cover. Torque the bolts to 106 inch lbs. (12 Nm).
   - Power steering pump bracket. Torque the bolts to 37 ft. lbs. (50 Nm).
   - Power steering pump. Torque the bolts to 18 ft. lbs. (25 Nm).
   - Power steering pump hoses and pulley
   - Coolant air bleed pipe
   - Intake manifold
   - Negative battery cable
9. Refill the cooling system.
10. Change the oil and filter.
11. Start the engine and check for leaks.

RIGHT SIDE

1. Before servicing the vehicle, refer to the precautions in the beginning of this section.
2. Drain the cooling system.
3. Remove or disconnect the following:
   - Negative battery cable
   - Coolant air bleed pipe
   - Intake manifold
   - Exhaust manifold
   - Rocker arm cover
   - Rocker arms, pedestals and push rods
   - Wiring clip bolt from the cylinder head
   - Tranmission dipstick tube bolt from the cylinder head
   - Cylinder head

To install:

4. Install the cylinder head with a new gasket.

➡Before installing the head bolts, be sure to mark the short bolts so the correct tightening specifications are used.

5. Lubricate the M11 bolts with clean engine oil, then install them.
6. Apply a 0.2 inch (5mm) bead of threadlock GM P/N 12345382 to the threads of the M8 cylinder head bolts.
7. Torque the cylinder head bolts in sequence as follows:
   a. M11 bolts 1–10: 22 ft. lbs. (30 Nm).
   b. M11 bolts 1–10: an additional 90 degrees.
   c. M11 bolts 1–8: an additional 90 degrees.
   d. M11 bolts 9–10: an additional 50 degrees.
   e. M8 bolts 11–15: 22 ft. lbs. (30 Nm). Begin at the center and alternate side-to-side working outward.
   - Wiring clip bolt to the cylinder head
   - Tranmission dipstick tube bolt to the cylinder head
   - Rocker arms, pedestals and push rods
   - Exhaust manifold
   - Rocker arm cover. Torque the bolts to 106 inch lbs. (12 Nm).

- Coolant air bleed pipe
- Intake manifold
- Negative battery cable

8. Refill the cooling system.
9. Change the oil and filter.
10. Start the engine and check for leaks.

Rocker Arms

REMOVAL & INSTALLATION

➡Be sure to keep all the components in the exact order of removal so they may be installed in their original locations. Coat the replacement rocker arm and ball with engine oil before installation.

1. Before servicing the vehicle, refer to the precautions in the beginning of this section.
2. Remove or disconnect the following:
   - Negative battery cable
   - Rocker arm cover
   - Rocker arm bolts and rocker arms
   - Retainer
   - Push rods

To install:

3. Lubricate the bearing surfaces with a thin coating of Molykote® or equivalent.
4. Install or connect the following:
5. The push rods, retainer, rocker arms and bolts.
6. On 3.8L engines, tighten the bolts to 11 ft. lbs. (15 Nm) plus an additional 90 degrees.
7. Adjust the rocker arm bolts for the 5.7L engine as follows:
   a. Step 1:Rotate the crankshaft until number 1 piston is at Top Dead Center (TDC).
   b. Step 2:Tighten exhaust rocker arms 1,2,7 and 8 to 22 ft. lbs. (30 Nm).

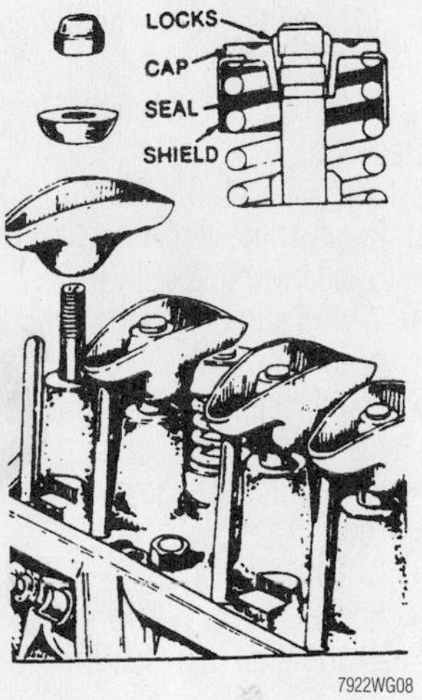

Exploded view of common rocker arm components and valve spring retainer

   c. Step 3:Tighten intake rocker arms 1,3,4 and 5 to 22 ft. lbs. (30 Nm).
   d. Step 4:Rotate the crankshaft 360 degrees.
   e. Step 5:Tighten exhaust rocker arms 3,4,5 and 6 to 22 ft. lbs. (30 Nm).
   f. Step 6:Tighten intake rocker arms 2,6,7 and 8 to 22 ft. lbs. (30 Nm).
8. Install the rocker arm cover and torque the bolts to 106 inch lbs. (12 Nm).
9. Install the negative battery cable.

Intake Manifold

REMOVAL & INSTALLATION

3.8L Engine

UPPER MANIFOLD

1. Before servicing the vehicle, refer to the precautions in the beginning of this section.
2. Relieve the fuel system pressure.
3. Drain the cooling system.
4. Remove or disconnect the following:
   - Negative battery cable
   - Intake Air Temperature (IAT) sensor electrical connector
   - Drive belt tensioner
   - Fuel pressure regulator vacuum and Evaporative Emission (EVAP) emission canister purge harness from the manifold vacuum source, purge solenoid valve and the fuel pressure regulator valve
   - Manifold vacuum source electrical connector
   - Manifold vacuum source screws
   - Manifold vacuum source from the upper intake manifold
   - Canister purge solenoid valve electrical connector and hose
   - Evaporative emissions canister purge solenoid valve
   - Idle Air Control (IAC) valve
   - Fuel lines from the fuel rail
   - EVAP pipe at the engine
   - Accelerator control cable bracket and control cables from the throttle body
   - Ignition control module and coil pack assembly
   - Brake booster hose from the upper manifold
   - Alternator brace
   - Throttle Position (TPS) sensor electrical connector
   - Mass Air Flow (MAF) sensor electrical connector
   - Wiring harness from the fuel rail clips

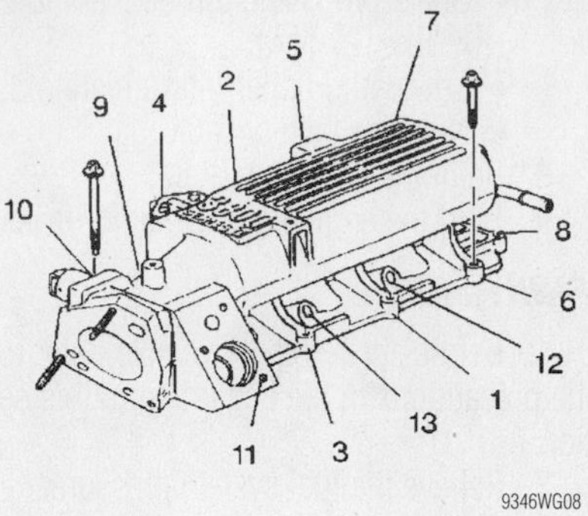

Upper intake manifold torque sequence—1998–99 3.8L engine

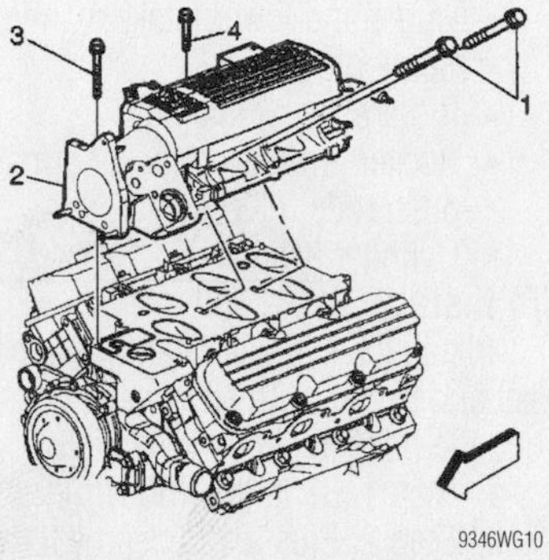

Upper intake manifold torque sequence—2000–01 3.8L engine

   - Throttle body
   - Electrical connectors from the fuel injectors
   - Fuel rail
   - Upper intake manifold

To install:

5. Install the 1998–99 models upper intake manifold using a new gasket. Torque the bolts as follows:
   a. Bolts 1–10, in sequence: 11 ft. lbs. (15 Nm).
   b. Coolant outlet bolts: 20 ft. lbs. (27 Nm).
   c. Side bolts: 22 ft. lbs. (30 Nm).
6. Install the 2000–01 models upper intake manifold using a new gasket. Torque the bolts as follows:
   a. Tighten the upper intake manifold bolts (1, 3 & 4): in sequence: 11 ft. lbs. (15 Nm).
   b. Tighten the nine upper intake manifold bolts (4): to 89 inch lbs. (10 Nm).
   c. Tighten the two upper intake manifold bolts (1) to 22 ft. lbs. (30 N).
7. Install the fuel rail as follows:
   a. Install new fuel injector O-rings lubricated with engine oil.
   b. Install the fuel injectors into the manifold bores.

c. Insert the fuel rail until the injectors are properly seated.

d. Torque the nuts to 75 inch lbs. (8.5 Nm).

8. Install or connect the following:
   - Fuel injector electrical connectors
   - Thermostat and thermostat housing. Torque the housing bolts to 20 ft. lbs. (27 Nm).
   - Throttle body and tighten the retainers to 89 inch lbs. (10 Nm)
   - Wiring harness to the fuel rail clips
   - Electrical connectors to the IAC, TPS, MAF and IAT sensors
   - Alternator brace
   - Brake booster hose
   - Ignition control module and coil pack assembly
   - Accelerator cable bracket and cruise control cables to the throttle body. Torque the bracket bolts to 12 ft. lbs. (16 Nm).
   - Accelerator and cruise control cables
   - Fuel lines to the fuel rail
   - IAC valve
   - Fuel pressure regulator
   - Canister purge harness
   - Evaporative emissions canister purge solenoid valve
   - MAP sensor and vacuum source
   - Drive belt tensioner. Torque the bolts to 37 ft. lbs. (50 Nm).
   - Air intake duct
   - IAT sensor connector
   - Drive belt
   - Negative battery cable
9. Refill the coolant system.
10. Start the engine and check for leaks.

LOWER MANIFOLD

1. Before servicing the vehicle, refer to the precautions in the beginning of this section.

➡There are 2 bolts that are hidden under the upper manifold. They are located in the front and left rear corners of the lower manifold. It is necessary to remove the upper manifold to service the lower manifold.

2. Remove or disconnect the following:
   - Negative battery cable
   - Upper intake manifold
   - Engine Coolant Temperature (ECT) sensor
   - Lower intake manifold

To install:

3. Install or connect the following:

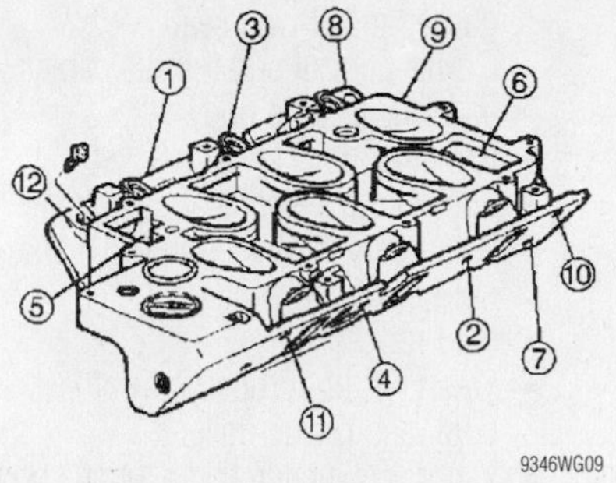

Lower intake manifold torque sequence—3.8L engine

   - New gaskets and seals
   - Lower manifold. Torque the manifold bolts in sequence to 11 ft. lbs. (15 Nm).
   - ECT sensor
   - Upper intake manifold
   - Negative battery cable

5.7L Engines

1998–98 MODELS

1. Before servicing the vehicle, refer to the precautions in the beginning of this section.
2. Relieve the fuel system pressure.
3. Drain the coolant system.
4. Remove or disconnect the following:
   - Negative battery cable
   - Transmission fluid level indicator tube bolt
   - Fuel lines from the fuel rail
   - Vapor line from the fuel vapor purge valve
   - Intake Air Temperature (IAT) sensor electrical connector
   - Mass Airflow (MAF) sensor electrical connection
   - Air inlet duct
   - Idle Air Control (IAC) valve and Throttle Position (TPS) sensor electrical connections from the throttle body
   - Drive belt
   - Exhaust Gas Recirculation (EGR) solenoid electrical connector
   - EGR valve pipe
   - Cruise control and throttle cables from the throttle body
   - Cruise control servo bracket
   - Fuel injector electrical connectors
   - Fuel rail
   - Manifold Absolute Pressure (MAP) sensor electrical connector
   - Vacuum hose from the vacuum port on the MAP sensor housing
   - Knock sensor harness electrical connector
   - Remaining electrical connections from the intake manifold
   - Fresh air tube from the throttle body and rocker cover
   - Vapor vent tube from the throttle body
   - Throttle body heater outlet hose from the throttle body
   - Evaporative emission (EVAP) canister purge tube
   - Canister purge valve and bracket from the intake manifold
   - Intake manifold bolts and the manifold

To install:

5. Install the intake manifold with new gaskets.
6. Torque the bolts in sequence as follows:

a. First pass in sequence: 44 inch lbs. (5 Nm)

b. Final pass in sequence to 89 inch lbs. (10 Nm).

7. Install or connect the following:
   - Canister purge valve and bracket to the intake manifold
   - EVAP canister purge tube
   - Throttle body heater outlet hose to the throttle body
   - Vapor vent tube to the throttle body
   - Fresh air tube to the throttle body and rocker cover
   - PCV valve pipe from the rocker cover
   - PCV pipe strap nut
   - Knock sensor harness electrical connector
   - MAP sensor electrical connector
   - Vacuum hose to the vacuum port on the MAP sensor housing
   - Fuel rail
   - Fuel injector electrical connectors
   - All remaining electrical connections
   - Cruise control servo bracket
   - Cruise control and throttle cables to the throttle body
8. Apply a light coating of clean engine oil to a NEW O-ring seal and install the seal onto the EGR valve pipe.
   - EGR valve pipe into the intake manifold. Finger start the EGR valve pipe to the intake manifold bolt and the EGR valve pipe to the cylinder head bolts.
9. Finger start a NEW EGR valve pipe exhaust manifold gasket and bolts and tighten as follows:

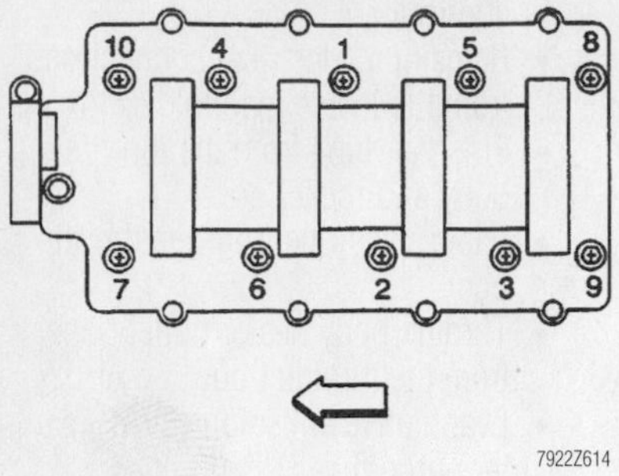

Intake manifold torque sequence—5.7L (VIN G) engine

a. Tighten the EGR valve pipe to the exhaust manifold bolts to 22 ft. lbs. (30 Nm).

b. Tighten the EGR valve pipe-to-cylinder head bolts to 37 ft. lbs. (50 Nm).

c. Tighten the EGR valve pipe-to-intake manifold bolt to 89 inch lbs. (10 Nm).

- EGR solenoid electrical connector
- Drive belt
- IAC valve and TPS sensor electrical connections to the throttle body
- Air inlet duct
- IAT and MAF sensor electrical connector
- Fuel lines to the fuel rail
- Vapor line to the fuel vapor purge valve
- Transmission fluid level indicator tube bolt
- Negative battery cable

2000–01 MODELS

1. Before servicing the vehicle, refer to the precautions in the beginning of this section.
2. Relieve the fuel system pressure.
3. Drain the coolant system.
4. Remove or disconnect the following:
   - Negative battery cable
   - Transmission fluid level indicator tube bolt
   - Fuel lines from the fuel rail
   - Vapor line from the fuel vapor purge valve
   - Intake Air Temperature (IAT) sensor electrical connector
   - Air inlet duct
   - Idle Air Control (IAC) valve and Throttle Position (TPS) sensor electrical connections from the throttle body
   - Drive belt
   - Exhaust Gas Recirculation (EGR) solenoid electrical connector
   - EGR valve pipe
   - O-ring seal from the EGR valve pipe. Discard the exhaust manifold gasket and O-ring seal.
   - Cruise control and throttle cables from the throttle body
   - Cruise control servo bracket
   - Fuel injector electrical connectors
   - Secondary Air Injection (AIR) solenoid electrical connector
   - Fuel rail
   - Fresh air tube from the throttle body and rocker cover
   - Vapor vent tube from the throttle body
   - Throttle body heater outlet hose from the throttle body
   - Evaporative emission (EVAP) canister purge tube
   - Canister purge valve and bracket from the intake manifold
   - Knock sensor harness electrical connector
   - Manifold Absolute Pressure (MAP) sensor electrical connector
   - Vacuum hose from the vacuum port on the MAP sensor housing
   - Remaining electrical connections from the intake manifold
   - Intake manifold bolts and the manifold

To install:

5. Install the intake manifold with new gaskets.
6. Torque the bolts in sequence as follows:

   a. First pass in sequence: 44 inch lbs. (5 Nm)

   b. Final pass in sequence to 89 inch lbs. (10 Nm).
7. Install or connect the following:
   - MAP sensor electrical connector
   - Vacuum hose from the vacuum port on the MAP sensor housing
   - Knock sensor harness electrical connector
   - Canister purge valve and bracket to the intake manifold
   - EVAP canister purge tube
   - Throttle body heater outlet hose to the throttle body
   - Vapor vent tube to the throttle body
   - Fresh air tube to the throttle body and rocker cover
   - Fuel rail
   - AIR solenoid electrical connector
   - Fuel injector electrical connectors
   - All remaining electrical connections
   - Cruise control servo bracket
   - Cruise control and throttle cables to the throttle body
8. Apply a light coating of clean engine oil to a NEW O-ring seal and install the seal onto the EGR valve pipe.
   - EGR valve pipe into the intake manifold. Finger start the EGR valve pipe to the intake manifold bolt and the EGR valve pipe to the cylinder head bolts.
9. Finger start a NEW EGR valve pipe exhaust manifold gasket and bolts and tighten as follows:

   a. Tighten the EGR valve pipe-to-intake manifold bolt to 89 inch lbs. (10 Nm).

   b. Tighten the EGR valve pipe-to-cylinder head bolts to 37 ft. lbs. (50 Nm).

   c. Tighten the EGR valve pipe to the exhaust manifold bolts to 22 ft. lbs. (30 Nm).
   - EGR solenoid electrical connector
   - Drive belt
   - IAC valve and TPS sensor electrical connections to the throttle body
   - Air inlet duct
   - IAT and MAF sensor electrical connector
   - Fuel lines to the fuel rail
   - Vapor line to the fuel vapor purge valve
   - Transmission fluid level indicator tube bolt
10. Refill the cooling system.
    - Negative battery cable

Exhaust Manifold

REMOVAL & INSTALLATION

3.8L Engine

LEFT SIDE

1. Before servicing the vehicle, refer to the precautions in the beginning of this section.
2. Remove or disconnect the following:
   - Negative battery cable
   - Catalytic converter pipe
   - Connector position assurance (CPA) retainer, if equipped
   - Oxygen Sensor (O_2S) electrical connector
   - O_2S sensor
   - Spark plug
   - Exhaust Gas Recirculation (EGR) valve adapter from the exhaust manifold, if equipped
   - 2 rear exhaust manifold studs and nut from the exhaust manifold
   - Exhaust manifold heat shields
   - Exhaust manifold and gasket

To install:

3. Install or connect the following:
   - Exhaust manifold with a new gasket. Tighten the 2 front exhaust manifold studs and nuts to the exhaust manifold to 11 ft lbs. (15

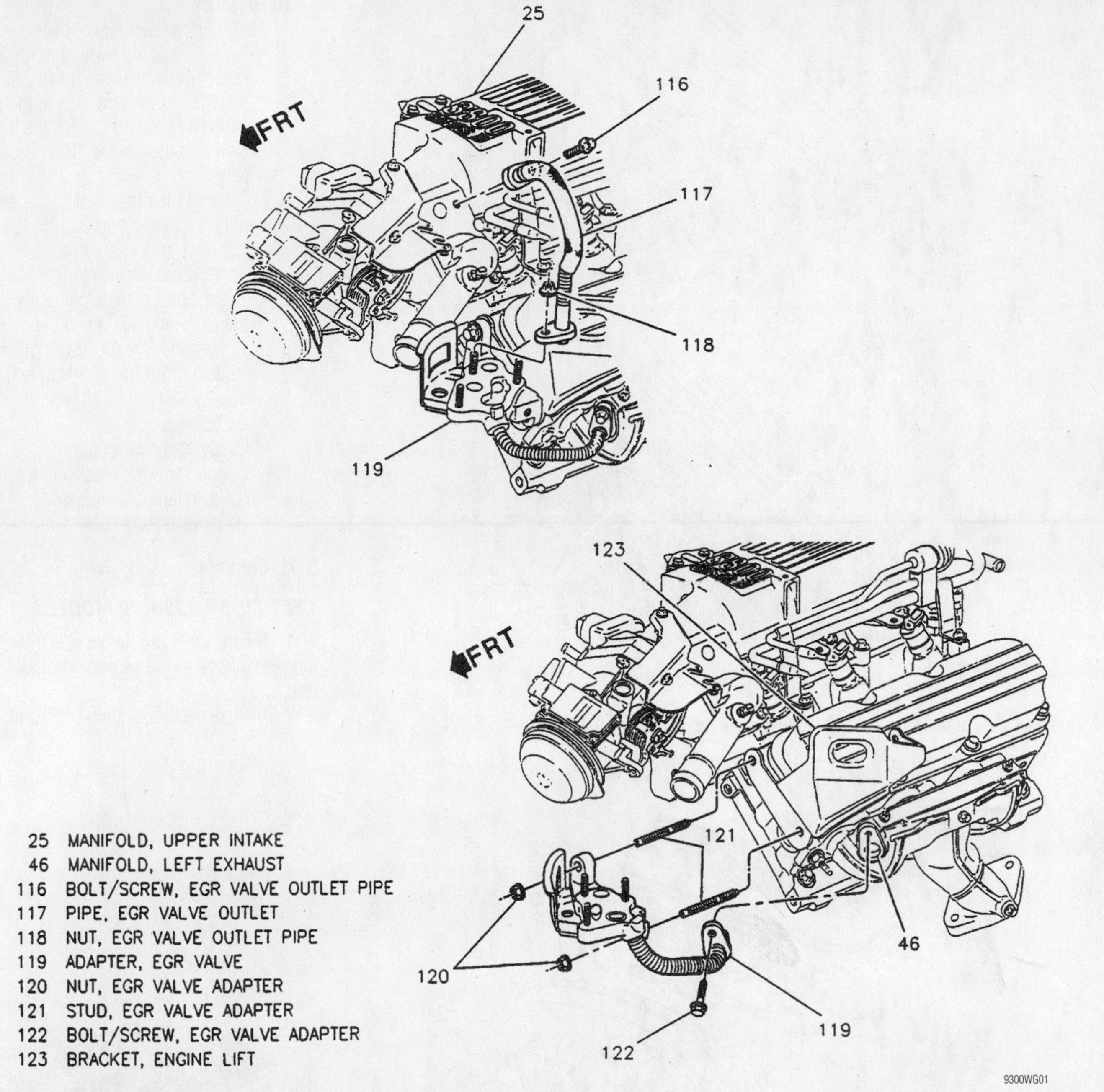

Exploded view of the EGR valve adapter and outlet pipe—3.8L engine

Nm) and the nuts to 13 ft. lbs. (18 Nm).

- Exhaust manifold heat shield and nuts. Tighten the to 89 inch lbs. (10 Nm).
- Spark plugs and wires
- 2 rear exhaust manifold studs and nut to the exhaust manifold. Tighten the 2 rear exhaust manifold studs to 11 ft lbs. (15 Nm) and the nuts to 13 ft. lbs. (18 Nm).
- O_2S sensor
- Catalytic converter pipe
- Connector position assurance (CPA) retainer, if equipped
- EGR valve adapter, if equipped and torque the bolt to 18 ft. lbs. (25 Nm)
- Negative battery cable

RIGHT SIDE

1. Before servicing the vehicle, refer to the precautions in the beginning of this section.
2. Remove or disconnect the following:
   - Negative battery cable
   - Connector position assurance (CPA) retainer, if equipped
   - Exhaust catalytic converter pipe
   - Oxygen Sensor (O_2S) electrical connector
   - Spark plugs and wires
   - 2 rear exhaust manifold studs and nut from the exhaust manifold
   - Exhaust manifold heat shields
   - Exhaust manifold and gasket

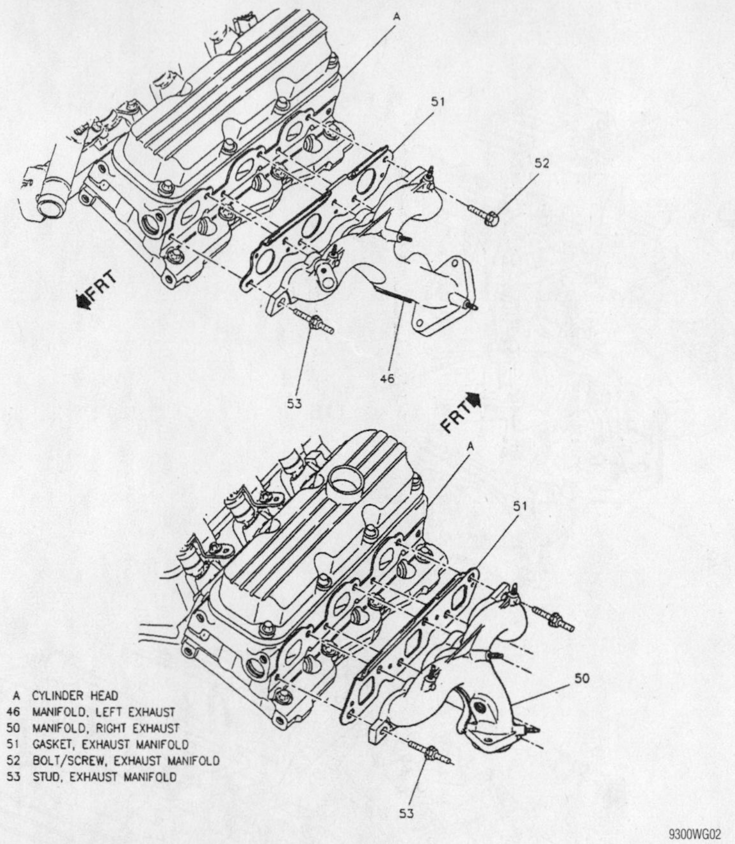

Exploded view of the right and left exhaust manifold mounting—3.8L engine

To install:

3. Install or connect the following:
   - Exhaust manifold with a new gasket. Tighten the 2 front exhaust manifold studs and nuts to the exhaust manifold to 11 ft lbs. (15 Nm) and the nuts to 13 ft. lbs. (18 Nm).
   - Exhaust manifold heat shield and nuts. Tighten the to 89 inch lbs. (10 Nm).
   - Spark plugs and wires
   - 2 rear exhaust manifold studs and nut to the exhaust manifold. Tighten the 2 rear exhaust manifold studs to 11 ft lbs. (15 Nm) and the nuts to 13 ft. lbs. (18 Nm).
   - O_2S sensor
   - Catalytic converter pipe
   - Connector position assurance (CPA) retainer, if equipped
   - Negative battery cable

5.7L Engines

LEFT SIDE—1998–99 MODELS

1. Before servicing the vehicle, refer to the precautions in the beginning of this section.
2. Relieve the fuel system pressure.

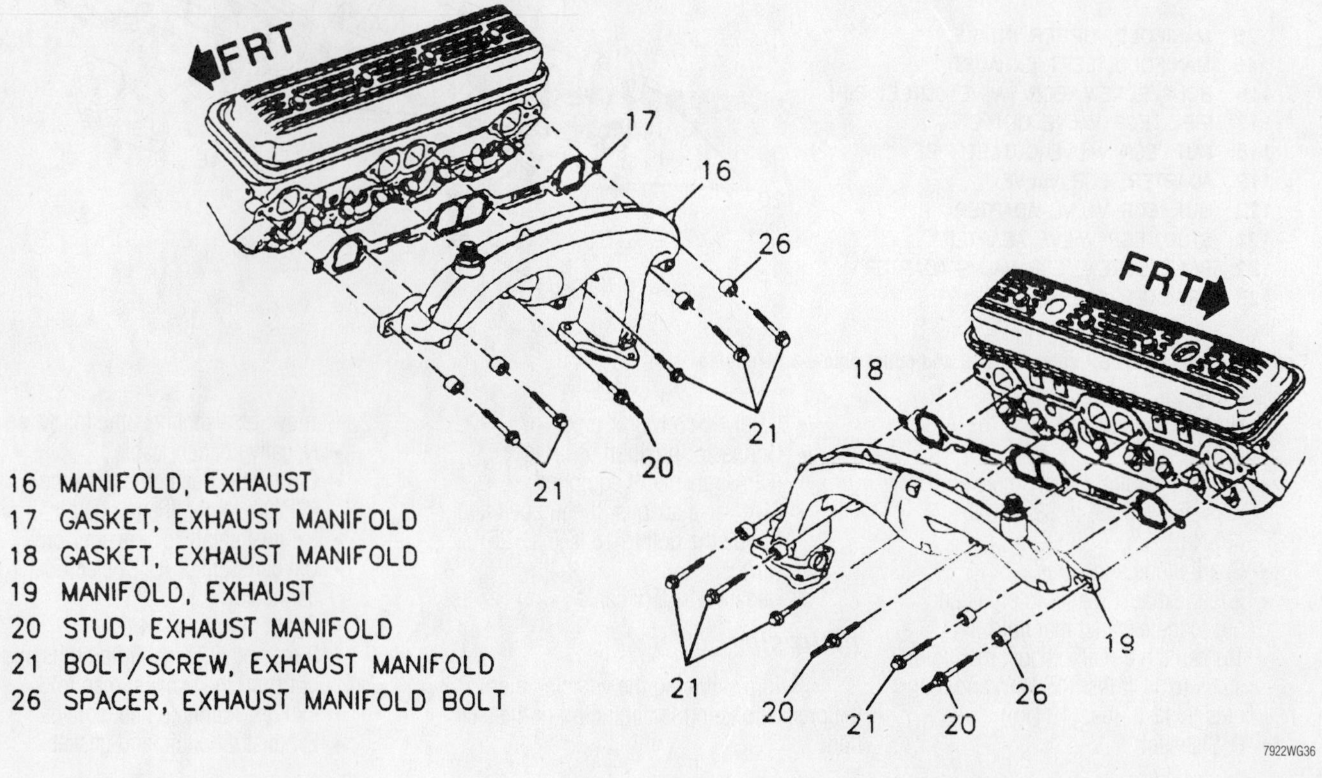

Exploded view of right and left exhaust manifold mounting—5.7L engines

3. Remove or disconnect the following:
   - Negative battery cable
   - Rocker arm cover
   - Fuel lines
   - Positive Crankcase Ventilation (PCV) hose
   - Spark plug wires
   - Spark plugs
   - Secondary Air Injection (AIR) hose
   - AIR pipe with check valve
   - Coolant sensor
   - Exhaust manifold bolts
   - Oxygen (O_2S) sensor
   - Exhaust manifold pipe nuts
   - Catalytic converter nuts
   - Exhaust manifold

To install:

Install or connect the following:
   - Exhaust manifold with a new gasket. Torque the catalytic converter nuts to 18 ft. lbs. (25 Nm).
   - Exhaust manifold pipe nuts. Torque the nuts to 26 ft. lbs. (35 Nm).
   - O_2S. Torque the sensor to 31 ft. lbs. (42 Nm).
   - Exhaust manifold bolts. Torque the bolts to 18 ft. lbs. (25 Nm).
   - Coolant sensor
   - AIR pipe. Torque the bolts to 15 ft. lbs. (20 Nm).
   - AIR hose
   - Spark plugs. Torque the spark plugs to 11 ft. lbs. (15 Nm).
   - Spark plug wires
   - PCV hose
   - Fuel lines
   - Negative battery cable

LEFT SIDE—2000–01 MODELS

1. Before servicing the vehicle, refer to the precautions in the beginning of this section.
2. Remove or disconnect the following:
   - Negative battery cable
   - Rocker arm cover
   - Spark plug wires
   - Spark plugs
   - Left side catalytic converter
   - Exhaust manifold bolts
   - Exhaust manifold and gasket

To install:

3. Apply a 0.2 inch (5mm)) wide band of threadlock GM P/N 12345493 to the threads of the exhaust manifold bolts.
4. Install a NEW exhaust manifold gasket and the exhaust manifold.
5. Install the exhaust manifold bolts and tighten as follows:
   a. Tighten the exhaust manifold bolts on first pass to 11 ft. lbs. (15 Nm). Tighten the manifold bolts beginning with the center 2 bolts. Alternate from side-to-side, and work toward the outside bolts.
   b. Tighten the exhaust manifold bolts on final pass to 18 ft. lbs. (25 Nm). Tighten the exhaust manifold bolts beginning with the center 2 bolts. Alternate from side-to-side, and work toward the outside bolts.
6. Using a flat punch, bend over the exposed edge of the exhaust manifold gasket at the rear of the cylinder head.

Install or connect the following:
   - Left side catalytic converter
   - Rocker arm cover
   - Spark plug wires
   - Spark plugs
   - Negative battery cable

RIGHT SIDE—1998–99 MODELS

1. Before servicing the vehicle, refer to the precautions in the beginning of this section.
2. Remove or disconnect the following:
   - Negative battery cable
   - Oil level indicator and tube
   - Exhaust manifold pipe nuts
   - Secondary Air Injection (AIR) hose
   - AIR pipe with check valve
   - Ignition coil main electrical connector
   - Spark plug wires
   - Spark plugs
   - Rocker arm cover
   - Exhaust Gas Recirculation (EGR) valve electrical connector
   - EGR valve pipe
   - Exhaust manifold

To install:

3. Install the exhaust manifold with a new gasket. Torque the bolts to 18 ft. lbs. (25 Nm).
4. Install the EGR valve pipe. Torque the bolts as follows:
   a. EGR valve pipe-to-exhaust manifold bolts to 18 ft. lbs. (25 Nm).
   b. EGR valve pipe-to-cylinder head bolt to 37 ft. lbs. (50 Nm).
   c. EGR valve pipe-to-throttle body bolt to 106 inch lbs. (12 Nm).
5. Install or connect the following:
   - EGR valve electrical connector
   - Rocker arm cover. Torque the bolts to 106 inch lbs. (12 Nm).
   - Spark plugs. Torque the spark plugs to 12 ft. lbs. (15 Nm).
   - Spark plug wires
   - Ignition coil main electrical connector
   - AIR pipe. Torque the bolts to 18 ft. lbs. (25 Nm).
   - AIR hose
   - Exhaust manifold pipe nuts. Torque the nuts to 26 ft. lbs. (35 Nm).
   - Oil level indicator and tube
   - Negative battery cable

RIGHT SIDE—2000–01 MODELS

1. Before servicing the vehicle, refer to the precautions in the beginning of this section.
2. Remove or disconnect the following:
   - Negative battery cable
   - Rocker arm cover
   - Spark plug wires
   - Spark plugs
   - Right side catalytic converter
   - Exhaust Gas Recirculation (EGR) valve electrical connector
   - EGR valve pipe bolts
   - EGR valve pipe up out of the intake manifold, using mild force
   - O-ring seal from the EGR valve pipe. Discard the exhaust manifold gasket and O-ring seal.
   - Exhaust manifold bolts
   - Exhaust manifold and gasket

To install:

3. Apply a 0.2 inch (5mm)) wide band of threadlock GM P/N 12345493 to the threads of the exhaust manifold bolts.
4. Install a NEW exhaust manifold gasket and the exhaust manifold.
5. Install the exhaust manifold bolts and tighten as follows:
   a. Tighten the exhaust manifold bolts on first pass to 11 ft. lbs. (15 Nm). Tighten the manifold bolts beginning with the center 2 bolts. Alternate from side-to-side, and work toward the outside bolts.
   b. Tighten the exhaust manifold bolts on final pass to 18 ft. lbs. (25 Nm). Tighten the exhaust manifold bolts beginning with the center 2 bolts. Alternate from side-to-side, and work toward the outside bolts.
6. Using a flat punch, bend over the exposed edge of the exhaust manifold gasket at the rear of the cylinder head.
7. Apply a light coating of clean engine oil to a NEW O-ring seal and install the seal onto the EGR valve pipe.
8. Install the EGR valve pipe into the intake manifold. Finger start the EGR valve pipe to the intake manifold bolt and the EGR valve pipe to the cylinder head bolts.

9. Finger start a NEW EGR valve pipe exhaust manifold gasket and bolts and tighten as follows:
   a. Tighten the EGR valve pipe-to-intake manifold bolt to 89 inch lbs. (10 Nm).
   b. Tighten the EGR valve pipe-to-cylinder head bolts to 37 ft. lbs. (50 Nm).
   c. Tighten the EGR valve pipe to the exhaust manifold bolts to 22 ft. lbs. (30 Nm).

   Install or connect the following:
   - EGR solenoid electrical connector
   - Right side catalytic converter
   - Rocker arm cover
   - Spark plug wires
   - Spark plugs
   - Negative battery cable

Camshaft and Valve Lifters

REMOVAL & INSTALLATION

3.8L Engine

1. Before servicing the vehicle, refer to the precautions in the beginning of this section.
2. Recover the refrigerant in the air conditioning system.
3. Drain the engine cooling system.
4. Remove or disconnect the following:
   - Negative battery cable
   - Radiator with the air conditioning condenser assembly
   - Upper intake manifold
   - Rocker arm covers
   - Rocker arms and the push rods
   - Lower intake manifold
   - Valve lifter guides
   - Valve lifters
   - Front cover
   - Camshaft sprocket and timing chain
   - Balance shaft gear
   - Camshaft thrust plate
   - Camshaft key
5. Install 3 5⁄16–18 x 4-inch bolts in the camshaft bolt holes.
6. Carefully rotate and pull the camshaft assembly out of the bearings.

To install:

7. Install or connect the following:
   - Camshaft lubricated with assembly lubricant
   - Camshaft key
   - Thrust plate. Torque the bolts to 11 ft. lbs. (15 Nm).
   - Balance shaft gear
   - Camshaft sprocket
   - Timing chain
   - Front cover
   - Valve lifters
   - Valve lifter guides
   - Lower intake manifold
   - Rocker arms and push rods

➡Make sure the push rods are seated in the lifter seat.

8. The engine must be on the No. 1 firing position before proceeding.
9. Tighten exhaust valves No. 1, 2 and 3; then intake valves No. 1, 5 and 6.
10. Loosen the adjusting nut until lash is felt at the push rod.
11. Tighten the adjusting nut until all lash is removed, then tighten the nut an additional 1 ½ turns to center the lifter plunger.
12. Rotate the crankshaft 1 revolution until the "0" timing mark is once again aligned.
13. Adjust exhaust valves No. 4, 5 and 6 and intake valves No. 2, 3 and 4.
14. Install or connect the following:
    - Rocker arm cover
    - Upper intake manifold
    - Radiator with the air conditioning condenser assembly
    - Negative battery cable
15. Evacuate and recharge the air conditioning system.
16. Refill the cooling system.
17. Start the engine and check for leaks.

5.7L Engines

1. Before servicing the vehicle, refer to the precautions in the beginning of this section.
2. Recover the refrigerant from the air conditioning system.
3. Drain the engine cooling system.
4. Remove or disconnect the following:
   - Negative battery cable
   - Intake manifold
   - Exhaust manifold
   - Rocker arm covers
   - Rocker arms and push rod assemblies
   - Cylinder head
   - Valve lifters
   - Radiator with the condenser
   - Front cover
   - Oil pan
   - Oil pump drive assembly
   - Timing chain and sprocket
   - Camshaft retainer

⁂ WARNING

All camshaft journals are the same diameter so care must be used when removing the camshaft to avoid damaging the bearings.

5. Install 3 5⁄16–18 x 4-inch bolts in the camshaft bolt holes.
6. Carefully rotate and pull the camshaft assembly out of the bearings.

To install:

➡When installing a new camshaft assembly, coat the camshaft lobes with "Molykote"® or equivalent. When installing a new camshaft, replace all valve lifters to ensure durability of the camshaft lobes and lifters.

7. Install or connect the following:
   - Camshaft
   - Camshaft retainer. Torque the bolts to 18 ft. lbs. (25 Nm).
   - Timing chain and camshaft sprocket
   - Oil pump drive assembly. Torque the bolt to 25 ft. lbs. (34 Nm).
   - Oil pan
   - Front cover
   - Radiator with the air conditioning condenser
   - Air conditioning receiver and dehydrator hose to the condenser
   - Air conditioning compressor and condenser hose to the condenser
   - Valve lifters
   - Cylinder head with a new gasket
   - Push rods and rocker arms

➡Make sure the push rods are seated in the lifter seat.

8. Place the engine on Top Dead Center (TDC) of the No. 1 firing position.
9. Tighten the exhaust valves No. 1, 2, 7 and 8; then intake valves No. 1, 3, 4 and 5 to 22 ft. lbs. (30 Nm).
10. Rotate the crankshaft 1 revolution until the "0" timing mark is once again aligned.
11. Adjust exhaust valves No. 3, 4, 5 and 6; then intake valves 2, 6, 7 and 8 to 22 ft. lbs. (30 Nm).
12. Install or connect the following:
    - Exhaust manifold
    - Rocker arm covers
    - Intake manifold with a new gasket
    - Negative battery cable
13. Recharge the air conditioning system.
14. Fill the engine cooling system.

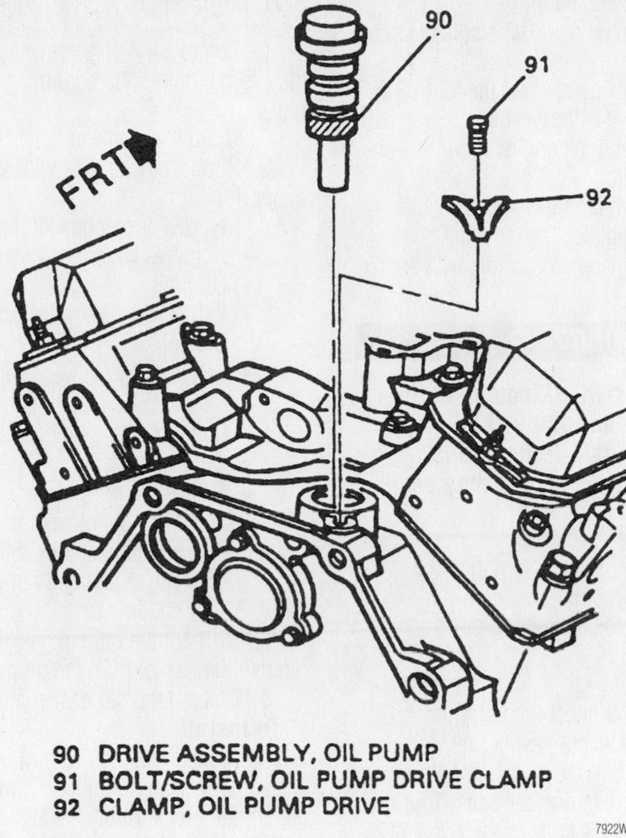

Unfasten the retaining bolt and clamp, then remove the oil pump drive—5.7L Engines

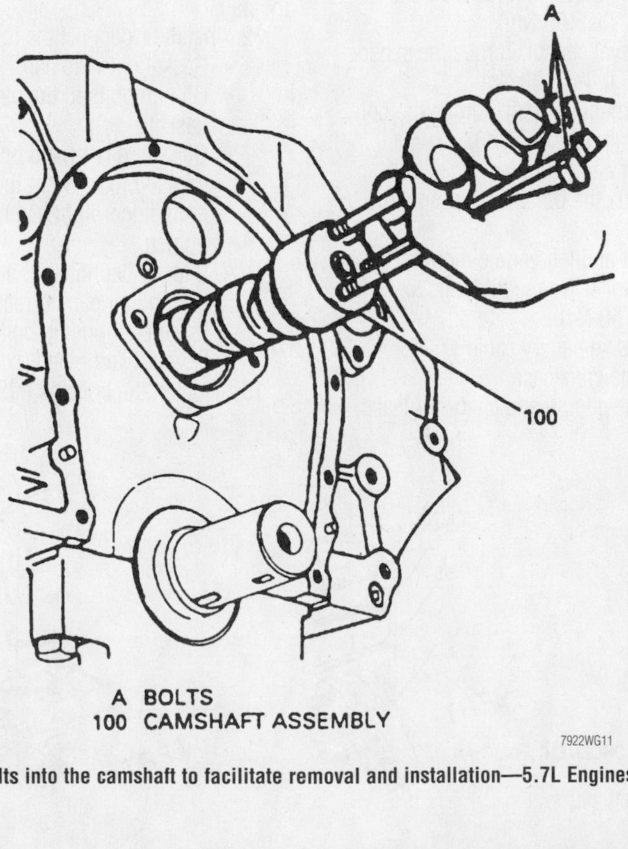

Thread 3 bolts into the camshaft to facilitate removal and installation—5.7L Engines

Valve Lash

ADJUSTMENT

➡The engines in this section utilize hydraulic valve lifters no adjustment is necessary.

Starter Motor

REMOVAL & INSTALLATION

3.8L Engine

1. Before servicing the vehicle, refer to the precautions in the beginning of this section.
2. Remove or disconnect the following:
   - Negative battery cable
   - Starter shield
   - Starter bolt and nut
   - Starter electrical connectors
   - Starter motor

To install:

Install or connect the following:

- Starter motor, do not install the bolts at this time
- Starter electrical connectors. Torque the lead nut to 17 inch lbs. (2 Nm) and the positive battery nut to 89 inch lbs. (10 Nm).
- Starter bolt and nut. Torque the starter motor bolt to 35 ft. lbs. (47 Nm) and the nut to 33 ft. lbs. (45 Nm).
- Starter shield
- Negative battery cable

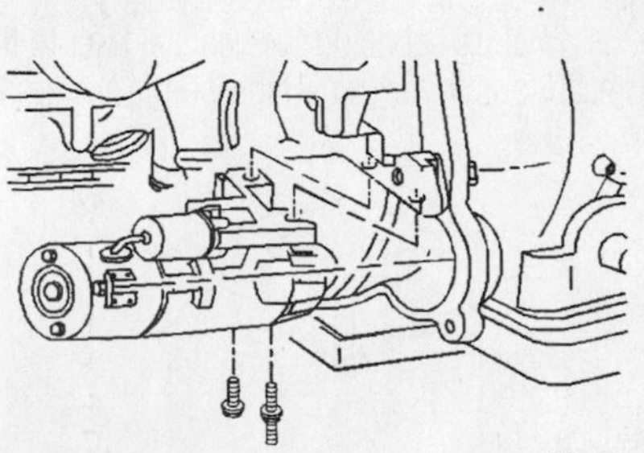

View of the starter—3.8L engine

5.7L Engine

1. Before servicing the vehicle, refer to the precautions in the beginning of this section.
2. Remove or disconnect the following:
   - Negative battery cable
   - Left side catalytic converter
   - Starter bolts
   - Starter electrical connectors
   - Starter motor

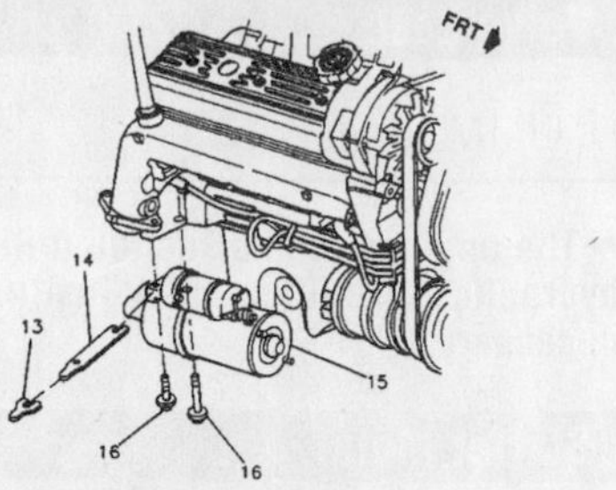

View of the starter—5.7L engine

To install:

Install or connect the following:

- Starter motor, do not install the bolts at this time
- Starter electrical connectors. Torque the lead nut to 18 inch lbs. (2 Nm) and the positive battery nut to 89 inch lbs. (10 Nm).
- Starter bolts. Torque the bolts to 37 ft. lbs. (50 Nm).
- Left side catalytic converter. Torque the bolts to 26 ft. lbs. (35 Nm).
- Negative battery cable

Oil Pan

REMOVAL & INSTALLATION

3.8L Engine

1. Before servicing the vehicle, refer to the precautions in the beginning of this section.

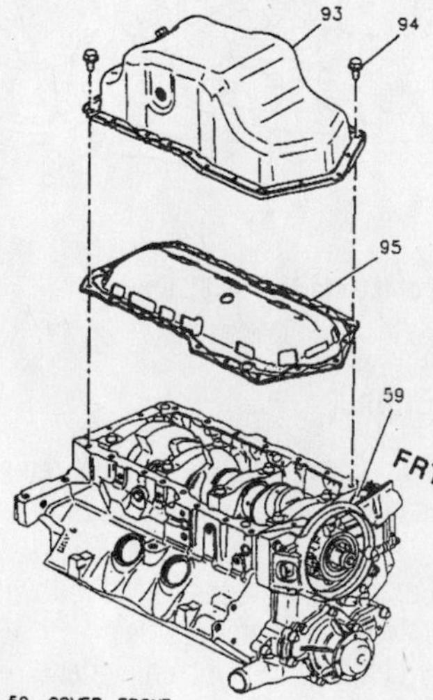

Exploded view of the oil pan and gasket—3.8L engine

2. Drain the engine oil.
3. Support the engine with suitable support fixture.
4. Remove or disconnect the following:
   - Negative battery cable
   - Right and left engine mount-to-cradle bolts
   - Right and left lower shock bolts
   - Intermediate steering shaft bolt
5. Support the engine cradle with appropriate equipment.

✲✲ WARNING

The oil level switch, located in the oil pan, must be removed before removing the pan. If the pan is removed first, the switch may be damaged.

   - Oil level sensor from the oil pan
   - Starter motor
   - Front crossmember bolts
   - Oil pan

To install:

6. Install or connect the following:
   - Oil pan with a new gasket. Torque the bolts to 10 ft. lbs. (14 Nm).
   - Front crossmember bolts. Torque the upper 4 bolts to 92 ft. lbs. (125 Nm) and the lower 2 bolts to 107 ft. lbs. (145 Nm).
   - Starter motor. Torque the bolts to 37 ft. lbs. (50 Nm).
   - Oil level sensor. Torque the sensor to 15 ft. lbs. (20 Nm).
   - Intermediate shaft. Torque the bolt to 35 ft. lbs. (47 Nm).
   - Right and left lower shock bolts. Torque the bolts to 48 ft. lbs. (65 Nm).
   - Right and left engine mount to cradle bolts. Torque the bolts to 43 ft. lbs. (58 Nm).
   - Negative battery cable
7. Refill the crankcase.
8. Start the engine and check for leaks.

5.7L Engines

1. Before servicing the vehicle, refer to the precautions in the beginning of this section.
2. Drain the crankcase.
3. Support the engine with suitable support fixture.
4. Remove or disconnect the following:
   - Negative battery cable
   - Oil filter
   - Left and right engine mount-to-cradle bolts
   - Left and right lower shock bolts
   - Intermediate steering shaft bolt
5. Support the engine cradle.
   - Oil level sensor
   - Starter motor
   - Left and right closeouts
6. Loosen the six cradle bolts.
7. Lower the cradle and remove the oil pan.
8. Drill out the oil pan gasket rivets and remove the gasket from the oil pan.
9. Discard the old gasket and rivets.

To install:

10. Apply a 0.2 inch (5mm) bead of sealant RTV sealer 0.8 inch (20mm) long to the engine block. Apply the sealant directly onto the tabs of the front cover gasket that protrudes into the oil pan surface
11. Pre-assemble the oil pan gasket to the pan.
12. Install or connect the following:
    - Gasket onto the pan
    - Oil pan bolts to the pan through the gasket
    - Oil pan, gasket and bolts to the engine block. Snug the oil pan bolts finger tight. Do not overtighten.
    - 2 lower bellhousing bolts to position the oil pan correctly. Snug the lower bellhousing bolt finger tight. Do not overtighten.
13. Tighten the bolts as follows:

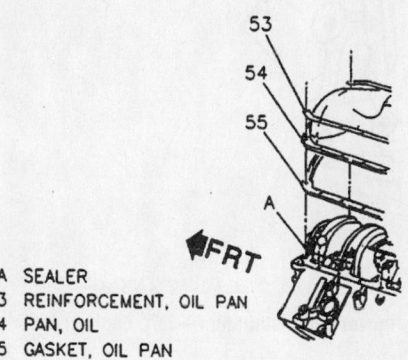

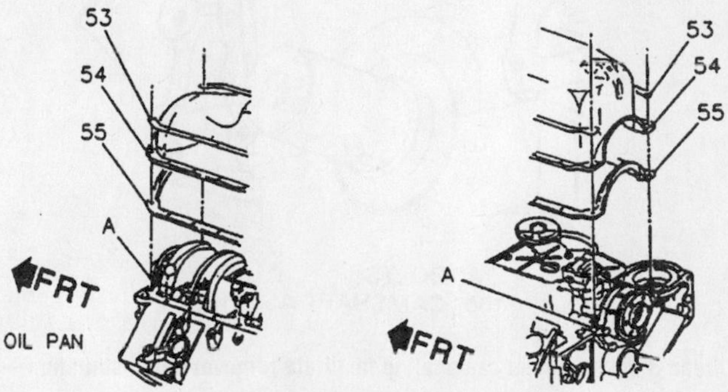

Oil pan sealer and gasket location—5.7L engines

a. Tighten the oil pan-to-block and oil pan-to-oil pan front cover bolts to 18 ft. lbs. (25 Nm).

b. Tighten the oil pan-to-rear cover bolts to 106 inch lbs. (12 Nm).

c. Tighten the bellhousing bolts to 37 ft. lbs. (50 Nm).

14. Install or connect the following:
    - Engine cradle. Torque the upper four bolts to 92 ft. lbs. (125 Nm) and the lower two bolts to 107 ft. lbs. (145 Nm).
    - Side closeouts and bolts. Tighten the bolts to 106 inch lbs. (12 Nm).
    - Starter motor. Torque the bolts to 37 ft. lbs. (50 Nm).
    - Oil level sensor. Torque it to 26 ft. lbs. (35 Nm).
    - Intermediate steering shaft bolt. Torque the bolt to 35 ft. lbs. (47 Nm).
    - Left and right lower shock bolts. Torque the bolts to 48 ft. lbs. (65 Nm).
    - Left and right engine mount-to-cradle bolts. Torque the bolts to 43 ft. lbs. (58 Nm).
    - Oil filter
    - Negative battery cable
15. Refill the crankcase.
16. Start the engine and check for leaks.

Oil Pump

REMOVAL & INSTALLATION

3.8L Engine

1. Before servicing the vehicle, refer to the precautions in the beginning of this section.

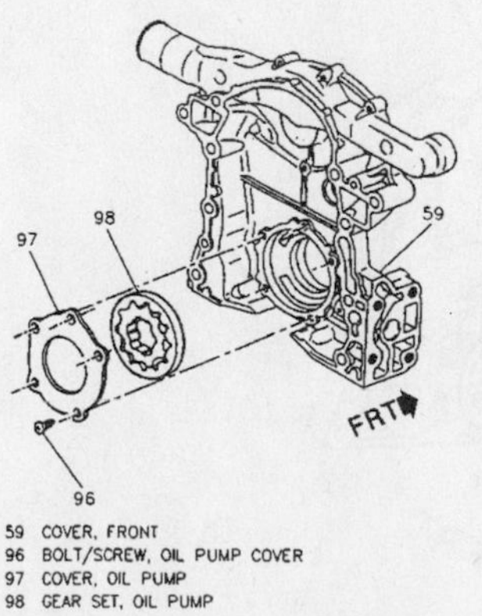

Exploded view of the oil pump gear set—3.8L engine

2. Remove or disconnect the following:
    - Negative battery cable
    - Front cover
    - Oil filter adapter, pressure valve and spring
    - Oil pump cover
    - Oil pump gear set

To install:

3. Lubricate the gear set with petroleum jelly.

4. Install the oil pump gear set into the front cover.

5. Pack the gear cavity with petroleum jelly for proper priming.

6. Install or connect the following:
    - Oil pump cover. Torque the screws to 98 inch lbs. (11 Nm).
    - Oil filter adapter with a new gasket, pressure valve and spring. Torque the bolts to 11 ft. lbs. (15 Nm).
    - Front cover. Torque the bolts to 11 ft. lbs. (15 Nm).
    - Negative battery cable

5.7L Engines

1. Before servicing the vehicle, refer to the precautions in the beginning of this section.
2. Drain the crankcase.
3. Remove or disconnect the following:
    - Front cover
    - Oil pan
    - Oil pump screen
    - Oil pan deflector
    - Oil pump and propeller shaft

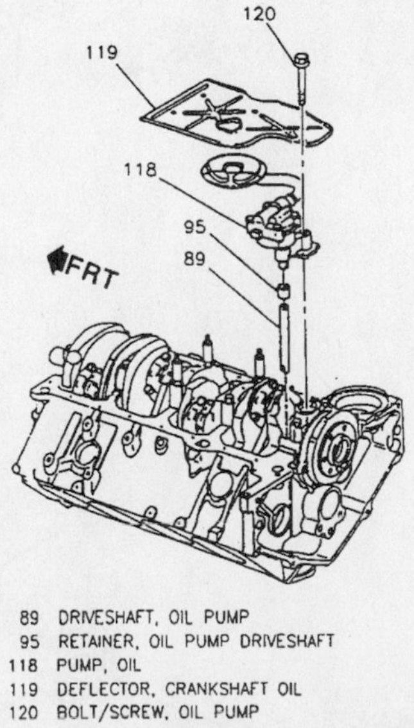

Exploded view of the oil pump and propeller shaft mounting—5.7L engine

To install:

4. Install or connect the following:
    - Oil pump and propeller shaft. Torque the bolts to 18 ft. lbs. (25 Nm).
    - Oil pan deflector. Torque the nuts to 18 ft. lbs. (25 Nm).
    - Oil pump screen. Torque the bolt to 106 inch lbs. (12 Nm).
    - Oil pan
    - Front cover. Torque the bolts to 18 ft. lbs. (25 Nm).
5. Refill the crankcase.

Rear Main Seal

REMOVAL & INSTALLATION

➡The rear main seal is a 1-piece unit. It can be removed or installed without removing the oil pan or crankshaft.

1. Before servicing the vehicle, refer to the precautions in the beginning of this section.
2. Remove or disconnect the following:
    - Transmission
    - Clutch and pressure plate, if equipped
    - Flywheel
    - Rear main seal
3. Inspect the crankshaft for nicks or burrs and correct as required.

To install:

4. Install a new rear seal lubricated with engine oil.
5. Install or connect the following:
    - Flywheel. Torque the bolts to 74 ft. lbs. (100 Nm) if equipped with a manual transmission, or to 44 ft. lbs. (60 Nm) if equipped with an automatic.
    - Clutch and pressure plate, if equipped
    - Transmission

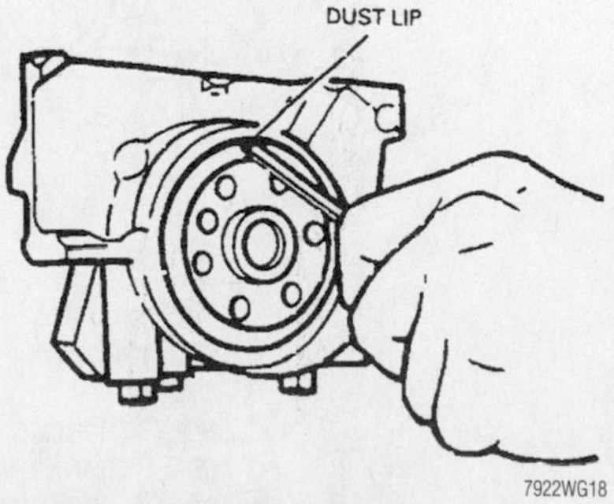

When prying the rear main seal out, be careful not to damage the crankshaft sealing surface or the seal bore

6. Check the fluid levels.
7. Start the engine and check for leaks.

Timing Chain, Sprockets, Front Cover and Seal

REMOVAL & INSTALLATION

3.8L Engine

1. Before servicing the vehicle, refer to the precautions in the beginning of this section.
2. Drain the crankcase and cooling system.
3. Remove or disconnect the following:
   - Negative battery cable
   - Air cleaner and intake air duct
   - Coolant pump pulley bolts, loosen
   - Drive belt tensioner

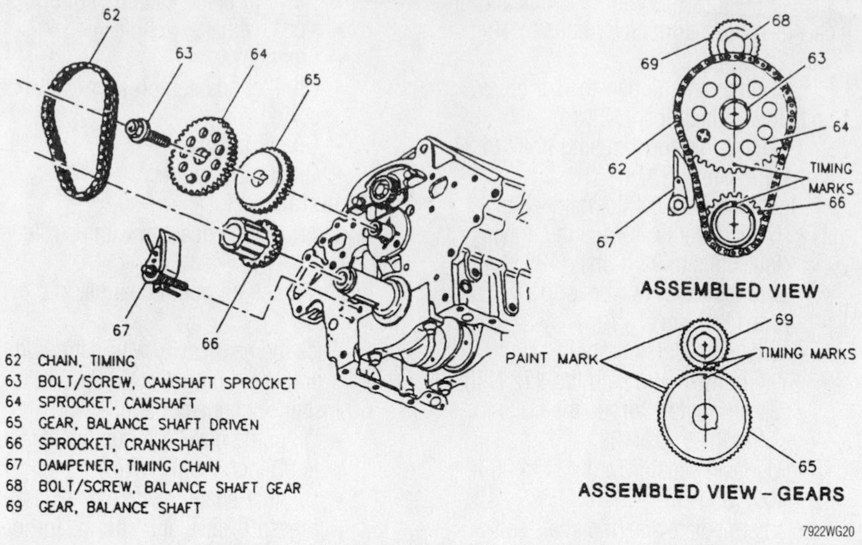

Exploded view of the timing chain and sprockets and timing mark alignment—3.8L engine

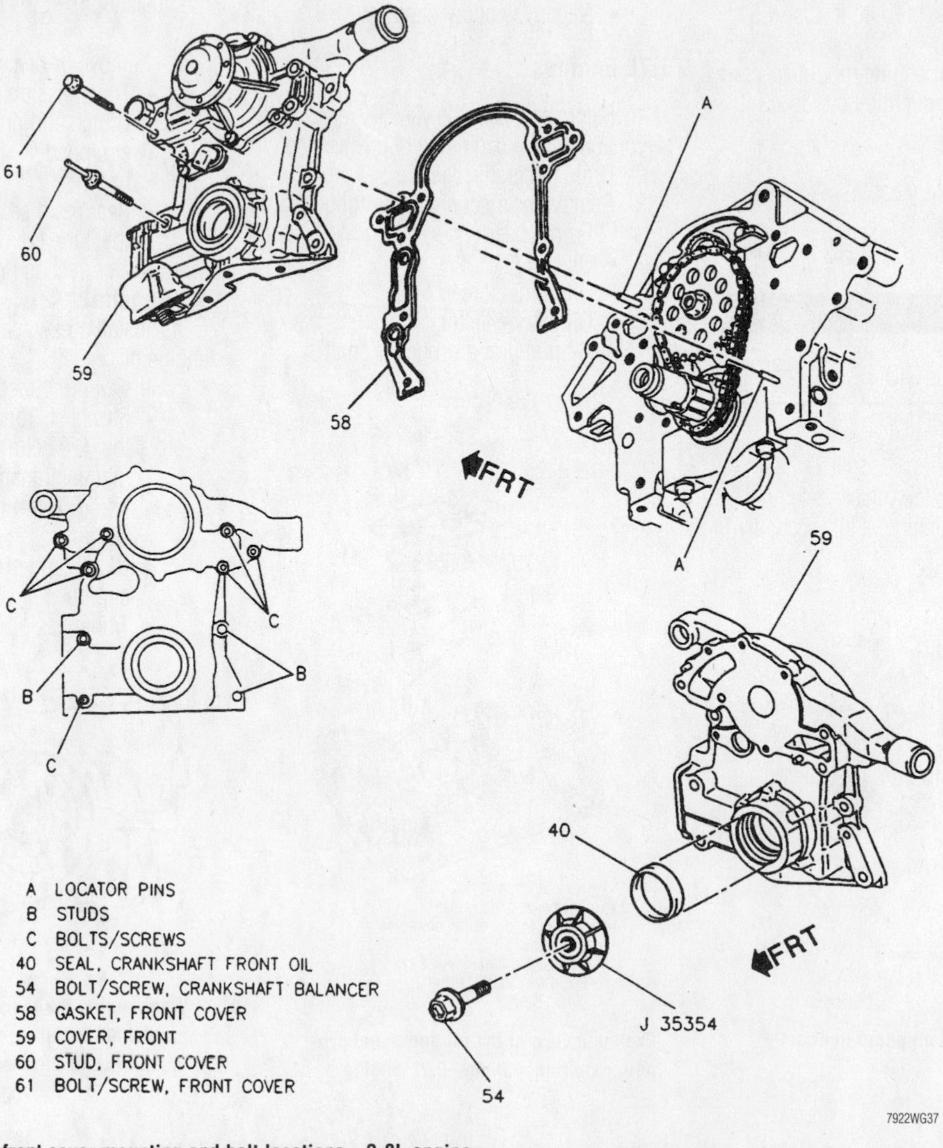

Exploded view of the front cover mounting and bolt locations—3.8L engine

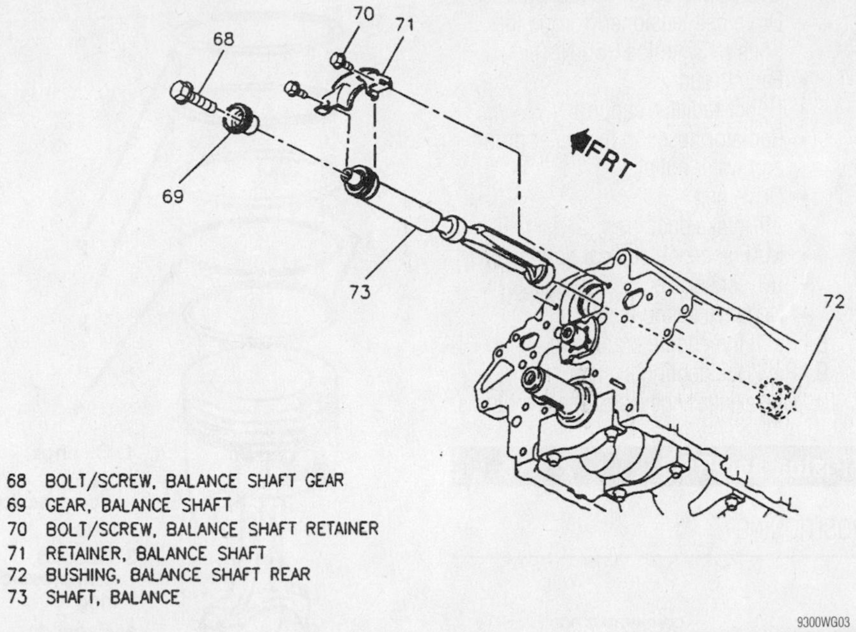

Exploded view of the balance shaft assembly—3.8L engine

- Power steering pump pulley, if necessary
- Inlet hose from the power steering pump, if necessary

4. Reposition the hose clamp on the return line, if necessary.
   - Reservoir hose return line from the power steering pump, if necessary
   - Power steering pump, if necessary
   - Crankshaft balancer
   - Crankshaft Position (CKP) sensor shield and the sensor
   - Oil pan-to-front cover bolts
   - Oil pan bolts, loosen and lower the pan slightly
   - Water pump
   - Radiator outlet hose from the front cover
   - Front cover bolts and studs
   - Cover and gasket
   - Crankshaft seal from the cover
5. Align the timing marks on the sprockets so they are as close together as possible.
   - Timing chain damper
   - Camshaft sprocket and timing chain
   - Crankshaft sprocket

To install:

6. If the crankshaft has been turned in the engine, perform the following:
   a. Turn the crankshaft so the No. 1 piston is at top Dead Center (TDC) of its compression stroke.
   b. Turn the camshaft so that, with the sprocket temporarily installed, the timing mark is straight down.
7. Install or connect the following:
   - Timing chain on the sprockets with the timing marks aligned
   - Timing chain and sprockets. Torque the bolt to 74 ft. lbs. (100 Nm) plus an additional 90 degrees.
   - Timing chain damper. Torque the bolt to 16 ft. lbs. (22 Nm).
8. Rotate the engine 2 revolutions, then check to be sure the timing marks are aligned.
   - Front cover with a new crankshaft oil seal. Torque the fasteners to 15 ft. lbs. (20 Nm) plus 40 degrees.
   - Radiator outlet hose to the front cover
   - Water pump
   - Oil pan-to-front cover bolts. Torque the bolts to 125 inch lbs. (14 Nm).
   - Oil pan bolts
   - CKP sensor and torque the fasteners to 15 ft. lbs. (20 Nm) plus 40 degrees
   - CKP sensor shield
   - Crankshaft balancer
   - Power steering pump, hoses and pulley, if necessary
   - Drive belt tensioner
   - Air cleaner assembly
   - Negative battery cable
9. Refill the crankcase and cooling system.
10. Start the engine and check for leaks.

5.7L Engines

1. Before servicing the vehicle, refer to the precautions in the beginning of this section.
2. Drain the crankcase and cooling system.
3. Remove or disconnect the following:
   - Negative battery cable
   - Mass Air flow (MAF) sensor electrical connector
   - Intake Air Temperature (IAT) sensor electrical connector
   - Air intake duct
   - Drive belt
   - Radiator hoses from the water pump and water outlet
   - Upper radiator support
   - Fan shroud
   - Drive belt tensioner
   - Overflow hose from the radiator
   - Throttle body heater hose from the radiator
   - Drive belt idler pulley
   - Water pump
   - Starter motor
   - Crankshaft balancer
   - Oil pan
   - Front cover
   - Oil pump
4. Rotate the crankshaft until the timing marks on the crankshaft and camshaft sprockets are aligned.
5. Remove or disconnect the following:
   - Camshaft sprocket and timing chain
   - Crankshaft sprocket

✻✻ WARNING

Do not turn the crankshaft after the timing chain has been removed to prevent damage to the pistons or valves.

To install:

6. Install or connect the following:

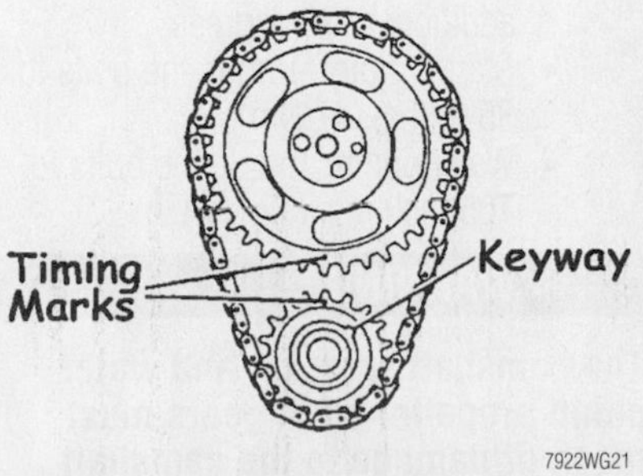

When assembled, be sure the marks on the gears are facing each other—5.7L engine

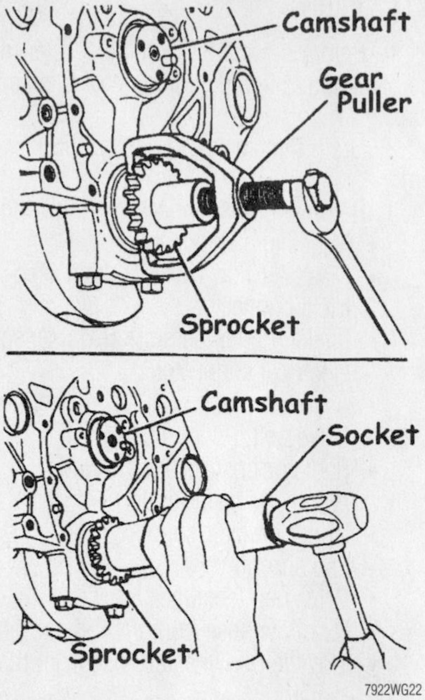

Use a gear puller to remove, and a large socket to install, the crankshaft sprocket—5.7L engine shown

- Crankshaft sprocket
- Camshaft sprocket and timing chain assembly. Torque the bolts to 26 ft. lbs. (35 Nm).
- Oil pump
- NEW crankshaft oil seal to the front cover

7. Apply a 0.20 inch (5mm bead of RTV sealant to the corner where the oil pan meets the engine block.

- Front cover. Torque the bolts to 18 ft. lbs. (25 Nm).
- Oil pan. Torque the oil pan-to-block and oil pan-to-front cover bolts to 18 ft. lbs. (25 Nm) and the oil pan-to-rear cover bolts to 106 inch lbs. (12 Nm).
- Crankshaft balancer. Torque the bolt to 37 ft. lbs. (50 Nm), then an additional 120 degrees.
- Starter motor. Torque the bolts to 35 ft. lbs. (47 Nm).
- Water pump. Torque the bolts to 105 inch lbs. (12 Nm).

⁂ WARNING

The camshaft sprocket and water pump propeller shaft gears must mesh or damage to the camshaft retainer may occur.

- Drive belt idler pulley. Torque the bolt to 37 ft. lbs. (50 Nm).
- Throttle body heater hose to the radiator
- Overflow hose to the radiator
- Drive belt tensioner. Torque the bolts to 37 ft. lbs. (50 Nm).
- Fan shroud
- Upper radiator support
- Radiator hoses to the water pump and water outlet
- Drive belt
- Air intake duct
- MAF sensor electrical connector
- IAT sensor electrical connector
- Negative battery cable

8. Refill the engine crankcase.
9. Refill the cooling system.
10. Operate the engine and check for leaks.

Piston and Ring

POSITIONING

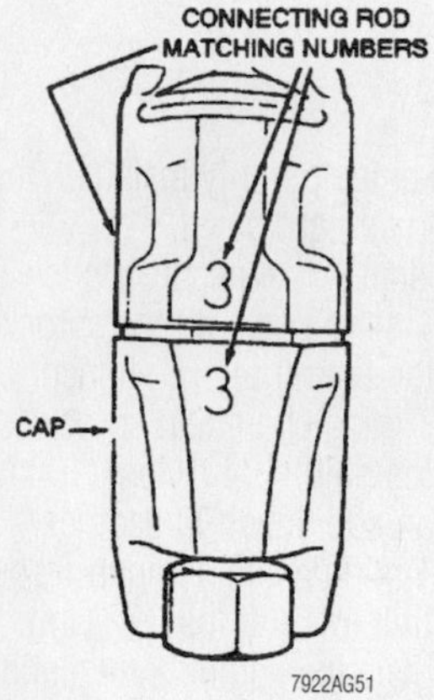

Engine connecting rod and cap installation. Be sure to matchmark the cap and rod prior to disassembly, as shown

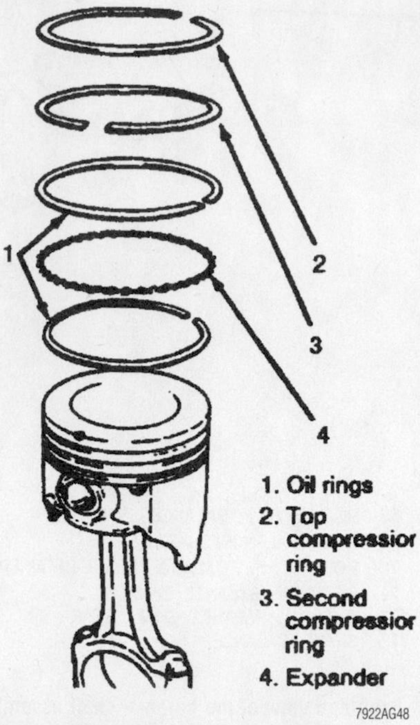

Piston ring positioning—3.8L engine

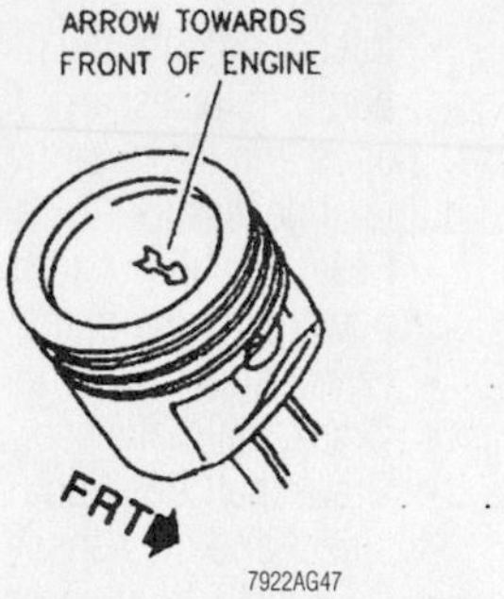

Piston positioning. Often the arrow is replaced by a notch, which also must face the front of the engine—3.8L engine

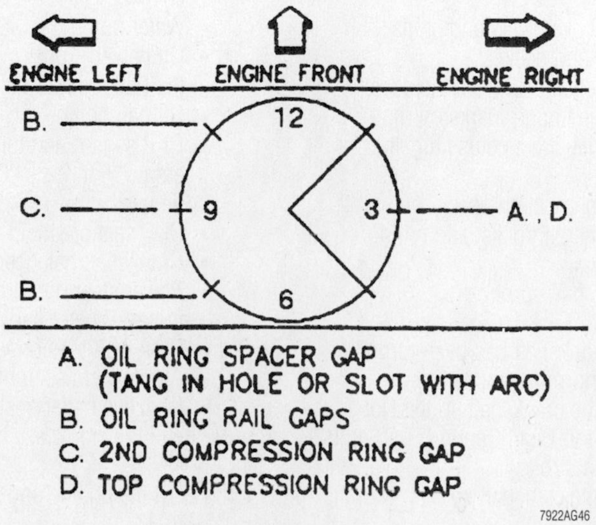

Piston ring end-gap spacing—3.8L engine

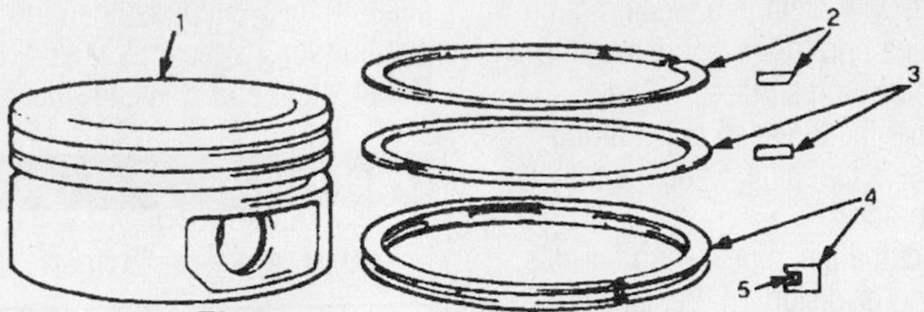

1. Piston
2. Upper compression piston ring
3. Lower compression piston ring
4. Oil control piston ring
5. Oil control ring spring w/spacer

7922AG43

Piston ring positioning—5.7L engine

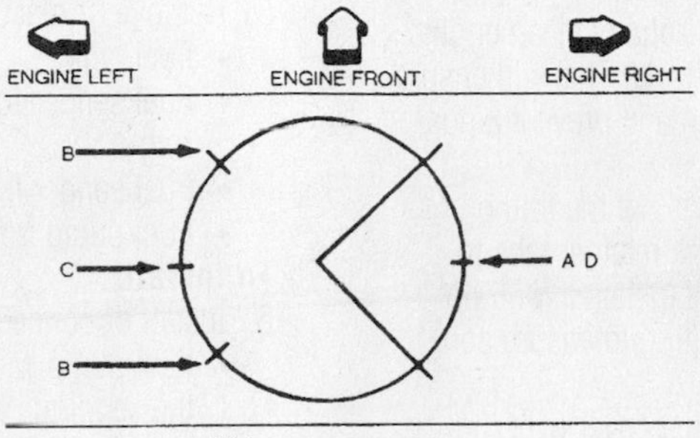

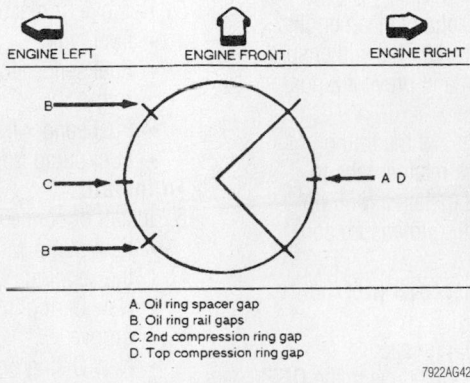

A. Oil ring spacer gap
B. Oil ring rail gaps
C. 2nd compression ring gap
D. Top compression ring gap

7922AG42

Piston ring end-gap spacing—5.7L engine

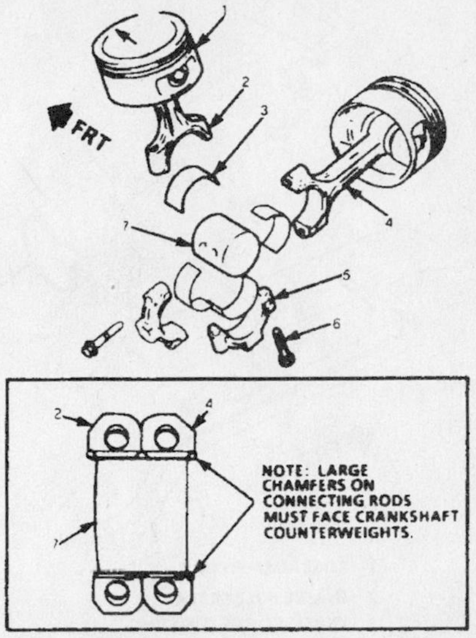

1. Piston
2. Connecting rod LH
3. Connecting rod bearing
4. Connecting rod RH
5. Connecting rod bearing cap
6. Connecting rod bearing cap bolt
7. Crankshaft

7922AG44

Piston and connecting rod assembly positioning—5.7L engine

FUEL SYSTEM

Fuel System Service Precautions

Safety is the most important factor when performing not only fuel system maintenance but any type of maintenance. Failure to conduct maintenance and repairs in a safe manner may result in serious personal injury or death. Maintenance and testing of the vehicle's fuel system components can be accomplished safely and effectively by adhering to the following rules and guidelines.

- To avoid the possibility of fire and personal injury, always disconnect the negative battery cable unless the repair or test procedure requires that battery voltage be applied.
- Always relieve the fuel system pressure prior to disconnecting any fuel system component (injector, fuel rail, pressure regulator, etc.), fitting or fuel line connection. Exercise extreme caution whenever relieving fuel system pressure, to avoid exposing skin, face and eyes to fuel spray. Please be advised that fuel under pressure may penetrate the skin or any part of the body that it contacts.
- Always place a shop towel or cloth around the fitting or connection prior to loosening to absorb any excess fuel due to spillage. Ensure that all fuel spillage (should it occur) is quickly removed from engine surfaces. Ensure that all fuel soaked cloths or towels are deposited into a suitable waste container.
- Always keep a dry chemical (Class B) fire extinguisher near the work area.
- Do not allow fuel spray or fuel vapors to come into contact with a spark or open flame.
- Always use a back-up wrench when loosening and tightening fuel line connection fittings. This will prevent unnecessary stress and torsion to fuel line piping.
- Always replace worn fuel fitting O-rings with new. Do not substitute fuel hose or equivalent, where fuel pipe is installed.

Fuel System Pressure

RELIEVING

1. Before servicing the vehicle, refer to the precautions in the beginning of this section.
2. Disconnect the negative battery cable to prevent fuel discharge if the key is accidentally turned to the **RUN** position.
3. Loosen the fuel filler cap to relieve the tank pressure and do not tighten until service has been completed.
4. Connect a fuel pressure gauge to the fuel pressure valve. Wrap a shop cloth around the fitting while connecting the gauge to avoid spillage.
5. Place the end of the bleed hose into a suitable container and open the valve to relieve the fuel system pressure.

Fuel Filter

REMOVAL & INSTALLATION

The inline fuel filter is located on the fuel feed pipe before the fuel injection system and mounted directly in front of the rear axle. The filter housing is constructed of steel with quick-connect inlet and threaded outlet fittings. The threaded fitting is sealed with an O-ring. In order to disengage quick-

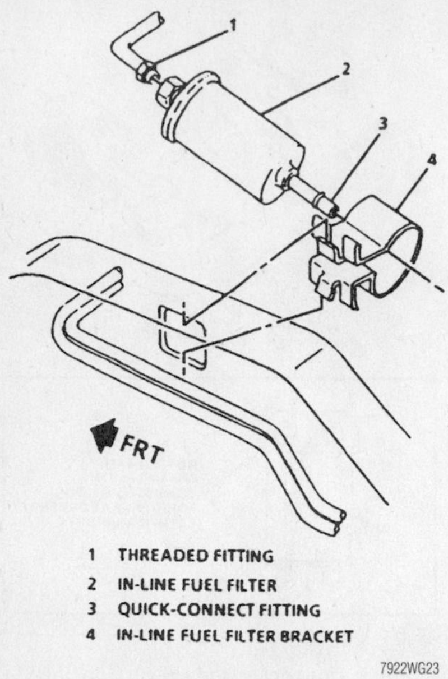

The inline fuel filter is mounted in a bracket located under the vehicle, directly in front of the rear axle

connect fittings, a fuel line quick-connect separator tool set, such as J-37088-A or equivalent, is required. There is no service interval for fuel filter replacement. We suggest replacing the filter every 30–40 thousand miles.

1. Before servicing the vehicle, refer to the precautions in the beginning of this section.
2. Relieve the fuel system pressure.
3. Clean both the inlet and the outlet fittings on the fuel filter.
4. Disengage the quick-connect fittings at the fuel filter inlet as follows:
   a. Slide the dust covers from the quick-connect fittings.
   b. Grasp both sides of the fitting. Twist the female connector ¼ turn in each direction to loosen any dirt within the fitting. Using compressed air and safety glasses, blow any accumulated dirt out of the fitting.
   c. If equipped with the plastic hand releasable fitting, squeeze the plastic retainer release tabs and pull the connection apart.
   d. If equipped with metal fittings, choose the correct size quick release tool and insert the tool into the female connector, then push inward to release the locking tabs. Pull the connector apart.
   e. Use a clean lint free rag to clean male pipe ends. Inspect both ends of the fitting for dirt and burrs. Clean or replace components as required. If it is necessary to remove rust or burrs from the fuel pipe, use emery cloth in a radial motion with the pipe end to prevent damage to the O-ring sealing surface.
5. Remove the threaded outlet fitting from the chassis fuel pipe. Slide the fuel filter from the bracket.
6. Inspect the fuel pipe O-ring for cuts, nicks, swelling or distortion. Replace if necessary.

To install:

7. Slide the fuel filter into the bracket.
8. Tighten the outlet fitting to the chassis fuel pipe. Torque the fitting to 22 ft. lbs. (30 Nm).
9. Engage the quick-connect inlet fitting as follows:
   a. Apply a few drops of clean engine oil to the male pipe end. This will ensure proper reconnection and prevent a possible fuel leak.
   b. Push both sides of the fitting together to cause the retainer tabs to snap in place. Once installed, pull on both sides of the fitting to ensure connection is secure.
   c. Reposition dust cover over the quick-connect fitting.
10. Tighten the fuel filler cap.
11. Confirm that the ignition is in the **OFF** position. Connect the negative battery cable.
12. Pressurize the fuel system by cycling the ignition without attempting to start the engine. Turn the ignition switch to the **ON** position for 2 seconds, then turn to the **OFF** position for 10 seconds. Again, turn to the **ON** position and check for fuel leaks.

Fuel Pump

REMOVAL & INSTALLATION

1. Before servicing the vehicle, refer to the precautions in the beginning of this section.
2. Relieve the fuel system pressure.
3. Drain the fuel tank.
4. Clean the area surrounding the sender assembly to prevent contamination of the fuel system.
5. Remove or disconnect the following:
   - Fuel tank
   - Fuel sender-to-fuel tank retaining ring
   - Fuel sender from the fuel tank
   - Fuel pump from the sending unit

To install:

6. Install or connect the following:
   - Fuel pump with new strainer onto the sending unit
   - New O-ring in the tank opening groove
   - New O-ring on the fuel sender feed tube

➡The fuel pump strainer must be in a horizontal position and must not block the float arm travel.

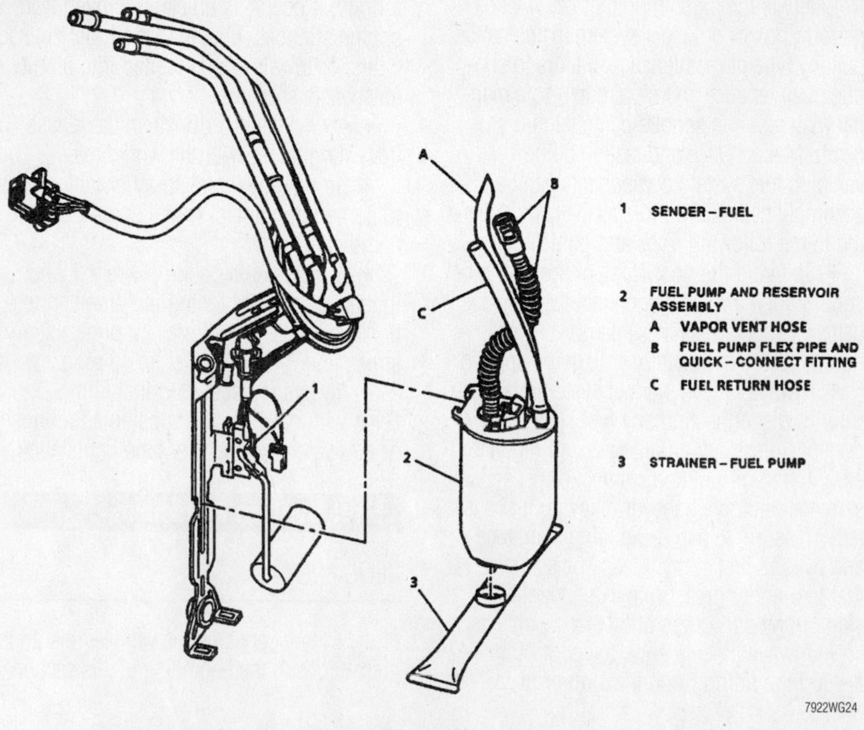

Fuel sender/pump assembly component identification

7. Fold the strainer over itself and slowly position the sending assembly in the tank so the strainer is not damaged or trapped by the sump walls.
   - Retaining ring. Torque the nuts to 63 inch lbs. (7 Nm).
   - Fuel tank. Torque the bolts to 24 ft. lbs. (33 Nm).
   - Fuel filler cap
   - Negative battery cable

8. Turn the ignition switch to the **ON** position for 2 seconds, **OFF** for 10 seconds, then back to the **ON** position. Check for fuel leaks.

Fuel Injector

REMOVAL & INSTALLATION

3.8L Engine

1. Before servicing the vehicle, refer to the precautions in the beginning of this section.
2. Relieve the fuel system pressure.
3. Remove or disconnect the following:
   - Fuel feed and return lines from the fuel rail
   - Fuel injector electrical connectors
   - Vacuum line from the fuel pressure regulator
   - Manifold Absolute Pressure (MAP) sensor electrical connector
   - Vacuum line from the vacuum switch at the fuel pipe bundle
   - Fuel injector harness fasteners
   - Fuel rail hold-down bolts
   - Fuel rail with fuel injectors
   - Fuel injectors by removing the retainer clips

To install:

⁂ WARNING

If the new O-rings are different colors (black and brown), install the black one in the upper portion and the brown one in the lower portion of the fuel injector.

4. Coat the injector O-rings with clean engine oil prior to installation.
5. Install or connect the following:
   - Fuel injectors with new O-rings to the fuel rail by installing the retainer clips
   - Fuel rail with fuel injectors. Torque the bolts to 89 inch lbs. (10 Nm).
   - Fuel injector harness fasteners
   - Vacuum line to the vacuum switch at the fuel pipe bundle
   - MAP sensor electrical connector
   - Vacuum line to the fuel pressure regulator
   - Fuel injector electrical connectors
   - Fuel feed and return lines to the fuel rail
   - Negative battery cable
6. Perform the following procedure in order to check for leaks:
   a. Turn the ignition switch ON for 2 seconds.
   b. Turn the ignition switch OFF for 10 seconds.
   c. Turn the ignition switch ON and check for fuel leaks.

5.7L Engine

1. Before servicing the vehicle, refer to the precautions in the beginning of this section.
2. Relieve the fuel system pressure.
3. Remove or disconnect the following:
   - Fuel feed hose from the fuel rail
   - Accelerator cable and cable bracket
   - Fuel injector electrical connectors
   - Electrical harness from the fuel rail
   - Fuel rail assembly
   - Fuel injectors from the fuel rail by spreading the injector retainer clips

To install:

4. Lubricate the new injector O-ring seals with clean engine oil prior to installation.
5. Install or connect the following:
   - New O-rings on the fuel injectors
   - Fuel injectors to the fuel rail using new retaining clips

➡Position the fuel injector electrical connector facing outward.

   - Fuel rail assembly. Torque the bolts to 89 inch lbs. (10 Nm).
   - Electrical harness to the fuel rail
   - Fuel injector electrical connectors
   - Accelerator cable and cable bracket
   - Fuel feed hose to the fuel rail
   - Negative battery cable
6. Perform the following procedure:
   a. Turn the ignition switch ON for 2 seconds.
   b. Turn the ignition switch OFF for 10 seconds.
   c. Turn the ignition switch ON and check for fuel leaks.

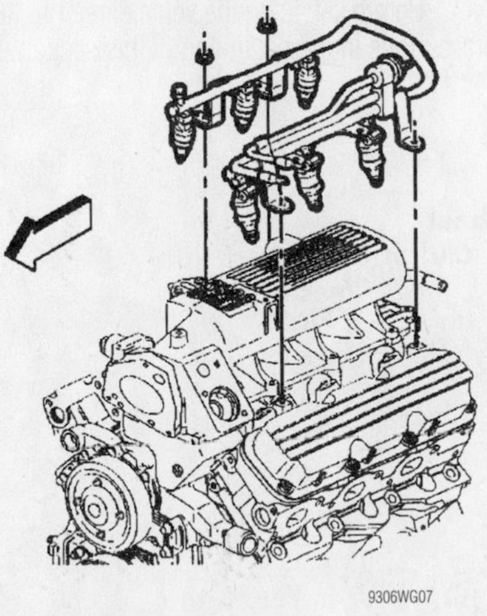

View of the fuel rail assembly—3.8L engine

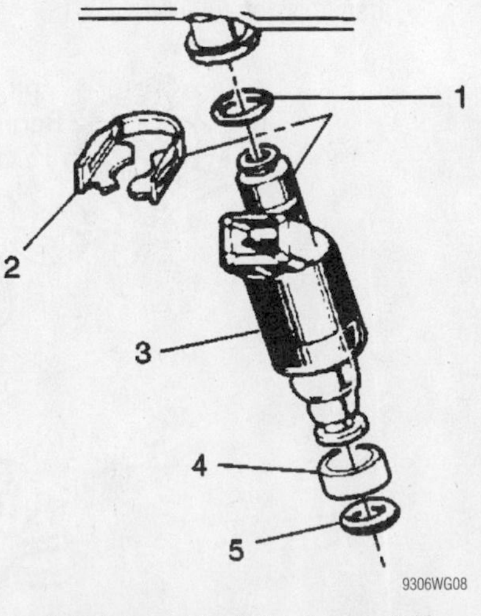

Exploded view of the fuel injector—3.8L engine

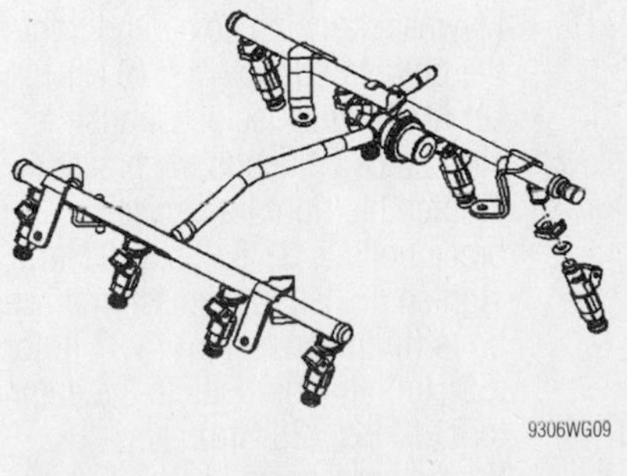

Exploded view of the fuel injector and fuel rail assembly—5.7L engine

DRIVE TRAIN

Transmission

REMOVAL & INSTALLATION

Manual

1. Before servicing the vehicle, refer to the precautions in the beginning of this section.
2. Drain the oil from the transmission.
3. Remove or disconnect the following:
   - Negative battery cable
   - Shift control lever boot assembly
   - Shift control lever
   - Propeller shaft
   - Rear axle torque arm
   - Right side catalytic converter
   - Electrical connectors from the transmission
   - Starter motor
   - Wiring harness and bracket from the transmission
   - Transmission brace bolts and the braces, on M49 transmissions
   - Flywheel housing cover
4. Depress the white circular release ring on the actuator hose and simultaneously pull on the master cylinder hose to disconnect.
5. Support the transmission.
6. Remove or disconnect the following:
   - Transmission support
   - Transmission mounting bolts
   - Transmission from the vehicle

To install:

7. Install or connect the following:
   - Transmission to the vehicle. Torque the bolts to 75 ft. lbs. (90 Nm) on M49 transmissions and 37 ft. lbs. (50 Nm) on MM6 transmissions.
   - Transmission support. Torque the bolts to 43 ft. lbs. (57 Nm).
   - Clutch actuator hose to the clutch master cylinder hose.
   - Flywheel housing cover and tighten the bolts to 80 inch lbs. (9 Nm)
   - On M49 transmissions Transmission brace bolts and the braces. Tighten the right transmission brace bolts to 37 ft. lbs. (50 Nm). Tighten the left transmission brace bolts (to transmission) to 37 ft. lbs. (50 Nm) and the bolts to the engine to 21 ft. lbs. (28 Nm).
   - Wiring harness and bracket
   - Electrical connectors
   - Starter motor
   - Right side catalytic converter. Torque the bolts to 18 ft. lbs. (25 Nm).
   - Rear axle torque arm. Torque the bolts to 37 ft. lbs. (50 Nm) and the nuts to 30 ft. lbs. (41 Nm).
   - Propeller shaft. Torque the bolts to 16 ft. lbs. (22 Nm).
   - Transmission oil drain plug. Torque the plug to 20 ft. lbs. (27 Nm).
8. Refill the transmission with oil.
9. Bleed the clutch hydraulic system.
10. Install or connect the following:
    - Shift control lever. Torque the bolts to 13 ft. lbs. (18 Nm).
    - Shift control lever boot assembly
    - Negative battery cable

Automatic

1. Before servicing the vehicle, refer to the precautions in the beginning of this section.
2. Remove or disconnect the following:
   - Negative battery cable
   - Intake Air Temperature (IAT) sensor electrical connector
   - Air intake duct
   - Catalytic converter
   - Range selector cable from the transmission
   - Propeller shaft
   - Rear axle torque arm
   - Torque converter cover
   - Torque converter bolts
   - Transmission oil cooler lines
   - Wiring harness from the transmission
   - Transmission 20-way connector
   - Vehicle Speed sensor (VSS) electrical connector
   - Transmission support
   - Transmission fill tube
   - Transmission nuts and bolts
   - Transmission from the vehicle

To install:

3. Install or connect the following:
   - Transmission to the vehicle. Torque the bolts and nuts to 70 ft. lbs. (95 Nm) on 3.8L engines and 37 ft. lbs. (50 Nm) on 5.7L engines
   - Transmission fill tube
   - Transmission support. Torque the bolts to 66 ft. lbs. (90 Nm).
   - VSS electrical connector
   - 20-way connector
   - Wiring harness clamp to the transmission. Torque the bolt to 22 inch lbs. (2.5 Nm).
   - Transmission oil cooler lines
   - Torque converter bolts. Torque the bolts to 47 ft. lbs. (63 Nm).
   - Torque converter cover
   - Rear axle torque arm. Torque the bolts to 37 ft. lbs. (50 Nm) and the nuts to 30 ft. lbs. (41 Nm).
   - Propeller shaft. Torque the bolts to 16 ft. lbs. (22 Nm).
   - Range selector cable
   - Catalytic converter. Torque the bolts to 18 ft. lbs. (25 Nm).
   - Air intake duct
   - IAT sensor electrical connector
   - Negative battery cable

Clutch

REMOVAL & INSTALLATION

1. Before servicing the vehicle, refer to the precautions in the beginning of this section.

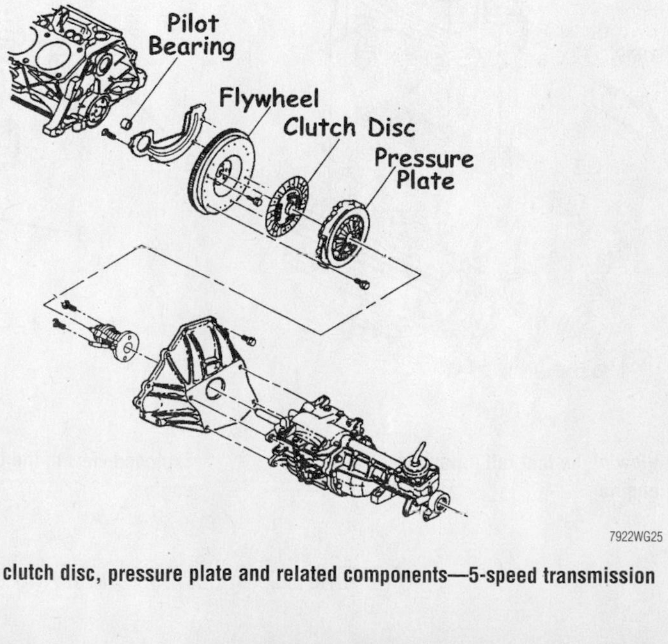

Exploded view of the clutch disc, pressure plate and related components—5-speed transmission

2. Remove or disconnect the following:
   - Negative battery cable
   - Transmission assembly
   - Flywheel housing cover
   - Flywheel housing

3. Install a clutch disc alignment tool through the center of the disc and into the pilot bearing to prevent the disc from falling when the pressure plate is removed.
4. Remove or disconnect the following:
   - Pressure plate retaining bolts
   - Pressure plate with clutch disc

To install:

5. Install or connect the following:
   - Clutch plate to the flywheel
   - Pressure plate and cover. Do not torque the bolts at this time.

6. Align the clutch plate with the pilot bearing and clutch pressure plate.
7. On 3.8L models, torque the clutch pressure plate and cover bolts in a star pattern to 15 ft. lbs. (20 Nm), plus an additional 45 degree turn.
8. On 5.7L models, torque the bolts using a star pattern torque sequence to 52 ft. lbs. (70 Nm).
9. Install or connect the following:
   - Flywheel housing
   - Flywheel housing cover
   - Transmission
   - Negative battery cable

Hydraulic Clutch System

BLEEDING

Bleeding air from the hydraulic clutch system is necessary whenever any part of the system has been disconnected or the fluid level (in the reservoir) has been allowed to fall so low that air has been drawn into the master cylinder.

⁂ WARNING

NEVER use fluid that has been bled from a clutch system to fill the master cylinder reservoir, as it may be aerated, contain excessive moisture and/or be contaminated in some other way.

1. Before servicing the vehicle, refer to the precautions in the beginning of this section.
2. Fill the clutch master cylinder reservoir with new hydraulic clutch fluid.
3. Attach a hose to the bleeder on the clutch actuator and submerge the other end of the hose in a container of hydraulic clutch fluid.
4. Have an assistant slowly depress and hold the clutch pedal.
5. Loosen the bleeder to purge air.
6. Tighten the bleeder.
7. Repeat the above 3 steps until all air is completely purged from the system.
8. Refill the clutch master cylinder reservoir.

Axle Shaft, Bearing and Seal

REMOVAL & INSTALLATION

1. Before servicing the vehicle, refer to the precautions in the beginning of this section.
2. Remove or disconnect the following:
   - Wheel
   - Brake rotor
   - Rear axle housing cover
   - Pinion gear shaft lock bolt (2)
   - Pinion gear shaft

3. Push the axle shaft into the housing in order to gain access to the rear axle shaft C-clip (1).
4. Remove the C-clip and remove the axle from the housing.

⁂ WARNING

Do not damage the rear wheel speed sensor reluctor wheel (3) during axle shaft removal.

5. Remove or disconnect the following:
   - Brake backing plate
   - Brake caliper mounting bracket
   - Oil seal, by prying it out
   - Rear wheel bearing with suitable puller

To install:

6. Lubricate the new wheel bearing with gear oil.
7. Install or connect the following:
   - Wheel bearing, using suitable installation tool
   - Oil seal, using suitable installation tool
   - Rear brake caliper mounting bracket
   - Brake backing plate
   - Axle shaft and C-clip

➡Be sure the splines on the end of the axle engage with the splines of the differential side gear.

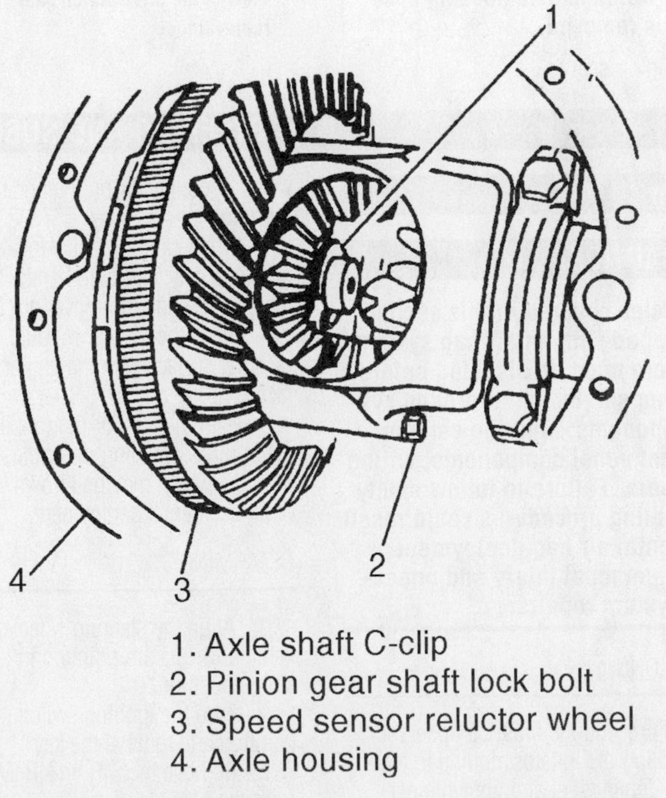

View of the rear axle housing with cover removed

- Pinion gear shaft
- Pinion gear shaft lock bolt. Torque the bolt to 27 ft. lbs. (36 Nm).
- Rear axle housing cover with new gasket. Torque the bolts in a star pattern to 22 ft. lbs. (30 Nm).
- Rear brake rotor.
- Wheel

Pinion Seal

REMOVAL & INSTALLATION

1. Before servicing the vehicle, refer to the precautions in the beginning of this section.
2. Matchmark the propeller shaft-to-drive pinion gear yoke, drive pinion gear and drive pinion yoke nut.
3. Remove or disconnect the following:
   - Propeller shaft
   - Drive pinion gear yoke nut using Pinion Flange Remover/Installer Tool J-8614-01 to hold the yoke and a socket wrench
   - Drive pinion gear yoke using Tools J-8614-01, J-8614-2 and J-8614-3

➡Use a container to catch the oil that may flow from the axle housing once the yoke is removed.

   - Pinion seal

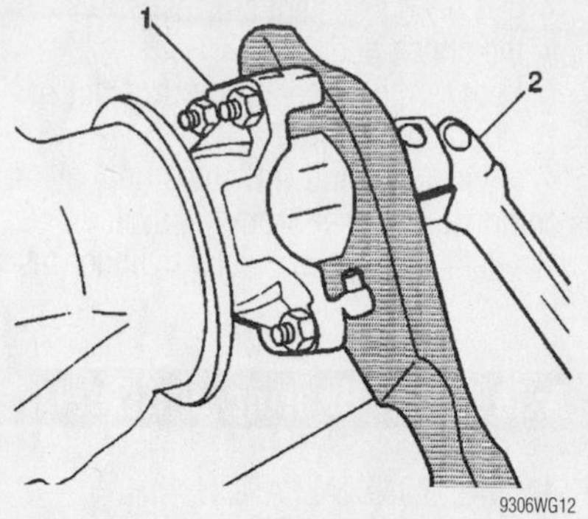

View of the drive pinion gear yoke nut removal tools

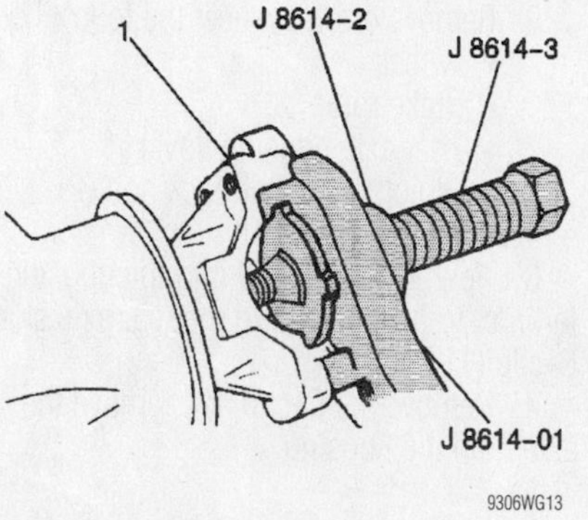

View of the drive pinion gear yoke removal tools

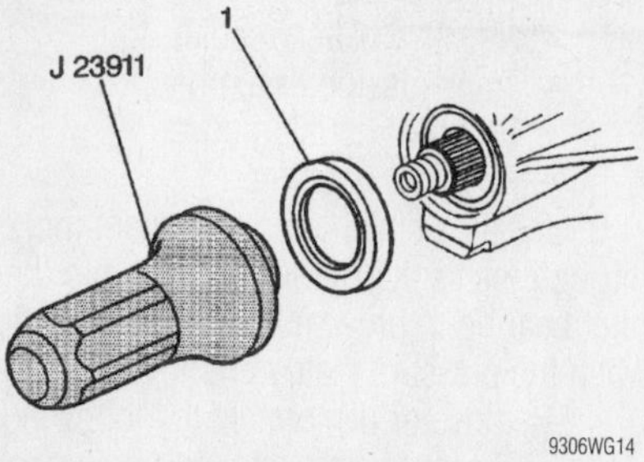

View of the pinion seal nut installer tool

➡It may be necessary to use a blunt chisel to drive the seal from the housing.

To install:

4. Install or connect the following:
   - New pinion seal, lubricated with chassis grease using pinion oil seal installer Tool J 8614-01 until it is flush with the housing
   - Drive pinion gear yoke using Tool J 8614-01 and a socket wrench
   - Drive pinion gear nut, tighten to 1/16 in. (1.59mm) beyond the alignment mark made earlier
   - Propeller shaft. Torque the propeller shaft-to-pinion yoke bolts to 16 ft. lbs. (22 Nm) and the propeller shaft-to-center support to 37 ft. lbs. (50 Nm) for 2-piece shaft.
5. Refill the axle housing.

STEERING AND SUSPENSION

Air Bag

⁂ CAUTION

The vehicles covered in this section are equipped with an air bag system. The system must be disabled before performing service on or around system components, steering column, instrument panel components, wiring and sensors. Failure to follow safety and disabling procedures could result in accidental air bag deployment, possible personal injury and unnecessary system repairs.

PRECAUTIONS

Several precautions must be observed when handling the inflator module to avoid accidental deployment and possible personal injury.

- Never carry the inflator module by the wires or connector on the underside of the module.
- When carrying a live inflator module, hold securely with both hands, and ensure that the bag and trim cover are pointed away.
- Place the inflator module on a bench or other surface with the bag and trim cover facing up.
- With the inflator module on the bench, never place anything on or close to the module, which may be thrown in the event of an accidental deployment.

DISARMING

1. Align the steering wheel so the vehicle wheels are pointing in the straight-ahead position.
2. Turn the ignition switch to the **LOCK** position and remove the key.
3. Remove the SIR or AIR BAG fuse from the fuse block.
4. Remove the left instrument panel insulator.
5. Remove the connector retainer, then disengage the yellow 2-way SIR wiring harness connector at the base of the steering column.
6. Remove the right instrument panel insulator.
7. Remove the connector retainer, then disengage the yellow 2-way SIR wiring harness connector located behind the IP compartment door.

ARMING

1. Engage the yellow 2-way connector at the base of the steering column and behind the IP compartment door, then install the connector retainer.
2. Install the instrument panel insulators.
3. Reinstall the SIR or AIR BAG fuse.
4. Turn the ignition switch to the **RUN** position.
5. Verify the SIR indicator light flashes 7–9 times, if not, inspect system for malfunction.

Power Rack and Pinion Steering Gear

REMOVAL & INSTALLATION

1. Before servicing the vehicle, refer to the precautions in the beginning of this section.
2. Remove or disconnect the following:
   - Serpentine belt
   - Air intake resonator
   - Front wheels
   - Outer tie rod ends from the knuckles
   - Alternator
   - Left side engine mount through bolt
   - Inlet and outlet hoses from the steering gear
   - Steering gear coupling shaft from the steering gear
3. Steering gear retainers
   - Steering gear from the vehicle

To install:

4. Install or connect the following:
   - Steering gear and its retainers
   - Steering gear to the crossmember. Torque the bolts to 63 ft. lbs. (85 Nm).
   - Steering gear to the coupling shaft. Torque the bolt to 35 ft. lbs. (47 Nm).
   - Inlet and outlet hoses to the steering gear. Torque the hoses to 21 ft. lbs. (28 Nm).
   - Left side engine mount through bolt
   - Alternator
   - Outer tie rod ends to the steering knuckle. Torque the nuts to 35 ft. lbs. (47 Nm).
   - Front wheels
   - Serpentine belt
   - Air intake resonator
5. Refill and bleed the power steering system.

Shock Absorber

REMOVAL & INSTALLATION

Front

1. Before servicing the vehicle, refer to the precautions in the beginning of this section.
2. If removing the driver's side spring and shock assembly, unbolt the brake mas-

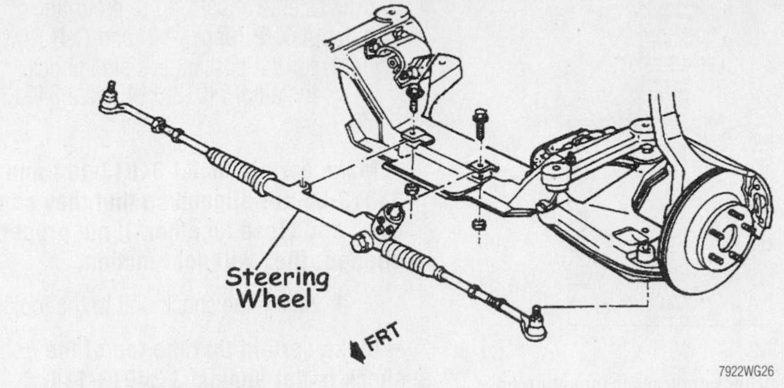

Exploded view of the power steering gear mounting

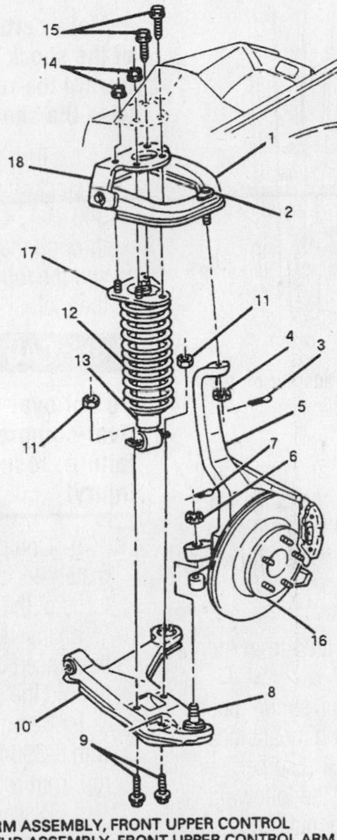

1 ARM ASSEMBLY, FRONT UPPER CONTROL
2 STUD ASSEMBLY, FRONT UPPER CONTROL ARM BALL
3 PIN, FRONT UPPER CONTROL ARM COTTER
4 NUT, FRONT UPPER CONTROL ARM, 53 N•m (39 LB. FT.)
5 KNUCKLE ASSEMBLY, STEERING
6 NUT, FRONT LOWER CONTROL ARM, 110 N•m (81 LB. FT.)
7 PIN, FRONT LOWER CONTROL ARM COTTER
8 STUD ASSEMBLY, FRONT LOWER CONTROL ARM BALL
9 BOLT/SCREW, FRONT SHOCK ABSORBER, 65 N•m (48 LB. FT.)
10 ARM ASSEMBLY, FRONT LOWER CONTROL
11 NUT, FRONT SHOCK ABSORBER LOWER BRACKET, 65 N•m (48 LB. FT.)
12 SPRING ASSEMBLY, FRONT
13 ABSORBER ASSEMBLY, FRONT SHOCK
14 NUT, FRONT SHOCK ABSORBER UPPER MOUNT, 43 N•m (32 LB. FT.)
15 BOLT/SCREW, FRONT SHOCK ABSORBER UPPER MOUNT, 50 N•m (37 LB. FT.)
16 HUB ASSEMBLY, FRONT WHEEL
17 MOUNT ASSEMBLY, FRONT UPPER SHOCK ABSORBER
18 SUPPORT, FRONT UPPER CONTROL ARM

7922WG27

Exploded view of the shock absorber unit and related suspension components

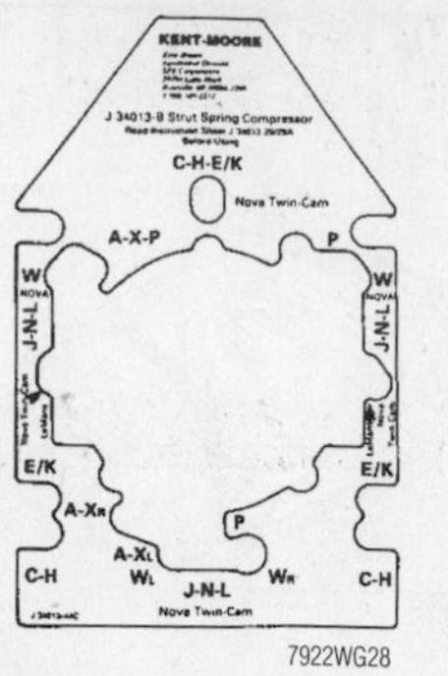

Shock absorber compressor mounting hole locations

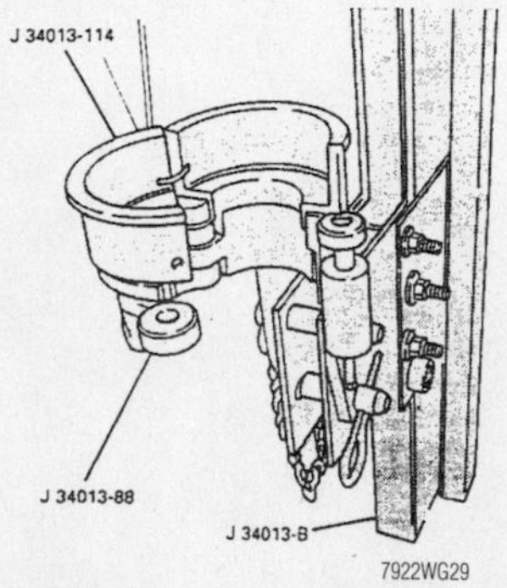

Install the strut compressor adapter on the spring compressor

ter cylinder from the booster and position it aside, but do not disconnect the brake lines.

3. Remove or disconnect the following:
   - Upper spring/shock mounting bolts and nuts
   - Front wheels
   - Stabilizer bar (shaft link) from the control arm

4. Matchmark the lower coil spring mount location to the upper coil spring mount location before removing the shock absorber.

5. Remove or disconnect the following:
   - Lower shock absorber nuts and bolts
   - Lower ball joint stud from the steering knuckle
   - Spring/shock absorber unit from the vehicle

6. Remove the spring from the shock absorber assembly by performing the following steps:

➡If using other than a GM spring compressor, follow the manufacturers instructions regarding the use of the specific tool you are using.

a. Assemble J-34013-B and J-34013-114 on the spring unit.

b. Use the wing nuts to secure the tool to mounting holes **C-H** (lower left corner) and **P** (upper right corner) for the driver's side shock and to mounting holes **A-X-P** (upper left) and **C-H** (lower right) for the passenger's side shock.

c. Install J-34013-114 and J 34013-88.

➡Make certain that J-34013-114 and J-34013-88 are aligned so that they can open and close together. If not properly aligned, they will not function.

d. Attach the shock unit to the tools.

➡Make certain that the top of the shock is flat against J-34013-114.

e. Close the tools and install the locking pin.

➡Make certain that the mounting ears of the shock are facing downward toward the rear of J-34013-B, otherwise the shock will not align properly.

f. Turn the screw of J-34013-B counterclockwise to raise the shock up to J-34013-114. Be sure that the studs go through the guide holes in J-34013-114 and the top of the shock is flat against the tool.

✲✲ CAUTION

Do not over-compress the spring! Over-compression can cause tool failure, resulting in severe bodily injury!

g. Compress the spring approximately ½ in. (13mm) or 3–4 complete turns of the screw on J-34013-114.

h. Insert J-39642-1 on the shock nut, then insert J-39642-2 through J-39642-1 to hold the shock absorber rod in place.

i. Remove the shock absorber nut with J-39642-1, while holding the shock rod from rotating with J-39642-2.

j. Discard the shock absorber nut.

k. Turn J-34013-B clockwise to fully relieve spring pressure and remove the spring from the shock.

To install:

7. Assemble the shock absorber/spring assembly by performing the following procedure:

a. Assemble J-34013-B spring compressor and J-34013-114 adapter on the spring unit. Use wing nuts to secure the tool to mounting holes **C-H** (lower left corner and **P** (upper right corner) for the driver's side shock and to mounting holes **A-X-P** (upper left) and **C-H** (lower right) for the passenger's side shock.

b. Install Tools J-34013-114 and J-34013-88.

c. Attach the shock unit to the tools.

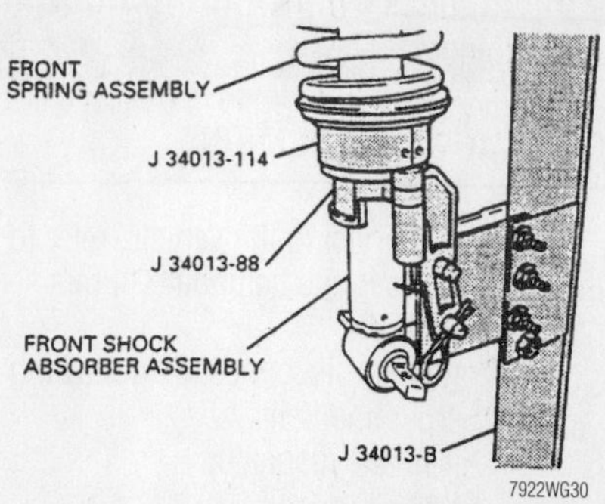

Installing the shock absorber on the compressor

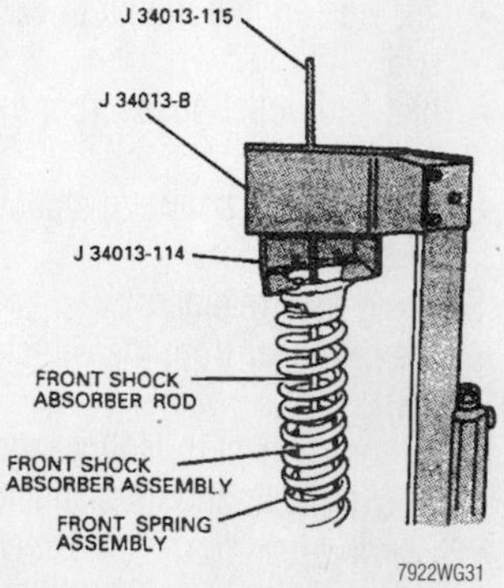

Inserting the shock assembly alignment rod

➡Make certain that the mounting ears of the shock are facing downward towards the rear of J 34013-B!

d. Close the tools and install the locking pin.

e. Be sure that the spring seats are positioned properly.

➡Make certain that the top of the shock is flat against J-34013-114.

f. Turn the screws of J-34013-B counterclockwise to raise the shock up to J-34013-114. Be sure that the studs go through the guide holes in J-34013-114 and the top of the shock is flat against the tool.

➡Turn the screw only enough to secure the shock. DO NOT compress the spring.

g. Place J-34013-115 down through the top of J-34013-B, through the top of the shock absorber and onto the rod.

➡Make certain that J-34013-115 is straight with the shock.

✲✲ CAUTION

Do not over-compress the spring! Over-compression can cause tool failure, resulting in severe bodily injury!

h. Turn the operating screw clockwise to compress the spring until the threaded portion of the rod is through the top of the shock. Remove J-34013-115.

i. Install a new shock absorber nut.

j. Install J-39642-1 on the nut, insert J-39642-2 through J-39642-1 and tighten the nut while holding J-39642-2.

8. Remove the shock/spring assembly from the tool.
9. Install or connect the following:
   - Shock/spring assembly on the lower control arm. Torque the lower nuts to 48 ft. lbs. (65 Nm).
   - Lower ball joint to the steering knuckle
   - Stabilizer shaft link
   - Front wheel
   - Upper shock/spring assembly. Torque the bolts to 37 ft. lbs. (50 Nm) and the nuts to 32 ft. lbs. (43 Nm).
   - Master cylinder

Rear

1. Before servicing the vehicle, refer to the precautions in the beginning of this section.
2. Place a support under the rear axle.

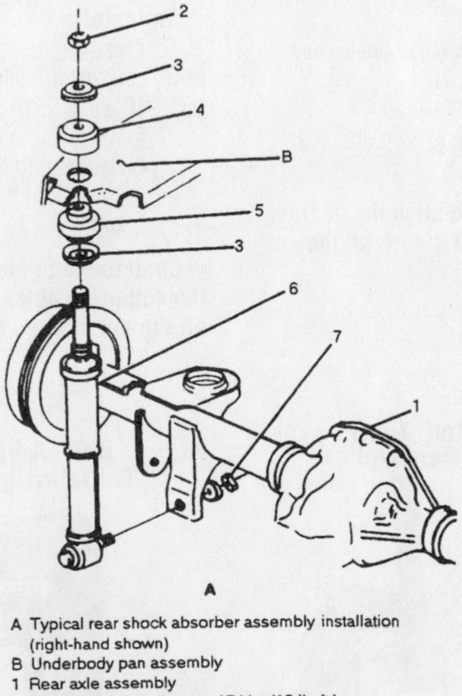

A Typical rear shock absorber assembly installation (right-hand shown)
B Underbody pan assembly
1 Rear axle assembly
2 Rear shock absorber nut 17 Nm (13 lb. ft.)
3 Rear shock absorber upper insulator retainer
4 Rear shock absorber upper insulator
5 Rear shock absorber lower insulator
6 Rear shock absorber assembly
7 Rear shock absorber nut 90 Nm (66 lb. ft.)

7922WG32

Exploded view of the rear shock absorber mounting

3. Remove or disconnect the following:
   - Rear seat back by folding it down
   - Quarter panel trim assembly
   - Carpet by folding it back

➡The rear axle must be supported before removing the upper mounting nut to avoid any possible damage to the brake hose lines, tie rods and propeller shaft.

   - Upper shock nut, retainer and upper insulator
   - Lower insulator and retainer
   - Lower shock-to-rear axle nut
   - Shock absorber from the vehicle

To install:

4. Install or connect the following:
   - Rear shock absorber. Torque the lower shock-to-axle nut to 66 ft. lbs. (90 Nm).
   - Lower insulator and retainer
   - Upper insulator and retainer
   - Upper shock mounting nut. Torque the nut to 13 ft. lbs. (17 Nm).

✲✲ WARNING

Turning the shock absorber while tightening the nut could damage the shock. To prevent damage, keep the shock absorber stationary when tightening the nut.

   - Carpet
   - Quarter panel trim assembly

Coil Spring

REMOVAL & INSTALLATION

Front

Refer to the from shock absorber removal and installation procedure in this manual for coils spring removal & installation.

Rear

1. Before servicing the vehicle, refer to the precautions in the beginning of this section.
2. Raise and safely support the vehicle by the frame so that the rear axle can be independently raised and lowered.
3. Support the rear axle with an adjustable support.
4. Remove or disconnect the following:
   - Brake hose brackets, if equipped, allowing the hoses to hang free

➡Perform the previous step only if the hoses would be stretched and damaged when the axle is lowered.

   - Lower shock absorber bolts
5. Lower the rear axle.
6. Remove or disconnect the following:
   - Upper insulator
   - Spring

➡The springs are painted with a protective coating. Take care to avoid damaging this coating. If the coating is chipped or damaged, paint the exposed spring to prevent rust.

To install:

7. Install or connect the following:
   - Spring on the axle with the open lower end facing forward
   - Upper insulator
8. Raise and safely support the rear axle.
   - Lower shock absorber nuts
   - Brake hose brackets, if removed

Upper Ball Joint

REMOVAL & INSTALLATION

1. Before servicing the vehicle, refer to the precautions in the beginning of this section.

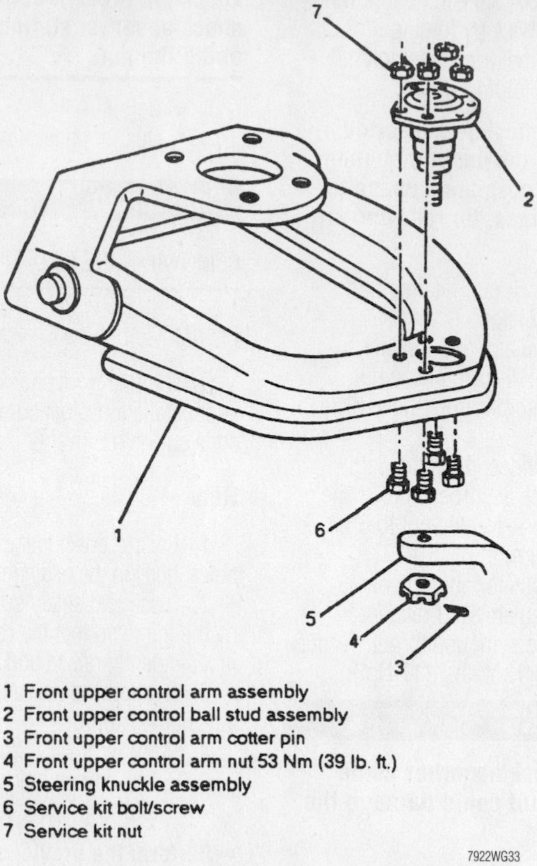

1 Front upper control arm assembly
2 Front upper control ball stud assembly
3 Front upper control arm cotter pin
4 Front upper control arm nut 53 Nm (39 lb. ft.)
5 Steering knuckle assembly
6 Service kit bolt/screw
7 Service kit nut

Position the new upper ball joint stud in the control arm and secure with the replacement nuts and bolts

2. Remove the front wheel.
3. Position a floor jack under the shock mount for support.

⁂ WARNING

The jack must remain in place for the entire duration of the procedure to hold the spring and lower control arm in proper position.

4. Remove or disconnect the following:
   - Cotter pin
   - Ball stud nut
   - 4 ball joint rivets by drilling them 0.25 in. (6mm) deep using a ⅛ in. (3.175mm) bit
   - Rivet heads by drilling them off using a ½ in. (12.7mm) bit
   - Rivets by using a small drift to punch them out
   - Upper ball joint from the control arm

To install:

5. Install or connect the following:
   - New upper ball joint to the control arm. Torque the nuts and bolts according to the specifications given in the replacement kit.
   - Ball joint to the steering knuckle. Torque the nut to 39 ft. lbs. (53 Nm).

➡Advance the nut to align the nearest cotter pin hole; NEVER back off the nut to align a hole!

   - New cotter pin
   - Front wheel

Lower Ball Joint

REMOVAL & INSTALLATION

➡To prevent component damage, an on-car ball joint press should be used.

1. Before servicing the vehicle, refer to the precautions in the beginning of this section.
2. Remove the front wheel.
3. Position a support under the shock mount for support.

⁂ WARNING

The support must remain in place for the entire duration of the procedure to hold the spring and lower control arm in proper position.

4. Remove or disconnect the following:
   - Lower ball joint cotter pin
   - Lower ball joint stud nut
   - Lower ball joint from the steering knuckle
5. Press the lower ball joint from the lower control arm using Tools J-9519-7, J-9519-18 and J-9519-23 as illustrated.

To install:

6. Press the new lower ball joint into the lower control arm using Tools J 9519-9, J 9519-18 and J 9519-23 as illustrated.
7. Install or connect the following:
   - Ball joint to the steering knuckle. Torque the nut to 81 ft. lbs. (110 Nm).

➡Continue to tighten the nut just until the cotter pin holes align; NEVER back off the nut to align the holes!

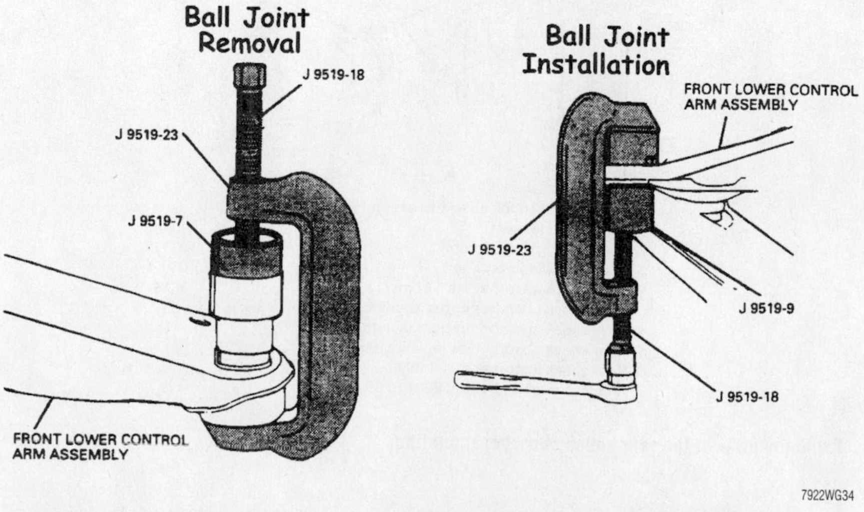

Lower ball joint replacement requires the use of special tools, such as the ones shown

- New cotter pin
- Front wheel

8. Check and/or adjust the alignment.

Upper Control Arm

REMOVAL & INSTALLATION

1. Before servicing the vehicle, refer to the precautions in the beginning of this section.
2. Remove or disconnect the following:
   - Brake master cylinder, if removing the drivers side control arm
   - Upper shock absorber mount bolts
   - Front wheel
   - Stabilizer shaft link
   - Lower shock absorber bolts
   - Upper ball joint nut
3. Support the steering knuckle.
   - Upper ball joint from the steering knuckle
   - Upper control arm with the shock absorber
   - Upper control arm from the shock absorber

To install:

4. Install or connect the following:
   - Upper control arm to the shock absorber with the bolts finger tight
   - Upper control arm and shock absorber to the vehicle. Torque the upper control arm-to-chassis bolts to 72 ft. lbs. (98 Nm).
   - Upper ball joint to the steering knuckle. Torque the upper ball joint nut to 39 ft. lbs. (53 Nm).

➡Never loosen the ball joint nut to align the cotter pin; the nut can be over torqued ⅙ of a turn maximum.

   - New cotter pin
5. Remove the steering knuckle support.
   - Lower shock absorber to the steering knuckle. Torque the bolts to 48 ft. lbs. (65 Nm).
   - Stabilizer shaft link. Torque the nuts to 17 ft. lbs. (23 Nm).
   - Front wheel
   - Upper shock absorber. Torque the nuts to 30 ft. lbs. (41 Nm) and the bolts to 37 ft. lbs. (50 Nm).
   - Brake master cylinder. Torque the nuts to 21 ft. lbs. (29 Nm).

CONTROL ARM BUSHING REPLACEMENT

1. Before servicing the vehicle, refer to the precautions in the beginning of this section.
2. Remove or disconnect the following:
   - Upper control arm
   - Upper control arm from the upper control arm support
3. Remove the bushing from the upper control arm by performing the following procedure:
   a. Thread the Upper Control Arm Screw Tool J-21474-19 through the Upper Control Arm Bushing Receiver/Installer Tool J-39930; the 3 tangs must be against the screw head.
   b. Thread the Upper Control Arm Screw Tool J-21474-19 through the upper control arm bushing.
   c. Run the smaller end of the Control Arm Bushing Receiver Tool J-21474-5 onto the Upper Control Arm Screw Tool J-21474-19; then, place the thrust washer onto Tool J-21474-19 with the seams facing Tool J-21474-5.
   d. Position the Half Moon Spacer Tool J-39872 around the outside of the bushing to prevent metal distortion during removal.
   e. Ensure that the tools are aligned; then, install the Upper Control Arm Nut Tool J-21474-18 on the Upper Control Arm Screw Tool J-21474-19.

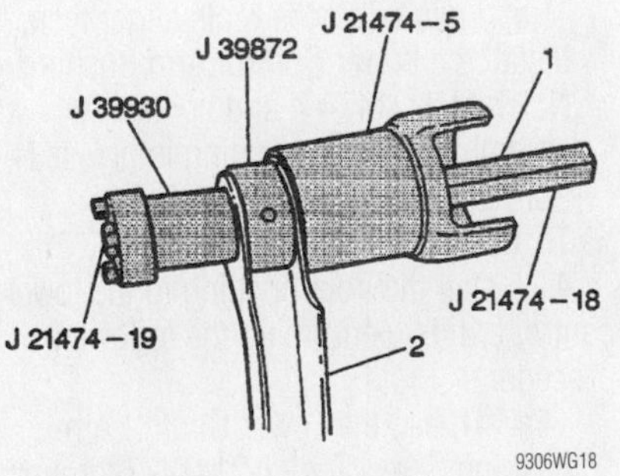

Removing the upper control arm bushing

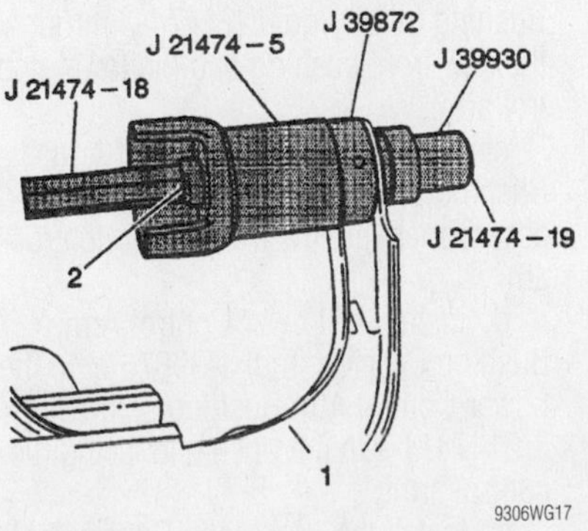

Installing the upper control arm bushing

   f. Tighen the Upper Control Arm Nut Tool J-21474-18 and Upper Control Arm Screw Tool J-21474-19 until the busing is pressed from the upper control arm.

To install:

4. Install the bushing to the upper control arm by performing the following procedure:
   a. Thread the Upper Control Arm Screw Tool J-21474-19 through the Upper Control Arm Bushing Receiver/Installer Tool J-39930; the 3 tangs must NOT be against the screw head.
   b. Install a new bushing onto the Upper Control Arm Screw Tool J-21474-19 with the 3 indentations facing the 3 tangs on the Upper Control Arm Bushing Receiver/Installer Tool J-39930.
   c. Install the threaded end of the Upper Control Arm Screw Tool J-21474-19 to the upper control arm from the outside.
   d. Install the Control Arm Bushing Receiver Tool J-21474-5 onto the Upper Control Arm Screw Tool J-21474-19 from the inside.
   e. Place the thrust washer onto Tool J-21474-19 with the seam facing the bushing.
   f. Position the Half Moon Spacer Tool J-39872 around the outside of the bushing to prevent metal distortion during installation.
   g. Install the Upper Control Arm Nut Tool J-21474-18 on the Upper Control Arm Screw Tool J-21474-19; ensure that the 3 tangs on the Upper Control Arm Bushing Receiver/Installer Tool J-39930 fit into the bushing indentations.
   h. Tighen the assembly until the busing is pressed into the upper control arm.
5. Install or connect the following:
   - Upper control arm to the upper control arm support. Torque the nuts/bolts to 72 ft. lbs. (98 Nm).
   - Upper control arm

Lower Control Arm

REMOVAL & INSTALLATION

1. Before servicing the vehicle, refer to the precautions in the beginning of this section.
2. Remove or disconnect the following:
   - Front wheel
   - Stabilizer shaft link
   - Outer tie rod end from the steering knuckle

- Lower shock absorber bolts
- Lower ball joint from the steering knuckle
- Lower control arm nuts
- Lower control arm

To install:

3. Install or connect the following:
   - Lower control arm.
   - On 1998–99 models, torque the lower control arm-to-crossmember horizontal nut to 74 ft. lbs. (100 Nm) and the vertical nut to 85 ft. lbs. (115 Nm)
   - On 2000–01 models, torque the lower control arm-to-crossmember horizontal nut 39 ft. lbs. (53 Nm).
   - Lower ball joint to the steering knuckle. Torque the ball joint nut to 81 ft. lbs. (110 Nm).

➡Never loosen the ball joint nut to align the cotter pin; the nut can be over torqued 1/6 of a turn maximum in order to align the slots for the cotter pin.

   - New lower ball joint cotter pin
   - Lower shock absorber. Torque the bolts to 48 ft. lbs. (65 Nm).
   - Outer tie rod end to the steering knuckle. Torque the nut to 35 ft. lbs. (47 Nm).
   - New outer tie rod end cotter pin
   - Stabilizer shaft link. Torque the nut to 17 ft. lbs. (23 Nm).
   - Front wheel

CONTROL ARM BUSHING REPLACEMENT

Front Bushing

1. Before servicing the vehicle, refer to the precautions in the beginning of this section.
2. Remove the lower control arm.
3. Remove the front bushing from the upper control arm by performing the following procedure:
   a. Install the Lower Control Arm Bushing Screw Tool J-21474-3 through the large end of the Control Arm Bushing Receiver Tool J-21474-6.
   b. Install the Lower Control Arm Bushing Screw Tool J-21474-3 through the front bushing and the open end of the Lower Control Arm Bushing Receiver Tool J-39876.
   c. Place the thrust washer onto the Lower Control Arm Bushing Screw Tool J-21474-3 so the seam faces the front bushing.
   d. Install the Half Moon Spacer Tool J-39875 around the outside of the bushing to avoid metal distortion during removal.

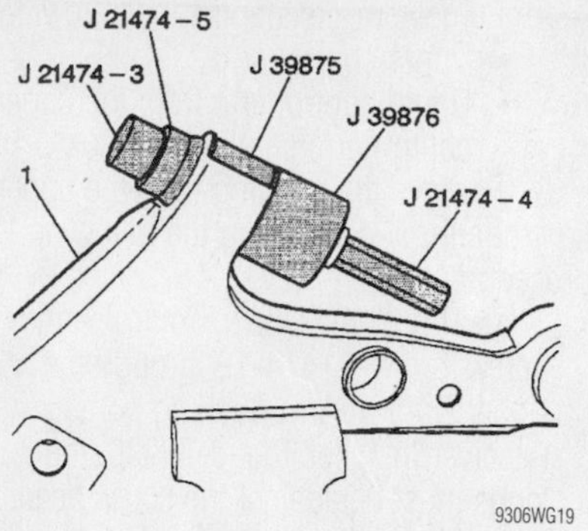

Removing the lower control arm front bushing

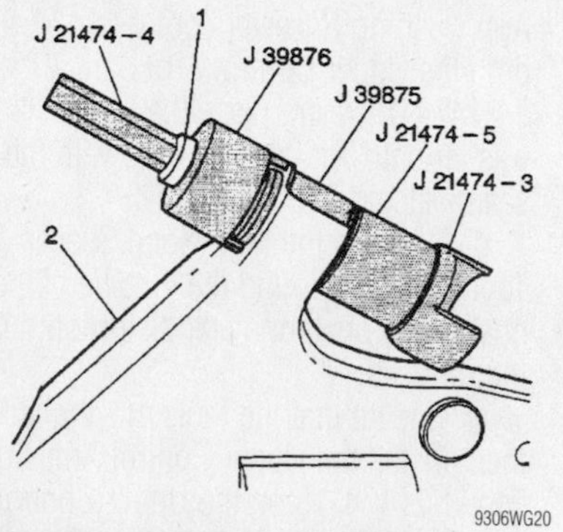

Installing the lower control arm front bushing

   e. Ensure that the tools align; then, install the Lower Control Arm Bushing Nut Tool J-21474-4 and tighten the assembly until the bushing is pressed from the control arm.

To install:

4. Install the front bushing to the lower control arm by performing the following procedure:
   a. Thread the Lower Control Arm Bushing Screw Tool J-21474-3 through the Control Arm Bushing Receiver Tool J-21474-5.
   b. Install the Lower Control Arm Bushing Screw Tool J-21474-3 through the new front bushing and the lower control arm.
   c. Thread the Lower Control Arm Bushing Screw Tool J-21474-3 and accessories into the front lower control arm.
   d. Install the Lower Control Arm Bushing Receiver Tool J-39876 onto the Lower Control Arm Bushing screw Tool J-21474-3 with the open end facing the control arm.
   e. Install the Half Moon Spacer Tool J-39875 around the outside of the bushing to avoid metal distortion during installation.
   f. Place the thrust washer onto the Lower Control Arm Bushing Screw Tool J-21474-3 threaded end with the seam facing the control arm.
   g. Ensure that the tools align; then, install the Lower Control Arm Bushing Nut and the Lower Control Arm Bushing Screw Tool J-21474-3.
   h. Tighten the assembly until the front bushing is flush with the lower control arm.
5. Install the lower control arm.

Rear Bushing

1. Before servicing the vehicle, refer to the precautions in the beginning of this section.
2. Remove the lower control arm.
3. Remove the rear bushing from the upper control arm by performing the following procedure:
   a. Install the Lower Control Arm Bushing Receiver Tool J-39874 onto the Lower Control Arm Bushing Screw Tool J-21474-3.
   b. Install the Lower Control Arm Bushing Screw Tool J-21474-3 through the rear bushing, the control arm and into the small end of the Lower Control Arm Bushing Receiver Tool J-39931.

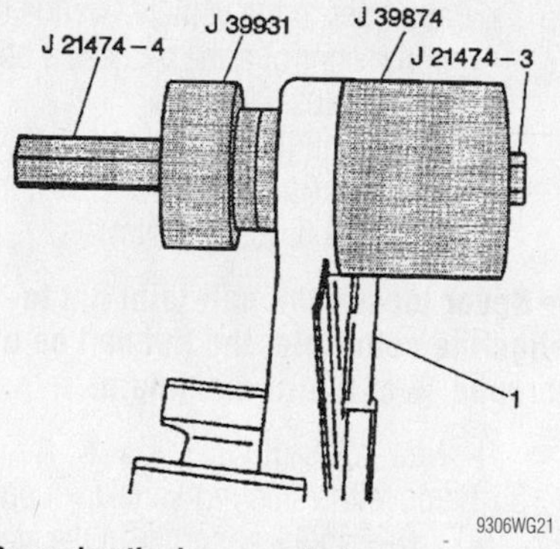

Removing the lower control arm rear bushing

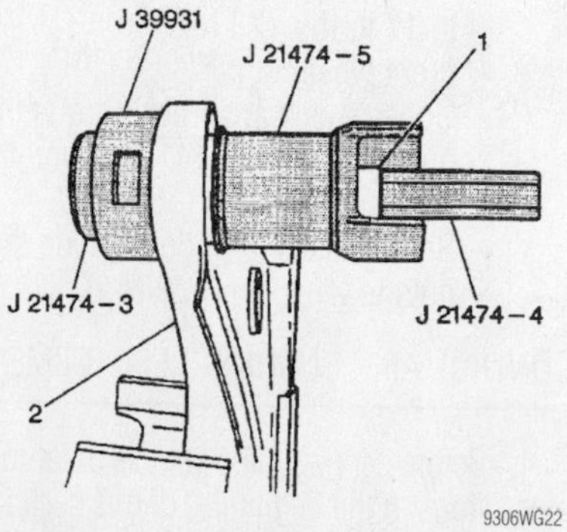

Installing the lower control arm rear bushing

c. Place the thrust washer onto the Lower Control Arm Bushing Screw Tool J-21474-3 so the seam faces the control arm.

d. Install the Lower Control Arm Bushing Nut Tool J-21474-4 onto the Lower Control Arm Bushing Screw Tool J-21474-3; then, ensure that the assembly aligns with the bushing.

e. Tighten the assembly until the bushing is pressed from the control arm.

To install:

4. Install the rear bushing to the lower control arm by performing the following procedure:

a. Thread the Lower Control Arm Bushing Screw Tool J-21474-3 through the small end of the Lower Control Arm Bushing Receiver Tool J-39931.

b. Install the Lower Control Arm Bushing Screw Tool J-21474-3 through the new rear bushing and the closed end of the lower control arm.

c. Place the small end of the Control Arm Bushing Receiver Tool J-21474-5 onto the Lower Control Arm Bushing Screw Tool J-21474-3.

d. Place the thrust washer onto the Lower Control Arm Bushing Screw Tool J-21474-3 with the seam facing the control arm.

e. Ensure that the tools align and the window on the Lower Control Arm Bushing Receiver Tool J-39931 faces straight up and is visible.

f. Install the Lower Control Arm Bushing Nut Tool J-21474-4 onto the Lower Control Arm Bushing Screw Tool J-21474-3; then, tighten the assembly until the rubber protrusion on the Lower Control Arm Bushing Receiver Tool J-39931 side bottoms against Tool J-39931.

5. Install the lower control arm.

Wheel Bearings

ADJUSTMENT

The front wheel bearing assembly used on these vehicles is a sealed, non-serviceable unit. No wheel bearing adjustments (front or rear) are necessary or possible.

REMOVAL & INSTALLATION

Front

1. Before servicing the vehicle, refer to the precautions in the beginning of this section.
2. Remove or disconnect the following:
   - Front wheel
   - Brake caliper
   - Brake rotor
   - Wheel speed sensor electrical connector
   - Hub bolts
   - Hub/bearing assembly by pulling it from the spindle

To install:

3. Install or connect the following:
   - Hub/bearing assembly on the spindle. Torque the bolts to 63 ft. lbs. (86 Nm).
   - Wheel speed sensor electrical connector

WARNING

Be sure the wheel speed sensor electrical connector is reattached to the sensor wire bracket and sensor or the wires could be damaged.

- Brake rotor
- Brake caliper
- Front wheel

Rear

1. Before servicing the vehicle, refer to the precautions in the beginning of this section.
2. Remove or disconnect the following:
   - Rear wheel
   - Brake rotor or brake drum and components, as equipped
3. Clean the axle housing cover and surrounding area to prevent dirt or contamination from entering the housing.
4. Remove the axle housing cover.
5. Install an Anti-lock Brake System (ABS) tone ring protector kit.
6. Remove or disconnect the following:
   - Rear axle pinion shaft lock screw
   - Pinion gear shaft
7. Push the flanged end of the axle shaft into the axle housing in order to access the rear axle shaft lock.
8. Remove or disconnect the following:
   - Shaft lock C-clip from the axle shaft
   - Axle shaft from the axle housing
9. Use a small suitable pry tool to

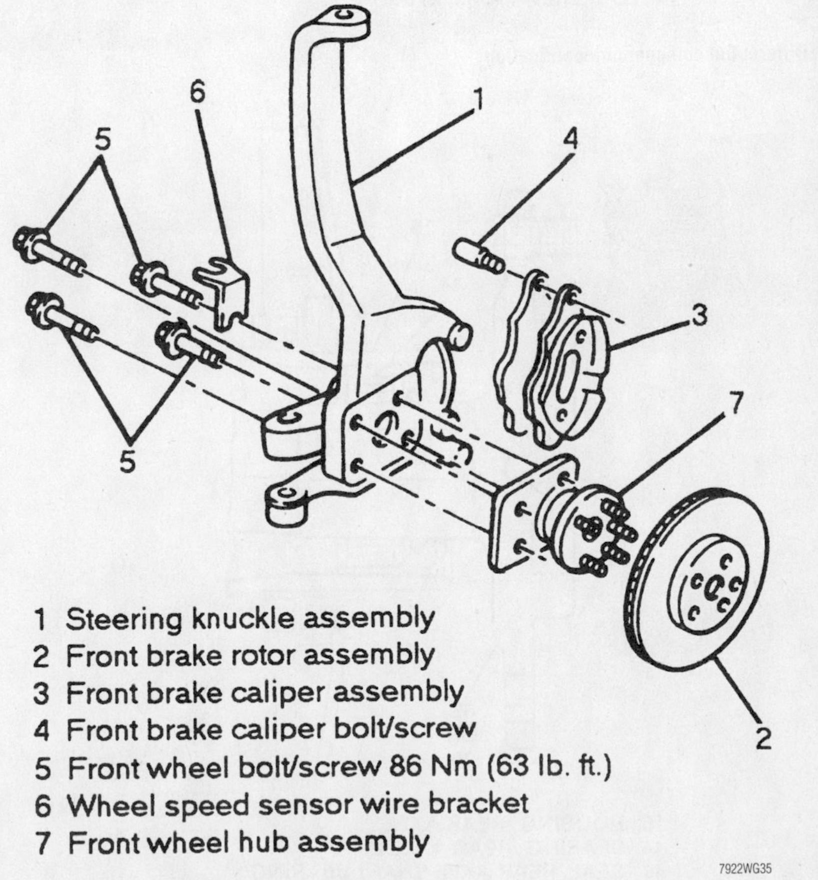

Exploded view of the wheel hub/bearing unit mounting

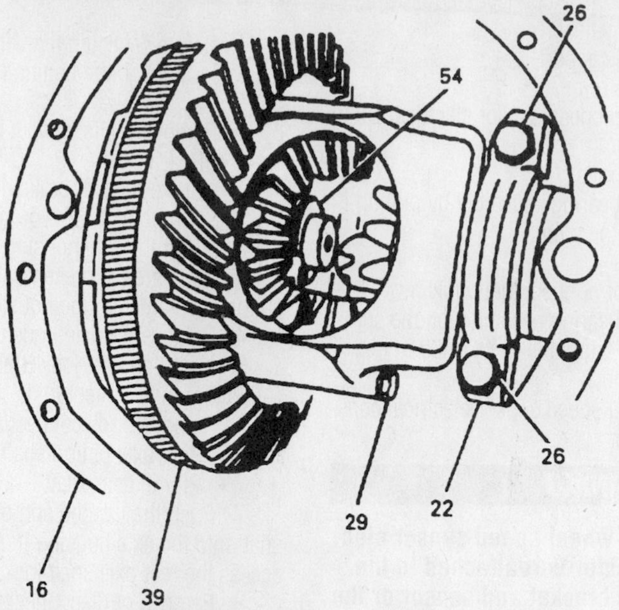

16 HOUSING, REAR AXLE
22 CAP, DIFFERENTIAL CARRIER BEARING
26 BOLT/SCREW, DIFFERENTIAL BEARING CAP
29 BOLT/SCREW, DIFFERENTIAL PINION GEAR SHAFT LOCK
39 WHEEL, REAR WHEEL SPEED SENSOR RELUCTOR
54 LOCK, REAR AXLE SHAFT

7922WG38

Differential component identification

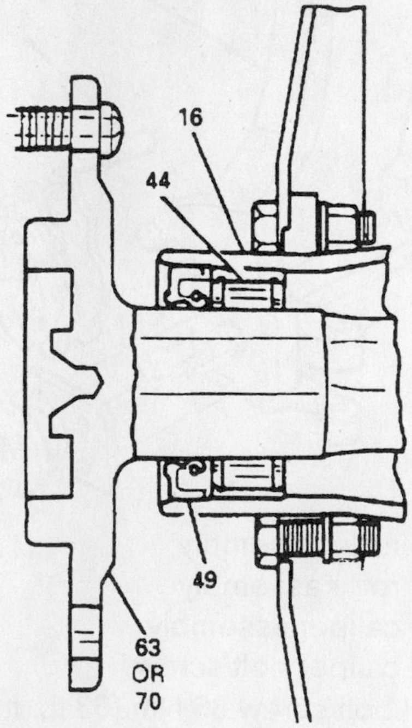

16 HOUSING, REAR AXLE
44 BEARING, REAR AXLE SHAFT
49 SEAL, REAR AXLE SHAFT BEARING
63 SHAFT, REAR AXLE (DRUM BRAKE ASSEMBLY)
70 SHAFT, REAR AXLE (DISC BRAKE ASSEMBLY)

7922WG39

Cut-away view of the rear axle bearing and seal

remove the oil seal from the axle housing. Be careful not to score or damage the housing.

10. Install an axle bearing remover into the bore of the axle housing and position it behind the bearing, ensure the tangs of the tool engage the outer race. Remove the bearing using a slide hammer.

To install:

11. Install a new bearing lubricated with gear oil with a suitable driver so the tool bottoms against the shoulder in the axle housing.

12. Position a new seal lubricated with gear oil on a suitable seal installer, then insert the seal into the housing bore.

13. Install or connect the following:
- Axle shaft
- C-clip so it seats in the axle side gear counterbore
- Pinion gear shaft through the differential case, thrust washer and pinion gears
- Shaft lock screw. Torque it to 27 ft. lbs. (36 Nm).
- New axle housing cover with gasket. Torque the bolts in a crosswise pattern to 22 ft. lbs. (30 Nm).

➡When refilling a limited slip differential rear axle with gear oil, 4 oz. (118ml) of limited slip additive should be added.

14. Refill the rear axle with SAE 80-90W GL-5 gear lubricant, then install the plug.
- Rear brake assemblies
- Rear wheels

GENERAL MOTORS CORPORATION—G-BODY

24

1998–01

Buick-Riviera • **Oldsmobile**-Aurora

PRECAUTIONS

Before servicing any vehicle, please be sure to read all of the following precautions, which deal with personal safety, prevention of component damage, and important points to take into consideration when servicing a motor vehicle:

- Never open, service or drain the radiator or cooling system when the engine is hot; serious burns can occur from the steam and hot coolant.
- Observe all applicable safety precautions when working around fuel. Whenever servicing the fuel system, always work in a well-ventilated area. Do not allow fuel spray or vapors to come in contact with a spark, open flame or excessive heat (a hot drop light, for example). Keep a dry chemical fire extinguisher near the work area. Always keep fuel in a container specifically designed for fuel storage; also, always properly seal fuel containers to avoid the possibility of fire or explosion. Refer to the additional fuel system precautions later in this section.
- Fuel injection systems often remain pressurized, even after the engine has been turned **OFF**. The fuel system pressure must be relieved before disconnecting any fuel lines. Failure to do so may result in fire and/or personal injury.
- Brake fluid often contains polyglycol ethers and polyglycols. Avoid contact with the eyes and wash your hands thoroughly after handling brake fluid. If you do get brake fluid in your eyes, flush your eyes with clean, running water for 15 minutes. If eye irritation persists, or if you have taken brake fluid internally, IMMEDIATELY seek medical assistance.
- The EPA warns that prolonged contact with used engine oil may cause a number of skin disorders, including cancer! You should make every effort to minimize your exposure to used engine oil. Protective gloves should be worn when changing oil. Wash your hands and any other exposed skin areas as soon as possible after exposure to used engine oil. Soap and water, or waterless hand cleaner should be used.
- All new vehicles are now equipped with an air bag system. The system must be disabled before performing service on or around system components, steering column, instrument panel components, wiring and sensors. Failure to follow safety and disabling procedures could result in accidental air bag deployment, possible personal injury and unnecessary system repairs.
- Always wear safety goggles when working with, or around, the air bag system. When carrying a non-deployed air bag, be sure the bag and trim cover are pointed away from your body. When placing a non-deployed air bag on a work surface, always face the bag and trim cover upward, away from the surface. This will reduce the motion of the module if it is accidentally deployed. Refer to the additional air bag system precautions later in this section.
- Clean, high quality brake fluid from a sealed container is essential to the safe and proper operation of the brake system. You should always buy the correct type of brake fluid for your vehicle. If the brake fluid becomes contaminated, completely flush the system with new fluid. Never reuse any brake fluid. Any brake fluid that is removed from the system should be discarded. Also, do not allow any brake fluid to come in contact with a painted surface; it will damage the paint.
- Never operate the engine without the proper amount and type of engine oil; doing so WILL result in severe engine damage.
- Timing belt maintenance is extremely important! Many models utilize an interference-type, non-freewheeling engine. If the timing belt breaks, the valves in the cylinder head may strike the pistons, causing potentially serious (also time-consuming and expensive) engine damage. Refer to the maintenance interval charts in the front of this manual for the recommended replacement interval for the timing belt, and to the timing belt section for belt replacement and inspection.
- Disconnecting the negative battery cable on some vehicles may interfere with the functions of the on-board computer system(s) and may require the computer to undergo a relearning process once the negative battery cable is reconnected.
- When servicing drum brakes, only disassemble and assemble one side at a time, leaving the remaining side intact for reference.
- Only an MVAC-trained, EPA-certified automotive technician should service the air conditioning system or its components.

ENGINE REPAIR

Alternator

REMOVAL

3.5L Engine

1. Before servicing the vehicle, refer to the precautions in the beginning of this section.
2. Remove or disconnect the following:
   - Negative battery cable
   - Drive belt
   - Engine cooling fan assembly
   - Alternator electrical connectors
   - Idler pulley
   - 2 upper alternator mounting bolts
   - A/C compressor
   - Alternator

3.8L Engine

1. Before servicing the vehicle, refer to the precautions in the beginning of this section.
2. Remove or disconnect the following:
   - Rear seat cushion
   - Negative battery cable
   - Accessory drive belt
   - Front cover
   - Rear alternator brace
   - Alternator electrical connectors
   - Alternator

4.0L Engine

1998–99 MODELS

1. Before servicing the vehicle, refer to the precautions in the beginning of this section.
2. Remove or disconnect the following:
   - Rear seat cushion
   - Negative battery cable
   - Upper radiator shroud
   - Air inlet duct
   - Both cooling fans
   - Accessory drive belts
   - Alternator electrical connectors
   - Lower splash shield
   - Alternator bolts, nuts and studs
   - Power steering hose from the reservoir
   - Alternator through the top left side of the engine compartment

2001 MODELS

1. Before servicing the vehicle, refer to the precautions in the beginning of this section.

2. Remove or disconnect the following:
   - Rear seat cushion
   - Negative battery cable
   - Radiator
   - Alternator electrical connectors
   - Alternator bolts
   - Alternator

INSTALLATION

3.5L Engine

1. Install or connect the following:
   - Alternator. Do not torque the bolts at this time.
   - A/C compressor. Torque the bolts and the front nut to 37 ft. lbs. (50 Nm) and the rear nut to 18 ft. lbs. (25 Nm).
   - Idler pulley. Torque the bolt to 37 ft. lbs. (50 Nm).
2. Torque the alternator bolts to 37 ft. lbs. (50 Nm).
3. Install or connect the following:
   - Alternator electrical connectors. Torque the positive battery terminal to 111 inch lbs. (12.5 Nm).
   - Engine cooling fan assembly. Torque the bolts to 53 inch lbs. (6 Nm).
   - Drive belt
   - Negative battery cable

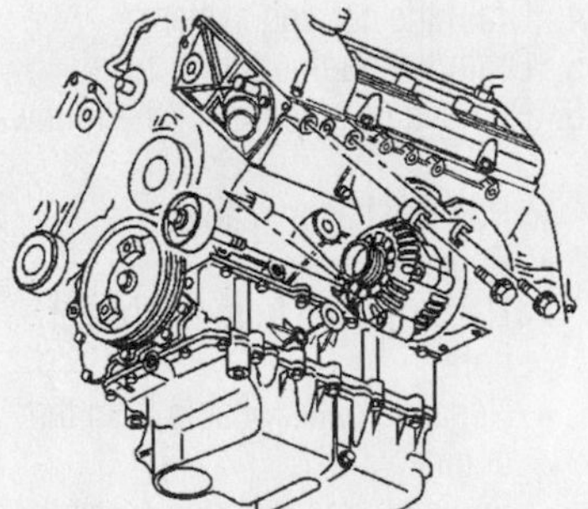

View of the alternator mounting—3.5L engine

3.8L Engine

Install or connect the following:
- Alternator. Torque the bolts to 37 ft. lbs. (50 Nm).
- Alternator electrical connectors. Torque the positive battery terminal to 15 ft. lbs. (20 Nm).
- Rear alternator brace. Torque the bolts to 22 ft. lbs. (30 Nm).
- Front cover. Torque the bolts to 11 ft. lbs. (15 Nm).
- Accessory drive belt
- Negative battery cable
- Rear seat cushion

4.0L Engine

1998–99 MODELS

1. Install or connect the following:
   - Alternator
   - Air inlet duct
   - Lower radiator hose
   - Power steering line bracket
   - Ground strap
   - Alternator through the top left side of the engine compartment
   - Power steering hose from the reservoir
   - Lower alternator bolts. Torque the bolts to 28 ft. lbs. (38 Nm).
   - Upper alternator bolts. Torque the bolts to 35 ft. lbs. (47 Nm).
   - Lower splash shield
   - Alternator electrical connectors
   - Alternator electrical connectors. Torque the positive battery terminal to 15 ft. lbs. (20 Nm).
   - Accessory drive belts
   - Both cooling fans. Torque the bolts to 53 inch lbs. (6 Nm).
   - Air inlet duct
   - Upper radiator shroud
   - Negative battery cable
   - Rear seat cushion

2001 MODELS

1. Install or connect the following:
   - Alternator
   - Alternator . Torque the bolts to 37 ft. lbs. (50 Nm).
   - Alternator electrical connectors
   - Radiator
   - Negative battery cable
   - Rear seat cushion

Ignition Timing

ADJUSTMENT

The engines are equipped with a Distributorless Ignition System (DIS). The system consists of 2 Crankshaft Position (CKP) sensors, crankshaft reluctor ring, Camshaft Position (CMP) sensor, Ignition Control Module (ICM), 4 ignition coils, spark plug wires and spark plugs, Knock Sensor (KS) and the Powertrain Control Module (PCM).

The PCM controls spark advance under all driving conditions. The PCM incorporates a permanent spark control override, which electronically lowers the base timing if spark knock (detonation) is encountered during normal operation due to the use of low octane fuel.

Engine Assembly

REMOVAL & INSTALLATION

3.5L Engine

1. Before servicing the vehicle, refer to the precautions in the beginning of this section.
2. Drain the crankcase and cooling system.
3. Remove or disconnect the following:
   - Negative battery cable
   - Accessory drive belt
   - Fuel injector sight shield
   - Power steering pump
   - Air intake duct
   - Fuel lines from the fuel supply rail
   - Fuel vapor line
   - Throttle and cruise control cables with the bracket from the throttle body
   - Range selector cable from the transmission
   - Brake booster vacuum hose from the engine
   - Air conditioning vacuum hose from the engine
   - Wiring harness from the engine and transmission
   - Upper radiator hose from the engine
   - Transmission fluid cooler lines from the radiator
   - Surge tank inlet hose
   - Heater hoses from the engine
   - Lower radiator air deflector
   - Secondary Air Injection (AIR) pipe from the AIR inlet valve
   - Battery cables from the retainers
   - Lower radiator hose from the engine
   - Alternator
   - Air conditioning compressor, move it aside without disconnecting the lines
   - Torque converter cover
   - Starter motor
   - Flexplate-to-torque converter bolts. Matchmark the bolts before removal.
   - Lower transmission-to-engine bolts

- Catalytic converter from the rear exhaust manifold
- Front wheels and splash shields
- Wheel speed sensor harness conduits from the lower control arms
- Tie rod ends from the steering knuckles
- Lower ball joints from the steering knuckles
- Intermediate shaft from the steering gear

4. Support the engine frame.

WARNING

Secure the front of the vehicle to the lift. The vehicle may become unstable as the engine and transmission assembly is removed from the vehicle.

5. Remove the frame-to-chassis bolts.

WARNING

Do not damage the air conditioning compressor or lines when removing the frame from the vehicle.

6. Lift the vehicle from the engine and transmission assembly.
7. Remove or disconnect the following:
   - Transmission-to-engine bracket
   - Transmission brace
   - Upper transmission-to-engine bolts
   - Transmission from the engine

To install:

8. Install or connect the following:
   - Transmission to the engine. Torque the bolts to 55 ft. lbs. (75 Nm).
   - Transmission brace. Torque the bolts to 32 ft. lbs. (43 Nm).
   - Transmission-to-engine bracket. Torque the bolts in sequence to 43 ft. lbs. (58 Nm).
   - Engine and transmission assembly under the vehicle

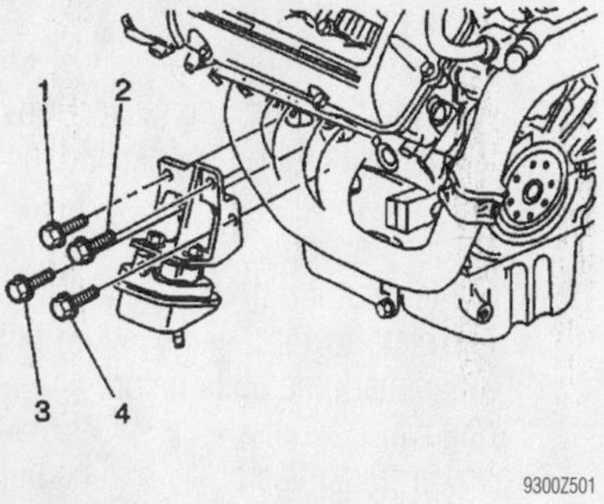

Engine mount bracket bolt tightening sequence—3.5L engine

9. Lower the vehicle onto the assembly.
10. Align the sub-frame on the vehicle using 2 bolts or drill bits, ¾ inches thick by 8 inches long through the alignment holes on the right side of the frame.
11. Install new frame-to-body bolts. Torque the bolts to 133 ft. lbs. (180 Nm) starting with the rear bolts and then the front bolts.
12. Raise the vehicle and remove the frame table.
13. Install or connect the following:
   - Intermediate shaft to the steering gear. Torque the bolt to 35 ft. lbs. (48 Nm).

➡Be sure the shaft is fully seated on the stub before installing the pinch bolt.

   - Lower ball joints to the steering knuckles. Torque the nuts to 41 ft. lbs. (55 Nm).
   - Tie rod ends to the steering knuckles. Torque the nuts to 55 ft. lbs. (75 Nm).
   - Wheel speed sensor conduits to the lower control arms
   - Splash shields and front wheels
   - Lower transmission-to-engine bolts. Torque the bolts to 55 ft. lbs. (75 Nm).
   - Catalytic converter to the rear exhaust manifold. Torque the bolts to 22 ft. lbs. (30 Nm).
   - Flexplate-to-torque converter by aligning the matchmarks. Torque the bolts to 47 ft. lbs. (63 Nm).
   - Starter motor. Torque the bolts to 37 ft. lbs. (50 Nm).
   - Torque converter cover. Torque the bolts to 11 ft. lbs. (15 Nm).
   - Air conditioning compressor. Torque the nuts and bolts to 37 ft. lbs. (50 Nm).
   - Alternator. Torque the bolts to 37 ft. lbs. (50 Nm).
   - Lower radiator hose
   - Battery cables in the retainers
   - AIR pipe
   - Lower radiator air deflector
   - Heater hoses
   - Surge tank hose
   - Transmission fluid cooler lines to the radiator
   - Upper radiator hose to the engine
   - Wiring harness connectors to the engine and transmission
   - Air conditioning vacuum hose to the engine
   - Brake booster hose to the engine
   - Range selector cable to the transmission

➡GM recommends you use a new throttle cable when replacing the engine assembly.

14. Remove the trim panel under the left instrument panel and detach the throttle cable from the top of the pedal, then squeeze the retainer and push the cable through the bulkhead to remove it.
15. Install or connect the following:
   - New throttle cable
   - Cruise control and throttle cables to the throttle body
   - Fuel vapor line
   - Fuel lines to the supply rail
   - Air intake duct
   - Power steering pump. Torque the bolts to 25 ft. lbs. (34 Nm).
   - Fuel injector sight shield
   - Drive belt
   - Negative battery cable
16. Refill the cooling system.
17. Install a new oil filter and refill the crankcase.
18. Start the engine and check for leaks.

3.8L Engine

1. Before servicing the vehicle, refer to the precautions in the beginning of this section.
2. Relieve the fuel system pressure.
3. Matchmark the hood hinges and remove the hood.
4. Drain the cooling system.
5. Drain the engine oil.
6. Remove or disconnect the following:
   - Positive battery cable
   - Radiator hoses
   - Heater hoses from the heater core
   - Negative battery cable from the engine
   - Engine harness connector at the bulkhead
   - Accessory drive belts
   - Power steering pump without disconnecting the power steering lines from the pump
   - Air intake duct
   - Throttle cable from the linkage bracket
   - All cables from the throttle body
   - Manifold Air Temperature (MAT) sensor electrical connector
   - Throttle Position (TPS) sensor electrical connector
   - Idle Air Control (IAC) valve electrical connector
   - Oxygen (O_2S) sensor electrical connector

- Air conditioning compressor electrical connector
- Oil pressure switch electrical connector
- Power steering cutout switch electrical connector
- Vehicle Speed Sensor (VSS) electrical connector
- Low oil level sensor electrical connector
- Ignition assembly ground strap from the inner fender panel
- Fuel feed and return lines from the fuel rail and the pressure regulator
- Evaporative Emissions (EVAP) canister hoses from the throttle body
- Vacuum lines from the power brake booster
- Heater control hoses
- Cruise control servo
- Exhaust pipe from the right side exhaust manifold
- Air conditioning compressor from the bracket. DO NOT disconnect the refrigerant lines.
- Right front engine-to-transmission brace
- Flexplate cover
- Starter motor
- Torque converter-to-flexplate bolts. Matchmark the bolts before removal.

7. Install an engine lifting device and slightly raise the engine to take the weight off the front engine mount.
8. Remove the torque axis engine mount.
9. Support the transmission.
10. Remove or disconnect the following:
    - Engine-to-transmission bolts
    - Engine

⁂ WARNING

Be sure all electrical and vacuum connections are disconnected.

To install:

11. Install the engine. Torque the engine-to-transmission bolts to 55 ft. lbs. (75 Nm).
12. Torque axis engine mount. Torque the mount-to-frame bolts to 64 ft. lbs. (87 Nm), the engine side through-bolt to 65 ft. lbs. (87 Nm) and the frame side through-bolt to 52 ft. lbs. (70 Nm).
13. Remove the engine lifting assembly and transmission support.
14. Install or connect the following:
    - Torque converter-to-flexplate bolts, aligning the matchmarks made earlier. Torque the bolts to 46 ft. lbs. (62 Nm).
    - Starter motor. Torque the bolts to 32 ft. lbs. (43 Nm).
    - Flexplate cover. Torque the screws to 88 inch lbs. (10 Nm).
    - Right side engine-to-transmission brace
    - Air conditioning compressor to the bracket. Torque the front mounting bolts to 44 ft. lbs. (60 Nm), the rear mounting nut and bolts to 18 ft. lbs. (25 Nm).
    - Exhaust pipe to the rear exhaust manifold. Torque the nuts to 18 ft. lbs. (25 Nm).
    - Cruise control servo
    - Power brake booster vacuum lines
    - Heater control hoses
    - EVAP canister hoses to the throttle body
    - Fuel feed and return lines to the fuel rail and the pressure regulator
    - Ignition assembly ground strap to the inner fender well
    - Low oil level sensor electrical connector
    - VSS electrical connector
    - Power steering cutout switch electrical connector
    - Oil pressure switch electrical connector
    - Air conditioning compressor electrical connector
    - O_2S electrical connector
    - IAC valve electrical connector
    - TPS sensor electrical connector
    - MAT sensor electrical connector
    - Control cables to the throttle body lever
    - Throttle cable to the mounting bracket
    - Air intake duct
    - Power steering pump. Torque the bolts to 20 ft. lbs. (27 Nm).
    - Accessory drive belts
    - Main engine harness at the bulkhead connector
    - Negative battery cable to the engine
    - Heater hoses to the heater core
    - Upper and lower radiator hoses
    - Battery cables
    - Hood
15. Refill the crankcase.
16. Pressurize the fuel system and verify no leaks.
17. Refill and bleed the cooling system.

4.0L Engine

1998–99 MODELS

1. Before servicing the vehicle, refer to the precautions in the beginning of this section.
2. Relieve the fuel system pressure.
3. Drain the cooling system and the crankcase.
4. Recover the air conditioning system refrigerant.
5. Remove or disconnect the following:
    - Air intake duct
    - Fuel feed and return lines from the fuel rail
    - Vacuum hoses from the rear of the intake manifold
    - Throttle cables from the throttle body
    - Engine harness connector at the bulkhead
    - Coolant hoses from the radiator surge tank
    - Upper radiator hose from the coolant crossover
    - Lower radiator hose from the thermostat housing
    - Upper transmission cooler line from the radiator
    - Lower transmission cooler line from the radiator
    - Heater hoses from the heater pipes
    - Vacuum supply line from the power brake booster
    - Range selector cable from the transmission
    - Vacuum reservoir
    - Electrical connectors and vacuum lines from the cruise control servo
6. Install an engine support fixture.
    - Positive battery cable at the junction block
    - Torque axis mount through-bolt
    - Front engine mount bolts
    - Negative battery cable from the engine
    - Both front wheels
    - Both inner fender well splash shields
    - Power steering rack electrical connector
    - Lower ball joints from the steering knuckles
    - Outer tie rod ends from the steering knuckles
    - Halfshafts from the wheel bearing and hub assemblies
    - Intermediate shaft
    - Exhaust system

- Engine oil cooler lines at the adapter
- Oil pan brace
- Front and rear engine-to-transmission braces
- Flexplate cover
- Torque converter-to-flexplate bolts. Matchmark the bolts before removal.
- Oil filter adapter from the engine
- A/C hose and muffler from the rear of the air conditioning compressor

7. Position a frame support table under the powertrain assembly.
   - Left transmission mount
   - Subframe bolts
8. Raise the vehicle off the engine table and verify no electrical or vacuum connections are still hooked to the engine.
   - Transmission and engine assembly
   - Engine from the transmission and subframe assembly

To install:

9. Install or connect the following:
   - Engine to transmission. Torque the bolts to 55 ft. lbs. (75 Nm).
   - Engine assembly onto the subframe. Torque the bolts to 55 ft. lbs. (75 Nm).
   - Powertrain assembly under the vehicle
10. Lower the vehicle and align the subframe. Torque the subframe bolts to 142 ft. lbs. (192 Nm).
11. Raise the vehicle and remove the engine dolly.
12. Install or connect the following:
    - Left side transmission mount
    - Air conditioning hose and muffler to the air conditioning compressor
    - Oil filter adapter to the engine. Torque the bolts to 12 ft. lbs. (16 Nm).
    - Torque converter-to-flexplate aligning the matchmarks made earlier. Torque the bolts to 35 ft. lbs. (47 Nm).
    - Flexplate cover
    - Rear engine-to-transmission brace. Torque the bolts to 44 ft. lbs. (60 Nm).
    - Front engine-to-transmission brace. Torque the bolts to 44 ft. lbs. (60 Nm).
    - Oil pan brace. Torque the bolts to 37 ft. lbs. (50 Nm).
    - Oil cooler hoses to the adapter. Torque the fittings to 12 ft. lbs. (18 Nm).
    - Exhaust system
    - Intermediate shaft to the rack and pinion. Torque the pinch bolt to 35 ft. lbs. (47 Nm).
    - Halfshafts to the hub and bearing assemblies. Torque the axle nuts to 107 ft. lbs. (148 Nm).
    - Tie rod ends to the steering knuckle. Torque the castle nuts to 41 ft. lbs. (55 Nm).

➡If necessary, tighten the castle nuts up to 60 degrees additional to align the cotter pin holes. NEVER loosen the castle nut and DO NOT exceed 52 ft. lbs. (70 Nm) to align the holes.

- New tie rod end cotter pins
- Ball joints to the steering knuckles. Torque the castle nuts to 88 inch lbs. (10 Nm) plus 120 degrees additional.

➡If necessary tighten the castle nuts up to 60 degrees additional to align the cotter pin holes. NEVER loosen the castle nuts to align the holes.

- New ball joint cotter pins
- Electrical connector to the rack and pinion assembly
- Both inner fender splash shields
- Front wheels. Torque the nuts to 100 ft. lbs. (140 Nm).
- Negative battery cable to the engine
- Front engine mount bracket-to-torque axis mount bolts
- Torque axis mount through-bolt
- Positive battery cable to the junction block

13. Remove the engine support fixture.
14. Install or connect the following:
    - Electrical and vacuum connectors to the cruise control servo
    - Vacuum reservoir
    - Range selector cable to the transmission
    - Vacuum hose to the power brake booster
    - Heater hoses
    - Transmission oil cooler lines
    - Lower radiator hose to the thermostat housing
    - Upper radiator hose to the coolant crossover
    - Coolant surge tank hoses
    - Main engine harness to the bulkhead connector
    - Throttle cable to the throttle body lever
    - Vacuum hoses to the rear of the intake manifold
    - Fuel feed and return lines
    - Air intake duct
    - Negative battery cable
15. Refill the cooling system, engine crankcase and recharge the air conditioning system.
16. Start the vehicle and check for leaks.

2001 MODELS

1. Before servicing the vehicle, refer to the precautions in the beginning of this section.
2. Relieve the fuel system pressure.
3. Drain the cooling system and the crankcase.
4. Recover the air conditioning system refrigerant.
5. Remove or disconnect the following:
   - Air intake duct
   - Intake manifold sight shield
   - Inlet radiator hose from the engine
   - Outlet radiator hose from the thermostat housing
   - Upper transmission cooler line retaining bolt from the fan shroud
   - Internal spring clip from the transmission oil cooler fitting
   - Upper transmission cooler line from the radiator
   - Lower transmission cooler line from the radiator
   - Surge tank inlet hose from the engine
   - Surge tank outlet hose from the heater pipe

⁂ CAUTION

In order to avoid possible injury or vehicle damage, always replace the accelerator control cable with a NEW cable whenever you remove the engine from the vehicle.

- Accelerator and cruise control cables from the throttle body
- Heater hoses from the engine
- Brake pipes from the master cylinder and plug the openings to avoid contamination
- Shift cable from the manual shift lever and the bracket
- Vacuum hose from the brake booster
- EVAP hose from the EVAP purge valve
- Fuel inlet and return quick-connect fittings at the fuel rail
- Engine ground cable from the frame
- Starter solenoid/knock sensor connector located near the power steering pump
- Electrical harness from the underhood fuse block
- Air deflector
- Wheels
- Powertrain Control Module (PCM) electrical connections

- 2 rear brake pipes at the rear of the engine frame
- Tie rod ends from the steering knuckles
- Lower ball joint from the steering knuckle
- Drive shafts
- Engine oil cooler lines at the adapter
- A/C hoses from the compressor
- Steering gear from the intermediate shaft
- Wheel speed and brake wear sensors at the right and left strut towers
- Heated Oxygen (HO_2S) sensor at the sensor pigtail
- Exhaust front pipe
- Oil pan brace
- Front and rear engine-to-transmission braces
- Flexplate cover
- Torque converter-to-flexplate bolts. Matchmark the bolts before removal.

6. Position a frame support table under the powertrain assembly.
   - Right engine mount-to-engine mount bracket
   - Left transmission mount
   - Subframe bolts
   - Nut securing the coil pack ground wire to the cylinder head
   - Bolt securing the crossover pipe to the cylinder head
   - Power steering pump
   - Transaxle braces
7. Raise the vehicle off the engine table and verify no electrical or vacuum connections are still hooked to the engine.
8. Remove or disconnect the following:
   - Transmission-to-engine bolts
   - Engine from the transmission and subframe assembly

To install:

9. Install or connect the following:
   - Engine to transmission. Torque the bolts to 55 ft. lbs. (75 Nm).
   - Engine assembly onto the subframe. Torque the bolts to 52 ft. lbs. (70 Nm).
   - Transaxle braces and torque the bolts to 37 ft. lbs. (50 Nm)
   - Boltt securing the crossover pipe to the cylinder head and tighten the bolt to 18 ft. lbs. (25 Nm)
   - Nut securing the coil pack ground wire to the cylinder head and tighten to 13 ft. lbs. (17 Nm)
   - Power steering pump
   - Powertrain assembly under the vehicle
10. Lower the vehicle and align the subframe. Torque the subframe bolts to 141 ft. lbs. (191 Nm).
    - Left and right side transmission mounts and tighten to 59 ft. lbs. (80 Nm)
11. Raise the vehicle and remove the engine dolly.
    - Torque converter-to-flexplate aligning the matchmarks made earlier. Torque the bolts to 44 ft. lbs. (60 Nm).
    - Flexplate cover
    - Oil pan-to-transaxle brace. Torque the bolts to 37 ft. lbs. (50 Nm).
    - Exhaust front pipe
    - HO_2S sensor at the sensor pigtail
    - Weel speed and brake wear sensors at the right and left strut towers
    - Intermediate shaft to the rack and pinion. Torque the pinch bolt to 33 ft. lbs. (45Nm).
    - Air conditioning hoses to the air conditioning compressor and tighten the fittings to 15 ft. lbs. (20 Nm)
    - Oil cooler hoses to the adapter. Torque the fittings to 12 ft. lbs. (18 Nm).
    - Halfshafts to the hub and bearing assemblies
    - Tie rod ends and ball joints to the steering knuckle
    - Real time dampening sensor links to the lower control arms
    - Right and left road sensing suspension electrical connectors
    - 2 rear brake pipes at the rear of the engine frame and tighten to 11 ft. lbs. (15 Nm)
    - Wheels
    - Air deflector
    - Engine ground cable to the frame
    - Starter solenoid/knock sensor connector located near the power steering pump
    - Electrical harness to the underhood fuse block
    - Shift cable to the manual shift lever and the bracket
    - Vacuum hose to the brake booster
    - EVAP hose to the EVAP purge valve
    - Fuel inlet and return quick-connect fittings to the fuel rail
    - Heater hoses to the engine
    - Heater pipes
    - 2 brake pipes to the master cylinder and tighten to 11 ft. lbs. (15 Nm)
12. Bleed the hydraulic brake system.
    - Surge tank outlet and inlet hoses

CAUTION

In order to avoid possible injury or vehicle damage, always replace the accelerator control cable with a NEW cable whenever you remove the engine from the vehicle.

- Accelerator and cruise control cables to the throttle body
- Lower transmission cooler line to the radiator and tighten to 26 ft. lbs. (35 Nm)
- Secure the internal spring clip to the fitting
- Upper transmission cooler line to the radiator by pushing until attached
- Upper transmission cooler line retaining bolt to the fan shroud
- Outlet radiator hose to the thermostat housing
- Inlet radiator hose to the engine
- Intake manifold sight shield
- Air intake duct
- Negative battery cable

13. Refill the cooling system, engine crankcase and recharge the air conditioning system.
14. Start the vehicle and check for leaks.

Water Pump

REMOVAL & INSTALLATION

3.5L Engine

1. Before servicing the vehicle, refer to the precautions in the beginning of this section.
2. Drain the cooling system.
3. Remove or disconnect the following:
   - Water pump pulley bolts, loosen but do not remove at this time
   - Drive belt
   - Idler pulley
   - Water pump pulley

➡The water pump is attached to the engine with both long and short bolts, be sure to note their locations.

- Water pump

To install:

4. Install or connect the following:
   - Water pump using a new gasket.

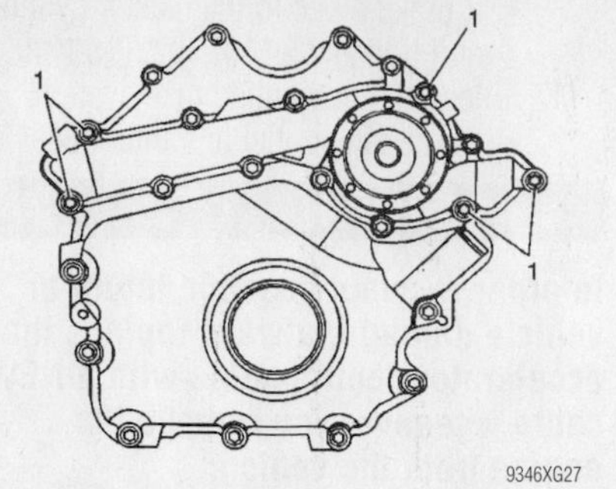

Be sure to install the 5 long water pump bolts (1) in the correct locations—3.5L engine

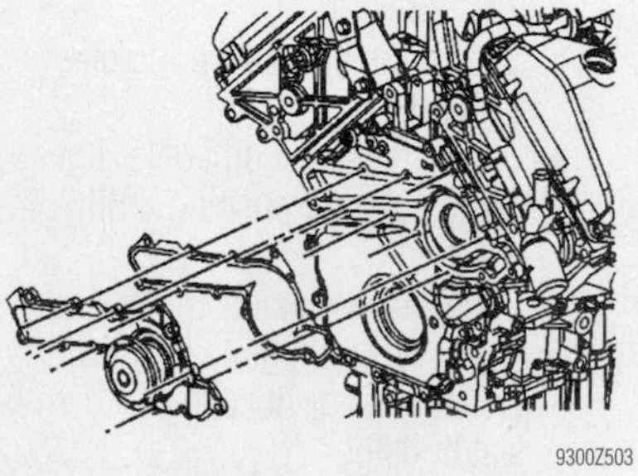

Water pump mounting—3.5L engine

Torque the bolts to 124 inch lbs. (14 Nm).
- Water pump pulley. Torque the bolts to 106 inch lbs. (12 Nm).
- Idler pulley. Torque the bolt to 37 ft. lbs. (50 Nm).
- Drive belt

5. Refill the cooling system.
6. Start the engine and check for leaks.

3.8L Engine

1. Before servicing the vehicle, refer to the precautions in the beginning of this section.
2. Drain the cooling system.
3. Remove or disconnect the following:
   - Negative battery cable
   - A/C compressor splash shield
   - Drive belts
   - Coil pack
   - Supercharger belt tensioner
   - Engine mount
   - Power steering pump
   - Engine mount bracket
   - Idler pulley
   - Water pump pulley
   - Water pump

To install:

4. Install or connect the following:
   - Water pump with a new gasket. Torque the bolts to 11 ft. lbs. (15 Nm), then to 22 ft. lbs. (30 Nm).
   - Water pump pulley
   - Idler pulley. Torque the bolt to 37 ft. lbs. (50 Nm).
   - Engine mount bracket. Torque the upper nut to 22 ft. lbs. (30 Nm) and the lower nut to 52 ft. lbs. (70 Nm).
   - Power steering pump. Torque the bolts to 20 ft. lbs. (27 Nm).
   - Engine mount. Torque the bolts to 58 ft. lbs. (78 Nm).
   - Supercharger belt tensioner. Torque the bolts to 37 ft. lbs. (50 Nm).
   - Coil pack
   - Drive belts
   - A/C compressor splash shield
   - Negative battery cable
5. Refill the cooling system.
6. Start the engine and check for leaks.

4.0L Engine

1. Before servicing the vehicle, refer to the precautions in the beginning of this section.
2. Drain the cooling system.
3. Remove or disconnect the following:
   - Negative battery cable
   - Air cleaner assembly
   - Secondary AIR injection control valve, if necessary
   - Oil level indicator tube nut, if necessary
   - Water pump drive belt cover
   - Water pump drive belt
   - Lower radiator hose
   - Bypass hose
   - Water pump cover
   - Water pump from the water pump housing, by turning the locking ring with a Water Pump Remover/Installer tool J-38816

To install:

4. Install or connect the following:
   - Water pump with new o-ring. Torque the pump to 73 ft. lbs. (100 Nm).
   - Water pump cover. Torque the bolts to 89 inch lbs. (10 Nm).
   - Lower radiator hose
   - Coolant bypass hose
   - Water pump drive belt
   - Water pump drive belt cover
   - Oil level indicator tube nut, if necessary
   - Secondary AIR injection control valve, if necessary
   - Air cleaner assembly
   - Negative battery cable
5. Refill and bleed the cooling system.

Cylinder Head

REMOVAL & INSTALLATION

3.5L Engine

LEFT SIDE (FRONT)

1. Before servicing the vehicle, refer to the precautions in the beginning of this section.
2. Drain the crankcase and the cooling system.
3. Remove or disconnect the following:
   - Negative battery cable
   - Intake manifold
   - Coolant crossover pipe
   - Left exhaust manifold
   - Camshaft cover
4. Install a holding tool on the camshafts to hold them in position.
5. Remove or disconnect the following:
   - Camshaft primary chain
   - Camshafts from the left cylinder head
   - Rocker arms and valve lifters

➡Be sure to keep the rocker arms and lifters in order so they can be installed in their original locations.

   - M6 bolts from the front of the cylinder head, note the location of the longer bolt
   - M11 cylinder head bolts
   - Cylinder head

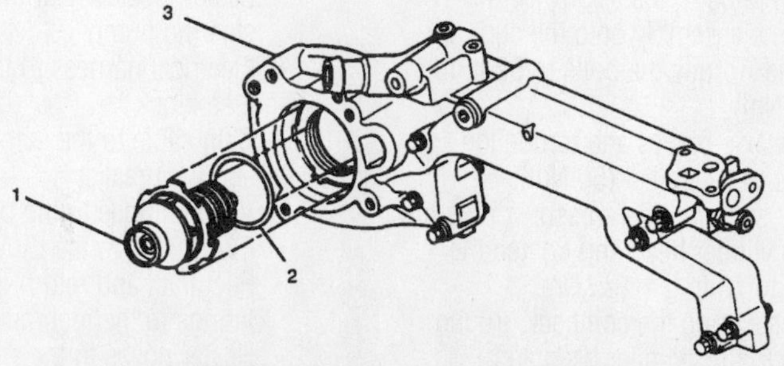

1 WATER PUMP ASM.
2 O-RING SEAL
3 WATER PUMP HOUSING ASM.

7922XG02

Exploded view of the water pump housing and water pump—4.0L engine

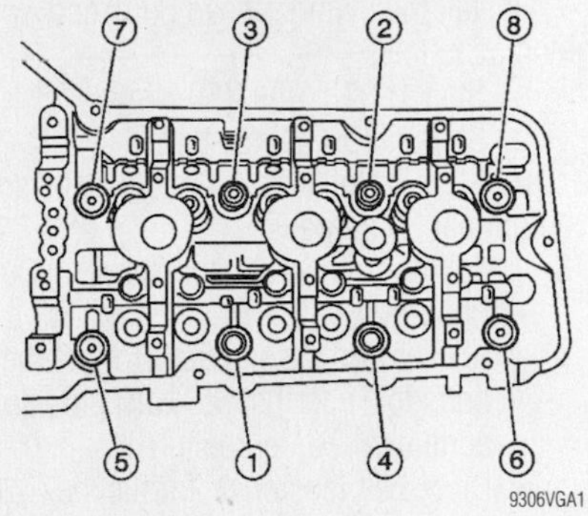

Front cylinder head bolt torque sequence—3.5L (VIN H) engine

To install:

6. Install the cylinder head with a new gasket.
7. Install new M11 bolts and M6 bolts .
8. Torque the M11 bolts in sequence to:
   a. Step 1: 22 ft. lbs. (30 Nm).
   b. Step 2: 60 degree turn.
   c. Step 3: 60 degree turn.
   d. Step 4: 60 degree turn.
9. Torque the long M6 bolt to 22 ft. lbs. (30 Nm).
10. Torque both shorter M6 bolts to 106 inch lbs. (12 Nm).
11. Install or connect the following:
   - Lifters
   - Rocker arms
   - Camshafts
   - Primary chain
12. Remove the timing chain holding tool.
13. Install or connect the following:
   - Camshaft covers. Torque the bolts to 80 inch lbs. (9 Nm).
   - Exhaust manifold. Torque the bolts to 22 ft. lbs. (30 Nm).
   - Coolant crossover pipe. Torque the bolts to 18 ft. lbs. (25 Nm).

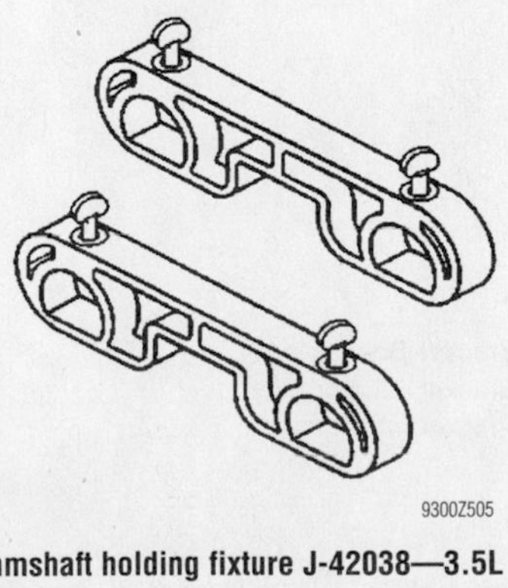

Camshaft holding fixture J-42038—3.5L engine

   - Intake manifold. Torque the bolts to 62 inch lbs. (7 Nm).
   - New oil filter
   - Negative battery cable
14. Refill the crankcase.
15. Refill the cooling system.
16. Start the engine and check for leaks.

RIGHT SIDE (REAR)

1. Before servicing the vehicle, refer to the precautions in the beginning of this section.
2. Drain the crankcase and the cooling system.
3. Remove or disconnect the following:
   - Negative battery cable
   - Intake manifold
   - Coolant crossover pipe

➡Do not remove the rear exhaust manifold. Detach it from the cylinder head and the connection from the front manifold; then, move it aside.

   - Rear exhaust manifold from the cylinder head and front manifold
   - Camshaft cover
4. Install a holding tool on the camshafts to hold them in position.
5. Remove or disconnect the following:
   - Primary camshaft chain
   - Right side camshafts
   - Rocker arms and lifters

➡Keep the rocker arms and lifters in order so they can be installed in their original positions.

   - Engine Coolant Temperature (ECT) sensor from the cylinder head
   - M6 bolts from the front of the cylinder head, note the location of the longer bolt
   - M11 cylinder head bolts
   - Cylinder head

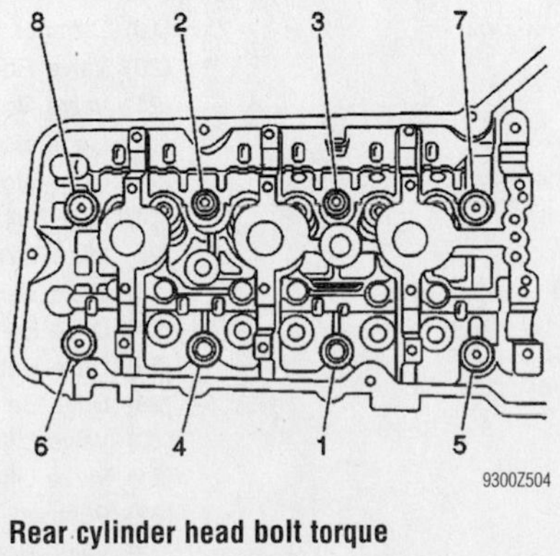

Rear cylinder head bolt torque sequence—3.5L (VIN H) engine

To install:

6. Install the cylinder head with a new gasket.
7. Install new M11 bolts .
8. Install M6 bolts in the front of the cylinder head.
9. Tighten the M11 bolts in sequence using the following steps:
   a. Step 1: 22 ft. lbs. (30 Nm).
   b. Step 2: 60 degree turn.
   c. Step 3: 60 degree turn.
   d. Step 4: 60 degree turn.
10. Torque the long M6 bolt to 22 ft. lbs. (30 Nm).
11. Torque both short M6 bolts to 106 inch lbs. (12 Nm).
12. Install or connect the following:
   - ECT sensor. Torque the sensor to 15 ft. lbs. (20 Nm).
   - Lifters and rocker arms in their original positions
   - Camshafts
   - Primary chain
   - Camshaft covers. Torque the bolts to 80 inch lbs. (9 Nm).
   - Exhaust manifold on the cylinder head. Torque the bolts to 22 ft. lbs. (30 Nm).
   - Coolant crossover pipe. Torque the bolts to 18 ft. lbs. (25 Nm).
   - Intake manifold. Torque the bolts to 62 inch lbs. (7 Nm).
   - New oil filter
   - Negative battery cable
13. Refill the crankcase.
14. Refill the cooling system.
15. Start the engine and check for leaks.

3.8L Engine

LEFT SIDE (FRONT)

1. Before servicing the vehicle, refer to the precautions in the beginning of this section.
2. Disconnect the negative battery cable.
3. Relieve the fuel system pressure.
4. Drain the cooling system
5. Remove or disconnect the following:
   - Accessory drive belts
   - Intake manifold
   - Left exhaust manifold
   - Ignition module
   - Spark plug wires
   - Alternator and bracket
   - Rocker arm covers
   - Air conditioning bracket-to-cylinder head bolt
   - Power steering pump
   - Drive belt tensioner

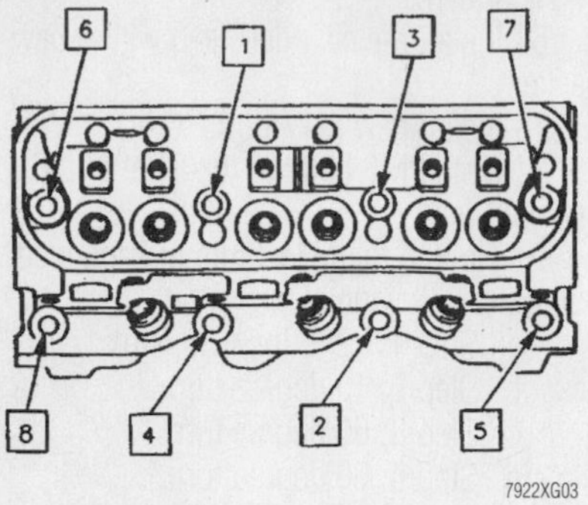

Cylinder head bolt torque sequence—3.8L (VIN 1and K) engines

- Fuel pipe heat shield
- Rocker arm assemblies and the pushrods
- Cylinder head

To install:

6. Install the cylinder head with a new gasket.

➡This engine uses special torque-to-yield head bolts. The procedure must be followed carefully and new bolts must be used whenever the head is removed. Total bolt torque should not exceed 60 ft. lbs. (81 Nm).

7. Install the cylinder head bolts and tighten as follows:
   a. Step 1: 37 ft. lbs. (50 Nm).
   b. Step 2: 130 degree turn.
   c. Step 3: Rotate the 4 center bolts an additional 30 degrees.
8. Install or connect the following:
   - Pushrods
   - Rocker arm assemblies. Torque the bolts to 19 ft. lbs. (25 Nm) plus an additional 70 degree turn.
   - Drive belt tensioner. Torque the bolts to 37 ft. lbs. (50 Nm).
   - Air conditioning compressor

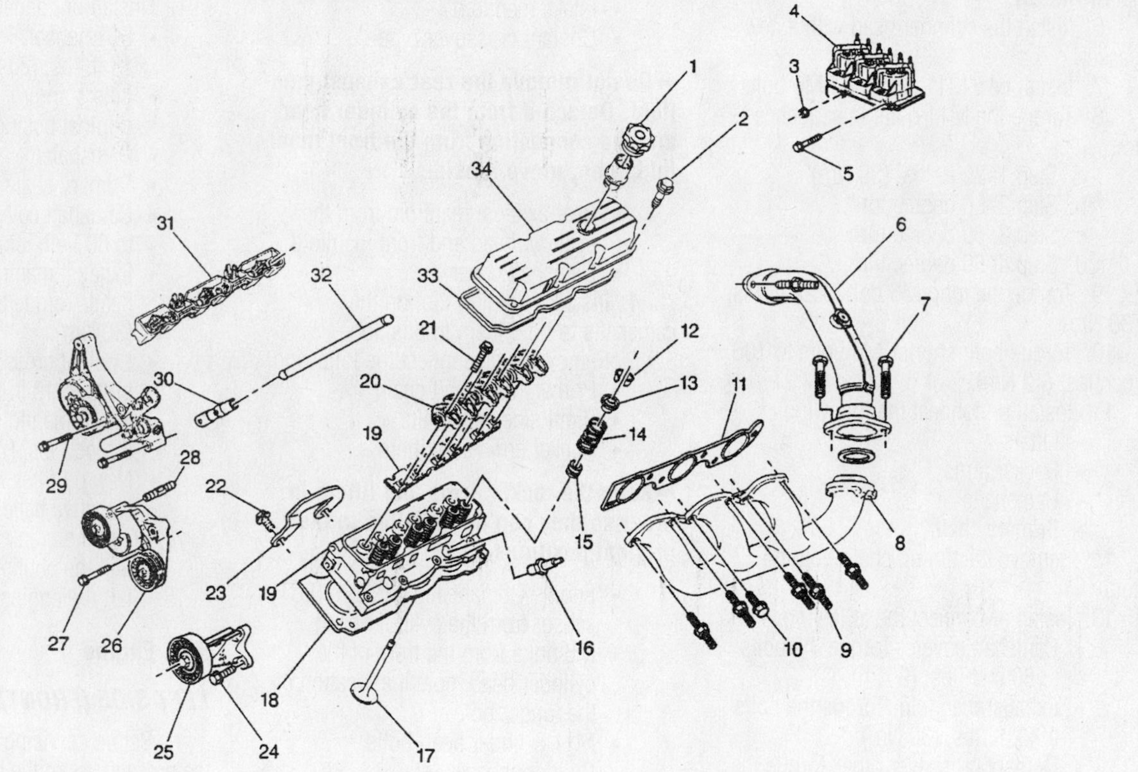

Legend

(1) Oil Fill Cap
(2) Valve Rocker Arm Cover Bolt
(3) ICM Bracket Nut
(4) ICM
(5) ICM Bracket Bolt
(6) Exhaust Crossover Pipe
(7) Exhaust Crossover Bolt
(8) Exhaust Manifold
(9) Exhaust Manifold Stud
(10) Exhaust Manifold Bolt
(11) Exhaust Manifold Gasket
(12) Valve Stem Key
(13) Valve Stem Cap
(14) Valve Spring
(15) Valve Seal
(16) Spark Plug
(17) Intake Valve
(18) Cylinder Head Gasket
(19) Cylinder Head
(20) Valve Rocker Arm
(21) Valve Rocker Arm Bolt
(22) Generator Brace Bolt
(23) Generator Brace
(24) Idler Pulley Bracket Bolt
(25) Idler Pulley
(26) Drive Belt Tensioner
(27) Drive Belt Tensioner Bracket Bolt
(28) Drive Belt Tensioner Bracket Stud
(29) Drive Belt Tensioner Bracket
(30) Valve Lifter
(31) Valve Lifter Guide
(32) Pushrod
(33) Valve Rocker Arm Cover Gasket
(34) Valve Rocker Arm Cover

9300XG03

Exploded view of the cylinder head and related components—3.8L (VIN 1) engine

bracket bolt. Torque the bolt to 52 ft. lbs. (70 Nm).

- Intake manifold. Torque the bolts to 11 ft. lbs. (15 Nm).
- Rocker arm covers. Torque the bolts to 89 inch lbs. (10 Nm).
- Alternator front mount bracket
- Alternator. Torque the bolts to 37 ft. lbs. (50 Nm).
- Ignition module
- Spark plug wires
- Exhaust manifold. Torque the bolts to 22 ft. lbs. (30 Nm).
- Accessory drive belts
- Power steering pump. Torque the bolts to 20 ft. lbs. (27 Nm).
- Fuel pipe heat shield
- Negative battery cable

9. Refill the cooling system.

RIGHT SIDE (REAR)

1. Before servicing the vehicle, refer to the precautions in the beginning of this section.
2. Disconnect the negative battery cable.
3. Relieve the fuel system pressure.
4. Drain the cooling system.
5. Remove or disconnect the following:
   - Intake manifold
   - Right side exhaust manifold
   - Accessory drive belt
   - Belt tensioner
   - Power steering pump and bracket
   - Fuel line heat shield
   - Spark plug wires
   - Rocker arm cover
   - Rocker arm assemblies and pushrods, keep everything in order for reinstallation
   - Cylinder head

To install:

6. Install the cylinder head with a new gasket.

➡This engine uses special torque-to-yield head bolts. The procedure must be followed carefully and new bolts must be used whenever the head is removed. Total bolt torque should not exceed 60 ft. lbs. (81 Nm).

7. Torque the cylinder head bolts in sequence as follows:
   a. Step 1: 37 ft. lbs. (50 Nm).
   b. Step 2: 130 degree turn.
   c. Step 3: Rotate 4 center bolts an additional 30 degree turn.
8. Install or connect the following:
   - Pushrods
   - Rocker arm assemblies. Torque the nut to 19 ft. lbs. (25 Nm) plus an additional 70 degree turn.
   - Rocker arm cover. Torque the bolts to 89 inch lbs. (10 Nm).
   - Spark plug wires
   - Fuel line heat shield
   - Power steering pump bracket. Torque the bolts to 35 ft. lbs. (47 Nm).
   - Power steering pump. Torque the bolts to 20 ft. lbs. (27 Nm).
   - Belt tensioner. Torque the bolts to 37 ft. lbs. (50 Nm).
   - Accessory drive belt
   - Exhaust manifold. Torque the bolts to 22 ft. lbs. (30 Nm).
   - Intake manifold. Torque the bolts to 11 ft. lbs. (15 Nm).
   - Negative battery cable
9. Refill the cooling system.

4.0L Engine

1998–99

➡The manufacturer recommends that the entire powertrain be removed from the vehicle before removing the cylinder heads.

1. Before servicing the vehicle, refer to the precautions in the beginning of this section.
2. Relieve the fuel system pressure.
3. Drain the cooling system.
4. Drain the engine oil.
5. Remove or disconnect the following:
   - Powertrain assembly
   - Intake manifold
   - Camshaft covers
   - Harmonic balancer
   - Front cover
   - Oil pump

➡Align all timing marks before performing the next step.

   - Timing chain tensioner for the cylinder head being removed
   - Camshaft sprockets from the head being removed

➡The timing chain remains in the chain case.

   - Timing chain guides

➡Access the retaining screws through the plugs at the front of the cylinder head.

   - Water crossover
   - Exhaust manifold
   - Cylinder head bolts a little at a time by reversing the torque sequence
   - Cylinder head

⁂ WARNING

With the camshafts remaining in the cylinder head, some valves will be open at all times. Do not rest the cylinder head on a flat service with the cylinder face down or valve damage will result.

6. Check the cylinder head for warpage using a straightedge and feeler gauge. Measure along each edge, at the center and across both ends.
7. If warpage is less than 0.002 in. (0.05mm), the cylinder head surface is usable. If warpage is 0.002–0.008 in. (0.05–0.2mm), the cylinder head must be

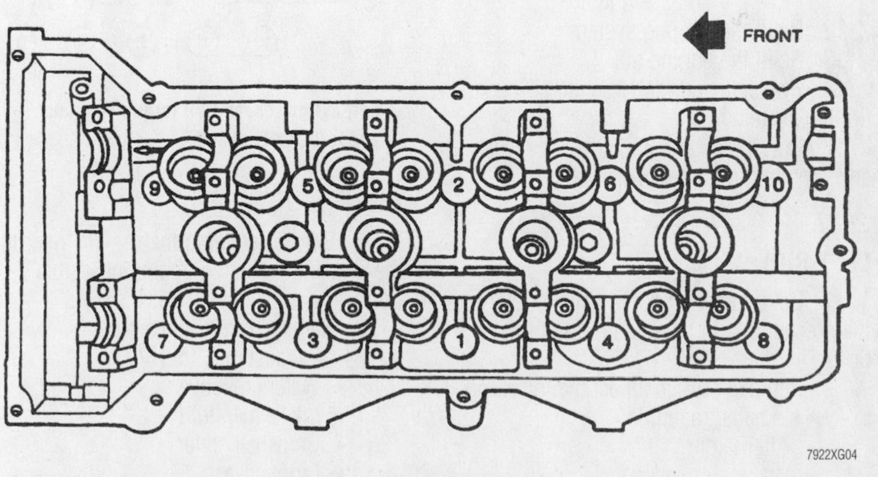

Cylinder head bolt tightening sequence—1998–99 4.0L engine

resurfaced. After resurfacing, the dimension between the combustion chamber gauge pad and the deck surface must be at least 10.5mm.

To install:

8. Install the cylinder head with a new gasket.

➡Lube the washer and the underside of the bolt head with engine oil prior to installation. New replacement head bolts are recommended.

9. Torque the M11 cylinder head bolts in sequence as follows:
   a. Step 1: 22 ft. lbs. (30 Nm).
   b. Step 2: 90 degree turn.
   c. Step 3: 75 degree turn.
10. Torque the M6 cylinder head bolts in sequence to 10 ft. lbs. (12 Nm).
11. Install or connect the following:
   - Exhaust manifold. Torque the nuts to 22 ft. lbs. (30 Nm) and the bolts to 18 ft. lbs. (25 Nm).
   - Water crossover. Torque the bolts to 18 ft. lbs. (25 Nm).
   - Timing chain guides. Torque the bolts to 18 ft. lbs. (25 Nm).
   - Camshaft sprockets. Torque the bolts to 90 ft. lbs. (120 Nm).
   - Timing chain tensioner. Torque the bolts to 18 ft. lbs. (25 Nm).
   - Oil pump
   - Front cover. Torque the bolts to 89 inch lbs. (10 Nm).
   - Harmonic balancer. Torque the bolt to 37 ft. lbs. (50 Nm), plus an additional 120 degree turn.
   - Camshaft cover. Torque the bolts to 89 inch lbs. (10 Nm).
   - Intake manifold . Torque the bolts to 89 inch lbs. (10 Nm).
   - Powertrain assembly
12. Refill the cooling system.
13. Refill the engine oil.
14. Evacuate and recharge the air conditioning system.
15. Run the engine and check for leaks and proper engine performance.

2001

LEFT SIDE

1. Before servicing the vehicle, refer to the precautions in the beginning of this section.
2. Remove or disconnect the following:
   - Exhaust manifold
   - Alternator
   - Water crossover
   - Intake manifold
   - Camshaft cover
   - Front cover
   - Secondary camshaft drive chain
   - Power steering hose retaining bolt from the head
   - Cylinder head

To install:

3. Install the cylinder head with a new gasket.
4. Install new M11 head bolts and M6 head bolts.
5. Torque the M11 bolts in sequence as follows:
   a. Step 1: 30 ft. lbs. (40 Nm).
   b. Step 2: 70 degree turn.
   c. Step 3: 60 degree turn.
   d. Step 4: 60 degree turn.
6. Torque the M6 bolts to 106 inch lbs. (12 Nm).
7. Install or connect the following:
   - Power steering hose retaining bolt. Torque the bolt to 37 ft. lbs. (50 Nm).
   - Secondary camshaft drive chain
   - Front cover. Torque the bolts to 89 inch lbs. (10 Nm).
   - Camshaft cover. Torque the bolts to 89 inch lbs. (10 Nm).
   - Intake manifold. Torque the bolts to 89 inch lbs. (10 Nm).
   - Water crossover. Torque the bolts to 18 ft. lbs. (25 Nm).
   - Alternator. Torque the bolts to 37 ft. lbs. (50 Nm).
   - Exhaust manifold. Torque the bolts to 18 ft. lbs. (25 Nm).

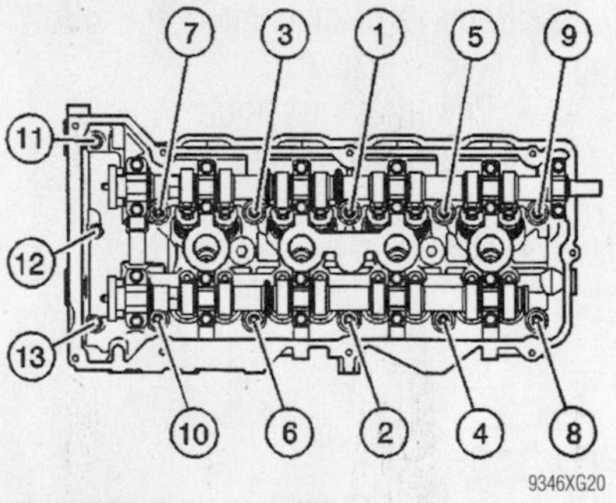

Left cylinder head bolt torque sequence—2001–01 4.0L models

RIGHT SIDE

8. Before servicing the vehicle, refer to the precautions in the beginning of this section.
9. Remove or disconnect the following:
   - Exhaust manifold
   - Water crossover
   - Intake manifold
   - Camshaft cover
   - Front cover
   - Secondary camshaft drive chain
   - Engine Coolant Temperature (ECT) sensor electrical connector
   - Ground wire from the head

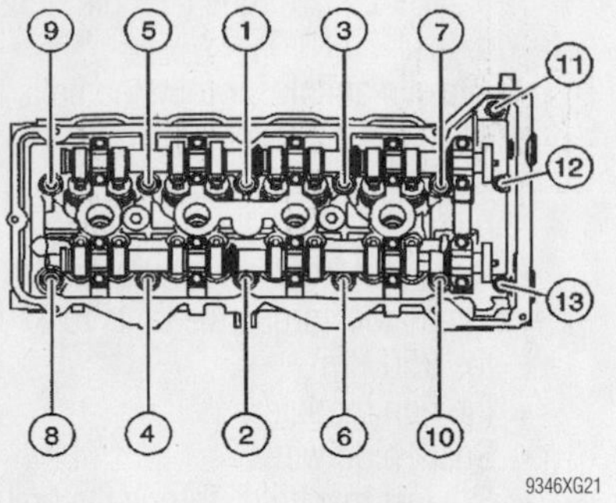

Right cylinder head bolt torque sequence—2001 4.0L models

   - Front and rear transmission braces from the head
   - Cylinder head

To install:

10. Install the cylinder head with a new gasket.
11. Install new M11 head bolts and M6 head bolts.
12. Torque the M11 head bolts in sequence as follows:
   a. Step 1: 30 ft. lbs. (40 Nm).
   b. Step 2: 70 degree turn.
   c. Step 3: 60 degree turn.
   d. Step 4: 60 degree turn.
13. Torque the M6 head bolts to 106 inch lbs. (12 Nm).
14. Install or connect the following:
   - Front and rear transmission braces to the head. Torque the bolts to 37 ft. lbs. (50 Nm).
   - ECT electrical connector
   - Ground wire to the head. Torque the nut to 13 ft. lbs. (17 Nm).
   - Secondary camshaft drive chain
   - Front cover. Torque the bolts to 18 ft. lbs. (25 Nm).
   - Camshaft cover. Torque the bolts to 89 inch lbs. (10 Nm).
   - Intake manifold. Torque the bolts to 89 inch lbs. (10 Nm).
   - Water crossover. Torque the bolts to 18 ft. lbs. (25 Nm).
   - Exhaust manifold. Torque the bolts to 22 ft. lbs. (30 Nm).

Rocker Arms

REMOVAL & INSTALLATION

➡All valve train components should be kept in the order that they were removed, so that they can be reinstalled in their original position.

3.5L Engine

Refer to the camshaft removal and installation procedure for rocker arm service.

3.8L Engine

LEFT SIDE (FRONT)

1. Before servicing the vehicle, refer to the precautions in the beginning of this section.
2. Remove or disconnect the following:
   - Negative battery cable
   - Spark plug wires
   - Rocker arm cover
   - Rocker arm bolt, pedestal and rocker arm
   - Pushrod

To install:

3. Install or connect the following:
   - Pushrod into the lifter
   - Rocker arm and pedestal. Torque the bolt to 11 ft. lbs. (15 Nm) plus a 90 degree turn.
   - Rocker arm cover with a new gasket
   - Spark plug wires
   - Negative battery cable

RIGHT SIDE (REAR)

1. Before servicing the vehicle, refer to the precautions in the beginning of this section.
2. Remove or disconnect the following:
   - Negative battery cable
   - Accessory drive belt
   - Alternator rear brace
   - Spark plug wires
   - Rocker arm cover
   - Rocker arm bolt, pedestal and rocker arm
   - Pushrod

To install:

3. Install or connect the following:
   - Pushrod into the lifter
   - Rocker arm and pedestal. Torque the bolt to 11 ft. lbs. (15 Nm) plus a 90 degree turn.
   - Rocker arm cover with a new gasket
   - Spark plug wires
   - Alternator rear brace
   - Accessory drive belt
   - Negative battery cable

4.0L Engine

The 4.0L engine is not equipped with rocker arms. The camshaft directly actuates the valves.

Supercharger

REMOVAL & INSTALLATION

3.8L Engine

1. Before servicing the vehicle, refer to the precautions in the beginning of this section.
2. Relieve the fuel system pressure.
3. Remove or disconnect the following:
   - Front cover
   - Drive belt from the supercharger pulley

➡It is not necessary to remove the drive belt from the remainder of the pulleys.

   - Right side spark plug wires from the ignition module
   - Alternator brace
   - Fuel injector electrical connectors
   - Manifold Absolute Pressure (MAP) sensor bracket
   - Fuel rail with the injectors
   - Boost control solenoid
   - Throttle body
   - Supercharger

To install:

4. Install or connect the following:
   - Supercharger with a new gasket. Torque the bolts to 17 ft. lbs. (23 Nm).
   - Throttle body to the supercharger. Torque the nuts to 88 inch lbs. (10 Nm).
   - Boost control solenoid
   - Fuel rail
   - MAP sensor bracket
   - Alternator brace
   - Right side spark plug wires
   - Supercharger belt
   - Front cover
   - Negative battery cable
5. Start the engine and check for proper operation.

Intake Manifold

REMOVAL & INSTALLATION

3.5L Engine

1. Before servicing the vehicle, refer to the precautions in the beginning of this section.
2. Partially drain the engine coolant.
3. Relieve the fuel system pressure.
4. Remove or disconnect the following:
   - Air duct from the throttle body
   - Fuel injector sight shield
   - Accelerator cable and bracket
   - Cruise control cable and bracket
   - Coolant hoses from the throttle body
   - Fuel lines from the fuel supply rail
   - Fuel vapor line from the Evaporative Emission (EVAP) canister purge solenoid
   - Brake booster vacuum hose
   - Air conditioning vacuum hose from the engine
   - Surge tank inlet pipe retainer from the fuel supply rail
   - Fuel injector electrical connectors
   - Throttle Position Sensor (TPS) electrical connector
   - Idle Air Control (IAC) valve electrical connector
   - Evaporative Emission (EVAP) canister purge solenoid connector
   - Manifold Absolute Pressure (MAP) sensor connector
   - Wiring harness from the camshaft covers
   - Vacuum hose from the fuel pressure regulator and throttle body
   - Positive Crankcase Ventilation (PCV) tubes from both camshaft covers and the intake manifold
   - Exhaust Gas Recirculation (EGR) valve outlet pipe
   - Fuel supply rail with injectors

➡Disengage the snap-lock retainers by pushing toward the camshaft covers and lifting.

   - Intake manifold

➡The manifold-to-cylinder head seals are reusable unless cut or damaged.

To install:

5. Install or connect the following:
   - Intake manifold with new gaskets (if necessary). Torque the bolts in a circular pattern, starting from the center to 62 inch lbs. (7 Nm).
   - New O-rings on the fuel injectors
   - Fuel supply rail with the injectors
   - EGR pipe. Torque the intake manifold bolts to 89 inch lbs. (10 Nm) and the coolant crossover bolt to 18 ft. lbs. (24 Nm).

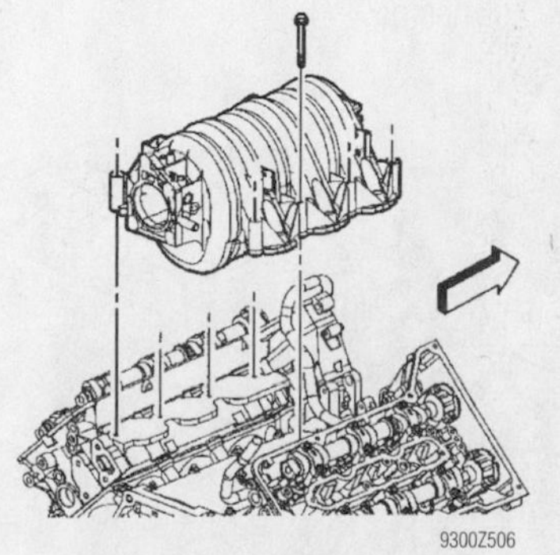

Intake manifold assembly—3.5L engine

- PCV valve and feed tubes
- Fuel pressure regulator vacuum hose
- Engine wiring harness to the camshaft covers. Torque the bolts to 89 inch lbs. (10 Nm).
- TPS electrical connector
- IAC valve electrical connector
- EVAP solenoid electrical connector
- MAP sensor electrical connector
- Fuel injector electrical connectors
- Surge tank pipe retainer to the fuel supply rail
- A/C vacuum hose
- Brake booster vacuum hose
- Vapor line to the EVAP canister purge solenoid
- Fuel lines to the fuel supply rail
- Coolant hoses to the throttle body
- Cruise control cable and bracket
- Accelerator cable and bracket
- Fuel injector sight shield
- Air duct to the throttle body
- Negative battery cable

6. Refill the cooling system.
7. Start the engine and check for leaks.

3.8L Engine

1. Before servicing the vehicle, refer to the precautions in the beginning of this section.
2. Relieve the fuel system pressure.
3. Drain the cooling system.
4. Remove or disconnect the following:
   - Supercharger
   - Thermostat housing
   - Exhaust Gas Recirculation (EGR) tube at the intake manifold
   - Temperature sensor electrical connector
   - Intake manifold

To install:

5. Install or connect the following:
   - Intake manifold with new gaskets. Torque the bolts in sequence to 11 ft. lbs. (15 Nm).
   - Temperature sensor electrical connector
   - EGR tube
   - Thermostat housing. Torque the bolts to 89 inch lbs. (10 Nm).
   - Supercharger
   - Negative battery cable
6. Refill the cooling system.
7. Start the vehicle and check for proper operation.

4.0L Engine

1998—99

1. Before servicing the vehicle, refer to the precautions in the beginning of this section.
2. Relieve the fuel system pressure.
3. Remove or disconnect the following:
   - Fuel injector sight shield
   - Air intake duct
   - Transmission vent hose
   - Range selector cable
   - Throttle Position Sensor (TPS) electrical connector
   - Idle Air Control valve (IAC) electrical connector
   - Accelerator cable
   - Cruise control cable
   - Front spark plug wires
   - Throttle body coolant hoses
   - Exhaust Gas Recirculation (EGR) pipe
   - Crankcase ventilation pipe
   - Brake booster vacuum hose
   - Fuel rail ground wire
   - Fuel lines at the fuel rail
   - Positive Crankcase Ventilation (PCV) hose
   - Fuel injector harness main connector
   - Intake manifold

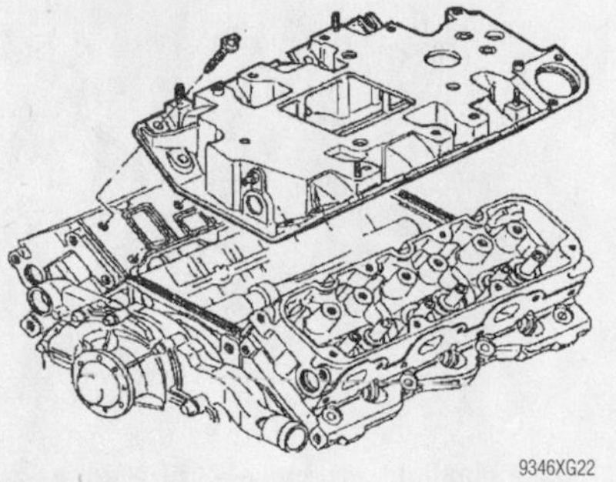

9346XG22

Intake manifold assembly—3.8L (VIN 1 and K) engines

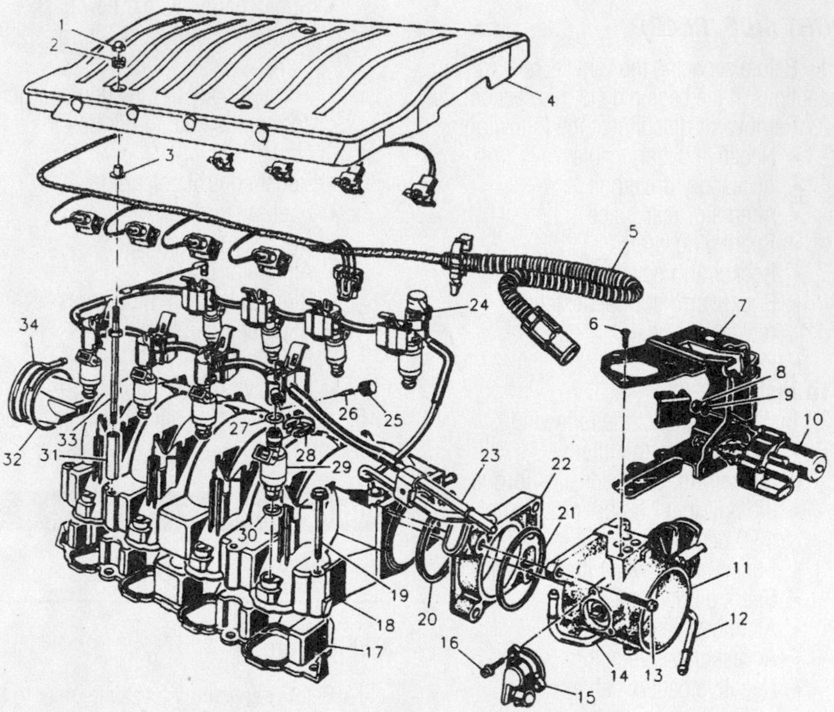

1 NUT, ATTACHING COVER
2 GROMMET, COVER
3 WASHER, COVER GROMMET
4 COVER, INTAKE MANIFOLD
5 WIRING HARNESS ASM
6 SCREW, ISC BRACKET ASM ATTACHING
7 BRACKET ASM, IDLE SPEED CONTROL (ISC)
8 NUT, HEX
9 WASHER, LOCK
10 ACTUATOR ASM, IDLE SPEED CONTROL (ISC)
11 BODY ASM, THROTTLE
12 TUBE, COOLANT OUTLET
13 BOLT, THROTTLE BODY ATTACHING
14 TUBE, COOLANT INLET
15 SENSOR, THROTTLE POSITION (TP)
16 SCREW ASM, TP SENSOR ATTACHING
17 GASKET, INTAKE MANIFOLD
18 HOUSING ASM, INTAKE MANIFOLD
19 BOLT, INTAKE MANIFOLD ATTACHING
20 SEAL, EGR TRANSFER SPACER TO INTAKE MANIFOLD
21 SEAL, THROTTLE BODY TO EGR TRANSFER SPACER
22 SPACER, EGR TRANSFER
23 RAIL ASM, FUEL
24 CARTRIDGE REGULATOR ASM, FUEL PRESSURE
25 CAP, FUEL PRESSURE CONNECTION
26 CORE ASM, FUEL PRESSURE CONNECTION VALVE
27 O-RING, MFI FUEL INJECTOR UPPER
28 CLIP, MFI FUEL INJECTOR RETAINER
29 INJECTOR ASM, MFI FUEL
30 O-RING, MFI FUEL INJECTOR LOWER
31 SPACER, INTAKE MANIFOLD
32 O-RING, PRESSURE RELIEF VALVE
33 STUD, INTAKE MANIFOLD/COVER ATTACHING
34 VALVE ASM, PRESSURE RELIEF

7922XG11

Exploded view of the intake manifold and related components—4.0L engine

To install:

4. Install or connect the following:
   - Intake manifold with a new gasket. Torque the bolts to 89 inch lbs. (10 Nm).
   - Fuel injector harness main connector
   - PCV hose
   - Fuel lines to the fuel rail
   - Fuel rail ground wire
   - Brake booster vacuum hose
   - EGR pipe. Torque the bolt to 21 ft. lbs. (28 Nm).
   - Crankcase ventilation pipe
   - Throttle body coolant hoses
   - Front spark plug wires
   - Accelerator cable
   - Cruise control cable
   - TPS electrical connector
   - IAC electrical connector
   - Transmission vent hose
   - Range selector cable
   - Air intake duct
   - Negative battery cable
5. Pressurize the fuel system and check for leaks.
6. Install the fuel injector sight shield.

2001

1. Before servicing the vehicle, refer to the precautions in the beginning of this section.
2. Relieve the fuel system pressure.
3. Remove or disconnect the following:
   - Fuel injector sight shield
   - Ignition module connectors
   - Positive Crankcase Ventilation (PCV) valve
   - PCV fresh air tube
   - Fuel regulator vacuum tube
   - Secondary Air Injection (AIR) solenoid vacuum tubes
   - Fuel lines at the fuel rail
   - Engine coolant heater wire
   - Fuel injector electrical connectors
   - Fuel rail with injectors
   - Plenum duct clamp
   - Intake manifold

To install:

4. Install or connect the following:
   - Intake manifold with a new gasket. Torque the bolts to 89 inch lbs. (10 Nm) and the plenum duct clamp to 20 inch lbs. (2.25 Nm).
   - Fuel rail with the injectors. Torque the bolts to 35 inch lbs. (4 Nm).
   - Fuel injector electrical connectors
   - Engine coolant heater wire
   - Fuel lines to the fuel rail
   - AIR solenoid vacuum tubes
   - Fuel regulator vacuum tube
   - PCV fresh air tube
   - PCV valve
   - Ignition module connectors
5. Pressurize the fuel system and check for leaks.
6. Install the fuel injector sight shield. Torque the bolts to 27 inch lbs. (3 Nm).

Exhaust Manifold

REMOVAL & INSTALLATION

3.5L Engine

LEFT

1. Before servicing the vehicle, refer to the precautions in the beginning of this section.
2. Drain the cooling system.
3. Remove or disconnect the following:
   - Negative battery cable
   - Fuel injector sight shield
   - Engine mount strut and bracket
   - Cooling fans
   - Radiator hoses
   - Transmission fluid lines from the radiator
   - Radiator
   - Alternator
   - Heat shield from the manifold
   - Oil level indicator tube
   - Secondary Air Injection (AIR) control valve assembly from the engine mount strut bracket
   - Exhaust manifold bolts
   - Exhaust manifold-to-crossover pipe studs
   - Exhaust manifold

To install:

4. Install or connect the following:
   - Exhaust manifold with a new gasket. Torque the bolts to 18 ft. lbs. (25 Nm).
   - AIR valve. Torque the pipe nut to 44 ft. lbs. (60 Nm) and the bolt to 80 inch lbs. (9 Nm).
   - Oil level indicator tube
   - Heat shield
   - Alternator. Torque the bolts to 37 ft. lbs. (50 Nm).
   - Radiator
   - Transmission fluid lines
   - Radiator hoses
   - Cooling fans
   - Engine mount strut
   - Fuel injector sight shield
   - Negative battery cable
5. Fill the cooling system.

RIGHT

1. Before servicing the vehicle, refer to the precautions in the beginning of this section.
2. Relieve the fuel system pressure.
3. Drain the cooling system.
4. Remove or disconnect the following:
   - Engine assembly
   - Crossover pipe from the front manifold
   - Exhaust Gas Recirculation (EGR) pipe from the crossover pipe
   - Right exhaust manifold from the engine

To install:

5. Install or connect the following:
   - Right exhaust manifold with a new gasket. Torque the bolts to 18 ft. lbs. (25 Nm).
   - Crossover pipe to the front exhaust manifold. Torque the bolts to 18 ft. lbs. (25 Nm).
   - EGR pipe to the crossover pipe. Torque the pipe nut to 44 ft. lbs. (60 Nm).
   - Engine assembly
6. Refill the cooling system.
7. Start the engine and check for leaks.

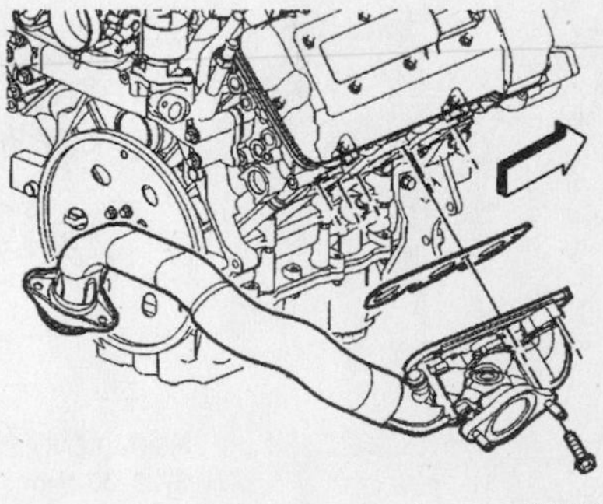

Exploded view of the right exhaust manifold—3.5L engine

3.8L Engines

LEFT SIDE (FRONT)

1. Before servicing the vehicle, refer to the precautions in the beginning of this section.
2. Remove or disconnect the following:
   - Negative battery cable
   - Exhaust manifold-to-crossover pipe bolts
   - Spark plug wires

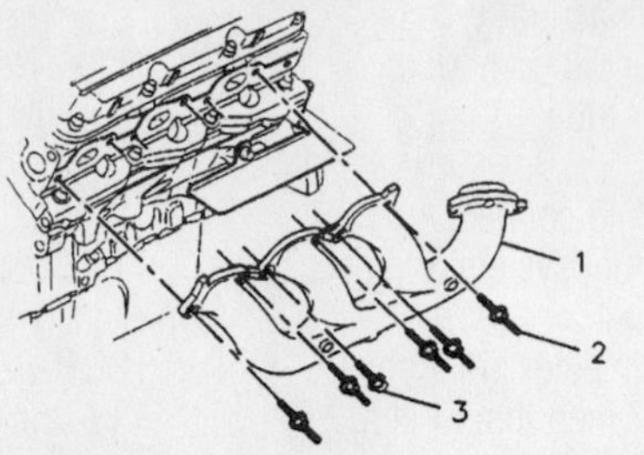

1 LEFT (FRONT) EXHAUST MANIFOLD
2 STUD 30 N•m (22 LB. FT.)
3 BOLT 30 N•m (22 LB. FT.)

7922XG30

Exploded view of the left exhaust manifold mounting—3.8L (VIN 1) engines

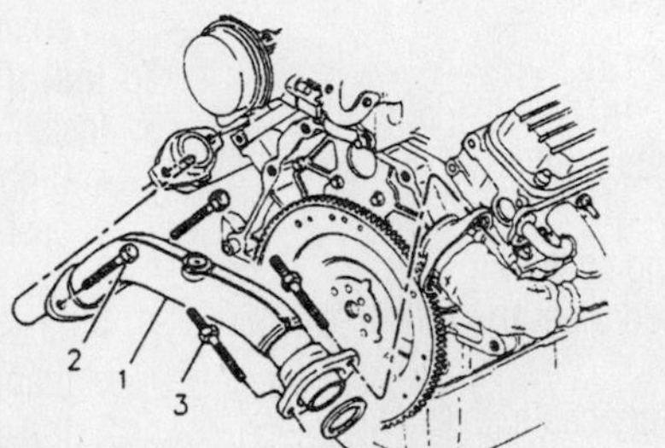

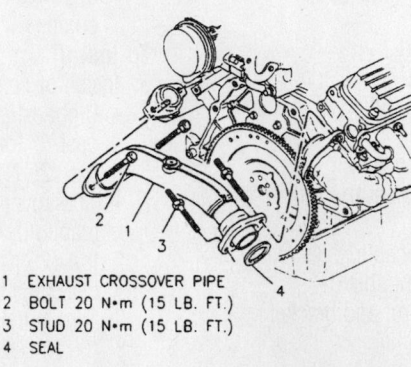

1 EXHAUST CROSSOVER PIPE
2 BOLT 20 N•m (15 LB. FT.)
3 STUD 20 N•m (15 LB. FT.)
4 SEAL

7922XG31

Exploded view of the crossover pipe mounting—3.8L (VIN 1) engines

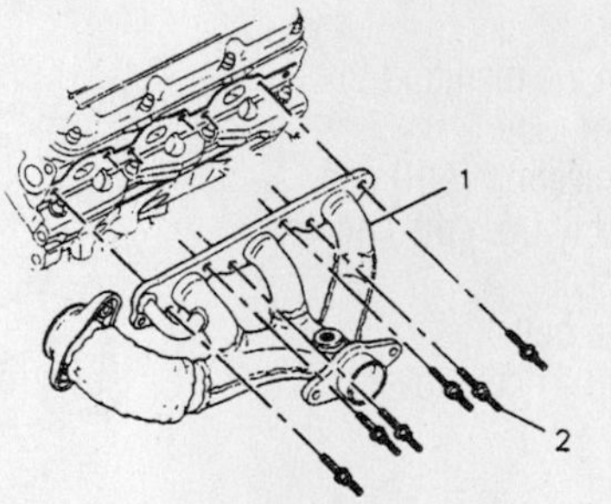

1 RIGHT (REAR) EXHAUST MANIFOLD
2 STUD 30 N•m (22 LB. FT.)

7922XG29

Exploded view of the right exhaust manifold mounting—3.8L (VIN 1) engines

- Oil level indicator tube
- Exhaust manifold

To install:

3. Install or connect the following:
   - Exhaust manifold with a new gasket. Torque the bolts to 22 ft. lbs. (30 Nm).
   - Oil level indicator tube
   - Spark plug wires
   - Exhaust manifold-to-crossover pipe. Torque the bolts to 15 ft. lbs. (20 Nm).
   - Negative battery cable
4. Start the vehicle and check for leaks.

RIGHT SIDE (REAR)

1. Before servicing the vehicle, refer to the precautions in the beginning of this section.
2. Remove or disconnect the following:
   - Negative battery cable
   - Spark plug wires
   - Transmission level indicator tube
   - Oxygen Sensor (O_2S) electrical connector
   - Exhaust manifold-to-crossover pipe bolts
   - Exhaust pipe from the manifold
   - Rear engine lift bracket
   - Exhaust manifold

To install:

3. Install or connect the following:
   - Exhaust manifold with a new gasket. Torque the bolts to 22 ft. lbs. (30 Nm).
   - Rear engine lift bracket
   - Exhaust pipe to the manifold. Torque the bolts to 15 ft. lbs. (20 Nm).
   - Right exhaust manifold-to-crossover pipe. Torque the bolts to 15 ft. lbs. (20 Nm).
   - Transmission level indicator tube
   - O_2S sensor electrical connector
   - Spark plug wires
   - Negative battery cable
4. Start the vehicle and check for leaks.

4.0L Engine

LEFT SIDE (FRONT)

1. Before servicing the vehicle, refer to the precautions in the beginning of this section.
2. Remove or disconnect the following:
   - Negative battery cable
   - Accessory drive belt
   - Alternator upper mounting bolt
   - Right inner fender well splash shield
   - Lower center air deflector
   - Alternator rear bracket
   - Alternator
   - Exhaust manifold-to-exhaust crossover pipe bolts
   - Oxygen Sensor (O_2S) electrical connector
   - Power steering line retainer bolts
   - Exhaust manifold

To install:

3. Install or connect the following:
   - Exhaust manifold with a new gasket. Torque the nuts starting in the center and working outward to 18 ft. lbs. (24 Nm).
   - O_2S electrical connector
   - Power steering line retainers. Torque the bolts to 10 ft. lbs. (14 Nm).
   - Alternator
   - Crossover pipe bolts. Torque the bolts to 37 ft. lbs. (50 Nm).
   - Alternator rear bracket
   - Lower center air deflector
   - Right inner fender well splash shield
   - Upper alternator bolt
   - Accessory drive belt
   - Negative battery cable
4. Start the vehicle and check for leaks.

RIGHT SIDE (REAR)

1. Before servicing the vehicle, refer to the precautions in the beginning of this section.
2. Remove or disconnect the following:

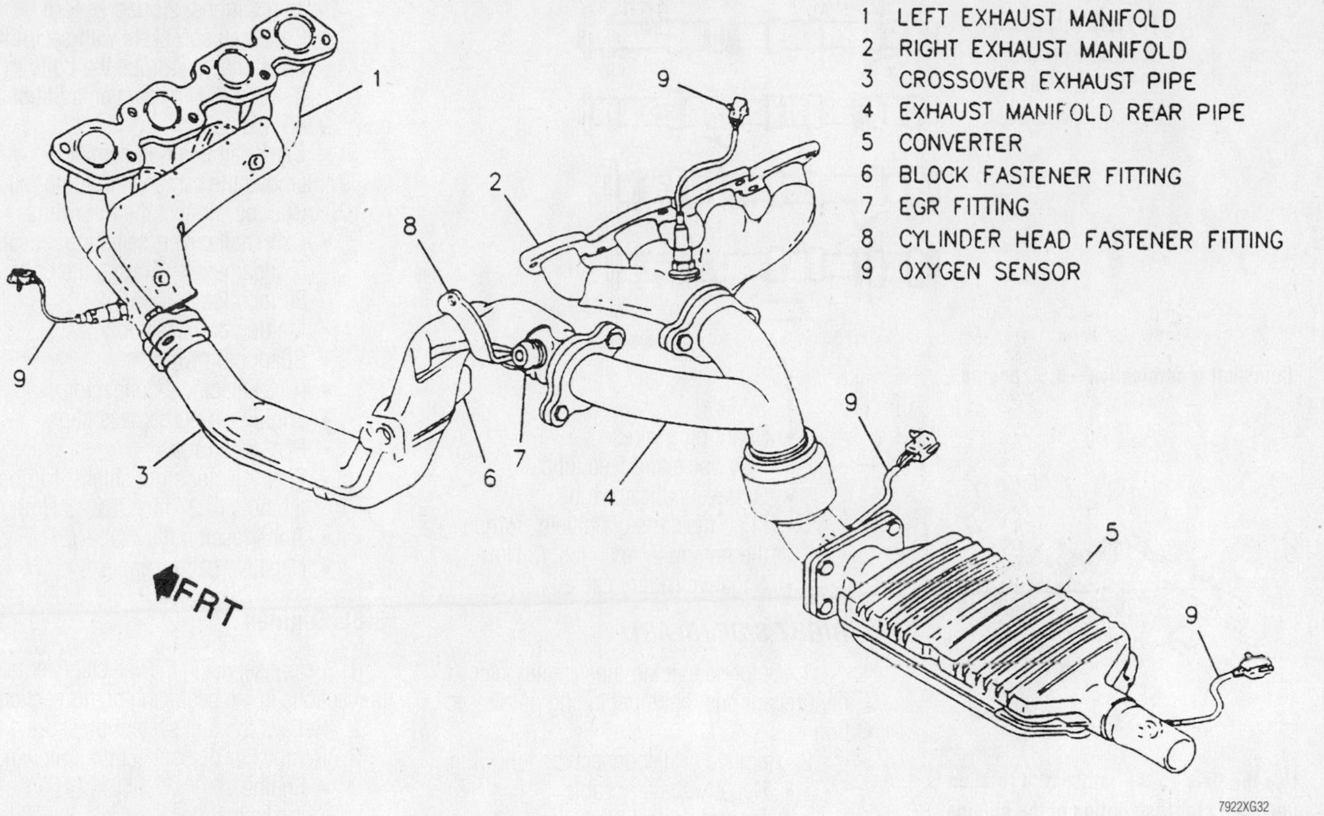

Exhaust system component identification—4.0L engine

- Negative battery cable
- Exhaust manifold rear pipe
- Crossover pipe
- Oxygen Sensor (O_2) electrical connector
- Exhaust manifold

To install:

3. Install or connect the following:
   - Exhaust manifold with a new gasket. Torque the nuts starting in the center and working outward, to 18 ft. lbs. (24 Nm).
   - O_2electrical connector
   - Crossover pipe. Torque the bolts to 30 ft. lbs. (40 Nm).
   - Exhaust manifold rear pipe. Torque the nuts to 18 ft. lbs. (25 Nm).
   - Negative battery cable
4. Start the vehicle and check for leaks.

Camshaft and Valve Lifters

REMOVAL & INSTALLATION

➡All valve train components should be kept in the order that they were removed, so that they can be reinstalled in their original position.

3.5L Engine

LEFT SIDE (FRONT)

1. Before servicing the vehicle, refer to the precautions in the beginning of this section.
2. Disconnect the negative battery cable.
3. Remove or disconnect the following:
   - Fuel injector sight shield
   - Oil level indicator tube
   - Positive Crankcase ventilation (PCV) valve
   - Spark plug wires
   - Ignition coil assembly
   - Fuel injector wiring harness
   - Engine wiring harness
   - Camshaft cover
4. Install the camshaft holding fixture (J 42038) on the camshafts.
5. Remove or disconnect the following:
   - Camshaft sprockets with the secondary drive chains
   - Camshaft bearing caps
   - Camshaft holding fixture
   - Camshafts
   - Rocker arms
   - Lifters

To install:

6. Coat the lifters with clean engine oil and place them into position in the engine.
7. Coat the rocker arms with clean engine oil and set them in place in the head.
8. Coat the camshafts with clean engine oil and place them in their proper position in the head.
9. Install or connect the following:
   - Camshaft holding fixture
   - Camshaft bearing caps. Torque the

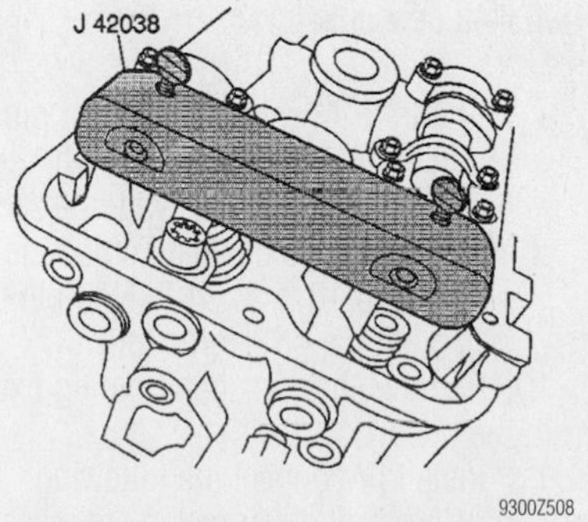

Camshaft holding fixture J-42038 installed on the camshafts—3.5L engine

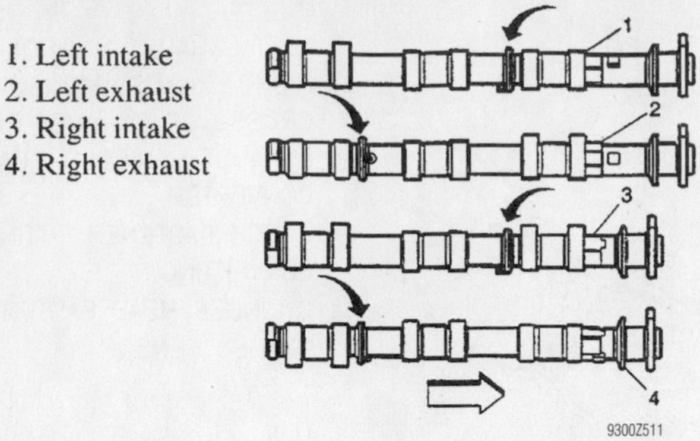

Camshaft identification—3.5L engine

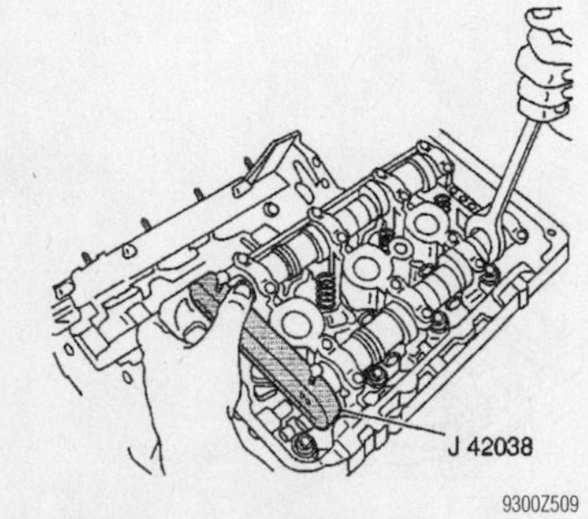

Use the flats on the camshaft if rotation is necessary for installation of the holding tool—3.5L engine

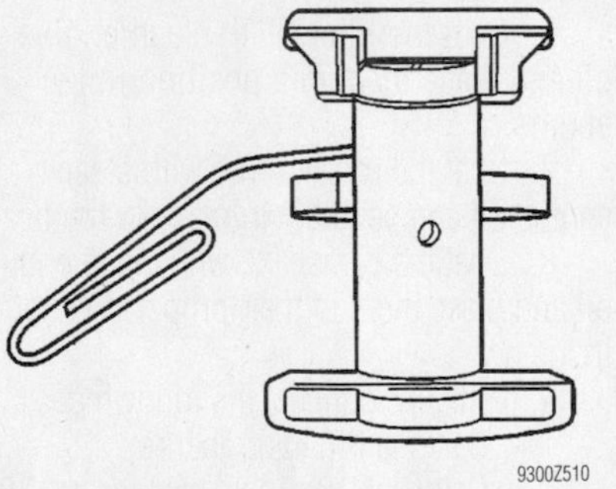

Before installation, compress the tensioner and lock it in place with a piece of wire—3.5L engine

bolts to 71 inch lbs. (8 Nm) plus an additional 22 degree turn.
- Camshaft sprockets with the secondary drive chains. Torque the bolts to 18 ft. lbs. (25 Nm) plus an additional 45 degree turn.

10. Remove the camshaft holding fixture from the head.
11. Install or connect the following:
- Camshaft cover (seals are reusable if they are not damaged). Torque the bolts to 80 inch lbs. (9 Nm).
- Engine wiring harness
- Fuel injector wiring harness
- Ignition coil assembly
- Spark plug wires
- PCV valve and feed tube
- Oil level indicator tube
- Fuel injector sight shield. Torque the nuts to 27 inch lbs. (3 Nm).
- Negative battery cable

RIGHT SIDE (REAR)

1. Before servicing the vehicle, refer to the precautions in the beginning of this section.
2. Remove or disconnect the following:
- Negative battery cable
- Transmission filler tube
- Fuel injector sight shield
- Positive Crankcase Ventilation (PCV) feed tube
- Engine wiring harness clips
- Oxygen (O_2S) sensor electrical connector
- Spark plug wires
- Ignition coil assembly
- Camshaft cover

3. Install the camshaft holding fixture (tool J-42038).
4. Remove or disconnect the following:
- Camshaft position sensor
- camshaft sprockets with the secondary drive chains
- Camshaft bearing caps
- Camshaft holding fixture
- Camshafts
- Rocker arms
- Lifters

To install:

5. Coat the lifters with clean engine oil and place them into position in the engine.
6. Coat the rocker arms with clean engine oil and set them in place in the head.
7. Coat the camshafts with clean engine oil and place them in position in the head.
8. Install or connect the following:
- Camshaft holding fixture
- Camshaft bearing caps. Torque the bolts to 71 inch lbs. (8 Nm). plus an additional 22 degree turn.
- Camshaft sprockets with secondary drive chains. Torque the bolts to 18 ft. lbs. (25 Nm) plus an additional 45 degree turn.
- Camshaft position sensor

9. Remove the camshaft holding fixture.
10. Install or connect the following:
- Camshaft cover (seals are reusable if undamaged). Torque the bolts to 80 inch lbs. (9 Nm).
- Ignition coil assembly
- Spark plug wires
- (O_2S) electrical connector
- Engine wiring harness clips
- PCV feed tube
- Fuel injector sight shield. Torque the nuts to 27 inch lbs. (3 Nm).
- Transmission filler tube
- Negative battery cable

3.8L Engines

1. Before servicing the vehicle, refer to the precautions in the beginning of this section.
2. Relieve the fuel system pressure.
3. Remove or disconnect the following:
- Engine
- Intake manifold
- Rocker arm covers
- Rocker arm assemblies and pushrods
- Lifter guide retainer
- Lifter guides
- Lifters
- Crankshaft balancer by pressing it from the crankshaft
- Crankshaft Position (CKP) sensor cover and sensor
- Front cover

WARNING

Align the camshaft and crankshaft sprocket timing marks to avoid burring the camshaft journals by the crankshaft.

- Camshaft sprocket and timing chain
- Camshaft thrust plate
- Camshaft

To install:

4. Coat the camshaft lobes and bearings with a suitable assembly lubricant prior to installation.
5. Install or connect the following:
- Camshaft
- Camshaft thrust plate. Torque the bolts to 11 ft. lbs. (15 Nm).
- Camshaft sprocket and timing chain. Torque the sprocket bolt to 74 ft. lbs. (100 Nm) plus an additional 90 degree turn.

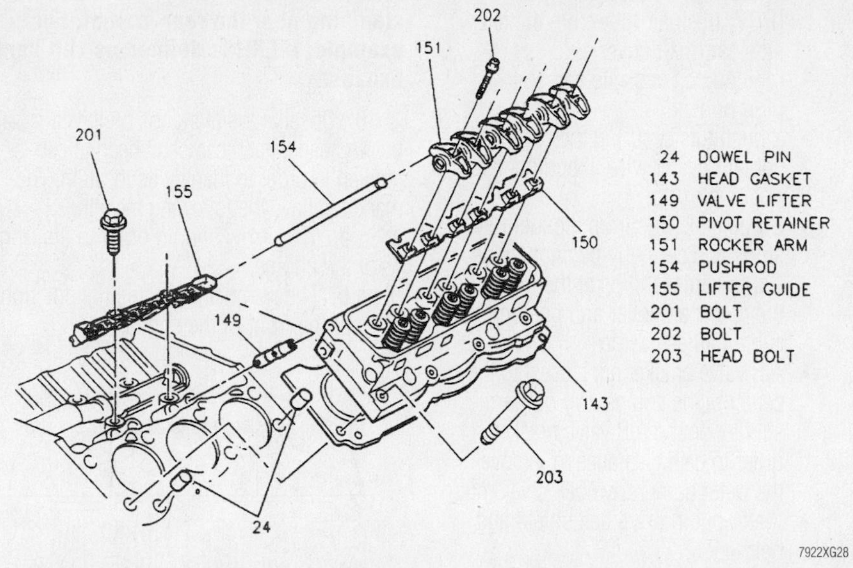

Exploded view of the cylinder head and valve train components—3.8L (VIN 1 and K) engines

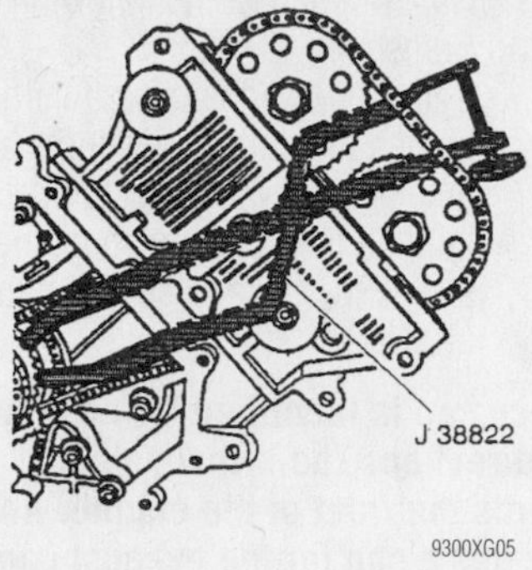

Camshaft chain holding tool J-38822—4.0L engine

- Front cover. Torque the bolts to 11 ft. lbs. (15 Nm) plus an additional 40 degree turn.
- CKP sensor and cover
- Crankshaft balancer. Torque the bolt to 111 ft. lbs. (150 Nm); then, an additional 76 degree turn.

⁂ WARNING

If the camshaft was replaced the lifters must also be replaced. The old lifters have developed a wear pattern and will cause the new camshaft to wear prematurely.

- Valve lifters, lubricated with clean engine oil
- Lifter guides and retainer. Torque the bolts to 27 ft. lbs. (37 Nm).
- Pushrods and rocker arms. Torque the bolts to 28 ft. lbs. (38 Nm).
- Rocker arm covers. Torque the bolts to 89 inch lbs. (10 Nm).
- Intake manifold. Torque the bolts to 89 inch lbs. (10 Nm).
- Engine
- Negative battery cable

6. Start the engine and check for leaks.

4.0L Engine

LEFT SIDE (FRONT)—1998–99 MODELS

1. Before servicing the vehicle, refer to the precautions in the beginning of this section.
2. Drain the cooling system.
3. Remove or disconnect the following:
   - Negative battery cable
   - Fuel injector sight shield
   - Oil level indicator tube
   - Upper radiator hose from the thermostat housing
   - Spark plug wires
   - Upper radiator support assembly
   - Positive Crankcase Ventilation (PCV) fresh air tube from the left side camshaft cover
   - Air inlet duct
   - Exhaust Gas Recirculation (EGR) outlet pipe
   - Water pump drive belt cover
   - Water pump drive belt and drive belt tensioner
   - Water pump pulley
   - Camshaft seal retainer and seal
   - Camshaft cover
4. Secure the camshaft sprocket to the timing chain by installing tie-wraps through the camshaft sprocket holes. Use 4 tie-wraps per sprocket.

➡**The sprocket and chain relationship must be maintained throughout this procedure or camshaft timing will be lost and require further engine disassembly to retime.**

5. Working behind the sprockets, install a chain holder so that it is positioned between the chain tensioner and chain guide. Apply tension to the tool by tightening the tension adjusting screw.
6. Remove or disconnect the following:
   - Camshaft sprockets. Note the relative location of the camshaft drive pins.
   - Camshaft bearing caps

➡**Alternately, loosen the cap bolts a few turns at a time, until all valve spring pressure has been released.**

   - Camshaft

To install:

7. Lubricate the camshaft lobes with assembly lubricant, and the camshaft journals with engine oil.

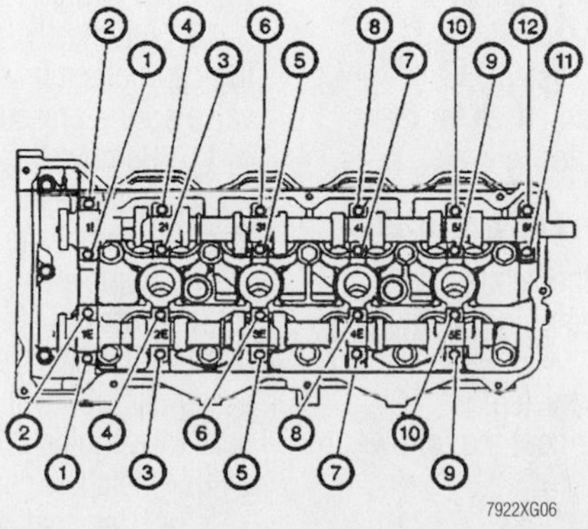

Left cylinder head camshaft bearing cap tightening sequence—4.0L (VIN C) engine

8. Install or connect the following:
   - Camshaft
   - Camshaft bearing caps. Alternately tighten the bolts, a few turns at a time against valve spring pressure until all the bolts are snug; then, torque the bolts to 108 inch lbs. (12 Nm).

➡Each cap is identified for position and direction. The arrow points towards the front of the engine. An "E" indicates a cap for the exhaust cam. An "I" indicates a cap for the intake cam. Position No. 1 is towards the front of the engine.

9. Using the hex cast into the camshaft, rotate the cams until the drive pins are in position to engage the camshaft sprockets over the cams and install the retaining bolts.
10. Install the camshaft sprockets. Torque the bolts to 90 ft. lbs. (120 Nm).
11. Remove the chain holder and the camshaft sprocket tie-wraps.
12. Install or connect the following:
   - Camshaft cover with new gaskets. Torque the bolts to 89 inch lbs. (10 Nm).
   - Camshaft seal retainer. Torque the bolts to 27 inch lbs. (3 Nm).
   - Water pump pulley. Torque the bolts to 106 inch lbs. (12 Nm).
   - Water pump belt tensioner and belt. Torque the bolts to 89 inch lbs. (10 Nm).
   - Water pump belt cover. Torque the bolts to 89 inch lbs. (10 Nm).
   - EGR outlet pipe
   - Air inlet duct
   - Upper radiator support
   - Spark plug wires
   - PCV hose to the camshaft cover
   - Upper radiator hose to the thermostat housing
   - Oil level indicator tube
   - Fuel injector sight shield
   - Negative battery cable
13. Refill the cooling system.
14. Run the engine and check for leaks and proper engine operation.

LEFT SIDE (FRONT)—2001 MODELS

1. Before servicing the vehicle, refer to the precautions in the beginning of this section.
2. Drain the cooling system.
3. Remove or disconnect the following:
   - Negative battery cable
   - Fuel injector sight shield
   - Inlet radiator hose from the water housing crossover
   - Positive Crankcase Ventilation (PCV) fresh air tube from the left side camshaft cover
   - Ignition coil cassette and spark plug boots
   - 2 pushnuts securing the engine coolant heater wire, if equipped and position aside
   - Surge tank pipe from the fuel rail studs and carefully position aside
   - Cable harness clips at the front of the camshaft cover and position the cable harness aside
   - AIR valve bracket nut closest to the center of the engine. Pry outward slightly on the AIR valve bracket in order to gain clearance to remove the water pump drive belt shield nut.
   - Water pump drive belt shield and belt
   - Water pump belt tensioner
   - 3 camshaft seal retainer bolts

➡DO NOT reuse the camshaft seal.

   - Camshaft seal
   - Camshaft cover
4. Rotate the crankshaft to Top dead Center (TDC) of the number 1 cylinders compression stroke, both camshaft sprocket drive pins should be at the top of their rotation.
5. Install the camshaft holding tool J44212 over the camshafts.
   - Camshaft sprocket bolts
6. Install the Secondary Drive Timing Chain/Sprocket Holding Fixture Tool J 44213 onto the front rail of the cylinder head.
7. Remove the secondary camshaft drive chain guide upper bolt access plug.
8. Loosen the secondary camshaft drive chain guide upper bolt ONLY two turns.
9. Slide the intake and exhaust camshaft sprockets onto the pins of the Secondary Drive Timing chain/Sprocket Holding Fixture Tool J 44213.
10. Alternately loosen the camshaft bearing cap bolts a few turns at a time until all valve spring pressure has been released.
11. Remove the camshaft bearing caps.
12. Remove the camshaft holding tool from the camshafts.
13. Remove the camshafts and followers.

To install:

14. Lubricate the camshaft lobes with assembly lubricant, and the camshaft journals with engine oil.
15. Install or connect the following:
   - Camshaft with the camshaft sprocket drive pins near the top of their rotation and the camshaft lobes in a neutral position.

➡The camshafts can be identified by a stamping near the rear journal. For example: L-EXH is defined as Left bank Exhaust.

16. Observe the markings on the camshaft bearing caps. Each camshaft bearing cap is marked in order to identify its location. The markings have the following meanings:
   a. The arrow should point to the front of the engine.
   b. The number indicates the position from the front of the engine.

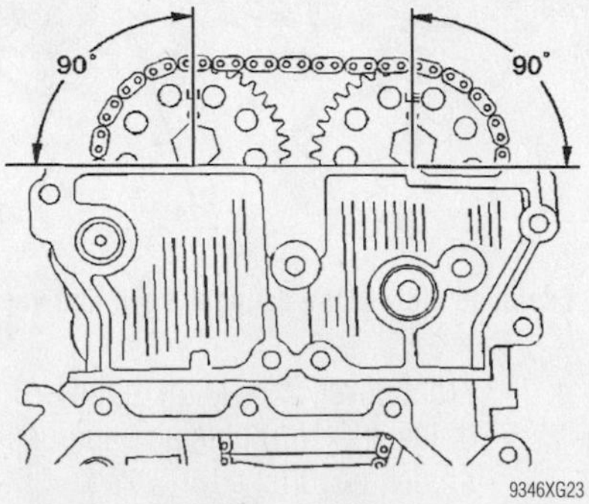

Rotate the crankshaft to Top Dead Center (TDC) of the number 1 cylinders compression stroke, both camshaft sprocket drive pins should be at the top of their rotation—2001 models

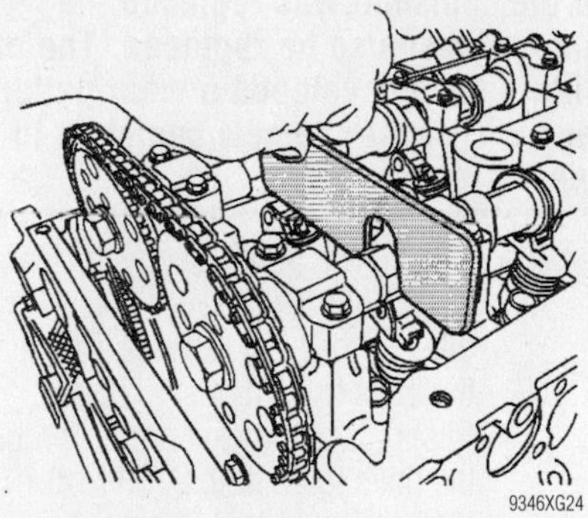

Install the camshaft holding tool J44212 over the camshafts—2001 models

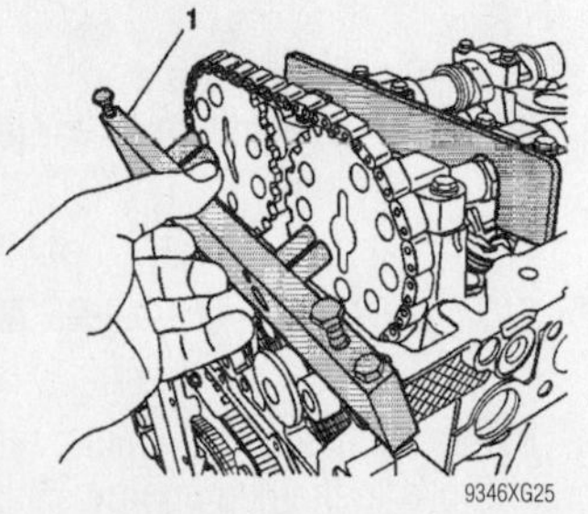

Install Secondary Drive Timing Chain/Sprocket Holding Fixture Tool J 44213 onto the front rail of the cylinder head —2001 models

c. The "E" indicates the exhaust camshaft.

d. The "I" indicates the Intake camshaft.

17. Apply a liberal amount of lubricant GM P/N 12345001 or equivalent to the camshaft bearing caps.

18. Install the camshaft bearing caps and bolts. Tighten in sequence as follows:

a. Alternately hand tighten the camshaft bearing cap bolts a few turns at a time until all caps are fully seated.

b. Tighten camshaft bearing cap bolts to 44 inch lbs. (5 Nm).

c. Tighten camshaft bearing cap bolts an additional 30 degrees.

19. Align the camshafts.

20. Install the camshaft holding fixture.

➡Ensure the camshaft sprockets properly engage the camshaft sprocket drive pins and camshafts.

21. Slide the intake and exhaust camshaft sprockets off the pins of the Secondary Drive Timing chain/Sprocket Holding Fixture Tool J 44213 and onto the pins of the camshafts.

22. Tighten the secondary camshaft drive chain guide upper bolt to 18 ft. lbs. (25 Nm).

23. Install the secondary camshaft drive chain guide upper bolt access plug and tighten to 39 inch lbs. (4.5Nm).

24. Remove the Secondary Drive Timing chain/Sprocket Holding Fixture Tool J 44213 while carefully holding the camshaft sprockets against the camshafts.

25. Install the camshaft sprocket bolts and tighten to 89 ft. lbs. (120 Nm).

26. Verify the camshaft sprocket alignment.

27. Remove the camshaft holding tool.

28. Install or connect the following:

- Camshaft cover and tighten the bolts to 89 inch lbs. (10 Nm)
- Camshaft seal lips lubricated with clean oil
- Camshaft seal retainers coated with sealer GM P/N 1052080 and tighten to 27 inch lbs. (3 Nm)
- Water pump belt tensioner
- Water pump drive belt and shield
- AIR valve bracket nut to 80 inch lbs. (9Nm)
- Cable harness clips at the front of the camshaft cover
- 2 pushnuts securing the engine coolant heater wire, if equipped
- Inlet radiator hose to the water housing crossover
- Fuel injector sight shield
- Surge tank pipe to the fuel rail studs
- Ignition coil cassette and spark plug boots
- PCV fresh air tube to the left side camshaft cover
- Fuel injector sight shield
- Negative battery cable

29. Refill the cooling system.

RIGHT SIDE (REAR)—1998–99 MODELS

1. Before servicing the vehicle, refer to the precautions in the beginning of this section.

2. Remove or disconnect the following:

- Negative battery cable
- Fuel injector sight shield
- Vacuum reservoir
- Cruise control servo
- Ignition assembly electrical connectors
- Spark plug wires
- Ignition coil assembly
- Positive Crankcase Ventilation (PCV) valve from the camshaft cover
- Evaporative Emission (EVAP) canister solenoid from the right camshaft cover
- Knock Sensor (KS) electrical connector
- Vehicle Speed Sensor (VSS) electrical connector
- Power steering pressure switch electrical connector
- Wiring harness retainers from the cover
- Camshaft cover

3. Secure the camshaft sprocket to the timing chain by installing tie-wraps through the camshaft sprocket holes. Use 4 tie-wraps per sprocket.

➡The sprocket and chain relationship must be maintained throughout this procedure or camshaft timing will be lost and require further engine disassembly to retime.

4. Working from behind the sprockets, install a chain holder so that it is positioned between the chain tensioner and chain guide. Apply tension to the tool by tightening the tension adjusting screw.

5. Remove or disconnect the following:

- Camshaft sprockets. Note the relative location of the camshaft drive pins.
- Camshaft bearing caps

➡Alternately loosen the cap bolts, a few turns at a time, until all valve spring pressure has been released.

- Camshaft

To install:

6. Lubricate the camshaft lobes with assembly lubricant. Lubricate the camshaft journals with clean engine oil.

7. Install or connect the following:

- Camshaft
- Camshaft bearing caps. Alternately tighten the bolts, a few turns at a time against valve spring pressure, until all the bolts are snug, then, torque the bolts to 108 inch lbs. (12 Nm).

➡Each cap is identified for position and direction. The arrow points towards the front of the engine. An "E" indicates a cap for the exhaust camshaft. An "I" indicates a cap for the intake camshaft. Position No. 1 is towards the front of the engine.

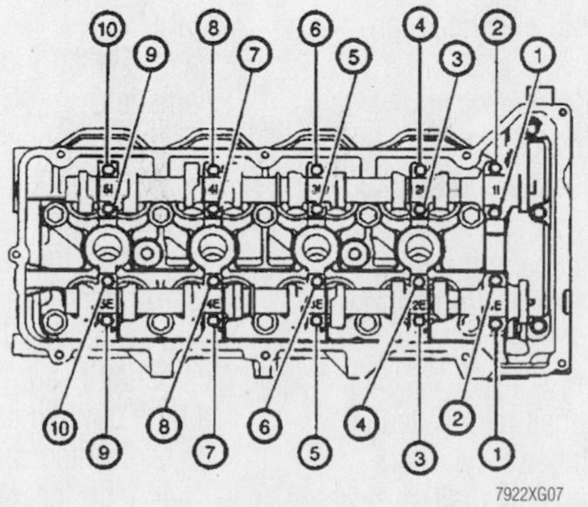

Right cylinder head camshaft bearing cap tightening sequence—4.0L engine

8. Using the hex cast into the camshaft, rotate the camshafts until the drive pins are in position to engage the camshaft sprockets over the camshafts and install the retaining bolts.

9. Install the camshaft sprockets. Torque the bolts to 90 ft. lbs. (120 Nm).

10. Remove the chain holder and the camshaft sprocket tie-wraps.

11. Install or connect the following:
- Camshaft cover with new gaskets. Torque the bolts to 89 inch lbs. (10 Nm).
- Wiring harness retainers to the camshaft cover
- KS, VSS and power steering pressure switch electrical connectors
- EVAP canister solenoid on the right side camshaft cover
- PCV valve
- Ignition coil assembly on the right side cover
- Spark plugs wires
- Electrical connector to the ignition assembly
- Cruise control servo
- Vacuum reservoir
- Fuel injector sight shield
- Negative battery cable

12. Run the engine and check for leaks and proper engine operation.

RIGHT SIDE (REAR)—2000–01 MODELS

1. Before servicing the vehicle, refer to the precautions in the beginning of this section.
2. Drain the cooling system.
3. Remove or disconnect the following:
- Negative battery cable
- Fuel injector sight shield
- Positive Crankcase Ventilation (PCV) valve from the camshaft cover
- Vacuum tubes from the AIR vent solenoid
- AIR vent solenoid electrical connector
- AIR control valve bracket
- Nut securing the AIR tube
- Ignition coil cassette and spark plug boots
- Cable harness clips at the front of the camshaft cover and position the cable harness aside
- Camshaft cover

4. Rotate the crankshaft to Top dead Center (TDC) of the number 1 cylinders compression stroke, both camshaft sprocket drive pins should be at the top of their rotation.

5. Install the camshaft holding tool J44212 over the camshafts.
- Camshaft sprocket bolts

6. Install the Secondary Drive Timing Chain/Sprocket Holding Fixture Tool J 44213 onto the front rail of the cylinder head.

7. Remove the secondary camshaft drive chain guide upper bolt access plug.

8. Loosen the secondary camshaft drive chain guide upper bolt ONLY two turns.

9. Slide the intake and exhaust camshaft sprockets onto the pins of the Secondary Drive Timing chain/Sprocket Holding Fixture Tool J 44213.

10. Alternately loosen the camshaft bearing cap bolts a few turns at a time until all valve spring pressure has been released.

11. Remove the camshaft bearing caps.

12. Remove the camshaft holding tool from the camshafts.

13. Remove the camshafts and followers.

To install:

14. Lubricate the camshaft lobes with assembly lubricant, and the camshaft journals with engine oil.

15. Install or connect the following:
- Camshaft with the camshaft sprocket drive pins near the top of their rotation and the camshaft lobes in a neutral position.

➡The camshafts can be identified by a stamping near the rear journal. For example: L-EXH is defined as Left bank Exhaust.

16. Observe the markings on the camshaft bearing caps. Each camshaft bearing cap is marked in order to identify its location. The markings have the following meanings:
   a. The arrow should point to the front of the engine.
   b. The number indicates the position from the front of the engine.
   c. The "E" indicates the exhaust camshaft.
   d. The "I" indicates the Intake camshaft.

17. Apply a liberal amount of lubricant GM P/N 12345001 or equivalent to the camshaft bearing caps.

18. Install the camshaft bearing caps and bolts. Tighten in sequence as follows:
   a. Alternately hand tighten the camshaft bearing cap bolts a few turns at a time until all caps are fully seated.
   b. Tighten camshaft bearing cap bolts to 44 inch lbs. (5 Nm).
   c. Tighten camshaft bearing cap bolts an additional 30 degrees.

19. Align the camshafts.

20. Install the camshaft holding fixture.

➡Ensure the camshaft sprockets properly engage the camshaft sprocket drive pins and camshafts.

21. Slide the intake and exhaust camshaft sprockets off the pins of the Secondary Drive Timing chain/Sprocket Holding Fixture Tool J 44213 and onto the pins of the camshafts.

22. Tighten the secondary camshaft drive chain guide upper bolt to 18 ft. lbs. (25 Nm).

23. Install the secondary camshaft drive chain guide upper bolt access plug and tighten to 39 inch lbs. (4.5Nm).

24. Remove the Secondary Drive Timing chain/Sprocket Holding Fixture Tool J 44213 while carefully holding the camshaft sprockets against the camshafts.

25. Install the camshaft sprocket bolts and tighten to 89 ft. lbs. (120 Nm).

26. Verify the camshaft sprocket alignment.

27. Remove the camshaft holding tool.

28. Install or connect the following:
- Camshaft cover and tighten the bolts to 89 inch lbs. (10 Nm)
- Cable harness clips at the front of the camshaft cover
- Ignition coil cassette and spark plug boots
- Nut securing the AIR tube
- AIR control valve bracket
- AIR vent solenoid electrical connector
- Vacuum tubes from the AIR vent solenoid
- Positive Crankcase Ventilation (PCV) valve to the camshaft cover
- Fuel injector sight shield
- Negative battery cable

29. Refill the cooling system.

Valve Lash

ADJUSTMENT

The valve lash in these models cannot be adjusted.

Starter Motor

REMOVAL & INSTALLATION

3.5L Engine

1. Before servicing the vehicle, refer to the precautions in the beginning of this section.
2. Remove or disconnect the following:
- Negative battery cable

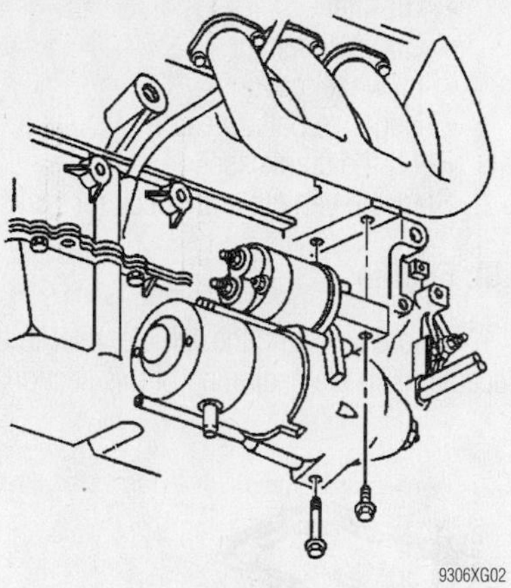

View of the starter—3.5L engine

- Lower front air deflector
- Torque converter cover
- Starter electrical connectors
- Starter motor

To install:

3. Install or connect the following:
   - Starter motor. Torque the bolts to 37 ft. lbs. (50 Nm).
   - Starter electrical connectors. Torque the positive battery terminal nut to 84 inch lbs. (10 Nm) and the "S" terminal nut to 30 inch lbs. (3.5 Nm).
   - Torque converter cover
   - Lower front air deflector
   - Negative battery cable

3.8L Engine

1. Before servicing the vehicle, refer to the precautions in the beginning of this section.
2. Remove or disconnect the following:
   - Negative battery cable
   - Splash shield
   - Flexplate cover
   - Starter electrical connectors
   - Starter motor

To install:

3. Install or connect the following:
   - Starter motor. Torque the bolts to 32 ft. lbs. (43 Nm).
   - Starter electrical connectors. Torque the positive battery terminal nut to 12 ft. lbs. (16 Nm) and the "S" terminal nut to 22 inch lbs. (2.5 Nm).
   - Flexplate cover. Torque the bolts to 22 inch lbs. (7 Nm).
   - Splash shield
   - Negative battery cable

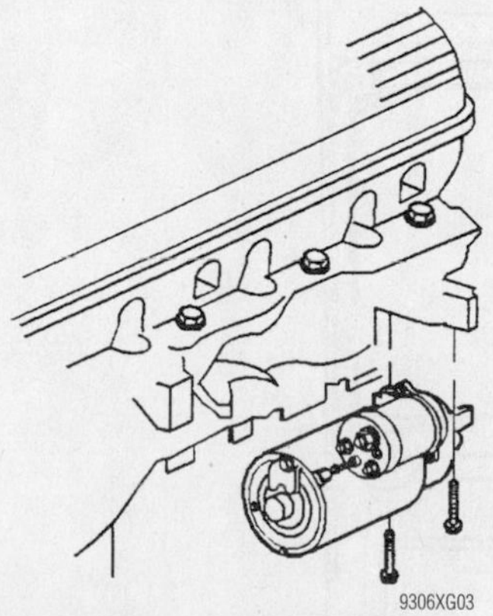

View of the starter—3.8L engine

4.0L Engine

1. Before servicing the vehicle, refer to the precautions in the beginning of this section.
2. Remove or disconnect the following:
   - Negative battery cable
   - Intake manifold
   - Starter electrical connectors
   - Starter motor

To install:

✻✻ WARNING

Before installing the starter, torque the inner solenoid and battery terminal nuts to 70 inch lbs. (8 Nm). If not properly tightened, the starter may fail, due to terminal or cap damage.

3. Install or connect the following:
   - Starter motor. Torque the bolts to 22 ft. lbs. (30 Nm).
   - Starter electrical connectors. Torque the positive battery terminal nut to 70 inch lbs. (8 Nm) and the "S" terminal nut to 26 inch lbs. (3 Nm).
   - Intake manifold
   - Negative battery cable

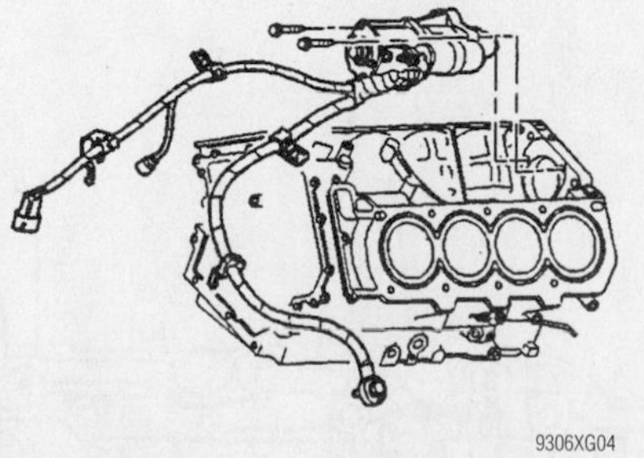

Exploded view of the starter—4.0L engine

Oil Pan

REMOVAL & INSTALLATION

3.5L Engine

1. Before servicing the vehicle, refer to the precautions in the beginning of this section.
2. Drain the crankcase.
3. Remove or disconnect the following:
   - Oil filter
   - Oil level sensor
   - Transmission brace
   - Oil pan

To install:

4. Install or connect the following:
   - Oil pan with a new gasket. Do not tighten the bolts at this time.
   - Transmission brace on the engine block only. Torque the bolts to 18 ft. lbs. (25 Nm).
   - Oil level sensor. Torque the bolt to 80 inch lbs. (9 Nm).
   - Brace-to-oil pan bolts, loosely install them
5. Align the rear of the oil pan flush with the rear of the engine block. Use a straight edge for reference.
6. Press the front of the oil pan against the transmission brace; then torque the brace-to-oil pan bolts to 18 ft. lbs. (25 Nm). Be sure to keep the rear of the pan flush with the rear of the engine.
7. Torque the oil pan bolts to 18 ft. lbs. (25 Nm).
8. Install or connect the following:
   - Brace-to-transmission bolts. Torque them to 32 ft. lbs. (43 Nm).
   - Oil level sensor electrical connector
   - Drain plug. Torque it to 15 ft. lbs. (20 m).

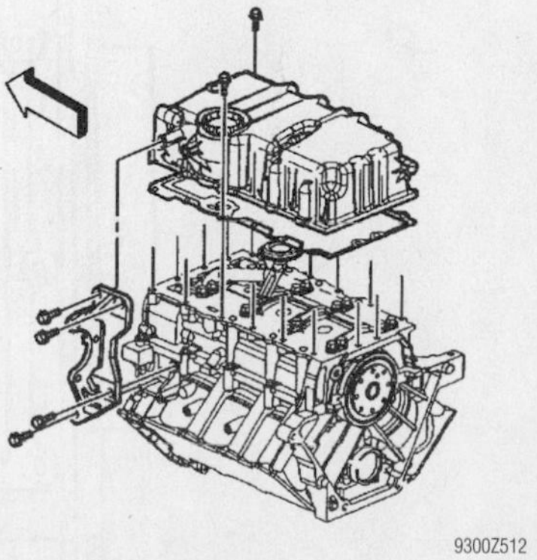

Exploded view of the oil pan—3.5L engine

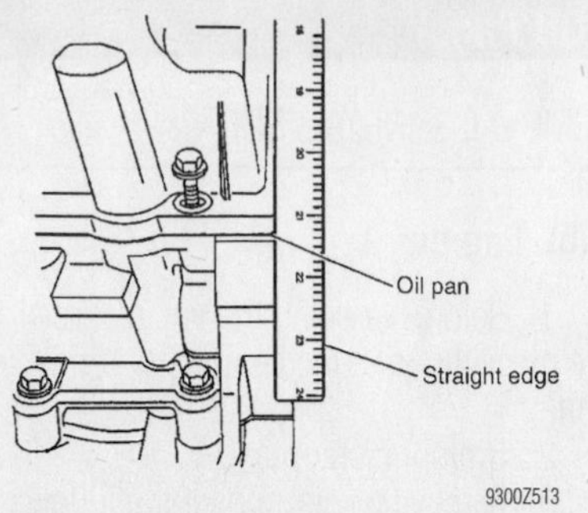

Use a straight edge to align the rear of the oil pan to the rear of the engine—3.5L engine

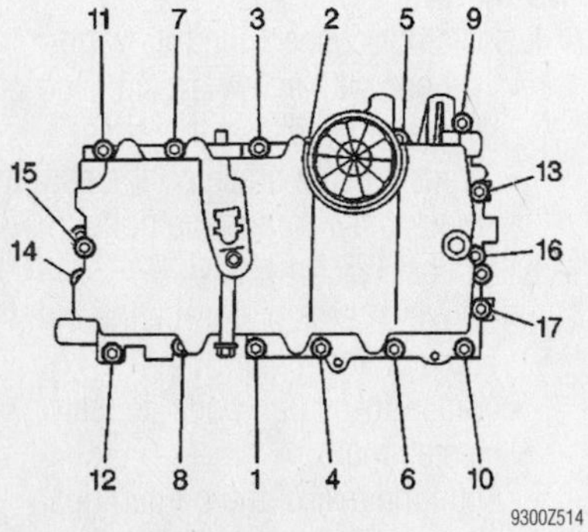

Oil pan mounting bolt tightening sequence—3.5L engine

- New oil filter. Torque the cap to 18 ft. lbs. (25 Nm).

9. Refill the crankcase.
10. Start the engine and inspect for leaks.

3.8L Engines

1. Before servicing the vehicle, refer to the precautions in the beginning of this section.
2. Disconnect the negative battery cable.
3. Drain the crankcase.
4. Remove or disconnect the following:
   - Flywheel cover
   - Oil level sensor
   - Oil filter
   - Oil pan

To install:

5. Install or connect the following:
   - Oil pan with a new gasket. Torque the bolts to 10 ft. lbs. (14 Nm).
   - Oil filter
   - Oil level sensor
   - Flywheel cover
   - Negative battery cable
6. Refill the crankcase.
7. Start the vehicle and check for leaks.

4.0L Engine

1. Before servicing the vehicle, refer to the precautions in the beginning of this section.

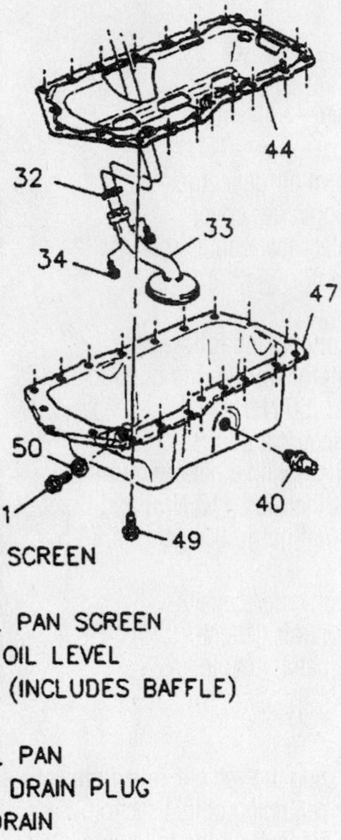

Exploded view of the oil pan mounting and related components—3.8L (VIN 1) engines

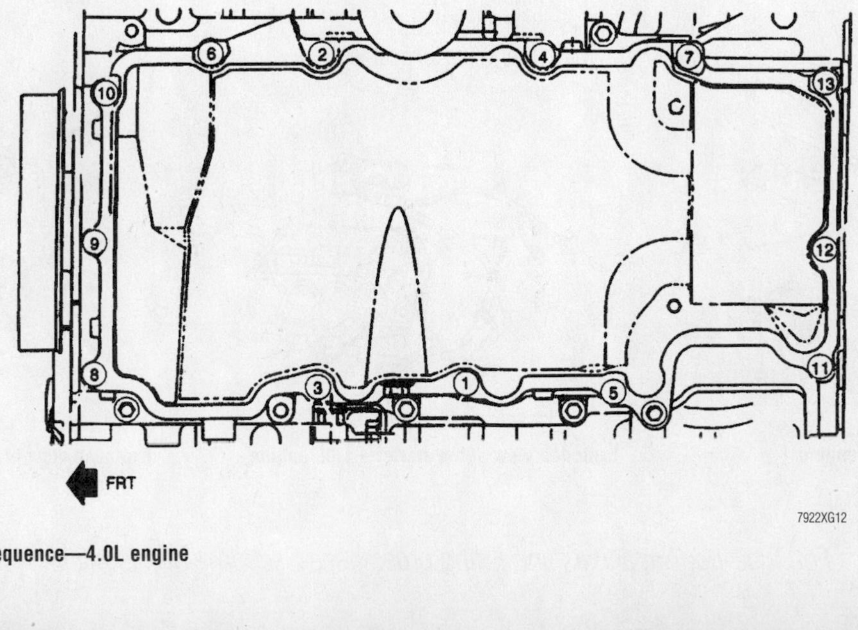

Oil pan bolt tightening sequence—4.0L engine

2. Raise the rear seat cushion to access the battery.
3. Disconnect the negative battery cable.
4. Drain the crankcase.
5. Remove or disconnect the following:
   - Transmission
   - Exhaust crossover pipe
   - Oil pan

➡The oil pan gasket is reusable unless it is damaged. Do not remove the gasket from the oil pan groove unless gasket replacement is required.

To install:

6. Install or connect the following:
   - Oil pan. Torque the bolts in sequence to 89 inch. lbs. (10 Nm).
   - Exhaust crossover pipe. Torque the bolts to 18 ft. lbs. (25 Nm).
   - Transmission
   - Oil pan drain plug. Torque it to 15 ft. lbs. (20 Nm).
   - Negative battery cable
   - Rear seat cushion
7. Refill the crankcase.

Oil Pump

REMOVAL & INSTALLATION

3.5L Engine

1. Before servicing the vehicle, refer to the precautions in the beginning of this section.
2. Remove or disconnect the following:
   - Front cover
   - Rocker arm covers
3. Install camshaft holding fixtures on both sets of camshafts.
4. Remove or disconnect the following:
   - Primary chain tensioner
   - Primary chain from the drive sprocket

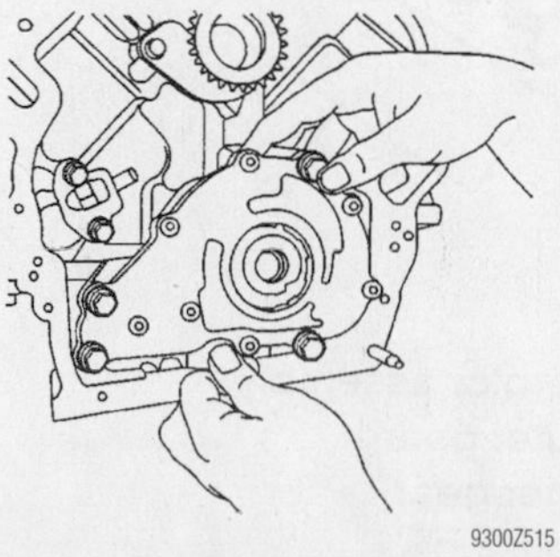

The oil pump is mounted on the front of the engine and driven by the crankshaft—3.5L engine

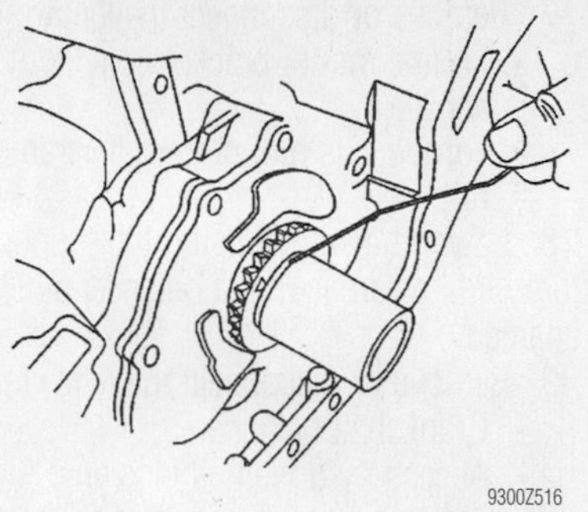

Correct position of the crankshaft sprocket when the oil pump is installed correctly—3.5L engine

   - Oil pan
   - Oil pump pipe and screen
   - Oil pump by sliding it off the crankshaft

➡The internal parts of the oil pump are not serviced separately. The oil pump may be opened for inspection. If damage or wear is noted, replace the entire pump as an assembly.

To install:

5. Pack the oil pump housing with white petroleum jelly to insure priming.
6. Install or connect the following:
   - Oil pump. Torque the bolts to 18 ft. lbs. (25 Nm).
   - Oil pump pipe and screen
   - Oil pan
   - Oil pump housing cover. Torque the bolts to 97 inch lbs. (11 Nm).
   - Primary chain on the sprocket

➡Be sure to maintain correct timing.

   - Chain tensioner
7. Remove the camshaft holding tools.
8. Install or connect the following:
   - Rocker arm covers. Torque the bolts to 80 inch lbs. (9 Nm).
   - Front cover. Torque the bolts to 124 inch lbs. (14 Nm).

3.8L Engines

1. Before servicing the vehicle, refer to the precautions in the beginning of this section.
2. Disconnect the negative battery cable.
3. Drain the crankcase.
4. Remove or disconnect the following:
   - Front cover
   - Oil filter adapter, pressure regulator valve and spring
   - Oil pump cover
   - Inner and outer pump gears
5. Make the following measurements and replace any components not within specification:

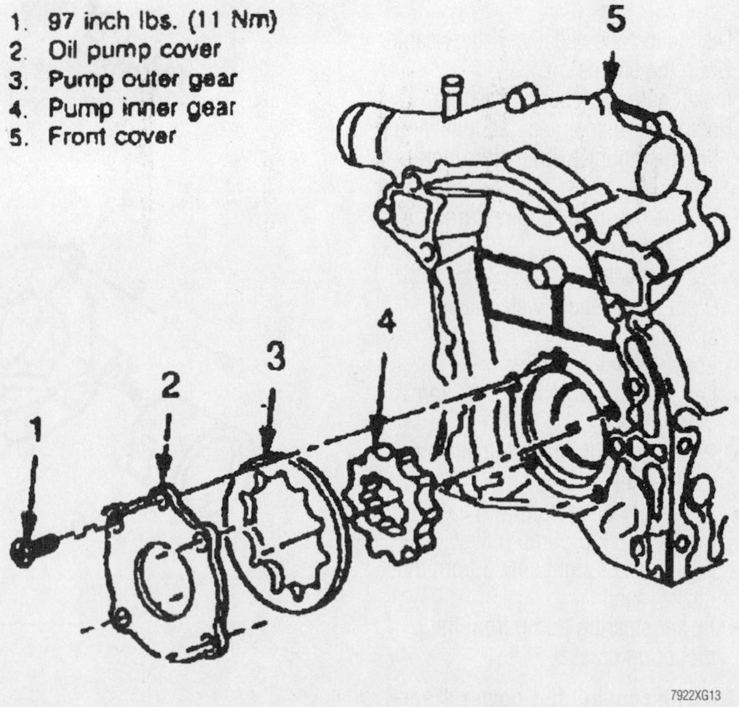

Exploded view of the oil pump assembly—3.8L (VIN 1 and K) engines

a. Gear pocket depth: 0.461–0.4265 in. (11.71–11.75mm).

b. Gear pocket diameter: 3.508–3.512 in. (89.10–89.20mm).

c. Inner gear tip clearance: 0.006 in. (0.152mm).

d. Outer gear diameter clearance: 0.008–0.015 in. (0.203–0.381mm).

e. Gear end clearance: 0.001–0.0035 in. (0.025–0.089mm).

To install:

6. Lubricate the gears with petroleum jelly and install the gears into the housing.
7. Pack the gear cavity with petroleum jelly after the gears have been installed in the housing.
8. Install or connect the following:
   - Oil pump cover. Torque the screws to 97 inch lbs. (11 Nm).
   - Oil filter adapter, pressure regulator valve and spring. Torque the bolts to 24 ft. lbs. (33 Nm).
   - Front cover. Torque the bolts to 11 ft. lbs. (15 Nm) plus an additional 40 degree turn.
   - Negative battery cable
9. Refill the crankcase.
10. Start the vehicle and verify no leaks and proper oil pressure.

4.0L Engine

1. Before servicing the vehicle, refer to the precautions in the beginning of this section.
2. Disconnect the negative battery cable.
3. Drain the engine oil.
4. Install an engine support fixture.
5. Remove or disconnect the following:
   - Engine mount-to-frame through-bolt and nut
   - Engine mount-to-engine through-bolt and nut
   - Front wheels
   - Right inner fender well splash shield
   - Lower center air deflector
   - Left transmission mount-to-frame through-bolt
   - Power steering line retainer from the bracket
   - Accessory drive belt from the power steering pump pulley
   - Fuel injector sight shield from the intake manifold
   - Power steering pump from the mounting bracket

➡DO NOT disconnect the power steering lines from the pump.

6. Raise the engine using a support fixture.
7. Remove or disconnect the following:
   - Engine mount bracket from the engine
   - Torque axis mount from the frame rail
8. Lower the engine using the support fixture, until clearance for a balancer puller is attained.
9. Remove or disconnect the following:
   - Crankshaft balancer
   - Accessory drive belt tensioner
   - Accessory drive belt idler pulley
   - Front cover

➡DO NOT discard the gasket, if it is undamaged it can be re-used.

   - Oil pump and drive spacer
10. If necessary, disassemble and inspect the pump as follows:

a. Remove the drive spacer from the pump housing.

b. Remove the 2 screws holding the pump housing halves together.

c. Remove the inner (drive) and outer (driven) rotors from the housing. Indicate the mating surfaces (dimples).

d. Remove the pressure relief valve.

e. If any components show signs of excessive wear or damage, replace the pump assembly.

To install:

11. If the pump was disassembled, reassemble as follows:

a. Install the inner and outer rotors to the pump cover in the same orientation as removed.

b. Install the pressure relief valve seat, spring and pilot in the pump housing.

c. Pack the pump housing halves with petroleum jelly to ensure pump priming.

d. Assemble the housing and cover over the locating dowel.

e. Insert a 9mm drill in the pump mounting hole on the opposite side to aid alignment of the housing and cover. Install the 2 screws and tighten to 108 inch lbs. (12 Nm).

12. Install or connect the following:
    - Oil pump drive spacer into the oil pump from the rear so the drive flat engages the pump rotor
    - Oil pump. Torque the bolts to 89 inch lbs. (10 Nm); then an additional 35 degree turn.
13. Place a small amount of RTV sealant at the split line of the upper and lower crankcases.
14. Install or connect the following:

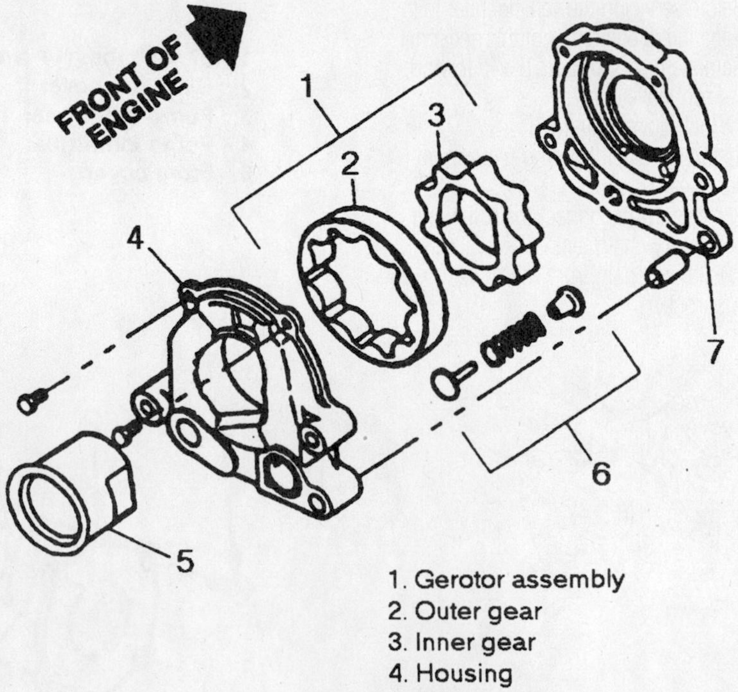

Exploded view of the oil pump—4.0L engine

- Front cover with a new gasket. Torque the bolts to 89 inch lbs. (10 Nm).
- Accessory drive belt idler pulley. Torque the bolt to 37 ft. lbs. (50 Nm).
- Accessory drive belt tensioner. Torque the bolt to 37 ft. lbs. (50 Nm).

15. Coat the seal contact area on the crankshaft balancer with engine oil.
16. Install the crankshaft balancer. Torque the bolt to 44 ft. lbs. (60 Nm); then an additional 120 degrees (⅔) turn.
17. Raise the engine with the support fixture.
18. Install or connect the following:
    - Torque axis mount on the body
    - Engine mount bracket on the engine. Torque the nuts to 30 ft. lbs. (40 Nm) and the bolts to 41 ft. lbs. (55 Nm).
19. Lower the engine support fixture until the engine is at its normal height.
20. Install or connect the following:
    - Power steering pump to the bracket
    - Fuel injector sight shield
    - Accessory drive belt
    - Power steering line retainer to the engine mount bracket
    - Left transmission mount-to-frame through-bolt. Torque the bolt to 63 ft. lbs. (85 Nm).
    - Lower center air deflector
    - Right inner fender well splash shield
    - Front wheels. Torque the nuts to 100 ft. lbs. (140 Nm).
    - Engine mount-to-engine bracket. Torque the through-bolt to 70 ft. lbs. (95 Nm).
    - Engine mount-to-frame. Torque the through-bolt to 37 ft. lbs. (50 Nm).
    - Negative battery cable
21. Remove the engine support fixture.
22. Replace the oil filter and refill the crankcase.
23. Run the engine and check for leaks and proper engine operation.

Rear Main Seal

REMOVAL & INSTALLATION

3.5L Engine

1. Before servicing the vehicle, refer to the precautions in the beginning of this section.

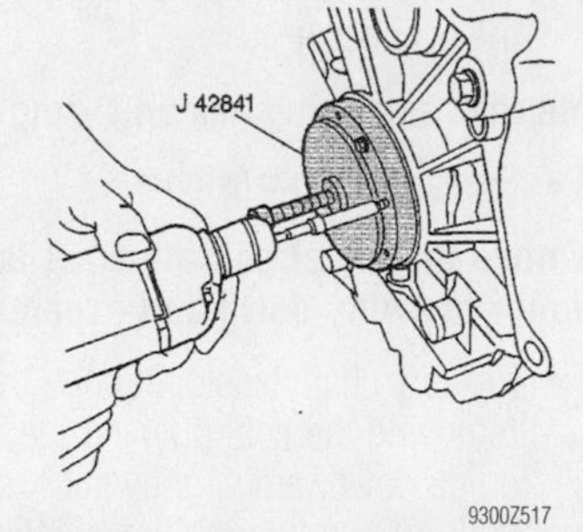

Use the guide holes in tool J-42841 to install the screws in the seal—3.5L engine

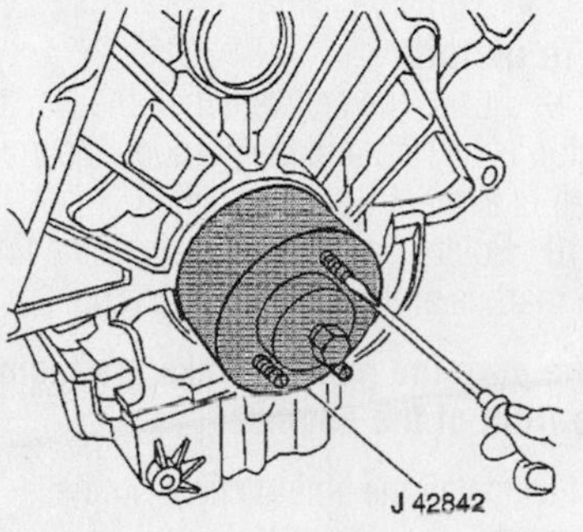

Use rear main seal installer J-42842 to install the seal—3.5L engine

2. Remove or disconnect the following:
    - Engine and transmission assembly
    - Transmission from the engine
    - Flexplate
3. Remove the seal as follows:
    a. Place Seal Removal tool J-42841 onto the crankshaft.
    b. Install 8, 1-inch self starting screws through the guide holes of the tool and into the seal. A variable speed drill will be helpful for installing the screws.
    c. Tighten the center screw of the tool to remove the seal.
4. Clean out the drain at the bottom of the seal bore with a piece of wire or pipe cleaner.

To install:

5. Apply a small amount of RTV gasket maker at the crankcase split line across the end of the upper and lower crankcase seal.
6. Coat the outer diameter of the rear crankshaft seal area with engine oil.
7. Wipe the outer diameter of the crankshaft flexplate flange with a lint-free cloth.
8. Lubricate the outer rubber surface of the seal with clean engine oil. Do not apply any oil to the green coating, pre-applied to the inner diameter of the seal.
9. Loosen the center bolt of the seal installer until the center hub protrudes about ½ inch past the outer plate.
10. Thread the 3 mounting bolts into the crankshaft flange until the tool is firmly mounted on the crankshaft.
11. Install the seal by tightening the center bolt until the tool bottoms against the crankshaft.
12. Remove the tool and verify that the seal is installed evenly.
13. Install or connect the following:
    - Flexplate. Torque the bolts to 11 ft. lbs. (15 Nm); then an additional 50 degree turn.
    - Transmission to the engine
    - Engine and transmission assembly

3.8L and 4.0L Engines

1. Before servicing the vehicle, refer to the precautions in the beginning of this section.
2. Remove or disconnect the following:
    - Negative battery cable
    - Transmission
    - Flexplate
3. Carefully pry the oil seal from the housing using a suitable prying tool, taking care not to damage the housing or the crankshaft sealing surface.

To install:

4. On 4.0L engines, place a small amount of GM Gasket Maker® at the top of the crankcase split line across the end of the upper and lower crankcase seal.
5. Apply clean engine oil to the inside and outside diameter of the oil seal.
6. Install or connect the following:
    - New seal
    - Flexplate
    - Transmission
    - Negative battery cable
7. Start the engine and check for leaks.

Timing Chain, Sprockets, Front Cover and Seal

REMOVAL & INSTALLATION

3.5L Engine

PRIMARY CHAIN

1. Before servicing the vehicle, refer to the precautions in the beginning of this section.
2. Remove or disconnect the following:
    - Negative battery cable
    - Rocker arm covers

3. Rotate the crankshaft so the No. 1 piston is at Top Dead Center (TDC) and the flats on the rear of the camshafts are parallel with the camshaft cover sealing surface.

4. Install camshaft holding fixtures on both sets of camshafts.

5. Drain the cooling system.

6. Remove or disconnect the following:

- Front diagonal brace
- Battery and tray
- Washer and coolant reservoirs
- Underhood accessory wiring junction block and move it aside
- Drive belt
- Power steering pump pulley
- Idler pulley
- Belt tensioner
- Water pump

7. Support the engine cradle

8. Remove or disconnect the following:

- Right side engine cradle bolts and lower the cradle
- Crankshaft balancer
- Front cover
- Lift bracket from the front of the engine
- Camshaft Position (CMP) sensor
- Sprocket bolt from the exhaust camshaft on the right cylinder head to allow for clearance of the chain guide
- 4 chain guide access plugs from the cylinder heads

➡Note that each plug has an O-ring.

- Primary chain tensioner

➡Remove the lower bolt allowing the tensioner to swing down and expand.

- Primary chain tensioner shoe by removing the bolt, pushing the guide downward slightly and pulling it up through the cylinder head
- Primary chain from the right camshaft, allowing it to fall into the oil pump area
- Primary chain

To install:

9. Rotate the crankshaft so the No. 1 piston is at TDC and the mark on the crankshaft is at the 4 o'clock position.

10. Rotate the balance shaft so the timing mark is at the 5 o'clock position.

➡Be sure the painted links are facing the front of the engine.

11. Install the timing chain on the sprockets.

12. Center the mark on the left intake camshaft sprocket between the 2 painted links.

13. Lift the chain onto the right intake

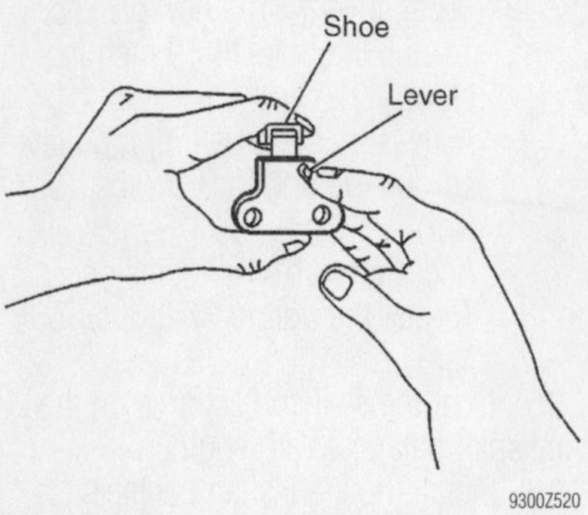

Compressing the primary chain tensioner—3.5L engine

camshaft sprocket. While doing this, align the marks on the balance shaft and crankshaft sprockets with the painted marks on the chain.

14. Verify that all of the timing marks are aligned.

15. Install the primary chain tensioner shoe. Torque the bolt to 22 ft. lbs. (30 Nm).

16. Compress the primary chain tensioner using the following steps:

a. Rotate the ratchet release lever counterclockwise and hold it.

b. Press the tensioner shoe in and hold it.

c. Release the ratchet lever and slowly release the pressure on the shoe.

d. Insert a pin through the hole in the lever as the lever moves to the first click. The ratchet should hold the shoe in the compressed position.

➡Be sure the lever on the tensioner is facing you when installed.

17. Install or connect the following:

- Primary chain tensioner. Torque the bolts to 18 ft. lbs. (25 Nm); then remove the chain tensioner pin.
- 4 chain guide access plugs. Torque the plugs to 44 inch lbs. (5 Nm).
- Front engine lift bracket. Torque the hex head bolt to 37 ft. lbs. (50 Nm) and the internal drive bolt to 18 ft. lbs. (25 Nm).
- CMP sensor. Torque the bolts to 80 inch lbs. (9 Nm).

18. Remove the camshaft holding tools.

19. Place a small bead of RTV sealant on the 3 areas indicated in the diagram.

20. Install or connect the following:

- Front cover with a new gasket. Torque the bolts to 124 inch lbs. (14 Nm) and the coolant drain plug to 89 inch lbs. (10 Nm).
- Crankshaft balancer. Torque the

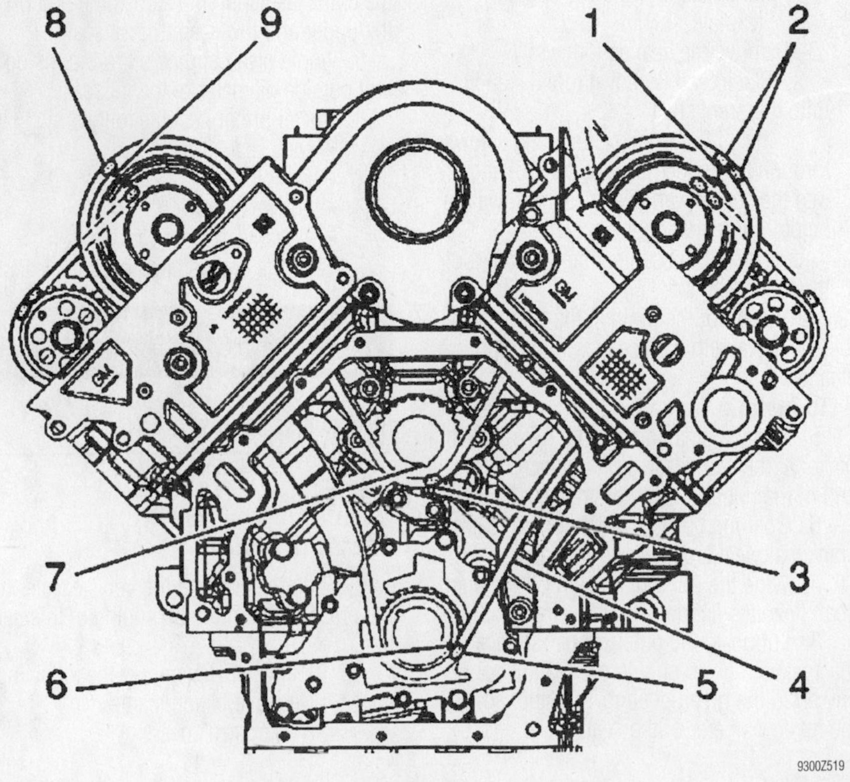

Primary timing chain alignment marks—3.5L engine

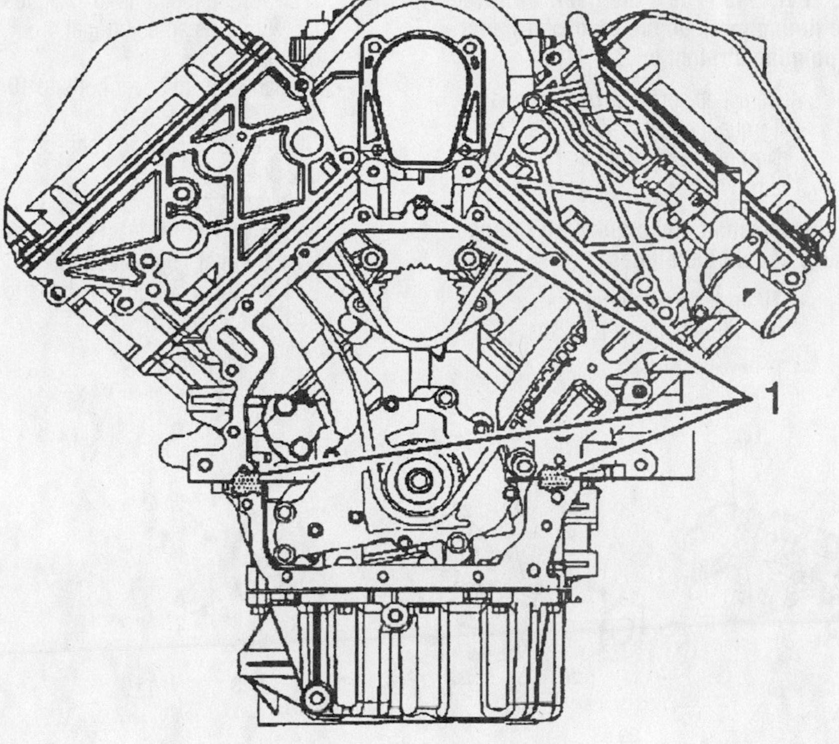

Apply RTV sealant to the 3 areas indicated before installing the front cover and gasket—3.5L engine

bolt to 37 ft. lbs. (50 Nm); then an additional 120 degree turn.

21. Raise the engine cradle and install new bolts. Torque the bolts to 133 ft. lbs. (180 Nm).
22. Coat the sub-frame bushings with rubber lubricant.
23. Install or connect the following:
    - Water pump with a new gasket. Torque the bolts to 124 inch lbs. (14 Nm).
    - Belt tensioner. Torque the bolts to 37 ft. lbs. (50 Nm).
    - Idler pulley. Torque the bolt to 37 ft. lbs. (50 Nm).
    - Power steering pump pulley
    - Drive belt
    - Underhood accessory wiring junction block
    - Washer and coolant reservoirs
    - Battery and tray
    - Front diagonal brace
    - Rocker arm covers. Torque the bolts to 80 inch lbs. (9 Nm).
    - Negative battery cable
24. Refill the cooling system.
25. Refill the crankcase.
26. Start the engine and verify no leaks.

SECONDARY TIMING CHAIN

1. Before servicing the vehicle, refer to the precautions in the beginning of this section.
2. Disconnect the negative battery cable.
3. Remove the rocker arm cover and install camshaft holding fixture J-42038.
4. Remove the camshaft sprocket bolts and install the timing chain holding fixture J-42042 on the cylinder head.
5. Remove or disconnect the following:

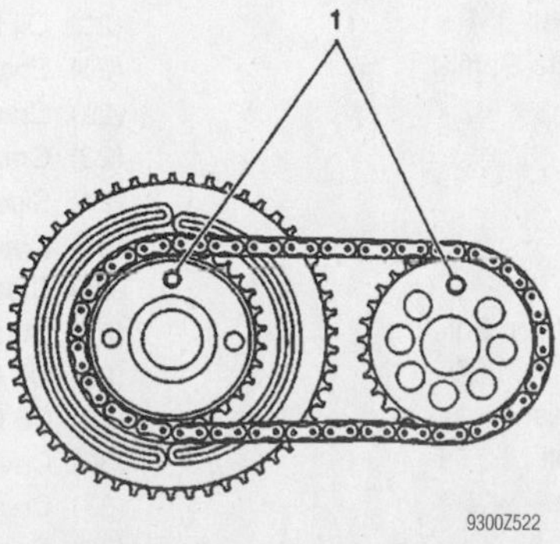

Correct sprocket alignment for the left secondary timing chain—3.5L engine

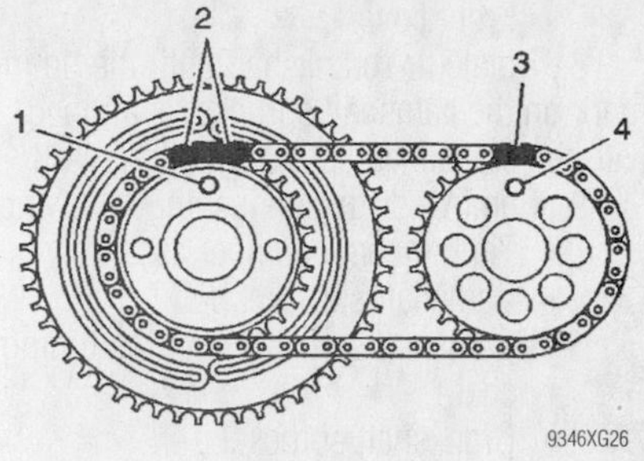

Correct sprocket alignment for the right secondary timing chain—3.5L engine

- Sprockets and chain
- Secondary timing sprocket and chain

To install:

6. Install or connect the following:
    - Secondary timing chain on the sprockets, with the drive pins at the 12 o'clock positions
    - Sprockets and chain assembly onto the camshafts, with the chain properly aligned on the tensioner
7. Remove the timing chain holding fixture
8. Install the sprocket bolts. Torque the bolts to 18 ft. lbs. (25 Nm); then an additional 45 degree turn.
9. Remove the camshaft holding fixture.
10. Install or connect the following:
    - Rocker arm cover
    - Negative battery cable

3.8L Engines

1. Before servicing the vehicle, refer to the precautions in the beginning of this section.
2. Disconnect the negative battery cable.
3. Drain the coolant system.
4. Remove the vacuum reservoir.
5. Install an engine support fixture.
6. Raise the engine so that the weight is removed from the torque axis mount.
7. Remove or disconnect the following:
    - Torque axis mount
    - Accessory drive belts
    - Engine mount bracket
    - Alternator
    - Drive belt tensioner
    - Crankshaft balancer
    - Crankshaft Position (CKP) sensor shield
    - CKP sensor electrical connector
    - Oil pan to the front cover bolts

- Front cover attaching bolts
- Front cover

8. Rotate the crankshaft until the timing mark on the camshaft sprocket is aligned with the crankshaft sprocket timing mark.
9. Remove or disconnect the following:
   - Timing chain damper assembly
   - Camshaft sprocket bolt
   - Camshaft sprocket with the timing chain
   - Crankshaft sprocket

To install:

10. Install or connect the following:
    - Crankshaft sprocket

➡It may be necessary to use a gear installer to fully seat the gear. Be sure the timing mark on the crankshaft gear is pointing straight up.

- Camshaft sprocket with the chain. Torque the bolt to 74 ft. lbs. (100 Nm) plus and additional 90 degree turn.

➡The camshaft and crankshaft timing marks should be aligned.

- Timing chain damper. Torque the mounting bolts to 16 ft. lbs. (22 Nm).
- Front cover with a new gasket and seal. Torque the bolts to 11 ft. lbs. (15 Nm) plus an additional 40 degrees.
- Oil pan-to-front cover bolts to 10 ft. lbs. (14 Nm)
- CKP sensor electrical connector and shield
- Crankshaft balancer. Torque the bolt to 111 ft. lbs. (150 Nm); then an additional 75 degree turn.
- Belt tensioner. Torque the bolts to 37 ft. lbs. (50 Nm).
- Alternator

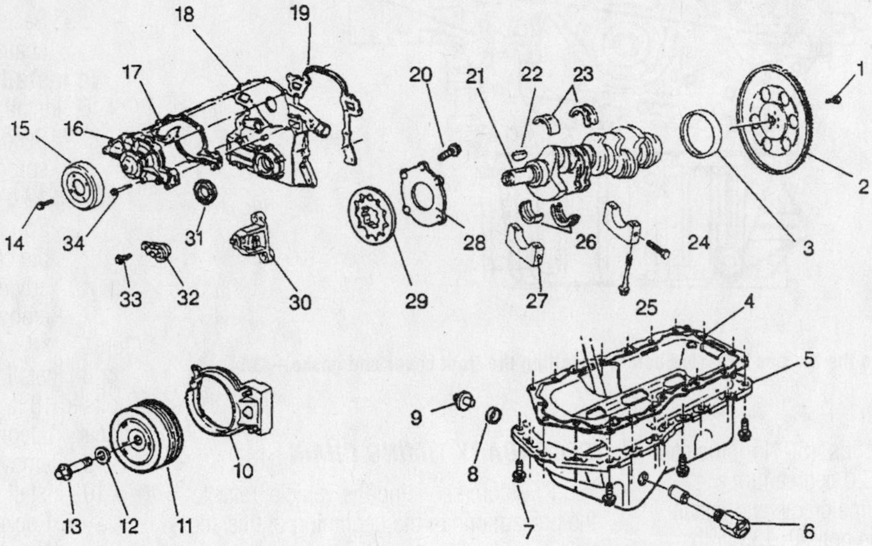

Legend
(1) Flywheel Bolt
(2) Flywheel
(3) Crankshaft Rear Oil Seal
(4) Oil Pan Gasket (Includes Baffle)
(5) Engine Oil Pan
(6) Oil Level Sensor
(7) Oil Pan Bolt
(8) Oil Pan Drain Plug
(9) Oil Pan Drain Gasket
(10) Crankshaft Position Sensor Shield
(11) Crankshaft Balancer
(12) Crankshaft Balancer Washer
(13) Crankshaft Balancer Bolt
(14) Water Pump Pulley Bolt
(15) Water Pump Pulley
(16) Water Pump
(17) Water Pump Gasket
(18) Engine Front Cover
(19) Engine Front Cover Gasket
(20) Oil Pump Cover Bolt
(21) Engine Crankshaft
(22) Crankshaft Balancer Key
(23) Crankshaft Upper Bearing
(24) Side Main Bolt
(25) Crankshaft Main Bearing Cap Bolt
(26) Crankshaft Lower Bearing
(27) Crankshaft Main Bearing Cap
(28) Oil Pump Cover
(29) Oil Pump Gear Set
(30) Crankshaft Position Sensor
(31) Crankshaft Front Oil Seal
(32) Camshaft Position Sensor
(33) Camshaft Position Sensor Bolt
(34) Water Pump Bolt

9300XG04

Exploded view of lower engine components—3.8L (VIN 1) engines

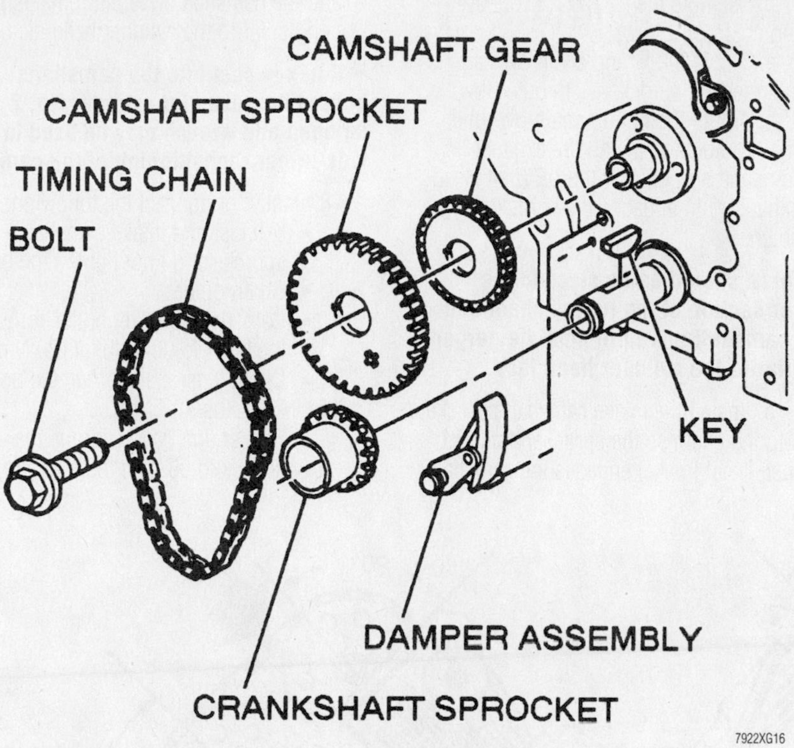

Exploded view of the timing chain and sprockets—3.8L (VIN 1) engines

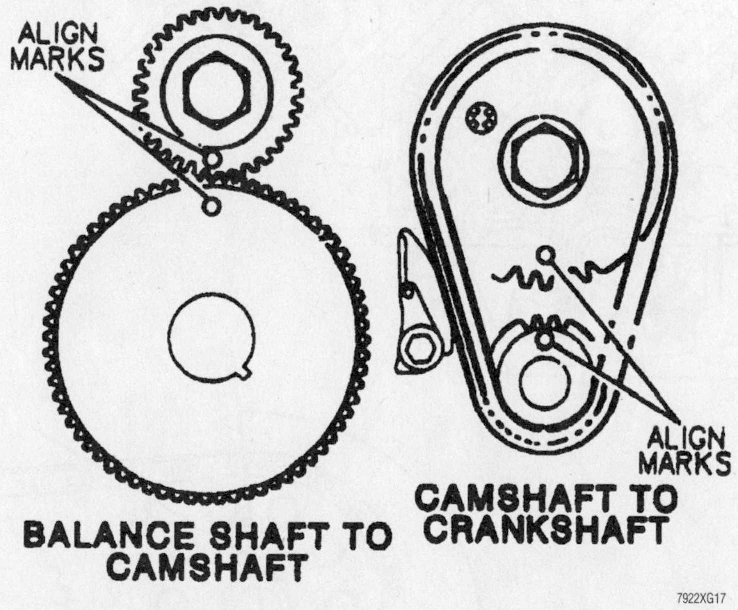

Balance shaft-to-camshaft and camshaft-to-crankshaft timing mark alignment—3.8L (VIN 1) engines

- Engine mount bracket. Torque the bolts to 65 ft. lbs. (87 Nm).
- Drive belt
- Torque axis mount. Torque the torque axis mount-to-frame bolts to 52 ft. lbs. (70 Nm) and the through-bolts to 65 ft. lbs. (87 Nm).

11. Remove the engine support fixture.
12. Install or connect the following:
    - Vacuum reservoir
    - Negative battery cable
13. Refill the cooling system.
14. Start the vehicle and check for leaks.
15. Road test the vehicle and ensure proper operation.

4.0L Engine

LEFT SIDE SECONDARY CHAIN

➡The secondary timing chains can be removed without removing the engine from the vehicle.

1. Before servicing the vehicle, refer to the precautions in the beginning of this section.
2. Remove or disconnect the following:
    - Negative battery cable
    - Drive belt
    - Tower to tower brace
    - Exhaust Y-pipe at the converter
    - Front wheels
    - Wheel well splash shields
    - Crankshaft balancer bolt
3. Support the engine cradle.
    - 3 right-side engine cradle bolts
    - Vehicle Speed Sensor (VSS) from the right control arm
4. Lower the cradle to gain access for the crankshaft damper puller.
5. Remove or disconnect the following:
    - Crankshaft balancer
    - Drive belt tensioner and idler pulley
    - Front cover and gasket

➡The front cover gasket is reusable as long as it is not damaged.

6. Partially drain the cooling system.
    - Upper radiator hose at the water crossover
    - Spark plug wires
    - Right side fan
    - Battery cable at the alternator
    - Spark plug wires
    - Wiring harness at the rocker arm cover
    - Positive Crankcase Ventilation (PCV) tube from the rocker arm cover
    - Right and left torque struts
    - Battery cable retainer at the front of the rocker arm cover
    - Ignition Control Module (ICM)
    - Rocker arm cover by pivoting it around the water pump drive shaft

➡Continue moving the cover upward and pivoting so that the edge of the cover closely follows the left edge of the intake manifold cover. The gasket

is reusable as long as it is not damaged.

- Left side secondary chain tensioner
- Left side chain guide

➡Access the upper chain guide mounting bolt through the hole in the cylinder head capped with the plastic plug.

- Secondary drive chain and sprockets

To install:

➡Correct timing exists when the crankshaft and intermediate shaft sprocket timing marks are in alignment and all 4 camshaft drive pins are perpendicular (90 degrees) to the cylinder head surface.

7. Assemble the left side secondary timing chain as follows:
   a. Route the timing chain over the intermediate sprocket teeth outer row.
   b. Route the timing chain over the chain guide and install the exhaust camshaft sprocket so the **LE** (Left Head Exhaust) pin engages the sprocket notch.

➡There should be no slack in the lower section of the timing chain and the camshaft drive pin must be perpendicular to the cylinder head face.

   c. Install the intake camshaft sprocket into the chain so the sprocket notch **LI** (Left Head Intake) engages the camshaft and the camshaft drive pin remains perpendicular to the cylinder head face.

➡A hex is cast into the camshafts behind the lobes for cylinder No. 2, so an open end wrench may be used to provide minor repositioning of the cams.

8. Install or connect the following:
   - Exhaust and intake camshaft sprockets, do not tighten the bolts
   - Chain guide
   - Camshaft sprocket bolts. Torque the bolts to 90 ft. lbs. (120 Nm).
   - Chain tensioner. Torque the bolts to 20 ft. lbs. (27 Nm).
   - Rocker arm cover. Torque the screws to 89 inch lbs. (10 Nm).

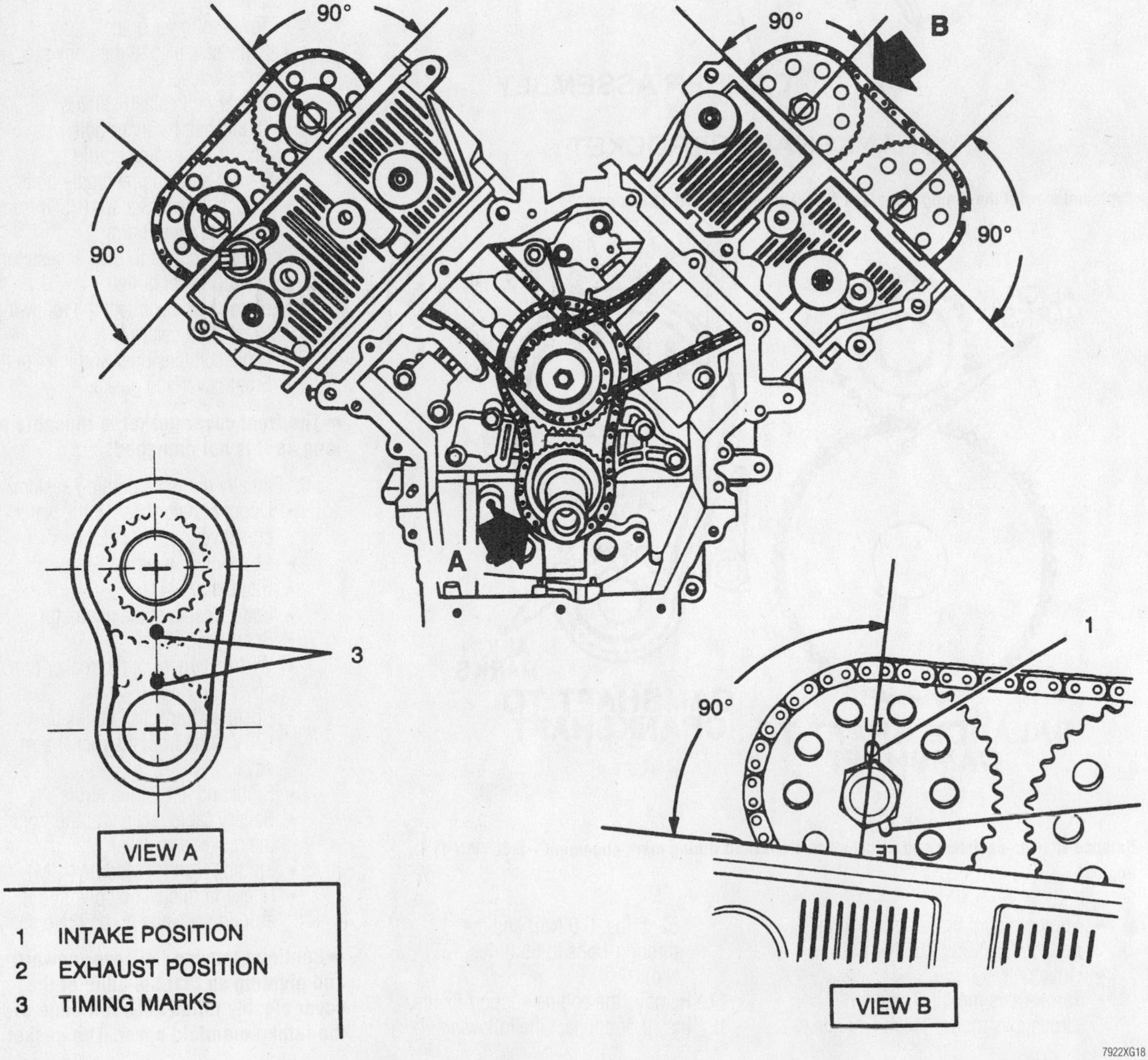

Primary and secondary timing mark alignment—4.0L (VIN C) engine

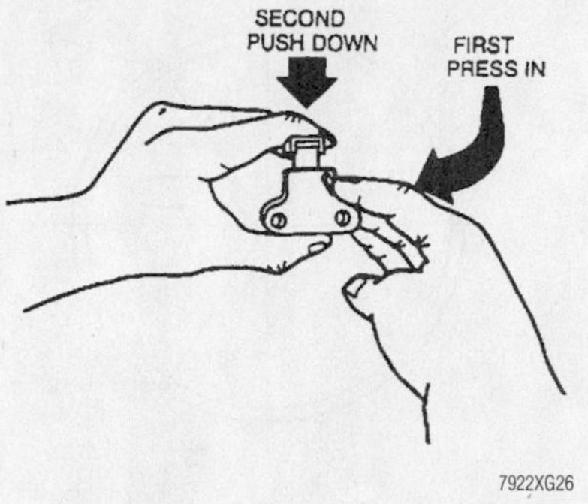

Rotating tensioner release lever—4.0L (VIN C) engine

- ICM
- Battery cable retainer to the front of the rocker arm cover

9. Install the right and left torque struts and torque the bolts as follows:

➡It is important during installation that the engine torque struts are not preloaded in their installed position. Adjustment is provided at the point the strut fastens to the core support bracket. Be sure this bolt is loose during assembly.

a. Step 1: Strut bracket-to-cylinder head (M10) bolt: 35 ft. lbs. (50 Nm).
b. Step 2: Strut bracket-to-water manifold (M8) bolts: 20 ft. lbs. (25 Nm).
c. Step 3: Strut-to-core support bracket bolt: 45 ft. lbs. (60 Nm).

10. Install or connect the following:
    - PCV fresh air tube to the rocker arm cover
    - Wiring harness to the cover
    - Spark plug wires
    - Battery cable at the alternator
    - Right side fan
    - Upper radiator hose to the water crossover
    - Front cover with a new gasket. Torque the bolts to 89 inch lbs. (10 Nm).
11. Apply a dab of RTV to the split line between the upper and lower crankcase assemblies.
12. Install or connect the following:
    - Drive belt idler pulley. Torque the bolt to 35 ft. lbs. (47 Nm).
    - Drive belt tensioner. Torque the nut to 35 ft. lbs. (47 Nm).
    - Crankshaft balancer. Lubricate the bolt threads with engine oil and torque the bolt to 44 ft. lbs. (60 Nm) then an additional 120 degree turn.

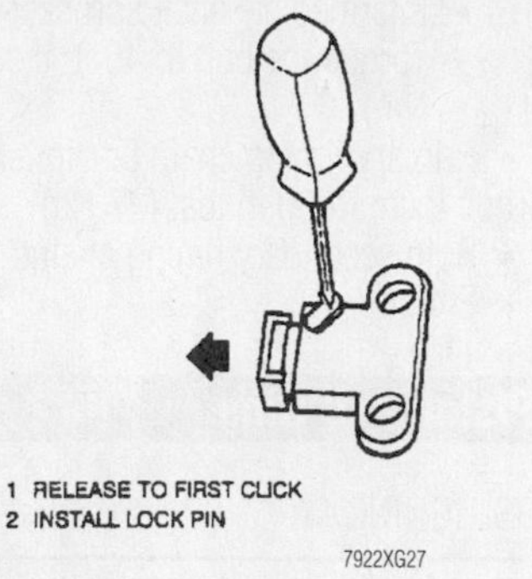

Locking the tensioner into position—4.0L (VIN C) engine

- VSS sensor
- Engine cradle. Torque the bolts to 75 ft. lbs. (102 Nm).
- Wheel well splash shields
- Wheels
- Exhaust Y-pipe to the converter. Torque the bolts to 20 ft. lbs. (25 Nm).
- Tower-to-tower brace
- Drive belt
- Negative battery cable

13. Refill the cooling system.
14. Start the engine and check for leaks.

RIGHT SIDE SECONDARY CHAIN

➡The secondary timing chains can be removed without removing the engine from the vehicle.

1. Before servicing the vehicle, refer to the precautions in the beginning of this section.
2. Remove or disconnect the following:
    - Exhaust Y-pipe from the converter
    - Tower-to-tower brace
    - Ignition Control Module (ICM)
    - Spark plug wires from the right bank
    - Positive Crankcase Ventilation (PCV) valve
    - Purge canister solenoid from the rear of the cover
    - Rocker arm cover
3. Safely support the front of the engine cradle.
4. Remove or disconnect the following:
    - 2 mounting bolts at the front of the cradle
    - Right and left torque struts
5. Lower the engine cradle (or raise the vehicle) to provide clearance at the rear of the engine compartment.
6. Remove or disconnect the following:
    - Left side secondary timing chain
    - Right side secondary chain tensioner
    - Right side chain guide

➡Access the upper chain guide mounting bolt through the hole in the cylinder head capped with the plastic plug.

- Right side camshaft sprocket bolts and camshaft sprockets
- Secondary drive chain

To install:

➡Correct timing exists when the crankshaft and intermediate shaft sprocket timing marks are in alignment and all 4 camshaft drive pins are perpendicular (90 degrees) to the cylinder head surface.

7. Install the secondary timing chain guide.
8. Assemble the right side secondary timing chain as follows:

a. Over the intermediate shaft sprocket inner row of teeth.
b. Over the chain guide.
c. Exhaust camshaft sprocket so the **RE** (Right Head Exhaust) pin engages the sprocket notch.

➡There should be no slack in the lower section of the timing chain and the camshaft drive pin must be perpendicular to the cylinder head face.

d. Intake camshaft sprocket into the chain so the sprocket notch **RI** (Right Head Intake) engages the camshaft and the camshaft drive pin remains perpendicular to the cylinder head face.

➡A hex is cast into the camshafts behind the lobes for cylinder No. 1, so an open end wrench may be used to provide minor repositioning of the cams.

9. Install or connect the following:
    - Exhaust and intake camshaft sprockets, do not tighten
    - Timing chain tensioner. Torque the bolts to 20 ft. lbs. (27 Nm).
    - Camshaft sprocket bolts. Torque the bolts to 90 ft. lbs. (120 Nm).
    - Left side secondary timing chain
    - Torque struts
    - Engine cradle. Torque the bolts to 75 ft. lbs. (102 Nm).
    - Rocker arm cover. Torque the screws to 84 inch lbs. (10 Nm).
    - Purge canister solenoid
    - PCV valve

- Spark plug wires
- ICM
- Tower to tower brace
- Exhaust Y-pipe to the converter. Torque the bolts to 20 ft. lbs. (25 Nm).

PRIMARY CHAIN

The engine must be removed from the vehicle and supported on an engine stand.

1. Before servicing the vehicle, refer to the precautions in the beginning of this section.
2. Remove or disconnect the following:
   - Engine
   - Both secondary timing chains
   - Primary timing chain tensioner
   - Intermediate shaft sprocket
   - Primary chain and sprocket assembly by sliding it off the shafts

To install:

3. Install or connect the following:
   - Crankshaft sprocket, intermediate shaft sprocket and primary timing chain assembly

➡If it is necessary to turn the crankshaft sprocket, the intermediate shaft sprocket will also have to be turned so the timing mark aligns with the crankshaft sprocket.

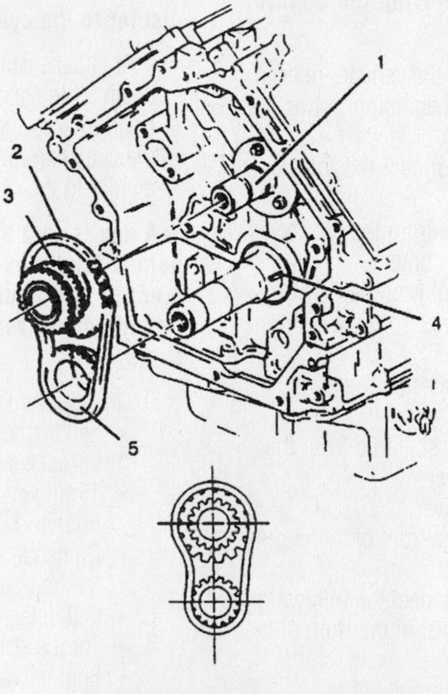

1 INTERMEDIATE SHAFT
2 PRIMARY CHAIN
3 INTERMEDIATE SHAFT SPROCKET
4 CRANKSHAFT SPROCKET KEY
5 SPROCKET

7922XG25

Primary drive chain components—4.0L (VIN C) engine

- Intermediate shaft sprocket bolt. Torque the bolt to 45 ft. lbs. (61 Nm).
- Primary timing chain tensioner bolts. Torque them to 20 ft. lbs. (27 Nm).
- Both secondary timing chains
- Engine

Piston and Ring

POSITIONING

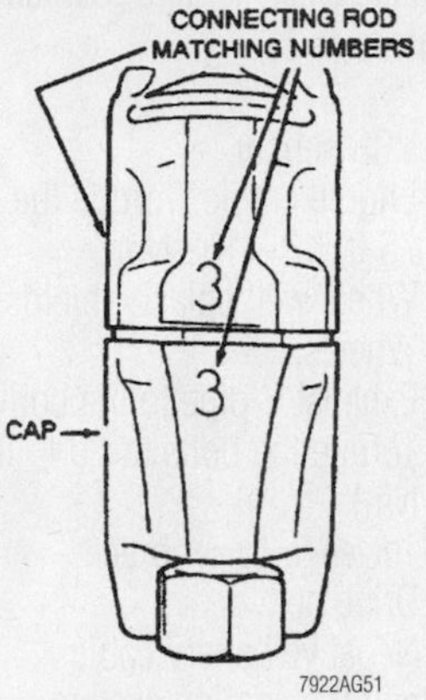

7922AG51

Connecting rod and cap installation. Be sure to matchmark the cap and rod prior to disassembly, as shown

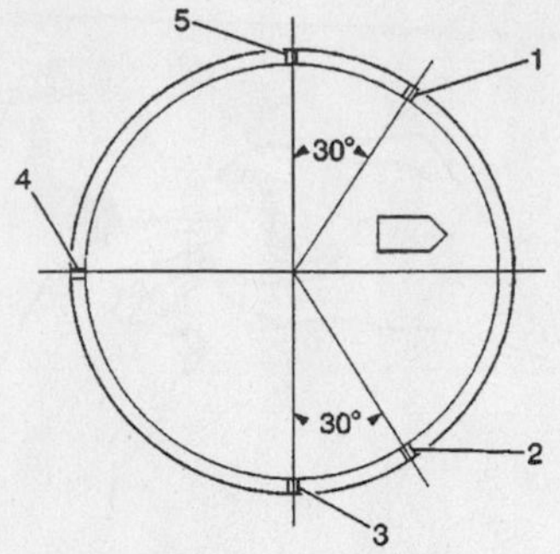

1. Lower oil control ring
2. Upper oil control ring
3. Top Ring
4. Oil control ring expander
5. Second ring

9306XG05

Piston ring end-gap positioning—3.5L engine

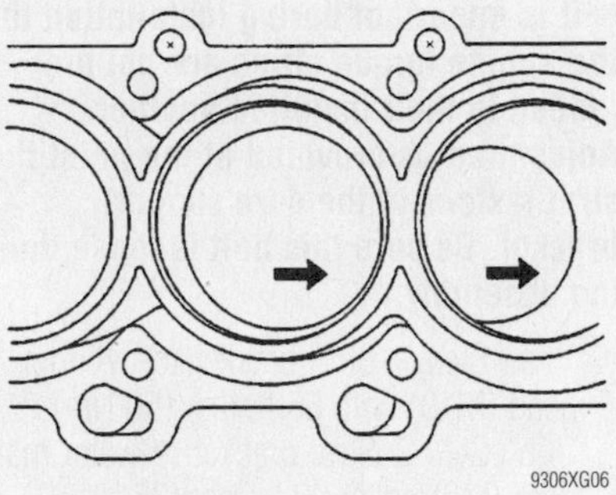

9306XG06

Piston positioning—3.5L engine

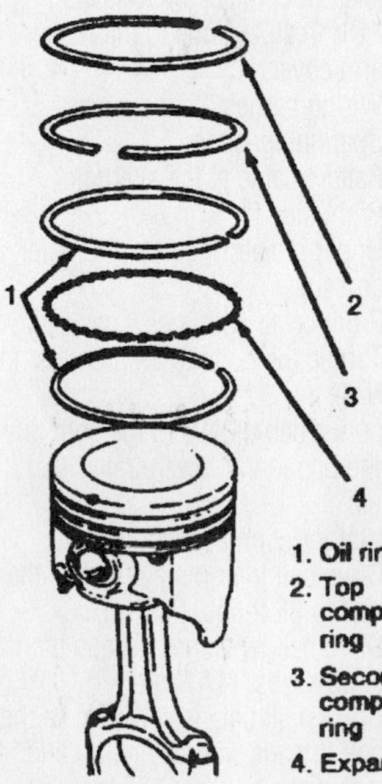

1. **Oil rings**
2. **Top compression ring**
3. **Second compression ring**
4. **Expander**

7922AG48

Piston ring positioning—3.5L and 3.8L engines

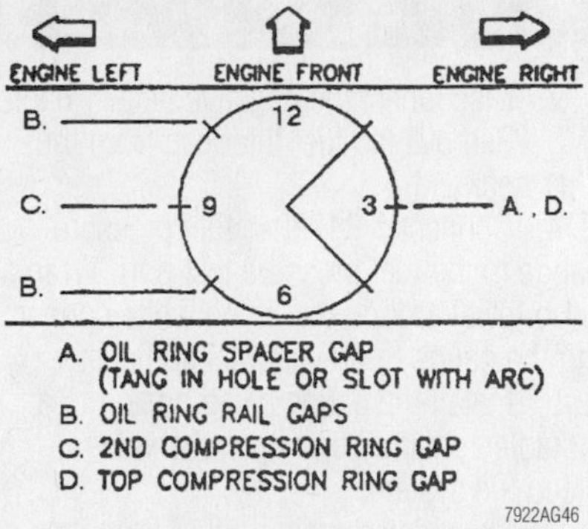

Piston ring end-gap spacing—3.8L engine

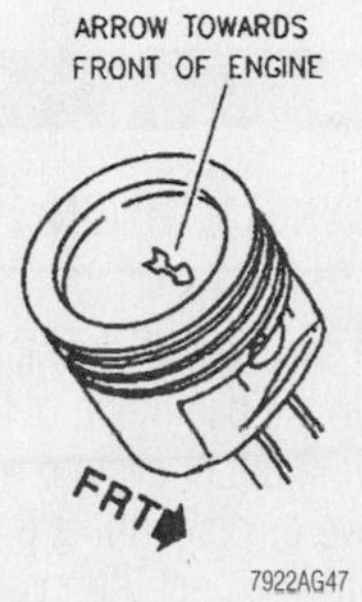

Piston positioning. Often the arrow is replaced by a notch, which also must face toward the front of the engine—3.8L engine

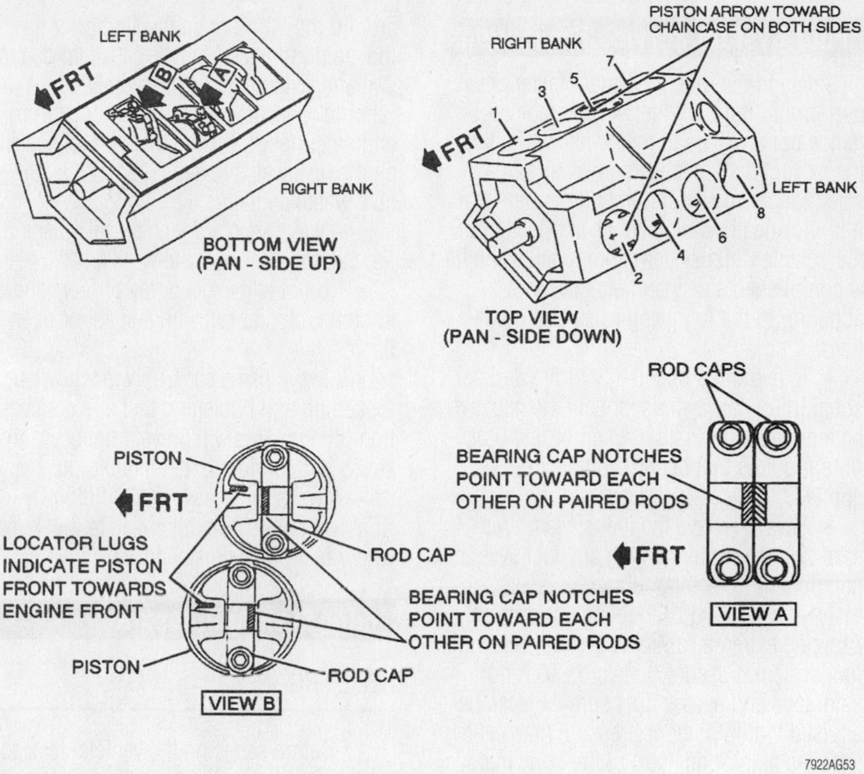

Piston and connecting rod assembly positioning—4.0L engine

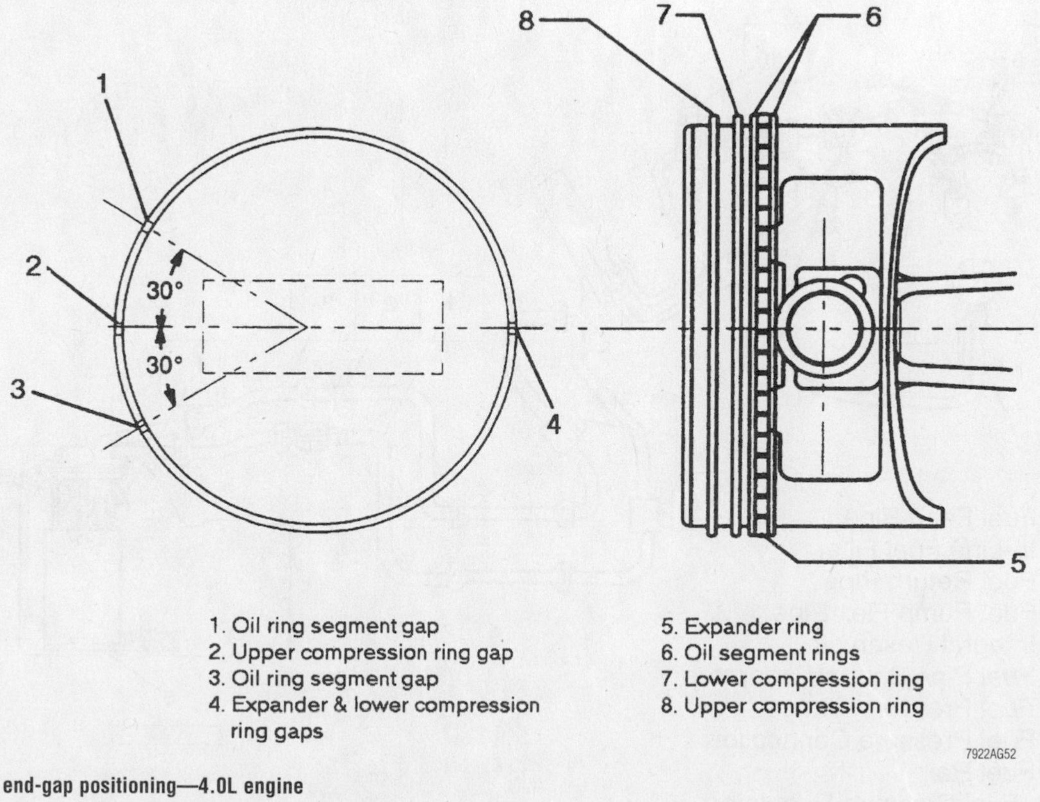

Piston ring and end-gap positioning—4.0L engine

FUEL SYSTEM

Fuel System Service Precautions

Safety is the most important factor when performing not only fuel system maintenance but any type of maintenance. Failure to conduct maintenance and repairs in a safe manner may result in serious personal injury or death. Maintenance and testing of the vehicle's fuel system components can be accomplished safely and effectively by adhering to the following rules and guidelines.

- To avoid the possibility of fire and personal injury, always disconnect the negative battery cable unless the repair or test procedure requires that battery voltage be applied.
- Always relieve the fuel system pressure prior to disconnecting any fuel system component (injector, fuel rail, pressure regulator, etc.), fitting or fuel line connection. Exercise extreme caution whenever relieving fuel system pressure, to avoid exposing skin, face and eyes to fuel spray. Please be advised that fuel under pressure may penetrate the skin or any part of the body that it contacts.
- Always place a shop towel or cloth around the fitting or connection prior to loosening to absorb any excess fuel due to spillage. Ensure that all fuel spillage (should it occur) is quickly removed from engine surfaces. Ensure that all fuel soaked cloths or towels are deposited into a suitable waste container.
- Always keep a dry chemical (Class B) fire extinguisher near the work area.
- Do not allow fuel spray or fuel vapors to come into contact with a spark or open flame.
- Always use a back-up wrench when loosening and tightening fuel line connection fittings. This will prevent unnecessary stress and torsion to fuel line piping.
- Always replace worn fuel fitting O-rings with new. Do not substitute fuel hose where fuel pipe is installed.

Fuel System Pressure

RELIEVING

1. Before servicing the vehicle, refer to the precautions in the beginning of this section.
2. Disconnect the negative battery cable.
3. Remove the fuel filler cap from the filler neck.
4. Connect J-34730-1 fuel pressure gauge to the fuel pressure test port. Wrap a shop towel around the fitting while connecting the gauge to prevent fuel spillage.
5. Install the gauge bleed hose into a suitable container and open the valve to bleed the system.
6. Drain any remaining fuel from the gauge into the container and remove the gauge from the test port.

Fuel Filter

REMOVAL & INSTALLATION

1. Before servicing the vehicle, refer to the precautions in the beginning of this section.
2. Relieve the fuel system pressure.
3. Remove or disconnect the following:
   - Quick connect fitting at the fuel filter inlet
   - Fuel filter outlet fitting from the fuel filter while holding the filter fitting with a back-up wrench

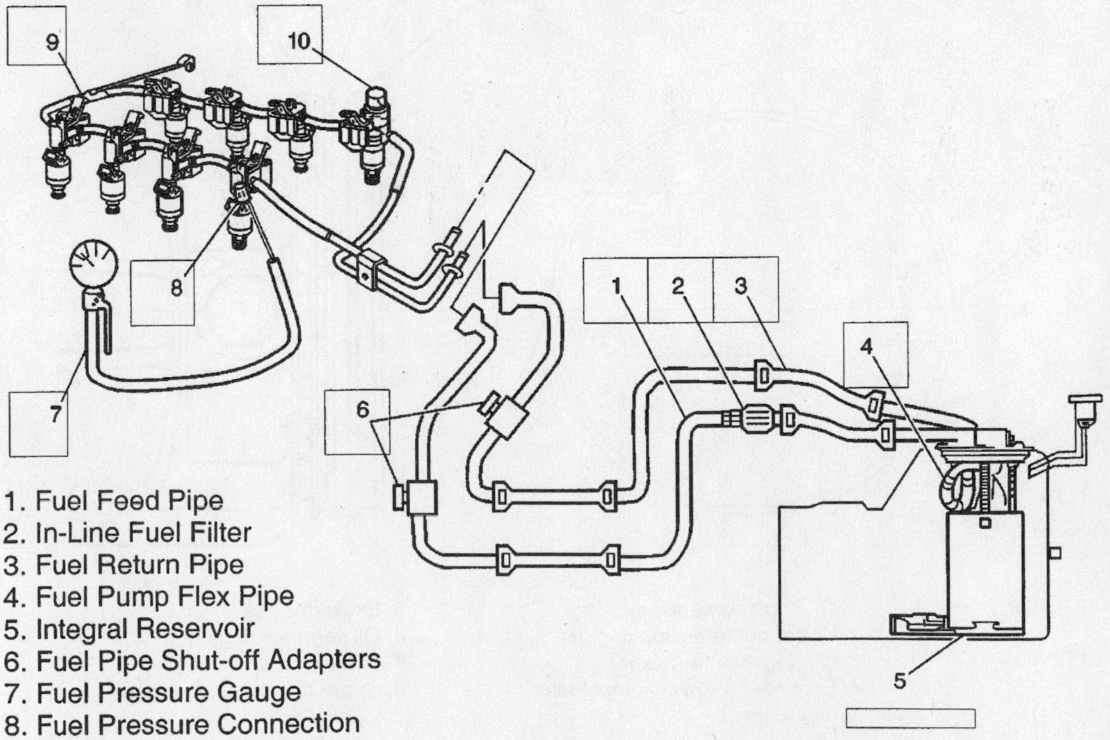

1. Fuel Feed Pipe
2. In-Line Fuel Filter
3. Fuel Return Pipe
4. Fuel Pump Flex Pipe
5. Integral Reservoir
6. Fuel Pipe Shut-off Adapters
7. Fuel Pressure Gauge
8. Fuel Pressure Connection
9. Fuel Rail
10. Fuel Pressure Regulator

9300XG07

Fuel system circuit showing the pressure test port on the supply rail—4.0L engine

- Fuel filter from the vehicle

To install:

4. Position the new fuel filter in the bracket.
5. Install a new plastic retainer on the fuel inlet line.
6. Apply a drop of oil on the fuel filter inlet fitting and snap the fitting onto the fuel filter.
7. Reconnect the fuel outlet line to the filter by holding the filter with a back-up wrench and tightening the line fitting to 22 ft. lbs. (30 Nm).
8. Pressurize the fuel system and verify no leaks.

Fuel Pump

REMOVAL & INSTALLATION

1. Before servicing the vehicle, refer to the precautions in the beginning of this section.
2. Relieve the fuel system pressure.
3. Drain the fuel tank.
4. Remove or disconnect the following:
   - Spare tire and jack
   - Floor trunk liner by pulling it back

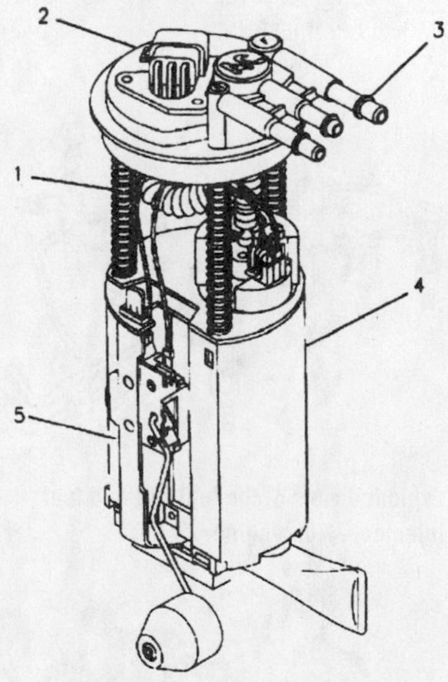

1 SUPPORT ASSEMBLY - FUEL SENDER
2 COVER ASSEMBLY - FUEL SENDER
3 FUEL PIPES (ABOVE COVER)
4 RESERVOIR - FUEL PUMP FUEL
5 SENSOR ASSEMBLY - FUEL LEVEL

7922XG19

Fuel pump and sending unit module assembly

   - Fuel sender access panel
   - Fuel sender assembly quick connect fittings
   - Fuel sender assembly electrical connector

⁂ WARNING

When the lock-ring is removed from the fuel sender, the sender assembly will spring up. Downward pressure should be kept on the assembly and slowly released to ensure the sender assembly does not get damaged.

   - Lock-ring from the fuel sender

⁂ CAUTION

The reservoir bucket on the fuel sender assembly will be full of fuel when it is removed from the tank. Be sure to have a catch pan nearby to drain the sender into.

   - Sender assembly from the tank

To install:

5. Install or connect the following:
   - New O-ring on top of the tank
   - Sender assembly into the tank
   - Retainer on top of the fuel tank, compress the sender until the retainer can be engaged; then lock it in place
   - Quick connect fittings to the fuel sender assembly
   - Sender assembly electrical connector
   - Negative battery cable
6. Pressurize the fuel system and verify no leaks.
7. Install or connect the following:
   - Fuel sender access panel
   - Trunk liner
   - Spare tire and jack

Fuel Injector

REMOVAL & INSTALLATION

3.5L Engine

1. Before servicing the vehicle, refer to the precautions in the beginning of this section.
2. Relieve the fuel system pressure.
3. Remove or disconnect the following:
   - Fuel injector sight shield
   - Fuel injector electrical connectors
   - Vacuum hose from the fuel pressure regulator
   - Positive Crankcase Ventilation (PCV) hose from the throttle body

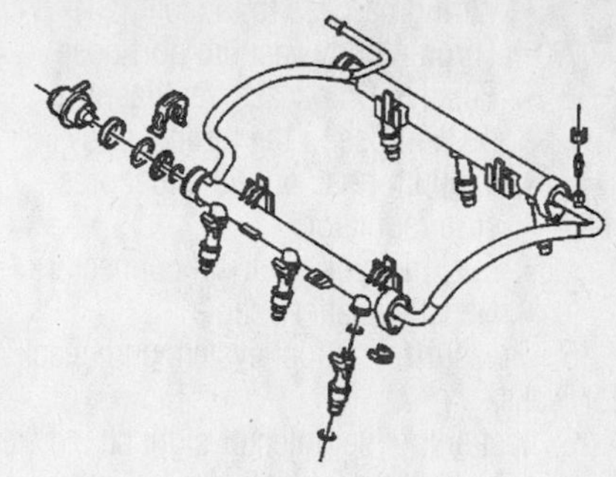

9306XG07

Exploded view of the fuel rail assembly–3.5L engine

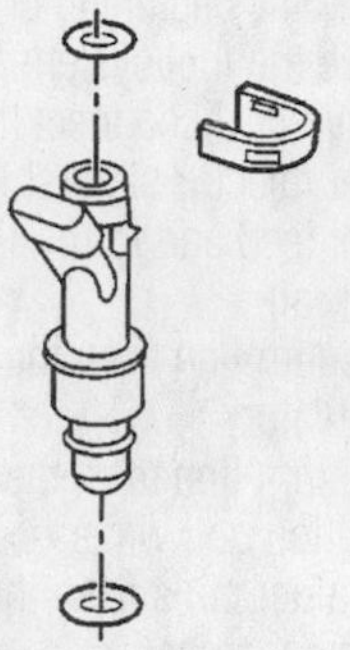

9306XG08

Exploded view of the fuel injector–3.5L engine

   - Throttle body vacuum port hose
   - Wiring harness from the fuel rail
   - Fuel rail from the intake manifold by pushing the locking tab away from the center of the intake manifold
   - Fuel injector from the fuel rail and by spreading the retainer clip to release it

To install:

⁂ WARNING

If the O-rings are different colors, install the black O-ring in the upper position and the brown O-ring in the lower position.

4. Install or connect the following:
   - New O-rings on the fuel injectors
   - Fuel injector on the fuel rail by pushing the retainer far enough to engage the clip

➡Position the fuel rail so the electrical connectors may be installed easily

   - Fuel rail onto the intake manifold by pushing the locking tab until it locks in place

- Wiring harness to the fuel rail
- Throttle body vacuum port hose
- Positive Crankcase Ventilation (PCV) hose to the throttle body
- Vacuum hose onto the fuel pressure regulator
- Fuel injector electrical connectors
- Negative battery cable

5. Pressurize the fuel system and check for leaks.
6. Install the fuel injector sight shield. Torque the bolts to 27 inch lbs. (3 Nm).

3.8L Engine

1. Before servicing the vehicle, refer to the precautions in the beginning of this section.
2. Relieve the fuel system pressure.
3. Remove or disconnect the following:
   - Fuel injector sight shield
   - Fuel feed and return lines from the fuel rail
   - Vacuum line from the fuel pressure regulator
   - Vacuum line from the throttle body
   - Ignition coil wires
   - Retainer clips from the top of the supercharger
   - Alternator and rear mount bracket
   - Fuel injector electrical connectors
   - Fuel rail hold-down bolts
   - Fuel Rail by applying equal force on both sides
   - Fuel injector-to-fuel rail retainer clips
   - Fuel injectors

To install:

⁂ WARNING

If the O-rings are different colors, install the black O-ring in the upper position and the brown O-ring in the lower position.

4. Install or connect the following:
   - New O-rings on the fuel injectors

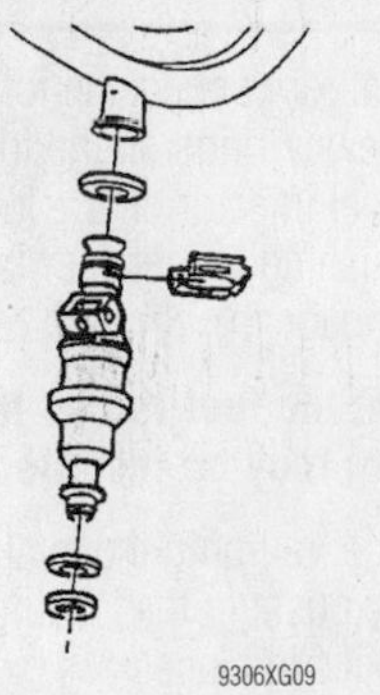

9306XG09

Exploded view of the fuel injector—3.8L engine

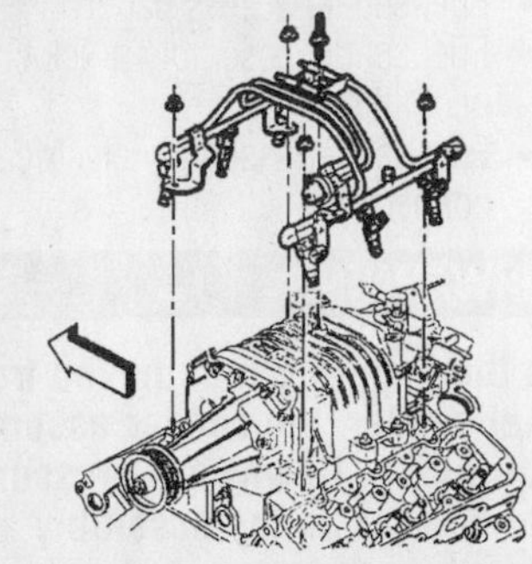

9306XG10

Exploded view of the fuel rail assembly—3.8L (VIN 1) engine

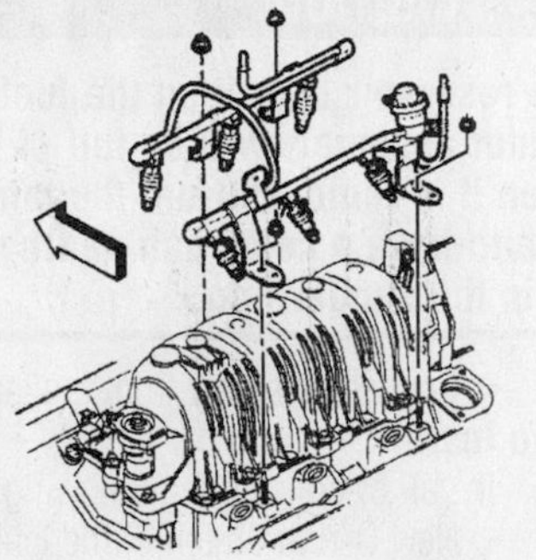

9306XG11

Exploded view of the fuel rail assembly—3.8L (VIN K) engine

   - Fuel injectors onto the fuel rail and secure with the retainers
   - Fuel rail. Torque the bolts to 7 ft. lbs. (10 Nm) and the stud to 18 ft. lbs. (25 Nm).
   - Alternator and rear mount bracket. Torque the bolts to 37 ft. lbs. (50 Nm).
   - Fuel injector electrical connectors
   - Ignition coil wires
   - Retainer clips to the top of the supercharger
   - Vacuum line to the fuel pressure regulator
   - Vacuum line to the throttle body
   - Fuel feed and return lines to the fuel rail by squeezing the tabs and pulling the lines apart
   - Negative battery cable
   - Rear seat cushion

5. Pressurize the fuel system and check for leaks.
6. Install the fuel injector sight shield. Torque the bolts to 18 inch lbs. (2 Nm).

4.0L Engine

1998–99 MODELS

1. Before servicing the vehicle, refer to the precautions in the beginning of this section.
2. Relieve the fuel system pressure.
3. Remove or disconnect the following:
   - Air intake duct
   - Exhaust Gas Recirculation (EGR) pipe at throttle body spacer
   - Brake booster vacuum hose at intake manifold vacuum fitting
   - Front bank spark plug wires
   - Manifold Absolute Pressure (MAP) sensor electrical connector from the Mass Air Flow (MAF) sensor
   - Fuel injector electrical connectors
   - Fuel rail ground wire bolt at the rear cylinder head
   - Fuel pressure regulator from the fuel rail
   - Crankcase ventilation vacuum lines
   - Quick-connect fittings at the fuel rail
   - Fuel rail end-point bracket retainer bolt
   - Throttle body mount bolts
   - Throttle body
   - Throttle body spacer
   - Intake manifold to throttle body spacer O-ring seal
   - Fuel rail-to-intake manifold locking tab, release it by pushing it toward the center of the intake manifold
   - Fuel injector from the fuel rail by using removal tool J 41081
   - Fuel injector

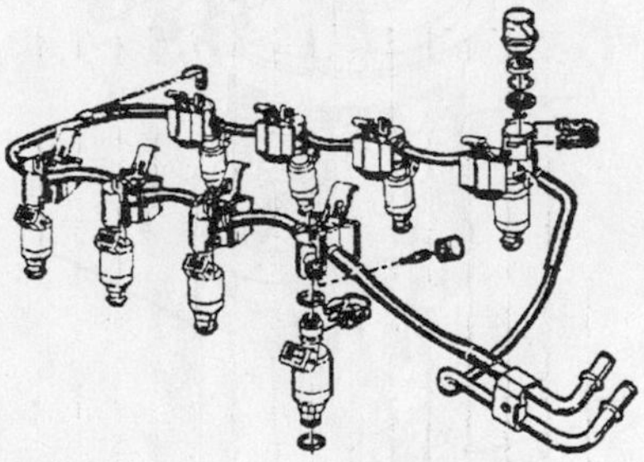

9306XG12

Exploded view of the fuel rail and fuel injector—4.0L engine

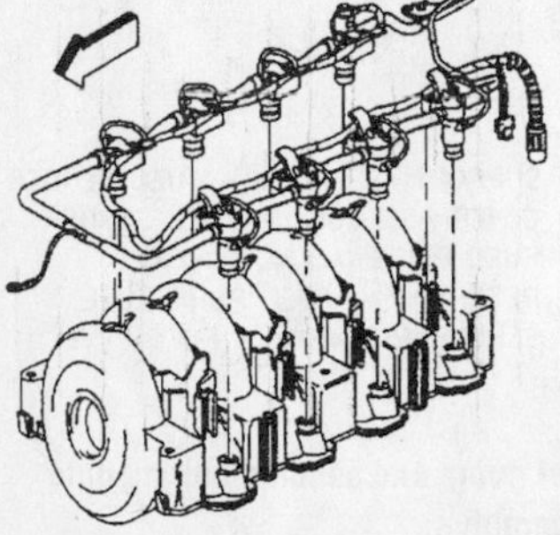

9306XG13

Exploded view of the fuel rail and intake manifold—4.0L engine

To install:

4. Lubricate the new O-rings with engine oil.
5. Install or connect the following:
   - Fuel injector using a new retainer clip and new O-rings
   - Fuel rail onto the intake manifold until the retainers latch
   - Fuel injector electrical connectors
   - Intake manifold to throttle body spacer O-ring seal
   - Throttle body spacer and tighten to 106 inch lbs. (12 Nm)
   - Throttle body
   - Throttle body mount bolts and tighten to 106 inch lbs. (12 Nm)
   - Fuel pressure regulator to the fuel rail
   - Quick-connect fittings at the fuel rail
   - Crankcase ventilation vacuum lines
   - Fuel rail ground wire bolt at the rear cylinder head
   - Fuel injector electrical connectors
   - MAP sensor electrical connector to the MAF sensor
   - Front bank spark plug wires
   - Brake booster vacuum hose at intake manifold vacuum fitting
   - EGR pipe at throttle body spacer
   - Fuel rail end-point bracket retainer bolt and tighten to 97 inch lbs. (11 Nm)
   - Air intake duct
   - Negative battery cable
6. Pressurize the system and check for leaks.

2001 MODELS

1. Before servicing the vehicle, refer to the precautions in the beginning of this section.
2. Relieve the fuel system pressure.
3. Remove or disconnect the following:
   - Air intake duct
   - Quick-connect fittings at the fuel rail
   - Fuel pressure regulator and Positive Crankcase ventilation Valve (PCV) vacuum hoses
   - Fuel rail end-point bracket retainer bolt
   - Fuel injector electrical connectors
   - Fuel rail bolts
   - Rail and injectors as an assembly from the manifold
   - Fuel injector from the fuel rail by spreading the retainer
   - Fuel injector

To install:

4. Lubricate the new O-rings with engine oil.
5. Install or connect the following:
   - Fuel injector using a new retainer clip and new O-rings
   - Rail and injectors as an assembly into the manifold until fully seated
   - Fuel rail bolts and tighten to 88 inch lbs. (10 Nm)
   - Fuel injector electrical connectors
   - Fuel rail end-point bracket retainer bolt and tighten to 88 inch lbs. (10 Nm)
   - Fuel pressure regulator and PCV vacuum hoses
   - Quick-connect fittings at the fuel rail
   - Air intake duct
   - Negative battery cable
6. Pressurize the system and check for leaks.

DRIVE TRAIN

Transmission Assembly

REMOVAL & INSTALLATION

Riviera

1. Before servicing the vehicle, refer to the precautions in the beginning of this section.
2. Remove or disconnect the following:
   - Negative battery cable
   - Air intake duct
   - Transmission electrical connector
   - Range selector cable and bracket
   - Top transmission bolts
   - Oxygen (O_2S) sensor electrical connector
   - Wheels
   - Front air deflector
   - Stabilizer links
   - Splash shield at the front crossmember
   - Wiring harness from the front crossmember
   - Intermediate shaft
   - Heat shield
   - Rack and pinion cooler pipe
   - Rack and pinion
   - Front and rear transmission mounts
   - Torque strut
3. Support the frame.
4. Remove the bolts and lower the frame.
5. Remove or disconnect the following:
   - Flexplate cover
   - Torque converter bolts
   - Input Speed Sensor (ISS) electrical connectors
   - Engine-to-transmission bracket
   - Transmission filler tube
   - Rear engine mount
   - Remaining (lower) transmission bolts
   - Transmission

To install:

6. Install or connect the following:
   - Transmission. Torque the bolts to 55 ft. lbs. (75 Nm).
   - Rear engine mount. Torque the bolts to 52 ft. lbs. (70 Nm).
   - Transmission filler tube. Torque the bolt to 15 ft. lbs. (20 Nm).
   - Transmission brace. Torque the bolts to 44 ft. lbs. (60 Nm).
   - ISS electrical connectors
   - Torque converter. Torque the bolts to 46 ft. lbs. (62 Nm).
   - Flexplate cover. Torque the bolts to 115 inch lbs. (13 Nm).
   - Engine frame. Torque the bolts to 66 ft. lbs. (90 Nm).
7. Remove the frame support fixture.
8. Install or connect the following:
   - Torque strut. Torque the bolt to 30 ft. lbs. (40 Nm).
   - Front and rear transmission mounts. Torque the bolts to 30 ft. lbs. (40 Nm).
   - Rack and pinion. Torque the bolts to 48 ft. lbs. (65 Nm).
   - Intermediate shaft. Torque the pinch bolt to 35 ft. lbs. (47 Nm).
   - Rack and pinion cooler pipe
   - Heat shield. Torque the bolts to 89 inch lbs. (10 Nm).
   - Wiring harness to the front crossmember
   - Splash shield to the front crossmember
   - Stabilizer links. Torque the nuts to 13 ft. lbs. (17 Nm).
   - Front air deflector
   - Wheels
   - (O_2S) electrical connector
   - Range selector cable and bracket
   - Transmission electrical connector

- Air intake duct
- Negative battery cable

9. Check and adjust the transmission fluid level.

10. Road test the vehicle and check for leaks.

Aurora

4T80-E TRANSMISSION

1. Before servicing the vehicle, refer to the precautions in the beginning of this section.
2. Remove or disconnect the following:
   - Negative battery cable
   - Air intake duct
   - Range selector cable from the transmission
   - Upper transmission oil cooler line from the radiator
   - Lower transmission oil cooler line from the transmission
   - Both heater tube retainers from the upper case side cover
   - Coolant temperature sensor electrical connector
   - Ground wire at the rear of the transmission
   - Left front Wheel Speed Sensor (WSS) electrical connector
   - Vehicle Speed Sensor (VSS) from the transmission
   - Rack and pinion electrical connector
   - Transmission vent hose
   - Upper transmission bolts
3. Install an engine support fixture.
4. Remove or disconnect the following:
   - Vacuum reservoir
   - Ball joints from the steering knuckles
   - Antilock Brake System (ABS) module
   - Secondary Air Injection (AIR) pumps
   - Left and right splash shields
   - Air deflector
   - Power steering line brackets at the frame
   - Front and rear transmission mount nuts
   - Rack and pinion from the frame
5. Support the engine frame.
6. Remove the 6 frame bolts and lower the frame from the body.
7. Remove or disconnect the following:
   - Left and right halfshafts
   - Exhaust heat shield
   - Transmission-to-engine brace
   - Front engine-to-transmission pencil brace
   - Three ground connections at the front of the transmission
   - Transmission main wiring harness
   - Engine to transmission bracket
   - Flexplate cover
   - Torque converter bolts
   - Right transmission bracket
8. Support the transmission.
9. Remove the lower transmission bolts and lower the transmission towards the left so it can clear the starter motor.

To install:

10. Install or connect the following:
    - Transmission. Torque the transmission-to-engine bolts to 55 ft. lbs. (75 Nm).
    - Right transmission bracket. Torque the bolt to 55 ft. lbs. (75 Nm).
    - Torque converter bolts. Torque the bolts to 47 ft. lbs. (63 Nm).
    - Flexplate cover. Torque the bolts to 20 ft. lbs. (27 Nm).
    - Transmission brace to the engine. Torque the bolts to 44 ft. lbs. (60 Nm).
    - Transmission main wiring harness
    - Three ground connections to the transmission
    - Front engine-to-transmission pencil brace
    - Engine-to-transmission bracket
    - Exhaust heat shield
    - Left and right halfshafts
    - Engine frame. Torque the bolts to 142 ft. lbs. (192 Nm).
11. Remove the frame support fixture.
12. Install or connect the following:
    - Rack and pinion. Torque the bolts to 48 ft. lbs. (65 Nm).
    - Front and rear transmission mounts. Torque the nuts to 48 ft. lbs. (65 Nm).
    - Power steering line brackets. Torque the bolts to 53 inch lbs. (6 Nm).
    - Air deflector
    - Left and right splash shields
    - AIR pumps
    - ABS module
    - Lower ball joints
13. Remove the engine support fixture.
14. Install or connect the following:
    - Vacuum reservoir
    - Upper transmission bolts. Torque the bolts to 55 ft. lbs. (75 Nm).
    - Transmission vent hose
    - Rack and pinion electrical connector
    - VSS to the transmission
    - WSS electrical connector
    - Ground wire at the rear of the transmission
    - Coolant temperature sensor electrical connector
    - Both heater tube retainers from the upper case side cover
    - Lower transmission oil cooler line
    - Upper transmission oil cooler line
    - Range selector cable and bracket
    - Air intake duct
    - Negative battery cable
15. Check and adjust the transmission fluid level.
16. Check the front end alignment.

4T65-E TRANSMISSION

1. Before servicing the vehicle, refer to the precautions in the beginning of this section.
2. Install a suitable engine support fixture
3. Remove or disconnect the following:
   - Negative battery cable
   - Air cleaner assembly
   - Transaxle electrical connector
   - Shift control cable bracket with cable attached
   - Upper transaxle case-to-engine bolts
   - Upper bolt from the rear transaxle mount
   - Oxygen (O_2S) sensor connector
   - Engine frame
   - Left and right drive axles from the transaxle
   - Torque converter cover
   - Flywheel-to-torque converter bolts
   - Input Speed Sensor (ISS) electrical connectors
   - Engine-to-transaxle brace
   - Transaxle fluid filler tube
   - Right transaxle mount
   - Remaining transaxle-to-engine bolts
   - Transaxle from vehicle

To install:

4. Install or connect the following:
   - Transmission. Torque the transmission-to-engine bolts to 55 ft. lbs. (75 Nm).
   - Right transmission bracket. Torque the bolt to 55 ft. lbs. (75 Nm).
   - Transaxle fluid filler tube
   - Engine-to-transaxle brace and tighten to 44 ft. lbs. (60 Nm)
   - ISS electrical connectors
   - Flywheel-to-torque converter bolts and tighten to 47 ft. lbs. (63 Nm)
   - Torque converter cover
   - Left and right drive axles from the transaxle
   - Engine frame and tighten the 6 frame insulator bolts to 142 inch lbs. (192 Nm)
   - O_2S sensor connector
   - Upper transaxle case-to-engine

bolts and tighten to 55 ft. lbs. (75 Nm)
- Upper bolt from the rear transaxle mount and tighten to 42 ft. lbs. (58 Nm)
- Shift control cable bracket with cable attached
- Transaxle electrical connector
- Air cleaner assembly
- Negative battery cable

5. Check and adjust the transmission fluid level.
6. Check the front end alignment.

Halfshaft

REMOVAL & INSTALLATION

1. Before servicing the vehicle, refer to the precautions in the beginning of this section.
2. Remove or disconnect the following:
   - Front wheel
   - Sway bar link kit
   - Lower ball joint
   - Halfshaft hub nut
   - Halfshaft from the hub

➡Once the halfshaft is clear of the knuckle, swing the strut assembly rearward.

- Halfshaft from the transmission using a slide hammer and adapter

To install:

3. If installing the right side halfshaft, install a tear away axle seal protector over the seal.
4. Install or connect the following:
   - Halfshaft into the transmission

➡Be sure the joint is properly seated by grasping the inboard joint and making sure it won't pull out of the transmission. DO NOT pull on the shaft itself or the inboard joint can become damaged.

- Halfshaft through the hub. Torque the nut to 107 ft. lbs. (145 Nm).
- Ball joint to the steering knuckle. Torque the castle nut to 41 ft. lbs. (55 Nm).

➡If necessary to align the cotter pin holes, rotate the nut up to an additional 60 degree turn. NEVER loosen the nut to align the holes.

- Sway bar link kit. Torque the bolt to 13 ft. lbs. (17 Nm).

5. Remove the tear away seal protector. Be sure no pieces of the protector remain in the transmission.
6. Install the front wheel and torque the nuts to 100 ft. lbs. (140 Nm).

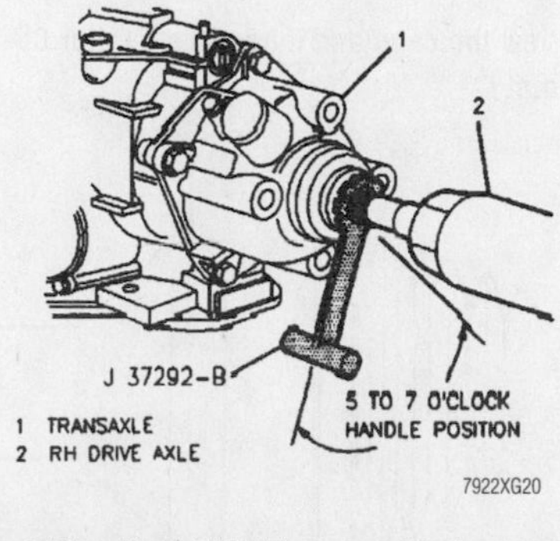

Installing the right side axle with the tear away seal protector

CV-Joint

REMOVAL & INSTALLATION

Inner (Tri-Pod) Joint

1. Before servicing the vehicle, refer to the precautions in the beginning of this section.
2. Remove or disconnect the following:
   - Front wheel
   - Halfshaft
   - Swage ring
   - Large CV-joint boot clamp
   - CV-joint boot by sliding it away from the tri-pod joint
   - Tri-pod housing from the tri-pod spider
   - Trilobal tri-pod bushing from the housing
   - Inboard spacer ring slide, it rearward on the shaft
   - Outboard retaining ring
   - Tri-pod joint spider assembly

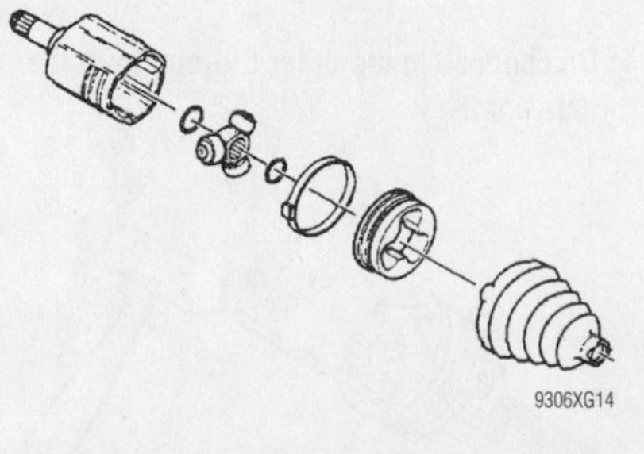

Exploded view of the inner (tri-pod) joint

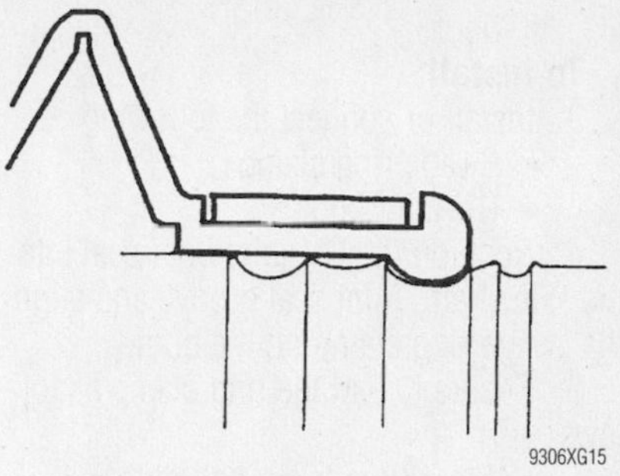

Positioning the inner CV-joint boot seal and swage ring—Inner (tri-pod) joint

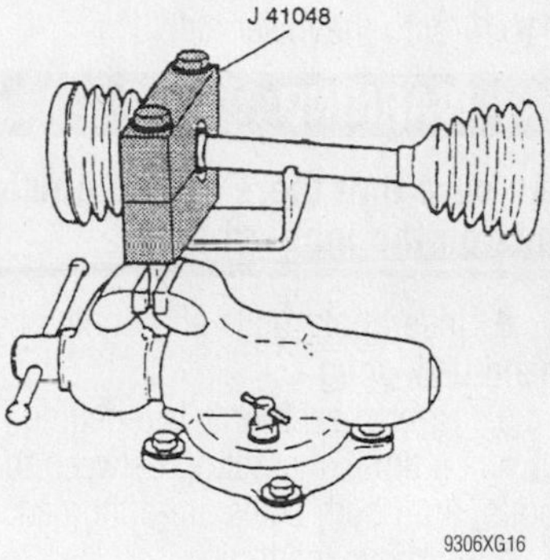

View of the swage ring crimping tool—Inner (tri-pod) joint

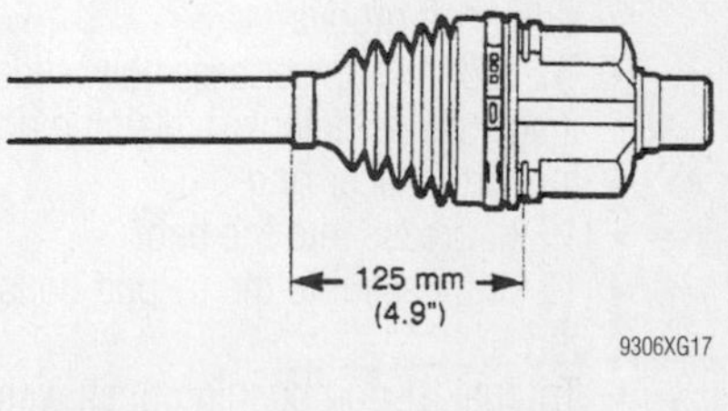

Boot measurement—Inner (tri-pod) joint

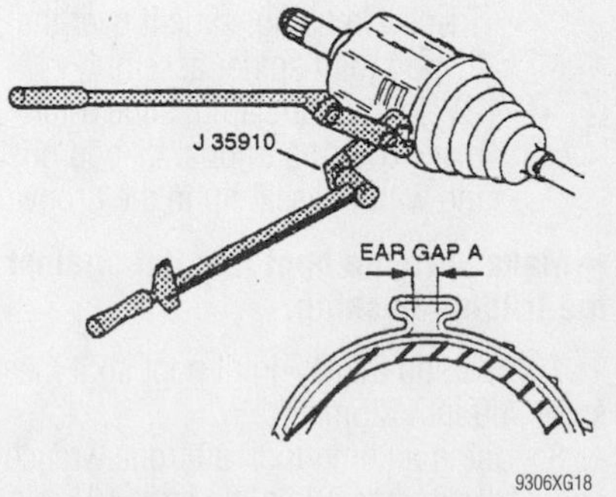

Crimping the large CV-joint boot ring—Inner (tri-pod) joint

- Inboard spacer ring and CV-joint boot

To install:

3. Install or connect the following:
   - Swage ring clamp
   - CV-joint boot

4. Position the CV-joint boot seal into the axle shaft's joint seal groove and align the swage ring clamp on the boot.

5. Secure the swage ring clamp as follows:

a. Mount the lower half of tool J-41048 in a vise.

b. Position the outboard of the halfshaft in the tool.

c. Position the upper end of tool J-41048 onto the lower half.

WARNING

Make sure that there are no pinch points on the inboard seal.

d. Insert both bolts and tighten by hand until snug.

e. Tighten each bolt 180 degree (½) turn at a time, alternating between the bolts, until both sides are bottomed.

f. Remove the tool.

6. Install or connect the following:
   - Inboard spacer ring slide it rearward on the shaft
   - Tri-pod joint spider assembly onto the shaft
   - Outboard retaining ring into the axle shaft groove
   - Tri-pod joint spider assembly, slide it against the outboard retaining ring
   - Inboard spacer ring
   - ½ kit grease into the boot
   - ½ kit grease into the tri-pod housing
   - Trilobal tri-pod bushing flush with the tri-pod housing face
   - New large seal clamp onto the CV-joint boot
   - Tri-pod housing, slide it over the tri-pod joint spider assembly
   - CV-joint boot/clamp, slide it into place over the trilobal tri-pod bushing with the seal lip in the groove

➡Make sure the boot lies flat against the trilobal bushing.

7. Position the CV-joint boot so it measures 4.9 in. (125mm).

8. Using a crimp tool, a torque wrench and a breaker bar, crimp the large CV-joint boot clamp to 130 ft. lbs. (176 Nm).

9. Install or connect the following:
   - Halfshaft
   - Front wheel

Outer Joint

1. Before servicing the vehicle, refer to the precautions in the beginning of this section.

2. Remove or disconnect the following:
   - Front wheel
   - Halfshaft
   - Swage ring
   - Large boot clamp
   - CV-joint boot, slide it away from the CV-joint
   - CV-joint assembly by spreading the inner race-to-axle shaft retaining ring ears
   - CV-joint boot from the axle shaft

3. Disassemble the chrome alloy balls from the CV-joint cage as follows:

a. Position a brass drift against the CV-joint cage and tap it with a hammer to tilt the cage.

b. Remove the 1st chrome alloy ball from the cage.

c. Tilt the cage in the opposite direction.

d. Remove the opposite chrome alloy ball.

e. Repeat the procedure until all 6 balls are removed.

4. Disassemble the CV-joint cage and inner race as follows:

a. Pivot the cage and race 90 degrees to the center line of the outer race.

b. Align the cage windows with outer race lands.

c. Remove the cage from the outer race.

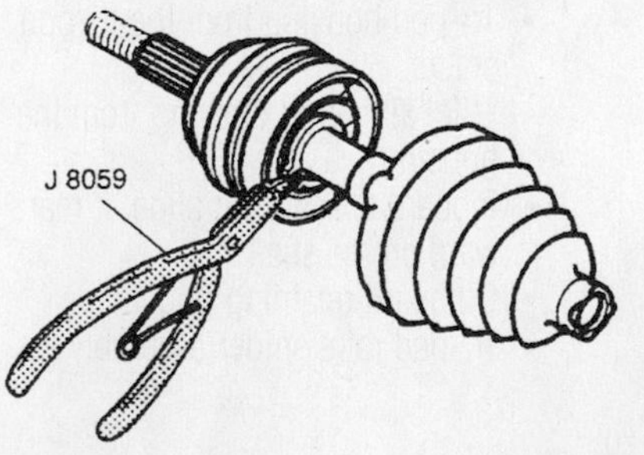

Disconnecting the outer CV-joint from the axle shaft

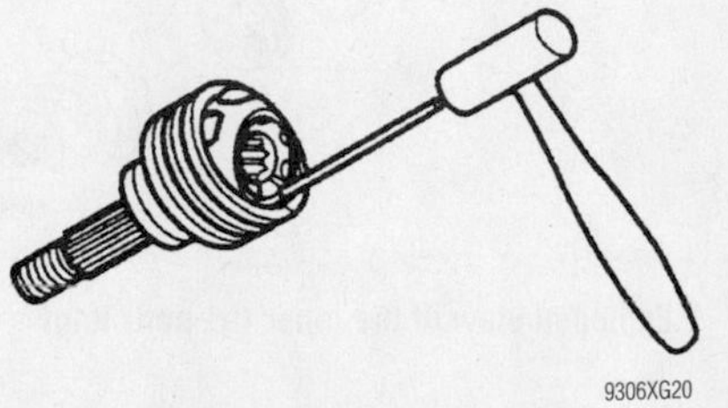

Tilting the cage—Outer CV-joint

d. Rotate the inner race upward and remove it from the cage.

To install:

5. Lubricate the parts with a light coat of grease.

6. Assemble the CV-joint cage and inner race, as follows:

a. Rotate the inner race 90 degrees to the cage centerline.

b. Align the cage windows with inner race lands.

c. Insert the inner race into the cage by rotating the inner race downward.

d. Insert the cage/inner race into the outer race.

7. Assemble the chrome alloy balls into the CV-joint cage, as follows:

a. Position a brass drift against the CV-joint cage and tap it with a hammer to tilt the cage.

b. Insert the 1st chrome alloy ball into the cage.

c. Tilt the cage in the opposite direction.

d. Insert the opposite chrome alloy ball.

e. Repeat the procedure until all 6 balls are inserted.

8. Install or connect the following:
   - ½ kit grease into the CV-joint boot
   - ½ kit grease into the CV-joint

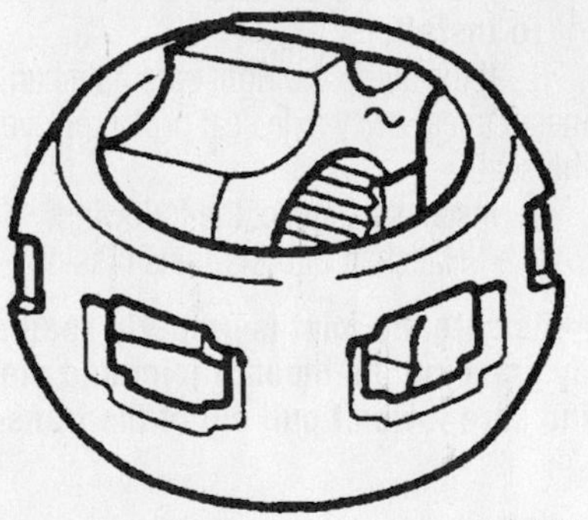

View the cage and inner race—Outer CV-joint

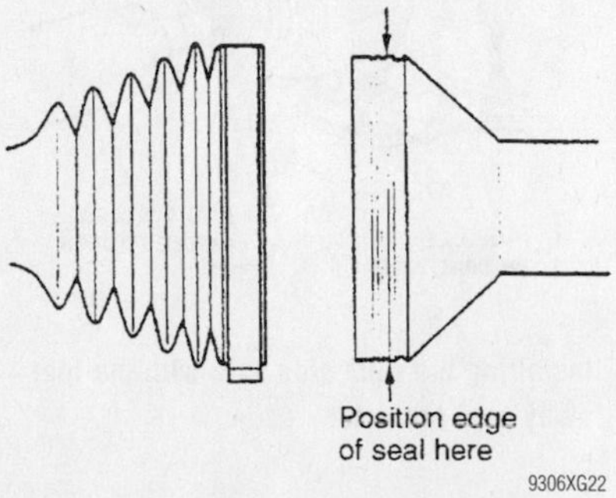

Positioning the boot—Outer CV-joint

- Swage ring clamp
- CV-joint boot
- CV-joint onto the axle shaft until the retaining ring seats into the groove

9. Position the CV-joint boot seal into the axle shaft's joint seal groove and align the swage ring clamp on the boot.

10. Secure the swage ring clamp as follows:

a. Mount the lower half of tool J-41048 in a vise.

b. Position the outboard of the halfshaft in the tool.

c. Position the upper end of tool J-41048 onto the lower half.

⁂ WARNING

Make sure that there are no pinch points on the inboard seal.

d. Insert both bolts and tighten by hand until snug.

e. Tighten each bolt 180 degree (½) turn at a time, alternating between the bolts, until both sides are bottomed.

f. Remove the tool.

11. Install or connect the following:

- New large seal clamp onto the CV-joint boot
- CV-joint boot/clamp, slide it into place over the outer race with the seal lip in the groove

➡Make sure the boot lies flat against the outer race.

12. Using a crimp tool, a torque wrench and a breaker bar, crimp the large CV-joint boot clamp to 130 ft. lbs. (176 Nm).

13. Install or connect the following:

- Halfshaft
- Front wheel

STEERING AND SUSPENSION

Air Bag

⁂ CAUTION

The vehicles are equipped with the Supplemental Inflatable Restraint (SIR) or air bag system.

The SIR system must be disabled before performing service on or around SIR system components, steering column, instrument panel components, wiring and sensors. Failure to follow safety and disabling procedures could result in accidental air bag deployment, possible personal injury and unnecessary SIR system repairs.

PRECAUTIONS

Several precautions must be observed when handling the inflator module to avoid accidental deployment and possible personal injury.

- Never carry the inflator module by the wires or connector on the underside of the module.
- When carrying a live inflator module, hold securely with both hands and ensure that the bag and trim cover are pointed away.
- Place the inflator module on a bench or other surface with the bag and trim cover facing up.
- With the inflator module on the bench, never place anything on or close to the module, which may be thrown in the event of an accidental deployment.

DISARMING

1. Turn the steering wheel so the vehicle wheels are pointing straight-ahead.
2. Turn the ignition key to the **LOCK** position and remove the key.
3. Remove the AIR BAG fuse from the fuse block.
4. Remove the left side sound insulator.
5. Detach the Connector Position Assurance (CPA) and yellow 2-way Supplemental Inflatable Restraint (SIR) connector at the multi-use bracket near the base of the steering column. The driver's side air bag is now disabled.
6. Remove the right side sound insulator.
7. Detach the Connector Position Assurance (CPA) and yellow 2-way SIR connector at the base of the steering wheel. The passenger's side air bag is now disabled.

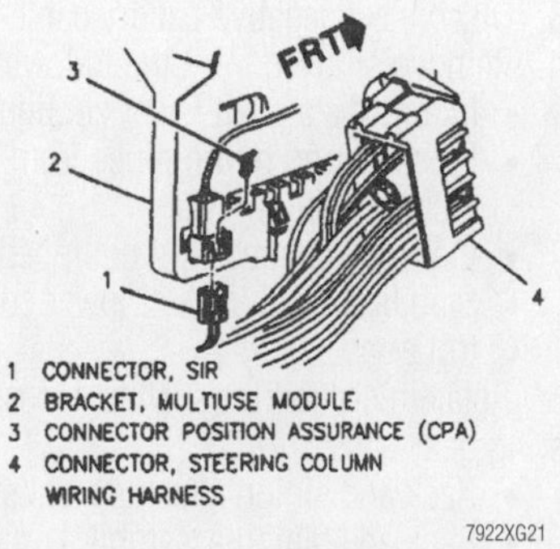

SRS 2-way connector location—driver's side

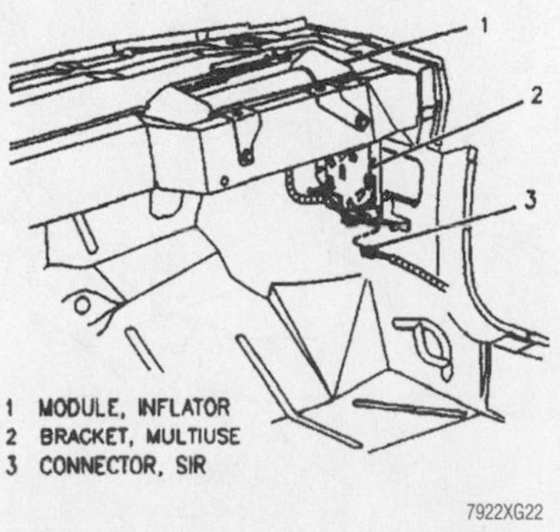

SRS 2-way connector location—passenger's side

ARMING

After necessary repairs are made, re-enable the air bag system as follows:

1. Be sure the ignition is locked and the key is removed.
2. Attach the yellow 2-way Supplemental Inflatable Restraint (SIR) connector and Connector Position Assurance (CPA) at the base of the steering wheel.
3. Install the right side sound insulator.
4. Attach the yellow 2-way SIR connector and Connector Position Assurance (CPA) at the multi-use bracket at the base of the column.
5. Install the left side sound insulator.
6. Install the AIR BAG fuse.
7. Turn the ignition switch to the **RUN** position and verify the AIR BAG light flashes 7 times, then shuts off.

Power Rack and Pinion Steering Gear

REMOVAL & INSTALLATION

Except Aurora

1. Before servicing the vehicle, refer to the precautions in the beginning of this section.

2. Lock the steering wheel in the straight-ahead position.
3. Remove or disconnect the following:
- Front wheel
- Outer tie rod end from the steering knuckle
- Exhaust pipe from the rear manifold
- Intermediate exhaust pipe hangers
- Intermediate shaft from the rack and pinion stub shaft
- Rack and pinion heat shield
- Magnasteer®electrical connector
- Power steering lines from the rack and pinion and cap them
- Rack and pinion-to-chassis bolts
4. Support the rear of the subframe.
5. Remove both rear subframe-to-chassis bolts.
6. Lower the frame enough to remove the rack and pinion assembly.
7. Remove the rack and pinion assembly through the right side wheel well.

To install:

8. Install or connect the following:
- Rack and pinion through the right side wheel well
- 3 rack and pinion-to-chassis bolts, do not tighten
- Subframe. Torque the bolts to 142 ft. lbs. (192 Nm).
9. Remove the subframe support.
10. Install or connect the following:
- Rack and pinion bolts. Torque the bolts in sequence to 48 ft. lbs. (65 Nm).

➡Start with the vertically installed bolt closest to the pinion housing and work toward the passenger side of the vehicle.

- Power steering lines to the rack and pinion. Torque the fittings to 20 ft. lbs. (27 Nm).
- Magnasteer®electrical connector
- Rack and pinion heat shield
- Intermediate shaft to the rack and pinion stub shaft. Torque the pinch bolt to 35 ft. lbs. (47 Nm).
- Exhaust pipe to the exhaust manifold
- Intermediate exhaust pipe hangers
- Tie rod ends to the steering knuckles. Torque the castle nuts to 35 ft. lbs. (47 Nm).

➡If necessary to align the cotter pin holes, tighten the nut up to an additional 60 degree turn. NEVER loosen the castle nut to align the cotter pin holes.

- Front wheel. Torque the nuts to 100 ft. lbs. (140 Nm).

➡Whenever the vehicle sub-frame is removed or lowered, the front wheel alignment should be checked.

11. Refill and bleed the power steering system. Check for leaks.
12. Check the front end alignment and adjust as necessary.

Aurora

1. Before servicing the vehicle, refer to the precautions at the beginning of this section.

CAUTION

Make sure the wheels of the vehicle are straight ahead and the steering column in the LOCK position before disconnecting the steering column or intermediate shaft from the steering gear. Failure to do so will cause the coil assembly in the steering column to become uncentered which will cause damage to the coil assembly.

2. Lock the steering column by installing the Steering Column Anti Rotation Pin tool J 42640 into the underside of the steering column.
3. Disconnect negative battery cable.
4. Remove or disconnect the following:
- Intermediate shaft lower coupling
- Pressure and return pipes from the rack and pinion gear
- Electrical connector, if equipped
- Stabilizer shaft links at lower control arms
5. Rotate the stabilizer shaft to access steering gear bolts.
- Rack and pinion attaching bolts
- Rack and pinion assembly

To install:

6. Install or connect the following:

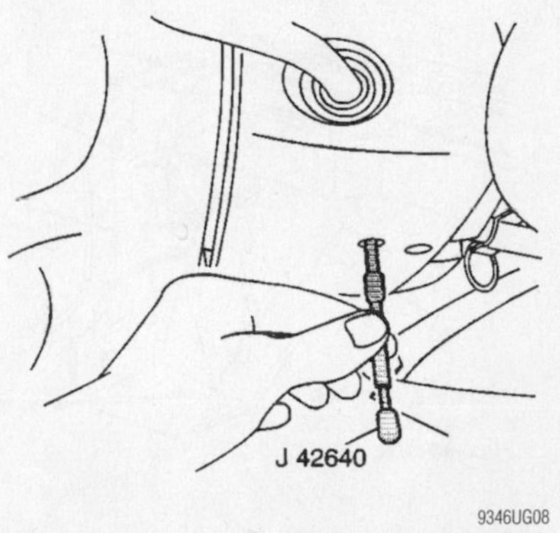

Lock the steering column by installing the Steering Column Anti Rotation Pin tool J 42640 into the underside of the steering column—Aurora

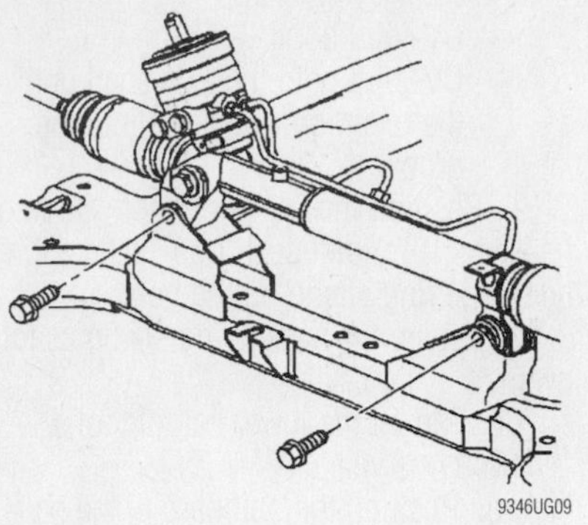

Steering gear mounting bolts—Aurora

- Rack and pinion assembly and tighten the bolts to 70 ft. lbs. (95 Nm)
- Electrical connector, if equipped
- Pressure and return lines to the steering gear and tighten the fittings to 22 ft. lbs. (30 Nm)
- Heat shield and bolts. Tighten the bolts to 7 ft. Lbs. (10 N)
7. Align stabilizer shaft.
- Links to lower control arms
- Intermediate shaft to the steering gear and tighten the pinch bolt to 33 ft. lbs. (45 Nm)
8. Remove the steering column anti rotation pin from the steering column.
9. Fill and bleed the power steering system.

Strut

REMOVAL & INSTALLATION

Front

1. Before servicing the vehicle, refer to the precautions in the beginning of this section.
2. Remove or disconnect the following:
- Negative battery cable
- Front wheel
- Anti-lock Brake System (ABS) wheel speed sensor electrical connector
- ABS speed sensor bracket from the strut
- Brake line bracket, if removing the left strut
3. Matchmark the lower strut bracket to the steering knuckle.
4. Support the steering knuckle.
5. Remove or disconnect the following:
- Lower strut bracket nut and through-bolts
- 3 upper strut plate mounting nuts and washers
- Strut from the vehicle

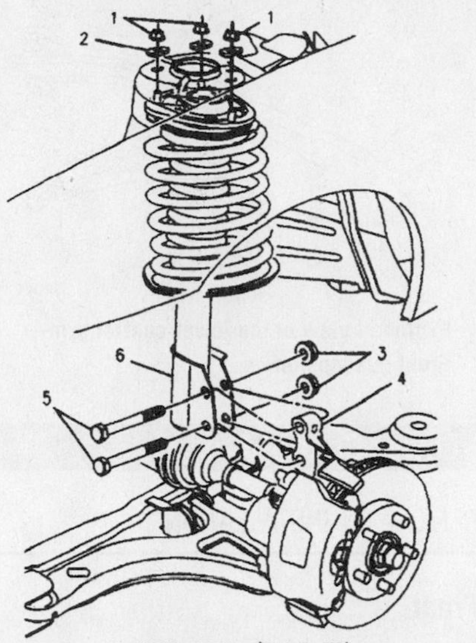

Exploded view of the upper and lower strut mounting components

To install:

6. Install or connect the following:
   - Strut, do not tighten the upper strut plate-to-body nuts
   - Lower strut through-bolts by aligning the matchmarks. Torque the nuts to 136 ft. lbs. (185 Nm).
7. Remove the steering knuckle support.
8. Install or connect the following:
   - Brake line bracket to the strut, if removed
   - ABS wheel speed sensor bracket on the strut
   - ABS sensor electrical connector
   - Front wheel. Torque the nuts to 100 ft. lbs. (140 Nm).
   - Upper strut plate nuts. Torque the nuts to 35 ft. lbs. (47 Nm).
   - Negative battery cable
9. Check the front end alignment and adjust as necessary.

Shock Absorber

REMOVAL & INSTALLATION

Rear

1. Before servicing the vehicle, refer to the precautions in the beginning of this section.
2. Support the lower control arm at such a height that the upper shock bolts will still be accessible.
3. Remove or disconnect the following:

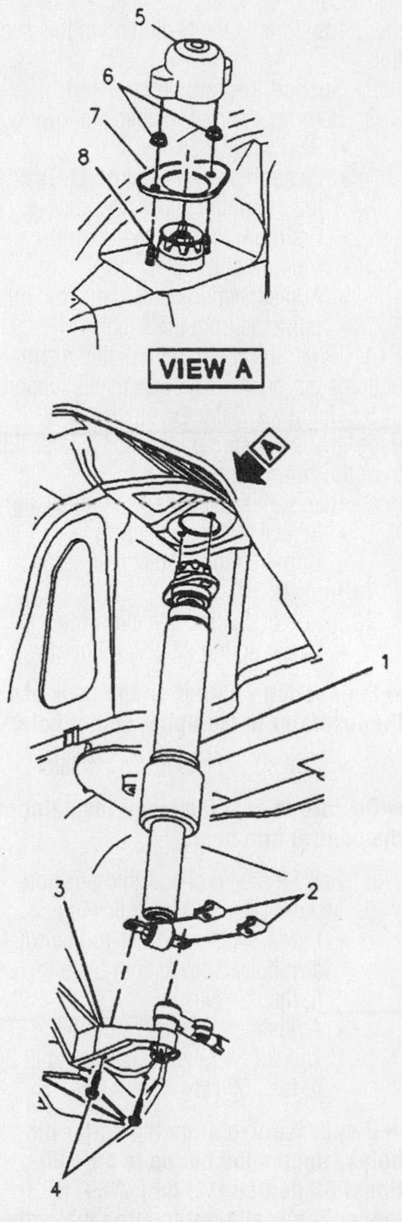

Exploded view of the rear shock mounting

- Rear wheel
- Electronic Level Control (ELC) air tube from the shock
- Both lower shock mount bolts
- Trunk trim panel to access the upper shock mount bolts
- Upper shock cap
- Both upper shock mount nuts and reinforcement
- Shock

To install:

4. Install or connect the following:
   - Shock
   - Reinforcement and upper shock mounting nuts. Torque the nuts to 15 ft. lbs. (20 Nm).
   - Upper shock cap and inner trunk trim
   - Lower shock bolts. Torque the bolts to 18 ft. lbs. (24 Nm).
   - ELC air tube to the shock
   - Rear wheel. Torque the nuts to 100 ft. lbs. (140 Nm).

Coil Spring

REMOVAL & INSTALLATION

Front

1. Remove the strut from the vehicle.
2. Disassemble the strut as follows:

a. Step 1: Place the strut assembly into compressor tool, to compress the coil spring.

b. Step 2: Compress the spring slightly.

c. Step 3: Hold the strut shaft from turning using a No. 50 Torx® socket and remove the 24mm nut on the top end of the strut.

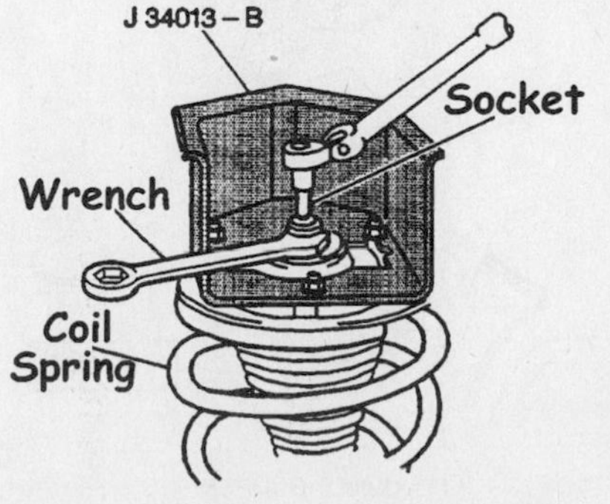

Use a Torx® socket to keep the piston rod from turning while removing the upper nut—front strut shown

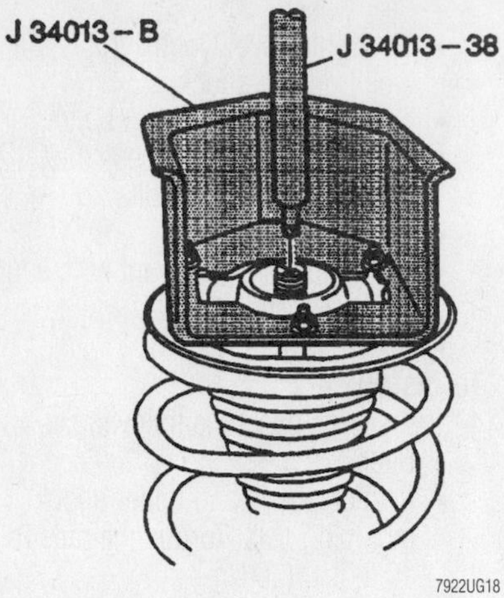

Install Rod J 34013-38 to help guide the strut shaft from the upper mount assembly—front strut shown

d. Step 4: Install Rod tool J 34013-38 to help guide the strut shaft from the upper mount assembly.

d. Step 5: Loosen the spring compressor tool until the coil spring and mount can be removed as an assembly. Remove the lower spring insulator, if equipped.

To install:

3. Assemble the strut as follows:
   a. Step 1: Place the strut in compressor tool.
   b. Step 2: Install the coil spring over the strut.
   c. Step 3: Compress the coil spring while guiding strut shaft through the top of the strut assembly.
   d. Step 4: Install the top strut nut. Torque the nut to 55 ft. lbs. (75 Nm).
   e. Step 5: Remove the strut from the compressor.

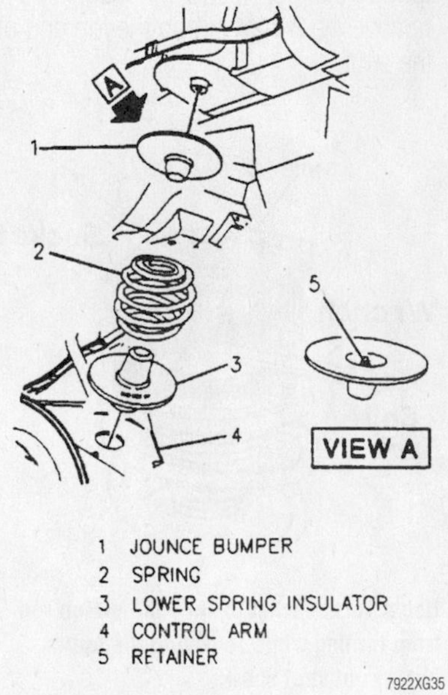

Exploded view of the rear coil spring and related components

Rear

1. Before servicing the vehicle, refer to the precautions in the beginning of this section.
2. Support the lower control arm.
3. Remove or disconnect the following:
   - Rear wheel
   - Electronic Level Control (ELC) air tube from the shock absorber
   - Both lower shock absorber-to-control arm bolts
   - Adjustment link outer ball stud nut
   - Ball stud from the knuckle
4. Lower the control arm until the arm bottoms out on the rear suspension support.
5. Using a suitable prying tool, pry under the lower spring insulator to unseat it from the control arm.
6. Remove or disconnect the following:
   - Insulator and coil spring
   - Upper spring insulator, if necessary

To install:

7. Install or connect the following:
   - Upper spring insulator, if removed

➡Engage the retainer on the back of the insulator in the upper mount hole.

   - Coil spring and lower insulator

➡Be sure to seat the lower insulator in the control arm hole.

8. Raise the control arm into position.
9. Install or connect the following:
   - Lower shock absorber-to-control arm bolts. Torque both bolts to 18 ft. lbs. (24 Nm).
   - Adjustment link ball stud to the knuckle. Torque the castle nut to 36 ft. lbs. (50 Nm).

➡If necessary to align the cotter pin holes, tighten the nut up to an additional 60 degree (1/6) turn. NEVER loosen the castle nut to align the cotter pin holes.

   - ELC air tube to the shock absorber
   - Rear wheel. Torque the nuts to 100 ft. lbs. (140 Nm).

Lower Ball Joint

REMOVAL & INSTALLATION

The ball joint is an integral part of the control arm. If defective the control arm must be replaced.

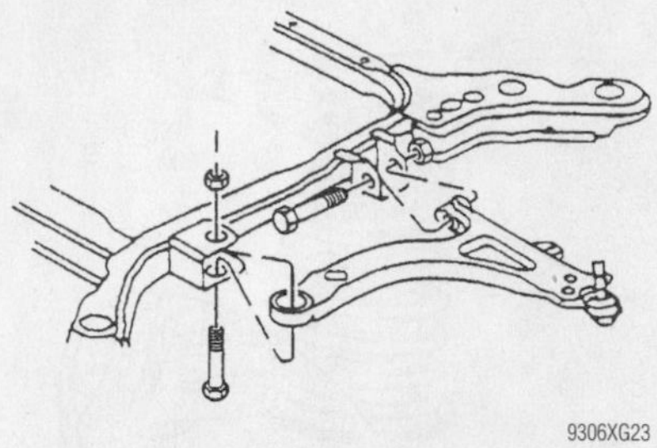

Exploded view of the lower control arm—Front suspension

Lower Control Arm

REMOVAL & INSTALLATION

Front

1. Before servicing the vehicle, refer to the precautions in the beginning of this section.
2. Remove or disconnect the following:
   - Front wheel
   - Sway bar link kit

➡Take note of the positions of the washers and insulators for installation purposes.

   - Lower ball joint from the steering knuckle

⁂ WARNING

Be careful not to overextend the halfshaft tri-pod joint.

   - Lower control arm-to-engine frame nuts and bolts
   - Lower control arm

To install:

3. Install or connect the following:
   - Lower control arm, do not tighten the nuts and bolts
   - Lower ball joint to the steering knuckle. Torque the castle nut to 41 ft. lbs. (55 Nm).

➡If necessary, tighten the castle nut up to an additional 60 degree (1/6 turn to align the cotter pin holes. NEVER loosen the nut to make the alignment.

   - Sway bar link kit. Torque the nut to 13 ft. lbs. (17 Nm).
   - Front wheel. Torque the nuts to 100 ft. lbs. (140 Nm).
4. Lower the vehicle in order to achieve proper trim height.
5. On all except Aurora, torque the lower control arm-to-frame rear bolt to 117 ft. lbs. (158 Nm) and the lower control arm-to-frame front nut to 93 ft. lbs. (126 Nm).

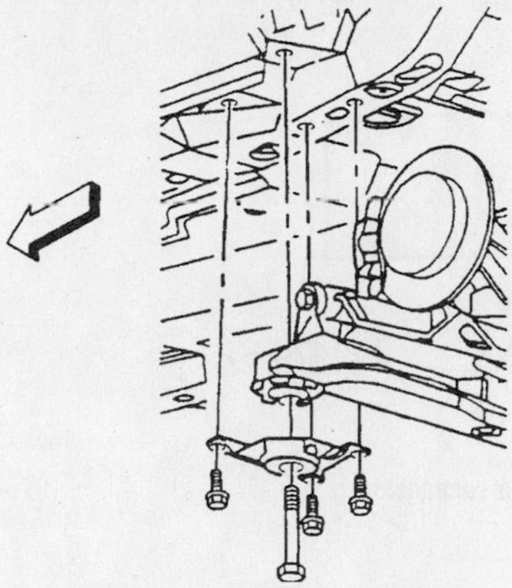

9306XG24

Exploded view of the support assembly—Rear suspension

6. On Aurora, torque the lower control arm-to-frame rear bolt to 117 ft. lbs. (158 Nm) and the lower control arm-to-frame front nut to 117 ft. lbs. (158 Nm).

Rear

1. Before servicing the vehicle, refer to the precautions in the beginning of this section.
2. Remove or disconnect the following:
   - Rear wheels
   - Exhaust system
   - Coil springs
   - Parking brake cable from the brake calipers
   - Brake calipers from the control arms
   - Parking brake cable from the rear suspension support assembly
   - Support assembly electrical connectors
   - Electronic Level Control (ELC) electrical connector and vent hose
   - ELC air tube from the air compressor
3. Support the rear suspension support assembly.
4. Remove or disconnect the following:
   - 3 support bracket-to-chassis bolts at each side
   - Both front and both rear support assembly bolts
   - Support assembly
   - ELC height sensor link from the left control arm
   - Stabilizer link bolt and nut
   - Anti-lock Brake System (ABS) electrical connector
   - Wheel/hub assembly
   - Lower control arm-to-rear suspension support assembly nuts and bolts
   - Lower control arm

To install:

5. Install or connect the following:
   - Lower control arm
   - Lower control arm-to-rear suspension support assembly nuts and bolts, do not tighten
   - Wheel/hub assembly
   - ABS electrical connector
   - Stabilizer link nut and bolt. Torque the nut to 11 ft. lbs. (15 Nm).
   - ELC height sensor link to the left control arm
   - Support assembly
   - Both front and both rear support assembly bolts. Torque the front support assembly bolts to 141 ft. lbs. (191 Nm) and the rear support assembly bolts to 122 ft. lbs. (165 Nm).
   - 3 support bracket-to-chassis bolts at each side. Torque the bolts to 63 ft. lbs. (86 Nm).
   - ELC air tube from the air compressor
   - ELC electrical connector and vent hose
   - Support assembly electrical connectors
   - Parking brake cable to the rear suspension support assembly
   - Brake calipers to the control arms
   - Parking brake cable to the brake calipers
   - Coil springs
   - Exhaust system
   - Rear wheels. Torque the nuts to 100 ft. lbs. (140 Nm).
6. Lower the vehicle in order to achieve proper trim height.

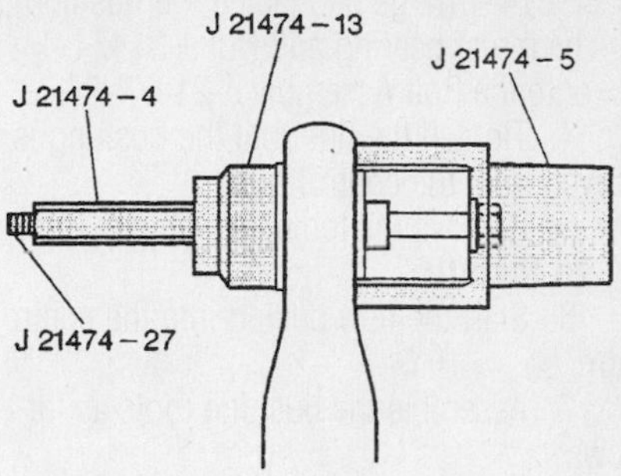

9306XG26

Installing the bushing to the lower control arm—Front suspension

7. Torque the lower control arm nuts to 78 ft. lbs. (106 Nm).

CONTROL ARM BUSHING REPLACEMENT

Front

EXCEPT AURORA—FRONT

The rear (horizontal) control bushing is replaced with the control arm and cannot be replace separately.

1. Before servicing the vehicle, refer to the precautions in the beginning of this section.
2. Remove or disconnect the following:
   - Front wheel
   - Lower control arm
3. Assemble the bushing tools as follows:
   a. Position the Bolt Assembly tool J-21474-27 with a washer, through the Bushing Receiver tool J-21474-5 with the larger diameter end over the bushing against the control arm.
   b. Lubricate the bolt threads with high pressure lubricant.
   c. Install the Bushing Remover tool J-

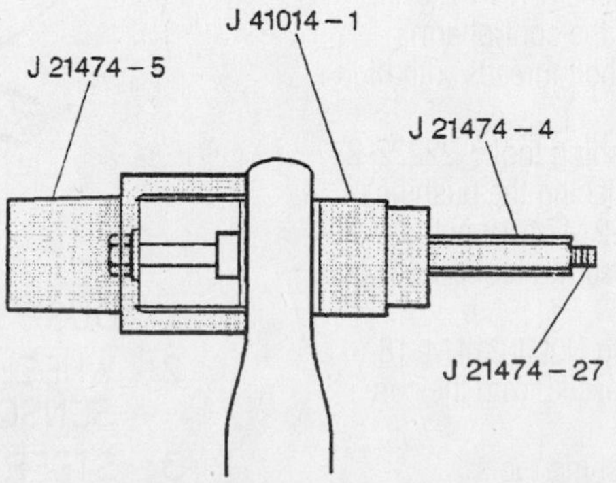

9306XG25

Removing the bushing from the lower control arm—Front suspension

41014-1 (large end facing the bushing), the thrust bearing and Nut J-21474-4 onto the Bolt Assembly J-21474-27.

4. Tighten the nut until the bushing is driven from the control arm.
5. Remove the tools.

To install:

6. Start the new busing into the control arm.
7. Assemble the bushing tools as follows:
   a. Position the Bolt Assembly tool J-21474-27 with a washer, through the Bushing Receiver tool J-21474-5 with the larger diameter end over the bushing against the control arm.
   b. Lubricate the bolt threads with high pressure lubricant.
   c. Install the Bushing Installer tool J-21474-13 (large end facing the bushing), the thrust bearing and Nut J-21474-4 onto the Bolt Assembly J-21474-27.
8. Tighten the nut until the bushing is fully seated in the control arm.
9. Remove the tools.
10. Install or connect the following:
    - Lower control arm
    - Front wheel

AURORA—REAR

The control arm bushings are not a serviceable component and if found to be defective the control arm assembly must be replaced.

1. Before servicing the vehicle, refer to the precautions in the beginning of this section.
2. Remove or disconnect the following:
   - Rear wheel
   - Lower control arm
3. Assemble the bushing tools as follows:
   a. Assemble the Puller Bolt/Thrust Bearing tool J-21474-19 through the Bushing Receiver tool J-14014-2 over the bushing against the control arm.
   b. Lubricate the bolt threads with high pressure lubricant.
   c. Install the Bushing tool J-22222-2 (with the small end facing the bushing) and the Long Nut J-21474-18 onto the Puller Bolt/Thrust Bearing tool J-21474-19.
4. Tighten the Long Nut J-21474-18 until the bushing is pressed from the control arm.
5. Remove the bushing tools.

To install:

6. Start the bushing onto the control arm.

➡Position the bushing flat vertical and rearward.

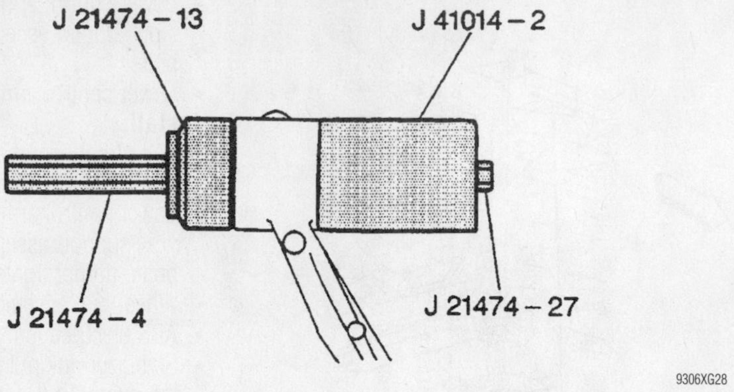

Installing the bushing to the lower control arm—Rear suspension

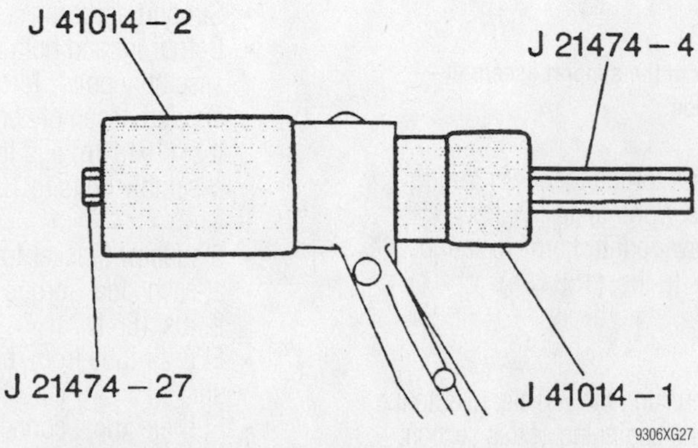

Removing the bushing from the lower control arm—Rear suspension

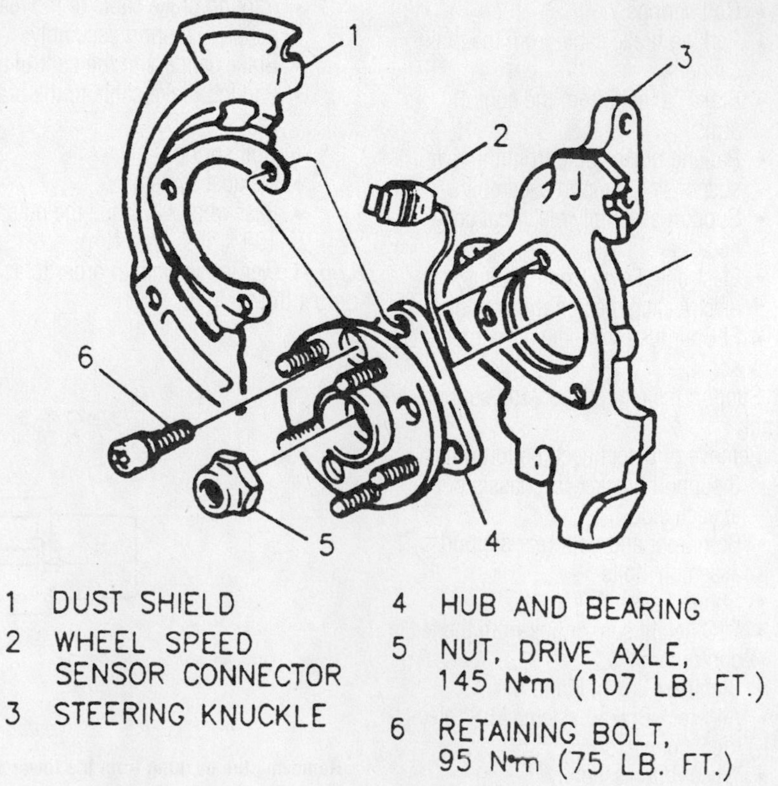

Exploded view of the front hub mounting and related components

7. Assemble the bushing tools as follows:
   a. Position the Puller Bolt/Thrust Bearing tool J-21474-19 through the Bushing Receiver tool J-14014-2 over the bushing against the control arm.
   b. Lubricate the bolt threads with high pressure lubricant.
   c. Install the Bushing Installer tool J-28685 (large end facing the bushing) and Long Nut J-21474-18 onto the Puller Bolt/Thrust Bearing tool J-21474-19.
8. Tighten the Puller Bolt/Thrust Bearing tool J-21474-19 until the bushing is fully seated in the control arm.
9. Remove the tools.
10. Install the lower control arm.
11. Install the front wheel.

Wheel Bearings

ADJUSTMENT

The wheel bearings are not adjustable. If the wheel bearings are defective, the hub and bearing assembly must be replaced.

REMOVAL & INSTALLATION

Front

The front wheel bearings are not serviced separately. If the front wheel bearings are defective, the hub and bearing assembly must be replaced.

1. Before servicing the vehicle, refer to the precautions in the beginning of this section.
2. Remove or disconnect the following:
   - Negative battery cable
   - Front wheel
3. Lubricate the threads on the halfshaft with clean engine oil.
4. Remove or disconnect the following:
   - Halfshaft hub nut
   - Caliper from the steering knuckle

⁂ WARNING

DO NOT allow the brake hose to support the weight of the caliper.

   - Brake rotor
   - Anti-lock Brake System (ABS) sensor from the backing plate
   - Hub and bearing assembly from the backing plate
   - Hub and bearing assembly from the halfshaft
   - Hub and bearing assembly

To install:

5. Install or connect the following:
   - Hub and bearing assembly onto the halfshaft
   - Hub and bearing assembly to the backing plate. Torque the 3 bolts alternately and evenly to 70 ft. lbs. (95 Nm).
   - ABS sensor
   - Brake rotor
   - Caliper on the steering knuckle. Torque the bolts to 38 ft. lbs. (51 Nm).
   - Halfshaft nut. Torque it to 107 ft. lbs. (145 Nm) on all Riviera and 1998–99 Aurora models or 118 ft. lbs. (160 Nm) on 2001 Aurora models.
   - Front wheel. Torque the nuts to 100 ft. lbs. (140 Nm).
   - Negative battery cable

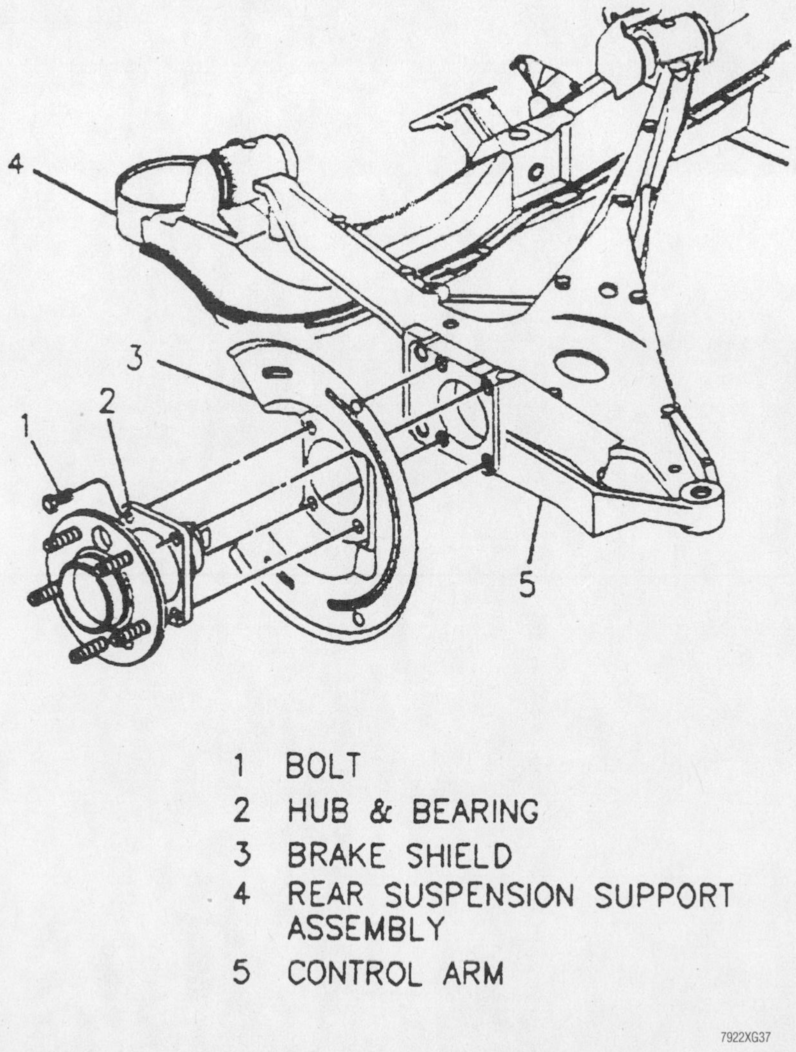

Exploded view of the rear hub mounting and related components

Rear

The rear wheel bearings are not serviced separately. If the rear wheel bearings are defective, the hub and bearing assembly must be replaced.

1. Before servicing the vehicle, refer to the precautions in the beginning of this section.
2. Remove or disconnect the following:
   - Negative battery cable
   - Rear wheel
   - Caliper and bracket assembly from the brake rotor

✻✻ WARNING

DO NOT allow the brake hose to support the weight of the caliper and bracket.

- Brake rotor
- Anti-lock Brake System (ABS) sensor electrical connector
- Hub and bearing assembly from the rear control arm

To install:

3. Install or connect the following:
   - Hub and bearing assembly onto the control arm, loosely install the bolts
   - ABS sensor electrical connector
   - Hub and bearing bolts. Torque the bolts alternately and evenly to 52 ft. lbs. (70 Nm).
   - Brake rotor
   - Caliper and bracket assembly. Torque the bolts to 35 ft. lbs. (48 Nm).
   - Rear wheel. Torque the nuts to 100 ft. lbs. (140 Nm).
   - Negative battery cable

GENERAL MOTORS CORPORATION—J-BODY

25

1998–01

Chevrolet-Cavalier • **Pontiac**-Sunfire

PRECAUTIONS

Before servicing any vehicle, please be sure to read all of the following precautions, which deal with personal safety, prevention of component damage, and important points to take into consideration when servicing a motor vehicle:

- Never open, service or drain the radiator or cooling system when the engine is hot; serious burns can occur from the steam and hot coolant.
- Observe all applicable safety precautions when working around fuel. Whenever servicing the fuel system, always work in a well-ventilated area. Do not allow fuel spray or vapors to come in contact with a spark, open flame, or excessive heat (a hot drop light, for example). Keep a dry chemical fire extinguisher near the work area. Always keep fuel in a container specifically designed for fuel storage; also, always properly seal fuel containers to avoid the possibility of fire or explosion. Refer to the additional fuel system precautions later in this section.
- Fuel injection systems often remain pressurized, even after the engine has been turned **OFF**. The fuel system pressure must be relieved before disconnecting any fuel lines. Failure to do so may result in fire and/or personal injury.
- Brake fluid often contains polyglycol ethers and polyglycols. Avoid contact with the eyes and wash your hands thoroughly after handling brake fluid. If you do get brake fluid in your eyes, flush your eyes with clean, running water for 15 minutes. If eye irritation persists, or if you have taken brake fluid internally, IMMEDIATELY seek medical assistance.
- The EPA warns that prolonged contact with used engine oil may cause a number of skin disorders, including cancer! You should make every effort to minimize your exposure to used engine oil. Protective gloves should be worn when changing oil. Wash your hands and any other exposed skin areas as soon as possible after exposure to used engine oil. Soap and water, or waterless hand cleaner should be used.
- All new vehicles are now equipped with an air bag system, often referred to as a Supplemental Restraint System (SRS) or Supplemental Inflatable Restraint (SIR) system. The system must be disabled before performing service on or around system components, steering column, instrument panel components, wiring and sensors. Failure to follow safety and disabling procedures could result in accidental air bag deployment, possible personal injury and unnecessary system repairs.
- Always wear safety goggles when working with, or around, the air bag system. When carrying a non-deployed air bag, be sure the bag and trim cover are pointed away from your body. When placing a non-deployed air bag on a work surface, always face the bag and trim cover upward, away from the surface. This will reduce the motion of the module if it is accidentally deployed. Refer to the additional air bag system precautions later in this section.
- Clean, high quality brake fluid from a sealed container is essential to the safe and proper operation of the brake system. You should always buy the correct type of brake fluid for your vehicle. If the brake fluid becomes contaminated, completely flush the system with new fluid. Never reuse any brake fluid. Any brake fluid that is removed from the system should be discarded. Also, do not allow any brake fluid to come in contact with a painted surface; it will damage the paint.
- Never operate the engine without the proper amount and type of engine oil; doing so WILL result in severe engine damage.
- Timing belt maintenance is extremely important! Many models utilize an interference-type, non-freewheeling engine. If the timing belt breaks, the valves in the cylinder head may strike the pistons, causing potentially serious (also time-consuming and expensive) engine damage. Refer to the maintenance interval charts in the front of this manual for the recommended replacement interval for the timing belt, and to the timing belt section for belt replacement and inspection.
- Disconnecting the negative battery cable on some vehicles may interfere with the functions of the on-board computer system(s) and may require the computer to undergo a relearning process once the negative battery cable is reconnected.
- When servicing drum brakes, only disassemble and assemble one side at a time, leaving the remaining side intact for reference.
- Only an MVAC-trained, EPA-certified automotive technician should service the air conditioning system or its components.

ENGINE REPAIR

Alternator

REMOVAL

2.2L Engines

1. Before servicing the vehicle, refer to the precautions in the beginning of this section.
2. Remove or disconnect the following:
   - Negative battery cable
   - Accessory drive belt
   - Alternator mounting bolts
   - Alternator electrical connectors
   - Alternator

2.4L Engines

1. Before servicing the vehicle, refer to the precautions in the beginning of this section.
2. Remove or disconnect the following:
   - Negative battery cable
   - Accessory drive belt
   - Alternator electrical connectors
   - Power steering line clip
   - Rear alternator brace
   - Alternator mounting bolts
   - Alternator

INSTALLATION

2.2L Engines

Install or connect the following:
- Alternator. Torque the bolts to 37 ft. lbs. (50 Nm).
- Alternator electrical connectors
- Accessory drive belt
- Negative battery cable

2.4L Engines

Install or connect the following:
- Alternator
- Alternator mounting bolts. Torque the bolts to 37 ft. lbs. (50 Nm).
- Rear alternator brace
- Power steering line clip
- Alternator electrical connectors
- Accessory drive belt
- Negative battery cable

Ignition Timing

ADJUSTMENT

Ignition timing is controlled by the Powertrain Control Module (PCM). No adjustment is necessary or possible.

Engine Assembly

REMOVAL & INSTALLATION

2.2L Engine

WITH MANUAL TRANSMISSION

1. Before servicing the vehicle, refer to the precautions in the beginning of this section.
2. Relieve the fuel system pressure.
3. Drain the cooling system.
4. Remove or disconnect the following:
   - Both battery cables
   - Throttle body air inlet duct
   - Upper radiator hose
   - Vacuum hose from the power brake booster
   - Idle Air Control (IAC) valve electrical connector
   - Alternator electrical connectors
   - Throttle Position (TPS) sensor electrical connector
   - Manifold Absolute Pressure (MAP) sensor electrical connector
   - EVAP emission solenoid electrical connection
   - Fuel injector electrical connectors
   - Exhaust Gas Recirculation (EGR) valve electrical connector
   - Engine Coolant Temperature (ECT) sender electrical connector
   - Oxygen (O_2S) sensor electrical connector
   - Engine grounds electrical connection
   - Accessory drive belt

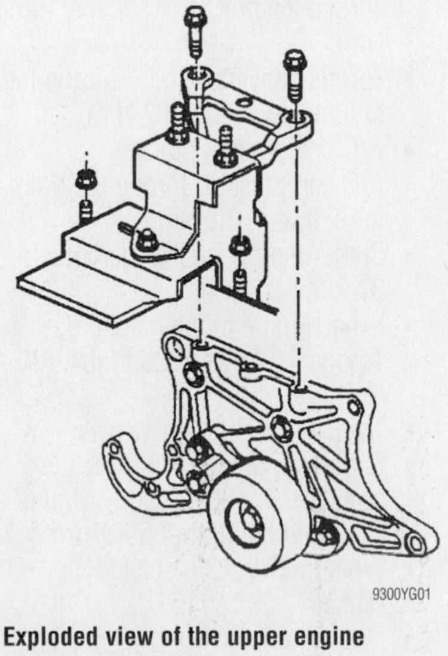

Exploded view of the upper engine mount—2.2L engine

   - Shift control cable
   - Coolant surge tank
   - Vacuum line near the master cylinder
5. Install an engine support fixture.
   - Lower radiator hose
   - Wheels
   - Wheel well splash shields
   - Exhaust pipe from the manifold
   - Engine mount strut
   - Wheel Speed Sensor (WSS) electrical connection
   - Ball joints from the steering knuckles
   - Tie rod ends from the steering knuckles
   - Brake lines from the suspension support
   - Air conditioning compressor. DO NOT disconnect the lines.
   - Ignition Control Module (ICM) electrical connector
   - Vehicle Speed Sensor (VSS) electrical connector
   - Cooling fan electrical connector
   - Starter electrical connector
   - Oil pressure sensor electrical connector
   - Power steering lines from the rack and pinion
   - Intermediate shaft
   - Accelerator and cruise control cables
   - Rack and pinion bolts
   - Suspension support extension brace
   - Stabilizer bar
   - Suspension support
   - Heater hoses
   - Halfshafts from the transmission
   - Fuel lines
   - Transmission mount
   - Engine mount
6. Remove the engine support fixture.
7. Remove or disconnect the following:

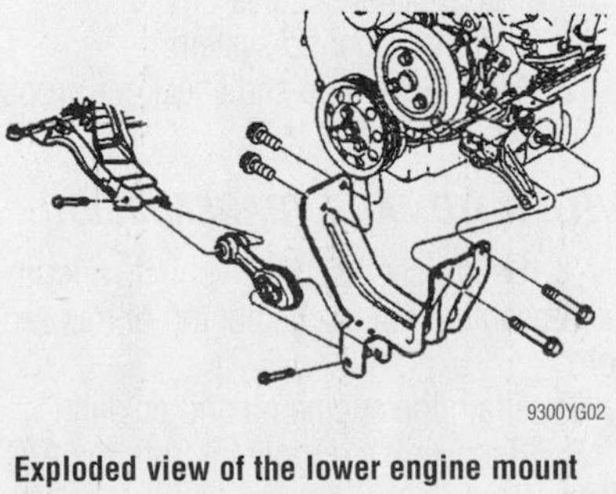

Exploded view of the lower engine mount and strut—2.2L engine

   - Engine and transmission assembly from the vehicle
   - Engine from the transmission

To install:

8. Install or connect the following:
   - Transaxle to the engine. Torque the bolts to 55 ft. lbs. (75 Nm).
   - Engine and transmission assembly to the vehicle
   - Engine support fixture
   - Engine mount. Torque the bolts to 49 ft. lbs. (66 Nm).
   - Transmission mount. Torque the bolts to 49 ft. lbs. (66 Nm).
   - Halfshafts
   - Heater hoses
   - Fuel lines
   - Suspension support. Torque the bolts to 71 ft. lbs. (110 Nm) plus an additional 90 degree turn.
   - Stabilizer bar. Torque the bolts to 49 ft. lbs. (66 Nm).
   - Suspension support extension brace. Torque the bolts to 53 ft. lbs. (72 Nm).
   - Rack and pinion bolts. Torque the bolts to 89 ft. lbs. (120 Nm).
   - Accelerator and cruise control cables
   - Intermediate shaft. Torque the pinch bolt to 30 ft. lbs. (41 Nm).
   - Power steering lines
   - Ignition module electrical connector
   - VSS electrical connector
   - Cooling fan electrical connector
   - Starter electrical connector
   - Oil pressure sensor electrical connector
   - A/C compressor. Torque the front bolts to 33 ft. lbs. (45 Nm) and the rear bolt to 48 ft. lbs. (65 Nm).
   - Brake lines to the suspension support
   - Tie rod ends to the steering knuckles. Torque the nuts to 7 ft. lbs. (10 Nm) plus an additional 120 degree turn.
   - Ball joints to the steering knuckles. Torque the nuts to 46 ft. lbs. (62 Nm).
   - WSS electrical connector
   - Engine mount strut. Torque the bolts to 74 ft. lbs. (100 Nm) plus an additional turn of 90 degrees.
   - Exhaust pipe to the manifold. Torque the bolts to 26 ft. lbs. (35 Nm).
   - Wheel well splash shields

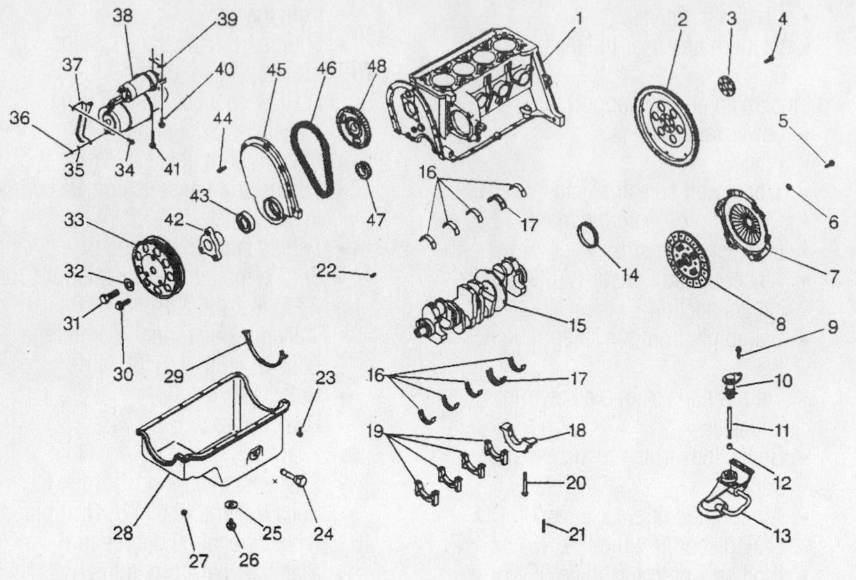

1 Cylinder Block
2 Flywheel
3 Flywheel Retainer
4 Flywheel Bolt
5 Clutch Pressure Plate Bolt
6 Clutch Pressure Plate Washer
7 Clutch Pressure Plate
8 Clutch Disc
9 Oil Pump Drive Bolt
10 Oil Pump Drive
11 Oil Pump Drive Shaft
12 Oil Pump Drive Shaft Retainer
13 Oil Pump
14 Crankshaft Rear Oil Seal
15 Crankshaft
16 Crankshaft Bearing
17 Crankshaft Thrust Bearing
18 Main Bearing Cap
19 Main Bearing Cap
20 Crankshaft Bearing Cap Bolt
21 Oil Pan Stud
22 Connecting Rod Nut
23 Oil Pan Bolt
24 Connecting Rod Nut
25 Oil Pan Drain Plug Gasket
26 Oil Pan Drain Plug
27 Oil Pan Nut
28 Oil Pan
29 Oil Pan Rear Seal
30 Crankshaft Pulley Bolt
31 Crankshaft Pulley Hub Bolt
32 Crankshaft Pulley Hub Bolt Washer
33 Crankshaft Pulley
34 Bracket Bolt
35 Starter Motor Bracket Nut
36 Starter Motor Bracket Nut
37 Starter Motor Bracket
38 Starter Motor
39 Starter Motor Shim
40 Starter Motor Bolt
41 Starter Motor Bolt
42 Crankshaft Pulley Hub
43 Crankshaft Front Oil Seal
44 Crankcase Front Cover Bolt
45 Crankcase Front Cover
46 Timing Chain
47 Crankshaft Sprocket
48 Camshaft Sprocket

9300YG06

Exploded view of the crankshaft and related components—2.2L engine

- Wheels
- Lower radiator hose

9. Remove the engine support fixture.
10. Install or connect the following:
    - Coolant surge tank. Torque the bolts to 89 inch lbs. (10 Nm).
    - Shift control cable
    - Accessory drive belt
    - O_2S electrical connector
    - ECT sender electrical connector
    - EGR valve electrical connector
    - Fuel injector electrical connectors
    - MAP sensor electrical connector
    - TPS sensor electrical connector
    - Alternator electrical connector
    - IAC electrical connector
    - Brake booster vacuum hose
    - Upper radiator hose
    - Throttle body air inlet duct
    - Both battery cables
11. Refill the cooling system.
12. Start the engine and check for proper operation.

WITH AUTOMATIC TRANSMISSION

1. Before servicing the vehicle, refer to the precautions in the beginning of this section.
2. Drain the engine oil and coolant.
3. Recover the refrigerant from the A/C system.
4. Matchmark the bolts and remove the hood.
5. Remove or disconnect the following:
    - Both battery cables
    - Air cleaner assembly
    - Brake booster vacuum line
    - Fuel injector sight shield
    - Accelerator and cruise control cables
    - Engine wiring harness
    - Cruise control module
    - A/C line from the accumulator
    - Accessory drive belt
    - Coolant surge tank
    - Upper and lower radiator hoses
    - Fuel feed and return lines from the engine
    - Coolant pipe assembly
    - Wheels
    - Wheel well splash shields
    - Engine mount strut
    - Torque converter dust cover
    - Exhaust pipe from the manifold
    - Starter motor
    - A/C compressor and bracket
    - Torque converter bolts
    - Transmission-to-engine support brace
    - Upper front engine mount
    - Transmission-to-engine bolts
    - Transmission from the engine
6. Lift the engine from the vehicle.

To install:

7. Position the engine in the vehicle, and hand tighten the upper transmission bolts.
8. Install or connect the following:
    - Transmission-to-engine bolts. Torque the bolts to 43 ft. lbs. (58 Nm).
    - Upper front engine mount. Torque the bolts to 55 ft. lbs. (75 Nm).
    - Transmission-to-engine brace. Torque the bolts to 71 ft. lbs. (96 Nm).
    - Torque converter bolts. Torque the bolts to 46 ft. lbs. (62 Nm).
    - A/C compressor bracket
    - A/C compressor. Torque the bolts to 37 ft. lbs. (50 Nm).
    - Starter motor. Torque the bolts to 37 ft. lbs. (50 Nm).
    - Exhaust pipe to the manifold. Torque the bolts to 26 ft. lbs. (35 Nm).
    - Torque converter dust cover
    - Engine mount strut. Torque the bolts to 74 ft. lbs. (100 Nm) plus an additional turn of 90 degrees.
    - Wheel well splash shields
    - Wheels
    - Coolant pipe assembly
    - Fuel feed and return lines
    - Upper and lower radiator hoses
    - Coolant surge tank. Torque the bolts to 89 inch lbs. (10 Nm).

- Accessory drive belt
- A/C line to the accumulator
- Cruise control module
- Engine wiring harness
- Accelerator and cruise control cables
- Brake booster vacuum line
- Air cleaner assembly
- Hood. Torque the bolts to 15 ft. lbs. (20 Nm).
- Battery cables

9. Evacuate and recharge the A/C system.
10. Refill the engine oil and coolant.
11. Start the engine and check for leaks.
12. Install the fuel injector sight shield.

2.4L Engine

WITH MANUAL TRANSMISSION

1. Before servicing the vehicle, refer to the precautions in the beginning of this section.
2. Discharge and recover the air conditioning refrigerant.
3. Properly drain the cooling system and engine oil.
4. Relieve the fuel system pressure.
5. Remove or disconnect the following:
   - Both battery cables
   - Clutch pushrod from the pedal assembly
   - Heater hose at the thermostat housing
   - Upper radiator hose
   - Air cleaner assembly
   - Coolant fan
   - Refrigerant hose assembly at the compressor
   - Both vacuum hoses from the front of the engine
   - Alternator electrical connector
   - Air conditioning compressor electrical connector
   - Fuel injector harness

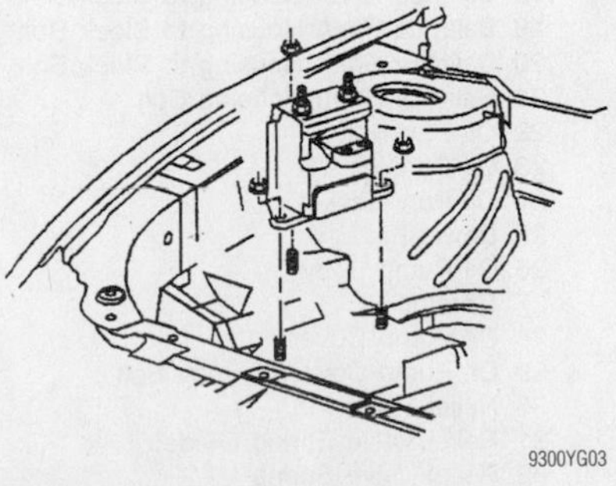

Exploded view of the upper engine mount—2.4L engine

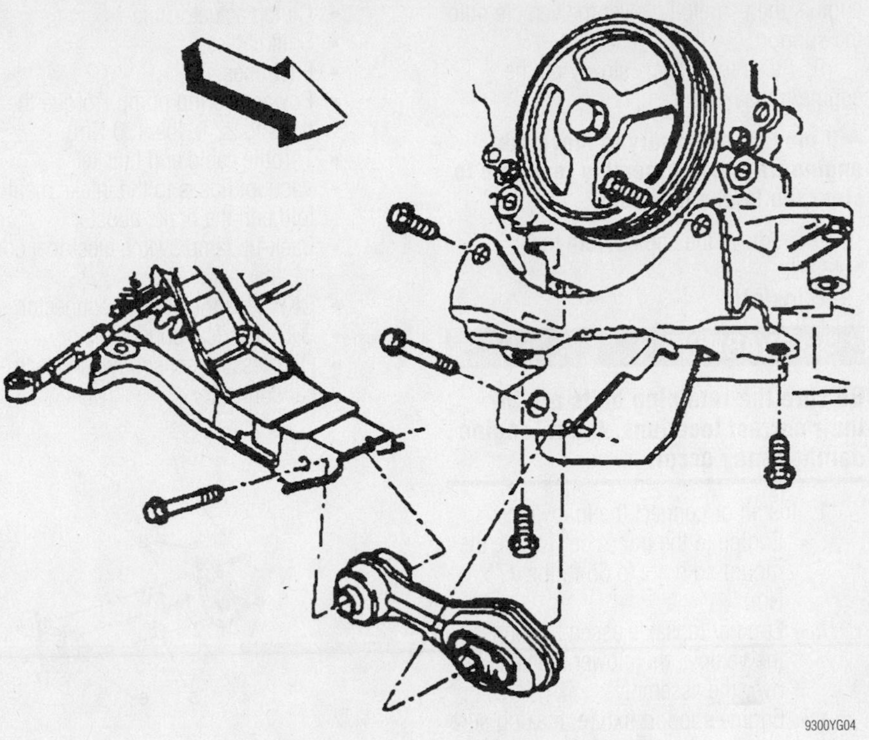

Exploded view of the lower engine mount and strut—2.4L engine

   - Idle Air Control (IAC) electrical connector
   - Throttle Position (TPS) sensor electrical connector
   - Manifold Absolute Pressure (MAP) sensor electrical connector
   - Intake Air Temperature (IAT) sensor electrical connector
   - Evaporative Emissions (EVAP) canister purge solenoid electrical connector
   - Negative battery cable from the transaxle
   - Electronic ignition module electrical connector
   - Engine Coolant Temperature (ECT) sensors electrical connectors
   - Oil pressure sensor/switch electrical connector
   - Oxygen (O_2S) sensor electrical connector
   - Crankshaft Position (CKP) sensor electrical connector
   - Back-up lamp switch electrical connector
   - Power brake vacuum hose from the throttle body
   - Throttle cable and bracket
   - Power steering pump and move it aside with the lines attached
   - Fuel lines
   - Shift cables
   - Clutch actuator line
   - Exhaust manifold and heat shield
   - Lower radiator hose
6. Install an engine support fixture.
7. Remove or disconnect the following:
   - Coolant recovery/surge tank, move it aside with the hoses attached
   - Engine strut mount
   - Front wheels
   - Right splash shield
   - Vehicle Speed Sensor (VSS) electrical connector
   - Knock Sensor (KS) electrical connector
   - Starter solenoid electrical connector
   - Both front Anti-lock Brake System (ABS) wheel speed sensor electrical connectors
   - Engine mount strut
   - Transmission mount
   - Ball joints from the steering knuckles
   - Suspension supports, crossmember and stabilizer shaft as an assembly
   - Heater outlet hose from the radiator outlet pipe
   - Halfshaft from the transmission and intermediate shaft
   - Flywheel housing cover
8. Position a suitable support below the

engine, then carefully lower the vehicle onto the support.

9. Raise the vehicle slowly off the engine/transaxle assembly.

➡It may be necessary to move the engine/transaxle assembly rearward to clear the intake manifold.

10. Remove the engine from the transaxle.

To install:

✲✲ WARNING

Be sure the retaining bolts are in their correct locations. If not, engine damage may occur.

11. Install or connect the following:
    - Engine to the transaxle. Torque the mounting bolts to 55 ft. lbs. (75 Nm).
    - Engine/transaxle assembly under the vehicle, then lower the vehicle over the assembly
    - Engine support fixture, making sure to adjust it to the previous setting
    - Halfshafts to the transaxle
    - Heater outlet hose to the radiator outlet pipe
    - Suspension supports, crossmember and stabilizer shaft assembly. Torque the bolts to 53 ft. lbs. (72 Nm).
    - Ball joints to the steering knuckles. Torque the nuts to 44 ft. lbs. (60 Nm) plus an additional turn of 180 degrees.
    - Engine mount assembly. Torque the bolts to 46 ft. lbs. (62 Nm).
    - Transmission mount. Torque the bolts to 66 ft. lbs. (90 Nm).
    - Both front ABS wheel speed sensor electrical connectors
    - Starter solenoid electrical connector
    - KS electrical connector
    - VSS electrical connector
    - Right splash shield
    - Front wheels
12. Carefully raise the vehicle off the support.
13. Install or connect the following:
    - Engine strut mount. Torque the bolts to 74 ft. lbs. (100 Nm).
    - Coolant recovery/surge tank
    - Lower radiator hose
    - Exhaust manifold. Torque the nuts to 31 ft. lbs. (42 Nm).
    - Exhaust manifold heat shield. Torque the bolts to 19 ft. lbs. (26 Nm).
    - Clutch actuator line
    - Shift cables
    - Fuel lines
    - Power steering pump. Torque the bolts to 22 ft. lbs. (30 Nm).
    - Throttle cable and bracket
    - Vacuum hoses to the intake manifold and the brake booster
    - Back-up lamp switch electrical connector
    - CKP sensor electrical connector
    - O_2S electrical connector
    - Oil pressure sensor/switch electrical connector
    - ECT sensors electrical connectors
    - Electronic ignition module electrical connector
    - Negative battery cable to the transaxle
    - EVAP canister purge solenoid electrical connector
    - IAT sensor electrical connector
    - MAP sensor electrical connector
    - TPS sensor electrical connector
    - IAC sensor electrical connector
    - Fuel injector harness
    - Air conditioning compressor electrical connector

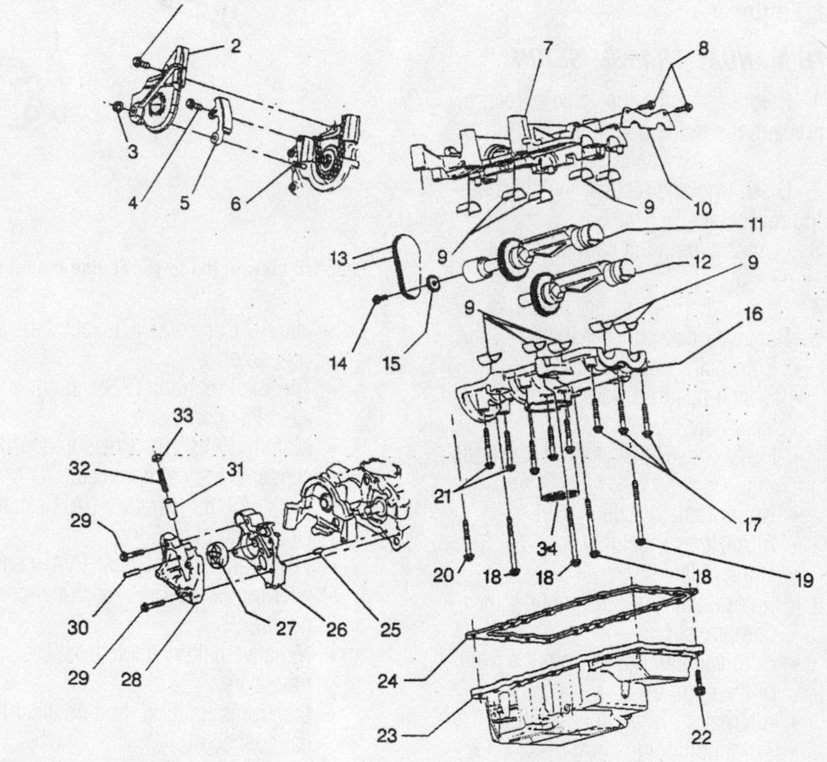

1 Balance Shaft Chain Cover Bolt
2 Balance Shaft Chain Cover
3 Balance Shaft Guide Adjuster Nut
4 Balance Shaft Chain Guide Adjuster Bolt
5 Balance Shaft Chain Adjuster Guide
6 Stud
7 Balance Shaft Upper Housing
8 Balance Shaft Thrust Plate Retainer Bolt
9 Balance Shaft Bearings
10 Balance Shaft Thrust Plate Retainer
11 Balance Shaft Drive
12 Balance Shaft Driven
13 Balance Shaft Drive Chain
14 Driven Sprocket Bolt-Note: Left Hand Thread
15 Balance Shaft Driven Sprocket
16 Balance Shaft Lower Housing
17 Balance Shaft Housing Bolt
18 Balance Shaft Housing to Block Bolt
19 Balance Shaft Housing to Block Bolt
20 Balance Shaft Housing to Block Bolt
21 Balance Shaft Housing Bolt
22 Oil Pan Bolt
23 Oil Pan
24 Oil Pan Gasket
25 Dowel Pin
26 Oil Pump Body
27 Gerotor
28 Oil Pump Cover
29 Oil Pump Cover to Body Bolt
30 Relief Valve Pin
31 Relief Valve Spring Guide
32 Relief Valve Spring
33 Relief Valve
34 Pickup Screen

9300YG09

Exploded view of the balance shafts and related components—2.4L engine

- Alternator electrical connector
- Refrigerant hose assembly to the compressor
- Coolant fan
- Air cleaner assembly
- Upper radiator hose
- Heater hose at the thermostat housing
- Clutch pushrod to the pedal assembly
- Both battery cables

14. Refill the transaxle and the crankcase.
15. Evacuate and recharge the air conditioning system.
16. Refill the cooling system.
17. Start the engine and check for leaks.

WITH AUTOMATIC TRANSMISSION

1. Before servicing the vehicle, refer to the precautions in the beginning of this section.
2. Matchmark the bolts and remove the hood.
3. Drain the engine coolant.
4. Recover the refrigerant.
5. Drain the engine oil.
6. Relieve the fuel system pressure.
7. Remove or disconnect the following:
   - Air cleaner assembly
   - Brake booster vacuum line
   - Fuel injector sight shield
   - Accelerator and cruise control cables
   - Engine wiring harness
   - Cruise control module
   - A/C line from the accumulator
   - Accessory drive belt
   - Coolant surge tank
   - Upper and lower radiator hoses
   - Fuel feed and return lines
   - Coolant pipe assembly
   - Wheels
   - Wheel well splash shields
   - Engine mount strut
   - Torque converter cover
   - Exhaust pipe from the manifold
   - Starter motor
   - A/C compressor
   - Torque converter bolts
   - Transmission-to-engine brace
   - Upper front engine mount
   - Transmission-to-engine bolts
8. Separate the engine from the transmission and lift the engine from the vehicle.

To install:

9. Position the engine in the vehicle.
10. Hand tighten the upper transmission bolts.
11. Install or connect the following:
    - Upper front engine mount. Torque the bolts to 46 ft. lbs. (62 Nm).
    - Lower transmission-to-engine bolts. Torque all bolts to 55 ft. lbs. (75 Nm).
    - Transmission-to-engine brace. Torque the bolts to 71 ft. lbs. (96 Nm).
    - Torque converter bolts. Torque the bolts to 46 ft. lbs. (62 Nm).
    - A/C compressor. Torque the bolts to 37 ft. lbs. (50 Nm).
    - Starter motor. Torque the bolts to 37 ft. lbs. (50 Nm).
    - Exhaust pipe to the manifold. Torque the bolts to 26 ft. lbs. (35 Nm).
    - Torque converter cover
    - Engine mount strut. Torque the bolts to 74 ft. lbs. (100 Nm).
    - Wheel well splash shields
    - Wheels
    - Coolant pipe assembly
    - Fuel feed and return lines
    - Upper and lower radiator hoses
    - Coolant surge tank
    - Accessory drive belt
    - A/C line to the accumulator
    - Cruise control module
    - Engine wiring harness
    - Accelerator and cruise control cables
    - Brake booster vacuum line
    - Air cleaner assembly
    - Both battery cables
12. Refill the crankcase and cooling system.
13. Evacuate and recharge the A/C system.
14. Install the hood.
15. Start engine and check for leaks.
16. Install the fuel injector sight shield.

Water Pump

REMOVAL & INSTALLATION

2.2L Engine

1. Before servicing the vehicle, refer to the precautions in the beginning of this section.
2. Drain the cooling system.
3. Remove or disconnect the following:
   - Negative battery cable
   - Water pump pulley bolts, loosen
   - Accessory drive belt
   - Water pump pulley
   - Surge tank hose

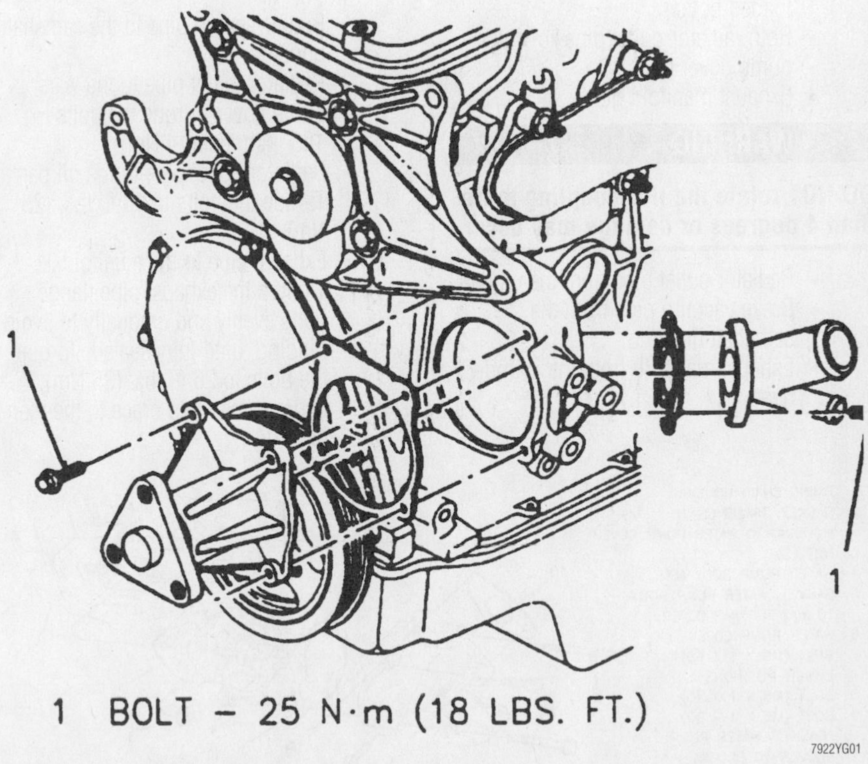

Exploded view of the water pump mounting—2.2L engine

Timing belt service is covered in Section 3 of this manual

- Water pump bolts, pump and gasket

To install:

4. Apply a thin bead of sealer around the outer edge of the water pump gasket seating area and place the gasket on the pump.
5. Install or connect the following:
   - Water pump. Torque the bolts to 18 ft. lbs. (25 Nm).
   - Water pump pulley and hand-tighten the bolts
   - Surge tank hose
   - Accessory drive belt
   - Water pump pulley bolts. Torque the bolts to 22 ft. lbs. (30 Nm).
6. Connect the negative battery cable.
7. Refill and bleed the cooling system.

2.4L Engine

1. Before servicing the vehicle, refer to the precautions in the beginning of this section.
2. Drain the cooling system.
3. Remove or disconnect the following:
   - Negative battery cable
   - Oxygen (O_2S) sensor electrical connector
   - Upper exhaust manifold heat shield
   - Exhaust manifold brace-to-manifold bolt
   - Manifold-to-exhaust pipe spring loaded bolts
   - Both radiator outlet pipe-to-water pump cover bolts
   - Exhaust manifold pipe

WARNING

DO NOT rotate the flex coupling more than 4 degrees or damage may occur.

- Radiator outlet pipe from the oil pan
- Brake vacuum pipe from the camshaft housing
- Exhaust manifold from the cylinder head
- Front cover
- Timing chain tensioner
- Water pump cover and water pump as an assembly

4. Separate the water pump from the water pump cover.

To install:

5. Install or connect the following:
   - Water pump with a new gasket to the cover and tighten the bolts finger-tight

➡Lubricate the splines with clean grease

   - Water pump and cover assembly using new gaskets and tighten the fasteners finger-tight
6. Torque the bolts in the following sequence;
   a. Pump assembly-to-chain housing nuts: 19 ft. lbs. (26 Nm).
   b. Pump cover-to-pump assembly: 124 inch lbs. (14 Nm).
   c. Cover-to-block, bottom bolt first: 19 ft. lbs. (26 Nm).
7. Install or connect the following:
   - Timing chain tensioner
   - Front cover. Torque the bolts to 97 inch lbs. (11 Nm).
   - Exhaust manifold with a new gasket. Torque the bolts to 31 ft. lbs. (42 Nm).
   - Brake vacuum pipe to the camshaft housing
   - Radiator outlet pipe to the water pump cover. Torque the bolts to 125 inch lbs. (14 Nm).
   - Radiator outlet pipe to the oil pan. Torque the bolts to 18 ft. lbs. (25 Nm).
   - Exhaust pipe to the manifold. Tighten the exhaust pipe flange bolts evenly and gradually to avoid binding, until fully seated. Torque the bolts to 26 ft. lbs. (35 Nm).
   - Exhaust manifold brace to the manifold. Torque the bolts to 41 ft. lbs. (56 Nm).
   - Upper heat shield. Torque the bolts to 125 inch lbs. (14 Nm).
   - O_2S electrical connector
   - Negative battery cable
8. Refill the cooling system until it comes out of the heater hose outlet; then connect the heater hose.
9. Start the engine. Run the vehicle until the thermostat opens, refill the radiator and recovery tank, then turn the engine **OFF**.
10. Once the vehicle has cooled, recheck the coolant level.

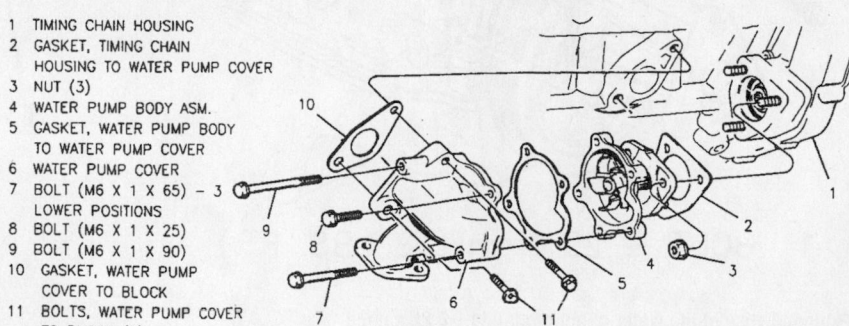

1 TIMING CHAIN HOUSING
2 GASKET, TIMING CHAIN HOUSING TO WATER PUMP COVER
3 NUT (3)
4 WATER PUMP BODY ASM.
5 GASKET, WATER PUMP BODY TO WATER PUMP COVER
6 WATER PUMP COVER
7 BOLT (M6 X 1 X 65) – 3 LOWER POSITIONS
8 BOLT (M6 X 1 X 25)
9 BOLT (M6 X 1 X 90)
10 GASKET, WATER PUMP COVER TO BLOCK
11 BOLTS, WATER PUMP COVER TO BLOCK (2)

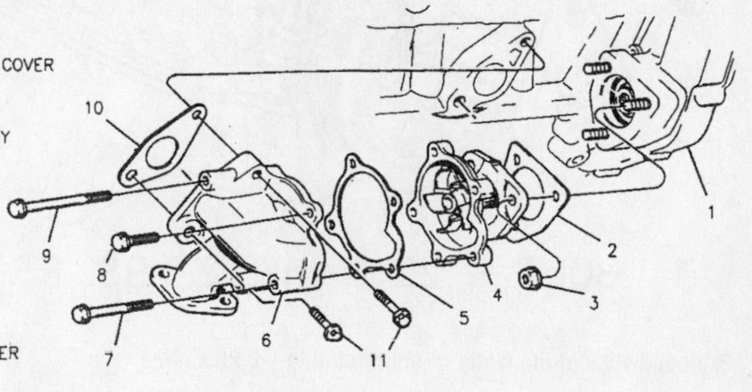

7922YG02

Exploded view of the water pump mounting and related components—2.4L engine

Cylinder Head

REMOVAL & INSTALLATION

2.2L Engine

1. Before servicing the vehicle, refer to the precautions in the beginning of this section.
2. Relieve the fuel system pressure.
3. Drain the cooling system.
4. Remove or disconnect the following:
   - Accessory drive belt
   - Air cleaner outlet duct
   - Engine mount
   - Intake manifold
   - Exhaust manifold
   - Engine Coolant Temperature (ECT) sensor electrical connector
   - Power steering pump
   - Radiator inlet pipe
   - Spark plug wires
   - Rocker arm cover

➡Whenever valve train components are removed, keep them in order for installation purposes.

   - Rocker arms and pushrods
   - Lifters
   - Alternator rear brace and the alternator
   - Power steering pump
   - Radiator inlet pipe
   - Ignition coil assembly
   - Accessory bracket
   - Spark plug wires
   - Cylinder head bolts

WARNING

Two sizes of bolts are used; note the location of each.

   - Cylinder head
5. Inspect the cylinder head and block surface for cracks, nicks, heavy scratches and flatness.

To install:

6. Install or connect the following:
   - Cylinder head with a new gasket and new cylinder head bolts. Torque the long bolts, in sequence to 46 ft. lbs. (63 Nm) and the short bolts to 43 ft. lbs. (58 Nm) plus an additional 90 degree turn on all the bolts.
   - Lifters
   - Pushrods and rocker arms. Torque nuts to 22 ft. lbs. (30 Nm).
   - Rocker arm cover. Torque the bolts to 89 inch lbs. (10 Nm).
   - Spark plug wires
   - Accessory bracket
   - Ignition coil assembly
   - Radiator inlet pipe
   - Power steering pump. Torque the bolts to 22 ft. lbs. (30 Nm).
   - ECT sensor electrical connector
   - Exhaust manifold. Torque the nuts to 118 inch lbs. (13 Nm).
   - Alternator. Torque the bolts to 37 ft. lbs. (50 Nm).
   - Intake manifold. Torque the bolts to 17 ft. lbs. (24 Nm).
   - Engine mount. Torque the bolts to 55 ft. lbs. (75 Nm).
   - Air cleaner outlet duct
   - Accessory drive belt
   - Negative battery cable
7. Refill and bleed the cooling system.
8. Start the vehicle and inspect for leaks.

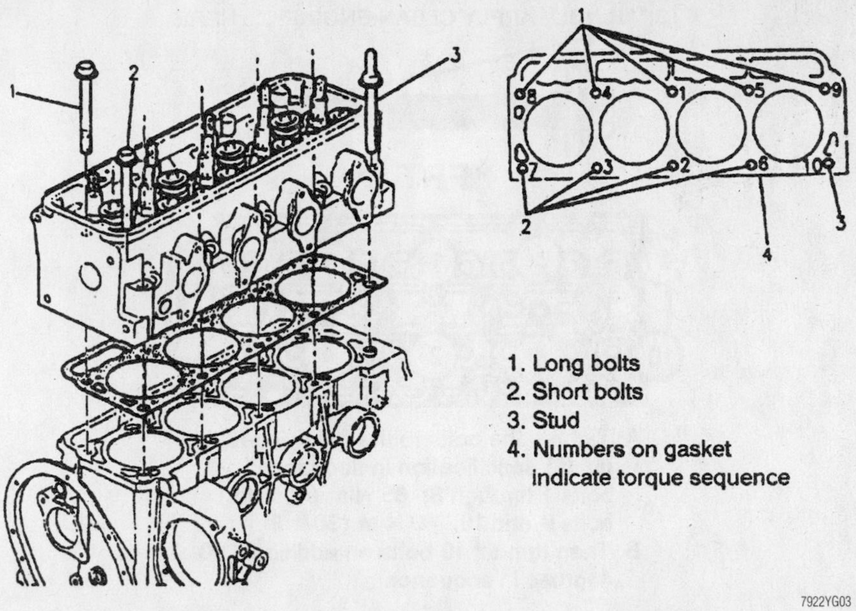

Cylinder head torquing sequence—2.2L engine

2.4L Engine

1. Before servicing the vehicle, refer to the precautions in the beginning of this section.
2. Relieve the fuel system pressure.
3. Drain the cooling system.
4. Remove or disconnect the following:
   - Throttle body-to-air cleaner duct
   - Heater inlet hose
   - Throttle body heater hose
   - Power brake vacuum hose from throttle body
   - Manifold Absolute Pressure (MAP) sensor electrical connector
   - Intake Air Temperature (IAT) sensor electrical connector
   - Evaporative Emission (EVAP) canister purge solenoid
   - Camshaft Position (CMP) sensor electrical connector
   - Intake manifold
   - Exhaust manifold
   - Ignition coil assembly
   - CMP sensor
   - Power steering pump

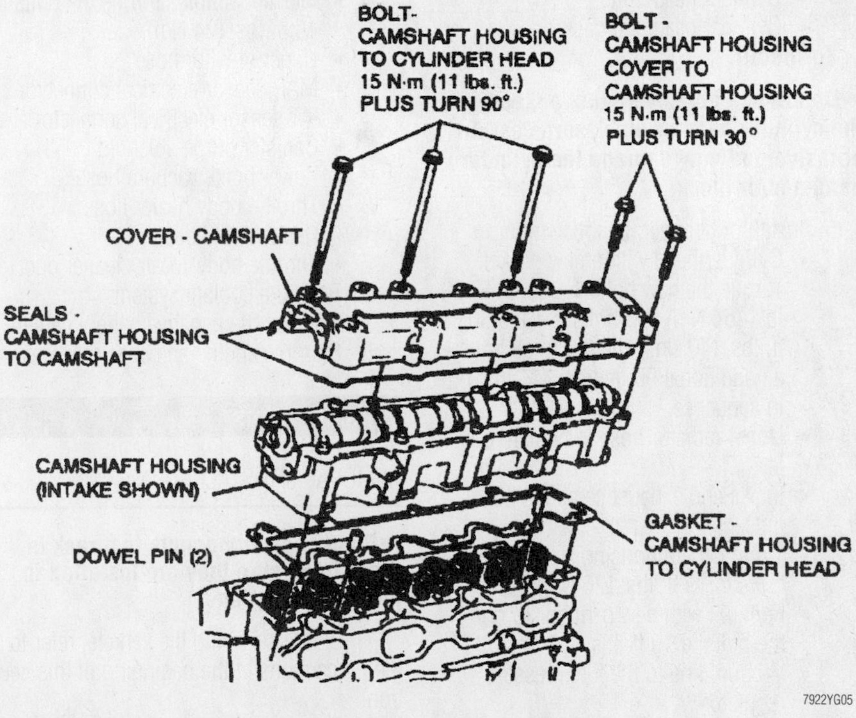

Exploded view of the camshaft housing cover mounting—2.4L engine

   - Vacuum line from the fuel pressure regulator
   - Fuel injector harness connector
   - Fuel rail
   - Timing chain housing from the intake camshaft housing. DO NOT remove it from the vehicle.
   - Oil pressure switch electrical connector
   - Transmission fluid level indicator tube
   - Intake and exhaust camshaft housing
   - Upper radiator hose
   - Coolant temperature sensor connector

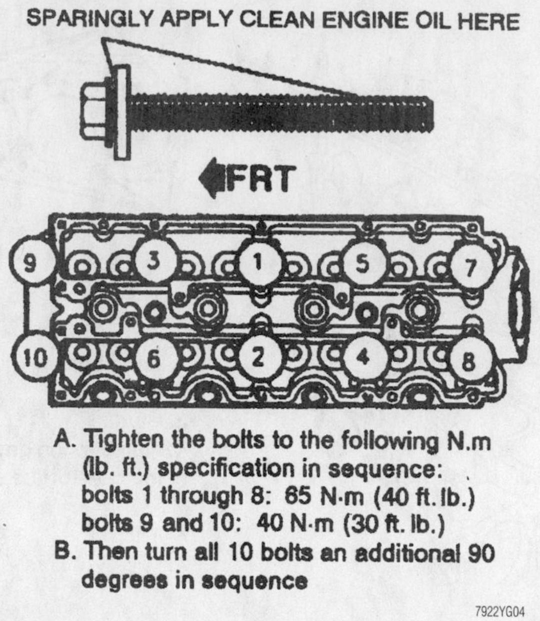

Head bolt torque sequence—2.4L engine

- Cylinder head bolts
- Cylinder head and gasket

To install:

➡Do not use abrasive pads to clean the cylinder head or block surfaces. An abrasive pad may damage the cylinder head and/or block.

5. Install or connect the following:
   - Cylinder head with a new gasket. Torque the new bolts 1-8 to 40 ft. lbs. (54 Nm) and bolts 9-10 to 30 ft. lbs. (40 Nm); then, turn all bolts an additional 90 degrees (¼ turn) in sequence.
   - Upper radiator hose to coolant outlet
   - Intake and exhaust camshaft housings
   - Timing chain housing. Torque the bolts to 19 ft. lbs. (26 Nm).
   - Fuel rail with new o-rings. Torque the bolts to 19 ft. lbs. (26 Nm).
   - Vacuum line to the fuel pressure regulator
   - Fuel injector electrical connectors
   - CMP
   - Power steering pump. Torque the bolts to 19 ft. lbs. (26 Nm).
   - Oil pressure switch electrical connector
   - Transmission fluid level indicator tube
   - Ignition coil assembly. Torque the bolts to 11 ft. lbs. (15 Nm) plus an additional 30 degree turn.
   - Exhaust manifold. Torque the bolts to 31 ft. lbs. (42 Nm).
   - Intake manifold. Torque the bolts to 18 ft. lbs. (24 Nm).
   - Upper radiator hose
   - MAP sensor electrical connector
   - IAT sensor electrical connector
   - Canister purge solenoid
   - Power brake vacuum hose
   - Throttle body heater hose
   - Heater inlet hose
   - Throttle body-to-air cleaner duct
6. Refill the coolant system.
7. Connect the negative battery cable.
8. Start the engine and check for leaks.

Rocker Arms

REMOVAL & INSTALLATION

➡Place the components in a rack in order to be sure they are installed in the same location.

1. Before servicing the vehicle, refer to the precautions in the beginning of this section.
2. Remove or disconnect the following:
   - Negative battery cable
   - Rocker arm cover(s)
   - Rocker arm nuts
   - Rocker arm pivot ball(s)
   - Rocker arm(s)
   - Pushrods

To install:

3. Install the pushrods.
4. Coat the bearing surfaces of the rocker arms and pivot balls with camshaft lubricant.
5. Install or connect the following:
   - Rocker arms. Torque the nuts to 22 ft. lbs. (30 Nm).
   - Rocker arm covers. Torque the bolts to 89 inch lbs. (10 Nm).
   - Negative battery cable

Intake Manifold

REMOVAL & INSTALLATION

2.2L Engine

1. Before servicing the vehicle, refer to the precautions in the beginning of this section.
2. Properly relieve the fuel system pressure.
3. Remove or disconnect the following:
   - Air cleaner inlet duct
   - Air inlet resonator and bracket
   - Throttle and cruise control cable from the throttle body
   - Manifold Absolute Pressure (MAP) sensor
   - Throttle Position (TPS) sensor
   - Idle Air Control (IAC) valve
   - Fuel supply line
   - Throttle body
   - Fuel inlet pipe
   - Intake manifold

To install:

4. Install or connect the following:
   - Intake manifold with a new gasket. Torque the bolts/nuts, in sequence, to 18 ft. lbs. (24 Nm).
   - Fuel inlet pipe. Torque the bolts to 18 ft. lbs. (24 Nm).
   - Throttle body. Torque the bolts to 89 inch lbs. (10 Nm).
   - Fuel supply line
   - MAP sensor
   - TPS sensor
   - IAC valve
   - Cruise control and throttle cables to the throttle body
   - Air inlet resonator bracket and resonator
   - Air cleaner inlet duct
   - Negative battery cable

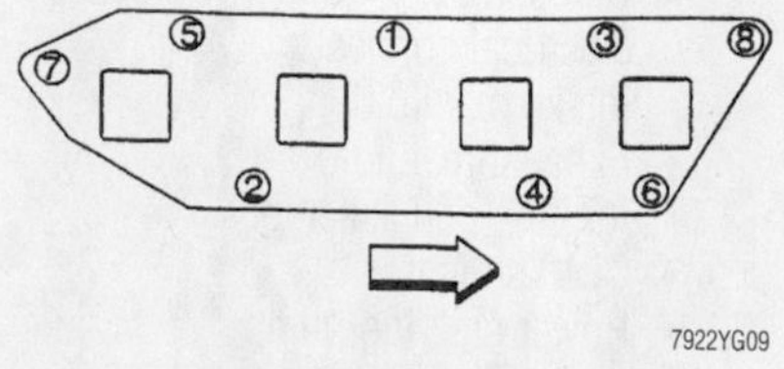

Intake manifold tightening sequence—2.2L engine

2.4L Engine

1. Before servicing the vehicle, refer to the precautions in the beginning of this section.
2. Properly relieve the fuel system pressure.
3. Drain the cooling system.
4. Remove or disconnect the following:
   - Manifold Absolute Pressure (MAP) sensor electrical connector
   - Intake Air Temperature (IAT) sensor electrical connector
   - Evaporative Emission (EVAP) canister purge solenoid electrical connector
   - Fuel injector harness
   - Fuel regulator vacuum hose
   - EVAP canister purge solenoid to canister vacuum hose
   - Air cleaner duct
   - Accelerator control cable bracket
   - Stud-ended alternator mount bolt
   - Exhaust Gas Recirculation (EGR) pipe from the EGR adapter
   - Intake manifold support brace
   - Intake manifold

To install:

5. Install or connect the following:
   - Intake manifold with a new gasket. Torque the bolts/nuts, in sequence, to 18 ft. lbs. (24 Nm).
   - EGR pipe to the adapter. Torque the fasteners to 19 ft. lbs. (26 Nm).
   - Stud-ended alternator bolt
   - Accelerator control cable bracket
   - Fuel injector electrical connectors
   - Vacuum hoses to the fuel regulator and EVAP canister purge solenoid
   - MAP sensor electrical connector
   - IAT sensor electrical connector
   - EVAP canister purge solenoid electrical connector
   - Air cleaner duct
   - Negative battery cable
6. Refill the cooling system.
7. Start the engine and inspect for leaks.

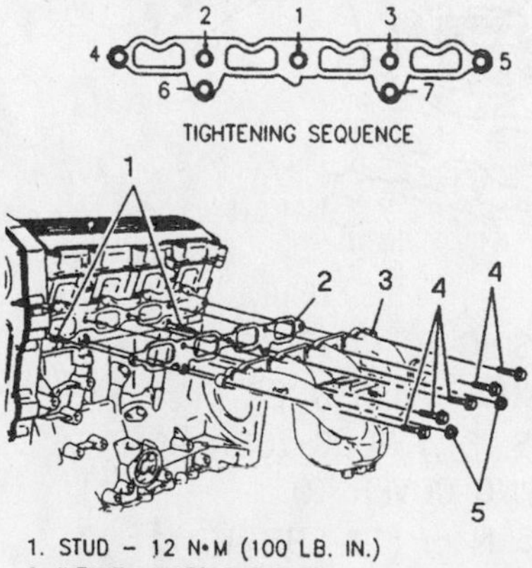

Intake manifold torque sequence—2.4L engine

Exhaust Manifold

REMOVAL & INSTALLATION

2.2L Engine

1. Before servicing the vehicle, refer to the precautions in the beginning of this section.
2. Drain the cooling system.
3. Remove or disconnect the following:
   - Negative battery cable
   - Oxygen (O_2S) sensor electrical connector
   - Accessory drive belt
   - Alternator
   - Radiator inlet pipe
   - Exhaust pipe-to-exhaust manifold bolts
   - Oil filler tube
   - Heater outlet hose assembly-to-exhaust manifold nut
   - Exhaust manifold

To install:

4. Install or connect the following:
   - Exhaust manifold with new gaskets. Torque the nuts to 115 inch lbs. (13 Nm).
   - Heater outlet hose assembly-to-exhaust manifold nut. Torque the nut to 18 ft. lbs. (25 Nm).
   - Oil filler tube
   - Exhaust pipe-to-exhaust manifold bolts. Torque the bolts to 26 ft. lbs. (35 Nm).
   - Radiator inlet pipe
   - Alternator. Torque the bolts to 37 ft. lbs. (50 Nm).
   - Accessory drive belt
   - O_2S sensor electrical connector
   - Negative battery cable
5. Refill the cooling system.
6. Start the engine and check for exhaust leaks.

2.4L Engine

1. Before servicing the vehicle, refer to the precautions in the beginning of this section.
2. Remove or disconnect the following:
   - Negative battery cable
   - Oxygen (O_2S) sensor electrical connector
   - Upper heat shield
   - Exhaust manifold brace-to-manifold bolt
   - Exhaust manifold-to-oil pan nuts

➡Do not bend the exhaust flex coupler more than necessary to remove it. Excessive movement will damage the flex coupler.

   - Exhaust pipe from the exhaust manifold
   - Exhaust manifold

To install:

3. Install or connect the following:
   - Exhaust manifold with new gaskets. Torque the nuts to 110 inch lbs. (12 Nm), in sequence.
   - Heat shield. Torque the bolts to 124 inch lbs. (14 Nm).
   - Exhaust manifold brace-to-manifold bolt and oil pan nuts. Torque the bolts to 41 ft. lbs. (56 Nm) and nuts to 19 ft. lbs. (26 Nm).
   - Manifold-to-flex coupler fasteners. Torque the bolts to 26 ft. lbs. (35 Nm).
   - O_2S sensor electrical connector
   - Negative battery cable
4. Start the engine and check for exhaust leaks.

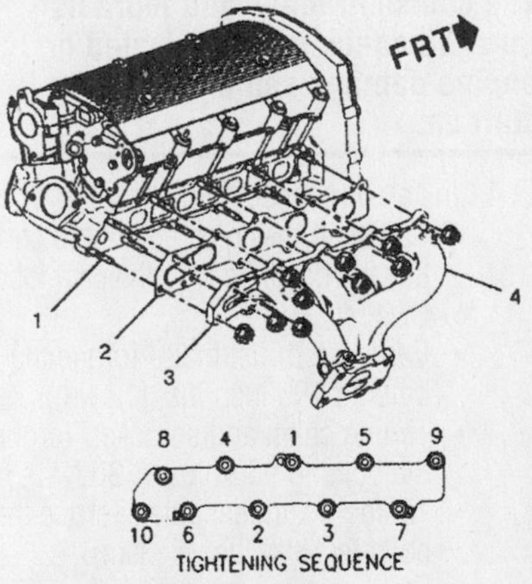

Exploded view of the exhaust manifold, showing the torque sequence—2.4L engine

Camshaft and Valve Lifters

REMOVAL & INSTALLATION

2.2L Engine

1. Before servicing the vehicle, refer to the precautions in the beginning of this section.
2. Drain the engine oil and coolant.
3. Remove or disconnect the following:
   - Engine
   - Accessory drive belt
   - Alternator
   - Power steering pump
   - Drive belt tensioner
   - Water pump pulley
   - Crankshaft pulley
   - Rocker arm cover
   - Pushrods, by pivoting the rocker arms to the sides, keeping them in order
   - Cylinder head
   - Valve lifters, keeping them in order
   - Front cover
   - Camshaft timing sprocket
   - Timing chain
   - Timing chain tensioner
   - Camshaft thrust plate
   - Camshaft Position (CMP) sensor
   - Camshaft from the block, being sure the lobes do not contact the bearings

To install:

WARNING

The camshaft lobes and journals must be adequately lubricated or engine damage could occur upon start up.

4. Install or connect the following:
   - Camshaft, being careful not to contact the bearings with the cam lobes
   - CMP sensor
   - Camshaft thrust plate. Torque the bolts to 106 inch lbs. (12 Nm).
   - Timing chain and sprocket. Torque the bolts to 96 ft. lbs. (130 Nm).
   - Timing chain tensioner. Torque the bolts to 18 ft. lbs. (24 Nm).
   - Front cover. Torque the bolts to 97 inch lbs. (11 Nm).
   - Valve lifters
   - Cylinder head. Torque the long bolts to 46 ft. lbs. (63 Nm) and the short bolts to 43 ft. lbs. (58 Nm), plus an additional 90 degree turn on all bolts.
   - Pushrods
   - Rocker arms
5. Adjust the valve lash after installing the engine.
6. Install or connect the following:
   - Rocker arm cover. Torque the bolts to 89 inch lbs. (10 Nm).
   - Crankshaft pulley. Torque the bolts to 37 ft. lbs. (50 Nm).
   - Water pump pulley. Torque the bolts to 22 ft. lbs. (30 Nm).
   - Drive belt tensioner. Torque the bolts to 37 ft. lbs. (50 Nm).
   - Power steering pump. Torque the bolts to 22 ft. lbs. (30 Nm).
   - Alternator. Torque the bolts to 37 ft. lbs. (50 Nm).
   - Accessory drive belt
   - Engine
7. Replace the oil filter and refill engine with new oil.
8. Refill the cooling system.

2.4L Engine

INTAKE CAMSHAFT

1. Before servicing the vehicle, refer to the precautions in the beginning of this section.
2. Relieve the fuel system pressure.
3. Remove or disconnect the following:
   - Ignition coil and module assembly
   - Camshaft Position (CMP) sensor electrical connector
   - Power steering pump. DO NOT disconnect the lines.
   - Fuel pressure regulator vacuum line
   - Fuel rail
   - Timing chain housing, DO NOT remove from the engine
   - Intake camshaft housing cover

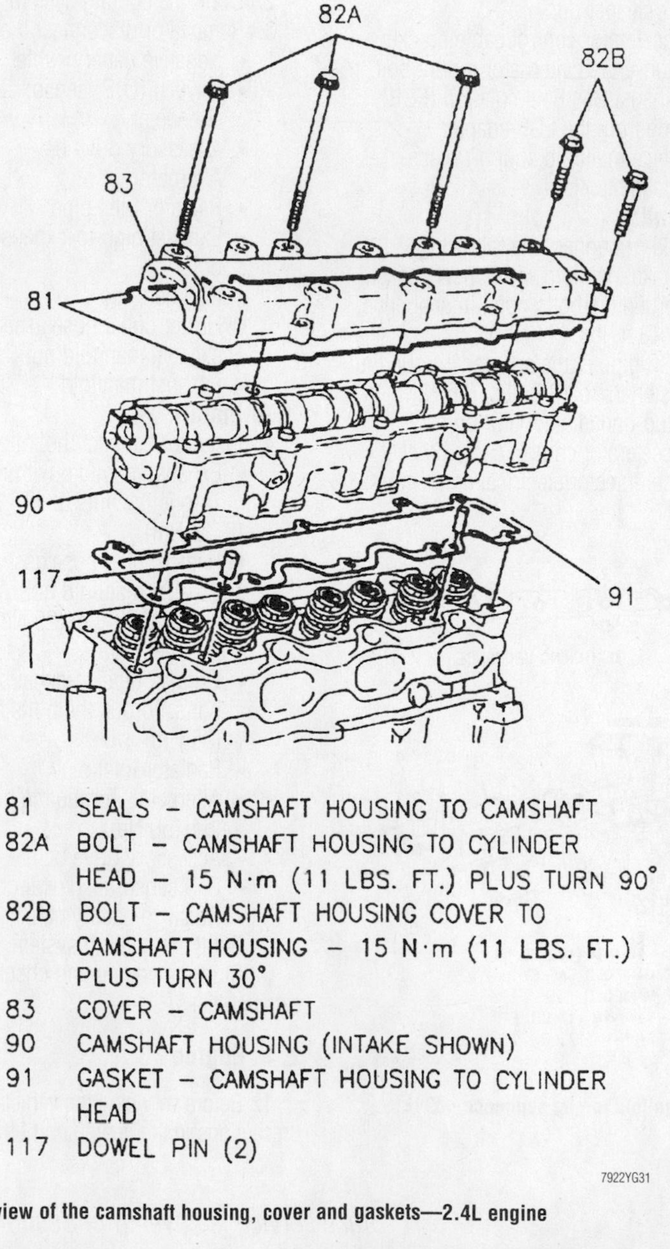

Exploded view of the camshaft housing, cover and gaskets—2.4L engine

- Intake camshaft housing, by reversing the torque sequence

➡**Leave 2 of the bolts loosely in place to hold the camshaft housing while separating the camshaft cover from the housing.**

- Intake camshaft oil seal from the camshaft
- Camshaft

➡**The seal must be replaced any time the housing and cover are separated.**

- Camshaft housing from the cylinder head

To install:

4. Install or connect the following:
   - Lifters into their bores

⁂ WARNING

The camshaft lobes and journals must be adequately lubricated or engine damage could occur upon start up.

- Camshaft
- Camshaft housing with a new gasket
- New intake camshaft seal lubricated with engine oil

➡**The timing chain sprocket dowel pin should be straight up and align with the centerline of the lifter bores.**

- Camshaft housing cover with a new gasket. Torque the bolts, in sequence, to 11 ft. lbs. (15 Nm) plus an additional 90 degree turn (except for the 2 rear fuel pipe-to-camshaft housing bolts). Torque the 2 rear bolts to 11 ft. lbs. (15 Nm), plus an additional 30 degree turn.
- Timing chain housing and the timing chain
- Fuel rail with new injector o-rings
- Vacuum line to the fuel pressure regulator
- Fuel injector harness connectors
- Power steering pump. Torque the bolts to 22 ft. lbs. (30 Nm).
- Camshaft position sensor
- Ignition coil/module assembly. Torque the bolts to 11 ft. lbs. (15 Nm) plus 30 degrees.
- Negative battery cable

5. Start the engine and check for leaks.

EXHAUST CAMSHAFT

1. Before servicing the vehicle, refer to the precautions in the beginning of this section.
2. Relieve the fuel system pressure.
3. Remove or disconnect the following:
   - Ignition coil/module assembly
   - Oil pressure switch electrical connector
   - Transaxle fluid level indicator tube assembly, (if equipped) from exhaust camshaft cover
   - Timing chain housing; do not remove from the engine
   - Exhaust camshaft cover
   - Exhaust camshaft housing by reversing of the torque sequence
4. Press the cover off the housing by threading 4 housing-to-head bolts into the camshaft housing cover tapped holes. Tighten the bolts evenly so the cover does not bind on the dowel pins.
5. Remove the camshaft housing cover.
6. Loosely install a camshaft housing-to-cylinder head bolt to retain the housing during camshaft and lifter removal.

➡**Note the position of the chain sprocket dowel pin for reassembly.**

7. Remove or disconnect the following:
   - Lifters
   - Camshaft
   - Camshaft housing from the cylinder head

To install:

8. Install or connect the following:
   - Camshaft housing with a new gasket on the cylinder head with 1 bolt loosely to hold it in place
   - Lifters

⁂ WARNING

The camshaft lobes and journals must be adequately lubricated or engine damage could occur upon start up.

- Camshaft

➡**The timing chain sprocket dowel pin should be straight up and align with the centerline of the lifter bores.**

➡**Apply thread locking compound to the camshaft housing cover bolt threads.**

- Camshaft housing cover with a new gasket. Torque the bolts, in sequence, to 11 ft. lbs. (15 Nm) plus an additional 90 degree turn (except for the 2 rear fuel pipe-to-camshaft housing bolts). Torque the 2 rear bolts to 11 ft. lbs. (15 Nm), plus an additional 30 degree turn.
- Timing chain housing
- Transaxle fluid level indicator tube assembly to the exhaust camshaft cover
- Oil pressure switch electrical connector
- Ignition coil/module assembly. Torque the bolts to 11 ft. lbs. (15 Nm) plus an additional 30 degrees.
- Negative battery cable

9. Start the engine and check for leaks.

Valve Lash

ADJUSTMENT

All of the engines are equipped with hydraulic valve lifters. There is no adjustment.

Starter Motor

REMOVAL & INSTALLATION

1. Before servicing the vehicle, refer to the precautions in the beginning of this section.
2. Remove or disconnect the following:
   - Negative battery cable
   - Air inlet duct
   - Top starter bolt
   - Lower starter bolt
   - Starter electrical connectors
   - Starter motor

To install:

3. Install or connect the following:
   - Starter motor. Torque the bolts to 66 ft. lbs. (90 Nm) (for 2.4L) or 37 ft. lbs. (50 Nm) (for the 2.2 L)
   - Starter electrical connectors
   - Air inlet duct to the throttle body
   - Negative battery cable

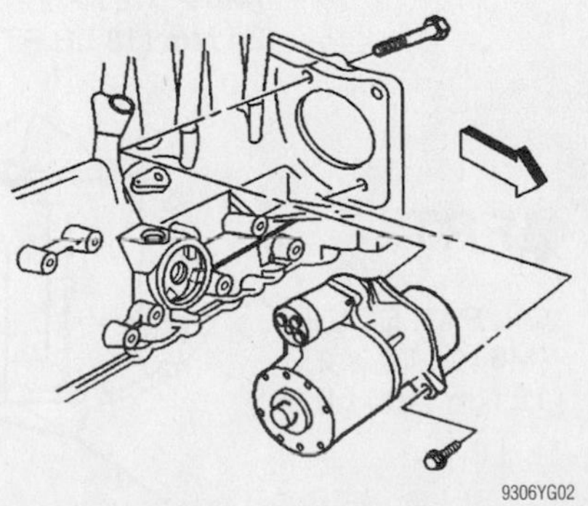

Exploded view of the starter motor—2.4L engine

Oil Pan

REMOVAL & INSTALLATION

2.2L Engine

1. Before servicing the vehicle, refer to the precautions in the beginning of this section.
2. Drain the engine oil.
3. Remove or disconnect the following:
   - Negative battery cable
   - Right front wheel
   - Right inner fender splash shield
   - Starter motor and bracket
   - Engine mount strut bracket
   - Oil pan

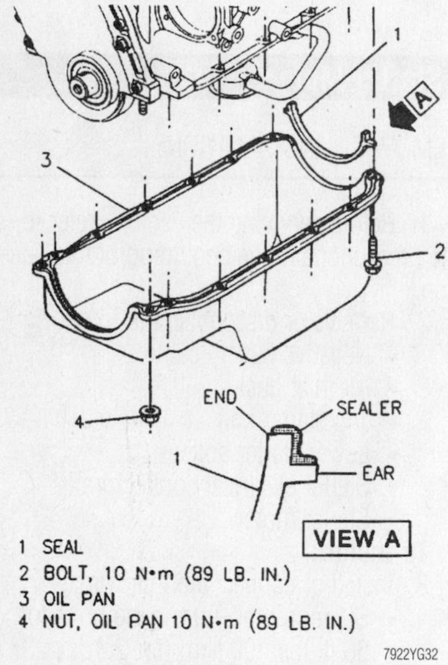

Exploded view of the oil pan mounting and related components—2.2L engine

To install:

4. Place a 2mm bead of RTV sealer to the oil pan sealing surface except at the rear seal mounting surface. Using a new oil pan rear seal, apply a thin coat of RTV sealer on the end down to the ears.
5. Install or connect the following:
   - Oil pan with a new gasket. Torque the nuts and bolts to 89 inch lbs. (10 Nm).
   - Engine mount strut bracket. Torque the bolts to 49 ft. lbs. (66 Nm).
   - Starter motor and bracket. Torque the bolts to 37 ft. lbs. (50 Nm).
   - Right fender splash shield
   - Right front wheel. Torque the nuts to 100 ft. lbs. (140 Nm).
   - Negative battery cable
6. Refill the crankcase.
7. Start the vehicle and verify no leaks.

2.4L Engine

1. Before servicing the vehicle, refer to the precautions in the beginning of this section.
2. Drain the engine oil.
3. Drain the cooling system.
4. Remove or disconnect the following:
   - Negative battery cable
   - Flywheel/converter cover
   - Right wheel
   - Right wheel well splash shield
   - Accessory drive belt
   - Air conditioning compressor lower bolts
   - Transmission-to-engine brace
   - Engine mount strut bracket
   - Radiator outlet pipe bolts
   - Radiator outlet pipe from the oil pan
   - Oil pan to the flywheel cover bolt and nut
   - Flywheel cover stud for clearance
   - Radiator outlet pipe from the lower radiator hose and oil pan
   - Oil level sensor connector
   - Oil pan

To install:

5. Inspect the oil pan gasket; it is reusable if not damaged.
6. Install or connect the following:
   - Oil pan with the gasket. Torque the M8 bolts to 18 ft. lbs. (24 Nm) and the M6 bolts to 106 inch lbs. (12 Nm).
   - Oil pan to the transmission nut
   - Oil level sensor connector
   - Radiator outlet pipe to the lower radiator hose and oil pan
   - Exhaust manifold brace
   - Radiator outlet pipe. Torque the bolts to 124 inch lbs. (14 Nm).
   - Engine mount strut bracket. Torque the bolts to 55 ft. lbs. (75 Nm).
   - Transmission to the engine brace
   - Air conditioning compressor lower bolts. Torque the bolts to 37 ft. lbs. (50 Nm).
   - Accessory drive belt
   - Right splash shield
   - Right front wheel
   - Flywheel/converter cover
   - Negative battery cable
7. Refill the crankcase.
8. Refill the cooling system.
9. Start the vehicle and verify no leaks.

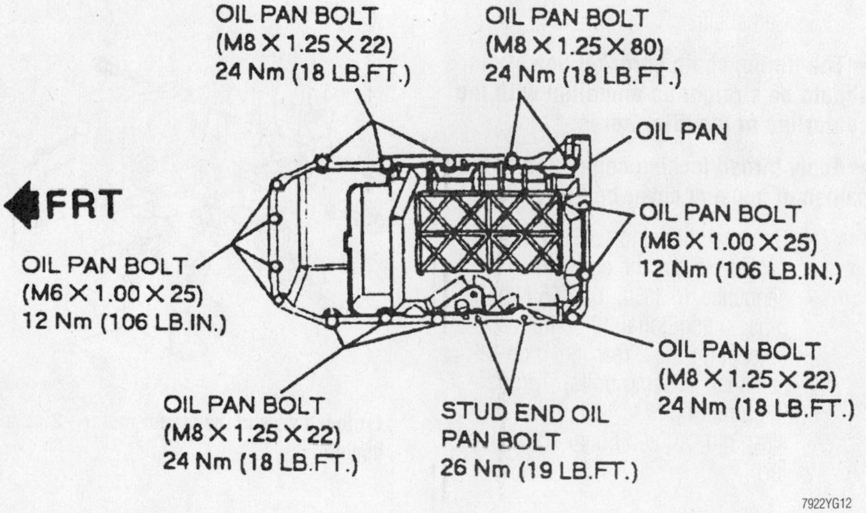

Oil pan fastener torque specifications—2.4L engine

Oil Pump

REMOVAL & INSTALLATION

2.2L Engine

1. Before servicing the vehicle, refer to the precautions in the beginning of this section.
2. Drain the crankcase.
3. Remove or disconnect the following:
   - Negative battery cable
   - Oil pan
   - Oil pump and extension shaft

To install:

**** WARNING**

A plastic extension shaft retainer connects the oil pump driveshaft to the oil pump. Heat the extension shaft retainer in hot water prior to assembly. Be sure the retainer does not crack upon installation.

4. Fill the oil pump cavities with petroleum jelly before installing the gears into the pump body.
5. Install or connect the following:

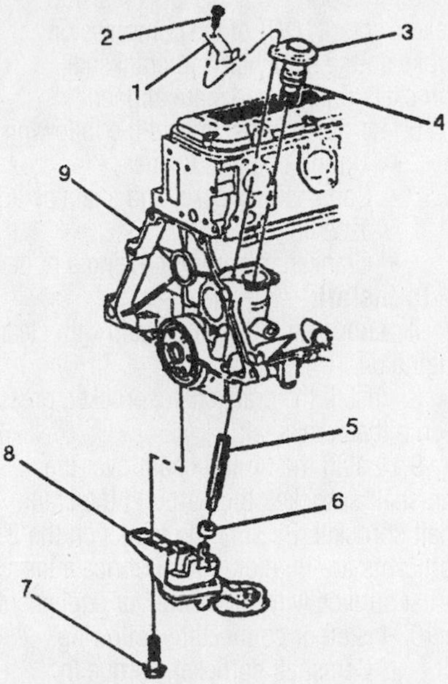

1 Bracket
2 Bolt
3 Oil pump drive assembly
4 O-ring
5 Shaft
6 Retainer; Heat and water soak prior to installation
7 Bolt
8 Oil pump
9 Cylinder block

7922YG13

Exploded view of the oil pump mounting to engine block—2.2L engine

- Extension shaft and oil pump. Torque the oil pump-to-bearing cap bolt to 32 ft. lbs. (43 Nm) and the upper oil pump drive bolt to 18 ft. lbs. (25 Nm).
- Oil pan
- Negative battery cable

6. Refill the crankcase.
7. Start the engine and check oil pressure and check for leaks.

2.4L Engine

1. Before servicing the vehicle, refer to the precautions in the beginning of this section.
2. Disconnect the negative battery cable.
3. Install an engine support fixture.
4. Drain the crankcase.
5. Remove or disconnect the following:
   - Oil pan
   - Balance shaft chain cover and tensioner
   - Oil pump
6. Disassemble the oil pump as follows:

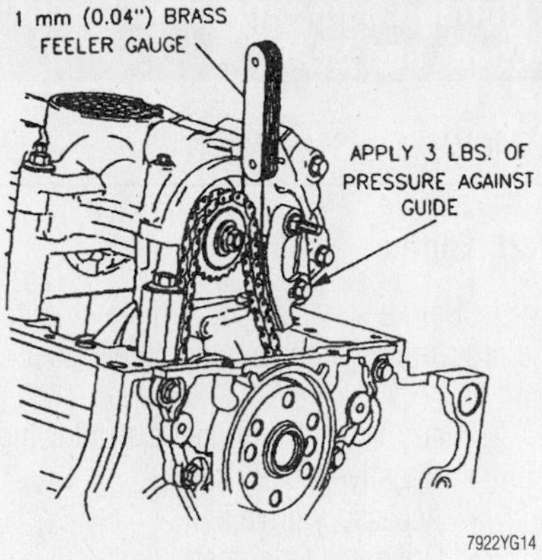

7922YG14

Using a feeler gauge to check the chain tension—2.4L engine

a. Remove the oil pump cover.
b. Remove the pump gear.
c. Remove the sub-assembly from the balance shaft assembly
d. Remove the gerotor from the oil pump housing.
e. Remove the oil pump from the balance shaft housing.
f. Remove the pressure relief valve.

To install:

7. Lubricate the gears with clean engine oil.
8. Assemble the oil pump as follows:
   a. Place the gerotor gear into the housing.
   b. Fill the oil pump cavities with petroleum jelly.
   c. Install the pressure relief valve using a 9⁄16 in. deep well socket to seat the valve.
   d. Connect the pump housing to the balance shaft assembly.
   e. Install the pump cover.
9. Install or connect the following:
   - Oil pump. Torque the bolts to 40 ft. lbs. (54 Nm).
   - Balance shaft chain tensioner and chain

➡**A brass feeler gauge must be used to ensure that correct measurements are obtained. If a steel gauge is used, it will not bend to conform to the guide and will allow for incorrect measurements.**

10. Adjust the chain tension as follows:
    a. Insert a 0.40 in. (1mm) brass feeler between the chain guide and the chain.
    b. Press the guide against the chain using about 3 pounds of force.
    c. Torque the chain tensioner fastener to 115 inch lbs. (13 Nm).
11. Install or connect the following:
    - Balance shaft chain cover. Torque the nut and bolt to 115 inch lbs. (13 Nm).
    - Oil pan
    - Negative battery cable.
12. Refill the crankcase.
13. Start the vehicle and verify oil pressure and no leaks.

Rear Main Seal

REMOVAL & INSTALLATION

2.2L Engine

1. Before servicing the vehicle, refer to the precautions in the beginning of this section.
2. Remove or disconnect the following:
   - Negative battery cable
   - Transaxle
   - Clutch/pressure plate assembly, if equipped with a manual transmission

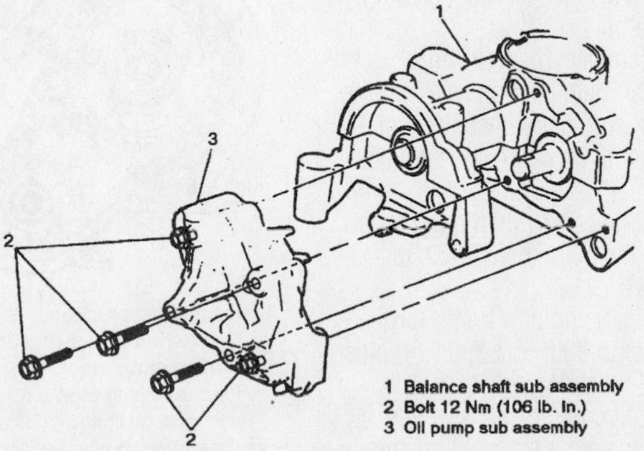

7922YG15

Exploded view of the oil pump assembly mounting—2.4L engine

- Flywheel
- Rear main bearing seal by prying it from the engine

➡**Be careful not to damage or scratch the seal mounting surfaces.**

To install:

3. Lubricate the new rear main bearing seal with engine oil.
4. Install or connect the following:
   - New rear main bearing seal using seal installer J-34686 until its flush with the block
   - Flywheel
   - Clutch/pressure plate assembly, if equipped with a manual transmission
   - Transaxle
   - Negative battery cable
5. Start the engine and check for leaks.

2.4L Engine

1. Before servicing the vehicle, refer to the precautions in the beginning of this section.
2. Remove or disconnect the following:
   - Negative battery cable
   - Transaxle
   - Clutch/pressure plate assembly, if equipped with a manual transmission
   - Flywheel
   - Seal housing and discard the gasket
   - Oil seal from the transaxle side of the seal housing

To install:

3. Clean and inspect the gasket mounting surfaces.
4. If necessary, add silicone sealer along the oil pan-to-cylinder block mating surface.
5. Lubricate the new rear main bearing seal with engine oil.
6. Install or connect the following:
   - New rear main bearing seal into the seal housing using installer tool J-36005 until its flush with the housing.
   - Oil seal housing with a new gasket. Torque the bolts to 106 inch lbs. (12 Nm).
   - Flywheel, using new bolts. Torque the bolts to 22 ft. lbs. (30 Nm) plus an additional 45 degree turn.
   - Clutch/pressure plate assembly, if equipped with a manual transmission
   - Transaxle
   - Negative battery cable
7. Start the engine and check for leaks.

Timing Chain, Sprockets, Front Cover and Seal

REMOVAL & INSTALLATION

2.2L Engine

1. Before servicing the vehicle, refer to the precautions in the beginning of this section.
2. Remove or disconnect the following:
   - Negative battery cable
   - Accessory drive belt
   - Drive belt tensioner
3. Install an engine support fixture.
4. Remove or disconnect the following:
   - Engine mount assembly
   - Alternator
   - Power steering pump and move it aside with the lines attached
   - Oil pan
   - Crankshaft pulley/hub
   - Front cover

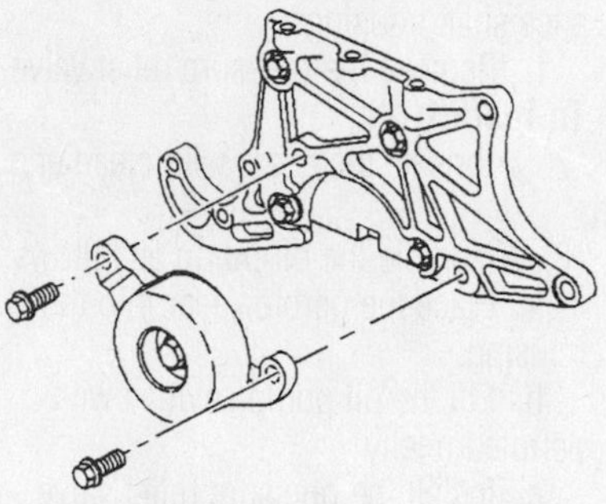

9300YG05

Exploded view of the accessory drive belt tensioner mounting—2.2L engine

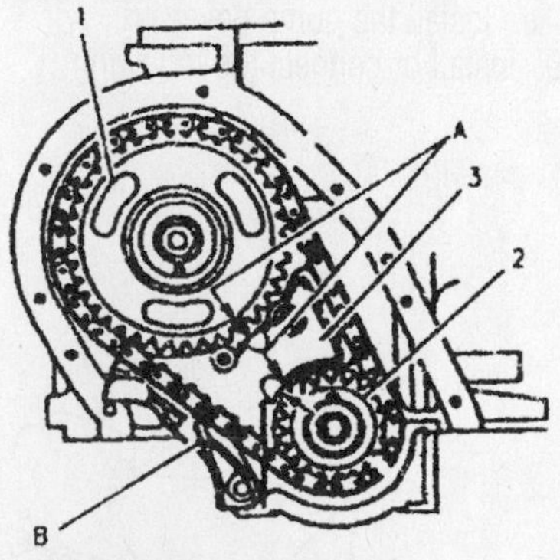

1. Camshaft sprocket
2. Crankshaft sprocket
3. Timing chain tensioner
A. Line up timing marks on sprockets with tabs on timing chain tensioner
B. Remove pin after timing chain is installed

7922YG16

Align the sprocket timing marks with the alignment tabs on the tensioner during timing chain installation—2.2L engine

5. Position the No. 1 piston at Top Dead Center (TDC) of the compression stroke so the camshaft and crankshaft sprockets timing marks are aligned.
6. Remove or disconnect the following:
   - Timing chain tensioner
   - Camshaft sprocket and chain as an assembly
   - Crankshaft sprocket, using a puller

To install:

7. Lubricate the timing chain with clean engine oil.
8. Install the crankshaft sprocket, press it onto the crankshaft.
9. Install the timing chain over the camshaft sprocket, then around the crankshaft sprocket. Be sure the marks on the 2 sprockets are in alignment. Lubricate the thrust surface with Molykote® or equivalent.
10. Install or connect the following:
    - Camshaft sprocket. Torque the bolts to 96 ft. lbs. (130 Nm).
11. Compress the timing chain tensioner spring. Insert a cotter pin or a nail into the hole in the tensioner to retain the timing chain tensioner shoe.
    - Chain tensioner. Torque the bolts to 18 ft. lbs. (24 Nm).
12. Remove the cotter pin or nail from the hole in the tensioner.
    - Front cover with a new gasket. Torque the bolts to 97 inch lbs. (11 Nm).
    - Crankshaft pulley/hub. Torque the pulley bolts to 37 ft. lbs. (50 Nm) and the hub bolt to 77 ft. lbs. (105 Nm).
    - Oil pan with new gaskets. Torque the nuts and bolts to 89 inch lbs. (10 Nm).
    - Power steering pump. Torque the bolts to 22 ft. lbs. (30 Nm).
    - Alternator. Torque the bolts to 37 ft. lbs. (50 Nm).
    - Engine mount assembly. Torque the bolts to 55 ft. lbs. (75 Nm).
13. Remove the engine support fixture.
14. Install or connect the following:
    - Accessory drive belt tensioner. Torque the bolts to 37 ft. lbs. (50 Nm).
    - Accessory drive belt
    - Negative battery cable

2.4L Engine

➡**It is recommended that the entire procedure be reviewed before attempting to service the timing chain.**

1. Before servicing the vehicle, refer to the precautions in the beginning of this section.
2. Disconnect the negative battery cable.

3. Drain the cooling system.
4. Install an engine support.
5. Remove or disconnect the following:
   - Coolant surge tank
   - Accessory drive belt
   - Upper fasteners from the front cover
   - Right engine mount and bracket
   - Right front wheel
   - Splash shield from the right wheel well
   - Crankshaft balancer
   - Front cover
6. Rotate the crankshaft clockwise, as viewed from front of engine (normal rotation), until the camshaft sprocket's timing dowel pin holes align with the holes in the timing chain housing. The mark on the crankshaft sprocket should align with the mark on the cylinder block. The crankshaft sprocket keyway should point upwards and align with the center line of the cylinder bores. This is the normal timed position.
7. Remove or disconnect the following:
   - Timing chain guides

➡Make sure that all of the slack in the timing chain is above the tensioner assembly.

The timing chain must be disengaged from any wear grooves in the tensioner shoe in order to remove the shoe. Slide a screwdriver blade under the timing chain while pulling the shoe outward.

   - Tensioner assembly
8. Mark the crankshaft sprocket and the timing chain outer surface.
   - Timing chain

To install:

9. Install the intake camshaft sprocket. Torque the bolt to 52 ft. lbs. (70 Nm).

➡Install the Special tool J 36008-A through the holes in the camshaft sprockets and into the holes in the timing chain housing. This positions the camshafts for correct timing.

10. If the camshafts are out of position and must be rotated more than ⅛ turn in order to install the alignment dowel pins, perform the following:
    a. Rotate the crankshaft 90 degrees clockwise off Top Dead Center (TDC) in order to give the valves adequate clearance to open.
    b. Once the camshafts are in position and the dowels installed, rotate the crankshaft counterclockwise back to TDC.

⁂ WARNING

Do not rotate the crankshaft clockwise to TDC or valve and piston damage may occur.

11. Install the timing chain over the exhaust camshaft sprocket, around the coolant pump sprocket and around the crankshaft sprocket.
12. Remove the alignment dowel pin from the intake camshaft. Using tool J 39579, rotate the intake camshaft sprocket counterclockwise enough to slide the timing chain over the intake camshaft sprocket. Release the camshaft sprocket wrench. The length of chain between the 2 camshaft sprockets will tighten. If properly timed, the intake camshaft alignment dowel pin should slide in easily. If the dowel pin does not fully index, the camshafts are not timed correctly and the procedure must be repeated.
13. Leave the alignment dowel pins installed.
14. With slack removed from the chain between the intake camshaft sprocket and the crankshaft sprocket, the crankshaft keyway and the cylinder block mark should be aligned. If not aligned, move the chain 1 tooth forward or rearward. Remove the slack and recheck marks.
15. Reload timing chain tensioner assembly to its **0** position as follows:
    a. Insert the tensioner plunger assembly into the tensioner housing.
    b. With the tensioner plunger fully extended, turn the complete assembly upside down on a flat surface.
    c. Press the bottom of the tensioner housing to compress the plunger into the housing until it is seated.
    d. Make sure that the plunger does not extend out of the tensioner housing more than 0.07 in. (1.7mm).
    e. Loosely install the tensioner assembly to the timing chain housing.
16. Install or connect the following:
    - Tensioner shoe on the stud.
    - Tensioner assembly by applying hand pressure on the timing chain tensioner shoe until the locking tab seats in the stud groove. Tighten the bolts to 89 inch lbs. (10 Nm).

⁂ WARNING

If the timing chain tensioner plunger is not released from the installation position, engine damage will occur upon start up.

17. Release the tensioner plunger by firmly pressing a flat blade tool against the plunger face.
18. Remove tool J 36008-A from the camshaft sprockets.
    a. Rotate crankshaft clockwise 2 full rotations. Align crankshaft keyway with mark on cylinder block and reinstall alignment dowel pins. Alignment dowel pins will slide in easily if engine is timed correctly.
19. Install or connect the following:
    - Timing chain guides
    - New seal lubricated with engine oil
    - Front cover using new gaskets. Tighten the nuts and bolts to 115 inch lbs. (13 Nm).
    - Crankshaft balancer. Torque the bolt to 129 ft. lbs. (175 Nm) plus a 90 degree turn.
    - Right front wheel well splash shield
    - Wheel
    - Right engine mount bracket. Torque the bolts to 44 ft. lbs. (60 Nm) plus an additional 60 degree turn.
    - Right engine mount. Torque the bolts to 55 ft. lbs. (75 Nm).
20. Remove the engine support.
21. Install or connect the following:
    - Accessory drive belt
    - Coolant surge tank
    - Negative battery cable
22. Refill the cooling system and check for leaks.

Piston and Ring

POSITIONING

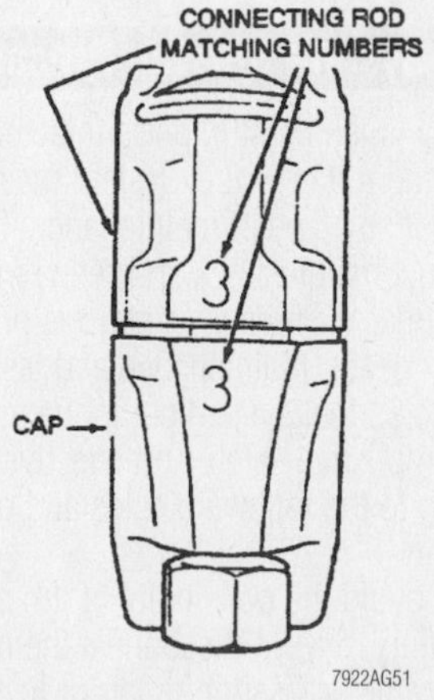

Connecting rod and cap installation. Be sure to matchmark the cap and rod prior to disassembly—2.2L and 2.4L engines

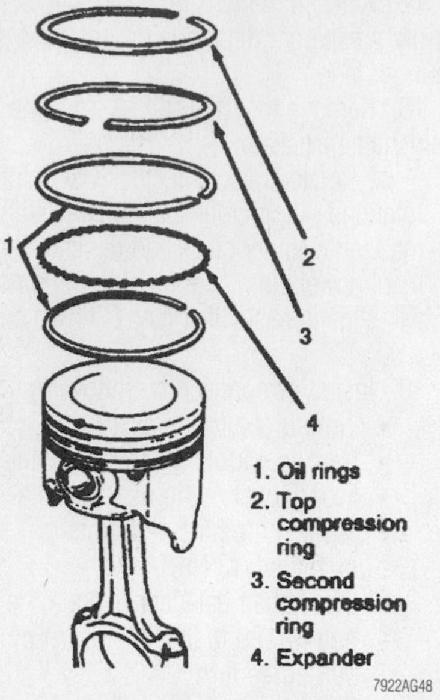

Piston ring positioning—2.2L engine

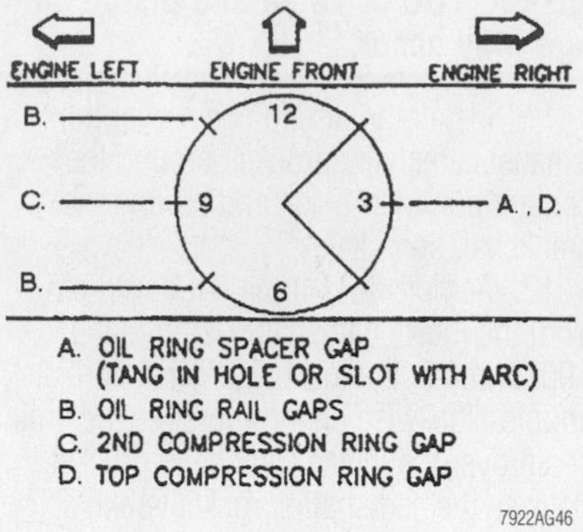

Piston ring end-gap spacing—2.2L engine

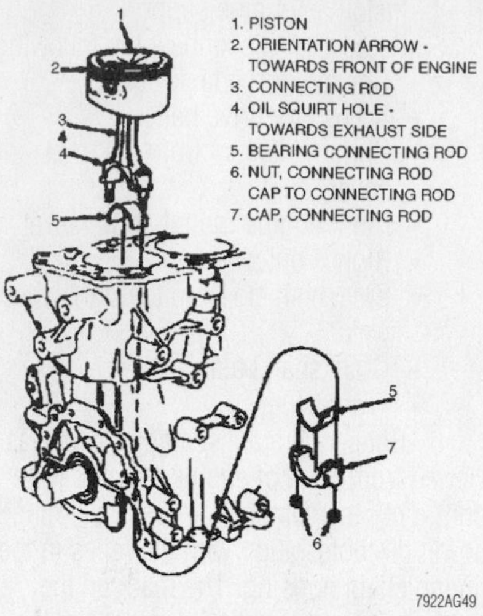

Piston and connecting rod assembly positioning—2.4L engine

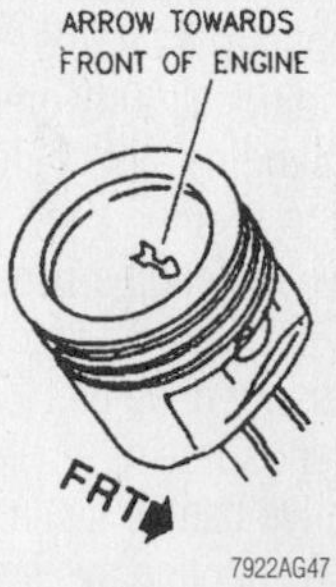

Piston positioning. Often the arrow is replaced by a notch, which also must face toward the front of the engine—2.2L engine

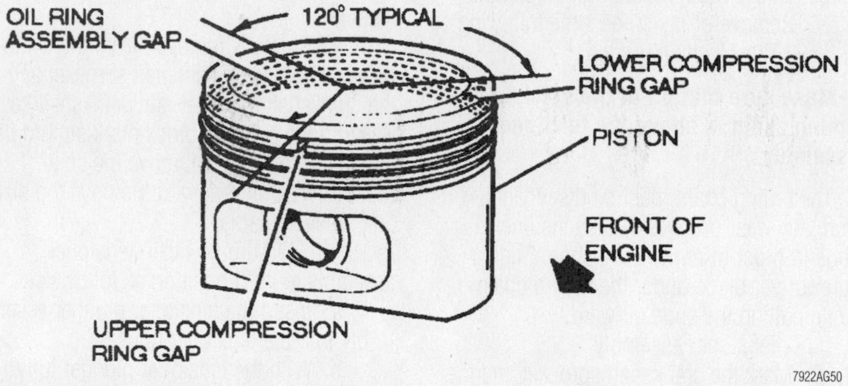

Piston ring end-gap spacing—2.4L engine

FUEL SYSTEM

Fuel System Service Precautions

Safety is the most important factor when performing not only fuel system maintenance but any type of maintenance. Failure to conduct maintenance and repairs in a safe manner may result in serious personal injury or death. Maintenance and testing of the vehicle's fuel system components can be accomplished safely and effectively by adhering to the following rules and guidelines.

- To avoid the possibility of fire and personal injury, always disconnect the negative battery cable unless the repair or test procedure requires that battery voltage be applied.
- Always relieve the fuel system pressure prior to disconnecting any fuel system component (injector, fuel rail, pressure regulator, etc.), fitting or fuel line connection. Exercise extreme caution whenever relieving fuel system pressure, to avoid exposing skin, face and eyes to fuel spray. Please be advised that fuel under pressure may penetrate the skin or any part of the body that it contacts.
- Always place a shop towel or cloth around the fitting or connection prior to loosening to absorb any excess fuel due to spillage. Ensure that all fuel spillage (should it occur) is quickly removed from engine surfaces. Ensure that all fuel soaked cloths or towels are deposited into a suitable waste container.
- Always keep a dry chemical (Class B) fire extinguisher near the work area.
- Do not allow fuel spray or fuel vapors to come into contact with a spark or open flame.
- Always use a back-up wrench when loosening and tightening fuel line connection fittings. This will prevent unnecessary stress and torsion to fuel line piping.
- Always replace worn fuel fitting O-rings with new. Do not substitute fuel hose or equivalent, where fuel pipe is installed.

Fuel System Pressure

RELIEVING

1. Before servicing the vehicle, refer to the precautions in the beginning of this section.
2. Loosen the fuel filler cap in order to relieve the pressure in the tank (do not tighten at this time).
3. Raise and safely support the vehicle.
4. Detach the fuel pump electrical connector.

5. Start and run the vehicle until it stalls, then engage the starter for an additional 3 seconds to ensure the relief of any remaining pressure.

6. Disconnect the negative battery cable.

7. Once the tests or repairs are completed, reattach the fuel pump electrical connector.

8. Connect the negative battery cable.

9. Lower the vehicle.

10. Tighten the fuel filler cap.

11. prime the fuel system by cycling the ignition switch **ON** for 2 seconds, **OFF** for 10 seconds, then **ON** again. Repeat, if necessary to build system pressure.

Fuel Filter

REMOVAL & INSTALLATION

1. Before servicing the vehicle, refer to the precautions in the beginning of this section.

2. Relieve the fuel system pressure.

3. Remove the fuel filter from the fuel line using a back-up wrench on the filter.

4. Grasp the filter and nylon connection line fitting. Twist the quick-connect fitting ¼ turn in each direction to loosen any dirt within the fitting. Disconnect the quick-connect fitting from the fuel filter.

5. Remove the fuel filter from the bracket.

To install:

6. Before installing a new filter, always apply a few drops of clean engine oil to the male tube end of the filter and to the fuel sending unit assembly connection. This will help ensure proper connection and prevent possible fuel leaks. During normal operation, the O-rings located in the female connector will swell and may prevent proper connection if not lubricated.

7. Install the fuel filter in the mounting bracket.

8. Connect the quick-connect fitting to the fuel filter using the following procedure:

a. Apply a few drops of clean engine oil to the male ends of the filter and the fuel sender assembly.

b. Push the connectors together to cause the retaining tabs/fingers to snap into place.

c. Once installed, pull on both ends of each connection to be sure they are secure.

9. Install or connect the following:

- New O-ring
- Fuel filter. Torque the fitting to 20 ft. lbs. (27 Nm).
- Negative battery cable

10. Pressurize the fuel system and verify no leaks.

Fuel Pump

REMOVAL & INSTALLATION

1. Before servicing the vehicle, refer to the precautions in the beginning of this section.

2. Relieve the fuel system pressure.

3. Drain the fuel tank.

4. Remove the fuel tank from the vehicle.

5. While holding the modular fuel sender assembly down, remove the snapring.

**** WARNING**

The modular fuel sender assembly may spring up from its position. When removing the modular fuel sender from the tank, be aware that the reservoir bucket is full of fuel. It must be tipped slightly during removal to avoid damage to the float.

6. Remove or disconnect the following:

- External fuel strainer
- Connector retainer from the fuel pump electrical connector

7. Gently release the tabs on the sides of the fuel sender at the cover assembly. Begin by squeezing the sides of the reservoir and releasing the tab opposite the fuel level sensor. Move clockwise to release the 2nd and 3rd tabs in the same manner.

8. Remove or disconnect the following:

- Fuel pump electrical connector
- Fuel pump baffle/pump assembly by rotating it counterclockwise from the retainer
- Fuel pump by sliding it from the slot
- Fuel pump outlet seal

To install:

9. Install or connect the following:

- Fuel pump outlet seal
- Fuel pump outlet, slide it into the reservoir cover slots
- Fuel pump/baffle assembly onto the reservoir retainer; rotate it clockwise until seated
- Lower retainer assembly partially

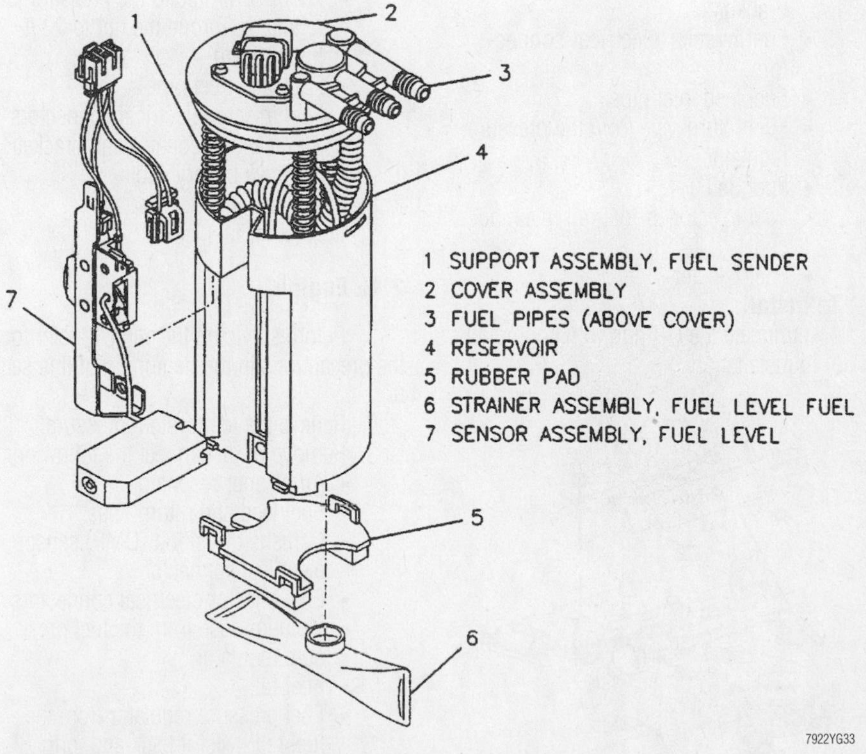

Modular fuel pump component identification

into the reservoir; align the 3 tabs and press the retainer onto the reservoir making sure all 3 tabs are firmly seated

➡Gently, pull on the fuel pump reservoir to assure it is secure to the retainer. If not secure, replace the entire fuel sender.

- Fuel pump electrical connector
- Connector retainer to the fuel sender cover
- New external fuel strainer
- Modular fuel sender
- Fuel tank in the vehicle
- Negative battery cable

10. Pressurize the fuel system and verify no leaks.

Fuel Injector

REMOVAL & INSTALLATION

2.2L Engine

1. Before servicing the vehicle, refer to the precautions in the beginning of this section.
2. Relieve the fuel system pressure.
3. Remove or disconnect the following:
   - Air cleaner resonator and bracket
   - Fuel injector electrical connectors
   - Fuel feed inlet pipe
   - Fuel return pipe from the pressure regulator
   - Fuel Rail
   - Fuel injector-to-fuel rail retaining clip
   - Fuel injector

To install:

4. Lubricate the O-rings with engine oil prior to installation.

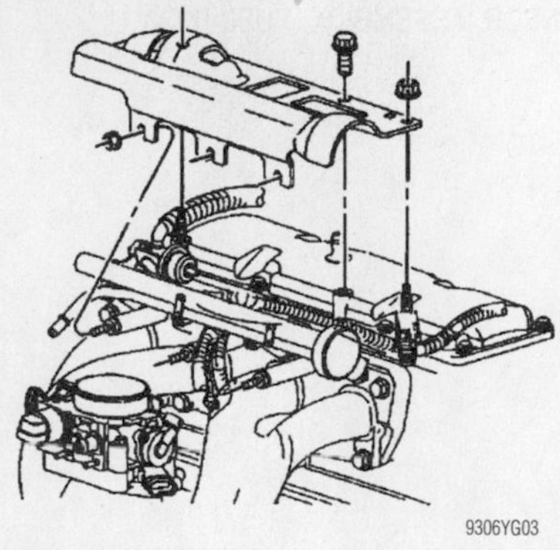

View of the fuel rail assembly—2.2L engine

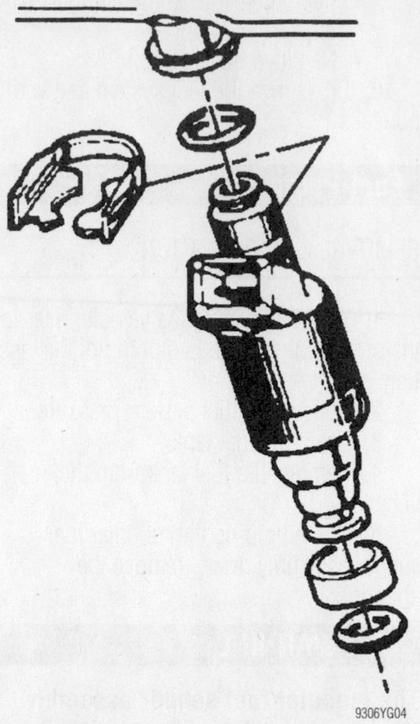

Exploded view of the fuel injector—2.2L engine

5. Install or connect the following:
   - Fuel injectors with new O-rings
   - Fuel Rail. Torque the bolts to 18 ft. lbs. (24 Nm).
   - Fuel return pipe to the pressure regulator. Torque the nut to 22 ft. lbs. (30 Nm).
   - Fuel feed inlet pipe
   - Fuel injector electrical connectors
   - Air cleaner resonator and bracket
   - Negative battery cable
6. Pressurize the fuel system.
7. Inspect for fuel leaks.

2.4L Engine

1. Before servicing the vehicle, refer to the precautions in the beginning of this section.
2. Relieve the fuel system pressure.
3. Remove or disconnect the following:
   - Air cleaner resonator
   - Fuel feed and return lines
   - Camshaft Position (CMP) sensor electrical connector
   - Fuel injector electrical connectors
   - Vacuum hose from the fuel pressure regulator
   - Fuel rail
   - Fuel pressure regulator from the fuel rail, twist it back and forth
   - Fuel injector-to-fuel rail retaining clip
   - Fuel injector

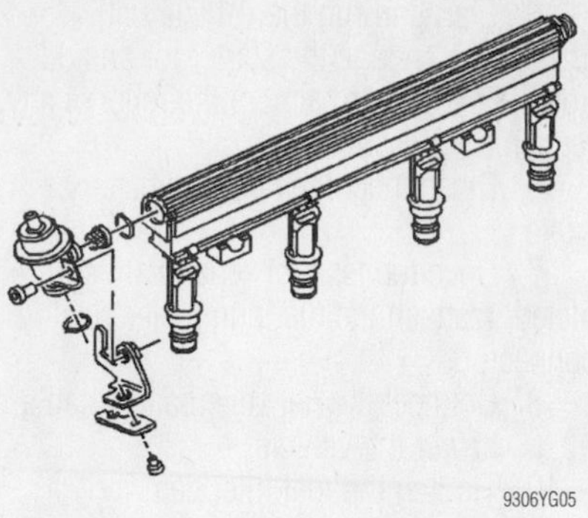

Exploded view of the fuel rail assembly—2.4L engine

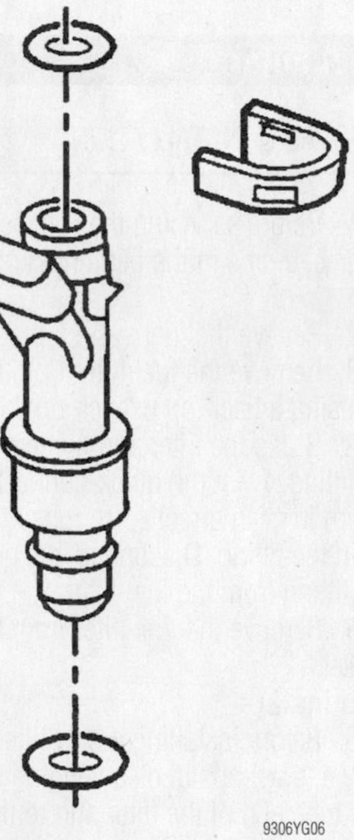

Exploded view of the fuel injector—2.4L engine

To install:

4. Lubricate the O-rings with engine oil prior to installation.
5. Install or connect the following:
   - Fuel injector(s) with new O-rings
   - Fuel rail. Torque the bolts to 18 ft. lbs. (24 Nm).
   - Fuel pressure regulator with a new O-ring to the fuel rail. Torque the bolt to 53 inch lbs. (6 Nm).
   - Fuel feed and return lines
   - Fuel injector electrical connectors
   - Vacuum hose to the fuel pressure regulator
   - CMP sensor electrical connector
   - Air cleaner resonator
   - Negative battery cable
6. Pressurize the fuel system.
7. Inspect for fuel leaks.

DRIVE TRAIN

Transmission Assembly

REMOVAL & INSTALLATION

Manual

1. Before servicing the vehicle, refer to the precautions in the beginning of this section.
2. Install and engine support fixture.
3. Raise the engine enough to take pressure off of the transaxle mounts.
4. Install an engine support fixture and raise the engine enough to take the pressure off the transaxle mounts.
5. Remove or disconnect the following:
   - Battery
   - Air cleaner and duct assembly
   - Pushrod from the clutch pedal
   - Wiring harness from the upper transmission mount bracket
   - Upper transaxle mount-to-transaxle bolts
   - Negative battery cable from the transmission
   - Starter motor
   - Pressure line from the clutch actuator cylinder
   - Back-up light switch connector
   - Rear transmission mount bolts
   - Vehicle Speed Sensor (VSS) electrical connector
   - Upper transmission mounting bolts
   - Shift cables
   - Cable bracket
   - Rack and pinion mounting bolts
   - Front wheels
   - Splash shields
   - Engine strut
   - Front exhaust pipe
   - Both front Anti-lock Brake System (ABS) wheel speed sensor harness
   - Intermediate steering shaft
   - Ball joints from the steering knuckles
   - Tie rod ends from the steering knuckles
   - Brake lines from the support frame
   - Power steering hoses from the rack and pinion
6. Support the suspension crossmember.
7. Remove or disconnect the following:
   - Left side suspension support bolts
   - Flywheel cover
   - Halfshafts from the transaxle
   - Front lower transaxle mount
8. Position a suitable jack under the transaxle.
9. Remove or disconnect the following:
   - Transaxle-to-engine mounting bolts (noting their location)
   - Transaxle from the engine

To install:

10. Install or connect the following:
    - Transaxle
    - Transaxle-to-engine mounting bolts. Torque the bolts to 55 ft. lbs. (75 Nm).
    - Front transaxle mount. Torque the bolt to 44 ft. lbs. (60 Nm).
    - Flywheel cover
    - Halfshafts into the transaxle
    - Suspension support. Torque the bolts to 81 ft. lbs. (110 Nm).
    - Power steering hoses to the rack and pinion
    - Brake lines to the support frame
    - Ball joints to the steering knuckle. Torque the nuts to 48 ft. lbs. (65 Nm).
    - Tie rod ends to the steering knuckles. Torque the nuts to 33 ft. lbs. (45 Nm).
    - Intermediate steering shaft. Torque the pinch bolt to 12 ft. lbs. (17 Nm).
    - ABS wheel speed sensor wiring harness connectors
    - Front exhaust pipe. Torque the bolts to 26 ft. lbs. (35 Nm).
    - Engine strut. Torque the bolts to 74 ft. lbs. (100 Nm) plus an additional turn of 90 degrees.
    - Inner splash shield
    - Front wheels
    - Rack and pinion mounting bolts. Torque the bolts to 89 ft. lbs. (120 Nm).
    - Starter motor. Torque the bolts to 37 ft. lbs. (50 Nm).
    - Upper transaxle bolts. Torque the bolts to 71 ft. lbs. (96 Nm).
    - Pressure line to clutch actuator cylinder
    - VSS electrical connector
    - Back-up light switch connector
    - Rear transaxle mount. Torque the bolts to 55 ft. lbs. (75 Nm).
    - Wiring harness to the mount bracket
11. Remove the engine support fixture.
12. Install or connect the following:
    - Shift cable clamp. Torque the nut to 89 inch lbs. (10 Nm).
    - Shift cables
    - Negative battery cable to the transmission
    - Air cleaner and duct assembly to the throttle body
    - Pushrod to the clutch pedal
    - Battery
13. Refill the transaxle.
14. Refill and bleed the power steering system.
15. Road test the vehicle and verify proper operation.

Automatic

1. Before servicing the vehicle, refer to the precautions in the beginning of this section.
2. Remove or disconnect the following:
   - Negative battery cable
   - Air intake duct
   - Throttle Valve (TV) cable
   - Shift cablc and bracket
   - Vacuum lines
   - Electrical connectors
   - Filler tube
3. Attach an engine support fixture.
4. Remove or disconnect the following:
   - Upper engine-to-transaxle bolts
   - Front wheels
   - Left side splash shield
   - Both front Anti-lock Brake System (ABS) wheel speed sensors and harness from left suspension support
   - Both lower ball joints
   - Stabilizer shaft links
   - Front air deflector
   - Both halfshafts
   - Engine-to-transaxle brace
   - Torque converter cover
   - Starter motor
   - Torque converter bolts
   - Transaxle oil cooler lines and brace
   - Ground wires from the transaxle
   - Engine-to-transaxle mount bolts
5. Support the transaxle with a jack.
6. Remove or disconnect the following:
   - Transaxle mount-to-body bolts
   - Heater core hose brace from transaxle
   - Engine-to-transmission bolts
   - Transmission

To install:

7. Install or connect the following:
   - Transaxle while installing right halfshaft
   - Lower engine-to-transaxle bolts.

Torque the bolts to 71 ft. lbs. (96 Nm).
- Transmission mount-to-body bolts. Torque the bolts to 49 ft. lbs. (66 Nm).
- Heater core hose brace
- Ground wires to the transaxle
- Oil cooler lines
- Torque converter. Torque the bolts to 46 ft. lbs. (62 Nm).
- Torque converter cover
- Starter motor. Torque the bolts to 37 ft. lbs. (50 Nm).
- Engine-to-transaxle brace. Torque the bolts to 32 ft. lbs. (43 Nm).
- Halfshafts
- Front air deflector
- Stabilizer links. Torque the bolts to 49 ft. lbs. (66 Nm).
- Lower ball joints. Torque the nuts to 45 ft. lbs. (61 Nm).
- Both ABS wheel speed sensors
- Left splash shield
- Front wheels
- Upper engine-to-transaxle bolts. Torque the bolts to 71 ft. lbs. (96 Nm).

8. Remove the engine support fixture.
9. Install or connect the following:
   - Filler tube
   - Electrical connectors
   - Vacuum lines
   - Shift cable and bracket
   - TV cable
   - Intake air duct
   - Negative battery cable

10. Refill the transaxle, start the vehicle and verify that there are no leaks.
11. Road test the vehicle.

Clutch

REMOVAL & INSTALLATION

1. Before servicing the vehicle, refer to the precautions in the beginning of this section.
2. Remove or disconnect the following:
   - Negative battery cable
   - Clutch master cylinder pushrod from the clutch pedal
   - Transaxle
3. If any of the parts are to be reused, mark the pressure plate assembly and the flywheel so they can be assembled in the same position.
4. Loosen the attaching bolts 1 turn at a time until spring tension is relieved.
5. Support the pressure plate and remove the bolts.

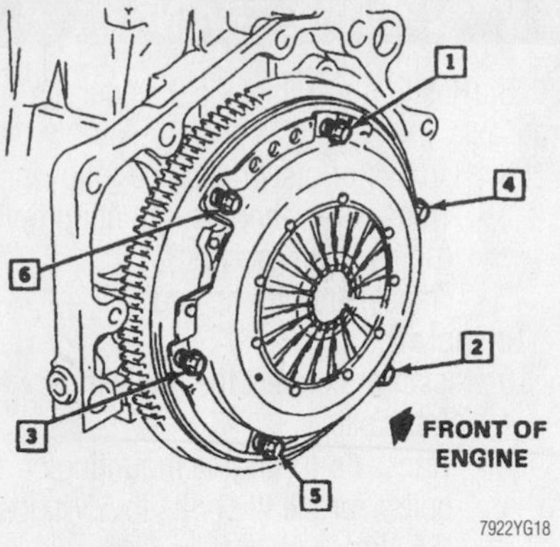

Clutch cover bolt tightening sequence

6. Remove the pressure plate and clutch disc as an assembly.

➡Do not disassemble the pressure plate assembly. Replace it if defective.

To install:

7. Install the pressure plate and clutch disc assembly, supporting it with a clutch aligning tool.
8. Torque the pressure plate-to-flywheel bolts, gradually, in a cross pattern, as follows:
   a. Step 1: Lightly seat all bolts.
   b. Step 2: Tighten the bolts to 16 ft. lbs. (20 Nm).
   c. Step 3: Tighten the bolts an additional 45 degrees in sequence
9. Lubricate the outside groove and the inside recess of the release bearing with high temperature grease.
10. Install or connect the following:
    - Release bearing
    - Transaxle
    - Clutch master cylinder pushrod to the clutch pedal and secure with the retaining clip
    - Negative battery cable
11. Bleed clutch system as necessary and road test vehicle.

Hydraulic Clutch System

BLEEDING

1. Before servicing the vehicle, refer to the precautions in the beginning of this section.
2. Attach a hose to the bleeder screw on the clutch actuator assembly and submerge the other end of the hose in a container of hydraulic clutch fluid.
3. Depress the clutch pedal slowly and hold.
4. Loosen the bleeder screw to purge air.
5. Tighten the bleeder screw to 18 inch lbs. (2 Nm).
6. Release the clutch pedal.
7. Repeat Steps 3 through 6 until all air is purged from the system.
8. Refill the reservoir with hydraulic clutch fluid.
9. Repeat this bleeding procedure if there is a grinding noise during the clutch spin down procedure.

Halfshaft

REMOVAL & INSTALLATION

1. Before servicing the vehicle, refer to the precautions in the beginning of this section.
2. Remove or disconnect the following:
   - Negative battery cable
   - Front wheel
   - Hub nut and washer
   - Lower ball joint
   - Wheel Speed Sensor (WSS) electrical connector

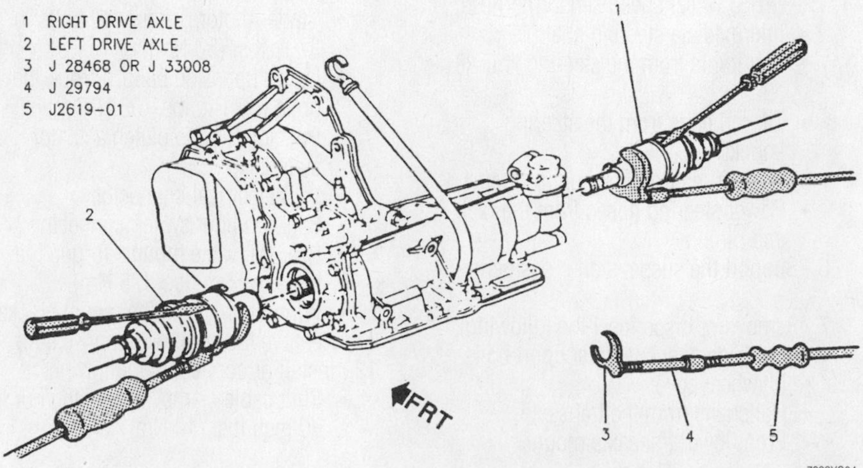

To prevent damaging the transaxle or halfshaft, use the tools as shown to remove the halfshafts

- Stabilizer bar link
- Halfshaft, press it from wheel bearing/hub assembly
- Halfshaft from the transaxle

⁂ WARNING

Do not pull the halfshaft by the CV-joint boot or on the joint itself.

To install:

3. Install the halfshaft into the transaxle (or intermediate shaft, if equipped) by placing a brass drift pin into the groove on the joint housing and tapping until seated. Be careful not to damage the axle seal or dislodge the seal garter spring when installing the axle.

➡Be sure the halfshaft is fully engaged in the transaxle. Verify that the halfshaft is seated by grasping the inner joint housing and pulling outward. Do not pull on the shaft or the boot, but on the inner joint housing only.

4. Install or connect the following:
   - Halfshaft into the hub/bearing assembly
   - Stabilizer link. Torque the bolts to 13 ft. lbs. (17 Nm).
   - WSS electrical connector
   - Lower ball joint to the steering knuckle. Torque the nut to 41–48 ft. lbs. (55–65 Nm).
   - Washer and a new hub nut. Torque the nut to 144 ft. lbs. (200 Nm).
   - Front wheel
   - Negative battery cable

CV-Joints

OVERHAUL

Outer CV-Joint

1. Before servicing the vehicle, refer to the precautions in the beginning of this section.
2. Remove or disconnect the following:
   - Front wheel
   - Halfshaft and position it in a vise
   - Large CV-joint boot clamp
   - Small CV-joint boot clamp
   - CV-joint boot and slide it back on the shaft
3. Perform the following:

   a. Choose a reference mark on the halfshaft.

   b. Measure the distance between the reference mark and the CV-joint inner race face; retain this measurement.

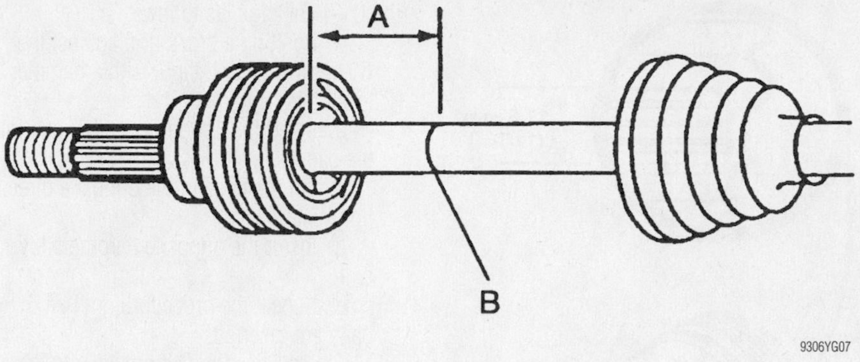

Measuring the halfshaft-to-inner race reference distance—Outer CV-joint

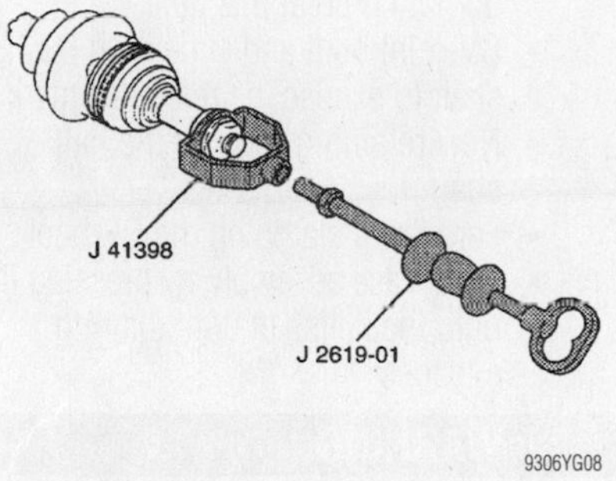

Removing the outer race from the halfshaft—Outer CV-joint

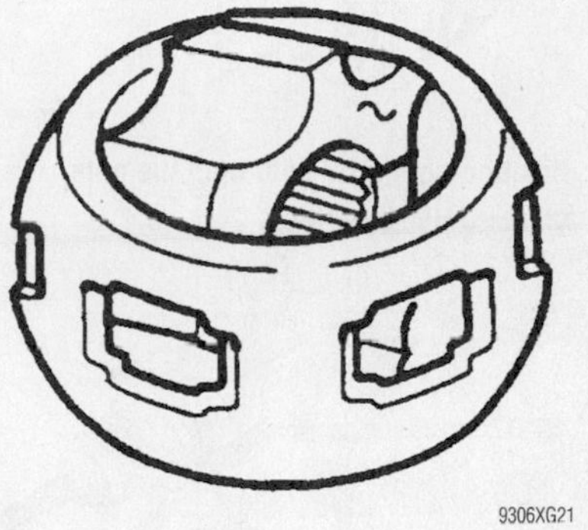

View the cage and inner race—Outer CV-joint

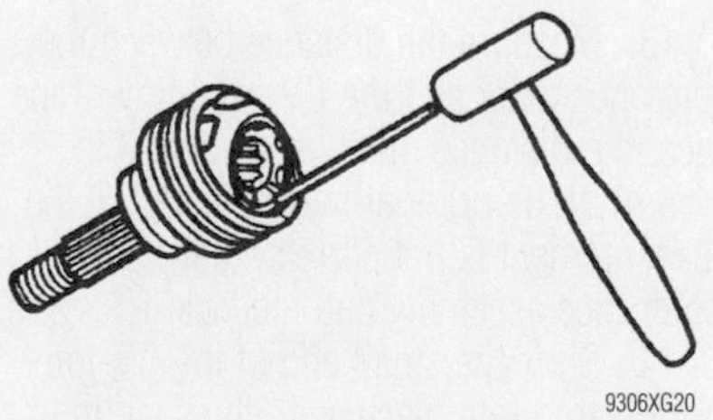

Tilting the cage—Outer CV-joint

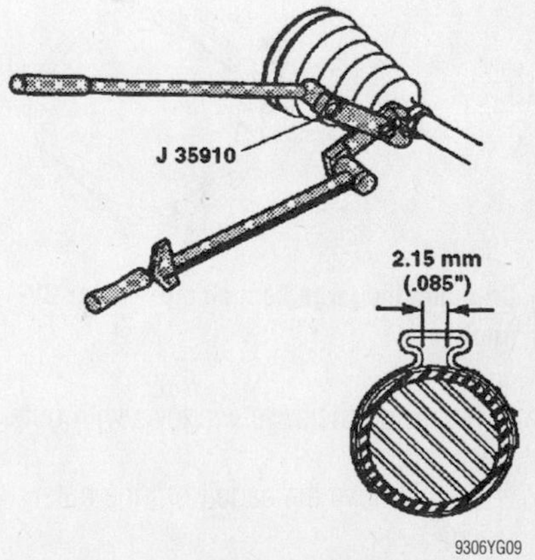

Crimping the small boot clamp—Outer CV-joint

4. Attach CV Puller tool J-41398 to the outer race threaded area.
5. Remove or disconnect the following:
   - CV-joint outer race from the halfshaft using a slide hammer puller
   - CV Puller tool J-41398 and slide hammer puller from the outer race
   - Retaining ring from the halfshaft
   - CV-joint boot from the halfshaft
6. Disassemble the chrome alloy balls from the CV-joint cage as follows:

   a. Position a brass drift against the CV-joint cage and tap it with a hammer to tilt the cage.

   b. Remove the 1st chrome alloy ball from the cage.

   c. Tilt the cage in the opposite direction.

   d. Remove the opposite chrome alloy ball.

   e. Repeat the procedure until all 6 balls are removed.
7. Disassemble the CV-joint cage and inner race as follows:

   a. Pivot the cage and race 90 degrees to the center line of the outer race.

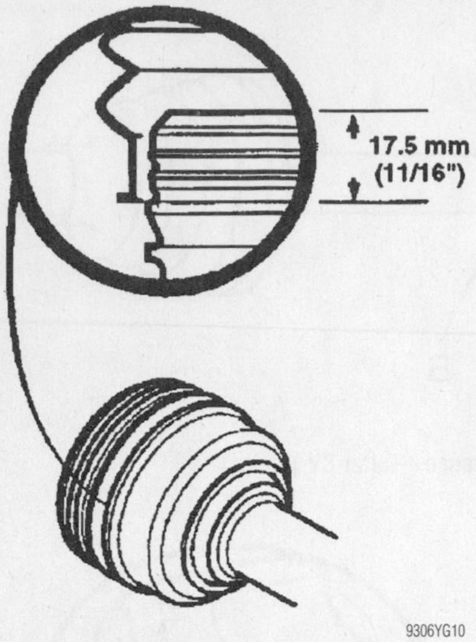

Positioning the CV boot onto the outer race—Outer CV-joint

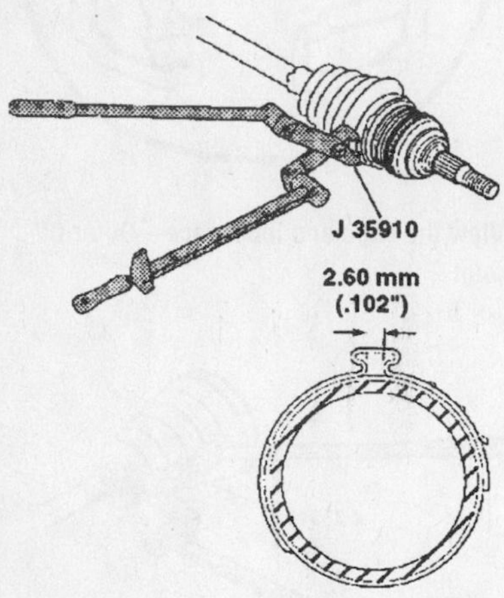

Crimping the large boot clamp—Outer CV-joint

b. Align the cage windows with outer race lands.

c. Remove the cage from the outer race.

d. Rotate the inner race upward and remove it from the cage.

To install:

8. Lubricate the parts with a light coat of grease.

9. Assemble the CV-joint cage and inner race, as follows:

a. Rotate the inner race 90 degrees to the cage centerline.

b. Align the cage windows with inner race lands.

c. Insert the inner race into the cage by rotating the inner race downward.

d. Insert the cage/inner race into the outer race.

10. Assemble the chrome alloy balls into the CV-joint cage, as follows:

a. Position a brass drift against the CV-joint cage and tap it with a hammer to tilt the cage.

b. Insert the 1st chrome alloy ball into the cage.

c. Tilt the cage in the opposite direction.

d. Insert the opposite chrome alloy ball.

e. Repeat the procedure until all 6 balls are inserted.

11. Install ½ of the kit grease into the CV-joint.

12. Install or connect the following:

- Small ring clamp on the CV boot, do not crimp at this time
- CV-joint boot and slide it up the shaft to expose the reference mark
- New retaining ring on the half-shaft
- Large ring clamp on the CV boot
- Outer race assembly by pressing it onto the halfshaft until the ring engages

WARNING

When installing the outer race assembly onto the halfshaft, do not exceed 4,000 lbs. pressure.

13. Measure the distance between the reference mark and the CV-joint inner race face; the distance must be +/- 0.039 in. (1mm) of the original measurement. If the measurement is not correct, repress the outer race assembly and recheck it.

14. Slide the small end of the CV-joint boot/clamp into place, with the seal lip in the halfshaft groove.

➡Make sure the boot lies flat against the halfshaft.

15. Using a Crimp tool, a torque wrench and a breaker bar, crimp the small CV-joint boot clamp to 100 ft. lbs. (136 Nm).

16. Check the clamp gap dimension; if it is not 0.085 in. (2.15mm), continue tightening the clamp until it is.

17. Install ½ of the kit grease into the CV-joint boot.

18. Slide the large end of the CV boot/clamp into place, with the seal lip in place over the outer race.

➡Make sure the boot lies flat against the outer race.

19. Using a Crimp tool, a torque wrench and a breaker bar, crimp the large CV-joint boot clamp to 130 ft. lbs. (176 Nm).

20. Check the clamp gap dimension; if it is not 0.102 in. (2.60mm), continue tightening the clamp until it is.

21. Install the halfshaft and the front wheel.

Inner (Tri-Pod) Joint

2.2L ENGINE

1. Before servicing the vehicle, refer to the precautions in the beginning of this section.

2. Remove or disconnect the following:

- Front wheel
- Halfshaft
- Small CV-joint boot clamp
- Large CV-joint boot clamp
- CV-joint boot by sliding it away from the tri-pod joint
- Inboard spacer ring
- Outboard retaining ring
- Tri-pod joint spider assembly by tapping it from the halfshaft with a brass drift
- Tri-pod spider retaining ring
- Trilobal tri-pod bushing from the housing
- CV-joint boot

To install:

3. Install or connect the following:

- Small boot clamp
- CV-joint boot
- New inboard spacer ring slide it rearward on the shaft past the 2nd groove
- Tri-pod joint spider assembly onto the shaft

4. Assemble the tri-pod spider assembly onto the halfshaft as follows:

a. Position the tri-pod spider assembly onto the shop press plate.

b. Position the halfshaft onto the tri-pod spider assembly, in the shop press.

c. Press the halfshaft into the tri-pod spider assembly until the spider assembly passes the 2nd groove.

WARNING

When assembling the tri-pod assembly onto the halfshaft, do not exceed 4,000 lbs. pressure.

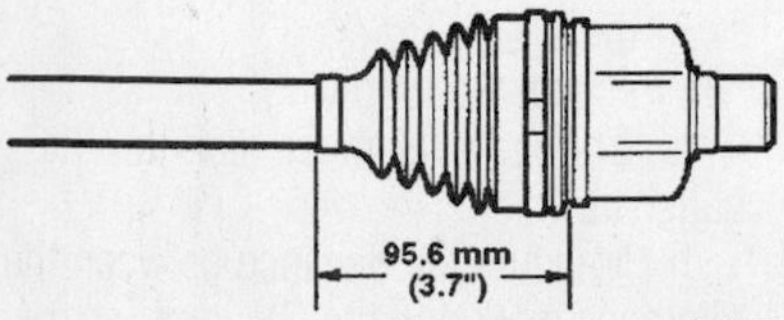

CV-joint boot measurement—Inner (tri-pod) joint—1998–01 2.2L engine

5. Remove the halfshaft from the shop press and place it in vise.

6. Install or connect the following:
- New outboard retaining ring into the axle shaft groove
- Tri-pod joint spider assembly, slide it against the outboard retaining ring using a brass drift

7. Seat the inboard spacer ring in the proper groove.

8. Slide the CV-joint boot into place and crimp the small boot clamp.

9. Install or connect the following:
- ½ of the kit grease into the boot
- ½ of the kit grease into the tri-pod housing
- Trilobal tri-pod bushing flush with the tri-pod housing face
- New large seal clamp onto the CV-joint boot
- Tri-pod housing, slide it over the tri-pod joint spider assembly
- CV-joint boot/clamp, slide it into place, over the trilobal tri-pod bushing with the seal lip in the groove

➡Make sure the boot lies flat against the trilobal bushing.

10. Position the CV-joint boot so it measures 4.9 in. (102mm) for 1998–99 or 3.7 in. (95.7) for 2000–01.

11. Using a Crimp tool, a torque wrench and a breaker bar, crimp the small CV-joint boot clamp to 100 ft. lbs. (136 Nm).

12. Crimp the large CV-joint boot clamp.

13. Install the halfshaft and the front wheel.

2.4L ENGINE

1. Before servicing the vehicle, refer to the precautions in the beginning of this section.

2. Remove or disconnect the following:
- Front wheel
- Halfshaft and place it in a vise
- Snapring from the stub shaft

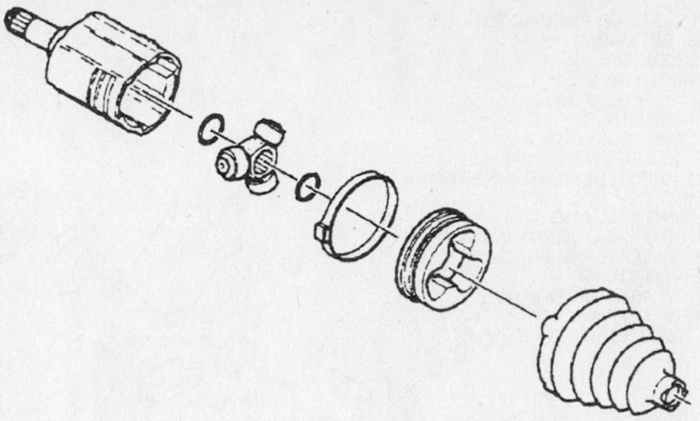

Exploded view of the inner (tri-pod) joint

- Small CV-joint boot clamp
- Large CV-joint boot clamp
- CV-joint boot by sliding it away from the tri-pod joint

3. Install a Stub Shaft Removal tool J-38868-A to the stub shaft snapring groove.

4. Using a slide hammer puller, pull the stub shaft from the tri-pod housing.

5. Remove or disconnect the following:
- Tri-pod housing from the tri-pod spider
- Inboard spacer ring slide it rearward on the shaft
- Outboard retaining ring
- Tri-pod joint spider assembly
- Inboard spacer ring
- CV-joint boot
- Trilobal tri-pod bushing from the housing

To install:

6. Install or connect the following:
- New snapring onto the stub shaft
- Small boot clamp
- CV-joint boot

7. Using a Crimp tool, a torque wrench and a breaker bar, crimp the small CV-joint boot clamp to 100 ft. lbs. (136 Nm).

8. Install or connect the following:
- Inboard spacer ring slide it rearward on the shaft, past the 2nd groove
- Tri-pod joint spider assembly onto the shaft until it passes the 2nd groove
- Outboard retaining ring into the axle shaft groove
- Tri-pod joint spider assembly, slide it against the outboard retaining ring
- Inboard spacer ring, seat it in the groove
- ½ kit grease into the boot
- ½ kit grease into the tri-pod housing
- Trilobal tri-pod bushing flush with the tri-pod housing face
- New large seal clamp onto the CV-joint boot
- Tri-pod housing, slide it over the tri-pod joint spider assembly
- CV-joint boot/clamp, slide it into place, over the trilobal tri-pod bushing with the seal lip in the groove

➡Make sure the boot lies flat against the trilobal bushing.

9. Position the CV-joint boot so it measures 4.9 in. (125mm).

10. Crimp the large CV-joint boot clamp.

11. Install the halfshaft and the front wheel.

STEERING AND SUSPENSION

Air Bag

⁂ CAUTION

Some vehicles are equipped with an air bag system. The system must be disabled before performing service on or around system components, steering column, instrument panel components, wiring and sensors. Failure to follow safety and disabling procedures could result in accidental air bag deployment, possible personal injury and unnecessary system repairs.

PRECAUTIONS

Several precautions must be observed when handling the inflator module to avoid accidental deployment and possible personal injury.

- Never carry the inflator module by the wires or connector on the underside of the module.

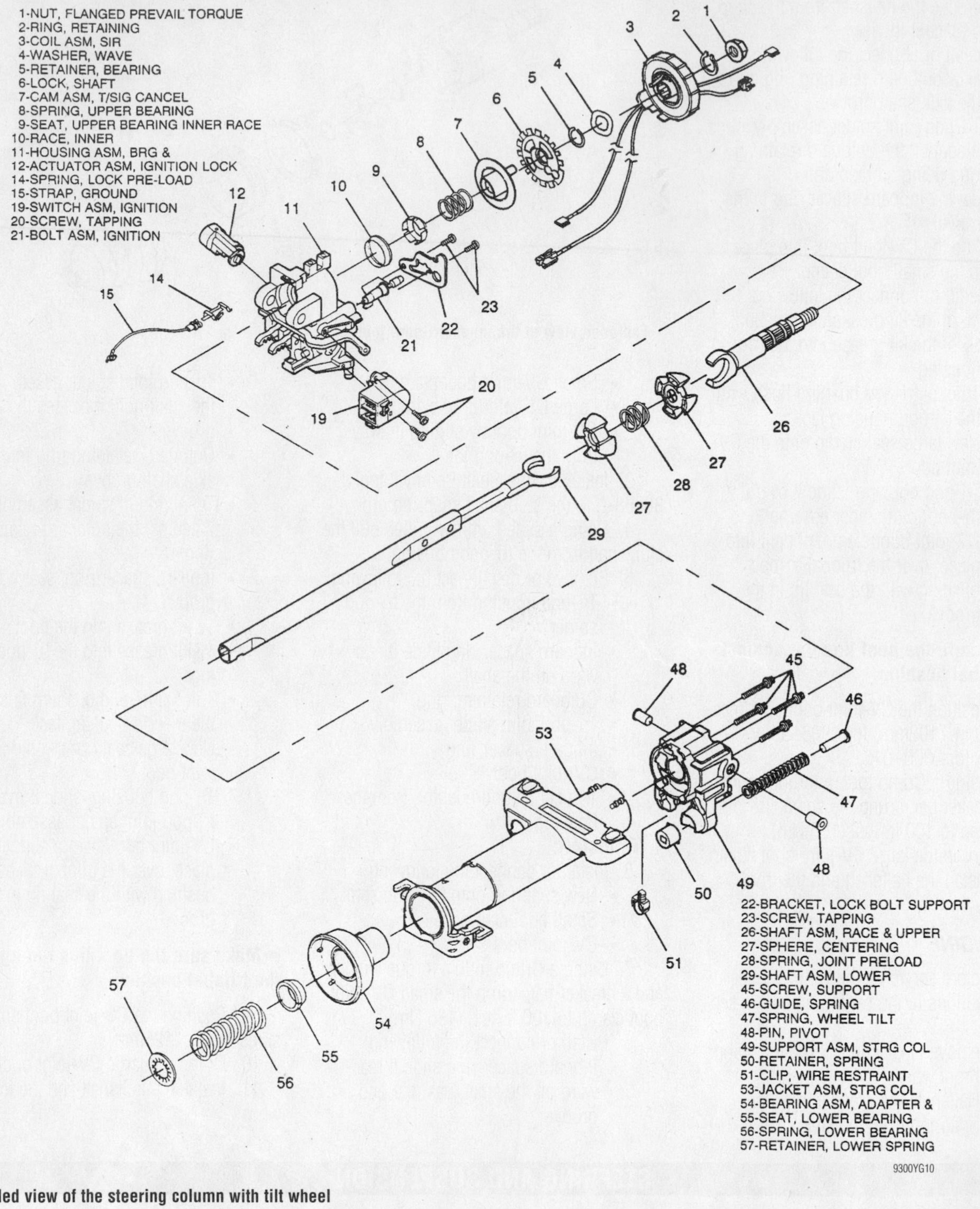

Exploded view of the steering column with tilt wheel

• When carrying a live inflator module, hold securely with both hands, and ensure that the bag and trim cover are pointed away.

• Place the inflator module on a bench or other surface with the bag and trim cover facing up.

• With the inflator module on the bench, never place anything on or close to the module which may be thrown in the event of an accidental deployment.

DISARMING

CAUTION

The Supplemental Inflatable Restraint (SIR) system must be disarmed before performing many in-vehicle service procedures. Failure to do so may cause accidental deployment of the air bag, resulting in unnecessary SIR system repairs and/or personal injury.

1. Turn the steering wheel so the vehicle's wheels are pointing straight-ahead.
2. Turn the ignition switch to the **LOCK** position.
3. Remove or disconnect the following:
   • Ignition key
   • AIR BAG fuse from the instrument panel fuse block

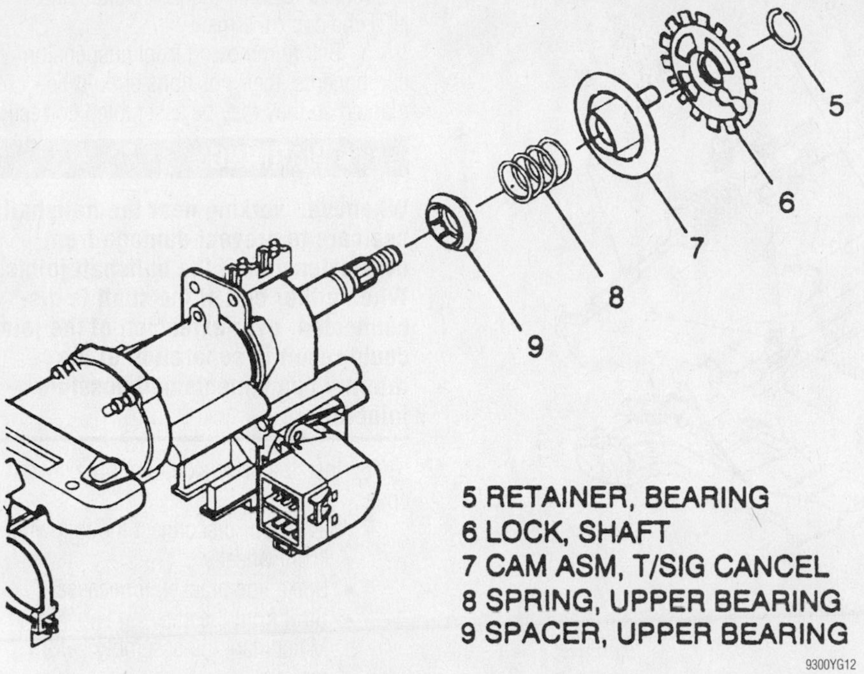

Exploded view of the upper steering column components

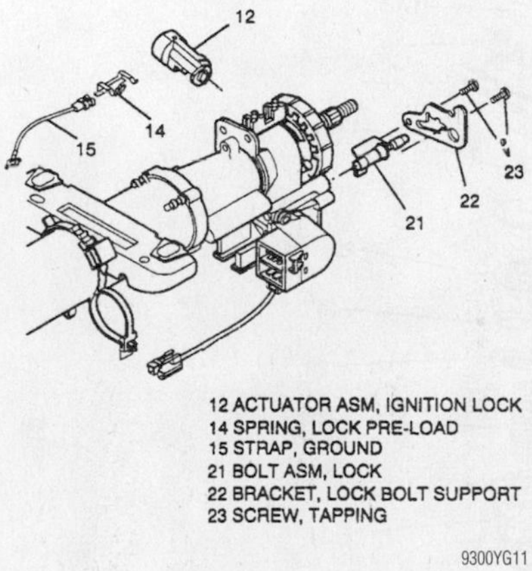

Exploded view of the ignition lock cylinder and related components

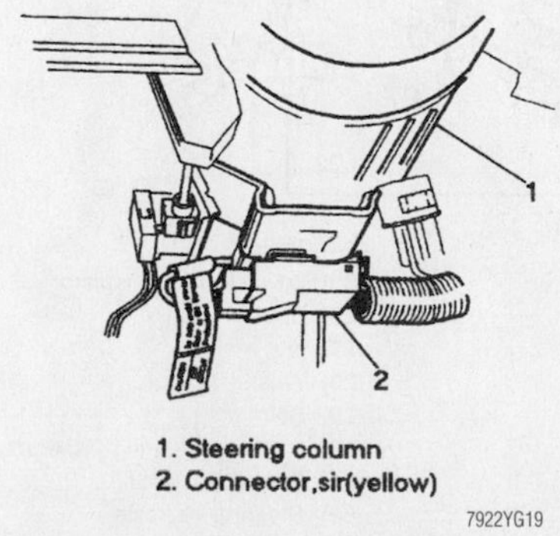

Driver's side Yellow 2-way air bag connector retainer location

- Left-hand sound insulator
- Connector retainer and yellow 2-way connector at the base of the steering column.
- Right-hand sound insulator, if equipped with passenger side air bags
- Connector retainer and yellow 2-way connector from the passenger inflator module pigtail, if equipped with passenger side air bags

ARMING

1. Turn the ignition switch to the **LOCK** position and remove the key.
2. Install or connect the following:
   - Connector retainer and yellow 2-way connector to the passenger inflator module pigtail, if equipped with passenger side air bags
   - Right-hand sound insulator, if equipped with passenger side air bags
   - Yellow 2-way connector and connector retainer at the base of the steering column
   - Left-hand insulator
   - AIR BAG fuse into the instrument panel fuse block
3. Turn the ignition switch to the **RUN** position and verify that the AIR BAG warning lamp flashes 7–9 times, then the light turns **OFF**.

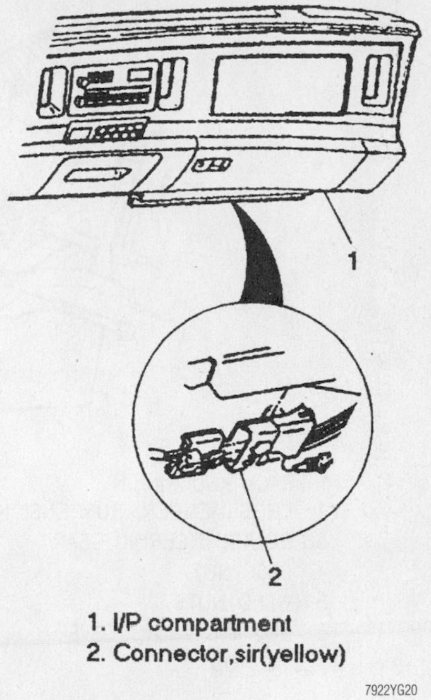

Passenger's side Yellow 2-way air bag connector location

Rack and Pinion Steering Gear

REMOVAL & INSTALLATION

1. Before servicing the vehicle, refer to the precautions in the beginning of this section.
2. Remove or disconnect the following:
   - Negative battery cable
   - Front wheels
   - Inlet and outlet hose assemblies from the rack and pinion
   - Tie rod ends from the steering knuckles
   - Intermediate shaft
   - Brake pipes from the retainers on the front suspension support
   - Crossmember bolts, loosen them to gain additional clearance
   - Rack and pinion mounting bolts
   - Rack and pinion

To install:

3. Install the rack and pinion. Torque the bolts to 89 ft. lbs. (120 Nm).
4. Install the crossmember bolts in the following order:
   a. Step 1: Left rear outboard to 71 ft. lbs. (110 Nm).
   b. Step 2: Right rear outboard 2nd to 71 ft. lbs. (110 Nm).

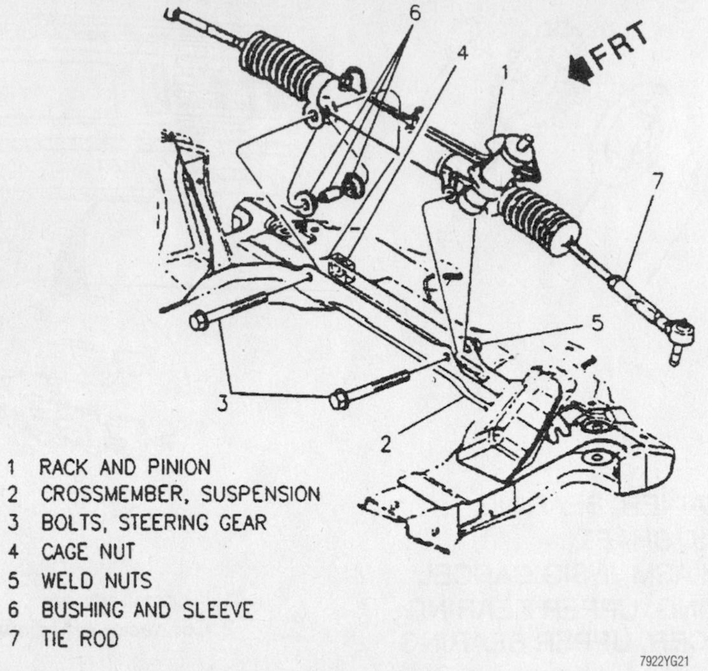

Exploded view of the rack and pinion steering gear mounting

c. Step 3: Front upper bolts to 71 ft. lbs. (110 Nm).

d. Step 4: Rear inboard bolts to 71 ft. lbs. (110 Nm).

5. Install or connect the following:
   - Intermediate shaft. Torque the pinch bolt to 30 ft. lbs. (41 Nm).
   - Brake pipes to the retainers on the front suspension crossmember
   - Tie rod ends. Torque the nuts to 44 ft. lbs. (60 Nm), use new cotter pins.
   - Inlet and outlet hoses to the rack and pinion. Torque the fittings to 20 ft. lbs. (27 Nm).
   - Wheels
   - Negative battery cable
6. Refill and bleed the power steering system.
7. Check and/or adjust the toe setting.
8. Road test vehicle and verify no leaks.

Strut

REMOVAL & INSTALLATION

Front

1. Before servicing the vehicle, refer to the precautions in the beginning of this section.
2. Remove the upper strut-to-body nuts and/or bolts.
3. Support the front crossmember with jack stands.
4. Lower the vehicle so the weight of the vehicle rests on the jack stands and NOT the control arms.
5. Before removing front suspension components, their positions should be marked so they may be assembled correctly.

WARNING

Whenever working near the halfshaft, use care to prevent damage from over-extension of the halfshaft joints. When either end of the shaft is disconnected, over-extension of the joint could result in separation of the internal components and possible joint failure.

6. Install a drive axle joint protective cover.
7. Remove or disconnect the following:
   - Front wheel
   - Brake line bracket, if necessary
   - Strut from the steering knuckle. Matchmark the assembly before removal.

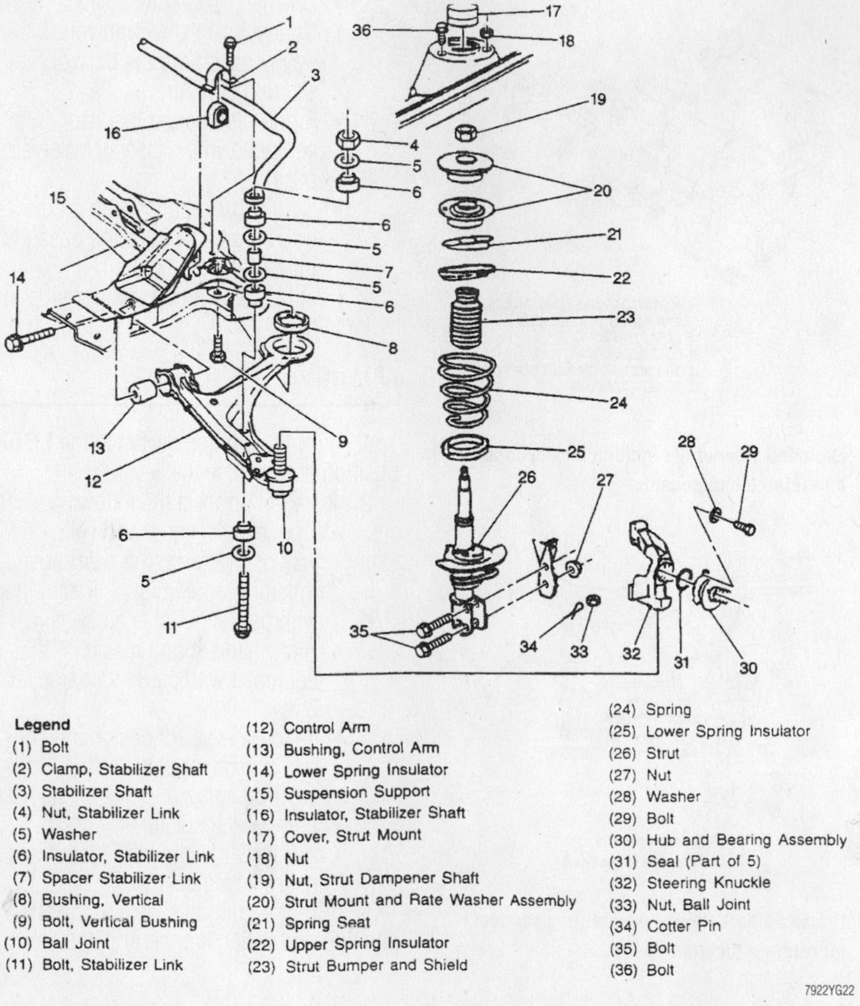

Exploded view of the front suspension

⁂ WARNING

The steering knuckle MUST be supported to prevent axle joint overextension.

To install:

8. Install or connect the following:
   - Strut to the steering knuckle, aligning the matchmarks made during removal. Torque the bolts and nuts to 133 ft. lbs. (180 Nm).
   - Brake line bracket
   - Upper strut to the body, Torque the nuts/bolts to 18 ft. lbs. (25 Nm).
9. Remove the jack stands from under the crossmember.
10. Install the front wheel. Torque the lug nuts to 100 ft. lbs. (140 Nm).

Shock Absorber

REMOVAL & INSTALLATION

Rear

1. Before servicing the vehicle, refer to the precautions in the beginning of this section.
2. Open the deck lid and position the carpet out of the way.
3. Raise and safely support the rear axle with jack stands.
4. Remove or disconnect the following:
   - Wheel
   - Upper mounting bolts
   - Lower mounting bolt
   - Shock absorber

To install:

5. Install or connect the following:
   - Shock
   - Lower mounting bolt. Torque it to 51 ft. lbs. (70 Nm).
   - Upper mounting bolts. Torque them to 18 ft. lbs. (25 Nm).
   - Wheel
   - Rear compartment carpet
6. Road test the vehicle.

Coil Spring

REMOVAL & INSTALLATION

Front

1. Before servicing the vehicle, refer to the precautions in the beginning of this section.
2. Remove the strut from the vehicle.
3. Mount the strut assembly into a strut compressor. Note that the strut compressor has strut mounting holes drilled for specific vehicle lines.
4. Compress the strut approximately ½ its height after initial contact with the top cap.

⁂ WARNING

Never bottom the spring or damper rod!

5. Remove the nut from the strut damper shaft and place an alignment/guiding rod on top of the damper shaft. Use the rod to guide the damper shaft straight down

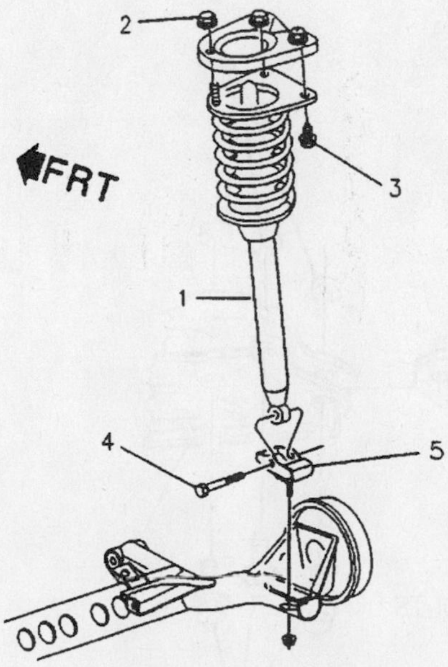

Exploded view of the rear shock mounting

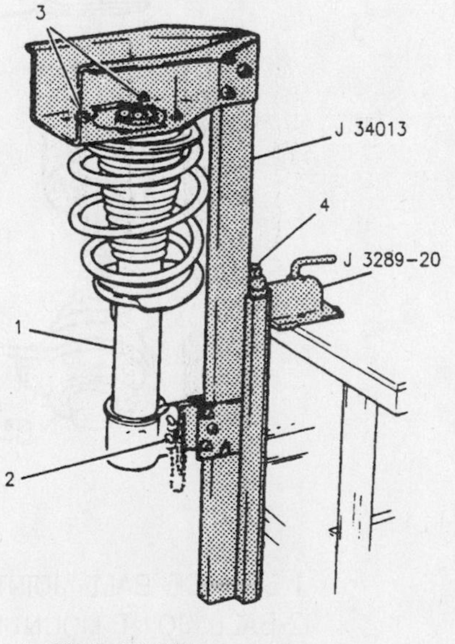

View of a typical strut assembly mounted in a compressor

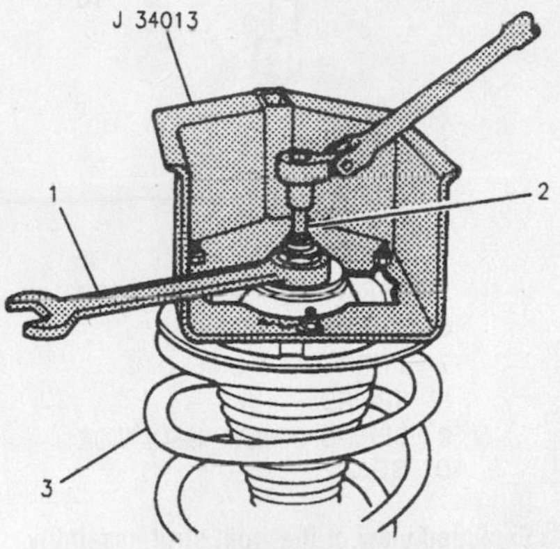

Use a socket and a wrench to remove the damper shaft nut spring cap while compressing the spring

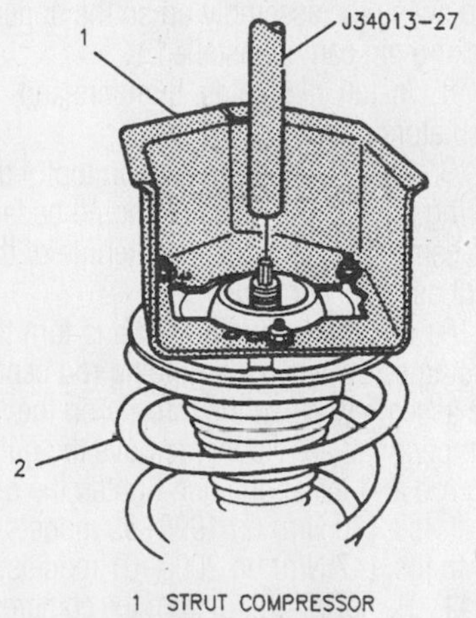

Install the rod to guide the damper shaft straight down through the spring cap while compressing the spring

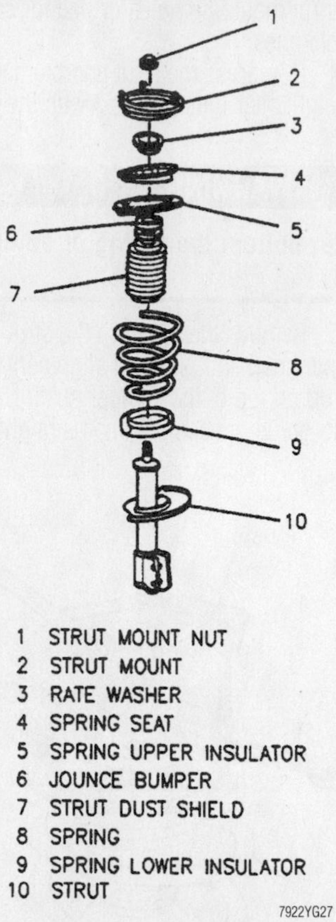

Exploded view of the front strut assembly

through the spring cap while compressing the spring. Remove the components.

To install:

6. Mount the strut assembly in the strut compressor, using the bottom locking pin only.
7. Install the spring over the damper and swing the assembly up so the upper locking pin can be installed.
8. Install all shields, bumpers and insulators on the spring seat.
9. Install the spring seat on top of the spring. The spring seat flat should be facing the same direction as the centerline of the strut assembly spindle.
10. Install the guiding rod and turn the forcing screw while the guiding rod centers the assembly. When the threads on the damper shaft are visible, remove the guiding rod and install the nut. tighten the nut to 52 ft. lbs. (70 Nm) on 1998–99 models and 34 ft. lbs. (47 Nm) on 2000–01 models.
11. Remove the strut from the compressor.
12. Install the strut to the vehicle.

Rear

1. Remove the shock absorber from the vehicle.
2. Place the shock absorber in a strut compressor tool using a shock adapter.
3. Compress the spring.
4. Remove the nut.
5. Slowly release the tension on the spring.
6. Remove the spring and shock absorber from the compressor.

To install:

7. Place the shock absorber in the strut compressor.
8. Slide the spring over the shock absorber.
9. Install the spring seat on top of the spring.
10. Compress the spring just until threads are showing through the spring seat.
11. Install the retaining nut. Torque the nut to 52 ft. lbs. (70 Nm).
12. Remove the shock from the compressor.
13. Install the shock/spring assembly onto the vehicle.

Lower Ball Joint

REMOVAL & INSTALLATION

1. Before servicing the vehicle, refer to the precautions in the beginning of this section.

✲✲ WARNING

Care must be exercised to prevent the axle shaft joints from being over-extended.

2. Remove or disconnect the following:
   - Front wheel
   - Stabilizer shaft link
   - Wiring harness from the lower control arm
   - Lower ball joint from the steering knuckle
3. Drill out the 3 ball joint-to-lower control arm rivets as follows:
   a. Use an ⅛ in. (3mm) drill bit to make a pilot hole through the rivets.
   b. Finish drilling rivets with a ½ in. (13mm) drill bit.
   c. Remove the rivets with a punch, if necessary.
4. Remove the ball joint from the steering knuckle.

To install:

5. Install or connect the following:
   - Ball joint into the control arm
   - 3 new nuts/bolts. Tighten the nuts/bolts to the specifications supplied with new ball joint
   - Ball joint to the steering knuckle.

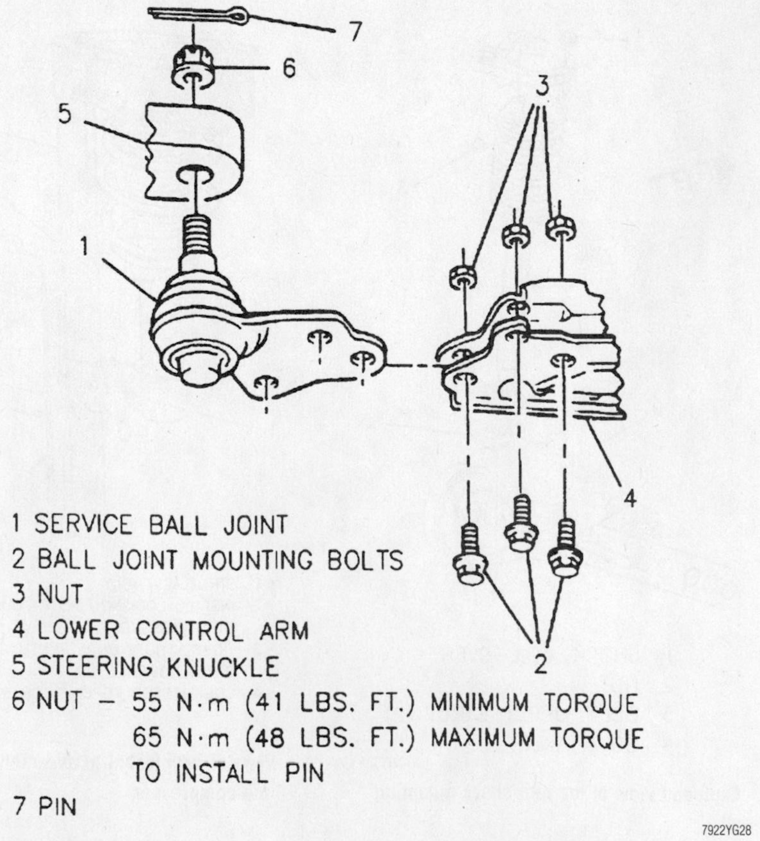

Exploded view of the ball joint mounting

Torque the nuts to 50 ft. lbs. (65 Nm), use a new cotter pin.
- Wiring harness to the lower control arm
- Stabilizer shaft link. Torque the nuts to 13 ft. lbs. (17 Nm).
- Front wheel

6. Check and/or adjust the front wheel alignment.

Lower Control Arm

REMOVAL & INSTALLATION

1. Before servicing the vehicle, refer to the precautions in the beginning of this section.
2. Remove or disconnect the following:
   - Front wheel
   - Stabilizer shaft link
   - Ball joint from the steering knuckle
   - Wiring harness from the lower control arm
   - Lower control arm

To install:

3. Install or connect the following:
   - Lower control arm to the crossmember. DO NOT torque the bolts at this time.
   - Wiring harness to the lower control arm
   - Ball joint to the steering knuckle. Torque the nut to 50 ft. lbs. (65 Nm), use a new cotter pin.
   - Stabilizer shaft link. Torque the nut to 13 ft. lbs. (17 Nm).
   - Front wheel
4. Lower the vehicle and allow it to assume proper trim height.
5. Torque the front control arm bolt to 79 ft. lbs. (107 Nm) and the rear bolt to 81 ft. lbs. (110 Nm) on 1998–98 models or 125 ft. lbs. (170 Nm) on 2000–01 models.
6. Check and/or adjust the front wheel alignment.

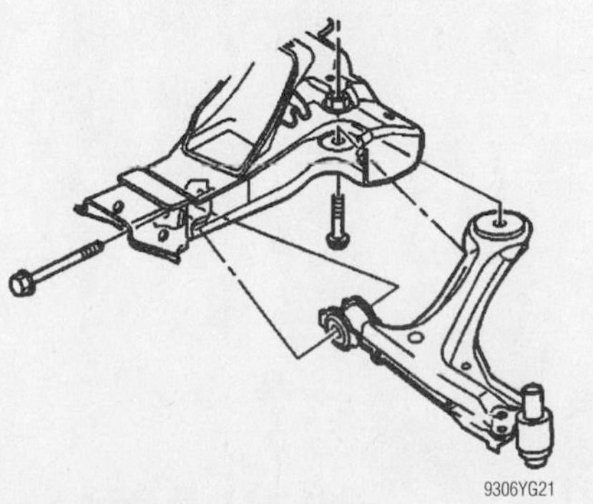

View of the lower control arm

CONTROL ARM BUSHING REPLACEMENT

Front Bushing

1. Before servicing the vehicle, refer to the precautions in the beginning of this section.
2. Remove or disconnect the following:
   - Front wheel
   - Lower control arm and place it in a vise
3. Lubricate the threads of tool J-21474-19 with High Pressure Lubricant J-23444-A.
4. Assemble 3/8 in. Bolt J-21474-19, Remover/Installer tool J-41397-1A, Remover/Installer tool J-41397-2A and 3/8 in. Nut J-21474-18 onto the lower control arm bushing.
5. Tighten the 3/8 in. Nut J-21474-18 to press the front bushing from the lower control arm.
6. Disassemble the tools.

To install:

7. Lubricate the outer case of the new front bushing.
8. Insert the new front bushing into the lower control arm.
9. Assemble 3/8 in. Bolt J-21474-19, Remover/Installer tool J-41397-1A, Remover/Installer tool J-41397-2A and 3/8 in. Nut J-21474-18 onto the lower control arm.
10. Tighten the 3/8 in. Nut J-21474-18 to press the bushing into the lower control arm.
11. Disassemble the tools.

Rear Bushing

1. Before servicing the vehicle, refer to the precautions in the beginning of this section.
2. Remove or disconnect the following:
   - Front wheel
   - Lower control arm and place it in a vise
3. Lubricate the threads of tool J-21474-27 with High Pressure Lubricant J-23444-A.
4. Assemble bolt J-21474-27, Remover/Installer tool J-41211-1, Remover/Installer tool J-41211-3 and 1/2 in. Nut J-21474-4.
5. Tighten the Bolt J-21474-27 to press the rear bushing from the lower control arm.
6. Disassemble the tools.

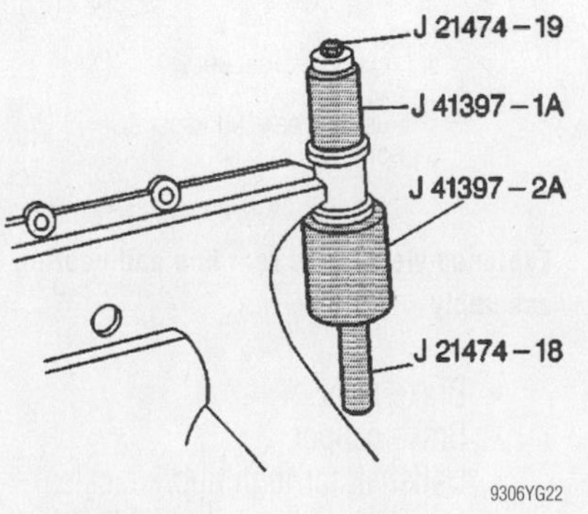

Removing the lower control arm front bushing

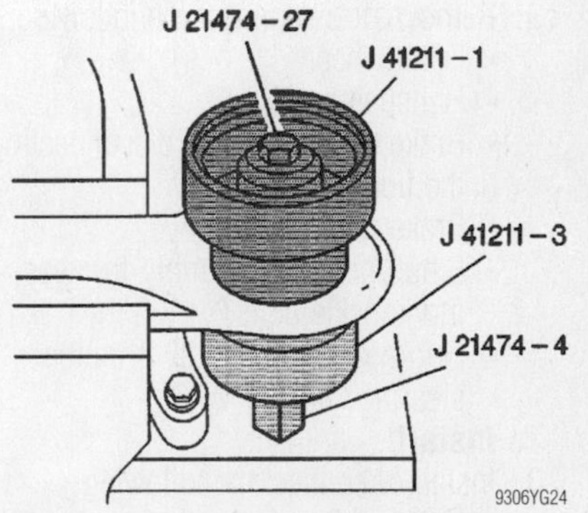

Removing the lower control arm rear bushing

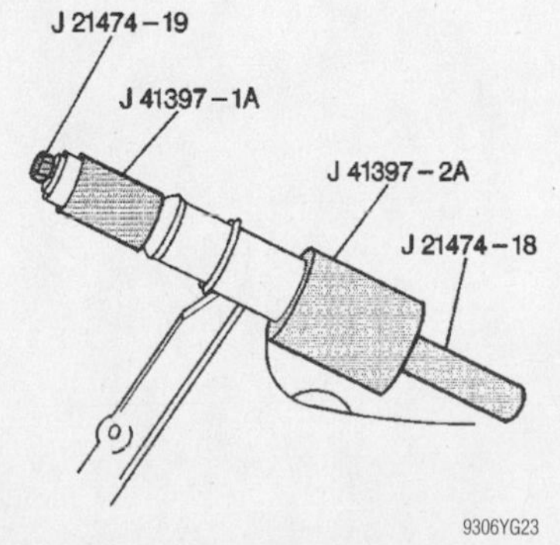

Installing the lower control arm front bushing

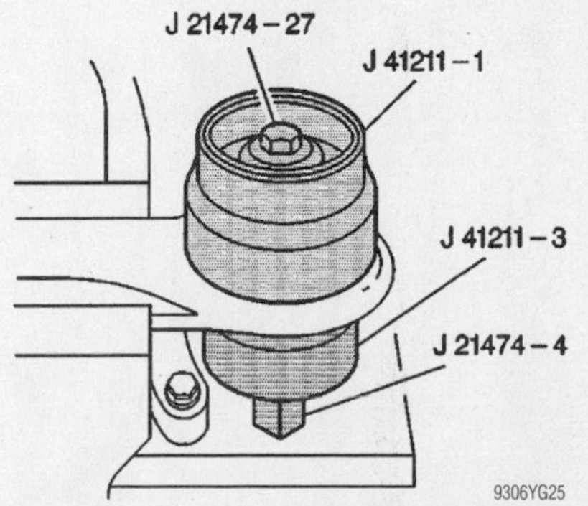

Installing the lower control arm rear bushing

Heater Core replacement is covered in Section 2 of this manual

To install:

7. Insert the rear lower control arm bushing.
8. Assemble bolt J-21474-27, Remover/Installer tool J-41211-1, Remover/Installer tool J-41211-3 and ½ in. Nut J-21474-4.
9. Tighten the Nut J-21474-4 to press the rear bushing into the lower control arm.
10. Disassemble the tools.

Wheel Bearings

ADJUSTMENT

These vehicles are equipped with sealed hub and bearing assemblies. The hub and bearing assemblies are non-serviceable. If the assembly is damaged, the complete unit must be replaced.

REMOVAL & INSTALLATION

Front

1. Before servicing the vehicle, refer to the precautions in the beginning of this section.
2. Remove or disconnect the following:
   - Front wheel
   - Halfshaft nut
   - Brake caliper without disconnecting the fluid line
   - Brake rotor
   - 3 hub/bearing assembly-to-steering knuckle bolts
   - Hub/bearing assembly from the steering knuckle

To install:

3. Install or connect the following:
   - Hub/bearing assembly to the steering knuckle. Torque the bolts to 70 ft. lbs. (95 Nm).

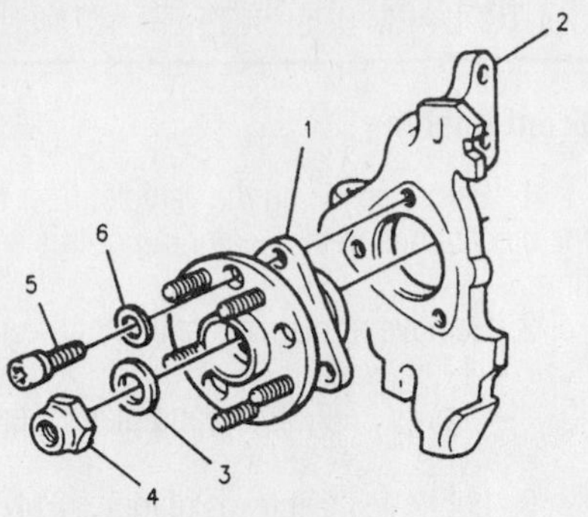

Exploded view of the front hub and bearing assembly

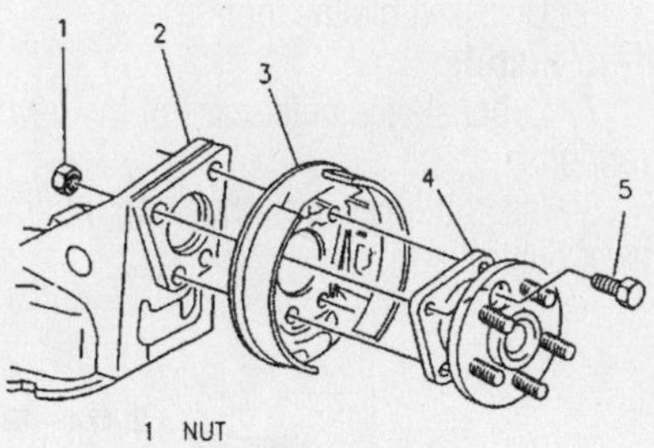

Exploded view of the rear hub and bearing assembly

   - Brake rotor
   - Brake caliper
   - Halfshaft through hub/knuckle assembly. Torque the nut to 144 ft. lbs. (200 Nm).
   - Front wheel

4. Road test the vehicle and verify proper operation.

Rear

A single-unit hub/bearing assembly is bolted to both ends of the rear axle assembly or rear knuckle assembly. The hub/bearing assembly is a sealed unit.

1. Before servicing the vehicle, refer to the precautions in the beginning of this section.
2. Remove or disconnect the following:
   - Rear wheel
   - Brake drum
   - Anti-lock Brake System (ABS) wheel speed sensor electrical connector
   - Hub/bearing assembly-to-rear axle assembly nuts/bolts

➡The top rear mounting bolt will not clear the brake shoe when removing the hub and bearing assembly. Partially remove the hub and bearing assembly prior to removing this bolt.

   - Hub/bearing assembly from the rear axle

To install:

3. Install or connect the following:
   - Hub/bearing assembly. Torque the nuts/bolts to 44 ft. lbs. (60 Nm).

➡Position the top rear mounting bolt in the hub and bearing assembly prior to installation to the axle housing.

   - Rear ABS wheel speed sensor wire connector
   - Brake drum
   - Rear wheel

4. Road test the vehicle.

GENERAL MOTORS CORPORATION—L/N BODY

26

1998–01

Chevrolet-Malibu • **Oldsmobile-**Cutlass

PRECAUTIONS

Before servicing any vehicle, please be sure to read all of the following precautions, which deal with personal safety, prevention of component damage, and important points to take into consideration when servicing a motor vehicle:

- Never open, service or drain the radiator or cooling system when the engine is hot; serious burns can occur from the steam and hot coolant.
- Observe all applicable safety precautions when working around fuel. Whenever servicing the fuel system, always work in a well-ventilated area. Do not allow fuel spray or vapors to come in contact with a spark, open flame or excessive heat (a hot drop light, for example). Keep a dry chemical fire extinguisher near the work area. Always keep fuel in a container specifically designed for fuel storage; also, always properly seal fuel containers to avoid the possibility of fire or explosion. Refer to the additional fuel system precautions later in this section.
- Fuel injection systems often remain pressurized, even after the engine has been turned **OFF**. The fuel system pressure must be relieved before disconnecting any fuel lines. Failure to do so may result in fire and/or personal injury.
- Brake fluid often contains polyglycol ethers and polyglycols. Avoid contact with the eyes and wash your hands thoroughly after handling brake fluid. If you do get brake fluid in your eyes, flush your eyes with clean, running water for 15 minutes. If eye irritation persists, or if you have taken brake fluid internally, IMMEDIATELY seek medical assistance.
- The EPA warns that prolonged contact with used engine oil may cause a number of skin disorders, including cancer! You should make every effort to minimize your exposure to used engine oil. Protective gloves should be worn when changing oil. Wash your hands and any other exposed skin areas as soon as possible after exposure to used engine oil. Soap and water, or waterless hand cleaner should be used.
- All new vehicles are now equipped with an air bag system, often referred to as a Supplemental Restraint System (SRS) or Supplemental Inflatable Restraint (SIR) system. The system must be disabled before performing service on or around system components, steering column, instrument panel components, wiring and sensors. Failure to follow safety and disabling procedures could result in accidental air bag deployment, possible personal injury and unnecessary system repairs.
- Always wear safety goggles when working with, or around, the air bag system. When carrying a non-deployed air bag, be sure the bag and trim cover are pointed away from your body. When placing a non-deployed air bag on a work surface, always face the bag and trim cover upward, away from the surface. This will reduce the motion of the module if it is accidentally deployed. Refer to the additional air bag system precautions later in this section.
- Clean, high quality brake fluid from a sealed container is essential to the safe and proper operation of the brake system. You should always buy the correct type of brake fluid for your vehicle. If the brake fluid becomes contaminated, completely flush the system with new fluid. Never reuse any brake fluid. Any brake fluid that is removed from the system should be discarded. Also, do not allow any brake fluid to come in contact with a painted surface; it will damage the paint.
- Never operate the engine without the proper amount and type of engine oil; doing so WILL result in severe engine damage.
- Timing belt maintenance is extremely important! Many models utilize an interference-type, non-freewheeling engine. If the timing belt breaks, the valves in the cylinder head may strike the pistons, causing potentially serious (also time-consuming and expensive) engine damage. Refer to the maintenance interval charts in the front of this manual for the recommended replacement interval for the timing belt, and to the timing belt section for belt replacement and inspection.
- Disconnecting the negative battery cable on some vehicles may interfere with the functions of the on-board computer system(s) and may require the computer to undergo a relearning process once the negative battery cable is reconnected.
- When servicing drum brakes, only disassemble and assemble one side at a time, leaving the remaining side intact for reference.

ENGINE REPAIR

Alternator

REMOVAL

2.4L Engine

1. Before servicing the vehicle, refer to the precautions in the beginning of this section.
2. Remove or disconnect the following:
   - Negative battery cable
   - Accessory drive belt
   - Alternator electrical connectors
   - Alternator

3.1L Engine

1. Before servicing the vehicle, refer to the precautions in the beginning of this section.
2. Remove or disconnect the following:
   - Negative battery cable
   - Accessory drive belt
   - Alternator electrical connectors
   - Power steering line clip
   - Alternator

INSTALLATION

2.4L Engine

1. Install or connect the following:
   - Alternator electrical connectors
   - Alternator. Torque the nuts/bolts to 37 ft. lbs. (50 Nm).
   - Accessory drive belt
   - Negative battery cable

3.1L Engine

1. Install or connect the following:
   - Alternator. Torque the nuts to 22 ft. lbs. (30 Nm) and the bolts to 37 ft. lbs. (50 Nm).
   - Alternator electrical connector
   - Power steering line clip
   - Accessory drive belt
   - Negative battery cable

Ignition Timing

ADJUSTMENT

Ignition timing is controlled by the Powertrain Control Module (PCM). No adjustment is necessary or possible.

Engine Assembly

REMOVAL & INSTALLATION

2.4L Engine

1. Before servicing the vehicle, refer to the precautions in the beginning of this section.
2. Properly drain the cooling system.
3. Relieve the fuel system pressure.
4. Drain the engine oil.
5. Remove or disconnect the following:
   - Fuel rail assembly
   - Air intake duct and bracket
   - Ignition coil assembly
   - Camshaft Position (CMP) sensor
   - Power steering pump. DO NOT remove the lines.
   - Oil pressure sending switch
   - Cruise control assembly
6. Install an engine support fixture.
   - Engine mount assembly
   - Fuel pressure regulator vacuum line
7. Raise the engine with the engine support fixture J-hook.
   - Engine mount bracket
   - Accessory drive belt
   - Front wheels
   - Right front splash shield
   - Crankshaft balancer
   - Manifold Absolute Pressure (MAP) sensor electrical connector
   - Intake Air Temperature (IAT) sensor electrical connector
   - Exhaust Gas Recirculation (EGR) sensor electrical connector
   - Evaporator canister electrical connector
   - Alternator
   - Accelerator cable and bracket
   - Starter motor
   - Exhaust manifold heat shield
   - Exhaust pipe from the manifold
   - Engine intake coolant pipe
   - Torque converter cover
   - Torque converter bolts
   - Oxygen (O_2S) sensor
   - A/C compressor. DO NOT remove the hoses
   - Oil pan-to-bell housing bolts
   - Transmission mount
   - Transmission bolts
8. Install an engine lifting device and a transmission support.
9. Raise the engine slightly and separate it from the transmission to remove it from the vehicle.

To install:

⁂ WARNING

Be sure the retaining bolts are in their correct locations. If not, engine damage may occur.

10. Install or connect the following:
    - Engine to the transaxle. Torque the bolts to 75 ft. lbs. (100 Nm).
    - Transmission mount. Torque the through bolt to 55 ft. lbs. (75 Nm).
    - Oil pan-to-bell housing bolts. Torque the bolts to 17 ft. lbs. (23 Nm).
    - A/C compressor. Torque the bolts to 37 ft. lbs. (50 Nm).
    - O_2S
    - Torque converter bolts. Torque the bolts to 46 ft. lbs. (62 Nm).
    - Torque converter cover. Torque the bolts to 115 inch lbs. (13 Nm).
    - Intake coolant pipe
    - Exhaust pipe to the manifold. Torque the bolts to 26 ft. lbs. (35 Nm).
    - Exhaust manifold heat shield. Torque the bolts to 124 inch lbs. (14 Nm).
    - Starter motor. Torque the bolts to 66 ft. lbs. (90 Nm).
    - Accelerator cable and bracket
    - Alternator. Torque the bolts to 37 ft. lbs. (50 Nm).
    - Fuel pressure regulator vacuum line
    - Evaporator canister electrical connector
    - EGR sensor electrical connector
    - IAT sensor electrical connector
    - MAP sensor electrical connector
    - Crankshaft balancer. Torque the bolt to 129 ft. lbs. (175 Nm).
    - Right front splash shield
    - Front wheels
    - Accessory drive belt
    - Engine mount bracket. Torque the bolts to 96 ft. lbs. (130 Nm).
    - Engine mount assembly. Torque the bolts to 49 ft. lbs. (66 Nm).
    - Cruise control assembly
    - Oil pressure sending switch
    - Power steering pump. Torque the bolts to 19 ft. lbs. (26 Nm).
    - CMP sensor
    - Ignition coil assembly
    - Air intake assembly
    - Fuel rail assembly
11. Refill the cooling system and engine oil.
12. Start the engine and check for proper operation.

3.1L Engine

1. Before servicing the vehicle, refer to the precautions in the beginning of this section.
2. Relieve the fuel system pressure.
3. Drain the engine coolant.
4. Remove or disconnect the following:
   - Air cleaner assembly
   - Hood
   - Accessory drive belt
   - Coolant inlet line from the coolant surge tank
   - Cruise control module
   - Upper engine wiring harness
   - Throttle and cruise control cables
   - Starter motor
   - A/C compressor. DO NOT remove the lines.
   - Lower engine wiring harness and position aside
   - Catalytic converter
   - Torque converter cover
   - Torque converter bolts
   - Engine splash shields
   - Transaxle-to-engine brace
   - 2 outer transmission mounting bolts
   - Upper and lower radiator hoses
   - Fuel lines
   - Brake booster vacuum line
   - Heater hoses
5. Install an engine support fixture.
6. Remove or disconnect the following:
   - Engine mount
   - Power steering pump
   - Transmission-to-engine bolts
   - Engine from the vehicle

To install:

7. Install or connect the following:
   - Transaxle to the engine. Torque the bolts to 55 ft. lbs. (75 Nm).
   - Power steering pump. Torque the bolts to 25 ft. lbs. (34 Nm).
   - Engine mount. Torque the bolts to 35 ft. lbs. (47 Nm).
8. Remove the engine support fixture.
9. Install or connect the following:
   - Heater hoses
   - Brake booster vacuum line
   - Fuel lines
   - Radiator hoses
   - Engine splash shields
   - Torque converter bolts. Torque the bolts to 46 ft. lbs. (62 Nm).
   - Torque converter cover. Torque the bolts to 89 inch lbs. (10 Nm).

- Catalytic converter
- Lower engine wiring harness
- A/C compressor. Torque the bolts to 37 ft. lbs. (50 Nm).
- Starter motor
- Throttle and cruise control cables
- Upper engine wiring harness
- Cruise control module
- Coolant surge tank hose
- Accessory drive belt
- Hood
- Air cleaner assembly

10. Refill the coolant.
11. Start the engine and check for proper operation.

Water Pump

REMOVAL & INSTALLATION

2.4L Engine

1. Before servicing the vehicle, refer to the precautions in the beginning of this section.
2. Drain the cooling system.
3. Remove or disconnect the following:
   - Negative battery cable
   - Exhaust manifold heat shield
   - Coolant inlet housing bolt through the exhaust manifold
   - Oxygen (O_2S) sensor electrical connector
   - Manifold-to-exhaust pipe bolts
   - Coolant inlet housing assembly-to-water pump cover bolt
   - Coolant inlet pipe
   - Exhaust manifold
   - Timing chain
   - Water pump and cover assembly
4. Separate the water pump from the cover.

To install:

5. Lubricate the water pump splines with clean grease.
6. Install or connect the following:
   - Water pump with a new gasket to the cover. Hand tighten the bolts.
   - Water pump using new gaskets to the engine and tighten the bolts and nuts finger-tight
   - Radiator outlet pipe O-ring, lubricated with antifreeze
   - Radiator outlet pipe onto the water pump cover and tighten the bolts finger-tight
7. With all gaps closed, torque the bolts, in the following sequence, to the proper values:
   a. Pump assembly-to-chain housing nuts to 19 ft. lbs. (26 Nm).

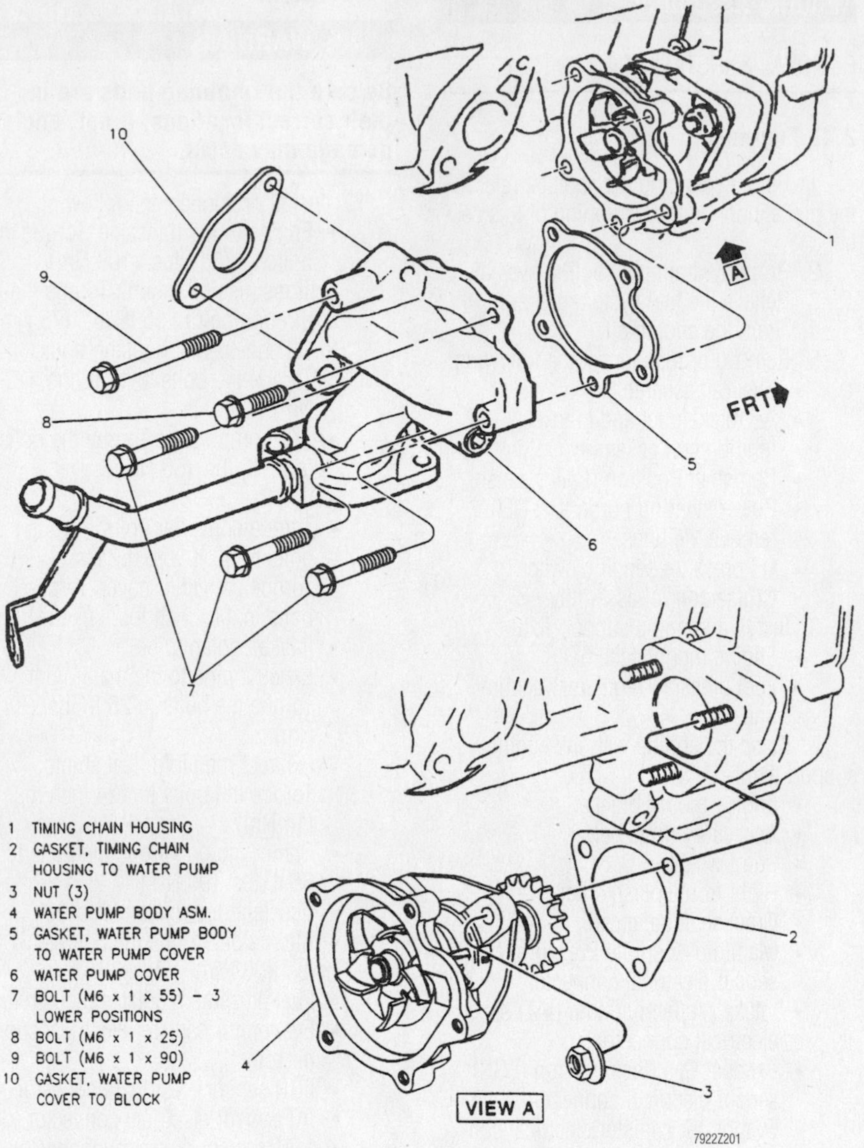

Exploded view of the water pump mounting—2.4L engine

   b. Pump cover-to-pump bolts to 124 inch lbs. (14 Nm).
   c. Cover-to-block bottom bolt to 19 ft. lbs. (26 Nm).
   d. Radiator outlet pipe assembly-to-pump cover bolt to 124 inch lbs. (14 Nm).
8. Install or connect the following:
   - Exhaust manifold with new gaskets. Torque the nuts to 12 ft. lbs. (16 Nm).
   - Exhaust pipe to the manifold. Torque the bolts to 26 ft. lbs. (35 Nm).
   - Radiator outlet pipe to the transaxle and oil pan. Torque the nuts to 19 ft. lbs. (26 Nm).
   - Timing chain and front cover
   - Manifold heat shield
   - O_2S electrical connector
   - Negative battery cable
9. Refill the cooling system until it comes out of the thermostat housing heater hose outlet; then connect the heater hose.
10. Start the engine. Run the vehicle until the thermostat opens, refill the radiator and recovery tank, then turn the engine **OFF**.
11. Once the vehicle has cooled, recheck the coolant level.

3.1L Engine

1. Before servicing the vehicle, refer to the precautions in the beginning of this section.
2. Drain the cooling system.
3. Remove or disconnect the following:
   - Negative battery cable
   - Water pump pulley bolts, loosen
   - Accessory drive belt

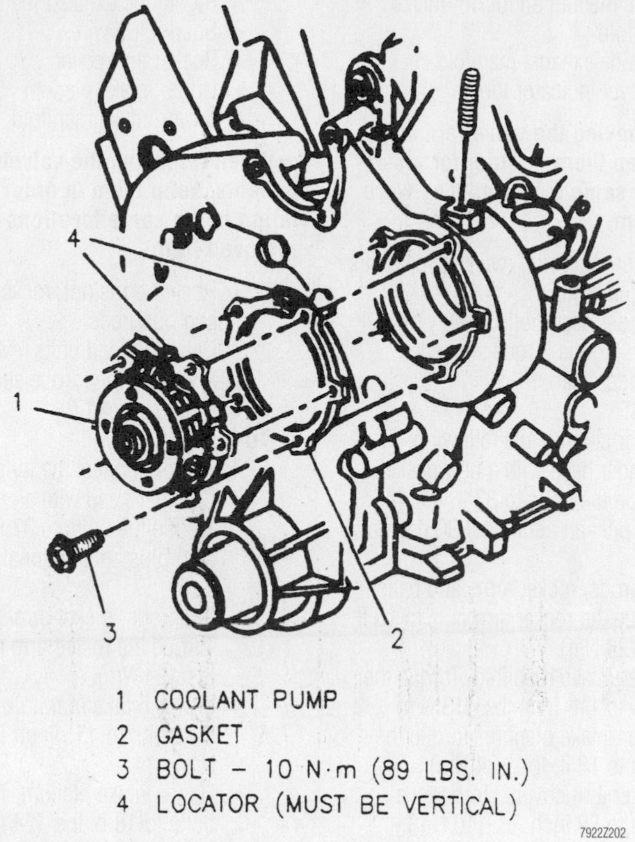

Exploded view of the water pump mounting—3.1L engine

- Water pump pulley
- Water pump bolts, pump and gasket

To install:

4. Apply a thin bead of sealer around the outside edge of the water pump along the gasket sealing area.
5. Install or connect the following:
   - Water pump with a new gasket. Torque the bolts to 89 inch lbs. (10 Nm).
   - Water pump pulley and finger-tighten the bolts
   - Accessory drive belt
   - Water pump pulley bolts and torque to 18 ft. lbs. (24 Nm)
   - Negative battery cable
6. Refill and bleed the cooling system.
7. Operate the engine and check for leaks.

Cylinder Head

REMOVAL & INSTALLATION

2.4L Engine

1. Before servicing the vehicle, refer to the precautions in the beginning of this section.
2. Relieve the fuel system pressure.
3. Drain the cooling system.
4. Remove or disconnect the following:
   - Air inlet duct
   - Heater inlet and throttle body hoses from the water outlet
   - Power brake vacuum hose from the throttle body
   - Manifold Absolute Pressure (MAP) sensor electrical connector
   - Evaporative Emission (EVAP) purge solenoid electrical connector
   - Camshaft Position (CMP) sensor electrical connector
   - Alternator stud-ended bolt
   - Intake manifold
5. Install the stud-ended alternator bolt back onto the engine.
6. Install an engine support fixture.
7. Remove or disconnect the following:
   - Exhaust manifold
   - Ignition coil and module assembly
   - Power steering pump and move it aside without disconnecting the lines
   - Vacuum line from the fuel pressure regulator

Cylinder head bolt torque sequence—2.4L engine

- Fuel line clamp from the bracket located on top of the intake camshaft housing
- Fuel injector electrical connectors
- Fuel line rail and move it aside with the fuel lines attached
- Timing chains and sprockets
- Water pump
- Intake camshaft housing
- Exhaust camshaft housing
- Lifters
- Oil pressure switch electrical connector
- Upper radiator hose from the coolant outlet
- Coolant temperature sensor electrical connectors
- Cylinder head by reversing the bolt torque sequence

**** WARNING**

Do not use abrasive pads to clean the cylinder head or block surfaces. An abrasive pad may damage the cylinder head and block. Use a razor or scraper to clean the gasket surfaces.

To install:

8. Install the cylinder head with a new gasket.

➡New replacement head bolts should be used. Coat the new cylinder head bolt threads with clean engine oil before installation.

9. Torque the new cylinder head bolts in sequence as follows:
   a. Step 1: Bolts 1 through 8: 40 ft. lbs. (65 Nm).
   b. Step 2: Bolts 9 and 10: 30 ft. lbs. (40 Nm).
   c. Step 3: All 10 bolts: an additional 90 degree turn.
10. Install or connect the following:
    - Lifters
    - Intake and exhaust camshaft housings. Torque the long bolts to 11 ft. lbs. (15 Nm) plus an additional 90 degree turn and the short bolts to

11 ft. lbs. (15 Nm) plus an additional 30 degree turn.
- Timing chains and sprockets
- Coolant temperature sensor electrical connector
- Upper radiator hose
- Oil pressure switch electrical connector
- Fuel rail on the intake manifold
- Fuel pressure regulator vacuum line
- Power steering pump. Torque the bolts to 19 ft. lbs. (26 Nm).
- Ignition coil assembly
- Exhaust manifold

11. Remove the engine support fixture.
12. Install or connect the following:
    - Intake manifold. Torque the bolts to 18 ft. lbs. (24 Nm).
    - MAP sensor electrical connector
    - EVAP purge solenoid electrical connector
    - CMP sensor electrical connector
    - Power brake vacuum hose
    - Heater inlet and throttle body hoses
    - Air inlet duct
    - Negative battery cable
13. Fill all fluids to their proper levels.

➡An oil and filter change is recommended.

14. Start the vehicle and check for leaks.

3.1L Engine

LEFT (FRONT)

1. Before servicing the vehicle, refer to the precautions in the beginning of this section.
2. Relieve the fuel system pressure.
3. Drain the cooling system.
4. Remove or disconnect the following:
    - Upper half of the air cleaner assembly
    - Air inlet duct
    - Exhaust crossover pipe heat shield and the crossover pipe
    - Spark plug wires
    - Rocker arm covers

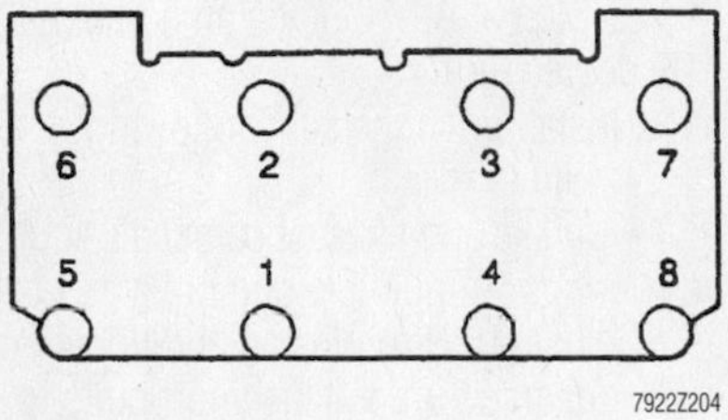

Cylinder head bolt torque sequence—3.1L engine

- Intake plenum and lower intake manifold
- Left side exhaust manifold
- Oil level indicator tube

➡When removing the valvetrain components, keep them in order for installation in the same locations they were removed from.

- Rocker arm nuts, rocker arms, balls and pushrods
- Cylinder head bolts, evenly, by reversing the torque sequence
- Cylinder head

To install:

5. Install or connect the following:
    - Cylinder head with a new gasket. Torque the bolts to 33 ft. lbs. (45 Nm) plus an additional 90 degree turn.
    - Pushrods, rocker arms and balls. Torque the rocker arm nuts to 18 ft. lbs. (24 Nm).
    - Lower intake manifold. Torque the bolts to 115 inch lbs. (13 Nm).
    - Upper intake plenum. Torque the bolts to 18 ft. lbs. (24 Nm).
    - Rocker arm covers. Torque the bolts to 89 inch lbs. (10 Nm).
    - Oil level indicator tube
    - Spark plug wires
    - Left side exhaust manifold. Torque the bolts to 12 ft. lbs. (16 Nm).
    - Exhaust crossover pipe. Torque the bolts to 18 ft. lbs. (24 Nm).
    - Crossover pipe heat shield. Torque the bolts to 89 inch lbs. (10 Nm).
    - Upper half of the air cleaner assembly
    - Air inlet duct
    - Negative battery cable
6. Refill the cooling system.

➡An oil change is recommended.

7. Start the vehicle and check for leaks.

RIGHT (REAR)

1. Before servicing the vehicle, refer to the precautions in the beginning of this section.
2. Relieve the fuel system pressure.
3. Drain the cooling system.
4. Remove or disconnect the following:
    - Upper half of the air cleaner assembly
    - Throttle body air inlet duct
    - Exhaust crossover pipe heat shield and the crossover pipe
    - Oxygen (O_2S) sensor
    - Exhaust pipe from the exhaust manifold
    - Right side exhaust manifold
    - Spark plug wires
    - Rocker arm cover
    - Upper intake plenum
    - Lower intake manifold

➡When removing the valvetrain components, keep them in order for installation in the same locations they were removed from.

- Rocker arms nut, rocker arms, balls and pushrods
- Cylinder head bolts, evenly, by reversing the torque sequence
- Cylinder head

To install:

5. Install or connect the following:
    - Cylinder head with a new gasket. Torque the bolts to 37 ft. lbs. (50 Nm) plus an additional 90 degree turn.
    - Pushrods, rocker arms and balls. Torque the rocker arm nuts to 18 ft. lbs. (24 Nm).
    - Lower intake manifold. Torque the bolts to 115 inch lbs. (13 Nm).
    - Upper intake plenum. Torque the bolts to 18 ft. lbs. (24 Nm).
    - Rocker arm covers. Torque the bolts to 89 inch lbs. (10 Nm).
    - Spark plug wires
    - Exhaust manifold. Torque the bolts to 12 ft. lbs. (16 Nm).
    - Exhaust pipe to the exhaust manifold. Torque the bolts to 26 ft. lbs. (35 Nm).
    - O_2S sensor
    - Exhaust crossover pipe. Torque the bolts to 18 ft. lbs. (24 Nm).
    - Heat shield. Torque the bolts to 89 inch lbs. (10 Nm).
    - Upper half of the air cleaner assembly
    - Throttle body air inlet duct
    - Negative battery cable
6. Refill the cooling system.
7. An oil and filter change is recommended.
8. Start the vehicle and check for leaks.

Rocker Arms

REMOVAL & INSTALLATION

2.4L Engine

The 2.4L engine is not equipped with rocker arms. The camshafts directly actuate the valves.

3.1L Engine

LEFT SIDE (FRONT)

1. Before servicing the vehicle, refer to the precautions in the beginning of this section.
2. Disconnect the negative battery cable.
3. Partially drain the cooling system to a level below the coolant pipe.
4. Remove or disconnect the following:
   - Rear ignition wire harness at the upper intake manifold and at the spark plugs
   - Coolant bypass hose clamp at the coolant tube
   - Coolant tube from the cylinder head and move it aside
   - Positive Crankcase Ventilation (PCV) valve from the rocker arm cover
   - Rocker arm cover
   - Rocker arms and pushrods

To install:

➡The intake pushrods are identified with yellow stripes and are 5 • inches long. The exhaust pushrods are identified with green stripes and are 6 inches long.

5. Coat all valvetrain components with engine oil.
6. Install or connect the following:
   - Pushrods and rocker arms. Torque the rocker arm bolts to 89 inch lbs. (10 Nm) plus an additional 30 degree turn.
   - Rocker arm cover with a new gasket. Torque the bolts to 89 inch lbs. (10 Nm).
   - PCV valve
   - Coolant tube and thermostat bypass hose. Torque the screw at the water pump to 106 inch lbs. (12 Nm) and the nut/bolt at the cylinder head to 18 ft. lbs. (24 Nm).
   - Rear ignition wire harness to the upper intake manifold and the spark plugs
   - Negative battery cable
7. Refill the cooling system.
8. Start the vehicle and verify no leaks.

RIGHT SIDE (REAR)

1. Before servicing the vehicle, refer to the precautions in the beginning of this section.
2. Remove or disconnect the following:

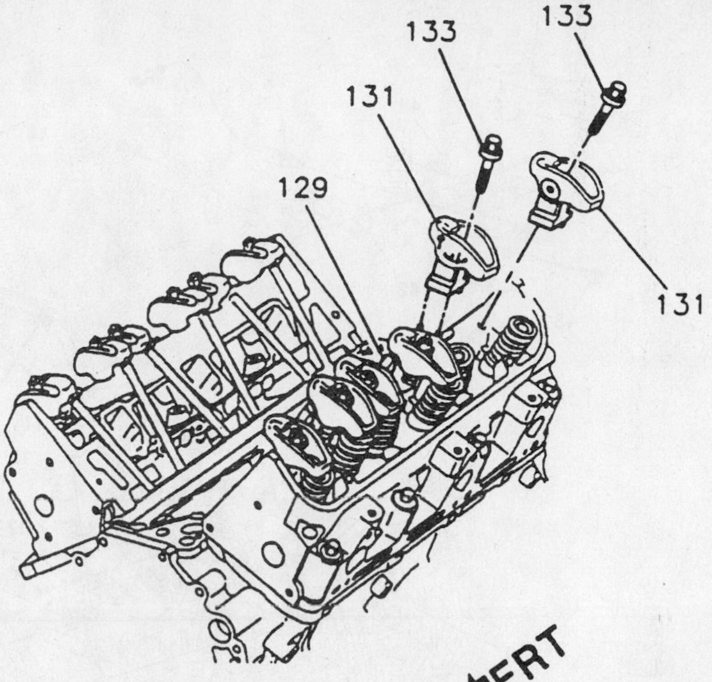

Rocker arm components—3.1L engine

   - Negative battery cable
   - Spark plug wires
   - Power brake booster vacuum pipe from the intake plenum
   - Accessory drive belt
   - Alternator
   - Ignition coil assembly
   - Evaporative Emission (EVAP) canister purge solenoid
   - Rocker arm cover
   - Rocker arms and pushrods

To install:

➡The intake pushrods are identified with yellow stripes and are 5 • inches long. The exhaust pushrods are identified with green stripes and are 6 inches long.

3. Coat all valvetrain components with engine oil.
4. Install or connect the following:
   - Pushrods and rocker arms. Torque the bolts to 89 inch lbs. (10 Nm) plus an additional 30 degree turn.
   - Rocker arm cover with a new gasket. Torque the bolts to 89 inch lbs. (10 Nm).
   - Ignition coil assembly
   - EVAP canister purge solenoid
   - Alternator. Torque the bolts to 18 ft. lbs. (24 Nm).
   - Accessory drive belt
   - Power brake booster vacuum pipe to the plenum
   - Spark plug wires
   - Negative battery cable

Intake Manifold

REMOVAL & INSTALLATION

2.4L Engine

1. Before servicing the vehicle, refer to the precautions in the beginning of this section.
2. Relieve the fuel system pressure.
3. Drain the cooling system to a level below the intake manifold.
4. Remove or disconnect the following:
   - Air cleaner intake duct
   - Manifold Absolute Pressure (MAP) sensor electrical connector and vacuum line
   - Intake Air Temperature (IAT) sensor electrical connector

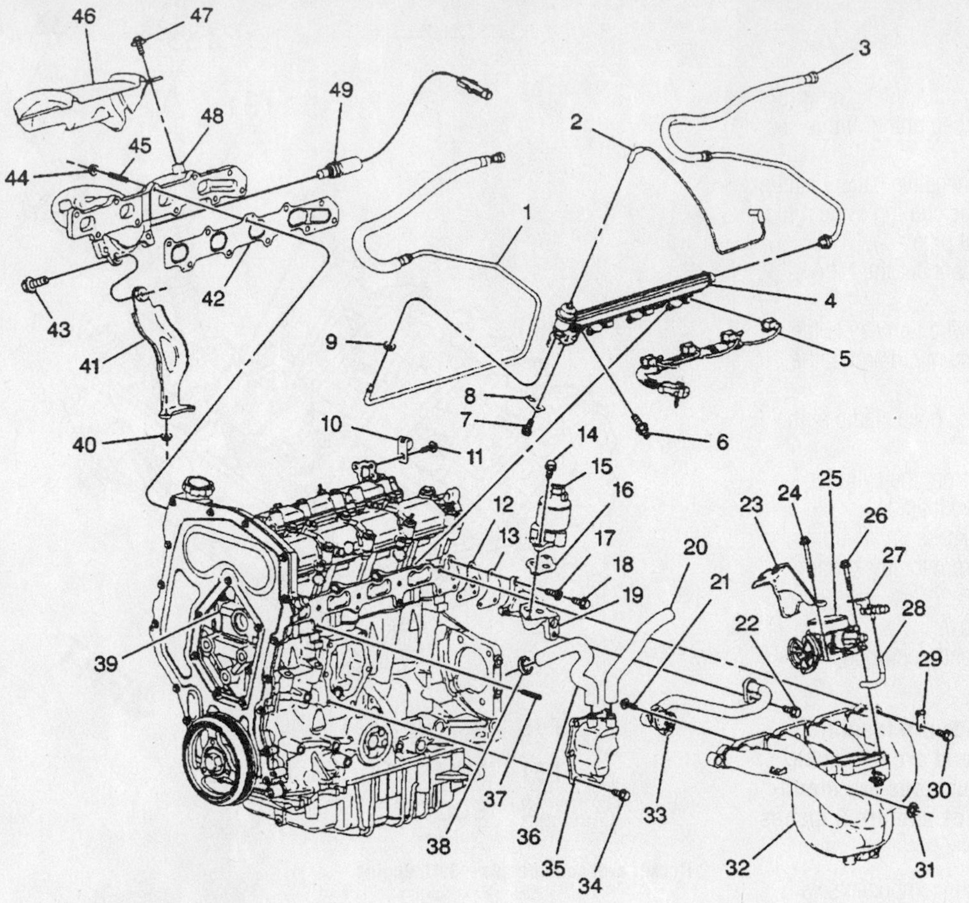

(1) Fuel Return Line
(2) Fuel Pressure Regulator Vacuum Harness
(3) Fuel Line
(4) Fuel Rail (with Injectors)
(5) Fuel Injector Harness
(6) Bolt, Fuel Rail to Cylinder Head
(7) Bolt, Fuel Line Retainer
(8) Fuel Line Retainer
(9) Fuel Line Seal
(10) Fuel Line Retainer
(11) Bolt, Fuel Line Retainer
(12) Gasket, EGR Adapter
(13) EGR Passage Cover (Export Only)
(14) Bolt, EGR Valve
(15) EGR Valve
(16) Gasket, EGR Valve
(17) Bolt, EGR Valve Adapter to Cylinder Head
(18) Bolt, EGR Passage Cover (Export Only)
(19) EGR Adapter
(20) Oil/Air Separator Fresh Air Tube
(21) Bolt, EGR Tube to Intake Manifold
(22) Bolt, EGR Tube to EGR Adapter
(23) Throttle Cable Bracket
(24) Bolt, Throttle Body to Intake Manifold
(25) Throttle Body
(26) Bolt, Throttle Body to Intake Manifold
(27) Manifold Absolute Pressure Sensor
(28) MAP Sensor Vacuum Line
(29) Fuel Rail Bracket
(30) Bolt, Intake Manifold to Cylinder Head
(31) Nut, Intake Manifold to Cylinder Head
(32) Intake Manifold
(33) EGR Pipe
(34) Bolt, Oil/Air Separator to Engine Block
(35) Oil/Air Separator
(36) Oil/Air Separator Foul Air Tube
(37) Stud, Intake Manifold to Cylinder Head
(38) Clamp, Oil/Air Separator Tube
(39) Cylinder Head
(40) Nut, Exhaust Manifold Brace to Oil Pan Stud
(41) Exhaust Manifold Brace
(42) Gasket, Exhaust Manifold
(43) Bolt, Exhaust Manifold Brace to Exhaust Manifold
(44) Nut, Exhaust Manifold to Cylinder Head
(45) Stud, Exhaust Manifold to Cylinder Head
(46) Exhaust Manifold Heat Shield
(47) Bolt, Exhaust Manifold Heat Shield to Exhaust Manifold
(48) Exhaust Manifold
(49) Oxygen Sensor

79227206

Exploded view of the intake and exhaust manifolds—2.4L engine

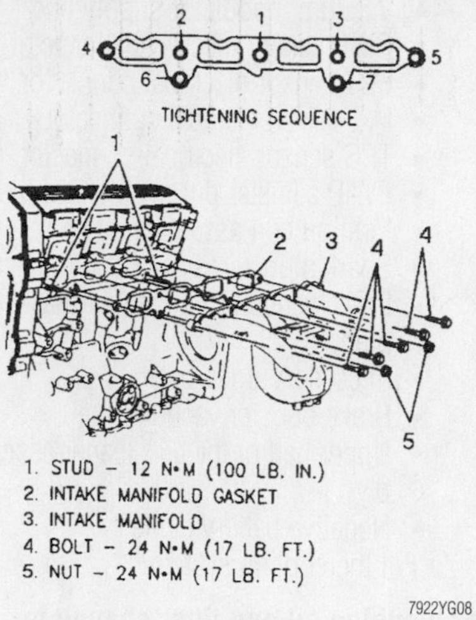

Intake manifold torque sequence—2.4L engine

- Evaporative Emissions (EVAP) canister purge solenoid electrical connector and vacuum line
- Fuel injector electrical connectors
- Fuel pressure regulator vacuum line
- Accelerator control cable bracket from the intake manifold
- Coolant lines from the throttle body
- Brake booster vacuum line
- Intake manifold

To install:

5. Install or connect the following:
   - Intake manifold using a new gasket. Torque the fasteners, in sequence, to 18 ft. lbs. (24 Nm).
   - Coolant lines to the throttle body
   - Brake booster vacuum line
   - Control cable bracket to the intake manifold
   - Fuel pressure regulator vacuum line
   - Fuel injector electrical connectors
   - EVAP canister purge solenoid electrical connector
   - IAT sensor electrical connector
   - MAP sensor electrical connector
   - Air cleaner intake duct
6. Refill the cooling system.
7. Connect the negative battery cable.

3.1L Engine

1. Before servicing the vehicle, refer to the precautions in the beginning of this section.
2. Drain the cooling system.
3. Relieve the fuel system pressure.
4. Remove or disconnect the following:
   - Upper half of the air cleaner assembly

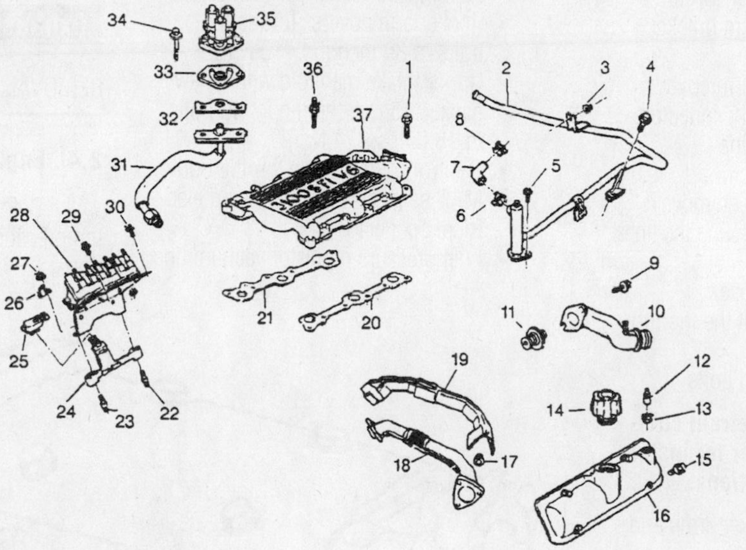

Exploded view of the upper intake and related components—3.1L engine

Heater Core replacement is covered in Section 2 of this manual

- Brake vacuum pipe at the intake plenum
- Fuel pressure regulator vacuum line
- Spark plug wires
- Ignition coil assembly
- Evaporative Emission (EVAP) canister purge solenoid
- Throttle Position (TPS) sensor electrical connector
- Idle Air Control (IAC) valve electrical connector
- Fuel Injector electrical connectors
- Engine Coolant Temperature (ECT) sensor electrical connector
- Manifold Absolute Pressure (MAP) sensor electrical connector and vacuum line
- Camshaft Position (CMP) sensor electrical connector
- Vacuum modulator supply hose
- Control cables from the throttle body and intake plenum bracket
- Upper intake manifold
- Left and right rocker arm covers
- Fuel injector electrical connectors
- Fuel feed and return line
- Fuel rail with the injectors
- Power steering pump and move it aside without disconnecting the lines
- Heater inlet pipe
- Thermostat bypass hose
- Upper radiator hose at the thermostat housing
- Lower intake manifold bolts

➡**When removing the valvetrain components, keep them in order for installation in their original locations.**

- Rocker arm nuts, rocker arms and pushrods
- Intake manifold

To install:

5. Place a 0.08–0.11 inch (8–12mm) bead of RTV on each ridge, where the front and rear of the intake manifold contact the block.
6. Install or connect the following:

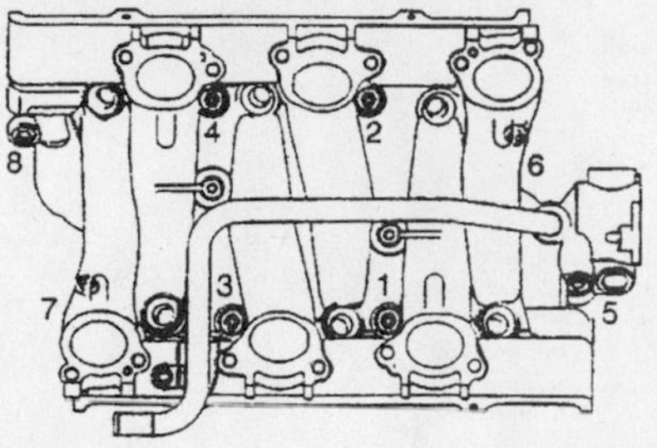

7922Z208

Lower intake manifold torque sequence—3.1L engine

- Lower intake manifold using a new gasket
- Pushrods and rocker arms

➡**Be sure the pushrods are properly seated in the valve lifters and rocker arms.**

- Rocker arm nuts. Torque them to 18 ft. lbs. (24 Nm).
- Lower intake manifold bolts, lubricated with sealant. Torque the bolts to 115 inch lbs. (13 Nm).
- Upper radiator hose at the thermostat housing
- Thermostat bypass hose
- Heater inlet pipe
- Power steering pump. Torque the bolts to 25 ft. lbs. (34 Nm).
- Fuel rail with the injectors
- Fuel feed and return lines
- Fuel injector electrical connectors
- Rocker arm covers. Torque the bolts to 89 inch lbs. (10 Nm).
- Upper intake manifold with a new gasket. Torque the mounting bolts to 18 ft. lbs. (24 Nm).
- Control cables to the throttle body
- MAP sensor vacuum line and electrical connector
- Fuel pressure regulator vacuum line
- Vacuum modulator supply line
- CMP sensor electrical connector
- ECT sensor electrical connector
- IAC valve electrical connector
- TPS sensor electrical connector
- EVAP canister purge solenoid
- Ignition coil assembly
- Spark plug wires
- Brake vacuum pipe at the intake plenum
- Accessory drive belt
- Brake booster vacuum line
- Upper half of the air cleaner assembly
- Negative battery cable

7. Fill the cooling system.

➡**An engine oil and filter change is recommended.**

8. Start the vehicle and verify no leaks.

Exhaust Manifold

REMOVAL & INSTALLATION

2.4L Engine

1. Before servicing the vehicle, refer to the precautions in the beginning of this section.

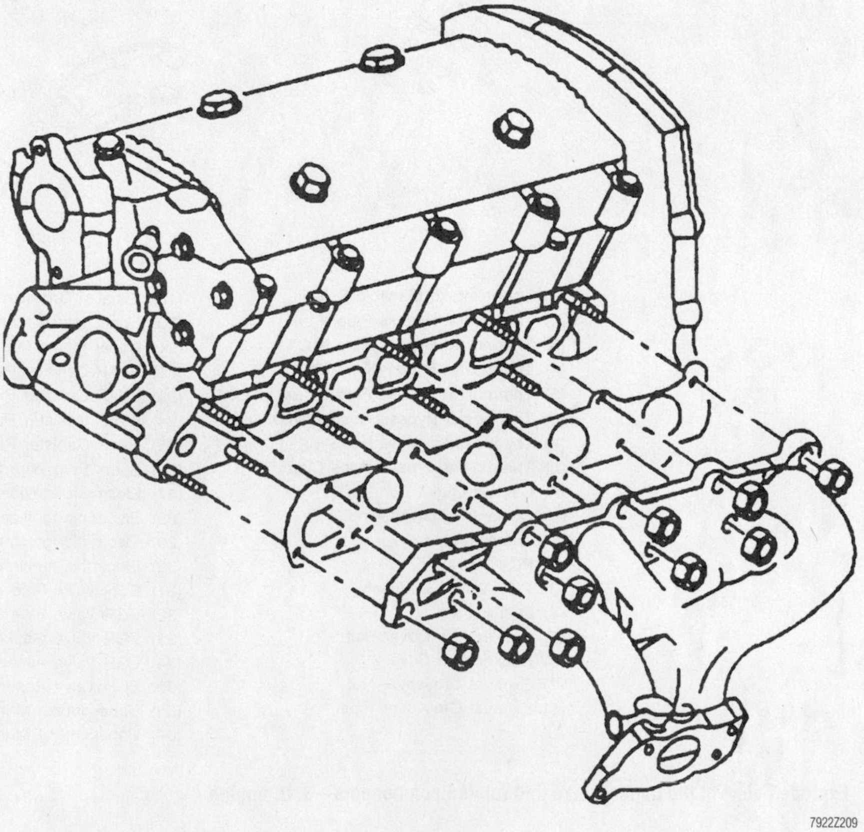

7922Z209

Exploded view of the exhaust manifold—2.4L engine

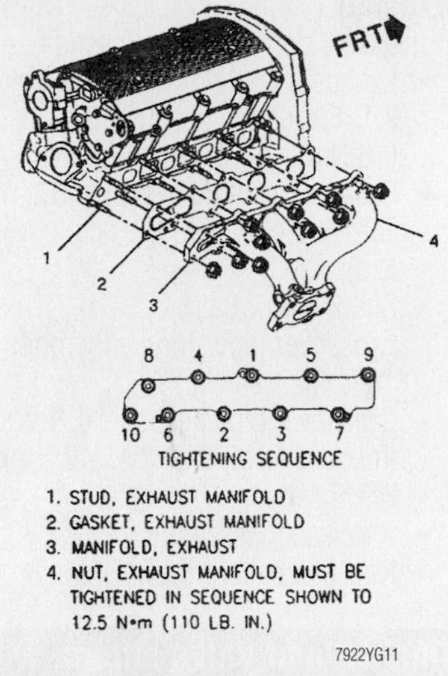

Exhaust manifold torque sequence—2.4L engine

2. Remove or disconnect the following:
   - Negative battery cable
   - Oxygen (O_2S) sensor
   - Heat shield
   - Exhaust manifold brace from the manifold
   - Exhaust manifold-to-exhaust pipe spring loaded bolts

➡The nuts should be loosened alternately and evenly to prevent the pipe from binding the nuts.

3. Pull down and back on the exhaust pipe to disengage it from the exhaust manifold.

⁂ WARNING

do not bend the exhaust flex coupler more than 3 degrees in any direction. Movement of more than 3 degrees will damage the flex coupler.

4. Remove the exhaust manifold.

To install:

5. Install or connect the following:
   - Exhaust manifold using a new gasket. Torque the nuts, in sequence, to 110 inch lbs. (13 Nm).
   - Exhaust manifold brace to the manifold. Tighten the bolt to 41 ft. lbs. (56 Nm) and the nuts to 19 ft. lbs. (26 Nm).
   - Heat shield. Torque the bolts to 89 inch lbs. (10 Nm).
   - Exhaust manifold-to-exhaust pipe bolts. Torque the bolts to 26 ft. lbs. (35 Nm).
   - O_2S sensor
   - Negative battery cable

6. Start the vehicle and verify no exhaust leaks.

3.1L Engine

LEFT SIDE (FRONT)

1. Before servicing the vehicle, refer to the precautions in the beginning of this section.
2. Partially drain the cooling system.
3. Remove or disconnect the following:
   - Negative battery cable
   - Air cleaner assembly
   - Throttle body duct
   - Exhaust crossover heat shield
   - Exhaust crossover pipe from the manifold
   - Upper radiator hose from the thermostat housing
   - Spark plug wires
   - Exhaust manifold heat shield
   - Exhaust manifold

To install:

4. Install or connect the following:
   - Exhaust manifold using a new gasket. Torque the nuts to 12 ft. lbs. (16 Nm).
   - Exhaust manifold heat shield. Torque the bolts to 89 inch lbs. (10 Nm).
   - Spark plug wires
   - Upper radiator hose
   - Exhaust crossover pipe to the manifold. Torque the bolts to 18 ft. lbs. (24 Nm).
   - Exhaust crossover pipe heat shield. Torque the bolts to 89 inch lbs. (10 Nm).
   - Throttle body duct
   - Air cleaner assembly
   - Negative battery cable
5. Fill the cooling system.

RIGHT SIDE (REAR)

1. Before servicing the vehicle, refer to the precautions in the beginning of this section.
2. Remove or disconnect the following:
   - Negative battery cable
   - Air cleaner assembly
   - Heated Oxygen (HO_2S) sensor
   - Crossover pipe heat shield
   - Crossover pipe
   - Spark plug wires
   - Exhaust manifold heat shield
   - Exhaust manifold

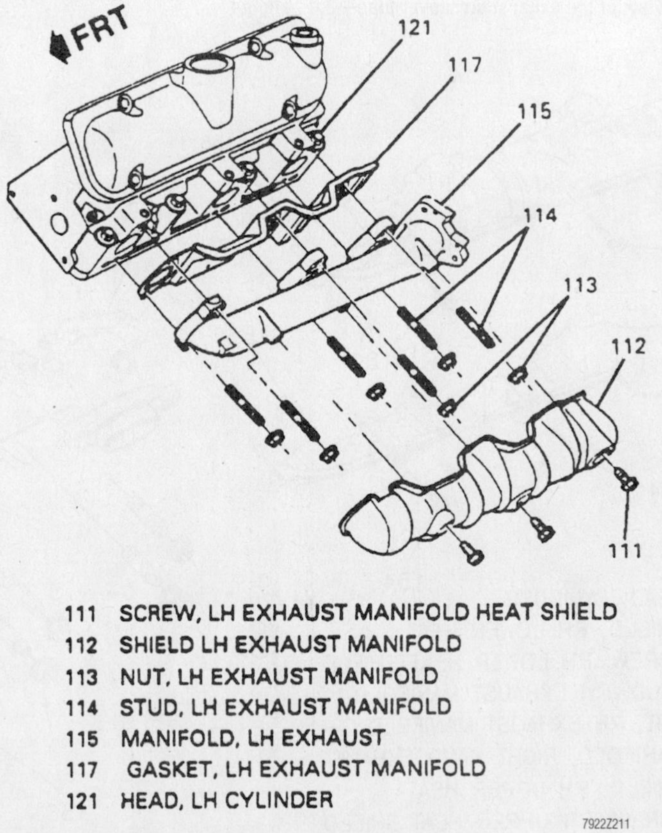

Exploded view of the left-hand exhaust manifold—3.1L engine

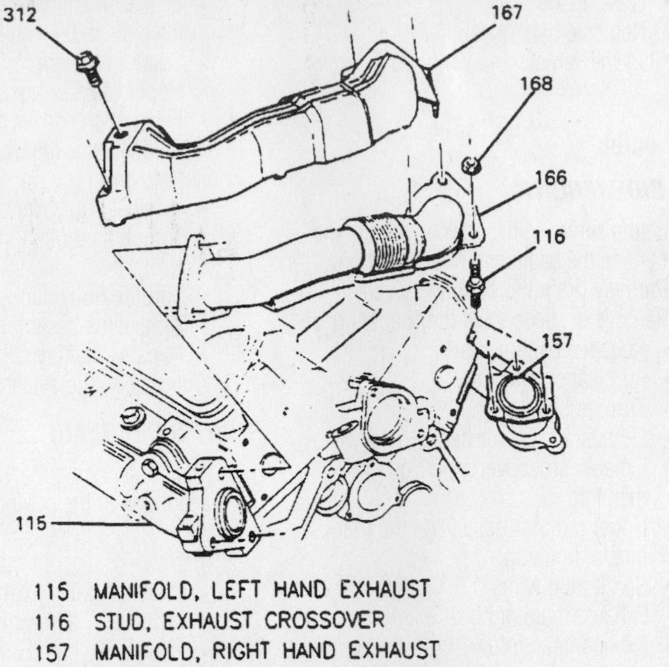

115 MANIFOLD, LEFT HAND EXHAUST
116 STUD, EXHAUST CROSSOVER
157 MANIFOLD, RIGHT HAND EXHAUST
166 CROSSOVER PIPE, EXHAUST
167 SHIELD, EXHAUST CROSSOVER UPPER HEAT
168 NUT, EXHAUST CROSSOVER
312 BOLT/SCREW, EXHAUST CROSSOVER UPPER HEAT SHIELD

79222212

Exploded view of the exhaust crossover pipe—3.1L engine

To install:

3. Install or connect the following:
   - Exhaust manifold with a new gasket. Torque the bolts to 12 ft. lbs. (16 Nm).
   - Exhaust manifold heat shield. Torque the bolts to 89 inch lbs. (10 Nm).
   - Spark plug wires
   - Crossover pipe. Torque the bolts to 18 ft. lbs. (24 Nm).
   - Crossover pipe heat shield. Torque the bolts to 89 inch lbs. (10 Nm).
   - HO_2S sensor
   - Air cleaner assembly
   - Negative battery cable

Camshaft and Valve Lifters

REMOVAL & INSTALLATION

2.4L Engine

➡Anytime the camshaft housing-to-cylinder head bolts are loosened or removed, the camshaft housing-to-cylinder head gasket must be replaced.

1. Before servicing the vehicle, refer to the precautions in the beginning of this section.

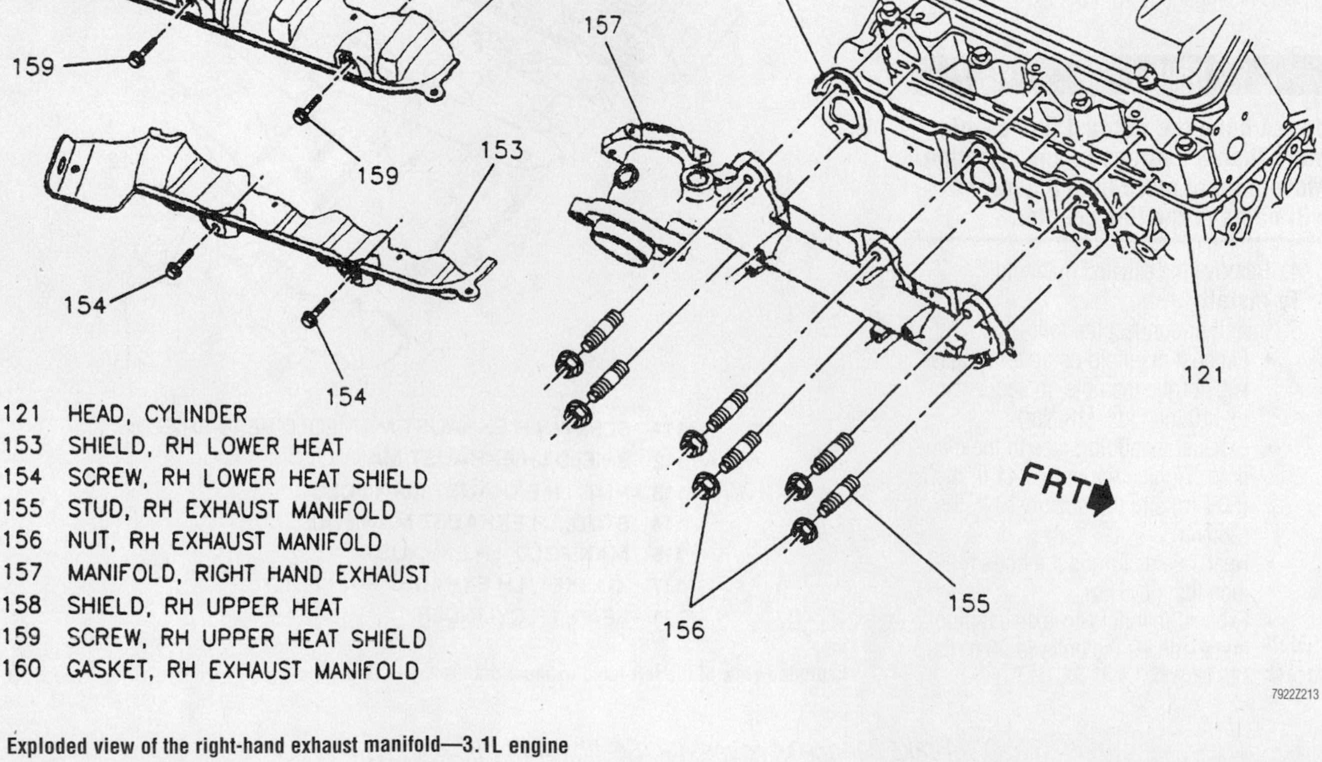

121 HEAD, CYLINDER
153 SHIELD, RH LOWER HEAT
154 SCREW, RH LOWER HEAT SHIELD
155 STUD, RH EXHAUST MANIFOLD
156 NUT, RH EXHAUST MANIFOLD
157 MANIFOLD, RIGHT HAND EXHAUST
158 SHIELD, RH UPPER HEAT
159 SCREW, RH UPPER HEAT SHIELD
160 GASKET, RH EXHAUST MANIFOLD

79222213

Exploded view of the right-hand exhaust manifold—3.1L engine

2. Remove or disconnect the following:
   - Negative battery cable
   - 11-pin connector from the ignition assembly cover
   - 4 ignition assembly cover-to-camshaft housing bolts and the assembly by pulling it straight up
   - Connector assemblies using a spark plug boot wire remover
3. On the exhaust camshaft side, remove or disconnect the following:
   - Transaxle fill tube
   - Oil pressure switch
4. On the intake camshaft side, remove or disconnect the following:
   - Camshaft Position (CMP) sensor and power steering pressure switch connector
   - Drive belt
   - Power steering pump and move it aside with the lines attached
   - Oil/air separator hoses from the oil fill tube, timing chain housing and intake manifold
   - Oil/air separator with the hoses attached
   - Vacuum line from the fuel pressure regulator
   - Fuel injector wiring harness
   - Fuel line clamps from the intake camshaft housing
   - Fuel rail-to-camshaft housing bolts
   - Fuel rail from the cylinder head and move it aside with the fuel lines attached.
   - Timing chain and camshaft sprockets
   - Timing chain housing bolts but do not the housing
   - Camshaft housing cover-to-camshaft housing bolts
   - Camshaft housing-to-cylinder head bolts by reversing the tightening sequence

➡Leave 2 bolts loosely in place to hold the camshaft housing while separating camshaft cover from housing.

5. Push the cover off the housing by threading 4 of the housing-to-cylinder head bolts into the tapped holes in the cam housing cover. Tighten the bolts evenly so the cover does not bind on the dowel pins.
6. Remove or disconnect the following:
   - Both loosely installed camshaft housing-to-cylinder head bolts
   - Camshaft housing cover and discard the gaskets

➡Note the position of the chain sprocket dowel pin for reassembly.

   - Camshaft, be careful do not damage the camshaft oil seal
   - Camshaft oil seal from camshaft and discard it

➡The oil seal must be replaced any time the housing and cover are separated.

➡The valve lifters must be kept in order for installation in the same locations they were removed.

   - Valve lifters from the camshaft housing
   - Camshaft carrier from the cylinder head and discard the gasket

To install:

7. Clean all gasket surfaces.
8. Install or connect the following:
   - New gasket on the cylinder head
   - Camshaft housing using 1 bolt loosely to hold it in place

➡If the camshaft is replaced, the valve lifters must also be replaced.

   - Valve lifters in their original bores

➡Lubricate the camshaft lobes, journals and lifters with camshaft and lifter prelube.

⁂ WARNING

The camshaft lobes and journals must be adequately lubricated or engine damage could occur upon start up.

   - Camshaft, so the timing chain sprocket dowel pin faces upward and aligns with the lifter bores centerline
   - New camshaft seals in the cover

➡The seals for the intake and exhaust covers are different; be sure the correct seals are used.

9. Remove the camshaft housing securing bolt
10. Lubricate the camshaft housing-to-cover bolts with thread locking compound.
11. Install or connect the following:
   - Camshaft housing cover. Torque the long bolts to 16 ft. lbs. (22 Nm) plus 90 degrees additional rotation and the short bolts to 16 ft. lbs. (22 Nm) plus 30 degrees additional rotation.
   - Timing chain housing bolts
   - Timing chain and sprockets
   - New O-rings on the fuel injectors, lubricated with engine oil

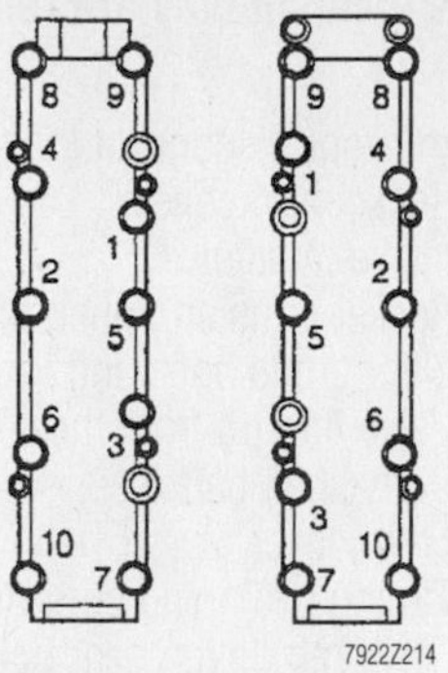

Camshaft housing bolt tightening sequence—2.4L engine

   - Fuel rail. Tighten the bolts to 19 ft. lbs. (26 Nm).
12. On the intake camshaft side, install or connect the following:
   - Fuel line clamp bolts on top of the intake camshaft housing
   - Vacuum line to the fuel pressure regulator
   - Fuel injector harness connector
   - Oil/air separator
   - Oil/air separator hoses to the oil fill tube, timing chain housing and intake manifold
   - New oil seal into the intake camshaft housing, lubricated with oil
   - Power steering pump pulley onto the intake camshaft
   - Power steering pump
   - Drive belt
   - CMP sensor and power steering pressure switch connectors
13. On the exhaust camshaft side, install or connect the following:
   - Transaxle fill tube
   - Oil pressure switch
14. Install or connect the following:
   - Ignition assembly on the camshaft housing. Tighten the bolts to 13 ft. lbs. (18 Nm).
   - 11-pin connector to ignition assembly
   - Negative battery cable
15. Start the vehicle and verify proper operation and no leaks.

3.1L Engine

1. Before servicing the vehicle, refer to the precautions in the beginning of this section.
2. Relieve the fuel system pressure.

➡When removing valvetrain components, marked them for installation in

the same location they are removed from.

3. Remove or disconnect the following:
   - Rocker arm covers
   - Intake manifold
   - Rocker arms and push rods
   - Lifter guide bolts and the guide
   - Valve lifter(s) from the lifter bores
   - Crankshaft balancer
   - Front cover
   - Timing chain and sprockets
   - Oil pump driven gear bolt and gear
   - Camshaft thrust plate
   - Camshaft

To install:

4. Install or connect the following:
   - Camshaft, lubricated with camshaft lubricant
   - Camshaft thrust plate. Torque bolts to 89 inch lbs. (10 Nm).
   - Oil pump driven gear. Torque the bolt to 27 ft. lbs. (36 Nm).
   - Timing chain and sprocket. Torque the bolt to 103 ft. lbs. (140 Nm).
   - Front cover. Torque the bolts to 35 ft. lbs. (47 Nm).
   - Crankshaft balancer. Torque the bolt to 76 ft. lbs. (103 Nm).
   - Valve lifters in their original locations
   - Lifter guide. Torque the bolts to 89 inch lbs. (10 Nm).
   - Pushrods, rocker arms, rocker balls and rocker arm nuts. Torque the rocker arm nuts to 89 inch lbs. (10 Nm) plus an additional 30 degrees.
   - Intake manifold. Torque the upper and lower intake bolts to 115 inch lbs. (13 Nm).
   - Rocker arm covers. Torque the bolts to 89 inch lbs. (10 Nm).
   - Negative battery cable
5. Start the engine and verify no oil leaks.

Valve Lash

ADJUSTMENT

The 2.4L and 3.1L engines are equipped with hydraulic valve lifters; no adjustment is necessary.

Starter Motor

REMOVAL & INSTALLATION

2.4L Engine

1. Before servicing the vehicle, refer to the precautions in the beginning of this section.
2. Remove or disconnect the following:
   - Negative battery cable
   - Air inlet duct from the throttle body
   - Starter bolts
   - Starter electrical connectors
   - Starter motor

To install:

3. Install or connect the following:
   - Starter motor. Torque the bolts to 66 ft. lbs. (90 Nm).
   - Starter electrical connectors
   - Air inlet duct to the throttle body
   - Negative battery cable

3.1L Engine

1. Before servicing the vehicle, refer to the precautions in the beginning of this section.
2. Remove or disconnect the following:
   - Negative battery cable
   - Lower closeout panel
   - Starter bolts
   - Starter electrical connectors
   - Starter motor

To install:

3. Install or connect the following:
   - Starter electrical connectors. Torque the solenoid cable to 106 inch lbs. (12 Nm).
   - Starter motor. Torque the bolts to 37 ft. lbs. (50 Nm).
   - Lower closeout panel
   - Negative battery cable

Oil Pan

REMOVAL & INSTALLATION

2.4L Engine

1. Before servicing the vehicle, refer to the precautions in the beginning of this section.
2. Disconnect the negative battery cable.
3. Drain the engine oil and cooling system.
4. Install an engine support fixture.
5. Remove or disconnect the following:
   - Engine mount and raise the engine approximately 1.5 inches (3.8cm)
   - Flywheel dust shield
   - Right front wheel
   - Right inner fender splash shield
   - Accessory drive belt
   - Air conditioning compressor and support it aside with the lines attached
   - Transmission-to-engine brace
   - Radiator outlet pipe
   - Bolt and the nut which secure the oil pan to the flywheel cover

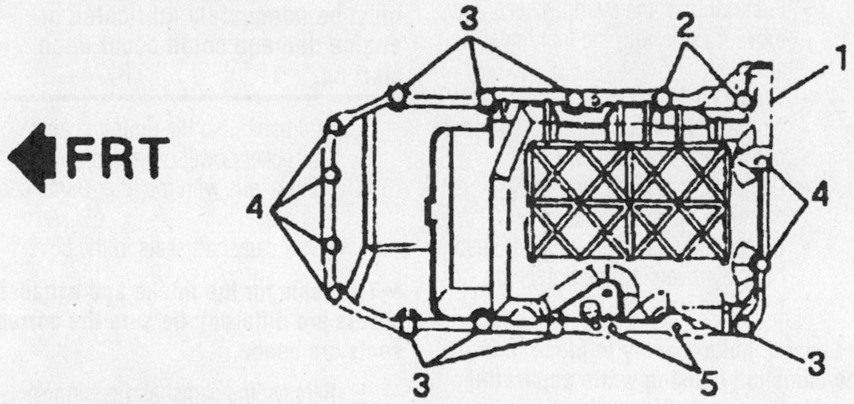

1 Oil pan
2 Oil pan bolt (M8 x 1.25 x 80) 24 Nm (18 lb. ft.)
3 Oil pan bolt (M8 x 1.25 x 22) 24 Nm (18 lb. ft.)
4 Oil pan bolt (M6 x 1.00 x 25) 12 Nm (106 lb. in.)
5 Stud end oil pan bolt 26 Nm (19 lb. ft.)

7922Z215

Oil pan mounting bolt locations—2.4L engine

- Flywheel cover stud for clearance
- Oil pan bolts and the pan

To install:

6. Install or connect the following:
   - Oil pan using a new gasket. Torque the chain housing-to-oil pan bolts to 106 inch lbs. (12 Nm) and the oil pan-to-block bolts to 18 ft. lbs. (24 Nm).
   - Oil pan-to-transaxle nut. Torque the nut to 41 ft. lbs. (56 Nm).
   - Radiator outlet pipe
   - Transmission-to-engine brace. Torque the bolts to 37 ft. lbs. (50 Nm).
   - Air conditioning compressor. Torque the bolts to 37 ft. lbs. (50 Nm).
   - Accessory drive belt
   - Right inner fender splash shield
   - Right front wheel
   - Flywheel dust shield
   - Engine mount. Torque the bolts to 49 ft. lbs. (66 Nm).
   - Negative battery cable
7. Fill the crankcase with oil.

➡A filter change is recommended.

8. Refill the crankcase.
9. Fill the cooling system.
10. Remove the engine support fixture.
11. Start the engine and check for leaks.

3.1L Engine

1. Before servicing the vehicle, refer to the precautions in the beginning of this section.
2. Evacuate and recover the A/C system.
3. Drain the engine oil.
4. Remove or disconnect the following:
   - Negative battery cable
   - Accessory drive belt
   - Upper and lower radiator hoses
   - Right front wheel
   - Inner fender splash shield
   - Antilock Brake System (ABS) Wheel Speed Sensor (WSS) electrical connector
   - Right front ball joint from the steering knuckle
   - Right side outer tie rod end
   - A/C compressor bolts and move the compressor aside

⁂ WARNING

DO NOT disconnect the refrigerant lines or allow the compressor to hang unsupported.

- Evaporator-to-accumulator A/C line
- Flywheel cover
- Engine cradle bolts
- Crankshaft balancer
- Starter motor
- Oil pan

To install:

5. Install or connect the following:
   - Oil pan using a new gasket

➡Apply silicone sealer to the portion of the pan that contacts the rear of the block.

- Oil pan bolts. Torque the bolts to 18 ft. lbs. (24 Nm) and the side bolts to 37 ft. lbs. (50 Nm).
- Starter motor. Torque the bolts to 37 ft. lbs. (50 Nm).
- Flywheel cover
- Crankshaft balancer. Torque the bolt to 76 ft. lbs. (103 Nm).
- A/C compressor. Torque the bolts to 37 ft. lbs. (50 Nm).
- Evaporator-to-accumulator line
- Engine cradle bolts. Torque the bolts to 89 ft. lbs. (120 Nm).
- Ball joint to the steering knuckle using a new cotter pin. Torque the nut 48 ft. lbs. (60 Nm).
- Outer tie rod end. Torque the nut to 33 ft. lbs. (45 Nm).
- ABS wheel speed sensor electrical connector
- Inner fender splash shield

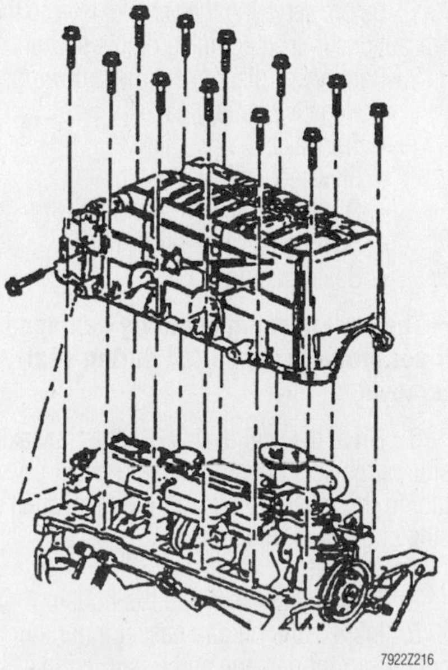

Exploded view of the oil pan mounting—3.1L engine

- Right front wheel
- Accessory drive belt
- Upper and lower radiator hoses
- Negative battery cable

6. Refill the crankcase.
7. Evacuate and recharge the A/C system.
8. Start the engine and check for leaks.

➡Whenever the vehicle subframe is removed or lowered, the wheel alignment should be checked.

Oil Pump

REMOVAL & INSTALLATION

2.4L Engine

➡Please note that the transaxle must be removed from the vehicle to service the oil pump.

1. Before servicing the vehicle, refer to the precautions in the beginning of this section.
2. Disconnect the negative battery cable.
3. Install an engine support fixture.
4. Drain the engine oil.
5. Remove or disconnect the following:
   - Oil pan
   - Transaxle
   - Flywheel
   - Balance shaft chain cover and chain guide
   - Oil pump cover bolts and the cover
   - Pump gear from the balance shaft by pulling the housing
   - Oil pump housing from the balance shaft

To install:

6. Clean all of the parts in suitable cleaning solvent. Remove all varnish sludge and dirt.
7. Lubricate the gears with clean engine oil.
8. Install or connect the following:
   - Gerotor gear into the housing; then, fill the oil pump cavities with petroleum jelly

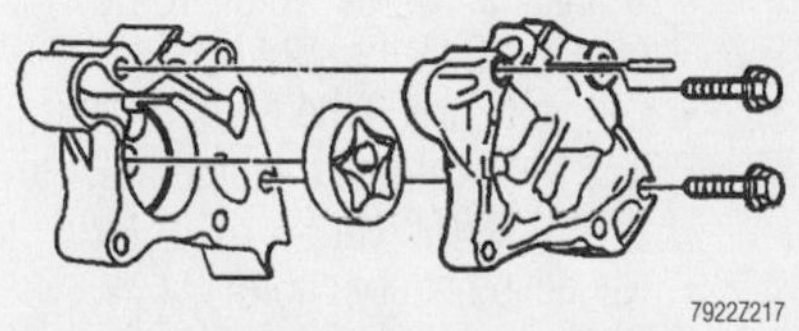

Exploded view of the oil pump components—2.4L engine

- Oil pump housing to the balance shaft
- Oil pump cover to the housing. Tighten the bolts to 40 ft. lbs. (54 Nm).
- Balance shaft chain guide and chain

9. Adjust the chain tension by performing the following procedure:

a. Insert a 0.40 in. (1mm) brass feeler between the chain guide and chain.

➡A brass feeler gauge must be used to ensure that correct measurements are obtained. If a steel gauge is used, it will not bend to conform to the guide and will allow for incorrect measurements.

b. Press the guide against the chain using about 3 lbs. (13.3 N) of force.

c. Tighten the chain tensioner fastener to 115 inch lbs. (13 Nm).

10. Install or connect the following:
    - Balance shaft chain cover. Tighten the nut and bolt to 115 inch lbs. (13 Nm).
    - Flywheel
    - Transaxle
    - Oil pan
    - Negative battery cable
11. Refill the crankcase.

➡A filter change is recommended.

12. Remove the engine support fixture.
13. Start the engine and verify oil pressure and no leaks.

3.1L Engine

1. Before servicing the vehicle, refer to the precautions in the beginning of this section.
2. Disconnect the negative battery cable.
3. Drain the engine oil.
4. Remove or disconnect the following:
    - Oil pan
    - Crankshaft oil deflector
    - Oil pump and pump driveshaft

To install:

5. Install or connect the following:
    - Oil pump and pump driveshaft. Torque the bolts to 30 ft. lbs. (41 Nm).
    - Crankshaft oil deflector. Torque the nuts to 18 ft. lbs. (24 Nm).
    - Oil pan. Torque the bolts to 18 ft. lbs. (24 Nm) and the side bolts to 37 ft. lbs. (50 Nm).
    - Negative battery cable
6. Refill the crankcase.

➡A filter change is recommended.

7. Start the engine, check the oil pressure and check for leaks.

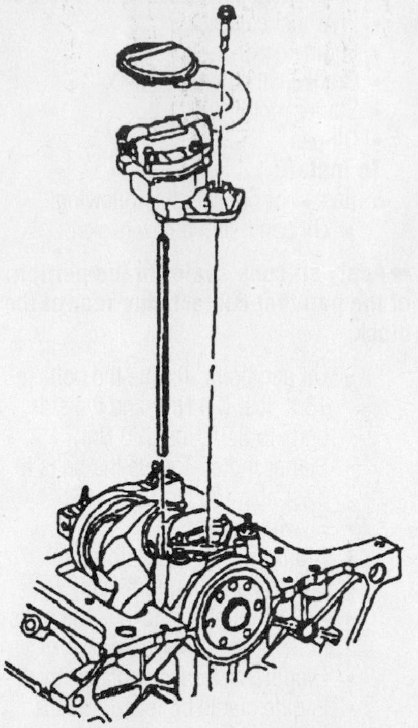

Exploded view of the oil pump mounting—3.1L engine

Rear Main Seal

REMOVAL & INSTALLATION

2.4L Engine

1. Before servicing the vehicle, refer to the precautions in the beginning of this section.
2. Remove or disconnect the following:
    - Negative battery cable
    - Transaxle
    - Flywheel
    - Oil pan-to-seal housing bolts
    - Seal housing-to-block bolts
    - Seal housing

➡The seal housing could be damaged if not properly supported during seal removal.

3. Drive the seal evenly out the transaxle side of the seal housing using a small pry tool in the relief grooves on the crankshaft side of the seal housing.

To install:

4. Press a new seal into the housing.
5. Inspect the oil pan gasket inner silicone bead for damage and repair using a silicone sealant, if necessary.
6. Lubricate the seal lip with clean engine oil.
7. Install or connect the following:
    - Seal housing with a new gasket. Torque the housing-to-engine bolts to 106 inch lbs. (12 Nm) and the oil pan-to-seal housing bolts to 106 inch lbs. (12 Nm).
    - Flywheel. Torque the bolts to 22 ft. lbs. (30 Nm) plus an additional 45 degree rotation.
    - Transaxle
    - Negative battery cable
8. Start the engine and check for leaks.

3.1L Engine

1. Before servicing the vehicle, refer to the precautions in the beginning of this section.
2. Support the engine.
3. Remove or disconnect the following:
    - Transaxle
    - Flywheel
    - Rear main seal by prying it from the housing

✲✲ WARNING

Use care not to damage the crankshaft seal surface with a prytool.

To install:

4. Lubricate the seal bore and new seal with engine oil.
5. Install the new seal by performing the following procedure:

a. Slide the new seal over the mandrel until the dust lip bottoms squarely against the tool collar.

b. Align the dowel pin of the tool with the dowel pin hole in the crankshaft and attach the tool to the crankshaft. Tighten the attaching screws to 24–60 inch lbs. (2.7–6.8 Nm).

c. Tighten the T-handle of the tool to push the seal into the bore. Continue until the tool collar is flush against the block.

d. Loosen the T-handle completely. Remove the attaching screws and the tool.

➡Check to see that the seal is squarely seated in the bore.

6. Install or connect the following:
    - Flywheel
    - Transaxle
7. Start the engine and check for leaks.

Timing Chain, Sprockets, Front Cover and Seal

REMOVAL & INSTALLATION

2.4L Engine

➡It is recommended that the entire procedure be reviewed before attempting to service the timing chain.

1. Before servicing the vehicle, refer to the precautions in the beginning of this section.

2. Disconnect the negative battery cable.
3. Drain the cooling system.
4. Install an engine support.
5. Remove or disconnect the following:
   - Coolant surge tank
   - Accessory drive belt
   - Upper front cover fasteners
   - Right engine mount and bracket
   - Right front wheel
   - Lower splash shield from the right wheel house
   - Crankshaft balancer
   - Lower front cover fasteners
   - Front cover
6. Rotate the crankshaft clockwise, as viewed from the front of the engine (normal rotation), until the camshaft sprocket's timing dowel pin holes align with the holes in the timing chain housing. The mark on the crankshaft sprocket should align with the mark on the cylinder block. The crankshaft sprocket keyway should point upwards and align with the center line of the cylinder bores. This is the normal timed position.
7. Remove or disconnect the following:
   - Timing chain guides
   - Timing chain tensioner
   - Timing chain tensioner shoe

⁂ CAUTION

The tensioner piston is spring loaded and could fly out causing personal injury.

   - Timing chain
   - Camshaft sprockets, using a 3 jaw puller

To install:

8. Install the intake camshaft sprocket. Torque the bolt to 52 ft. lbs. (70 Nm).

➡Install the Special Tool J 36008-A through the holes in the camshaft sprockets into the holes in the timing chain housing. This positions the camshafts for correct timing.

9. If the camshafts are out of position and must be rotated more than 1/8 turn in order to install the alignment dowel pins, perform the following:
   a. Rotate the crankshaft 90 degrees clockwise off Top Dead Center (TDC) in order to give the valves adequate clearance to open.
   b. Once the camshafts are in position and the dowels installed, rotate the crankshaft counterclockwise back to TDC.

⁂ WARNING

Do not rotate the crankshaft clockwise to TDC or valve and piston damage may occur.

10. Install the timing chain over the exhaust camshaft sprocket, around the coolant pump sprocket and around the crankshaft sprocket.
11. Remove the alignment dowel pin from the intake camshaft. Rotate the intake camshaft sprocket counterclockwise enough to slide the timing chain over the intake camshaft sprocket. If properly timed, the intake camshaft alignment dowel pin should slide in easily. If the dowel pin does not fully index, the camshafts are not timed correctly and the procedure must be repeated.
12. Leave the alignment dowel pins installed.
13. With slack removed from the chain between the intake camshaft sprocket and the crankshaft sprocket, the crankshaft keyway and the cylinder block mark should be aligned. If not aligned, move the chain 1 tooth forward or rearward. Remove the slack and recheck the marks.
14. Tighten the chain housing-to-block stud. The stud is installed under the timing chain. Torque the stud to 19 ft. lbs. (26 Nm).
15. Reload timing chain tensioner assembly to its **0** position as follows:
   a. Insert the tensioner plunger assembly into the tensioner housing.
   b. With the tensioner plunger fully extended, turn the complete assembly upside down on a flat surface.
   c. Press the bottom of the tensioner housing to compress the plunger into the housing until it is seated.
   d. Make sure that the plunger does not extend out of the tensioner housing more than 0.07 in. (1.7mm).
   e. Loosely install the tensioner assembly to the timing chain housing.
16. Install or connect the following:
   - Tensioner shoe on the stud
   - Tensioner assembly by applying hand pressure on the timing chain tensioner shoe until the locking tab seats in the stud groove. Torque the bolts to 89 inch lbs. (10 Nm).

⁂ WARNING

If the timing chain tensioner plunger is not released from the installation position, engine damage will occur upon start up.

17. Release the tensioner plunger by firmly pressing a flat blade tool against the plunger face.

⁂ WARNING

If the tensioner plunger cannot be depressed, it is not properly reset and the resetting procedure must be repeated.

18. Rotate the crankshaft clockwise 2 full rotations. Align the crankshaft keyway with the mark on the cylinder block and reinstall alignment dowel pins. The alignment dowel pins will slide in easily if engine is timed correctly.
19. Install or connect the following:
   - Timing chain guides
   - Front cover using new gaskets. Torque the nuts and bolts to 106 inch lbs. (12 Nm).
   - Crankshaft balancer. Torque the bolt to 129 ft. lbs. (175 Nm) plus an additional 90 degree rotation.
   - Right front lower splash shield
   - Wheel
   - Right engine mount bracket. Torque the bolts to 81 ft. lbs. (110 Nm) plus an additional 90 degree rotation.
   - Right engine mount. Torque the bolt to 49 ft. lbs. (66 Nm).
20. Remove the engine support.
21. Install or connect the following:
   - Accessory drive belt
   - Coolant surge tank
   - Negative battery cable
22. Refill the cooling system and check for leaks.

3.1L Engine

1. Before servicing the vehicle, refer to the precautions in the beginning of this section.
2. Disconnect the negative battery cable.
3. Drain the cooling system.
4. Drain the engine oil.
5. Recover the A/C system refrigerant.
6. Install an engine support fixture.
7. Remove or disconnect the following:
   - Right engine mount assembly
   - Accessory drive belt
   - Air cleaner assembly
   - Throttle body tube
   - Power steering line at the pump
   - Alternator and bracket
   - Right front wheel
   - Right inner fender well splash shield

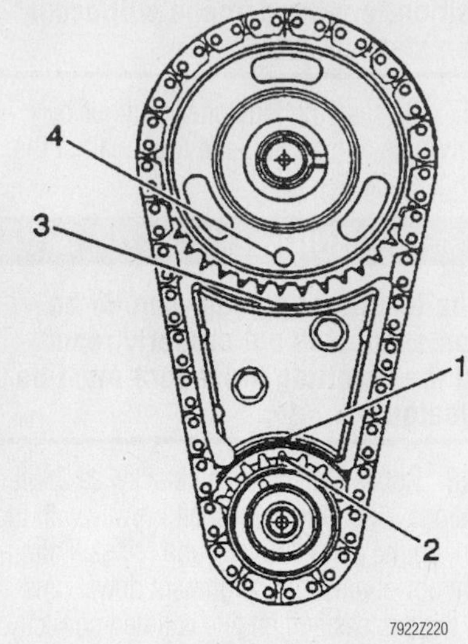

Timing chain and sprocket timing mark alignment—3.1L engine

- Right engine mount bracket
- Crankshaft balancer
- Drive belt tensioner
- Oil pan
- Crankshaft Position (CKP) sensor
- Coolant bypass pipe from the water pump and the intake manifold
- Lower radiator hose from the front cover outlet
- Front cover

8. Rotate the crankshaft until the timing marks on the camshaft and crankshaft sprockets are in alignment.

9. Remove or disconnect the following:
   - Camshaft sprocket and timing chain
   - Crankshaft sprocket
   - Timing chain damper

To install:

10. Install or connect the following:
    - Timing chain damper. Torque the bolts to 15 ft. lbs. (21 Nm).
    - Crankshaft sprocket

11. Be sure the crankshaft sprocket timing mark is pointing straight up.

12. Install the timing chain over the camshaft sprocket and hold the sprocket in such a way, that the timing mark is pointing down, and the timing chain is hanging down off the sprocket.

13. Loop the timing chain under the crankshaft sprocket and install the camshaft sprocket on the camshaft. The sprocket will only fit on the camshaft if the dowel on the camshaft aligns with the hole in the sprocket.

14. Verify that the marks are aligned (the camshaft sprocket will be at the 6 o'clock position and the crankshaft sprocket will be in the 12 o'clock position).

15. Torque the camshaft sprocket mounting bolt to 103 ft. lbs. (140 Nm).

16. Lubricate the timing chain components with engine oil.

17. Apply a thin bead of sealer around the gasket sealing area of the front cover.

18. Install or connect the following:
    - Front cover with a new seal and gasket. Torque the small bolts to 15 ft. lbs. (21 Nm) and the large bolts to 35 ft. lbs. (47 Nm).
    - Radiator hose to the coolant outlet
    - Coolant bypass pipe to the water pump and intake manifold
    - CKP sensor
    - Oil pan. Torque the bolts to 18 ft. lbs. (24 Nm) and the side bolts to 37 ft. lbs. (50 Nm).
    - Crankshaft balancer. Torque the bolt to 76 ft. lbs. (103 Nm).
    - Accessory drive belt tensioner. Torque the bolt to 40 ft. lbs. (54 Nm).
    - Right engine mount bracket. Torque the bolts to 96 ft. lbs. (130 Nm).
    - Right inner fender well splash shield
    - Wheel

19. Remove the engine support fixture.

20. Install or connect the following:
    - Alternator. Torque the front bolt to 37 ft. lbs. (50 Nm) and the rear bolt to 18 ft. lbs. (24 Nm).
    - Power steering line
    - Throttle body tube
    - Air cleaner assembly
    - Accessory drive belt
    - Negative battery cable

21. Refill the cooling system.

22. Check the engine oil level.

➡An oil and filter change is recommended.

23. Start the engine and verify that there are no leaks.

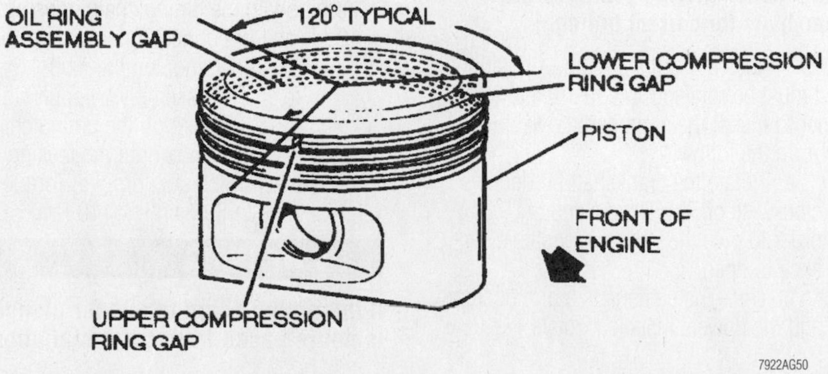

Piston ring end-gap spacing—2.4L engine

Piston and Ring

POSITIONING

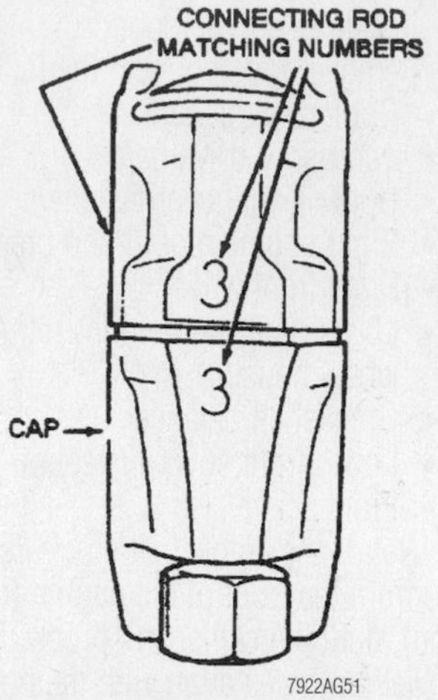

Connecting rod and cap installation. Be sure to matchmark the cap and rod prior to disassembly, as shown

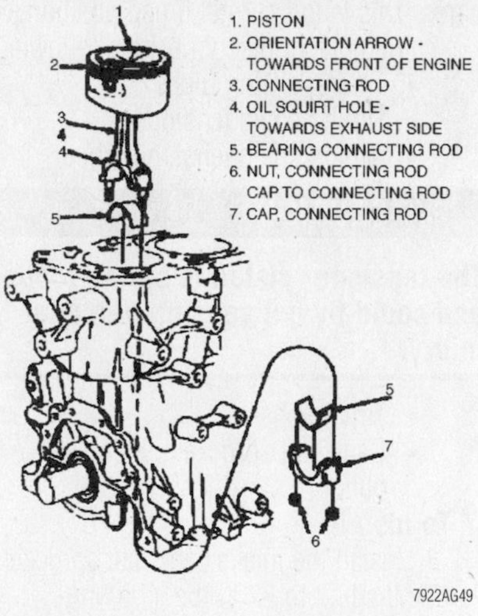

Piston and connecting rod assembly positioning—2.4L engine

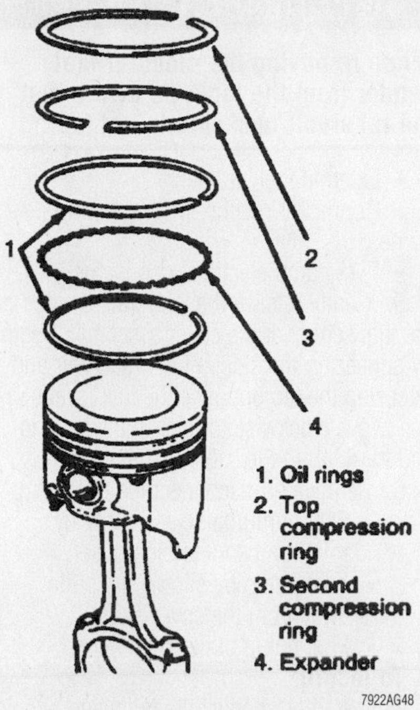

Piston ring positioning—3.1L engine

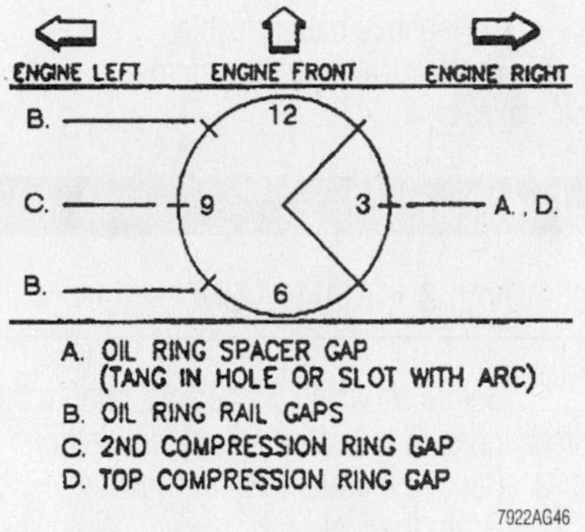

Piston ring end-gap spacing—3.1L engine

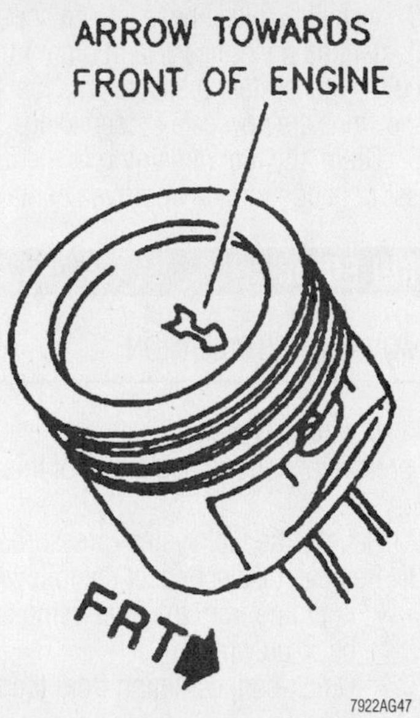

Piston positioning. Often the arrow is replaced by a notch, which also must face toward the front of the engine—3.1L engine

FUEL SYSTEM

Fuel System Service Precautions

Safety is the most important factor when performing not only fuel system maintenance but any type of maintenance. Failure to conduct maintenance and repairs in a safe manner may result in serious personal injury or death. Maintenance and testing of the vehicle's fuel system components can be accomplished safely and effectively by adhering to the following rules and guidelines.

- To avoid the possibility of fire and personal injury, always disconnect the negative battery cable unless the repair or test procedure requires that battery voltage be applied.
- Always relieve the fuel system pressure prior to disconnecting any fuel system component (injector, fuel rail, pressure regulator, etc.), fitting or fuel line connection. Exercise extreme caution whenever relieving fuel system pressure, to avoid exposing skin, face and eyes to fuel spray. Please be advised that fuel under pressure may penetrate the skin or any part of the body that it contacts.
- Always place a shop towel or cloth around the fitting or connection prior to loosening to absorb any excess fuel due to spillage. Ensure that all fuel spillage (should it occur) is quickly removed from engine surfaces. Ensure that all fuel soaked cloths or towels are deposited into a suitable waste container.
- Always keep a dry chemical (Class B) fire extinguisher near the work area.
- Do not allow fuel spray or fuel vapors to come into contact with a spark or open flame.
- Always use a back-up wrench when loosening and tightening fuel line connection fittings. This will prevent unnecessary stress and torsion to fuel line piping. Always follow the proper torque specifications.
- Always replace worn fuel fitting O-rings with new. Do not substitute fuel hose, where fuel pipe is installed.

Fuel System Pressure

RELIEVING

2.4L Engine

1. Before servicing the vehicle, refer to the precautions in the beginning of this section.
2. Loosen the fuel filler cap in order to relieve the pressure in the tank (do not tighten at this time).
3. Detach the fuel pump electrical connector.
4. Start and run the vehicle until it stalls, then engage the starter for an additional 3 seconds to ensure the relief of any remaining pressure.
5. Disconnect the negative battery cable.
6. Once the tests or repairs are completed, reattach the fuel pump electrical connector.
7. Connect the negative battery cable.
8. Tighten the fuel filler cap.
9. prime the fuel system by cycling the ignition switch **ON** for 2 seconds, **OFF** for 10 seconds, then **ON** again. Repeat, if necessary to build system pressure.

3.1L Engine

1. Before servicing the vehicle, refer to the precautions in the beginning of this section.
2. Disconnect the negative battery cable in order to avoid possible fuel discharge if an accidental attempt is made to start the engine.
3. Loosen the fuel tank filler cap in order to relieve fuel tank pressure.
4. Connect a fuel pressure gauge to the fuel pressure test port connection. Wrap a towel around the fuel pressure connection when installing the fuel pressure gauge in order to avoid fuel spillage.

5. Install the bleed hose into an approved container and open the valve in order to bleed the fuel system pressure. The fuel pipe connections are now safe for servicing.
6. Drain any fuel remaining in the fuel pressure gauge into an approved container.

Fuel Filter

REMOVAL & INSTALLATION

1. Before servicing the vehicle, refer to the precautions in the beginning of this section.
2. Relieve the fuel system pressure.
3. Remove or disconnect the following:
   - Fuel line from the filter using a back-up wrench
   - Quick-connect fitting from the fuel filter
   - Fuel filter from the mounting bracket

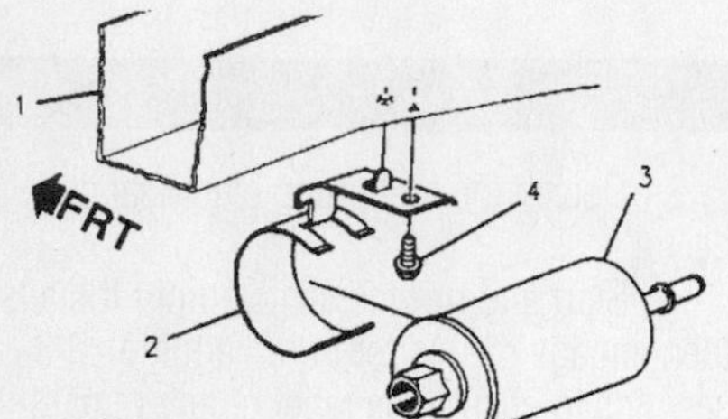

1 BODY
2 FUEL FILTER BRACKET
3 FUEL FILTER
4 SCREW - FULLY DRIVEN, SEATED AND NOT STRIPPED

79222221

Exploded view of the fuel filter mounting

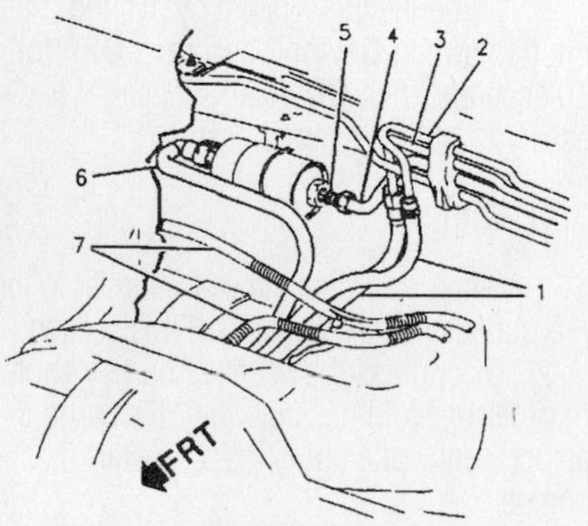

1 HOSE, PART OF FUEL SENDER
2 FUEL VAPOR PIPE
3 FUEL RETURN PIPE
4 FUEL FEED PIPE
5 FUEL FEED PIPE NUT 27 N.m (20 LBS. FT.)
6 HOSE, PART OF FUEL SENDER
7 ABS AND FUEL SENDER HARNESS

79222222

Fuel filter mounting location and component identification

To install:

4. Install or connect the following:
   - Fuel filter to the mounting bracket
   - Fuel line. Torque the fuel line fitting to 20 ft. lbs. (27 Nm).
   - Quick-connect fitting to the fuel filter
   - Negative battery cable
5. Pressurize the fuel system and verify no leaks.

Fuel Pump

REMOVAL & INSTALLATION

1. Before servicing the vehicle, refer to the precautions in the beginning of this section.
2. Relieve the fuel system pressure.
3. Drain the fuel tank.
4. Remove or disconnect the following:
   - Fuel tank
   - Modular fuel sender assembly-to-fuel tank snapring by pressing the assembly downward, on 1998–99 models or on 2000–01 models, use fuel sender lock nut Tool J-39765, press down and rotate the cam lock ring until free of the fuel sender retaining tabs.

➡**When removing the modular fuel sender from the tank, be aware that it may spring upward.**

   - Modular fuel sender assembly

⁂ CAUTION

When removing the modular fuel sender from the tank, be aware that the reservoir bucket is full of fuel.

   - External fuel strainer
   - Connector retainer from the wiring harness
   - Fuel pump electrical connector

5. Gently release the tabs on the sides of the fuel sender at the cover assembly. Begin by squeezing the sides of the reservoir and releasing the tab opposite the fuel level sensor. Move clockwise to release the second and third tab in the same manner.
6. Remove or disconnect the following:
   - Fuel pump/baffle assembly by rotating it counterclockwise
   - Fuel pump by sliding the outlet away from the cover slots
   - Fuel pump outlet seal

To install:

7. Install or connect the following:
   - Fuel pump outlet seal
   - Fuel pump outlet by sliding it into the reservoir cover slots
   - Fuel pump/baffle assembly onto the reservoir retainer by rotating it clockwise until seated
   - Lower retainer assembly by aligning the 3 sleeve tabs and pressing the retainer onto the reservoir until tabs are firmly seated

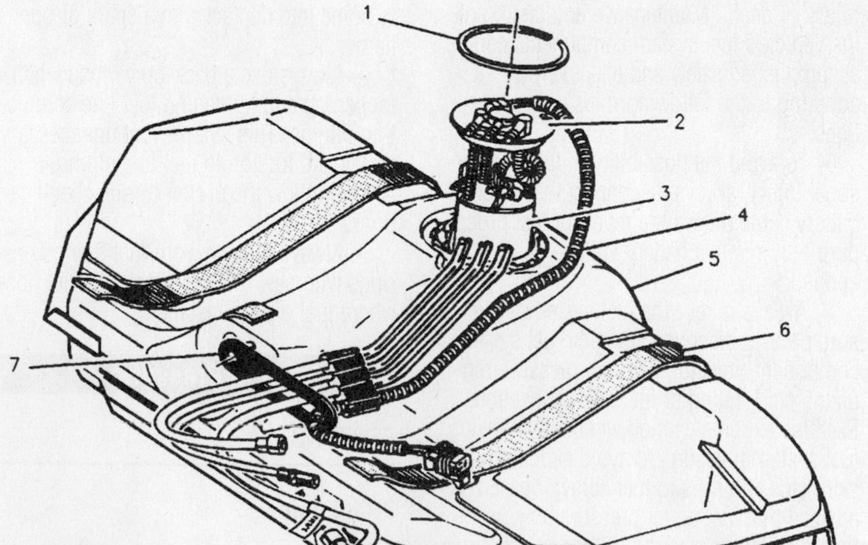

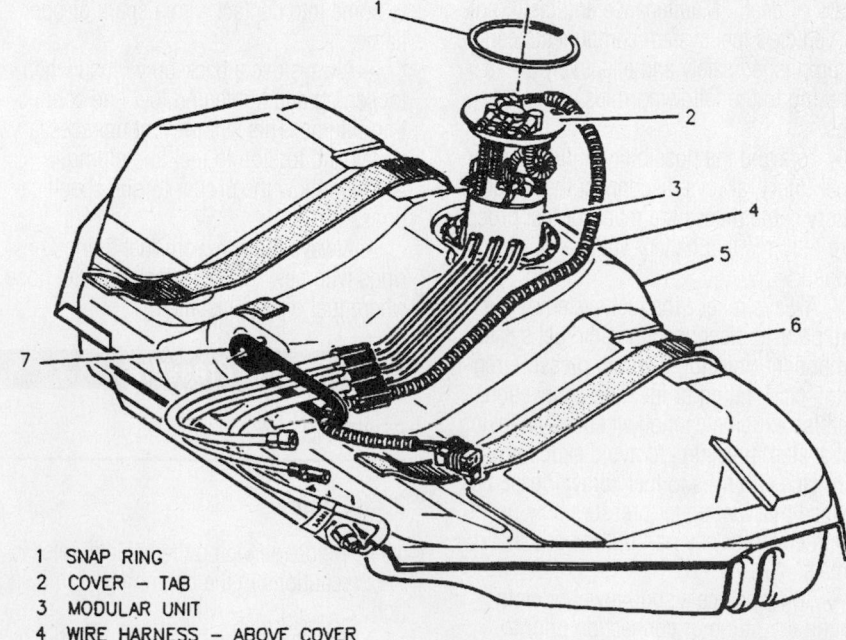

1 SNAP RING
2 COVER - TAB
3 MODULAR UNIT
4 WIRE HARNESS - ABOVE COVER
5 FUEL TANK
6 TANK ISOLATION STRIPS (3)
7 RUBBER ISOLATOR

79222223

Exploded view of the fuel sender assembly mounting to the tank

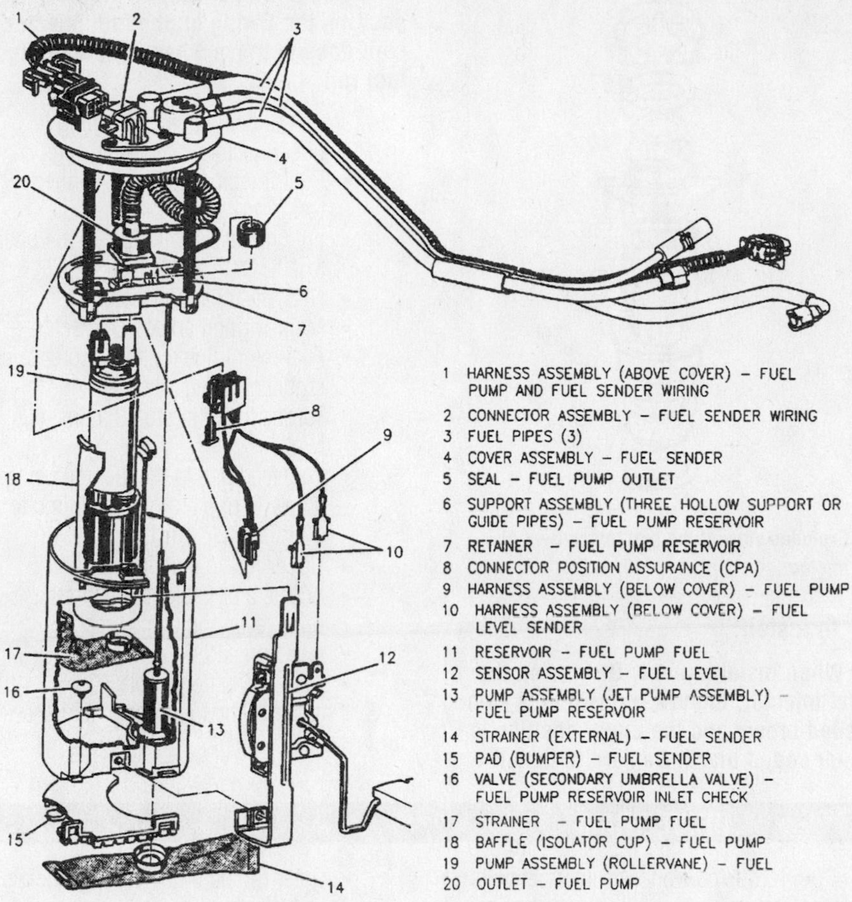

Exploded view of the fuel pump assembly

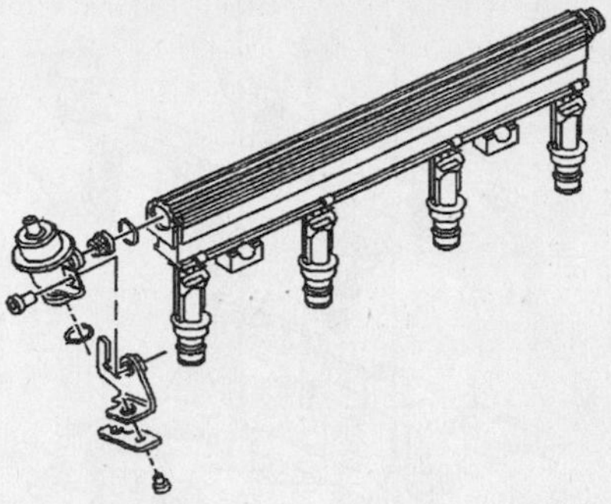

Exploded view of the fuel rail assembly—2.4L engine

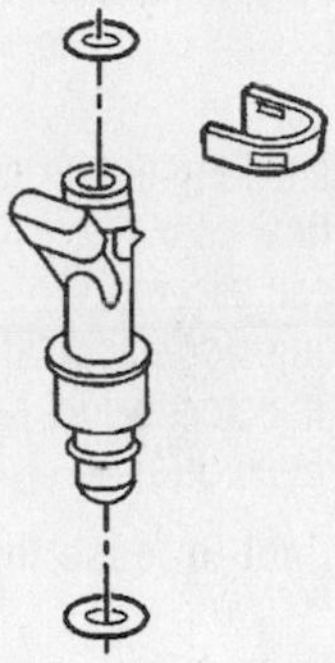

Exploded view of the fuel injector—2.4L engine

- Fuel pump electrical connector
- Connector retainer to the wiring harness
- External fuel strainer
- Modular fuel sender assembly
- Modular fuel sender assembly-to-fuel tank snapring by pressing the assembly downward on 1998–99 models or lock the cam ring using Tool J-39765 on 2000–01 models
- Fuel tank
- Negative battery cable

8. Refill the fuel tank.
9. Pressurize the fuel system and verify no leaks.

Fuel Injector

REMOVAL & INSTALLATION

2.4L Engine

1. Before servicing the vehicle, refer to the precautions in the beginning of this section.
2. Relieve the fuel system pressure.
3. Remove or disconnect the following:
   - Air cleaner assembly
   - Fuel pressure regulator vacuum line
   - Camshaft Position (CMP) sensor electrical connector
   - Fuel injector electrical connectors
   - Fuel line from the fuel pressure regulator
   - Fuel rail
   - Fuel pressure regulator screw
   - Fuel pressure regulator from the fuel rail twisting it back and forth
   - Fuel injector-to-fuel rail retaining clip
   - Fuel injector and discard the O-rings

To install:

4. Install or connect the following:
   - Fuel injector(s) with new O-rings
   - Fuel injector retaining clip(s)
   - New fuel pressure regulator O-ring, lubricated with engine oil
   - Fuel pressure regulator to the fuel rail. Torque the screw to 97 inch lbs. (11 Nm).
   - Fuel rail
   - Fuel line to the fuel pressure regulator
   - Fuel injector electrical connectors
   - Camshaft position sensor
   - Fuel pressure regulator vacuum line
   - Negative battery cable
5. Pressurize the fuel system and check for leaks.

3.1L Engine

1. Before servicing the vehicle, refer to the precautions in the beginning of this section.
2. Relieve the fuel system pressure.
3. Remove or disconnect the following:
   - Accelerator cable from the throttle body lever and cable bracket
   - Upper intake manifold
   - Fuel feed line from the fuel rail and discard the O-ring
   - Fuel return line from the fuel pressure regulator
   - Main wiring harness connectors located near the alternator

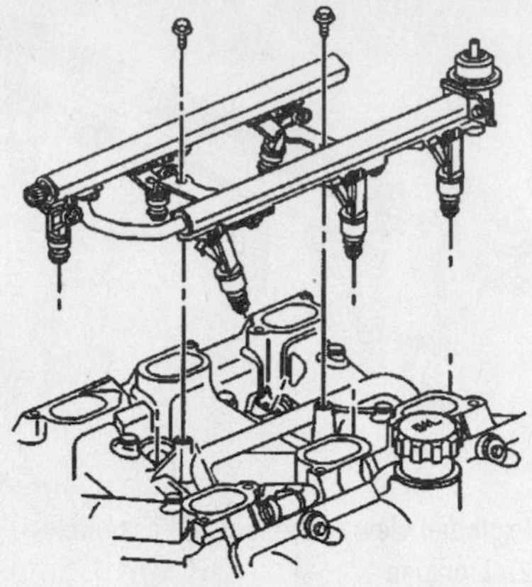

Exploded view of the fuel rail assembly—3.1L engine

- Coolant Temperature Sensor (CTS) electrical connector
- Fuel rail assembly
- Fuel injector electrical connectors
- Fuel injector-to-fuel rail clips
- Fuel injector(s)

➡Be careful not to loose the O-ring backups.

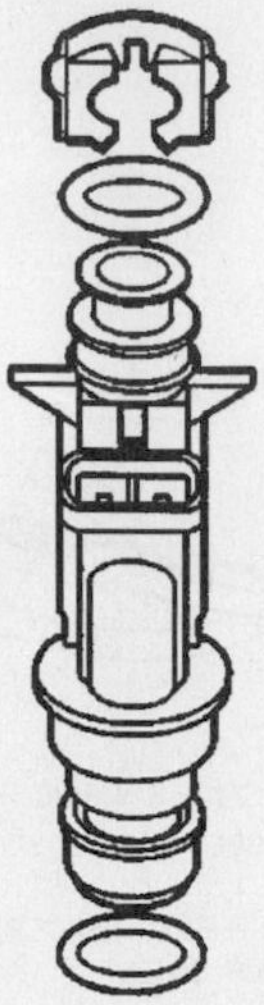

Exploded view of the fuel injector—3.1L engine

To install:

➡When installing new O-rings on the fuel injector, the lower position is color coded brown and the upper position is color coded black. Be sure to install the nylon O-ring backup to properly position the O-ring on the fuel injector so it doesn't move when installing the fuel rail.

4. Install or connect the following:
   - Fuel injector with new O-rings
   - Fuel injector electrical connectors
   - Fuel rail assembly. Torque the bolts to 7 ft. lbs. (10 Nm).
   - CTS electrical connector
   - Main wiring harness connectors
   - Fuel return line to the fuel pressure regulator using a new O-ring. Torque the fitting to 13 ft. lbs. (17 Nm).
   - Fuel feed line to the fuel rail using a new O-ring. Torque the fitting to 13 ft. lbs. (17 Nm).
   - Upper intake manifold
   - Accelerator cable to the throttle body lever and cable bracket
   - Fuel cap and tighten it
   - Negative battery cable
5. Pressurize the fuel system and check for leaks.

DRIVE TRAIN

Transmission Assembly

REMOVAL & INSTALLATION

1. Before servicing the vehicle, refer to the precautions in the beginning of this section.
2. Install an Engine Support Fixture.
3. Drain the transmission.
4. Remove or disconnect the following:
   - Negative battery cable
   - Air cleaner assembly
   - Front transmission mount bolts
   - Front wheels
   - Left and right splash shields
   - Shift linkage from the transaxle
   - Wiring harness connection from the transaxle
   - Ground cables from the engine block
   - Park Neutral Position (PNP) switch electrical connector
   - Lower radiator and condenser support
   - Front transmission mount bracket
   - Anti-lock Brake System (ABS) wheel speed sensors and electrical harnesses from the suspension supports
   - Brake modulator assembly
   - Vehicle Speed Sensor (VSS)
   - Torque converter cover
   - Torque converter-to-flywheel bolts
   - Ball joints from the steering knuckles
   - Halfshafts from the transaxle
   - ABS module
   - Outer tie rod ends from the steering knuckles
   - Pressure line from the rack and pinion
   - Transmission fluid cooler lines
   - Brake hose bracket from the body
   - Intermediate shaft
   - Transmission-to-engine bolts
   - Transaxle from the engine

To install:

5. Apply a thin film of grease on the torque converter pilot hub.

⁂ WARNING

Be sure to properly seat the torque converter in the pump.

6. Install or connect the following:
   - Transaxle. Torque the transmission-to-engine bolts to 66 ft. lbs. (90 Nm).
   - Intermediate shaft. Torque the pinch bolt to 15 ft. lbs. (20 Nm).
   - Brake hose bracket to the body
   - Transaxle fluid cooler lines
   - Pressure line to the rack and pinion
   - Outer tie rod ends. Torque the nuts to 14 ft. lbs. (20 Nm) plus an additional 180 degree turn.
   - ABS module
   - Halfshafts to the transaxle
   - Ball joints. Torque the nuts to 48 ft. lbs. (65 Nm).
   - Torque converter. Torque the bolts to 46 ft. lbs. (62 Nm).
   - Torque converter cover
   - VSS electrical connector
   - Brake modulator assembly
   - Front ABS wheel speed sensors and harnesses to the suspension support
   - Front transmission mount bracket
   - Lower radiator and condenser support
   - PNP switch electrical connector
   - Ground cables to the engine block
   - Transmission electrical connections
   - Shift cable bracket. Torque the bolt to 18 ft. lbs. (24 Nm) and the nut to 37 ft. lbs. (50 Nm).
   - Air cleaner assembly
   - Left and right splash shields
   - Front wheels
7. Remove the engine support fixture.
8. Install or connect the following:
   - Shift linkage
   - Negative battery cable

9. Refill the transmission.
10. Apply the brakes and start the engine.
11. Shift the transaxle from **R** to **D** and back to **P**.
12. Recheck the fluid level.

Halfshaft

REMOVAL & INSTALLATION

1. Before servicing the vehicle, refer to the precautions in the beginning of this section.
2. Remove or disconnect the following:
   - Wheel
   - Tie rod from the steering knuckle
   - Halfshaft hub nut and washer
   - Stabilizer link
   - Lower ball joint from the steering knuckle
   - Halfshaft from the hub/bearing assembly using a puller
   - Halfshaft from the transaxle

To install:

3. Install the halfshaft into the transaxle by placing a tool in the joint housing groove and tapping until seated.

⁂ WARNING

Be careful not to damage the axle seal or dislodge the transaxle seal garter spring when installing the halfshaft.

4. Verify that the halfshaft is seated in the transaxle by grasping on the housing and pulling outward.
5. Install or connect the following:
   - Halfshaft into the hub/bearing assembly
   - Tie rod to the steering knuckle. Torque the nut to 33 ft. lbs. (45 Nm).
   - Lower ball joint to the steering knuckle. Torque the nut to 48 ft. lbs. (65 Nm).
   - New cotter pin to the lower ball joint
   - Stabilizer link. Torque the nut to 13 ft. lbs. (17 Nm).
   - New halfshaft nut. Torque it to 284 ft. lbs. (385 Nm).
   - Wheel

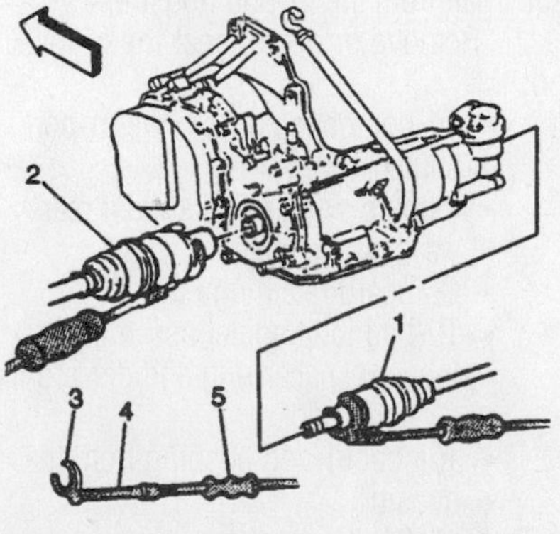

(1) Drive Axle, Right Side
(2) Drive Axle, Left Side
(3) J 28468 or J 33008
(4) J 29794
(5) J 2619-01

7922Z225

Removing the left and right halfshafts

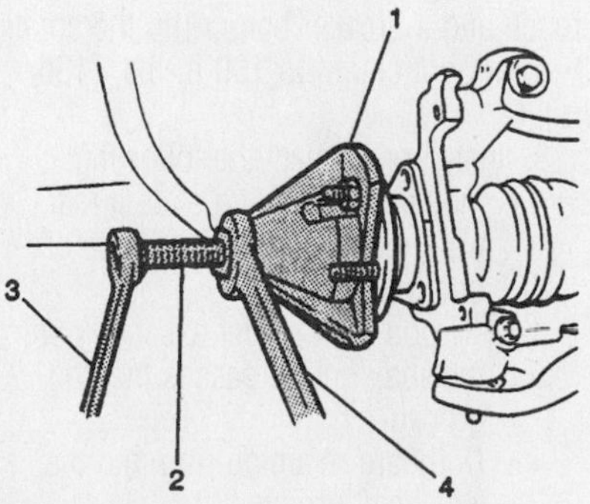

(1) J 28733A
(2) Forcing Screw
(3) Turn Box Wrench
(4) Hold Wrench

7922Z226

Removing the halfshaft from the hub utilizing the appropriate tools

CV-Joint

OVERHAUL

Outer CV-Joint

1. Before servicing the vehicle, refer to the precautions in the beginning of this section.
2. Remove or disconnect the following:
   - Front wheel
   - Halfshaft and position it in a vise
   - Large CV-joint boot clamp
   - Small CV-joint boot clamp
   - CV-joint boot and slide it back on the shaft
   - Outer race from the halfshaft by spreading the outer race-to-halfshaft retaining ring
   - Retaining ring from the halfshaft
   - CV-joint boot from the halfshaft

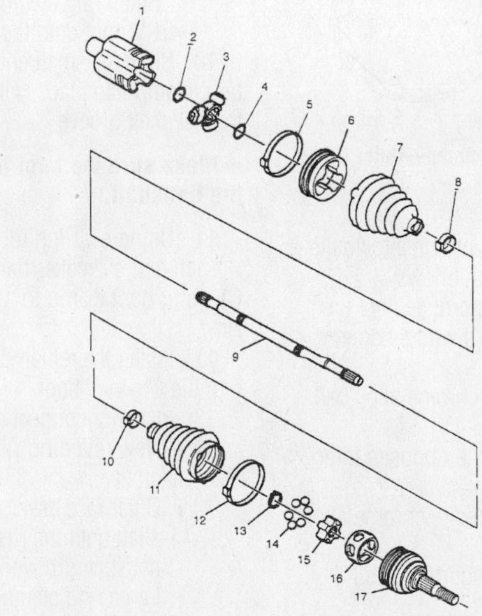

1 Retainer and Housing Assembly
2 Shaft Retaining Ring
3 Tripot Joint Spider Assembly
4 Spacer Ring
5 Seal Retaining Clamp
6 Tripot Trilobal Bushing
7 Drive Axle Inboard Seal
8 Seal Retaining Clamp
9 Axle Shaft
10 Seal Retaining Clamp
11 Drive Axle Outboard Seal
12 Seal Retaining Clamp
13 Race Retaining Ring
14 Chrome Alloy Ball
15 CV Joint Inner Race
16 CV Joint Cage
17 CV Joint Outer Race

9306ZG01

Exploded view of the halfshaft assembly

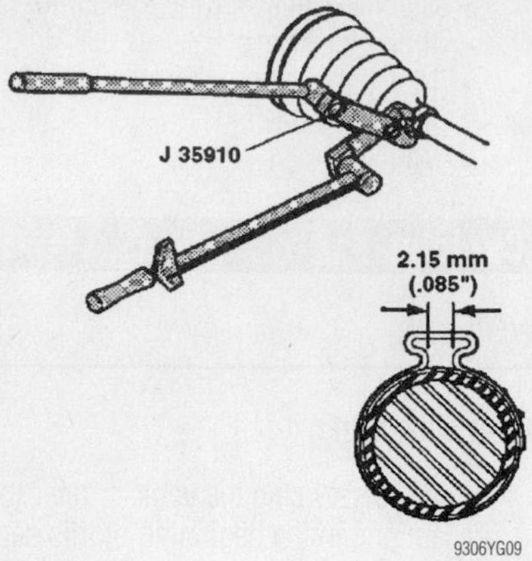

Crimping the small boot clamp—Outer CV-joint

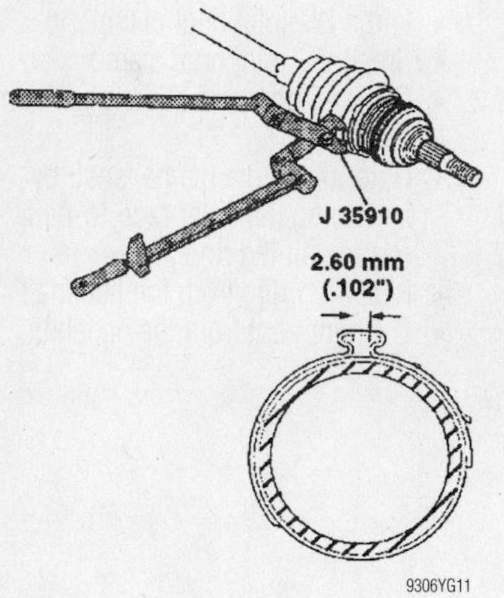

Crimping the large boot clamp—Outer CV-joint

3. Disassemble the chrome alloy balls from the CV-joint cage as follows:
   a. Position a brass drift against the CV-joint cage and tap it with a hammer to tilt the cage.
   b. Remove the 1st chrome alloy ball from the cage.
   c. Tilt the cage in the opposite direction.
   d. Remove the opposite chrome alloy ball.
   e. Repeat the procedure until all 6 balls are removed.
4. Disassemble the CV-joint cage and inner race as follows:
   a. Pivot the cage and race 90 degrees to the center line of the outer race.
   b. Align the cage windows with outer race lands.
   c. Remove the cage from the outer race.
   d. Rotate the inner race upward and remove it from the cage.

To install:

5. Lubricate the parts with a light coat of grease.
6. Assemble the CV-joint cage and inner race, as follows:
   a. Rotate the inner race 90 degrees to the cage centerline.
   b. Align the cage windows with inner race lands.
   c. Insert the inner race into the cage by rotating the inner race downward.
   d. Insert the cage/inner race into the outer race.
7. Assemble the chrome alloy balls into the CV-joint cage, as follows:
   a. Position a brass drift against the CV-joint cage and tap it with a hammer to tilt the cage.
   b. Insert the 1st chrome alloy ball into the cage.
   c. Tilt the cage in the opposite direction.
   d. Insert the opposite chrome alloy ball.
   e. Repeat the procedure until all 6 balls are inserted.
8. Install ½ of the grease provided into the CV-joint.
9. Install or connect the following:
   - Small ring clamp on the CV boot
   - CV boot onto the halfshaft
10. Slide the small end of the CV-joint boot/clamp into place, with the seal lip in the halfshaft groove.

➡Make sure the boot lies flat against the halfshaft.

11. Using a Crimp tool, a torque wrench and a breaker bar, crimp the small CV-joint boot clamp to 100 ft. lbs. (136 Nm).
12. Install the remainder of the grease into the CV-joint boot.
13. Install or connect the following:
    - New retaining ring on the halfshaft
    - Outer race assembly onto the halfshaft until the ring engages the halfshaft groove
    - Large ring clamp on the CV boot
14. Using a Crimp tool, a torque wrench and a breaker bar, crimp the large CV-joint boot clamp to 130 ft. lbs. (176 Nm).
15. Install the halfshaft and the front wheel.

Inner (Tri-Pod) Joint

1. Before servicing the vehicle, refer to the precautions in the beginning of this section.

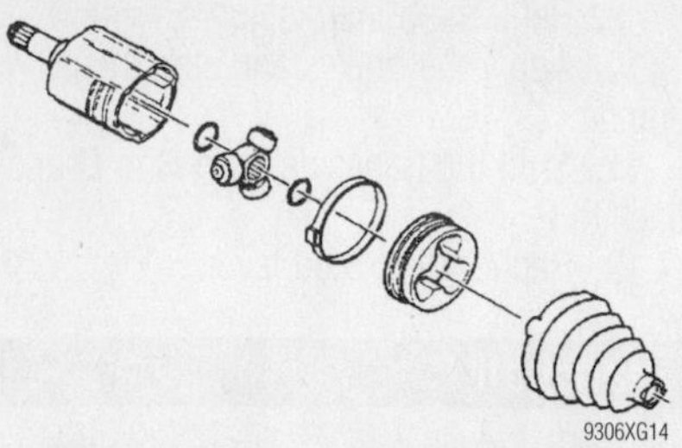

Exploded view of the inner (tri-pod) joint

2. Remove or disconnect the following:
   - Front wheel
   - Halfshaft and place it in a vise
   - Snapring from the stub shaft
   - Small CV-joint boot clamp
   - Large CV-joint boot clamp
   - CV-joint boot by sliding it away from the tri-pod joint
3. Install a Stub Shaft Removal Tool J-38868-A to the stub shaft snapring groove.
4. Using a slide hammer puller, pull the stub shaft from the tri-pod housing.
5. Remove or disconnect the following:
   - Tri-pod housing from the tri-pod spider
   - Inboard spacer ring slide it rearward on the shaft
   - Outboard retaining ring
   - Tri-pod joint spider assembly
   - Inboard spacer ring and discard it
   - CV-joint boot
   - Trilobal tri-pod bushing from the housing

To install:

6. Install or connect the following:
   - New snapring onto the stub shaft
   - Small boot clamp
   - CV-joint boot
7. Using a Crimp tool, a torque wrench and a breaker bar, crimp the small CV-joint boot clamp to 100 ft. lbs. (136 Nm).
8. Install or connect the following:
   - Inboard spacer ring slide it rearward on the shaft beyond the second groove
   - Tri-pod joint spider assembly onto the shaft until it passes the 2nd groove
   - Outboard retaining ring into the axle shaft groove
   - Tri-pod joint spider assembly, slide it against the outboard retaining ring
   - Inboard spacer ring, seat it in the groove
   - ½ of the grease provided into the boot

- The remaining grease into the tri-pod housing
- Trilobal tri-pod bushing flush with the tri-pod housing face
- New large seal clamp onto the CV-joint boot
- Tri-pod housing, slide it over the tri-pod joint spider assembly
- CV-joint boot/clamp, slide it into place, over the trilobal tri-pod bushing with the seal lip in the groove

➡**Make sure the boot lies flat against the trilobal bushing.**

9. Using a Crimp tool, a torque wrench and a breaker bar, crimp the large CV-joint boot clamp to 130 ft. lbs. (176 Nm).
10. Check the clamp gap dimension; if it is not 0.085 in. (2.16mm), continue tightening the clamp until it is.
11. Install the halfshaft and the front wheel.

STEERING AND SUSPENSION

Air Bag

⁂ CAUTION

Some vehicles are equipped with an air bag system, also known as the Supplemental Inflatable Restraint (SIR) or Supplemental Restraint System (SRS). The system must be disabled before performing service on or around system components, steering column, instrument panel components, wiring and sensors. Failure to follow safety and disabling procedures could result in accidental air bag deployment, possible personal injury and unnecessary system repairs.

PRECAUTIONS

Several precautions must be observed when handling the inflator module to avoid accidental deployment and possible personal injury.

- Never carry the inflator module by the wires or connector on the underside of the module.
- When carrying a live inflator module, hold securely with both hands, and ensure that the bag and trim cover are pointed away.
- Place the inflator module on a bench or other surface with the bag and trim cover facing up.
- With the inflator module on the bench, never place anything on or close to the module, which may be thrown in the event of an accidental deployment.

DISARMING

⁂ CAUTION

The Supplemental Restraint System (SRS) must be disarmed before performing service procedures around the air bag or SRS wiring. Failure to do so may cause accidental deployment of the air bag, resulting in unnecessary SRS repairs and/or personal injury.

1. Disconnect the negative battery cable.
2. Turn the steering wheel so the vehicle's wheels are pointing straight-ahead.
3. Turn the ignition switch to the **LOCK** position.
4. Remove or disconnect the following:
   - Key
   - **AIR BAG** fuse from the fuse block
   - Left sound insulator
   - Connector retainer clip from the yellow 2-way connector at the base of the steering column
   - Connector retainer and yellow 2-way connector from the passenger air bag lead, if equipped with a passenger's side air bag

ARMING

1. Turn the ignition switch to the **LOCK** position and remove the key.
2. Install or connect the following:
   - Yellow 2-way connector at the base of steering column and the connector retainer clip
   - Yellow 2-way connector at the passenger air bag lead and secure it with the connector retainer clip, if equipped with a passenger's side air bag
   - Left sound insulator
   - **AIR BAG** fuse in the fuse block
3. Turn the ignition switch to the **RUN** position and verify that the **AIR BAG** warning lamp flashes 7 times, then turns **OFF**.
4. Connect the negative battery cable.

Power Rack and Pinion Steering Gear

REMOVAL & INSTALLATION

1. Before servicing the vehicle, refer to the precautions in the beginning of this section.
2. Disconnect the negative battery cable.
3. Siphon the power steering fluid from the reservoir.
4. Remove or disconnect the following:
   - Left front wheel
   - Intermediate shaft assembly
   - Tie rod ends from the steering knuckles
   - Stabilizer shaft links
   - Steering gear mounting bolts
5. Support the rear of the support frame.
6. Remove or disconnect the following:
   - Transaxle-to-crossmember mounting bolt
   - Rear crossmember-to-body bolts
   - Front crossmember bolts, loosen them
7. Lower the support frame for access.
8. Remove or disconnect the following:
   - Power steering hoses from the steering gear
   - Steering gear through the left wheel opening

To install:

9. Install or connect the following:
   - Steering gear through the left wheel opening
   - Power steering hoses to the steering gear
10. Raise the support frame.
11. Install or connect the following:
    - Front crossmember bolts. Torque them to 71 ft. lbs. (110 Nm) plus an additional 90 degree turn.
    - Rear crossmember bolts. Torque them to 71 ft. lbs. (110 Nm) plus an additional 90 degree turn.
    - Transaxle mount-to-crossmember bolt. Torque it to 49 ft. lbs. (66 Nm).
    - Steering gear bolts. Torque them to 89 ft. lbs. (120 Nm).
    - Stabilizer shaft links. Torque the nuts to 13 ft. lbs. (17 Nm).
    - Tie rod ends to the steering knuckle. Torque the nut to 15 ft.

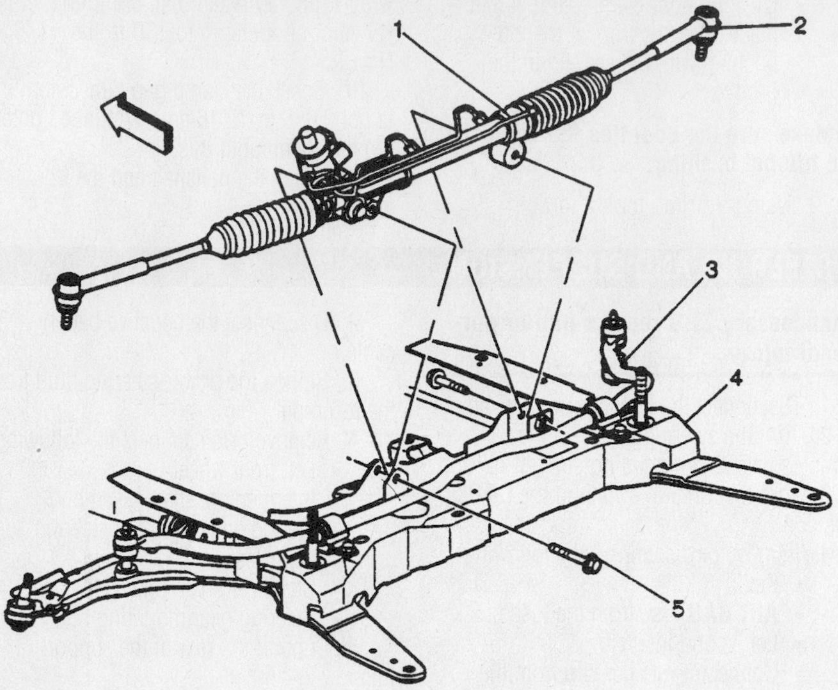

(1) Gear Assembly, Power Steering
(2) Tie Rod
(3) Shaft, Front Stabilizer
(4) Frame Assembly, Drivetrain and Front Crossmember
(5) Bolt, Power Steering Gear Assembly

79222227

Exploded view of the power rack and pinion steering gear mounting

- lbs. (20 Nm) plus an additional 180 degree turn.
- Intermediate shaft. Torque the pinch bolt to 16 ft. lbs. (22 Nm).
- Left front wheel
- Negative battery cable

12. Refill and bleed the power steering system.
13. Check and/or adjust the front end alignment.

Strut

REMOVAL & INSTALLATION

Front

1. Before servicing the vehicle, refer to the precautions in the beginning of this section.
2. Support the front crossmember.
3. Matchmark the front strut-to-steering knuckle location.
4. Remove or disconnect the following:
   - Front wheel
   - Brake line bracket from the strut
   - Strut lower mounting bracket nut and through-bolts
   - Upper strut-to-chassis nuts
   - Strut

To install:

5. Install or connect the following:
   - Strut and finger-tighten the upper strut-to-chassis nuts
   - Strut to the steering knuckle. Torque the through-bolts/nuts to 133 ft. lbs. (180 Nm) with the reference marks aligned.
   - Brake line bracket to the strut. Torque the bolt to 10 ft. lbs. (14 Nm).
   - Wheel
6. Torque the upper strut fasteners to 18 ft. lbs. (24 Nm).
7. Check and/or adjust the front end alignment.

Rear

1. Before servicing the vehicle, refer to the precautions in the beginning of this section.
2. Matchmark the strut-to-knuckle position.
3. Remove or disconnect the following:
   - Rear wheels
   - Upper strut-to-chassis nuts
   - Strut-to-knuckle bolts
   - Strut

To install:

4. Install or connect the following:

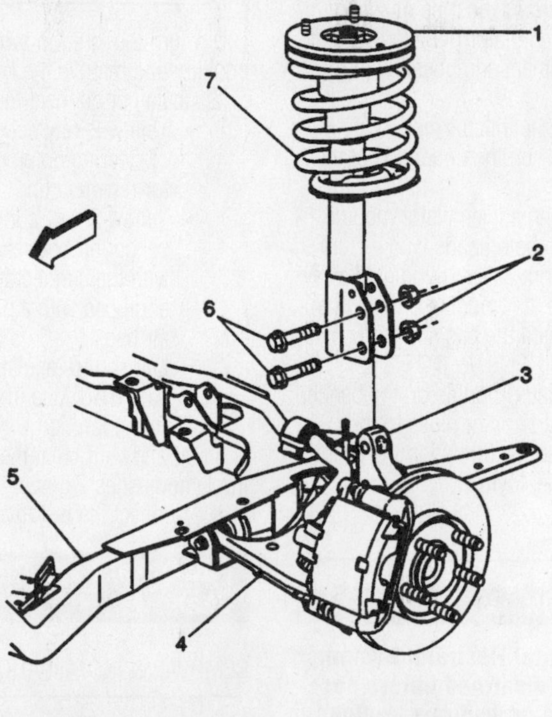

(1) Bearing, Strut Mount
(2) Nuts, Strut Mount
(3) Knuckle
(4) Arm, Control
(5) Crossmember
(6) Bolts, Strut Mount
(7) Spring, Strut Coil

79222228

Exploded view and component identification of the front strut

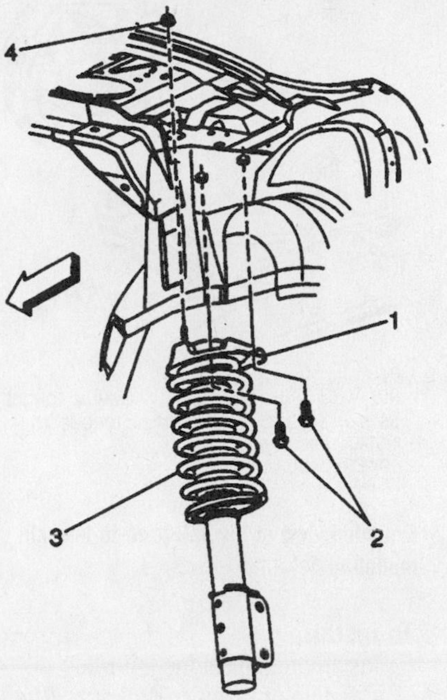

(1) Mount, Strut Upper (3) Spring, Coil
(2) Bolt, Strut Mount (4) Nut

7922Z229

Exploded view of the upper strut mounting

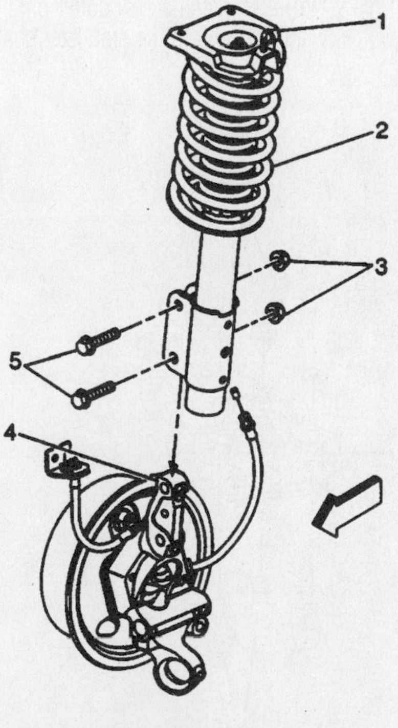

(1) Mount, Strut Upper
(2) Spring, Coil
(3) Nuts, Strut Bolt
(4) Knuckle
(5) Bolt, Strut Mount

7922Z230

Exploded view of the lower strut mounting

- Strut and finger-tighten the strut-to-knuckle bolts
- Upper strut-to-chassis nuts. Torque the nuts to 18 ft. lbs. (24 Nm).

➡Align the matchmarks to ensure proper alignment.

- Strut-to-knuckle bolts. Torque them to 89 ft. lbs. (120 Nm).
- Rear wheels

5. Check and/or adjust the alignment.

Coil Spring

REMOVAL & INSTALLATION

Front and Rear

1. Before servicing the vehicle, refer to the precautions in the beginning of this section.
2. Remove the strut.
3. Mount the strut in a strut compressor.

➡The strut compressor has strut mounting holes drilled for specific vehicle lines.

4. Compress the strut approximately ½ of its height after initial contact with the top cap.

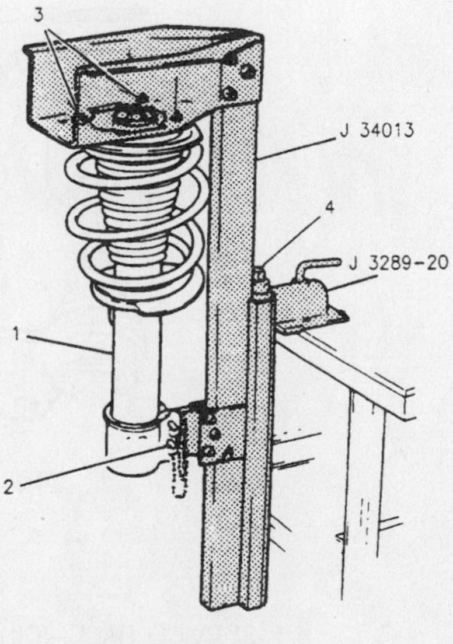

1 STRUT ASSEMBLY
2 INSTALL LOCKING PINS THROUGH STRUT ASSEMBLY
3 TIGHTEN NUTS UNTIL FLUSH WITH STRUT COMPRESSOR
4 COMPRESSOR FORCING SCREW

7922Z231

View of the strut mounted in the compressor

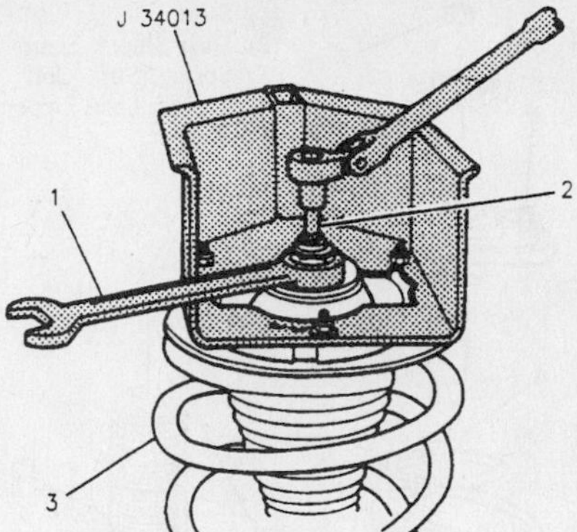

1 WRENCH
2 SOCKET
3 STRUT ASSEMBLY

7922Z232

Use a socket and a wrench to remove the damper shaft nut spring cap while compressing the spring

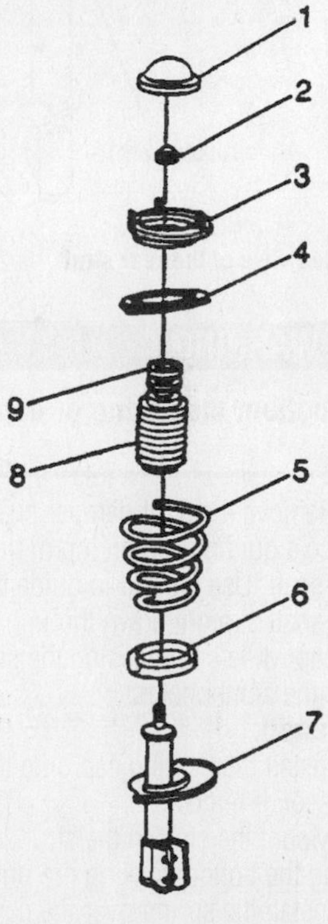

(1) Cap
(2) Nut, Strut Mount
(3) Bearing, Strut Mount
(4) Insulator, Spring (Upper)
(5) Spring, Strut Coil
(6) Insulator, Spring (Lower)
(7) Strut
(8) Bumper, Jounce
(9) Shield, Strut Dust

7922Z233

Exploded view of the front strut

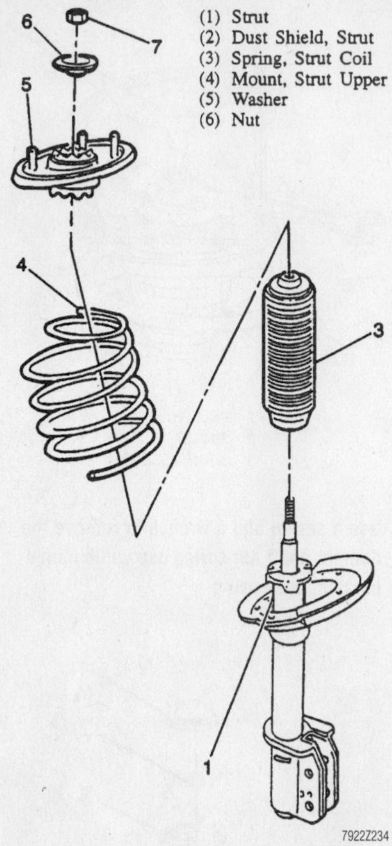

Exploded view of the rear strut

⁂ WARNING

Never bottom the spring or damper rod.

5. Remove the strut damper shaft nut and place a guiding rod on top of the damper shaft. Use the rod to guide the damper shaft straight down through the spring cap while compressing the spring. Remove the components.

To install:

6. Install the bearing cap onto the strut compressor, if removed.
7. Mount the strut in the strut compressor using the bottom locking pin only.
8. Install the spring over the damper and swing the assembly up so the upper locking pin can be installed.
9. Install all shields, bumpers and insulators on the spring seat. Install the spring seat on top of the spring. Be sure the flat on the upper spring seat is facing in the proper direction. The spring seat flat should be facing the same direction as the centerline of the strut assembly spindle.
10. Install the guiding rod and turn the forcing screw while the guiding rod centers the assembly. When the threads on the damper shaft are visible, remove the guiding rod and install the nut. Torque the nut to 34 ft. lbs. (47 Nm).
11. Remove the assembly from the compressor.
12. Install the strut on the vehicle.

Lower Ball Joint

REMOVAL & INSTALLATION

1. Before servicing the vehicle, refer to the precautions in the beginning of this section.
2. Remove or disconnect the following:
   - Wheel

➡Care must be exercised to prevent the halfshaft joints from being over-extended.

   - Wiring harness from the control arm
   - Stabilizer shaft link
3. Separate the ball joint from the steering knuckle.
4. Drill out the 3 ball joint-to-lower control arm rivets. Use an ⅛ in. (3mm) drill bit to make a pilot hole through the rivets; then, finish drilling the rivets with a ½ in. (13mm) drill bit.
5. Remove the ball joint from the control arm.

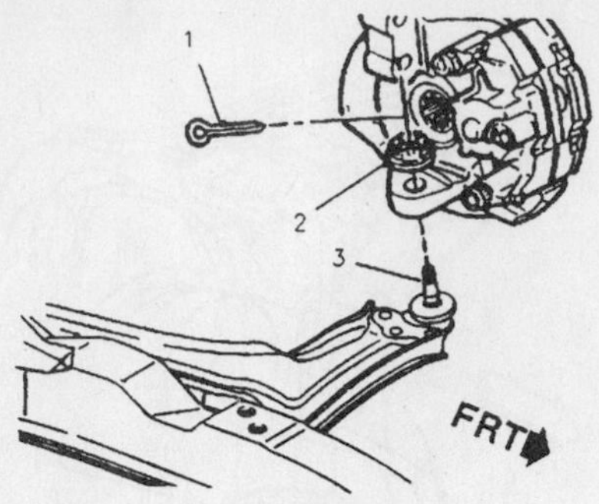

Exploded view of the ball joint-to-knuckle mounting

To install:

6. Install or connect the following:
   - Ball joint to the control arm. Torque the 3 new nuts/bolts (supplied with new ball joint) to the specifications included with the kit.
   - Ball joint to the steering knuckle. Torque the nut to 48 ft. lbs. (65 Nm) and install a new cotter pin.
   - Stabilizer link to the stabilizer shaft.

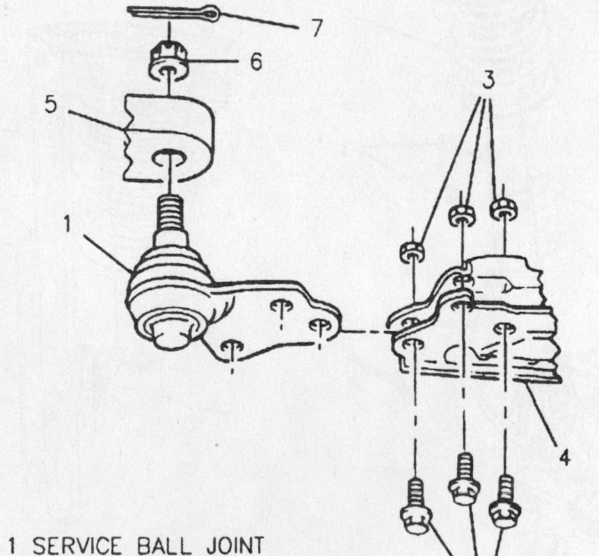

Exploded view of the ball joint mounting

Torque the nut to 13 ft. lbs. (17 Nm).
- Wiring harness to the control arm
- Wheel

7. Check and/or adjust the front wheel alignment.

Lower Control Arm

REMOVAL & INSTALLATION

1. Before servicing the vehicle, refer to the precautions in the beginning of this section.
2. Remove or disconnect the following:
   - Front wheel
   - Stabilizer shaft link
   - Lower ball joint from the steering knuckle
   - Lower control arm-to-suspension crossmember bolts
   - Lower control arm

To install:

3. Install or connect the following:
   - Lower control arm-to-suspension crossmember and hand-tighten the bolts
   - Lower ball joint to the steering knuckle. Torque the nut to 48 ft. lbs. (65 Nm).
   - Stabilizer shaft link. Torque the nut to 13 ft. lbs. (17 Nm).
   - Front wheel
4. Position the vehicle at curb height.
5. On 1998–99 models, torque the bolts as follows:
   a. Front lower control arm-to-suspension crossmember bolt to 79 ft. lbs. (107 Nm).
   b. Rear lower control arm-to-suspension crossmember bolt to 74 ft. lbs. (100 Nm).
6. On 2000–01 models, torque the bolts as follows:
   a. Front lower control arm-to-suspension crossmember bolt to 84 ft. lbs. (115 Nm) plus an additional 120 degrees.
   b. Rear lower control arm-to-suspension crossmember bolt to 180 ft. lbs. (245 Nm) plus an additional 180 degrees.
7. Check and/or adjust the front alignment.

CONTROL ARM BUSHING REPLACEMENT

Front Bushing

1. Before servicing the vehicle, refer to the precautions in the beginning of this section.
2. Remove the lower control arm and place it in a vise.
3. Lubricate the threads of Screw J-21474-19 with high pressure lubricant.
4. Assemble Tools Screw J-21474-19, Remover/Installer J-41397-1A, Receiver J-41397-2A and J-21474-18 onto the front control arm bushing.
5. Tighten Tool J-21474-18 until the front bushing is pressed from the control arm.
6. Disassemble the tools.

To install:

7. Lubricate the new front bushing outer casing.
8. Insert the new bushing into the control arm.
9. Assemble Tools Screw J-21474-19, Remover/Installer J-41397-1A, Receiver J-41397-2A and J-21474-18 onto the front control arm bushing.
10. Tighten Screw J-21474-19 until the front bushing is pressed into the control arm.
11. Disassemble the tools.
12. Install the lower control arm.

Rear Bushing

1. Before servicing the vehicle, refer to the precautions in the beginning of this section.
2. Remove the lower control arm and place it in a vise.
3. Assemble Tools Screw J-21474-27, Remover/Installer J-41211-1, Receiver J-

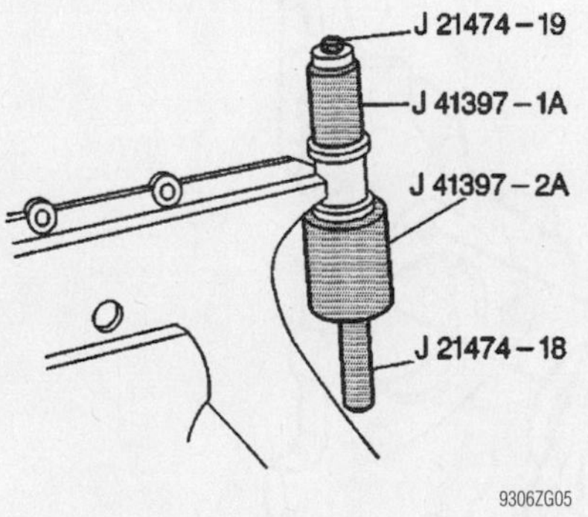

Removing the front bushing from the lower control arm

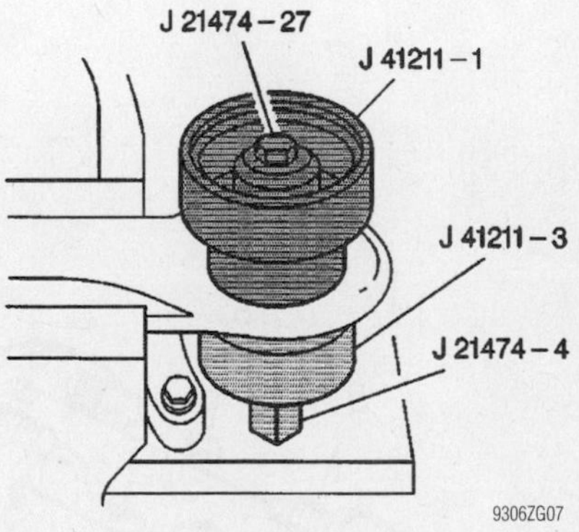

Removing the rear bushing from the lower control arm

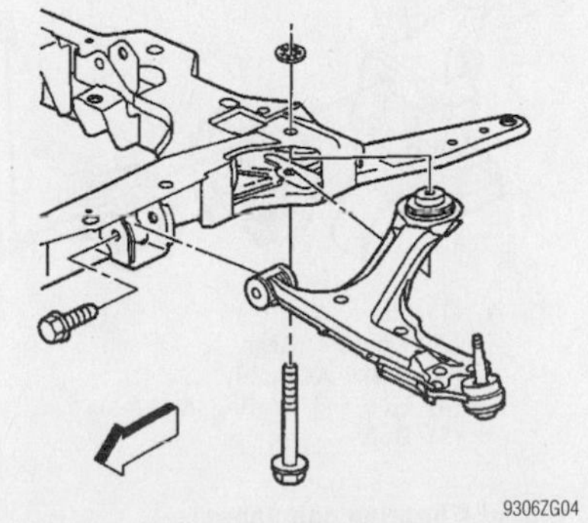

Exploded view of the lower control arm and related components

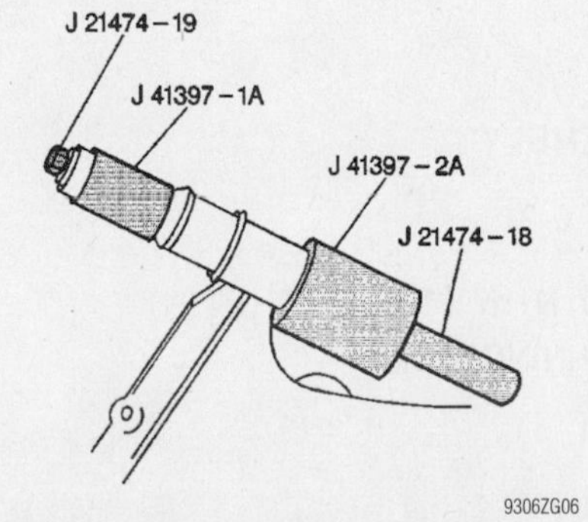

Installing the front bushing to the lower control arm

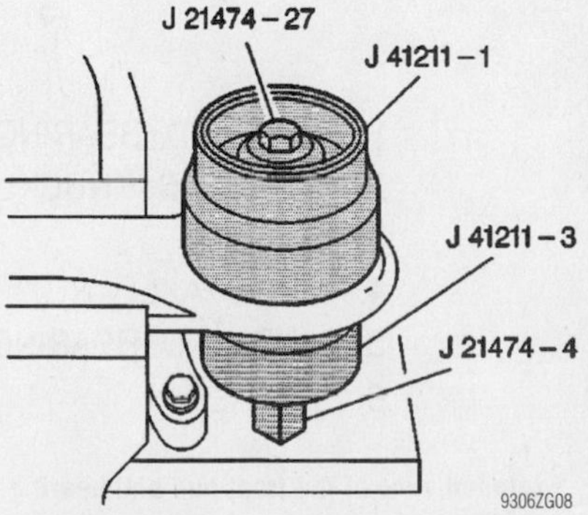

Installing the rear bushing to the lower control arm

41211-3 and J-21474-4 onto the rear control arm bushing.

4. Tighten Tool J-21474-27 until the rear bushing is pressed from the control arm.
5. Disassemble the tools.

To install:

6. Insert the new bushing into the control arm.
7. Assemble Tools Screw J-21474-27, Remover/Installer J-41211-1, Receiver J-41211-3 and J-21474-4 onto the rear control arm bushing.
8. Tighten Screw J-21474-4 until the rear bushing is pressed into the control arm.
9. Disassemble the tools.
10. Install the lower control arm.

Wheel Bearings

ADJUSTMENT

These vehicles are equipped with sealed hub and bearing assemblies. The hub and bearing assemblies are non-serviceable. If the assembly is damaged, the complete unit must be replaced.

REMOVAL & INSTALLATION

Front

1. Before servicing the vehicle, refer to the precautions in the beginning of this section.
2. Remove or disconnect the following:
   - Front wheel
   - Halfshaft nut
   - Caliper from the steering knuckle and support it aside

✻✻ WARNING

DO NOT allow the caliper to hang unsupported from the brake hose.

   - Brake rotor
   - 3 hub bearing-to-steering knuckle bolts
   - Halfshaft from the hub/bearing assembly
   - Hub/bearing assembly

To install:

3. Install or connect the following:
   - Hub/bearing assembly on the halfshaft
   - 3 hub/bearing-to-steering knuckle bolts. Torque the bolts to 70 ft. lbs. (95 Nm).
   - Brake rotor
   - Caliper onto the steering knuckle. Torque the bolts to 38 ft. lbs. (51 Nm).
   - Halfshaft nut. Torque it to 284 ft. lbs. (385 Nm).
   - Front wheel. Torque the nuts to 100 ft. lbs. (140 Nm).

Rear

1. Before servicing the vehicle, refer to the precautions in the beginning of this section.
2. Remove or disconnect the following:
   - Wheel
   - Brake drum
   - 4 hub/bearing assembly-to-knuckle nuts
   - Anti-lock Brake System (ABS) speed sensor wire from the hub/bearing assembly
   - Hub/bearing assembly

To install:

3. Install or connect the following:
   - Hub/bearing assembly
   - ABS wheel speed sensor
   - Hub/bearing-to-knuckle nuts. Torque the nuts to 70 ft. lbs. (95 Nm) on 1998–99 models or 62 ft. lbs. (85 Nm) on 2000–01 models.
   - Brake drum
   - Wheel. Torque the bolts to 100 ft. lbs. (140 Nm).

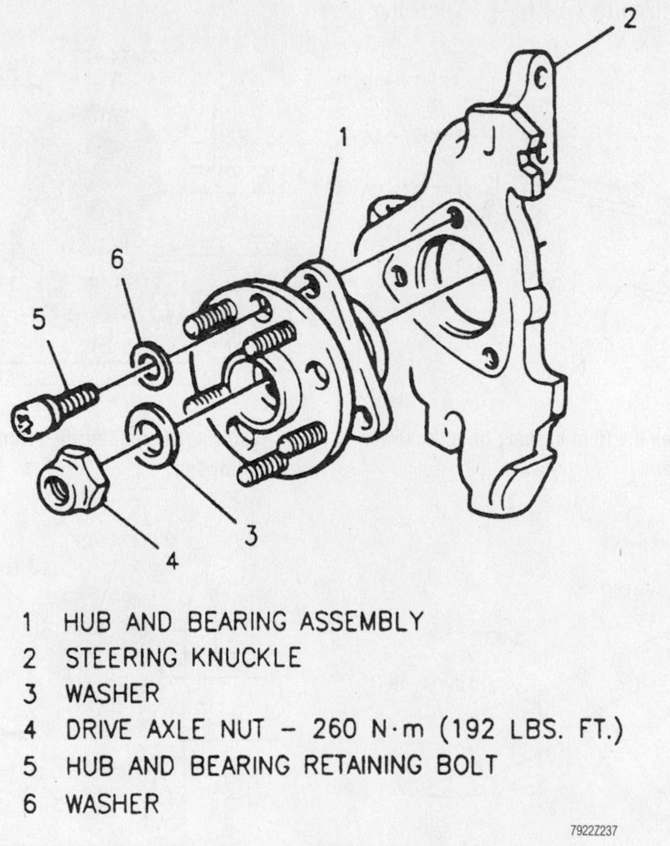

Exploded view of the front hub and bearing assembly

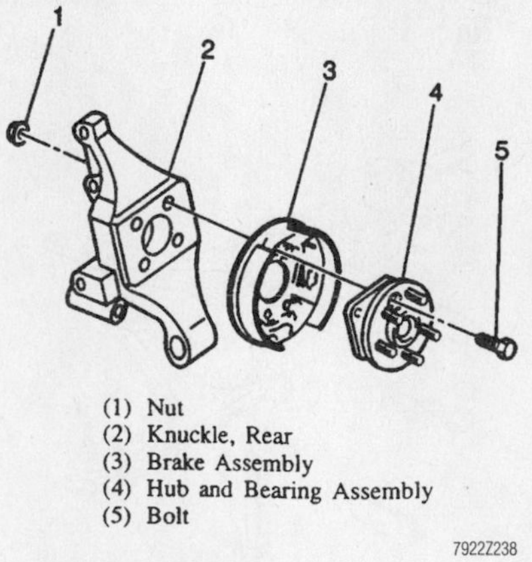

Hub and bearing components

GENERAL MOTORS CORPORATION—N-BODY

27

1998–01

Buick-Skylark • **Oldsmobile-**Achieva • Alero • **Pontiac-**Grand Am

PRECAUTIONS

Before servicing any vehicle, please be sure to read all of the following precautions, which deal with personal safety, prevention of component damage, and important points to take into consideration when servicing a motor vehicle:

- Never open, service or drain the radiator or cooling system when the engine is hot; serious burns can occur from the steam and hot coolant.
- Observe all applicable safety precautions when working around fuel. Whenever servicing the fuel system, always work in a well-ventilated area. Do not allow fuel spray or vapors to come in contact with a spark, open flame or excessive heat (a hot drop light, for example). Keep a dry chemical fire extinguisher near the work area. Always keep fuel in a container specifically designed for fuel storage; also, always properly seal fuel containers to avoid the possibility of fire or explosion. Refer to the additional fuel system precautions later in this section.
- Fuel injection systems often remain pressurized, even after the engine has been turned **OFF**. The fuel system pressure must be relieved before disconnecting any fuel lines. Failure to do so may result in fire and/or personal injury.
- Brake fluid often contains polyglycol ethers and polyglycols. Avoid contact with the eyes and wash your hands thoroughly after handling brake fluid. If you do get brake fluid in your eyes, flush your eyes with clean, running water for 15 minutes. If eye irritation persists, or if you have taken brake fluid internally, IMMEDIATELY seek medical assistance.
- The EPA warns that prolonged contact with used engine oil may cause a number of skin disorders, including cancer! You should make every effort to minimize your exposure to used engine oil. Protective gloves should be worn when changing oil. Wash your hands and any other exposed skin areas as soon as possible after exposure to used engine oil. Soap and water, or waterless hand cleaner should be used.
- All new vehicles are now equipped with an air bag system. The system must be disabled before performing service on or around system components, steering column, instrument panel components, wiring and sensors. Failure to follow safety and disabling procedures could result in accidental air bag deployment, possible personal injury and unnecessary system repairs.
- Always wear safety goggles when working with, or around, the air bag system. When carrying a non-deployed air bag, be sure the bag and trim cover are pointed away from your body. When placing a non-deployed air bag on a work surface, always face the bag and trim cover upward, away from the surface. This will reduce the motion of the module if it is accidentally deployed. Refer to the additional air bag system precautions later in this section.
- Clean, high quality brake fluid from a sealed container is essential to the safe and proper operation of the brake system. You should always buy the correct type of brake fluid for your vehicle. If the brake fluid becomes contaminated, completely flush the system with new fluid. Never reuse any brake fluid. Any brake fluid that is removed from the system should be discarded. Also, do not allow any brake fluid to come in contact with a painted surface; it will damage the paint.
- Never operate the engine without the proper amount and type of engine oil; doing so WILL result in severe engine damage.
- Timing belt maintenance is extremely important! Many models utilize an interference-type, non-freewheeling engine. If the timing belt breaks, the valves in the cylinder head may strike the pistons, causing potentially serious (also time-consuming and expensive) engine damage. Refer to the maintenance interval charts in the front of this manual for the recommended replacement interval for the timing belt, and to the timing belt section for belt replacement and inspection.
- Disconnecting the negative battery cable on some vehicles may interfere with the functions of the on-board computer system(s) and may require the computer to undergo a relearning process once the negative battery cable is reconnected.
- When servicing drum brakes, only disassemble and assemble one side at a time, leaving the remaining side intact for reference.

ENGINE REPAIR

Alternator

REMOVAL

2.4L Engine

1. Before servicing the vehicle, refer to the precautions in the beginning of this section.
2. Remove or disconnect the following:
   - Negative battery cable
   - Accessory drive belt
   - Alternator mounting bolts
   - Alternator electrical connectors
   - Alternator

3.1L Engine

1. Before servicing the vehicle, refer to the precautions in the beginning of this section.
2. Remove or disconnect the following:
   - Negative battery cable
   - Accessory drive belt
   - Alternator electrical connectors
   - Power steering-to-alternator line clip
   - Rear alternator brace
   - Alternator air intake connector
   - Alternator-to-bracket nut/bolt
   - Alternator

3.4L Engine

1. Before servicing the vehicle, refer to the precautions in the beginning of this section.
2. Remove or disconnect the following:
   - Negative battery cable
   - Accessory drive belt
   - Alternator electrical connectors
   - Power steering-to-alternator line clip
   - Alternator mounting nuts and bolts
   - Alternator

INSTALLATION

2.4L Engine

Install or connect the following:

- Alternator
- Alternator electrical connectors
- Alternator mounting bolts. Torque the bolts to 37 ft. lbs. (50 Nm).
- Accessory drive belt
- Negative battery cable

3.1L Engine

Install or connect the following:

- Alternator. Torque the nut to 37 ft. lbs. (50 Nm) and the bolt to 18 ft. lbs. (25 Nm).

- Alternator air intake connector
- Rear alternator brace. Torque the nut/bolt to 18 ft. lbs. (25 Nm).
- Power steering-to-alternator line clip
- Alternator electrical connectors
- Accessory drive belt
- Negative battery cable

3.4L Engine

Install or connect the following:

- Alternator. Torque the nuts to 22 ft. lbs. (30 Nm) and the bolts to 37 ft. lbs. (50 Nm).
- Power steering-to-alternator line clip
- Alternator electrical connectors
- Accessory drive belt
- Negative battery cable

Ignition Timing

ADJUSTMENT

The ignition timing is not adjustable, and is set electronically according to engine demand.

Engine Assembly

REMOVAL & INSTALLATION

2.4L Engine

1998 MODELS

The engine and transaxle are removed as a unit from under the vehicle.

1. Before servicing the vehicle, refer to the precautions in the beginning of this section.
2. Relieve the fuel system pressure.
3. Drain the cooling system.
4. Discharge and recover the refrigerant.
5. Remove or disconnect the following:
   - Left sound insulator
   - Clutch pushrod from the pedal assembly, manual transaxle only
   - Heater and upper radiator hoses from the thermostat housing
   - Air cleaner assembly
   - Cooling fan assembly
   - Compressor/condenser hose at the compressor
   - Vacuum hoses from the front of the engine
   - Alternator electrical connector
   - A/C compressor electrical connector
   - Fuel injector electrical connectors
   - Throttle Position (TPS) sensor electrical connector
   - Manifold Absolute Pressure (MAP) sensor electrical connector
   - Idle Air Control (IAC) valve electrical connector
   - Intake Air Temperature (IAT) sensor electrical connector
   - Evaporative Emissions (EVAP) control solenoid electrical connector
   - Starter electrical connector
   - Ground connections from the engine
   - Negative battery cable from the transmission
   - Ignition coil and module electrical connections
   - 2 Engine Coolant Temperature (ECT) sensors electrical connectors
   - Oil pressure sensor electrical connector
   - Oxygen (O_2S) sensor electrical connector
   - Crankshaft Position (CKP) sensor electrical connector
   - Backup lamp switch electrical connector
   - Camshaft Position (CMP) sensor electrical connector
   - Power brake booster vacuum lines
   - Throttle cables and the cable bracket
   - Power steering pump without disconnecting the lines
   - Fuel lines from the fuel rail
   - Shift cable and the Throttle Valve (TV) cable from the transmission (on automatic transaxle)
   - Exhaust manifold and heat shield
   - Lower radiator hose
   - Coolant recovery tank and move it aside with the hoses attached
6. Install an engine support fixture.
7. Remove or disconnect the following:
   - Engine mount
   - Front wheels
   - Right wheel well splash shield
   - Vehicle Speed Sensor (VSS) electrical connector
   - Knock sensor electrical connector
   - Anti-lock Brake System (ABS) Wheel Speed sensors (WSS) electrical connectors
   - Engine mount strut
   - Transmission mount
   - Lower ball joints from the steering knuckles
   - Stabilizer shaft
   - Heater outlet hose from the radiator outlet pipe
   - Halfshafts from the transaxle
   - Intermediate shaft, if equipped with a manual transaxle
8. Position an engine table under the powertrain and lower the vehicle until the powertrain is resting on the table.
9. Remove the suspension crossmember bolts.
10. Remove the engine support fixture.
11. Raise the vehicle until it is clear of the engine.
12. Separate the transaxle from the engine.

To install:

13. Install or connect the following:
    - Transaxle onto the engine. Torque the bolts to 55 ft. lbs. (75 Nm).
    - Powertrain assembly and the engine table under the vehicle
    - Suspension crossmember bolts. Torque the bolts to 84 ft. lbs. (115 Nm) plus an additional 120 degree turn.
    - Transaxle mounts. Torque the bolts to 55 ft. lbs. (75 Nm).
    - Halfshafts to the transaxle
    - Heater outlet hose to the radiator outlet pipe
    - Stabilizer shafts. Torque the nuts to 22 ft. lbs. (30 Nm).
    - Lower ball joints to the steering knuckles. Torque the nuts to 41 ft. lbs. (55 Nm).
    - Engine mount strut. Torque the bolts to 74 ft. lbs. (100 Nm) plus an additional 90 degree turn.
    - ABS and WSS electrical connections
    - Knock sensor electrical connector
    - VSS electrical connector
    - Front wheels and the right splash shield
    - Right engine mount. Torque the bolts to 96 ft. lbs. (130 Nm).
14. Remove the engine Support Fixture.
15. Install or connect the following:
    - Coolant recovery tank
    - Lower radiator hose
    - Exhaust manifold. Torque the nuts to 110 inch lbs. (13 Nm).
    - Exhaust manifold heat shield. Torque the bolts to 89 inch lbs. (10 Nm).
    - Shift cable to the transaxle and the TV cable, on automatic transaxle
    - Fuel lines to the fuel rail
    - Power steering pump. Torque the bolts to 19 ft. lbs. (26 Nm).

- Throttle cables and the cable bracket
- Power brake booster vacuum lines
- CMP sensor electrical connector
- Backup light switch electrical connector
- CKP electrical connector
- O_2S electrical connector
- Oil pressure sensor electrical connector
- 2 ECT sensor electrical connectors
- Negative battery cable to the transmission
- Ground connections to the engine
- Ignition coil and module electrical connectors
- Starter electrical connectors
- EVAP solenoid electrical connectors
- IAT sensor electrical connector
- MAP sensor electrical connector
- IAC solenoid electrical connector
- TPS electrical connector
- Fuel injector electrical connectors
- A/C compressor electrical connector
- Alternator electrical connectors
- Vacuum hoses to the front of the engine
- Compressor/condenser hose, using new O-rings
- Cooling fan assembly
- Air cleaner assembly
- Heater and upper radiator hoses to the thermostat housing
- Clutch pushrod to the pedal assembly, manual transaxle only
- Left sound insulator
- Negative battery cable

16. Refill the cooling system.
17. Refill the crankcase.
18. Recharge the air conditioning system.
19. Start the vehicle and verify no leaks.
20. Check and/or adjust the wheel alignment.

EXCEPT ALL 1998 AND 2000–01 MODELS WITH A MANUAL TRANSMISSION

1. Before servicing the vehicle, refer to the precautions in the beginning of this section.
2. Properly drain the cooling system.
3. Relieve the fuel system pressure.
4. Drain the engine oil.
5. Remove or disconnect the following:
   - Fuel rail assembly
   - Air intake duct and bracket
   - Ignition coil assembly
   - Camshaft Position (CMP) sensor
   - Power steering pump. DO NOT remove the lines.
   - Oil pressure sending switch
   - Cruise control assembly
6. Install an engine support fixture.
   - Engine mount assembly
   - Fuel pressure regulator vacuum line
7. Raise the engine with the engine support fixture J-hook.
   - Engine mount bracket
   - Accessory drive belt
   - Front wheels
   - Right front splash shield
   - Crankshaft balancer
   - Manifold Absolute Pressure (MAP) sensor electrical connector
   - Intake Air Temperature (IAT) sensor electrical connector
   - Exhaust Gas Recirculation (EGR) sensor electrical connector
   - Evaporator canister electrical connector
   - Alternator
   - Accelerator cable and bracket
   - Starter motor
   - Exhaust manifold heat shield
   - Exhaust pipe from the manifold
   - Engine intake coolant pipe
   - Torque converter cover
   - Torque converter bolts
   - Oxygen (O_2S) sensor
   - A/C compressor. DO NOT remove the hoses.
   - Oil pan-to-bell housing bolts
   - Transmission mount
   - Transmission bolts
8. Install an engine lifting device and a transmission support.
9. Raise the engine slightly and separate it from the transmission to remove it from the vehicle.

To install:

⁂ WARNING

Be sure the retaining bolts are in their correct locations. If not, engine damage may occur.

10. Install or connect the following:
    - Engine to the transaxle. Torque the bolts to 75 ft. lbs. (100 Nm).
    - Transmission mount. Torque the through bolt to 55 ft. lbs. (75 Nm).
    - Oil pan-to-bell housing bolts. Torque the bolts to 17 ft. lbs. (23 Nm).
    - A/C compressor. Torque the bolts to 37 ft. lbs. (50 Nm).
    - O_2S
    - Torque converter bolts. Torque the bolts to 46 ft. lbs. (62 Nm).
    - Torque converter cover. Torque the bolts to 115 inch lbs. (13 Nm).
    - Intake coolant pipe
    - Exhaust pipe to the manifold. Torque the bolts to 26 ft. lbs. (35 Nm).
    - Exhaust manifold heat shield. Torque the bolts to 124 inch lbs. (14 Nm).
    - Starter motor. Torque the bolts to 66 ft. lbs. (90 Nm).
    - Accelerator cable and bracket
    - Alternator
    - Fuel pressure regulator vacuum line
    - Evaporator canister electrical connector
    - EGR sensor electrical connector
    - IAT sensor electrical connector
    - MAP sensor electrical connector
    - Crankshaft balancer. Torque the bolt to 129 ft. lbs. (175 Nm).
    - Right front splash shield
    - Front wheels
    - Accessory drive belt
    - Engine mount bracket. Torque the bolts to 96 ft. lbs. (130 Nm).
    - Engine mount assembly. Torque the bolts to 49 ft. lbs. (66 Nm).
    - Cruise control assembly
    - Oil pressure sending switch
    - Power steering pump. Torque the bolts to 19 ft. lbs. (26 Nm).
    - CMP sensor
    - Ignition coil assembly
    - Air intake assembly
    - Fuel rail assembly
11. Refill the cooling system and engine oil.
12. Start the engine and check for proper operation.

2000–01 MODELS WITH MANUAL TRANSMISSION

1. Before servicing the vehicle, refer to the precautions in the beginning of this section.
2. Properly drain the cooling system.
3. Relieve the fuel system pressure.
4. Drain the engine oil.
5. Discharge and recover the refrigerant.
6. Remove or disconnect the following:
   - Air cleaner duct
   - Air cleaner
   - Underhood fuse block
   - Air cleaner bracket
   - Shift cables at the shift control
   - Shift cables from the bracket and remove the bracket
   - Back up lamp switch electrical connection
   - Vehicle Speed Sensor (VSS) electrical connection

- Engine controls harness
- Vacuum lines
- Fuel line quick disconnect fittings
- Radiator hoses
- Heater hoses from the core
- Slave cylinder hydraulic line
- Evaporative (EVAP) solenoid
- Ground cables at the rear of the engine block
- Power steering pump with lines attached and position the pump aside
- Electrical connector from the A/C compressor and the Crankshaft Position (CKP) sensor and position the harness aside
- Starter with the wires attached and position it aside
- Surge tank bypass hose from the engine
- Cruise and accelerator cables from the throttle body
- Cruise control module

7. Tie the radiator to the hood latch panel with mechanics wire
8. Raise and support the vehicle. Safety strap the front of the vehicle to the hoist.
   - Oil filter
   - Front splash shields
   - Front closeout panel
   - Lower radiator support
   - Right front brake hose
   - Retaining nut from the Brake Pressure Modulator Valve (BPMV) at the mounting bracket
   - Wheel Speed Sensors (WSS)
   - WSS harnesses from the control arm retainers and the frame retainers and position them aside
   - Driveshafts
   - Ball joints from the control arms
   - Outer tie rod ends from the control arms
   - Catalytic converter from the exhaust manifold
   - A/C compressor hose from the A/C compressor
   - Bolt from the power steering pressure line retainer
9. Lower the vehicle until the front suspension crossmember rests on the support table. Position a three inch block of wood between the front of the oil pan and the crossmember.
   - Front engine mount-to-bracket bolts
   - Front suspension crossmember retaining bolts
10. Carefully raise the vehicle off of the engine/transaxle assembly. Install the engine hoist to the engine/transmission assembly.
   - Front and rear transmission mount thru-bolts
   - 2 side transmission mount lower nuts
   - Engine/transaxle assembly off of the front suspension crossmember
   - Transaxle from the engine
   - Clutch drive plate and clutch driven plate
   - Flywheel
11. Mount the engine on a suitable engine stand.

To install:

12. Remove the engine from the engine stand
13. Install or connect the following:
   - Flywheel
   - Clutch driven plate and clutch drive plate
   - Transaxle to the engine
14. Lower the engine/transaxle assembly on to the front suspension crossmember.
   - 2 side transmission mount lower nuts and tighten the nuts to 60 ft. lbs. (44 Nm)
   - Front and rear transmission mount thru-bolts. Tighten the thru-bolts to 75 ft. lbs. (55 Nm).
15. Remove the engine hoist from the engine/transaxle assembly.
16. Carefully lower the vehicle on to the engine/transaxle assembly.
   - Front suspension crossmember retaining bolts
   - Front engine mount-to-bracket bolts
   - Power steering pressure line retainer bolt.
   - A/C compressor hose to the A/C compressor
   - Catalytic converter to the exhaust manifold
   - Outer tie rod ends to the control arms
   - Ball joints to the control arms
   - Driveshafts
   - WSS harnesses to the frame retainers and the control arm retainers
   - Wheel speed sensors
   - Retaining nut to the BPMV at the mounting bracket
   - Right front brake hose to the vehicle
   - Lower radiator support
   - Front closeout panel and the panel fasteners
   - Front splash shields
   - New oil filter
17. Remove the safety straps from the front of the vehicle and the hoist. Lower the vehicle. Untie the radiator from the hood latch panel.
   - Cruise control module
   - Cruise and accelerator cables to the throttle body
   - Surge tank bypass hose to the engine
   - Starter
   - Connectors to the A/C compressor and the CKP sensor
   - Power steering pump
   - Ground cables at the rear of the engine block
   - EVAP solenoid
   - Slave cylinder hydraulic line
   - Heater hoses to the heater core
   - Radiator hoses
   - Fuel line fittings
   - Vacuum lines
   - Engine controls harness
   - VSS
   - Back up lamp switch
   - Shift cable bracket and the shift cables to the bracket
   - Shift cables at the shift control
   - Air cleaner bracket
   - Underhood fuse block
   - Air cleaner
   - Air cleaner duct
   - Negative battery cable
18. Refill the cooling system.
19. Refill the crankcase.
20. Recharge the A/C system.
21. Start the vehicle and verify no leaks.
22. Check and/or adjust the wheel alignment.

3.1L Engines

Please note that the engine and transaxle are removed as an assembly from under the vehicle.

1. Before servicing the vehicle, refer to the precautions in the beginning of this section.
2. Relieve the fuel system pressure.
3. Drain the cooling system.
4. Drain the engine oil.
5. Remove or disconnect the following:
   - Air cleaner assembly
   - Upper and lower radiator hoses
   - Coolant inlet line from the coolant surge tank
   - Vacuum hoses from the Evaporative Emissions (EVAP) canister purge

valve, vacuum modulator and power brake booster
- Heater outlet hose from the water pump
- Accessory drive belt
- Control cables from the throttle body lever and intake manifold bracket
- Ignition coil assembly electrical connectors
- Heated Oxygen (HO_2S) sensor electrical connector
- Fuel injector electrical connectors
- Idle Air Control (IAC) valve electrical connector
- Throttle Position (TPS) sensor electrical connector
- Engine Coolant Temperature (ECT) sensor electrical connector
- Park/neutral position switch electrical connector
- Exhaust Gas Recirculation (EGR) valve electrical connector
- Ground wire from the transmission
- Alternator
- Power steering lines from the power steering pump
- Fuel lines from the fuel rail
- Cooling fan assembly
- Shift control cable from the transaxle shift lever and cable bracket
- Transaxle vent tube
- Vacuum hose from the vacuum reservoir

6. Install an engine support fixture.
7. Loosen, but do not remove the top 2 A/C compressor mounting bolts.
8. Remove or disconnect the following:
   - Front wheels
   - Right and left inner fender splash shields
   - Engine mount strut
   - Anti-lock Brake System (ABS) sensor wires from the Wheel Speed Sensors (WSS)
   - Lower ball joints from the steering knuckles
   - Suspension support assembly
   - Halfshafts from the transaxle
   - Oil filter and oil filter adapter
   - Flywheel cover
   - Starter motor
   - Heater hoses from the heater core
   - Knock sensor electrical connector
   - 2 Crankshaft Position (CKP) sensor electrical connectors
   - Oil level sensor electrical connector
   - Vehicle Speed Sensor (VSS) electrical connector
   - Ground wires from the transmission
   - A/C compressor and move it aside with the refrigerant lines connected
   - Vacuum reservoir tank
   - Exhaust pipe from the exhaust manifold
   - Engine mount strut bracket from the engine
   - Transaxle cooler lines from the radiator
   - Transaxle oil fill tube
9. Lower the vehicle until the powertrain assembly is resting on an engine table.
10. Remove or disconnect the following:
    - Transaxle mount-to-body bolts
    - Intermediate bracket from the right engine mount
    - Engine support fixture
11. Raise the vehicle leaving the powertrain assembly on the engine table.
12. Separate the engine from the transaxle.

To install:

13. Install the transaxle to the engine. Torque the bolts to 55 ft. lbs. (75 Nm).
14. Lower the vehicle over the powertrain assembly.
15. Install or connect the following:
    - Engine support fixture
    - Intermediate bracket to the right engine mount
    - Transaxle mount-to-body bolts. Torque the bolts to 49 ft. lbs. (66 Nm).
    - Transaxle oil fill tube
    - Transaxle cooler lines to the radiator
    - Engine mount strut bracket to the engine. Torque the bolts to 44 ft. lbs. (60 Nm).
    - Exhaust pipe to the exhaust manifold. Torque the bolts to 18 ft. lbs. (25 Nm).
    - Vacuum reservoir tank
    - A/C compressor. Torque the bolts to 37 ft. lbs. (50 Nm).
    - Ground wires to the transmission
    - VSS electrical connector
    - Oil level sensor electrical connector
    - CKP electrical connectors
    - Knock sensor electrical connector
    - Heater hoses to the heater core
    - Starter motor. Torque the bolts to 32 ft. lbs. (43 Nm).
    - Flywheel cover. Torque the bolts to 115 inch lbs. (13 Nm).
    - Oil filter and oil filter adapter
    - Halfshafts to the transaxle
    - Suspension support assembly. Torque the bolts to 89 ft. lbs. (120 Nm).
    - Lower ball joints to the steering knuckles, using new cotter pins. Torque the nuts to 41 ft. lbs. (55 Nm).
    - WSS electrical connectors
    - Engine mount strut. Torque the bolts to 44 ft. lbs. (60 Nm) plus an additional 90 degree turn.
    - Right and left inner fender splash shields
    - Front wheel
    - Top 2 A/C compressor mounting bolts. Torque the bolts to 37 ft. lbs. (50 Nm).
16. Remove the engine support fixture.
17. Install or connect the following:
    - Vacuum hose to the vacuum reservoir
    - Transaxle vent tube to the transaxle
    - Shift control cable to the transaxle shift lever and cable bracket
    - Cooling fan assembly
    - Fuel lines to the fuel rail
    - Power steering lines to the power steering pump
    - Alternator. Torque the nut to 37 ft. lbs. (50 Nm).
    - Ground wire to the transmission
    - EGR valve electrical connector
    - Park/neutral position switch electrical connector
    - ECT electrical connector
    - TPS electrical connector
    - IAC electrical connector
    - Fuel injector electrical connectors
    - HO_2S sensor electrical connector
    - Ignition coil assembly electrical connector
    - Control cables to the throttle body lever and intake manifold bracket
    - Accessory drive belt
    - Heater outlet hose to the water pump
    - Vacuum hoses to the EVAP canister purge valve, vacuum modulator and power brake booster
    - Coolant inlet line to the coolant surge tank
    - Upper and lower radiator hoses
    - Air cleaner assembly
    - Negative battery cable
18. Check and fill all the engine fluids as necessary.
19. Start the vehicle and bleed the power steering system.

3.4L Engine

1. Before servicing the vehicle, refer to the precautions in the beginning of this section.
2. Disconnect the negative battery cable.
3. Drain the engine coolant.

4. Remove or disconnect the following:
   - Air cleaner assembly
   - Hood
   - Accessory drive belt
   - Hoses from the surge tank
   - Cruise control module
   - Upper wiring harness from the engine
   - Throttle and cruise control cables
   - Starter motor
   - A/C compressor and move it aside with the lines attached
   - Lower wiring harness from the engine components
   - Catalytic converter from the rear exhaust manifold
   - Torque converter cover
   - Torque converter-to-flywheel bolts
   - Engine splash shields
   - Transaxle-to-engine brace and the 2 outer transaxle mounting bolts
   - Upper and lower radiator hoses
   - Fuel lines from the engine
   - Vacuum hose from the power brake booster
   - Heater hoses
5. Install an engine hoist and raise the engine slightly.
6. Remove or disconnect the following:
   - Engine mount and bracket
   - Power steering pump
   - Transaxle-to-engine bolts
   - Engine from the vehicle

To install:

7. Install or connect the following:
   - Engine. Torque the upper transaxle-to-engine bolts to 66 ft. lbs. (90 Nm).
   - Engine support fixture and remove the engine hoist
   - Power steering pump. Torque the bolts to 25 ft. lbs. (34 Nm).
   - Engine mount bracket and mount. Torque the bolts to 43 ft. lbs. (58 Nm) and the nuts to 35 ft. lbs. (47 Nm).
   - Heater hoses
   - Fuel lines
   - Vacuum hose to the power brake booster
   - Upper and lower radiator hoses
   - Both outer transaxle to engine bolts. Torque the bolts to 66 ft. lbs. (90 Nm).
   - Transaxle to the engine brace. Torque the bolts to 32 ft. lbs. (43 Nm).
   - Engine splash shields
   - Flexplate to the torque converter. Torque the bolts to 46 ft. lbs. (62 Nm).
   - Torque converter cover. Torque the bolts to 89 inch lbs. (10 Nm).
   - Catalytic converter to the exhaust manifold. Torque the bolts to 25 ft. lbs. (34 Nm).
   - A/C compressor. Torque the bolts to 37 ft. lbs. (50 Nm).
   - Starter motor. Torque the bolts to 32 ft. lbs. (43 Nm).
   - Lower and upper wiring harnesses

✱✱ CAUTION

To avoid personal injury and/or damage to the vehicle, replace the throttle cable with a new one any time the engine has been removed from the vehicle.

   - Cruise control and throttle cables
   - Cruise control module
   - Accessory drive belt
   - Hoses to the surge tank
   - Hood. Torque the bolts to 13 ft. lbs. (17 Nm).
   - Air cleaner assembly
8. Refill the cooling system.
9. Refill the crankcase.
10. Start the engine and inspect for coolant and/or oil leaks.
11. Stop the engine and recheck the fluid levels after the engine has cooled.

Water Pump

REMOVAL & INSTALLATION

2.4L Engine

1. Before servicing the vehicle, refer to the precautions in the beginning of this section.
2. Disconnect the negative battery cable.
3. Drain the cooling system.
4. Remove or disconnect the following:
   - Oxygen (O_2S) sensor electrical connector
   - Exhaust manifold heat shield
   - Coolant inlet housing bolt through the exhaust manifold
   - Exhaust manifold brace-to-manifold bolt
   - Manifold-to-exhaust pipe studs
   - Coolant inlet housing-to-water pump cover bolt
   - Exhaust pipe from the exhaust manifold by pulling it downward

✱✱ WARNING

Do not rotate the flex coupling more than 4 degrees or damage may occur.

   - Coolant inlet pipe from the oil pan
   - Brake vacuum pipe from the camshaft housing
   - Exhaust manifold from the cylinder head

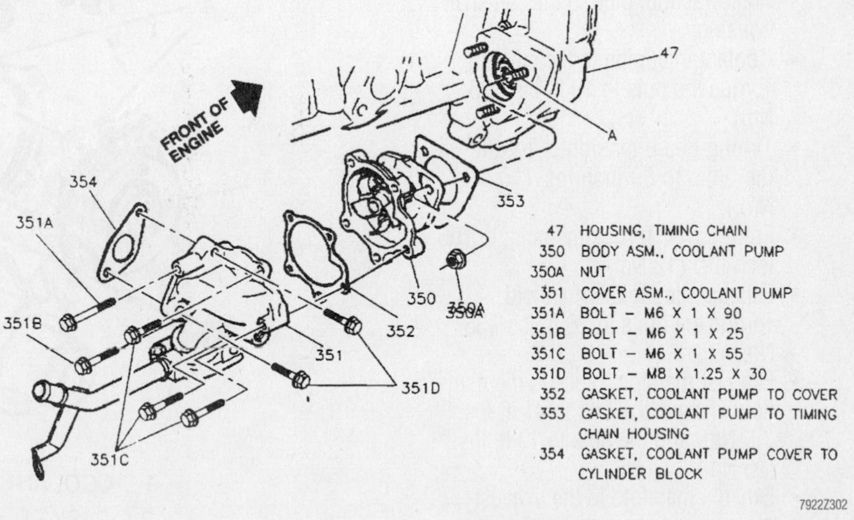

Exploded view of the water pump mounting—2.4L engine

- Heater hose from the heater outlet pipe
- Timing chain cover and tensioner
- Water pump cover-to-engine bolts
- 3 water pump-to-timing chain housing nuts
- Water pump and cover assembly
- Water pump cover from the pump

To install:

5. Install or connect the following:
   - Water pump with a new gasket to the cover and tighten the bolts finger-tight
   - Water pump cover-to-engine bolts and tighten finger-tight
   - Water pump-to-timing chain housing nuts and tighten finger-tight
   - Coolant inlet pipe into the water pump cover and tighten the bolts finger-tight

6. With all gaps closed, torque the bolts, in the following sequence, to the proper values:

   a. Pump assembly-to-chain housing nuts to 19 ft. lbs. (26 Nm).

   b. Pump cover-to-pump assembly to 124 inch lbs. (14 Nm).

   c. Water pump cover-to-engine bolts to 19 ft. lbs. (26 Nm).

   d. Coolant inlet pipe assembly-to-water pump cover bolts to 124 inch lbs. (14 Nm).

7. Install or connect the following:
   - Exhaust manifold to the cylinder head. Torque the nuts to 11 ft. lbs. (15 Nm).
   - Brake vacuum pipe to the camshaft housing
   - Coolant inlet pipe to the oil pan. Torque the nuts to 19 ft. lbs. (25 Nm).
   - Timing chain tensioner. Torque the bolts to 89 inch lbs. (10 Nm.).
   - Front cover. Torque the bolts to 106 inch lbs. (12 Nm).
   - Exhaust pipe to the manifold. Torque the bolts to 26 ft. lbs. (35 Nm).
   - Exhaust manifold brace to the manifold. Torque the bolt to 41 ft. lbs. (56 Nm) and the nuts to 19 ft. lbs. (26 Nm).
   - Exhaust manifold to the exhaust pipe. Torque the nuts to 26 ft. lbs. (35 Nm).
   - Heater hose to the heater pipe
   - Exhaust manifold heat shield. Torque the bolts to 124 inch lbs. (14 Nm).
   - O_2S electrical connector
   - Negative battery cable

8. Refill the cooling system.
9. Start the engine and check for leaks.

3.1L and 3.4L Engines

1. Before servicing the vehicle, refer to the precautions in the beginning of this section.
2. Disconnect the negative battery cable.
3. Drain the cooling system.
4. Remove or disconnect the following:
   - Accessory drive belt
   - Water pump pulley
   - Water pump

To install:

5. Install or connect the following:
   - Water pump with a new gasket. Torque the bolts to 89 inch lbs. (10 Nm).
   - Water pump pulley. Torque the bolts to 18 ft. lbs. (25 Nm).
   - Accessory drive belt
   - Negative battery cable
6. Refill and bleed the cooling system.
7. Test run the engine and check for leaks.

Cylinder Head

REMOVAL & INSTALLATION

2.4L Engine

1. Before servicing the vehicle, refer to the precautions in the beginning of this section.
2. Relieve the fuel system pressure.
3. Drain the cooling system.
4. Remove or disconnect the following:
   - Throttle body air intake duct
   - Heater inlet and throttle body heater hoses from the water outlet
   - Power brake vacuum hose from the throttle body
   - Manifold Absolute Pressure (MAP) sensor electrical connector
   - Intake Air Temperature (IAT) sensor electrical connector
   - Evaporative Emissions (EVAP) purge solenoid electrical connector
   - Camshaft Position (CMP) sensor electrical connector
   - Stud-ended bolt from the alternator
   - Intake manifold

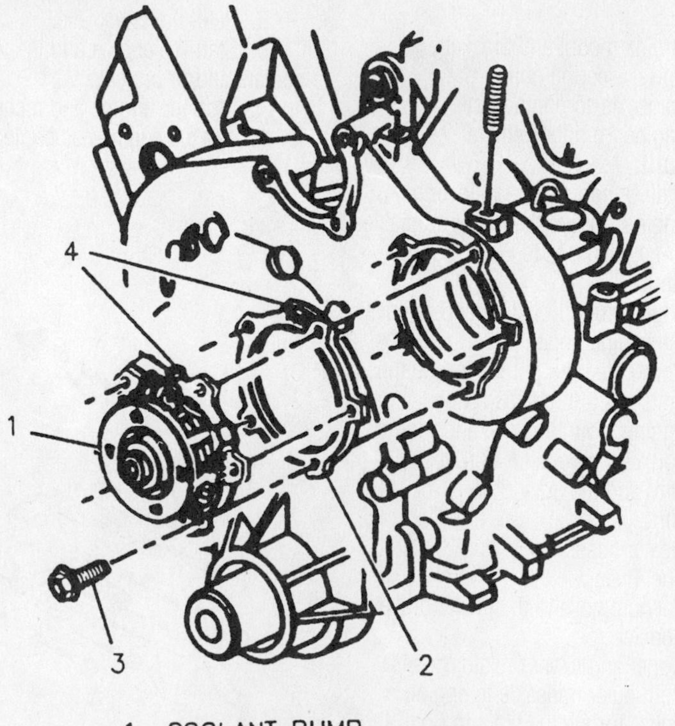

Exploded view of the water pump mounting—3.1L and 3.4L engines

Cylinder head bolt tightening sequence—2.4L engine

5. Install the stud-ended alternator bolt back into the engine.
6. Install an engine support fixture.
7. Remove or disconnect the following:
   - Exhaust manifold
   - Ignition coil and module assembly
   - Power steering pump
   - Vacuum line from the fuel pressure regulator
   - Fuel injector electrical connectors
   - Fuel line clamp from the intake camshaft housing bracket
   - Fuel rail without disconnecting the fuel lines
   - Timing chain and sprocket
   - Water pump
   - Electrical connection from the oil switch
   - Transaxle fluid level indicator tube assembly from the exhaust camshaft cover and move it aside, if equipped with an automatic transaxle

➡Any time the camshaft housing-to-cylinder head bolts are loosened or removed, the camshaft housing-to-cylinder head gasket must be replaced.

   - Intake camshaft housing

➡Turn the camshaft housing upside down as soon as it is removed from the cylinder head; otherwise, the lifters may fall out.

   - Exhaust camshaft housing
   - Upper radiator hose
   - Engine Coolant Temperature (ECT) sensor electrical connector
   - Cylinder head bolts by reversing of the tightening sequence
   - Cylinder head

To install:

8. Install or connect the following:
   - Cylinder head with a new gasket
   - New cylinder head bolts lubricated with clean engine oil
9. Torque the cylinder head bolts, in sequence, as follows:
   a. Step 1: Bolts 1 through 8 to 40 ft. lbs. (65 Nm).
   b. Step 2: Bolts 9 and 10 to 30 ft. lbs. (40 Nm).
   c. Step 3: All bolts an additional 90 degree (¼) turn.
10. Install the intake and exhaust camshaft housings and torque the bolts as follows:
   a. Long bolts: 11 ft. lbs. (15 Nm) plus an additional 90 degree turn.
   b. Short bolts: 11 ft. lbs. (15 Nm) plus an additional 30 degree turn.
11. Install or connect the following:
   - Water pump. Torque the pump-to-cover bolts to 124 inch lbs. (14 Nm) and the water pump cover-to-block bolts to 19 ft. lbs. (26 Nm).
   - Timing chain and sprocket. Torque the sprocket bolt to 52 ft. lbs. (70 Nm).
   - Upper radiator hose to the water outlet
   - ECT sensor connector
   - Fuel rail
   - Fuel injector electrical connectors
   - Fuel pressure regulator vacuum line
   - MAP sensor electrical connector
   - IAT sensor electrical connector
   - EVAP purge solenoid electrical connector
   - CMP sensor electrical connector
   - Oil switch electrical connector
   - Exhaust manifold. Torque the nuts to 110 inch lbs. (13 Nm).
   - Intake manifold. Torque the bolts to 18 ft. lbs. (24 Nm).
   - Power steering pump. Torque the bolts to 19 ft. lbs. (26 Nm).
   - Throttle body air intake duct
   - Oil fill tube into the engine
   - Heater inlet and throttle body heater hoses to the water outlet
   - Power brake vacuum hose
   - Ignition coil and module assembly
12. Remove the engine support fixture.
13. Fill all fluids to their proper levels.

➡An oil and filter change is recommended.

14. Connect the negative battery cable. Start the vehicle and verify no leaks.

3.1L and 3.4L Engines

LEFT (FRONT) CYLINDER HEAD

1. Before servicing the vehicle, refer to the precautions in the beginning of this section.
2. Relieve the fuel system pressure.

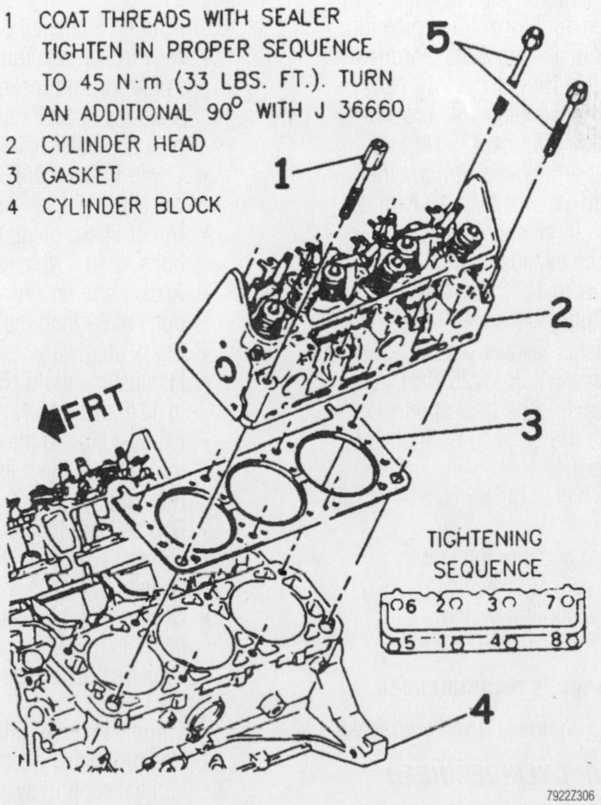

Cylinder head bolt tightening sequence—3.1L and 3.4L engines

Heater Core replacement is covered in Section 2 of this manual

3. Drain the crankcase.
4. Drain the cooling system.
5. Remove or disconnect the following:
   - Upper half of the air cleaner assembly
   - Throttle body air inlet duct
   - Exhaust crossover pipe heat shield
   - Crossover pipe
   - Spark plug wires from the spark plugs
   - Rocker arm covers
   - Upper intake plenum and lower intake manifold
   - Left side exhaust manifold
   - Oil level indicator tube

➡**When removing the valvetrain components, keep them in order for installation purposes.**

   - Rocker arms and pushrods
   - Cylinder head

To install:

6. Install the cylinder head with a new gasket.
7. Torque the cylinder head bolts to 33 ft. lbs. (45 Nm) plus an additional 90 degree turn.
8. Install or connect the following:
   - New intake manifold gasket
   - Pushrods and rocker arms. Torque the bolts to 89 inch lbs. (10 Nm) plus an additional 30 degree turn.
   - Lower intake manifold. Torque the bolts to 115 inch lbs. (13 Nm).
   - Upper intake plenum. Torque the bolts to 18 ft. lbs. (25 Nm).
   - Rocker arm covers. Torque the bolts to 89 inch lbs. (10 Nm).
   - Oil level indicator tube
   - Left side exhaust manifold. Torque the nuts to 12 ft. lbs. (16 Nm).
   - Spark plug wires
   - Exhaust crossover pipe. Torque the bolts to 18 ft. lbs. (25 Nm).
   - Crossover pipe heat shield. Torque the bolts to 89 inch lbs. (10 Nm).
   - Upper half of the air cleaner assembly
   - Throttle body air inlet duct
   - Negative battery cable
9. Refill the cooling system.
10. Refill the crankcase.

➡**A filter change is recommended.**

11. Start the engine and verify no leaks.

RIGHT (REAR) CYLINDER HEAD

1. Before servicing the vehicle, refer to the precautions in the beginning of this section.
2. Relieve the fuel system pressure.
3. Drain the crankcase.
4. Drain the cooling system.
5. Remove or disconnect the following:
   - Upper half of the air cleaner assembly
   - Throttle body air inlet duct
   - Exhaust crossover pipe heat shield
   - Crossover pipe
   - Oxygen (O_2S) sensor electrical connector
   - Exhaust pipe from the exhaust manifold
   - Right side exhaust manifold
   - Spark plug wires from the spark plugs
   - Rocker arm covers
   - Upper intake plenum
   - Lower intake manifold

➡**When removing the valvetrain components keep them in order for installation purposes.**

   - Rocker arms and pushrods
   - Cylinder head

To install:

6. Install the cylinder head with a new gasket.
7. Torque the cylinder head bolts to 33 ft. lbs. (45 Nm) plus an additional 90 degree turn.
8. Install or connect the following:
   - New intake manifold gasket
   - Pushrods and rocker arms. Torque the bolts to 89 inch lbs. (10 Nm) plus an additional 30 degree turn.
   - Lower intake manifold. Torque the bolts to 115 inch lbs. (13 Nm).
   - Upper intake plenum. Torque the bolts to 18 ft. lbs. (25 Nm).
   - Rocker arm covers. Torque the bolts to 89 inch lbs. (10 Nm).
   - Spark plug wires
   - Exhaust manifold. Torque the nuts to 12 ft. lbs. (16 Nm).
   - Exhaust pipe to the exhaust manifold. Torque the nuts to 33 ft. lbs. (45 Nm).
   - O_2S sensor electrical connector
   - Exhaust crossover pipe. Torque the bolts to 18 ft. lbs. (25 Nm).
   - Crossover pipe heat shield. Torque the nuts to 89 inch lbs. (10 Nm).
   - Upper half of the air cleaner assembly
   - Throttle body air inlet duct
   - Negative battery cable
9. Refill the cooling system.
10. Refill the crankcase.

➡**An oil filter change is recommended.**

11. Start the engine and verify no leaks.

Rocker Arms

REMOVAL & INSTALLATION

2.4L Engine

The 2.4L engine is not equipped with rocker arms. The camshafts directly actuate the valves.

3.1L and 3.4L Engines

LEFT SIDE

1. Before servicing the vehicle, refer to the precautions in the beginning of this section.
2. Drain the cooling system to a level below the coolant pipe on the front of the engine.
3. Remove or disconnect the following:
   - Negative battery cable
   - Spark plug wires
   - Heater bypass pipe
   - Positive Crankcase Ventilation (PCV) valve and hose
   - Rocker arm cover

➡**Keep the pushrods in order. Intake pushrods are 5¾inches long and exhaust pushrods are 6 inches long.**

   - Rocker arms and pushrods

To install:

4. Lubricate all the valvetrain components with engine oil.
5. Install or connect the following:
   - Pushrods and the rocker arms. Torque the bolts to 89 inch lbs. (10 Nm) plus an additional 30 degree turn.
   - Rocker arm cover using a new gasket. Torque the rocker cover bolts to 89 inch lbs. (10 Nm).
   - PCV valve and hose
   - Heater bypass pipe. Torque the screw at the water pump to 106 inch lbs. (12 Nm), the bolt at the cylinder head corner to 18 ft. lbs. (25 Nm) and the nut to 18 ft. lbs. (25 Nm).
   - Spark plug wires
   - Negative battery cable
6. Refill the cooling system.
7. Start the vehicle and verify no leaks.

RIGHT SIDE

1. Before servicing the vehicle, refer to the precautions in the beginning of this section.
2. Remove or disconnect the following:
   - Negative battery cable
   - Spark plug wires from the spark

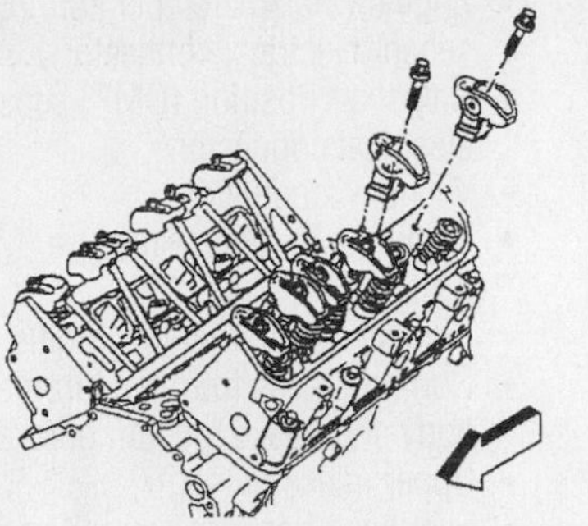

Rocker arm components—3.1L and 3.4L engines

plugs and the upper intake plenum wire retainer
- Power brake booster vacuum pipe from the intake plenum
- Accessory drive belt
- Alternator
- Ignition coil assembly and Evaporative Emissions (EVAP) canister purge solenoid as an assembly
- Rocker arm cover

➡**Keep the pushrods in order. Intake pushrods are 5¾inches long and exhaust pushrods are 6 inches long.**

- Rocker arms and pushrods

To install:

3. Lubricate all the valvetrain components with engine oil.
4. Install or connect the following:
   - Pushrods and the rocker arms. Torque the bolts to 89 inch lbs. (10 Nm) plus an additional 30 degree turn.
   - Rocker arm cover using a new gasket. Torque the rocker cover bolts to 89 inch lbs. (10 Nm).
   - Ignition coil and EVAP solenoid assembly
   - Alternator. Torque the bolts to 37 ft. lbs. (50 Nm).
   - Accessory drive belt
   - Power brake booster vacuum pipe to the plenum
   - Spark plug wires
   - Negative battery cable
5. Start the vehicle and verify no leaks.

Intake Manifold

REMOVAL & INSTALLATION

These vehicles were filled at the factory with an antifreeze/coolant called GM Goodwrench DEX-COOL®. When adding coolant to vehicles, it is important that you use GM Goodwrench DEX-COOL (orange-colored, silicate-free) coolant. **Propylene glycol is not recommended for use in GM vehicles**. A 50/50 mixture of DEX-COOL and clean water will provide all the recommended protection. **DO NOT mix DEX-COOL with any other type of antifreeze**.

2.4L Engine

1. Before servicing the vehicle, refer to the precautions in the beginning of this section.
2. Relieve the fuel system pressure.
3. Drain the cooling system to a level below the intake manifold.
4. Remove or disconnect the following:
   - Air cleaner duct
   - Manifold Absolute Pressure (MAP) sensor electrical connector
   - Intake Air Temperature (IAT) sensor electrical connector
   - Evaporative Emissions (EVAP) canister purge solenoid electrical connector

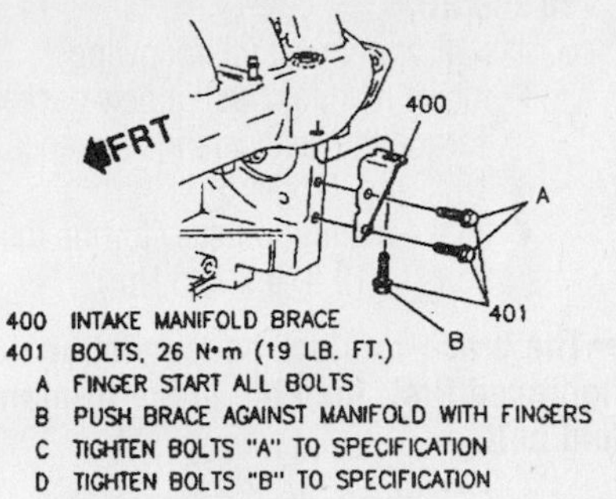

Exploded view of the intake manifold brace mounting—2.4L engine

   - Fuel injector electrical connectors
   - Vacuum hoses from the intake manifold, fuel pressure regulator and EVAP canister purge solenoid
   - Accelerator control cable bracket
   - Coolant lines from the throttle body
   - Stud-end generator mount bolt
   - Exhaust Gas Recirculation (EGR) pipe from the EGR adapter
   - Intake manifold support brace
   - Intake manifold

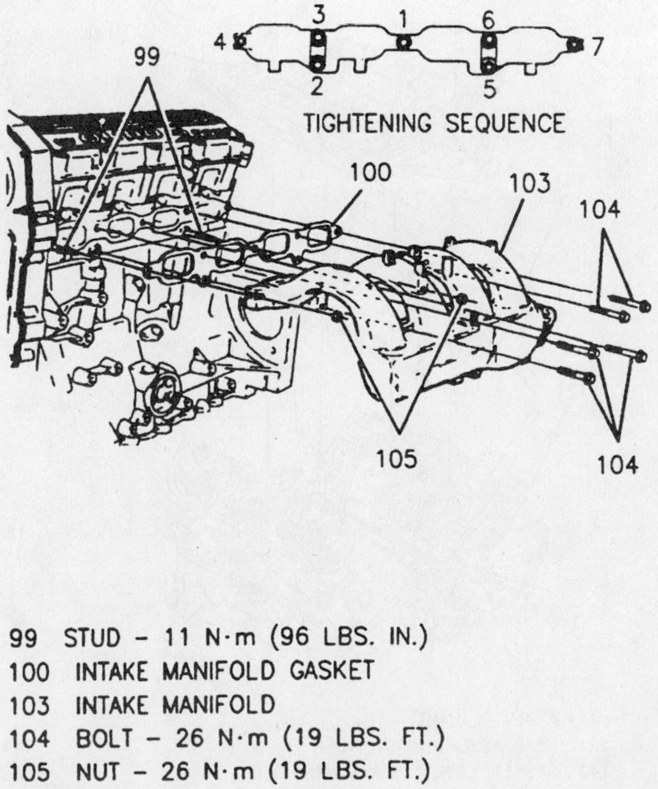

Intake manifold tightening sequence—2.4L engine

To install:

5. Install or connect the following:
   - Intake manifold with a new gasket. Torque the fasteners in sequence to 18 ft. lbs. (25 Nm).
   - Intake manifold brace. Torque the bolts to 19 ft. lbs. (26 Nm).

➡The brace-to-block bolts must be tightened first, then the brace-to-manifold bolt.

   - EGR pipe to the EGR adapter
   - Stud-end generator mount bolt
   - Coolant lines to the throttle body
   - Accelerator control cable bracket
   - Vacuum hoses to the intake manifold, fuel pressure regulator and EVAP canister purge solenoid
   - Electrical connectors to the fuel injectors
   - MAP sensor electrical connector
   - IAT sensor electrical connector
   - EVAP canister purge solenoid electrical connector
   - Air cleaner duct
6. Refill the cooling system.
7. Connect the negative battery cable.

3.1L and 3.4L Engines

1. Before servicing the vehicle, refer to the precautions in the beginning of this section.
2. Relieve the fuel system pressure.
3. Drain the cooling system.
4. Remove or disconnect the following:
   - Air cleaner assembly
   - Accessory drive belt
   - Exhaust Gas Recirculation (EGR) valve
   - Brake vacuum pipe at the intake plenum
   - Fuel pressure regulator vacuum line
   - Spark plug wires from the spark plugs and the intake plenum retainers
   - Ignition coil assembly and the Evaporative Emissions (EVAP) canister purge solenoid as an assembly
   - Throttle Position (TPS) sensor electrical connector
   - Idle Air Control (IAC) sensor electrical connector
   - Fuel injector electrical connectors
   - Engine Coolant Temperature (ECT) sensor electrical connector
   - Camshaft Position (CMP) sensor electrical connector
   - Vacuum modulator
   - Manifold Absolute Pressure (MAP) sensor
   - Coolant hoses from the throttle body
   - Control cables from the throttle body and intake plenum bracket
   - Upper intake plenum
   - Fuel lines from the fuel rail and fuel line bracket
   - Fuel rail with the injectors
   - Inlet cooling pipe from the outlet housing
   - Heater bypass hose from the water pump and the cylinder head
   - Upper radiator hose at the thermostat housing
   - Thermostat housing
   - Both rocker arm covers
   - Lower intake manifold

➡When removing the valvetrain components, keep them in order for installation purposes.

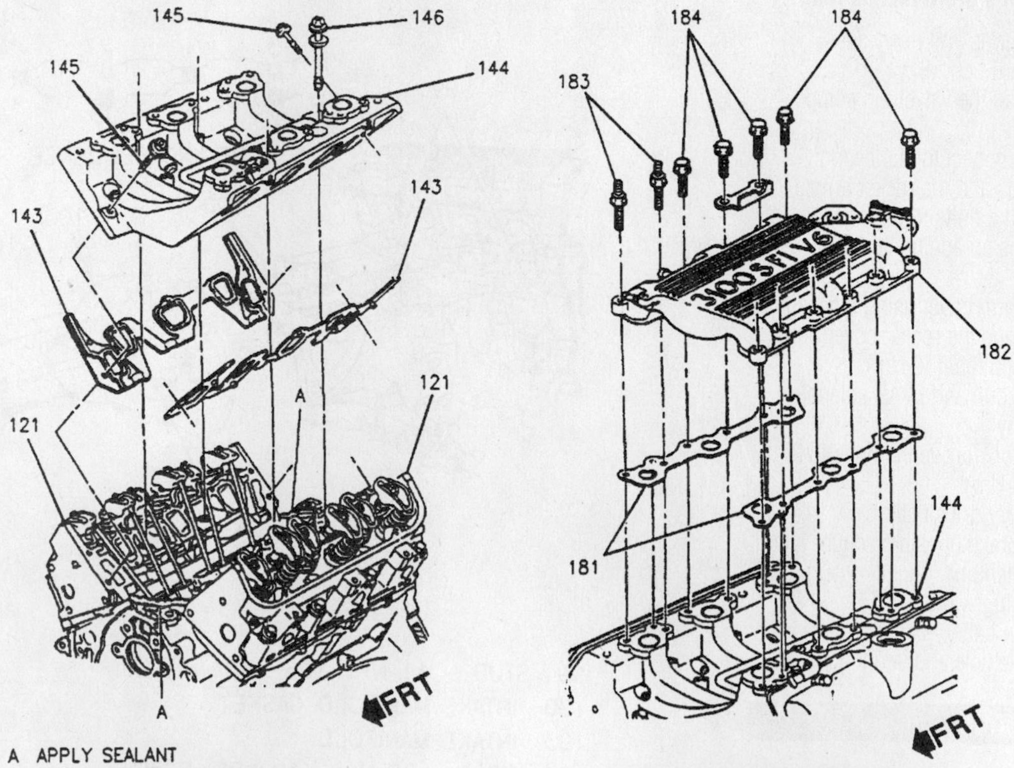

A APPLY SEALANT
121 HEAD ASSEMBLY, CYLINDER
143 GASKET, LOWER INTAKE MANIFOLD
144 BOLT, LOWER INTAKE
145 BOLT, LOWER INTAKE MANIFOLD
146 BOLT, LOWER INTAKE MANIFOLD

144 MANIFOLD, LOWER INTAKE
181 GASKET, UPPER INTAKE MANIFOLD
182 MANIFOLD, UPPER INTAKE
183 STUD, UPPER INTAKE MANIFOLD
184 BOLT, UPPER INTAKE MANIFOLD

79222310

Exploded view of the upper and lower intake—3.1L and 3.4L engines

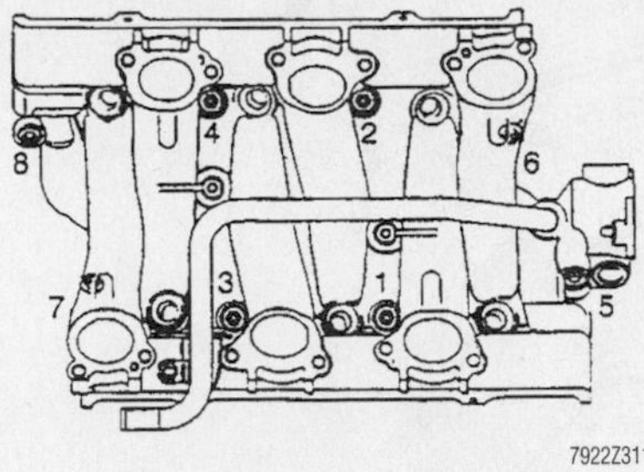

Lower intake manifold tightening sequence—3.1L and 3.4L engines

- Rocker arms and pushrods

To install:

5. Place a 3mm bead of RTV, on each ridge, where the front and rear of the intake manifold contact the block.
6. Install or connect the following:
   - New intake manifold gasket
   - Pushrods and rocker arms. Torque the bolts to 89 inch lbs. (10 Nm) plus an additional 30 degree turn.
   - Lower intake manifold. Apply sealant to the threads of the bolts and torque the bolts to 115 inch lbs. (13 Nm).

WARNING

In order to prevent oil leaks at the intake manifold, tighten the vertical bolts before the diagonal bolts.

   - Both rocker arm covers. Torque the bolts to 89 inch lbs. (10 Nm).
   - Thermostat housing with a new gasket. Torque the bolts to 19 ft. lbs. (26 Nm).
   - Upper radiator hose at thermostat housing
   - Coolant bypass hose to the water pump and the cylinder head
   - Coolant inlet pipe to coolant outlet housing
   - Fuel rail with the injectors
   - Fuel lines to the fuel rail and fuel line bracket
   - Upper intake manifold with a new gasket. Torque the bolts to 18 ft. lbs. (25 Nm).
   - Coolant hoses to the throttle body
   - MAP sensor
   - Vacuum modulator
   - CMP sensor electrical connector
   - ECT sensor electrical connector
   - Fuel injector electrical connectors
   - IAC sensor electrical connector
   - TPS sensor electrical connector
   - Ignition coil assembly and the EVAP canister purge solenoid as an assembly
   - Spark plug wires
   - Fuel pressure regulator vacuum line
   - EGR valve. Torque the bolts to 22 ft. lbs. (30 Nm).
   - Control cables to the throttle body and intake plenum bracket
   - Brake vacuum pipe at the intake plenum
   - Accessory drive belt
   - Air cleaner assembly
   - Negative battery cable
7. Refill the cooling system.

➡An engine oil and filter change is recommended.

8. Start the vehicle and verify no leaks.

Exhaust Manifold

REMOVAL & INSTALLATION

2.4L Engine

1. Before servicing the vehicle, refer to the precautions in the beginning of this section.
2. Remove or disconnect the following:
   - Negative battery cable
   - Oxygen (O_2S) sensor electrical connector
   - Exhaust manifold brace
   - Exhaust manifold-to-exhaust pipe spring loaded bolts
   - Exhaust pipe by pulling it down and back from the exhaust manifold

➡Do not bend the exhaust flex coupler more than 3 degrees in any direction, for it may damage the flex coupler.

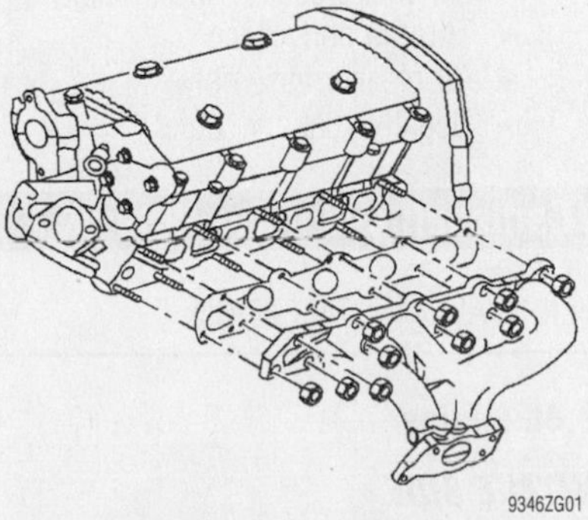

Exploded view of the exhaust manifold assembly mounting— 2.4L engine

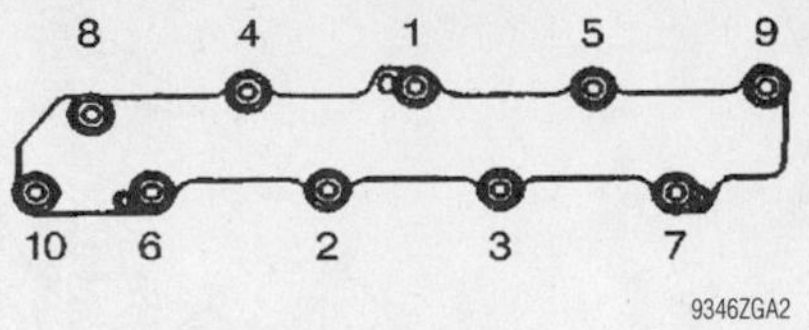

Exhaust manifold assembly torque sequence— 2.4L engine

   - Exhaust manifold

To install:

3. Install or connect the following:
   - Exhaust manifold with new gaskets. Torque the nuts, in sequence, to 110 inch lbs. (13 Nm).
   - Exhaust manifold brace. Torque the bolt to 41 ft. lbs. (56 Nm) and the nuts to 19 ft. lbs. (26 Nm).
   - Exhaust pipe. Torque the bolts evenly to 26 ft. lbs. (35 Nm).
   - O_2S sensor electrical connector
   - Negative battery cable
4. Start the vehicle and verify no exhaust leaks.

3.1L and 3.4L Engines

LEFT SIDE

1. Before servicing the vehicle, refer to the precautions in the beginning of this section.
2. Partially drain the cooling system.
3. Remove or disconnect the following:
   - Negative battery cable
   - Air cleaner assembly
   - Throttle body duct
   - Exhaust crossover heat shield
   - Exhaust crossover pipe from the manifold
   - Radiator hose from the thermostat housing
   - Spark plug wires
   - Exhaust manifold heat shield
   - Exhaust manifold

To install:

4. Install or connect the following:
   - Exhaust manifold with a new gasket. Torque the nuts to 12 ft. lbs. (16 Nm).
   - Exhaust manifold heat shield. Torque the nuts to 89 inch lbs. (10 Nm).
   - Exhaust crossover pipe. Torque the bolts to 18 ft. lbs. (25 Nm).
   - Crossover pipe heat shield. Torque the bolts to 89 inch lbs. (10 Nm).
   - Spark plug wires
   - Radiator hose to the coolant outlet housing

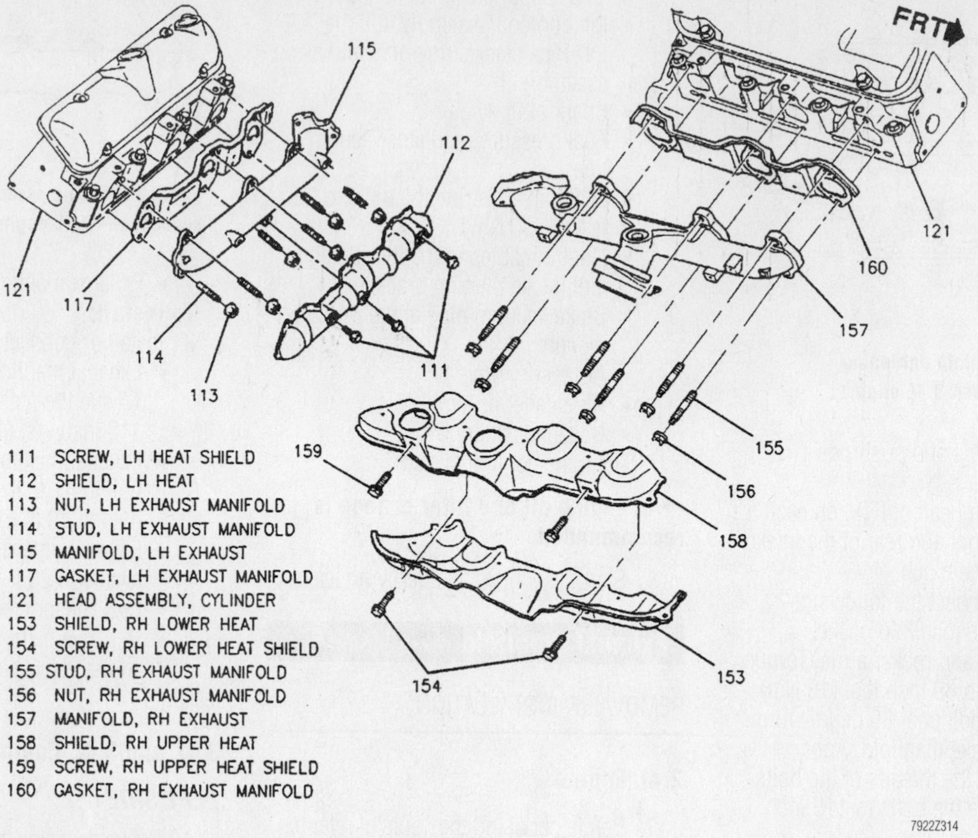

Exploded view of the exhaust manifold mounting—3.1L and 3.4L engines

- Air cleaner assembly
- Throttle body duct
- Negative battery cable

RIGHT SIDE

1. Before servicing the vehicle, refer to the precautions in the beginning of this section.
2. Remove or disconnect the following:
   - Negative battery cable
   - Air cleaner assembly
   - Throttle body duct
   - Heated Oxygen (HO_2S) sensor
   - Exhaust crossover heat shield
   - Exhaust Gas Recirculation (EGR) pipe from the exhaust manifold
   - Exhaust crossover pipe
   - Exhaust manifold heat shield
   - Transaxle oil fill tube
   - Front exhaust pipe from the exhaust manifold
   - Wires from the spark plugs
   - Exhaust manifold

To install:

3. Install or connect the following:
   - Exhaust manifold with a new gasket. Torque the nuts to 12 ft. lbs. (16 Nm).
   - Exhaust manifold heat shield. Torque the nuts to 89 inch lbs. (10 Nm).
   - Exhaust crossover pipe to the manifold. Torque the bolts to 18 ft. lbs. (25 Nm).
   - Exhaust crossover pipe heat shield. Torque the bolts to 89 inch lbs. (10 Nm).
   - Exhaust pipe to the exhaust manifold. Torque the bolts to 33 ft. lbs. (45 Nm).
   - Transaxle oil level indicator and fill tube assembly
   - HO_2S sensor
   - EGR pipe to the exhaust manifold
   - Exhaust crossover heat shield
   - Throttle body duct
   - Air cleaner assembly
   - Negative battery cable

Camshaft and Valve Lifters

REMOVAL & INSTALLATION

2.4L Engine

INTAKE SIDE

➡Anytime the camshaft housing-to-cylinder head bolts are loosened or removed, the camshaft housing-to-cylinder head gasket must be replaced.

1. Before servicing the vehicle, refer to the precautions in the beginning of this section.
2. Remove or disconnect the following:
   - Negative battery cable
   - Ignition assembly electrical connector
   - 4 ignition assembly-to-camshaft housing bolts and the assembly by pulling it straight up

➡Use a spark plug boot wire remover to remove connector assemblies.

   - Camshaft Position (CMP) sensor electrical connector
   - Power steering pump and move it aside without disconnecting the lines
   - Vacuum line from fuel pressure regulator
   - Fuel injector wiring harness
   - Both fuel line-to-intake camshaft housing clamps
   - Fuel rail-to-camshaft housing bolts
   - Fuel rail from the cylinder head and move it aside with the fuel lines attached
   - Timing chain and camshaft sprockets
   - Timing chain housing bolts but do not remove from the engine

- Camshaft housing cover-to-camshaft housing bolts
- Camshaft housing-to-cylinder head bolts by reversing of the tightening sequence

➡Leave 2 bolts loosely in place to hold the camshaft housing while separating the camshaft cover from housing.

3. Press the cover off the housing by threading 4 of the housing-to-cylinder head bolts into the tapped camshaft housing cover holes. Tighten the bolts in evenly so the cover does not bind on the dowel pins.
4. Remove the camshaft housing cover and discard the gaskets.
5. Note the position of the chain sprocket dowel pin for reassembly.
6. Remove or disconnect the following:
   - Camshaft
   - Camshaft oil seal from camshaft and discard it

➡The camshaft seal must be replaced any time the housing and cover are separated.

➡Store the valve lifters in order so that they may be installed in the their locations.

- Valve lifters from the camshaft housing
- Camshaft carrier from the cylinder head and discard the gasket

To install:

7. Clean all the gasket surfaces completely.
8. Install or connect the following:
   - New camshaft housing gasket
   - Camshaft housing

➡Install 1 bolt loosely to hold the housing in place.

➡If the camshaft was replaced the valve lifters must also be replaced.

9. Install the lifters into their original bores.
10. Lubricate the camshaft lobes, journals and lifters with camshaft and lifter prelube. The camshaft lobes and journals must be adequately lubricated or engine damage could occur upon start up.
11. Install or connect the following:
   - Camshaft in its original position with the timing chain sprocket dowel pin straight up and aligned with the centerline of the lifter bores.
   - New Green camshaft housing cover seal

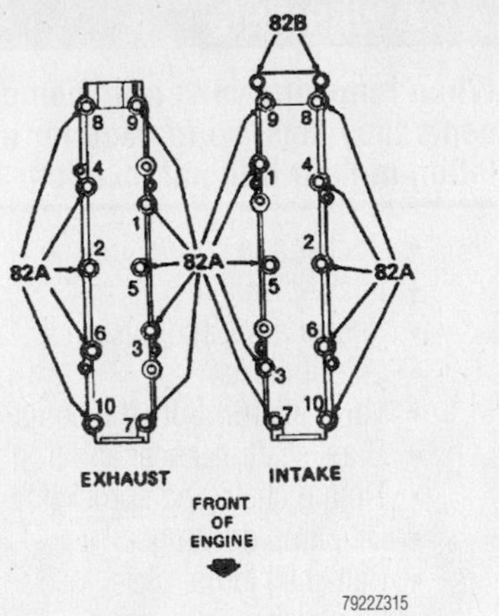

Camshaft housing bolt tightening sequence—2.4L engine

➡The seals for the intake and exhaust covers are different and the correct seals must be used.

12. Remove the bolt holding the housing in place.
13. Apply thread locking compound to the camshaft housing and cover bolt threads.
14. Install or connect the following:
   - Camshaft housing cover. Tighten the bolts, in sequence, to 16 ft. lbs. (22 Nm) plus an additional 90 degree turn (long bolts) and 16 ft. lbs. (22 Nm) plus an additional 30 degree turn (short bolts).
   - Timing chain housing bolts
   - Timing chain and sprockets
   - New fuel injector O-ring seals lubricated with engine oil
   - Fuel rail. Tighten the bolts to 19 ft. lbs. (26 Nm).
   - Ignition assembly on the camshaft housing. Tighten the bolts to 11 ft. lbs. (15 Nm), plus an additional 30 degree turn.
   - Ignition assembly electrical connector
   - Negative battery cable
15. Start the vehicle and verify proper operation and no leaks.

EXHAUST SIDE

➡Anytime the camshaft housing-to-cylinder head bolts are loosened or removed, the camshaft housing-to-cylinder head gasket must be replaced.

1. Before servicing the vehicle, refer to the precautions in the beginning of this section.
2. Remove or disconnect the following:
   - Negative battery cable
   - Ignition assembly electrical connector
   - 4 ignition assembly-to-camshaft housing bolts and the assembly by pulling it straight up

➡Use a spark plug boot wire remover to remove connector assemblies.

- Oil pressure switch electrical connector
- Transaxle fill tube
- Timing chain and camshaft sprockets
- Timing chain housing bolts but do not remove from the engine
- Camshaft housing cover-to-camshaft housing bolts
- Camshaft housing-to-cylinder head bolts by reversing of the tightening sequence

➡Leave 2 bolts loosely in place to hold the camshaft housing while separating the camshaft cover from housing.

3. Press the cover off the housing by threading 4 of the housing-to-cylinder head bolts into the tapped camshaft housing cover holes. Tighten the bolts in evenly so the cover does not bind on the dowel pins.
4. Remove the camshaft housing cover and discard the gaskets.
5. Note the position of the chain sprocket dowel pin for reassembly.
6. Remove or disconnect the following:
   - Camshaft
   - Camshaft oil seal from camshaft and discard it

➡The camshaft seal must be replaced any time the housing and cover are separated.

➡Store the valve lifters in order so that they may be installed in the their locations.

- Valve lifters from the camshaft housing
- Camshaft carrier from the cylinder head and discard the gasket

To install:

7. Clean all the gasket surfaces completely.
8. Install or connect the following:
   - New camshaft housing gasket
   - Camshaft housing

➡Install 1 bolt loosely to hold the housing in place.

➡If the camshaft was replaced the valve lifters must also be replaced.

9. Install the lifters into their original bores.
10. Lubricate the camshaft lobes, journals and lifters with camshaft and lifter prelube. The camshaft lobes and journals must be adequately lubricated or engine damage could occur upon start up.
11. Install or connect the following:
   - Camshaft in it original position with the timing chain sprocket dowel pin straight up and aligned with the centerline of the lifter bores.
   - New Orange camshaft housing cover seal

➡The seals for the intake and exhaust covers are different and the correct seals must be used.

12. Remove the bolt holding the housing in place.
13. Apply thread locking compound to the camshaft housing and cover bolt threads.
14. Install or connect the following:
   - Camshaft housing cover. Tighten the bolts, in sequence, to 16 ft. lbs. (22 Nm) plus an additional 90 degree turn (long bolts) and 16 ft. lbs. (22 Nm) plus an additional 30 degree turn (short bolts).
   - Timing chain housing mounting bolts
   - Timing chain and sprockets
   - New fuel injector O-ring seals lubricated with engine oil
   - Fuel rail. Tighten the bolts to 19 ft. lbs. (26 Nm).
   - Transaxle fill tube
   - Oil pressure switch electrical connector
   - Ignition assembly on the camshaft housing. Tighten the bolts to 11 ft. lbs. (15 Nm), plus an additional 30 degree turn.
   - Ignition assembly electrical connector
   - Negative battery cable
15. Start the vehicle and verify proper operation and no leaks.

3.1L and 3.4L Engines

1. Before servicing the vehicle, refer to the precautions in the beginning of this section.
2. Relieve the fuel system pressure.
3. Remove or disconnect the following:
   - Engine assembly

⁂ WARNING

When removing valvetrain components they must be marked for installation in their original location.

   - Rocker arm covers
   - Intake manifold
   - Rocker arms and pushrods
   - Lifter guides
   - Valve lifter(s) from the bores
   - Crankshaft balancer and front cover
   - Timing chain and sprockets
   - Oil pump driven gear
   - Camshaft thrust plate
   - Camshaft

⁂ WARNING

Avoid damaging the camshaft bearing surfaces.

To install:

4. Coat the camshaft with camshaft lubricant.
5. Install or connect the following:
   - Camshaft
   - Camshaft thrust plate. Torque the bolts to 89 inch lbs. (10 Nm).
   - Oil pump driven gear. Torque the bolt to 27 ft. lbs. (36 Nm).
   - Timing chain and sprocket. Torque the sprocket bolt to 103 ft. lbs. (140 Nm).
   - Front cover. Torque the bolts to 15 ft. lbs. (21 Nm).
   - Crankshaft balancer. Torque the bolt to 76 ft. lbs. (103 Nm).
6. Lubricate the bearing surfaces with Molykote®.
7. Install or connect the following:
   - Lifters in their original locations
   - Lifter guides. Torque the guide bolts to 89 inch lbs. (10 Nm).
   - Pushrods and rocker arms. Torque the nuts to 89 inch lbs. (10 Nm) plus an additional 30 degree turn.
   - Intake manifold
   - Rocker arm covers. Torque the bolts to 89 inch lbs. (10 Nm).
   - Engine assembly
   - Negative battery cable
8. Adjust the valves, as required. Start the engine and verify no oil leaks.

Valve Lash

ADJUSTMENT

The engines are equipped with hydraulic valve lifters that do not require periodic valve lash adjustment. Adjustment to zero lash is maintained automatically by hydraulic pressure in the lifters.

Starter Motor

REMOVAL & INSTALLATION

2.4L Engine

1. Before servicing the vehicle, refer to the precautions in the beginning of this section.
2. Remove or disconnect the following:
   - Negative battery cable
   - Air inlet duct from the throttle body
   - Upper starter bolt
   - Lower closeout panel
   - Lower starter bolt
   - Starter electrical connectors
   - Starter motor

To install:

3. Install or connect the following:
   - Starter electrical connectors
   - Starter motor. Torque the bolts to 66 ft. lbs. (90 Nm).
   - Lower closeout panel
   - Air inlet duct to the throttle body
   - Negative battery cable

3.1L Engine

1. Before servicing the vehicle, refer to the precautions in the beginning of this section.
2. Remove or disconnect the following:
   - Negative battery cable
   - Starter electrical connectors
   - Starter motor

To install:

3. Install or connect the following:
   - Starter motor. Torque the bolts to 32 ft. lbs. (43 Nm).
   - Starter electrical connectors
   - Negative battery cable

3.4L Engine

1. Before servicing the vehicle, refer to the precautions in the beginning of this section.
2. Remove or disconnect the following:
   - Negative battery cable
   - Flywheel inspection cover
   - Starter electrical connectors
   - Starter motor

To install:

3. Install or connect the following:
   - Starter motor. Torque the bolts to 32 ft. lbs. (43 Nm).
   - Starter electrical connectors
   - Flywheel inspection cover. Torque the bolts to 89 inch lbs. (10 Nm).
   - Negative battery cable

Oil Pan

REMOVAL & INSTALLATION

2.4L Engine

1. Before servicing the vehicle, refer to the precautions in the beginning of this section.
2. Drain the engine oil.
3. Drain the cooling system.
4. Remove or disconnect the following:
   - Negative battery cable
   - Flywheel/converter cover
   - Right wheel
   - Right wheel well splash shield
   - Accessory drive belt
   - Air conditioning compressor lower bolts
   - Transmission-to-engine brace
   - Engine mount strut bracket
   - Radiator outlet pipe bolts
   - Radiator outlet pipe from the oil pan
   - Oil pan to the flywheel cover bolt and nut
   - Flywheel cover stud for clearance
   - Radiator outlet pipe from the lower radiator hose and oil pan
   - Oil level sensor connector
   - Oil pan

To install:

5. Inspect the oil pan gasket; it is reusable if not damaged.
6. Install or connect the following:
   - Oil pan with the gasket. Torque the M8 bolts to 18 ft. lbs. (24 Nm) and the M6 bolts to 106 inch lbs. (12 Nm).
   - Oil pan to the transmission nut
   - Oil level sensor connector
   - Radiator outlet pipe to the lower radiator hose and oil pan
   - Exhaust manifold brace
   - Radiator outlet pipe. Torque the bolts to 124 inch lbs. (14 Nm).
   - Engine mount strut bracket. Torque the bolts to 55 ft. lbs. (75 Nm).
   - Transmission to the engine brace
   - Air conditioning compressor lower bolts. Torque the bolts to 37 ft. lbs. (50 Nm).
   - Accessory drive belt
   - Right splash shield
   - Right front wheel
   - Flywheel/converter cover
   - Negative battery cable
7. Refill the crankcase.
8. Refill the cooling system.
9. Start the vehicle and verify no leaks.

3.1L and 3.4L Engine

1. Before servicing the vehicle, refer to the precautions in the beginning of this section.
2. Drain the engine oil.
3. Evacuate the A/C system.
4. Remove or disconnect the following:
   - Negative battery cable
   - Accessory drive belt
   - Right front wheel
   - Right splash shield
   - Anti-lock Brake System (ABS) Wheel Speed (WSS) sensor from the right subframe
   - Right lower ball joint from the steering knuckle
   - Right outer tie rod end
   - A/C compressor without disconnecting the lines
   - Evaporator-to-accumulator A/C line
   - Flywheel cover
   - Right side engine cradle bolts
   - Crankshaft balancer
   - Starter motor
   - Oil pan

To install:

5. Apply silicone sealer to the portion of the pan that contacts the rear of the block.
6. Install or connect the following:
   - Oil pan with a new gasket. Torque

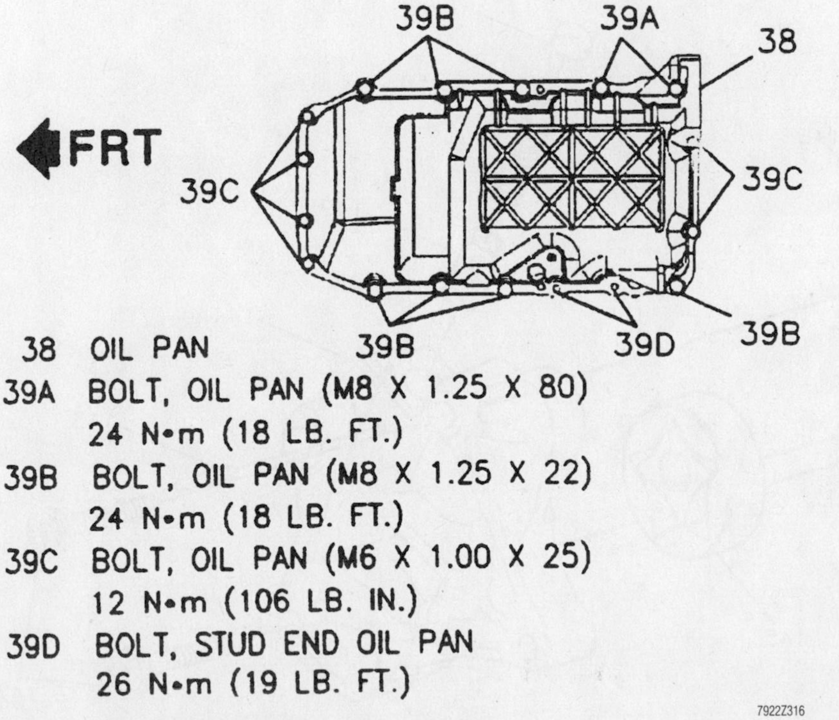

Oil pan mounting bolt locations—2.4L engine

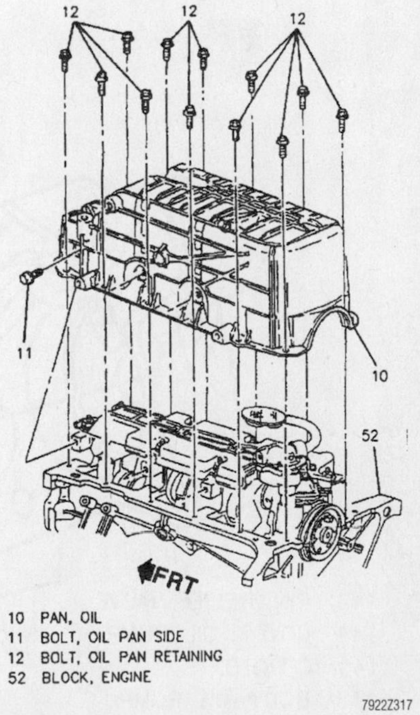

Exploded view of the oil pan mounting—3.1L engine

the flange bolts to 18 ft. lbs. (25 Nm) and the side bolts to 37 ft. lbs. (50 Nm).
- Starter motor. Torque the bolts to 32 ft. lbs. (43 Nm).
- Flywheel cover. Torque the bolts to 89 inch lbs. (10 Nm).
- Crankshaft balancer. Torque the bolt to 76 ft. lbs. (103 Nm).
- Engine cradle bolts. Torque the bolts to 84 ft. lbs. (115 Nm) plus an additional 120 degree turn.
- A/C compressor. Torque the bolts to 37 ft. lbs. (50 Nm).
- Evaporator-to-accumulator line
- Ball joint. Torque the nut to 48 ft. lbs. (60 Nm).
- Outer tie rod end. Torque the nut to 44 ft. lbs. (60 Nm).
- WSS electrical connector
- Wheel well splash shield
- Right front wheel
- Accessory drive belt
- Negative battery cable

7. Fill the crankcase.
8. Evacuate and recharge the A/C system.
9. Start the engine and check for leaks.

➡Whenever the vehicle subframe is removed or lowered, the wheel alignment should be checked.

10. Check and/or adjust the front end alignment.

Oil Pump

REMOVAL & INSTALLATION

2.4L Engine

1. Before servicing the vehicle, refer to the precautions in the beginning of this section.
2. Disconnect the negative battery cable.
3. Install an engine support fixture.
4. Properly drain the engine oil.
5. Remove or disconnect the following:
   - Oil pan
   - Balance shaft chain cover
   - Balance shaft chain tensioner
   - Oil pump cover
   - Oil pump assembly from the balance shaft assembly, by pulling the housing to disconnect the pump gear from the balance shaft

To install:

6. Lubricate the gears with clean engine oil.
7. Assemble the geroter gear into the housing.

➡Fill the oil pump cavities with petroleum jelly prior to installation. This seals the pump and acts like a "prime" so the pump will draw oil as soon as the engine begins to turn. This will ensure that there is oil pressure immediately on start-up and will prevent engine damage.

8. Install or connect the following:
   - Oil pump to the balance shaft assembly. Torque the bolts to 40 ft. lbs. (54 Nm).
   - Oil pump cover. Torque the bolts to 40 ft. lbs. (54 Nm).
   - Balance shaft chain tensioner with the bolts finger tight
9. Adjust the chain tension inserting a 0.40 in. (1mm) brass feeler, between the chain guide and the chain.

➡A brass feeler gauge must be used to ensure that correct measurements are obtained. If a steel gauge is used, it will not bend to conform to the guide and will allow for incorrect measurements.

10. Press the guide against the chain using about 10 lbs. of force. Torque the chain tensioner fastener to 115 inch lbs. (13 Nm).

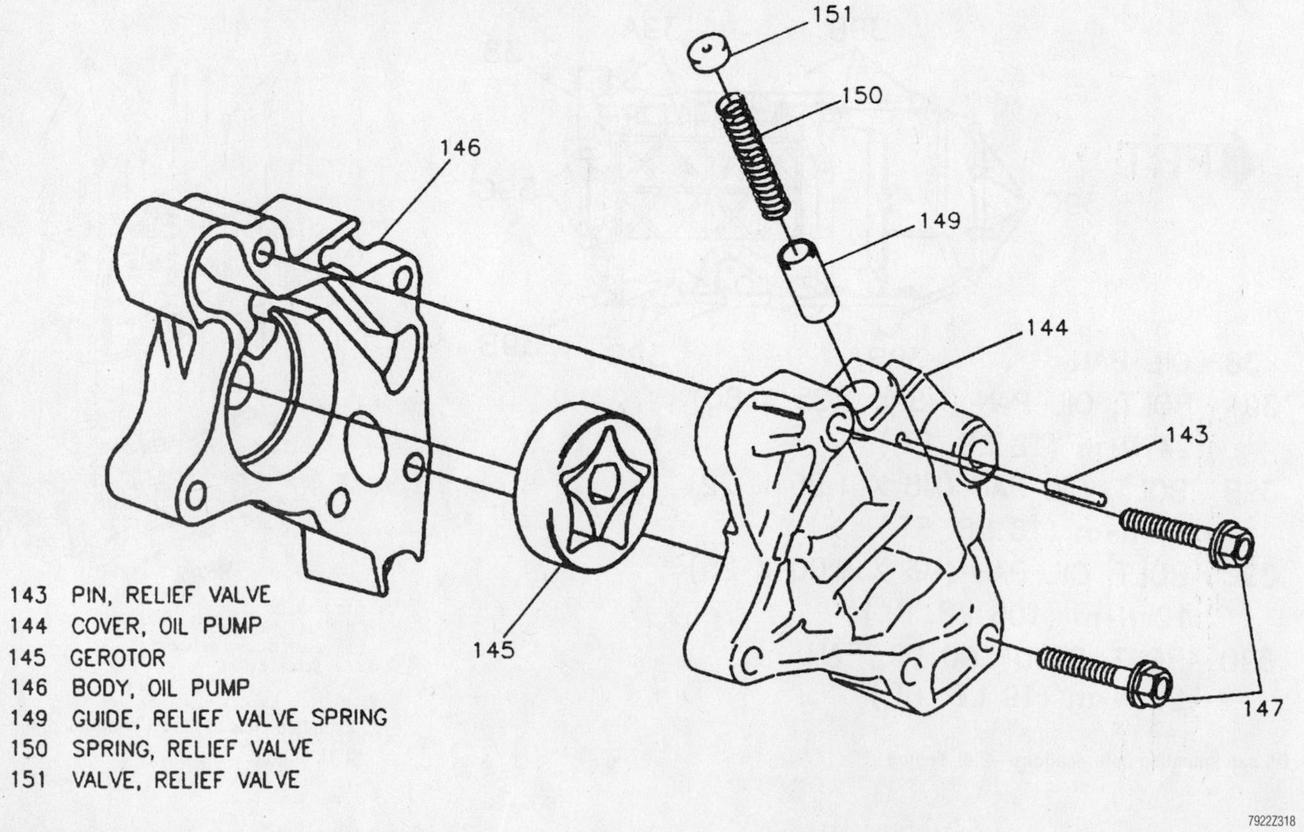

Exploded view of the oil pump components—2.4L engine

11. Install or connect the following:
   - Balance shaft chain cover. Torque the nut/bolt to 10 ft. lbs. (13 Nm).
   - Oil pan. Torque the bolts to 18 ft. lbs. (24 Nm).
   - Negative battery cable
12. Fill the crankcase.

➡An oil filter change is recommended.

13. Start the engine and verify oil pressure and no leaks.

3.1L and 3.4L Engines

1. Before servicing the vehicle, refer to the precautions in the beginning of this section.
2. Disconnect the negative battery cable.
3. Drain the engine oil.
4. Remove or disconnect the following:
   - Oil pan
   - Oil pump and pump driveshaft

To install:

5. Install or connect the following:
   - Oil pump by engaging the oil pump driveshaft. Torque the oil pump bolts to 30 ft. lbs. (41 Nm).
   - Oil pan. Torque the bolts to 18 ft. lbs. (25 Nm).
   - Negative battery cable
6. Fill the crankcase.

➡An oil filter change is recommended.

7. Start the engine, check the oil pressure and check for leaks.

Rear Main Seal

REMOVAL & INSTALLATION

2.4L Engine

1. Before servicing the vehicle, refer to the precautions in the beginning of this section.
2. Remove or disconnect the following:
   - Negative battery cable
   - Transaxle
   - Pressure plate and clutch disc, if equipped
   - Flywheel
   - Oil pan-to-seal housing bolts
   - Seal housing
   - Rear main seal from the housing

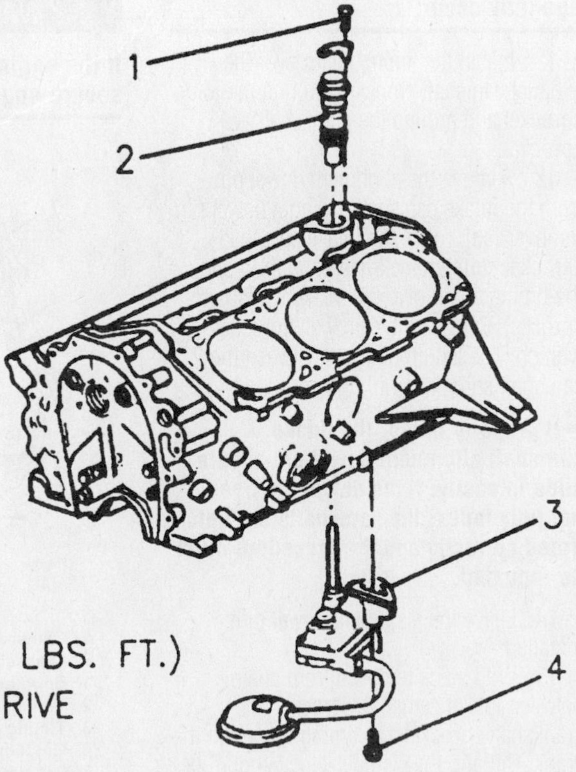

Exploded view of the oil pump mounting—3.1L and 3.4L engines

⁂ WARNING

Be careful not to damage the seal housing sealing surface; damage may result in an oil leak.

To install:

3. Install the new rear main seal into the housing.
4. Inspect the oil pan gasket inner silicone bead for damage and repair using a silicone sealant, if necessary.
5. Lubricate the lip of the seal with clean engine oil.
6. Install or connect the following:
   - Seal housing with a new gasket. Torque the bolts to 106 inch lbs. (12 Nm).
   - Flywheel. Torque the bolts to 22 ft. lbs. (30 Nm) plus an additional 45 degree turn.
   - Clutch, pressure plate and clutch cover assembly, if equipped with a manual transaxle
   - Transaxle
   - Negative battery cable
7. Start the engine and check for leaks.

3.1L and 3.4L Engines

1. Before servicing the vehicle, refer to the precautions in the beginning of this section.
2. Support the engine.
3. Remove or disconnect the following:
   - Transaxle
   - Flywheel
   - Rear main seal using a prytool

⁂ WARNING

Be careful not to damage the crankshaft sealing surface.

To install:

4. Lubricate the seal and bore with engine oil.
5. Install the new seal by sliding it over the mandrel until the dust lip bottoms squarely against the tool collar.
6. Align the dowel pin of the tool with the dowel pin hole in the crankshaft and attach the tool to the crankshaft. Tighten the attaching screws to 24–60 inch lbs. (2.7–6.8 Nm).
7. Tighten the tool T-handle to press the seal into the bore until the tool collar is flush against the engine.
8. Remove the installation tool.

➡Make sure that the seal is squarely seated in the bore.

9. Install or connect the following:
   - Flywheel. Tighten the flywheel-to-crankshaft bolts to 52 ft. lbs. (71 Nm).
   - Transaxle
10. Start the engine and check for leaks.

Timing Chain, Sprockets, Front Cover and Seal

REMOVAL & INSTALLATION

2.4L Engine

➡It is recommended that the entire procedure be reviewed before attempting to service the timing chain.

1. Before servicing the vehicle, refer to the precautions in the beginning of this section.
2. Drain the cooling system.
3. Remove or disconnect the following:
   - Negative battery cable
   - Coolant surge tank
   - Accessory drive belt
   - Alternator
4. Install an engine support.
5. Remove or disconnect the following:
   - Upper cover fasteners
   - Front cover vent hose
   - Right engine mount and bracket
   - Right front wheel
   - Right lower splash shield
   - Crankshaft balancer
   - Lower cover fasteners
   - Front cover
6. Rotate the crankshaft clockwise, as viewed from front of engine (normal rotation) until the camshaft sprocket's timing dowel pin holes align with the timing chain housing holes. The crankshaft sprocket mark should align with the engine mark. The crankshaft sprocket keyway should point upward and align with the cylinder bores centerline. This is the normal timed position.
7. Remove the timing chain guides.
8. Remove the timing chain tensioner.

➡Be sure all the slack in the timing chain is above the tensioner assembly when removing it.

⁂ CAUTION

The tensioner plunger is spring loaded and could fly out causing personal injury.

9. Remove or disconnect the following:
   - Timing chain
   - Camshaft sprockets

To install:

10. Install or connect the following:
    - Camshaft sprockets. Torque the bolts to 52 ft. lbs. (70 Nm).
    - Camshaft sprocket alignment pin through the camshaft sprockets holes into the timing chain housing holes to position the camshafts for timing.
11. If the camshafts are out of position and must be rotated more than ⅛turn in order to install the alignment dowel pins, perform the following:
    a. Rotated the crankshaft 90 degrees clockwise off Top Dead Center (TDC) in order to give the valves adequate clearance to open.
    b. Once the camshafts are positioned and the dowels installed, rotate the crankshaft counterclockwise back to TDC.

⁂ WARNING

Do not rotate the crankshaft clockwise to TDC or valve and piston damage may occur.

12. Install the timing chain over the exhaust camshaft sprocket, around the idler sprocket and around the crankshaft sprocket.
13. Remove the alignment dowel pin from the intake camshaft. Using a dowel pin remover tool, rotate the intake camshaft sprocket counterclockwise enough to slide the timing chain over the intake camshaft sprocket. Release the camshaft sprocket wrench. The length of chain between the 2 camshaft sprockets will tighten.

➡If properly timed, the intake camshaft alignment dowel pin should slide in easily. If the dowel pin does not fully index, the camshafts are not timed correctly and the procedure must be repeated.

14. Leave the alignment dowel pins installed.
15. With slack removed from chain between intake camshaft sprocket and crankshaft sprocket, the timing marks on the crankshaft and the cylinder block should be aligned. If marks are not aligned, move the chain 1 tooth forward or rearward, remove slack and recheck the marks.
16. Tighten the chain housing to engine stud. The stud is installed under the timing chain. Torque it to 19 ft. lbs. (26 Nm).
17. Reload the timing chain tensioner as follows:
    a. Form a keeper from heavy gauge wire.
    b. Slightly, compress the shoe plunger and insert a small screwdriver into the access hole.
    c. Release the ratchet pawl and compress the plunger completely into the hole.
    d. Insert the keeper between the access hole and the blade.
18. Install or connect the following:
    - Tensioner assembly to the chain housing. Torque the bolts to 89 inch lbs. (10 Nm).

➡Recheck plunger assembly installation. It is correctly installed when the long end is toward the crankshaft.

   - Tensioner shoe and retainer. Torque the bolts to 89 inch lbs. (10 Nm).

19. Remove the alignment dowel pins. Rotate crankshaft clockwise 2 full rotations. Align the crankshaft timing mark with mark on cylinder block and reinstall alignment dowel pins. Alignment dowel pins will slide in easily if engine is timed correctly.

⁂ WARNING

If the engine is not correctly timed, severe engine damage could occur.

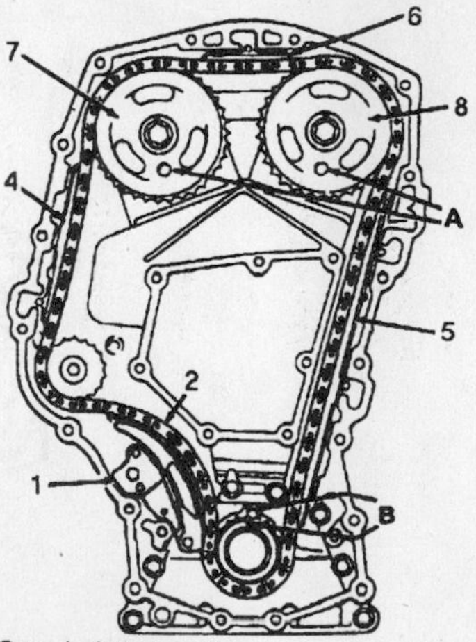

A. Camshaft timing alignment pin locations
B. Crankshaft gear timing marks
1. Shoe asm. timing chain tensioner
2. Timing chain
3. Timing chain tensioner
4. R.H. timing chain guide
5. L.H. timing chain guide
6. Upper timing chain guide
7. Exhaust camshaft sprocket
8. Intake camshaft sprocket

79227321

Timing chain and sprocket alignment positions—2.4L DOHC engine

20. Install or connect the following:
    - Timing chain guides
    - New seal into the front cover by lubricating the seal lip and tapping it into place
    - Front cover and gaskets. Torque the nuts and bolts to 106 inch lbs. (12 Nm).
    - Crankshaft balancer. Torque the bolt to 129 ft. lbs. (175 Nm).
    - Right front lower splash shield
    - Front wheel. Torque the nuts to 100 ft. lbs. (140 Nm).
    - Right engine mount bracket. Torque the bolts to 81 ft. lbs. (110 Nm) plus an additional 90 degree turn.
    - Right engine mount. Torque the bolt to 49 ft. lbs. (60 Nm).
    - Upper cover vent hose
21. Remove the engine support.
22. Install or connect the following:
    - Alternator. Torque the bolts to 37 ft. lbs. (50 Nm).
    - Accessory drive belt
    - Coolant surge tank
    - Negative battery cable
23. Refill the cooling system.
24. Start the engine and check for leaks.

3.1L and 3.4L Engines

1. Before servicing the vehicle, refer to the precautions in the beginning of this section.
2. Disconnect the negative battery cable.
3. Drain the cooling system.
4. Drain the engine oil.
5. Discharge and recover the A/C refrigerant.
6. Install an engine support fixture.
7. Remove or disconnect the following:
    - Front engine mount and bracket
    - Accessory drive belt
    - Air cleaner assembly
    - Air intake duct
    - 2 upper air conditioning compressor mounting bolts
    - Power steering pump
    - Alternator and bracket
    - Right front wheel
    - Right wheel well splash shield
    - Crankshaft balancer
    - Drive belt tensioner
    - Right Wheel Speed sensor (WSS) harness at the suspension support
    - Lower ball joint
    - Stabilizer bar from the control arm and suspension support
    - Suspension support
    - A/C compressor-to-oil pan bolts
    - Oil filter and adapter
    - Starter motor
    - Oil pan
    - Crankshaft Position (CKP) sensor
    - Lower front cover bolts
    - Coolant bypass hose
    - Upper radiator hose
    - Engine front cover
8. Place the number one piston at Top Dead Center (TDC).
9. Remove or disconnect the following:
    - Camshaft sprocket
    - Timing chain
    - Crankshaft sprocket

To install:

10. Install the crankshaft sprocket.

➡Be sure the timing mark on the crankshaft sprocket is pointing toward the mark on the chain damper.

11. Place the timing chain over the camshaft sprocket and hold the sprocket so the timing mark is pointing down and the timing chain is hanging off the sprocket.
12. Loop the timing chain under the crankshaft sprocket and install the camshaft sprocket on the camshaft.
13. Verify that the marks are aligned; the camshaft sprocket will be at the 6 o'clock position and the crankshaft sprocket at the 12 o'clock position.

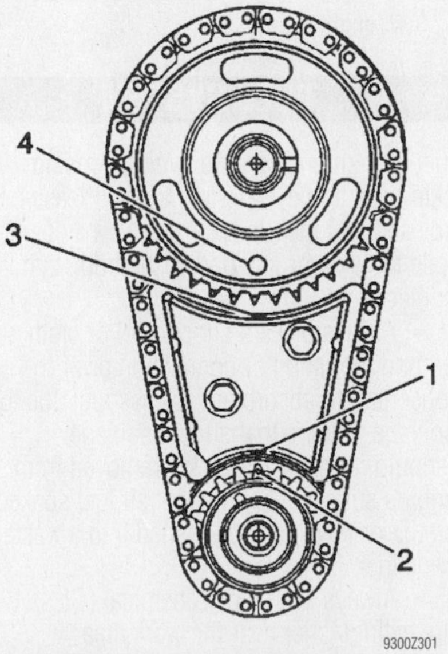

Be sure to align the damper mark (1) with the crankshaft mark (2) and the damper mark (3) with the camshaft sprocket mark (4)—3.1L and 3.4L engines

➡The No. 1 piston will be at TDC and the No. 4 piston will also be at TDC but on the compression stroke.

14. Torque the camshaft sprocket bolt to 103 ft. lbs. (140 Nm).
15. Install or connect the following:
    - Front cover, using a new gasket and seal. Torque the small bolts to 15 ft. lbs. (21 Nm) and the long bolts to 35 ft. lbs. (47 Nm).
    - Upper radiator hose
    - Coolant bypass hose
    - CKP sensor
    - Oil pan. Torque the bolts to 18 ft. lbs. (25 Nm).
    - Starter motor. Torque the bolts to 32 ft. lbs. (43 Nm).
    - Oil filter adapter and filter
    - A/C compressor-to-oil pan bolts. Torque the bolts to 37 ft. lbs. (50 Nm).
    - Suspension support. Torque the bolts to 61 ft. lbs. (82 Nm).
    - Stabilizer bar. Torque the bolts to 13 ft. lbs. (17 Nm).
    - Ball joint. Torque the nuts to 41 ft. lbs. (55 Nm).
    - WSS electrical connector
    - Drive belt tensioner. Torque the bolts to 33 ft. lbs. (45 Nm).
    - Crankshaft balancer. Torque the bolt to 76 ft. lbs. (103 Nm).
    - Right wheel well splash shield
    - Right front wheel
    - Alternator. Torque the front bolt to 37 ft. lbs. (50 Nm) and the rear bolt to 18 ft. lbs. (25 Nm).
    - 2 upper air conditioning compressor mounting bolts. Torque the bolts to 37 ft. lbs. (50 Nm).
    - Power steering pump. Torque the bolts to 25 ft. lbs. (34 Nm).
    - Air intake duct
    - Air cleaner assembly
    - Accessory drive belt
    - Front engine mount and bracket. Torque the 8mm bolts to 15 ft. lbs. (20 Nm) and the 12mm bolts to 30 ft. lbs. (40 Nm).
    - Negative battery cable
16. Evacuate and recharge the A/C system.
17. Refill the engine oil and coolant.
18. Start the engine and check for leaks.

Piston and Ring

POSITIONING

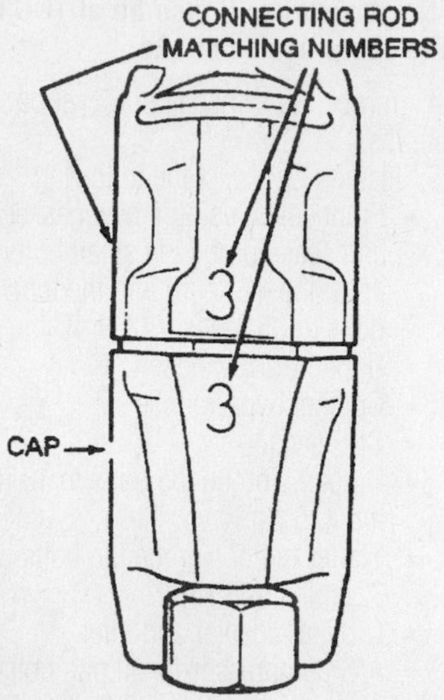

Connecting rod and cap installation. Be sure to matchmark the cap and rod prior to disassembly, as shown—All engines

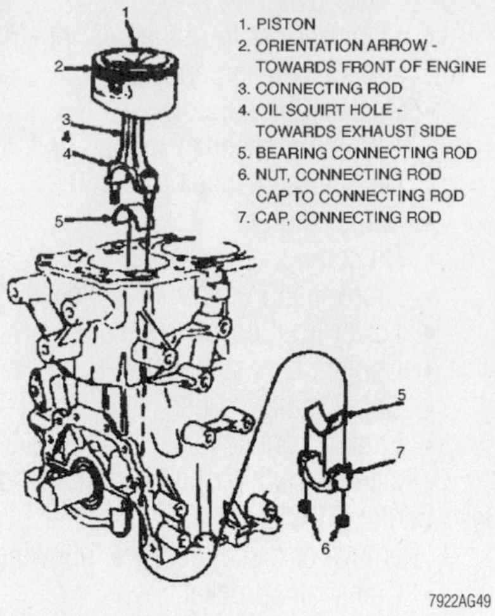

Piston and connecting rod assembly positioning—2.4L engine

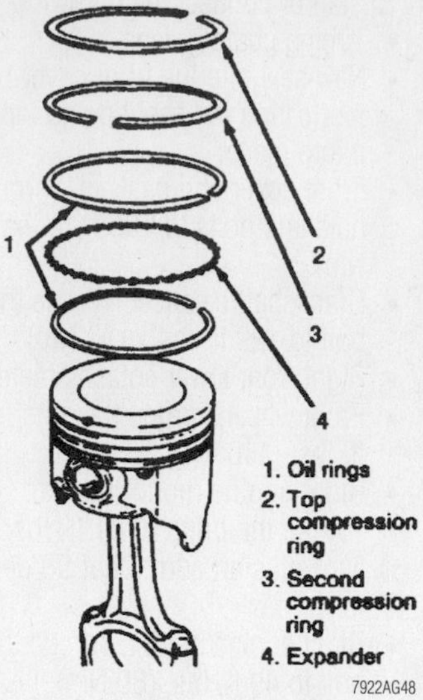

Piston ring postioning—3.1 and 3.4L engines

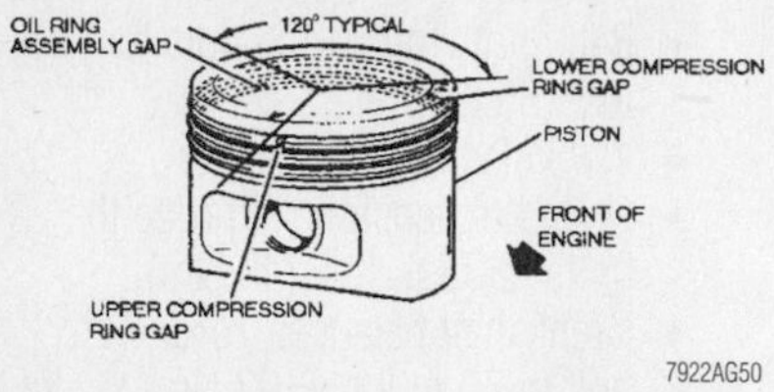

Piston ring end-gap spacing—2.4L engine

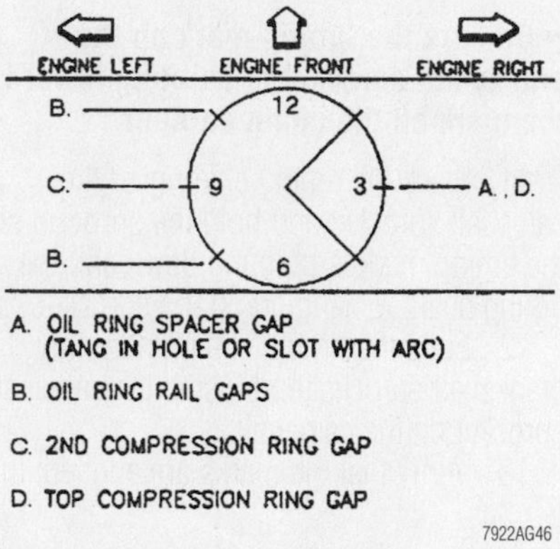

Piston ring end-gap spacing—3.1 and 3.4L engines

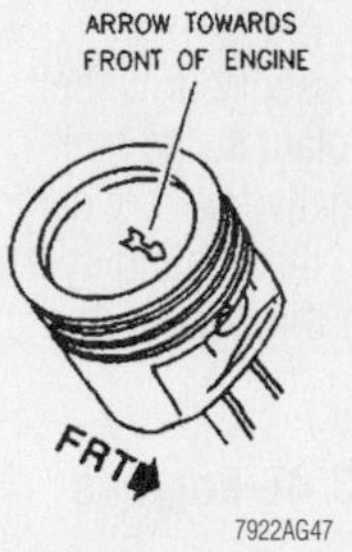

Piston positioning. Often the arrow is replaced by a notch, which also must face the front of the engine—3.1 and 3.4L engines

FUEL SYSTEM

Fuel System Service Precautions

Safety is the most important factor when performing not only fuel system maintenance but any type of maintenance. Failure to conduct maintenance and repairs in a safe manner may result in serious personal injury or death. Maintenance and testing of the vehicle's fuel system components can be accomplished safely and effectively by adhering to the following rules and guidelines.

- To avoid the possibility of fire and personal injury, always disconnect the negative battery cable unless the repair or test procedure requires that battery voltage be applied.
- Always relieve the fuel system pressure prior to disconnecting any fuel system component (injector, fuel rail, pressure regulator, etc.), fitting or fuel line connection. Exercise extreme caution whenever relieving fuel system pressure, to avoid exposing skin, face and eyes to fuel spray. Please be advised that fuel under pressure may penetrate the skin or any part of the body that it contacts.
- Always place a shop towel or cloth around the fitting or connection prior to loosening to absorb any excess fuel due to spillage. Ensure that all fuel spillage (should it occur) is quickly removed from engine surfaces. Ensure that all fuel soaked cloths or towels are deposited into a waste container.
- Always keep a dry chemical (Class B) fire extinguisher near the work area.
- Do not allow fuel spray or fuel vapors to come into contact with a spark or open flame.
- Always use a backup wrench when loosening and tightening fuel line connection fittings. This will prevent unnecessary stress and torsion to fuel line piping. Always follow the proper torque specifications.
- Always replace worn fuel fitting O-rings with new. Do not substitute fuel hose or equivalent, where fuel pipe is installed.

Fuel System Pressure

RELIEVING

2.4L Engine

1. Before servicing the vehicle, refer to the precautions in the beginning of this section.
2. Loosen the fuel filler cap in order to relieve the pressure in the tank (do not tighten at this time).
3. Detach the fuel pump electrical connector.
4. Start and run the vehicle until it stalls, then engage the starter for an addi-

tional 3 seconds to ensure the relief of any remaining pressure.

5. Disconnect the negative battery cable.
6. Once the tests or repairs are completed, reattach the fuel pump electrical connector.
7. Connect the negative battery cable.
8. Tighten the fuel filler cap.
9. prime the fuel system by cycling the ignition switch **ON** for 2 seconds, **OFF** for 10 seconds, then **ON** again. Repeat, if necessary to build system pressure.

3.1L and 3.4L Engines

1. Before servicing the vehicle, refer to the precautions in the beginning of this section.
2. Disconnect the negative battery cable in order to avoid possible fuel discharge if an accidental attempt is made to start the engine.
3. Loosen the fuel tank filler cap in order to relieve fuel tank pressure.
4. Connect a fuel pressure gauge (with bleed hose) to the fuel pressure test port connection. Wrap a towel around the fuel pressure connection when installing the fuel pressure gauge in order to avoid fuel spillage.
5. Install the bleed hose into an approved container and open the valve in order to bleed the fuel system pressure. The fuel pipe connections are now safe for servicing.
6. Drain any fuel remaining in the fuel pressure gauge into an approved container.

Fuel Filter

REMOVAL & INSTALLATION

1. Before servicing the vehicle, refer to the precautions in the beginning of this section.
2. Relieve the fuel system pressure.
3. Remove or disconnect the following:
   - Fuel line from the filter, using a backup wrench
   - Quick-connect fitting from the fuel filter by compressing the tabs while pulling outward on the line
   - Fuel filter from the mounting bracket

To install:

4. Install or connect the following:
   - Fuel filter to the mounting bracket
   - Fuel line using a backup wrench. Torque the fitting to 20 ft. lbs. (27 Nm).

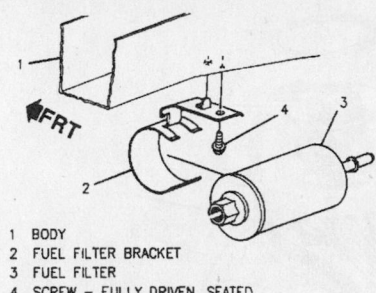

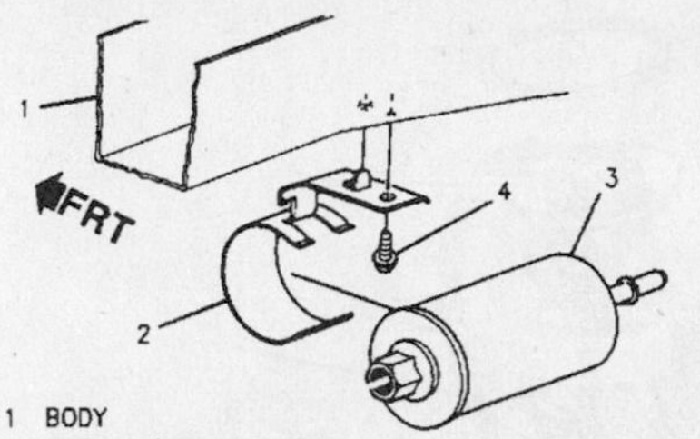

Exploded view of the fuel filter mounting

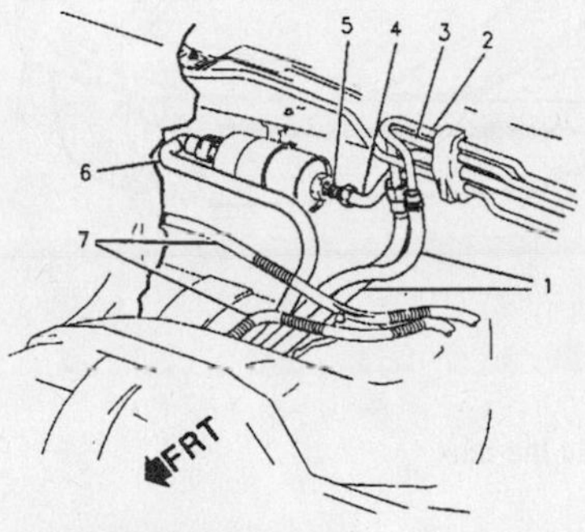

Fuel filter mounting location and component identification

   - Quick-connect fitting to the fuel filter
   - Negative battery cable
5. Pressurize the fuel system and verify no leaks.

Fuel Pump

REMOVAL & INSTALLATION

1. Before servicing the vehicle, refer to the precautions in the beginning of this section.
2. Relieve the fuel system pressure.
3. Drain the fuel tank.
4. Remove or disconnect the following:
5. While holding the modular fuel sender assembly down, remove the snapring, on 1998–99 models or use Fuel Sender Lock Nut Tool J-39765 in order to press down and rotate the cam lock ring, then remove the lock ring on 2000–01 models.

✻✻ WARNING

If the modular fuel sender assembly is retained by a snapring, it may spring up from its position. When removing the modular fuel sender from the tank, be aware that the reservoir bucket is full of fuel. It must be tipped slightly during removal to avoid damage to the float.

   - Fuel sender assembly
   - External fuel strainer
   - Connector retainer from the wiring harness and the fuel pump
6. Gently release the tabs on the sides of the fuel sender at the cover assembly. Begin by squeezing the sides of the reservoir and releasing the tab opposite the fuel level sensor. Move clockwise to release the second and third tab in the same manner.
7. Remove or disconnect the following:
   - Fuel pump electrical connection by lifting the cover assembly
   - Baffle and pump assembly from the retainer by rotating the fuel pump baffle counterclockwise
   - Fuel pump outlet by sliding it out of slot
   - Fuel pump outlet seal

To install:

8. Install or connect the following:
   - Fuel pump outlet with a new seal by sliding it in the reservoir cover slots
   - Fuel pump and baffle assembly onto the reservoir retainer by rotating it clockwise until seated
   - Lower retainer assembly partially into the reservoir by aligning all 3 sleeve tabs and pressing the retainer onto the reservoir making sure all 3 tabs are firmly seated

➡Gently pull on the fuel pump reservoir to assure it is secure. If not secure, replace the entire fuel sender.

   - Connector retainer to the wiring harness and the fuel pump
   - External fuel strainer
   - Snapring to the retainer slots while holding the modular fuel sender assembly down on 1998–99 models or use Fuel Sender Lock Nut Tool J-39765 in order to install the cam lock ring on 2000–01 models

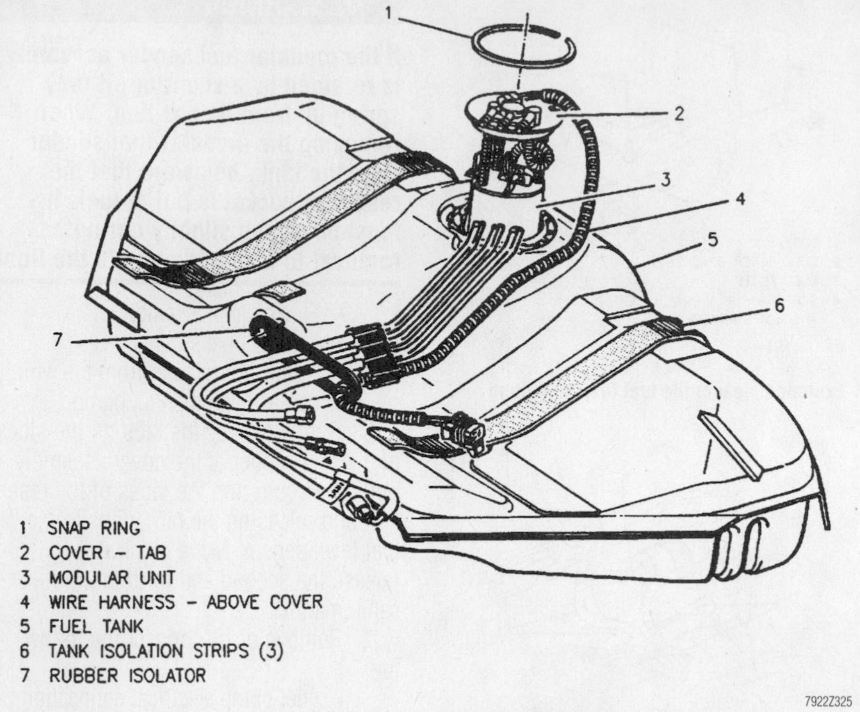

Exploded view of the fuel sender assembly mounting to the tank

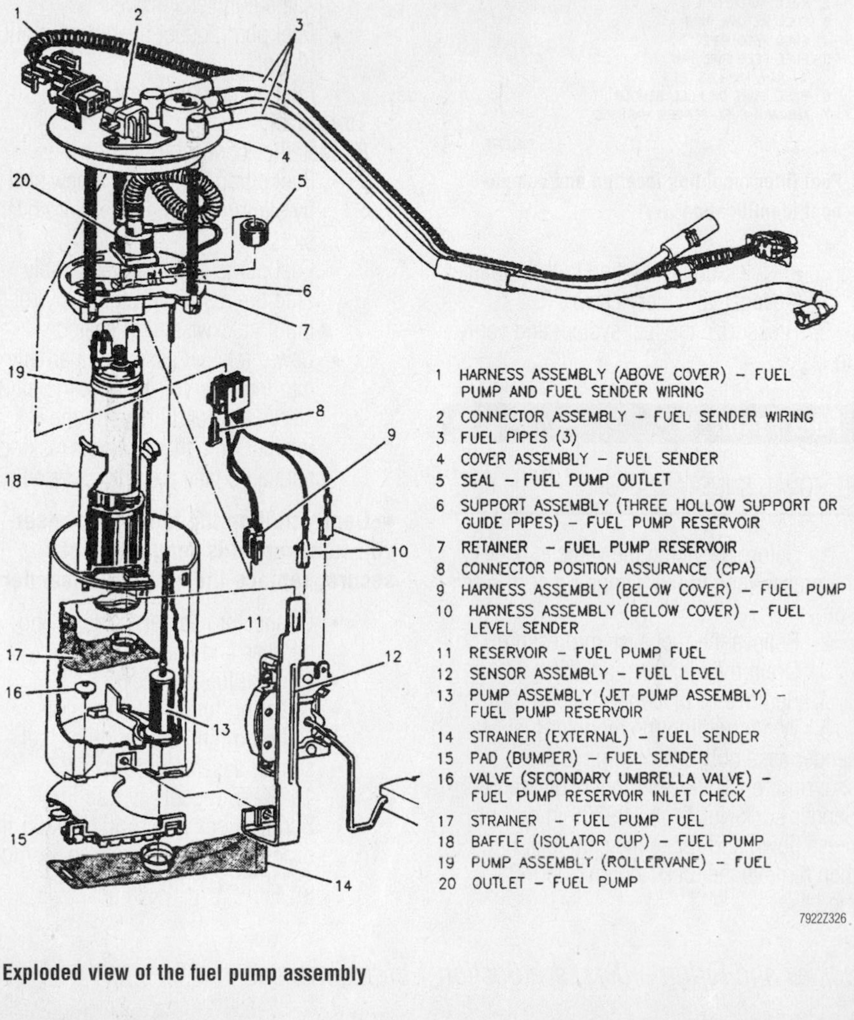

Exploded view of the fuel pump assembly

- Fuel tank
- Negative battery cable

9. Pressurize the fuel system and verify no leaks.

Fuel Injector

REMOVAL & INSTALLATION

2.4L Engine

1. Before servicing the vehicle, refer to the precautions in the beginning of this section.
2. Relieve the fuel system pressure.
3. Remove or disconnect the following:
   - Air cleaner outlet resonator
   - Fuel pressure regulator vacuum hose
   - Camshaft Position (CMP) sensor electrical connector
   - Fuel injector electrical connectors
   - Fuel inlet pipe at the fuel rail
   - Fuel return pipe bracket screw and separate the fuel pipe from the pressure regulator
   - Fuel rail assembly
   - Fuel injector retaining clip
   - Fuel injector

To install:

4. Lubricate the O-rings with engine oil prior to installation.
5. Install or connect the following:
   - Fuel injector with new O-rings
   - Fuel injector retaining clip
   - Fuel rail assembly. Torque the bolts to 19 ft. lbs. (26 Nm).
   - New fuel pipe O-rings lubricated with engine oil
   - Fuel inlet pipe. Torque the fitting to 22 ft. lbs. (30 Nm).

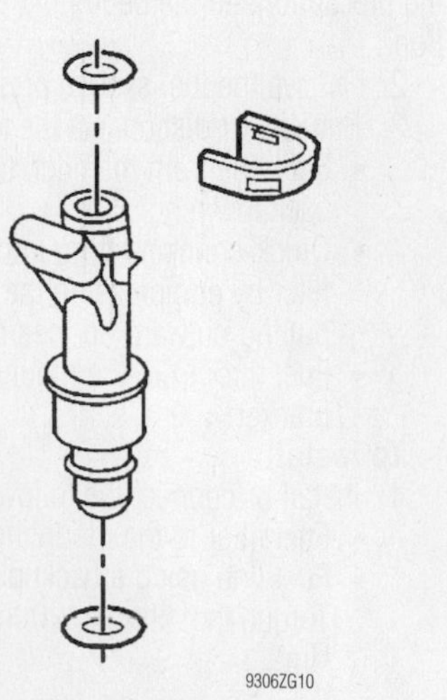

Exploded view of the fuel injector—2.4L engine

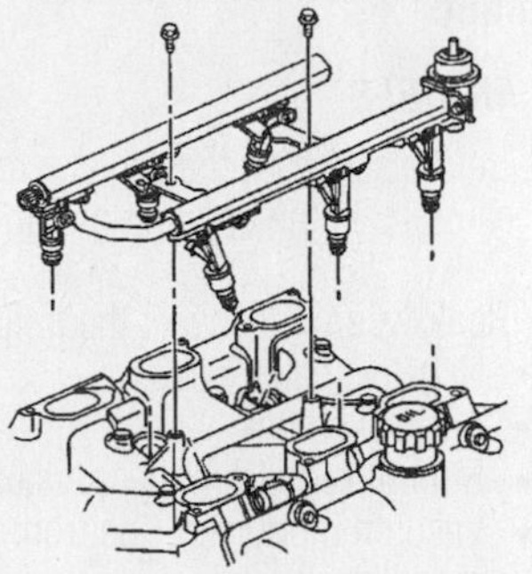

9306ZG11

Exploded view of the fuel rail assembly—3.4L engine—3.1L engine is similar

- Fuel return pipe to the pressure regulator. Torque the screw to 53 inch lbs. (6 Nm).
- Fuel injector electrical connectors
- CMP sensor electrical connector
- Fuel pressure regulator vacuum hose
- Air cleaner outlet resonator
- Negative battery cable

6. Re-pressurize the fuel system and check for leaks.

3.1L and 3.4L Engines

1. Before servicing the vehicle, refer to the precautions in the beginning of this section.

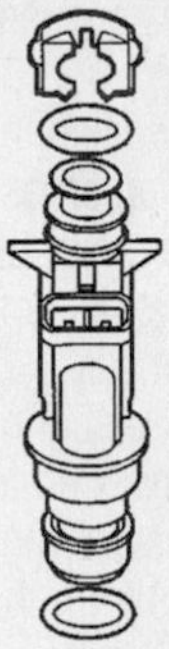

9306ZG12

Exploded view of the fuel injector—3.4L engine—3.1L engine is similar

2. Relieve the fuel system pressure.
3. Remove or disconnect the following:
   - Upper intake manifold
   - Fuel feed pipe
   - Fuel return pipe from the pressure regulator
   - Main fuel injector wiring harness electrical connector
   - Engine Coolant Temperature (ECT) sensor electrical connector
   - Fuel rail assembly
   - Fuel injector electrical connectors
   - Fuel injector-to-fuel rail clip
   - Fuel injector

To install:

WARNING

If the fuel injector O-rings are color coded, install the black O-ring in the upper position and the brown O-ring in the lower position.

4. Lubricate the O-rings with engine oil prior to installation.
5. Install or connect the following:
   - New fuel injector O-rings
   - Fuel injector
   - Fuel injector retaining clip
   - Fuel injector electrical connectors
   - Fuel rail assembly. Torque the bolts to 89 inch lbs. (10 Nm).
   - ECT sensor electrical connector
   - Main fuel injector wiring harness electrical connector
   - New fuel pipe O-rings lubricated with engine oil
   - Fuel feed pipe at the fuel rail. Torque the pipe nut to 13 ft. lbs. (17 Nm).
   - Fuel return pipe to the pressure regulator. Torque the pipe nut to 13 ft. lbs. (17 Nm).
   - Upper intake manifold
6. Re-pressurize the fuel system and check for leaks.

DRIVE TRAIN

Transaxle

REMOVAL & INSTALLATION

Manual

1998 MODELS

1. Before servicing the vehicle, refer to the precautions in the beginning of this section.
2. Disconnect the negative battery cable.
3. Drain the transaxle fluid.
4. Install an engine support fixture and use the fixture to take the weight off the engine mounts.
5. Remove or disconnect the following:
   - Left side sound insulator
   - Clutch master cylinder pushrod from the clutch pedal
   - Air cleaner and duct work from the throttle body
   - Shift cable from the manual shift lever on the transaxle
   - Wiring harness from the transaxle mount bracket
   - Upper transaxle mount
   - Clutch master cylinder from the slave cylinder
   - Ground wires from the transaxle
   - Backup light switch connector
   - Transaxle vent tube
   - Rear transaxle mount
6. Lower the powertrain assembly with the support fixture.
   - Front wheels
   - Left inner fender well splash shield
   - Anti-lock Brake System (ABS) Wheel Speed sensor (WSS) electrical connectors
   - Flywheel housing cover
   - Vehicle Speed Sensor (VSS) from the transaxle
   - Ball joints from the steering knuckles
   - Left side stabilizer link
   - Left side U-bolt from the sway bar
   - Left side suspension support
   - Halfshafts from the transaxle
   - Front lower transaxle mount
7. Support the transaxle with a jack.
8. Remove the transaxle from the vehicle.

To install:

9. Install or connect the following:
   - Transaxle. Torque the bolts to 55 ft. lbs. (75 Nm).
   - Front transaxle mount. Torque the bolts to 55 ft. lbs. (75 Nm).
   - Flywheel housing cover. Torque the bolts to 89 inch lbs. (10 Nm).
   - Halfshafts to the transaxle
   - Left side suspension support. Torque the bolts to 89 ft. lbs. (120 Nm).
   - Left side U-bolt to the sway bar. Torque the nuts to 22 ft. lbs. (30 Nm).
   - Ball joints. Torque the nuts to 45 ft. lbs. (61 Nm).
   - Stabilizer link. Torque the nuts to 22 ft. lbs. (30 Nm).
   - WSS electrical connectors
   - Left inner fender well splash shield

- Front wheels
- Vehicle Speed Sensor (VSS) to the transaxle. Torque the bolt to 80 inch lbs. (9 Nm).
- Rear transaxle mount. Torque the bolts to 44 ft. lbs. (60 Nm).
- Ground wires to the transaxle
- Transaxle vent tube
- Backup light switch connector
- Upper transaxle mount. Torque the bolts to 55 ft. lbs. (75 Nm).
- Clutch master cylinder to the slave cylinder
- Wiring harness to the transaxle mount bracket

10. Remove the engine support fixture.
11. Install or connect the following:
    - Shift cable to the manual shift lever on the transaxle. Torque the nut to 89 inch lbs. (10 Nm).
    - Air cleaner and duct work to the throttle body
    - Clutch master cylinder pushrod to the clutch pedal
    - Left side sound insulator
    - Negative battery cable
12. Refill the transaxle.

2000–01 MODELS

1. Before servicing the vehicle, refer to the precautions in the beginning of this section.
2. Remove or disconnect the following:
    - Negative battery cable
    - Air cleaner assembly
    - Clutch slave cylinder line and bracket
    - Shifter cables from the transaxle
    - Cable bracket
    - Vehicle Speed Sensor (VSS) electrical connector
    - Backup lamp switch electrical connector
    - Starter motor
3. Tie the radiator to the upper hood latch panel.
4. Install an engine support fixture.
5. Remove or disconnect the following:
    - Transaxle bolts
    - Lower closeout panel
    - Wheels
    - Both wheel well splash shields
    - Lower radiator support
    - Wheel Speed sensor (WSS) electrical connectors
    - Both outer tie rod ends
    - Both ball joints
    - Both stabilizer shaft links
    - Front transaxle mount bolts
    - Side transaxle mount nuts
    - Rear transaxle mount stud
    - Right brake hose
6. Support the suspension crossmember.
7. Remove or disconnect the following:
    - Suspension crossmember bolts
    - Rack and pinion bolts
8. Raise the vehicle off of the front suspension crossmember.
9. Remove or disconnect the following:
    - Front transaxle mount
    - Rear transaxle mount
    - Halfshafts from the vehicle
10. Lower the engine and transaxle assembly enough to clear the left side inner body panel.
11. Support and remove the transmission from the vehicle.

To install:

12. Install or connect the following:
    - Transaxle. Torque the bolts to 74 ft. lbs. (100 Nm).
    - Halfshafts
    - Rear transaxle mount
    - Front transaxle mount
    - Rear transaxle mount bolt. Torque the bolt to 81 ft. lbs. (110 Nm).
    - Front transaxle mount bolts. Torque the bolts to 84 ft. lbs. (115 Nm).
    - Rack and pinion bolts. Torque the bolts to 81 ft. lbs. (110 Nm).
    - Suspension crossmember bolts. Torque the rear bolts to 180 ft. lbs. (245 Nm) and the front bolts to 84 ft. lbs. (115 Nm).
    - Right front brake hose
    - Both stabilizer shaft links. Torque the nuts to 22 ft. lbs. (30 Nm).
    - Both outer tie rod ends. Torque the nuts to 15 ft. lbs. (20 Nm) plus an additional 180 degree turn.
    - Both ball joints. Torque the nuts to 45 ft. lbs. (61 Nm).
    - WSS electrical connectors
    - Lower radiator support
    - Both wheel well splash shields
    - Front wheels
    - Lower closeout panel
13. Remove the engine support fixture.
14. Untie the radiator support.
15. Install or connect the following:
    - Starter motor. Torque the bolts to 66 ft. lbs. (90 Nm) on 2.4L engines and 32 ft. lbs. (43 Nm) on 3.4L engines.
    - Backup lamp switch electrical connector
    - VSS electrical connector
    - Shifter cables
    - Clutch slave cylinder bracket and line
    - Air cleaner assembly
    - Negative battery cable

Automatic

1998 MODELS

1. Before servicing the vehicle, refer to the precautions in the beginning of this section.
2. Remove or disconnect the following:
    - Negative battery cable
    - Air intake duct
    - Shift linkage from the transaxle
    - Vacuum modulator line from the modulator
    - Torque Converter Clutch (TCC) electrical connector
    - Park/Neutral position switch electrical connector
    - Shift solenoid electrical connector
3. Install an engine support fixture.
4. Remove or disconnect the following:
    - Upper (2) transaxle-to-engine bolts
    - Rubber hose from the transaxle vent pipe
    - Remaining upper transaxle-to-engine bolts
    - Both front wheels
    - Right and left engine splash shields
    - Anti-lock Brake System (ABS) Wheel Speed Sensor (WSS) connectors
    - Both ball joints from the control arms
    - Left side stabilizer shaft link pin bolt
    - Left side stabilizer shaft frame bushing clamp nuts
    - Left side suspension support assembly
    - Both halfshafts
    - Engine-to-transaxle brace
    - Starter motor
    - Transaxle converter cover
    - Heater core hose pipe brace-to-transaxle nut and bolt
    - Torque converter-to-flywheel bolts

➡Using a scribe mark the flywheel-to-torque converter relationship to assure proper reassembly.

- Oil level indicator and fill tube
- Transaxle cooler lines
- Vehicle Speed Sensor (VSS)
- Vacuum reservoir tank

5. Position a jack under the transaxle.
6. Remove or disconnect the following:
    - Transaxle mount-to-body bolts
    - Remaining engine-to-transaxle bolts
    - Transaxle

➡Transaxle cooler and lines should be flushed whenever the transaxle has been removed for overhaul or replacement.

To install:

➡Be sure to properly seat the torque converter in the pump, if removed.

7. Install or connect the following:
   - Transaxle
   - Lower transaxle-to-engine bolts. Torque the bolts to 66 ft. lbs. (90 Nm).
   - Transaxle mount-to-body bolts. Torque the bolts to 66 ft. lbs. (90 Nm).
   - Transaxle cooler lines
   - Torque converter-to-flywheel bolts. Torque the bolts to 46 ft. (62 Nm).
   - Oil level indicator and fill tube
   - Vacuum reserve tank
   - Starter motor. Torque the bolts to 32 ft. lbs. (43 Nm).
   - Transaxle converter cover. Torque the bolts to 115 inch lbs. (13 Nm).
   - VSS electrical connector
   - Halfshafts
   - Left side suspension support assembly. Torque the bolts to 89 ft. lbs. (120 Nm).
   - Left stabilizer shaft frame bushing nuts. Torque the nuts to 22 ft. lbs. (30 Nm).
   - Stabilizer shaft link pin bolt. Torque the bolts to 22 ft. lbs. (30 Nm).
   - Engine-to-transaxle brace.
   - Ball joints to the control arms. Torque the nuts to 45 ft. lbs. (61 Nm).
   - WSS harness and connectors
   - Right and left side splash shields
   - Heater core pipe brace-to-transaxle nut and bolt
   - Wheels
   - Upper transaxle-to-engine bolts. Torque the bolts to 66 ft. lbs. (90 Nm).
   - Shift solenoid electrical connector
   - Shift linkage to the transaxle
   - Torque converter clutch electrical connector
   - Park/neutral switch electrical connector
   - Backup lamp switch electrical connector
8. Remove the engine support fixture.
9. Install or connect the following:
   - Rubber hose to the vent pipe
   - Vacuum line to the modulator
   - Air intake duct
   - Negative battery cable
10. Refill the transaxle.
11. Verify proper shift linkage adjustment. Start the engine and check for leaks.
12. Road test the vehicle and verify proper operation.
13. Check the transaxle fluid level.
14. Check and/or adjust the wheel alignment.

1999–01 MODELS

1. Before servicing the vehicle, refer to the precautions in the beginning of this section.
2. Disconnect the negative battery cable.
3. Drain the transaxle.
4. Install an engine support fixture.
5. Remove or disconnect the following:
   - Front transmission mount bolts
   - Splash shield
   - Air cleaner and duct assembly
   - Upper transaxle mount
   - Shifter cable
6. Secure the radiator and condenser to the upper radiator support.
7. Remove or disconnect the following:
   - Ground cables from the engine
   - Park/Neutral Position (PNP) switch electrical connectors
   - Front suspension crossmember
   - Front wheels and splash shields
   - Lower radiator and condenser support
   - Front transaxle mount and bracket
   - Both front Antilock Brake System (ABS) Wheel Speed sensor (WSS) electrical connectors
   - Brake modulator assembly
   - Vehicle Speed Sensor (VSS)
   - Torque converter bolts
   - Ball joints from the control arms
   - Halfshafts
   - Front transaxle mount
   - ABS brake control module
   - Tie rod ends from the steering knuckles
   - Power steering pressure line from the steering rack
   - Brake lines from the retainer under the steering rack
   - Steering column intermediate shaft from the steering gear
8. Place a jack under the transaxle.

✲✲ CAUTION

The torque converter may fall out of the transaxle if it is tilted downward.

9. Remove or disconnect the following:
   - Transaxle-to-engine bolts
   - Transaxle

To install:

10. Install the transaxle onto the engine and torque the bolts to 66 ft. lbs. (90 Nm).
11. Remove the jack from the transaxle.
12. Install or connect the following:
   - Intermediate shaft to the steering gear. Torque the pinch bolt to 30 ft. lbs. (41 Nm).
   - Brake lines in the retainer under the steering gear
   - Power steering pressure line to the steering gear
   - Tie rod ends to the steering knuckles. Torque the nuts to 15 ft. lbs. (20 Nm) plus an additional 180 degree turn.
   - ABS control module
   - Front transaxle mount bracket. Torque the bolts to 96 ft. lbs. (130 Nm).
   - Front transaxle mount. Torque the bolt to 55 ft. lbs. (75 Nm).
   - Halfshafts
   - Ball joints to the control arms. Torque the nuts to 41 ft. lbs. (55 Nm).
   - Torque converter bolts. Torque them to 46 ft. lbs. (62 Nm).
   - VSS
   - Brake modulator assembly. Torque the nut to 22 ft. lbs. (30 Nm) and the fluid lines to 18 ft. lbs. (24 Nm).
   - WSS electrical connectors
   - Lower radiator and condenser support
   - Splash shields and the front wheels
   - Front suspension crossmember. Torque the bolts to 81 ft. lbs. (110 Nm).
   - PNP switch electrical connector
   - Ground wiring to the engine
   - Shifter cable and the upper transaxle mount
   - Upper transaxle mount. Torque the bolts to 89 ft. lbs. (120 Nm).
   - Air cleaner and duct assembly
   - Splash shield
   - Front transaxle mount. Torque the bolts to 89 ft. lbs. (120 Nm).
13. Remove the engine support fixture.
14. Connect the negative battery cable.
15. Refill the transaxle with fluid.
16. Refill and bleed the power steering system.

Clutch

REMOVAL & INSTALLATION

1. Before servicing the vehicle, refer to the precautions in the beginning of this section.
2. Remove or disconnect the following:
   - Negative battery cable
   - Hydraulic line from the clutch actuator (slave cylinder)
   - Transaxle from the vehicle
3. If any clutch components are to be reused, use the following procedure:
   a. Matchmark the clutch pressure plate to the flywheel. This is to retain the balance of the original parts. If all parts are to be replaced with new, this step is not necessary.
   b. If the pressure plate is to be reused, loosen the pressure plate mounting bolts by turning each bolt 1 full turn until all the spring pressure is removed. This helps avoid warping the pressure plate.
   c. Remove the clutch disc and pressure plate.

To install:

4. Apply a small amount of high-temperature grease to the pilot bearing as well

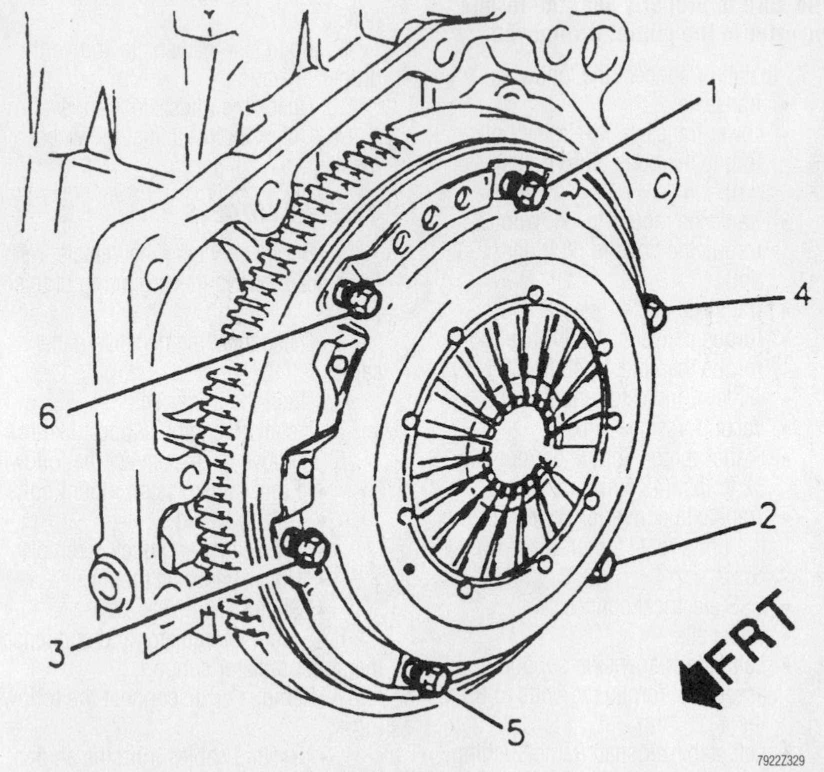

Clutch cover tightening sequence

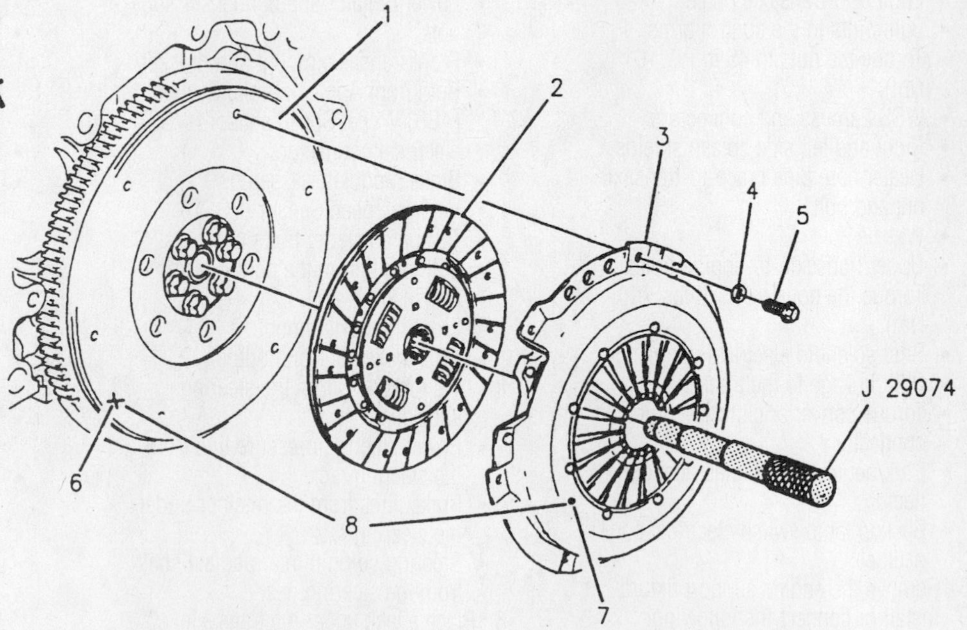

Exploded view of the clutch components—showing the clutch disk alignment tool

as the tip of the transaxle input shaft and the clutch splines.

5. Install or connect the following:
   - Clutch alignment tool into the flywheel
   - Clutch disc onto the tool
   - Pressure plate

➡New pressure plate-to-flywheel replacement bolts are recommended.

6. On 1998–99 models, torque the pressure plate-to-flywheel bolts, in sequence, as follows:
   a. 2.4L engines: bolts to 15 ft. lbs. (20 Nm) plus an additional 30 degrees.
   b. 3.1L engines: bolts to 15 ft. lbs. (20 Nm) plus an additional 45 degrees.
7. On 2000–01 models, torque the pressure plate-to-flywheel bolts, in sequence, as follows:
   a. Step 1: 12 ft. lbs. (16 Nm).
   b. Step 2: 15 ft. lbs. (20 Nm).
   c. Step 3: 30 degree turn if equipped with a 3.1L engine or 45 degree turn if equipped with the 2.4L engine.
8. Remove the clutch alignment tool.
9. Lubricate the inside diameter of the actuator (slave cylinder/throw out bearing) with clutch bearing lubricant.
10. Install or connect the following:
    - Transaxle
    - Hydraulic line to the actuator (slave cylinder)
    - Negative battery cable
11. Bleed the clutch hydraulic system.
12. Road test the vehicle to verify correct operation and easy shifting.

Hydraulic Clutch System

BLEEDING

With Bleeder Screw

1. Before servicing the vehicle, refer to the precautions in the beginning of this section.
2. Be sure the reservoir is kept full throughout this procedure.
3. Depress the clutch pedal.
4. Loosen the bleed screw, located on the actuator cylinder body next to the inlet connection.
5. Torque the bleeder screw to 17 inch lbs. (2 Nm).
6. Repeat previous 3 steps until all air is removed from the system.
7. Refill the fluid reservoir.
8. To check the system, start the engine and wait 10 seconds.
9. Depress the clutch pedal and shift into **R**. If there is any gear clash, air may still be present.

Without Bleeder Screw

1. Before servicing the vehicle, refer to the precautions in the beginning of this section.
2. Remove or disconnect the following:
   - Actuator cylinder from the transaxle
   - Loosen the master cylinder attaching nuts to the ends of the studs
   - Reservoir cap and diaphragm
3. Depress the actuator cylinder pushrod about ¾ in. into its bore and hold the position.
4. Install the reservoir diaphragm and cap while holding the actuator pushrod.
5. Release the pushrod when the diaphragm and cap are properly installed.
6. With the actuator lower than the master cylinder, hold the actuator vertically with the pushrod end facing the ground.
7. Press the actuator pushrod into its bore with ½ in. strokes. Check the reservoir for bubbles. Continue until no bubbles enter the reservoir.
8. Install the master cylinder and actuator. Refill the fluid reservoir.
9. To check the system, start the engine and wait 10 seconds.
10. Depress the clutch pedal and shift into reverse. If there is any gear clash, air may still be present.

Halfshaft

REMOVAL & INSTALLATION

Left and Right Halfshafts

1. Before servicing the vehicle, refer to the precautions in the beginning of this section.
2. Remove or disconnect the following:
   - Negative battery cable
   - Wheel
   - Hub nut and washer
   - Lower ball joint from the steering knuckle
   - Anti-lock Brake System (ABS) electrical connector
   - Sway bar link kit
3. Press the halfshaft from the hub/bearing assembly.
4. Pull the halfshaft from the transmission using a slide hammer and axle removal tool.

To install:

5. Install or connect the following:
   - Halfshaft into the transaxle or intermediate shaft using a brass drift positioned in the inboard joint groove and tap the joint in until it is seated

➡Verify the joint is seated properly by grasping the inboard joint and pulling on it firmly.

⁂ WARNING

DO NOT pull on the halfshaft or damage to the inner joint may result.

   - Halfshaft into the hub assembly
   - Lower ball joint to the steering knuckle. Torque the nut to 44 ft. lbs. (60 Nm).

➡If necessary, tighten the nut up to 60 degree (⅙) turn additional rotation to align the cotter pins. NEVER loosen the nut to make the holes align.

   - New cotter pin
   - Washer and hub nut. Torque the hub nut to 74 ft. lbs. (100 Nm) plus an additional 40 degree turn for 1998 models, or to 284 ft. lbs. (385 Nm), for 1999–01 models.
   - Sway bar link kit. Torque the nut to 13 ft. lbs. (17 Nm).
   - Wheel. Torque the nuts to 100 ft. lbs. (140 Nm).
   - Negative battery cable
6. Check the transaxle fluid level and top off as necessary.

Intermediate Shaft

1. Before servicing the vehicle, refer to the precautions in the beginning of this section.
2. Install an engine support fixture.
3. Remove or disconnect the following:
   - Right side wheel
   - Sway bar link kit
   - Lower ball joint from the steering knuckle
   - Halfshaft from the intermediate shaft
   - Intermediate shaft support bracket-to-engine bolts
   - Intermediate shaft from the transaxle

To install:

4. Install or connect the following:
   - Intermediate shaft into the transaxle. Torque the intermediate shaft support bracket-to-engine bolts to 49 ft. lbs. (66 Nm). Coat the splines of the intermediate shaft with chassis grease.

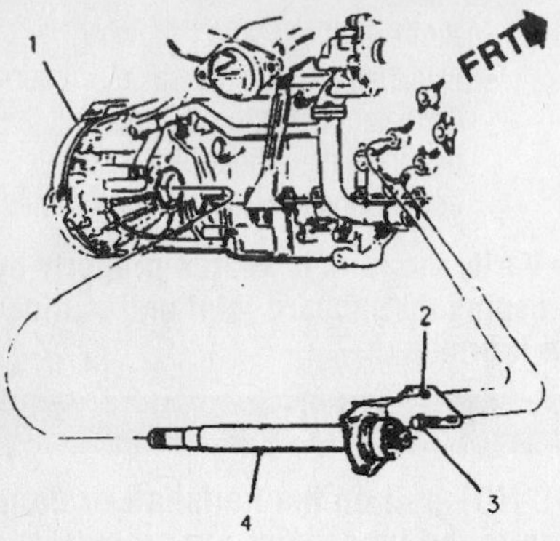

Intermediate shaft components

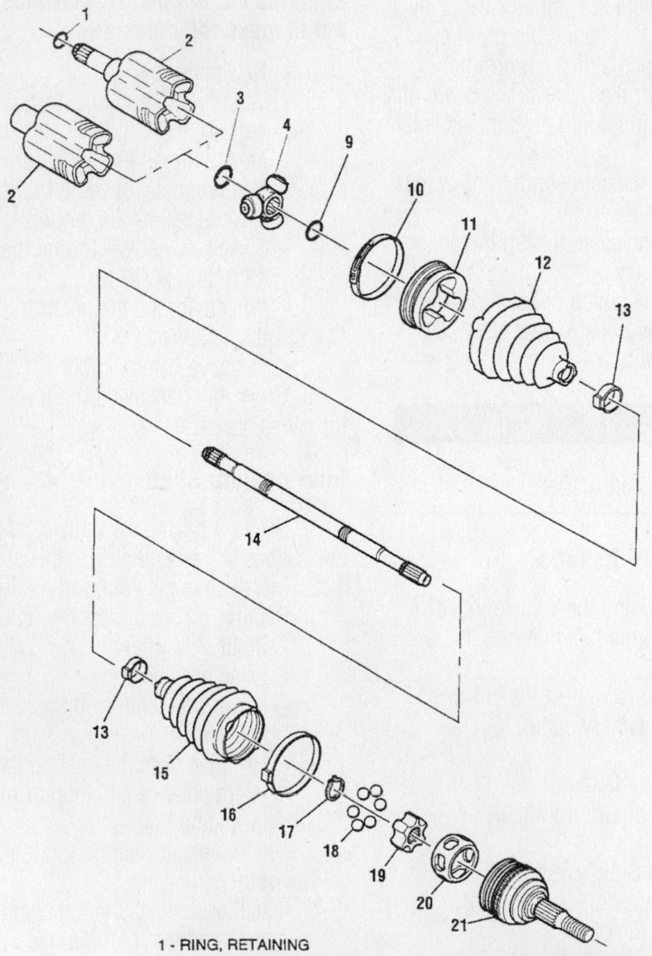

Exploded view of the halfshaft assembly

- Halfshaft to the intermediate shaft
- Lower ball joint to the steering knuckle. Torque the castle nut to 44 ft. lbs. (60 Nm).

➡If necessary, tighten the nut up to 60 degrees (1/6) turn additional rotation to align the cotter pins. NEVER loosen the nut to make the holes align.

- New cotter pin
- Sway bar link kit. Torque the nut to 13 ft. lbs. (17 Nm).
- Wheel. Torque the nuts to 100 ft. lbs. (140 Nm).

5. Remove the engine support fixture.
6. Connect the negative battery cable.
7. Check the transaxle fluid level and top off as necessary.

CV-Joints

OVERHAUL

Outer CV-Joint

1. Before servicing the vehicle, refer to the precautions in the beginning of this section.
2. Remove or disconnect the following:
   - Front wheel
   - Halfshaft, position it in a vise
   - Large CV-joint boot clamp
   - Small CV-joint boot clamp
   - CV-joint boot and slide it back on the shaft
   - Outer race from the halfshaft, by spreading retaining ring
   - Retaining ring from the halfshaft
   - CV-joint boot from the halfshaft and discard it if damaged
3. Disassemble the chrome alloy balls from the CV-joint cage as follows:
   a. Position a brass drift against the CV-joint cage and tap it with a hammer to tilt the cage.
   b. Remove the 1st chrome alloy ball from the cage.
   c. Tilt the cage in the opposite direction.
   d. Remove the opposite chrome alloy ball.
   e. Repeat the procedure until all 6 balls are removed.
4. Disassemble the CV-joint cage and inner race as follows:
   a. Pivot the cage and race 90 degrees to the center line of the outer race.
   b. Align the cage windows with outer race lands.
   c. Remove the cage from the outer race.
   d. Rotate the inner race upward and remove it from the cage.

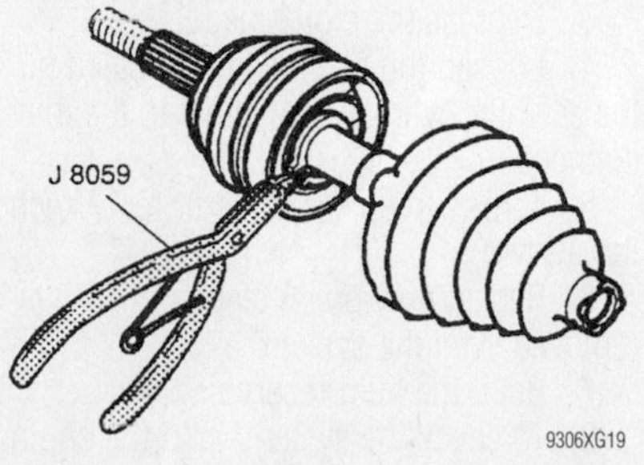

Disconnecting the outer CV-joint from the axle shaft

To install:

5. Lubricate the parts with a light coat of grease.

6. Assemble the CV-joint cage and inner race, as follows:

a. Rotate the inner race 90 degrees to the cage centerline.

b. Align the cage windows with inner race lands.

c. Insert the inner race into the cage by rotating the inner race downward.

d. Insert the cage/inner race into the outer race.

7. Assemble the chrome alloy balls into the CV-joint cage, as follows:

a. Position a brass drift against the CV-joint cage and tap it with a hammer to tilt the cage.

b. Insert the 1st chrome alloy ball into the cage.

c. Tilt the cage in the opposite direction.

d. Insert the opposite chrome alloy ball.

e. Repeat the procedure until all 6 balls are inserted.

8. Install ½ of the grease provided, into the CV-joint.

9. Install or connect the following:

- Small ring clamp on the CV boot
- New retaining ring on the halfshaft
- Large ring clamp on the CV boot
- Outer race assembly onto the halfshaft until the ring engages the halfshaft groove

10. Slide the small end of the CV-joint boot/clamp into place, with the seal lip in the halfshaft groove.

➡Make sure the boot lies flat against the halfshaft.

11. Using a crimp tool, a torque wrench and a breaker bar, crimp the small CV-joint boot clamp to 100 ft. lbs. (136 Nm).

12. Check the clamp gap dimension; if it is not 0.085 in. (2.15mm), continue tightening the clamp until it is.

13. Install ½ kit grease into the CV-joint boot.

14. Measure approximately 0.687 in. (17.5mm) up from the bottom edge of the outer CV-joint assembly.

15. Slide the large end of the CV boot/clamp into place, with the seal lip in place over the outer race.

➡Make sure the boot lies flat against the outer race.

16. Using a crimp tool, a torque wrench and a breaker bar, crimp the large CV-joint boot clamp to 130 ft. lbs. (176 Nm).

17. Check the clamp gap dimension; if it is not 0.102 in. (2.60mm), continue tightening the clamp until it is.

18. Install the halfshaft and the front wheel.

Inner (Tri-Pod) Joint

1. Before servicing the vehicle, refer to the precautions in the beginning of this section.

2. Remove or disconnect the following:

- Front wheel
- Halfshaft and place it in a vise
- Small CV-joint boot clamp
- Large CV-joint boot clamp
- CV-joint boot by sliding it away from the tri-pod joint
- Tri-pod housing from the tri-pod spider
- Inboard spacer ring and slide it rearward on the shaft, using snapring pliers
- Outboard retaining ring
- Tri-pod joint spider assembly by tapping it from the halfshaft with a brass drift
- Inboard tri-pod spider retaining ring
- Trilobal tri-pod bushing from the housing
- CV-joint boot

3. Thoroughly clean and inspect all parts.

To install:

4. Install or connect the following:

- Small boot clamp
- CV-joint boot
- New tri-pod spider retaining ring onto the halfshaft, slide it past the 2nd ring groove

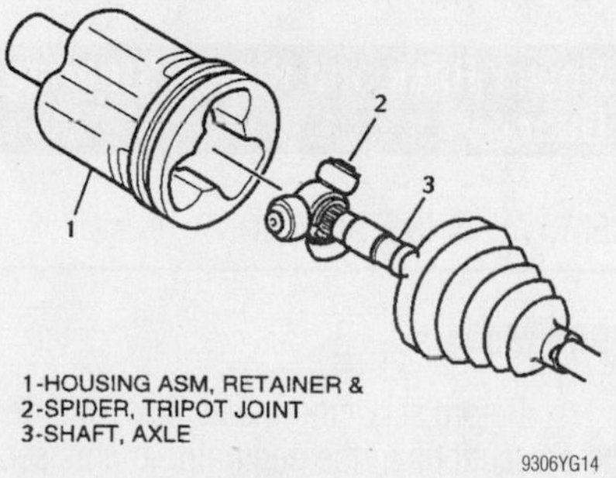

Exploded view of the inner (tri-pod) joint

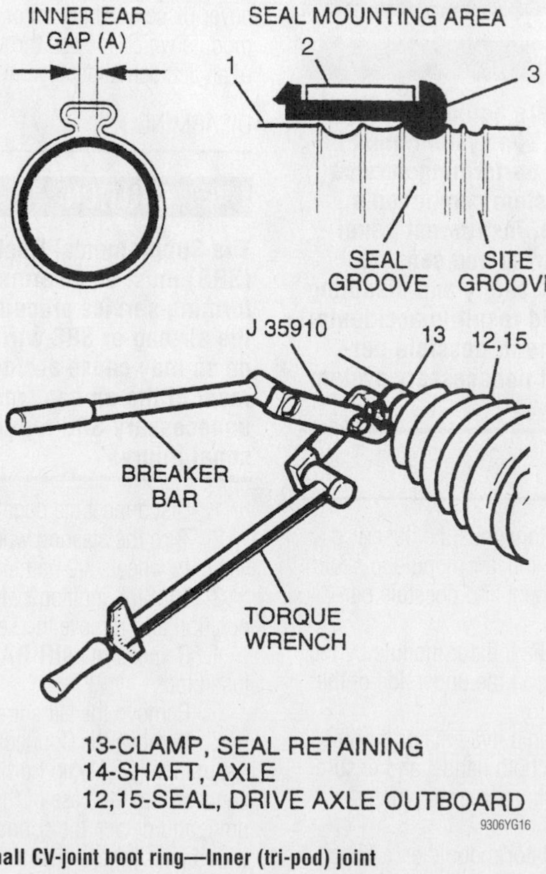

Crimping the small CV-joint boot ring—Inner (tri-pod) joint

Heater Core replacement is covered in Section 2 of this manual

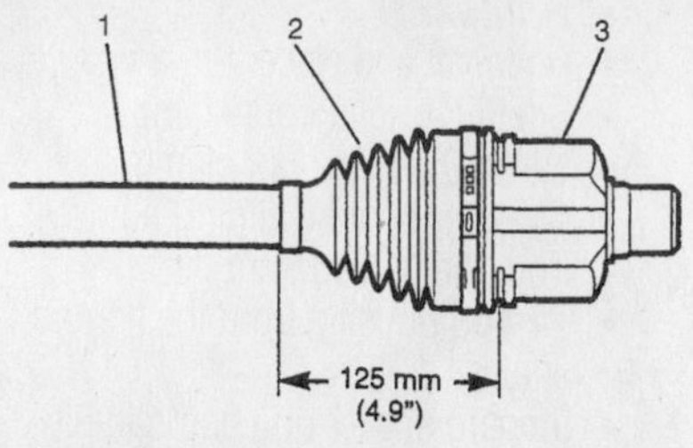

CV-joint boot measurement—Inner (tri-pod) joint

5. Assemble the tri-pod spider assembly onto the halfshaft as follows:
   a. Position the tri-pod spider assembly onto the shop press plate.
   b. Position the halfshaft onto the tri-pod spider assembly, in the shop press.
   c. Press the halfshaft into the tri-pod spider assembly until it passes the 2nd ring groove.

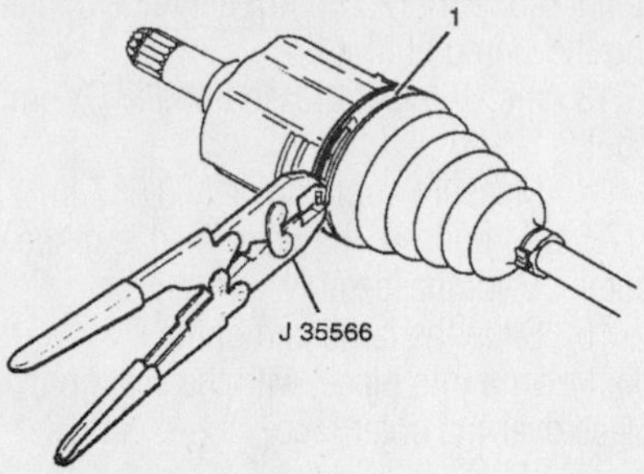

Latching the large CV-joint boot ring—Inner (tri-pod) joint

6. Remove the halfshaft from the shop press.
7. Install or connect the following:
   - Outboard retaining ring into the axle shaft groove using snapring pliers
   - Tri-pod joint spider assembly, slide it against the outboard retaining ring
   - Inboard spacer ring, seat it in the groove
   - ½ of the kit grease into the boot
   - ½ of the kit grease into the tri-pod housing
   - Trilobal tri-pod bushing flush with the tri-pod housing face
   - New large seal clamp onto the CV-joint boot
   - Tri-pod housing, slide it over the tri-pod joint spider assembly
   - CV-joint boot/clamp, slide it into place, over the trilobal tri-pod bushing with the seal lip in the groove

➡Make sure the boot lies flat against the trilobal bushing.

8. Position the CV-joint boot so it measures 4.9 in. (125mm).
9. Using a crimp tool, a torque wrench and a breaker bar, crimp the small CV-joint boot clamp to 100 ft. lbs. (136 Nm).
10. Using a crimp tool, latch the large CV-joint boot clamp.
11. Install the halfshaft and the front wheel.

STEERING AND SUSPENSION

Air Bag

⁂ CAUTION

Some vehicles are equipped with an air bag system. The system must be disabled before performing service on or around system components, steering column, instrument panel components, wiring and sensors. Failure to follow safety and disabling procedures could result in accidental air bag deployment, possible personal injury and unnecessary system repairs.

PRECAUTIONS

Several precautions must be observed when handling the inflator module to avoid accidental deployment and possible personal injury.

- Never carry the inflator module by the wires or connector on the underside of the module.
- When carrying a live inflator module, hold securely with both hands, and ensure that the bag and trim cover are pointed away.
- Place the inflator module on a bench or other surface with the bag and trim cover facing up.
- With the inflator module on the bench, never place anything on or close to the module which may be thrown in the event of an accidental deployment.

DISARMING

⁂ CAUTION

The Supplemental Restraint System (SRS) must be disarmed before performing service procedures around the air bag or SRS wiring. Failure to do so may cause accidental deployment of the air bag, resulting in unnecessary SRS repairs and/or personal injury.

1. Disconnect the negative battery cable.
2. Turn the steering wheel so the vehicle's wheels are pointing straight ahead.
3. Turn the ignition switch to the **LOCK** position and remove the key.
4. Remove the **AIR BAG** fuse from the fuse block.
5. Remove the left sound insulator.
6. Remove the Connector Position Assurance (CPA) clip from the yellow 2-way connector at the base of the steering column, and detach the connector. If equipped with a passenger's side air bag, remove the CPA and detach the yellow 2-way connector from the passenger air bag lead.

REARMING

1. Turn the ignition switch to the **LOCK** position and remove the key.
2. Attach the yellow 2-way connector at the base of steering column and secure it with the Connector Position Assurance (CPA) clip. If equipped with a passenger's side air bag, attach the yellow 2-way connector at the passenger air bag lead and secure it with the CPA clip.
3. Install the left sound insulator.
4. Install the **AIR BAG** fuse in the fuse block.
5. Turn the ignition switch to the **RUN** position and verify that the **AIR BAG** warning lamp flashes 7 times, then turns **OFF**.
6. Connect the negative battery cable.

Power Rack and Pinion Steering Gear

REMOVAL & INSTALLATION

1998 Models

1. Before servicing the vehicle, refer to the precautions in the beginning of this section.
2. Remove or disconnect the following:
   - Left sound insulator from under the dashboard

- Power steering line retainer
- Power brake booster

3. Pull the booster away from the firewall and position away from the rack and pinion mounting clamp. DO NOT remove the master cylinder or allow the brake lines to kink.
4. Remove or disconnect the following:
   - Inner tie rod bolts, lock plates and washers
   - Left and right mounting clamps
   - Power steering lines from the rack and pinion unit
5. Pull the rack away from the firewall and separate the steering shaft from the rack and pinion.
6. Remove or disconnect the following:
   - Column seal from the rack and pinion housing
   - Rack and pinion unit

To install:

7. If when removing the rack and pinion mounting clamp nuts the studs came out, remove the nuts from the studs and install the studs back into the firewall. Torque the studs to 15 ft. lbs. (20 Nm).
8. Install or connect the following:
   - Rack and pinion assembly
   - Column seal
   - Steering shaft coupling to the stub shaft. Torque the pinch bolt to 30 ft. lbs. (41 Nm).
   - Power steering lines to the rack and pinion assembly. Torque the fittings to 20 ft. lbs. (27 Nm).
   - Rack and pinion clamps. Torque the left side clamps first, then the right to 22 ft. lbs. (30 Nm).
   - Inner tie rod(s). Torque the bolts to 65 ft. lbs. (90 Nm).
   - Power brake booster. Torque the nuts to 20 ft. lbs. (27 Nm).
   - Power steering line retainer
   - Left side sound insulator
9. Refill and bleed the power steering system.
10. Check and/or adjust the front end alignment.

1999–01 Models

1. Before servicing the vehicle, refer to the precautions in the beginning of this section.
2. Remove or disconnect the following:
   - Negative battery cable
   - Both front wheels
   - Stabilizer shaft links from the control arms
   - Tie rod ends from the steering knuckles
   - Intermediate shaft lower pinch bolt
   - Through-bolt from the rear transmission mount
   - Power steering line bracket from the crossmember
   - Stabilizer bar
3. Support the rear of the sub-frame with jacks.
4. Remove the rear sub-frame mounting bolts and loosen the front ones.
5. Lower the sub-frame about 3 inches using the jacks.
6. Remove or disconnect the following:
   - Hoses from the steering gear
   - Steering gear through the left wheel opening

To install:

7. Install or connect the following:
   - Steering gear. Torque the bolts to 81 ft. lbs. (110 Nm).
   - Hoses to the steering gear
   - Sub-frame. Torque the bolts in the following order: left rear, right rear, left front and right front to 71 ft. lbs. (110 Nm).
   - Stabilizer bar bracket. Torque the bolts to 49 ft. lbs. (66 Nm).
   - Power steering line bracket
   - Rear transaxle mount. Torque the bolt to 89 ft. lbs. (120 Nm).
   - Intermediate shaft. Torque the pinch bolt to 15 ft. lbs. (20 Nm).
   - Tie rod ends. Torque the nuts to 15 ft. lbs. (20 Nm) plus an additional 180 degree turn.
   - Stabilizer bar links. Torque the nuts to 13 ft. lbs. (17 Nm).
   - Negative battery cable
8. Verify all steering hose fittings are tight.
9. Install the front wheels.
10. Fill the steering reservoir.
11. Install an adapter cap on the fluid reservoir with a vacuum pump attached to it.
12. Apply about 20 inches of vacuum to the system and wait 5 minutes. Typical vacuum drop is 2–3 inches. If the vacuum drop is greater, there may be a leak in the system allowing air to enter.
13. Remove the tools and install the reservoir cap.
14. Start the engine and allow it to idle.

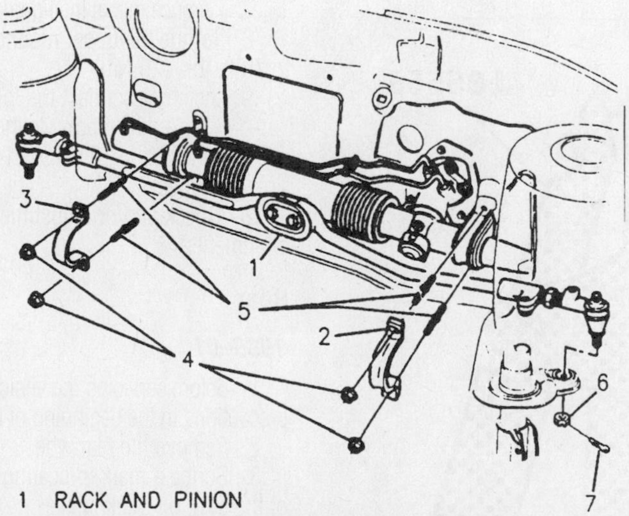

Rack and pinion mounting components–1998 models

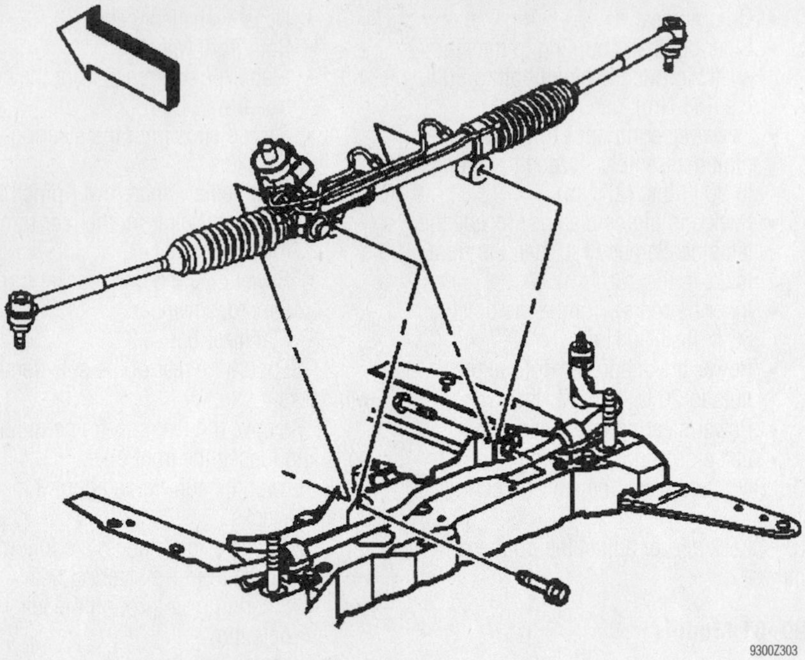

Power steering gear mounting—1999–01 models

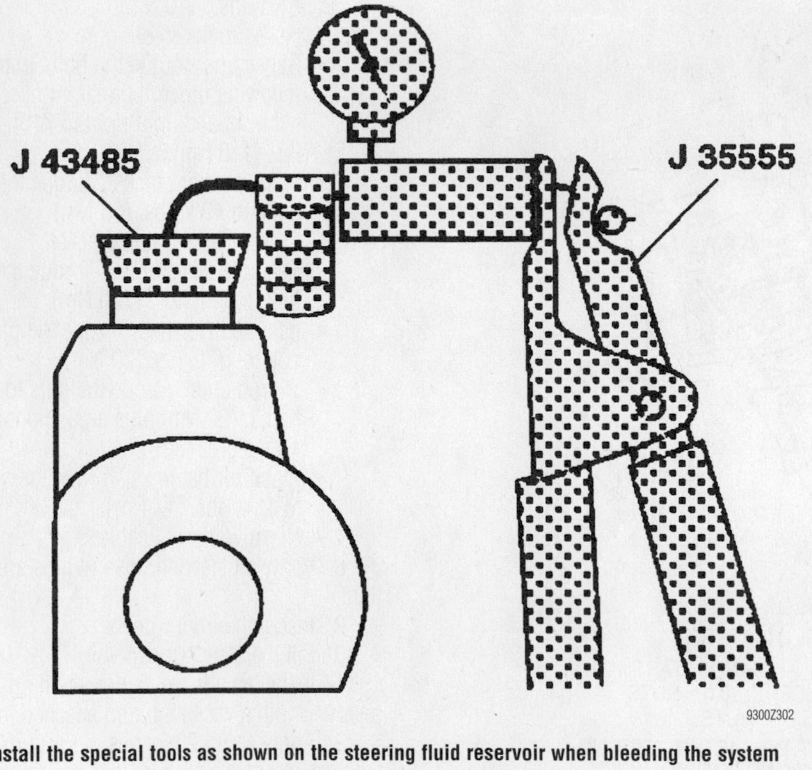

Install the special tools as shown on the steering fluid reservoir when bleeding the system

15. Turn the engine **OFF** and check the fluid level. Do this until the fluid level stabilizes.

16. Start the engine and allow it to idle.

17. Turn the steering wheel in both directions 180–360 degrees 5 times.

18. Turn the engine **OFF** and check the fluid level.

19. Install an adapter cap on the fluid reservoir with a vacuum pump attached to it once again.

20. Apply about 20 inches of vacuum to the system and wait 5 minutes.

21. Remove the tools and check the fluid level. Install the cap.

Strut

REMOVAL & INSTALLATION

Front

1. Before servicing the vehicle, refer to the precautions in the beginning of this section.
2. Support the front crossmember.
3. Remove or disconnect the following:
   - Upper strut mounting fasteners
   - Front wheel
   - Brake line bracket from the strut
4. Scribe reference marks on the front strut and steering knuckle for installation purposes.
5. Remove the strut lower mounting bracket nut and through-bolts.
6. Remove the strut from the vehicle.

To install:

7. Install or connect the following:
   - Strut assembly
   - Upper strut plate mounting nuts finger tight
   - Lower strut bracket to the steering knuckle. Torque the through-bolt/nuts to 133 ft. lbs. (180 Nm) with the reference marks in alignment.
8. Torque the upper mounting fasteners to 18 ft. lbs. (25 Nm).
9. Install or connect the following:
   - Brake line bracket to the strut. Torque the bolt to 10 ft. lbs. (14 Nm).
   - Wheel
10. Check and/or adjust the front end alignment.

Rear

1999–01

1. Before servicing the vehicle, refer to the precautions in the beginning of this section.
2. Remove the rear wheel.
3. Scribe a mark indicating the position of the strut on the knuckle.
4. Remove or disconnect the following:
   - Strut nuts from inside the trunk
   - Strut bolts from the wheel well area
   - Strut-to-knuckle bolts
   - Strut from the vehicle

To install:

5. Install or connect the following:
   - Strut. Torque the upper bolts and nuts to 18 ft. lbs. (25 Nm).
   - Strut on the knuckle by aligning the scribe marks made earlier. Torque the bolts to 89 ft. lbs. (120 Nm).
   - Wheel
6. Check and/or adjust the wheel alignment.

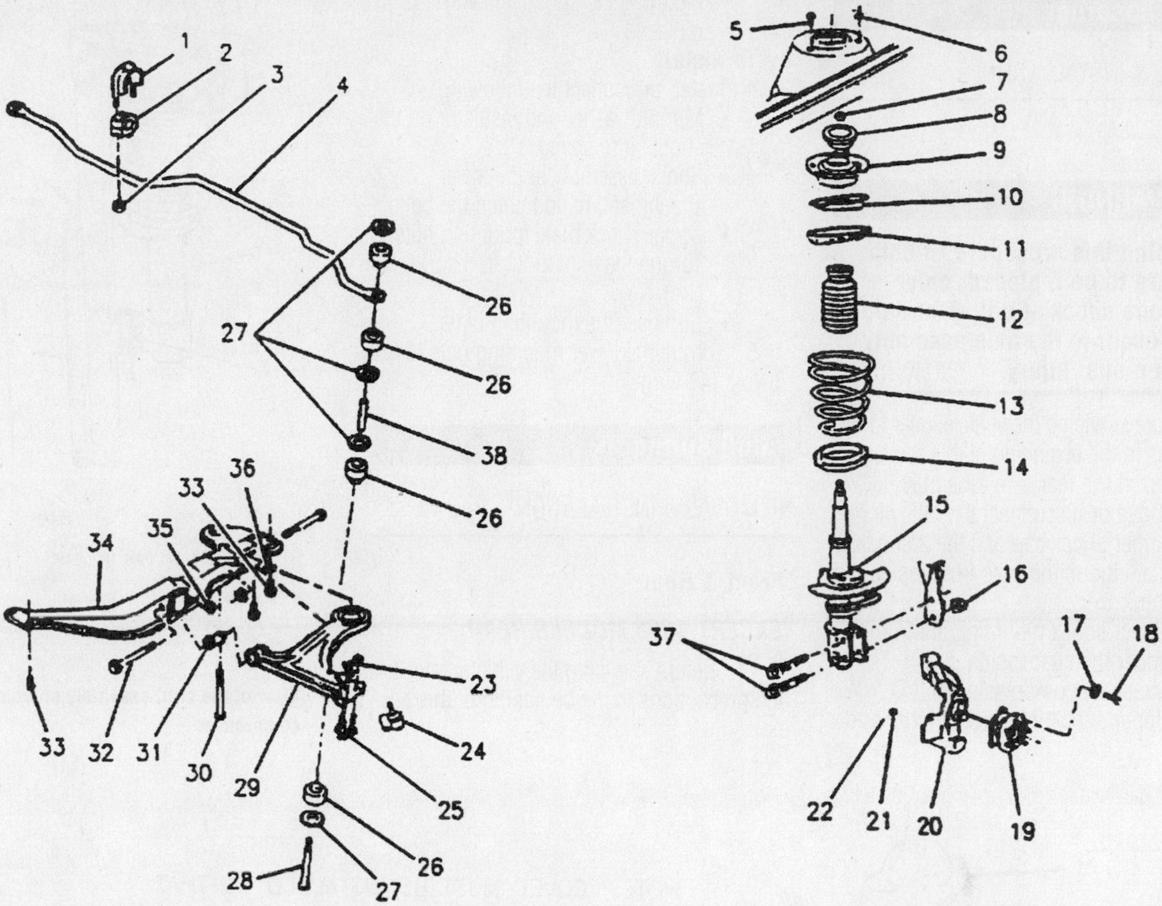

1 CLAMP, STABILIZER SHAFT
2 INSULATOR, STABILIZER SHAFT
3 NUT
4 STABILIZER SHAFT
5 BOLT
6 NUT
7 NUT, STRUT DAMPENER SHAFT
8 RATE WASHER
9 STRUT MOUNT
10 UPPER SPRING SEAT
11 UPPER SPRING INSULATOR
12 DUST TUBE ASSEMBLY
13 SPRING
14 LOWER SPRING INSULATOR
15 STRUT
16 NUT
17 WASHER
18 BOLT
19 HUB AND BEARING ASSEMBLY
20 STEERING KNUCKLE
21 NUT, BALL JOINT
22 COTTER PIN
23 NUT
24 BALL JOINT
25 BOLT
26 INSULATOR, STABILIZER LINK
27 WASHER, STABILIZER LINK
28 BOLT, STABILIZER LINK
29 CONTROL ARM
30 BOLT
31 BUSHING, CONTROL ARM
32 BOLT
33 BOLT
34 SUSPENSION SUPPORT
35 NUT
36 WASHER
37 BOLT
38 SPACER, STABILIZER LINK

7922Z332

Exploded view of the front suspension

Shock Absorber

REMOVAL & INSTALLATION

1998

**** WARNING**

When doing this procedure, if both shocks are to be replaced, only remove one shock at a time or damage can occur to the axle assembly and/or personal injury.

1. Before servicing the vehicle, refer to the precautions in the beginning of this section.
2. Support the rear axle assembly.
3. Remove or disconnect the following:
   - Upper shock cap and the 2 shock plate mounting nuts from inside the trunk
   - Lower shock mounting nut from under the suspension
   - Shock by compressing it
   - Upper shock insulator, washer and mount, if the shock is being replaced

To install:

4. Install or connect the following:
   - Mount, washer and insulator on the shock, if removed.
   - Shock assembly at the lower attachment, hand tighten the bolts
   - Upper shock plate mounting nuts. Torque the nuts to 21 ft. lbs. (29 Nm).
   - Cap over the mounting plate
5. Torque the lower mounting nuts to 35 ft. lbs. (47 Nm).

Coil Spring

REMOVAL & INSTALLATION

Front & Rear

EXCEPT 1998 MODELS REAR

1. Before servicing the vehicle, refer to the precautions in the beginning of this section.

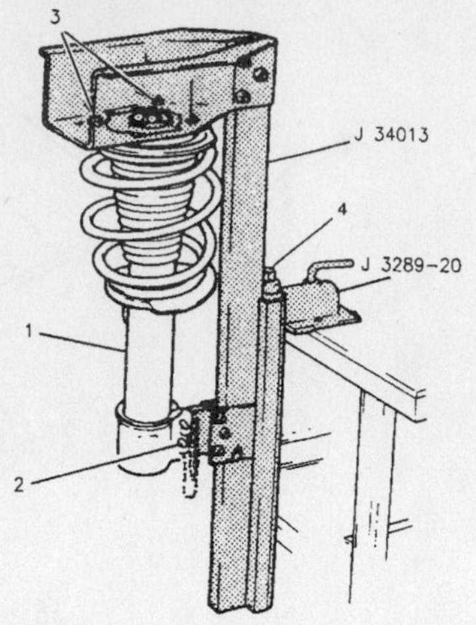

1 STRUT ASSEMBLY
2 INSTALL LOCKING PINS THROUGH STRUT ASSEMBLY
3 TIGHTEN NUTS UNTIL FLUSH WITH STRUT COMPRESSOR
4 COMPRESSOR FORCING SCREW

7922Z334

View of the strut assembly mounted in a compressor

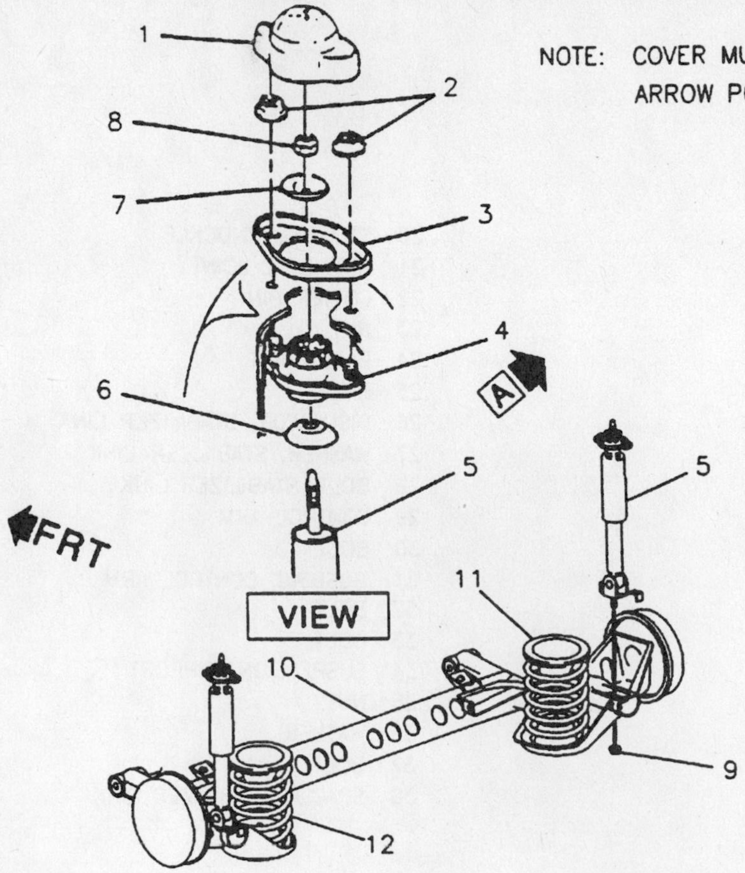

NOTE: COVER MUST BE INSTALLED SO THAT ARROW POINTS TO LEFT SIDE OF VEHICLE

1 COVER (MUST BE INSTALLED WITH ARROW POINTING TO LEFT OF VEHICLE)
2 NUT
3 REINFORCEMENT
4 UPPER SHOCK ABSORBER MOUNT
5 SHOCK ABSORBER
6 WASHER
7 WASHER
8 NUT
9 NUT
10 REAR AXLE; SPRING-ON-CENTER TYPE
11 INSULATOR, COIL SPRING
12 COIL SPRING

7922Z333

Exploded view of the rear suspension—1998 models

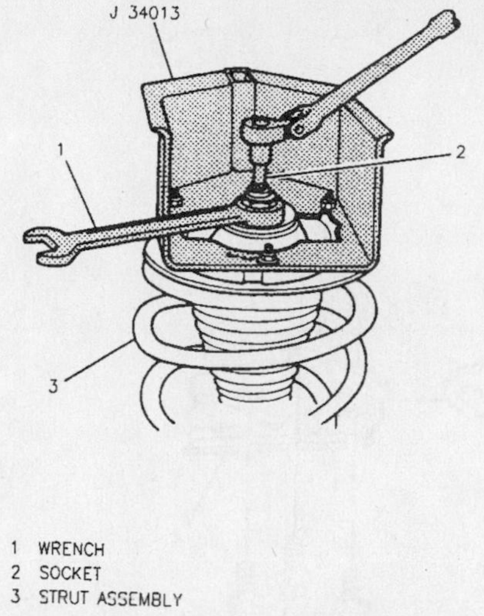

Use a socket and a wrench to remove the damper shaft nut spring cap while compressing the spring

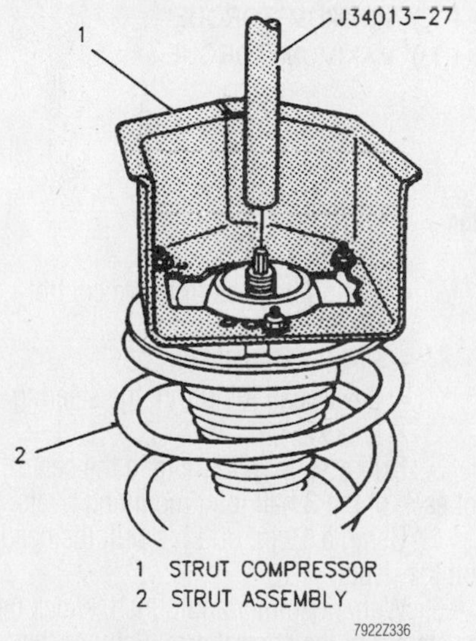

Install the rod to guide the damper shaft straight down through the spring cap while compressing the spring

2. Remove the strut assembly from the vehicle.

3. Mount the strut compressor in a holding fixture.

4. Mount the strut assembly into the compressor. Note that the strut compressor has strut mounting holes drilled for specific vehicle lines.

5. Compress the strut approximately ½ its height after initial contact with the top cap.

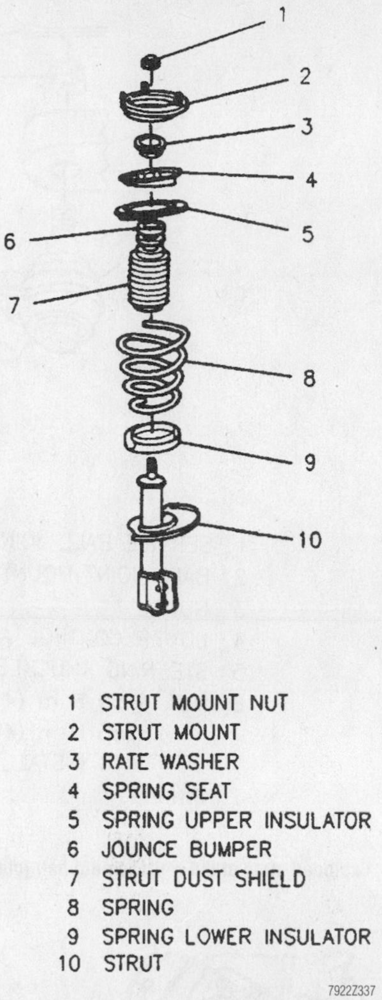

Exploded view of the front strut assembly

WARNING

Never bottom the spring or damper rod.

6. Remove the nut from the strut damper shaft and place alignment/guiding rod J-34013-27 on top of the damper shaft. Use the rod to guide the damper shaft straight down through the spring cap while compressing the spring. Remove the components.

To install:

7. Install the bearing cap into the strut compressor, if removed.

8. Mount the strut assembly in strut compressor, using bottom locking pin only. Extend the damper shaft and install clamp J-34013-20 on the damper shaft.

9. Install the spring over the damper and swing the assembly up so the upper locking pin can be installed.

10. Install all shields, bumpers and insulators on the spring seat. Install the spring seat on top of the spring. Be sure the flat on the upper spring seat is facing in the proper direction. The spring seat flat should be facing the same direction as the centerline of the strut assembly spindle.

11. Install the guiding rod and turn the forcing screw while the guiding rod centers the assembly. When the threads on the damper shaft are visible, remove the guiding rod and install the nut. tighten the nut to 52 ft. lbs. (70 Nm) on 1998–99 models or 34 ft. lbs. (47 Nm) on 2000–01 models. Use a crow's foot line wrench while holding the damper shaft with a socket.

12. Remove the clamp.

1998 MODELS REAR

1. Before servicing the vehicle, refer to the precautions in the beginning of this section.

2. Support the rear axle with a jack.

3. Remove the lower shock absorber mounting nuts. Remove the right and left brake line mounting bolts from the floor of the vehicle.

4. Slowly lower the rear axle assembly until the tension is removed from the coil springs. Remove the coil springs and insulators.

To install:

5. Install the springs and insulator into the vehicle. The ends of the lowest coil must be within 9⁄16 in. of the spring stop.

6. Raise the rear axle assembly until the lower shock nuts can be installed. Torque the nuts to 35 ft. lbs. (47 Nm).

7. Connect the brake line brackets to the frame and torque the mounting bolts to 97 inch lbs. (11 Nm).

Torsion Bars

REMOVAL & INSTALLATION

1. Before servicing the vehicle, refer to the precautions in the beginning of this section.

2. Remove or disconnect the following:
   - Torsion bar-to-knuckle bolt, washer and bushing
   - Torsion bar-to-chassis bolt
   - Torsion bar from the chassis

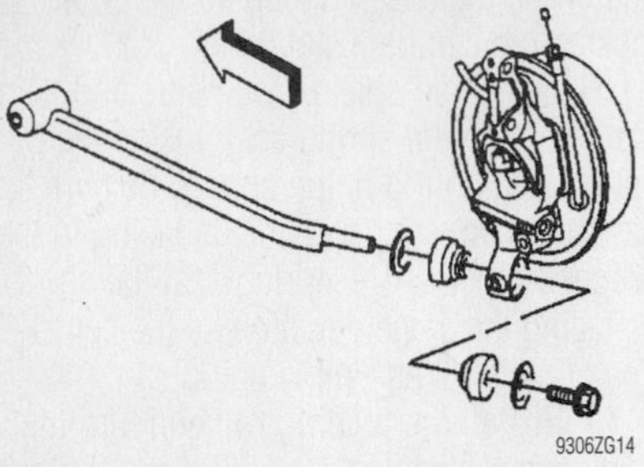

Exploded view of the torsion bar

To install:

3. Install or connect the following:
   - Torsion bar to the chassis. Torque the nut/bolt to 48 ft. lbs. (65 Nm) plus an additional 120 degree turn.
   - Torsion bar to the knuckle. Torque the bolt to 51 ft. lbs. (69 Nm).

Lower Ball Joint

REMOVAL & INSTALLATION

1. Before servicing the vehicle, refer to the precautions in the beginning of this section.
2. Remove or disconnect the following:
   - Wheel

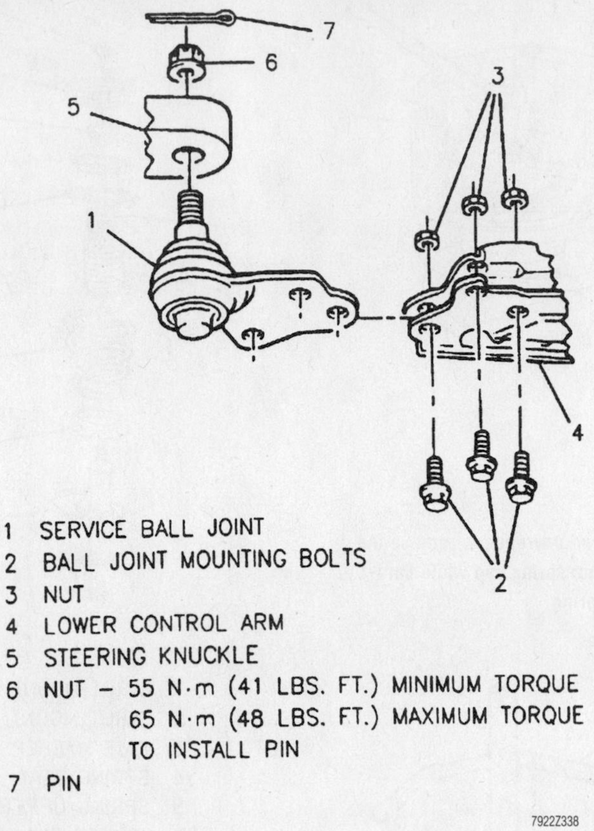

Exploded view of the replacement ball joint mounting

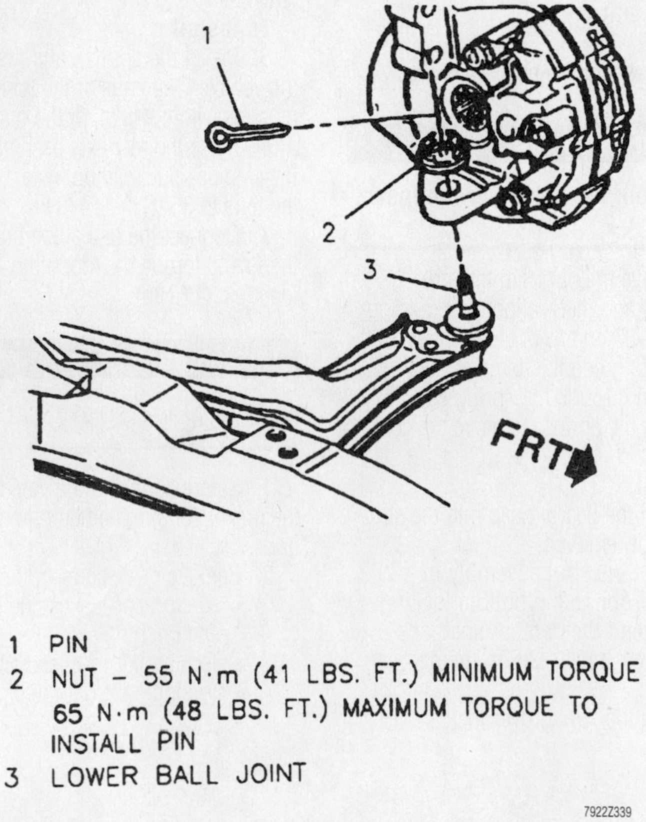

Exploded view of the ball joint-to-knuckle mounting

   - Wiring harness from the control arm
   - Stabilizer shaft link
   - Lower ball joint from the steering knuckle

3. Drill a ⅛ in. pilot hole in the center of each of the 3 ball joint mounting rivets.
4. Using a ½ in. drill bit, drill the heads off the rivets.
5. With a hammer and punch, knock the rivets out of the control arm. Remove the sway bar link kit.
6. Pull the control arm down so the ball stud clears the steering knuckle. Slide the ball joint out of the control arm.

To install:

7. Position the new ball joint into the lower control arm and install the bolts. The nuts must be on top of the control arm. Tighten the mounting bolts to the specification provided with the ball joint service kit.
8. Connect the ball joint to the steering knuckle and torque the castle nut to 41 ft. lbs. (55 Nm).

➡If necessary to align the cotter pin holes tighten the nut up to 60 degree (⅙) turn additional rotation. NEVER loosen the nut to make the holes align.

9. Install or connect the following:

- New cotter pin
- Stabilizer shaft link. Torque the nut to 22 ft. lbs. (30 Nm).
- Wiring harness to the control arm
- Wheel. Torque the nuts to 100 ft. lbs. (140 Nm).

Lower Control Arm

REMOVAL & INSTALLATION

Front

1. Before servicing the vehicle, refer to the precautions in the beginning of this section.
2. Remove or disconnect the following:
   - Front wheel
   - Stabilizer link
   - Lower ball joint from the steering knuckle
   - Lower control arm-to-suspension crossmember bolts
   - Lower control arm

To install:

3. Install or connect the following:
   - Lower control arm-to-suspension crossmember and hand-tighten the bolts
   - Lower ball joint to the steering knuckle. Torque the nut to 45 ft. lbs. (61 Nm).
   - Stabilizer link. Torque the nut to 22 ft. lbs. (30 Nm).
   - Front wheel
4. Position the vehicle at curb height.
5. Tighten the bolts on 1998–99 models, with the vehicle at curb height as follows:
   a. Center bolts: 89 ft. lbs. (120 Nm).
   b. Front bolts: 89 ft. lbs. (120 Nm).
   c. Rear bolts: 89 ft. lbs. (120 Nm).
   d. Control arm-to-suspension support bolts at the front bushing: 89 ft. lbs. (120 Nm).
   e. Control arm-to-suspension support bolts at the rear vertical bushing: 125 ft. lbs. (170 Nm).
6. Tighten the bolts as follows on 2000–01 models, with the vehicle at curb height as follows:
   a. Front lower control arm-to-suspension crossmember bolt to 84 ft. lbs. (115 Nm), plus an additional 120 degree turn.
   b. Rear lower control arm-to-suspension crossmember bolt to 180 ft. lbs. (245 Nm), plus an additional 180 degree turn.
7. Check and/or adjust the front alignment.

Rear—1998 models

1. Before servicing the vehicle, refer to the precautions in the beginning of this section.
2. Remove or disconnect the following:
   - Wheels
   - Brake line from the body
   - Parking brake cable from the body
   - Control arm from the vehicle

To install:

3. Install or connect the following:
   - Control arm to the vehicle with the nuts finger tight
   - Parking brake cable to the body
   - Brake line to the body. Torque the screws to 97 inch lbs. (11 Nm).
   - Wheels
4. Lower the vehicle to curb height and torque the control arm nuts to 59 ft. lbs. (80 Nm) plus an additional 120 degree turn.

CONTROL ARM BUSHING REPLACEMENT

Front Bushing

1. Before servicing the vehicle, refer to the precautions in the beginning of this section.
2. Remove the lower control arm and place it in a vise.
3. Lubricate the threads of a bushing driver with high pressure lubricant.
4. Assemble the tool onto the front control arm bushing.
5. Tighten the tool until the front bushing is pressed from the control arm.
6. Disassemble the tools.

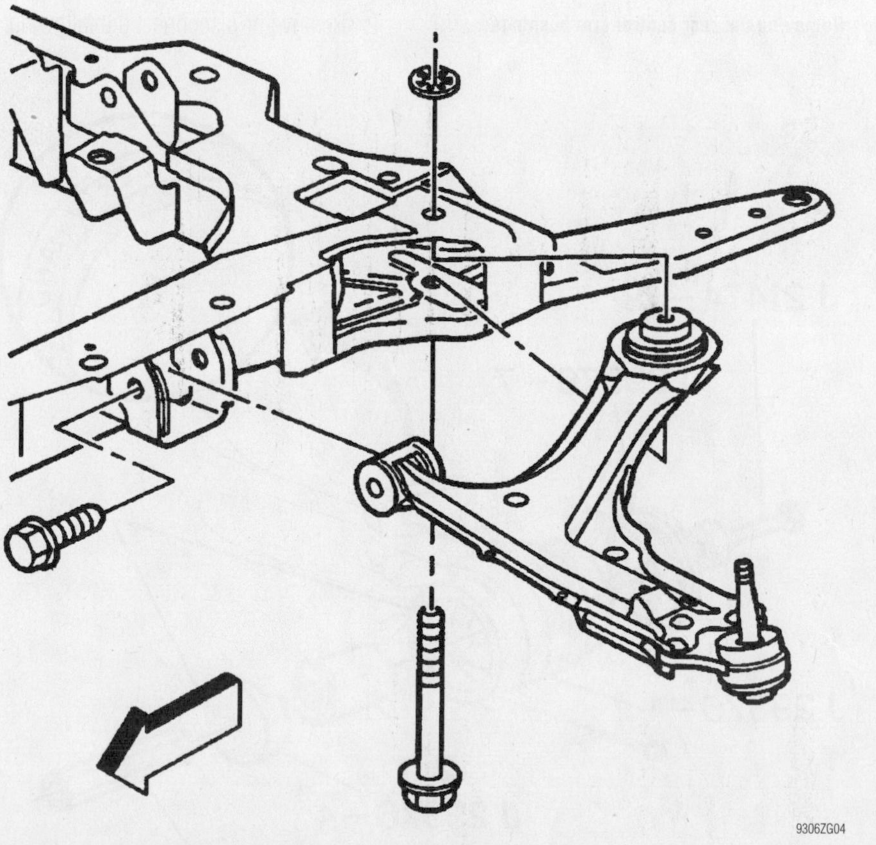

Exploded view of the lower control arm and related components

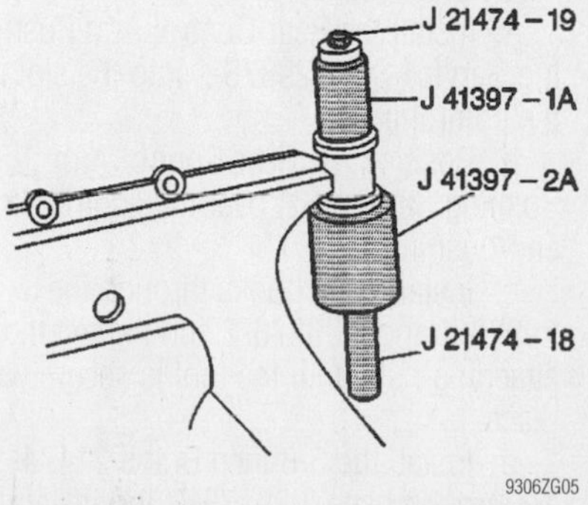

Removing the front bushing from the lower control arm

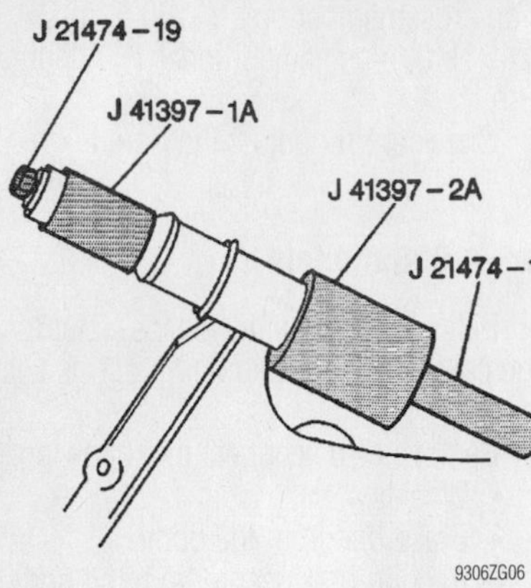

Installing the front bushing to the lower control arm

To install:

7. Lubricate the new front bushing outer casing.
8. Insert the new bushing into the control arm.
9. Assemble a bushing driver/installer set onto the front control.
10. Tighten the screw until the front bushing is pressed into the control arm.
11. Disassemble the tools.
12. Install the lower control arm.

Rear Bushing

1. Before servicing the vehicle, refer to the precautions in the beginning of this section.
2. Remove the wheels.
3. If removing the right bushings, disconnect the brake lines from the body.
4. If removing the left bushings, disconnect the brake line bracket from the body and the parking brake cable from the hook guide on the body.
5. Remove the nut, bolt, and washer from the control arm and underbody attachment.
6. Rotate the control arm downward.
7. Remove the bushing using the following tools and procedure:
   a. Install the Rear Control Arm Bushing Service Set J 29376-2 into the slot in the control arm.
   b. Position the Rear Control Arm Bushing Service Set over the control arm end/bushing.
   c. Install the J hooks through the J 29376-1 and J 29376-2 and tighten the attaching nuts until the tool is securely in place.
   d. Install the 3/8 inch Bolt J 21474-19 through plate J 29376-7 and install into the J 29376-1.
   e. Place the J 29376-6A remover into position on the bushing.
   f. Install the 3/8 inch Nut J 21474-18 onto the J 21474-19 bolt.
   g. Remove the bushing from the control arm by turning the bolt.

To install:

8. Install the bushing as follows:
   a. Install the J 29376 receiver on the control arm.
   b. Install the J 21474-19 bolt through plate J 29376-7 and into the receiver.
   c. Install the bushing on the bolt and position into the housing. Align the bushing installer arrow with the arrow on the receiver for proper indexing of the bushing. A high-pressure lubricant may be necessary for assembly.

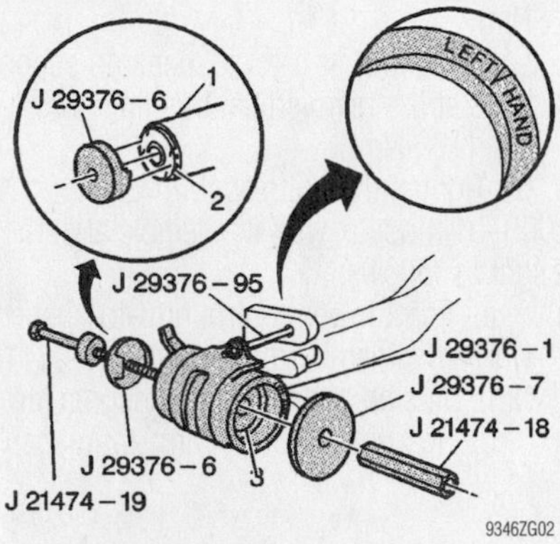

Removing the rear control arm bushings

   d. Install J 21474-18 (3/8 inch) nut onto the J 21474-19 (3/8 inch) bolt.
   e. Press the bushing into the control arm by turning the bolt. When the bushing is in the proper position, the end flange will be flush against the face of the control arm.

➡**The washer and nut must be installed on the outboard side.**

9. Install the control arm.

Wheel Bearings

ADJUSTMENT

These vehicles are equipped with sealed hub and bearing assemblies. The hub and bearing assemblies are non-serviceable. If the assembly is damaged, the complete unit must be replaced.

REMOVAL & INSTALLATION

Front

1. Before servicing the vehicle, refer to the precautions in the beginning of this section.
2. Remove or disconnect the following:

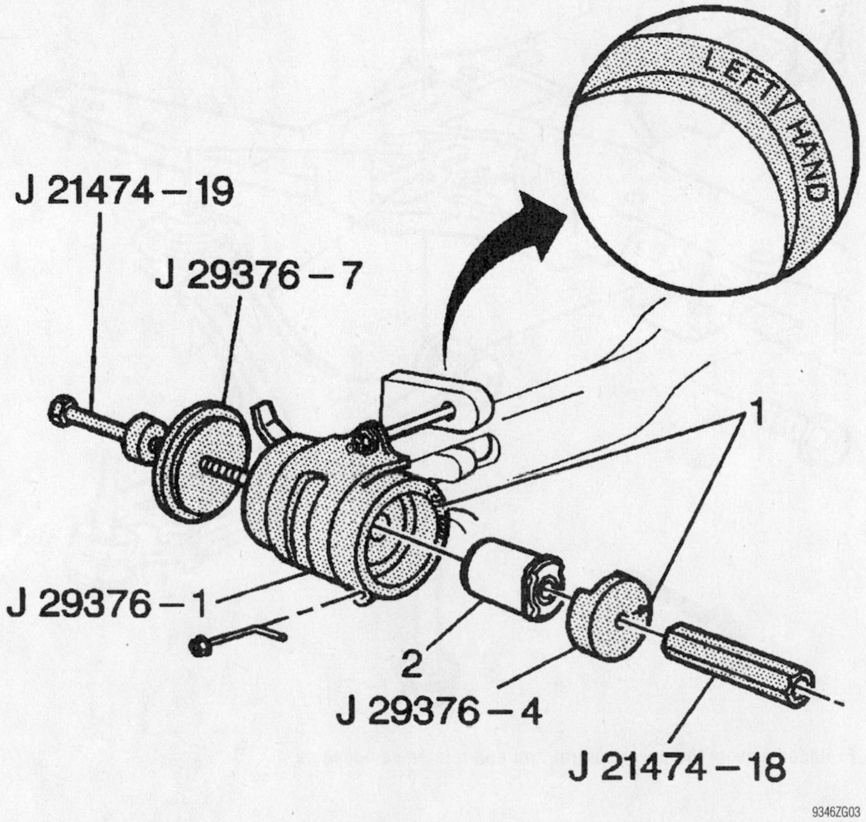

Installing the rear control arm bushings

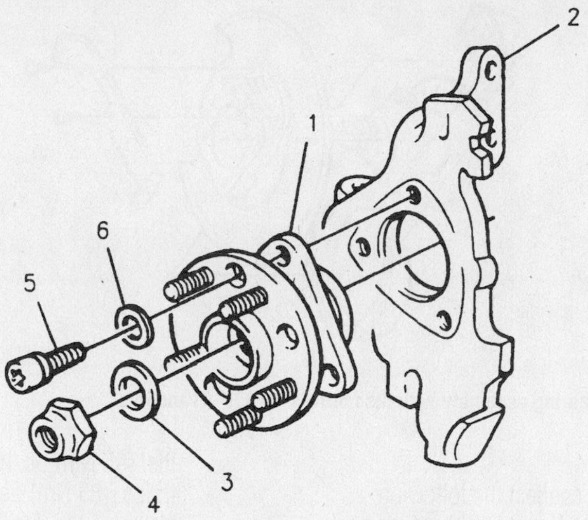

1 HUB AND BEARING ASSEMBLY
2 STEERING KNUCKLE
3 WASHER
4 DRIVE AXLE NUT – 260 N·m (192 LBS. FT.)
5 HUB AND BEARING RETAINING BOLT
6 WASHER

79222340

Exploded view of the front hub/bearing assembly

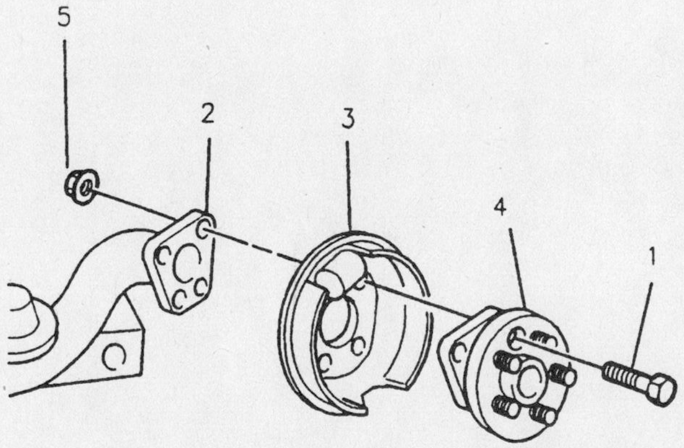

1 BOLT
2 REAR AXLE ASSEMBLY
3 BACKING PLATE
4 HUB AND BEARING ASSEMBLY
5 LOCKNUT

79222341

Hub and bearing components

- Front wheel
- Halfshaft nut and washer
- Caliper from the steering knuckle and support it aside

⁂ WARNING

DO NOT allow the caliper to hang unsupported from the brake hose.

- Brake rotor
- 3 hub/bearing assembly bolts
- Halfshaft from the hub/bearing assembly
- Hub/bearing assembly

To install:

3. Install or connect the following:
   - Hub/bearing assembly onto the halfshaft, making sure the splines engage smoothly
   - Hub/bearing assembly to the steering knuckle. Torque the bolts to 70 ft. lbs. (95 Nm).
   - Brake rotor
   - Caliper onto the steering knuckle. Torque the bolts to 38 ft. lbs. (51 Nm).
   - Halfshaft nut. Torque the hub nut to 74 ft. lbs. (100 Nm) plus an additional 40 degree turn for 1998 models, or to 284 ft. lbs. (385 Nm), for 1999–01 models.
   - Front wheel. Torque the nuts to 100 ft. lbs. (140 Nm).

Rear

DRUM BRAKE

1. Before servicing the vehicle, refer to the precautions in the beginning of this section.
2. Remove or disconnect the following:
   - Wheel
   - Brake drum
   - 4 hub/bearing assembly to knuckle nuts

➡The top rear bolt will not clear the brake shoes and must be removed with the bearing assembly.

- Anti-lock Brake System (ABS) speed sensor wire from the hub/bearing assembly
- Hub/bearing assembly

To install:

3. Install or connect the following:
   - Hub/bearing assembly on the knuckle
   - ABS wheel speed sensor

- Hub/bearing nuts. Torque the nuts to 44 ft. lbs. (60 Nm).
- Brake drum
- Wheel. Torque the nuts to 100 ft. lbs. (140 Nm).

DISC BRAKE

1. Before servicing the vehicle, refer to the precautions in the beginning of this section.
2. Remove or disconnect the following:
   - Wheel
   - Brake rotor
   - Parking brake cable from the lever
   - Wheel Speed Sensor (WSS) electrical connector
   - Hub/bearing assembly bolts from the knuckle
   - Torx® bolts from the rear of the hub/bearing assembly
   - Hub/bearing assembly from the backing plate

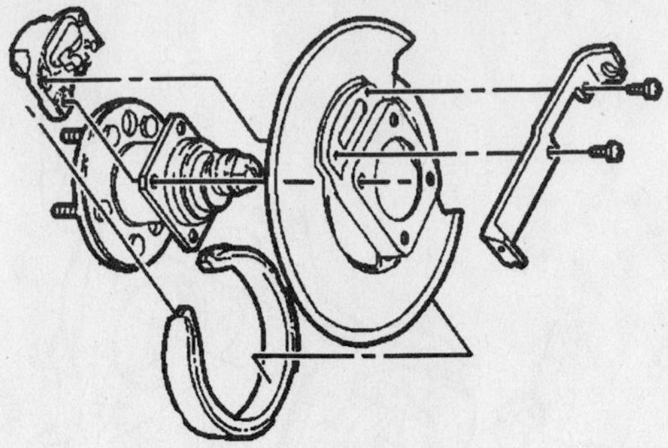

Rear wheel bearing assembly with disc brakes—1999–01 models

To install:

3. Install or connect the following:
   - Hub/bearing assembly to the backing plate. Tighten the Torx®bolts to 89 inch lbs. (10 Nm).
   - Hub/bearing assembly on the knuckle. Torque the bolts to 70 ft. lbs. (95 Nm) on 1999 models or 62 ft. lbs. (85 Nm) on 2000–01 models.
   - WSS electrical connector
   - Parking brake cable to the lever
   - Brake rotor
   - Wheel assembly

GENERAL MOTORS CORPORATION—V-BODY

1998–01
Cadillac-Catera

28

PRECAUTIONS

Before servicing any vehicle, please be sure to read all of the following precautions, which deal with personal safety, prevention of component damage, and important points to take into consideration when servicing a motor vehicle:

- Never open, service or drain the radiator or cooling system when the engine is hot, serious burns can occur from the steam and hot coolant.
- Observe all applicable safety precautions when working around fuel. Whenever servicing the fuel system, always work in a well-ventilated area. Do not allow fuel spray or vapors to come in contact with a spark, open flame or excessive heat (a hot drop light, for example). Keep a dry chemical fire extinguisher near the work area. Always keep fuel in a container specifically designed for fuel storage; also, always properly seal fuel containers to avoid the possibility of fire or explosion. Refer to the additional fuel system precautions later in this section.
- Fuel injection systems often remain pressurized, even after the engine has been turned **OFF**. The fuel system pressure must be relieved before disconnecting any fuel lines. Failure to do so may result in fire and/or personal injury.
- Brake fluid often contains polyglycol ethers and polyglycols. Avoid contact with the eyes and wash your hands thoroughly after handling brake fluid. If you do get brake fluid in your eyes, flush your eyes with clean, running water for 15 minutes. If eye irritation persists, or if you have taken brake fluid internally, seek medical assistance IMMEDIATELY.
- The EPA warns that prolonged contact with used engine oil may cause a number of skin disorders, including cancer! You should make every effort to minimize your exposure to used engine oil. Protective gloves should be worn when changing oil. Wash your hands and any other exposed skin areas as soon as possible after exposure to used engine oil. Soap and water, or waterless hand cleaner should be used.
- All new vehicles are now equipped with an air bag system. The system must be disabled before performing service on or around system components, steering column, instrument panel components, wiring and sensors. Failure to follow safety and disabling procedures could result in accidental air bag deployment, possible personal injury and unnecessary system repairs.
- Always wear safety goggles when working with, or around, the air bag system. When carrying a non-deployed air bag, be sure the bag and trim cover are pointed away from your body. When placing a non-deployed air bag on a work surface, always face the bag and trim cover upward, away from the surface. This will reduce the motion of the module if it is accidentally deployed. Refer to the additional air bag system precautions later in this section.
- Clean, high quality brake fluid from a sealed container is essential to the safe and proper operation of the brake system. You should always buy the correct type of brake fluid for your vehicle. If the brake fluid becomes contaminated, completely flush the system with new fluid. Never reuse any brake fluid. Any brake fluid that is removed from the system should be discarded. Also, do not allow any brake fluid to come in contact with a painted surface; it will damage the paint.
- Never operate the engine without the proper amount and type of engine oil; doing so WILL result in severe engine damage.
- Timing belt maintenance is extremely important! Many models utilize an interference-type, non-freewheeling engine. If the timing belt breaks, the valves in the cylinder head may strike the pistons, causing potentially serious (also time-consuming and expensive) engine damage. Refer to the maintenance interval charts in the front of this manual for the recommended replacement interval for the timing belt, and to the timing belt section for belt replacement and inspection.
- Disconnecting the negative battery cable on some vehicles may interfere with the functions of the on-board computer system(s) and may require the computer to undergo a relearning process once the negative battery cable is reconnected.
- When servicing drum brakes, only disassemble and assemble one side at a time, leaving the remaining side intact for reference.

ENGINE REPAIR

➡Disconnecting the negative battery cable on some vehicles may interfere with the operation of the on board computer system. The computer may undergo a relearning process once the negative battery cable is reconnected.

Alternator

REMOVAL

1. Before servicing the vehicle, refer to the precautions in the beginning of this section.
2. Remove or disconnect the following:
   - Negative battery cable
   - Intake air resonator
   - Drive belt
   - Coolant heater, if equipped
   - Alternator cooling duct
   - Alternator electrical connectors
   - Alternator

INSTALLATION

1. Install or connect the following:
   - Alternator. Torque the lower nut to 26 ft. lbs. (35 Nm) and the upper nut to 30 ft. lbs. (40 Nm).
   - Cooling duct
   - Alternator electrical connectors
   - Coolant heater, if equipped
   - Drive belt
   - Intake air resonator
   - Negative battery cable
2. Perform a charging system test and verify that the system is operating properly.

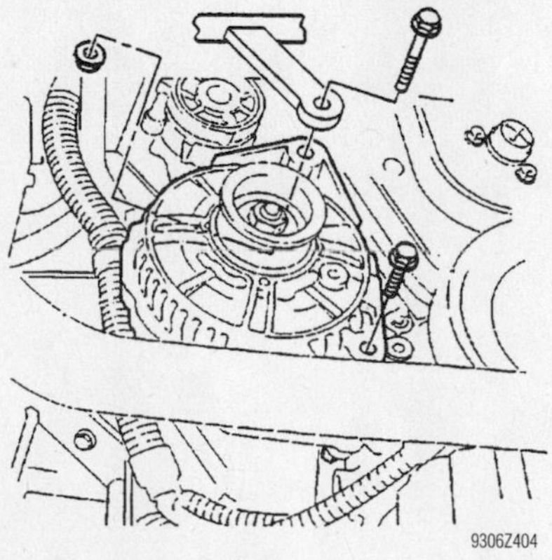

Alternator mounting hardware

Ignition Timing

ADJUSTMENT

➡The 3.0L DOHC engine used in the Catera utilizes a Distributorless Ignition System (DIS). No ignition timing adjustment is possible.

Engine Assembly

REMOVAL & INSTALLATION

1. Before servicing the vehicle, refer to the precautions in the beginning of this section.
2. Drain the cooling system.
3. Drain the engine oil.
4. Recover the A/C refrigerant.
5. Relieve the fuel system pressure.
6. Remove or disconnect the following:
   - Battery
   - Wiper arms
   - Left and right air inlet grilles
   - Hood
   - Intake air resonator
   - Air filter housing
   - Red, white and black wiring harness connectors
   - Body ground cable
   - Power supply wires at the battery cable end
   - Power steering hoses from the power steering pump
   - Brake booster vacuum lines
   - Electronic Control Module (ECM) and harness from the electrical center
   - Relays and wiring harness from electrical center box
   - Cruise control and accelerator cables

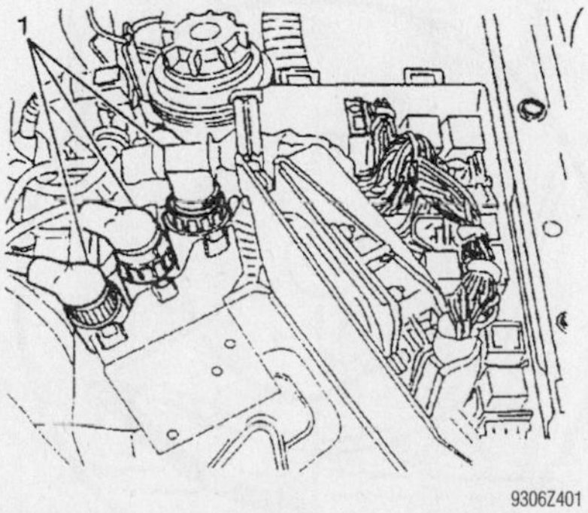

Red, white and black harness connectors (1)

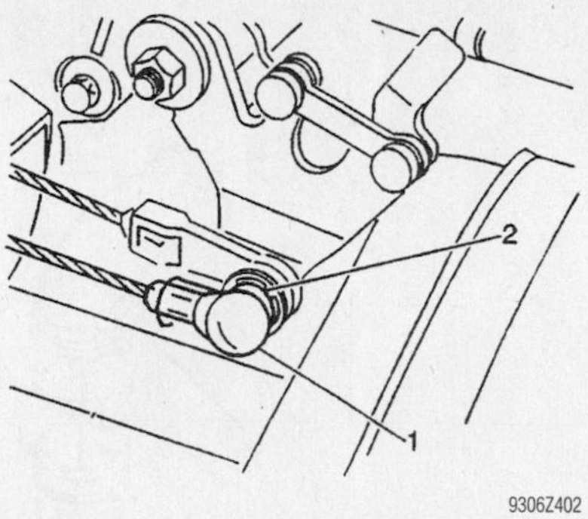

Accelerator (1) and cruise control (2) cables

   - Fuel lines from the fuel rail. Use a backup wrench to prevent damage to the fuel rail.
   - Coolant hose from the throttle body
   - Vacuum hose from the purge valve on the engine ventilation chamber
   - Vacuum hose from the heater control valve
   - Coolant reservoir hose from the coolant inlet pipe
   - Electric water pump with hoses attached
   - Coolant hoses from the heater core
   - Radiator
   - Refrigerant line from the A/C compressor and compressor bracket
   - Condenser line from the A/C condenser
   - A/C quick-connect hose near the low pressure service valve
   - Splash shield
   - A/C compressor electrical connector

➡Support the engine whenever the transmission is removed. The motor mounts are silicone filled and do not provide sufficient rigidity to support the engine with the transmission removed.

7. Remove the transmission from the vehicle.
8. Attach a hoist to the engine lifting eyes. Raise the engine slightly and remove the engine support fixture.
9. Remove the engine mount nuts and lift the engine out of the vehicle. Raise the engine slowly after being certain all wiring, cables and hoses have been removed.

To install:

10. Install the engine.
11. Install the engine mount. Torque the upper motor mount nuts to 30 ft. lbs. (40 Nm). Torque the lower nuts to 41 ft. lbs. (55 Nm).
12. Install an engine support fixture and remove the chain hoist.
13. Install or connect the following:
   - Transmission. Torque the mounting bolts to 44 ft. lbs. (60 Nm).
   - A/C compressor electrical connector
   - Splash shield and tighten securely
   - Quick connects for the high and low pressure fittings
   - A/C condenser line to the condenser
   - A/C compressor line
   - Radiator
   - Coolant hoses to the heater core
   - Electric water pump
   - Coolant reservoir hose to the coolant inlet pipe
   - Vacuum hose to the heater control valve
   - Vacuum hose for the purge valve on the engine ventilation chamber
   - Coolant hose to the throttle body
   - Fuel return and supply lines. Torque the lines to 11 ft. lbs. (15 Nm).
   - Cruise control and accelerator cables to the throttle body
   - Relays and wiring harness to the electrical center box
   - ECM
   - Brake booster vacuum lines
   - Power steering hoses to the steering pump. Torque the discharge hose fasteners to 21 ft. lbs. (28 Nm).
   - Power supply wires at the battery cable ends
   - Body ground cable
   - Red, white and black wiring harness connectors
   - Air filter housing
   - Intake air resonator
   - Hood
   - Left and right air inlet grilles
   - Wiper arms
   - Battery
   - Negative battery cable
14. Fill and bleed the power steering system.
15. Refill the engine oil.

➡An oil filter change is recommended.

16. Fill and bleed the coolant system.

➡When refilling the coolant system add 2 crushed engine coolant supplement sealant pellets (PN 3634621) into the reservoir.

17. Recharge the A/C system and check for leaks.

18. Start the vehicle and inspect for leaks.

Water Pump

REMOVAL & INSTALLATION

1. Before servicing the vehicle, refer to the precautions in the beginning of this section.
2. Drain the coolant.
3. Remove or disconnect the following:
   - Negative battery cable
   - Intake air resonator
   - Water pump pulley bolts, loosen only
   - Front timing belt cover
   - Water pump pulley
   - Water pump

To install:

4. Install or connect the following:
   - Water pump with a new O-ring. Torque the bolts to 18 ft. lbs. (25 Nm).
   - Water pump pulley
   - Front timing belt cover
   - Intake air resonator
   - Negative battery cable
5. Fill the cooling system through the reservoir tank.

➡When refilling the cooling system, add 2 crushed engine coolant supplement sealant pellets (PN 3634621) into the coolant reservoir.

6. Start the vehicle and inspect the coolant systems for leaks.

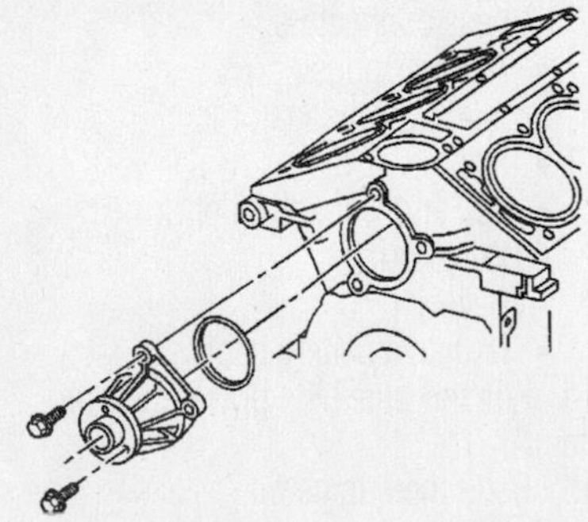

Exploded view of the water pump mounting

Cylinder Head

REMOVAL & INSTALLATION

1. Before servicing the vehicle, refer to the precautions in the beginning of this section.

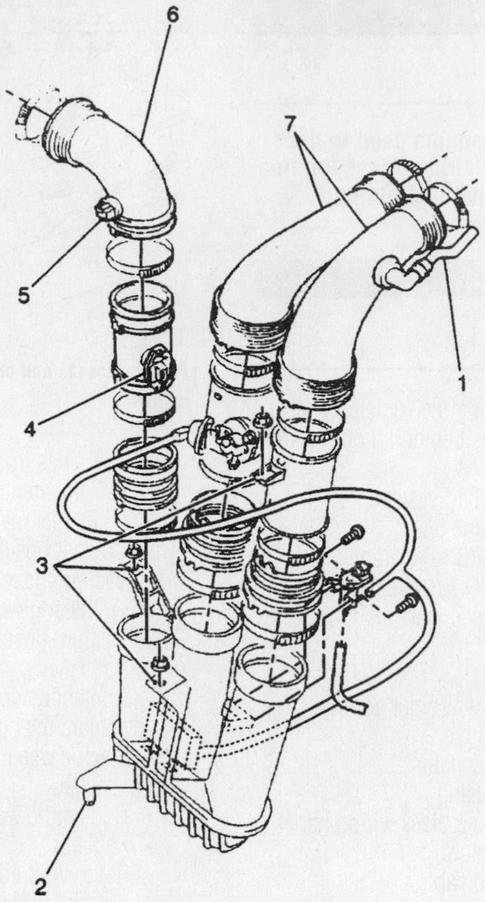

(1) Idle Air Control (IAC) Inlet Hose
(2) Resonance Chamber Guide Pin
(3) Resonance Chamber Nut
(4) Mass Air Flow (MAF) Sensor
(5) Intake Air Temperature (IAT) Sensor
(6) Resonance Chamber Air Intake Hose
(7) Intake Plenum Air Inlet Hose

79222406

Intake air resonator and related components

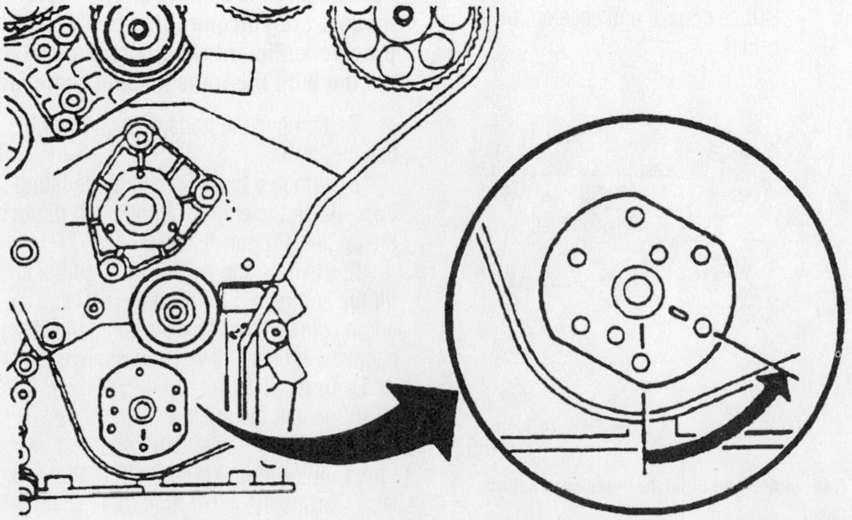

Before removing the timing belt, be sure to turn the crankshaft 60 degrees BTDC to avoid valve-to-piston contact and subsequent engine damage.

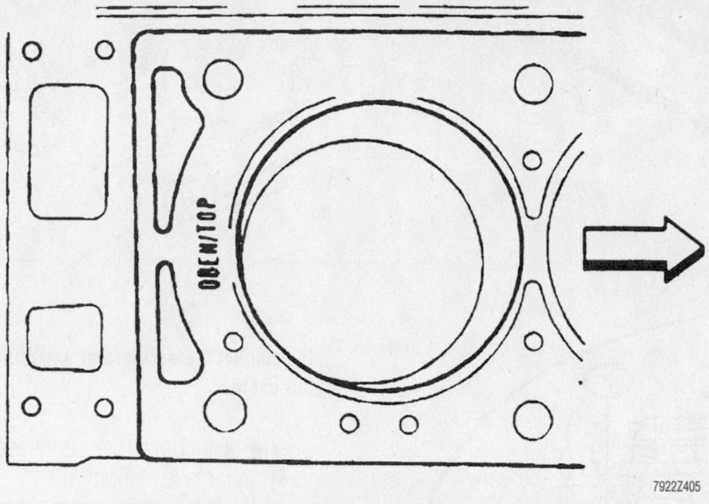

Proper head gasket installation position—right cylinder head

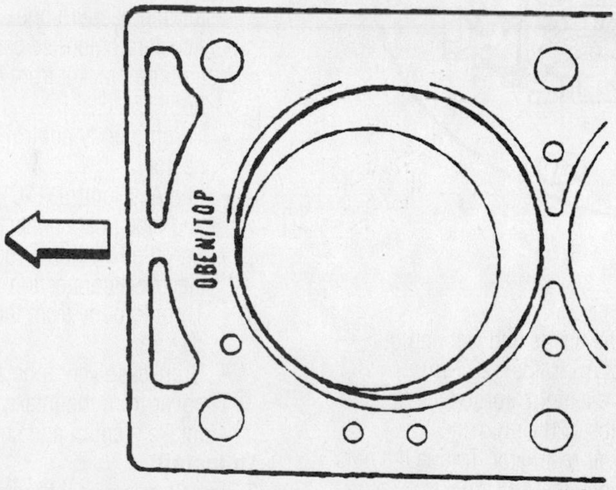

Proper head gasket installation position—left cylinder head

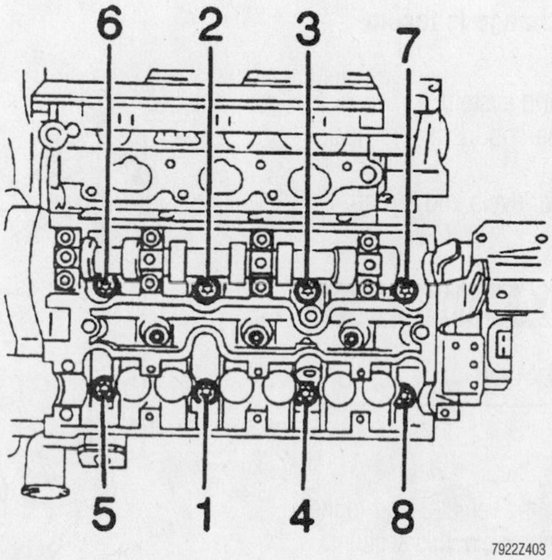

Cylinder head torque sequence—left cylinder head

2. Drain the engine coolant.
3. Relieve the fuel system pressure.
4. Remove or disconnect the following:
   - Intake plenum
   - Intake air resonator
   - Intake manifold and spacer
   - Both Engine Coolant Temperature (ECT) sensor electrical connectors
   - Coolant crossover
   - Ignition coil assembly
   - Camshaft cover
   - Front timing belt cover
5. Position the crankshaft 60 degrees Before Top Dead Center (BTDC) to avoid contact between the valves and the pistons.
6. Remove or disconnect the following:
   - Timing belt
   - Timing belt tensioner bracket
   - Four camshaft gears
   - Water pump
   - Rear timing belt cover
   - Camshaft Position (CMP) sensor electrical connector
   - Exhaust camshaft
   - Coolant pipe/engine lift bracket from the cylinder head
   - Oil level indicator tube
   - Upper radiator hose from the coolant pipe
   - Coolant pipe
   - Exhaust manifold from the cylinder head
   - Cylinder head

To install:

➡Be sure to use new cylinder head bolts when installing the cylinder head. The old bolts have been stretched and are not reusable.

7. Install a new cylinder head gasket.
8. Install the cylinder head.
9. Install new cylinder head bolts and tighten the bolts as follows:
   a. Step 1: Torque the bolts to 18 ft. lbs. (25 Nm).
   b. Step 2: An additional 90 degrees.
   c. Step 3: An additional 90 degrees.
   d. Step 4: An additional 90 degrees.
   e. Step 5: An additional 15 degrees.
10. Install or connect the following:
    - Exhaust manifold using a new gasket. Torque the nuts to 15 ft. lbs. (20 Nm).
    - Coolant pipe with new O-rings
    - Oil level indicator tube
    - Coolant pipe/engine lift bracket. Torque the bolt to 15 ft. lbs. (20 Nm).

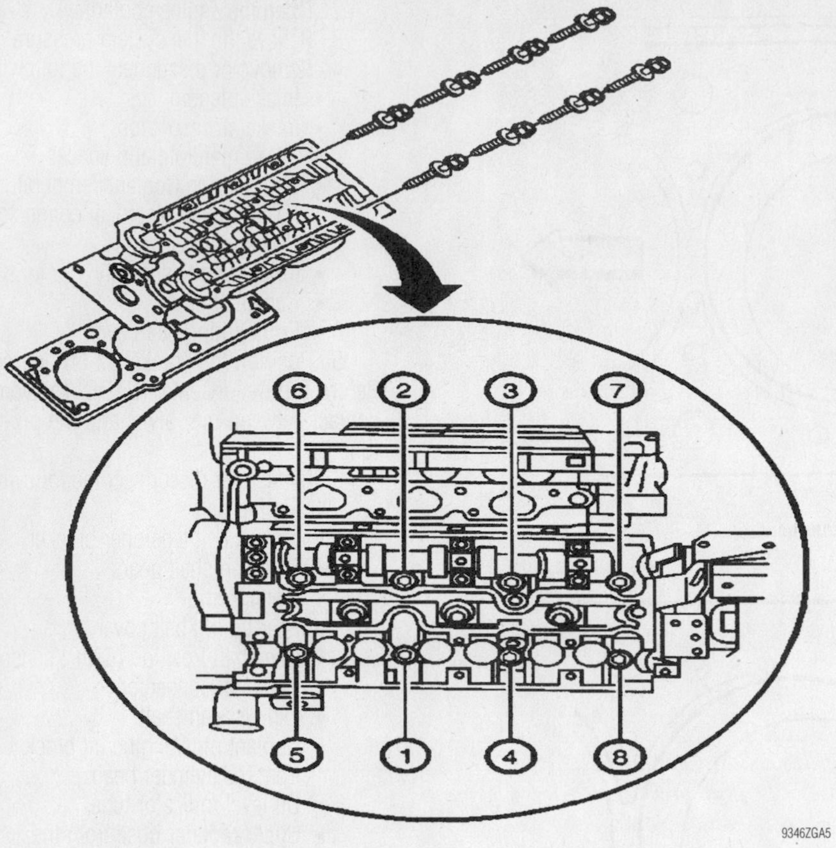

Cylinder head torque sequence—right cylinder head

- Upper radiator hose to the coolant pipe
- Exhaust camshaft. Torque the bearing cap bolts to 71 inch lbs. (8 Nm).
- CMP electrical connector
- Rear timing belt cover. Torque the bolts to 71 inch lbs. (8 Nm).
- Water pump. Torque the bolts to 18 ft. lbs. (25 Nm).
- Camshaft gears. Torque the bolts to 37 ft. lbs. (50 Nm) plus a 60 degree turn, plus another 15 degree turn.
- Timing belt tensioner bracket. Torque the bolts to 30 ft. lbs. (40 Nm).
- Timing belt
- Front timing belt cover. Torque the bolts to 71 inch lbs. (8 Nm).
- Left camshaft cover with new O-rings and gaskets. Torque the fasteners to 71 inch lbs. (8 Nm).
- Ignition coil assembly
- Coolant crossover. Torque the bolts to 22 ft. lbs. (30 Nm).
- Both ECT sensor electrical connectors
- Intake manifold spacer. Torque the bolts in a spiral direction, starting from the inside and working outward to 15 ft. lbs. (20 Nm).
- Intake manifold. Torque the bolts to 15 ft. lbs. (20 Nm).
- Intake air resonator. Torque the nuts to 27 inch lbs. (3 Nm).
- Intake plenum. Torque the bolts to 71 inch lbs. (8 Nm).
- Negative battery cable

➡An oil and filter change is recommended.

11. Refill the cooling system.
12. Start the engine and inspect for leakage.
13. Inspect all fluid levels and top off, if necessary.

Intake Manifold Assembly

REMOVAL & INSTALLATION

Intake Plenum

1. Before servicing the vehicle, refer to the precautions in the beginning of this section.
2. Remove or disconnect the following:
   - Negative battery cable
   - Intake plenum air inlet hoses from the throttle body

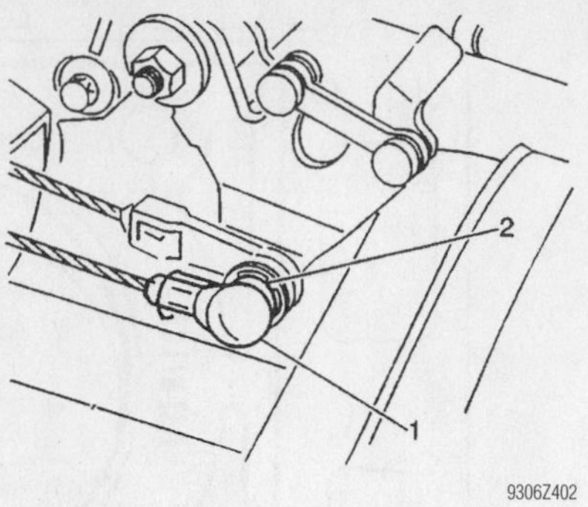

Disconnect the accelerator and cruise control cables

   - Brake booster vacuum hose from the intake plenum
   - Wiring harness channel from the plenum
   - Switch over valve electrical connector and vacuum line
   - Accelerator, cruise control cables and the bracket from the throttle body
   - Throttle body control electrical connector
   - Idle Air Control (IAC) inlet hose and electrical connector
   - Throttle Position Sensor (TPS) electrical connection
   - Throttle body from the intake plenum
   - Crankcase vent tube adapter and cover from the intake plenum
   - Intake plenum and O-rings

To install:

3. Install or connect the following:
   - Intake plenum with new O-rings. Torque the bolts to 71 inch lbs. (8 Nm).

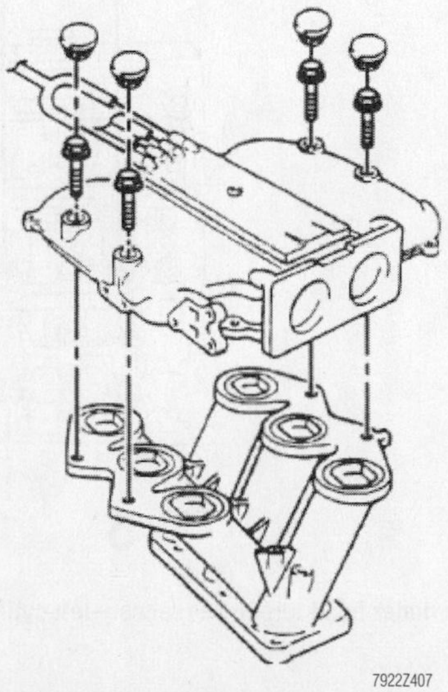

Intake plenum mounting bolt locations

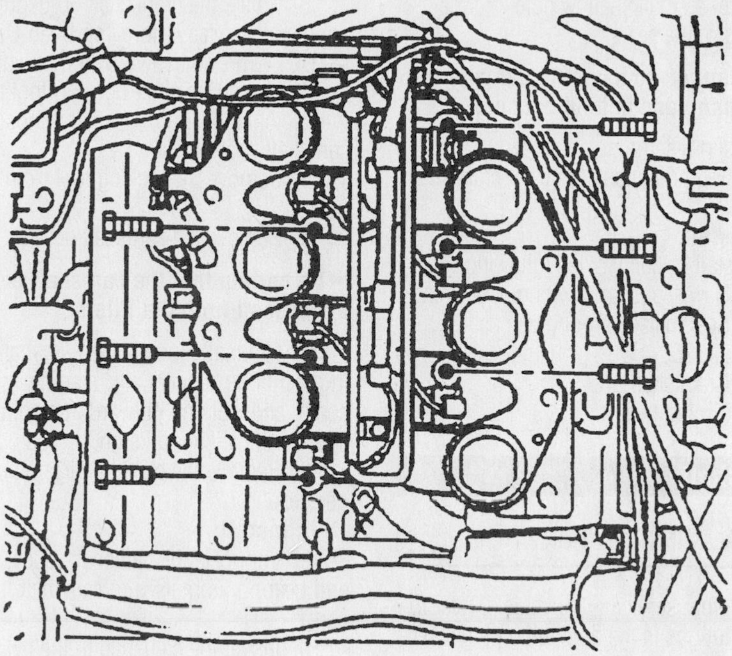

Intake manifold mounting bolt locations

- Crankcase vent tube adapter to the intake plenum. Torque the fastener to 71 inch lbs. (8 Nm).
- Throttle body to the intake plenum. Torque the fasteners to 106 inch lbs. (12 Nm).
- TPS electrical connector
- IAC electrical connector and vacuum line
- Throttle body control electrical connector
- Accelerator and cruise control bracket and cables. Torque the bolts to 71 inch lbs. (8 Nm).
- Electrical connector and vacuum hose to the intake plenum switchover valve solenoid
- Wiring channel to the intake plenum. Torque the bolts to 71 inch lbs. (8 Nm).
- Threaded brake booster vacuum hose to the intake plenum

➡Be certain not to strip the threads on the brake booster vacuum hose.

- Intake plenum air inlet hoses to the throttle body
- Negative battery cable

4. Start the engine and check for proper performance.

Intake Manifold

1. Before servicing the vehicle, refer to the precautions in the beginning of this section.
2. Relieve the fuel system pressure.
3. Remove or disconnect the following:
   - Intake plenum
   - Fuel supply/return lines from the fuel rail by loosening the fittings
   - Fuel injector electrical harness connector
   - Fuel pressure regulator vacuum line
   - Intake manifold bolts

➡Cover the open ports to the intake manifold spacer. Severe damage may occur if foreign material enters the engine.

   - Intake manifold

To install:

4. Install or connect the following:
   - Intake manifold with new gaskets. Torque the bolts to 15 ft. lbs. (20 Nm).
   - Fuel pressure regulator vacuum line

➡The fuel injector connectors are numbered. Be sure to attach the correct connector to the proper fuel injector.

   - Fuel injector electrical harness connectors
   - Fuel supply/return hoses. Torque the fittings to 11 ft. lbs. (15 Nm).
   - Intake plenum
   - Negative battery cable
5. Start the vehicle and inspect for leaks.

Exhaust Manifold

REMOVAL & INSTALLATION

Left Side

1. Before servicing the vehicle, refer to the precautions in the beginning of this section.
2. Drain the engine coolant.
3. Remove or disconnect the following:
   - Engine assembly
   - Exhaust manifold lower/upper heat shields
   - Secondary Air Injection (AIR) pipe from the exhaust manifold
   - Coolant pipe and engine lift bracket from the cylinder head
   - Oil level indicator tube
   - Exhaust manifold

To install:

4. Install or connect the following:
   - Exhaust manifold with a new gasket. Torque the nuts to 15 ft. lbs. (20 Nm).
   - Oil level indicator tube. Torque the bolt to 15 ft. lbs. (20 Nm).
   - Engine lifting bracket and coolant pipe. Torque the bolt to 15 ft. lbs. (20 Nm).
   - Air injection pipe to the exhaust manifold. Torque the bolts to 15 ft. lbs. (20 Nm).
   - Exhaust manifold upper/lower heat shields. Torque the bolts to 71 inch lbs. (8 Nm).
   - Engine assembly
5. Refill the engine coolant.
6. Start the vehicle and inspect for any leaks.

Right Side

1. Before servicing the vehicle, refer to the precautions in the beginning of this section.

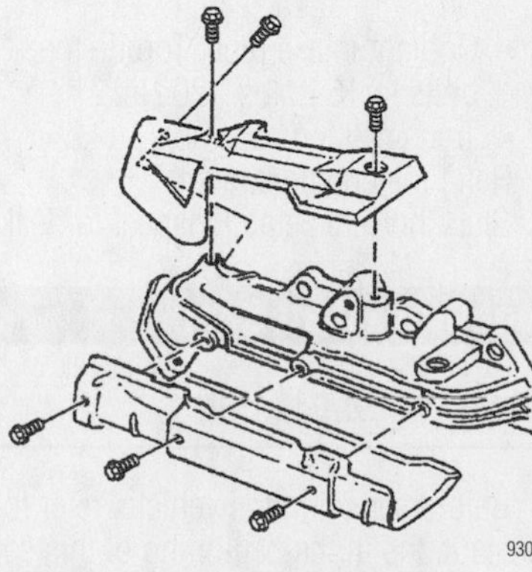

Exhaust manifold heat shield bolts

2. Drain the engine coolant.
3. Remove or disconnect the following:
   - Transmission assembly
   - Coolant intake pipe
4. Remove or disconnect the following:
   - Exhaust manifold lower heat shield
   - Exhaust manifold lower rear nuts
   - Exhaust manifold upper rear heat shield bolt
   - Catalytic converter nuts from the exhaust manifold
   - Drive belt tensioner
   - Exhaust manifold front two nuts
   - Exhaust manifold upper heat shield bolts

➡It is not necessary to remove the upper heat shield.

   - Secondary Air Injection (AIR) injection pipe from the exhaust manifold
   - Exhaust manifold

To install:

5. Install or connect the following:
   - Exhaust manifold with a new gasket. Torque the 2 upper nuts to 15 ft. lbs. (20 Nm).
   - AIR injection pipe to the exhaust manifold. Torque the bolts to 15 ft. lbs. (20 Nm).
   - Upper heat shield bolts. Torque the bolts to 71 inch lbs. (8 Nm).
   - Exhaust manifold upper heat shield bolts. Torque the bolts to 71 inch lbs. (8 Nm).
   - Exhaust manifold lower front nuts. Torque the nuts to 15 ft. lbs. (20 Nm).
   - Drive belt tensioner. Torque the bolts to 30 ft. lbs. (40 Nm).
   - Catalytic converter nuts. Torque the nuts to 15 ft. lbs. (20 Nm).
   - Exhaust manifold lower rear nuts. Torque the nuts to 15 ft. lbs. (20 Nm).
   - Exhaust manifold lower heat shield. Torque the bolts to 71 inch lbs. (8 Nm).
   - Coolant intake pipe. Torque the bolts to 15 ft. lbs. (20 Nm).
   - Transmission assembly
6. Refill the engine coolant.
7. Start the vehicle and inspect for leaks.

Front Crankshaft Seal

REMOVAL & INSTALLATION

1. Before servicing the vehicle, refer to the precautions in the beginning of this section.
2. Remove the timing belt and crankshaft gear.
3. Drill a small shallow hole into the steel ring of the seal.

➡Use caution so as not to damage the area around and behind the seal.

4. Insert a self-tapping screw.
5. Using pliers, pull the front crankshaft seal out.

To install:

6. Install or connect the following:
   - New seal coated with grease using Seal Installer Tool (such as J35268-A)
   - Crankshaft gear
   - Timing belt

Camshaft

REMOVAL & INSTALLATION

1. Before servicing the vehicle, refer to the precautions in the beginning of this section.
2. Remove or disconnect the following:
   - Negative battery cable
   - Intake plenum
   - Intake air resonator
   - Camshaft cover
   - Front timing belt cover
   - Timing belt
3. Rotate the crankshaft 60 degrees counterclockwise Before Top Dead Center (BTDC) to prevent valve/piston contact.
4. Use a Camshaft Locking tool to hold the camshaft gears in place and loosen the camshaft gear bolt.
5. Remove the locking tool from the gears.
6. Remove the camshaft gear(s).

➡Be certain that the camshaft is not under load from the lifters.

7. Gradually loosen the camshaft bearing cap bolts sequentially, starting in the center and working outward in a spiral. Note the identification marks on the caps.
8. Remove the camshaft from the cylinder head.

To install:

9. Lubricate the camshaft lobes, lifters and bearing journals with camshaft lubricant.
10. Install the camshaft to the cylinder head.
11. To reduce the lifter load, position the pin on the front of the camshaft as follows:
   - 1 o'clock position: right exhaust camshaft
   - 11 o'clock position: right intake camshaft

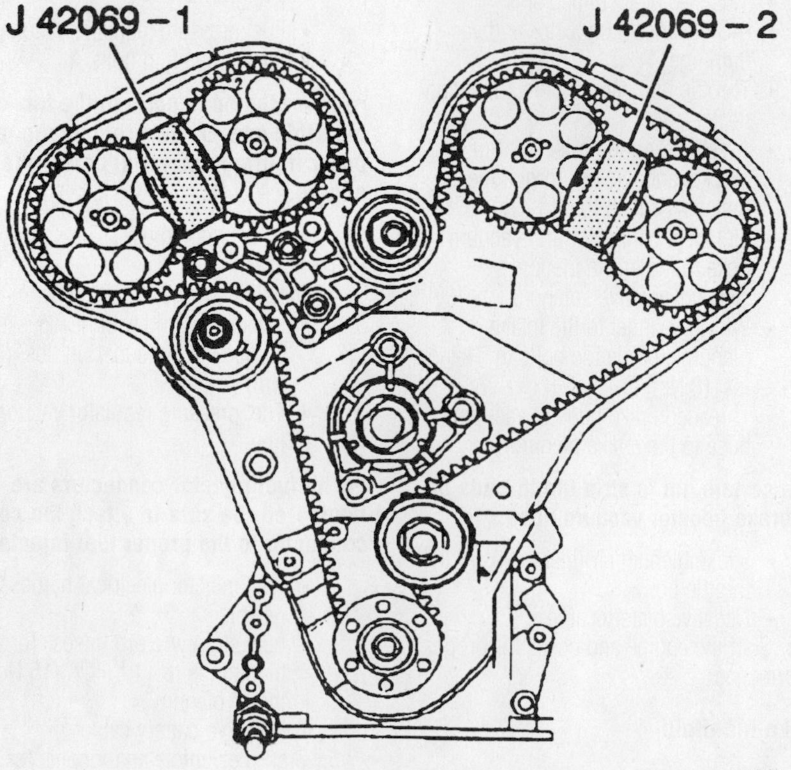

Camshaft gear holding tools must be installed before attempting to remove the gears from the camshafts

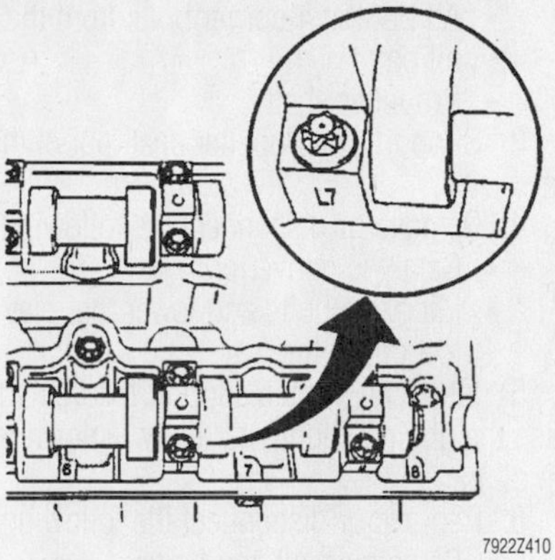

The camshaft bearing caps have identification marks stamped on the side

- 12 o'clock position: left exhaust camshaft
- 7 o'clock position: left intake camshaft

12. Place a small amount of Locktite® on the edge of the front bearing cap to ensure a good seal between the cap and surface of the cylinder head. Do not allow the sealer to get into the oil journal of the cap.

WARNING

The bearing caps must be installed in their original positions.

13. Install or connect the following:
    - Camshaft bearing caps. Torque the bolts in sequence starting in the center and working outwards to 71 inch lbs. (8 Nm).
    - New camshaft seal lubricated with engine oil

➡Make certain that the camshaft seal is fully seated.

- Camshaft gear. Tighten the new bolt to 37 ft. lbs. (50 Nm), plus a 60 degree turn, plus a 15 degree turn.

WARNING

A new camshaft gear bolt must be installed. The required tightening method will stretch the bolt making the original bolt unusable.

- Timing belt and adjust as needed
- Front timing belt cover. Torque the bolts to 71 inch lbs. (8 Nm).
- Camshaft cover. Torque the bolts to 71 inch lbs. (8 Nm).
- Intake plenum. Torque the bolts to 71 inch lbs. (8 Nm).
- Intake air resonator

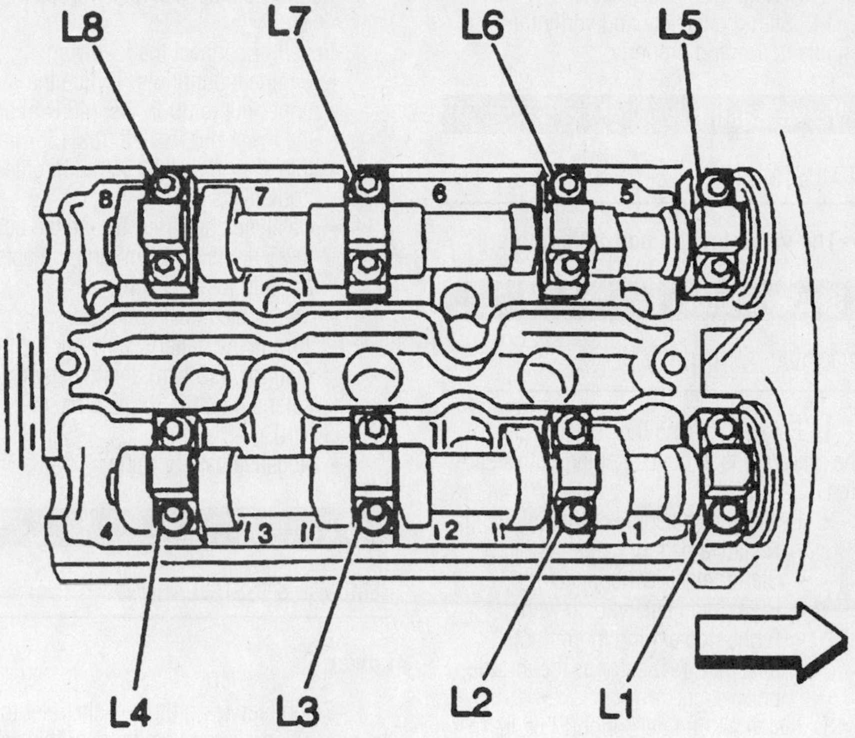

Camshaft bearing cap locations-right cylinder head

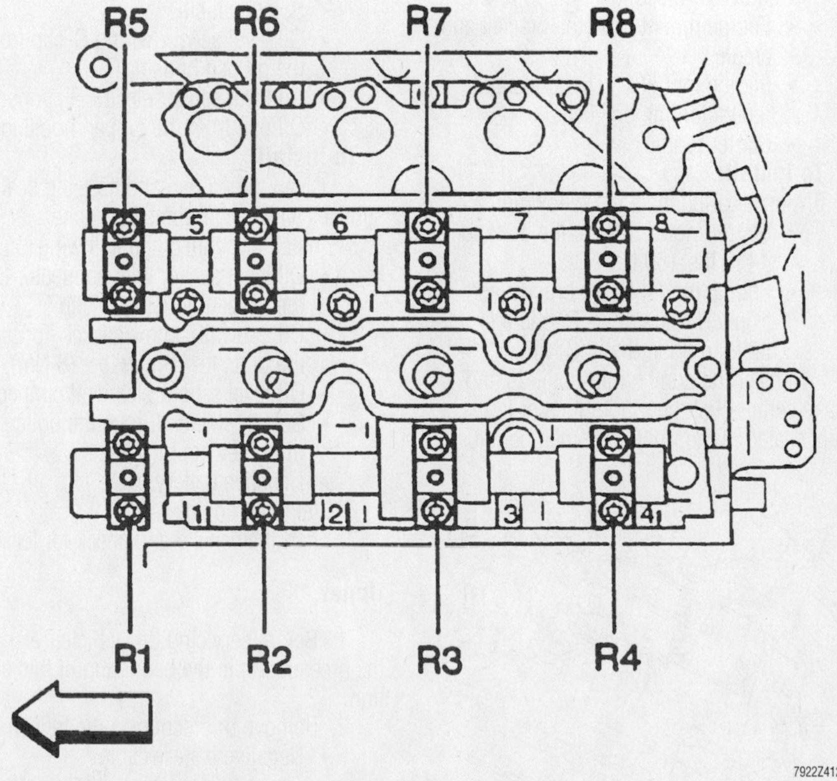

Camshaft bearing cap locations—left cylinder head

Heater Core replacement is covered in Section 2 of this manual

- Negative battery cable

14. Start the vehicle and verify the engine is running properly.

Valve Lash

ADJUSTMENT

➡**The valve lash is non-adjustable.**

Starter Motor

REMOVAL & INSTALLATION

1. Before servicing the vehicle, refer to the precautions in the beginning of this section.
2. Remove or disconnect the following:
   - Negative battery cable
   - Starter electrical connectors
   - Right hand catalytic converter
   - Right side engine mount nuts
   - Intake air resonator to the throttle body ducts
3. Install an Engine Support Fixture to the right side of the engine.
4. Raise the right side of the engine approximately 1.5 inches (38mm) to gain access to the starter motor bolts.
5. Remove or disconnect the following:
   - Engine mount from the bracket and cradle
   - Engine mount bracket bolts and reposition the bracket
   - Starter motor

To install:

6. Install or connect the following:
   - Starter motor. Torque the bolts to 44 ft. lbs. (60 Nm).
   - Engine mount to the bracket and front crossmember. Torque the bolts to 30 ft. lbs. (40 Nm).
7. Lower the engine into position and make certain that the mount locator tab is fully seated in the front crossmember slot.
8. Remove the special tools from the engine.
9. Install or connect the following:
   - Engine mount nuts. Torque the upper nut to 30 ft. lbs. (40 Nm) and the lower nut to 41 ft. lbs. (55 Nm).
   - Intake air resonator to the throttle body ducts
   - Catalytic converter. Torque the bolt to 25 ft. lbs. (34 Nm) and the nuts to 18 ft. lbs. (25 Nm).
   - Starter electrical connectors. Torque the battery cable nut to 115 inch lbs. (13 Nm) and the starter solenoid nut to 35 inch lbs. (4 Nm).
   - Negative battery cable

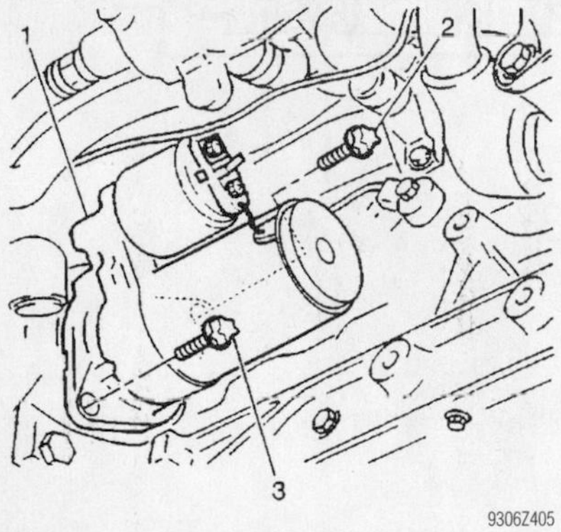

Remove the starter motor and bolts

Oil Pan

REMOVAL & INSTALLATION

Lower

1. Before servicing the vehicle, refer to the precautions in the beginning of this section.
2. Drain the engine oil.
3. Remove or disconnect the following:
   - Negative battery cable
   - Splash shield
   - Oil level sensor wiring C-clip from the oil pan housing
   - Oil level sensor electrical connector
   - Oil pan from the oil pan housing

To install:

4. Clean the oil pan and housing sealing surfaces with a non-abrasive cleaner.
5. Install or connect the following:
   - Oil level sensor wire connector to the oil pan housing C-clip
   - Oil pan with a new gasket. Torque the bolts to 71 inch lbs. (8 Nm).
   - Oil level sensor electrical connector
   - Splash shield. Torque the bolts until they are fully seated.
   - Negative battery cable
6. Refill the engine oil.
7. Start the vehicle and check for leaks.

Upper

1. Before servicing the vehicle, refer to the precautions in the beginning of this section.
2. Remove or disconnect the following:
   - Negative battery cable
   - Lower oil pan
   - Engine mount lower nuts
   - A/C compressor hose strap from the oil pan
   - Transmission bolts from the oil pan
   - All but the 4 corner bolts from the oil pan
   - Propeller shaft
3. Support the propeller shaft out of the way.
4. Remove or disconnect the following:
   - Catalytic converter
   - Idler arm bolts and lower the relay rod out of the way
5. Install an engine support fixture and raise the engine slightly to allow removal of the oil pan.
6. Remove or disconnect the following:
   - Remaining oil pan bolts
   - Oil intake pipe
   - Oil pan from the vehicle

To install:

7. Apply a bead of silicone sealer in the bottom of the upper oil pan groove.
8. Install a new rubber seal into the groove in the pan.
9. Apply a 3 mm (0.12 in) bead of silicone sealant on the OUTSIDE edge of the seal, at the front of the upper oil pan and apply an identical bead on the INSIDE edge of the seal at the rear of the upper oil pan. The beads will overlap at the middle of the upper oil pan.
10. Install or connect the following:
    - Upper oil pan
    - Oil intake pipe. Torque the bolt to 71 inch lbs. (8 Nm).
    - Four oil pan corner bolts finger tight
11. Lower the engine into place.
12. Install or connect the following:
    - Remaining oil pan bolts. Torque the bolts to 11 ft. lbs. (15 Nm).
    - Transmission-to-oil pan bolts. Torque the bolts to 30 ft. lbs. (40 Nm).
    - Idler arm bolts. Torque the bolts to 44 ft. lbs. (60 Nm).
    - Catalytic converter. Torque the nuts to 18 ft. lbs. (25 Nm).
    - Propeller shaft. Torque the bolts to 70 ft. lbs. (95 Nm).
    - A/C compressor hose strap. Torque the bolt to 71 inch lbs. (8 Nm).
    - Engine mount lower nuts. Torque the nuts to 41 ft. lbs. (55 Nm).
    - Lower oil pan
13. Refill the engine oil.

Oil Pump

REMOVAL & INSTALLATION

1. Before servicing the vehicle, refer to the precautions in the beginning of this section.
2. Drain the engine oil.

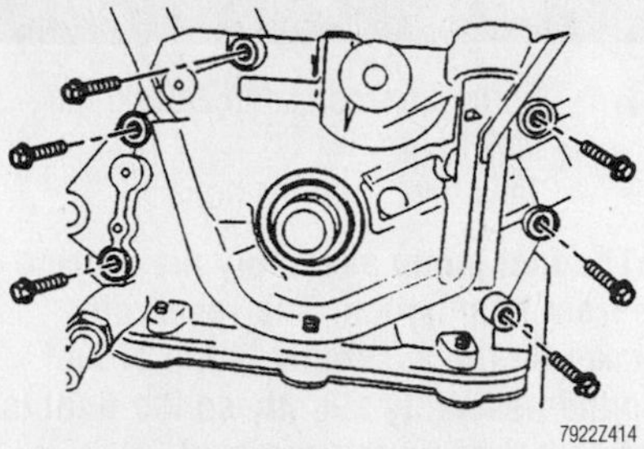

7922Z414

Oil pump mounting bolt locations

3. Drain the engine coolant.
4. Discharge and recover the A/C system.
5. Remove or disconnect the following:
   - Negative battery cable
   - Intake air resonator
   - Front timing belt cover
   - Intake plenum
   - Timing belt
   - Rear timing belt cover
   - A/C compressor
   - A/C compressor and power steering pump bracket and position it out of the way of the oil pump housing
   - Lower alternator mounting bolt and loosen the upper mounting bolt then, position the alternator aside
   - Lower and upper oil pans
   - Crankshaft drive gear
   - Oil pump
   - Front main oil seal and collar

To install:

6. Coat the pump side of the oil pump gasket with a thin layer of sealant. Do not cover or restrict the gasket openings with sealant.
7. Install or connect the following:
   - Oil pump collar
   - New front main oil seal
   - Oil pump with a new gasket by aligning the guide pins. Torque the bolts to 9 ft. lbs. (12 Nm).
   - Crankshaft gear. Torque the bolt to 184 ft. lbs. (250 Nm) plus an additional 60 degree turn.
   - Upper oil pan. Torque the bolts to 11 ft. lbs. (15 Nm).
   - Lower oil pan. Torque the bolts to 71 inch lbs. (8 Nm).
   - Alternator. Torque the upper and lower bolts to 30 ft. lbs. (40 Nm).
8. Re-tighten the oil pump bolts to 15 ft. lbs. (20 Nm).
9. Install or connect the following:
   - A/C compressor and power steering pump bracket. Torque the fasteners to 30 ft. lbs. (40 Nm).
   - A/C compressor. Torque the bolts to 30 ft. lbs. (40 Nm).
   - Rear timing belt cover. Torque the bolts to 71 inch lbs. (8 Nm).
   - Timing belt
   - Front timing belt cover. Torque the bolts to 71 inch lbs. (8 Nm).
   - Intake plenum. Torque the bolts to 71 inch lbs. (8 Nm).
   - Intake air resonator
   - Negative battery cable
10. Fill the engine oil.
11. Fill the coolant.
12. Recharge the A/C system.
13. Start the vehicle and checks for leaks.

Rear Main Seal

REMOVAL & INSTALLATION

1. Before servicing the vehicle, refer to the precautions in the beginning of this section.
2. Remove or disconnect the following:
   - Negative battery cable.
   - Flywheel
3. Center punch the steel ring of the rear main seal.
4. Drill a small hole into the steel ring.

➡Make certain not to damage any engine components on the opposite side of the seal.

5. Install a small self tapping screw.
6. Remove the rear main oil seal.

To install:

7. Clean all sealing surfaces with a non-abrasive cleaner.
8. Install or connect the following:
   - New seal lubricated with chassis grease

7922Z416

Thread a self-tapping screw into the metal part of the seal and remove the seal

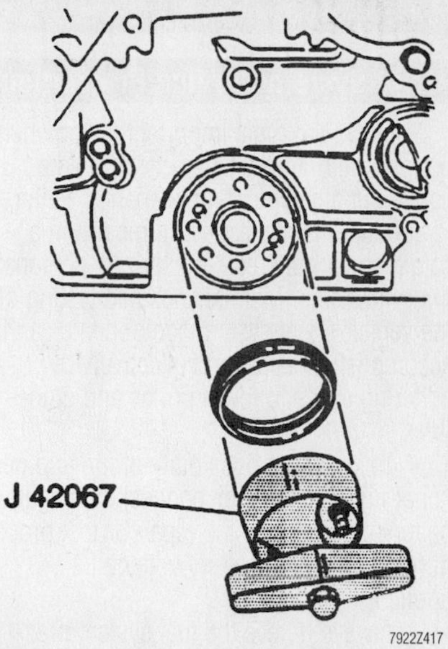

7922Z417

Press the rear main seal into position using a threaded seal installer like J 42067

   - Rear main oil seal using an Oil Seal Installer tool J 42067
   - Flywheel. Torque the new bolts to 48 ft. lbs. (65 Nm), plus a 30 degree turn, plus a 15 degree turn.
   - Negative battery cable

Piston and Ring

POSITIONING

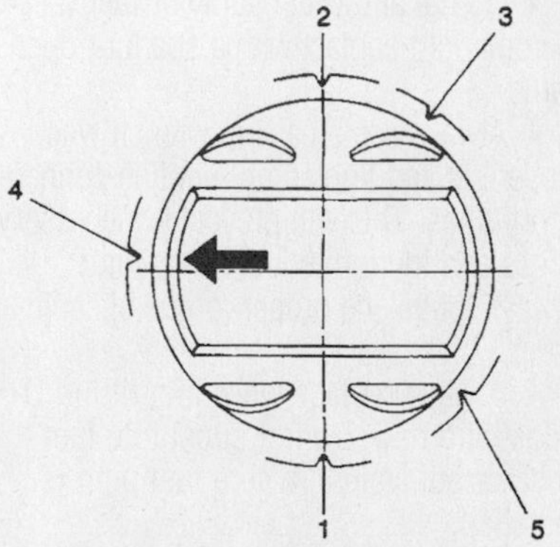

(1) 1st Compression Ring End Gap Location
(2) 2nd Compression Ring End Gap Location
(3) Oil Control Ring Upper Ring End Gap Location
(4) Oil Control Ring Spacer End Gap Location
(5) Oil Control Ring Lower Ring End Gap Location

7922AG55

Piston ring end-gap spacing—GM 3.0L engine

FUEL SYSTEM

Fuel System Service Precautions

Safety is the most important factor when performing not only fuel system maintenance but any type of maintenance. Failure to conduct maintenance and repairs in a safe manner may result in serious personal injury or death. Maintenance and testing of the vehicle's fuel system components can be accomplished safely and effectively by adhering to the following rules and guidelines.

- To avoid the possibility of fire and personal injury, always disconnect the negative battery cable unless the repair or test procedure requires that battery voltage be applied.
- Always relieve the fuel system pressure prior to disconnecting any fuel system component (injector, fuel rail, pressure regulator, etc.), fitting or fuel line connection. Exercise extreme caution whenever relieving fuel system pressure, to avoid exposing skin, face and eyes to fuel spray. Please be advised that fuel under pressure may penetrate the skin or any part of the body that it contacts.
- Always place a shop towel or cloth around the fitting or connection prior to loosening to absorb any excess fuel due to spillage. Ensure that all fuel spillage (should it occur) is quickly removed from engine surfaces. Ensure that all fuel soaked cloths or towels are deposited into a suitable waste container.
- Always keep a dry chemical (Class B) fire extinguisher near the work area.
- Do not allow fuel spray or fuel vapors to come into contact with a spark or open flame.
- Always use a backup wrench when loosening and tightening fuel line connection fittings. This will prevent unnecessary stress and torsion to fuel line piping. Always follow the proper torque specifications.
- Always replace worn fuel fitting O-rings with new. Do not substitute fuel hose or equivalent, where fuel pipe is installed.

Fuel System Pressure

RELIEVING

1. Before servicing the vehicle, refer to the precautions in the beginning of this section.
2. Loosen the fuel filler cap to relieve the tank pressure.
3. Remove or disconnect the following:
   - Negative battery cable
   - Intake manifold top cover
4. Install a Fuel Pressure Gauge to the fuel pressure fitting. Wrap a shop towel around the fitting while installing the gauge.
5. Connect a bleed hose into an approved container and open the valve to bleed the system.
6. Close the valve and disconnect the gauge.
7. Drain any remaining fuel from the gauge into the approved container.

Fuel Filter

REMOVAL & INSTALLATION

1. Before servicing the vehicle, refer to the precautions in the beginning of this section.
2. Loosen the fuel filler cap to relieve pressure in the tank.
3. Relieve the fuel system pressure.
4. Remove or disconnect the following:
   - Fuel feed and return lines
   - Fuel filter from the bracket

To install:

5. Clean the bolts and fittings on both fuel lines.
6. Install or connect the following:
   - Filter in the retaining strap making certain that the flow is in the proper direction
   - Both fuel lines. Torque the fittings to 18 ft. lbs. (25 Nm).
   - Fuel filter bracket mounting bolt. Torque the bolt to 13 ft. lbs. (18 Nm).
   - Fuel filler cap
7. Crank the engine for a few seconds and check for leakage.

Fuel Pump

REMOVAL & INSTALLATION

1. Before servicing the vehicle, refer to the precautions in the beginning of this section.
2. Relieve the fuel system pressure.
3. Drain the fuel tank.
4. Remove or disconnect the following:
   - Fuel tank
   - Fuel tank pressure sensor
   - Fuel tank connection cover electrical connector
   - Spring loaded clamp around the tank boot
   - Tank boot from the housing

➡The fuel pump assembly may spring up from its position. The reservoir bucket on the assembly is full of fuel. Tip the assembly slightly so the float is not damaged during removal.

   - Fuel sender locking nut from the tank using a Fuel Tank Sender Wrench (J 42219)
   - Fuel pump electrical connectors
   - Fuel tank connection cover
   - Hose from the pump
   - Fuel pump from the tank by pushing the tabs inward
   - Locking ring by pushing the tabs in and pushing the locking ring from the housing simultaneously
   - Fuel pump from the housing
   - Lip seal from the cover by sliding it downward, past the reservoir and over the float arm and discard it
5. Drain the remaining fuel from the reservoir into an approved container.

To install:

6. Install or connect the following:
   - New lip seal lubricated with engine oil
   - Fuel pump into the housing
   - Locking ring into the housing
   - New fuel inlet screen
   - Fuel pump assembly into the tank
   - Fuel line to the pump using a new clamp
   - Fuel pump electrical connectors
   - Fuel tank connection cover
   - Fuel sender locking nut. Torque the nut to 37 ft. lbs. (50 Nm).

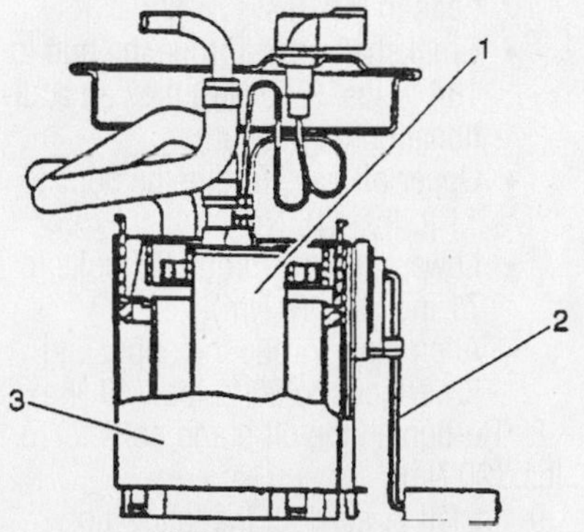

(1) Fuel pump
(2) Fuel gauge float arm
(3) Fuel reservoir

79222419

Fuel pump module components

- Fuel tank boot to the housing
- Fuel tank pressure sensor. Torque the fastener to 18 inch lbs. (4 Nm).
- Air reference hose to the pressure sensor
- Fuel feed and return lines using new clamps
- Fuel tank pressure sensor and connection cover electrical connectors
- Fuel tank to the vehicle. Torque the bolts to 22 ft. lbs. (30 Nm).
- Negative battery cable

7. Refill the fuel tank.
8. Start the vehicle and inspect for leaks.

Fuel Injector

REMOVAL & INSTALLATION

CAUTION

Remove the fuel rail assembly carefully to prevent damage to the injector electrical connector terminals and spray tips. Support the fuel rail after it is removed in order to avoid damaging the fuel rail components. Cap the fittings and plug the holes when servicing the fuel system to prevent debris from entering open ports.

Fuel injection systems often remain pressurized, even after the engine has been turned **OFF**. The fuel system pressure must be relieved before disconnecting any fuel lines. Failure to do so may result in fire and/or personal injury.

1. Before servicing the vehicle, refer to the precautions in the beginning of this section.
2. Relieve fuel system pressure.
3. Remove or disconnect the following:
   - Intake plenum
   - Fuel rail supply and return lines from the fuel rail
   - Fuel injector electrical connectors
   - Harness from the fuel rail
   - Fuel rail attaching bolts from the intake manifold
   - Fuel pressure regulator vacuum lines
   - Fuel rail assembly
   - Injectors from the fuel rail and discard the O-rings

To install:

4. Install or connect the following:
   - New O-rings lubricated with clean engine oil onto the fuel injectors
   - Fuel injectors with new retaining clips
   - Fuel rail to the intake manifold. Torque the bolt to 71 inch lbs. (8 Nm).
   - Fuel pressure regulator vacuum line
   - Fuel injector electrical connectors
   - Harness to the fuel rail
   - Fuel supply/return lines to the fuel rail. Torque the fasteners to 11 ft. lbs. (15 Nm).
   - Fuel line bracket. Torque the bolts to 37 ft. lbs. (50 Nm).
   - Intake plenum. Torque the bolts to 71 inch lbs. (8 Nm).
   - Negative battery cable
5. Crank the engine several times to pressurize the system.
6. Check for leaks and repair, if necessary.

DRIVE TRAIN

Transmission Assembly

REMOVAL & INSTALLATION

1. Before servicing the vehicle, refer to the precautions in the beginning of this section.
2. Remove or disconnect the following:
   - Negative battery cable
   - Shift linkage from the transmission
   - Propeller shaft coupling bolts
   - Propeller shaft coupling from the drive flange
   - Access holes plugs from the oil pan and bell housing

➡Mark the flywheel to the torque converter to ease the installation procedure.

   - Flywheel-to-torque converter bolts
   - Oil cooler inlet and outlet pipes
   - Oxygen (O_2S) sensor connections
   - Catalytic converters
   - Transmission housing-to-oil pan bolts
3. Support the transmission.
4. Remove or disconnect the following:
   - Transmission crossmember to mount nuts
   - Transmission crossmember
   - Vent hose
   - Electrical connector from the transmission control selector switch
   - Electrical connector from the adapter case
   - Electrical connector from the main case
   - Electrical connector from the transmission speed sensor
5. Support the front of the engine.
6. Lower the transmission to gain access to the upper housing bolts.
7. Remove the upper transmission housing bolts.
8. Carefully remove the transmission from the vehicle.

To install:

WARNING

The transmission fluid cooler and lines must be flushed out prior to installing a new or reconditioned transmission.

9. Apply thread locking compound to the transmission-to-engine mounting bolts.
10. Install or connect the following:
    - Transmission. Torque the transmission-to-engine bolts to 44 ft. lbs. (60 Nm).
    - Electrical connector to the transmission control selector switch
    - Electrical connector to the adapter case
    - Electrical connector to the main case
    - Electrical connector to the speed sensor
    - Vent hose
    - Crossmember. Torque the bolts to 33 ft. lbs. (45 Nm).
    - Crossmember-to-mount and torque the nuts to 15 ft. lbs. (20 Nm)
11. Remove the transmission support.
12. Install or connect the following:
    - Transmission housing-to-oil pan bolts and torque the bolts to 15 ft. lbs. (20 Nm)
    - Catalytic converters. Torque the bolts to 18 ft. lbs. (25 Nm).
    - O_2S connectors
    - Oil cooler inlet and outlet pipes to the center pipes
    - Torque converter to the flywheel by aligning the matchmarks. Torque the new bolts to 22 ft. lbs. (30 Nm).

➡Be sure the torque converter weld nuts are flush with the flywheel.

- Oil pan access hole plugs
- Bell housing access hole plugs
- Propeller shaft coupling-to-drive flange an torque the bolts to 70 ft. lbs. (95 Nm)
- Shift linkage to the transmission
- Negative battery cable

13. Adjust the shift lever rod as follows:
a. Place the shift control lever in the **P** position.
b. Loosen the adjusting bolt on the rod.
c. Hold the selector lever on the transmission toward the rear stop to eliminate any free-play.
d. Tighten the adjusting bolt to 71 inch lbs. (8 Nm).
14. Refill the transmission. Tighten the plug to 33 ft. lbs. (45 Nm).
15. Start the vehicle and check for leaks. Be sure the vehicle will only start in the **P** and **N** positions.

Axle Shaft, Bearing and Seal

REMOVAL & INSTALLATION

1. Before servicing the vehicle, refer to the precautions in the beginning of this section.
2. Place the gear selector in **N**.
3. Remove or disconnect the following:
   - Rear wheel
   - Drive axle flange bolts
   - Outer end of the drive axle from the wheel bearing/hub inner flange

➡Make certain that the drive axle separator tool is properly aligned to the differential and drive axle. The tool is clearly marked "Differential Side". If installed incorrectly, damage to the Anti-lock Brake System (ABS) sensor reluctor ring is possible.

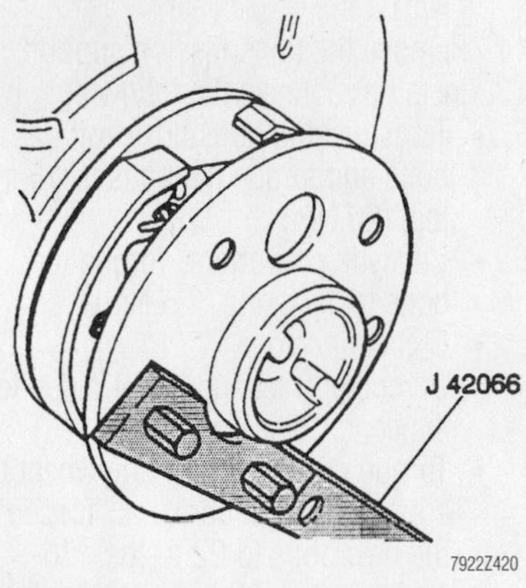

Prevent the outer hub from turning by installing a tool such as Hub Holding Tool J 42066

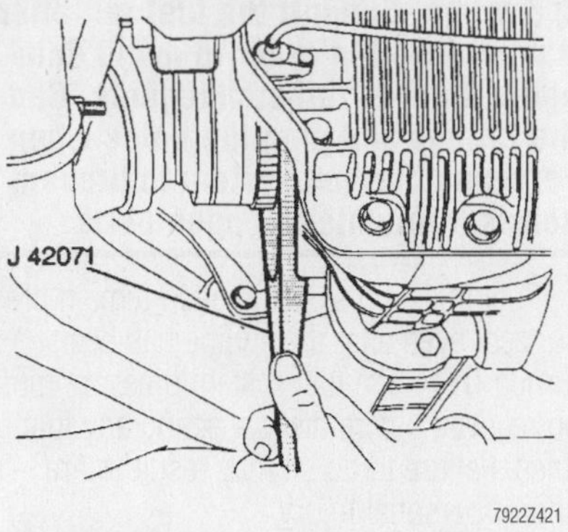

Pry the halfshaft out of the differential

4. Remove the axle shaft.
5. Use a suitable pry tool to remove the axle seal from the differential.

To install:

6. Use a Seal Installation Tool (such as J 26234) to install the new seal.
7. Lubricate the splines of the halfshaft and the sealing surfaces with differential oil.
8. Install or connect the following:
   - Axle shaft into the differential bore by aligning the splines

➡If necessary use a soft faced tool, such as a rubber mallet, to seat the halfshaft fully in the differential bore.

   - Outer end of the axle shaft into the wheel bearing/hub inner flange. Torque the bolts to 37 ft. lbs. (50 Nm) plus an additional 67 degree turn.
   - Rear wheel
9. Road test the vehicle.

CV-Joints

INNER & OUTER JOINT

1. Before servicing the vehicle, refer to the precautions in the beginning of this section.
2. Disconnect the negative battery cable.
3. Remove the axle shaft and place it in a vise.
4. Remove the large boot retaining clamp.
5. Remove the small boot retaining clamp.
6. Slide the boot up the shaft away from the joint.
7. With a brass drift and a hammer tap on the inner race of the cross grove.
8. Remove the CV joint retaining ring.
9. Remove the CV boot.

➡The joint is not serviceable, if damaged it must be replaced.

To install:

10. Pack the CV joint with ½ of the grease provided in the service kit.
11. Install new small and large boot retaining clamps on the ends of the boot
12. Install a new boot on the halfshaft.
13. Install a new joint retaining ring onto the halfshaft.
14. Place halfshaft into a shop press and press the joint into place.
15. Slide the CV boot over the joint .
16. Using a crimping tool a breaker bar and a torque wrench, torque the small retaining clamp to 100 ft. lbs. (140 Nm).
17. Place the remaining grease into the CV boot.
18. Using a crimping tool a breaker bar and a torque wrench, torque the large retaining clamp to 130 ft. lbs. (176 Nm).
19. Install the halfshaft in the vehicle.

Pinion Seal

REMOVAL & INSTALLATION

1. Before servicing the vehicle, refer to the precautions in the beginning of this section.
2. Place the transmission in **N**.
3. Remove or disconnect the following:
   - Underbody heat shields
   - Bolts from the propeller shaft center bearing bracket
   - Propeller shaft from the front and rear propeller shaft couplings
   - Rear propeller shaft coupling
4. Mark the position of the gear yoke, gear and nut for correct bearing pre-load.
5. Remove or disconnect the following:
   - Drive pinion flange nut
   - Drive pinion flange using a gear puller
   - Pinion seal from the rear axle housing by using a blunt chisel to drive the seal out

To install:

6. Apply lubricant to the outer diameter of the gear yoke and the outer lip of the new seal.
7. Install or connect the following:
   - New seal lubricated with chassis grease
   - Drive pinion flange
   - Drive pinion gear nut and hand-tighten

8. Tighten the nut to the position marked in the removal procedure. Tighten the nut an additional 0.062 inch (1.59mm).
9. Install or connect the following:
 - Rear propeller shaft coupling
 - Propeller shaft center bearing bracket. Torque the bolts to 15 ft. lbs. (20 Nm).
 - Propeller shaft to the rear coupling. Torque the bolts to 70 ft. lbs. (95 Nm).
 - Propeller shaft to the front coupling. Torque the bolts to 70 ft. lbs. (95 Nm).
 - Underbody heat shields. Torque the bolts to 18 inch lbs. (2 Nm).
10. Road test the vehicle.

STEERING AND SUSPENSION

Air Bag

CAUTION

These vehicles are equipped with an air bag system, known as a Supplemental Inflatable Restraint (SIR) system. The system must be disabled before performing service on or around system components, steering column, instrument panel components, wiring and sensors. Failure to follow safety and disabling procedures could result in accidental air bag deployment, possible personal injury and unnecessary system repairs.

PRECAUTIONS

Several precautions must be observed when handling the inflator module to avoid accidental deployment and possible personal injury.

- Never carry the inflator module by the wires or connector on the underside of the module.
- When carrying a live inflator module, hold securely with both hands, and ensure that the bag and trim cover are pointed away.
- Place the inflator module on a bench or other surface with the bag and trim cover facing up.
- With the inflator module on the bench, never place anything on or close to the module which may be thrown in the event of an accidental deployment.

DISARMING

1. Turn the steering wheel to align the wheels in the straight-ahead position.
2. Turn the ignition switch to the **LOCK** position and remove the key.
3. Wait 1 minute until the capacitors in the Sensing and Diagnostic Module (SDM) discharge.
4. Disconnect the negative battery cable.

WARNING

If major electrical or mechanical repairs are being performed on the vehicle, including painting, body part replacement or welding, the SDM module must first be disarmed.

REARMING

1. Turn the steering wheel so that the wheels are pointing straight ahead.
2. Turn the ignition switch to the **LOCK** position and remove the key.
3. Reconnect the negative battery cable.
4. Wait at least 1 minute for the capacitors to recharge.
5. While staying away from the inflator modules, turn the key to the **RUN** position. The air bag warning lamp should turn ON for 3–4 seconds, then turn OFF.

Recirculating Ball Power Steering Gear

REMOVAL & INSTALLATION

1. Before servicing the vehicle, refer to the precautions in the beginning of this section.
2. Center the front wheels, then turn the ignition key to the **LOCK** position.
3. Drain the coolant.
4. Evacuate the A/C system.
5. Siphon the power steering fluid from the reservoir.
6. Siphon the brake fluid from the reservoir.
7. Remove or disconnect the following:
 - Negative battery cable
 - Windshield wiper assembly
 - 3 body harness electrical connectors

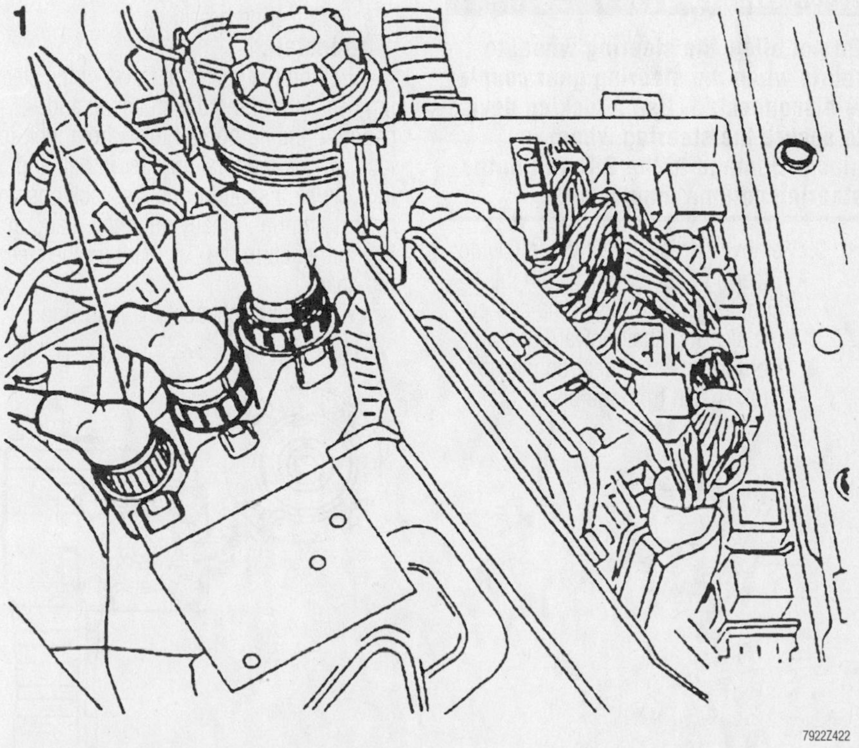

Body harness electrical connector locations (1)

For Accessory Drive Belt illustrations, see Section 1 of this manual

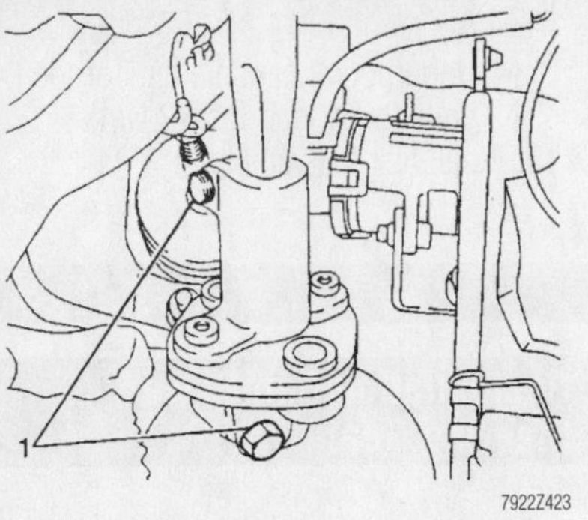

Matchmark the coupler to the steering gear before removing the coupler bolts (1)

- Electronic Control Module (ECM) from the electrical box
- Upper radiator hose
- Evaporative line extension bolt
- Power steering fluid reservoir and bracket
- Brake booster vacuum connection from the intake plenum
- Brake pipes from the master cylinder
- Electrical connector from the master cylinder reservoir cap
- Sound insulator

8. Matchmark the steering coupler and input shaft of the steering gear.

⁂ WARNING

Do not allow the steering wheel to rotate when the steering gear coupler is disconnected. Use a locking devise to secure the steering wheel in place. Damage to the SIR coil in the steering column may occur.

9. Remove or disconnect the following:
   - Steering coupler bolts from the coupler. Spread the clamp and pull the steering shaft upward and away from the steering gear connection.
   - Brake pedal from the power brake booster link rod by removing the retaining clip from the pin and driving the pin out of the linkage rod
   - Instrument panel drivers side knee bolster energy absorber
   - Relay panel. Move the fuse and relay panel away from the brake booster.
   - Brake booster nuts from the inward side of the cowling
   - Brake booster and master cylinder as one unit
   - A/C evaporator line quick connect fitting
   - Power steering hoses from the steering gear
   - Electronic Brake Traction Control Module (EBTCM)
   - Heat shield upper fastening bolt
   - Pitman arm washer and nut. Mark the alignment of the pitman arm to the steering gear for ease of installation.
   - Pitman arm using a suitable puller
   - Lower heat shield mounting nuts
   - Steering gear lower washers, bolts and nuts
   - Heat Shield
   - Electrical connector from the power steering fluid flow control valve actuator
   - Upper steering gear shims and bolt
   - Steering gear

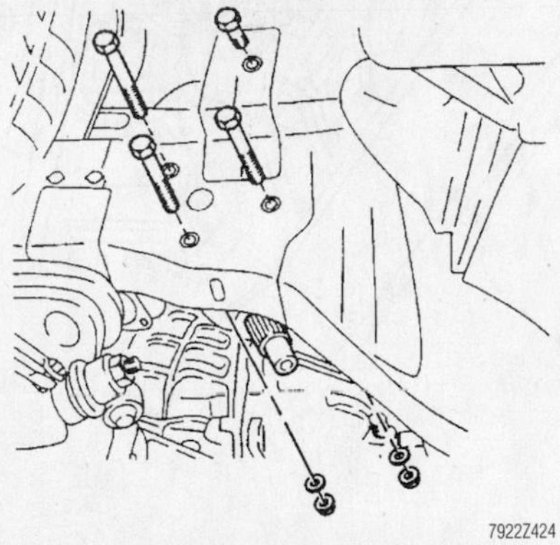

Steering gear mounting bolt locations

To install:

10. Find the center of travel of the steering gear by turning the stubshaft and counting the number of turns from lock-to-lock. Divide the total number in half and turn the stub shaft from either lock position by this amount. Finally align the mark on the stub shaft to the "V" mark on the steering gear.

11. Install or connect the following:

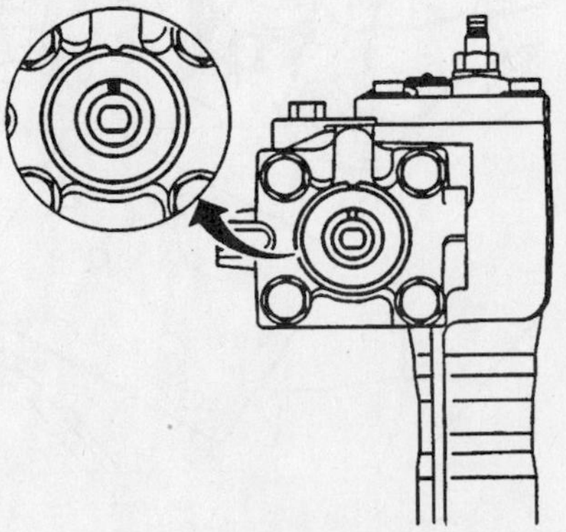

Align the mark on the stub shaft with the "V" mark on the steering gear mounting bolt locations

- Steering gear
- Steering gear shims and bolt. Torque the bolt snugly at this time. Do not tighten the bolt to proper specification at this time.
- Electrical connector to the power steering flow control valve actuator
- Heat shield with the upper mounting bolt. Torque the bolt to 71 inch lbs. (8 Nm).
- Lower steering gear washers, bolts and nuts. Torque all fasteners to 30 ft. lbs. (40 Nm).
- Heat shield lower mounting bolt. Torque the bolt to 11 ft. lbs. (15 Nm).
- Pitman arm to the steering gear. Make certain that the steering gear shaft protector is in place.
- Pitman arm washer and nut. Torque the fastener to 118 ft. lbs. (160 Nm).
- EBTCM assembly
- Inlet and outlet hoses to the steering gear. Torque the hoses to 21 ft. lbs. (28 Nm).
- A/C evaporator line quick connect with a new O-ring lubricated with mineral oil
- Vacuum booster/master cylinder assembly. Make certain the vacuum booster nuts are facing the inward side of the cowl. Torque the fasteners to 15 ft. lbs. (20 Nm).
- Fuse and relay panels and tighten the fasteners securely
- Drivers side instrument panel knee bolster energy absorber
- Brake pedal to the power booster by driving the pin into the linkage rod
- Retaining clip to the pin
- Steering coupler to the gear by aligning the matchmarks
- Coupler bolts. Torque the bolts to 16 ft. lbs. (22 Nm).
- Sound insulator
- Brake pipes to the master cylinder. Torque the fasteners to 12 ft. lbs. (16 Nm).
- Electrical connector to the master cylinder reservoir cap
- Brake booster vacuum connection to the intake plenum
- Power steering fluid reservoir and clamp. Torque the clamp to 62 inch lbs. (7 Nm).
- New O-ring lubricated with mineral oil to the A/C evaporator line extension
- Evaporator line to the cowl.

Torque the bolt to 15 ft. lbs. (20 Nm).
- Upper radiator hose
- ECM to the electrical box
- Body harness electrical connectors
- Wiper assembly
- Negative battery cable

12. Fill and bleed the brake fluid.
13. Fill and bleed the power steering system.
14. Fill the coolant system.
15. Recharge the A/C system.
16. Start the vehicle and check all fluid system for leaks.
17. Check all fluid levels and top off, if necessary.
18. Road test the vehicle.
19. Bleed the systems again, if necessary.

Strut

REMOVAL & INSTALLATION

Front

1. Before servicing the vehicle, refer to the precautions in the beginning of this section.
2. Remove or disconnect the following:
   - Front wheel
   - Wheel Speed Sensor (WSS) wiring from the strut
   - Brake wear indicator wires from the strut
   - Brake flex hose clip from the strut
   - Brake caliper and bracket from the steering knuckle
   - Stabilizer shaft link
   - Lower strut mounting bolts from the steering knuckle
   - Upper support plate protective cap
   - Upper support plate
   - Strut from the wheel well

To install:

3. Install or connect the following:
   - Strut into the strut tower
   - Upper support plate and nut. Torque the nut to 41 ft. lbs. (55 Nm).
   - Cap to the upper support nut
   - Steering knuckle to the strut
   - New bolts for the steering knuckle. Torque the lower ball joint pinch bolt to 74 ft. lbs. (100 Nm).

➡The steering knuckle-to-strut bolts will not be tightened to specifications until after camber corrections have been made.

   - Stabilizer shaft link to the strut. Torque the nut to 48 ft. lbs. (65 Nm).

➡Before installing the bolts for the caliper, run an M12 x 1.5 tap through the holes and coat the new bolts with a thread locking compound.

   - Brake caliper and bracket to the steering knuckle. Torque the bracket bolts to 70 ft. lbs. (95 Nm) plus a 37 degree turn.
   - Brake flex hose to the strut and retain it with the clip
   - WSS wiring to the strut
   - Brake wear indicator wires to the strut
   - Front wheel

4. Adjust the wheel alignment.
5. Torque the steering knuckle-to-strut bolts to 52 ft. lbs. (70 Nm).
6. Road test the vehicle.

Shock Absorber

REMOVAL & INSTALLATION

Rear

1. Before servicing the vehicle, refer to the precautions in the beginning of this section.
2. Position the rear seat backs forward to allow access to the upper shock absorber mounting.
3. Remove or disconnect the following:
   - Protective cap and nut from shock tower
   - Upper mounting nut, washer and grommet

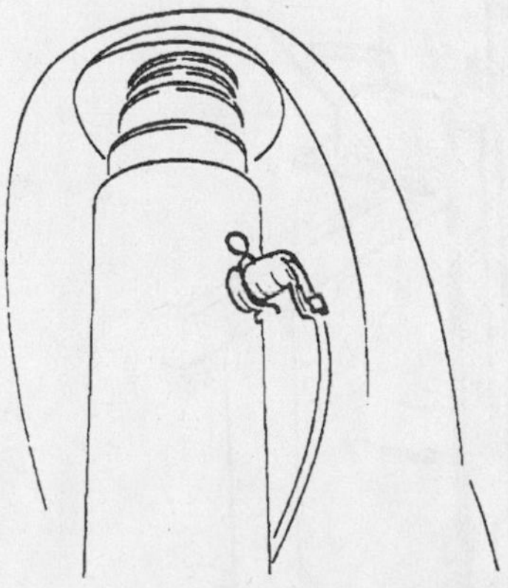

Air line connection to the shock absorber for Automatic Level Control (ALC)

   - Automatic Level Control (ALC) air line connection from the shock
   - Lower shock mounting bolt
   - Shock absorber

To install:

4. Install or connect the following:
   - Lower rubber mounting grommet and washer
   - Shock into the tower
   - Upper shock mounting grommet, washer and nut. Torque the nut 15 ft. lbs. (20 Nm).
   - Protective cap to the shock tower
   - Lower shock mounting bolt. Torque the bolt 81 ft. lbs. (110 Nm).

➡It may be necessary to support the lower control arm for ease of installation of the lower nut.

   - ALC air line connection to the shock

5. Position the rear seats in their proper position.
6. Road test the vehicle and verify a smooth ride.

Coil Spring

REMOVAL & INSTALLATION

Front

1. Before servicing the vehicle, refer to the precautions in the beginning of this section.
2. Remove the strut from the vehicle.
3. Using a Strut Compressor Holding tool J 3289-20 place the strut in a Compressor tool J 34013-A along with an Adapter J 3413-88
4. Remove or disconnect the following:
   - Compress the spring
   - Upper bearing support nut
   - Bearing and plate assembly
5. Decompress the spring.
6. Remove or disconnect the following:
   - Upper spring support plate
   - Upper insulator
   - Strut bumper
   - Cover from the strut
   - Spring from the strut
   - Lower insulator from the strut

To install:

7. Before servicing the vehicle, refer to the precautions in the beginning of this section.
   - Lower insulator to the strut
   - Spring on the strut
   - Upper spring support plate

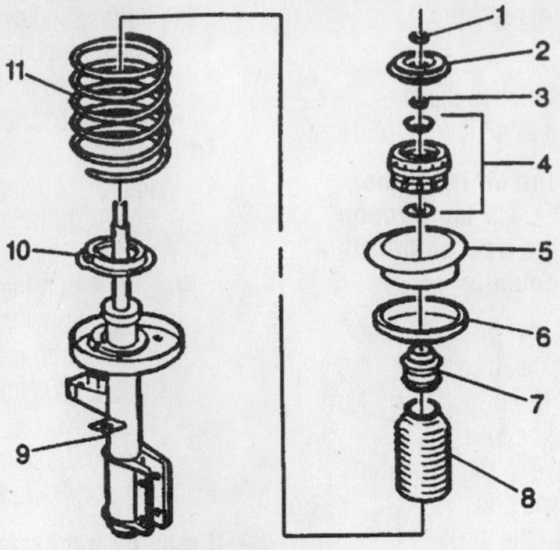

(1) Upper Support Plate Nut
(2) Upper Support Plate
(3) Upper Bearing Support Nut
(4) Bearing and Bearing Plate Assembly
(5) Upper Spring Support Plate
(6) Upper Insulator
(7) Strut Bumper
(8) Strut Cover
(9) Strut
(10) Lower Insulator
(11) Spring

Exploded view of the strut assembly

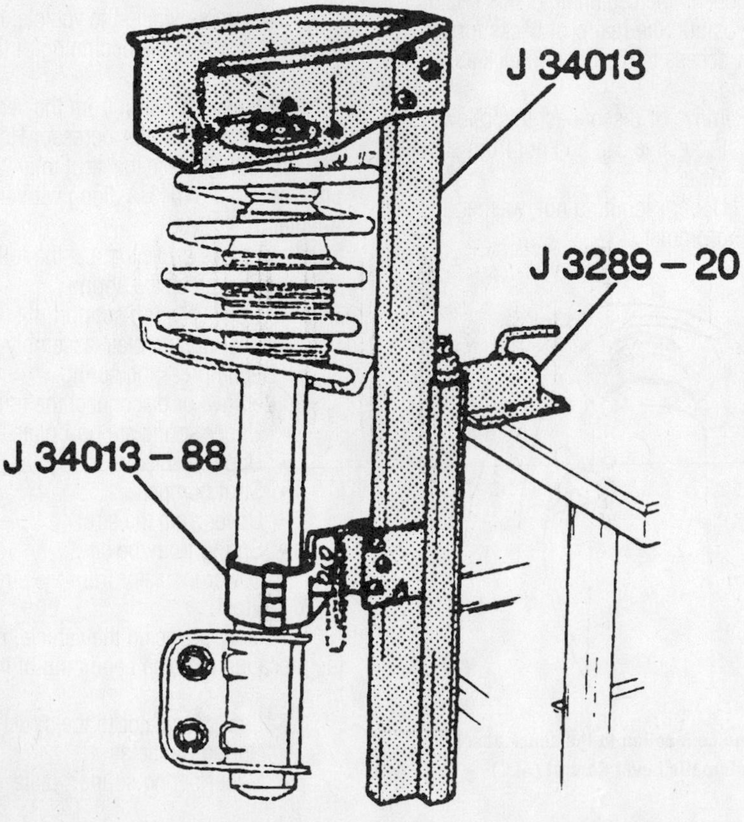

Be sure to compress the coil spring before removing the upper bearing support nut

- Upper insulator
- Strut bumper
- Cover

8. Align the spring between the upper and lower rubber isolation rings.
9. Compress the spring.
10. Install or connect the following:
    - Bearing and plate assembly
    - Upper bearing support nut. Torque the nut to 52 ft. lbs. (70 Nm).
11. Release the tension from the spring.
12. Remove the strut compressor tools from the strut.
13. Install the strut to the vehicle.
14. Road test the vehicle and verify a smooth ride and no abnormal noises from the strut and spring.

Rear

1. Before servicing the vehicle, refer to the precautions in the beginning of this section.
2. Remove or disconnect the following:
    - Retainers from the lower control arms
    - Brake pipes from the lower control arms without disconnect the brake fitting
    - Stabilizer shaft link bolts
    - Stabilizer shaft link from the lower control arms
    - Rubber exhaust insulators from the hangers. Support the exhaust system so no damage occurs to the pipes or gaskets.
    - Rear wheel speed sensor electrical connections
    - Shock lower mounting bolt after properly supporting the lower control arms
    - Support from the lower control arms and place on the rear differential
    - Rear axle cradle mounts to the body bolts
3. Lower the rear differential so that coil spring can be removed.
4. Remove or disconnect the following:
    - Rear springs with the seats on the spring
    - Seats from the spring

To install:

5. Install or connect the following:
    - Seats to the spring
    - Rear springs to the lower control arm and body. Make certain that the spring leg is aligned properly.
    - Protective shields
    - Rear differential with the cradle mounting bolts. Torque the bolts to 48 ft. lbs. (65 Nm).

- Shock absorber. Torque the bolt to 81 ft. lbs. (110 Nm).
- Rear wheel speed sensor electrical connections
- Rubber exhaust insulators to the hangers
- Stabilizer shaft link to the lower control arm. Torque the bolts to 15 ft. lbs. (20 Nm).
- Brake pipes to the lower control arm and secure with the retainers

6. Check and/or adjust the rear toe.

Lower Ball Joint

REMOVAL & INSTALLATION

1. Before servicing the vehicle, refer to the precautions in the beginning of this section.
2. Remove or disconnect the following:
   - Front wheel
   - Lower control arm
3. Using a 0.5 inch (13mm) drill, drill out the 3 rivet heads that attach the ball stud to the lower control arm.
4. Remove the ball stud from the lower control arm.

To install:

5. Install or connect the following:
   - Ball stud to the lower control arm
   - Bolts through the upper side of the lower control arm. Torque the new bolts to 26 ft. lbs. (35 Nm).
   - Lower control arm

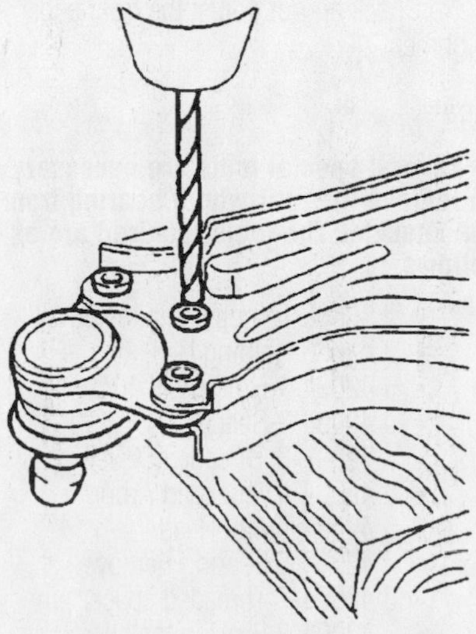

7922Z430

Drill out the rivets to remove the ball joint from the lower control arm

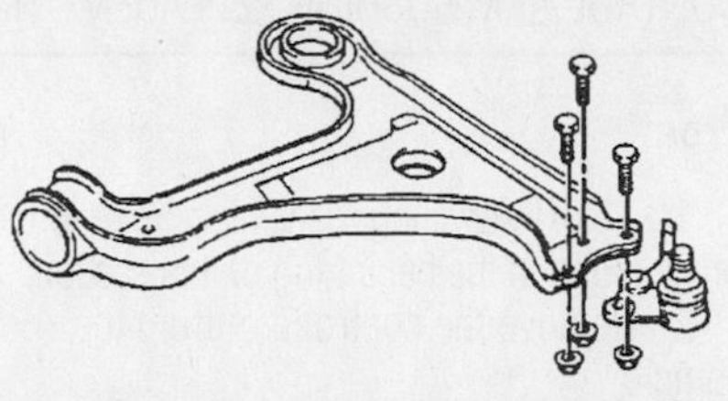

7922Z431

The replacement ball joint will be bolted to the lower control arm with the bolts supplied in the kit

- Front wheel

6. Check and/or adjust the front end alignment.

Lower Control Arm

REMOVAL & INSTALLATION

Front

1. Before servicing the vehicle, refer to the precautions in the beginning of this section.
2. Remove or disconnect the follow-ing:
   - Front wheel
   - Brake pad wear indicator rubber sleeve
   - Wheel Speed Sensor (WSS)
   - Brake hose from the strut bracket
   - Brake caliper from the steering knuckle

➡Do not open the hydraulic brake system.

- Outer tie rod from the steering knuckle
- Stabilizer shaft nut from the shaft
- Steering knuckle from the strut
- Lower control arm ball stud pinch bolt
- Steering knuckle from the lower control arm ball stud. The rotor and wheel hub will remain attached.

3. Loosen the lower control arm horizontal/vertical bolts.
4. Remove or disconnect the following:
   - Horizontal bolts
   - Vertical bolts by rotating the lower control arm from the support bracket
   - Lower control arm

To install:

5. Install or connect the following:
   - Lower control arm into the support bracket
   - Vertical bolt and hand-tighten at this time
   - Horizontal bolt by inserting it from the rear side of the vehicle. Torque all bolts to 103 ft. lbs. (140 Nm).

➡Be certain that the lower control arm is in the horizontal position before tightening the bolts to prevent the bushings from binding.

- Steering knuckle to the lower control arm ball stud. Torque the bolt to 74 ft. lbs. (100 Nm).
- Steering knuckle to the strut with new bolts. Hand tighten the bolts at this time.
- Stabilizer shaft link to the shaft. Torque the nut to 48 ft. lbs. (65 Nm).
- Outer tie rod to the steering knuckle
- New outer tie rod nut. Torque the nut 44 ft. lbs. (60 Nm).

➡Clean the bolt holes for the brake caliper by running a M12 x 105 tap through the holes. Use new bolts for the caliper to steering knuckle connection after applying a thread locking compound.

- Caliper to the steering knuckle. Torque the new bolts to 70 ft. lbs. (95 Nm) plus a 37 degree turn.
- Brake hose to the strut bracket with a new clip
- WSS. Torque the bolt to 71 inch lbs. (8 Nm).
- Brake pad wear indicator sleeve to the strut
- Front wheel

6. Adjust the wheel alignment and tighten the lower strut mounting bolts to 52 ft. lbs. (70 Nm).
7. Road test the vehicle.

Rear

1. Before servicing the vehicle, refer to the precautions in the beginning of this section.
2. Remove or disconnect the following:
   - Wheel
   - Driveshaft from the rear wheel hub flange
   - Brake pipe from the lower control arm
   - Brake caliper
   - Brake rotor
   - Parking brake cable from the actuator bracket
   - Rear hub and flange

- Wheel bearing
- Brake backing plate
- Exhaust system from the rubber mounts
- Rear outer tie rod from the rear control arm
- Stabilizer shaft link connection at the lower control arm
- Rear axle cradle mounting bolts

➡**Make certain that the rear differential is properly supported.**

- Lower shock mounting bolts
- Rear coil spring
- Underbody heat shield
- Propeller shaft center bearing bracket bolts
- Rear support flange bolts
- Rear support bushing bolt
- Lower control arm

To install:

3. Install or connect the following:
   - Lower control arm. Torque the bolts to 74 ft. lbs. (100 Nm).
   - Rear support bushing bolt. Torque the bolt to 92 ft. lbs. (125 Nm).
   - Rear support flange bolts. Torque the bolts to 48 ft. lbs. (65 Nm).
   - Propeller shaft center bearing bracket. Torque the bolts to 15 ft. lbs. (20 Nm).
   - Underbody heat shield. Torque the bolts to 18 inch lbs. (2 Nm).
   - Rear coil spring
   - Rear axle cradle mount. Torque the body bolts to 48 ft. lbs. (65 Nm).
   - Shock absorber. Torque the lower mount bolt to 81 ft. lbs. (110 Nm).
   - Stabilizer shaft link to the lower control arm. Torque the bolt to 15 ft. lbs. (20 Nm).
   - Outer tie rod. Torque the nut to 44 ft. lbs. (60 Nm).
   - Exhaust system to the rubber mounts
   - Brake backing plate
   - Wheel bearing
   - Hub and flange
   - Parking brake cable and route it through the bracket in the rear lower control arm
   - Brake rotor and set screw. Torque the screw to 35 inch lbs. (4 Nm).
   - Brake caliper. Torque the bolts to 59 ft. lbs. (80 Nm).
   - Brake pipe and clip to the lower control arm
   - Driveshaft. Torque the bolts to 37 ft. lbs. (50 Nm) plus an additional 70 degree turn.
   - Wheel
4. Check and/or adjust the rear toe.

CONTROL ARM BUSHING REPLACEMENT

Front

1. Before servicing the vehicle, refer to the precautions in the beginning of this section.
2. Remove the control arm from the vehicle.
3. Press the bushings from the control arm.

To install:

4. Press the new bushings into the control arm.
5. Install the control arm in the vehicle.

Rear

1. Before servicing the vehicle, refer to the precautions in the beginning of this section.
2. Remove or disconnect the following:
   - Wheel
   - Rear axle lower control arm
   - Collar for the inboard and outboard lower control arm bushings by cutting it off
3. Press the bushings out of the control arm.

To install:

4. Press the inboard side bushing into the control arm with the collar toward the rear differential.
5. Press the outboard side bushing into the control arm.
6. Install or connect the following:
   - Lower control arm
   - Wheel

Wheel Bearings

ADJUSTMENT

The wheel bearings are not adjustable.

REMOVAL & INSTALLATION

Front

1. Before servicing the vehicle, refer to the precautions in the beginning of this section.
2. Remove or disconnect the following:
   - Front wheel
   - Brake pad wear indicator wire from the strut bracket
   - Brake hose from the strut bracket
   - Caliper bolts from the steering knuckle
   - Brake caliper. The brake system should remain closed.
   - Brake rotor and setscrew
   - Dust cap from the hub
   - Hub with the bearing
   - Bearing by pressing it from the hub

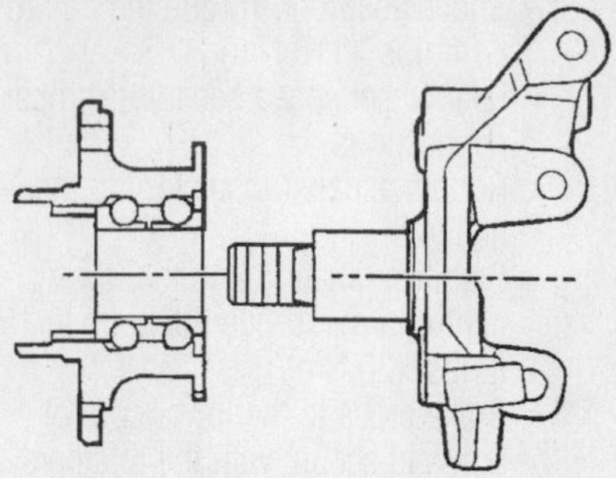

79222429

Cut-away view of the hub/wheel bearing assembly and steering knuckle

To install:

3. Install or connect the following:
   - Bearing onto the spindle and press the outer bearing ring into the hub
   - Hub nut. Torque the nut to 236 ft. lbs. (320 Nm).
   - Dust cap to the hub
   - Brake rotor and setscrew. Torque the screw to 35 inch lbs. (4 Nm).
   - Brake caliper to the steering knuckle after cleaning the bolt holes with an M12 x 1.5 tap and coating the new bolts with a thread locking compound. Torque the bolts to 70 ft. lbs. (95 Nm), plus a 37 degree turn.
   - Brake hose to the strut bracket with a retaining clip
   - Brake pad wear indicator wire to the strut
   - Front wheel
4. Road test the vehicle.
5. Check and/or adjust the front end alignment.

Rear

➡**Several special tools are necessary to remove the rear wheel bearing from the knuckle. The tools required are as follows:**

- J 36660 Torque angle meter
- J 42066 Holding tool
- J 42094-1 Holding Fixture
- J 42094-2 Spacer
- J 42094-3 Threaded Driver
- J 42094-4 Threaded Arbor
- J 42094-5 Ball Head
- J 42094-6 Bearing Remover
- J 42094-7 Threaded spacer pin
- J 42094-8 Bearing Installer
- J 42094-9 Hub Installer
- J 42094-10 Thrust Bearing
- J 42072 Triple Hex Head (10mm) deep socket

The tool numbers mentioned are Kent-

Moore tools used by GM. Equivalent tools may be available from other sources.

1. Before servicing the vehicle, refer to the precautions in the beginning of this section.

2. Remove the rear wheel.

3. Remove the outer tie rod from the knuckle.

4. Use a Holding tool J 42066 and a 0.5 inch breaker bar to counter hold the rear wheel hub.

5. Remove or disconnect the following:

- Driveshaft from the rear hub
- Brake pipe and retaining clip from the lower control arm
- Rear brake caliper and support it aside
- Set screw and brake rotor
- Three of the four brake backing plate bolts approximately 9 revolutions by installing a deep well socket
- Rear wheel hub nut by installing a

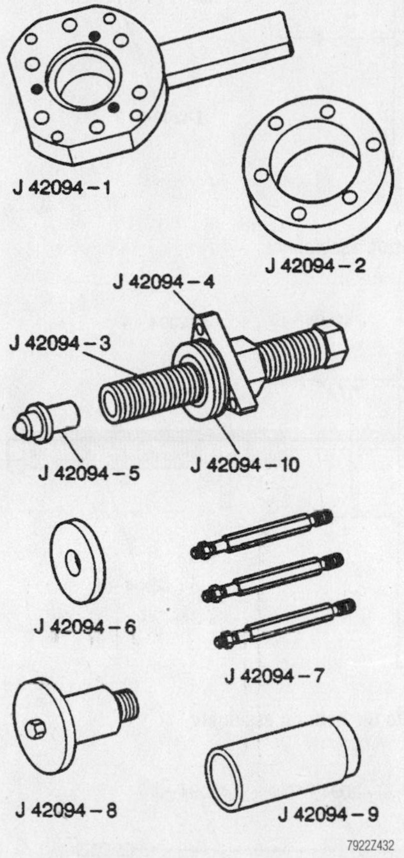

Several tools make up the J 42094 tool kit for rear wheel bearing removal and installation

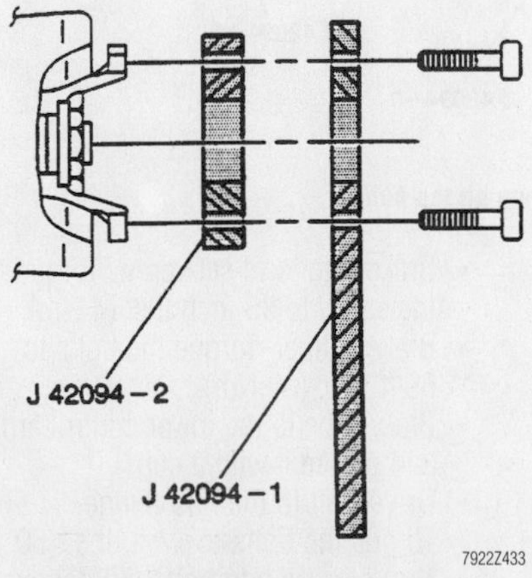

Attach the spacer and holding fixture to the flange before removing the hub retaining nut

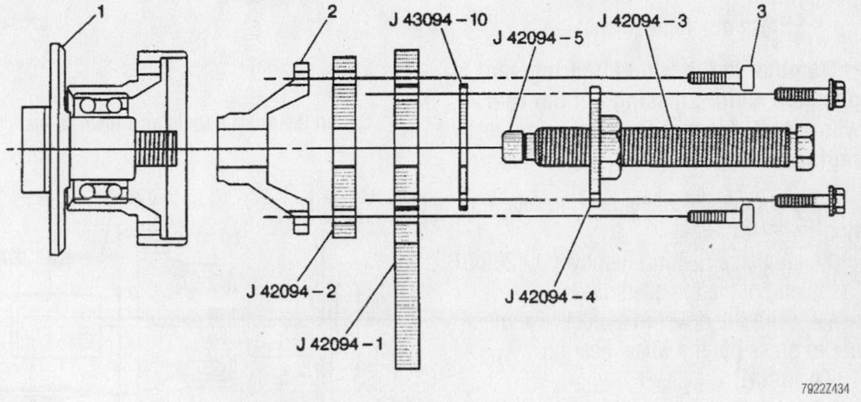

Set up the special tools as shown to remove the flange

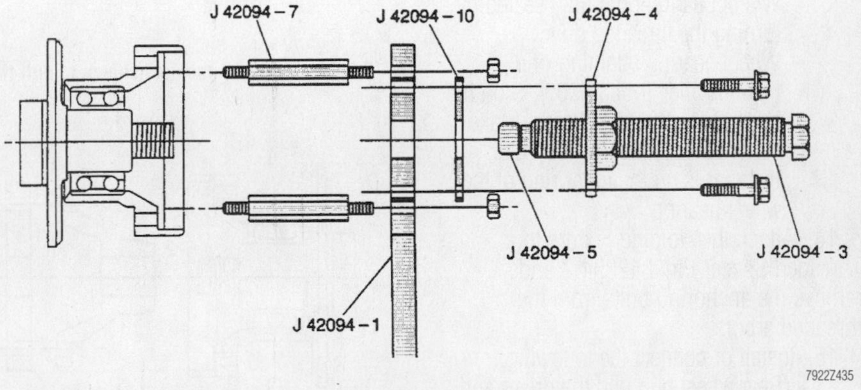

Set up the special tools as shown to remove the hub

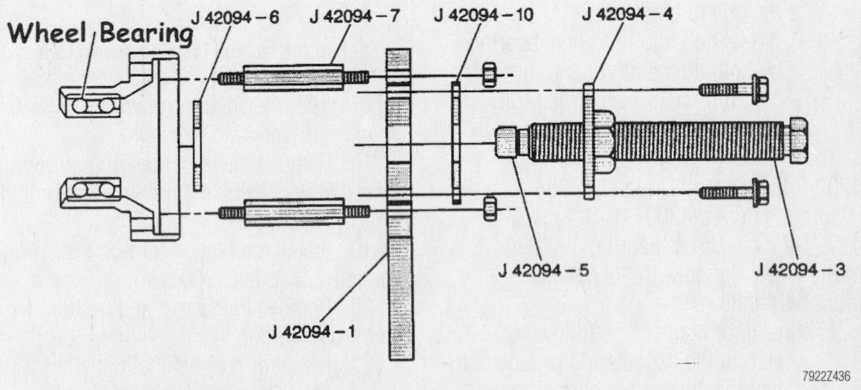

Set up the special tools as shown to remove the bearing assembly

Spacer J 42094-2 and Holding Fixture J 42094-1 to the rear wheel hub

6. Use a Thrust Bearing J 42094-10 as a spacer to attach a Threaded Arbor J42094-4 to the holding Fixture J 42094-7 with 3 bolts.

7. Attach a Threaded Driver into the threaded arbor.

8. Screw the bolts into the backing plate and attach the holding fixture with the stem upward.

9. Press out the rear wheel hub.

➡Damage to the wheel bearing seal is possible while pressing out the rear wheel hub. Inspect the seal and replace if needed.

10. Remove the wheel bearing retaining ring.

11. Attach a Bearing Remover J 42094-6 to the end of the threaded driver.

12. Turn the driver in a clockwise manner to press out the wheel bearing.

To install:

13. Install or connect the following:
- Wheel Bearing Installer Tool J 42094-8 through the wheel bearing and on to the threaded driver
- Wheel bearing until fully seated by turning the threaded driver
- Wheel bearing retaining ring
- Hub Installer Tool J 42094-9 on the shaft of the threaded driver. Make certain that the rear wheel hub installer is on the inner ring of the wheel bearing

14. Attach the Holding Fixture to a Threaded Spacer Pin J 42094-7 and remove the anchoring bolts from the threaded arbor.

15. Install or connect the following:
- Rear wheel hub into the driver and make certain the driver is seated properly on the wheel bearing. If not centered properly, it may cause the hub to bind
- Wheel hub into the wheel bearing by holding the driver and turning the arbor clockwise. When fully seated, remove the tools

16. Connect the Threaded Arbor J 42094-4, Threaded Driver J 42094-3, Holding Fixture J 42094-1, Spacer J 42094-2, and Thrust Bearing J 42094-10 to the wheel flange with the halfshaft mounting bolts.

17. Install or connect the following:
- Hub to the flange and make certain that the splines are aligned properly
- Flange fully onto the hub by turning the arbor clockwise while counter holding it with the driver

18. Remove the thrust bearing, arbor and driver while leaving the holding fixture and spacer attached to the hub.

19. Install the rear wheel hub nut. Torque the nut 221 ft. lbs. (300 Nm).

20. Remove the spacer and holding fixture from the hub.

21. Install or connect the following:
- Retaining washer to the hub
- Backing plate bolts. Torque the bolts to 37 ft. lbs. (50 Nm) plus a 40 degree turn.
- Brake rotor and set screw. Torque the screw to 35 inch lbs. (4 Nm).
- Brake caliper. Torque the bolts to 59 ft. lbs. (80 Nm).
- Brake pipe to the lower control arm and secure it with a clip
- Driveshaft to the hub flange. Torque the bolts to 37 ft. lbs. (50 Nm) plus an additional 70 degree turn.
- Tie rod end. Torque the nut to 44 ft. lbs. (60 Nm).
- Rear wheel

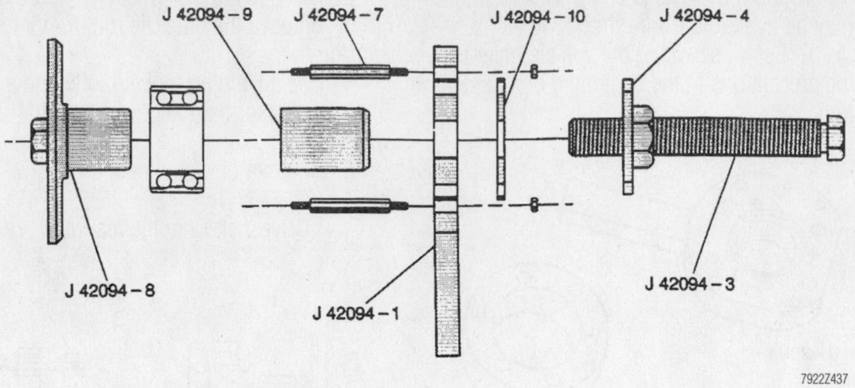

Set up the special tools as shown to install the bearing assembly

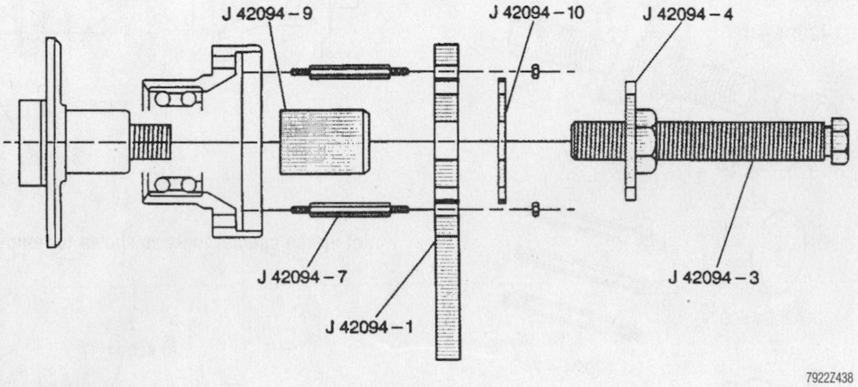

Set up the special tools as shown to pull the hub into the bearing assembly

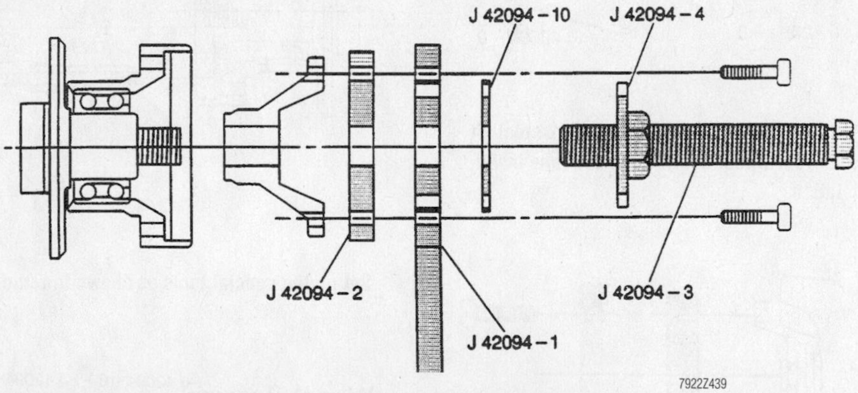

Set up the special tools as shown to install the flange on the hub

GENERAL MOTORS CORPORATION—W-BODY

29

1998–01

Buick-Century • Regal • **Chevrolet-**Impala • Lumina • Monte Carlo
Oldsmobile-Intrigue • **Pontiac-**Grand Prix

PRECAUTIONS

Before servicing any vehicle, please be sure to read all of the following precautions, which deal with personal safety, prevention of component damage, and important points to take into consideration when servicing a motor vehicle:

- Never open, service or drain the radiator or cooling system when the engine is hot; serious burns can occur from the steam and hot coolant.
- Observe all applicable safety precautions when working around fuel. Whenever servicing the fuel system, always work in a well-ventilated area. Do not allow fuel spray or vapors to come in contact with a spark, open flame or excessive heat (a hot drop light, for example). Keep a dry chemical fire extinguisher near the work area. Always keep fuel in a container specifically designed for fuel storage; also, always properly seal fuel containers to avoid the possibility of fire or explosion. Refer to the additional fuel system precautions later in this section.
- Fuel injection systems often remain pressurized, even after the engine has been turned **OFF**. The fuel system pressure must be relieved before disconnecting any fuel lines. Failure to do so may result in fire and/or personal injury.
- Brake fluid often contains polyglycol ethers and polyglycols. Avoid contact with the eyes and wash your hands thoroughly after handling brake fluid. If you do get brake fluid in your eyes, flush your eyes with clean, running water for 15 minutes. If eye irritation persists, or if you have taken brake fluid internally, seek medical assistance IMMEDIATELY.
- The EPA warns that prolonged contact with used engine oil may cause a number of skin disorders, including cancer! You should make every effort to minimize your exposure to used engine oil. Protective gloves should be worn when changing oil. Wash your hands and any other exposed skin areas as soon as possible after exposure to used engine oil. Soap and water, or waterless hand cleaner should be used.
- All new vehicles are now equipped with an air bag system. The system must be disabled before performing service on or around system components, steering column, instrument panel components, wiring and sensors. Failure to follow safety and disabling procedures could result in accidental air bag deployment, possible personal injury and unnecessary system repairs.
- Always wear safety goggles when working with, or around, the air bag system. When carrying a non-deployed air bag, be sure the bag and trim cover are pointed away from your body. When placing a non-deployed air bag on a work surface, always face the bag and trim cover upward, away from the surface. This will reduce the motion of the module if it is accidentally deployed. Refer to the additional air bag system precautions later in this section.
- Clean, high quality brake fluid from a sealed container is essential to the safe and proper operation of the brake system. You should always buy the correct type of brake fluid for your vehicle. If the brake fluid becomes contaminated, completely flush the system with new fluid. Never reuse any brake fluid. Any brake fluid that is removed from the system should be discarded. Also, do not allow any brake fluid to come in contact with a painted surface; it will damage the paint.
- Never operate the engine without the proper amount and type of engine oil; doing so WILL result in severe engine damage.
- Timing belt maintenance is extremely important! Many models utilize an interference-type, non-freewheeling engine. If the timing belt breaks, the valves in the cylinder head may strike the pistons, causing potentially serious (also time-consuming and expensive) engine damage. Refer to the maintenance interval charts in the front of this manual for the recommended replacement interval for the timing belt, and to the timing belt section for belt replacement and inspection.
- Disconnecting the negative battery cable on some vehicles may interfere with the functions of the on-board computer system(s) and may require the computer to undergo a relearning process once the negative battery cable is reconnected.
- When servicing drum brakes, only disassemble and assemble one side at a time, leaving the remaining side intact for reference.

ENGINE REPAIR

Distributor

The engines in this section all utilize a Distributorless Ignition System (DIS).

Alternator

REMOVAL

3.1L Engine

1. Before servicing the vehicle, refer to the precautions in the beginning of this section.
2. Remove or disconnect the following:
   - Negative battery cable
   - Drive belt from the alternator
   - Coolant recovery reservoir and place it away from the alternator
   - Alternator electrical connector
   - Alternator bolts and the alternator

3.4L Engine

1. Before servicing the vehicle, refer to the precautions in the beginning of this section.
2. Remove or disconnect the following:
   - Negative battery cable
   - Engine compartment cross brace
   - Accessory drive belt
   - Coolant recovery reservoir and place it aside
   - Alternator bolts
   - Alternator electrical connector
   - Alternator

3.5L Engine

1. Before servicing the vehicle, refer to the precautions in the beginning of this section.
2. Drain the cooling system.
3. Remove or disconnect the following:
   - Battery and tray
   - Drive belt
   - Engine cooling fan assembly
   - Thermostat housing and radiator hose
   - Outboard/inboard alternator bolts
   - Idler pulley bolt and pulley
   - Alternator electrical connectors
   - Alternator bolts and the alternator

3.8L Engine

1. Before servicing the vehicle, refer to the precautions in the beginning of this section.
2. Remove or disconnect the following:
   - Negative battery cable
   - Drive belt
   - Alternator electrical connectors
   - Rear alternator brace
   - Alternator bolts and the alternator

INSTALLATION

3.1L Engine

Install or connect the following:

- Alternator
- Alternator output BAT terminal and torque the nut to 15 ft. lbs. (20 Nm)
- Protective boot from the output terminal
- Electrical connector
- Alternator bolts and torque them to 37 ft. lbs. (50 Nm)
- Coolant recovery reservoir
- Drive belt
- Negative battery cable

3.4L Engine

Install or connect the following:

- Alternator
- Alternator output BAT wire and torque the nut to 15 ft. lbs. (20 Nm)
- Alternator electrical connector
- Alternator bolts and torque them to 37 ft. lbs. (50 Nm)
- Coolant recovery reservoir
- Drive belt
- Negative battery cable

3.5L Engine

1. Install or connect the following:
   - Alternator
   - Alternator electrical connectors. Torque the positive battery terminal to 15 ft. lbs. (20 Nm).
   - Idler pulley and torque the bolt to 37 ft. lbs. (50 Nm)
   - Alternator bolts and torque them to 37 ft. lbs. (50 Nm)
   - Thermostat housing with radiator hose and torque the bolts to 80 inch lbs. (9 Nm)

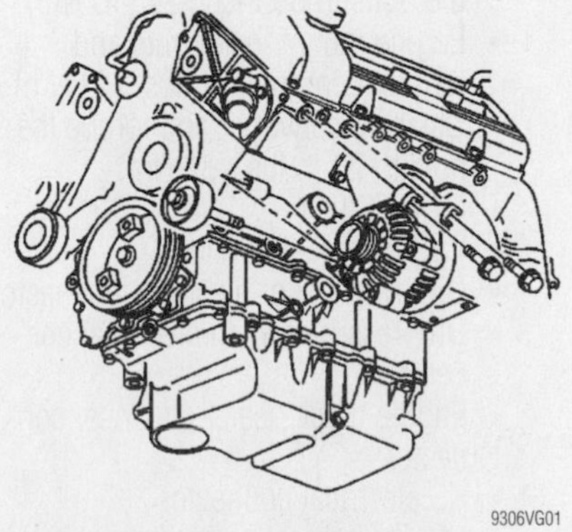

View of the alternator—3.5L engine

   - Engine cooling fan assembly
   - Drive belt
   - Battery tray and torque the bolts to 44 inch lbs. (5 Nm)
   - Battery
2. Refill the cooling system.

3.8L Engine

1. Install or connect the following:
   - Alternator
   - Alternator bolts and torque them to 37 ft. lbs. (50 Nm)
   - Alternator electrical connectors and torque the positive battery terminal to 15 ft. lbs. (20 Nm)
   - Rear alternator brace and torque the bolts to 37 ft. lbs. (50 Nm)
   - Accessory drive belt
   - Negative battery cable

Ignition Timing

ADJUSTMENT

The ignition timing is not adjustable.

Engine Assembly

REMOVAL & INSTALLATION

3.1L Engine

1. Before servicing the vehicle, refer to the precautions in the beginning of this section.
2. Drain the engine coolant.
3. Drain the engine oil.
4. Relieve the fuel system pressure.
5. Remove or disconnect the following:
   - Negative battery cable
   - Hood
   - Throttle body air inlet duct
   - Air cleaner
   - Engine mount struts
   - Drive belt
   - Knock Sensor (KS) electrical connector
   - Heated Oxygen (HO_2S) sensor electrical connector
   - Camshaft Position (CMP) sensor electrical connector
   - Crankshaft Position (CKP) sensor electrical connector
   - Manifold Air Pressure (MAP) sensor electrical connector
   - Exhaust Gas Recirculation (EGR) valve electrical connector
   - Evaporative (EVAP) emissions canister purge solenoid valve
   - Throttle Position (TPS) sensor electrical connector
   - Idle Air Control (IAC) valve electrical connector
   - Starter motor electrical connector
   - Alternator electrical connector
   - Ignition coil electrical connector
   - Wiring harness ground connections
   - Two body wiring harness-to-engine harness connectors
   - Oil filter
   - Catalytic converter pipe from the rear exhaust manifold
   - Engine mount lower nuts
   - Torque converter cover
   - A/C compressor
   - Right side splash shield
   - Torque converter bolts
   - Transaxle brace
   - Radiator outlet hose from the engine
   - Accelerator and cruise control cables
   - Heater inlet and outlet hoses
   - Vacuum hoses from the upper intake manifold
   - Fuel feed and return hoses
   - Power steering pump and reposition
   - Radiator inlet and outlet hoses
6. Attach an engine lifting device.
7. Remove the transmission-to-engine bolts.
8. Remove the engine from the vehicle.

To install:

9. Install or connect the following:
   - Engine to the vehicle
   - Transmission-to-engine bolts and torque them to 55 ft. lbs. (75 Nm)
10. Remove the engine lifting device.
    - Power steering pump and torque the bolts to 25 ft. lbs. (34 Nm)
    - Power steering lines to the pump and torque the fasteners to 20 ft. lbs. (27 Nm)
    - Radiator inlet hose
    - Heater inlet and outlet hoses
    - Fuel feed and return hoses
    - Vacuum hoses to the upper intake manifold

➡In order to avoid possible injury or vehicle damage, always replace the accelerator control cable with a NEW cable whenever you remove the engine from the vehicle.

11. Remove the trim panel under the left instrument panel and detach the throttle cable from the top of the pedal then squeeze the retainer and push the cable through the bulkhead to remove it.
12. Install or connect the following:
   - New accelerator cable
   - Cruise control cable
   - Radiator outlet hose
   - Transmission brace and torque the transmission bolts to 32 ft. lbs. (43 Nm) and the engine bolts to 46 ft. lbs. (63 Nm)
   - Right side splash shield and torque the fasteners to 44 inch lbs. (5 Nm)
   - Torque converter bolts and torque them to 47 ft. lbs. (63 Nm)
   - A/C compressor and torque the bolts to 37 ft. lbs. (50 Nm)
   - Starter motor and torque the bolts to 32 ft. lbs. (43 Nm)
   - Torque converter cover and torque the bolts to 89 inch lbs. (10 Nm)
   - Engine mount lower nuts and torque them to 32 ft. lbs. (43 Nm)
   - Catalytic converter pipe to the rear exhaust manifold and torque the fasteners to 26 ft. lbs. (35 Nm)
   - Oil Filter
   - Two body wiring to engine harness
   - Wiring harness grounds
   - Ignition coil electrical connector
   - Alternator electrical connector
   - A/C compressor electrical connector
   - Starter motor electrical connector
   - IAC valve electrical connector
   - TPS sensor electrical connector
   - EVAP canister purge solenoid valve electrical connector
   - EGR valve electrical connector
   - MAP sensor electrical connector
   - CKP sensor electrical connector
   - CMP sensor electrical connector
   - HO_2sensor electrical connector
   - KS sensor electrical connector
   - Drive belt
   - Engine mount struts and torque the fasteners to 35 ft. lbs. (48 Nm)
   - Hood and torque the bolts to 18 ft. lbs. (25 Nm)
   - Throttle body air inlet duct
   - Air cleaner
   - Negative battery cable
13. Fill the engine with clean oil.
14. Fill the coolant system.
15. Start the vehicle and check for leaks, repair if necessary.

3.4L Engine

1. Before servicing the vehicle, refer to the precautions in the beginning of this section.
2. Drain the cooling system.
3. Drain the engine oil.
4. Relieve the fuel system pressure.
5. Remove or disconnect the following:
   - Hood
   - Cross vehicle brace
   - Engine mount struts
   - Drive belt
   - Brake booster vacuum hose
   - Heated Oxygen (HO_2S) sensor electrical connector
   - Secondary Air Injection (AIR) check valve electrical connector
   - Exhaust Gas Recirculation (EGR) valve electrical connector
   - Evaporative Emissions (EVAP) canister purge solenoid valve electrical connector
   - Throttle Position (TPS) sensor electrical connector
   - Idle Air Control (IAC) valve electrical connector
   - Alternator electrical connector
   - Ignition coil electrical connector
   - Wiring harness grounds
   - Two engine wiring harness connectors
   - Lower radiator air baffle
   - Engine splash shields
   - Oil filter
   - Vehicle Speed (VSS) sensor electrical connector
   - Oil level sensor electrical connector
   - Oil pressure switch electrical connector
   - Engine block heater (if equipped)
   - Knock (KS) sensor electrical connector
   - Crankshaft Position (CKP) sensor
   - A/C compressor electrical connector
   - Catalytic converter
   - Engine mount lower nuts
   - Torque converter cover
   - Starter motor
   - A/C compressor without disconnecting the lines
   - Torque converter bolts
   - Transaxle brace
   - Radiator outlet hose
   - Accelerator and cruise control cables
   - Vacuum hoses from the upper intake manifold
   - Fuel feed and return hoses
   - Power steering pump without disconnecting the hoses
   - Heater inlet and outlet hoses
   - Radiator inlet hose
6. Attach an engine lifting device.
7. Remove the transmission bolts.
8. Remove the engine from the vehicle.

To install:

9. Install or connect the following:
   - Engine to the transaxle and torque the bolts to 55 ft. lbs. (75 Nm)
   - Radiator inlet hose
   - Heater inlet and outlet hoses
   - Power steering pump and torque the bolts to 25 ft. lbs. (34 Nm)
   - Fuel feed and return hoses
   - Vacuum hoses to the upper intake manifold

➡In order to avoid possible injury or vehicle damage, always replace the accelerator control cable with a NEW cable whenever you remove the engine from the vehicle.

10. Remove the trim panel under the left instrument panel and detach the throttle cable from the top of the pedal then squeeze the retainer and push the cable through the bulkhead to remove it.
11. Install or connect the following:
   - New accelerator cable
   - Cruise control cable
   - Radiator outlet hose
   - Transmission brace and torque the transmission bolts to 32 ft. lbs. (43 Nm) and the engine bolts to 46 ft. lbs. (63 Nm)
   - Torque converter-to-flywheel bolts and torque them to 47 ft. lbs. (63 Nm)
   - A/C compressor and torque the bolts to 37 ft. lbs. (50 Nm)
   - Starter motor and torque the bolts to 32 ft. lbs. (43 Nm)
   - Torque converter cover and torque the bolts to 89 inch lbs. (10 Nm)
   - Engine mount lower nuts and torque them to 32 ft. lbs. (43 Nm)
   - Catalytic converter and torque the nuts to 24 ft. lbs. (32 Nm)
   - New oil filter
   - VSS electrical connector
   - Oil level sensor electrical connector
   - Oil pressure switch electrical connector
   - Engine block heater electrical connector
   - KS electrical connector
   - HO_2S electrical connector
   - CKP sensor electrical connector
   - A/C compressor electrical connector
   - Wiring harness grounds

- Lower radiator air baffle and torque the bolts to 15 ft. lbs. (20 Nm)
- Engine splash shield and torque the fasteners to 18 inch lbs. (2 Nm)
- AIR check valve solenoid electrical connector
- EGR valve electrical connector
- EVAP canister purge solenoid valve electrical connector
- TPS electrical connector
- IAC valve electrical connector
- Alternator electrical connector
- Ignition coil electrical connector
- Wiring harness grounds
- Engine wiring harness connectors
- Drive belt
- Engine mount strut and torque the bolt to 35 ft. lbs. (48 Nm)
- Throttle body air inlet duct
- Cross vehicle brace and torque the nuts to 29 ft. lbs. (40 Nm)
- Hood and torque the bolts to 18 ft. lbs. (25 Nm)
- Negative battery cable

12. Fill the engine oil.
13. Fill the engine coolant.
14. Inspect the transmission fluid level and top off if necessary.
15. Turn the ignition to the **ON** position several times to pressurize the fuel system.
16. Start the engine and inspect for any leaks, repair if necessary.
17. Check and top off the fluid levels if required.

3.5L Engine

1. Before servicing the vehicle, refer to the precautions in the beginning of this section.
2. Relieve the fuel system pressure.
3. Drain the engine oil.
4. Drain the engine cooling system.
5. Remove or disconnect the following:
   - Drive belt
   - Fuel injector sight shield
   - Power steering pump and move it aside
   - Air intake duct from the throttle body
   - Engine mount strut from the bracket
   - Fuel lines from the fuel supply rail
   - Fuel vapor line
   - Throttle and cruise control cables with the bracket from the throttle body
   - Range selector cable from the Park/Neutral Position (PNP) switch
   - Brake booster vacuum hose from the engine
   - Air conditioning vacuum hose from the engine
   - Wiring harness from the engine and transaxle
   - Radiator inlet hose from the engine
   - Transaxle fluid cooler lines from the radiator
   - Thermostat housing
   - Surge tank inlet hose
   - Heater hoses from the engine
   - Lower radiator air deflector
   - Secondary Air Injection (AIR) pipe from the AIR inlet valve
   - Battery cables from the retainers
   - Radiator outlet hose from the engine
   - Alternator
   - A/C compressor, move it aside without disconnecting the lines
   - Torque converter cover
   - Starter motor
   - Flexplate-to-torque converter bolts
   - Catalytic converter from the rear exhaust manifold
   - Lower transaxle-to-engine bolts
   - Front wheels and splash shields
   - Fog lamp electrical connectors
   - Wheel Speed Sensor (WSS) harness from the lower control arms
   - Tie rod ends from the steering knuckles
   - Lower ball joints from the steering knuckles
   - Halfshafts
   - Intermediate shaft from the steering rack

WARNING

Secure the front of the vehicle to the lift. The vehicle may become unstable as the engine/transaxle assembly is removed from the vehicle.

6. Position a frame table under the vehicle.
7. Lower the vehicle so the frame is resting on the frame table.
8. Remove the frame-to-chassis bolts.

WARNING

Do not damage the air conditioning compressor or lines when removing the frame from the vehicle.

9. Lift the vehicle from the engine/transaxle assembly.

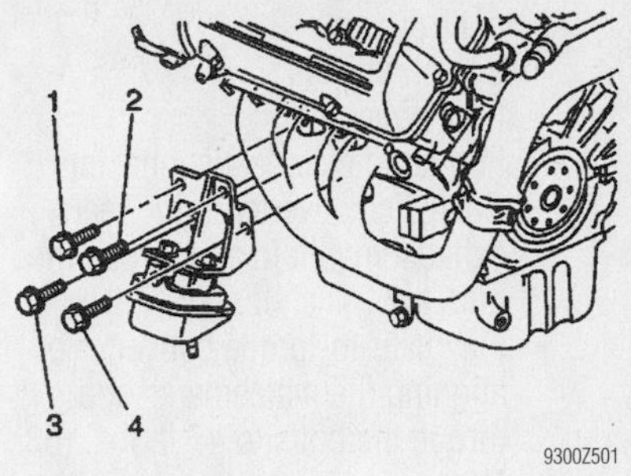

Engine mount bracket bolt tightening sequence—3.5L engine

10. Remove or disconnect the following:
    - Remove the engine mount bracket bolts
    - Remove the transmission brace
    - Remove the engine from the transaxle

To install:

11. Install or connect the following:
    - Transaxle to the engine and torque the bolts to 55 ft. lbs. (75 Nm)
    - Transaxle brace and torque the bolts to 32 ft. lbs. (43 Nm)
    - Engine mount bracket to the front of the engine and torque the bolts to 43 ft. lbs. (58 Nm)
    - Engine/transaxle assembly under the vehicle
12. Coat the sub-frame bushings with rubber lubricant.
13. Lower the vehicle onto the assembly. Align the sub-frame on the vehicle using 2 bolts or drill bits, ¾ inches thick by 8 inches long through the alignment holes on the right side of the frame.
14. Install new frame-to-body bolts and torque them to 133 ft. lbs. (180 Nm) starting with the rear bolts and then the front bolts
15. Raise the vehicle and remove the frame table.
16. Install or connect the following:
    - Intermediate shaft to the steering rack and torque the pinch bolt to 35 ft. lbs. (48 Nm)

➡Be sure the shaft is fully seated on the stub before installing the pinch bolt.

- Halfshafts
- Lower ball joints and torque the nuts to 40 ft. lbs. (50 Nm)
- Tie rod ends and torque the nuts to 22 ft. lbs. (30 Nm) plus an additional 120 degree turn

- WSS wiring harness to the lower control arm
- Splash shields
- Front wheels
- Fog lamp electrical connectors
- Catalytic converter to the rear exhaust manifold and torque the nuts to 53 inch lbs. (6 Nm)
- Flexplate-to-torque converter by aligning the matchmarks and torque the bolts to 47 ft. lbs. (63 Nm)
- Starter motor and torque the bolts to 37 ft. lbs. (50 Nm)
- Torque converter cover and torque the bolts to 89 inch lbs. (10 Nm)
- A/C compressor and torque the bolts to 37 ft. lbs. (50 Nm)
- Alternator and torque the bolts to 37 ft. lbs. (50 Nm)
- Lower radiator hose to the engine
- Battery cables in the retainers
- AIR pipe
- Lower radiator air deflector and torque the bolts to 15 ft. lbs. (20 Nm)
- Thermostat housing and torque the bolts to 80 inch lbs. (9 Nm)
- Heater hoses
- Surge tank hose
- Transaxle fluid cooler lines to the radiator
- Upper radiator hose to the engine
- Wiring harness connectors to the engine/transaxle
- A/C vacuum hose to the engine
- Brake booster hose to the engine
- Shifter cable to the transaxle PNP switch

➡In order to avoid possible injury or vehicle damage, always use a new accelerator cable when replacing the engine assembly.

17. Remove the trim panel under the left instrument panel and detach the throttle cable from the top of the pedal then squeeze the retainer and push the cable through the bulkhead to remove it.
18. Install or connect the following:
    - New throttle cable
    - Cruise control cable to the throttle body
    - Fuel vapor line
    - Fuel lines to the supply rail
    - Engine mount strut and torque the bolt to 35 ft. lbs. (48 Nm)
    - Air duct to the throttle body
    - Power steering pump and torque the bolts to 25 ft. lbs. (34 Nm)
    - Fuel injector shield and torque the bolts to 27 inch lbs. (3 Nm)
    - Drive belt
    - Negative battery cable
19. Refill the cooling system.
20. Install a new oil filter and refill the engine with new oil.
21. Start the engine and check for leaks, repair if necessary.

3.8L Engine

1. Before servicing the vehicle, refer to the precautions in the beginning of this section.
2. Relieve the fuel system pressure.
3. Drain the cooling system.
4. Drain the engine oil.
5. Remove or disconnect the following:
    - Hood
    - Fuel injector sight shield
    - Throttle body air inlet duct
    - Air cleaner
    - Radiator inlet/outlet hoses
    - A/C compressor without disconnecting the lines
    - Starter motor wiring harness
    - Accelerator and cruise control cables
    - Wiring harness from the top of the engine
    - Fuel lines from the fuel rail
    - Fuel vapor line
    - Power brake booster vacuum hose
    - A/C vacuum hose
    - Heater hoses
    - Engine mount struts from the brackets
    - Power steering pump and move it aside
    - Torque converter cover
    - Starter motor
    - Flywheel-to-torque converter bolts
    - Engine mount lower nuts
    - Transaxle brace
    - Catalytic converter pipe from the rear exhaust manifold
    - Lower transaxle to engine bolts
6. Lower the vehicle and install a suitable engine lifting device.
7. Remove the upper transaxle to engine bolts.
8. Remove the engine from the vehicle.

To install:

9. Install the engine to the vehicle.
10. Torque the transmission-to-engine bolts to 55 ft. lbs. (75 Nm).
11. Remove the engine lifting device.
12. Install or connect the following:
    - Catalytic converter to the rear exhaust manifold. Torque the nuts to 26 ft. lbs. (35 Nm).
    - Transaxle brace and torque the transaxle bolts to 32 ft. lbs. (43 Nm) and the engine bolts to 46 ft. lbs. (63 Nm)
    - Engine mount lower nuts and torque them to 35 ft. lbs. (47 Nm)
    - Flywheel-to-torque converter bolts and torque them to 47 ft. lbs. (63 Nm)
    - Starter motor and torque the bolts to 32 ft. lbs. (43 Nm)
    - Torque converter cover and torque the bolts to 89 inch lbs. (10 Nm)
    - Power steering pump and torque the bolt to 25 ft. lbs. (34 Nm)
    - Engine mount struts and torque the bolts to 35 ft. lbs. (48 Nm)
    - Heater hoses
    - A/C vacuum hose
    - Power brake booster vacuum hose
    - Fuel vapor line
    - Fuel lines
    - Wiring harness to the top of the engine

➡In order to avoid possible injury or vehicle damage, always replace the accelerator control cable with a NEW cable whenever you remove the engine from the vehicle.

13. Remove the trim panel under the left instrument panel and detach the throttle cable from the top of the pedal then squeeze the retainer and push the cable through the bulkhead to remove it.
14. Install or connect the following:
    - New throttle cable
    - Cruise control cable to the throttle body
    - Starter motor wiring harness
    - A/C compressor and torque the bolts to 23 ft. lbs. (31 Nm) and the nuts to 74 ft. lbs. (100 Nm)
    - Inlet and outlet radiator hoses
    - Throttle body air inlet duct
    - Air cleaner
    - Hood and torque the bolts to 20 ft. lbs. (27 Nm)
    - Fuel injector sight shield
    - Negative battery cable
15. Fill the engine with new oil.

➡A filter change is recommended.

16. Fill the cooling system.
17. Start the engine and check for leaks, repair if necessary.

Water Pump

REMOVAL & INSTALLATION

3.1L and 3.4L Engines

1. Before servicing the vehicle, refer to the precautions in the beginning of this section.
2. Drain the cooling system.
3. Remove or disconnect the following:
   - Negative battery cable
   - Accessory drive belt guard
   - Water pump pulley bolts, loosen
   - Accessory drive belt
   - Water pump pulley
   - Water pump

To install:

4. Install or connect the following:
   - Water pump with a new gasket and torque the bolts to 89 inch lbs. (10 Nm)
   - Water pump pulley and torque the bolts to 18 ft. lbs. (25 Nm)
   - Accessory drive belt
   - Drive belt guard
5. Fill the cooling system.
6. Start the engine and check for leaks, repair if necessary.

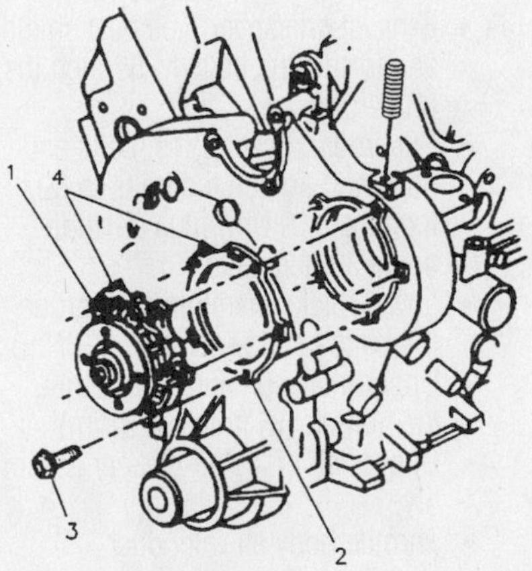

1 WATER PUMP
2 GASKET
3 10 N·m (89 LB. IN.)
4 LOCATOR – MUST BE VERTICAL

7924LG01

Water pump assembly mounting—3.1L and 3.4L engines

3.5L Engine

1. Before servicing the vehicle, refer to the precautions in the beginning of this section.
2. Partially drain the cooling system.

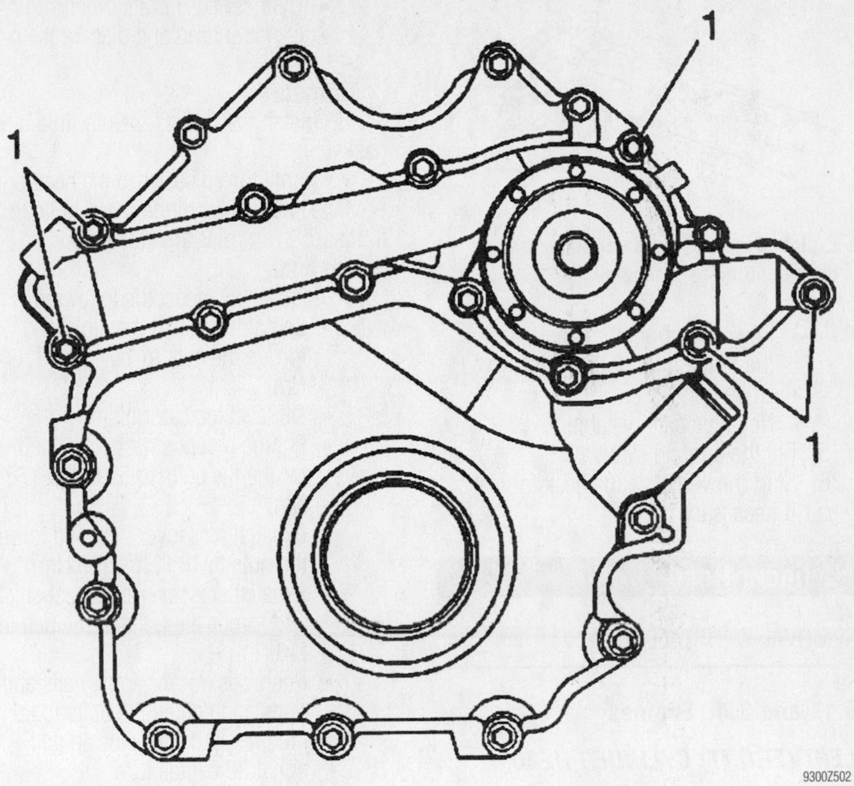

9300Z502

Be sure to install the 5 long water pump bolts in the correct locations—3.5L engine

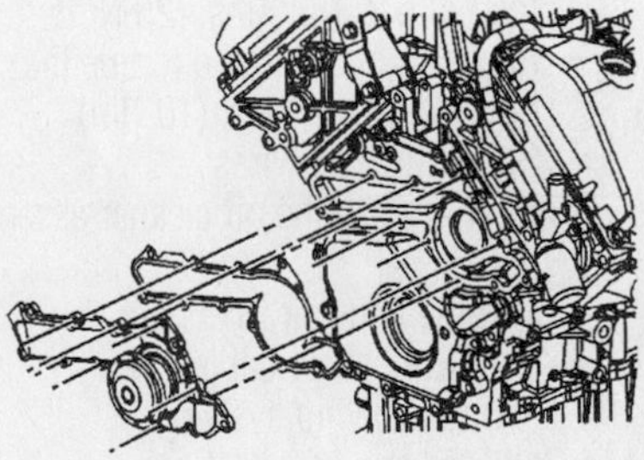

9300Z503

Water pump mounting—3.5L engine

3. Remove or disconnect the following:
   - Surge tank outlet hose
   - Drive belt
   - Idler pulley
   - Water pump pulley

➡The water pump is attached to the engine with both long and short bolts, be sure to note their locations.

   - Water pump

To install:

4. Install or connect the following:
   - Water pump using a new gasket and torque the bolts to 124 inch lbs. (14 Nm)
   - Water pump pulley and torque the bolts to 18 ft. lbs. (25 Nm)
   - Idler pulley and torque the bolt to 115 inch lbs. (13 Nm)
   - Drive belt
   - Surge tank hose
5. Refill the cooling system.
6. Start the engine and check for leaks.

3.8L Engines

1. Before servicing the vehicle, refer to the precautions in the beginning of this section.
2. Drain the engine coolant.
3. Remove or disconnect the following:
   - Negative battery cable.
   - Drive belt
   - Water pump pulley
   - Power steering pump and move it aside
   - Water pump

To install:

4. Install or connect the following:
   - Water pump using a new gasket and torque the long bolts to 15 ft. lbs. (20 Nm) plus a 40 degree turn and the short bolts to 11 ft. lbs. (15 Nm) plus an additional 80 degree turn
   - Water pump pulley and torque the bolts to 115 inch lbs. (13 Nm)

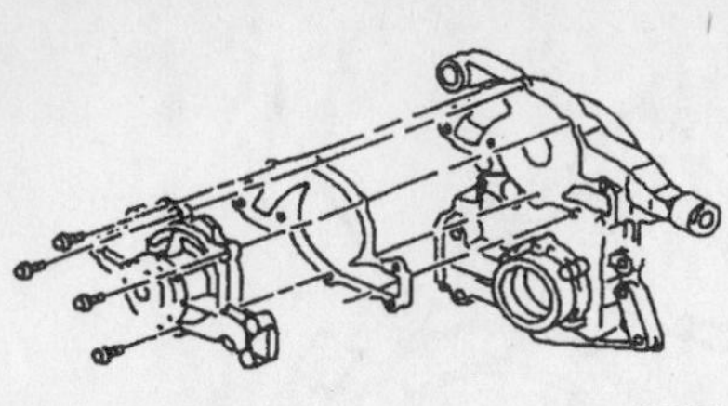

7922Z504

Exploded view of the water pump assembly mounting—3.8L engine

- Power steering pump and torque the bolts to 25 ft. lbs. (34 Nm)
- Drive belt
- Negative battery cable

5. Fill the cooling system.
6. Start the vehicle and check for leaks, repair if necessary.

Cylinder Head

REMOVAL & INSTALLATION

3.1L and 3.4L Engines

LEFT (FRONT) CYLINDER HEAD

1. Before servicing the vehicle, refer to the precautions in the beginning of this section.
2. Relieve the fuel system pressure.
3. Drain the engine oil.
4. Drain the cooling system.
5. Remove or disconnect the following:
   - Upper half of the air cleaner
   - Throttle body air inlet duct
   - Upper intake manifold
   - Spark plug wires from the spark plugs
   - Rocker arm covers
   - Lower intake manifold
   - Rocker arm bolt, rocker arms, balls and pushrods
   - Exhaust crossover pipe
   - Engine mount strut bracket from the cylinder head
   - Left side exhaust manifold
   - Oil level indicator tube

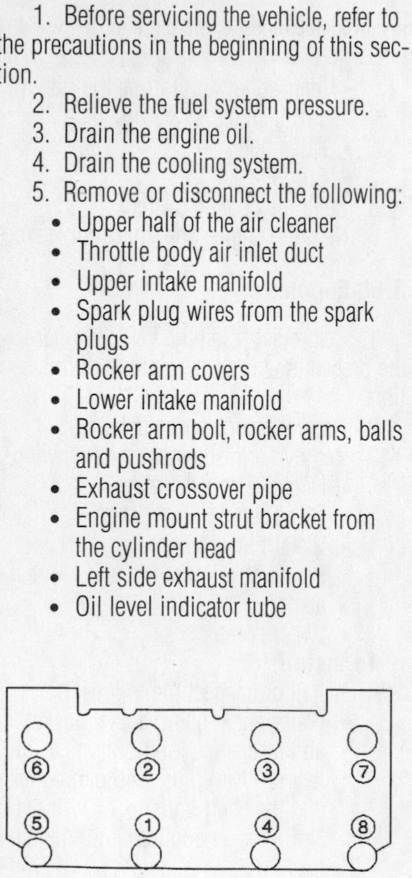

9346ZGA9

Tighten the cylinder head bolts using the following sequence—31L and 3.4L engines

   - Cylinder head bolts evenly
   - Cylinder head and discard the gasket

To install:

6. Install the cylinder head with a new gasket.
7. Apply thread sealer to the head bolts.
8. Tighten the cylinder head bolts to 37 ft. lbs. (50 Nm) plus an additional 90 degree turn.
9. Install or connect the following:
   - Left side exhaust manifold and torque the nuts to 12 ft. lbs. (16 Nm)
   - Oil level indicator tube
   - Engine mount strut bracket and torque the bolts to 52 ft. lbs. (70 Nm)
   - Exhaust crossover pipe and torque the nuts to 18 ft. lbs. (25 Nm)
   - Exhaust crossover pipe heat shield and torque the bolts to 89 inch lbs. (10 Nm)
   - Pushrods, rocker arms, balls and bolts and tighten the bolts to 89 inch lbs. (10 Nm) plus an additional 30 degree turn
   - Lower intake manifold and torque the bolts to 115 inch lbs. (13 Nm)
   - Upper intake plenum and torque the bolts to 18 ft. lbs. (25 Nm)
   - Rocker arm cover and torque the bolts to 89 inch lbs. (10 Nm)
   - Spark the plug wires
   - Upper half of the air cleaner assembly
   - Throttle body air inlet duct
   - Negative battery cable
10. Refill the cooling system.
11. Refill the engine with clean oil.

➡A filter change is also recommended.

12. Start the engine and check for leaks, repair if necessary.

RIGHT (REAR) CYLINDER HEAD

1. Before servicing the vehicle, refer to the precautions in the beginning of this section.
2. Relieve the fuel system pressure.
3. Drain the engine oil.
4. Drain the cooling system.
5. Remove or disconnect the following:
   - Upper half of the air cleaner
   - Throttle body air inlet duct
   - Upper intake plenum
   - Lower intake manifold
   - Spark plug wires from the spark plugs
   - Rocker arm covers
   - Exhaust crossover pipe heat shield
   - Crossover pipe
   - Right side exhaust manifold
   - Fuel line bracket
   - Alternator and bracket
   - Rocker arms bolt, rocker arms, balls and pushrods
   - Cylinder head bolts evenly
   - Cylinder head

To install:

6. Install the cylinder head with a new gasket.
7. Apply thread sealer to the head bolts.
8. Tighten the cylinder head bolts to 37 ft. lbs. (50 Nm) plus an additional 90 degree turn.
9. Install or connect the following:
   - Pushrods, rocker arms, balls and rocker arm bolts and tighten the bolts to 89 inch lbs. (10 Nm) plus an additional 30 degrees
   - Exhaust manifold and torque the nuts to 12 ft. lbs. (16 Nm)
   - Alternator bracket and torque the bolts to 37 ft. lbs. (50 Nm)
   - Alternator and torque the bolts to 37 ft. lbs. (50 Nm)
   - Rocker arm cover and torque the bolts to 89 inch lbs. (10 Nm)
   - Spark plug wires
   - Exhaust crossover pipe and torque the nuts to 18 ft. lbs. (25 Nm)
   - Exhaust crossover pipe heat shield and torque the bolts to 89 inch lbs. (10 Nm)
   - Pushrods, rocker arms, balls and bolts and tighten the bolts to 89 inch lbs. (10 Nm) plus an additional 30 degree turn
   - Lower intake manifold and torque the bolts to 115 inch lbs. (13 Nm)
   - Upper intake plenum and torque the bolts to 18 ft. lbs. (25 Nm)
   - Upper half of the air cleaner assembly
   - Throttle body air inlet duct
   - Negative battery cable
10. Refill the cooling system.
11. Refill the engine with clean oil.

➡An oil filter change is recommended.

12. Start the engine and check for leaks, repair if necessary.

3.5L Engine

FRONT

1. Before servicing the vehicle, refer to the precautions in the beginning of this section.
2. Drain the engine oil.
3. Drain the cooling system.
4. Remove or disconnect the following:

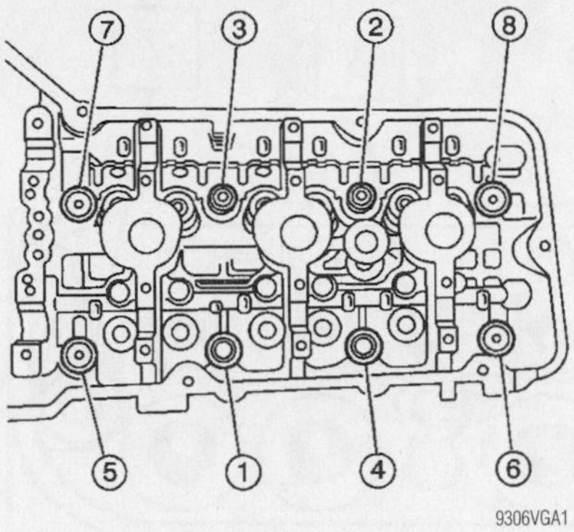

Front cylinder head bolt torque sequence—3.5L engine

- Negative battery cable
- Intake manifold
- Water outlet housing
- Engine mount strut bracket
- Coolant crossover pipe
- Front exhaust manifold
- Camshaft cover

5. Install a holding tool on the camshafts to hold them in position.
6. Remove or disconnect the following:
    - Camshaft primary chain
    - Camshafts from the front cylinder head
    - Rocker arms and valve lifters

➡Be sure to keep the arms and lifters in order so they can be installed the their original locations.

- M6 bolts from the front of the cylinder head
- M11 cylinder head bolts and discard
- Cylinder head

To install:

7. Be sure the dowels are securely mounted in the engine block.
8. Install or connect the following:
    - New gasket
    - Cylinder head
    - New M11 bolts
    - M6 bolts in the front of the cylinder head
9. Torque the M11 bolts in sequence to:
    a. Step 1: 22 ft. lbs. (30 Nm).
    b. Step 2: 60 degree turn.
    c. Step 3: 60 degree turn.
    d. Step 4: 60 degree turn.
10. Torque the long M6 bolt to 22 ft. lbs. (30 Nm).
11. Torque both shorter M6 bolts to 106 inch lbs. (12 Nm).
12. Install or connect the following:
    - Lifters and rocker arms in their original positions

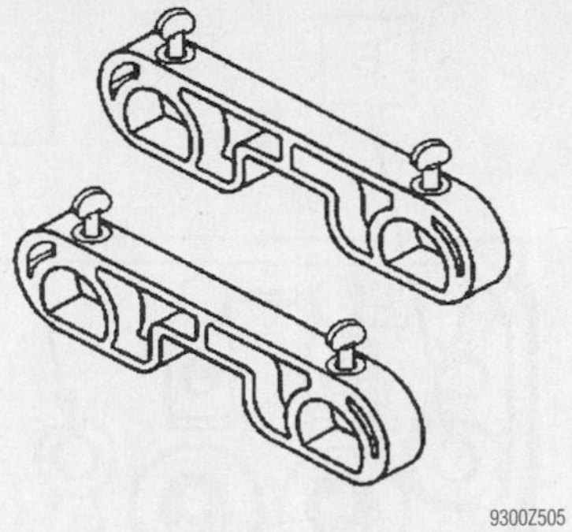

Camshaft holding fixture J-42038—3.5L engine

- Camshafts and torque the bearing cap bolts to 71 inch lbs. (8 Nm) plus an additional 22 degree turn
- Primary camshaft drive chain
- Camshaft cover and torque the bolts to 80 inch lbs. (9 Nm)
- Exhaust manifold and torque the bolts to 18 ft. lbs. (25 Nm)
- Coolant crossover pipe and torque the bolts to 18 ft. lbs. (25 Nm)
- Engine mount strut bracket and torque the bolts to 37 ft. lbs. (50 Nm)
- Water outlet housing and torque the bolts to 80 inch lbs. (9 Nm)
- Intake manifold and torque the bolts to 62 inch lbs. (7 Nm)
- New oil filter
- Negative battery cable

13. Refill the engine with clean oil.
14. Refill the cooling system.
15. Start the engine and check for leaks, repair if necessary.

REAR

1. Before servicing the vehicle, refer to the precautions in the beginning of this section.
2. Drain the engine oil.
3. Drain the cooling system.
4. Remove or disconnect the following:
    - Negative battery cable
    - Intake manifold

➡Do not remove the rear exhaust manifold. Detach it from the cylinder head and the connection from the front manifold; then, move it aside.

- Exhaust manifold
- Coolant crossover pipe
- Camshaft cover

5. Install a holding tool on the camshafts to hold them in position.
6. Remove or disconnect the following:
    - Primary camshaft drive chain
    - Camshafts

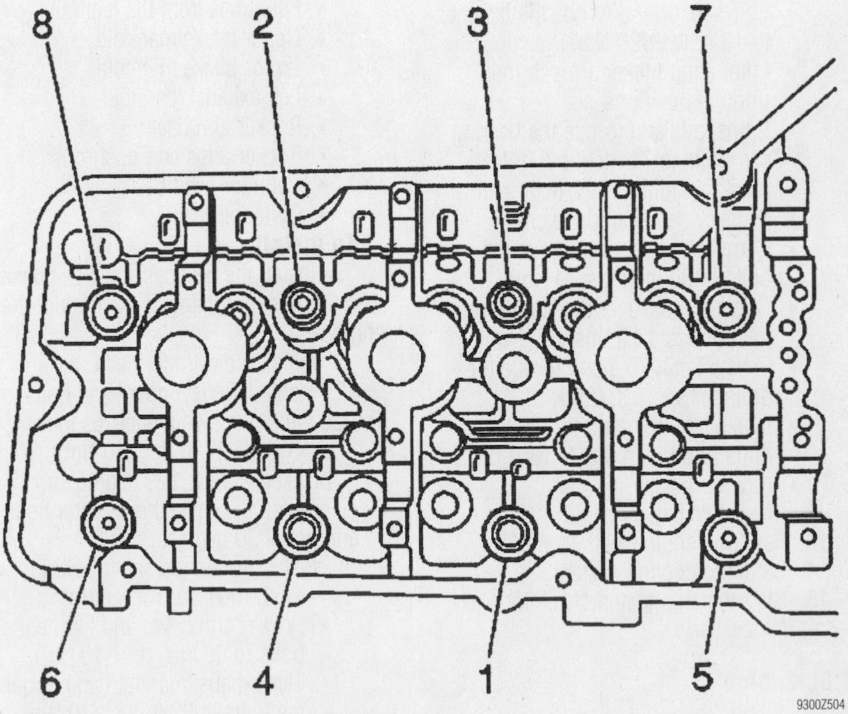

Rear cylinder head bolt torque sequence—3.5L engine

- Rocker arms and valve lifters

➡Keep the valve train parts in order so they can be installed in their original positions.

- Engine Coolant Temperature (ECT) sensor from the cylinder head
- M6 bolts from the front of the cylinder head, note the location of the longer bolt
- M11 cylinder head bolts and discard the bolts
- Cylinder head

To install:

7. Be sure the dowels are securely mounted in the engine block.
8. Install or connect the following:
   - New gasket
   - Cylinder head
   - New M11 bolts
   - M6 bolts in the front of the cylinder head
9. Torque the M11 bolts in sequence as follows:
   a. Step 1: 22 ft. lbs. (30 Nm).
   b. Step 2: 60 degree turn.
   c. Step 3: 60 degree turn.
   d. Step 4: 60 degree turn.
10. Torque the long M6 bolt to 22 ft. lbs. (30 Nm).
11. Torque both short M6 bolts to 106 inch lbs. (12 Nm).
12. Install or connect the following:
   - ECT sensor and torque the fastener to 15 ft. lbs. (20 Nm)
   - Lifters and rocker arms in their original positions
   - Camshafts and torque the bearing cap bolts to 71 inch lbs. (8 Nm) plus an additional 22 degree turn
   - Primary camshaft drive chain
   - Camshaft cover and torque the bolts to 80 inch lbs. (9 Nm)
   - Coolant crossover pipe and torque the bolts to 18 ft. lbs. (25 Nm)
   - Exhaust manifold and torque the bolts to18 ft. lbs. (25 Nm)
   - Intake manifold and torque the bolts to 62 inch lbs. (7 Nm)
   - New oil filter
   - Negative battery cable
13. Refill the engine with clean oil.
14. Refill the cooling system.
15. Start the engine and check for leaks, repair if necessary.

3.8L Engine

LEFT SIDE

1. Before servicing the vehicle, refer to the precautions in the beginning of this section.

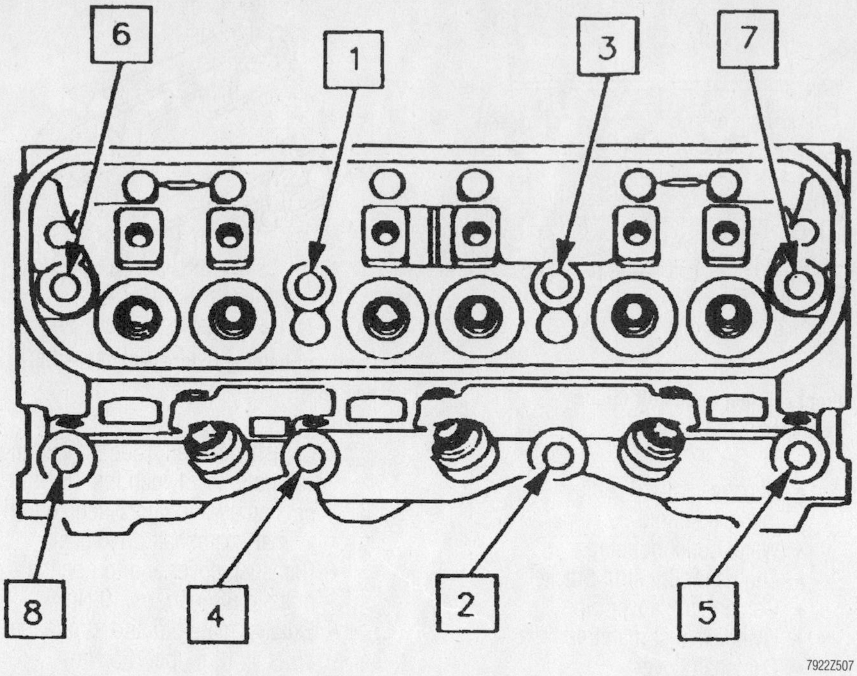

Cylinder head bolt torque sequence—3.8L engine

2. Relieve the fuel system pressure.
3. Drain the cooling system.
4. Remove or disconnect the following:
   - Fuel injector sight shield
   - Throttle body air inlet duct
   - Right and left engine mount strut brackets
   - Fuel lines from the fuel rail
   - Upper intake manifold
   - Lower intake manifold
   - Left exhaust manifold
   - Rocker arm cover
   - Rocker arms and pushrods
   - Cylinder head bolts
   - Cylinder head

To install:

5. Install the new cylinder head gasket with the arrow pointing to the front of the engine.
6. Install the cylinder head.
7. Install NEW cylinder head bolts. Torque the bolts in sequence, as follows:
   a. Step 1: 37 ft. lbs. (50 Nm).
   b. Step 2: Plus 120 degree turn.
   c. Step 3: Rotate the 4 center bolts an additional 30 degrees.
8. Install or connect the following:
   - Push rods and rocker arms
   - Rocker arm cover and torque the bolts to 89 inch lbs. (10 Nm)
   - Left exhaust manifold and torque the bolts to 22 ft. lbs. (30 Nm)
   - Lower intake manifold and torque the bolts to 11 ft. lbs. (15 Nm)
   - Upper intake manifold and torque the bolts to 89 inch lbs. (10 Nm)
   - Left and right engine mount strut brackets and torque the bolts to 37 ft. lbs. (50 Nm)
   - Throttle body air inlet duct
   - Fuel injector sight shield and torque the bolts to 27 inch lbs. (3 Nm)
   - Negative battery cable
9. Refill the cooling system.
10. Start the engine and check for leaks.

RIGHT SIDE

1. Before servicing the vehicle, refer to the precautions in the beginning of this section.
2. Relieve the fuel system pressure.
3. Drain the cooling system.
4. Remove or disconnect the following:
   - Throttle body air inlet duct
   - Fuel injector sight shield
   - Accessory drive belt
   - Alternator
   - Catalytic converter from the exhaust manifold
   - Engine mount struts
5. Place the transmission in neutral and rotate the engine forward for access.
6. Remove or disconnect the following:
   - Drive belt tensioner
   - Power steering pump without disconnecting the lines
   - Heater hoses from the engine
   - Spark plug wires
   - Oxygen (O_2S) sensor electrical connector
   - Throttle and cruise control cables

- Fuel lines from the fuel rail
- Fuel rail
- Exhaust Gas Recirculation (EGR) valve heat shield
- EGR valve
- EGR valve outlet pipe
- EGR valve adapter
- Upper intake manifold
- Lower intake manifold
- Exhaust crossover pipe
- Right exhaust manifold
- Rocker arm cover
- Rocker arms and pushrods
- Cylinder head

To install:

7. Install a new cylinder head gasket with the arrow pointing to the front of the engine.

8. Install the cylinder head and secure with NEW head bolts. Torque the bolts, in sequence, as follows:

a. Step 1: 37 ft. lbs. (50 Nm).
b. Step 2: Plus a 120 degree turn.
c. Step 3: Rotate the 4 center bolts an additional 30 degrees.

9. Install or connect the following:
- Push rods and rocker arms
- Rocker arm cover and torque the bolts to 89 inch lbs. (10 Nm)
- Right exhaust manifold and torque the bolts to 22 ft. lbs. (30 Nm)
- Exhaust crossover pipe and torque the nuts to 18 ft. lbs. (25 Nm)
- Lower intake manifold and torque the bolts to 11 ft. lbs. (15 Nm)
- Upper intake manifold and torque the bolts to 89 inch lbs. (10 Nm)
- EGR valve adapter and torque the bolts to 37 ft. lbs. (50 Nm)
- EGR valve outlet pipe and torque the bolts to 21 ft. lbs. (29 Nm)
- EGR valve and torque the nuts to 21 ft. lbs. (29 Nm)
- EGR valve heat shield and torque the bolts to 89 inch lbs. (10 Nm)
- Fuel rail
- Fuel lines to the fuel rail
- Throttle and cruise control cables
- O_2S electrical connector
- Spark plugs and torque them to 20 ft. lbs. (27 Nm)
- Spark plug wires
- Heater hoses
- Power steering pump and torque the bolts to 25 ft. lbs. (34 Nm)
- Drive belt tensioner and torque the bolts to 37 ft. lbs. (50 Nm)

10. Carefully return the engine to its proper position.

11. Place the transmission in park.

12. Install or connect the following:
- Throttle body air inlet duct
- Engine mount struts and torque the bolts to 35 ft. lbs. (48 Nm)
- Catalytic converter and torque the bolts to 26 ft. lbs. (35 Nm)
- Alternator and torque the bolts to 37 ft. lbs. (50 Nm)
- Accessory drive belt
- Fuel injector sight shield
- Negative battery cable

13. Refill the cooling system.

14. Start the engine and check for leaks.

Rocker Arms

REMOVAL & INSTALLATION

3.1L and 3.4L Engines

LEFT SIDE

1. Before servicing the vehicle, refer to the precautions in the beginning of this section.
2. Disconnect the negative battery cable.
3. Drain the cooling system.
4. Remove or disconnect the following:
- Spark plug wires
- Engine mount strut
- Secondary Air Injection (AIR) check valve pipe, if equipped
- Thermostat bypass hose and pipe
- Positive Crankcase Ventilation (PCV) valve from the rocker arm cover
- Rocker arm cover

➡Keep the pushrods in order. Intake pushrods are 5¾ inches long and exhaust pushrods are 6 inches long.

- Rocker arm bolts
- Rocker arms
- Pushrods

To install:

5. Lubricate all the valvetrain components with engine oil.

6. Install or connect the following:
- Pushrods and the rocker arms and tighten the bolts to 89 inch lbs. (10 Nm) plus an additional 30 degree turn on 3.1L engines or 14 ft. lbs. (19 Nm) plus 30 degrees on 3.4L engines
- Rocker arm cover using a new gasket and tighten the rocker cover bolts to 89 inch lbs. (10 Nm)
- PCV valve to the rocker arm cover
- Coolant tube with the thermostat bypass hose and tighten the bolt to 98 inch lbs. (11 Nm) and the nut to 18 ft. lbs. (25 Nm)
- AIR check valve pipe and torque the adapters to 22 ft. lbs. (30 Nm)
- Engine mount strut and torque the bolt to 35 ft. lbs. (48 Nm)
- Spark plug wires
- Negative battery cable

7. Refill the cooling system.
8. Start the vehicle and verify no leaks.

RIGHT SIDE

1. Before servicing the vehicle, refer to the precautions in the beginning of this section.
2. Remove or disconnect the following:
- Negative battery cable
- Accessory drive belt
- Alternator
- Alternator Bracket
- Spark plug wires
- Ignition coil and Evaporative (EVAP) emissions canister purge solenoid as an assembly
- Power brake booster vacuum pipe from the intake plenum
- Rocker arm cover
- Rocker arm bolts, balls, rocker arms and pushrods

To install:

3. Lubricate all the valve train components with engine oil.
4. Install or connect the following:
- Pushrods and the rocker arms and tighten the bolts to 89 inch lbs. (10 Nm) plus an additional 30 degree turn on 3.1L engines or 14 ft. lbs. (19 Nm) plus 30 degrees on 3.4L engines
- Rocker arm cover using a new gas-

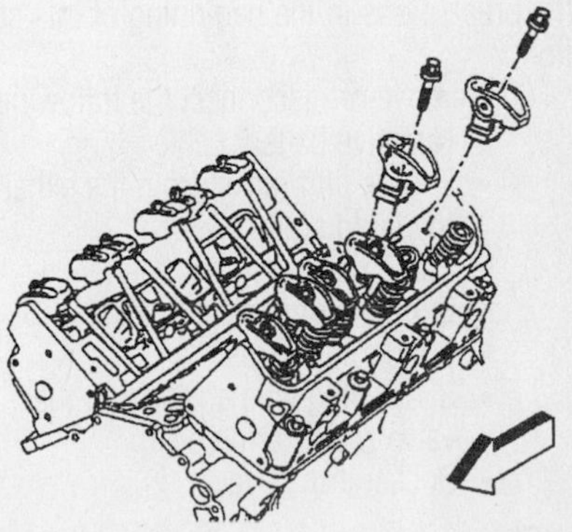

Rocker arm components—3.1L and 3.4L engines

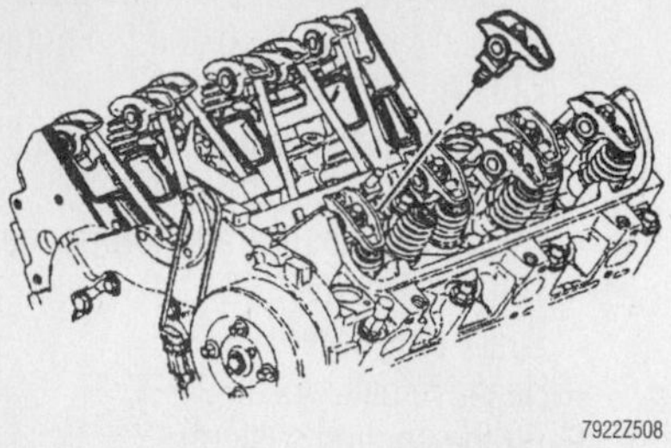

Rocker arm mounting—3.1L engine

ket and tighten the rocker cover bolts to 89 inch lbs. (10 Nm)
- Power brake booster vacuum pipe to the plenum
- EVAP solenoid and ignition coil assembly
- Spark plug wires
- Alternator bracket and torque the bolts to 37 ft. lbs. (50 Nm)
- Alternator and torque the bolts to 37 ft. lbs. (50 Nm)
- Accessory drive belt
- Negative battery cable

5. Start the vehicle and verify no leaks.

3.5L Engine

Refer to the camshaft removal and installation procedure for rocker arm service.

3.8L Engine

WARNING

The rocker arm bolts have been permanently stretched during installation. New bolts must be used each time the rocker arm assemblies are removed.

LEFT SIDE (FRONT) ROCKER ARMS

1. Before servicing the vehicle, refer to the precautions in the beginning of this section.
2. Remove or disconnect the following:
   - Negative battery cable
   - Engine lift bracket from the exhaust manifold studs
   - Fuel injector sight shield
   - Engine mount strut bracket
   - Spark plug wires
   - Spark plug wire cover from the valve rocker arm cover
   - Rocker arm cover

➡The rocker arms, pushrods, pedestals and bolts must be kept in order for installation in the same locations they were removed from.

- Rocker arm bolts, pedestals, rocker arms and pushrods

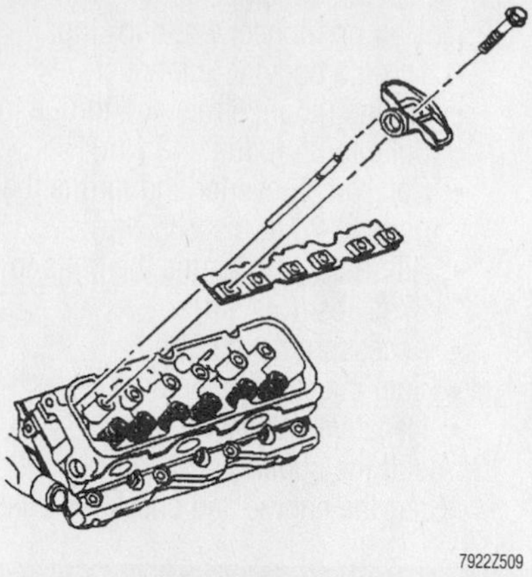

Exploded view of the rocker arm, pushrod and pushrod guide plate assembly—3.8L engine

To install:

3. Apply thread locking compound to the rocker arm bolts.
4. Install or connect the following:
   - Pushrod in the lifter
   - Rocker arm, pedestal and mounting bolt and tighten the mounting bolt to 11 ft. lbs. (15 Nm) plus an additional 90 degrees
   - Rocker arm cover with a new gasket and torque the mounting bolts to 89 inch lbs. (10 Nm)
   - Spark plug wire cover to the rocker arm cover
   - Spark plug wires
   - Engine mount strut bracket and torque the bolt to 75 ft. lbs. (102 Nm)
   - Fuel injector sight shield
   - Engine lift bracket to the exhaust manifold and torque the bolt to 22 ft. lbs. (30 Nm)
   - Negative battery cable
5. Start the vehicle and check for leaks, repair if necessary.

RIGHT SIDE (REAR) ROCKER ARMS

1. Before servicing the vehicle, refer to the precautions in the beginning of this section.
2. Remove or disconnect the following:
   - Negative battery cable
   - Fuel injector sight shield
   - Accessory drive belt
   - Alternator
   - Drive belt tensioner
   - Engine mount struts
   - Throttle body air inlet duct
3. Rotate the engine forward for proper access.
   - Evaporative (EVAP) emissions purge solenoid and brace
   - Right side spark plug wires
   - Fuel injector sight shield bracket
   - Rocker arm cover

➡The rocker arms, pushrods, pedestals and bolts must be kept in order for installation in the same locations they were removed from.

- Rocker arm bolts, pedestals, rocker arms and pushrods

To install:

4. Seat the pushrod in the lifter and install the rocker arm, pedestal and mounting bolt. Tighten the mounting bolt to 11 ft. lbs. (15 Nm) plus an additional 90 degrees.
5. Install or connect the following:
   - Rocker arm cover using a new gasket and torque the bolts to 89 inch lbs. (10 Nm)
   - Fuel injector sight shield bracket and torque the bolts to 22 ft. lbs. (30 Nm)
   - EVAP purge solenoid and brace
   - Right side spark plug wires
6. Return the engine back to its original position.
   - Throttle body air inlet duct
   - Engine mount struts and torque the bolt to 35 ft. lbs. (48 Nm)
   - Drive belt tensioner and torque the bolts to 37 ft. lbs. (50 Nm)
   - Alternator and torque the bolt to 37 ft. lbs. (50 Nm)
   - Fuel injector sight shield
   - Negative battery cable
7. Start the vehicle and check for leaks, repair if necessary.

Super Charger

REMOVAL & INSTALLATION

3.8L (VIN 1) Engine

1. Before servicing the vehicle, refer to the precautions in the beginning of this section.
2. Relieve the fuel system pressure.
3. Remove or disconnect the following:
   - Fuel injector sight shield
   - Exhaust Gas Recirculation (EGR) valve heat shield
   - Supercharger drive belt
   - Right side spark plug wires from the ignition module
   - Alternator brace
   - Fuel injector electrical connectors
   - Manifold Absolute Pressure (MAP) sensor bracket
   - Fuel lines
   - Fuel rail with the injectors

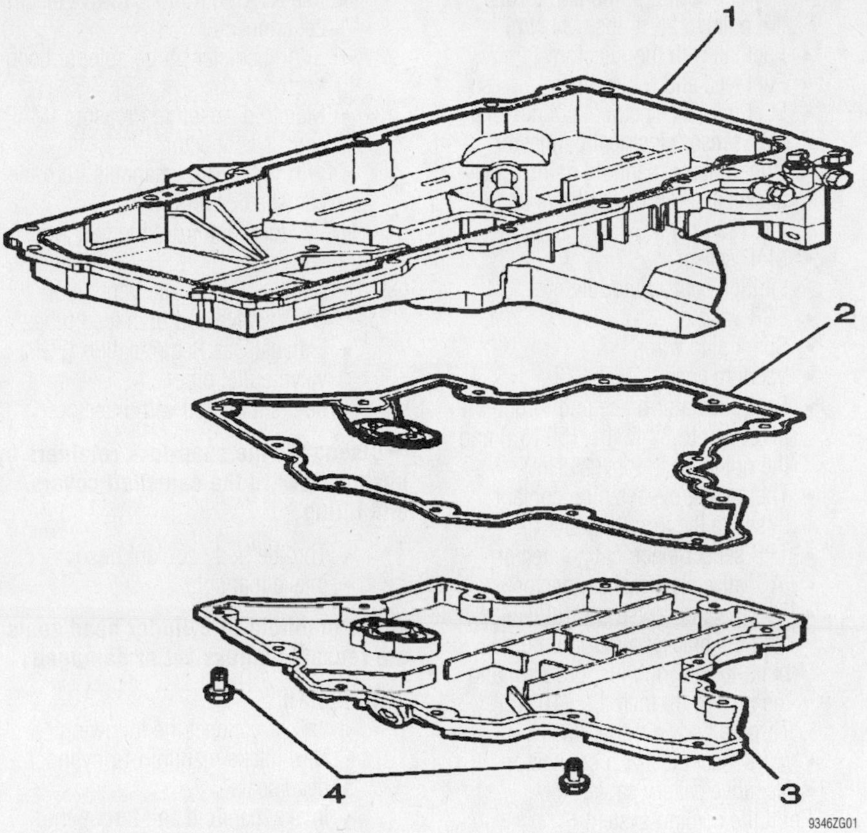

Supercharger bolt locations–3.8L (VIN 1) engine

- Boost control solenoid
- Electrical and vacuum connections as necessary
- Throttle and cruise control cables with the bracket
- Throttle body air inlet duct
- Supercharger

To install:

4. Install or connect the following:
   - Supercharger with new a new gasket and O-rings and torque the bolts to 17 ft. lbs. (23 Nm)
   - Throttle body air inlet duct
   - Throttle and cruise control cables with the bracket and torque the bolts to 142 inch lbs. (16 Nm)
   - Electrical and vacuum connectors
   - Boost control solenoid and torque the nut to 72 inch lbs. (8 Nm)
   - Fuel rail with fuel injectors and torque the hold-down bolts to 7 ft. lbs. (10 Nm) and the stud to 18 ft. lbs. (25 Nm)
   - Fuel lines
   - MAP sensor bracket
   - Fuel injector electrical connectors
   - Alternator brace and torque the bolt and nut to 37 ft. lbs. (50 Nm)
   - Spark plug wires
   - Supercharger drive belt
   - EGR valve heat shield
   - Fuel injector sight shield
   - Negative battery cable
5. Start the engine and ensure proper operation.

Intake Manifold

REMOVAL & INSTALLATION

3.1L and 3.4L Engines

1. Before servicing the vehicle, refer to the precautions in the beginning of this section.
2. Relieve the fuel system pressure.
3. Drain the cooling system.
4. Remove or disconnect the following:
   - Intake Air Temperature (IAT) sensor electrical connector
   - Throttle body air inlet duct
   - Throttle and cruise control cables with bracket
   - Throttle Position (TPS) sensor electrical connector
   - Idle Air Control (IAC) valve electrical connector

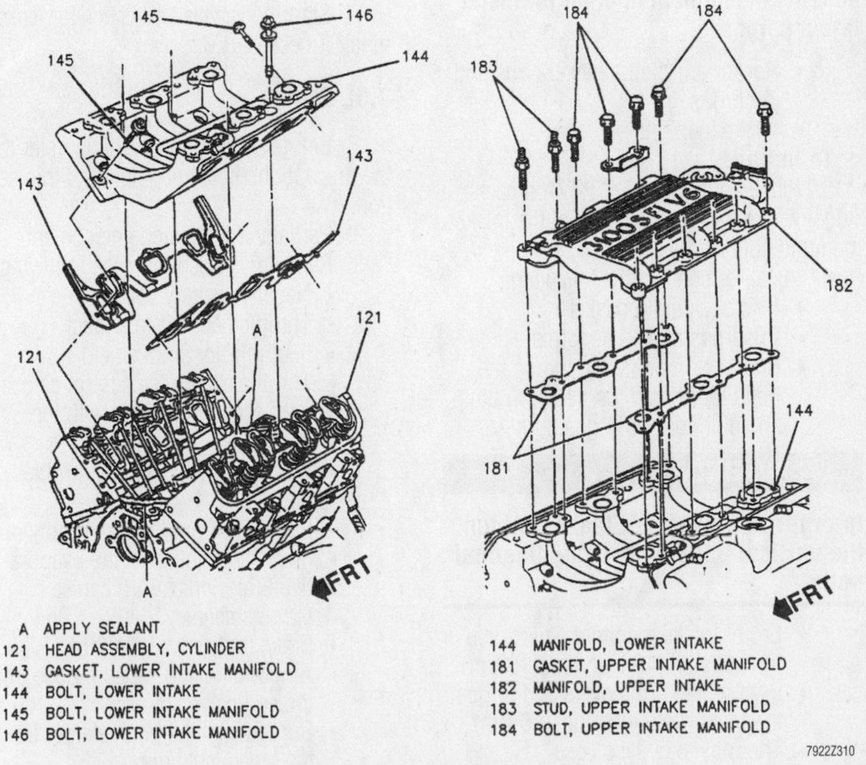

Lower intake manifold torque sequence—3.1L and 3.4L engines

- Camshaft Position (CMP) sensor wiring harness from the intake
- Spark plug wires from the intake
- Thermostat bypass hoses from the throttle body
- Rear alternator brace
- Vacuum lines from the upper intake manifold
- Manifold Absolute Pressure (MAP) sensor
- Exhaust Gas Recirculation (EGR) valve
- Spark plug wires
- Ignition control module
- Upper intake manifold
- Engine Coolant Temperature (ECT) sensor electrical connector
- Fuel injector electrical connectors
- Fuel lines from the fuel rail and fuel line bracket
- Fuel rail with the injectors
- Power steering pump and move it aside without disconnecting the lines
- Coolant inlet pipe from the coolant outlet housing
- Coolant bypass hose
- Upper radiator hose at thermostat housing
- Both rocker arm covers
- Lower intake manifold bolts

➡When removing the valvetrain components, keep them in order for installation purposes.

- Rocker arm bolts, rocker arms and pushrods
- Intake manifold

To install:

5. Place a 3mm bead of RTV, on each ridge, where the front and rear of the intake manifold contact the block.
6. Install or connect the following:
   - Intake manifold gasket
   - Pushrods and rocker arms
   - Rocker arm bolts and torque the bolts to 89 inch lbs. (10 Nm) plus an additional 30 degree turn

⁂ WARNING

In order to prevent oil leaks, tighten the vertical bolts before the diagonal bolts.

- Lower intake manifold and torque the bolts to 115 inch lbs. (13 Nm)
- Both rocker arm covers and torque the bolts to 89 inch lbs. (10 Nm)
- Thermostat bypass hose
- Upper radiator hose
- Heater inlet pipe and heater hose to the lower intake
- Power steering pump and torque the bolts to 25 ft. lbs. (34 Nm)
- Fuel fail with the injectors
- Fuel feed and return lines
- Fuel injector electrical connectors
- ECT sensor electrical connector
- Upper intake manifold using a new gasket and torque the bolts to 18 ft. lbs. (25 Nm)
- MAP sensor
- Ignition control module
- EGR valve
- Spark plug wires
- Vacuum lines
- Rear alternator brace and torque the pivot bolt to 37 ft. lbs. (50 Nm) and the nut to 18 ft. lbs. (25 Nm)
- Thermostat bypass pipe coolant hoses to the throttle body
- TPS sensor electrical connector
- IAC valve electrical connector
- Throttle and cruise control cables with the bracket and torque the bolts to 115 inch lbs. (13 Nm) and the nuts to 89 inch lbs. (10 Nm)
- Throttle body air inlet duct
- IAT sensor electrical connector
- Negative battery cable

7. Refill the cooling system.

➡An engine oil and filter change is recommended.

8. Start the vehicle and check for leaks, repair if necessary.

3.5L Engine

1. Before servicing the vehicle, refer to the precautions in the beginning of this section.
2. Partially drain the engine coolant.
3. Remove or disconnect the following:
   - Negative battery cable
   - Throttle body air inlet duct
   - Fuel injector sight shield
   - Throttle and cruise control cables from the throttle body with the bracket
   - Coolant hoses from the throttle body
   - Fuel lines from the fuel supply rail
   - Fuel vapor line from the Evaporative Emission (EVAP) canister purge solenoid
   - Brake booster vacuum hose
   - Air conditioning vacuum hose from the engine
   - Surge tank inlet pipe retainer from the fucl supply rail
   - Fuel injector electrical connectors
   - Throttle Position Sensor (TPS) electrical connectors
   - Idle Air Control (IAC) valve electrical connector
   - EVAP canister purge solenoid connector
   - Manifold Absolute Pressure (MAP) sensor connector
   - Wiring harness channels from the camshaft covers
   - Vacuum tube from the fuel pressure regulator
   - Positive Crankcase Ventilation (PCV) valve and both feed tubes
   - Exhaust Gas Recirculation (EGR) valve outlet pipe
   - Fuel supply rail with injectors

➡Disengage the snap-lock retainers by pushing toward the camshaft covers and lifting.

- Throttle body coolant hose
- Intake manifold

➡The manifold-to-cylinder head seals are reusable unless cut or damaged.

To install:

4. Install or connect the following:
   - New intake manifold-to-cylinder head seals
   - Intake manifold and torque the bolts, in a circular pattern, starting from the center to 62 inch lbs. (7 Nm)
   - Throttle body heater hose
   - New O-rings on the fuel injectors
   - Fuel supply rail with injectors
   - EGR pipe and torque the intake manifold bolt to 89 inch lbs. (10 Nm) and the coolant crossover bolt to 18 ft. lbs. (24 Nm)
   - PCV valve and related tubing
   - Brake booster vacuum hose
   - Fuel pressure regulator vacuum hose
   - Air conditioning vacuum hose

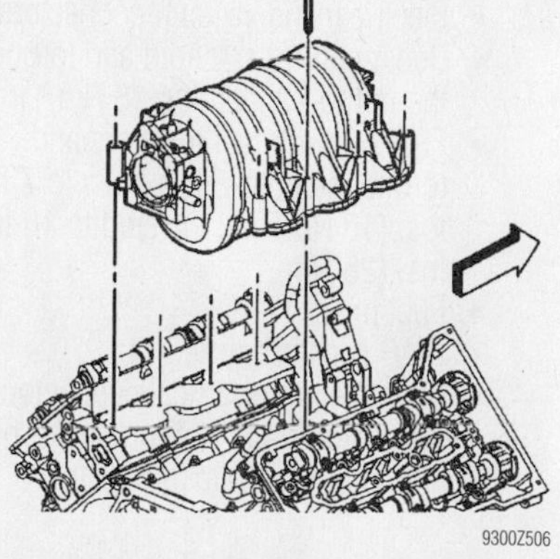

Intake manifold assembly—3.5L engine

- Engine wiring harness with channel to the camshaft covers and torque the bolts to 89 inch lbs. (10 Nm)
- Surge tank pipe retainer to the fuel supply rail
- TPS electrical connector
- IAC electrical connector
- EVAP solenoid electrical connector
- MAP sensor electrical connector
- Fuel injector electrical connectors
- Coolant hoses to the throttle body
- Throttle and cruise control cables to the throttle body
- Fuel lines to the fuel supply rail
- Fuel vapor line to the purge solenoid
- Fuel injector sight shield
- Throttle body air inlet duct
- Negative battery cable

5. Refill the cooling system.
6. Start the engine and check for leaks, repair if necessary.

3.8L Engine—Not Equipped With A Supercharger

➡There are 2 bolts hidden beneath the upper intake manifold. These bolts are located in the right front and left rear corners of the lower intake manifold. It is necessary to remove the upper intake manifold to service the lower intake manifold.

1. Before servicing the vehicle, refer to the precautions in the beginning of this section.
2. Relieve the fuel system pressure.
3. Drain the cooling system.
4. Remove or disconnect the following:
   - Negative battery cable
   - Fuel injector sight shield and air inlet duct
   - Spark plug wires from the right side of the engine
   - Evaporative emissions (EVAP) canister purge solenoid electrical connector
   - Fuel feed and return lines
   - Vacuum lines from the throttle body
   - Accessory drive belt
   - Alternator and bracket
   - Fuel injector electrical connectors
   - Fuel pressure regulator vacuum line
   - Fuel rail hold down bolts
   - Fuel rail with injectors
   - Brake booster hose from the manifold
   - Exhaust Gas Recirculation (EGR) valve heat shield

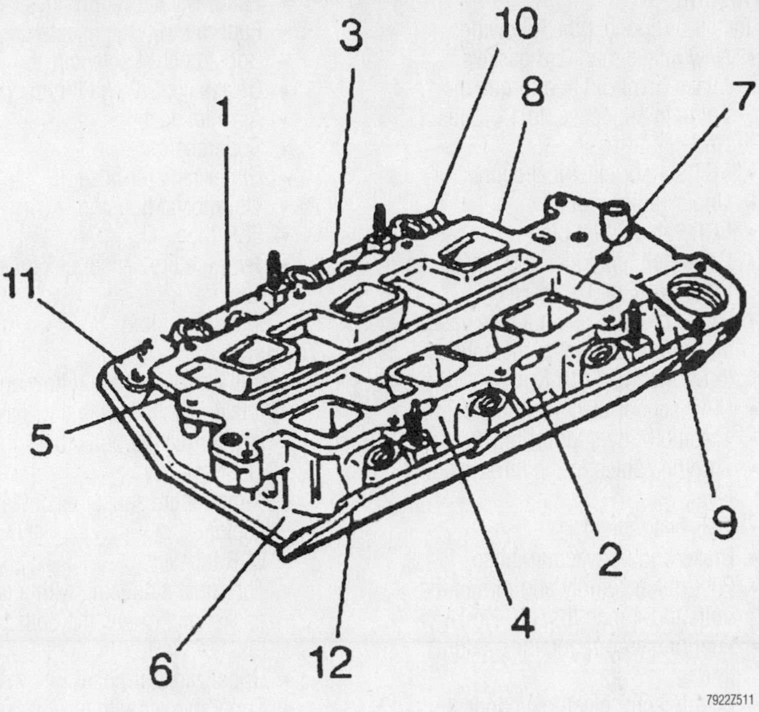

Always tighten the lower intake manifold bolts in sequence—3.8L (VIN 1 and K) engines

   - Throttle cables from the throttle body lever
   - Throttle body support bracket
   - Manifold Absolute Pressure (MAP) sensor electrical connector
   - Upper intake plenum
   - Drive belt tensioner
   - EGR valve outlet pipe
   - Upper radiator hose
   - Engine Coolant Temperature (ECT) sensor electrical connector
   - Lower intake manifold

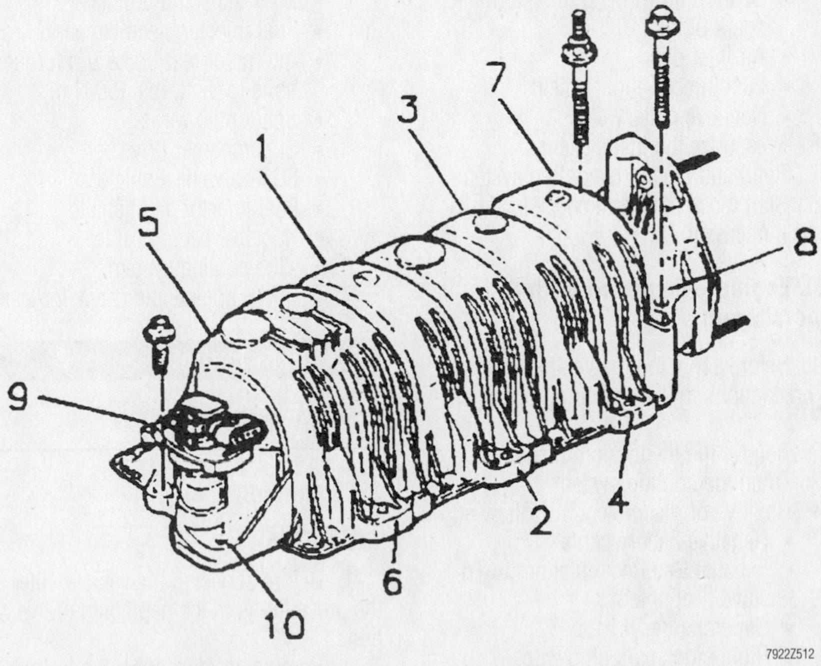

Upper intake manifold bolt tightening sequence—3.8L (VIN 1 and K) engines

For Accessory Drive Belt illustrations, see Section 1 of this manual

To install:

5. Install or connect the following:
   - New intake manifold gaskets
   - Intake manifold and torque the bolts, in sequence, to 11 ft. lbs. (15 Nm)
   - ECT sensor electrical connector
   - Upper radiator hose
   - EGR valve outlet pipe
   - Drive belt tensioner and torque the bolts to 37 ft. lbs. (50 Nm)
   - Intake plenum using a new gasket and torque the bolts, in sequence, to 89 inch lbs. (10 Nm)
   - MAP sensor electrical connector
   - Throttle body support bracket
   - Throttle cables to the throttle body lever
   - EGR heat shield
   - Brake booster vacuum hose
   - Fuel rail assembly and torque the bolts to 84 inch lbs. (10 Nm)
   - Fuel pressure regulator vacuum line
   - Fuel injector electrical connectors
   - Alternator bracket and torque the nut and bolt to 37 ft. lbs. (50 Nm)
   - Alternator and torque the bolts to 37 ft. lbs. (50 Nm)
   - Accessory drive belt
   - Vacuum lines to the throttle body
   - Fuel feed and return lines
   - EVAP canister purge solenoid electrical connector
   - Spark plug wires to the rear bank spark plugs
   - Air inlet duct
   - Fuel injector sight shield
   - Negative battery cable
6. Pressurize the fuel system.
7. Refill and bleed the cooling system.
8. Start the vehicle and check for leaks, repair if necessary.

3.8L Engine—Equipped With A Supercharger

1. Before servicing the vehicle, refer to the precautions in the beginning of this section.
2. Relieve the fuel system pressure.
3. Drain the cooling system.
4. Remove or disconnect the following:
   - Negative battery cable
   - Exhaust Gas Recirculation (EGR) valve heat shield
   - Supercharger drive belt
   - Right side spark plug wires
   - Alternator rear brace
   - Fuel injector electrical connectors
   - Manifold Absolute Pressure (MAP) sensor and bracket
   - Fuel feed and return lines
   - Fuel rail with the injectors
   - Boost control solenoid
   - Cruise control and throttle cables
   - Air inlet duct
   - Supercharger
   - Upper radiator hose
   - Thermostat housing
   - EGR tube
   - Temperature sensor electrical connector
   - Intake manifold

To install:

5. Install or connect the following:
   - Intake manifold with new gaskets and torque the bolts to 11 ft. lbs. (15 Nm)
   - Temperature sensor electrical connector
   - EGR tube
   - Thermostat housing with a new gasket and torque the bolts to 20 ft. lbs. (27 Nm)
   - Upper radiator hose
   - Supercharger with new gaskets and torque the bolts to 17 ft. lbs. (23 Nm)
   - Air inlet duct
   - Cruise control and throttle cables
   - Boost control solenoid and torque the bolts to 72 inch lbs. (8 Nm)
   - Fuel rail and torque the hold down bolts to 7 ft. lbs. (10 Nm)
   - Fuel feed and return lines
   - MAP sensor and bracket
   - Fuel injector electrical connectors
   - Alternator rear brace and torque the bolts to 37 ft. lbs. (50 Nm)
   - Spark plug wires
   - Supercharger drive belt
   - EGR valve heat shield
   - Fuel injector sight shield
   - Negative battery cable
6. Fill the cooling system.
7. Start the engine and check for leaks.

Exhaust Manifold

REMOVAL & INSTALLATION

3.1L Engine

LEFT SIDE

1. Before servicing the vehicle, refer to the precautions in the beginning of this section.
2. Remove or disconnect the following:
   - Negative battery cable
   - Throttle body air inlet duct
   - Right engine mount strut bracket
   - Crossover pipe heat shield

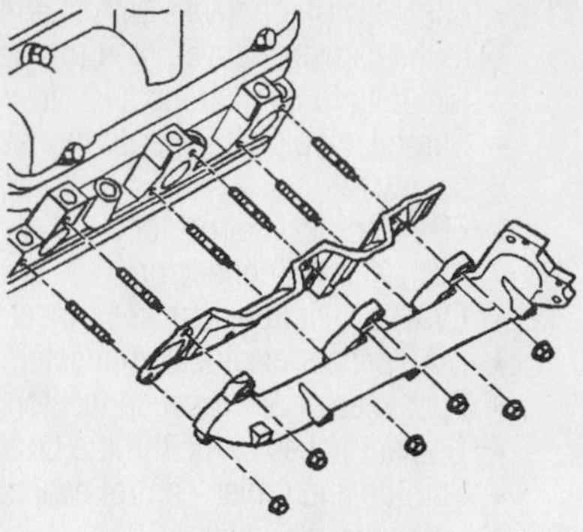

Exploded view of the left exhaust manifold mounting–3.1L engine

   - Crossover pipe nuts to the left exhaust manifold
   - Exhaust manifold heat shield
   - Exhaust manifold

To install:

3. Install or connect the following:
   - Exhaust manifold with a new gasket and torque the nuts to 12 ft. lbs. (16 Nm)
   - Exhaust manifold heat shield and torque the nuts to 89 inch lbs. (10 Nm)
   - Exhaust crossover pipe to the manifold and torque the bolt to 18 ft. lbs. (25 Nm)
   - Crossover pipe heat shield and torque the bolt to 89 inch lbs. (10 Nm)
   - Left side engine mount strut bracket and torque the bolt to 37 ft. lbs. (50 Nm)
   - Throttle body air inlet duct
   - Negative battery cable

RIGHT SIDE

1. Before servicing the vehicle, refer to the precautions in the beginning of this section.
2. Remove or disconnect the following:
   - Negative battery cable
   - Throttle body air inlet duct
   - Exhaust crossover pipe heat shield
   - Exhaust crossover pipe-to-right exhaust manifold nuts
   - Heated Oxygen (HO_2S) sensor electrical connector
   - Catalytic converter pipe from the exhaust manifold
   - Exhaust Gas Recirculation (EGR) pipe from the exhaust manifold
   - Exhaust manifold heat shields
   - Exhaust manifold

To install:

3. Install or connect the following:
   - Exhaust manifold with a new gasket

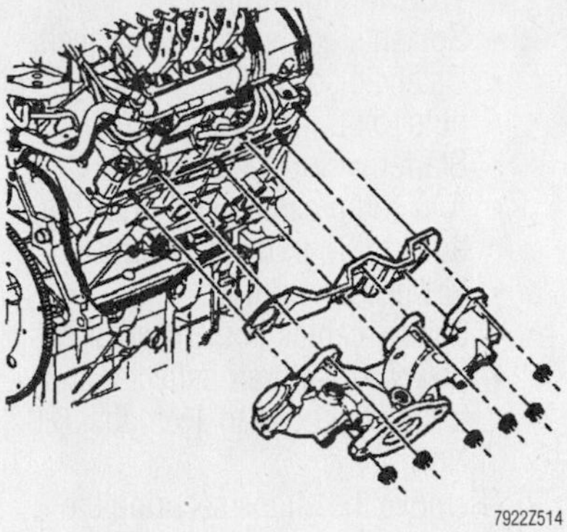

Exploded view of the right exhaust manifold mounting—3.1L engine

and torque the nuts to 12 ft. lbs. (16 Nm)
- Exhaust manifold heat shields and torque the bolts to 89 inch lbs. (10 Nm)
- EGR pipe to the exhaust manifold and torque the bolt to 22 ft. lbs. (30 Nm)
- Catalytic converter pipe with a new gasket and torque the nuts to 26 ft. lbs. (35 Nm).
- HO_2S sensor electrical connector
- Exhaust crossover pipe and torque the bolts to 18 ft. lbs. (25 Nm)
- Exhaust crossover pipe heat shield and torque the bolts to 89 inch lbs. (10 Nm)
- Throttle body air inlet duct
- Negative battery cable

3.4L Engine

LEFT SIDE

1. Before servicing the vehicle, refer to the precautions in the beginning of this section.
2. Remove or disconnect the following:
   - Negative battery cable
   - Air cleaner
   - Throttle body air inlet duct
   - Ignition wires to the front spark plugs
   - Secondary Air Injection (AIR) pipe, if equipped
   - Thermostat bypass pipe
   - Exhaust crossover pipe heat shield
   - Exhaust crossover pipe nuts to the left exhaust manifold
   - Exhaust manifold heat shield
   - Exhaust manifold

To install:

3. Install or connect the following:
   - Exhaust manifold with a new gasket and torque the bolts to 12 ft. lbs. (16 Nm)
   - Exhaust manifold heat shield and torque the bolts to 89 inch lbs. (10 Nm)
   - Crossover pipe and torque the bolts to 18 ft. lbs. (25 Nm)
   - Crossover pipe heat shield and torque the bolts to 89 inch lbs. (10 Nm)
   - Thermostat bypass pipe
   - Ignition wires to the front spark plugs
   - AIR pipe, if equipped
   - Negative battery cable
   - Throttle body air inlet duct
   - Air cleaner
4. Start the vehicle and check for leaks, repair if necessary.

RIGHT SIDE

1. Before servicing the vehicle, refer to the precautions in the beginning of this section.
2. Remove or disconnect the following:
   - Negative battery cable
   - Throttle body air inlet duct
   - Secondary Air Injection (AIR) pipe, if equipped
   - Exhaust crossover heat shield
   - Exhaust crossover pipe from the manifold
   - Heated Oxygen (HO_2S) sensor electrical connector
   - Catalytic converter from the manifold
   - Exhaust Gas Recirculation (EGR) valve from the intake manifold
   - Exhaust manifold heat shields
   - Exhaust manifold

To install:

3. Install or connect the following:
   - Exhaust manifold with a new gasket and torque the nuts to 12 ft. lbs. (16 Nm)
   - Exhaust manifold heat shields and torque the bolts to 89 inch lbs. (10 Nm)
   - EGR valve and torque the bolts to 22 ft. lbs. (30 Nm)
   - Catalytic converter and torque the nuts to 24 ft. lbs. (32 Nm)
   - HO_2S electrical connector
   - Crossover pipe to the manifold and torque the bolts to 18 ft. lbs. (25 Nm)
   - Crossover pipe heat shield and torque the bolts to 89 inch lbs. (10 Nm)
   - AIR pipe and torque the bolt to 15 ft. lbs. (20 Nm), if equipped
   - Throttle body air inlet duct
   - Negative battery cable
4. Start the vehicle and check for leaks, repair if necessary.

3.5L Engine

LEFT

1. Before servicing the vehicle, refer to the precautions in the beginning of this section.
2. Drain the cooling system.
3. Remove or disconnect the following:
   - Negative battery cable
   - Fuel injector sight shield
   - Engine mount strut and bracket
   - Cooling fans
   - Upper and lower radiator hoses
   - Transaxle oil cooler lines from the radiator
   - Radiator
   - Alternator
   - Exhaust manifold heat shield
   - Oil level indicator tube
   - Secondary Air Injection (AIR) control valve assembly from the engine mount strut bracket
   - Exhaust manifold-to-crossover pipe studs
   - Exhaust manifold

To install:

4. Install or connect the following:
   - Manifold to the crossover pipe using a new gasket and torque the studs to 18 ft. lbs. (25 Nm)
   - Exhaust manifold with a new gasket and torque the bolts to 18 ft. lbs. (25 Nm)
   - AIR valve and torque the pipe nut to 44 ft. lbs. (60 Nm) and the bolt to 80 inch lbs. (9 Nm)
   - Oil level indicator tube and torque the bolt to 80 inch lbs. (9 Nm)
   - Exhaust manifold heat shield and torque the bolts to 80 inch lbs. (9 Nm)
   - Alternator and torque the bolts to 37 ft. lbs. (50 Nm)
   - Radiator and torque the bolts to 89 inch lbs. (10 Nm)
   - Transaxle oil cooler lines and torque the fitting to 17 ft. lbs. (23 Nm)
   - Upper and lower radiator hoses
   - Cooling fans and torque the bolts to 89 inch lbs. (10 Nm)
   - Engine mount strut bracket and

torque the bolts to 37 ft. lbs. (50 Nm)
- Engine mount strut and torque the bolts to 35 ft. lbs. (48 Nm)
- Fuel injector sight shield and torque the nuts to 27 inch lbs. (3 Nm)
- Negative battery cable

5. Fill the cooling system.
6. Start the vehicle and check for leaks, repair if necessary.

RIGHT

1. Before servicing the vehicle, refer to the precautions in the beginning of this section.
2. Relieve the fuel system pressure.
3. Drain the cooling system.
4. Remove or disconnect the following:
   - Negative battery cable
   - Fuel injector cover
   - Air intake duct
   - Engine mount strut
   - Fuel lines from the supply rail
   - Cruise control and accelerator cables from the throttle body
   - Transaxle selector range cable and cable brackets
   - Transaxle shift cable from the shift module
   - Brake booster vacuum hose from the engine
   - Wiring harness connectors from the engine and transaxle
   - Upper radiator hose
   - Transaxle oil cooler lines from the radiator
   - Surge tank inlet hose
   - Heater hoses from the engine
   - Lower radiator air deflector
   - Battery cables from the retainers
   - Lower radiator hose
   - A/C compressor without disconnecting the lines and move it aside
   - Starter wiring
   - Catalytic converter from the manifold
   - Front wheels and splash shields
   - Wheel speed sensor (WSS) wiring from the lower control arms
   - Tie rod ends from the steering knuckles
   - Lower ball joints from the knuckles
   - Halfshafts
   - Intermediate shaft from the steering rack
5. Secure the vehicle to the lift in preparation for engine removal.
6. Position an engine/frame support table under the vehicle and lower the vehicle to meet the table.
7. Remove or disconnect the following:

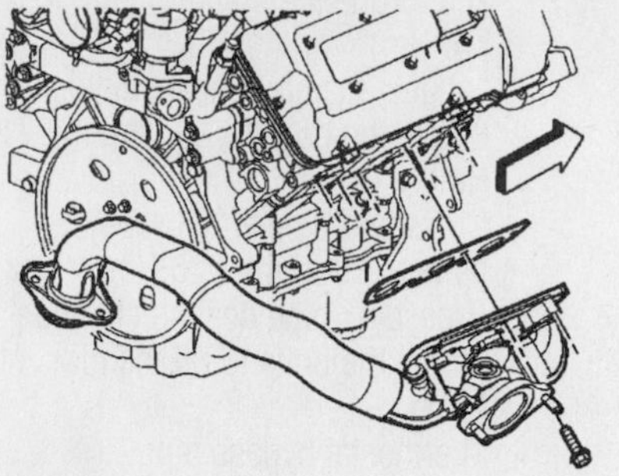

Exploded view of the right exhaust manifold—3.5L engine

- Frame-to-body bolts
- Crossover pipe from the front manifold
- Exhaust Gas Recirculation (EGR) pipe from the crossover pipe
- Right exhaust manifold from the engine

To install:

8. Install or connect the following:
   - Right exhaust manifold with a new gasket and torque the bolts to 18 ft. lbs. (25 Nm)
   - Crossover pipe to the front exhaust manifold and torque the bolts to 18 ft. lbs. (25 Nm)
   - EGR pipe to the crossover pipe and torque the pipe nut to 44 ft. lbs. (60 Nm)
9. Position the engine/transaxle assembly under the vehicle.
10. Coat the sub-frame bushings with rubber lubricant.
11. Lower the vehicle onto the assembly. Align the sub-frame on the vehicle using 2 bolts or drill bits, ¾ inches thick by 8 inches long through the alignment holes on the right side of the frame.
12. Install new frame-to-body bolts. Torque the bolts to 133 ft. lbs. (180 Nm) starting with the rear bolts and then the front bolts.
13. Raise the vehicle and remove the frame table.
14. Install or connect the following:
    - Intermediate shaft to the steering rack and torque the bolts to 35 ft. lbs. (48 Nm)

➡Be sure the shaft is fully seated on the stub before installing the pinch bolt.

- Halfshafts
- Ball joints and torque the nuts to 40 ft. lbs. (55 Nm)
- Tie rod ends and torque the nuts to 22 ft. lbs. (30 Nm) plus an additional 120 degree turn
- WSS wiring harness
- Splash shields and front wheels
- Catalytic converter and torque the nuts to 53 inch lbs. (6 Nm)
- Starter wiring
- A/C compressor and torque the bolts to 37 ft. lbs. (50 Nm)
- Lower radiator hose
- Battery cables in their retainers
- Lower radiator air deflector and torque the bolts to 15 ft. lbs. (20 Nm)

15. Remove the straps securing the vehicle to the lift.
    - Heater hoses
    - Surge tank hose
    - Transaxle oil cooler lines to the radiator and torque the fittings to 17 ft. lbs. (23 Nm)
    - Upper radiator hose
    - Wiring harness connectors to the engine and transaxle
    - Transaxle shift cable to the shift module and bracket
    - Transaxle range selector cable
    - Cruise control and accelerator cables to the throttle body
    - Fuel lines to the supply rail
    - Engine mount strut and torque the bolts to 35 ft. lbs. (48 Nm)
    - Air inlet duct
    - Fuel injector sight shield
    - Negative battery cable
16. Refill the cooling system.
17. Start the engine and check for leaks, repair if necessary.

3.8L Engines

LEFT SIDE

1. Before servicing the vehicle, refer to the precautions in the beginning of this section.
2. Remove or disconnect the following:
   - Negative battery cable
   - Fuel injector sight shield
   - Engine mount struts
   - Exhaust crossover from the manifold
   - Spark plug wires from the plugs
   - Oil level indicator tube
   - Exhaust manifold heat shield
   - Engine lift hook
   - Exhaust manifold

To install:

3. Install or connect the following:
   - Exhaust manifold with a new gasket and torque the bolts to 22 ft. lbs. (30 Nm)
   - Engine lift hook and torque the bolts to 22 ft. lbs. (30 Nm)
   - Exhaust manifold heat shield and torque the bolts to 15 ft. lbs. (20 Nm).

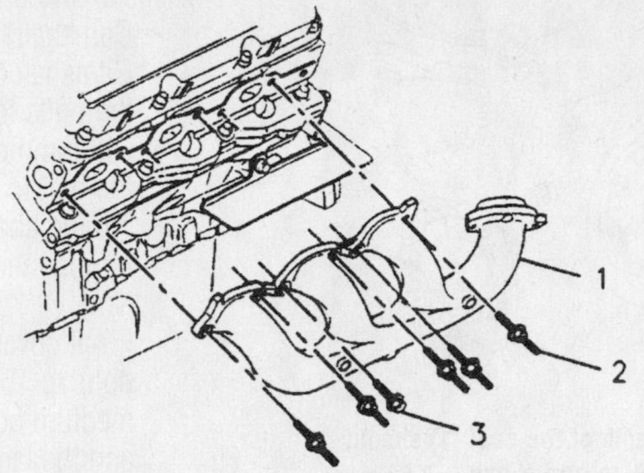

Exploded view of the left exhaust manifold mounting—3.8L engine

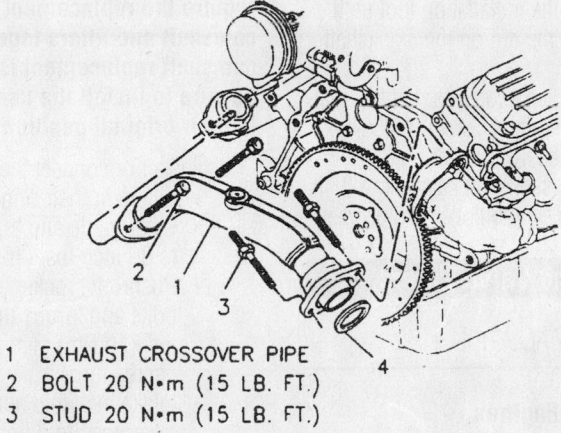

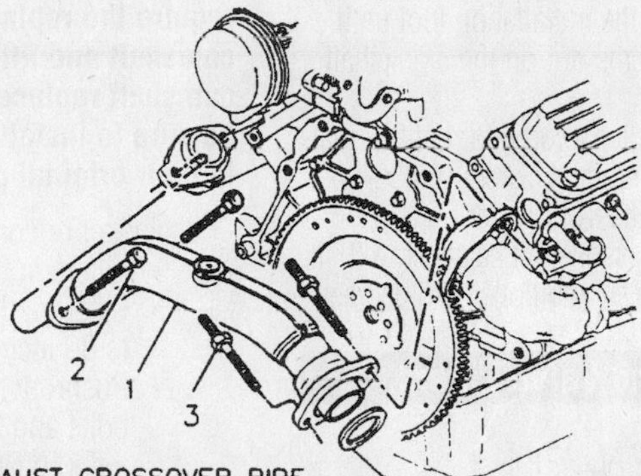

Exploded view of the crossover pipe mounting—3.8L engine

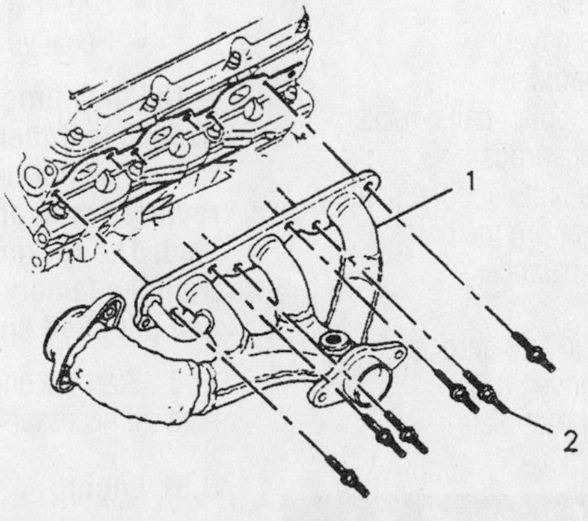

Exploded view of the right exhaust manifold mounting—3.8L engine

- Oil level indicator tube and torque the nut to 14 ft. lbs. (19 Nm)
- Spark plug wires
- Exhaust crossover and torque the bolts to 15 ft. lbs. (20 Nm)
- Engine mount struts and torque the bolts to 35 ft. lbs. (48 Nm)
- Fuel injector sight shield
- Negative battery cable

4. Start the vehicle and check for leaks, repair if necessary.

RIGHT SIDE

1. Before servicing the vehicle, refer to the precautions in the beginning of this section.
2. Remove or disconnect the following:
   - Negative battery cable
   - Fuel injector sight shield
   - Exhaust crossover pipe from the manifold
   - Exhaust Gas Recirculation (EGR) adapter pipe from the manifold
   - Catalytic converter
   - Spark plug wires from the plugs
   - Engine lift bracket
   - Fuel injector sight shield bracket
   - Oxygen Sensor (O_2S) electrical connector
   - Exhaust manifold

To install:

3. Install or connect the following:
   - Exhaust manifold with a new gasket and torque the bolts to 22 ft. lbs. (30 Nm)
   - O_2S electrical connector
   - Fuel injector sight shield bracket and torque the bolts to 22 ft. lbs. (30 Nm)
   - Engine lift bracket and torque the bolts to 22 ft. lbs. (30 Nm)
   - Spark plug wires
   - Catalytic converter and torque the bolts to 26 ft. lbs. (35 Nm)
   - EGR adapter pipe and torque the bolt to 21 ft. lbs. (29 Nm)
   - Exhaust crossover pipe and torque the bolts to 13 ft. lbs. (18 Nm)
   - Fuel injector sight shield
   - Negative battery cable
4. Start the vehicle and check for leaks, repair if necessary.

Front Crankshaft Seal

REMOVAL & INSTALLATION

Except 3.5L Engines

1. Before servicing the vehicle, refer to the precautions in the beginning of this section.

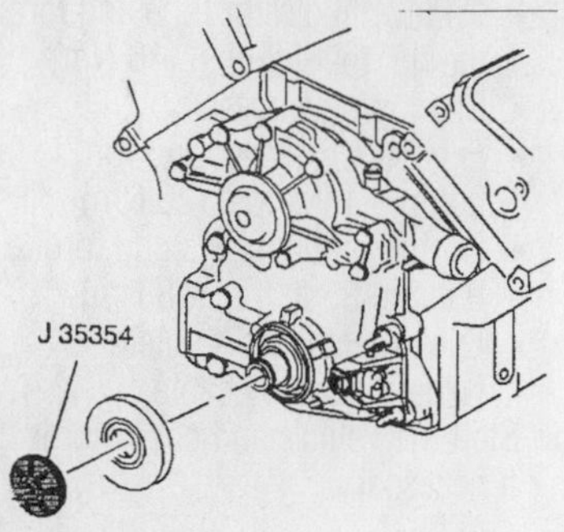

Installing the front crankshaft seal—except 3.5L engines

2. Remove the crankshaft balancer using an appropriate puller.
3. Pry the front crankshaft seal out.

To install:

4. Coat the new seal with engine oil to ease installation.
5. Install the front crankshaft seal using seal installer J-35468 on 3.1L and 3.4L engines. On 3.8L engines use tool J 35354. Make sure that the crankshaft front oil seal lip faces the engine.
6. Make sure the seal is flush with the engine when installed.
   - Crankshaft balancer and torque the bolt to 76 ft. lbs. (103 Nm)

3.5L Engines

1. Before servicing the vehicle, refer to the precautions in the beginning of this section.
2. Remove the crankshaft balancer using an appropriate puller.
3. Pry the front crankshaft seal out.

To install:

4. Coat the new seal with engine oil to ease installation.
5. Place the front seal onto the mandrel of the installation tool J-42041.
6. Slide the mandrel onto the screw of the installation tool.

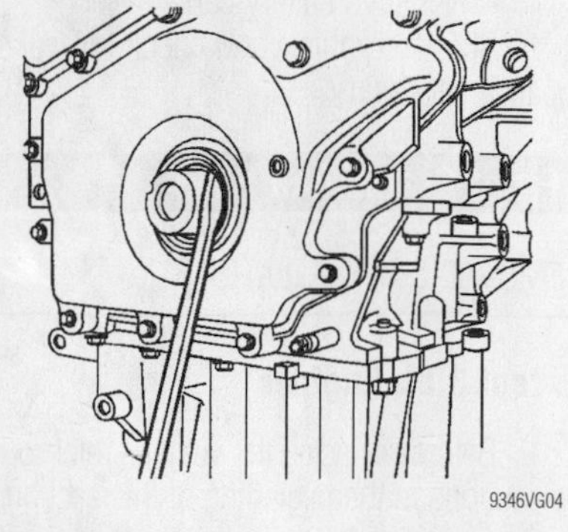

Remove the front oil seal, using a suitable prying tool

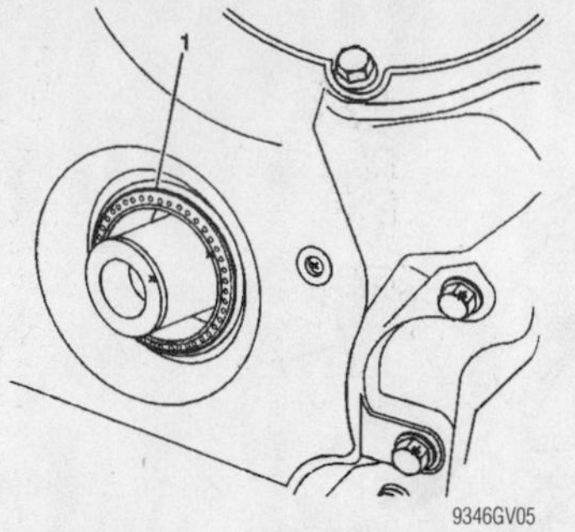

The installed depth of the seal (1) should protrude 1-2 mm (0.039-0.078 in.) from the front cover—3.5L engine

7. Thread the installation tool in the crankshaft. Ensure you engage at least ten threads of the tool before pressing the seal in place.
8. Push the seal into position by tightening the nut on the installation tool until the mandrel bottoms out on the crankshaft sprocket.
9. Check the installed depth of the seal, which should protrude 1-2 mm (0.039-0.078 in.), from the front cover.
10. Install the crankshaft balancer to 37 ft. lbs. (50 Nm) plus an additional 125 degrees.

Camshaft and Valve Lifters

REMOVAL & INSTALLATION

3.1L and 3.4L Engines

1. Before servicing the vehicle, refer to the precautions in the beginning of this section.
2. Relieve the fuel system pressure.
3. Remove or disconnect the following:
   - Engine assembly
   - Rocker arm covers
   - Intake manifold
   - Rocker arm bolts, balls, rocker arms and pushrods
   - Lifter guide bolts and the guide
   - Valve lifters from the bores
   - Crankshaft balancer
   - Front cover
   - Timing chain and sprockets
   - Oil pump driven gear
   - Camshaft thrust plate
   - Camshaft

WARNING

Avoid damaging the camshaft bearing surfaces.

To install:

4. Coat the camshaft with assembly lubricant.
5. Install or connect the following:
   - Camshaft
   - Camshaft thrust plate and torque the bolts to 89 inch lbs. (10 Nm)
   - Oil pump driven gear and torque the bolt to 27 ft. lbs. (36 Nm)
   - Timing chain and sprocket and torque the bolt to 103 ft. lbs. (140 Nm)
   - Front cover and torque the small bolts to 15 ft. lbs. (21 Nm), the medium bolts to 35 ft. lbs. (47 Nm) and the large bolts to 41 ft. lbs. (55 Nm)
   - Crankshaft balancer and torque the bolt to 76 ft. lbs. (103 Nm)
6. Lubricate the bearing surfaces with Molykote®.

➡Installation of a new camshaft or a wear pattern on the old valve lifter will require the replacement of the camshaft and lifters together. If camshaft replacement is not necessary, be sure to install the used valve lifters in their original position.

7. Install or connect the following:
   - Lifters in their original locations
   - Lifter guide and torque guide bolts to 89 inch lbs. (10 Nm)
   - Pushrods, rocker arms, balls and bolts and torque the nuts to 89 inch lbs. (10 Nm) plus an additional 30 degree turn
   - Intake manifold and torque the lower manifold bolts to 115 inch lbs. (13 Nm) and the upper manifold bolts to 18 ft. lbs. (25 Nm)
   - Rocker arm covers and torque the bolts to 89 inch lbs. (10 Nm)
   - Engine assembly
   - Negative battery cable

➡The only time valve adjustment in needed is if there was a valve job performed or the rocker studs have been replaced with an adjustable rocker arm stud. The rocker arm stud installed from the factory should be shouldered and not need any adjustment.

8. Start the engine and check for leaks, repair if necessary.

3.5L Engine

1. Before servicing the vehicle, refer to the precautions in the beginning of this section.
2. Drain the cooling system.
3. Remove or disconnect the following:
   - Negative battery cable
   - Fuel injector sight shield

- Thermostat housing for clearance when installing the camshaft Holding Fixture
- Camshaft cover

4. Rotate the crankshaft so the camshaft flats are parallel to the camshaft's sealing surface, then install a camshaft holding fixture.
5. Remove the camshaft sprocket bolts.
6. Install a timing chain/sprocket holding fixture J 42042.
7. Evenly slide the camshaft sprocket and chain from the camshafts onto the holding tool.

➡The camshaft bearing caps are marked. Be sure the raised portion of the cap faces the outside of the engine. They must always be installed in their original positions.

8. Remove or disconnect the following:
   - Camshaft bearing caps
   - Camshaft holding fixture
   - Camshafts

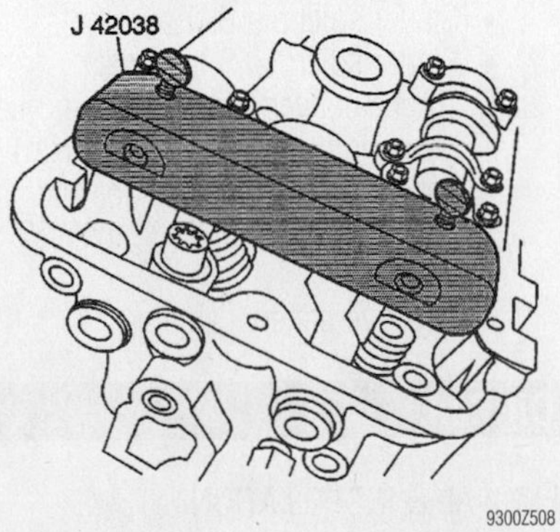

Camshaft holding fixture installed on the camshafts—3.5L engine

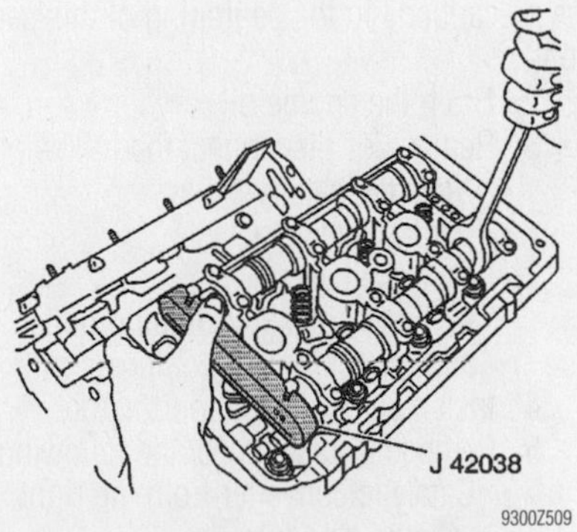

Timing chain/sprocket holding fixture installed on the cylinder head; use the flats on the camshaft if rotation is necessary for installation of the holding tool—3.5L engine

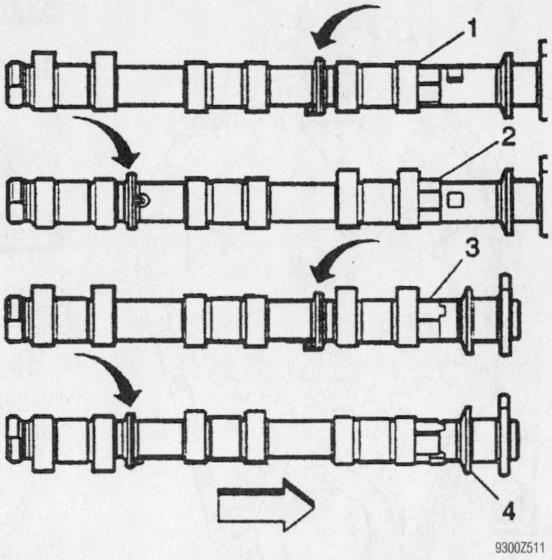

Camshaft identification—3.5L engine

To install:

9. Coat the rocker arms with engine oil and place them in their original positions.

➡Be sure to install the rounded end on the lash adjuster and the flat end on the tip of the valve.

10. Install or connect the following:
    - Camshafts, lubricated with engine oil, with the sprocket drive pin notch located at the top
    - Bearing caps and torque the bolts to 71 inch lbs. (8 Nm) plus an additional 22 degree turn
    - camshaft Holding Fixture tool onto the camshaft(s) at the rear of the cylinder head

➡Use the camshaft flats to turn the camshaft.

11. Compress the secondary timing chain tensioner by hand and insert a wire into the access hole to lock it in place.

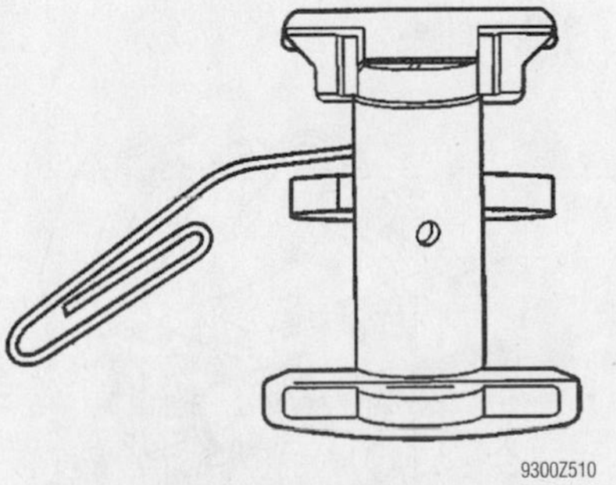

Before installation, compress the tensioner and lock it in place with a piece of wire—3.5L engine

12. Slide the camshaft sprockets/timing chain off the tool and onto the camshafts. Be sure to align the drive pins.
13. Remove the timing chain/sprocket holder from the front of the cylinder head. Torque the sprocket bolts to 18 ft. lbs. (25 Nm); then, an additional 45 degree turn.
14. Remove the wire from the chain tensioner and allow the tensioner to apply pressure to the chain.
15. Remove the camshaft holding fixture.
16. Install or connect the following:
    - Camshaft cover and torque the bolts to 80 inch lbs. (9 Nm)
    - Thermostat housing, if removed
    - Fuel injector sight shield and torque the nuts to 27 inch lbs. (3 Nm)
    - Negative battery cable
17. Refill the cooling system.
18. Start the vehicle and check for leaks, repair if necessary.

3.8L Engine

1. Before servicing the vehicle, refer to the precautions in the beginning of this section.
2. Discharge the A/C system.
3. Drain the cooling system.
4. Remove or disconnect the following:
   - Negative battery cable
   - Radiator with the air conditioning condenser assembly
   - Rocker arm cover

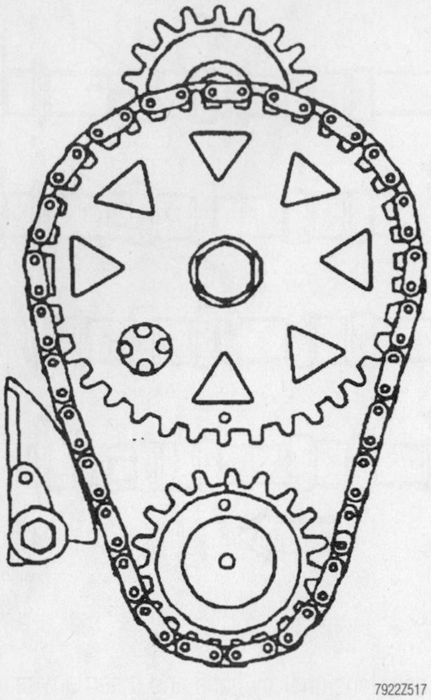

The timing marks should face each other if the chain and gears are installed properly—3.8L engine

- Valve lifters
- Front cover
- Camshaft sprocket and timing chain
- Camshaft thrust plate

5. Remove the camshaft assembly, as follows:
   a. Step 1:Install 1/2–20 x 6 inch bolt in the camshaft front bolt hole
   b. Step 2:Carefully rotate and pull the camshaft assembly out of the bearings.
6. Inspect the camshaft for damage and replace if necessary.

To install:

7. Install or connect the following:
   - Camshaft lubricated with assembly lubricant
   - Thrust plate and torque the bolts to 11 ft. lbs. (15 Nm)
   - Front cover and torque the bolts to 15 ft. lbs. (20 Nm) plus an additional 40 degree turn
   - Valve lifters
   - Rocker arm cover and torque the bolts to 89 inch lbs. (10 Nm)
   - Radiator with the air conditioning condenser and torque the bolts to 18 ft. lbs. (25 Nm)
   - Negative battery cable
8. Evacuate and recharge the A/C system.
9. Refill the cooling system.
10. Start the engine and check for leaks, repair if necessary.

Valve Lash

ADJUSTMENT

The valve clearance cannot be adjusted on these engines.

Starter Motor

REMOVAL & INSTALLATION

1. Before servicing the vehicle, refer to the precautions in the beginning of this section.
2. Remove or disconnect the following:
   - Negative battery cable
   - Radiator lower air deflector
   - Torque converter cover
   - Electrical connections from the starter
   - Starter motor

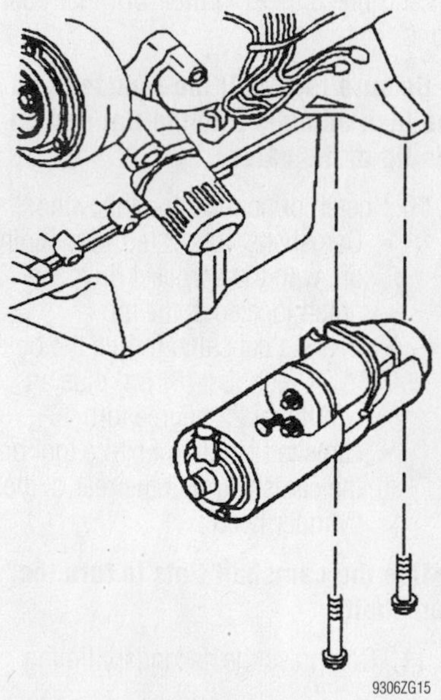

Exploded view of the starter—3.1L and 3.4L engines

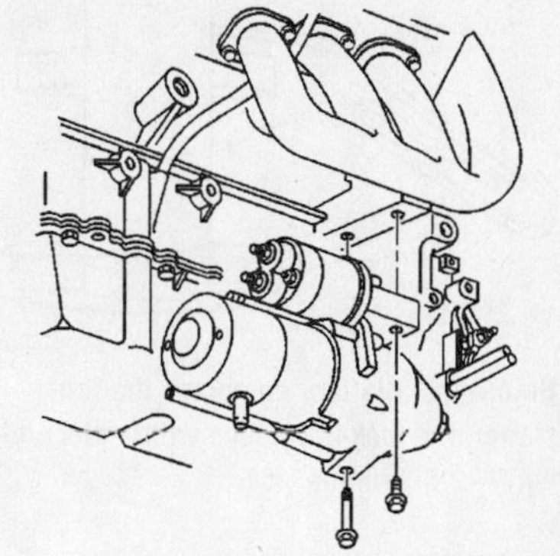

Exploded view of the starter—3.5L engine

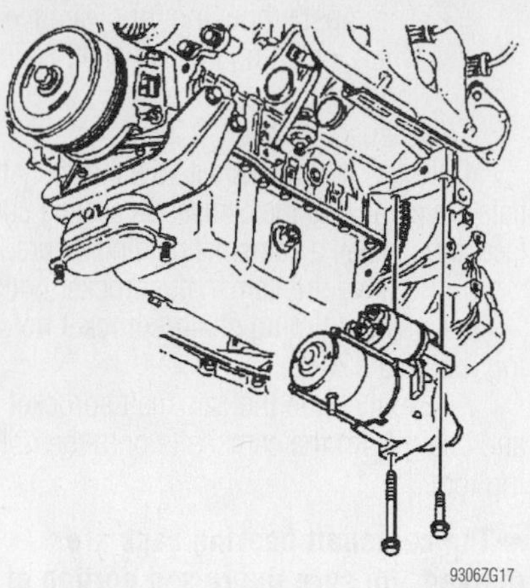

Exploded view of the starter—3.8L engine

To install:

3. Install or connect the following:
   - Starter motor and torque the bolts to 32 ft. lbs. (43 Nm) on all models except Intrigue. On Intrigue models, tighten the bolts to 37 ft. lbs. (50 Nm).
   - Starter solenoid BAT terminal
   - Starter solenoid S terminal
   - Torque converter cover and torque the bolts to 89 inch lbs. (10 Nm)
   - Radiator lower air deflector and torque the bolts to 15 ft. lbs. (20 Nm)
   - Negative battery cable

Oil Pan

REMOVAL & INSTALLATION

3.1L and 3.4L Engines

1. Before servicing the vehicle, refer to the precautions in the beginning of this section.
2. Drain the engine oil.
3. Remove or disconnect the following:
   - Negative battery cable
   - Engine mount struts
   - Drive belt
   - A/C compressor mounting bolts and reposition the compressor
4. Install an engine support fixture.
5. Remove or disconnect the following:
   - Catalytic converter from the right exhaust manifold
   - Oil level sensor electrical connection
   - Starter motor
   - Transaxle brace from the oil pan
   - Transaxle mount lower nuts
   - Engine mount lower nuts
6. Raise the engine to gain access to the oil pan.

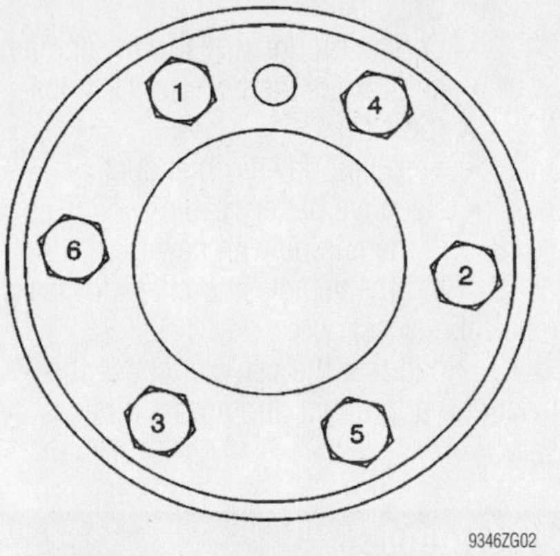

Exploded view of the oil pan side bolts—3.1L and 3.4L engines

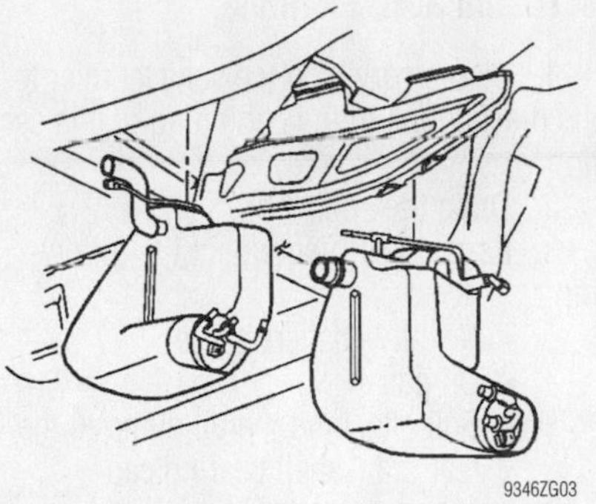

Exploded view of the oil pan retaining bolts—3.1L and 3.4L engines

7. Remove the engine mount from the oil pan.
8. Remove the oil pan.

To install:

9. Apply a small amount of sealer on both sides of the rear main bearing cap between the gasket and the engine.
10. Install or connect the following:
    - Oil pan with a new gasket and torque the retaining bolts to 18 ft. lbs. (25 Nm)
    - Front and rear oil pan bolts and torque the bolts to 37 ft. lbs. (50 Nm)
    - Engine mount and bracket to the oil pan and torque the bolts to 43 ft. lbs. (58 Nm)
11. Lower the engine into position.
12. Install or connect the following:
    - Engine mount lower nuts and torque the nuts to 32 ft. lbs. (43 Nm)
    - Transaxle mount lower nuts and torque the nuts to 35 ft. lbs. (47 Nm)
    - Transaxle brace to the oil pan and torque the bolts to 46 ft. lbs. (63 Nm)
    - Starter motor and torque the bolts to 32 ft. lbs. (43 Nm)
    - Oil level sensor electrical connection
    - Catalytic converter to the rear manifold and torque the nuts to 26 ft. lbs. (35 Nm)
13. Remove the engine support fixture.
    - A/C compressor and torque the bolts to 37 ft. lbs. (50 Nm)
    - Drive belt
    - Engine mount struts and torque the bolts to 35 ft. lbs. (48 Nm)
    - Negative battery cable
14. Fill the engine with new oil.
15. Start the vehicle and check for leaks, repair if necessary.

3.5L Engine

1. Before servicing the vehicle, refer to the precautions in the beginning of this section.
2. Drain the engine oil.
3. Remove or disconnect the following:
    - Negative battery cable
    - Oil filter cap and filter
    - Oil level sensor electrical connector
    - Transaxle brace
    - Oil pan

To install:

4. Install or connect the following:
    - Oil pan with a new gasket, Do not tighten the bolts
    - Transaxle brace on the engine block only and torque the bolts to 18 ft. lbs. (25 Nm)

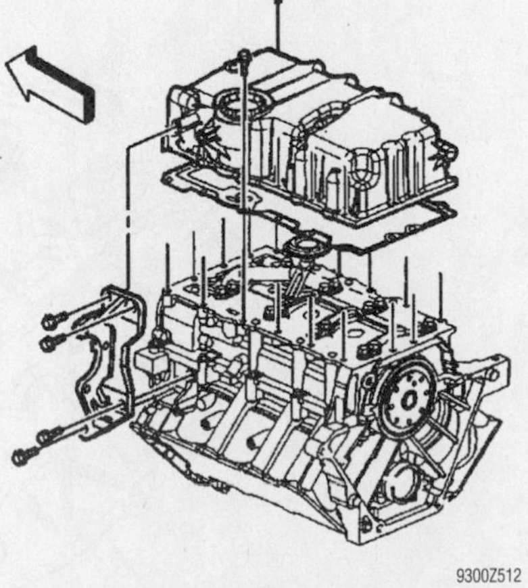

Exploded view of the oil pan—3.5L engine

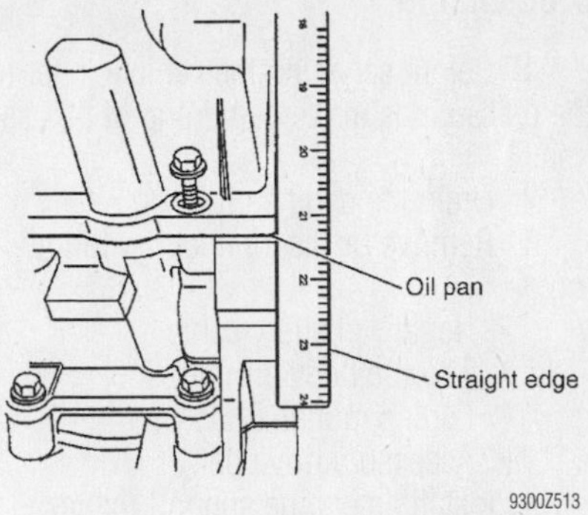

Use a straight-edge to align the rear of the oil pan to the rear of the engine—3.5L engine

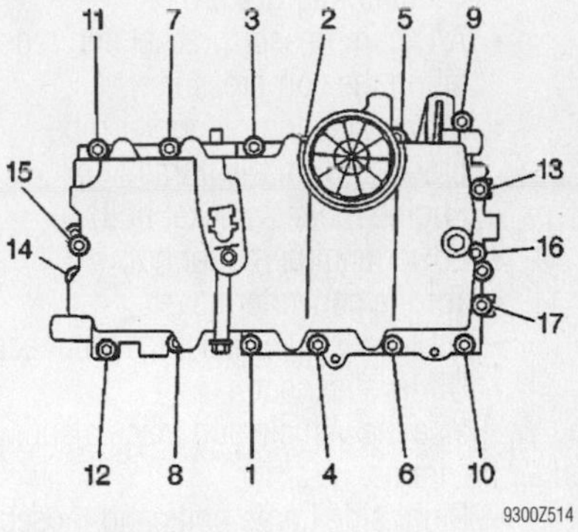

Oil pan mounting bolt tightening sequence—3.5L engine

- Brace-to-oil pan bolts, loosely install them

5. Align the rear of the oil pan flush with the rear of the engine block. Use a straight edge for reference.
6. Press the front of the oil pan against the transaxle brace; then, torque the brace-to-oil pan bolts to 18 ft. lbs. (25 Nm). Be sure to keep the rear of the pan flush with the rear of the engine.
7. Torque the oil pan bolts in sequence to 18 ft. lbs. (25 Nm).
8. Torque the brace-to-transaxle bolts to 32 ft. lbs. (43 Nm).
9. Install or connect the following:
    - Oil level sensor electrical connection
    - Drain plug and torque it to 15 ft. lbs. (20 m)
    - New oil filter and torque the cap to 18 ft. lbs. (25 Nm)
    - Negative battery cable
10. Refill the engine with clean oil.
11. Start the vehicle and check for leaks, repair if necessary.

3.8L Engine

1. Before servicing the vehicle, refer to the precautions in the beginning of this section.
2. Drain the engine oil.
3. Remove or disconnect the following:
   - Negative battery cable
   - Throttle body air inlet duct
   - Engine mount struts
   - Accessory drive belt
4. Install an engine support fixture.
   - Catalytic converter from the right exhaust manifold
   - Right front wheel
   - Right side splash shield
   - Oil filter and discard
   - A/C compressor bracket and reposition the compressor
   - Power steering oil cooler pipe brackets from the frame
   - Engine mount bracket bolts
   - Lower engine mount nuts
   - Torque converter cover
   - Oil level sensor electrical connector
   - Oil level sensor
5. Raise the vehicle and place a support under the frame.
   - Right side frame bolts and loosen the left side bolts
   - Engine mount and bracket
   - Oil pan
   - Oil pump pipe and screen

To install:

6. Install or connect the following:
   - New oil pan gasket and the oil pump pipe and screen assembly and torque the bolts to 11 ft. lbs. (15 Nm)

➡Do not over tighten the oil pan bolts or damage to the pan may occur.

   - Oil pan and torque the bolts to 125 inch lbs. (14 Nm)
   - Engine mount and bracket and torque the bolts to 50 ft. lbs. (68 Nm)
   - New right side frame bolts and torque all the bolts to 133 ft. lbs. (180 Nm)
   - Lower engine mount and torque the nuts to 35 ft. lbs. (47 Nm)
   - Oil level sensor and torque the fastener to 15 ft. lbs. (20 Nm)
   - Oil level wire connector
   - Torque converter cover and torque the bolts to 89 inch lbs. (10 Nm)
   - Power steering oil cooler pipe brackets to the frame
   - A/C compressor and torque the upper bolt to 37 ft. lbs. (50 Nm) and the lower bolt to 59 ft. lbs. (80 Nm)
   - New oil filter
   - Drain plug and torque the plug to 22 ft. lbs. (30 Nm)
   - Splash shield
   - Right front wheel
   - Catalytic converter pipe to the right exhaust manifold and torque the bolts to 26 ft. lbs. (35 Nm)
7. Remove the engine support fixture.
   - Drive belt
   - Engine mount struts to the engine and torque the bolts to 41 ft. lbs. (56 Nm)
   - Throttle body air inlet duct
   - Negative battery cable
8. Fill the engine with new oil.
9. Start the vehicle and check for leaks, repair if necessary.
10. Road test the vehicle, check the front end alignment and adjust if necessary.

Oil Pump

REMOVAL & INSTALLATION

3.1L and 3.4L Engines

1. Before servicing the vehicle, refer to the precautions in the beginning of this section.
2. Drain the engine oil.
3. Remove or disconnect the following:
   - Negative battery cable
   - Oil pan
   - Bolt attaching the oil pump to the rear crankshaft bearing cap
   - Oil pump and driveshaft

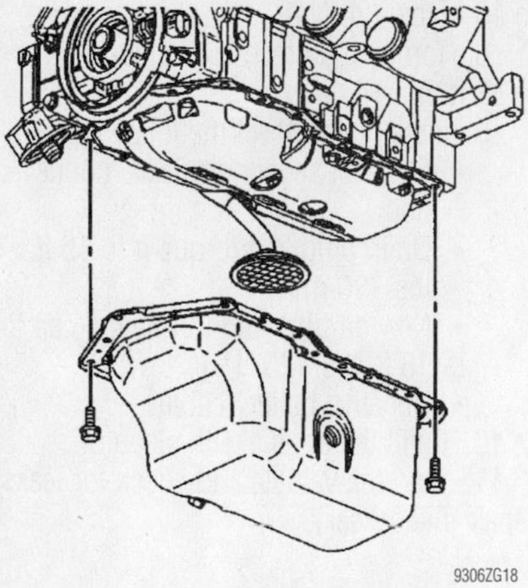

Exploded view of the oil pan mounting—3.8L engine

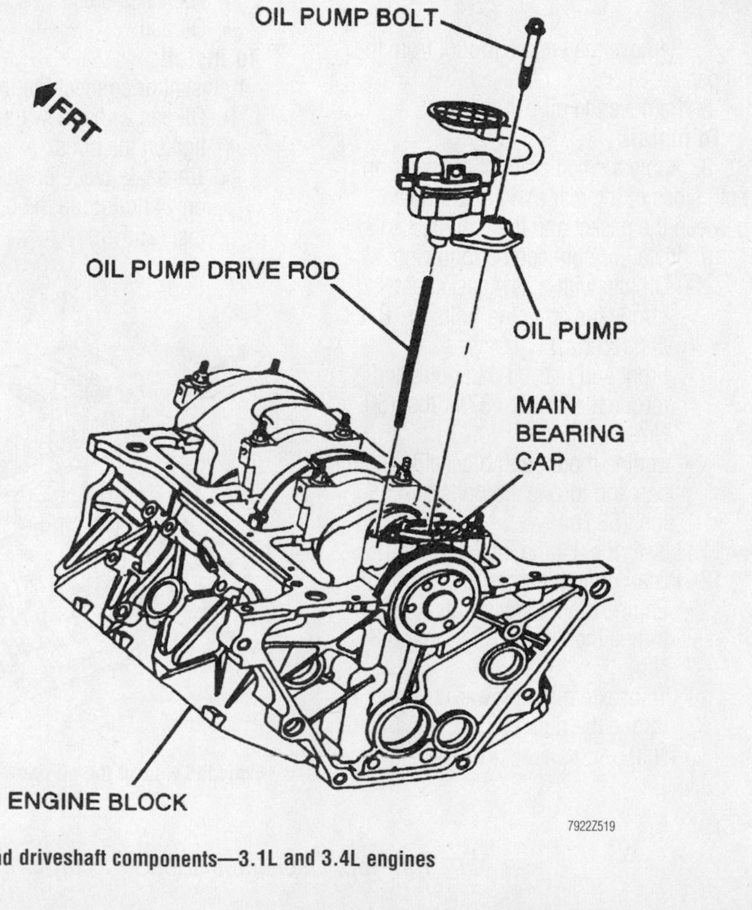

Oil pump and driveshaft components—3.1L and 3.4L engines

To install:

➡Rotate the driveshaft as required to obtain the proper engagement with the oil pump drive unit.

4. Install or connect the following:
   - Oil pump and driveshaft and torque the bolt to 30 ft. lbs. (41 Nm)
   - Oil pan
   - Negative battery cable
5. Fill the engine with new oil.
6. Start the vehicle and check for leaks, repair if necessary.

3.5L Engine

1. Before servicing the vehicle, refer to the precautions in the beginning of this section.
2. Drain the engine oil.
3. Remove or disconnect the following:
   - Negative battery cable
   - Front cover
   - Rocker arm covers
4. Install camshaft holding fixtures J 42038 on both sets of camshafts. Turn the hex portion of the camshaft to align them for tool installation. When installed, the flats on the rear of the camshafts will be parallel with the camshaft cover sealing surface.
5. Remove or disconnect the following:
   - Oil pan
   - Oil pump pipe and screen
   - Primary chain tensioner
   - Primary chain from the drive sprocket
   - Oil pump by sliding it off the crankshaft

To install:

6. Pack the oil pump housing with white petroleum jelly to insure priming.
7. Align the oil pump sprocket with the crankshaft and install the pump on the engine until a positive stop is felt. When installed properly, the sprocket will protrude slightly from the oil pump and the face of the sprocket will be behind the machined step in the crankshaft.
8. Install or connect the following:
   - Oil pump and torque the bolts to 18 ft. lbs. (25 Nm)
   - Primary chain on the sprocket

➡Be sure to maintain correct timing.

   - Chain tensioner
   - Oil pump pipe with the screen and torque the nut and bolt to 89 inch lbs. (10 Nm)
   - Oil pan and torque the bolts to 18 ft. lbs. (25 Nm)
9. Remove the camshaft holding tools.
10. Install or connect the following:
    - Camshaft covers and torque the bolts to 80 inch lbs. (99 Nm)
    - Engine front cover and torque the bolts to 124 inch lbs. (14 Nm)
    - Negative battery cable
11. Fill the engine with new oil.
12. Start the vehicle and check for leaks, repair if necessary.

3.8L Engines

1. Before servicing the vehicle, refer to the precautions in the beginning of this section.
2. Drain the engine oil.
3. Remove or disconnect the following:
   - Negative battery cable
   - Front cover
   - Oil pump cover
   - Oil pump gear set

To install:

4. Lubricate the oil pump gears with petroleum jelly and install the gears into the housing.
5. Pack the gear cavity with petroleum jelly after the gears have been installed.
6. Install or connect the following:

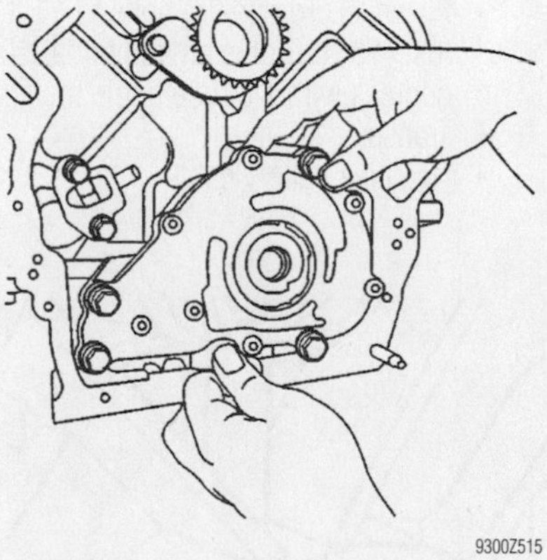

The oil pump is mounted on the front of the engine and driven by the crankshaft—3.5L engine

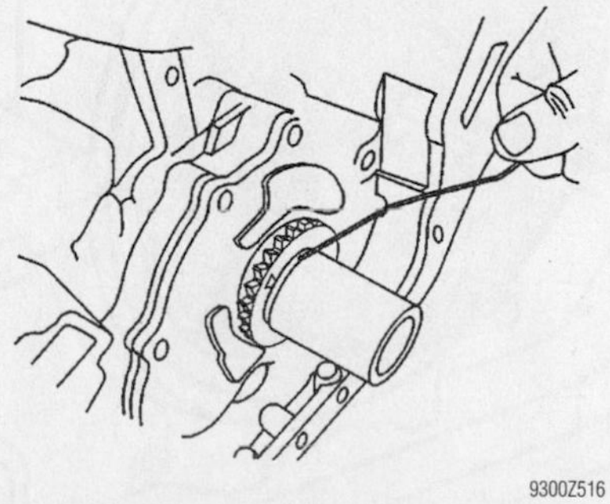

Correct position of the crankshaft sprocket when the oil pump is installed correctly—3.5L engine

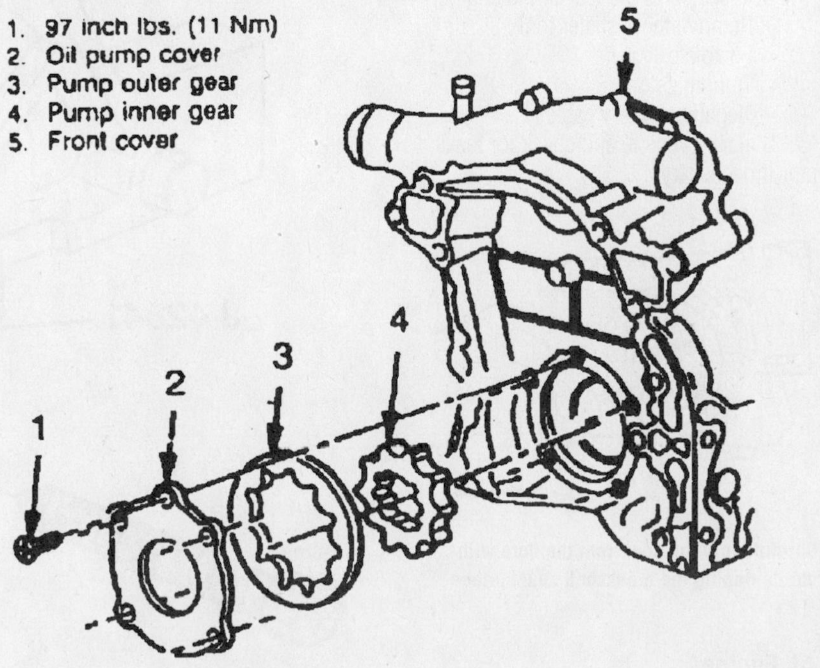

Oil pump assembly—3.8L engine

- Oil pump cover and torque the screws to 98 inch lbs. (11 Nm)
- Front cover
- Negative battery cable

7. Fill the engine oil. A new oil filter is recommended.
8. Start the vehicle and check for leaks, repair if necessary.

Rear Main Seal

REMOVAL & INSTALLATION

Except 3.5L Engine

1. Before servicing the vehicle, refer to the precautions in the beginning of this section.
2. Remove or disconnect the following:
   - Negative battery cable
   - Transmission
   - Flexplate
   - Rear main seal by prying it out

To install:

3. Lubricate the lip and the outer edge of the new seal with clean engine oil.
4. Install or connect the following:
   - Position the new seal on the mandrel until the back of the seal is flush against the collar of the tool
   - Seal installer tool to the rear of the crankshaft with the 2 mounting bolts. Turn the handle until the seal is seated in the rear of the engine. Remove the installer tool
   - Flexplate
   - Transmission
   - Negative battery cable
5. Start the vehicle and check for leaks, repair if necessary.

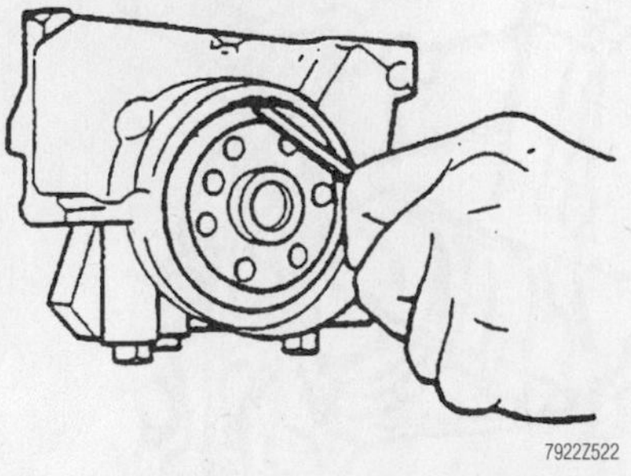

Carefully pry the seal from the bore without damaging the crankshaft seal surface

3.5L Engine

1. Before servicing the vehicle, refer to the precautions in the beginning of this section.
2. Remove or disconnect the following:

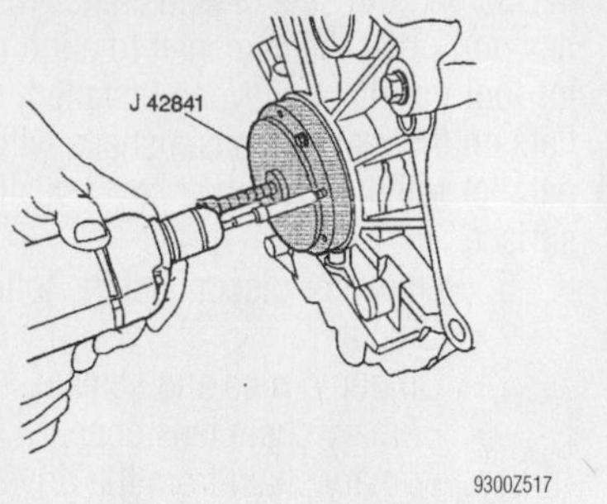

Use the guide holes in Tool J 42841 to install the screws in the seal—3.5L engine

   - Negative battery cable
   - Transaxle
   - Engine flywheel
3. Place Rear Seal Remover Tool J 42841 on the crankshaft with retaining bolts.
4. Install eight one-inch self starting screws through the guide holes of the tool. Tighten the screws.
5. Install the two retaining bolts.
6. Install a center forcing screw into the removal tool and pull the seal off the end of the crankshaft.

To install:

7. Clean debris from the crankshaft rear seal drain. The seal may leak if the drain is not properly cleaned.
8. Place a small amount of gasket maker to the crankcase split line across the end of the upper and lower crankcase seal.
9. Coat the outer diameter of the block with clean engine oil.
10. Clean the outer diameter of the flywheel flange with a lint-free cloth.

**** CAUTION**

Do not apply any oil on the green coating of the new seal.

11. Loosen the center bolt of the seal installer tool until the hub protrudes past the outer plate (approximately inch).
12. Install or connect the following:
    - Three mounting bolts into the crankshaft flange until the tool is fully seated on the crankshaft
    - New seal by tightening the center bolt until the tool bottoms out against the crankshaft
13. Remove the removal/installer tool and make certain the seal is installed properly
14. Install or connect the following:
    - Flywheel. Torque the bolts to 11 ft. lbs. (15 Nm) plus an additional 50 degrees with a torque angle meter.
    - Transaxle
    - Negative battery cable

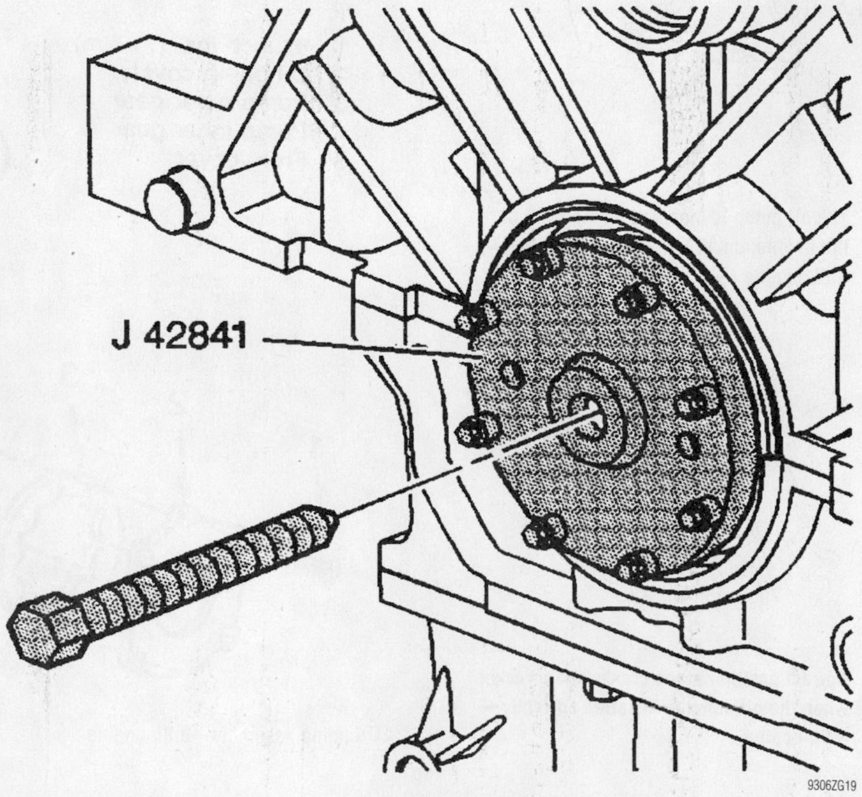

Install a center forcing screw into the removal tool—3.5L engine

15. Top off the engine oil if needed.
16. Start the vehicle and check for leaks, repair if necessary.

Timing Chain, Sprockets, Front Cover and Seal

REMOVAL & INSTALLATION

3.1L and 3.4L Engines

1. Before servicing the vehicle, refer to the precautions in the beginning of this section.
2. Drain the cooling system.
3. Drain the engine oil.
4. Remove or disconnect the following:
   - Negative battery cable
   - Coolant reservoir
   - Accessory drive belt
   - Crankshaft balancer
   - Drive belt tensioner
   - Power steering pump, DO NOT disconnect the lines
   - Thermostat bypass pipe from the front cover
   - Upper radiator hose
   - Water pump pulley
   - Lower Crankshaft Position (CKP) sensor
   - Front cover
5. Rotate the crankshaft until the timing marks on the camshaft and crankshaft sprockets are in alignment (facing each other).
   - Camshaft sprocket bolt, sprocket and timing chain
   - Crankshaft sprocket
   - Timing chain damper bolts and damper, if necessary

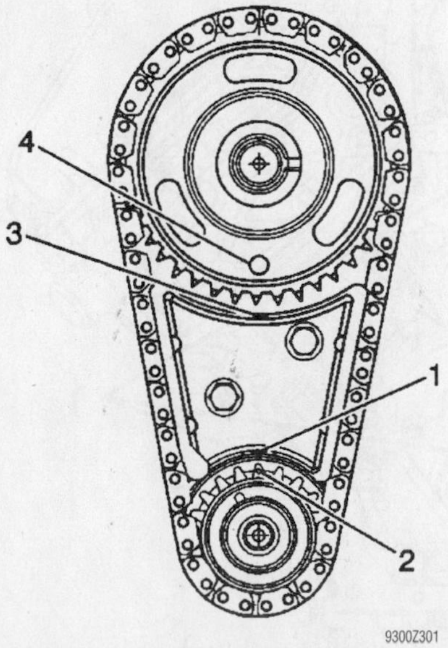

Be sure to align the damper mark (1) with the crankshaft mark (2) and the damper mark (3) with the camshaft sprocket mark (4)—3.1L and 3.4L engines

To install:

6. Install or connect the following:
   - Timing chain damper (if removed) and torque the bolts to 15 ft. lbs. (21 Nm)
   - Crankshaft sprocket

➡Be sure the timing mark on the crankshaft sprocket is pointing toward the mark on the chain damper.

   - Timing chain over the camshaft sprocket
7. Loop the timing chain under the crankshaft sprocket and install the camshaft sprocket on the camshaft.
8. Verify that the marks are aligned; the camshaft sprocket will be at the 6 o'clock position and the crankshaft sprocket at the 12 o'clock position.

➡The No. 1 piston will be at Top Dead Center (TDC) and the No. 4 piston will also be at TDC but on the compression stroke.

9. Tighten the camshaft sprocket bolt to 103 ft. lbs. (140 Nm).
10. Lubricate the timing chain components with engine oil.
11. Install or connect the following:
    - New front cover seal
    - Front cover using a new gasket and torque the small bolts to 15 ft. lbs. (21 Nm) and the large bolts to 35 ft. lbs. (47 Nm)
    - Water pump pulley and torque the bolts to 18 ft. lbs. (25 Nm)
    - Upper radiator hose
    - Thermostat bypass pipe
    - Power steering pump and torque the bolts to 25 ft. lbs. (34 Nm)
    - Drive belt tensioner and torque the bolt to 37 ft. lbs. (50 Nm)
    - CKP sensor
    - Crankshaft balancer and torque the bolt to 76 ft. lbs. (103 Nm)
    - Accessory drive belt
    - Coolant reservoir
    - Negative battery cable
12. Refill the fluids.
13. Start the engine and check for leaks, repair if necessary.

3.5L Engine

PRIMARY CHAIN

1. Before servicing the vehicle, refer to the precautions in the beginning of this section.
2. Drain the engine oil.
3. Drain the cooling system.
4. Disconnect the negative battery cable.
5. Remove the camshaft covers.
6. Rotate the crankshaft so the No. 1 piston is at Top Dead Center (TDC) and the flats on the rear of the camshafts are parallel with the camshaft cover sealing surface.
7. Install camshaft holding fixtures on both sets of camshafts. Turn the hex portion of the camshaft to align them for tool installation.

➡When installed, the flats on the rear of the camshafts will be parallel with the camshaft cover sealing surface.

8. Remove or disconnect the following:
   - Right diagonal brace
   - Battery and tray
   - Coolant reservoir
   - Underhood accessory wiring junction block, move it aside
   - Drive belt
   - Power steering pump pulley
   - Idler pulley and belt tensioner
   - Water pump pulley
   - Water pump drive belt shield
   - Water pump
9. Support the engine cradle.
10. Remove the right side engine cradle bolts.
11. Lower the cradle.
12. Remove or disconnect the following:
    - Crankshaft balancer
    - Front cover
    - Engine lift bracket from the front of the engine
    - Camshaft Position (CMP) sensor
    - Sprocket bolt from the exhaust camshaft on the right cylinder head to allow for clearance of the chain guide
    - Four chain guide access plugs from the cylinder heads

➡Note that each plug has an O-ring.

    - Primary chain tensioner

➡Remove the lower bolt allowing the tensioner to swing down and expand.

    - Primary chain tensioner shoe, by removing the bolt, pushing the guide downward slightly and

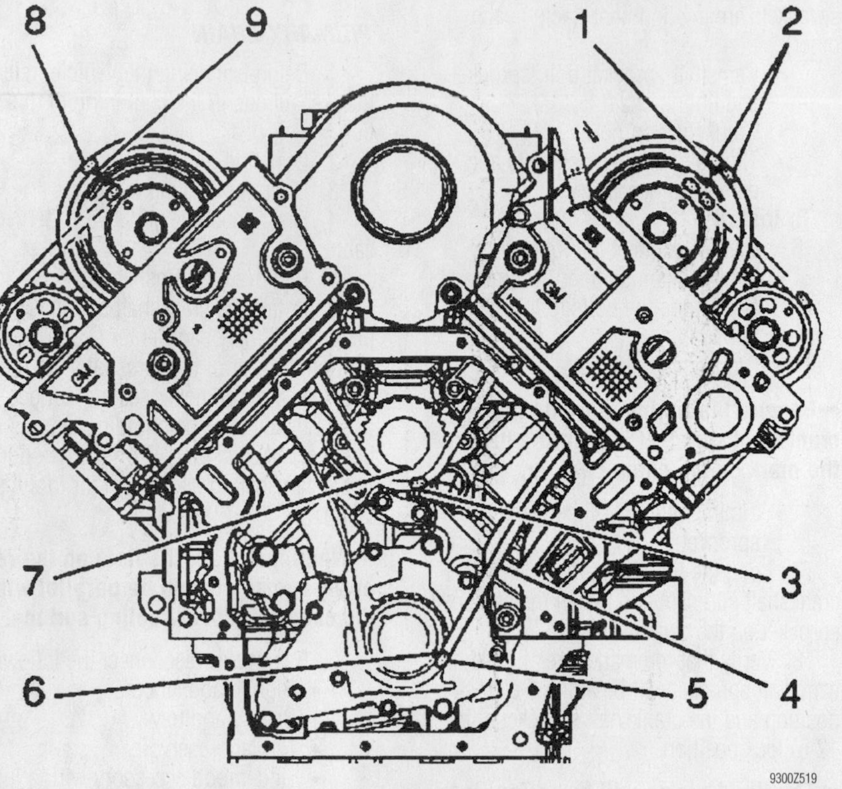

Primary timing chain alignment marks—3.5L engine

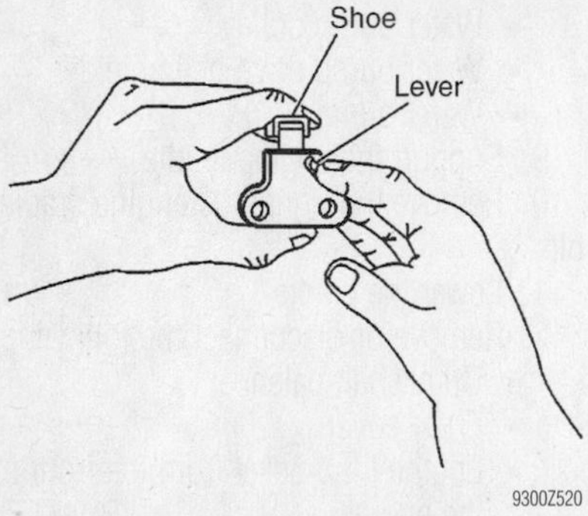

Compressing the primary chain tensioner—3.5L engine

pulling it up through the cylinder head
- Primary chain from the right camshaft, allowing it to fall into the oil pump area
- Primary chain

To install:

13. Rotate the crankshaft so the No. 1 piston is at Top Dead Center (TDC) and the mark on the crankshaft is at the 4 o'clock position.

14. Rotate the balance shaft so the timing mark is at the 5 o'clock position.

➡Be sure the painted links are facing the front of the engine.

15. Install the timing chain on the sprockets.

16. Center the mark on the left intake camshaft sprocket between the 2 painted links.

17. Verify that all of the timing marks are aligned.

18. Install the primary chain tensioner shoe. Torque the bolt to 22 ft. lbs. (30 Nm).

19. Compress the primary chain tensioner using the following sub-steps:

a. Step 1:Rotate the ratchet release lever counterclockwise and hold it.

b. Step 2:Press the tensioner shoe in and hold it.

c. Step 3:Release the ratchet lever and slowly release the pressure on the shoe.

d. Step 4:Insert a pin through the hole in the lever as the lever moves to the first click. The ratchet should hold the shoe in the compressed position.

➡Be sure the lever on the tensioner is facing you when installed.

20. Install or connect the following:
- Primary chain tensioner and torque the bolts to 18 ft. lbs. (25 Nm); then, remove the chain tensioner pin
- Four chain guide access plugs and torque the plugs to 44 inch lbs. (5 Nm)

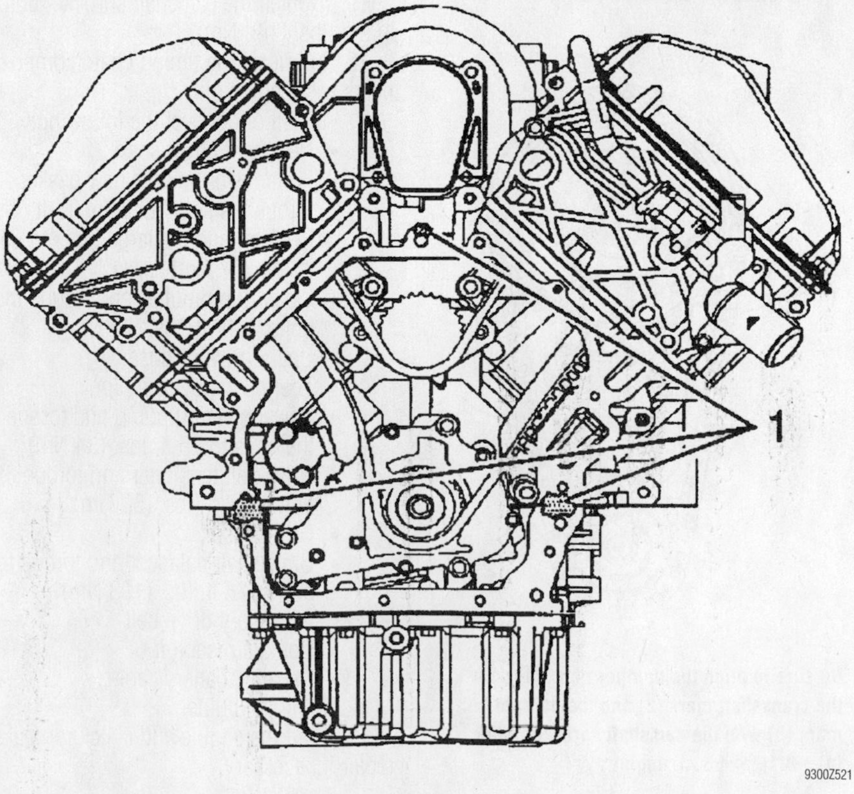

Apply RTV sealant to the 3 areas indicated before installing the front cover and gasket—3.5L engine

- Front engine lift bracket and torque the hex head bolt to 37 ft. lbs. (50 Nm) and the internal drive bolt to 18 ft. lbs. (25 Nm)
- CMP sensor and torque the bolts to 80 inch lbs. (9 Nm)

21. Remove the camshaft holding tools.
22. Install the rocker arm covers. Torque the bolts to 80 inch lbs. (9 Nm).
23. Place a small bead of RTV sealant on the 3 areas indicated in the diagram.
24. Install or connect the following:
    - Front cover with a new gasket and torque the bolts to 124 inch lbs. (14 Nm) and the coolant drain plug to 89 inch lbs. (10 Nm)
    - Crankshaft balancer and torque the bolt to 37 ft. lbs. (50 Nm) plus an additional 120 degree turn
25. Raise the engine cradle and install new bolts loosely.
26. Coat the sub-frame bushings with rubber lubricant.
27. Lower the vehicle onto the assembly. Align the sub-frame on the vehicle using 2 bolts or drill bits, ¾ inches thick by 8 inches long through the alignment holes on the right side of the frame.
28. Install or connect the following:
    - New frame-to-body bolts and torque the bolts to 133 ft. lbs. (180 Nm)
    - Water pump with a new gasket and torque the bolts to 124 inch lbs. (14 Nm)
    - Water pump pulley and torque the bolts to 106 inch lbs. (12 Nm)
    - Belt tensioner and torque the bolt to 37 ft. lbs. (50 Nm)
    - Power steering pump pulley
    - Drive belt
    - Underhood accessory wiring junction
    - Coolant reservoir and torque the nuts to 30 inch lbs. (3 Nm)
    - Battery and tray
    - Right diagonal brace and torque the bolts to 35 ft. lbs. (47 Nm)
    - Camshaft covers and torque the bolts to 80 inch lbs. (9 Nm)
    - Negative battery cable
29. Refill the engine cooling system.
30. Refill the engine with new oil.
31. Start the vehicle and check for leaks, repair if necessary.

SECONDARY TIMING CHAIN

1. Before servicing the vehicle, refer to the precautions in the beginning of this section.
2. Drain the cooling system.

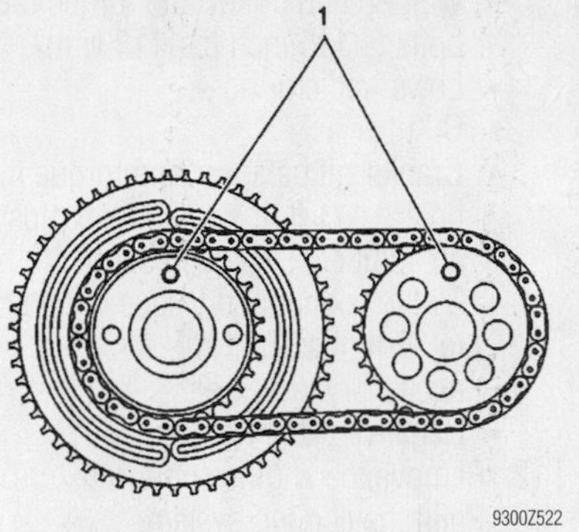

Correct sprocket alignment for secondary timing chain—3.5L engine

3. Remove or disconnect the following:
    - Negative battery cable
    - Thermostat housing for clearance (when working on the front cylinder head)
    - Rocker arm cover and install a Camshaft Holding Fixture
    - Camshaft position (CMP) sensor
    - Camshaft sprocket bolts
4. Install the timing chain/sprocket holding fixture on the cylinder head.
5. Evenly slide the secondary drive chain and sprockets off the camshafts.

To install:

6. Install the secondary timing chain on the sprockets, with the drive pins at the 12 o'clock positions.
7. Install the sprockets/chain assembly onto the camshafts, with the chain properly aligned on the tensioner.
8. Remove the sprocket holding fixture from the cylinder head.
9. Install the sprocket bolts and torque the bolts to 18 ft. lbs. (25 Nm) plus an additional 45 degree turn.
10. Remove the camshaft holding fixture.
11. Install or connect the following:
    - Rocker arm cover and torque the bolts to 80 inch lbs. (9 Nm)
    - Thermostat housing (if removed) and torque the bolts to 80 inch lbs. (9 Nm)
    - CMP sensor and torque the bolts to 80 inch lbs. (9 Nm)
    - Negative battery cable
12. Refill the cooling system.
13. Start the vehicle and check for leaks, repair if necessary.

3.8L Engines

1. Before servicing the vehicle, refer to the precautions in the beginning of this section.
2. Drain the coolant system.
3. Drain the engine oil.
4. Install an Engine Support Fixture.
5. Raise the engine so that the weight is removed from the engine mount.
6. Remove or disconnect the following:
    - Negative battery cable
    - Accessory drive belt(s)
    - Drive belt tensioner
    - Crankshaft balancer
    - Crankshaft Position (CKP) sensor shield
    - Engine mount bracket
    - Front oil pan-to-front cover bolts
    - Oil filter
    - Lower radiator hose
    - Water pump pulley
    - CKP sensor
    - Front cover
7. Rotate the crankshaft until the timing mark on the camshaft sprocket is aligned with the crankshaft sprocket timing mark.
    - Timing chain damper assembly
    - Camshaft sprocket bolt
    - Camshaft sprocket with the timing chain
    - Crankshaft sprocket

To install:

8. Install the crankshaft sprocket.

➡It may be necessary to use a gear installer to fully seat the gear. Be sure the timing mark on the crankshaft gear is pointing straight up.

9. Install the camshaft gear with the timing chain.

➡Hold the sprocket with the timing mark facing downward and the chain hanging down off the sprocket; then, loop the chain under the crankshaft sprocket.

10. If the marks are not in alignment perform the following:
    a. Step 1:Remove the camshaft sprocket and timing chain.
    b. Step 2:Install the camshaft sprocket onto the camshaft and rotate the camshaft until the camshaft and crankshaft marks are aligned.
    c. Step 3:Remove the camshaft sprocket.
    d. Step 4:Reinstall the assembly.
11. Install or connect the following:
    - Camshaft sprocket bolt and torque the bolt to 74 ft. lbs. (100 Nm) plus an additional 90 degree turn
    - Timing chain damper and torque the mounting bolts to 16 ft. lbs. (22 Nm)

- Front cover seal lubricated with engine oil, using the appropriate seal driver
- New front cover gasket
- Front cover and torque the bolts to 15 ft. lbs. (20 Nm) plus an additional 40 degree turn
- Oil pan-to-front cover bolts and torque the bolts to 125 inch lbs. (14 Nm)
- CKP sensor and torque the bolts to 21 ft. lbs. (28 Nm)
- CKP sensor shield
- Engine mount bracket and torque the bolts to 65 ft. lbs. (87 Nm)
- Water pump pulley and torque the bolts to 115 inch lbs. (13 Nm)
- Lower radiator hose
- Oil filter
- Crankshaft balancer and torque the bolt to 111 ft. lbs. (150 Nm) plus an additional 75 degree turn
- Belt tensioner and torque the bolts to 37 ft. lbs. (50 Nm)
- Accessory drive belt
- Negative battery cable

12. Remove the engine support fixture.
13. Refill the cooling system.
14. Refill the engine oil.
15. Start the engine and check for leaks, repair if necessary.

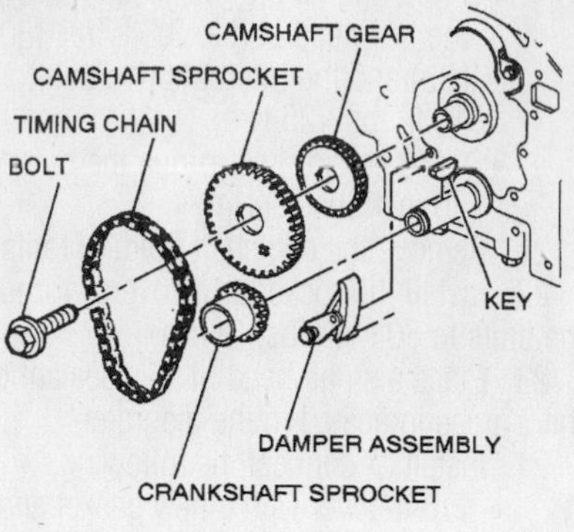

Exploded view of the timing chain and sprockets—3.8L engines

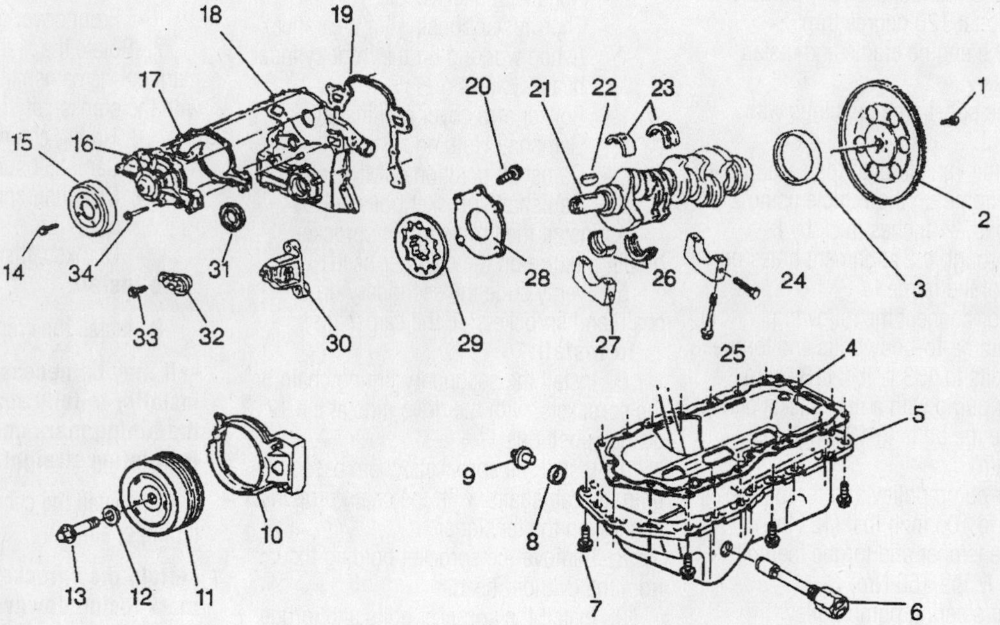

Legend

(1) Flywheel Bolt
(2) Flywheel
(3) Crankshaft Rear Oil Seal
(4) Oil Pan Gasket (Includes Baffle)
(5) Engine Oil Pan
(6) Oil Level Sensor
(7) Oil Pan Bolt
(8) Oil Pan Drain Plug
(9) Oil Pan Drain Gasket
(10) Crankshaft Position Sensor Shield
(11) Crankshaft Balancer
(12) Crankshaft Balancer Washer
(13) Crankshaft Balancer Bolt
(14) Water Pump Pulley Bolt
(15) Water Pump Pulley
(16) Water Pump
(17) Water Pump Gasket
(18) Engine Front Cover
(19) Engine Front Cover Gasket
(20) Oil Pump Cover Bolt
(21) Engine Crankshaft
(22) Crankshaft Balancer Key
(23) Crankshaft Upper Bearing
(24) Side Main Bolt
(25) Crankshaft Main Bearing Cap Bolt
(26) Crankshaft Lower Bearing
(27) Crankshaft Main Bearing Cap
(28) Oil Pump Cover
(29) Oil Pump Gear Set
(30) Crankshaft Position Sensor
(31) Crankshaft Front Oil Seal
(32) Camshaft Position Sensor
(33) Camshaft Position Sensor Bolt
(34) Water Pump Bolt

9300XG04

Exploded view of lower engine components—3.8L engine

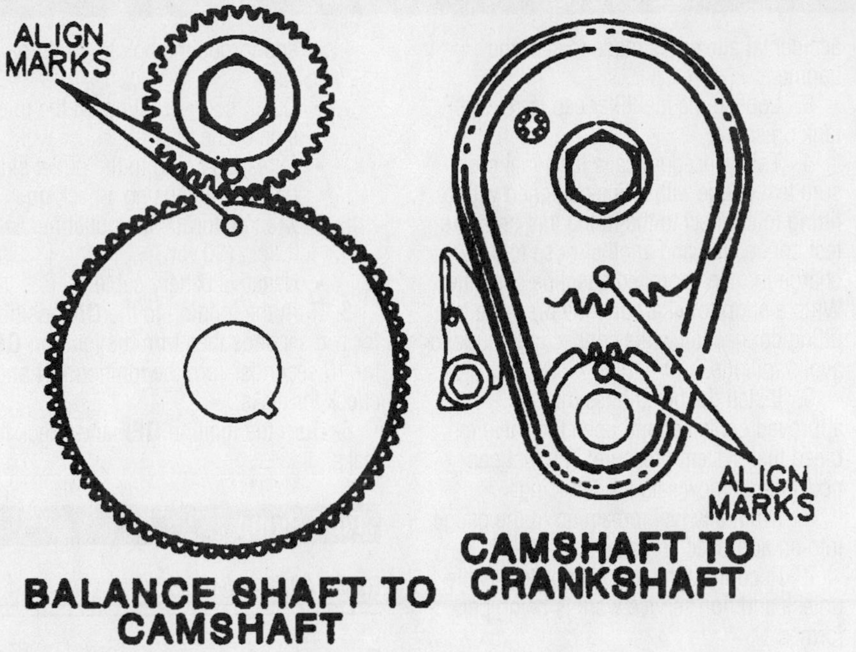

Balance shaft-to-camshaft and camshaft-to-crankshaft timing mark alignment—3.8L engine

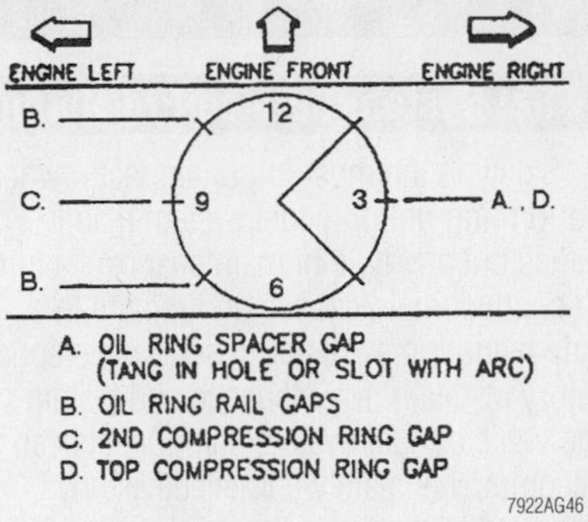

Piston ring end-gap spacing—3.1L, 3.4L, 3.8L engines

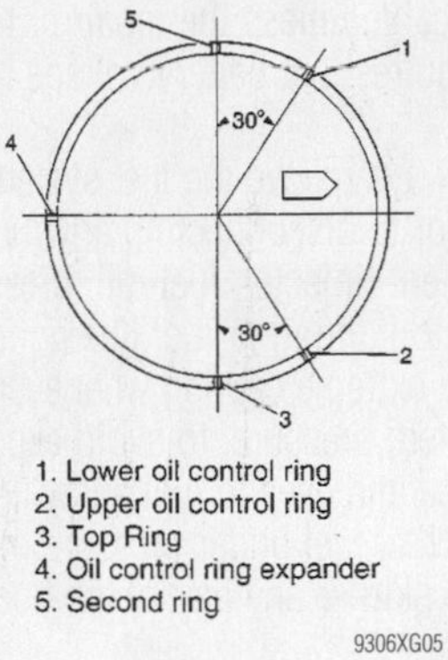

Piston ring end-gap positioning—3.5L engine

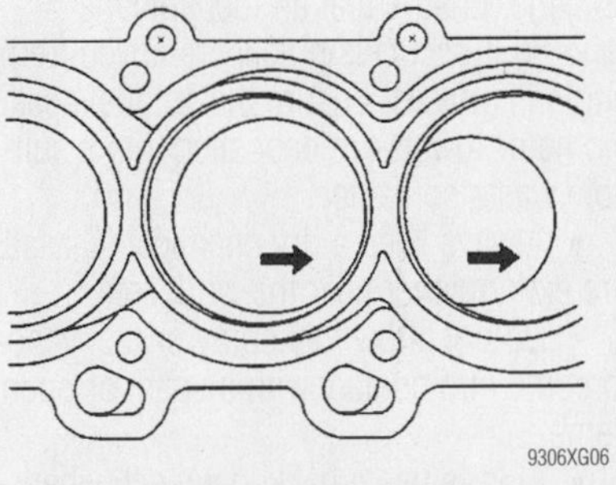

Piston positioning—3.5L engine

Piston and Ring

POSITIONING

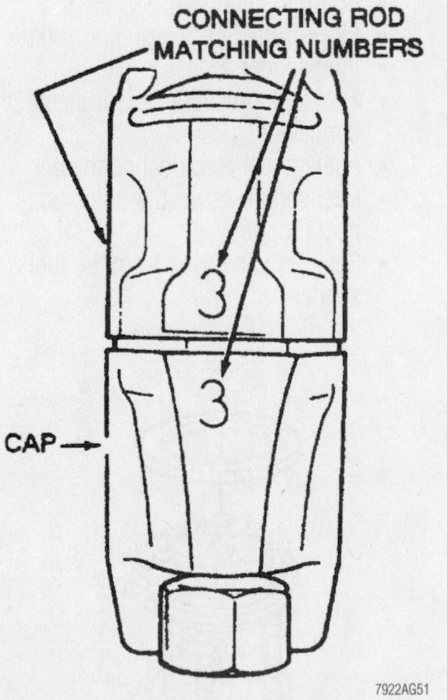

Connecting rod and cap installation. Be sure to matchmark the cap and rod prior to disassembly, as shown

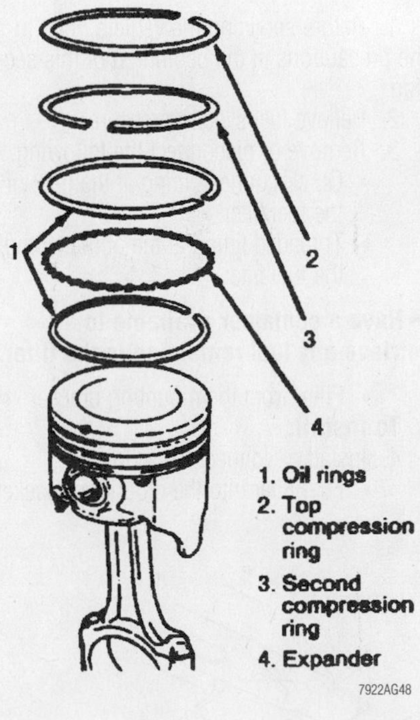

Piston ring positioning—3.1L, 3.4L, 3.5L and 3.8L engines

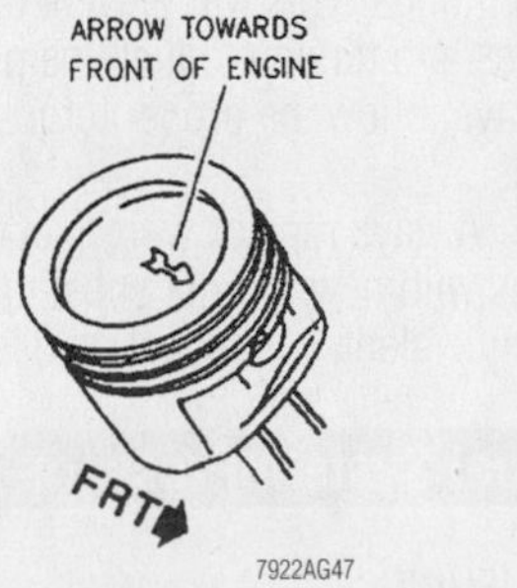

Piston positioning. Often the arrow is replaced by a notch, which also must face toward the front of the engine—3.1L, 3.4L and 3.8L engines

Heater Core replacement is covered in Section 2 of this manual

FUEL SYSTEM

Fuel System Service Precautions

Safety is the most important factor when performing not only fuel system maintenance but any type of maintenance. Failure to conduct maintenance and repairs in a safe manner may result in serious personal injury or death. Maintenance and testing of the vehicle's fuel system components can be accomplished safely and effectively by adhering to the following rules and guidelines.

- To avoid the possibility of fire and personal injury, always disconnect the negative battery cable unless the repair or test procedure requires that battery voltage be applied.
- Always relieve the fuel system pressure prior to disconnecting any fuel system component (injector, fuel rail, pressure regulator, etc.), fitting or fuel line connection. Exercise extreme caution whenever relieving fuel system pressure, to avoid exposing skin, face and eyes to fuel spray. Please be advised that fuel under pressure may penetrate the skin or any part of the body that it contacts.
- Always place a shop towel or cloth around the fitting or connection prior to loosening to absorb any excess fuel due to spillage. Ensure that all fuel spillage (should it occur) is quickly removed from engine surfaces. Ensure that all fuel soaked cloths or towels are deposited into a suitable waste container.
- Always keep a dry chemical (Class B) fire extinguisher near the work area.
- Do not allow fuel spray or fuel vapors to come into contact with a spark or open flame.
- Always use a backup wrench when loosening and tightening fuel line connection fittings. This will prevent unnecessary stress and torsion to fuel line piping. Always follow the proper torque specifications.
- Always replace worn fuel fitting O-rings with new. Do not substitute fuel hose or equivalent, where fuel pipe is installed.

Fuel System Pressure

RELIEVING

1. Before servicing the vehicle, refer to the precautions in the beginning of this section.
2. Disconnect the negative battery cable to prevent possible discharge of fuel if an accidental attempt is made to start the engine.
3. Loosen the fuel filler cap to relieve tank pressure.
4. This procedure calls for a fuel pressure test gauge with a line equipped with a fitting to connect to the to the fuel pressure test connection and another hose to discharge into an approved gasoline container. Wrap a shop towel around the pressure test fitting connection while connecting gauge to avoid spillage.
5. Install the bleed hose into an approved container and open the valve to bleed fuel system pressure. The fuel connections are now safe for servicing.
6. Drain any fuel remaining in the gauge into an approved container.
7. Reconnect the negative battery cable unless addition service work is being performed.

Fuel Filter

REMOVAL & INSTALLATION

All Models

1. Before servicing the vehicle, refer to the precautions in the beginning of this section.
2. Relieve fuel system pressure.
3. Remove or disconnect the following:
   - Quick-connect fitting at the inlet of the fuel filter
   - Threaded fitting at the outlet side of the fuel filter.

➡Have a container available to retrieve any fuel remaining in the filter.

   - Filter from the mounting bracket

To install:

4. Install or connect the following:
   - Fuel filter into the mounting bracket
   and torque the bolt to 15 ft. lbs. (20 Nm)
   - Quick-connect fitting on the inlet side of the filter
   - Threaded fitting to the outlet side of the filter and using a back up wrench, torque the outlet nut to 22 ft. lbs. (30 Nm)
   - Negative battery cable
5. Turn the ignition to the **ON** position for two seconds then turn the ignition **OFF** for 10 seconds. Turn the ignition **ON** and check for leaks.
6. Turn the ignition **OFF** and check for leaks.

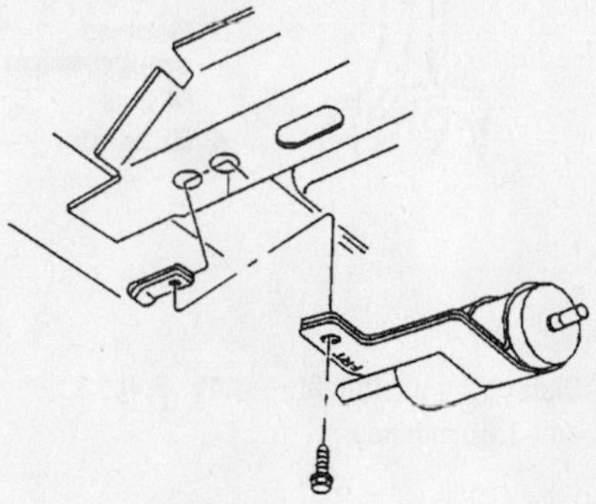

9306ZG20

Common fuel filter mounting

Fuel Pump

REMOVAL & INSTALLATION

Except Lumina

1. Before servicing the vehicle, refer to the precautions in the beginning of this section.
2. Relieve the fuel system pressure.
3. Drain the fuel tank with a hand held siphon until the level is less than ¼ full.
4. Remove or disconnect the following:
   - Negative battery cable
   - Spare tire and jack
   - Floor trunk liner by pulling it back
   - Fuel sender access panel
   - Fuel tank pressure sensor electrical connector
   - Fuel sender electrical connector
   - Fuel sender assembly quick connect fittings
   - Retaining snapring from the fuel sender

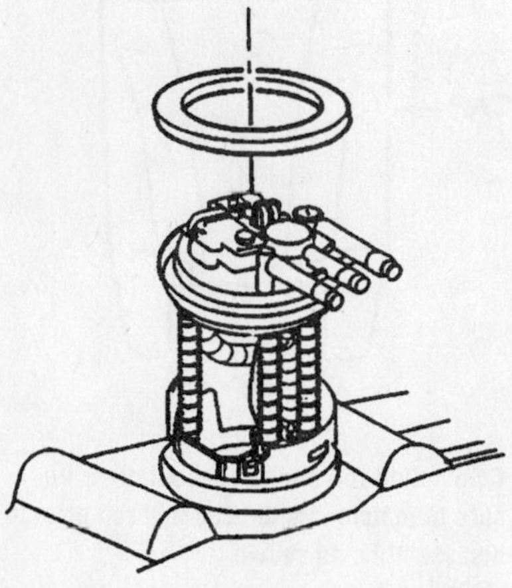

7922Z528

Remove the fuel pump from the tank after removing the locking ring

WARNING

When the Lockring is removed from the fuel sender, the sender assembly will spring up. Downward pressure should be kept on the assembly and slowly released to ensure the sender assembly does not get damaged.

- Modular fuel sender assembly

To install:

5. Install or connect the following:
   - New O-ring on the fuel tank
   - Fuel sender
   - Snapring on the fuel sender
   - Fuel sender electrical connector
   - Fuel tank pressure sensor electrical connector
   - Quick connect fittings at the fuel sender
   - Negative battery cable
6. Add a small amount of fuel to the fuel tank.
7. Turn the ignition to the **ON** position for 2 seconds then turn the ignition **OFF** for 10 seconds. Turn the ignition **ON** and check for leaks.
8. Turn the ignition **OFF** and check for leaks.
9. Install or connect the following:
   - Fuel sender access panel and

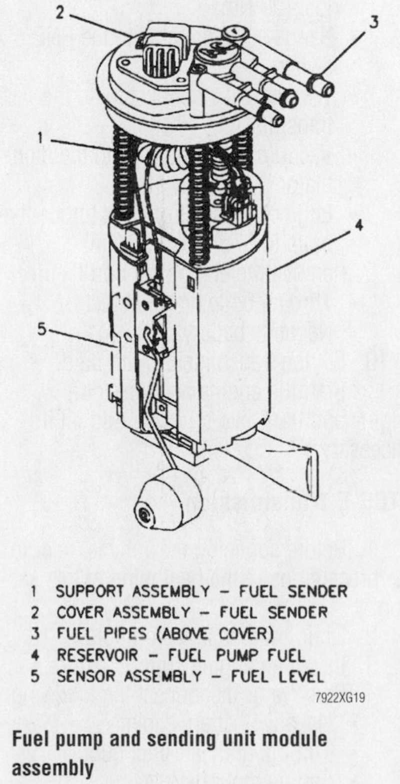

Fuel pump and sending unit module assembly

torque the nuts to 89 inch lbs. (10 Nm)
   - Trunk liner
   - Spare tire, jack and spare tire cover
10. Refill the fuel tank.

Lumina

1. Before servicing the vehicle, refer to the precautions in the beginning of this section.
2. Relieve the fuel system pressure.
3. Drain the fuel tank.
4. Disconnect the negative battery cable.
5. Remove the fuel tank from the vehicle.
6. Clean all of pipe and hose connections. Also clean all areas around the connections.
7. Disconnect the quick-connect fittings at the fuel sender assembly.
8. Remove the fuel sender assembly retaining snap ring, the fuel sender assembly and the O-ring from the fuel tank. Discard the fuel sender assembly O-ring.

To install:

9. Position the new fuel sender assembly O-ring on the fuel tank.

➡Make sure not to fold over or twist the fuel pump strainer when installing the fuel sender assembly, as this will restrict fuel flow. Also, ensure that the fuel pump strainer does not block full travel of float arm.

10. Install the fuel sender assembly and the fuel sender assembly retaining snap ring.
11. Connect the quick-connect fittings at the fuel sender assembly.
12. Install the fuel tank and add fuel.
13. Pressurize the system as follows to check for leaks:
    a. Turn the ignition switch ON for 2 seconds.
    b. Turn the ignition switch OFF for 10 seconds.
    c. Turn the ignition switch to the ON.
    d. Check for fuel leaks.

Fuel Injector

REMOVAL & INSTALLATION

3.1L and 3.4L Engines

1. Before servicing the vehicle, refer to the precautions in the beginning of this section.

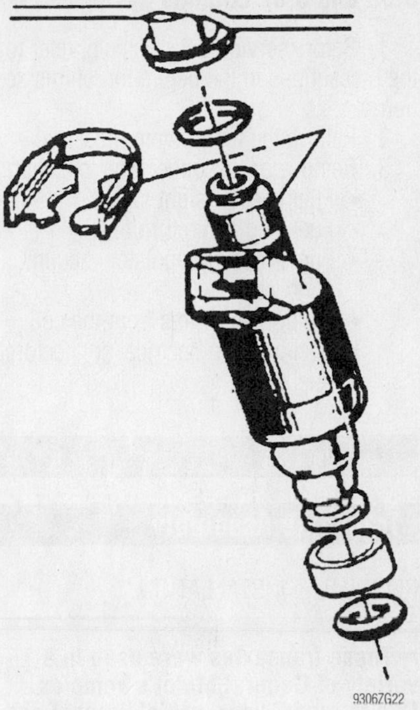

Exploded view of the fuel injector—3.1L and 3.4L engines

2. Relieve the fuel system pressure.
3. Remove or disconnect the following:
   - Upper intake manifold
   - Fuel inlet and return lines
   - Fuel injector electrical connectors
   - Coolant temperature sensor electrical connector
   - Fuel rail
   - Fuel injector retaining clips
   - Fuel injectors

To install:

4. Coat the new fuel injector O-rings with clean engine oil.
5. Install or connect the following:
   - Lower backup O-ring
   - Upper O-ring
   - Lower O-ring
   - Fuel injector
   - Fuel injector retaining clips
   - Fuel rail and torque the bolts to 7 ft. lbs. (10 Nm)
   - Coolant temperature sensor electrical connector
   - Fuel injector electrical connectors
   - Fuel inlet and return lines with new O-rings and torque the fittings to 13 ft. lbs. (17 Nm)
   - Upper intake manifold and torque the bolts to 18 ft. lbs. (25 Nm)
   - Negative battery cable
6. Start the vehicle and check for leaks, repair if necessary.

3.5L and 3.8L Engines

1. Before servicing the vehicle, refer to the precautions in the beginning of this section.
2. Relieve the fuel system pressure.
3. Remove or disconnect the following:
   - Fuel injector sight shield
   - Fuel feed and return lines
   - Fuel pressure regulator vacuum connection
   - Ignition coil wires from the coil
   - Fuel injector electrical connectors
   - Fuel rail
   - Fuel injector retaining clips
   - Fuel injectors

To install:

4. Coat the new fuel injector O-rings with clean engine oil.
5. Install or connect the following:
   - Lower backup O-ring
   - Upper O-ring
   - Lower O-ring
   - Fuel injector
   - Fuel injector retaining clips
   - Fuel rail and torque the bolts to 7 ft. lbs. (10 Nm) if equipped with bolt retainers. If equipped with snap-lock tabs retainers, push down until the retainers snap into place.
   - Fuel injector electrical connectors
   - Ignition coil wires to the coil
   - Fuel pressure regulator vacuum connection
   - Fuel feed and return lines
   - Negative battery cable
6. Start the vehicle and check for leaks, repair if necessary.
7. Install the fuel injector sight shield and torque the nuts to 27 inch lbs. (3 Nm).

DRIVE TRAIN

Transaxle Assembly

REMOVAL & INSTALLATION

➡These transaxles were used in a variety of General Motor's vehicles. Due to model year, vehicle model and installed options, the removal and installation procedures may vary slightly. The procedures given here should suffice for most all vehicles using these transaxles.

4T60-E Transmission

1. Before servicing the vehicle, refer to the precautions in the beginning of this section.
2. Drain the transmission fluid.
3. Install an engine support fixture.
4. Remove or disconnect the following:
   - Negative battery cable
   - Throttle body air inlet duct
   - Engine mount struts
   - Wire harness connectors from the transmission
   - Vacuum hose and pipe from the modulator
   - Range selector cable from the Park/Neutral Position (PNP) switch
   - PNP switch
   - Fluid filler tube
   - Upper transmission bolts
   - Wire harness grounds
   - Both front wheels
   - Engine splash shields
   - Both tie rod ends from the steering knuckles
   - Power steering gear from the frame and secure it to the body of the vehicle
   - Power steering cooler line clamps
   - Engine mount lower nuts
   - Lower ball joints from the steering knuckles
   - Torque converter cover
   - Starter motor
   - Torque converter bolts
   - Oil cooler hoses
   - Drive axles and secure them to the steering knuckles and struts
   - Wheel Speed Sensor (WSS) connectors
   - Vehicle Speed Sensor (VSS) connectors
5. Use a transmission table to support the transmission.
   - Transmission brace
   - Lower transmission bolts
   - Frame-to-body bolts
6. Lower the frame from the vehicle.
7. Remove the transmission from the frame.

To install:

8. Install or connect the following:
   - Transmission
   - Lower transmission bolts and torque them to 55 ft. lbs. (75 Nm)
   - New frame to body bolts and torque them to 125 ft. lbs. (170 Nm)
   - Transmission brace and torque the bolts to 35 ft. lbs. (47 Nm)
   - VSS electrical connector
   - WSS electrical connectors
   - Drive axles to the transmission
   - Oil cooler hoses
   - Torque converter bolts and torque them to 46 ft. lbs. (63 Nm)
   - Starter motor and torque the bolts to 32 ft. lbs. (43 Nm)
   - Torque converter cover and torque the bolts to 89 inch lbs. (10 Nm)
   - Lower ball joints to the steering knuckle and torque the nuts to 40 ft. lbs. (55 Nm)
   - Engine mount lower nuts and torque them to 32 ft. lbs. (43 Nm)
   - Power steering cooler line clamps to the frame
   - Power steering gear to the frame and torque the bolts to 59 ft. lbs. (80 Nm)
   - Tie rod ends to the steering knuckles and torque the nuts to 63 ft. lbs. (85 Nm)
   - Engine splash shields
   - Front wheels
   - Fluid filler tube
   - Upper transaxle bolts and torque them to 55 ft. lbs. (75 Nm)
   - PNP switch and torque the bolts to 18 ft. lbs. (25 Nm)
   - Range selector cable with the bracket and torque the nut to 15 ft. lbs. (20 Nm)
   - Range selector cable to the PNP switch
   - Wire harness connectors to the transmission
   - Vacuum hose and pipe to the modulator
   - Engine mount struts and torque the bolts to 37 ft. lbs. (50 Nm)
9. Remove the engine support fixture.
   - Throttle body air inlet duct
   - Negative battery cable
10. Fill the transmission with fluid.
11. Start the engine and check the engine and transaxle oil level. Add oil if necessary.

4T65-E Transmission

1. Before servicing the vehicle, refer to the precautions in the beginning of this section.
2. Drain the transmission fluid.
3. Install an engine support fixture.
4. Remove or disconnect the following:
   - Negative battery cable
   - Throttle body air inlet duct
   - Engine mount struts
   - Wire harness connectors from the transmission

- Range selector cable from the Park Neutral Position (PNP) switch
- Range selector cable and bracket
- PNP switch
- Power steering gear to frame retaining bolts
- Fluid filler tube
- Upper transmission bolts
- Wire harness grounds
- Both front wheels
- Engine splash shields
- Both tie rod ends from the steering knuckles
- Power steering gear from the frame and secure it to the body of the vehicle
- Power steering cooler line clamps
- Engine mount lower nuts
- Lower ball joints from the steering knuckles
- Torque converter cover
- Starter motor
- Torque converter bolts
- Oil cooler hoses
- Drive axles and secure them to the steering knuckles
- Wheel Speed Sensor (WSS) electrical connectors
- Vehicle Speed Sensor (VSS) electrical connectors

5. Use a transmission table to support the transmission.
- Engine mount lower nuts
- Transmission brace
- Lower transmission bolts
- Frame-to-body bolts

6. Separate the transmission from the engine.

7. Lower the transmission and frame from the vehicle.

8. Remove the transmission.

To install:

9. Install or connect the following:
- Transmission to the frame and raise the assembly into position
- New frame-to-body bolts and torque them to 133 ft. lbs. (180 Nm)
- Lower transmission-to-engine bolts and torque them to 55 ft. lbs. (75 Nm)
- Transmission brace and torque the transmission bolts to 32 ft. lbs. (43 Nm) and the engine bolts to 46 ft. lbs. (63 Nm)
- VSS electrical connector
- WSS electrical connectors
- Drive axles to the transmission
- Oil cooler hoses
- Torque converter bolts and torque them to 46 ft. lbs. (63 Nm)
- Starter motor and torque the bolts to 32 ft. lbs. (43 Nm)
- Torque converter cover and torque the bolts to 89 inch lbs. (10 Nm).
- Lower ball joints to the steering knuckle and torque the nuts to 40 ft. lbs. (55 Nm)
- Engine mount lower nuts and torque them to 35 ft. lbs. (47 Nm)
- Power steering cooler line clamps to the frame
- Power steering gear to the frame and torque the bolts to 59 ft. lbs. (80 Nm)
- Tie rod ends to the steering knuckles and torque the nuts to 63 ft. lbs. (85 Nm)
- Engine splash shields
- Front wheels
- Fluid filler tube
- Upper transmission bolts and torque them to 55 ft. lbs. (75 Nm)
- PNP switch and torque the bolts to 18 ft. lbs. (25 Nm)
- Range selector cable with the bracket and torque the nut to 15 ft. lbs. (20 Nm)
- Range selector cable to the PNP switch
- Wire harness connectors to the transmission
- Engine mount strut and torque the bolts to 37 ft. lbs. (50 Nm)

10. Remove the engine support fixture.
- Throttle body air inlet duct
- Negative battery cable

11. Fill the transmission with fluid.

12. Adjust the wheel alignment.

13. Start engine and check the engine and transaxle oil levels. Add oil if necessary.

Halfshaft

REMOVAL & INSTALLATION

Lumina and 1998–99 Monte Carlo

➡If equipped with Antilock Brake System (ABS) brakes, use care to avoid damage to the ABS toothed ring. Damage to the ring may cause the self-diagnostic feature of the ABS system to set a system fault code.

1. Before servicing the vehicle, refer to the precautions in the beginning of this section.

2. Remove or disconnect the following:
- Front wheel
- Drive shaft nut
- Brake caliper and bracket assembly
- Brake rotor
- Wheel Speed Sensor (WSS) electrical connector
- Wheel hub/bearing
- Axle shaft using an appropriate puller

3. Remove the halfshaft/bearing assembly through the steering knuckle.

To install:

4. Install or connect the following:
- Halfshaft/bearing assembly through the knuckle and into the transaxle
- Bearing-to-knuckle bolts, loosely
- Halfshaft into the transaxle

5. Verify that the snapring is seated properly by tapping on the inner groove with a prying tool. Grasp the inner housing of the axle shaft and pull outward. Do not pull on the axle shaft. If the snapring is properly seated, the axle will remain in place.

6. Install or connect the following:
- WSS electrical connector
- Front wheel driveshaft bearing and torque the bolts to 52 ft. lbs. (70 Nm)
- New driveshaft nut and torque it to 159 ft. lbs. (215 Nm)
- Brake rotor
- Brake caliper and attaching bracket and torque the bracket bolts to 148 ft. lbs. (200 Nm) and the caliper bolts to 80 ft. lbs. (108 Nm)
- Front wheel
- Negative battery cable

7. Check the transaxle fluid level.

8. Check the front alignment and adjust if necessary.

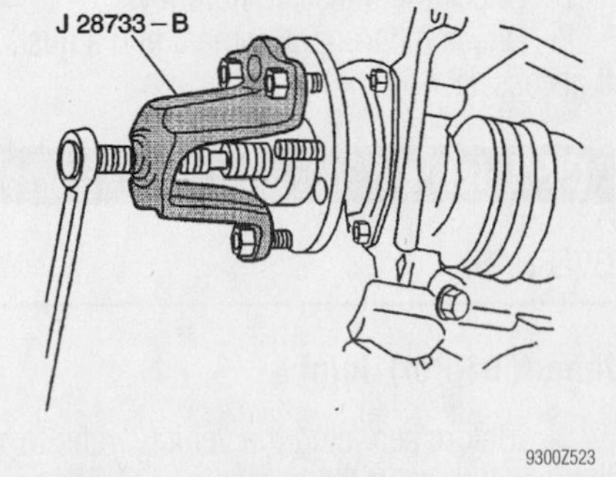

Use a puller to press the axle shaft through the hub/bearing assembly

Century, Regal, Impala, Intrigue, Gran Prix and 2000–01 Monte Carlo

1. Before servicing the vehicle, refer to the precautions in the beginning of this section.
2. Remove or disconnect the following:
   - Front wheel
   - Stabilizer shaft link
   - Drive shaft nut
   - Outer tie rod end from the steering knuckle
   - Ball joint from the steering knuckle
3. Press the axle shaft through the hub.

WARNING

To prevent damage to the inner CV-joint, do not pull on the axle shaft to remove it from the transaxle.

4. Place a drain pan under the transaxle to catch any transaxle fluid that leaks out when the axle shaft is removed.
5. Remove the axle shaft from the transaxle by prying between the transaxle and the inner CV-joint housing.

To install:

6. Install or connect the following:
   - Axle shaft in the transaxle. Verify that it is seated by pulling on the housing
   - Axle shaft through the hub/bearing assembly
   - Ball joint and torque the nut to 40 ft. lbs. (55 Nm)
   - Tie rod end and torque the nut to 22 ft. lbs. (30 Nm) plus an additional 120 degree turn
   - New drive shaft nut and torque it to 118 ft. lbs. (160 Nm) on all except Impala models or 159 ft. lbs. (215 Nm) on Impala models
   - Stabilizer shaft link and torque the nut to 17 ft. lbs. (23 Nm)
   - Front wheel
7. Check the transaxle fluid level.
8. Check the front alignment and adjust, if necessary.

CV-Joint

OVERHAUL

Inner (Tri-Pot) Joint

1. Before servicing the vehicle, refer to the precautions in the beginning of this section.
2. Remove or disconnect the following:
   - Front wheel
   - Halfshaft
   - Swage ring using a hand grinder
   - Large CV-joint boot clamp, cut and discard it
   - CV-joint boot by sliding it away from the tri-pot joint
   - Tri-pot housing from the tri-pot spider
   - Trilobal tri-pot bushing from the housing
   - Inboard spacer ring slide it rearward on the shaft using Snapring Pliers Tool J-8059
   - Outboard retaining ring using Snapring Pliers Tool J-8059
   - Tri-pot joint spider assembly
   - Inboard spacer ring and CV-joint boot
3. Throughly clean and inspect all parts.

To install:

4. Install or connect the following:

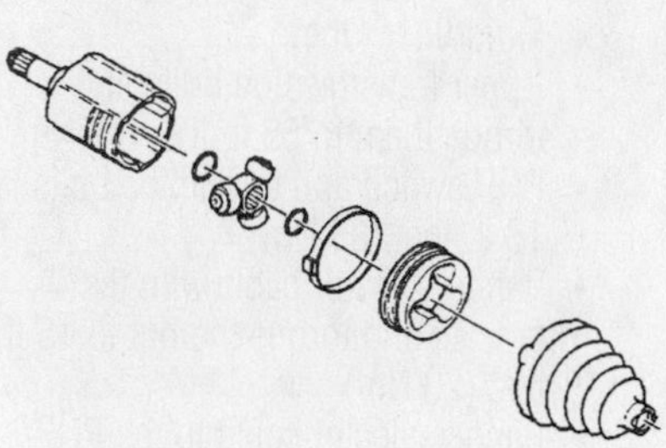
9306XG14

Exploded view of the inner (tri-pot) joint

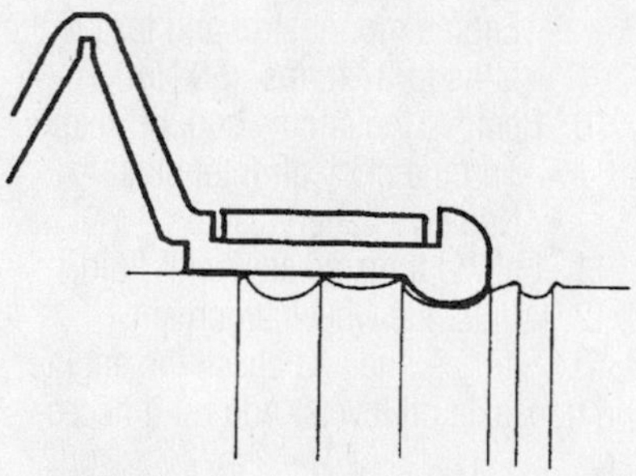
9306XG15

Positioning the inner CV-joint boot seal and swage ring—Inner (tri-pot) joint

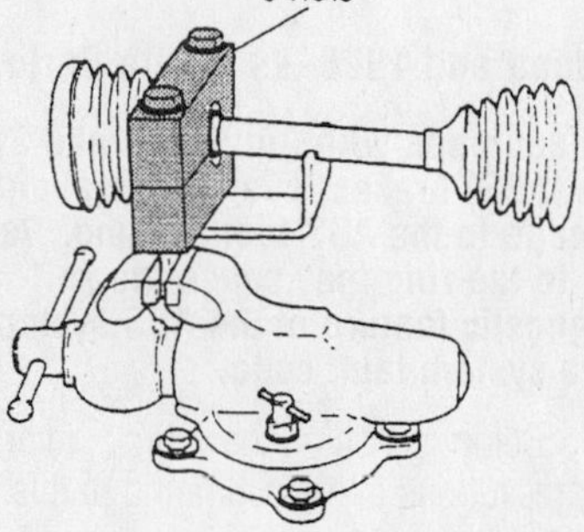

9306XG16

View of the swage ring crimping tool—Inner (tri-pot) joint

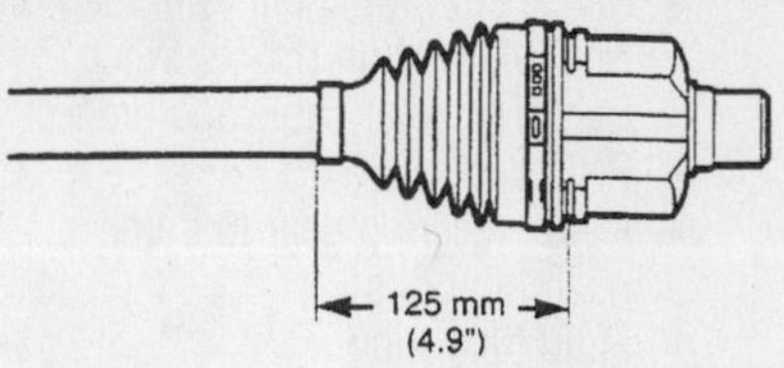

9306XG17

Boot measurement—Inner (tri-pot) joint

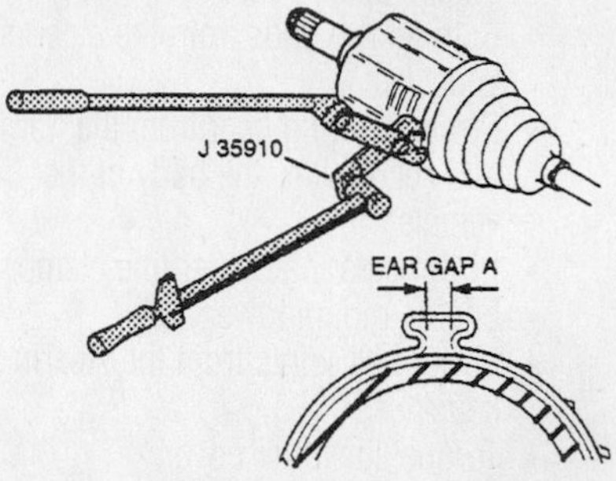

9306XG18

Crimping the large CV-joint boot ring—Inner (tri-pot) joint

   - Swage ring clamp
   - CV-joint boot
5. Position the CV-joint boot seal into the axle shaft's joint seal groove and align the swage ring clamp on the boot.
6. Secure the swage ring clamp as follows:
   a. Mount the lower half of Tool J-41048 in a vise.
   b. Position the outboard of the half-shaft in the tool.
   c. Position the upper end of Tool J-41048 onto the lower half.

WARNING

Make sure that there are no pinch points on the inboard seal.

   d. Insert both bolts and tighten by hand until snug.
   e. Tighten each bolt 180 degree (½ turn) at a time, alternating between the bolts, until both sides are bottomed.
   f. Remove the tool.
7. Install or connect the following:
   - Inboard spacer ring, slide it rearward on the shaft using Snapring Pliers Tool J-8059
   - Tri-pot joint spider assembly onto the shaft
   - Outboard retaining ring into the axle shaft groove using Snapring Pliers Tool J-8059
   - Tri-pot joint spider assembly, slide it against the outboard retaining ring

- Inboard spacer ring, seat it in the groove
- ½ kit grease into the boot
- ½ kit grease into the tri-pot housing
- Trilobal tip-pot bushing flush with the tri-pot housing face
- New large seal clamp onto the CV-joint boot
- Tri-pot housing, slide it over the tri-pot joint spider assembly
- CV-joint boot/clamp, slide it into place, over the trilobal tri-pot bushing with the seal lip in the groove

➡Make sure the boot lies flat against the trilobal bushing.

8. Position the CV-joint boot so it measures 4.9 in. (125mm).
9. Using the Crimp Tool J-35910, a torque wrench and a breaker bar, crimp the large CV-joint boot clamp to 130 ft. lbs. (176 Nm).
10. Install or connect the following:
    - Halfshaft
    - Front wheel

Outer Joint

1. Before servicing the vehicle, refer to the precautions in the beginning of this section.
2. Remove or disconnect the following:
    - Front wheel
    - Halfshaft
    - Swage ring using a hand grinder
    - Large boot clamp, cut and discard it
    - CV-joint boot, slide it away from the CV-joint
    - CV-joint assembly by spreading the inner race-to-axle shaft retaining ring ears using Snapring Pliers Tool J-8059
    - CV-joint boot from the axle shaft and discard it
3. Disassemble the chrome alloy balls from the CV-joint cage as follows:
    a. Position a brass drift against the CV-joint cage and tap it with a hammer to tilt the cage.
    b. Chrome alloy ball from the cage.
    c. Tilt the cage in the opposite direction.
    d. Remove the opposite chrome alloy ball.
    e. Repeat the procedure until all 6 balls are removed.

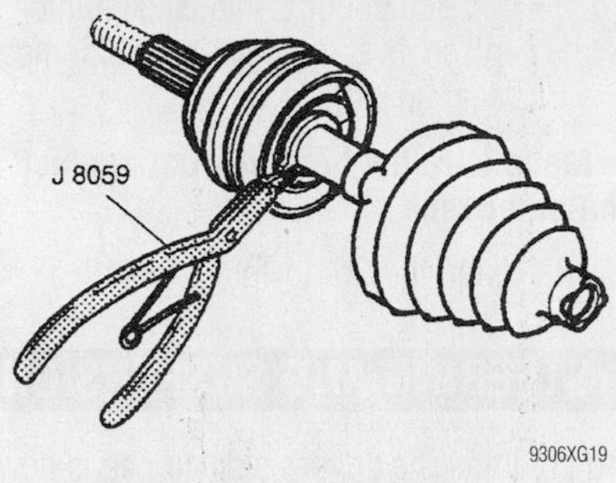

Disconnecting the outer CV-joint from the axle shaft

Tilting the cage—Outer CV-joint

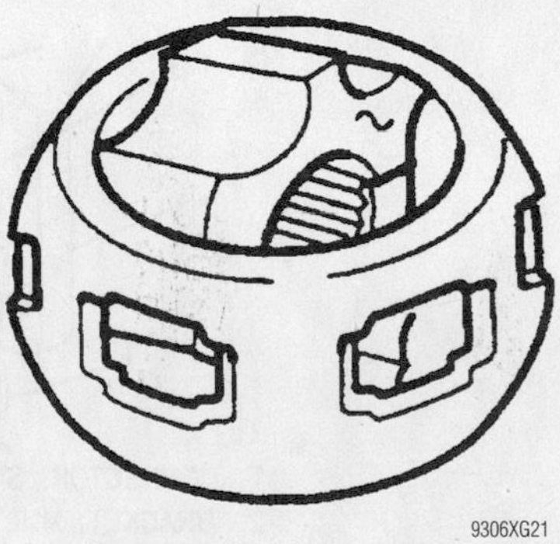

View the cage and inner race—Outer CV-joint

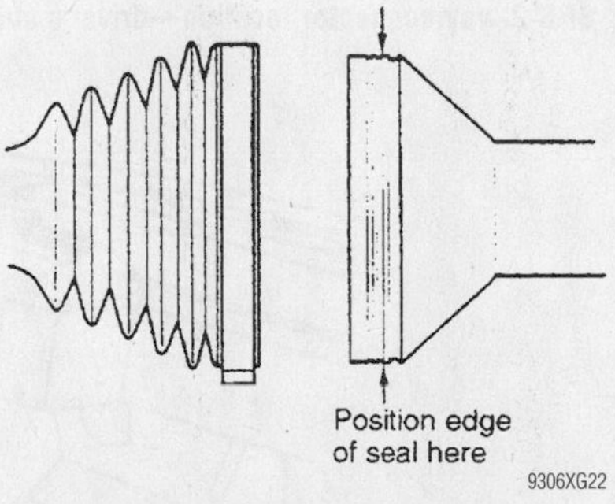

Positioning the boot—Outer CV-joint

4. Disassemble the CV-joint cage and inner race as follows:
    a. Pivot the cage and race 90 degrees to the center line of the outer race.
    b. Align the cage windows with outer race lands.
    c. Remove the cage from the outer race.
    d. Rotate the inner race upward and remove it from the cage.
5. Throughly clean and inspect all parts.

To install:

6. Lubricate the parts with a light coat of grease.
7. Assemble the CV-joint cage and inner race, as follows:
    a. Rotate the inner race 90 degrees to the cage centerline.
    b. Align the cage windows with inner race lands.
    c. Insert the inner race into the cage by rotating the inner race downward.
    d. Insert the cage/inner race into the outer race.
8. Assemble the chrome alloy balls into the CV-joint cage, as follows:
    a. Position a brass drift against the CV-joint cage and tap it with a hammer to tilt the cage.
    b. Insert the 1st chrome alloy ball into the cage.
    c. Tilt the cage in the opposite direction.
    d. Insert the opposite chrome alloy ball.
    e. Repeat the procedure until all 6 balls are inserted.
9. Install or connect the following:
    - ½ kit grease into the CV-joint boot
    - ½ kit grease into the CV-joint
    - Swage ring clamp
    - CV-joint boot
    - CV-joint onto the axle shaft until the retaining ring seats into the groove
10. Position the CV-joint boot seal into the axle shaft's joint seal groove and align the swage ring clamp on the boot.
11. Secure the swage ring clamp as follows:
    a. Mount the lower half of tool J-41048 in a vise.
    b. Position the outboard of the halfshaft in the tool.
    c. Position the upper end of tool J-41048 onto the lower half.

⁂ WARNING

Make sure that there are no pinch points on the inboard seal.

    d. Insert both bolts and tighten by hand until snug.

e. Tighten each bolt 180 degree (½) turn at a time, alternating between the bolts, until both sides are bottomed.

f. Remove the tool.

12. Install or connect the following:
- New large seal clamp onto the CV-joint boot
- CV-joint boot/clamp, slide it into place, over the outer race with the seal lip in the groove

➡Make sure the boot lies flat against the outer race.

13. Using the Crimp Tool J-35910, a torque wrench and a breaker bar, crimp the large CV-joint boot clamp to 130 ft. lbs. (176 Nm).

14. Install or connect the following:
- Halfshaft
- Front wheel

STEERING AND SUSPENSION

Air Bag

⁂ CAUTION

The vehicles are equipped with the Supplemental Inflatable Restraint (SIR) or air bag system. The SIR system must be disabled before performing service on or around SIR system components, steering column, instrument panel components, wiring and sensors. Failure to follow safety and disabling procedures could result in accidental air bag deployment, possible personal injury and unnecessary SIR system repairs.

PRECAUTIONS

Several precautions must be observed when handling the inflator module to avoid accidental deployment and possible personal injury.

- Never carry the inflator module by the wires or connector on the underside of the module.
- When carrying a live inflator module, hold securely with both hands, and ensure that the bag and trim cover are pointed away.
- Place the inflator module on a bench or other surface with the bag and trim cover facing up.
- With the inflator module on the bench, never place anything on or close to the module which may be thrown in the event of an accidental deployment.

DISARMING

1. Turn the steering wheel so the vehicle wheels are pointing straight-ahead.
2. Turn the ignition key to the **LOCK** position and remove the key.
3. Remove the AIR BAG fuse from the fuse block.
4. Remove the left side sound insulator.
5. Detach the Connector Position Assurance (CPA) and yellow 2-way Supplemental Inflatable Restraint (SIR) connector at the multi-use bracket near the base of the steering column. The driver's side air bag is now disabled.
6. Remove the right side sound insulator.
7. Detach the CPA and yellow 2-way SIR connector at the base of the steering wheel. The passenger's side air bag is now disabled.

ARMING

After necessary repairs are made, re-enable the air bag system as follows:

1. Be sure the ignition is locked and the key is removed.
2. Attach the yellow 2-way Supplemental Inflatable Restraint (SIR) connector and CPA at the base of the steering wheel.
3. Install the right side sound insulator.
4. Attach the yellow 2-way SIR connector and Connector Position Assurance (CPA) at the multi-use bracket at the base of the column.
5. Install the left side sound insulator.
6. Install the AIR BAG fuse.
7. Turn the ignition switch to the **RUN** position and verify the AIR BAG light flashes 7 times, then shuts off.

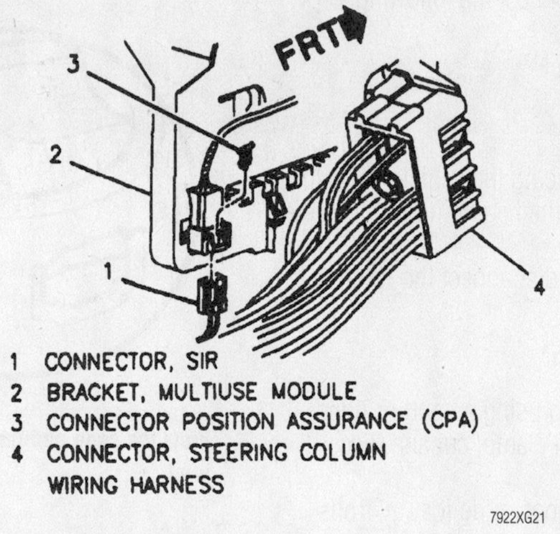

SRS 2-way connector location—driver's side

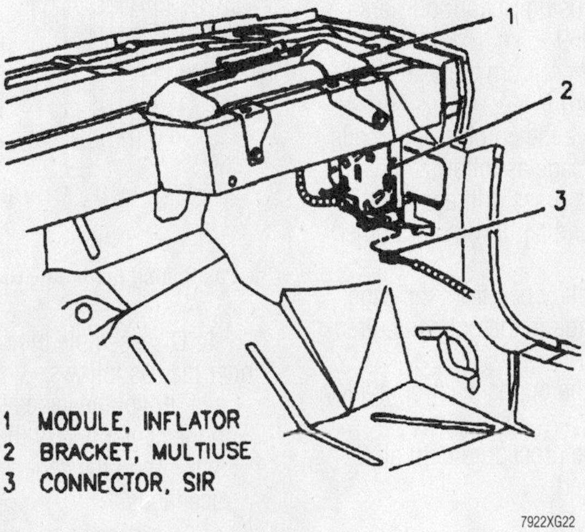

SRS 2-way connector location—passenger's side

Power Rack and Pinion Steering Gear

REMOVAL & INSTALLATION

1. Before servicing the vehicle, refer to the precautions in the beginning of this section.
2. Remove or disconnect the following:
   - Negative battery cable
   - Front wheels

CAUTION

Failure to disconnect the intermediate shaft from the rack and pinion stub shaft may result in damage to the steering gear. This damage may cause a loss of steering control and may cause personal injury.

➡Set the steering shaft so that the block tooth on the upper steering shaft is at the 12 o'clock position. The wheels should be straight ahead. Set the ignition key lock to the LOCK position. Failure to follow these procedures could result in damage to the SIR coil assembly.

3. Remove or disconnect the following:

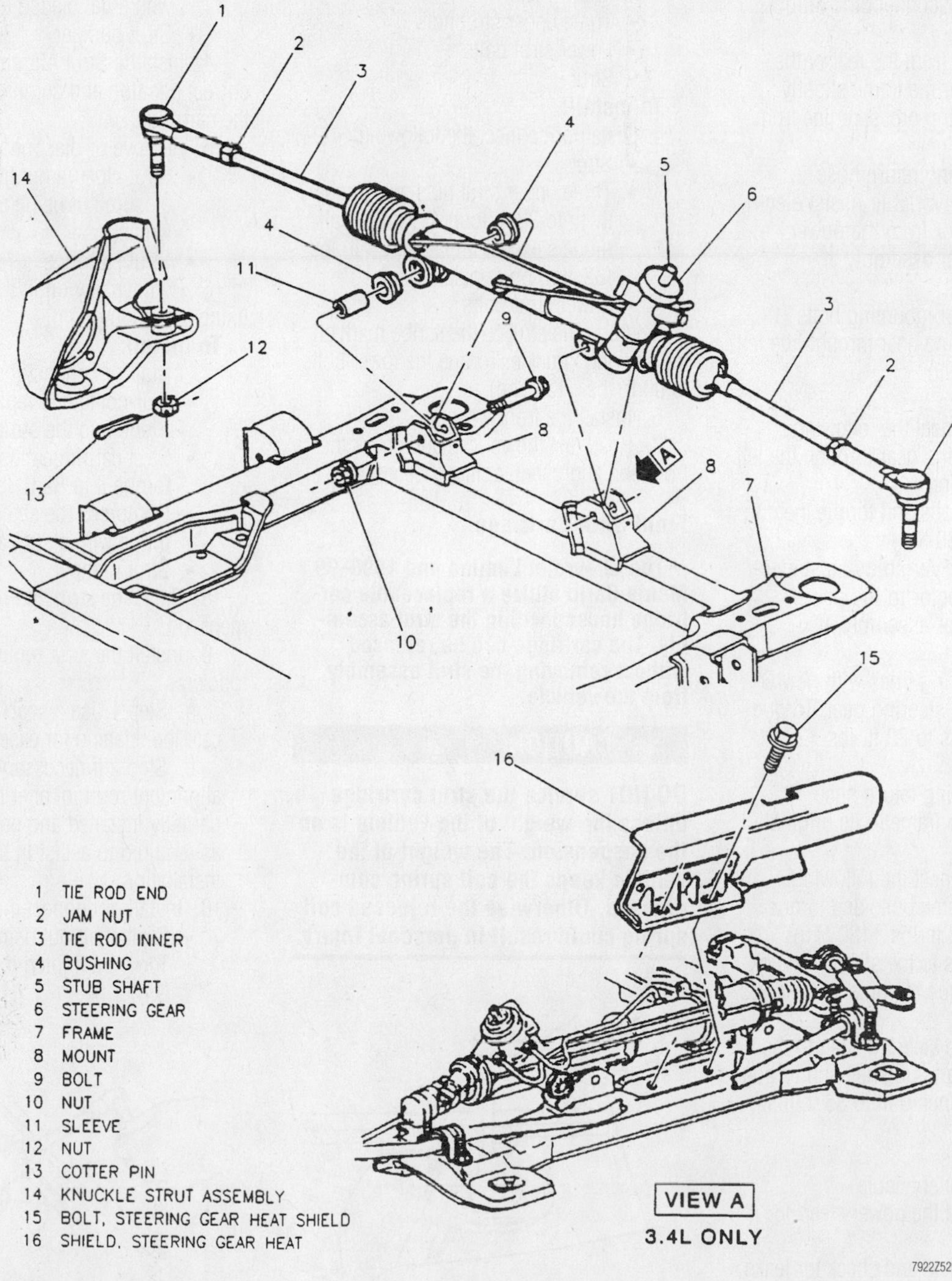

Common rack and pinion steering gear mounting

- Intermediate steering shaft lower pinch bolt from the steering gear stub shaft
- Intermediate steering shaft
- Both tie rod ends from the steering knuckles

4. Support the frame at the center rear.

➡**DO NOT lower the frame too far. Engine components near the firewall may be damaged.**

5. Remove or disconnect the following:
   - Frame bolts from the rear of the frame. Lower the frame slightly
   - Power steering pressure line from the steering gear
   - Power steering return hose
   - Magnasteer Variable Assist electrical connector from the power steering gear assembly, if equipped
   - Steering gear mounting bolts
   - Power steering gear through the left wheel opening

To install:

6. Install or connect the following:
   - Power steering gear through the left wheel opening
   - Mounting bolts and torque them to 59 ft. lbs. (80 Nm)
   - Magnasteer Variable Assist electrical connector to the power steering gear assembly, if equipped
   - Power steering lines with new O-rings to the steering gear. Torque the fasteners to 20 ft. lbs. (27 Nm).
   - Power steering return hose
7. Raise the rear frame to its original position.
8. Install or connect the following:
   - New rear frame bolts and torque them to 133 ft. lbs. (180 Nm)
   - Tie rod ends to the steering knuckles and torque the nuts to 22 ft. lbs. (30 Nm)
   - Intermediate steering shaft to the steering gear stub shaft and torque the lower pinch bolt to 35 ft. lbs. (48 Nm)
   - Front wheels
   - Negative battery cable
9. Fill and bleed the power steering system.
10. Start the vehicle and check for leaks, repair if necessary.
11. Check the front end alignment and adjust as needed.

Strut

REMOVAL & INSTALLATION

Front Strut

1. Before servicing the vehicle, refer to the precautions in the beginning of this section.
2. Scribe the strut to the steering knuckle for proper installation.
3. Remove or disconnect the following:
   - Front wheel
   - Three upper strut nuts
   - Lower strut bolts
   - Strut

To install:

4. Install or connect the following:
   - Strut
   - Three upper strut nuts and torque them to 30 ft. lbs. (41 Nm) on all models except Impala or 24 ft. lbs. (33 Nm) on Impala models.
   - Lower strut bolts
5. Align the strut to the scribe mark on the steering knuckle. Torque the lower bolts to 90 ft. lbs. (123 Nm).
6. Install the front wheel.
7. Road test the vehicle and check the front end alignment, adjust if necessary.

Front Strut Cartridge

➡**The Chevrolet Lumina and 1998–99 Monte Carlo utilize a replaceable cartridge housed within the strut assembly. The cartridge can be replaced without removing the strut assembly from the vehicle.**

✱✱ CAUTION

DO NOT service the strut cartridge unless the weight of the vehicle is on the suspension. The weight of the vehicle keeps the coil spring compressed. Otherwise the released coil spring could result in personal injury.

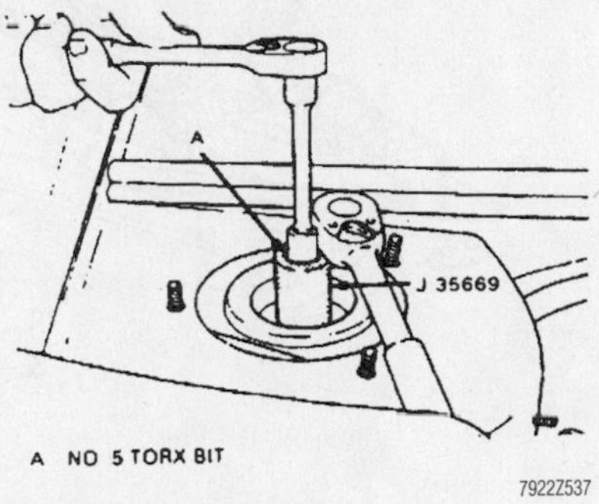

Strut shaft nut removal

1. Before servicing the vehicle, refer to the precautions in the beginning of this section.
2. Scribe the strut cover to body to assure proper camber adjustment.
3. Remove or disconnect the following:
   - Wheel
   - Strut cover by removing the three cover nuts
   - Strut shaft nut by using a No. 50 Torx® bit
   - Strut mount insulator by prying with a flat bladed tool
   - Strut bumper
4. Install a Strut Alignment tool in the correct position and compress the strut into the cartridge.
5. Remove or disconnect the following:
   - Strut closure nut by unscrewing the closure nut using a Strut Cap Nut wrench
   - Strut cartridge
6. Remove any oil in the strut housing using a suction pump.

To install:

7. Install or connect the following:
   - Self-contained replacement cartridge into the strut housing
   - Strut cartridge closure nut and torque it to 82 ft. lbs. (110 Nm)
   - Compress the shaft into the cartridge with a strut alignment tool
   - Strut bumper
8. Raise the strut and remove the alignment rod.
9. Install the strut mount insulator as follows:
   a. Step 1:Use a soap solution to lubricate the bushing for ease of installation.
   b. Step 2:If necessary, install an alignment rod tool after the bushing is partially installed and position the strut as required to assist in the bushing installation.
10. Install or connect the following:
    - Strut shaft nut using the No. 50 Torx® bit tighten it to 59 ft. lbs. (80 Nm)

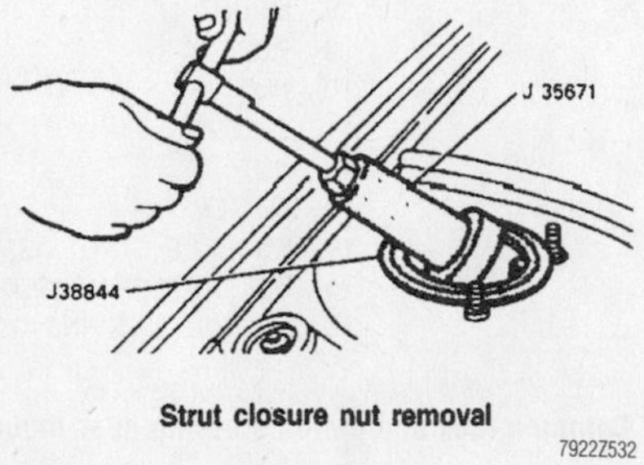

Strut closure nut removal

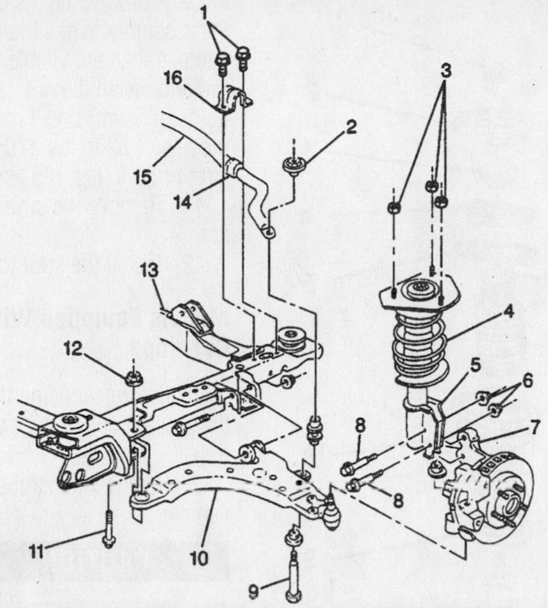

(1) Front Stabilizer Shaft Insulator Clamp Bolt/screw
(2) Front Stabilizer Shaft Link Nut
(3) Front Suspension Strut Mount Nut
(4) Front Suspension Spring
(5) Front Suspension Strut
(6) Strut To Knuckle Nut
(7) Front Steering Knuckle
(8) Strut To Knuckle Bolt/screw
(9) Front Stabilizer Shaft Link]
(10) Front Lower Control Arm
(11) Front Lower Control Arm Bolt/screw
(12) Front Lower Cotrol Arm Nut
(13) Frame
(14) Front Stabilizer Shaft Insulator
(15) Front Stabilizer Shaft
(16) Front Stabilizer Shaft Clamp

7922Z530

Exploded view of the front suspension without replaceable strut cartridge

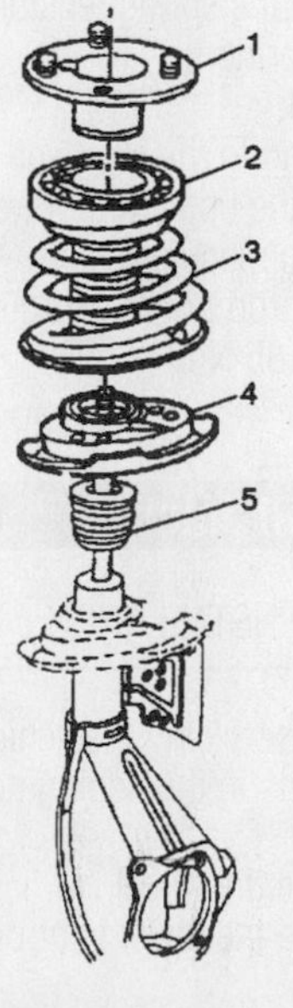

(1) Front Suspension Strut Mount Retainer
(2) Front Spring Upper Insulator
(3) Front Spring
(4) Front Spring Seat
(5) Front Suspension Strut Bumper

7922Z531

Knuckle and strut assembly with replaceable strut cartridge

- Strut cover while aligning scribe marks and torque the bolts to 24 ft. lbs. (33 Nm)

11. A 4-wheel alignment is recommended after any steering/suspension repairs are performed.

Rear Strut

EXCEPT LUMINA

1. Before servicing the vehicle, refer to the precautions in the beginning of this section.
2. Remove or disconnect the following:
   - Negative battery cable
   - Three strut to body nuts
   - Rear tire and wheel
   - Stabilizer shaft link from the strut
3. Scribe the strut to the knuckle.

➡The knuckle must be retained after the strut to knuckle bolts have been removed. Damage may occur to the ball joint or drive axle if the knuckle is not retained.

4. Remove or disconnect the following:
   - Strut to knuckle bolts
   - Strut

To install:

5. Install or connect the following:
   - Strut
   - Strut to knuckle bolts and torque the bolts to 90 ft. lbs. (122 Nm)
   - Stabilizer shaft link to the strut
   - Rear tire and wheel
   - Three strut to body mount nuts and torque them to 30 ft. lbs. (41 Nm)
6. Road test the vehicle and adjust the rear wheel alignment if needed.

LUMINA

1. Before servicing the vehicle, refer to the precautions in the beginning of this section.
2. Remove or disconnect the following:
   - Negative battery cable
   - Wheel
   - Brake hose bracket at the strut
   - Strut mount to body nuts
   - Strut/stabilizer shaft bracket from the knuckle
   - Strut

To install:

3. Install or connect the following:
   - Strut mount-to-body nuts and tighten to 37 ft. lbs. (50 Nm)
   - Strut/stabilizer shaft bracket to the knuckle
   - Strut-to-knuckle bolts and nuts. Tighten the nuts to 82 ft. lbs. (112 Nm).
   - Brake hose bracket
   - Wheel

Coil Spring

REMOVAL & INSTALLATION

Except Models Equipped With A Front Strut Cartridge

1. Before servicing the vehicle, refer to the precautions in the beginning of this section.
2. Remove the strut from the vehicle.
3. Mount the strut assembly into a strut compressor. Note that the strut compressor has strut mounting holes drilled for specific vehicle lines.
4. Compress the strut approximately ½ its height after initial contact with the top cap.

⁂ WARNING

Never bottom the spring or damper rod!

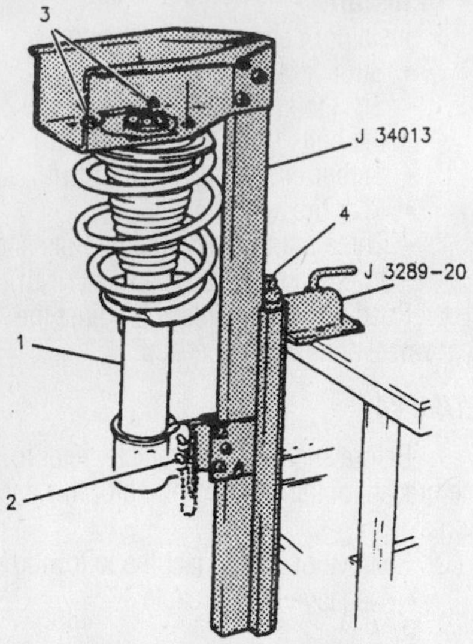

1 STRUT ASSEMBLY
2 INSTALL LOCKING PINS THROUGH STRUT ASSEMBLY
3 TIGHTEN NUTS UNTIL FLUSH WITH STRUT COMPRESSOR
4 COMPRESSOR FORCING SCREW

7922YG24

View of a typical strut assembly mounted in a compressor

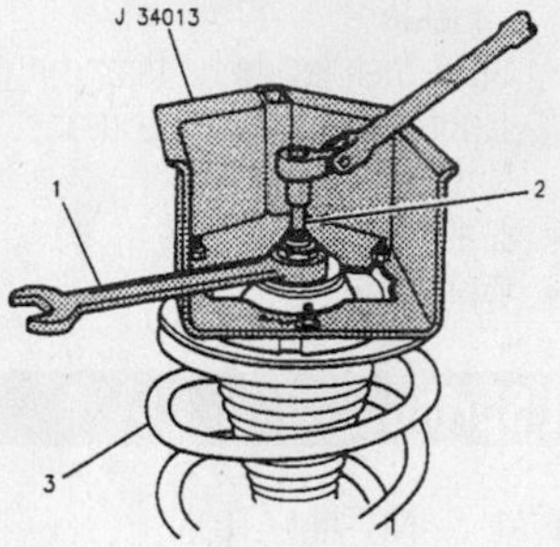

1 WRENCH
2 SOCKET
3 STRUT ASSEMBLY

7922YG25

Use a socket and a wrench to remove the damper shaft nut spring cap while compressing the spring

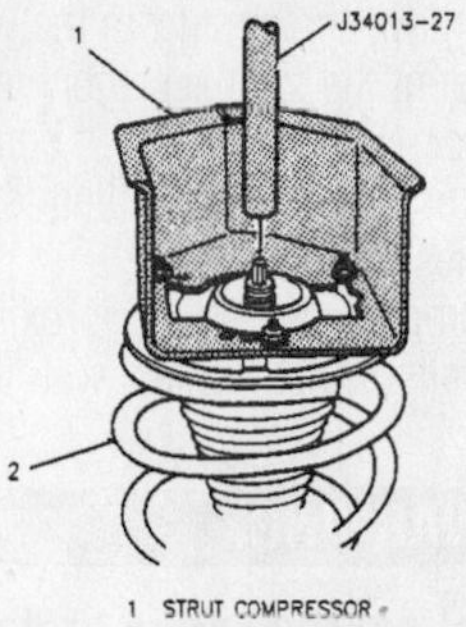

1 STRUT COMPRESSOR
2 STRUT ASSEMBLY

7922YG26

Install the rod to guide the damper shaft straight down through the spring cap while compressing the spring

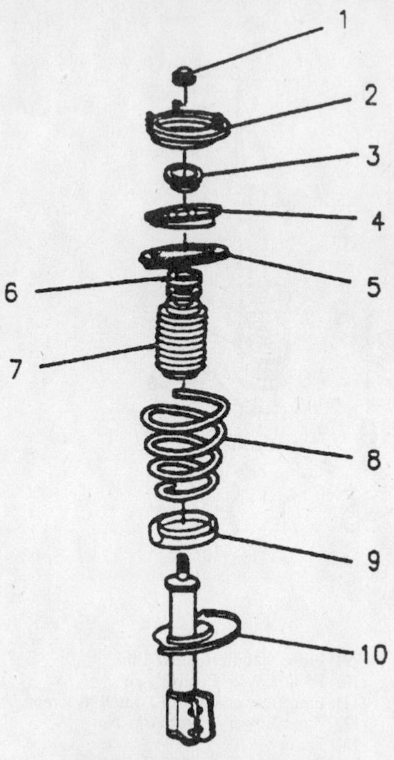

1 STRUT MOUNT NUT
2 STRUT MOUNT
3 RATE WASHER
4 SPRING SEAT
5 SPRING UPPER INSULATOR
6 JOUNCE BUMPER
7 STRUT DUST SHIELD
8 SPRING
9 SPRING LOWER INSULATOR
10 STRUT

7922YG27

Exploded view of the front strut assembly

5. Remove the nut from the strut damper shaft and place an alignment/guiding rod on top of the damper shaft. Use the rod to guide the damper shaft straight down through the spring cap while compressing the spring. Remove the components.

To install:

6. Mount the strut assembly in the strut compressor, using the bottom locking pin only.
7. Install the spring over the damper and swing the assembly up so the upper locking pin can be installed.
8. Install all shields, bumpers and insulators on the spring seat.
9. Install the spring seat on top of the spring. The spring seat flat should be facing the same direction as the centerline of the strut assembly spindle.
10. Install the guiding rod and turn the forcing screw while the guiding rod centers the assembly. When the threads on the damper shaft are visible, remove the guiding rod and install the nut. tighten the nut to 63 ft. lbs. (85 Nm) on all models except Impala models or 52 ft. lbs. (70 Nm), for a front strut or 55 ft. lbs. (75 Nm) (for a rear strut)
11. Remove the strut from the compressor.
12. Install the strut to the vehicle.

Models Equipped With A Front Strut Cartridge

1. Before servicing the vehicle, refer to the precautions in the beginning of this section.
2. Remove the strut assembly from the vehicle.

**** CAUTION**

Do not over compress the coil spring. Only compress the spring until it comes away from the seat.

3. Compress the coil spring approximately ½inch (13mm) using a spring compressor.
4. Remove or disconnect the following:
   - Strut rod nut with the proper socket while not allowing the rod to rod to rotate. Discard the nut
5. Relieve the spring tension and remove the spring.

To install:

- Spring to the strut and make certain that the upper and lower seats are positioned correctly
- Strut rod nut and torque to 63 ft. lbs. (85 Nm)
- Strut

Lower Ball Joint

REMOVAL & INSTALLATION

1. Before servicing the vehicle, refer to the precautions in the beginning of this section.
2. Remove the wheel.
3. Remove the lower control arm from the vehicle.
4. Remove the ball joint from the lower control arm by drilling out the 4 rivets retaining the ball joint to the control arm. Use an ⅛ in. drill bit to make a pilot hole through the rivets. Finish drilling the rivets using a ½ in. drill bit.
5. Remove the ball joint.

To install:

6. Install or connect the following:

- Ball joint to the control arm
- Bolts with the heads facing down and torque them to 50 ft. lbs. (68 Nm)
- Lower control arm to the vehicle
- Wheel

➡A 4-wheel alignment is recommended after any steering/suspension repairs are performed.

Lower Control Arm

REMOVAL & INSTALLATION

1. Before servicing the vehicle, refer to the precautions in the beginning of this section.
2. Remove or disconnect the following:
   - Front wheel
   - Antilock Brake System (ABS) wheel speed sensor connector and jumper harness from the retainer
   - Stabilizer shaft link
   - Cotter pin from the ball stud and loosen the nut
3. Install a ball joint removal tool over the ball joint and lower control arm. Rotate the ball stud nut counterclockwise to separate the ball stud from the steering knuckle.
4. Remove the lower control arm.

To install:

5. Install the lower control arm and bolts. Do not tighten the nuts at this time.

➡Align the ball stud cotter pin hole parallel to the knuckle to ease the cotter pin installation.

6. Install or connect the following:
   - Ball stud to the knuckle and torque the nut to 40 ft. lbs. (55 Nm) on Monte Carlo, Grand Prix, Century and Regal models. On Impala models, tighten to 15 ft. lbs. (20 Nm) plus an additional 120 degrees on Impala models. On Lumina models tighten to 63 ft. lbs. (85 Nm).
   - New cotter pin and bend the ends. Make certain the ends do not make contact with the ABS wheel speed sensor
   - Stabilizer shaft link
   - ABS wheel speed sensor wire harness to the retainer clips
   - ABS wheel speed sensor connector
   - Lower control arm nuts and torque them to 83 ft. lbs. (113 Nm) on all models except Lumina. On Lumina models tighten the control arm-to-frame nuts to 52 ft. lbs. (70 Nm).
   - Front wheel

CONTROL ARM BUSHING REPLACEMENT

1. Before servicing the vehicle, refer to the precautions in the beginning of this section.
2. Remove or disconnect the following:
   - Front wheel
   - Lower control arm
3. Lubricate the threads of a bushing driver with high pressure lubricant.
4. Assemble the tool onto the control arm bushing.
5. Tighten the tool until the bushing is pressed from the control arm.
6. Disassemble the tools.

To install:

7. Lubricate the new bushing outer casing.
8. Install a new bushing into the control arm.
9. Assemble a bushing driver/installer set onto the control arm.

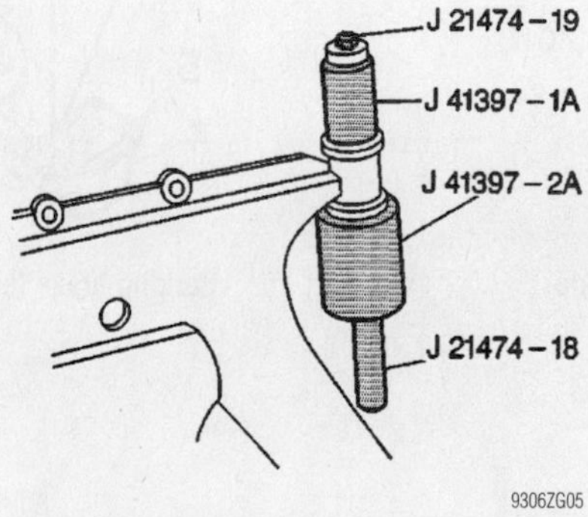

Removing the bushing from the lower control arm

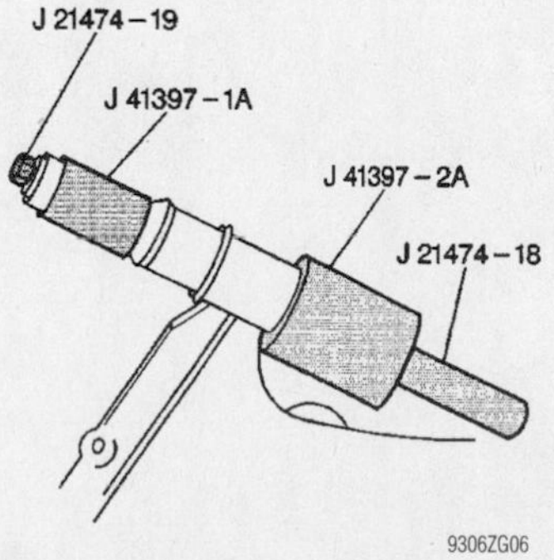

Installing the bushing to the lower control arm

10. Tighten the screw until the bushing is pressed into the control arm.
11. Disassemble the tools.
12. Install or connect the following:
    - Lower control arm
    - Front wheel

Wheel Bearings

ADJUSTMENT

The wheel bearings are not adjustable. If a wheel bearing is out of specification, it must be replaced. Using a dial indicator, check for looseness. If it exceeds 0.005 in. (0.127mm) on drum or disc brakes the bearing wear is excessive and the hub and bearing should be replaced.

REMOVAL & INSTALLATION

Front

1. Before servicing the vehicle, refer to the precautions in the beginning of this section.
2. Remove or disconnect the following:
   - Front wheel
   - Wheel Speed Sensor (WSS) electrical connector
   - Brake caliper and bracket
   - Rotor
   - Driveshaft nut
3. Install a front hub removal tool to the wheel bearing/hub assembly with three wheel nuts. Use the tool to push the driveshaft out of the wheel bearing/hub.
4. Remove the wheel bearing/hub assembly and discard the bolts.

To install:

5. Install or connect the following:
   - Wheel bearing/hub assembly using new bolts and torque them to 96 ft. lbs. (130 Nm)
   - New drive shaft nut and torque it to 118 ft. lbs. (160 Nm) on all except Impala models or 159 ft. lbs. (215 Nm) on Impala models
   - Brake rotor
   - Front caliper with the bracket and torque the bracket bolts to 137 ft. lbs. (185 Nm) and the caliper mounting bolts to 63 ft. lbs. (85 Nm)
   - WSS electrical connector
   - Front wheel

Rear

The rear wheel bearing/hub is integrated into one unit. The unit is non-serviceable. If the hub or bearing is damaged, the complete hub and bearing unit must be replaced.

1. Before servicing the vehicle, refer to the precautions in the beginning of this section.
2. Remove or disconnect the following:
   - Rear wheel
   - Brake drum, if equipped
   - Rear caliper and bracket, if equipped
   - Brake rotor, if equipped
   - Antilock Brake System (ABS) Wheel Speed Sensor (WSS) electrical connector
   - Rear wheel hub to knuckle bolts
   - Parking brake lever bracket and parking brake actuator
   - Rear wheel hub from the knuckle

To install:

3. Install or connect the following:
   - Parking Brake lever bracket and actuator
   - Hub and bearing assembly and torque the bolts to 55 ft. lbs. (75 Nm)
   - WSS electrical connector
   - Brake rotor, if equipped
   - Brake caliper with the bracket and torque the bracket bolts to 92 ft. lbs. (125 Nm) and the caliper bolts to 32 ft. lbs. (44 Nm) , if equipped
   - Brake drum, if equipped
   - Rear wheel
4. A 4 wheel alignment is recommended after any steering/suspension repairs have been performed.

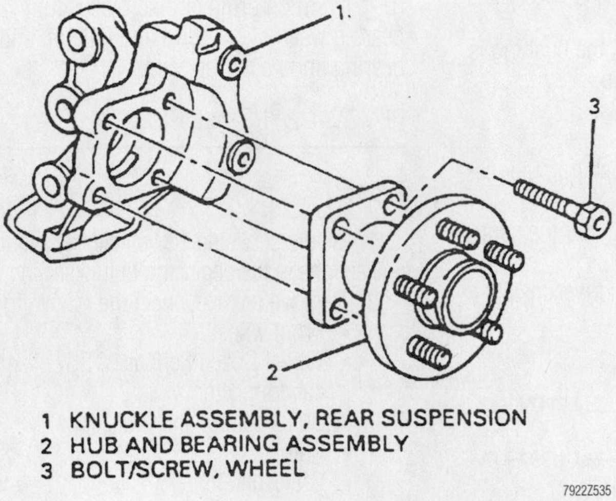

The rear hub/bearing assembly is bolted to the knuckle

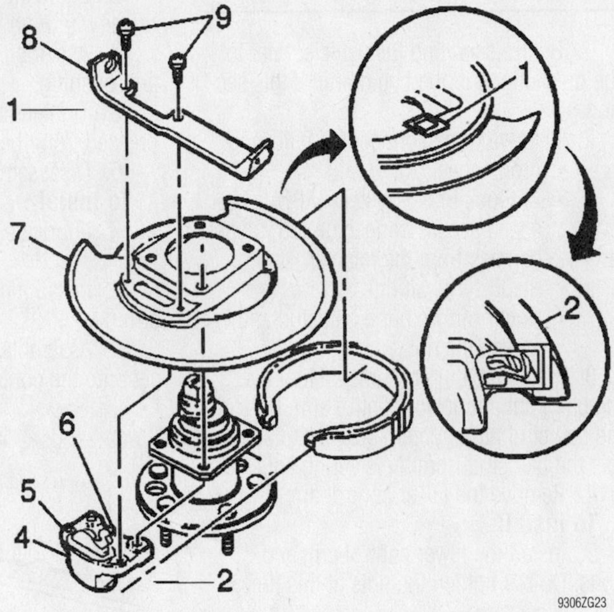

Parking brake lever bracket (1) and parking brake actuator (2)

GENERAL MOTORS CORPORATION—Y-BODY

30

1998–01
Chevrolet-Corvette

PRECAUTIONS

Before servicing any vehicle, please be sure to read all of the following precautions, which deal with personal safety, prevention of component damage, and important points to take into consideration when servicing a motor vehicle:

- Never open, service or drain the radiator or cooling system when the engine is hot; serious burns can occur from the steam and hot coolant.
- Observe all applicable safety precautions when working around fuel. Whenever servicing the fuel system, always work in a well-ventilated area. Do not allow fuel spray or vapors to come in contact with a spark, open flame or excessive heat (a hot drop light, for example). Keep a dry chemical fire extinguisher near the work area. Always keep fuel in a container specifically designed for fuel storage; also, always properly seal fuel containers to avoid the possibility of fire or explosion. Refer to the additional fuel system precautions later in this section.
- Fuel injection systems often remain pressurized, even after the engine has been turned **OFF**. The fuel system pressure must be relieved before disconnecting any fuel lines. Failure to do so may result in fire and/or personal injury.
- Brake fluid often contains polyglycol ethers and polyglycols. Avoid contact with the eyes and wash your hands thoroughly after handling brake fluid. If you do get brake fluid in your eyes, flush your eyes with clean, running water for 15 minutes. If eye irritation persists, or if you have taken brake fluid internally, seek medical assistance IMMEDIATELY.
- The EPA warns that prolonged contact with used engine oil may cause a number of skin disorders, including cancer! You should make every effort to minimize your exposure to used engine oil. Protective gloves should be worn when changing oil. Wash your hands and any other exposed skin areas as soon as possible after exposure to used engine oil. Soap and water, or waterless hand cleaner should be used.
- All new vehicles are now equipped with an air bag system. The system must be disabled before performing service on or around system components, steering column, instrument panel components, wiring and sensors. Failure to follow safety and disabling procedures could result in accidental air bag deployment, possible personal injury and unnecessary system repairs.
- Always wear safety goggles when working with, or around, the air bag system. When carrying a non-deployed air bag, be sure the bag and trim cover are pointed away from your body. When placing a non-deployed air bag on a work surface, always face the bag and trim cover upward, away from the surface. This will reduce the motion of the module if it is accidentally deployed. Refer to the additional air bag system precautions later in this section.
- Clean, high quality brake fluid from a sealed container is essential to the safe and proper operation of the brake system. You should always buy the correct type of brake fluid for your vehicle. If the brake fluid becomes contaminated, completely flush the system with new fluid. Never reuse any brake fluid. Any brake fluid that is removed from the system should be discarded. Also, do not allow any brake fluid to come in contact with a painted surface; it will damage the paint.
- Never operate the engine without the proper amount and type of engine oil; doing so WILL result in severe engine damage.
- Timing belt maintenance is extremely important! Many models utilize an interference-type, non-freewheeling engine. If the timing belt breaks, the valves in the cylinder head may strike the pistons, causing potentially serious (also time-consuming and expensive) engine damage. Refer to the maintenance interval charts in the front of this manual for the recommended replacement interval for the timing belt, and to the timing belt section for belt replacement and inspection.
- Disconnecting the negative battery cable on some vehicles may interfere with the functions of the on-board computer system(s) and may require the computer to undergo a relearning process once the negative battery cable is reconnected.
- When servicing drum brakes, only disassemble and assemble one side at a time, leaving the remaining side intact for reference.

ENGINE REPAIR

Distributor

All 5.7L (VIN G) engines use a Direct Ignition System.

Alternator

REMOVAL

1. Before servicing the vehicle, refer to the precautions in the beginning of this section.
2. Remove or disconnect the following:
   - Negative battery cable
   - Regulator connector and battery to alternator terminal from the rear of the alternator
   - Accessory drive belt
   - Rear mounting bracket
   - Alternator mounting bolts
   - Alternator

To install:

3. Install or connect the following:
   - Alternator
   - Alternator mounting bolts and torque them to 37 ft. lbs. (50 Nm)
   - Alternator rear mounting bracket and torque the bolts to 37 ft. lbs. (50 Nm)
   - Drive belt
   - Regulator connector and battery to

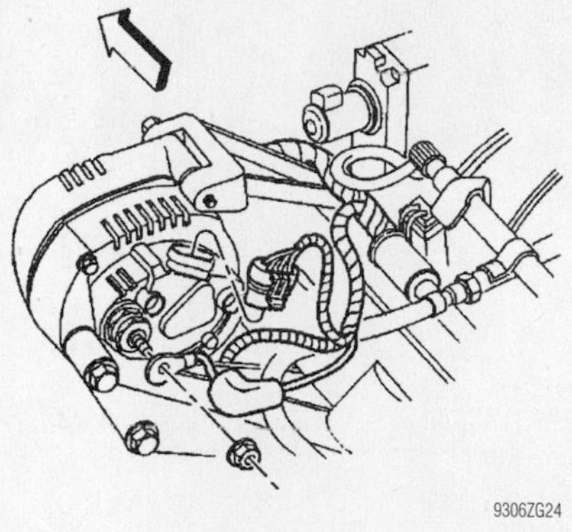

Exploded view of the alternator—5.7L engine

alternator terminals and torque the alternator terminal nut to 10 inch lbs. (13 Nm)
- Negative battery cable

Engine Assembly

REMOVAL & INSTALLATION

1. Before servicing the vehicle, refer to the precautions in the beginning of this section.

The following tools will be required in addition to the basic hand tools:
- Transverse spring compressor and adapters
- Ball joint separator
- Engine support table
- Driveshaft support strap
- Fuel pressure gauge

2. Relieve the fuel system pressure.
3. Drain the cooling system.
4. Drain the engine oil.
5. Evacuate the A/C system.
6. Remove or disconnect the following:

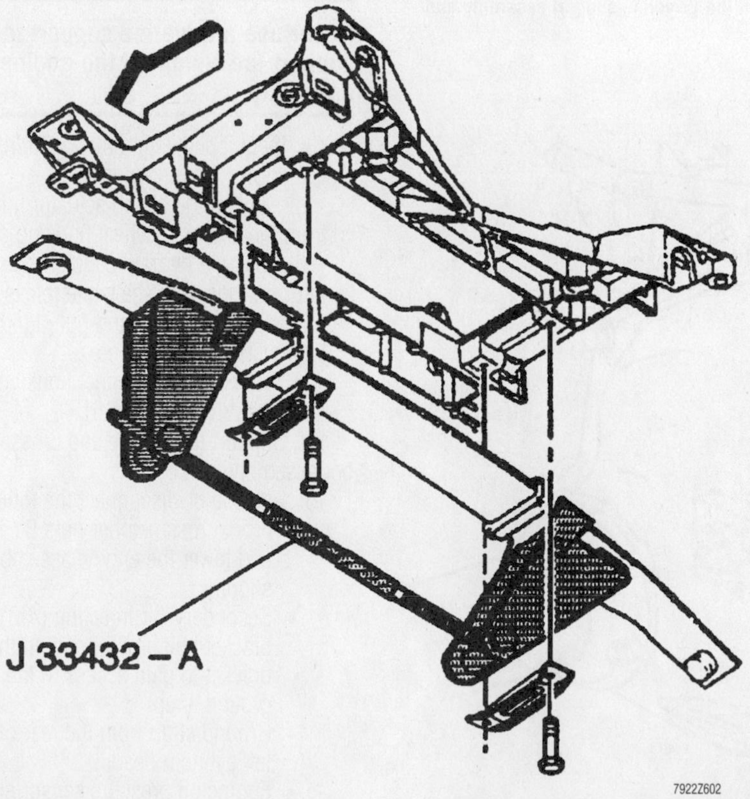

Use a compressor such as J 33432-A to compress the front transverse spring—5.7L (VIN G) engine

- Intake Air Temperature (IAT) sensor electrical connector
- Mass Air Flow (MAF) sensor electrical connector
- Air intake duct and air cleaner
- Upper radiator support
- Radiator
- Electronic Brake Control Traction Module/Brake Pressure Modulator Valve (EBTCM/BPMV) and bracket
- Brake pipes
- Accessory drive belt
- Fuel line from the connector at the front of the dash

➡Cap and plug all openings for the fuel system to prevent contaminants from entering the system.

- Fuel rail covers
- Fuel line at the fuel rail
- Radiator hoses from the water pump
- Heater hoses from the water pump
- Fuel injector electrical connectors
- Ignition coil main harness connector
- Evaporative Emission (EVAP) solenoid electrical connector
- Electric throttle motor electrical connector
- Throttle Position Sensor (TPS) electrical connector
- Engine Coolant Temperature (ECT) sensor electrical connector
- A/C compressor electrical connector
- Alternator electrical connectors
- Alternator
- Brake booster vacuum hose
- Intermediate shaft to steering gear bolt
- Intermediate shaft from the steering gear
- Secondary Air Injection (AIR) hose from the left exhaust manifold
- Both front wheels
- Heated Oxygen (HO_2S) sensor connectors from the intermediate exhaust pipes
- Intermediate exhaust pipes
- Close-out panel
- Starter motor electrical connectors
- Starter motor
- Oil level sensor connector
- Crankshaft Position (CKP) sensor electrical connector
- A/C compressor hose assembly
- Engine oil temperature sensor electrical connector
- Left Heated Oxygen (HO_2S) sensor
- Front stabilizer shaft
- Connectors from the cooling fans
- Cooling fan assembly
- Tie rod ends from the steering knuckles
- Antilock Brake System (ABS) electrical connectors from the crossmember, if equipped
- Electronic Variable Orifice (EVO) connector clips from the crossmember, if equipped
- Real Time Damping (RTD) connector clips from the crossmember, if equipped
- Shock absorber lower mounting bolts
- Front transverse leaf spring
- Automatic transmission cooler pipes at the flywheel housing junction
- Automatic transmission cooler pipe clamps from the engine oil pan
- Automatic transmission cooler pipe from the radiator

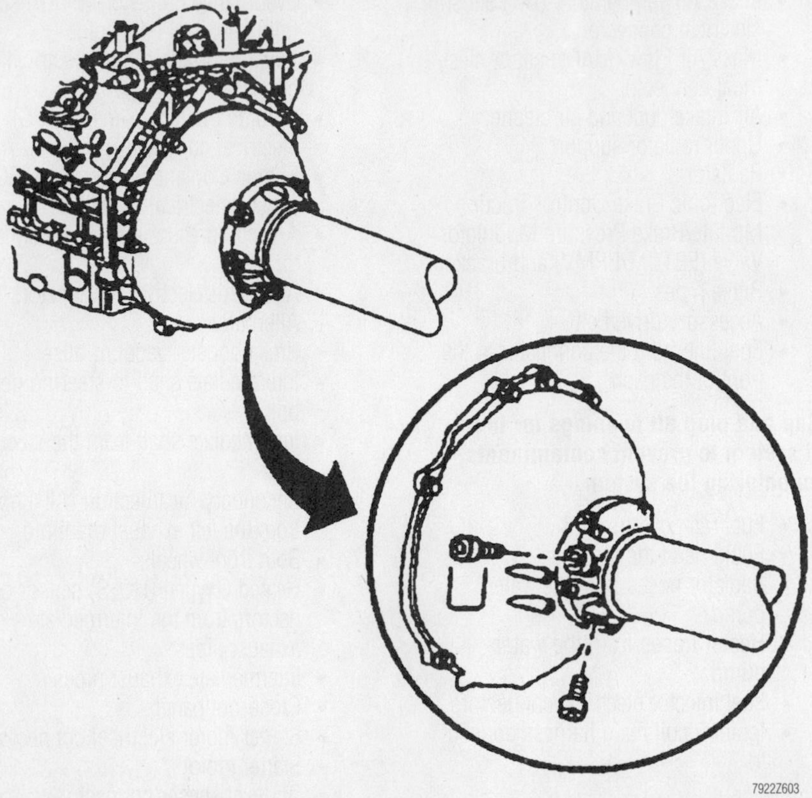

On automatic transmission vehicles, remove the plugs from the driveline support assembly and install 2 M10 x 1.5 bolts into the plug holes

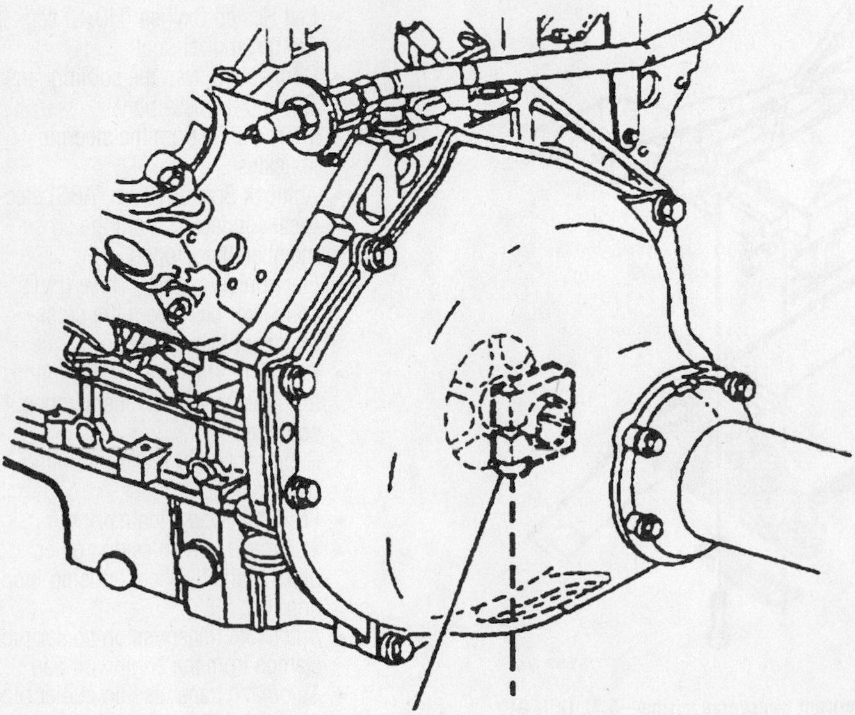

Loosen the bolt on the flywheel hub collar after turning it for access

CAUTION

Failure to use the minimum fastener length specified will prevent the proper retention of the propeller shaft during assembly.

7. For vehicles with an automatic transmission, perform the following steps:
   a. Remove the two plug bolts in the driveline support assembly.
   b. Install an M10 x 55 bolt, or longer, in each plug location. Torque the bearing support bolts to 26 ft. lbs. (35 Nm).
8. On vehicles with automatic transmission, remove the bell housing inspection cover, then turn the flywheel hub collar to access the bolt and loosen it.
9. Remove the engine flywheel housing access plug.
10. Rotate the automatic transmission collar so that the bolt is facing down.
11. Loosen the bolt and unclip the wire harness from the engine and reposition it to the driveline.
12. Install a driveline support tool to the close out panel flange.

CAUTION

Never use a driveline support tool to support the weight of the engine assembly.

13. If equipped with a manual transmission, perform the following steps:
   a. Unclip the clutch actuator hose from the engine flywheel housing clip.
   b. Using a hydraulic clutch line separator tool, depress the white release ring on the actuator hose and pull lightly on the master cylinder hose.
   c. Remove the flywheel housing bolts from the driveline support.
14. Support the engine and crossmember assembly.
15. Remove or disconnect the following:
   - Front crossmember nuts BY HAND and lower the engine assembly slightly
   - Secondary Air Injection (AIR) tube bracket bolt and reposition the bracket to gain access to the ground strap
   - Ground strap from the rear of the left cylinder head
   - Engine oil pressure sensor electrical connector
   - Camshaft Position (CMP) sensor electrical connector
   - Manifold Absolute Pressure (MAP) sensor electrical connector

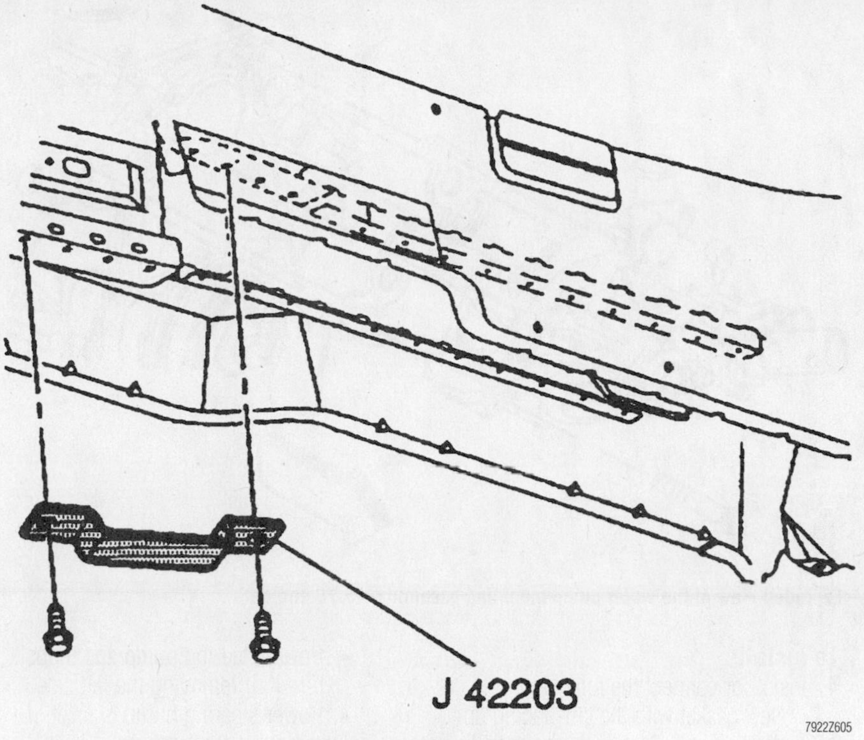

Install a Driveline Support tool J 42203 to the under cover flange

- Knock Sensor (KS) electrical connector
- Front driveline support assembly bolts
- Engine from the driveline, by placing a flat blade tool between edge of the driveline and the flywheel housing

16. Pull the engine away from the propeller shaft.
17. Raise the vehicle off of the engine and crossmember.
18. Remove the engine mount-to-crossmember nuts.
19. Remove the engine from the crossmember.

To install:

20. Install the engine on the crossmember and torque the nuts to 40 ft. lbs. (54 Nm).
21. Lower the vehicle onto the engine/crossmember assembly.
22. Install or connect the following:
    - Power steering pump and reservoir on the engine and torque the bolts to 18 ft. lbs. (25 Nm)
    - Ground strap to the rear of the left cylinder head and torque the bolt to 24 ft. lbs. (32 Nm)
    - Engine oil pressure sensor electrical connector
    - CMP sensor electrical connector
    - MAP sensor electrical connector
    - KS electrical connector
    - Install or connect the following:
    - Front driveline support bolts and torque them to 37 ft. lbs. (50 Nm)
    - Air tube and bracket and torque the bolt to 37 ft. lbs. (50 Nm)
    - New crossmember nuts and torque them to 81 ft. lbs. (110 Nm)
    - Flywheel housing bolts and torque them to 37 ft. lbs. (50 Nm)
    - Clutch actuator hose to the flywheel housing (if equipped)
    - Master cylinder hose to the clutch actuator hose (if equipped)
    - Wire harness to the rear of the engine
23. Remove the driveline support tool.
24. Remove the M10 x 55mm bolts from the driveline support assembly.
    - Two plugs in the driveline support assembly after removing the M10 x 55 bolts and torque them to 37 ft. lbs. (50 Nm)
    - Automatic transmission oil cooler pipes at the flywheel housing junction and torque them to 20 ft. lbs. (27 Nm) (if equipped)
    - Automatic transmission cooler pipes to the front and rear of the oil pan
    - Front transverse spring and torque the bolts to 46 ft. lbs. (62 Nm)
    - Shock absorber lower mounting nuts and torque them to 21 ft. lbs. (28 Nm)
    - Front stabilizer shaft and torque the bolts to 53 ft. lbs. (72 Nm)
    - ABS electrical connector
    - EVO electrical connector
    - RTD electrical connector
    - Tie rods to the steering knuckle and torque the nuts to 33 ft. lbs. (45 Nm).
    - Radiator
    - Cooling fan assembly to the radiator and attach the electrical connectors to the fans
    - Ground wires to the left side of the engine
    - HO_2S sensor connectors
    - Engine oil temperature sensor connector
    - A/C compressor hoses to the compressor
    - A/C compressor line retaining bolt and torque it to 26 ft. lbs. (35 Nm)
    - CKP sensor electrical connector
    - Oil level sensor electrical connector
    - Wire harness ground wires to the right side of the engine
    - Starter motor and torque the bolts to 37 ft. lbs. (50 Nm)
    - Driveline close out panel and torque the bolts to 106 inch lbs. (12 Nm)
    - Intermediate exhaust pipe and hangers and torque the bolts to 37 ft. lbs. (50 Nm) and the nuts to 15 ft. lbs. (20 Nm)
    - Front wheels
    - Automatic transmission cooler pipes to the radiator (if equipped) and torque the fittings to 26 ft. lbs. (35 Nm)
    - AIR hose to the left exhaust manifold
    - Intermediate steering shaft to the steering gear and torque the bolt to 35 ft. lbs. (48 Nm)
    - Brake booster vacuum hose
    - Alternator bracket and torque the bolts to 37 ft. lbs. (50 Nm)
    - Alternator and torque the bolts to 37 ft. lbs. (50 Nm)
    - Alternator connectors and torque the nuts to 10 ft. lbs. (13 Nm)
    - Fuel injector electrical connectors
    - Ignition coil main electrical connectors

- EVAP solenoid electrical connectors
- Throttle motor electrical connector
- ECT sensor electrical connector
- TPS electrical connector
- A/C compressor electrical connector
- Heater hoses to the water pump
- Radiator hoses to the water pump
- Fuel line to the fuel rail
- Fuel rail covers
- Drive belt
- EBTCM/BPMV bracket and torque the bolts to 20 ft. lbs. (27 Nm)
- EBTCM/BPMV to the bracket and torque the nuts to 89 inch lbs. (10 Nm)
- Upper radiator hose
- Upper radiator support
- Air cleaner and air intake duct
- IAT sensor electrical connector
- MAF sensor electrical connector
- Negative battery cable

25. Fill the engine with new oil.
26. Fill the cooling system.
27. Evacuate and recharge the A/C system.
28. On vehicles with an automatic transmission, start the engine and allow it to idle for 10 minutes. Torque the hub collar bolt to 96 ft. lbs. (130 Nm) and install the flywheel inspection plug.
29. A wheel alignment is recommended when the engine and crossmember have been removed from the vehicle.
30. Check the vehicle for leaks and repair if necessary.

Water Pump

REMOVAL & INSTALLATION

1. Before servicing the vehicle, refer to the precautions in the beginning of this section.
2. Drain the cooling system.
3. Remove or disconnect the following:
   - Negative battery cable
   - Intake Air Temperature (IAT) sensor electrical connector
   - Mass Air Flow (MAF) sensor electrical connector
   - Air intake duct
   - Accessory drive belt
   - Inlet and outlet hoses from the water pump
   - Heater hoses from the water pump
   - Water pump pulley bolts and pulley (if equipped)
   - Six water pump retaining bolts
   - Water pump

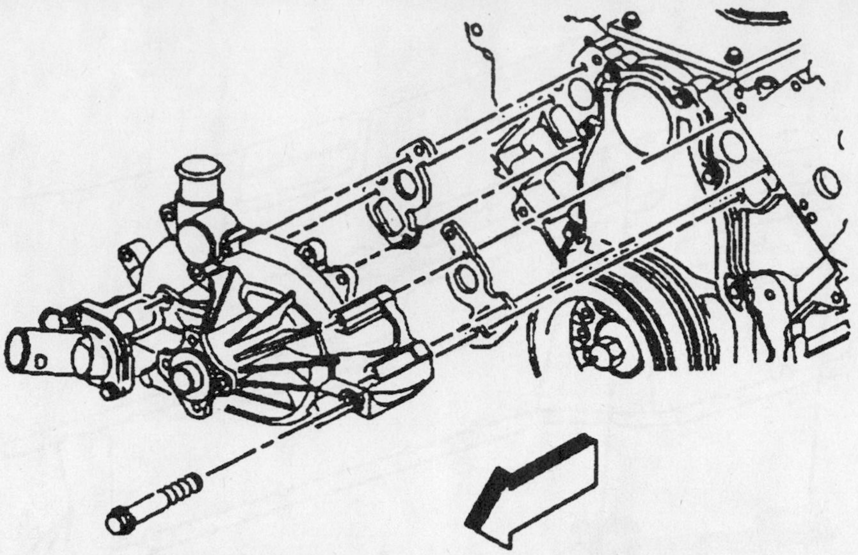

Exploded view of the water pump mounting assembly—5.7L engine

To install:

4. Install or connect the following:
   - New gasket with the tabs facing up
   - Water pump. Torque the bolts to 11 ft. lbs. (15 Nm) and then to 22 ft. lbs. (30 Nm).
   - Water pump pulley (if equipped). Torque the bolts to 89 inch lbs. (10 Nm) and then to 18 ft. lbs. (25 Nm).
   - Heater hoses to the water pump
   - Radiator inlet and outlet hoses to the water pump
   - Accessory drive belt
   - Air intake duct
   - MAF sensor electrical connector
   - IAT sensor electrical connector
   - Negative battery cable
5. Fill the cooling system.
6. Start the vehicle and check for leaks, repair if necessary.

Cylinder Head

REMOVAL & INSTALLATION

1. Before servicing the vehicle, refer to the precautions in the beginning of this section.
2. Drain the engine coolant.
3. Relieve the fuel system pressure.
4. Remove or disconnect the following:
   - Negative battery cable
   - Rocker arm cover
   - Exhaust manifold from the cylinder head
   - Intake manifold
   - Vapor vent pipe
   - Power steering pump pulley, if removing the left side
   - Power steering pump and reposition it, if removing the left side
   - Power steering pump bracket, if removing the left side
   - Ground wire bolt from the rear of the cylinder head, if removing the left side
   - Cylinder head

To install:

➡New M11 cylinder head bolts must be used when installing the cylinder head assembly.

5. Install the new cylinder head gasket onto the locating pins.
6. Place the cylinder head on the engine and tighten the bolts in sequence to:
   a. M11 bolts: 22 ft. lbs. (30 Nm).
   b. M11 bolts: plus 90 degrees.
   c. M11 bolts 1–8: plus 90 degrees.
   d. M11 bolts 9–10: plus 50 degrees.
   e. M8 bolts (11–15): 22 ft. lbs. (30 Nm).
7. Install or connect the following:
   - Ground wire to the rear of the cylinder head and torque the bolt to 24 ft. lbs. (32 Nm), if removed
   - Power steering pump bracket and torque the bolts to 37 ft. lbs. (50 Nm) , if removed
   - Power steering pump and torque the bolts to 18 ft. lbs. (25 Nm) , if removed
   - Power steering pump pulley, if removed
   - Exhaust manifold and torque the bolts, in sequence, to 11 ft. lbs. (15 Nm), then to 18 ft. lbs. (25 Nm)
   - Valve rocker arm cover and torque

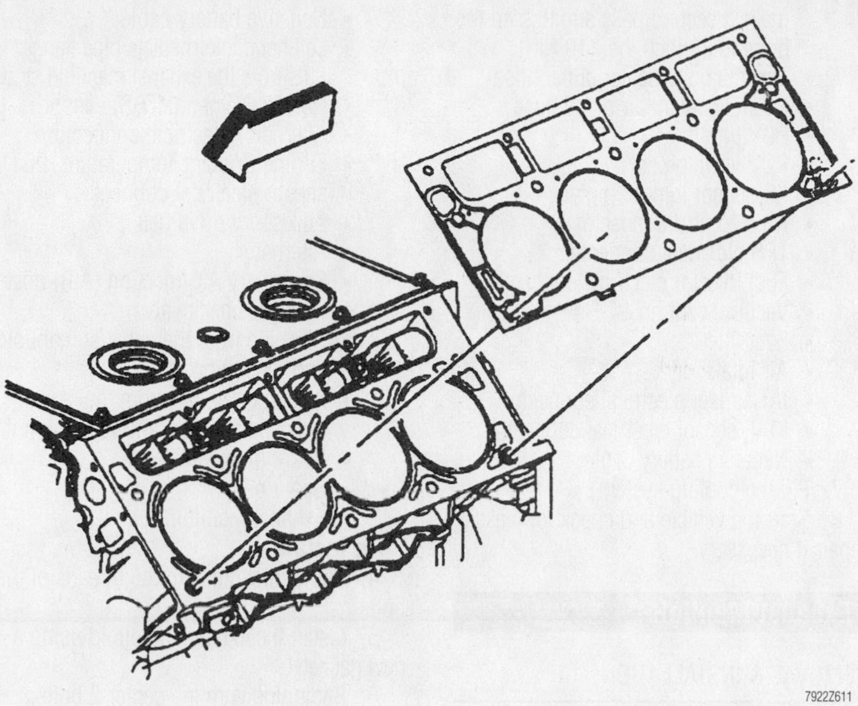

Be sure the tab on the edge of the gasket is closer to the front of the engine when installed

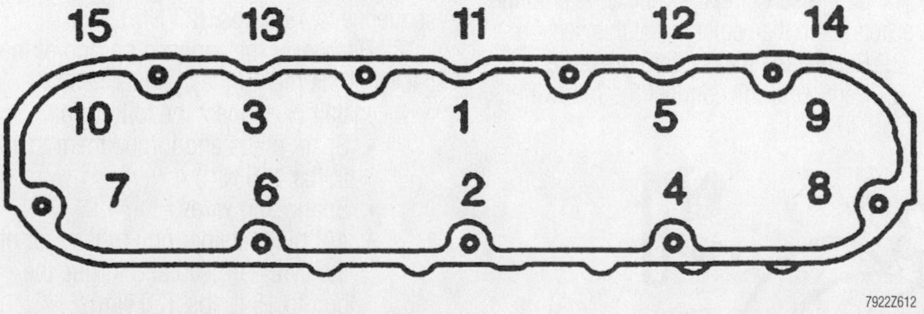

Cylinder head bolt torque sequence—5.7L engine

the bolts to 106 inch lbs. (12 Nm)

- Intake manifold and torque the bolts in sequence to 44 inch lbs. (5 Nm) then to 89 inch lbs. 910 Nm)
- Vapor vent pipe and torque the bolt to 106 inch lbs. (12 Nm)
- Negative battery cable

8. Fill the cooling system.
9. Start the vehicle and check for leaks, repair if necessary.

Rocker Arms

REMOVAL & INSTALLATION

➡Always keep the rocker arms and pushrods in order so they can be installed in their original positions.

1. Before servicing the vehicle, refer to the precautions in the beginning of this section.
2. Relieve the fuel system pressure.
3. Remove or disconnect the following:
   - Rocker arm cover
   - Rocker arm
   - Rocker arm, push rods and pedestals
   - Rocker arm pivot support, if necessary

To install:

4. If removed, install the rocker arm pivot support.
5. Lubricate the rocker arms and pushrods with clean engine oil.
6. Lubricate the flange and washer surface of the rocker arm mounting bolts.
7. Install the rocker arms but do not tighten the bolts at this time.
8. Rotate the crankshaft so the No. 1 piston is at Top Dead Center (TDC).
9. With the engine in this position, tighten the exhaust valve rocker arm bolts on cylinders No. 1, 2, 7 and 8 to 22 ft. lbs. (30 Nm), then, tighten the intake valve rocker arm bolts on cylinders No. 1, 3, 4 and 5 to 22 ft. lbs. (30 Nm).
10. Rotate the crankshaft 1 revolution (360 degrees).
11. With the engine in this position, tighten the exhaust valve rocker arm bolts on cylinders No. 3, 4, 5 and 6 to 22 ft. lbs. (30 Nm), then, tighten the intake valve rocker arm bolts on cylinders No. 2, 6, 7 and 8 to 22 ft. lbs. (30 Nm).
12. Install the rocker arm cover using a new gasket and torque the bolts to 106 inch lbs. (12 Nm).
13. Negative battery cable.
14. Start the vehicle and check for leaks, repair if necessary.

Intake Manifold

REMOVAL & INSTALLATION

➡The intake manifold, throttle body, fuel injectors and rail may be removed from the engine as an assembly.

1. Before servicing the vehicle, refer to the precautions in the beginning of this section.
2. Drain the cooling system.
3. Relieve the fuel system pressure.
4. Remove or disconnect the following:
   - Intake Air Temperature (IAT) sensor electrical connector
   - Mass Air Flow (MAF) sensor electrical connector
   - Air intake duct
   - Fuel line from the fuel rail
   - Vacuum ventilation hose
   - Fuel injector electrical connectors
   - Throttle Position Sensor (TPS) electrical connector
   - Knock Sensor (KS) connector
   - KS jumper pigtail from the fuel rail stop bracket
   - Positive Crankcase Ventilation (PCV) valve pipes from the rocker arm covers
   - PCV tube from the throttle body
   - Engine cooling air bleed hose from the throttle body
   - Throttle body heater outlet hose
   - Intake manifold and fuel stop bracket

To install:

5. Apply a 0.20 in. (5mm) bead of threadlock to the threads of the intake manifold bolts.

Timing belt service is covered in Section 3 of this manual

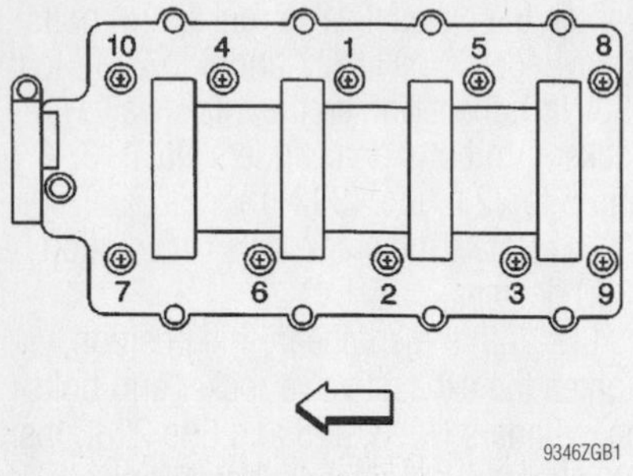

Intake manifold torque sequence—5.7L engine

CAUTION

The fuel rail stop bracket must be installed on the intake manifold in its original position. Failure to reinstall this part may result in fuel spray causing personal injury.

6. Install or connect the following:
   - Intake manifold with a new gasket
   - Fuel stop bracket
   - Intake manifold bolts and torque the bolts first to 44 inch lbs. (5 Nm) in the sequence shown, then make a second pass and tighten the bolts to 89 inch lbs. (10 Nm)
   - Throttle body heater outlet hose
   - Engine coolant air bleed hose
   - PCV tube
   - PCV valve pipes
   - KS sensor jumper pigtail
   - KS electrical connector
   - TPS electrical connector
   - Fuel injector electrical connectors
   - Vacuum vent hose
   - Fuel line
   - Air intake duct
   - IAT sensor electrical connector
   - MAF sensor electrical connector
   - Negative battery cable
7. Fill the cooling system.
8. Start the vehicle and check for leaks, repair if necessary.

Exhaust Manifold

REMOVAL & INSTALLATION

Left Side

1. Before servicing the vehicle, refer to the precautions in the beginning of this section.
2. Relieve the fuel system pressure.
3. Remove or disconnect the following:
   - Negative battery cable
   - Left hand intermediate pipe flange nuts from the exhaust manifold studs
   - Heated Oxygen (HO_2S) sensor
   - Alternator electrical connector
   - Engine Coolant Temperature (ECT) sensor electrical connector
   - Accessory drive belt
   - Alternator
   - Secondary Air Injection (AIR) hose from the check valve
   - AIR pipe from the exhaust manifold and reposition it
   - AIR pipe retainer from the rear of the cylinder head and reposition it
   - Spark plug wires
   - Spark plugs
   - Exhaust manifold

To install:

4. Apply threadlock to the threads of the bolts.
5. Install the exhaust manifold using a new gasket.
6. Beginning with the center 2 bolts, torque the bolts to 11 ft. lbs. (15 Nm) alternating from side-to-side until all the bolts are tight. Torque the bolts again in the same sequence to 18 ft. lbs. (25 Nm).
7. Bend over the exposed portion of the gasket at the rear of the cylinder head.
8. Install or connect the following:
   - Spark plugs and torque them to 11 ft. lbs. (15 Nm)
   - Spark plug wires
   - AIR pipe retainer bolt to the rear of the cylinder head and torque the bolt to 15 ft. lbs. (20 Nm)
   - AIR pipe with a new gasket and torque the bolts to 15 ft. lbs. (20 Nm)
   - AIR hose to the check valve and torque it to 15 ft. lbs. (20 Nm)
   - Alternator and torque the bolts to 37 ft. lbs. (50 Nm)
   - Accessory drive belt
   - Alternator electrical connector
   - ECT sensor electrical connectors
   - HO_2S sensor and torque it to 30 ft. lbs. (42 Nm)
   - Fuel lines
   - Intermediate pipe flange-to-exhaust manifold nuts and torque them to 15 ft. lbs. (20 Nm)
   - Negative battery cable
9. Start the vehicle and check for leaks, repair if necessary.

Right Side

1. Before servicing the vehicle, refer to the precautions in the beginning of this section.

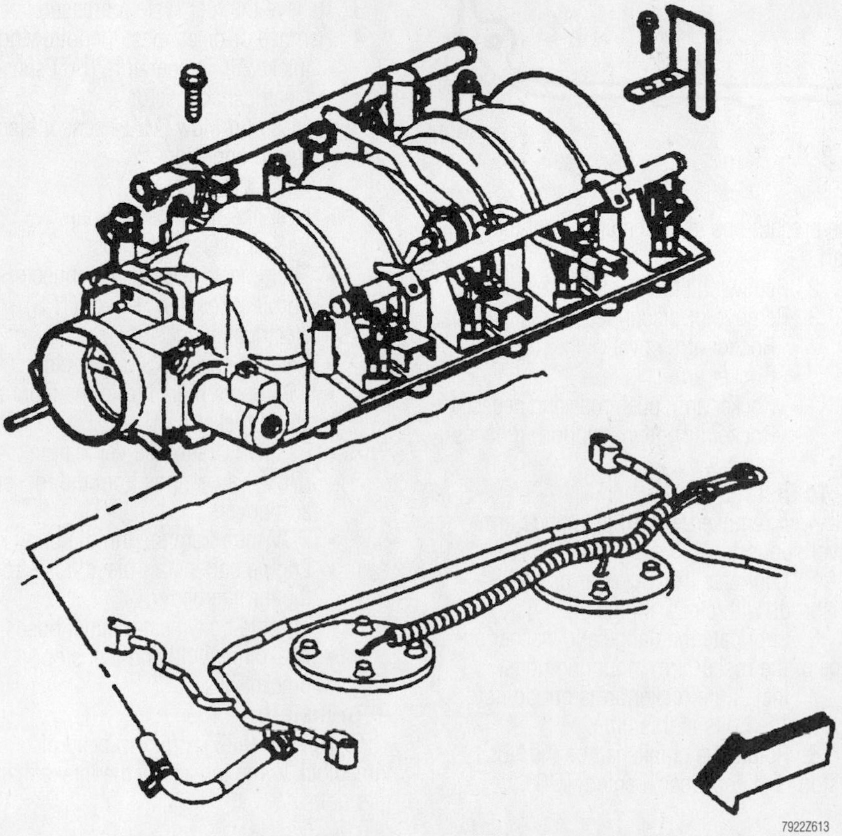

Be sure to reinstall the fuel stop bracket when installing the intake manifold

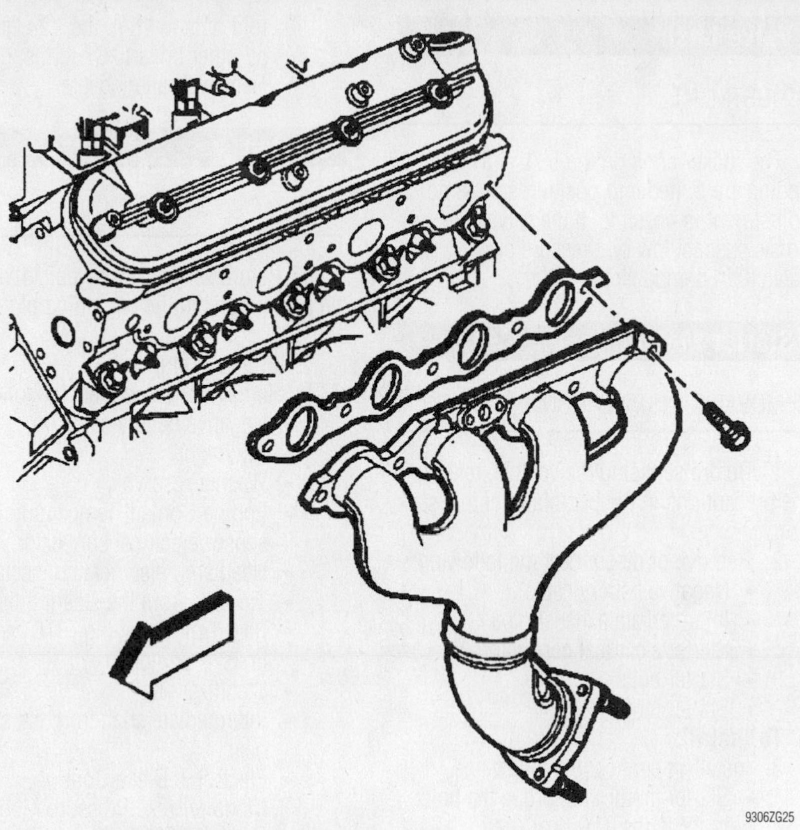

Exploded view of the exhaust manifold—5.7L engine

2. Remove or disconnect the following:
   - Negative battery cable
   - Right side intermediate exhaust pipe nuts from the manifold studs
   - Heated Oxygen (HO_2S) sensor
   - Secondary Air Injection (AIR) pipe from the exhaust manifold
   - Spark plug wires
   - Spark plugs
   - Oil level indicator and tube
   - Exhaust manifold

To install:

3. Apply threadlock to the threads of the bolts.
4. Install the exhaust manifold using a new gasket.
5. Beginning with the center 2 bolts, tighten the bolts to 11 ft. lbs. (15 Nm) alternating from side-to-side until all the bolts are tight. Then, tighten the bolts again in the same sequence to 18 ft. lbs. (25 Nm).
6. Bend over the exposed portion of the gasket at the rear of the cylinder head.
7. Install or connect the following:
   - Oil level indicator tube and torque the bolt to 12 ft. lbs. (16 Nm)
   - Spark plugs and torque them to 11 ft. lbs. (15 Nm)
   - Spark plug wires
   - AIR pipe using a new gasket and torque the bolts to 15 ft. lbs. (20 Nm)
   - HO_2S sensor in the manifold and torque it to 30 ft. lbs. (42 Nm)
   - Intermediate pipe flange-to-exhaust manifold nuts and torque them to 15 ft. lbs. (20 Nm)
   - Negative battery cable
8. Start the vehicle and check for leaks, repair if necessary.

Camshaft

REMOVAL & INSTALLATION

1. Before servicing the vehicle, refer to the precautions in the beginning of this section.
2. Drain the cooling system.
3. Relieve the fuel system pressure.
4. Remove or disconnect the following:
   - Fuel lines from the fuel rail
   - Rocker arm covers
   - Rocker arms and pushrods
   - Both cylinder heads
   - Valve lifter guides and the lifters
   - Radiator

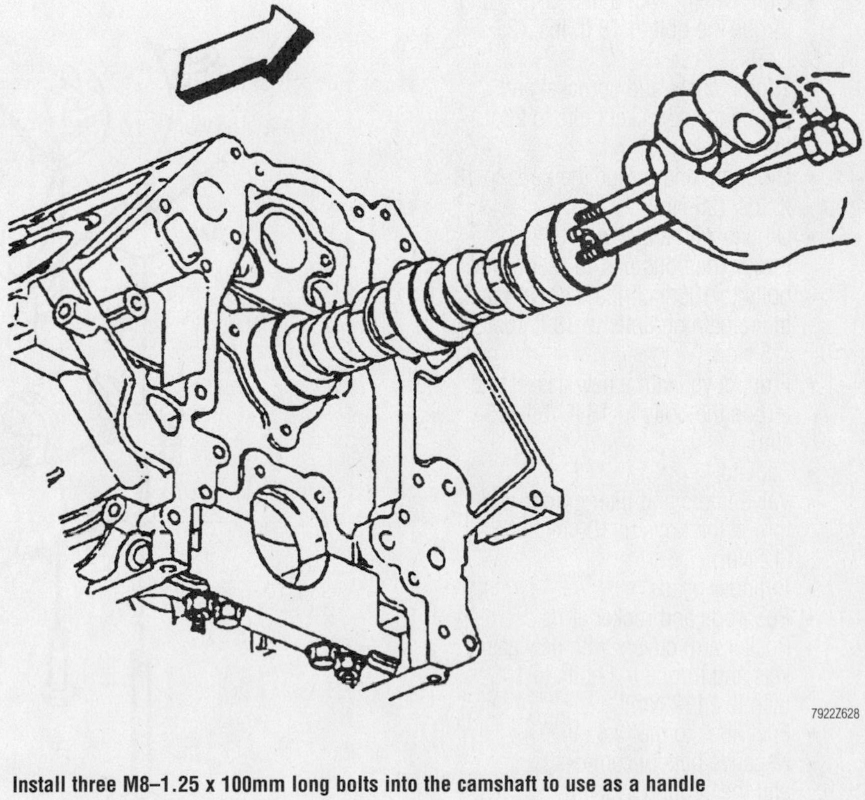

Install three M8–1.25 x 100mm long bolts into the camshaft to use as a handle

- Engine front cover
- Oil pan
- Oil pump
- Camshaft sprocket and timing chain
- Camshaft Position (CMP) sensor bolt and sensor
- Camshaft retainer plate

WARNING

All camshaft bearings are the same diameter, extreme care must be taken during camshaft removal or installation to avoid damaging them.

5. Install 3 M8 x 100mm long bolts in the front of the camshaft to use as a handle.
6. Carefully remove the camshaft from the engine, be sure not to damage the camshaft bearings.

To install:

7. Lubricate the camshaft with clean engine oil.
8. Install or connect the following:
   - Three M8–1.25 x 100mm long bolts in the front of the camshaft to use as a handle
   - Camshaft into the engine and remove the 3 bolts
   - Camshaft retainer using a new gasket and torque the bolts to 18 ft. lbs. (25 Nm)
   - CMP sensor with a new O-ring and torque the bolt to 18 ft. lbs. (25 Nm)
   - Timing chain and sprocket and torque the sprocket bolts to 26 ft. lbs. (35 Nm)
   - Oil pump and torque the bolts to 18 ft. lbs. (25 Nm)
   - Oil pan with a new gasket and torque the bolts pan-to-rear cover bolts to 106 inch lbs. (12 Nm) and the remaining bolts to 18 ft. lbs. (25 Nm)
   - Front cover with a new gasket and torque the bolts to 18 ft. lbs. (25 Nm)
   - Radiator
   - Valve lifters and lifter guides and torque the bolts to 106 inch lbs. (12 Nm)
   - Cylinder heads
   - Pushrods and rocker arms
   - Rocker arm covers with new gaskets and torque the bolts to 106 inch lbs. (12 Nm)
   - Fuel lines to the fuel rail
   - Negative battery cable
9. Refill the cooling system.
10. Start the engine and check for leaks, repair if necessary.

Valve Lash

ADJUSTMENT

The rocker arms on the 5.7L (VIN G) engine are bolted into position so that no adjustment is possible. If the valves are noisy, suspect low oil pressure or worn valve train components.

Starter Motor

REMOVAL & INSTALLATION

1. Before servicing the vehicle, refer to the precautions in the beginning of this section.
2. Remove or disconnect the following:
   - Negative battery cable
   - Intermediate exhaust pipe
   - Starter electrical connectors
   - Starter bolts
   - Starter motor

To install:

3. Install or connect the following:
   - Starter motor and torque the bolts to 37 ft. lbs. (50 Nm)
   - Starter electrical connectors
   - Intermediate exhaust pipe and torque intermediate pipe-to-manifold nuts to 15 ft. lbs. (20 Nm) and all other bolts to 37 ft. lbs. (50 Nm)
   - Negative battery cable

Oil Pan

REMOVAL & INSTALLATION

1. Before servicing the vehicle, refer to the precautions in the beginning of this section.
2. Drain the engine oil.
3. Remove or disconnect the following:
   - Negative battery cable
   - Alternator
   - Washer reservoir
   - Engine Coolant Temperature (ECT) sensor electrical connector
   - Headlamp electrical connector
   - Tie rods from the steering linkage
   - Real Time Damping (RTD) sensor links, if equipped
   - Stabilizer shaft
   - Intermediate shaft from the steering gear
   - Electronic Brake Control Module/Brake Pressure Modulator Valve (EBCM/BPMV) bracket and reposition
   - Power steering gear

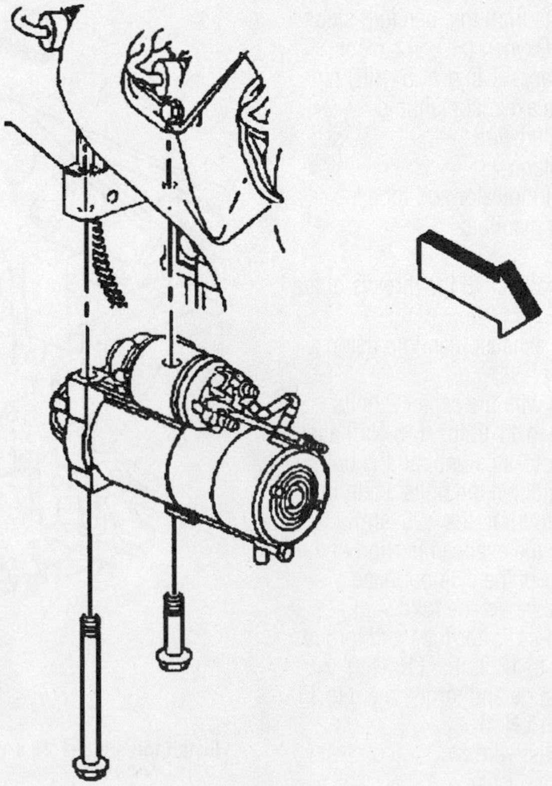

Exploded view of the starter assembly—5.7L engine

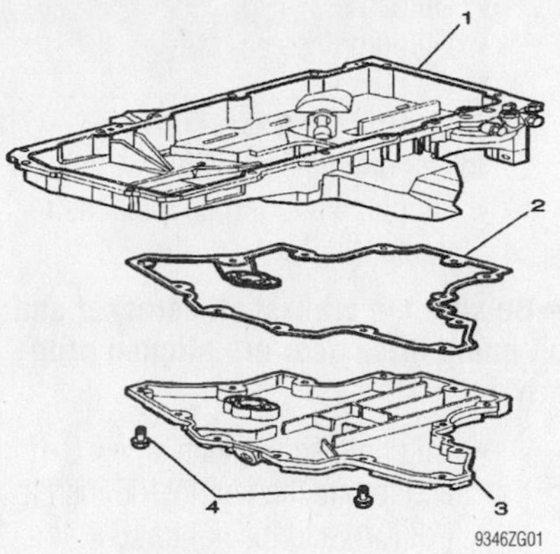

Separate the upper (1) and lower (3) oil pans

- Lower shock absorber mounting bolts
- Transverse leaf spring using a suitable compressor
- Lower control arm bolts from the suspension crossmember
- Engine mount lower nuts
- Wheel Speed Sensor (WSS) wiring harness from the crossmember
- Brake line from the crossmember

4. Support the suspension crossmember.
   - Crossmember mounting nuts
   - Front crossmember
   - Oil filter
   - Automatic transmission cooler line front and rear retaining clamp bolts (if equipped)
   - Engine flywheel housing-to-oil pan bolts
   - Engine flywheel close out bolts
   - Engine oil level sensor
   - Engine oil temperature sensor electrical connector
   - Oil pan assembly
   - Lower oil pan bolts and separate the pans (if necessary)

➡To remove the oil pan gasket it is necessary to drill out 3 rivets securing it to the pan.

To install:

5. If separated, connect the upper and lower oil pans with a new gasket and torque the bolts to 106 inch lbs. (12 Nm).
6. Apply a 0.20 in. (5mm) bead of RTV sealant directly on the tabs of the front and rear cover gaskets that extend onto the oil pan mounting surface.
7. Install or connect the following:
   - New gasket on the oil pan. It is not necessary to rivet the new gasket on the oil pan.

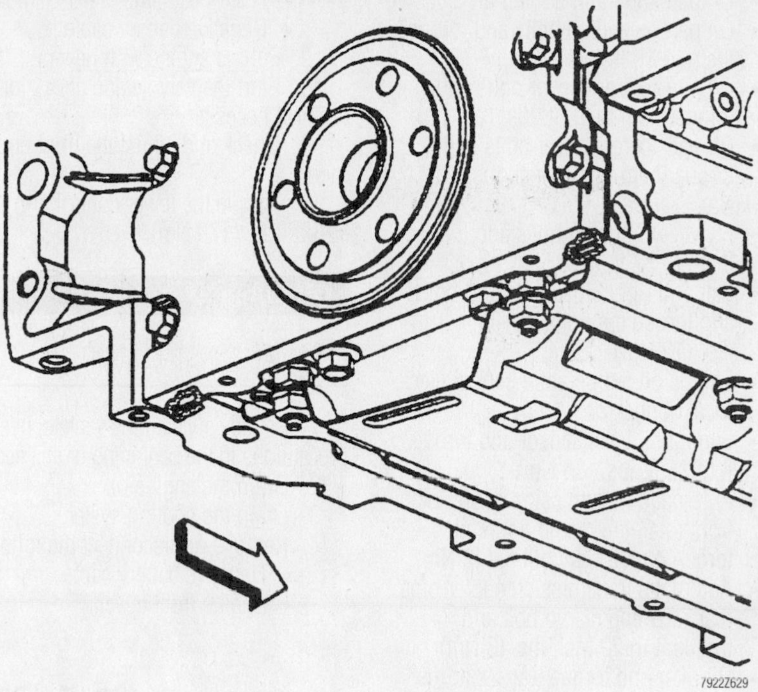

Apply sealant to the areas where the front and rear covers attach to the engine block—5.7L engine

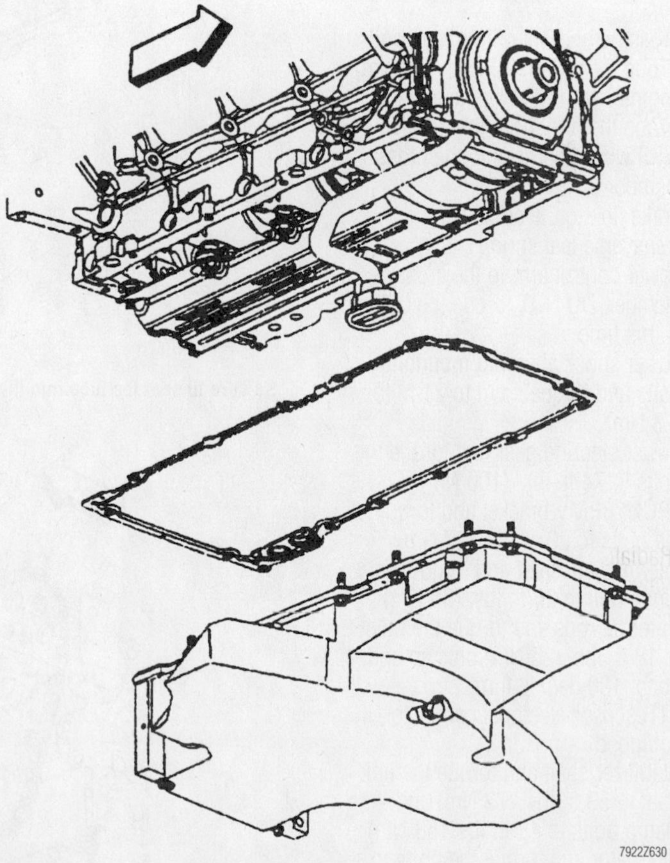

Exploded view of the oil pan mounting—5.7L engine

- Oil pan and hand tighten the bolts
- Oil pan-to-engine bolts and torque them to 18 ft. lbs. (25 Nm)
- Oil pan-to-front cover bolts and torque them to 18 ft. lbs. (25 Nm)
- Oil pan-to-rear cover bolts and torque them to 106 inch lbs. (12 Nm)
- Flywheel housing bolts and torque them to 37 ft. lbs. (50 Nm)
- Left and right close outs and bolts and torque the bolts to 106 inch lbs. (12 Nm)
- Engine oil temperature sensor electrical connector
- Engine oil level sensor and torque it to 26 ft. lbs. (35 Nm)
- Automatic transmission cooler pipe front retaining clamp bolt and torque it to 106 inch lbs. (12 Nm)
- Automatic transmission cooler pipe rear retaining clamp bolt and torque it to 22 inch lbs. (3 Nm)
- Oil filter and torque it to 22 ft. lbs. (30 Nm)
- Oil drain plug and torque it to 18 ft. lbs. (25 Nm)

8. Raise the suspension crossmember into position.
   - Crossmember mounting nuts and torque them to 81 ft. lbs. (110 Nm)
   - Engine mount lower nuts and torque them to 48 ft. lbs. (65 Nm)
   - WSS wiring harness to the crossmember
   - Brake line to the crossmember
   - Transverse leaf spring
   - Lower control arm to the crossmember, DO NOT torque the bolts at this time
   - Lower shock absorber mounting bolts and torque them to 21 ft. lbs. (28 Nm)
   - Power steering gear and torque the bolts to 74 ft. lbs. (100 Nm)
   - EBCM/BPMV bracket and torque the bolts to 20 ft. lbs. (27 Nm)
   - Intermediate shaft and torque the pinch bolt to 25 ft. lbs. (35 Nm)
   - Outer tie rods and torque the nuts to 18 ft. lbs. (25 Nm) plus an additional 180 degree turn
   - RTD sensor electrical connector, if equipped
   - Stabilizer shaft and torque the link nuts to 53 ft. lbs. (72 Nm) and the clamp bolts to 43 ft. lbs. (58 Nm)
   - Alternator and torque the bolts to 37 ft. lbs. (50 Nm)
   - Washer reservoir and torque the nuts to 66 inch lbs. (7.5 Nm)
   - ECT sensor electrical connector
   - Front headlamp electrical connector
   - Negative battery cable
9. Fill the engine with new oil.
10. Start the vehicle and check for leaks, repair if necessary.
11. Check and adjust the front end alignment.
12. Torque the lower control arm bolts to 125 ft. lbs. (170 Nm).

Oil Pump

REMOVAL & INSTALLATION

1. Before servicing the vehicle, refer to the precautions in the beginning of this section.
2. Drain the engine oil.
3. Drain the cooling system.
4. Remove or disconnect the following:
   - Negative battery cable
   - Engine front cover
   - Oil pan
   - Oil pump screen
   - Oil pump

To install:

5. Install or connect the following:
   - Oil pump and torque the bolts to 18 ft. lbs. (25 Nm)

➡Be sure the crankshaft sprocket and oil pump drive gear are aligned properly.

   - Oil pump screen with a new O-ring and torque the bolt to 106 inch lbs. (12 Nm) and the nut to 18 ft. lbs. (25 Nm)
   - Oil pan
   - Engine front cover
   - Negative battery cable
6. Fill the engine with new oil.
7. Fill the cooling system.
8. Start the vehicle and check for leaks, repair if necessary.

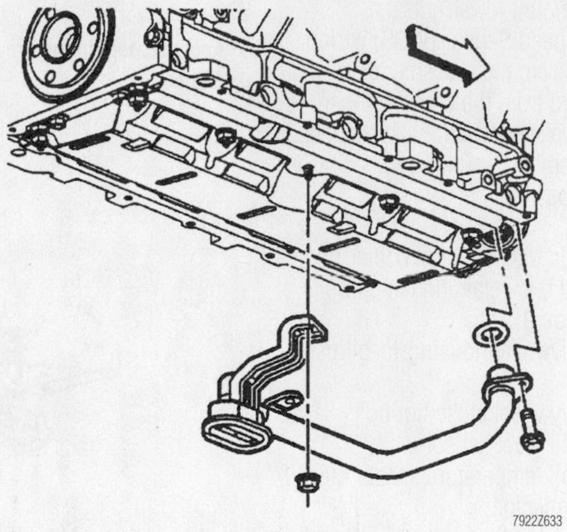

Be sure to seat the tube into the pump before installing the bolt

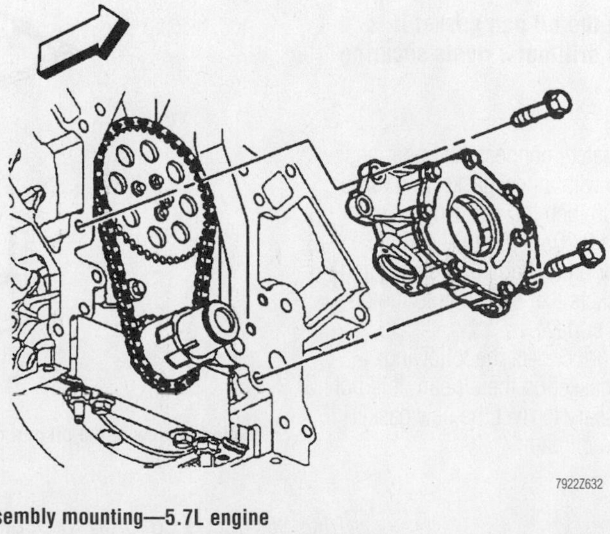

Oil pump assembly mounting—5.7L engine

Rear Main Seal

REMOVAL & INSTALLATION

➡The rear main seal is a 1-piece unit. It can be removed or installed without removing the oil pan or crankshaft.

1. Before servicing the vehicle, refer to the precautions in the beginning of this section.
2. Remove or disconnect the following:
   - Transmission
   - Clutch and pressure plate, if equipped with a manual transmission
   - Flywheel inspection cover
   - Flywheel
3. Using a pry tool, carefully pry the old seal out.

To install:

4. Inspect the crankshaft for nicks or burrs, correct as required.
5. Clean the area and coat the seal and crankshaft with engine oil.
6. Install or connect the following:
   - Seal, using a seal installer
   - Flywheel and torque the bolts in sequence to 15 ft. lbs. (20 Nm) then to 37 ft. lbs. (50 Nm) and finally to 74 ft. lbs. (100 Nm)
   - Flywheel inspection cover and torque the bolts 18 ft. lbs. (25 Nm)
   - Clutch and pressure plate if equipped with a manual transmission
   - Transmission
7. Start the vehicle and check for leaks, repair if necessary.

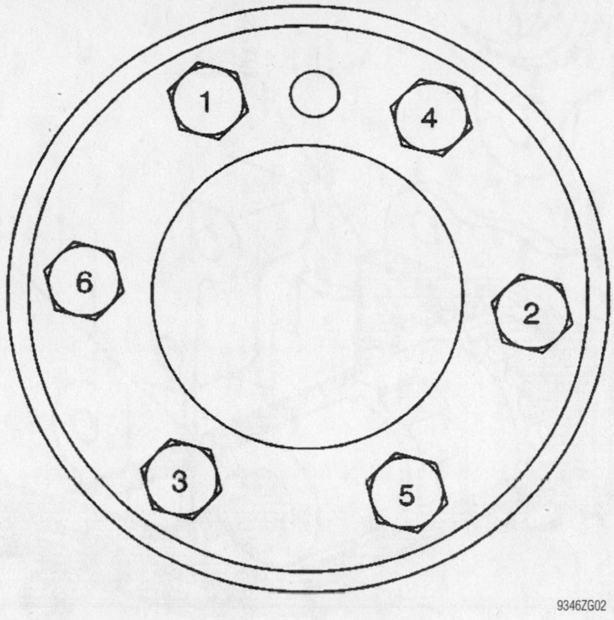

Flywheel torque sequence

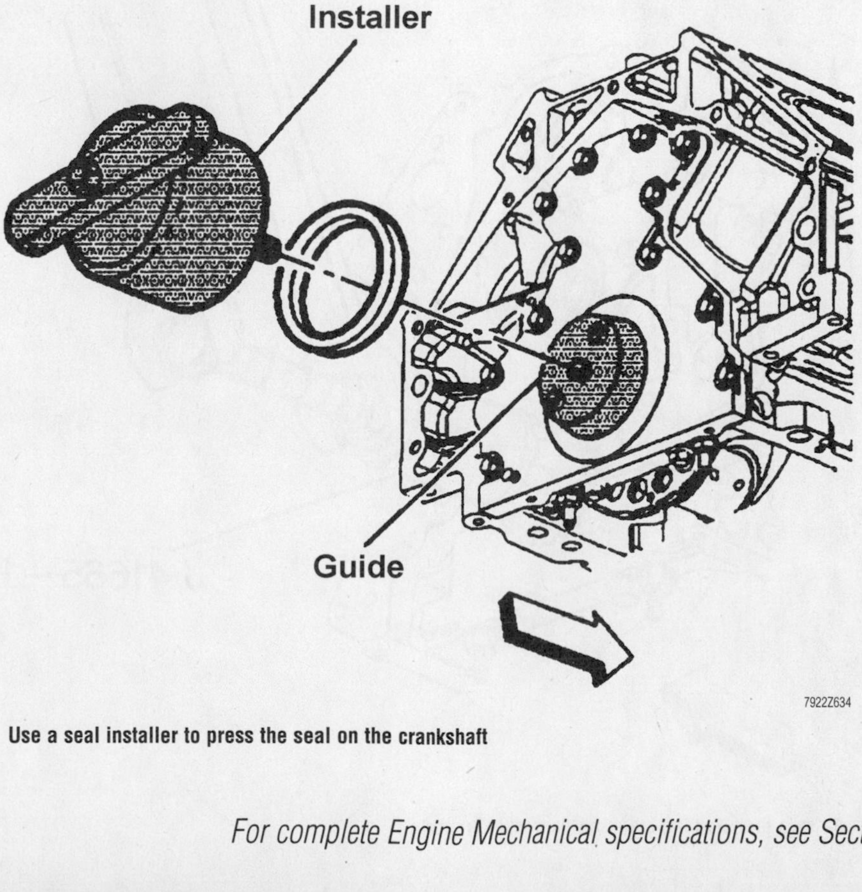

Use a seal installer to press the seal on the crankshaft

Timing Chain, Sprockets, Front Cover and Seal

REMOVAL & INSTALLATION

1. Before servicing the vehicle, refer to the precautions in the beginning of this section.
2. Drain the cooling system.
3. Drain the engine oil.
4. Remove or disconnect the following:
   - Negative battery cable
   - Engine front cover
   - Oil pan
   - Oil pump
5. Rotate the crankshaft until the timing marks are aligned. The marks should face each other. Remove the camshaft sprocket bolts.

➡Do not turn the crankshaft after the timing chain has been removed. Damage to the pistons or valves may occur.

6. Remove or disconnect the following:
   - Camshaft sprocket and timing chain
   - Crankshaft sprocket using a suitable puller

To install:

7. Be sure the timing mark on the crankshaft sprocket is at the 12 o'clock position. If not rotate the crankshaft to the correct position.

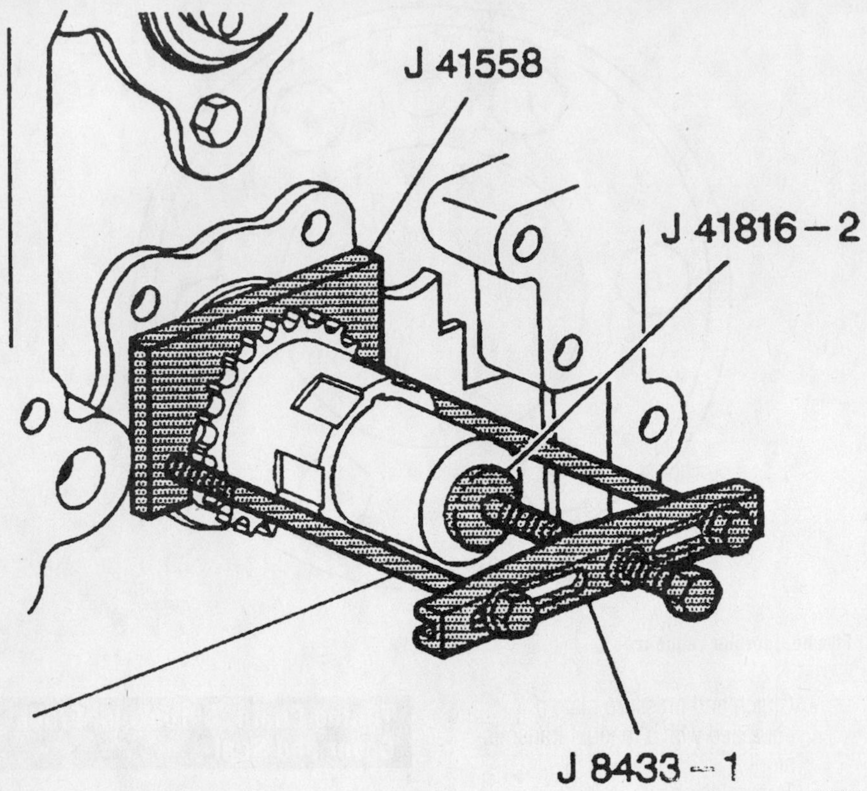

A puller is needed to remove the crankshaft sprocket

8. Install or connect the following:
   - Crankshaft sprocket with a Crankshaft Balancer and Sprocket Installer tool
   - Timing chain on the camshaft sprocket.
   - Timing chain and sprocket assembly so the timing marks are aligned and torque the camshaft sprocket bolts to 26 ft. lbs. (35 Nm)
   - Oil pump and torque the bolts to 18 ft. lbs. (25 Nm)
   - Oil pan with a new gasket
   - Front cover with a new gasket and seal and hand tighten the bolts at this time

9. Install the oil pan to cover bolts and hand-tighten them at this time. Make certain that the front cover is properly aligned and torque the oil pan-to-front cover bolts to 18 ft. lbs. (25 Nm) then the remaining bolts to 18 ft. lbs. (25 Nm).

10. Connect the negative battery cable.

11. Fill the cooling system.

12. Fill the engine with new oil.

13. Start the vehicle and check for leaks, repair if necessary.

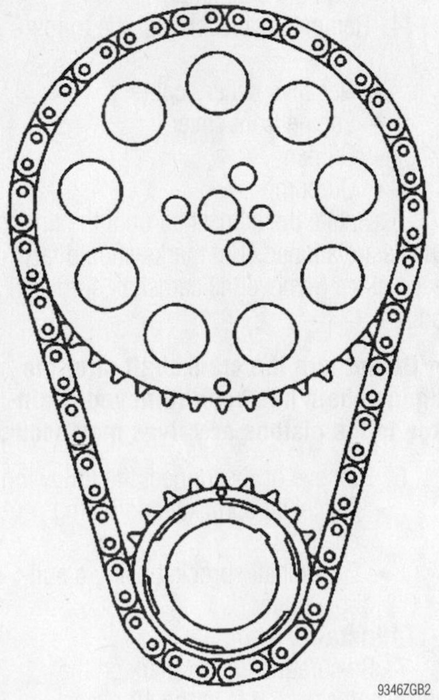

Timing chain alignment—5.7L engine

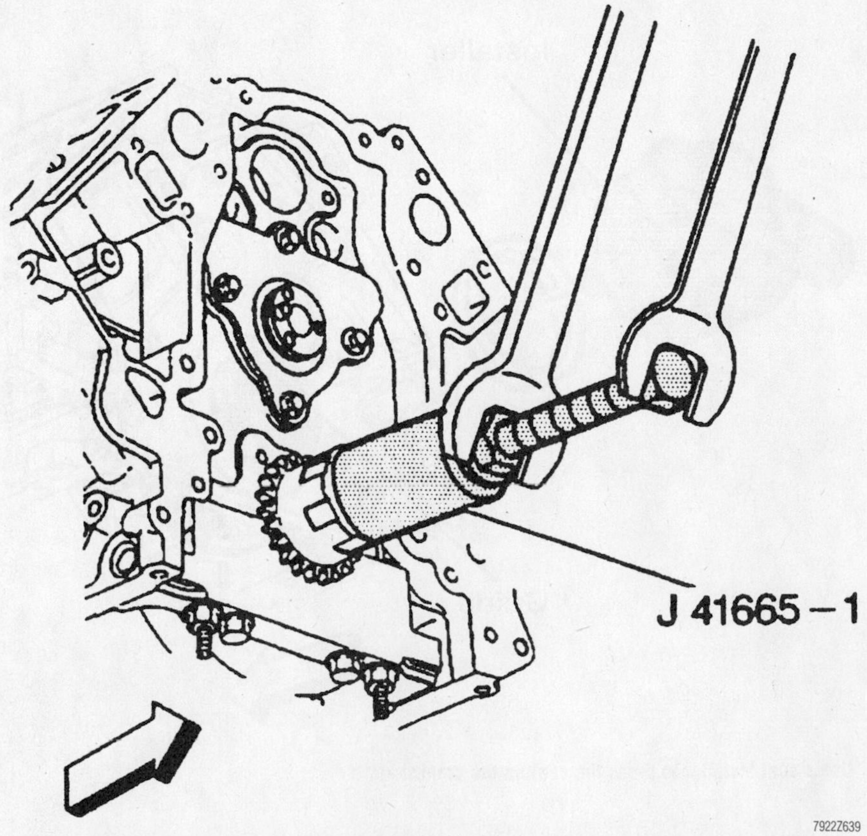

Be sure to install the crankshaft sprocket properly

Piston and Ring

POSITIONING

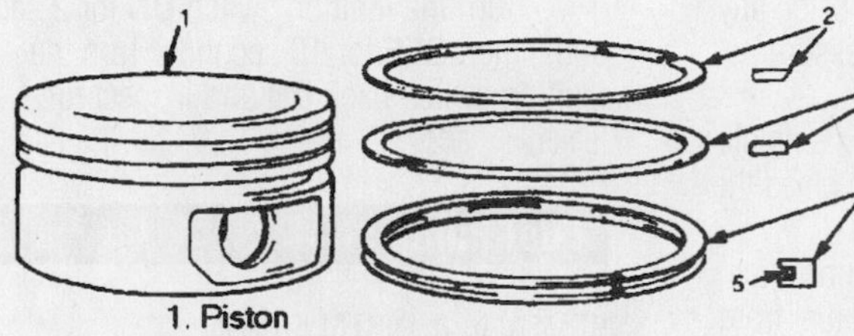

1. Piston
2. Upper compression piston ring
3. Lower compression piston ring
4. Oil control piston ring
5. Oil control ring spring w/spacer

7922AG43

Piston ring positioning—5.7L engine

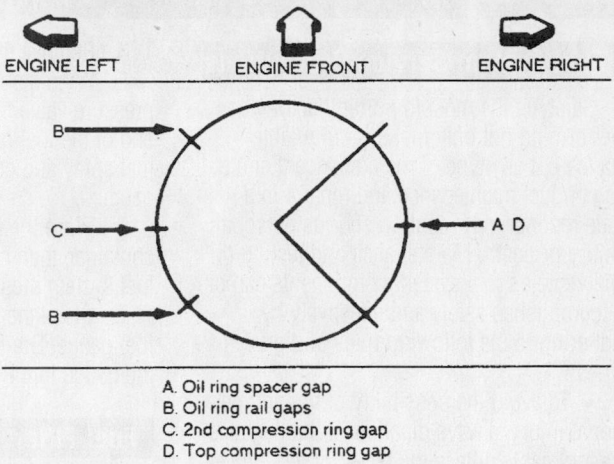

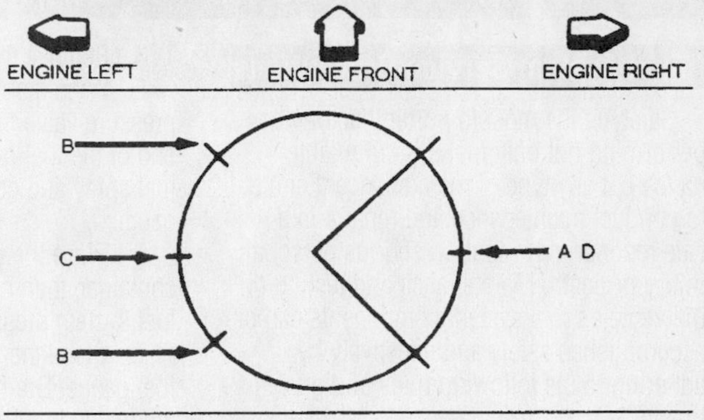

A. Oil ring spacer gap
B. Oil ring rail gaps
C. 2nd compression ring gap
D. Top compression ring gap

7922AG42

Piston ring end-gap spacing—5.7L engine

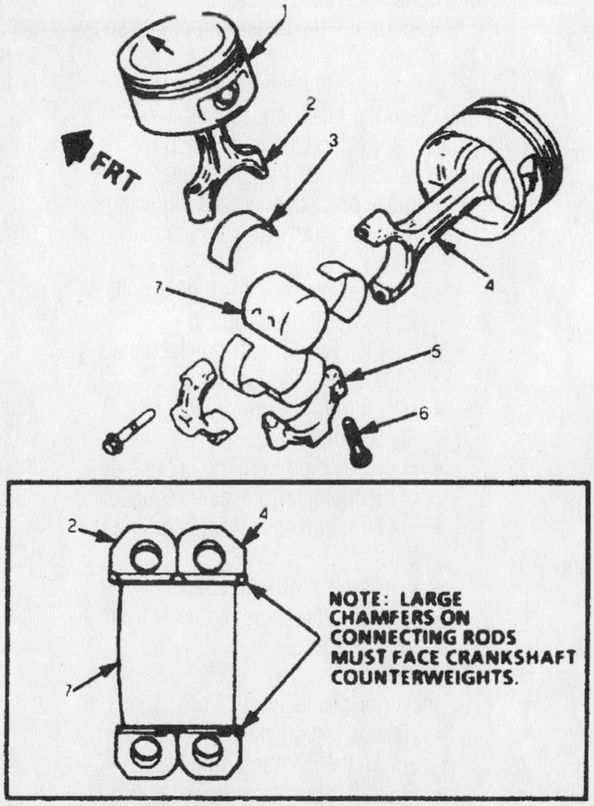

1. Piston
2. Connecting rod LH
3. Connecting rod bearing
4. Connecting rod RH
5. Connecting rod bearing cap
6. Connecting rod bearing cap bolt
7. Crankshaft

7922AG44

Piston and connecting rod positioning—5.7L engine

FUEL SYSTEM

Fuel System Service Precautions

Safety is the most important factor when performing not only fuel system maintenance but any type of maintenance. Failure to conduct maintenance and repairs in a safe manner may result in serious personal injury or death. Maintenance and testing of the vehicle's fuel system components can be accomplished safely and effectively by adhering to the following rules and guidelines.

- To avoid the possibility of fire and personal injury, always disconnect the negative battery cable unless the repair or test procedure requires that battery voltage be applied.
- Always relieve the fuel system pressure prior to disconnecting any fuel system component (injector, fuel rail, pressure regulator, etc.), fitting or fuel line connection. Exercise extreme caution whenever relieving fuel system pressure, to avoid exposing skin, face and eyes to fuel spray. Please be advised that fuel under pressure may penetrate the skin or any part of the body that it contacts.
- Always place a shop towel or cloth around the fitting or connection prior to loosening to absorb any excess fuel due to spillage. Ensure that all fuel spillage (should it occur) is quickly removed from engine surfaces. Ensure that all fuel soaked cloths or towels are deposited into a suitable waste container.
- Always keep a dry chemical (Class B) fire extinguisher near the work area.
- Do not allow fuel spray or fuel vapors to come into contact with a spark or open flame.
- Always use a back-up wrench when loosening and tightening fuel line connection fittings. This will prevent unnecessary stress and torsion to fuel line piping. Always follow the proper torque specifications.
- Always replace worn fuel fitting O-rings with new. Do not substitute fuel hose or equivalent, where fuel pipe is installed.

Fuel System Pressure

RELIEVING

1. Before servicing the vehicle, refer to the precautions in the beginning of this section.
2. Disconnect the negative battery cable.
3. Loosen the fuel filler cap to relieve the tank pressure.
4. Remove the left fuel rail cover.
5. Wrap a shop towel around the fuel pressure valve fitting (located on the side or end of the fuel rail assembly) to catch any fuel spray and connect a fuel pressure gauge.
6. Place the bleed hose into a suitable container, then open the valve to bleed the fuel system pressure.
7. Close the valve and disconnect the fuel gauge. Drain any remaining fuel from the gauge into the bleed container.

Fuel Filter

REMOVAL & INSTALLATION

1. Before servicing the vehicle, refer to the precautions in the beginning of this section.
2. Relieve the fuel system pressure.
3. Clean the filter connections, then depress the locking tabs and detach the quick-connect fittings from the filter.
4. Remove or disconnect the following:
   - Rear stabilizer bar from the rear cradle
   - Intermediate pipe-to-muffler bolts and lower the left muffler
   - Fuel feed and return quick connect fittings from the fuel filter
   - Fuel filter/pressure regulator bracket mount nut
   - Fuel feed pipe from the outlet side of the fuel filter/pressure regulator
   - Fuel filter/pressure regulator and bracket
   - Fuel system ground strap
   - Fuel filter/pressure regulator from the bracket

To install:

5. Install or connect the following:
   - New plastic quick connector retainers on the inlet and outlet tubes
   - Fuel filter/pressure regulator into the bracket
   - Fuel system ground strap
   - Fuel feed pipe to the outlet side of the fuel filter/pressure regulator
   - Fuel filter/pressure regulator and bracket to the mounting stud and torque the nut to 40 inch lbs. (4.5 Nm)
   - Fuel return and feed quick connect fittings
   - Muffler and torque the hanger nuts to 12 ft. lbs. (16 Nm) and the intermediate pipe bolts to 37 ft. lbs. (50 Nm)
   - Stabilizer shaft and torque the nuts to 70 ft. lbs. (95 Nm)
6. Connect the negative battery cable.
7. Turn the ignition switch **ON** for 2 seconds, then **OFF** for 10 seconds. Turn the ignition switch back **ON** and inspect for leaks.

Fuel Pump

REMOVAL & INSTALLATION

These models have 2 fuel tanks (right and left). The fuel pump is part of the fuel sender assembly in the left fuel tank. The right fuel tank contains a fuel sender assembly with a siphon jet pump which supplies fuel to the left tank through the fuel sender feed pipe. Although there are 2 sending units, the removal and installation procedure is the same for both.

1. Before servicing the vehicle, refer to the precautions in the beginning of this section.
2. Properly relieve the fuel system pressure.
3. Drain the fuel tank(s).
4. Remove or disconnect the following:
   - Both rear wheels
   - Fuel tank shield(s)
5. Clean the area around the fuel sender assembly.
6. Mark each fuel line to help identify them during installation.
7. Remove or disconnect the following:
   - Quick connect fittings from the fuel sender
   - Fuel sender electrical connector
   - Fuel tank strap and properly support the fuel tank
   - Fuel sender attaching bolts and discard them

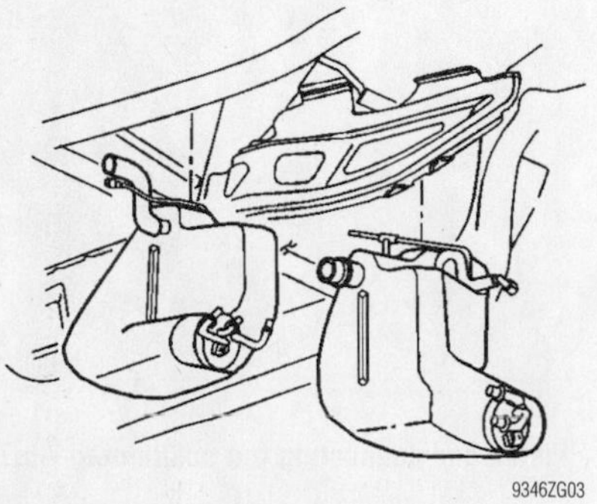

These models have 2 fuel tanks connected by a fuel sender feed pipe

- Float arm retaining clip and the float arm (left side sender only)
- Fuel sender and gasket

To install:

➡Always install a new fuel strainer before installing the fuel sender assembly. A strainer that has been exposed to fuel will not unfold completely and may interfere with the full travel of the float arm.

8. Install a new gasket on the fuel sender.

⁂ WARNING

Do not damage the float arm during installation.

9. Fold the long strainer over itself and hold the strainer in this position.
10. Pinch both strainers upward toward each other.
11. Install the float arm through the fuel tank opening with the folded strainers.

➡It may be necessary to rotate the sender for proper installation.

12. Look into the fuel tank opening and make certain that the long strainer is visible. If it is not visible, rotate the fuel sender until the strainer is free
13. Align the fuel sender gasket tab with the fuel sender cover mark and align the cover mark with the fuel tank mark.
14. Install the new break away head attaching bolts to the fuel sender and hand tighten them.

⁂ CAUTION

The upper hex head portion of the fuel sender attaching bolts is designed to shear off the lower section of the bolt when the proper torque is reached. Do not tighten the bolts after the head is sheared off.

15. Tighten the new break away attaching bolts in sequence until the hex head shears off of the lower section.
16. Install or connect the following:
    - Fuel feed and return pipes
    - Fuel sender electrical connector
    - Fuel tank strap and torque the bolts to 18 ft. lbs. (25 Nm)
    - Fuel tank shield and torque the bolts to 18 ft. lbs. (25 Nm)
    - Both rear wheels
    - Negative battery cable

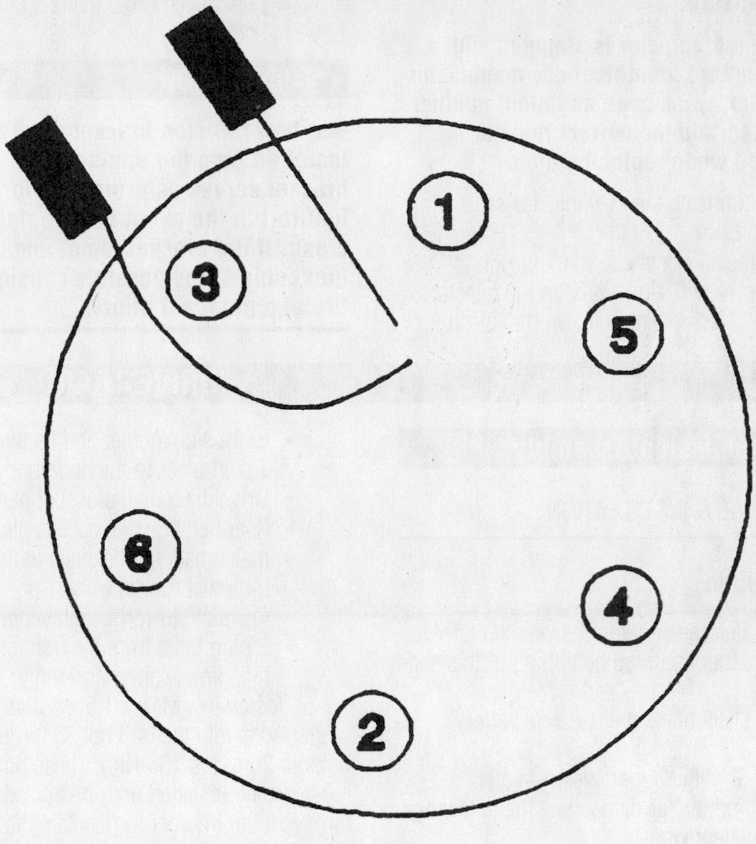

7922Z641

Tighten the fuel pump mounting bolts in the sequence shown—5.7L engine

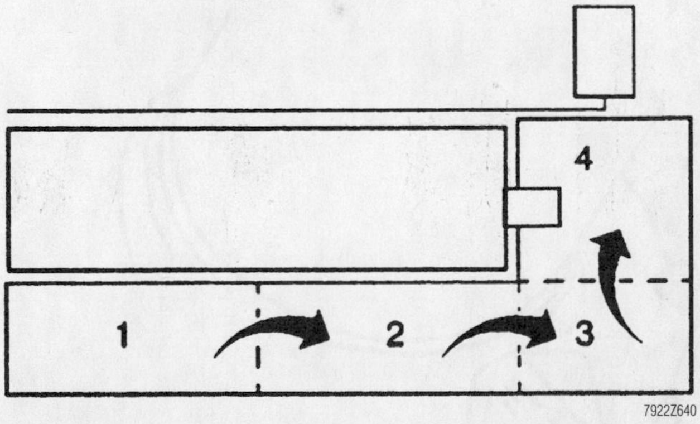

7922Z640

Fold the strainer on itself 3 times so it will fit into the opening in the fuel tank

17. Fill the fuel tank and install the fuel filler cap.
18. Turn the ignition **ON** for 2 seconds, **OFF** for 10 seconds, then **ON** again and inspect the system for leaks.

Fuel Injector

REMOVAL & INSTALLATION

1. Before servicing the vehicle, refer to the precautions in the beginning of this section.
2. Relieve the fuel system pressure.
3. Remove or disconnect the following:
    - Both fuel rail covers
    - Fuel feed hose from the fuel rail
    - Fuel injector electrical connectors
    - Fuel rail mounting bolts
    - Fuel rail
4. Spread the injector clip to release the injector from the fuel rail.
5. Fuel injector by spreading the retaining clip.

To install:

➡**The fuel injector is stamped with a part number identification, manufacturing date, week code and plant number. Make certain the correct injector is ordered when replacing them.**

6. Lubricate the new injector seals with clean oil.
7. Install or connect the following:
   - New O-ring seals on the injectors
   - New retainer clip on the injector
   - Fuel injector into the fuel rail socket

⁂ CAUTION

The fuel rail stop bracket must be installed onto the engine. The bracket serves as protection for the fuel rail in the event of a frontal crash. If the bracket is not installed, fuel could spray possibly causing a fire and personal injury.

   - Fuel rail and ground strap to the intake manifold and torque the fuel rail attaching bolts to 89 inch lbs. (10 Nm)
   - Electrical connectors to the injectors
   - Fuel feed hose to the fuel rail
   - Negative battery cable
   - Left and right fuel rail covers
8. Turn the ignition **ON** for 2 seconds, **OFF** for 10 seconds, then **ON** again and inspect the system for leaks.

DRIVE TRAIN

Transmission Assembly

REMOVAL & INSTALLATION

Automatic

1. Before servicing the vehicle, refer to the precautions in the beginning of this section.
2. Disconnect the negative battery cable.
3. Shift the transmission into **N**.
4. Remove or disconnect the following:
   - Rear wheels
   - Intermediate exhaust pipe
   - Right side muffler and tie the left side muffler to the underbody
   - Driveline tunnel closeout panel
   - Rear bell housing access plug and matchmark the flexplate to the torque converter
   - Flexplate-to-torque converter bolts
   - 2 plug bolts from the front of the driveline support assembly
5. Install two M10 x 1.5 x 55mm or longer bolts into the bolt holes. Torque the bolts to 26 ft. lbs. (35 Nm). These bolts must remain installed until instructed to remove them in order to maintain the position of the input shaft bearing.
6. Remove or disconnect the following:
   - Engine flywheel housing access plug
   - Propeller shaft hub clamp bolt
   - Shift cable bracket nuts
   - Shift control cable from the shift lever
   - Rear transverse spring and support the lower control arms
   - Outer tie rod ends from the knuckle
   - Shock absorber lower mounting bolts
   - Lower ball joints from the knuckles
7. Install a transmission support fixture to the transmission.
   - Wiring harness and brake pipe clip retainers from the rear crossmember
   - Lower transmission-to-differential nut
   - Transaxle mount-to-rear crossmember nut
   - Rear crossmember retaining nuts while supporting the crossmember
   - Crossmember
   - Transaxle mount bracket
8. Separate the axle shafts from the differential and tie them to the underbody.
9. Release the retainer securing the wiring harness from the "L" shaped brackets along the driveline support and move the harness out of the way.
10. Lower the driveline slightly and tilt it to gain access to the electrical connectors.
11. Remove or disconnect the following:
    - Vehicle Speed Sensor (VSS) electrical connector
    - Wire harness retainer from the differential rear cover stud
    - Wire harness retainer clip from the top of the differential
    - Transmission harness 20 way connector
    - Park Neutral Position (PNP) switch electrical connectors

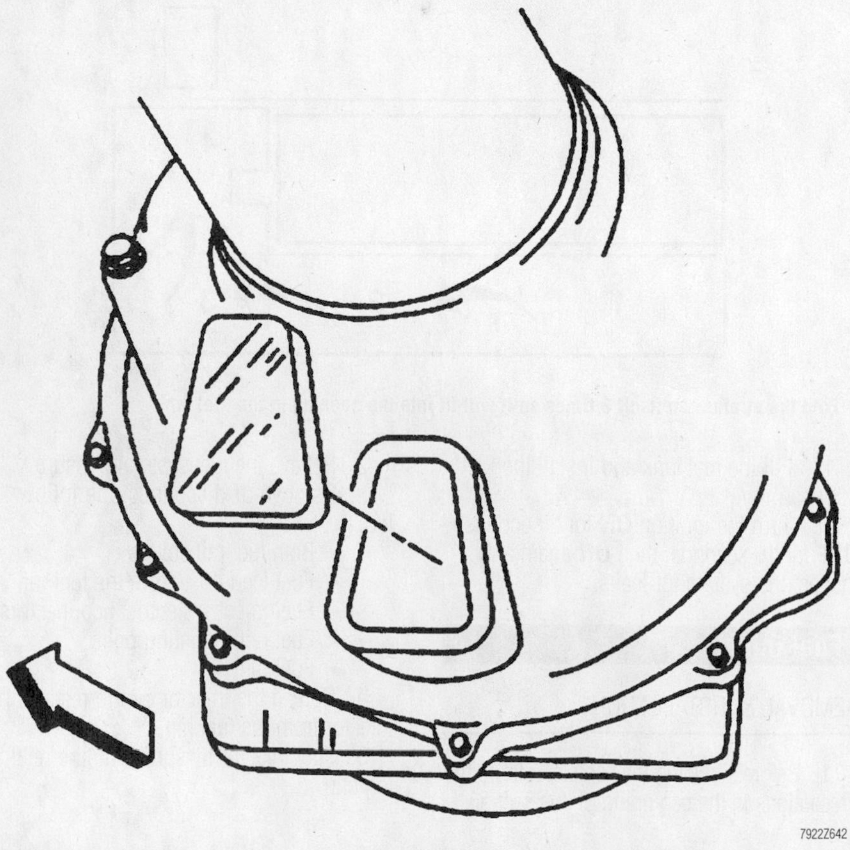

Remove the inspection plug, then remove the flexplate-to-torque converter bolts

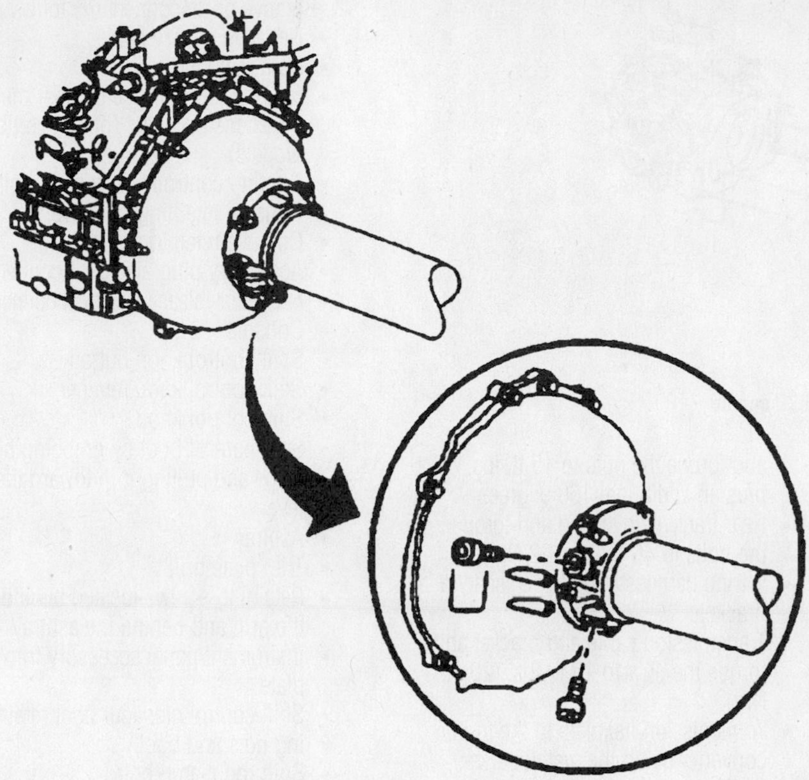

Remove the plugs and install two M10 x 1.5 x 55mm or longer bolts into the bolt holes to secure the bearing—5.7L engine

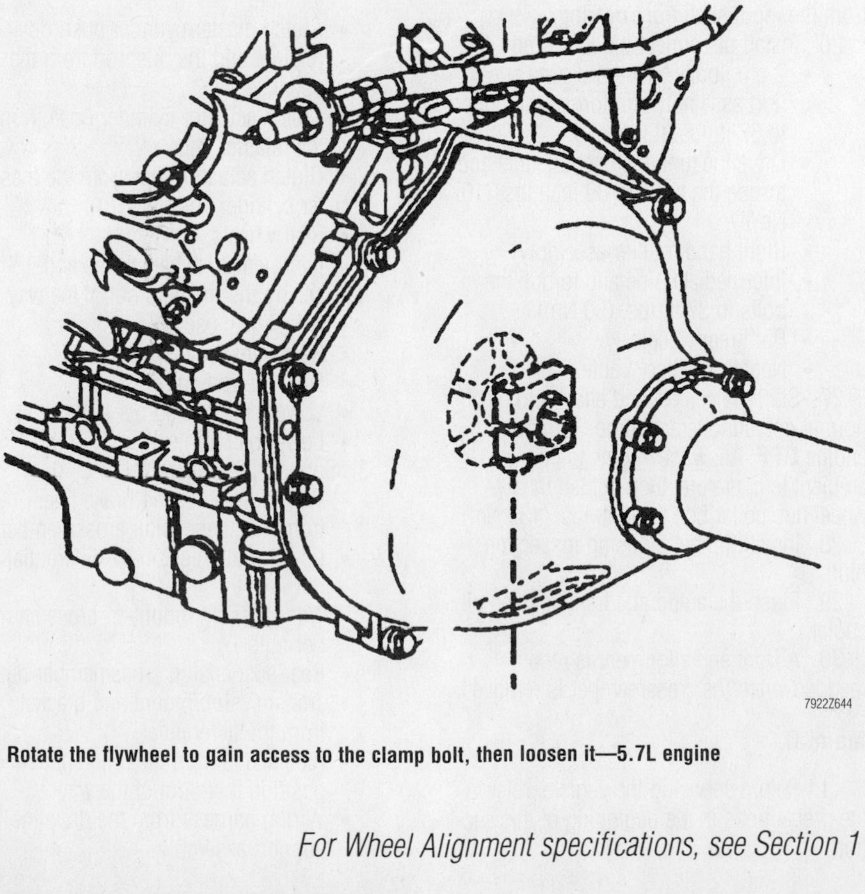

Rotate the flywheel to gain access to the clamp bolt, then loosen it—5.7L engine

- Bolt securing the wire harness to the left side of the case

12. Lower the driveline and angle it while observing the top rear of the differential and the lowest part of the rear compartment floor panel. The engine Positive Crankcase Ventilation (PCV) pipes will most likely contact the dash panel. Make certain not to damage the dash.
13. Remove or disconnect the following:
    - Wire harness from the retainer along the top of transmission
    - Transmission rear oil cooler pipes from the junction fittings at the flywheel housing
    - 5 driveline to flywheel housing bolts after supporting the rear of the engine
14. Bend the wiring harness bracket away from the driveline, toward the tunnel wall to have access to remove the driveline.
15. Separate the driveline from the engine by prying them apart with a flat blade tool.
16. Lower the driveline and tilt it away from the engine until the input shaft clears the flywheel housing.
17. Remove or disconnect the following:
    - Driveline
    - Transmission oil cooler rear pipes from the fittings
    - Transmission to driveline bolts
    - Separate the transmission from the driveline with a flat blade tool
18. Using a lifting device, place the assembly on a bench.
    - Rear transmission oil cooler fittings from the transmission
    - Transmission-to-driveline assembly support bolts
    - Driveline from the transmission while supporting the torque converter
    - Differential-to-transmission bolts
    - Differential from the transmission

To install:

19. Install or connect the following:
    - Differential to the transmission and torque the bolts to 37 ft. lbs. (50 Nm)
    - Transmission to the driveline support and torque the bolts to 37 ft. lbs. (50 Nm)
    - Rear transmission oil cooler fittings to the transmission and torque the fittings to 30 ft. lbs. (40 Nm)
20. Using a chain hoist, place the assembly on a transmission jack.
21. Carefully raise the assembly into the

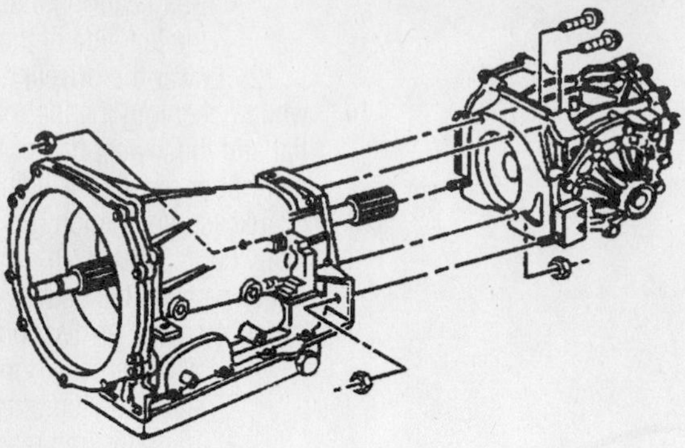

Transmission-to-differential mounting bolt locations—5.7L engine

vehicle while placing the wiring harness loosely into the harness retaining slots.

22. Align the assembly for installation into the engine. The driveline will slide into the rear of the engine as long as the angles are the same.

23. Install or connect the following:
   - Driveline assembly in the vehicle. Reposition the wiring harness bracket to align with the appropriate hole in the driveline assembly.
   - Driveline support-to-flywheel housing bolts and torque them to 37 ft. lbs. (50 Nm)
   - Oil cooler lines and torque the fittings to 20 ft. lbs. (27 Nm)
   - Wiring harness to the left side of the transmission and torque the bolt to 22 inch lbs. (2.5 Nm)
   - PNP switch electrical connector
   - Transmission harness 20 way connector
   - Wiring harness clip to the top of the differential
   - VSS electrical connector

24. Raise the transmission to installation height.
   - Axle shafts
   - Transmission mount to the crossmember and torque the nuts to 37 ft. lbs. (50 Nm)
   - Suspension crossmember and torque the nuts to 81 ft. lbs. (110 Nm)
   - Lower transmission-to-differential nut and torque it to 37 ft. lbs. (50 Nm)
   - Wiring harness and brake pipe clip retainers to the crossmember
   - Lower ball joints to the suspension knuckles and torque the nuts to 52 ft. lbs. (70 Nm)
   - Shock absorber lower mounting bolts and torque them to 162 ft. lbs. (220 Nm)
   - Outer tie rod ends to the knuckles and torque the nuts to 15 ft. lbs. plus an additional 160 degrees
   - Rear transverse spring and torque the bolts to 46 ft. lbs. (62 Nm)
   - Wiring harness into the "L" shaped brackets
   - Transmission cable and bracket and torque the nuts to 15 ft. lbs. (20 Nm)
   - Transmission flexplate to the torque converter using the matchmarks made during removal and torque the bolts to 47 ft. lbs. (63 Nm)
   - Rear bell housing access plug
   - Propeller shaft hub bolt until it is finger tight

25. Remove the two M10 x 55mm bolts from the input shaft front bearing.

26. Install or connect the following:
   - 2 plug bolts to the driveline support assembly and torque the bolts to 37 ft. lbs. (50 Nm)
   - Driveline tunnel closeout panel and torque the bolts to 89 inch lbs. (10 Nm)
   - Right hand muffler assembly
   - Intermediate pipe and torque the bolts to 37 ft. lbs. (50 Nm)
   - Both rear wheels
   - Negative battery cable

27. Start the vehicle and allow it to reach normal operating temperature. Turn the engine **OFF**. Allow the engine to cool to ambient temperature, then tighten the flywheel hub collar bolt to 96 ft. lbs. (130 Nm).

28. Install the bell housing inspection plug.

29. Flush the automatic transmission oil cooler.

30. A front end alignment is recommended when the crossmember is removed.

Manual

1. Before servicing the vehicle, refer to the precautions in the beginning of this section.
2. Remove or disconnect the following:
   - Negative battery cable
   - Both rear wheels
   - Folding top stowage compartment lid extension panel (on convertible models)
   - Traction control/ride control switch
   - Console retaining nut covers
   - Console retaining nuts
   - Accessory plug electrical connector
   - Fuel door release electrical connector
   - Console
   - Shift control knob button
   - Shift control knob retainer
   - Shift control knob
   - Shift control boot by grabbing both sides and pulling it in toward the lever
   - Ashtray
   - Trim plate grill
   - Retaining screws located behind the grill and behind the ashtray
   - Instrument panel accessory trim plate
   - Shift control closeout boot retaining nuts and boot
   - Shift rod clamp bolt
   - Shift control mounting bolts
   - Shift control assembly
   - Courtesy lamp
   - Left side lower closeout panel while guiding the courtesy lamp through the hole
   - Clutch master cylinder pushrod retainer and the pushrod from the pedal
   - Clutch actuator cylinder hose from the retainer clip
   - Clutch actuator hose from the master cylinder hose
   - Rear wheels
   - Intermediate exhaust pipe and secure the mufflers out of the way
   - Driveline closeout panel
   - Rear transverse spring
   - Outer tie rod ends
   - Lower shock absorber bolts
   - Lower ball joints
3. Support the transmission
   - Wiring harness and brake lines from the suspension crossmember
   - Lower transmission-to-differential nut
   - Transmission mount-to-crossmember nuts
   - Rear suspension crossmember nuts
   - Transmission mount and bracket from the differential
   - Axle shafts from the differential and position them out of the way
   - Wiring harness from the driveline support assembly

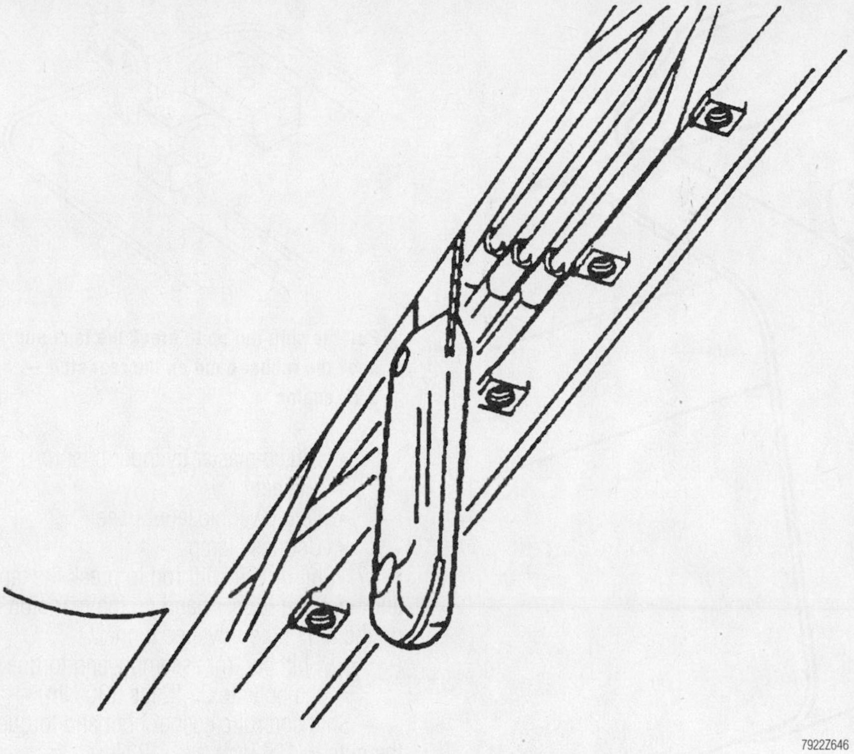

Insert a flat bladed tool between the shifter bracket and brake liner retainer before lowering the transmission assembly—5.7L engine

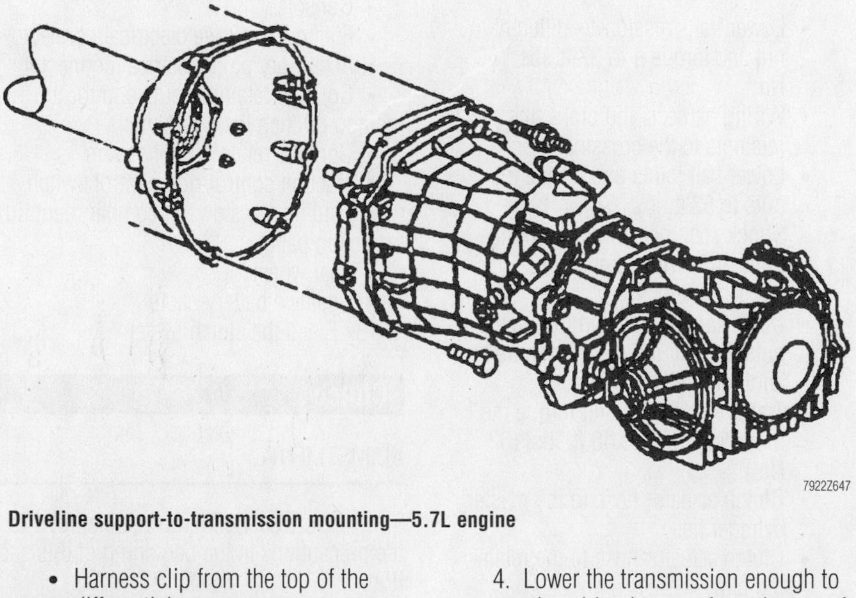

Driveline support-to-transmission mounting—5.7L engine

- Harness clip from the top of the differential

⁂ WARNING

Do not lower the top rear portion of the differential past the bottom of the storage compartment or the PCV pipe will hit the dash panel possibly causing damage.

4. Lower the transmission enough to remove the wiring harness from the top of the assembly.
5. Remove or disconnect the following:
   - Vehicle Speed Sensor (VSS) electrical connector
   - Backup lamp switch electrical connector
   - Reverse lockout solenoid electrical connector
   - Gear select (skip shift) solenoid electrical connector
   - Transmission fluid temperature sensor electrical connector (if equipped)
6. Support the rear of the engine
   - Driveline support-to-bell housing bolts
   - Driveline support assembly out of the bell housing while moving the assembly rearward
7. Carefully lower the assembly away from the vehicle while simultaneously adjusting the angle.
8. Attach a chain hoist to the assembly, then remove it from the jack and place it on a workbench.
9. Remove or disconnect the following:
   - Driveline support-to-transmission mounting bolts. Carefully pry the driveline assembly away from the transmission while guiding the shift rod through the opening in the driveline support
   - Transmission shift rod from the transmission
   - Transmission-to-differential mounting bolts and separate the differential from the transmission

To install:

10. Install or connect the following:
    - Differential to the transmission and torque the bolts to 37 ft. lbs. (50 Nm)
    - Transmission shift rod
    - Driveline support to the transmission and torque the bolts to 37 ft. lbs. (50 Nm)
11. To aid in later installation place a rubber band around the shift rod then tape the shift rod to the driveline support assembly.
12. Place the assembly on the transmission jack with a chain hoist.
13. Begin to raise the assembly into the vehicle while loosely installing the wiring harness along the driveline assembly retaining slots.
14. Have an assistant guide the front of the driveline support to the bell housing.
15. Be sure the driveline assembly is at the same angle as the engine before trying to install it.
16. Install or connect the following:
    - Input propeller shaft into the clutch disc
    - Wiring harness bracket to align

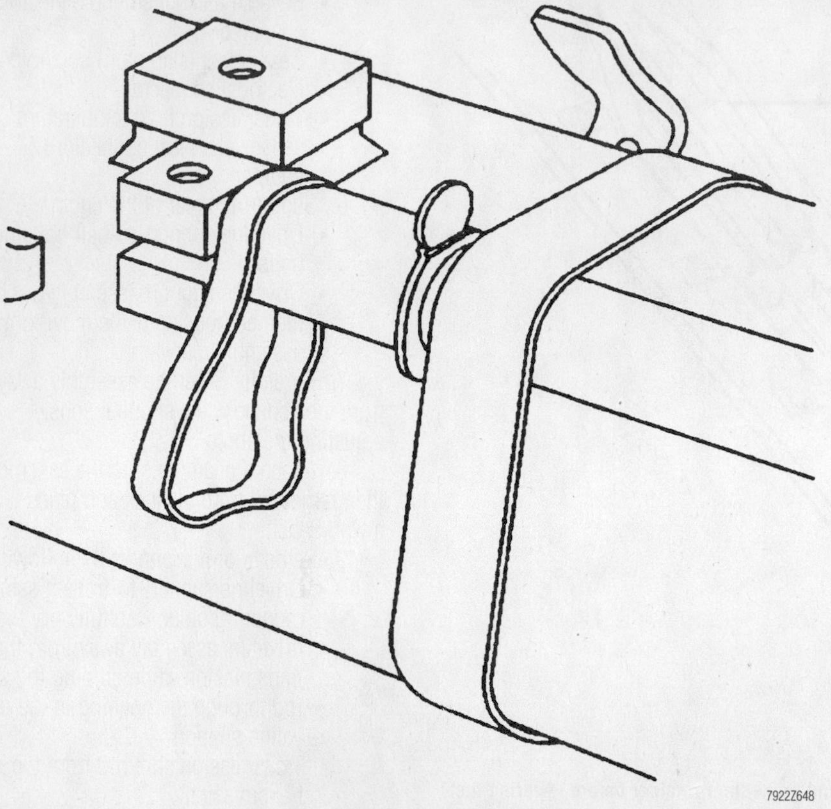

Place a rubber band on the shift rod, then tape the rod to the driveline support assembly—5.7L engine

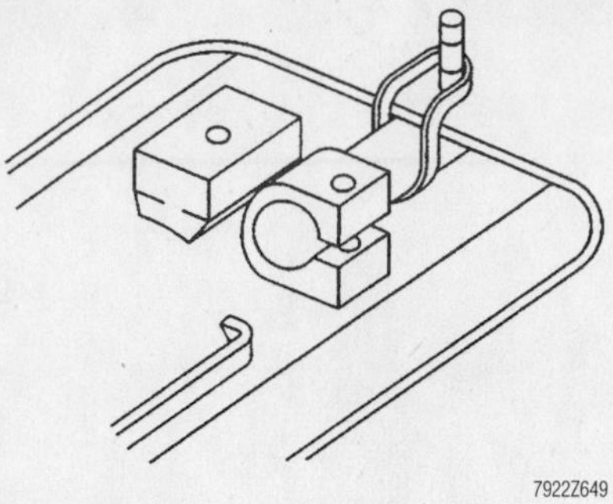

Pull the shift rod up to break the tape and hook the rubber band on the rear stud—5.7L engine

with the appropriate hole in the driveline assembly

- Driveline support-to-flywheel housing bolts and torque them to 37 ft. lbs. (50 Nm)
- Wiring harness to the retainer on the top of the transmission and attach the connectors
- Transmission fluid temperature sensor electrical connector (if equipped)
- Gear select/skip shift solenoid electrical connector
- Reverse lockout solenoid electrical connector
- Backup lamp switch electrical connector
- Wiring harness to the top of the differential
- VSS electrical connector
- Axle shafts in the differential
- Transmission mount on the differential and torque the bolts to 37 ft. lbs. (50 Nm)
- Rear suspension crossmember and torque the nuts to 81 ft. lbs. (110 Nm)
- Transmission mount-to-crossmember nuts and torque them to 37 ft. lbs. (50 Nm)
- Lower transmission-to-differential nut and torque it to 37 ft. lbs. (50 Nm)
- Wiring harness and brake line retainers to the crossmember
- Lower ball joints and torque the nuts to 52 ft. lbs. (70 Nm)
- Shock absorber lower mounting bolts and torque them to 162 ft. lbs. (220 Nm)
- Outer tie rod ends and torque the nuts to 15 ft. lbs. (20 Nm) plus an additional 160 degrees
- Transverse spring and torque the mounting bolts to 46 ft. lbs. (62 Nm)
- Clutch actuator hose to the master cylinder hose
- Clutch actuator hose to the retaining clip
- Driveline tunnel closeout panel and torque the bolts to 89 inch lbs. (10 Nm)
- Muffler assemblies
- Intermediate exhaust pipe and torque the bolts to 37 ft. lbs. (50 Nm)
- Rear wheels
- Clutch master cylinder pushrod to the clutch pedal
- Clutch master cylinder pushrod retainer
- Left lower closeout panel
- Courtesy lamp

17. Pull up the shift rod to break the tape and hook the rubber band on the rear stud on the top of the driveline tunnel.

- Shift control assembly and torque the bolts to 22 ft. lbs. (30 Nm)
- Shift control closeout boot and torque the nuts to 106 inch lbs. (12 Nm)
- Instrument panel accessory trim plate
- Shift control boot
- Shift control knob
- Shift control knob retainer and button
- Console
- Fuel door release electrical connector
- Accessory plug electrical connector
- Console retaining nuts and torque them to 89 inch lbs. (10 Nm)
- Console retaining nut covers
- Traction control/ride control switch
- Folding top stowage compartment lid extension panel
- Rear wheels
- Negative battery cable

18. Bleed the clutch system.

Clutch

ADJUSTMENT

1. Before servicing the vehicle, refer to the precautions in the beginning of this section.
2. Remove the flywheel inspection cover.
3. Have an assistant depress the clutch pedal until the tension is released from the stepped adjusting ring.
4. Using 2 screwdrivers, rotate the stepped adjusting ring counterclockwise until fully adjusted out.
5. Continue to hold them in this position and have the assistant release the clutch pedal.

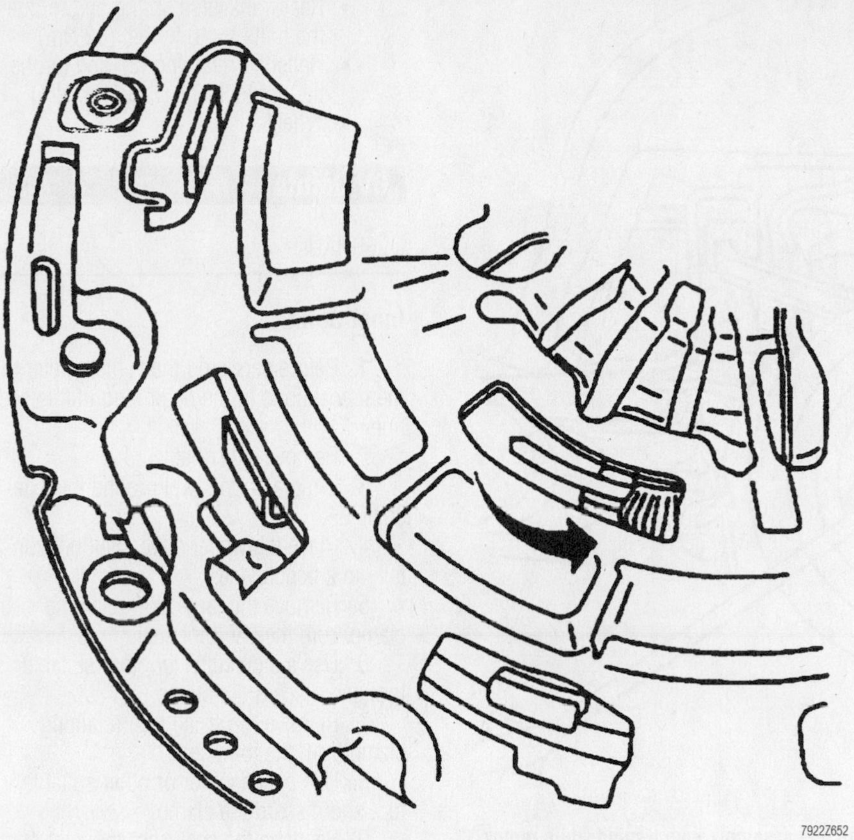

Rotate the stepped adjusting ring counterclockwise to compress the springs

6. Remove the screwdrivers.
7. Install the inspection cover. Torque the bolts to 18 ft. lbs. (25 Nm).

REMOVAL & INSTALLATION

1. Before servicing the vehicle, refer to the precautions in the beginning of this section.
2. Remove or disconnect the following:

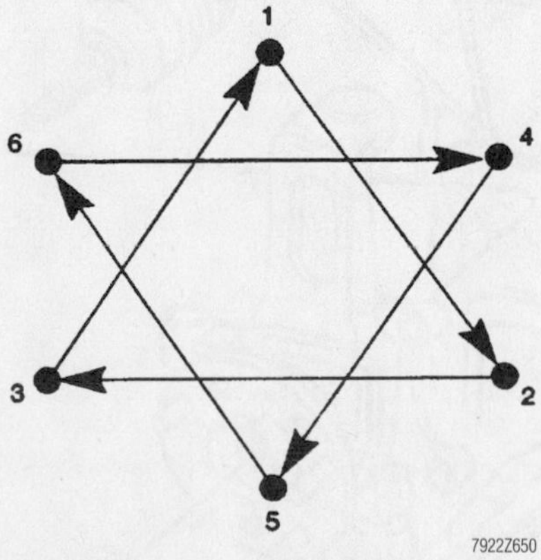

Clutch assembly torque sequence

- Negative battery cable
- Exhaust system
- Driveline support and transmission as an assembly
- Flywheel inspection cover
- Pressure plate bolts. It will be necessary to rotate the flywheel to access all the bolts
- Pressure plate and disc

To install:

3. Install or connect the following:
   - Clutch disc and pressure plate on the flywheel
   - Clutch disc alignment tool through the disc to keep it in place
   - Pressure plate bolts finger-tight. Turn the flywheel to access all the bolt holes
4. Torque the bolts in the sequence shown, using 3 steps to 52 ft. lbs. (70 Nm).
   - Inspection cover and torque the bolts to 18 ft. lbs. (25 Nm)
   - Driveline support and transmission assembly
   - Exhaust system
   - Negative battery cable
5. Bleed the clutch system.

Hydraulic Clutch System

BLEEDING

1. Before servicing the vehicle, refer to the precautions in the beginning of this section.
2. Fill the clutch master cylinder with clean clutch hydraulic fluid.
3. Raise and safely support the vehicle with an assistant in it.
4. Remove the intermediate exhaust pipe.
5. Remove the driveline tunnel cover.
6. Have the assistant depress and hold the clutch pedal down.
7. Loosen the bleeder screw on the actuator cylinder to release the air, then tighten the screw.

⁂ WARNING

Do not allow the clutch pedal to be released until the bleeder screw is closed or air will be drawn into the system.

8. Repeat the prior two steps until the air has been purged from the system. Check the master cylinder fluid level and refill as needed.
9. Install the driveline tunnel cover.
10. Install the intermediate exhaust pipe.
11. Lower the vehicle.

Halfshaft

REMOVAL & INSTALLATION

1. Before servicing the vehicle, refer to the precautions in the beginning of this section.
2. Apply the parking brake.
3. Remove or disconnect the following:
   - Wheel
   - Axle nut
   - Rear transverse spring
   - Outer tie rod end from the knuckle
   - Antilock Brake System (ABS) Wheel Speed Sensor (WSS) electrical connector
   - Park brake cable from the lever and bracket

➡Be sure to support the halfshaft until it is removed. Do not let it hang by the CV-joint.

4. Attach a puller on the wheel studs and start to push the axle shaft into the hub

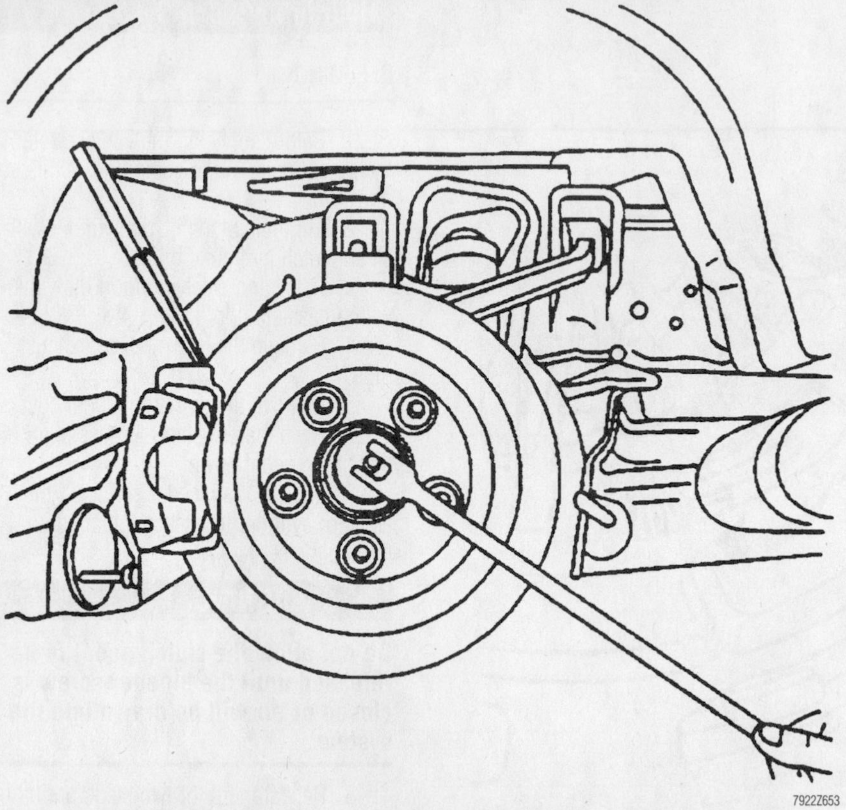

Insert a large drift through the cooling fins to keep the hub assembly from turning while removing the retaining nut—5.7L engine

assembly. This will provide clearance for the ball joint to be separated from the knuckle assembly.

5. Remove or disconnect the following:
   - Ball joint from the knuckle
   - Axle shaft from the hub
   - Halfshaft assembly from the differential by inserting a suitable tool between the CV-joint and differential and prying them apart

To install:

6. Install or connect the following:
   - Halfshaft on the differential output shaft. Use light force to be sure it is fully seated
   - Halfshaft through the hub assembly but do not install completely. This will provide clearance for installing the ball joint
   - Ball joint to the knuckle and torque the nut to 41 ft. lbs. (55 Nm)
   - Halfshaft through the hub assembly completely
   - Park brake cable to the bracket and the lever
   - WSS electrical connector
   - Tie rod end to the knuckle and torque the nut to 15 ft. lbs. (20 Nm) plus an additional 160 degrees
   - Rear transverse spring and torque the bolts to 46 ft. lbs. (62 Nm)
   - Halfshaft retaining nut and torque the nut to 118 ft. lbs. (160 Nm)
   - Wheel

CV-Joints

OVERHAUL

Inner Joint

1. Before servicing the vehicle, refer to the precautions in the beginning of this section.
2. Remove axle shaft.
3. Wrap a shop towel around the axle shaft.
4. Place the wheel drive shaft horizontally in a bench vise.
5. Remove the large seal retaining clamp from the CV joint seal.
6. Use a side cutter or other suitable tool and discard the clamp.
7. Remove the small seal retaining clamp from the joint seal.
8. Use a side cutter or other suitable tool and discard the clamp.
9. Separate the seal from the joint outer race at the large diameter end.

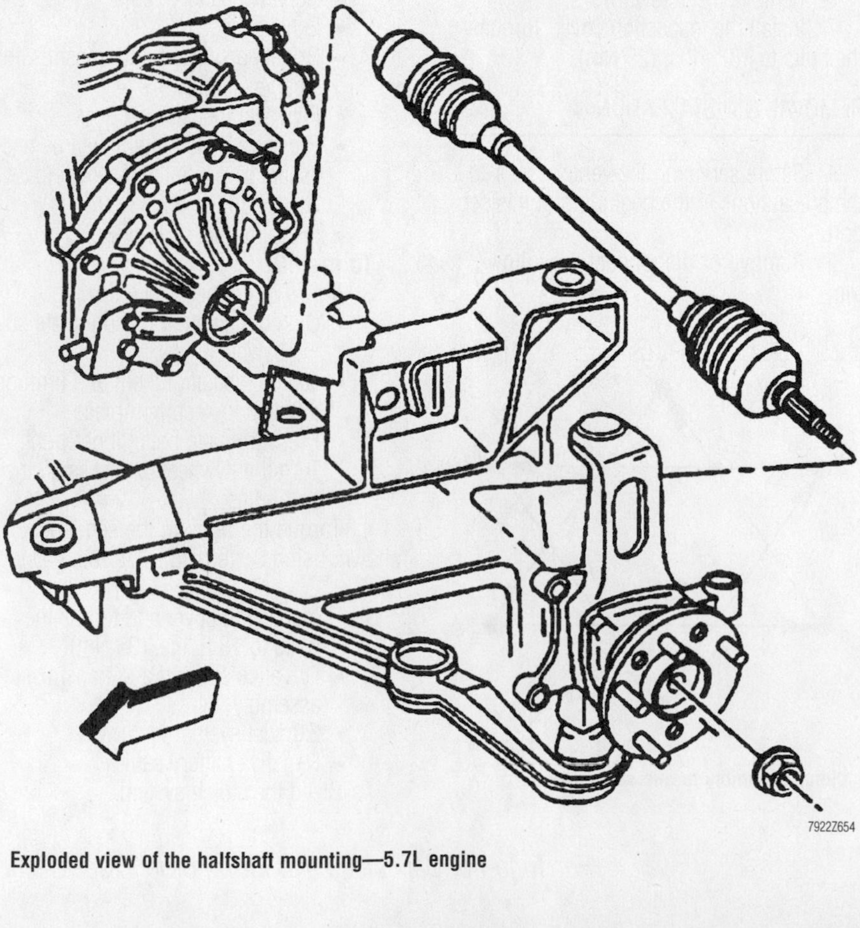

Exploded view of the halfshaft mounting—5.7L engine

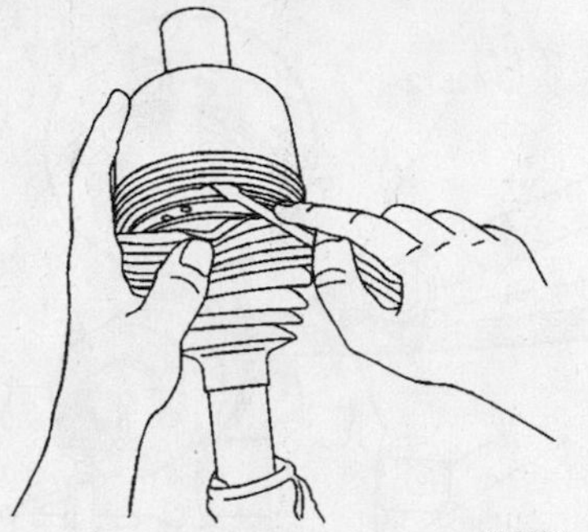

9346ZGB3

Separate the seal from the joint outer race at the large diameter end–inner joint

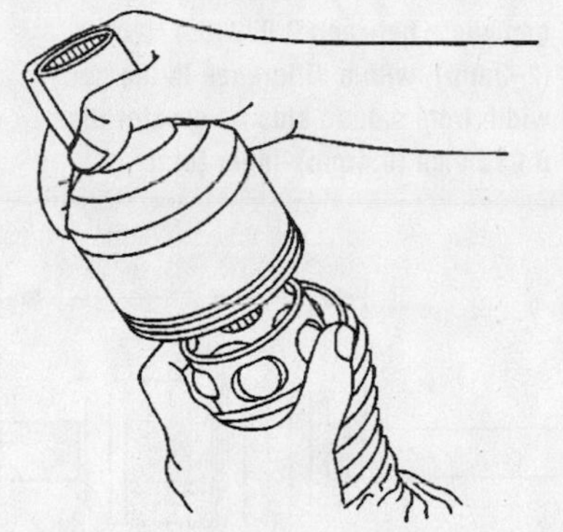

9346ZGB4

Remove the outer race from the axle shaft–inner joint

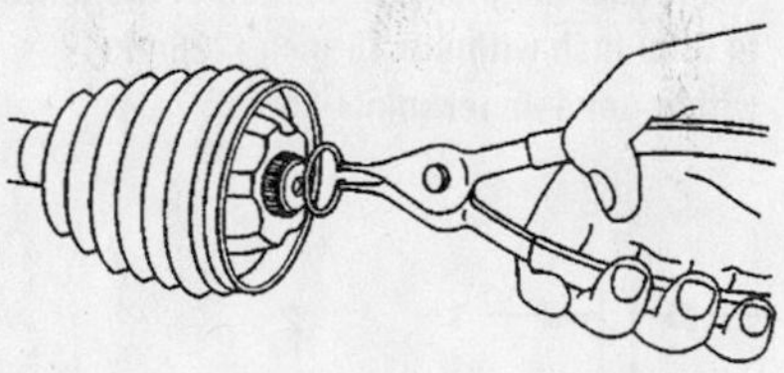

9346ZGB5

Remove the snap ring from the axle shaft–inner joint

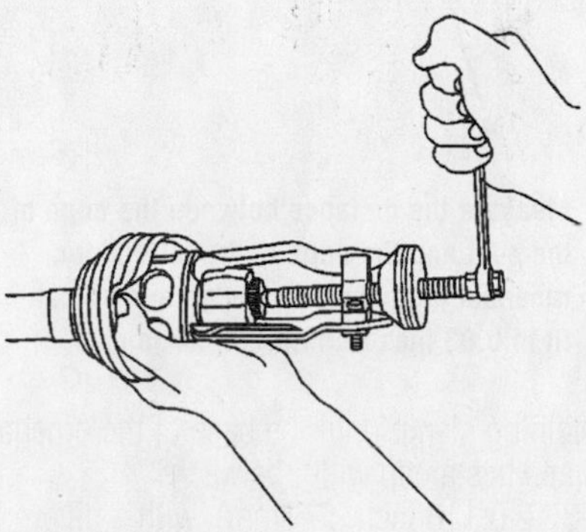

9346ZGB6

Remove the inner race from the axle shaft using a three jaw puller–inner joint

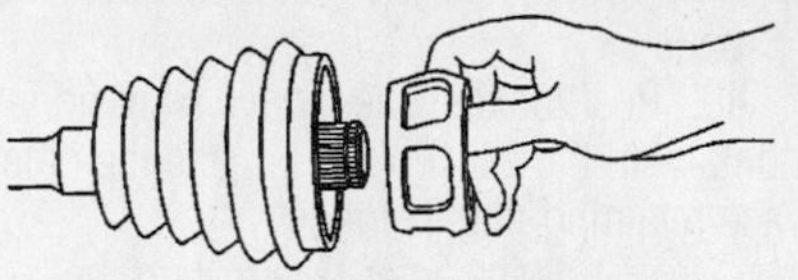

9346ZGB7

Remove the cage from the axle shaft–inner joint

10. Position the seal behind the joint face.
11. Position the wheel drive shaft vertically in the bench vise so the inner joint is up.
12. Slide the joint outer race down toward the vise.
13. Disengage the outer race retaining ring as follows:
    a. Insert a small flat bladed screwdriver between the retaining ring and the outer race.
    b. Pry the retaining ring from the outer race.
    c. Position the retaining ring along the axle shaft away from the outer race.

➡The balls may fall out of the cage and inner race when the outer race is removed.

14. Remove the outer race from the axle shaft as follows:
    a. Use the seal to catch any balls which are not retained by grease.
    b. Lift the outer race off the axle shaft.
15. Remove any remaining balls from the cage and inner race.
16. Remove any balls caught by the seal.
17. Position the wheel drive shaft horizontally in the bench vise.
18. Remove the outer race retaining ring from the axle shaft.
19. Remove the snap ring from the axle shaft.
20. Align the cage lands with the inner race ball tracks.
21. Reposition the cage along the axle shaft away from the inner race.
22. Wipe the grease from the inner race.
23. Remove the inner race from the axle shaft using a three jaw puller.
24. Remove the cage from the axle shaft.
25. Remove the seal from the axle shaft.
26. Remove the wheel drive shaft from the bench vise.

➡All traces of old grease and any contaminates must be removed.

27. Clean al parts thoroughly with clean solvent.
28. Thoroughly air dry all the parts.

To install:

➡Protect the wheel drive shaft boots, seals and clamps from sharp objects when servicing on or near the wheel drive shaft(s). Damage to the boot(s), the seal(s) or the clamp(s) may cause lubricant to leak from the joint and lead to increased noise and possible failure of the wheel drive shaft.

29. Wrap a shop towel around the axle shaft.
30. Place the wheel drive shaft horizontally in a bench vise.
31. Install a new small seal retaining clamp onto the wheel drive shaft.
32. Install the seal onto the axle shaft.
33. Install the cage onto the axle shaft so the smaller diameter end faces the vise.

➡The inner race spline relief must face away from the end of the axle shaft.

34. Install the inner race onto the axle shaft as follows:
    a. Engage the inner race splines onto the axle shaft splines. Be sure to install the inner race spline relief side onto the axle shaft first.
    b. Position a wood block squarely over the end of the inner race.
    c. Use a hammer to begin to drive the inner race onto the axle shaft.
    d. Reposition the wood block along the face of the inner race to avoid the axle shaft.
    e. Work evenly around the inner race and continue to drive the inner race, until you feel the inner race seat fully onto the axle shaft.
35. Inspect to be sure that the axle shaft snap ring groove is exposed.
36. Install the snap ring to the axle shaft.
37. Position the cage so the cage lands align with the inner race ball tracks.
38. Install the cage onto the inner race.
39. Position the cage windows to align with the inner race ball tracks.
40. Insert approximately 60 percent of the grease from the service kit into the outer race.
41. Position the wheel drive shaft vertically in the bench vise so the inner joint end is up.
42. Apply a small amount of the grease from the service kit to the cage windows and inner race ball tracks.
43. Insert the remaining grease from the service kit into the seal.

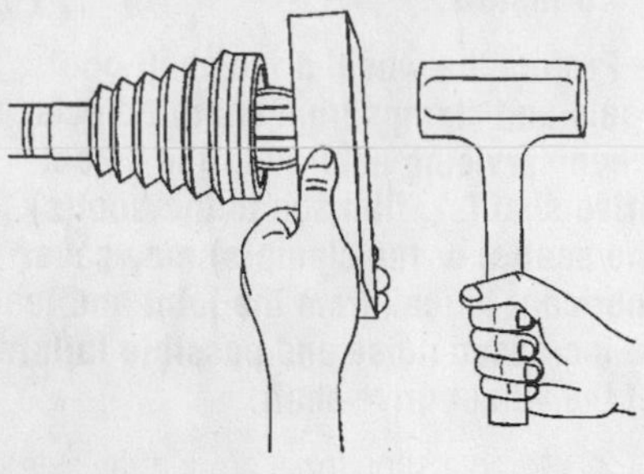

9346ZGB8

Use a hammer and a block of wood to drive the inner race onto the axle shaft–inner joint

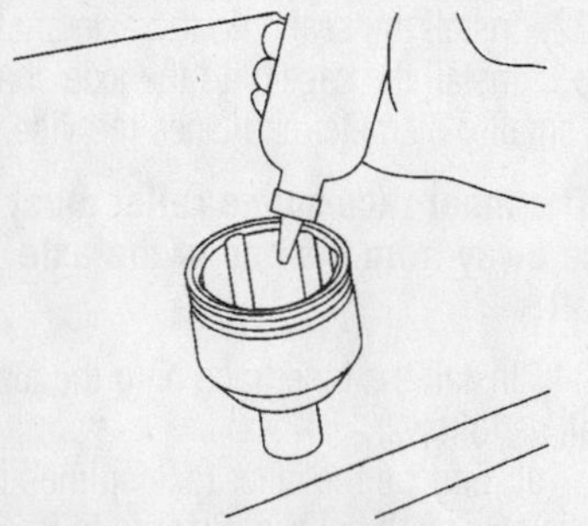

9346ZGB9

Insert approximately 60 percent of the grease from the service kit into the outer race–inner joint

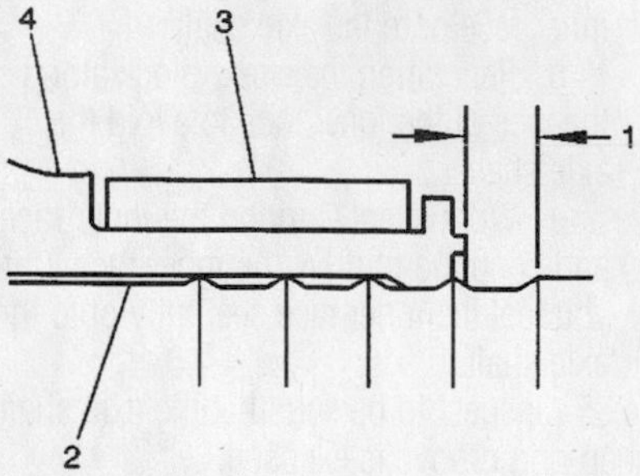

9346ZGC1

Measure the distance between the edge of the seal and the edge of the last axle shaft groove closing edge and adjust fit to 0.10 inch (2.5mm)–inner joint

44. Install the outer race retaining ring onto the axle shaft.
45. Position the retaining ring below the cage, toward the vise.
46. Install the balls through the cage windows to the inner race ball tracks.
47. Use the seal to keep the balls in position if necessary.
48. Install the outer race onto the axle shaft as follows:
   a. Be careful not to allow the grease in the outer race to leak out.
   b. Align the outer race ball tracks to the balls.
   c. Slide the outer race down over the balls.
49. Position the wheel drive shaft horizontally in the bench vise. Engage the outer race retaining ring as follows:
   a. Slide the outer race toward the vise.
   b. Insert the outer race retaining ring into the groove along the outer edge of the outer race.
   c. Position the outer race retaining ring so the opening in the ring aligns with an outer race land (not a ball track).
50. Position the large diameter end of the seal onto the outer race.
51. Position the small seal retaining clamp onto the neck of the seal.
52. Position the seal and small retaining clamp to the axle shaft.
53. Measure the distance between the edge of the seal and the edge of the last axle shaft groove closing edge and adjust fit to 0.10 inch (2.5mm).

➡The seal retaining clamp must not be over-tightened or under-tightened.

54. Crimp the small seal retaining clamp using tool J 42572. Tighten the small seal retaining clamp until the base of the omega shape has a gap width between 0.079–0.118 inch (2–3mm), with a difference in the gap width from side to side no greater than 0.016 inch (0.4mm). The clamping hold time must be no less than 2 seconds.
55. Measure the distance between the end of the seal and the end of the joint outer race and adjust the plunging motion of the joint to 8.90 inch within 0.16 inch (226mm within 4mm).
56. Position the large seal retaining clamp onto the seal.
57. Position the seal and large retaining clamp to the joint outer race. Measure the distance between the edge of the seal and the edge of the joint outer race last groove closing edge and adjust fit to 0.03 inch (0.8mm).

➡The seal must not be dimpled, stretched or out of shape in any way.

58. Inspect the seal for proper shape. If the seal is NOT shaped correctly, equalize the pressure in the seal and shape the seal properly by hand. Inspect the seal for damage. If the seal has been cut or punctured during assembly, you must discard and replace the seal.

➡The seal retaining clamp must not be over-tightened or under-tightened.

59. Crimp the large seal retaining clamp using tool J 42572. Tighten the small seal retaining clamp until the base of the omega shape has a gap width between 0.079–0.118 inch (2–3mm), with a difference in the gap width from side to side no greater than 0.016 inch (0.4mm). The clamping hold time must be no less than 2 seconds.
60. Remove the wheel drive shaft from the bench vise.

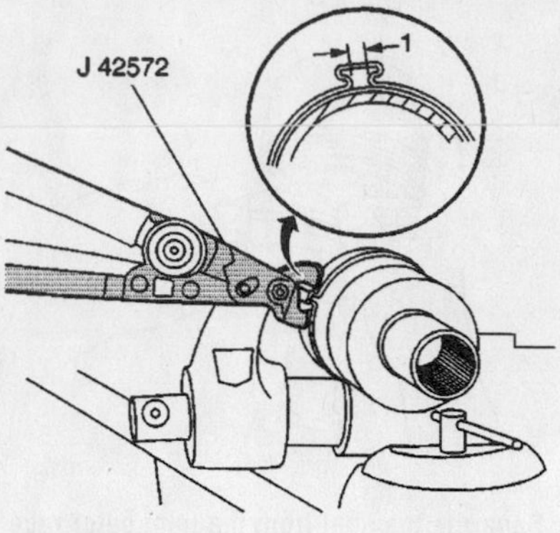

9346ZGC2

Tighten the small seal retaining clamp until the base of the omega shape has a gap width between 0.079–0.118 inch (2–3mm), with a difference in the gap width from side to side no greater than 0.016 inch (0.4mm)–inner joint

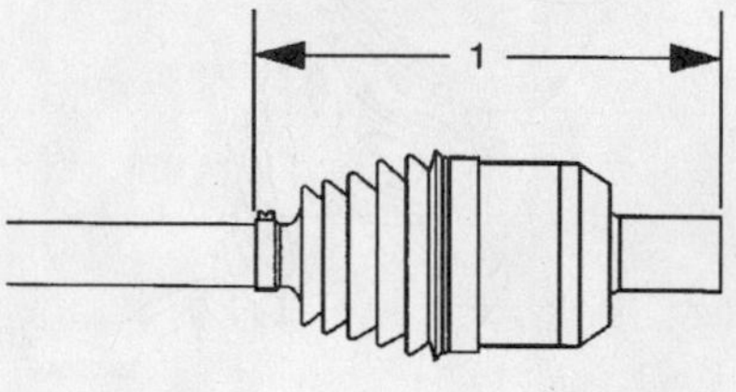

9346ZGC3

Measure the distance between the end of the seal and the end of the joint outer race and adjust the plunging motion of the joint to 8.90 inch within 0.16 inch (226mm within 4mm–inner joint

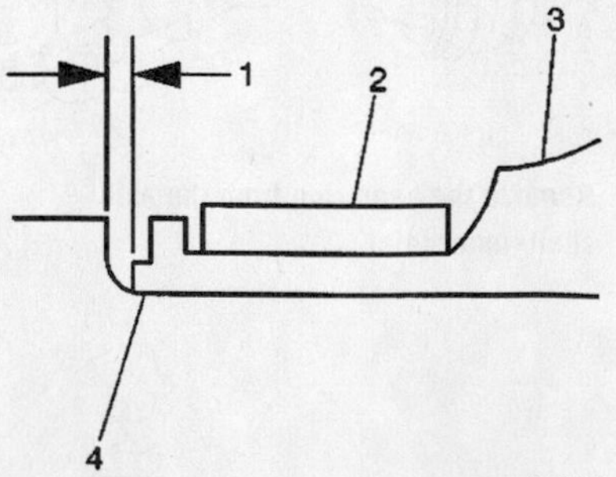

9346ZGC4

Measure the distance between the edge of the seal and the edge of the joint outer race last groove closing edge and adjust fit to 0.03 inch (0.8mm)–inner joint

61. Distribute the grease within the inner CV joint.

62. Plunge the joint back and forth four or five times.

63. Inspect the inner CV joint and wheel drive shaft for smooth operation as follows:

a. Hold the wheel drive shaft vertically, with the outer joint at the bottom.

b. Rotate the wheel drive shaft four or five times in a circular motion.

64. Install the axle shaft.

Outer Joint

1. Before servicing the vehicle, refer to the precautions in the beginning of this section.

2. Remove or disconnect the following:

- Axle shaft from the vehicle
- Large CV boot retaining clamp
- Small CV boot retaining clamp
- CV boot from the joint

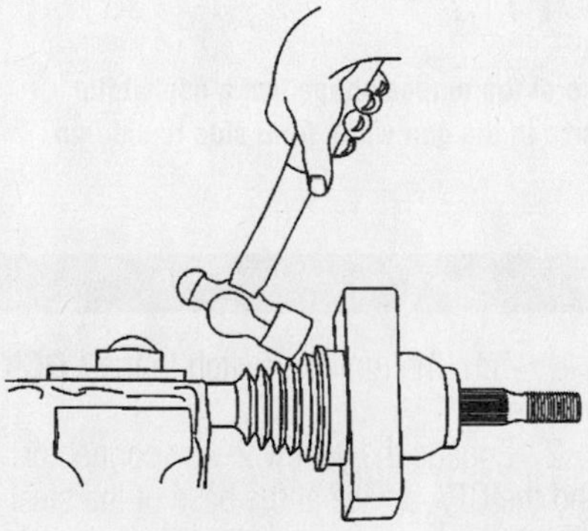

9346ZGC5

Separate the outer joint from the axle shaft using a hammer and wood block–outer joint

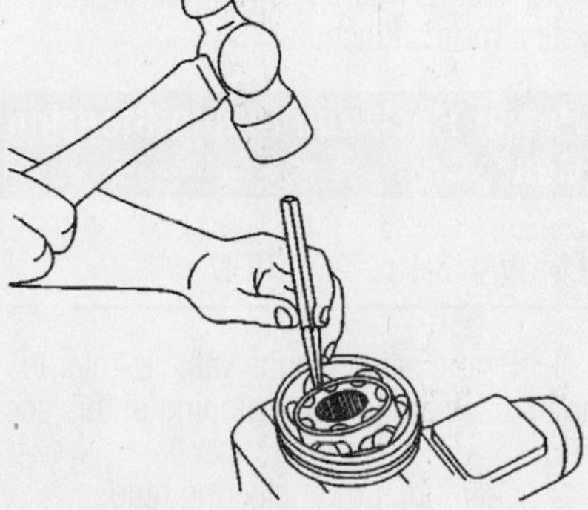

9346ZGC6

Position a brass drift against the CV-joint cage and tap it with a hammer to tilt the cage–outer joint

- Outer joint from the axle shaft using a hammer and wood block
- Axle shaft retaining ring
- CV boot

3. Disassemble the chrome alloy balls from the CV-joint cage as follows:

a. Position a brass drift against the CV-joint cage and tap it with a hammer to tilt the cage.

b. Remove the 1st chrome alloy ball from the cage.

c. Tilt the cage in the opposite direction.

d. Remove the opposite chrome alloy ball.

e. Repeat the procedure until all 6 balls are removed.

4. Disassemble the CV-joint cage and inner race as follows:

a. Pivot the cage and race 90 degrees to the center line of the outer race.

b. Align the cage windows with outer race lands.

c. Remove the cage from the outer race.

d. Rotate the inner race upward and remove it from the cage.

To install:

5. Lubricate the parts with a light coat of grease.

6. Assemble the CV-joint cage and inner race, as follows:

a. Rotate the inner race 90 degrees to the cage centerline.

b. Align the cage windows with inner race lands.

c. Insert the inner race into the cage by rotating the inner race downward.

d. Insert the cage/inner race into the outer race.

7. Assemble the chrome alloy balls into the CV-joint cage, as follows:

a. Position a brass drift against the CV-joint cage and tap it with a hammer to tilt the cage.

b. Insert the 1st chrome alloy ball into the cage.

c. Tilt the cage in the opposite direction.

d. Insert the opposite chrome alloy ball.

e. Repeat the procedure until all 6 balls are inserted.

8. Install ½of the grease provided, into the CV-joint.

9. Install or connect the following:

- Small CV boot retaining ring
- CV boot on the halfshaft
- New retaining ring on the half-shaft
- Large ring clamp on the CV boot
- Outer joint onto the axle shaft

10. Compress the retaining ring using a

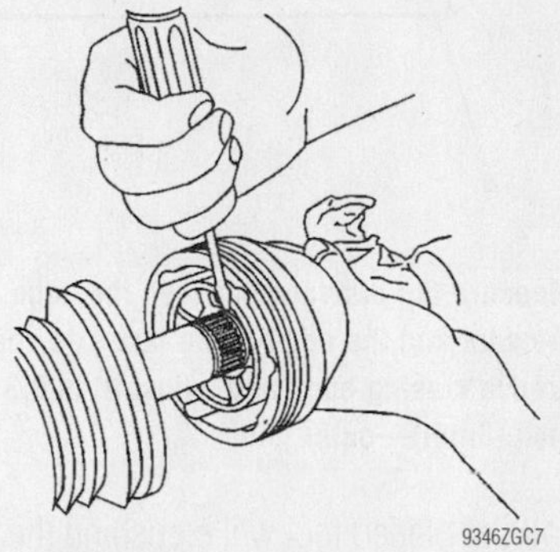

9346ZGC7

Install the axle shaft retaining ring–outer joint

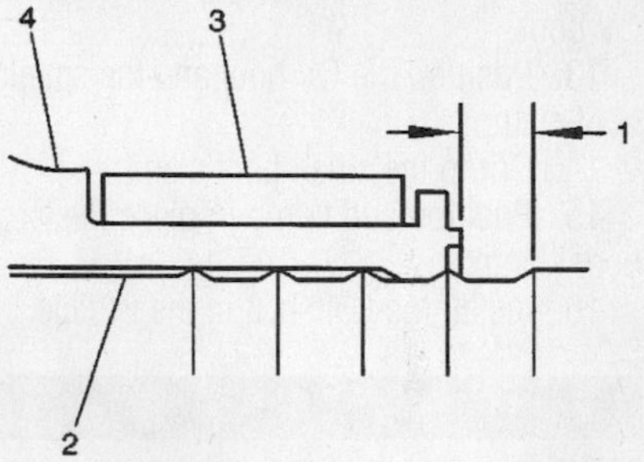

9346ZGC8

Measure the distance between the edge of the seal and the edge of the last axle shaft groove closing edge and adjust fit to 0.10 inch (2.5mm)

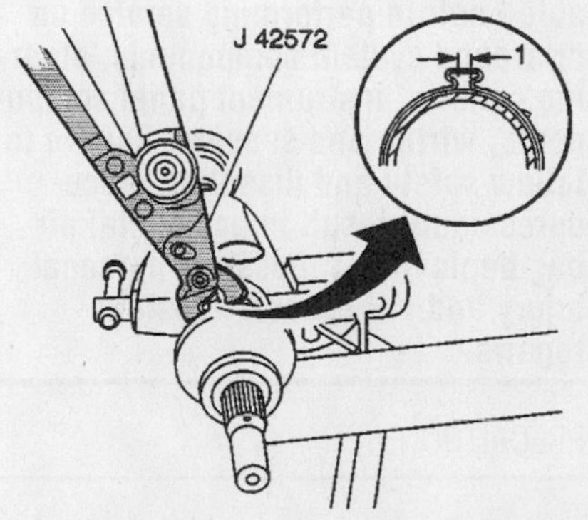

9346ZGC9

Tighten the small seal retaining clamp until the base of the omega shape has a gap width between 0.079–0.118 inch (2–3mm), with a difference in the gap width from side to side no greater than 0.016 inch (0.4mm)–outer joint

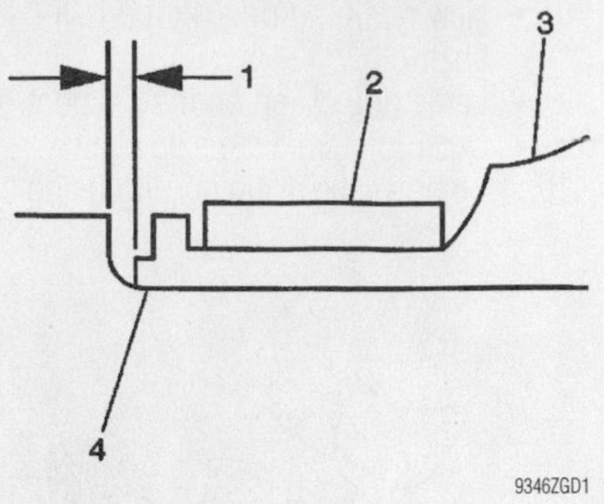

Measure the distance between the edge of the seal and the edge of the last axle shaft groove closing edge and adjust fit to 0.3 inch (8mm)—outer joint

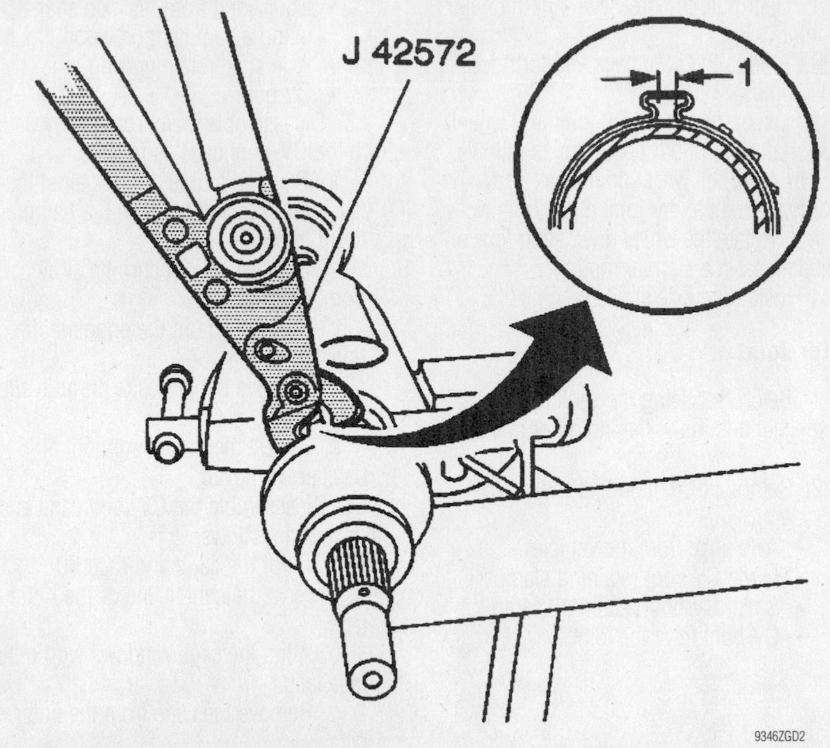

Tighten the small seal retaining clamp until the base of the omega shape has a gap width between 0.079–0.118 inch (2–3mm), with a difference in the gap width from side to side no greater than 0.016 inch (0.4mm)–outer joint

small flat bladed tool while pushing the outer joint onto the axle shaft.

11. Seat the outer joint on the shaft using a hammer and block of wood.
12. Install the remaining grease into the CV boot.
13. Position the CV boot and the small boot clamp.
14. Crimp the small boot clamp.
15. Position and crimp in place the large boot clamp.
16. Install the Halfshaft in the vehicle.

STEERING AND SUSPENSION

Air Bag

CAUTION

All vehicles are equipped with an air bag system. The system must be disabled before performing service on or around system components, steering column, instrument panel components, wiring and sensors. Failure to follow safety and disabling procedures could result in accidental air bag deployment, possible personal injury and unnecessary system repairs.

PRECAUTIONS

Several precautions must be observed when handling the inflator module to avoid accidental deployment and possible personal injury.

1. Never carry the inflator module by the wires or connector on the underside of the module.
2. When carrying a live inflator module, hold securely with both hands, and ensure that the bag and trim cover are pointed away.
3. Place the inflator module on a bench or other surface with the bag and trim cover facing up.
4. With the inflator module on the bench, never place anything on or close to the module which may be thrown in the event of an accidental deployment.

DISARMING

1. Before servicing the vehicle, refer to the precautions in the beginning of this section.
2. Turn the steering wheel to align the wheels in the straight-ahead position.
3. Turn the ignition switch to the **LOCK** position.
4. Remove the AIR BAG fuse from the fuse block.
5. Remove the left side lower trim panel, then unplug the Connector Position Assurance (CPA) device and the yellow 2-way SIR harness wire connector at the base of the steering column.

ARMING

After the necessary repairs have been made, re-enable the air bag system as follows:

1. Turn the ignition switch to the **LOCK** position.
2. Engage the yellow 2-way connector and the CPA device at the base of the steering column.
3. Install the left side lower trim panel, then install the SIR fuse to the fuse block.
4. Turn the ignition switch to the **RUN** position.
5. Verify the SIR indicator light flashes 7–9 times, then turns OFF. If not, inspect system for malfunction.

Power Rack and Pinion Steering Gear

REMOVAL & INSTALLATION

1. Before servicing the vehicle, refer to the precautions in the beginning of this section.
2. Drain the power steering fluid.
3. Remove or disconnect the following:
   - Negative battery cable
   - Brake Pressure Modulator Valve (BPMV) bracket, if equipped
   - Both front wheels
   - Intermediate shaft shield

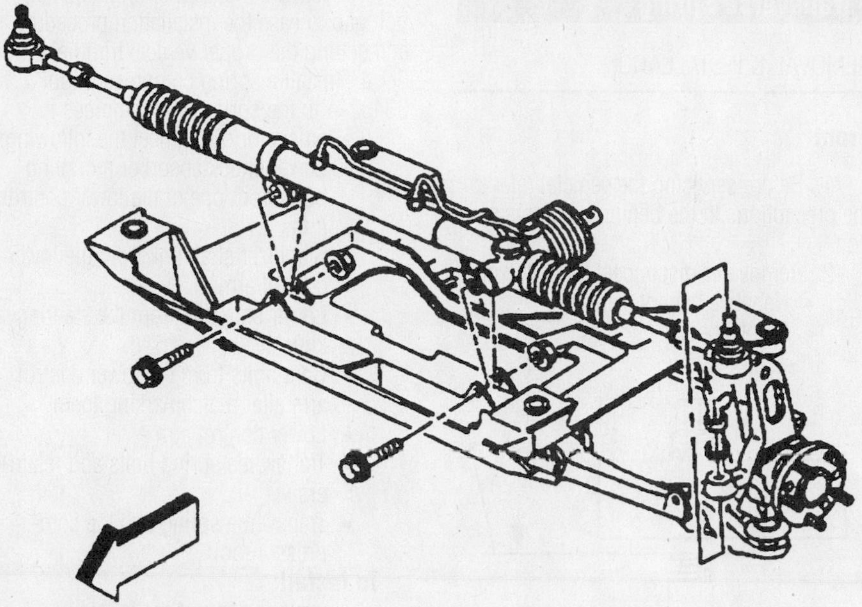

Exploded view of the power steering gear mounting—5.7L engine

- Intermediate shaft lower coupling from the power steering gear
- Power steering inlet hose from the steering gear
- Steering cooler pipe from the steering gear
- Power steering cooler
- Both outer tie rod ends from the knuckles
- Magnasteer electrical connector
- Stabilizer shaft from the crossmember
- Wire harness clips from the crossmember
- Brake pipe from the crossmember
- Steering gear mounting bolts and loosen the crossmember mounting nuts
- Power steering gear through the left wheel opening

To install:

4. Install or connect the following:
   - Steering gear to the crossmember and torque the nuts to 74 ft. lbs. (100 Nm)

5. Torque the crossmember mounting nuts 81 ft. lbs. (110 Nm)
   - Brake pipe to the crossmember
   - Stabilizer shaft
   - Both outer tie rod ends to the steering knuckle and torque the nuts to 18 ft. lbs. (25 Nm) plus and additional 180 degrees
   - Intermediate shaft lower steering coupling to the steering gear and torque the bolt to 25 ft. lbs. (34 Nm)
   - Magnasteer electrical connector
   - Wire harness clips to the crossmember
   - Cooler to the crossmember
   - Cooler pipe to the steering gear and torque the fitting to 20 ft. lbs. (27 Nm)
   - Inlet hose to the steering gear and torque the fitting to 20 ft. lbs. (27 Nm)
   - Intermediate shaft shield and torque the clamp to 32 inch lbs. (3.5 Nm)
   - Both front wheels
   - Negative battery cable

6. Refill and bleed the power steering system.
7. Check and adjust the front end alignment.

Shock Absorber

REMOVAL & INSTALLATION

Front

1. Before servicing the vehicle, refer to the precautions in the beginning of this section.
2. Remove or disconnect the following:

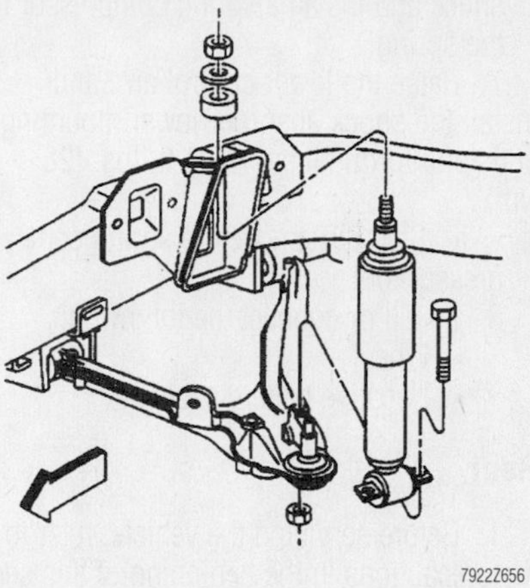

Exploded view of the front shock absorber mounting—5.7L engine

   - Negative battery cable
   - Wheel
   - Real Time Damper (RTD) electrical connector
   - Upper mounting nut retainer and insulator
   - Shock absorber lower mounting bolts
   - Shock absorber from the upper tower

3. If equipped with heavy duty shocks (FE3) perform the following steps:
   a. Compress the shock absorber from the bottom upward.
   b. Install a shock support tool to the shock while it is compressed.
   c. Remove the shock from the vehicle.
   d. Remove the tool from the shock.
4. Remove the shock absorber.

To install:

5. Install or connect the following:
   - Retainer and insulator to the shock absorber
   - Shock absorber to the upper shock tower
   - Upper insulator, retainer and nut and torque the nut to 19 ft. lbs. (26 Nm)
   - Lower mounting bolts and torque them to 21 ft. lbs. (28 Nm)
   - RTD electrical connector
6. If equipped with heavy duty shocks (FE3) perform the following steps:
   a. Install a shock support tool to the shock absorber.
   b. Install the shock to the vehicle.
   c. Install the upper insulator, retainer and nut and torque the nut to 19 ft. lbs. (26 Nm).
   d. Remove the support tool from the

shock and install a spring compressor to the spring.

7. Raise the lower control arm and install the shock absorber lower mounting bolts and torque them to 21 ft. lbs. (28 Nm).

a. Step 6: Remove the spring compressor tool.

8. Install or connect the following:
- Wheel
- Negative battery cable

Rear

1. Before servicing the vehicle, refer to the precautions in the beginning of this section.
2. Remove or disconnect the following:
- Negative battery cable
- Wheel
- Rear position sensor electrical connector, if equipped
- Lower mounting bolt
- Upper mounting bolts
- Shock absorber from the lower control arm and shock tower
- Upper insulator and retainer from the shock absorber

To install:

3. Install or connect the following:
- Upper insulator and retainer to the shock absorber, if removed
- Shock absorber to the shock tower and lower control arm
- Upper mounting bolts and torque them to 22 ft. lbs. (30 Nm)
- Lower shock absorber mounting bolt and torque it to 162 ft. lbs. (220 Nm)
- Rear position sensor electrical connector
- Wheel
- Negative battery cable

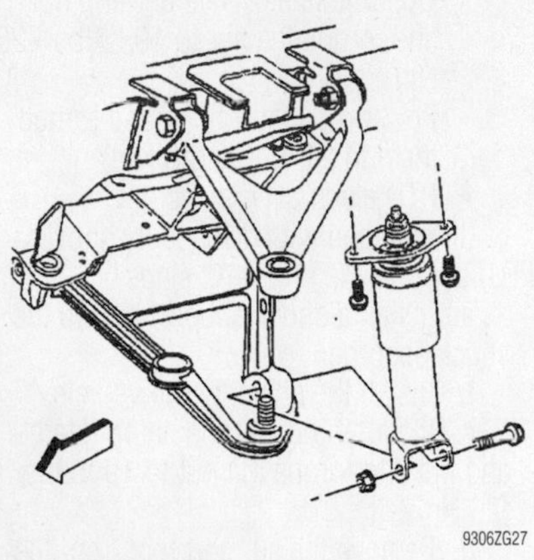

Exploded view of the rear shock absorber—5.7L engine

Transverse Spring

REMOVAL & INSTALLATION

Front

1. Before servicing the vehicle, refer to the precautions in the beginning of this section.
2. Remove or disconnect the following:
- Negative battery cable
- Front wheels

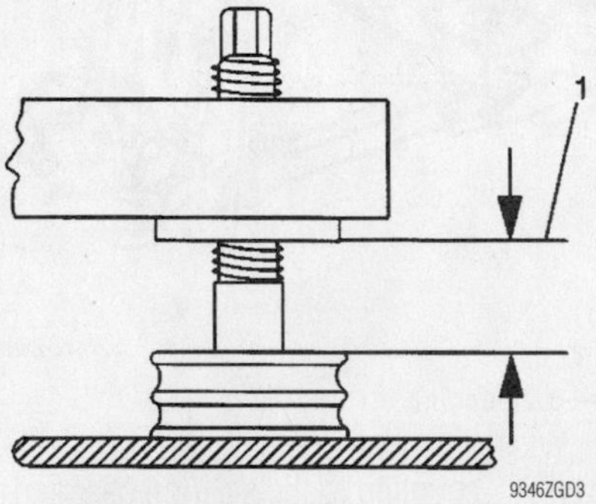

Measure the front spring adjuster bolt gap to ease the installation procedure and setting the proper vehicle trim height—front assembly

3. Measure the front spring adjuster bolt gap to ease the installation procedure and setting the proper vehicle trim height.
4. Install a spring compressor tool J 33432-A to the spring and compress it.
5. Remove or disconnect the following:
- Lower shock absorber mounting bolts from one of the lower control arms
- Stabilizer shaft link from the lower control arm
- Lower ball joint from the steering knuckle
- Cam bolts from the lower control arm after matchmarking them
- Lower control arm
- Transverse spring bolts and retainers
- Transverse spring and the compressor tool

To install:

6. Install or connect the following:
- Spring to the crossmember
- New spring retainers and bolts to the crossmember and torque them to 46 ft. lbs. (62 Nm)
- Lower control arm to the crossmember
- Cam bolts to the matchmarks made during the removal procedure.

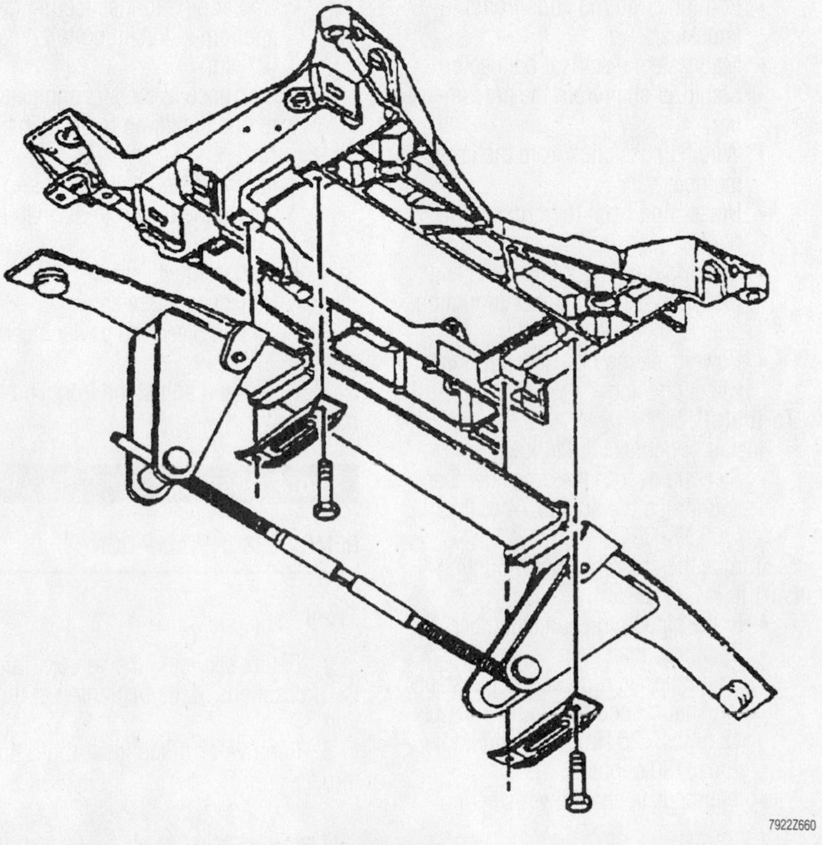

Exploded view of the front transverse spring—5.7L engine

Hand tighten the cam bolts at this time
- Lower control arm ball joint stud to the steering knuckle and torque the nut to 15 ft. lbs. (20 Nm) plus an additional 210 degrees
- Shock absorber
- Shock absorber lower mounting bolts and torque them to 21 ft. lbs. (28 Nm)
- Stabilizer shaft link to the lower control arm and torque the nut to 53 ft. lbs. (72 Nm)

7. Remove the spring compressor tool from the transverse spring and move the lower control arm supports.

8. Install or connect the following:
- Front wheels
- Negative battery cable

9. Adjust the front trim height.

10. Perform a front end alignment.

11. Torque the control arm cam bolts to 125 ft. lbs. (170 Nm)

Rear

1. Before servicing the vehicle, refer to the precautions in the beginning of this section.

2. Remove or disconnect the following:
- Negative battery cable
- Rear wheels

3. Measure the transverse spring stud height to ease the installation procedure and setting the proper vehicle trim height.

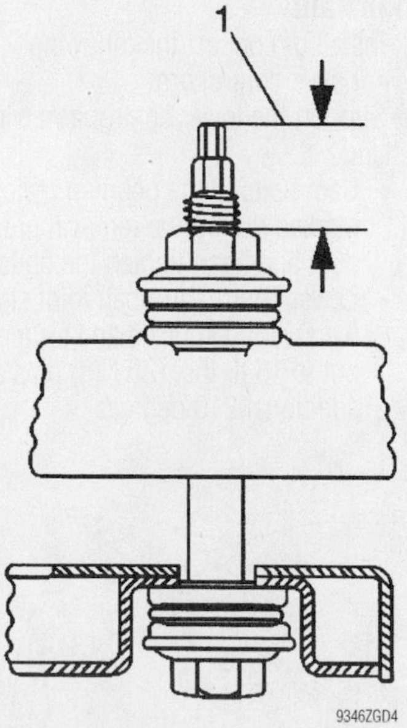

Measure the transverse spring stud height to ease the installation procedure and setting the proper vehicle trim height—rear assembly

4. Install a spring compressor tool J 33432-A to the spring and compress it.

5. Remove the spring mounting bolts, spacers and insulators from the crossmember and control arm.

6. Remove the spring from the vehicle and remove the compressor tool.

To install:

7. Install or connect the following:
- Spring compressor to the transverse spring
- Spring to the vehicle
- Spring spacers, insulators and mounting bolts to the crossmember and torque the bolts to 46 ft. lbs. (62 Nm)
- Spring to the lower control arm
- Lower control arm to spring bolts and insulators and release the compressor tool
- Rear wheel
- Negative battery cable

8. Adjust the trim height and place the retainers on the control arm bolts.

9. Check and adjust the alignment, if needed.

Upper Ball Joint

REMOVAL & INSTALLATION

Front

1. Before servicing the vehicle, refer to the precautions in the beginning of this section.

2. Remove or disconnect the following:
- Wheel
- Brake caliper
- Brake rotor
- Stabilizer shaft link from the lower control arm
- Wheel Speed Sensor (WSS) electrical connector
- Outer tie rod from the steering knuckle
- Upper ball joint from the control arm
- Steering knuckle from the vehicle

3. Press the ball joint from the steering knuckle.

To install:

4. Install or connect the following:
- Ball joint to the steering knuckle
- Upper ball joint to the control arm and torque the nut to 15 ft. lbs. (20 Nm) plus an additional 250 degree turn
- Lower ball joint to the steering knuckle and torque the nut to 15 ft. lbs. (20 Nm) plus an additional 210 degree turn
- Outer tie rod to the steering knuckle and torque the nut to 18 ft. lbs. (25 Nm) plus an additional 180 degree turn
- Stabilizer shaft link to the control arm and torque the nut to 53 ft. lbs. (72 Nm)
- WSS electrical connector
- Brake rotor
- Brake caliper and torque the mounting bracket bolts to 125 ft. lbs. (175 Nm)
- Wheel

5. Check and adjust the front alignment.

Rear

1. Before servicing the vehicle, refer to the precautions in the beginning of this section.

2. Remove or disconnect the following:
- Wheel
- Wheel Speed Sensor (WSS) electrical connector
- Real Time Damping (RTD) position sensor, if equipped
- Brake caliper
- Brake rotor
- Shock absorber solenoid electrical connector, if equipped
- Outer tie rod from the suspension knuckle
- Axle shaft nut
- Upper control arm from the knuckle
- Lower ball joint from the knuckle
- Suspension knuckle from the vehicle

3. Press the upper ball joint from the knuckle.

To install:

4. Press the ball joint into the suspension knuckle.

5. Install or connect the following:
- Knuckle to the vehicle
- Lower ball joint to the knuckle and torque the nut to 41 ft. lbs. (55 Nm)
- Upper control arm to the knuckle and torque the nut to 15 ft. lbs. (20 Nm) plus an additional 250 degree turn
- Axle shaft nut and torque it to 118 ft. lbs. (160 Nm)
- Outer tie rod to the knuckle and torque the nut to 15 ft. lbs. (20 Nm) plus an additional 160 degree turn
- Brake rotor

- Brake caliper and torque the mounting bracket bolts to 125 ft. lbs. (175 Nm)
- WSS electrical connector
- Shock absorber solenoid electrical connector, if equipped
- RTD position sensor, if equipped
- Wheel

6. Check and adjust the rear wheel alignment.

Lower Ball Joint

REMOVAL & INSTALLATION

Front and Rear

1. Before servicing the vehicle, refer to the precautions in the beginning of this section.
2. Remove the lower control arm from the vehicle.
3. Press the ball joint from the control arm.

To install:

4. Press the ball joint into the control arm.
5. Install the control arm in the vehicle.

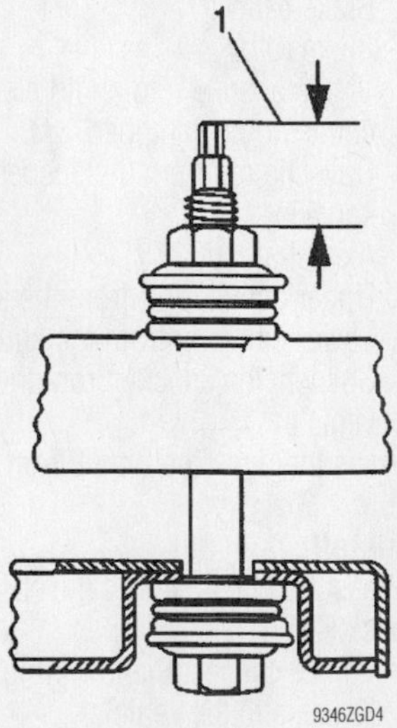

Ball joint removal and installation tool set–lower ball joint

Upper Control Arm

REMOVAL & INSTALLATION

Front

1. Before servicing the vehicle, refer to the precautions in the beginning of this section.
2. Remove or disconnect the following:
   - Negative battery cable
   - Front wheel
   - Real Time Damping (RTD) sensor link and support the lower control arm
   - Upper ball joint from the upper control arm
   - Upper control arm bolts and shims
   - Upper control arm

To install:

3. Install or connect the following:
   - Upper control arm
   - Upper control arm shims and bolts and torque the bolts to 48 ft. lbs. (65 Nm)
   - Ball joint stud into the upper control arm and torque the nut to 15 ft. lbs. (20 Nm) plus an additional 250 degrees
   - RTD sensor link
   - Front wheel
   - Negative battery cable
4. Check and adjust the front wheel alignment.

Rear

1. Before servicing the vehicle, refer to the precautions in the beginning of this section.
2. Remove or disconnect the following:
   - Negative battery cable
   - Rear wheel
   - Wheel Speed Sensor (WSS) electrical connector
   - Real Time Damping (RTD) position sensor link, if equipped
3. Support the lower control arm with a jack stand.
   - Ball joint stud from the control arm
   - Upper control arm bolts
   - Upper control arm

To install:

4. Install or connect the following:
   - Upper control arm and torque the bolts to 81 ft. lbs. (110 Nm)
   - Suspension knuckle ball joint stud nut into the upper control arm. It may be necessary to use an Allen wrench to keep the ball joint stud from spinning while tightening the ball joint stud nut. Torque the nut to 15 ft. lbs. (20 Nm) plus an additional 250 degrees
   - WSS electrical connector
   - RTD position sensor link, if equipped
   - Rear wheel
   - Negative battery cable
5. Check and adjust the rear wheel alignment.

CONTROL ARM BUSHING REPLACEMENT

The manufacturer does not provide a bushing replacement procedure. The bushing is replaced when a new control arm is installed.

Lower Control Arm

REMOVAL & INSTALLATION

Front

1. Before servicing the vehicle, refer to the precautions in the beginning of this section.
2. Remove or disconnect the following:
   - Front wheel
   - Front transverse spring
   - Wheel Speed Sensor (WSS) electrical connector
   - Real Time Damping (RTD) electrical connector, if equipped
3. Support the lower control arm with a jack stand
   - Shock absorber from the lower control arm
   - Stabilizer shaft link from the lower control arm
   - Ball joint stud from the knuckle
   - Cam bolts, washers and nuts after matchmarking them
   - Lower control arm from the vehicle

To install:

4. Install or connect the following:
   - Lower control arm
5. Support the lower control arm with a jack stand
   - Cam bolts to the position match marked during the removal procedure and hand tighten the bolts
   - Lower control arm ball joint stud to the steering knuckle and torque the nut to 15 ft. lbs. (20 Nm) plus an additional 210 degrees

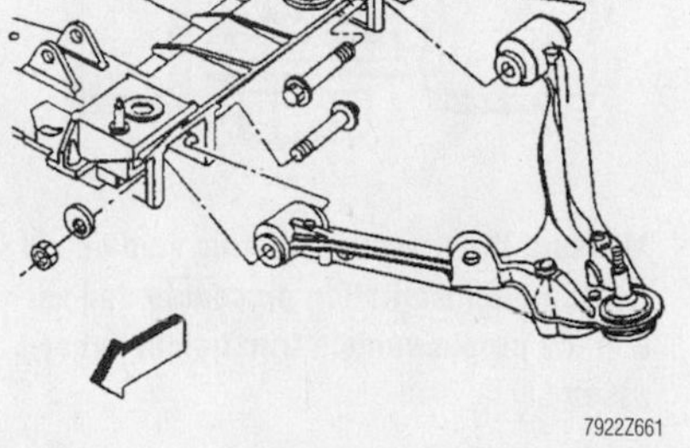

Lower control arm mounting bolts—5.7L engine

- Stabilizer shaft link to the lower control arm and torque the nut to 53 ft. lbs. (72 Nm)
- Transverse spring and remove
- Shock absorber lower mounting bolts and torque them to 21 ft. lbs. (28 Nm)
- RTD electrical connector, if equipped
- WSS electrical connector
- Front wheel
- Negative battery cable

6. Perform a front wheel alignment.
7. Tighten the lower control arm nuts to 125 ft. lbs. (170 Nm).

Rear

1. Before servicing the vehicle, refer to the precautions in the beginning of this section.
2. Remove or disconnect the following:
   - Negative battery cable
   - Rear wheel
   - Transverse spring from the lower control arm
   - Shock absorber from the lower control arm
   - Lower ball joint stud nut from the knuckle
   - Stabilizer shaft link from the lower control arm
   - Cam bolts, washers and nuts after matchmarking them
   - Lower control arm from the vehicle

To install:

3. Install or connect the following:
   - Lower control arm
   - Cam bolts to the position match marked during the removal procedure and hand tighten the bolts
   - Lower ball joint stud to the steering knuckle and torque the nut to 15 ft. lbs. (20 Nm) plus an additional 210 degrees
   - Stabilizer shaft link to the lower control arm and torque the link nut to 53 ft. lbs. (72 Nm)
   - Shock absorber lower mounting bolts and torque the bolts to 162 ft. lbs. (220 Nm)
   - Transverse spring to the lower control arm
   - Rear wheel
   - Negative battery cable
4. Perform a rear wheel alignment.
5. Torque the lower control arm front cam bolt to 107 ft. lbs. (145 Nm) and the rear cam bolt to 70 ft. lbs. (95 Nm).

CONTROL ARM BUSHING REPLACEMENT

The manufacturer does not provide a bushing replacement procedure. The bushing is replaced when a new control arm is installed.

Wheel Bearings

ADJUSTMENT

No periodic wheel bearing adjustment is necessary. The wheel bearings are a sealed unit that must be replaced if loose or noisy.

REMOVAL & INSTALLATION

Front

1. Before servicing the vehicle, refer to the precautions in the beginning of this section.
2. Remove or disconnect the following:
   - Negative battery cable
   - Front wheel
   - Wheel Speed Sensor (WSS) electrical connector
   - Brake caliper
   - Brake rotor
   - Stabilizer shaft link from the lower control arm and support the lower control arm
   - Separate the outer tie rod ball stud from the steering knuckle
   - Lower ball joint stud from the steering knuckle
   - Wheel hub mounting bolts
   - Wheel hub/bearing assembly from the knuckle

To install:

3. Install or connect the following:
   - Hub/bearing assembly to the steering knuckle. Make certain that the speed sensor cable connection is facing rearward and torque the mounting bolts to 96 ft. lbs. (130 Nm)
   - Lower control arm ball stud to the steering knuckle and torque the nut to 55 ft. lbs. (75 Nm)
   - Outer tie rod to the steering knuckle and torque the nut to 33 ft. lbs. (45 Nm)
   - WSS electrical connector

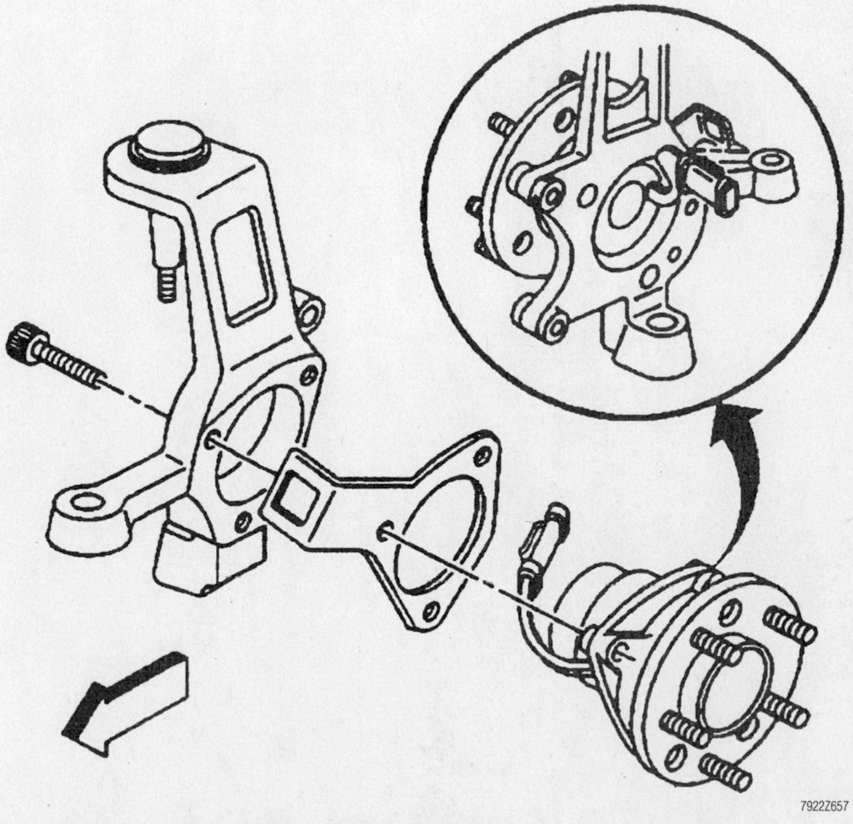

Exploded view of the front hub/wheel bearing and knuckle assembly—5.7L engine

- Stabilizer shaft link to the lower control arm and torque the nut to 55 ft. lbs. (75 Nm)
- Brake rotor
- Brake caliper and torque the bracket mounting bolts to 125 ft. lbs. (175 Nm)
- Front wheel
- Negative battery cable

Rear

1. Before servicing the vehicle, refer to the precautions in the beginning of this section.
2. Remove or disconnect the following:
   - Negative battery cable
   - Rear wheel
   - Wheel Speed Sensor (WSS) electrical connector
   - Real Time Damping (RTD) position sensor link, if equipped
   - Brake caliper
   - Brake rotor
   - Shock absorber solenoid electrical connector, if equipped
   - Outer tie rod end from the suspension knuckle

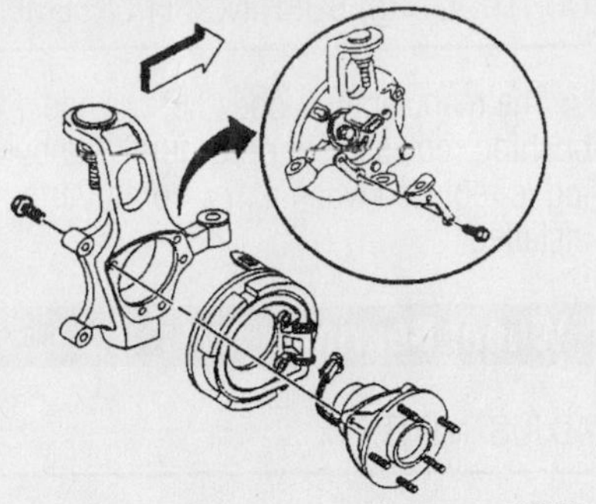

Exploded view of the rear hub/wheel bearing and knuckle assembly—5.7L engine

   - Axle nut
   - Separate the upper control arm from the suspension knuckle
   - Steering knuckle from the lower control arm ball joint stud
   - Suspension knuckle
   - Hub/bearing bolts
   - Hub/bearing from the knuckle

To install:

3. Install or connect the following:
   - Hub/bearing assembly to the steering knuckle. Make certain that the speed sensor cable connection is facing rearward and torque the hub mounting bolts to 96 ft. lbs. (130 Nm)
   - Upper control arm to the steering knuckle and torque the bolt to 49 ft. lbs. (65 Nm)
   - Lower control arm ball stud to the suspension knuckle and torque the nut to 55 ft. lbs. (75 Nm)
   - Axle nut and torque it to 118 ft. lbs. (160 Nm)
   - Outer tie rod ball stud to the steering knuckle and torque the nut to 33 ft. lbs. (45 Nm)
   - Brake rotor
   - Brake caliper and torque the mounting bracket bolts to 125 ft. lbs. (175 Nm)
   - WSS electrical connector
   - Shock absorber solenoid electrical connector
   - RTD position sensor link, if equipped
   - Rear wheel
   - Negative battery cable

GEO/CHEVROLET

31

1998–01

Metro • Prizm

PRECAUTIONS

Before servicing any vehicle, please be sure to read all of the following precautions, which deal with personal safety, prevention of component damage, and important points to take into consideration when servicing a motor vehicle:

• Never open, service or drain the radiator or cooling system when the engine is hot; serious burns can occur from the steam and hot coolant.

• Observe all applicable safety precautions when working around fuel. Whenever servicing the fuel system, always work in a well-ventilated area. Do not allow fuel spray or vapors to come in contact with a spark, open flame or excessive heat (a hot drop light, for example). Keep a dry chemical fire extinguisher near the work area. Always keep fuel in a container specifically designed for fuel storage; also, always properly seal fuel containers to avoid the possibility of fire or explosion. Refer to the additional fuel system precautions later in this section.

• Fuel injection systems often remain pressurized, even after the engine has been turned **OFF**. The fuel system pressure must be relieved before disconnecting any fuel lines. Failure to do so may result in fire and/or personal injury.

• Brake fluid often contains polyglycol ethers and polyglycols. Avoid contact with the eyes and wash your hands thoroughly after handling brake fluid. If you do get brake fluid in your eyes, flush your eyes with clean, running water for 15 minutes. If eye irritation persists, or if you have taken brake fluid internally, IMMEDIATELY seek medical assistance.

• The EPA warns that prolonged contact with used engine oil may cause a number of skin disorders, including cancer! You should make every effort to minimize your exposure to used engine oil. Protective gloves should be worn when changing oil. Wash your hands and any other exposed skin areas as soon as possible after exposure to used engine oil. Soap and water, or waterless hand cleaner should be used.

• All new vehicles are now equipped with an air bag system. The system must be disabled before performing service on or around system components, steering column, instrument panel components, wiring and sensors. Failure to follow safety and disabling procedures could result in accidental air bag deployment, possible personal injury and unnecessary system repairs.

• Always wear safety goggles when working with, or around, the air bag system. When carrying a non-deployed air bag, be sure the bag and trim cover are pointed away from your body. When placing a non-deployed air bag on a work surface, always face the bag and trim cover upward, away from the surface. This will reduce the motion of the module if it is accidentally deployed. Refer to the additional air bag system precautions later in this section.

• Clean, high quality brake fluid from a sealed container is essential to the safe and proper operation of the brake system. You should always buy the correct type of brake fluid for your vehicle. If the brake fluid becomes contaminated, completely flush the system with new fluid. Never reuse any brake fluid. Any brake fluid that is removed from the system should be discarded. Also, do not allow any brake fluid to come in contact with a painted surface; it will damage the paint.

• Never operate the engine without the proper amount and type of engine oil; doing so WILL result in severe engine damage.

• Timing belt maintenance is extremely important! Many models utilize an interference-type, non-freewheeling engine. If the timing belt breaks, the valves in the cylinder head may strike the pistons, causing potentially serious (also time-consuming and expensive) engine damage. Refer to the maintenance interval charts in the front of this manual for the recommended replacement interval for the timing belt, and to the timing belt section for belt replacement and inspection.

• Disconnecting the negative battery cable on some vehicles may interfere with the functions of the on-board computer system(s) and may require the computer to undergo a relearning process once the negative battery cable is reconnected.

• When servicing drum brakes, only disassemble and assemble one side at a time, leaving the remaining side intact for reference.

ENGINE REPAIR

Distributor

REMOVAL

1. Before servicing the vehicle, refer to the precautions in the beginning of this section.
2. Remove or disconnect the following:
   - Negative battery cable
   - Distributor electrical connector
   - Spark plug wires from the distributor cap
   - Distributor cap
3. Mark the position of the distributor rotor in relation to the distributor body and mark the position of the distributor body in relation to the cylinder head.
4. Remove the distributor hold-down bolts.
5. Remove the distributor from the engine.

INSTALLATION

Timing Not Disturbed

1. Align the reference marks made during removal. Position the distributor carefully and be sure the drive gear engages properly within the slot.
2. Install or connect the following:
   - Hold-down bolts but do not tighten
   - Distributor electrical connector
   - Spark plug wires
   - Distributor cap
   - Negative battery cable
3. Check and/or adjust the ignition timing.
4. Tighten the distributor hold-down bolts to 11 ft. lbs. (15 Nm).

Timing Disturbed

1. Remove the No. 1 spark plug.
2. Place a thumb over the spark plug hole. Have someone rotate the engine by hand using a wrench on the crankshaft pulley, until compression is felt.
3. Align the timing mark on the crankshaft pulley with the **0** degrees mark on the timing scale attached to the front of the engine. This places the engine at Top Dead Center (TDC) on the compression stroke.
4. Turn the distributor shaft until the rotor points to the No. 1 spark plug tower on the cap.
5. Install or connect the following:
   - Distributor, by aligning the distributor housing-to-cylinder head marks made during the removal procedure

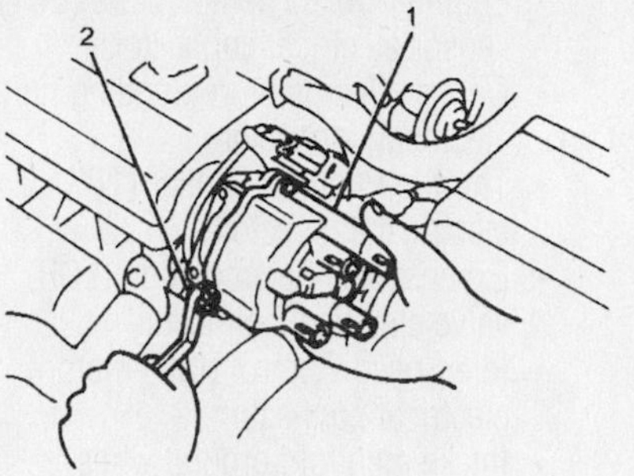

1. Distributor
2. Distributor mounting bolts

7922Z701

Be sure to tighten the distributor mounting bolts to prevent the timing from changing—Prizm

- Spark plug wires and electrical wires
- Distributor cap
- Negative battery cable

6. Check and/or adjust ignition timing.
7. Tighten the distributor hold-down bolts to 11 ft. lbs. (15 Nm).

Alternator

REMOVAL

Metro

1. Before servicing the vehicle, refer to the precautions in the beginning of this section.
2. Remove or disconnect the following:
   - Negative battery cable
   - Air cleaner assembly
   - Alternator electrical connectors
   - Accessory drive belt
   - Upper and lower mounting bolts
   - Alternator

Prizm

1. Before servicing the vehicle, refer to the precautions in the beginning of this section.
2. Remove or disconnect the following:
   - Negative battery cable
   - Accessory drive belt
   - Alternator electrical connectors
   - Alternator

INSTALLATION

Metro

Install or connect the following:

- Alternator and the bolts, do not tighten the bolts at this time

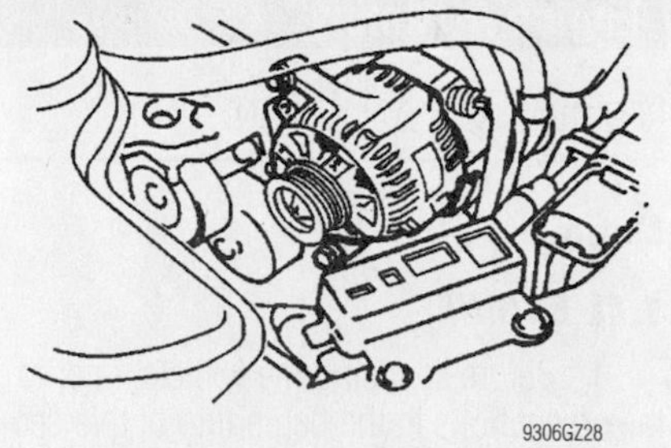

9306GZ28

Exploded view of the alternator—Metro and Prizm models

- Accessory drive belt and adjust as necessary
- Alternator electrical connectors, then torque the alternator bolts to 17 ft. lbs. (23 Nm)
- Air cleaner assembly
- Negative battery cable

Prizm

1. Install the alternator and hand-tighten the bolts.
2. Install the alternator belt and adjust the tension until a deflection of 0.24–0.31 inch (6–8mm) deflection with 22 lbs. (10 kg) of pressure exerted when installing a used belt. On a new belt the belt deflection should be 0.20–0.27 inch (5–7mm) with 22 lbs. (10 kg) of pressure exerted.
3. Install or connect the following:
   - Alternator bolts to 17 ft. lbs. (23 Nm)
   - Alternator electrical connectors
   - Accessory drive belt
   - Negative battery cable

Ignition Timing

ADJUSTMENT

➡The ignition timing on the Prizm is not adjustable.

Metro

1. Before servicing the vehicle, refer to the precautions in the beginning of this section.
2. Run the engine until it reaches normal operating temperature. Stop the engine, but keep the ignition switch in the **ON** position for approximately 5 seconds. Start the engine again.
3. Run the engine at 2000 RPM for approximately 5 minutes. After 5 minutes, allow it to run at idle speed.
4. Be sure all accessories are turned **OFF**.

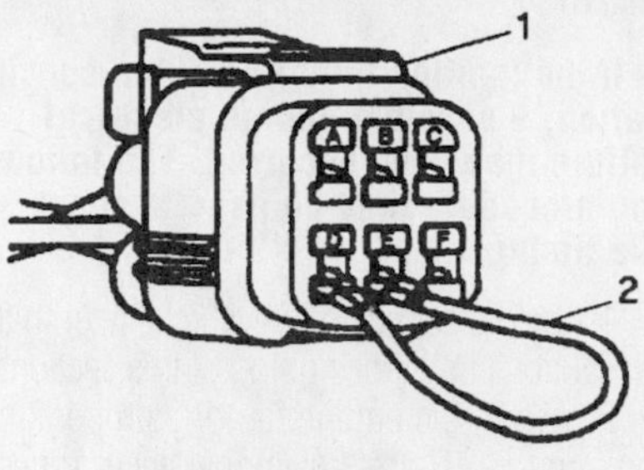

A. Blank (cavity 1)
B. Diagnostic request terminal (cavity 2)
C. Diagnostic output terminal (cavity 3)
D. Ground terminal(cavity 4)
E. Test switch terminal(cavity 5)
F. Duty check terminal(cavity 6)
1. Duty check DLC
2. Jumper

7922Z702

Jumping terminals in the 6 terminal DLC—Metro models

5. Fully engage the parking brake.
6. Be sure the shift lever is placed in the **NEUTRAL** or **PARK** depending on transaxle type.
7. Connect a tachometer to the negative terminal of the ignition coil. Connect a timing light to the No. 1 spark plug wire. Check the engine idle speed and adjust, if needed.
8. Refer to the vehicle emission control information label (located under the hood) for ignition timing specifications.
9. Remove the cap from the diagnostic check connector, located next to the ignition coil or left side strut tower and insert a fused jumper wire between appropriate the terminals.
10. On 4 terminal connectors, hold the connector with the locking tab at the top and jump the lower 2 terminals (**C** and **D**).
11. On 6 terminal connectors, hold the connector with the locking tab at the top and jump the 2 terminals at the lower left (**D** and **E**).
12. Loosen the distributor hold-down bolt and rotate the distributor until the correct timing marks are aligned.
13. Tighten the distributor hold-down bolt to 11–15 ft. lbs. (15–20 Nm) and recheck the timing. Be sure the timing advances according to engine speed.
14. Remove the diagnostic check jumper. Remove the timing light and tachometer.

Prizm

➡If the ignition timing is out of specification, a possible engine electrical failure may have occurred. The following procedure may be used to check the timing:

1. Before servicing the vehicle, refer to the precautions in the beginning of this section.
2. Warm the engine to normal operating temperature. Turn all electrical accessories OFF.
3. Connect a tachometer or engine scan tool and check the engine idle speed and be sure it is 650–750 rpm.
4. Connect a timing light. Remove the cap on the Data Link Connector (DLC). Using a fused jumper wire, connect terminals **E1** and **TE1**.
5. Start the engine and check timing. With the jumper wire connected, the timing should be at 10 degrees Before Top Dead Center (BTDC)..
6. Remove the jumper wire from the DLC and reinstall the cap.
7. Shut the engine **OFF** and disconnect all test equipment.

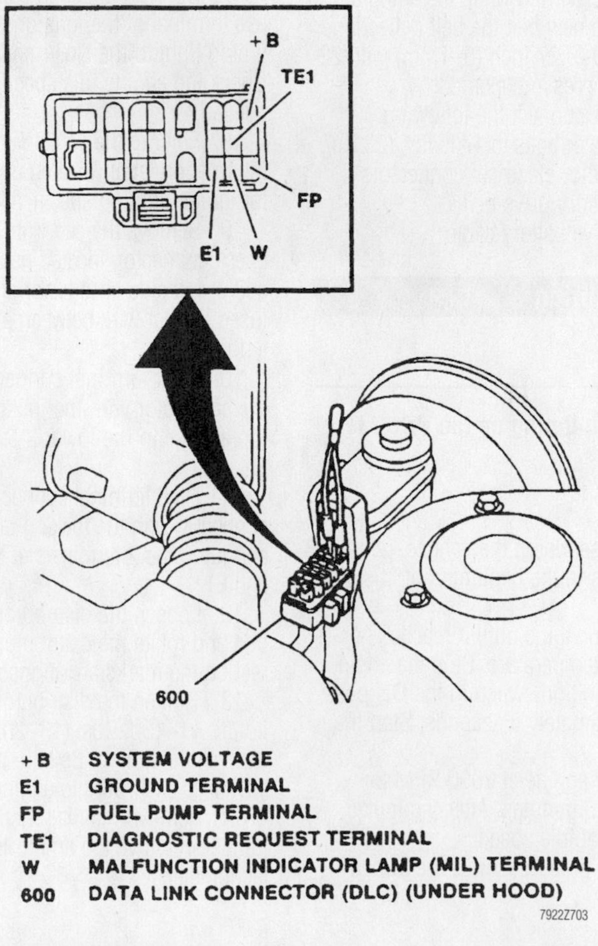

DLC pin locations—Prizm models

Engine Assembly

REMOVAL & INSTALLATION

Metro

1.0L ENGINE

1. Before servicing the vehicle, refer to the precautions in the beginning of this section.
2. Properly relieve the fuel system pressure.
3. Drain the cooling system.
4. Drain the engine oil.
5. Drain the transmission fluid, if equipped with an automatic transmisson.
6. Remove or disconnect the following:
   - Hood after scribing the hood hinge to the hood
   - Air cleaner
   - Coil wire from the distributor
   - Radiator and cooling fan assembly
   - Evaporative emissions (EVAP) purge solenoid electrical connector
   - Distributor electrical connector
   - Engine Coolant Temperature (ECT) sensor electrical connector
   - Coolant Temperature Sending unit electrical connector
   - Throttle Position Sensor (TPS) electrical connector
   - Exhaust Gas Recirculation (EGR) valve electrical connector
   - Idle Speed Control (ISC) motor electrical connector
   - Intake manifold ground wires
   - Manifold Absolute Pressure (MAP) sensor electrical connector
   - Heated Oxygen (HO_2S) sensor electrical connector
   - Throttle Body Injection (TBI) unit electrical connector
   - Early Fuel Evaporation (EFE) heater electrical connector, if equipped
   - Oil pressure switch electrical connector
   - Alternator connectors
   - Starter solenoid electrical connector
   - A/C compressor clutch coil electrical connector, if equipped
   - EVAP canister vacuum hose
   - MAP sensor vacuum hose
   - Brake booster hose
   - Accelerator cable from the throttle body
   - Clutch cable, if equipped
   - Gear select cable for automatic transmissions
   - Speedometer cable
   - Heater inlet and outlet hoses
   - Fuel feed and return hoses
7. Support the engine.
   - Exhaust pipe from the manifold
   - Gearshift control shaft and extension rod from the transmission
   - Backup lamp switch electrical connector
   - Crankshaft Position (CKP) sensor electrical connector
   - Both halfshafts

➡It is not necessary to disconnect the halfshaft from the knuckle.

   - A/C compressor without disconnecting the hoses
   - Power steering hoses from the pump, if equipped
8. Attach a lifting device to the engine.
   - Left side transmission mount
   - Right side and rear engine mount
   - Engine/transmission assembly

To install:

9. Install or connect the following:
   - Engine/transmission assembly
   - Left transmission mount and torque the nut to 41 ft. lbs. (55 Nm)

- Right and rear engine mounts and torque the bolts to 41 ft. lbs. (55 Nm)

10. Remove the engine lifting device and properly support the engine assembly.
    - Exhaust pipe to the manifold and torque the bolts to 33 ft. lbs. (45 Nm)
    - Gearshift control shaft and extension
    - Backup lamp switch
    - Halfshafts
    - Power steering hoses to the pump
    - A/C compressor and torque the bolts to 37 ft. lbs. (50 Nm)
    - Speedometer cable
    - Gear select cable, if equipped
    - Clutch cable, if equipped
    - Accelerator cable
    - Heater inlet/outlet hoses
    - Engine vacuum lines
    - Alternator and torque the bolts to 17 ft. lbs. (23 Nm)
    - Fuel return and feed lines
    - Distributor electrical connector
    - Starter solenoid electrical connector
    - CKP sensor electrical connector
    - MAP sensor electrical connector
    - EVAP purge solenoid electrical connector
    - ECT sensor connectors
    - Coolant temperature sending unit electrical connector
    - TPS electrical connector
    - EGR valve electrical connector
    - ISC motor electrical connector
    - Fuel injector connectors
    - HO_2S sensor connector
    - Intake manifold ground wires
    - EFE heater electrical connector
    - Oil pressure switch electrical connector
    - Alternator electrical connector
    - A/C Compressor clutch coil electrical connector, if equipped
    - Starter solenoid electrical connector
    - MAP sensor electrical connector
    - EVAP canister vacuum hose
    - Map sensor vacuum hose
    - Brake booster vacuum hose
    - Accelerator cable
    - Clutch cable, if equipped
    - Speedometer cable
    - Radiator and cooling fan and torque the bolts to 89 inch lbs. (10 Nm)
    - Air cleaner assembly
    - Hood and torque the bolts to 20 ft. lbs. (27 Nm)
    - Negative battery cable
11. Adjust the clutch pedal free-play, gear select cable and accelerator cable play, if equipped.
12. Fill the engine with clean oil.
13. Fill the cooling system.
14. Fill the transmission.
15. Fill the power steering reservoir.
16. Start the engine and check for leaks, repair if necessary.

1.3L ENGINE

1. Before servicing the vehicle, refer to the precautions in the beginning of this section.
2. Properly relieve the fuel system pressure.
3. Drain the cooling system.
4. Evacuate and recover the A/C system.
5. Drain the engine oil.
6. Drain the transmission fluid.
7. Remove or disconnect the following:
    - Positive and negative battery cables
    - Hood
    - Radiator and cooling fan assembly
    - Air cleaner assembly
    - Engine Coolant Temperature (ECT) sensor electrical connector
    - Throttle Position (TPS) sensor electrical connector
    - Evaporative emissions (EVAP) purge solenoid electrical connector
    - Camshaft Position (CMP) sensor electrical connector
    - Crankshaft Position (CKP) sensor electrical connector
    - Idle Speed Control (ISC) motor electrical connector
    - Heated Oxygen (HO_2S) sensor electrical connector
    - Fuel injector electrical connectors
    - Manifold Absolute Pressure (MAP) sensor electrical connector
    - Power steering pump pressure switch electrical connector, if equipped
    - Oil pressure switch electrical connector
    - A/C compressor clutch electrical connector, if equipped
    - Park/Neutral Position (PNP) switch electrical connector
    - Ignition coil connectors
    - Alternator electrical connector
    - Starter solenoid electrical connector
    - Backup lamp switch electrical connector
    - Vehicle Speed (VSS) sensor electrical connector
    - Direct clutch and second brake solenoid electrical connector
    - Brake booster hose
    - Accelerator cable
    - Clutch cable, if equipped
    - Shift select cable, if equipped
    - Speedometer cable
    - Heater inlet and outlet hoses
    - Fuel feed and return hoses
8. Install and engine support fixture.
    - Right side engine mount
    - Left transmission mount
    - Rear engine mount
    - Exhaust pipe from the manifold
    - Gearshift control shaft and extension rod, if equipped
    - Both halfshafts
    - Power steering pump, if equipped
    - A/C compressor, if equipped
9. Place an engine table under the vehicle.
10. Remove the engine support fixture.
11. Raise the vehicle from the engine/transmission assembly.

To install:

12. Place the engine/transmission assembly on the table, under the vehicle.
13. Lower the vehicle onto the engine/transmission.
14. Install or connect the following:
    - Left transmission mount and torque the bolts to 41 ft. lbs. (55 Nm)
    - Right side engine mount and torque the bolts to 41 ft. lbs. (55 Nm)
    - Rear engine mount and torque the bolts to 41 ft. lbs. (55 Nm)
    - Power steering pump and torque the bolts to 25 ft. lbs. (35 Nm)
    - A/C compressor and torque the bolts to 18 ft. lbs. (25 Nm)
    - Both halfshafts
    - Gearshift control shaft and extension rod
    - Exhaust pipe to the manifold and torque the bolts to 37 ft. lbs. (50 Nm)
15. Remove the engine support fixture.
    - Fuel feed and return hoses
    - Heater inlet and outlet hoses
    - Accelerator cable
    - Clutch cable, if equipped
    - Shift select cable, if equipped
    - Speedometer cable
    - Brake booster hose

- ECT sensor electrical connector
- TPS sensor electrical connector
- EVAP purge solenoid electrical connector
- CMP sensor electrical connector
- CKP sensor electrical connector
- ISC motor electrical connector
- HO_2S sensor electrical connector
- Fuel injector electrical connectors
- MAP sensor electrical connector
- Power steering pump pressure switch electrical connector, if equipped
- Oil pressure switch electrical connector
- A/C compressor clutch electrical connector, if equipped
- PNP switch electrical connector
- Ignition coil connectors
- Alternator electrical connector
- Starter solenoid electrical connector
- Backup lamp switch electrical connector
- VSS sensor electrical connector
- Direct clutch and second brake solenoid electrical connector
- Radiator and cooling fan assembly and torque the bolts to 89 inch lbs. (10 Nm)
- Air cleaner assembly
- Hood
- Positive and negative battery cables

16. Fill the cooling system.
17. Fill the engine oil.
18. Evacuate and recharge the A/C system.
19. Fill the transmission fluid.
20. Start the engine and check for proper operation.

Prizm

1. Before servicing the vehicle, refer to the precautions in the beginning of this section.
2. Properly relieve the fuel system pressure.
3. Drain the cooling system.
4. Drain the engine oil.
5. Drain the transmission fluid.
6. Evacuate the A/C system, if equipped.
7. Remove or disconnect the following:
   - Hood after scribing the hood hinge to the hood
   - Ignition coil electrical connectors
   - Accessory drive belt
   - A/C compressor, if equipped
   - Alternator
   - Crankshaft pulley, if equipped with manual transmission
   - Drive belt tensioner, if equipped with manual transmission
   - Accelerator cable
   - Intake Air Temperature (IAT) sensor electrical connector
   - Air cleaner assembly
   - Upper radiator hose
   - Windshield washer reservoir
   - Oxygen (O_2S) sensor
   - Engine wiring harness and place aside
   - Lower radiator hose from the thermostat housing
   - Splash shields
   - Both halfshafts
   - Starter motor electrical connectors
   - Starter motor
   - Torque converter bolts, if equipped with an automatic transmission
   - Power steering pump, do not disconnect the hoses
   - Exhaust pipe from the manifold
   - Transmission-to-engine bolts
   - Manifold Absolute Pressure (MAP) sensor hose
   - Brake booster hose
   - MAP sensor
   - Data Link Connector (DLC) and A/C pressure switch wiring harnesses
   - Ground wires from the intake manifold and fenders
   - Heater inlet and outlet hoses
   - Fuel feed and return hoses
   - Shift select cable
8. Attach an engine hoist to the engine.
9. Remove or disconnect the following:
   - Front and rear transmission mounts
   - Right engine mount
   - Engine assembly

To install:

10. Install or connect the following:
    - Engine assembly
    - Engine-to-transmission bolts and torque them to 47 ft. lbs. (64 Nm)
    - Right side engine mount and torque the bolts to 40 ft. lbs. (54 Nm)
    - Front and rear transmission mounts and torque the front mount bolts to 47 ft. lbs. (64 Nm) and the through-bolt to 64 ft. lbs. (87 Nm). Torque the rear transmission nuts to 42 ft. lbs. (52 Nm) and the through-bolt to 64 ft. lbs. (87 Nm).
11. Remove the engine hoist and support the engine.
    - Both halfshafts
    - Front exhaust pipe and torque the bolts to 46 ft. lbs. (62 Nm)
    - O_2S sensor and torque the nuts to 32 ft. lbs. (40 Nm)
    - Power steering pump and torque the bolts to 27 ft. lbs. (37 Nm)
    - Torque converter bolts and tighten them to 26 ft. lbs. (35 Nm)
    - Starter motor and torque the bolts to 22 ft. lbs. (30 Nm)
    - Engine splash shields and torque the bolts to 89 inch lbs. (10 Nm)
    - Lower radiator hose
    - Engine wiring harness
    - Alternator and torque the bolts to 17 ft. lbs. (23 Nm)
    - Windshield washer reservoir and torque the bolts to 89 inch lbs. (10 Nm)
    - Upper radiator hose
    - Air cleaner assembly and torque the bolts to 89 inch lbs. (10 Nm)
    - IAT sensor electrical connector
    - Accelerator cable to the throttle lever
    - Accessory drive belt
    - A/C compressor and torque the bolts to 18 ft. lbs. (25 Nm)
    - Alternator and torque the upper bolt to 18 ft. lbs. (25 Nm) and the lower bolt to 40 ft. lbs. (54 Nm)
    - Drive belt tensioner (if equipped with manual transmission) and torque the bolt to 51 ft. lbs. (69 Nm) and the nut to 21 ft. lbs. (29 Nm)
    - Crankshaft pulley, if equipped with manual transmission, and toque the bolt to 105 ft. lbs. (142 Nm)
    - Ignition coil electrical connectors
    - Shift cable
    - Fuel return hose
    - Fuel feed pipe to the fuel rail, using new gaskets and torque the bolt to 22 ft. lbs. (29 Nm)
    - Heater inlet and outlet hoses
    - Engine ground wires to the intake manifold and fender
    - Radiator and cooling fan and torque the bolts to 9 ft. lbs. (13 Nm)
    - Hood and torque the bolts to 20 ft. lbs. (27 Nm)
    - Negative battery cable
12. Adjust the clutch pedal free-play, gear select cable and accelerator cable play, if equipped.
13. Recharge the A/C system.
14. Fill the engine with clean oil.
15. Fill the cooling system.
16. Fill the transmission.
17. Fill the power steering reservoir.
18. Start the engine and check for leaks, repair if necessary.

Water Pump

REMOVAL & INSTALLATION

Metro

1. Before servicing the vehicle, refer to the precautions in the beginning of this section.
2. Drain the cooling system.
3. Remove or disconnect the following:
   - Negative battery cable
   - Air cleaner assembly
   - A/C compressor suction pipe bracket, if equipped
   - Water pump pulley bolts, loosen but do not remove the bolts
   - Right side lower splash shield
   - A/C compressor drive belt, if equipped
   - Lower alternator cover plate
   - Water pump/alternator drive belt
   - Crankshaft pulley
   - Water pump pulley
   - Timing belt
   - Oil level indicator tube
   - Upper alternator bracket from the water pump
   - Water pump bolts and nuts
   - Water pump

To install:

4. Install or connect the following:
   - Water pump with a new gasket and torque the bolts to 115 inch lbs. (13 Nm)
   - New rubber seals
   - Upper alternator adjusting bracket and torque the bolt to 17 ft. lbs. (23 Nm)
   - Oil level indicator tube
   - Timing belt
   - Crankshaft pulley and torque the bolts to 12 ft. lbs. (16 Nm)
   - Water pump/alternator drive belt
   - Lower alternator cover plate and torque the bolts to 89 inch lbs. (10 Nm)
   - A/C compressor drive belt, if equipped
   - Right side lower splash shield
   - Water pump pulley and torque the bolts to 18 ft. lbs. (24 Nm)
5. Adjust the water pump drive belt tension and torque the alternator adjustment bolt to 17 ft. lbs. (23 Nm).
   - A/C compressor suction pipe bracket, if equipped
   - Air cleaner assembly
   - Negative battery cable
6. Refill the cooling system.
7. Start the engine and check for leaks, repair if necessary.

Prizm

1. Before servicing the vehicle, refer to the precautions in the beginning of this section.
2. Drain the engine coolant.
3. Remove or disconnect the following:
   - Negative battery cable
   - Accessory drive belt
   - Right side lower splash shield
   - Water pump bolts
   - Water pump

To install:

4. Install or connect the following:
   - Water pump with a new O-ring and torque the bolts to 8 ft. lbs. (11 Nm)
   - Right side lower splash shield
   - Accessory drive belt
   - Negative battery cable
5. Fill the cooling system.
6. Start the vehicle and check for leaks, repair if necessary.

Cylinder Head

REMOVAL & INSTALLATION

Metro

1. Before servicing the vehicle, refer to the precautions in the beginning of this section.
2. Remove or disconnect the following:
   - Negative battery cable
   - Intake manifold
   - Exhaust manifold
   - Rocker arm cover
   - Timing belt and tensioner
   - Camshaft
   - Valve lifters
   - Distributor
   - Cylinder head bolts
   - Cylinder head

To install:

3. Install or connect the following:
   - Cylinder head with a new gasket and torque the bolts in 3 even

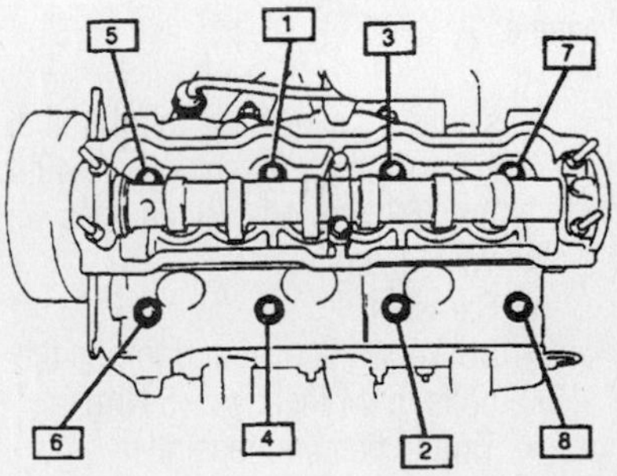

7922Z709

Cylinder head torque sequence—1.0L engine

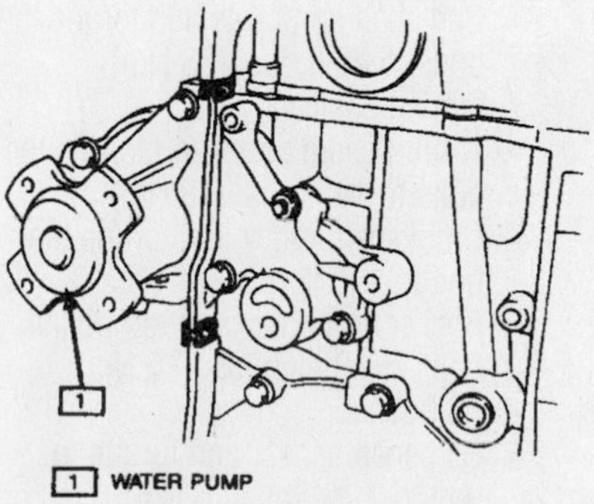

7922Z704

To ensure a tight seal, be sure gasket surfaces are properly prepared—Metro

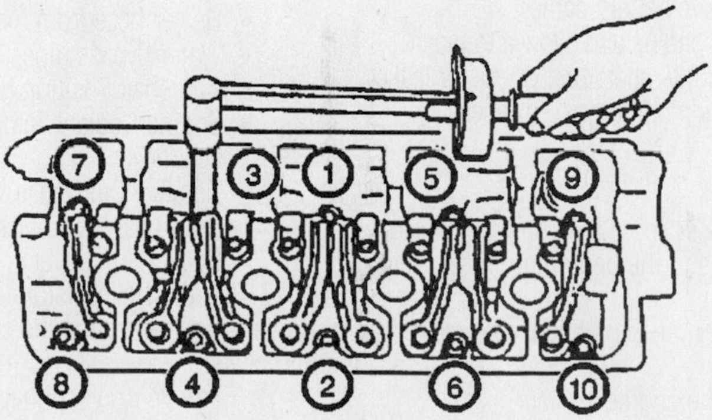

9346ZGD7

Cylinder head torque sequence—1.3L engine

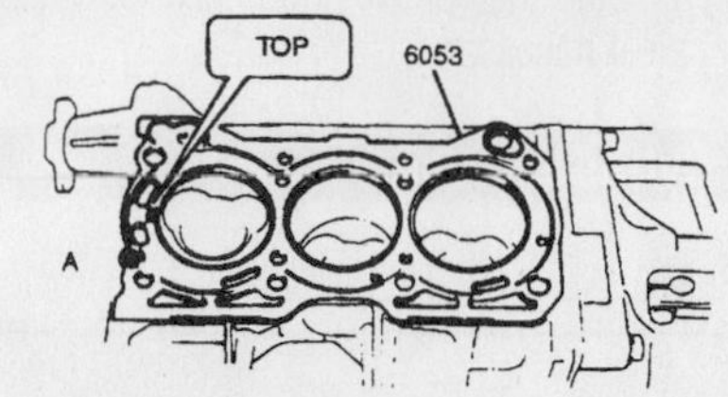

A CRANKSHAFT PULLEY SIDE
6053 CYLINDER HEAD GASKET

7922Z708

Cylinder head gasket positioning—1.0L engine

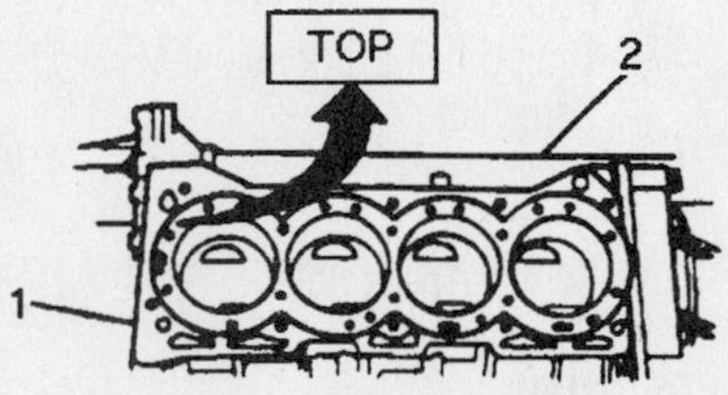

1. Cylinder head gasket
2. Cylinder block

7922Z799

Cylinder head gasket positioning—1.3L engine

stages, in sequence, to 54 ft. lbs. (73 Nm) on 1.0L engines or 49 ft. lbs. (68 Nm) on 1.3L engines

- Valve lifters
- Camshaft
- Rocker arm cover and torque the bolts to 44 inch lbs. (5 Nm)
- Timing belt and tensioner
- Distributor
- Intake manifold and torque the bolts to 17 ft. lbs. (23 Nm)
- Exhaust manifold and torque the bolts to 17 ft. lbs. (23 Nm)
- Negative battery cable

4. Start the engine and allow it to reach normal operating temperature. Check for leaks and adjust the ignition timing, if necessary.

Prizm

1. Before servicing the vehicle, refer to the precautions in the beginning of this section.
2. Properly relieve the fuel system pressure.
3. Drain the engine coolant.
4. Drain the engine oil.
5. Remove or disconnect the following:
   - Negative battery cable
   - Washer tank and pump
   - Accessory drive belt
   - Alternator
   - Accelerator cable from the throttle body
   - Intake Air Temperature (IAT) sensor electrical connector
   - Air cleaner hose
   - Throttle Position (TPS) sensor electrical connector
   - Idle Air Control (IAC) valve electrical connector
   - Manifold Absolute Pressure (MAP) sensor electrical connector
   - Coolant bypass hoses
   - Fuel injector electrical connectors
   - Fuel injector harness from the intake manifold
   - Intake manifold support bracket
   - Oxygen (O_2S) sensor
   - Exhaust pipe from the manifold
6. Install an engine support fixture.
   - A/C receiver pinch clamp and move the receiver aside
   - Right side engine mount
   - Ignition coil connectors
   - Fuel line at the fuel rail
   - Spark plug wires
   - Ignition coils
   - Fuel rail
   - Ground wires from the cylinder head
   - Heater hose from the cylinder head
   - Water bypass pipe
   - Camshaft Position (CMP) sensor
   - Engine Coolant Temperature (ECT) sensor
   - Positive Crankcase Ventilation (PCV) valve and hoses
   - Rocker arm cover
7. Set the No. 1 piston to Top Dead Center (TDC) and align the camshaft timing sprockets.
8. Remove or disconnect the following:
   - Power steering oil pressure switch connector
   - Right side lower engine splash shield
   - Power steering pump
   - Crankshaft pulley
   - Crankshaft Position (CKP) sensor
   - Drive belt tensioner
   - Right side engine mount bracket
   - Timing chain tensioner
   - Timing chain cover
   - Crank sensor reluctor
   - Timing chain slipper
   - Crankshaft sprocket
   - Timing chain
   - Camshaft sprockets
   - Camshaft bearing caps
   - Camshafts
   - Valve lifters
   - Intake manifold
   - Exhaust manifold
   - Cylinder head

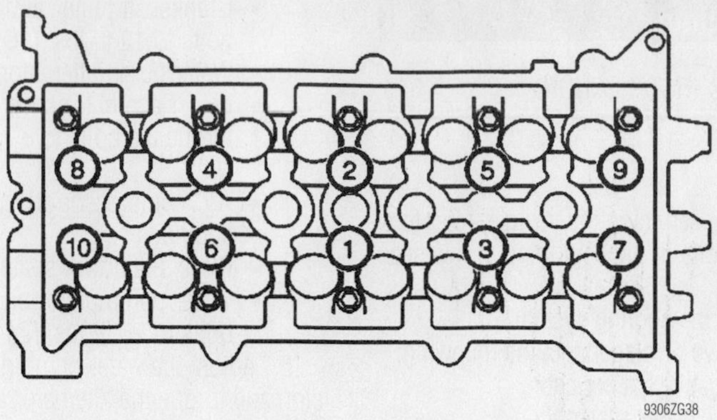

9306ZG38

Cylinder head bolt torque sequence—Prizm

To install:

9. Install the cylinder head with a new gasket.
10. Torque the cylinder head bolts in the following sequence:
    a. 18 ft. lbs. (25 Nm).
    b. 36 ft. lbs. (49 Nm).
    c. Plus an additional 90 degrees.
11. Install or connect the following:
    - Camshafts and torque the bearing cap bolts to 10 ft. lbs. (13 Nm)
    - Camshaft sprockets and torque the bolts to 40 ft. lbs. (55 Nm)
    - Timing chain
    - Timing chain cover and torque the bolts to 14 ft. lbs. (18 Nm)
    - Crankshaft pulley and torque the bolt to 105 ft. lbs. (142 Nm)
    - Rocker arm cover and torque the bolts to 89 inch lbs. (10 Nm)
    - ECT sensor
    - Camshaft sensor and torque the bolt to 11 ft. lbs. (15 Nm)
    - Water bypass pipe and torque the bolt to 11 ft. lbs. (15 Nm)
    - Heater hoses
    - Ground wires to the cylinder head and torque the bolts to 11 ft. lbs. (15 Nm)

- Fuel rail and torque the bolts to 10 ft. lbs. (14 Nm)
- Fuel line to the rail
- Ignition coils and torque the bolts to 10 ft. lbs. (14 Nm)
- Right side engine mount and torque the bolts to 40 ft. lbs. (54 Nm)
- A/C receiver pinch clamp

12. Remove the engine support fixture.
    - Exhaust manifold support bracket and torque the bolt to 26 ft. lbs. (35 Nm)
    - Exhaust pipe with a new gasket and torque the bolts to 46 ft. lbs. (62 Nm)
    - O_2sensor and torque the nuts to 30 ft. lbs. (41 Nm)
    - Fuel injector harness brackets
    - Vacuum hoses
    - Intake manifold support bracket and torque the bolts to 37 ft. lbs. (50 Nm)
    - Fuel injector harness to the intake manifold and torque the bolts to 106 inch lbs. (12 Nm)
    - Upper radiator hose
    - Fuel injector connectors
    - Coolant bypass hoses
    - MAP sensor electrical connector
    - IAC valve electrical connector
    - TPS sensor electrical connector
    - IAT sensor electrical connector
    - Accelerator cable to the throttle body
    - Alternator and torque the bolts to 17 ft. lbs. (23 Nm)
    - Washer pump and torque the bolt to 89 inch lbs. (10 Nm)
    - Accessory drive belt
    - Negative battery cable
13. Fill the engine with new oil.
14. Fill and bleed the cooling system.
15. Start the vehicle and check for leaks, repair if necessary. Road test and check for proper operation.

Rocker Arms/Shafts

REMOVAL & INSTALLATION

The 1.0L and 1.8L engines do not use rocker arms/shafts, the camshaft directly actuates the valves.

1.3L Engine

1. Before servicing the vehicle, refer to the precautions in the beginning of this section.
2. Remove or disconnect the following:
    - Both battery cables
    - Battery
    - Rocker arm cover
    - Timing belt
    - Camshaft sprocket
    - Camshaft Position (CMP) sensor
    - Camshaft
    - Rocker shaft retaining bolts. Lift out the exhaust and intake rocker arm shafts, with the springs and rocker arms attached.
    - If necessary, remove the rocker arms, washers and springs from the rocker shafts. Note the order in which the components are removed; they must be re-assembled in the same order and position.

➡The intake and exhaust rocker arm shafts are NOT the same. Both can be distinguished by looking at the ends of the shafts, which are different. Install the intake rocker arm shaft with the stepped end toward the distributor side.

To install:

3. If disassembled earlier, assemble the springs, washers and rocker arms on to the rocker shafts.
4. Install or connect the following:
    - Lubricate the rocker arms and shafts with clean engine oil. Position the intake and exhaust rocker arm shafts in their original positions and torque the bolts, in the correct sequence, to 97 inch lbs. (11 Nm).
    - Camshaft
    - CMP sensor and torque the bolts to 97 inch lbs. (11 Nm)
    - Camshaft sprocket and torque the bolt to 43 ft. lbs. (60 Nm)
    - Timing belt
    - Rocker arm cover and torque the bolts to 44 inch lbs. (5 Nm)
    - Battery
    - Both battery cables

Intake Manifold

REMOVAL & INSTALLATION

Metro

1.0L ENGINE

1. Before servicing the vehicle, refer to the precautions in the beginning of this section.
2. Relieve the fuel system pressure.
3. Drain the cooling system.
4. Remove or disconnect the following:
    - Air cleaner
    - Engine Coolant Temperature (ECT) sensor electrical connector
    - ECT sending unit connector
    - Throttle Position (TPS) sensor electrical connector
    - Exhaust Gas Recirculation (EGR) valve electrical connector
    - Idle Speed Control (ISC) motor electrical connector
    - Ground wires from the intake manifold
    - Throttle body electrical connector
    - Manifold Absolute Pressure (MAP) sensor electrical connector
    - Early Fuel Evaporative (EFE) heater electrical connector, if equipped
    - Fuel feed and return hoses from the throttle body
    - Coolant hoses from the throttle body
    - EGR Modulator
    - EGR valve

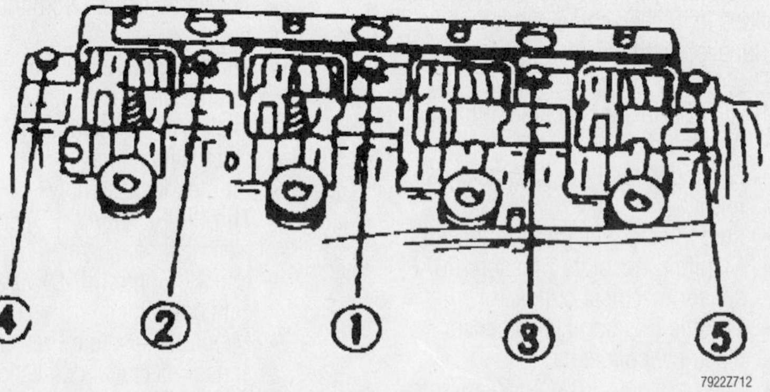

Rocker arm shaft retaining bolts tightening sequence—1.3L engines

- Evaporative Emissions (EVAP) canister vacuum hose
- MAP sensor hose from the intake manifold
- Brake booster vacuum hose
- Positive Crankcase Ventilation (PCV) vacuum hose
- Accelerator cable from the throttle body
- Intake manifold/throttle body unit

To install:

5. Install or connect the following:
   - Intake manifold with a new gasket and torque the bolts to 17 ft. lbs. (23 Nm)
   - PCV hose
   - EVAP canister vacuum hose
   - MAP sensor vacuum hose
   - Brake booster vacuum hose
   - Coolant hoses
   - Fuel feed and return hoses
   - ECT sensor electrical connector
   - Coolant temperature sending unit electrical connector
   - EFE heater electrical connector, if equipped
   - ECT switch electrical connector
   - TPS sensor electrical connector
   - EGR solenoid valve electrical connector
   - ISC motor electrical connector
   - Ground wires to the intake manifold
   - Throttle body electrical connector
   - Wiring harness to the retaining clamps
   - Accelerator cable to the throttle body and adjust as needed
   - Air cleaner assembly
   - Negative battery cable
6. Fill the cooling system.
7. Start the vehicle and check for leaks, repair if necessary.

1.3L ENGINE

1. Before servicing the vehicle, refer to the precautions in the beginning of this section.
2. Relieve the fuel system pressure.
3. Drain the cooling system.
4. Remove or disconnect the following:
   - Air cleaner
   - Ground wires from the intake manifold
   - Fuel injector electrical connectors
   - Manifold Absolute Pressure (MAP) sensor electrical connector
   - Throttle Position (TPS) sensor electrical connector
   - Idle Air Control (IAC) valve electrical connector
   - Evaporative Emissions (EVAP) canister purge valve electrical connector
   - Wiring harness from the retaining clamps
   - Fuel feed and return hoses
   - Coolant hoses from the throttle body
   - Brake booster hose
   - Canister purge hose
   - Accelerator cable
   - Intake manifold bolts
   - Intake manifold

To install:

5. Install or connect the following:
   - Intake manifold with a new gasket and torque the nuts and bolts to 17 ft. lbs. (23 Nm)
   - Canister purge hose
   - Brake booster hose
   - Coolant hoses
   - Fuel feed and return hoses
   - Wiring harness to the retaining clamps
   - Ground wires to the intake manifold
   - Fuel injector electrical connectors
   - MAP sensor electrical connector
   - TPS sensor electrical connector
   - IAC valve electrical connector
   - EVAP canister purge valve electrical connector
   - Accelerator cable
   - Air cleaner assembly
   - Negative battery cable
6. Fill the cooling system.
7. Start the vehicle and check for leaks, repair as necessary.

Prizm

1. Before servicing the vehicle, refer to the precautions in the beginning of this section.
2. Relieve the fuel system pressure.
3. Drain the cooling system.
4. Remove or disconnect the following:
   - Accelerator cable from the throttle body
   - Throttle valve cable from the throttle body, if equipped
   - Intake Air Temperature (IAT) sensor electrical connector
   - Air cleaner assembly
   - Throttle Position (TPS) sensor electrical connector
   - Idle Air Control (IAC) valve electrical connector
   - Manifold Absolute Pressure (MAP) sensor electrical connector
   - Coolant bypass hoses
   - Fuel injector electrical connectors
   - Wiring harness from the intake manifold
   - Intake manifold support bracket
   - Upper radiator hose and support bracket
   - Brake booster vacuum hose
   - Fuel feed and return hoses
   - Intake manifold

To install:

5. Install or connect the following:
   - Intake manifold with a new gasket and torque the bolts to 13 ft. lbs. (18 Nm)
   - Brake booster vacuum hose
   - Fuel inlet pipe to the fuel rail and torque the fitting to 22 ft. lbs. (29 Nm)
   - Fuel return hose
   - Intake manifold support bracket and torque the bolts to 13 ft. lbs. (18 Nm)
   - Upper radiator hose and support bracket and torque the bolt to 13 ft. lbs. (18 Nm)
   - Engine wire harness to the intake manifold and torque the bolt to 9 ft. lbs. (13 Nm)
   - Fuel injector connectors to the fuel injectors
   - Coolant bypass hoses
   - MAP sensor electrical connector
   - IAC valve electrical connector
   - TPS sensor electrical connector
   - Air cleaner assembly
   - IAT sensor electrical connector
   - Accelerator cable and torque the bolts to 98 inch lbs. (11 Nm)
   - Throttle valve cable, if equipped and torque the bolts to 11 ft. lbs. (15 Nm)
   - Negative battery cable
6. Fill the cooling system.
7. Start the vehicle and check for leaks, repair if necessary.

Exhaust Manifold

REMOVAL & INSTALLATION

Metro

1.0L ENGINE

1. Before servicing the vehicle, refer to the precautions in the beginning of this section.
2. Remove or disconnect the following:
   - Negative battery cable
   - Front pipe from the exhaust manifold
   - Heated Oxygen (HO_2S) sensor
   - Exhaust manifold heat shield
   - Exhaust manifold

To install:

3. Install or connect the following:
   - Exhaust manifold with a new gasket and torque the bolts to 17 ft. lbs. (23 Nm)
   - Exhaust manifold heat shield and torque the bolts to 11 ft. lbs. (15 Nm)
   - Front pipe to the exhaust manifold with a new seal and torque the bolts to 33 ft. lbs. (45 Nm)
   - HO_2S and torque the bolts to 33 ft. lbs. (45 Nm)
   - Negative battery cable
4. Start the vehicle and check for exhaust leaks, repair If necessary.

1.3L ENGINE

1. Before servicing the vehicle, refer to the precautions in the beginning of this section.
2. Remove or disconnect the following:
   - Negative battery cable
   - Front pipe from the exhaust manifold
   - Exhaust manifold stiffener bracket
   - Oxygen (O_2S) sensor electrical connector
   - Exhaust manifold heat shield
   - Low side A/C line bracket and position aside, if equipped
   - Exhaust manifold

To install:

3. Install or connect the following:
   - Exhaust manifold with a new gasket and torque the bolts to 24 ft. lbs. (32 Nm)
   - Exhaust manifold heat shield and torque the bolts to 11 ft. lbs. (15 Nm)
   - Low side A/C line bracket, if equipped and torque the bolts to 11 ft. lbs. (15 Nm)
   - Exhaust manifold stiffener bracket and torque the bolts to 37 ft. lbs. (50 Nm)
   - Front pipe and torque the bolts to 37 ft. lbs. (50 Nm)
   - O_2S electrical connector
   - Negative battery cable

Prizm

1. Before servicing the vehicle, refer to the precautions in the beginning of this section.
2. Remove or disconnect the following:
   - Negative battery cable
   - Heated Oxygen (HO_2S) sensor and sub-oxygen sensor
   - Front exhaust pipe from the exhaust manifold
   - Exhaust manifold support bracket
   - Exhaust manifold upper heat shield
   - Exhaust manifold
   - Lower heat shield from the exhaust manifold, if necessary

To install:

3. Install or connect the following:
   - Lower heat shield, if removed, and torque the bolts to 11 ft. lbs. (15 Nm)
   - Exhaust manifold with a new gasket and torque the bolts to 36 ft. lbs. (48 Nm)
   - Exhaust manifold upper heat shield and torque the bolts to 11 ft. lbs. (15 Nm)
   - Exhaust manifold support bracket and torque the bolts to 24 ft. lbs. (32 Nm)
   - Front pipe to the exhaust manifold and torque the bolts to 46 ft. lbs. (62 Nm)
   - HO_2sensor and torque the fasteners to 30 ft. lbs. (41 Nm)
   - Negative battery cable
4. Start the vehicle and check for exhaust leaks, repair if necessary.

Front Crankshaft Seal

REMOVAL & INSTALLATION

Metro

1. Before servicing the vehicle, refer to the precautions in the beginning of this section.
2. Drain the engine oil.
3. Remove or disconnect the following:
   - Negative battery cable
   - Air cleaner assembly
   - Right side engine splash shield
   - Lower alternator cover
   - Accessory drive belt
   - Crankshaft pulley
   - Water pump pulley
   - Timing belt cover
4. Rotate the Crankshaft to align the timing marks.
   - Timing belt
   - Crankshaft oil seal

To install:

5. Lubricate the lip of the new seal with clean engine oil.

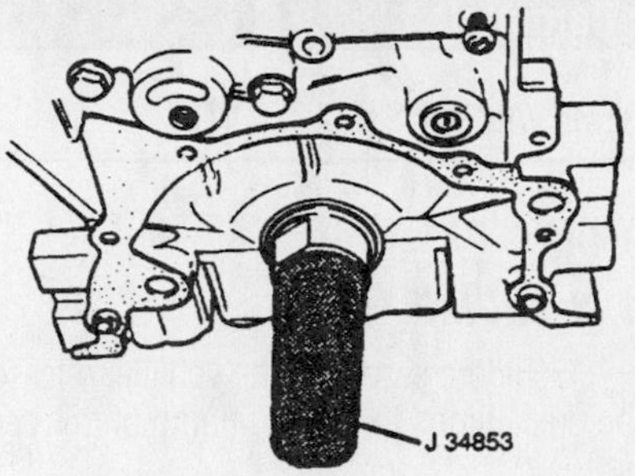

To prevent damage to the lip of the seal use an oil seal protector as shown—Metro

6. Install or connect the following:
   - New oil seal
   - Timing belt and torque the tensioner bolt to 20 ft. lbs. (27 Nm)
   - Timing belt cover and torque the bolts to 97 inch lbs. (11 Nm)
   - Water pump pulley and torque the bolts to 18 ft. lbs. (24 Nm)
   - Crankshaft pulley and torque the bolts to 12 ft. lbs. (16 Nm)
   - Accessory drive belt
   - Alternator cover plate and torque the bolts to 49 inch lbs. (5.5 Nm)
   - Right side splash shield
   - Air cleaner assembly
   - Negative battery cable
7. Fill the engine with new oil.
8. Start the vehicle and check for leaks, repair if necessary.

Prizm

1. Before servicing the vehicle, refer to the precautions in the beginning of this section.
2. Remove or disconnect the following:
   - Negative battery cable
   - Accessory drive belt
   - Right side splash shield
   - Crankshaft pulley
   - Crankshaft front seal

To install:

3. Apply clean engine oil to the new oil seal and lubricate the lip of the oil seal with multi-purpose grease.
4. Install or connect the following:
   - New oil seal using a seal driver
   - Crankshaft pulley and torque the bolt to 105 ft. lbs. (142 Nm)
   - Right side splash shield
   - Accessory drive belt
   - Negative battery cable
5. Start the vehicle and check for leaks, repair if necessary.

Camshaft

REMOVAL & INSTALLATION

Metro

1.0L ENGINE

1. Before servicing the vehicle, refer to the precautions in the beginning of this section.
2. Remove or disconnect the following:
   - Negative battery cable
   - Rocker arm cover
   - Distributor and housing
   - Timing belt
   - Camshaft timing belt sprocket. Lock the camshaft with a 0.39 in. (10mm) rod inserted into the hole in the camshaft, before loosening the sprocket retaining bolt.

WARNING

The mating surface of the cylinder head and cover must not be damaged during this procedure. Place a clean shop cloth between the rod and mating surfaces and use care not to bump the rod when loosening.

3. Turn the crankshaft until the crankshaft sprocket timing mark is 60 degrees to the left of the arrow mark on the oil pump case.
4. Remove or disconnect the following:
   - Camshaft caps from the cylinder head
   - Camshaft
   - Valve lifters

To install:

5. Pour engine oil through the camshaft journal oil holes and check that engine oil comes out from the oil holes in the valve lifter bores.
6. Install the valve lifters in their original positions, if being re-used.
7. Install the camshaft in the cylinder head. After applying engine oil to the camshaft journal and all around the cam, position the camshaft so the camshaft timing sprocket pin hole in the camshaft is at the lower position.
8. Apply clean engine oil to the sliding surface of each bearing cap against the camshaft journal.
9. Apply silicone sealant to the mating surface of the No. 1 and No. 3 bearing cap, which will mate with the cylinder head.
10. There are marks provided on each camshaft bearing cap indicating the position and direction for installation. Install the bearing cap as indicated by the marks.

➡**Camshaft bearing cap No. 1 is installed first. It retains the camshaft in the proper position and thrust direction.**

11. Apply clean engine oil to the retaining bolts and torque them to 97 inch lbs. (11 Nm).
12. Install or connect the following:
   - Camshaft oil seal after applying engine oil to the seal lip
   - Camshaft sprocket and torque the bolt to 44 ft. lbs. (60 Nm)
   - Valve cover using a new gasket and torque the bolts to 44 inch lbs. (5 Nm)
   - Distributor housing
   - Distributor
   - Timing belt
   - Negative battery cable
13. Start the engine and check the ignition timing. Adjust the timing as necessary.

➡**When the engine is started, if air is trapped in the Hydraulic Valve Lash (HVL) adjuster, the valve may make a tapping sound when the engine is operated. In such a case, run the engine at 2000 rpm until the air is purged and the tapping sound ceases.**

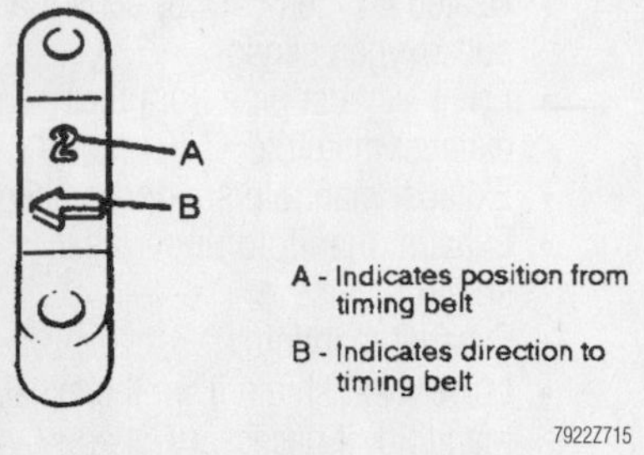

Directional markings for the camshaft bearing caps—Metro with 1.0L engine

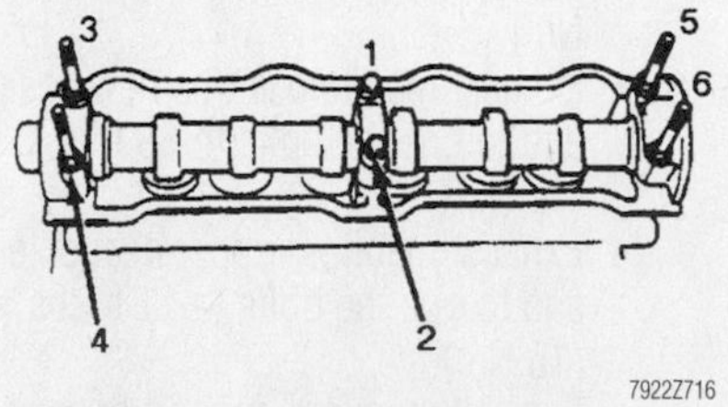

Torque sequence for the camshaft bearing cap bolts—Metro with 1.0L engine

1.3L ENGINE

1. Before servicing the vehicle, refer to the precautions in the beginning of this section.
2. Remove or disconnect the following:
   - Negative battery cable
   - Rocker arm cover
   - Timing belt
   - Camshaft timing belt pulley
   - Camshaft Position (CMP) sensor and housing
3. Loosen all valve adjusting screws to allow the rocker arms to move freely.
   - Camshaft housings
   - Camshaft

To install:

4. Apply clean engine oil to the camshaft, oil journals and the oil seal.
5. Install or connect the following:
   - Camshaft
   - Camshaft housings and torque the bolts in sequence to 97 inch lbs. (11 Nm)
   - New camshaft oil seal

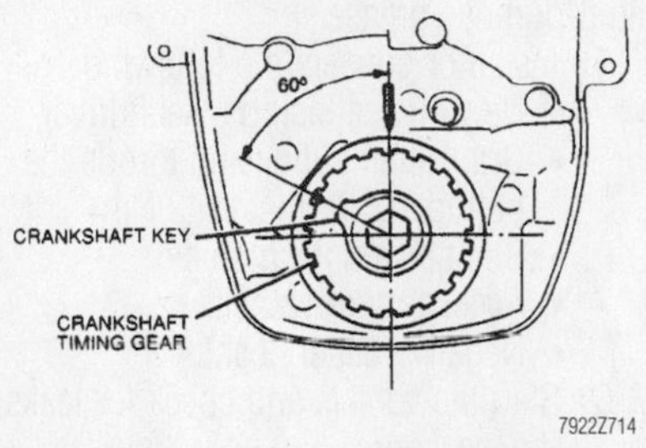

Position the crankshaft as shown before removing the camshaft—Metro with 1.0L engines

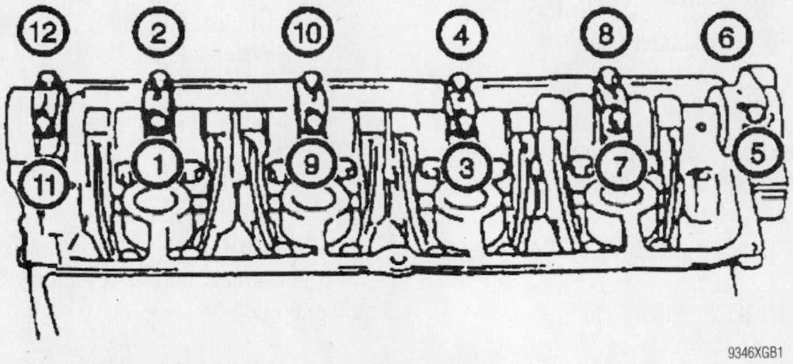

Camshaft housing torque sequence

- Camshaft timing sprocket and torque the bolt to 43 ft. lbs. (60 Nm)
- Timing belt

6. Perform the following steps to adjust the valve lash.

a. Set the number one cylinder to top dead center.

b. Ensure the camshaft is in proper phase.

c. Adjust valves 1, 2, 8, and 6.

d. Rotate the crankshaft one revolution (360°).

e. Adjust valves 3, 4, 7, and 5.

7. Install or connect the following:

- CMP sensor housing and torque the bolts to 97 inch lbs. (11 Nm)
- CMP sensor
- Rocker arm cover and torque the bolts to 97 inch lbs. (11 Nm)
- Negative battery cable

Prizm

➡The procedure is the same for the exhaust camshaft and the intake camshaft.

1. Before servicing the vehicle, refer to the precautions in the beginning of this section.
2. Remove or disconnect the following:

- Negative battery cable
- Timing chain
- Camshaft sprocket
- Camshaft bearing cap bolts, in sequence, by starting at the ends and working towards the center of the camshaft
- Camshaft from the cylinder head

To install:

3. Install or connect the following:

- Camshaft to the cylinder head
- Camshaft bearing caps in their proper location and direction and torque the bolts in sequence to 10 ft. lbs. (13 Nm). Torque the front bearing cap bolts to 17 ft. lbs. (23 Nm).
- Sprocket to the camshaft and torque the bolt to 40 ft. lbs. (55 Nm)
- Timing chain
- Negative battery cable

4. Check the valve clearance and adjust if necessary.

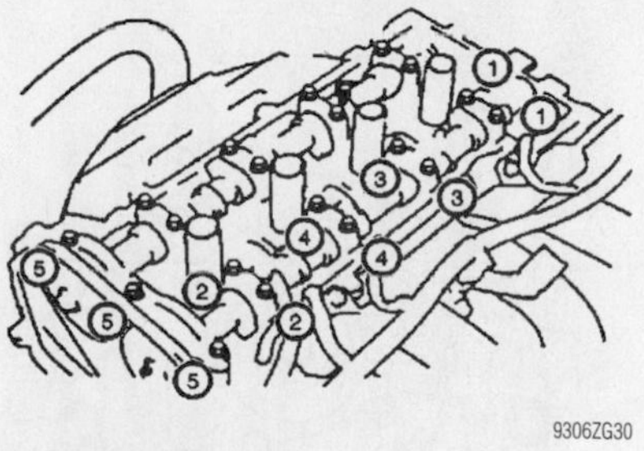

Camshaft bearing cap tightening sequence—Prizm

Valve Lash

ADJUSTMENT

1.0L Metro

Lash adjusters, located between the camshaft and valve stems, are used to adjust the valve clearance to 0 lash automatically at all times. Adjustment is not required or possible.

Prizm and 1.3L Metro

1. Before servicing the vehicle, refer to the precautions in the beginning of this section.
2. Remove the negative battery cable.
3. Disconnect the ignition wires.
4. Remove the rocker arm cover.
5. Turn the crankshaft until the piston in No. 1 cylinder is at Top Dead Center (TDC) on the compression stroke. Align the groove in the crankshaft pulley with the **0** mark on the timing belt cover. Be sure the camshaft gears are aligned.
6. The intake valve clearance should be 0.006–0.010 in. (0.15–0.25mm) and the exhaust valve clearance should be 0.010–0.014 in. (0.25–0.35mm).
7. Using a feeler gauge, measure the clearance between the camshaft and valve lifter shim at the No. 1 cylinder intake and exhaust valves, the No. 2 cylinder intake valves and the No. 3 cylinder exhaust valves. Record the clearance measurements for all valves that are not within specification, in order to determine the required replacement shims.

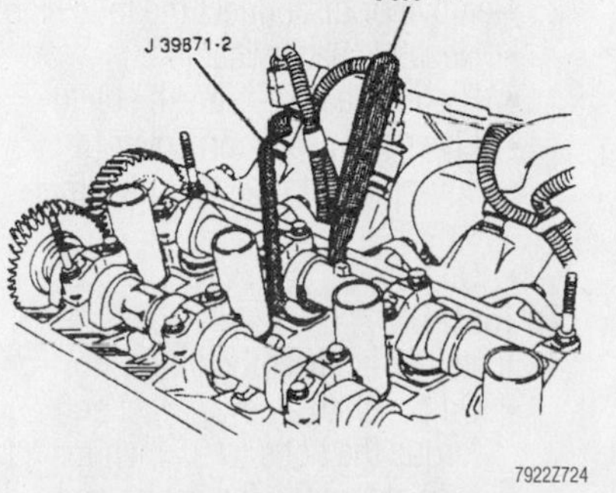

Using the valve clearance adjustment tool set to hold the lifter

8. Rotate the crankshaft pulley 1 full turn (360 degrees) and check the clearance at the No. 2 cylinder exhaust valves, the No. 3 cylinder intake valves and the No. 4 cylinder intake and exhaust valves. Record the clearance measurements for all valves that are not within specification, in order to determine the required replacement shims.
9. For valves requiring adjustment, proceed as follows:

a. Be sure the base of the camshaft lobe is directly over the valve (camshaft lobe pointing away from the valve).

b. Insert tool J-3987-1 between the camshaft and lifter adjustment shim, to compress the valve spring and push the lifter down.

c. Insert tool J-39871-2 between the camshaft and the lifter, to hold the lifter away from the camshaft. Position the bottom edge of the tool on the lifter.

d. Using a small screwdriver and a magnet, remove the adjustment shim from the top of the lifter.

e. Use the micrometer to measure the thickness of the removed shim. Determine the thickness of the new shim using the formula below. For the purposes of the following formula, T = Thickness of the shim removed; A = Valve clearance measured; N = Thickness of the required new shim.

10. For the intake camshaft valves: N = T + A—0.008 in. (0.20mm).
11. For the exhaust camshaft valves: N = T + A—0.010 in. (0.25mm).

a. Select a shim closest to the calculated thickness. Shims are available in 16 sizes, in increments of 0.002 in. (0.050mm), from 0.1004 in. (2.55mm) to 0.1299 in. (3.30mm).

b. Install the shim on the valve lifter and remove tool J-39871-2. Recheck the valve clearance.

12. Install the rocker arm cover, using a new gasket and torque the nuts to 53 inch lbs. (6 Nm).
13. Connect the ignition wires.
14. Connect the negative battery cable.
15. Run the engine and check operation.

Starter Motor

REMOVAL & INSTALLATION

Metro

1. Before servicing the vehicle, refer to the precautions in the beginning of this section.
2. Remove or disconnect the following:
   - Negative battery cable
   - Starter solenoid electrical connectors
   - Starter motor

To install:

3. Install or connect the following:
   - Starter motor and torque the bolts to 17 ft. lbs. (23 Nm)
   - Starter solenoid electrical connectors
   - Negative battery cable

Prizm

1. Before servicing the vehicle, refer to the precautions in the beginning of this section.
2. Remove or disconnect the following:
   - Negative battery cable
   - Upper mounting bolt
   - Right side splash shield
   - Starter motor electrical connectors
   - Starter motor

To install:

3. Install or connect the following:
   - Starter motor and torque the bolts to 27 ft. lbs. (37 Nm)
   - Starter motor electrical connectors
   - Right side splash shield
   - Negative battery cable

Oil Pan

REMOVAL & INSTALLATION

Metro

1. Before servicing the vehicle, refer to the precautions in the beginning of this section.
2. Drain the engine oil.
3. Remove or disconnect the following:
   - Negative battery cable
   - Front pipe from the exhaust manifold and separate it from the center/resonator pipe
   - Catalytic converter
   - Crankshaft Position (CKP) sensor
   - Flywheel inspection cover
   - Oil pan
   - Oil pump strainer and bracket
4. Clean the mating surface of the oil pan and the cylinder block.

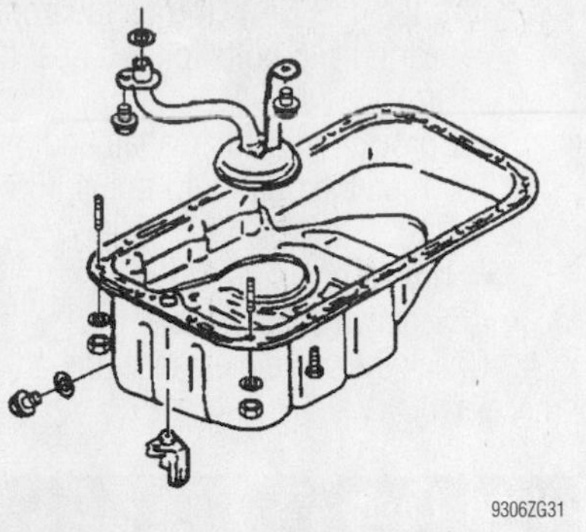

Exploded view of the oil pan and strainer—Metro

5. Clean the oil pan and oil pump strainer.

To install:

6. Install or connect the following:
   - New oil pump strainer seal
   - Oil pump strainer to the cylinder block and torque the bolts to 97 inch lbs. (11 Nm)
   - Oil pan to the cylinder block after applying a continuous bead of sealant around the oil pan. Torque the bolts to 97 inch lbs. (11 Nm) starting in the center and working outward.
   - CKP sensor
   - Flywheel inspection cover
   - Front pipe to the exhaust manifold
   - Center pipe to the front pipe
   - Catalytic converter
   - Negative battery cable
7. Fill the engine with new oil.
8. Start the vehicle and check for leaks, repair if necessary.

Prizm

1. Before servicing the vehicle, refer to the precautions in the beginning of this section.
2. Drain the engine oil.
3. Remove or disconnect the following:
   - Negative battery cable
   - Right side lower splash shield
   - Flywheel inspection cover (if equipped with manual transmission)
   - Oil pan

To install:

4. Install or connect the following:
   - Oil pan with a new gasket and torque the bolts to 97 inch lbs. (11 Nm) starting in the center and working outward
   - Flywheel inspection cover, if removed
   - Right side lower splash shield
   - Negative battery cable

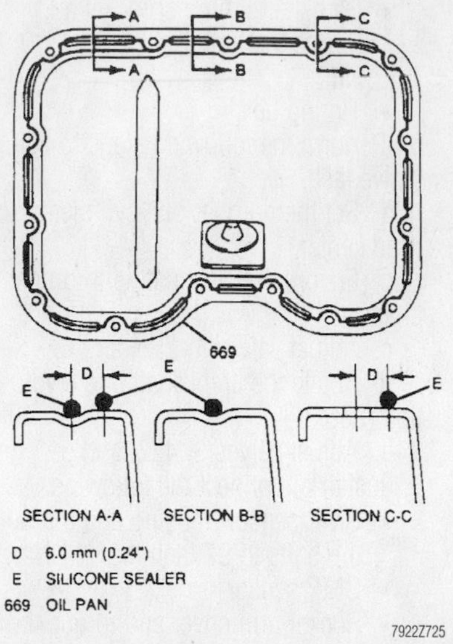

To ensure a leak-free seal, apply sealer as shown—Prizm 1.8L engine

5. Start the vehicle and check for leaks, repair if necessary.

Oil Pump

REMOVAL & INSTALLATION

Metro

1. Before servicing the vehicle, refer to the precautions in the beginning of this section.
2. Drain the engine oil.
3. Remove or disconnect the following:
   - Negative battery cable
   - Oil pan
   - Timing belt and tensioner
   - A/C compressor, if equipped and lock the crankshaft
   - Timing belt guide
   - Oil pump and discard the gasket
   - Rotor plate pins from the oil pump

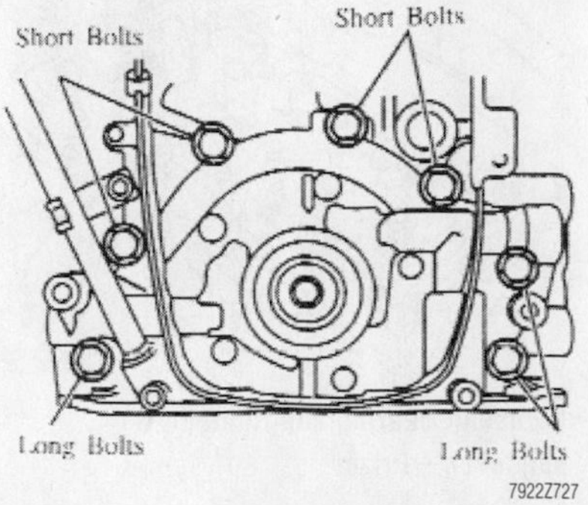

Oil pump mounting bolt identification—Metro

To install:

4. Install or connect the following:
   - Rotor plate pins to the oil pump
   - Oil seal over the crankshaft and make certain that the lip is not turned out
   - Apply Loctite® sealant to the threads of the short mounting bolts and install the oil pump. Torque the bolts to 97 inch lbs. (11 Nm).
   - Rubber seal between the oil pump and the water pump
   - Timing belt guide and lock the crankshaft
   - A/c compressor and bracket, if equipped
   - Oil pan
   - Timing belt and tensioner
   - Guide tube to the oil pump. Torque the bolt to 97 inch lbs. (11 Nm).
   - Negative battery cable
5. Fill the engine with new oil.
6. Start the vehicle and check for leaks, repair if necessary.

Prizm

1. Before servicing the vehicle, refer to the precautions in the beginning of this section.
2. Drain the engine oil.
3. Remove or disconnect the following:
   - Negative battery cable
   - Timing chain
   - Oil pump

To install:

4. Install or connect the following:
   - Oil pump with a new gasket and torque the bolts to 97 inch lbs. (11 Nm)
   - Timing chain
   - Negative battery cable
5. Fill the engine with new oil.
6. Start the vehicle and check for leaks, repair if necessary.

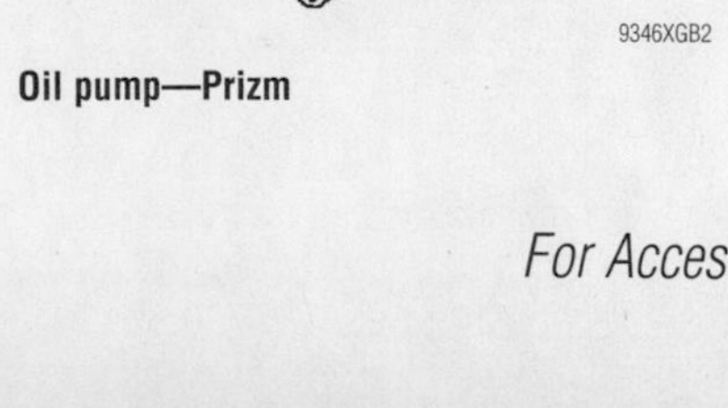

9346XGB2

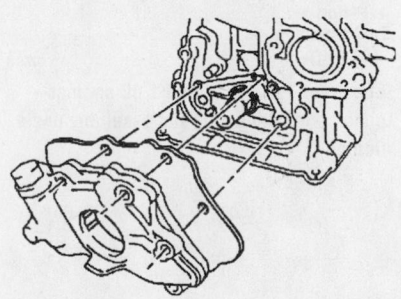

Oil pump—Prizm

Rear Main Seal

REMOVAL & INSTALLATION

Metro

1. Before servicing the vehicle, refer to the precautions in the beginning of this section.
2. Remove or disconnect the following:
   - Negative battery cable
   - Transmission from the vehicle
   - Pressure plate and clutch disc, for manual transmission
3. Matchmark the flywheel-to-engine position.
   - Flywheel from the crankshaft
   - Crankshaft position (CKP) sensor
   - Oil pan
   - Rear crankshaft seal housing
   - Rear crankshaft seal

To install:

4. Lubricate the inside and outside edges of the rear crankshaft seal.
5. Install or connect the following:
   - Rear crankshaft seal housing with a new seal and torque the bolts to 97 inch lbs. (11 Nm)
   - Oil pan and torque the bolts to 97 inch lbs. (11 Nm)
   - CKP sensor and torque the bolt to 97 inch lbs. (11 Nm)
6. Apply Loctite® sealant to the flywheel retaining bolt threads.
   - Flywheel and torque the bolts to 55 ft. lbs. (75 Nm)
   - Pressure plate and clutch disc, for manual transmission
   - Transmission
   - Negative battery cable
7. Check the oil level and top off if necessary.
8. Start the vehicle and check for leaks, repair if necessary.

Prizm

1. Before servicing the vehicle, refer to the precautions in the beginning of this section.
2. Remove or disconnect the following:
   - Negative battery cable
   - Transmission
   - Pressure plate and clutch disc, for manual transmission
3. Matchmark the flywheel-to-engine position.
   - Flywheel from the crankshaft
   - Rear crankshaft seal

To install:

4. Lubricate the lip of the new rear oil seal.
5. Install or connect the following:
   - Rear oil seal and tap it into place
   - Flywheel and torque the bolts to 67 ft. lbs. (83 Nm)
   - Pressure plate and clutch disc, for manual transmission
   - Transmission
   - Negative battery cable
6. Check the oil level and top off, if necessary.
7. Start the vehicle and check for leaks, repair if necessary.

Timing Chain, Sprockets, Front Cover & Seal

REMOVAL & INSTALLATION

Prizm

1. Before servicing the vehicle, refer to the precautions in the beginning of this section.
2. Remove or disconnect the following:
3. Drain the cooling system.
   - Negative battery cable
   - Windshield washer reservoir
   - Drive belt
   - Alternator and install an engine support fixture
   - Right side engine mount
   - Cylinder head cover

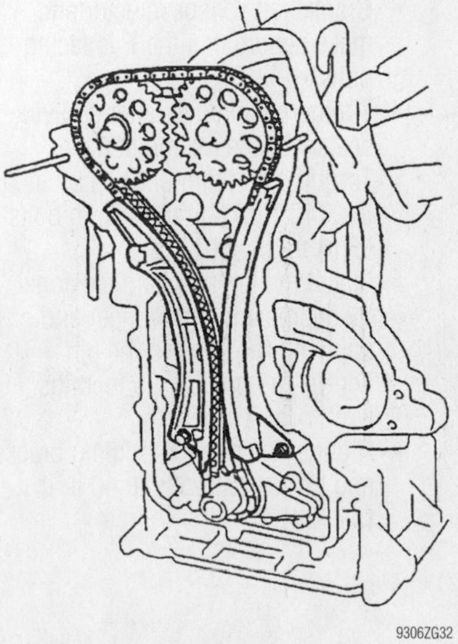

9306ZG32

Exploded view of the timing chain and sprockets–1.8L Prizm

4. Set the No. 1 piston to the Top Dead Center (TDC) position on the compression stroke and align the camshaft timing sprockets.

5. Remove or disconnect the following:
- Power steering oil pressure switch connector
- Power steering pump
- Crankshaft pulley
- Crankshaft Position (CKP) sensor
- Drive belt tensioner
- Timing chain cover
- Crankshaft sensor reluctor
- Timing chain damper and shoe
- Crankshaft sprocket
- Timing chain
- Front seal
- Camshaft sprocket bolt
- Camshaft sprocket

To install:

6. Install or connect the following:
- Camshaft sprocket. Use a wrench to turn the camshafts to align the timing marks
- Crankshaft bolt and turn it until the keyway faces upward. When aligned properly, torque the sprocket bolts to 33 ft. lbs. (44 Nm).
- Front seal and tap it into position until the surface is flush against the retainer edge
- Timing chain
- Crankshaft timing sprocket
- Timing chain shoe and torque the bolts to 89 inch lbs. (10 Nm)
- Timing chain damper and torque the bolt to 14 ft. lbs. (18 Nm)
- Crankshaft sensor reluctor and make certain that the **F** is facing outward
- Timing chain cover after applying a sealant to the mating surface. Torque the 10mm bolts to 89 inch lbs. (10 Nm) and the 12mm bolts to 14 ft. lbs. (18 Nm).
- Install the timing chain tensioner by depressing the plunger and applying the hook to the pin. Torque the bolts to 89 inch lbs. (10 Nm).
- Right side engine mounting bracket and torque the bolts to 40 ft. lbs. (54 Nm)

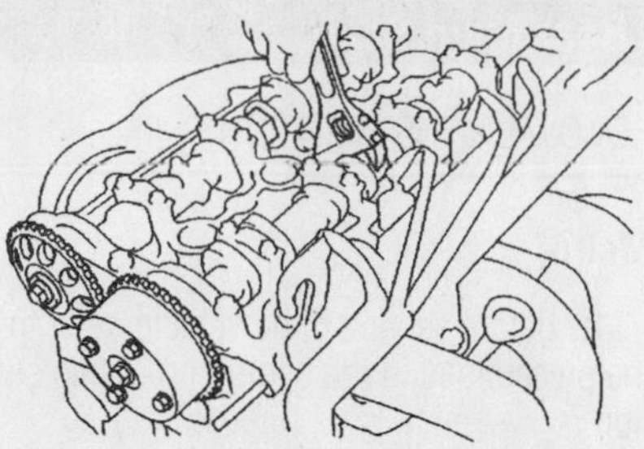

9346ZGD8

Use a wrench to turn the camshafts to align the timing marks—Prism

- Drive belt tensioner and torque the bolt to 51 ft. lbs. (69 Nm) and the nut to 21 ft. lbs. (29 Nm)
- CKP sensor and torque the bolt to 106 inch lbs. (12 Nm)
- Crankshaft pulley and torque the pulley to 105 ft. lbs. (142 Nm)
- Power steering pump and torque the bolts to 32 ft. lbs. (43 Nm)
- Power steering oil pressure switch

7. Rotate the crankshaft clockwise 1 revolution and verify that the plunger on the timing chain tensioner has released. If the plunger does not release perform the following steps:

a. Step 1: Press the timing chain dampener into the tensioner.

b. Step 2: Release the hook from the pin.

c. Step 3: Verify proper timing chain alignment after the plunger is released.

8. Install or connect the following:
- Cylinder head cover
- Right side engine mount and insulator. Torque the bolts to 40 ft. lbs. (54 Nm).
- Alternator

9. Remove the engine support fixture.

10. Install or connect the following:
- Windshield washer reservoir
- Negative battery cable

11. Fill the cooling system.

12. Start the vehicle and check for leaks, repair if necessary.

Piston & Ring

POSITIONING

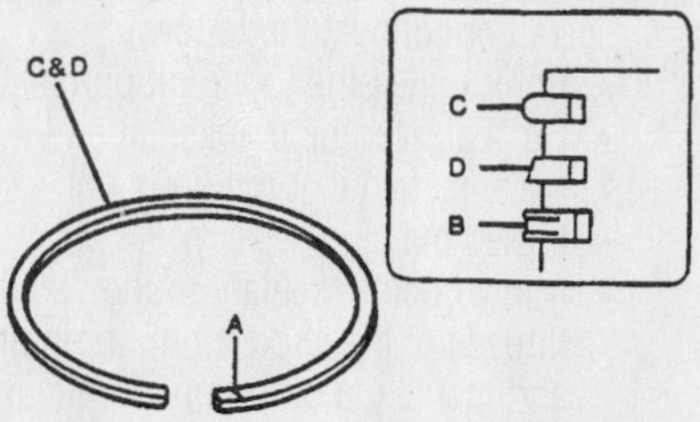

A "R" OR "T" MARK C UPPER COMPRESSION RING
B OIL RING D LOWER COMPRESSION RING

7922AG40

Chevrolet 1.0L, 1.3L and 1.8L engines—piston ring positioning

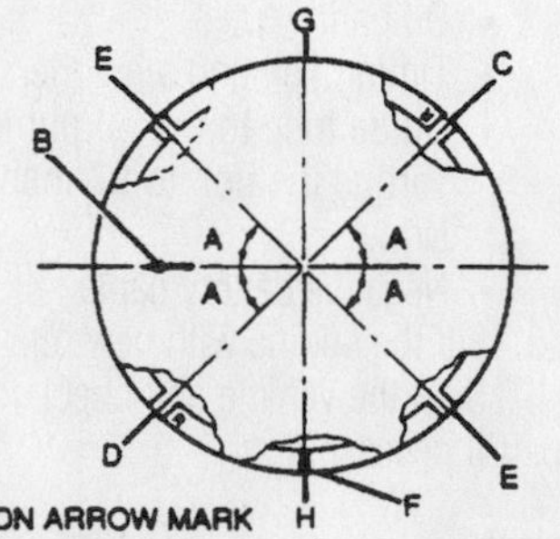

A 45°
B PISTON ARROW MARK
C FIRST RING END GAP
D SECOND RING END GAP G INTAKE SIDE
E OIL RING END GAPS H EXHAUST SIDE
F OIL RING SPACER GAP

7922AG39

Chevrolet 1.0L, 1.3L and 1.8L engines—piston ring end-gap spacing

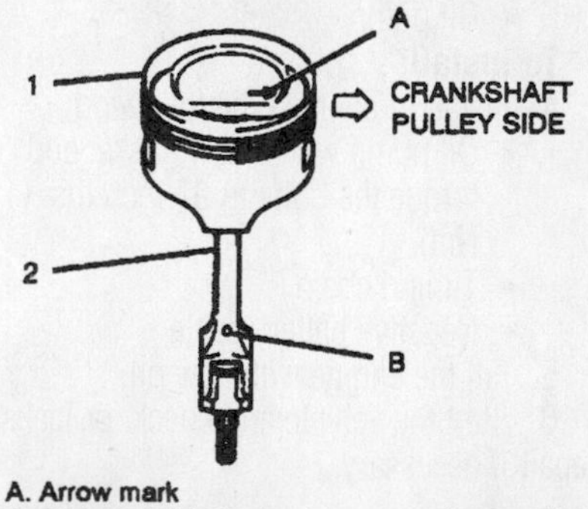

A. Arrow mark
B. Oil hole (oil hole should come on intake side)
1. Piston
2. Connecting rod

7922AG41

Chevrolet 1.0L, 1.3L and 1.8L engines—piston and connecting rod assembly positioning

FUEL SYSTEM

Fuel System Pressure

RELIEVING

1. Before servicing the vehicle, refer to the precautions in the beginning of this section.
2. Remove the fuel filler cap.
3. On Metro, perform the following:
   a. Remove the control relay box cover from the relay box.
   b. Disconnect the fuel pump relay from the relay box connector.
4. Remove the circuit opening relay above the left side lower kick panel.
5. Attempt to start the engine, if it starts, let it run until the engine stalls due to lack of fuel.
6. Engage the starter for a few seconds to assure relief of remaining fuel pressure.

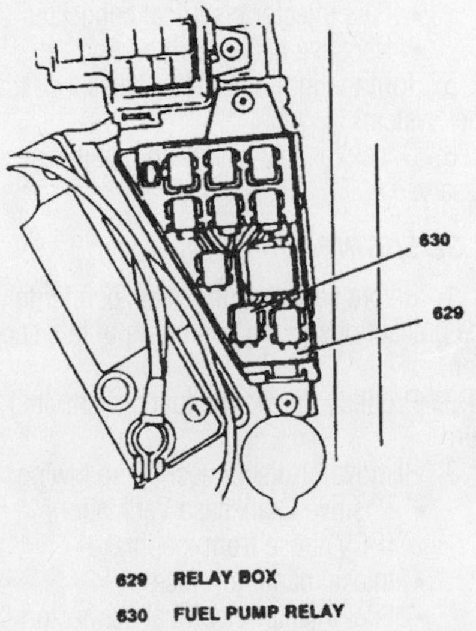

View of the relay box showing the location of the fuel pump relay—Metro

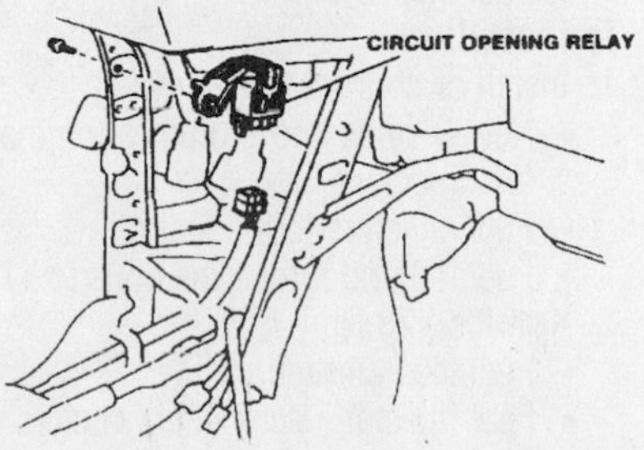

Leave the circuit opening relay unplugged during fuel system servicing—Prizm

7. Disconnect the negative battery cable.
8. Continue with the required service procedure(s).

Fuel Filter

REMOVAL & INSTALLATION

The fuel filter is an integral part of the fuel sender assembly which is located in the fuel tank. The fuel filter does not require replacement at regular intervals.

Fuel Pump

REMOVAL & INSTALLATION

Metro

1. Before servicing the vehicle, refer to the precautions in the beginning of this section.
2. Properly relieve the fuel system pressure.
3. Drain the fuel tank by pumping the fuel out through the filler neck into a suitable container.
4. Remove or disconnect the following:
   - Rear seat cushion
   - Fuel pump electrical connector
   - Muffler and exhaust pipe assembly
   - Fuel filler hose
   - Fuel feed and return clamps and hoses from the pump
   - Fuel tank from the vehicle
   - Screws from the pump assembly
   - Fuel sender assembly
5. Disassemble the fuel sending unit by removing or disconnecting the following:
   - Fuel level sensor by depressing the tab
   - Lower assembly end cap
   - Fuel pump cushion
   - Fuel sender sub assembly electrical connectors
   - Fuel pump from the sender assembly
   - Fuel tube and 2 grommets from the fuel sender sub assembly
6. The sender sub assembly contains the fuel filter and must be serviced as an assembly.

To install:

7. Install or connect the following:
   - Fuel tube and 2 new grommets from the fuel sender sub assembly
   - Fuel pump to the sender assembly
   - Fuel sender sub assembly electrical connectors
   - New fuel pump cushion
   - Lower assembly end cap
   - Fuel level sensor
   - Fuel sender assembly with a new gasket in the fuel tank and torque the screws to 89 inch lbs. (10 Nm)
   - Fuel feed and return hoses and clamps on the fuel pump
   - Fuel sender electrical connectors
   - Fuel tank and torque the bolts to 18 ft. lbs. (25 Nm)
   - Fuel filler hose
   - Muffler and exhaust pipe assembly and torque the bolts to 26 ft. lbs. (35 Nm)
   - Rear seat cushion
   - Negative battery cable
8. Refill the fuel tank.
9. Turn the ignition key **ON** and allow the fuel system to pressurize.
10. Start the engine and check for leaks, repair if necessary.

Prizm

1. Before servicing the vehicle, refer to the precautions in the beginning of this section.
2. Properly relieve the fuel system pressure.
3. Drain the fuel tank into a suitable container.
4. Remove or disconnect the following:
   - Rear seat cushion to gain access to the service panel
   - Access panel
   - Fuel sender electrical connectors

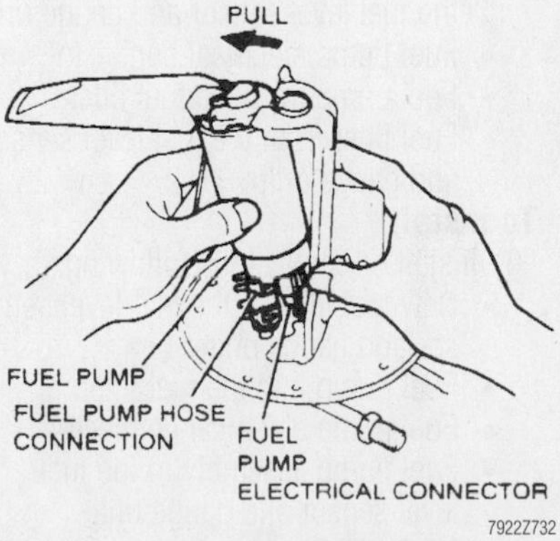

Remove the fuel pump and the sender assembly from the fuel tank slowly to keep from splashing fuel—Prizm

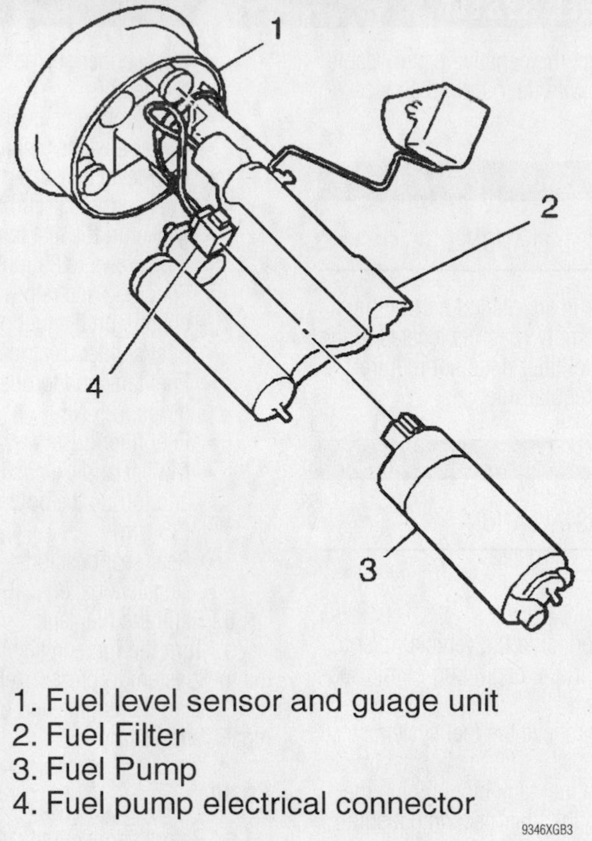

Exploded view of the fuel sender assembly unit

- Fuel line retaining clip
- Fuel line
- Fuel sender from the tank

5. Disassemble the Fuel sender assembly by removing or disconnecting the following:
- End cap
- Fuel pressure regulator
- Fuel pump strainer retaining clip
- Fuel pump strainer
- Fuel sender assembly housing from the fuel level sensor and gauge unit
- Fuel pump electrical connector
- Fuel pump from the fuel filter
- Fuel filter from the fuel level sensor and gauge unit

To install:

6. Install or connect the following:
- New fuel filter to the fuel level sensor and gauge unit
- Fuel pump into the fuel filter
- Fuel pump electrical connector
- Fuel pump assembly to the fuel level sensor and gauge unit
- Fuel pressure regulator
- Fuel pump strainer
- End cap
- Fuel sender assembly to the tank using a new gasket and torque the bolts to 35 inch lbs. (4 Nm)
- Fuel line and retaining clip
- Fuel sender electrical connector
- Negative battery cable

7. Refill the fuel tank.

8. Turn the ignition **ON** to pressurize the fuel system.

9. Check for fuel leaks and repair if necessary.

10. Install the service access panel and rear seat cushion.

Fuel Injector

REMOVAL & INSTALLATION

Metro

1.0L ENGINE

1. Before servicing the vehicle, refer to the precautions in the beginning of this section.

2. Properly relieve the fuel system pressure.

3. Remove or disconnect the following:
- Air cleaner assembly
- Air cleaner stud bracket from the Throttle Body Injection (TBI) unit
- Fuel injector cover
- Fuel injector electrical connectors
- Fuel injector

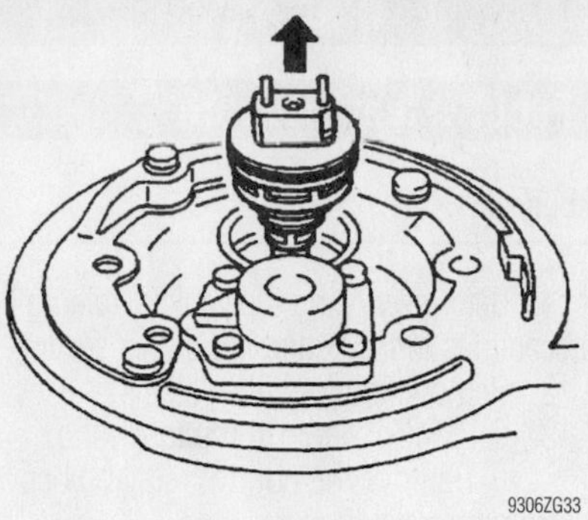

Remove the fuel injector from the Throttle Body Unit (TBI)—Metro

To install:

4. Install or connect the following:
- New O-rings coated with clean engine oil to the injector
- Injector into the TBI unit
- Fuel injector cover
- Air cleaner stud bracket and torque the bolts to 89 inch lbs. (10 Nm)
- Fuel injector electrical connector
- Negative battery cable

5. Turn the ignition **ON** to pressurize the fuel system.

6. Check for fuel leaks and repair if necessary.

1.3L ENGINE

1. Before servicing the vehicle, refer to the precautions in the beginning of this section.

2. Properly relieve the fuel system pressure.

3. Remove or disconnect the following:
- Positive Crankcase Ventilation (PCV) hose from the intake
- Intake manifold brace
- Fuel injector electrical connectors
- Fuel pressure regulator vacuum hose
- Fuel feed and return lines
- Fuel rail
- Fuel injectors

To install:

4. Install or connect the following:
- New O-rings and grommets on the fuel injectors
- Fuel injectors to the fuel rail
- Fuel rail and torque the bolts to 17 ft. lbs. (23 Nm)
- Fuel feed and return lines
- Fuel injector electrical connectors
- Fuel pressure regulator vacuum hose
- Intake manifold brace and torque the bolts to 22 ft. lbs. (30 Nm)
- PCV hose
- Negative battery cable

5. Turn the ignition **ON** to pressurize the fuel system.
6. Check for fuel leaks and repair if necessary.

Prizm

1. Before servicing the vehicle, refer to the precautions in the beginning of this section.
2. Remove the engine cover.
3. Properly relieve the fuel system pressure.
4. Remove or disconnect the following:
   - Fuel feed hose quick connect coupling cover
   - Fuel feed hose
   - Throttle Valve (TV) cable and bracket, if equipped
   - Fuel injector electrical connectors
   - Fuel rail from the cylinder head
   - Fuel injectors from the fuel rail

To install:

5. Apply clean engine oil to the new O-rings and grommets.
6. Install or connect the following:
   - New O-rings and grommets to the fuel injectors
   - Fuel injectors to the fuel rail
   - Fuel rail to the cylinder head and torque the bolts to 13 ft. lbs. (18.5 Nm)
   - TV cable and bracket and torque the bolts to 77 inch lbs. (9 Nm)
   - Fuel feed hose quick connect fitting
   - Fuel feed hose coupling cover
   - Fuel injector electrical connectors
   - Negative battery cable
7. Turn the ignition **ON** to pressurize the fuel system.
8. Check for fuel leaks and repair if necessary.

DRIVE TRAIN

Transaxle Assembly

REMOVAL & INSTALLATION

Manual

METRO

1. Before servicing the vehicle, refer to the precautions in the beginning of this section.
2. Drain the transmission fluid.
3. Remove or disconnect the following:
   - Negative battery cable
   - Both front wheels
   - Backup lamp switch electrical connector
   - Clutch release lever from the clutch release shaft
   - Clutch cable bracket and move it aside
   - Negative battery cable and hanger from the transmission
   - Gearshift guide case bolts and set the engine wire harness and bracket aside
   - Speedometer cable
   - Starter motor
   - Two right side upper case-to-engine bolts
4. Install an engine support fixture.
   - Both drive axles after properly supporting the transmission
   - Flywheel cover
   - Control joint shaft through bolt and separate the control shaft from the control joint
   - Gearshift control lever guide plate and separate the extension rod from the lever guide plate
5. Support the transmission.
   - Bolts from the rear transmission mount
   - Left side mounting bracket bolts
   - Right side case-to-engine mounting bolts
   - Transmission through the left side of the engine compartment and make certain that the input shaft clears the clutch pressure plate

To install:

6. Install or connect the following:
   - Transmission to the engine and torque the bolts to 44 ft. lbs. (60 Nm)
   - Left side mounting bracket and torque the bolts to 44 ft. lbs. (60 Nm)
   - Rear transmission through bolt and torque it to 44 ft. lbs. (60 Nm)
   - Starter motor and torque the bolts to 21 ft. lbs. (28 Nm)
   - Extension rod to the mounting stud on the transmission and torque the nut to 29 ft. lbs. (40 Nm)
   - Extension rod to the gearshift control lever guide plate and torque the bolts to 89 inch lbs. (10 Nm)
   - Gearshift control shaft and joint and torque the joint through bolt to 15 ft. lbs. (20 Nm)
   - Flywheel cover and torque the bolts to 15 ft. lbs. (20 Nm)
   - Both driveshafts
7. Remove the transmission support.
   - Speedometer cable to the speedometer driven gear case
   - Speedometer cable retaining clip
   - Engine wire harness and bracket to the gearshift guide case and torque the bolts to 106 inch lbs. (12 Nm)
   - Negative battery cable and hanger and torque the bolt to 106 inch lbs. (12 Nm)
   - Clutch cable and bracket and torque the bolts to 21 ft. lbs. (28 Nm)
   - Clutch release shaft to the clutch release lever and torque the bolt to 17 ft. lbs. (23 Nm)
   - Backup lamp switch electrical connector
   - Both wheels
   - Negative battery cable
8. Fill the transmission with clean fluid.
9. Start the vehicle and check for leaks, repair if necessary.

PRIZM

1. Before servicing the vehicle, refer to the precautions in the beginning of this section.
2. Drain the transmission fluid.
3. Remove or disconnect the following:
   - Negative battery cable
   - Battery and tray
   - Both front wheels
   - Air cleaner
   - Clutch release cylinder and line
   - Clutch line bracket
   - Backup switch electrical connector
   - Vehicle Speed Sensor (VSS) electrical connectors
   - Ground cable from the transmission
   - Control cables from the bracket and move them to the side
   - Upper transmission to engine bolts
4. Install an engine support fixture.
   - Left engine mount and bracket
   - Splash shields
   - Both axle shafts
   - Heat insulator

- Front suspension crossmember
- Starter motor

5. Support the transaxle.
   - Lower transmission mounting bolts
   - Transmission from the engine

To install:

6. Install or connect the following:
   - Transmission to the engine and torque the lower bolts to 46 ft. lbs. (34 Nm) and the left side bolts to 11 ft. lbs. (23 Nm)
   - Starter motor and torque the bolts to 29 ft. lbs. (39 Nm)
   - Starter electrical connectors
   - Front suspension crossmember and torque the bolts in the following sequence:

a. Main crossmember to underbody bolts to 152 ft. lbs. (206 Nm).

b. Lower "A" frame to underbody bolt to 161 ft. lbs. (218 Nm).

c. Lower "A" frame to underbody bolts to 109 ft. lbs. (147 Nm).

d. Radiator support bolts to 45 ft. lbs. (61 Nm).

7. Remove the transmission support.
8. Install or connect the following:
   - Heat insulator and torque the bolt and nut to 15 ft. lbs. (20 Nm)
   - Both axle shafts
   - Both splash shields
   - Left engine mounting bracket and torque the bolts to 21 ft. lbs. (28 Nm)
   - Left transmission mount and torque the bolts to 64 ft. lbs. (87 Nm)
9. Remove the engine support.
   - Control cables
   - Ground cable
   - VSS electrical connector
   - Backup light switch electrical connector
   - Clutch release cylinder and line and torque the bolts to 106 inch lbs. (12 Nm)
   - Clutch line bracket and torque the bolts to 106 inch lbs. (12 Nm)
   - Both front wheels
   - Air cleaner
   - Battery and cables
10. Fill the transmission with clean fluid.
11. Start the vehicle and check for leaks, repair if necessary.

Automatic

METRO

1. Before servicing the vehicle, refer to the precautions in the beginning of this section.
2. Drain the transmission fluid.
3. Remove or disconnect the following:
   - Negative battery cable
   - Hood
   - Battery
   - Transmission electrical connectors
   - Speedometer cable
   - Throttle Valve (TV) cable from the accelerator cable
   - Shift select cable from the manual select cable joint and bracket
   - Accelerator cable from the bracket on top of the transmission case
   - Vehicle Speed Sensor (VSS) electrical connector
   - Shift solenoid electrical connectors
   - Park Neutral Position (PNP) switch electrical connector
   - Starter motor
   - Fluid cooler hoses
   - Upper rear transmission-to-bulkhead bolt
   - 2 upper transmission-to-engine bolts
   - Negative battery cable from the transmission
4. Install an engine support fixture.
   - Both front wheels
   - Splash shields
   - Control arms from the steering knuckles
   - Both axle shafts
   - Rear engine torque rod
   - Flywheel cover
   - Flywheel-to-torque converter bolts
   - Muffler mounting bracket from the rear of the engine
   - Rear engine mount bracket from the transmission
5. Properly support the transmission.
   - Left transmission mount
   - Remaining engine-to-transmission bolts
   - Transmission

To install:

6. Use grease to lubricate the cup around the center of the torque converter.

7. Measure the distance between the torque converter and the edge of the transaxle housing; it should be at least 0.85 in. (21.4mm). If the distance is less than specified, the torque converter is improperly installed. Remove and reinstall it.

8. Install or connect the following:
   - Transmission to the engine and torque the lower transaxle-to-engine bolt to 40 ft. lbs. (55 Nm)
   - Left transmission mount and bracket and torque the bolts to 40 ft. lbs. (55 Nm)
   - Rear engine mount bracket to the transmission and torque the bolts to 40 ft. lbs. (55 Nm)
   - Rear engine mount-to-bulkhead bolt and torque it to 40 ft. lbs. (55 Nm)
   - Muffler mount to the rear exhaust hanger
   - Flywheel-to-torque converter bolts and torque them to 14 ft. lbs. (19 Nm)
   - Flywheel inspection cover and torque the bolts to 89 inch lbs. (10 Nm)
   - Both driveshafts
   - Rear engine torque rod assembly and torque the bolts to 40 ft. lbs. (55 Nm)
   - Control arms to the steering knuckles and torque the bolts 44 ft. lbs. (60 Nm)
   - Splash shields
   - Both front wheels
9. Remove the engine support fixture.
   - Negative battery cable to the transmission and torque the bolt to 11 ft. lbs. (15 Nm)
   - Upper transmission-to-engine mounting bolts and torque them to 40 ft. lbs. (55 Nm)
   - Through-bolt to the rear engine mount and torque it to 40 ft. lbs. (55 Nm)
   - Starter motor and torque the bolts to 17 ft. lbs. (23 Nm)
   - Inlet and outlet fluid cooler lines with new hose clamps
   - Speedometer cable into the gear case
   - VSS electrical connector
   - Shift solenoid electrical connector
   - PNP switch electrical connector
   - Engine wire harness bracket to the rear of the transmission and torque the bolts 12 ft. lbs. (16 Nm)
   - Accelerator and shift select cables into the bracket on top of the transmission
   - Shift select cable end to the manual select cable joint and adjust as necessary
   - TV cable to the accelerator cable
   - Negative battery cable
10. Fill the transmission with new oil.
11. Flush the oil cooler lines.
12. Start the engine and verify the proper gear select range.
13. Check the fluid level and adjust, as needed.

PRIZM

1. Before servicing the vehicle, refer to the precautions in the beginning of this section.
2. Drain the transmission fluid.
3. Remove or disconnect the following:
   - Both battery cables
   - Battery and tray

- Intake Air Temperature (IAT) sensor electrical connector
- Air cleaner
- Park Neutral Position (PNP) switch electrical connector
- Throttle Valve (TV) cable from the linkage and the bracket
- Shift select cable from the manual lever and the TV cable bracket guide
- Upper transmission to engine bolts
- Starter motor

4. Install an engine support fixture.
   - Left transmission mounting bracket
   - Both splash shields
   - Cooler hoses from the pipes at the transmission
   - Both front wheels
   - Both axle shafts
   - Front transmission mount
   - Rear transmission mount
   - Front crossmember
5. Support the transmission.
   - Engine reinforcement brace and lower it
   - Flywheel access cover
   - Flywheel-to-torque converter bolts
   - Lower transmission-to-engine bolts
   - Transmission

To install:

6. Before installing the transmission assembly, perform the following:

a. Apply grease around the pilot shaft at the center of the torque converter.

b. Measure the distance between the outside edge of the torque converter housing and the torque converter lug. The distance should be more than 0.906 in. (23mm). If it is less than 0.906 in. (23mm), the torque converter is improperly installed. Remove and properly seat on the input shaft.

7. Install or connect the following:
   - Transmission and torque the two lower transmission-to-engine bolts to 47 ft. lbs. (64 Nm)
   - Flywheel-to-torque converter bolts and torque them to 14 ft. lbs. (19 Nm)
   - Flywheel access cover
   - Lower engine reinforcement brace and torque the bolts to 47 ft. lbs. (64 Nm)
8. Remove the transmission support fixture.
   - Front suspension crossmember and torque the 10 front bolts to 152 ft. lbs. (206 Nm) and the two center bolts to 45 ft. lbs. (61 Nm)
   - Rear transmission mount and torque the bolts to 42 ft. lbs. (57 Nm)
   - Front transmission mount and torque the bolts to 47 ft. lbs. (64 Nm)
   - Both driveshafts
   - Both front wheels
   - Left transmission mounting bracket and torque the bolts to 41 ft. lbs. (56 Nm)
   - Oil cooler hoses to the pipes and secure them with new clamps
   - Both splash shields and torque the bolts to 44 inch lbs. (5 Nm)
   - Starter motor and torque the bolts to 27 ft. lbs. (37 Nm)

⁂ WARNING

The differential portion of the 3-speed transmission is separated from the rest of the transmission and must be drained and refilled separately. The differential cannot be drained or refilled through the transmission drain plug or filler tube.

9. On 3-speed transmission, refill the differential with 1½ qts. (1.4 L) of DEXRON®III automatic transmission fluid, into the differential filler plug hole. The fluid level should be even with the bottom of the differential filler plug hole. Install the filler plug and tighten to 29 ft. lbs. (39 Nm).
10. Install or connect the following:
    - Left side transmission mounting bracket and torque the bolt to 41 ft. lbs. (56 Nm)
    - Left transmission bracket reinforcement and torque the bolts to 15 ft. lbs. (21 Nm)
    - Upper transmission-to-engine bolts and torque them to 47 ft. lbs. (64 Nm)
    - TV guide cable bracket onto the transmission and torque the bolt to 71 inch lbs. (8 Nm)
    - Shift select cable into the bracket at the transaxle and secure with clip
    - Shift select cable to the manual lever and torque the nut to 106 inch lbs. (12 Nm)
    - VSS electrical connector
    - TV cable to the throttle linkage and bracket
11. Adjust the TV cable as follows:

a. Be sure the throttle valve is fully closed.

b. Measure the distance between the end of the outer cable boot and the end of the TV cable stopper, it should be 0–0.04 in. (0–1mm).

c. If the distance is greater than specified, loosen the TV cable locknut and tighten the adjust nut until within specification.

d. If the distance is less than specified, loosen the TV cable adjust nut. Torque the locknut to 71 inch lbs. (8 Nm).

12. Install or connect the following:
    - Ground cable and torque the bolt to 115 inch lbs. (13 Nm)
    - PNP electrical connector
    - Air cleaner and torque the bolts to 106 inch lbs. (12 Nm)
    - IAT electrical connector
    - Battery
    - Both battery cables
13. Fill the transmission with new oil.
14. Flush the oil cooler lines.
15. Start the engine and verify the proper gear select range.
16. Check the fluid level and adjust, as needed.

Clutch

ADJUSTMENT

Clutch Pedal Height

Inspect the clutch pedal height. If it is not within 5.45–5.84 in. (138.4–148.4mm), adjust the pedal height as follows:

1. Before servicing the vehicle, refer to the precautions in the beginning of this section.
2. Remove the lower instrument panel and air duct.
3. Loosen the locknut.
4. Turn the stopper bolt until the pedal height is within 5.12–5.51 in. (130–140mm).
5. Tighten the locknut and install the air duct and lower instrument panel.

Clutch Pedal Free Play

Inspect the clutch pedal free play. If it is not within 0.197–0.521 in. (5.0–15.0mm), adjust the pedal free-play as follows:

1. Before servicing the vehicle, refer to the precautions in the beginning of this section.
2. Remove the lower instrument panel and air duct.
3. Depress the pedal until resistance is felt.

4. Loosen the locknut.
5. Turn the pushrod until the free play is within 0.197–0.591 in. (5.0–15.0mm).
6. Tighten the locknut and install the air duct and lower instrument panel.

REMOVAL & INSTALLATION

1. Before servicing the vehicle, refer to the precautions in the beginning of this section.
2. Remove or disconnect the following:
   - Negative battery cable
   - Transmission assembly and matchmark the flywheel to the engine
   - On Prizm, unfasten the release fork bearing clips and remove the release bearing, fork and boot
   - Clutch pressure plate cover by slowly and evenly loosening the bolts until the spring tension is released
   - Clutch disc

To install:

3. Clean the flywheel mating surfaces of all oil, grease and metal deposits.
4. Check the diaphragm spring and pressure plate for wear or damage. If the spring or plate is excessively worn, replace the clutch cover assembly.
5. Check the pilot bearing for smooth operation. If the bearing does not spin freely, replace it.
6. Inspect the disc, pressure plate and flywheel for damage and wear using a caliper to measure depth and width and a dial indicator to measure run-out.
7. The minimum clutch disc rivet head depth: 0.02 in. (0.5mm).
8. The maximum clutch disc run-out: 0.031 in. (0.8mm).
9. The maximum pressure plate spring depth: 0.024 in. (0.6mm).
10. The maximum pressure plate spring width: 0.197 in. (5.0mm).
11. The maximum flywheel run-out: 0.004 in. (0.1mm).
12. When reassembling, apply a thin coating of multi-purpose grease to the release bearing hub and release fork contact points. Also, pack the groove inside the

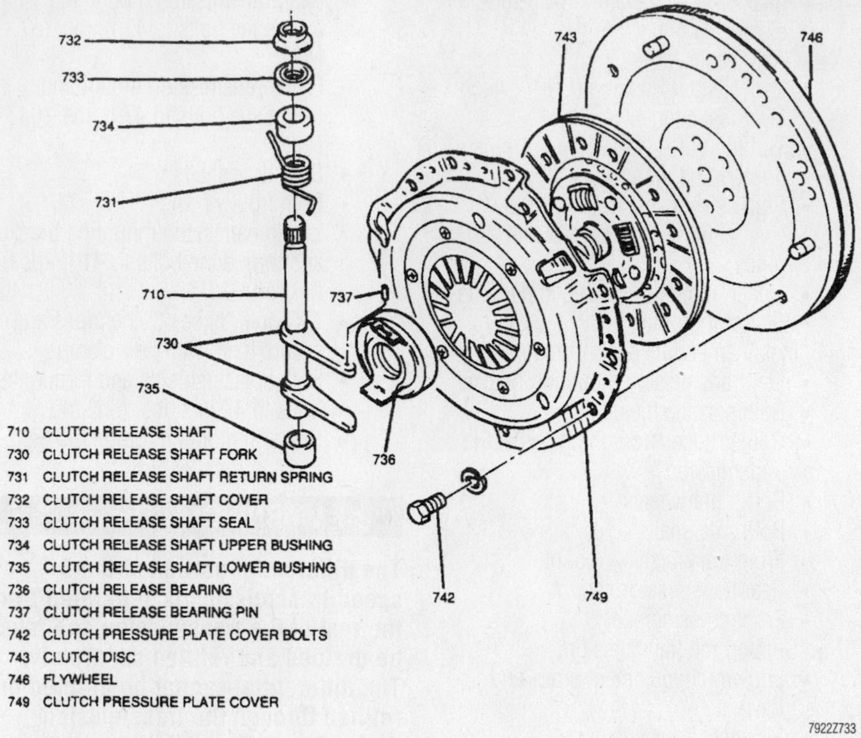

Exploded view of the clutch assembly—Metro

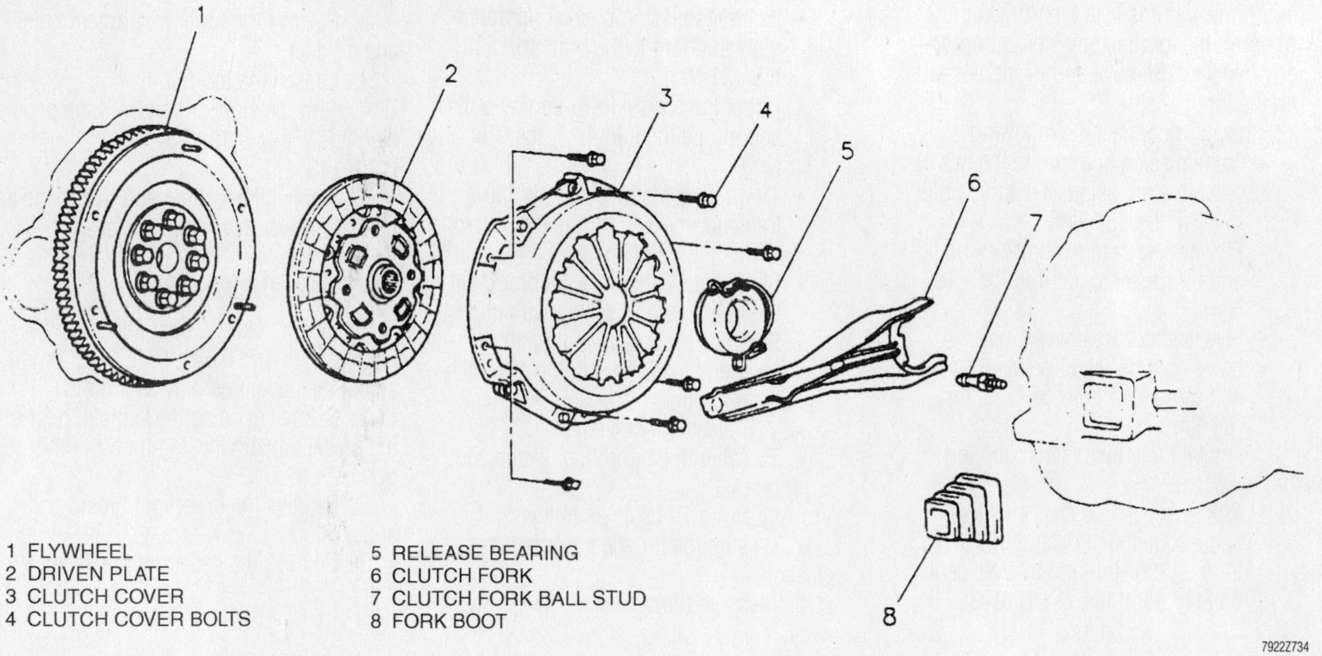

Exploded view of the clutch assembly—Prizm

clutch hub with multi-purpose grease and lubricate the pivot points of the release fork.

13. Install or connect the following:
    - Clutch disc and clutch cover
    - Clutch cover bolts and torque them to 17 ft. lbs. (23 Nm)
    - Release bearing, fork, and boot on the Prizm
    - Transmission
    - Negative battery cable
14. Adjust the clutch cable, if needed.

Hydraulic Clutch System

BLEEDING

1. If any maintenance on the clutch system was performed or the system is suspected of containing air, bleed the system as follows:
   a. Fill the clutch reservoir with brake fluid. Check the reservoir level frequently and add fluid as needed.
   b. Connect one end of a vinyl tube to the bleeder plug on the slave cylinder and submerge the other end into a clear container half-filled with clean brake fluid.
   c. Slowly pump the clutch pedal several times.
   d. Repeat Steps 2 and 3 until all of the air bubbles are removed from the system.
   e. Tighten the bleeder screw to 97 inch lbs. (11 Nm).
   f. Refill the master cylinder to the proper level.
   g. Check the system for leaks.

Halfshaft

REMOVAL & INSTALLATION

Metro

1. Before servicing the vehicle, refer to the precautions in the beginning of this section.
2. Drain the transmission fluid.
3. Remove or disconnect the following:
   - Negative battery cable
   - Front wheel
   - Anti-lock Brake System (ABS) speed sensor from the steering knuckle, if equipped
   - Unstake and remove the halfshaft nut
   - Ball joint from the steering knuckle

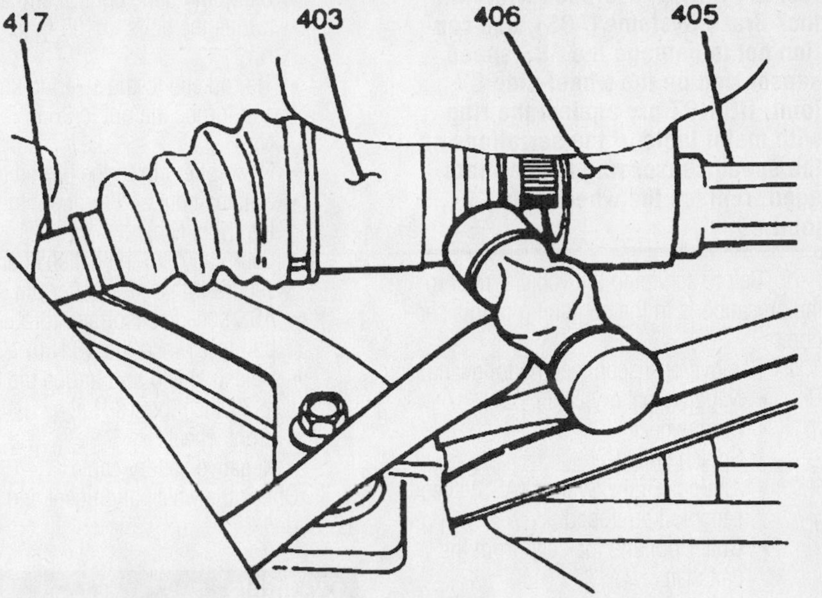

403 DIFFERENTIAL-SIDE JOINT HOUSING
405 RIGHT INNER DRIVE AXLE SHAFT
406 RIGHT INNER DRIVE AXLE SUPPORT ARBOR
417 RIGHT DRIVE AXLE ASSEMBLY

79222735

Tapping out the right inner halfshaft from the transaxle—4-door models

➡On 4-door sedan, the right side halfshaft is stabilized by a center support bearing between the transmission and the inner Tri-pod joint. It is not necessary to remove the right inner drive axle and center support bearing assembly. The right side halfshaft assembly can be separated from the center support bearing by lightly tapping with a plastic mallet.

4. Using a suitable pry tool, pry on the inboard joints of the halfshaft to detach the snapring lock.

➡On 4-door sedans, separate the right side halfshaft assembly from the center support bearing by tapping lightly with a plastic mallet.

   - Halfshaft from the steering knuckle

To install:

5. Inspect the CV-joint boots for tears or deterioration and replace, if necessary.
6. Install or connect the following:
   - Halfshaft into the steering knuckle
   - Halfshaft into the transmission, or intermediate shaft
   - Control arm to the steering knuckle and torque the ball stud bolt to 44 ft. lbs. (60 Nm)
   - Halfshaft nut and washer and torque the fastener to 129 ft. lbs. (175 Nm) and stake the nut
   - ABS speed sensor to the steering knuckle and torque the bolt to 89 inch lbs. (10 Nm)
   - Front wheel
   - Negative battery cable
7. Fill the transmission with clean fluid.
8. Road test the vehicle.

Prizm

⁂ WARNING

Care must be exercised to prevent the differential-side CV-joint from being over-extended. Over-extension of the CV-joint could result in separation of internal components and possible joint failure. If the vehicle is to be lowered and moved, the front wheel bearings must be supported using a 9/16 in. bolt, 1¾ in. washer, 2 in. washer and a 9/16 in. nut, assembled through the shaft opening in the hub. Tighten the nut and bolt to 40 ft.

lbs. (54 Nm). If equipped with Antilock Brake Systems (ABS), use caution not to damage the ABS speed sensor ring on the wheel-side CV-joint. DO NOT pry against the ring with metal tools. If the serrations on the speed sensor ring appear damaged, replace the wheel-side CV-joint.

1. Before servicing the vehicle, refer to the precautions in the beginning of this section.
2. Remove or disconnect the following:
   - Negative battery cable
   - Front wheel
   - Splash shield
   - Antilock Brake System (ABS) speed sensor, if equipped
   - Cotter pin and lock cap from the halfshaft
   - Halfshaft nut
   - Tie rod end from the steering knuckle
   - Ball joint from the control arm
   - Outer CV-joint from the steering knuckle
   - Inner CV-joint from the transmission by gently prying the joint out
   - Halfshaft from the vehicle

To install:

3. Inspect the CV-joint boots for tears or deterioration. Replace as necessary.
4. Inspect the front wheel bearing inner and outer oil seals for damage or deterioration. Replace as necessary.
5. Inspect the halfshaft fluid seal at the transmission for leakage or damage. Replace as necessary.
6. Install or connect the following:
   - Inner CV-joint into the transmission
   - Outer CV-joint into the steering knuckle
   - Ball joint to the control arm and torque the bolts to 105 ft. lbs. (142 Nm)
   - Tie rod end to the steering knuckle and torque the nut to 36 ft. lbs. (49 Nm)
   - New cotter pin to the tie rod end
   - Halfshaft nut and torque it to 166 ft. lbs. (225 Nm)
   - Lock cap onto the halfshaft and secure with a new cotter pin
   - ABS speed sensor and torque the bolt to 71 inch lbs. (8 Nm)
   - Splash shield and torque the bolts to 71 inch lbs. (8 Nm)
   - Front wheel
   - Negative battery cable
7. Check the wheel alignment and adjust as needed.

CV-Joints

OVERHAUL

Inner

METRO

1. Before servicing the vehicle, refer to the precautions in the beginning of this section.
2. Remove the halfshaft.
3. Place a mark on the inner joint and the halfshaft for proper assembly, prior to removal.
4. Remove or disconnect the following:
   - Inner and outer boot bands
   - Inner joint housing from the tri-pod joint spider. Place a mark on the tri-pod joint spider and the halfshaft for proper assembly, prior to removal.
   - Snap ring from the halfshaft
   - Tri-pod joint spider
   - CV boot

To install:

5. Install or connect the following:
   - Inner boot
   - Inner boot band
   - Tri-pod joint spider aligning the marks made earlier
   - Snap ring on the halfshaft
6. Pack the inner joint housing with 2.8–3.5 oz. (80–100 g) of the grease provided with the service kit.
   - Inner joint housing onto the tri-pod joint spider aligning the marks made earlier
   - Outer boot band
   - Halfshaft

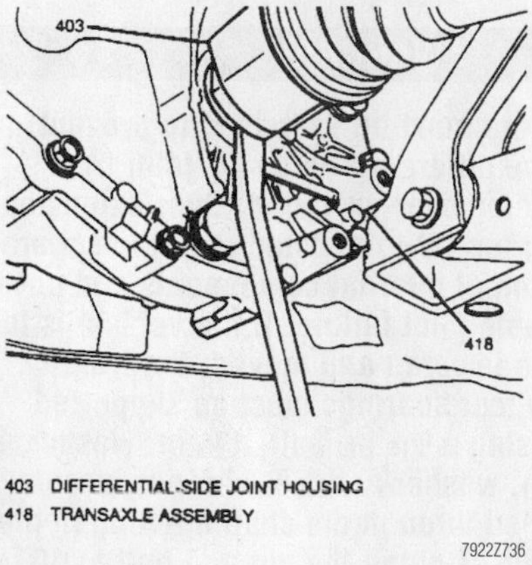

Prying out the inner halfshaft from the transaxle—Prizm

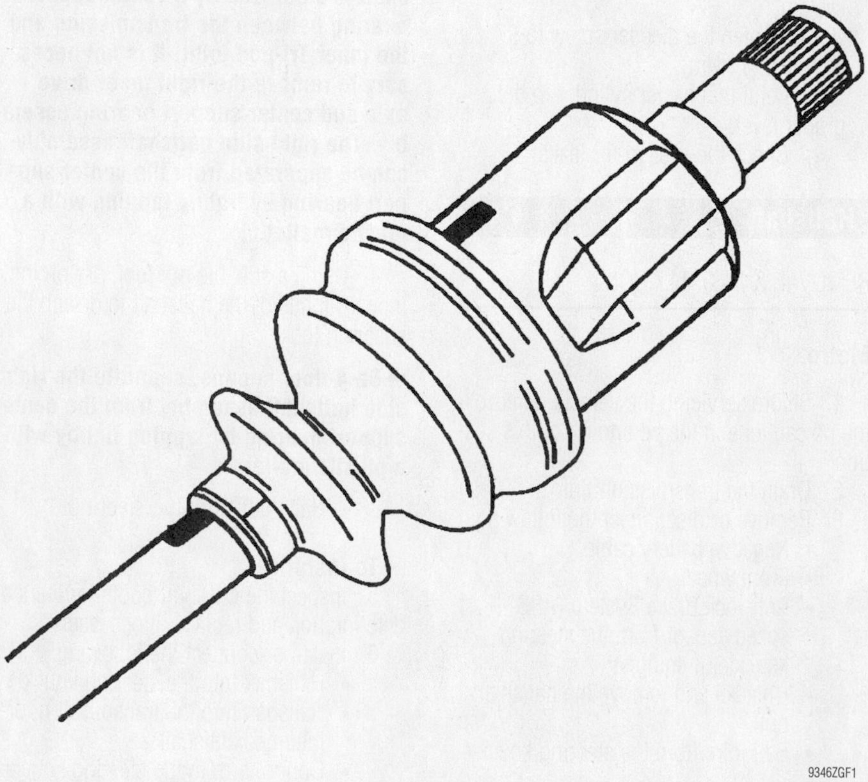

Place a mark on the inner joint and the halfshaft for proper assembly, prior to removal–inner joint Metro

PRISM

1. Before servicing the vehicle, refer to the precautions in the beginning of this section.
2. Remove the halfshaft.
3. Remove the small side inner joint boot clamp.
4. Slide the boot towards the center of the wheel drive shaft.
5. Place index marks on the inner joint housing and the wheel drive shaft to ensure correct assembly.

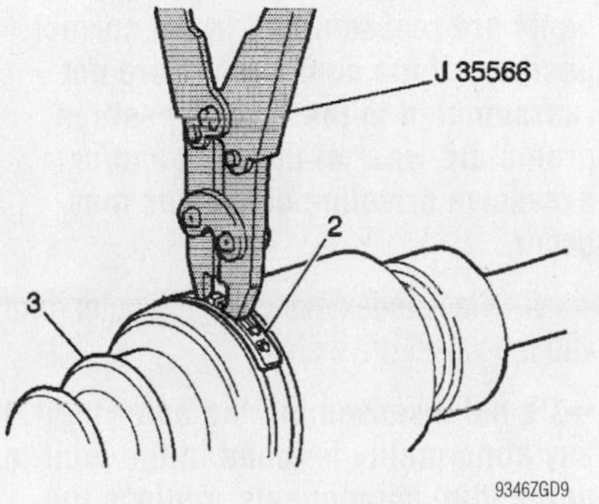

Removing the joint bands–inner joint Prism

6. Remove the inner joint housing from the tripot joint spider.
7. Place index marks on the tripot joint spider and the wheel drive shaft.
8. Remove the snap ring from the wheel drive shaft.
9. Use a brass drift and a hammer to remove the tripot joint from the wheel drive shaft. Place the drift on the tripot body, not the roller.
10. Remove the tripot joint from the wheel drive shaft.

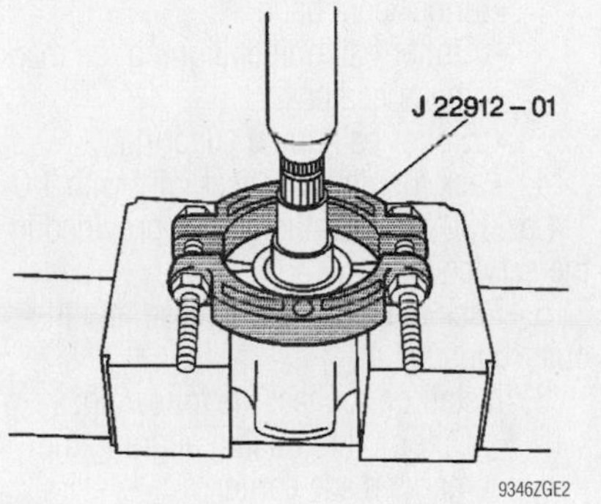

Remove the inner joint seal using J-22912-01 and a press–inner joint Prism

➡Do not wash the Tri-Pot joint spider in solvent or degreaser. Washing the Tri-Pot joint spider in solvent or degreaser will remove all of the lubrication in the spider's needle bearings. Use only a clean, dry, solvent-free rag to clean the Tri-Pot spider assembly.

11. Clean the tripot joint with a clean, dry, solvent free cloth. Inspect the tripot joint for excessive wear or damage. If any excessive wear, damage, or abnormality is found, replace the tripot joint as an assembly.
12. Remove the inner boot from the wheel drive shaft.

➡Do not clean the wheel drive shaft boot(s) in solvent. Cleaning the wheel drive shaft boot(s) in degreaser or other solvents can cause the boot(s) to deteriorate. Use only a clean, dry, solvent-free cloth to clean the wheel drive shaft boot(s).

13. Clean the inner boot with a clean, dry, solvent free cloth. Inspect the inner boot for tears, damage or fatigue. Replace as necessary.
14. Inspect the dynamic dampener on the right wheel drive shaft for damage or distortion. Replace as necessary.
15. Remove the inner joint housing seal using the Bearing Puller Tool J-22912-01 and a press.

To install:

16. Install the inner joint seal using a press.
17. If the damper was removed, install the damper to the wheel drive shaft before installing the inner joint.
18. Install the damper assembly on the right side wheel drive shaft only, to the wheel drive shaft. The damper should be located on the groove on the wheel drive shaft. The damper centerline should be 16.980 inch (431mm), plus or minus 0.118 inch (3mm) from the outer joint.

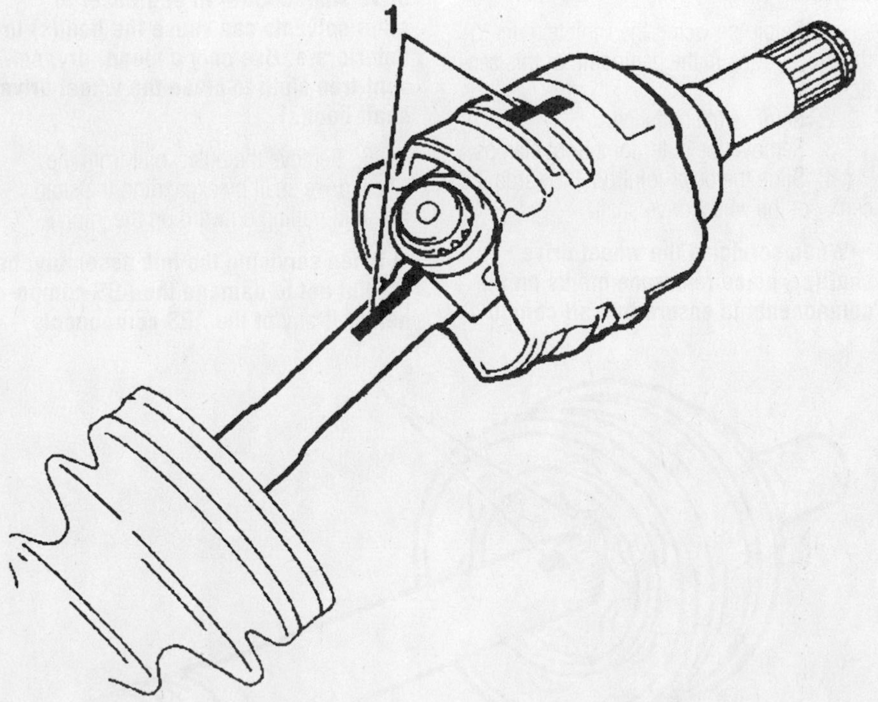

Place a mark on the inner joint and the halfshaft for proper assembly, prior to removal–inner joint Prism

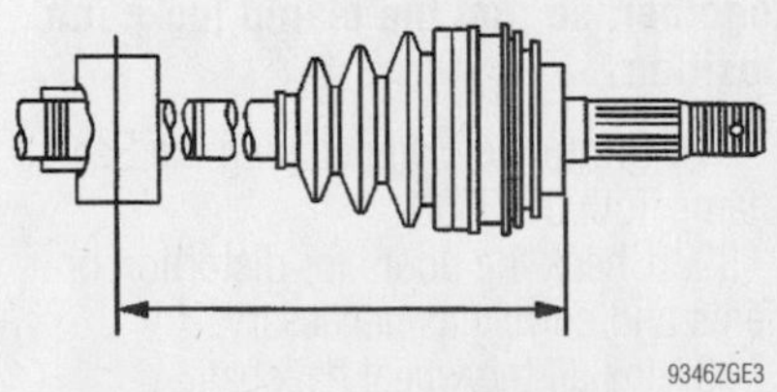

The damper centerline should be 16.980 inch (431mm), plus or minus 0.118 inch (3mm) from the outer joint–inner joint Prism

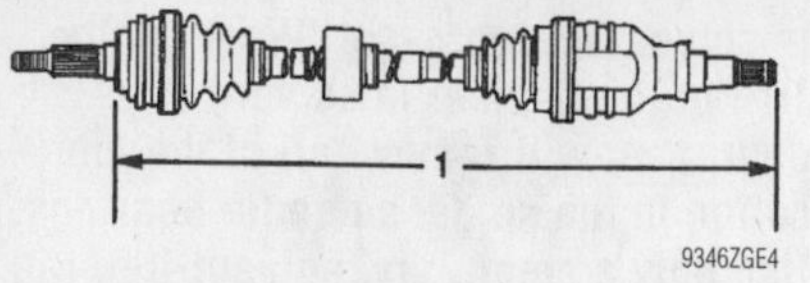

Measure the wheel drive shaft length to ensure that the boots are not stretched or excessively contracted when the wheel drive shaft is at standard length–inner joint Prism

19. Crimp the damper clamp.
20. Install the inner joint housing boot and clamps to the wheel drive shaft. Do not crimp the clamps at this time.
21. Align the index marks on the tripot joint and the wheel drive shaft.

➡Place the side of the tripot joint with the bevelled splines (1) away from the transaxle.

22. Install the tripot joint onto the wheel drive shaft.
23. Install the snap ring.
24. Pack the inner joint housing with 5.8–6.4 oz. (165–185 grams) of grease supplied with inner joint boot kit.
25. Align the index marks on the inner joint housing and the tripot joint.
26. Install the inner joint housing to the tripot joint.
27. Install the inner joint boot and clamps to the inner joint housing.
28. Measure the wheel drive shaft length to ensure that the boots are not stretched or excessively contracted when the wheel drive shaft is at standard length.
29. The right wheel drive shaft standard length is 33.756 inch (857.4mm), plus or minus 0.197 inch (5mm).
30. The left wheel drive shaft standard length is 21.268 inch (540.2mm), plus or minus 0.197 inch (5mm).

➡Use the Seal Clamp Tool J-35566 to draw the closing hooks of the clamp together, so that the clamp locks into position.

31. Crimp the clamps using the Seal Clamp Tool J-35566.
32. Check the boots for distortion or dents and correct as necessary.
33. Install the wheel halfshaft.

Outer

METRO

1. Before servicing the vehicle, refer to the precautions in the beginning of this section.
2. Remove or disconnect the following:
   - Halfshaft. Place a mark on the outer joint and the halfshaft for proper assembly, prior to removal.
   - Inner and outer boot bands
   - Snapring and outer joint housing from the halfshaft
   - Roller ball guide snapring. Place a mark on the roller ball guide and the halfshaft for proper assembly, prior to removal.
   - Roller ball guide from the halfshaft
   - Outer boot

To install:

3. Install or connect the following:
   - Outer boot
   - Inner boot band
   - Roller ball guide aligning the mark made earlier
   - Roller ball guide snapring
4. Pack the double offset joint with 1.0-1.4 oz. (30-40 g) of the grease provided in the service kit.
5. Pack the remaining grease into the outer housing.
6. Install or connect the following:
   - Outer joint housing aligning the marks made earlier
   - Outer housing snapring
   - Outer boot band
   - Halfshaft

PRISM

1. Before servicing the vehicle, refer to the precautions in the beginning of this section.
2. Remove the halfshaft.
3. Remove the outer joint boot clamps.
4. Slide the outer joint boot towards the center of the wheel drive shaft.

➡When servicing the wheel drive shaft(s), place reference marks on all components to ensure that all components are reassembled in the correct position. If the components are not reassembled in the correct position, premature wear of the joint and/or excessive driveline vibrations may occur.

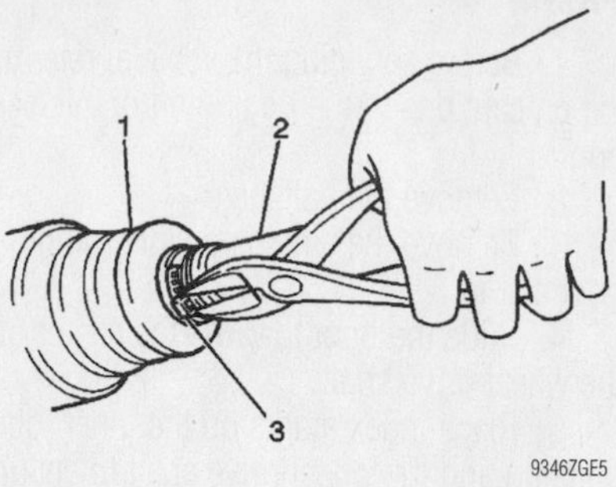

Remove the outer band clamps–outer joint Prism

5. Place index marks on the outer joint and the wheel drive shaft.

➡Do not disassemble the outer joint. If any abnormality is found in the joint, or in the joint components, replace the joint as an assembly.

➡Do not clean the wheel drive shaft boot(s) in solvent. Cleaning the wheel drive shaft boot(s) in degreaser or other solvents can cause the boot(s) to deteriorate. Use only a clean, dry, solvent-free cloth to clean the wheel drive shaft boot(s).

6. Remove the outer joint from the wheel drive shaft by expanding the snap ring and pulling outward on the joint.

➡When servicing the hub assembly, be careful not to damage the ABS components. If any of the ABS components

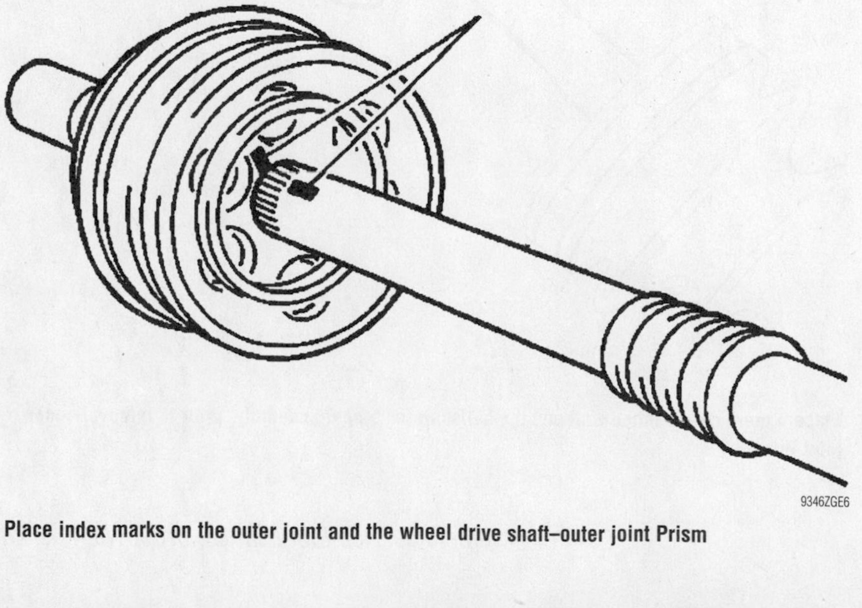

Place index marks on the outer joint and the wheel drive shaft–outer joint Prism

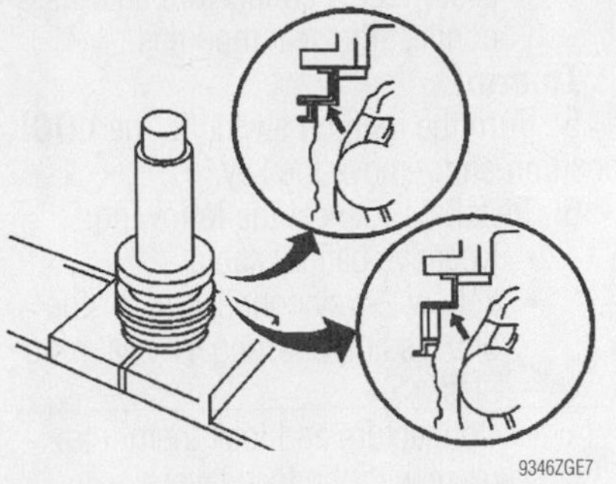

Install the outer joint seal using a suitable sleeve and a press–outer joint Prism

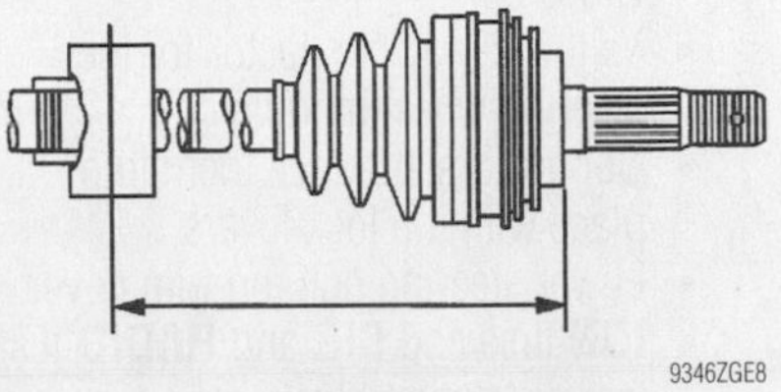

The damper centerline should be 16.980 inch (431mm), plus or minus 0.118 inch (3mm) from the outer joint–outer joint Prism

are damaged, the ABS system may not function properly and the damaged component will need to be replaced.

7. Mount the wheel drive shaft in a soft jaw vise.
8. Remove the outer joint seal.
9. Clean the outer boot with a clean, dry, solvent free cloth.
10. Inspect the outer boot for tears damage or fatigue. Replace as necessary.
11. Inspect the outer joint for excessive wear or damage. If any excessive wear, damage or abnormality is found, replace the outer joint as an assembly.

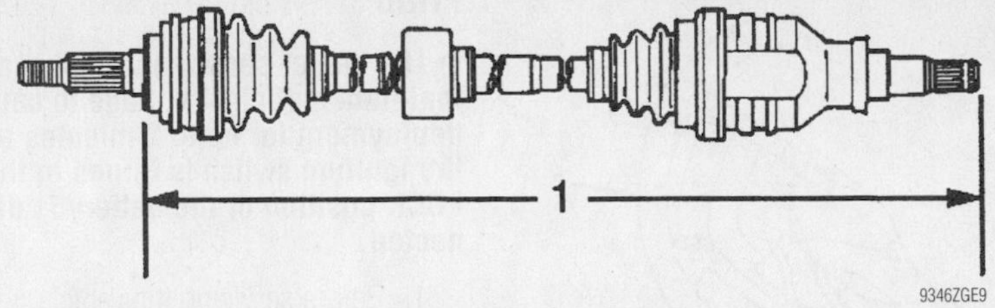

Measure the wheel drive shaft length to ensure that the boots are not stretched or excessively contracted when the wheel drive shaft is at standard length–outer joint Prism

12. Inspect the dynamic dampener on the right wheel drive shaft for damage or distortion. Replace as necessary.

To install:

13. Install the outer joint seal using a suitable sleeve and a press.
14. If the damper was removed, install the damper to the wheel drive shaft before installing the outer joint.
15. Install the damper assembly on the right side wheel drive shaft only, to the wheel drive shaft.
16. The damper should be located on the groove on the wheel drive shaft.
17. The damper centerline should be 16.980 inch (431mm), plus or minus 0.118 inch (3mm) from the outer joint.
18. Crimp the damper clamp.
19. Install the outer boot and clamps to the wheel drive shaft. Do not crimp the clamps at this time.
20. Align the index marks on the outer joint and the wheel drive shaft.

➡Make sure the snap ring is securely seated in the groove in the wheel drive shaft.

21. Expand the snap ring and slide the outer joint onto the wheel drive shaft.
22. Pack the inner joint housing with 5.8–6.4 oz. (165–185 grams) of grease supplied with inner joint boot kit.
23. Install the outer joint large boot clamp.
24. Measure the wheel drive shaft length to ensure that the boots are not stretched or excessively contracted when the wheel drive shaft is at standard length.
25. The right wheel drive shaft standard length is 33.756 inch (857.4mm), plus or minus 0.197 inch (5mm).
26. The left wheel drive shaft standard length is 21.268 inch (540.2mm), plus or minus 0.197 inch (5mm).

➡Use the Seal Clamp Tool J-35566 to draw the closing hooks of the clamp together, so that the clamp locks into position.

27. Crimp the clamps using the Seal Clamp Tool J-35566.
28. Check the boots for distortion or dents and correct as necessary.
29. Install the wheel halfshaft.

STEERING AND SUSPENSION

Air Bag

⁂ CAUTION

All vehicles are equipped with an Air Bag system. The system must be disabled before performing service on or around system components, steering column, instrument panel components, wiring and sensors. Failure to follow safety and disabling procedures could result in accidental Air Bag deployment, possible personal injury and unnecessary system repairs.

PRECAUTIONS

Several precautions must be observed when handling the inflator module to avoid accidental deployment and possible personal injury.

- Never carry the inflator module by the wires or connector on the underside of the module.
- When carrying a live inflator module, hold securely with both hands, and ensure that the bag and trim cover are pointed away.
- Place the inflator module on a bench or other surface with the bag and trim cover facing up.
- With the inflator module on the bench, never place anything on or close to the module which may be thrown in the event of an accidental deployment.

DISARMING

Metro

1. Before servicing the vehicle, refer to the precautions in the beginning of this section.
2. Turn the steering wheel so the wheels are pointing straight-ahead.
3. Turn the ignition switch to the **LOCK** position and remove the key.

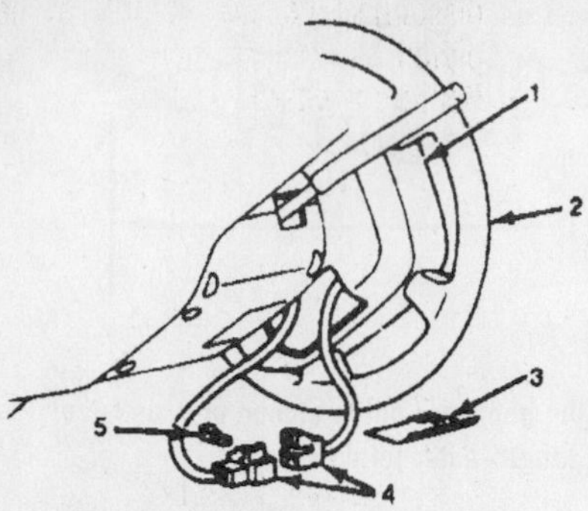

Location of the SIR harness connector—Metro

4. Remove or disconnect the following:
   - Negative battery cable
   - **AIR BAG-IG** fuse from the junction block near the base of the steering column
   - Steering wheel side cap, Connector Position Assurance (CPA) and yellow 2-way connector for the driver's side Air Bag (inflator) module
   - Glove box while pushing in on the stoppers from the left and the right sides. Disconnect the yellow connector for the passenger Air Bag (inflator) module.

To arm:

5. Install or connect the following:
   - Negative battery cable
   - Turn the ignition switch to the **LOCK** position and remove the key
   - Yellow connector for the passenger's side Air Bag (inflator) module and the yellow connector for the driver's air bag (inflator) module. Be sure to lock each connector with the lock lever
   - Glove box
   - Left side steering wheel side cap
   - **AIR BAG-IG** fuse to the junction block
6. Staying away from both air bags, turn the ignition switch to the **ON** position and verify that the **AIR BAG** warning lamp flashes 7 times, then turns OFF. If the system does not operate as described, diagnosis and repairs to the air bag system are necessary.

Prizm

➡The center sensor assembly can maintain sufficient voltage to cause deployment for up to 2 minutes after the ignition switch is turned to the LOCK position or the battery is disconnected.

1. Before servicing the vehicle, refer to the precautions in the beginning of this section.
2. Turn the steering wheel so the front wheels are in the straight-ahead position.
3. Turn the ignition switch to **LOCK**.
4. Remove or disconnect the following:
   - Negative battery cable
   - Lower steering column trim cover
   - **IGN** fuse and **CIG** and **RADIO** fuse from junction block 1
   - Connector Position Assurance (CPA) and disconnect the yellow 2-way connector at the base of the steering column
   - Inflatable restraint steering wheel module coil connector
   - Glove box
   - Unlock the passenger's side module connector and disconnect the pigtail
   - Unlock the left and right hand side seat module connectors and disconnect the seat modules

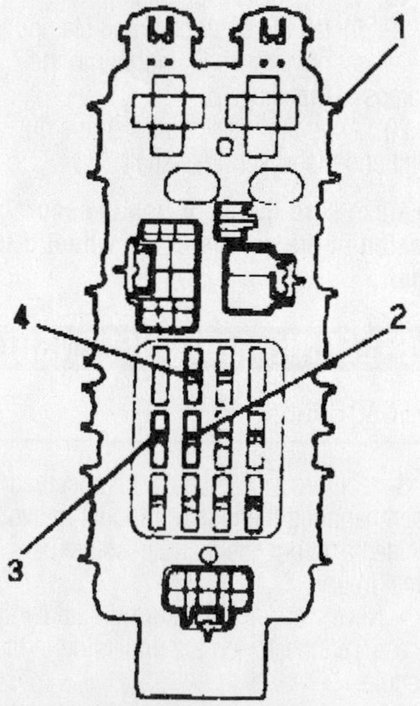

Connector view and terminal identification of junction block 1—Prizm

To arm:

5. Turn the ignition switch to the **LOCK** position and remove the key.
6. Install or connect the following:
   - Negative battery cable
   - Yellow 2-way connectors for the driver's and passenger's seat modules
   - Connectors and lock them into place with the lock levers
   - Yellow 2-way connector to the passenger's side inflatable restraint pigtail. Connect the lock with the lever
   - Glove box
   - Yellow 2-way connector for the steering wheel module coil
   - Connectors and lock them into place with the lock levers
   - Lower steering column trim cover
   - **IGN** fuse and **CIG** and **RADIO** fuse to the junction block
7. Staying away from both air bags, turn the ignition **ON** and verify that the **AIR BAG** warning lamp flashes 7 times, then turns OFF. If the system does not operate as described, diagnosis and repairs to the air bag system are necessary.

Rack and Pinion Steering Gear

REMOVAL & INSTALLATION

Manual

METRO

1. Before servicing the vehicle, refer to the precautions in the beginning of this section.
2. Remove or disconnect the following:
   - Negative battery cable
   - Forward section of driver's side carpet
   - Steering shaft joint cover
   - Loosen the steering shaft upper joint bolt but do not remove
   - Steering shaft lower joint bolt
   - Lower joint from the pinion
   - Both front wheels
   - Tie rod ends from the steering knuckles
   - Exhaust pipe
   - Inlet and outlet pipes
   - Steering gear mounting bolts and brackets
   - Steering gear from the vehicle

To install:

3. Install or connect the following:
   - Steering gear through the front wheel opening

- Steering gear mounting brackets and torque the bolts to 18 ft. lbs. (25 Nm)
- Inlet and outlet pipes and torque the fittings to 28.5 ft. lbs. (40 Nm), if removed
- Exhaust pipe and torque the bolts to 37 ft. lbs. (50 Nm), if removed
- Right and left tie rod ends to the steering knuckles and torque the nuts to 32 ft. lbs. (43 Nm)
- New cotter pins
- Both front wheels
- Steering shaft to the steering gear and torque the upper and lower steering shaft joint bolts to 18 ft. lbs. (25 Nm)
- Steering shaft joint cover
- Driver's side floor carpet
- Negative battery cable

4. Check and adjust the front wheel alignment.

PRIZM

1. Before servicing the vehicle, refer to the precautions in the beginning of this section.
2. The front wheels must be straight and the steering column in the **LOCK** position before disconnecting the intermediate shaft from the steering gear.
3. Remove or disconnect the following:
   - Negative battery cable
   - Steering column upper cover
   - Steering shaft lower coupling
   - Outlet pipe heat shield
   - Inlet and outlet pipes from the steering gear
   - Heated Oxygen (HO_2S) sensor
4. Install an engine support fixture.
   - Both splash shields
   - Both front wheels
   - Exhaust manifold heat shield
   - Outer tie rods
   - Front crossmember brace
   - Front suspension crossmember
   - Rear transmission mount and bracket
   - Exhaust pipe from the exhaust manifold
   - Steering gear boot heat shield
   - Steering gear mounting clamps
   - Steering gear

To install:

5. Install or connect the following:
   - Steering gear
   - Steering gear clamps and torque the bolts to 52 ft. lbs. (71 Nm)
   - Steering gear boot heat shield and torque the bolt to 48 inch lbs. (5 Nm)
   - Exhaust manifold pipe and torque the bolts to 46 ft. lbs. (62 Nm)
   - Rear transmission mount bracket and torque the bolts to 57 ft. lbs. (77 Nm)
   - Rear transmission mount and torque the bolt to 64 ft. lbs. (87 Nm)
   - Front suspension crossmember and torque the bolts to 45 ft. lbs. (60 Nm)
   - Front crossmember brace and torque the bolts to 51 ft. lbs. (69 Nm)
   - Outer tie rods and torque the nuts to 36 ft. lbs. (49 Nm)
   - Front wheels
   - Exhaust manifold heat shield and torque the fasteners to 48 inch lbs. (5 Nm)
   - Both splash shields and torque the bolts to 89 inch lbs. (10 Nm)
   - HO_2S sensor and torque the nuts to 32 ft. lbs. (44 Nm)
   - Inlet and outlet pipes and torque the bolts to 108 inch lbs. (13 Nm)
   - Inlet and outlet clips and torque the bolts to 48 inch lbs. (5 Nm)
   - Outlet pipe heat shield and torque the bolt to 48 inch lbs. (5 Nm)
   - Steering shaft lower coupling and torque the bolt to 26 ft. lbs. (35 Nm)
   - Steering column upper cover and torque the bolts to 48 inch lbs. (5 Nm)
   - Negative battery cable
6. Refill and bleed the power steering system.
7. Check and adjust the front end alignment as needed.

Power

METRO

1. Before servicing the vehicle, refer to the precautions in the beginning of this section.
2. Remove or disconnect the following:
   - Negative battery cable
   - Slide the driver's seat back, pull off the front part of the floor carpet on the driver's side and remove the steering shaft joint cover
   - Loosen the steering shaft upper joint bolt but do not remove
   - Steering shaft lower joint bolt and disconnect the lower joint from the pinion
   - Both front wheels
   - Tie rod ends
   - Exhaust pipe at the manifold
   - Separate the shift linkage and extension rod from the transmission, for manual transmission
   - Engine rear torque rod with the bracket from the transmission, for automatic transmission
   - Power steering lines from the rack and pinion assembly and plug the openings to prevent system contamination
   - Rack and pinion from the vehicle

To install:

3. Install or connect the following:
   - Steering rack, brackets and mounting bolts to the bulkhead. When aligned properly, torque the bolts to 18 ft. lbs. (25 Nm).
   - Power steering lines to the rack and pinion. Torque the inlet and outlet fluid lines to 25 ft. lbs. (35 Nm) and the remaining lines to 18 ft. lbs. (25 Nm).
   - Shift linkage and extension rod to the manual transmission
   - Engine rear torque rod and bracket to the automatic transmission
   - Front exhaust pipe to the manifold using a new seal and torque the bolts to 37 ft. lbs. (50 Nm)
   - Tie rod ends to the steering knuckles and torque the castle nuts to 32 ft. lbs. (43 Nm)
   - New cotter pins
   - Both front wheels
   - Lower steering shaft-to-steering gear and torque the bolts to 18 ft. lbs. (25 Nm)
   - Upper steering shaft and torque the bolt to 18 ft. lbs. (25 Nm)
   - Steering joint cover
   - Negative battery cable
4. Place the driver's side floor carpet back into the original position.
5. Bleed the steering system and check for leaks, repair if necessary.
6. Check and adjust the front wheel alignment.

PRIZM

1. Before servicing the vehicle, refer to the precautions in the beginning of this section.
2. The front wheels must be straight and the steering column in the **LOCK** position

before disconnecting the intermediate shaft from the steering gear.

3. Remove or disconnect the following:
   - Negative battery cable
   - Steering column upper cover
   - Steering shaft lower coupling
   - Inlet and outlet pipes from the steering gear
   - Oxygen (O_2S) sensor and install an engine support fixture
   - Both splash shields
   - Both front wheels
   - Exhaust manifold heat shield
   - Outer tie rods
   - Front crossmember brace
   - Front suspension crossmember, transmission support, both control arms and front stabilizer shaft as a complete unit
   - Rear transmission mount and bracket
   - Steering gear boot heat shield
   - Steering gear

To install:

4. Install or connect the following:
   - Steering gear and clamps and torque the bolts to 52 ft. lbs. (71 Nm)
   - Stedering gear boot heat shield and torque the bolt to 48 inch lbs. (5 Nm)
   - Exhaust manifold pipe and torque the bolts to 46 ft. lbs. (62 Nm)
   - Rear transmission mount bracket and torque the bolts to 57 ft. lbs. (77 Nm)
   - Rear transmission mount and torque the bolt to 64 ft. lbs. (87 Nm)
   - Front suspension crossmember, transmission support, control arms and stabilizer shaft
   - Front crossmember and torque the bolts to 51 ft. lbs. (69 Nm)
   - Outer tie rods
   - Front wheels
   - Exhaust manifold heat shield and torque the bolt to 48 inch lbs. (5 Nm)
   - Both splash shields and torque the bolts to 89 inch lbs. (10 Nm)
   - O_2sensor and remove the engine support fixture
   - Inlet and outlet pipes and torque the bolts to 108 inch lbs. (13 Nm)
   - Inlet and outlet clips and torque the bolts to 48 inch lbs. (5 Nm)
   - Outlet pipe heat shield and torque the bolt to 48 inch lbs. (5 Nm)
   - Steering shaft lower coupling and torque the bolt to 26 ft. lbs. (35 Nm)
   - Steering column upper cover and torque the bolts to 48 inch lbs. (5 Nm)
   - Negative battery cable

5. Refill and bleed the power steering system.

6. Check and adjust the front end alignment as needed.

Strut

REMOVAL & INSTALLATION

Front

METRO

1. Before servicing the vehicle, refer to the precautions in the beginning of this section.

2. Remove or disconnect the following:
   - Negative battery cable
   - Upper strut support nuts from the tower
   - Front wheel
   - Wheel Speed Sensor (WSS) harness from the strut, if equipped
   - Brake hose
   - Strut from the steering knuckle
   - Strut from the vehicle

To install:

3. Install or connect the following:
   - Strut to the vehicle and torque the upper support nuts to 21 ft. lbs. (28 Nm)
   - Strut to steering knuckle and torque the bolts to 59 ft. lbs. (80 Nm)

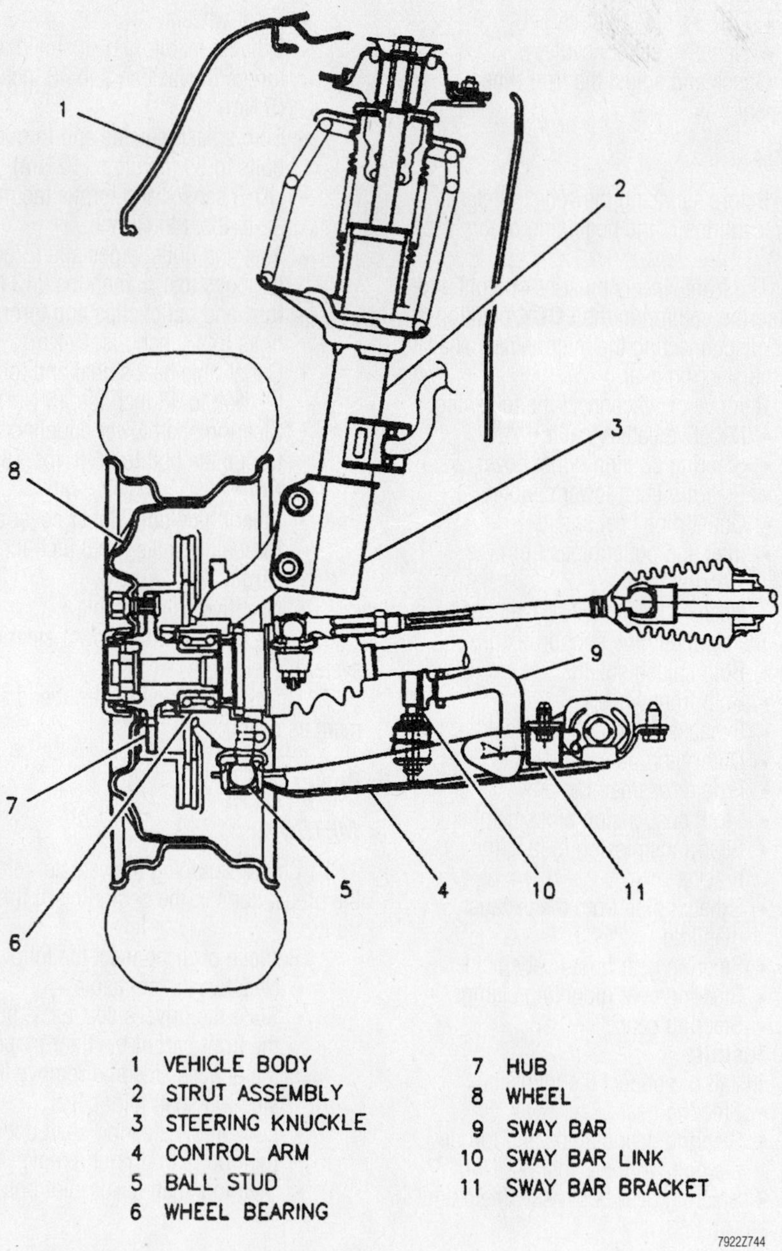

Cut-away view of the front suspension—Metro

- Brake hose to the strut bracket and secure with a new E-ring
- WSS to the strut, if equipped and torque the bolt to 89 inch lbs. (10 Nm)
- Front wheel
- Negative battery cable

4. Check the front alignment and adjust if necessary.

PRIZM

1. Before servicing the vehicle, refer to the precautions in the beginning of this section.
2. Properly support the front suspension crossmember.
3. Remove or disconnect the following:
   - Negative battery cable
   - Front wheel and make certain that the weight of the vehicle is resting on the crossmember and not the control arms
   - Wheel Speed Sensor (WSS) electrical connector, if equipped
   - Brake hose from the strut
   - Strut assembly from the strut tower
   - Lower strut bolts
   - Strut from the vehicle

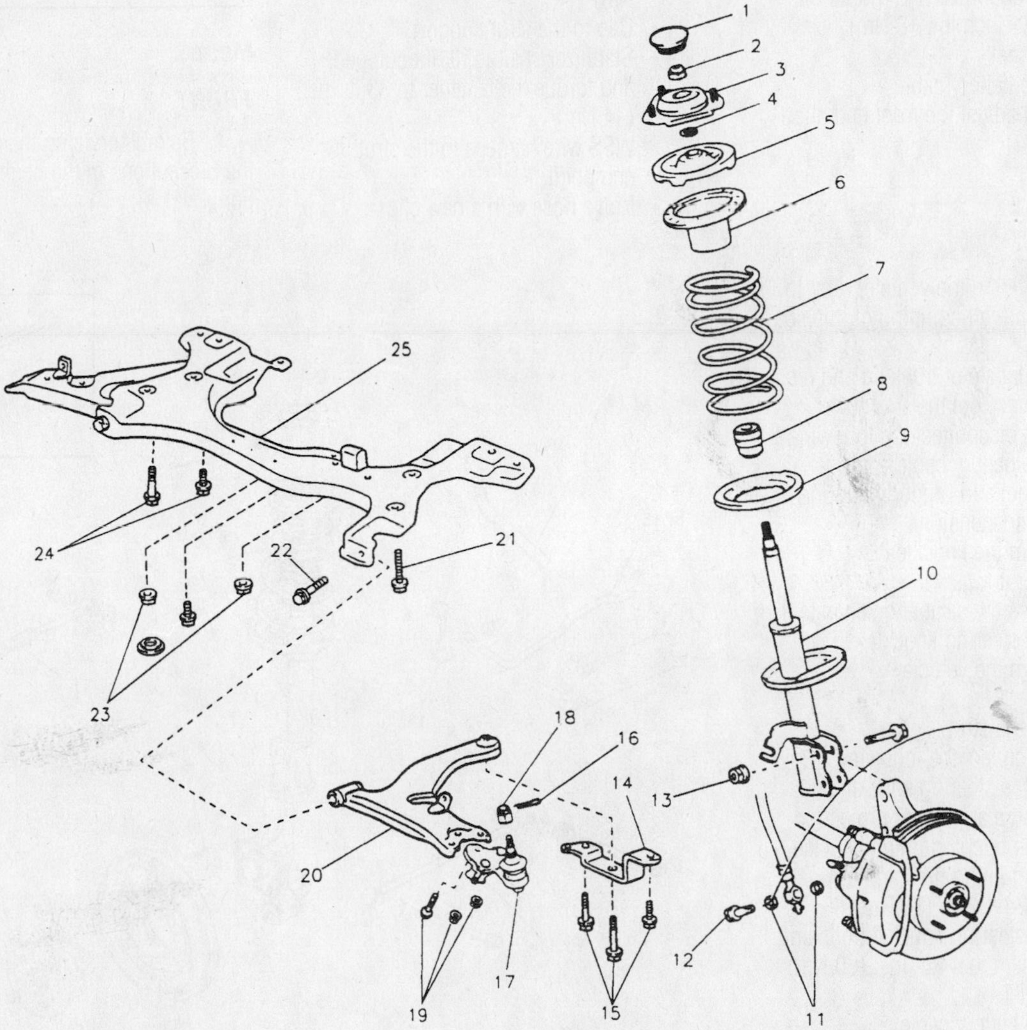

1 DUST CAP
2 STRUT ROD PISTON NUT
3 STRUT SUPPORT
4 DUST SEAL
5 SPRING SEAT
6 UPPER INSULATOR
7 COIL SPRING
8 SPRING BUMPER
9 LOWER INSULATOR
10 STRUT
11 BRAKE LINE GASKETS
12 BRAKE LINE-TO-CALIPER BOLT
13 STRUT MOUNTING NUT AND BOLT
14 LOWER CONTROL ARM RETAINING BRACKET
15 LOWER CONTROL ARM RETAINING BRACKET BOLTS
16 COTTER PIN
17 BALL JOINT
18 BALL JOINT CASTLE NUT
19 BALL JOINT MOUNTING NUT AND BOLT
20 CONTROL ARM
21 CROSSMEMBER MOUNTING BOLTS
22 CROSSMEMBER-TO-CONTROL ARM BOLT
23 CROSSMEMBER MOUNTING NUTS
24 CROSSMEMBER MOUNTING BOLTS
25 SUSPENSION CROSSMEMBER

79222739

Exploded view of the front suspension assembly—Prizm

Heater Core replacement is covered in Section 2 of this manual

To install:

4. Install or connect the following:
   - Strut to the lower mount and hand-tighten the bolts
   - Strut to the strut tower and torque the strut tower nuts to 29 ft. lbs. (39 Nm) and the lower mounting bolts to 203 ft. lbs. (274 Nm)
   - Brake hose to the strut and torque the bolt to 22 ft. lbs. (29 Nm)
   - WSS, if equipped and torque the bolt to 71 inch lbs. (8 Nm)
   - Front wheel
   - Negative battery cable

5. Check and adjust the front end alignment.

Rear

METRO

1. Before servicing the vehicle, refer to the precautions in the beginning of this section.
2. Open the lift gate or trunk lid and reposition the trim cover from the strut tower.
3. Remove or disconnect the following:
   - Negative battery cable
   - Rear wheel and properly support the rear suspension
   - Strut from the knuckle
   - Upper strut nuts
   - Compress the strut and separate it from the steering knuckle
   - Strut from the vehicle

To install:

4. Compress the strut.
5. Install or connect the following:
   - Strut to the steering knuckle and position the alignment projection inside the opening in the knuckle
   - Upper mounting nuts and torque them to 24 ft. lbs. (30 Nm)
   - Strut-to-steering knuckle bolts and torque them to 44 ft. lbs. (60 Nm)
   - Rear wheel
   - Negative battery cable
6. Reposition the rear trim cover.

PRIZM

1. Before servicing the vehicle, refer to the precautions in the beginning of this section.
2. Support the rear suspension.
3. Remove or disconnect the following:
   - Negative battery cable
   - Rear seat cushions
   - Rear wheel
   - Wheel Speed Sensor (WSS) harness from the strut, if equipped
   - Brake hose from the strut
   - Stabilizer shaft joint from the strut, if equipped
   - Cap from the strut support
   - Upper strut mounting fasteners
   - Strut from the suspension knuckle
   - Strut from the vehicle

To install:

4. Install or connect the following:
   - Strut to the rear knuckle and hand-tighten the nuts at this time
   - Strut to the support and torque the nuts to 29 ft. lbs. (39 Nm) and the knuckle bolts to 105 ft. lbs. (142 Nm)
   - Cap to the strut support
   - Stabilizer shaft joint, if equipped and torque the fastener to 33 ft. lbs. (44 Nm)
   - WSS wire harness to the strut, if equipped
   - Brake hose with a new clip
   - Rear wheel
   - Rear seat cushions
   - Negative battery cable
5. Fill and bleed the brake system.
6. Check and adjust the rear wheel alignment as needed.

Coil Spring

REMOVAL & INSTALLATION

Metro

FRONT

1. Before servicing the vehicle, refer to the precautions in the beginning of this section.

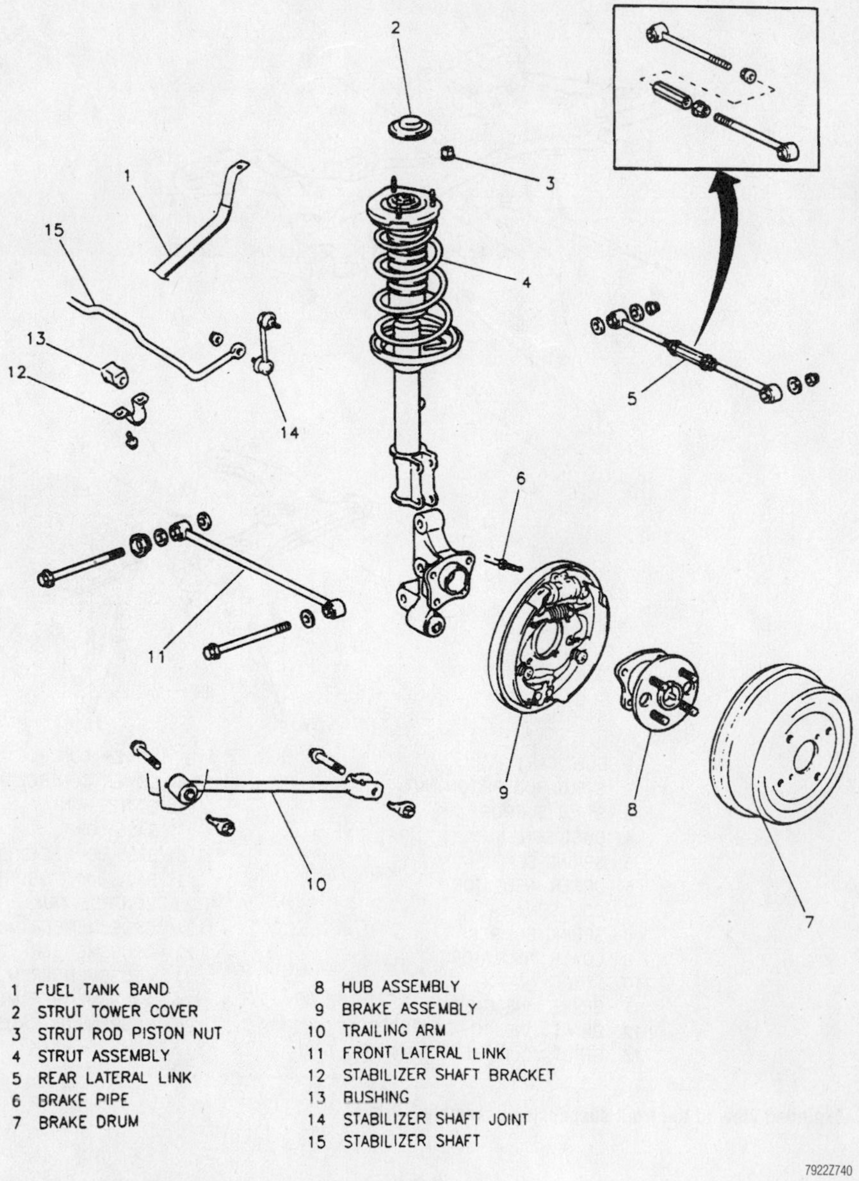

Exploded view of the rear strut assembly—Prizm

2. Remove or disconnect the following:
   - Negative battery cable
   - Rear wheel
   - Strut

3. Mount a strut compressor tool into a holding fixture and install the strut using an adapter.
4. Compress the coil spring.
5. Remove or disconnect the following:
   - Strut rod piston nut and carefully release the compressed spring
   - Strut from the compressor tool
   - Strut cap, nuts, insulator and coil spring

To install:

6. Install or connect the following:
   - Coil spring, insulator nuts and strut cap
   - Mount the strut into a compressor tool and compress the coil spring

⁂ CAUTION

Do not over compress the spring. Excess overloading may result in tool failure and could result in bodily injury.

   - Strut rod piston nut and torque it to 37 ft. lbs. (50 Nm)
   - Release the coil spring and remove the compressor tool
   - Strut
   - Rear wheel
   - Negative battery cable

REAR

1. Before servicing the vehicle, refer to the precautions in the beginning of this section.
2. Remove or disconnect the following:
   - Negative battery cable
   - Rear wheel
   - Control rod from the knuckle

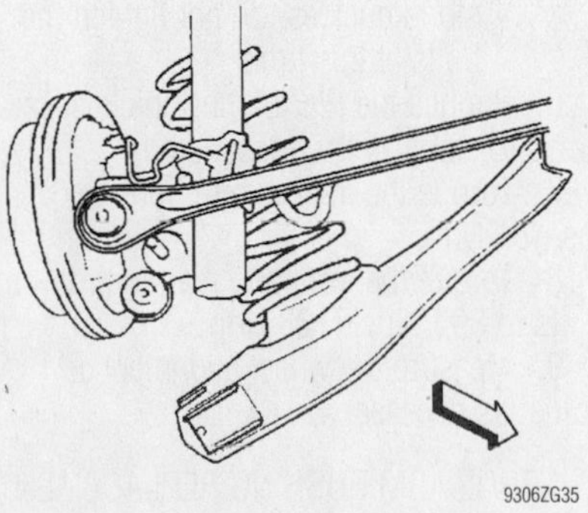

Remove the coil spring from the rear suspension—Metro

   - Wheel Speed Sensor (WSS) electrical connector from the control arm, if equipped
   - Stabilizer link from the control arm and properly support the rear suspension
   - Knuckle from the suspension arm
   - Coil spring from the suspension arm and upper seat

To install:

3. Install or connect the following:
   - Spring to the suspension arm with the larger diameter on the bottom
   - Suspension arm to steering knuckle and torque the bolt to 29 ft. lbs. (40 Nm)
   - WSS to the control arm, if equipped
   - Stabilizer link to the control arm and torque the nut to 18 ft. lbs. (25 Nm)
   - Control rod to the knuckle stud and torque the nut to 59 ft. lbs. (80 Nm)
   - Rear wheel
   - Negative battery cable

Prizm

1. Before servicing the vehicle, refer to the precautions in the beginning of this section.
2. Remove or disconnect the following:
   - Negative battery cable
   - Rear wheel
   - Strut

3. Mount a strut compressor tool into a holding fixture and install the strut using an adapter.
4. Compress the coil spring.
5. Remove or disconnect the following:
   - Strut rod piston nut and carefully release the compressed spring
   - Strut from the compressor tool
   - Strut cap, nuts, insulator and coil spring

To install:

6. Install coil spring, insulator nuts and strut cap.
7. Mount the strut into a compressor tool and compress the coil spring.

⁂ CAUTION

Do not over compress the spring. Excess overloading may result in tool failure and could result in bodily injury.

8. Install the strut rod piston nut and torque it to 34 ft. lbs. (47 Nm).
9. Release the coil spring and remove the compressor tool.
10. Install the strut.
11. Install the rear wheel.
12. Connect the negative battery cable.

Lower Ball Joint

REMOVAL & INSTALLATION

Metro

The lower ball joint is an integral part of the lower control arm assembly and is not serviceable as an individual component. If the lower ball joint is defective, the entire lower control arm must be replaced.

Prizm

1. Before servicing the vehicle, refer to the precautions in the beginning of this section.
2. Remove or disconnect the following:
   - Negative battery cable
   - Front wheel

3. Position a jack stand under the suspension crossmember.
   - Cotter pin from the ball joint nut
   - Ball joint from the steering knuckle
   - Ball joint from the control arm

To install:

4. Install or connect the following:
   - Ball joint to the steering knuckle and torque the new castle nut to 91 ft. lbs. (124 Nm)
   - New cotter pin
   - Ball joint to the control arm and torque the bolt and nuts to 105 ft. lbs. (142 Nm)
   - Front wheel
   - Negative battery cable

5. Check the front wheel alignment and adjust as needed.

Lower Control Arm

REMOVAL & INSTALLATION

Metro

FRONT

1. Before servicing the vehicle, refer to the precautions in the beginning of this section.
2. Remove or disconnect the following:
   - Negative battery cable
   - Stabilizer shaft link from the control arm

- Ball joint nut
- Ball joint from the steering knuckle
- Control arm front mounting bracket bolts
- Control arm rear mounting bracket bolts
- Control arm from the vehicle

To install:

3. Install or connect the following:
   - Control arm to the vehicle
   - Front and rear mounting brackets, do not tighten the bolts at this time
   - Ball joint to the steering knuckle, do not tighten the nut at this time
   - Stabilizer shaft link to the lower control arm
   - Front wheel
4. Bounce the front end a few times to stabilize the suspension.
5. Torque the rear control arm bolts to 32 ft. lbs. (43 Nm).
6. Torque the front control arm bolts to 66 ft. lbs. (90 Nm).
7. Torque the ball joint nut to 44 ft. lbs. (60 Nm).
8. Torque the stabilizer link nut to 20 ft. lbs. (28 Nm).
9. Connect the negative battery cable.
10. Check and adjust the front end alignment, if needed.

REAR

1. Before servicing the vehicle, refer to the precautions in the beginning of this section.
2. Remove or disconnect the following:
   - Coil spring
   - Suspension arm front mounting bolts
   - Suspension arm rear bolt
   - Suspension arm from the vehicle

To install:

3. Install or connect the following:
   - Suspension arm to the vehicle
   - Rear mounting bolt and torque it to 29 ft. lbs. (40 Nm)
   - Front mounting bolts and torque them to 33 ft. lbs. (45 Nm)
   - Coil spring

Prizm

FRONT

If removing the left side lower control arm from a vehicle with an automatic transmission, remove the crossmember, transmission support, both lower control arms and the front stabilizer bar as a complete unit.

1. Install an engine support fixture.
2. Before servicing the vehicle, refer to the precautions in the beginning of this section.

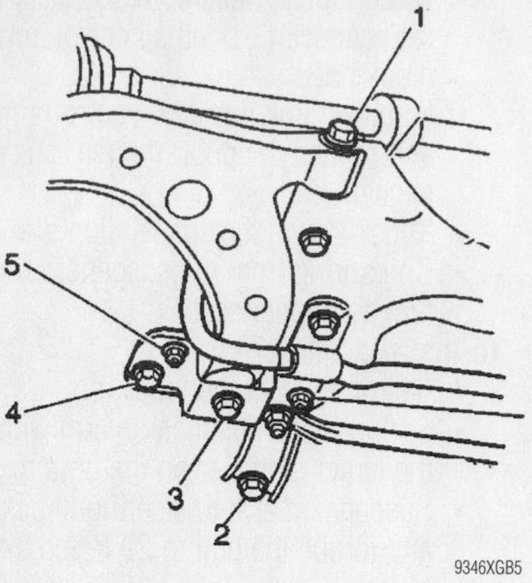

Disconnect the hardware as shown (1–5) in order to remove the control arm–Prizm

3. Remove or disconnect the following:
   - Negative battery cable
   - Front wheel
   - Suspension support brace
   - Stabilizer shaft link, if equipped
   - Mounting bolts (1—5) from the lower control arm
   - Lower control arm from the vehicle

To install:

4. Install or connect the following:
   - Lower control arm and hand-tighten the bolts
   - Stabilizer shaft link to the lower control arm and hand-tighten the nut, if equipped
   - Lower control arm bracket and torque the nut to 14 ft. lbs. (19 Nm)
   - Control to the steering knuckle and torque the fasteners to 105 ft. lbs. (142 Nm)
   - Suspension support brace and torque the nuts to 51 ft. lbs. (69 Nm)
   - Front wheel
5. Bounce the front end several times to stabilize the suspension.
6. Torque the bolts in the following sequence:

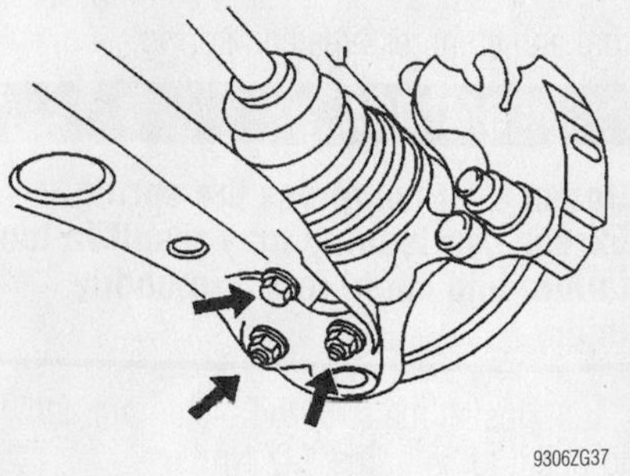

Install the two bolts and nut to the lower control arm–Prizm

   a. Bolt No. 3: 129 ft. lbs. (175 Nm).
   b. Bolt No. 1: 158 ft. lbs. (215 Nm).
   c. Bolt No. 5: 167 ft. lbs. (225 Nm).
   d. Bolt No. 4: 109 ft. lbs. (147 Nm).
   e. Bolt No. 2: 91 ft. lbs. (123 Nm).
   f. Stabilizer shaft nut to 33 ft. lbs. (44 Nm).
7. Remove the engine support fixture and connect the negative battery cable.
8. Check and adjust the front end alignment, if needed.

REAR

➡**The rear suspension contains a front and rear control arm.**

1. Before servicing the vehicle, refer to the precautions in the beginning of this section.
2. Remove or disconnect the following:
   - Negative battery cable
   - Both trailing arms from the suspension knuckles
   - Both rear suspension rear control arms
3. Support the rear suspension crossmember.
   - Suspension crossmember mounting bolts
   - Both rear suspension front control arms from the crossmember

To install:

4. Install or connect the following:
   - Both rear suspension front control arms to the crossmember, do not tighten the bolts at this time
   - Rear suspension crossmember and torque the outer bolts to 55 ft. lbs. (74 Nm) and the inner bolts to 14 ft. lbs. (19 Nm)
5. Remove the crossmember support.
   - Both rear suspension rear control arms, do not tighten the bolts at this time
   - Both trailing arms to the suspension knuckles, do not tighten the bolts at this time
6. Bounce the rear of the vehicle several times to stabilize the suspension.
7. Torque the trailing arm bolts to 67 ft. lbs. (91 Nm).
8. Torque the front and rear control arm bolts to 89 ft. lbs. (125 Nm).
9. Measure the wheel alignment and adjust as necessary.

CONTROL ARM BUSHING REPLACEMENT

Metro

1. Before servicing the vehicle, refer to the precautions in the beginning of this section.

2. Remove or disconnect the following:
   - Negative battery cable
   - Control arm
3. Cut the flange off of the front bushing.
   - Control arm rear bushing from the control arm with Tool J 34865
   - Control arm bracket
   - Front bushing from the control arm with Tool J 34865

To install:

4. Lubricate the circumference of the bushings with soapy water to ease installation.
5. Install or connect the following:
   - Center the bushing in the control arm and using tool J 34865 install the front bushing to the control arm
   - Rear bushing to the control arm with tool J 34865
   - Control arm bracket. Torque the nut to 92 ft. lbs. (125 Nm).
   - Control arm
   - Negative battery cable

Prizm

➡The lower control arm bushings are service as part of the control arm and cannot be serviced separately.

Wheel Bearings

ADJUSTMENT

The wheel bearings are sealed units and cannot be adjusted.

REMOVAL & INSTALLATION

Front

METRO

1. Before servicing the vehicle, refer to the precautions in the beginning of this section.
2. Remove or disconnect the following:
   - Negative battery cable
   - Front wheel
   - Brake caliper
   - Brake caliper bracket
   - Brake rotor
   - Drive axle nut
   - Hub/bearing from the steering knuckle using a slide hammer
3. Inspect the wheel bearings for excessive wear, corrosion or contamination by dirt or water; replace as necessary.

To install:

4. Install or connect the following:
   - Hub/bearing to the steering knuckle using a piece of wood and a hammer
   - Drive axle nut and torque it to 129 ft. lbs. (175 Nm)
   - Brake rotor
   - Brake caliper bracket and torque the bolts to 37 ft. lbs. (50 Nm)
   - Brake caliper and torque the bolts to 22 ft. lbs. (30 Nm)
   - Front wheel
   - Negative battery cable
5. Check the wheel alignment and adjust, if necessary.

PRIZM

1. Before servicing the vehicle, refer to the precautions in the beginning of this section.
2. Remove or disconnect the following:
   - Negative battery cable
   - Front wheel
   - Wheel Speed Sensor (WSS), if equipped
   - Brake hose from the strut
   - Brake caliper
   - Brake caliper bracket
   - Brake rotor
   - Steering knuckle and clamp it in a soft face vise
   - Inner seal shield from the knuckle
   - Inner oil seal from the knuckle
   - Wheel bearing retainer from the knuckle
   - Hub and brake shield from the knuckle
   - Outer seal
3. Press the front wheel bearing outside inner race from the knuckle.
4. Remove the wheel bearing.

To install:

5. Remove the inner and outer races from the new bearing.
6. Press a new bearing into the steering knuckle.
7. Install or connect the following:
   - New races into the bearing
   - Drive the outer seal into the steering knuckle with a Hub Seal Installer tool
   - Brake shield to the knuckle and torque the bolts to 74 inch lbs. (8 Nm)
   - Press the hub into the knuckle
   - Front wheel bearing retainer to the knuckle
   - Inner seal to the knuckle
8. If equipped with Anti-lock Brake System (ABS), align the holes for the speed sensor in the inner shield and the knuckle.
9. Install or connect the following:
   - Inner seal shield to the knuckle
   - Steering knuckle to the vehicle
   - Brake rotor
   - Brake caliper bracket and torque the bolts to 65 ft. lbs. (88 Nm)
   - Brake caliper and torque the bolts to 25 ft. lbs. (34 Nm)
   - Brake hose to the strut and torque the bolt to 22 ft. lbs. (29 Nm)
   - WSS, if equipped and torque the bolt to 70 inch lbs. (8 Nm)
   - Front wheel
   - Negative battery cable
10. Check the front wheel alignment and adjust if necessary.

Rear

METRO

1. Before servicing the vehicle, refer to the precautions in the beginning of this section.
2. Remove or disconnect the following:
   - Negative battery cable
   - Rear wheel
   - Spindle nut dust cap
   - Unstake the spindle nut and washer
   - If not equipped with a wheel hub, tap on the brake drum/hub assembly to loosen and remove the assembly from the spindle
   - Inner bearing from the outboard side of the assembly, using a hammer and punch

➡It is necessary to move the bearing spacer from side-to-side to work the punch around the perimeter of the inner bearing.

   - Outer bearing from the inboard side of the assembly, using a hammer and punch

To install:

➡The outer bearing is smaller in diameter than the inner one.

3. Install or connect the following:
   - Outer bearing in the drum/hub assembly using a Bearing Installation Tool J-34842 and driver handle J-7079-2
   - Bearing spacer in the brake drum hub with the inner lip toward the outer bearing

➡Be sure the bearing spacer is properly installed. The assembly will not fit on the spindle if the spacer lip is toward the inner bearing.

- Install the inner bearing in the drum/hub assembly using a Bearing Installation Tool J-34842 and driver handle J-7079-2
- Brake drum/hub assembly to the spindle
- Washer and a new spindle nut and torque the nut to 120 ft. lbs. (170 Nm) if equipped with a wheel hub or to 74 ft. lbs. (100 Nm) without a wheel hub
- Stake the spindle nut and install the dust cap
- Rear wheel
- Negative battery cable

PRIZM

1. Before servicing the vehicle, refer to the precautions in the beginning of this section.
2. Remove or disconnect the following:
   - Negative battery cable
   - Rear wheel
   - Brake drum
   - Wheel Speed Sensor (WSS), if equipped
   - Axle hub mounting bolts
   - Axle hub
   - O-ring from the backing plate

To install:

3. Install or connect the following:
   - New O-ring onto the backing plate
   - Hub to the knuckle and torque the bolts to 59 ft. lbs. (80 Nm)
   - WSS, if equipped and torque the bolt to 70 inch lbs. (8 Nm)
   - Brake drum
   - Rear wheel
   - Negative Battery cable
4. Check the rear wheel alignment and adjust, if necessary.

SATURN

32

1998–01

LW • SC1 • SC2 • SL • SL1 • SL2 • SW1 • SW2

PRECAUTIONS

Before servicing any vehicle, please be sure to read all of the following precautions, which deal with personal safety, prevention of component damage, and important points to take into consideration when servicing a motor vehicle:

- Never open, service or drain the radiator or cooling system when the engine is hot; serious burns can occur from the steam and hot coolant.
- Observe all applicable safety precautions when working around fuel. Whenever servicing the fuel system, always work in a well-ventilated area. Do not allow fuel spray or vapors to come in contact with a spark, open flame or excessive heat (a hot drop light, for example). Keep a dry chemical fire extinguisher near the work area. Always keep fuel in a container specifically designed for fuel storage; also, always properly seal fuel containers to avoid the possibility of fire or explosion. Refer to the additional fuel system precautions later in this section.
- Fuel injection systems often remain pressurized, even after the engine has been turned **OFF**. The fuel system pressure must be relieved before disconnecting any fuel lines. Failure to do so may result in fire and/or personal injury.
- Brake fluid often contains polyglycol ethers and polyglycols. Avoid contact with the eyes and wash your hands thoroughly after handling brake fluid. If you do get brake fluid in your eyes, flush your eyes with clean, running water for 15 minutes. If eye irritation persists, or if you have taken brake fluid internally, IMMEDIATELY seek medical assistance.
- The EPA warns that prolonged contact with used engine oil may cause a number of skin disorders, including cancer! You should make every effort to minimize your exposure to used engine oil. Protective gloves should be worn when changing oil. Wash your hands and any other exposed skin areas as soon as possible after exposure to used engine oil. Soap and water, or waterless hand cleaner should be used.
- All new vehicles are now equipped with an air bag system. The system must be disabled before performing service on or around system components, steering column, instrument panel components, wiring and sensors. Failure to follow safety and disabling procedures could result in accidental air bag deployment, possible personal injury and unnecessary system repairs.
- Always wear safety goggles when working with, or around, the air bag system. When carrying a non-deployed air bag, be sure the bag and trim cover are pointed away from your body. When placing a non-deployed air bag on a work surface, always face the bag and trim cover upward, away from the surface. This will reduce the motion of the module if it is accidentally deployed. Refer to the additional air bag system precautions later in this section.
- Clean, high quality brake fluid from a sealed container is essential to the safe and proper operation of the brake system. You should always buy the correct type of brake fluid for your vehicle. If the brake fluid becomes contaminated, completely flush the system with new fluid. Never reuse any brake fluid. Any brake fluid that is removed from the system should be discarded. Also, do not allow any brake fluid to come in contact with a painted surface; it will damage the paint.
- Never operate the engine without the proper amount and type of engine oil; doing so WILL result in severe engine damage.
- Timing belt maintenance is extremely important! Many models utilize an interference-type, non-freewheeling engine. If the timing belt breaks, the valves in the cylinder head may strike the pistons, causing potentially serious (also time-consuming and expensive) engine damage. Refer to the maintenance interval charts in the front of this manual for the recommended replacement interval for the timing belt, and to the timing belt section for belt replacement and inspection.
- Disconnecting the negative battery cable on some vehicles may interfere with the functions of the on-board computer system(s) and may require the computer to undergo a relearning process once the negative battery cable is reconnected.
- When servicing drum brakes, only disassemble and assemble one side at a time, leaving the remaining side intact for reference.
- Only an MVAC-trained, EPA-certified automotive technician should service the air conditioning system or its components.

ENGINE REPAIR

Ignition Timing

ADJUSTMENT

The 1.9L, 2.2L and 3.0L engines utilize a Distributorless Ignition System (DIS), no adjustment is possible.

Alternator

REMOVAL

1.9L Engines

1. Before servicing the vehicle, refer to the precautions in the beginning of this section.
2. Remove or disconnect the following:

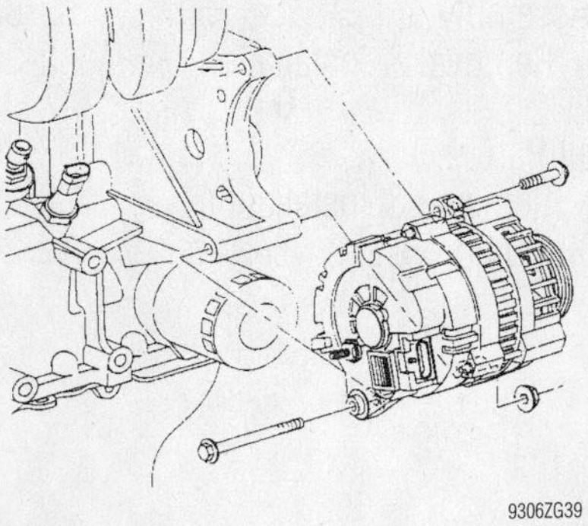

Exploded view of the alternator–1.9L engines

- Negative battery cable
- Accessory drive belt
- Right front wheel
- Right front wheel well splash shield
- Alternator electrical connectors
- Alternator bolts
- Alternator through the wheel well opening

2.2L Engine

1. Before servicing the vehicle, refer to the precautions in the beginning of this section.
2. Remove or disconnect the following:
   - Negative battery cable
   - Throttle body air duct
   - Accessory drive belt
   - Alternator electrical connectors
   - Alternator bolts
   - Alternator

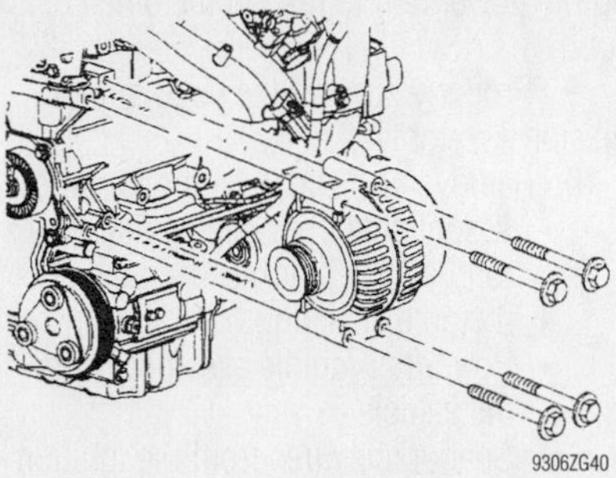

Exploded view of the alternator–2.2L engine

3.0L Engine

1. Before servicing the vehicle, refer to the precautions in the beginning of this section.
2. Remove or disconnect the following:
   - Negative battery cable
   - Accessory drive belt and tensioner
   - Upper alternator bolts
   - Alternator lower bolts
   - Alternator electrical connectors
   - Alternator

INSTALLATION

1.9L Engines

➡Make certain the alternator shield is in place before installation.

Install or connect the following:
- Alternator to the cylinder block bracket and torque the bolts to 24 ft. lbs. (32 Nm)
- Alternator electrical connectors
- Splash shield
- Right front wheel
- Accessory drive belt
- Negative battery cable

2.2L Engine

Install or connect the following:
- Alternator and torque the bolts to 15 ft. lbs. (20 Nm)
- Alternator electrical connectors
- Accessory drive belt
- Throttle body air duct
- Negative battery cable

3.0L Engine

Install or connect the following:
- Alternator electrical connectors
- Alternator and torque the bolts to 30 ft. lbs. (40 Nm)
- Drive belt tensioner and torque the bolts to 30 ft lbs. (40 Nm)
- Accessory drive belt
- Negative battery cable

Engine Assembly

REMOVAL & INSTALLATION

1.9L Engines

➡The manufacturer recommends that the engine and transmission be removed as a complete unit. Disconnect the cradle and lower the entire assembly instead of lifting the assembly out of the vehicle. Both the Single Over Head Camshaft (SOHC) and Dual Over Head Camshaft (DOHC) engines are removed or installed in the same manner. We have found that it is often possible, however, to remove the engine or transmission alone on some models with automatic transmissions by raising it up and out of the engine compartment.

1. Before servicing the vehicle, refer to the precautions in the beginning of this section.
2. Properly disable the Supplemental Inflatable Restraint (SIR) system.
3. Properly relieve the fuel system pressure.
4. Drain the engine coolant.
5. Drain the engine oil.
6. Remove or disconnect the following:
   - Both battery cables
   - Air cleaner and intake duct assembly
   - Engine Coolant Temperature (ECT) sensor electrical connector
   - Oxygen (O_2S) sensor and clip
   - Idle Air Control (IAC) valve
   - Ignition coil connectors
   - Throttle Position (TPS) sensor
   - Manifold Absolute Pressure (MAP) sensor
   - Exhaust Gas Recirculation (EGR) solenoid
   - Brake booster hose
   - Ground connectors at the rear of the cylinder block
   - Fuel injector connectors
   - Neutral Safety Selector (NSS) switch, if equipped
   - Valve body actuator connector, if equipped
   - Transmission temperature sensor, if equipped
   - Turbine speed sensor, if equipped
   - Back-up light switch on manual transmissions
   - Accelerator cable
   - Fuel pressure and return lines
   - Drive belt
   - Upper radiator hose and de-aeration hose
   - A/C compressor without removing the hoses
   - Transmission cooler lines, if equipped
   - Automatic transmission shifter cable, if equipped
   - Manual transmission shifter cables
   - Hydraulic clutch actuator, if equipped
   - Tie the radiator, condenser and fan module to the crossbar
   - Front wheels
   - Splash shields
   - Brake caliper brackets and secure the calipers to the shock tower
   - Struts from the knuckles
   - Lower radiator hose
   - Heater inlet and outlet hoses
   - Steering shaft
   - Power steering pressure switch electrical connector, if equipped
   - Front exhaust pipe from the exhaust manifold
   - Powertrain stiffening bracket
   - Flywheel cover
   - Torque converter bolts, if equipped
   - Starter motor electrical connectors
   - Alternator electrical connector
   - Oil pressure sensor electrical connector
   - Knock Sensor (KS) electrical connector
   - Crankshaft Position (CKP) sensor electrical connector
   - Electronic Variable Orifice (EVO) solenoid electrical connector, if equipped
   - Vehicle Speed Sensor (VSS) electrical connector
   - Evaporative Emissions (EVAP) purge solenoid electrical connector
   - Powertrain Control Module (PCM) electrical connector
   - Antilock Brake System (ABS) connectors, if equipped
   - Brake lines from the rear of the cradle
   - Electrical harness from the engine and position it out of the way
7. Place a block of wood between the torque strut and cradle.
8. Remove the 3 right side upper engine torque axis-to-front cover nuts and the 2 mount-to-midrail bracket nuts, allow-

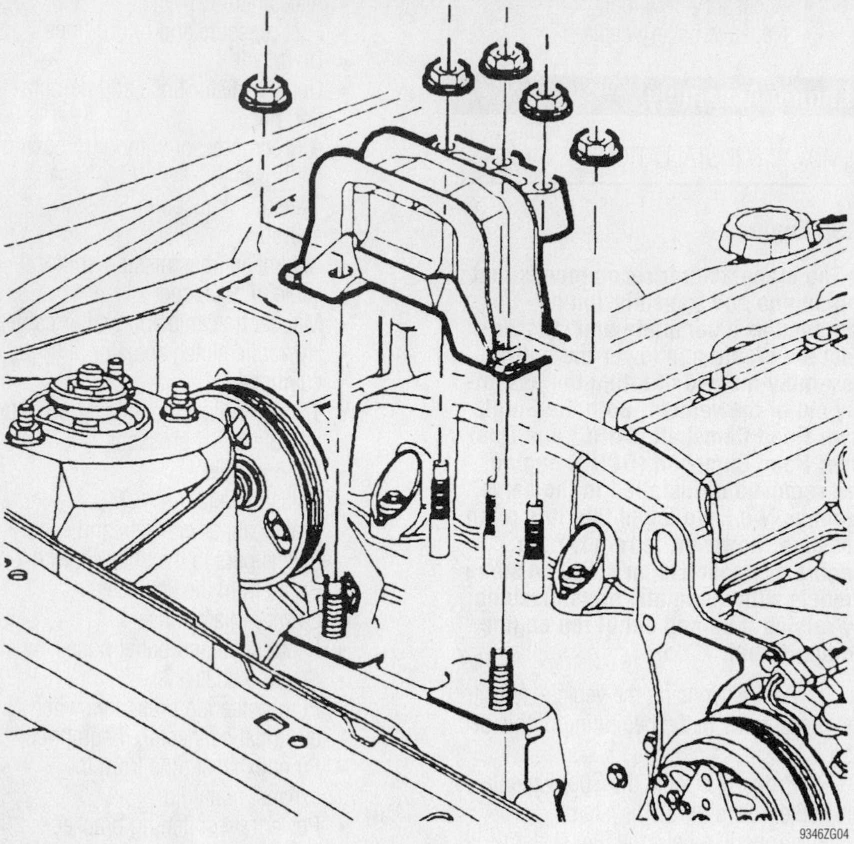

Right hand upper engine torque axis–1.9L engines

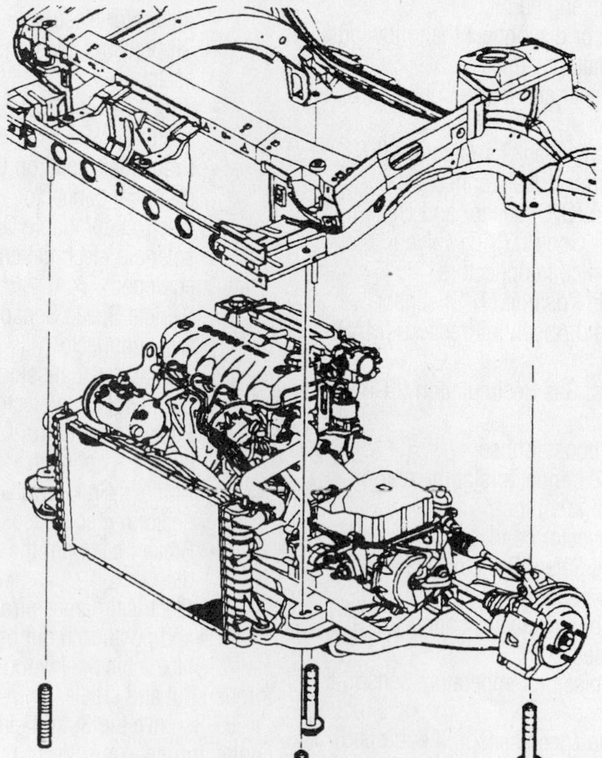

Exploded view of the cradle to body bolts–1.9L engines

ing the powertrain to rest on the block of wood.

9. Properly support the engine/transmission assembly.

10. Remove or disconnect the following:
- 2 right side front engine mount torque strut bracket-to-cradle nuts
- 4 cradle attaching bolts
- Powertrain/cradle assembly from the vehicle
- Spark plug wires from the ignition module
- Power steering pump and bracket, if equipped

11. Install an engine lifting device to the service support brackets.
- Front mount assembly
- Intake manifold support brace, DOHC engines only
- 3 axle shaft bracket support bolts and allow the bracket to rotate rearward
- Transmission attaching bolts
- Engine from the transmission

To install:

12. Install or connect the following:
- Engine to the transmission and torque the lower transmission bolts to 96 ft. lbs. (130 Nm) and the upper transmission bolts to 66 ft. lbs. (90 Nm)
- Front engine mount and torque the bolts to 41 ft. lbs. (55 Nm)
- Engine mount torque strut bracket and torque the bolts to 52 ft. lbs. (70 Nm)
- Engine strut-to-engine bracket and torque the bolts to 52 ft. lbs. (70 Nm)
- Axle shaft bracket and torque the bolts to 41 ft. lbs. (55 Nm)
- Starter bracket and torque the bolts to 80 inch lbs. (9 Nm)
- Engine/transmission assembly to the chassis and torque the cradle-to-body bolts to 151 ft. lbs. (205 Nm)
- Brake lines to the cradle
- Steering shaft and torque the bolts to 35 ft. lbs. (47 Nm)
- Starter motor electrical connectors
- Alternator electrical connectors
- Oil pressure sensor electrical connector
- KS sensor electrical connector
- CKP sensor electrical connector
- EVAP canister purge valve electrical connector
- Wiring harness PCM ground
- Wiring harness to the transmission/engine block and torque the fasteners to 18 ft. lbs. (25 Nm)

- EVO solenoid electrical connector, if equipped
- VSS electrical connector
- ABS wheel sensor electrical connectors, if equipped
- Front exhaust pipe to the manifold and torque the bolts to 23 ft. lbs. (31 Nm)
- Exhaust pipe-to-stiffening bracket and torque the bolts to 35 ft. lbs. (47 Nm)
- Exhaust pipe-to-support bracket and torque the bolts to 23 ft. lbs. (31 Nm)
- Heater hoses
- Lower radiator hose
- Steering knuckle to the strut and torque the bolts to 148 ft. lbs. (200 Nm)
- Brake caliper and torque the bolts to 81 ft. lbs. (110 Nm)
- Hydraulic clutch slave cylinder and torque the fasteners to 19 ft. lbs. (25 Nm), if equipped
- Shift cables
- Transmission oil cooler lines, if equipped
- A/C compressor bracket and torque the bolts to 22 ft. lbs. (30 Nm)
- A/C compressor and torque the front bolt to 40 ft. lbs. (54 Nm) and the rear bolt to 22 ft. lbs. (30 Nm)
- Accessory drive belt
- Engine mount-to-midrail bracket and torque the bolts to 37 ft. lbs. (50 Nm)
- Engine mount to the front cover and torque the bolts to 37 ft. lbs. (50 Nm)
- Torque converter-to-flexplate bolts and torque them to 52 ft. lbs. (70 Nm)
- Dust cover and torque the fasteners to 89 inch lbs. (10 Nm)
- Splash shields
- Both front wheels
- ECT sensor electrical connector
- O_2S sensor electrical connector
- IAC valve electrical connector
- Fuel injector connectors
- TPS sensor electrical connectors
- MAP sensor electrical connector
- EGR valve electrical connector
- A/C compressor electrical connector, if equipped
- Brake booster vacuum hose
- Ground connections
- NSS switch connectors
- PNP switch connectors
- Valve body actuator connector
- Back-up light switch connector, if equipped
- Temperature sensor electrical connector
- Accelerator cable
- Fuel feed and return lines
- Upper radiator and de-aeration hoses

➡Check the upper cooling module grommets for misalignment. The cooling module must be able to move freely.

- Air cleaner and intake duct assembly
- Both battery cables

13. Fill the engine with new oil.
14. Fill the cooling system.
15. Prime the fuel system by cycling the ignition **ON**for 5 seconds and **OFF** for 10 seconds a few times without cranking the engine. Start the engine and check for leaks.
16. Perform a short road test and check the engine again for leaks. Be sure the cooling system is filled to the surge tank FULL COLD line.

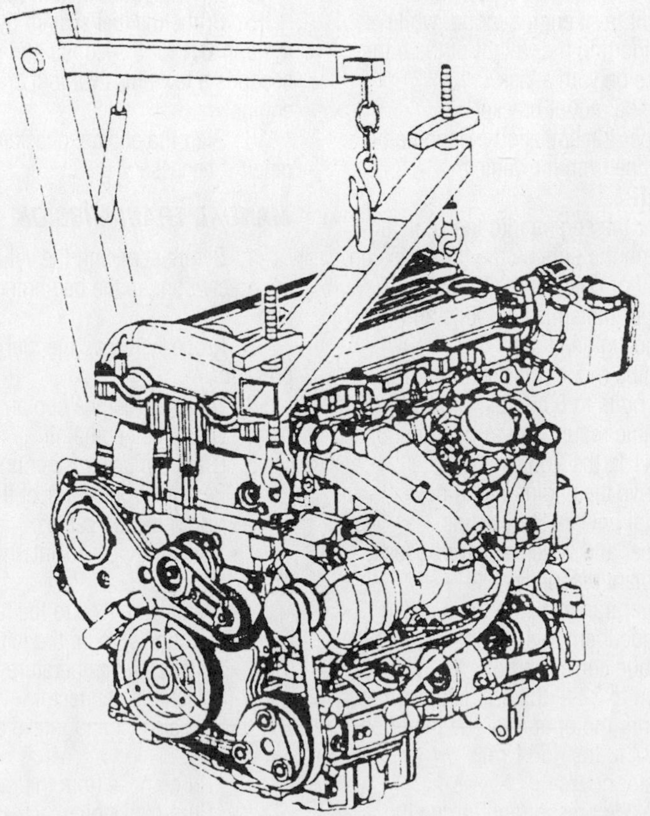

Remove the engine with the lifting hoist–2.2L engine

2.2L Engine

AUTOMATIC TRANSMISSION

1. Before servicing the vehicle, refer to the precautions in the beginning of this section.
2. Properly relieve the fuel system pressure.
3. Drain the engine coolant.
4. Drain the engine oil.
5. Drain the power steering fluid.
6. Remove or disconnect the following:
   - Both battery cables
   - Battery
   - Main wire feed from the fuse block
   - Intake Air Temperature (IAT) sensor electrical connector
   - Air cleaner and intake duct assembly
   - Purge hose from the throttle body
   - Purge solenoid electrical connector
   - Rear Heated Oxygen (HO_2S) sensor electrical connector
   - Main harness connector by the master cylinder
   - Alternator electrical connector
   - A/C pressure switch electrical connector

- Transmission electrical connectors
- Right front Wheel Speed Sensor (WSS) electrical connector
- Cowl cover
- Powertrain Control Module (PCM) boot from the cowl
- PCM electrical connectors
- Engine harness connectors from the fuse block
- Battery tray and fuse block tray
- Cruise control and throttle cables
- Brake assist vacuum line from the throttle body
- Fuel feed and return
- Evaporative Emissions (EVAP) purge hose at the solenoid
- Starter motor
- Torque converter bolts
- Exhaust system from the catalytic converter to the exhaust manifold
- A/C compressor without removing the hoses
- Lower engine-to-transmission bolts
- Heater hoses at the lower cowl
- Coolant reservoir hose
- Upper and lower radiator hoses
- Metal power steering line

7. Attach an engine lifting hoist to the lifting eyes.
   - Right front engine mount while supporting the weight of the transmission with a jack
   - Engine mount bracket
   - Upper transmission-to-engine bolts
   - Engine from the vehicle

To install:

8. Lower the engine into the vehicle and align it on the pins on the transmission.
9. Install or connect the following:
   - Upper transmission bolts and hand-tighten them
   - Engine mount bracket and torque the bolts to 66 ft. lbs. (90 Nm)
   - Engine mount and torque the bolts to 41 ft. lbs. (55 Nm)
10. Remove the engine lifting device.
    - Metal power steering line
    - Upper and lower radiator hoses
    - Coolant reservoir hose
    - Lower transmission bolts and torque them to 48 ft. lbs. (65 Nm)
    - Torque converter bolts and torque them to 33 ft. lbs. (45 Nm)
    - Starter motor and torque the bolts to 37 ft. lbs. (50 Nm)
    - Heater hoses
    - A/C compressor and torque the bolts to 18 ft. lbs. (25 Nm)
    - Accessory drive belt
    - Exhaust from the catalytic converter to the exhaust manifold and torque the bolts to 12 ft. lbs. (16 Nm)
11. Torque the upper bell housing bolts to 48 ft. lbs. (65 Nm).
    - EVAP purge solenoid electrical connector and hose
    - Fuel feed and return lines
    - Brake booster vacuum hose
    - Cruise control and throttle cables
    - Fuse block and battery trays
    - Engine harness to the fuse block
    - Main engine harness under the fuse block panel
    - PCM electrical connector
    - PCM boot to the cowl
    - Cowl cover
    - Main harness connector near the master cylinder
    - Rear HO_2S sensor electrical connector
    - Purge solenoid electrical connector
    - Purge hose to the throttle body
    - Air cleaner and intake duct assembly
    - IAT sensor electrical connector
    - Ground wire to the left fender well
    - Main wire feed to the fuse block
    - Battery and both cables
12. Fill the engine with coolant.
13. Fill the engine with new oil.
14. Fill the power steering fluid.
15. Prime the fuel system by cycling the ignition **ON** for 5 seconds and **OFF** for 10 seconds a few times without cranking the engine.
16. Start the engine, check for leaks, and repair if necessary.

MANUAL TRANSMISSION

1. Before servicing the vehicle, refer to the precautions in the beginning of this section.
2. Properly relieve the fuel system pressure.
3. Drain the engine coolant.
4. Drain the engine oil.
5. Drain the power steering fluid.
6. Remove or disconnect the following:
   - Both battery cables
   - Steering gear-to-intermediate shaft pinch bolt
   - Main wire feed to the fuse block
   - Ground wire on the left fender well
   - Intake Air Temperature (IAT) sensor electrical connector
   - Air cleaner and intake duct assembly
   - Purge hose from the throttle body
   - Purge solenoid electrical connector
   - Rear Heated Oxygen (HO_2S) sensor electrical connector
   - Back-up switch electrical connector
   - Front dash cover
   - Powertrain Control Module (PCM) boot from the front of the dash
   - Main harness connector by the master cylinder
   - PCM electrical connectors
   - Engine harness connectors from the fuse block
   - Battery, tray and fuse block tray
   - Cruise control and throttle cables
   - Brake booster vacuum hose
   - Control assembly rod-to-control shaft lever pinch bolt
   - Retaining clips connecting the control shaft lever to the transmission
   - Control shaft lever assembly
   - Heater hoses at the lower cowl
   - Upper and lower radiator hoses
   - Fuel feed and return lines
7. Secure the radiator to the upper radiator support.
   - Both front wheels
   - Right front splash shield
   - Left front wheel liner push pin from the frame
8. Install wood blocks between the transmission case and frame and the crank pulley and frame.
   - Right side engine mount
   - Left side transmission mount
   - Exhaust system from the catalytic converter to the exhaust manifold
   - A/C compressor without removing the hoses
   - Tie rods from the steering knuckles
   - Stabilizer bar links from the struts
   - Lower ball joints from the steering knuckles
   - Suspension support assemblies
   - Suspension support cage nuts from the body
9. Properly support the powertrain and frame assembly.
   - Frame-to-body attachment fasteners
   - Powertrain/frame assembly from the vehicle

To install:

10. Raise the powertrain into the vehicle.
11. Install or connect the following:
    - Powertrain/frame assembly
    - New frame-to-body bolts do not tighten at this time
    - Suspension supports with new bolts and torque the suspension support and frame bolts to 74 ft. lbs. (100 Nm) plus 45 degrees
    - Ball joint to the steering knuckle and torque the nuts to 74 ft. lbs. (100 Nm)
    - Stabilizer links to the struts and torque the fasteners to 50 ft. lbs. (65 Nm)

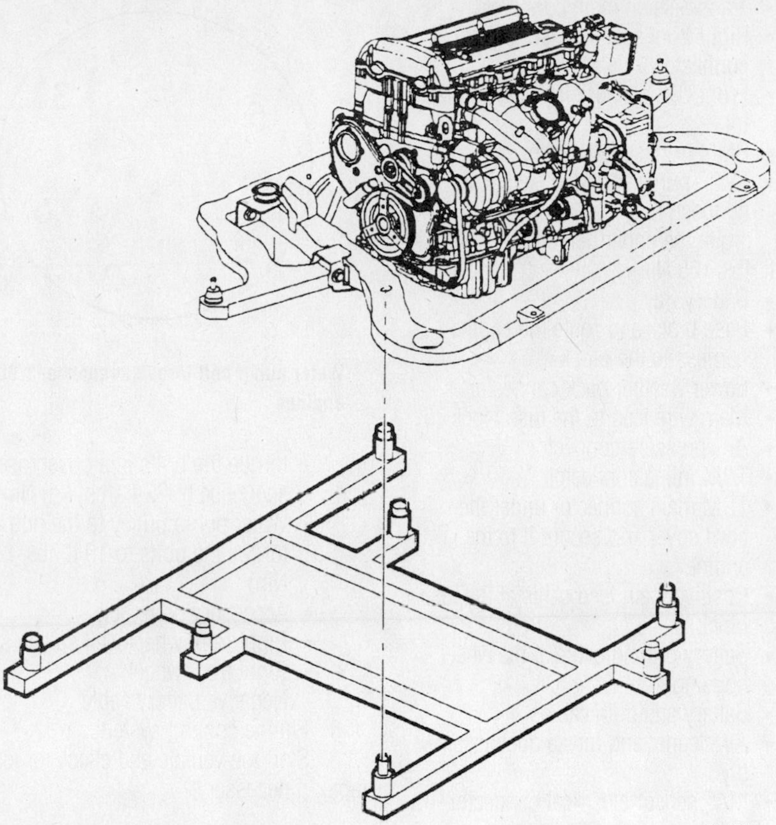

Powertrain/frame assembly with lifting table–2.2L engine

- Tie rods to the steering knuckles and torque the new nuts to 45 ft. lbs. (65 Nm)
- Exhaust system to the exhaust manifold and torque the bolts to 12 ft. lbs. (16 Nm)
- A/C compressor and torque the bolts to 18 ft. lbs. (25 Nm)
- Rear HO_2S sensor electrical connector
- Right side engine mount and torque the bolts to 41 ft. lbs. (55 Nm)
- Left side transmission mount and torque the bolts to 41 ft. lbs. (55 Nm)

12. Remove the wood blocks.
    - Left front wheel liner push pins
    - Right front splash shield
    - Both front wheels
    - Fuel feed and return lines
    - Upper and lower radiator hoses
    - Heater hoses
    - Brake booster vacuum hose
    - Shift control rod
13. Adjust the shift control rod as follows:

a. Install the shift rod into the shift linkage assembly.

b. Rotate the transmission shift control shaft clockwise and depress the spring loaded transmission shift linkage lock pin on the case.

c. Pull up the shift lever boot.

d. Move the shift lever to the 5 (reverse gate) and use a 3/8 inch punch to hold it in place.

e. Tighten the pinch bolt to 9 ft. lbs. (12 Nm) plus 180 degrees

14. Install or connect the following:
    - Cruise control and throttle cables
    - Fuse block and battery trays
    - Fuse block
    - Engine harness connector to the bottom of the fuse block
    - Fuse block
    - PCM electrical connector
    - PCM boot to the cowl
    - Cowl cover
    - EVAP purge solenoid electrical connector and hose
    - Main harness connector near the master cylinder
    - Transmission back-up switch electrical connector
    - Purge hose to the throttle body
    - Air cleaner and intake duct assembly
    - IAT sensor electrical connector
    - Ground wire to the left fender well
    - Main wire feed to the fuse block
    - Battery and both cables
    - Steering gear-to-intermediate shaft and torque the pinch bolt to 22 ft. lbs. (30 Nm)
15. Fill the engine with coolant.
16. Fill the engine with new oil.
17. Fill the power steering fluid.
18. Prime the fuel system by cycling the ignition **ON** for 5 seconds and **OFF** for 10 seconds a few times without cranking the engine.
19. Start the engine, check for leaks, and repair if necessary.

3.0L Engine

1. Before servicing the vehicle, refer to the precautions in the beginning of this section.
2. Properly relieve the fuel system pressure.
3. Drain the engine coolant.
4. Drain the engine oil.
5. Drain the power steering fluid.
6. Remove or disconnect the following:
    - Both battery cables
    - Air cleaner assembly
    - Mass Air Flow (MAF) sensor electrical connector
    - Battery ground cable from the fuse block
    - Transmission Control Module (TCM) main connector from under the cowl cover
    - TCM inline connector by the brake master cylinder
    - A/C pressure connector
    - Black engine harness connector from under the fuse block panel
    - Lower weather pack connector from inside the fuse block and secure the engine harness to the engine
    - Fuse block
    - Battery tray
    - Evaporative Emissions (EVAP) purge connector
    - Right front speed sensor connector
    - Front Oxygen (O_2S) sensor from the down pipe
    - Transmission ground electrical connector
    - Transmission main electrical connector
    - Transmission shift control electrical connector
    - Rear connector on the Engine Control Module (ECM)

- Brake booster vacuum hose
- Fuel lines
- EVAP purge hose and solenoid
- Starter motor
- Torque converter bolts
- Exhaust system from the catalytic converter to the exhaust manifold
- Transmission nose bracket
- A/C compressor without removing the hoses
- Heater hoses at the lower cowl
- Lower engine-to-transmission bolts
- Upper and lower radiator hoses
- coolant reservoir hose
- Metal power steering line
- Power steering reservoir from the radiator support

7. Attach an engine lifting device to the engine lifting eyes.
- Right front engine mount and bracket
- Upper transmission bolts
- Engine from the vehicle

To install:

8. Lower the engine into the vehicle and align it on the pins on the transmission.
9. Install or connect the following:
- Upper transmission bolts and hand-tighten them
- Engine mount bracket and torque the bolt to 30 ft. lbs. (40 Nm)
- Engine mount and torque the upper bolt to 30 ft. lbs. (40 Nm) and the lower bolt to 41 ft. lbs. (55 Nm)
- Transmission nose bracket and torque the bolts to 30 ft. lbs. (40 Nm)

10. Remove the engine lifting device.
- Metal power steering line
- Power steering reservoir to the radiator support
- Upper and lower radiator hoses
- Coolant reservoir hose
- Lower transmission bolts and torque them to 48 ft. lbs. (65 Nm)
- Torque converter and torque them to 48 ft. lbs. (65 Nm)
- Starter motor and torque the bolts to 30 ft. lbs. (40 Nm)
- Heater hoses at the lower cowl
- A/C compressor and torque the bolts to 30 ft. lbs. (40 Nm)
- Exhaust system to the exhaust manifold and torque the nuts to 15 ft. lbs. (20 Nm)
- EVAP purge solenoid and hose
- Fuel feed and return lines
- Brake booster vacuum hose
- ECM rear connector
- Transmission shift control electrical connector
- Transmission ground connector
- Transmission main connector
- Right front speed sensor electrical connector
- Front O_2S sensor electrical connector
- Right front speed sensor
- EVAP purge solenoid electrical connector

11. Torque the upper bell housing bolts to 48 ft. lbs. (65 Nm).
- Battery tray
- Fuse block and route the engine harness to the block
- Lower weather pack connector
- Main wire feed to the fuse block
- A/C pressure connector
- TCM inline connector
- TCM main connector under the cowl cover and secure it to the engine
- Positive main feed cable at the fuse block
- Battery ground cable at the wheel housing
- Battery and both cables
- Air cleaner and intake duct assembly
- MAF sensor electrical connector

12. Fill the engine with coolant.
13. Fill the engine with new oil.
14. Fill the power steering fluid.
15. Prime the fuel system by cycling the ignition **ON** for 5 seconds and **OFF** for 10 seconds a few times without cranking the engine.
16. Start the engine, check for leaks, and repair if necessary.

Water Pump

REMOVAL & INSTALLATION

1.9L Engines

1. Before servicing the vehicle, refer to the precautions in the beginning of this section.
2. Drain the cooling system.
3. Remove or disconnect the following:
- Negative battery cable
- Accessory drive belt
- Right front wheel
- Inner wheel well splash shield
- Water pump pulley bolts and allow the pulley to hang freely on the water pump hub
- Water pump flange bolts
- Water pump

To install:

4. Install or connect the following:
- Water pump with a new gasket and torque the bolts in a crisscross sequence to 22 ft. lbs. (30 Nm)

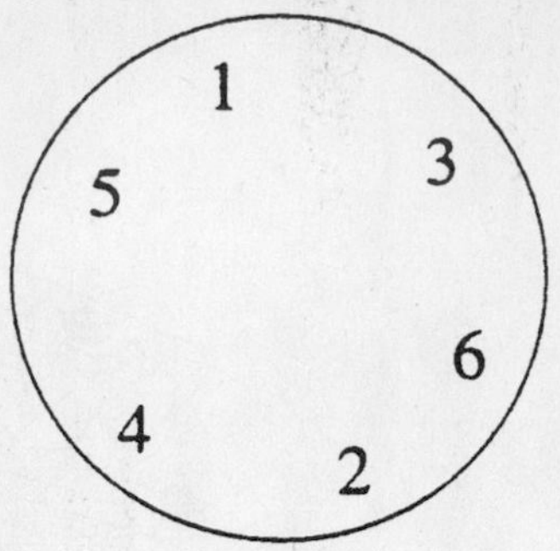

Water pump bolt torque sequence–1.9L engines

- Water pump pulley to the hub and torque the bolts to 19 ft. lbs. (25 Nm)
- Accessory drive belt
- Right front wheel well splash shield
- Right front wheel
- Negative battery cable

5. Fill the cooling system.
6. Start the vehicle and check for leaks, repair if necessary.

2.2L Engine

1. Before servicing the vehicle, refer to the precautions in the beginning of this section.
2. Drain the cooling system.
3. Remove or disconnect the following:
- Negative battery cable
- Air cleaner assembly
- Exhaust manifold heat shield

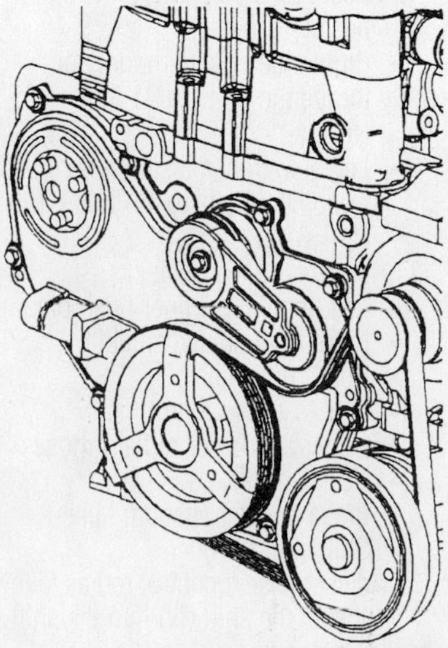

Water pump holding tool J43651–2.2L engine

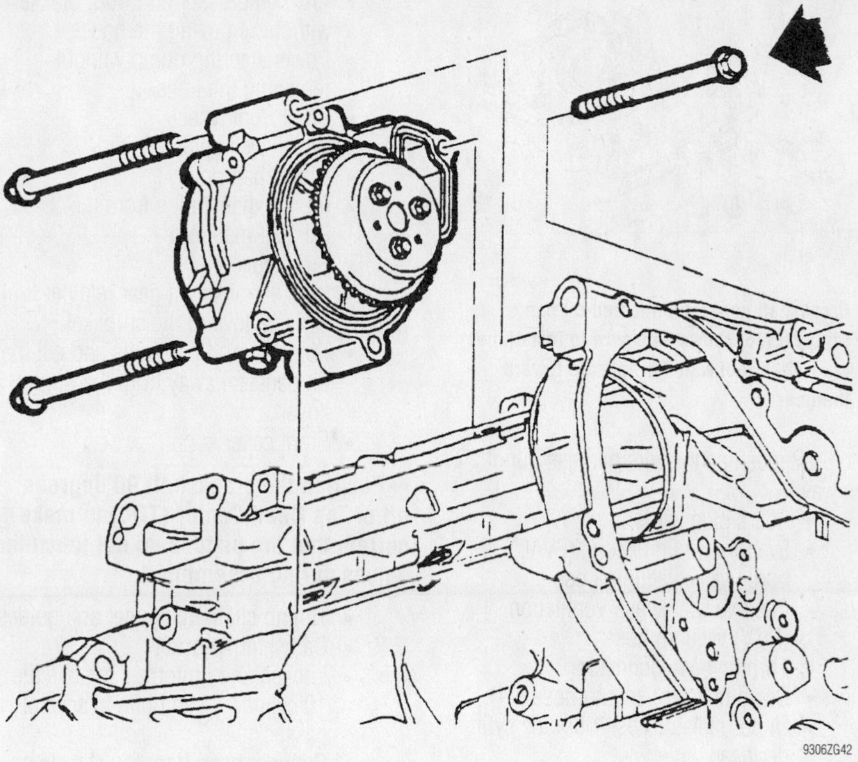

Exploded view of the water pump–2.2L engine

- Thermostat housing
- Water pump feed pipe
- Water pump sprocket access plate from the front cover

4. Install a Water Pump Holding Tool J43651.
   - Water pump retaining bolts
   - Water pump

To install:

5. Install or connect the following:
   - Water pump and torque the bolts to 18 ft. lbs. (25 Nm)
   - Water pump sprocket and torque the bolts to 89 inch lbs. (10 Nm)
   - Water pump sprocket access plate
   - Water feed tube after lubricating the O-ring
   - Thermostat housing and torque the bolts to 89 inch lbs. (10 Nm)
   - Exhaust manifold heat shield
   - Air cleaner assembly
   - Negative battery cable
6. Fill the cooling system.
7. Start the vehicle and check for leaks, repair if necessary.

3.0L Engine

1. Before servicing the vehicle, refer to the precautions in the beginning of this section.
2. Drain the cooling system.
3. Remove or disconnect the following:
   - Negative battery cable
   - Air cleaner assembly
   - Left front wheel
   - Wheel well splash shield
   - Loosen the water pump pulley and power steering pulley bolts, but do not remove them
4. Install an engine support fixture.
   - Right front engine mount
   - Accessory drive belt
   - Release the retaining tabs on the wiring harness channel and remove the front cover
   - Wiring harness from the channel
   - Water pump pulley
   - Power steering pump pulley
   - Drive belt tensioner
   - Front timing belt cover
   - Water pump

To install:

5. Install or connect the following:
   - Water pump with a new O-ring and torque the bolts to 18 ft. lbs. (25 Nm)
   - Front timing belt cover and torque the bolts to 71 inch lbs. (8 Nm)

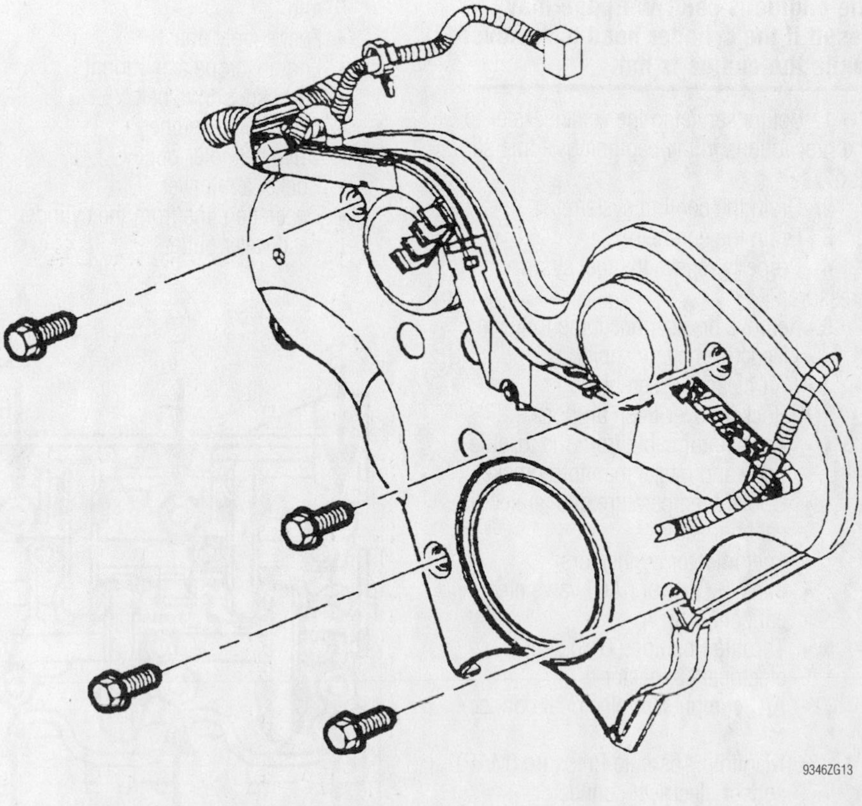

Timing belt cover bolt locations–3.0L engine

Heater Core replacement is covered in Section 2 of this manual

- Drive belt tensioner and torque the bolts to 30 ft. lbs. (40 Nm)
- Power steering pump and torque the bolts to 71 inch lbs. (8 Nm)
- Water pump pulley's and torque the bolts to 15 ft. lbs. (20 Nm)
- Wiring harness into the channel
- Wiring harness channel front cover
- Accessory drive belt

6. Remove the engine support fixture.
   - Wheel well splash shield and torque the bolts to 44 inch lbs. (5 Nm)
   - Front wheel
   - Air cleaner and intake duct
   - Negative battery cable
7. Fill the cooling system.
8. Start the vehicle and check for leaks, repair if necessary.

Cylinder Head

REMOVAL & INSTALLATION

1.9L Engines

WARNING

Only remove the cylinder head when the engine is cold. Warpage may result if the cylinder head is removed while the engine is hot.

1. Before servicing the vehicle, refer to the precautions in the beginning of this section.
2. Drain the cooling system.
3. Drain the engine oil.
4. Properly relieve the fuel system pressure.
5. Remove or disconnect the following:
   - Negative battery cable
   - Air cleaner assembly
   - Rocker cover fresh air hose
   - Accelerator cable from the throttle body and intake manifold bracket
   - Coolant temperature gauge electrical connector
   - Fuel injector connectors
   - Idle Air Control (IAC) valve electrical connector
   - Throttle Position (TPS) sensor electrical connector
   - A/C compressor electrical connector
   - Manifold Absolute Pressure (MAP) sensor electrical connector
   - Oxygen (O_2S) sensor electrical connector
   - Exhaust Gas Recirculation (EGR) solenoid

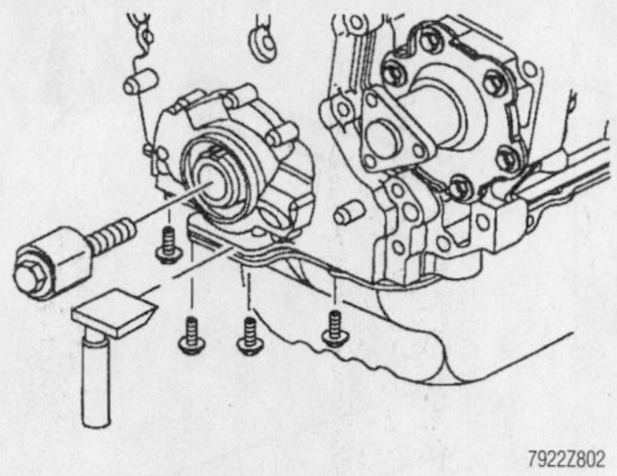

Crankshaft gear retaining and oil pan removal tool shown—be sure to install the crankshaft tool with the flat side toward the gear

6. Reposition the wire harness out of the way.
   - Spark plug wires
   - Evaporative Emissions (EVAP) purge valve vacuum hose
   - Positive Crankcase Ventilation (PCV) vacuum hose
   - Throttle body connector
   - Brake booster vacuum hose
   - Upper radiator hose from the cylinder head outlet
   - Deaeration hose from the intake manifold
   - Heater hose from the intake manifold
   - Fuel supply line
   - Engine torque axis mount
   - Accessory drive belt
   - Drive belt tensioner
   - Drive belt idler pulley
   - Rocker arm cover
   - Deaeration line from the cylinder head water outlet
   - A/C compressor and front bracket without removing the hoses
   - Power steering pump without removing the lines
   - Right front wheel
   - Wheel well splash shield
   - Crankshaft pulley
   - Front exhaust pipe from the exhaust manifold
   - Front cover
7. Install a crankshaft gear retainer tool with the flat side toward the sprocket.
   - Front four oil pan bolts and cut the oil pan seal away from the front cover
   - Front cover

➡Position the crankshaft 90 degrees off of Top Dead Center (TDC) to make certain that the pistons do not touch the valves during assembly.

- Timing chain, tensioner and guides
- Camshaft sprocket
- Loosen and uniformly remove the 10 cylinder head bolts in several passes
- Cylinder head from the dowels on the engine block

To install:

➡Before installing the cylinder head, it should be cleaned and inspected for excessive wear or damage.

8. Clean the gasket mating surfaces. Be careful not to damage the aluminum components and be sure the block bolt holes are clean of any residual sealer, oil or foreign matter.
9. Using a dial gauge, check that the

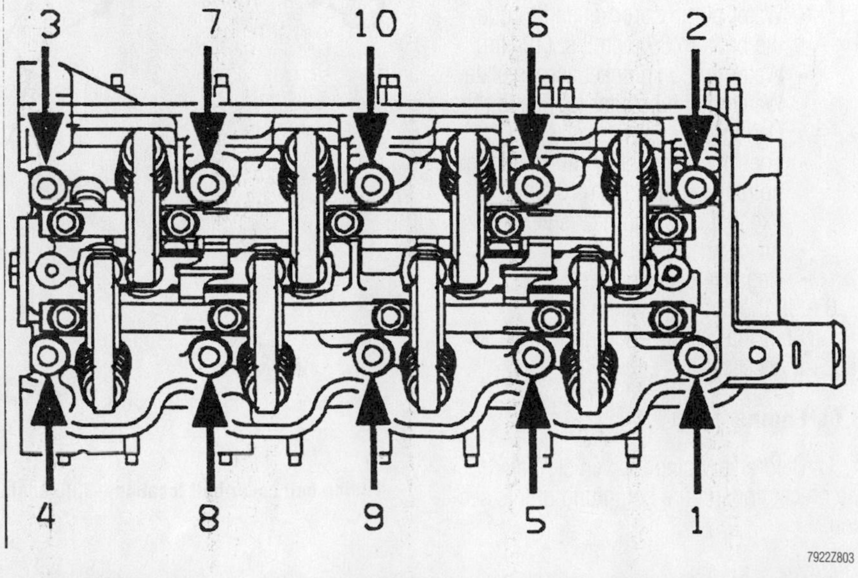

Gradually remove the cylinder head bolts in the sequence shown to prevent warping the head

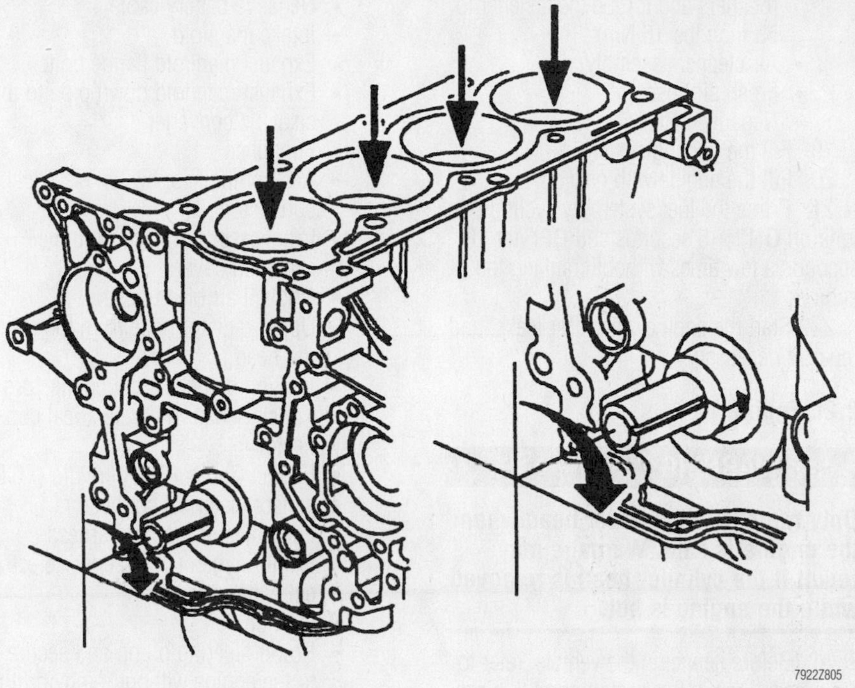

Rotate the crankshaft clockwise to 90 degrees past TDC (the crankshaft sprocket timing mark will be at 3 o'clock) to prevent valve damage during assembly

cylinder liners are flush or do not deviate more than 0.0005 in. (0.013mm).

10. Be sure the crankshaft is still 90 degrees past TDC and that the camshaft(s) are properly positioned with the dowel pin(s) at the 12 o'clock position to prevent valve damage.

11. Install the cylinder head gasket and carefully guide the head into place over the dowels.

12. If the head bolts and/or the block were replaced, install the bolts and tighten in sequence to 48 ft. lbs. (65 Nm) to insure proper clamp load; then, remove the bolts.

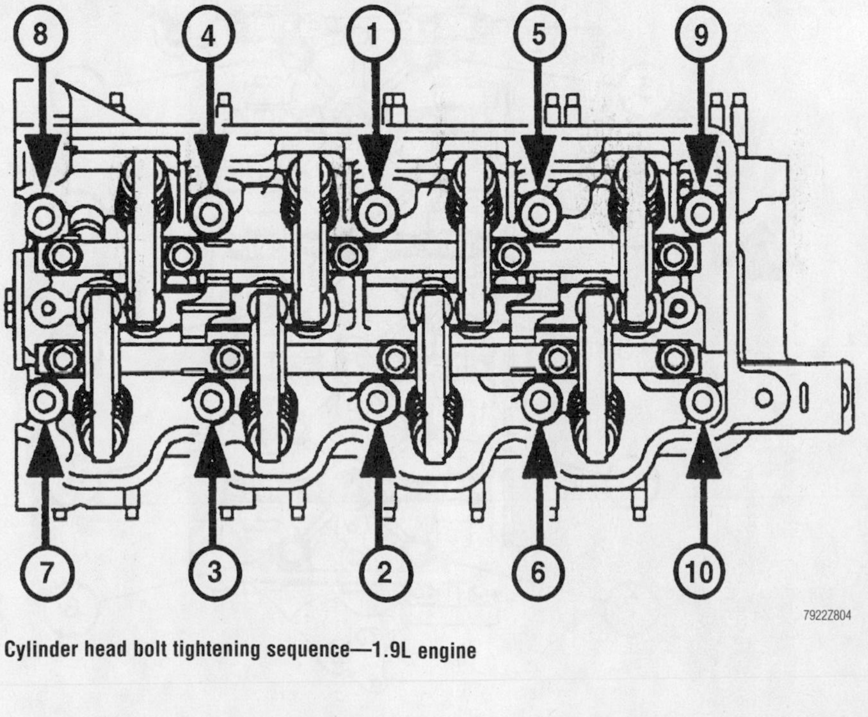

Cylinder head bolt tightening sequence—1.9L engine

13. Coat the cylinder head bolts with clean engine oil and thread the bolts by hand until finger-tight.

14. Install the cylinder head bolts, in sequence, as follows:

a. Tighten the bolts to 22 ft. lbs. (30 Nm).

b. Tighten the bolts to 33 ft. lbs. (45 Nm) for SOHC engines or to 37 ft. lbs. (50 Nm) for DOHC engines.

c. Tighten each bolt an additional 90 degree turn.

15. Install the timing chain, sprockets, guides and tensioners, using the following procedure :

a. Verify that the crankshaft is positioned 90 degrees clockwise past TDC. The crankshaft key way sprocket timing mark must be aligned with the cylinder block main bearing cap split line to prevent piston and valve damage.

b. If required, rotate the camshaft up to No. 1 TDC position.

c. Rotate the crankshaft 90 degrees counterclockwise to No. 1 TDC position. The crankshaft sprocket timing mark must align with the cylinder block timing mark.

16. Install or connect the following:

- Timing chain over the camshaft sprocket and under the crankshaft sprocket. Slide the camshaft sprocket onto the camshaft with the letters **FRT**facing away from the cylinder head
- Camshaft sprocket timing pin
- Camshaft washer and bolt and torque the bolt to 74 ft. lbs. (100 Nm)
- Fixed chain guide and torque the fastener to 19 ft. lbs. (26 Nm). The timing chain should be snug against the fixed guide.
- Pivoting chain guide making certain there is clearance between the block and head and torque the bolts to 19 ft. lbs. (26 Nm)
- Timing chain tensioner and torque the bolts to 168 inch lbs. (19 Nm)
- Crankshaft timing gear retainer tool to align the gerotor oil pump to the front cover
- Front cover and torque the perimeter bolts to 22 ft. lbs. (30 Nm) and the lower center bolt to 89 inch lbs. (10 Nm)
- Oil pan and torque the bolts to 80 inch lbs. (9 Nm)
- Water pump pulley and torque the

bolts to 19 ft. lbs. (25 Nm) (if removed)

17. Remove the crankshaft timing gear retaining tool.

18. Apply RTV across the cylinder head and front cover T-joints and install a new rocker cover gasket

- Crankshaft pulley and torque the bolt to 159 ft. lbs. (215 Nm)
- Rocker cover and torque the bolts to 22 ft. lbs. (30 Nm)
- Drive belt tensioner and torque the fastener to 22 ft. lbs. (30 Nm)
- Drive belt idler pulley and torque the fastener to 20 ft. lbs. (27 Nm)
- Accelerator linkage bracket and the throttle cable. Make certain that the cable is routed properly and is not binding and torque the bolts to 18 ft. lbs. (25 Nm).
- Alternator and torque the bolts to 24 ft. lbs. (32 Nm) (if removed)
- Power steering pump and torque the bolts to 22 ft. lbs. (30 Nm)
- A/C compressor and torque the rear bracket bolts to 18 ft. lbs. (25 Nm) and the front bracket bolts to 35 ft. lbs. (47 Nm)
- Accessory drive belt
- Upper mount and torque the bracket nuts to 37 ft. lbs. (50 Nm) and then torque the mount to cover bolts to 37 ft. lbs. (50 Nm)
- Wheel well splash shield
- Wheel
- EVAP canister purge hose
- Throttle body vacuum harness hose
- PCV hose
- Brake booster vacuum hose
- Upper radiator hose
- Heater hose
- Ground wire electrical connectors
- O_2S sensor electrical connector
- Fuel injector electrical connectors
- IAC valve electrical connector
- EGR valve electrical connector
- TPS sensor electrical connector
- EVAP canister purge valve connectors
- MAP sensor connectors
- Alternator electrical connectors
- Starter motor electrical connectors
- A/C compressor electrical connectors
- Intake manifold support bracket and torque the bolts to 22 ft. lbs. (30 Nm)
- Deaeration line and torque the cylinder head nut to 80 inch lbs. (9 Nm) and the rear lift bracket nut to 97 inch lbs. (11 Nm)
- Fuel feed and return lines with new retainers and torque the fastener to 53 inch lbs. (6 Nm)
- Air cleaner assembly
- Fresh air hose
- Negative battery cable

19. Fill the cooling system.

20. Fill the engine with new oil.

21. Prime the fuel system by cycling the ignition **ON** for 5 seconds and **OFF** for 10 seconds a few times without cranking the engine.

22. Start the engine, check for leaks, and repair if necessary.

2.2L Engine

⁂ WARNING

Only remove the cylinder head when the engine is cold. Warpage may result if the cylinder head is removed while the engine is hot.

1. Before servicing the vehicle, refer to the precautions in the beginning of this section.
2. Drain the cooling system.
3. Drain the engine oil.
4. Properly relieve the fuel system pressure.
5. Remove or disconnect the following:

- Negative battery cable
- Intake manifold
- Exhaust manifold flange bolts
- Exhaust manifold down pipe to the catalytic converter
- Fuel lines
- A/C compressor switch electrical connector
- Crankcase vent hose from the camshaft cover
- Coolant air bleed hose
- Upper radiator hose from the cylinder head
- Ignition coil electrical connectors
- Knock Sensor (KS) electrical connector
- Engine Coolant Temperature (ECT) sensor electrical connector
- Fuel injector jumper harness
- Front Oxygen (O_2S) sensor electrical connector
- Ignition coil
- Power steering pump and secure it to the engine without removing the lines
- Camshaft cover ground strap
- Camshaft cover
- Upper timing chain guide
- Upper timing chain tensioner
- Camshaft sprocket bolts

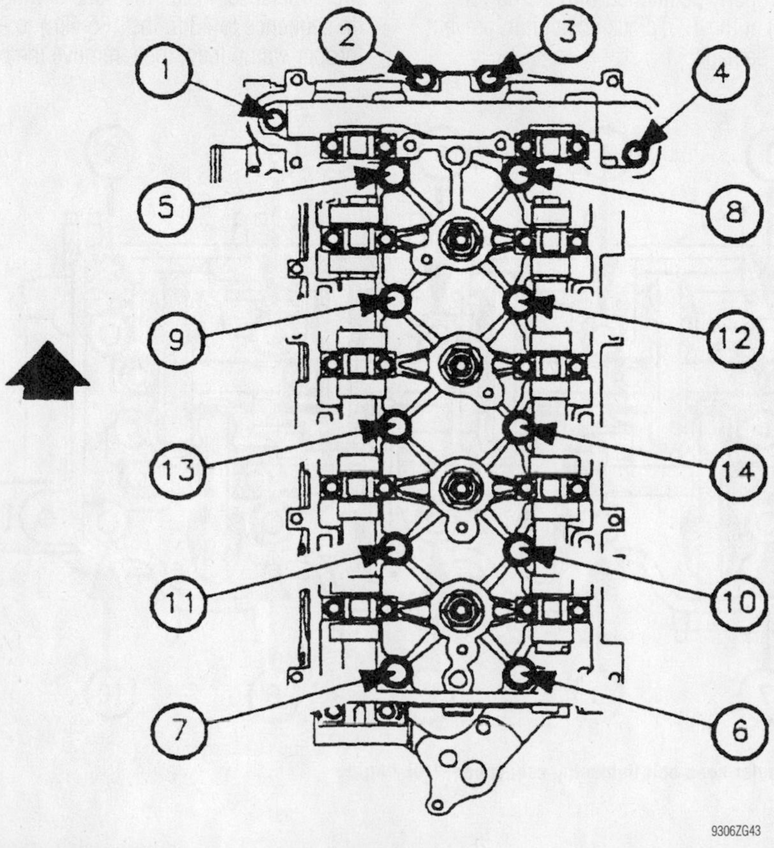

Cylinder head bolt tightening sequence—2.2L engine

- Camshaft sprockets
- Plug to gain access to the fixed timing chain guide
- Fixed timing chain guide upper bolt only
- Cylinder head bolts using the proper sequence
- Cylinder head

➡The manufacturer recommends to perform the remaining steps to complete this procedure. It may be not be necessary to perform the remaining steps if the timing chain can be supported without slipping or moving on the crankshaft pulley.

- Right front wheel and splash shield
- Drive belt and tensioner
- A/C suction line retaining clips and move the line away to allow clearance to the crankshaft pulley bolt

6. Position the crankshaft 60 degrees Before Top Dead Center (BTDC) and install a crankshaft pulley holder tool.
   - Crankshaft pulley
   - Front cover bolts except for the one blocked by the engine mount bracket
   - Loosen water pump bolt but do not remove it
7. Support the engine and remove the right side engine mount.
   - Engine mount bracket
   - Upper front cover bolt which was blocked by the mount bracket
   - Water pump bolt
   - Front cover
   - Adjustable timing chain guide
   - Fixed timing chain guide lower bolt

To install:

➡Set the crankshaft to 60 degrees Before Top Dead Center (BTDC) or after Top Dead Center (TDC) to prevent contact between the pistons and valves.

8. Install or connect the following:
   - New cylinder head gasket with the side imprinted **OBEN**facing up
   - Cylinder head and align it on the dowels
   - New cylinder head bolts and torque them in sequence to 22 ft. lbs. (30 Nm) plus 155 degrees
   - Front 4 cylinder head bolts coated with Loctite® and torque them to 18 ft. lbs. (22 Nm)
9. Position the exhaust camshaft with the offset slot in the 2 o'clock position and the intake camshaft with the offset slot in the 11 o'clock position.
   - Timing chain around the intake camshaft sprocket with the copper link aligned with the **INT**diamond timing mark
   - Sprocket to the camshaft and align it with the offset slot
   - New camshaft sprocket bolt but do not tighten
   - Timing chain around the crankshaft sprocket and align the silver link to the timing mark
   - Adjustable timing chain guide through the opening on top of the cylinder head and torque the bolt to 89 inch lbs. (10 Nm)
   - Timing chain around the exhaust camshaft sprocket with the silver link aligned with the offset slot. Install but do not tighten a new sprocket bolt

⁂ CAUTION

Make certain that all timing marks and colored links are aligned properly before proceeding to the next step. If the timing chain is not aligned properly, severe engine damage may occur.

10. Torque the intake and exhaust camshaft bolts to 63 ft. lbs. (85 Nm) plus a 30 degree turn.
11. Install or connect the following:
    - Fixed timing guide and torque the bolt to 89 inch lbs. (10 Nm)
    - Fixed timing guide bolt access plug after applying Loctite® to the threads and torque it to 30 ft. lbs. (40 Nm)
    - Timing chain tensioner and torque the bolts 44 ft. lbs. (60 Nm)
12. Tap the top of the timing chain between the camshaft sprockets to engage the tensioner.
    - Upper timing chain guide and torque the bolts to 89 inch lbs. (10 Nm)
    - Front cover with a new gasket and torque the bolts to 18 ft. lbs. (25 Nm)
    - Water pump bolt and torque it to 18 ft. lbs. (25 Nm)
    - Drive belt tensioner and torque the bolts to 30 ft. lbs. (40 Nm)
    - Engine mount bracket and torque the bolts to 66 ft. lbs. (90 Nm)
    - Engine mount to the side rail and torque the fasteners to 41 ft. lbs. (55 Nm)
    - Engine mount to the bracket and torque the bolts to 41 ft. lbs. (55 Nm)
    - Crankshaft damper and torque the bolt to 74 ft. lbs. (100 Nm)
    - Accessory drive belt
    - A/C line clamps
    - Right wheel splash shield
    - Right front wheel
    - Camshaft cover and torque the bolts to 89 inch lbs. (10 Nm)
    - Camshaft cover ground strap and torque the bolt to 89 inch lbs. (10 Nm)
    - Power steering pump and torque the bolts to 18 ft. lbs. (25 Nm)
    - Ignition coil/module assembly and torque the bolts to 71 inch lbs. (8 Nm)
    - Intake manifold with a new gasket and torque the bolts to 89 inch lbs. (10 Nm)
    - Wiring harness to the intake manifold bracket
    - Engine wiring harness and torque the fasteners to 89 inch lbs. (10 Nm)
    - All component electrical connections
    - Upper radiator hose
    - Crankcase ventilation hose
    - Fuel pressure regulator hose
    - Brake booster vacuum hose
    - EVAP purge hose
    - Fuel lines
    - Fuel lines to the bracket and torque the bracket bolt to 89 inch lbs. (10 Nm)
    - Exhaust down pipe to the exhaust manifold and torque the bolts to 25 ft. lbs. (30 Nm)
    - Exhaust pipe to the resonator and torque the bolts to 15 ft. lbs. (20 Nm)

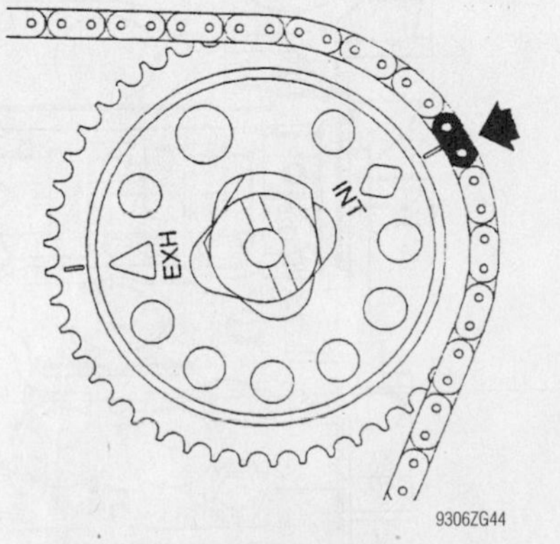

Align the copper link on the timing chain with the INT diamond timing mark

- Exhaust manifold heat shield and torque the bolts to 18 ft. lbs. (25 Nm)
- Air cleaner assembly
- Negative battery cable

13. Fill the engine with clean oil.
14. Fill the cooling system.
15. Prime the fuel system by cycling the ignition **ON** for 5 seconds and **OFF** for 10 seconds a few times without cranking the engine.
16. Start the engine, check for leaks, and repair if necessary.

3.0L Engine

FRONT

⁂ WARNING

Only remove the cylinder head when the engine is cold. Warpage may result if the cylinder head is removed while the engine is hot.

1. Before servicing the vehicle, refer to the precautions in the beginning of this section.
2. Drain the cooling system.
3. Drain the engine oil.
4. Properly relieve the fuel system pressure.
5. Remove or disconnect the following:
   - Negative battery cable
   - Air cleaner assembly
   - Intake plenum
   - Intake manifold
   - Intake manifold spacer
   - Coolant bridge
   - Upper radiator hose from the coolant extension housing
6. Properly support the powertrain assembly.
   - Front transmission mount through bolt
   - Extension housing over the coolant module
   - Oil level indicator tube
   - Front camshaft cover
   - Ground wires from the lift bracket
   - Oxygen (O_2S) sensor electrical connector
   - Down pipe from the exhaust manifold
   - Front timing belt cover
   - Timing belt
   - Timing belt tensioner bracket
   - Rear timing belt cover
   - Camshaft position (CMP) sensor electrical connector
   - Exhaust camshaft
   - Loosen the cylinder head bolts in stages as shown

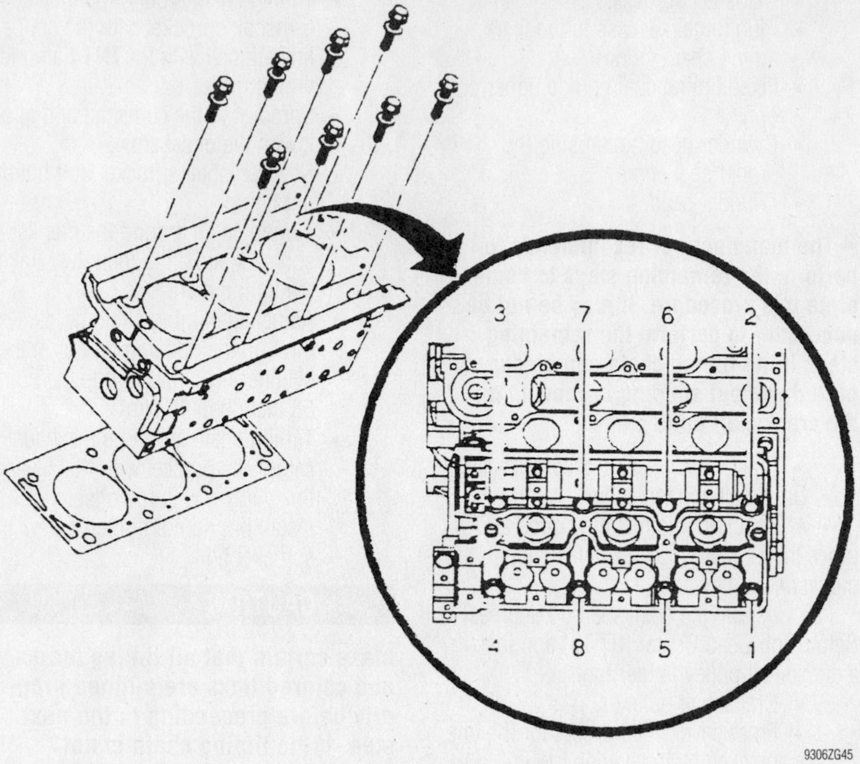

Cylinder head bolt removal for the front cylinder head–3.0L engine

- Cylinder head
- Exhaust manifold from the cylinder head (if necessary)

To install:

7. Install or connect the following:
   - Exhaust manifold with a new gasket and torque the bolts to 15 ft. lbs. (20 Nm)
   - New cylinder head gasket with the part number imprint facing the top of the engine
   - Cylinder head

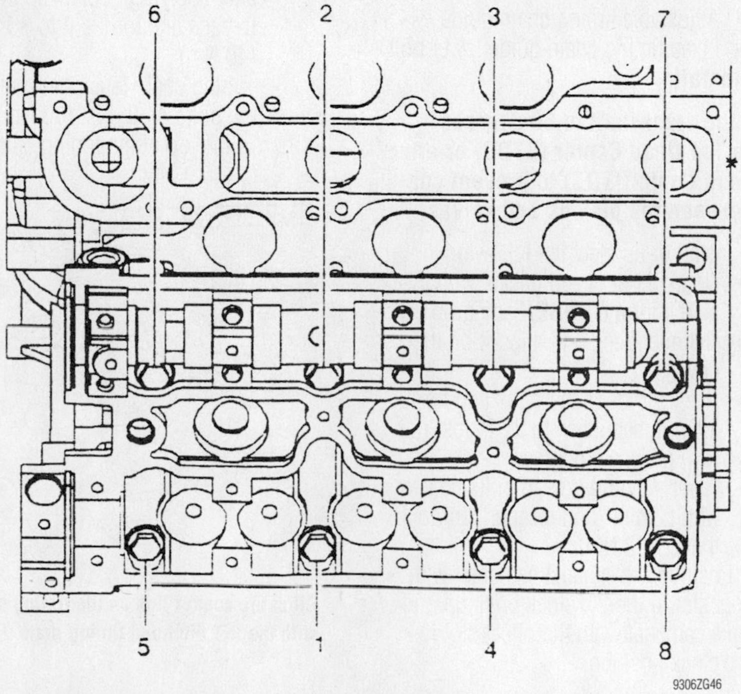

Front cylinder head bolt tightening sequence—3.0L engine

8. Torque the new cylinder head bolts, in sequence, as follows:
   a. 18 ft. lbs. (25 Nm).
   b. 90 degree turn.
   c. 90 degree turn.
   d. 90 degree turn.
   e. 15 degree turn.
9. Install or connect the following:
   - Coolant pipe with new sealing rings
   - Engine lift bracket bolt and torque it to 15 ft. lbs. (20 Nm)
   - Upper radiator hose to the coolant pipe
   - Front transmission mount through bolt
   - Exhaust camshaft
   - CMP sensor electrical connector
   - Rear timing belt cover and torque the bolts to 71 inch lbs. (8 Nm)
   - Camshaft gears
   - Timing belt tensioner bracket
   - Timing belt
   - Front timing belt cover and torque the bolts to 71 inch lbs. (8 Nm)
   - Down pipe to the exhaust manifold
   - O_2S electrical connector
   - Front camshaft cover and torque the bolts to 71 inch lbs. (8 Nm)
   - Coolant bridge and torque the bolt to 22 ft. lbs. (33 Nm)
   - Intake manifold spacer and torque the bolts in a spiral direction from the inside and working out to 15 ft. lbs. (20 Nm)
   - Intake manifold and torque the bolts to 15 ft. lbs. (20 Nm)
   - Intake plenum and torque the bolts to 71 inch lbs. (8 Nm)
   - Air cleaner assembly
   - Negative battery cable
10. Fill the engine with clean oil.
11. Fill the cooling system.
12. Prime the fuel system by cycling the ignition **ON** for 5 seconds and **OFF** for 10 seconds a few times without cranking the engine.
13. Start the engine, check for leaks, and repair if necessary.

REAR

WARNING

Only remove the cylinder head when the engine is cold. Warpage may result if the cylinder head is removed while the engine is hot.

1. Before servicing the vehicle, refer to the precautions in the beginning of this section.
2. Drain the cooling system.
3. Drain the engine oil.
4. Properly relieve the fuel system pressure.
5. Remove or disconnect the following:
   - Negative battery cable
   - Air cleaner assembly
   - Intake plenum
   - Intake manifold
   - Intake manifold spacer
   - Coolant bridge
   - Engine ventilation chamber
   - Rear camshaft cover
   - Front timing belt cover
   - Timing belt
   - Timing belt tensioner bracket
   - Camshaft gears
   - Rear timing belt cover
   - Exhaust manifold pipe heat shield
   - Front exhaust manifold pipe-to-rear exhaust manifold pipe fasteners
   - Rear exhaust manifold pipe nuts, pull the manifold pipe down and discard the gasket
   - Exhaust Gas Recirculation (EGR)-to-exhaust manifold pipe
   - Exhaust camshaft
   - Cylinder head bolts
   - Cylinder head
   - Exhaust manifold from the cylinder head

To install:

6. Install or connect the following:
   - Exhaust manifold with a new gasket and torque the bolts to 15 ft. lbs. (20 Nm), if removed
   - New cylinder head gasket with the part number imprint facing the top of the engine
   - Cylinder head
7. Torque the new cylinder head bolts, in sequence, as follows:
   a. 18 ft. lbs. (25 Nm).
   b. 90 degree turn.
   c. 90 degree turn.
   d. 90 degree turn.
   e. 15 degree turn.
8. Install or connect the following:
   - Exhaust camshaft
   - Rear timing belt cover and torque the bolts to 71 inch lbs. (8 Nm)
   - Camshaft gears and torque the bolts to 37 ft. lbs. (50 Nm) plus a 60 degree turn plus another 15 degree turn
   - Timing belt tensioner bracket and torque the bolts to 30 ft. lbs. 940 Nm)
   - Timing belt
   - Front timing belt cover and torque the bolts to 71 inch lbs. (8 Nm)

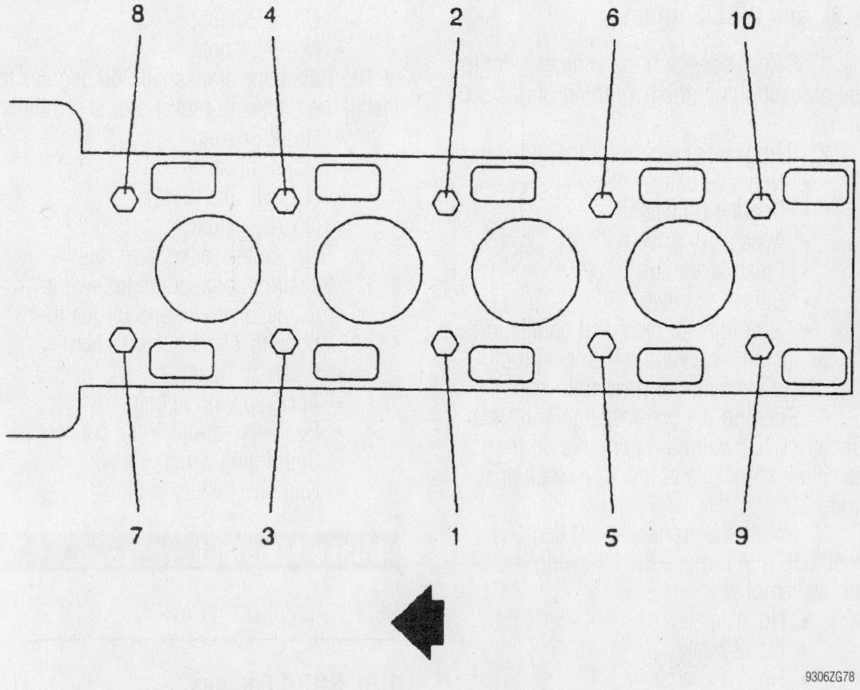

Rear cylinder head bolt tightening sequence—3.0L engine

- Rear camshaft cover and torque the bolts to71 inch lbs. (8 Nm)
- Exhaust manifold pipe to the manifold
- Exhaust manifold pipe heat shield
- EGR-to-exhaust manifold pipe and torque the nut to 18 ft. lbs. (25 Nm)
- Coolant bridge and torque the bolt to 22 ft. lbs. (33 Nm)
- Engine ventilation chamber and torque the bolts to 71 inch lbs. (8 Nm)
- Intake manifold spacer and torque the bolts to 15 ft. lbs. (20 Nm)
- Intake manifold and torque the bolts to 15 ft. lbs. (20 Nm)
- Intake plenum and torque the bolts to 71 inch lbs. (8 Nm)
- Air cleaner assembly
- Negative battery cable

9. Fill the engine with clean oil.
10. Fill the cooling system.
11. Prime the fuel system by cycling the ignition **ON** for 5 seconds and **OFF** for 10 seconds a few times without cranking the engine.
12. Start the engine, check for leaks, and repair if necessary.

Rocker Arms/Shafts

REMOVAL & INSTALLATION

1.9L and 2.2L Engines

1. Before servicing the vehicle, refer to the precautions in the beginning of this section.
2. Remove or disconnect the following:
   - Negative battery cable
   - Spark plug wires
   - Accessory drive belt
   - Fresh air hose
   - Camshaft cover
   - Fuel injector electrical connectors
3. Install a rocker arm removal tool J43223 to the cylinder head rail.
4. Position the crankshaft at Top Dead Center (TDC) with the pip marks on the camshaft sprockets at the 12 o'clock position.
5. Rotate the crankshaft 90 degrees past TDC and remove the following rocker arm assemblies:
   - No. 1 Intake
   - No. 2 Intake
   - No. 1 Exhaust
   - No. 3 Exhaust
6. Rotate the crankshaft 360 degrees to remove the following rocker arm assemblies:

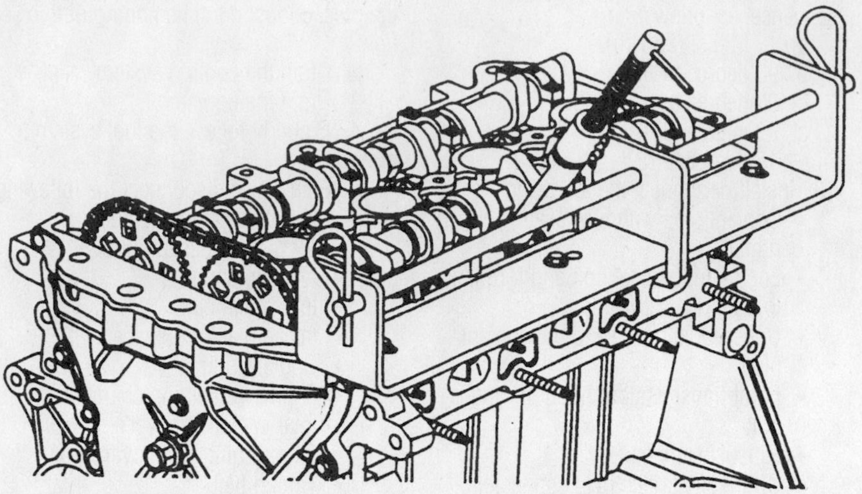

Exploded view of the valve springs compressed–1.9L DOHC engines

   - No. 2 Exhaust
   - No. 4 Intake
   - No. 3 Intake
   - No. 4 Exhaust
7. Compress the valve springs.
8. Remove the rocker arm and/or valve lifters.

To install:

9. Rotate the crankshaft 90 degrees past TDC and install the following rocker arm assemblies:
   - No. 1 Intake
   - No. 2 Intake
   - No. 1 Exhaust
   - No. 3 Exhaust
10. Rotate the crankshaft 360 degrees to install the following rocker arm assemblies:
    - No. 2 Exhaust
    - No. 4 Intake
    - No. 3 Intake
    - No. 4 Exhaust
11. Remove the rocker arm tool.
12. Install or connect the following:
    - Camshaft cover and torque the bolts to 71 inch lbs. (8 Nm)
    - Fresh air hose
    - Accessory drive belt
    - Fuel injector electrical connectors
    - Spark plug wires
    - Negative battery cable

Intake Manifold

REMOVAL & INSTALLATION

1.9L SOHC Engines

1. Before servicing the vehicle, refer to the precautions in the beginning of this section.
2. Properly relieve the fuel system pressure.
3. Drain the cooling system.
4. Remove or disconnect the following:
   - Negative battery cable
   - Air cleaner assembly
   - Positive Crankcase Ventilation (PCV) valve hose
   - Fuel line from the rail
   - Throttle cable from the throttle body
   - Throttle cable bracket nuts
   - Fuel injector electrical connectors
   - Throttle Position (TPS) sensor electrical connectors
   - Idle Air Control (IAC) valve electrical connectors
   - Manifold Absolute Pressure (MAP) sensor electrical connectors
   - Exhaust Gas Recirculation (EGR) valve electrical connectors
   - Heater hose from the intake manifold outlet
5. Position the wiring harness over the brake master cylinder.
   - Intake manifold support bracket bolt
   - Accessory drive belt
   - Power steering pump without removing the lines
   - Upper intake manifold nuts
   - Evaporative Emissions (EVAP) canister purge valve solenoid vacuum hose
   - Brake booster vacuum hose
   - Lower intake manifold nuts
   - Intake manifold

To install:

6. Install or connect the following:
   - Intake manifold with a new gasket and torque the nuts in sequence to 22 ft. lbs. (30 Nm)

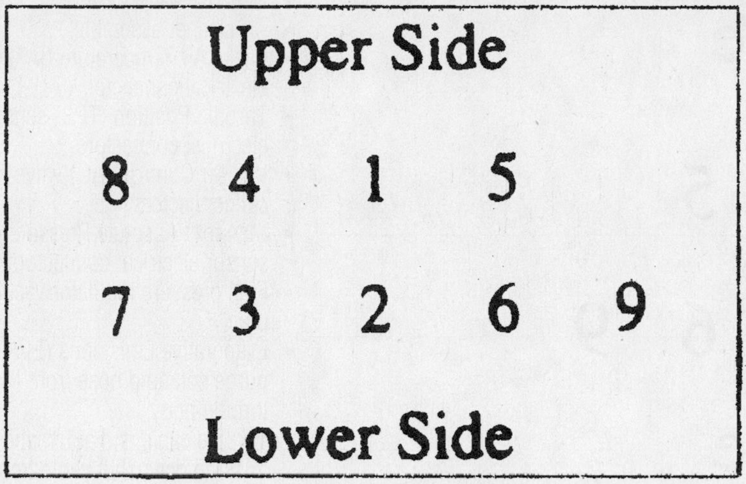

Intake manifold bolt torque sequence—1.9L SOHC engine

- Power steering pump and torque the fasteners to 27 ft. lbs. (35 Nm)
- Accessory drive belt
- Heater hose
- Intake manifold support bracket bolt and torque it to 22 ft. lbs. (30 Nm)
- Fuel supply and return lines
- Fuel line(s) in the retaining bracket and torque the mounting screw to 36 inch lbs. (4 Nm)
- Throttle cable to the throttle body and accelerator cable bracket and torque the bolts to 22 ft. lbs. (30 Nm)
- IAC valve electrical connector
- TPS sensor electrical connector
- MAP sensor electrical connector
- Fuel injector electrical connectors
- EGR valve electrical connector
- EVAP purge solenoid vacuum hose
- PCV valve hose
- Air cleaner assembly
- Negative battery cable

7. Fill the cooling system.
8. Prime the fuel system by cycling the ignition **ON** for 5 seconds and **OFF** for 10 seconds a few times without cranking the engine.
9. Start the engine, check for leaks, and repair if necessary.

1.9L DOHC Engines

1. Before servicing the vehicle, refer to the precautions in the beginning of this section.
2. Properly relieve the fuel system pressure.
3. Drain the cooling system.
4. Remove or disconnect the following:
   - Negative battery cable
   - Air cleaner fresh air hose
   - Positive Crankcase Ventilation (PCV) valve hose
   - Fuel line from the rail
   - Throttle cable from the throttle body
   - Throttle cable bracket nuts
   - Fuel injector electrical connectors
   - Throttle Position (TPS) sensor electrical connector
   - Idle Air Control (IAC) valve electrical connector
   - Manifold Absolute Pressure (MAP) sensor electrical connector
   - Exhaust Gas Recirculation (EGR) valve electrical connector
   - Heater hose from the intake manifold outlet
   - Power steering support brace
   - Fuel rail
   - EGR pipe from the cylinder head
   - Brake booster vacuum hose
   - Accessory drive belt
   - Power steering pump without removing the lines
   - Upper axis torque mount nuts and midrail bracket nuts. Allow the powertrain assembly to rest on a support fixture.
   - Resonator and air cleaner box
   - Intake Air Temperature (IAT) sensor electrical connector
   - Transmission strut-to-midrail bolt and rotate the engine
   - Intake manifold

To install:

5. Install or connect the following:
   - Intake manifold with a new gasket and torque the nuts in sequence to 22 ft. lbs. (30 Nm) for 1998–99 1.9L DOHC engines or to 115 inch lbs. (13 Nm) for 2000–01 1.9L DOHC engines
   - EGR pipe with a new gasket and torque the fasteners to 18 ft. lbs. (25 Nm)
   - Heater hose
   - Fuel rail and torque the bolts to 106 inch lbs. (12 Nm)
   - IAC valve electrical connector
   - TPS sensor electrical connector
   - MAP sensor electrical connector
   - Fuel injector electrical connector
   - EGR valve electrical connector

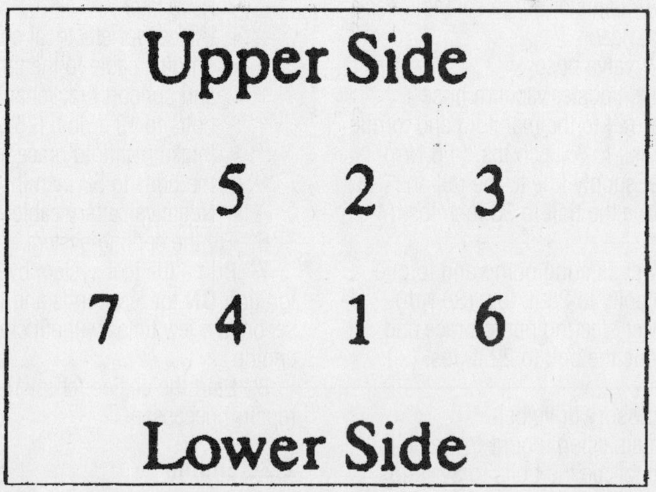

Intake manifold bolt torque sequence—1998 1.9L DOHC engine

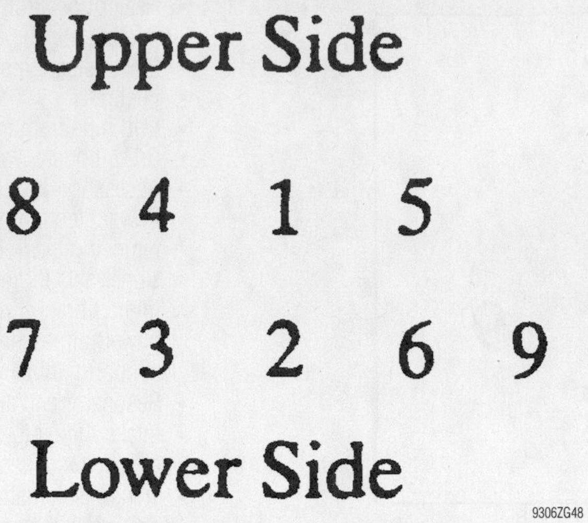

Intake manifold bolt torque sequence—1999 1.9L DOHC engine

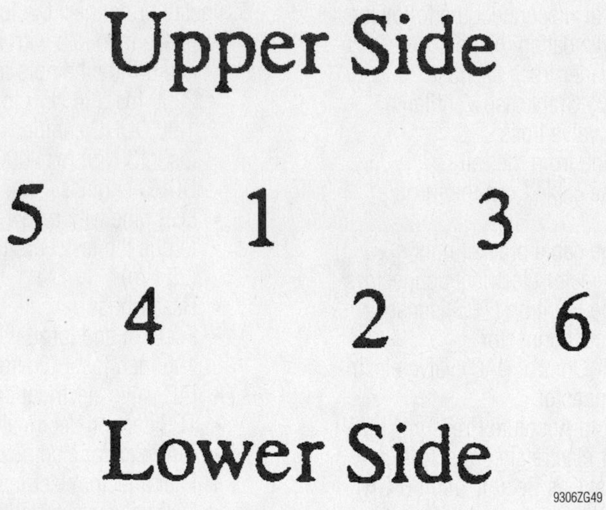

Intake manifold bolt torque sequence—2000–01 1.9L DOHC engine

- EVAP canister purge solenoid vacuum hose
- PCV valve hose
- Brake booster vacuum hose
- Fuel rail to the manifold and torque the nut to 36 inch lbs. (4.5 Nm)
- Fuel supply line to the rail and torque the bolt to 36 inch lbs. (4 Nm)
- Power steering pump and torque the bolts to 28 ft. lbs. (38 Nm)
- Power steering pump brace and torque the bolt to 22 ft. lbs. (30 Nm)
- Accessory drive belt
- Transmission mount-to-frame rail and torque the bolts to 52 ft. lbs. (70 Nm)
- Engine mount-to-midrail nuts and torque them to 37 ft. lbs. (50 Nm)
- Air cleaner box and torque the bolts to 89 inch lbs. (10 Nm)
- Resonator
- IAT sensor electrical connector
- Throttle cable to the throttle body and support bracket and torque the bolts to 19 ft. lbs. (25 Nm)
- Intake manifold brace and torque the bolts to 58 inch lbs. (6.5 Nm)
- Negative battery cable

6. Fill the cooling system.

7. Prime the fuel system by cycling the ignition **ON** for 5 seconds and **OFF** for 10 seconds a few times without cranking the engine.

8. Start the engine, check for leaks, and repair if necessary.

2.2L Engine

1. Before servicing the vehicle, refer to the precautions in the beginning of this section.

2. Remove or disconnect the following:
   - Negative battery cable
   - Air cleaner assembly
   - Intake Air Temperature (IAT) sensor electrical connector
   - Throttle Position (TPS) sensor electrical connectors
   - Idle Air Control (IAC) valve electrical connectors
   - Manifold Absolute Pressure (MAP) sensor electrical connectors
   - Fuel pressure regulator vacuum pipe
   - Evaporative Emissions (EVAP) purge solenoid hose from the throttle body
   - Throttle cable and automatic transmission downshift cable from the throttle body
   - Throttle cable bracket nuts
   - Brake booster vacuum hose
   - Oil level indicator tube
   - Throttle body
   - Intake manifold

To install:

3. Install or connect the following:
   - Intake manifold with a new gasket and torque the nuts to 89 inch lbs. (10 Nm) starting from the center and working outward
   - Throttle body to the intake manifold and torque the bolts to 19 ft. lbs. (25 Nm)
   - Throttle cable bracket to the throttle body studs
   - Oil level indicator tube and torque the fastener to 89 inch lbs. (10 Nm)
   - Brake booster vacuum pipe
   - Throttle cable to the throttle body and install the support bracket. Torque the bolts to 19 ft. lbs. (25 Nm)
   - EVAP canister purge solenoid vacuum hose
   - Fuel pressure regulator vacuum pipe to the throttle body
   - IAC valve electrical connector
   - TPS sensor electrical connector
   - MAP sensor electrical connector
   - Air inlet hose to the throttle body
   - Crankcase vent hose to the camshaft cover
   - IAT sensor electrical connector
   - Air cleaner assembly
   - Negative battery cable

4. Start the engine, check for leaks, and repair if necessary.

3.0L Engine

1. Before servicing the vehicle, refer to the precautions in the beginning of this section.

2. Properly relieve the fuel system pressure.
3. Remove or disconnect the following:
   - Negative battery cable
   - Air cleaner assembly
   - Throttle body electrical connector
   - Fuel pressure regulator vacuum hose
   - Both intake runners
   - Intake plenum
   - Fuel supply and return lines
   - Fuel injector electrical connectors
   - Fuel rail
   - Intake manifold
   - Intake manifold spacer and O-ring gaskets

To install:

4. Install or connect the following:
   - Intake manifold spacer with new seals and torque the bolts to 15 ft. lbs. (20 Nm) starting from the center and working outward
   - Intake manifold with a new gasket and torque the bolts to 15 ft. lbs. (20 Nm)
   - Fuel rail
   - Fuel injector electrical connectors
   - Fuel supply and return hoses and torque the fastener to 11 ft. lbs. (15 Nm)
   - Intake plenum and torque the bolts to 71 inch lbs. (8 Nm)
   - Intake manifold runners and torque the bolts to 71 inch lbs. (8 Nm)
   - Negative battery cable
5. Prime the fuel system by cycling the ignition **ON** for 5 seconds and **OFF** for 10 seconds a few times without cranking the engine.
6. Start the engine, check for leaks, and repair if necessary.

Exhaust Manifold

REMOVAL & INSTALLATION

1.9L Engines

1. Before servicing the vehicle, refer to the precautions in the beginning of this section.
2. Remove or disconnect the following:
   - Negative battery cable
   - Front exhaust pipe-to-engine support bracket mounting fasteners
   - Pipe-to-manifold nuts and lower the pipe
   - A/C compressor and bracket from the engine without removing the lines
   - Oxygen (O_2S) sensor electrical connector
   - Exhaust manifold

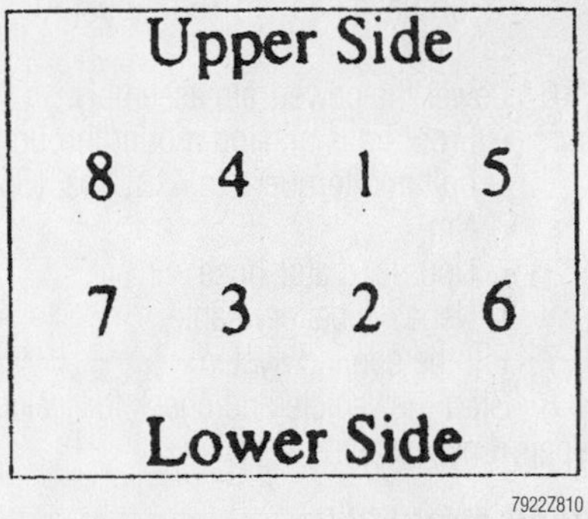

Exhaust manifold bolt torque sequence—1.9L SOHC engines

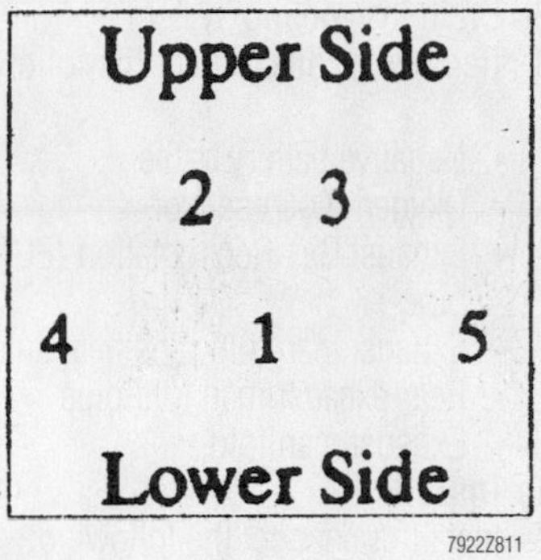

Exhaust manifold bolt torque sequence—1.9L DOHC engines

To install:

3. Install or connect the following:
   - Exhaust manifold with a new gasket and torque the nuts in sequence to 16 ft. lbs. (22 Nm) for the SOHC engine or to 150 inch lbs. (17 Nm) for the DOHC engine
   - O_2S sensor electrical connector
   - A/C compressor and bracket and torque the front bracket-to-compressor to 40 ft. lbs. (54 Nm) and all remaining fasteners to19 ft. lbs. (25 Nm)
   - Pipe-to-manifold with a new gasket and torque the fasteners in a crosswise pattern to 23 ft. lbs. (31 Nm)
   - Exhaust pipe-to-engine support bracket and torque the mounting fasteners to 23 ft. lbs. (31 Nm)
   - Negative battery cable
4. Start the vehicle and check for leaks, repair if necessary.

2.2L Engine

1. Before servicing the vehicle, refer to the precautions in the beginning of this section.
2. Remove or disconnect the following:
   - Negative battery cable
   - Exhaust pipe from the manifold
   - Exhaust manifold heat shield
   - Oxygen (O_2S) sensor from the manifold
   - Exhaust manifold

To install:

3. Install or connect the following:
   - Exhaust manifold with a new gasket and torque the bolts, starting from the center and working outward, to 13 ft. lbs. (18 Nm)

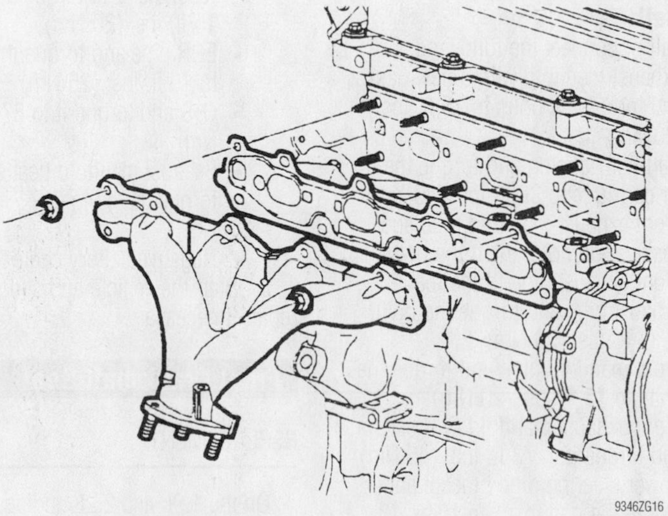

Remove the exhaust manifold and gasket.

- O_2S sensor and torque it to 33 ft. lbs. (45 Nm)
- Exhaust manifold heat shield and torque the bolts to 18 ft. lbs. (25 Nm)
- Exhaust pipe to the manifold with a new gasket and torque the nuts to 22 ft. lbs. (30 Nm)
- Negative battery cable

4. Start the vehicle and check for leaks, repair if necessary.

3.0L Engine

FRONT MANIFOLD

1. Before servicing the vehicle, refer to the precautions in the beginning of this section.
2. Drain the cooling system.
3. Remove or disconnect the following:
   - Negative battery cable
   - Exhaust manifold Oxygen (O_2S) sensor
   - Upper radiator hose from the coolant extension housing
4. Install and engine support fixture.
   - Front transmission through bolt and raise the powertrain assembly
   - Oil level indicator tube and coolant extension tube bolt
   - Coolant extension housing
   - Power steering pipe bracket bolt
   - Upper exhaust manifold nuts
   - Front exhaust manifold pipe from the manifold
   - Font exhaust manifold pipe from the rear exhaust manifold pipe
   - Oil filter housing
   - Lower exhaust manifold nuts
   - Exhaust manifold

To install:

5. Install or connect the following:
   - Exhaust manifold with a new gasket and torque the bolts to 15 ft. lbs. (20 Nm)
   - Oil filter housing and torque the filter cartridge to 33 ft. lbs. (45 Nm)
   - Front exhaust manifold pipe and gaskets, do not tighten the bolts
   - Front exhaust pipe to the rear exhaust pipe and torque the bolts to 15 ft. lbs. (20 Nm)
   - Front manifold pipe and torque the bolts to 15 ft. lbs. (20 Nm)
   - Front exhaust manifold O_2S sensor and torque it to 73 ft. lbs. (45 Nm)
   - Power steering pipe bracket and torque the bolt to 71 inch lbs. (8 Nm)
   - New O-rings to the coolant extension housing
   - Coolant extension housing and torque the bolt to 15 ft. lbs. (20 Nm)
6. Lower the powertrain assembly.
   - Front transmission mount through bolt and torque it to 41 ft. lbs. (55 Nm)
   - Upper radiator hose
   - Negative battery cable
7. Fill the cooling system.
8. Start the vehicle and check for leaks, repair if necessary.

REAR MANIFOLD

1. Before servicing the vehicle, refer to the precautions in the beginning of this section.
2. Drain the cooling system.
3. Remove or disconnect the following:
   - Negative battery cable
   - Oxygen (O_2S) sensor
   - Exhaust Gas Recirculation (EGR) pipe
   - Exhaust manifold pipe heat shield
   - Rear exhaust manifold pipe
   - Exhaust manifold

To install:

4. Install or connect the following:
   - Exhaust manifold with a new gasket and torque the bolts to 15 ft. lbs. (20 Nm)
   - Rear exhaust manifold pipe and torque the nuts to 25 ft. lbs. (30 Nm)
   - Front exhaust manifold pipe to the rear exhaust manifold pipe and torque the bolts to 15 ft. lbs. (20 Nm)
   - Rear exhaust manifold pipe to the resonator and torque the bolts to 15 ft. lbs. (20 Nm)
   - EGR pipe and torque the fasteners to 19 ft. lbs. (25 Nm)
   - O_2S and torque it to 37 ft. lbs. (50 Nm)
   - Exhaust manifold heat shield and torque the bolts to 71 inch lbs. (8 Nm)
   - Negative battery cable
5. Start the vehicle and check for leaks, repair if necessary.

Front Crankshaft Seal

REPLACEMENT

On the 1.9L and 2.2L engines the front crankshaft seal is located in the timing chain front cover. Refer to the timing chain procedure for information about removing the front cover and replacing the seal.

3.0L Engine

1. Before servicing the vehicle, refer to the precautions in the beginning of this section.
2. Remove or disconnect the following:
   - Negative battery cable
   - Timing belt
   - Crankshaft gear
3. Drill a small pilot hole into the steel ring of the seal.
4. Screw in a self taping screw.
5. Use pliers to pull out the oil seal.

To install:

6. Coat the lip of the new oil seal with engine oil.
7. Install the oil seal using a suitable seal installer.
8. Install the crankshaft gear and torque the bolt to 184 ft. lbs. (250 Nm) plus an additional 45 degrees then another 15 degrees.
9. Install the timing belt.

Camshaft and Lifters

REMOVAL & INSTALLATION

1.9L (SOHC) Engines

1. Before servicing the vehicle, refer to the precautions in the beginning of this section.
2. Remove or disconnect the following:
   - Both battery cables
   - Battery and tray
   - Timing chain front cover
   - Timing chain and camshaft sprocket
   - Rocker arm/shaft assemblies
   - Lifters and label or position them for assembly in their original locations
3. Drive the camshaft plug inward, then remove it from the cylinder head with a magnet.
4. Carefully pull the camshaft from the rear of the cylinder head through the oversized camshaft plug hole. Turn the camshaft back and forth slowly while withdrawing to help prevent journal or bearing damage.

To install:

5. Clean and inspect all parts prior to installation. Lubricate the camshaft and carefully insert it through the hole at the rear of the cylinder head.
6. Install or connect the following:
   - New rear cylinder head plug coated with Loctite® 242
   - Valve lifters into their original bores, or if the camshaft has been replaced, install new lifters

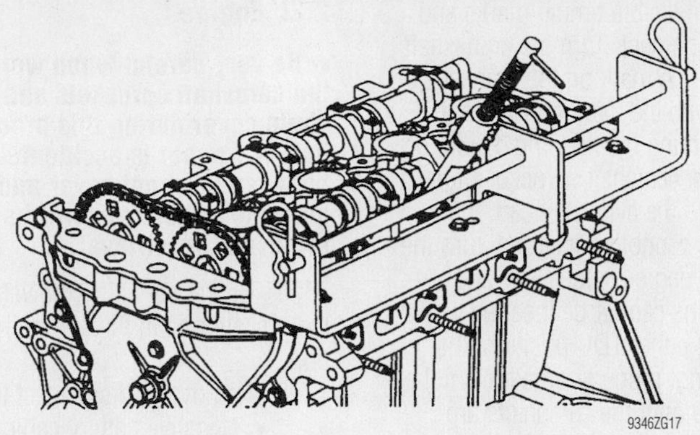

Remove the rocker arm/shaft assemblies

- Rocker arm/shaft assemblies
- Timing chain and camshaft sprocket
- Timing chain front cover
- Battery and tray and torque the battery hold-down nut and screw to 80 inch lbs. (9 Nm)
- Both battery cables

7. Start the engine and check for leaks, repair if necessary.

1.9L (DOHC) Engines

➡Be careful when working around the camshaft sprockets and timing chain cover during this procedure. If a bolt or washer is accidentally dropped between the front cover and engine assembly, the cover will have to be removed for retrieval.

1. Before servicing the vehicle, refer to the precautions in the beginning of this section.
2. Remove or disconnect the following:
   - Negative battery cable
   - Accessory drive belt
   - Spark plug wires
   - Positive Crankcase Ventilation (PCV) fresh air hose
   - Cam cover
3. Turn the crankshaft clockwise until the mark on the crankshaft pulley is in alignment with the pointer on the front cover and the No. 1 cylinder is at Top Dead Center (TDC) of the compression stroke. Both camshaft dowel pins will be at the 12 o'clock position and the timing pin holes will be aligned when the No. 1 cylinder is at TDC. If necessary, the right wheel and splash shield can be removed to help observe the timing marks.
4. Remove the camshaft sprocket retaining bolts. Use a 7/8 in. (21mm) open-end wrench to hold the camshaft from turning while removing the bolts.
5. Position a front angled support fixture in front of the camshaft sprockets.
6. Attach the camshaft sprocket adapters to the end of each camshaft using the pilot bolts, but do not tighten the bolts. The front angled support should come between the sprocket adapters and camshaft sprockets.
7. Remove or disconnect the following:
   - Upper timing chain guide
   - Both front camshaft bearing caps
8. Secure the support fixture using 7/8 in. bolts/blocks and align the 2 holes in each camshaft sprocket, adapter and the front support fixture. Install the 4 nuts, but do not tighten.

➡The steel blocks should be installed against the rearward side of the camshaft sprocket.

9. Tighten the sprocket pilot bolts to 19 ft. lbs. (25 Nm) while holding the camshafts from turning with an open end wrench.
10. Move each camshaft sprocket off the end of the camshaft by rocking the sprocket forward or by carefully prying between the end of the camshaft and the sprocket. Then, tighten the 4 nuts and bolts with blocks from the side of the support fixture to 19 ft. lbs. (25 Nm).
11. Install the 2 bolts retaining the support fixture to the engine front cover and tighten the bolts to 89 inch lbs. (10 Nm). Then, remove each camshaft sprocket pilot bolt while holding the camshafts with a wrench.
12. Carefully pry between the sprocket and the end of the camshaft to move the camshaft rearward. Pry only enough to

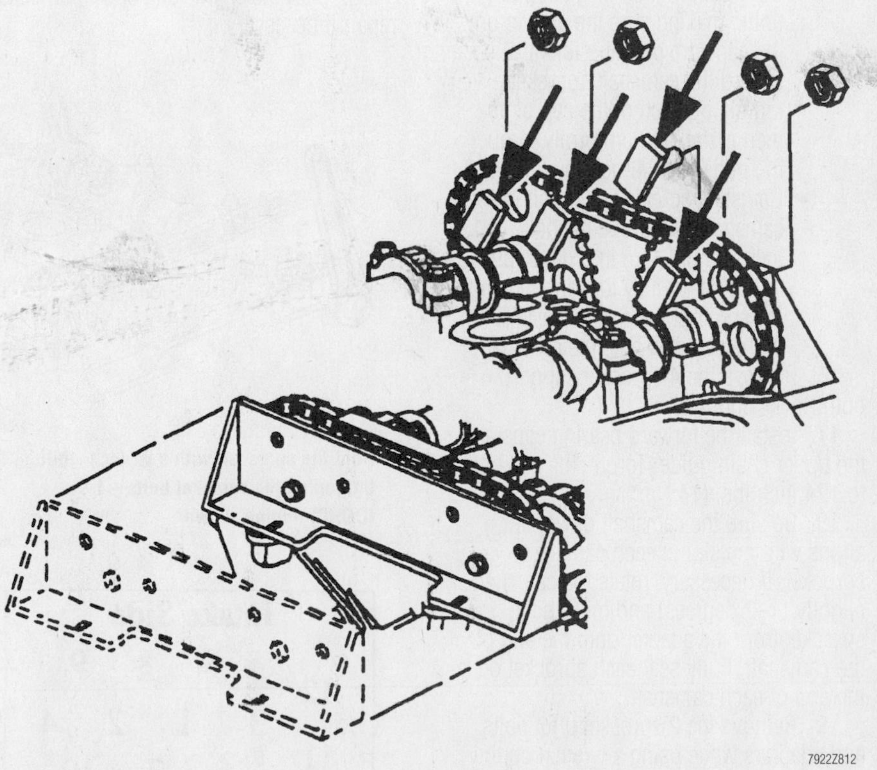

Install the camshaft support fixture to securely hold the camshafts in position—1.9L (DOHC) engines remove its end from inside the sprocket pilot otherwise camshaft or lifter damage may occur.

13. Remove or disconnect the following:
   - Uniformly loosen and remove the remaining camshaft bearing cap bolts. To prevent bolt/cap damage, do not use power tools and make several passes.
   - Camshafts
   - Pull the lifters out to remove them, always place them in the order in which they were removed and with the camshaft side facing down. Oil will drain out of the lifter if it is placed valve side down.

To install:

14. Clean and inspect all parts prior to installation. Oil the camshaft and install with the **IN** camshaft on the intake side and **EX** camshaft on the exhaust side.

➡The dowel pin in each camshaft must be located at the 12 o'clock position during installation to prevent valve and piston damage.

15. Install or connect the following:
   - Lifters in their original locations
   - Bearing caps, except for the forward pair, in their original positions, making sure the arrows on the caps are pointing forward toward the camshaft sprockets. Lightly oil each of the cap bolts, then install and uniformly torque the bolts to 124 inch lbs. (14 Nm).
   - Camshaft sprocket pilot bolt in each camshaft and torque the bolts to 124 inch lbs. (14 Nm) in order to pull the camshaft fully forward and align the sprocket support for installation of the sprocket onto the camshaft

16. Remove the 4 sprocket support bolt/blocks and nuts.

17. Install the forward bearing caps and the upper chain guide. Torque the cap bolts to 124 inch lbs. (14 Nm).

18. Be sure the camshaft dowel pin aligns with the slot in each camshaft sprocket. If necessary, rotate the camshaft slightly (1–2 degrees) and move each sprocket from the adapter onto the end of the camshaft. Fully seat each sprocket on the end of each camshaft.

19. Remove the 2 sprocket pilot bolts and adapters while using a wrench on the camshaft flats to assure the camshaft cannot move.

20. Remove the support angle fixture.

21. Install the camshaft sprocket retaining bolts and washers. Hold the camshafts and torque the bolts to 76 ft. lbs. (103 Nm).

22. Verify all visible timing marks and holes are in alignment. Turn the crankshaft clockwise until the mark on the crankshaft pulley aligns with the mark on the front cover. Check timing by inserting 3⁄16 in. drill bits through the camshaft sprocket alignment holes, into the cylinder head. If the alignment pins cannot be inserted, turn the crankshaft 360 degrees clockwise and repeat. If the pins cannot be inserted within 1–2 degrees of either TDC position, the camshafts are not properly timed. Do not start the engine until the camshafts are timed.

23. Apply a small drop of RTV across the cylinder head and front cover T-joints. Inspect the old camshaft cover gasket and replace if damaged. Install the gasket and the camshaft cover. Torque the fasteners in proper sequence to 89 inch lbs. (10 Nm).

24. Install or connect the following:
   - Right splash shield and wheel, if removed to observe the timing marks
   - PCV and fresh air hoses and the spark plug wires
   - Accessory drive belt
   - Negative battery cable

25. Start the engine and check for leaks, repair if necessary.

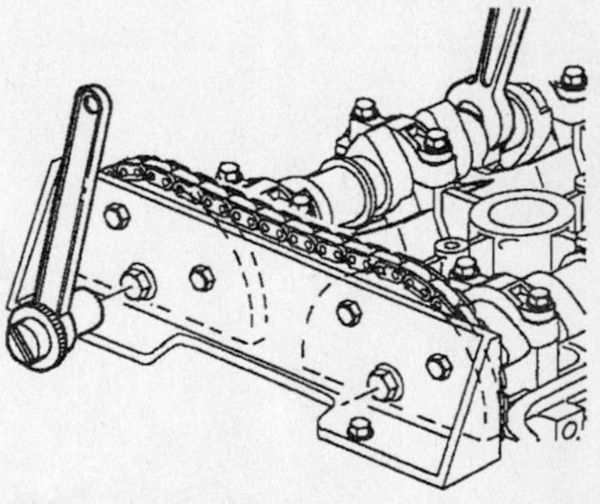

7922Z813

Hold the camshaft with a wrench while tightening the sprocket bolts—1.9L (DOHC) engine shown

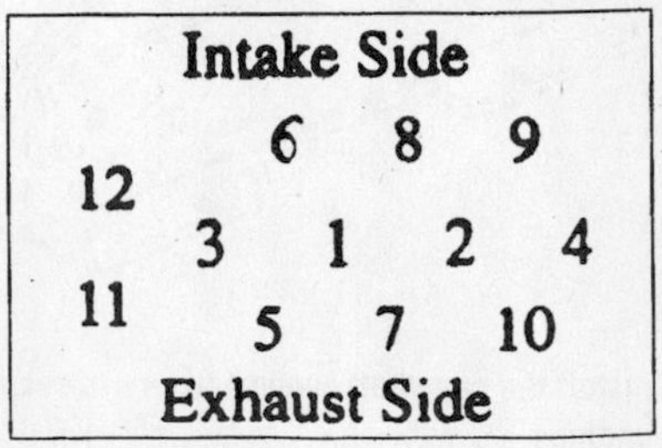

7922Z817

Camshaft cover bolt tightening sequence—1.9L DOHC engines

2.2L Engine

➡Be very careful when working around the camshaft sprockets and timing chain cover during this procedure. If a bolt or washer is accidentally dropped between the front cover and engine assembly, the cover will have to be removed for retrieval.

1. Before servicing the vehicle, refer to the precautions in the beginning of this section.

2. Remove or disconnect the following:
   - Negative battery cable
   - Ignition coil
   - Ground strap
   - Positive Crankcase Ventilation (PCV) fresh air hose
   - Cam cover
   - Fuel line
   - Wire harness bracket
   - Power steering pump without removing the lines

➡To avoid valve piston contact, the No. 1 cylinder piston must be positioned at 60 degrees Before Top Dead Center (BTDC). The pistons are properly aligned when the diamond shaped hole on the intake camshaft sprocket is located at 12 o'clock.

3. Remove the upper timing chain guide and front camshaft caps.

4. Install a camshaft sprocket holding tool through the sprocket holes from the timing chain side. Align the guide pins into the slot on the support head. Torque the pins to 89 inch lbs. (10 Nm).

5. Hold each camshaft in place with a 24mm open end wrench and remove the camshaft timing sprocket retaining bolts and washers. Discard the bolts.

6. Uniformly loosen and remove the remaining camshaft bearing caps.

7. Slide the camshaft sprockets away from the camshafts and remove the camshaft.

To install:

8. Lubricate the camshaft bearing journals with clean engine oil.

9. Install both camshafts and all bearing caps except the front cap on each camshaft.

10. Torque the bearing caps uniformly, except for the front caps and the rear intake cap, to 89 inch lbs. (10 Nm).

➡Make certain that the alignment notches are properly positioned with the notches in the camshaft sprockets before final torque is applied. Also, be sure that the timing chain is properly aligned on the fixed guide.

Remove the camshaft and bearing caps–2.2L engine

11. Slide the camshaft sprockets and timing chain on the guide pins toward the camshafts. Rotate the camshafts with a 24mm open end wrench to align the camshaft and sprocket.
12. Install new camshaft sprocket bolts and torque them to 63 ft. lbs. (85 Nm) plus 30 degrees.
13. Remove the camshaft sprocket holding tool.
14. Install or connect the following:
    - Front camshaft bearing caps and torque the bolts to 89 inch lbs. (10 Nm)
    - Upper timing chain guide and apply Loctite® to the bolts
    - Rear intake camshaft bearing cap and torque the bolts 19 ft. lbs. (25 Nm)
    - Power steering pump and torque the bolts to 19 ft. lbs. (25 Nm)
    - Ignition coil
    - PCV fresh air hose
    - Ground strap
    - Fuel line
    - Wire harness bracket
    - Cam cover
    - Negative battery cable
15. Start the vehicle and check for leaks, repair if necessary.

3.0L Engine

This engine is equipped with front and rear camshafts.

The front camshaft bearing caps for the cylinder head are marked R1–R8 and the rear cylinder head bearing caps are marked L1–L8.

➡Be very careful when working around the camshaft sprockets and timing chain cover during this procedure. If a bolt or washer is accidentally dropped between the front cover and engine assembly, the cover will have to be removed for retrieval.

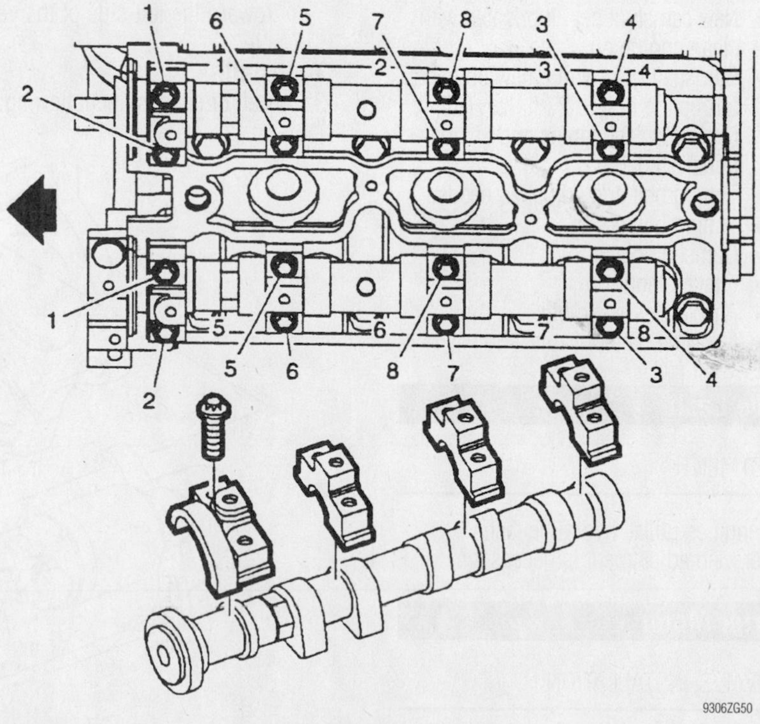

Front camshaft bearing cap removal sequence–3.0L engine

1. Before servicing the vehicle, refer to the precautions in the beginning of this section.
2. Remove or disconnect the following:
    - Negative battery cable
    - Intake plenum
    - Air cleaner assembly
    - Camshaft cover
    - Front timing belt cover
    - Timing belt

➡Rotate the crankshaft counterclockwise to 60 degrees Before Top Dead Center (BTDC) to prevent valve to piston contact.

3. Install a Camshaft Gear Locking Tool into the camshaft gears.
4. Remove or disconnect the following:
    - Loosen the camshaft gear bolt, remove the holding tool
    - Camshaft gear bolt and discard it
    - Camshaft gear
    - Loosen the camshaft bearing caps sequentially starting in the center and working outward in a spiral direction
    - Camshaft bearing caps
    - Camshaft

To install:

5. Lubricate all bearing surfaces with clean engine oil.
6. Install or connect the following:

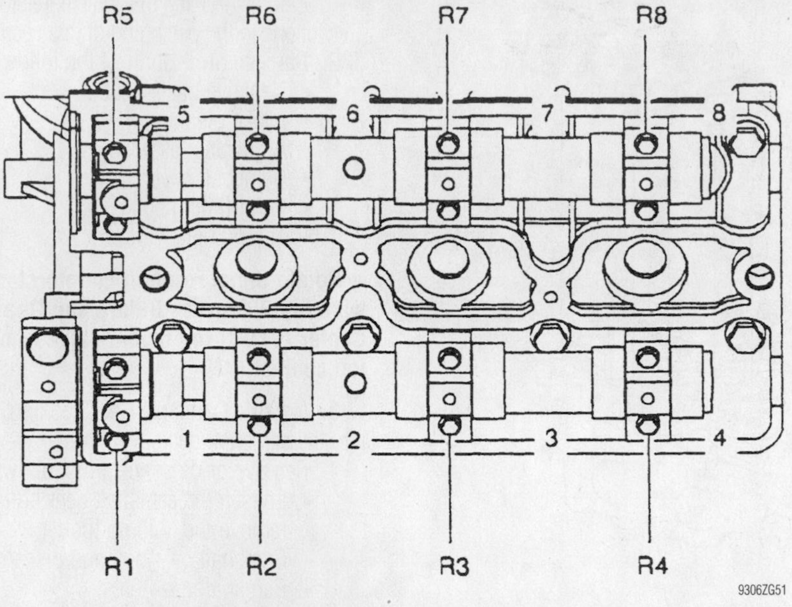

Front camshaft bearing cap installation sequence–3.0L engine

- Camshaft to the cylinder head and make sure that the pin on the exhaust camshaft is in the 12 o'clock position or that the pin on the intake camshaft is in the 7 o'clock position
- Camshaft bearing caps in their proper position and torque the bolts, starting in the center and working outward, to 71 inch lbs. (8 Nm)
- New camshaft seal lubricated with clean engine oil
- Camshaft gear with a new bolt and torque the bolt to 27 ft. lbs. (50 Nm) plus 60 degrees and an additional 15 degrees
- Timing belt and adjust as needed
- Timing belt cover
- Camshaft cover
- Intake plenum
- Air cleaner assembly
- Negative battery cable

Valve Lash

ADJUSTMENT

All engines utilize hydraulic lash adjusters; no adjustment is necessary.

Starter Motor

REMOVAL & INSTALLATION

1.9L and 2.2L Engines

1. Before servicing the vehicle, refer to the precautions in the beginning of this section.
2. Remove or disconnect the following:
   - Negative battery cable

➡**Spray the starter solenoid electrical connectors with penetrating oil before removal.**

   - Starter electrical connections
   - Starter bolts
   - Starter assembly by pulling it toward the left side of the vehicle

To install:

3. Install or connect the following:

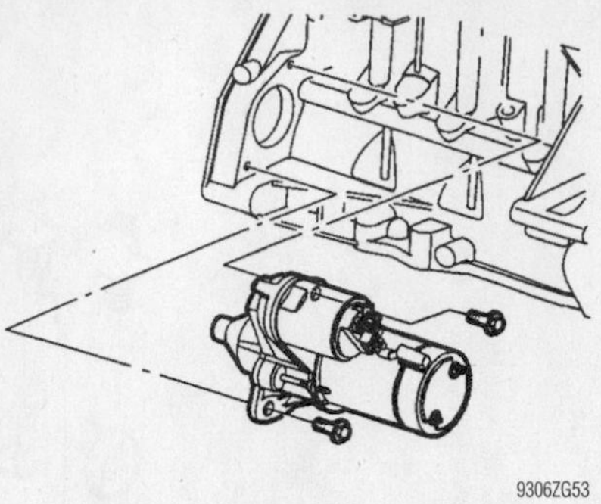

Starter assembly removal–1.9L and 2.2L engines

   - Starter to the flywheel housing and torque the bolts to 27 ft. lbs. (37 Nm)
   - Starter electrical connectors and torque the solenoid ignition wire to 44 inch lbs. (5 Nm) and the positive battery cable to 89 inch lbs. (10 Nm)
   - Negative battery cable

3.0L Engine

1. Before servicing the vehicle, refer to the precautions in the beginning of this section.
2. Remove or disconnect the following:
   - Negative battery cable
   - Right front wheel
   - Starter electrical connections
   - Loosen the fastener securing the electrical harness bracket to the engine

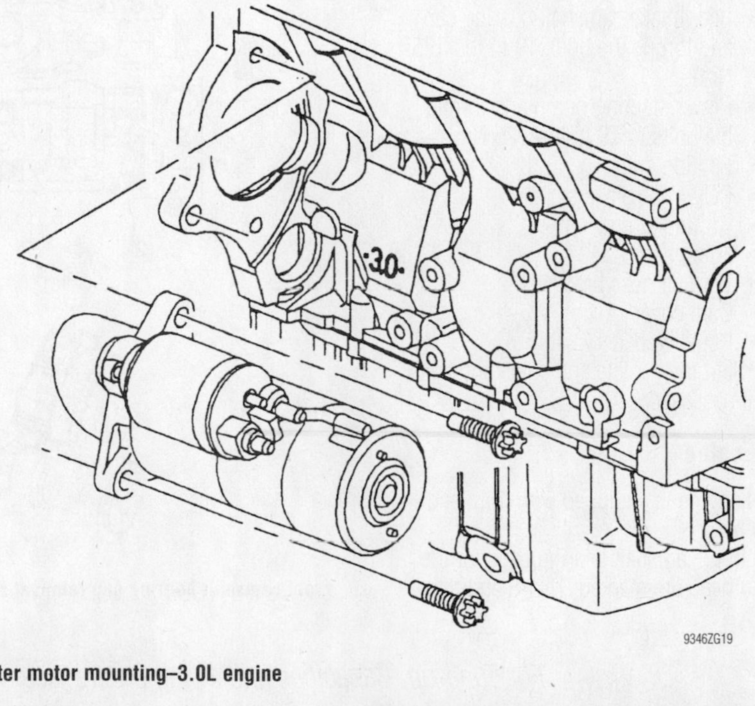

Starter motor mounting–3.0L engine

- Starter bolts
- Starter assembly by pulling it toward the left side of the vehicle

To install:

3. Install or connect the following:
   - Starter to the flywheel housing and torque the bolts to 30 ft. lbs. (40 Nm)
   - Starter electrical connections and tighten the electrical harness bracket bolt
   - Right front wheel
   - Negative battery cable

Oil Pan

REMOVAL & INSTALLATION

1.9L Engines

1. Before servicing the vehicle, refer to the precautions in the beginning of this section.
2. Drain the oil from the engine.
3. Remove or disconnect the following:
   - Negative battery cable
   - Front exhaust pipe
   - Front stiffening bracket
   - Flywheel cover
   - Right front wheel
   - Wheel well splash shield
4. Loosen the 4 front motor mount bolts approximately ½ in. (12mm).
   - Oil pan bolts

➡If equipped with a manual transaxle, an 8mm flex socket may be used to access the rear oil pan bolts located next to the flywheel.

5. Using an RTV cutter tool, separate the oil pan from the engine. Drive the tool around the pan to shear the RTV seam, then tap the pan sideways with a rubber mallet to loosen.
6. Pry the engine mount away from the engine as necessary and remove the oil pan. Be careful not to damage or score component surfaces when prying.

To install:

7. Apply a 0.16 in. (4mm) bead of RTV sealer to the pan flange. Be sure the RTV is applied to the inner side of the flange from the bolt holes.
8. Install or connect the following:
   - Oil pan and torque the bolts to 80 inch lbs. (9 Nm)
   - Front engine mount bolts and torque them to 37 ft. lbs. (50 Nm)
   - Right splash shield and wheel

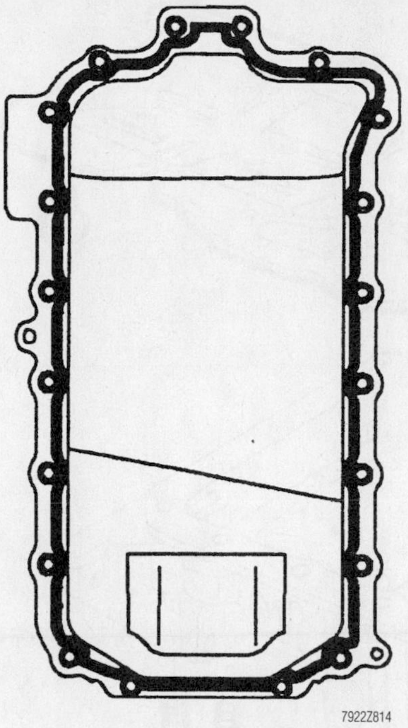

Apply a 0.16 in (4mm) bead of RTV to the oil pan flange to the inner side of the bolt holes–1.9L engines

   - Engine stiffening bracket
   - Flywheel cover
   - Exhaust pipe and torque the pipe-to-manifold nuts in a crosswise pattern to 23 ft. lbs. (31 Nm) and the pipe to converter bolts to 33 ft. lbs. (45 Nm)
   - Negative battery cable
9. Fill the engine with clean oil.
10. Start the vehicle and check for leaks, repair if necessary.

2.2L Engine

1. Before servicing the vehicle, refer to the precautions in the beginning of this section.
2. Drain the engine oil.
3. Remove or disconnect the following:
   - Negative battery cable
   - Oil pan bolts
4. Using a flat blade tool, pry the oil pan from the engine block.

To install:

5. Apply a 0.16 in. (4mm) bead of RTV sealer to the pan flange. Be sure the RTV is applied to the inner side of the flange.
6. Install or connect the following:
   - Oil pan and torque the bolts in the proper sequence to 18 ft. lbs. (25 Nm)
   - Negative battery cable
7. Fill the engine with clean oil.
8. Start the vehicle and check for leaks, repair if necessary.

3.0L Engine

1. Before servicing the vehicle, refer to the precautions in the beginning of this section.
2. Drain the engine oil.

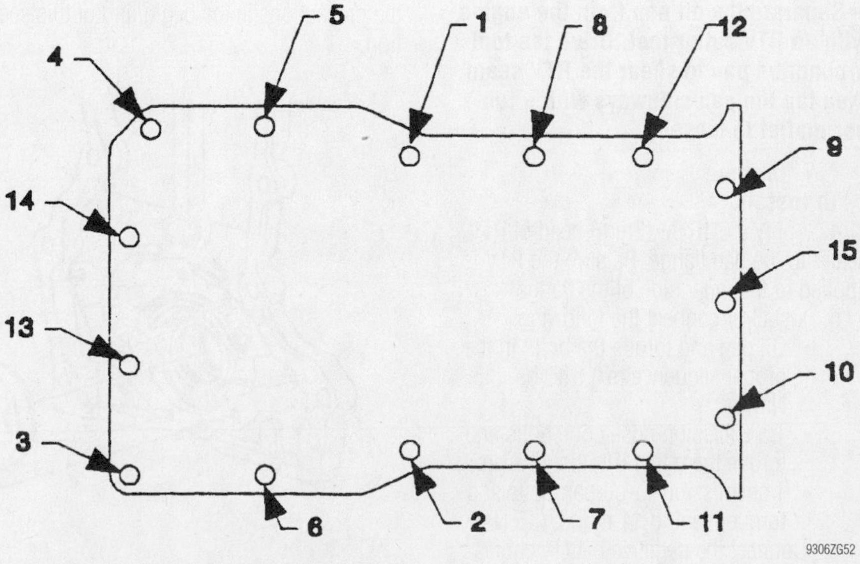

Oil pan bolts torque sequence–2.2L engine

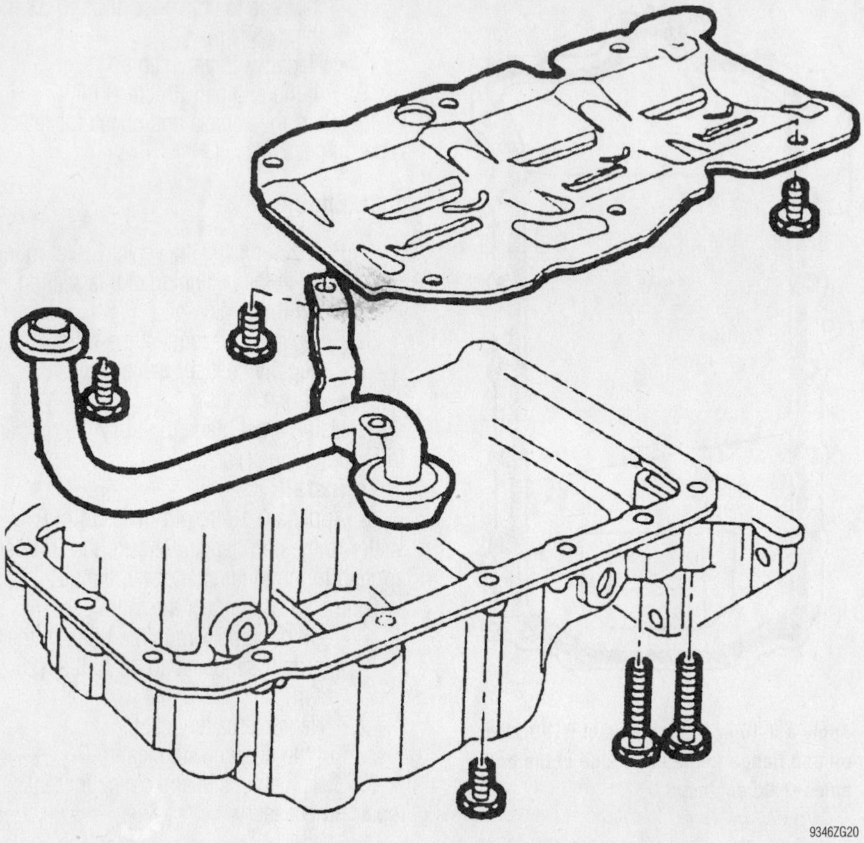

Oil pan and related components–3.0L engine

3. Remove or disconnect the following:
   - Negative battery cable
   - Nose cone bracket bolts from the oil pan
   - Lower transmission flange-to-oil pan bolts
   - Oil pan bolts

➡**Separate the oil pan from the engine with an RTV cutter tool. Drive the tool around the pan to shear the RTV seam, then tap the pan sideways with a rubber mallet to loosen.**

   - Oil pan

To install:

4. Apply a 0.10 in. (2mm) bead of RTV sealer to the pan flange. Be sure the RTV is applied to the inner side of the flange.
5. Install or connect the following:
   - Oil pan and torque the bolts in the proper sequence to 11 ft. lbs. (15 Nm)
   - Transmission nose cone bolts and torque them to 11 ft. lbs. (15 Nm)
   - Transmission-to-oil pan bolts and torque them to 11 ft. lbs. (15 Nm)
6. Connect the negative battery cable.
7. Fill the engine with clean oil.
8. Start the vehicle and check for leaks, repair if necessary.

Oil Pump

REMOVAL & INSTALLATION

1.9L Engines

1. Before servicing the vehicle, refer to the precautions in the beginning of this section.

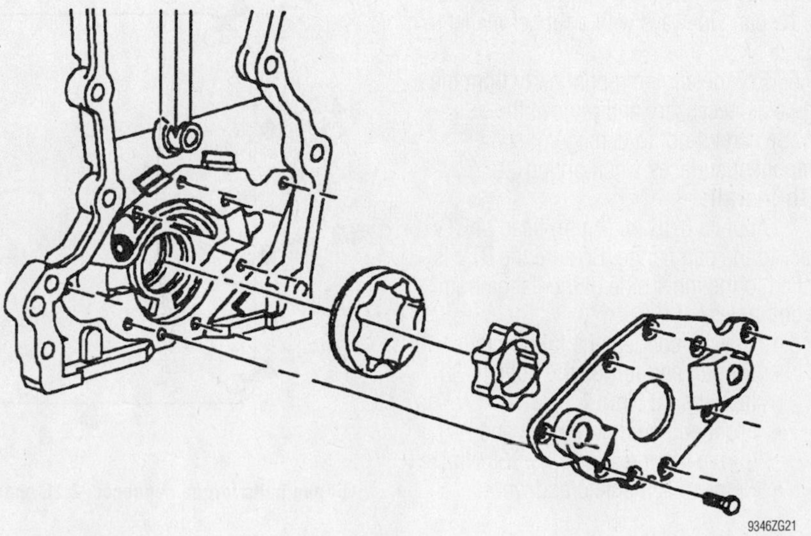

Oil pump drive rotor and driven rotor

2. Drain the engine oil.
3. Remove or disconnect the following:
   - Negative battery cable
   - Right front wheel and splash shield
   - Accessory drive belt
   - Front crankshaft vibration damper/pulley
   - Drive belt idler pulley
   - Power steering pump, SOHC only
   - Belt tensioner
   - Camshaft cover, DOHC only
   - Right side engine mount to front cover nuts and the engine mount to midrail bracket nuts
   - Front 4 oil pan bolts
   - Front cover bolts
   - Drive rotor and driven rotor
4. If necessary, remove the relief valve. Because the puller jaws will damage the relief valve sealing seat, the valve cannot be used again when removed.

To install:

5. Install or connect the following:
   - New relief valve into the cover bore, if removed. Coat the valve with clean engine oil and tap it into the bore.

➡**Whenever the oil pump is installed, the assembly must be packed with petroleum jelly in order to prime the pump.**

   - Drive and driven rotors into the pump with the chamfer toward the front oil seal
   - Oil pump body cover using new bolts and torque them to 97 inch lbs. (11 Nm)
   - New oil pressure and suction seals in the cylinder block
   - Front oil pan bolts and torque them to 80 inch lbs. (9 Nm)

- Front cover and torque the perimeter and center bolts to 19 ft. lbs. (25 Nm) and the lower center bolt to 89 inch lbs. (10 Nm)
- Front oil pan bolts and torque them to 80 inch lbs. (9 Nm)
- Camshaft cover
- Drive belt tensioner and torque the fasteners to 26 ft. lbs. (35 Nm)
- Power steering pump, if removed
- Drive belt and torque the pulley bolts to 33 ft. lbs. (45 Nm)
- Engine mount to midrail bracket nuts first and the engine mount to front cover nuts last and torque all nuts to 37 ft. lbs. (50 Nm)
- Right front splash shield and wheel
- Negative battery cable

6. Fill the engine with clean oil and replace the oil filter.
7. Start the vehicle and check for leaks, repair if necessary.

2.2L Engine

1. Before servicing the vehicle, refer to the precautions in the beginning of this section.
2. Drain the engine oil.
3. Remove or disconnect the following:
   - Negative battery cable
   - Air cleaner assembly
   - Right front wheel and splash shield
   - Accessory drive belt
   - Crankshaft damper pulley
   - Belt tensioner
4. Install an engine support fixture.
   - Right front engine mount
   - Front cover bolts and the 13mm bolt under the water pump
   - Front cover
   - Oil pump cover plate
   - Drive rotor and driven rotor
   - Pressure relief valve

To install:

5. Install or connect the following:
   - New relief valve into the cover bore, if removed. Coat the valve with clean engine oil and tap it into the bore. Torque the plug to 30 ft. lbs. (40 Nm).

➡Whenever the oil pump is installed, the assembly must be packed with petroleum jelly in order to prime the pump.

   - Drive and driven rotors into the pump with the chamfer toward the front oil seal
   - Oil pump body cover using new bolts and torque the bolts to 53 inch lbs. (6 Nm)
   - Front cover with a new oil seal and torque the perimeter and center bolts to 19 ft. lbs. (25 Nm) and the lower center bolt to 89 inch lbs. (10 Nm)
   - Right side engine mount and torque the bolts to 41 ft. lbs. (55 Nm)
6. Remove the engine support fixture.
   - Drive belt tensioner and torque the bolts 37 ft. lbs. (50 Nm)
   - Crankshaft damper pulley and torque the bolt to 74 ft. lbs. (100 Nm) plus 75 degrees
   - Accessory drive belt
   - Right front splash shield and wheel
   - Air cleaner assembly
   - Negative battery cable
7. Fill the engine with clean oil and replace the oil filter.
8. Start the vehicle and check for leaks, repair if necessary.

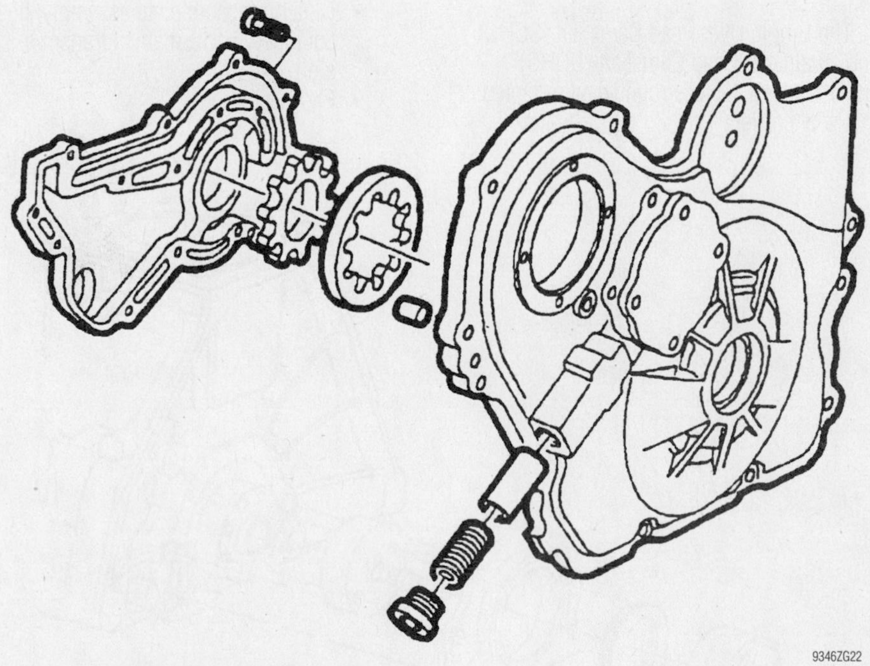

Front cover and oil pump assembly–2.2L engine

3.0L Engine

1. Before servicing the vehicle, refer to the precautions in the beginning of this section.
2. Drain the engine oil.
3. Drain the cooling system.
4. Remove or disconnect the following:
   - Negative battery cable
   - Air cleaner assembly
   - Front timing belt cover
   - Timing belt
   - Rear timing belt cover
   - A/C compressor and power steering pump bracket and move them away from the oil pump housing
   - Alternator bolts and move the alternator out of the way
   - Oil pan
5. Mount a crank hub holding tool to the crankshaft drive gear and remove the drive gear.
   - Oil pump bolts
   - Oil pan housing bolts that thread into the oil pump
   - Oil pump
   - Front main oil seal and collar
   - Pressure relief valve
   - Oil pump cover
   - Drive rotor and driven rotor

To install:

6. Install the new relief valve into the cover bore (if removed) and torque the plug to 30 ft. lbs. (40 Nm).

➡Whenever the oil pump is installed, the new gasket must be coated with a thin bead of sealing Loctite 518®.

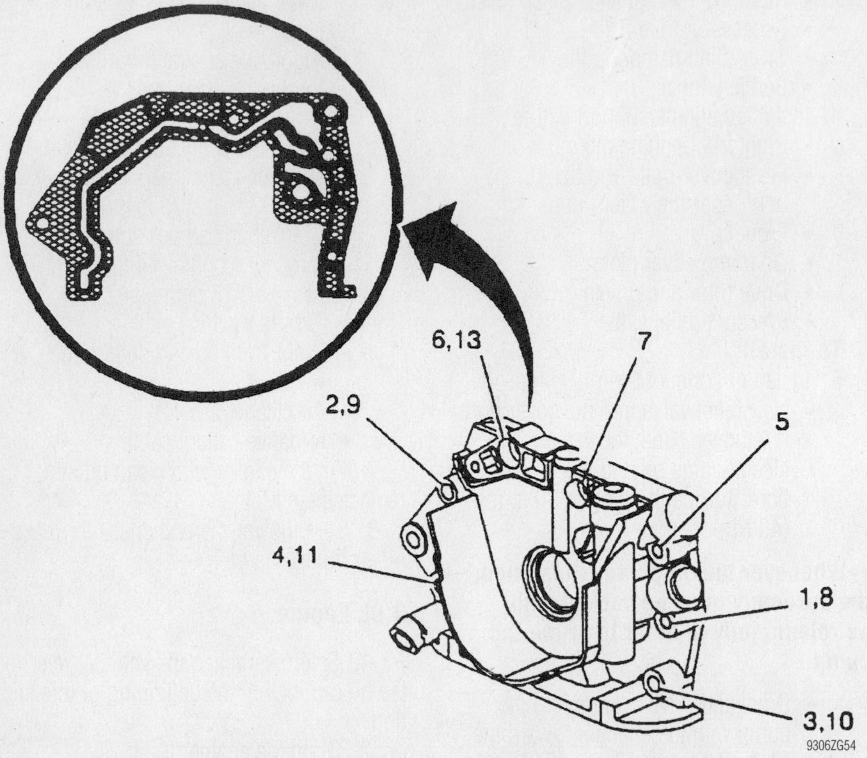

Oil pump bolt tightening sequence–3.0L engine

7. Install or connect the following:
   - Drive and driven rotors into the pump with the chamfer toward the front oil seal
   - Oil pump body cover using new bolts and torque them to 89 inch lbs. (10 Nm)
   - Drive gear and torque the new bolt to 184 ft. lbs. (250 Nm) plus 45 degrees then an additional 15 degrees
   - Oil pan and torque the bolts to 11 ft. lbs. (15 Nm)
   - Alternator and torque the bolts to 30 ft. lbs. (40 Nm)
   - A/C compressor and power steering pump bracket. Torque the bolts to 30 ft. lbs. (40 Nm).
   - Rear timing belt cover—Refer to section 4 for the timing belt procedure.
   - Drive belt idler pulley—Refer to section 4 for the timing belt procedure.
   - Timing belt
   - Front cover with a new oil seal
   - Air cleaner assembly
   - Negative battery cable
8. Fill the engine with clean oil. An oil filter replacement is also recommended.
9. Fill the cooling system.
10. Start the vehicle and check for leaks, repair if necessary.

Rear Main Seal

REMOVAL & INSTALLATION

1.9L Engines

The Single Over Head Camshaft (SOHC) and Dual Over Head Camshaft (DOHC) engines use a 1-piece round seal mounted in a seal carrier.

1. Before servicing the vehicle, refer to the precautions in the beginning of this section.
2. Drain the engine oil.
3. Disconnect the negative battery cable.
4. Remove the oil pan.
5. Use the prying tangs provided in the carrier to remove the seal with a small suitable pry bar and hammer. Be careful not to damage the crankshaft oil seal lip contact surface.

To install:

6. Clean the carrier and crankshaft with solvent and a rag to prevent seal lip damage during installation. Check for scores or damage to the sealing surfaces.
7. Apply a light coat of clean engine oil to the seal lip and the carrier inner diameter.
8. Install a new rear main seal, using a Seal Installer tool.
9. Install the oil pan.
10. Connect the negative battery cable.
11. Fill the engine with clean oil.
12. Start the vehicle and check for leaks, repair if necessary.

2.2L Engine

1. Before servicing the vehicle, refer to the precautions in the beginning of this section.
2. Remove or disconnect the following:
   - Negative battery cable
   - Transmission
   - Clutch/pressure plate assembly, if equipped with a manual transmission
   - Flywheel

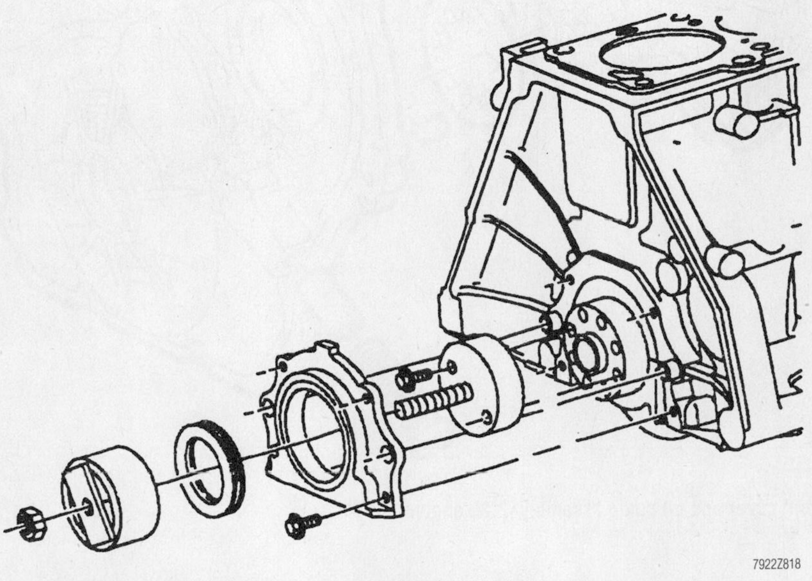

Exploded view of the rear main seal installation–1.9L engines

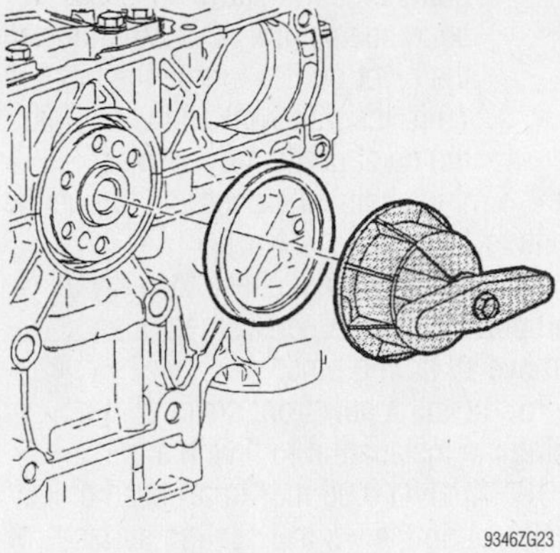

Rear oil seal and installation tool J42067–2.2L engine

- Rear main bearing seal by prying it from the engine

➡**Be careful not to damage or scratch the seal mounting surfaces.**

To install:

3. Lubricate the new rear main bearing seal with engine oil.
4. Install or connect the following:
   - New rear main seal using a Rear Main Bearing Oil Seal Installer Tool J42067 until it is flush with the block
   - Flywheel
   - Clutch/pressure plate assembly, if equipped with a manual transmission
   - Transmission
   - Negative battery cable
5. Start the engine and check for leaks, repair if necessary.

3.0L Engine

1. Before servicing the vehicle, refer to the precautions in the beginning of this section.
2. Remove or disconnect the following:
   - Negative battery cable
   - Transmission
   - Clutch/pressure plate assembly, if equipped with a manual transmission
   - Flywheel
3. Center punch the steel ring of the oil seal.
4. Drill a small hole into the steel ring.
5. Install a self-tapping screw and using pliers, pull out the rear main oil seal.

➡**Be careful not to damage or scratch the seal mounting surfaces.**

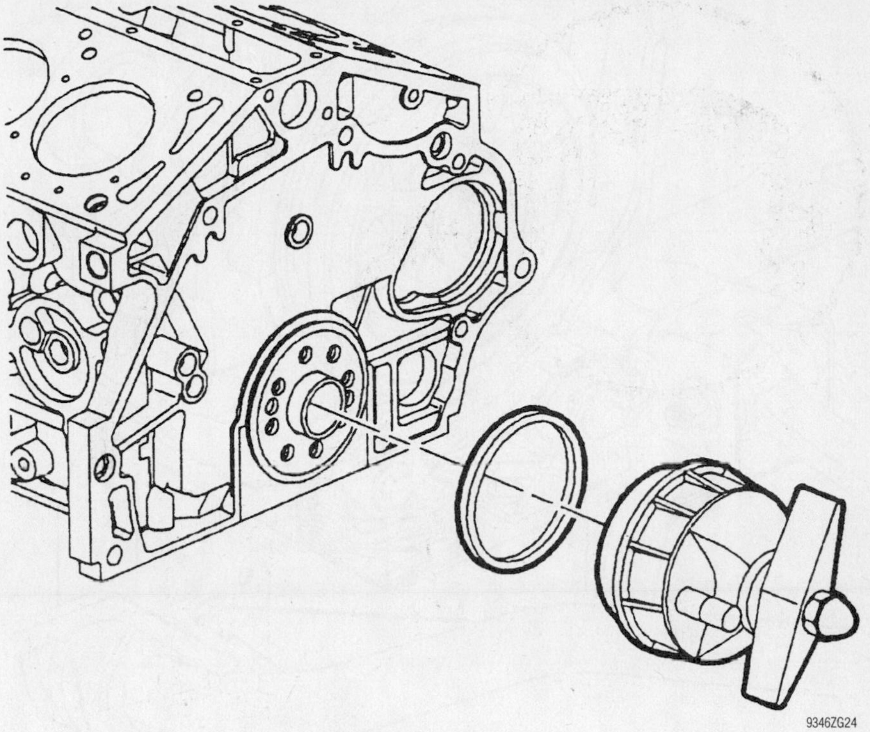

Rear main oil seal and installation tool J42067–3.0L engine

To install:

6. Lubricate the new rear main oil seal with engine oil.
7. Install or connect the following:
   - New rear main seal using a Rear Main Bearing Oil Seal Installer Tool SA9121E until it is flush with the block
   - Flywheel
   - Clutch/pressure plate assembly, if equipped with a manual transmission
   - Transmission
   - Negative battery cable
8. Start the engine and check for leaks, repair if necessary.

Timing Chain, Sprockets, Front Cover and Seal

REMOVAL & INSTALLATION

1.9L SOHC Engines

1. Before servicing the vehicle, refer to the precautions in the beginning of this section.
2. Drain the engine oil.
3. Remove or disconnect the following:
   - Negative battery cable
   - Right front wheel and splash shield

➡**Place a 1 in. X 1 in. X 2 in. long block of wood between the torque strut and cradle to ease removal and installation of the torque engine mount.**

   - 3 right side upper engine torque axis-to-front cover nuts and the 2 mount to midrail bracket nuts, allowing the powertrain to rest on the block of wood
   - Drive belt, tensioner and pulley
   - Power steering pump attaching bolts and set the pump to the side with the lines still attached
   - A/C compressor from the bracket and set aside with the lines attached
   - Camshaft cover
   - Crankshaft damper/pulley assembly from the crankshaft
4. Install the special oil seal replacement tool SA9104E, to be sure the front crankshaft timing sprocket is held firmly in place and prevent guide damage. Install with the flat side towards the crankshaft sprocket.
5. Remove or disconnect the following:
   - Front 4 oil pan bolts and cut the seal away from the front cover
   - Front cover bolts and carefully pry the cover away from the cylinder block

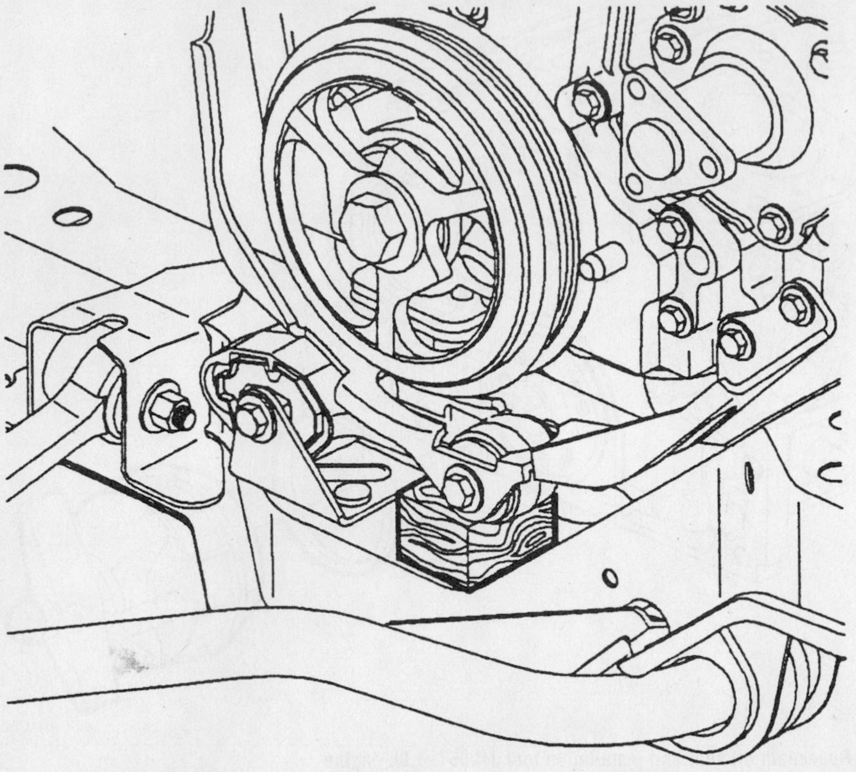

7922Z819

Place a 1 x 1 x 2 in. (25 x 25 x 51mm) piece of wood between the torque strut and cradle before removal of the torque engine mount—1.9L (SOHC) engines

- Front cover oil seal from the cover

6. Carefully rotate the crankshaft clockwise so the timing mark on the crankshaft sprocket and keyway align with the main bearing cap split line (90 degrees past Top Dead Center [TDC]).
7. Remove or disconnect the following:
   - Timing chain guides and tensioner
   - Camshaft sprocket bolt
   - Timing chain and camshaft sprocket
   - Crankshaft sprocket

To install:

8. Inspect the chain for wear and damage. Check the inside diameter of the chain, it should be no more than 16.77 in. (426mm). Inspect the chain guides for wear or cracks and the timing sprockets for teeth or key wear. Replace components as necessary.
9. Verify that the crankshaft keyway is positioned 90 degrees clockwise past TDC (keyway at 3 o'clock). The keyway should align with the split between the bearing cap and engine block.
10. Bring the camshaft up to No. 1 TDC by loosely installing the sprocket and rotating the sprocket until the timing pin can be inserted. The camshaft contains wrench flats to assist in turning the shaft. The dowel pin should be at 12 o'clock when the camshaft is at TDC.
11. Install the crankshaft sprocket, then rotate the crankshaft counterclockwise 90 degrees up to No. 1 TDC (keyway at 12 o'clock).
12. Position the chain under the crankshaft sprocket and over the camshaft sprocket. The timing chain should be positioned so that 1 silver link plate aligns with the reference mark on the camshaft sprocket and the other aligns with the downward tooth (at the 6 o'clock position) on the crankshaft sprocket. The letters FRT on the camshaft sprocket must face forward, away from the cylinder head and excess chain slack should be located on the tensioner side of the block.
13. Install or connect the following:
   - Timing pin to verify proper alignment of the camshaft and sprocket
   - Camshaft sprocket and torque the sprocket bolt to 75 ft. lbs. (102 Nm)

➡Do not allow the camshaft retaining bolt to torque against the timing pin or cylinder head damage will result.

   - Timing chain guides with the words FRONT facing out. Install the fixed guide first and verify the chain is snug against the guide, then install the pivot guide. Torque the bolts to 19 ft. lbs. (26 Nm) and verify that the pivot guide moves freely.
   - Timing chain tensioner and torque the bolts to 14 ft. lbs. (19 Nm)

14. Make one final check to verify all components are properly timed, then remove all timing pins.
15. Install a new front cover oil seal using the installation tool with a press.
16. Apply a 0.08 in. (2mm) bead of RTV sealer along the vertical sealing surfaces of the front cover to the inside of the bolt holes and to the front of the oil pan. Extra sealer is necessary at the oil pan and cylinder head joints. Be sure to assemble the front cover to the engine within 3 minutes of RTV application.
17. Install or connect the following:
   - Crankshaft sprocket retaining tool SA9104E, to align the oil pump and crankshaft during cover installation
   - Front cover to the engine and torque the perimeter bolts starting at the center and working outwards on both sides to 19 ft. lbs. (25 Nm)
   - Front cover center or inner bolts and torque them to 89 inch lbs. (10 Nm), except for the upper inside bolt, which should be tightened to 22 ft. lbs. (30 Nm)
   - Oil pan front bolts and torque them to 80 inch lbs. (9 Nm)
18. After front cover installation, spray 6–12 squirts of oil through the front oil seal drain back hole to verify that it is not plugged.
19. Apply a thin film of RTV between the

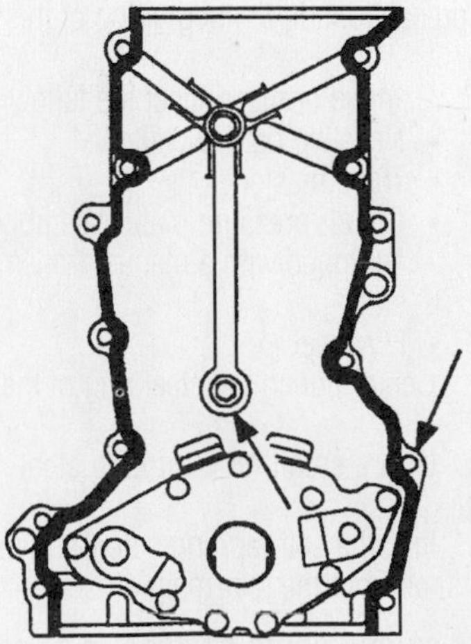

7922Z820

Apply a 0.08 in. (2mm) bead of RTV sealer along the vertical sealing surfaces of the front cover—1.9L (SOHC) engines

damper/pulley assembly flange and washer only; the washer and bolt head flange are designed to prevent oil leakage.

20. Remove the crankshaft retaining tool SA9104E.

21. Install the crankshaft damper/pulley assembly and torque the bolt to 159 ft. lbs. (215 Nm).

22. Apply a small drop of RTV across the cylinder head and front cover T-joints. Inspect the old camshaft cover gasket and replace if damaged. Install the gasket and the camshaft cover and torque the fasteners uniformly to 22 ft. lbs. (30 Nm).

23. Install or connect the following:

- A/C compressor assembly and/or the power steering pump assembly
- Drive belt idler pulley
- Drive belt tensioner
- Drive belt
- Engine mount-to-midrail bracket nuts and torque them to 37 ft. lbs. (50 Nm)
- 3 mount-to-front cover nuts and torque them uniformly to 37 ft. lbs. (50 Nm) in order to prevent front cover damage
- Splash shield and the wheel assembly
- Negative battery cable

24. Fill the engine with clean oil.

25. Start the engine and check for leaks, repair if necessary.

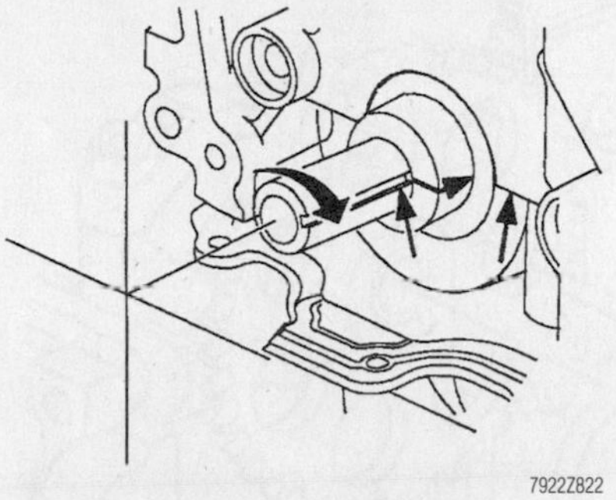

At 90 degrees past TDC, the crankshaft sprocket keyway will align with the main bearing cap split line—1.9L (SOHC) engines

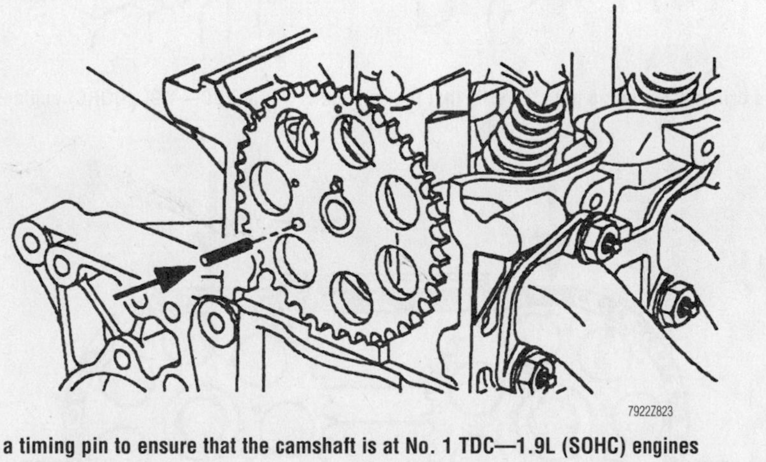

Insert a timing pin to ensure that the camshaft is at No. 1 TDC—1.9L (SOHC) engines

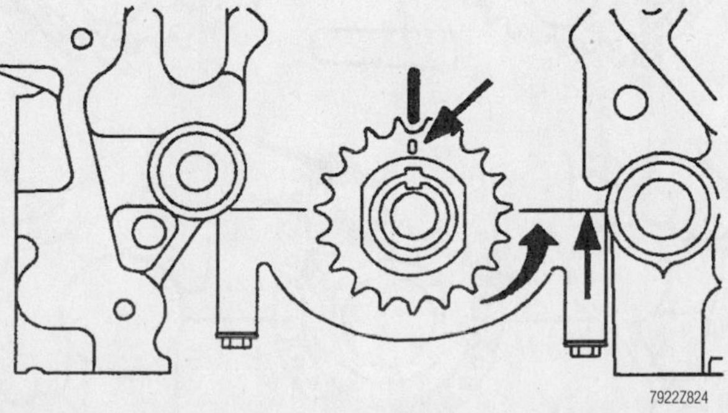

When the camshaft is at TDC, rotate the crankshaft counterclockwise 90 degrees to achieve TDC—1.9L (SOHC) engines

1.9L (DOHC) Engines

1. Before servicing the vehicle, refer to the precautions in the beginning of this section.

2. drain the engine oil.

3. Remove or disconnect the following:

- Negative battery cable
- Right front wheel and splash shield

➡Place a 1 X 1 X 2 in. (25 X 25 X 51mm) block of wood between the torque strut and cradle to ease removal and installation of the torque engine mount.

- 3 right side upper engine torque axis-to-front cover nuts and the 2 mount-to-midrail bracket nuts, allowing the powertrain to rest on the block of wood
- Accessory drive belt
- Drive belt tensioner and pulley
- Power steering pump attaching bolts and set the pump to the side with the lines still attached
- A/C compressor from the bracket and set it to the side with the lines attached
- Camshaft cover
- Crankshaft damper/pulley

4. Install an Oil Seal Replacement Tool SA9104E to be sure the front crankshaft timing sprocket is held firmly in place and prevent guide damage. Install with the flat side towards the crankshaft sprocket.

- Front 4 oil pan bolts and cut the seal away from the front cover
- Front cover bolts and carefully pry the cover away from the cylinder block
- Front cover oil seal from the cover

5. Carefully rotate the crankshaft clockwise so the timing mark on the crankshaft sprocket and keyway align with the main bearing cap split line (90 degrees past Top Dead Center (TDC).

- Timing guides and tensioner
- Camshaft sprocket bolt
- Timing chain and camshaft sprocket
- Crankshaft sprocket

To install:

6. Inspect the chain for wear and damage. Check the inside diameter of the chain, it should be no more than 23.15 in. (588mm). Inspect the chain guides for wear or cracks and the timing sprockets for teeth or key wear. Replace components as necessary.

7. Verify that the crankshaft keyway is positioned 90 degrees clockwise past TDC

(keyway at 3 o'clock). The keyway should align with the split between the bearing cap and engine block.

8. Install the crankshaft sprocket.

9. Rotate the crankshaft counterclockwise 90 degrees up to No. 1 TDC (keyway at 12 o'clock).

10. Position the chain under the crankshaft sprocket and over the camshaft sprocket. The timing chain should be positioned so that 1 silver link plate aligns with the reference mark on the camshaft sprocket and the other aligns with the downward tooth (at the 6 o'clock position) on the crankshaft sprocket. The letters FRT on the camshaft sprocket must face forward, away from the cylinder head and excess chain slack should be located on the tensioner side of the block. Torque the camshaft sprocket bolt to 75 ft. lbs. (102 Nm).

11. Verify that the crankshaft reference mark aligns with the cylinder block mark at 12 o'clock.

12. Install or connect the following:

- Timing chain fixed guide and torque the bolts to 21 ft. lbs. (28 Nm)
- Pivoting chain guide and torque the bolt to 19 ft. lbs. (26 Nm) and make certain the guide moves freely
- Two forward bearing caps and the upper timing chain guide and torque the retaining bolts to 124 inch lbs. (14 Nm)
- Timing chain tensioner and torque the bolts to 14 ft. lbs. (19 Nm)

13. Make one final check to verify all components are properly timed, then remove all timing pins.

- Seat a new front cover oil seal using the installation tool with a press

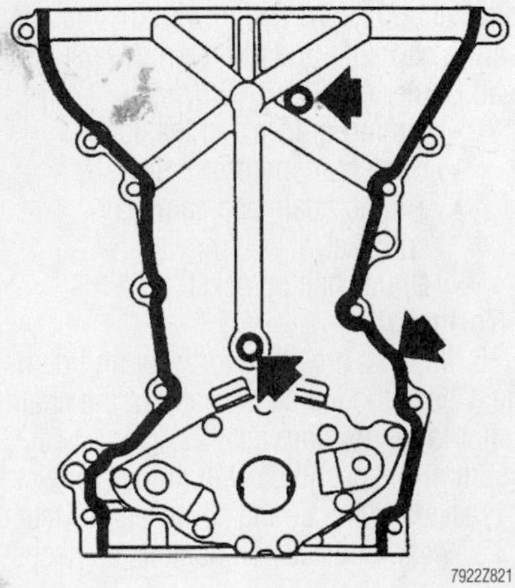

Apply a 0.08 in. (2mm) bead of RTV sealer along the vertical sealing surfaces of the front cover and the front of the oil pan—1.9L (DOHC) engines

- If the engine front cover casting or assembly is replaced, the 3 torque axis mount studs should also be replaced. Torque the new studs to 19 ft. lbs. (25 Nm).

➡Apply a 0.08 in. (2mm) bead of RTV sealer along the vertical sealing surfaces of the front cover to the inside of the bolt holes and to the front of the oil pan. Extra sealer is necessary at the oil

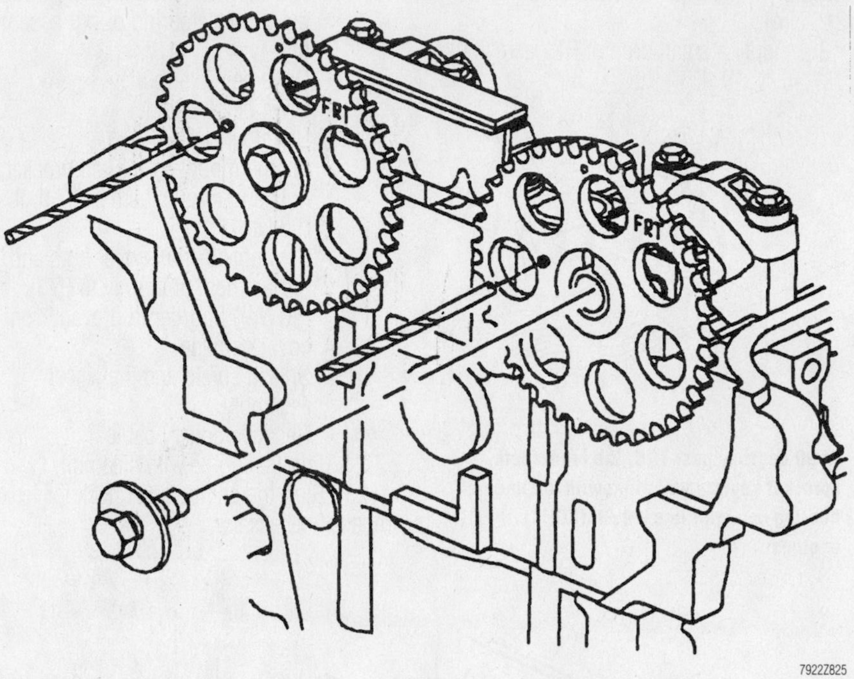

Use drill bits as timing pins to verify that the camshafts are at TDC—1.9L (DOHC) engines

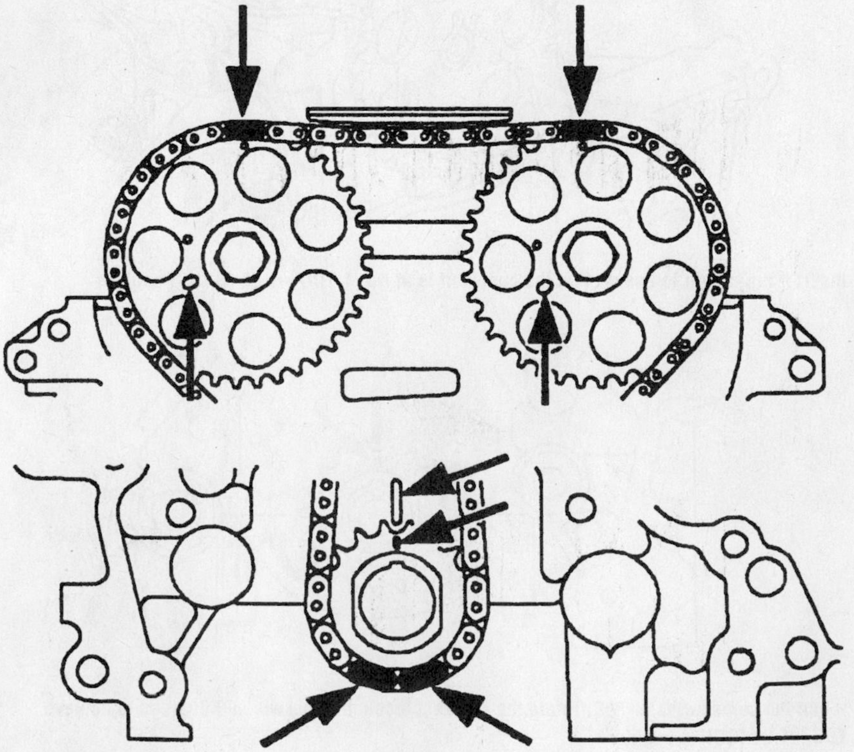

Be sure that the silver link plates and reference marks are all in alignment as shown—1.9L (DOHC) engines

pan and cylinder head joints. Be sure to assemble the front cover to the engine within 3 minutes of RTV application.

- Crankshaft sprocket retaining tool to align the oil pump and crankshaft during cover installation
- Front cover to the engine and torque the perimeter bolts starting at the center and working outwards on both sides to 22 ft. lbs. (30 Nm)
- Front cover center or inner bolts to 89 inch lbs. (10 Nm) except for the upper inside bolt which should be tightened to 22 ft. lbs. (30 Nm)
- 4 oil pan front bolts and torque them to 80 inch lbs. (9 Nm)

14. After front cover installation, spray 6–12 squirts of oil through the front oil seal drain back hole to verify that it is not plugged.
15. Apply a thin film of RTV between the damper/pulley assembly flange and washer only; the washer and bolt head flange are designed to prevent oil leakage.
16. Position the crankshaft damper/pulley assembly, then secure using the wood or strap wrench (as accomplished during removal). Torque the bolt to 159 ft. lbs. (215 Nm).
17. Apply a small drop of RTV across the cylinder head and front cover T-joints. Inspect the old camshaft cover gasket and replace if damaged. Install the gasket and the camshaft cover. Torque the fasteners uniformly to 89 inch lbs. (10 Nm).
18. Remove the crankshaft retaining tool.
19. Install or connect the following:
    - A/C compressor assembly and/or the power steering pump assembly
    - Drive belt idler pulley
    - Drive belt tensioner
    - Accessory drive belt
    - 2 engine mount-to-midrail bracket nuts and torque them to 37 ft. lbs. (50 Nm)
    - 3 mount-to-front cover nuts and torque them uniformly to 37 ft. lbs. (50 Nm) in order to prevent front cover damage
    - Splash shield and the wheel
    - Negative battery cable
20. Fill the engine with clean oil.
21. Start the engine and check for leaks, repair if necessary.

2.2L Engine

1. Before servicing the vehicle, refer to the precautions in the beginning of this section.
2. Drain the engine oil.
3. Remove or disconnect the following:
    - Negative battery cable
    - Front cover
    - Camshaft cover
    - Timing chain tensioner
    - Upper timing chain guide
    - Exhaust camshaft sprocket
    - Adjustable timing chain guide
    - Fixed timing chain guide access plug and guide
    - Intake camshaft sprocket
    - Timing chain through the top of the cylinder head
    - Timing chain sprocket from the crankshaft
    - Timing chain oiling nozzle

To install:

4. Inspect the chain guides for wear and damage. Replace the guides if wear exceeds 0.045 inch (1.12mm). Inspect the timing chain shoe. Replacc the shoe if wear exceeds 0.045 inch (1.12mm).
5. Install or connect the following:
    - Timing chain sprocket to the crankshaft. Rotate the crankshaft so that the mark on the sprocket is at the 5 o'clock position.
    - Timing chain to the intake camshaft sprocket alignment copper link to the "INT" diamond timing mark on the camshaft sprocket

➡When lowering the timing chain, rotate the assembly 90 degrees to allow the chain to fall between the cylinder block bosses. Rotate the chain back so that the camshaft sprocket is facing forward.

- Chain through the housing opening on top of the cylinder head. Make certain that the chain goes around both sides of the bosses
- Timing chain around the crankshaft sprocket and align the silver link to the timing mark
- Intake camshaft sprocket loosely on the camshaft and hand-tighten the new bolt
- Adjustable timing chain guide and torque the bolt to 89 inch lbs. (10 Nm)
- Exhaust camshaft sprocket loosely on the camshaft with the timing mark on the sprocket aligned with the silver link on the chain and hand-tighten the new bolt at this time

6. Align the camshaft sprocket to

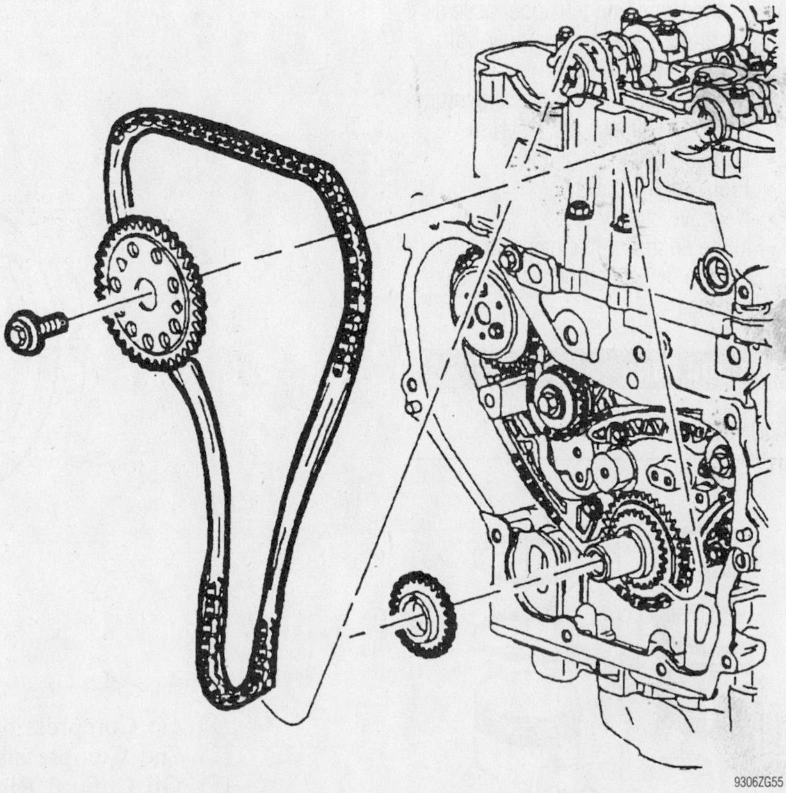

Remove the timing chain through the top of the cylinder head–2.2L engine

camshaft and tighten the bolt using the following procedure:

a. Make certain that the sprocket timing mark is at the 5 o'clock position.

b. Rotate the intake camshaft using a 24mm wrench on flats of the camshaft until the sprocket to camshaft alignment notch seats.

c. When seated properly hand-tighten the bolt.

d. Rotate the exhaust camshaft using a 24mm wrench on flats of the camshaft until the sprocket to camshaft alignment notch seats.

e. When seated properly hand-tighten the bolt.

7. Verify that all colored links are aligned with the proper marks on the camshaft and crankshaft sprockets.

8. Install or connect the following:

- Fixed timing chain guide bolt and torque it to 89 inch lbs. (10 Nm)
- Fixed timing chain guide bolt access plug and torque it to 30 ft. lbs. (40 Nm)
- Upper timing chain guide and torque the bolts to 89 inch lbs. (10 Nm)
- Intake and exhaust camshaft sprocket bolts and torque them to 63 ft. lbs. (85 Nm) plus 30 degrees
- Sealing ring and tensioner assembly and torque the tensioner bolts to 44 ft. lbs. (60 Nm)
- Timing chain oiling nozzle and torque the bolt to 89 inch lbs. (10 Nm)
- Camshaft cover
- Front engine cover
- Negative battery cable

9. Fill the engine with clean oil.

10. Start the vehicle and check for leaks, repair if necessary.

Piston and Ring

POSITIONING

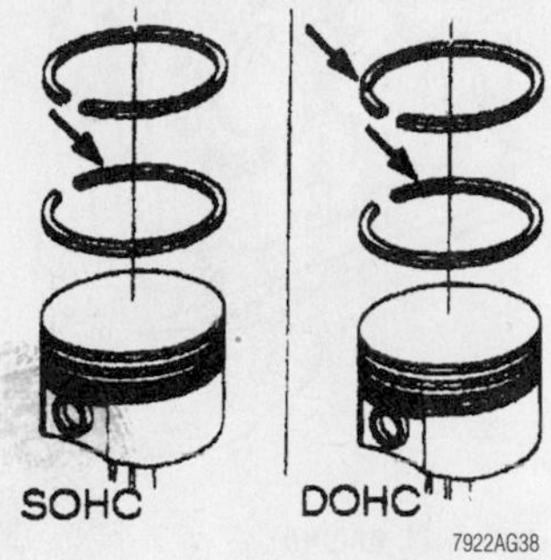

Both upper rings (DOHC) or the second ring (SOHC only) can be installed either way—1.9L engines

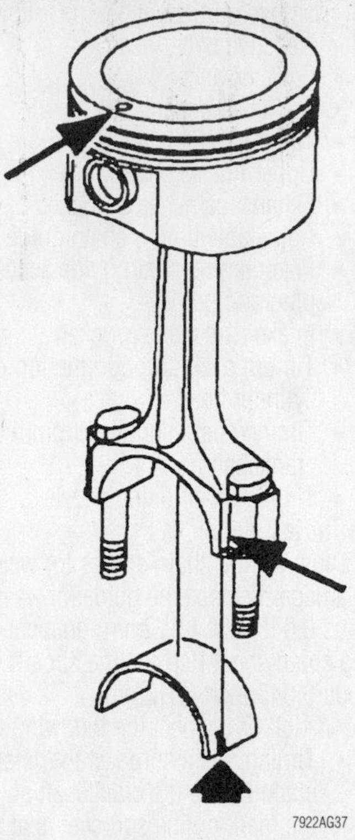

Piston and connecting rod assembly positioning mark locations. The mark on the piston and rod bearing tang slots must face the front of the engine—1.9L engines

Piston ring positioning—2.2L engine

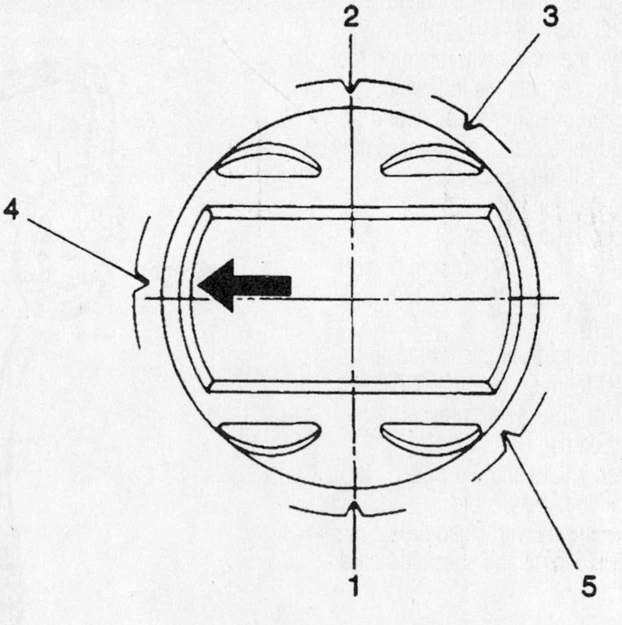

(1) 1st Compression Ring End Gap Location
(2) 2nd Compression Ring End Gap Location
(3) Oil Control Ring Upper Ring End Gap Location
(4) Oil Control Ring Spacer End Gap Location
(5) Oil Control Ring Lower Ring End Gap Location

7922AG55

Piston ring positioning—3.0L engine

FUEL SYSTEM

Fuel System Service Precautions

Safety is the most important factor when performing not only fuel system maintenance but any type of maintenance. Failure to conduct maintenance and repairs in a safe manner may result in serious personal injury or death. Maintenance and testing of the vehicle's fuel system components can be accomplished safely and effectively by adhering to the following rules and guidelines.

- To avoid the possibility of fire and personal injury, always disconnect the negative battery cable unless the repair or test procedure requires that battery voltage be applied
- Always relieve the fuel system pressure prior to disconnecting any fuel system component (injector, fuel rail, pressure regulator, etc.), fitting or fuel line connection. Exercise extreme caution whenever relieving fuel system pressure, to avoid exposing skin, face and eyes to fuel spray. Please be advised that fuel under pressure may penetrate the skin or any part of the body that it contacts
- Always place a shop towel or cloth around the fitting or connection prior to loosening to absorb any excess fuel due to spillage. Ensure that all fuel spillage (should it occur) is quickly removed from engine surfaces. Ensure that all fuel soaked cloths or towels are deposited into a suitable waste container
- Always keep a dry chemical (Class B) fire extinguisher near the work area
- Do not allow fuel spray or fuel vapors to come into contact with a spark or open flame
- Always use a back-up wrench when loosening and tightening fuel line connection fittings. This will prevent unnecessary stress and torsion to fuel line piping
- Always replace worn fuel fitting O-rings with new. Do not substitute fuel hose or equivalent, where fuel pipe is installed

Fuel System Pressure

RELIEVING

1. Before servicing the vehicle, refer to the precautions in the beginning of this section.
2. Unless battery voltage is necessary for testing, disconnect the negative battery cable. This will prevent the fuel pump from running and causing a fuel spill through the disconnected components if the ignition key is accidentally turned **ON**.
3. Remove the air cleaner assembly, for access.
4. Wrap a shop rag around the fuel test port fitting, located at the lower rear of the engine, then remove the cap and connect a fuel pressure gauge.
5. Install the bleed hose from the pressure gauge into an approved container and open the valve to bleed the system pressure.
6. After the pressure is bled, remove the gauge from the test port and recap it.
7. Install the air cleaner assembly.
8. After servicing the vehicle, connect the negative battery cable and prime the fuel system as follows:
   a. Turn the ignition **ON** for 5 seconds, then **OFF** for 10 seconds.
   b. Repeat the **ON/OFF** cycle 2 more times.
   c. Crank the engine until it starts.
   d. If the engine does not readily start, repeat sub-steps A–C.
9. Run the engine and check for leaks.

Fuel Filter

The fuel filter and fuel pressure regulator are one integral component of the new anti-return fuel injection system, and is located underneath the vehicle at the forward edge of the left side of the fuel tank.

REMOVAL & INSTALLATION

1.9L Engines

1. Before servicing the vehicle, refer to the precautions in the beginning of this section.
2. Properly relieve the fuel system pressure.
3. Remove or disconnect the following:
   - Negative battery cable
   - Fuel filter/pressure regulator bracket screws
   - Fuel line bundle retaining clip on the left side of the fuel tank

⁂ WARNING

Exercise extreme care when opening the retaining clip. The fuel lines must be retained in this clip; if damaged, the fuel tank assembly must be replaced, since the fitting is not serviced separately.

   - Evaporative Emissions (EVAP) purge line at the 90 degree quick connect fitting
   - Outlet of the fuel filter/regulator from the support on the fuel tank bracket and remove the fuel feed line
   - Fuel feed and return line quick-connect fittings on the fuel filter/regulator
   - Fuel filter

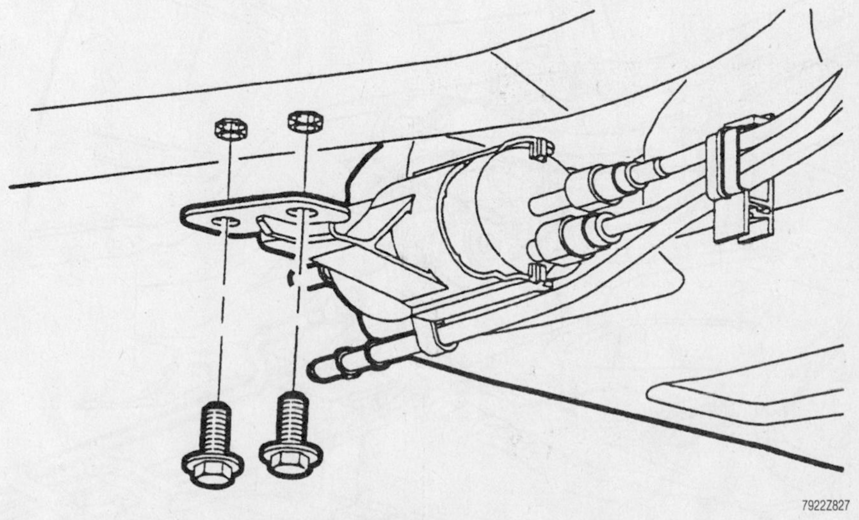

The fuel filter/regulator bracket is held to the frame with 2 bolts–1.9L engines

➡It is not necessary to separate the fuel filter/regulator from the mounting bracket, since both items are serviced as an assembly.

To install:

4. Install or connect the following:
   - Fuel filter
   - New fuel line retainers (3) into the female portion of the quick-connect fuel line fittings
   - Fuel feed and return lines onto the fuel filter/regulator and snap them closed
   - Fuel feed, return, and EVAP purge lines into the fuel tank's retaining clip

⁂ WARNING

Be sure to route the chassis fuel feed and purge lines above the parking brake cable. The parking brake cable must be firmly secured to the underbody to support the fuel and purge lines.

   - 90 degree EVAP purge line quick-connect fitting to the purge line and snap it closed
   - Fuel feed outlet pipe of the fuel filter/regulator into the retaining clip on the fuel tank bracket
   - Fuel filter/regulator bracket mounting screws and torque them to 71 inch lbs. (8 Nm)
   - Negative battery cable
5. Prime the fuel system as follows:
   a. Turn the ignition **ON** for 5 seconds, then **OFF** for 10 seconds.
   b. Repeat the **ON/OFF** cycle 2 more times.
   c. Crank the engine until it starts.
   d. If it does not start, repeat the 3 above steps.
6. Start the engine and check for leaks, repair if necessary.

2.2L And 3.0L Engines

1. Before servicing the vehicle, refer to the precautions in the beginning of this section.
2. Properly relieve the fuel system pressure.
3. Remove or disconnect the following:
   - Negative battery cable
   - Fuel filter bracket screw
   - Fuel feed and return lines from the filter
   - Fuel filter from the bracket

To install:

4. Install or connect the following:
   - New fuel filter into the bracket
   - New fuel line retainers to the female portion of the quick connect fittings
   - Fuel feed and return lines
   - Fuel filter bracket attaching screw and torque it to 35 inch lbs. (4 Nm)
   - Negative battery cable
5. Prime the fuel system as follows:
   a. Turn the ignition **ON** for 5 seconds, then **OFF** for 10 seconds.
   b. Repeat the **ON/OFF** cycle 2 more times.
   c. Crank the engine until it starts.
   d. If it does not start, repeat the 3 above steps.
6. Start the engine and check for leaks, repair if necessary.

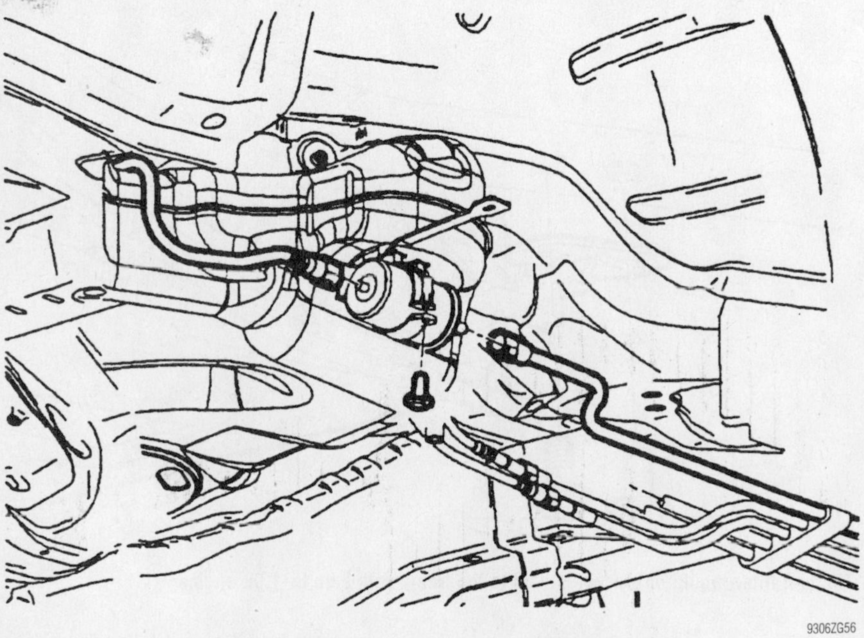

Remove the fuel filter bracket attaching screw 2.2L and 3.0L engines

Fuel Pump

REMOVAL & INSTALLATION

1.9L Engines

To prevent excessive fuel spillage, whenever the tank is removed from the vehicle it should be no more than ½ full. Removal of the fuel pump module assembly requires the removal of the fuel tank.

1. Before servicing the vehicle, refer to the precautions in the beginning of this section.
2. Properly relieve the fuel system pressure.
3. Remove or disconnect the following:
   - Negative battery cable
   - Fuel feed and return lines from the filter/pressure regulator
   - Fuel pump vapor line from the fuel tank vent pipe
   - Fuel pump module retaining ring with service Tool SA9156E. A ½ inch breaker bar will loosen the lockring

➡To prevent bending of the sending unit float arm, lift the pump module up slightly to disengage the orientation tabs in the tank. Rotate the module 90 degrees clockwise until the fuel lines are facing the 1 o'clock position.

   - Fuel pump module from the fuel tank until the bottom of the pump module is close to the fuel tank opening
   - Tilt the pump module approximately 45 degrees to the right hand side of the tank and lift the pump from the fuel tank
   - Fuel pump to tank seal and discard it

To install:

4. Install or connect the following:
   - Fuel pump module into the fuel tank by orientating the float to face the right side and tilting the module approximately 45 degrees
   - Fuel pump into the tank and rotate the assembly 90 degrees counterclockwise to align the modulc tabs with the slots in the tank

➡The fuel pump retaining ring cannot be properly installed if the flange loca-

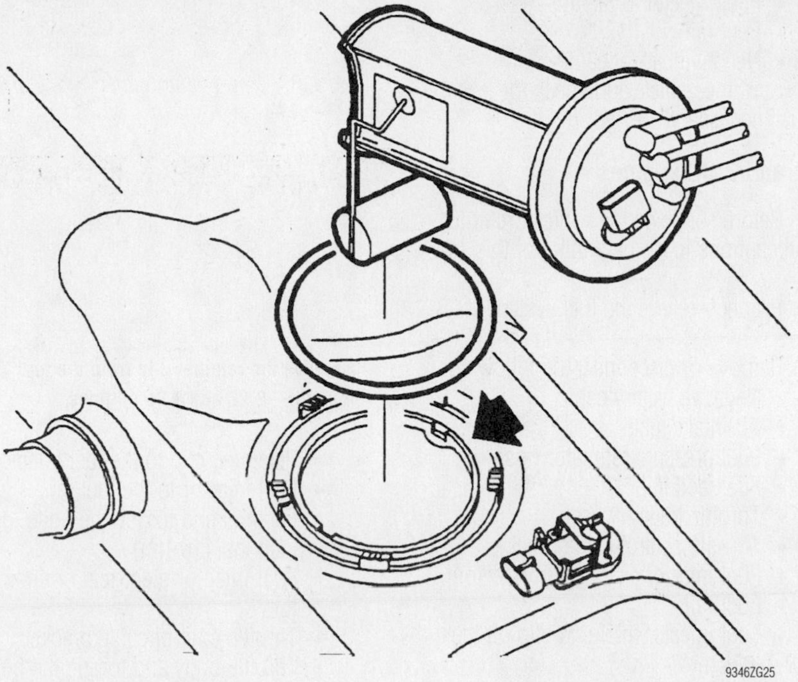
Installing the fuel sending unit and float–1.9L engines

tor tabs are not aligned with the slots in the fuel tank.

- Retaining lockring with Tool SA9156E
- Vapor line to the fuel tank vent pipe
- Fuel feed and return lines to the filter/pressure regulator
- Fuel feed, return and EVAP canister purge line in the fuel tank retaining clip
- Fuel tank
- Negative battery cable

5. Start the vehicle and check for leaks, repair if necessary.

2.2L and 3.0L Engines

1. Before servicing the vehicle, refer to the precautions in the beginning of this section.
2. Properly relieve the fuel system pressure.
3. Remove or disconnect the following:
   - Negative battery cable
   - Fuel tank
   - Fuel lines from the fuel pump module cover
   - Fuel pump module retaining ring with a Sending Unit Wrench, J43827
   - Pull the retaining clip toward the float arm and lift up
   - Fuel pump straight up from the fuel tank
   - Fuel pump tank seal and discard the seal
   - Fuel feed line from the bottom of the fuel pump cover with Clamp Pliers J43914
   - Fuel pump electrical connector

To install:

4. Install or connect the following:
   - Fuel pump feed line to the cover
   - Fuel pump electrical connector
   - Fuel pump to the new seal
   - Fuel pump cover lockring with Tool J-43827
   - Fuel lines and wiring harness
   - Fuel tank
   - Negative battery cable
5. Start the vehicle and check for leaks, repair if necessary.

Fuel Injector

REMOVAL AND INSTALLATION

1.9L Engines

1. Before servicing the vehicle, refer to the precautions in the beginning of this section.
2. Properly relieve the fuel system pressure.
3. Remove or disconnect the following:
   - Negative battery cable
   - Fuel rail
   - Fuel injector retaining clip off the injector
   - Fuel injector
   - Fuel injector O-rings and discard them

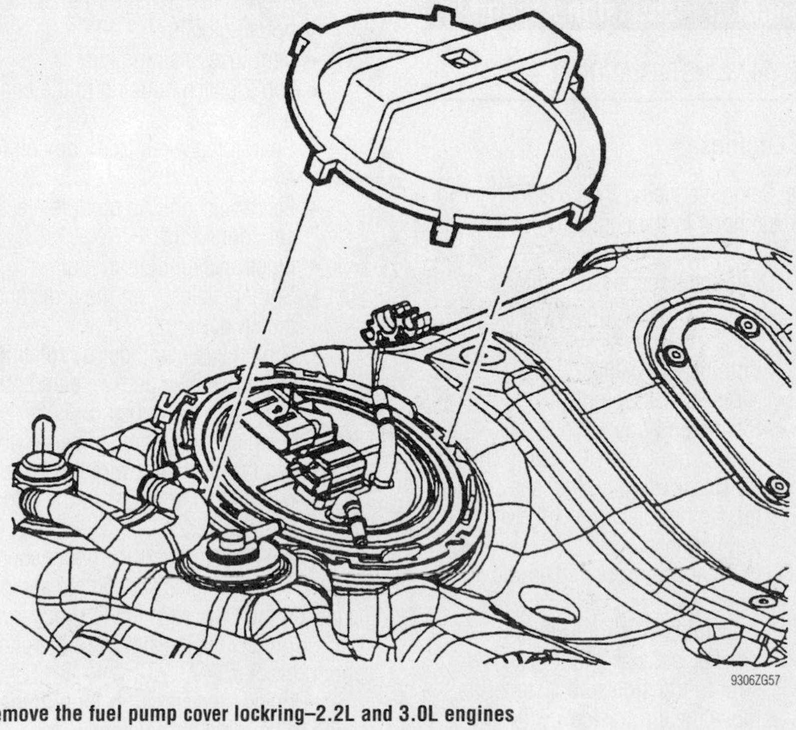
Remove the fuel pump cover lockring–2.2L and 3.0L engines

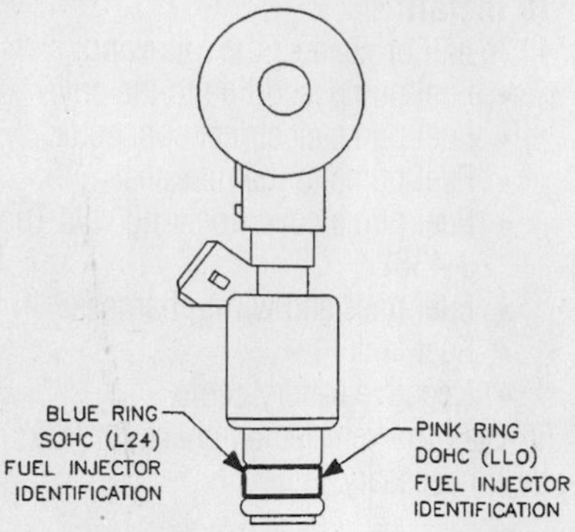

Fuel injector identification–1.9L engine

To install:

➡The SOHC and DOHC fuel injectors are not interchangeable. The injectors are identified by a color coded plastic ring near the bottom of the injector. The SOHC injector has a blue ring near the O-ring that mates to the intake manifold and the DOHC has a pink ring.

4. Lubricate the new fuel injector O-ring with clean engine oil.
5. Install or connect the following:
   - New O-ring seals on the fuel injector
   - Retaining clip to the fuel injector
   - Fuel injector to the fuel rail
   - Fuel rail
   - Negative battery cable
6. Start the vehicle and check for leaks, repair if necessary.

2.2L and 3.0L Engines

1. Before servicing the vehicle, refer to the precautions in the beginning of this section.
2. Properly relieve the fuel system pressure.
3. Remove or disconnect the following:
   - Negative battery cable
   - Air intake tube
   - Fuel pressure regulator hose
   - Fuel feed line
   - Throttle body
   - Throttle control cable bracket
   - Fuel injector electrical connectors
   - Fuel rail
   - Fuel injector retaining clip off the injector
   - Fuel injector
   - Fuel injector O-rings

To install:

4. Lubricate the new fuel injector O-ring with clean engine oil.
5. Install or connect the following:
   - New O-ring seals on the fuel injector

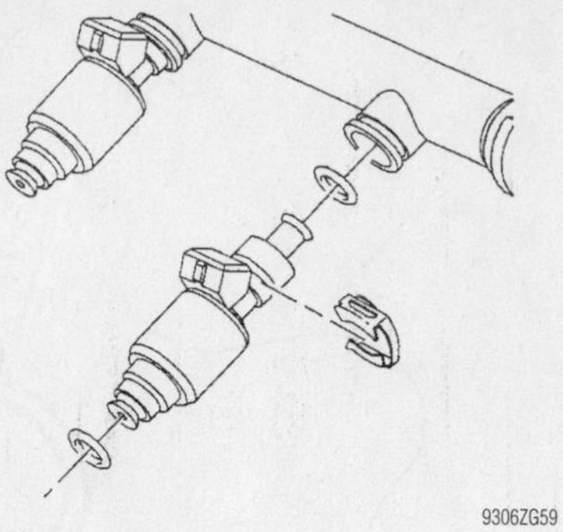

Remove the retainer clip from the fuel injector—2.2L and 3.0L engines

   - Retaining clip to the fuel injector
   - Fuel injector to the fuel rail
   - Fuel rail and torque the bolts to 89 inch lbs. (10 Nm)
   - Fuel injector electrical connectors
   - Throttle control cable bracket
   - Throttle body and torque the bolts to 89 inch lbs. (10 Nm)
   - Fuel feed line
   - Fuel pressure regulator hose
   - Air intake tube
   - Negative battery cable
6. Start the vehicle and check for leaks, repair if necessary.

DRIVE TRAIN

Manual Transmission Assembly

REMOVAL & INSTALLATION

1.9L Engines

1. Before servicing the vehicle, refer to the precautions in the beginning of this section.
2. Drain the transmission fluid.
3. Remove or disconnect the following:
   - Both battery cables
   - Air cleaner assembly
   - Battery and tray
   - Powertrain Control Module (PCM) **J2** (black 28-way) harness connector. Do not disconnect the **J1** (80-way) connector.
   - PCM attaching bolts and move it aside
   - Transmission strut to midrail bracket fastener and loosen the strut to transmission fastener and move the strut aside
   - Back-up lamp electrical connector
   - Vehicle Speed Sensor (VSS) electrical connector, if equipped
   - Ground terminals from the clutch housing bolts
   - Vent tube retaining clip
   - Top 2 clutch housing to the engine bolts
   - Spark plug wires from the coil towers
   - Electronic ignition module electrical connectors
   - Electronic ignition module
   - Shifter cables from the arms and clutch housing
   - Clutch slave cylinder by rotating it ¼turn counterclockwise and pushing it into the housing
   - 2 clutch hydraulic damper to housing fasteners and wire the actuator cylinder and damper to the upper radiator hose
4. Wire the radiator to the upper support.
5. Install an engine support bar and place the feet on the outer edge of the shock tower.
6. Connect the bar hooks to the engine support bracket.
7. Position the stabilizer foot on the engine to the right of the dipstick tube.
8. Adjust the hooks and stabilizer to remove any looseness.
9. Remove or disconnect the following:
   - Both front wheels and inner splash shields
   - Engine splash shield
   - Engine strut cradle bracket fasteners
   - Transmission mount-to-cradle fasteners
   - Front exhaust pipe-to-manifold and catalytic converter fasteners
   - Front exhaust pipe
   - Steering gear to cradle fasteners
   - Brake pipe bracket push pin
   - Engine-to-transmission stiffening brace
   - Dust cover
   - Cotter pin from the lower ball joints
   - Ball joint from the lower control arm with Separator Tool SA9132S
   - Separate the left axle from the transmission and install an axle seal protector
   - Separate the right axle from the intermediate shaft
   - Intake bracket-to-intake manifold bolt on DOHC engines
   - Intake bracket to intermediate shaft support bolt

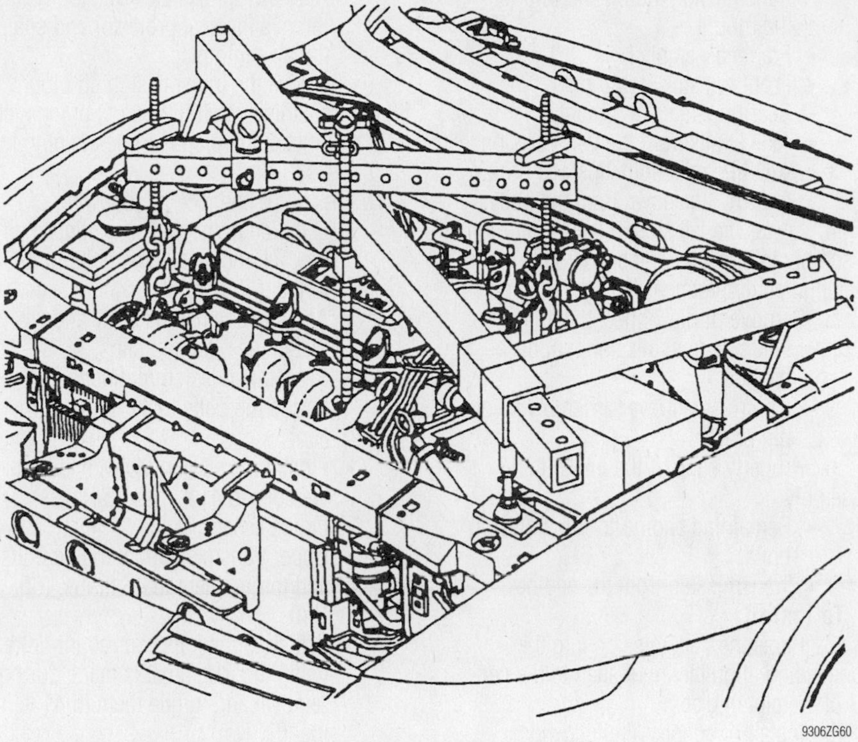

Install an Engine Support tool

- Support bracket
- Intermediate shaft from the transmission
- 4 cradle-to-body bolts and lower the cradle on a support dolly
- Bottom clutch housing bolts and install a guide bolt into the rear housing bolt hole

10. Separate the transmission from the engine.

To install:

11. Place the transmission assembly securely onto a jack and position under the vehicle. Install axle seal protectors into seals on both sides.

12. Raise the transmission into the vehicle to align the input shaft to the center of the clutch. Rotate the transmission back and forth to align the input shaft splines to the clutch disc.

13. Install or connect the following:
- 2 lower clutch housing-to-engine bolts and torque them to 103 ft. lbs. (140 Nm)
- Axle Seal Protector tool to the inside seal and remove the transmission jack
- Left side axle to the transmission. When the splines clear the seal protector, remove the tool.
- Intermediate shaft to the transmission
- Intermediate shaft and torque the bolts to 40 ft. lbs. (54 Nm)
- Starter bracket to the intermediate shaft and torque the bolts to 22 ft. lbs. (30 Nm)
- Intake manifold support bracket (on DOHC) and torque the bolts to 22 ft. lbs. (30 Nm)
- Intake bracket to the intermediate shaft support and torque the bolt to 40 ft. lbs. (54 Nm)
- Right side axle to the intermediate shaft

14. Raise the cradle up on a support dolly and place the ball joints into the knuckles. Verify the correct positioning of the lower control arm bar studs to the knuckles, the cooling module support bushings, the engine strut bracket and the transaxle mount.

15. Insert 9⁄16 in. round steel rods into the cradle-to-body alignment holes near the front cradle to body fastener holes. Guide the cradle into position making sure all mount studs are properly guided into their holes.

16. Be sure the washers are in place, then install the 2 rear cradle to body bolts. Verify proper cradle positioning and install the 2 front cradle bolts, then torque the 4 cradle bolts to 155 ft. lbs. (210 Nm).

17. Remove the support dolly and lower the vehicle sufficiently for underhood access. Remove the engine support bar assembly.

18. Install or connect the following:
- Transmission strut-to-cradle bracket through-bolt and nut and torque them to 40 ft. lbs. (54 Nm)
- Strut cradle bracket-to-cradle bolt and torque the it to 52 ft. lbs. (70 Nm)
- Ignition module and torque the bolts to 71 inch lbs. (8 Nm)
- 2 top transmission housing-to-engine studs and torque them to 74 ft. lbs. (100 Nm)
- VSS electrical connector
- Back-up lamp electrical connector
- Ground terminals to the clutch housing studs
- Vent hose clip to the housing
- Shifter cables to the shift arms and clutch housing
- Hydraulic clutch system to the clutch housing and torque the fasteners to 18 ft. lbs. (25 Nm)
- Actuator to the clutch housing. Push in and rotate the actuator 1⁄4 turn
- Battery tray and torque the bolts to 89 inch lbs. (10 Nm)
- PCM and torque the bolts to 53 inch lbs. (6 Nm)
- Connect the PCM **J2** (28-way) harness
- Battery and hold down retainer and torque the fastener to 80 inch lbs. (9 Nm)
- Air cleaner assembly
- Transmission lower mount-to-cradle fastener and torque it to 37 ft. lbs. (50 Nm)
- 2 engine strut cradle bracket-to-cradle fasteners and torque them to 37 ft. lbs. (50 Nm)
- Steering gear-to-cradle fasteners and torque them to 37 ft. lbs. (50 Nm)
- Brake pipe-to-cradle retainer
- Dust cover and torque the bolts to 8 ft. lbs. (11 Nm)
- Engine-to-transmission stiffening brace and torque the bolts to 40 ft. lbs. (54 Nm)
- Exhaust manifold pipe and torque the bolts to 23 ft. lbs. (31 Nm)
- Intermediate exhaust pipe to the catalytic converter and torque the bolts to 18 ft. lbs. (25 Nm)

- Front exhaust pipe support to the stiffening bracket and torque the fasteners to 89 inch lbs. (10 Nm)
- Both front wheels
- Both battery cables

19. Fill the transmission to the proper level.

20. Warm the engine and check the transmission fluid. Check and adjust vehicle alignment, as necessary.

2.2L and 3.0L Engines

1. Before servicing the vehicle, refer to the precautions in the beginning of this section.
2. Drain the transmission fluid.
3. Remove or disconnect the following:
   - Both battery cables
   - Battery feed to the underhood fuse block
   - Release the retaining tabs on the fuse block cover and remove it
   - Engine 68-way electrical connector
   - Forward lamp 68-way connector
   - Forward lamp 2-way (white) connector
   - Black, green and brown instrument panel 2-way connectors
   - Fuse block from the case
   - Engine, forward lamp and instrument panel harness from the fuse block case
   - Case from the battery tray
   - Battery tray and battery
   - Loosen the pinch bolt securing the control assembly rod to the control shaft lever

➡The control shaft lever retaining pin has a spring loaded locking feature to keep the pin in place. The spring must be depressed before attempting to remove it.

- Pin retaining the control shaft lever to shift control shaft
- Retaining clips holding the control shaft lever to the transmission and frame brackets
- Control shaft lever
- Back-up lamp switch electrical connector
- Wire harness from the transmission
- Spring loaded locking pin from the hole on top of the transmission, if equipped
- Clutch hydraulic fitting from the transmission
- Upper transmission to engine bolts and install and engine support bar, SA9105E, and Adapter Kit J43405
- Matchmark the position of the left transmission mount and remove the mount
- Frame assembly
- Front transmission mount
- Rear transmission mount
- Axle shafts from the transmission
- A/C line retaining clips and drop the A/C line down from the body

4. Lower the left side of the powertrain assembly to allow the transmission to clear the engine compartment rail.
5. Remove or disconnect the following:
   - Left side transmission mount bracket
   - Control shaft lever assembly pivot pin bracket
6. Properly support the powertrain assembly
   - Remaining engine-to-transmission bolts
   - Transmission from the engine

To install:

7. Raise the transmission into the vehicle and align the input shaft to the center of the clutch disc.
8. Install or connect the following:
   - Lower transmission to engine bolts and torque them to 48 ft. lbs. (65 Nm)
   - Control shaft lever assembly pivot pin bracket and torque the fastener to 18 ft. lbs. (24 Nm)
   - Axle shafts to the transmission. After the splines clear the tool, remove the seal protector and snap the axle into place.
9. Remove the transmission fill plug, fill the transmission with 2 quarts of manual fluid and install the plug. Torque the plug to 22 ft. lbs. (30 Nm).
10. Install or connect the following:
    - Left side transmission mount bracket and torque the bolts to 41 ft. lbs. (55 Nm)
    - A/C line against the body and install the retaining clips
    - Rear transmission mount and torque the bolts to 41 ft. lbs. (55 Nm)
    - Front transmission mount and torque the bolts to 41 ft. lbs. (55 Nm)
    - Frame assembly
    - Upper transmission to engine bolts and torque them to 48 ft. lbs. (65 Nm)
    - Left side transmission mount bolts using the matchmarks made during removal and torque them to 41 ft. lbs. (55 Nm)
    - Back-up lamp switch electrical connector
    - Control shaft lever assembly onto the pivot pin brackets
    - Retaining clips holding the control shaft lever to the transmission and frame brackets

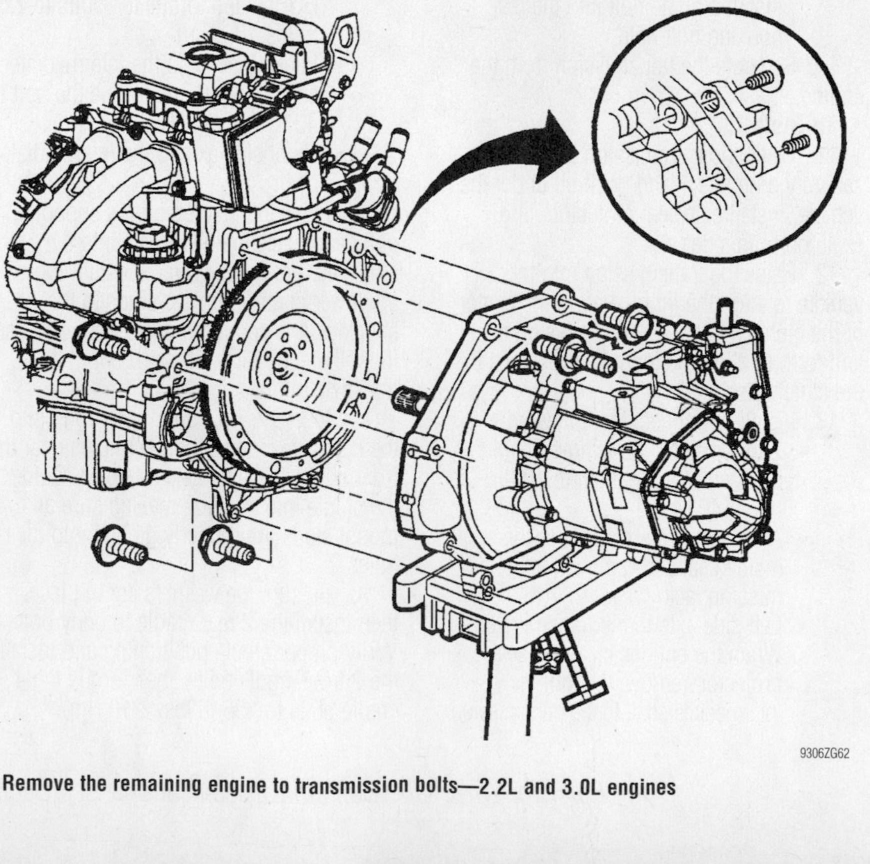

Remove the remaining engine to transmission bolts—2.2L and 3.0L engines

- Pin to retain the control shaft lever to the shift control shaft
- Rotate the shift control shaft clockwise and install the shift linkage lock pin
- Move the shift lever to the 5th gear/reverse gate and install a 3/8 inch punch
- Pinch bolt securing the control assembly rod to the control shaft lever and torque it to 9 ft. lbs. (12 Nm) plus 180 degrees
- Shift lever hole plug
- Clutch hydraulic fitting to the transmission and bleed the hydraulic system
- Battery tray and torque the fasteners to 11 ft. lbs. (15 Nm)
- Underhood fuse block to the battery tray and torque the fasteners to 80 inch lbs. (9 Nm)
- Wire harness to the fuse case
- Snap the fuse block onto the hinges
- Engine 68-way connectors
- Forward lamp 68-way connectors
- Instrument panel 68-way connectors
- Forward 2-way (white) connector
- Black, green and brown 2-way connectors
- Snap the fuse block to the case
- Coolant hose to the fuse block
- Battery feed to the fuse block and torque the fastener to 12 inch lbs. (16 Nm)
- Battery tray and hold-down bracket and torque the fastener to 15 ft. lbs. (20 Nm)
- Battery
- Fan control module to the battery hold down bracket
- Both battery cables

11. Fill the transmission to the proper level.

12. Warm the engine and check the transmission fluid. Check and adjust vehicle alignment, as necessary.

Automatic Transmission Assembly

REMOVAL & INSTALLATION

1.9L Engines

1. Before servicing the vehicle, refer to the precautions in the beginning of this section.
2. Drain the transmission fluid.
3. Remove or disconnect the following:
   - Both battery cables
   - Air cleaner assembly
   - Battery and tray
   - Powertrain Control Module (PCM) **J2** (black 28-way) harness connector. Do not disconnect the **J1** (80-way) connector.
   - PCM attaching bolts and move aside
   - Transmission strut to midrail bracket fastener and loosen the strut to transmission fastener and move the strut aside
   - Transmission solenoid harness connector
   - Vehicle speed and input sensor harness connectors
   - Transmission fluid temperature sensor connector
   - Range switch harness connector
   - Ground terminals from the converter housing
   - Ground wire from the range switch
   - Spark plug wires from the coil towers
   - Electronic ignition module electrical connectors
   - Electronic ignition module
4. Wire the radiator to the upper support.
5. Install an engine support bar and place the feet on the outer edge of the shock tower.
6. Connect the bar hooks to the engine support bracket.
7. Position the stabilizer foot on the engine to the right of the dipstick tube.
8. Adjust the hooks and stabilizer to remove any looseness.
9. Remove or disconnect the following:
   - Both front wheels and inner splash shields
   - Engine splash shield
   - Engine strut cradle bracket fasteners
   - Transmission mount to cradle fasteners
   - Front exhaust pipe to manifold and catalytic converter fasteners
   - Front exhaust pipe
   - Steering gear to cradle fasteners
   - Brake pipe bracket push pin

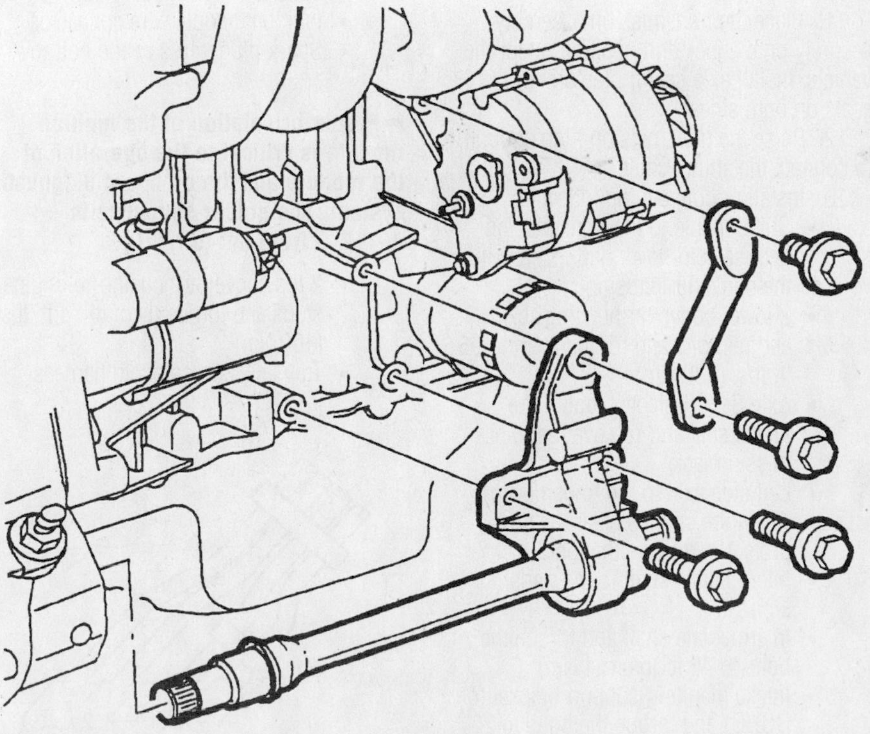

9346ZG26

Remove the intake-to-intermediate shaft support bracket and the intermediate shaft–1.9L engines

- Engine-to-transmission stiffening brace
- Dust cover
- Torque converter-to-flywheel bolts
- Ball joint from the lower control arm
- Separate the left axle from the transmission and install an axle seal protector
- Separate the right axle from the intermediate shaft
- Intake bracket-to-intake manifold bolt (on DOHC engines)
- Intake bracket-to-intermediate shaft support bolt
- Support bracket
- Intermediate shaft from the transmission
- Transmission cooler lines
- 4 cradle to body bolts and lower the cradle on a support dolly
- Lower torque converter bolts
- Separate the transmission from the engine
- Shifter cable from the converter housing
- Transmission from the vehicle

To install:

10. Flush the transmission oil cooler lines.

11. Place the transmission assembly securely onto a jack and position under the vehicle. Install axle seal protectors into seals on both sides.

12. Raise the transmission high enough to connect the shifter cable.

13. Install or connect the following:
   - Shifter cable to the transmission gear selector lever and insert it in the converter housing
   - 2 lower converter housing-to-engine bolts and torque them to 96 ft. lbs. (130 Nm)
   - Axle Seal Protector tool to the inside seal and remove the transmission jack
   - Left side axle to the transmission. When the splines clear the seal protector, remove the tool.
   - Intermediate shaft to the transmission
   - Intermediate shaft and torque the bolts to 40 ft. lbs. (54 Nm)
   - Intake manifold support bracket (on DOHC) and torque the bolts to 22 ft. lbs. (30 Nm)
   - Intake bracket to the intermediate shaft support and torque the bolt to 40 ft. lbs. (54 Nm)
   - Right side axle to the intermediate shaft
   - Cooler lines to the transmission

14. Raise the cradle up on a support dolly and place the ball joints into the knuckles. Verify the correct positioning of the lower control arm bar studs to the knuckles, the cooling module support bushings, the engine strut bracket and the transaxle mount.

15. Insert 9⁄16 in. round steel rods into the cradle-to-body alignment holes near the front cradle to body fastener holes. Guide the cradle into position making sure all mount studs are properly guided into their holes.

16. Be sure the washers are in place, then install the 2 rear cradle to body bolts. Verify proper cradle positioning and install the 2 front cradle bolts, then torque the 4 cradle bolts to 155 ft. lbs. (210 Nm).

17. Remove the support dolly and lower the vehicle sufficiently for underhood access. Remove the engine support bar assembly.

18. Install or connect the following:
   - Transmission strut-to-cradle bracket through-bolt and nut and torque them to 40 ft. lbs. (54 Nm)
   - Strut midrail bracket and torque the bolt to 52 ft. lbs. (70 Nm)
   - Ignition module and torque the bolts to 71 inch lbs. (8 Nm)
   - Ignition module wire connector
   - Spark plug wires to the coil towers

➡Proper orientation of the ignition module is critical to the operation of the module and the on board diagnostic system. The proper sequence is 4–1–2–3 from left to right.

   - 2 top converter housing-to-engine studs and torque them to 74 ft. lbs. (100 Nm)
   - Transmission solenoid harness connector and torque the fastener to 22 inch lbs. (2.5 Nm)
   - Vehicle speed and input sensor connectors
   - Transmission fluid temperature sensor connectors
   - Transmission range switch harness connectors
   - Ground terminals to the top 2 converter housing bolts and torque the bolts to 19 ft. lbs. (25 Nm)
   - Ground wire to the neutral selector switch
   - Adjust the control cable
   - Air cleaner assembly
   - Battery tray and torque the bolts to 89 inch lbs. (10 Nm)
   - PCM and torque the bolts to 53 inch lbs. (6 Nm)
   - Connect the PCM **J2** (28-way) harness
   - Battery and hold down retainer and torque the fastener to 80 inch lbs. (9 Nm)
   - Transmission lower mount to cradle fastener and torque it to 37 ft. lbs. (50 Nm)
   - 2 engine strut cradle bracket-to-cradle fasteners and torque them to 37 ft. lbs. (50 Nm)
   - Steering gear to cradle fasteners and torque them to 37 ft. lbs. (50 Nm)
   - Brake pipe to cradle retainer
   - Torque converter to flexplate and torque the bolts to 52 ft. lbs. (70 Nm)
   - Dust cover and torque the bolts to 8 ft. lbs. (11 Nm)
   - Engine to transmission stiffening brace and torque the bolts to 40 ft. lbs. (54 Nm)
   - Exhaust manifold pipe and torque the bolts to 23 ft. lbs. (31 Nm)
   - Intermediate exhaust pipe to the catalytic converter and torque the bolts to 18 ft. lbs. (25 Nm)
   - Front exhaust pipe support to the stiffening bracket and torque the fasteners to 23 ft. lbs. (31 Nm)
   - Lower ball joint and torque the nut to 55 ft. lbs. (75 Nm)
   - New cotter pins
   - Splash shields
   - Both front wheels
   - Both battery cables

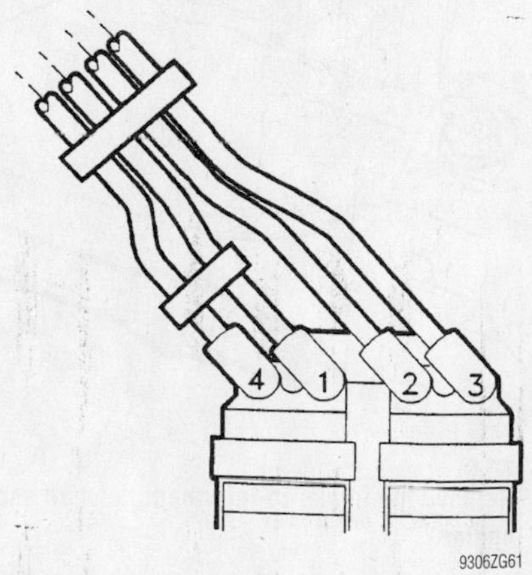

Electronic ignition module wire sequence 4–1–2–3–1.9L engines

19. Fill the transmission to the proper level.

20. Warm the engine and check the transmission fluid. Check and adjust vehicle alignment, as necessary.

2.2L And 3.0L Engines

1. Before servicing the vehicle, refer to the precautions in the beginning of this section.
2. Drain the transmission fluid.
3. Remove or disconnect the following:
   - Both battery cables
   - Move the fan control module up and out of the way
   - Battery feed to the underhood fuse block
   - Coolant hose from the underhood fuse block
   - Release the retaining tabs on the fuse block cover and remove it
   - Engine 68-way electrical connector
   - Forward lamp 68-way connector
   - Forward lamp 2-way (white) connector
   - Black, green and brown instrument panel 2-way connectors
   - Fuse block from the case
   - Engine, forward lamp and instrument panel harness from the fuse block case
   - Case from the battery tray
   - Battery tray and battery
   - Control cable from the range switch lever
   - Control cable bracket from the rear powertrain mount
   - Transmission electrical connector and ground wire
   - Remaining wire harness connectors from the transmission
   - Transmission range switch electrical connector
   - Upper engine to transmission bolts
   - Drive belt from the alternator, for 3.0L engine
   - Drive belt tensioner, for 3.0L engine
   - Alternator, for 3.0L engine
   - Output speed sensor electrical connector
4. Install an Engine Support Fixture.
   - Frame
   - Axle shafts from the transmission
   - Engine to transmission bracket
   - Starter solenoid electrical connector
   - Starter motor

➡The wire harness mounting bracket on the rear of the engine must be removed on the 3.0L.

   - Torque converter-to-flexplate bolts through the starter opening
   - Transmission oil cooler lines
   - 2 lower engine to transmission bolts
   - Left side transmission mount bolts
5. Slide a hydraulic lifting table under the transmission and remove the remaining mounting bolts.
6. Separate the transmission from the engine.

To install:

7. With the transmission on a hydraulic lift, raise the transmission to the engine.
8. Install or connect the following:
   - Rear wire harness bracket, for 3.0L engine
   - Engine to transmission mounting bolts and torque them to 48 ft. lbs. (60 Nm)
   - Left side transmission mount bolts and torque them to 41 ft. lbs. (55 Nm)
   - Bottom 2 engine to transmission mounting bolts and torque them to 48 ft. lbs. (65 Nm)
   - Transmission oil cooler lines after lubricating them with automatic transmission fluid and torque the fasteners to 71 inch lbs. (8 Nm)
   - Loosely install the flex plate to torque converter bolts. When aligned properly, torque the bolts to 33 ft. lbs. (45 Nm).
   - Starter motor and torque the bolts to 30 ft. lbs. (40 Nm)
   - Starter solenoid electrical connectors
   - Engine to transmission bracket and torque the bolts to 26 ft. lbs. (35 Nm)
   - New axle shaft retaining rings on the end of the output shafts and install the axle shafts to the transmission
   - Frame assembly
   - Output speed sensor electrical connector and secure it to the mounting stud
   - Alternator and torque the bolts to 30 ft. lbs. (40 Nm), for 3.0L engine
   - Drive belt tensioner and torque the bolts to 30 ft. lbs. (40 Nm), for 3.0L engine
   - Drive belt, for 3.0L engine
   - Remaining transmission to engine mounting bolts and torque them to 48 ft. lbs. (65 Nm)
   - Transmission electrical connector and ground wire and torque the fasteners to 18 ft. lbs. (25 Nm)
   - Wire harness attachments to the transmission side cover and torque the fasteners to 15 ft. lbs. (20 Nm)
   - Control cable mounting bracket and

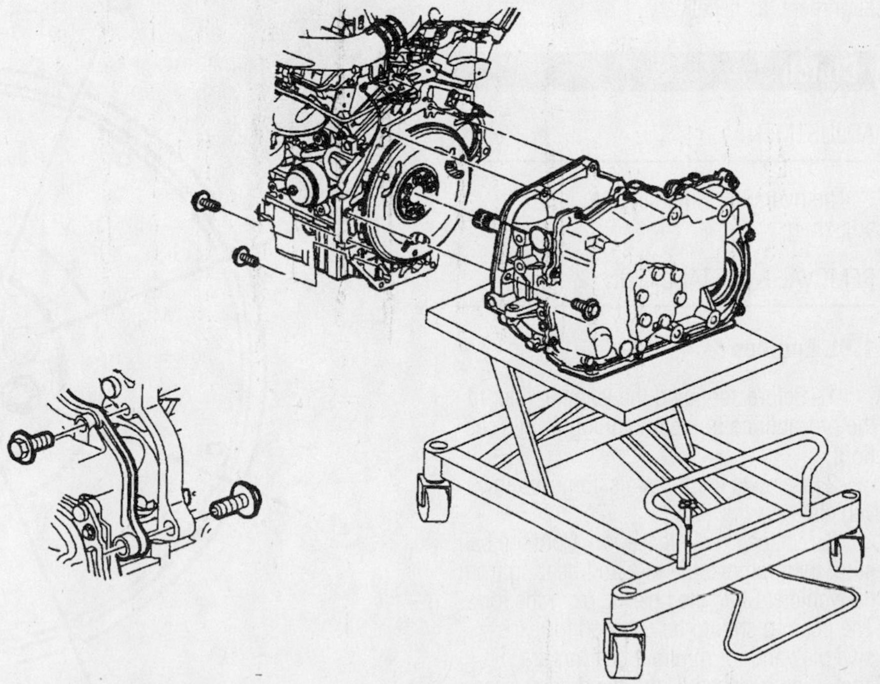

Lifting table mounting points–2.2L and 3.0L engines

torque the bolts to 15 ft. lbs. (20 Nm)

- Control cable to the range switch lever
- Battery tray and torque the fasteners to 11 ft. lbs. (15 Nm)
- Underhood fuse block case to the battery tray. Torque the bolt to 80 inch lbs. (9 Nm) and snap the block onto the hinges.
- Engine 68-way connectors
- Forward lamp 68-way connectors
- Instrument panel 68-way connectors
- Forward 2-way (white) connector
- Black, green and brown 2-way connectors
- Snap the fuse block to the case
- Coolant hose to the fuse block
- Battery feed to the fuse block and torque the fastener to 12 inch lbs. (16 Nm)
- Battery tray and hold-down bracket and torque the fastener to 15 ft. lbs. (20 Nm)
- Battery
- Fan control module to the battery hold down bracket
- Both battery cables

9. Fill the transmission to the proper level.

10. Warm the engine and check the transmission fluid. Check and adjust vehicle alignment, as necessary.

Clutch

ADJUSTMENT

The hydraulic clutch system is self-adjusting.

REMOVAL & INSTALLATION

1.9L Engines

1. Before servicing the vehicle, refer to the precautions in the beginning of this section.

2. Remove the transmission from the vehicle.

3. Unsnap the release fork from the ball stud, then remove the fork and bearing from the vehicle. Slide the bearing from the fork. The bearing should be checked for excessive play and for minimal bearing drag. It should be replaced if no/little drag or excessive play is found.

➡The release bearing is packed with grease and should not be washed with solvent.

4. Using a feeler gauge, measure the distance between the pressure plate and flywheel surfaces in order to determine clutch face thickness. Replace the clutch disc if it is not within 0.205–0.287 in. (5.2–7.3mm) for 1998–00 or to 0.18 –0.28 in. (4.65mm–7.10mm) for 2001.

5. Remove the pressure plate-to-flywheel bolts in a progressive crisscross pattern to prevent warping the cover, then remove the pressure plate and clutch disc.

6. Inspect the pressure plate, as follows:

a. Check for excessive wear, chatter marks, cracks or overheating (indicated by a blue discoloration). Black random spots on the friction surface of the pressure plate is normal.

b. Check the plate for warpage using a straightedge and a feeler gauge; the maximum allowable warpage is 0.006 in. (0.15mm).

c. Replace the plate, if necessary.

7. Inspect the clutch disc, as follows:

a. Check the disc face for oil or burnt spots.

b. Check the disc for loose damper springs, hub or rivets.

c. Replace the disc, if necessary.

8. Check the flywheel, as follows:

a. Check the ring gear for wear or damage.

b. Check the friction surface for excessive wear, chatter marks, cracks or overheating.

c. Check flywheel thickness; the minimum allowable is 1.102 in. (28mm).

d. Measure flywheel run-out using a dial indicator, positioned for at least 2 flywheel revolutions. Push the crankshaft forward to take up thrust bearing clearance. Maximum flywheel run-out is 0.006 in. (0.15mm).

e. Check the flywheel for warpage using a straight-edge and a feeler gauge; the maximum allowable warpage is 0.006 in. (0.15mm).

f. Replace the flywheel, if necessary.

9. If necessary, remove the flywheel retaining bolts and remove the flywheel from the crankshaft.

To install:

10. If removed, install the flywheel and torque the bolts in a crisscross sequence to 59 ft. lbs. (80 Nm).

11. Install or connect the following:

- Clutch disc and pressure plate with the yellow dot on the pressure plate aligned as close as possible to the mark on the flywheel. The clutch

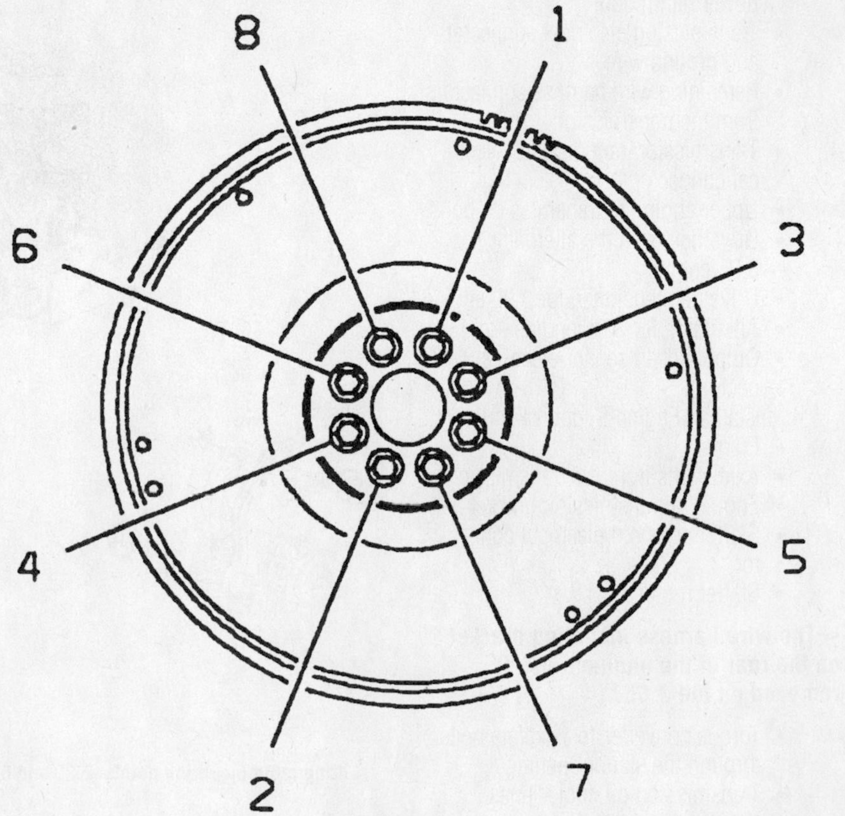

Flywheel mounting bolt tightening sequence—1.9L engines

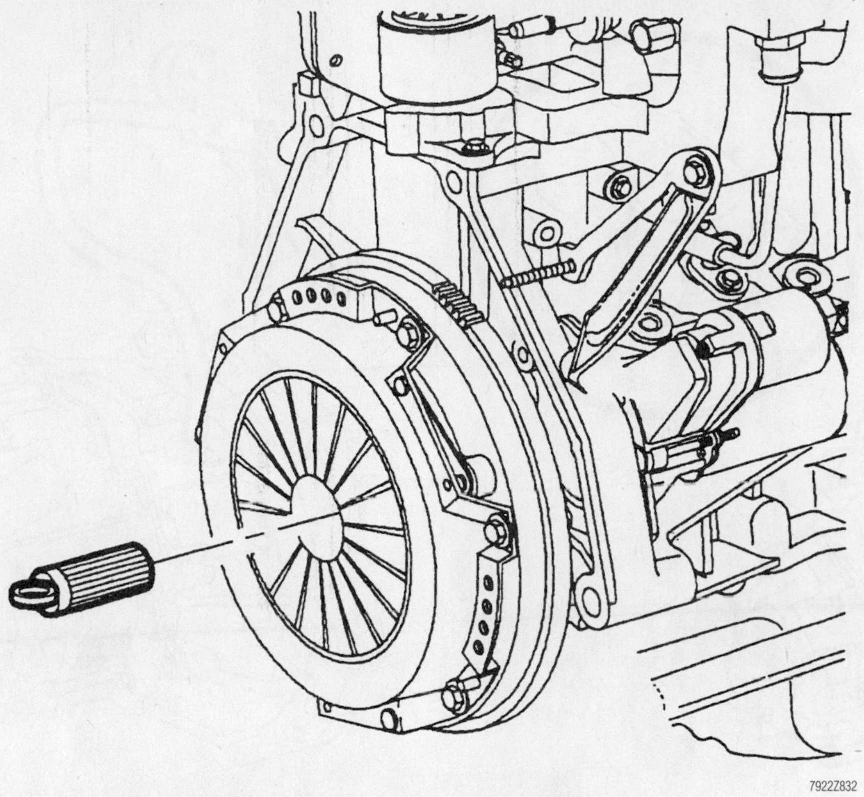

Use a proper clutch alignment tool before tightening the pressure plate retaining bolts—1.9L engines

disc is labeled FLYWHEEL SIDE in order to help correctly position the disc. Loosely install the pressure plate bolts.
- Clutch alignment tool in the clutch disc and install the release bearing to the fork and torque the pressure plate bolts to 18 ft. lbs. (25 Nm)
- Snap the release bearing and fork onto the ball stud
- Lubricate the splines of the input shaft lightly with a high temperature grease
- Transmission assembly
- Negative battery cable

2.2L and 3.0L Engines

1. Before servicing the vehicle, refer to the precautions in the beginning of this section.
2. Remove the transmission from the vehicle.
3. Check the slave cylinder release bearing minimal bearing drag. Replace the slave cylinder if little or drag is found.
4. Remove the pressure plate and clutch disc.
5. Inspect the pressure plate, as follows:
   a. Check for excessive wear, chatter marks, cracks or overheating (indicated by a blue discoloration). Black random spots on the friction surface of the pressure plate is normal.
   b. Check the plate for warpage using a straightedge and a feeler gauge; the maximum allowable warpage is 0.006 in. (0.15mm).
   c. Replace the plate, if necessary.
6. Inspect the clutch disc, as follows:
   a. Check the disc face for oil or burnt spots.
   b. Check the disc for loose damper springs, hub or rivets.
   c. Replace the disc, if necessary.
7. Check the flywheel, as follows:
   a. Check the ring gear for wear or damage.
   b. Check the friction surface for excessive wear, chatter marks, cracks or overheating.
   c. Check flywheel thickness; the minimum allowable is 1.102 in. (28mm).
   d. Measure flywheel run-out using a dial indicator, positioned for at least 2 flywheel revolutions. Push the crankshaft forward to take up thrust bearing clearance. Maximum flywheel run-out is 0.006 in. (0.15mm).
   e. Check the flywheel for warpage using a straight-edge and a feeler gauge; the maximum allowable warpage is 0.006 in. (0.15mm).
   f. Replace the flywheel, if necessary.
8. If necessary, remove the flywheel retaining bolts and remove the flywheel from the crankshaft.

To install:

9. Install or connect the following:
   - Flywheel (if removed) and torque the bolts in a crisscross pattern to 39 ft. lbs. (53 Nm) plus 25 degrees
   - Clutch disc and pressure plate and loosely install the pressure plate bolts
   - Clutch alignment tool in the clutch disc, and push in until it bottoms out in the crankshaft
10. Tighten the pressure plate bolts using multiple passes of a crisscross sequence to 11 ft. lbs. (15 Nm) and remove the alignment tool.
11. Lubricate the splines of the input shaft lightly with a high temperature grease.
12. Install the transmission assembly.
13. Connect the negative battery cable.

Hydraulic Clutch System

BLEEDING

1.9L Engines

The clutch hydraulic assembly has been filled with fluid and bled of air at the factory. Do not attempt to bleed the hydraulic system. While the unit does not require periodic checking, it must be serviced, when necessary, as a complete assembly. The system is full when the reservoir is half full.

Only DOT 3 brake fluid should be added to the system. If the fluid level drops, inspect the system, including the slave cylinder, for leakage. A slight wetting of the slave cylinder surface is normal. Fill the clutch master cylinder reservoir with brake fluid. Be careful not to spill brake fluid on the painted surface of the vehicle.

2.2L and 3.0L Engines

TRANSMISSION IN VEHICLE

This procedure outlines how to bleed the hydraulic clutch with the transmission in the vehicle. Only **DOT 3**brake fluid should be added to the system.

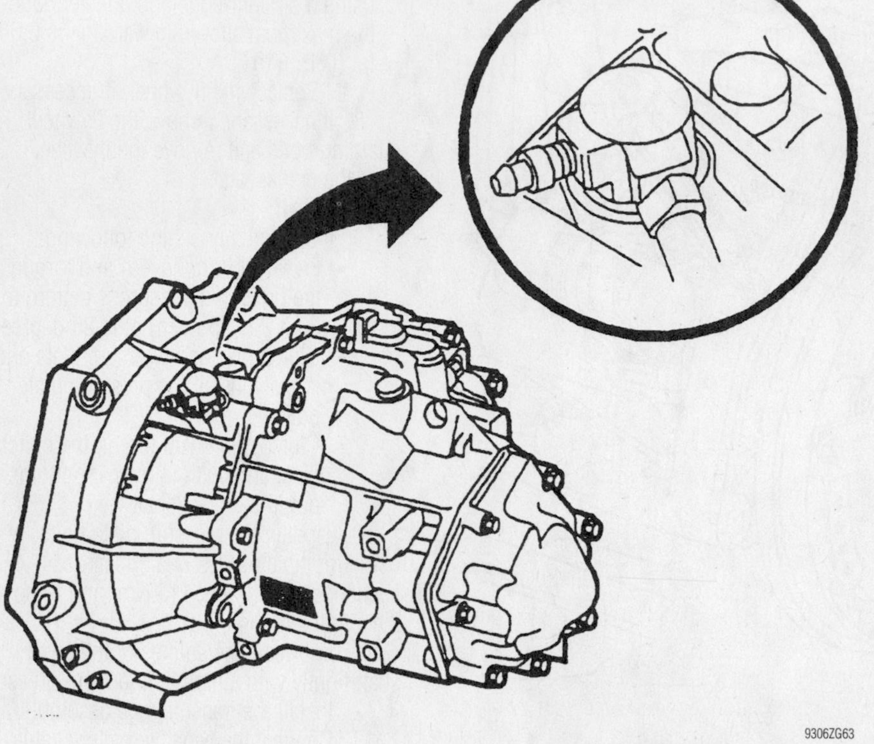

Loosen the clutch bleeder screw on the hydraulic fitting

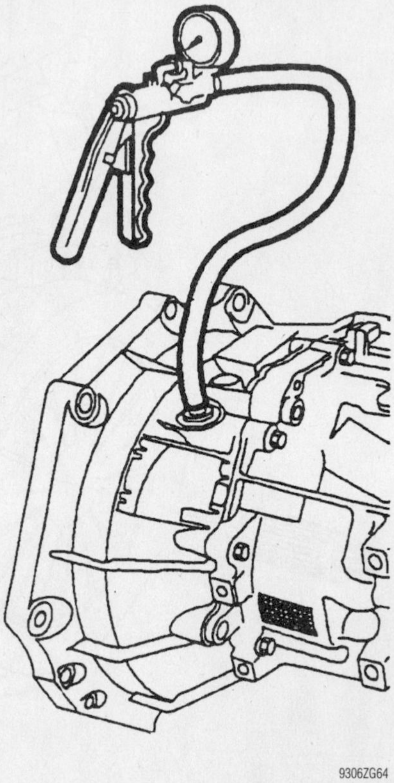

Install a vacuum pump and pressure adapter to bleed the slave cylinder.

1. Before servicing the vehicle, refer to the precautions in the beginning of this section.
2. Remove or disconnect the following:
3. Remove the reservoir cap and fill the reservoir with new brake fluid.
4. Install a Bleeder Adapter Tool J43915, to the reservoir and connect a pressure bleeder to the adapter.
5. Charge the pressure bleeder to 20–25 psi (138–172 kPa).
6. Attach a transparent hose over the clutch bleeder screw nipple and submerge the opposite end of the hose in a container of brake fluid.
7. Loosen the bleeder screw on the transmission hydraulic fitting.
8. Bleed the system until no air bubbles are seen in the hose.
9. Tighten the bleeder screw.
10. Check the clutch pedal for a spongy feel. If the pedal feels soft, repeat the bleeding procedure.
11. Remove the bleeder tools and top off the fluid level if necessary.

TRANSMISSION OUT OF VEHICLE

This procedure outlines how to bleed the clutch slave cylinder while the transmission is out of the vehicle. Only **DOT 3**brake fluid should be added to the system.

1. Before servicing the vehicle, refer to the precautions in the beginning of this section.
2. Remove or disconnect the following:
3. Connect a long piece of transparent hose to the slave cylinder fitting.
4. Depress the release bearing on the slave cylinder towards the clutch housing and release it.
5. The release bearing will spring back, but the piston will remain depressed next to the clutch housing.
6. Fill the hose with 13.8 inches (350mm) of DOT 3 brake fluid.
7. Connect a Vacuum Pump SA9180NE with a Pressure Adapter J35555-92, to the top of the hose.
8. Apply pressure to the hose to force the brake fluid into the slave cylinder. Pressure in the gauge will increase when the slave cylinder piston reaches the end of its travel.
9. Remove the vacuum pressure pump from the hose and depress the release bearing. Air bubbles should appear in the hose.
10. Repeat this procedure until no air bubbles are visible in the hose.
11. Drain and remove the hose..

Halfshafts

REMOVAL & INSTALLATION

1.9L Engines

1. Before servicing the vehicle, refer to the precautions in the beginning of this section.
2. Remove the wheel cover or the center cap for access to the halfshaft nut. Have an assistant depress the brake pedal and loosen the front halfshaft nut.
3. Remove or disconnect the following:
   - Wheel and splash shields
   - 2 push pins
   - Lower shield to cradle molded-in fasteners at the cradle
   - Drain the transmission fluid if replacing the left side halfshaft
   - Drive axle nut and washer and discard them
   - Lower control arm to steering knuckle cotter pin and discard it
   - Loosen the lower control arm to steering knuckle castle nut so that the top of the nut is even with the top of the ball stud

WARNING

The outer CV-joint, if equipped with Anti-lock Brake System (ABS), contains a speed sensor ring. Use of an incorrect tool to separate the control arm from the knuckle may result in damage and loss of the ABS system.

- Lower control arm from the steering knuckle by using a Lower Control Arm Separator Tool SA9132S
- Cotter pin and castle nut and discard them, if equipped
- Torque prevailing nut and discard it, if equipped
- Separate the tie rod end from the steering knuckle using a Tie Rod Separator Tool, SA91100C
- Lower control arm to steering knuckle castle nut
- Position a pry bar at the proper cradle and front stabilizer and place a cloth over the ball stud boot
- Separate the lower control arm from the steering knuckle and pull the knuckle away from the ball stud
- While pulling the knuckle/strut away, pull the outer end of the halfshaft out of the wheel hub and properly support the halfshaft

4. For right side halfshafts, remove the halfshaft from the intermediate shaft by tapping the axle with a block of wood and hammer. Separate the halfshaft from the intermediate shaft.
5. For left side halfshafts, remove the halfshaft by installing a prybar and prying the halfshaft from the transmission.

To install:

➡Be careful not to damage the halfshaft oil seal when installing the halfshaft into the transmission.

6. Install or connect the following:
   - Transaxle Seal Protector Tool SA91112T, when installing the left side halfshaft
   - Halfshaft into the transmission. After the splines have safely passed the oil seal, remove the protector tool
   - Insert the inner end of the halfshaft onto the outer end of the intermediate shaft for the right side halfshaft. Push on the axle firmly to engage the retaining ring
   - Outer end of the halfshaft into the wheel hub. Do not install any hardware at this time
   - Lower control arm ball stud into the steering knuckle. Hand-tighten the castle nut at this time
   - Attach the tie rod end to the steering knuckle, if equipped. Torque the nut to 33 ft. lbs. (45 Nm) and install a new cotter pin
   - If equipped with a torque prevailing nut, attach the tie rod to steering knuckle with a new nut. Do not allow the ball stud to turn while tightening the prevailing nut. Torque the nut to 41 ft. lbs. (55 Nm).
   - Axle to hub washer and new nut. Torque the nut to 148 ft. lbs. (200 Nm).

➡When tightening the axle nut, have an assistant depress the brake pedal to prevent axle rotation.

   - Splash shields
   - Align the molded in shield fasteners with the holes in the cradle and install push pins
   - Wheel

7. Top off the transmission with the proper fluid.
8. Check and adjust the front end alignment as necessary.

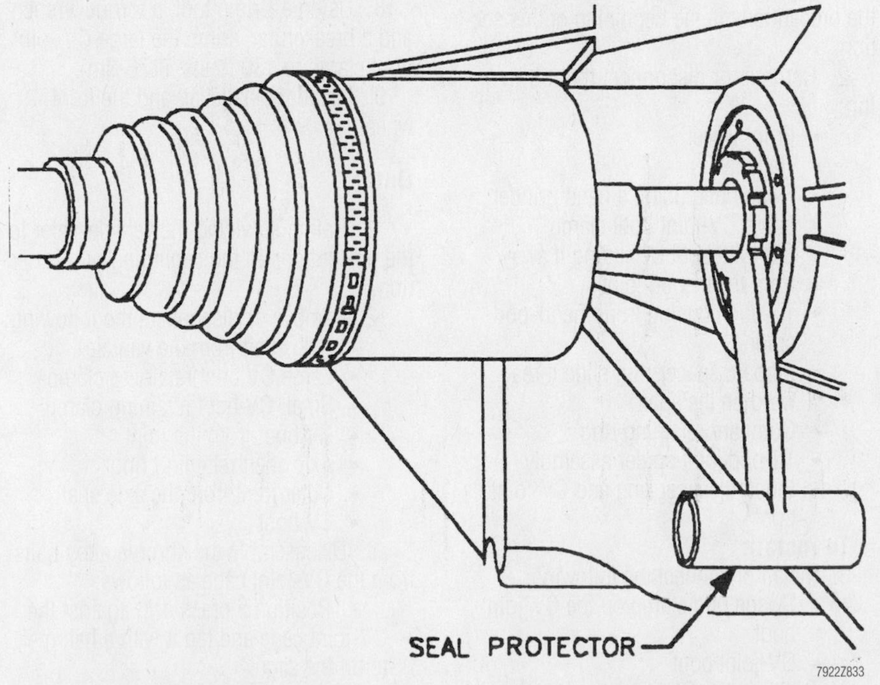

Failure to use a seal protector may allow the halfshaft splines to damage the transaxle seal

2.2L and 3.0L Engines

1. Before servicing the vehicle, refer to the precautions in the beginning of this section.
2. Remove the wheel cover or the center cap for access to the halfshaft nut. Have an assistant depress the brake pedal and loosen the front halfshaft nut.
3. Remove or disconnect the following:
   - Wheel
   - Tie rod end torque prevailing nut and discard it
   - Tie rod end from the steering knuckle, using a Tie Rod Separator Tool SA91100C
   - Lower control arm to steering knuckle

➡Do not allow the steering knuckle to contact the ball stud seal. Contact may cause the seal to rip and the ball stud will need replacement.

   - Pull the steering knuckle/strut assembly away from the vehicle and pull the halfshaft out of the hub
   - Properly support the halfshaft and remove the halfshaft from the transmission
   - Shaft retaining ring and discard it

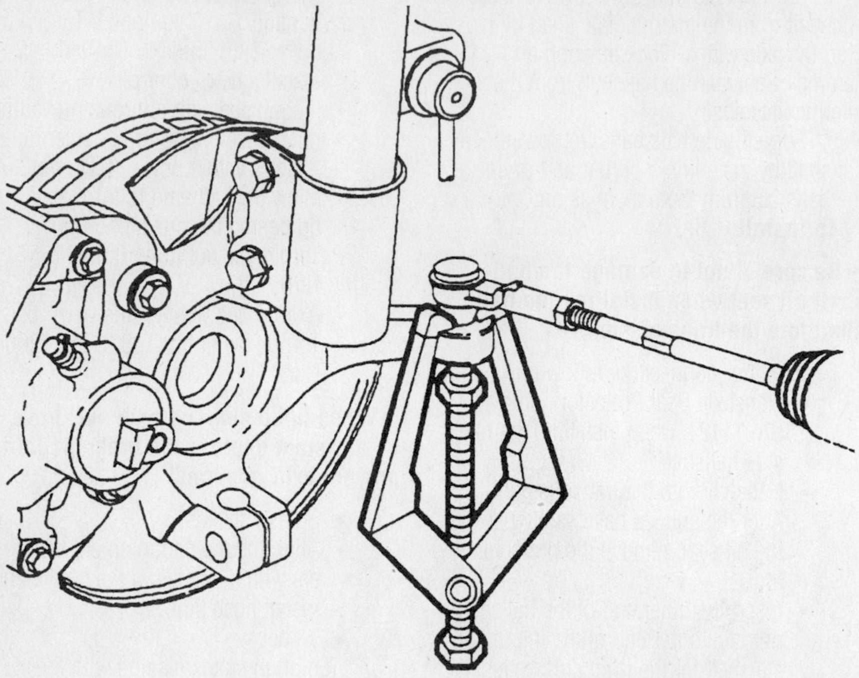

Separate the tie rod end from the steering knuckle with a Tie Rod Separator tool

To install:

4. Apply Output Shaft Lubricant 7847638, to the splines of the output shaft, if equipped with an automatic transmission.

5. Install or connect the following:
   - New stub shaft retaining ring
   - Halfshaft to the transmission after installing a Seal Protector Tool SA91112T

6. Remove the seal protector tool after the splines have passed the oil seal.

7. Install or connect the following:
   - Fully seat the halfshaft into the transmission
   - Outer end of the halfshaft to the wheel hub with a new washer and nut
   - Lower control arm ball stud to the steering knuckle. Torque the fastener to 75 ft. lbs. (100 Nm).
   - Tie rod end to the steering knuckle. When seated properly, torque the fastener to 45 ft. lbs. (60 Nm).
   - Wheel
   - Halfshaft to wheel nut. Torque the nut to 74–118 ft. lbs. (100–160 Nm), release the nut until it turns freely and torque the nut to 15 ft. lbs. (20 Nm) plus 90 degrees.
   - Cotter pin

8. Top off the transmission with the proper fluid.

9. Check and adjust the front end alignment as necessary.

CV-Joints

OVERHAUL

Inner

1. Before servicing the vehicle, refer to the precautions in the beginning of this section.

2. Remove or disconnect the following:
   - Front wheel
   - Halfshaft
   - Swage ring using a hand grinder
   - Large CV-joint boot clamp
   - CV-joint boot by sliding it away from the tri-pod joint
   - Tri-pod housing from the tri-pod spider
   - Inboard spacer ring slide it rearward on the shaft
   - Outboard retaining ring
   - Tri-pod joint spider assembly
   - Inboard spacer ring and CV-joint boot

To install:

3. Install or connect the following:
   - Swage ring clamp on the CV-joint boot
   - CV-joint boot

4. Position the CV-joint boot seal into the axle shaft's joint seal groove and align the swage ring clamp on the boot.

⁂ WARNING

Make sure that there are no pinch points on the inboard seal.

5. Crimp the swage ring

6. Install or connect the following:
   - Inboard spacer ring, slide it rearward on the shaft
   - Tri-pod joint spider assembly onto the shaft
   - Outboard retaining ring into the axle shaft groove
   - Tri-pod joint spider assembly, slide it against the outboard retaining ring
   - Inboard spacer ring, seat it in the groove
   - ½ kit grease into the boot
   - ½ kit grease into the tri-pod housing
   - New large seal clamp onto the CV-joint boot
   - Tri-pod housing, slide it over the tri-pod joint spider assembly
   - CV-joint boot/clamp, slide it into place, over the trilobal tri-pod bushing with the seal lip in the groove

➡Make sure the boot lies flat against the trilobal bushing.

7. Position the CV-joint boot so it measures 4.9 in. (125mm).

8. Using a Crimp tool, a torque wrench and a breaker bar, crimp the large CV-joint boot clamp to 130 ft. lbs. (176 Nm).

9. Install the halfshaft and the front wheel.

Outer

1. Before servicing the vehicle, refer to the precautions in the beginning of this section.

2. Remove or disconnect the following:
   - Axle shaft from the vehicle
   - Large CV boot retaining clamp
   - Small CV boot retaining clamp
   - CV boot from the joint
   - Axle shaft retaining ring
   - Outer joint from the axle shaft
   - CV boot

3. Disassemble the chrome alloy balls from the CV-joint cage as follows:

   a. Position a brass drift against the CV-joint cage and tap it with a hammer to tilt the cage.

   b. Remove the 1st chrome alloy ball from the cage.

   c. Tilt the cage in the opposite direction.

d. Remove the opposite chrome alloy ball.

e. Repeat the procedure until all 6 balls are removed.

4. Disassemble the CV-joint cage and inner race as follows:

a. Pivot the cage and race 90 degrees to the center line of the outer race.

b. Align the cage windows with outer race lands.

c. Remove the cage from the outer race.

d. Rotate the inner race upward and remove it from the cage.

To install:

5. Lubricate the parts with a light coat of grease.

6. Assemble the CV-joint cage and inner race, as follows:

a. Rotate the inner race 90 degrees to the cage centerline.

b. Align the cage windows with inner race lands.

c. Insert the inner race into the cage by rotating the inner race downward.

d. Insert the cage/inner race into the outer race.

7. Assemble the chrome alloy balls into the CV-joint cage, as follows:

a. Position a brass drift against the CV-joint cage and tap it with a hammer to tilt the cage.

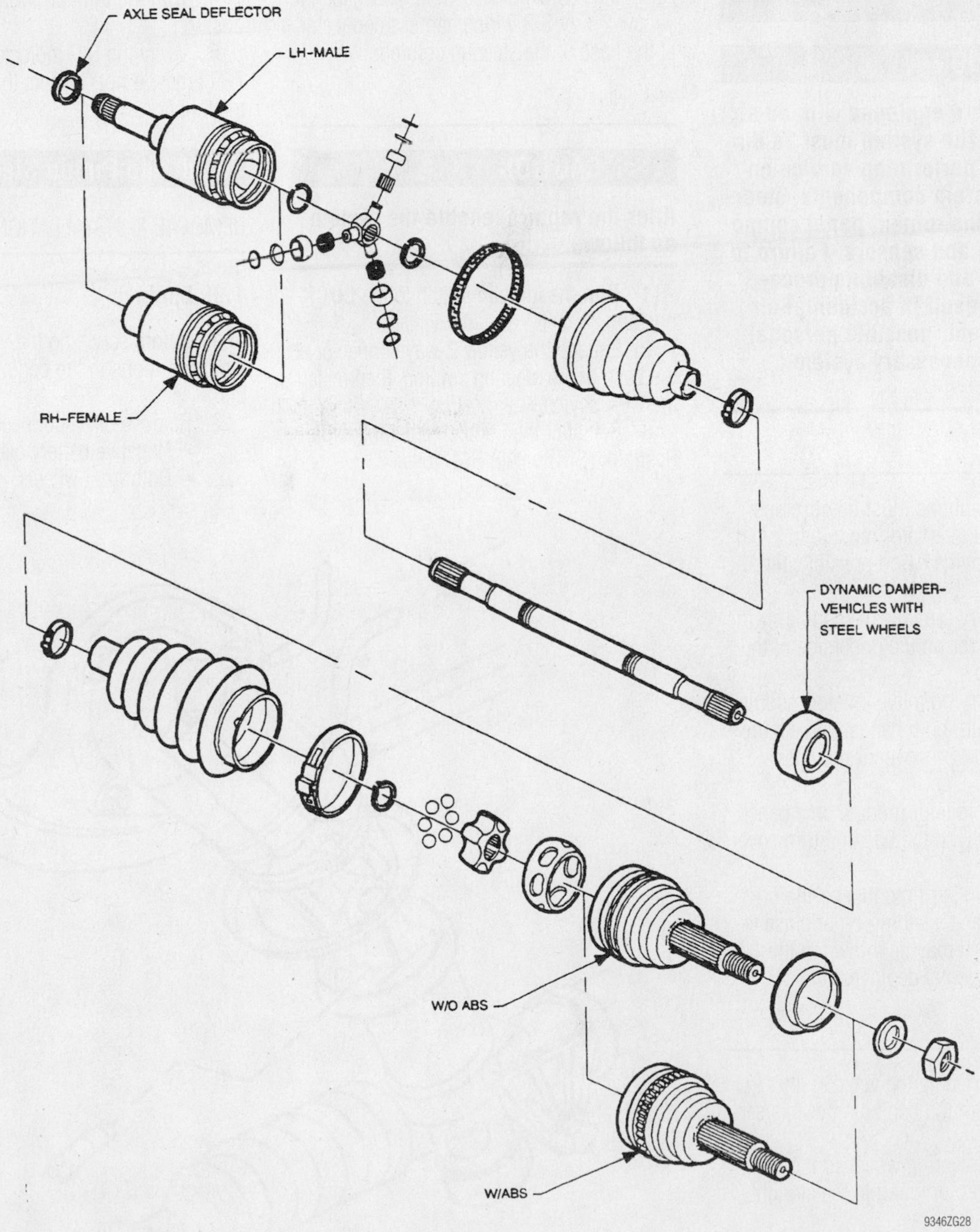

Exploded view of axle shaft—S series shown, L series is similar

b. Insert the 1st chrome alloy ball into the cage.

c. Tilt the cage in the opposite direction.

d. Insert the opposite chrome alloy ball.

e. Repeat the procedure until all 6 balls are inserted.

8. Install ½of the grease provided, into the CV-joint.

9. Install or connect the following:
- Small CV boot retaining ring
- CV boot on the halfshaft
- New retaining ring on the halfshaft
- Large ring clamp on the CV boot
- Outer joint onto the axle shaft
- Retaining ring into the outer race

10. Install the remaining grease into the CV boot.

11. Position the CV boot and the small boot clamp.

12. Crimp the small boot clamp.

13. Position and crimp in place the large boot clamp.

14. Install the Halfshaft in the vehicle.

STEERING AND SUSPENSION

Air Bag

✻✻ CAUTION

All vehicles are equipped with an air bag system. The system must be disabled before performing service on or around system components, steering column, instrument panel components, wiring and sensors. Failure to follow safety and disabling procedures could result in accidental air bag deployment, possible personal injury and unnecessary system repairs.

PRECAUTIONS

Several precautions must be observed when handling the inflator module to avoid accidental deployment and possible personal injury.

1. Never carry the inflator module by the wires or connector on the underside of the module.
2. When carrying a live inflator module, hold securely with both hands, and ensure that the bag and trim cover are pointed away.
3. Place the inflator module on a bench or other surface with the bag and trim cover facing up.
4. With the inflator module on the bench, never place anything on or close to the module which may be thrown in the event of an accidental deployment.

DISARMING

1. Before servicing the vehicle, refer to the precautions in the beginning of this section.
2. Align the steering wheel so the vehicle wheels are pointing in the straight-ahead position.
3. Turn the ignition switch to the **LOCK** position.
4. Remove the SIR or AIR BAG fuse from the fuse block.
5. Remove the Connector Position Assurance (CPA) device, then disengage the yellow 2-way SIR wiring harness connector at the base of the steering column.

ARMING

✻✻ CAUTION

After the repairs, enable the system as follows:

1. Turn the ignition switch to the **LOCK** position.
2. Engage the yellow 2-way connector at the base of the steering column, then install the CPA device.
3. Reinstall the Supplemental Inflatable Restraint (SIR) or AIR BAG fuse.
4. Turn the ignition switch to the **RUN** position.
5. Verify the SIR indicator light flashes 7–9 times, if not, inspect the system for malfunction.

Rack and Pinion Steering Gear

REMOVAL & INSTALLATION

1.9L Engines

1. Before servicing the vehicle, refer to the precautions in the beginning of this section.
2. Remove or disconnect the following:
   - Negative battery cable
   - Both front wheels

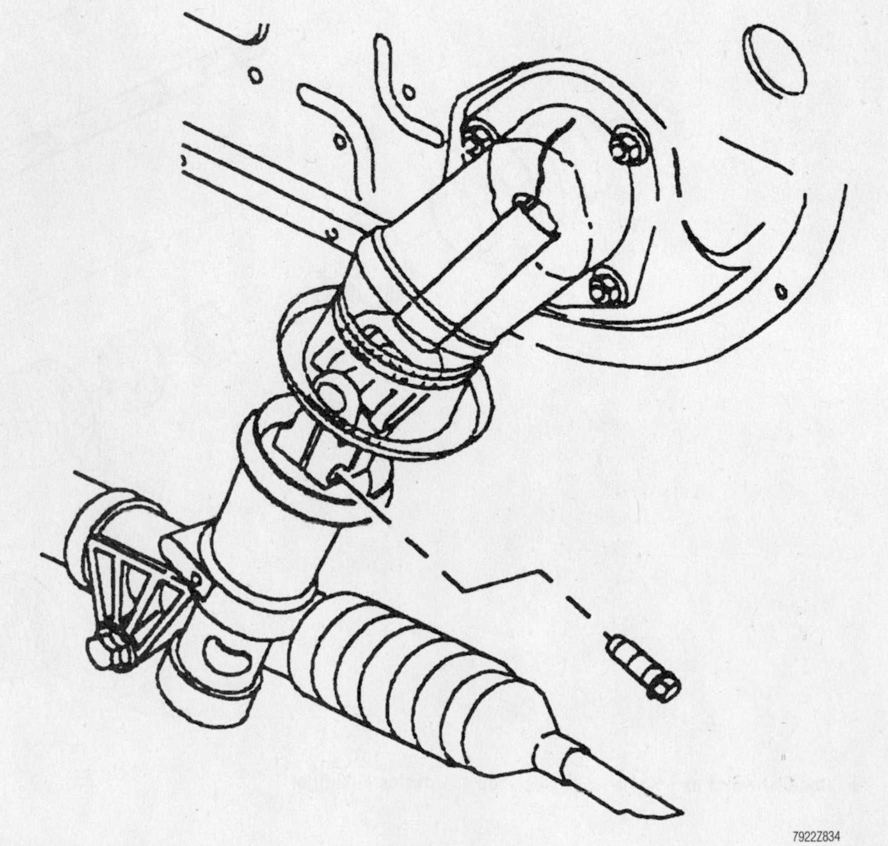

The pinch bolt is located under the intermediate shaft cover—1.9L engines

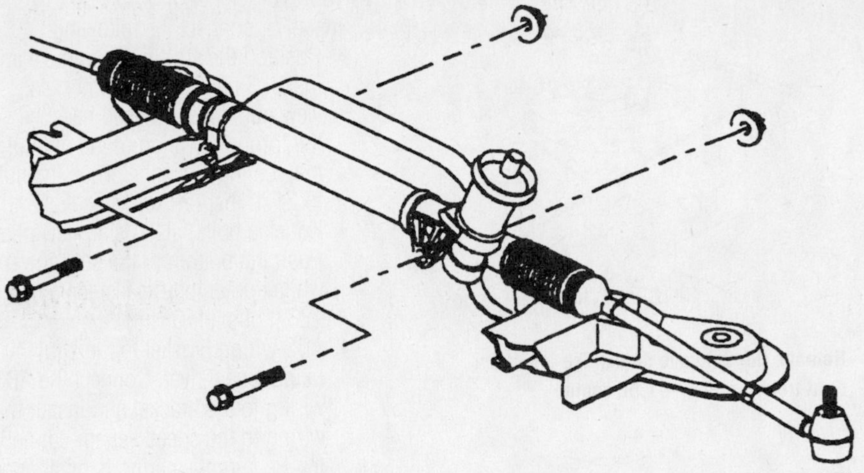

7922Z835

Remove the steering gear-to-cradle fasteners, then remove the gear through the left fender well—1.9L engines

- Outer tie rod from the steering knuckle
- Left side splash shield
- Loosen the intermediate shaft cover from the steering gear and move it aside
- Intermediate shaft pinch bolt

3. On vehicles with power steering, place a suitable container under the steering assembly. Disconnect the pressure and return lines at the steering gear and allow the system to drain.
4. Remove the steering gear to cradle fasteners.
5. Remove the steering gear through the left fender well.

To install:

6. Install or connect the following:
   - Steering gear to cradle with new nuts and torque the nuts to 37 ft. lbs. (50 Nm)
   - Intermediate shaft to the steering gear and torque the pinch bolt to 35 ft. lbs. (47 Nm)
   - Power steering pressure and return lines and torque the fittings to 20 ft. lbs. (28 Nm)
   - Left side splash shield
   - Tie rod ends to the steering knuckle and torque the castle nuts to 41 ft. lbs. (55 Nm)
   - New cotter pins to the castle nuts
   - Front wheels
   - Negative battery cable
7. Check the alignment and adjust as necessary.
8. On vehicles with power steering, bleed the power steering system.

2.2L and 3.0L Engines

1. Before servicing the vehicle, refer to the precautions in the beginning of this section.
2. Remove or disconnect the following:
   - Negative battery cable
   - Both front wheels
   - Intermediate shaft from the steering gear
   - Rear exhaust manifold pipe heat shield (on 3.0L engine)
   - Rear transmission mount through bolt
   - Transmission mount-to-frame bolt
   - Power steering pressure and return hoses
   - Front wheels
   - Right front lower splash shield
   - Exhaust manifold pipe (on 2.2L engine with manual transmission and all 3.0L engines)
   - Remaining rear transmission mount-to-frame bolts
   - Tie rods from the steering knuckles
   - Steering gear-to-frame bolts
   - Steering gear heat shield (on 2.2L engine with automatic transmission)
   - Stabilizer bar links from the struts
   - Front suspension support assemblies
   - Loosen the remaining frame to body fasteners until there is clearance to remove the steering gear
   - Steering gear through the left side wheel opening

To install:

3. One at a time remove and replace the frame bolts and nuts. Torque retention of the old fasteners may not be sufficient.
4. Install or connect the following:
   - Steering gear through the left wheel opening
   - Raise the frame assembly to the undercarriage and torque the fasteners to 66 ft. lbs. (90 Nm) plus a 46–60 degree turn
   - Suspension supports with new bolts and torque the bolts to 66 ft. lbs. (90 Nm) plus a 45–60 degree turn
   - Steering gear to the frame with new bolts and torque the bolts to 35 ft. lbs. (45 Nm) plus 45–60 degrees
   - Steering gear heat shield (on 2.2L engine with an automatic transmission) and torque the bolts to 35 inch lbs. (4 Nm)
   - Tie rod ends to the steering knuckle and torque the nut to 44 ft. lbs. (60 Nm)
   - Stabilizer bar links to the strut and torque the nuts to 48 ft. lbs. (65 Nm)
   - Rear transmission mount-to-frame bolts and torque them to 44 ft. lbs. (60 Nm)
   - Power steering pressure and return lines and torque the fittings to 20 ft. lbs. (27 Nm)
   - Exhaust manifold pipe (2.2L engines with a manual transmission and all 3.0L engines) to the manifold and torque the nuts to 22 ft. lbs. (30 Nm)
   - Exhaust manifold pipe to the exhaust pipe and torque the bolts to 15 ft. lbs. (20 Nm)
   - Front wheels
   - Rear transmission mount-to-frame bolt and torque it to 45 ft. lbs. (60 Nm)
   - Rear transmission mount through bolt and torque it to 66 ft. lbs. (90 Nm)
   - Exhaust manifold pipe heat shield (on 3.0L engine) and torque the bolts to 71 inch lbs. (8 Nm)
   - Intermediate shaft to the steering gear and torque the new pinch bolt to 21 ft. lbs. (28 Nm)
   - Negative battery cable
5. Check the alignment and adjust if necessary.
6. Bleed the power steering system.

Timing belt service is covered in Section 3 of this manual

POWER STEERING SYSTEM BLEEDING

1. Fill the power steering fluid reservoir.
2. Raise the front of the vehicle just sufficiently so that the drive wheels are off the ground, then safely support the vehicle.
3. Bleed the system by turning the wheels from side-to-side without hitting the stops. It may take several cycles to bleed the system.

➡**Maintain the reservoir at the FULL mark during this procedure.**

4. Lower the vehicle.
5. Start the engine and, with the transmission in **P** (automatic transmission) or **N** (manual transmission), check the fluid level with the engine idling. If necessary, add fluid to bring the level to the FULL mark.
6. Road test the vehicle and check for proper operation. Recheck the fluid level and make sure it is at or slightly above the FULL mark after the system has stabilized at normal operating temperature.

Strut

REMOVAL & INSTALLATION

Front

1.9L ENGINES

1. Before servicing the vehicle, refer to the precautions in the beginning of this section.
2. If equipped with an Anti-lock Brake System (ABS), disconnect the negative battery cable, then raise and support the vehicle safely. Be sure the vehicle is at a height where underhood access is still possible.
3. Remove or disconnect the following:
   - Front wheel
   - Unplug and disconnect the ABS wire from the strut wiring bracket. Note the wiring position for assembly purposes, then place the ABS wiring out of the way to prevent damage. If the strut is being replaced, drill the rivet head retaining the ABS wiring bracket to the strut and remove the bracket
   - Loosen the 2 steering knuckle-to-strut housing bolts, but do not remove them at this time. For reassembly purposes, matchmark the strut position to the steering knuckle
   - 3 upper strut-to-body nuts and discard them
   - Place a rag over the CV-joint seal to protect it from damage, then remove the 2 steering knuckle-to-strut housing bolts
   - Strut assembly from the vehicle

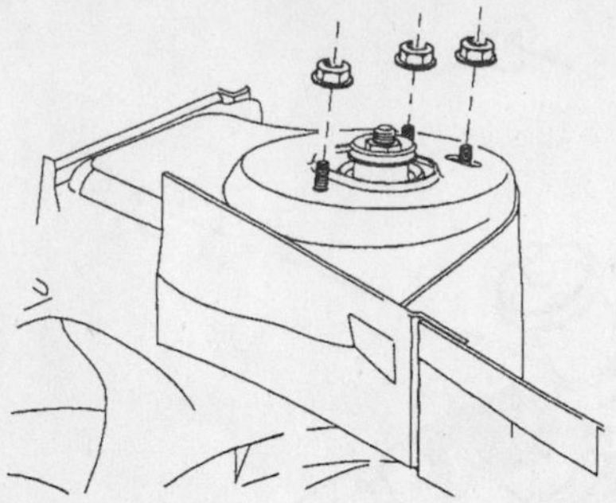
7922Z836

Remove the 3 nuts to detach the top of the strut from the body—1.9L engine

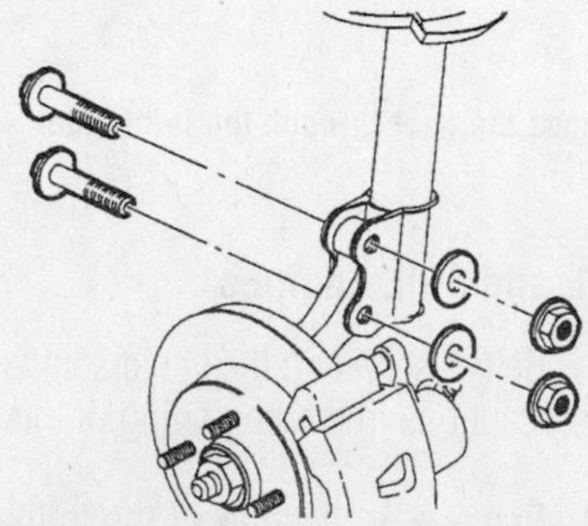
7922Z837

Matchmark, then remove the 2 bolts that secure the strut to the knuckle assembly—1.9L engine

To install:

4. Install or connect the following:
   - Position the strut in the vehicle and install 3 new upper mount nuts. New nuts must be used because the torque retention of the old nut may be insufficient. Torque the nuts to 21 ft. lbs. (29 Nm).
   - Knuckle bolts, also using new nuts. Push the bottom of the strut inward while tightening the fasteners to 126 ft. lbs. (170 Nm)
   - ABS wiring bracket to the strut using a new rivet. Connect the ABS wiring to the bracket and install the wiring to the speed sensor connector. Be sure the wiring is positioned as noted during removal
   - Wheel assembly, remove the supports and lower the vehicle.
   - Negative battery cable
5. Check and adjust the alignment if necessary.

2.2L AND 3.0L ENGINES

1. Before servicing the vehicle, refer to the precautions in the beginning of this section.
2. If equipped with an Anti-lock Brake System (ABS), disconnect the negative bat-

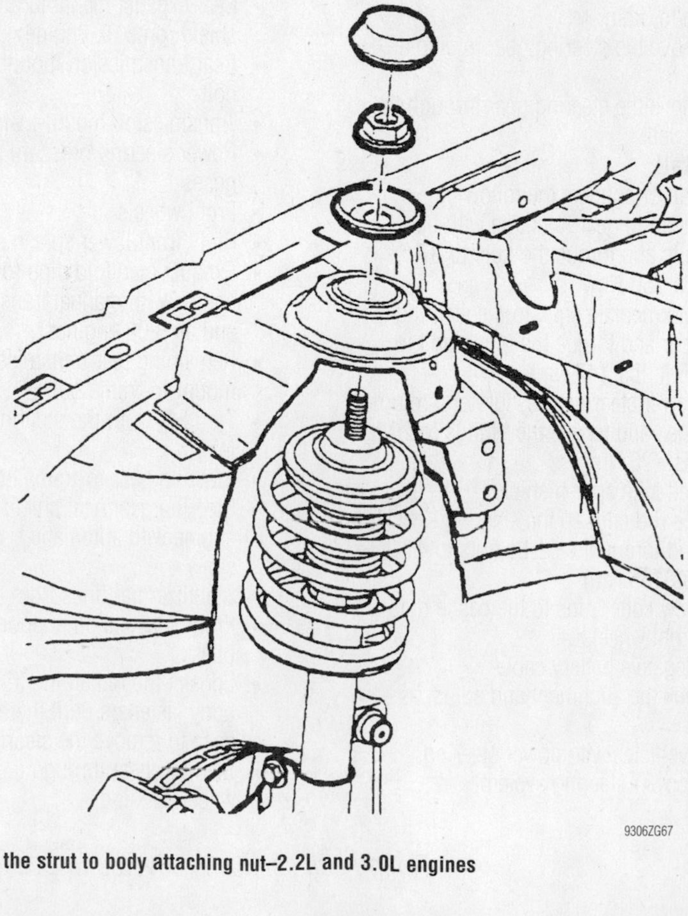
9306ZG67

Remove the strut to body attaching nut–2.2L and 3.0L engines

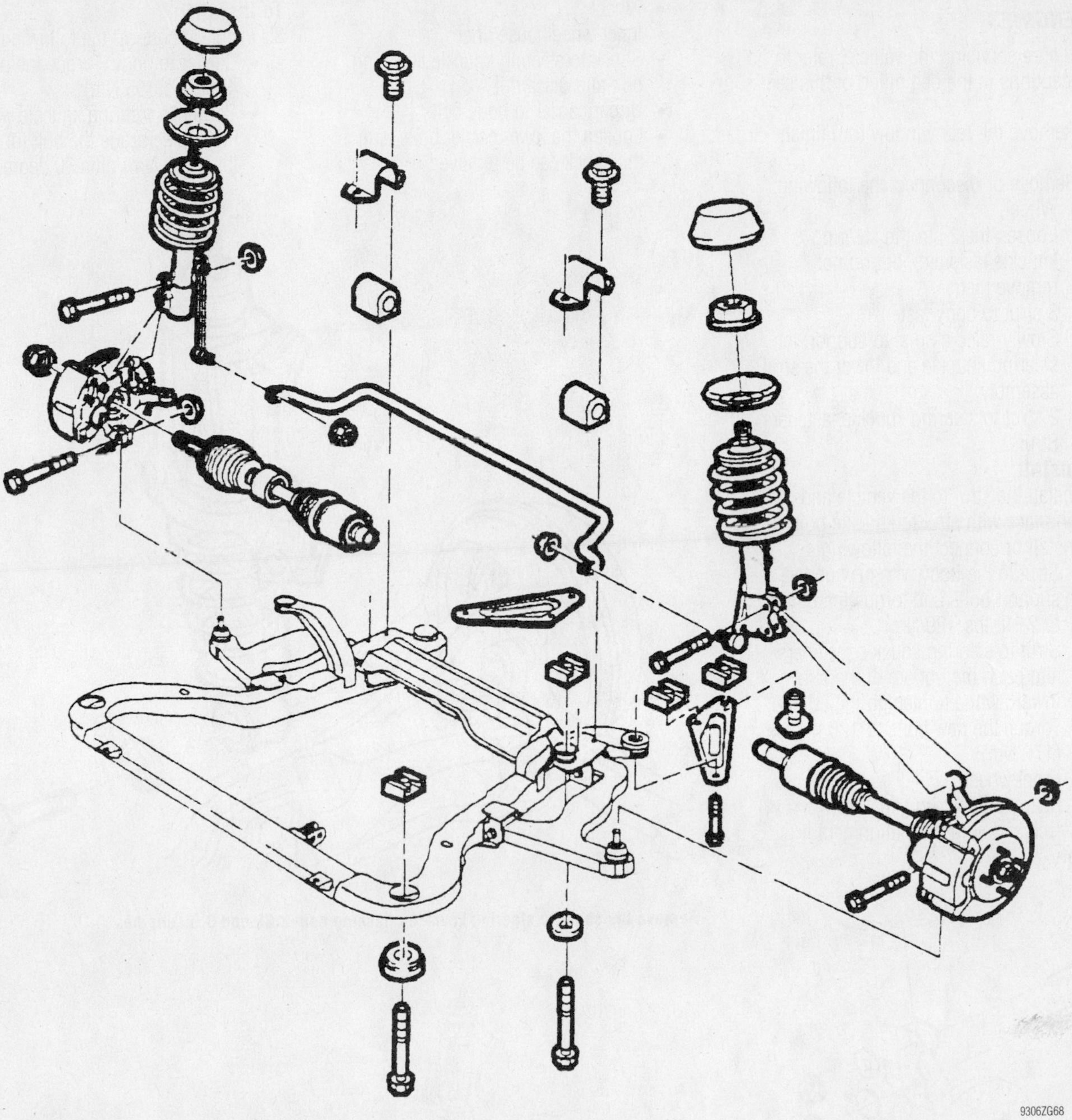

Exploded view of the front suspension–2.2L and 3.0L engines

tery cable, then raise and support the vehicle safely. Be sure the vehicle is at a height where underhood access is still possible.

3. Remove or disconnect the following:
   - Front wheel
   - Brake hose and ABS harness from the strut assembly
   - Loosen the steering knuckle to strut fasteners, but do not remove them
   - Stabilizer bar link to the strut assembly attaching nut and move it toward the rear of the vehicle
   - Strut to body attaching nuts
   - Place a rag over the CV-joint seal to protect it from damage, then remove the 2 steering knuckle-to-strut housing bolts
   - Steering knuckle to strut fasteners
   - Strut assembly from the vehicle

To install:

4. Install or connect the following:
   - Strut to the body and torque the new attaching nut to 40 ft. lbs. (55 Nm)
   - Strut to the steering knuckle and torque the new fasteners to 37 ft. lbs. (50 Nm) then to 66 ft. lbs. (90 Nm) plus a 45 degree turn
   - Stabilizer bar link to the strut and torque the fastener to 50 ft. lbs. (65 Nm)
   - Brake hose and ABS wiring to the strut
   - Front wheel
   - Negative battery cable

5. Check and adjust the alignment as necessary.

Heater Core replacement is covered in Section 2 of this manual

Rear

1.9L ENGINES

1. Before servicing the vehicle, refer to the precautions in the beginning of this section.
2. Remove the rear window trim finish panel.
3. Remove or disconnect the following:
   - Wheel
   - Loosen the 2 strut to steering knuckle fasteners, but do not remove them
   - 3 strut to body fasteners
   - Slowly raise a hoist to support the steering knuckle and lower the strut assembly
   - 2 strut to steering knuckle fasteners
   - Strut

To install:

4. Install the strut to the vehicle and hold it in place with strut to knuckle bolts.
5. Install or connect the following:
   - Strut to the body with new upper support bolts and torque the bolts to 21 ft. lbs. (29 Nm)
   - Strut to steering knuckle fasteners and push the bottom of the strut inward while tightening the bolts. Torque the new bolts to 126 ft. lbs. (170 Nm).
   - Front wheel
   - Rear window trim panel
6. Check and adjust the alignment if necessary.

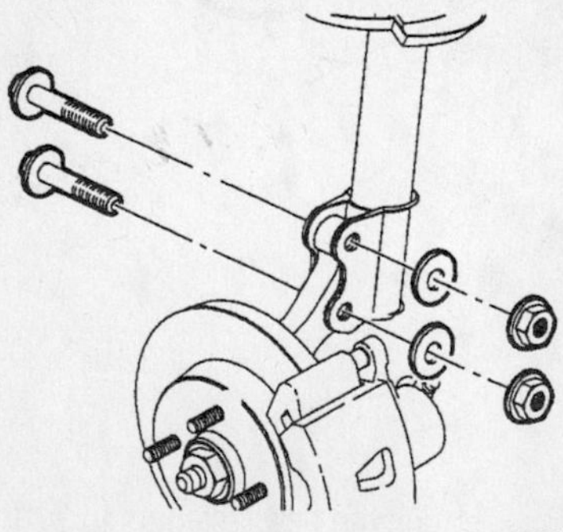

7922Z837

Remove the two strut to steering knuckle fasteners–1.9L engine

Shock Absorber

REMOVAL & INSTALLATION

2.2L and 3.0L Engines

1. Before servicing the vehicle, refer to the precautions in the beginning of this section.
2. Remove or disconnect the following:
   - Wheel
   - Inner wheelhouse liner
   - Shock to steering knuckle attaching bolt and discard it
   - Upper carrier to body bolts
   - Loosen the lower carrier bolts until the shock can be removed
   - Shock

To install:

3. Install or connect the following:
   - Carrier to body. Torque the bolts to 40 ft. lbs. (55 Nm).
   - Carrier to steering knuckle with a new bolt. Torque the bolt to 110 ft. lbs. (150 Nm) plus 30 degrees

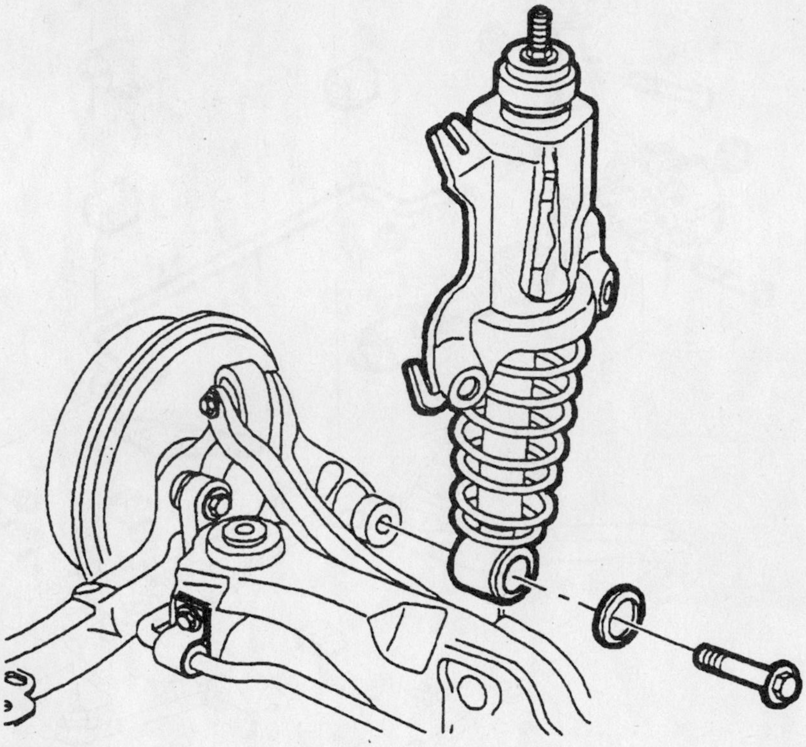

9306ZG69

Remove the shock to steering knuckle attaching bolt–2.2L and 3.0L engines

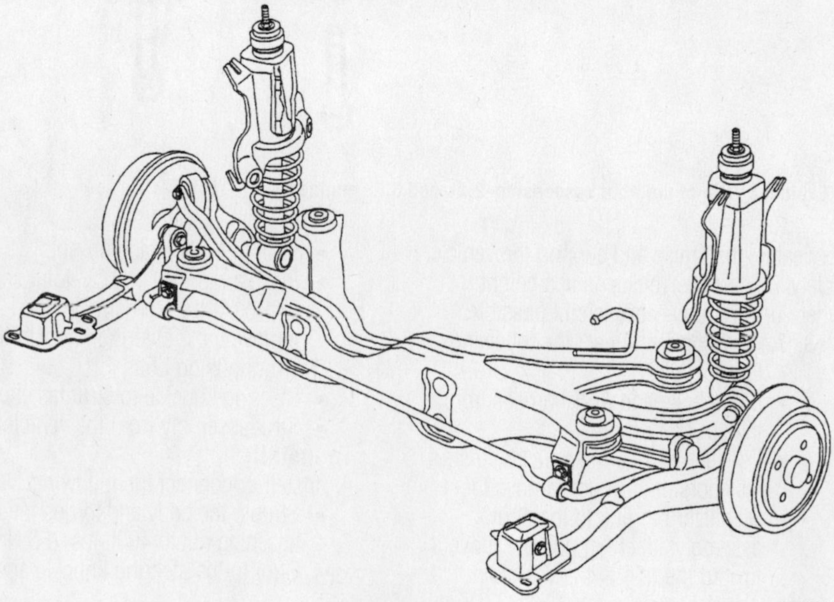

9306ZG70

Exploded view of the rear suspension–2.2L and 3.0L engines

- Inner wheelhouse liner
- Wheel

Coil Spring

REMOVAL & INSTALLATION

Front

1.9L ENGINES

1. Before servicing the vehicle, refer to the precautions in the beginning of this section.
2. Remove or disconnect the following:
   - Strut from the vehicle
   - Mount the strut in a bench vise, then attach a spring compressor/holding fixture; be sure that the strut component is firmly secured
   - Compress the spring sufficiently to completely unload the upper strut mount
   - Shaft nut while holding the strut stationary with a Torx® head socket wrench
   - Upper spring support and inspect the rubber for cracks or deterioration
   - Spring from the strut and inspect the spring for damage
   - Dust shield assembly and inspect for cracks or deterioration
   - Strut from the vise or applicable holding fixture and retract the strut shaft, checking for smooth, even resistance
3. If replacing the coil spring, carefully release the spring compressor.

To install:

4. Secure the strut in the bench vise or applicable holding fixture.
5. Extend the strut shaft to the limit of its travel.
6. Install or connect the following:
   - Dust shield assembly onto the strut, then install the spring with the compressor tool installed
   - Spring isolator and the strut mount to the top of the assembly
   - Guide the strut shaft through the upper strut mount assembly. Compress the coil until the washer and shaft nut can be installed to the end of the shaft, but do not over-compress and damage the spring
   - Tighten the shaft to the nut using a Torx® head socket wrench and a torque wrench, while holding the nut steady with an open end wrench. Tighten the fastener to 37 ft. lbs. (50 Nm).
   - Release the spring compressor tool and remove the strut from the fixture
   - Strut assembly in the vehicle.

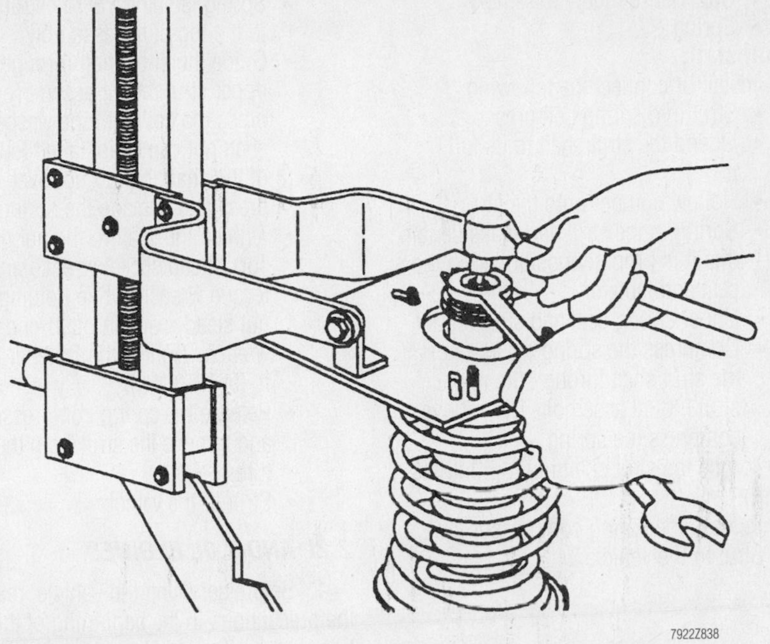

Remove the strut shaft nut while the coil spring is compressed to unload the upper strut mount–1.9L engines

2.2L AND 3.0L ENGINES

1. Before servicing the vehicle, refer to the precautions in the beginning of this section.
2. Remove or disconnect the following:
   - Strut from the vehicle
   - Place the strut into a spring compressor. Fasten the assembly with a strut to steering knuckle bolt through the lower mounting hole
   - Compress the spring enough to completely unload the upper strut mount
   - Strut shaft nut while holding the shaft stationary with a TORX® socket
   - Release the spring compressor and tilt the strut outward

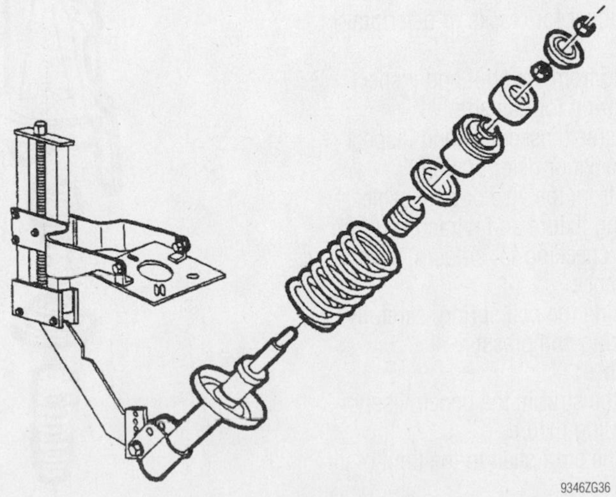

Exploded view of the coil spring and related components–2.2L and 3.0L engines

- Upper strut mount assembly
- Spring

To install:

3. Install or connect the following:
   - Strut into spring compressor
   - Extend the strut shaft to its full travel
   - Hollow bumper onto the strut
   - Spring to the strut and make certain that it is properly positioned in the seat and isolator
   - Upper spring seat and strut mount
   - Compress the spring while guiding the strut shaft through the upper strut mount assembly. Do not over compress the spring
4. Torque the strut shaft nut to 40 ft. lbs. (55 Nm).
5. Release the spring compressor tool.
6. Strut to the vehicle.

Rear

1.9L ENGINES

1. Before servicing the vehicle, refer to the precautions in the beginning of this section.
2. Remove or disconnect the following:
   - Strut assembly from the vehicle
   - Mount the strut in a bench vise, then attach a suitable spring compressor/holding fixture; be sure that the strut component is firmly secured
   - Compress the spring sufficiently to completely unload the upper strut mount
   - Strut shaft nut while holding the strut stationary with a Torx® head socket wrench
   - Upper spring support and inspect the rubber for cracks or deterioration
   - Spring from the strut and inspect the spring for damage
   - Dust shield assembly and inspect for cracks or deterioration
   - Strut from the vise or applicable holding fixture and retract the strut shaft, checking for smooth, even resistance
3. If replacing the coil spring, carefully release the spring compressor.

To assemble:

4. Secure the strut in the bench vise, or applicable holding fixture.
5. Extend the strut shaft to the limit of its travel.
6. Install or connect the following:
   - Dust shield assembly onto the strut, then install the spring with the compressor tool installed
   - Spring isolator and the strut mount to the top of the assembly
   - Guide the strut shaft through the upper strut mount assembly. Compress the coil until the washer and shaft nut can be installed to the end of the shaft, but do not over-compress and damage the spring
   - Tighten the shaft to the nut using a Torx® head socket wrench and a torque wrench, while holding the nut steady with an open-end wrench. Tighten the fastener to 37 ft. lbs. (50 Nm).
   - Release the spring compressor tool and remove the strut from the fixture
   - Strut to the vehicle

2.2L AND 3.0L ENGINES

1. Before servicing the vehicle, refer to the precautions in the beginning of this section.
2. Remove or disconnect the following:
   - Shock absorber assembly from the vehicle
   - Mount the carrier assembly Spring Compressor SA9155S, in a holding fixture
   - Mount the shock assembly to the spring compressor using an adapter, SA9155–3
   - Compress the spring enough to unload the upper spring supports
   - Shock absorber shaft nut while holding the shaft securely with a TORX® socket
   - Release the spring compressor and remove the carrier assembly
   - Rear suspension support and inspect the inner and outer bumpers for cracks or deterioration, replace if necessary
   - Rear spring upper insulator
   - Extend and retract the shock absorber and check for smooth resistance

To install:

3. Install or connect the following:
   - Shock into a spring compressor
   - Extend the shock to its full limit of travel
   - Inner and outer bumpers and make certain that the springs are properly positioned in their support seats
   - Compress the spring while guiding the shock absorber shaft through the upper mount until the washer and nut can be installed. Torque the shaft nut to 15 ft. lbs. (20 Nm)

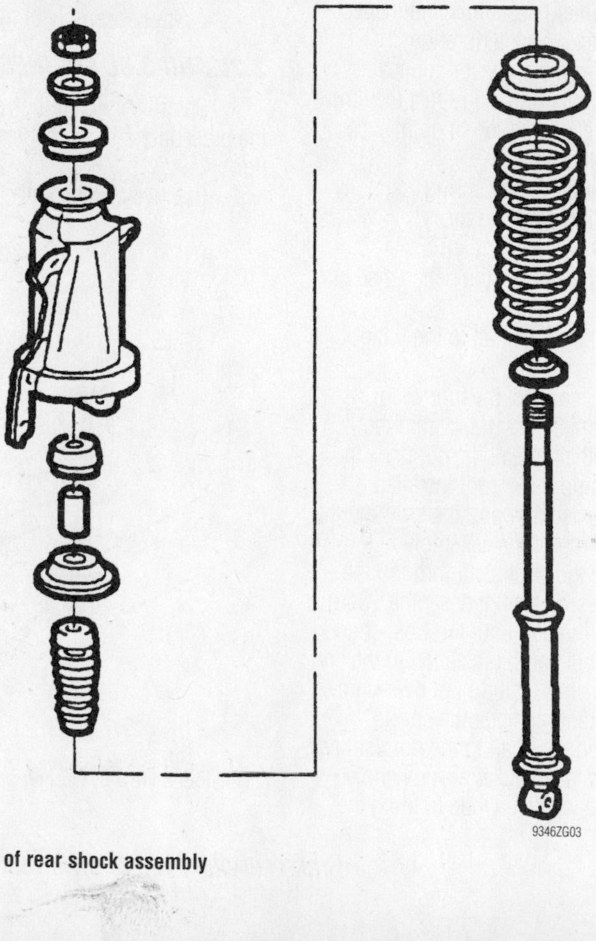

Exploded view of rear shock assembly

while holding the shaft nut with a TORX® socket.
- Release the compressor tool and remove the carrier from the compressor
- Shock absorber to the vehicle

Lower Ball Joint

REMOVAL & INSTALLATION

1.9L Engines

The ball joint is part of the lower control arm and must be replaced as a unit.

2.2L and 3.0L Engines

1. Before servicing the vehicle, refer to the precautions in the beginning of this section.
2. Install or connect the following:
   - Lower control arm from the vehicle
   - Rivets retaining the ball joint to the control arm using a 13 mm (½ in.) drill bit
   - Ball joint from the control arm

To install:

3. Install or connect the following:
   - Ball joint into the control arm
   - Nuts and bolts (included with new ball joint kit) as shown and torque them to 25 ft. lbs. (35 Nm)
   - Control arm to the vehicle

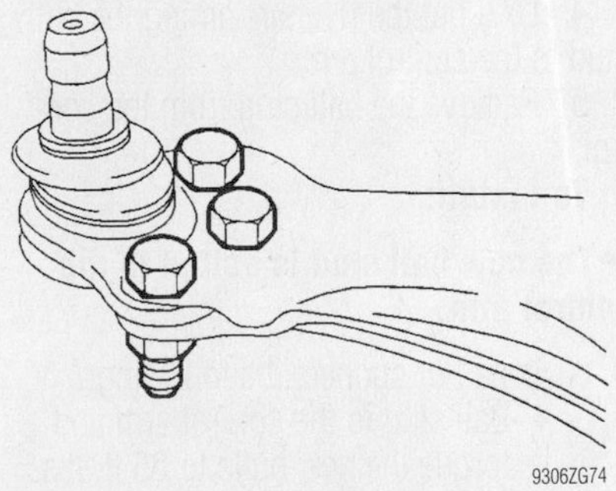
9306ZG74

The new ball stud is bolted into the control arm–2.2L and 3.0L engines

Lower Control Arm

REMOVAL & INSTALLATION

Front

1.9L ENGINES

1. Before servicing the vehicle, refer to the precautions in the beginning of this section.
2. Remove or disconnect the following:
   - Wheel
   - Lower control arm ball stud cotter pin and discard the pin
   - Loosen the castle nut until it is level with the top of the ball stud

➡Use caution not do damage the Anti-lock Braking System (ABS) speed sensor ring. If the ring is damaged a malfunction of the ABS system is possible.

3. Separate the lower control arm from the steering knuckle, using a Lower Control Arm Ball Stud Separator Tool SA9132S
4. Remove or disconnect the following:
   - Lower control arm castle nut
   - Front inner fender splash shield and remove the push pins
   - Front section of the shield first on the left hand side and the rear section first on the right side of the vehicle
   - Lower control arm to cradle fasteners
   - front stabilizer bar nut at the lower control arm
   - Lower control arm

To install:

5. Install or connect the following:
   - Control arm to the front stabilizer bar
6. Position the end of the control arm into the cradle. Torque the bolt to 92 ft. lbs. (125 Nm) and the nut to 74 ft. lbs. (100 Nm).
   - Control arm to front stabilizer bar nut and torque the nut to 106 ft. lbs. (144 Nm)
   - Ball stud to the steering knuckle and torque the castle nut to 55 ft. lbs. (75 Nm) and install a new cotter pin
   - Rear section of the shield first on the left side and the front section first on the right side of the vehicle
   - Inner fender splash shield
   - Align the molded in fasteners with the cradle holes and push them straight in
   - Wheel
7. Check and adjust the alignment if necessary.

2.2L AND 3.0L ENGINES

1. Before servicing the vehicle, refer to the precautions in the beginning of this section.

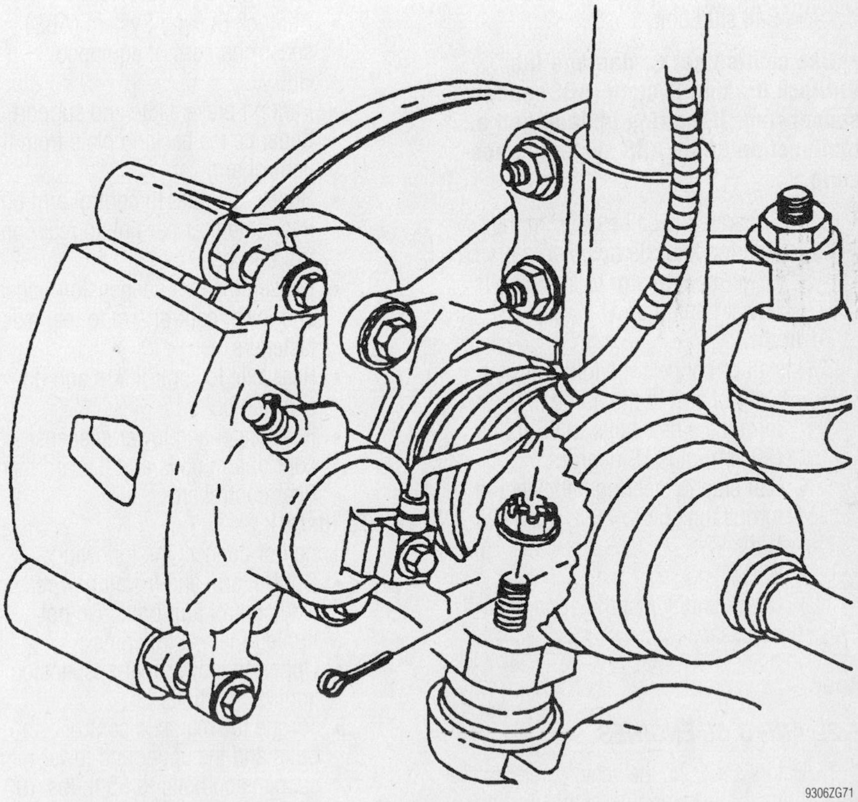
9306ZG71

Connect the ball stud to the steering knuckle–1.9L engines

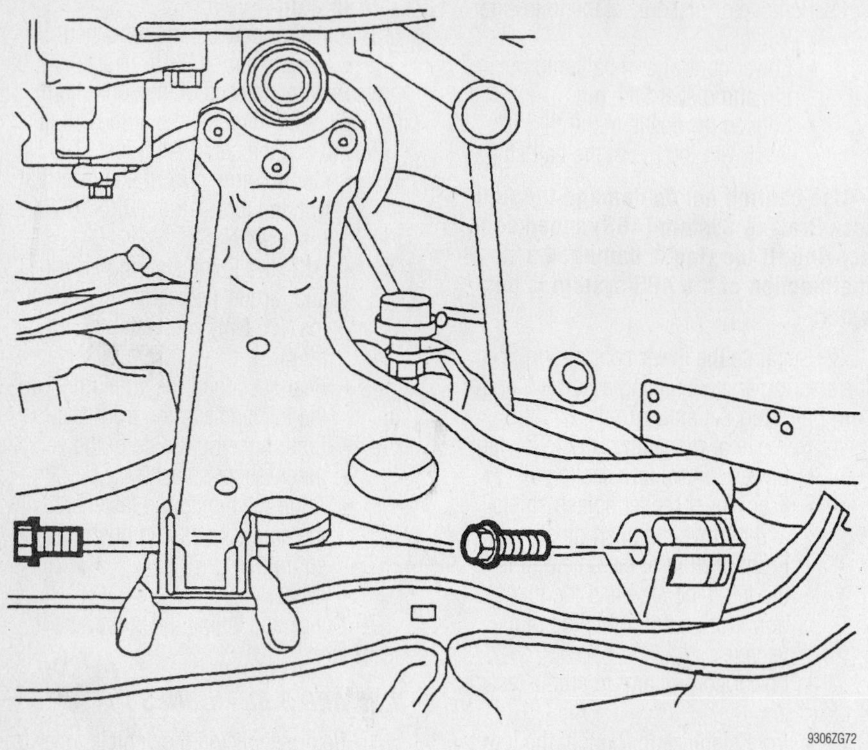

Remove the lower control arm to frame bolts–2.2L and 3.0L engines

2. Remove or disconnect the following:
   - Wheel
   - Ball stud bolt

➡Use caution not do damage the Antilock Braking System (ABS) speed sensor ring. If the ring is damaged a malfunction of the ABS system is possible.

   - Separate the ball stud from the steering knuckle by using a pry bar
   - Lower control arm to frame bolts
   - Lower control arm

To install:

3. Install or connect the following:
   - Control arm to the frame and torque the new bolts to 65 ft. lbs. (90 Nm) plus 75 degrees
   - Ball stud to steering knuckle and torque the bolt to 75 ft. lbs. (100 Nm)
   - Wheel
   - Check and adjust the alignment, if necessary.

Rear

2.2L AND 3.0L ENGINES

1. Before servicing the vehicle, refer to the precautions in the beginning of this section.
2. Remove or disconnect the following:
   - Rear wheel
   - Rear brake hose and bracket from the control arm
   - Caliper, if equipped
   - Brake drum/rotor
   - Antilock Braking System (ABS) sensor harness, if equipped
   - Hub
   - Parking brake cable and support
   - Separate the backing plate from the control arm
   - Shock absorber to control arm bolt
   - Rear stabilizer bar link to rear control arm bolt
   - Loosen the rear suspension upper and lower control arm to rear axle fasteners
   - Rear axle to control arm and discard the bolts
   - Rear upper and lower suspensions control arm bolts and discard them
   - Rear control arm

To install:

3. Install or connect the following:
   - Control arm and install new rear axle control arm bolts. Do not tighten them at this time
   - Upper and lower rear suspension control arm bolts
   - Torque the rear axle control arm bolts and the upper and lower rear suspension bolts to 65 ft. lbs. (90 Nm) plus 30 degrees
   - Shock absorber to rear axle control arm and torque the new bolt to 110 ft. lbs. (150 Nm) plus 30 degrees
   - Stabilizer bar link to the control arm. And torque the bolt 41 ft. lbs. (55 Nm)
   - Backing plate and hub assembly and torque the bolts to 37 ft. lbs. (50 Nm) plus 30 degrees
   - Rear drum/disc and torque the bolts to 35 inch lbs. (4 Nm)
   - Rear brake caliper and torque the bolts to 59 ft. lbs. (80 Nm), if equipped
   - Parking brake cable and support bracket and torque the bolts to 71 inch lbs. (8 Nm)
   - Brake line bracket to the control arm and torque the bolts to 71 inch lbs. (8 Nm)
   - Rear wheel
4. Check and adjust the alignment, as needed.

LOWER CONTROL ARM BUSHING REPLACEMENT

2.2L and 3.0L Engines

FRONT

1. Before servicing the vehicle, refer to the precautions in the beginning of this section.
2. Remove or disconnect the following:
   - Control arm
3. Press out the front bushing by using Tools KM-508–3 and KM508–1.
4. Drill out the rivets retaining the ball stud to the control arm.
5. Remove the ball stud from the control.

To install:

➡The new ball stud is bolted to the control arm.

6. Install or connect the following:
   - Ball stud to the control arm and torque the new bolts to 25 ft. lbs. (35 Nm)

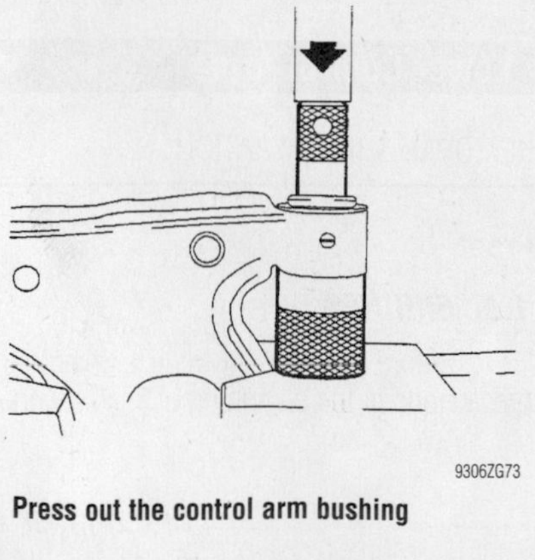

Press out the control arm bushing

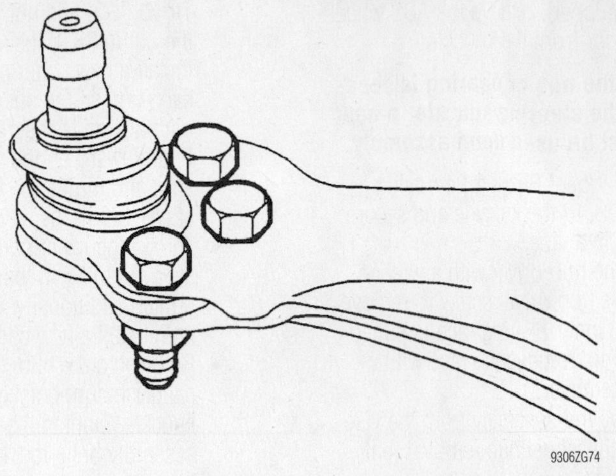

The new ball stud is bolted into the control arm–2.2L and 3.0L engines

7. Press in the new control arm bushing using Tools KM508–1, KM508–2 and KM508–3.
   - Control arm to the frame and torque the new bolts to 65 ft. lbs. (90 Nm) plus 75 degrees
   - Ball stud to the steering knuckle and torque the bolts to 75 ft. lbs. (100 Nm)

REAR

1. Before servicing the vehicle, refer to the precautions in the beginning of this section.
2. Remove or disconnect the following:
   - Control arm and place it in a vise
   - Rear axle control arm bolt and discard it
3. Press out the rear axle upper and lower control arm bushings with Bushing Removal Tools KM-671—KM-906–62—and KM-906–61.
4. Press out the rear axle control arm front bushing with Bushing Removal Tools KM-671—KM-906–41 and KM-906–44.

To install:

5. Install or connect the following:

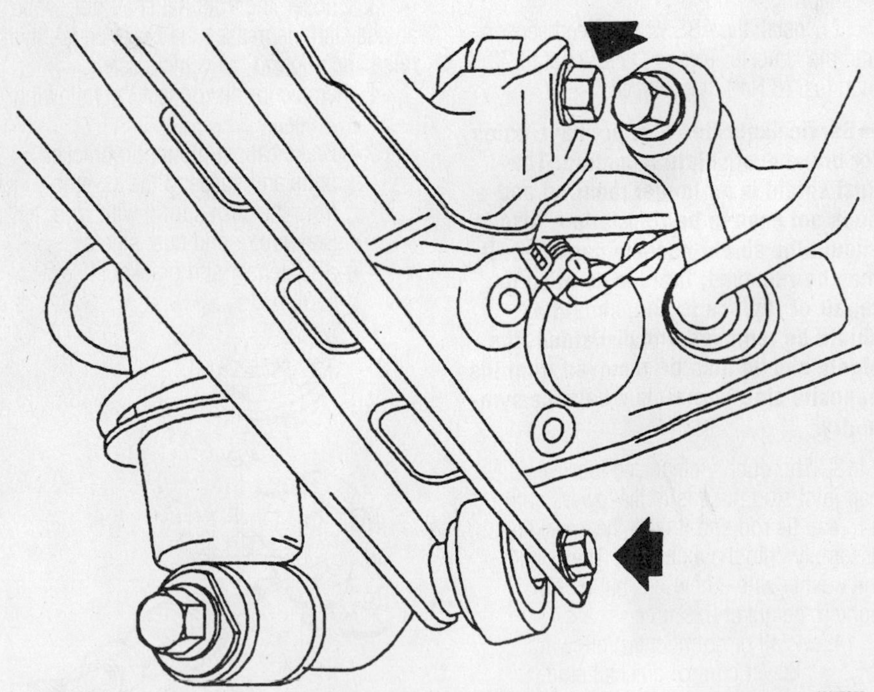

Remove the rear axle control arm upper and lower bushings

6. Press in the new rear axle front control arm bushing using tools KM-671, KM-906-42 and KM-906-43.
7. Press in the rear axle upper and lower control arm bushings with Bushing Removal Tools KM-671, KM-906-64 and KM-906-631.
   - Rear axle control arm bracket. Torque the new bolt to 65 ft. lbs. (90 Nm) plus 60 degrees

Wheel Bearings

ADJUSTMENT

The wheel bearing are sealed at the factory and do not require any adjustment or maintenance.

REMOVAL & INSTALLATION

Front

1.9L ENGINES

1. Before servicing the vehicle, refer to the precautions in the beginning of this section.
2. If equipped with an Antilock Braking System (ABS), disconnect the negative battery cable.
3. Loosen the front halfshaft nut, while an assistant depresses the brake pedal, then raise and support the vehicle safely.
4. Remove or disconnect the following:
   - Wheel
   - Brake caliper mounting bracket bolts and suspend the assembly from the strut spring with wire
   - Loosen the strut-to-knuckle bolts, but do not remove at this time
   - Rotor, axle nut and washer
   - Cotter pin from the lower control arm ball joint. Back the ball joint

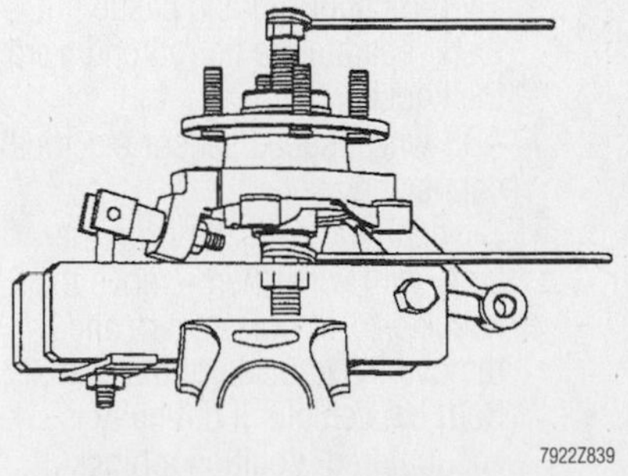

Tighten the hub driver screw to remove the hub while the assembly is held firmly in a vise–1.9L engine

For Accessory Drive Belt illustrations, see Section 1 of this manual

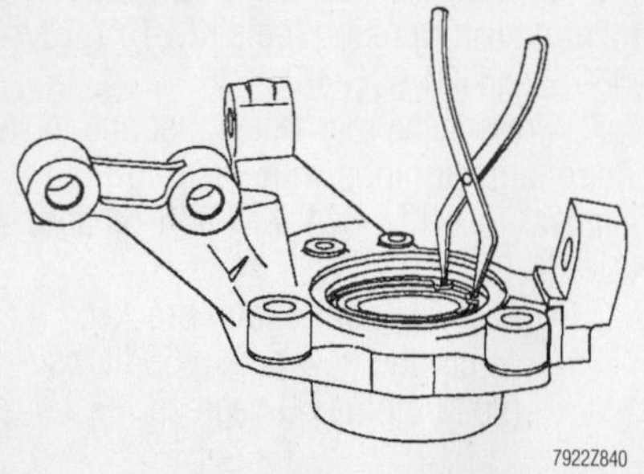

Remove the bearing retainer snapring before pressing out the bearing–1.9L engine

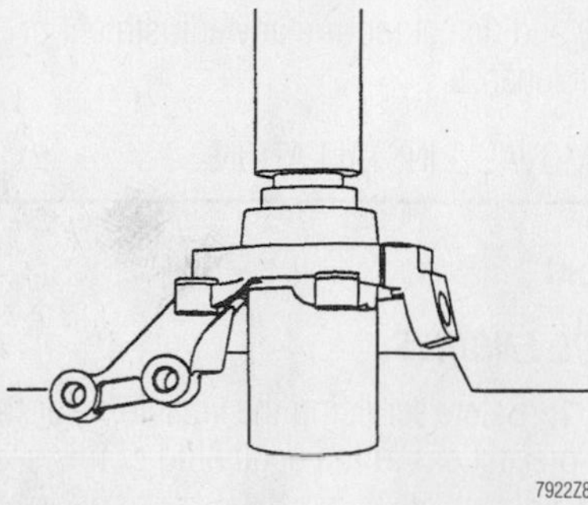

Carefully press the bearing out of the knuckle after removing the snapring–1.9L engine

nut until the top of the nut is even with the top of the threads
- Separate the lower control arm from the steering knuckle, then remove the nut. Do not use a wedge tool or seal damage may occur

➡The outer CV-joint for vehicles equipped with ABS contains a speed sensor ring. Use of an incorrect tool to separate the control arm from the knuckle may result in damage and loss of the ABS system.

- Tie rod cotter pin and castle nut, then separate the tie rod end from the knuckle
- ABS wheel speed sensor electrical connector
- Suspend the halfshaft from the body with wire, then remove the 2 knuckle-to-strut fasteners and remove the knuckle/hub assembly from the vehicle. If difficulty is encountered, position a block of wood on the end of the halfshaft and tap on the wood with a hammer to free the hub assembly

5. Disassemble the knuckle hub assembly as follows:

a. If equipped, remove the ABS wheel speed sensor from the knuckle.

➡Any time the hub or bearing is separated from the steering knuckle, a new bearing must be used upon assembly.

6. Install Wheel Bearing Removing Tools SA9159S, to the knuckle and secure the assembly in a vise.

7. Hold the hub driver with a wrench and tighten the hub driver screw to remove the hub. If the inner bearing race is pulled out with the hub, remove the race with a bearing race remover.

8. Remove the assembly from the vise and separate the wheel hub removal tool from the knuckle.

9. Remove the bearing retainer snapring.

10. Position the knuckle in a shop press on a knuckle support tube and press the bearing from the knuckle with a small driver.

To install:

11. If necessary, assemble the knuckle hub assembly as follows:

a. Use a large driver and press in the new bearing until seated.

b. Use the small driver and the knuckle support tube to press in the hub assembly. The small driver must be used to support the bearing inner race with its small (pilot) side facing towards the press and away from the bearing.

c. Install the bearing retainer snapring.

12. Install the ABS wheel speed sensor into the knuckle. Torque the fastener to 72 inch lbs. (8 Nm), if equipped.

➡Service knuckles may not have holes for brake dust shield mounting. The dust shield is no longer required and does not have to be reinstalled. Also, should the shield become damaged, it may be removed; there is no need to repair or replace it. But, should a shield be removed and discarded, the shield should also be removed from the opposite side to maintain balance/symmetry.

13. Thoroughly clean and lubricate the ball joint stud threads of the lower control arm and tie rod end. Install the knuckle/hub assembly onto the axle shaft. Then, install the washer with a new nut, but do not tighten the nut at this time.

14. Install or connect the following:
- Lower control arm ball stud through the knuckle bore and install the nut, but do not tighten at this time
- Steering knuckle-to-strut fasteners, but do not tighten at this time
- Tie rod end and nut, then torque the nut to 33 ft. lbs. (45 Nm) and install a new cotter pin. If necessary, tighten the nut additionally, do not back off to insert the cotter pin
- Push inward on the bottom of the strut and torque the knuckle fasteners to 126 ft. lbs. (170 Nm).
- Torque the lower control arm ball stud nut to 55 ft. lbs. (75 Nm), tighten additionally, if necessary, and install a new cotter pin
- Rotor onto the hub, then install the caliper mount bracket onto the knuckle. Tighten the mount bracket assembly bolts to 81 ft. lbs. (110 Nm).
- While an assistant depresses the brake pedal, tighten the halfshaft nut to 103 ft. lbs. (140 Nm).
- Wheel assembly and lower the vehicle
- Negative battery cable

15. Check and adjust the alignment, if necessary.

2.2L AND 3.0L ENGINES

1. Before servicing the vehicle, refer to the precautions in the beginning of this section.

2. If equipped with an Antilock Braking System (ABS), disconnect the negative battery cable.

3. Loosen the front halfshaft nut, while an assistant depresses the brake pedal, then raise and support the vehicle safely.

4. Remove or disconnect the following:
- Wheel
- Brake caliper mounting bracket bolts and suspend the assembly from the strut spring with wire
- Brake rotor and dust shield
- ABS sensor and bracket, if equipped

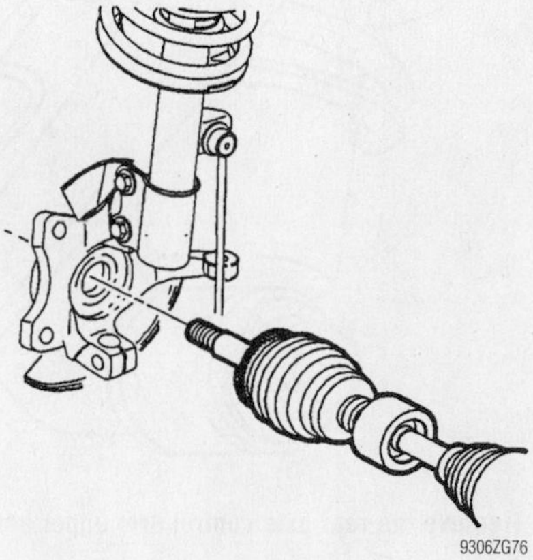

Separate the axle from the wheel hub–2.2L and 3.0L engines

- Brake rotor, if equipped
- Caliper to the steering knuckle, if equipped. Torque the bolts to 63 ft. lbs. (85 Nm).
- Rear wheel
- Negative battery cable

2.2L AND 3.0L ENGINES

1. Before servicing the vehicle, refer to the precautions in the beginning of this section.
2. Remove or disconnect the following:
   - Negative battery cable, if equipped with an Antilock Braking System (ABS)
   - Rear wheel
   - Brake line to rear axle control arm attaching clip, if equipped with rear disc brakes
   - Caliper to steering knuckle bolts and move the caliper aside, if equipped with rear disc brakes
   - Brake drum/rotor
   - ABS electrical connector, if equipped
   - Hub to control arm nuts and discard them
   - Hub

To install:

3. Install or connect the following:
   - Hub to control arm and torque the new nuts to 35 ft. lbs. (50 Nm) plus 30 degrees
   - ABS electrical connector, if equipped
   - Brake drum/rotor attaching screw and torque the screw to 35 inch lbs. (4 Nm)
   - Brake caliper, if equipped and torque the bolts to 59 ft. lbs. (80 Nm)
   - Brake line to the rear axle control arm bracket, if equipped. Torque the bolt to 71 inch lbs. (8 Nm).
   - Rear wheel
   - Negative battery cable

- Loosen the strut-to-knuckle bolts, but do not remove at this time
- Axle to hub nut and discard it
- Tie rod end nut and discard it

5. If difficulty is encountered, position a block of wood on the end of the halfshaft and tap on the wood with a hammer to free the hub assembly.
   - Steering knuckle/hub assembly
6. To remove the wheel bearing proceed, as follows:
   a. Install Wheel Bearing Removing Tools SA9159S, to the knuckle and secure the assembly in a vise.
   b. Hold the hub driver with a wrench and tighten the hub driver screw to remove the hub. If the inner bearing race is pulled out with the hub, remove the race with a bearing race remover.
   c. Remove the assembly from the vise and separate the wheel hub removal tool from the knuckle.
   d. Remove the bearing retainer snapring.
   e. Position the knuckle in a shop press on a knuckle support tube and press the bearing from the knuckle with a small driver.

To install:

7. If necessary, assemble the knuckle hub assembly, as follows:
   a. Use a suitable large driver and press in the new bearing until seated.
   b. Use the small driver and the knuckle support tube to press in the hub assembly. The small driver must be used to support the bearing inner race with its small (pilot) side facing towards the press and away from the bearing.
   c. Install the bearing retainer snap ring.
8. Thoroughly clean and lubricate the ball joint stud threads of the lower control arm and tie rod end.
9. Install or connect the following:
   - Knuckle/hub assembly onto the axle shaft. Then, install the washer with a new nut, but do not tighten the nut at this time
   - Lower control arm ball stud through the knuckle bore and install the nut, but do not tighten at this time
   - Steering knuckle-to-strut fasteners. Torque the fasteners to 40 ft. lbs. (50 Nm); then, to 65 ft. lbs. (90 Nm) plus 45 degrees
10. Torque the ball stud fastener to 75 ft. lbs. (100 Nm).
   - Tie rod end into the steering knuckle using a Linkage Installer Tool J44015. Torque the fastener to 35 ft. lbs. (45 Nm).
   - ABS sensor and mounting bracket, if equipped and torque the bolts to 71 inch lbs. (8 Nm).
   - Dust shield and torque the bolts to 35 in lbs. (4 Nm)
   - Rotor and screw and torque the screw to 27 inch lbs. (3.5 Nm)
   - Caliper mount bracket onto the knuckle and torque the bolts to 70 ft. lbs. (90 Nm)
   - Stabilizer bar link to the strut and torque the bolts to 50 ft. lbs. (65 Nm)
   - Wheel
   - Negative battery cable
11. Depress the brake pedal and torque the axle to hub nut, in the following sequence:
   - Step 1: 85 ft. lbs. (115 Nm) to seat the bearing.
   - Step 2: Back the nut off
   - Step 3: 15 ft. lbs. (20 Nm).
   - Step 4: Turn the nut an additional 90 degrees, using a torque angle gauge
12. Install the cotter pin.
13. Check and adjust the alignment

Rear

1.9L ENGINES

Unlike the front wheel bearings, which may be removed from the hub for replacement, the rear wheel hub and bearing assembly is not serviceable. If damaged or worn, the hub and bearing assembly must be replaced as a unit.

1. Before servicing the vehicle, refer to the precautions in the beginning of this section.
2. Remove or disconnect the following:
   - Negative battery cable, if equipped with an Antilock Braking System (ABS)
   - Rear wheel
   - ABS speed sensor connector, if equipped
   - Caliper assembly-to-knuckle mounting bolts and support it with a wire from the strut, then remove the rotor, if equipped with disc brakes
   - Brake drum, if equipped with drum brakes
   - 4 hub/bearing-to-knuckle bolts, then remove the assembly from the vehicle

To install:

3. Install or connect the following:
   - Brake backing plate, hub/bearing assembly and retaining bolts and torque the bolts to 63 ft. lbs. (85 Nm)
   - Brake drum or rotor and caliper assembly and torque the caliper retaining bolts to 63 ft. lbs. (85 Nm), if equipped
   - ABS speed sensor connector, if equipped
   - Brake drum, if equipped

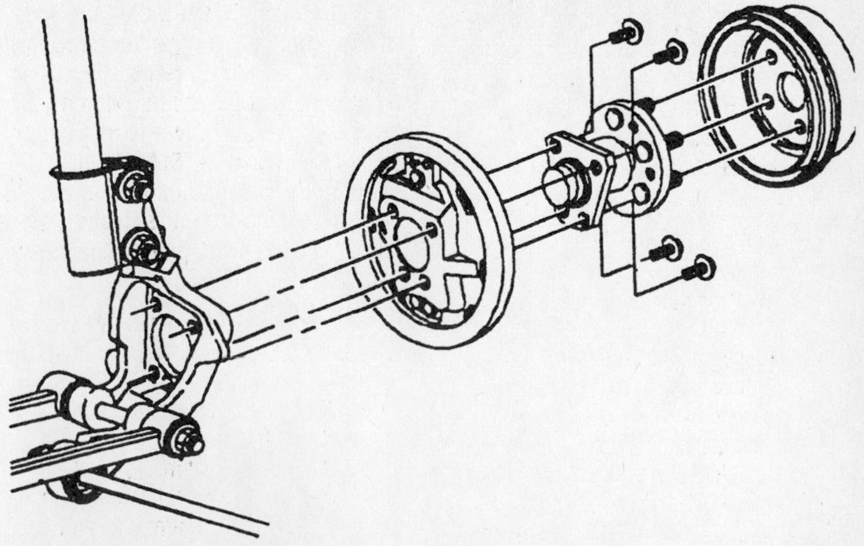

Exploded view of the rear hub/bearing assembly—drum brake set-up shown—1.9L engine

GLOSSARY

ABS: Anti-lock braking system. An electro-mechanical braking system which is designed to minimize or prevent wheel lock-up during braking.

ABSOLUTE PRESSURE: Atmospheric (barometric) pressure plus the pressure gauge reading.

ACCELERATOR PUMP: A small pump located in the carburetor that feeds fuel into the air/fuel mixture during acceleration.

ACCUMULATOR: A device that controls shift quality by cushioning the shock of hydraulic oil pressure being applied to a clutch or band.

ACTUATING MECHANISM: The mechanical output devices of a hydraulic system, for example, clutch pistons and band servos.

ACTUATOR: The output component of a hydraulic or electronic system.

ADVANCE: Setting the ignition timing so that spark occurs earlier before the piston reaches top dead center (TDC).

ADAPTIVE MEMORY (ADAPTIVE STRATEGY): The learning ability of the TCM or PCM to redefine its decision-making process to provide optimum shift quality.

AFTER TOP DEAD CENTER (ATDC): The point after the piston reaches the top of its travel on the compression stroke.

AIR BAG: Device on the inside of the car designed to inflate on impact of crash, protecting the occupants of the car.

AIR CHARGE TEMPERATURE (ACT) SENSOR: The temperature of the airflow into the engine is measured by an ACT sensor, usually located in the lower intake manifold or air cleaner. ALDL (assembly line diagnostic link): **Electrical connector for scanning ECM/PCM/TCM input and output devices.**

AIR CLEANER: An assembly consisting of a housing, filter and any connecting ductwork. The filter element is made up of a porous paper, sometimes with a wire mesh screening, and is designed to prevent airborne particles from entering the engine through the carburetor or throttle body.

AIR INJECTION: One method of reducing harmful exhaust emissions by injecting air into each of the exhaust ports of an engine. The fresh air entering the hot exhaust manifold causes any remaining fuel to be burned before it can exit the tailpipe.

AIR PUMP: An emission control device that supplies fresh air to the exhaust manifold to aid in more completely burning exhaust gases.

AIR/FUEL RATIO: The ratio of air-to-gasoline by weight in the fuel mixture drawn into the engine.

ALIGNMENT RACK: A special drive-on vehicle lift apparatus/measuring device used to adjust a vehicle's toe, caster and camber angles.

ALL WHEEL DRIVE: Term used to describe a full time four wheel drive system or any other vehicle drive system that continuously delivers power to all four wheels. This system is found primarily on station wagon vehicles and SUVs not utilized for significant off road use.

ALTERNATING CURRENT (AC): Electric current that flows first in one direction, then in the opposite direction, continually reversing flow.

ALTERNATOR: A device which produces AC (alternating current) which is converted to DC (direct current) to charge the car battery.

AMMETER: An instrument, calibrated in amperes, used to measure the flow of an electrical current in a circuit. Ammeters are always connected in series with the circuit being tested.

AMPERAGE: The total amount of current (amperes) flowing in a circuit.

AMPLIFIER: A device used in an electrical circuit to increase the voltage of an output signal.

AMP/HR. RATING (BATTERY): Measurement of the ability of a battery to deliver a stated amount of current for a stated period of time. The higher the amp/hr. rating, the better the battery.

AMPERE: The rate of flow of electrical current present when one volt of electrical pressure is applied against one ohm of electrical resistance.

ANALOG COMPUTER: Any microprocessor that uses similar (analogous) electrical signals to make its calculations.

ANODIZED: A special coating applied to the surface of aluminum valves for extended service life.

ANTIFREEZE: A substance (ethylene or propylene glycol) added to the coolant to prevent freezing in cold weather.

ANTI-FOAM AGENTS: Minimize fluid foaming from the whipping action encountered in the converter and planetary action.

ANTI-WEAR AGENTS: Zinc agents that control wear on the gears, bushings, and thrust washers.

ANTI-LOCK BRAKING SYSTEM: A supplementary system to the base hydraulic system that prevents sustained lock-up of the wheels during braking as well as automatically controlling wheel slip.

ANTI-ROLL BAR: See stabilizer bar.

ARC: A flow of electricity through the air between two electrodes or contact points that produces a spark.

ARMATURE: A laminated, soft iron core wrapped by a wire that converts electrical energy to mechanical energy as in a motor or relay. When rotated in a magnetic field, it changes mechanical energy into electrical energy as in a generator.

ATDC: After Top Dead Center.

ATF: Automatic transmission fluid.

ATMOSPHERIC PRESSURE: The pressure on the Earth's surface caused by the weight of the air in the atmosphere. At sea level, this pressure is 14.7 psi at 32°F (101 kPa at 0°C).

ATOMIZATION: The breaking down of a liquid into a fine mist that can be suspended in air.

AUXILIARY ADD-ON COOLER: A supplemental transmission fluid cooling device that is installed in series with the heat exchanger (cooler), located inside the radiator, to provide additional support to cool the hot fluid leaving the torque converter.

AUXILIARY PRESSURE: An added fluid pressure that is introduced into a regulator or balanced valve system to control valve movement. The auxiliary pressure itself can be either a fixed or a variable value. (See balanced valve; regulator valve.)

AWD: All wheel drive.

AXIAL FORCE: A side or end thrust force acting in or along the same plane as the power flow.

AXIAL PLAY: Movement parallel to a shaft or bearing bore.

AXLE CAPACITY: The maximum load-carrying capacity of the axle itself, as specified by the manufacturer. This is usually a higher number than the GAWR.

AXLE RATIO: This is a number (3.07:1, 4.56:1, for example) expressing the ratio between driveshaft revolutions and wheel revolutions. A low numerical ratio allows the engine to work easier because it doesn't have to turn as fast. A high numerical ratio means that the engine has to turn more rpm's to move the wheels through the same number of turns.

BACKFIRE: The sudden combustion of gases in the intake or exhaust system that results in a loud explosion.

BACKLASH: The clearance or play between two parts, such as meshed gears.

BACKPRESSURE: Restrictions in the exhaust system that slow the exit of exhaust gases from the combustion chamber.

BAKELITE®: A heat resistant, plastic insulator material commonly used in printed circuit boards and transistorized components.

BALANCED VALVE: A valve that is positioned by opposing auxiliary hydraulic pressures and/or spring force. Examples include mainline regulator, throttle, and governor valves. (See regulator valve.)

BAND: A flexible ring of steel with an inner lining of friction material. When tightened around the outside of a drum, a planetary member is held stationary to the transmission/transaxle case.

BALL BEARING: A bearing made up of hardened inner and outer races between which hardened steel balls roll.

BALL JOINT: A ball and matching socket connecting suspension components (steering knuckle to lower control arms). It permits rotating movement in any direction between the components that are joined.

BARO (BAROMETRIC PRESSURE SENSOR): Measures the change in the intake manifold pressure caused by changes in altitude.

BAROMETRIC MANIFOLD ABSOLUTE PRESSURE (BMAP) SEN-

SOR: Operates similarly to a conventional MAP sensor; reads intake manifold pressure and is also responsible for determining altitude and barometric pressure prior to engine operation.

BAROMETRIC PRESSURE: (See atmospheric pressure.)

BALLAST RESISTOR: A resistor in the primary ignition circuit that lowers voltage after the engine is started to reduce wear on ignition components.

BATTERY: A direct current electrical storage unit, consisting of the basic active materials of lead and sulfuric acid, which converts chemical energy into electrical energy. Used to provide current for the operation of the starter as well as other equipment, such as the radio, lighting, etc.

BEAD: The portion of a tire that holds it on the rim.

BEARING: A friction reducing, supportive device usually located between a stationary part and a moving part.

BEFORE TOP DEAD CENTER (BTDC): The point just before the piston reaches the top of its travel on the compression stroke.

BELTED TIRE: Tire construction similar to bias-ply tires, but using two or more layers of reinforced belts between body plies and the tread.

BEZEL: Piece of metal surrounding radio, headlights, gauges or similar components; sometimes used to hold the glass face of a gauge in the dash.

BIAS-PLY TIRE: Tire construction, using body ply reinforcing cords which run at alternating angles to the center line of the tread.

BI-METAL TEMPERATURE SENSOR: Any sensor or switch made of two dissimilar types of metal that bend when heated or cooled due to the different expansion rates of the alloys. These types of sensors usually function as an on/off switch.

BLOCK: See Engine Block.

BLOW-BY: Combustion gases, composed of water vapor and unburned fuel, that leak past the piston rings into the crankcase during normal engine operation. These gases are removed by the PCV system to prevent the buildup of harmful acids in the crankcase.

BOOK TIME: See Labor Time.

BOOK VALUE: The average value of a car, widely used to determine trade-in and resale value.

BOOST VALVE: Used at the base of the regulator valve to increase mainline pressure.

BORE: Diameter of a cylinder.

BRAKE CALIPER: The housing that fits over the brake disc. The caliper holds the brake pads, which are pressed against the discs by the caliper pistons when the brake pedal is depressed.

BRAKE HORSEPOWER(BHP): The actual horsepower available at the engine flywheel as measured by a dynamometer.

BRAKE FADE: Loss of braking power, usually caused by excessive heat after repeated brake applications.

BRAKE HORSEPOWER: Usable horsepower of an engine measured at the crankshaft.

BRAKE PAD: A brake shoe and lining assembly used with disc brakes.

BRAKE PROPORTIONING VALVE: A valve on the master cylinder which restricts hydraulic brake pressure to the wheels to a specified amount, preventing wheel lock-up.

BREAKAWAY: Often used by Chrysler to identify first-gear operation in D and 2 ranges. In these ranges, first-gear operation depends on a one-way roller clutch that holds on acceleration and releases (breaks away) on deceleration, resulting in a freewheeling coast-down condition.

BRAKE SHOE: The backing for the brake lining. The term is, however, usually applied to the assembly of the brake backing and lining.

BREAKER POINTS: A set of points inside the distributor, operated by a cam, which make and break the ignition circuit.

BRINNELLING: A wear pattern identified by a series of indentations at regular intervals. This condition is caused by a lack of lube, overload situations, and/or vibrations.

BTDC: Before Top Dead Center.

BUMP: Sudden and forceful apply of a clutch or band.

BUSHING: A liner, usually removable, for a bearing; an anti-friction liner used in place of a bearing.

CALIFORNIA ENGINE: An engine certified by the EPA for use in California only; conforms to more stringent emission regulations than Federal engine.

CALIPER: A hydraulically activated device in a disc brake system, which is mounted straddling the brake rotor (disc). The caliper contains at least one piston and two brake pads. Hydraulic pressure on the piston(s) forces the pads against the rotor.

CAPACITY: The quantity of electricity that can be delivered from a unit, as from a battery in ampere-hours, or output, as from a generator.

CAMBER: One of the factors of wheel alignment. Viewed from the front of the car, it is the inward or outward tilt of the wheel. The top of the tire will lean outward (positive camber) or inward (negative camber).

CAMSHAFT: A shaft in the engine on which are the lobes (cams) which operate the valves. The camshaft is driven by the crankshaft, via a belt, chain or gears, at one half the crankshaft speed.

CAPACITOR: A device which stores an electrical charge.

CARBON MONOXIDE (CO): A colorless, odorless gas given off as a normal byproduct of combustion. It is poisonous and extremely dangerous in confined areas, building up slowly to toxic levels without warning if adequate ventilation is not available.

CARBURETOR: A device, usually mounted on the intake manifold of an engine, which mixes the air and fuel in the proper proportion to allow even combustion.

CASTER: The forward or rearward tilt of an imaginary line drawn through the upper ball joint and the center of the wheel. Viewed from the sides, positive caster (forward tilt) lends directional stability, while negative caster (rearward tilt) produces instability.

CATALYTIC CONVERTER: A device installed in the exhaust system, like a muffler, that converts harmful byproducts of combustion into carbon dioxide and water vapor by means of a heat-producing chemical reaction.

CENTRIFUGAL ADVANCE: A mechanical method of advancing the spark timing by using flyweights in the distributor that react to centrifugal force generated by the distributor shaft rotation.

CENTRIFUGAL FORCE: The outward pull of a revolving object, away from the center of revolution. Centrifugal force increases with the speed of rotation.

CETANE RATING: A measure of the ignition value of diesel fuel. The higher the cetane rating, the better the fuel. Diesel fuel cetane rating is roughly comparable to gasoline octane rating.

CHECK VALVE: Any one-way valve installed to permit the flow of air, fuel or vacuum in one direction only.

CHOKE: The valve/plate that restricts the amount of air entering an engine on the induction stroke, thereby enriching the air/fuel ratio.

CHUGGLE: Bucking or jerking condition that may be engine related and may be most noticeable when converter clutch is engaged; similar to the feel of towing a trailer.

CIRCLIP: A split steel snapring that fits into a groove to hold various parts in place.

CIRCUIT BREAKER: A switch which protects an electrical circuit from overload by opening the circuit when the current flow exceeds a pre-determined level. Some circuit breakers must be reset manually, while most reset automatically.

CIRCUIT: Any unbroken path through which an electrical current can flow. Also used to describe fuel flow in some instances.

CIRCUIT, BYPASS: Another circuit in parallel with the major circuit through which power is diverted.

CIRCUIT, CLOSED: An electrical circuit in which there is no interruption of current flow.

CIRCUIT, GROUND: The non-insulated portion of a complete circuit used as a common potential point. In automotive circuits, the ground is composed of metal parts, such as the engine, body sheet metal, and frame and is usually a negative potential.

CIRCUIT, HOT: That portion of a circuit not at ground potential. The hot circuit is usually insulated and is connected to the positive side of the battery.

CIRCUIT, OPEN: A break or lack of contact in an electrical circuit, either intentional (switch) or unintentional (bad connection or broken wire).

CIRCUIT, PARALLEL: A circuit having two or more paths for current flow with common positive and negative tie points. The same voltage is applied to each load device or parallel branch.

CIRCUIT, SERIES: An electrical system in which separate parts are connected end to end, using one wire, to form a single path for current to flow.

CIRCUIT, SHORT: A circuit that is accidentally completed in an electrical path for which it was not intended.

CLAMPING (ISOLATION) DIODES: Diodes positioned in a circuit to prevent self-induction from damaging electronic components.

CLEARCOAT: A transparent layer which, when sprayed over a vehicle's paint job, adds gloss and depth as well as an additional protective coating to the finish.

CLUTCH: Part of the power train used to connect/disconnect power to the rear wheels.

CLUTCH, FLUID: The same as a fluid coupling. A fluid clutch or coupling performs the same function as a friction clutch by utilizing fluid friction and inertia as opposed to solid friction used by a friction clutch. (See fluid coupling.)

CLUTCH, FRICTION: A coupling device that provides a means of smooth and positive engagement and disengagement of engine torque to the vehicle powertrain. Transmission of power through the clutch is accomplished by bringing one or more rotating drive members into contact with complementing driven members.

COAST: Vehicle deceleration caused by engine braking conditions.

COEFFICIENT OF FRICTION: The amount of surface tension between two contacting surfaces; identified by a scientifically calculated number.

COIL: Part of the ignition system that boosts the relatively low voltage supplied by the car's electrical system to the high voltage required to fire the spark plugs.

COMBINATION MANIFOLD: An assembly which includes both the intake and exhaust manifolds in one casting.

COMBINATION VALVE: A device used in some fuel systems that routes fuel vapors to a charcoal storage canister instead of venting them into the atmosphere. The valve relieves fuel tank pressure and allows fresh air into the tank as the fuel level drops to prevent a vapor lock situation.

COMBUSTION CHAMBER: The part of the engine in the cylinder head where combustion takes place.

COMPOUND GEAR: A gear consisting of two or more simple gears with a common shaft.

COMPOUND PLANETARY: A gearset that has more than the three elements found in a simple gearset and is constructed by combining members of two planetary gearsets to create additional gear ratio possibilities.

COMPRESSION CHECK: A test involving removing each spark plug and inserting a gauge. When the engine is cranked, the gauge will record a pressure reading in the individual cylinder. General operating condition can be determined from a compression check.

COMPRESSION RATIO: The ratio of the volume between the piston and cylinder head when the piston is at the bottom of its stroke (bottom dead center) and when the piston is at the top of its stroke (top dead center).

COMPUTER: An electronic control module that correlates input data according to prearranged engineered instructions; used for the management of an actuator system or systems.

CONDENSER:

An electrical device which acts to store an electrical charge, preventing voltage surges.

2. A radiator-like device in the air conditioning system in which refrigerant gas condenses into a liquid, giving off heat.

CONDUCTOR: Any material through which an electrical current can be transmitted easily.

CONNECTING ROD: The connecting link between the crankshaft and piston.

CONSTANT VELOCITY JOINT: Type of universal joint in a halfshaft assembly in which the output shaft turns at a constant angular velocity without variation, provided that the speed of the input shaft is constant.

CONTINUITY: Continuous or complete circuit. Can be checked with an ohmmeter.

CONTROL ARM: The upper or lower suspension components which are mounted on the frame and support the ball joints and steering knuckles.

CONVENTIONAL IGNITION: Ignition system which uses breaker points.

CONVERTER: (See torque converter.)

CONVERTER LOCKUP: The switching from hydrodynamic to direct mechanical drive, usually through the application of a friction element called the converter clutch.

COOLANT: Mixture of water and anti-freeze circulated through the engine to carry off heat produced by the engine.

CORROSION INHIBITOR: An inhibitor in ATF that prevents corrosion of bushings, thrust washers, and oil cooler brazed joints.

COUNTERSHAFT: An intermediate shaft which is rotated by a mainshaft and transmits, in turn, that rotation to a working part.

COUPLING PHASE: Occurs when the torque converter is operating at its greatest hydraulic efficiency. The speed differential between the impeller and the turbine is at its minimum. At this point, the stator freewheels, and there is no torque multiplication.

CRANKCASE: The lower part of an engine in which the crankshaft and related parts operate.

CRANKSHAFT: Engine component (connected to pistons by connecting rods) which converts the reciprocating (up and down) motion of pistons to rotary motion used to turn the driveshaft.

CURB WEIGHT: The weight of a vehicle without passengers or payload, but including all fluids (oil, gas, coolant, etc.) and other equipment specified as standard.

CURRENT: The flow (or rate) of electrons moving through a circuit. Current is measured in amperes (amp).

CURRENT FLOW CONVENTIONAL: Current flows through a circuit from the positive terminal of the source to the negative terminal (plus to minus).

CURRENT FLOW, ELECTRON: Current or electrons flow from the negative terminal of the source, through the circuit, to the positive terminal (minus to plus).

CV-JOINT: Constant velocity joint.

CYCLIC VIBRATIONS: The off-center movement of a rotating object that is affected by its initial balance, speed of rotation, and working angles.

CYLINDER BLOCK: See engine block.

CYLINDER HEAD: The detachable portion of the engine, usually fastened to the top of the cylinder block and containing all or most of the combustion chambers. On overhead valve engines, it contains the valves and their operating parts. On overhead cam engines, it contains the camshaft as well.

CYLINDER: In an engine, the round hole in the engine block in which the piston(s) ride.

DATA LINK CONNECTOR (DLC): Current acronym/term applied to the federally mandated, diagnostic junction connector that is used to monitor ECM/PC/TCM inputs, processing strategies, and outputs including diagnostic trouble codes (DTCs).

DEAD CENTER: The extreme top or bottom of the piston stroke.

DECELERATION BUMP: When referring to a torque converter clutch in the applied position, a sudden release of the accelerator pedal causes a forceful reversal of power through the drivetrain (engine braking), just prior to the apply plate actually being released.

DELAYED (LATE OR EXTENDED): Condition where shift is expected but does not occur for a period of time, for example, where clutch or band engagement does not occur as quickly as expected during part throttle or wide open throttle apply of accelerator or when manually downshifting to a lower range.

DETENT: A spring-loaded plunger, pin, ball, or pawl used as a holding device on a ratchet wheel or shaft. In automatic transmissions, a detent mechanism is used for locking the manual valve in place.

DETENT DOWNSHIFT: (See kickdown.)

DETERGENT: An additive in engine oil to improve its operating characteristics.

DETONATION: An unwanted explosion of the air/fuel mixture in the combustion chamber caused by excess heat and compression, advanced timing, or an overly lean mixture. Also referred to as "ping".

DEXRON®: A brand of automatic transmission fluid.

DIAGNOSTIC TROUBLE CODES (DTCs): A digital display from the control module memory that identifies the input, processor, or output device circuit that is related to the powertrain emission/driveability malfunction detected. Diagnostic trouble codes can be read by the MIL to flash any codes or by using a handheld scanner.

DIAPHRAGM: A thin, flexible wall separating two cavities, such as in a vacuum advance unit.

DIESELING: The engine continues to run after the car is shut off; caused by fuel continuing to be burned in the combustion chamber.

DIFFERENTIAL: A geared assembly which allows the transmission of motion between drive axles, giving one axle the ability to rotate faster than the other, as in cornering.

DIFFERENTIAL AREAS: When opposing faces of a spool valve are acted upon by the same pressure but their areas differ in size, the face with the larger area produces the differential force and valve movement. (See spool valve.)

DIFFERENTIAL FORCE: (See differential areas)

DIGITAL READOUT: A display of numbers or a combination of numbers and letters.

DIGITAL VOLT OHMMETER: An electronic diagnostic tool used to measure voltage, ohms and amps as well as several other functions, with the readings displayed on a digital screen in tenths, hundredths and thousandths.

DIODE: An electrical device that will allow current to flow in one direction only.

DIRECT CURRENT (DC): Electrical current that flows in one direction only.

DIRECT DRIVE: The gear ratio is 1:1, with no change occurring in the torque and speed input/output relationship.

DISC BRAKE: A hydraulic braking assembly consisting of a brake disc, or rotor, mounted on an axle shaft, and a caliper assembly containing, usually two brake pads which are activated by hydraulic pressure. The pads are forced against the sides of the disc, creating friction which slows the vehicle.

DISPERSANTS: Suspend dirt and prevent sludge buildup in a liquid, such as engine oil.

DOUBLE BUMP (DOUBLE FEEL): Two sudden and forceful applies of a clutch or band.

DISPLACEMENT: The total volume of air that is displaced by all pistons as the engine turns through one complete revolution.

DISTRIBUTOR: A mechanically driven device on an engine which is responsible for electrically firing the spark plug at a pre-determined point of the piston stroke.

DOHC: Double overhead camshaft.

DOUBLE OVERHEAD CAMSHAFT: The engine utilizes two camshafts mounted in one cylinder head. One camshaft operates the exhaust valves, while the other operates the intake valves.

DOWEL PIN: A pin, inserted in mating holes in two different parts allowing those parts to maintain a fixed relationship.

DRIVELINE: The drive connection between the transmission and the drive wheels.

DRIVE TRAIN: The components that transmit the flow of power from the engine to the wheels. The components include the clutch, transmission, driveshafts (or axle shafts in front wheel drive), U-joints and differential.

DRUM BRAKE: A braking system which consists of two brake shoes and one or two wheel cylinders, mounted on a fixed backing plate, and a brake drum, mounted on an axle, which revolves around the assembly.

DRY CHARGED BATTERY: Battery to which electrolyte is added when the battery is placed in service.

DVOM: Digital volt ohmmeter

DWELL: The rate, measured in degrees of shaft rotation, at which an electrical circuit cycles on and off.

DYNAMIC: An application in which there is rotating or reciprocating motion between the parts.

EARLY: Condition where shift occurs before vehicle has reached proper speed, which tends to labor engine after upshift.

EBCM: See Electronic Control Unit (ECU).

ECM: See Electronic Control Unit (ECU).

ECU: Electronic control unit.

ELECTRODE: Conductor (positive or negative) of electric current.

ELECTROLYSIS: A surface etching or bonding of current conducting transmission/transaxle components that may occur when grounding straps are missing or in poor condition.

ELECTROLYTE: A solution of water and sulfuric acid used to activate the battery. Electrolyte is extremely corrosive.

ELECTROMAGNET: A coil that produces a magnetic field when current flows through its windings.

ELECTROMAGNETIC INDUCTION: A method to create (generate) current flow through the use of magnetism.

ELECTROMAGNETISM: The effects surrounding the relationship between electricity and magnetism.

ELECTROMOTIVE FORCE (EMF): The force or pressure (voltage) that causes current movement in an electrical circuit.

ELECTRONIC CONTROL UNIT: A digital computer that controls engine (and sometimes transmission, brake or other vehicle system) functions based on data received from various sensors. Examples used by some manufacturers include Electronic Brake Control Module (EBCM), Engine Control Module (ECM), Powertrain Control Module (PCM) or Vehicle Control Module (VCM).

ELECTRONIC IGNITION: A system in which the timing and firing of the spark plugs is controlled by an electronic control unit, usually called a module. These systems have no points or condenser.

ELECTRONIC PRESSURE CONTROL (EPC) SOLENOID: A specially designed solenoid containing a spool valve and spring assembly to control fluid mainline pressure. A variable current flow, controlled by the ECM/PCM, varies the internal force of the solenoid on the spool valve and resulting mainline pressure. (See variable force solenoid.)

ELECTRONICS: Miniaturized electrical circuits utilizing semiconductors, solid-state devices, and printed circuits. Electronic circuits utilize small amounts of power.

ELECTRONIFICATION: The application of electronic circuitry to a mechanical device. Regarding automatic transmissions, electrification is incorporated into converter clutch lockup, shift scheduling, and line pressure control systems.

ELECTROSTATIC DISCHARGE (ESD): An unwanted, high-voltage electrical current released by an individual who has taken on a static charge of electricity. Electronic components can be easily damaged by ESD.

ELEMENT: A device within a hydrodynamic drive unit designed with a set of blades to direct fluid flow.

ENAMEL: Type of paint that dries to a smooth, glossy finish.

END BUMP (END FEEL OR SLIP BUMP): Firmer feel at end of shift when compared with feel at start of shift.

END-PLAY: The clearance/gap between two components that allows for expansion of the parts as they warm up, to prevent binding and to allow space for lubrication.

ENERGY: The ability or capacity to do work.

ENGINE: The primary motor or power apparatus of a vehicle, which converts liquid or gas fuel into mechanical energy.

ENGINE BLOCK: The basic engine casting containing the cylinders, the crankshaft main bearings, as well as machined surfaces for the mounting of other components such as the cylinder head, oil pan, transmission, etc..

ENGINE BRAKING: Use of engine to slow vehicle by manually downshifting during zero-throttle coast down.

ENGINE CONTROL MODULE (ECM): Manages the engine and incorporates output control over the torque converter clutch solenoid. (Note: Current designation for the ECM in late model vehicles is PCM.)

ENGINE COOLANT TEMPERATURE (ECT) SENSOR: Prevents converter clutch engagement with a cold engine; also used for shift timing and shift quality.

EP LUBRICANT: EP (extreme pressure) lubricants are specially formulated for use with gears involving heavy loads (transmissions, differentials, etc.).

ETHYL: A substance added to gasoline to improve its resistance to knock, by slowing down the rate of combustion.

ETHYLENE GLYCOL: The base substance of antifreeze.

EXHAUST MANIFOLD: A set of cast passages or pipes which conduct exhaust gases from the engine.

FAIL-SAFE (BACKUP) CONTROL: A substitute value used by the PCM/TCM to replace a faulty signal from an input sensor. The temporary value allows the vehicle to continue to be operated.

FAST IDLE: The speed of the engine when the choke is on. Fast idle speeds engine warm-up.

FEDERAL ENGINE: An engine certified by the EPA for use in any of the 49 states (except California).

FEEDBACK: A circuit malfunction whereby current can find another path to feed load devices.

FEELER GAUGE: A blade, usually metal, of precisely predetermined thickness, used to measure the clearance between two parts.

FILAMENT: The part of a bulb that glows; the filament creates high resistance to current flow and actually glows from the resulting heat.

FINAL DRIVE: An essential part of the axle drive assembly where final gear reduction takes place in the powertrain. In RWD applications and north-south FWD applications, it must also change the power flow direction to the axle shaft by ninety degrees. (Also see axle ratio).

FIRING ORDER: The order in which combustion occurs in the cylinders of an engine. Also the order in which spark is distributed to the plugs by the distributor.

FIRM: A noticeable quick apply of a clutch or band that is considered normal with medium to heavy throttle shift; should not be confused with harsh or rough.

FLAME FRONT: The term used to describe certain aspects of the fuel explosion in the cylinders. The flame front should move in a controlled pattern across the cylinder, rather than simply exploding immediately.

FLARE (SLIPPING): A quick increase in engine rpm accompanied by momentary loss of torque; generally occurs during shift.

FLAT ENGINE: Engine design in which the pistons are horizontally opposed. Porsche, Subaru and some old VW are common examples of flat engines.

FLAT RATE: A dealership term referring to the amount of money paid to a technician for a repair or diagnostic service based on that particular service versus dealership's labor time (NOT based on the actual time the technician spent on the job).

FLAT SPOT: A point during acceleration when the engine seems to lose power for an instant.

FLOODING: The presence of too much fuel in the intake manifold and combustion chamber which prevents the air/fuel mixture from firing, thereby causing a no-start situation.

FLUID: A fluid can be either liquid or gas. In hydraulics, a liquid is used for transmitting force or motion.

FLUID COUPLING: The simplest form of hydrodynamic drive, the fluid coupling consists of two look-alike members with straight radial varies referred to as the impeller (pump) and the turbine. Input torque is always equal to the output torque.

FLUID DRIVE: Either a fluid coupling or a fluid torque converter. (See hydrodynamic drive units.)

FLUID TORQUE CONVERTER: A hydrodynamic drive that has the ability to act both as a torque multiplier and fluid coupling. (See hydrodynamic drive units; torque converter.)

FLUID VISCOSITY: The resistance of a liquid to flow. A cold fluid (oil) has greater viscosity and flows more slowly than a hot fluid (oil).

FLYWHEEL: A heavy disc of metal attached to the rear of the crankshaft. It smoothes the firing impulses of the engine and keeps the crankshaft turning during periods when no firing takes place. The starter also engages the flywheel to start the engine.

FOOT POUND (ft. lbs., lbs. ft. or sometimes, ft. lb.): The amount of energy or work needed to raise an item weighing one pound, a distance of one foot.

FREEZE PLUG: A plug in the engine block which will be pushed out if the coolant freezes. Sometimes called expansion plugs, they protect the block from cracking should the coolant freeze.

FRICTION: The resistance that occurs between contacting surfaces. This relationship is expressed by a ratio called the coefficient of friction (CL).

FRICTION, COEFFICIENT OF: The amount of surface tension between two contacting surfaces; expressed by a scientifically calculated number.

FRONT END ALIGNMENT: A service to set caster, camber and toe-in to the correct specifications. This will ensure that the car steers and handles properly and that the tires wear properly.

FRICTION MODIFIER: Changes the coefficient of friction of the fluid between the mating steel and composition clutch/band surfaces during the engagement process and allows for a certain amount of intentional slipping for a good "shift-feel".

FRONTAL AREA: The total frontal area of a vehicle exposed to air flow.

FUEL FILTER: A component of the fuel system containing a porous paper element used to prevent any impurities from entering the engine through the fuel system. It usually takes the form of a canister-like housing, mounted in-line with the fuel hose, located anywhere on a vehicle between the fuel tank and engine.

FUEL INJECTION: A system replacing the carburetor that sprays fuel into the cylinder through nozzles. The amount of fuel can be more precisely controlled with fuel injection.

FULL FLOATING AXLE: An axle in which the axle housing extends through the wheel giving bearing support on the outside of the housing. The front axle of a four-wheel drive vehicle is usually a full floating axle, as are the rear axles of many larger (1 ton and over) pick-ups and vans.

FULL-TIME FOUR-WHEEL DRIVE: A four-wheel drive system that continuously delivers power to all four wheels. A differential between the front and rear driveshafts permits variations in axle speeds to control gear wind-up without damage.

FULL THROTTLE DETENT DOWNSHIFT: A quick apply of accelerator pedal to its full travel, forcing a downshift.

FUSE: A protective device in a circuit which prevents circuit overload by breaking the circuit when a specific amperage is present. The device is constructed around a strip or wire of a lower amperage rating than the circuit it is designed to protect. When an amperage higher than that stamped on the fuse is present in the circuit, the strip or wire melts, opening the circuit.

FUSIBLE LINK: A piece of wire in a wiring harness that performs the same job as a fuse. If overloaded, the fusible link will melt and interrupt the circuit.

FWD: Front wheel drive.

GAWR: (Gross axle weight rating) the total maximum weight an axle is designed to carry.

GCW: (Gross combined weight) total combined weight of a tow vehicle and trailer.

GARAGE SHIFT: initial engagement feel of transmission, neutral to reverse or neutral to a forward drive.

GARAGE SHIFT FEEL: A quick check of the engagement quality and responsiveness of reverse and forward gears. This test is done with the vehicle stationary.

GEAR: A toothed mechanical device that acts as a rotating lever to transmit power or turning effort from one shaft to another. (See gear ratio.)

GEAR RATIO: A ratio expressing the number of turns a smaller gear will make to turn a larger gear through one revolution. The ratio is found by dividing the number of teeth on the smaller gear into the number of teeth on the larger gear.

GEARBOX: Transmission

GEAR REDUCTION: Torque is multiplied and speed decreased by the factor of the gear ratio. For example, a 3:1 gear ratio changes an input torque of 180 ft. lbs. and an input speed of 2700 rpm to 540 Ft. lbs. and 900 rpm, respectively. (No account is taken of frictional losses, which are always present.)

GEARTRAIN: A succession of intermeshing gears that form an assembly and provide for one or more torque changes as the power input is transmitted to the power output.

GEL COAT: A thin coat of plastic resin covering fiberglass body panels.

GENERATOR: A device which produces direct current (DC) necessary to charge the battery.

GOVERNOR: A device that senses vehicle speed and generates a hydraulic oil pressure. As vehicle speed increases, governor oil pressure rises.

GROUND CIRCUIT: (See circuit, ground.)

GROUND SIDE SWITCHING: The electrical/electronic circuit control switch is located after the circuit load.

GVWR: (Gross vehicle weight rating) total maximum weight a vehicle is designed to carry including the weight of the vehicle, passengers, equipment, gas, oil, etc.

HALOGEN: A special type of lamp known for its quality of brilliant white light. Originally used for fog lights and driving lights.

HARD CODES: DTCs that are present at the time of testing; also called continuous or current codes.

HARSH(ROUGH): An apply of a clutch or band that is more noticeable than a firm one; considered undesirable at any throttle position.

HEADER TANK: An expansion tank for the radiator coolant. It can be located remotely or built into the radiator.

HEAT RANGE: A term used to describe the ability of a spark plug to carry away heat. Plugs with longer nosed insulators take longer to carry heat off effectively.

HEAT RISER: A flapper in the exhaust manifold that is closed when the engine is cold, causing hot exhaust gases to heat the intake manifold providing better cold engine operation. A thermostatic spring opens the flapper when the engine warms up.

HEAVY THROTTLE: Approximately three-fourths of accelerator pedal travel.

HEMI: A name given an engine using hemispherical combustion chambers.

HERTZ (HZ): The international unit of frequency equal to one cycle per second (10,000 Hertz equals 10,000 cycles per second).

HIGH-IMPEDANCE DVOM (DIGITAL VOLT-OHMMETER): This styled device provides a built-in resistance value and is capable of limiting circuit current flow to safe milliamp levels.

HIGH RESISTANCE: Often refers to a circuit where there is an excessive amount of opposition to normal current flow.

HORSEPOWER: A measurement of the amount of work; one horsepower is the amount of work necessary to lift 33,000 lbs. one foot in one minute. Brake horsepower (bhp) is the horsepower delivered by an engine on a dynamometer. Net horsepower is the power remaining (measured at the flywheel of the engine) that can be used to turn the wheels after power is consumed through friction and running the engine accessories (water pump, alternator, air pump, fan etc.)

HOT CIRCUIT: (See circuit, hot; hot lead.) hot lead: **A wire or conductor in the power side of the circuit. (See circuit, hot.)**

HOT SIDE SWITCHING: The electrical/electronic circuit control switch is located before the circuit load.

HUB: The center part of a wheel or gear.

HUNTING (BUSYNESS): Repeating quick series of up-shifts and downshifts that causes noticeable change in engine rpm, for example, as in a 4-3-4 shift pattern.

HYDRAULICS: The use of liquid under pressure to transfer force of motion.

HYDROCARBON (HC): Any chemical compound made up of hydrogen and carbon. A major pollutant formed by the engine as a by-product of combustion.

HYDRODYNAMIC DRIVE UNITS: Devices that transmit power solely by the action of a kinetic fluid flow in a closed recirculating path. An impeller energizes the fluid and discharges the high-speed jet stream into the turbine for power output.

HYDROMETER: An instrument used to measure the specific gravity of a solution.

HYDROPLANING: A phenomenon of driving when water builds up under the tire tread, causing it to lose contact with the road. Slowing down will usually restore normal tire contact with the road.

HYPOID GEARSET: The drive pinion gear may be placed below or above the centerline of the driven gear; often used as a final drive gearset.

IDLE MIXTURE: The mixture of air and fuel (usually about 14:1) being fed to the cylinders. The idle mixture screw(s) are sometimes adjusted as part of a tune-up.

IDLER ARM: Component of the steering linkage which is a geometric duplicate of the steering gear arm. It supports the right side of the center steering link.

IMPELLER: Often called a pump, the impeller is the power input (drive) member of a hydrodynamic drive. As part of the torque converter cover, it acts as a centrifugal pump and puts the fluid in motion.

INCH POUND (inch lbs.; sometimes in. lb. or in. lbs.): One twelfth of a foot pound.

INDUCTANCE: The force that produces voltage when a conductor is passed through a magnetic field.

INDUCTION: A means of transferring electrical energy in the form of a magnetic field. Principle used in the ignition coil to increase voltage.

INITIAL FEEL: A distinct firmer feel at start of shift when compared with feel at finish of shift.

INJECTOR: A device which receives metered fuel under relatively low pressure and is activated to inject the fuel into the engine under relatively high pressure at a predetermined time.

INPUT: In an automatic transmission, the source of power from the engine is absorbed by the torque converter, which provides the power input into the transmission. The turbine drives the input(turbine)shaft.

INPUT SHAFT: The shaft to which torque is applied, usually carrying the driving gear or gears.

INTAKE MANIFOLD: A casting of passages or pipes used to conduct air or a fuel/air mixture to the cylinders.

INTERNAL GEAR: The ring-like outer gear of a planetary gearset with the gear teeth cut on the inside of the ring to provide a mesh with the planet pinions.

ISOLATION (CLAMPING) DIODES: Diodes positioned in a circuit to prevent self-induction from damaging electronic components.

IX ROTARY GEAR PUMP: Contains two rotating members, one shaped with internal gear teeth and the other with external gear teeth. As the gears separate, the fluid fills the gaps between gear teeth, is pulled across a crescent-shaped divider, and then is forced to flow through the outlet as the gears mesh.

IX ROTARY LOBE PUMP: Sometimes referred to as a gerotor type pump. Two rotating members, one shaped with internal lobes and the other with external lobes, separate and then mesh to cause fluid to flow.

JOURNAL: The bearing surface within which a shaft operates.

JUMPER CABLES: Two heavy duty wires with large alligator clips used to provide power from a charged battery to a discharged battery mounted in a vehicle.

JUMPSTART: Utilizing the sufficiently charged battery of one vehicle to start the engine of another vehicle with a discharged battery by the use of jumper cables.

KEY: A small block usually fitted in a notch between a shaft and a hub to prevent slippage of the two parts.

KICKDOWN: Detent downshift system; either linkage, cable, or electrically controlled.

KILO: A prefix used in the metric system to indicate one thousand.

KNOCK: Noise which results from the spontaneous ignition of a portion of the air-fuel mixture in the engine cylinder caused by overly advanced ignition timing or use of incorrectly low octane fuel for that engine.

KNOCK SENSOR: An input device that responds to spark knock, caused by over advanced ignition timing.

LABOR TIME: A specific amount of time required to perform a certain repair or diagnostic service as defined by a vehicle or after-market manufacturer .

LACQUER: A quick-drying automotive paint.

LATE: Shift that occurs when engine is at higher than normal rpm for given amount of throttle.

LIGHT-EMITTING DIODE (LED): A semiconductor diode that emits light as electrical current flows through it; used in some electronic display devices to emit a red or other color light.

LIGHT THROTTLE: Approximately one-fourth of accelerator pedal travel.

LIMITED SLIP: A type of differential which transfers driving force to the wheel with the best traction.

LIMP-IN MODE: Electrical shutdown of the transmission/ transaxle output solenoids, allowing only forward and reverse gears that are hydraulically energized by the manual valve. This permits the vehicle to be driven to a service facility for repair.

LIP SEAL: Molded synthetic rubber seal designed with an outer sealing edge (lip) that points into the fluid containing area to be sealed. This type of seal is used where rotational and axial forces are present.

LITHIUM-BASE GREASE: Chassis and wheel bearing grease using lithium as a base. Not compatible with sodium-base grease.

LOAD DEVICE: A circuit's resistance that converts the electrical energy into light, sound, heat, or mechanical movement.

LOAD RANGE: Indicates the number of plies at which a tire is rated. Load range B equals four-ply rating; C equals six-ply rating; and, D equals an eight-ply rating.

LOAD TORQUE: The amount of output torque needed from the transmission/transaxle to overcome the vehicle load.

LOCKING HUBS: Accessories used on part-time four-wheel drive systems that allow the front wheels to be disengaged from the drive train when four-wheel drive is not being used. When four-wheel drive is desired, the hubs are engaged, locking the wheels to the drive train.

LOCKUP CONVERTER: A torque converter that operates hydraulically and mechanically. When an internal apply plate (lockup plate) clamps to the torque converter cover, hydraulic slippage is eliminated.

LOCK RING: See Circlip or Snapring

MAGNET: Any body with the property of attracting iron or steel.

MAGNETIC FIELD: The area surrounding the poles of a magnet that is affected by its attraction or repulsion forces.

MAIN LINE PRESSURE: Often called control pressure or line pressure, it refers to the pressure of the oil leaving the pump and is controlled by the pressure regulator valve.

MALFUNCTION INDICATOR LAMP (MIL): Previously known as a check engine light, the dash-mounted MIL illuminates and signals the driver that an emission or driveability problem with the powertrain has been detected by the ECM/PCM. When this occurs, at least one diagnostic trouble code (DTC) has been stored into the control module memory.

MANIFOLD ABSOLUTE PRESSURE (MAP) SENSOR: Reads the amount of air pressure (vacuum) in the engine's intake manifold system; its signal is used to analyze engine load conditions.

MANIFOLD VACUUM: Low pressure in an engine intake manifold formed just below the throttle plates. Manifold vacuum is highest at idle and drops under acceleration.

MANIFOLD: A casting of passages or set of pipes which connect the cylinders to an inlet or outlet source.

MANUAL LEVER POSITION SWITCH (MLPS): A mechanical switching unit that is typically mounted externally to the transmission/transaxle to inform the PCM/ECM which gear range the driver has selected.

MANUAL VALVE: Located inside the transmission/transaxle, it is directly connected to the driver's shift lever. The position of the manual valve determines which hydraulic circuits will be charged with oil pressure and the operating mode of the transmission.

MANUAL VALVE LEVER POSITION SENSOR (MVLPS): The input from this device tells the TCM what gear range was selected.

MASS AIR FLOW (MAF) SENSOR: Measures the airflow into the engine.

MASTER CYLINDER: The primary fluid pressurizing device in a hydraulic system. In automotive use, it is found in brake and hydraulic clutch systems and is pedal activated, either directly or, in a power brake system, through the power booster.

MacPherson STRUT: A suspension component combining a shock absorber and spring in one unit.

MEDIUM THROTTLE: Approximately one-half of accelerator pedal travel.

MEGA: A metric prefix indicating one million.

MEMBER: An independent component of a hydrodynamic unit such as an impeller, a stator, or a turbine. It may have one or more elements.

MERCON: A fluid developed by Ford Motor Company in 1988. It contains a friction modifier and closely resembles operating characteristics of Dexron.

METAL SEALING RINGS: Made from cast iron or aluminum, their primary application is with dynamic components involving pressure sealing circuits of rotating members. These rings are designed with either butt or hook lock end joints.

METER (ANALOG): A linear-style meter representing data as lengths; a needle-style instrument interfacing with logical numerical increments. This style of electrical meter uses relatively low impedance internal resistance and cannot be used for testing electronic circuitry.

METER(DIGITAL): Uses numbers as a direct readout to show values. Most meters of this style use high impedance internal resistance and must be used for testing low current electronic circuitry.

MICRO: A metric prefix indicating one-millionth (0.000001).

MILLI: A metric prefix indicating one-thousandth (0.001).

MINIMUM THROTTLE: The least amount of throttle opening required for upshift; normally close to zero throttle.

MISFIRE: Condition occurring when the fuel mixture in a cylinder fails to ignite, causing the engine to run roughly.

MODULE: Electronic control unit, amplifier or igniter of solid state or integrated design which controls the current flow in the ignition primary circuit based on input from the pick-up coil. When the module opens the primary circuit, high secondary voltage is induced in the coil.

MODULATED: In an electronic-hydraulic converter clutch system (or shift valve system), the term modulated refers to the pulsing of a solenoid, at a variable rate. This action controls the buildup of oil pressure in the hydraulic circuit to allow a controlled amount of clutch slippage.

MODULATED CONVERTER CLUTCH CONTROL (MCCC): A pulse width duty cycle valve that controls the converter lockup apply pressure and maximizes smoother transitions between lock and unlock conditions.

MODULATOR PRESSURE (THROTTLE PRESSURE): A hydraulic signal oil pressure relating to the amount of engine load, based on either the amount of throttle plate opening or engine vacuum.

MODULATOR VALVE: A regulator valve that is controlled by engine vacuum, providing a hydraulic pressure that varies in relation to engine torque. The hydraulic torque signal functions to delay the shift pattern and provide a line pressure boost. (See throttle valve.)

MOTOR: An electromagnetic device used to convert electrical energy into mechanical energy.

MULTIPLE-DISC CLUTCH: A grouping of steel and friction lined plates that, when compressed together by hydraulic pressure acting upon a piston, lock or unlock a planetary member.

MULTI-WEIGHT: Type of oil that provides adequate lubrication at both high and low temperatures.

needed to move one amp through a resistance of one ohm.

MUSHY: Same as soft; slow and drawn out clutch apply with very little shift feel.

MUTUAL INDUCTION: The generation of **Current from one wire circuit to another by movement of the magnetic field surrounding a current-carrying circuit as its ampere flow increases or decreases.**

NEEDLE BEARING: A bearing which consists of a number (usually a large number) of long, thin rollers.

NITROGEN OXIDE (NOx): One of the three basic pollutants found in the exhaust emission of an internal combustion engine. The amount of NOx usually varies in an inverse proportion to the amount of HC and CO.

NONPOSITIVE SEALING: A sealing method that allows some minor leakage, which normally assists in lubrication.

O2 SENSOR: Located in the engine's exhaust system, it is an input device to the ECM/PCM for managing the fuel delivery and ignition system. A scanner can be used to observe the fluctuating voltage readings produced by an O2 sensor as the oxygen content of the exhaust is analyzed.

O-RING SEAL: Molded synthetic rubber seal designed with a circular cross-section. This type of seal is used primarily in static applications.

OBD II (ON-BOARD DIAGNOSTICS, SECOND GENERATION): Refers to the federal law mandating tighter control of 1996 and newer vehicle emissions, active monitoring of related devices, and standardization of terminology, data link connectors, and other technician concerns.

OCTANE RATING: A number, indicating the quality of gasoline based on its ability to resist knock. The higher the number, the better the quality. Higher compression engines require higher octane gas.

OEM: Original Equipment Manufactured. OEM equipment is that furnished standard by the manufacturer.

OFFSET: The distance between the vertical center of the wheel and the mounting surface at the lugs. Offset is positive if the center is outside the lug circle; negative offset puts the center line inside the lug circle.

OHM'S LAW: A law of electricity that states the relationship between voltage, current, and resistance. Volts = amperes x ohms

OHM: The unit used to measure the resistance of conductor-to-electrical flow. One ohm is the amount of resistance that limits current flow to one ampere in a circuit with one volt of pressure.

OHMMETER: An instrument used for measuring the resistance, in ohms, in an electrical circuit.

ONE-WAY CLUTCH: A mechanical clutch of roller or sprag design that resists torque or transmits power in one direction only. It is used to either hold or drive a planetary member.

ONE-WAY ROLLER CLUTCH: A mechanical device that transmits or holds torque in one direction only.

OPENCIRCUIT: A break or lack of contact in an electrical circuit, either intentional (switch) or unintentional (bad connection or broken wire).

ORIFICE: Located in hydraulic oil circuits, it acts as a restriction. It slows down fluid flow to either create back pressure or delay pressure buildup downstream.

OSCILLOSCOPE: A piece of test equipment that shows electric impulses as a pattern on a screen. Engine performance can be analyzed by interpreting these patterns.

OUTPUT SHAFT: The shaft which transmits torque from a device, such as a transmission.

OUTPUT SPEED SENSOR (OSS): Identifies transmission/transaxle output shaft speed for shift timing and may be used to calculate TCC slip; often functions as the VSS (vehicle speed sensor).

OVERDRIVE: (1.) A device attached to or incorporated in a transmission/transaxle that allows the engine to turn less than one full revolution for every complete revolution of the wheels. The net effect is to reduce engine rpm, thereby using less fuel. A typical overdrive gear ratio would be .87:1, instead of the normal 1:1 in high gear. (2.) A gear assembly which produces more shaft revolutions than that transmitted to it.

OVERDRIVE PLANETARY GEARSET: A single planetary gearset designed to provide a direct drive and overdrive ratio. When coupled to a three-speed transmission/transaxle configuration, a four-speed/overdrive unit is present.

OVERHEAD CAMSHAFT (OHC): An engine configuration in which the camshaft is mounted on top of the cylinder head and operates the valve either directly or by means of rocker arms.

OVERHEAD VALVE (OHV): An engine configuration in which all of the valves are located in the cylinder head and the camshaft is located in the cylinder block. The camshaft operates the valves via lifters and pushrods.

OVERRUNCLUTCH: Another name for a one-way mechanical clutch. Applies to both roller and sprag designs.

OVERSTEER: The tendency of some vehicles, when steering into a turn, to over-respond or steer more than required, which could result in excessive slip of the rear wheels. Opposite of under-steer.

OXIDATION STABILIZERS: Absorb and dissipate heat. Automatic transmission fluid has high resistance to varnish and sludge buildup that occurs from excessive heat that is generated primarily in the torque converter. Local temperatures as high as 6000F (3150C) can occur at the clutch plates during engagement, and this heat must be absorbed and dissipated. If the fluid cannot withstand the heat, it burns or oxidizes, resulting in an almost immediate destruction of friction materials, clogged filter screen and hydraulic passages, and sticky valves.

OXIDES OF NITROGEN: See nitrogen oxide (NOx).

OXYGEN SENSOR: Used with a feedback system to sense the presence of oxygen in the exhaust gas and signal the computer which can use the voltage signal to determine engine operating efficiency and adjust the air/fuel ratio.

PARALLEL CIRCUIT: (See circuit, parallel.)

PARTS WASHER: A basin or tub, usually with a built-in pump mechanism and hose used for circulating chemical solvent for the purpose of cleaning greasy, oily and dirty components.

PART-TIME FOUR WHEEL DRIVE: A system that is normally in the two wheel drive mode and only runs in four-wheel drive when the system is manually engaged because more traction is desired. Two or four wheel drive is normally selected by a lever to engage the front axle, but if locking hubs are used, these must also be manually engaged in the Lock position. Otherwise, the front axle will not drive the front wheels.

PASSIVE RESTRAINT: Safety systems such as air bags or automatic seat belts which operate with no action required on the part of the driver or passenger. Mandated by Federal regulations on all vehicles sold in the U.S. after 1990.

PAYLOAD: The weight the vehicle is capable of carrying in addition to its own weight. Payload includes weight of the driver, passengers and cargo, but not coolant, fuel, lubricant, spare tire, etc.

PCM: Powertrain control module.

PCV VALVE: A valve usually located in the rocker cover that vents crankcase vapors back into the engine to be reburned.

PERCOLATION: A condition in which the fuel actually "boils," due to excessive heat. Percolation prevents proper atomization of the fuel causing rough running.

PICK-UP COIL: The coil in which voltage is induced in an electronic ignition.

PINION GEAR: The smallest gear in a drive gear assembly. piston: **A disc or cup that fits in a cylinder bore and is free to move. In hydraulics, it provides the means of converting hydraulic pressure into a usable force. Examples of piston applications are found in servo, clutch, and accumulator units.**

PING: A metallic rattling sound produced by the engine during acceleration. It is usually due to incorrect ignition timing or a poor grade of gasoline.

PINION: The smaller of two gears. The rear axle pinion drives the ring gear which transmits motion to the axle shafts.

PISTON RING: An open-ended ring which fits into a groove on the outer diameter of the piston. Its chief function is to form a seal between the piston and cylinder wall. Most automotive pistons have three rings: two for compression sealing; one for oil sealing.

PITMAN ARM: A lever which transmits steering force from the steering gear to the steering linkage.

PLANET CARRIER: A basic member of a planetary gear assembly that carries the pinion gears.

PLANET PINIONS: Gears housed in a planet carrier that are in constant mesh with the sun gear and internal gear. Because they have their own independent rotating centers, the pinions are capable of rotating around the sun gear or the inside of the internal gear.

PLANETARY GEAR RATIO: The reduction or overdrive ratio developed by a planetary gearset.

PLANETARY GEARSET: In its simplest form, it is made up of a basic assembly group containing a sun gear, internal gear, and planet carrier. The gears are always in constant mesh and offer a wide range of gear ratio possibilities.

PLANETARY GEARSET(COMPOUND): Two planetary gearsets combined together.

PLANETARY GEARSET(SIMPLE): An assembly of gears in constant mesh consisting of a sun gear, several pinion gears mounted in a carrier, and a ring gear. It provides gear ratio and direction changes, in addition to a direct drive and a neutral.

PLY RATING: A. rating given a tire which indicates strength (but not necessarily actual plies). A two-ply/four-ply rating has only two plies, but the strength of a four-ply tire.

POLARITY: Indication (positive or negative) of the two poles of a battery.

PORT: An opening for fluid intake or exhaust.

POSITIVE SEALING: A sealing method that completely prevents leakage.

POTENTIAL: Electrical force measured in volts; sometimes used interchangeably with voltage.

POWER: The ability to do work per unit of time, as expressed in horsepower; one horsepower equals 33,000 ft. lbs. of work per minute, or 550 ft. lbs. of work per second.

POWER FLOW: The systematic flow or transmission of power through the gears, from the input shaft to the output shaft.

POWER-TO-WEIGHT RATIO: Ratio of horsepower to weight of car.

POWERTRAIN: See Drivetrain.

POWERTRAIN CONTROL MODULE(PCM): Current designation for the engine control module (ECM). In many cases, late model vehicle control units manage the engine as well as the transmission. In other settings, the PCM controls the engine and is interfaced with a TCM to control transmission functions.

Ppm: Parts per million; unit used to measure exhaust emissions.

PREIGNITION: Early ignition of fuel in the cylinder, sometimes due to glowing carbon deposits in the combustion chamber. Preignition can be damaging since combustion takes place prematurely.

PRELOAD: A predetermined load placed on a bearing during assembly or by adjustment.

PRESS FIT: The mating of two parts under pressure, due to the inner diameter of one being smaller than the outer diameter of the other, or vice versa; an interference fit.

PRESSURE: The amount of force exerted upon a surface area.

PRESSURE CONTROL SOLENOID (PCS): An output device that provides a boost oil pressure to the mainline regulator valve to control line pressure. Its operation is determined by the amount of current sent from the PCM.

PRESSURE GAUGE: An instrument used for measuring the fluid pressure in a hydraulic circuit.

PRESSURE REGULATOR VALVE: In automatic transmissions, its purpose is to regulate the pressure of the pump output and supply the basic fluid pressure necessary to operate the transmission. The regulated fluid pressure may be referred to as mainline pressure, line pressure, or control pressure.

PRESSURE SWITCH ASSEMBLY (PSA): Mounted inside the transmission, it is a grouping of oil pressure switches that inputs to the PCM when certain hydraulic passages are charged with oil pressure.

PRESSURE PLATE: A spring-loaded plate (part of the clutch) that transmits power to the driven (friction) plate when the clutch is engaged.

PRIMARY CIRCUIT: The low voltage side of the ignition system which consists of the ignition switch, ballast resistor or resistance wire, bypass, coil, electronic control unit and pick-up coil as well as the connecting wires and harnesses.

PROFILE: Term used for tire measurement (tire series), which is the ratio of tire height to tread width.

PROM (PROGRAMMABLE READ-ONLY MEMORY): The heart of the computer that compares input data and makes the engineered program or strategy decisions about when to trigger the appropriate output based on stored computer instructions.

Pulse generator: A two-wire pickup sensor used to produce a fluctuating electrical signal. This changing signal is read by the controller to determine the speed of the object and can be used to measure transmission/transaxle input speed, output speed, and vehicle speed.

PSI: Pounds per square inch; a measurement of pressure.

PULSE WIDTH DUTY CYCLE SOLENOID (PULSE WIDTH MODULATED SOLENOID): A computer-controlled solenoid that turns on and off at a variable rate producing a modulated oil pressure; often referred to as a pulse width modulated (PWM) solenoid. Employed in many electronic automatic transmissions and transaxles, these solenoids are used to manage shift control and converter clutch hydraulic circuits.

PUSHROD: A steel rod between the hydraulic valve lifter and the valve rocker arm in overhead valve (OHV) engines.

PUMP: A mechanical device designed to create fluid flow and pressure buildup in a hydraulic system.

QUARTER PANEL: General term used to refer to a rear fender. Quarter panel is the area from the rear door opening to the tail light area and from rear wheel well to the base of the trunk and roof-line.

RACE: The surface on the inner or outer ring of a bearing on which the balls, needles or rollers move.

RACK AND PINION: A type of automotive steering system using a pinion gear attached to the end of the steering shaft. The pinion meshes with a long rack attached to the steering linkage.

RADIAL TIRE: Tire design which uses body cords running at right angles to the center line of the tire. Two or more belts are used to give tread strength. Radials can be identified by their characteristic sidewall bulge.

RADIATOR: Part of the cooling system for a water-cooled engine, mounted in the front of the vehicle and connected to the engine with rubber hoses. Through the radiator, excess combustion heat is dissipated into the atmosphere through forced convection using a water and glycol based mixture that circulates through, and cools, the engine.

RANGE REFERENCE AND CLUTCH/BAND APPLY CHART: A guide that shows the application of clutches and bands for each gear, within the selector range positions. These charts are extremely useful for understanding how the unit operates and for diagnosing malfunctions.

RAVIGNEAUX GEARSET: A compound planetary gearset that features matched dual planetary pinions (sets of two) mounted in a single planet carrier. Two sun gears and one ring mesh with the carrier pinions.

REACTION MEMBER: The stationary planetary member, in a planetary gearset, that is grounded to the transmission/transaxle case through the use of friction and wedging devices known as bands, disc clutches, and one-way clutches.

REACTION PRESSURE: The fluid pressure that moves a spool valve against an opposing force or forces; the area on which the opposing force acts. The opposing force can be a spring or a combination of spring force and auxiliary hydraulic force.

REACTOR, TORQUE CONVERTER: The reaction member of a fluid torque converter, more commonly called a stator. (See stator.)

REAR MAIN OIL SEAL: A synthetic or rope-type seal that prevents oil from leaking out of the engine past the rear main crankshaft bearing.

RECIRCULATING BALL: Type of steering system in which recirculating steel balls occupy the area between the nut and worm wheel, causing a reduction in friction.

RECTIFIER: A device (used primarily in alternators) that permits electrical current to flow in one direction only.

REDUCTION: (See gear reduction.) regulator valve: **A valve that changes the pressure of the oil in a hydraulic circuit as the oil passes through the valve by bleeding off (or exhausting) some of the volume of oil supplied to the valve.**

REFRIGERANT 12 (R-12) or 134 (R-134): The generic name of the refrigerant used in automotive air conditioning systems.

REGULATOR: A device which maintains the amperage and/or voltage levels of a circuit at predetermined values.

RELAY: A switch which automatically opens and/or closes a circuit.

RELAY VALVE: A valve that directs flow and pressure. Relay valves simply connect or disconnect interrelated passages without restricting the fluid flow or changing the pressure.

RELIEF VALVE: A spring-loaded, pressure-operated valve that limits oil pressure buildup in a hydraulic circuit to a predetermined maximum value.

RELUCTOR: A wheel that rotates inside the distributor and triggers the release of voltage in an electronic ignition.

RESERVOIR: The storage area for fluid in a hydraulic system; often called a sump.

RESIN: A liquid plastic used in body work.

RESIDUAL MAGNETISM: The magnetic strength stored in a material after a magnetizing field has been removed.

RESISTANCE: The opposition to the flow of current through a circuit or electrical device, and is measured in ohms. Resistance is equal to the voltage divided by the amperage.

RESISTOR SPARK PLUG: A spark plug using a resistor to shorten the spark duration. This suppresses radio interference and lengthens plug life.

RESISTOR: A device, usually made of wire, which offers a preset amount of resistance in an electrical circuit.

RESULTANT FORCE: The single effective directional thrust of the fluid force on the turbine produced by the vortex and rotary forces acting in different planes.

RETARD: Set the ignition timing so that spark occurs later (fewer degrees before TDC).

RHEOSTAT: A device for regulating a current by means of a variable resistance.

RING GEAR: The name given to a ring-shaped gear attached to a differential case, or affixed to a flywheel or as part of a planetary gear set.

ROADLOAD: grade.

ROCKER ARM: A lever which rotates around a shaft pushing down (opening) the valve with an end when the other end is pushed up by the pushrod. Spring pressure will later close the valve.

ROCKER PANEL: The body panel below the doors between the wheel opening.

ROLLER BEARING: A bearing made up of hardened inner and outer races between which hardened steel rollers move.

ROLLER CLUTCH: A type of one-way clutch design using rollers and springs mounted within an inner and outer cam race assembly.

ROTARY FLOW: The path of the fluid trapped between the blades of the members as they revolve with the rotation of the torque converter cover (rotational inertia).

ROTOR: (1.) The disc-shaped part of a disc brake assembly, upon which the brake pads bear; also called, brake disc. (2.) The device mounted atop the distributor shaft, which passes current to the distributor cap tower contacts.

ROTARY ENGINE: See Wankel engine.

RPM: Revolutions per minute (usually indicates engine speed).

RTV: A gasket making compound that cures as it is exposed to the atmosphere. It is used between surfaces that are not perfectly machined to one another, leaving a slight gap that the RTV fills and in which it hardens. The letters RTV represent room temperature vulcanizing.

RUN-ON: Condition when the engine continues to run, even when the key is turned off. See dieseling.

SEALED BEAM: A automotive headlight. The lens, reflector and filament from a single unit.

SEATBELT INTERLOCK: A system whereby the car cannot be started unless the seatbelt is buckled.

SECONDARY CIRCUIT: The high voltage side of the ignition system, usually above 20,000 volts. The secondary includes the ignition coil, coil wire, distributor cap and rotor, spark plug wires and spark plugs.

SELF-INDUCTION: The generation of voltage in a current-carrying wire by changing the amount of current flowing within that wire.

SEMI-CONDUCTOR: A material (silicon or germanium) that is neither a good conductor nor an insulator; used in diodes and transistors.

SEMI-FLOATING AXLE: In this design, a wheel is attached to the axle shaft, which takes both drive and cornering loads. Almost all solid axle passenger cars and light trucks use this design.

SENDING UNIT: A mechanical, electrical, hydraulic or electromagnetic device which transmits information to a gauge.

SENSOR: Any device designed to measure engine operating conditions or ambient pressures and temperatures. Usually electronic in nature and designed to send a voltage signal to an on-board computer, some sensors may operate as a simple on/off switch or they may provide a variable voltage signal (like a potentiometer) as conditions or measured parameters change.

SERIES CIRCUIT: (See circuit, series.)

SERPENTINE BELT: An accessory drive belt, with small multiple v-ribs, routed around most or all of the engine-powered accessories such as the alternator and power steering pump. Usually both the front and the back side of the belt comes into contact with various pulleys.

SERVO: In an automatic transmission, it is a piston in a cylinder assembly that converts hydraulic pressure into mechanical force and movement; used for the application of the bands and clutches.

SHIFT BUSYNESS: When referring to a torque converter clutch, it is the frequent apply and release of the clutch plate due to uncommon driving conditions.

SHIFT VALVE: Classified as a relay valve, it triggers the automatic shift in response to a governor and a throttle signal by directing fluid to the appropriate band and clutch apply combination to cause the shift to occur.

SHIM: Spacers of precise, predetermined thickness used between parts to establish a proper working relationship.

SHIMMY: Vibration (sometimes violent) in the front end caused by misaligned front end, out of balance tires or worn suspension components.

SHORT CIRCUIT: An electrical malfunction where current takes the path of least resistance to ground (usually through damaged insulation). Current flow is excessive from low resistance resulting in a blown fuse.

SHUDDER: Repeated jerking or stick-slip sensation, similar to chuggle but more severe and rapid in nature, that may be most noticeable during certain ranges of vehicle speed; also used to define condition after converter clutch engagement.

SIMPSON GEARSET: A compound planetary gear train that integrates two simple planetary gearsets referred to as the front planetary and the rear planetary.

SINGLE OVERHEAD CAMSHAFT: See overhead camshaft.

SKIDPLATE: A metal plate attached to the underside of the body to protect the fuel tank, transfer case or other vulnerable parts from damage.

SLAVE CYLINDER: In automotive use, a device in the hydraulic clutch system which is activated by hydraulic force, disengaging the clutch.

SLIPPING: Noticeable increase in engine rpm without vehicle speed increase; usually occurs during or after initial clutch or band engagement.

SLUDGE: Thick, black deposits in engine formed from dirt, oil, water, etc. It is usually formed in engines when oil changes are neglected.

SNAP RING: A circular retaining clip used inside or outside a shaft or part to secure a shaft, such as a floating wrist pin.

SOFT: Slow, almost unnoticeable clutch apply with very little shift feel.

SOFTCODES: DTCs that have been set into the PCM memory but are not present at the time of testing; often referred to as history or intermittent codes.

SOHC: Single overhead camshaft.

SOLENOID: An electrically operated, magnetic switching device.

SPALLING: A wear pattern identified by metal chips flaking off the hardened surface. This condition is caused by foreign particles, overloading situations, and/or normal wear.

SPARK PLUG: A device screwed into the combustion chamber of a spark ignition engine. The basic construction is a conductive core inside of a ceramic insulator, mounted in an outer conductive base. An electrical charge from the spark plug wire travels along the conductive core and jumps a preset air gap to a grounding point or points at the end of the conductive base. The resultant spark ignites the fuel/air mixture in the combustion chamber.

SPECIFIC GRAVITY (BATTERY): The relative weight of liquid (battery electrolyte) as compared to the weight of an equal volume of water.

SPLINES: Ridges machined or cast onto the outer diameter of a shaft or inner diameter of a bore to enable parts to mate without rotation.

SPLIT TORQUE DRIVE: In a torque converter, it refers to parallel paths of torque transmission, one of which is mechanical and the other hydraulic.

SPONGY PEDAL: A soft or spongy feeling when the brake pedal is depressed. It is usually due to air in the brake lines.

SPOOLVALVE: A precision-machined, cylindrically shaped valve made up of lands and grooves. Depending on its position in the valve bore, various interconnecting hydraulic circuit passages are either opened or closed.

SPRAG CLUTCH: A type of one-way clutch design using cams or contoured-shaped sprags between inner and outer races. (See one-way clutch.)

SPRUNG WEIGHT: The weight of a car supported by the springs.

SQUARE-CUT SEAL: Molded synthetic rubber seal designed with a square- or rectangular-shaped cross-section. This type of seal is used for both dynamic and static applications.

SRS: Supplemental restraint system

STABILIZER (SWAY) BAR: A bar linking both sides of the suspension. It resists sway on turns by taking some of added load from one wheel and putting it on the other.

STAGE: The number of turbine sets separated by a stator. A turbine set may be made up of one or more turbine members. A three-element converter is classified as a single stage.

STALL: In fluid drive transmission/transaxle applications, stall refers to engine rpm with the transmission/transaxle engaged and the vehicle stationary; throttle valve can be in any position between closed and wide open.

STALL SPEED: In fluid drive transmission/transaxle applications, stall speed refers to the maximum engine rpm with the transmission/transaxle engaged and vehicle stationary, when the throttle valve is wide open. (See stall; stall test.)

STALL TEST: A procedure recommended by many manufacturers to help determine the integrity of an engine, the torque converter stator, and certain clutch and band combinations. With the shift lever in each of the forward and reverse positions and with the brakes firmly applied, the accelerator pedal is momentarily pressed to the wide open throttle (WOT) position. The engine rpm reading at full throttle can provide clues for diagnosing the condition of the items listed above.

STALL TORQUE: The maximum design or engineered torque ratio of a fluid torque converter, produced under stall speed conditions. (See stall speed.)

STARTER: A high-torque electric motor used for the purpose of starting the engine, typically through a high ratio geared drive connected to the flywheel ring gear.

STATIC: A sealing application in which the parts being sealed do not move in relation to each other.

STATOR (REACTOR): The reaction member of a fluid torque converter that changes the direction of the fluid as it leaves the turbine to enter the impeller vanes. During the torque multiplication phase, this action assists the impeller's rotary force and results in an increase in torque.

STEERING GEOMETRY: Combination of various angles of suspension components (caster, camber, toe-in); roughly equivalent to front end alignment.

STRAIGHT WEIGHT: Term designating motor oil as suitable for use within a narrow range of temperatures. Outside the narrow temperature range its flow characteristics will not adequately lubricate.

STROKE: The distance the piston travels from bottom dead center to top dead center.

SUBSTITUTION: Replacing one part suspected of a defect with a like part of known quality.

SUMP: The storage vessel or reservoir that provides a ready source of fluid to the pump. In an automatic transmission, the sump is the oil pan. All fluid eventually returns to the sump for recycling into the hydraulic system.

SUN GEAR: In a planetary gearset, it is the center gear that meshes with a cluster of planet pinions.

SUPERCHARGER: An air pump driven mechanically by the engine through belts, chains, shafts or gears from the crankshaft. Two general types of supercharger are the positive displacement and centrifugal type, which pump air in direct relationship to the speed of the engine.

SUPPLEMENTAL RESTRAINT SYSTEM: See air bag.

SURGE: Repeating engine-related feeling of acceleration and deceleration that is less intense than chuggle.

SWITCH: A device used to open, close, or redirect the current in an electrical circuit.

SYNCHROMESH: A manual transmission/transaxle that is equipped with devices (synchronizers) that match the gear speeds so that the transmission/transaxle can be downshifted without clashing gears.

SYNTHETIC OIL: Non-petroleum based oil.

TACHOMETER: A device used to measure the rotary speed of an engine, shaft, gear, etc., usually in rotations per minute.

TDC: Top dead center. The exact top of the piston's stroke.

TEFLON SEALING RINGS: Teflon is a soft, durable, plastic-like material that is resistant to heat and provides excellent sealing. These rings are designed with either scarf-cut joints or as one-piece rings. Teflon sealing rings have replaced many metal ring applications.

TERMINAL: A device attached to the end of a wire or cable to make an electrical connection.

TEST LIGHT, CIRCUIT-POWERED: Uses available circuit voltage to test circuit continuity.

TEST LIGHT, SELF-POWERED: Uses its own battery source to test circuit continuity.

THERMISTOR: A special resistor used to measure fluid temperature; it decreases its resistance with increases in temperature.

THERMOSTAT: A valve, located in the cooling system of an engine, which is closed when cold and opens gradually in response to engine heating, controlling the temperature of the coolant and rate of coolant flow.

THERMOSTATIC ELEMENT: A heat-sensitive, spring-type device that controls a drain port from the upper sump area to the lower sump. When the transaxle fluid reaches operating temperature, the port is closed and the upper sump fills, thus reducing the fluid level in the lower sump.

THROTTLE POSITION (TP) SENSOR: Reads the degree of throttle opening; its signal is used to analyze engine load conditions. The ECM/PCM decides to apply the TCC, or to disengage it for coast or load conditions that need a converter torque boost.

THROTTLE PRESSURE/MODULATOR PRESSURE: A hydraulic signal oil pressure relating to the amount of engine load, based on either the amount of throttle plate opening or engine vacuum.

THROTTLE VALVE: A regulating or balanced valve that is controlled mechanically by throttle linkage or engine vacuum. It sends a hydraulic signal to the shift valve body to control shift timing and shift quality. (See balanced valve; modulator valve.)

THROW-OUT BEARING: As the clutch pedal is depressed, the throwout bearing moves against the spring fingers of the pressure plate, forcing the pressure plate to disengage from the driven disc.

TIE ROD: A rod connecting the steering arms. Tie rods have threaded ends that are used to adjust toe-in.

TIE-UP: Condition where two opposing clutches are attempting to apply at same time, causing engine to labor with noticeable loss of engine rpm.

TIMING BELT: A square-toothed, reinforced rubber belt that is driven by the crankshaft and operates the camshaft.

TIMING CHAIN: A roller chain that is driven by the crankshaft and operates the camshaft.

TIRE ROTATION: Moving the tires from one position to another to make the tires wear evenly.

TOE-IN (OUT): A term comparing the extreme front and rear of the front tires. Closer together at the front is toe-in; farther apart at the front is toe-out.

TOP DEAD CENTER (TDC): The point at which the piston reaches the top of its travel on the compression stroke.

TORQUE: Measurement of turning or twisting force, expressed as foot-pounds or inch-pounds.

TORQUE CONVERTER: A turbine used to transmit power from a driving member to a driven member via hydraulic action, providing changes in drive ratio and torque. In automotive use, it links the driveplate at the rear of the engine to the automatic transmission.

TORQUE CONVERTER CLUTCH: The apply plate (lockup plate) assembly used for mechanical power flow through the converter.

TORQUE PHASE: Sometimes referred to as slip phase or stall phase, torque multiplication occurs when the turbine is turning at a slower speed than the impeller, and the stator is reactionary (stationary). This sequence generates a boost in output torque.

TORQUE RATING (STALL TORQUE): The maximum torque multiplication that occurs during stall conditions, with the engine at wide open throttle (WOT) and zero turbine speed.

TORQUE RATIO: An expression of the gear ratio factor on torque effect. A 3:1 gear ratio or 3:1 torque ratio increases the torque input by the ratio factor of 3. Input torque (100 ft. lbs.)x 3 = output torque (300 ft. lbs.)

TRACTION: The amount of usable tractive effort before the drive wheels slip on the road contact surface.

TORSION BAR SUSPENSION: Long rods of spring steel which take the place of springs. One end of the bar is anchored and the other arm (attached to the suspension) is free to twist. The bars' resistance to twisting causes springing action.

TRACK: Distance between the centers of the tires where they contact the ground.

TRACTION CONTROL: A control system that prevents the spinning of a vehicle's drive wheels when excess power is applied.

TRACTIVE EFFORT: The amount of force available to the drive wheels, to move the vehicle.

TRANSAXLE: A single housing containing the transmission and differential. Transaxles are usually found on front engine/front wheel drive or rear engine/rear wheel drive cars.

TRANSDUCER: A device that changes energy from one form to another. For example, a transducer in a microphone changes sound energy to electrical energy. In automotive air-conditioning controls used in automatic temperature systems, a transducer changes an electrical signal to a vacuum signal, which operates mechanical doors.

TRANSMISSION: A powertrain component designed to modify torque and speed developed by the engine; also provides direct drive, reverse, and neutral.

TRANSMISSION CONTROL MODULE (TCM): Manages transmission functions. These vary according to the manufacturer's product design but may include converter clutch operation, electronic shift scheduling, and mainline pressure.

TRANSMISSION FLUID TEMPERATURE (TFT)SENSOR: Originally called a transmission oil temperature (TOT) sensor, this input device to the ECM/PCM senses the fluid temperature and provides a resistance value. It operates on the thermistor principle.

TRANSMISSION INPUT SPEED (TIS) SENSOR: Measures turbine shaft (input shaft) rpm's and compares to engine rpm's to determine torque converter slip. When compared to the transmission output speed sensor or VSS, gear ratio and clutch engagement timing can be determined.

TRANSMISSION OIL TEMPERATURE (TOT) SENSOR: (See transmission fluid temperature (TFT) sensor.)

TRANSMISSION RANGE SELECTOR (TRS) SWITCH: Tells the module which gear shift position the driver has chosen. turbine: **The output (driven) member of a fluid coupling or fluid torque converter. It is splined to the input (turbine) shaft of the transmission.**

TRANSFER CASE: A gearbox driven from the transmission that delivers power to both front and rear driveshafts in a four-wheel drive system. Transfer cases usually have a high and low range set of gears, used depending on how much pulling power is needed.

TRANSISTOR: A semi-conductor component which can be actuated by a small voltage to perform an electrical switching function.

TREAD WEAR INDICATOR: Bars molded into the tire at right angles to the tread that appear as horizontal bars when 1/16 in. of tread remains.

TREAD WEAR PATTERN: The pattern of wear on tires which can be "read" to diagnose problems in the front suspension.

TUNE-UP: A regular maintenance function, usually associated with the replacement and adjustment of parts and components in the electrical and fuel systems of a vehicle for the purpose of attaining optimum performance.

TURBOCHARGER: An exhaust driven pump which compresses intake air and forces it into the combustion chambers at higher than atmospheric pressures. The increased air pressure allows more fuel to be burned and results in increased horsepower being produced.

TURBULENCE: The interference of molecules of a fluid (or vapor) with each other in a fluid flow.

TYPE F: Transmission fluid developed and used by Ford Motor Company up to 1982. This fluid type provides a high coefficient of friction.

TYPE 7176: The preferred choice of transmission fluid for Chrysler automatic transmissions and transaxles. Developed in 1986, it closely resembles Dexron and Mercon. Type 7176 is the recommended service fill fluid for all Chrysler products utilizing a lockup torque converter dating back to 1978.

U-JOINT (UNIVERSAL JOINT): A flexible coupling in the drive train that allows the driveshafts or axle shafts to operate at different angles and still transmit rotary power.

UNDERSTEER: The tendency of a car to continue straight ahead while negotiating a turn.

UNIT BODY: Design in which the car body acts as the frame.

UNLEADED FUEL: Fuel which contains no lead (a common gasoline additive). The presence of lead in fuel will destroy the functioning elements of a catalytic converter, making it useless.

UNSPRUNG WEIGHT: The weight of car components not supported by the springs (wheels, tires, brakes, rear axle, control arms, etc.).

UPSHIFT: A shift that results in a decrease in torque ratio and an increase in speed.

VACUUM: A negative pressure; any pressure less than atmospheric pressure.

VACUUM ADVANCE: A device which advances the ignition timing in response to increased engine vacuum.

VACUUM GAUGE: An instrument used for measuring the existing vacuum in a vacuum circuit or chamber. The unit of measure is inches (of mercury in a barometer).

VACUUM MODULATOR: Generates a hydraulic oil pressure in response to the amount of engine vacuum.

VALVES: Devices that can open or close fluid passages in a hydraulic system and are used for directing fluid flow and controlling pressure.

VALVE BODY ASSEMBLY: The main hydraulic control assembly of the transmission/transaxle that contains numerous valves, check balls, and other components to control the distribution of pressurized oil throughout the transmission.

VALVE CLEARANCE: The measured gap between the end of the valve stem and the rocker arm, cam lobe or follower that activates the valve.

VALVE GUIDES: The guide through which the stem of the valve passes. The guide is designed to keep the valve in proper alignment.

VALVE LASH (clearance): The operating clearance in the valve train.

VALVE TRAIN: The system that operates intake and exhaust valves, consisting of camshaft, valves and springs, lifters, pushrods and rocker arms.

VAPOR LOCK: Boiling of the fuel in the fuel lines due to excess heat. This will interfere with the flow of fuel in the lines and can completely stop the flow. Vapor lock normally only occurs in hot weather.

VARIABLE DISPLACEMENT (VARIABLE CAPACITY) VANE PUMP: Slipper-type vanes, mounted in a revolving rotor and contained within the bore of a movable slide, capture and then force fluid to flow. Movement of the slide to various positions changes the size of the vane chambers and the amount of fluid flow. Note: **GM refers to this pump design as variable displacement, and Ford terms it variable capacity.**

VARIABLE FORCE SOLENOID (VFS): Commonly referred to as the electronic pressure control (EPC) solenoid, it replaces the cable/linkage style of TV system control and is integrated with a spool valve and spring assembly to control pressure. A variable computer-controlled current flow varies the internal force of the solenoid on the spool valve and resulting control pressure.

VARIABLE ORIFICE THERMAL VALVE: Temperature-sensitive hydraulic oil control device that adjusts the size of a circuit path opening. By altering the size of the opening, the oil flow rate is adapted for cold to hot oil viscosity changes.

VARNISH: Term applied to the residue formed when gasoline gets old and stale.

VCM: See Electronic Control Unit (ECU).

VEHICLE SPEED SENSOR (VSS): Provides an electrical signal to the computer module, measuring vehicle speed, and affects the torque converter clutch engagement and release.

VESPEL SEALING RINGS: Hard plastic material that produces excellent sealing in dynamic settings. These rings are found in late versions of the 4T60 and in all 4T60-E and 4T80-E transaxles.

VISCOSITY: The ability of a fluid to flow. The lower the viscosity rating, the easier the fluid will flow. 10 weight motor oil will flow much easier than 40 weight motor oil.

VISCOSITY INDEX IMPROVERS: Keeps the viscosity nearly constant with changes in temperature. This is especially important at low temperatures, when the oil needs to be thin to aid in shifting and for cold-weather starting. Yet it must not be so thin that at high temperatures it will cause excessive hydraulic leakage so that pumps are unable to maintain the proper pressures.

VISCOUS CLUTCH: A specially designed torque converter clutch apply plate that, through the use of a silicon fluid, clamps smoothly and absorbs torsional vibrations.

VOLT: Unit used to measure the force or pressure of electricity. It is defined as the pressure

VOLTAGE: The electrical pressure that causes current to flow. Voltage is measured in volts (V).

VOLTAGE, APPLIED: The actual voltage read at a given point in a circuit. It equals the available voltage of the power supply minus the losses in the circuit up to that point.

VOLTAGE DROP: The voltage lost or used in a circuit by normal loads such as a motor or lamp or by abnormal loads such as a poor (high-resistance) lead or terminal connection.

VOLTAGE REGULATOR: A device that controls the current output of the alternator or generator.

VOLTMETER: An instrument used for measuring electrical force in units called volts. Voltmeters are always connected parallel with the circuit being tested.

VORTEX FLOW: The crosswise or circulatory flow of oil between the blades of the members caused by the centrifugal pumping action of the impeller.

WANKEL ENGINE: An engine which uses no pistons. In place of pistons, triangular-shaped rotors revolve in specially shaped housings.

WATER PUMP: A belt driven component of the cooling system that mounts on the engine, circulating the coolant under pressure.

WATT: The unit for measuring electrical power. One watt is the product of one ampere and one volt (watts equals amps times volts). Wattage is the horsepower of electricity (746 watts equal one horsepower).

WHEEL ALIGNMENT: Inclusive term to describe the front end geometry (caster, camber, toe-in/out).

WHEEL CYLINDER: Found in the automotive drum brake assembly, it is a device, actuated by hydraulic pressure, which, through internal pistons, pushes the brake shoes outward against the drums.

WHEEL WEIGHT: Small weights attached to the wheel to balance the wheel and tire assembly. Out-of-balance tires quickly wear out and also give erratic handling when installed on the front.

WHEELBASE: Distance between the center of front wheels and the center of rear wheels.

WIDE OPEN THROTTLE (WOT): Full travel of accelerator pedal.

WORK: The force exerted to move a mass or object. Work involves motion; if a force is exerted and no motion takes place, no work is done. Work per unit of time is called power. Work = force x distance = ft. lbs. 33,000 ft. lbs. in one minute = 1 horsepower

ZERO-THROTTLE COAST DOWN: A full release of accelerator pedal while vehicle is in motion and in drive range.

Commonly Used Abbreviations

2

2WD	Two Wheel Drive

4

4WD	Four Wheel Drive

A

A/C	Air Conditioning
ABDC	After Bottom Dead Center
ABS	Anti-lock Brakes
AC	Alternating Current
ACL	Air cleaner
ACT	Air Charge Temperature
AIR	Secondary Air Injection
ALCL	Assembly Line Communications Link
ALDL	Assembly Line Diagnostic Link
AT	Automatic Transaxle/Transmission
ATDC	After Top Dead Center
ATF	Automatic Transmission Fluid
ATS	Air Temperature Sensor
AWD	All Wheel Drive

B

BAP	Barometric Absolute Pressure
BARO	Barometric Pressure
BBDC	Before Bottom Dead Center
BCM	Body Control Module
BDC	Bottom Dead Center
BPT	Backpressure Transducer
BTDC	Before Top Dead Center
BVSV	Bimetallic Vacuum Switching Valve

C

CAC	Charge Air Cooler
CARB	California Air Resources Board
CAT	Catalytic Converter
CCC	Computer Command Control
CCCC	Computer Controlled Catalytic Converter
CCCI	Computer Controlled Coil Ignition
CCD	Computer Controlled Dwell
CDI	Capacitor Discharge Ignition
CEC	Computerized Engine Control
CFI	Continuous Fuel Injection
CIS	Continuous Injection System
CIS-E	Continuous Injection System - Electronic
CKP	Crankshaft Position
CL	Closed Loop
CMP	Camshaft Position
CPP	Clutch Pedal Position
CTOX	Continuous Trap Oxidizer System
CTP	Closed Throttle Position
CVC	Constant Vacuum Control
CYL	Cylinder

D

DBC	Dual Bed Catalyst
DC	Direct Current
DFI	Direct Fuel Injection
DIS	Distributorless Ignition System
DLC	Data Link Connector
DMM	Digital Multimeter
DOHC	Double Overhead Camshaft
DRB	Diagnostic Readout Box
DTC	Diagnostic Trouble Code
DTM	Diagnostic Test Mode
DVOM	Digital Volt/Ohmmeter

E

EBCM	Electronic Brake Control Module
ECM	Engine Control Module
ECT	Engine Coolant Temperature
ECU	Engine Control Unit or Electronic Control Unit
EDIS	Electronic Distributorless Ignition System
EEC	Electronic Engine Control
EEPROM	Electrically Erasable Programmable Read Only Memory
EFE	Early Fuel Evaporation
EGR	Exhaust Gas Recirculation
EGRT	Exhaust Gas Recirculation Temperature
EGRVC	EGR Valve Control
EPROM	Erasable Programmable Read Only Memory
EVAP	Evaporative Emissions
EVP	EGR Valve Position

F

FBC	Feedback Carburetor
FEEPROM	Flash Electrically Erasable Programmable Read Only Memory
FF	Flexible Fuel
FI	Fuel Injection
FT	Fuel Trim
FWD	Front Wheel Drive

G

GND	Ground

H

HAC	High Altitude Compensation
HEGO	Heated Exhaust Gas Oxygen sensor
HEI	High Energy Ignition
HO2 Sensor	Heated Oxygen Sensor

I

IAC	Idle Air Control
IAT	Intake Air Temperature
ICM	Ignition Control Module
IFI	Indirect Fuel Injection
IFS	Inertia Fuel Shutoff
ISC	Idle Speed Control
IVSV	Idle Vacuum Switching Valve

Commonly Used Abbreviations

K

KOEO	Key On, Engine Off
KOER	Key ON, Engine Running
KS	Knock Sensor

M

MAF	Mass Air Flow
MAP	Manifold Absolute Pressure
MAT	Manifold Air Temperature
MC	Mixture Control
MDP	Manifold Differential Pressure
MFI	Multiport Fuel Injection
MIL	Malfunction Indicator Lamp or Maintenance
MST	Manifold Surface Temperature
MVZ	Manifold Vacuum Zone

N

NVRAM	Nonvolatile Random Access Memory

O

O2 Sensor	Oxygen Sensor
OBD	On-Board Diagnostic
OC	Oxidation Catalyst
OHC	Overhead Camshaft
OL	Open Loop

P

P/S	Power Steering
PAIR	Pulsed Secondary Air Injection
PCM	Powertrain Control Module
PCS	Purge Control Solenoid
PCV	Positive Crankcase Ventilation
PIP	Profile Ignition Pick-up
PNP	Park/Neutral Position
PROM	Programmable Read Only Memory
PSP	Power Steering Pressure
PTO	Power Take-Off
PTOX	Periodic Trap Oxidizer System

R

RABS	Rear Anti-lock Brake System
RAM	Random Access Memory
ROM	Read Only Memory
RPM	Revolutions Per Minute
RWAL	Rear Wheel Anti-lock Brakes
RWD	Rear Wheel Drive

S

SBC	Single Bed Converter
SBEC	Single Board Engine Controller
SC	Supercharger
SCB	Supercharger Bypass
SFI	Sequential Multiport Fuel Injection
SIR	Supplemental Inflatible Restraint
SOHC	Single Overhead Camshaft
SPL	Smoke Puff Limiter
SPOUT	Spark Output
SRI	Service Reminder Indicator
SRS	Supplemental Restraint System
SRT	System Readiness Test
SSI	Solid State Ignition
ST	Scan Tool
STO	Self-Test Output

T

TAC	Thermostatic Air Cleaner
TBI	Throttle Body Fuel Injection
TC	Turbocharger
TCC	Torque Converter Clutch
TCM	Transmission Control Module
TDC	Top Dead Center
TFI	Thick Film Ignition
TP	Throttle Position
TR Sensor	Transaxle/Transmission Range Sensor
TVV	Thermal Vacuum Valve
TWC	Three-way Catalytic Converter

V

VAF	Volume Air Flow, or Vane Air Flow
VAPS	Variable Assist Power Steering
VRV	Vacuum Regulator Valve
VSS	Vehicle Speed Sensor
VSV	Vacuum Switching Valve

W

WOT	Wide Open Throttle
WU-TWC	Warm Up Three-way Catalytic Converter

ENGLISH TO METRIC CONVERSION: TORQUE

To convert foot-pounds (ft. lbs.) to Newton-meters (Nm), multiply the number of ft. lbs. by 1.36

To convert Newton-meters (Nm) to foot-pounds (ft. lbs.), multiply the number of Nm by 0.7376

ft. lbs.	Nm	ft. lbs.	Nm	ft. lbs.	Nm	ft. lbs.	Nm
0.1	0.1	34	46.2	76	103.4	118	160.5
0.2	0.3	35	47.6	77	104.7	119	161.8
0.3	0.4	36	49.0	78	106.1	120	163.2
0.4	0.5	37	50.3	79	107.4	121	164.6
0.5	0.7	38	51.7	80	108.8	122	165.9
0.6	0.8	39	53.0	81	110.2	123	167.3
0.7	1.0	40	54.4	82	111.5	124	168.6
0.8	1.1	41	55.8	83	112.9	125	170.0
0.9	1.2	42	57.1	84	114.2	126	171.4
1	1.4	43	58.5	85	115.6	127	172.7
2	2.7	44	59.8	86	117.0	128	174.1
3	4.1	45	61.2	87	118.3	129	175.4
4	5.4	46	62.6	88	119.7	130	176.8
5	6.8	47	63.9	89	121.0	131	178.2
6	8.2	48	65.3	90	122.4	132	179.5
7	9.5	49	66.6	91	123.8	133	180.9
8	10.9	50	68.0	92	125.1	134	182.2
9	12.2	51	69.4	93	126.5	135	183.6
10	13.6	52	70.7	94	127.8	136	185.0
11	15.0	53	72.1	95	129.2	137	186.3
12	16.3	54	73.4	96	130.6	138	187.7
13	17.7	55	74.8	97	131.9	139	189.0
14	19.0	56	76.2	98	133.3	140	190.4
15	20.4	57	77.5	99	134.6	141	191.8
16	21.8	58	78.9	100	136.0	142	193.1
17	23.1	59	80.2	101	137.4	143	194.5
18	24.5	60	81.6	102	138.7	144	195.8
19	25.8	61	83.0	103	140.1	145	197.2
20	27.2	62	84.3	104	141.4	146	198.6
21	28.6	63	85.7	105	142.8	147	199.9
22	29.9	64	87.0	106	144.2	148	201.3
23	31.3	65	88.4	107	145.5	149	202.6
24	32.6	66	89.8	108	146.9	150	204.0
25	34.0	67	91.1	109	148.2	151	205.4
26	35.4	68	92.5	110	149.6	152	206.7
27	36.7	69	93.8	111	151.0	153	208.1
28	38.1	70	95.2	112	152.3	154	209.4
29	39.4	71	96.6	113	153.7	155	210.8
30	40.8	72	97.9	114	155.0	156	212.2
31	42.2	73	99.3	115	156.4	157	213.5
32	43.5	74	100.6	116	157.8	158	214.9
33	44.9	75	102.0	117	159.1	159	216.2

METRIC TO ENGLISH CONVERSION: TORQUE

To convert foot-pounds (ft. lbs.) to Newton-meters (Nm), multiply the number of ft. lbs. by 1.36
To convert Newton-meters (Nm) to foot-pounds (ft. lbs.), multiply the number of Nm by 0.7376

Nm	ft. lbs.	Nm	ft. lbs.	Nm	ft. lbs.	Nm	ft. lbs.	Nm	ft. lbs.
0.1	0.1	34	25.0	76	55.9	118	86.8	160	117.6
0.2	0.1	35	25.7	77	56.6	119	87.5	161	118.4
0.3	0.2	36	26.5	78	57.4	120	88.2	162	119.1
0.4	0.3	37	27.2	79	58.1	121	89.0	163	119.9
0.5	0.4	38	27.9	80	58.8	122	89.7	164	120.6
0.6	0.4	39	28.7	81	59.6	123	90.4	165	121.3
0.7	0.5	40	29.4	82	60.3	124	91.2	166	122.1
0.8	0.6	41	30.1	83	61.0	125	91.9	167	122.8
0.9	0.7	42	30.9	84	61.8	126	92.6	168	123.5
1	0.7	43	31.6	85	62.5	127	93.4	169	124.3
2	1.5	44	32.4	86	63.2	128	94.1	170	125.0
3	2.2	45	33.1	87	64.0	129	94.9	171	125.7
4	2.9	46	33.8	88	64.7	130	95.6	172	126.5
5	3.7	47	34.6	89	65.4	131	96.3	173	127.2
6	4.4	48	35.3	90	66.2	132	97.1	174	127.9
7	5.1	49	36.0	91	66.9	133	97.8	175	128.7
8	5.9	50	36.8	92	67.6	134	98.5	176	129.4
9	6.6	51	37.5	93	68.4	135	99.3	177	130.1
10	7.4	52	38.2	94	69.1	136	100.0	178	130.9
11	8.1	53	39.0	95	69.9	137	100.7	179	131.6
12	8.8	54	39.7	96	70.6	138	101.5	180	132.4
13	9.6	55	40.4	97	71.3	139	102.2	181	133.1
14	10.3	56	41.2	98	72.1	140	102.9	182	133.8
15	11.0	57	41.9	99	72.8	141	103.7	183	134.6
16	11.8	58	42.6	100	73.5	142	104.4	184	135.3
17	12.5	59	43.4	101	74.3	143	105.1	185	136.0
18	13.2	60	44.1	102	75.0	144	105.9	186	136.8
19	14.0	61	44.9	103	75.7	145	106.6	187	137.5
20	14.7	62	45.6	104	76.5	146	107.4	188	138.2
21	15.4	63	46.3	105	77.2	147	108.1	189	139.0
22	16.2	64	47.1	106	77.9	148	108.8	190	139.7
23	16.9	65	47.8	107	78.7	149	109.6	191	140.4
24	17.6	66	48.5	108	79.4	150	110.3	192	141.2
25	18.4	67	49.3	109	80.1	151	111.0	193	141.9
26	19.1	68	50.0	110	80.9	152	111.8	194	142.6
27	19.9	69	50.7	111	81.6	153	112.5	195	143.4
28	20.6	70	51.5	112	82.4	154	113.2	196	144.1
29	21.3	71	52.2	113	83.1	155	114.0	197	144.9
30	22.1	72	52.9	114	83.8	156	114.7	198	145.6
31	22.8	73	53.7	115	84.6	157	115.4	199	146.3
32	23.5	74	54.4	116	85.3	158	116.2	200	147.1
33	24.3	75	55.1	117	86.0	159	116.9	201	147.8

ENGLISH/METRIC CONVERSION: TEMPERATURE

To convert Fahrenheit (F°) to Celsius (C°), take F° temperature and subtract 32, multiply the result by 5 and divide the result by 9
To convert Celsius (C°) to Fahrenheit (F°), take C° temperature and multiply it by 9, divide the result by 5 and add 32

F°	C°	F°	C°	C°	F°	C°	F°
-40	-40.0	150	65.6	-38	-36.4	46	114.8
-35	-37.2	155	68.3	-36	-32.8	48	118.4
-30	-34.4	160	71.1	-34	-29.2	50	122
-25	-31.7	165	73.9	-32	-25.6	52	125.6
-20	-28.9	170	76.7	-30	-22	54	129.2
-15	-26.1	175	79.4	-28	-18.4	56	132.8
-10	-23.3	180	82.2	-26	-14.8	58	136.4
-5	-20.6	185	85.0	-24	-11.2	60	140
0	-17.8	190	87.8	-22	-7.6	62	143.6
1	-17.2	195	90.6	-20	-4	64	147.2
2	-16.7	200	93.3	-18	-0.4	66	150.8
3	-16.1	205	96.1	-16	3.2	68	154.4
4	-15.6	210	98.9	-14	6.8	70	158
5	-15.0	212	100.0	-12	10.4	72	161.6
10	-12.2	215	101.7	-10	14	74	165.2
15	-9.4	220	104.4	-8	17.6	76	168.8
20	-6.7	225	107.2	-6	21.2	78	172.4
25	-3.9	230	110.0	-4	24.8	80	176
30	-1.1	235	112.8	-2	28.4	82	179.6
35	1.7	240	115.6	0	32	84	183.2
40	4.4	245	118.3	2	35.6	86	186.8
45	7.2	250	121.1	4	39.2	88	190.4
50	10.0	255	123.9	6	42.8	90	194
55	12.8	260	126.7	8	46.4	92	197.6
60	15.6	265	129.4	10	50	94	201.2
65	18.3	270	132.2	12	53.6	96	204.8
70	21.1	275	135.0	14	57.2	98	208.4
75	23.9	280	137.8	16	60.8	100	212
80	26.7	285	140.6	18	64.4	102	215.6
85	29.4	290	143.3	20	68	104	219.2
90	32.2	295	146.1	22	71.6	106	222.8
95	35.0	300	148.9	24	75.2	108	226.4
100	37.8	305	151.7	26	78.8	110	230
105	40.6	310	154.4	28	82.4	112	233.6
110	43.3	315	157.2	30	86	114	237.2
115	46.1	320	160.0	32	89.6	116	240.8
120	48.9	325	162.8	34	93.2	118	244.4
125	51.7	330	165.6	36	96.8	120	248
130	54.4	335	168.3	38	100.4	122	251.6
135	57.2	340	171.1	40	104	124	255.2
140	60.0	345	173.9	42	107.6	126	258.8
145	62.8	350	176.7	44	111.2	128	262.4

LENGTH CONVERSION

To convert inches (in.) to millimeters (mm), multiply the number of inches by 25.4
To convert millimeters (mm) to inches (in.), multiply the number of millimeters by 0.04

Inches	Millimeters	Inches	Millimeters	Inches	Millimeters	Inches	Millimeters
0.0001	0.00254	0.005	0.1270	0.09	2.286	4	101.6
0.0002	0.00508	0.006	0.1524	0.1	2.54	5	127.0
0.0003	0.00762	0.007	0.1778	0.2	5.08	6	152.4
0.0004	0.01016	0.008	0.2032	0.3	7.62	7	177.8
0.0005	0.01270	0.009	0.2286	0.4	10.16	8	203.2
0.0006	0.01524	0.01	0.254	0.5	12.70	9	228.6
0.0007	0.01778	0.02	0.508	0.6	15.24	10	254.0
0.0008	0.02032	0.03	0.762	0.7	17.78	11	279.4
0.0009	0.02286	0.04	1.016	0.8	20.32	12	304.8
0.001	0.0254	0.05	1.270	0.9	22.86	13	330.2
0.002	0.0508	0.06	1.524	1	25.4	14	355.6
0.003	0.0762	0.07	1.778	2	50.8	15	381.0
0.004	0.1016	0.08	2.032	3	76.2	16	406.4

ENGLISH/METRIC CONVERSION: LENGTH

To convert inches (in.) to millimeters (mm), multiply the number of inches by 25.4

To convert millimeters (mm) to inches (in.), multiply the number of millimeters by 0.04

Inches		Millimeters	Inches		Millimeters	Inches		Millimeters
Fraction	Decimal	Decimal	Fraction	Decimal	Decimal	Fraction	Decimal	Decimal
1/64	0.016	0.397	11/32	0.344	8.731	11/16	0.688	17.463
1/32	0.031	0.794	23/64	0.359	9.128	45/64	0.703	17.859
3/64	0.047	1.191	3/8	0.375	9.525	23/32	0.719	18.256
1/16	0.063	1.588	25/64	0.391	9.922	47/64	0.734	18.653
5/64	0.078	1.984	13/32	0.406	10.319	3/4	0.750	19.050
3/32	0.094	2.381	27/64	0.422	10.716	49/64	0.766	19.447
7/64	0.109	2.778	7/16	0.438	11.113	25/32	0.781	19.844
1/8	0.125	3.175	29/64	0.453	11.509	51/64	0.797	20.241
9/64	0.141	3.572	15/32	0.469	11.906	13/16	0.813	20.638
5/32	0.156	3.969	31/64	0.484	12.303	53/64	0.828	21.034
11/64	0.172	4.366	1/2	0.500	12.700	27/32	0.844	21.431
3/16	0.188	4.763	33/64	0.516	13.097	55/64	0.859	21.828
13/64	0.203	5.159	17/32	0.531	13.494	7/8	0.875	22.225
7/32	0.219	5.556	35/64	0.547	13.891	57/64	0.891	22.622
15/64	0.234	5.953	9/16	0.563	14.288	29/32	0.906	23.019
1/4	0.250	6.350	37/64	0.578	14.684	59/64	0.922	23.416
17/64	0.266	6.747	19/32	0.594	15.081	15/16	0.938	23.813
9/32	0.281	7.144	39/64	0.609	15.478	61/64	0.953	24.209
19/64	0.297	7.541	5/8	0.625	15.875	31/32	0.969	24.606
5/16	0.313	7.938	41/64	0.641	16.272	63/64	0.984	25.003
21/64	0.328	8.334	21/32	0.656	16.669	1/1	1.000	25.400
			43/64	0.672	17.066			